2026 IEEE International Solid-State Circuits Conference (ISSCC 2026)

San Francisco, California, USA
15-19 February 2026

Pages 342-691

IEEE Catalog Number: CFP26ISS-POD
ISBN: 979-8-3315-8937-0

**Copyright © 2026 by the Institute of Electrical and Electronics Engineers, Inc.
All Rights Reserved**

Copyright and Reprint Permissions: Abstracting is permitted with credit to the source. Libraries are permitted to photocopy beyond the limit of U.S. copyright law for private use of patrons those articles in this volume that carry a code at the bottom of the first page, provided the per-copy fee indicated in the code is paid through Copyright Clearance Center, 222 Rosewood Drive, Danvers, MA 01923.

For other copying, reprint or republication permission, write to IEEE Copyrights Manager, IEEE Service Center, 445 Hoes Lane, Piscataway, NJ 08854. All rights reserved.

*** This is a print representation of what appears in the IEEE Digital Library. Some format issues inherent in the e-media version may also appear in this print version.

IEEE Catalog Number: CFP26ISS-POD
ISBN (Print-On-Demand):   979-8-3315-8937-0
ISBN (Online):   979-8-3315-8936-3
ISSN: 0193-6530

Additional Copies of This Publication Are Available From:

Curran Associates, Inc
57 Morehouse Lane
Red Hook, NY 12571 USA
Phone: (845) 758-0400
Fax: (845) 758-2633
E-mail: curran@proceedings.com
Web: www.proceedings.com

Pagination in this book matches the original digital media

TABLE OF CONTENTS

REFLECTIONS..4
FOREWORD..5
AWARDS...23

PAPER SESSIONS

1 Plenary - Invited Papers..8

2 Processors..40

3 Wearable and Wireless Biomedical Systems...................62

4 Analog Techniques & Amplifiers...............................74

5 Sub-THz and mm-Wave Phased Arrays and Beamformers...........88

6 Exploratory Receiver Architectures from GHz to THz.....................98

7 Image Sensors and Ranging..110

8 Die-to-Die and High-Speed Electrical Transceivers.....................132

9 Wireless Power..156

10 Digital Processing and Circuit Techniques...................166

11 Pipeline and Ultra-High-Speed Data Converters.........................188

12 Frequency Synthesizers and VCOs..........................206

13 Circuits for AI and AI for Circuits...........................226

14 Unusual Interconnects and Other Uses for Light.....................238

15 DRAM, SRAM, and Non-Volatile Memories.....................252

16 Energy Harvesting, Piezo and Chargers.....................274

17 Highlighted Chip Releases for AI.............................296

18 Technology and Circuits for Domain-Specific Accelerators.........306

19 High-Voltage, Isolated and Display Power..................318

20 RF Transceiver Subsystems from cm-Wave to THz....................338

21 Sensor Interfaces...362

22 Circuits in Extreme Environments.............................382

23 Next-Generation Optical Transceivers.......................394

24 Displays..408

25 Hardware Security..422

26 Compute Power and Supply Modulators.....................444

27 Frequency Generators, Multipliers, and Modulators...................464

28 Innovations from Outside the (ISSCC) Box..................486

29 Biochemical Sensors for Life Sciences and Agriculture...............496

30 Compute-in-Memory...510

31 AI Accelerators..530

32 Low-Power Noise-Shaping ADCs..............................550

33 Time-Varying Circuit Techniques from RF to mm-Wave..............560

34 Integrated Radar and UWB Transceivers.....................574
      from Microwave to Sub-THz

35 Low Power Wireless Transceivers.............................586
      for Localization and Communications

36 Neural and Biomedical Interfaces.............................598

37 Memory Interface...622

TUTORIALS

TUTORIALS 1-10...643

FORUMS

F1 Power Efficient Circuits and Systems for...................646
      Next-Gen Agentic and Robotics

F2 Electrical and Optical Links Towards 400G+ Connectivity.........648

F3 Powering the Future of AI, HPC, and Chiplet Architectures:......650
      From Dies to Package and Rack

F4 The Race for 6G FR3 (7-24GHz): From Network Deployment....652
      to System Integration and Breakthrough Technology

F5 Analog for AI and AI for Analog: What the Analog/RF..............654
      People Can Do and Leverage in the AI Era

F6 Calibration and Dynamic Matching Techniques.................656
      for High-Performance Data Converters

SPECIAL EVENTS

EE1 Student Research Preview: 658
      Short Presentations with Poster Session

      CAREER PANEL: Chip In for Change: Career Insights...........663
      for Young Designers in an AI-First, Eco-Conscious World

EE2 Generative AI for Silicon Design: Mastering Complexity,......665
      Democratizing Design, and Building Trust

EE3 The Augmented Human – Will Chips in Our Brain...............667
      Enhance Our Cognitive Abilities?

SHORT COURSE

SC Circuits for Optical Subsystems:...........................670
      Communications and Beyond

EXECUTIVE COMMITTEE..680

INTERNATIONAL TECHNICAL PROGRAM COMMITTEE..................681

ITPC EWAA SUBCOMMITTEE...685

ITPC APAC SUBCOMMITTEE...686

ITPC AM SUBCOMMITTEE..687

2027 CALL FOR PAPERS...688

CONFERENCE TIMETABLE..689

CONFERENCE SPACE LAYOUT...690

INDEX TO AUTHORS

ISSCC 2026 / SESSION 20 / RF TRANSCEIVER SUBSYSTEMS FROM cm-WAVE TO THz / 20.2

20.2 A High Back-off Efficiency Unequal-Stacked Doherty Power Amplifier Achieving 16.7dBm P_{avg} in a 22nm FDSOI CMOS Technology for 5G FR2 Applications

Seungwon Park, Jooseok Lee, Seungjae Baek, Taewan Kim, Yifei Chen, Sehyug Jeon, Sung-gi Yang

Samsung Electronics, Seoul, Korea

Abstract

This paper proposes an unequal-stacked Doherty power amplifier. The proposed Doherty structure consists of a common-source topology for the carrier amplifier and an N-stacked topology for the peaking amplifier with a theoretically determined quarter-wave line. Thanks to the proposed Doherty structure, the back-off efficiency is enhanced. With a 5G NR 64-QAM OFDM 100MHz signal, the P_{avg} and PAE_{avg} exceed 15.8dBm and 16.9%, respectively, from 24 to 28GHz with the EVM of -25dB.

Phase arrays are essential for millimeter-wave (mm-wave) cellular systems, enabling high equivalent isotropic radiated power (EIRP) for wide coverage. Since a significant number of power amplifiers (PAs) are integrated into a phased-array IC, it is crucial that these PAs are compact, highly efficient, and have high output power to minimize overall power consumption and chip area [1]. Numerous studies have been reported on these aspects employing various structures [2-5]. Among these, Doherty PAs (DPAs) have been extensively researched to achieve high efficiency for high peak-to-average-power-ratio (PAPR) signals. However, a conventional DPA has a theoretical back-off range of 6dB, which is insufficient to meet the high PAPR (> 10dB) of 5G FR2 signals. To overcome this, alternative load-modulated PA architectures have been proposed and implemented, such as an N-way DPA [6], a distributed efficient PA (DEPA) [7], a load-modulated balanced PA (LMBA) [8], and an asymmetric Doherty PA (ADPA) [9].

Figure 20.2.1 (top) shows the block diagrams of an N-way DPA, a DEPA, and an LMBA. These architectures achieve high efficiency in the deep back-off region but require multiple amplifiers or large passive components. This leads to a large chip size, making them unsuitable for highly integrated phased-array systems. In contrast, the ADPA uses only two amplifiers, making it attractive for compact phased-array systems. However, achieving high output power with the ADPA remains challenging. A common approach to increase output power is to adopt a stacked topology, which increases output voltage swing. However, due to parasitic components, efficiency degrades as the number of transistors in a stack increases.

Figure 20.2.1 (bottom) shows the simulated drain efficiency (DE) of common-source (CS) and stacked amplifiers as a function of back-off power. It can be seen that the DE of stacked amplifiers decreases as the number of transistors in the stack increases. Furthermore, the efficiency degradation is much larger when the load impedance of the stacked amplifier is twice the optimum value, which is the situation for the carrier amplifier in a DPA. Therefore, while a stacked-based ADPA can achieve high output power, it exhibits relatively lower back-off efficiency compared to a CS-based ADPA.

CS-based ADPAs, on the other hand, maintain higher back-off efficiency without stacking. Yet, delivering high output power requires an extremely large transistor in the peaking amplifier. Hence, the CS-based ADPA exhibits high 9dB back-off efficiency only in narrow frequency bands due to the large parasitic capacitance of the peaking amplifier as shown in Fig. 20.2.1. Consequently, a new design approach is needed to realize a high-power deep back-off DPA while maintaining compact size.

In this work, we propose an unequal-stacked DPA that employs different stacked topologies for the carrier and peaking amplifiers in order to improve deep back-off efficiency in a compact die area while achieving high output power. The proposed DPA exhibits an average output power (P_{avg}) of 16.4dBm and an average power-added efficiency (PAE_{avg}) of 19.3% at 25GHz using 5G NR OFDM 100MHz.

Figure 20.2.2 (left) illustrates the diagram and operation principle of the proposed unequal-stacked DPA. It consists of a CS-based carrier amplifier, an N-stacked peaking amplifier with a quarter-wave line (TL_Q), and inductors (L_{OUT}, $N \cdot L_{OUT}$) adopted to resonate out the capacitances of both amplifiers. The impedance seen looking into the load from each amplifier (R_C, R_P) and the output power of each amplifier (P_C, P_P) are listed in the table of Fig. 20.2.2. R_{OPT} and P_{OUT} are the optimum impedance and saturated output power of the CS topology, respectively. N is the number of transistors in the stack in the peaking amplifier. Considering the table in Fig. 20.2.2, the back-off range is calculated as $20 \log(N+1)$. Therefore, the back-off range increases as N increases, and, further, the output power of the DPA increases with N. This paper employs a 3-stacked topology in the peaking amplifier, achieving a 12dB back-off range, which is higher than that of a conventional DPA, while delivering high output power. Furthermore, the back-off efficiency is high due to the CS in the carrier amplifier, and the bandwidth is not degraded since the stacked topology in the peaking amplifier has a relatively lower output capacitance compared to the CS topology. Figure 20.2.2 (bottom right) shows the simulated 9dB back-off efficiency of the conventional CS- and 3-stacked-based ADPAs and the proposed unequal-stacked DPA. In the simulation, the passive components of the DPA structure are assumed to be lossless.

Therefore, when implemented using high-loss technologies such as CMOS, especially at mm-wave frequencies, the efficiency and bandwidth enhancement of the proposed structure become limited. Even with these considerations, the proposed structure demonstrates relatively high back-off efficiency while achieving wide bandwidth.

However, connecting amplifiers with different stacked topologies may cause reliability issues due to differences in maximum voltage swings. To avoid this, the drain voltage swing of the carrier amplifier must be 1/N of the peaking amplifier. Thus, TL_Q must be properly designed to ensure the reliability of the carrier amplifier. Furthermore, TL_Q parameters should also be determined so that both amplifiers deliver maximum output power. For maximum output power at saturation, R_C and R_P must be R_{OPT} and $N R_{OPT}$, respectively. Considering the nature of the DPA, R_P becomes $N R_{OPT}$ with a load impedance (R_L) of $N^2 R_{OPT}$ over N+1. With the R_L, the impedance seen into the load from TL_Q (R_C') is $N^2 R_{OPT}$. To achieve optimal R_M and maximum carrier output power, TL_Q must be set as $N R_{OPT}$ and transform $N^2 R_{OPT}$ to R_{OPT}, so the voltage swing decreases by 1/N from load to carrier. This guarantees reliability and maximum voltage swing for both amplifiers, regardless of the number of stacks in the peaking amplifier.

A schematic of the proposed DPA is shown in Fig. 20.2.3. The PA consists of carrier and peaking amplifiers at the power stage and a single driver. The driver and carrier amplifiers use a differential CS topology, while the peaking amplifier employs a differential 3-stacked topology. A single-transformer-based Doherty matching network at the output stage [10] and a differential-breaking phase-offset (DBPO) technique at the inter stage [11] are utilized for a compact design. In addition, operational amplifiers (OPAs) stabilize and supply the gate bias of the 3-stacked structure [12].

Figure 20.2.3 (bottom left) shows the simulated drain voltage of the carrier and peaking amplifiers. The drain voltage swing of the carrier amplifier remains below $2V_{DD}$ (2.4V), which is the breakdown voltage swing of a transistor, even though a 3-stacked topology is used in the peaking amplifier. Therefore, the theoretically designed quarter-wave network ensures the reliability of the proposed unequal-stacked DPA.

In developing PAs for wireless communication systems, another key topic is minimizing performance variations caused by changes in antenna impedance. Figure 20.2.4 (top left) shows a diagram of phased-array wireless communication systems. Since PAs are directly connected to the antenna through passive components such as a matching network or a switch, PA performance is highly dependent on the antenna input impedance. Power delivered from a PA to an antenna is inversely proportional to the $|1-S_{22} \times \Gamma_{ant}|^2$ [13], where Γ_{ant} and S_{22} are the input reflection coefficient of antenna and the output reflection coefficient of the PA, respectively. Therefore, a low S_{22} can alleviate power reduction and promotes stable amplifier operation.

Figure 20.2.4 (bottom left) shows the simulated PAE load-pull and optimum matching circles of the CS and stacked topologies. When two circles are close, an amplifier can achieve high PAE with low S_{22} simultaneously. The circles are relatively closer for the CS topology than for stacked topologies. As a result, the CS topology can demonstrate high efficiency with better matching performance compared to stacked topologies. Since the proposed unequal-stacked DPA employs a CS topology for the carrier, it achieves superior output matching compared to other load-modulated PAs that utilize stacked topologies. Figure 20.2.4 (right) presents the measured output reflection coefficient (S_{22}) and simulated PAE performance based on the -10dB return loss circle for both the conventional 3-stacked DPA and proposed DPA. The proposed DPA presents an S_{22} that is 27dB lower than that of the conventional DPA at 28GHz. In addition, under a large-signal S-parameter simulation in the saturation region, the S_{22} of the proposed DPA remains 9.7dB lower than that of the conventional DPA at 28GHz. Furthermore, the proposed DPA exhibits only a 9% variation in the 9dB back-off efficiency under load impedance variation, whereas the conventional stacked DPA shows a 30% variation. Consequently, the proposed DPA exhibits improved load insensitivity while maintaining high back-off efficiency and high output power.

The proposed DPA was fabricated in a 22nm FD-SOI CMOS technology. Figure 20.2.7 shows the fabricated die micrograph of the DPA. The OPA-based bias network is embedded below

342 • 2026 IEEE International Solid-State Circuits Conference

979-8-3315-8937-0/26 $31.00 © 2026 IEEE

the ground metal. The chip size is $0.70 \times 0.49\text{mm}^2$ including probing pads, and the core area is $0.47 \times 0.27\text{mm}^2$. Figure 20.2.5 shows the measured performance of the proposed DPA from 24 to 28GHz. With a 1-tone continuous wave, the saturated output power (P_{sat}) and 9dB back-off PAE (PAE_{9dB}) are 23.2 to 24.0dBm and 14.7 to 16.6%, respectively. Compared to the simulated PAE of the Class-AB-biased PA, the proposed DPA shows 57% higher PAE at 9dB back-off. A 5G NR FR2 100MHz 1-CC 64-QAM signal (PAPR > 10dB) is applied to the DPA without any additional digital pre-distortion. Over the frequency range from 24 to 28GHz, the DPA exhibits P_{avg} of 15.8 to 16.7dBm with the EVM under -25dB, and the PAE_{avg} is higher than 16.9%.

For comparison, the performance of the proposed unequal-stacked DPA and recent PAs is summarized in Fig. 20.2.6. Thanks to the unequal-stacked structure, which employs a CS topology in the carrier amplifier and a stacked topology in the peaking amplifier, the proposed design achieves not only high P_{avg} and PAE_{avg} but also excellent output matching performance. Moreover, the proposed PA achieves the smallest core size while maintaining high PAE_{avg} and P_{avg} among the compared PAs.

References:
[1] S. Baek et al., "A Large-Scale, Low-Power, Compact 5G mm-Wave Phased-Array Transceiver in 45 nm RFSOI CMOS", *IEEE TMTT*, vol. 73, no. 4, pp. 2097-2110, Apr. 2025. https://doi.org/10.1109/TMTT.2025.3544620
[2] T.-W. Li et al., "A Continuous-Mode Harmonically Tuned 19-to-29.5GHz Ultra-Linear PA Supporting 18Gb/s at 18.4% Modulation PAE and 43.5% Peak PAE", *ISSCC*, pp. 410-412, Feb. 2018. https://doi.org/10.1109/ISSCC.2018.8310358
[3] B. Rabet et al., "A High-Efficiency 28GHz Outphasing PA with 23dBm Output Power Using a Triaxial Balun Combiner", *ISSCC*, pp. 174-176, Feb. 2018. https://doi.org/10.1109/ISSCC.2018.8310240
[4] M. Pashaeifar et al., "A Millimeter-Wave CMOS Series-Doherty Power Amplifier with Post-Silicon Inter-Stage Passive Validation", *IEEE JSSC*, vol. 57, no. 10, pp. 2999-3013, Oct. 2022. https://doi.org/10.1109/JSSC.2022.3175685
[5] E. Liu et al., "A Ka-Band Doherty-Like Non-Load Modulated Power Amplifier", *IEEE JSSC*, vol. 60, no. 5, pp. 1584-1593, May 2025. https://doi.org/10.1109/JSSC.2025.3532578
[6] A. K. Kumaran et al., "A Single-Supply Balun-First Three-Way mm-Wave Doherty PA", *IEEE TMTT*, vol. 72, no. 5, pp. 2757-2772, May 2024. https://doi.org/10.1109/TMTT.2024.3365697
[7] P. Saad et al., "A 1.8–3.8-GHz Power Amplifier With 40% Efficiency at 8-dB Power Back-Off", *IEEE TMTT*, vol. 66, no. 11, pp. 4870-4882, Nov. 2018. https://doi.org/10.1109/TMTT.2018.2867426
[8] V. Qunaj et al., "A Doherty-Like Load-Modulated Balanced Power Amplifier Achieving 15.5dBm Average P_{out} and 20% Average PAE at a Data Rate of 18Gb/s in 28nm CMOS", *ISSCC*, pp. 356-358, Feb. 2021. https://doi.org/10.1109/ISSCC42613.2021.9365966
[9] X. Zhang et al., "A 24-to-29GHz Compact Transmit/Receive Front-End Module Featuring an Asymmetric Doherty Power Amplifier and 0.22mm² Area", *ISSCC*, pp. 464-466, Feb. 2025. https://doi.org/10.1109/ISSCC49661.2025.10904541
[10] H.-C. Park et al., "Single Transformer-Based Compact Doherty Power Amplifiers for 5G RF Phased-Array ICs", *IEEE JSSC*, vol. 57, no. 5, pp. 1267-1279, May 2022. https://doi.org/10.1109/JSSC.2022.3148044

[11] H. Oh et al., "A 24.25-to-29.5GHz Extremely Compact Doherty Power Amplifier with Differential-Breaking Phase Offset Achieving 23.7% PAE_{avg} for 5G Base-Station Transceivers", *ISSCC*, pp. 522-524, Feb. 2024. https://doi.org/10.1109/ISSCC49657.2024.10454406
[12] J. Lee et al., "A 22nm FDSOI CMOS-Based Compact 3-Stack Doherty Power Amplifier with a Stacked OPA-Based Bias Scheme Achieving >16.5dBm P_{avg} for 5G FR2 Applications", *ISSCC*, pp. 96-98, Feb. 2025. https://doi.org/10.1109/ISSCC49661.2025.10904681
[13] M. Eleraky et al., "An Ultra-Compact Wideband Load-Insensitive Complex-Cascode LC-Neutralized Power Amplifier for 4:1-VSWR-Resilient Operations in Large-Scale Phased Arrays", *ISSCC*, pp. 98-100, Feb. 2025. https://doi.org/10.1109/ISSCC49661.2025.10904694
[14] E. Liu et al., "An Ultra-Compact 28GHz Doherty Power Amplifier with an Asymmetrically-Coupled-Transformer Output Combiner", *ISSCC*, pp. 536-538, Feb. 2024. https://doi.org/10.1109/ISSCC49657.2024.10454274

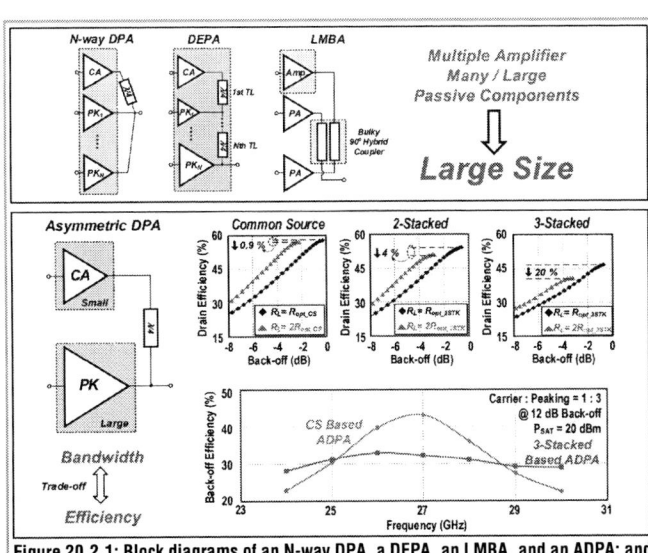

Figure 20.2.1: Block diagrams of an N-way DPA, a DEPA, an LMBA, and an ADPA; and efficiency of conventional ADPAs.

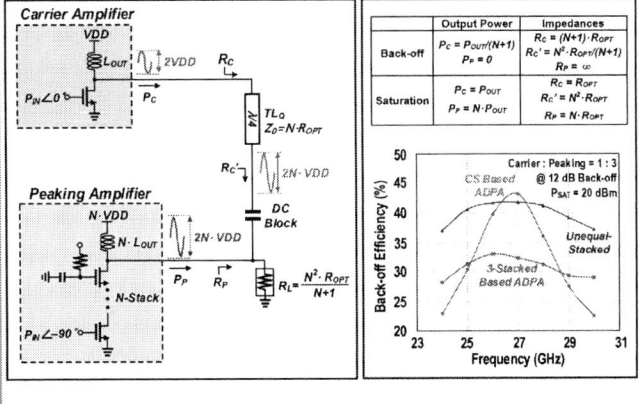

Figure 20.2.2: Operation principle of the unequal-stacked DPA, and performance comparison with conventional ADPAs.

ISSCC 2026 / SESSION 20 / RF TRANSCEIVER SUBSYSTEMS FROM CM-WAVE TO THz / 20.2

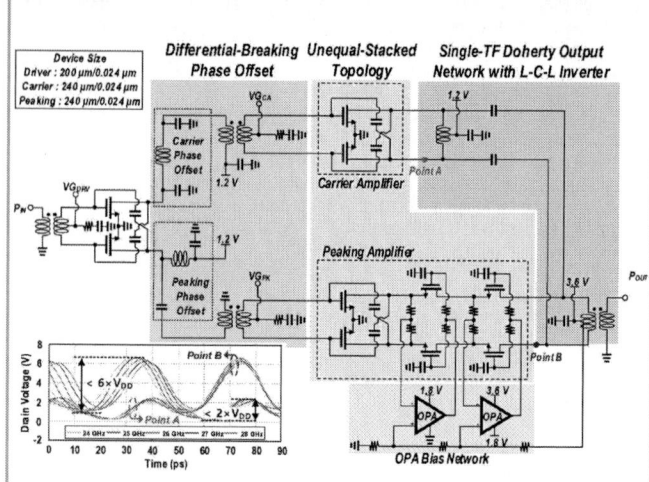

Figure 20.2.3: Schematic of the proposed DPA and simulated drain-voltage swing of the carrier and peaking amplifiers.

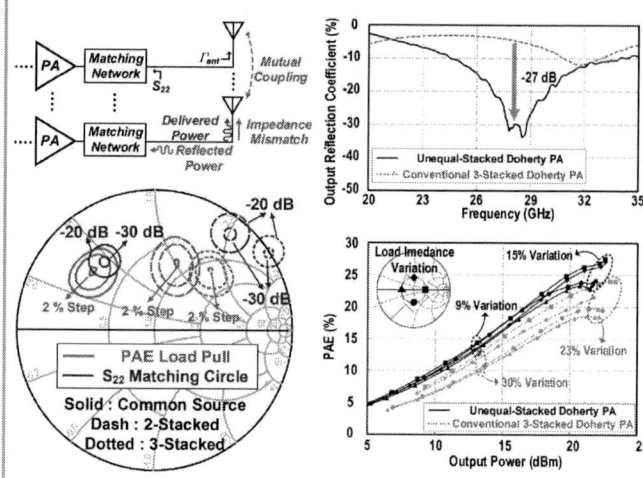

Figure 20.2.4: Simulated PAE load-pull and matching circle, measured S_{22}, and simulated large-signal performances.

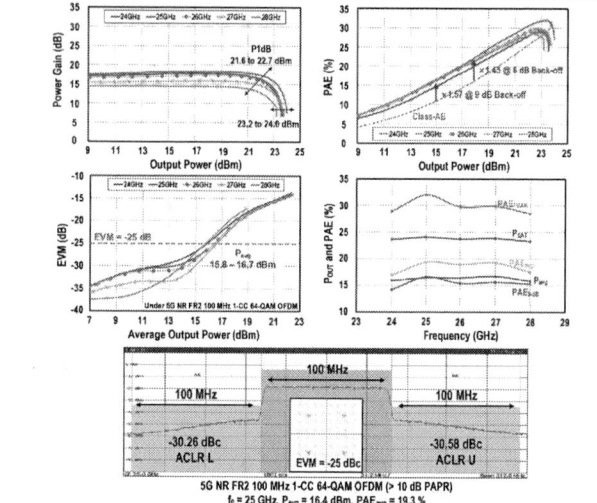

Figure 20.2.5: Measured performances of the unequal-stacked DPA under CW and 100MHz 1-CC 5G NR 64-QAM OFDM signals.

Figure 20.2.6: Performance comparison table of recently reported mm-wave PAs.

Ref.	This work	JSSC '22 Pashaeifar [4]	JSSC '25 Liu [5]	TMTT '24 Kumaran [6]	JSSC '22 Park [10]	ISSCC '24 Oh [11]	ISSCC '25 Lee [12]	ISSCC '24 Liu [14]
Architecture	Unequal-Stacked Doherty	Doherty	Non-Load Modulated	N-Way Doherty	Doherty	Doherty	Doherty	Doherty
Technology	22nm FD-SOI CMOS	40 bulk CMOS	45nm SOI CMOS	40nm bulk CMOS	28nm bulk CMOS	45nm SOI CMOS	22nm FD-SOI CMOS	45nm SOI CMOS
Freq. (GHz)	24.0 to 28.0	25.0 to 30.0	26.0 to 40.0	24.0 to 30.0	24.5 to 29.5	24.25 to 29.5	24.0 to 29.0	26.0 to 30.0
Supply (V)	3.6 / 1.2	1.8	2.0	2.4	1.8	2.2	3.6	2.0
Gain (dB)	13.7 to 16.8	> 17.0*	13.7 to 14.9	14.0 to 20.0*	12.0 to 17.3*	19.2 to 22.3	16.7 to 20.7	15.7 to 16.7
S22 (dB)	-31.1 to -8.6	< -6.2	-8.0 to -5.0*	< -8.5	-10.0 to -5.0*	-7.0 to -5.0*	-7.0 to -3.0*	-4.0 to -2.0*
Psat (dBm)	23.2 to 24.0	> 20.2*	19.1 to 22.6	> 20.0	18.3 to 18.8	20.3 to 22.0	23.0 to 24.1	> 21.0
PAEpeak (%)	28.4 to 32.0	> 42.0*	20.9 to 37.6	> 33.0†	26.0 to 30.5	32.5 to 42.3	22.3 to 24.7	31.7 to 38.0
PAE6dB (%)	20.1 to 22.9	> 29.0	14.0 to 21.0*	> 22.0†	18.0 to 22.0	21.5 to 27.0	> 17.0	26.5 to 29.3
PAE9dB (%)	14.7 to 16.6	> 22.0*	10.4 to 18.2	> 15.0†#	12.0 to 12.5*	-	-	18.6 to 21.5
Core Area (mm²)	0.13	0.37	1.07	0.77	0.20	0.14	0.15	0.15
Modulation	64-QAM OFDM	64-QAM OFDM	64-QAM OFDM	64-QAM OFDM	64-QAM OFDM	64-QAM OFDM	64-QAM OFDM	64-QAM OFDM
PAPR (dB)	> 10	9.7	9.6	9.7	> 10	> 10	> 10	> 9
Bandwidth (MHz)	100	400	200	400	100	100	100	100
Pavg (dBm)	15.8 to 16.7	> 7.3	8.1 to 10.4	9.3 to 9.8	> 12.0	13.2 to 14.6	16.5 to 17.1	9.2 to 12.1
PAEavg (%)	16.9 to 19.3	> 11.3	7.4 to 13.2	3.0 to 15.0†	> 17.5	20.0 to 25.6	15.9 to 16.8	14.4 to 21.0
EVM (dB)	< -25.0	< -24.5	< -25.0	< -24.1	< -25.0	< -25.0	< -25.0	< -25.0

*Graphically estimated, †Drain Efficiency, #9.5 dB PBO.

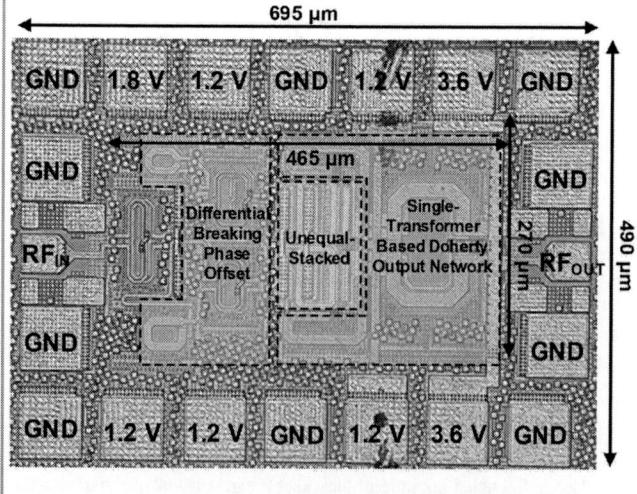

Figure 20.2.7: Die micrograph of the proposed unequal-stacked DPA.

• 2026 IEEE International Solid-State Circuits Conference

ISSCC 2026 / SESSION 20 / RF TRANSCEIVER SUBSYSTEMS FROM CM-WAVE TO THz / 20.3

20.3 A mm-Wave Doherty Power Amplifier in a Single-Path Footprint Using Compact Reciprocal Doherty Networks

Lianbo Liu[1], Yidong Fang[1], Qiang Zhou[2], Taiyun Chi[2], Sensen Li[1]

[1]University of Texas, Austin, TX, [2]Rice University, Houston, TX

Abstract

This paper presents a compact, single-path-footprint Doherty PA with minimized passive network areas. A systematic analysis for its realization is developed based on the proposed theory of reciprocal Doherty networks employing a single transformer. Implemented with only three transformer footprints and a single driver, the design occupies a core area of 0.12mm², comparable to that of a linear PA, while supporting wideband, high-order QAM OFDM signals with high average efficiency.

Beamforming phased arrays have become indispensable for wireless communication and sensing at upper mid-band and millimeter-wave (mm-wave) spectra, where they provide access to abundant bandwidth while compensating for severe path loss. By integrating multiple RF chains, phased-array transceivers enable highly directional communication that improves link quality, while the adoption of dual polarization further doubles the number of RF paths, thereby increasing chip area and implementation complexity [1-3]. Among the RF front-end blocks, power amplifiers (PAs) dominate the transceiver power consumption and efficiency. Modern communication standards rely heavily on orthogonal frequency-division multiplexing (OFDM), which inherently exhibits a high peak-to-average power ratio (PAPR). This forces PAs to operate under significant output power back-off (PBO), where conventional linear PAs suffer from poor efficiency. The Doherty PA, with its load modulation principle, has emerged as the preferred architecture for enhancing efficiency under PBO, while maintaining compatibility with high-order QAM OFDM signals and supporting RF-in/RF-out operation [4-6]. However, conventional Doherty implementations typically occupy a larger footprint than linear PAs, posing challenges for integration in multi-channel array systems. Consequently, there is a pressing need to develop compact Doherty PA solutions that simultaneously deliver high efficiency under back-off and minimize area overhead—making them ideally suited for energy-efficient, large-scale array transceivers at higher spectrum.

Doherty PAs typically require an impedance inverting network (IIN) and an impedance scaling network (ISN) to achieve load-pull matching with a 90° phase shift, thus enabling correct load modulation for efficiency enhancement [7]. The regular Doherty PAs often utilize two transformers to implement IIN and ISN, which makes the overall design quite bulky [8-12]. To address this challenge, recent efforts toward compact Doherty PA implementations, as illustrated in Fig. 20.3.1 (top), have focused on merging the IIN and ISN within approximately a single transformer footprint [13-16]. However, these techniques have been primarily applied on the output side. Other passive networks—such as input IQ generation and interstage matching—still occupy additional area compared to their conventional Class-AB counterparts and, therefore, must also be minimized to realize a truly compact design. Furthermore, a systematic analysis and theoretical framework for realizing a fully compact Doherty PA is still lacking. In this work, we present a compact Doherty PA implemented within a single-path footprint by employing a reciprocal Doherty network, achieving a core area of only 0.12mm².

Two types of reciprocal Doherty networks are introduced: the reciprocal parallel network (RPN) and the reciprocal series network (RSN), as illustrated in Fig. 20.3.1 (bottom left). When signals propagate from ports 1 and 2 to port 3, the RPN and RSN function as parallel and series Doherty combiners, respectively. Conversely, when the signal travels in the opposite direction, they split the signal with a 90° phase shift, effectively operating as quadrature hybrids. These reciprocity characteristics are exploited to repurpose the single-transformer-footprint Doherty combiner as a 90° hybrid with built-in impedance matching. This approach significantly reduces the input-side area of the Doherty PA and minimizes the total number of passive footprints. Since both RPN and RSN can function as the input quadrature hybrid and the output Doherty combiner, three possible variants of Doherty implementation can be realized using these networks (Fig. 20.3.1, bottom right). To ensure correct Doherty operation, the following criteria must be satisfied: 1) the output powers of the main and auxiliary amplifiers must combine in phase when both are on; and 2) the auxiliary amplifier must not load the network when it is off. According to this rubric, neither the type-1 nor the type-2 variant achieves proper Doherty operation: in type-1, the signals combine destructively due to a 180° phase difference, while in type-2, the auxiliary path undesirably loads either the input or the output network. To address these issues, it is necessary to place one ISN and one IIN on the same main or auxiliary path to satisfy the phase alignment requirement, and to provide a short termination for the RSN and an open termination for the RPN when the auxiliary path is off, so that the network is not loaded. Based on this analysis, the type-3 variants are proven to support correct Doherty operation. Depending on whether the RPN or RSN is used as the output Doherty combiner, a parallel or series Doherty PA can be realized, and their theoretical bandwidth and linearity performance are compared (Fig. 20.3.1, bottom right). The series Doherty PA offers a wider bandwidth for active load modulation and reduced AM-AM/AM-PM distortion. Consequently, the final implementation adopts an asymmetric RPN+RSN variant, requiring only three transformer footprints and a single driver amplifier, enabling a compact and fully integrated single-path footprint solution.

Figure 20.3.2 (top) shows the 3D EM models of the proposed RSN and RPN, each realized within a single-transformer footprint. The networks are implemented using the top three back-end-of-line (BEOL) metal layers and folding multiple coils to achieve a compact layout while supporting Doherty load modulation or quadrature signal generation. The RSN comprises four coils vertically interleaved across the three layers: the two larger, outer coils are optimized for strong coupling to implement the ISN (coupling factor k=0.8), while the two smaller, centrally placed coils provide the lower coupling required for the IIN (k=0.4). This layout also limits unintended mutual coupling between the inner and outer coils to k<0.15, rendering its impact on the intended load modulation negligible. At 0dB PBO, both the main and auxiliary amplifiers see a wideband differential 50Ω impedance (R_{opt}) centered around 28GHz, as shown in Fig. 20.3.2 (bottom). At 6dB PBO, the impedance presented to the main amplifier increases to 91Ω—nearly twice R_{opt}—as a result of effective Doherty active load modulation. EM simulation of the Doherty output network demonstrates a peak passive efficiency of 76.1% at 6dB PBO, optimized for enhanced efficiency in the back-off region. In addition to quadrature signal generation, the RPN implemented at the inputs of the main and auxiliary amplifiers inherently provides impedance transformation. When matched to their designated impedances (i.e., the inputs of the main and auxiliary amplifiers and the output of the driver amplifier), the equivalent reflection coefficients of the three RPN ports exhibit good matching behavior across the band of interest. As a result, no additional interstage matching networks are required, further reducing the overall area. The final layout sizes of the RSN and RPN are only 0.032mm² and 0.033mm², respectively.

The top-level schematic, along with the EM models of the single-path passive networks, is shown in Fig. 20.3.3. A compact multi-turn input balun matches the driver amplifier input impedance to 50Ω, achieving amplitude imbalance below 0.3dB and phase imbalance under 5° across a 10GHz bandwidth. It also provides wideband input matching, with S_{11} less than −20dB from 23 to 31GHz. A single common-source (CS) driver amplifier is implemented to feed the RPN-based quadrature hybrid, enabling a compact footprint. The main and auxiliary amplifiers employ a cascode topology with capacitive neutralization. Activation of the auxiliary amplifier is controlled via adaptive biasing [17,18], with the bias signal routed on the top metal layer to minimize parasitic resistance and capacitance and thus improve switching speed. This approach yields a sharp turn-on of the auxiliary amplifier, as verified by its DC-power–versus–RF-power characteristic in Fig. 20.3.3 (bottom right). As a result, effective Doherty active load modulation is achieved through the RSN-based compact output network.

The proposed compact Doherty PA prototype is fabricated in a 22nm CMOS FDSOI process. Continuous-wave (CW) measurement results are summarized in Fig. 20.3.4. The PA achieves a peak gain of 16.8dB and a 3dB gain-variation bandwidth of 6.2GHz. Input matching, represented by S_{11}, remains better than −20dB within the band of interest and better than −10dB from 20 to 34GHz. At 25/27GHz, the PA demonstrates OP_{1dB} of 19.8/19.9dBm, P_{sat} of 20.4/20.5dBm, PAE_{0dB} of 30.7/29.2%, and PAE_{6dB} of 22.0/22.5%, corresponding to 1.43×/1.54× efficiency enhancement compared to a conventional Class-B counterpart. Over frequency, the PA maintains flat and consistent large-signal performance, with PAE and output-power variations smaller than 1.5% and 0.8dB, respectively, across 24 to 28GHz. When evaluated with 5G NR modulation signals (Fig. 20.3.5), the proposed compact Doherty PA demonstrates the capability to support wideband, high-order QAM with high average efficiency. For 64-QAM OFDM signals with 100/400/800MHz bandwidths, the PA achieves P_{avg} of 10.2/9.7/9.1dBm and PAE_{avg} of 14.0/12.9/11.0%, with EVM_{rms} of −25.3/−24.8/−26.3dB and ACPR of −30.9/−33.6/−27.8dBc. The PA was further characterized under 256-QAM OFDM signals to validate its capability with higher-order modulation. For 100/200/400MHz bandwidths, the PA delivers P_{avg} of 9.3/8.6/8.9dBm and PAE_{avg} of 11.1/9.6/10.2%, with EVM_{rms} of −28.0/−27.8/−27.8dB and ACPR of −36.4/−35.3/−32.8dBc.

Figure 20.3.6 presents a comparison with prior-art compact Doherty PAs [13-16], conventional Doherty and load-modulated PAs [10,19,20], and linear PAs [21,22] operating at similar frequencies. The proposed prototype demonstrates CW and modulation performance comparable to that of other Doherty and load-modulated PAs in Fig. 20.3.6, particularly in terms of bandwidth and its ability to support high-order QAM OFDM 5G NR signals. Notably, this Doherty PA occupies the smallest reported core area of only 0.12mm². The die micrograph is shown in Fig. 20.3.7. Owing to the proposed reciprocal Doherty networks, the design is realized in a truly single-path footprint, with a core area comparable to that of a linear PA. Consequently, the proposed compact Doherty PA is well suited for integration in large-scale beamforming or MIMO arrays to enable high-speed, advanced high-order QAM OFDM transmission.

Acknowledgement:
The authors would like to thank GlobalFoundries for chip fabrication; MediaTek for supporting this project; Keysight Technologies and National Instruments (part of Emerson) for providing measurement equipment; and all members of the UT-ACE Lab for their valuable technical discussions.

References:
[1] J. D. Dunworth et al., "A 28GHz Bulk-CMOS Dual-Polarization Phased-Array Transceiver with 24 Channels for 5G User and Basestation Equipment," *ISSCC*, pp. 70-72, Feb. 2018. https://doi.org/10.1109/ISSCC.2018.8310188
[2] J. Pang et al., "A 28-GHz CMOS Phased-Array Beamformer Utilizing Neutralized Bi-Directional Technique Supporting Dual-Polarized MIMO for 5G NR," *IEEE JSSC*, vol. 55, no. 9, pp. 2371-2386, Sep. 2020. https://doi.org/10.1109/JSSC.2020.2995039
[3] B. Sadhu et al., "A 24-to-30GHz 256-Element Dual-Polarized 5G Phased Array with Fast Beam-Switching Support for >30,000 Beams," *ISSCC*, pp. 436-438, Feb. 2022. https://doi.org/10.1109/ISSCC42614.2022.9731778
[4] X. Zhang et al., "A 47GHz 4-way Doherty PA with 23.7dBm P1dB and 21.7% / 13.1% PAE at 6 / 12dB Back-off Supporting 2000MHz 5G NR 64-QAM OFDM," *ISSCC*, pp. 520-522, Feb. 2024.10454571. https://doi.org/10.1109/ISSCC49657.2024.10454571
[5] M. Mortazavi et al., "A Four-Way Series Doherty Digital Polar Transmitter at mm-Wave Frequencies," *IEEE JSSC*, vol. 57, no. 3, pp. 803-817, Mar. 2022. https://doi.org/10.1109/JSSC.2021.3133861
[6] X. Zhang et al., "A 24-to-29GHz Compact Transmit/Receive Front-End Module Featuring an Asymmetric Doherty Power Amplifier and 0.22mm² Area," *ISSCC*, pp. 464-465, Feb. 2025. https://doi.org/10.1109/ISSCC49661.2025.10904541
[7] N. S. Mannem et al., "Broadband Active Load-Modulation Power Amplification Using Coupled-Line Baluns: A Multifrequency Role-Exchange Coupler Doherty Amplifier Architecture," *IEEE JSSC*, vol. 56, no. 10, pp. 3109-3122, Oct. 2021. https://doi.org/10.1109/JSSC.2021.3078322
[8] M. Pashaeifar et al., "A Millimeter-Wave CMOS Series-Doherty Power Amplifier with Post-Silicon Inter-Stage Passive Validation," *IEEE JSSC*, vol. 57, no. 10, pp. 2999-3013, Oct. 2022. https://doi.org/10.1109/JSSC.2022.3175685
[9] F. Wang et al., "A Highly Linear Super-Resolution Mixed-Signal Doherty Power Amplifier for High-Efficiency mm-Wave 5G Multi-Gb/s Communications," *ISSCC*, pp. 88-90, Feb. 2019. https://doi.org/10.1109/ISSCC.2019.8662497
[10] Z. Zong et al., "A 28-GHz SOI-CMOS Doherty Power Amplifier with a Compact Transformer-Based Output Combiner," *IEEE TMTT*, vol. 69, no. 6, pp. 2795-2808, June 2021. https://doi.org/10.1109/TMTT.2021.3064022
[11] S. Hu et al., "A 28-/37-/39-GHz Linear Doherty Power Amplifier in Silicon for 5G Applications," *IEEE JSSC*, vol. 54, no. 6, pp. 1586-1599, June 2019. https://doi.org/10.1109/JSSC.2019.2902307

[12] N. Rostomyan et al., "28 GHz Doherty Power Amplifier in CMOS SOI with 28% Back-Off PAE," *IEEE MWCL*, vol. 28, no. 5, pp. 446-448, May 2018. https://doi.org/10.1109/LMWC.2018.2813882
[13] J. Lee et al., "A 22nm FDSOI CMOS-Based Compact 3-Stack Doherty Power Amplifier with a Stacked OPA-Based Bias Scheme Achieving >16.5dBm Pavg for 5G FR2 Applications," *ISSCC*, pp. 96-98, Feb. 2025. https://doi.org/10.1109/ISSCC49661.2025.10904681
[14] E. Liu and H. Wang, "An Ultra-Compact 28GHz Doherty Power Amplifier with an Asymmetrically-Coupled-Transformer Output Combiner," *ISSCC*, pp. 536-538, Feb. 2024. https://doi.org/10.1109/ISSCC49657.2024.10454274
[15] H. Oh et al., "A 24.25-to-29.5GHz Extremely Compact Doherty Power Amplifier with Differential-Breaking Phase Offset Achieving 23.7% PAE_avg for 5G Base-Station Transceivers," *ISSCC*, pp. 522-524, Feb. 2024. https://doi.org/10.1109/ISSCC49657.2024.10454406
[16] H.-C. Park et al., "Single Transformer-Based Compact Doherty Power Amplifiers for 5G RF Phased-Array ICs," *IEEE JSSC*, vol. 57, no. 5, pp. 1267-1279, May 2022. https://doi.org/10.1109/JSSC.2022.3148044
[17] H. Yu et al., "A Blocker-Tolerant mm-Wave Low-Noise Amplifier Utilizing Doherty Active Load Modulation for Linearity Enhancement," *ISSCC*, pp. 110-112, Feb. 2025. https://doi.org/10.1109/ISSCC49661.2025.10904630
[18] X. Zhang et al., "A Millimeter-Wave Three-Way Doherty Power Amplifier for 5G NR OFDM," *IEEE JSSC*, vol. 58, no. 5, pp. 1256-1270, May 2023. https://doi.org/10.1109/JSSC.2023.3238766
[19] Z. Ma et al., "A Compact Doherty-Like Load-Modulated Coupled Amplifier (LMCA) for 5G Phased-Array Applications," *IEEE TMTT*, vol. 73, no. 6, pp. 3479-3490, June 2025. https://doi.org/10.1109/TMTT.2024.3495692
[20] T.-Y. Huang, "A 26-to-60GHz Continuous Coupler-Doherty Linear Power Amplifier for Over-An-Octave Back-Off Efficiency Enhancement," *ISSCC*, pp. 354-356, Feb. 2021. https://doi.org/10.1109/ISSCC42613.2021.9365858
[21] J. Dong et al., "An Ultra-Compact Wideband-Linearized Power Amplifier Achieving 0.24° AM-PM Distortion and Supporting 64-/256-/1024-/4096-QAM," *IEEE CICC*, pp. 1-3, Apr. 2025. https://doi.org/10.1109/CICC63670.2025.10983528
[22] W. Zeng et al., "A 19.7-to-43.8GHz Power Amplifier with Broadband Linearization Technique in 28nm Bulk CMOS," *ISSCC*, pp. 372-374, Feb. 2023. https://doi.org/10.1109/ISSCC42615.2023.10067840

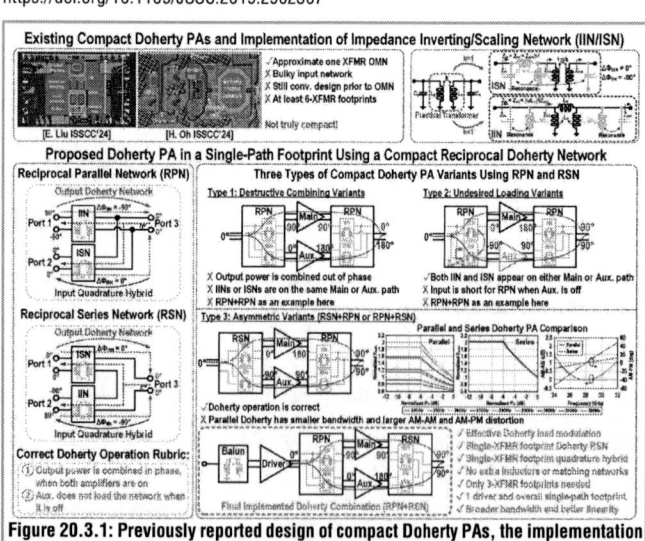

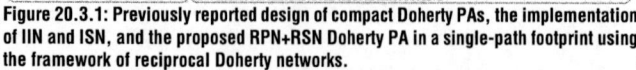

Figure 20.3.1: Previously reported design of compact Doherty PAs, the implementation of IIN and ISN, and the proposed RPN+RSN Doherty PA in a single-path footprint using the framework of reciprocal Doherty networks.

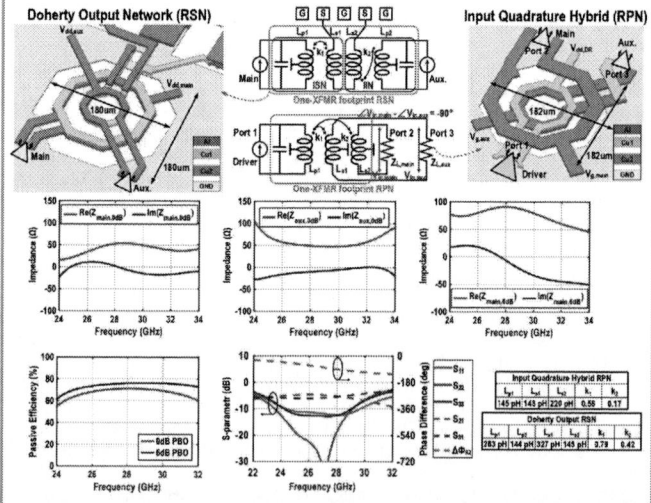

Figure 20.3.2: 3D models of the single-transformer-footprint Doherty output network using RSN and the input quadrature hybrid using RPN with their EM simulation results.

ISSCC 2026 / SESSION 20 / RF TRANSCEIVER SUBSYSTEMS FROM cm-WAVE TO THz / 20.3

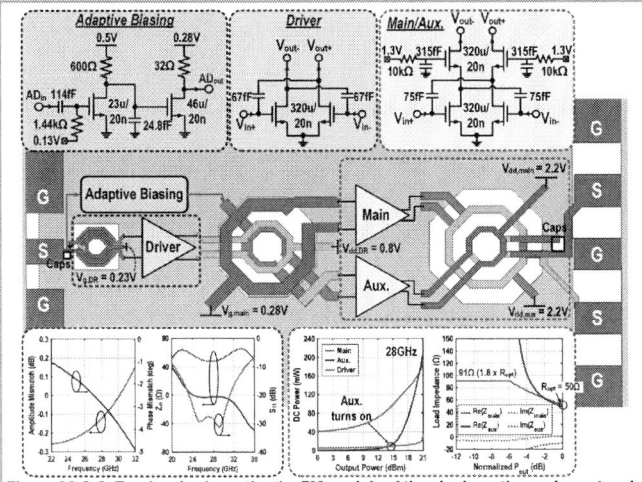

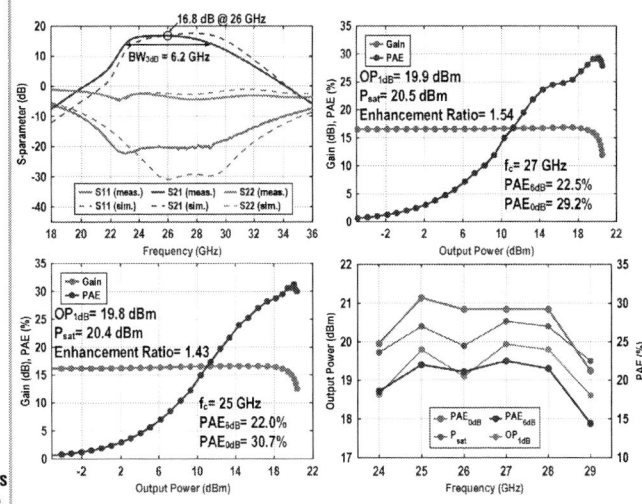

Figure 20.3.3: Top-level schematic, the EM models of the single-path passive networks and the EM simulation results of the input balun matching and balance performance, the achieved Doherty load modulation, and the simulated DC power versus P_{out} at 28GHz.

Figure 20.3.4: CW measurement results: small-signal S-parameters, large-signal power sweep at 25GHz and 27GHz, and large-signal performance across frequencies.

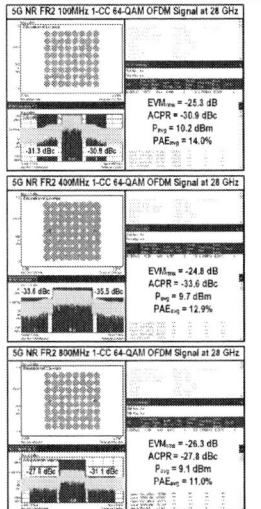

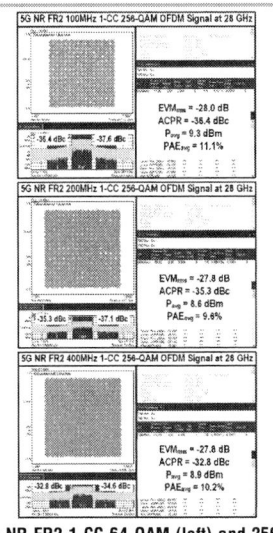

Figure 20.3.5: Modulation test results: 5G NR FR2 1-CC 64-QAM (left) and 256-QAM (right) OFDM signals over bandwidths from 100 to 800 MHz.

Figure 20.3.6: Performance summary and comparison with prior-art compact Doherty PAs, conventional Doherty, load modulated PAs, and linear PAs at similar frequencies.

Ref.	This work		Compact Doherty PAs				Regular Doherty and Load Modulated PAs			Linear mm-Wave PAs	
			J. Lee ISSCC'25	E. Liu ISSCC'24	H.Oh ISSCC'24	H. Park JSSC'22	Z. Ma TMTT'25	Z. Zong TMTT'21	T. Huang ISSCC'21	J. Dong CICC'25	W. Zeng ISSCC'23
Technology	22nm SOI		22nm SOI	45nm SOI	45nm SOI	28nm CMOS	65nm CMOS	22nm SOI	45nm SOI	65nm CMOS	28nm CMOS
Architectures	Compact Doherty with Reciprocal Parallel and Series Networks		Stacked Doherty	Parallel Doherty	Parallel Doherty	Parallel Doherty	Load Modulated Coupled Amplifier	Stacked Doherty	Continuous Coupler Doherty	Tunable Inductor Linearizer	Two-stage PA with Feedback Linearizer
BW$_{3dB}$ (GHz)	23 to 29.2 (18.9%)		24 to 29 (18.9%)	26 to 33 (23.7%)	24.3 to 29.5 (19.3%)	24.5 to 28.5* (18.5%)	24 to 29* (18.9%)	26 to 30 (14.3%)	26 to 60***	21 to 28.8 (31.3%)	19.7 to 43.8 (78%)
Gain (dB)	16.8		20.7	18.7	23.3	16.5	15	26.1	11	18.2	20.5
Freq. (GHz)	25	27	27	28	28	27	27	28	28	28	28
OP$_{1dB}$ (dBm)	19.8	19.9	21.5*	20.4	19.7*	17.5	14.5	21.1	18*	16.7	19
P$_{sat}$ (dBm)	20.4	20.5	23.5*	21	20.8*	18.8	15.2	22.5	20.8	18.5	20.3
PAE$_{6dB}$ (%)	30.7	29.2	22.5*/24.2**	36	33*	30.1**	28	26.6	23.3***	24.1	29.7
PAE$_{0dB}$ (%)	22	22.5	13**/17**	28.3	22.9*	22**	21.6	22.1	13.4***	7*	12*
# of Inductor Footprint	3		7	6	7	4	5	11	7	3	4
Core Area (mm²)	0.12		0.131	0.154	0.14	0.16	0.18/0.14##	0.2	0.62	0.06	0.108
Linearization Techniques	No		No	-	-	-	No	No	No	Yes	Yes
Adaptive Biasing	Yes		No	Yes	No	No	No	Yes	Yes	No	No
Modulation Scheme	5G NR FR2 1-CC 64-QAM OFDM	5G NR FR2 1-CC 256-QAM OFDM	64-QAM OFDM	64-256-QAM OFDM	64-256-QAM OFDM	64-QAM OFDM	64-QAM	64-256-QAM	64-QAM OFDM	Single-carrier 64-QAM	64-QAM OFDM
Carrier Freq. (GHz)	28		28	28	28	28	27	28	32.5	28	28
Modulation BW (Hz)	100M	400M	800M	100M	200M	400M	8x130M	200M	100M/8x100M	100M/8x190M	500M	400M/100M	200M	800M	200M
P$_{avg}$ (dBm)	10.2	9.7	9.1	9.3	8.6	8.9	16.2*	11.1/8.4	12.4/12.9	12.4/11.4	6.18	10.9/10.2	9.6	13.9	10.9
Power Backoff (dB)###	9.6	10.1	10.7	10.5	11.2	10.9	3.6*	9.3/12	6.6* @ 28GHz	5.1/6.1	9.32	10.2/10.9	12	2.8	8.1
PAE$_{avg}$ (%)	14.0	12.9	11	11.1	9.6	8.9	15.8*	18.4/12.9	18.3*/20.8	20.2/16.1	14.6	9.29	15.5	14.8	8.8
EVM$_{dB}$ (dB)	-25.3	-24.9	-26.3	-28	-27.8	-27.8	<-25	-25/-36	-25/-26.8	-25	-25.1	-25.1/-30	-25.4	-25.1	-25.1

* Graphically Estimated
** PAE$_{6dB/sat}$ is referred from P_{sat} rather than OP$_{1dB}$
*** Large-signal sweeping frequency range, not S$_{21}$ BW$_{3dB}$

\# Drain efficiency of the last stage only
\## Exclude the output balun area
\### Power Backoff = OP$_{1dB}$ − P$_{avg}$

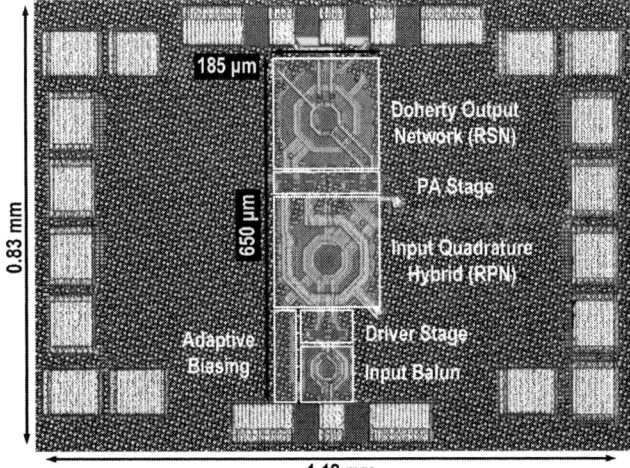

Figure 20.3.7: Die micrograph with a core area of only 0.12mm².

• 2026 IEEE International Solid-State Circuits Conference

ISSCC 2026 / SESSION 20 / RF TRANSCEIVER SUBSYSTEMS FROM cm-WAVE TO THz / 20.4

20.4 An Ultra-Compact, Inherently Resilient FR3 Source-Follower Power Amplifier with 4:1 VSWR Resilience and RIMD Immunity for Large-Scale SATCOM and 6G Phased Arrays

Mohamed Eleraky, Jianping Zeng, Hua Wang

ETH Zurich, Zurich, Switzerland

Abstract

This paper presents a source-follower power amplifier (PA) designed to combat high VSWR and reverse intermodulation distortion (RIMD) in phased arrays. Its inherently low output impedance enables conjugate matching and robust resilience to load variations, mitigating degradation from antenna mismatches up to 4:1 VSWR. The design maintains strong large-signal performance without compromise, proving its suitability for compact, calibration-free integration in large-scale SATCOM and 6G systems.

The fast-paced growth of satellite-communication (SATCOM) systems, together with the transition toward sixth-generation (6G) wireless connectivity, is driving an increasing demand for large-scale phased-array architectures, particularly those operating in the FR3 band (7 to 24GHz) [1,2]. In such dense arrays, one of the primary challenges is the impedance mismatch caused by mutual coupling between closely spaced antennas. These impedance variations depend on element positioning and are exacerbated by wide bandwidths and wide scan angles, causing voltage standing wave ratio (VSWR) to reach as high as 4:1. Addressing these conditions requires compact PAs that provide moderate output power, which is enhanced by the array gain, while most critically maintaining resilience to severe VSWR variations. Figure 20.4.1 shows the simulated impedance variation of a 4×4 patch antenna array at frequencies of f_c and $1.05 \times f_c$ and for scan angles of $\theta = 0°$, 40°, and 80°. As depicted, the impedance fluctuates within a region corresponding to a 4:1 VSWR at the PA stage. Such elevated VSWR levels significantly degrade PA performance by impairing key metrics, including saturated output power (P_{sat}), power-added efficiency (PAE), and power-gain (PG) flatness. Moreover, the linearity of the PA is adversely affected by signal coupling from nearby antennas. When this coupled signal is injected into the PA output, it mixes with the fundamental frequency, generating reverse-intermodulation (RIMD) products that degrade performance, a mechanism analogous to third-order intermodulation (IMD3).

Several strategies have been proposed to mitigate high VSWR effects in PAs. Conventional non-reciprocal devices, such as circulators and isolators, decouple a PA from its load, but their bulk, insertion loss, and the incompatibility of their ferrite materials with CMOS preclude on-chip integration; even recent CMOS-compatible implementations suffer from prohibitive area overhead [3]. Alternatively, reconfigurable PA architectures and tunable matching networks dynamically adjust the PA load line to accommodate varying VSWR conditions [4]. Although these techniques can recover performance, they introduce substantial complexity, additional loss, and require sophisticated calibration algorithms, which limit their practicality in large-scale phased array systems. Contrarily, balanced PAs (BPAs), employs 90° couplers to cancel reflected waves [5,6]. This technique provides only partial mismatch compensation, as the antenna reflection coefficient is mapped with opposing phases onto the constituent amplifiers, causing one device to operate into a low impedance (higher power, lower PAE) while the other sees a high impedance (lower power, higher PAE). This results in only partial isolation, which, coupled with the BPA inherent insertion loss and large footprint, renders the architecture impractical for dense array configurations.

Alternatively, designing PAs with a low output reflection coefficient (S_{22}) has been shown to improve resilience against load variations [7]. As shown in the signal-flow graph of Fig. 20.4.1(middle), the power delivered to the antenna is inversely proportional to $|1-S_{22}\Gamma_{ant}|$. Also, an antenna mismatch ($\Gamma_{ant} \neq 0$) not only degrades delivered power but also introduces phase distortion, which in turn alters the beam pattern. Moreover, as established in [8], the severity of RIMD is directly proportional to the PA output matching. Therefore, minimizing S_{22} is critical to mitigate power degradation, RIMD, and phase distortion under varying load conditions.

Achieving a low S_{22} in conventional PAs is challenging, especially in load-pull optimized designs where S_{22} is typically suboptimal, a difficulty that is exacerbated at frequencies far below the transistor f_t and f_{max}. As illustrated in Fig. 20.4.1, several techniques can be employed to decrease the output impedance (R_{out}). For instance, applying RC negative feedback from the output to the input reduces R_{out} but introduces stability concerns and degrades large-signal performance due to the lossy feedback path. Cascode drain-source neutralization can also decrease R_{out} by increasing the neutralization capacitance (C_N) [9]; however, the requisite larger C_N increases the output capacitance and the output quality factor (Q_L), which elevates the loss of the matching network and consequently degrades large-signal performance.

Alternatively, using the cascode LC-neutralization technique, which is effective at high mm-wave frequencies [7], is unsuitable for lower frequency bands, where the output impedance of even a common-source (CS) topology remains elevated. To address this challenge, this work proposes employing a source-follower architecture in the final stage [10,11]. The low output impedance of the source follower significantly improves S_{22} over the targeted frequency range, enhancing the PA resilience under high-VSWR conditions without compromising efficiency or output power.

Figure 20.4.2 shows the schematic of the proposed two-stage PA. To achieve good S_{22} and ensure VSWR resilience, a source follower is used as the final stage, while the first stage employs CS with capacitive neutralization. To stabilize the source-follower PA, C_{gs} neutralization is employed to boost power gain and ensure differential stability. While the source follower provides effective power amplification, its voltage gain is constrained to be less than or equal to 1, much like the CS transistor in a cascode PA. The output power and PAE load-pull contours are simulated for the source follower PA, as shown in Fig. 20.4.2. The large coverage of these contours on the Smith chart proves its resilience against the load mismatches. Transformers are utilized for stage-to-stage matching. However, the optimization of the output matching network is hindered by parasitic ground and inter-winding capacitances, which are not captured in ideal transformer models. A more accurate coupled-line-based balun model is employed to predict performance. For Ku-band operation, which requires a multi-coil structure, the model is implemented by introducing a cut at the midpoint of the second coil (Fig. 20.4.2). A 3D view of the output matching network and the corresponding model is shown, along with the simulated passive efficiency, indicating an insertion loss of 0.8dB at 12GHz.

As a proof of concept, a two-stage PA is implemented using a GlobalFoundries (GF) 22nm CMOS SOI technology. In large-scale phased arrays, the cumulative current from hundreds of PAs induces significant IR drop across the power delivery network. This voltage drop is a critical bottleneck with the low ~0.8V supplies of advanced nodes, while local LDO regulation is precluded by efficiency penalties. This work utilizes the GF 22nm Extended Drain P-well (EDP-well) transistor, which supports a higher 1.2V supply [12]. The low knee voltage of this device further enhances its suitability for an efficient PA design.

The die micrograph of the proposed PA, shown in Fig. 20.4.7, reveals a compact core area of 0.166mm². The PA achieves a peak small-signal gain of 26.22dB at 12GHz, with a 3dB bandwidth ranging from 11.5 to 13.8GHz (Fig. 20.4.3). Additionally, the PA demonstrates an S_{22} value better than -10dB from 11.5 to 25GHz. The stability factors K_f, μ, and μ_p, consistently exceed 1 across the entire frequency range. The large-signal performance at 12.25GHz achieves a P_{sat} of 16.9dBm, an OP1dB of 15.2dBm, and a PAE of 41.2% (Fig. 20.4.3). Furthermore, the large-signal performance across the frequency range is illustrated in Fig. 20.4.3. The PA shows peak P_{sat}, OP1dB, and PAE of 17.8dBm, 17.45dBm, 41.2%, respectively. The PA maintains a P_{sat} greater than 14.7dBm from 11.25 to 14.5GHz, with the OP1dB exceeding 12.3dBm in this range, and peak efficiency surpassing 27.5%.

The RIMD measurement configuration is shown in Fig. 20.4.3. The PA was operated at its OP1dB using a 12.5GHz tone (Fig. 20.4.3). To emulate antenna coupling, an interfering signal, generated by a power signal generator at a power level 4 to 6dB below the PA output and a frequency of 12.6GHz (100MHz spacing), was fed into port 1 of the circulator. Port 3 of the circulator was connected to a power meter to monitor the PA output power. The injected signal was monitored via the isolated port of the coupler, while the resulting fundamental tone and the generated RIMD3 products were measured at the coupled port using a spectrum analyzer. At 12.5GHz, the upper RIMD3 was below -48.25dBc and the lower RIMD3 was below -27.5dBc. The RIMD3 performance across the 12.5 to 14GHz frequency range is also reported.

To evaluate the PA performance under various VSWR conditions, a Maury load tuner is employed to accurately replicate antenna-impedance mismatches. The load tuner is calibrated for both frequency and angular variations (VSWR angle) using the PNA-X and the power meter. The PA is tested under the VSWR conditions of 2:1, 3:1, and 4:1 across a range of VSWR angles from 0° to 315° in 45° increments. The large-signal performance of the PA is compared to the 50Ω reference condition, and the PG deviation, P_{sat} deviation, and PAE deviation are reported in Fig. 20.4.4. The worst-case power gain deviation from the 50Ω matched case for the VSWR conditions of 2:1, 3:1, and 4:1 is less than 1.3/2/2.25dB, respectively, across the 12.5 to 14.5GHz frequency range. Meanwhile, the OP1dB deviation remains below 1.94/2.97/3.5dB, and the deviation in the peak PAE is below 5.3%, 7.8%, and 11.7% from the matched case.

346 • 2026 IEEE International Solid-State Circuits Conference

979-8-3315-8937-0/26 $31.00 © 2026 IEEE

The PA VSWR resilience was further tested using a 64-QAM signal with a peak-to-average power ratio (PAPR) of 8.5dB at the carrier frequencies of 12.5, 13.5, and 14.5GHz, and at the VSWR values of 3:1 and 4:1. The symbol rate was 100MHz, limited by the bandwidth of the load tuner, as shown in Fig. 20.4.5. While maintaining the error vector magnitude (EVM$_{rms}$) within a range of -24.5 and -27dB, the PA achieves an average output power (P$_{avg}$) greater than 10/10.5/8.5dBm and an average PAE (PAE$_{avg}$) exceeding 18.5/14.7/11%, respectively, at 12.5/13.5/14.5GHz. While maintaining a constant EVM$_{rms}$, the PA power gain deviated from its matched-case performance by less than 2.5dB when subjected to 3:1 and 4:1 VSWR load mismatches across the 12.5 to 14.5GHz range. The proposed PA demonstrates competitive performance compared to both VSWR and non-VSWR-resilient designs (Fig. 20.4.6).

The presented PA employs a source-follower output stage to simultaneously satisfy load-pull and conjugate-match conditions in the lower mm-wave range, providing inherent tolerance to VSWR up to 4:1 and suppressing RIMD without compromising large-signal performance. The PA features an ultra-compact footprint, the highest reported power density among VSWR-resilient PAs in Fig. 20.4.6, and it requires no reconfigurability or calibration, making it well suited for large-scale phased-array systems.

Acknowledgement:
The authors would like to thank the GlobalFoundries University Program for chip fabrication. This work was in part sponsored by the Swiss State Secretariat for Education, Research, and Innovation (SERI) under the SwissChips initiative, an ETH Zurich grant, the HORIZON-JU-SNS-2023 "6G-REFERENCE" project under Project 101139155, and HORIZON-JU-SNS-2023 "X-TREME 6G" project under Project 101192681.

References:
[1] N. Rostomyan et al., "15 GHz Doherty Power Amplifier with RF Predistortion Linearizer in CMOS SOI," *IEEE TMTT*, vol. 66, no. 3, pp. 1339-1348, Mar. 2018. https://doi.org/10.1109/TMTT.2017.2772785
[2] M. Ghorbanpoor et al., "12 GHz Stacked Power Amplifier with 22.9 dBm Psat and 44.9% PAE$_{sat}$ for 6G in 22nm FDSOI," *EuMIC*, pp. 62-65, Sep. 2024. https://doi.org/10.23919/EuMIC61603.2024.10732645
[3] M. Pashaeifar et al., "A Millimeter-Wave Power Amplifier with an Integrated CMOS Isolator/Circulator/Receiver," *IEEE JSSC*, Early Acces, 2025. https://doi.org/10.1109/JSSC.2025.3570656
[4] N. S. Mannem et al., "A Reconfigurable Series/Parallel Quadrature-Coupler-Based Doherty PA in CMOS SOI with VSWR Resilient Linearity and Back-Off PAE for 5G MIMO Arrays," *ISSCC*, pp. 364-366, Feb. 2020. https://doi.org/10.1109/ISSCC19947.2020.9062944
[5] M. Pashaeifar et al., "A 25.2dBm P$_{SAT}$, 35-to-43GHz VSWR-Resilient Chain-Weaver Eight-Way Balanced PA with an Embedded Impedance/Power Sensor," *ISSCC*, pp. 532-534, Feb. 2024. https://doi.org/10.1109/ISSCC49657.2024.10454427
[6] D. You et al., "A Small-Satellite-Mounted 256-Element Ka-Band CMOS Phased-Array Transmitter Achieving 63.8dBm EIRP Under 26.6W Power Consumption Using Single/Dual Circular Polarization Active Coupler," *ISSCC*, pp. 298-300, Feb.. 2023. https://doi.org/10.1109/ISSCC42615.2023.10067451

[7] M. Eleraky et al., "An Ultra-Compact Wideband Load-Insensitive Complex-Cascode LC-Neutralized Power Amplifier for 4:1-VSWR-Resilient Operations in Large-Scale Phased Arrays," *ISSCC*, pp. 98-99, Feb. 2025. https://doi.org/10.1109/ISSCC49661.2025.10904694
[8] A. N. Atanasov et al., "Reverse Intermodulation in Multi-Tone Array Transmitters," *IEEE BCICTS*, pp. 1-4, Nov. 2020. https://doi.org/10.1109/BCICTS48439.2020.9392972
[9] S. V. Thyagarajan et al., "A 60 GHz Drain-Source Neutralized Wideband Linear Power Amplifier in 28 nm CMOS," *IEEE TCAS-I*, vol. 61, no. 8, pp. 2253-2262, Aug. 2014. https://doi.org/10.1109/TCSI.2014.2333682
[10] Y. Chang et al., "A K-Band High-OP1dB Common-Drain Power Amplifier with Neutralization Technique in 90-nm CMOS Technology," *IEEE MWCL*, vol. 29, no. 12, pp. 795-797, Dec. 2019. https://doi.org/10.1109/LMWC.2019.2947247
[11] A. F. Aref et al., "Class-0: A highly Linear Class of Power Amplifiers in 0.13µm CMOS for WCDMA/LTE Applications," *ISSCC*, pp. 40-41, Feb. 2015. https://doi.org/10.1109/ISSCC.2015.7062915
[12] J. Xu et al., "A Compact Doubly Neutralized Ku-Band Power Amplifier With 39% Peak PAE and 23-dBm Output Power in 22FDX+ EDMOS for 6G FR3," *IEEE MWTL*, vol. 35, no. 6, pp. 856-859, June 2025. https://doi.org/10.1109/LMWT.2025.3566279
[13] G. Diverrez et al., "A 22-44 GHz 28nm FD-SOI CMOS 5G Doherty Power Amplifier with Wideband PAE$_{6dBPBO}$ Enhancement and 3:1 VSWR Resiliency," *IEEE RFIC*, pp. 131-134, June 2024. https://doi.org/10.1109/RFIC61187.2024.10600014
[14] N. Rostomyan et al., "15 GHz Doherty Power Amplifier with RF Predistortion Linearizer in CMOS SOI," *IEEE TMTT*, vol. 66, no. 3, pp. 1339-1348, Mar. 2018. https://doi.org/10.1109/TMTT.2017.2772785
[15] N. Rostomyan et al., "15 GHz 25 dBm Multigate-Cell Stacked CMOS Power Amplifier with 32 % PAE and ≥ 30 dB Gain for 5G Applications," *EuMIC*, pp. 265-268, Sep. 2016. https://doi.org/10.1109/EuMIC.2016.7777541

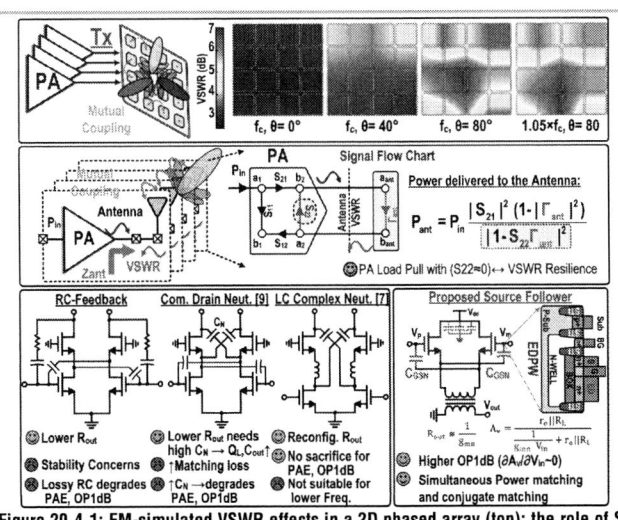

Figure 20.4.1: EM-simulated VSWR effects in a 2D phased array (top); the role of S$_{22}$ (middle); and a comparison of the conventional low-impedance solutions with the proposed PA.

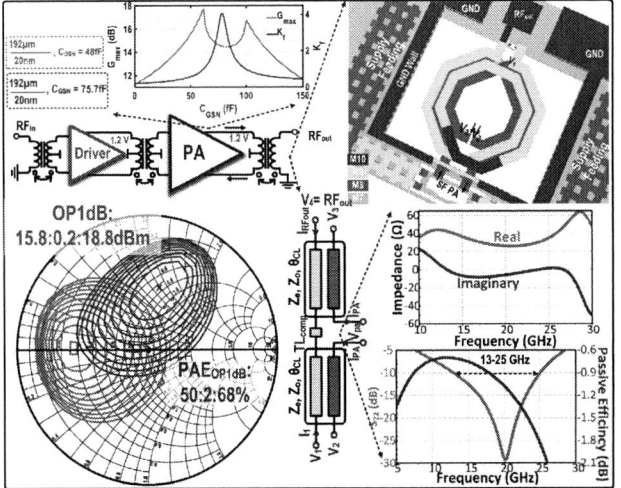

Figure 20.4.2: PA and driver schematic (top); load-pull of the PA showing VSWR resilience and the output-matching coupled-line model with its wideband S$_{22}$ (< -10dB) and low loss (bottom).

ISSCC 2026 / SESSION 20 / RF TRANSCEIVER SUBSYSTEMS FROM cm-WAVE TO THz / 20.4

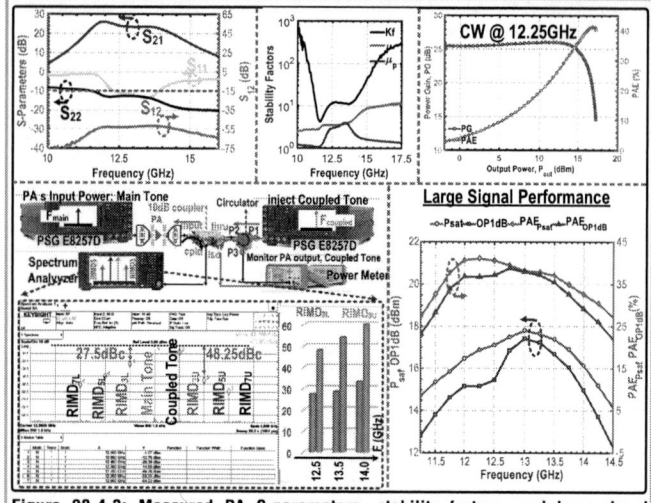

Figure 20.4.3: Measured PA S-parameters, stability factors, and large-signal performance at 12.25GHz (top), and across the frequency (bottom right); and the RIMD conceptual measurement setup and performance across frequency (bottom left).

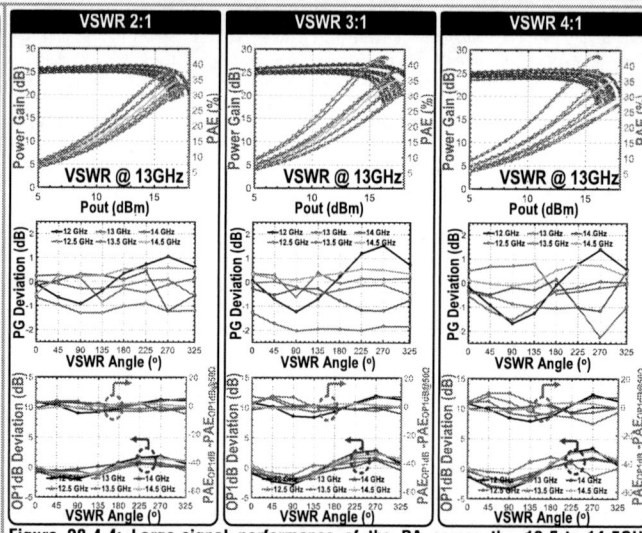

Figure 20.4.4: Large-signal performance of the PA across the 12.5-to-14.5GHz frequency range under three VSWR conditions: 2:1 (left), 3:1 (center), and 4:1 (right).

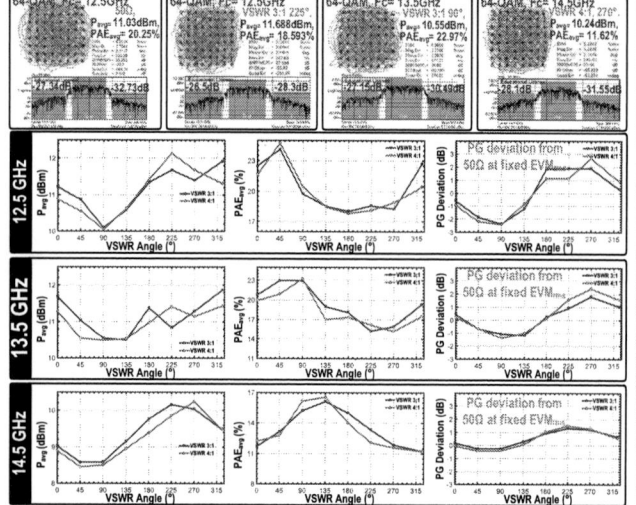

Figure 20.4.5: Modulation performance summary at the carrier frequencies of 12.5, 13.5, and 14.5GHz under the VSWR conditions of 3:1 and 4:1.

Figure 20.4.6: Performance summary of the proposed source-follower PA and a comparison with prior-art VSWR-resilient and non-VSWR-resilient PAs.

	VSWR Resilience PA					Non VSWR Resilience PA		
	This work	M. Elkeraky ISSCC 2025 [7]	G. Diverrez RFIC 2024 [13]	M. Pashaeifar ISSCC 2024 [5]	N.M. Sasikanth ISSCC 2020 [4]	N. Rostomyan TMTT 2018 [14]	N. Rostomyan EuMC 2016 [15]	M. Ghorbanpoor EuMC 2024 [2]
Technology	22nm SOI	22nm SOI	28nm SOI	28nm SOI	45nm SOI	45nm SOI	45nm SOI	22nm SOI
PA Architecture	Source-Follower	Complex LC Cascode	Inductive Doherty	Chain Weaver 8-Way Balanced	Reconfigurable Series/ Parallel Doherty	2-Stage Doherty 4-Stack	3-Stage 4-Stack	1-Stage 3-Stack
VDD (V)	1.2	1.6*/1.96	2	2	2	4.8	5	2.4
Frequency (GHz)	11.5 to 13.6	37 to 43	22 to 42	34 to 44	38.5 to 47	15	12 to 15	12
Gain (dB)	26.2	22.7	22	29.9	12.4	23	30	25 17
S₂₂ (dB)	< -19 (11.5 to 25)	<-10 (32 to 62)	<-10	<-20	<-9	<-10 (13 to 15)	<-10 (13.5 to 20)	>-2
Core Area (mm²)	0.166	0.093	0.82	2.08	1.2	1.0		0.195
OP₁dB (dBm)	17.44	17.3	19.8	22.7	20.2	23	24.5	22.1
Psat (dBm)	17.81	17.6	25.2	25.2	20.8	25.7	25.1	22.9
PAEOP1dB (%)	38.8	37	34*	N.R.	32.2	28*	31	42.2
Peak PAE (%)	41.15	37.5	34.4	16.2	33.2	31 20	32.4	44.8
Power Density (W/mm²)	0.36	0.60	0.13	0.16	0.10	0.75	0.58*	1.00
VSWR =	(2:1) / (3:1) / (4:1)	(2:1) / (3:1) / (4:1)	(3:1)*	(1.5:1) / (3:1)	3:1	N.A.	N.A.	N.A.
VSWR Freq. Range (GHz)	12 to 14.5	37 to 43	24 to 30	37 to 40	39	N.A.	N.A.	N.A.
Gain Deviation (dB)	1.3 / 2.7 / 2.88	1.2 / 1.5 / 1.5	N.R.	0.7	1	N.A.	N.A.	N.A.
OP1dB Deviation (dBm)	1.94 / 2.97 / 3.5	1.6 / 2.4 / 3	2.1	0.8	1.7	N.A.	N.A.	N.A.
PAEOP1dB (%)	5.3 / 7.8 / 11.2	7/10/11	N.R.	N.R.	N.R.	N.A.	N.A.	N.A.
Modulation Scheme	50Ω, VSWR 3:1, and 4:1	VSWR 3:1 (4:1)	50Ω	50Ω	VSWR = 3:1	50Ω	50Ω	50Ω
	Single Carrier 64-QAM (6.5dB PAPR)	Single Carrier 64-QAM	NA	OFDM 64-QAM	Single Carrier 64-QAM	Single Carrier 64-QAM	Single Carrier 64-QAM	Single Carrier 64-QAM
Frequency (GHz)	12.5 13.5 14.5	39 to 43	30	39	39	15		12
Data Rate (MHz)	100	100	200	2000	100	200		400
EVMrms (dB)	-27 to -24.5	< -24.6	-22.3	-25.00	< -22	-25.1	N.R.	-25
Pavg (dBm)	10→-12.2 10.5→-10.9 8.3→-9.2	7→-9.8 8.5→-6)	8.3	15.80	>11.2	16.4	N.R.	14.9
PAEavg (%)	18.5 -24 14.7 -23 11 -16.8	9.7→-17 (9→-13)	12.3	3.3	>9.6	15.2	N.R.	20.0

*No VSWR angles shown, *Estimated From reported figures, N.R. Not Reported, N.A. Not Applicable

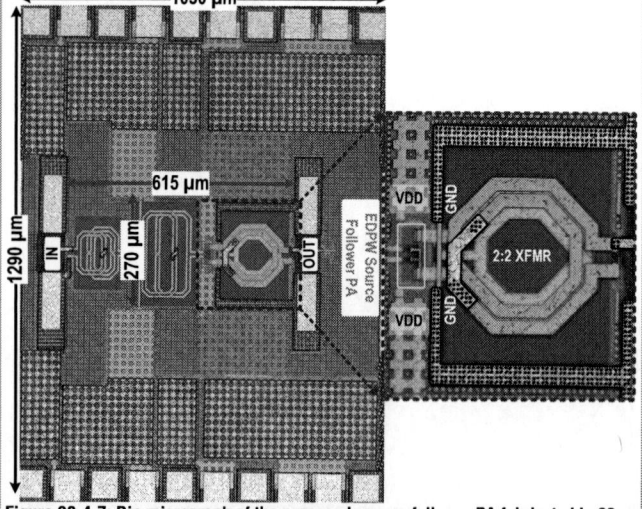

Figure 20.4.7: Die micrograph of the proposed source-follower PA fabricated in 22nm CMOS.

- 2026 IEEE International Solid-State Circuits Conference

979-8-3315-8937-0/26 $31.00 © 2026 IEEE

ISSCC 2026 / SESSION 20 / RF TRANSCEIVER SUBSYSTEMS FROM cm-WAVE TO THz / 20.5

20.5 A 24-to-27.5GHz Self-Adaptive Load-Modulated Balanced Amplifier for Integrated Communication, Sensing, and Power Transfer Scenarios

Luqi Yu*[1], Peng Chen*[1], Yucheng Yu[1], Gaojing Zhang[1], Xiaoyu Lu[1], Xiang Zhang[2], Jixin Chen[1], Xiaowei Zhu[1], Chao Yu[1], Wei Hong[1]

[1]Southeast University, Nanjing, China, [2]China Academy of Information and Communications Technology, Beijing, China
*Equally Credited Authors (ECAs)

Abstract

A 24-to-27.5GHz self-adaptive load-modulated balanced amplifier (SALMBA) is proposed for 6G integrated communication, sensing, and power-transfer scenarios. Based on the proposed self-adaptive power divider, the power divide ratio can dynamically change according to the input power levels, and the back-off PAE and saturated performance of the SALMBA can be significantly improved. The SALMBA achieves the saturated PAE of 28.5 to 33%, and 10dB OPBO PAE of 15.3 to 22.1% over 24 to 27.5GHz.

With the emerging sixth-generation (6G) millimeter wave (mm-wave) wireless communication, multi-functional convergence scenarios, such as vehicle-to-everything (V2X) and Internet of Things (IoT), are expected to deliver smarter and more efficient living environments. Consequently, future multifunctional front-ends will extend beyond mere communication capabilities to support sensing and power-transfer functions. There is no doubt that the design of integrated communication, sensing, and power transfer (ICSPT) can largely empower system functionalities and significantly reduce operation costs [1-4]. Figure 20.5.1 shows the multi-type signals used in the ICSPT scenarios, such as a modulated signal with a high peak-to-average power ratio (PAPR) for the communication scenario, which requires the front-ends to support high output-power back-off (OPBO) efficiency. Typically, a wideband frequency-modulated continuous-wave (FMCW) signal is used for the sensing scenario, while a single-tone signal is used for the power-transfer scenario, which both rely on high saturated power, efficiency, and gain. As the most energy-consuming devices in the transmitter, power amplifiers (PAs) are required to operate efficiently in both at OPBO and saturation states for ICSPT applications.

Multi-way PAs, such as balanced PAs (BPAs), Doherty PAs (DPAs) [5], and load-modulated balanced amplifiers (LMBAs) [6], are commonly adopted for high-efficiency PA designs. Generally, PAs used in communication systems [7-18] are typically optimized for OPBO performance, while others applied to mm-wave sensing [19,20] or power transfer are mainly targeted for saturated performance. As is well-investigated, a BPA suffers from low OPBO efficiency, whereas a DPA is limited in bandwidth [13-18]. In contrast to a BPA and a DPA, an LMBA offers the advantages of enhanced efficiency, high output power, and bandwidth extension. In view of this, it is of great interest and significance to propose an LMBA for the ICSPT scenarios that can achieve high OPBO performance for communication and high saturation performance for sensing and power transfer. In the conventional symmetric LMBA architecture, a Class-AB control amplifier (CA) saturates first to provide high OPBO efficiency, and two Class-C balanced amplifiers (BAs) turn on to provide a large OPBO range. As shown in Fig. 20.5.1, the main drawback of a conventional symmetric LMBA is that its CA becomes overdriven between CA saturation and LMBA saturation, making operation at the overall saturation point unreliable, especially in sensing and power-transfer scenarios. Most existing approaches address this issue by employing asymmetric couplers or asymmetric BAs biases, which induces load modulation on the CA to mitigate the overdrive [11,12]. However, these asymmetric architectures deteriorate transistor saturation performance and introduce large gain compression.

In this work, we proposed a self-adaptive LMBA (SALMBA) based on a self-adaptive power divider (SAPD) for ICSPT scenarios (Fig. 20.5.1). The power divide ratio of the SAPD can dynamically change according to the input power levels. With the assistance of the SAPD, the power provided to the CA decreases after CA saturation, which alleviates the CA overdriven state and maintains high efficiency. Simultaneously, the power provided to the BAs increases after CA saturation, which not only drives the BAs to saturation more quickly for higher saturated efficiency but also alleviates gain compression. This indicates that the SAPD-based SALMBA exhibits better potential in ICSPT scenarios, providing both high OPBO and saturated performance.

The SAPD consists of a conventional Lange coupler, two diodes, and a part of matching networks, as shown in Fig. 20.5.2. Diode-1 is placed at the coupled port of the Lange coupler to behave as a limiter, while diode-2 is placed at the isolated port of the Lange coupler to alter the reflection coefficient. Before CA saturation, diode-1 remains in the off-state and exhibits high impedance, which has negligible effects on the transmission coefficient. With the use of diode-2, barely any power enters the isolated port, and the port impedance is approximately 50Ω. After CA saturation, diode-1 gradually turns on limiting additional power from entering the CA and alleviating its overdrive state. Moreover, it reflects a portion of the power to the isolated port. At the same time, diode-2 increases the reflection coefficient at the isolated port so that part of the power is reflected to the through port of the Lange coupler. In this way, the proposed SAPD can dynamically change the power divide ratio according to the input power levels. In a 0.15μm gallium nitride (GaN) process, the different sizes of diode-1 exhibit different turn-on power threshold, and smaller diodes feature lower turn-on power threshold, as shown in Fig. 20.5.2. In this work, the proposed SAPD adopts a 2×25μm diode-1 and a 2×25μm diode-2, and its dynamic S-parameters are compared to those using two different fixed power dividers. Here, the power divide ratio represents the power ratio between the CA and BA branches. Figure 20.5.2 also shows the schematic of the 3-stage monolithic microwave integrated circuit (MMIC) SALMBA. A 2×75μm device is used in the driver amplifier 1, and the gate and drain voltages of the driver amplifier 1 are biased at -1.6V and 20V, respectively. The CA driver stage and power stage adopt 2×50μm and 6×50μm devices, while the BA driver stage and power stage use 6×50μm and 6×75μm devices, respectively. The gate and drain voltages of the CA branch are biased at -1.7V and 16V, while the gate and drain voltages of the BA branches are supplied at -2.5V and 26V, respectively.

The simulated power-added efficiency (PAE) and gain of the CA and BA branches are compared when using the SAPD and two different fixed power dividers, as shown in Fig. 20.5.3. After the CA-branch saturation, with the assistance of the SAPD, the overdriven state of the CA branch can be alleviated, enabling the CA branch to maintain high efficiency with higher input power. Meanwhile, as more input power is allocated to the BA branch, the efficiency and gain of the BA branch increase more rapidly. Figure 20.5.3 also illustrates the overall simulated PAE and gain of the LMBA with different power divide ratios versus output power at 25.5GHz. The LMBA with a fixed 1:1 power divide ratio sacrifices the saturated performance for high OPBO efficiency. On the contrary, the LMBA with a fixed 1:3 power divide ratio achieves higher saturated performance at the expense of a lower OPBO efficiency. Therefore, a fixed power divide ratio is not suitable for ICSPT scenarios, while the proposed LMBA with a SAPD exhibits better performance both at OPBO and saturation. The performance comparison of the LMBA with different power divide ratios versus frequency is shown in Fig. 20.5.3. It can be concluded that the proposed SAPD can effectively enhance both the OPBO and saturation performance of the LMBA, making it more suitable for ICSPT applications.

Based on the proposed SAPD, a MMIC SALMBA is demonstrated in a 0.15μm GaN process with a die area of 2.515×4.8mm² (Fig. 20.5.7). Figure 20.5.4 shows the simulated and measured results of the SAPD, where the measured results of S_{31} and S_{41} are in good agreement with the simulated results. Figure 20.5.4 shows small-signal S-parameters and large-signal measurement results of the proposed SALMBA. The SALMBA achieves the saturated output power of 34.3 to 35.9dBm, saturated PAE of 28.5 to 33%, saturated gain of 16 to 18.3 dB, and 10dB OPBO PAE of 15.3 to 22.1% over 24 to 27.5GHz. The measured results using FMCW signals are also shown in Fig. 20.5.4. It shows the SALMBA output signal spectra under the excitation of a 500MHz FMCW signal at 24.5GHz and a 1800MHz FMCW signal at 25.5GHz, respectively. Under the 500MHz FMCW signal, the SALMBA achieves the average output power (P_{ave}) of 34.3 to 35.9dBm and average PAE (PAE_{ave}) of 27.5 to 32.7%, and the gain fluctuations are <1.5dB.

To characterize the linearity in communication scenario, the SALMBA is further tested with a 400MHz 9dB PAPR 5G new radio (NR) signal with the decomposed-vector-rotation (DVR)-based digital predistortion (DPD) [21]. The constellation and power spectrum density (PSD) at 25.5GHz are provided in Fig. 20.5.5. It achieves P_{ave} of 25.5 to 26.6dBm and PAE_{ave} of 15.5 to 23.3% from 24GHz to 27.5GHz. The error vector magnitude (EVM) of the SALMBA is improved from <-14.3dB to <-31dB with the DPD over 24 to 27.5GHz. The adjacent-channel-leakage-ratio (ACLR) levels are lower than -20dBc before applying the DPD, while they are improved to better than -39dBc after applying the DPD over 24 to 27.5GHz.

Figure 20.5.6 summarizes the performance comparisons between the proposed SALMBA and other mm-wave PAs. Compared to those PAs based on CMOS technologies, the proposed SALMBA provides a much higher output power and PAE. In contrast to others based on GaN technologies, it achieves a wider bandwidth with competitive output power and PAE. Thus, the proposed SALMBA exhibits good OPBO and saturated performance, which is a promising PA architecture for the future ICSPT scenarios.

Acknowledgement:
This work was supported by National Key R&D Program of China (2022YFB2902403). Corresponding authors: Chao Yu and Wei Hong ({chao.yu, weihong}@seu.edu.cn).

References:

[1] L. Xu et al., "A 210×340×50μm Integrated CMOS System for Micro-Robots with Energy Harvesting, Sensing, Processing, Communication and Actuation," *ISSCC*, pp. 194-195, Feb. 2022. https://doi.org/10.1109/ISSCC42614.2022.9731743

[2] L. Lou et al., "An Early Fusion Complementary RADAR-LiDAR TRX in 65nm CMOS Supporting Gear-Shifting Sub-cm Resolution for Smart Sensing and Imaging," *ISSCC*, pp. 220-222. Feb. 2021. https://doi.org/10.1109/ISSCC42613.2021.9365756

[3] Y. Yu et al., "Digital Predistortion of Millimeter-Wave GaN Power Amplifiers for 6G Integrated Communication, Sensing, and Power Transfer Scenarios," *IEEE TMTT*, vol. 73, no. 1, pp. 26-37, Jan. 2025. https://doi.org/10.1109/TMTT.2024.3452555

[4] L. Yu et al., "Optimization of High-Efficiency GaN Load Modulated Balanced Amplifier for Integrated Sensing and Communication Applications," *IEEE TCADICS*, vol. 43, no. 12, pp. 4361-4372, Dec. 2024. https://doi.org/10.1109/TCAD.2024.3416442

[5] W. H. Doherty, "A New High Efficiency Power Amplifier for Modulated Waves," *Proc. IRE*, vol. 24, no. 9, pp. 1163-1182, Sept. 1936. https://doi.org/10.1109/JRPROC.1936.228468

[6] D. J. Shepphard et al., "An Efficient Broadband Reconfigurable Power Amplifier Using Active Load Modulation," *IEEE MWCL*, vol. 26, no. 6, pp. 443-445, June 2016. https://doi.org/10.1109/LMWC.2016.2559503

[7] W. Zhu et al., "A 27.8-to-38.7GHz Load-Modulated Balanced Power Amplifier with Scalable 7-to-1 Load-Modulated Power-Combine Network Achieving 27.2dBm Output Power and 28.8%/23.2%/16.3%/11.9% Peak/6/9/12dB Back-Off Efficiency," *ISSCC*, pp. 534-536, Feb. 2024. https://doi.org/10.1109/ISSCC49657.2024.10454540.

[8] V. Qunaj and P. Reynaert, "A Doherty-Like Load-Modulated Balanced Power Amplifier Achieving 15.5dBm Average Pout and 20% Average PAE at a Data Rate of 18Gb/s in 28nm CMOS," *ISSCC*, pp. 356-358, Feb. 2021. https://doi.org/10.1109/ISSCC42613.2021.9365966

[9] C. R. Chappidi et al., "Load Modulated Balanced mm-Wave CMOS PA with Integrated Linearity Enhancement for 5G applications," *IEEE IMS*, pp. 1101-1104, Aug. 2020. https://doi.org/10.1109/IMS30576.2020.9224038

[10] H. Jia et al., "A 26-GHz GaN MMIC Load-Modulated Balanced Amplifier with Miniaturized Dual-Loop Coupler," *IEEE TMTT*, vol. 73, no. 1, pp. 530-539, Jan. 2025. https://doi.org/10.1109/TMTT.2024.3421941

[11] Z. -M. Zhao et al., "A GaN MMIC Load-Modulated Balanced Amplifier with Modified Output Coupler for Efficiency Enhancement Over a Larger Power Back-Off Range," *IEEE TCAS-II*, vol. 70, no. 9, pp. 3373-3377, Sept. 2023. https://doi.org/10.1109/TCSII.2023.3265778

[12] L. Yu et al., "A GaN MMIC Dual-Asymmetrical-Lange-Coupler-Based Load-Modulated Balanced Amplifier for Back-Off Efficiency Enhancement," *IEEE MWTL*, vol. 34, no. 8, pp. 1031-1034, Aug. 2024. https://doi.org/10.1109/LMWT.2024.3420946

[13] A. Piacibello et al., "3-Way Doherty Power Amplifiers: Design Guidelines and MMIC Implementation at 28 GHz," *IEEE TMTT*, vol. 71, no. 5, pp. 2016-2028, May 2023. https://doi.org/10.1109/TMTT.2022.3225316

[14] Z. Ma et al., "A 28GHz Compact 3-Way Transformer-Based Parallel-Series Doherty Power Amplifier With 20.4%/14.2% PAE at 6/12-dB Power Back-off and 25.5dBm P_{SAT} in 55nm Bulk CMOS," *ISSCC*, pp. 320-322, Feb. 2022. https://doi.org/10.1109/ISSCC42614.2022.9731564

[15] J. Lee et al., "A 22nm FDSOI CMOS-Based Compact 3-Stack Doherty Power Amplifier with a Stacked OPA-Based Bias Scheme Achieving >16.5dBm P_{avg} for 5G FR2 Applications," *ISSCC*, pp. 96-98, Feb. 2025. https://doi.org/10.1109/ISSCC49661.2025.10904681

[16] A. Piacibello et al., "A 3-Way GaN Doherty Power Amplifier for 28 GHz 5G FR2 Operation," *IEEE IMS*, pp. 327-330, June 2023. https://doi.org/10.1109/IMS37964.2023.10187996

[17] R. Liu et al., "A Linearity-Improved 24–29-GHz GaN MMIC Doherty Power Amplifier with Reconfigurable Self-Adaptive Peaking Gate Bias Network," *IEEE TMTT*, vol. 72, no. 12, pp. 6845-6856, Dec. 2024. https://doi.org/10.1109/TMTT.2024.3409944

[18] P. Chen et al., "Simplified Emulation of Active Load Modulation for a Millimeter-Wave GaN MMIC Doherty Power Amplifier Design," *IEEE TMTT*, vol. 72, no. 1, pp. 149-159, Jan. 2024. https://doi.org/10.1109/TMTT.2023.3284258

[19] T. Dinc et al., "High-Efficiency Class-E Power Amplifiers for mmWave Radar Sensors: Design and Implementation," *IEEE JSSC*, vol. 57, no. 5, pp. 1291-1299, May 2022. https://doi.org/10.1109/JSSC.2022.3147723

[20] S. Wang et al., "A Low-Power 23–25.5-GHz FMCW Radar Transceiver in 65-nm CMOS for AIOT Applications," *IEEE TMTT*, vol. 72, no. 4, pp. 2560-2576, April 2024. https://doi.org/10.1109/TMTT.2024.3372289

[21] A. Zhu, "Decomposed Vector Rotation-Based Behavioral Modeling for Digital Predistortion of RF Power Amplifiers," *IEEE TMTT*, vol. 63, no. 2, pp. 737-744, Feb. 2015. https://doi.org/10.1109/TMTT.2014.2387853

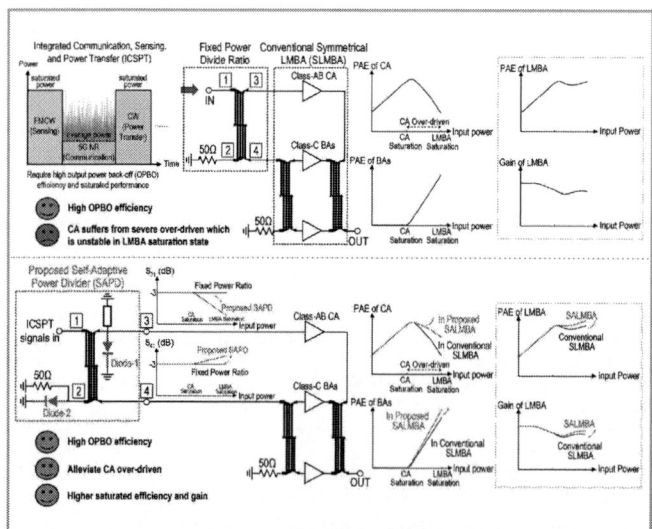

Figure 20.5.1: Conventional symmetrical LMBA and the proposed self-adaptive LMBA (SALMBA) based on self-adaptive power divider (SAPD) in ICSPT scenarios.

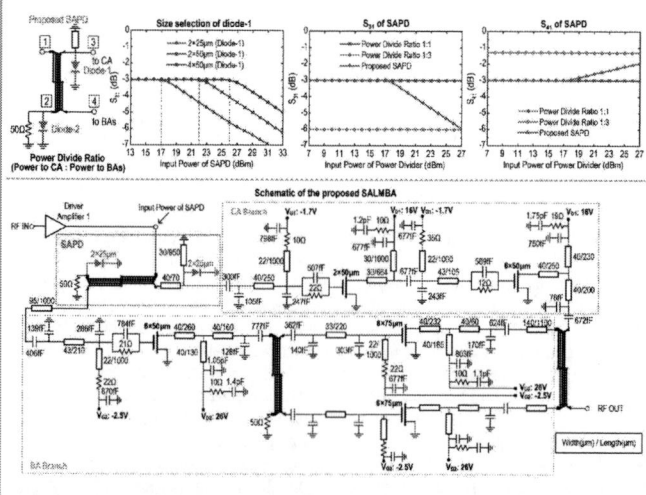

Figure 20.5.2: Size selection of the diode-1, the dynamic S-parameters of the proposed SAPD, and the schematic of the SALMBA.

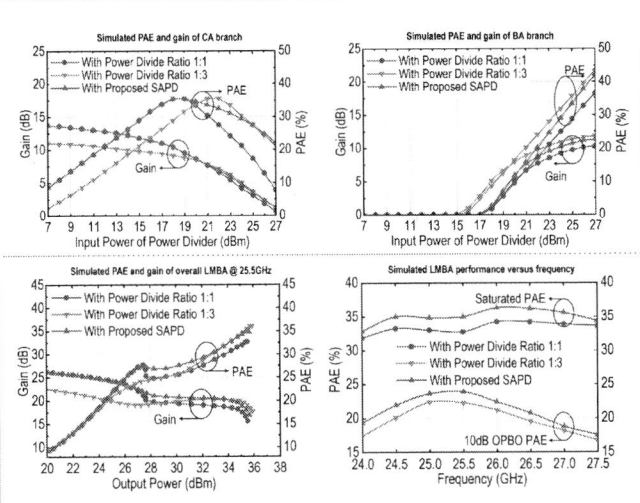

Figure 20.5.3: Simulated PAE and gain of the CA and BA branches, and the overall LMBA performance versus output power and frequency with different power dividers.

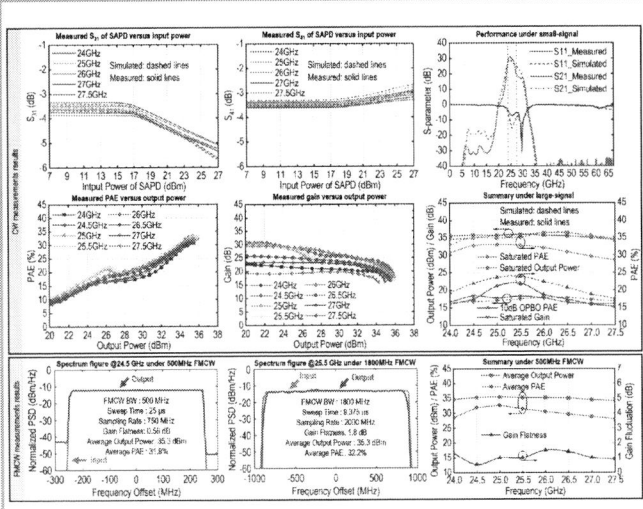

Figure 20.5.4: Measured results of the SAPD, and measured results of the SALMBA under the excitations of small-signal, large-signal CW, and FMCW.

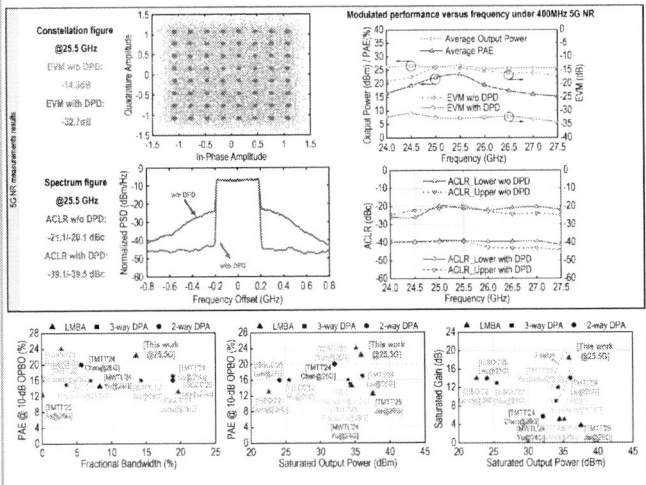

Figure 20.5.5: Measured constellation, spectrum, and summary figures under 400MHz 5G NR signal, and comparison with recent mm-wave PAs.

Figure 20.5.6: Performance summary and comparison with previously published mm-wave PAs.

	This work	ISSCC'24 Zhu [7]	ISSCC'21 Qunaj [8]	IMS'20 Chappidi [9]	TMTT'25 Jia [10]	TCASII'23 Zhao [11]	TMTT'23 Placibelo [13]	ISSCC'22 Ma [14]	ISSCC'25 Lee [15]	JSSC'22 Dinc [19]	TMTT'24 Wang [20]
Architecture	3-stage LMBA	3-stage 7-way LMBA	2-stage LMBA	2-stage LMBA	1-stage LMBA	2-stage LMBA	3-stage LMBA	2-stage 3-way DPA	2-stage 2-way DPA	2-stage class-E PA	2-stage class-AB PA
Process	0.15μm GaN	65nm CMOS	28nm CMOS	65nm bulk CMOS	0.12μm GaN	0.15μm GaN	0.15μm GaN	55nm Bulk CMOS	22nm FDSOI CMOS	0.13μm BiCMOS	65nm CMOS
Application	ICSPT	Communi-cation	Communi-cation	Communi-cation	Communi-cation	Communi-cation	Communi-cation	Communi-cation	Communi-cation	Sensing	Sensing
Bandwidth(GHz)	24.0 to 27.5	27.8 to 38.7	32 to 39	30 to 40	26	24.7 to 25.7	26 to 29	26 to 30	24 to 29	70 to 86	23.0 to 25.5
P_{sat} (dBm)	34.3 to 35.9	26.2 to 27.2	20.5 to 22.6	18.5 to 20.0	37.6	34.0 to 35.2	34.0 to 34.3	25.0 to 25.5	23.0 to 24.1	15.8 to 17.0	12.2 to 12.6†
P_{1dB} (dBm)	24.0 to 31.5	22.5 to 24.1	19.6@38G	N.A.	31	N.A.	N.A.	24.3@28G	N.A.	15.8@79G	N.A.
PAE@ P_{sat} (%)	28.5 to 33.9	25.4 to 28.8	30 to 32	30 to 38‡	21.3	20 to 25‡	20 to 22	22.0 to 25.2	22.3 to 24.7	25.0 to 30.5	23.0 to 28.5†
Gain@ P_{sat} (dB)	16.0 to 18.3	8 to 13*	10 to 13*	7 to 8*	3.7	4 to 5*	8 to 12	13*@28G	12 to 14*	10*@79G	N.A.
PAE@10-dB OPBO (%)	15.3 to 22.1	13.5 to 15.5*	11 to 14*	10*	12*	20 to 24	13 to 15	16*@28G	13*@25G	10*@79G	N.A.
Modulation Scheme	64-QAM	64-QAM	64-QAM	64-QAM	64-QAM	64-QAM	64-QAM	64-QAM	64-QAM	N.A.	N.A.
Data Rate(Gb/s)	2.4	4.5	18	1	0.6	0.6	0.24	1.5	4.8	N.A.	N.A.
Frequency(GHz)	25.5	28	36	33	25	25	28.5	28	27	N.A.	N.A.
EVM (dB)	-32.7#	-26.5	-25.1	-26.4	-33.2^	N.A.	-26	-25.2	-25	N.A.	N.A.
P_{avg} (dBm)	26.6	17.3	15.5	10.6	28	26.2	15	17.7	16.7	N.A.	N.A.
PAE_{avg} (%)	23.3	13.9	20	12.1	N.A.	25.2	8	17.5	15.5	N.A.	N.A.

*: Graphically Estimated;
#: EVM After DPD;
†: PA in Transmitter;
‡: Drain Efficiency

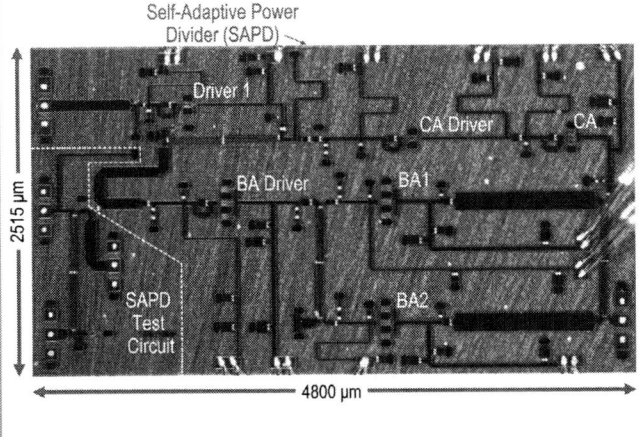

Figure 20.5.7: Die micrograph.

• 2026 IEEE International Solid-State Circuits Conference

ISSCC 2026 / SESSION 20 / RF TRANSCEIVER SUBSYSTEMS FROM cm-WAVE TO THz / 20.6

20.6 A 330-to-344GHz GaN Power Amplifier with Maximum-Available-Gain-Boosting Technique and Compact Tandem Coupler Achieving 86mW Output Power at 340GHz

Weibo Wang*[1,2], Wenhua Chen*[3], Kenan Xie*[4], Chenjie Luo[2], Yibin Zhang[2], Dechun Shang[2], Ziwei Jiang[2], Guodong Yu[2], Yan Chen[2], Keping Wang[4]

[1]National Key Laboratory of Solid-State Microwave Devices and Circuits, Nanjing, China, [2]Nanjing Electronic Device Institute, Nanjing, China, [3]Tsinghua University, Beijing, China, [4]Tianjin University, Tianjin, China
*Equally Credited Authors (ECAs)

Abstract

This paper demonstrates a 330-to-344GHz GaN power amplifier (PA), fabricated using a 35nm GaN HEMT process with a 15-stage cascaded architecture. The maximum-available-gain-boosting technique enhances the high frequency gain, while the miniaturized Tandem coupler reduces port reflection and combining losses. The PA achieves >21dB gain, >18dBm saturated output power (P_{SAT}), and >0.6% power-added efficiency. The peak P_{SAT} reaches 19.3dBm at 340GHz, with a power density exceeding 0.925W/mm.

The 340GHz frequency range, serving as a critical "atmospheric window," plays an indispensable role in systems such as high-frequency communications and radar imaging. Traditional silicon-based (CMOS and SiGe) and III-V (GaAs and InP) power amplifiers (PAs) have exhibited marginal improvements in saturated output power (P_{out}) within this frequency band [1-3]. Gallium Nitride (GaN), with its wide bandgap and high breakdown voltage, has emerged as a promising technology for enhancing the P_{out} performance of 340GHz PAs. However, realizing this potential faces significant challenges. Beyond increasing the device operating frequency, additional efforts are required to unlock the GaN potential. Although circuit design techniques like capacitance neutralization and g_m-boosting can enhance circuit gain [4-7], they typically rely on multi-layer interconnect architectures to mitigate parasitic effects in coupling loops, thereby limiting their application in III-V chips. Moreover, output-load-mismatch effects exert a disproportionately severe degradation on the performance of a 340GHz PA [8,9]. While conventional Lange couplers can theoretically provide a degree of impedance isolation [10], their transmission loss and isolation metrics are often insufficient, necessitating alternative approaches.

Figure 20.6.1 shows the circuit schematic of the presented 340GHz GaN PA. It utilizes four identical power amplifier units combined on-chip to boost the final P_{out}. The input and output networks employ two-stage cascaded quadrature power splitter/combiner. Circuit matching is implemented using a coplanar waveguide with ground (CPWG) topology. Each individual power-amplifier unit comprises a 15-stage common-source cascaded core to deliver the overall gain. To overcome the technical challenges, the following techniques are implemented to achieve an in-band P_{out} greater than 18dBm within the 330-to-344GHz band: First, a transistor maximum-available-gain-boosting (MAG-boosting) technique is proposed by introducing a microstrip line of optimized length at the device source, shifting the second-order inflection-point frequency of the HEMT MAG to around 340GHz, thus extending high-frequency gain; Second, a Tandem coupler is adopted for quadrature power splitting/combining, mitigating the load-mismatch degradation induced by non-ideal input/output port characteristics and preserving P_{out}. Additionally, an AlGaN back-barrier structure and a 35nm T-gate process are concurrently utilized to further improve the gain and frequency characteristics of the GaN HEMT devices.

Although scaling down the gate length of HEMT devices can raise operating frequency, the short-channel effects and reduced breakdown voltage degrade P_{out} and PAE. To secure high gain around 340GHz in GaN HEMTs for PAs avoiding gate length scaling, an effective strategy to boost MAG is imperative. While the double-G_{max}-core technique [6] can improve gain in CMOS technology, it collapses in compound semiconductors at 340GHz due to the extremely short microstrip line required for feedback, rendering input and output ports interconnection unfeasible. Similarly, the dual-peak G_{max}-core technique [7], though beneficial for gain enhancement in CMOS, proves impractical for compound semiconductor chips at 340GHz because of the need for fF-level capacitors and the low Q-factor of microstrip lines. As illustrated in Fig. 20.6.2 (top-right), a GaN HEMT MAG-boosting technique is realized by inserting a length-optimized microstrip line in series with the source. This relocates the second-order inflection point of the MAG, yielding a pronounced gain boost near 340GHz through deterministic microstrip-length tuning. Figure 20.6.2 (top-left) depicts the equivalent circuit of this structure. When the PA operates below the first MAG inflection frequency, the microstrip realizes feedback as conventional source-degenerating inductor. Its reactance reduces voltage division across the intrinsic capacitor C_{gs}, thereby decreasing V_{gs} and thus $g_m V_{gs}$, which degrades amplification. When operating above the first MAG inflection frequency, the source microstrip line L_s forms a resonant circuit with both the C_{gs} and the C_{gd}. As frequency increases, the input and output port impedances progressively shift towards the low-impedance region. This shift causes a greater portion of the power incident upon the input/output ports to be reflected, thereby enhancing the isolation between the output and input ports. Consequently, the MAG is significantly boosted near the resonant frequency. Figure 20.6.2 (bottom left) presents the MAG equation derived from a simplified equivalent circuit. The inflection points of the MAG, where its slope changes abruptly, can be determined by solving for the frequencies at which the stability factor K equals unity. The first (f_1) and second (f_2) inflection frequencies are distributed on either side of the resonant frequency f_0 of L_s and the combined capacitance ($C_{gs} + C_{ds}$). Figure 20.6.2 (bottom-right) compares the proposed MAG-boosting technique with those in [6] and [7] at 340GHz, demonstrating a simpler structure and superior gain enhancement.

At 340GHz, the GaN PA exhibits hypersensitivity to output-load impedance variations. Hence, a quadrature power combining architecture is adopted to desensitize the output match. While classical Lange couplers perform adequately at lower frequencies, their performance degrades significantly at high frequencies (e.g., 340GHz) due to thinned substrates and diminished ground-plane spacing, which weaken coupling efficiency. Consequently, achieving adequate balance between direct and coupled paths in power splitting/combining becomes formidable with Lange couplers at these frequencies. To address these limitations, a Tandem coupler for power combining is proposed. Tighter coupling is first established through multiple metallic cross-over air bridges, subsequently refined by reducing the distance between coupling lines and the ground plane for precise coupling control. This yields a compact design that simultaneously enhances port isolation and suppresses transmission loss compared to traditional Lange couplers. Figure 20.6.3 summarizes the simulated insertion loss, isolation, and other structural parameters of the proposed Tandem coupler. A comparison between the traditional Lange and Tandem couplers (Fig. 20.6.3, bottom-right) demonstrates the Tandem coupler superior performance at 340GHz, with an average transmission loss of approximately 0.7dB and port isolation greater than 16dB. Furthermore, Fig. 20.6.3 (top-left, top-right, bottom-left) illustrates the influence of key geometric parameters on Tandem coupler performance. To suppress undesirable internal coupling, a shielded-metal grounding technique is employed, markedly weakening adverse coupling effects.

This paper first presents a two-way power-combining cell formed by two identical amplifier units with a Tandem coupler; subsequently, another Tandem coupler coherently combines two such cells to form a four-way power-combining architecture, delivering higher P_{out} while further enhancing port isolation. To counteract the inherently low large-signal gain of 340GHz GaN HEMT devices, a CPWG structure minimizes losses in matching circuits. To ensure adequate power drive at each stage, a 1:1 drive ratio is maintained between successive stages. A 15-stage cascade amplifier using 2×20µm HEMT devices compensates for matching circuit losses, enhances gain, and reduces drive power requirements. Figure 20.6.4 provides a comparative overview of small-signal gain, P_{out}, PAE, and input/output power sweep measurements for two GaN PAs. Figure 20.6.4 (top-left) shows measured P_{out} and PAE. Under identical bias conditions and the input power (P_{in}) of 6dBm within 330 to 344GHz, the two-way combined amplifier achieves P_{out}>16dBm, with maximum in-band P_{out} of 17.5dBm at 340GHz; the four-way combined amplifier achieves P_{out}>18dBm, with maximum 19.2dBm at 340GHz. The gain difference of 2.5dB is primarily due to an additional Tandem coupler in the four-way amplifier, indicating approximately 0.75dB loss for each Tandem coupler. Maximum PAE is 0.8% for the two-way and 0.6% for the four-way amplifiers, consistent with earlier simulations. Figure 20.6.4 (bottom-left) displays the small-signal gain. Under V_{gs} = -0.5V, V_{ds} = 8V, P_{in} = -20dBm within 330 to 340GHz, the two-way gain >25.5dB and four-way gain >24dB. The 1.5dB difference, attributed the extra Tandem coupler, confirms a nearly 0.75dB loss for each coupler. Figure 20.6.4 (top-right) shows input/output power sweeps at 340GHz. Both amplifiers exhibit significant gain compression at P_{in} = -12dBm, with the two-way configuration showing 1.8dB compression and the four-way showing 1dB compression. This indicates that cascaded amplification, combined with short-channel effects in 35nm GaN HEMTs, collectively exacerbate gain compression. Figure 20.6.4 (bottom-right) presents input/output return losses (S_{11} and S_{22}), which are generally below -10dB for both amplifiers.

Figure 20.6.5 presents measured performance of the realized four-way combined GaN PA, including small-signal gain, P_{out}, PAE, and input/output power sweeps under different bias voltages. With V_{gs} fixed at -0.5V and V_{ds} at 6V, 7V, and 8V, corresponding quiescent drain currents (I_{dq}) are 1.25A, 1.33A, and 1.37A, respectively. Figure 20.6.5 (top-left) shows small-signal gain generally increases with drain voltage, with maximum improvement of about 1.8dB at 340GHz when V_{ds} increases from 6 to 8V. Figure 20.6.5 (bottom-left) indicates significant P_{out} enhancement with increasing drain voltage, reaching approximately 2dB of the maximum increase at 340GHz. Figure 20.6.5 (top-right) displays input/output power sweep at V_{ds} = 8V. When P_{in} >-10dBm, P_{out} remains >10dBm across 330 to 344GHz; for P_{in} >0dBm, P_{out} >17dBm; saturation occurs at P_{in} >6dBm. Total output loss is conservatively estimated at nearly 2.5dB, primarily from the two-stage Tandem coupler (≥1.5dB) and the output matching network of each single amplifier path (about 1dB). Thus, equivalent maximum P_{out} of the HEMT with the total gate width of 4×2×20µm is estimated

350 • 2026 IEEE International Solid-State Circuits Conference

979-8-3315-8937-0/26 $31.00 © 2026 IEEE

(19.2 + 2.5)dBm at 340GHz, yielding output power density of 0.925W/mm² at 340GHz for this 35nm GaN HEMT under V_{ds} = 8V. Figure 20.6.5 (bottom-right) shows PAE improves noticeably with drain voltage increase from 6 to 8V, with absolute maximum improvement roughly 0.18%, representing 40% relative enhancement.

Figure 20.6.6 benchmarks the fabricated four-way GaN PA against the prior art. The developed GaN PA demonstrates excellent saturated output power (P_{SAT}) and power density. Within 330 to 344GHz, our PA achieves small-signal gain >21dB, P_{SAT} >18dBm, and PAE >0.6%. P_{SAT} reaches 86mW (19.3dBm) at 340GHz, with output power density >0.925W/mm². Comparing to other PAs in Fig. 20.6.6, the maximum P_{SAT} of our PA surpasses any previously reported PA fabricated in InP HBT, GaAs mHEMT, or SiGe technologies by more than 400%, highlighting the significant power amplification potential of GaN HEMTs around 340GHz. Figure 20.6.7 shows the die microphotograph of the fabricated GaN PA.

Acknowledgement:
This work was supported in part by the National Natural Science Foundation of China under Grants U24B20163. Corresponding author: Keping Wang

References:
[1] J. Yu et al., "A 211-to-263-GHz Dual-LC-Tank-Based Broadband Power Amplifier With 14.7-dBm P_{SAT} and 16.4-dB Peak Gain in 130-nm SiGe BiCMOS," *IEEE JSSC*, vol. 58, no. 2, pp. 332-344, Feb. 2023. https://doi.org/10.1109/JSSC.2022.3192043
[2] L. John et al., "Highly-Compact 20-mW, 270-320-GHz InGaAs mHEMT Power Amplifier MMIC," *IEEE IMS*, pp. 970-973, June 2024. https://doi.org/10.1109/IMS40175.2024.10600274
[3] C.-G. Choi et al., "H-Band Differential Cascode Power Amplifier Achieving 9.5-dBm OP1dB at 260 GHz in 250-nm InP DHBT Process," *IEEE IMS*, pp. 1046-1049, June 2023. https://doi.org/10.1109/IMS37964.2023.10188092
[4] D. Simic et al., "Analysis and Design of Lossy Capacitive Over-Neutralization Technique for Amplifiers Operating Near f_{MAX}," *IEEE TCAS-I*, vol. 68, no. 5, pp. 1945-1955, May 2021. https://doi.org/10.1109/TCSI.2021.3060662
[5] F. He et al., "Analyze the Loss Utilization in Near- f_{max} Embedded Amplifiers Using Uniform 3-D Gain Space: The Super-Gain-Boosting Technique," *IEEE TMTT*, vol. 72, no. 5, pp. 2745-2756, May 2024. https://doi.org/10.1109/TMTT.2024.3365088
[6] D.-W. Park et al., "A 247 and 272 GHz Two-Stage Regenerative Amplifiers in 65 nm CMOS with 18 and 15 dB Gain Based on Double-G_{max} Gain Boosting Technique," *IEEE VLSI*, pp. 1-2, June 2020. https://doi.org/10.1109/VLSICircuits18222.2020.9162862
[7] D.-W. Park et al., "A 230-260-GHz Wideband and High-Gain Amplifier in 65-nm CMOS Based on Dual-Peak G_{max}-Core," *IEEE JSSC*, vol. 54, no. 6, pp. 1613-1623, June 2019. https://doi.org/10.1109/JSSC.2019.2899515
[8] A. Suárez et al., "Stability Analysis of Power Amplifiers under Mismatching Effects," *IEEE IMS*, pp. 1-3, June 2013. https://doi.org/10.1109/MWSYM.2013.6697625
[9] E. Zenteno et al., "Output Impedance Mismatch Effects on the Linearity Performance of Digitally Predistorted Power Amplifiers," *IEEE TMTT*, vol. 63, no. 2, pp. 754-765, Feb. 2015. https://doi.org/10.1109/TMTT.2014.2387060
[10] S. Sinha et al., "Balanced Power Amplifier Protection Against Load Mismatch," *IEEE MWCL*, vol. 28, no. 2, pp. 165-167, Feb. 2018. https://doi.org/10.1109/LMWC.2018.2792692

[11] L. John et al., "Broadband 300-GHz Power Amplifier MMICs in InGaAs mHEMT Technology," *IEEE TTST*, vol. 10, no. 3, pp. 309-320, May 2020. https://doi.org/10.1109/TTHZ.2020.2965808
[12] Z. Griffith et al., "A 6-10 mW Power Amplifier at 290-307.5 GHz in 250 nm InP HBT," *IEEE MWCL*, vol. 25, no. 9, pp. 597-599, Sept. 2015. https://doi.org/10.1109/LMWC.2015.2451360
[13] A. S. H. Ahmed et al., "A 190-210GHz Power Amplifier with 17.7-18.5dBm Output Power and 6.9-8.5% PAE," *IEEE IMS*, pp. 787-790, June 2021. https://doi.org/10.1109/IMS19712.2021.9574925
[14] A. S. H. Ahmed et al., "A compact H-band Power Amplifier with High Output Power," *IEEE RFIC*, pp. 123-126, June 2021. https://doi.org/10.1109/RFIC51843.2021.9490426
[15] J. Kim et al., "H-Band Power Amplifier Integrated Circuits Using 250-nm InP HBT Technology," *IEEE TTST*, vol. 5, no. 2, pp. 215-222, March 2015. https://doi.org/10.1109/TTHZ.2014.2387259
[16] Y. Kumazaki et al., "High Output Power and Efficiency 300-GHz Band InP-Based MOS-HEMT Power Amplifiers with Composite-Channel and Double-Side Doping," *IEEE JEDS*, vol. 12, pp. 965-973, Oct. 2024. https://doi.org/10.1109/JEDS.2024.3483305
[17] W. Wang et al., "A 216-to-226GHz Watt-Level GaN Solid-State Power Amplifier with Multiband Large-Signal Impedance Correction and Circuit-Package Co-Design Technique," *ISSCC*, pp. 544-546, Feb 2025. https://doi.org/10.1109/ISSCC49661.2025.10904715

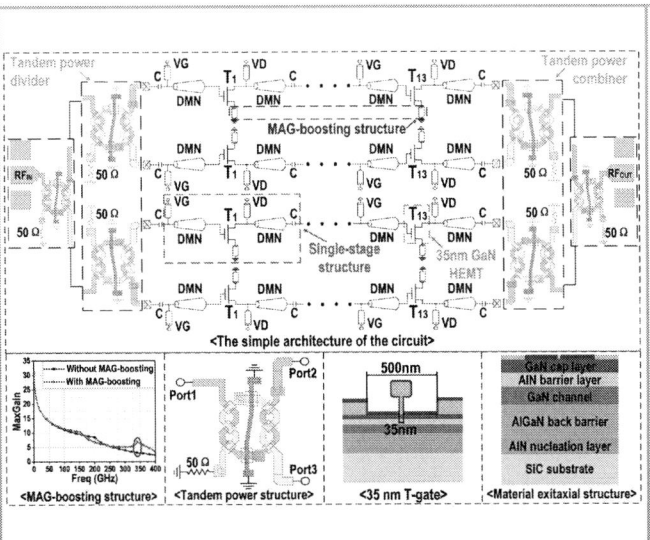

Figure 20.6.1: Simple circuit architecture and typical characteristics of devices.

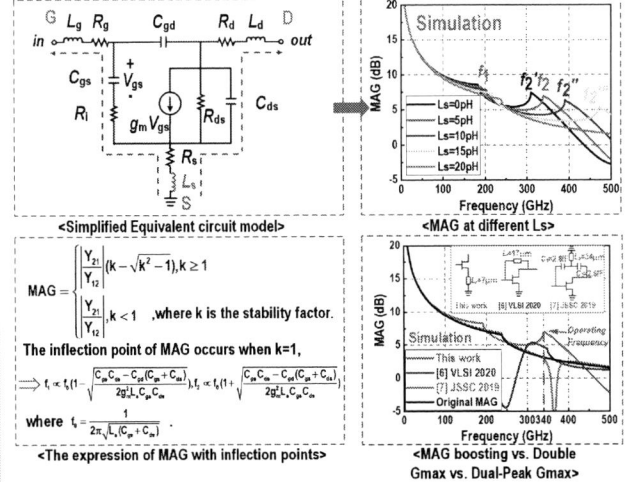

Figure 20.6.2: Simplified equivalent circuit model of a unit GaN HEMT, MAG at different L_s, the expression of MAG with inflection points, and MAG boosting vs. double Gmax vs. dual-peak Gmax.

ISSCC 2026 / SESSION 20 / RF TRANSCEIVER SUBSYSTEMS FROM CM-WAVE TO THz / 20.6

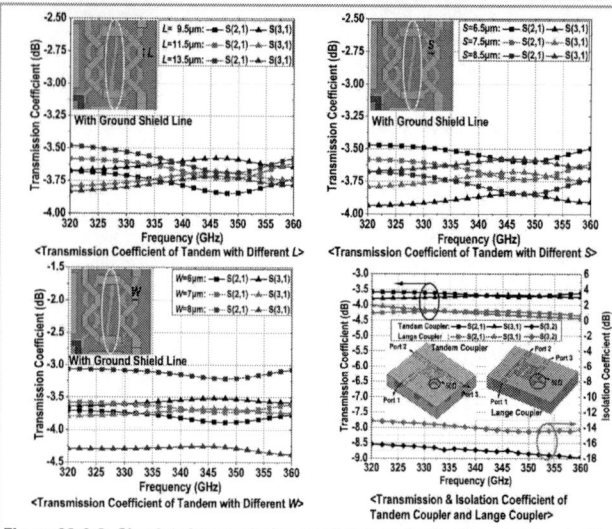

Figure 20.6.3: Simulated transmission coefficient of the Tandem coupler at different parameters.

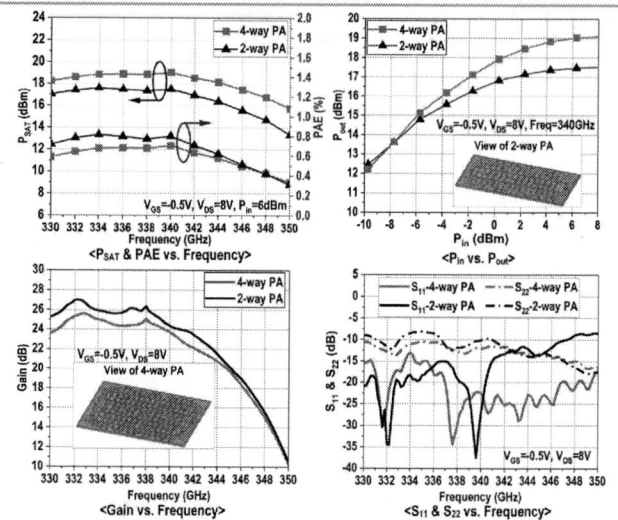

Figure 20.6.4: Measured P_{SAT}, PAE, P_{in} vs. P_{out}, Gain, and S-parameters of the 4-way and 2-way PAs.

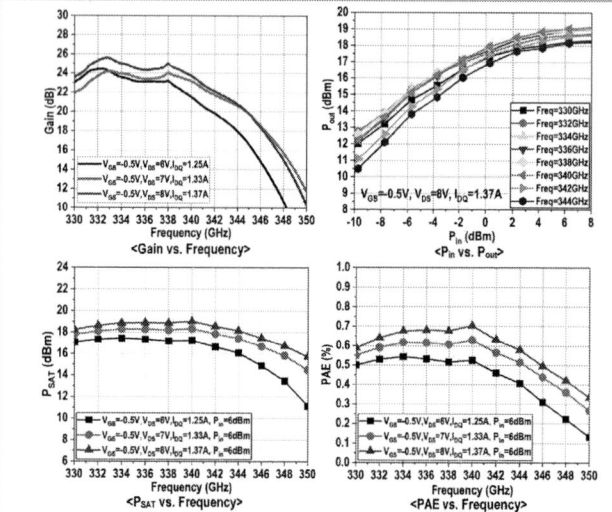

Figure 20.6.5: Measured Gain, P_{in} vs. P_{out}, P_{SAT}, and PAE of the 4-way PA at different conditions.

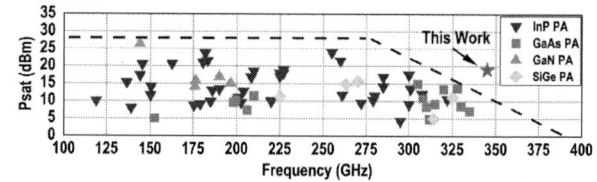

	Technology	Combining Ways	Frequency [GHz]	Gain [dB]	Max. P_{SAT} [dBm] @freq.	PAE_{MAX} [%]	P_{DC} [W]	Area [mm²]	Power Density [mW/mm]
This work	35nm GaN HEMT	2	330 to 344	25.5	17.6@334GHz	0.8	7.04	7.03	925
		4	330 to 344	24	19.3@340GHz	0.6	11.38	11.88	
[2] IMS '2024	35nm InGaAs mHEMT	8	270 to 320	20	13.4@290GHz	2.5	/	0.33	156**
[3] IMS '2023	0.25µm InP HBT	2	254 to 279	19.5	10.2@260GHz	3.1	0.33	0.50	/
[11] TTST '2020	35nm InGaAs mHEMT	2	280 to 328	26	13.7@300GHz	2.4	/	0.38*	67
[12] MWCL '2015	0.25µm InP HBT	2	290 to 307.5	23.5	10@297.5GHz	1.1	0.85	0.64	310
[13] IMS '2021	0.25µm InP HBT	4	190 to 210	16.8	18.5@194GHz	8.5	0.81	1.14	737**
[14] RFIC '2021	0.25µm InP HBT	4	266 to 285	10.9	16.8@270GHz	4	1.09	0.83	499**
[15] TTST '2015	0.25µm InP HBT	4	297 to 310	15.4	13.5@301GHz	/	/	0.46	233**
[16] JEDS '2024	75nm InP HEMT	16	250 to 300	26.6	16.9@270GHz	2.2	/	5.71	153**
[17] ISSCC '2025	50nm GaN HEMT	2	190 to 230	22.4	20.4@216GHz	2.3	2.30	9.35	1350

* Core area ** Estimated based on transistor size and output power

Figure 20.6.6: Comparison of P_{SAT}, performance summary, and results from prior works.

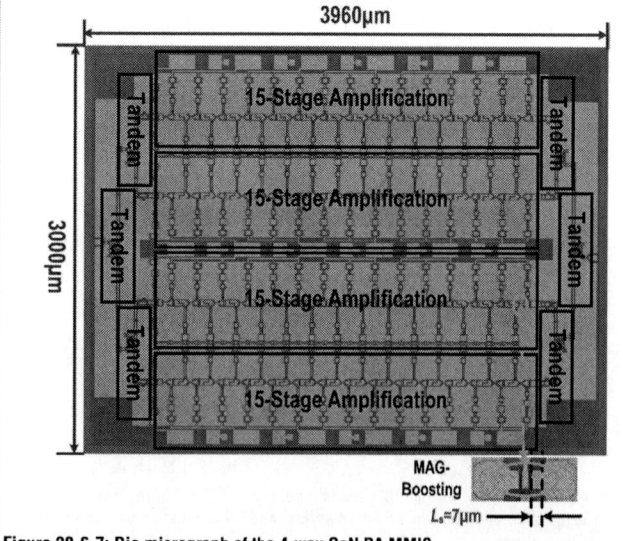

Figure 20.6.7: Die micrograph of the 4-way GaN PA MMIC.

• 2026 IEEE International Solid-State Circuits Conference

ISSCC 2026 / SESSION 20 / RF TRANSCEIVER SUBSYSTEMS FROM CM-WAVE TO THz / 20.7

20.7 An Ultra-Compact D-Band Transceiver Front-End Based on a Common-Gate Bidirectional Amplifier Achieving 11.2dBm TX P_{sat} and 8.2dB RX Average NF for an Area-Constrained 2D Beamformer AiP

Syed Mohammad Ashab Uddin[1,2], Wooram Lee[1]

[1]Pennsylvania State University, University Park, PA, [2]Texas Instruments, Dallas, TX

Abstract

An ultra-compact D-band transceiver based on a common-gate bi-directional amplifier is presented in a 45nm RFSOI process for an area-constrained 2D beamformer antenna-in-package. By employing a switchless symmetric inter-stage matching network and compact, low-loss reconfigurable input/output networks, the transceiver front-end achieves a TX peak P_{sat} of 11.2dBm at 135GHz and an average RX NF of 8.2dB over 129 to 145GHz. The active chip area, excluding pads, is only 0.08mm².

The continuous growth in demand for wireless data necessitates the utilization of the D-band spectrum (110 to 170GHz). However, the limited transistor performance of silicon-based RFICs and the severe path loss at such high frequencies require the development of large-scale D-band phased arrays. For a compact phased-array transceiver module, the transmitter (TX) and receiver (RX) circuits can share an antenna, typically through single-pole double-throw (SPDT) switches operating in time-division duplex (TDD) mode. However, the high insertion loss of silicon-based SPDT switches at D-band frequencies degrades the TX output power and efficiency, as well as the RX noise figure (NF) [1,2]. Moreover, for a 2D uniform array with an antenna spacing of ~λ/2, the available space for TRX circuits is limited to (λ/2)². Hence, at D-band (λ/2 ≈1.07mm@140GHz), a beamformer IC with separate TX and RX circuits is not optimal for integration with the area-constrained antenna-in-package (AiP) (Fig. 20.7.1). These challenges can be addressed by a compact bidirectional amplifier integrated with a bidirectional phase shifter [3], which shares silicon area for both TX and RX operation modes (Fig. 20.7.1). As a phased array transceiver front-end, the bidirectional amplifier must exhibit high output power in the TX mode and low noise figure (NF) in the RX mode. High linearity is also desirable in the RX mode to maintain a high dynamic range by mitigating the increased integrated noise within a wide signal bandwidth. The first D-band bidirectional amplifier was reported in [4] and exploited the device symmetry of a common-gate amplifier with a symmetric interstage matching network. The design exhibited limited output power (OP1dB of -5.1dBm), and with a reconfigurable T/R switch network, which used a tunable transmission line at the input and output terminals, was not area-efficient.

To significantly improve the TX mode output power, while further reducing chip area, for a D-band phased-array transceiver front-end, this paper presents an ultra-compact high-power bidirectional amplifier. A power amplifier (PA) and low-noise amplifier (LNA) are designed within a single amplifier footprint to minimize chip area and antenna TRX switching loss by utilizing a compact, switchless interstage matching network and low-loss, reconfigurable input/output networks. The proposed transceiver achieves a TX peak saturation power P_{sat} of 11.2dBm at 135GHz and an average RX NF of 8.2dB over the 129-to-145GHz frequency range. The RX IP1dB ranges from -11.6 to -6.4dBm, demonstrating superior RX linearity. The active chip area, excluding pads, is only 0.08mm². Additionally, the proposed transceiver features phase-invariant gain tuning, which is crucial for independent amplitude and phase control in a large-scale phased array [5,6].

The proposed bidirectional amplifier exploits the source-drain symmetry of MOSFETs in a common-gate (CG) amplifier with switchable S/D voltage potential (Fig. 20.7.1). In this work, the design methodology reported in [4] for a single-ended CG amplifier is further developed to implement a differential CG amplifier to increase the output power in the TX mode and linearity in the RX mode. A differential topology offers additional advantages, such as providing a virtual ground at the node of the supply switch, which minimizes the effect of the switch resistance on the small-signal gain. The capacitive neutralization technique is applied to the differential pair for unconditional stability. For the design of a symmetric interstage matching network (Fig. 20.7.2), a common series inductor L_{CM} is placed between the two adjacent stages to move the drain impedance of the first stage Z_A and the source impedance of the next stage Z_C to the same constant conductance circle (Z_B and Z_D) on the Smith chart. Then, shunt components L_1 and C_1 are added to resonate out the imaginary parts of Z_B and Z_D, which move to Z_E and Z_F, respectively, on the real axis of the Smith chart, completing impedance matching. The interstage matching network achieves bidirectional symmetry, as L_1 and C_1 are interchangeable parallel components, along with a common series inductor, L_{CM}. For a compact layout of the interstage matching network, four L_{CM} inductors, two L_1 inductors, and two C_1 capacitors can be transformed into a single differential transformer (Fig. 20.7.2). Half of the designed differential interstage matching network can be represented as a T network equivalent to a single-ended transformer, where a mutual inductor M provides an equivalent impedance at the design frequency to the parallel combination of shunt capacitor C_1 and inductor L_1. The self-inductance (L) and coupling factor (k) of transformers are selected based on L_{CM} and M.

A reconfigurable TRX antenna interface is designed by adding only a single series switch to a balun, resulting in minimal extra space and insertion loss (Fig. 20.7.3). In the PA mode, the antenna impedance needs to be transformed to the optimum load impedance for the transistors of the last stage to generate the maximum output power. In the LNA mode, the antenna impedance needs to be transformed to meet the noise and power matching conditions through a reconfigurable matching network. To meet these requirements for the PA and LNA modes simultaneously, while minimizing extra insertion loss and chip area, the bidirectional differential pair, a balun, and a series switch M_2 are tightly co-designed for a reconfigurable matching network, eliminating the need for any additional components. The on-state switch M_2 in the PA mode exhibits a small resistance (~6Ω), resulting in minimal impact on the insertion loss and moving Z_{AP} to Z_X on the Smith chart. The balun is designed (1) to move Z_X to Z_{Load}, which is close to the optimum output load impedance of the bidirectional differential pair Z_{OPT} in the PA mode, and (2) to move the input impedance of the bidirectional differential pair Z_{IN} to Z_Y, where Z_Y is moved close to 50Ω after adding the off-state switch on the Smith chart in the LNA mode (Fig. 20.7.3). The off-state switch M_2 can be modeled with a series capacitor C_{off} between the two terminals and two shunt capacitors C_{gnd} between each terminal and the ground. Note that the optimum noise matching condition is close to the conjugate matching condition, resulting in the simulated NF only 0.2dB lower than the minimum achievable NF, NF_{min}. Simulations show that the series switch adds only 0.2dB of loss in the RX mode and 0.4dB of loss in the TX mode at 140GHz, compared to ideal matching networks without a switch. A similar design approach is applied to the other port, which may be used as an interface to a bidirectional phase shifter. Note that the pad capacitance and interconnect transmission line are absorbed into a part of the matching network.

A six-stage bidirectional amplifier is designed using the proposed symmetric interstage matching networks and reconfigurable input and output matching networks (Fig. 20.7.4). For higher gain, ADNFETs, recently developed by GlobalFoundries for power amplifier design, are utilized for the first four stages, comprising M_0-M_7 [7]. In our transistor breakout measurement, it is found that the ADNFETs exhibit a ~1.8dB higher maximum available gain (G_{max}) at 140GHz in a common-gate configuration than the regular NFETs for a 24µm/40nm transistor size and a 0.3mA/µm bias condition. However, the simulated minimum achievable NF (NF_{min}) of ADNFETs is ~1.4dB higher than that of the regular NFET at 140GHz. Hence, the regular NFETs are used for the first two stages of the LNA (the last two stages of the PA), which comprises the transistors M_8-M_{11}. The six-stage bidirectional amplifier is fabricated using a GlobalFoundries 45RFSOI process, and the active core area, excluding pads, is only 0.08mm² (Fig. 20.7.7). The fabricated bidirectional amplifiers are characterized on a probe station using Keysight N5242A PNA-X, 5292A mmWave test controller, and N5262BW06 D-band VNAX modules. The input power levels applied to the IC from the VNAX module are calibrated using a VDI-Erickson PM5B power meter. The power supply voltages for the LNA and PA measurements are 1.0V and 1.0V/1.2V, respectively. The bias currents are adjusted separately for the LNA and PA modes. The measured peak gain in the LNA mode is 15dB at 137GHz with a 3dB bandwidth of 16GHz, while consuming a total DC power of 67mW (Fig. 20.7.4). The measured peak gain in the PA mode is 16.5dB at 137GHz with a 3dB bandwidth of 17GHz, while consuming total DC power of 114mW (Fig. 20.7.4). The phase-invariant gain tuning feature via bias current control is measured in the LNA mode, showing a maximum phase variation of 2 degrees for a gain tuning range of 10dB at 140GHz (Fig. 20.7.4). The relative phase variation for 10dB gain tuning is less than 5.625 degrees (=1/2 LSB for a 5-bit phase shifter) at frequencies from 138 to 147GHz (Fig. 20.7.4). The fabricated bidirectional amplifier exhibits unconditional stability with a µ-factor above unity across the measured frequency range of 110 to 170GHz in both the PA and LNA modes. The measured OP1dB, P_{sat}, and peak power-added efficiency (PAE) at 135GHz in the PA mode are 6.3dBm, 11.2dBm, and 8.4%, respectively (Fig. 20.7.5). The OP1dB and P_{sat} in the PA mode vary within 2dB over the 3dB gain bandwidth. The RX NF is measured based on the Y-factor method using a VDI WR6.5NS D-band noise source, Keysight N9029BV06 D-band SAX module, and Keysight N9042B signal analyzer. The measured minimum NF and average NF over the 3dB gain bandwidth are 7.3dB and 8.2dB, respectively, in the LNA mode (Fig. 20.7.5). The IP1dB in the LNA mode varies from -6.4 to -11.6dBm (Fig. 20.7.5).

Figure 20.7.6 summarizes the performance of the proposed bidirectional amplifier as compared with prior works in a similar frequency range. The proposed bidirectional amplifier achieves high saturation output power in the TX mode and low NF in the RX mode, while occupying a significantly smaller chip area. This work also demonstrates superior RX linearity, quantified by a Figure of Merit (FoM) defined as Gain × IP1dB (mW) normalized by DC power consumption, P_{DC} (mW). Moreover, this work reports phase-invariant gain tunability, a feature not previously reported for other D-band TRX designs in Fig. 20.7.6.

Acknowledgement:
This work was supported by the National Science Foundation under Award 2235336 and Award 2151190. The authors would like to thank the GlobalFoundries University Partnership Program for fabrication support.

References:
[1] W. T. Khan et al., "A D-band (110 to 170 GHz) SPDT switch in 32 nm CMOS SOI," *IEEE IMS*, pp. 1-3, June 2015. https://doi.org/10.1109/MWSYM.2015.7167061
[2] A. Ahmed et. al, "140-GHz 2-D Scalable On-Grid 8×8-Element Transmit–Receive Phased Arrays with Up/Down Converters Demonstrating a 5.2-m Link at 16 Gbps," *IEEE TMTT*, vol. 72, no. 5, pp. 2852-2868, May 2024. https://doi.org/10.1109/TMTT.2023.3346407
[3] M. Abbasi et. al, "A Low-Loss Passive D-Band Phase Shifter for Calibration-Free, Precise Phase Control," *IEEE JSSC*, vol. 59, no. 5, pp. 1371-1380, May 2024. https://doi.org/10.1109/JSSC.2024.3357738
[4] S. M. A. Uddin el. al, "Design and Implementation of a D-Band Bidirectional Common-Gate Amplifier in 45-nm RFSOI," *IEEE JSSC*, vol. 60, no. 7, pp. 2500-2510, July 2025. https://doi.org/10.1109/JSSC.2024.3504358
[5] J. Park et. al, "A 28-GHz Four-Channel Beamforming Front-End IC With Dual-Vector Variable Gain Phase Shifters for 64-Element Phased Array Antenna Module," *IEEE JSSC*, vol. 58, no. 4, pp. 1142-1159, Apr. 2023. https://doi.org/10.1109/JSSC.2022.3214436
[6] B. Sadhu et. al., "A 28-GHz 32-Element TRX Phased-Array IC With Concurrent Dual-Polarized Operation and Orthogonal Phase and Gain Control for 5G Communications," *IEEE JSSC*, vol. 52, no. 12, pp. 3373-3391, Dec. 2017. https://doi.org/10.1109/JSSC.2017.2766211
[7] S. H. Jain et al., "Novel mmWave NMOS Device for High Pout mmWave Power Amplifiers in 45RFSOI," *ESSDERC*, pp. 199-202, Sep. 2021. https://doi.org/10.1109/ESSDERC53440.2021.9631775
[8] X. Tang et.al, "Design and Analysis of a 140-GHz T/R Front-End Module in 22-nm FD-SOI CMOS," *IEEE JSSC*, vol. 57, no. 5, pp. 1300-1313, May 2022. https://doi.org/10.1109/JSSC.2021.3139359
[9] Y. Yamazaki et al., "A 150 GHz High-Power-Density Phased-Array Transceiver in 65nm CMOS for 6G UE Module," *IEEE Symp. VLSI Circuits*, pp. 1-3, June 2025. https://doi.org/10.23919/VLSITechnologyandCir65189.2025.11074823
[10] A. Karakuzulu et.al, "Full D-Band Transmit–Receive Module for Phased Array Systems in 130-nm SiGe BiCMOS," *IEEE SSCL*, vol. 4, pp. 40-43, Jan. 2021. https://doi.org/10.1109/LSSC.2021.3054512
[11] Y. Qi et al., "A 110–150 GHz Broadband TDD Front End in SiGe BiCMOS," *IEEE MWTL*, vol. 35, no. 8, pp. 1250-1253, Aug. 2025. https://doi.org/10.1109/LMWT.2025.3563580

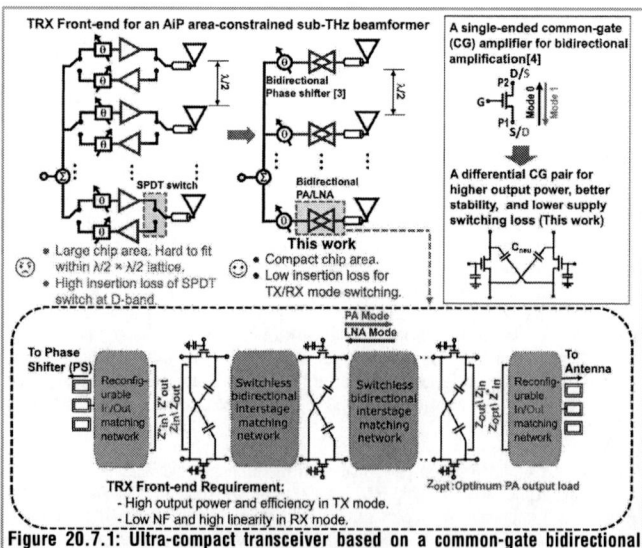

Figure 20.7.1: Ultra-compact transceiver based on a common-gate bidirectional amplifier for an area-constrained sub-THz beamformer.

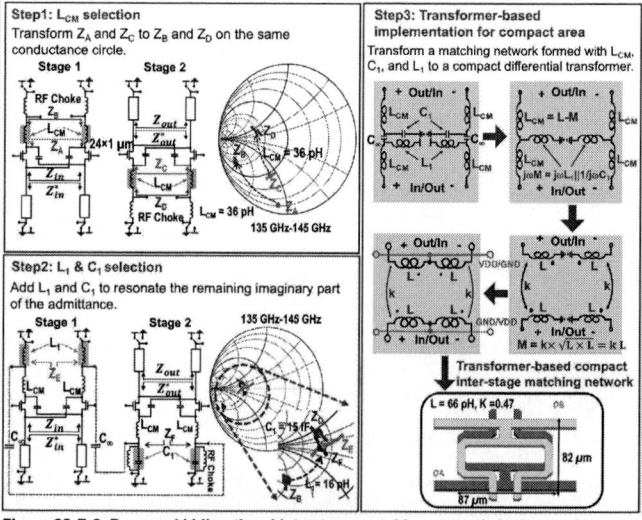

Figure 20.7.2: Proposed bidirectional interstage-matching-network design methodology and transformer-based implementation for a compact chip area.

ISSCC 2026 / SESSION 20 / RF TRANSCEIVER SUBSYSTEMS FROM cm-WAVE TO THz / 20.7

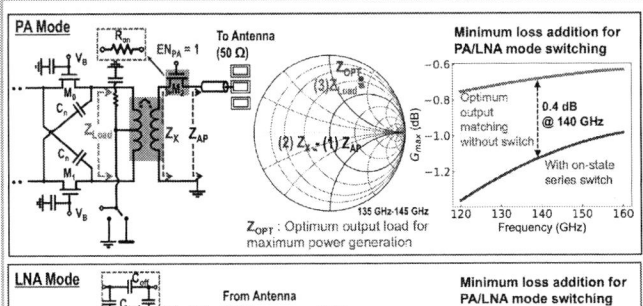

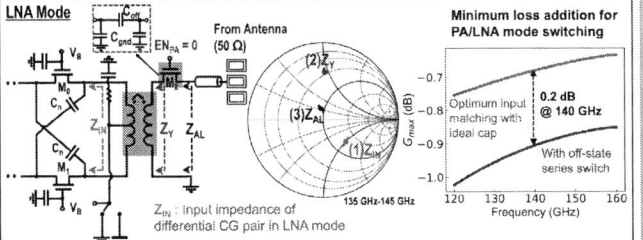

Figure 20.7.3: Reconfigurable low-loss, compact input (LNA)/output (PA) matching network design.

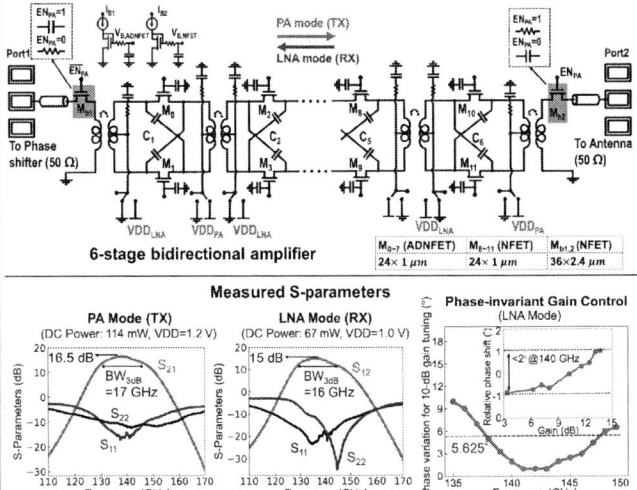

Figure 20.7.4: Schematic of the proposed transceiver front-end, measured S-parameters, and phase-invariant gain tunability.

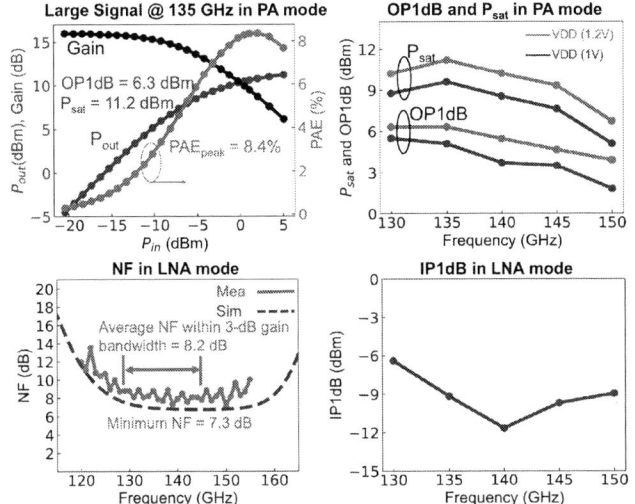

Figure 20.7.5: Measured large-signal behavior in the PA mode and NF and linearity in the LNA mode.

	This Work	JSSC 2025 [4]	JSSC 2022 [8]	VLSI 2025 [9]	SSCL 2021 [10]	MWTL 2025 [11]
Topology	Bidirectional Amplifier	Bidirectional Amplifier	PA/LNA/TRX Switch	PA/LNA/TRX Switch	PA/LNA/TRX Switch	PA/LNA/TRX Switch
Technology	45RFSOI	45RFSOI	22FDX	65nm CMOS	0.13μm SiGe	0.13μm SiGe
Active Area (mm²)	0.08	0.28	0.8	0.75#	2.04*	1.4*
PA						
Gain (dB)	16.5	14	33.6	22.7	22.4	20
$BW_{3dB\ Gain}$ (GHz)	17	21	30	22	57	40
Frequency (GHz)	130 to 147	124 to 145	140	142 to 164	113 to 170	110 to 150
Peak P_{sat} (dBm)	11.2	-2.1	12.5	9.5	9.5	12.8
Peak OP1dB (dBm)	6.3	-5.1	9.4	N/A	7	9.2
PAE_{peak} (%)	8.4	N/A	10.8	N/A	1.7(DE)	5.5
P_{DC} (mW)	114	28.5	152	150§	560	185
LNA						
Gain (dB)	15	14	20	17.3	28.3	20
$BW_{3dB\ Gain}$ (GHz)	16	21	12	22	60	37
Frequency (GHz)	129 to 145	124 to 145	140	142 to 164	110 to 170	111 to 148
IP1dB (dBm)	-11.6 to -6.4	-18	-24	N/A	-30	-18
Minimum NF /NF_{avg} (dB)	7.3/8.2	N/A/7	9.2/N/A	10.7/N/A	9/10.7	7/10
P_{DC} (mW)	57	28.5	20	93§	430	43
$\frac{Gain \times IP1dB\ (mW)}{P_{DC}(mW)} \times 100$	3.3 to 10.8	1.4	2	N/A	0.16	3.7
Phase-invariant gain tunability						
Phase variation for 10-dB gain tuning(°)	<5.625° (138 to 147 GHz)	N/A	N/A	N/A	N/A	N/A

*including pads, # Approximated from layout image, § The DC power includes a sub-harmonic mixer, IF buffer, and LO generation.

Figure 20.7.6: Performance summary and comparison table.

Figure 20.7.7: Die micrograph.

ISSCC 2026 / SESSION 20 / RF TRANSCEIVER SUBSYSTEMS FROM CM-WAVE TO THz / 20.8

20.8 A 16-to-256QAM G-Band Subharmonic Phase-Modulating Transmitter for Beyond-5G Communications

Jia Zhou[1], Jieqiong Du[1], Chao-Jen Tien[1], Jhih-Wei Chen[1], Ruei-Chen Soong[1], Francisco Cardenas Beltran[1], Lachlan Cuskelly[1], Christopher Chen[1], Minji Zhu[1], Adrian Tang[1,2], Sai-Wang Tam[3], Mau-Chung Frank Chang[1]

[1]University of California, Los Angeles, CA, [2]California Institute of Technology, Los Angeles, CA, [3]NXP Semiconductors, San Jose, CA.

Abstract

This paper presents a highly integrated G-band digital transmitter for beyond-5G communications featuring: 1) a subharmonic phase-modulating architecture with time-invariant phase tripling, 2) a GS/s segmented 9-bit digital-to-phase converter (DTPC) with a sub-degree rms error, and 3) a carrier-saturating and multiplying chain for direct QAM synthesis. It achieves 51.2Gbps/16QAM, 24Gbps/64QAM, and 8Gbps/256QAM with -2dBm peak output power while consuming 280-to-350mW DC power.

Sub-THz spectra, particularly at G-band and above, offer much increased bandwidth for beyond-5G (B5G) wireless communications. However, existing analog upconverting transmitters (TX) remain insufficient for achieving desired data rate and spectrum/energy efficiency due to their dependence on high-resolution power-hungry DACs and the back-off required by power amplifiers for linearity concerns [1,2]. Prior digital TX designs at sub-THz were also limited to low-order modulations, such as OOK [3], BPSK [4], and QPSK [5] due to the limited dynamic operation range and resolution of existing circuits. While high modulation-order TXs have been demonstrated at E-band [6], it is nontrivial to extend a similar concept to the sub-THz frequency regime owing to excessive phase-modulator losses [7,8]. Harmonic outphasing approaches have emerged recently to help alleviate such issues but are constrained to low-order constellations of PSK [9] and STAR-QAM [10] due to the lack of high-speed and fine-resolution phase modulators. To overcome such design challenges, a subharmonic phase-modulating transmitter (SHPM-TX) is presented to achieve high-speed modulations (51.2Gbps/16QAM, 24Gbps/64QAM, and 8Gbps/256QAM) with -2dBm peak output power, while consuming only 280-to-350mW DC power and occupying a 0.36mm² silicon real estate.

Shown in Fig. 20.8.1, the SHPM-TX starts by splitting its input at subharmonic $f_c/3$ (60 to 75GHz) to create dual phasors in parallel with a wideband balun, followed by digital-to-time/phase converters (DTPC) made of switched-capacitor-controlled differential transmission lines (TL). On-chip SRAM, serializers, and binary-to-thermometer (B2T) decoders are integrated as high-speed controllers for the DTPC. A subsequent carrier conditioning chain saturates a frequency tripler with input at $f_c/3$ to upconvert DTPC modulation to the carrier frequency f_c, consequently tripling the phase modulation in a time-invariant manner. An output buffer is inserted afterwards to suppress subharmonic spurs and simultaneously create isolation. Symbol synthesis is attained by summing dual constant-envelope phasor voltages through a wideband transformer. This subharmonic modulation scheme eliminates the need for separate baseband, clocking paths, and mixers, and enhances system integration and efficiency compared to prior art. The implementation of subharmonic DTPC at $f_c/3$ reduces the required phase modulation range by 3×, preserves the switched-capacitor quality factor, and lowers traveling-wave loss and propagation delay. The switched-capacitor TL, similar to DiCAD [11], also delivers finer phase resolution with a smaller footprint than the reflection-type phase shifters [12].

As shown in Fig. 20.8.2, the subharmonic DTPC employs a segmented 9-bit design composed of a 1-bit polarity flip, 3-bit thermometer-coded coarse and fine units, and two calibration bits, which cover roughly half and a quarter of a fine unit delay, respectively. The polarity flip is formed by a double-balanced mixer and also serves as an input buffer. The coarse unit integrates 52pH inductors with 30fF capacitors, while the fine unit uses 4.5pH inductors and 5.5fF capacitors; the calibration bits are realized with 4fF and 2fF capacitors. Coarse and fine paths are controlled by individual 3-to-7b B2T decoders to reduce propagation delays, while the calibration bits and polarity flip are both driven by the serializers. Figure 20.8.2 shows the simulated DTPC characteristics at 70GHz. The coarse path provides an ~80° range with ~11.5° steps (rms error ~1.2°), while the fine path adds ~10° tuning with ~1.3° steps (rms error ~0.2°). The switched-capacitor transmission line is terminated with 100Ω provided by wideband transformers to satisfy input and output matching while preserving system gain. Insertion loss remains flat at -5dB with only 0.7dB variation across all codes, enabling a wide phase shift range with fine resolution.

As shown in Fig. 20.8.3, the subharmonic carrier conditioning chain comprises a cascaded buffer, a frequency tripler, and an output isolation stage. The buffer employs a 3-stage topology: an initial gain amplifier compensates the 5dB insertion loss of the DTPC, followed by two driver stages that saturate the amplitude for driving the frequency tripler. The tripler is configured with a capacitor-neutralized common-source differential pair biased near the threshold, leveraging enhanced nonlinearity for third-harmonic generation. A final unity-gain buffer provides load isolation and suppresses the residual subharmonic leakage. At the output, a 50Ω pad ensures broadband matching, while a transformer-based combiner with 80Ω input impedance is used to synthesize the output signal. Post-layout simulations of the subharmonic carrier chain are summarized in Fig. 20.8.3. The tripler achieves a peak output power of -3dBm across 210 to 215GHz, with a DC power consumption of 205 to 220mW.

The measurement setup is shown in Fig. 20.8.3. A V-band ×4 multiplier, driven by a ~17.5GHz LO, generates a 70GHz input carrier at 0dBm for the SHPM-TX under test. The 210GHz TX is probed and downconverted by a G-band ×12 harmonic mixer, driven by an ~18.33GHz LO, to an intermediate frequency (IF) centered at 10GHz. The IF waveform is captured by an oscilloscope and downloaded for offline post-processing, where baseband symbol recovery is performed. Output power is measured through a power meter in a separate branch. There are two challenges in the measurement process: phase noise (PN) and inter-symbol interference (ISI) due to waveguide and coaxial channel reflections. Low-PN LOs are synchronized to a common 10MHz reference, and a pilot sequence is embedded in the random bit stream to track high-frequency phase noise. To reduce ISI, linear channel equalization is applied in the post-processing. To demonstrate 16/64/256-QAM modulations, static calibration is applied by mapping all measurable SHPM-TX output phasors to generate a desired static look-up table (LUT) for selected digital control bits that yield the lowest estimated EVMs. The LUT is first created at 50MBaud to reduce impairments due to PN and ISI and can retain its accuracy for the following linear modulation tests.

Figure 20.8.4 summarizes the measured peak continuous-wave output and DC power (P_{OUT} and P_{DC}, respectively), across the 195-to-225GHz band, along with a detailed power breakdown at 210GHz. The left plot shows that the peak CW P_{OUT} stays between -6dBm (195GHz) and -2dBm (225GHz), while P_{DC} decreases from 265mW (200GHz) to 245mW (225GHz). The bar chart shows the power distribution among various circuit blocks at 210GHz: the total chip consumes 280 to 350mW, dominated by the subharmonic multiplier chain (215mW), followed by the mixers (40mW) and the digital controls (ranging from 25mW at 3.2GBaud to 95mW at 12.8GBaud). Figure 20.8.4 also shows the mapped TX static phase-shifting steps at 210GHz across coarse, fine, and calibration states, mapped on the I/Q plane with full-circle coverage. The coarse steps (black circles) define the primary phase spacing, while the fine steps (gray circles) provide the intermediate resolution. The calibration states (red dots) further refine the phase accuracy. Phase-shift steps are measured across settings to capture systematic code dependencies. Coarse steps provide 23.3° resolution with a 2.25° rms error, while fine steps achieve 3.1° resolution with a 0.31° rms error. Calibration steps achieve sub-degree accuracy, with step-1 at 1.6° (0.14° rms error) and step-0 at 0.71° (0.17° rms error). A 20% range reduction is observed compared to simulation results, attributed to modeling inaccuracies.

The measured EVM versus symbol-rate trend is plotted in Fig. 20.8.5. The measured EVMs for 1GBaud/256QAM, 4GBaud/64QAM, 10GBaud/16QAM, and 12.8GBaud/16QAM are -30.37dB, -26.05dB, -29.48dB, and -16.37dB, respectively. Corresponding data rates are 8Gbps/256QAM, 24Gbps/64QAM, 40Gbps/16QAM, and 51.2Gbps/16QAM. The measured symbol rates for 256/64QAM are currently limited by the phase noise (corner rotations), whereas the symbol rate of 16QAM (10 vs 12.8GBaud) is limited by the 16.8GHz oscilloscope bandwidth. The IF spectra of the corresponding modulations and symbol rates are captured with sinc shapes using rectangular pulses. The average P_{OUT} for linear modulation measurements is around -10dBm.

A table to benchmark this work versus prior sub-THz transmitters and phase shifters is given in Fig. 20.8.6. With fully integrated digital baseband and carrier conditioning circuitry in 28nm CMOS, the SHPM-TX operates from 195 to 225GHz, achieving a peak P_{OUT} of -6 to -2dBm. It features fine phase granularity (down to 0.7° steps and <0.2° rms error) and realizes up to 51.2Gbps/16QAM, 24Gbps/64QAM, and 8Gbps/256QAM, while consuming a DC power of 280 to 350mW and occupying a silicon area of 0.36mm². Compared with prior transmitters in Fig. 20.8.6 at 200GHz and beyond, the SHPM-TX achieves scalable high-order QAM transmissions at higher data rates in a more integrated design.

Figure 20.8.7 shows the die micrograph of a prototype SHPM-TX, occupying 2.79mm × 1.13mm. The chip integrates SRAM, nine 32:1 serializers with B2T decoders, subharmonic DTPCs, gain and driving amplifiers, a frequency tripler, and a buffer that uses a transformer-based combiner. The zoomed-in view of the die highlights the DTPC core with seven coarse, seven fine, and two calibration units, enabling wide-range phase control with sub-degree resolutions.

354 • 2026 IEEE International Solid-State Circuits Conference

979-8-3315-8937-0/26 $31.00 © 2026 IEEE

Acknowledgement:
This work is supported by TSMC University Shuttle Program, NXP Semiconductors, and Semiconductor Research Corporation.

References:
[1] K. Takano et al., "A 105Gb/s 300GHz CMOS Transmitter," *ISSCC*, pp. 308-309, Feb. 2017. http://doi.org/10.1109/ISSCC.2017.7870384

[2] C. Wang et al., "A 236-to-266GHz 4-Element Amplifier-Last Phased-Array Transmitter in 65nm CMOS," *ISSCC*, pp. 415-417, Feb. 2024. http://doi.org/10.1109/ISSCC49657.2024.10454273

[3] B. Hadidian et al., "A 220-GHz Energy-Efficient High-Data-Rate Wireless ASK Transmitter Array," *IEEE JSSC*, vol. 57, no. 6, pp. 1623-1634, June 2022. http://doi.org/10.1109/JSSC.2021.3133512

[4] L. Steinweg et al., "8.0-pJ/bit BPSK Transmitter with LO Phase Steering and 52-Gbps Data Rate Operating at 246 GHz," *IEEE TMTT*, vol. 71, no. 7, pp. 3217-3226, July 2023. http://doi.org/10.1109/TMTT.2023.3239792

[5] S. Kang et al., "A 240 GHz Fully Integrated Wideband QPSK Transmitter in 65 nm CMOS," *IEEE JSSC*, vol. 50, no. 10, pp. 2256-2267, Oct. 2015. http://doi.org/10.1109/JSSC.2015.2467179

[6] J. Zhou et al., "A 71-86GHz 1024QAM Direct-Carrier Phase-Modulating Transmitter with Digital-to-Phase Converters and Constant-Envelope Phasors," *IEEE RFIC*, pp. 3-6, June 2025. http://doi.org/10.1109/RFIC61188.2025.11082945

[7] L. Piotto et al., "A 125-to-170GHz Power-Efficient Phase Shifter in SiGe BiCMOS with Outphasing Gain and Phase Corrections," *ISSCC*, pp. 546-548, Feb. 2025. http://doi.org/10.1109/ISSCC49661.2025.10904520

[8] K. Kuliabin et al., "A 100–300 GHz Attenuator-Based Ultrawideband Vector Modulator," *IEEE Trans. Terahertz Sci. Technol.*, vol. 14, no. 3, pp. 404-413, May 2024. http://doi.org/10.1109/TTHZ.2024.3371670

[9] J. S.-C. Chien and J. F. Buckwalter, "A 110-120-GHz, 12.2% Efficiency, 16.2-dBm Output Power Multiplying Outphasing Transmitter in 22-nm FDSOI," *IEEE ASSCC*, pp. 1-3, Nov. 2022. http://doi.org/10.1109/A-SSCC56115.2022.9980747

[10] A. Standaert and P. Reynaert, "A 390-GHz Outphasing Transmitter in 28-nm CMOS," *IEEE JSSC*, vol. 55, no. 10, pp. 2703-2713, Oct. 2020. http://doi.org/10.1109/JSSC.2020.3006433

[11] T. LaRocca et al., "Millimeter-Wave CMOS Digital Controlled Artificial Dielectric Differential Mode Transmission Lines for Reconfigurable ICs," *IEEE IMS*, pp. 181-184, June 2008. http://doi.org/10.1109/MWSYM.2008.4633133.

[12] B. A. Abdelmagid et al., "A Wideband Bidirectional Calibration-Free Frequency/Switching-Staggering 360° D-Band Phase Shifter with Frequency-Invariant Codes Achieving <2.38°/0.63dB RMS-Errors Over 24% Bandwidth," *ISSCC*, pp. 548-550, Feb. 2025. http://doi.org/10.1109/ISSCC49661.2025.10904552

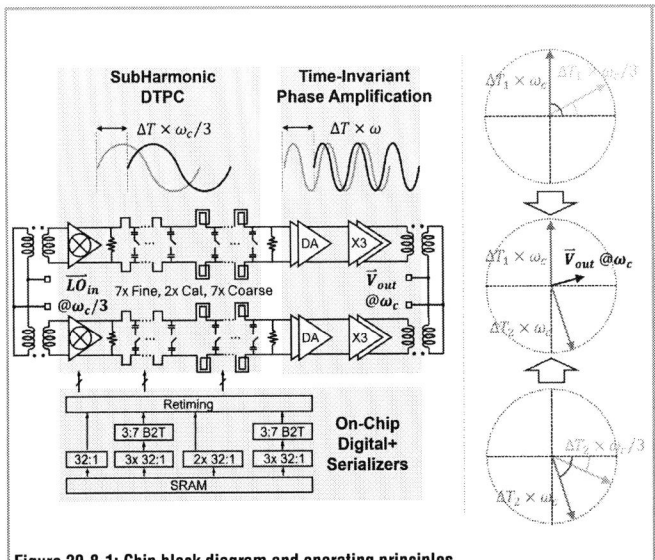

Figure 20.8.1: Chip block diagram and operating principles.

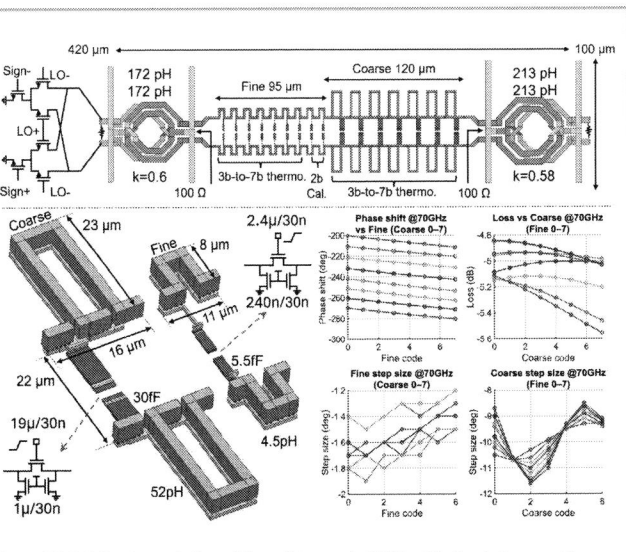

Figure 20.8.2: Implementation of the subharmonic DTPC with simulation results.

ISSCC 2026 / SESSION 20 / RF TRANSCEIVER SUBSYSTEMS FROM CM-WAVE TO THz / 20.8

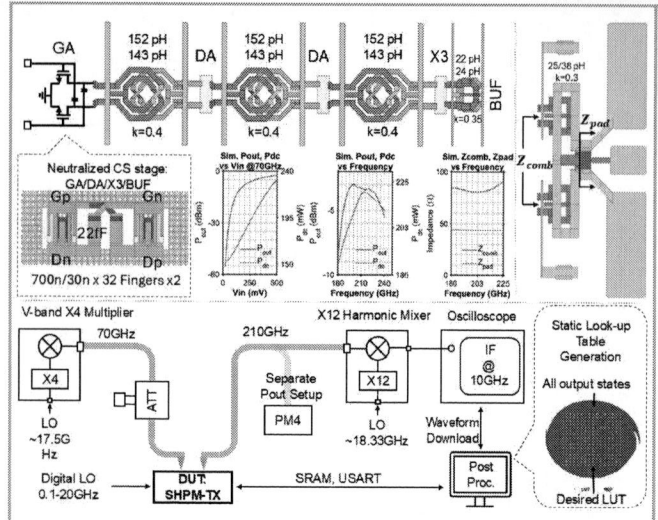

Figure 20.8.3: Implementation of the carrier conditioning chain with simulation results and measurement setup.

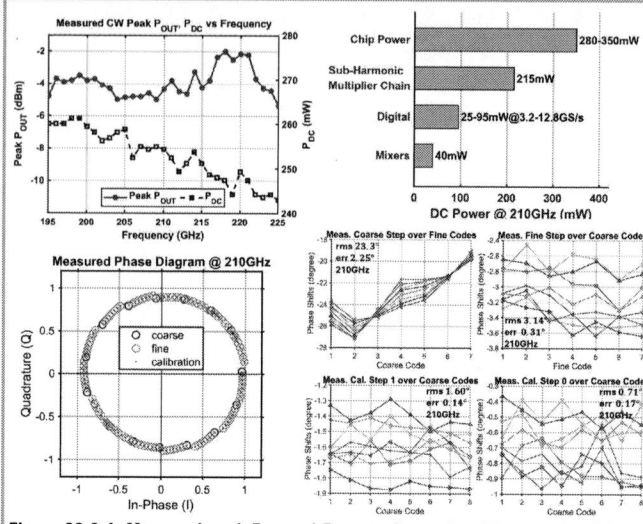

Figure 20.8.4: Measured peak P_OUT and P_DC over frequency; DC power breakdown; Measured static phase shift ranges, rms resolutions and errors.

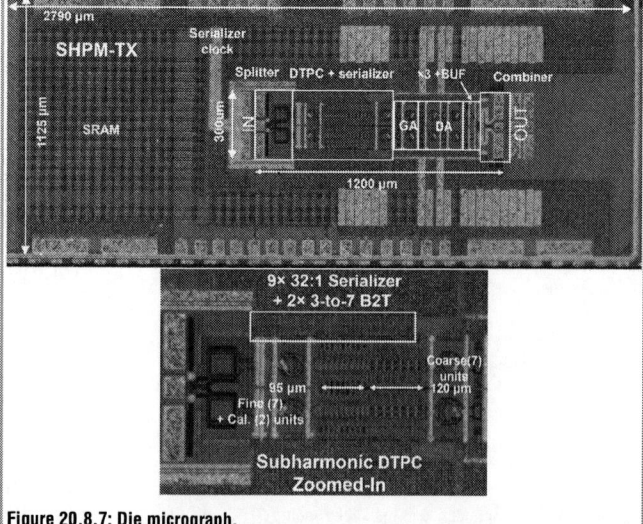

Figure 20.8.5: Measured constellations and IF spectra at 16, 64, 256QAM at various symbol rates.

	This work	ASSCC'22 [9]	JSSC' 20 [10]	JSSC' 22 [3]	TMTT' 23[4]	ISSCC' 24 [2]	ISSCC' 25 [7]
Technology	28nm CMOS	22nm SOI	28nm CMOS	55nm SiGe	0.13μm SiGe	65nm CMOS	55nm SiGe
Frequency (GHz)	195 to 225	110	390	225	246	236-266	125 to 170
Architecture	Subharmonic Phase-Modulating Transmitter	Harmonic Outphasing Transmitter	Harmonic Outphasing Transmitter	Amplitude Modulation Transmitter	Phase Modulation Transmitter	Analog Upconverter for Phased Array	Outphasing Phase Shifter
Modulator Type	Digital Switched-Capacitor Transmission Line	Analog RTPS	Digital Switched-Capacitor Delay Line	Digital Bias On-Off Switch	Digital Mixer	N/A#	Digital Switched-Capacitor Transmission Line
Digital Control Bits	2 × 9b	N/A	2 × 3b	1	1	N/A	2 × 4b
RMS Phase shift Step (°)	23.3 (coarse) 3.1 (fine) 1.6 (Cal. 1) 0.7 (Cal. 0)	N/A	45	N/A	N/A	2.4	9
RMS Phase Shift Step Error (°)	2.25 (coarse) 0.31 (fine) 0.14 (Cal. S1) 0.17 (Cal. S0)	N/A	N/A	N/A	N/A	N/A	5
Data Rate (Gbps) and Modulation	51.2@16QAM 24@64QAM 8@256QAM	10@QPSK 13.5@8PSK	6@QPSK 1@STAR-16 QAM	20@OOK	52@BPSK	108@16QAM 95@32QAM	N/A
EVM (dB)	-30.3@256QAM -26@64QAM -16.37@16QAM	-13.9@QPSK -14.5@8PSK	-11.6@QPSK -16.9@STAR-16QAM	N/A	N/A	-19.6@32QAM -16.5@16QAM	N/A
Peak Pout (dBm)	-6 to -2	16.2	-16	-3*	3.5	-3.4	2 to 3.5
Power Dissipation (mW)	280@3.2GS/s 350@12.8GS/s	334	1100	165.2	414	209**	31
Core Area (mm²)	0.36	0.54	2	0.89	1.18	2.47	0.2

* Simulated # Off-chip AWG ** Small-signal power

Figure 20.8.6: Comparison with prior art.

Figure 20.8.7: Die micrograph.

• 2026 IEEE International Solid-State Circuits Conference

ISSCC 2026 / SESSION 20 / RF TRANSCEIVER SUBSYSTEMS FROM cm-WAVE TO THz / 20.9

20.9 A Compact 26/38GHz-Reconfigurable Dual-Band Low-Noise Amplifier Using Transformer-Based Pole-Zero-Inversion Image-Rejection Technique Achieving >39/41dB IRR for 5G Multi-Band Applications

Jiaming Luo, Jincai Wen, Yuxia Wu, Xu Wang, Lingling Sun

Hangzhou Dianzi University, Hangzhou, China

Abstract

This paper presents a compact 26/38GHz reconfigurable dual-band low-noise amplifier (LNA) for a narrow-LO dual-side mixing. By adopting a transformer-based pole-zero-inversion image-rejection technique, the LNA achieves 20dB and 19dB of gain, 3.5dB and 3.9dB noise figures at 26GHz and 38GHz, respectively. The image rejection ratios are better than 39dB and 41dB with an intermediate frequency of 6GHz.

The fifth-generation (5G) communication technology utilizes millimeter-wave (mm-wave) spectrum to meet high-speed data transmission requirements. These 5G Frequency Range 2 (FR2) bands typically include the n257/n258/n261 and n259/n260 bands, which can be classified as a mm-wave low-frequency (LF) band (24.25 to 29.5GHz) and high-frequency (HF) band (37 to 43.5GHz). A large-scale multi-channel transceiver technology is the key to expanding mm-wave transmission distance and communication capacity. However, the number of transceiver channels is limited by the operating frequency bands and antenna size. Therefore, multi-band mm-wave circuits that can significantly reduce chip size and power consumption are both highly practical and technically challenges.

The traditional solution is to use two independent transceivers operating in the LF and HF frequency bands, but this significantly increases chip cost and power consumption. Another solution is to implement a wideband transceiver that covers all 5G FR2 frequency bands [1,2]. However, this approach is limited by suboptimal radio-frequency (RF) performance within the operating frequency bands and cross-band interference. To further suppress interference signals in the receiver, in-phase and quadrature (I/Q) structures have been employed in [1], and high-pass filters are used in [2]. Nevertheless, these methods have complex circuit structures and require a wideband local oscillator (LO). To solve the demand for broadband LO, dual-side (high-side and low-side) mixing scheme is a promising solution (Fig. 20.9.1). Compared to traditional single-side mixing method, the LO bandwidth can be significantly reduced to 7.25GHz with an intermediate frequency (IF) of 6GHz, greatly simplifying the design difficulty of the LO source. However, this configuration produces mutual image locations for the LF and HF signal bands; therefore, it is necessary to efficiently suppress the two image signals that are mutually mirrored around the LO frequency. Furthermore, to avoid using complex IQ mixing architectures in narrow-LO mixing reception [3-6], one feasible approach is to implement image rejection (IR) in the low-noise amplifier (LNA) stage [7-10]. A dual-band LNA introduces transmission zero by magneto-electric hybrid coupling to suppress image interference and support LF and HF bands, but it still requires a broadband LO source [7].

This paper presents a compact 26/38GHz dual-band high-IR LNA for narrow-LO dual-side-mixing reception. Through the proposed transformer-based pole-zero-inversion (PZI) IR technique, the dual-band LNA achieves 20dB and 19dB of gain, 3.5dB and 3.9dB noise figures (NF) at 26GHz and 38GHz, respectively. The image-rejection ratios (IRR) are better than 39dB and 41dB with the IF of 6GHz. The proposed dual-band LNA can be combined with a simplified non-IQ downconversion mixer to form a dual-band receiver, which is suitable for 5G multi-band multi-channel communication systems.

The implementation of the PZI IR structure is the key to achieving dual-band high IR LNA. The proposed PZI notch-filter structure, as shown in Fig. 20.9.2, consists of three groups of mutually coupled transformer (XFMR) coils, with the primary and secondary coils decomposed into two series-connected inductors (L_{P1}, L_{P2} and L_{S1}, L_{S2}), and the inductances are changed by respective switches (SW) to adapt to the LF and HF operating modes. The third group of coils consists of an LC series tank, and the inductor is split into two separate coils (L_T) that are respectively coupled to the primary and secondary coils, simplifying the coupling relationship and facilitating the layout implementation. It can be observed that in the LF mode (SW OFF) and the HF mode (SW ON), this structure has one pole and one zero each and exhibits low-pass and high-pass characteristics, respectively. Moreover, it is found that the pole-to-zero ratio is independent of the capacitance C_T and inductance L_T. The pole and zero can be shifted simultaneously by changing C_T or L_T, demonstrating the unique pole-zero-tracking capability of the proposed XFMR-based PZI IR structure. Therefore, by adjusting C_T (from $C_{T,HF}$ to $C_{T,LF}$), the pole and zero of the LF mode are simultaneously moved downwards, so that the pole of the LF mode is equal to the zero of the HF mode ($\omega_{p,LF}=\omega_{z,HF}$), and the zero in the LF mode can also be aligned with the pole in the HF mode ($\omega_{z,LF}=\omega_{p,HF}$), thereby achieving the reversal of pole and zero and suppressing interference signals in the mutual image signals of the dual bands.

As verification, a 26/38GHz reconfigurable dual-band LNA is presented that uses a narrow-LO mixing scheme (Fig. 20.9.3). By integrating the proposed XFMR-based PZI IR structure into the inter-stage matching network, the impact on noise-performance degradation is minimized. In order to maximize the quality factor of the coil and minimize losses, all XFMR coils are implemented with a top thick-metal layer. To generate the required coupling relationship and flexibly adjust the coupling coefficient (relationship), the PZI IR notch-filtering structure is constructed in an inner and outer quadrilateral style. The design process of the PZI IR matching structures can be divided into three steps. Step1: To simplify the design and quickly establish the prototype, consider the case where the port impedance is 50Ω and SW is ideal, then the inductance of the primary and secondary coils (L_{P1}, L_{P2}, L_{S1}, L_{S2}) can be determined for LF and HF matching requirements. Next, k_{P1T} is prioritized to address layout implementation and image notch depth. Furthermore, the parameters k_{P1T}, k_{P1S2}, L_T, and the tunable capacitance (C_T) are derived based on the pole-zero relationships. Step 2: In the case of inter-stage matching, both active amplification devices of the two stages exhibit complex port impedances, and their reactive part (capacitance) generates new poles in the passband, thereby widening the bandwidth of the operating frequency band without affecting the position of the zero. Step 3: Consider the complete design with the non-ideal SW and tunable capacitor (C_T). In the LF mode, a zero shift caused by SW parasitic capacitance can be corrected by reducing internal coil size and adjusting k_{P1S2} (k_{P2S1}) and L_{S2} (L_{P2}). It is worth noting that due to the combined effects of complex port impedance, non-ideal SW, and XFMR coils, the inter-stage matching network introduces high-frequency poles in the LF mode. Since these pole frequencies are relatively high, they do not affect the operating frequency bands of the circuit.

The proposed 26/38GHz dual-band LNA was implemented in a 65nm CMOS technology. The die micrograph is shown in Fig. 20.9.7. The LNA occupies a core area of 640μm × 220μm and draws 21.2mA from a supply voltage of 1V. Figure 20.9.4 shows the measured and simulated results of the proposed LNA. In the LF mode, the measured gain exhibits a maximum value of 20.6dB with a 3dB bandwidth of 25.3 to 28.6GHz. The measured IRR has a maximum value of 39.9dB at 26GHz and is greater than 30dB from 24.25 to 29GHz with the IF of 6GHz. In the HF mode, the measured gain shows a maximum value of 19dB with a 3dB bandwidth of 36.5 to 47.2GHz. The measured IRR has a maximum value of 41.6dB at 38.5GHz and is greater than 40dB from 36.8 to 40.3GHz with the IF of 6GHz. In addition, as expected, changing the IF can further boost the IRR in both the LF and HF ranges, resulting in good IR performance over a wide operating bandwidth. In the LF band, the NF ranges from 3.5 to 3.9dB, and input 1dB compression point (IP1dB) varies between -18.8 and -13dBm. In the HF band, the measured NF is between 3.9 and 5.5dB, and the IP1dB lies between -17 and -13.5dBm.

In addition, to further validate image-rejection capability and estimate the impact of image on the quality of the 5G NR FR2 signal, both the desired modulated signal (P_{RF}) at 26/38GHz band and a 38/26GHz one-tone image interference signal (P_{IM}) are applied the LNA through a wideband power combiner, with a single-balanced downconversion mixer capable of dual-side mixing connected behind the LNA through bonding wires. The measured error vector magnitude (EVM) deteriorates by just 1.0/1.5dB for the 400MHz 64-QAM signal in the 26/38GHz band when the image increases to 8/5dB higher than the power level of the desired signal. The EVMs under their respective interference signals are still less than -33dB, as shown in Fig. 20.9.5. It demonstrates that the proposed LNA features excellent image rejection for multi-band receiver application.

Figure 20.9.6 summarizes and compares the performance of recent mm-wave dual-band LNAs with IR functions. Compared to the notch filters used in both inter-stage and output matching in [7], this design uses the proposed image-rejection technique in inter-stage matching, achieving comparable IRR. Compared to other LNAs, this work exhibits superior IRR performance while having similar gain and noise characteristics. Overall, the proposed LNA achieves switchable dual-band operation with a compact size and high image-rejection capability. It demonstrates the ability to adapt to narrow-LO mixing scheme for multi-band 5G communication.

Acknowledgement:
This work was supported by the Zhejiang Provincial Natural Science Foundation of China under Grant LD25F040005. Corresponding author: Jincai Wen.

References:

[1] M.-Y. Huang et al., "A 24.5–43.5-GHz Ultra-Compact CMOS Receiver Front End with Calibration-Free Instantaneous Full-Band Image Rejection for Multiband 5G Massive MIMO," *IEEE JSSC*, vol. 55, no. 5, pp. 1177-1186, May 2020. http://doi.org/10.1109/JSSC.2019.2959495

[2] L. Gao and G. M. Rebeiz, "A 20–42-GHz IQ Receiver in 22-nm CMOS FD-SOI with 2.7–4.2-dB NF and –25-dBm IP1dB for Wideband 5G Systems," *IEEE TMTT*, vol. 69, no. 11, pp. 4951-4960, Nov. 2021. http://doi.org/10.1109/TMTT.2021.3095944

[3] S. Mondal et al., "A Reconfigurable 28/37GHz Hybrid-Beamforming MIMO Receiver with Inter-Band Carrier Aggregation and RF-Domain LMS Weight Adaptation," *ISSCC*, pp. 72-74, Feb. 2018. http://doi.org/10.1109/ISSCC.2018.8310189

[4] Z. Deng et al., "A 23–40-GHz Phased-Array Receiver Using 14-Bit Phase-Gain Manager and Wideband Noise-Canceling LNA," *IEEE JSSC*, vol. 58, no. 3, pp. 647-661, Mar. 2023. http://doi.org/10.1109/JSSC.2022.3223373

[5] Y. Yu et al., "A 26/28/39-GHz Reconfigurable Phased-Array Receiver Front-End with Built-In Calibration Technique for 5G New Radio," *IEEE JSSC*, vol. 60, no. 2, pp. 382-393, Feb. 2025. http://doi.org/10.1109/JSSC.2024.3425889

[6] D. Cheng et al., "A 28/39 GHz Concurrent/Band-Switching LNA with Three-Winding Transformer and Common-Gate-Based Multiplexer Supporting Multistream and Multiband 5G FR2 Communication," *IEEE TMTT*, vol. 73, no. 4, pp. 1924-1937, Apr. 2025. http://doi.org/10.1109/TMTT.2025.3529990

[7] N.-Z. Sun et al., "A Compact Millimeter-Wave Reconfigurable Dual-Band LNA with Image-Rejection in 28-nm Bulk CMOS for 5G Applications," *IEEE JSSC*, vol. 59, no. 10, pp. 3406-3416, Oct. 2024. http://doi.org/10.1109/JSSC.2024.3400952

[8] F. Zhao et al., "A Band-Shifting Millimeter-Wave T/R Front-End Using Inductance-Mutation Transformer Technique for Multiband Phased-Array Transceivers," *IEEE JSSC*, vol. 59, no. 5, pp. 1323-1336, May 2024. http://doi.org/10.1109/JSSC.2024.3353220

[9] H. Lin et al., "Design of 22.6-29.5/ 30.4-43.5 GHz Dual-Band Low Power LNA with 2.6-3.8 dB NF for Millimeter-Wave 5G Applications in 28-nm CMOS," *IEEE RFIC*, pp. 127-130, June 2025. http://doi.org/10.1109/RFIC61188.2025.11082893

[10] D. Cheng et al., "A Compact 28/39 GHz Dual-Band Concurrent/Band-Switching LNA for 5G Multi-Band Multi-Stream Applications," *IEEE RFIC*, pp. 315-318, June 2024. http://doi.org/10.1109/RFIC61187.2024.10600006

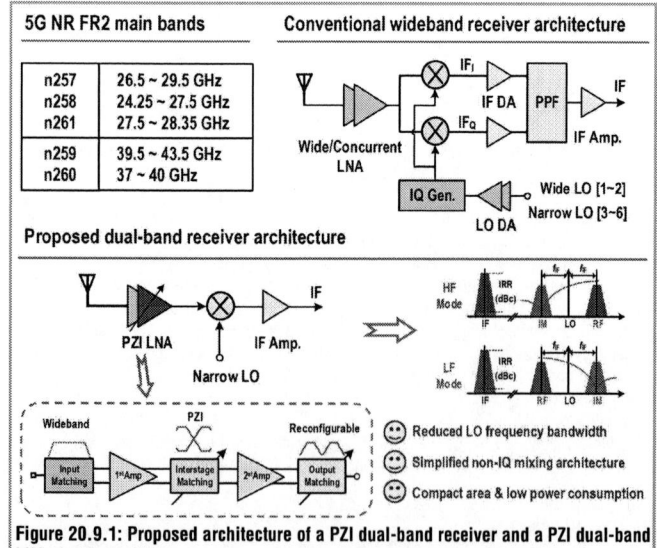

Figure 20.9.1: Proposed architecture of a PZI dual-band receiver and a PZI dual-band LNA circuit structure.

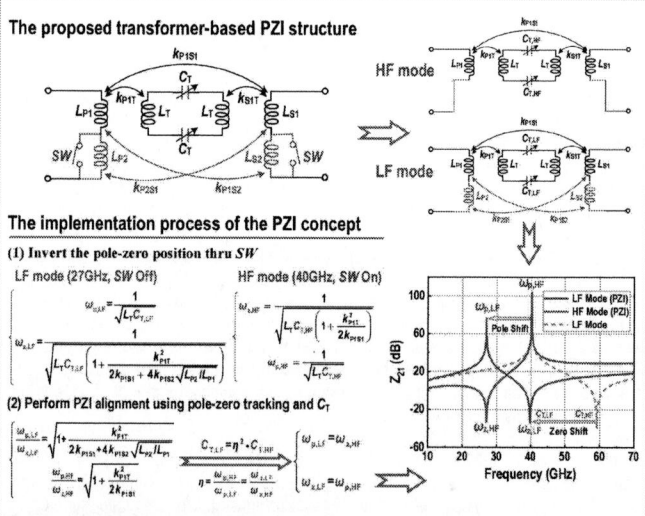

Figure 20.9.2: Proposed transformer-based PZI structure, and the implementation process of dual-band pole-zero inversion.

Proposed PZI dual-band LNA and the down mixer

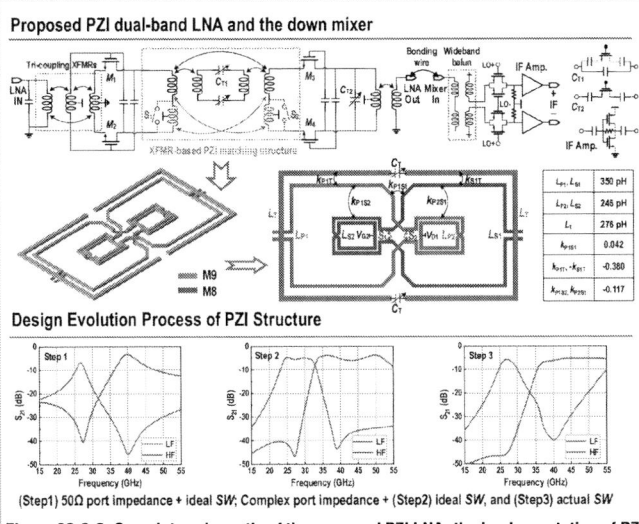

Figure 20.9.3: Complete schematic of the proposed PZI LNA, the implementation of PZI image-rejection matching structure, and design process.

Measurement Results

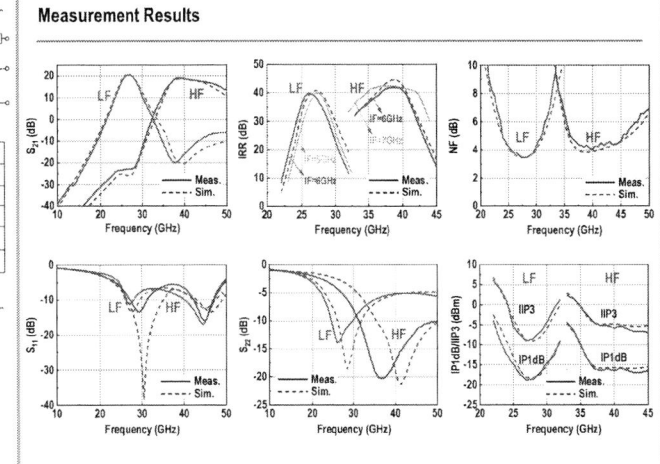

Figure 20.9.4: Measured results of the PZI dual-band LNA (S-parameters, gain, NF, IRR, IP1dB and IIP3).

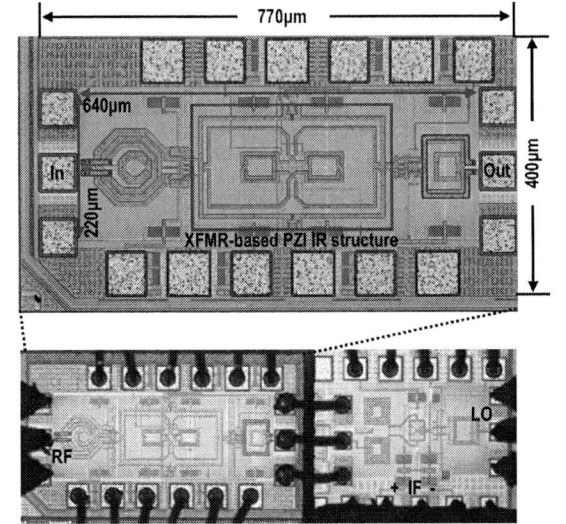

26GHz (EVM=-33.6 dB, P_{IM}/P_{RF}=8 dB)   38GHz (EVM=-33.9 dB, , P_{IM}/P_{RF}=5 dB)

Figure 20.9.5: EVM measurement results of the proposed dual-band RX with input signal consisting of a modulated signal and a single-tone image interference.

	This Work		JSSC 2024 [7]		JSSC 2024 [8]		RFIC 2025 [9]		RFIC 2024 [10]	
Technology	65nm CMOS		28nm CMOS		28nm CMOS		28nm CMOS		65nm CMOS	
Wide LO Requirement (GHz)#	NO (30.25 to 37.5)		YES (16.25 to 35.5)		NO (29 to 35.5)		NO (28 to 38.5)		NO (29.75 to 38)	
Freq. (GHz)	25.3 to 28.6	36.5 to 47.2	23.8 to 33.5	34.4 to 41.4	23.9 to 35.3	29.2 to 43.7	22.6 to 29.5	30.4 to 43.5	21.8 to 30.3	32.6 to 45.1
Peak Gain (dB)	20.6	19	18.1	18.9	21.3	18.3	19.7	19.2	26.3	25.5
NF (dB)	3.5 to 3.9	3.9 to 5.5	2.5 to 3.5	2.8 to 3.5	5.1	5.8	2.6 to 3.6	2.9 to 3.8	3.6	3.8
IRR (dBc)	39.9	41.6	>40	>38	>18		>20	>16	10.3 to 23.2	15.7 to 30.4
IP1dB (dBm)	-18.8 to -13	-17 to -13.5	-19 to -15.1	-18.5 to -15.2	/	/	-21 to -18.7	-22.9 to -18.3	/	/
IIP3 (dBm)	-8.5 to -5.5	-5.4 to -3.9	-9.5 to -6.8	-11.1 to -8	-10.8	-10.1	/	/	-18.5	-16.4
P_{DC} (mW)	21.2		14		19.7	19.8	10		19.2	
Core Area (mm²)	0.14		0.09		0.26*		0.09		0.1	

#:Assume that their operating frequency bands are all 5G FR2 (24.25 to 29.5GHz and 37 to 43.5GHz).
*: With power amplifier.

Figure 20.9.6: Performance summary and comparison with prior-art dual-band LNAs.

Figure 20.9.7: Die micrograph (top), photo of the receiver.

ISSCC 2026 / SESSION 20 / RF TRANSCEIVER SUBSYSTEMS FROM CM-WAVE TO THz / 20.10

20.10 A 214-to-242GHz Miniaturized Co-Packaged PA-Antenna Array with 29dBm Lens-less EIRP in a 0.13µm SiGe Process

Qianqi Meng, Zhihua Wang, Peigen Zhou, Dawei Tang, Siyuan Tang, Junyue Xiao, Rui Zhou, Jinben Li, Zhe Chen, Jixin Chen, Hao Gao, Wei Hong

Southeast University, Nanjing, China

Abstract

This work presents a miniaturized, high-power, and wideband THz PA-antenna array, realizing 29dBm lens-less peak EIRP from 214 to 242GHz. To improve the BW and P_{sat} of the PA, a folded 10th-order power-splitting/combining and staggered-matching method is proposed, achieving 17.4dB peak gain and 16.5dBm peak P_{sat} from 160 to 248GHz. To address the efficiency and BW limitation of the AoC, a co-packaged ULP slot-patch antenna is integrated, extending the radiation efficiency to 66.7% and BW to 36GHz.

The next generation of mobile and edge devices envision the convergence of ultra-high-resolution sensing and imaging, integrated seamlessly into compact form factors such as smartphone bezels. This ambition necessitates a trend toward the terahertz (THz) band, where more spectrum resources could be allocated. Existing solutions, relying on bulky lenses, off-chip waveguides, or narrowband oscillator arrays, fail to meet the co-design requirements in a compact footprint, while the reported lens-less systems remain constrained by the limited radiation power, bandwidth, and efficiency of solid-state power amplifiers (PAs) [1-6] and antennas, particularly in silicon-based technologies. To address these challenges, this work investigates a miniaturized, high-power, and wideband THz PA-antenna array leveraging silicon process technologies, realizing 29dBm lens-less peak effective isotropic radiated power (EIRP) from 214 to 242GHz. To improve the bandwidth and output power of the PA, a folded 10th-order power splitting/combining and staggered-matching network is adopted, achieving a 17.4dB peak gain from 135 to 241GHz and 16.5dBm peak output power from 160 to 248GHz. To address the efficiency and bandwidth limitation of an on-chip antenna, a co-packaged ultra-low-profile (ULP) slot-patch antenna is integrated, extending the radiation efficiency to 66.7% and a matching bandwidth between 204 and 240GHz. These features extend radar detection range to tens of meters and resolution to the millimeter level, all within an ultra-compact system with a form factor < 0.01cm³.

Figure 20.10.1 shows the system architecture of the PA-antenna hybrid power-combining array front-end (FE), integrating three key components: a wideband high-output-power PA, an ULP slot-patch antenna array, and a 6× multiplier chain. The PA implements a quad-channel differential power-combining topology that employs multi-stage equivalent LC-network-based power-splitting and -combining techniques at both the input and output, achieving a saturated output power (P_{sat}) of 16.5dBm at 222GHz with a 3dB small-signal bandwidth exceeding 110GHz. The antenna array, co-designed with the 6× multiplier-PA chains, utilizes optimized slot-loaded elements in a low-profile configuration with a λ/2 inter-element spacing, achieving superior radiation characteristics including a peak array gain of 9.1dBi and >15dB sidelobe suppression. The 6× multiplier chain incorporates integrated harmonic-rejection filter and power amplification stages to provide >4dBm driving power for the PA-antenna array.

Figure 20.10.2 details the wideband, high-power PA implementation, employing a three-stage inductively gm-boosted cascode architecture with four-way hybrid power combining. The design features multi-stage staggered matching to achieve flat gain response across the operating band. Traditional zero-degree power combiners using extended microstrip routing encounter significant area consumption and insertion loss due to skin effect and conductor loss in a THz frequency band. Although Marchand and transformer baluns can partially mitigate these issues, the inherent limitations of THz transistor performance, combined with gm-boosting-induced impedance degradation, render second-order baluns insufficient for full-band matching. To address these challenges, the design implements two key innovations: a compact 6th-order differential power combiner/splitter employing a microstrip-balun equivalent topology, modeled as an equivalent C-L-C network. This structure integrates a 6th-order matching network and leverages multiple metal layers with vertical-crossing layout techniques to minimize electromagnetic interference while enabling a folded configuration. A 4th-order low-loss zero-degree LC-ladder-based combiner is utilized, absorbing the pad capacitance and preceding matching elements, and maintaining a <1dB combining loss. The adaptability of this topology allows a similar implementation as the input matching network. The synthesized 10th-order input/output network achieves >100GHz matching bandwidth.

Figure 20.10.3 presents schematic and measurement results of the 6× frequency multiplier chain and the measurement results of the PA. The frequency multiplier chain combines a frequency tripler (3×FM) and a push-push doubler (PPFM) utilizing a two-way power combination structure with multi-conductor-coupling technique. Measurement results demonstrate a 3dB output power bandwidth spanning 158 to 248GHz, delivering over 4dBm driving power to a subsequent PA stage. Measurement results of the PA demonstrate a peak gain of 17.4dB at 215GHz with a 106GHz 3dB bandwidth (135 to 241GHz). Combined with the frequency multiplier chain, the PA delivers 16.5dBm maximum P_{sat} at 223GHz, maintaining <3dB power roll-off across 160 to 248GHz. Since the system focuses on the maximum EIRP, the linearity of the PA is not a concern.

Figure 20.10.4 presents the ULP slot-patch antenna of the radiation array FE. To address the challenges in system integration, a broadside radiation configuration is adopted with enhanced thermal management. Conventional broadside radiation antennas on silicon substrates (ε_r=11.9) exhibit limited performance. Our solution introduces a periodic slot structure to generate additional resonance modes. Bandwidth expansion is achieved by aligning slot resonance frequencies within the operational passband. Further improvements incorporate low-profile stacked layers comprising 35µm copper and 60µm Megtron6 dielectric (ε_r=3.71) above the on-chip antenna. This configuration achieves 204-to-240GHz impedance bandwidth (S11<-10dB) with 66.7% peak radiation efficiency and a 9.1dBi array gain for a 4×1 antenna configuration. Due to intrinsic limitation of the silicon process, the bandwidth of on-chip antenna is hard to further increase. However, a heterogeneously integrated ULP slot-patch antenna can be utilized to accommodate the PA bandwidth.

Fabricated in a 0.13µm SiGe BiCMOS process, the 214-to-242GHz lens-less PA-antenna hybrid power-combining array FE achieves a compact chip size of 2.54mm×1.85mm (including all bonding pads). The four-way PA and ULP slot-patch antenna array employ co-optimized horizontal dimension to maximize chip area utilization. As shown in Fig. 20.10.5, the PA-antenna array was directly wire-bonded to a test PCB for characterization. EIRP measurements were conducted in a terahertz anechoic chamber using the following configuration: The PA-antenna array was mounted on a carrier stage, with a one-sixth subharmonic input signal delivered through a coaxial probe. The receiver subsystem integrates a 12× LO chain to facilitate signal detection. Measurement results demonstrate that the integrated lens-less PA-antenna array achieves a 29dBm peak EIRP at 232GHz with 16.35% EIRP/PDC efficiency, exhibiting a 3dB bandwidth from 214 to 242GHz. The measured radiation pattern at 220GHz reveals a beamwidth of 25°. Due to the variation of the antenna gain, the overall efficiency fluctuates considerably.

Figure 20.10.6 presents performance comparisons with prior PAs and FEs operating around 220GHz and implemented in silicon technologies. The PA achieves a 3dB bandwidth of 106GHz (135 to 241GHz) with 16.5dBm peak output power and P_{sat} > 15dBm, demonstrating competitive performance in figures of merit (FOM₁ and FOM₂). The lower section of Fig. 20.10.6 highlights the performance metrics of the hybrid power-combining array FE. By combining the high-output-power PAs with gain- and efficiency-enhanced antennas, the PA-antenna array achieves a maximum EIRP of 29dBm at 232GHz without lens. This implementation simultaneously attains a peak EIRP/PDC efficiency of 16.35%. The die micrograph is shown in Fig. 20.10.7.

A proof-of-concept radar system was developed for verification, incorporating three essential components: a commercial frequency-modulated continuous-wave (FMCW) signal source, a custom-designed receiver chain, and a baseband processor for sensing/imaging applications. Experimental results successfully demonstrate resolution of adjacent targets and achieve 5.1mm range resolution over a 28GHz operational bandwidth. Long-range evaluation confirms effective detection capability beyond 26.2meters, validating the co-packaged PA-antenna-array far-field operation performance.

Acknowledgement:

This work was supported in part by Major Project of Jiangsu Province under Grant BG2024034, in part by the National Natural Science Foundation of China under Grant No. 62188102, and in part by the ZTE Industry-University-Institute Cooperation Funds. (Corresponding authors: Peigen Zhou, Dawei Tang, Wei Hong)

References:

[1] M. H. Eissa and D. Kissinger, "A 13.5dBm Fully Integrated 200-to-255GHz Power Amplifier with a 4-Way Power Combiner in SiGe:C BiCMOS," *ISSCC*, pp. 82-84, Feb. 2019. https://doi.org/10.1109/ISSCC.2019.8662424

[2] T. Bücher et al., "A Broadband 300 GHz Power Amplifier in a 130 nm SiGe BiCMOS Technology for Communication Applications," *IEEE JSSC*, vol. 57, no. 7, pp. 2024-2034, July 2022. https://doi.org/10.1109/JSSC.2022.3162079

[3] J. Yu et al., "A 211-to-263-GHz Dual-LC-Tank-Based Broadband Power Amplifier With 14.7-dBm P_{SAT} and 16.4-dB Peak Gain in 130-nm SiGe BiCMOS," *IEEE JSSC*, vol. 58, no. 2, pp. 332-344, Feb. 2023. https://doi.org/10.1109/JSSC.2022.3192043

[4] G. Park et al., "A 15.7-dBm 164–270 GHz Power Amplifier with Asymmetric Slotline-Based Series-Parallel Combiner in 130-nm SiGe BiCMOS Technology," *IEEE RFIC*, pp. 195-198, June 2024. https://doi.org/10.1109/RFIC61187.2024.10600010

[5] S. Fu et al., "A 6.3 dBm 258-314 GHz Power Amplifier using a Broadband 8-way SQWL Power Combiner in 130-nm SiGe BiCMOS Technology," 2024 *ESSERC*, pp. 713-716, Sep. 2024. https://doi.org/10.1109/ESSERC62670.2024.10719572

[6] J. Zhang et al., "A 124-to-152-GHz Power Amplifier Exploiting Chebyshev-Type Two-Section Wideband and Low-Loss Power-Combining Technique in 28-nm CMOS," *IEEE TMTT*, vol. 71, no. 5, pp. 1852-1865, May 2023. https://doi.org/10.1109/TMTT.2022.3231599

[7] C. Wang et al., "A 236-to-266GHz 4-Element Amplifier-Last Phased-Array Transmitter in 65nm CMOS," *ISSCC*, pp. 415-417, Feb. 2024. https://doi.org/10.1109/ISSCC49657.2024.10454273

[8] J. Wang et al., "A 232-to-260GHz CMOS Amplifier-Multiplier Chain with a Low-Cost, Matching-Sheet-Assisted Radiation Package and 11.1dBm Total Radiated Power," *ISSCC*, pp. 542-544, Feb. 2025. https://doi.org/10.1109/ISSCC49661.2025.10904783

[9] S.-Y. Tang et al., "A 200-GHz Phased Array Transmitter with Element-Level Scanning Antenna for ± 45° Scanning Range with 0.71 λ_0 Antenna Pitch," *IEEE RFIC*, pp. 171-174, June 2025. https://doi.org/10.1109/RFIC61188.2025.11082962

[10] X. Yi et al., "A 220-to-320-GHz FMCW Radar in 65-nm CMOS Using a Frequency-Comb Architecture," *IEEE JSSC*, vol. 56, no. 2, pp. 327-339, Feb. 2021. https://doi.org/10.1109/JSSC.2020.3020291

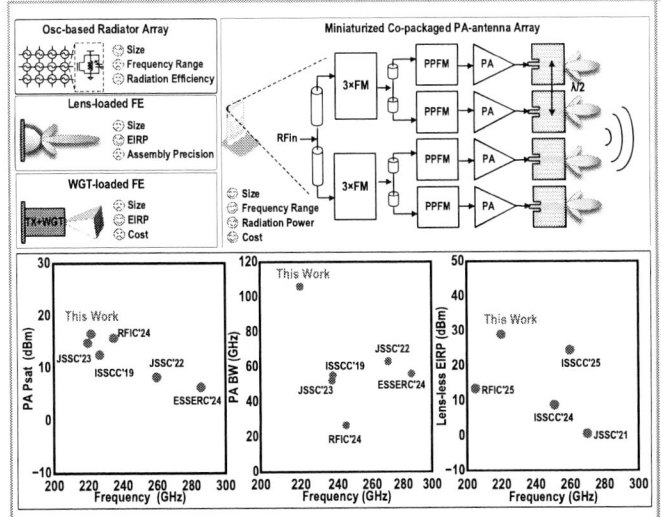

Figure 20.10.1: THz system architecture, detailed diagram of the miniaturized co-packaged PA-antenna array, and performance comparison.

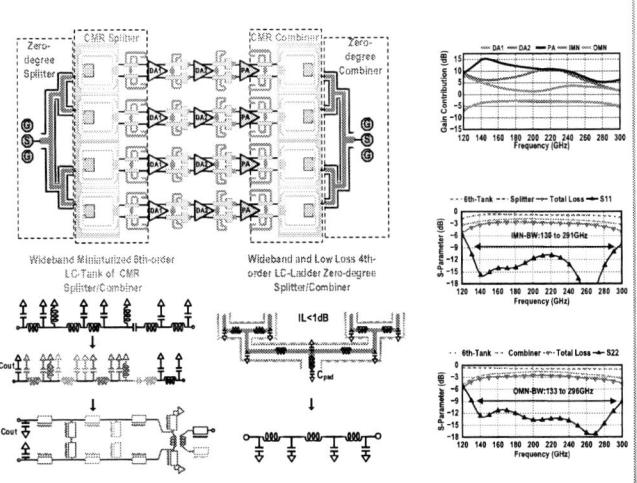

Figure 20.10.2: Schematic of the PA, gain-staggered technique, and an equivalent wideband miniaturized input and output matching network.

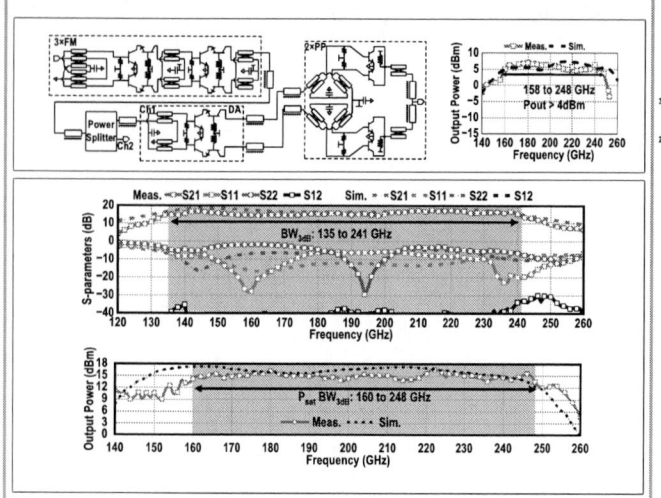

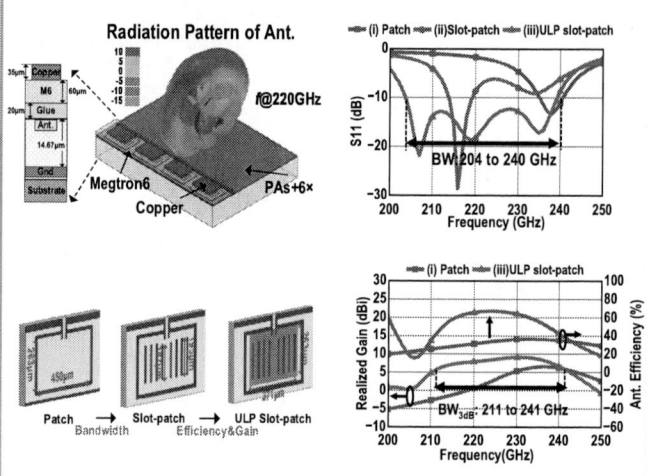

Figure 20.10.3: Schematic of 6× multiplier chain and measured output power, measured S-parameters, and P_sat of the PA.

Figure 20.10.4: Details of the co-packaged ULP slot-patch antenna.

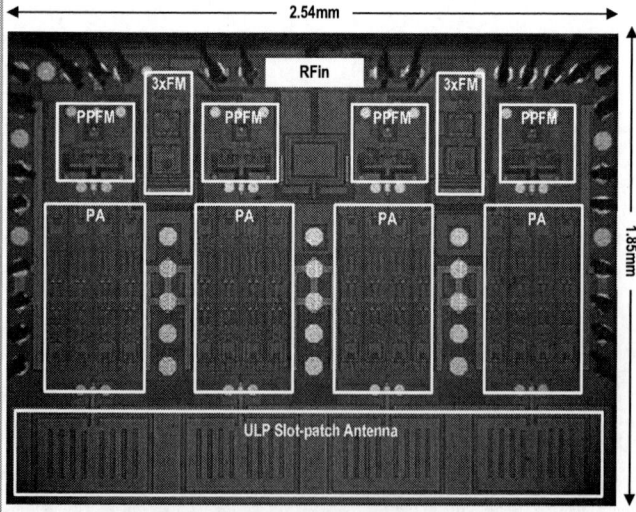

Figure 20.10.5: EIRP measurement setup, the measured EIRP, normalized gain, and EIRP/DC.

References	This PA	[1]ISSCC'19	[2]JSSC'22	[3]JSSC'23	[4]RFIC'24	[5]ESSERC'24	[6]TMTT'23
Technology	0.13μm SiGe	0.13μm SiGe	0.13μm SiGe	0.13μm SiGe	0.13μm SiGe	0.13μm SiGe	28nm CMOS
f_{oNF}(GHz)	450	500	650	450	450	450	300
Frequency(GHz)	135 to 241 (56%)	200 to 255 (23%)	239 to 302 (23%)	211 to 263 (22%)	231.5 to 258* (11%)	258 to 314 (20%)	124 to 152 (20%)
BW_{2dB}(GHz)	106	55	63	52	26.5 (78^8)	56	28
Peak Gain(dB)	17.4	12.5	23.0^8/20.1	16.4	20.1	15	22.6
Peak P_{sat}(dBm)	16.5	12.5*	8.2	14.7	15.7	6.3	16.2
Peak PAE(%)	3.11	1	1.38	3.13	2.12	0.8	8.6
P_{sat} BW_{3dB}(GHz)	88 (160 to 248)	47 (211 to 258)*	94 (223 to 317)	45 (210 to 255)*	106 (164 to 270)	40 (260 to 300)*	38* (122 to 160)*
Area(mm²)	0.71	0.83	0.26	0.83	0.61	0.97	0.33
FoM1	6.37	0.14	0.23	0.21	0.16	0.19	0.24
FoM2	54.6	43.1	49.2	46	43.6	38.4	48.9

FoM1=$\sqrt[4]{G(f_l f_{low}/f_h)}^2 (BW_{rel})^2$ FoM2=P_{sat}[dBm]+20lg(BW[GHz])+20lg(Freq/f_{max})-20lg(Area[mm²]))
*Estimated by the figures. ^§5dB-bandwidth. ^8Without balun loss

References	This Work	[7]ISSCC'25	[8]ISSCC'25	[9]RFIC'25	[10]JSSC'21
Architecture	Co-packaged PA-Ant. Array	PA-Ant. Array	Mul.-Ant. Array	Mul.-Ant. Array	Lens-Loaded Multiplier-Ant
Technology	0.13μm SiGe	65nm CMOS	Intel16 CMOS	0.13μm SiGe	65nm CMOS
Element Num.	4	4	30	4	5
Frequency(GHz)	214 to 242	236 to 266	232 to 260	191 to 205	220 to 320
Bandwidth(GHz)	28	30	28	14	20*5
Ant. Gain(dBi)	9.1	6	N.A.	N.A.	0
Radiation Eff.(%)	66.7	87	41.8	N.A.	25
Core EIRP(dBm)	29.0	8.8	24.5	13.5	0.6
Core EIRP/PDC(%)	16.35	0.90	5.12	0.56	0.14
DC Power(W)	4.86	0.84	5.5	4	0.84
Beam Steering Capability	No	Yes	No	Yes	No
Chip Area(mm²)	4.7	9.9	10.42	7.71	5
Form Factor(cm³)	<0.01	<0.01	<0.01	<0.01	>10 (w/ Lens)

Figure 20.10.6: Comparison with prior-art silicon-based PAs and FE arrays operating around 200GHz.

Figure 20.10.7: Die micrograph of the PA-antenna array.

• 2026 IEEE International Solid-State Circuits Conference

ISSCC 2026 / SESSION 20 / RF TRANSCEIVER SUBSYSTEMS FROM cm-WAVE TO THz / 20.11

20.11 A 215GHz 8×8 Radiator-Oscillator Array with Robust Coupling Achieving 25.5dBm EIRP and 12.9% FTR

Ahmed Elmenshawi[1], Sriram Muralidharan[2], Mona M. Hella[1]

[1]Rensselaer Polytechnic Institute, Troy, NY, [2]Oso Semiconductor, Berkeley, CA

Abstract

A 215GHz 8×8 radiator-oscillator array with robust coupling achieving 25.5dBm average EIRP over a 12.9% continuous frequency tuning range. The proposed coupling scheme, based on distributed coupling through multi-functional networks, achieves robust 2D-scalable coupling over a wide frequency range with sub-nanosecond synchronization. A high-directivity horn-antenna radiation package with a glass superstrate is proposed to achieve a high EIRP and radiation efficiency.

THz waves are key enablers for novel sensing and imaging solutions for which wide bandwidth, high power, and low phase-noise are required. However, low-cost silicon-based THz radiators fall short of delivering the needed performance due to the limited f_T/f_{MAX} of transistors and the low Q-factor of passives. In addition, the weak non-linearity of CMOS FETs leads to low power generation in harmonic radiators. Coupled-oscillator arrays are good candidates to overcome these limitations. For a coupled-oscillator array of N-oscillators, the generated power and phase noise (PN) are 10log(N) dB better compared to a single oscillator. In addition, wideband high-Q-factor resonators can be utilized to achieve a wide frequency tuning range [1].

Coupled-oscillator arrays, however, come with their own challenges. First, the coupling mechanism between the oscillators must ensure a single stable mode of oscillation. Second, the coupling network must be low loss to avoid tank de-Q and must not introduce significant reactive loading to maintain a wide frequency tuning range. Third, efficient power combining and radiation with high directivity is required to achieve a high EIRP. While the use of an antenna array can boost the directivity by 10log(N) dB, the congested layout of coupled-oscillator arrays may make it difficult to integrate a full-size antenna array with optimal spacing. Moreover, on-chip antennas on silicon substrates typically have low efficiency and directivity. High resistivity silicon lenses have been widely adopted [2-7]. However, the benefits of the lens are diminished when fed by an antenna array due to the undesirable beam-steering caused by antenna elements displacement from the lens axis [4]. In this work, a 2D scalable coupling scheme based on multi-functional networks (MFN) is presented to achieve a wide frequency tuning range. Two-level power combining, with local on-antenna and global quasi-optical power combining, is employed to overcome the limitations of antenna array integration. Furthermore, a low-cost horn-antenna package with a fused-silica superstrate is proposed for high-efficiency, high-directivity radiation.

The standalone VCO, shown in Fig. 20.11.1 (top), is discussed first. The standalone VCO utilizes a multi-section switch-loaded coupled-transmission-line resonator, which can achieve a wide frequency tuning range with a high Q factor without using any varactors [1]. One terminal of the resonator is connected to the VCO cross-coupled pair $M_{1,2}$, while the other terminal must be AC grounded. This is achieved in push-push operation by connecting the nodes out+ and out-, presenting a virtual ground at f_0. Capacitive degeneration with capacitors C_1 is employed at the source to boost high-frequency gain, while TL_S provides DC biasing. Inductor L_1 resonates with C_1 in the common mode, presenting a low impedance to the source of $M_{1,2}$ at the 2nd harmonic to boost the 2nd-harmonic power generation. In this work, the drain/gate and source MFNs are modified to allow for scalable 2D coupling. The proposed coupled-oscillator-array unit cell and the coupling scheme are shown in Fig. 20.11.1 (bottom).

Coupled-oscillator arrays are typically comprised of self-sustaining unit oscillators that are synchronized through coupling networks, which introduce parasitics and losses that limit the array performance. In addition, the locking range is typically narrow, leading to a reduced robustness to variations as well as limiting the size of the array [7]. Instead of direct cross-coupling of the gates and drains of the oscillator core, the drains and gates are magnetically coupled. The drain TL is broadside-coupled to the gate resonator, leading to a compact layout. Unit cells 1 and 2 are coupled at the drain/gate nodes d± and g±. In the desired differential mode, these nodes are virtual grounds, through which the 2nd harmonic is extracted, and DC bias is provided. To suppress common-mode oscillations, a small metal resistor R_{CM} is connected in [8] between nodes d± and g± to de-Q the tank. However, this also required using large decoupling capacitors at nodes d±. This is not applicable in our case since the 2nd harmonic is extracted at these nodes. If decoupling capacitors are not used, the bias/matching network loads the resonator in the common mode, pushing the resonance frequency to low frequency, where the gain is high and a small R_{CM} is not enough to suppress the common-mode oscillation. Instead, a large R_{CM} (~1kΩ) is used to completely disconnect the RF path between nodes d and g and only provide DC biasing to the gates. The other coupling path through the gate/drain coupled transmission lines is weak at low frequencies, since the electrical length of the coupled transmission line is shorter at lower frequencies. A 1D loop array could be constructed by coupling through the drain/gate MFN, which could be folded to create a pseudo-2D array. However, a loop with a large number of elements is at a higher risk of de-synchronization, limiting the size of the array. For robust synchronization and true 2D scalability, the source network is modified to allow for coupling. The degeneration capacitor is replaced by a transmission line TL_C. In the common mode between U_1 and U_3, nodes c± are virtual open and TL_C appears capacitive, boosting the gain and negative resistance of the cross-coupled core. In the differential mode, on the other hand, TL_C appears inductive, leading to a reduction in the gain. The core is carefully designed to ensure that the gain is low enough in the differential mode and oscillations can only be sustained in the common mode. TL_P and TL_S provide DC biasing for the source, while TL_P controls the impedance seen by the source at the 2nd harmonic to boost power generation.

The proposed topology of the oscillator array and startup are shown in Fig. 20.11.2. The loops of 8 oscillators are formed by coupling through the drain/gate MFNs. 8 such loops are coupled together through the source MFNs, creating an 8×8 coupled-oscillator array. Each oscillator is thus coupled to three neighboring oscillators. The proposed distributed coupling topology eliminates end effects and leads to a fast settling time, since all the unit cells oscillate collectively and only a single mode is supported. The array synchronizes in less than 100 cycles, two orders of magnitude lower than injection-locked arrays of the same size [9]. A two-tier power combining scheme is employed in this work, where local on-antenna and global quasi-optical power combining are both utilized. A compact dual-feed folded-slot antenna is proposed [10,11]. The 2nd-harmonic currents from two unit cells add in phase on the antenna. The 8×8 oscillator array feeds an 8×5 antenna array. The top and bottom rows are terminated with an LC termination to present the same impedance to the antenna. The antenna array radiates, and powers add in phase spatially. However, the high contrast between the permittivities of the silicon substrate and air generates surface waves and reduces radiation efficiency. High-resistivity silicon lenses are commonly used to reduce surface waves and achieve high directivity. However, their performance degrades when fed by a large antenna array and is sensitive to misalignments [4]. In this work, a custom pyramidal horn antenna is proposed as a low-cost alternative to silicon lenses. The package assembly is shown in Fig. 20.11.3. The input of the horn has dimensions comparable to the chip size, rather than standard waveguide dimensions. Consequently, the horn is fabricated with a standard CNC drilling process with low cost. The folded-slot antenna array and the horn are co-designed to excite the TE_{10} mode. The chip is attached with conductive epoxy to an FR4 PCB, and the PCB top copper layer underneath the chip acts as a reflector, directing radiation to the top side. DC bias and control is provided through wirebonds between the PCB and the pads on the chip top side, avoiding the need for flip-chipping. The horn is mounted on the PCB, enclosing the chip and the wirebonds. A 200μm fused silica glass superstrate (ε_r=3.8) is attached with non-conductive epoxy to the top side of the chip and operates as a matching layer between silicon and air, reducing surface waves and improving efficiency [12,13].

The chip, shown in Fig. 20.11.7, is fabricated in a 22nm FDSOI process. The chip has an area of 2.1mm×2.0mm, including pads. The chip is packaged as described previously. The measurement setup and radiation pattern are shown in Fig. 20.11.4. Standard VDI WR-05 conical horn and WR-03 diagonal horn are used on the receiver side. A VDI WR-05 SAX is used for spectrum measurements and an Erickson PM5 is used for power measurements. All measurements were performed at a 20cm distance, well in the far-field region of both the radiator and receiver. The radiator is mounted on a 2-axis motorized stage for radiation-pattern measurements. The radiator aperture center is offset by ~5cm from the mechanical rotation center. The true rotation angles and distance are corrected accordingly in post-processing. More importantly, there is a small angular offset (<10° in the measurement range) between the receiver horn boresight and the radiator-under-test, leading to a reduction in the receiver directivity as seen by the radiator. The directivity of the receiver horn as a function of the offset angle is calculated using the receiver horn datasheet and assuming a Gaussian beam. We note that this post-processing step is only required for radiation pattern measurements and does not affect peak EIRP measurements. The directivity of the radiator is calculated to be 24.8dBi by integrating the 2D radiation pattern over the measurement range. The control voltages v4:v1 are sequentially tuned from 0 to 0.8V and a continuous frequency tuning range of 27.7GHz (12.9%) at 215GHz center frequency is measured, as shown in Fig. 20.11.5. The EIRP is measured across the frequency tuning range for 0.8V and 0.9V supply voltages, with and without the glass superstrate. An average EIRP of 25.5dBm is measured with the glass superstrate and VDD=0.9V, with a peak EIRP of 29.5dBm at 201.7GHz. The use of the glass superstrate significantly boosts the EIRP by more than 3.5dB. The measured EIRP and directivity translates to an average radiated power of 0.7dBm, with a peak of 4.7dBm. The PN is also measured across the frequency tuning range. An average PN of -87.75dBc/Hz and -113.10dBc/Hz are measured at 1MHz and 10MHz offsets, respectively. The best-case PN is -91.7dBc/Hz at a 1MHz offset and -118.3dBc/Hz at a 10M offset, both measured at the lower end of the frequency band. The chip draws 1.22 to 1.44A from a 0.9V supply. The array is compared with the prior art in Fig. 20.11.6. The array simultaneously achieves a high EIRP and a wide frequency tuning range.

Acknowledgement:
The authors thank GlobalFoundries for chip fabrication and Analog Devices Inc. for support.

References:
[1] A. Elmenshawi et al., "A 224GHz 19.9% TR Varactor-Less VCO Utilizing a Multi-Section Switch-Loaded Coupled-Line Resonator," *ISSCC*, pp. 550-551, Feb. 2025. https://doi.org/10.1109/ISSCC49661.2025.10904538
[2] D. F. Filipovic et al., "Double-Slot Antennas on Extended Hemispherical and Elliptical Silicon Dielectric Lenses," *IEEE TMTT*, vol. 41, no. 10, pp. 1738-1749, Oct. 1993, doi: 10.1109/22.247919. https://doi.org/10.1109/22.247919
[3] S. Razavian and A. Babakhani, "A Highly Power Efficient 2×3 PIN-Diode-Based Intercoupled THz Radiating Array at 425GHz with 18.1dBm EIRP in 90nm SiGe BiCMOS," *ISSCC*, pp. 156-157, Feb. 2022. https://doi.org/10.1109/ISSCC42614.2022.9731731
[4] H. Jalili and O. Momeni, "A 0.46-THz 25-Element Scalable and Wideband Radiator Array with Optimized Lens Integration in 65-nm CMOS," *IEEE JSSC*, vol. 55, no. 9, pp. 2387-2400, Sep. 2020. https://doi.org/10.1109/JSSC.2020.2989897
[5] R. Jain et al., "A 0.42THz 9.2dBm 64-Pixel Source-Array SoC with Spatial Modulation Diversity for Computational Terahertz Imaging," *ISSCC*, pp. 440-442, Feb. 2020. https://doi.org/10.1109/ISSCC19947.2020.9063025
[6] R. Han and E. Afshari, "A CMOS High-Power Broadband 260-GHz Radiator Array for Spectroscopy," *IEEE JSSC*, vol. 48, no. 12, pp. 3090-3104, Dec. 2013. https://doi.org/10.1109/JSSC.2013.2272864
[7] H. Saeidi et al., "A 4×4 Distributed Multi-Layer Oscillator Network for Harmonic Injection and THz Beamforming with 14dBm EIRP at 416GHz in a Lensless 65nm CMOS IC," *ISSCC*, pp. 256-258, Feb. 2020. https://doi.org/10.1109/ISSCC19947.2020.9063076
[8] H. Jia et al., "A 53.6-to-60.2GHz Many-Core Fundamental Oscillator with Scalable Mesh Topology Achieving -136.0dBc/Hz Phase Noise at 10MHz Offset and 190.3dBc/Hz Peak FoM in 65nm CMOS," *ISSCC*, pp. 154-156, Feb. 2022. https://doi.org/10.1109/ISSCC42614.2022.9731581
[9] R. He and Y. Tousi, "Analysis of Stable Modes of a Scalable Coupled Oscillator Array," *IEEE TCAS-II*, vol. 68, no. 2, pp. 647-651, Feb. 2021. https://doi.org/10.1109/TCSII.2020.3014181
[10] T. M. Weller et al., "Single and Double Folded-Slot Antennas on Semi-Infinite Substrates," in *IEEE TAP*, vol. 43, no. 12, pp. 1423-1428, Dec. 1995. https://doi.org/10.1109/8.475932
[11] T. Chi et al., "A Multifeed Antenna for High-Efficiency On-Antenna Power Combining," *IEEE TAP*, vol. 65, no. 12, pp. 6937-6951, Dec. 2017. https://doi.org/10.1109/TAP.2017.2764119

[12] J. M. Edwards and G. M. Rebeiz, "High-Efficiency Elliptical Slot Antennas with Quartz Superstrates for Silicon RFICs," *IEEE TAP*, vol. 60, no. 11, pp. 5010-5020, Nov. 2012. https://doi.org/10.1109/TAP.2012.2207353
[13] J. Wang et al., "A 232-to-260GHz CMOS Amplifier-Multiplier Chain with a Low-Cost, Matching-Sheet-Assisted Radiation Package and 11.1dBm Total Radiated Power," *ISSCC*, pp. 542-544, Feb. 2025. https://doi.org/10.1109/ISSCC49661.2025.10904783
[14] H. Jalili and O. Momeni, "A 0.34-THz Wideband Wide-Angle 2-D Steering Phased Array in 0.13-μm SiGe BiCMOS," *IEEE JSSC*, vol. 54, no. 9, pp. 2449-2461, Sep. 2019. https://doi.org/10.1109/JSSC.2019.2925523

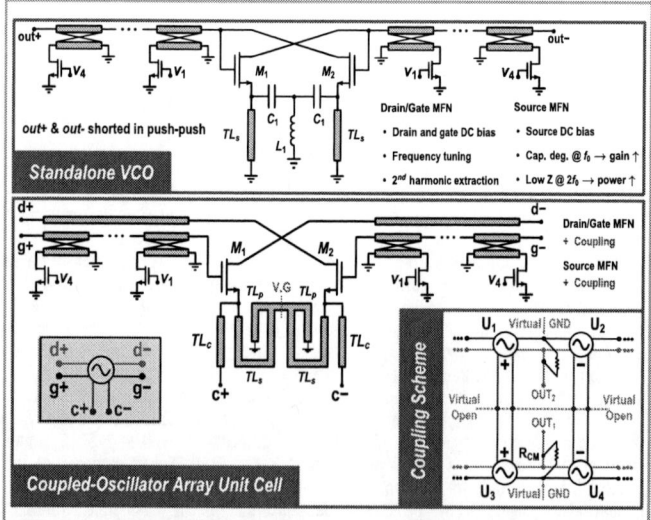

Figure 20.11.1: Standalone VCO schematic (top); and coupled-oscillator-array unit-cell schematic and coupling scheme (bottom).

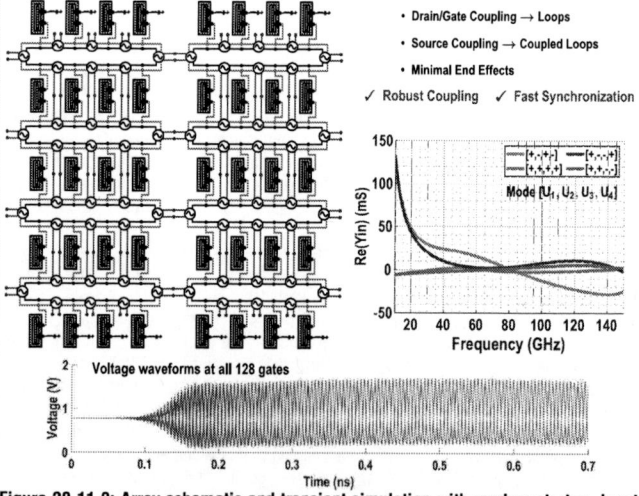

Figure 20.11.2: Array schematic and transient simulation with random startup. Input conductance at the gate for different modes (right). Only [+,-,+,-] mode can sustain oscillation.

ISSCC 2026 / SESSION 20 / RF TRANSCEIVER SUBSYSTEMS FROM cm-WAVE TO THz / 20.11

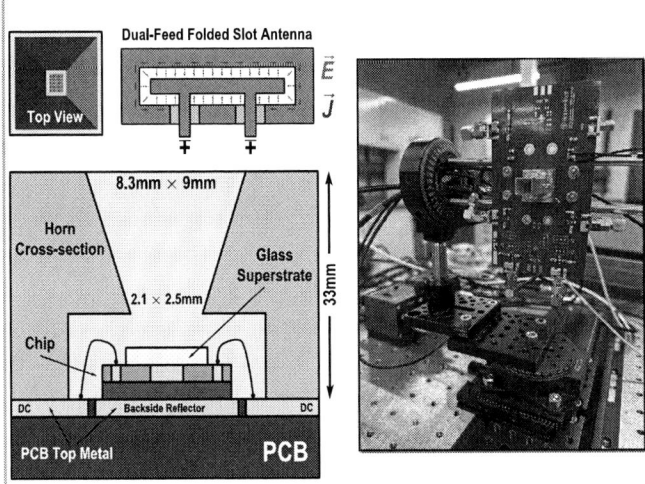

Figure 20.11.3: Horn antenna radiation package with glass superstrate. Dual-feed folded-slot antenna element (top).

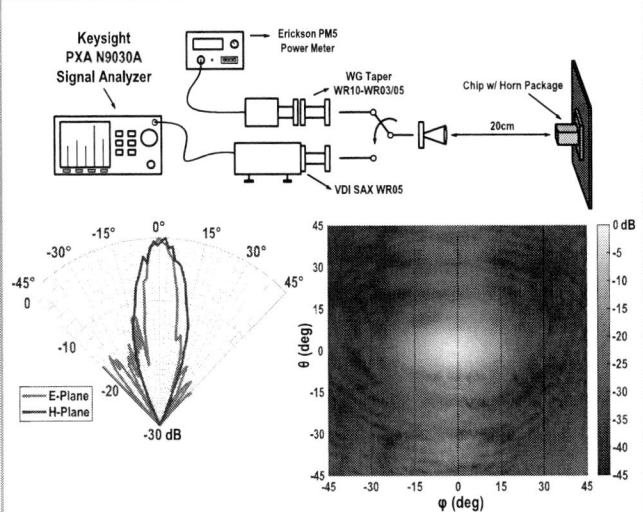

Figure 20.11.4: Radiation measurement setup and results.

Figure 20.11.5: Array measurements results: frequency tuning, EIRP and phase noise. Far field region > 13cm.

Figure 20.11.6: Comparison with the prior art. This work simultaneously achieves high EIRP and wideband frequency tuning.

Ref.	This Work	S. Razavian ISSCC'22 [3]	H. Jalili JSSC'20 [4]	R. Jain ISSCC'20 [5]	H. Jalili JSSC'19 [14]	R. Han JSSC'13 [6]
Process	22nm FDSOI	90nm SiGe	65nm CMOS	0.13μm SiGe	0.13μm SiGe	65nm CMOS
Harmonic	2	5	4	2[#]	4	2
Supply	0.9	2.5	0.9	2.1	0.8	1.5
Freq.	215.0	425	460	420	344	260
TR (GHz)	27.7	62.1	40.9	2.9	51.9	24.7
TR (%)	12.9	14.6	8.9	0.7	15.1	9.5
Radiation	Horn + Glass	Si Lens	Si Lens	Si Lens	Patch	Si Lens
Array Size	8×8	2×3	25	8×8	2×2	8
Steering	N	N	N	N	128/53°	N
EIRP (dBm)	29.5[*]	18.1	19.3	32.8	4.9	15.7
Prad (dBm)	4.7[**]	-5.1	-1.8	9.2	-6.8	0.5
P_{DC} (mW)	1098 to 1296	400	1470	6900	450	800
Efficiency (%)	0.269	0.08	0.045	0.19	0.046	0.14
PN (dBc/Hz)	-118.3[***] @ 10 MHz	-104 @ 10 MHz	-100.6 @ 10 MHz	-	-78.3 @ 1 MHz	-93.1 @ 10 MHz
Area (mm²)	4.2	0.98	3.94	12.6	1.2	2.3

[*] Peak EIRP, avg. is 25.5dBm   [**] Peak Prad, avg. is 0.7dBm
[#] Fundamental VCO + doubler   [***] Best-case PN, avg. is -113.1dBc/Hz

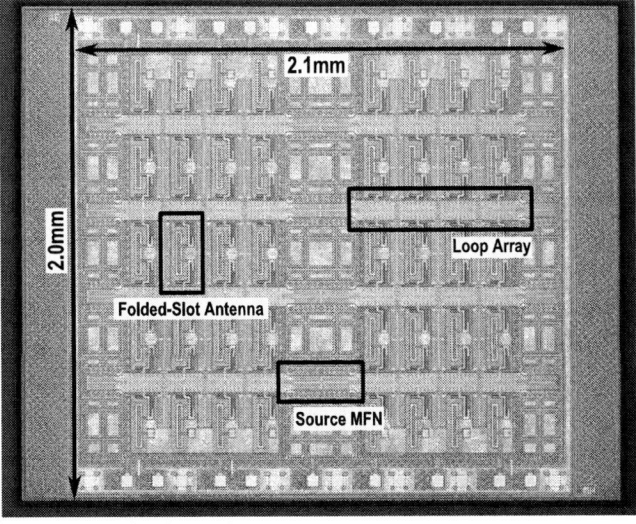

Figure 20.11.7: Die micrograph.

• 2026 IEEE International Solid-State Circuits Conference

979-8-3315-8937-0/26 $31.00 © 2026 IEEE

ISSCC 2026 / SESSION 21 / SENSOR INTERFACES / OVERVIEW

Session 21 Overview: *Sensor Interfaces*
ANALOG SUBCOMMITTEE

Session Chair: Drew Hall
University of California at San Diego
San Diego, CA

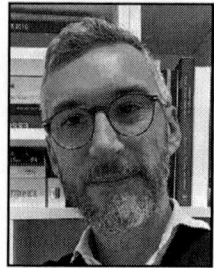

Session Co-Chair: Edoardo Bonizzoni
University of Pavia
Pavia, Italy

Advances in precision circuits and calibration techniques are enabling unprecedented levels of absolute accuracy in sensor interfaces, with demonstrated applications spanning automotive, wearable, and biosensing domains. The first paper in this session focuses on automotive systems, where a bipolar-sampling ADC enables accurate detection of busbar faults. The second and eighth papers extend these advances to biosensing and wearable platforms, improving readout time and dynamic range, respectively. The third, seventh, and ninth papers present state-of-the-art current sensors, while the sixth introduces a sampler with an automatic power-hold buffer that enhances SNDR. Finally, the remaining two papers feature temperature sensors that achieve best-in-class power efficiency and temperature accuracy.

1:30 PM

21.1 Sub-1mV-Accuracy, 24-CH Synchronous Battery Monitoring IC with Bipolar-Sampling ADCs and Calibration Engines
Jeongwon Han, Korea Advanced Institute of Science and Technology, Daejeon, Korea, Autosilicon, Daejeon, Korea
In Paper 21.1, KAIST and Autosilicon present a high-accuracy 24-channel battery-monitoring IC for fault detection in electric vehicles. The resulting system achieves sub-1mV accuracy across -2V and +5V and -40°C to +130°C.

1:55 PM

21.2 A Fully Integrated GMR Biosensor with On-Chip Coils and Sensors Achieving 605 Resolution FoM for Multiplexed PoC Diagnostics
Mengze Wu, University of California, San Diego, CA
In Paper 21.2, UC San Diego and Allegro MicroSystems report a fully integrated giant magnetoresistive (GMR) biosensor chip for point-of-care *in vitro* diagnostics using a magnetic immunoassay. The system achieves $120nT_{rms}$ input-referred noise, 0.38ppm sensitivity, <0.2% mismatch, and a resolution FoM of 605.

2:20 PM

21.3 A Temperature- and Aging-Compensated TMR Current Sensor with ±0.13% Sensitivity Variation from -40°C to 120°C
Tianxiang Qu, Fudan University, Shanghai, China
In Paper 21.3, Fudan University shows a TMR-based contactless current sensor that mitigates sensitivity drift due to temperature and aging. The adopted techniques enable ±0.13% sensitivity variation from -40°C to 120°C and reduce the aging drift to 0.1% at 25°C.

ISSCC 2026 / February 17, 2026 / 1:30 PM

2:45 PM

21.4 A Background-Calibrated NPN-Based Temperature Sensor with 0.05°C (3σ) Inaccuracy from -70°C to 125°C

Nandor G. Toth, TU Delft, Delft, The Netherlands

In Paper 21.4, TU Delft presents an NPN-based temperature sensor featuring a continuous-calibration scheme to correct all current-domain errors in its front-end. It achieves a 1-point trimmed inaccuracy of 0.05°C (3σ) from -70°C to +125°C, and a power sensitivity equal to 0.004°C/V.

3:00 PM

21.5 A 0.6V 625um² Fully Stacked RC-Based Temperature Sensor Using Low TCR Metal Resistor Achieving 0.017nJ·%²-Accuracy FoM in 2nm Gate-All-Around Process

Haejung Choi, Samsung Electronics, Yongin, Korea

In Paper 21.5, Samsung Electronics introduces an RC-based temperature sensor fabricated with a 2nm gate-all-around process that stacks a low TCR metal resistor and a ring-oscillator TDC. The conversion time is 12μs and the achieved accuracy-FoM is 0.017nJ%².

3:35 PM

21.6 A ±60mA-Inaccuracy Low-Side Average Current Sensor with Operating-Conditions-Insensitive Control Supporting 0.1-to-3A Load Range and Sub-100ns Sample Time for Automotive USB Charge Application

Jian-Jun Kuang, University of Electronic Science and Technology of China, Chengdu, China
SouthChip Semiconductor Technology, Chengdu, China

In Paper 21.6 UESTC and SouthChip Semiconductor Technology present a low-side average current sensor with operating-conditions-insensitive control. It enables ±60mA inaccuracy, duty cycles down to 100ns off-time, and load currents from 0.1 to 3A.

4:00 PM

21.7 A Battery-Free Wireless Electrochemical-Interface SoC Featuring 143dB Dynamic Range for Multimodal Wearables

Weixiao Wang, Zhejiang University, Hangzhou, China

In Paper 21.7 Zhejiang University introduces a battery-free wireless electrochemical interface SoC demonstrated in a sweatband prototype. The measured dynamic range is 143dB with a 23.3μW power consumption.

21

4:25 PM

21.8 A CMOS Hybrid Common-Gate Current-Integrating Sampler with >37dB SNDR Across 51GHz BW in a 128GS/s Front-End

Jun Dai, Tsinghua University, Beijing, China

In Paper 21.8 Tsinghua University describes a CMOS hybrid common-gate current-integrating sampler. The circuit achieves 48dB SFDR and 38dB SNDR near its 52GHz -3dB bandwidth.

4:50 PM

21.9 A -82.3dB THD+N 60V Fully Integrated Shunt-Resistor-Based In-Line Current Sensor with DLL-Assisted Dynamic Body-Biasing Technique

Heng Ma, TU Delft, Delft, The Netherlands

In Paper 21.9, TU Delft University describes a DLL-assisted dynamic body-biasing sensor applied to a shunt resistor-based current sensor. The circuit supports 60V common-mode, rejects PWM up to 2MHz, and achieves a peak THD+N of -82.3dB.

DIGEST OF TECHNICAL PAPERS • 363

979-8-3315-8937-0/26 $31.00 © 2026 IEEE

ISSCC 2026 / SESSION 21 / SENSOR INTERFACES / 21.1

21.1 Sub-1mV-Accuracy, 24-CH Synchronous Battery Monitoring IC with Bipolar-Sampling ADCs and Calibration Engines

Jeongwon Han[1,2], Sunsik Woo[2], Kwang-Seok Oh[2], Donghyeon Kim[2], Youngwoon Ko[2], Won-Jong Choi[1,2], Seungjun Han[2], Kyungdam Park[2], Juncheol Choi[2], Jinseop Lee[2], Jin-Yong Jeon[2], Young-Suk Son[2], Sang-Gug Lee[1], Kyeongha Kwon[1]

[1]Korea Advanced Institute of Science and Technology, Daejeon, Korea, [2]Autosilicon, Daejeon, Korea

Abstract

This paper presents a 24-CH BMIC with sub-1mV accuracy and enhanced fault detection for EVs. Dedicated bipolar-sampling ADCs enable simultaneous measurements across -2V to +5V input ranges, supporting busbar fault detection. Bipolar operation uses only 12V-rated MOS through active body biasing and clock doubling. Calibration engines correct for CH-dependent gain variations and input offsets. Fabricated in 0.13μm BCD, the BMIC achieves sub-1mV accuracy across -2V to +5V and -40°C to +130°C.

In battery packs of electric vehicles (EVs), multiple battery-monitoring ICs (BMICs) perform cell voltage/temperature measurements, cell balancing, and fault detection to diagnose the battery status (Fig. 21.1.1). LiFePO4 batteries, widely used in EVs, present monitoring challenges with their flat voltage profiles exhibiting ~1mV/% variation in the 20-to-80% state-of-charge (SOC) range [1], demanding ~1mV accuracy. Prior-art BMICs use multiplexed architectures that share a single ADC with a level shifter (LS) across multiple channels (CHs) for sequential measurement [2]. This requires high-voltage (HV) MOS rated for the full battery stack voltage (V_{TOP} = 70V for 14 CHs) and suffers from a latency-accuracy trade-off. The automotive industry's shift to advanced, higher-power systems demands BMICs capable of monitoring 24+ cells simultaneously with enhanced fault detection. Conventional multiplexed architectures face fundamental limitations with their scaling constraints. Scaling to 24 CH would require prohibitive HV ratings (>120V), imposing severe process constraints. EVs present noisy environments where high-current transients from motors and converters induce voltage noise across cell series resistances [3]. Sequential measurement divides the measurement window among CHs, resulting in shorter averaging periods that increase noise susceptibility and inter-channel delays that introduce time-correlated errors. Synchronous measurement enables extended averaging for all CHs simultaneously, eliminating inter-channel latency. Enhanced fault detection also requires bipolar sampling to monitor busbar degradation, where aging increases resistance and generates negative voltages down to -1.4V [4]. This work presents a 24-CH BMIC with dedicated HV bipolar-sampling ADCs and on-chip calibration engines, enabling simultaneous and negative voltage measurements. Validation using battery simulator equipment across -40°C to +130°C (AEC-Q100 Grade 1 compliant) demonstrates ±0.8mV total measurement error (TME) at the nominal EV battery voltage (3.5V [1]) and maintains TME within ±1mV across -2V to 5V input range, while EV battery measurements under room temperature (RT) achieve 28.23μV measurement precision over 1,000 consecutive samples.

Figure 21.1.2 shows the proposed 24-CH BMIC architecture featuring dedicated ADCs per CH. The system comprises measurement paths, cell balancer (CB) with open-wire (OW) diagnosis, digital controller (calibration engine), power supply and reference generation, and external RC filters. Dual-path configuration with primary and secondary paths ensures functional safety compliance. Each path incorporates 24 dedicated HV bipolar samplers (HVSPL) with sigma-delta ADCs, totaling 48 ADCs. Additionally, an auxiliary low-voltage (LV) ADC per path monitors cell temperature through external thermistors. This configuration enables simultaneous voltage measurement across all CHs with comparison-based fault detection. The HVSPL (Fig. 21.1.2; bottom) implements a chopper architecture using HV PMOS ($SW_{P1,2}$) and NMOS ($SW_{N1,2}$) switches at the sampler input terminals (V_{IN+} and V_{IN-}), controlled by dedicated SW_P and SW_N drivers that provide two-phase ($\Phi_{1,2}$) switching for bipolar input sampling. All switches employ 12V-rated MOS to withstand input differentials exceeding 5V during surge or ESD events. The secondary path incorporates CB and OW diagnosis circuits. CB performs passive balancing by activating discharge switches (SW_{CB}) for high-SOC cells, dissipating excess energy through external balancing resistors (R_B). OW diagnosis verifies connection integrity by monitoring voltage-ratio changes when SW_{OW} switches on/off, identifying OW when ratios deviate from the expected value $R_{OW2}/(2R_{OW1} + R_{OW2})$. The digital controller performs sampler clock (SCLK) generation, decimation filtering of ADC outputs, and ADC gain/offset calibration. To enhance accuracy across varying input voltages and CH configurations, two additional calibration engines address systematic error sources: common-mode voltage (V_{CM}) calibration corrects for voltage-dependent ADC gain variations, while input impedance (R_{IN}) calibration corrects for input current-induced offsets (V_{OS}). The reference generation circuit integrates a 3.2V Zener reference with 2-point trimming and curvature correction across -40°C to +130°C.

Figure 21.1.3 shows the HVSPL schematic. Conventional HV switches either use LV MOS with unipolar operation [5-6] or V_{TOP}-rated HV MOS for bipolar capability [2,7-8]. This work achieves bipolar capability using 12V-rated MOS through active body biasing and clock doubling implemented within driver circuits. Each driver incorporates a charge pump (CP), LS, a regulator (REG), and two clock doublers (CD_1 and CD_2), generating bias voltages for HV switches. Cross-coupled CP generates driver supply rails: $V_{IN+} + V_{CP}$ and $V_{IN-} - V_{CP}$. LS translates LV control signals ($\Phi_{1,2}$: ground-to-V_{CC} swing) to HV domain ($\Phi_{1,2,HV}$: input terminal-to-V_{CP} swing) through HV MOM capacitors, controlling CD internal switches for proper gate biasing. REG defines CD supply rails ($V_{IN+} + V_{REG}$ and $V_{IN-} - V_{REG}$) and maintains source-body voltage of HV switches (V_{SB}) one diode forward voltage drop (V_F) below the supply rails ($|V_{SB}| = V_{REG} - V_F$), ensuring V_{IN}-independent operation across PVT variations. CD pre-charges C_{VGS} capacitors to V_{REG} and switches their terminal connections according to $\Phi_{1,2,HV}$ to generate V_{GS} for HV switches. During Φ_1 = '1' ($\Phi_{1,HV}$ = '1'), the SW_P driver operates as follows: CD_1 connects $C_{VGS,1}$ across V_{IN+} and SW_{P1} gate ($V_{G,P1}$), satisfying $V_{SG} > |V_{TH}|$. CD_2 charges $C_{VGS,2}$ to V_{REG} and connects $V_{G,P2}$ to CD supply rails ($V_{IN+} + V_{REG}$), ensuring V_{SG}, $V_{DG} < |V_{TH}|$. Alternating $CD_{1,2}$ operation per $\Phi_{1,2}$ achieves voltage doubling, generating $\pm V_{REG}$ swing referenced to the input terminals. Fig. 21.1.3 (bottom) illustrates driver internal node voltages versus V_{IN} during Φ_1 = '1' (SW_{P1} on, SW_{P2} off). $SW_{P1,2}$ operation requires V_{SB}, $V_{DB} < V_F$ to prevent forward-biased body diode conduction, and V_{SG}, $V_{DG} < |V_{TH}|$ to prevent channel conduction. While a positive V_{IN} inherently satisfies these requirements, a negative V_{IN} causes the off-state $V_{D,P2}$ (= V_{OUT-}) to exceed $V_{S,P}$ (= V_{IN+}). Active body biasing through CP and REG elevates $V_{B,P}$ by the maximum negative V_{IN} magnitude, ensuring V_{SB}, $V_{DB} < V_F$ to prevent body diode conduction even when $V_{D,P2}$ exceeds $V_{S,P}$. Clock doubling maintains off-state SW_{P2} gate ($V_{G,P2} = V_{IN+} + V_{REG}$) to satisfy V_{SG}, $V_{DG} < |V_{TH}|$ and prevent channel conduction.

Switched-capacitor ADCs with DC blocking capacitors (C_S) have been adopted in BMICs to enable LV MOS integration [5,6] but suffers from systematic errors that scale with CH count and voltage range. These errors arise from i) channel-dependent ADC gain variations due to varying voltages ($V_{CAP,n}$) across C_S, and ii) input offset errors ($V_{OS,n}$) from finite R_{IN}-induced input current ($I_{IN,n}$) at n[th] channel (CH_n). These channel-dependent variations become amplified as channel count (n) increases in multi-CH BMICs, degrading measurement uniformity. Figure 21.1.4 presents calibration methods within the proposed calibration engine based on the CH_n ADC output ($D_{OUT,n}$). ADC gain variations occur because different CHs experience varying V_{CM} across C_S that increase with CH position, while feedback capacitors (C_{FB}) remain within the LV V_{CC} domain. This generates gain errors following the voltage coefficients ($VC_{1,2}$) of HV MOM capacitors: $(VC_1 \cdot V_{CM} + VC_2 \cdot V_{CM}^2) \cdot (C_S/C_{FB})$, where higher-order terms persist as residual error after conventional first-order ADC gain/offset calibration. V_{CM} calibration calculates the nth-channel V_{CM} ($V_{CM,n}$) through the cumulative summation of $D_{OUT,n}$ ($D_{CM,n} = D_{OUT,1} + D_{OUT,2} + \ldots + D_{OUT,n}$) and applies polynomial mapping [$f(D_{CM,n})$] for correction. During RT measurements across 4 chips (Fig. 21.1.4; top right), input voltage (V_{IN}) swept from 0 to 5V, where CH_1 operates with $V_{CM,1}$ of 0 to 5V while CH_{24} experiences $V_{CM,24}$ of 0 to 120V (= $24 \times V_{IN}$). Without V_{CM} calibration, CH_{24} exhibits substantially higher error than CH_1, but V_{CM} calibration reduces CH_{24} error to CH_1 levels. Offset errors ($V_{OS,n}$) occur because C_S exhibits finite R_{IN}, resulting in nonzero $I_{IN,n}$ that flow through RC filter resistance (R_F) and creates voltage offsets, $V_{OS,n} = I_{IN,n} \cdot R_F$. Adjacent CHs at identical voltages ($V_{IN,n+1} = V_{IN,n}$) naturally cancel these currents ($I_{IN,n+1} = I_{IN,n}$); however, voltage differences from cell faults or busbar measurements produce uncanceled currents ($\Delta I_{IN,n} = I_{IN,n+1} - I_{IN,n} \neq 0$), causing $V_{OS,n}$ proportional to the voltage difference and R_F value. R_{IN} calibration mitigates this effect by estimating $V_{OS,n}$ through analysis of adjacent CH ADC output differences ($\Delta D_{OUT,n} = D_{OUT,n+1} - D_{OUT,n}$), and applying the correction factor, $(\Delta D_{OUT,n-1} - \Delta D_{OUT,n}) \cdot (R_F/R_{IN})$. RT measurements with all CHs at 3.5V (nominal EV battery voltage) except CH_4 swept from -2V to 5V show these offset issues: CH_2 (identical adjacent voltages; $V_{IN,1} = V_{IN,2} = V_{IN,3}$) maintains constant measurement error, while CH_3, CH_4, and CH_5 (differing adjacent voltages; $V_{IN,3} = V_{IN,5} \neq V_{IN,4}$) exhibit error scaling with $V_{IN,4}$ variation. R_{IN} calibration resolves these issues and achieves uniform error across all CHs.

Measurement using battery cell simulator equipment evaluated TME across -40°C to +130°C under various input conditions (Fig. 21.1.5), while EV battery measurements at RT assessed inter-channel variation and noise (Fig. 21.1.6; top). The simulator utilizes series-connected isolated power supplies providing V_{IN} of -2, 0, 3.5, and 5V across 24 CHs (Fig. 21.1.5). For positive inputs (0, 3.5, and 5V), 8 chips yielded 192 samples (8 chips × 24CH). For negative input (-2V), a split configuration applies 5V to CH1-8 and -2V to CH9-24 (yielding 128 samples across 8 chips; 8 chips × 16-CH), ensuring all device terminals remain above ground while creating the target -2V input. The BMICs achieve ±0.8 mV TME at the nominal EV battery voltage (V_{IN} = 3.5V) and maintain TME within ±1mV across all test conditions (V_{IN} = -2 to 5V). EV battery validation employed a 1P24S configuration using 21700 cells (3.5 V, 5,000mAh) at RT (Fig. 21.1.6; top). The synchronous measurement architecture captures all 24 CHs within 1ms, providing a temporal margin for increased averaging. With 16× averaging, the system demonstrates 28.23μV standard deviation over 1,000

979-8-3315-8937-0/26 $31.00 © 2026 IEEE

consecutive measurements, totaling 24,000 samples. The comparison with state-of-the-art BMICs (Fig. 21.1.6; bottom) shows the prototype achieves sub-1mV accuracy at nominal EV battery voltage (3.5 V) across 24 CHs over AEC-Q100 Grade 1 temperature range, with ±1mV TME over the -2 to 5V input range. The integrated DC-DC converter enables power-efficient operation, consuming only 0.056mA per ADC CH at V_{TOP} = 84V (V_{IN} = 3.5V). Figure 21.1.7 presents the chip micrograph, fabricated in a 0.13μm BCD process. Although capable of handling V_{TOP} up to 120V, the design uses only 12V-rated MOS, while integrating 48 ADCs across dual measurement paths.

Acknowledgement:
K. Kwon acknowledges support from the National Research Foundation of Korea (NRF) grant funded by the Korea government (MSIT; RS-2023-00278995 and BK21) and the Institute of Information & Communications Technology Planning & Evaluation (IITP) under the Graduate School of Artificial Intelligence Semiconductor (IITP-2025-RS-2023-00256472) grant funded by the Korea government (MSIT).

References:
[1] Simolka, Matthias, et al. "Influence of Cycling Profile, Depth of Discharge and Temperature on Commercial LFP/C Cell Ageing: Cell Level Analysis with ICA, DVA and OCV Measurements." *Journal of The Electrochemical Society* 167.11: 110502, June 2020. http://doi.org/10.1149/1945-7111/ab9cd1
[2] J. -K. Lee et al., "ASIL-D Compliant Battery Monitoring IC with High Measurement Accuracy and Robust Communication," *ISSCC*, pp. 322-324, Feb. 2023. http://doi.org/10.1109/ISSCC42615.2023.10067607
[3] Amamra, Sid-Ali, et al. "Electric Vehicle Battery Performance Investigation Based on Real World Current Harmonics." *Energies*, 13.2: 489, 2020. http://doi.org/10.3390/en13020489
[4] Park, Sangjun, et al. "Fault Diagnosis for Electric Vehicle Battery Pack Interconnection System Using Real-World Driving Data." *IEEE TIE* (2025), vol. 72, no. 8, pp. 8583-8591, Aug. 2025. http://doi.org/10.1109/TIE.2024.3522461
[5] X. Zhang et al., "A 14-Cell Battery Monitoring AFE with 1mV Total Measurement Error and Integrated Electrochemical Impedance Spectroscopy," *CICC*, pp. 1-3, April 2025. http://doi.org/10.1109/CICC63670.2025.10982745.
[6] K. Deng et al., "An Area-Efficient High-Precision Analog Front End for Battery Management System," in *IEEE TCAS-II*: Express Briefs, vol. 72, no. 5, pp. 783-787, May 2025. http://doi.org/10.1109/TCSII.2025.3551150
[7] C. van Vroonhoven, "A 0-to-60V-Input VCM Coulomb Counter with Signal-Dependent Supply Current and ±0.5% Gain Inaccuracy from –50°C to 125°C," *ISSCC*, pp. 348-350, Feb. 2020. http://doi.org/10.1109/ISSCC19947.2020.9063066
[8] G. Zhu et al., "Multi-Cell Battery Sensing and Protection IC With Integrated Low-Temperature-Drift Reference for Series Battery Pack Management," in *IEEE TCAS-I*: Regular Papers (Early Access), March 2025. http://doi.org/10.1109/TCSI.2025.3551284
[9] Analog Devices, "ADBMS6830B: 16-Channel Multicell Battery Monitor," Datasheet, Rev. 0, Jan. 2024. https://www.analog.com/en/products/adbms6830b.html
[10] Texas Instruments, "BQ79616-Q1, BQ79614-Q1, BQ79612-Q1 Functional Safety-Compliant Automotive 16S/14S/12S Battery Monitor, Balancer and Integrated Hardware Protector," Datasheet, SLUSE81E, Rev. E, Nov. 2023. https://www.ti.com/product/BQ79616-Q1

[11] Infineon Technologies AG, "TLE9018DQK: Li-Ion Battery Monitoring and Balancing IC," Datasheet, Rev. 1.0, May 2025. https://www.infineon.com/part/TLE9018DQK
[12] K. Kadirvel et al., "A Stackable, 6-Cell, Li-Ion, Battery Management IC for Electric Vehicles with 13, 12-bit ΣΔ ADCs, Cell Balancing, and Direct-Connect Current-Mode Communications," in *IEEE JSSC*, vol. 49, no. 4, pp. 928-934, April 2014. http://doi.org/10.1109/JSSC.2014.2300861
[13] V. B. Vulligaddala et al., "A 7-Cell, Stackable, Li-Ion Monitoring and Active/Passive Balancing IC With In-Built Cell Balancing Switches for Electric and Hybrid Vehicles," in *IEEE TII*, vol. 16, no. 5, pp. 3335-3344, May 2020. http://doi.org/10.1109/TII.2019.2953939

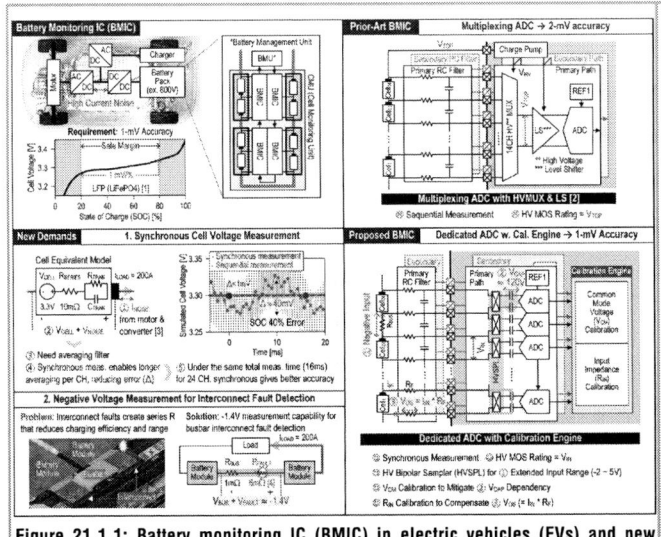

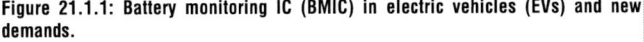

Figure 21.1.1: Battery monitoring IC (BMIC) in electric vehicles (EVs) and new demands.

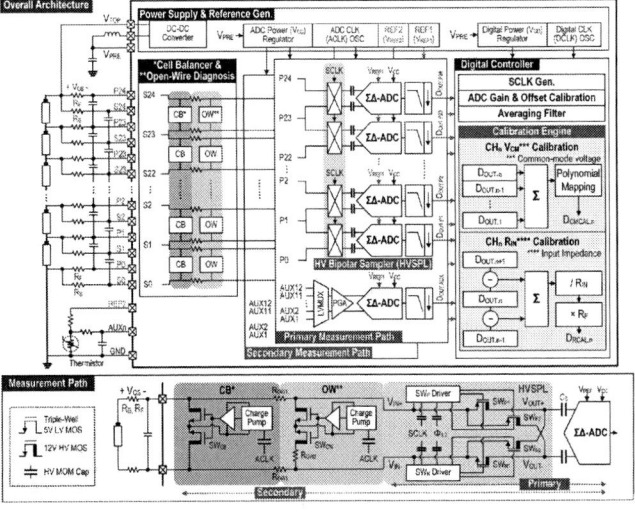

Figure 21.1.2: Overall architecture of proposed 24-channel BMIC.

ISSCC 2026 / SESSION 21 / SENSOR INTERFACES / 21.1

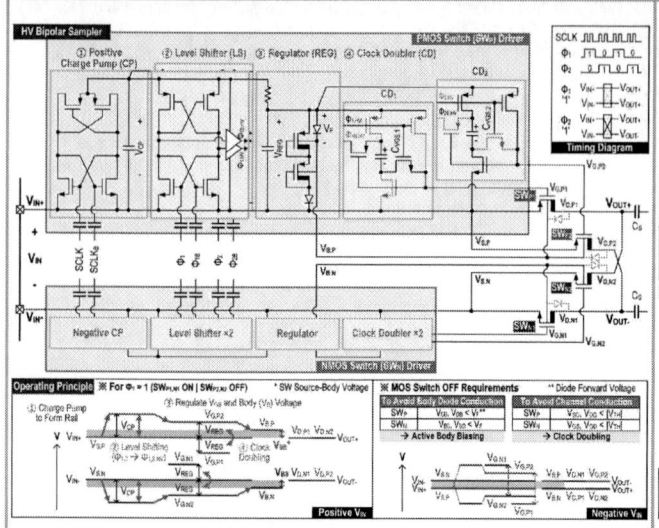

Figure 21.1.3: Schematic and operating principle of HV bipolar sampler.

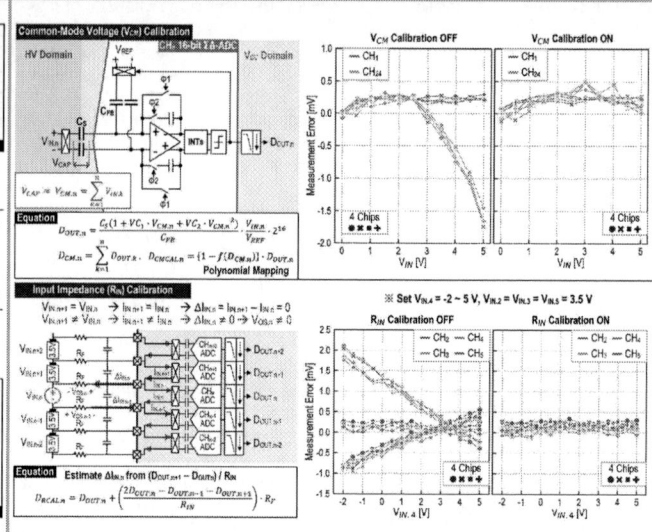

Figure 21.1.4: Common-mode voltage and input-impedance calibration.

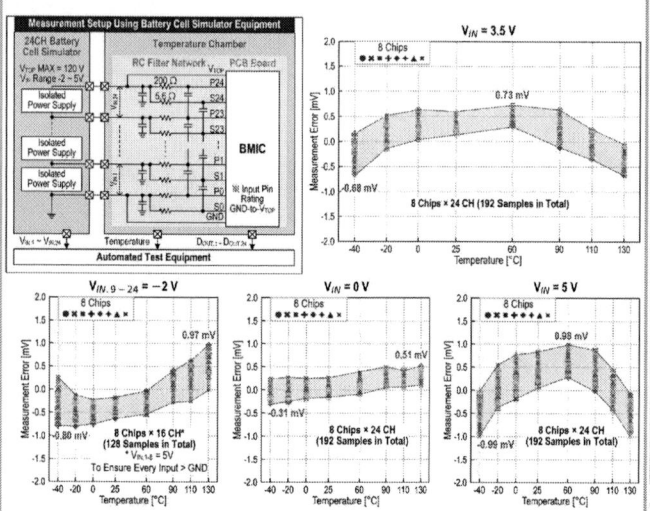

Figure 21.1.5: Measurement results using battery cell simulator equipment.

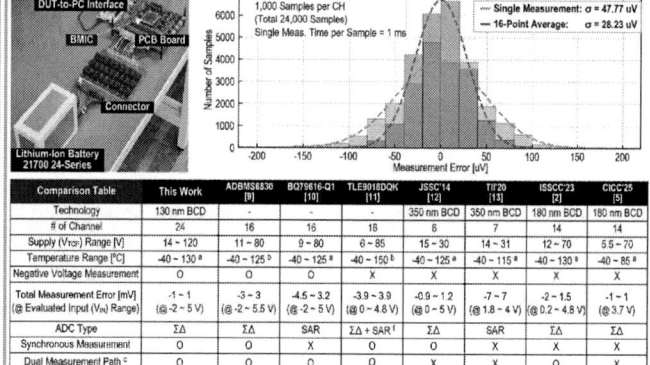

Comparison Table	This Work	ADBMS6830 [9]	BQ79616-Q1 [10]	TLE9018DQK [11]	JSSC'14 [12]	TI'20 [13]	ISSCC'23 [2]	CICC'25 [5]
Technology	130 nm BCD	-	-	-	350 nm BCD	350 nm BCD	180 nm BCD	180 nm BCD
# of Channel	24	16	16	18	6	7	14	14
Supply (V_TOP) Range [V]	14 ~ 120	11 ~ 80	9 ~ 80	6 ~ 85	15 ~ 30	14 ~ 31	12 ~ 70	5.5 ~ 70
Temperature Range [°C]	-40 ~ 130 [a]	-40 ~ 125 [b]	-40 ~ 125 [a]	-40 ~ 150 [b]	-40 ~ 125 [a]	-40 ~ 115 [a]	-40 ~ 130 [a]	-40 ~ 85 [a]
Negative Voltage Measurement	O	O	X	X	X	X	X	X
Total Measurement Error [mV] (@ Evaluated Input (V_IN) Range)	-1 ~ 1 (@ -2 ~ 5 V)	-3 ~ 3 (@ -2 ~ 5.5 V)	-4.5 ~ 3.2 (@ -2 ~ 5 V)	-3.9 ~ 3.9 (@ 0 ~ 4.8 V)	-0.9 ~ 1.2 (@ 0 ~ 5 V)	-7 ~ 7 (@ 1.8 ~ 4 V)	-2 ~ 1.5 (@ 0.2 ~ 4.8 V)	-1 ~ 1 (@ 3.7 V)
ADC Type	ΣΔ	ΣΔ	ΣΔ	ΣΔ + SAR	ΣΔ	SAR	ΣΔ	ΣΔ
Synchronous Measurement	O	O	X	X	O	X	X	X
Dual Measurement Path [c] (# of Cell ADC [d])	O (48)	O (32)	O (2)	O (19) [f]	X (6)	X (1)	O (2)	X (2)
Quiescent Current [uA]	14	4	16	3.4	15	17	14	-
Current Consumption During Cell Measurement [mA per ADC]	0.056 [e]	0.1375	0.4	-	-	-	-	1.15

[a] Ambient Temp. [b] Junction Temp. [c] Primary and secondary path configuration for functional safety compliance [d] ADCs for cell voltage measurement
[e] Integrated DCDC converter (@ V_TOP = 84 V) [f] Multiplexing architecture with SAR ADC for secondary measurement

Figure 21.1.6: Measurement results using EV battery and comparison with prior works.

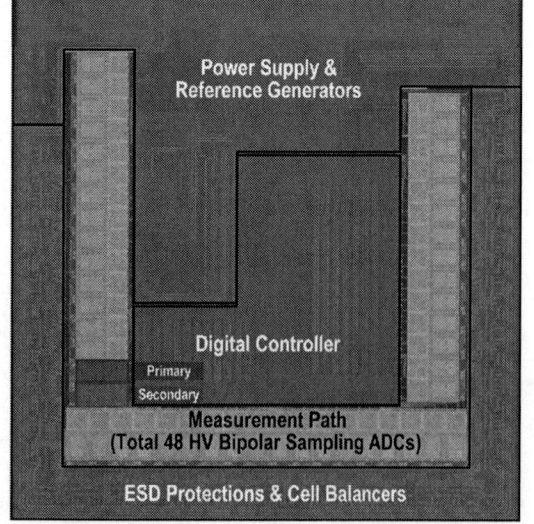

Figure 21.1.7: Chip micrograph.

• 2026 IEEE International Solid-State Circuits Conference

ISSCC 2026 / SESSION 21 / SENSOR INTERFACES / 21.2

21.2 A Fully Integrated GMR Biosensor with On-Chip Coils and Sensors Achieving 605 Resolution FoM for Multiplexed PoC Diagnostics

Mengze Wu[1], Remy Lassalle-Balier[2], Pablo Aguirre[3], Florencia Ferrer[3], Sina Haji Alizad[4], Shekhar Kummari[1], Alex Latham[4], Drew A. Hall[1]

[1]University of California, San Diego, CA, [2]Allegro MicroSystems, Chavanod, France, [3]Allegro MicroSystems, Montevideo, Uruguay, [4]Allegro MicroSystems, Manchester, NH

Abstract

This paper presents a fully integrated GMR biosensor chip for point-of-care *in vitro* diagnostics featuring on-chip sensors, excitation coils, and a digital back-end. The system achieves $120nT_{rms}$ noise, 0.38ppm sensitivity, <0.2% mismatch, and a resolution FoM of 605, outperforming prior GMR designs with 21× faster readout and 10× lower baseline with full integration. Real-time IL-6 immunoassay results demonstrate high specificity and multiplexing, enabling compact, high-throughput PoC testing.

Point-of-care (PoC) diagnostics are increasingly vital for rapid, decentralized healthcare, enabling timely decision-making in settings ranging from clinics to resource-limited environments. The COVID-19 pandemic accelerated their adoption worldwide, underscoring the value of portable, cost-effective tools that deliver laboratory-quality results anywhere. Simple colorimetric methods (*e.g.*, lateral flow assays commonly used for pregnancy tests) are inexpensive, portable, and easy to use but have limited analytical sensitivity. Enzyme-linked immunosorbent assays (ELISAs) offer significantly higher sensitivity but require bulky optics and can be compromised by sample matrix interference. Magnetic immunoassays overcome these limitations because biological samples exhibit no magnetic background, allowing for highly sensitive detection. However, realizing the full potential of magnetic biosensing requires sensitive electronic readout of giant magnetoresistive (GMR) sensors – thin-film quantum mechanical sensors that convert a change in magnetic field into a corresponding change in resistance through spin-dependent scattering events. While these devices are CMOS-compatible and thus well-suited for low-cost, scalable integration, the binding events generate minuscule resistance changes, creating significant challenges for analog front-end design.

The magnetic nanoparticle (MNP)-induced signal is very small ($\mu\Omega$) – orders of magnitude smaller than the sensor's nominal resistance ($k\Omega$) – requiring a high dynamic range (>120dB) front-end [1]. Prior work, shown in Fig. 21.2.1, has addressed this in two main ways. In [2-5], a magnetic double modulation scheme is used to spectrally separate part of the baseline from the signal, with the remaining baseline canceled via pseudo-differential sensing using a reference sensor. In [6-8], a time-domain scheme based on magnetorelaxation (MRX) is employed, where the magnetic field, H_{ext}, is rapidly removed and the temporal relaxation response is observed. However, the fast relaxation time requires a high-bandwidth front-end, incurring a significant noise penalty. Magnetic correlated double sampling can mitigate this penalty, but it requires long readout times to achieve comparable noise levels. Moreover, all prior work relies heavily on external components (*e.g.*, sensor arrays, Helmholtz coils, power amplifiers, etc.), which limits their portability. This work presents a fully integrated, monolithic GMR biosensor chip and achieves $280.8nT_{rms}$ input-referred noise, $23.2\mu T$ input-referred baseline, and a state-of-the-art resolution FoM of 605. Key contributions include: (1) a lock-in detection architecture with a Wheatstone bridge to relax the front-end design requirements; (2) an interdigitated sensor layout ensuring <0.2% mismatch; (3) a thin/thick passivation layer strategy enabling pseudo-differential signaling within the bridge; and (4) on-chip sensors, excitation coils, and a digital back-end for a complete, compact PoC system.

The chip architecture, shown in Fig. 21.2.2, consists of a 32-pixel sensor array, a floating bridge driver, a coil driver, an analog front-end (AFE), a relaxation oscillator, and a digital signal processor (DSP). A top-level digital controller generates the phase-aligned clocks to drive the electromagnet, producing an AC magnetic bias field that modulates the sensor output at f_H. Unlike double-modulation schemes, which bias both the sensor and the magnetic field at different frequencies, this system biases the sensors with a DC voltage while driving only the magnetic coil with an AC signal. This approach eliminates the common-mode swing at the amplifier, relaxing the subsequent circuit requirements. Double modulation was also used in prior work to mitigate electromagnetic interference (EMI) induced in inductive components (*e.g.*, wire bonds, long interconnects) from external AC fields [2], by shifting the EMI to a separate frequency. In this design, a highly localized AC field is generated beneath the pixel; therefore, the bridge structures and routing convert any residual EMI into common-mode (CM) noise, allowing for the use of a simpler lock-in detection method. Two AFE channels are time-division multiplexed across the array. A programmable-gain amplifier (PGA) amplifies and downconverts the modulated signal to DC, which is then digitized by an incremental delta-sigma ADC. The DSP processes and transmits the data off-chip via an I2C interface. All analog references, clocks, and power-management circuits are integrated on-chip, enabling operation from a single 3.3V supply with a simple 5-pin interface.

The pixels are paired to amortize the coil area and power overhead. Each pixel pair includes two GMR bridges, both biased by a shared coil located directly beneath the sensors, which is realized by the topmost metal layer, utilizing all the lower metals for a return path. From simulation, a 58mA coil current generates a 2.9mT in-plane field with less than 0.68% variation across the sensor. Each bridge consists of four GMR elements in a Wheatstone configuration, where the elements are arranged in an interdigitated fashion to reduce mismatch. To make the bridge sensitive to changes in the local magnetic field induced by binding events, a thin/thick passivation strategy is employed, where the reference sensors are covered with a thick (>200nm) passivation layer, whereas the active sensors have a thin (<70nm) passivation layer. This scheme ensures that the sensors see the same environment (magnetic, temperature, fabrication, etc.), but the bridge exhibits a differential response to MNPs tethered to the sensor surface, as MNPs on the thin layer generate measurable signals, while those on the thick layer remain below the noise floor. This interdigitated design with ultra-low mismatch (<0.2%) suppresses most of the sensor baseline and enables inherent temperature cancelation without the need for an algorithmic approach [3].

The signal path for a single GMR bridge readout is illustrated in Fig. 21.2.3. The bridge is DC biased by a bridge driver, implemented by a differential transimpedance amplifier, that keeps the sensor-bridge outputs centered around the PGA's input common-mode voltage. The bridge output is routed to a PGA via an analog multiplexer. The PGA is implemented using a differential-difference-amplifier (DDA) with a variable resistive feedback network to ensure a high input impedance (>5MΩ). Chopper switches embedded in the DDA down-modulate the signal to DC, effectively suppressing the upstream flicker noise and offset. The amplifier provides a tunable gain range of 21 to 210V/V, which is configured through a 3b resistive feedback ladder. The DDA utilizes an NPN input pair for high transconductor efficiency and low input-referred noise (<10nV/$\sqrt{Hz}$) without requiring a large bias current. The amplifier is a two-stage op-amp with Miller compensation and a Class-AB output buffer to achieve high loop gain with resistive feedback. A 3b capacitor bank sets the compensation capacitance, ensuring >2MHz bandwidth across the entire gain range. Dedicated common-mode feedback (CMFB) circuits for each stage ensure stability over a wide range of tuning conditions. The down-converted signal is digitized using a continuous-time incremental $\Delta\Sigma$ ADC (IADC). The IADC is implemented by a 2nd-order cascade-of-integrators feedback (CIFB) loop with an oversampling ratio (OSR) of 256 and a sampling frequency of 32MHz. The loop filter uses active RC integrators with a 9-level flash ADC and a 9-level IDAC. Dynamic element rotation is employed to scramble the DAC mismatch and preserve the linearity. The digitized output is processed by a decimation filter followed by an averaging filter to enhance the SNR. Both the amplifier chopping frequency and the ADC reset frequency are aligned with the notches of the averaging filter to suppress artifacts and prevent distortion from chopping and reset transients.

The chip was fabricated in a custom 0.18μm CMOS BCD process, occupying an area of 9.96mm². Figure 21.2.4 shows electrical measurements where the magnetic field generated by the on-chip coil is sensed indirectly using the on-chip GMR sensor array. To do this, the sensor was calibrated using an external Helmholtz coil and then used to characterize the on-chip coil. The measured field follows the expected trend until the coil driver saturates due to the parasitic coil and routing resistance. The bridge resistance (R_0) and sensitivity (S_0) mismatch were characterized, showing R_0 mismatch better than 0.2% and S_0 mismatch less than 0.8% across the array – substantially lower than the ~10% mismatch reported for external GMR sensor arrays [4]. The sensitivity can be increased by adjusting the filter's averaging window to improve the SNR, with performance ultimately limited by the residual flicker noise. The design achieves an input-referred noise of $120nT_{rms}$, corresponding to a sensitivity of 0.38ppm. The total system-level power consumption is 115.99mW, dominated by the on-chip coil driver.

Figure 21.2.5 illustrates the process for a magnetic immunoassay, where the sensors are first functionalized with capture probes specific to the target biomarker (*e.g.*, DNA, RNA, protein). Next, the sample is added (Step 2) and incubated. Finally, detection antibodies tagged with MNPs, which assemble to form a sandwich complex (Steps 3 and 4), are added [10]. The tethered MNPs perturb the local magnetic field (Step 5), which is detected by the underlying GMR sensors. In practice, the sensors are read out in real-time, not only at the end, as illustrated here, to observe real-time binding kinetics. To demonstrate magnetic immunoassay capability, the array was functionalized with: human interleukin-6 (IL-6) capture antibody (12 sensors), biotinylated bovine serum albumin (Biotin-BSA) as a positive control (6 sensors), bovine serum albumin (BSA) as a negative control (6 sensors), and epoxy as an inert negative control (8 sensors). Figure 21.2.5 shows measured real-time magnetic binding curves when 1nM of IL-6 target in binding buffer (0.1% BSA, 0.1% Tween-

366 • 2026 IEEE International Solid-State Circuits Conference

979-8-3315-8937-0/26 $31.00 © 2026 IEEE

20 in 1× PBS) was added, clearly differentiating the target and control groups. Figure 21.2.6 shows a table comparing this work to prior work on GMR biosensors [3-6]. The reported system achieves a 21× faster readout time and a 10× lower input-referred baseline. While [6] reports a lower input-referred noise and a higher resolution FoM, it lacks key features such as temperature robustness and integration of the sensor array with an on-chip coil. Relative to prior Hall sensor work with integrated coils and sensor arrays [7,8], this design shows a 99× improvement in resolution FoM. The GMR biosensor chip features a fully integrated on-chip coil and an ultra-low-mismatch sensor array, combining high sensitivity with the fastest reported readout time among the devices in the comparison table, enabling high-throughput, multiplexed, point-of-care assays. A die photo is shown in Fig. 21.2.7.

Acknowledgment:
This work is supported by Allegro Microsystems.

References:
[1] Hua Wang, Yan Chen, A. Hassibi, A. Scherer and A. Hajimiri, "A Frequency-Shift CMOS Magnetic Biosensor Array with Single-Bead Sensitivity and No External Magnet," *IEEE ISSCC*, San Francisco, CA, USA, 2009, pp. 438-439,439a.
https://doi.org/10.1109/isscc.2009.4977496
[2] S. -J. Han, H. Yu, B. Murmann, N. Pourmand and S. X. Wang, "A High-Density Magnetoresistive Biosensor Array with Drift-Compensation Mechanism," 2007 *IEEE ISSCC*, San Francisco, CA, USA, 2007, pp. 168-594.
https://doi.org/10.1109/isscc.2007.373347
[3] D. A. Hall, R. S. Gaster, K. A. A. Makinwa, S. X. Wang and B. Murmann, "A 256 Pixel Magnetoresistive Biosensor Microarray in 0.18 μm CMOS," in *IEEE JSSC*, vol. 48, no. 5, pp. 1290-1301, May 2013. https://doi.org/10.1109/jssc.2013.2245058
[4] T. Costa, F. A. Cardoso, J. Germano, P. P. Freitas and M. S. Piedade, "A CMOS Front-End with Integrated Magnetoresistive Sensors for Biomolecular Recognition Detection Applications," in *IEEE BCS*, vol. 11, no. 5, pp. 988-1000, Oct. 2017.
https://doi.org/10.1109/tbcas.2017.2743685
[5] X. Zhou, M. Sveiven and D. A. Hall, "A Fast-Readout Mismatch-Insensitive Magnetoresistive Biosensor Front-End Achieving Sub-ppm Sensitivity," 2019 *IEEE ISSCC*, San Francisco, CA, USA, 2019, pp. 196-198. https://doi.org/10.1109/isscc.2019.8662440
[6] X. Zhou et al., "A 9.7-nT_{rms}, 704-ms Magnetic Biosensor Front-End for Detecting Magneto-Relaxation," in *IEEE JSSC*, vol. 56, no. 7, pp. 2171-2181, July 2021.
https://doi.org/10.1109/jssc.2020.3043669
[7] P. P. Liu et al., "Magnetic Relaxation Detector for Microbead Labels," in *IEEE JSSC*, vol. 47, no. 4, pp. 1056-1064, April 2012. https://doi.org/10.1109/jssc.2012.2185339
[8] S. Gambini, K. Skucha, P. P. Liu, J. Kim and R. Krigel, "A 10 kPixel CMOS Hall Sensor Array with Baseline Suppression and Parallel Readout for Immunoassays," in *IEEE JSSC*, vol. 48, no. 1, pp. 302-317, Jan. 2013. https://doi.org/10.1109/jssc.2012.2224531
[9] O. Y. Galkin, O. B. Besarab, M. O. Pysmenna, Y. V. Gorshunov, and O. M. Dugan, "Modern Magnetic Immunoassay: Biophysical and biochemical aspects," *Regulatory Mechanisms in Biosystems*, vol. 9, no. 1, pp. 47–55, Nov. 2017.
https://doi.org/10.15421/021806

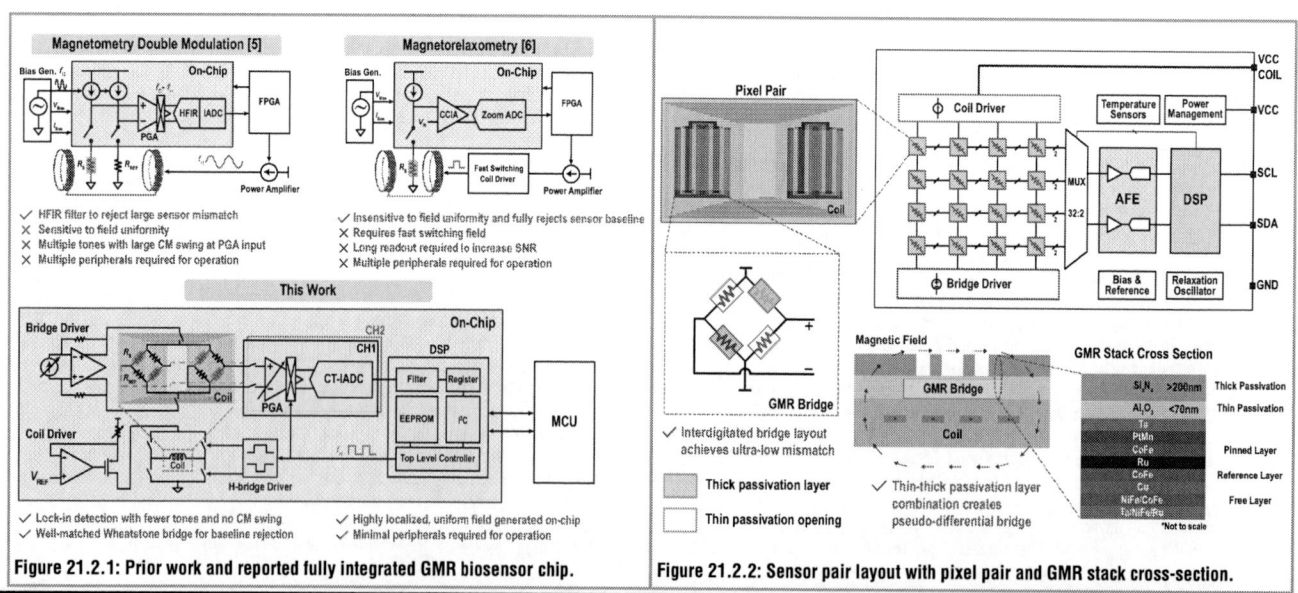

Figure 21.2.1: Prior work and reported fully integrated GMR biosensor chip.

Figure 21.2.2: Sensor pair layout with pixel pair and GMR stack cross-section.

ISSCC 2026 / SESSION 21 / SENSOR INTERFACES / 21.2

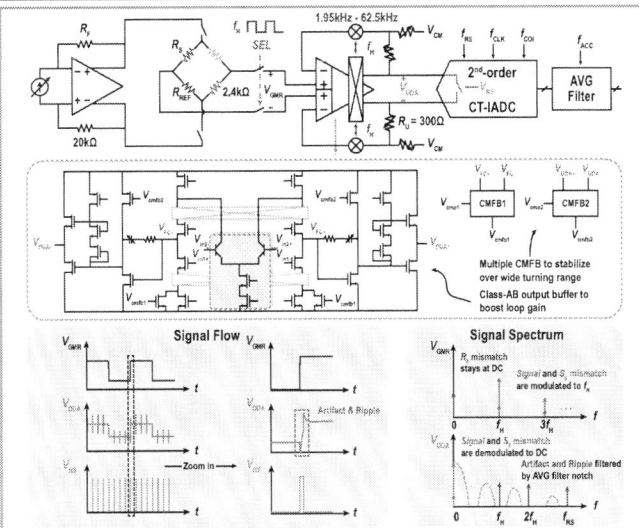

Figure 21.2.3: Schematic of the differential difference amplifier and signal flow diagram.

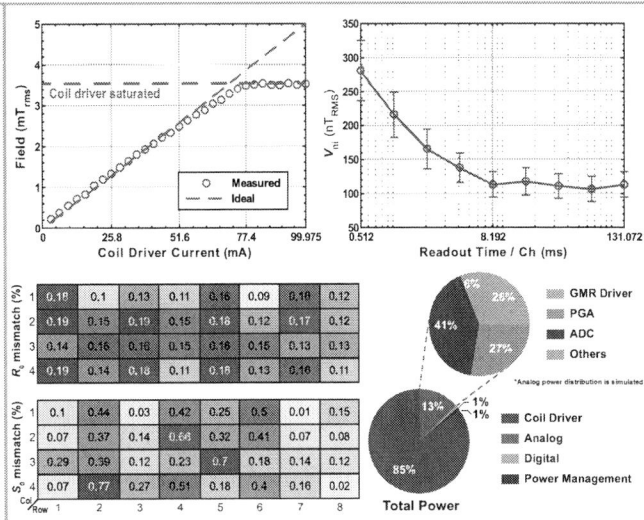

Figure 21.2.4: Measured magnetic field of the on-chip coil (top left), input-referred noise (top right), sensor mismatch (bottom left), and power distribution (bottom right).

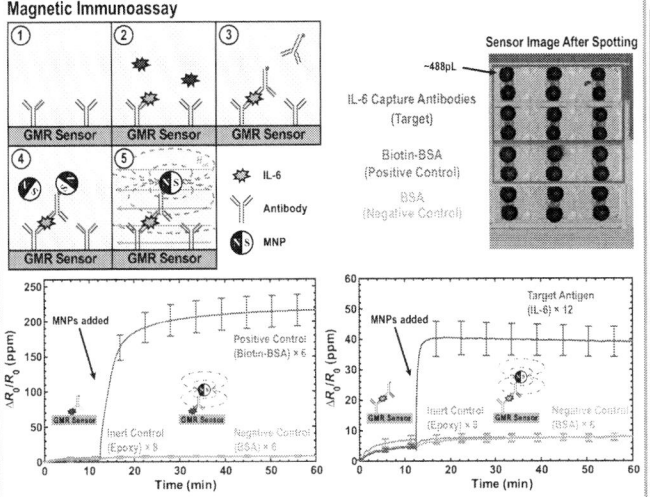

Figure 21.2.5: Magnetic immunoassay overview (top left), photograph after spotting (top right), and measured real-time IL-6 magnetic immunoassay results (bottom).

	JSSC '13 [3]	TBCAS '17 [4]	ISSCC '19 [5]	JSSC '21 [6]	JSSC '12 [7]	JSSC '13 [8]	This work
Sensor Type	GMR	GMR	GMR	GMR	Hall	Hall	GMR
Technology (μm)	0.18	0.35	0.18	0.18	0.18	0.18	0.18
Sensing Method	Magneto	Magneto	Magneto	MRX	MRX	MRX	Magneto
No. of Sensor	256	192	80	80	64	10,240	32
Input-referred Integrated Noise (nT_rms)	49	11.5§	460	9.7	15§	1,207.5	280.8
Readout Time/Ch. (ms)	250	1,000	11	704	64,000	50	0.512
Power/Ch. (mW)*	3.15	4.9⁴	1.39	4.32	6.2§	0.825	14.995 / 115.993^
Temperature Robustness	Yes	No	Yes	No	Yes	Yes	Yes
On-Chip Sensor	No	Yes	No	No	Yes	Yes	Yes
On-Chip Coil	No	No	No	No	Yes	Yes	Yes
Off-Chip Components	Power amplifier, Function generator	Power amplifier, Function generator, ADC	Power amplifier, Function generator	Fast switching coil driver, Function generator, DSP	ADC DSP	None	None
Input-referred Baseline (μT)	7,090	1,840	<235	3,000	<1	7	23.2
Resolution FoM (nT²·mJ)	1,891 N/A§	648§ N/A§	3,235 N/A^	266 N/A^	89,280§ 1,071,360^§,^	60,143 278,622§	695 4681^

*Power/Ch does not include sensor bias and magnetic field generator § Does not include ADC & Include on-chip coil driver and sensor bias FoM = Resolution² × Energy/Conversion

Figure 21.2.6: Comparison table.

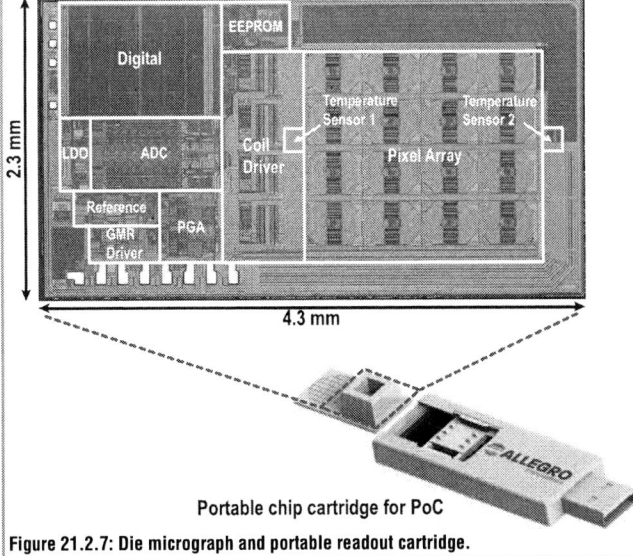

Figure 21.2.7: Die micrograph and portable readout cartridge.

• 2026 IEEE International Solid-State Circuits Conference

ISSCC 2026 / SESSION 21 / SENSOR INTERFACES / 21.3

21.3 A Temperature- and Aging-Compensated TMR Current Sensor with ±0.13% Sensitivity Variation from -40°C to 120°C

Tianxiang Qu, Nan Wang, Kaiwen Zhou, Jiayao Liu, Longxi Xiang, Mingqian Sun, Zhiliang Hong, Xiaoyang Zeng, Jiawei Xu

Fudan University, Shanghai, China

Abstract

This paper presents a TMR-based contactless current sensor that mitigates sensitivity drift due to temperature and aging. The proposed sensitivity stabilization loop continuously adjusts the TMR sensitivity, making residual drift dependent only on the on-chip curvature-corrected current reference (I_{REF}). A built-in stress sensor is used to further compensate the aging-induced drift of I_{REF}. These techniques achieve ±0.13% sensitivity variation from -40°C to 120°C and reduce the aging drift to 0.1% at 25°C.

High-precision contactless current sensors have gained increasing adoption in power inverters, electricity metering, and battery management systems owing to their inherent safety, reliability, and galvanic isolation. A low-noise, low-offset, and linear sensor interface with excellent energy efficiency is an essential prerequisite. For industrial applications, even more important is that sensors must not only ensure minimal sensitivity drift (<1%) across a wide temperature range, but also mitigate long-term sensitivity drift caused by aging effects. From a system perspective, convenient and practical calibration that allows uninterrupted current sensing is also crucial.

Conventional shunt-based current sensors [1,2] offer low sensitivity variation but lack galvanic isolation and can dissipate considerable power, a limitation that magnetic current sensors can overcome. Closed-loop magnetic sensors [3-5] have excellent linearity and temperature stability as the feedback mechanism helps to maintain a near-zero core flux density, thereby minimizing non-linearities arising from the core material. However, driving the feedback coil consumes significant power (≥10mW) [3,4]. Open-loop Hall sensors [6], while CMOS compatible, suffer from low sensitivity, large offset, and require relatively high bias current (in the mA range). In contrast, tunnel magnetoresistance (TMR) sensors [7] stand out for their high sensitivity to magnetic fields, particularly at low bias currents, while ever-decreasing manufacturing costs make them a suitable candidate for precision current sensing. However, the sensitivity of TMR sensors is notably susceptible to temperature variations and aging, posing a severe challenge for long-term accuracy and reliability. First-order temperature compensation [6,7] can be achieved by adjusting the temperature coefficient (TC) of the TMR bias current. However, due to the inherent nonlinear temperature characteristics of the TMR sensor and limitations in trimming step size, residual sensitivity drift often exceeds 1%. In addition, since existing calibration schemes [7] operate only in the foreground, aging-induced drift can accumulate over time and eventually dominate the sensitivity error.

This paper presents a TMR contactless current sensor that addresses the sensitivity drift due to temperature and aging, even with the TMR sensor not integrated on-chip. First, the real-time sensitivity stabilization loop (SSL) in Fig. 21.3.1 (top right) uses an on-chip generated reference magnetic field (B_{REF}) to dynamically adjust the TMR bias voltage ($V_{B,TMR}$ in Fig. 21.3.1) to counteract the TMR's sensitivity drift. Second, a built-in stress sensor (Fig. 21.3.1 top left) is introduced to monitor the local mechanical deformation due to aging, quantifies its effect on the reference current (I_{REF}) and the resulting B_{REF}, and applies digital-domain correction to further suppress long-term sensitivity drift. These techniques reduce system sensitivity variation to ±0.13% over the –40°C-to-120°C range and limit long-term drift to 0.1%. Finally, the TMR offset is nulled by a delta-sigma modulator (DSM) DAC (Fig. 21.3.1 bottom left) with a shared reference voltage ($V_{B,TMR}$), reducing the system offset to 6μV with low drift.

The TMR sensor in Fig. 21.3.1 (top right) is modeled as a voltage-biased Wheatstone bridge, in which the resistance variation ΔR is modulated by the surrounding magnetic field, producing an output voltage proportional to both the TMR bias voltage ($V_{B,TMR}$) and the current under test. The readout circuit (Fig. 21.3.1 bottom) consists of a chopper instrumentation amplifier (IA) followed by a 14-bit DSM ADC (Fig. 21.3.1, bottom), an SSL for continuous correction of TMR sensitivity (Fig. 21.3.1, top right), and a current reference to generate the B_{REF} for the SSL (Fig. 21.3.1, top left). This current reference is also temperature- and stress-compensated for long-term accuracy. It generates a 160kHz out-of-band magnetic field to ensure it does not interfere with the DC-to-20kHz current being measured. The corresponding output voltage at 160kHz is demodulated to DC, amplified by G_{SSL} (210×), and integrated to adjust $V_{B,TMR}$ in real time. The unity-gain bandwidth (BW) of the SSL is set to below 1kHz, which effectively suppresses the upmodulated input signals and TMR offset to prevent SSL saturation. Operating entirely in the background, the SSL continuously corrects TMR sensitivity without interrupting current sensing, marking a key distinction from previous offline calibration [1,7].

The temperature and aging stability of both the SSL and current reference are equally important in determining the overall sensitivity variation. Once the SSL settles, as illustrated in the formula in Fig. 21.3.1 (upper left corner), the TMR sensitivity depends only on the programmed reference voltage V_{SET}, the reference magnetic field B_{REF} (set by I_{REF}), and the SSL amplifier gain (G_{SSL}). As long as these parameters are kept constant, the TMR sensitivity remains the same and invariant to temperature and aging. In this work, V_{SET} is derived from the on-chip ADC reference ($V_{REF,ADC}$) via a resistive divider, establishing a ratio-metric readout scheme. The gain G_{SSL} is defined by resistor ratios and thus is insensitive to temperature and process variations. Furthermore, chopping suppresses offsets of IA_{SSL} to a negligible level compared to the B_{REF}-induced output signal. Owing to these combined factors, the SSL exhibits excellent long-term stability.

The current reference (I_{REF}), however, must rely on dedicated compensation techniques to ensure temperature and aging stability. In Fig. 21.3.2 (top), a PTAT current and a CTAT current are summed to achieve first-order temperature compensation. To extend correction accuracy over a wide temperature range, segmented curvature compensation [8,9] is implemented by injecting three pairs of currents (I_{Hn} and I_{Ln}) into high- and low-temperature regions, respectively. The combined current is then amplified by a current amplifier and chopped to generate I_{REF}=20mA at 160kHz. To suppress 1/f noise, amplifiers A1 to A3 are cyclically rotated with an auxiliary amplifier (Fig. 21.3.2 bottom left), allowing one amplifier to autozero while the other three remain in operation, thereby also reducing output ripple. The current sources utilize dynamic element matching (DEM) to mitigate both 1/f noise and mismatch, with the resulting ripple filtered by the decoupling capacitor (C_{DEP}).

Even with the temperature compensation described above, package stress drift induced by aging can still cause variations in I_{REF}, leading to residual sensitivity drift. To address this, an on-chip stress sensor (Fig. 21.3.2, bottom right) [10] is employed to monitor stress fluctuations in the vicinity of the current reference, enabling 1st-order polynomial compensation in the digital domain. This sensor exploits the difference in stress coefficients between N-diff and P-diff resistors, oriented along the X and Y axes, respectively, to generate a differential output voltage. Its readout path comprises a programmable gain amplifier (PGA) and a 12-bit SAR ADC, with the sensor offset trimmed by a DAC to prevent saturation at large gains. Since the TMR sensor is incompatible with CMOS processes, direct integration of stress sensors around the TMR element is not feasible. Nevertheless, the SSL can still continuously correct the sensitivity of the off-chip TMR by adjusting its bias voltage, ensuring robustness against both temperature- and stress-induced drift.

The TMR readout (Fig. 21.3.1, bottom) uses a pseudo-differential IA with embedded chopping to achieve high input impedance and thus low gain error. Compared with the CCIA [13], this IA eliminates the intermodulation distortion caused by input coupling capacitors and the TMR resistance during the chopping switching. To improve noise efficiency and minimize gain error, the core amplifier A_{IA} (Fig. 21.3.1) adopts a three-stage amplifier topology (Fig. 21.3.3, top) with a current-reuse input stage, resulting in a closed-loop gain error well below 0.1% from –40°C to 120°C (Fig. 21.3.3, top right). The IA's BW is set to 40kHz, filtering most of the 160kHz reference magnetic field signal. Both the residual signal (160kHz) and the chopping ripple (80kHz) coincide with the zeros of the digital decimation filter, providing additional suppression. The 14-bit 2nd-order DSM ADC (Fig. 21.3.1 bottom right) adopts an RC integrator at the input. To match the input resistor R_{INT}, a resistive DAC (RDAC) is used, instead of a current DAC [11,12], to reduce gain error and maintain current sensing accuracy.

To compensate the TMR sensor offset, a 4-bit DSM DAC (Fig. 21.3.3 bottom left) is engaged to provide an offset-nulling current (I_{cmop} in Fig. 21.3.1 bottom). Unlike conventional current-source approaches [7], this DAC generates the compensation current through a resistor array (Fig. 21.3.3 bottom left) and shares the same bias voltage ($V_{B,TMR}$) as the TMR sensor, ensuring robust offset nulling even if $V_{B,TMR}$ varies. In Fig. 21.3.3 (bottom right), the measured system offset of 5 chips drops from a maximum of 45mV to 6μV, approaching the intrinsic offset of the IA (Fig. 21.3.3 top right). Moreover, the offset drift remains below ±11μV over temperature (Fig. 21.3.3 bottom right).

The IC is implemented in a 0.18μm CMOS process (Fig. 21.3.7), while the TMR sensor is a separately packaged chip. The reference current-carrying trace is routed on the top PCB layer, and the measured current-carrying trace is on the second PCB layer. Figure 21.3.4 (top) shows the output FFT plot of the readout-only circuit, achieving a thermal-noise-limited SNR of 84.6dB with a 70mV$_{pk}$ sine-wave input. When the TMR sensor, SSL, and offset-trim

DAC are enabled, the input-referred noise (BW=20kHz) increases from 2.5µV$_{rms}$ to 4.7µV$_{rms}$ (Fig. 21.3.4 bottom), dominated by the TMR sensor rather than the readout. Figure 21.3.5 (left) indicates that the overall current sensor exhibits a sensitivity drift of ~-1000ppm/°C when the TMR bias voltage is fixed at 1.8V. Enabling the SSL allows the TMR bias voltage to increase with temperature, compensating for sensitivity drift. Across 5 test chips, the maximum TC of I$_{REF}$ is 15.5ppm/°C and the maximum sensitivity error is 0.26% (±0.13%). Figure 21.3.5 (right) shows that both I$_{REF}$ and the stress sensor output correlate linearly with applied stress, enabling digital compensation of I$_{REF}$ drift. After moisture expansion and accelerated aging, with a 200mA test current at 25°C, the SSL reduces the original 2.3% drift to 0.3%, and stress compensation further suppresses it below 0.1%.

Figure 21.3.6 summarizes the performance of this work in comparison with state-of-the-art current sensors. By incorporating the SSL with a current reference featuring segmented TC and stress compensation, the proposed current sensor not only implements real-time sensitivity compensation of the off-chip TMR sensor and achieves minimal sensitivity variation from −40°C to 120°C, but also substantially reduces the long-term drift to 0.1% at 25°C. In addition, this work also delivers competitive DR of 82dB and power efficiency of FoM=153dB within a 20kHz BW.

Acknowledgement:
This work is supported by NSFC (No. 62504049) and the State Key Laboratory of Integrated Chips and Systems, Fudan University.

References:
[1] Z. Tang et al., "A 40A Shunt-Based Current Sensor with ±0.2% Gain Error from -40°C to 125°C and Self-Calibration," *ISSCC*, pp. 348-350, Feb. 2023. https://doi.org/10.1109/ISSCC42615.2023.10067304
[2] Z. Tang et al., "A ±25A Versatile Shunt-Based Current Sensor with 10kHz Bandwidth and ±0.25% Gain Error from -40°C to 85°C Using 2-Current Calibration," *ISSCC*, pp. 66-68, Feb. 2022. https://doi.org/10.1109/ISSCC42614.2022.9731777
[3] M. Kashmiri et al., "A 200kS/s 13.5b Integrated-Fluxgate Differential-Magnetic-to-Digital Converter with an Oversampling Compensation Loop for Contactless Current Sensing," *ISSCC*, pp. 1- 3, Feb. 2015. https://doi.org/10.1109/ISSCC.2015.7063140
[4] P. Garcha et al., "A 770 kS/s Duty-Cycled Integrated-Fluxgate Magnetometer for Contactless Current Sensing," *ISSCC*, pp. 80-82, Feb. 2021. https://doi.org/10.1109/ISSCC42613.2021.9365837
[5] I. Akita et al., "A 4mW 45pT/√Hz Magnetoimpedance-Based ΔΣ Magnetometer with Background Gain Calibration and Short-Time CDS Techniques," *ISSCC*, pp. 62-64, Feb. 2024. https://doi.org/10.1109/ISSCC49657.2024.10454549
[6] A. Jouyaeian et al., "A 51A Hybrid Magnetic Current Sensor with a Dual Differential DC Servo Loop and 43mArms Resolution in a 5MHz Bandwidth," *ISSCC*, pp. 22-24, Feb. 2023. https://doi.org/10.1109/ISSCC42615.2023.10067677
[7] T. Qu et al., "A 2 MHz Bandwidth TMR-Based Contactless Current Sensor with Ping-Pong Auto-Zeroing and SAR-Assisted Offset Calibration," *IEEE JSSC*, vol. 60, no. 5, pp. 1708-1718, May. 2025. https://doi.org/10.1109/JSSC.2024.3468955
[8] H. -M. Chen et al., "A Sub-1 ppm/°C Precision Bandgap Reference with AdjustedTemperature-Curvature Compensation," *IEEE TCAS-I*, vol. 64, no. 6, pp. 1308-1317, June 2017. https://doi.org/10.1109/TCSI.2017.2658186

[9] K. Chen et al., "A 1.16-V 5.8-to-13.5-ppm/°C Curvature-Compensated CMOS Bandgap Reference Circuit with a Shared Offset-Cancellation Method for Internal Amplifiers," *IEEE JSSC*, vol. 56, no. 1, pp. 267-276, Jan. 2021. https://doi.org/10.1109/JSSC.2020.3033467
[10] U. Ausserlechner et al., "Compensation of the Piezo-Hall Effect in Integrated Hall Sensors on (100)-Si," *IEEE Sensors Journal*, vol. 7, no. 11, pp. 1475-1482, Nov. 2007. https://doi.org/10.1109/JSEN.2007.907039
[11] C. Lee et al., "A Miniaturized Wireless Neural Implant with Body-Coupled Data Transmission and Power Delivery for Freely Behaving Animals," *ISSCC*, pp. 1-3, Feb. 2022. https://doi.org/10.1109/ISSCC42614.2022.9731733
[12] Y. Li et al., "A 6.4GΩ-Input-Impedance 104.5dB-CMRR 96dB-DR DD-AFE with Tri-Level IDAC for Small-Diameter Dry-Electrode Interface," *IEEE TbioCAS*. https://doi.org/10.1109/TBCAS.2025.3558094
[13] H. Jiang et al., "An Energy-Efficient 3.7nv/√Hz Bridge-Readout IC with a Stable Bridge Offset Compensation Scheme," *ISSCC*, pp. 172-173, Feb. 2017. https://doi.org/10.1109/ISSCC.2017.7870316

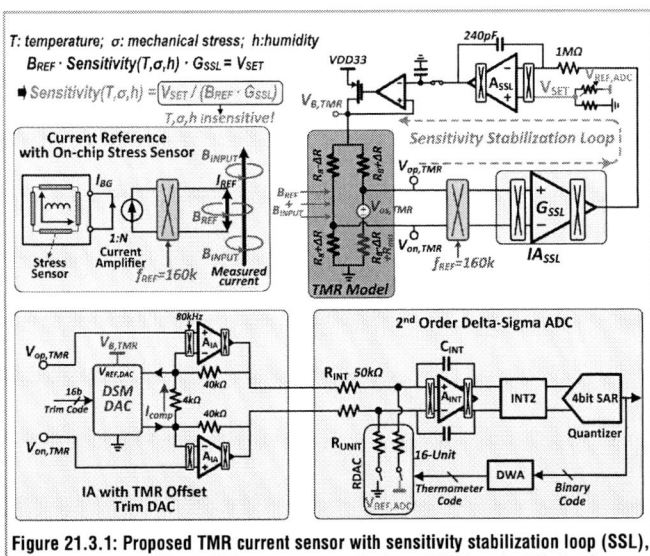

Figure 21.3.1: Proposed TMR current sensor with sensitivity stabilization loop (SSL), current reference with a built-in stress sensor (top), and TMR readout (bottom).

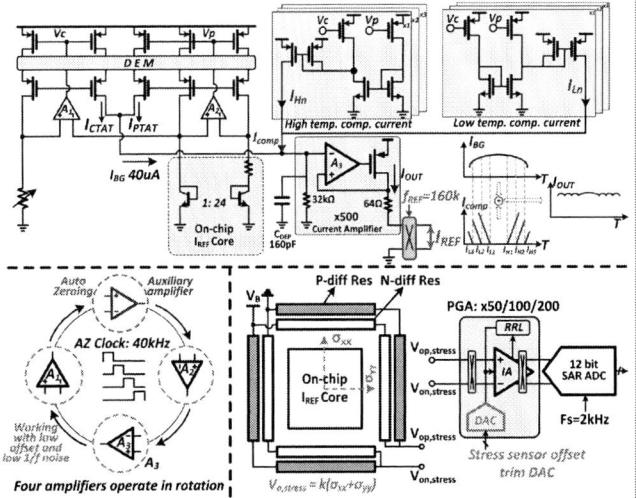

Figure 21.3.2: Current reference with segmental curvature compensation (top), and stress sensor with its readout circuit (bottom right).

ISSCC 2026 / SESSION 21 / SENSOR INTERFACES / 21.3

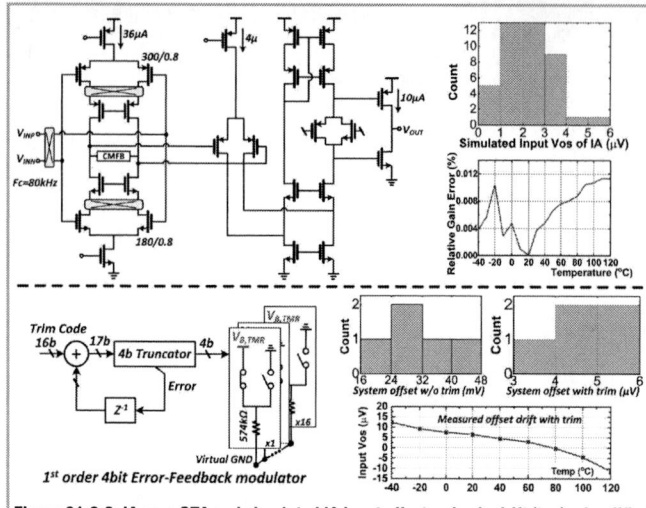

Figure 21.3.3: IA core OTA and simulated IA input offset and gain drift (top), simplified circuit diagram of the TMR offset-nulling DAC (bottom left), measured offset and offset drift (bottom right).

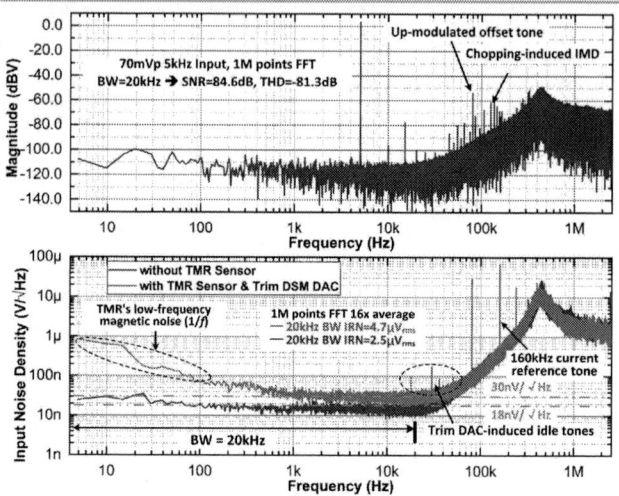

Figure 21.3.4: Measured spectrum of the TMR readout-only circuit with a 70mVpk sine wave input (top), and input-referred noise w/wo TMR sensor and offset trim DAC (bottom).

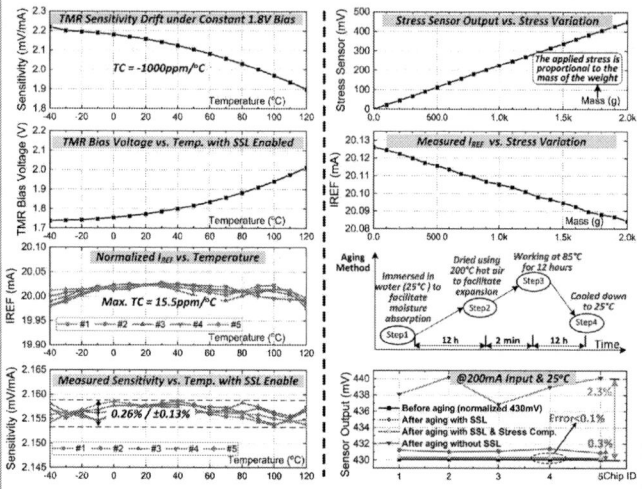

Figure 21.3.5: Measured sensitivity, I_{REF} and TMR bias voltage over Temp.(left), stress sensor output and I_{REF} as stress changes (top right), system output after accelerated aging (bottom right).

	ISSCC'21 [4]	ISSCC'15 [3]	ISSCC'24 [5]	ISSCC'23 [1]	ISSCC'22 [2]	ISSCC'23 [6]	JSSC'25 [7]	This work
Sensor Type	Fluxgate		MI	Shunt		Hall+Coil		TMR
Detection Mode	Closed-Loop			Open-Loop				
Process (μm)	0.25	0.6	0.18	0.18	0.18	0.18	0.18	0.18
Area (mm²)	4	9.8	5.71	0.38	0.36	3.9	0.97	2.99
Supply Voltage (V)	1.8	5	1.5	1.8	1.8	1.8	1.8	3.3/1.8
Power (mW)	12.96	280	3.96	0.5 *1	0.48	12.78	3.1	1.65 *1
Sensitivity Variation vs. Temp.	--	--	--	±0.2% (-40~125°C)	±0.25% (-40~85°C)	±2.7% (-40~85°C)	±1.25% (-40~125°C)	±0.13% (-40~120°C)
Nonlinearity	±0.2%	±0.1%	±0.6%	--	--	--	0.24%	±0.3%
Bandwidth (Hz)	125k	75k	10k	10k	10k	5M	2M	20k
Resolution	829nT	212nT	4.5nT *2	1.7mA	5.3mA	43mA	206nT	9.4nT *2 (50μA)
Dynamic Range (dB)	74 *3	82	82	87	73.5	61.5	58	82 *5
Offset	--	900nT	--	4μV	6μV	73μT	35μV	6μV (12nT)
Noise Density	--	--	45pT/√Hz	14nV/√Hz	85nV/√Hz	--	5.4nV/√Hz	30nV/√Hz
Aging Drift Compensation	No	No	No	No	No	No	No	Yes
Self-calibration	No	No	Yes (Off-line)	Yes (Off-line)	No	No	No	Yes (On-line)
*4 FoM (dB)	149	136	146	160 *1	147	147	146	153 *1

*1 The 20mA reference current is not included; *2 Estimated from noise density and bandwidth; *3 Estimated from rms noise and input range;
*4 FoM=DR+10lg(BW/Power); *5 The maximum input current of 800mA yields a 60mV TMR output with linearity <±0.3%.

Figure 21.3.6: Performance summary and comparison table.

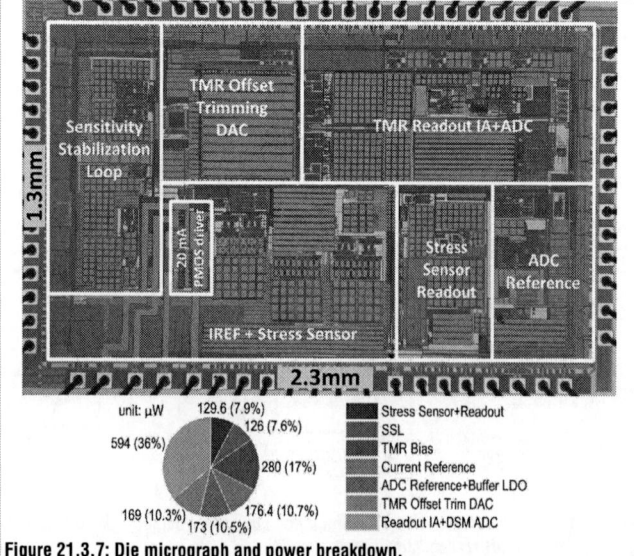

Figure 21.3.7: Die micrograph and power breakdown.

• 2026 IEEE International Solid-State Circuits Conference

979-8-3315-8937-0/26 $31.00 © 2026 IEEE

ISSCC 2026 / SESSION 21 / SENSOR INTERFACES / 21.4

21.4 A Background-Calibrated NPN-Based Temperature Sensor with 0.05°C (3σ) Inaccuracy from -70°C to 125°C

Nandor G. Toth, Kofi A. A. Makinwa

TU Delft, Delft, The Netherlands

Abstract

In this work, an NPN-based temperature sensor is presented that introduces a background-calibration scheme to correct all current-domain errors in its front-end. It achieves an inaccuracy of 0.05°C (3σ) from -70°C to 125°C (RIA=0.05%) after a one-point trim, which represents the state-of-the-art energy efficiency (100fJ·K^2) and power supply sensitivity (PSS) (0.004°C/V).

After a 1-point trim, BJT-based temperature sensors can be very accurate, with many designs achieving better than 0.2°C inaccuracy from -55°C to 125°C [1]. However, achieving sub-0.1°C accuracy has proven extremely challenging, having been reported only once before according to a comprehensive survey [1]. This is mainly due to the difficulty in completely correcting for errors caused by the decrease in BJT current gain (β) at low temperatures and the exponential increase of leakage currents at high temperatures. This paper proposes an NPN-based temperature sensor that uses a background calibration technique to correct such errors over a wide temperature range. It achieves a 1-point trimmed inaccuracy of 0.05°C (3σ) from -70°C to 125°C (RIA=0.05%), which represents the state of the art. In addition, it also achieves state-of-the-art energy efficiency (100fJ·K^2) and power supply sensitivity (PSS) (0.004°C/V). Compared to [2], it is 78× more energy efficient, and, compared to more recent designs [3-5], it is 2× more accurate and has 2.5× better PSS.

A simplified block diagram of the proposed temperature sensor is shown in Fig. 21.4.1. A PTAT current (I_{PTAT}) is generated by forcing the difference in the base-emitter voltages (ΔV_{BE}) of two NPNs ($Q_{1,2}$) across a resistor (R_1). This current is then mirrored (I_1) to the virtual ground of the 1st integrator of a ΔΣ-modulator, which is also used to generate a CTAT current (I_2) by forcing V_{BE} across a resistive DAC (R_2). The modulator's bitstream output then drives the RDAC such that, on average, I_2 exactly balances I_1. As a result, the bitstream average (μ_1) will be a function of temperature [3].

In the proposed sensor, a significant source of I_1-error is the spread in the R_2/R_1 ratio. In [6], this was partially corrected with the help of a room-temperature (RT) calibration. However, this does not correct for its temperature dependence, e.g., due to the finite R_{ON} of the RDAC switch. Another source of I_1-error is the base current (I_{B1}) of Q_1, which is most problematic at low temperatures. Although this can be mitigated by the β-compensation technique proposed in [6], the dependence of β on collector current density still causes a residual error related to $\Delta\beta = \beta_{02} - \beta_{Q1}$. The net leakage current (I_L) of the junction-isolated source/drain regions of the PMOS current mirrors and the collectors of the NPNs is yet another source of error, especially at high temperatures. In consequence, I_1 can be expressed as $(1+\varepsilon)\cdot I_{PTAT}$, where ε is a temperature-dependent parameter that contains all sources of error (Fig. 21.4.1).

In this work, the error component $(1+\varepsilon)$ is extracted by a background-calibration (B-CAL) technique. This involves performing a calibration conversion in which the 1st integrator's virtual ground is switched to ΔV_{BE}, instead of V_{BE}, so that $I_{2,cal}=\Delta V_{BE}/R_2$ (Fig. 21.4.1, bottom). Since $I_{2,cal} < I_1$, charge balancing is performed by using a current DAC (IDAC) to switch I_1, while constantly enabling the RDAC to keep $I_{2,cal}$ constant. As a result, the modulator's bitstream average (μ_2) will now be equal to $1/(R_2/R_1\cdot(1+\varepsilon))$. This can then be used to digitally trim μ_1 by simply computing $\mu_{TRIM} = \mu_1\cdot\mu_2$, which only depends on $\Delta V_{BE}/V_{BE}$ and is thus insensitive to any I_1-error over PVT.

A simplified block diagram of the proposed sensor is shown in Fig. 21.4.2. Q_1 and Q_2 are biased at a current ratio of 1 and an area ratio of 7, which, in combination with R_1 (=130kΩ), results in I_{PTAT}=390nA at room temperature (RT). A feedback loop containing transistor M_1 and a base-current mirror sets the collector of Q_2 to $V_{GS,M1}$ (~V_{BE}), while an amplifier (A_1) drives the PMOS collector-current mirrors to equalize both collector voltages. To save power, I_{PTAT} is mirrored to the ADC with a 2:1 ratio. In this work, R_2/R_1=13, and so μ_1=6.5·$\Delta V_{BE}/V_{BE}$, which ranges from 0.24 to 0.92 from -70°C to 125°C. Since the RDAC is quite large (R_2=1.7MΩ), it is segmented to minimize switching transients due to the charge stored in its parasitic capacitance [3]. During calibration, when I_1 is switched, switching transients are minimized by dumping unused current into 2·R_1, thus maintaining the IDAC output at ~ΔV_{BE}. Lastly, the mismatch of the NPNs and PMOS current mirrors is mitigated by barrel-shifting their unit elements at $f_s/320$ and $f_s/160$, respectively.

For a maximum calibration error of 0.01°C (3σ) after an RT trim, the resolution of μ_2 must be less than 30ppm (σ). Furthermore, the errors in $I_{2,cal}$, e.g. due to DAC switch leakage, must be less than 1.2pA (σ) at 125°C. During calibration, $\mu_2 \approx 1/6.5 = 0.15$, which means that the noise of I_1 is averaged for only 15% of the total conversion time. Since the modulator enforces charge balancing, more noise averaging, and thus, more resolution can be obtained by increasing $I_{2,cal}$ (= $\mu_2\cdot I_1$). To accomplish this, R_2 is folded into 2 equal segments that are connected in parallel (Fig. 21.4.2). This increases $I_{2,cal}$ by 4× and proportionally increases the averaging of I_1's noise. This, in turn, results in a 4× reduction in the conversion time needed to achieve a given resolution, and in the sensitivity of $I_{2,cal}$ to DAC switch leakage. It is also very accurate; a segment mismatch of 0.12% (σ) results in only a small additional error ($\Delta R_2/R_2 \sim$ 1ppm (σ)) after folding.

The RDAC consists of 4 resistor segments and 5 HVT NMOS switches. The R_{ON} of these switches, as well as the R_{ON} of the NPN DEM switches, will introduce temperature-dependent errors in the R_2/R_1 ratio. To reduce these errors, the switches are driven by boosted clocks generated by level-shifters (LS) connected to a high-voltage rail (V_H). As shown in Fig. 21.4.2 (right), V_H is generated by a regulated charge-pump (RCP), which consists of two cascaded voltage doublers that boost the V_{GS} (~0.95V) of an I_{PTAT}-biased diode-connected HVT NMOS by 4×. V_H thus tracks the HVT NMOS process corner, making R_{ON} robust to PVT variations. This allows the use of smaller DEM and DAC switches, which reduces their leakage and charge injection. In simulations, the RCP reduces R_{ON}-spread by about 2.5× while consuming only 85nA (at RT).

The ΔΣ-modulator employs a 2nd-order CIFF loop filter, which consists of a continuous-time 1st integrator and a compact switched-capacitor 2nd integrator. It is clocked at f_s = 50kHz. The 1st integrator is based on a current-assisted two-stage amplifier (A_2) consisting of a folded cascode input stage and a source-follower output stage [3]. Its high DC gain and GBW (>78dB and >1.3MHz) ensure a stable virtual ground, reducing errors in I_{CTAT} and $I_{2,cal}$. To achieve less than 30ppm (σ) error during calibration, the offset of A_2 must be less than 1μV (σ) at -70°C. To ensure this, nested chopping is applied, with the inner choppers clocked at f_s, while the outer choppers are clocked at $f_s/320$. During calibration, the bias current of A_2 is doubled to increase its GBW (> 2.5MHz) and reduce settling errors. In simulation, such errors correspond to equivalent errors of <1mK (σ) during a regular conversion and <20ppm (σ) during calibration.

The sensor was fabricated in a 0.18μm CMOS process and occupies 0.063mm^2 (Fig. 21.4.7). It draws 1.9uA (61% front-end, 33% ADC, 6% digital) from a supply ranging from 1.4V to 2.2V. Two identical sensors were placed on each die to enable differential temperature measurements, which cancels the effect of ambient temperature drift. The sinc^2 decimation filter was implemented off-chip for flexibility.

To measure the sensor's resolution, a long bitstream (2^{21} bits) is decimated by a sinc^2 filter while the modulator is operated in free-running mode. As shown in (Fig. 21.4.3, top left), the sensor becomes thermal-noise limited for T_{conv} > 30ms and remains immune to drift and $1/f$ noise up to T_{conv} < 5s. With T_{conv} = 38.4ms, the sensor achieves 1mK (σ) resolution, resulting in a resolution FoM of 100fJ·K^2. Figure 21.4.3 (top right) shows the resolution during calibration expressed in ppm (σ). As expected, folding R_2 reduces the conversion time required to achieve 30ppm (σ) by 4×: from 76.8ms to 19.2ms, thus significantly reducing the calibration overhead. If calibration is performed after every conversion, which is not strictly necessary, the resolution FoM increases to 2pJ·K^2 due to the increased noise and conversion time.

30 sensors on 15 dies in ceramic DIL packages were characterized from -70°C to 125°C in a temperature-controlled oven. To suppress ambient temperature drift, the samples were placed in good thermal contact with a large aluminum block. The decimated data of a regular conversion (μ_1) before and after B-CAL is shown in Fig. 21.4.3 (bottom left). The results of calibration ($1/\mu_2$) are shown in Fig. 21.4.3 (bottom right) with β-compensation [6] both enabled and disabled. It can be observed that (Δ)β-errors at low temperatures and I_L-errors at high temperatures cause significant deviations in $1/\mu_2$ from the expected value of 6.5.

After a batch calibration, the master curves in Fig. 21.4.3 (bottom left) are mapped to a fixed 5th-order polynomial. After an RT trim and an RT calibration, the resulting temperature spread is 0.4°C (3σ) (Fig. 21.4.4, top left), which is mainly limited by β-error at low temperatures. This improves to 0.1°C (3σ) with β-compensation (Fig. 21.4.4, top right), but is still limited by Δβ and I_L-errors. Employing the proposed B-CAL technique results in a state-of-the-art 0.05°C (3σ) inaccuracy (Fig. 21.4.4, bottom left). Similar results are obtained with β-compensation, demonstrating its redundancy after applying B-CAL.

370 • 2026 IEEE International Solid-State Circuits Conference

979-8-3315-8937-0/26 $31.00 © 2026 IEEE

Another low-cost calibration technique is voltage calibration (V-CAL) [7]. This obviates the need for calibrating at a known temperature by leveraging the accuracy of ΔV_{BE}. In this design, V-CAL can be implemented by connecting an external voltage reference (V_{EXT}) to the virtual ground of the 1st integrator (Fig. 21.4.2). As a result, the bitstream average is $6.5 \cdot \Delta V_{BE}/V_{EXT}$, from which ΔV_{BE}, and thus the die temperature, can be determined. For the proposed sensor, V-CAL results in an inaccuracy of 0.08°C (3σ), which is 2× better than previously reported [2,8,9].

The measured PSS of the proposed sensor is shown in Fig. 21.4.5. With V_H fixed at $2 \cdot V_{DD}$ and without B-CAL, the PSS is 0.43°C/V from 1.4V to 2.2V (top left). B-CAL reduces the temperature dependence of R_2/R_1 and thus improves the PSS to 0.15°C/V (top right). Using the RCP improves the PSS significantly, to 0.025°C/V without B-CAL (bottom left) and even to 0.004°C/V with B-CAL (bottom right), which is the lowest reported in [1].

In Fig. 21.4.6, the performance of this sensor is benchmarked against state-of-the-art BJT-based temperature sensors. It achieves the lowest 1-point trimmed inaccuracy (0.05°C (3σ) from -70°C to 125°C), the lowest voltage calibrated inaccuracy (0.08°C 3σ from -70°C to 125°C), and the lowest PSS (0.004°C/V), while its area (0.063mm²) and energy efficiency (100fJ·K²) are in line with the state-of-the-art. This performance is enabled by the combination of an accurate current-mode front-end and the proposed background calibration technique.

References:
[1] K. A. A. Makinwa. Smart Temperature Sensor Survey. Accessed: Aug. 2024. [Online]. Available: http://ei.ewi.tudelft.nl/docs/TSensor_survey.xls. https://doi.org/10.1109/ISSCC49661.2025.10904777
[2] B. Yousefzadeh et al., "A BJT-Based Temperature-to-Digital Converter with ±60mK (3σ) Inaccuracy from −55°C to +125°C in 0.16-µm CMOS" *IEEE JSSC*, vol. 52, no. 4, pp. 1044-1052, April 2017. https://doi.org/10.1109/JSSC.2016.2638464
[3] N. G. Toth, K. A. A. Makinwa, "A BJT-Based Temperature Sensor with an 80fJ.K² Resolution FoM," *ISSCC*, pp. 476-478, Feb. 2025. https://doi.org/10.1109/ISSCC49661.2025.10904777
[4] Z. Tang et al., "A Sub-1 V Capacitively Biased BJT-Based Temperature Sensor with an Inaccuracy of ±0.15 °C (3σ) from -55 °C to 125 °C" *IEEE JSSC*, vol. 58, no. 12, pp. 3433-3441, Dec. 2023. https://doi.org/10.1109/JSSC.2023.3308554
[5] S. H. Shalmany et al., "A 620µW BJT-Based Temperature-to-Digital Converter with 0.65mK Resolution and FoM of 190fJ·K²," *ISSCC*, pp. 70-71, Feb. 2020. https://doi.org/10.1109/ISSCC19947.2020.9063007
[6] N. G. Toth, K. A. A. Makinwa, "A β-Compensated NPN-Based Temperature Sensor with ±0.1 °C (3σ) Inaccuracy from -55°C to 125°C and 200fJ·K²] Resolution FoM," *IEEE JSSC*, vol. 59, no. 12, pp. 4068-4076, Dec. 2024. https://doi.org/10.1109/JSSC.2024.3440071
[7] M. A. P. Pertijs et al., "Low-Cost Calibration Techniques for Smart Temperature Sensors," *IEEE Sensors Journal*, vol. 10, pp. 1098-1105, June 2010. https://doi.org/10.1109/JSEN.2010.2040730
[8] N. G. Toth et al., "A PNP-Based Temperature Sensor with Continuous-Time Readout and ±0.1 °C (3σ) Inaccuracy from -55 °C to 125 °C," *IEEE JSSC*, vol. 60, no. 2, pp. 593-602, Feb. 2025. https://doi.org/10.1109/JSSC.2024.3402131

[9] K. Souri et al., "A CMOS Temperature Sensor with a Voltage-Calibrated Inaccuracy of ± 0.15° C (3σ) from − 55° C to 125° C," *IEEE JSSC*, vol. 48, no. 1, pp. 292-301, Jan. 2013. https://doi.org/10.1109/JSSC.2012.2214831
[10] B. Wang et al., "A BJT-Based CMOS Temperature Sensor Achieving an Inaccuracy of ±0.45°C(3σ) from -50°C to 180°C and a Resolution-FoM of 7.2pJ·K² at 150°C," *ISSCC*, pp. 72-73, Feb. 2022. https://doi.org/10.1109/ISSCC42614.2022.9731647

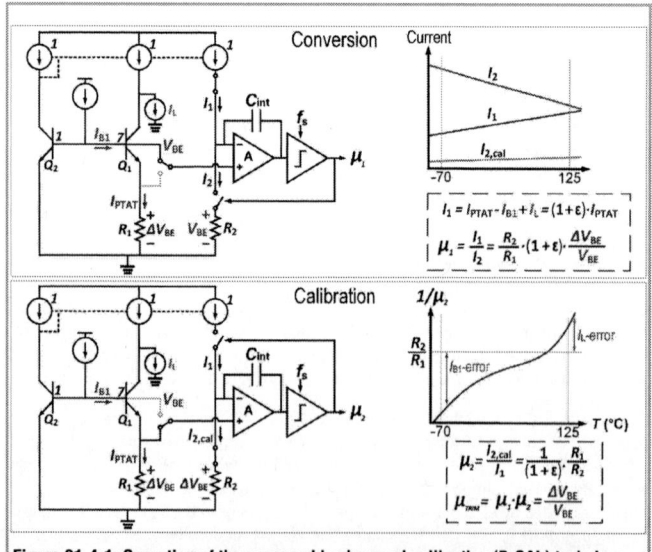

Figure 21.4.1: Operation of the proposed background calibration (B-CAL) technique.

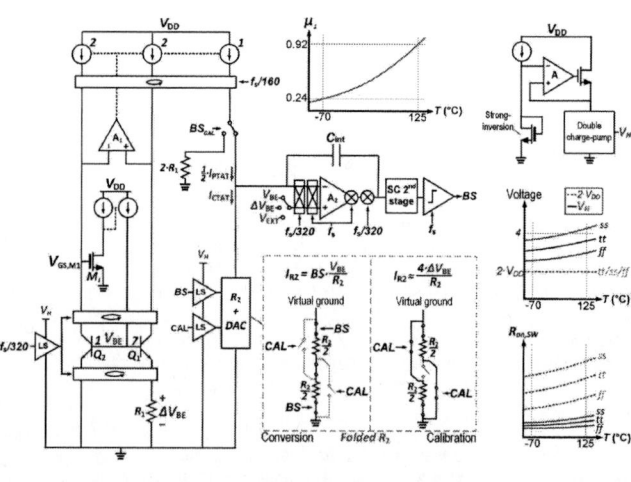

Figure 21.4.2: Block diagram of the proposed sensor (left), the RCP and its effect on the R_{ON} of the HVT NMOS switches (right).

ISSCC 2026 / SESSION 21 / SENSOR INTERFACES / 21.4

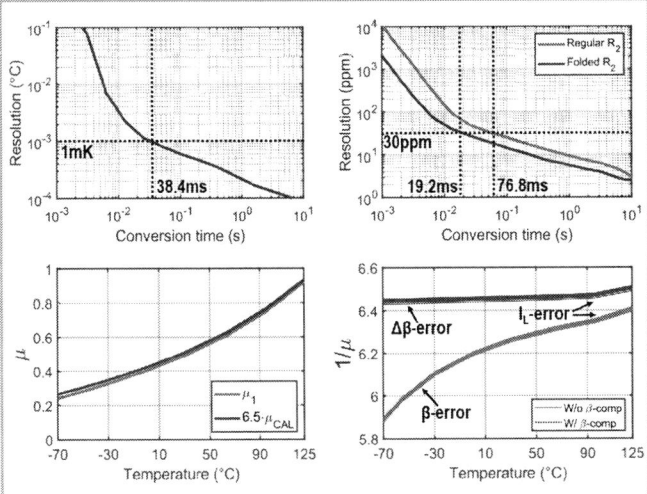

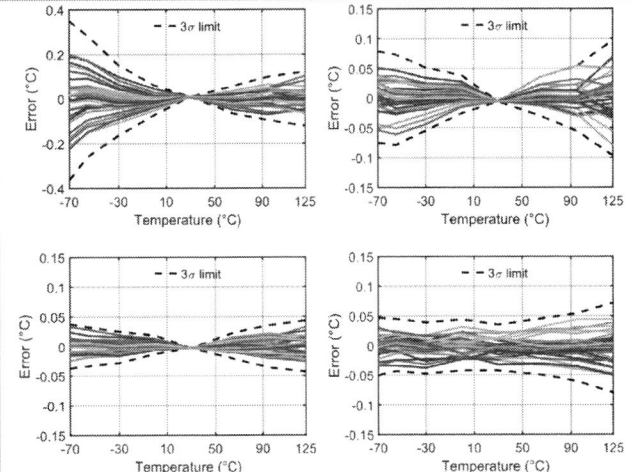

Figure 21.4.3: Resolution vs. conversion time during conversion (top left) and calibration (top right), and the decimated data over temperature during conversion (bottom left) and calibration (bottom right).

Figure 21.4.4: Measured temperature error of 30 samples with an RT trim and RT calibration (top left), with an RT trim, RT calibration and β-compensation (top right), with an RT trim and B-CAL (bottom left), with a voltage calibration and B-CAL (bottom right).

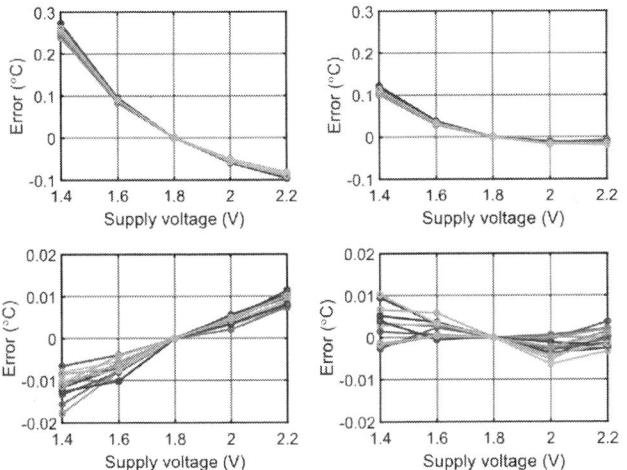

Figure 21.4.5: PSS without B-CAL and RCP (top left), with B-CAL and without RCP (top right), without B-CAL and with RCP (bottom left), and with B-CAL and RCP (bottom right).

	Wang JSSC'22 [10]	Shalmany ISSCC'20 [5]	Tang JSSC'23 [4]	Yousefzadeh JSSC'17 [2]	Toth ISSCC'25 [3]	This work
Sensor type	PNP	NPN	PNP	PNP	NPN	**NPN**
Technology	0.18μm	0.11μm	0.18μm	0.16μm	0.18μm	**0.18μm**
Chip area [mm²]	0.42	0.2	0.25	0.16	0.05	**0.063**
Supply current [μA]	2.5	550	0.85	4.6	1.8	**1.9**
Supply voltage [V]	1.5 to 2	1.125	0.95 to 1.4	1.5 to 2	1.4 to 2.2	**1.4 to 2.2**
Supply sensitivity [°C/V]	0.44	-	0.2	0.01	0.025	**0.004**
Temperature range	-50°C to 180°C	-35°C to 95°C	-55°C to 125°C	-55°C to 125°C	-70°C to 125°C	**-70°C to 125°C**
3σ Inaccuracy [°C] after a 1-pt trim	±0.45	-	±0.15	±0.06	±0.1	**±0.05**
Relative inaccuracy [%]	0.39	-	0.17	0.07	0.11	**0.05**
3σ Inaccuracy [°C] after voltage calibration	-	-	-	±0.17[2]	-	**±0.08**
Resolution [mK]	17.6	0.72	1.8	15	0.79	**1**
Conversion time [ms]	8.3	0.65	128	5	51	**38.4**
Resolution FoM[1] [fJ·K²]	9700	190	340	7800	80	**100**

[1] FoM = Energy/Conversion · Resolution²
[2] From a single batch, estimated from Fig. 13

Figure 21.4.6: Performance summary and comparison table.

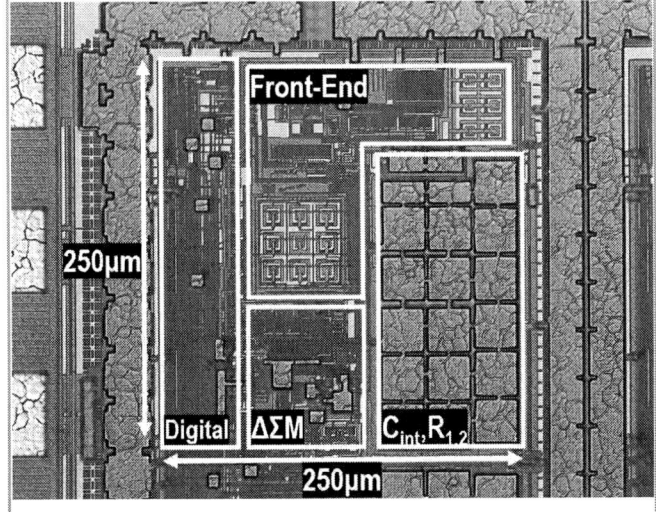

Figure 21.4.7: Die micrograph of the sensor.

• 2026 IEEE International Solid-State Circuits Conference

ISSCC 2026 / SESSION 21 / SENSOR INTERFACES / 21.5

21.5 A 0.6V 625um² Fully Stacked RC-Based Temperature Sensor Using Low TCR Metal Resistor Achieving 0.017nJ·%²-Accuracy FoM in 2nm Gate-All-Around Process

Haejung Choi, Jooseong Kim, Woojoong Jung, Sungmin Yoo, Jun-Hyeok Yang, Sunghyuck Lee, Jihye Park, Michael Choi, Ben Rhew

Samsung Electronics, Yongin, Korea

Abstract

The proposed RC-based temperature sensor, fabricated on a 2nm gate-all-around process, minimizes silicon area by fully stacking low temperature coefficient of resistance (TCR) metal resistors and a ring-oscillator time-to-digital converter. The conversion time is 12µs by using a time-offset compression technique to overcome the disadvantage of a low TCR metal resistor while having ±0.5°C (3σ) inaccuracy after 2-point calibration with a range of -40 to 125°C, resulting in accuracy-FoM of 0.017nJ·%².

As technology scaling continues to drive higher integration density and operating speed of SoCs, on-chip thermal management becomes essential. To provide effective thermal management, temperature sensors require high accuracy and fast conversion for optimizing chip performance, and a small size for placing sensors adjacent to on-chip hotspots. While traditional BJT-based temperature sensors have been widely used for high accuracy [6,7], they suffer from a significant area penalty as they require complex circuits such as bandgap references. While remote-probe techniques mitigate area limitation by placing only the BJT in a hotspot, this approach requires time-multiplexing a central sensor; hence, it is a trade-off between area and conversion time. Furthermore, optimizing BJT devices is facing significant challenges in advanced gate-all-around (GAA) processes [10]. To address these limitations, various alternatives have been suggested. While MOS-type sensors are advantageous from an area perspective, these sensors are sensitive to supply variations [8,9]. To overcome the limitations of both BJT and MOS-type sensors, resistor-based solutions have been actively researched. The work in [1] achieves a considerably small area, yet its clock-counting architecture features a conversion time of 1ms, which is too slow for real-time thermal tracking. In contrast, although the work in [2] implements a fast 12µs conversion time by adopting a SAR-DLL scheme, it comes at the cost of a large footprint from its extensive delay-line array and the area inefficiency of routing metal as the sensing element. This work employs a standard process design kit (PDK)-provided metal resistor to achieve high area efficiency. The proposed time-offset compression technique resolves the trade-off between resolution and conversion time caused by the low temperature coefficient of resistance (TCR) of a metal resistor. Furthermore, an area-efficient ring-oscillator (RO)-based time-to-digital converter (TDC), composed of only a few inverter cells, replaces the large delay line of [2] to reduce silicon area. Consequently, the proposed sensor achieves both a significant area reduction and a fast 12µs conversion time.

Previous works have typically used routing metal for its high TCR [1-3]. High TCR means a large change in resistance with temperature (ΔR), enabling high resolution which is a prerequisite for high accuracy. However, ΔR is proportional not only to TCR but also to the initial resistance(R_0), as described by $\Delta R = R_0 \cdot TCR \cdot \Delta T$ (Fig. 21.5.1). This implies that even for a low TCR, a large R_0 enables a PDK metal resistor to achieve a sufficient ΔR. Using the metal resistors has three key advantages over routing metal. First, the metal resistor offers 518 times higher sheet resistance than routing metal. For the same trace width, it requires 1/100 of the area to achieve the same ΔR. Second, being implemented in upper metal layers, it allows for active circuits underneath, eliminating front-end-of-line (FEoL) area overhead. In contrast, routing metal spans from lower to upper layers to achieve sufficient resistance, restricting circuit placement underneath, and thus wasting FEoL area. Third, as a standard PDK-provided device, the characteristics and variations of the metal resistor are well-characterized, which is a crucial advantage for ensuring performance in mass-production. While the metal resistor offers an area advantage, using a large R_0 introduces a trade-off between conversion time and resolution required for high accuracy. RC-based temperature sensors detect temperature through the temperature-dependent variation of an RC time constant ($\Delta \tau = \Delta RC$) (Fig. 21.5.1, middle). To achieve high accuracy, a large $\Delta \tau$ is required, as $\Delta \tau$ determines resolution of the sensor. A large R_0 increases $\Delta \tau$, thereby enhancing resolution, while slowing the overall conversion time. Conversely, a small R_0 enables fast conversion while yielding a small $\Delta \tau$, which degrades the achievable accuracy (Fig. 21.5.1, bottom middle). To resolve this trade-off, this work proposes a time-offset compression technique (Fig. 21.5.1, bottom right). Given that the temperature information is captured by $\Delta \tau$, the initial part of the RC-filter charging period is a time-offset (T_{OS}) which does not contain temperature information. This technique minimizes this redundant T_{OS} by rapidly charging the RC-filters with a small resistor for a precisely controlled duration that does not corrupt the subsequent $\Delta \tau$ measurement window. Subsequently, the RC-filters switch to a large resistor, slowing the voltage slope of VP and VN to generate a sufficiently large $\Delta \tau$ for high accuracy.

The proposed sensor structure, shown in Fig. 21.5.2, consists of a sensor core, a TDC, and a CTRL logic. The sensor core is composed of RC filters, where the resistance is switched between 0.1R and 1R based on the Φ_{OC} signal. The TDC converts the temperature-dependent time (T_{PTAT}) from the sensor core and a reference time (T_{REF}) into digital codes of D_{PTAT} and D_{REF}, respectively. The CTRL logic controls the entire operation and calculates the temperature code (TSCODE). The sensor operates in two modes (Fig. 21.5.2, middle). In TSENSE mode, the TDC generates D_{PTAT} by adjusting the delay generator's output (DLY) to match T_{PTAT} using a successive-approximation-register (SAR) search algorithm. The proposed time-offset compression technique alters the T_{PTAT}. The T_{PTAT} is determined in two stages: an initial T_{OS} period with the small resistor (0.1R), and a subsequent charging period with the large resistor (1R) with a duration of $RC \cdot \ln(2) - 10 \cdot T_{OS}$. Therefore, T_{PTAT} is $RC \cdot \ln(2) - 9 \cdot T_{OS}$. In normalization mode [2], the TDC measures T_{REF} to generate D_{REF}, which compensates for PVT variations of the unit delay in TDC (T_{DLY}). Since D_{PTAT}, D_{REF}, and the code corresponding to T_{OS} (D_{OS}) are all quantized by the same TDC, PVT-induced variations in T_{DLY} are canceled out. The D_{OS} value, predetermined by an auto-time-offset optimization technique described later, is arithmetically eliminated in the CTRL logic. The effects of comparator delay and offset are minimized through a dynamic offset cancellation and a hold-and-compare scheme. Furthermore, input-referred noise of the comparator is minimized to ensure sufficient accuracy for the temperature sensor. Consequently, the TSCODE is a ratio-metric value dependent only on the RC characteristics and the designed reference time, T_{REF}, ensuring robust operation (Fig. 21.5.2, bottom).

Figure 21.5.3 illustrates the proposed circuit techniques. The delay-line-based DLL in [2] required a vast number of delay cells and thus occupied a large area. To address this area issue, this work adopts an RO delay generator, which utilizes RO output (FRO) count (CNT) and tapped delays within the RO (uDLY) for coarse and fine bits, respectively (Fig. 21.5.3, top). As a result, the delay generator area decreases by 99.1% (from ~2500µm² in [2] to 22.6µm²). However, this method can cause non-monotonicity in the TDC output due to transition jitter at the CNT switching points. To overcome this issue, this work proposes a phase-interleaved counting technique. This technique decouples the sampling of fine bits from the transition edges of the coarse counter which are subject to jitter. Fine bits from the first half-period of the FRO are paired with the count from the falling-edge (CNTb), while those from the second half-period are paired with the count from the rising-edge (CNT). This method ensures a timing margin equivalent to a half FRO period for all TDC outputs, effectively preventing non-monotonicity. The bottom of Fig. 21.5.3 shows the proposed time offset auto-optimization technique. Determining an appropriate T_{OS} is crucial to ensure proper operation of the sensor. If T_{OS} is too long, the crossing point occurs within the T_{OS}, distorting the temperature data. If T_{OS} is too short, the crossing may not occur within the measurement window defined by the Φ_0 high period. Furthermore, the optimal T_{OS} varies with process variations from sample to sample. Lastly, precisely controlling T_{OS} (typical 50ns) of the proposed sensor requires a GHz-level clock, which imposes a system-level burden. This work proposes an auto-optimization technique that automatically finds the optimal T_{OS}. The same RO, introduced to create the area-efficient delay generator, also provides the 1.6GHz clock required for this technique, resolving the need for a dedicated high-frequency clock. The sensor first finds the crossing time with the fast-charging resistor (0.1R) enabled, using this time as a reference. T_{OS} is then set to a slightly smaller value than the reference time, with a margin to account for temperature-induced variations. The corresponding TDC output code (D_{OS}) is then recorded in the CTRL logic. Even if the temperature-induced RC variation exceeds the initial margin, the CTRL logic can automatically adjust it by comparing D_{PTAT} and D_{OS}. The normalization mode corrects for PVT-induced variations in the RO. This optimization step, performed at startup or during wafer-level testing, ensures PVT-robust sensing without increasing the conversion time in normal operation.

The proposed temperature sensor is fabricated in a 2nm gate-all-around (GAA) process. Measurement results from 20 samples over -40°C to 125°C show the output code varies from 300 to 3000 on average, corresponding to a sensitivity of 16.4 LSB/°C (Fig. 21.5.4, left). The inaccuracy after 1-point calibration is ±1.9°C (p2p), while the 3σ inaccuracy is ±3.9°C. A 2-point calibration improves the inaccuracy to ±0.5°C (3σ) and the p2p inaccuracy is ±0.8°C, over the temperature range, without additional curve-fitting to reduce inaccuracy. The sensor consumes 15µW from 0.6V supply at room temperature.

Figure 21.5.5 compares the proposed fully stacked RC-based temperature sensor with state-of-the-art works. The sensor operates from 0.6V to 0.77V, with a measured supply sensitivity of only 0.13°C (±1LSB) (Fig. 21.5.5, Top). This robustness is an inherent feature of the RC-based principle. The proposed sensor overcomes the trade-off between conversion time-resolution needed for high accuracy of low-TCR resistors and utilizes an improved RO-based TDC readout. As a result, it achieves a 12µs conversion time, 83 times faster than [1] and comparable to [2], while attaining the smallest area among RC-based sensors (Fig. 21.5.5, bottom left). Moreover, the use of a well-characterized PDK-provided device contributes to the high ±0.5°C inaccuracy. This combination of metrics results in an improved accuracy-FoM of 0.017nJ·%² (Fig. 21.5.5, bottom right). The resolution-FoM is 0.7pJ·%².

The die photo in Fig. 21.5.7 shows the physical implementation of the sensor. The proposed RC-based temperature sensor features a fully-stacked architecture where the capacitor, metal resistor, and circuits are vertically integrated, resulting in a total area of 625µm². Figure 21.5.6 summarizes the final performance achieved through this implementation and compares it with the state-of-the-arts.

References:
[1] Bei-Shing Lien et al., "A 0.65V 900µm² BEoL RC-Based Temperature Sensor with ±1°C Inaccuracy from -25°C to 125°C," *ISSCC*, pp. 68-70, Feb. 2024. https://doi.org/10.1109/ISSCC49657.2024.10454423
[2] J. Park et al., "A 0.65V 1316µm² Fully Synthesizable Digital Temperature Sensor Using Wire Metal Achieving 0.16nJ·%² -Accuracy FoM in 5nm FinFET CMOS," *ISSCC*, pp. 220-221, Feb. 2022. https://doi.org/10.1109/ISSCC42614.2022.9731766
[3] Dan Shi et al., "A 4,100µm² Wire-Metal-Based Temperature Sensor with a Fractional-Discharge FLL and a Time-Domain Amplifier with ±0.2°C Inaccuracy (3σ) from –40 to 125°C and 45fJ·K² Resolution FoM in 28nm CMOS," *ISSCC*, pp. 478-480, Feb. 2025. https://doi.org/10.1109/ISSCC49661.2025.10904592
[4] Jan A. Angevare et al., "A Highly Digital 2210µm² Resistor-Based Temperature Sensor with a 1-Point Trimmed Inaccuracy of ±1.3°C (3σ) from -55°C to 125°C in 65nm CMOS," *ISSCC*, pp. 76-78, Feb. 2021. https://doi.org/10.1109/ISSCC42613.2021.9365995
[5] Amr Khashaba et al., "A 0.0088mm² Resistor-Based Temperature Sensor Achieving 92fJ·K² FoM in 65nm CMOS," *ISSCC*, pp. 60-62, Feb. 2020. https://doi.org/10.1109/ISSCC19947.2020.9062956
[6] Nandor G. Toth et al., "A β-Compensated NPN-Based Temperature Sensor with ±0.1°C (3σ) Inaccuracy from -55°C to 125°C and a 200fJ·K² Resolution FoM," *ISSCC*, pp. 66-68, Feb. 2024. https://doi.org/10.1109/ISSCC49657.2024.10454408
[7] Nandor G. Toth et al., "A BJT-Based Temperature Sensor with ±0.1°C (3σ) Inaccuracy from -55°C to 125°C and a 0.85pJ·K² Resolution FoM Using Continuous-Time Readout," *ISSCC*, pp. 358-360, Feb. 2023. https://doi.org/10.1109/ISSCC42615.2023.10067457
[8] Vinshtok-Melnik et al., "Ultra Miniature 1850 µm² Ring Oscillator Based Temperature Sensor," *IEEE Access*, pp. 91415-91423, May. 2020. https://doi.org/10.1109/ACCESS.2020.2994326
[9] Martin Cochet et al., "A 225 µm² Probe Single-Point Calibration Digital Temperature Sensor Using Body-Bias Adjustment in 28 nm FD-SOI CMOS," *ISSCL*, pp. 14-17, Jan. 2018. https://doi.org/10.1109/LSSC.2018.2797427
[10] J Deng et al., "Enhancing Power-Performance-Area Scaling of Std-Cell, SRAM, and Analog Designs Through Design-Technology Co-Optimization," *IEEE VLSI Technology and Circuits*, pp. 1-3, June. 2025. https://doi.org/10.23919/VLSITechnologyandCir65189.2025.11074818

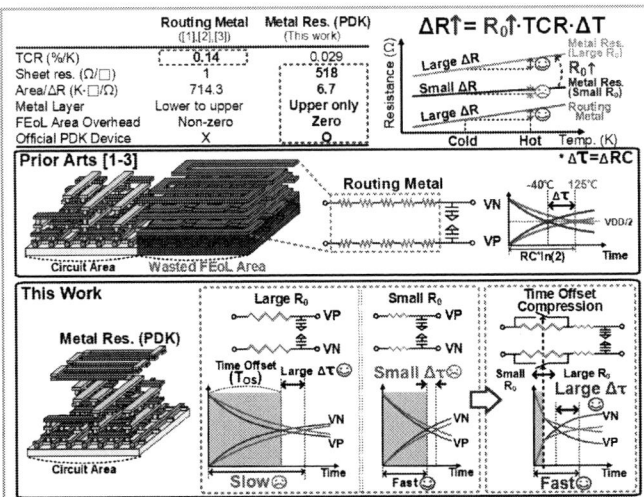

Figure 21.5.1: Temperature sensing device selection, comparison of routing metal and metal resistor (top), architecture of prior arts (middle), the proposed RC-based temperature sensor, (R=800kΩ, C=1.5pF) (bottom).

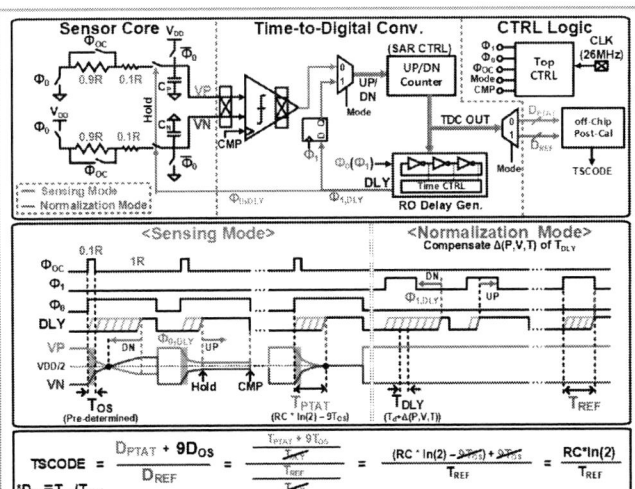

Figure 21.5.2: Overall structure of the proposed temperature sensor, block diagram(top), timing diagram (middle), temperature code generation equation (bottom).

ISSCC 2026 / SESSION 21 / SENSOR INTERFACES / 21.5

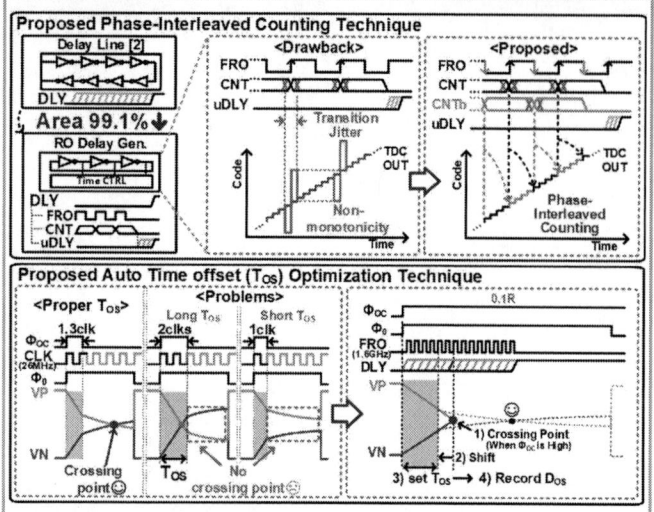

Figure 21.5.3: Details of proposed circuit techniques including phase-interleaved-counting (top), auto time offset optimization (bottom).

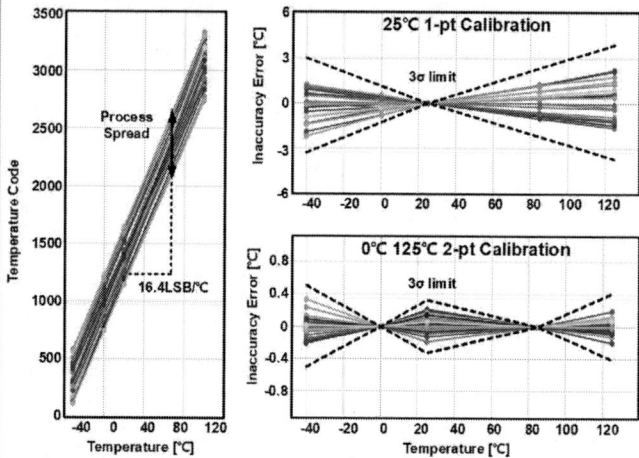

Figure 21.5.4: Measurement results showing temperature code without calibration (left), inaccuracy error after 1-point calibration (top right), inaccuracy error after 2-point calibration (bottom right).

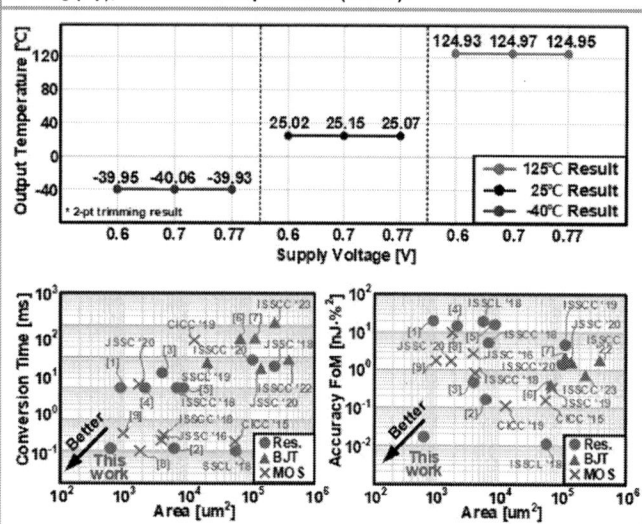

Figure 21.5.5: Measurement results showing supply sensitivity across 0.6 to 0.77V (top), conversion time and accuracy-FoM versus area (bottom).

	This Work		ISSCC '24 [1]	ISSCC '22 [2]	ISSCC '25 [3]		ISSCC '21 [4]	ISSCC '20 [5]
Technology	2nm Gate-All-Around		3nm finfet	5nm finfet	28nm		65nm	65nm
Sensor Type	Metal Res.		Routing metal	Routing metal	Routing metal		Polysilicon Res.	Polysilicon Res.
Official PDK Device	O		X	X	X		O	O
Required # of Resistors	1		2[a]	1	1		1	1
Required Metal Layer for Resistor	Upper only		Lower to Upper	Lower	Lower to Upper		-	-
Wasted FEoL Area [μm²]	0		670[b]	510[b]	730[b]		1334[b]	3690[b]
Area [μm²]	625[c]		900[c]	1316[c] 6350	4100[c]		2210[c]	8800[b]
Read-Out Type	Ring-OSC TDC		Counter	SAR DLL	FLL		ΔΣ ADC	FLL
Min. Supply Voltage [V]	0.6		0.65	0.65	0.7		0.9	-
Power [μW]	15		20	15	10.5		28	45
Conversion Time [ms]	0.012		1	0.012	3		1	1
Energy/Conversion [nJ]	0.18		20	0.18	31.5		28	45
Temperature Range [°C]	-40 to 125		-25 to 125	-40 to 150	-40 to 125		-55 to 125	-30 to 90
Calibration Points	1	2	1	1	1	2	1	1
Inaccuracy(3σ) [°C]	3.9	0.5	1.5	-	1.5	0.2	1.3	-
Inaccuracy(p2p) [°C]	1.9	0.8	1	2.7	1.8	-	-	0.72
Resolution [mK]	61		22.1	114.1	1.2		12.8	1.43
Relative Inaccuracy [%]	0.30		1	0.95	0.12		0.72	0.6
Accuracy FoM[e] [nJ-%²]	0.017		20	0.16	0.46		14.6	16.2
Resolution FoM[f] [pJ-%²]	0.7		9.8	2.3	0.05		4.6	0.09

a) Sensor and reference resistor   b) Estimated RC filter area
c) Excluding off-chip post-calculation block
d) Excluding off-chip post-calculation and read-out circuit
e) Accuracy FoM = energy/conversion x (relative inaccuracy)²
f) Resolution FoM = energy/conversion x (resolution)²

Figure 21.5.6: Summary and comparison table.

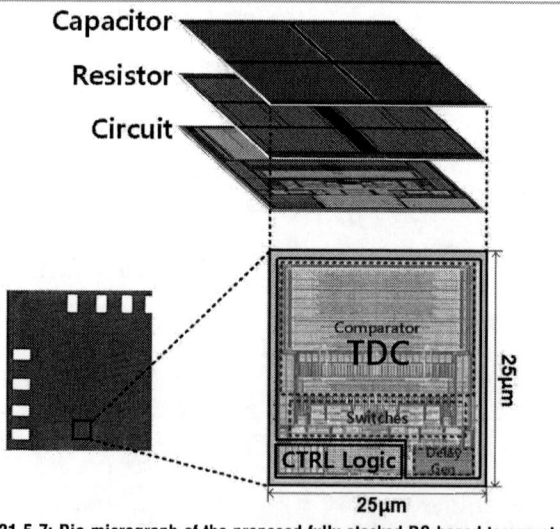

Figure 21.5.7: Die micrograph of the proposed fully-stacked RC-based temperature sensor.

ISSCC 2026 / SESSION 21 / SENSOR INTERFACES / 21.6

21.6 A ±60mA-Inaccuracy Low-Side Average Current Sensor with Operating-Conditions-Insensitive Control Supporting 0.1-to-3A Load Range and Sub-100ns Sample Time for Automotive USB Charge Application

Jian-Jun Kuang[1,2], Xin Ming[1], Xin-Ce Gong[1], Tian-Chen Lang[2], Xiang Geng[2], Bo Zhang[1]

[1]University of Electronic Science and Technology of China, Chengdu, China, [2]SouthChip Semiconductor Technology, Chengdu, China

Abstract

In Paper 21.6, UESTC and SouthChip Semiconductor Technology present a low-side average current sensor for automotive USB charging, which supports 0.1-to-3A load range and sub-100ns sample time with ±60mA inaccuracy. This is achieved by 1) an adaptive settling helper to reduce the settling delay, 2) a DC level-shift bias to improve the light-load current sense accuracy, and 3) an anti-duty resistor to suppress the drift of -3dB bandwidth.

Optimizing in-vehicle USB (universal serial bus) charging lets users enjoy the convenience of charging [1,2]. High CC (constant current) and CV (constant voltage) accuracy are crucial for efficient and reliable power delivery. The top left of Fig. 21.6.1 gives the accuracy requirement of the common charging protocols including USB PPS (programmable power supply) and UFCS (universal fast-charging specification), where PPS is an important function of USB PD (power delivery). Targeting to achieve high-precision CC and CV charging, a fully integrated USB charging solution with 33W (11V, 3A) maximum output power is proposed (top right of Fig. 21.6.1), which mainly consists of an integrated Buck converter and a protocol controller [1]. This Buck converter adopts peak-current-mode (PCM) control and works in FCCM (forced CCM) mode.

In general, high CV accuracy is easy to realize, but achieving high CC accuracy over a wide load current (I_{BUS}) range (i.e., 0.1A to 3A) with duty-cycle (D) variation is challenging, since the feedback information of I_{BUS} cannot be obtained through a simple resistor voltage divider as with V_{BUS} (output voltage) in the CV loop. Thus, the CC loop usually needs an average current sensor (ACS) to sense I_{BUS} and generate V_{CS}. Sense gain A_{SEN} is defined as $\Delta V_{CS}/\Delta I_{BUS}$. V_{CS} and the CC-loop reference (V_{REFI}) are input to the CC loop error amplifier (CC_EA) for loop regulation. ACS can be implemented with both off-chip and on-chip schemes. The main concern of off-chip ACS is that it will result in extra BOM (bill of material) cost and chip-pins consumption [2,3]. [4,5] used on-chip full-cycle ACS. However, since high-side and low-side sense-FETs, and the circuits related to them, are all needed, these schemes are complicated and occupy large chip area. Low-side (LS) ACS is an attractive on-chip scheme. Except for sense-FETs, all other circuits can be implemented with low-voltage transistors [6], leading to compact chip area, simple implementation and easy timing control, which is fairly suitable for high voltage applications. The bottom left of Fig. 21.6.1 shows the conventional LS ACS. It has two operation phases (sample phase T_1 and hold phase T_2). In T_1 (V_{LG} = Hi, M_L turned on), the source of sense-FET M_{S1} is clamped to V_{SSP} (power GND, V_{SSA} is analog GND) with amplifier A_1 and M_1. The current of M_{S1} is proportionally injected into R_{CS} (generating V_{C1}). Meanwhile, switch S_1 is on, so V_{CS} is filtered (sampled) by R_F and C_F. In T_2 (V_{LG} = Lo, M_L turned off), S_1 is off and V_{CS} is held. With this sample-hold (S/H) operation, V_{CS} will stabilize at the average value of V_{C1} within T_1 in steady state. Assuming both response time and mismatch are zero, V_{C1} can precisely reflect I_L during T_1, hence V_{CS} senses the average I_L as well as I_{BUS}. Note that the function of LS ACS relies on the average I_L during the sample phase (i.e., T_1) equaling I_{BUS}. Thus, FCCM mode (allowing $I_L < 0$) is usually adopted for LS ACS. It can provide continuous I_L even at light I_{BUS} to meet the above condition.

However, conventional LS ACS has the following three issues with operating conditions (D and I_{BUS}) changes. (1) In T_2, $V_A \approx 0V$ (M_1 off), $V_{C1} \approx 0V$. Thus, from T_2 to T_1, A_1 (and M_1) and V_{C1} must recover from no current state, and the settling delay (proportional to I_L) will greatly deteriorate the accuracy of A_{SEN} at large D (short T_1) and heavy I_{BUS} case (high V_{C1} at T_1). (2) The current of M_1 can only flow out to V_{SW} and can't flow back to V_{DD}. With small I_{BUS}, I_L will be < 0A at some time of a switching cycle. Once $I_L < 0A$ ($V_{SW} > 0V$), A_1 will drive M_1 into a turn-off state, and accuracy of A_{SEN} decreases. Notably, the drift of A_{SEN} due to mismatch between devices and process variation is not severe, since A_{SEN} can be trimmed under DC I_L. (3) From an AC perspective, the ACS's -3dB bandwidth (termed ω_{ACS}) is decided by R_F, C_F and D, where D ≈ $T_2/(T_1+T_2)$. R_F with S_1 is a duty-cycled resistor [7], and ω_{ACS} can be expressed as $\omega_{ACS} = 1/(R_F \times C_F \times 1/(1-D))$ [7], where ω_{ACS} will vary significantly with the change of D.

Bottom right of Fig. 21.6.1 shows the block diagram of the proposed LS ACS. For achieving high sense accuracy with operating conditions (D and I_{BUS}) variation, three schemes are proposed to orderly deal with the above three issues (1-3), including an adaptive settling helper (ASH), DC level-shift Bias (DLSB) and an anti-duty resistor (ADR). The top of Fig. 21.6.2 shows the operation states of this ACS in steady state. It also works in an S/H pattern.

ASH is proposed to adaptively minimize the settling delay. The core idea of ASH is to pre-bias the signal path (from A_1's output (V_A) to node V_{C1}) during T_2 with a proper level close to its initial value in T_1. Based on this consideration, as shown in the top left of Fig. 21.6.2, ASH reuses V_{CS} (corresponding to the average I_L) to keep A_1 active and $V_{C1}=V_{CS}$ in T_2 (via an I_{ASH} generator, two R_H and S_{3-4}). The level of V_{HD} (and V_B, function of V_B will be explained later) is relatively low ($V_{HD} \approx 60mV$ at $V_{CS} = 0.8V$) and doesn't cause the obvious DC shift on V_A in T_2. Thus, the variation range of V_A from T_2 to T_1 is also faint. Notably, since V_{CS} is reused for pre-bias in T_2, ASH can adapt to I_{BUS} change. As shown in the bottom left of Fig. 21.6.2 (stable state), from T_2 to T_1, with ASH, V_A is pre-set, and V_{C1} only needs to rise from the average value to the peak value (corresponding to I_L). Compared to the case without ASH, the settling delay with ASH is greatly reduced and the accuracy loss is slight even when operating with large D and heavy load (i.e., minimum off-time case with a 3A load).

DLSB is introduced to sense negative I_L at the sample phase (T_1). The top right of Fig. 21.6.2 shows the operation of this ACS in T_1, where S_{1-2} are on and S_{3-4} are off. With DLSB, V_{CR} is not tied to V_{SSP} and is lifted up with a DC current I_{LS} injecting into sense-FET M_{S2}. V_{CR} has a >0V DC level ($V_B \approx 27mV$), thus M_1 can remain active when $I_L < 0A$. As shown in the bottom right of Fig. 21.6.2, sense gain A_{SEN} is set to 200mΩ. I_{LS} and M_{S2} are designed to support up to -1A I_L (i.e., adding 200mV DC shift on V_{CS}). By choosing the same transistor type for M_{S2}, M_L and M_{S1} (M_{S1} and M_{S2} are placed near to M_L in layout), and setting I_{LS} and R_{CS} (low TC poly resistor) have same temperature coefficient (TC), the 200mV DC shift is robust to temperature variation. In addition, to calibrate both the drift of A_{SEN} and the 200mV DC shift magnitude over process and mismatch, M_{6T} and I_{LS} are trimmed in sequence under DC I_L (bottom of Fig. 21.6.2).

ADR is proposed to suppress the drift of ω_{ACS} (the -3dB bandwidth of ACS) with D variation. In fact, ω_{ACS} contributes a pole within the CC loop, and its drift will affect the stability of the CC loop. Implementation of ADR is given in the top right of Fig. 21.6.3. Since this Buck works in FCCM mode, $V_{BUS}/V_{IN} \approx D$ is met over the full load range. Based on this, we reduce the value of R_F in stages as V_{BUS}/V_{IN} increases to compensate the term of $1/(1-D)$. With ADR, the drift of ω_{ACS} is ±44% with D variation. Correspondingly, as shown in the bottom right of Fig. 21.6.3, the bandwidth (BW) drift of the CC loop with ADR is ±33% (10.7kHz to 21.3kHz, phase margin > 60°), while the minimal BW of the CC loop without ADR is only 3.5kHz with 30° phase margin.

Moreover, the bottom left of Fig. 21.6.3 shows the schematic of A_1. The BJT (high beta)-based 1st stage is a wide-BW pre-amplifier with low offset, whose gain can attenuate the 2nd-stage offset and increase the BW of A_1. Total 3σ offset of A_1 is ~0.4mV (this offset will cause non-linear gain error on A_{SEN}). According to the AC simulation (bottom right of Fig. 21.6.3), the A_1-based clamp loop at T_1 has >18MHz unity-gain BW (UGB) when $I_L > 1A$, and >50° phase margin. V_{CS} can also be used for line-drop compensation (LDC), which is realized with a V to I current of V_{CS} (I_C) flowing through the V_{BUS} feedback resistor (R_1). The bottom middle of Fig. 21.6.3 shows the schematic of the I_C generator and V_{REFI} generator. The I_C generator reuses the V to I current of ASH, and the V_{REFI} generator is an 8-bits R-string DAC. They are all added with the corresponding DC offset to match the 200mV DC shift of V_{CS}.

This charging IC with ACS is fabricated in a 0.18μm 40V BCD process (Fig. 21.6.7). M_H and M_L are implemented with high-voltage NLDMOS, to support wide V_{IN} range up to 40V under load-dump environments. Measured transient performance of the CC loop for a typical sample is shown in Fig. 21.6.4. Since the bandgap, CC_EA and V_{REFI} generator are respectively trimmed, the accuracy of the CC loop is mainly determined by the proposed ACS. The top of Fig. 21.6.4 shows the result by sweeping V_{REFI}, where the current error is less than 20mA from 0.1A to 3A. The bottom left of Fig. 21.6.4 shows the stable waveform with minimum off-time operation. The on time of M_L is shorter than 100ns, and CC error is less than 30mA. The bottom right of Fig. 21.6.4 shows the transient behavior of the CC loop under V_{BUS} on-line change, where V_{BUS}/V_{IN} changes from 42.4% to 92.4% and vice versa. The CC loop is always stable with ADR. Figure 21.6.5 provides CC accuracy from 0.1A to 3A for 7 samples under different V_{IN} and V_{BUS} conditions (up to 16V V_{IN}, matching the maximum steady voltage of a car battery). As can be seen, the CC inaccuracy (from 0.1A to 3A) is less than ±60mA even with minimum off-time operation.

The table in Fig. 21.6.6 summarizes performance of the proposed charging solution and compares it with other works that consist of a DC-DC converter with a CC loop (or average current sensor). This work shows the effort to design a simple LS ACS (0.1mm²) with high accuracy supporting the full range of duty cycle (including minimum off-time cases, M_L on <100ns) and load current (0.1A ~ 3A, meeting the requirements of UFCS). The die micrograph and charging test are shown in Fig. 21.6.7.

374 • 2026 IEEE International Solid-State Circuits Conference

979-8-3315-8937-0/26 $31.00 © 2026 IEEE

Acknowledgement:

This work was supported by the National Key Research and Development Program of China (2022YFB3604204). Corresponding Author: Xin Ming.

References:

[1] Monolithic Power System, MPQ4241-AEC1 Datasheet, "MPQ424165W, Fully Integrated USB PD Solution with 36V Transient Input Buck and PD3.1 DFP controller AEC-Q100 Qualified", June 2022, accessed on June, 4, 2025, https://www.monolithicpower.com/en/mpq4241-aec1.html

[2] Texas Instruments, TPS25762-Q1 Datasheet, "TPS25762-Q1, TPS25762-Q1 Automotive USB Type-C® Power Delivery Controller with Buck-Boost Regulator", Dec, 2022, accessed on June, 4, 2025. https://www.ti.com/lit/ds/symlink/tps25762-q1.pdf

[3] Yong-Seong Roh et al, "A Multiphase Synchronous Buck Converter with a Fully Integrated Current Balancing Scheme", *IEEE Trans. Power Electronics*, vol. 30, no. 9, pp. 5159-5169, Sept, 2015. https://doi.org/10.1109/TPEL.2014.2368130

[4] Sachin Rao et al, "A 1.2A Buck-Boost LED Driver with On-Chip Error Averaged SenseFET-Based Current Sensing Technique", *IEEE JSSC*, vol. 46, no. 12, pp. 2772-2783, Dec, 2011. https://doi.org/10.1109/JSSC.2011.2162921

[5] Vratislav Michal, "Absolute Value, 1% Linear and Lossless Current-Sensing Circuit for the Step-Down DC-DC Converters with Integrated Power Stage", *IEEE JSSC*, vol 49, no. 5, pp. 1256-1270, May, 2014. https://doi.org/10.1109/JSSC.2014.2309696

[6] Lin Cheng et al, "A 6.78-MHz Single-Stage Wireless Charger with Constant-Current Constant-Voltage Charging Technique", *IEEE JSSC* vol. 55, no. 4, pp. 999-1010, April, 2020. https://doi.org/10.1109/JSSC.2019.2961852

[7] Hariprasad Chandrakumar and Dejan Markovic, "A High Dynamic-Range Neural Recording Chopper Amplifier for Simultaneous Neural Recording and Stimulation", *IEEE JSSC*, vol. 52, no. 3, pp. 645-656, March, 2017. https://doi.org/10.1109/JSSC.2016.2645611

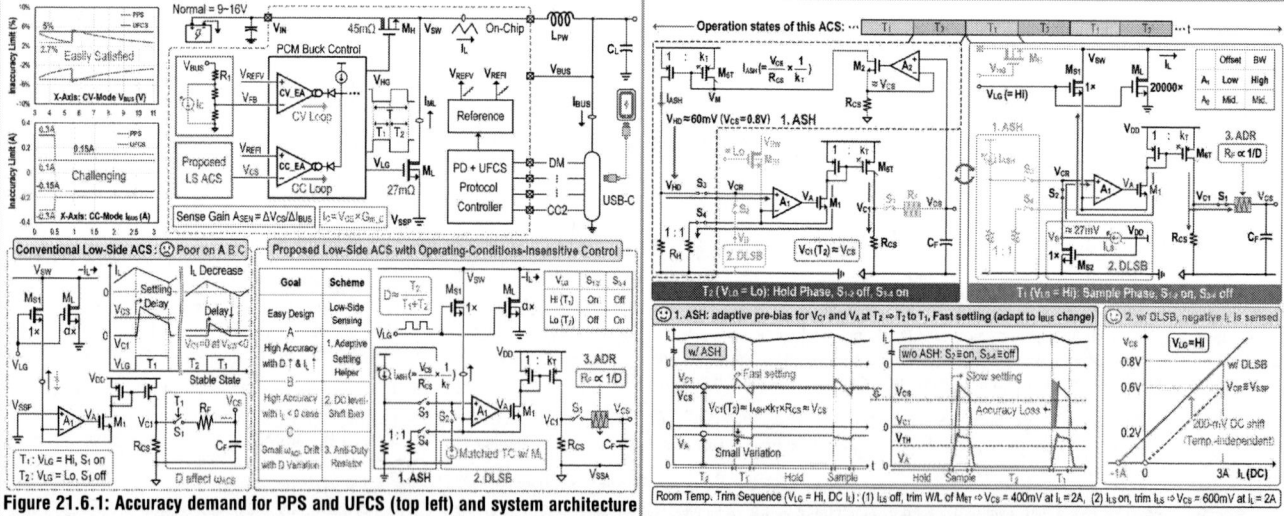

Figure 21.6.1: Accuracy demand for PPS and UFCS (top left) and system architecture of the proposed charging solution (dashed box of top). Block diagrams of conventional and the proposed low-side ACS, respectively (bottom).

Figure 21.6.2: Operation states of the proposed ACS (top). Technical advantages of ASH and DLSB, and trimming strategy for this ACS at room temperature (bottom).

ISSCC 2026 / SESSION 21 / SENSOR INTERFACES / 21.6

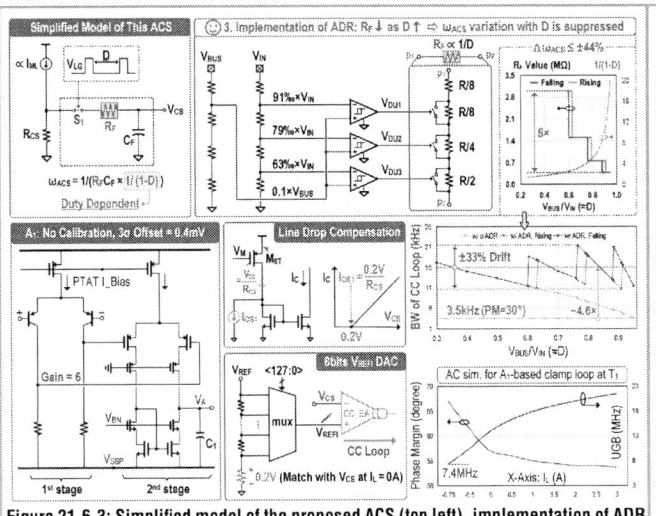

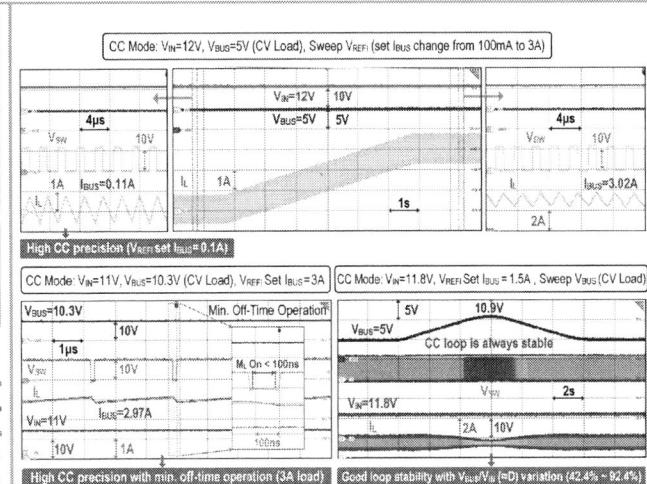

Figure 21.6.3: Simplified model of the proposed ACS (top left), implementation of ADR (top right). Schematic of A_1 (bottom left), LDC and V_{REFI} generator (bottom middle). AC simulations for BW of CC loop and A_1-based clamp loop (bottom right).

Figure 21.6.4: Measured CC-loop transient waveforms for sweeping V_{REFI}, steady state with minimum off-time operation and sweeping V_{BUS} (F_{SW}=420kHz, L_{PW}=4.7μH).

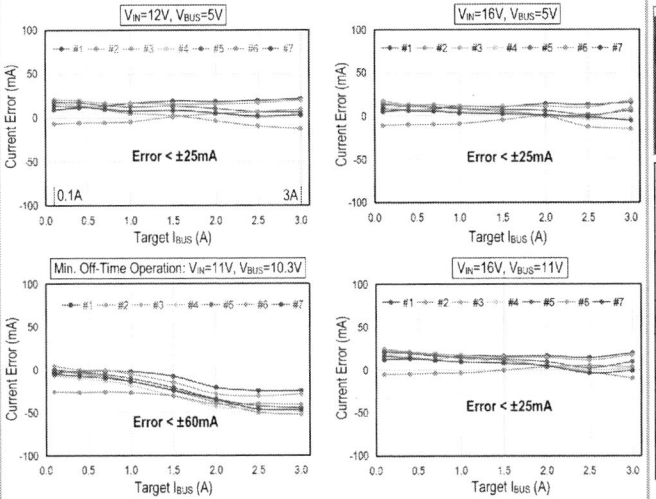

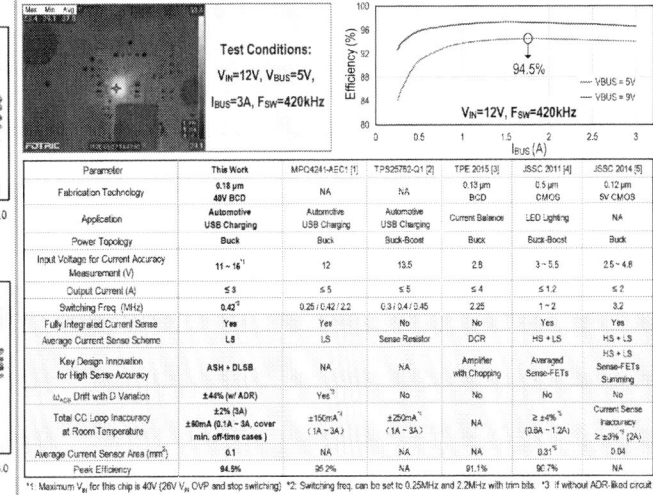

Figure 21.6.5: Measured current inaccuracy at different V_{IN} and V_{BUS} (buck works in CC mode, F_{SW}=420kHz, L_{PW}=4.7μH).

Figure 21.6.6: Performance summary and comparison with other prior works.

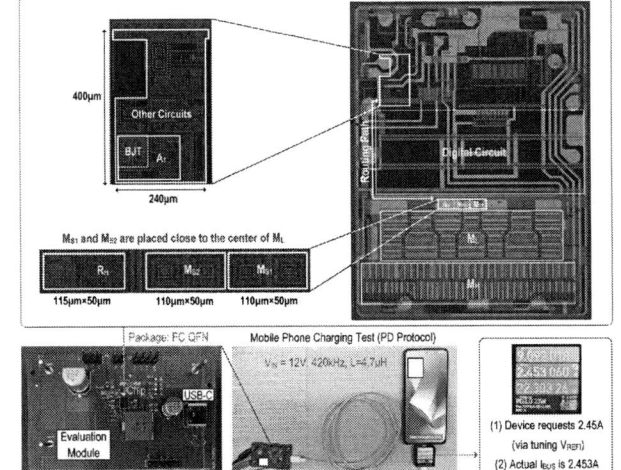

Figure 21.6.7: Layout of the proposed ACS (top left) and die micrograph of this charging solution (top right). PD charging test for this charging solution (CC mode), device requests 2.45A and the actual I_{BUS} is 2.453A (3mA current error) (bottom).

• 2026 IEEE International Solid-State Circuits Conference

ISSCC 2026 / SESSION 21 / SENSOR INTERFACES / 21.7

21.7 A Battery-Free Wireless Electrochemical-Interface SoC Featuring 143dB Dynamic Range for Multimodal Wearables

Weixiao Wang, Yili Shen, Tianyi Cai, Qijing Xiao, Yuxuan Luo, Bo Zhao

Zhejiang University, Hangzhou, China

Abstract

A battery-free wireless electrochemical-interface SoC is implemented and demonstrated in a sweatband prototype for physiological monitoring in body sweat, where the proposed techniques improve the dynamic range with a high linearity and low power consumption:

1) A self-adaptive stepwise-sinking technique for an OTA-less potentiostat extends the dynamic range to 143dB with 23.3μW power. 2) A frequency-digital-hybrid structure helps to ensure a high linearity of R^2=0.99997 across the 143dB range.

Electrochemical sensors can be used to detect various physiological parameters on the human body, such as glucose, lactate, alcohol, uric acid, etc. Due to the low fabrication cost, electrochemical sensors are potentially widely-adopted in disposable wearables, e.g. contact lenses [1], wristbands [2], bandages [3], and various patches [4,5]. However, various substances on the body surface require different types of electrochemical sensors, presenting a sensor current ranging from sub-pA to mA-level, which is determined by enzyme concentration, sample concentration, electrode-sample contact area, and other factors. To enable the disposable applications in various wearables, it's necessary for an electrochemical chip to cover a high dynamic range with a high linearity, and then a single chip can apply to a great variety of wearable sensors in mass production, cutting down the fabrication cost. As given in Fig. 21.7.1(top, bottom-left), there have been several methods to improve the dynamic range in prior arts: 1) Programmable current conveying. The current-conveyer ratio was digitally controlled to enlarge the dynamic range to 114dB [6], while the tuning could only be manual and multiple feedback amplifiers led to potential instability. 2) Time-domain closed-loop scaling. The dynamic range was expanded to 129.7dB by the resolution scaling through a differential current digital-to-analog converter (IDAC), a time-to-digital converter (TDC), and a current-to-frequency converter (I-to-F), which consumed high power in sensing a mA-level current [7]. 3) Coarse self-cancelling. A $\Sigma\Delta$ current detector detected the sensor current, which controlled an IDAC to cancel the coarse part of sensor current and realized a 136.5dB dynamic range [8]. However the power-hungry $\Sigma\Delta$ current detector increased the chip power to 425μW. In addition, most prior work required an OTA to form a negative-feedback potentiostat [6-9], leading to increased power consumption and potential instability issues. Therefore, a digital multi-loop potentiostat [10] simplified the implementation of potentiostats with only comparators, counters and IDACs. Although the stability of a single closed-loop has been verified, it requires simultaneous operation of three closed loops, still leading to instability. In addition, various sensor currents and electrolytes can cause significant decrease in counter electrode (CE) voltage, leading to output current distortion of the IDAC and nonlinearity of the potentiostat. Moreover, the dynamic range was limited by a fixed biasing current, which can only be controlled manually.

In this work, a battery-free wireless electrochemical-interface SoC is implemented and demonstrated in a sweatband prototype (Fig. 21.7.1, bottom-right): 1) The proposed self-adaptive stepwise-sinking technique for the OTA-less potentiostat improves the dynamic range to 143dB with 23.3uW power. 2) Within the dynamic range, the proposed frequency-digital-hybrid structure maintains a high linearity of R^2=0.99997 across the 143dB range.

Figure 21.7.2 shows the block diagram of the electrochemical-sensing system based on the proposed wireless battery-free electrochemical-interface chip. To miniaturize the wearable devices such as a sweatband, the chip eliminates the battery by RF power harvesting and backscatter telemetry. The antenna and AC-DC rectifier harvest the wireless power transmitted by an RF source, and then the power-management unit (PMU) provides a current reference and power supplies through bandgap and LDOs. In the OTA-less frequency-digital-hybrid potentiostat circuitry, the voltage of the reference electrode (RE) can be stabilized to the preset voltage V_{RE-REF} by the feedback loop. Meanwhile, the sensor current I_{SENS} is directly quantified by both the primary IDAC group and secondary IDAC group. The differential current I_{OSC} is extracted from the secondary IDAC group, which is then converted to a frequency signal F_{COV} by a current-calibration current-to-frequency converter (CC I-to-F). The IDAC control word D_{COV} and frequency signal F_{COV} (from CC I-to-F) are mixed and transmitted to the reader through backscatter, indicating the accurate value of sensor current I_{SENS}.

The self-adaptive stepwise-sinking technique eliminates the OTA in a potentiostat and extends the dynamic range (Fig. 21.7.3, left). The RE voltage is quantized by a 1-bit quantizer, feeding to the control logic. With dynamic element matching (DEM), the IDAC control word D_{COV} is conveyed to the primary/secondary IDAC groups, which sink the I_{SENS} from the CE. As RE voltage varies versus the current extracted by IDACs, the self-adaptive stepwise-sinking loop maintains the RE voltage at the preset voltage V_{RE-REF}. With the decreased unit current from IDAC L_1 to L_4, the current sunk by primary/secondary IDAC groups automatically converges in stepwise fashion, to match I_{SENS} ranging from sub-nA to mA-level, thereby directly quantifying the sensor current and achieving a high dynamic range with low power.

The frequency-digital-hybrid potentiostat maintains a high linearity across a high dynamic range of sensor currents and various electrolytes by extending the compliance voltage (Fig. 21.7.3, right). The compliance voltage (V_{CL}) refers to the maximum voltage that a potentiostat can provide across working electrode (WE) and CE to keep a preset voltage difference between WE and RE. To ensure distortion-free current output, the voltage difference between the WE and CE (V_{WE-CE}) must keep within the compliance voltage range of the potentiostat. In case of high-impedance electrolyte or a high sensor current, the electrochemical reaction requires a large V_{WE-CE}, which decreases V_{CE} at a fixed V_{WE}. In a conventional IDAC-based structure, the measured current I_{FIELD} obtained only by D_{COV} will be higher than the actual sensor current I_{SENS}. In the proposed frequency-digital-hybrid structure, the two inputs of the differential IDAC (I_{SENS} & I_{OSC}) are connected to CE and CC I-to-F, respectively. The output frequency signal F_{COV} of CC I-to-F is proportional to both I_{OSC} and D_{COV}. Compared to a conventional IDAC-based structure, the sum of these two currents ($I_{SUM} = I_{OSC} + I_{SENS}$) in the proposed frequency-digital-hybrid structure changes less versus the decrease in V_{CE}. In this way, the actual sensor current I_{SENS} can be accurately read out as $I_{SUM} - I_{OSC}$, even at a lower V_{CE}. Hence, the precise magnitude of sensor current can be derived from the combination of F_{COV} and D_{COV}, which ensures a high linearity of the potentiostat for various sensor currents and electrolytes.

Figure 21.7.4 presents the measurement setup and electrical performance of the proposed battery-free electrochemical-interface SoC. The far-field wireless power transfer (WPT) is measured at different reader-to-chip ranges over the air (Fig. 21.7.4, top-right). As a 1.5V DC voltage is required by the core circuits, the 30dBm reader power results in a maximal WPT range of 2.2m. Compared to the four sensor current samples at a low V_{CE} of 0.35V, the measured currents derived solely from D_{COV} exhibit an error of 7.5%, while the proposed frequency-digital-hybrid structure reduces the error of measured currents to only 0.96% (Fig. 21.7.4, bottom-left). Figure 21.7.4 (bottom-right) also shows the measured currents versus a sensor current range of 82pA-1.16mA, demonstrating a 143dB dynamic range with an R^2 linearity of 0.99997.

Figure 21.7.5 presents the structure diagram of glucose and lactate sensor electrodes as well as the in-vitro testing results of the sensor-and-chip system. The sensor electrodes are fabricated on Au substrate, where polyaniline and glutaral immobilizes the corresponding enzymes. The performance of the glucose and lactate sensors was tested in the corresponding solutions with the same concentration ranges as in body sweat, i.e., 10uM to 200uM for glucose and 5mM to 20mM for lactate [11]. The i-t curves show that the linear ranges of two sensors cover the actual concentration ranges of the corresponding analytes in body sweat. The proposed electrochemical-interface SoCs are connected to bare-electrode sensor, glucose sensor, and lactate sensor, respectively, to form the sensor-and-chip systems. Then, the sensors are immersed in the corresponding potassium ferrocyanide ($K_4[Fe(CN)_6]$), glucose, and lactate solutions, where the electrochemical-interface SoC extracts the sensor current signals. The sensor current signals are processed and transmitted to a remote reader in a battery-free way. The measured results in $K_4[Fe(CN)_6]$, glucose, and lactate solutions indicate an R^2 linearity of 0.9994, 0.9990 and 0.9965, respectively, which is only limited by the linearity of the self-designed sensors.

Figure 21.7.6(top) demonstrates the battery-free sweatband prototype on the human body. The glucose and lactate sensors are integrated with the proposed electrochemical-interface SoCs, which are assembled on the same sweatband. The reader is placed at the handlebars of the exercise apparatus, which emits RF energy to power the sweatband. Meanwhile, the sweatband backscatters the glucose and lactate sensing data to the reader, which can be displayed on the screen. The demonstration is conducted throughout the entire fitness process involving both exercise and food intake (Fig. 21.7.6, bottom). The lactate concentration in the sweatband significantly increases from warm-up to intense-exercise states, while returning to baseline after subsequent relaxation. The glucose concentration in the sweatband rises sharply right after a meal, and then returns to a normal level during an extended exercise.

Figure 21.7.7 shows the die micrograph, power breakdown, and performance summary compared with state-of-the-art electrochemical-interface chips. Fabricated in a 65nm CMOS process, the chip takes a total area of 1.18mm² including I/O frame. Consuming only 23.3uW power, the chip can be read battery-free by a 2.2m-range-away reader. The performance comparison shows that: 1) The proposed self-adaptive stepwise-sinking technique achieves the highest dynamic range of 143dB with only 23.3uW. 2) The proposed frequency-digital-hybrid structure realizes the highest R^2 linearity of 0.99997 across the 143dB dynamic range. 3) This may be the only electrochemical-interface chip that has been demonstrated on a human body, where the sweatband reflects glucose and lactate levels during the body fitness process.

Acknowledgement:
This work was supported by the National Key R&D Program of China (2023YFF1203600) and the National Natural Science Foundation of China (62534008). Corresponding Author: Bo Zhao (zhaobo@zju.edu.cn).

References:
[1] C. Jeon et al., "A Smart Contact Lens Controller IC Supporting Dual-Mode Telemetry with Wireless-Powered Backscattering LSK and EM-Radiated RF Transmission Using a Single-Loop Antenna," *IEEE JSSC*, vol. 55, no. 4, pp. 856-867, April. 2020. https://doi.org/10.1109/JSSC.2019.2959493
[2] Gao, Wei, et al. "Fully Integrated Wearable Sensor Arrays for Multiplexed In Situ Perspiration Analysis." *Nature* 529.7587 (2016): 509-514. https://doi.org/10.1038/nature16521
[3] Ciui, Bianca, et al. "Wearable Wireless Tyrosinase Bandage and Microneedle Sensors: Toward Melanoma Screening." *Advanced healthcare materials* 7.7 (2018): 1701264. https://doi.org/10.1002/adhm.201701264
[4] Sempionatto, Juliane R., et al. "An Epidermal Patch for the Simultaneous Monitoring of Haemodynamic and Metabolic Biomarkers." *Nature Biomedical Engineering* 5.7 (2021): 737-748. https://doi.org/10.1038/s41551-021-00685-1
[5] Wang, Minqiang, et al. "A Wearable Electrochemical Biosensor for The Monitoring of Metabolites and Nutrients." *Nature Biomedical Engineering* 6.11 (2022): 1225-1235. https://doi.org/10.1038/s41551-022-00916-z
[6] Q. Lin et al., "A 22 µW Peak Power Multimodal Electrochemical Sensor Interface IC for Bioreactor Monitoring," *ISSCC*, pp. 314-316, Feb. 2023. https://doi.org/10.1109/ISSCC42615.2023.10067298
[7] S. -Y. Lu et al., "A Wireless Multimodality System-on-a-Chip with Time-Based Resolution Scaling Technique for Chronic Wound Monitoring," *ISSCC*, pp. 282-284, Feb. 2021. https://doi.org/10.1109/ISSCC42613.2021.9365992
[8] S. Choi et al., "A Wide Dynamic Range Multi-Sensor ROIC for Portable Environmental Monitoring Systems with Two-Step Self-Optimization Schemes," in *IEEE TCAS-I*, vol. 68, no. 6, pp. 2432-2443, June. 2021. https://doi.org/10.1109/TCSI.2021.3065503
[9] W. Wang et al., "A Wireless Battery-Free Cerebral-Oxygen-Monitoring Micro-System Featuring 0.21-mmHg Sensing Resolution," *IEEE JSSC*, vol. 60, no. 10, pp. 3538-3550, Oct. 2025. https://doi.org/10.1109/JSSC.2025.3577123
[10] M. A. Akram, A. Aberra, S. -J. Kweon and S. Ha, "An Amplifier-Less CMOS Potentiostat IC Consuming 3.7nW Power all Over 129.5dB Dynamic Range for Electrochemical Biosensing," *ISSCC*, pp. 64-66, Feb. 2024. https://doi.org/10.1109/ISSCC49657.2024.10454401
[11] Xu, Jing, Yunsheng Fang, and Jun Chen. "Wearable Biosensors for Non-Invasive Sweat Diagnostics." *Biosensors* 11.8 (2021): 245. https://doi.org/10.3390/bios11080245

[12] J. -C. Chien, H. T. Soh and A. Arbabian, "A Cell-Capacitance-Insensitive CMOS Sample-and-Hold Chronoamperometric Sensor for Real-Time Measurement of Small Molecule Drugs in Whole Blood," *ISSCC*, pp. 406-408, Feb. 2020. https://doi.org/10.1109/ISSCC19947.2020.9063036
[13] A. Manickam et al., "A CMOS Biosensor Array with 1024 3-Electrode Voltammetry Pixels and 93dB Dynamic Range," *ISSCC*, pp. 192-194, Feb. 2019. https://doi.org/10.1109/ISSCC.2019.8662507

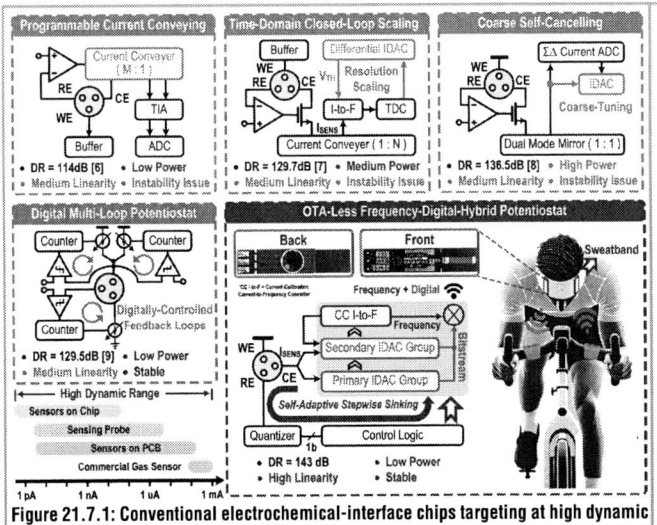

Figure 21.7.1: Conventional electrochemical-interface chips targeting at high dynamic ranges (DR) (top, bottom-left); Proposed battery-free electrochemical-interface SoC (143dB DR) for wearables (including sweatband demonstration) (bottom-right).

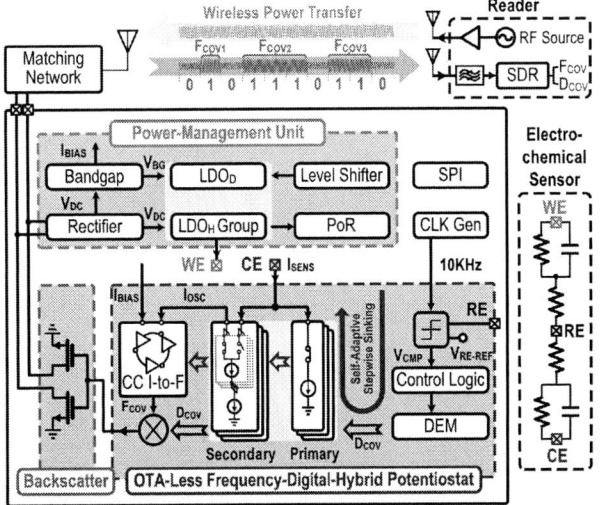

Figure 21.7.2: Complete wireless electrochemical sensing system including block diagram of proposed wireless battery-free electrochemical-interface chip.

ISSCC 2026 / SESSION 21 / SENSOR INTERFACES / 21.7

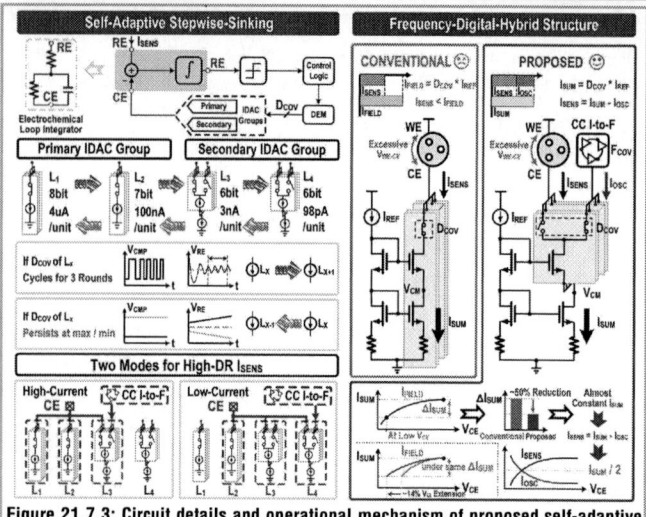

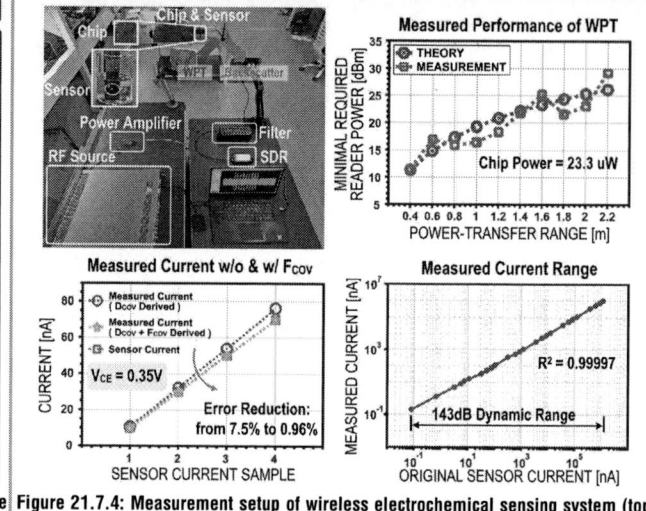

Figure 21.7.3: Circuit details and operational mechanism of proposed self-adaptive stepwise-sinking technique (left); Circuit details of proposed frequency-digital-hybrid structure (right).

Figure 21.7.4: Measurement setup of wireless electrochemical sensing system (top-left); Measured WPT performance (top-right); Measured current of frequency-digital-hybrid structure (bottom-left); Measured current versus sensor current (bottom-right).

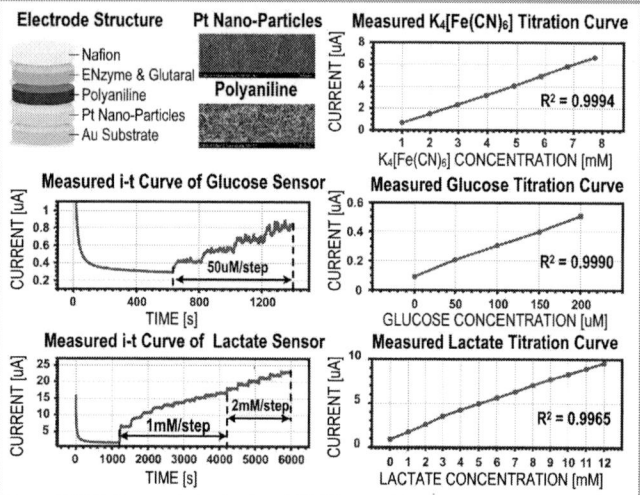

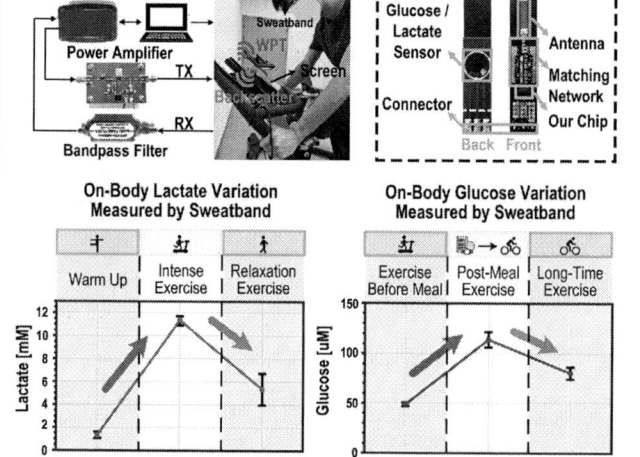

Figure 21.7.5: Sensor electrodes (top-left); Measured i-t curve of glucose (middle-left) and lactate sensors (bottom-left) by electrochemical workstation; In-vitro measurement of sensor-and- chip systems in $K_4[Fe(CN)_6]$, glucose, and lactate titration (right).

Figure 21.7.6: On-body sweatband demonstration (top); On-body lactate (bottom-left) and glucose (bottom-right) measured by sweatband including proposed chip.

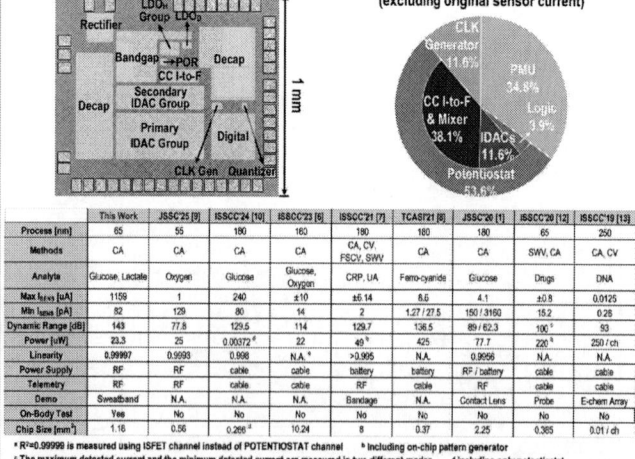

	This Work	JSSC'25 [9]	ISSCC'24 [10]	ISSCC'23 [6]	ISSCC'21 [7]	TCAS'21 [8]	JSSC'20 [1]	ISSCC'20 [12]	ISSCC'19 [13]
Process [nm]	65	55	180	180	180	180	180	65	250
Methods	CA	CA	CA	CA	CA, CV, FSCV, SWV	CA	SWV, CA	CA, CV
Analyte	Glucose, Lactate	Oxygen	Glucose	Glucose, Oxygen	CRP, UA	Ferro-cyanide	Glucose	Drugs	DNA
Max I_{SENS} [µA]	1159	1	240	±10	±6.14	8.6	4.1	±0.8	0.0125
Min I_{SENS} [pA]	82	129	80	14	2	1.27 / 27.5	150 / 3160	15.2	0.26
Dynamic Range [dB]	143	77.8	129.5	114	129.7	136.5	89 / 62.3	100 [c]	93
Power [µW]	23.3	25	0.00372 [d]	22	49 [b]	425	77.7	220 [c]	250 / ch
Linearity	0.99997	0.9993	0.998	N.A. [a]	>0.995	N.A.	0.9056	N.A.	N.A.
Power Supply	RF	RF	cable	cable	battery	battery	RF / battery	cable	cable
Telemetry	RF	RF	cable	cable	RF	cable	RF	cable	cable
Demo	Sweatband	N.A.	N.A.	N.A.	Bandage	N.A.	Contact Lens	Probe	E-chem Array
On-Body Test	Yes	No	No	No	No	No	No	No	No
Chip Size [mm²]	1.18	0.56	0.266 [d]	10.24	8	0.37	2.25	0.385	0.01 / ch

* R^2=0.99998 is measured using ISFET channel instead of POTENTIOSTAT channel   [b] Including on-chip pattern generator
[c] The maximum detected current and the minimum detected current are measured in two different modes   [d] including only potentiostat

Figure 21.7.7: Die micrograph (top-left) and power breakdown (top-right); Performance summary compared with state-of-the-art electrochemical-interface chips (bottom).

2026 IEEE International Solid-State Circuits Conference

ISSCC 2026 / SESSION 21 / SENSOR INTERFACES / 21.8

21.8 A CMOS Hybrid Common-Gate Current-Integrating Sampler with >37dB SNDR Across 51GHz BW in a 128GS/s Front-End

Jun Dai[1], Yi Zhong[1], Yunsong Tao[1], Aodong Zhang[1], Mingtao Zhan[1], Hao Zhang[2], Lu Jie[1], Nan Sun[1]

[1]Tsinghua University, Beijing, China, [2]Magnichip, Nanjing, China

Abstract

This work proposes a CMOS hybrid common-gate current integrating sampler to address linearity, BW, and jitter limitations in prior wideband ADC front-ends. The front-end employs a hybrid common-gate V-I converter, transformer-coupled inductive peaking, and an automatic power-gating hold buffer, enabling 48.1dB SFDR and 38.6dB SNDR near its 52GHz BW$_{-3dB}$. This work demonstrates superior SNDR$_{BW}$ (measured near BW$_{3dB}$), outperforming prior samplers/ADCs reported in ISSCC/JSSC/VLSI with BW>30GHz.

The rapid expansion of AI applications greatly increases the demand for wireline links. Long-reach receivers often adopt the ADC-DSP architecture, where ADCs enable flexible digital equalization. The ADC needs to have wide BW, low noise, and high linearity, all of which are limited by the front-end sample-and-hold (S/H) circuit [1-8]. Figure 21.8.1 top-left shows a conventional CMOS S/H design. Despite using a T-coil, the high loss of the CMOS substrate limits the BW to 30GHz [4-7]. Adding CTLE can extend the BW beyond 40GHz, but it causes significant distortions (SNDR≤30dB at 50GHz) [1,2,8]. Also, the high-frequency SNR of the switched-capacitor S/H is limited by jitter [1-8]. To increase the BW and reduce jitter sensitivity, the SiGe current integrating sampler (CIS) can be used [9-13] (Fig. 21.8.1 top-right). Nonetheless, it suffers from poor G_m linearity (SNDR≤34dB at 50GHz) [11-13]. In addition, the use of SiGe prevents monolithic integration with the CMOS quantizer and consumes high power (>1W) [10-13]. Hence, both prior CMOS and SiGe front-ends face significant challenges.

This work proposes a wideband and linear CMOS hybrid common-gate (CG) CIS front-end, which overcomes prior limitations with 3 key techniques: (1) a hybrid CG input stage to convert voltage to current linearly; (2) transformer-coupled inductive peaking to boost BW; and (3) an automatic power-gating hold buffer to boost performance. Equipped with the proposed techniques, a prototype 128GS/s CMOS CIS front-end achieves 52GHz BW, 48.1dB SFDR, and 38.6dB SNDR at 51GHz. Its high-frequency SNDR outperforms wideband samplers and ADCs with BW above 30GHz.

The proposed CMOS CIS front-end is shown at the bottom of Fig. 21.8.1. It uses a 2-stage sampling architecture. In the 1st stage, a broadband high-linearity V-I converter converts the input voltage signal to current, which is then passed to 4-way 32GS/s interleaved switches to be integrated on the sampling capacitor C_s. The sampled signal from the 1st stage is buffered to the 2nd stage, which consists of 32-way interleaved S/H circuits running at 4GS/s. Since the 2nd stage samples the held signal from the 1st stage, its requirements for BW, jitter, and timing skew are significantly relaxed. Thus, a simple voltage-domain switched capacitor sampler is employed in the 2nd stage to reduce power. The 4-phase 32GHz sampling clocks required by the 1st-stage CIS are generated by injecting a 32GHz differential clock into an on-chip injection-locking VCO. The 32-phase 4GHz clocks are generated by the 2nd-stage S/H are generated by passing the 4-phase 32GHz clock through a 1/8 divider (1 CML and 2 C²MOS divider stages). Compared with the conventional CMOS S/H front-end (Fig. 21.8.1 top-left), the proposed front-end design has several merits: 1) despite using the 28nm CMOS process, it enables 48dB SFDR with a 51GHz input; 2) the use of CIS reduces the impact of jitter by 2.5dB at 50GHz [10]; 3) the proposed CIS front-end requires only 4 clock phases to have low jitter and skew. By contrast, prior wideband CMOS S/H typically requires 16 or 32 critical clock phases. This simplifies the clock path matching and skew calibration.

Figure 21.8.2 shows the schematic of the proposed wideband and linear CMOS CIS front-end. The input resistor R_{in} (40Ω) and the CG transistor together provide the 50Ω termination, as well as performing the conversion of the input voltage to current. The low CG input impedance (10Ω) ensures that the majority of the input voltage drops across the passive resistor, resulting in high linearity. Traditional CIS designs use a common-source (CS) stage for V-I conversion. Although it can also use a large degeneration resistor for linearization, the drain-to-source voltage (V_{ds}) of the CS transistor exhibits large variation, thereby causing large distortion due to severe g_{ds} modulation. By contrast, the source and drain voltages are in phase for the proposed CG stage, thereby greatly reducing V_{ds} variation and the g_{ds} modulation effect, leading to 10dB SFDR improvement (Fig. 21.8.2 top-right). Although the required 50Ω input impedance matching limits the V/I conversion gain, additional gain of 3dB is provided by adding an AC-coupled CS stage at the CG source node. This auxiliary CS stage is not degenerated, but it does not cause nonlinearity due to low input swing (from the CG source node) and sufficient V_{ds} headroom (no degeneration resistor). To widen the BW, the ESD diodes (160fF parasitic capacitance) are placed at the low-impedance CG source node. More importantly, L_i and L_d are inserted into the main signal path to form series inductive peaking. Furthermore, L_b and L_g are added to form 2 sets of transformers, which enhance the inductive peaking effects by providing magnetic coupling, without increasing area overhead. As the signal amplitude across L_i and L_d increases with frequency, the magnetic-coupled signal amplitude across L_b and L_g also increases at high frequencies.

This compensates for the high-frequency gain loss in the circuit and thereby increases the CIS BW. Without transformer coupling, the BW is limited to only 20GHz. The BW can be increased to 40GHz with TF$_1$ and 60GHz with TF$_1$ & TF$_2$, which also compensates for the BW degradation due to the sinc transfer function of an ideal CIS (56GHz BW$_{-3dB}$ for 128GS/s [13]). The interleaved CIS switches adopt the timing scheme of [13]. This work uses NMOS CML switches and PMOS reset switches. Different from [13], this work adds differential PMOS reset switches, which significantly reduce the size of reset transistors, thereby obviating the degradation in integrating gain from the large parasitic capacitor.

Figure 21.8.3 shows the proposed hold buffer with intrinsic power-gating capability. It combines a PMOS source follower (SF) with a PMOS CS amplifier. This design takes advantage of the variation in the CIS output common-mode (CM) voltage, i.e., the hold buffer input CM voltage, $V_{i,cm}$ during different clock phases. During the reset phase (RST1 & RST2), $V_{i,cm}$ is pulled to V_{DD}, and thus, automatically shuts down the hold buffer for power saving. During the CIS integration and hold phase (INTE & HOLD), $V_{i,cm}$ gradually decreases, and thus, automatically turns on the hold buffer. This provides sufficient drivability for the 2nd-stage S/H circuits during the hold phase. Conventionally, power-gating of the hold buffer has to be realized by adding extra switches on the power rails. However, it degrades the BW and increases the design complexity, as the generation of 32GHz switch control signals is a significant burden. By contrast, the proposed hold buffer achieves automatic power-gating without any cost. Compared with a static hold buffer, the proposed buffer can be biased in a stronger inversion region with the same power consumption, thus achieving higher drivability, wider BW and higher linearity when buffering the 32GS/s held signal (Fig. 21.8.3 right). In addition, the combination of an SF and a CS amplifier also increases the voltage gain and slew rate of the hold buffer.

The prototype chip in a 28nm CMOS process occupies a total area of 0.11mm² (Fig. 21.8.7), including all active and passive components. The input signal and 32GHz clock signal are probed into the chip pads using a probe station. To facilitate the testing, a single output is extracted from the 32-way 4GS/s 2nd-stage S/Hs, and then passed through a 125MS/s low-speed S/H for 32 decimation. The final decimated output is buffered by a 50Ω driver and measured using an oscilloscope. This testing method does not capture the spurs due to mismatches among the interleaving channels (e.g., offset, gain, and timing skew), but previous studies have shown that these interleaving spurs can be effectively suppressed via calibration [2,4,6,14]. The primary factors that ultimately determine the high-frequency performance, including clock jitter, signal-path BW, and harmonic distortions, are accurately measured using this scheme [7,15,16]. Since this prototype implements the full 128GS/s CIS front-end, it ensures that the crosstalk among different channels and the performance degradation due to supply noise in a real-world deployment are properly evaluated. The input signal amplitude fed to the chip pads is 800mV$_{ppd}$, after compensating for the losses of the balun, cables, and probes. The non-idealities like noise and distortions from decimator and instruments are included. Figure 21.8.4 shows that the chip achieves 48.1dB SFDR and 38.6dB SNDR at 51GHz, while maintaining above 45dB SFDR and 37dB SNDR across the BW$_{-3dB}$ of 52GHz. Figure 21.8.5 shows the consistency of the measurement results across multiple chips. The BW variation is within 1GHz. The SNDR fluctuation is within 2dB under ±10% V_{DD} variations. The measured S_{11} remains below -12dB from DC to 67GHz. The total power consumption is 311mW. As shown in Fig. 21.8.7, this work achieves 38.6dB SNDR$_{BW}$ (measured near BW-3dB) which outperforms prior samplers/ADCs reported in ISSCC/JSSC/VLSI with BW>30GHz. Compared with other CMOS designs, for comparable BW of 50GHz or above, this work achieves 11dB increase in SNDR$_{BW}$ (from 27dB to 38dB) [2]; while for comparable SNDR$_{BW}$ of 38dB or above, this work increases the BW by 26GHz (from 26GHz to 52GHz) [6].

Acknowledgement:
This work is supported in part by the NSFC under Grant 62090042, 62374098, 62434004 and 62434005, National Key R&D Program of China (No. 2019YFB2205003), the Beijing National Research Center for Information Science and Technology, the Beijing Innovation Center for Future Chip. The corresponding author is Yi Zhong (zhongyi1104@tsinghua.edu.cn).

References:

[1] D. Pfaff et al., "A 224 Gb/s 3 pJ/bit 40 dB Insertion Loss Transceiver in 3-nm FinFET CMOS," in *IEEE JSSC*, vol. 60, no. 1, pp. 9-22, Jan. 2025. http://doi.org/10.1109/JSSC.2024.3466092.

[2] R. L. Nguyen et al., "A 200GS/s 8b 20fJ/c-s Receiver with >60GHz AFE Bandwidth for 800Gb/s Optical Coherent Communications in 5nm FinFET," *ISSCC*, San Francisco, CA, USA, 2024, pp. 344-346. http://doi.org/10.1109/ISSCC49657.2024.10454385.

[3] J. Heel, H. S. Bindra, S. Louwsma, A. Dezzani and B. Nauta, "A 12.8GS/s Sub-Sampling ADC Front-End With 38GHz Input Bandwidth and >39dB SNDR for 1 to 32GHz in 22nm FDSOI," *ISSCC*, San Francisco, CA, USA, 2025, pp. 76-78. http://doi.org/10.1109/ISSCC49661.2025.10904786.

[4] Y. Zhang, M. Zhang, Z. Wu, Y. Zhu, R. P. Martins and C. -H. Chan, "A 72GS/s 9b Time-Interleaved Pipeline-SAR ADC Achieving 55.3/49.3dB SFDR at 20GHz/Nyquist Inputs in 16nm FinFET," *ISSCC*, San Francisco, CA, USA, 2025, pp. 436-438. http://doi.org/10.1109/ISSCC49661.2025.10904672.

[5] C. Nani et al., "A 5nm 60GS/s 7b 64-Way Time Interleaved Partial Loop Unrolled SAR ADC Achieving 34dB SNDR up to 32GHz," *IEEE VLSI Technology and Circuits*), Honolulu, HI, USA, 2024, pp. 1-2. http://doi.org/10.1109/VLSITechnologyandCir46783.2024.10631522.

[6] L. Kull et al., "A 10-Bit 20–40 GS/S ADC with 37 dB SNDR at 40 GHz Input Using First Order Sampling Bandwidth Calibration," *IEEE Symposium on VLSI Circuits*, Honolulu, HI, USA, 2018, pp. 275-276. http://doi.org/10.1109/VLSIC.2018.8502268.

[7] Z. Wu, C. -H. Chan, Y. Zhu, R. P. Martins and M. Zhang, "An Inverter-Based Sampling Front-End Achieving >46-dB SFDR at 50-GHz Input," in *IEEE Solid-State Circuits Letters*, vol. 8, pp. 81-84, 2025. http://doi.org/10.1109/LSSC.2025.3552520.

[8] G. Tretter, D. Fritsche, M. M. Khafaji, C. Carta and F. Ellinger, "A 55-GHz-Bandwidth Track-and-Hold Amplifier in 28-nm Low-Power CMOS," in *IEEE TCAS II: Express Briefs*, vol. 63, no. 3, pp. 229-233, March 2016. http://doi.org/10.1109/TCSII.2015.2503579.

[9] X. -Q. Du, M. Grözing and M. Berroth, "A 25.6-GS/s 40-GHz 1-dB BW Current-Mode Track and Hold Circuit with more than 5-ENOB," *2018 IEEE BiCMOS and Compound Semiconductor Integrated Circuits and Technology Symposium (BCICTS)*, San Diego, CA, USA, 2018, pp. 56-59. http://doi.org/10.1109/BCICTS.2018.8550855.

[10] X. -Q. Du et al., "A 112-GS/s 1-to-4 ADC Front-End with More than 35-dBc SFDR and 28-dB SNDR up to 43-GHz in 130-nm SiGe BiCMOS," *2019 IEEE RFIC*, Boston, MA, USA, 2019, pp. 215-218. http://doi.org/10.1109/RFIC.2019.8701786.

[11] L. Wu and J. C. Scheytt, "Analysis and Design of a Charge Sampler with 70-GHz 1-dB Bandwidth in 130-nm SiGe BiCMOS," in *IEEE TCAS I: Regular Papers*, vol. 68, no. 9, pp. 3668-3681, Sept. 2021. http://doi.org/10.1109/TCSI.2021.3094428.

[12] P. Thomas, T. Tannert, M. Grözing, M. Berroth, Q. Hu and F. Buchali, "1-to-4 Analog Demultiplexer with up to 128 GS/s for Interleaving of Bandwidth-Limited Digitizers in Wireline and Optical Receivers," in *IEEE JSSC*, vol. 56, no. 9, pp. 2611-2623, Sept. 2021. http://doi.org/10.1109/JSSC.2021.3100677.

[13] S. Niu et al., "A 200–256-GS/s Current-Mode 4-Way Interleaved Sampling Front-End with over 67-GHz Bandwidth Using a Slew-Rate Insensitive Clocking Scheme," in *IEEE JSSC*, vol. 60, no. 1, pp. 244-259, Jan. 2025. http://doi.org/10.1109/JSSC.2024.3416528.

[14] G. Li et al., "A 600Gb/s DP-QAM64 Coherent Optical Transceiver Frontend with 4x105GS/s 8b ADC/DAC in 16nm CMOS," *ISSCC*, San Francisco, CA, USA, 2024, pp. 338-340. http://doi.org/10.1109/ISSCC49657.2024.10454499.

[15] K. N. Madsen, T. D. Gathman, S. Daneshgar, T. C. Oh, J. C. Li and J. F. Buckwalter, "A High-Linearity, 30 GS/s Track-and-Hold Amplifier and Time Interleaved Sample-and-Hold in an InP-on-CMOS Process," in *IEEE JSSC*, vol. 50, no. 11, pp. 2692-2702, Nov. 2015. http://doi.org/10.1109/JSSC.2015.2472642.

[16] A. Zandieh, N. Weiss, T. Nguyen, D. Haranne and S. P. Voinigescu, "128-GS/s ADC Front-End with over 60-GHz Input Bandwidth in 22-nm Si/SiGe FDSOI CMOS," *2018 IEEE BiCMOS and Compound Semiconductor Integrated Circuits and Technology Symposium (BCICTS)*, San Diego, CA, USA, 2018, pp. 271-274. http://doi.org/10.1109/BCICTS.2018.8550842.

[17] P. Thomas, J. Finkbeiner, M. Grözing and M. Berroth, "Time-Interleaved Switched Emitter Followers to Extend Front-End Sampling Rates to up to 200 GS/s," in *IEEE JSSC*, vol. 57, no. 9, pp. 2599-2610, Sept. 2022. http://doi.org/10.1109/JSSC.2022.3192546.

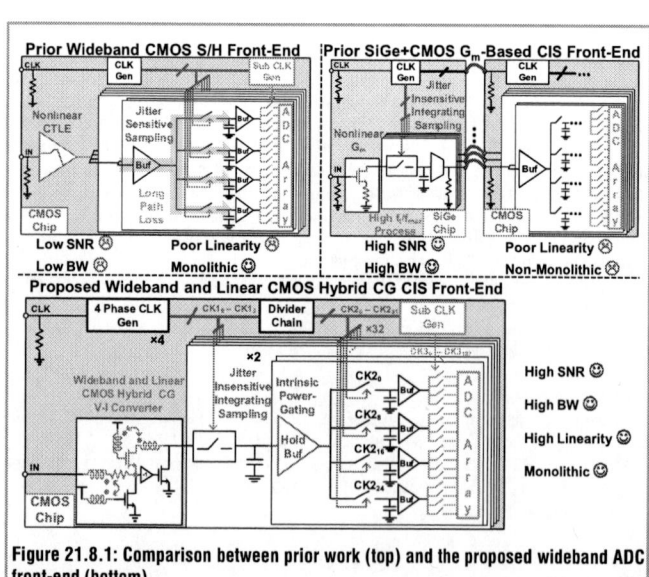

Figure 21.8.1: Comparison between prior work (top) and the proposed wideband ADC front-end (bottom).

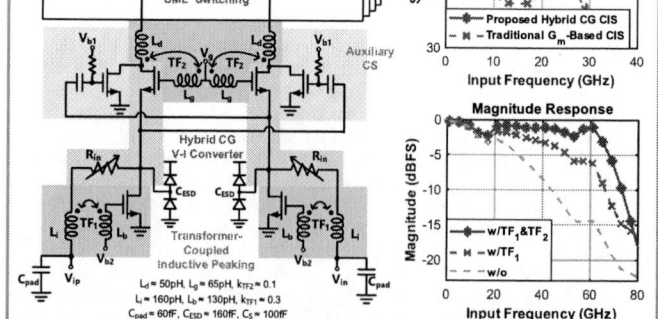

Figure 21.8.2: The proposed CMOS hybrid CG CIS (left) and the simulation results of the proposed techniques (right).

ISSCC 2026 / SESSION 21 / SENSOR INTERFACES / 21.8

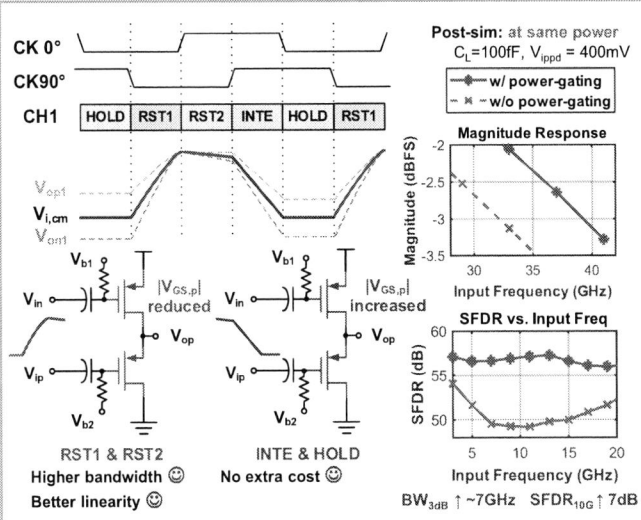

Figure 21.8.3: Operation principle of the proposed automatic power-gating hold buffer (left) and its simulation results (right).

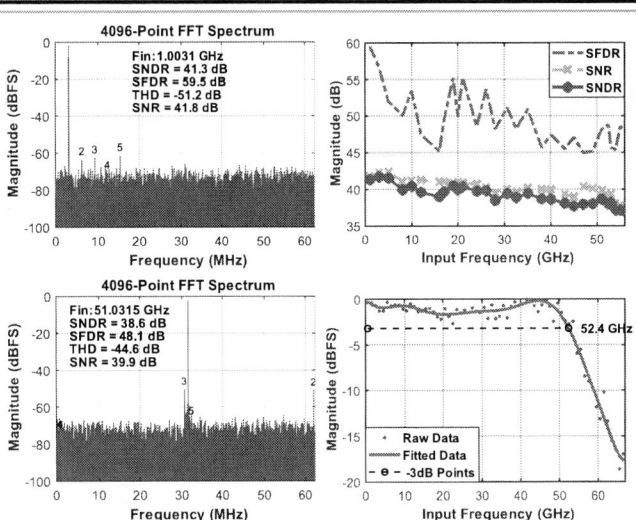

Figure 21.8.4: Measured output spectra at 1GHz and 51GHz input frequencies, SFDR/SNR/SNDR versus input frequency, and fitted magnitude response.

Figure 21.8.5: Measured multi-chip magnitude response, SFDR/SNDR variation under different analog V_{DD} variation, S_{11} performance, and power breakdown.

Figure 21.8.6: Performance summary and comparison with state-of-the-art samplers and sampler-limited ADCs.

	This work	ISSCC'25 Heel [3]	JSSC'25 Niu [13]	JSSC'22 Thomas [17]	JSSC'21 Thomas [12]	ISSCC '25 Zhang [4]	JSSC'25 Pfaff [1]	ISSCC '24 Nguyen [2]
Type	TI Front-End	TI Front-End	TI Front-End	TI Front-End	TI Front-End	Full ADC	Full ADC	Full ADC
Sampler Topology	Hybrid CG Based CIS + SC S/H	SC S/H	G_m-Based CIS	SEF	G_m-Based CIS	SC S/H	SC S/H	SC S/H
Channels	4/32	1/4	4	4	4	16/64	8/80	16/256
Measured Channels	1 of 32	4 of 4	4 of 4	4 of 4	4 of 4	64 of 64	80 of 80	256 of 256
Required Calibration[a]	O, G, T, B	O, G	O, G, T, B	O, G, T, B	O, G, T, B	O, G, T, B	O, G, T, B	O, G, T, B
Technology	28nm CMOS	22nm FDSOI	130nm SiGe	90nm SiGe	90nm SiGe	16nm FinFET	3nm FinFET	5nm FinFET
Input Swing [mV_{ppd}]	800	700	500	1000	500	500	n.a.	n.a.
Power [mW]	311	87	1300	3500	3600	227[b]	n.a.	n.a.
Area [mm²]	0.11	0.12	n.a.	0.2	n.a.	n.a.	n.a.	n.a.
fs [GS/s]	128	12.8	200	128	128	72	112	200
BW [GHz] (-3dB)	52	38	>67	>57	36	32	>50[c]	60
SFDR [dB]	>45 1-52GHz	>40[c] 1-38GHz	>90[c] 1-50GHz	>20[c] 1-50GHz	>35[c],[e] 1-50GHz	>47[c] 1-40GHz	>30[c] 1-50GHz	>28 1-47GHz
SNDR [dB]	>37 1-52GHz	>35[c] 1-38GHz	>27[c] 1-50GHz	>10[c],[d] 1-50GHz	>22 1-50GHz	>35[c] 1-40GHz	>25 1-50GHz	>26 1-47GHz

a: TI Calibration Requirement. O: Offset, G: Gain, T: Timing-Skew, B: Bandwidth.
b: Power consumption includes both the front-end and the clock generator for fair comparison.
c: Data are extracted from the figures in the referenced papers.
d: SNR is limited by the jitter performance of the measurement instruments.
e: SFDR is represented by HD3 due to significant mismatch in the differential signal path.

Figure 21.8.7: Chip micrograph and SNDR comparison with other ISSCC/JSSC/VLSI-published samplers/ADCs (BW≥20GHz, measured near BW$_{-3dB}$).

* The reported SNDR$_{BW}$ is measured when the input frequency is near the sampler/ADC BW$_{-3dB}$

• 2026 IEEE International Solid-State Circuits Conference

ISSCC 2026 / SESSION 21 / SENSOR INTERFACES / 21.9

21.9 A -82.3dB THD+N 60V Fully Integrated Shunt-Resistor-Based In-Line Current Sensor with DLL-Assisted Dynamic Body-Biasing Technique

Heng Ma, Huajun Zhang, Qinwen Fan

TU Delft, Delft, The Netherlands

Abstract

Floating-G_m-based current sensors are effective in rejecting PWM common-mode voltage. But their linearity drops at high voltages due to substrate-current-induced body effect in LDMOS devices. To address this issue, this work introduces a DLL-assisted dynamic body biasing sensor in 0.18μm BCD, improving HD2 by 31dB and THD+N by 14dB. And a peak THD+N of -82.3dB is achieved in the meanwhile. It supports 60V common-mode, rejects PWM up to 2MHz, and enables high-linearity sensing for high power audio and motor-driver systems.

High-linearity current sensors are essential for high-performance audio systems and precision motor drivers, etc. In particular, in-line shunt-resistor-based current sensors are gaining popularity due to their potentially high linearity [1]. To achieve high power efficiency, such systems often rely on switched power stages—e.g., power inverters or Class-D amplifiers—, controlled by pulse-width-modulated (PWM) signals to drive the load. The inline current sensor, placed between the power stage and the load, is then directly exposed to fast, high-voltage (HV), common-mode (CM) PWM pulses at the output of the power stages, which can induce severe glitches at the sensor output and significantly degrade linearity [2]. To mitigate this issue, a floating G_m stage can be used to effectively reject the HV fast PWM voltage [3,4], followed by an HV LDMOS buffer stage that transfers the signal from the floating voltage domain to the low-voltage (LV) domain. High linearity together with PWM rejection up to 2MHz was demonstrated in [3], but its low input voltage range (14.4V) makes it incompatible with many switching systems that operate under the increasingly common 48V supply standard and above [2]. Increasing the PWM voltage will lead to higher V_{DS} stress on the LDMOS transistors, which can cause severe body leakage current—a phenomenon known as substrate current-induced body effect (SCBE) shown in Fig. 21.9.1 top [5]. This can, surprisingly, but significantly increase the second-order harmonic distortion (HD2). To mitigate this, a DLL-assisted dynamic-body-biasing (DBB) technique is proposed, which yields up to 31dB improvement in HD2 and 14dB improvement in THD+N. Implemented in a 180nm BCD process, the proposed current sensor achieves a peak THD+N of -82.3dB with a 4A, 1kHz input signal, while rejecting HV PWM signal up to at least 2MHz. It offers a 60V input voltage range with a DC CMR of 132dB, while consuming 1.7mA current.

The proposed inline current sensor is shown in Fig. 21.9.1 bottom, which consists of an on-chip shunt resistor R_S, a floating G_m (FG$_m$) stage powered by the bootstrap voltage V_{BST} (commonly available in high-power switched amplifiers for high-side gate driving), an HV LDMOS current buffer ($M_{D1,2}$), and an LV (5V) transimpedance amplifier (TIA) built around A1. 70V LDMOS are chosen to allow 60V input CM range with extra safe margin. A resistor thermal compensation scheme is employed to mitigate nonlinearity due to R_S self-heating [3]. A DLL-assisted timing generator monitors the PWM voltage V_{SW} and generates control signals for the DBB circuit, which dynamically adjusts the LDMOS body voltage during V_{SW} transitions, effectively suppressing SCBE-induced distortion.

Figure 21.9.2 illustrates the proposed DBB scheme and compares it with conventional approaches. Conventionally, the body of $M_{D1,2}$ is either connected to the supply FVDD (BTF) or to its own source (BTS). In the former configuration (Fig. 21.9.2 top-left), where FVDD is 5V above V_{SW}, a body current I_{SCBE} is generated due to SCBE, which is dependent on V_{DS}: the higher V_{DS}, the higher I_{SCBE}. Since V_{DS} changes with V_{SW}, which is a PWM-modulated version of I_{SIG}, I_{SCBE} becomes signal dependent. At first glance, this appears to be a CM error. However, I_{SCBE} also depends on I_{SIG} directly, due to the fact that higher I_{SIG} induces more impact ionization—the origin of I_{SCBE} [5]. Consequently, I_{SCBE} contains a differential component proportional to $\sim I_{SIG}^2$, which can result in an HD2 distortion ~-67dB with a 4A current in this design. In the BTS configuration (Fig. 21.9.2, top-middle), I_{SCBE} would circulate inside the LDMOS, leaving the drain current undistorted. However, HV LDMOS devices are typically fabricated directly on the substrate (in non-SOI processes) without an additional isolation well, resulting in a parasitic body-to-substrate capacitance C_P. During high slew rate PWM transitions, the C_P charging and discharging current I_{LEAK} (up to mA$_p$ level) can lead to large voltage glitches on V_{CAS} (Fig. 21.9.1 bottom) and can drive the circuit shortly into a nonlinearity region and degrade performance. This effect becomes more severe with increasing PWM frequency (f_{PWM}): a trend that has become prevalent in many switching power systems to reduce system bulk.

To address the limitations discussed above, DBB is proposed, which combines BTS with a body-to-source buffer (BTB) path and is shown in Fig. 21.9.2 bottom with its control timing diagram. When V_{SW} is high, the BTS path is selected, preventing additional I_{SCBE} from being injected into the drain. During the PWM transitions, the body is switched to the BTB path, which allows I_{LEAK} to be sourced from V_{S_BUF}, where V_{S_BUF} is a buffered version of the source voltage. The source buffer is implemented using a floating Class-AB source follower to ensure fast settling during PWM transients. Native NMOS devices are used to reduce the voltage difference between V_S and V_{S_BUF}, allowing smooth switching transition and

preventing additional glitches on V_{CAS}. It should be noted that while it is crucial to stay in BTS configuration during V_{SW} is high to minimize I_{SCBE} injection into the drain, it is much less of an issue when V_{SW} is low, as I_{SCBE} is significantly lower. As a result, a BTB configuration is chosen when V_{SW} is low for simplicity in control as shown in Fig. 21.9.2 bottom-right. Simulations in Fig. 21.9.2 top-right show that with a 48V, 2MHz PWM signal under a slew rate ranging between 1 and 10V/ns, the proposed DBB can effectively reduce the V_{CAS} glitch from up to 2.3V to 0.5V compared to BTS.

To realize the proposed DBB operation, the control signal φ_0 (Fig. 21.9.2 bottom-right) must switch on immediately after the rising edge and shortly before the falling edge of the V_{SW}, so that BTS is active as long as possible when V_{SW} is high, while BTB is immediately active before each PWM switching. The former switching operation can be realized by timing the V_{SW} rising edge using an inverter buffer; the latter switching operation, however, requires the knowledge of the incoming V_{SW} falling edge in advance, which can be tricky for standalone current sensors since the internal PWM clocks of the switching power stage may not be accessible. To solve this issue, a DLL-based timing generator is employed to predict the PWM falling edge, which is shown in Fig. 21.9.3. The input V_{SW} is first attenuated by a 1/11 voltage divider to limit its swing within 5V. A hysteresis digital buffer then reshapes the PWM waveform and suppresses ringing due to parasitic inductance and high di/dt during transients, producing a clean digital signal DIN as the DLL input. The DLL consists of a phase-frequency detector (PFD), a charge pump (CP), a filter capacitor C_{FILT}, a voltage-controlled delay line (VCDL), and fixed delay cells t_{DELAY}. The PFD compares DIN with the feedback signal DFB, and the resulting phase error is integrated by the CP and C_{FILT}. Once locked, DIN and DFB are phase-aligned and the intermediate node DOUT, located before the fixed delay cell, leads DIN by a constant delay t_{DELAY}. Finally, by logically combining DOUT and a delayed version of DIN (DIN$_{DELAY}$) using an AND gate, the body control signal φ_0 is generated. t_{DELAY} should be long enough to ensure the LDMOS switch to BTB before the falling edge of V_{SW}, while tolerating slew rate variations of V_{SW} and phase errors due to the finite DLL bandwidth. However, a long t_{DELAY} will shorten the BTS duration when V_{SW} is high, producing un-negligible I_{SCBE} and HD2. This effect becomes more pronounced at higher PWM frequencies f_{PWM}, since a shorter PWM period makes t_{DELAY} contribute proportionally more I_{SCBE}. To minimize phase-lock error such that a shorter t_{DELAY} can be allowed especially for higher f_{PWM}, the DLL bandwidth is set as high as possible to extend the gain at the current signal frequency (f_{SIG}), but is still kept $<\sim0.1 \cdot f_{PWM}$ for stability [6]. Finally, t_{DELAY} is chosen to be ~50ns to accommodate a range of V_{SW} slew rate (>~1V/ns) and f_{PWM} (384kHz to 2MHz) such that the proposed current sensor is compatible with a wide range of commercial Class-D audio amplifiers. Figure 21.9.3 bottom presents measured results at $f_{PWM} = 38 \cdot f_{SIG}$ and $200 \cdot f_{SIG}$, respectively, where f_{SIG} is 10kHz and the current amplitude is 4A. In both cases, the φ_0 falling edge is successfully locked to lead DIN by t_{DELAY}. It's worth noting that the proposed scheme can also be applied to other use cases with different f_{SIG} and f_{PWM} ranges as long as $f_{PWM} >> f_{SIG}$, which is often the case.

The sensor was fabricated in a standard 0.18μm BCD process and occupies an area of 2mm² (Fig. 21.9.7). It is measured with a commercial Class-D audio amplifier TAS6684, which allows a wide range of internal f_{PWM} (384kHz to 2MHz) to demonstrate the versatility of the proposed design, but its maximum supply voltage (48V) and unadjustable slew rate also limit the setup. Figure 21.9.4 top-left shows the measured spectrum of V_{OUT} under a 4A, 1kHz input current and 1MHz f_{PWM}. Compared to BTB where I_{SCBE} is added to I_{SIG} similar to BTF, the proposed DBB improves HD2 by 27dB. The measured HD2 with 1kHz signal frequency versus signal amplitude (6mA to 6A) is shown in Fig. 21.9.4 top-middle and the HD2 with 4A signal amplitude versus signal frequency (20Hz to 10kHz) is shown in Fig. 21.9.4 top-right, both demonstrating significant HD2 improvements using DBB. V_{OUT} spectrum with 4A, 1kHz signal current is measured with various f_{PWM} (384kHz to 2MHz) and is shown in Fig. 21.9.4 left-bottom, proving the robustness and versatility of the proposed DBB. THD+N with 1kHz signal frequency versus signal amplitude and with 4A signal amplitude versus signal frequency are shown in Fig. 21.9.4 bottom-middle and right, demonstrating 14dB improvement with DBB. Figure 21.9.5 top-left shows that the measured THD+N versus signal amplitude of 3 samples varies only by <2dB. Figure 21.9.5 bottom-left illustrates the improved THD+N with a 4A, 1kHz signal varies negligibly over temperatures (-40°C to 85°C), using DBB. Figure 21.9.5 top-right shows the transient response of V_{OUT} under a 2MHz, 48V V_{SW}. It can be seen that DBB effectively reduces the

980 • 2026 IEEE International Solid-State Circuits Conference

979-8-3315-8937-0/26 $31.00 © 2026 IEEE

PWM switching glitches compared to BTS. The difference is expected to be more significant if a higher slew-rate power stage is used. Figure 21.9.5 bottom-right shows the measured DC CMR of the current sensor, demonstrating a 0V-to-60V input CM operating range and a 132dB CMR.

Figure 21.9.6 summarizes the sensor performance and compares it to state-of-the-art designs. Compared to [3], the proposed design extends the supply voltage capability from 14.4V to 60V while maintaining similar PWM rejection frequency and THD+N. Compared to other prior works with >40V input range, it achieves at least 16dB better THD+N, 16× higher PWM rejection frequency and 10dB higher DR, while consuming comparable current.

References:
[1] M. Mauerer et al., "Voltage/current measurement performance and power supply rejection in all-digital Class-D power amplifiers," *IECON*, pp. 666-673, Oct. 2016. http://doi.org/10.1109/IECON.2016.7793041
[2] Texas Instruments, *INA253 High Voltage, Bidirectional, Zero-Drift, Current-Shunt Monitor with Integrated, 2-mΩ, Precision, Low Inductive Shunt Resistor*, [Online]. Available: https://www.ti.com/product/INA253. https://www.ti.com/product/INA253
[3] H. Ma et al., "A -87.2 dB THD+N 89.1 dB DR Fully-Integrated Shunt-Resistor-Based In-Line Current Sensor with up to 2 MHz 14.4 V PWM Rejection," *IEEE Symp. VLSI Circuits*, pp. 1-2, June 2025. http://doi.org/10.23919/VLSITechnologyandCir65189.2025.11075038
[4] C. Larsen et al., "High Common Mode Rejection Ratio (CMRR) Current Monitoring Circuit Using Floating Supplies," U.S. Patent US11296666B1, Apr. 2022. https://patents.google.com/patent/US11296666B1
[5] S. Salahuddin et al., *BSIM4 4.8.3 MOSFET Model User's Manual*, [Online]. Available: https://bsim.berkeley.edu/models/bsim4/. https://bsim.berkeley.edu/models/bsim4/
[6] B. Razavi, "The Delay-Locked Loop [A Circuit for All Seasons]," *IEEE Solid-State Circuits Magazine*, vol. 10, no. 3, pp. 9-15, Aug. 2018. http://doi.org/10.1109/MSSC.2018.2844615
[7] V. Binet et al., "A Fully Integrated Class-D Amplifier in 40nm CMOS with Dynamic Cascode Bias and Load Current Sensing," *ESSCIRC*, pp. 319-322, Sep. 2014. http://doi.org/10.1109/ESSCIRC.2014.6942086

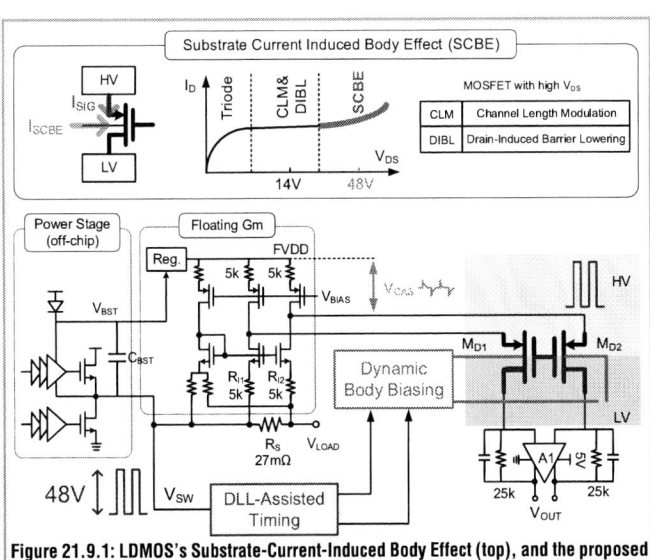

Figure 21.9.1: LDMOS's Substrate-Current-Induced Body Effect (top), and the proposed current sensor with DLL-assisted dynamic body-biasing technique (bottom).

Figure 21.9.2: Different body biasing schemes: BTF (top-left); BTS (top-middle); the simulated V_{CAS} glitch (top-right); the proposed DBB (bottom-left); and its control timing (bottom-right).

ISSCC 2026 / SESSION 21 / SENSOR INTERFACES / 21.9

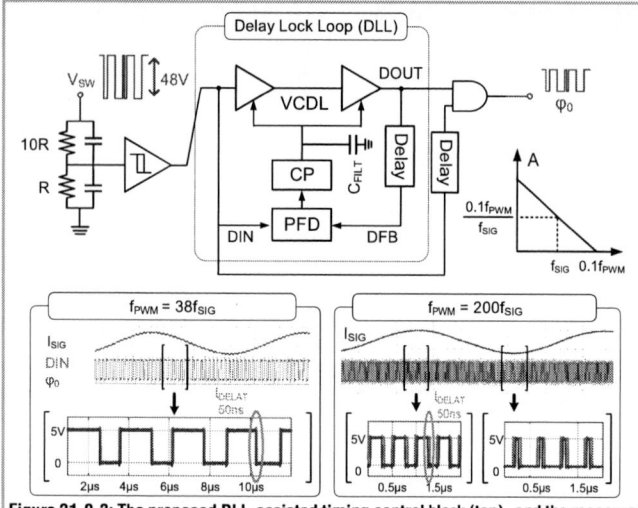

Figure 21.9.3: The proposed DLL-assisted timing control block (top), and the measured transient waveform under different PWM frequencies and different pulse duty cycles (bottom).

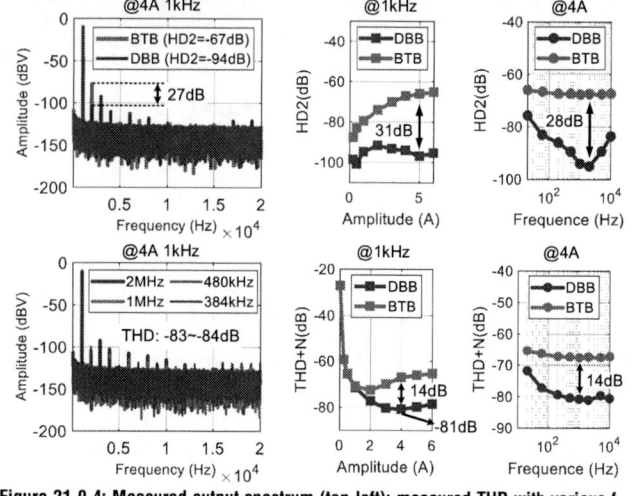

Figure 21.9.4: Measured output spectrum (top-left); measured THD with various f_{PWM} (bottom-left); measured HD2 and THD+N versus current amplitude and frequency (middle and right).

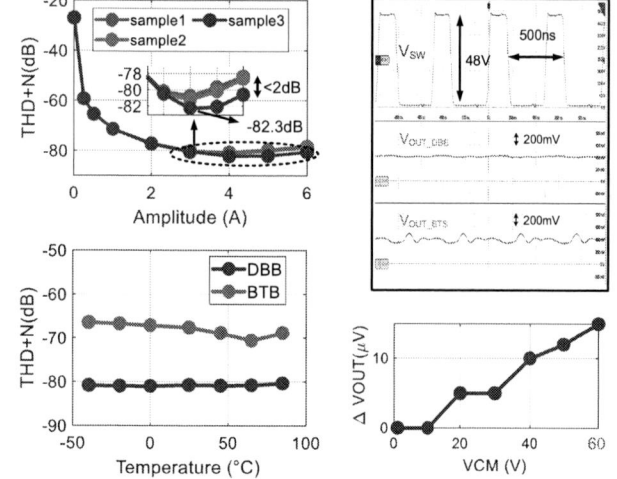

Figure 21.9.5: Measured THD+N with different samples (top-left) and temperatures (bottom-left); the transient waveform with DBB versus BTS (top-right); measured CMR (bottom-right).

	This work	INA253	INA241	VLSI'25 [3]	MAX98388	ESSCIRC'14 [7]
Fully Integrated?	**Yes**	Yes	No	Yes	Yes	Yes
In-line sensing?	**Yes**	Yes	Yes	Yes	/	No
Input CM voltage	60V	80V	110V	14.4V	10V	5V
PWM rejection frequency	2MHz	/	125kHz	2MHz	/	/
Peak THD+N @1kHz	-82.3dB	-66.4dB[1,2]	/	-83.1dB	-63dB[1]	-58dB[1]
DR[3] A	87dB	/	/	89.1dB	/	/
DR[3] Un	84.5dB	74.5dB[1]	/	86.9dB	73dB	77.5dB
Current Consumption	1.7mA	1.8mA	2.5mA	1.2mA	/	0.8mA[1]
Current Range	±6A	±15A[4]	/	±6A	±3A	±0.5A[1]

[1]Estimated from data/graph   [2]Measured at 90% full scale
[3]A-weighted (A) or unweighted (un)   [4]DC current

Figure 21.9.6: Comparison to the state-of-the-art current sensors for switched power systems.

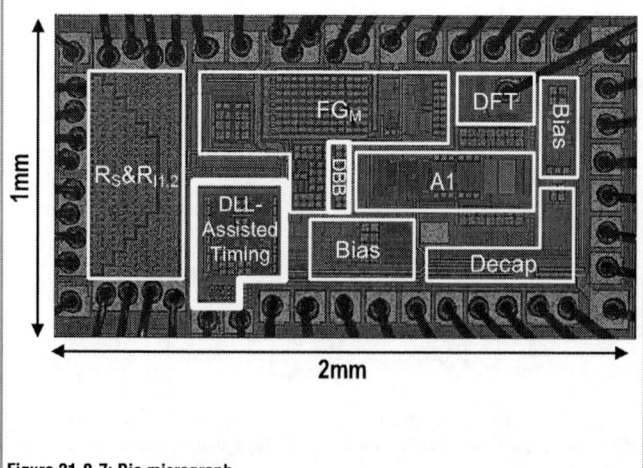

Figure 21.9.7: Die micrograph.

ISSCC 2026 / SESSION 22 / CIRCUITS IN EXTREME ENVIRONMENTS / OVERVIEW

Session 22 Overview: *Circuits in Extreme Environments*
TECHNOLOGY DIRECTIONS SUBCOMMITTEE

Session Chair: Giorgio Ferrari
Politecnico di Milano
Milano, Italy

Session Co-Chair: Joseph Bardin
University of Massachusetts Amherst,
Amherst, MA
and
Google, Goleta, CA

CMOS chips continue to serve a wider range of applications both at extremely low temperatures or in harsh radiation environments. The first three papers describe cryo-CMOS chips operating below 12K to control superconducting qubits and color-center qubits. The last two papers present a CMOS imager and a WiFi receiver that can sustain extreme radiation doses as expected in space and in nuclear power plants.

ISSCC 2026 / February 17, 2026 / 1:30 PM

1:30 PM

22.1 A Cryo-CMOS Color-Center Quantum Controller with Diamond Waveguide Micro-Chiplet Integration

Jinchen Wang, Massachusetts Institute of Technology, Cambridge, MA

In Paper 22.1, MIT shows a 16nm cryo-CMOS controller for color-center qubits in diamond. The chip drives an array of on-chip coils at 2.87GHz to drive the magnetic field in a diamond waveguide array on a micro-chiplet pick-and-placed on top of the CMOS die.

1:55 PM

22.2 A 16-Channel Low-Power Cryo-CMOS Flux Control Pulse Generator ASIC in 14nm FinFET Technology

Kevin Tien, IBM T. J. Watson Research Center, Yorktown Heights, NY

In Paper 22.2, IBM presents a cryogenic 14nm ASIC comprising a 16-channel flux-pulse generator ASIC for superconducting qubit control, consuming 7mW/channel. The chip is used to demonstrate 2-qubit control with error rates as low as 0.0017.

2:20 PM

22.3 A Multi-Qubit Cryo-CMOS SoC with Polar-Based Electron-Spin and PDM-Based Nuclear-Spin Controllers for Color Centers in Diamond

Niels Fakkel, TU Delft, Delft, The Netherlands

In Paper 22.3, TU Delft describes a cryo-CMOS 40nm SoC able to drive both electron-spin and nuclear-spin qubits in nitrogen-vacancy (NV) centers in diamond. The circuit can perform electron/nuclear quantum gates with high fidelity (99.3% and 99.8%, respectively) and high Rabi frequency (2.31MHz/1.93kHz) using multi-band (2.5-to-3.2GHz/1.9-to-2.1MHz) large-current (70mA/38mA) excitations.

2:45 PM

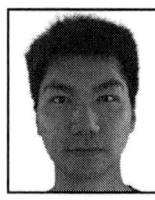

22.4 A Radiation-Hardened Self-Healing CMOS Imager with Online Pixel/Logic Annealing and Tile-Adaptive Compression for Space Applications

Quan Cheng, Southern University of Science and Technology, Shenzhen, China, Kyoto University, Kyoto, Japan

In Paper 22.4, Southern University of Science and Technology, Kyoto University, and Seoul National University present a 180nm CMOS self-healing imager for space applications with localized online thermal annealing for both pixels and logic. The chip shows 4.82×/3.97× reductions in pixel/logic leakage and nearly full image recovery after 20kGy X-ray exposure.

3:00 PM

22.5 A 500kGy Radiation-Hardened 2.4GHz Wi-Fi Receiver for Innovative Nuclear Power Plant Decommissioning

Yasuto Narukiyo, Institute of Science Tokyo, Tokyo, Japan

In Paper 22.5, the Institute of Science Tokyo, and KEK introduce a 2.4GHz Wi-Fi receiver in 65nm CMOS that is designed with radiation-hardened techniques to sustain operation under a total ionizing dose up to 500kGy to be used in nuclear power plant decommissioning.

22

DIGEST OF TECHNICAL PAPERS • 383

979-8-3315-8937-0/26 $31.00 © 2026 IEEE

ISSCC 2026 / SESSION 22 / CIRCUITS IN EXTREME ENVIRONMENTS / 22.1

22.1 A Cryo-CMOS Color-Center Quantum Controller with Diamond Waveguide Micro-Chiplet Integration

Jinchen Wang, Yuyang Han, Isaac B. Harris, Eunseok Lee, Yong Hu, Di Liu, Dirk Englund, Ruonan Han

Massachusetts Institute of Technology, Cambridge, MA

Abstract

We present a scalable cryo-CMOS controller for color-center-based quantum processors. A diamond waveguide micro-chiplet with NVs is pick-and-placed on CMOS with a 3D-printed prism for scalable photonic readout. A serial qubit driver array using grid inductors and pulse-width modulators generates a programmable 2.87GHz magnetic field for each qubit with low power and strong confinement. Up to 144 qubits can be operated on a single chip, with 204µW power consumption per qubit under active control.

Color centers in diamond are solid-state defects that confine charge carriers in optically accessible quantum states. They have shown great potentials for multi-modal quantum sensing, long-distance entanglement [1], and hybrid integration with CMOS [2,3]. Due to the long coherence time of nitrogen vacancy (NV) centers in diamond and high operating temperature (4K), the feasibility of using CMOS circuits to control the NV electron spin and the nearby C13 nuclear spin for quantum processors has been discussed [4]. In 2024, a cryo-CMOS controller for color-center-based quantum processors was reported [3]. At present, there are three major hurdles for color-center-based quantum processors: first, the qubit states of most color centers require a microwave driving field, thus mandating a large physical separation (>1mm in [3]) to avoid crosstalk. This dramatically limits the scalability of the color-center-based quantum processors with thousands of physical qubits and beyond. Secondly, most quantum information processing protocols require multiple qubits to be controlled simultaneously. However, all previously envisioned architectures require an individual RF driver for each qubit, consuming large DC power (~16mW/qubit in [3,5]) and space, while a standard 4K cryostat has a total cooling power budget of only ~100mW and limited space. Lastly, small qubit pitch requires a photonic readout scheme that enables not only high collection efficiency of the qubit photoluminescence, but also high separation of different optical signals. Such requirements are difficult to meet using the previously adopted bulk diamond scheme [2,3] due to the total internal reflection inside diamond and high background fluorescence from the CMOS chip. In [6], one promising solution for high-efficiency, multiplexed readout is reported where color centers are created within a dense array of diamond optical waveguides. Integration of such photonic structures with a custom-designed CMOS controller has not been demonstrated previously.

In this paper, we introduce a scalable cryo-CMOS controller for color-center-based quantum processors to address the above challenges. Shown in Fig. 22.1.1, a diamond waveguide array micro-chiplet is pick-and-placed on top of the CMOS die, allowing the built-in color center to emit into the guided mode. As the 2.87GHz magnetic field launcher under each micro-chiplet, a grid-shaped inductor is adopted for field confinement. Every 12 inductors are connected in series through switches controlled by on-chip pulse-width modulators (PWM), to form an array that shares a single RF driver. In addition, an on-chip photonic prism based on 3D nano-printing is added to redirect the color center photoluminescence from the ends of waveguides into freespace, enabling high-density optical readouts outside the cryostat. A maximum of 144 qubits can be operated on a single chip, with a per-qubit power consumption of 204µW under active control.

For high qubit density, the magnetic field launched by each inductor should only exist in the region right above the inductor, while rapidly decaying in regions nearby. As shown in Fig. 22.2.1, compared to a conventional RF loop inductor, the grid configuration of the inductor provides more space to accommodate the diamond within the grid and simultaneously offers field self-cancellation outside the grid. The crosstalk is further reduced by positioning each qubit (e.g. Qubit2 in Fig. 22.2.1) along the center lines of other adjacent grid inductors. For even further qubits (e.g. Inductor1 to Qubit3 in Fig. 22.1.1), although the magnetic field is only partially cancelled, sufficient suppression is still ensured by the larger distance. Fig. 22.1.2 presents the model and the simulated magnetic field of the grid inductor at 4K. A crosstalk suppression of 60dB is achieved at 20µm from the inductor core, while a conventional loop inductor needs a distance of 400µm. A suppression of 60dB implies that the system would require at least one thousand operating periods to accumulate at most a single error, which is sufficient for quantum operations. Considering the footprint of the local switches and resistors in each qubit driver unit, an inductor pitch of 34-to-42µm is adopted, leading to much higher array density compared to prior RF-coil-based schemes.

The series cascade connection of the inductors (Fig. 22.1.1) is based on a general fact that, the low impedance of small sized inductors prevents efficient power matching and leads to high DC power if each inductor is driven by an individual RF driver. 12 grid inductors in series effectively increase the impedance and power-delivery efficiency, allowing a single driver to power the entire line of inductors. To allow the inductors to be turned on and off independently, we employ a set of buried differential coaxial cables (BDCX) and PWM-controlled switch array, as shown in Fig. 22.1.2. When a pulse is applied to a unit, the inductor is activated, and when no RF pulse is present, a BDCX-based by-pass path is

activated. The BDCX, of which the side view is shown in Fig. 22.1.2, is located in the bottom layer and is driven by a differential RF signal for maximum crosstalk suppression. The simulated field profile of the BDCX structure at 4K is also shown in Fig. 22.1.2: when the BDCX is activated, its field leakage to the inductor within the same unit is below -60dB. As a result, the proposed circuitry allows simultaneous control of all qubits. Note that it is also important to ensure equal phase shifts between the inductor and the BDCX to avoid phase errors caused by switching different units on and off, since a phase error of 0.18° has a similar effect to a crosstalk of -60dB. The BDCX is intentionally designed to match the phase shift of the grid inductor, with a simulated phase discrepancy of only 0.021° at 2.87 GHz and 4K (Fig. 22.1.2). The error introduced by switching loss can be calibrated at the algorithm level or using on-chip strain controllers (not used in this project).

Figure 22.1.3 shows the schematic of the full chip. The 12 rows of driver circuits are arranged in an interleaved pattern, and the digital array is divided into two sides due to limited routing space. To realize quantum processing, the sinusoidal pulse area and phase must be controlled to operate θ and Φ on the Bloch sphere, respectively. Considering the trade-off between logic circuit area and functional implementation, we design the PWM logic unit using two down-counters and shift-register-based SPIs, as shown in Fig. 22.1.3. A 4-bit down-counter is used to control the pulse width, and another 4-bit down-counter is used to control the pulse delay. Both have a resolution of 15 clock periods. Each unit can store up to four instructions (e.g., $\pi/4$, $\pi/2$, and π) in write mode. In read mode, the high-speed addressing SPI retrieves the correct instruction to generate the pulse. The control of Φ and matching of the RF frequency are achieved by modulating the RF input with a chirp and time-division multiplexing using the controllable delay on-chip. Since the pulses are generated at different frequencies, their phases also differ. All qubits can be simultaneously and individually controlled, since each qubit has its own addressing code and instruction storage.

The chip was fabricated using the Intel16 CMOS process and has a die size of 5.4mm² (Fig. 22.1.7). The chip was first etched with a depth of 9µm, after which the prism array was 3D nano-printed with an UpNano NanoOne printer. Wire bonding was then performed, followed by pick-and-place of diamond micro-chiplet. Each nano-fabricated diamond micro-chiplet contains single-mode waveguides with multiple NV centers in them. Figure 22.1.4 shows the microscopic close-up of the controller assembly stack. A 532nm laser is used to excite off-resonantly a single NV center at a time within the waveguide, and the photoluminescence propagates through the waveguide before being totally internally reflected to the external detector at the top by the nano-printed dielectric prism. Figure 22.1.4 also presents the simulated far-field dipole emission profile from the color center via the prism using total internal reflection, which validates our design. For future use, we also implemented an optional on-chip programmable strain-controller array that enables integration of diamond micro-chiplets with compliant waveguides, which can be mechanically bent to fine-tune photon-emission frequencies for interference and entanglement [6].

The chip was attached and mounted on the 4K cold finger of a Montana cryostat. The measurements were conducted at 4.9K using the measurement setup shown in Fig. 22.1.4, which also shows the measured PWM output at a local test point. DATAI and CLKI denote the observed data and clock signals injected into the PWM logic unit from the last unit. These measurements verify the circuit's ability to implement programmable pulse-width and delay control. Figure 22.1.5 summarizes the measured temperature and power at 4.9K. During circuit operation, toggling the laser on and off does not change the cryostat temperature, implying that the absorbed laser power is <10µW. The total power consumption of the system is 29.3mW when all circuits are active. We also measured the optically detected magnetic resonance (ODMR) from the single NV center within the diamond micro-chiplet, as shown in Fig. 22.1.5, where the Zeeman splitting is due to the environment magnetic field. To quantitatively assess the suppression of the magnetic-field crosstalk at different positions on our chip, another chip sample is prepared with a uniform coating of nano diamond particles containing NV centers, as shown in Fig. 22.1.5. With all other conditions held constant, the dependence of ODMR contrast on the RF drive current at 4.9K is measured and compiled with a corresponding calibration table. The ODMR at different spatial positions is also measured (Fig. 22.1.6). From this series of measurements at 4.9K, the distance dependence of the magnetic-field strength is obtained. dB scale is not used because the measurement accuracy is insufficient to yield meaningful data for crosstalk below ~40dB. Nevertheless, the data indicate no detectable magnetic-field interference at a distance of 40µm from the driven inductor, which verifies our design. Figure 22.1.6 also shows a comparison with prior arts. The chip supports up to 144 qubits with individual RF control and photonic readout, and consumes 204µW of power per qubit under active control (all qubit drivers are turned on). The peak RF current for each qubit is only 1.64mA in our design, which is significantly smaller than that reported in [4]. This improvement is partly due to the 9µm etching depth of our chip, which reduces the vertical distance between the color center and the inductor core to about 300nm. Consequently, an RF magnetic field of approximately 1.5 gauss is achieved at qubit position with I_{RF}=1.64mAp. Because the CMOS geometry in this system is nonstandard, and the number of color centers in diamond microchiplet is far lower than ensembles in bulk diamond, additional architecture—spanning the photonics, sample, microscopy, and cryostat subsystems—must be developed to enable Rabi-oscillation measurements of the microchiplet and other algorithmic demonstrations in future work. The system can also be used to control group-IV color centers (e.g., SnV⁻) in diamond, which serve as more promising candidates for quantum computing and networking due to their smaller spectral diffusion and larger Debye-Waller factor [7].

Acknowledgement:
The chip is fabricated through the Intel University Shuttle Program. This work is supported by NSF under Grant FuSe2-2425611. The authors would like to thank Alessandro Buzzi at MIT for technical discussions and support.

References:
[1] C. M. Knaut *et al.*, "Entanglement of nanophotonic quantum memory nodes in a telecom network," *Nature*, pp. 573–578, May. 2024.
https://doi.org/10.1038/s41586-024-07252-z
[2] D. Kim *et al.*, "A CMOS-integrated quantum sensor based on nitrogen–vacancy centres," *Nat Electron*, pp. 284–289, Jul. 2019.
https://doi.org/10.1038/s41928-019-0275-5
[3] L. Enthoven *et al.*, "A Cryo-CMOS Controller with Class-DE Driver and DC Magnetic-Field Tuning for Color-Center-Based Quantum Computers," *ISSCC*, pp. 472–474, Feb. 2024. https://doi.org/10.1109/ISSCC49657.2024.10454348
[4] R. Ishihara *et al.*, "3D Integration Technology for Quantum Computer based on Diamond Spin Qubits," *IEDM*, pp. 14.5.1–14.5.4, Dec. 2021.
https://doi.org/10.1109/IEDM19574.2021.9720552
[5] J. Wang *et al.*, "A CMOS-Integrated Color Center Pulse-Sequence Control and Detection System," *CICC*, pp. 1–2, Apr. 2024.
https://doi.org/10.1109/CICC60959.2024.10529078
[6] N. H. Wan *et al.*, "Large-scale integration of artificial atoms in hybrid photonic circuits," *Nature*, pp. 226–231, Jul. 2020. https://doi.org/10.1038/s41586-020-2441-3
[7] I. B. W. Harris *et al.*, "Coherence of Group-IV Color Centers," *Phys. Rev. B*, 085414, Feb. 2024. https://doi.org/10.1103/PhysRevB.109.085414

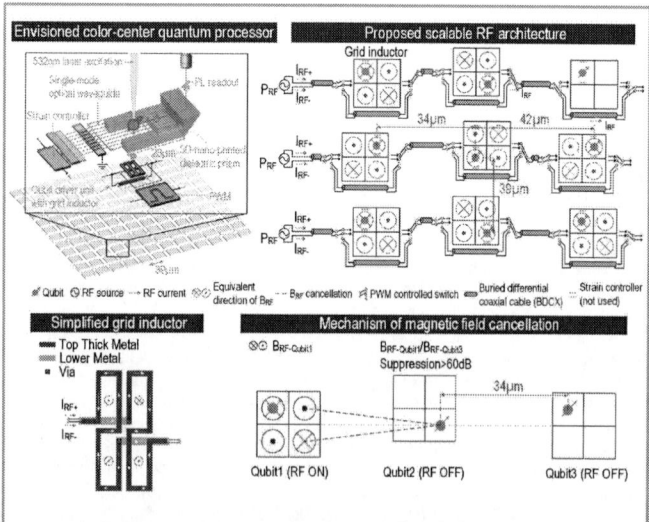

Figure 22.1.1: Proposed scalable architecture, including the structure of a simplified grid inductor and its cancellation mechanism.

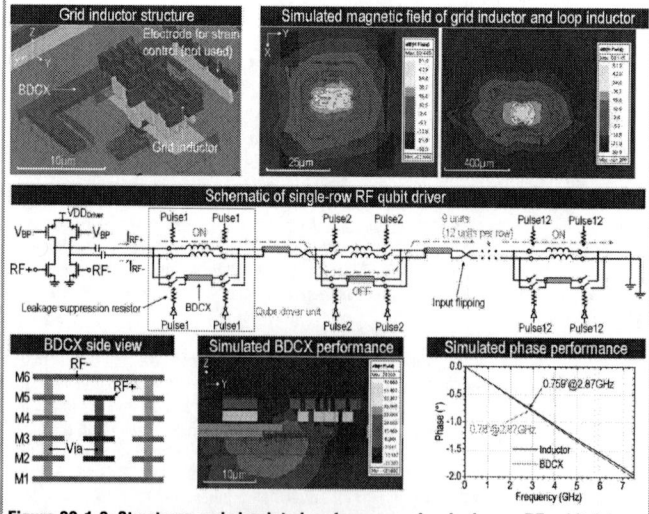

Figure 22.1.2: Structures and simulated performance of a single-row RF qubit driver, including its grid inductor and BDCX.

ISSCC 2026 / SESSION 22 / CIRCUITS IN EXTREME ENVIRONMENTS / 22.1

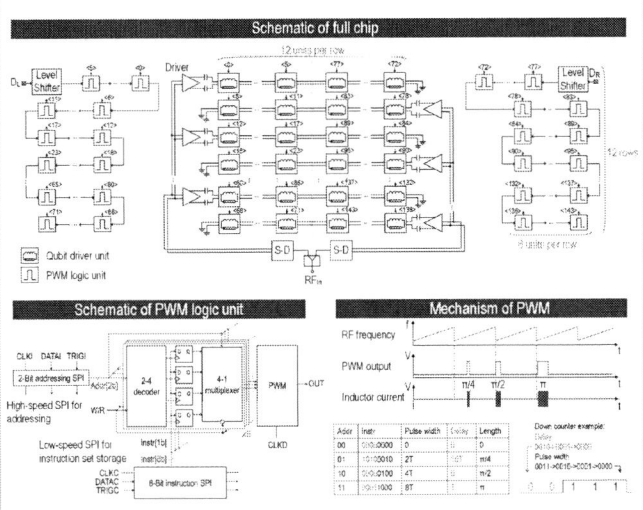

Figure 22.1.3: Schematic of the full chip and the PWM logic unit with its operating mechanism.

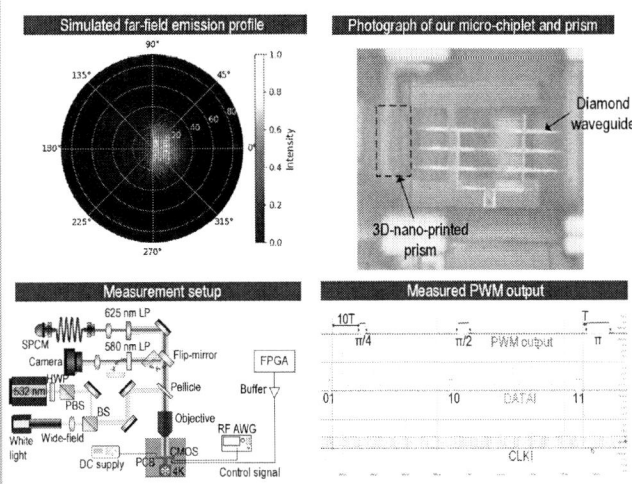

Figure 22.1.4: Design and photograph of the micro-chiplet and prism, the measurement setup, and measured on-chip PWM results.

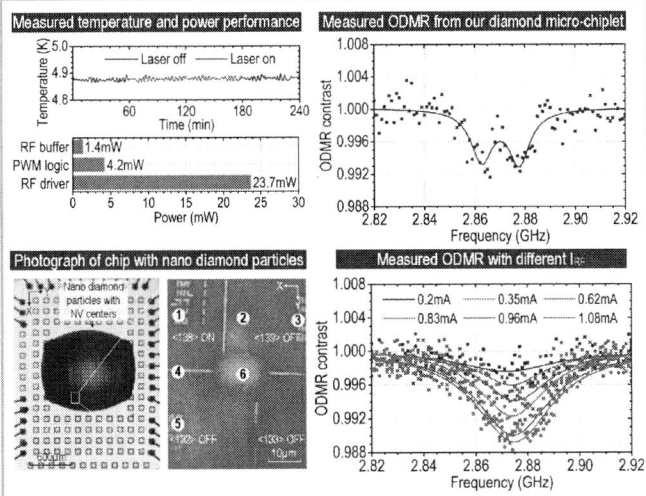

Figure 22.1.5: Measured performance of the chip and micro-chiplet, and a photograph and ODMR measurement of the chip using NV nano diamond particles.

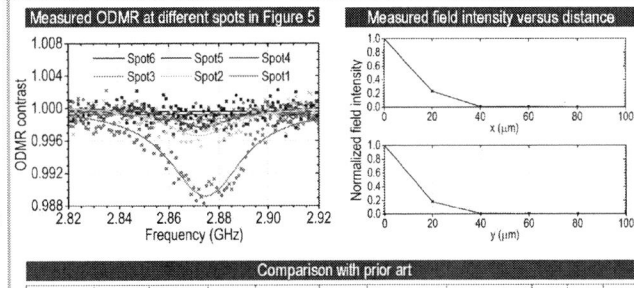

Figure 22.1.6: Magnetic-field crosstalk suppression measured using nano diamond particles, with comparison to prior-art CMOS NV controllers.

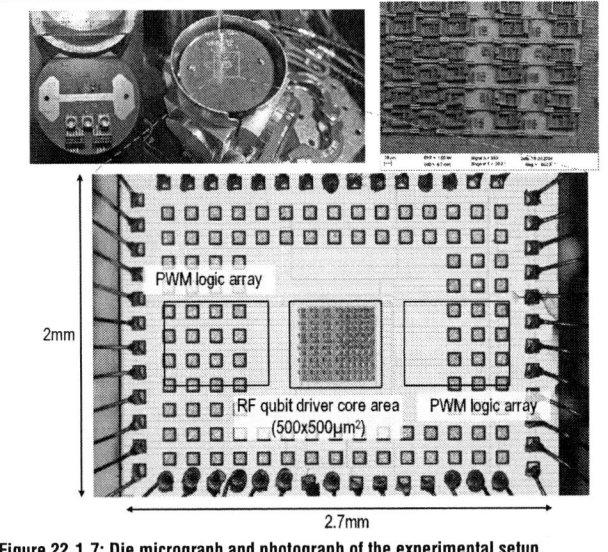

Figure 22.1.7: Die micrograph and photograph of the experimental setup.

• 2026 IEEE International Solid-State Circuits Conference

ISSCC 2026 / SESSION 22 / CIRCUITS IN EXTREME ENVIRONMENTS / 22.2

22.2 A 16-Channel Low-Power Cryo-CMOS Flux Control Pulse Generator ASIC in 14nm FinFET Technology

Kevin Tien[1], David J. Frank[1], John F. Bulzacchelli[1], Pat Rosno[2], Daniel Moertl[2], John Timmerwilke[1], Ari Noori[1], Devin Underwood[1], Ken Inoue[1], Subhajit Ray[1], Daniel Ramirez[2], Timothy J. Schmerbeck[2], Bryce Snell[2], Jeremy Ekman[2], Ryan Black[2], Mark Yeck[1], Tom Haselhorst[2], Emma Erickson[2], Kevin Demsky[2], Christian W. Baks[1], Jonathan Kaus[2], Andrea Ruffino[3], Pier Andrea Francese[3], Andy Davies[2], Sudipto Chakraborty[1], Scott Lekuch[1], Brian P. Gaucher[1], Bodhisatwa Sadhu[1], Scott M. Willenborg[2], Daniel J. Friedman[1]

[1]IBM T. J. Watson Research Center, Yorktown Heights, NY, [2]IBM Systems, Rochester, MN, [3]IBM Zurich Research Laboratory, Rueschlikon, Switzerland

Abstract

This work presents a 16-channel flux pulse generator ASIC for qubit control in 14nm FinFET technology consuming < 7mW per channel. Multiple 10b sub-DACs operating at different rates (up to 1GS/s) are used to realize a pulse generator with 14b level resolution and IIR precompensation. A custom SERDES designed for < 1mW power dissipation enables off-chip 2Gb/s (up) and 250Mb/s (down) communications. Multi-qubit control is demonstrated, and measured error-per-2-qubit-gate is as low as 0.0017.

Recent successful demonstrations of quantum processing units built on superconducting (SC) qubits have required two types of control signals: RF pulses for microwave-based state manipulation, and broadband (~DC-to-250MHz) flux control pulses (FCPs) for magnetic flux-based control [1]. Scaling the qubit control system to support the $O(10^6)$ physical qubits anticipated for full quantum error correction (QEC) [2], however, presents a major engineering challenge. While most prior work has focused on scaling RF pulse generation [3-9], FCP generation poses an additional scaling bottleneck, as recent quantum processing unit (QPU) architecture proposals require as many as 3-to-4 FCP generators per qubit [2,10].

FCPs, implemented as precisely shaped current pulses, generate magnetic flux to shift the energy eigenstate(s) of individual or coupled qubits, resulting in a frequency shift or phase rotation in the qubit state(s). Both room temperature (RT) [10-12] and cryogenic [3] approaches for FCP generation have been successfully demonstrated. Cryogenic implementations, however, offer key system scaling advantages such as reducing cryostat input wire count and reducing critical signal path length, though at the cost of tighter power constraints. Early demonstrations of cryo-CMOS-based broadband pulse generation, while promising, have suffered from limited configurability [3] or have been developed for non-SC QPUs [6]. A cryo-CMOS FCP generator that can be used as a building block to realize a QEC-capable quantum computer must meet superconducting-QPU-driven performance requirements and also support communication and coordination with the system stack, yet must also operate within the power constraints set by the cryostat cooling capacity. Noise/resolution requirements are set by limiting flux pulse height uncertainty at its peak to 100ppm, consistent with supporting gate error rates ≤10^{-5} [3]. As this uncertainty combines resolution and noise in quadrature, a specification of 14b resolution (+/- ~30ppm) with a 95ppm RMS noise budget meets the objective. Power targets can be set by considering long-term projected cryogenic cooling power of cryo-plant architectures using pumped liquid helium for systems with $O(10^6)$ qubits [3,13] and allocating half of a hypothetical 50mW/qubit budget to FCP generation, corresponding to an ~8mW/FCP generator target for the 3 FCP generators/qubit case. The cryo-CMOS FCP generator must also support integration within an end-to-end system stack without exceeding this power target.

This work presents the design and qubit control measurements of a first-of-a-kind 16-channel flux control array (FCA) chip built in 14nm FinFET technology which meets the above requirements. The FCA architecture is shown in Fig. 22.2.1, and an annotated die photo of the 30.4mm² test chip in Fig. 22.2.7. The FCA delivers precision flux control signals to the qubit plane; the signal generated by each channel is governed by dedicated digital circuitry and coordination with the higher-level system stack is mediated by global communications infrastructure. The FCA is comprised of four quadrants of four channels each, arrayed around a digital spine. Each channel contains a microprocessor and a multi-part current-steering DAC. Each quadrant also contains a block of circuitry for local cryogenic reference generation [14]/distribution and local test management. A chip-level support block provides additional redundant references and hierarchical test management. For off-chip communications, the chip includes a low-speed serial interface for bring-up and a custom asymmetric high-speed SERDES link for integration into a general-purpose qubit control software stack.

The per-channel microprocessors (Fig. 22.2.2) enable semi-autonomous sequence generation leveraging on-chip memories, increasing FCA flexibility and reducing off-chip communications needs. They operate at 62.5MHz, with waveform data serialized up to DAC sample rates. The ISA (32-bit instructions) comprises 40 general-purpose instructions (including branching and fixed-point math facilities) and 2 special instructions for playing waveforms, including constant waveforms. The special instructions are queued in a FIFO to support delivering sequences of waveforms without constraining waveform duration due to the processor's low frequency.

As 100ppm generator accuracy is only needed at the FCP peak [1,3], achieving 14b resolution at the full Nyquist-Shannon sampling rate is unnecessary. Instead, we enable significant power savings by tailoring our solution to the FCP requirements and distributing key elements of the composite DAC functionality across three lower-resolution sub-DACs operating at different rates, as shown in Fig. 22.2.2. Each is identically implemented using 10-bit differential current steering architectures similar to [5] (6b thermometer, 4b binary). A 1GS/s main DAC generates the main FCP shape, and static offset adjustment is provided by an offset DAC, with $LSB_{main}=LSB_{offset}$. A 250MS/s fine DAC sets the fine 14b resolution by providing an $LSB_{fine,effective}=LSB_{main}/16$, implemented by using 1/4 the main DAC reference current and attenuating the resultant output by an additional factor of 4 before combination with the main and offset DAC currents to form the aggregate DAC output.

Due to the distinct sampling rates, the aggregate DAC has 15b static resolution and 14b resolution at 250MS/s. This approach is sufficient for peak control, and the aggregate DAC consumes only 3.38mW in hardware. To correct for dispersive effects in the channel response $H(s)$ between the FCA and qubit plane, the data stream feeding the fine DAC also includes predistortion information generated by an IIR filter realizing an approximate channel inversion $1/(1+\tilde{H}(z))$, where $\tilde{H}(z)$ is derived by impulse invariance from a measured $H(s)$ (Fig. 22.2.3). In practice, the observed time constants characterizing $H(s)$ can span up to 6 orders of magnitude, so the IIR filter includes configurable downsampling on a per-model-component basis, allowing for 1) power optimization through appropriate sampling rate choice, and 2) hardware savings through precision scaling, since the impulse invariance conversion scales the time constants by the sampling rate.

Even with the use of per-channel microprocessors and local on-chip memories, high-speed SERDES circuits (Fig. 22.2.4) are needed for real-time management of the 16-channel FCA. QEC systems may need to send 1-to-2 instructions per QEC cycle per channel to trigger program starts, requiring a downlink (DL) minimum data rate of 64 bits/µs/channel ≈ 1Gb/s to the chip; to support other functions and to provide data bandwidth margin, a 2Gb/s downlink data rate has been chosen. Even state-of-the-art medium/long-reach < 5pJ/bit SERDES approaches do not meet power efficiency targets for the cryo-CMOS chip. A key insight for our application is that these links can be made asymmetric between RT and the cryo-CMOS ASIC by shifting core functionality such as clock-to-data alignment and channel equalization to RT. Transmission line terminations are only placed at the RT ends of the links; the power dissipation of a matched resistive load is eliminated by making the input resistors (Rin) of the DL RX high impedance (5kΩ). These input resistors are biased via a voltage DAC (VDAC) to set the input common-mode voltage (VCM). DL equalization is primarily provided by a 3-tap FFE in a commercial FPGA I/O. The DL RX uses a minimal 1-tap DFE with a half-rate speculative architecture. To save power, the DFE employs decision-gated clocking [15], activating only the slicer in the speculative path selected by the previous bit decision. The slicers are differential dynamic comparators with configurable thresholds for offset compensation and H1 tap adaptation. Simulated DL RX power is < 1mW at 2Gb/s.

The uplink (UL) TX cannot achieve similarly low power levels with CML or SST drivers, as multiple mA of current need to be delivered to a $Z_0=50\Omega$ cable. This work introduces a custom low-power differential UL TX powered from a RT current source. The output driver is an NMOS switch pair with < 10Ω individual on-resistances. If the UL switch is open, the applied current (2-8mA) flows through the termination resistor R_{term}, generating a high voltage, and if the UL switch is closed, the current is shunted to ground through the UL cable, lowering the voltage across R_{term}. Since $R_{term}=Z_0$, reflections are absorbed at the RT end of the link. TX power dissipation is limited to switch conduction losses and the dynamic switch driver power; the simulated UL TX consumes < 50µW at 250Mb/s.

To enable qubit test, the FCA chip was flip-chip bonded to a 225 mm² LGA laminate and socket-mounted on a PCB attached to a cold stage (5-to-10K, load dependent) of a dilution refrigerator (DF). The packaged chip is cooled through socket pins and a piston that thermally sinks the chip to the cold stage through a copper strap. FCP generator outputs are routed via flex cabling to a QPU payload in the 10mK stage of the DF, with no additional interposing circuitry. Flux control enables controlled-Z (CZ) gates between fixed-frequency data qubits, while RT electronics provides single-qubit control, readout, digitization, and qubit state discrimination.

During qubit control cycles, the FCA chip consumes ~ 7mW/channel, of which the microprocessor consumes 3.6mW. The active chip temperature measured using on-chip diode sensors is ~12K. The payload's qubits were pre-characterized using state-of-the-art RT electronics, with measured values of T1 ≈ 200µs and T2 ≈ 140µs. All 16 channels of the FCA were successfully validated through standalone measurements. Randomized benchmarking (RB) [16] of 2-qubit CZ gates (64ns duration) was used to evaluate flux pulse fidelity over a 4-qubit/3-coupler QPU slice; the resulting error per two-qubit gate is as low as 0.0017 (Fig. 22.2.5).

Cryogenic validation of the precompensation filter was done with before/after step response comparison (Fig. 22.2.5), as measured *in situ* with a qubit [11]. Standalone noise measurements in a distinct cryostat (Fig. 22.2.2) yield integrated noise (10µHz to 20MHz) at or below 68 ppm with no offset active. Cryogenic validation of the SERDES yields a measured bathtub curve of the downlink with 2Gb/s PRBS31 data that shows a horizontal eye opening of 63% at a BER=10⁻⁹, and the measured uplink eye diagram at the output of the RT amplifier is observed to be wide open at a data rate of 250Mb/s (Fig. 22.2.4); the SERDES was then used in all qubit experiments.

Figure 22.2.6 provides a comparison to prior work. This 16-channel flux control chip has been successfully demonstrated with a state-of-the-art QPU requiring flux control, in a full quantum computing context where the cryo-CMOS chip functions as part of a complex system with heterogeneous cryogenic and RT control, all while consuming only 7mW/channel during active operation. This reported 16-channel FCA represents a major step forward in proving the suitability of cryo-CMOS electronics for superconducting-QPU-based quantum computers.

Acknowledgement:
The authors wish to acknowledge S. Frei, M. B. Rothwell, S. Dhawan, S. Trcka, N. Young, B. Hinrichsen, T. Timpane, A. Carter, and M. Boraas.

References:
[1] J. Stehlik *et al.,* "Tunable Coupling Architecture for Fixed-Frequency Transmon Superconducting Qubits." *Physical review letters* vol. 127,8 (2021): 080505. https://doi.org/10.1103/PhysRevLett.127.080505
[2] S. Bravyi *et al.,* "High-threshold and low-overhead fault-tolerant quantum memory," *Nature* 627, 778–782 (2024). https://doi.org/10.1038/s41586-024-07107-7
[3] J. Yoo *et al.,* "Design and Characterization of a <4-mW/Qubit 28-nm Cryo-CMOS Integrated Circuit for Full Control of a Superconducting Quantum Processor Unit Cell," *IEEE Journal of Solid-State Circuits*, vol. 58, no. 11, pp. 3044-3059, Nov. 2023. http://doi.org/10.1109/JSSC.2023.3309317
[4] D. L. Underwood *et al.,* "Using Cryogenic CMOS Control Electronics to Enable a Two-Qubit Cross-Resonance Gate," *PRX Quantum*, vol. 5, 2024. https://doi.org/10.1103/PRXQuantum.5.010326
[5] S. Chakraborty *et al.,* "A Cryo-CMOS Low-Power Semi-Autonomous Transmon Qubit State Controller in 14-nm FinFET Technology," *IEEE Journal of Solid-State Circuits*, vol. 57, no. 11, pp. 3258-3273, Nov. 2022. http://doi.org/10.1109/JSSC.2022.3201775
[6] J. Park *et al.,* "A Fully Integrated Cryo-CMOS SoC for State Manipulation, Readout, and High-Speed Gate Pulsing of Spin Qubits," *IEEE Journal of Solid-State Circuits*, vol. 56, no. 11, pp. 3289-3306, Nov. 2021. http://doi.org/10.1109/JSSC.2021.3115988

[7] J. P. G. Van Dijk *et al.,* "A Scalable Cryo-CMOS Controller for the Wideband Frequency-Multiplexed Control of Spin Qubits and Transmons," *IEEE Journal of Solid-State Circuits*, vol. 55, no. 11, pp. 2930-2946, Nov. 2020. http://doi.org/10.1109/JSSC.2020.3024678
[8] K. Kang *et al.,* "A Cryo-CMOS Controller IC With Fully Integrated Frequency Generators for Superconducting Qubits," *IEEE International Solid-State Circuits Conference (ISSCC)*, Feb. 2022, pp. 362-364. http://doi.org/10.1109/ISSCC42614.2022.9731574
[9] B. Sadhu *et al.,* "Cryogenic CMOS Circuits for Future Scaled Quantum Computing Systems: Challenges and Solutions," *IEEE Custom Integrated Circuits Conference (CICC)*, 2025, pp. 1-3. http://doi.org/10.1109/CICC63670.2025.10982971
[10] F. Arute *et al.,* "Quantum supremacy using a programmable superconducting processor," *Nature* 574, 505–510 (2019). https://doi.org/10.1038/s41586-019-1666-5
[11] M. A. Rol *et al.,* "Time-domain characterization and correction of on-chip distortion of control pulses in a quantum processor," *Appl. Phys. Lett.* 3 February 2020; 116 (5): 054001. https://doi.org/10.1063/1.5133894
[12] I.N. Moskalenko *et al.,* "High fidelity two-qubit gates on fluxoniums using a tunable coupler," *npj Quantum Inf* 8, 130 (2022). https://doi.org/10.1038/s41534-022-00644-x
[13] M. A. Green, "The Cost of Helium Refrigerators and Coolers for Superconducting Devices as a Function of Cooling at 4 K," *Advances in Cryogenic Engineering*, vol. 985, 2008. https://doi.org/10.1063/1.2908683
[14] S. Ray *et al.,* "A 5.6-100K, 128ppm/K Cryo-CMOS Current Reference," *Symposium on VLSI Technology and Circuits*, 2025, pp. 1-3. http://doi.org/10.23919/VLSITechnologyandCir65189.2025.11075183
[15] E. Sacco *et al.,* "A 5Gb/s 7.1fJ/b/mm 8× multi-drop on-chip 10mm data link in 14nm FinFET CMOS SOI at 0.5V," *Symposium on VLSI Circuits*, 2017, pp. C54-C55. http://doi.org/10.23919/VLSIC.2017.8008545
[16] J. Emerson *et al.,* "Scalable noise estimation with random unitary operators," *J. Opt. B: Quantum Semiclass.* Opt. 7 S347, 2005. http://doi.org/10.1088/1464-4266/7/10/021

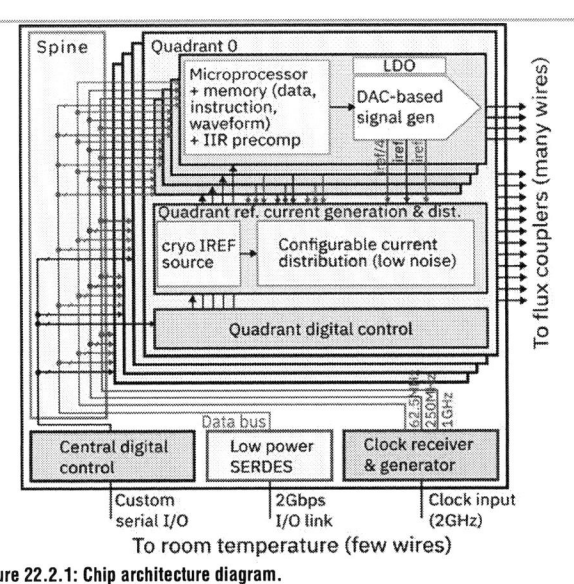

Figure 22.2.1: Chip architecture diagram.

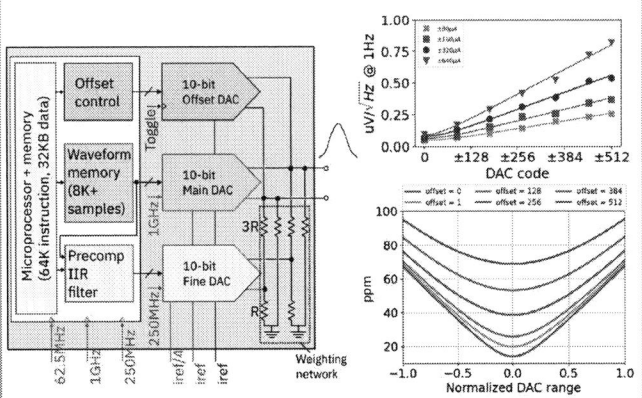

Figure 22.2.2: Channel-level block diagram (left) and DAC characterization measurements in cryogenic setup, including measured differential noise over a range of full-scale settings (top right) and integrated output noise (10µHz-to-20MHz) in ppm of pulse peak over offset DAC settings (bottom right).

ISSCC 2026 / SESSION 22 / CIRCUITS IN EXTREME ENVIRONMENTS / 22.2

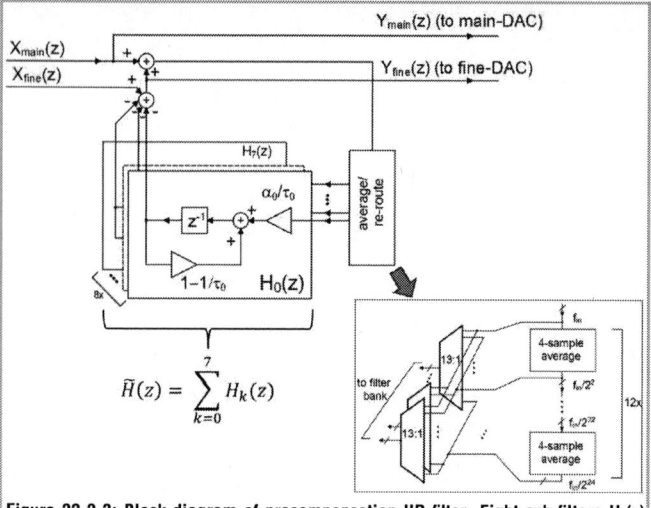

Figure 22.2.3: Block diagram of precompensation IIR filter. Eight sub-filters $H_k(z)$ (characterized by measured amplitude α_k and a time constant τ_k) are placed in parallel to form the channel model.

$$\widetilde{H}(z) = \sum_{k=0}^{7} H_k(z)$$

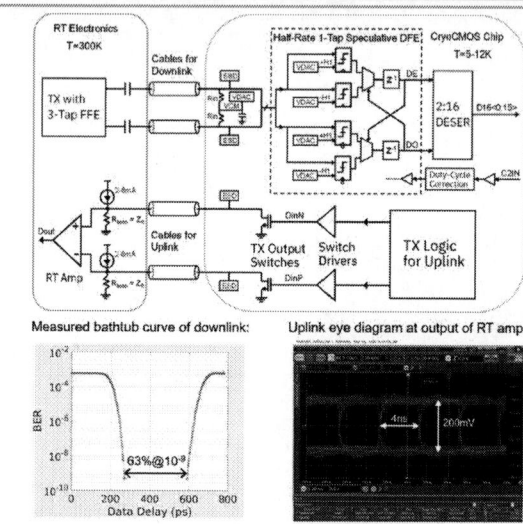

Figure 22.2.4: SERDES architectures (top); measurements (bottom): DL bathtub curve (2Gb/s PRBS31 data) (left), measured UL eye diagram (250Mb/s PRBS31 data) (right).

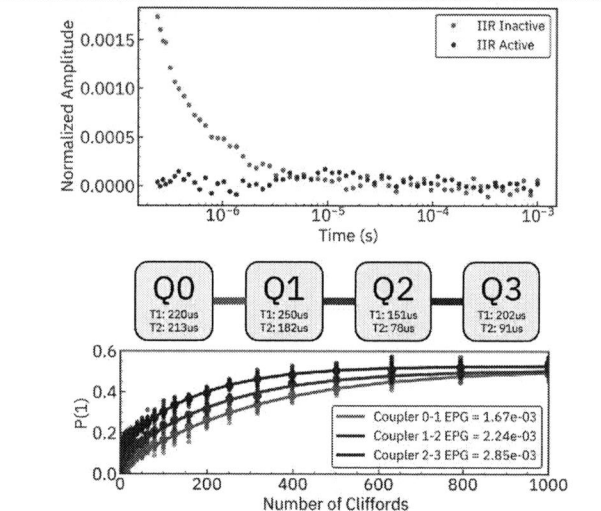

Figure 22.2.5: Falling step response tail (top); two-qubit RB on 3 couplers connecting 4 qubits, illustrating |1⟩ state probability vs. gate sequence length (bottom).

		This work	[3]	[6]*
Technology	Target qubit tech	Transmon	Transmon	Spin
	CMOS tech	14nm FinFET	28nm planar	22nm FinFET
Architecture	Channels	16 (flux)	3 (flux)	22 (gate driver)
	ISA complexity	Full, general-purpose + dedicated instructions	Limited, only instructions for waveform construction	Limited, only instructions for waveform construction
	On-chip digital capability	Per-channel µprocessor	Local sequencer	Global µcontroller, local sequencer
	Resolution	10-bit/14-bit/15-bit	3-bit/14-bit	11-bit
	Sampling rate	1GSps/250MSps/DC	1GSps/DC	not reported
	Interface	SERDES, serial interface	SPI	SPI, GPIO, UART, JTAG
	Output range	-640µA to 640µA	(0 to 500µA)**	-0.4V to 0.4V
Performance	Power/channel	6.93 mW / 3.55 mW (digital) / 3.38 mW (analog)	0.9 mW	2.3 mW
	Qubit test results	0.17% 2Q-gate error (RB, qubit-limited)	0.5% 2Q-gate error (XEB***, qubit-limited)	not tested with qubits
	Integrated RMS noise	< 100 ppm (10µHz-20MHz)	not reported	250 ppm (BW not reported)

*this work does not present a FCP generator directly; the gate driver circuit could be used to generate FCPs. **specification claim; no hardware data provided. ***cross-entropy benchmarking

Figure 22.2.6: Comparison table.

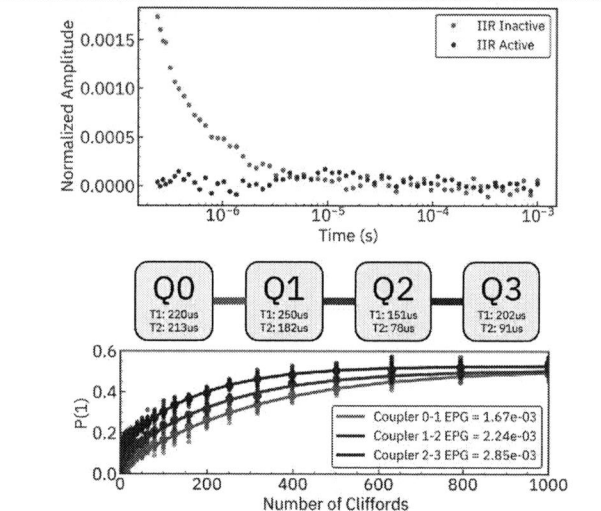

Figure 22.2.7: Annotated die micrograph.

• 2026 IEEE International Solid-State Circuits Conference

ISSCC 2026 / SESSION 22 / CIRCUITS IN EXTREME ENVIRONMENTS / 22.3

22.3 A Multi-Qubit Cryo-CMOS SoC with Polar-Based Electron-Spin and PDM-Based Nuclear-Spin Controllers for Color Centers in Diamond

Niels Fakkel*, Luc Enthoven*, Mohamed Abdelrahman Elbadry, Hendrik Benjamin van Ommen, Margriet van Riggelen, Jiwon Yun, Tim Hugo Taminiau, Fabio Sebastiano**, Masoud Babaie**

TU Delft, Delft, The Netherlands
*Equally Credited Authors (ECAs)

Abstract

Co-integrating a cryo-CMOS SoC with nitrogen-vacancy (NV) centers in diamond enables a scalable quantum platform. This work introduces a combined Class-DE RFDAC and class-D PDM driver for multi-qubit electron- and nuclear-spin control. A switch allows shared coil driving enabling multi-band 2.5-3.2GHz(1.9-2.1MHz), large-current 70mA(38mA), high-Rabi frequency 2.31MHz(1.93kHz) and high-fidelity 99.34(3)%(99.78(2)%) electron(nuclear) quantum logic gates with decoupled coherence times >50ms.

Quantum bits (qubits) based on nitrogen-vacancy (NV) centers in diamond combine high-fidelity electron-spin gates with long-coherence nuclear-spin memories, making them strong candidates for exploring quantum computation and networking. Electron spins enable fast GHz control and photon-mediated entanglement for distributed quantum architectures [1,2], while coupled nuclear spins provide stable memories [3] and localized quantum error correction [4]. NV centers operate near 4K [1], a temperature where cryostats offer significant cooling power. Such qubits can then be directly co-integrated with cryo-CMOS controllers, removing interconnect bottlenecks and enabling scalable quantum processors [5–7].

Figure 22.3.1 shows a quantum processing unit (QPU) composed of many NV-center unit cells, each hosting one electron and several nuclear spins in a static magnetic field B_0. The cells, spaced by 1mm, are interconnected via optical waveguides for readout and entanglement, while cryo-CMOS chips above each cell generate GHz/MHz pulses, control the optics, and read out single-photon detectors.

The B_0 field splits the energy levels of the electron spin, defining its Larmor frequency ($f_{0,e}$~2.5GHz), with slight variations due to the B_0 non-uniformity across the QPU. Hyperfine coupling to local Nitrogen-14 (^{14}N) and Carbon-13 (^{13}C) nuclear spins shifts $f_{0,e}$ by ±2.5MHz, creating multiple resonance peaks. Thus, to achieve high-fidelity single-qubit electron gates, the controller must generate fast (<200ns), broadband (>5MHz), and spectrally flat pulses, requiring ~100mAp of GHz current due to the ~20μm coil-to-electron-spin spacing.

Each unit cell also hosts several randomly positioned ^{13}C nuclear spins with distinct hyperfine couplings to the electron spin and unique Larmor frequencies ($f_{0,c}$). Due to their smaller gyromagnetic ratio, ^{13}C spins are limited to a lower Rabi frequency, which then requires long (>100ms) pulses with large amplitude (>40mAp).

The prior-art cryo-CMOS NV controller [7] demonstrated electron-spin driving, but had key limitations: it relied on a dedicated DC magnetic field regulator and a dedicated coil to align $f_{0,e}$ across units, degrading system scalability due to the extra coil, and lowering qubit coherence due to the regulator noise. Also, its limited phase resolution and amplitude shaping affected gate fidelity. Moreover, it did not demonstrate nuclear spin control of ^{13}C spins [8], essential for realizing a scalable NV-based quantum computer.

The proposed electron and nuclear spin controller (Fig. 22.3.1(top-right)) overcomes those limitations by introducing 1) a digital-intensive intermediate-frequency (IF) generation with single-sideband up-conversion, allowing each cell to operate at its unique Larmor frequency due to B_0 inhomogeneity, eliminating the need for DC field regulation. Setting the LO (f_{LO}) 20MHz above the Larmor frequencies minimizes undesired qubit rotations from LO leakage. 2) An efficient digital architecture for precise amplitude and phase; 3) A class-D Pulse-Density Modulator (PDM) to efficiently generate accurate MHz pulses for nuclear spins, using a DS modulator for quantization noise shaping and ADC feedback to reduce sensitivity to supply and process spread; 4) A switchable output passive network that allows a single coil for both GHz and MHz driving, improving coil-to-qubit coupling and scalability.

For the electron-spin controller (Fig. 22.3.2), the digital polar architecture comprises the RFDAC for amplitude modulation (AM) and a single-sideband upconverter for frequency shifting and phase modulation (PM). In the PM path, a numerically controlled oscillator (NCO) generates the required IF tone [9] and a digital rotational CORDIC [10] converts the truncated triangular phase input into four sinusoidal quadrature streams. Unlike prior art [11–13] that rely on ~1.6kb sinusoidal lookup tables (LUTs) to achieve <0.2° phase accuracy, the CORDIC uses only 8 LUT entries for storing arccot(2^1, 2^2, ..., 2^8) phase steps (88 bits total) and performs rotations with bit shifts and summations. This reduces the area by 10× and the simulated power consumption by 6×, while also enabling direct phase word (PW) inputs as required for instant phase updates in qubit gates.

Four 6-bit resistive unary-weighted DACs then generate the quadrature IF currents, which are up-converted by single-sideband passive mixers to f_{LO}-f_{IF}. Since the DAC output common-mode voltage is approximately ~V_{DD}/2, a level shifter raises the LO voltages to 0.5–1.5V_{DD} to minimize mixer size and on-resistance. The mixers' output currents are

converted to voltage by a transimpedance amplifier (TIA), whose low input impedance ensures current-mode DAC operation and suppresses voltage swings at the mixer terminals, improving linearity. Five inverter stages convert the TIA's sinusoidal output into frequency and phase-modulated square wave signals for RFDAC.

The RFDAC consists of 255 unary-weighted cells, each individually addressable to modulate the driver on-resistance and enable arbitrary envelope shaping. As the RFDAC requires 25% duty-cycle input clocks to operate in class-DE switching mode, each cell includes four (N)AND gates to convert the 50% duty-cycle quadrature inputs into the 25% pulses. Its matching network includes: (i) a series coil (L_{coil}) magnetically coupled to the electron spin, (ii) shunt capacitors (C_X) absorbing the parasitics of both GHz and MHz controllers for zero-voltage/current switching, and (iii) a series capacitor (C_S) incorporating the parasitics from the matching network switch transistors. Due to the voltage division between the coil parasitic resistance and the AM-dependent on-resistance class-DE driver, the amplitude and phase of the coil current do not scale linearly with the AM codeword. To correct this, two on-chip predistortion LUTs are used: one in the AM path and one before the PM data enters the CORDIC. Finally, since the AM and PM data experience different paths, a controllable delay aligns them before being combined in the RFDAC.

The nuclear-spin controller (Fig. 22.3.3) adopts a class-D output stage for high linearity and power efficiency while driving large currents. For the signal generation, 10 different 24-bit NCOs track the frequency (f_{nco}) and phase with an accuracy of 3Hz and 2e-5°, respectively, to address 10 carbon spins in the 200kHz bandwidth and enable driving of 2 tones simultaneously. The sawtooth NCO output is folded into a triangular waveform to suppress the even harmonics of f_{nco}, while avoiding another CORDIC or LUT. The amplitude of the triangular waveform is modulated using SRAM data, truncated to 10bit, and fed into a digital 1bit DS modulator that shapes the quantization noise for optimal resolution. The DS modulator allows 1) digital feedback from the 1bit quantizer output for minimum noise, power dissipation, and high linearity, and 2) analog feedback, which uses a SAR ADC sensing L+, to improve the robustness to supply variations and process spread, e.g., in the on-resistance of the output-bridge devices. Operating at 50MHz, the 2nd-order feed-forward loop filter uses zero compensation to counteract the analog-feedback path delay and a programmable resonator coefficient g to optimally place the resonator notch for minimum in-band quantization noise. The class-D output stage is implemented in a half-bridge structure with large devices to achieve a low on-resistance R_{ON}, allowing the use of low supplies (±V_{sup}) to reduce the power consumption for the same coil current, while also filtering high-frequency current noise thanks to the large L_{coil}/R_{ON} time constant. However, as excessively large devices would introduce more parasitic capacitance that cannot be completely absorbed in the matching network of the GHz driver, this design limits R_{ON} to 2Ω to enable a 40mAp output current with ±100mV supplies.

Figure 22.3.3 also shows the schematic of the matching-network switch, with series thick-oxide transistors M1-M2 designed to withstand large voltage swings when off during GHz operation. In MHz mode, EN is high, turning on M1 and M2 to form a low-impedance path to ground. In GHz mode, the transistors act as back-to-back diodes, limiting M2's AC gate voltage to keep it off and ensuring the gate-oxide voltages of both transistors remain within the foundry-specified limits.

The chip (Fig. 22.3.7) was fabricated in 40nm CMOS and characterized at room (RT) and cryogenic temperatures (CT). An alternate GHz driver with on-chip driving coil (L_{coil}) was tested in a cryogenic probe station using an on-chip pick-up coil with results shown in Fig. 22.3.4(left). At 2.5GHz(3.0GHz), it achieves >80(70)mAp coil current and >270(600)% output-stage current efficiency, while drawing 30(15)mA from a 1.1(0.85)V supply at RT (CT), and showing SNR= 53(52)dB over 50MHz bandwidth. The controller can generate arbitrary waveforms with 3ns-3μs pulse length, and <10kHz and <0.5° frequency and phase accuracy across 25MHz f_{IF} bandwidth. The MHz driver, measured in a liquid-helium bath with an on-chip buffer (Fig. 22.3.4(right)), achieves 57(58)dB SNDR in single-tone tests and 47(51)dB SFDR in two-tone tests over 1.9-to-2.1MHz.

To demonstrate the quantum controller's performance (Fig. 22.3.5), the cryo-CMOS was bonded to an NV-center sample with a gold-patterned stripline (L_{coil}=1.1nH) and placed in a cryostat at ~6K. RT electronics and optics were used for (^{14}N, ^{13}C, electron) spin initialization and readout. A permanent magnet (B_0=1980 Gauss) induces $f_{0,e}$=2.49GHz for electron spin and $f_{0,C0}$=2.46, $f_{0,C1}$=2.03MHz for two ^{13}C spins. The cryo-CMOS driver achieves a maximum Rabi frequency $f_{R,e}$=2.31MHz for the electron spin and $f_{R,C0}$=1.93kHz, $f_{R,C1}$=1.16kHz for the nuclear spins. The qubit characteristic chevron patterns were demonstrated for the electron spin by sweeping the f_{IF} frequency of the GHz NCO over the range f_{LO}-20±5MHz and for the nuclear spin by sweeping the MHz NCO over the range 2.025±0.1MHz. Coherence time T_2 could be extended from 0.8ms to 50ms by increasing N_{Dec} decoupling pulses in an echo experiment. For the electron {nuclear} spin, an average gate fidelity of 99.34(3)% {99.78(2)%} was found by gate set tomography (GST) with depth L=128 {L=64} [14].

The cryo-CMOS consumes an average of 45mW, 17mW and 12mW for the GHz and MHz drivers and in idle mode, respectively. Compared to other works (Fig. 22.3.6), this paper shows the first cryo-CMOS high-fidelity gate control of both nuclear and electron spins, proving co-integration of color centers with cryo-CMOS as a scalable candidate for quantum computing.

Acknowledgement:
We would like to thank Fujitsu Limited for funding; Z. Y. Chang from TU Delft; N. Alberts, T. Hiep, J. Mensingh; O. Benninghof from Qutech; R. Vollmer from TNO.

References:
[1] S. Pezzagna and J. Meijer, "Quantum computer based on color centers in diamond," *Applied Physics Reviews*, vol. 8, no. 1, p. 011308, Feb. 2021. https://doi.org/10.1063/5.0007444
[2] A. J. Stolk et al., "Metropolitan-scale heralded entanglement of solid-state qubits," *Science Advances*, vol. 10, no. 25, eadl4863, Jun. 2024. https://doi.org/10.1126/sciadv.adl4863
[3] C. E. Bradley et al., "A ten-qubit solid-state spin register with quantum memory up to one minute," *Physical Review X*, vol. 9, no. 3, 031045, Sep. 2019. https://doi.org/10.1103/PhysRevX.9.031045
[4] M. H. Abobeih et al., "Fault-tolerant operation of a logical qubit in a diamond quantum processor," *Nature*, vol. 606, pp. 884–889, Jun. 2022. https://doi.org/10.1038/s41586-022-04884-5
[5] R. Ishihara et al., "3D integration technology for quantum computer based on diamond spin qubits," *Proc. IEEE Int. Electron Devices Meeting (IEDM)*, Dec. 2021, pp. 1–4. https://doi.org/10.1109/IEDM19574.2021.9720552
[6] L. Enthoven, M. Babaie, and F. Sebastiano, "Optimizing the electrical interface for large-scale color-center quantum processors," *IEEE Transactions on Quantum Engineering*, vol. 5, pp. 1–17, 2024. https://doi.org/10.1109/TQE.2024.3416836
[7] N. Fakkel et al., "A cryo-CMOS controller with class-DE driver and DC magnetic-field tuning for quantum computers based on color centers in diamond," *IEEE Journal of Solid-State Circuits*, vol. 59, no. 11, pp. 3627–3643, Nov. 2024. https://doi.org/10.1109/JSSC.2024.3459392
[8] T. H. Taminiau et al., "Detection and control of individual nuclear spins using a weakly coupled electron spin," *Physical Review Letters*, vol. 109, 137602, Sep. 2012. https://doi.org/10.1103/PhysRevLett.109.137602

[9] Analog Devices, Inc., "A technical tutorial on digital signal synthesis," Application Note, 1999, pp. 1–122. https://www.analog.com/media/en/training-seminars/tutorials/450968421DDS_Tutorial_rev12-2-99.pdf
[10] J. E. Volder, "The CORDIC trigonometric computing technique," *IRE Transactions on Electronic Computers*, vol. EC-8, no. 3, pp. 330–334, Sep. 1959. https://doi.org/10.1109/TEC.1959.5222693
[11] D. J. Frank et al., "A cryo-CMOS low-power semi-autonomous qubit state controller in 14 nm FinFET technology," *IEEE Int. Solid-State Circuits Conf. (ISSCC)*, Feb. 2022, pp. 360–361. https://doi.org/10.1109/ISSCC42614.2022.9731538
[12] J. S. Park et al., "A fully integrated cryo-CMOS SoC for qubit control in quantum computers capable of state manipulation, readout and high-speed gate pulsing of spin qubits in Intel 22 nm FFL FinFET technology," *IEEE Int. Solid-State Circuits Conf. (ISSCC)*, Feb. 2021, pp. 208–209. https://doi.org/10.1109/ISSCC42613.2021.9365762
[13] B. Patra et al., "A scalable cryo-CMOS 2-to-20 GHz digitally-intensive controller for 4 × 32 frequency-multiplexed spin qubits/transmons in 22 nm FinFET technology for quantum computers," *IEEE Int. Solid-State Circuits Conf. (ISSCC) Dig. Tech. Papers*, Feb. 2020, pp. 304–306. https://doi.org/10.1109/ISSCC19947.2020.9063109
[14] H. P. Bartling et al., "Universal high-fidelity quantum gates for spin qubits in diamond," *Physical Review Applied*, vol. 23, 034052, Mar. 2025. https://doi.org/10.1103/PhysRevApplied.23.034052
[15] J. Yoo et al., "A 28-nm bulk-CMOS IC for full control of a superconducting quantum processor unit-cell," *IEEE Int. Solid-State Circuits Conf. (ISSCC)*, Feb. 2023, pp. 506–508. https://doi.org/10.1109/ISSCC42615.2023.10067292
[16] Y. Guo et al., "A cryo-CMOS quantum computing unit interface chipset in 28 nm bulk CMOS with phase-detection based readout and phase-shifter based pulse generation," *IEEE Int. Solid-State Circuits Conf. (ISSCC)*, Feb. 2024, pp. 476–478. https://doi.org/10.1109/ISSCC49657.2024.10454392
[17] L. Guevel, C. Wang, and J. C. Bardin, "A 22 nm FD-SOI <1.2 mW/active-qubit AWG-free cryo-CMOS controller for fluxonium qubits," *IEEE Int. Solid-State Circuits Conf. (ISSCC)*, Feb. 2024, pp. 1–3. https://doi.org/10.1109/ISSCC49657.2024.10454522

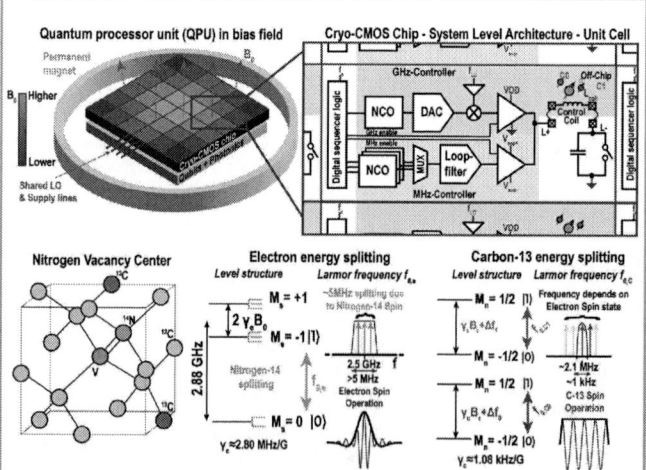

Figure 22.3.1: Quantum processor unit (QPU) based on nitrogen-vacancy (NV) centers and integrated cryo-CMOS SoC (top). The NV-center energy splitting and corresponding Larmor frequencies for electron and nuclear driving (bottom).

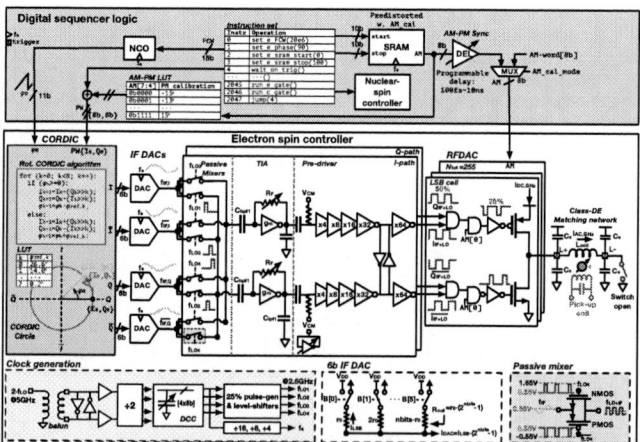

Figure 22.3.2: Electron spin controller comprising digital sequencer controller, SRAM, rotational CORDIC with f_{IF} DACs, quadrature mixers and pre-drivers for up-conversion to f_{LO}-f_{IF}, and Class-DE RFDAC output stage to drive the off-chip coil.

ISSCC 2026 / SESSION 22 / CIRCUITS IN EXTREME ENVIRONMENTS / 22.3

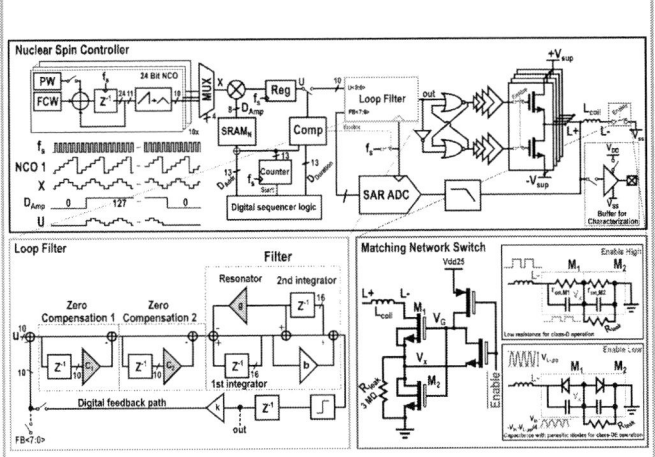

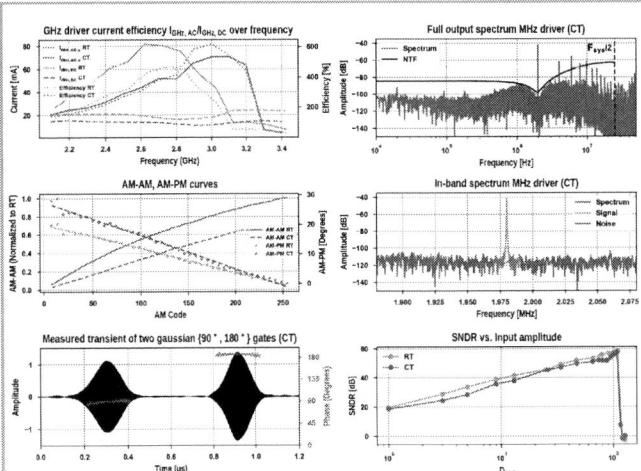

Figure 22.3.3: Nuclear spin controller comprising a class-D output stage with low supplies (±V_{sup}) and 10 NCOs for phase tracking (top); loop filter and thick oxide matching network switch using transistors to share L_{coil} (bottom).

Figure 22.3.4: Measured GHz-controller driver current efficiency, AM-AM and AM-PM, and phase control pulses(left); MHz-driver out-of-band PSD, in-band spectrum and SNDR using the analog feedback (right).

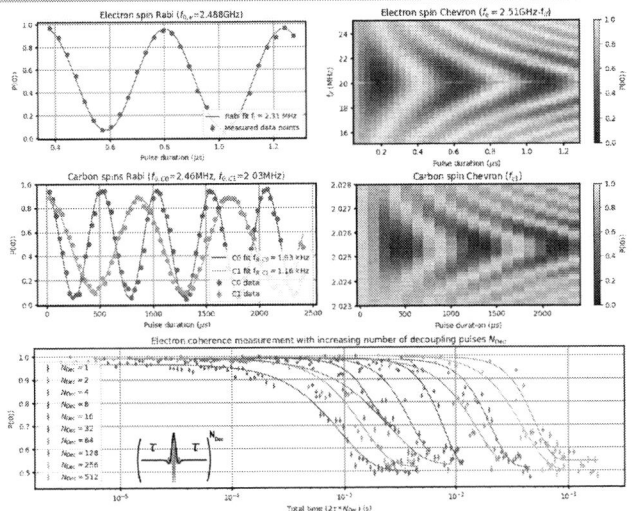

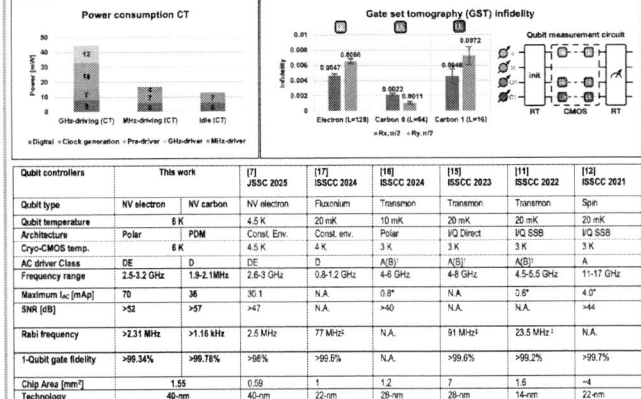

Figure 22.3.5: Rabi oscillation and chevron pattern with NCO sweep for electron spin (top), and for nuclear spins (middle); Coherence measurement (bottom).

Figure 22.3.6: Summary of the electrical and quantum-control performance and benchmark with prior art.

Qubit controllers	This work		[7] JSSC 2025	[17] ISSCC 2024	[16] ISSCC 2024	[15] ISSCC 2023	[11] ISSCC 2022	[12] ISSCC 2021
Qubit type	NV electron	NV carbon	NV electron	Fluxonium	Transmon	Transmon	Transmon	Spin
Qubit temperature	6 K		4.5 K	20 mK	10 mK	20 mK	20 mK	20 mK
Architecture	Polar	PDM	Const. Env.	Const. env.	Polar	I/Q Direct	I/Q SSB	I/Q SSB
Cryo-CMOS temp.	6 K		4.5 K	3 K	3 K	3 K	3 K	3 K
AC driver Class	DE	D	DE	D	A(B)†	A(B)†	A(B)†	A
Frequency range	2.5-3.2 GHz	1.9-2.1 MHz	2.6-3 GHz	0.8-1.2 GHz	4-6 GHz	4-8 GHz	4.5-5.5 GHz	11-17 GHz
Maximum I_{AC} [mAp]	70	36	30.1	N.A.	0.8*	N.A.	0.6*	4.0*
SNR [dB]	>52	>57	>47	N.A.	>40	N.A.	N.A.	>44
Rabi frequency	>2.31 MHz	>1.16 kHz	2.5 MHz	77 MHz‡	N.A.	91 MHz‡	23.5 MHz‡	N.A.
1-Qubit gate fidelity	>99.34%	>99.78%	>98%	>99.6%	N.A.	>99.6%	>99.2%	>99.7%
Chip Area [mm²]	1.55		0.59	4	1.2	7	1.5	≈4
Technology	40-nm		40-nm	22-nm	28-nm	28-nm	14-nm	22-nm
Power per qubit under active control	45 mW	17 mW	16.8 mW	1.2 mW	13.7 mW	<4 mW	23 mW	90 mW≈

† Derived from schematic; * Assuming P_{out} with 50Ω; ‡ Based on reported gate duration; ≈Estimated

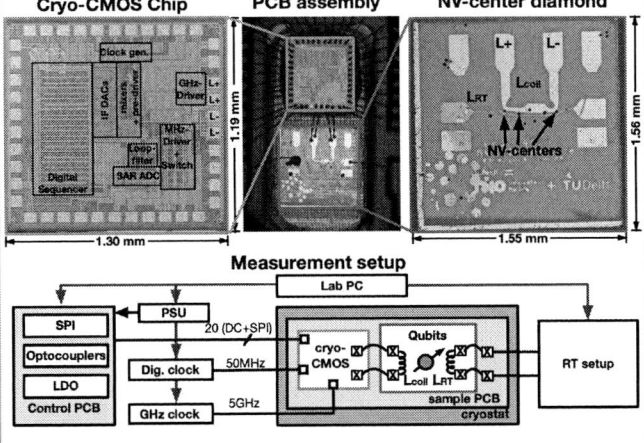

Figure 22.3.7: Qubit measurement assembly of the cryo-CMOS chip and NV-center sample. Two striplines are patterned close to the NVs; L_{coil} is bonded to the cryo-CMOS; L_{RT} is used with the RT setup as experimental reference.

- 2026 IEEE International Solid-State Circuits Conference

ISSCC 2026 / SESSION 22 / CIRCUITS IN EXTREME ENVIRONMENTS / 22.4

22.4 A Radiation-Hardened Self-Healing CMOS Imager with Online Pixel/Logic Annealing and Tile-Adaptive Compression for Space Applications

Quan Cheng*[1,2], Zhengke Yang*[1], Haoyuan Li[2], Qiufeng Li[1], Zhen Kong[1], Gaoqiang Niu[1], Yuan Liang[1], Jiamin Li[1], Jerald Yoo[3], Masanori Hashimoto[2], Longyang Lin[1]

[1]Southern University of Science and Technology, Shenzhen, China, [2]Kyoto University, Kyoto, Japan, [3]Seoul National University, Seoul, Korea
*Equally Credited Authors (ECAs)

Abstract

CMOS imagers in space suffer radiation-induced faults and downlink bandwidth limits. This work presents an 180nm self-healing imager with localized online thermal annealing for both pixels and logic, as well as adaptive ROI-lossless compression for 75.5% data reduction. Measurements show 4.82×/3.97× reductions in pixel/logic leakage and nearly full image recovery after 20kGy X-ray exposure, enabling continuous, robust space imaging.

CMOS imagers are indispensable in space missions for Earth observation, planetary exploration, and autonomous navigation [1-4]. However, long-duration deployment exposes them to harsh radiation, inducing transient and permanent faults in pixel arrays and logic, which manifest as hot pixels, leakage, and timing errors [3-8], ultimately degrading image quality and mission reliability. At the same time, massive onboard image data often exceeds available satellite downlink bandwidth [9], demanding imagers with both radiation tolerance and efficient in-sensor compression (Fig. 22.4.1).

Prior radiation-hardened techniques such as enclosed layout transistors (ELTs) [4–6], wide-bandgap photodiodes [7], and voltage compensation [8] improve radiation tolerance but cannot repair accumulated damage (i.e., hard errors). Redundancy-based self-healing [10] can replace damaged circuits with spares but introduces large area overhead. In contrast, thermal annealing can restore device performance by releasing trapped charges caused by Total Ionizing Dose (TID) and repairing lattice defects induced by Displacement Damage Dose (DDD). A prior space imager [3] employed spike-based architecture with on-chip thermal annealing for pixel recovery. However, this approach interrupts imaging during annealing, does not address degradation in logic circuits, and lacks scalability due to its spike-based processing. Therefore, there is no scalable, continuous-operation space imager that can heal both pixels and logic while also mitigating the downlink bottleneck.

This work presents a radiation-hardened self-healing imager that integrates localized online thermal annealing for both pixels and processing logic, enabling autonomous recovery from radiation-induced degradation. Unlike prior designs, the proposed architecture supports online annealing without interrupting imaging, allowing continuous operation during self-healing. In addition, a tile-based adaptive compression engine applies per-tile variable compression ratios, dynamically allocating bandwidth according to local image complexity. A 180nm test chip demonstrates successful online self-healing for nearly full image quality recovery, with a 4.82× reduction in pixel dark current and a 3.97× reduction in logic leakage after 20kGy X-ray exposure. The proposed compression reduces transmitted data volume by 75.5% without loss of information in regions of interest (ROI), while consuming 26.48mW at 100fps.

Illustrated in Fig. 22.4.2, the proposed imager integrates a 128×128 PWM-based pixel array with localized thermal annealing, column-parallel time-to-digital converters for pixel readout with support for online annealing via pixel masking, scan-assisted faulty-path detection and logic annealing, and a tile-based adaptive image compression engine. During normal imaging, the compression pipeline applies tile-based adaptive compression guided by edge complexity, preserving key scene content while reducing bandwidth. Radiation damage detection is performed intermittently: hot pixels are identified by shuttered readout compared to a threshold, while logic faults are detected by monitoring error flags from Razor flip-flops. Localized thermal annealing is then applied on individual damaged pixels, while faulty logic paths are healed by an internally generated high-frequency clock applied only to the selected path (Fig. 22.4.2, bottom).

Figure 22.4.3 (top) shows the proposed 6T PWM-based self-healing pixel design, where the photodiode (PD) operates in reverse bias for readout or forward bias for localized thermal annealing. During normal operation, the pixel is first reset by turning on M2 ($RSTN=0$) and M1 ($V_{RAMP}=V_{RST}$) for threshold variation cancellation (TVC), similar to [11]. After reset, M2 is turned off and a ramp signal is applied to V_{RAMP}, while the in-pixel comparator (M1 and M3) with gain stage M4 generates a pulse whose width is proportional to light intensity. Unlike [11], which omits the row selection transistor, the proposed design includes M5 to suppress leakage currents from unselected rows on the column lines. Radiation-damaged pixels are detected by temporarily closing the shutter to capture a dark frame; if any pixel value exceeds a predefined threshold, it is flagged as damaged. The healing process is then triggered by switching the corresponding pixel into forward bias through M6 and activating a column-shared annealing driver (1.5-to-2.0V) [3], which initiates localized thermal annealing. This selective activation minimizes power overhead while enabling per-pixel recovery. Continuous imaging with online pixel annealing is achieved through pixel masking (Fig. 22.4.4, left). During annealing, the driver occupies the selected column, making its pixel readouts unavailable. These missing values are interpolated from the average of the adjacent horizontal neighbors. Furthermore, annealing is temporarily suspended when the readout reaches the annealed row, yielding an effective annealing window of 127 out of 128 row periods per readout cycle.

Figure 22.4.3 (bottom) shows the principle of localized thermal annealing in digital logic using scan-assisted faulty-path detection and selection. Razor-based flip-flops [12], which support error detection, are augmented with a triple-modular redundancy (TMR) structure for immediate error correction and an additional latch (L1 in Fig. 22.4.3) to store error flags, enabling asynchronous readout via a scan chain (SC) without interrupting normal logic operation. Within the TMR, intentional delays are inserted in both the clock and data paths, introducing slight temporal skew among the three flip-flops to enhance single event transient (SET) tolerance and enable sensing of delay-induced timing violations by cumulative radiation effects (e.g., TID). A persistent error flag observed on the same path therefore serves as an indicator of a hard error, and the SC is reused to select this faulty path for annealing (e.g., via SC flip-flop F1 in Fig. 22.4.3). When annealing is enabled (HE=1), the selected path (SQ=1) is toggled at high frequency (>1.5GHz) by a digital annealer, inducing localized Joule heating (Fig. 22.4.3, bottom-right). The annealer generates a local clock with a compact ring oscillator (delay cell, inverter, and multiplexer) and maximizes switching activity on the faulty path by inverting the flip-flop input each cycle. To accommodate intermittent logic annealing, a short healing phase is inserted after normal operations (image readout, novelty detection, and image compression), resulting in only a slight frame-rate reduction (~5%). Meanwhile, the Razor-TMR flip-flops, designed with intentional delay skew, sustain normal operation under radiation-induced errors, as any error in one flip-flop can be detected and corrected by the majority voter leveraging the other two.

Figure 22.4.4 (right) shows the proposed tile-based adaptive image compression. Each frame is partitioned into tiles, where edge magnitudes are extracted using a Sobel operator and inter-frame edge differences set compression levels from 0 (lossless) to 15 (highly compressed). These levels control an adaptive blurrer that attenuates high-frequency content in less critical regions, with blur strength N mapped from the tile values. A lightweight pre-quantization then aligns pixels to coarse gray-level grids to reduce symbol diversity. The processed image is passed to a LOCO-I-based predictor and context modeler [13], with configurable near-lossless tolerance, and prediction errors are encoded via a Golomb encoder. This approach preserves key scene details while enabling efficient transmission under tight bandwidth and compute constraints.

Test-chip temperature maps captured by an infrared thermal camera confirm localized thermal annealing in both the pixel array (~167μW/μm² at 1.8V) and digital logic (~13μW/μm² at 1.8V), as shown in Fig. 22.4.5 (top-left). After exposure to 20kGy X-ray irradiation, the dark current in the pixel array and the leakage current in the digital logic increase by 181.13× and 4.13×, respectively, and the captured image quality degrades severely, indicating radiation-induced damage. After four annealing rounds (10 min/round for pixel; 1 hr/round for logic), leakage currents across 8 chips are reduced by 4.82× and 3.97×, respectively, with image quality progressively restored to nearly full recovery (Fig. 22.4.5, middle). The digital logic exhibits no timing errors under X-ray irradiation up to 20kGy, attributed to the relatively low operating frequency (≤80MHz) that provides ample timing margin. Hence, the healing effect in digital logic is primarily evidenced by reduced leakage currents. Shown in Fig. 22.4.5 (right), the adaptive image compression processes 128×128 frames (8-bit, 16 kB) via Sobel edge detection and tile-based object detection on 64 tiles, where tile values determine compression ratios. Detail-critical tiles (e.g., around objects) use low or lossless compression as required, while background tiles are heavily compressed. After LOCO-I prediction, context modeling, and Golomb encoding, the bitstream is reduced to 3.92kB (75.5%) while preserving full detail in the regions of interest.

Comparison with prior radiation-hardened imagers (Fig. 22.4.6) shows that the 180nm test chip (Fig. 22.4.7) uniquely supports both pixel and logic localized online thermal annealing together with ROI-lossless compression at up to 400fps, enabling continuous and robust operation for long-duration space missions. Unlike [3], which employs a spiking-pixel, fully parallel architecture with poor scalability and no logic healing, the proposed PWM-based architecture scales efficiently to large arrays while providing continuous self-healing and compression, retaining full radiation-hardening capability.

ISSCC 2026 / February 17, 2026 / 2:45 PM

Acknowledgement:
This work is partially supported by the National Natural Science Foundation of China under Grant 62274081, Guangdong Projects (Grant 2023QN10X177, Grant 2025B0101180002), and the Grant-in-Aid for Scientific Research (S) from the Japan Society for the Promotion of Science (JSPS) under Grant 24H00073.

References:
[1] A. A. Omar, M. M. Farag and R. A. Alhamad, "Artifical Intelligence: New Paradigm in Deep Space Exploration," *14th International Conference on Developments in eSystems Engineering (DeSE)*, Sharjah, United Arab Emirates, 2021, pp. 438-442. http://doi.org/10.1109/DeSE54285.2021.9719425
[2] I. Kostavelis, L. Nalpantidis, E. Boukas *et al.*, "Spartan: Developing a vision system for future autonomous space exploration robots". *Journal of Field Robotics*, 2014, 31(1): 107-140. http://doi.org/10.1002/rob.21484
[3] Q. Cheng *et al.*, "A Radiation-Hardened Neuromorphic Imager with Self-Healing Spiking Pixels and Unified Spiking Neural Network for Space Robotics," *Symposium on VLSI Technology and Circuits (VLSI Technology and Circuits)*, 2025, pp. 1-3. http://doi.org/10.23919/VLSITechnologyandCir65189.2025.11075180
[4] V. Goiffon *et al.*, "Total Ionizing Dose Effects on a Radiation-Hardened CMOS Image Sensor Demonstrator for ITER Remote Handling," *IEEE Transactions on Nuclear Science*, vol. 65, no. 1, pp. 101-110, Jan. 2018. http://doi.org/10.1109/TNS.2017.2765481
[5] C. Virmontois *et al.*, "Total Ionizing Dose Versus Displacement Damage Dose Induced Dark Current Random Telegraph Signals in CMOS Image Sensors," *IEEE Transactions on Nuclear Science*, vol. 58, no. 6, pp. 3085-3094, Dec. 2011. http://doi.org/10.1109/TNS.2011.2171005
[6] J. Bogaerts, B. Dierickx, G. Meynants and D. Uwaerts, "Total dose and displacement damage effects in a radiation-hardened CMOS APS," *IEEE Transactions on Electron Devices*, vol. 50, no. 1, pp. 84-90, Jan. 2003. http://doi.org/10.1109/TED.2002.807251
[7] M. Tsutsumi, T. Meguro, A. Takeyama, T. Ohshima, Y. Tanaka and S. -I. Kuroki, "Integrated 4H-SiC Photosensors With Active Pixel Sensor-Type Circuits for MGy-Class Radiation Hardened CMOS UV Image Sensor," *IEEE Electron Device Letters*, vol. 44, no. 1, pp. 100-103, Jan. 2023. http://doi.org/10.1109/LED.2022.3226494
[8] B. R. Dean *et al.*, "Back-Biasing Strategy to Mitigate TID-Induced Threshold Voltage Shift in 22nm FDSOI Transistors," *IEEE Transactions on Nuclear Science*. http://doi.org/10.1109/TNS.2025.3613463
[9] O. Kodheli *et al.*, "Satellite Communications in the New Space Era: A Survey and Future Challenges," *IEEE Communications Surveys & Tutorials*, vol. 23, no. 1, pp. 70-109, Firstquarter 2021. http://doi.org/10.1109/COMST.2020.3028247
[10] P. Balasubramanian and D. L. Maskell, "A Self-Healing Redundancy Scheme for Mission/Safety-Critical Applications," *IEEE Access*, vol. 6, pp. 69640-69649, 2018. http://doi.org/10.1109/ACCESS.2018.2880763
[11] A. Y. -C. Chiou and C. -C. Hsieh, "An ULV PWM CMOS Imager With Adaptive-Multiple-Sampling Linear Response, HDR Imaging, and Energy Harvesting," *IEEE Journal of Solid-State Circuits*, vol. 54, no. 1, pp. 298-306, Jan. 2019. http://doi.org/10.1109/JSSC.2018.2870559
[12] M. Fojtik *et al.*, "Bubble Razor: An architecture-independent approach to timing-error detection and correction," *IEEE International Solid-State Circuits Conference*, San Francisco, Feb. 2012, pp. 488-490. http://doi.org/10.1109/ISSCC.2012.6177103

[13] M. J. Weinberger, G. Seroussi and G. Sapiro, "The LOCO-I lossless image compression algorithm: principles and standardization into JPEG-LS," *IEEE Transactions on Image Processing*, vol. 9, no. 8, pp. 1309-1324, Aug. 2000. http://doi.org/10.1109/83.855427

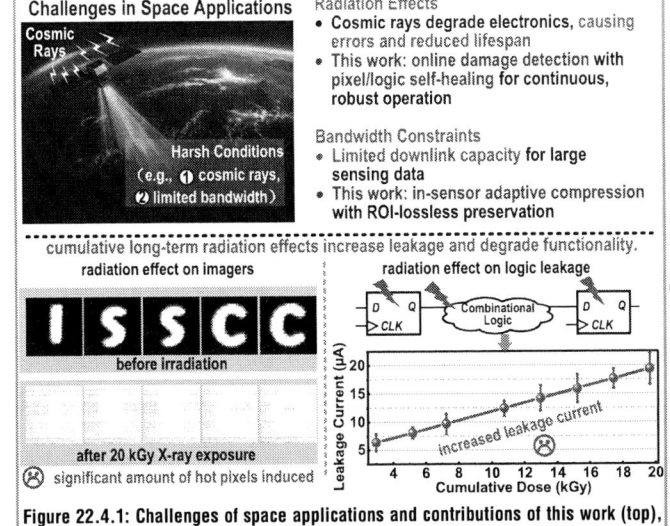

Figure 22.4.1: Challenges of space applications and contributions of this work (top), and cumulative long-term radiation effects on imagers and logic (bottom).

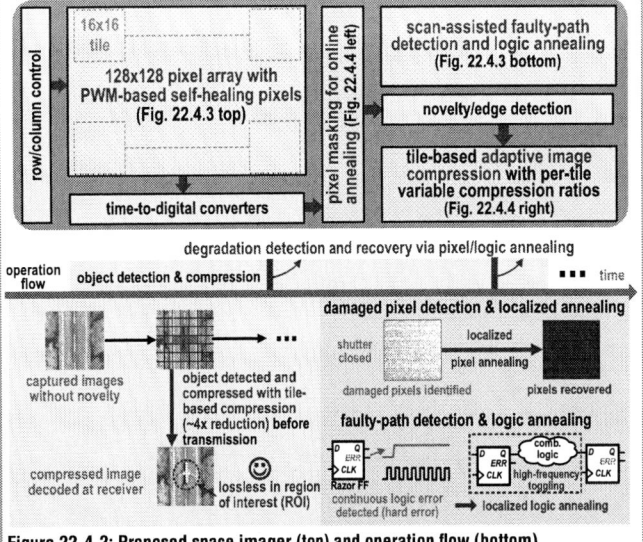

Figure 22.4.2: Proposed space imager (top) and operation flow (bottom).

DIGEST OF TECHNICAL PAPERS • 391

ISSCC 2026 / SESSION 22 / CIRCUITS IN EXTREME ENVIRONMENTS / 22.4

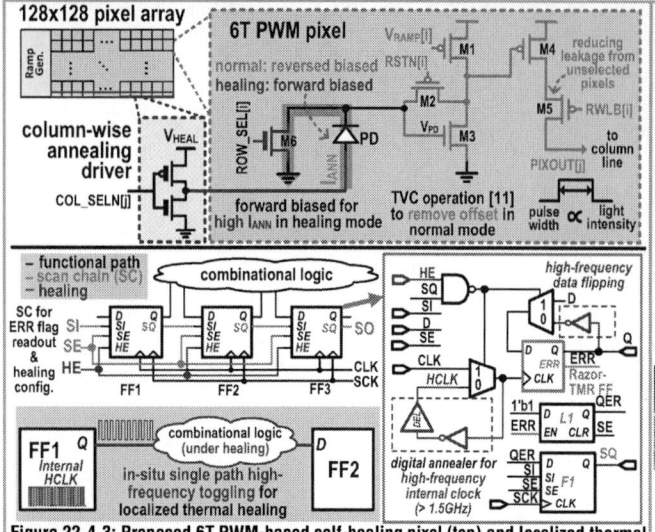

Figure 22.4.3: Proposed 6T PWM-based self-healing pixel (top) and localized thermal annealing of digital logic using scan-assisted path selection (bottom).

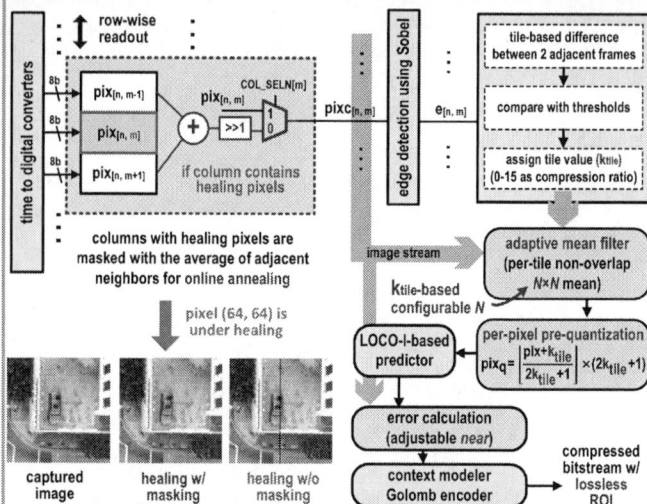

Figure 22.4.4: Proposed pixel masking for online annealing (left) and tile-based adaptive image compression (right).

Figure 22.4.5: Measurement results of X-ray irradiation and self-healing with near full image quality recovery, and tile-based adaptive image compression.

	[4] TNS'18	[3] VLSI'25	This work
Technology	180nm CIS	180nm CMOS	180nm CMOS
Architecture	imager only	imager, SNN	imager, novelty/edge detection, adaptive data compression
Supply voltage (V)	analog: 3.3 digital: 1.8	0.6-0.8 (normal) 1.5-2.5 (healing)	0.6-1.0 (analog), 1.3-1.8 (digital) 1.5-2.0 (healing)
Area (mm²)	not reported	28.275	18.243
Pixel array	256x256	73x73	128x128
Pixel size (µm²)	10x10	22x21	20x20
Pixel structure	3T	14T spike	6T PWM
Frame/event rate	25 fps	1-5000 event/s	100-400 fps
Image compression ratio	no compression	no compression	~ 4x (2.3x higher compared to LOCO-I)
Power consumption	not reported	76.3 µW @ 0.6V	31.80 µW @ 0.6V, 100fps (analog) 26.48 mW @1.3V, 100fps (digital)
Radiation hardening	ELT, gate-overlap PD	pixel annealing, NW-guarded PD	pixel/logic online annealing, Razor-TMR FF, NW-guarded PD
Online healing	no	no	yes
*Dark current increase	>100x @20kGy X-ray	~20x (after healing) @14kGy Alpha Am-241	~38x (after healing) @20kGy X-ray
Power density for thermal annealing	N/A	114 - 465 µW/µm² (pixel only)	93 - 221 µW/µm² (pixel) 7 - 18 µW/µm² (logic)
Dark/leakage current reduction after healing	N/A	6.25x (pixel only) @ 14kGy Alpha Am-241	4.82x @ 20kGy X-ray (pixel) 3.97x @ 20kGy X-ray (logic)

*the rise in pixel dark current after irradiation compared to that of the non-irradiated reference device.

Figure 22.4.6: Comparison with state-of-the-art radiation-hardened imagers.

Technology	180nm CMOS
Chip architecture	imager, novelty/edge detection, adaptive data compression
Chip size	4.83mm x 3.78mm (pixel array: 7.03mm²)
Power supply	0.6-1.0 (analog) 1.3-1.8 (digital) 1.5-2.0 (healing)
Pixel array	128 x 128
Pixel size (µm²)	20x20 (10x10 for PD)
Pixel structure	6T PWM
Frame rate	100-400 fps
Bit depth	8-bit
Compression	16 tile-based compression options (adaptive)
Power	31.80 µW @ 0.6V 100fps (analog) 26.48 mW @ 1.3V 100fps (digital)
Radiation hardening	online pixel annealing (93-221 µW/µm²), online logic annealing (7-18 µW/µm²), NW-guarded PD, Razor-TMR FF
Leakage current reduction ratio w/ self-healing	4.82x @ 20kGy X-ray (pixel) 3.97x @ 20kGy X-ray (logic)

Figure 22.4.7: Die micrograph and chip summary.

• 2026 IEEE International Solid-State Circuits Conference

ISSCC 2026 / SESSION 22 / CIRCUITS IN EXTREME ENVIRONMENTS / 22.5

22.5 A 500kGy Radiation-Hardened 2.4GHz Wi-Fi Receiver for Innovative Nuclear Power Plant Decommissioning

Yasuto Narukiyo[1], Sena Kato[1], Kotaro Yanaka[1], Yuya Takahashi[1], Masaya Miyahara[2], Jill Mayeda[1], Atsushi Shirane[1]

[1]Institute of Science Tokyo, Tokyo, Japan, [2]High Energy Accelerator Research Organization, Ibaraki, Japan

Abstract

This work presents a 2.4GHz Wi-Fi receiver in 65nm CMOS, designed with radiation-hardened techniques to sustain operation under total ionizing dose to 500kGy. Such tolerance addresses the requirements of nuclear power plant decommissioning, where conventional studies have considered doses of only several Gy. Measurement results confirm that even at 500kGy, the proposed receiver maintains acceptable performance showing the feasibility of wireless systems in extreme radiation environments.

Decommissioning operations at nuclear power plants have been steadily progressing; however, work in highly radiation environments, such as inside the nuclear reactor, remains extremely challenging for human operators. Consequently, the deployment of remotely operated robotic systems is indispensable. Currently, wired control using LAN cables is the predominant method for robot controls, but it imposes several limitations. Cables act as physical obstacles within the workspace, significantly reducing mobility and operational flexibility, and the number of simultaneously controllable robots is constrained, limiting overall efficiency. Wireless remote operation offers a viable solution to these challenges, potentially improving both operational efficiency and safety in decommissioning tasks.

However, the wireless transceivers deployed inside the reactor are exposed to intense gamma radiation emitted from fuel debris. In this environment, cumulative total ionizing dose (TID) levels are expected to reach up to 500kGy per year, posing an extremely high risk of severe performance degradation.

Figure 22.5.1 shows TID mechanism. TID effects induce charge trapping and defect generation at the Si/SiO_2 interface, resulting in threshold voltage shifts, increased leakage currents in MOSFETs [1]. Below 1MGy, the impact of threshold voltage variation is minimal; instead, the primary contributor to device degradation is the emergence of drain-source leakage paths, which significantly reduce the drain-source resistance (R_{DS}) in a 65nm CMOS process [2]. These effects are particularly pronounced in analog circuits and can critically impair wireless communication performance.

Notably, the 500kGy cumulative dose far exceeds the TID levels typically encountered in space environments. Conventional studies on space-grade semiconductor devices have mainly focused on doses ranging from 1Gy to 30kGy. In space, TID effect of high energy electrons and protons is substantially reduced by shielding with materials such as aluminum. For instance, cumulative TID experienced by instruments aboard geostationary or low-Earth-orbit satellites generally remain at the order of a few Gy when shielding is considered [3 - 5]. In contrast, the reactor environment is dominated by highly penetrative gamma radiation, which cannot be effectively attenuated by shielding such as aluminum. As a result, semiconductor devices in reactor environments are directly exposed to extremely high TID, potentially reaching 500kGy over an order of magnitude higher than space-grade TID. This extreme exposure exceeds the design assumptions of conventional space applications, necessitating new radiation-hardened design approaches beyond traditional space-grade guidelines.

This work targets the realization of a Wi-Fi receiver capable of stable operation under 500kGy TID conditions. Specifically, a 2.4GHz Wi-Fi receiver chip was developed with radiation-hardened circuit design strategies to minimize performance degradation.

Figure 22.5.2 shows the block diagram of the proposed radiation-hardened 2.4GHz Wi-Fi receiver. The architecture comprises an LNA for signal amplification, followed by a VGA for amplitude control, and an RFAMP mediated operational voltage for TIA. The signal is then directly down-converted to baseband via a mixer and finally amplified by a TIA stage to produce four BB channels delivered to the digital processing system.

Under gamma ray irradiation, TID effects manifest across all blocks. Specifically, the LNA exhibits gain reduction, S11 deterioration, and NF increase. The VGA shows a reduced gain tuning range, the RFAMP experiences P1dB and conversion gain degradation, and the mixer shows increased I/Q imbalance. To mitigate these effects, radiation-hardened strategies were applied at each circuit level.

This paper uses three ways to mitigate TID effects [1-2]. First is to reduce the number of MOSFETs, reducing the physical area of exposed oxide layers, decreasing the total trapped charge and mitigating TID effect. The second way is replacing PMOS with passive components, such as inductors. PMOS has higher TID susceptibility than NMOS. The third way is to increase the channel length and width (especially length). Short-channel MOSFETs are highly sensitive to interface defects and STI effects, leading to pronounced R_{DS} degradation and threshold shifts under gamma irradiation, whereas long-channel devices exhibit a more distributed electric field, mitigating TID effects. Gate length and width were therefore optimized for all key MOSFETs in the LNA, VGA, RFAMP, and mixer, minimizing TID effects.

Figure 22.5.3 summarizes the radiation-hardening strategies implemented for each block. For the radiation-hardened LNA, an increase in gate length generally results in a reduction of the gain, NF, S11, however, gate length and width were selected such that the performance degradation remained within an acceptable range, while prioritizing enhanced reliability under TID conditions. As a result, degradation in Gain, S11, and NF by TID is mitigated. Similarly to the LNA, in the radiation-hardened VGA, long and wide channel devices were adopted, and the tail transistor for gain control was replaced with an inductor. The tail inductor stabilizes the output resistance against TID variation, maintaining gain range stability. While the gain range is slightly reduced, overall radiation tolerance is improved. Typical RFAMPs use a cascoded PMOS, which exhibits higher output resistance degradation due to R_{DS} variation under TID than NMOS. In the radiation-hardened RFAMP, PMOS was replaced with an inductor to stabilize output impedance. To suppress I/Q imbalance degradation in the radiation-hardened Mixer, Gate length and width were increased in the MOSFETs to reduce R_{DS} variations under TID.

These strategies directly mitigate TID circuit-level degradation, maintaining the electrical stability and performance of the overall receiver.

Figure 22.5.5 shows TID test setup for the proposed receiver. Co-60 in the silver pillar served as gamma radiation source. Dose rate was 2kGy (SiO_2) per hour. Measurements were conducted after cumulative TID exposures of 300kGy and 500kGy to evaluate circuit performance degradation.

Figure 22.5.4 shows the receiver measurement results at 0kGy, 300kGy, and 500kGy. With an input tone of 2.41GHz at −60 to -10dBm and a 2.4GHz LO, the proposed receiver down-converts the signal to a 10MHz baseband output. The conversion gain of this system was 38.3dB at -60dBm, and IP1dB was −48.7dBm. After 300kGy gamma irradiation, the conversion gain decreased to 37.1dB at -60dBm, and the IP1dB increased to -47.8dBm. After 500kGy gamma irradiation, the conversion gain additionally decreased to 36.9dB at -60dBm, and the IP1dB additionally increased to −47.7dBm. The Pin vs Pout curve shows a decrease in gain within the linear region as the total ionizing dose increases. Due to the gain degradation in the linear region, the conversion gain decreased by 1.4dB, whereas the IP1dB increased by 1dB. The Gain control range of the VGA was 20.4dB from 200-to-600mV bias range. After 500kGy gamma irradiation, the range was 20.3dB. The gain control range degraded by 0.1dB after exposure to 500kGy TID.

Figure 22.5.4 also shows NF and S11 measurements, and at 20MHz the NF was 5.1dB and the maximum NF degradation was 1.26dB from 3.8dB at 0kGy to 5.4dB at 10.5MHz after 500kGy. The increase in NF is attributed to gain degradation induced by TID. S11 degradation showed a 0.7dB increase from −8.06dB at 0kGy to −7.4dB at 2.4GHz.

Figure 22.5.5 shows the degradation of power consumption with an input tone of 2.41GHz at −60 to -10dBm and a 2.4GHz LO. The degradation was improved 2.1mW from 65.3mW at 0kGy to 63.2mW at 500kGy at −60 dBm. Figure 22.5.5 also shows EVM tests using IEEE 802.11n 16QAM signals achieved an EVM of 4.0% at a bandwidth of 20MHz and 6.5% at a bandwidth of 40MHz, and 64QAM signals also achieved an EVM of 3.9% with a bandwidth of 20MHz and 6.5% with a 40MHz bandwidth even after 500kGy gamma irradiation.

Figure 22.5.6 shows the comparison between this work and the state-of-the-art Wi-Fi receiver and radiation hardened transceivers [3-7]. The proposed receiver exhibits only 1.4dB gain degradation, and 1.3dB NF increase at 500kGy, outperforming prior work. Even under 500kGy, the proposed receiver maintains sufficient performance for Wi-Fi reception.

The die micrograph is shown in Fig. 22.5.7. The proposed radiation hardened receiver is fabricated in a Low-power 65nm CMOS process and 0.9mm×1.5mm die size.

ISSCC 2026 / February 17, 2026 / 3:00 PM

Acknowledgement:
This work is partially supported by JAEA (JPJA24P24020833), MIC/SCOPE (#192203002, #192103003, and JP235003011), MIC (JPJ000254), and VDEC in collaboration with Cadence Design Systems, Inc., Mentor Graphics, Inc., and Keysight Technologies Japan, Ltd.

References:
[1] H. J. Barnaby, "Total-Ionizing-Dose Effects in Modern CMOS Technologies," *IEEE Transactions on Nuclear Science*, vol. 53, no. 6, pp. 3103-3121, Dec. 2006. https://doi.org/10.1109/TNS.2006.885952
[2] A. Nikolaou *et al.*, "Extending a 65nm CMOS process design kit for high total ionizing dose effects," *7th International Conference on Modern Circuits and Systems Technologies (MOCAST)*, pp. 1-4, May. 2018. https://doi.org/10.1109/MOCAST.2018.8376561.
[3] D. C. Howard *et al.*, "Mitigation of Total Dose Performance Degradation in an 8–18 GHz SiGe Reconfigurable Receiver," *IEEE Transactions on Nuclear Science*, vol. 61, no. 6, pp. 3226-3235, Dec. 2014. https://doi.org/10.1109/TNS.2014.2364580.
[4] X. Fu *et al.*, "A 3.4mW/element Radiation-Hardened Ka-Band CMOS Phased-Array Receiver Utilizing Magnetic-Tuning Phase Shifter for Small Satellite Constellation," *IEEE International Solid-State Circuits Conference (ISSCC)*, pp. 90-92, Feb. 2022. https://doi.org/10.1109/ISSCC42614.2022.9731557.
[5] A. Kawaguchi, J. Pang, Z. Li, K. Yanagisawa, A. Shirane and K. Okada, "Total Ionizing Dose Effects on 28GHz CMOS Bi-Directional Transceiver for 5G Non-Terrestrial Networks," *20th European Conference on Radiation and Its Effects on Components and Systems (RADECS)*, pp. 1-4, Oct. 2020. https://doi.org/10.1109/RADECS50773.2020.9857699.
[6] A. Kale, S. Popuri, M. Koeberle, J. Sturm and V. S. R. Pasupureddi, "A –40 dB EVM, 77 MHz Dual-Band Tunable Gain Sub-Sampling Receiver Front End in 65-nm CMOS," *IEEE Transactions on Circuits and Systems I: Regular Papers*, vol. 66, no. 3, pp. 1166-1179, March 2019. https://doi.org/10.1109/TCSI.2018.2878342.
[7] Y. -H. Chung *et al.*, "A dual-band 802.11abgn/ac transceiver with integrated PA and T/R switch in a digital noise controlled SoC," *IEEE Custom Integrated Circuits Conference (CICC)*, pp. 1-8, Sep. 2015. https://doi.org/10.1109/CICC.2015.7338361.

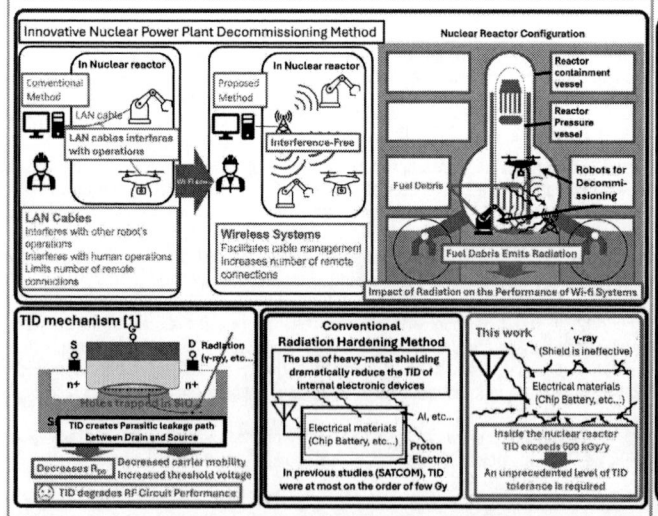

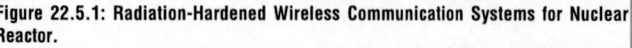

Figure 22.5.1: Radiation-Hardened Wireless Communication Systems for Nuclear Reactor.

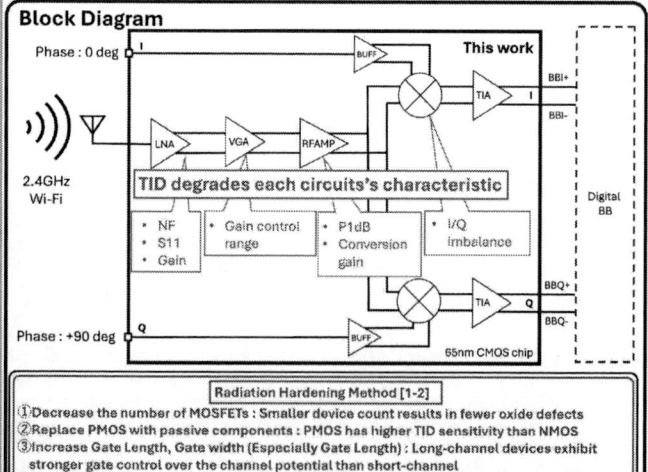

Figure 22.5.2: Block Diagram of Proposed Radiation Hardened 2.4GHz WLAN Receiver and Radiation Hardening Method.

DIGEST OF TECHNICAL PAPERS • 393

979-8-3315-8937-0/26 $31.00 © 2026 IEEE

ISSCC 2026 / SESSION 22 / CIRCUITS IN EXTREME ENVIRONMENTS / 22.5

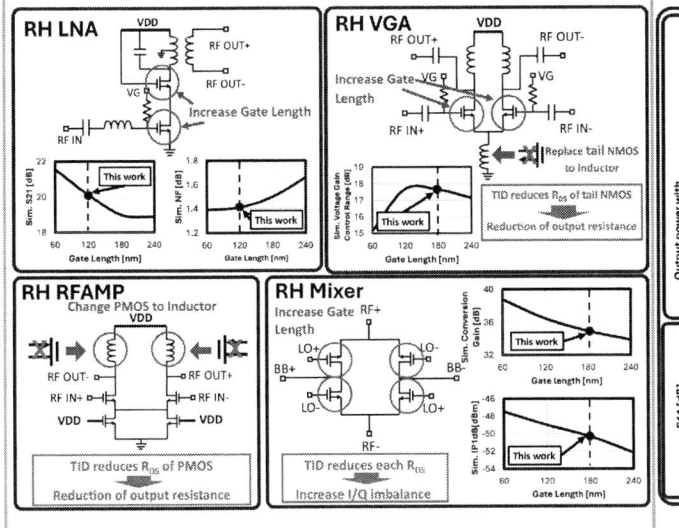

Figure 22.5.3: Proposed Each Radiation Hardened Circuit Configuration.

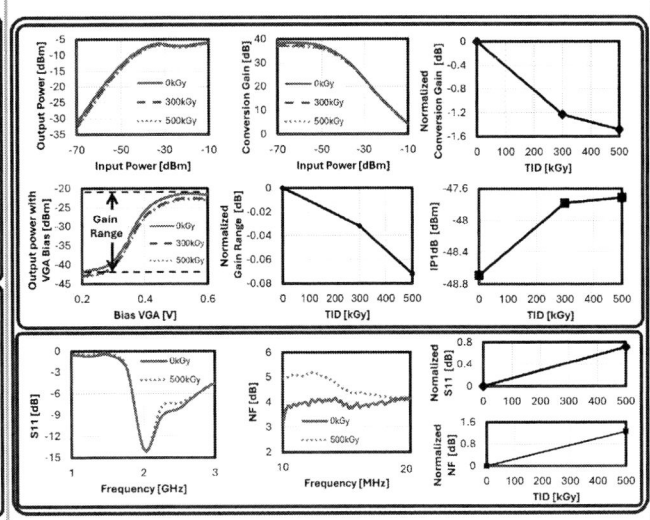

Figure 22.5.4: Proposed RX Measurement Results.

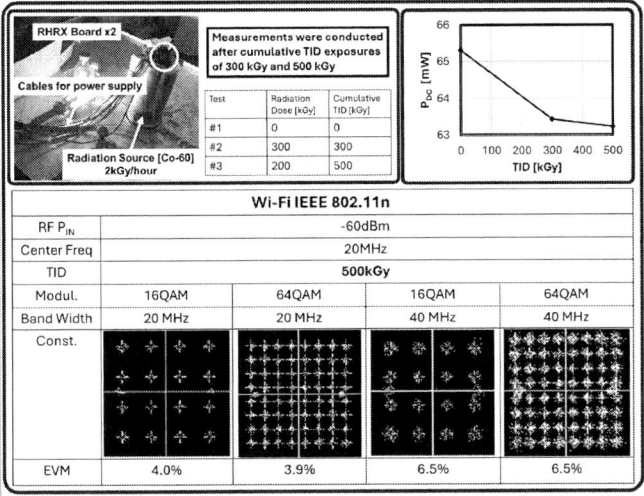

Figure 22.5.5: TID Test Setup and Measured RX EVM at 500kGy.

Figure 22.5.6: Performance Comparison with State-of-the-art Works.

	This work	TCS2019[6]	CICC2015[7]	TNS2014[3]	ISSCC2022[4]	RADECS2020[5]
Process	CMOS 65nm	CMOS 65nm	CMOS 55nm	SiGe BiCMOS 180nm	CMOS 65nm	CMOS 65nm
Application	Wi-Fi	Wi-Fi	Wi-Fi	SATCOM	SATCOM	SATCOM
Wi-Fi Standard	1X1 11n	1X1 11ac	1X1 11abg	N/A	N/A	N/A
Frequency	2.4GHz	2.4/5GHz	2.4/5GHz	10/16GHz	27.5 - 30GHz	28GHz
Support Band Width	20/40 MHz	20/40/77 MHz	20/40/80 MHz	N/A	N/A	N/A
Supply Voltage	1.2V	2.5V	N/A	4V	0.8	1V
P_{DC}	65.5mW	27.8mW	84mW	1.0mW	3.4mW	88mW
Max Gain	38.3dB	41dB	N/A	32.4dB	23dB	1.5dB
NF	4.2dB@20MHz	12dB	4.2dB	8.1dB	3.8dB	5.4dB
Modulation	64QAM	64QAM	256QAM	N/A	256APSK	64QAM
Area	1.4mm²	0.4mm²	3.4mm²	10.2mm²	0.2mm²	0.5mm²
TID Gain Degradation	-1.4dB @500kGy	N/A	N/A	-1dB @60kGy	-0.2dB @30kGy	-1.5dB @10kGy
TID NF Degradation	+1.3dB @500kGy	N/A	N/A	+0.3dB @60Gy	N/A	N/A

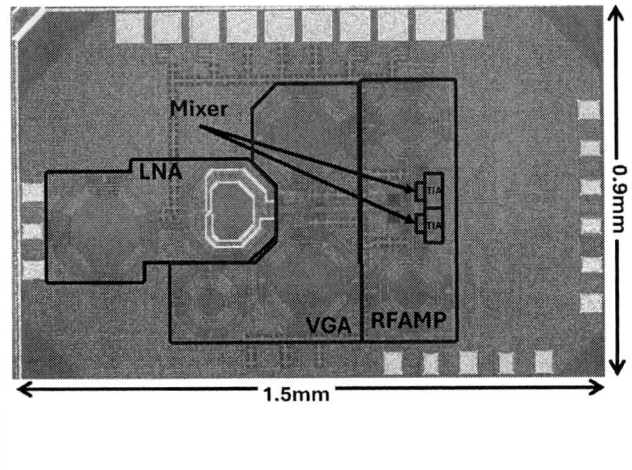

Figure 22.5.7: RHRX Die micrograph.

• 2026 IEEE International Solid-State Circuits Conference

979-8-3315-8937-0/26 $31.00 © 2026 IEEE

ISSCC 2026 / SESSION 23 / NEXT-GENERATION OPTICAL TRANSCEIVERS / OVERVIEW

Session 23 Overview: *Next-Generation Optical Transceivers*
WIRELINE SUBCOMMITTEE

Session Chair: Ahmed Mostafa
Marvell
Santa Clara, CA

Session Co-Chair: Quan Pan
Southern University of Science and Technology
Shenzhen, China

Machine learning-driven computing requires tight connections among components in a system. They require high bandwidth, energy efficiency, and low latency. High-density interconnects using co-packaged silicon photonics are suitable for chiplet-level and system-level connectivity in a data center. These technologies are transforming switch architectures for scalability and power efficiency. Long reach, up to a few tens of km, coherent or coherent-lite optical communication is another featured domain of this session. The first paper from Nvidia presents a 32Gb/s/λ DWDM link with 9 wavelengths, achieving 256Gb/s/fiber, 2.78pJ/b energy efficiency, and 1.33Tb/s/mm² area efficiency at <1e-11 BER. The second paper from Marvell presents a 5nm FinFET transceiver for 800Gb/s DP-16QAM links over 2-40km, optimized for campus interconnects. The third paper from Huazhong University of Science and Technology presents a 1Tb/s DWDM PAM-4 transceiver in 45nm CMOS SOI with 100Gb/s per channel and <2.5pJ/b efficiency. In the fourth paper, Broadcom presents a 6.4Tb/s Co-packaged optics ASIC with 7nm FinFET, achieving 4.2pJ/b efficiency and -11dBm TIA sensitivity. In the fifth paper, University of California at Berkeley presents a 212Gb/s QAM-16 OTX in 45nm CMOS-SOI with 0.91pJ/b efficiency and 5× area reduction. The sixth paper from Imec - Ghent University presents a burst-mode receiver with SiGe BiCMOS achieving 112Gb/s NRZ and −25.8dBm OMA sensitivity.

3:35 PM

23.1   A 32Gb/s/λ 256Gb/s/Fiber Half-Rate Bandpass-Filtered Clock-Forwarding DWDM Optical Link in a 3D-Stacked 7nm EIC/65nm PIC Technology

Sanquan Song, Nvidia, Santa Clara, CA

In Paper 23.1 Nvidia presents a 32Gbps/λ dense wavelength division multiplexing bandpass-filtered clock-forwarding link featuring nine 200GHz-spaced wavelengths (8 for data and 1 for clk). Fabricated in a 3D-stacked technology with electronic IC in 7nm process and photonic IC in 65nm SiPh process, the implementation achieves a throughput of 256Gb/s/fiber, delivering an energy efficiency of approximately 2.78pJ/b and an area efficiency of 1.33Tb/s/mm² while maintaining 0.46UI wide eye openings at <1e-11 BER.

394 • 2026 IEEE International Solid-State Circuits Conference

979-8-3315-8937-0/26 $31.00 © 2026 IEEE

4:00 PM

23.2 A 2-Channel 800Gb/s Transceiver for Coherent-Lite Applications with <300ns Latency in 5nm FinFET

Enrico Monaco, Marvell, Pavia, Italy

In Paper 23.2, Marvell presents a full transceiver suitable for coherent-lite optical communications over 2-to-40km distances. The low-latency transceiver combines the advantages of advanced coherent modulation with the benefits and manufacturing scale of O-band optics to deliver a power- and cost-effective module. Fabricated in 5nm FinFET, it comprises two 400Gb/s/λ transmitters and receivers to realize a 2-channel DP-16QAM 800Gb/s link tailored to the market of distributed campus data center interconnects.

4:25 PM

23.3 A 2×500Gb/s Monolithic Silicon-Photonic DWDM PAM-4 Transceiver in 45nm CMOS SOI

Ziang Xu, Huazhong University of Science and Technology, Wuhan, China

In Paper 23.3, Huazhong University of Science and Technology presents a 1Tb/s monolithic silicon photonic DWDM PAM-4 transceiver in 45nm CMOS SOI, integrating a 2×5-channel architecture supporting 100Gb/s PAM-4 per channel. V-groove couplers reduce insertion loss. Innovations include a low-noise TIA with Qtamed CTLE, high-speed MRM drivers, and PWM-based wavelength locking. It achieves <2.5pJ/b efficiency, 100Gb/s per channel, and high stability under thermal variations.

4:50 PM

23.4 A 6.4Tb/s 4.2pJ/b Co-Packaged Optics ASIC with Direct-Drive Integrated TIA and Retimed Segmented Mach-Zehnder Modulator Driver in 7nm FinFET

Mahdi Kashmiri, Broadcom, San Jose, CA

In Paper 23.4, Broadcom presents a 6.4Tb/s co-packaged optics ASIC, which integrates 64 ingress path direct-drive TIAs and 64 egress path retimed segmented Mach-Zehnder modulator drivers in 7nm FinFET. The 7nm ASIC is 3D packaged with a photonics IC (PIC) and achieves an energy efficiency of 4.2pJ/b. At 106.25Gb/s, its PAM-4 TIA has a sensitivity better than -11dBm and a BER floor better than 1E-9. The transmitter's PAM-4 optical eye achieves a TDEQ of 1.48dB and an extinction ratio of 4.57dB.

5:05 PM

23.5 A 212Gb/s/λ 0.91pJ/b Direct-Drive O-Band Monolithic Coherent Transmitter Based on Carrier-Injection-Mode Mach-Zehnder Modulator

Antroy Roy Chowdhury, University of California, Berkeley, CA

In Paper 23.5, University of California at Berkeley presents a 212Gb/s QAM-16 coherent direct-drive optical transmitter (OTX) in 45nm CMOS-SOI using a compact 0.4mm monolithically integrated and highly efficient forward-biased PIN-MZM. A multistage linear driver with shunt peaking, passive equalization and TX-FIR enables 106Gb/s PAM-4 IMDD and 212Gb/s QAM-16 coherent signaling. The OTX achieves 0.91pJ/bit electrical power efficiency with a 2.76× lower laser power and 5× area reduction versus a conventional reverse-biased TW-MZM fabricated on the same die.

5:20 PM

23.6 A 112Gb/s NRZ Heterodyne Detection 23ns Settled Burst-Mode RX with CD Suppression and Envelope Demodulation for 100G PON

Cheng Wang, imec - Ghent University, Ghent, Belgium

In Paper 23.6, Imec - Ghent University presents a burst-mode receiver based on an intensity modulation heterodyne detection scheme. SSB quadrature IF generation and envelope detection enhance baseband signal integrity and chromatic dispersion tolerance. Pre/post-demodulator loop separation allows parallel AGC/AOC in burst-mode. The SiGe BiCMOS front-end, co-integrated with an O-band silicon photonic PIC, achieves 112Gb/s NRZ with −25.8dBm OMA (B2B), 23ns settling, and −23.7/−22.9dBm OMA over 20km SMF in 1290/1329nm.

ISSCC 2026 / SESSION 23 / NEXT-GENERATION OPTICAL TRANSCEIVERS / 23.1

23.1 A 32Gb/s/λ 256Gb/s/Fiber Half-Rate Bandpass-Filtered Clock-Forwarding DWDM Optical Link in a 3D-Stacked 7nm EIC/65nm PIC Technology

Sanquan Song[1], Nandish Mehta[1], Nikola Nedovic[1], Angad Rekhi[1], Georgios Kalogerakis[1], Li Xu[1], Brian Zimmer[1], Stephen G. Tell[2], Yoshi Nishi[1], Xi Chen[1], Ward Lopes[1], Benjamin G. Lee[3], Thomas H. Greer III[2], C. Thomas Gray[2]

[1]Nvidia, Santa Clara, CA, [2]Nvidia, Durham, NC, [3]Nvidia, Ridgefield, CT

Abstract

We present a 32Gb/s/λ dense wavelength division multiplexing bandpass-filtered clock-forwarding link featuring nine 200GHz-spaced wavelengths (8 for data and 1 for clk). Fabricated in a 3D-stacked technology with an electronic IC in a 7nm process and a photonic

IC in a 65nm SiPh process, our implementation achieves a throughput of 256Gb/s/fiber, delivering an energy efficiency of approximately 2.78pJ/b and an area efficiency of 1.33Tbps/mm²—while maintaining 0.46UI wide eye openings at <1E-11 BER.

The rapid growth of data centers, driven by increasing demand from artificial intelligence training and inference, has highlighted the limitations of traditional copper cables in terms of density, reach, and energy efficiency. Photonic links, especially dense wavelength division multiplexing (DWDM) links [1-10], offer the capability to break these barriers and expand the all-to-all connection domain to thousands of processors. DWDM links leverage multiple wavelengths, each of which transmits data at a modest rate (typically in the lower tens of gigabits per second). This strategy eliminates the need for ultra high-speed signaling (such as 100+Gb/s), improving the area and energy efficiency as well as reducing system complexity.

High-density, low-power electrical chip-to-chip (C2C) links [11-14] are a promising way to connect hosts to DWDM optical engines (OE), commonly using forwarded clocking (FC) due to its excellent jitter tracking bandwidth. There is a strong interest in adapting the FC scheme to DWDM links, aiming to preserve its advantages. However, DWDM links are predominantly impacted by transimpedance amplifier (TIA) thermal noise-induced jitter. Simply forwarding a clock could substantially increase the jitter power, potentially doubling it at the point of data slicing. Consequently, embedded clocking (EC) techniques [1-4] are widely adopted, which come with additional Clock and Data Recovery (CDR) challenges and exposure to jitter beyond its limited bandwidth, such as the transmitter phase-locked loop (TXPLL) induced jitter and supply-induced jitter.

We propose a half-rate FC DWDM link architecture for co-packaged optics (CPO), shown in Fig. 23.1.1, where eight wavelengths carry data and one wavelength carries a half-rate FWDCLK, which is band-pass filtered at the receiver (RX) for data recovery. A band-pass filter with a bandwidth of 1 to 2GHz rejects much of TIA thermal noise induced uncorrelated random jitter on the clock because the noise bandwidth (BW) is comparable to the TIA BW. In contrast, it tracks most of the correlated jitter on the data, such as TXPLL jitter and supply noise-induced jitter, since its power is primarily concentrated at low frequencies. This enables clock forwarding with reduced noise, and thus combines the main benefits of both schemes, as shown in Fig. 23.1.2 [15].

The transmitter (TX) is shown in Fig. 23.1.3. A ring oscillator (RO) based TXPLL under a regulated supply domain is shared among nine identical TX lanes with four clock phases distributed at half rate. Each TX lane includes a phase interpolator to adjust for the half unit interval (UI) phase shift between the FWDCLK and data to align the RX de-serializer (DES) sampling clock to the center of the eye diagram. Furthermore, this compensates for per-lane skews caused by routing mismatch and chromatic dispersion. It offers improved clock and data path delay matching over the RX side PI scheme and streamlines RX clocking by eliminating the need for quadrature phase generation and distribution. A 16-to-1 serializer is followed by a single-ended-to-differential converter (S2D), TX driver, and a broadband TX level shifter. The latter biases the ring cathode at high voltage, nominally 1.5×VDDA, and the anode at low voltage, nominally 0.5×VDDA. Compared with a conventional bias-T level shifter, this TX broadband level shifter does not have a lower cutoff frequency, posing no restrictions on data encoding; also, it supports microring modulator (MRM) leakage current up to 300µA [16]. By setting the input parallel data pattern as 0101, any TX lane can transmit the FWDCLK.

The RX is shown in Fig. 23.1.4. An RX phase-locked loop (RXPLL) generates the reference voltage for the TIA regulator and RX regulator, ensuring the supplies for the TIA and other RX circuits track speed and process-voltage-temperature (PVT) variations [11]. This RXPLL acts as a reference generator as its clocks are not distributed by default. At the expense of the RXPLL and regulator overhead, the bandwidth of the TIA remains relatively constant across PVT corners, with sensitivity only to resistor variation. All TIAs share the same supply domain, isolated from the rest of the RX circuits to minimize supply noise over the TIA due to its higher sensitivity. The inverter-based TIA consists of a shunt-feedback amplifier followed by two Cherry-Hooper stages. An injection-locked oscillator (ILO) stage is inserted between the outputs of the 4th and 2nd inverters with tunable natural frequency, which is ON only if it receives FWDCLK. Together with the clock distribution ILO (see below), this structure forms a bandpass filter described above that significantly suppresses the wideband uncorrelated jitter while tracking low-frequency correlated jitter. Following the TIA, two delay paths are employed: one directs the buffered output to the 1:16 DES upon

data reception, while the other routes the signal to the RX clock distribution upon FWDCLK reception.

For DWDM systems, a fixed TX-RX ring mapping (e.g., TX ringₙ to RX ringₙ) can lead to high heater power consumption in the presence of large wafer-to-wafer ring resonance frequency variation [8]. To minimize the heater power, the system should support a flexible TX-RX ring mapping, which causes ambiguity in RX clock distribution as the FWDCLK can be received by any RX ring. Instead of a 9-to-1 mux at the source of RX clock distribution, which limits bandwidth and introduces an extra hurdle for clock to data matching, an ILO structure is embedded into the RX clock distribution. As shown in Fig. 23.1.4 (bottom), one lane drives the distributed net and others oscillate at the operation frequency or stay OFF. This allows the FWDCLK to be sourced from any lane and provides additional bandpass filtering. To reduce parasitic capacitance, the RX clock distribution is split into two groups that are both connected to the 2:1 input clock mux in all DESs, with only one group active during normal operation. Besides the jitter filtering benefit, this ILO-based RX clock distribution scheme also improves the resilience of the system to potential laser reliability issues and provides the possibility to assign desired wavelengths to FWDCLK for performance optimization. Compared with the quarter-rate FWDCLK scheme [8], this approach avoids multi-phase generation at the RX side, as such circuitry usually leads to unwanted mismatch between the data and clock paths.

Both TX and RX rings are tuned thermally to lock to the target wavelength with on-chip mixed-signal loops. The TX thermal tuning (TT) loop locks to a dialed-in drop point photodetector (PD) current level, determined by a pre-characterization step prior to transmitting live data. The RX thermal tuning loop taps the TIA DC-loop to maximize the RX PD DC current. As shown in Fig. 23.1.3 bottom (right), for either loop, the current reading is first digitized by an 8b temperature-insensitive switched cap-based combined integrator/successive approximation register analog-to-digital converter (SAR ADC). It is then processed digitally (either with proportional-integral control [TX] or a peak-finding algorithm [RX]) to form a pulse density modulated (PDM) signal to drive a metal heater co-integrated with the ring for maximum energy efficiency.

This architecture employs a 7nm electronic integrated circuit (EIC) stacked atop a 65nm silicon photonics (SiPh) photonic integrated circuit (PIC), interconnected through hybrid bonding technology to minimize interface parasitics and maximize receiver sensitivity, as shown in Fig. 23.1.7. Each TX PHY includes nine data/FWDCLK lanes, one TXPLL lane, one regulator lane for the TXPLL, and an extra lane for decoupling capacitors. Each RX PHY has nine data/FWDCLK lanes, one RXPLL lane, and two regulator lanes for the TIA and other RX circuits respectively. Each lane measures 80.94×80.94 µm² with an extra 18.012mm-wide routing corridor for parallel data, control, and status bits. The EIC is connected to the PIC via 9µm-pitch hybrid bonding pads, minimizing parasitics by shortening the routing distance from the TX driver to the TX ring and RX PD to the RX TIA. This is achieved through positioning each ring with PD directly below the corresponding electrical lane. The PIC features multiple ~5µm radius rings along the waveguide for DWDM experiments, with grating couplers used for fiber connections. The TX ring modulator quality factor Q is ~4.5K, and RX filter Q is 4K, selected using an in-house link modeling tool, and validated by device test chip measurements. We place the RX rings in radius-descending order to help partially cancel optical crosstalk [17].

A PIC TX bus spectrum sweep is shown on Fig. 23.1.5 top (left). All 9 rings are spread around 1310nm with about 200GHz spacing. The TT loop response to the input laser wavelength steps is shown in Fig. 23.1.5 top (right). The heater code follows the laser wavelength shift properly, confirming the functionality of the TT loop. Generated by a tunable laser source, a single wavelength is modulated by one TX ring with on-die generated PRBS7 pattern. The eye diagram of the modulated NRZ signal is shown in Fig. 23.1.5 bottom (left), indicating a robust TX. With one data lane and one FWDCLK lane on, the RX sensitivity is measured at -16.0dBm OMA at BER<1E-11 level using a PRBS7 pattern. A multi-wavelength laser source is used for the link measurement, with all TX lanes active at 32Gb/s using PRBS31 patterns (varied seeds), and the FWDCLK lane (TX2) set to 16GHz. Thermal loops lock all rings to their corresponding wavelengths. One RX ring (RX6) receives the FWDCLK, filters the jitter with the TIA with ILO enabled and sends it to all the lanes with the RX clock

396 • 2026 IEEE International Solid-State Circuits Conference

979-8-3315-8937-0/26 $31.00 © 2026 IEEE

distribution ILO ON. The bathtub of the bit error rate (BER) is captured by sweeping the TX PI code and querying RX PRBS checker for each lane. As shown in Fig. 23.1.5 bottom (right), with all lanes simultaneously ON, the aggregate eye opening is 0.46 UI wide at BER<1E-11 level. This work is summarized together with recent related publications in Fig. 23.1.6. Compared with the state-of-the-art DWDM links, it achieves ~6× higher shoreline and ~20× higher area density, important for future CPO applications in AI datacenters. Assuming 10% fiber-coupled laser wall-plug efficiency [18,19], laser energy efficiency is 0.76pJ/b. Combined with 0.76pJ/b energy for TX+RX ring heaters and 1.26pJ/b energy for TX, RX, clocking and TT circuits, it achieves 2.78pJ/b overall energy efficiency.

Acknowledgement:
The authors extend their sincere gratitude to NVIDIA's MSD team for their support with EIC/PIC design, layout, and testing; to the ATG team for CAD and tapeout assistance; to the Israel CPO team for their contributions to packaging, FAU, and substrate design; and to the NBU team for their contributions to PCB design.

References:
[1] K. Hosseini *et al.*, "5.12 Tbps Co-Packaged FPGA and Silicon Photonics Interconnect I/O," *2022 IEEE Symposium on VLSI Technology and Circuits (VLSI Technology and Circuits)*, Honolulu, HI, USA, 2022, pp. 260-261. https://doi.org/10.1109/VLSITechnologyandCir46769.2022.9830221
[2] C. Sun et al., "TeraPHY: An O-Band WDM Electro-Optic Platform for Low Power, Terabit/s Optical I/O," *2020 IEEE Symposium on VLSI Technology and Circuits (VLSI Technology and Circuits)*, Honolulu, HI, USA, 2020, pp. 1-2. https://doi.org/10.1109/VLSITechnology18217.2020.9265012
[3] M. Raj *et al.*, "A 0.96pJ/b 7 × 50Gb/s-per-Fiber WDM Receiver with Stacked 7nm CMOS and 45nm Silicon Photonic Dies," *2023 IEEE International Solid-State Circuits Conference (ISSCC)*, San Francisco, CA, USA, 2023, pp. 11-13. https://doi.org/10.1109/ISSCC42615.2023.10067617
[4] C. S. Levy et al., "8-λ × 50 Gbps/λ Heterogeneously Integrated Si-Ph DWDM Transmitter," in *IEEE Journal of Solid-State Circuits*, vol. 59, no. 3, pp. 690-701, March 2024. https://doi.org/10.1109/JSSC.2023.3344072
[5] A. Netherton et al., "Short-Reach Silicon Photonic Interconnects with Quantum Dot Mode Locked Laser Comb Sources," *2024 IEEE International Solid-State Circuits Conference (ISSCC)*, San Francisco, CA, USA, 2024, pp. 422-424. https://doi.org/10.1109/ISSCC49657.2024.10454400
[6] Daudlin, S., Rizzo, A., Lee, S. et al. "Three-dimensional photonic integration for ultra-low-energy, high-bandwidth interchip data links". *Nat. Photon.* 19, 502–509 (2025). https://doi.org/10.1038/s41566-025-01633-0
[7] H. Li *et al.*, "A 4×50 Gb/s All-Silicon Ring-based WDM Transceiver with CMOS IC," *2021 European Conference on Optical Communication (ECOC)*, Bordeaux, France, 2021, pp. 1-3. https://doi.org/10.1109/ECOC52684.2021.9605947
[8] P. -H. Chang et al., "A 3D Integrated Energy-Efficient Transceiver Realized by Direct Bond Interconnect of Co-Designed 12 nm FinFET and Silicon Photonic Integrated Circuits," in *Journal of Lightwave Technology*, vol. 41, no. 21, pp. 6741-6755, 1 Nov.1, 2023. https://doi.org/10.1109/JLT.2023.3291704

[9] Z. Xuan et al., "A 256 Gbps Heterogeneously Integrated Silicon Photonic Microring-based DWDM Receiver Suitable for In-Package Optical I/O," *2023 IEEE Symposium on VLSI Technology and Circuits (VLSI Technology and Circuits)*, Kyoto, Japan, 2023, pp. 1-2. https://doi.org/10.23919/VLSITechnologyandCir57934.2023.10185280
[10] N. Qi *et al.*, "A Monolithically Integrated DWDM Si-Photonics Transceiver for Chiplet Optical I/O," in *IEEE Journal of Solid-State Circuits*, vol. 60, no. 10, pp. 3613-3625, Oct. 2025. https://doi.org/10.1109/JSSC.2025.3585584
[11] J. W. Poulton et al., "A 1.17-pJ/b, 25-Gb/s/pin Ground-Referenced Single-Ended Serial Link for Off- and On-Package Communication Using a Process- and Temperature-Adaptive Voltage Regulator," in *IEEE Journal of Solid-State Circuits*, vol. 54, no. 1, pp. 43-54, Jan. 2019. https://doi.org/10.1109/JSSC.2018.2875092
[12] Y. Nishi et al., "A 0.297-pJ/Bit 50.4-Gb/s/Wire Inverter-Based Short-Reach Simultaneous Bi-Directional Transceiver for Die-to-Die Interface in 5-nm CMOS," in *IEEE Journal of Solid-State Circuits*, vol. 58, no. 4, pp. 1062-1073, April 2023. https://doi.org/10.1109/JSSC.2022.3232024
[13] Y. Wei et al., "NVLink-C2C: A Coherent Off Package Chip-to-Chip Interconnect with 40Gbps/pin Single-ended Signaling," *2023 IEEE International Solid-State Circuits Conference (ISSCC)*, San Francisco, CA, USA, 2023, pp. 160-162. https://doi.org/10.1109/ISSCC42615.2023.10067395
[14] K. Seong *et al.*, "A 4nm 32gb/s 8Tb/s/mm Die-to-Die Chiplet Using NRZ Single-Ended Transceiver With Equalization Schemes And Training Techniques," *2023 IEEE International Solid-State Circuits Conference (ISSCC)*, San Francisco, CA, USA, 2023, pp. 114-116. https://doi.org/10.1109/ISSCC42615.2023.10067477
[15] N. Mehta *et al.*, "A microring resonator-based clock-forwarded DWDM optical interconnect in a monolithic silicon photonics process," *49th European Conference on Optical Communications (ECOC 2023)*, Hybrid Conference, Glasgow, UK, 2023, pp. 1496-1499. https://doi.org/10.1049/icp.2023.2604
[16] N. Nedovic *et al.*, "Carrier Photogeneration Impact on Ring Dynamics," *2025 IEEE Photonics Society Summer Topicals Meeting Series (SUM)*, Berlin, Germany, 2025, pp. 1-2. https://doi.org/10.1109/SUM65312.2025.11121786
[17] N. Nedovic, "Optical Interchannel Interference in Dense Wavelength Division Multiplexing Systems," *IEEE Photonics Society Summer Topicals Meeting Series (SUM)*, Bridgetown, Barbados, 2024, TuE3.1. https://doi.org/10.1109/SUM60964.2024.10614498
[18] Shekhar, S., Bogaerts, W., Chrostowski, L. et al. "Roadmapping the next generation of silicon photonics," *Nat Commun* 15, 751, 2024. https://doi.org/10.1038/s41467-024-44750-0
[19] C. -H. Chen et al., "A comb laser-driven DWDM silicon photonic transmitter with microring modulator for optical interconnect," *2015 Conference on Lasers and Electro-Optics (CLEO)*, San Jose, CA, USA, 2015, pp. 1-2. https://doi.org/10.1364/CLEO_SI.2015.STu4F.1

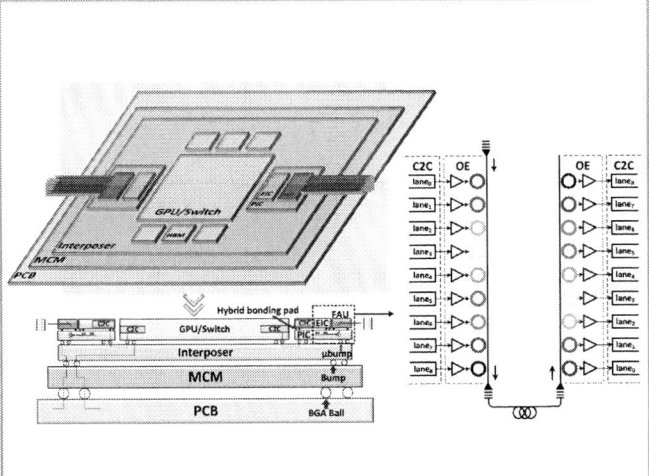

Figure 23.1.1: The proposed clock-forwarding DWDM links for CPO. One of the nine wavelengths, each spaced at 200GHz, is allocated to the FWDCLK, while the remaining eight are designated for data transmission.

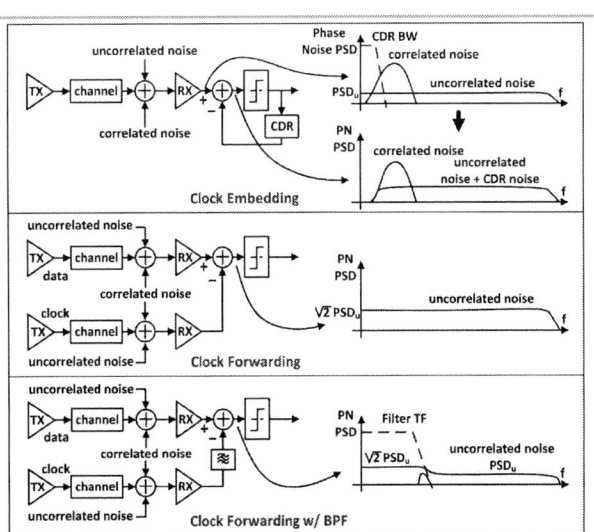

Figure 23.1.2: Phase noise comparison between the clock embedding (top), clock forwarding (middle), and clock forwarding with band-pass filtering (bottom).

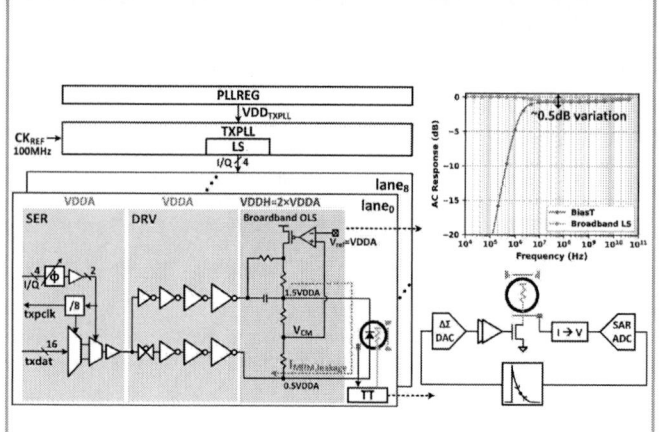

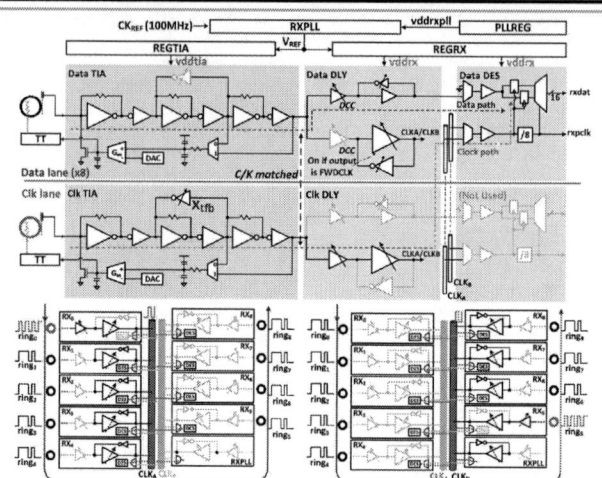

Figure 23.1.3: Left: TX PHY architecture; right: (top) the AC transfer functions of the proposed broadband level shifter vs a typical bias-T based output level shifter; (bottom) thermal tuning loop architecture.

Figure 23.1.4: Top: RX PHY architecture. Feedback inverter x_{tfb} ON for FWDCLK reception. Bottom: (left) RX clock distribution when $ring_0$ receives the FWDCLK. $Ring_0$ path drives CLK_A while $ring_{1-4}$ injection locking; (right) when $Ring_5$ receives the FWDCLK.

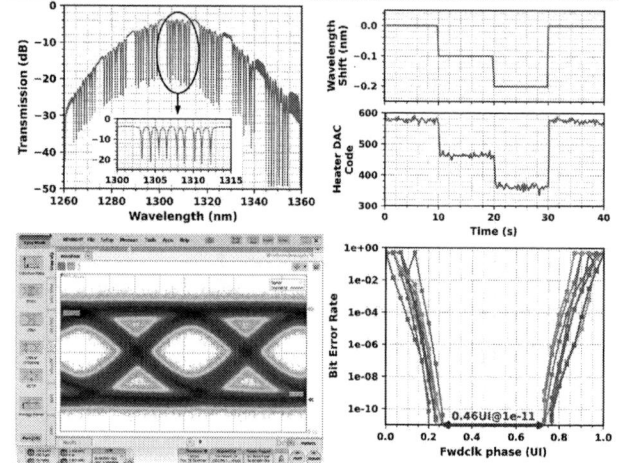

Figure 23.1.5: Top: (left) Frequency sweep of TX bus; (right) Laser wavelength shift and TX thermal tuning loop response. Bottom: (left) DCA captured TX output eye diagram for one λ at 32 Gb/s, PRBS7 pattern; (right) Bathtub curves with all lanes on simultaneously, PRBS31 pattern, TX2-RX6 pair for FWDCLK.

		This work	VLSI'22 [1]	VLSI'23 [9]	ISSCC'23 [3]
Process	EIC	7nm FinFET	45nm SOI CMOS	28nm CMOS	7nm CMOS
	PIC	65nm SOI SiPh		300mm SOI SiPh	45nm SOI SiPh
Integration		Hybrid	Monolithic	Hybrid	Hybrid
Architecture		FC	EC	EC	EC
Number of wavelengths		9	8	8	7
Per-λ Data rate <Gbps>		32	16	32	50
Per-Fiber Data rate <Gbps>		256	128	256	350
TX+RX Lane size <μm²>		0.013	-	~1	0.031 (RX only)
EIC shoreline density[a] <Tbps/mm>		0.8	~0.14	~0.064	
EIC area density[b] <Tbps/mm²>		1.33	~0.064	~0.032	
RX sensitivity OMA <dBm>		-16.0[c]	-	-10	-10.1 to -11.4
BER		<1E-11	<6.9E-12	<1E-12	<1E-12
Energy efficiency <pJ/b>	Laser	0.76[d]	-	-	-
	Thermal	0.76[e]|0.49[f]	<0.6	-	-
	TRX + Clock + TT circuits	1.26	4.9-5	3.8 (RX only)	0.96 (RX only)
	Total	2.78|2.51			

[a] Bi-directional, 1-PHY deep. [b] Bi-directional. [c] Measured with one data lane and one FWDCLK ON, PRBS7 pattern. [d] Assuming 10% fiber-coupled laser WPE. [e] Measured with the bench laser. [f] Estimated based on the measured heater efficiency, covering 3σ of ring and modeled laser variation.

Figure 23.1.6: Comparison with state-of-the-art DWDM links of NRZ modulation.

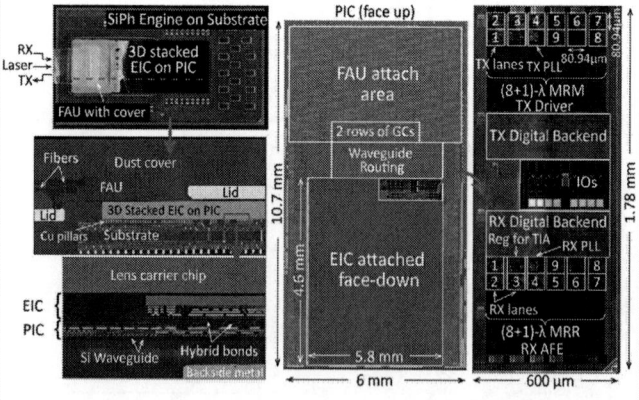

Figure 23.1.7: Optical engine on substrate, assembly cross-section, and die micrograph of EIC and PIC chips.

ISSCC 2026 / SESSION 23 / NEXT-GENERATION OPTICAL TRANSCEIVERS / 23.2

23.2 A 2-Channel 800Gb/s Transceiver for Coherent-Lite Applications with <300ns Latency in 5nm FinFET

Marco Sosio[1], Claudio Nani[1], Enrico Monaco[1], Nicola Ghittori[1], Domenico Albano[1], Alessio Di Pasquo[1], Alessandro Bosi[1], Travis Lovitt[2], Gabriele Gira[1], Fulvio Martinelli[1], Victor Karam[2], Devrishi Khanna[2], Mehdi N. Khiarak[2], Sasan Cyrusian[3], Mehdi Davoodi[3], Marco Garampazzi[1], Nimesh N. Miral[1], Fabio Giunco[1], Ivan Fabiano[1], Nicola Codega[1], Claudio Asero[1], Daniel L. Herbas[1], Enrico Temporiti[1], Shawn Scouten[2], Kishore Kota[3], Yang Fu[4], Ruibin Jin[3], Josef Mueller[4], Michael Leung[3], Arash Farhoodfar[4], Stephen Jantzi[3], Ken Chang[4]

[1]Marvell, Pavia, Italy, [2]Marvell, Ottawa, Canada, [3]Marvell, Irvine, CA, [4]Marvell, Santa Clara, CA

Abstract

A full transceiver suitable for coherent-lite optical communications over 2-to-20km distances is presented. The low-latency transceiver combines the advantages of advanced coherent modulation with the benefits and manufacturing scale of O-band optics to deliver a power- and cost-effective module. Fabricated in 5nm FinFET, it comprises two 400Gb/s/λ transmitters and receivers to realize a 2-channel DP-16QAM 800Gb/s link tailored to the market of distributed campus data center interconnects.

The rapid growth of AI and cloud services is increasing data center complexity, leading to challenges in physical space and power delivery. Data-center architectures are thus shifting from large-scale facilities to distributed campus-based architectures, increasing the need for optimized interconnects in the 2-to-40km range.

Currently, intensity modulation direct detection (IMDD) is the predominant technology in data centers, while coherent interconnects are used for regional connectivity (Fig. 23.2.1). However, IMDD hits a performance limit as reach and link-speed requirements increase. Concurrently, traditional coherent solutions become extremely power-inefficient at shorter distances. The coherent-lite solution introduced in this paper combines the benefits of IMDD and coherent technologies [1,2], enabling low-latency and power-optimized 2-40km communications at 400Gb/s per wavelength with DP-16QAM modulation. By leveraging the favorable trade-off between chromatic dispersion and optical attenuation in the O-band to limit DSP equalization complexity, the presented architecture achieves over 10× lower latency than conventional coherent systems and 2× higher power efficiency, positioning it as a compelling candidate for next-generation AI interconnects. The transceiver also integrates a software-reconfigurable high-swing optical modulator driver, reducing module power consumption and component count. Two channels are combined in a single die, resulting in an 800Gb/s aggregate data rate, suitable for integration into 400G/800G/1.6T pluggable modules. The device is fabricated using a 5nm FinFET process.

Figure 23.2.1 shows the block diagram of a 400Gb/s coherent-lite channel. A single clocking section (CLK) drives four line transmitter (LTX) or line receiver (LRX) lanes in a quad configuration. Each quad supports a 400Gb/s data-stream at the electrical interface, handling horizontal and vertical polarizations (X, Y) and in-phase and quadrature-phase modulations (I, Q). DP-16QAM modulation and demodulation are performed in the optical domain, utilizing highly integrated silicon photonics O-band optics. In the LRX path, external transimpedance amplifiers (TIAs) provide low noise amplification and maximize receiver sensitivity.

The quad LTX block diagram (Fig. 23.2.2) includes a fractional-N PLL, generating a 31.25GHz half-rate clock, and per-lane phase interpolators (PI). The PI output clocks are distributed across the four lanes (XI, XQ, YI, YQ). To ensure the synchronization required for coherent modulation, a skew-alignment calibration system compensates on-die and off-die skews by operating on the PI. The TX DSP includes convolution interleaving and inner FEC encoding. XI, XQ, YI, YQ data streams are generated by the DP-16QAM mapper and pilot symbol insertion. A 7-tap FIR filter enables signal equalization and mitigation of reflection artifacts [3]. The block diagram of an LTX lane is shown in Fig. 23.2.2. The 7b × 64 data stream from the DSP is fed to the serializer, which performs the 64:1 parallel to serial conversion. A full-speed pre-driver delivers the signal to a 7b DAC-based driver to transmit data in PAM-4 format. To address clock distribution impairments, a duty cycle correction (DCC) mechanism is adopted. Power efficiency and optimized jitter performance across PVT are achieved through local automatic voltage scaling (AVS), dynamically adjusting the supply levels.

A Class-AB architecture [4,5], driving both the NMOS and PMOS sections of the driver and pushing and pulling current into a differential termination, has been chosen for the pseudo-differential DAC driver to improve power efficiency (Fig. 23.2.2). The combination of AC-coupler-based level shifters and differential latches, used inside each bit within the DAC, enables consistent performance at high frequency and immunity to wandering effects [6]. Double cascodes provide low-impedance current collection extending the bandwidth and shielding the DAC from high-voltage swings. Dynamic biasing ensures long-term reliability [7]. The DAC driver slices feed into a differential termination (RTERM) and TCOIL. The RTERM is reconfigurable to optimize performance with a variety of Mach-Zehnder modulator (MZM) designs.

The DAC driver is reconfigurable as either an integrated silicon photonics (SiPho) driver, delivering $4V_{ppd}$ to a 66Ω differential load, or a standard (STD) driver, delivering $1V_{ppd}$ to a 100Ω load. In the former case, the integrated circuit directly drives a SiPho MZM modulator, while in the latter an external high-swing driver needs to be added at the module level, reducing the overall efficiency. The driver can be reconfigured to STD mode by operating on PMOS-bias and PMOS-NMOS cascode blocks and reducing LTX voltage supply.

Figure 23.2.3 shows the LRX architecture which operates with an 8/7 oversampling ratio using a half rate 35.7GHz clock. The LRX PLL architecture is the same as the LTX PLL, except for a dedicated VCO design. The PLL output clock is distributed to the XI, XQ, YI, YQ lanes. A coarse skew detection and alignment scheme ensures clock and data synchronization across lanes [8]. Residual skews are then compensated in the digital equalization engine.

The LRX lane, illustrated in Fig. 23.2.3, has been designed to achieve an SNDR in excess of 32dB up to 30GHz. An external TIA converts current into voltage and drives the analog front end. The internal 100Ω termination is DC coupled directly to the input pads, while the VGA input is AC coupled. The VGA is implemented as a single-stage PMOS-based CML amplifier utilizing programmable MOS transistors for resistive degeneration [3]. Bandwidth extension is achieved through cross-coupled MOM capacitors, implementing a negative capacitance at the VGA input, and inductive peaking. The gain is programmable between -2.5dB and 6dB. The VGA drives a 71.4GS/7b ADC with a 64-way time-interleaved architecture based on a partial-loop-unrolled SAR structure [9]. It adopts a hierarchical scheme, where 8 track-and-hold units each feeds 8 sub-ADCs. A dedicated clock generation block (CLK GEN) produces the required 8Ts and 64Ts sampling phases from the PLL clock. This block also includes delay lines to fine tune the track-and-hold units, achieving timing alignment with sub-100fs precision.

The RX DSP (Fig. 23.2.3) implements a low-power architecture suitable for low-latency coherent-lite modules. A 4×4 real multiple-input-multiple-output (MIMO) equalizer handles bandwidth and fiber dispersion effects, as well as carrier and polarization rotations introduced by the coherent optical modulator/demodulator and the optical fiber. It is implemented as a poly-phase filter with 8/7 oversampled inputs and symbol-rate outputs. This architecture is suitable for O-band designs with low chromatic dispersion tolerance requirements and, compared to coherent DSPs [10], provides lower latency and better immunity to skew, gain and quadrature phase imbalances between the four input lanes, as well as lower power dissipation. A carrier frequency/phase recovery block uses known pilot symbols, spaced every 64 symbols, to handle LO frequency offset and laser linewidth effects. A timing recovery block, controlling the PLL, generates a local LRX sampling clock that is synchronous to the transmitter clock. A post-processing block improves robustness to optical impairments and is key to the design of low-cost and low-power coherent-lite optical modules. It features adaptive level tracking and skew adjustment filters to provide robustness to transmit gain, skew, offset and quadrature phase imbalance impairments. A high-performance, low-power and low-latency forward error-correction (FEC) decoder is used to recover the transmit bits. The FEC concatenates an inner BCH(126,110) code with an outer RS(544,514) code, which is used in IEEE 802.3 compliant chip-to-module (C2M) interfaces. The inner code is added without termination of the RS code in either the TX or the RX paths to minimize latency. The inner code is decoded using a simple soft-input/hard-output Chase decoder [11], which results in an input BER threshold of $1×10^{-2}$, to achieve $<2.2×10^{-4}$ BER at the C2M interface and $<1×10^{-15}$ BER after RS FEC decoding in the host device. The latency from ADC outputs to the FEC decoder input is <20ns.

A photograph of the 5nm FinFET die is shown in Fig. 23.2.7. It features 8×100Gb/s PAM-4 TX and RX lanes on the host side and 2×400Gb/s coherent-lite channels on the line side.

Figure 23.2.4 presents the 125Gb/s PAM-4 eye-diagrams of the LTX lane for both STD and SiPho modes. The post-processed SNDR/bandwidth are 30.8dB/31GHz and 28dB/35GHz respectively. For the integrated SiPho mode, the driver is connected to the 100Ω oscilloscope load, though optimized for 66Ω. The measured random jitter of the TX path in both STD and SiPho mode is $<120fs_{rms}$.

The THD, SNR and SNDR of the LRX lane with -1dBFS signal are shown in Fig. 23.2.4 as a function of frequency. SNDR_nocal with de-activation of the calibration of sampling time interleaving errors is also reported to replicate the condition of loopback measurements.

Both electrical and optical loopback measurements have been performed with an LTX transmitting to an LRX (Fig. 23.2.5). The DP-16QAM constellations for X and Y polarizations measured in a 1.6T OSFP-XD module running without any optical amplification over 40km of standard single-mode fiber are shown in Fig. 23.2.5. The constellations are recovered at the LRX DSP in an O-band optical setup, with the transmit signal generated using the

398 • 2026 IEEE International Solid-State Circuits Conference

979-8-3315-8937-0/26 $31.00 © 2026 IEEE

integrated driver and a SiPho MZM suitable for small-form-factor optical modules. A better than 5×10⁻⁴ BER at the input of the FEC decoder is achieved, demonstrating substantial margin to the $1×10^{-2}$ decoder BER threshold. The overall round-trip latency, including the C2M interface, is <300ns, far lower than traditional coherent DSP modules (such as 400ZR [12]) which have latencies >3μs. The low power dissipation of the implementation is confirmed by the ability to build and cool a 1.6T OSFP-XD module. The device demonstrates the ability to tolerate high levels of mismatch between lanes (e.g. up to 3dB TX I-Q gain imbalance, substantially higher than the 1dB specification in 400ZR) and laser linewidths as high as 1MHz (compared to the <500kHz requirement in 400ZR), allowing integration of low-cost DFB lasers, as commonly used in IMDD systems. This increases module yields and lowers cost to be competitive with IMDD designs.

A comparison with state-of-the-art complete transceivers adopting coherent technologies is reported in Fig. 23.2.6. The proposed coherent-lite transceiver achieves twice the analog power efficiency compared to state-of-the-art conventional coherent systems, representing the optimal trade-off to address the growing market of interconnects for distributed campus-scale data center architectures. Furthermore, it is the first to enable the integration of a large-swing driver compatible with silicon photonics MZM.

Acknowledgement:
The authors would like to thank the Accelink team for the optical module assembly and optical measurements. They would like also to thank the entire team who contributed to this project and especially the analog layout and validation teams.

References:
[1] X. Zhou et al., "State-of-the-Art 800G/1.6T Datacom Interconnects and Outlook for 3.2T", Optical Fiber Communication Conference and Exhibition, March 2023, Digital Object Identifier: 10.1364/OFC.2023.W3D.1. http://doi.org/10.1364/OFC.2023
[2] Xiaoyan Ye et al., "Simplified Transceivers for Short-Reach Coherent-Lite Systems", Journal Of Lightwave Technology, VOL. 43, NO. 13, JULY 1, 2025, Digital Object Identifier: 10.1109/JLT.2025.3581121. http://doi.org/10.1109/JLT.2025.3581121
[3] F. Giunco et al., "An Eight-Lane 800-Gb/s Transceiver for PAM-4 Optical Direct-Detection Applications in 5-nm FinFET Process", Journal of Solid-State Circuits, Volume: 60, Issue: 4, pp: 1210-1222, 2025, Digital Object Identifier: 10.1109/JSSC.2025.3541174. http://doi.org/10.1109/JSSC.2025.3541174
[4] J. He et al., "A 56-Gb/s Reconfigurable Silicon-Photonics Transmitter Using High-Swing Distributed Driver and 2-Tap In-Segment Feed-Forward Equalizer in 65-nm CMOS", Transactions on Circuits and Systems I, vol. 69, pp. 1159-1170, March 2022, Digital Object Identifier: 10.1109/TCSI.2021.3127723. http://doi.org/10.1109/TCSI.2021.3127723
[5] N. Codega et al., "A Monolithic 400Gbps Electro-Optical Retimer with Integrated TIA and Class-AB Silicon-Photonics/VCSEL Driver in 5nm FinFET", 2025 Symposium on VLSI Technology and Circuits, Digital Object Identifier: 10.23919/VLSITechnologyandCir65189.2025.11074826. http://doi.org/10.23919/VLSITechnologyandCir65189.2025.11074826
[6] H. Li et al., "A 3D-Integrated Microring-Based 112Gb/s PAM-4 Silicon-Photonic Transmitter with Integrated Nonlinear Equalization and Thermal Control", International Solid-State Circuits Conference, Feb. 2020, pp. 208-209, Digital Object Identifier: 10.1109/ISSCC19947.2020.9063122. http://doi.org/10.1109/ISSCC19947.2020.9063122

[7] H. Sepehrian, S. Alie, P. Madeira, and D. Tonietto, "106 Gb/s PAM-4 Transmitter With 2.1 Vppd Swing in 7nm FinFET Process", IEEE Symposium on VLSI Circuits, 2021, Digital Object Identifier: 10.23919/VLSICircuits52068.2021.9492349. http://doi.org/10.23919/VLSICircuits52068.2021.9492349
[8] R. L. Nguyen et al., "A Highly Reconfigurable 40-97GS/s DAC and ADC with 40GHz AFE Bandwidth and Sub-35fJ/conv-step for 400Gb/s Coherent Optical Applications in 7nm FinFET", International Solid-State Circuit Conference, Feb. 2021, pp. 136-137, Digital Object Identifier: 10.1109/ISSCC42613.2021.9365746. http://doi.org/10.1109/ISSCC42613.2021.9365746
[9] C. Nani et al., "A 5nm 60GS/s 7b 64-way Time Interleaved Partial Loop Unrolled SAR ADC Achieving 35.2dB SNDR Up To 32GHz", Journal of Solid-State Circuits, Volume: 60, Issue: 4, pp: 1210-1222, 2025, Digital Object Identifier: 10.1109/JSSC.2024.3517333. http://doi.org/10.1109/JSSC.2024.3517333
[10] S. J. Savory, "Digital Coherent Optical Receivers: Algorithms and Subsystems", Journal of Selected Topics in Quantum Electronics, vol. 16, no. 5, pp. 1164-1179, Sept.-Oct. 2010, Digital Object Identifier: 10.1109/JSTQE.2010.2044751. http://doi.org/10.1109/JSTQE.2010.2044751
[11] D. Chase, "Class of algorithms for decoding block codes with channel measurement information", Transactions on Information Theory, vol. 18, no. 1, pp. 170-182, January 1972, Digital Object Identifier: 10.1109/TIT.1972.1054746. http://doi.org/10.1109/TIT.1972.1054746
[12] Optical Internetworking Forum (OIF). Implementation Agreement 400ZR. OIF-400ZR-03.0. October 2024. https://www.oiforum.com/technical-work/implementation-agreements-ias
[13] G. Li et al., "A 600Gb/s DP-QAM64 coherent optical transceiver frontend with 4×105GS/s 8b ADC/DAC in 16nm CMOS", International Solid-State Circuits Conference, Feb. 2024, pp. 338–340, Digital Object Identifier: 10.1109/ISSCC49657.2024.10454499. http://doi.org/10.1109/ISSCC49657.2024.10454499
[14] J. Cao et al., "A Transmitter and Receiver for 100Gb/s Coherent Networks with Integrated 4×64GS/s 8b ADCs and DACs in 20nm CMOS", International Solid-State Circuits Conference, Feb. 2017, pp. 484-485, Digital Object Identifier: 10.1109/ISSCC.2017.7870472. http://doi.org/10.1109/ISSCC.2017.7870472

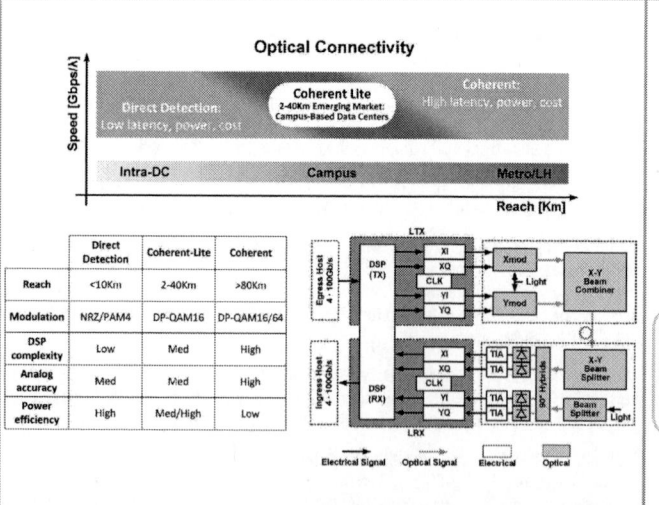

Figure 23.2.1: Coherent-lite application. Block diagram of the 400Gb/s coherent-lite transceiver.

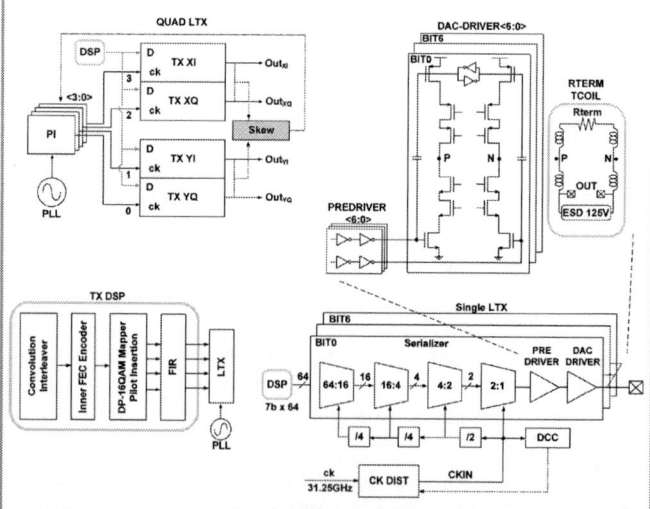

Figure 23.2.2: LTX quad block diagram. Single LTX block diagram with class-AB driver details. LTX DSP.

ISSCC 2026 / SESSION 23 / NEXT-GENERATION OPTICAL TRANSCEIVERS / 23.2

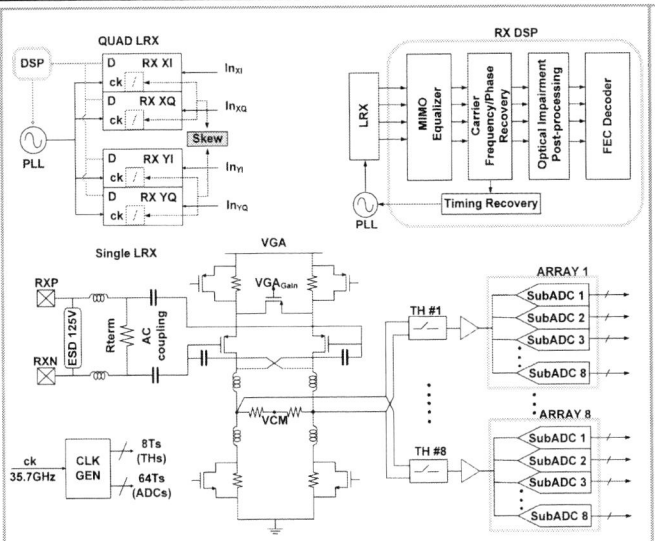

Figure 23.2.3: LRX quad block diagram. Single LRX block diagram. LRX DSP.

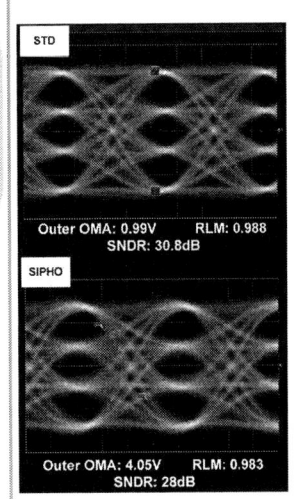

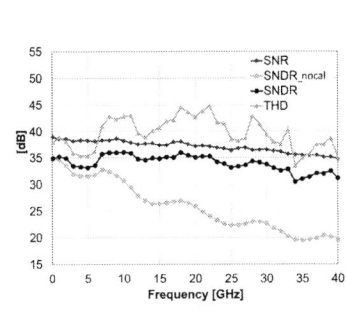

Figure 23.2.4: Measured 125Gb/s LTX eye diagrams (STD and SiPho mode). Measured LRX performance vs input frequency.

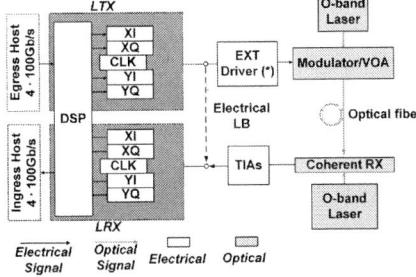

	This Work	[8]	[13]	[14]
Channel Datarate [Gb/s]	400	400	600	100
Modulation	DP-16QAM	PS-16QAM	DP-64QAM	DP-QPSK
Process	5n Finfet	7n Finfet	16nm Finfet	20nm
RX and TX resolution [bits]	7	8	8	8
RX Speed [GS/s]	71.4	97	105	64
RX SNDR low freq. [dB]	35	41.1	42.8	39.7
RX SNDR high freq. [dB]	31@35.7G	32@30G	39.2@20.9G	32.5@19G
TX speed [GS/s]	62.5	97	105	64
Optical driver support	SiPho	No	No	No
TX Swing SiPho/STD [Vppd]	4/1	No/1.1	No/1.2	No/0.7
TX Linear Fit SNDR SiPho/STD [dB]	28/30.8	No/NA	No/NA	No/NA
Analog Power SiPho [W]	2.2@400G	No	No	No
Analog Power STD [W]	1.49@400G	3@400G	4.5*@600G	6.3*@100G
Analog Power Efficiency SiPho [pJ/b]	5.5	No	No	No
Analog Power Efficiency STD [pJ/b]	3.72	7.5	7.5*	63*

* TX and RX only

Figure 23.2.6: Comparison table.

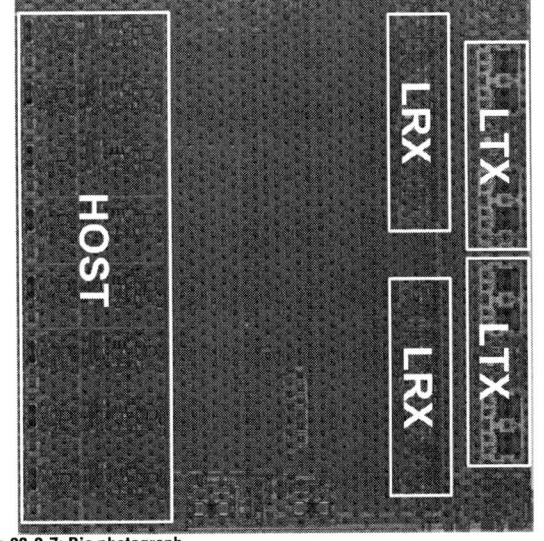

Figure 23.2.5: Loopback setup with (*) EXT driver bypassed in SiPho mode. Measured DP-16QAM constellations.

Figure 23.2.7: Die photograph.

• 2026 IEEE International Solid-State Circuits Conference

ISSCC 2026 / SESSION 23 / NEXT-GENERATION OPTICAL TRANSCEIVERS / 23.3

23.3 A 2×500Gb/s Monolithic Silicon-Photonic DWDM PAM-4 Transceiver in 45nm CMOS SOI

Ziang Xu*, Yalong Lin*, Jinxuan Jin*, Zhenkai Ye*, Dacheng Xu, Aolin Xu, Chaodi Sheng, Binwen Hong, Xiaojun Bi

Huazhong University of Science and Technology, Wuhan, China
*Equally Credited Authors (ECAs)

Abstract

This work presents a monolithic 2×5λ DWDM microring transceiver in 45nm CMOS SOI delivering 2×500Gb/s (100Gb/s PAM-4/channel). The transceiver features a low-noise TIA with Q-tamed CTLE, a high-speed MRM driver, and a PWM-based IL-adjustable wavelength-locking circuit. V-groove and a PSR-to-CRR front end enable low coupling loss and tight λ-spacing. Through electro-photonic co-design, the transceiver achieves an efficiency <2.5pJ/b at 100Gb/s per channel in end-to-end loopback.

The rapid growth of AI and machine learning (ML) is driving unprecedented bandwidth demands in hyperscale datacenters [1], necessitating short-reach, high-density optical interconnects with improved energy efficiency. Silicon photonics offers a scalable solution through monolithic integration of photonic and electronic devices, which eliminates pad and wire-bonding parasitics and enables direct fiber-to-chip coupling. Prior monolithic microring transceivers have demonstrated the potential of this approach [2-5], though channel data rates remain limited to 64Gb/s NRZ [1-2] and 40Gb/s PAM-4 [3], which are insufficient for 800G and future 1.6T modules [6]. While PAM-4 microring transceivers [3], [7], [8] improve bandwidth density and energy efficiency, several challenges persist:

1) The microring modulator (MRM) needs to balance bandwidth, linearity, and optical modulation amplitude (OMA) when operating at different insertion loss (IL) points, to address dynamic application requirements [9-11].

2) With denser wavelength division multiplexing (WDM) requiring narrower wavelength spacing, sharp microring resonator (MRR) filters can effectively reduce crosstalk but at the expense of limiting the demultiplexed bandwidth.

3) Thermal drift of MRR resonance requires active stabilization and monitor photodiode (MPD) current mismatches exist between channels in multiple channel wavelength locking (WLL) situations.

4) A fundamental trade-off exists among power consumption, compensation capability, and frequency response in the continuous-time linear equalizer (CTLE) required in the transceiver link.

These limitations motivate the need for advanced co-design techniques to further enhance performance.

As seen in Fig. 23.3.1, this work presents a monolithic 2×500Gb/s dense wavelength division multiplexing (DWDM) microring transceiver in 45nm CMOS SOI. A co-designed electro-photonic architecture implements a 2×5λ transceiver supporting 100Gb/s PAM-4 per channel. V-groove couplers are employed to achieve lower fiber-to-chip coupling loss [12]. In the TX, microrings with a radius of 7.5μm and a pre-alignment of 0.054μm exhibit a free spectral range (FSR) of 9.267nm and a channel spacing of 1.853nm (≈ 324GHz). At the RX input, V-groove couplers, on-chip polarization splitter rotator (PSR), coupled-ring resonator (CRR) filters and dual-detector PD are deployed to enhance crosstalk suppression and enable tighter channel spacing, while the increased RX bandwidth helps maintain signal integrity. Furthermore, each channel integrates an on-chip WLL circuit, ensuring both MRMs and CRRs operate at their optimal wavelength points without requiring extra off-chip control logic.

Figure 23.3.2 depicts the implemented transimpedance amplifier (TIA), which employs a low-noise bandwidth-enhanced architecture for 112Gb/s PAM-4 operation. The transimpedance stage (TIS) utilizes a large feedback resistor to achieve high front-end gain and low input-referred noise, which intensifies the design constraints for subsequent equalization. Conventional CTLE solutions face critical limitations: source-degenerated topologies sacrifice gain for peaking generation with poor scalability beyond 30GHz, while inductive peaking techniques suffer from peaking amplitude versus response shape trade-offs – high peaking amplitude reduces peaking frequency and elevates Q-factor, degrading time-domain performance. To overcome these challenges, a cascaded Q-tamed CTLE is introduced. The CTLE adopts a 2-stage Gm-ZT topology with series inductive peaking (CTLE1 + CTLE2), while a Q-tamed path (QTP) is designed for frequency response shaping. The QTP is implemented through a hybrid passive low-Q inductor and Gm-based active inductor between stages. As shown in Fig. 23.3.2, the QTP splits the resonance peaks of CTLE1 and CTLE2 (originally coincident at 41GHz), boosts mid-band gain (10~30GHz), and suppresses the overall Q-factor. The Q-tamed CTLE achieves 6.1dB baseline gain with 12.4dB peaking at 41GHz. Under the same output voltage swing, a 60% improvement in eye height and a 30% improvement in eye width were achieved by time-domain simulation. The output buffer after the CTLE leverages inverter-based architecture for 50Ω matching. The TIA achieves 62dBΩ gain, 43.5GHz bandwidth, and 3.14μA$_{rms}$ input-referred noise while consuming 67mW, validating the design as a breakthrough solution reconciling high-frequency peaking, low-Q response, and power efficiency in >100G receiver front ends.

Figure 23.3.3 shows the architecture of the MRM driver, consisting of a three-stage preamplifier, a level shifter, and a main driver. The preamplifier converts single-ended MSB/LSB inputs into differential signals, provides equalization and gain, and employs shunt peaking and negative capacitance to achieve a bandwidth of >30GHz. The level shifter generates full-swing outputs in two voltage domains (0–VDDL and VDDL–VDDH) using stacked inverters for high swing capability. The main driver then combines the two NRZ signals into PAM-4. A series-peaking inductor with a tunable parallel resistor forms an R+L CTLE, which reduces swing to 2.5V$_{ppd}$ while extending output bandwidth. An on-chip bias-T provides DC biasing and enables direct connection to the MRM anode and cathode.

The architecture concept diagram of the WLL circuit is shown in Fig. 23.3.4. At the front end, an MPD current (I$_{ph}$) amplifier is designed, which is automatically adjusted by a 6b digital signal to regulate the I$_{ph}$ gain, eliminating mismatches between channels in the WDM system and handling variations in incoming optical power and system optical path losses. This ensures the best matching of I$_{ph}$ with the circuit's dynamic input range. A 1-bit bang-bang ADC is used for A/D conversion, thereby reducing hardware overhead compared to a traditional SAR ADC. The locking with adjustable IL scheme is implemented as follows: the circuit first adjusts the gain of the I$_{ph}$ amplifier in a SWEEP & ADJUST mode to match the peak values of I$_{ph}$ and IL reference DAC, the latter of which is initialized to V$_{max}$. Subsequently, by changing the DAC input code to V$_{IL}$ via the on-chip SPI interface, where Desired IL = 1-(V$_{IL}$/V$_{max}$), the circuit achieves wavelength locking precisely to the target IL point by tuning the microring so that the resulting voltage of I$_{ph}$ matches V$_{IL}$. This scheme is compatible with WLL applications for both TX MRM and RX CRR and addresses the bandwidth, linearity, and OMA trade-offs presented by MRM at different IL points to meet dynamic application requirements. The effective number of bits of the thermal control DAC is extended to 14b by PWM, which further reduces system area and complexity, with a size advantage compared to the previously used delta-sigma modulation module. The output driver stage achieves thermal-tuning current exceeding 15mA.

Figure 23.3.5 shows eye diagrams for the 50GBaud PAM-4 TX and 56GBaud PAM-4 RX. The TX achieves a 4.82dB extinction ratio and an RLM of 0.99 at 100Gb/s PAM-4 under a –6.2dB IL operating point, with MSB/LSB skew applied. The figure also plots BER versus received optical power for the five RX channels at 112Gb/s PAM-4, demonstrating a sensitivity of –5.3dBm at the BER limit of 2.4e-4. The end-to-end TRX attains an energy efficiency below 2.5pJ/b. The test results for WLL are also illustrated in Fig. 23.3.5, showing the SWEEP & ADJUST mode clearly after system reset, followed by locking and bang-bang thermal tuning processes. By configuring different IL points, the microring resonance can be locked to various IL values, successfully validating operation at 6dB, 3dB, 1.5dB, and near the resonance peak. Thermal stabilization was verified under an environmental thermal disturbance of approximately 40°C. The measured thermal tuning efficiency is 0.3nm/mW for TX MRM, and is 0.42nm/mW for CRR, with corresponding tuning ranges are 5.5nm and over 7nm, respectively. The total power consumption of the entire WLL circuit is 26.7mW, with the control part consuming only 7.9mW.

Figure 23.3.6 shows the performance comparison between this work and previously reported MRM TRXs. Compared with other monolithic integrations, this work achieves the highest single-channel data rate of 100Gb/s and an energy efficiency of <2.5pJ/b under PAM-4 modulation. It also demonstrates leading performance in terms of bandwidth density and total bandwidth per fiber.

Figure 23.3.7 shows the monolithic 2×500 Gb/s microring transceiver, fabricated in 45nm CMOS SOI, integrating ten-channel drivers and TIAs, optical components, and V-groove couplers. The area of a single TX channel is 400μm × 550μm, and that of a single RX channel is 400μm × 535μm. In the TX/RX, the area of a single WLL is 0.063/0.15mm², respectively, and the overall energy efficiency of TX/RX is 1.82pJ/b and 0.6pJ/b, respectively.

Acknowledgement:

The authors would like to thank M.M. Zhang, S. Zheng, Z.Y. Zhou, and Z.H. Sun from Huazhong University of Science and Technology for their assistance with the design and deployment of the optical components. This work was supported in part by the National Key Research and Development Program of China under Grant 2022YFB2802600, and in part by the National Natural Science Foundation of China under Grant 62571210. Corresponding author: Xiaojun Bi

ISSCC 2026 / February 17, 2026 / 4:25 PM

References:

[1] R. Nagarajan, I. Lyubomirsky, and O. Agazzi, "Low power DSP-based transceivers for data center optical fiber communications (invited tutorial)," *Journal of Lightwave Technology*, vol. 39, no. 16, pp. 5221–5231, Aug. 2021.
https://doi.org/10.1109/JLT.2021.3089901

[2] N. Qi *et al.*, "A monolithically integrated DWDM si-photonics transceiver for chiplet optical I/O," *IEEE Journal of Solid-State Circuits*, pp. 1–13, 2025.
https://doi.org/10.1109/JSSC.2025.3585584

[3] A. Sadr and A. Chan Carusone, "A monolithic microring modulator-based transmitter with a multiobjective thermal controller," *IEEE Open Journal of the Solid-State Circuits Society*, vol. 4, pp. 340–350, 2024. https://doi.org/10.1109/OJSSCS.2024.3507754

[4] K. Omirzakhov, H. Hao, A. Pirmoradi, and F. Aflatouni, "Energy efficient monolithically integrated 256 gb/s optical transmitter with autonomous wavelength stabilization in 45 nm CMOS SOI," *IEEE Journal of Solid-State Circuits*, pp. 1–10, 2024.
https://doi.org/10.1109/JSSC.2024.3511673

[5] K. Omirzakhov and F. Aflatouni, "Monolithically integrated sub-63 fJ/b 8-channel 256Gb/s optical transmitter with autonomous wavelength locking in 45nm CMOS SOI," in *2024 IEEE International Solid-State Circuits Conference (ISSCC)*, Feb. 2024, pp. 218–220. https://doi.org/10.1109/ISSCC49657.2024.10454519

[6] P. Dong *et al.*, "Silicon photonics for 800G and beyond: [invited]," in *2022 Optical Fiber Communications Conference and Exhibition (OFC)*, Mar. 2022, pp. 01–03. Accessed: Jul. 31, 2025. [Online]. https://doi.org/10.1364/OFC.2022.M4H.1

[7] S. Ma *et al.*, "A 4λ×128Gb/s PAM-4 si-photonic transmitter with micro-ring modulator and co-designed linear driver for chiplet optical I/O," in *Optical Fiber Communication Conference (OFC) 2025 (2025)*, paper M2H.2, Optica Publishing Group, Mar. 2025, p. M2H.2. https://doi.org/10.1364/OFC.2025.M2H.2

[8] H. Li *et al.*, "A 3D-Integrated Microring-Based 112Gb/s PAM-4 Silicon-Photonic Transmitter with Integrated Nonlinear Equalization and Thermal Control," in *2020 IEEE International Solid-State Circuits Conference - (ISSCC)*, Feb. 2020, pp. 208–210.
https://doi.org/10.1109/ISSCC19947.2020.9063122

[9] H. Li *et al.*, "A 112 gb/s PAM4 silicon photonics transmitter with microring modulator and CMOS driver," *Journal of Lightwave Technology*, vol. 38, no. 1, pp. 131–138, Jan. 2020. https://doi.org/10.1109/JLT.2019.2938731

[10] S. Agarwal, M. Ingels, M. Pantouvaki, M. Steyaert, P. Absil, and J. Van Campenhout, "Wavelength locking of a si ring modulator using an integrated drop-port OMA monitoring circuit," *IEEE Journal of Solid-State Circuits*, vol. 51, no. 10, pp. 2328–2344, Oct. 2016.
https://doi.org/10.1109/JSSC.2016.2592691

[11] C. S. Levy *et al.*, "8-λ × 50 gbps/λ heterogeneously integrated si-ph DWDM transmitter," *IEEE Journal of Solid-State Circuits*, vol. 59, no. 3, pp. 690–701, Mar. 2024.
https://doi.org/10.1109/JSSC.2023.3344072

[12] W. Kocon, Y. Bian, K. Giewont, and T. Letavic, "Key technologies and performance aspects for electrical and optical interconnects," in *2025 Symposium on VLSI Technology and Circuits (VLSI Technology and Circuits)*, Jun. 2025, pp. 1–3.
doi: 10.23919/VLSITechnologyandCir65189.2025.11075044.
https://doi.org/10.23919/VLSITechnologyandCir65189.2025.11075044

[13] Z. Xuan et al., "A 256 gbps heterogeneously integrated silicon photonic microring-based DWDM receiver suitable for in-package optical I/O," in 2023 IEEE Symposium on VLSI Technology and Circuits (VLSI Technology and Circuits), Kyoto, Japan: IEEE, June 2023, pp. 1–2.

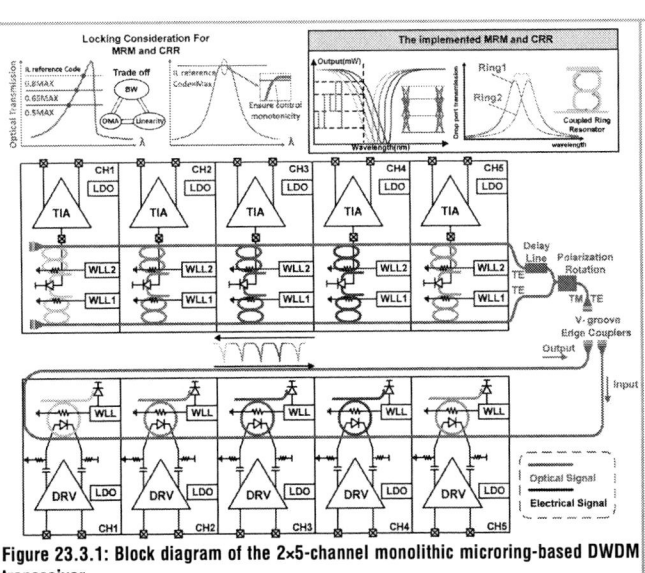

Figure 23.3.1: Block diagram of the 2×5-channel monolithic microring-based DWDM transceiver.

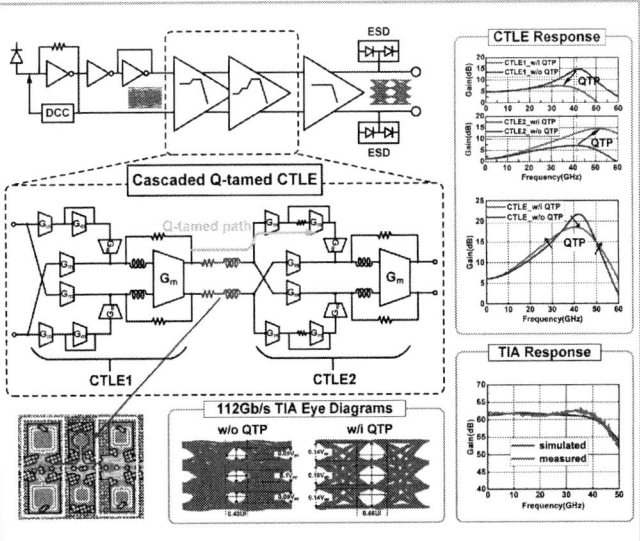

Figure 23.3.2: Architecture of cascaded Q-tamed CTLE-based TIA.

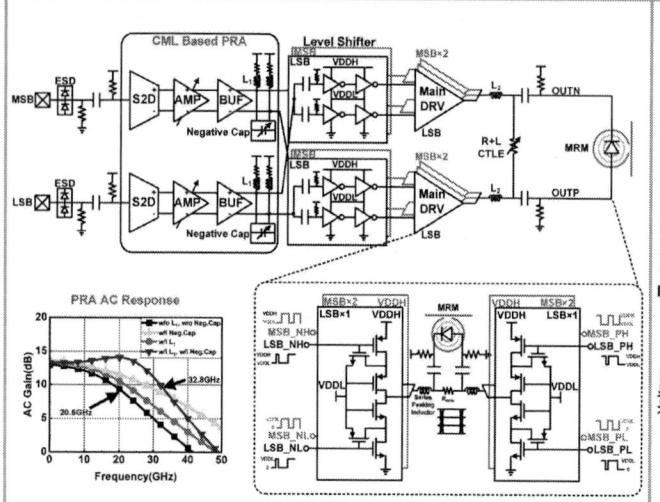

Figure 23.3.3: Architecture of the MRM driver.

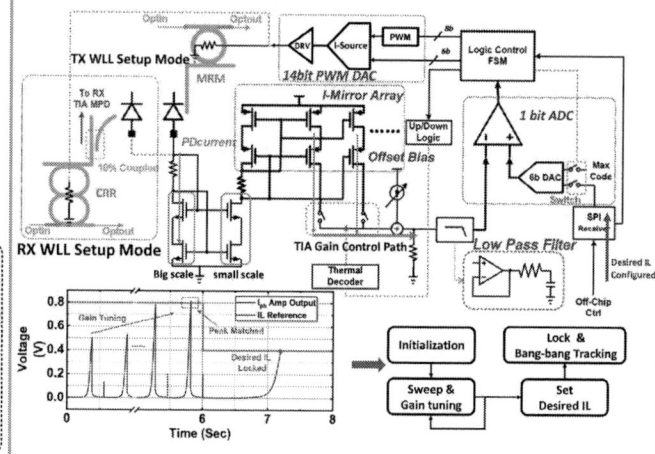

Figure 23.3.4: Schematic and concept of the wavelength-tuning circuit.

Figure 23.3.5: Measurement results.

Figure 23.3.6: Performance summary and comparison.

	JSSC'24[11](TX) VLSI'23[13](RX)	OJSSC'24[3]	ISSCC'24[5]	JSSC'25[2]	This Work
Integration approach	Flip-chip	Monolithic	Monolithic	Monolithic	Monolithic
Technology	28nm CMOS	45nm CMOS SOI	45nm CMOS SOI	45nm CMOS SOI	45nm CMOS SOI
Locking Method	IL Based	RLM, ER, OMA	IL Based	IL Based (Off-Chip DSP)	IL Based (Adjustable)
Wavelength Tuning Range	11.5nm	N/R	N/R	N/R	5.5nm(MRM) >7nm(CRR)
Wavelength Tuning Resolution	14b DSM	12b	5b Capacitive + 7b Thermal	12b	14b PWM
Channel Speed (Gb/s)	50(TX) 32(RX)	40(TX) PAM-4	32(TX)	64(TX) 64(RX)	100(TX) 112(RX)
Total BW (Gb/s)	400(TX) 256(RX)	N/R	256(TX)	256(TX) 256(RX)	1000(TX) 1120(RX)
BW Density (Tb/s/mm²)	0.03(TX+RX)	N/R	13.25(TX)	0.226(TX+RX)	0.551*/0.745(TX+RX)
Power Efficiency (pJ/bit)	2.5*(TX) 3.8*(RX)	1.7*(TX)	0.328*(TX)	1.6(TX) 1.25(RX)	1.82/1.90*(TX)@100Gb/s 0.67/0.83*(RX)@100Gb/s 0.60/0.74*(RX)@112Gb/s

* Includes the wavelength locking circuit

Figure 23.3.7: Micrograph of the chip and the TRX power breakdown.

ISSCC 2026 / SESSION 23 / NEXT-GENERATION OPTICAL TRANSCEIVERS / 23.4

23.4 A 6.4Tb/s 4.2pJ/b Co-Packaged Optics ASIC with Direct-Drive Integrated TIA and Retimed Segmented Mach-Zehnder Modulator Driver in 7nm FinFET

Mahdi Kashmiri, Ajay Yadav, Jin Namkoong, Behrooz Nakhkoob, Tony Kao, Shen Shen, Vaibhav Pandey, Kuan-Chang Chen, Jinho Han, Sudheer Gaddam, Shayan Kazemkhani, Sang Young Kim, Simar Maangat, Bo Nguyen, Mike Robinson, Hiva Hedayati

Broadcom, San Jose, CA

Abstract

A 6.4Tb/s co-packaged optics ASIC integrates 64 ingress path direct-drive TIAs and 64 egress path retimed segmented Mach-Zehnder modulator drivers in 7nm FinFET. The 7nm ASIC is 3D packaged with a photonics IC (PIC) and achieves an energy efficiency of 4.2pJ/b.

At 106.25Gb/s, its PAM-4 TIA has a sensitivity better than -11dBm and a BER floor better than 1E-9. The transmitted PAM-4 optical eye achieves a TDEQ of 1.48dB and an extinction ratio of 4.57dB.

Optical interconnects are gaining importance for the high-bandwidth communication needs of the scale-out AI clusters, replacing the pluggable modules. Co-packaged optics (CPO) is a heterogeneous integration technology to address the bandwidth and power consumption challenges by enabling scaling through integration. The shortened chip-to-chip interconnects between a photonic integrated IC (PIC) and a networking ASIC, reduces complexity, power consumption and signal integrity issues of otherwise lengthy interconnects between optics and electronics. This work presents a 6.4Tb/s retimer ASIC for CPO applications. Figure 23.4.1 shows the monolithic integration of 64 egress and ingress lanes of 106.25Gb/s PAM-4 channels. These lanes directly interface the line side on a PIC driving 64 segmented Mach-Zehnder modulators (MZM) [1] for TX and interface 64 photodiodes (PD) for RX [2]. An AFE followed by a CDR retimes and deserializes the host egress path incoming data. The CDR runs off the 4t clock from the TX PLL. The deserialized data bus feeds the MZM driver through the TX DSP, where the thermometer coding of the PAM-4 signal allows the concurrent drive of two segments with the MSB data. The ingress path implements a DSP-adapted TIA. The TIA output directly drives the host ASIC through an organic substrate. Furthermore, 320 monitoring photodiode readouts and 320 heater DACs are integrated within the ASIC for the optical sense and control of the PIC.

The host interface in the egress path is shown in Fig. 23.4.2. The input signal is capacitively coupled to the inputs of a continuous-time linear equalizer (CTLE) through a calibrated termination, which utilizes bandwidth extension t-coils for the input ESD. The CTLE, whose input common-mode is set by a programmable DAC, drives a variable-gain amplifier (VGA), which then drives the 4× time-interleaved sampler. The CTLE and VGA both utilize resistor calibration and t-coils for bandwidth extension.

The 4 sampling phases of the CDR are generated off the TX PLL clk4t clocks using a clkgen block. Within the clkgen block, a digitally adapted DLL-based quadrature clock generator (IQgen) drives a phase interpolator (PI) followed by a second IQgen producing 0, 90, 180 and 270 degrees sampling phases. The 4 slicer banks feed data and error information to a de-serializer, which in turn produces 128 data and 64 error bits at clk64t update rate to the TX DSP. The TX DSP contains CDR adaptation where the clock recovery loop is implemented. It eventually feeds the PAM-4 MSB and LSB data to the line side. A slicer level adaptation algorithm within the TX DSP adjusts the 4 threshold levels of the slicers within each sampler bank using sixteen 9b DACs.

The line side interface of the egress path is shown in Fig. 23.4.3. The serializer first translates the 64b retimed parallel MSB and LSB data from the TX DSP into 8b data streams. These data buses in the 8t clock domain are then passed to 3 driver segments, with the first segment assigned to the LSB and the last two to the MSB. The TXPLL 4t clock is distributed along the 3 MZM driver segments. Each segment's local clock generator corrects for clock distortion and generates 4 phases by means of a quadrature clock generation block. The 8t and 4t calibrated clocks from the IQ gen drive the last two serializer stages of the segment driver. The 4:1 mux output is translated into a 1.5V NRZ pattern by means of a two-stage level shifter pre-driver that drives the PMOS and NMOS gates of the stacked output stage. Pre-driver and cascode voltage levels are generated within the segment by means of level generation LDOs and DACs, respectively.

The ingress path consists of a direct drive TIA as shown in Fig. 23.4.4. On the line side, the PD anode interfaces a current-to-voltage converter (i2v). An input coil L_{in} extends the bandwidth by resonating the input parasitic capacitance. The i2v core is built around an inverter with a feedback resistor R_{fb} and regulated bias current. The i2v stage drives a first VGA1 pseudo-differentially after which the signal path through VGA2, 3 and the output driver remain fully-differential. The VGAs fan out towards the driver and incorporate bandwidth extension by means of t-coils and make use of 8b adjustable calibrated load resistors. The gain adjustment is through a digital gain adjustment code. The output driver uses 5b termination calibration and t-coils for bandwidth extension and driving the ESD protection and output bumps. The TIA differential output drives the host switching ASIC directly through an organic substrate. Each lane's TIA is digitally adapted by means of a local DSP, which drives the VGA gain control word for amplitude regulation, as well as a DC cancellation DAC connected to the TIA input. Adaptation information is provided to the DSP by means of an adaptation frontend. Each VGA is capable of ~6dB gain with gain

programmability steps of 0.2dB. The TIA has a nominal input current noise density of 15pA/√Hz while consuming less than 50mW of power. Furthermore, the ingress path integrates local PTAT bias and supply regulation within each TIA channel.

The CPO ASIC is fabricated in a 7nm FinFET technology (Fig. 23.4.7). The monolithic die integrates 64 egress and ingress lanes, TXPLLs, auxiliary common circuits, ADC and DAC components. The 6.4Tb/s optical engines are built by 3D packaging of the 7nm ASIC and the PIC silicon. Multiple engines are co-packaged with a host networking ASIC for scalable throughputs. The performance of the direct drive TIA ingress path is shown in Fig. 23.4.5, where the TIA output has been equalized by the host networking ASIC. The BER vs input OMA curves across 512-channels of a 51.2Tb/s CPO system (8×6.4Tb/s optical engines) is shown across the 4 utilized wavelengths. The TIA achieves an average sensitivity better than -11dBm (IEEE802.3 BER of 2.4E-4 for 106.25Gb/s PAM-4) and an average BER floor better than 1E-9. At 106.25Gb/s PAM-4 data rate, the egress path has a pre-FEC BER of better than 1E-9 where more than 99.98% of channels are error free. An eye-scan of the egress path's CDR, captured by the TX DSP, shows an open PAM-4 eye in Fig. 23.4.6. The optical PAM-4 eye (Fig. 23.4.6) for a PRBS-13 pattern is measured by a DCA applying an IEEE 802.3bs reference RX equalizer with a 5-tap FFE. The TX achieves a transmitter and dispersion eye closure quaternary (TDEQ) of 1.48dB and an outer extinction-ratio (ER) of 4.57dB and a ratio-level mismatch (RLM) of 0.958. Comparison to state-of-the-art [3 – 6] optical interface transceivers (Fig. 23.4.6) shows the 6.4Tb/s CPO ASIC achieves the highest level of integration and throughput with ×64 106.25Gb/s PAM-4 lanes at a competitive energy efficiency of 4.2pJ/b.

References:

[1] A. Hashemi, et al., "A 2.4pJ/b 100Gb/s 3D-Integrated PAM-4 Optical Transmitter with Segmented SiP MOSCAP Modulators and a 2-Channel 28nm CMOS Driver," *IEEE ISSCC*, pp.284 – 285, Feb. 2022. https://doi.org/10.1109/ISSCC42614.2022.9731563

[2] K. Lakshmikumar, et al., "A 7 pA/√Hz Asymmetric Differential TIA for 100Gb/s PAM-4 links with -14dBm Optical Sensitivity in 16nm CMOS," *IEEE ISSCC*, pp.206 – 207, Feb. 2023. https://doi.org/10.1109/ISSCC42615.2023.10067483

[3] V. Gurumoorthy, et al., "A 212Gb/s PAM-4 Retimer with Integrated High-Swing Optical Driver and Chip-to-Module Long Reach Capability of 40dB in 5nm FinFET," *IEEE ISSCC*, pp.602 – 603, Feb. 2025. https://doi.org/10.1109/ISSCC49661.2025.10904515

[4] S. Krishnamurthy. et. al., "A 0.9pJ/b 108Gb/s PAM-4 VCSEL-Based Direct-Drive Optical Engine," *IEEE ISSCC*, pp.592 – 593, Feb. 2025. https://doi.org/10.1109/ISSCC49661.2025.10904635

[5] J. Q. Wang, et al., "A 2.69pJ/b 212Gb/s DSP-Based PAM-4 Transceiver for Optical Direct-Detect Application in 5nm FinFET," *IEEE ISSCC*, pp.124 – 125, Feb. 2024. https://doi.org/10.1109/ISSCC49657.2024.10454275

[6] S. Mondal, et al., "A 4×64Gb/s NRZ 1.3pJ/b Co-Packaged and Fiber-Terminated 4-Ch VCSEL-Based Optical Transmitter," *IEEE ISSCC*, pp.340 – 341, Feb. 2024. https://doi.org/10.1109/ISSCC49657.2024.10454455

ISSCC 2026 / February 17, 2026 / 4:50 PM

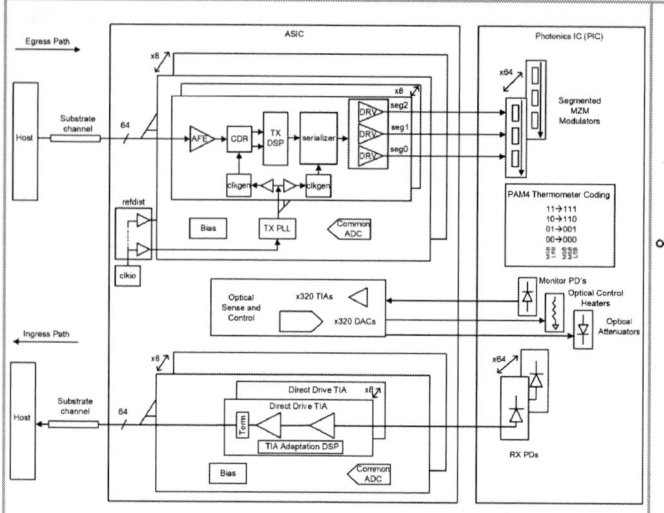

Figure 23.4.1: 64Tb/s ASIC and ingress and egress paths to the host and photonics IC (PIC).

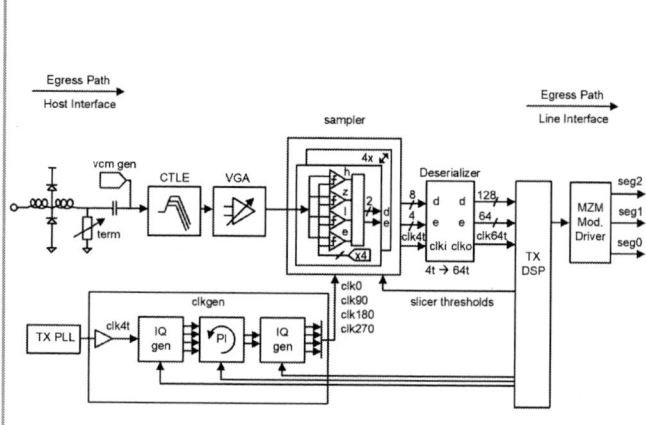

Figure 23.4.2: Egress path host interface AFE, CDR sampler and deserializer.

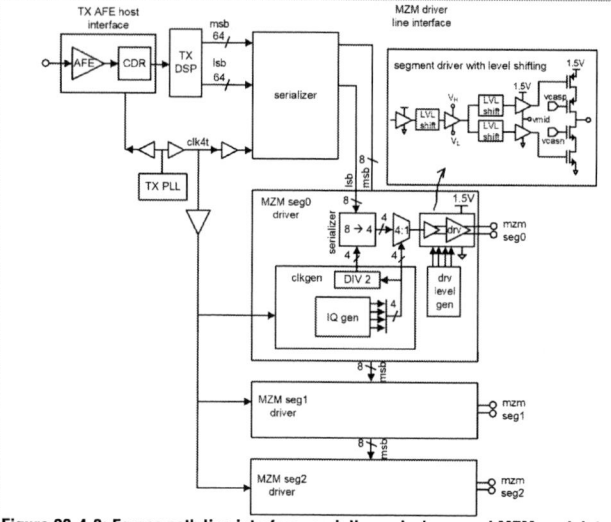

Figure 23.4.3: Egress path line interface, serializer, clock gen and MZM modulator driver.

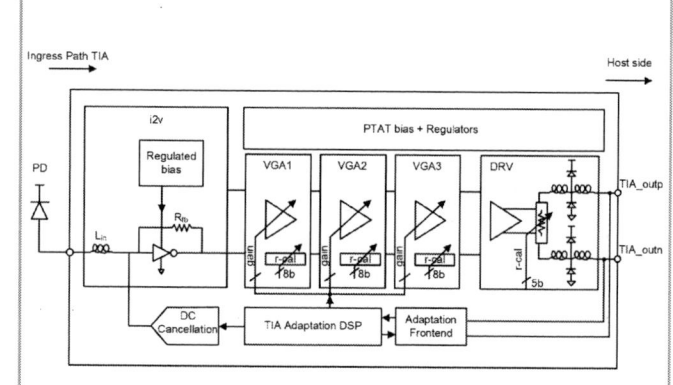

Figure 23.4.4: Ingress path direct drive TIA.

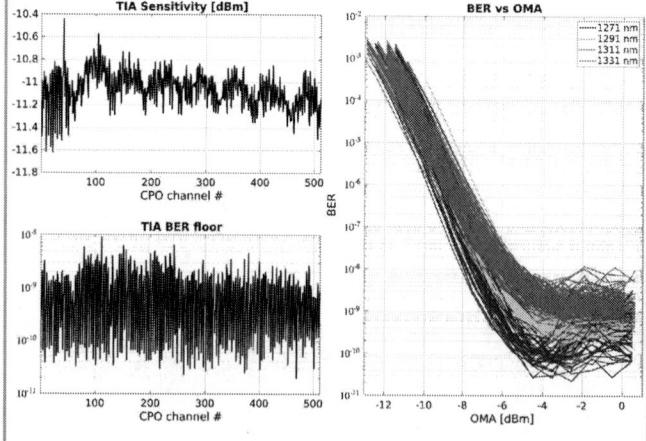

Figure 23.4.5: Direct drive TIA 106.25Gb/s PAM-4 sensitivity, BER floor and BER vs input OMA across 512 channels of a 51.2Tb/s CPO system comprising eight 6.4Tb/s optical engines.

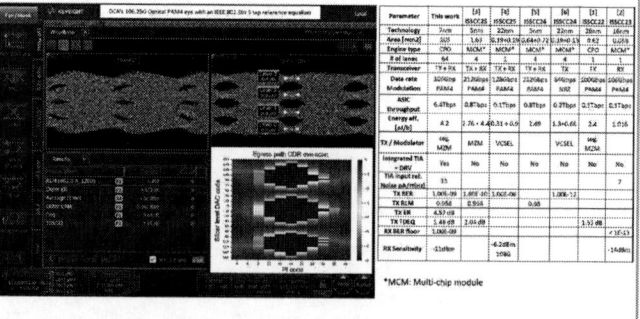

Figure 23.4.6: TX PAM-4 optical eye at 106.25Gb/s, egress CDR eye-scan and comparison with state-of-the-art.

DIGEST OF TECHNICAL PAPERS • 403

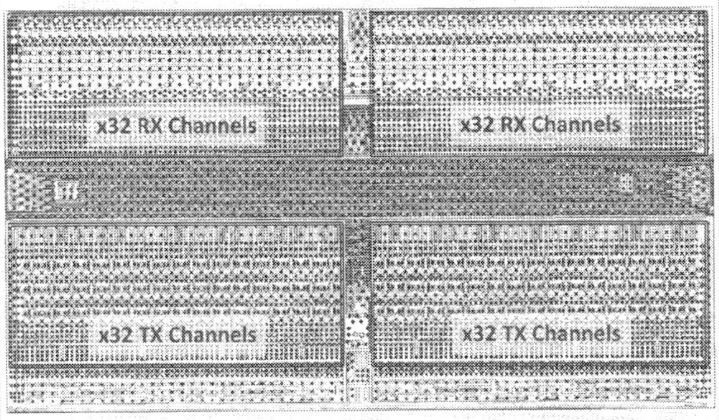

Figure 23.4.7: 6.4Tb/s co-packaged optics ASIC in 7nm FinFET.

979-8-3315-8937-0/26 $31.00 © 2026 IEEE

ISSCC 2026 / SESSION 23 / NEXT-GENERATION OPTICAL TRANSCEIVERS / 23.5

23.5 A 212Gb/s/λ 0.91pJ/b Direct-Drive O-Band Monolithic Coherent Transmitter Based on Carrier-Injection-Mode Mach-Zehnder Modulator

Antroy Roy Chowdhury[1], Wahid Rahman[1], Ahmed ElShater[2], Tamer Ali[2], Vladimir Stojanovic[1,3]

[1]University of California, Berkeley, CA, [2]MediaTek, Irvine, CA, [3]Ayar Labs, San Jose, CA

Abstract

This work presents an O-band coherent direct-drive optical transmitter (OTX) in 45nm CMOS-SOI using a compact 0.4mm monolithically integrated forward-biased PIN-MZM. A multistage linear driver with shunt peaking, passive equalization and TX-FIR enables 106Gb/s PAM-4 IMDD and 212Gb/s QAM16 coherent signaling. The OTX achieves 0.91pJ/b electrical power efficiency with a 2.76× lower laser power and 5× area reduction versus a conventional reverse-biased TW-MZM fabricated on the same die.

To continue the bandwidth density scaling of short-reach intra-data-center optical links, recently there has been a significant focus on coherent signaling [1,2], such as QPSK and QAM-16, compared to the conventional intensity modulation direct-detection (IMDD) approaches such as on-off keying and PAM-4. With the evolution of co-packaged optics (CPO), emerging direct-drive optical transceivers [3] prove to have better power efficiency, higher performance, and a more simplified design than pluggable solutions. While power consumption of the optical transmitter (OTX) including the laser source remains a dominant part of the overall link power, the choice of efficient electro-optic (EO) modulators can significantly improve the OTX power consumption. Mach-Zehnder modulators (MZMs) are thermally robust, less sensitive to process variation and exhibit better linearity when compared with microring modulators (MRMs), which primarily suffer from thermal instability and EO nonlinearity [4]. However, silicon photonic MZMs based on conventional reverse-biased p-n junction phase modulators are very long (>3mm) due to poor modulation efficiency $V_\pi L_\pi$ (voltage-length product for π phase shift), often exceeding 10V·mm. Hence, they are often realized either by a traveling-wave (TW) topology consuming a large die area and driver power [5], or a segmented topology requiring separate MSB and LSB data for PAM-4 operation [6,7]. MZMs based on forward-biased p-i-n phase modulators (PIN-MZM) [8,9], on the other hand, exhibit significantly better $V_\pi L_\pi$ (<0.05V·mm at DC) leading to a more compact and laser-power-efficient OTX design. However, a significantly lower EO bandwidth (~200-300MHz) and frequency-dependent resistive load of the forward-biased diode impose specific challenges on the design of the linear electrical driver for 100+Gb/s operation of PIN-MZM which has not been addressed in the literature before. This work presents a monolithically integrated 212Gb/s QAM-16 coherent direct-drive OTX based on PIN-MZM. The proposed linear multi-stage driver architecture exploits several bandwidth extension techniques along with low impedance driving suitable for PIN-MZM to achieve 106Gb/s PAM-4 electrical data rate.

Figure 23.5.1 shows the block diagram of the proposed coherent OTX. The incoming laser is equally split into the MZMs of I and Q paths. In each MZM, the PIN phase shifters are used to modulate the incoming light and low-power thermal undercut phase shifters (TUPS) are used to set the bias of the MZM at the null point for coherent operation and at the quadrature point for IMDD operation. Additional TUPS are deployed to adjust the relative phase shift between the I and Q paths for coherent operation. On chip optical power monitors are used to determine the required bias voltages for all TUPS. The MZM is driven in a push-pull fashion by monolithically integrated multistage linear electrical driver using the data generated by the host SerDes. The physical cross-section and simplified electrical model [10] of the forward-biased PIN phase shifter are shown in Fig. 23.5.1. At low frequencies, the equivalent load impedance consists of the contact resistance R_S in series with the forward-biased diode resistance, R_F. At high frequency, R_S dominates the net load impedance. The parasitic component values, $V_\pi L_\pi$ and optical loss are highly dependent on the forward-bias current density, I_{DEN} [10]. With increasing I_{DEN} and phase shifter length, L, the phase shift improves, whereas the optical loss also increases [9]. Design choice of L=0.4mm and I_{DEN} = 1.5mA/mm has a 0.6dB optical loss resulting in optimal optical modulation amplitude (OMA) with a compact footprint. Although the low frequency $V_\pi L_\pi$ degrades at higher frequencies due to the large diffusion capacitance ($C_{0,1}$ > 20pF/mm) leading to a low EO bandwidth, it still remains better than a conventional reverse-biased PN phase shifter even at a very high frequency as shown in Fig. 23.5.1.

The electrical data path of the OTX is designed to equalize the intrinsic PIN-MZM EO response while maintaining its superior modulation efficiency. Figure 23.5.2 summarizes the proposed approach. Cascading the PIN phase shifter with a passive RC high-pass filter de-emphasizes the low frequency component. By proper choice of R_E and C_E, a flat EO response can be obtained with higher bandwidth. For a 0.4mm long PIN phase shifter with a 0.6mA bias current, R_E = 600Ω requires C_E = 3.76pF to obtain a flat EO response with a 4.5GHz bandwidth. Scaling the R_E, C_E values while keeping the $R_E C_E$ product fixed maintains the flatness of the EO response while scaling the bandwidth, as shown in Fig. 23.5.2. An effective EO bandwidth of 28GHz can be obtained for R_E = 3.6kΩ and C_E = 627fF, thus fully equalizing the intrinsic EO response of the PIN MZM. However, very strong RC equalization also degrades the effective $V_\pi L_\pi$ by reducing the EO response at the Nyquist frequency (28GHz for 112Gb/s PAM-4 operation). TX pre-emphasis (PE) or TX FIR, on the other hand, provides unity gain at the Nyquist frequency while extending the bandwidth. Employing TX

PE along with an RC equalizer restores the effective $V_\pi L_\pi$. Nonetheless, excessive TX PE increases group delay variation and creates ripple in the flat band. As Fig. 23.5.2 shows, by combining 0.56× PE (post-cursor tap) with 20dB RC equalization, a 1.6× improvement in transmit OMA is obtained as compared to the fully RC-equalized case.

Figure 23.5.3 shows the proposed electrical data path consisting of the RC equalizer, two-stage linear driver and passive input stage. The tunable RC filter provides a nominal 20dB equalization across process variation. The equivalent load impedance of the RC equalized PIN-MZM drops from 635Ω at low frequencies to 8Ω at high frequencies necessitating a low-output-impedance driver which is realized by a complementary source follower stage. The source follower provides a low drive impedance and achieves very high input and output linearity, while the complementary topology reduces the required bias current. To compensate for the sub-unity gain (-1.9dB) and MZM load-dependent high-frequency loss (-3.7dB) of the driver, a pre-driver stage is used. Realized by a tailless complementary CML topology, the pre-driver is AC-coupled to the main driver. It utilizes a higher supply voltage and cascode devices to achieve high output linearity. Additionally, 30Ω passive resistors are used to set its DC gain along with 300pH shunt peaking inductors to achieve 4dB high-frequency peaking. The resulting two-stage driver shows an electrical bandwidth of 40GHz and maximum linear swing of 1.15V_{ppd} assuming a -30dB THD definition. Finally, the input stage utilizes 200pH inductors to decouple the load capacitances arising from the pre-driver, Cu-pillar bump and ESD diodes. The inductors extend the input bandwidth to 65GHz while achieving good input return loss (S_{DD11} = -16dB, S_{CC11} = -18dB).

The monolithic coherent transceiver consisting of 0.4mm-long custom-designed PIN-MZM is fabricated in a GlobalFoundries 45nm CMOS-SOI (45SPCLO) process. The coherent TX core (electronics + photonics) uses 0.612mm² die area (Fig. 23.5.7). Light is coupled into the monolithic chip using silicon-nitride (SiN) V-groove edge couplers and spot-size converters (SSC), whereas a 135μm-spaced Cu-pillar bump array is used for electrical signals. Figure 23.5.7 also shows the floorplan of the TX through its physical layout. The monolithic die is flip-chip bonded to an organic substrate and co-packaged with the host SerDes die (Fig. 23.5.5) acting as the high-speed data generator. A 1310nm DFB laser source sends light into the chip through 4m long polarization-maintaining (PM) fiber. In IMDD mode, the transmitted optical signal is converted to electrical using an optical receiver (RXM25AF) configured to have a conversion gain of 500V/W with a 25GHz electrical bandwidth. In coherent mode, the transmitted optical signal and the forwarded LO from the DFB laser go to a Quantifi Photonics IQRX coherent receiver consisting of optical hybrid and balanced PDs.

Figure 23.5.4 shows the measured IMDD raw eyes obtained with a laser power of 9dBm. A complete ISI-free 26Gb/s NRZ eye with an OMA of 2dBm is obtained with just a post-cursor TX FEE tap (0.56× PE) substantiating the efficacy of the proposed bandwidth extension approach based on RC equalization and TX PE. 106Gb/s IMDD PAM-4 operation is demonstrated with a 4-tap TX FFE and aided by the scope FFE feature of a Keysight UXR0334B real-time oscilloscope. The OMA drops from 2.04dBm at 26Gb/s NRZ to -2.5dBm at 106.25Gb/s PAM-4. For coherent operation, data through a single lane (I or Q) is transmitted at a time and the output current of the passive coherent RX is directly measured with UXR0334B. Although the received data is not stable for a long duration due to the absence of carrier phase recovery and phase tracking loop in the receiver, the obtained real-time data is post-processed to get eye diagrams containing 10.6k UIs as shown in Fig. 23.5.5. The same TX-FFE and scope FFE configurations are used as in the 106.25Gb/s IMDD PAM-4 case.

To compare the performance of the proposed forward-biased PIN-MZM based OTX with conventional reverse-biased MZMs, a lumped PN-MZM based OTX and a traveling-wave PN-MZM (TW-MZM) based OTX are also fabricated on the same die and are compared in Fig. 23.5.4. For identical transmit OMA at 106.25Gb/s, due to the higher modulation efficiency of PIN-MZM, it consumes 2.5× and 2.76× lower laser power compared to the lumped PN-MZM and TW-MZM respectively. The total energy efficiency (electrical + laser) of the PIN-MZM lane is 1.63pJ/b, which is 1.5× better than the PN-MZM and 1.45× better than the conventional TW-MZM. Moreover, it is 5× more area efficient compared to the TW-MZM due to elimination of the transmission line. Figure 23.5.6 compares the performance

404 • 2026 IEEE International Solid-State Circuits Conference

979-8-3315-8937-0/26 $31.00 © 2026 IEEE

of the proposed OTX with prior silicon photonic MZM based OTXs. This work reports the shortest MZM in O-band with the lowest energy cost per bit of 0.91pJ/b compared to the prior art. Moreover, coherent operation doubles the per wavelength data rate paving the way for future high-bandwidth-density and energy-efficient optical interconnects.

References:

[1] A. E. Abdelrahman, M. B. Younis, M. O. Selim, M. S. Aly, M. A. Khalil and P. K. Hanumolu, "A 1.54pJ/b 64Gb/s 16-QAM Intradyne Coherent Optical Receiver in 28nm CMOS," *2025 IEEE International Solid-State Circuits Conference (ISSCC)*, San Francisco, CA, USA, 2025, pp. 598-600.
https://doi.org/10.1109/ISSCC49661.2025.10904665

[2] A. R. Chowdhury, W. Rahman and V. Stojanovic, "Electronic-Photonic Co-Optimization of Linear Drive Laser-Forwarded Coherent Silicon Photonic Transmitters for Co-Packaged Optical (CPO) Links," in Journal of Lightwave Technology, vol. 43, no. 9, pp. 4338-4351, 1 May1, 2025. https://doi.org/10.1109/JLT.2025.3532994

[3] S. Krishnamurthy *et al.*, "A 0.9pJ/b 108Gb/s PAM-4 VCSEL-Based Direct-Drive Optical Engine," *2025 IEEE International Solid-State Circuits Conference (ISSCC)*, San Francisco, CA, USA, 2025, pp. 592-594. https://doi.org/10.1109/ISSCC49661.2025.10904635

[4] C. Sun et al., "A 45 nm CMOS-SOI Monolithic Photonics Platform With Bit-Statistics-Based Resonant Microring Thermal Tuning," in IEEE Journal of Solid-State Circuits, vol. 51, no. 4, pp. 893-907, April 2016. https://doi.org/10.1109/JSSC.2016.2519390

[5] E. Sentieri et al., "A 4-Channel 200Gb/s PAM-4 BiCMOS Transceiver with Silicon Photonics Front-Ends for Gigabit Ethernet Applications," 2020 IEEE International Solid-State Circuits Conference - (ISSCC), San Francisco, CA, USA, 2020, pp. 210-212. https://doi.org/10.1109/ISSCC19947.2020.9062992

[6] Q. Liao et al., "A 50-Gb/s PAM-4 Silicon-Photonic Transmitter Incorporating Lumped-Segment MZM, Distributed CMOS Driver, and Integrated CDR," in IEEE Journal of Solid-State Circuits, vol. 57, no. 3, pp. 767- 780, March 2022.
https://doi.org/10.1109/JSSC.2021.3134874

[7] A. H. Talkhooncheh et al., "A 2.4pJ/b 100Gb/s 3D-integrated PAM-4 Optical Transmitter with Segmented SiP MOSCAP Modulators and a 2-Channel 28nm CMOS Driver," 2022 IEEE International Solid-State Circuits Conference (ISSCC), San Francisco, CA, USA, 2022, pp. 284-286. https://doi.org/10.1109/ISSCC42614.2022.9731563

[8] S. Tanaka et al., "Ultralow-power (1.59 mW/Gb/s), 56-Gb/s PAM4 operation of Si photonic transmitter integrating segmented PIN Mach–Zehnder modulator and 28-nm CMOS driver," J. Lightw. Technol., vol. 36, no. 5, pp. 1275–1280, Mar. 1, 2018. https://doi.org/10.1109/JLT.2018.2799965

[9] A. R. Chowdhury et al., "A Carrier Injection Mode O-Band Mach-Zehnder Modulator with 0.02 V·mm VπL in 45nm Monolithic CMOS-SOI Process," in CLEO 2025, Technical Digest Series (Optica Publishing Group, 2025), paper SS187_3.
https://doi.org/10.1364/CLEO_SI.2025.SS187_3

[10] S. Tanaka, T. Usuki and Y. Tanaka, "Accurate SPICE Model of Forward- Biased Silicon PIN Mach–Zehnder Modulator for an Energy-Efficient Multilevel Transmitter," in Journal of Lightwave Technology, vol. 36, no. 10, pp. 1959-1969, 15 May 15, 2018.
https://doi.org/10.1109/JLT.2018.2797184

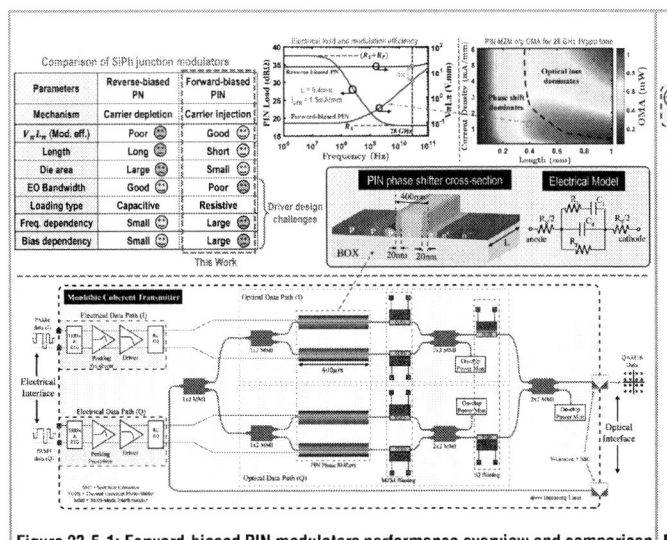

Figure 23.5.1: Forward-biased PIN modulators performance overview and comparison with conventional PN modulators (top); proposed coherent TX architecture (bottom).

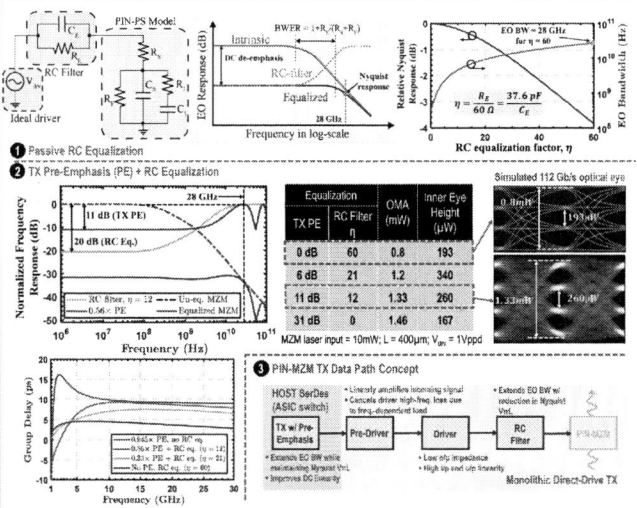

Figure 23.5.2: Enabling high-speed operation of PIN-MZM through passive RC equalization and TX pre-emphasis.

ISSCC 2026 / SESSION 23 / NEXT-GENERATION OPTICAL TRANSCEIVERS / 23.5

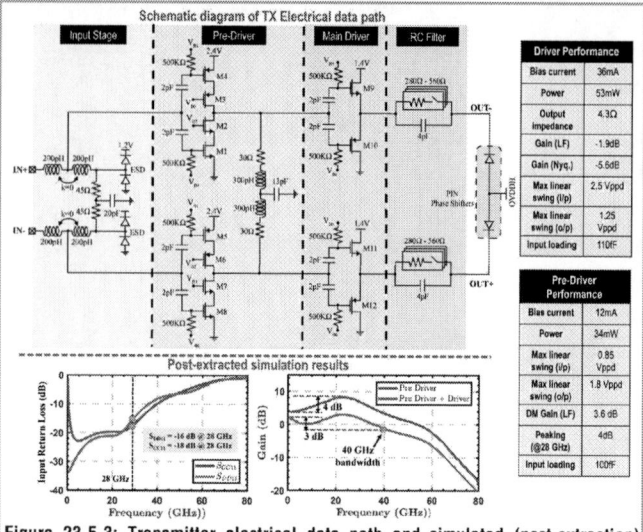

Driver Performance

Bias current	36mA
Power	53mW
Output impedance	4.3Ω
Gain (LF)	-1.9dB
Gain (Nyq.)	-5.6dB
Max linear swing (I/p)	2.5 Vppd
Max linear swing (o/p)	1.25 Vppd
Input loading	110fF

Pre-Driver Performance

Bias current	12mA
Power	34mW
Max linear swing (I/p)	0.85 Vppd
Max linear swing (o/p)	1.8 Vppd
DM Gain (LF)	3.6 dB
Peaking (@28 GHz)	4dB
Input loading	100fF

Figure 23.5.3: Transmitter electrical data path and simulated (post-extraction) performance.

Comparison (IMDD) with reverse-biased MZM-based OTXs fabricated on the same die

Parameters	Lumped PIN-MZM	Lumped PN-MZM	TW PN-MZM
Data rate (Gb/s)	106.25	106.25	106.25
OMA (dBm)	-2.5	-2.5	-2.5
TX FIR	4-taps	6-taps	6-taps
Scope FFE	On	On	On
MZM length (mm)	0.4	1	1.85
Active area* (mm²)	0.306	0.28	1.542
Monolithic driver	Yes	Yes	No
Electrical shoreline density (Gb/s/mm)	417	417	177
Laser EE** (pJ/bit)	0.75	1.88	2.07
Electrical EE[a] (pJ/bit)	0.88	0.56	0.3[b]
Total EE (pJ/bit)	1.63	2.44	2.37

*both photonic and electrical area **Laser wall-plug efficiency of 10% assumed
[a]Thermal phase shifter heater power included [b]excluding driver power

Figure 23.5.4: Measured IMDD transmit eye received using RXM25AF optical receiver (left); comparison with other fabricated MZM lanes for similar transmit eyes (right).

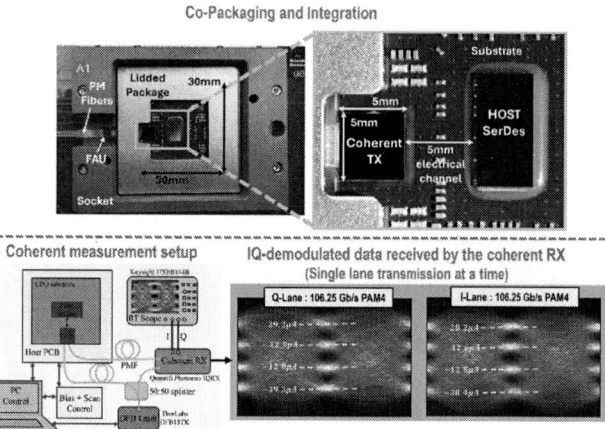

Figure 23.5.5: System integration and co-packaging of the monolithic TX with host SerDes (top); coherent measurement (bottom).

References	[7] ISSCC'22	[6] JSSC'22	[5] ISSCC'20	[8] JLT'18	This Work
Technology	28nm CMOS	40nm CMOS	55nm BiCMOS	28nm CMOS	45nm CMOS-SOI
Integration	3D	Wire-bond	3D	3D	Monolithic
Modulator type	Si MOSCAP	PN (depletion)	PN (depletion)	PIN (injection)	PIN (injection)
Modulator Structure	Lumped-MZM	Segmented-MZM	TW-MZM	Segmented-MZM	Lumped-MZM
Modulator length	0.62 mm	3 mm	6 mm	0.75 mm	0.4 mm
Optical modulation	IMDD	IMDD	IMDD	IMDD	Coherent
Electrical signaling	PAM4	PAM4	PAM4	PAM4	PAM4
Data rate (/λ)	100 Gb/s	50 Gb/s	53 Gb/s	56 Gb/s	212 Gb/s
Driver type	Switching	Switching	Linear	Switching	Linear
Operating wavelength	1550 nm	1550 nm	1310 nm	1550 nm	1310 nm
Input laser power	10 dBm	19 dBm	13.5 dBm	-	9 dBm[1]
Fiber coupling	Grating coupler	Grating coupler	Grating coupler	Edge coupler	V-grooves + SSC
Optical OMA	1.4 dBm	ER = 9.8 dB	0.4 dBm	-3.6 dBm	-2.5 dBm[2]
Electrical power efficiency	2.4 pJ/bit	13.64 pJ/bit	5.2 pJ/bit	1.59 pJ/bit	0.91 pJ/bit
Laser power efficiency[3]	1 pJ/bit	15.8 pJ/bit	4.2 pJ/bit	-	0.75 pJ/bit
Die active area (mm²)	0.62	4.92	44	0.4	0.612[4]

[1]Considering IMDD operation [2]Loss not de-embedded [3]Laser wall-plug efficiency of 10% assumed
[4]Includes both photonic and electrical area, excluding V-grooves

Figure 23.5.6: Performance comparison with prior silicon photonic MZM-based transmitters.

Figure 23.5.7: Die micrograph and floorplan of the monolithic TX.

• 2026 IEEE International Solid-State Circuits Conference

ISSCC 2026 / SESSION 23 / NEXT-GENERATION OPTICAL TRANSCEIVERS / 23.6

23.6 A 112Gb/s NRZ Heterodyne Detection 23ns Settled Burst-Mode RX with CD Suppression and Envelope Demodulation for 100G PON

Cheng Wang, Ye Gu, Gertjan Coudyzer, Shengpu Niu, Jing Zhang, Xin Yin

imec - Ghent University, Ghent, Belgium

Abstract

This work presents a burst-mode receiver based on an intensity modulation heterodyne detection scheme. SSB quadrature IF generation and envelope detection enhance baseband signal integrity and chromatic dispersion tolerance. Pre/post-demodulator loop separation allows parallel AGC/AOC in burst-mode. The SiGe BiCMOS front-end, co-integrated with an O-band silicon photonic PIC, achieves 112Gb/s NRZ with −25.8dBm OMA (B2B), 23ns settling, and −23.7/−22.9dBm OMA over 20km SMF in 1290/1329nm.

With the development of AI infrastructure and 6G communication, the demand for higher data rates in passive optical networks (PONs) is rapidly increasing. While 50G PON has been standardized [1], scaling to 100G PON presents fundamental challenges in transceiver designs, particularly in intensity modulation direct detection (IMDD) links. Conventional approaches using APD and SOA-PIN receivers [2-3] suffer from device gain-bandwidth (GBW) product limitations and severe chromatic dispersion (CD), especially for 20km fiber reach in PONs. At 100Gb/s, CD-induced phase distortion becomes ~4× worse than at 50Gb/s, requiring complex equalization techniques such as a combination of 21-tap FFE and maximum-likelihood sequence estimation (MLSE) to achieve a bit-error rate (BER) of 1e-2 for 12km fiber [3]. Coherent modulation and detection (Fig. 23.6.1(b)) offer the potential for high rates and robust CD tolerance via high order modulation and advanced digital signal processing (DSP) [4]. But its cost, power and complexity can be prohibitive for PON deployment.

This work presents a burst-mode receiver front-end architecture based on intensity modulation heterodyne detection (IMHD), offering a compelling scheme that combines low complexity, good sensitivity and strong CD tolerance (Fig. 23.6.1(c)). The proposed integrated receiver generates quadrature intermediate frequency (IF) signals via heterodyne detection and exploits the flat frequency response of the transimpedance amplifier (TIA) below the 3dB BW to carry single-sideband (SSB) amplitude-modulated (AM) I/Q signals. This approach filters out CD-distorted upper sidebands and hence preserves phase information, significantly reducing the need for CD compensation. After squaring AM I/Q signals and then adding the squared outputs, the demodulated baseband signal is enhanced while the corresponding content around the 2× center frequency (f_0) of the IF signal is suppressed through destructive cancellation. Similar to the quasi-coherent approach [5], the proposed scheme eliminates the need for complex coherent DSP, such as phase locking or carrier phase recovery. Critically, the IMHD architecture also enables highly efficient burst-mode (BM) operation, essential for upstream reception in time-division multiplexed PONs. BM optical signals are down converted to BM SSB IF signals at the TIA front end, eliminating abrupt DC drifts at the TIA input, and allowing the automatic offset control (AOC) to be relocated after the envelope demodulator. The common mode interference due to BM signals is suppressed by the fully differential TIA. As a result, coarse automatic gain control (CAGC) and AOC loops can operate in parallel rather than sequentially, significantly reducing the 2R (re-amplify and re-shape) settling time. Fabricated in 0.13μm SiGe BiCMOS and integrated with an O-band silicon photonic integrated circuit (PIC), the receiver achieves 112Gb/s NRZ operation with a BER below 2.4e-4 and a BM settling time of 23ns, over 20km SMF in the 1290–1329nm (limited by instrument) wavelength range.

The top half of Fig. 23.6.2 shows the circuit architecture of the proposed RX front end, including a PIC and an electronic integrated circuit (EIC). PIC and EIC are integrated by wire bonds with ~330pH inductance per line. The PIC consists of an O-band 90-degree hybrid 2×4 MMI and two pairs of common cathode balanced photodiodes (CC-BPDs). The signal path of the EIC contains three cascaded parts:

Part 1 (for both I and Q) – linear transimpedance amplifier stages including shunt feedback TIA front ends followed by two stages of variable-gain amplifiers (VGAs), combined with continuous time linear equalizers (CTLEs).

Part 2 – envelope demodulator realized by two gilbert multipliers and one resistor adder.

Part 3 – post linear amplifier and hybrid equalizers.

The total capacitance of the CC-BPD is half of a single PD, which benefits creating a high resonance frequency due to wire bond inductance and BPD at the TIA input. Tunable CTLEs boost BW linearly, minimizing the group delay variation and maximizing the flatness of the frequency response. The simulated group delay variation of part 1 is <2ps from 1GHz to 40GHz. Center frequency(f_0) is chosen at 40GHz for filtering the upper sideband of the IF signal. In this way, the SSB I/Q IF signal is carried from −40GHz to 40GHz. Due to the physical channel pitch limitation of the PIC, the gilbert multipliers of the I and Q path are separated from each other by 750μm. The resistor adder combines the squared signals of the I and Q path travelling along transmission lines. Source termination resistors are used

for absorbing the reflections due to unmatched impedance caused by the resistor adder. In part 3, passive equalization by differential series inductance and CTLEs cooperate to enhance BW of post amplifier to 67GHz in simulation. The simulated optical-electrical (O-E) BW including PIC and EIC under large signal is 60GHz. In both the I and Q path of part 1, there is an input DC current control (IDCC) loop for sinking common mode DC current generated by the LO on BPD and a DC offset cancellation (DCOC) loop to balance the DC level due to mismatch. The bottom half of Fig. 23.6.2 shows the simulated output of the proposed RX and DD RX at 112Gb/s NRZ after 20km SMF with 1.6ps/(nm*km) CD parameter without considering noise. By using a 3-tap FFE in DSP, CD-induced phase distortion can be compensated in the proposed RX, whereas the output of a DD RX is severely distorted and only achieves <40mV eye opening after using a 21-tap FFE.

For fast and accurate BM settling, proper timing and loop BW control is implemented as shown in Fig. 23.6.3(a). As the start of burst (SOB) rises to high, the on-chip oscillator works, and the number of clock cycles is counted. Depending on the counting value, the state machine generates BM control signals, which switch the BW of the BM control loops. The additional fine AGC (FGC) loop is used for adaptation to variant extinction ratio (ER) from different TXs of ONUs. During S1 and S2, CAGC and AOC loops work in parallel at high-BW mode and converge at the end of S2. The FGC loop adjusts quickly in S2 and S3. After the settlement of the FGC loop until S5, the AOC loop is switched to low-BW mode in S6 for less power penalty. At the end of a burst, the AOC loop is reset to high BW mode for responding to the next burst quickly. The timing sequence is programmable by setting the clock speed and state value in SPI. The simulated BM settling of critical points A, B and C in signal path (Fig. 23.6.2) are shown in Fig. 23.6.3(a). The CAGC and AOC loops converge in parallel and take 6ns to settle. The FGC loop takes another 10ns to settle. A high-speed peak detector captures both differential-mode and common-mode voltage values of the signal. The relative difference between the detected peak value and the reference peak voltage is generated by a current mode subtractor, maximizing detection speed and minimizing detection error.

The EIC in this work is fabricated in 0.13μm SiGe BiCMOS technology and the PIC is manufactured in iSiPP50G technology. The PD has a responsivity of 0.9A/W. Figure 23.6.3 (b) shows the measurement setup for continuous and burst mode reception. The optical signal TX includes an ECL with λ1, an MZM and an SOA. Channel one of AWG drives the MZM directly by sending continuous NRZ data with a pattern of PRBS 2^{13}-1. For continuous mode transmission, the SOA is biased by a DC voltage from a function generator (FG). For burst mode, channel two of the AWG triggers the FG to generate two synchronized pulse signals. One is used to switch the SOA between on and off states, creating burst optical packets. The other is the SOB signal for the RX. A 20km length of ITU-T G.652.D single-mode fiber (SMF) is inserted into the optical link for emulating transmission in PONs. No special data pattern is used as preamble sequence for BM. A 100ns guard time between two bursts is chosen according to the 50G PON standard [1]. A variable optical attenuator (VOA) controls the signal power for sensitivity measurement. An optical switch selects the signal flowing into either the RX or a 70GHz BW PD for the input signal check on a digital sampling oscilloscope (DSO). Another ECL with λ2 is used as an LO, which boosts the optical signal through the PIC. The center frequency of the IF signals is chosen at 40GHz by setting wavelength offset between λ1 and λ2. The output of the RX is probed and can be either connected to the RTO for BER check by counting bit errors or be connected to DSO for monitoring eye diagram.

Figure 23.6.4 illustrates the measured eye diagrams and BER versus optical modulated amplitude (OMA) of continuous-mode reception for an NRZ signal. Adhering to the LDPC FEC BER limit of 1e-2 [1] for 50G PON, without an equalizer, the RX achieves sensitivity of -22.9dBm by using 4.5dBm in waveguide LO power at 112Gb/s NRZ in back-to-back(B2B) case. After implementing a 3-tap FFE, the OMA is improved to -25.8dBm. A 128Gb/s NRZ pattern with BER 9.5e-5 is also obtained at -12.9dBm OMA. After 20km fiber transmission, the highly distorted optical signal (checked on 70GHz BW PD) due to CD can be recovered by the proposed RX for 112Gb/s NRZ. By adopting 3-tap FFE, the measured OMA is -23.7dBm and -22.9dBm for 1290nm and 1329nm respectively. The FFE is implemented by a built-in linear equalizer math function on scope.

Figure 23.6.5 shows the measured BM settling waveforms and BER curve under loud-loud (LL) and loud-soft (LS) burst packets. Among different powers of burst packets, the longest convergence time is 23ns with -25.8dBm input OMA. Both LL and LS packets can be detected under BER 2.4e-4 and achieve >20.8dB dynamic range at BER 1e-2. Compared with B2B continuous mode (CM) reception, power penalty (PP) for BM reception is obtained. Without FFE, at a BER of 1e-2, the PP is 0.5dB for LS burst packets. At a BER level of 1e-3, 0.2dB and 0.6dB PP are observed for LL and LS packets, respectively. After using a 3-tap FFE, the PP can be reduced to 0 and 0.1dB at a BER level 1e-3 for LL and LS packets respectively. Compared to 20km SMF CM transmission at 1290nm and 1329nm, LL packets have no PP at a BER level 1e-2 and 1e-3.

To characterize O-E BW and the quality of the demodulated signal, Figure 23.6.6 gives the measured O-E large signal frequency response and signal to noise and distortion ratio (SNDR) of the RX. Corresponding to the best BER value in Fig. 23.6.4, LO power is set at 4.5dBm and the received OMA of a single sine waveform (modulated by MZM) is -11.9dBm for different frequency values. The center frequency of IF signals is chosen as 40GHz. The RX is set at fixed gain. The output sine waveforms of the RX are captured by RTO. 62GHz O-E BW is obtained (limited by the BW of RTO). The measured SNDR keeps higher than 11dB from 1GHz to 60GHz. Comparison table is listed at the end. This work achieves 112Gb/s BM NRZ signal detection with 23ns settling time, >20.8dB dynamic range and -25.8dBm received OMA. For 20km SMF transmission at 1290/1329nm (limited by wavelength range of SOA), -23.7/ -22.9dBm OMA are measured after using an off-chip 3-tap FFE which can be embedded in a fast-locked BM CDR [10].

Acknowledgement:
This work is funded by FWO(G041420N). The authors thank Jasper Jans and Joris Van Kerrebrouck from IDLab-design group for PIC-EIC assembly.

References:
[1] ITU-T, 50-Gigabit-capable passive optical networks(50GPON): PMD layer specification. ITU-T G.9804.3(2021) Amd.1, 03/2024. https://www.itu.int/rec/T-REC-G.9804.3-202403-I!Amd2/en. http://doi.org
[2] G. Coudyzer et al., "100 Gbit/s PAM-4 Linear Burst-Mode Transimpedance Amplifier for Upstream Flexible Passive Optical Networks," in Journal of Lightwave Technology, vol. 41, no. 12, pp. 3652-3659, 15 June15, 2023. http://doi.org/10.1109/JLT.2023.3262319
[3] C. Füllner et al., "First 100G NRZ-OOK PON Demonstration with >31 dB Loss Budget and Coexistence Study over Field-Deployed Fiber," 2025 Optical Fiber Communications Conference and Exhibition (OFC), San Francisco, CA, USA, 2025, pp. 1-3. http://doi.org
[4] T. A. Eriksson et al., "Real-time bidirectional coherent point-to-multipoint passive optical network," 49th European Conference on Optical Communications (ECOC 2023), Hybrid Conference, Glasgow, UK, 2023, pp. 487-490. http://doi.org/10.1049/icp.2023.2112
[5] C. Wang et al., "Low-Complexity Integrated Optical-Electrical Quasi-Coherent RX Front End Supporting 5 km SSMF Transmission in C-Band for 50G PON," in Journal of Lightwave Technology, vol. 43, no. 11, pp. 5149-5155, 1 June1, 2025. http://doi.org/10.1109/JLT.2025.3548403
[6] M. Sieben, J. Conradi and D. E. Dodds, "Optical single sideband transmission at 10 Gb/s using only electrical dispersion compensation," in Journal of Lightwave Technology, vol. 17, no. 10, pp. 1742-1749, Oct. 1999. http://doi.org/10.1109/50.793744

[7] G. Coudyzer, P. Ossieur, J. Bauwelinck and X. Yin, "A 25Gbaud PAM-4 Linear Burst-Mode Receiver With Analog Gain- and Offset Control in 0.25μm SiGe:C BiCMOS," in IEEE Journal of Solid-State Circuits, vol. 55, no. 8, pp. 2206-2218, Aug. 2020. http://doi.org/10.1109/JSSC.2020.2987680
[8] A. Rylyakov et al., "A 25Gb/s burst-mode receiver for rapidly reconfigurable optical networks," 2015 IEEE International Solid-State Circuits Conference - (ISSCC) Digest of Technical Papers, San Francisco, CA, USA, 2015, pp. 1-3. http://doi.org/10.1109/ISSCC.2015.7063095.
[9] N. TANAKA, D. UMEDA, Y. SUGIMOTO, T. FUNADA, K. TANAKA and S. OGITA, "25.78-Gbit/s Burst-mode Receiver for 50G-EPON OLT," 2020 Optical Fiber Communications Conference and Exhibition (OFC), San Diego, CA, USA, 2020, pp. 1-3. http://doi.org
[10] B. Zhang et al., "A 50Gb/s Burst-Mode NRZ Receiver with 5-Tap FFE, 7-Tap DFE and 15ns Lock Time in 28nm CMOS for Symmetric 50G-PON," *2025 IEEE International Solid-State Circuits Conference (ISSCC)*, San Francisco, CA, USA, 2025, pp. 1-3. http://doi.org/10.1109/ISSCC49661.2025.10904530

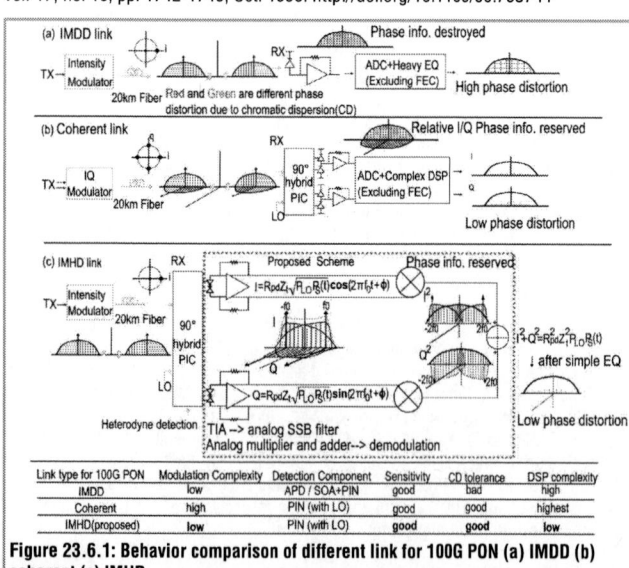

Figure 23.6.1: Behavior comparison of different link for 100G PON (a) IMDD (b) coherent (c) IMHD.

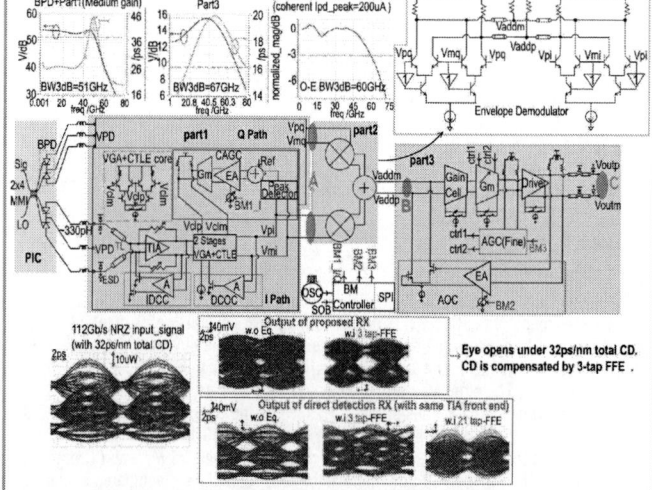

Figure 23.6.2: Architecture of proposed RX and simulation results under a distorted input signal.

ISSCC 2026 / SESSION 23 / NEXT-GENERATION OPTICAL TRANSCEIVERS / 23.6

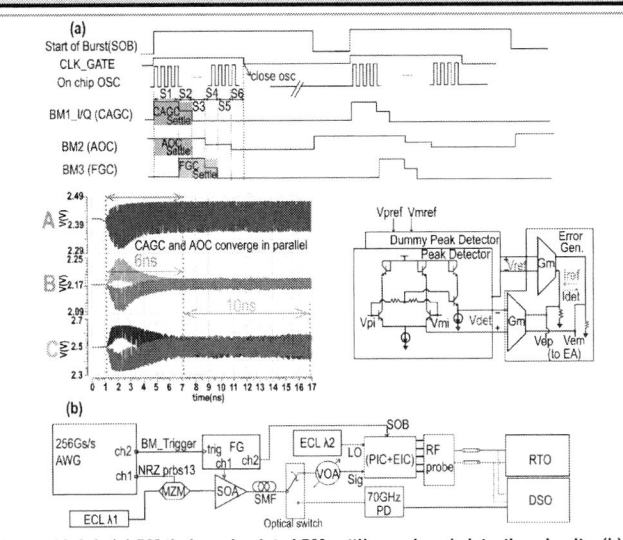

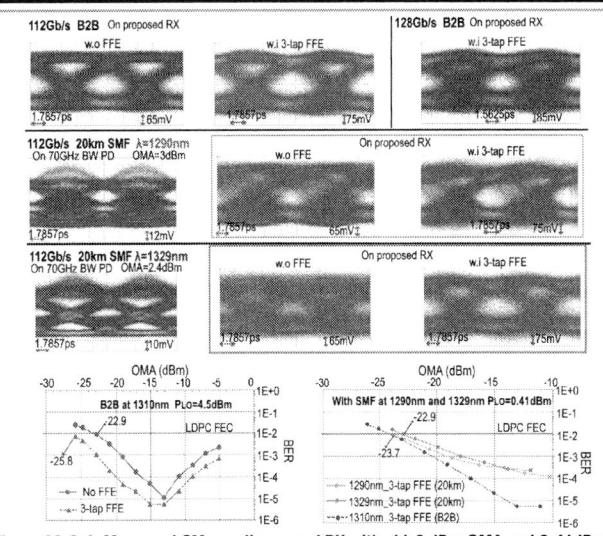

Figure 23.6.3: (a) BM timing, simulated BM settling and peak detection circuits; (b) measurement setup.

Figure 23.6.4: Measured CM eye diagram of RX with -11.9 dBm OMA and 0.41dBm LO and BER curve.

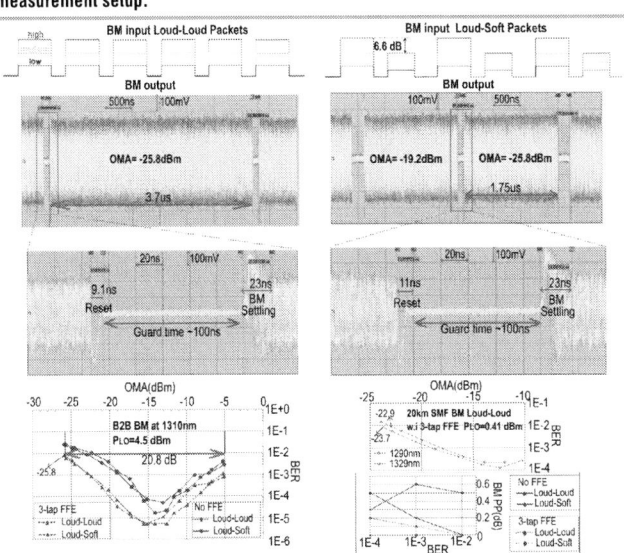

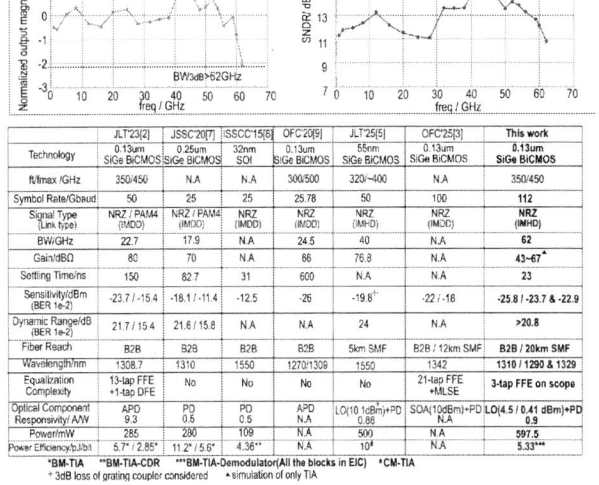

Figure 23.6.5: Measured burst-mode output, BER curve and BM power penalty.

Figure 23.6.6: O-E signal response and comparison table.

Technology	JLT'23[2] 0.13um SiGe BiCMOS	JSSC'20[7] 0.25um SiGe BiCMOS	ISSCC'15[8] 32nm SOI	OFC'20[9] 0.13um SiGe BiCMOS	JLT'25[5] 55nm SiGe BiCMOS	OFC'25[3] 0.13um SiGe BiCMOS	This work 0.13um SiGe BiCMOS
ft/fmax /GHz	350/450	N.A	N.A	300/500	320/~400	N.A	350/450
Symbol Rate/Gbaud	50	25	25	25.78	50	100	112
Signal Type (Link type)	NRZ / PAM4 (IMDD)	NRZ / PAM4 (IMDD)	NRZ (IMDD)	NRZ (IMDD)	NRZ (IMI-D)	NRZ (IMDD)	NRZ (IMHD)
BW/GHz	22.7	17.9	N.A	24.5	40	N.A	62
Gain/dBΩ	80	70	N.A	66	76.8	N.A	43~67 [▲]
Settling Time/ns	150	82.7	31	600	N.A	N.A	23
Sensitivity/dBm (BER 1e-2)	-23.7 / -15.4	-18.1 / -11.4	-12.5	-26	-19.8 [**]	-22 / -18	-25.8 / -23.7 & -22.9
Dynamic Range/dB (BER 1e-2)	21.7 / 15.4	21.6 / 15.8	N.A	N.A	24	N.A	>20.8
Fiber Reach	B2B	B2B	B2B	B2B	5km SMF	B2B / 12km SMF	B2B / 20km SMF
Wavelength/nm	1308.7	1310	1550	1270/1309	1550	1342	1310 / 1290 & 1329
Equalization Complexity	13-tap FFE +1-tap DFE	No	No	No	No	21-tap FFE +MLSE	3-tap FFE on scope
Optical Component Responsivity/ A/W	APD 9.3	PD 0.5	PD 0.5	APD N.A	LO(10.1dBm)+PD 0.86	SOA(10dBm)+PD N.A	LO(4.5 / 0.41 dBm)+PD 0.9
Power/mW	285	280	109	N.A	500	N.A	597.5
Power Efficiency/pJ/bit	5.7* / 2.85*	11.2* / 5.6**	4.3**	N.A	10[#]	N.A	5.33***

*BM-TIA **BM-TIA-CDR ***BM-TIA-Demodulator(All the blocks in EIC) +CM-TIA
[▲] 3dB loss of grating coupler considered [▲] simulation of only TIA

Figure 23.6.7: PIC+EIC photo.

ISSCC 2026 / SESSION 24 / DISPLAYS / OVERVIEW

Session 24 Overview: *Displays*
IMAGE SENSORS AND DISPLAYS SUBCOMMITTEE

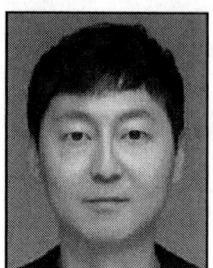

Session Chair: Seunghoon Ko
Kwangwoon University
Seoul, Korea

Session Co-Chair: Sanshiro Shishido
Panasonic
Osaka, Japan

The evolution of mobile display technology requires more accurate and faster drivers and sensors, while reducing power consumption. The first two papers describe compact CMOS backplane circuits for micro-LED for AR applications. The third paper presents a 3D display processor that converts the focal stack to tensor display. Following this, two papers disclose 10b source drivers that can achieve a horizontal time of <1µs. The final paper showcases a touch-embedded display-driver IC to suppress display-induced noise affecting the touch signals.

3:35 PM

24.1 A 28nm CMOS 1-Chip In-Pixel Memory Backplane Circuit with Pixel-Level Sensing and Compensation for 6652-PPI MicroLED AR Glasses

Sunkwon Kim, Samsung Electronics, Suwon, Korea

In Paper 24.1, Samsung Electronics presents a 28nm CMOS 1-chip in-pixel memory backplane for a 6652-PPI micro-LED AR display with integrated bias tuning and variation control. System techniques such as self-refresh, partial update, and foveated rendering reduce bandwidth by 59.2% and cut power by up to 40.1%.

4:00 PM

24.2 A 4042-PPI 10b 240Hz Digital-PWM CMOS Backplane for Micro-LED-on-Silicon Displays with a Shared, Unified Memory-in-Pixel Architecture

Kihyun Kim, KAIST, Daejeon, Korea

In Paper 24.2, KAIST describes a 10b 240Hz digital-PWM CMOS backplane for µLED-on-Silicon with a shared, unified memory-in-pixel structure. The architecture adopts a hierarchical time-divided PWM with a shuffled rolling-shutter scheme to time-multiplex the LSB engine across 1×6 pixel clusters.

4:25 PM

24.3 A 144mW 161Mpixels/s Tensor Display Processor for 3D Virtual Reality

Ching-Yen Lee, National Taiwan University, Taipei, Taiwan

In Paper 24.3, National Taiwan University demonstrates the first energy-efficient focal stack to 3D tensor display processor, which achieves a 2.6-to-2.8× higher throughput with 5-to-18× lower normalized energy consumption than the prior work.

4:50 PM

24.4 A 512ch 10b Source-Driver IC with Current-Mode Auto-Zeroing Scheme Achieving 1.4mV DVO and 600ns 1-Horizontal Time for Mobile OLED Displays

Yousung Park, KAIST, Daejeon, Korea

In Paper 24.4, KAIST discloses a 10b source-driver IC for ultra-fast, uniform-luminance mobile OLED displays. A current-mode auto-zeroing scheme transparently cancels output-buffer offset, lowering DVO with zero timing penalty. Fabricated in 180nm CMOS with an 11.37µm channel pitch, the 512ch SD-IC achieves 600ns 1-H time and a 1.4mV DVO.

24

5:05 PM

24.5 A 10b Display Driver IC with a Pivoting Translinear-Loop DAC-in-Buffer Enabling 1µs 1-H Scan Time and 1722µm² per Channel

Hyeong-Joon Kim, KAIST, Daejeon, Korea

In Paper 24.5, KAIST introduces 10b display driver IC for high-resolution OLEDs that features a novel pivoting trans-linear-loop DAC-in-buffer that overcomes the area, speed, and linearity trade-offs. It achieves a fast 1-H scan time of 1µs and a compact 1722µm² per channel in 180nm CMOS.

5:20 PM

24.6 A Touch-Embedded OLED Display-Driver IC with Display Noise Referencing and Display Coupling Noise Prediction Based on Dedicated Neural Networks for Mobile Applications with CoE Display

Yun-Rae Jo, Samsung Electronics, Hwaseong, Korea

In Paper 24.6, Samsung Electronics reveals a touch-embedded display driver IC to overcome the interference issue between touch and display electrodes. In addition, a novel analog front-end circuit is disclosed to suppress coupling noise from display, and a dedicated neural networks engine is used to improve interference based on displayed image.

ISSCC 2026 / SESSION 24 / DISPLAYS / 24.1

24.1 A 28nm CMOS 1-Chip In-Pixel Memory Backplane Circuit with Pixel-Level Sensing and Compensation for 6652-PPI MicroLED AR Glasses

Yongil Kwon, Sunkwon Kim, Youngkil Choi, Taehyun Kwon, Seongyoung Ryu, Sewhan Na, Kangjoo Kim, Uijong Song, Hyun-Wook Lim, Jongwoo Lee, Jae-Yeol Lee

Samsung Electronics, Suwon, Korea

Abstract

This work presents a 28nm CMOS 1-chip in-pixel memory backplane for a 6652-PPI microLED AR display. Pixel-level sensing with pBIST enables accurate calibration and compensation of current variation. The 2.7μm-pitch array integrates an 8b shifter with MSB latch for PWM. System techniques including self-refresh, partial update, and foveated rendering reduce MIPI bandwidth by 59.2% and power by up to 40.1%. Measured DNL <±0.25 LSB and flicker SNR −72 dB validate circuit and system effectiveness.

Augmented reality (AR) glasses are emerging as a key wearable platform, requiring compact high-resolution displays integrated into the optical path without sacrificing comfort. For outdoor use, high luminance and a wide color gamut are essential, making micro light-emitting diodes (microLEDs) suitable due to their brightness, efficiency, and long lifetime [1,2]. Limited battery capacity makes low-power operation critical. Process variations cause luminance non-uniformity (mura) and pixel brightness differences, degrading image quality. Addressing these challenges requires circuit-level solutions such as pixel bias tuning, built-in self-test (pBIST), and sensing for precise compensation. This work presents a 28nm CMOS LED-on-Silicon (LEDoS) backplane with in-pixel memory, integrating these functions to reduce variation, improve uniformity, and lower system power, making it suitable for AR glasses.

The system-level architecture in Fig. 24.1.1(a) integrates an application processor (AP), LEDoS display, and waveguide. The AP connects to a smartphone via Wi-Fi and to LEDoS through MIPI (image) and SPI (control). It also interfaces with a camera, audio, and speaker, enabling AR applications. The LEDoS system integrates flash memory and a PMIC on the same substrate as the silicon backplane (Si-BP), and couples its optical output into the waveguide for see-through display. As compared in Fig. 24.1.1(b), the 1-chip Si-BP solution offers compactness and high pixel density over the 3-chip approach, motivating its adoption in this work. The proposed stack in Fig. 24.1.1(c) integrates a micro-lens array, RGB microLEDs, hybrid Cu bonding (HCB), and a 28nm CMOS Si-BP. HCB ensures high alignment accuracy and low-resistance interconnects, enabling ultra-fine pitch and high pixel density [3].

The proposed Si-BP adopts a 2.7μm-pitch RGBG subpixel structure array with a resolution of 1800×1350. As shown in Fig. 24.1.2(a), the pixel array is surrounded by a row driver on the left and a column driver at the bottom, with four temperature sensors placed at the array corners. An AHB bus serves as the main interconnect, linking memory and peripheral blocks. MIPI input data passes through gamma correction, dithering, and a display controller before driving the column and row drivers to control the microLED array. An SPI interface provides a side channel for Si-BP control. Figure 24.1.2(b) details the row and column driver circuits. The row driver generates PCLK, DCLK, and CSEL, where DCLK clocks pixel data during addressing, PCLK drives the PWM generator during emission, and CSEL selects between them to supply the pixel. The column driver delivers pixel data synchronized with these clocks. The simplified 8b in-pixel memory (iPxM) schematic is shown in Fig. 24.1.2(c). Each pixel integrates an 8b shift register with MSB latch (8b-SRM), comprising an 8b shift register with a feedback bit for self-refresh and an additional MSB shuffling latch, thereby providing full 8b depth within the 2.7μm pitch constraint. Pixel bias currents for R, G, and B are independently generated using separate current mirror circuits, enabling fine per-color current tuning. Figure 24.1.2(d) presents the timing diagram for the PWM driving scheme. DCLK[7:0] sequentially clocks binary data into pixels during addressing, while PCLK[7:0] drives the PWM generator to produce binary emission patterns during emission. FB_SEL selects the feedback path for self-refresh operation. P_EN controls the pixel current path, enabling emission when asserted. The final binary PWM signal, denoted as P_SIG, is generated by ANDing P_EN with the output of the 8b SRM. P_SIG produces pulse widths of $T/2^1$ to $T/2^8$ through sequential clock shifting of PCLK[7:0], realizing 8b luminance control via binary PWM.

Figure 24.1.3 illustrates the power-reduction techniques implemented in the Si-BP. Figure 24.1.3(a-1) illustrates the block-gating scheme, where in normal mode the MIPI, DSC, and image-processing blocks remain active every frame, while in self-refresh mode only the pixel array stays active until new data arrive, with all upstream blocks disabled for power reduction. As shown in Fig. 24.1.3(a-2), the iPxM employs an 8b-SRM for in-pixel data storage and MSB shuffling. Depending on the FB_SEL signal, new data are shifted into the 8b SRM for normal operation, or the data retained in the 8b SRM are fed back for self-refresh mode. Figure 24.1.3(a-3) compares the timing of normal and self-refresh modes, showing that in self-refresh the emission of the next frame is driven by the stored data without incoming MIPI data, reducing system power. The partial-update scheme in Fig. 24.1.3(b) enables region-specific activation by transmitting only updated regions of the image, such as a small navigation icon or clock digits, while the rest of the frame is maintained via self-refresh. This approach minimizes MIPI burst activity and downstream logic switching, which are the dominant contributors to dynamic power consumption in display pipelines. The fast-update mode of Fig. 24.1.3(c) accelerates data-write by optimizing row-sequencing and driver timing, enabling emission to complete within the same frame. A steep-slope drive option allows emission to be aligned across the panel, reducing perceived motion blur and maximizing the low-power idle interval of the MIPI interface. Figure 24.1.3(d) illustrates foveated upscaling with compression, where the AP transmits the high-resolution (HA) region at full detail, while the low-resolution (LA) region is downscaled and compressed for bandwidth reduction. The Si-BP decompresses and upscales the LA before merging it with the HA [4]. By concentrating bandwidth only where the human eye perceives detail, this method reduces the required MIPI throughput from 8.4 to 3.43Gb/s (59.2% reduction) compared to full-resolution transmission, while maintaining perceptual image quality.

PWM-based driving often suffers from noticeable flicker and contour artifacts when large MSB pulses dominate the emission period. These issues are mitigated by decomposing the MSB into multiple smaller segments and distributing them across the frame, as shown in Fig. 24.1.4(a). For instance, 128 gray levels can be split into 64, 32, and 32 levels, emitted sequentially across early, middle, and late sub-periods, thus reducing temporal luminance fluctuation. To further suppress artifacts such as false contours and flicker, the system supports sub-frame driving, as illustrated in Fig. 24.1.4(b). Each frame can be divided into up to 16 sub-frames, with adjustable duty ratio, count, and inter-frame spacing. This flexibility allows optimization for both image quality and power efficiency. Pixel-to-pixel current variation due to transistor mismatch in the current mirror structure is compensated through 2b bias calibration and tuning (2BCT), shown in Fig. 24.1.4(c). Each pixel incorporates a 2BCT for 4-step bias tuning, along with an additional NMOS in the current path to implement a supply-voltage-insensitive bias (SVIB). The system integrates a pBIST to support verification and calibration, as shown in Fig. 24.1.4(d). The pBIST detects iPxM shifter functionality using simple logic and incorporates a high-resolution on-chip ADC that measures each pixel's current and the forward voltage of the microLED, enabling accurate electrical-to-optical correlation and pixel-level compensation. The display pipeline includes a 12b gamma correction block, described in Fig. 24.1.4(e). An 8b input is remapped using a 12b LUT to represent a nonlinear gamma curve up to 2.2, enhancing gradation precision and brightness control. To preserve 12b image quality with 8b PWM driving, spatial and temporal dithering techniques are combined, as shown in Figs. 24.1.4(f-1) and 24.1.4(f-2). Spatial dithering employs a 4×4 pixel mask to distribute errors across neighboring pixels, while temporal dithering modulates pixel intensity across frames, together achieving fractional gray levels beyond the native bit depth. Dead pixels caused by bonding or microLED fabrication defects are compensated using a 3×3 kernel-based interpolation, demonstrated in Fig. 24.1.4(g). Dead pixel data are redistributed to neighboring sub-pixels to preserve the local average luminance [5].

Figure 24.1.5(a) shows the fabricated 1800×1350 microLED display with an RGBG subpixel pattern, demonstrating a tiger image. In Fig. 24.1.5(b-1), a full-frame update displays the complete image, while Fig. 24.1.5(b-2) illustrates a partial update example, where only specific rows (e.g., rows 400 to 700) are refreshed, significantly reducing the active transmission period over MIPI. Figure 24.1.5(c) presents the measured gray-level response, showing a highly linear relationship between input code (0 to 255) and relative luminance, with a DNL of less than ±0.25LSB. Figure 24.1.5(d) shows that applying 2BCT reduces the standard deviation from 0.15 to 0.08μA (approximately 47.7%) while maintaining the mean current near 1.0μA. As shown in Figure 24.1.5(e), flicker SNR improves with increasing sub-frame count, and at 12 sub-frames the measured value reaches −72dB, exceeding the −60dB threshold where flicker becomes imperceptible. Figure 24.1.5(f-1) compares measured and simulated pixel current, showing less than 4.7% deviation. Figure 24.1.5 (f-2) shows the measured pixel current distribution, with 1σ variation of 6.5% at 2μA and 12.9% at 100nA, within 4.6% of simulation. Figure 24.1.5(f-3) summarizes these results in tabular form for different current levels. As shown in Fig. 24.1.5(g), an SVIB circuit maintains a constant V_{gs} against supply variation, reducing the average pixel current variation from 8.78% down to 0.96%. In Fig. 24.1.5(h), measured power consumption at 60Hz is 37.4mW in normal mode, 28.5mW in partial update mode, and 22.4mW in self-refresh mode. Compared to normal operation, partial update and self-refresh reduce power consumption by 23.8% and 40.1%, respectively, with results within 3% of simulation. All pixel current measurements were obtained using the integrated pBIST function.

410 • 2026 IEEE International Solid-State Circuits Conference

979-8-3315-8937-0/26 $31.00 © 2026 IEEE

ISSCC 2026 / February 17, 2026 / 3:35 PM

The comparison in Fig. 24.1.6 highlights advantages of this work over prior art [6-10]. While previous designs achieved PPIs ranging from 3400 to 5000 using larger technology nodes and limited grayscale control, this work demonstrates a 6652 PPI microLED display implemented in 28nm CMOS, enabling an 8b iPxM circuit at a 2.7μm pitch. In addition to the compact pixel implementation, the system integrates partial update, per-pixel current sensing, and bias tuning, achieving low DNL (<0.25LSB) and power savings at 60Hz (22.4mW self-refresh vs. 37.4mW normal). These features collectively provide the highest pixel density and among the compared works. Fig. 24.1.7 shows the die micrograph of the 28nm CMOS Si-BP and bonded module with compact integration.

References:

[1] Y. Huang et al., "Mini-LED, Micro-LED and OLED displays: present status and future perspectives," *Light: Sci. Appl.*, vol. 9, art. 105, 2020. http://doi.org/10.1038/s41377-020-0341-9

[2] J. H. Xiong et al., "Augmented reality and virtual reality displays: emerging technologies and future perspectives," *Light: Science & Applications*, vol. 10, art. 216, 2021. http://doi.org/10.1038/s41377-021-00658-8

[3] S. Lee et al., "A study on memory stack process by hybrid copper bonding (HCB) technology," *IEEE ECTC*, pp. 1085–1089, 2022. http://doi.org/10.1109/ECTC51906.2022.00175

[4] L. Wang et al., "Foveated Rendering: A State-of-the-Art Survey," *Comput. Visual Media*, vol. 9, pp. 195–228, 2023. http://doi.org/10.1007/s41095-022-0306-4

[5] T. Kimpe et al., "Defective pixels in medical LCD displays: problem analysis and fundamental solution," *J. Digit. Imaging*, vol. 19, no. 1, pp. 76–84, 2006. http://doi.org/10.1007/s10278-005-9239-6

[6] J. Seong et al., "CMOS backplane pixel circuit with leakage and voltage drop compensation for a micro-LED display achieving 5000 PPI or higher," *IEEE Access*, vol. 8, pp. 49467–49476, 2020. http://doi.org/10.1109/ACCESS.2020.2979883

[7] X. X. Ji et al., "3400 PPI active-matrix monolithic blue and green micro-LED display," *IEEE Trans. Electron Devices*, vol. 70, no. 9, pp. 4689–4695, 2023. http://doi.org/10.1109/TED.2023.3295764

[8] M.-C. Wu et al., "High pixel density 960×540 flip-chip AlGaInP red micro-LED display," *IEEE Trans. Electron Devices*, vol. 69, no. 12, pp. 6206–6212, Dec. 2022. http://doi.org/10.1109/TED.2022.3209134

[9] P.-Y. Lai et al., "A 10-Bit 1280 × 720 Micro-LED Display Driver With 2-Transistor Pixel Circuits and Current-Mode Pulse Width Modulation," *IEEE Solid-State Circuits Lett.*, vol. 5, pp. 234–237, 2022. http://doi.org/10.1109/LSSC.2022.3174137

[10] S.-S. Cheong et al., "A new SRAM-embedded pixel circuit that modulates accurate gray level for PWM-driven micro-LED displays," *IEEE Solid-State Circuits Lett.*, vol. 6, pp. 408–411, 2023. http://doi.org/10.1109/LSSC.2023.3282904

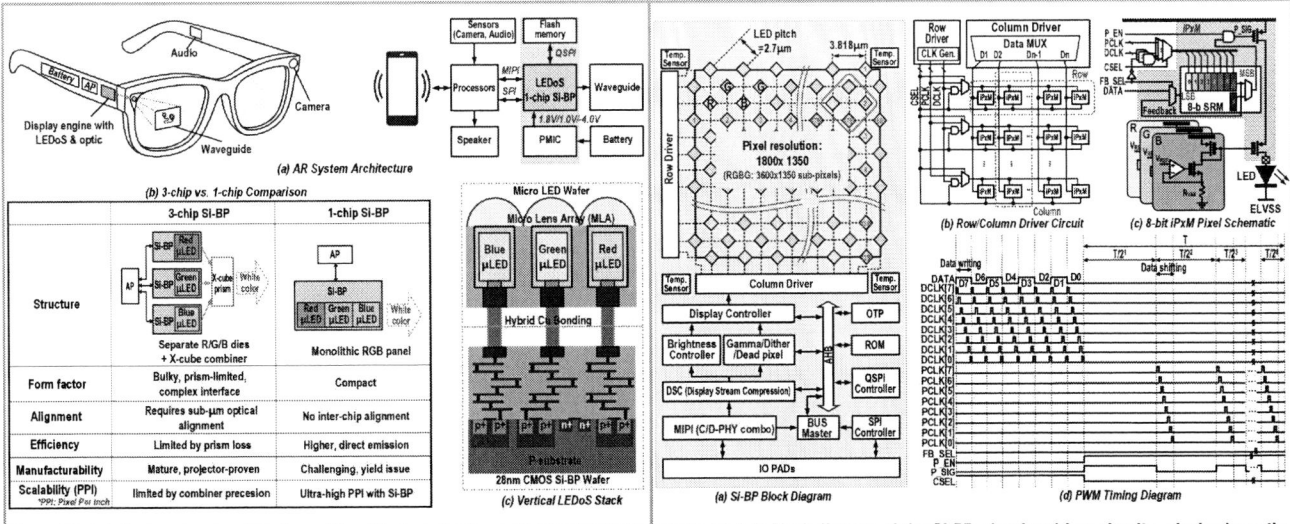

Figure 24.1.1: System-level architecture of the AR glass display, including 3-chip vs. 1-chip Si-BP comparison and the vertical LEDoS stack with CMOS backplane.

Figure 24.1.2: Block diagram of the Si-BP, showing driver circuits, pixel schematic, and the timing diagram of the PWM scheme.

DIGEST OF TECHNICAL PAPERS • 411

979-8-3315-8937-0/26 $31.00 © 2026 IEEE

ISSCC 2026 / SESSION 24 / DISPLAYS / 24.1

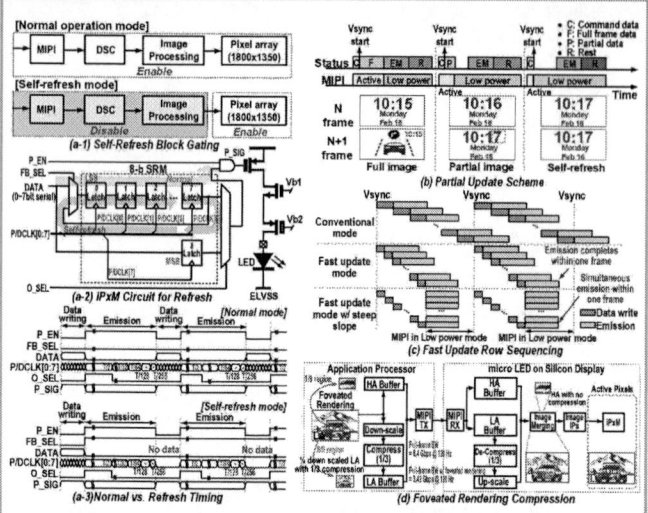

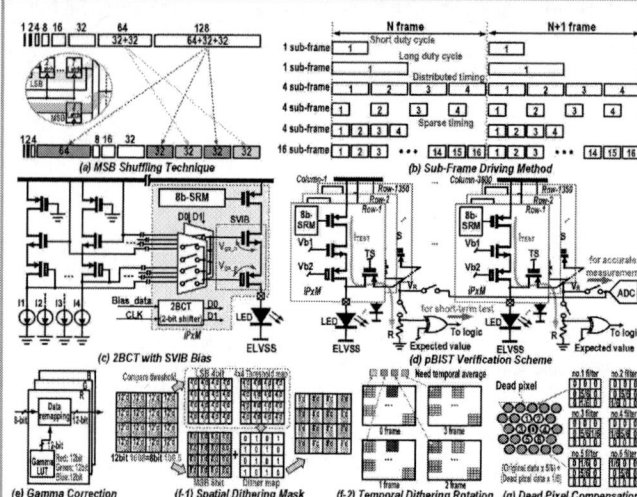

Figure 24.1.3: Power-saving techniques including self-refresh gating, partial update, fast update with row sequencing, and foveated rendering with compression.

Figure 24.1.4: Compensation and enhancement methods with MSB shuffling, sub-frame driving, 2BCT and SVIB, pBIST, gamma correction, dithering, and dead-pixel repair.

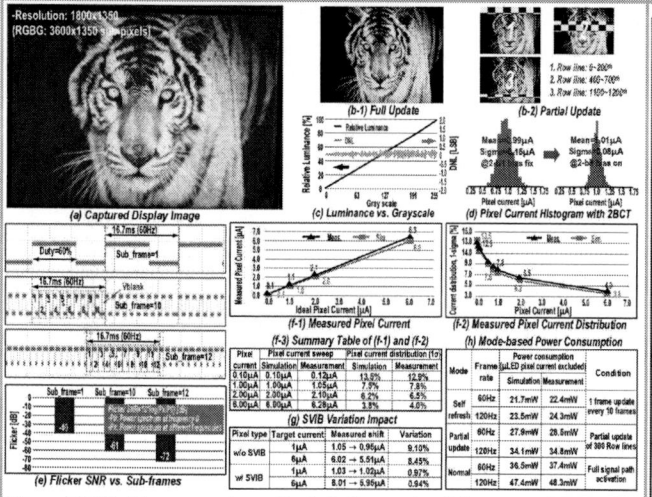

Figure 24.1.5: Measurement results including captured image, update modes, luminance vs. grayscale, pixel current, flicker SNR, SVIB impact, and power consumption.

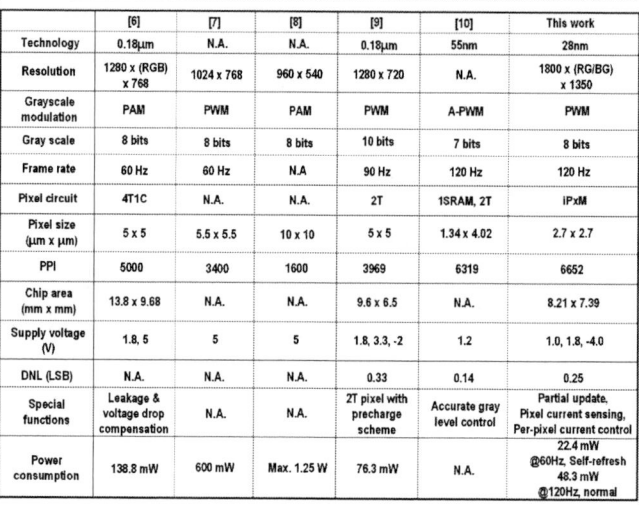

Figure 24.1.6: Summary and comparison of prior LEDoS backplanes and this work, highlighting node, resolution, pixel pitch, modulation, and power consumption.

	[6]	[7]	[8]	[9]	[10]	This work
Technology	0.18μm	N.A.	N.A.	0.18μm	55nm	28nm
Resolution	1280 x (RGB) x 768	1024 x 768	960 x 540	1280 x 720	N.A.	1800 x (RG/BG) x 1350
Grayscale modulation	PAM	PWM	PAM	PWM	A-PWM	PWM
Gray scale	8 bits	8 bits	8 bits	10 bits	7 bits	8 bits
Frame rate	60 Hz	60 Hz	N.A	90 Hz	120 Hz	120 Hz
Pixel circuit	4T1C	N.A.	N.A.	2T	1SRAM, 2T	IPxM
Pixel size (μm x μm)	5 x 5	5.5 x 5.5	10 x 10	5 x 5	1.34 x 4.02	2.7 x 2.7
PPI	5000	3400	1600	3969	6319	6652
Chip area (mm x mm)	13.8 x 9.68	N.A.	N.A.	9.6 x 6.5	N.A.	8.21 x 7.39
Supply voltage (V)	1.8, 5	5	5	1.8, 3.3, -2	1.2	1.0, 1.8, -4.0
DNL (LSB)	N.A.	N.A.	N.A.	0.33	0.14	0.25
Special functions	Leakage & voltage drop compensation	N.A.	N.A.	2T pixel with precharge scheme	Accurate gray level control	Partial update, Pixel current sensing, Per-pixel current control
Power consumption	138.8 mW	600 mW	Max. 1.25 W	76.3 mW	N.A.	22.4 mW @60Hz, Self-refresh 48.3 mW @120Hz, normal

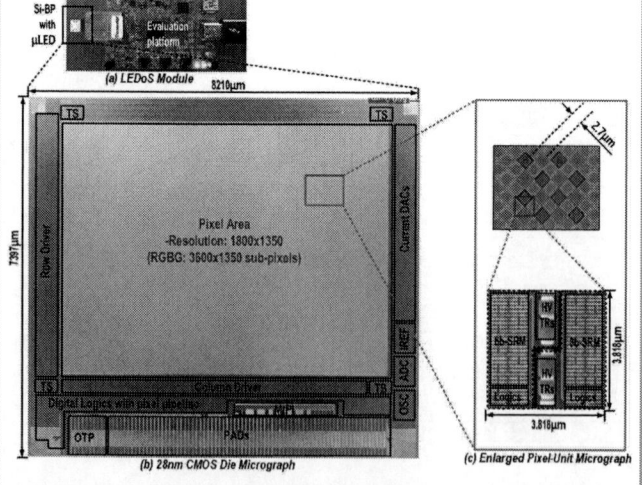

Figure 24.1.7: Fabricated hardware with bonded LEDoS module and evaluation platform, die micrograph of the 28nm CMOS Si-BP, and enlarged pixel-unit micrograph.

- 2026 IEEE International Solid-State Circuits Conference

979-8-3315-8937-0/26 $31.00 © 2026 IEEE

ISSCC 2026 / SESSION 24 / DISPLAYS / 24.2

24.2 A 4042-PPI 10b 240Hz Digital-PWM CMOS Backplane for Micro-LED-on-Silicon Displays with a Shared, Unified Memory-in-Pixel Architecture

Kihyun Kim, Jun-Gi Lee, Hyun-Sik Kim

KAIST, Daejeon, Korea

Abstract

This paper presents a 10b 240Hz digital-PWM CMOS backplane for µLED-on-Silicon with a shared, unified memory-in-pixel structure. A unified D-FF chain (UDC) merges the in-pixel register, counter, and comparator into one reconfigurable unit. Hierarchical time-divided PWM (HTD-PWM) with shuffled rolling-shutter (SRS) time-multiplexes an LSB engine across 1×6 pixel clusters. Fabricated in 28nm CMOS, the 120×120 backplane achieves 4042-PPI and an 88% active-area, with DNL/INL of 0.9/7.1LSBs.

Micro-light-emitting diode (µLED) is a promising technology for next-generation AR/VR head-mounted displays (HMDs) [1–4]. However, achieving high-quality, cost-effective µLED-on-silicon (µLEDoS) displays requires a CMOS backplane that meets several key requirements (Fig. 24.2.1, top). First, constant-current driving is essential for color stability, as the µLED emission wavelength is current-density dependent [5,6]. Second, high pixels-per-inch (PPI) demands an extremely compact pixel circuit. To maximize cost-effectiveness in advanced CMOS processes, a high active-area ratio, X/(X+Y)—where X is the pixel-array area and Y is the peripheral area—is crucial. Finally, the backplane must support pre- and post-µLED-bond testing to ensure high manufacturing yield.

Gray levels in µLEDoS backplanes are typically generated with pulse-width modulation (PWM), which is either analog or digital (Fig. 24.2.1, bottom). In analog PWM [7–9], the column-driver DAC converts image data to an analog V_{DATA} stored on a per-pixel capacitor; a row-driven ramp V_{RAMP} is compared to V_{DATA} to set the pulse width. Although the pixel circuit can be compact, large peripheral DAC/ramp blocks increase the peripheral area (Y) and reduce the active-area ratio, X/(X+Y). In addition, capacitor leakage and comparator offset/slew degrade low-gray linearity. To address these issues, memory-in-pixel (MIP) digital PWM has been explored [1,10–13]. Here, each pixel receives digital data directly (no DAC) and latches it with CLK_W; an in-pixel binary counter clocked by CLK_P (LSB interval T_{LSB}) counts, and a code comparator asserts when the count equals the stored data, defining the pulse width. Fully digital operation removes leakage-drift paths, sharpens transitions, and eliminates large analog driver blocks, improving linearity and X/(X+Y). The remaining challenge is the area: per-pixel register, counter, and code comparator consume substantial silicon, limiting achievable PPI.

In this work, we present a CMOS backplane with 10b memory-in-pixel (MIP) digital PWM for 120×120, 240Hz µLEDoS displays, addressing the above challenges through four innovations: (i) a unified D-flip-flop (D-FF) chain (UDC) that merges the register, counter, and comparator into a single D-FF chain; (ii) hierarchical time-divided PWM (HTD-PWM) enabling time-multiplexed sharing of PWM logic across multiple pixels; (iii) a shuffled rolling-shutter (SRS) scan that reduces routing complexity during sharing; and (iv) built-in self-test (BIST). Using these digital-intensive schemes, the µLEDoS backplane achieves 4042 PPI and an 88% active-area ratio.

Figure 24.2.2 shows the MIP digital-PWM pixel circuit. The in-pixel PWM generator is a unified D-FF chain (UDC) that operates in two modes (Fig. 24.2.2, bottom-left): memory during data load and a down-counter during LED emission. In the write phase, UDC cells form a shift register; the 10b code $D_{DATA}<9:0>$ is shifted MSB-to-LSB on D_{IN} with CLK_W until stored. With SCAN = low, the UDC is reconfigured as a binary down-counter preloaded with $D_{DATA}<9:0>$ and clocked by CLK_P. On underflow (rollover from 0), a tail underflow detector—a toggle flip-flop (T-FF)—toggles, switching the PWM output from ON to OFF; the µLED emission time is proportional to $D_{DATA}<9:0>$. Reusing the UDC for memory and counting minimizes per-pixel logic. For R-bit resolution, a conventional digital PWM uses ~$2R$ D-FFs (R-bit register + R-bit counter) plus an R-bit code comparator. The design uses R reconfigurable UDC cells (D-FFs) plus one underflow detector (T-FF), totaling (R+1) flip-flops, with minimal control gating. Each UDC cell includes four reconfiguration switches to form either a serial shift-register loop or a parallel down-counter. Pulse-width termination is implicit in the underflow event, eliminating the multi-bit comparator. Overall, the flip-flop count is halved and the comparator removed, reducing memory/logic area by >50% without sacrificing PWM resolution or linearity.

To further reduce area (improve PPI), we propose hierarchical time-divided PWM (HTD-PWM), splitting the 10b PWM into two sequential phases (Fig. 24.2.2, bottom-right): an N-bit LSB phase and an M-bit MSB phase ($N+M = R = 10$; $N = M = 5$ in this work). During the LSB phase, the UDC down-counter runs with lower 5b at CLK_{PL} to produce PWM_L. After completion, the MSB phase counts the remaining upper 5b at CLK_{PM} to output PWM_M. $CLK_{PM} = CLK_{PL}/2^N$ (32× in this design), so one MSB tick spans the entire LSB interval; the counter thus processes higher-weight bits with a period scaled by 2^N. A MASK signal distinguishes the LSB and MSB phases for control gating. This two-step PWM scheme preserves 10b grayscale while activating the LSB hardware only early in the frame. In conventional PWM, all counter bits toggle across the full frame (0→1023); here, once LSB

pulses are issued, LSB-associated UDC cells stop toggling and only the MSB portion continues. The resulting LSB idle time enables time-multiplexing of the LSB memory/counter across multiple pixels in non-overlapping slots. The HTD-PWM therefore enables the clustering and sharing described next, reusing the LSB engine multiple times per frame and substantially reducing total hardware in each pixel, thereby drastically improving the PPI.

Figure 24.2.3 illustrates pixel clustering and shuffled rolling shutter (SRS) scanning for LSB-UDC sharing among pixels. HTD-PWM enables LSB-logic sharing, but adjacent pixels must do so without long routes or timing collisions. We satisfy both constraints using 1×6 pixel clusters and an SRS scan. In each cluster, every pixel retains one LSB D-FF; the six per-pixel LSB D-FFs collectively implement a shared 5b LSB-UDC (five UDC cells + one underflow detector) that is time-multiplexed across the six pixels. Each pixel also keeps its own six D-FFs (MSB-UDC) for the MSB phase. A naive "no-overlap" rule fails under conventional row-by-row rolling shutter scanning because neighboring rows often overlap in their data-write or LSB-emission intervals, limiting sharing to distant pixels—impractical due to long interconnects and parasitics. To enable local sharing of LSB-UDC hardware, we reorder the scan in a "cluster-wise" interleaved pattern: activate the 1st-row pixel of every cluster sequentially across the panel, then the 2nd-row of every cluster, and so on. By the time the driver returns to the next pixel within a given cluster, that cluster's prior LSB phase has completed, so adjacent pixels never contend for the shared LSB-UDC. Data loading for one pixel occurs while its neighbor is in the MSB phase or idle, enabling conflict-free time-multiplexing of the shared LSB engine. The shuffled scan preserves the standard per-row line time—only the order changes—so the frame time and refresh rate are unchanged. In our prototype (20 clusters × 6 pixels/cluster = 120 rows), each cluster's 5b LSB-UDC with underflow detection is realized from the six per-pixel LSB D-FFs (one per pixel) with lightweight steering logic, enabling local time-multiplexing without contention. This removes dedicated per-pixel 5b LSB memory/counters (each pixel retains only a single LSB D-FF) and yields ≈40% pixel-circuit area reduction. Combined with HTD-PWM, this clustering plus SRS strategy preserves precise digital memory-in-pixel driving while significantly shrinking pixel size.

Figure 24.2.4 shows the on-chip built-in self-test (BIST) that employs a current-controlled-oscillator (CCO)-based ADC [14]. To improve yield and reliability, the CMOS backplane verifies each pixel's current-drive in two modes: pre-bond (backplane-only) by applying a test voltage to the electrode to emulate the LED load, and post-bond (µLED attached) by driving the actual LED. In both cases, the pixel current is digitized by the current-mode 15b ADC and compared against programmable limits to flag out-of-range behavior. Each pixel includes four switches (S_1–S_4) to enter BIST: S_1 ties the test-voltage source to the electrode for pre-bond emulation; S_2 opens to isolate the µLED during test (or to disconnect a faulty LED post-bond); S_3/S_4 act as row/column selectors, routing the selected pixel's current to a global sense line and onward to the ADC. During BIST, scan logic steps through all addresses, enabling one pixel at a time. The ADC code is checked on-chip; violations (weak/leaky drivers or dead/misaligned LEDs) are logged. After all 120×120 pixels are tested, a FINISH flag asserts, and a PASS/FAIL result is produced based on a programmable threshold for bad pixels. This built-in flow enables early screening of backplane defects prior to µLED transfer and post-bond verification afterward, reducing external test time and cost while boosting overall yield.

The 10b digital-PWM CMOS backplane for a 120×120 µLED-on-silicon display was fabricated in a 28nm CMOS process (Fig. 24.2.7). Of the total die area of 0.79mm², the effective pixel array occupies 0.7mm², achieving an active-area ratio of 88%. Figure 24.2.5 (top) shows measured waveforms and pixel current (via a sense resistor) for various grayscale codes under a frame rate of 240Hz. Note that CLK_W and CLK_P are distinct signals; for measurement, they are observed at a single port by logically OR-ing them in the measured waveform. Because the 5b LSB engine is shared by six pixels, the LSB PWM clock (CLK_{PL}) runs six times per frame—once per pixel in each 1×6 cluster. The measured on-times (LED emission times) under the proposed HTD-PWM follow the expected $D_{DATA}/1024$ of the 4.17ms frame—e.g., $D_{DATA} = 48$ (low gray) → 195.3µs; $D_{DATA} = 520$ (mid gray) → 2.116ms; $D_{DATA} = 963$ (high gray) → 3.918ms—confirming correct PWM operation. Figure 24.2.5 (bottom-left) plots the frame-averaged pixel current (i.e., PWM current averaged over one 240Hz frame) versus input code $D_{DATA}<9:0>$, showing measured

ISSCC 2026 / February 17, 2026 / 4:00 PM

linearity: maximum DNL of +0.9LSB and INL of −7.1LSB at 10b resolution, with no degradation at low gray levels—a behavior that commonly degrades in analog PWM schemes. Figure 24.2.5 (bottom-right) compares the input image data with the reconstructed display image from the on-chip BIST, successfully demonstrating both accurate pixel driving and full-array BIST readout capability.

Figure 24.2.6 provides a performance comparison and shows the pixel layout. Each pixel measures 6.08×6.48µm², yielding a pixel density of approximately 4042 PPI, which is the highest reported for a MIP-based digital µLEDoS backplane at this bit depth in the comparison table. All pixel, cluster-control, and peripheral circuits (except for a small portion of BIST) were synthesized and auto-placed using standard cells, demonstrating the scalability of the highly digital-intensive approach.

Acknowledgement:
The chip fabrication in a 28nm CMOS process was supported by Samsung Foundry.

References:
[1] J. Seong *et al.*, "Multi-bit MIP(Memory-in-Pixel)-based Pixel Circuit of CMOS Backplane for Micro-LED Display," *SID Symp.*, Aug. 2020, pp. 359–362. http://doi.org/10.1002/sdtp.13878
[2] P.-Y. Lee *et al.*, "A 10-Bit 1280 × 720 Micro-LED Display Driver with 2-Transistor Pixel Circuits and Current-Mode Pulse Width Modulation," *IEEE SSC-L*, vol. 5, pp. 134–137, May. 2022. http://doi.org/10.1109/LSSC.2022.3174137
[3] L. Wu *et al.*, "An AMLED Microdisplay Driver SoC with Built-In 1.25-Mb/s VLC Transmitter," *IEEE Symp. VLSI Circuits*, June 2015, pp. C328–C329. http://doi.org/10.1109/VLSIC.2015.7231310
[4] J. Jiang *et al.*, "A Fully-Integrated Micro-Display System with Hybrid Voltage Regulator," *IEEE JETCAS*, vol. 13, no. 2, pp. 605–616, June 2023. http://doi.org/10.1109/JETCAS.2023.3270291
[5] Z. G *et al.*, "Size-Dependent Light Output, Spectral Shift, and Self-Heating of 400 nm InGaN Light-Emitting Diodes," *J. Appl. Phys.*, vol. 107, no. 1, pp. 0131031–0131036, Nov. 2010. http://doi.org/10.1063/1.3276156
[6] C.-C. Lin *et al.*, "Ultra-Fine Pitch Thin-Film Micro LED Display for Indoor Applications," *SID Symp.*, May 2018, pp. 780–785. http://doi.org/10.1002/sdtp.12373
[7] T. Blalock *et al.*, "True color 1024×768 Microdisplay with Analog In-Pixel Pulsewidth Modulation and Retinal Averaging Offset Correction," *IEEE JSSC*, vol. 36, no. 5, pp. 838–845, May 2001. http://doi.org/10.1109/4.918923
[8] P.-A. Zou *et al.*, "A New Analog PWM Pixel Circuit with Metal Oxide TFTs for Micro-LED Displays," *IEEE Trans. Electron Dev. (T-ED)*, vol. 69, no. 8, pp. 4306–4311, Aug. 2022. http://doi.org/10.1109/TED.2022.3178363
[9] S.-S. Cheng and P. Chao, "A New SRAM-Embedded Pixel Circuit That Modulates Accurately Gray Level for PWM-Driven Micro-LED Displays," *IEEE SSC-L*, vol. 6, pp. 157–160, June 2023. http://doi.org/10.1109/LSSC.2023.3282904
[10] H.-A. Ahn *et al.*, "An Active Matrix Micro-Pixelated LED Display Driver for High Luminance Uniformity Using Resistance Mismatch Compensation Method," *IEEE TCAS-II*, vol. 65, no. 6, pp. 724–728, June 2018. http://doi.org/10.1109/TCSII.2018.2790412
[11] H.-A. Ahn *et al.*, "A Driving and Compensation Method for AMLED Displays Using Adaptive Reference Generator for High Luminance Uniformity," *IEEE TCAS-II*, vol. 67, no. 10, pp. 1725–1729, Oct. 2020. http://doi.org/10.1109/TCSII.2019.2946207

[12] W. Jeon *et al.*, "Active-Matrix Pixelated-LED Control System for Automotive Headlamps," *IEEE Access*, vol. 10, pp. 45553–45561, April 2022. http://doi.org/10.1109/ACCESS.2022.3170113
[13] W. Jeon *et al.*, "A Pixel Circuit with Leakage Current Compensation Using PWM-Based Data Cycling for AMLED Displays," *IEEE TCAS-II*, vol. 70, no. 1, pp. 61–65, Jan. 2023. http://doi.org/10.1109/TCSII.2022.3203045
[14] P. Prabha *et al.*, "A Highly Digital VCO-Based ADC Architecture for Current Sensing Applications," *IEEE JSSC*, vol. 50, no. 8, pp. 1785–1795, Aug. 2015. http://doi.org/10.1109/JSSC.2015.2414428

24

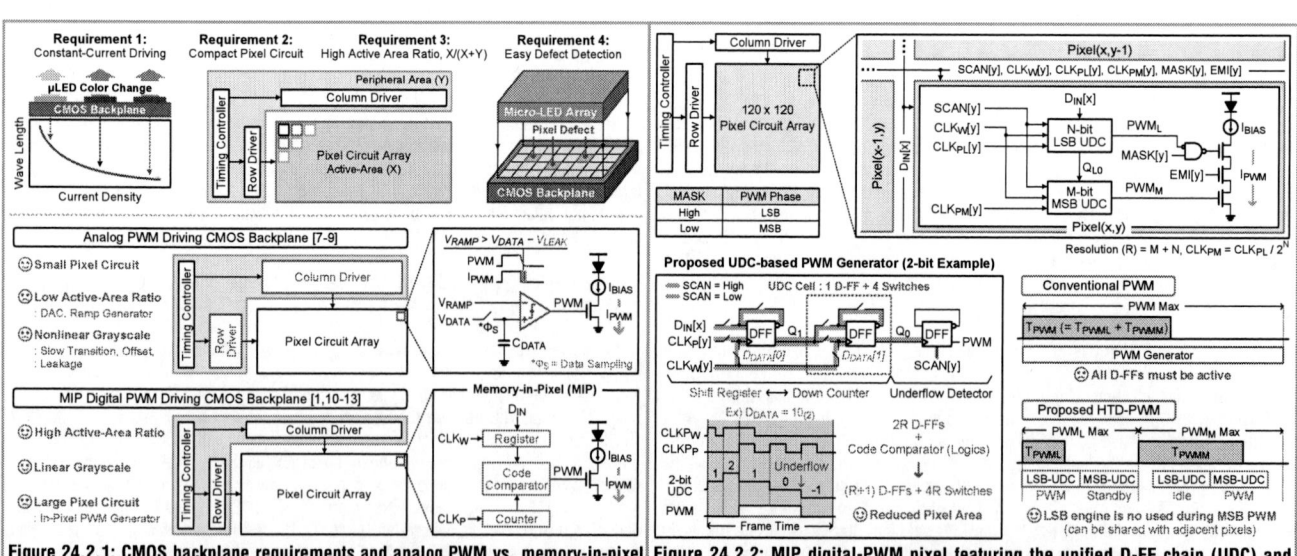

Figure 24.2.1: CMOS backplane requirements and analog PWM vs. memory-in-pixel (MIP) digital PWM for micro-LED-on-silicon (µLEDoS) displays.

Figure 24.2.2: MIP digital-PWM pixel featuring the unified D-FF chain (UDC) and hierarchical time-divided PWM (HTD-PWM) enabling time-multiplexed sharing of LSBs.

DIGEST OF TECHNICAL PAPERS • 413

ISSCC 2026 / SESSION 24 / DISPLAYS / 24.2

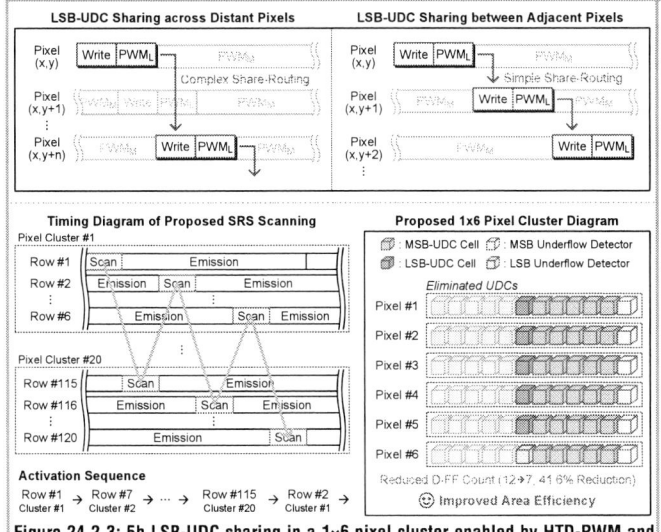

Figure 24.2.3: 5b LSB-UDC sharing in a 1×6 pixel cluster enabled by HTD-PWM and the shuffled rolling-shutter (SRS) scanning technique.

Figure 24.2.4: On-chip built-in-self-test (BIST) using a CCO-based current-mode 15b ADC, and a pre-/post-μLED-bond test flow to detect defective pixels.

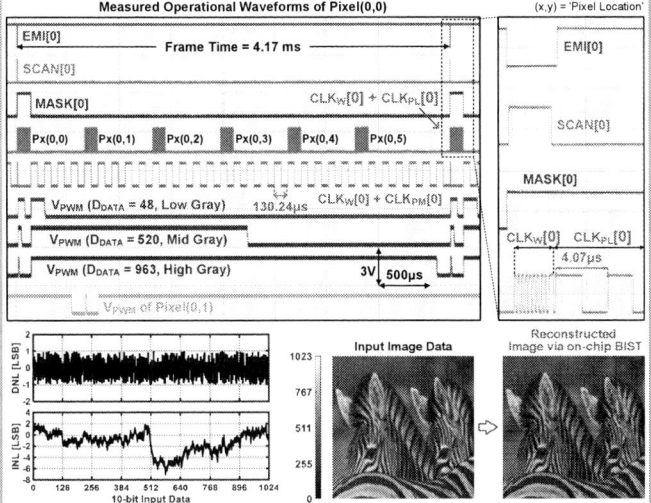

Figure 24.2.5: Measured PWM waveforms, PWM linearity (DNL/INL), and the reconstructed display image acquired via the on-chip BIST.

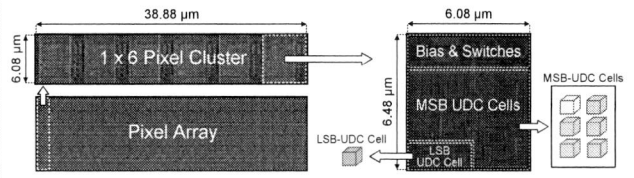

	IEEE JETCAS'23 [4]	IEEE SSC-L'23 [9]	SID'20 [1]	IEEE TCAS-II'23 [13]	This Work
Technology	180-nm CMOS	55-nm CMOS	180-nm CMOS	180-nm CMOS	28-nm CMOS
Pixel Array	64 × 36	1 Pixel	400RGB × 240	2 × 16	120 × 120
PMW Type	Digital	Analog	Digital MIP	Digital MIP	Digital MIP
Power	N/A	N/A	9.67 mW	2.12 mW	1.9 mW
Grayscale	4-bit	7-bit	10-bit	8-bit	10-bit
Frame Rate	100 Hz	120 Hz	120 Hz	60 / 120 Hz	240 Hz
Pixel Dimension (PPI)	40 μm × 40 μm (635)	1.34 μm × 4.02 μm (8477)	30 μm × 30 μm (847)	110 μm × 110 μm (231)	6.08 μm × 6.48 μm (4042)
Active-Area Ratio	*85%	N/A	*75%	*48%	88%
DNL/INL	N/A	0.14 / 4.16 LSB	N/A	N/A	0.9 / -7.1 LSB
BIST	No	No	No	No	Yes

*Active-area ratio is estimated from die photograph

Figure 24.2.6: Pixel layout and performance summary.

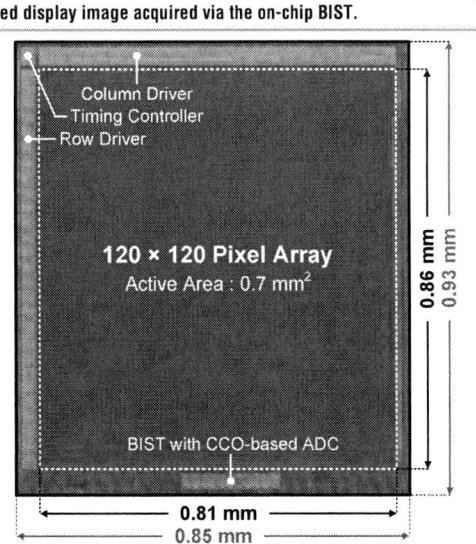

Figure 24.2.7: Die micrograph of the 4042-PPI 120×120 μLEDoS backplane (in 28nm CMOS).

ISSCC 2026 / SESSION 24 / DISPLAYS / 24.3

24.3 A 144mW 161Mpixels/s Tensor Display Processor for 3D Virtual Reality

Ching-Yen Lee, Yueh-Feng Tsai, Ying-Sheng Lin, Chia-Hsiang Yang

National Taiwan University, Taipei, Taiwan

Abstract

This work presents a 3D display processor that converts the focal stack to tensor display. Algorithm-architecture co-optimization is applied to reduce the computational complexity and hardware complexity. It achieves the maximum throughput of 161 Mpixels/s (equivalent to 87 fps at HD resolution) at a clock frequency of 200 MHz, with normalized energy consumption of 0.9nJ/pixel, and a PSNR of 23.7 to 24.8dB,

The metaverse is progressively becoming technically feasible through integration of sensing technology, virtual reality (VR), and 3D displays [1,2], as illustrated in Fig. 24.3.1. For 3D displays, tensor display is a promising solution to creating 3D effects by showing the scenes on multiple image layers [3]. Given multiple light-field (LF) images (captured by an LF camera), the image layers can be generated through non-negative matrix factorization (NMF) [3,4]. This process is highly time-consuming because a large volume of LF data needs to be processed; it takes 33.4 seconds to process two 1280×720 image layers on a CPU. To address this issue, the 3D display processor in [5] adopts half-block-based factorization and sparse sampling to reduce both the computational complexity and external memory access. However, a more efficient way is to preserve essential information while reducing the input data volume by converting the LF images to focal stack (FS) layers first [6]. The FS layers can be generated by accumulating scaled LF pixel values for respective focal depths, along a direction determined by a reference plane. The LF-to-FS conversion can be performed offline since the LF images are captured in advance. In such a system, the image layers for the tensor display are generated in an iterative way by employing NMF [6]. Starting from the initialized image layers, latent image layers are generated, and the generated image layers are used to update the image layers. The process is terminated once the 3D image quality meets the specification or a maximum number of iterations is reached. Despite the reduced amount of data, dedicated FS-based tensor display processors have not been explored in the open literature. In response, this work presents a 3D display processor that converts the focal stack to tensor display. Algorithm-architecture co-optimization is applied to reduce the computational and hardware complexity.

Figure 24.3.2 illustrates the algorithm-architecture co-optimization techniques: correlative data movement (CDM), resolution adaptation (RA), and partial layer reuse (PLR). According to our analysis, the computational complexity is dominated by NMF. CDM is applied to reduce the computations by exploiting data redundancy and approximate computing. An image layer is divided into blocks with 8×16 pixels and the computations are conducted at the block level. In the baseline design, blocks are processed image layer by image layer, and they are repeatedly accessed from the memory. In this work, blocks are processed across image layers and the partial terms can be reused for the subsequent block to save computations. Since the mean value of full block is required for scaling, a subset, instead of all, pixels in a block can be used to approximate the computations. CDM reduces the computational complexity by 49%. RA is applied to skip computations since the data distributions of the image layers remain the same for feasible resolutions. The difference between the image layers also decreases during the iterative process [7]. The image resolution can be reduced, and an image pyramid, consisting of image layers with various (8×, 4×, 2×, 1×) downsampling factors, is built to approximate the distributions with a lower computational cost. Starting from an image layer with the lowest resolution, the resolution of the image layer increases when the difference between the image layers drops below a predefined threshold. The process continues until the two consecutive image layers converge and the resolution of the final image layer reaches the original one. The computational complexity is further reduced by 67% by applying RA. PLR is employed to reduce computational complexity by employing the spatial correlation between adjacent pixels in a block. In the direct-mapped implementation, updating data for a specific image layer requires referencing multiple blocks across image layers. For 3D scenes, similar blocks in different image layers only have slight positional shifts. The positions of the blocks from the previous layer can be reused with a negligible performance loss, reducing the computational complexity by 25%. Overall, the computational complexity can be reduced by up to 87% by the proposed techniques (CDM, RA, and PLR), and the image quality, measured in peak signal-to-noise ratio (PSNR), only decreases by 0.5dB.

Figure 24.3.3 shows the system architecture of the tensor display processor, which includes a tensor display engine, a layer calculator, a layer updater, a block handler, and a memory bank. The tensor display engine is designed to perform compute-intensive multiply-accumulate (MAC) operations for NMF. It includes 30 processing element (PE) arrays, each with 5×5 MAC units, and a data distributor. The blocks for an image layer are distributed to the MAC units by the data distributor. The layer calculator is used to accumulate partial terms of the MAC array and to compute the scaled image layers. The layer updater adjusts the resolution and to obtain the positions of the blocks in the corresponding image layers. The block handler is designed to retrieve the required blocks across multiple image layers

from the memory bank. The memory bank stores essential data for FS images and image layers. Only the difference between the FS images and part of the image layers are required to minimize the memory storage.

Figure 24.3.4 shows the design details of the tensor display engine and the layer calculator. For the tensor display engine, the input blocks of a size of 8×16 pixels are retrieved sequentially and sent to the PE arrays and the subsequent blocks are preloaded. Overlaps occur at the block boundaries, so the associated computations can be reduced by 75% by skipping the repeated pixels. The PE array is configurable to support operations from image layers along different dimensions (height, width, and channel). The hardware utilization is maximized by dynamically allocating blocks to multiple MAC units. For the layer calculator, a partial sum accumulator is designed to support three operating modes for the required flexibility. In Mode 1, the data is accumulated for an image layer whenever the PE array outputs are available. In Mode 2, partial sums from the overlapped regions are accumulated according to the configuration of the PE array. In Mode 3, the PE array outputs can be reused for the succeeding computations. A high dynamic range is required to accommodate possible partial sums. A dataflow with a fixed 20b width suffices, but at a cost of high hardware cost. In our design, pixels and partial sums are represented by 8b and 16b, respectively. The latent images are represented by 20b after scaling. The area is reduced by 30% with a negligible PSNR loss, when compared to the baseline with a fixed bit width.

Figure 24.3.5 shows the design details of the layer updater. Iterations for updating image layers are terminated when the difference between the image layers becomes small or the maximum iteration count is reached. According to our experiments, it usually takes 5 to 8 iterations for processing the image layers with a specific resolution. The maximum number of iterations can be set to 8 to achieve the target PSNR. An input-stationary dataflow is employed to reuse the pixels according to their positions. The memory allocation is optimized to reduce the memory access. Since the FS layers exhibit high similarity, the sparsity extracted from their differences can be leveraged to reduce data access. A group of 16b is stored for accessing a block, increasing the memory bandwidth utilization. Memory access is reduced by 42% by applying the proposed memory allocation scheme.

Figure 24.3.6 shows the verification results and performance comparison. Fabricated in a 40nm CMOS technology, the tensor display processor integrates 2.6M logic gates in core area of 2.15mm² and includes 55.3KB of on-chip SRAM. The functionality of the chip is verified using the commonly used datasets [8,9] for performance evaluation. The reconstructed 3D scenes achieve a PSNR of 22.6 to 28.9dB using the LF images as the ground truth. The chip dissipates 20 to 144mW at a clock frequency of 60 to 200MHz from a supply voltage of 0.60 to 1.05V. It achieves the maximum throughput of 161Mpixels/s (equivalent to 87fps at HD resolution) at a clock frequency of 200MHz, with normalized energy consumption of 0.9nJ/pixel. Compared to the state-of-the-art 3D display processor [5], which supports HD resolution at 31 to 33 fps on 3D displays with a PSNR of 23.7 to 24.8 dB, the chip achieves a 2.6 to 2.8× higher throughput and 5 to 18× lower normalized energy, with higher PSNR on the same dataset. This work provides an energy-efficient solution for high throughput 3D display processing on wearable devices. Figure 24.3.7 shows the chip micrograph and summary.

Acknowledgement:
This work is supported by National Science and Technology Council (NSTC) of Taiwan and Intelligent & Sustainable Medical Electronics Research Fund in National Taiwan University. The authors also thank Taiwan Semiconductor Research Institute (TSRI) for technical support on chip design and fabrication.

References:
[1] J.-W. Jo *et al.*, "Progress in 3D Display Technologies for Immersive Visual Experiences," *IEEE Open Journal on Immersive Displays*, vol. 1, pp. 155-164, Sep. 2024. http://doi.org/10.1109/OJID.2024.3457495
[2] H. Wang *et al.*, "A Survey on the Metaverse: The State-of-the-Art, Technologies, Applications, and Challenges," *IEEE Internet of Things Journal*, vol. 10, no. 16, pp. 14671-14688, Aug. 2023. http://doi.org/10.1109/JIOT.2023.3278329

[3] G. Wetzstein *et al.*, "Tensor displays: compressive light field synthesis using multilayer displays with directional backlighting," *ACM Transactions on Graphics*, no. 4, pp. 1-11, July 2012. https://doi.org/10.1145/2185520.218557

[4] D. Lanman *et al.*, "Content-adaptive parallax barriers: optimizing dual-layer 3D displays using low-rank light field factorization," *ACM Transactions on Graphics*, no. 163, pp. 1-10, Dec. 2010. https://doi.org/10.1145/1882261.1866164

[5] L.-Q. Weng et al., "A HD 31fps 7x7-View Light-Field Factorization Processor for Dual-Layer 3D Factored Display," *ISSCC*, pp. 508-509, Feb. 2022. http://doi.org/10.1109/ISSCC42614.2022.9731661

[6] K. Takahashi *et al.*, "From Focal Stack to Tensor Light-Field Display," *IEEE Transactions on Image Processing*, vol. 27, no. 9, pp. 4571-4584, Sep. 2018. http://doi.org/10.1109/TIP.2018.2839263

[7] X. Cao *et al.*, "Accelerating decomposition of light field video for compressive multi-layer display," *Opt. Express*, vol. 23, pp. 34007-34022, 2015. http://doi.org/10.1364/OE.23.034007

[8] G. Wetzstein, "Synthetic Light Field Archive," https://www.media.mit.edu/~gordonw/SyntheticLightFields/, 2018. https://www.media.mit.edu/~gordonw/SyntheticLightFields/

[9] K. Honauer *et al.*,"A dataset and evaluation methodology for depth estimation on 4D light fields," *Asian Conference on Computer Vision*, pp. 19-34, Nov. 2016. https://doi.org/10.1007/978-3-319-54187-7_2

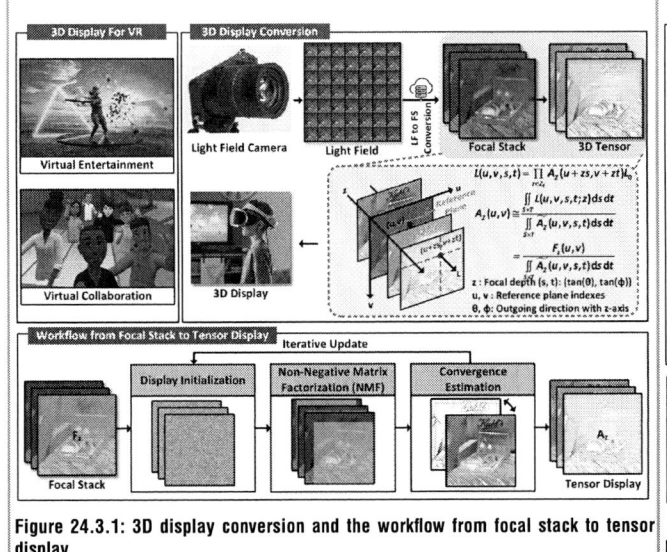

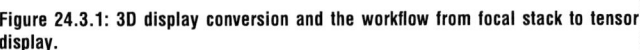

Figure 24.3.1: 3D display conversion and the workflow from focal stack to tensor display.

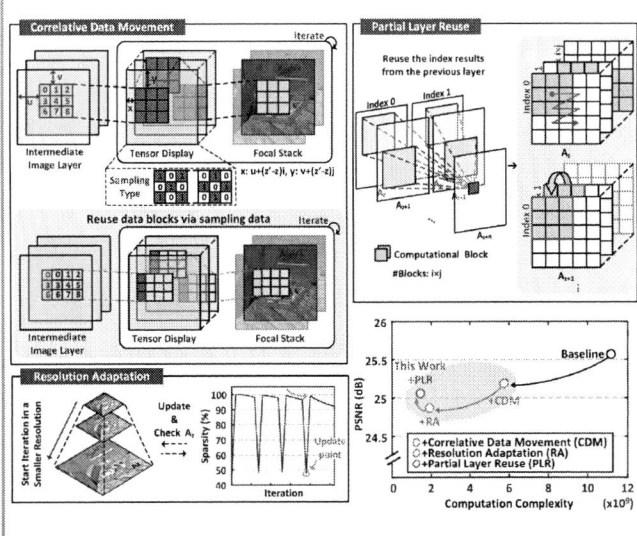

Figure 24.3.2: Algorithm-architecture co-optimization and design techniques.

ISSCC 2026 / SESSION 24 / DISPLAYS / 24.3

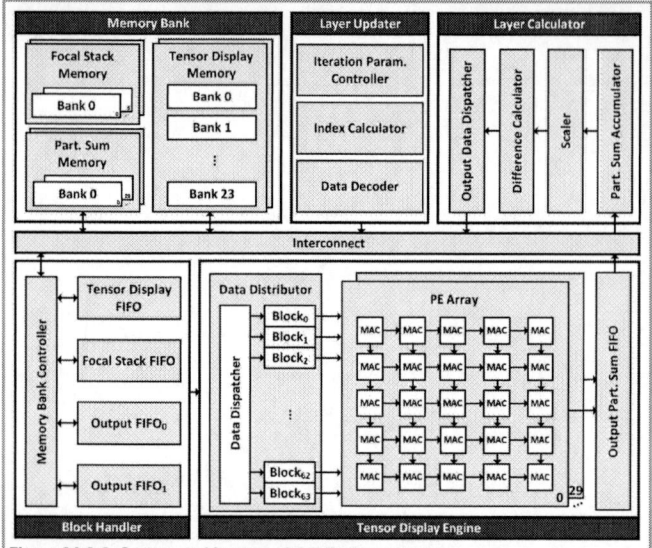

Figure 24.3.3: System architecture of the display processor.

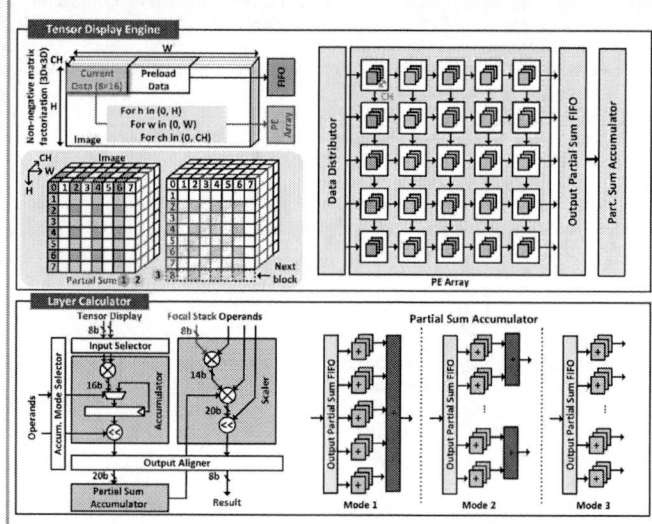

Figure 24.3.4: Architecture design of the tensor display engine and layer calculator.

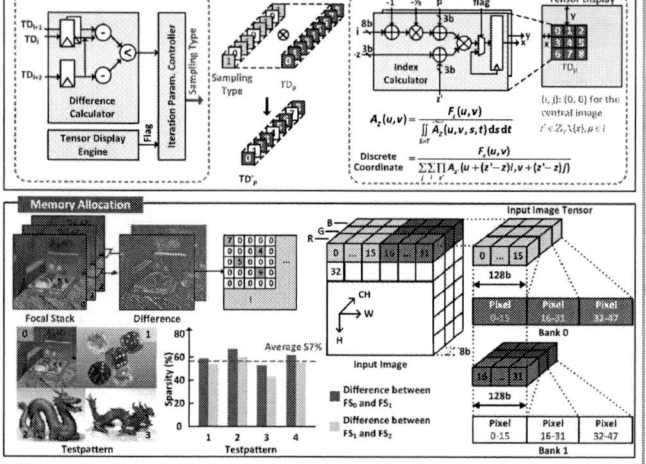

Figure 24.3.5: Design details of layer updater and optimized memory allocation.

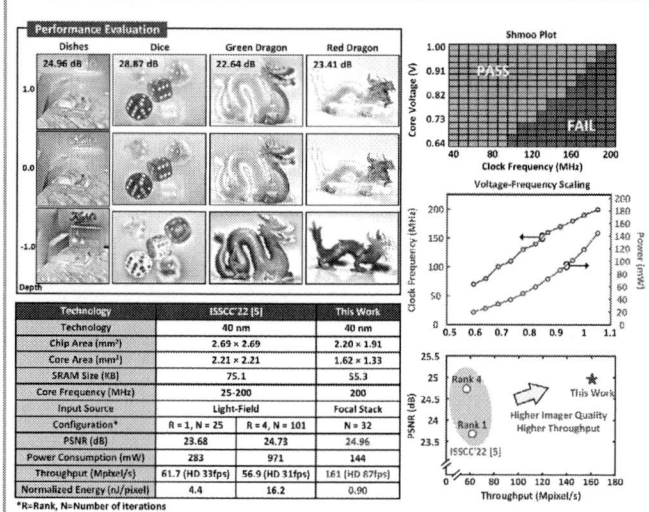

Figure 24.3.6: Experimental verification and performance comparison.

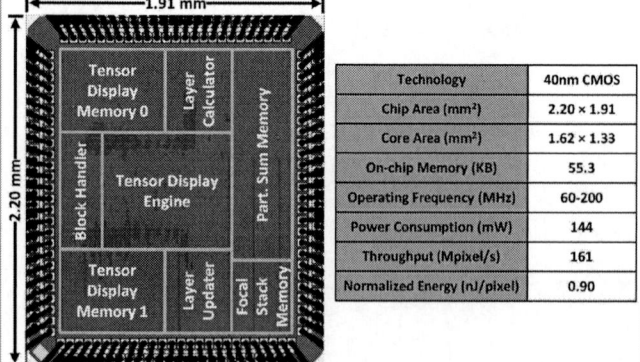

Figure 24.3.7: Chip micrograph and summary.

Technology	40nm CMOS
Chip Area (mm²)	2.20 × 1.91
Core Area (mm²)	1.62 × 1.33
On-chip Memory (KB)	55.3
Operating Frequency (MHz)	60-200
Power Consumption (mW)	144
Throughput (Mpixel/s)	161
Normalized Energy (nJ/pixel)	0.90

• 2026 IEEE International Solid-State Circuits Conference

ISSCC 2026 / SESSION 24 / DISPLAYS / 24.4

24.4 A 512ch 10b Source-Driver IC with Current-Mode Auto-Zeroing Scheme Achieving 1.4mV DVO and 600ns 1-Horizontal Time for Mobile OLED Displays

Yousung Park, Gyu-Wan Lim, Hyun-Sik Kim

KAIST, Daejeon, Korea

Abstract

This paper presents a 10b source-driver IC (SD-IC) for ultra-fast, uniform-luminance mobile OLED displays. A current-mode auto-zeroing (CM-AZ) scheme transparently cancels output-buffer offset—lowering DVO with zero timing penalty—while simultaneously suppressing AC-coupling errors in the LSB-stacked DAC. Fabricated in 180nm CMOS with an 11.37µm channel pitch, the 512ch SD-IC achieves a 600ns 1-H time and a 1.4mV DVO, successfully demonstrated on an LED display panel.

The growing demand for mobile OLED displays with ultra-high frame rates and superior luminance uniformity drives advances in source-driver ICs (SD-ICs) [1,2]. Pushing the frame rate beyond 240Hz—well above the legacy 60Hz—in high-resolution panels shortens the one-horizontal (1-H) interval, defined as 1-H time = 1 / (frame rate × number of scan lines). Within this 1-H period, the SD-IC must complete digital-to-analog conversion (DAC) and program each OLED pixel through its buffer amplifier. Display uniformity, meanwhile, depends on minimizing the deviation of voltage outputs (DVO) across hundreds of source channels—a metric dominated by the offset-voltage (V_{OS}) dispersion of the output buffers.

Frame-by-frame chopping [3–6] has been explored to lower DVO by alternately inverting the V_{OS} polarity every frame; however, modern variable-refresh-rate (VRR) displays can throttle the frame rate down to 1Hz to save power [7–9], and at such ultra-low rates, the luminance ripple become visible flicker. Conventional voltage-mode auto-zeroing (AZ) [10–13] cancels the buffer's V_{OS} by sampling it onto a capacitor, but it suffers two major drawbacks: (1) the dedicated V_{OS}-sampling phase consumes precious 1-H timing margin, and (2) the AZ capacitor is directly coupled to the output. This coupling allows sample-and-hold (S/H) errors to appear at the output, ultimately limiting the attainable DVO. Foreground calibration [14,15] is an option, but requires a lengthy start-up, is constrained by the calibration resolution, and cannot track temporal V_{OS} drift during normal operation.

This paper proposes a current-mode auto-zeroing (CM-AZ) technique. Unlike classical AZ, the CM-AZ removes buffer offsets transparently during normal data driving with zero timing overhead. Leveraging this scheme, we implemented a 512-channel, 10b SD-IC that achieves both a low DVO and a fast 1-H time.

The top-left of Fig. 24.4.1 shows a simplified model of a typical unity-gain buffer. The 1st stage is a folded-cascode amplifier that sets the transconductance (G_m) and boosted output resistance (r_{o1}). It comprises a differential input-pair (DIP) biased by a tail current of $2 \cdot I_B$, bias current sources of ($I_B + I_S$), and a cascode stage biased by $2 \cdot I_S$ that acts as a current conveyor. The 2nd-stage amplifier (-A) provides additional gain. An input-referred offset voltage (V_{OS}), caused primarily by transistor mismatch in the DIP, results in a final output of $V_O = V_{IN} + V_{OS}$, rather than the ideal $V_O = V_{IN}$. Conceptually partitioning the combined bias ($I_B + I_S$) into individual I_B and I_S sources reveals a zero-current bridge (ZCB) between them (top-right of Fig. 24.4.1). This structure presents a key opportunity to correct the DIP's V_{OS}. If an offset current (I_{OS}) that matches V_{OS} (as $I_{OS} = G_m \cdot V_{OS}$) could be intentionally injected into the left side of the ZCB, the offset would be cancelled, yielding the ideal $V_O = V_{IN}$. In a steady state, the ZCB carries no DC current and thus behaves like a virtual open circuit. These observations—that a purposefully introduced I_{OS} can compensate V_{OS}, and that the ZCB virtually isolates the V_{OS}-corrected DIP from the cascode stage—led to the conception of our proposed current-mode auto-zeroing (CM-AZ) technique.

The bottom of Fig. 24.4.1 presents the CM-AZ architecture, which transparently cancels V_{OS} in the background while the buffer drives V_O. The original I_B sources are replaced by current-mode S/Hs positioned to the left side of the ZCB, and an auxiliary DIP is added to its right side. Operation proceeds in two overlapping phases. During the initial drive—or sampling—interval (T_{SAMP}), as new data arrives at V_{IN}, the auxiliary DIP steers V_O to slew toward V_{IN}. Concurrently, the main DIP is isolated from the bridge by shorting both its inputs to V_{IN}, allowing it to produce the V_{OS}-matched offset current I_{OS} (= $G_m \cdot V_{OS}$). The S/H then samples a current of $I_B \pm I_{OS}$. After T_{SAMP}, the S/H switches to hold mode, and control of the buffer is returned to the main DIP to complete the settling of V_O. Because the sampled $\pm I_{OS}$ actively cancels V_{OS} from the main DIP, any offset component is prevented from propagating into the cascode stage even at $V_O = V_{IN}$. Consequently, as V_O settles to V_{IN}, it remains free of offset, and the bridge current becomes zero ($I_{ZCB} = 0$). The DC-floating nature of the ZCB enables a seamless hand-off between the (temporary-drive) auxiliary DIP and the (now offset-compensated) main DIP. This achieves in-operation ("zero-wait") offset cancellation without a dedicated sampling slot, meaning the CM-AZ technique incurs no speed penalty.

Figure 24.4.2 depicts the schematic of the CM-AZ buffer amplifier. During T_{SAMP} ($t_0 < t < t_1$), when $<\Phi_S, \Phi_H> = <high, low>$, the negative input node of the main DIP is connected to V_{IN} via MUX$_1$. Concurrently, the current-mode S/H (M$_{SH}$ and C$_{SH}$) samples the main DIP's currents, I_{IN1} (= $I_B + I_{OS}$) and I_{IN2} (= $I_B - I_{OS}$). For a rising step at V_{IN}, the auxiliary DIP temporarily supplies slewing currents, I_{SUM+} (= $2 \cdot I_B - I_B$) and I_{SUM-} (= $0 - I_B$), to the folded-

cascode stage, driving the initial positive transition of V_O. After T_{SAMP} (during $t_1 < t < t_2$), with $<\Phi_S, \Phi_H> = <low, high>$, MUX$_{2-3}$ hand control of the buffer amplifier back from the auxiliary DIP to the main DIP while the S/H pair holds the sampled currents. Because V_O is still slewing, the main DIP now delivers $I_{SUM+} = I_B - I_{OS}$ and $I_{SUM-} = -(I_B - I_{OS})$ to the following stage. The nearly unchanged magnitude of $I_{SUM\pm}$ before and after the DIP hand-over ensures a smooth transition. As the buffer amplifier enters its fine-settling phase, the I_{OS} term stored in the S/H cancels (absorbs) the main DIP's offset V_{OS}, driving the bridge current $I_{ZCB\pm}$ toward true-zero and forcing V_O to accurately track V_{IN}, free of V_{OS}.

In each 1-H cycle, the S/H circuits (M$_{SH}$ and C$_{SH}$) capture a nearly constant $I_B \pm I_{OS}$, as V_{OS} changes negligibly from one cycle to the next. Thus, a long sampling time (T_{SAMP}) is not required to achieve high accuracy, making the CM-AZ offset cancellation insensitive to the exact T_{SAMP}. Because a high sampling rate—set by $g_{m,SH}/C_{SH}$—is not needed, M$_{SH}$ can be designed with a low $g_{m,SH}$. A key benefit of this low $g_{m,SH}$ is its ability to effectively attenuate error voltages on C_{SH}, which in turn allows C_{SH} to be kept small ($\leq$40fF). Moreover, since the V_{OS} is cancelled from the instantaneous V_{IN} level, the CM-AZ continuously tracks V_{OS} variations caused by DC shifts in V_{IN}—an advantage over foreground-calibrations. The auxiliary DIP adds only negligible overhead; with the offset effectively suppressed, both the main and auxiliary DIPs can be designed with minimum area. Although CM-AZ cannot correct the mismatch current between M$_{P1}$ (M$_{N1}$) and M$_{P2}$ (M$_{N2}$) of the folded-cascode stage, that mismatch is attenuated by the main DIP's G_m. It appears only as a negligible input-referred offset—less than 2% of the total V_{OS} in our design.

Figure 24.4.3 describes the in-operation LSB-stacking DAC scheme. In our SD-IC design, we adopt the LSB-stacking DAC of [12]: a 6b resistor-string DAC (R-DAC) first performs a coarse conversion, producing V_H and V_L (= V_{MSB}). Subsequently, a 4b switched-capacitor (SC) DAC interpolates V_{LSB} between V_H and V_L and then "stacks" this fine voltage onto V_{MSB}, resulting in the 10b $V_O = V_{MSB}$ (6b) + V_{LSB} (4b). However, [12] suffers from an AC-coupling error (top-left of Fig. 24.4.3). If V_{MSB} is driven to the output V_O while stacking V_{LSB} via C_{LSB}, the parasitic C_{IN} of the buffer's negative input (V_{IN-}) introduces an error voltage, V_E [= $C_{IN}/C_{LSB} \cdot (V_{MSB} - V_X)$], where V_X is the buffer's V_{IN-} voltage at the V_{LSB}-stacking instant ($t = t_1$). To avoid V_E, driving V_{MSB} to V_O and stacking V_{LSB} must be time-separated, lengthening the total DAC interval. The CM-AZ buffer can remove this constraint (right of Fig. 24.4.3). During the I_{OS}-sampling phase ($t_0 < t < t_1$), MUX$_1$ (with $\Phi_H = low$) ties the main DIP's V_{IN-} directly to V_{IN+} (= V_{MSB}); this pre-charges C_{IN} to V_{MSB} (i.e., $V_X = V_{MSB}$) prior to the V_{LSB}-stacking event, thereby suppressing V_E and obviating time-separation of the two operational phases. Thus, the CM-AZ buffer inherently enables faster operation of the LSB-stacking DAC.

The top of Fig. 24.4.4 shows the architecture of the 512-channel SD-IC. Each channel processes serial data through latches and level shifters, generating a 10b voltage (V_O) using a combined 6b R-DAC and 4b LSB-stacking SC-DAC. To drive the highly capacitive data lines of a high-resolution OLED panel, the CM-AZ buffer amplifier is augmented with a high-slew-rate (HSR) assist stage (top-right of Fig. 24.4.4). With V_{IN} and V_O as inputs, the HSR circuit detects rising or falling transitions by comparing the transient current via M$_{P(N)<8:6>}$ with a reference I_{CMP}. Once the detection threshold is exceeded, switch M$_{N(P)9}$ turns on, allowing an assist current I_{SR}—much larger than the nominal bias I_B—to be pulled from or pushed to the buffer's intermediate node $V_{X(Y)}$, thereby accelerating the output slew rate.

The 512-channel 10b SD-IC was fabricated in a 180nm CMOS (Fig. 24.4.7). The bottom-left of Fig. 24.4.4 presents the output (V_O) waveforms measured with various 10b input codes (D<9:0>) under a display panel load of 30kΩ and 30pF. The HSR buffer achieves a slew rate of 31.6V/µs—a 4.78× improvement over the non-HSR setup—without increasing the quiescent current ($I_Q = 1.7$µA). The V_O settled with the LSB-stacking DAC exhibits no AC-coupling error ($V_E = 0$). By contrast, if AC-coupling issue were not addressed, V_O would be severely distorted. This work achieves the minimum 1-H time of 600ns while guaranteeing 10b accuracy. The bottom-center of Fig. 24.4.4 confirms that the CM-AZ incurs no speed penalty; the V_O-settling curves with and without CM-AZ are essentially identical, demonstrating transparent offset cancellation during the buffer's normal operation. The bottom-right of Fig. 24.4.4 shows that the CM-AZ performance remains stable for different offset-sampling windows ($T_{SAMP} = 200$ns and 250ns).

The top of Fig. 24.4.5 presents demonstration photos from a 128×128 red/green LED display panel driven by the SD-IC. As confirmed by the middle of Fig. 24.4.5, the CM-AZ effectively suppresses offsets. Among 128 channels, the maximum DVO of 1.4mV was measured, which is a 7.7× reduction over the non-CM-AZ version, achieved under identical silicon area, 1-H time, and power budget constraints. The bottom of Fig. 24.4.5 shows the measured linearity of the 10b DAC. The maximum DNL was 0.27LSB. Applying 7 voltage-taps (used for gamma-correction) to the global 6b R-string improved the maximum INL to 0.37LSB. Figure 24.4.6 presents a performance comparison to prior art and the 1-channel layout. Compared to prior art, the proposed chip achieves the smallest DVO (1.4mV) and the fastest 1-H time (600ns), all while maintaining a competitive size and a wide rail-to-rail output range.

Acknowledgement:
This work was supported by LX Semicon.

References:
[1] Y.-R. Jo *et al.*, "An OLED Display Driver IC with High-Gain Fast-Slew Circuit and On-the-Fly Self-Repair Technique for High-Resolution Display," *J. SID*, vol. 32, no. 5, pp. 415–425, May 2024. http://doi.org/10.1002/jsid.1307

[2] Y. Park *et al.*, "A 10-bit Source-Driver IC with Charge-Modulation DAC for Enhanced Frame-Rate Mobile OLED Displays," *IEEE JSSC*, vol. 59, no. 11, pp. 3511–3524, Nov. 2024. http://doi.org/10.1109/JSSC.2024.3442248

[3] J.-S. Kang *et al.*, "10-bit Driver IC Using 3-bit DAC Embedded Operational Amplifier for Spatial Optical Modulators (SOMs)," *IEEE JSSC*, vol. 42, no. 12, pp. 2913–2922, Dec. 2007. http://doi.org/10.1109/JSSC.2007.908690

[4] H.-M. Lee *et al.*, "An Area and Power Efficient Interpolation Scheme Using Variable Current Control for 10-Bit Data Drivers in Mobile Active-Matrix LCDs," *IEEE Trans. Consum. Electron.*, vol. 65, no. 2, pp. 253–262, May 2019. http://doi.org/10.1109/TCE.2019.2900512

[5] J.-S. Na *et al.*, "A Highly Linear 10-Bit DAC of Data Driver IC Using Source Degeneration Load for Active Matrix Flat-Panel Displays," *IEEE TCAS-II*, vol. 67, no. 11, pp. 2312–2316, Nov. 2020. http://doi.org/10.1109/TCSII.2020.2972370

[6] K. Ryu *et al.*, "A Source-Driver IC Including Power-Switching Fast-Slew-Rate Buffer and 8Gb/s Effective 3-Tap DFE Receiver Achieving 4.9mV DVRMS and 17V/μs Slew Rate for 8K Displays and Beyond," *ISSCC*, Feb. 2023, pp. 382–383. http://doi.org/10.1109/ISSCC42615.2023.10067592

[7] Y. Kim *et al.*, "A Highly Uniform Luminance and Low-Flicker Pixel Circuit and Its Driving Methods for Variable Frame Rate AMOLED Displays," *IEEE Access*, vol. 11, pp. 74301–74311, July 2023. http://doi.org/10.1109/ACCESS.2023.3296787

[8] Y. Kim *et al.*, "A Pixel Circuit with Improved Luminance Uniformity and Flicker for AMOLED Displays With a Wide VRR Range of 15 Hz to 360 Hz," *IEEE Electron Device Lett.*, vol. 46, no. 4, pp. 600–603, April 2025. http://doi.org/10.1109/LED.2025.3535717

[9] D. Sim *et al.*, "Improvement of Flicker Phenomenon at Low Frequencies in AMOLED Displays by Applying Compensation Scheme of Variable Reset Voltage," *Sci. Rep.*, vol. 15, no. 1, April 2025. http://doi.org/10.1038/s41598-025-96847-1

[10] C.-W. Lu, "A Rail-To-Rail Class-AB Amplifier with an Offset Cancellation for LCD Drivers," *IEEE JSSC*, vol. 44, no. 2, pp. 525–537, Feb. 2009. http://doi.org/10.1109/JSSC.2008.2010995

[11] M.-W. Hsu and C.-L. Chen, "A Cost-Effective Offset Cancellation Structure for LCD Source Driver," *ISCAS*, June 2014, pp. 2317–2320. http://doi.org/10.1109/ISCAS.2014.6865635

[12] G.-W. Lim *et al.*, "An Area-Efficient 10-Bit Source-Driver IC With LSB-Stacked LV-to-HV-Amplify DAC for Mobile OLED Displays," *IEEE JSSC*, vol. 58, no. 11, pp. 3164–3178, Nov. 2023. http://doi.org/10.1109/JSSC.2023.3289503

[13] J. Ahn *et al.*, "A Fully Nonlinear Compact 10b Source Driver with Low-Voltage Gamma Slope DAC and Data/Phase Dependent Current Modulation Achieving 2411μm²/Channel for Mobile OLED Displays," *ISSCC*, Feb. 2024, pp. 434–435. http://doi.org/10.1109/ISSCC49657.2024.10454504

[14] C. Kim *et al.*, "Effective 10-Bit OLED Driver IC with 11-Bit DAC, Double Capacitor-Coupled Adder, and Offset Calibration for Enhanced Panel Driving," *J. SID*, vol. 33, no. 5, pp. 644–652, April 2025. http://doi.org/10.1002/jsid.2068

[15] J. Oh *et al.*, "A 10b Source-Driver IC with All-Channel Automatic Offset Calibration and Slew-Rate-Enhanced Amplifier Achieving 2273pm²/Channel and 1.9mV DVO for 6285-PPI OLED-on-Silicon Displays," *ISSCC*, Feb. 2025, pp. 126–127. http://doi.org/10.1109/ISSCC49661.2025.10904712

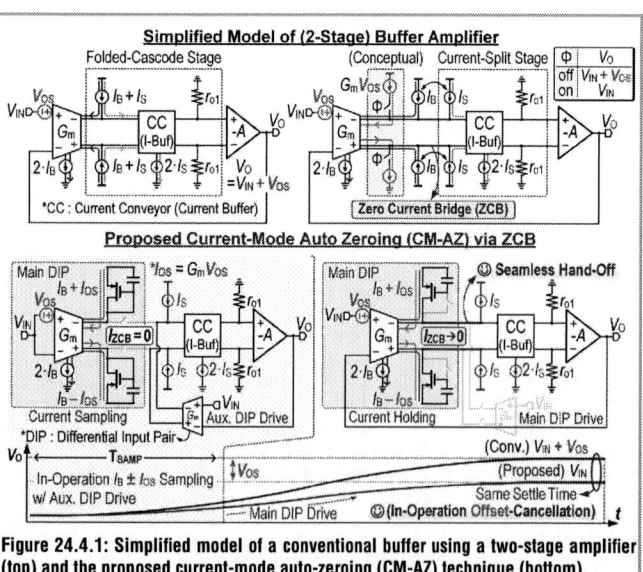

Figure 24.4.1: Simplified model of a conventional buffer using a two-stage amplifier (top) and the proposed current-mode auto-zeroing (CM-AZ) technique (bottom).

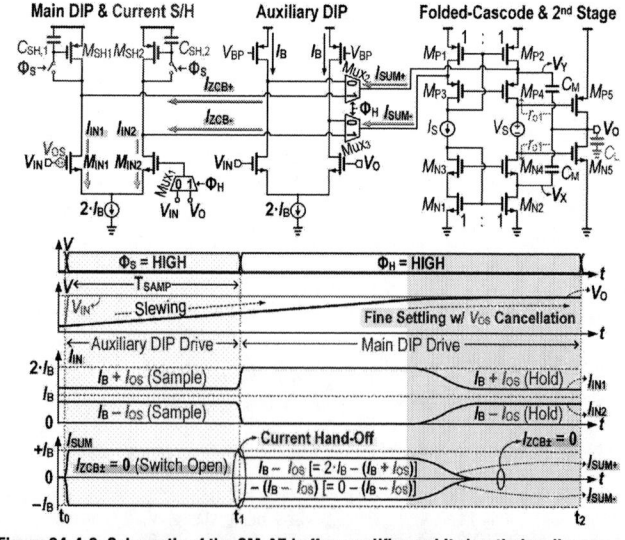

Figure 24.4.2: Schematic of the CM-AZ buffer amplifier and its key timing diagram.

ISSCC 2026 / SESSION 24 / DISPLAYS / 24.4

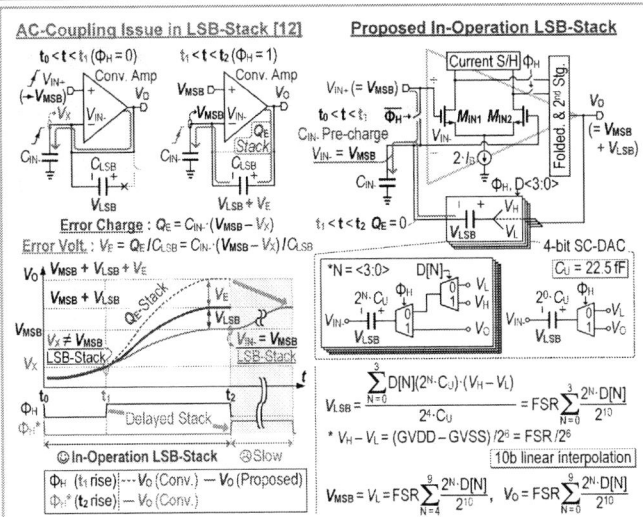

Figure 24.4.3: Mitigation of the AC-coupling error in LSB-stacking SC-DAC using the inherent C_{IN} pre-charge of the CM-AZ technique.

Figure 24.4.4: Top architecture of the 512-ch 10b SD-IC (top-left), schematic of the proposed high slew-rate stage (top-right), and measured key waveforms (bottom).

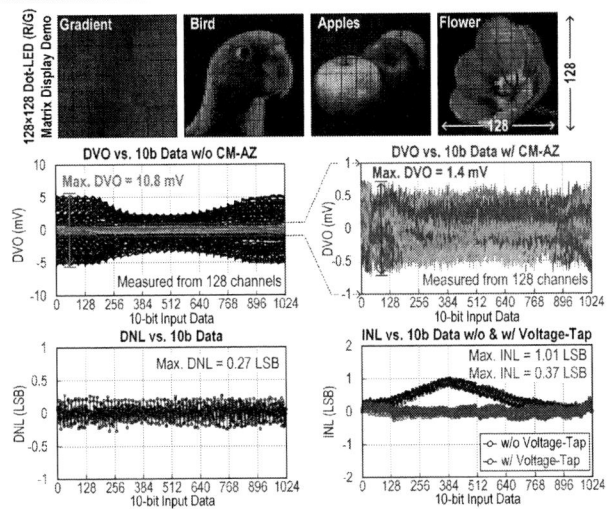

Figure 24.4.5: Photograph of the LED display demonstration driven by the SD-IC, measured deviation of voltage outputs (DVO), and linearity (DNL/INL).

Proposed 10-bit SD-IC Channel Layout

	ISSCC'23 [6]	JSSC'23 [12]	JSSC'24 [2]	ISSCC'25 [15]	This Work
CMOS Process	180-nm	130-nm	180-nm	65-nm	180-nm
Gray Scale (DAC Architecture)	– (R-DAC)	10-bit (SC-multiplier + LSB stack-up)	10-bit (QM-DAC)	10-bit (DAC-embedded Amplifier)	10-bit (SC-DAC)
Output Range	0.2 to 17.8 V	0.3 to 4.5 V	0.25 to 4.85 V	2.1 to 3.1 V*	0.2 to 4.8 V
Static Current (I_Q/ch)	19 μA	1.8 μA	2.4 μA	5.4 μA	1.7 μA
DNL/INL	–	0.39 / 0.9 LSB	0.21 / 0.41 LSB	0.27 / 0.59 LSB	0.27 / 0.37 LSB
Offset Cancellation Method	Per-Frame Chopping	Auto-Zeroing	None	Foreground Calibration	CM-AZ
Overhead	Visible Flicker	Long 1-H	High DVO	In-situ V_{OS} Drift	None
Max. DVO	4.9 mV	4.82 mV	11.5 mV	1.9 mV	1.4 mV
1-Horizontal Time***	1** μs (1.67 ×)	8.2 μs (13.67 ×)	1.5 μs (2.5 ×)	0.69 μs (1.15 ×)	0.6 μs (1 ×)
Slew Rate @ R_L & C_L	17 V/μs @ 5 kΩ & 350 pF	8.1** V/μs @ 30 kΩ & 30 pF	22 V/μs @ 3 kΩ & 100 pF	6.8 V/μs @ 10 kΩ & 10 pF	31.6 V/μs @ 30 kΩ & 30 pF
DAC + Buffer Area per Channel	10 × 560** μm² (2.47 ×)	16 × 144 μm² (1.02 ×)	15 × 221 μm² (1.46 ×)	16 × 123.8 μm² (0.87 ×)	11.37 × 199 μm² (1 ×)

* Output range for OLEDoS application ** Estimation from die micrograph or measured waveforms
*** 1-H time = The time required for the analog output to settle within 10b accuracy after a digital input

Figure 24.4.6: Performance summary with 1-channel layout of the SD-IC.

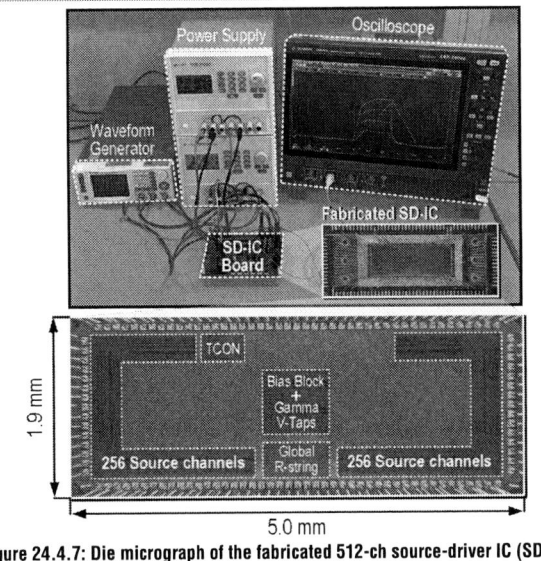

Figure 24.4.7: Die micrograph of the fabricated 512-ch source-driver IC (SD-IC).

• 2026 IEEE International Solid-State Circuits Conference

ISSCC 2026 / SESSION 24 / DISPLAYS / 24.5

24.5 A 10b Display Driver IC with a Pivoting Translinear-Loop DAC-in-Buffer Enabling 1μs 1-H Scan Time and 1722μm² per Channel

Hyeong-Joon Kim, Yousung Park, Kihyun Kim, Seunghwa Shin, Hyun-Sik Kim

KAIST, Daejeon, Korea

Abstract

This paper presents a 10b display driver IC (DDI) for high-resolution OLEDs. It features a novel pivoting translinear-loop (PTL) DAC-in-buffer that overcomes the area, speed, and linearity trade-offs. A key enabler is a floating voltage-mirror (FVM) that decouples coarse/fine domains, allowing an area-efficient 5b+5b architecture via true-DC interpolation while maintaining high linearity. Fabricated in 180nm CMOS, the DDI achieves a fast 1-H scan time of 1μs and a compact 1722μm²/channel.

As mobile OLED displays push toward ever-higher spatial resolutions, the display driver IC (DDI) must accommodate a growing number of sourcing channels, forcing extremely tight area budgets. To meet these constraints, most high-density DDIs adopt the two-stage digital-to-analog converter (DAC) architecture (Fig. 24.5.1, top-left). In each channel, an M-bit decoder selects two adjacent tap voltages, V_H and V_L, from a globally shared resistor-string (R-string) that provides 2^M levels between V_{REFH} and V_{REFL}. An in-channel N-bit sub-DAC then interpolates between V_H and V_L, producing the full ($M+N$)-bit output, which is driven to the OLED pixel through an output buffer amplifier. For a 10b system, area efficiency tends to peak when the code space is split evenly ($M:N = 5:5$) rather than in a combination where $M > N$ [1]. Lowering M widens the differential $\Delta V_{HL} [= V_H - V_L = (V_{REFH} - V_{REFL})/2^M]$. Unfortunately, a large ΔV_{HL} imposes a linearity burden on the sub-DAC, especially when it is realized with a DAC-embedded amplifier [2–7]. Under large-signal differential inputs, the amplifier's small-signal transconductance (g_m)—the basis for its voltage interpolation—becomes highly nonlinear. This ΔV_{HL}-induced g_m distortion degrades sub-DAC linearity, compelling designers to choose $M \gg N$ at the expense of optimal area efficiency. A true-DC interpolative buffer [1] remains highly linear even for large ΔV_{HL}, but its topology inherently supports only 1b interpolation ($N = 1$). Resistor-resistor (R-R) DACs [8–10] address this by implementing the sub-DAC with a secondary R-string inside each channel (Fig. 24.5.1, top-center). Operating in the true-DC domain, this approach preserves good linearity even with a wide ΔV_{HL}, allowing a near-optimal bit split ($M \approx N$). However, the in-channel R-string—and the two extra isolation buffers required to decouple it from the global R-string—greatly inflate channel area, rendering R-R DACs unsuitable for high-density DDIs. Numerous alternatives [1,11–15] have attempted to shrink channel size without sacrificing linearity, but these gains come at the cost of a longer D/A-conversion time (T_{DAC}). As high-frame-rate displays shorten the one-horizontal (1-H) scan period, DACs with a lengthy T_{DAC} are becoming untenable. In summary (Fig. 24.5.1, top-right), existing DAC topologies cannot simultaneously achieve compact area, high linearity, and a fast T_{DAC}.

To break through this three-way trade-off, we propose a pivoting translinear-loop (PTL)-DAC-in-buffer (Fig. 24.5.1, bottom). Integrating the multi-bit sub-DAC inside the output buffer lets the design share circuit resources and shrink the per-channel area. Because DAC and output driving occur simultaneously through the PTL-DAC-in-buffer, the architecture also meets the stringent timing of a fast 1-H scan rate. The true-DC translinear loop, which performs the entire multi-bit interpolation, guarantees excellent linearity while achieving a minimal T_{DAC}.

The 10 ($M+N$)-bit PTL-DAC-in-buffer adopts an area-optimal bit combination of $M = 5b$ (coarse) and $N = 5b$ (fine). The upper 5b, $D<9:5>$, selects two taps, V_H and V_L, from the global main R-string; these taps are interpolated within the buffer to form a center voltage $V_{MID} = (V_H + V_L)/2$. The remaining 5b, $D<4:0>$, generates a signed differential voltage $\Delta V_F (= V_{F+} - V_{F-})$ through a 4b decoder and a 1b polarity inverter. The translinear loop of the amplifier's input stage then adds half of this signed ΔV_F to V_{MID}, pivoting around the center to produce the final output: $V_0 = V_{MID} + (\text{signed}) \Delta V_F/2$. To guarantee monotonic steps for a 10b resolution, where $V_{LSB} = (V_{REFH} - V_{REFL})/2^{10}$, the voltage across each segment of the main R-string is fixed at $2^5 \cdot V_{LSB}$. Correspondingly, the minimum step of ΔV_F is set to $2V_{LSB}$. A crucial feature of the PTL-DAC is that the common-mode level of the fine-voltage pair (V_{F+} and V_{F-}), which produces ΔV_F, is entirely independent of the coarse voltages (V_H and V_L). This powerful decoupling is enabled by a floating voltage-mirror (FVM), proposed in this work. The FVM behaves like an electrical transformer: it isolates its input and output stages while transmitting only the differential component (ΔV_F). Because the FVM removes any DC-level dependency between the coarse and fine domains, the 16-level (4b) secondary R-string for the fine-voltage pair (V_{F+} & V_{F-}) can be eliminated from the individual channels, as its functionality is now subsumed by one segment (which is composed of 16×R) of the globally shared main R-string. In contrast to R-R DACs, obviating the per-channel secondary R-string and its isolation buffers yields a substantial area reduction.

The top of Fig. 24.5.2 depicts the principle of the FVM, which copies only the differential component ($V_{F+} - V_{F-}$) to its output. The FVM's differential input-pair (M_1 and M_2) accepts V_{F+} and V_{F-}, encoding their voltage difference into currents by steering their tail current ($I_{BIAS,N}$). These currents are mirrored to the output side via V_{P1} and V_{P2}, and ΔV_F appears across the differential output-pair (M_3 and M_4) as the difference of their gate-source voltages:

$\Delta V_F = (V_{CM} - V_{gs,4}) - (V_{CM} - V_{gs,3}) = V_{gs,3} - V_{gs,4} = V_{F+} - V_{F-}$, provided ($M_1$ & M_2) and (M_3 & M_4) are identically sized. Notably, in the FVM operation, ΔV_F remains fully independent of the common-mode voltages at V_{F+}, V_{F-}, and V_{CM}.

The bottom of Fig. 24.5.2 details the pivoting translinear-loop (PTL) at the buffer amplifier's input stage. Built upon a typical buffer design, the differential output-pair (M_3 & M_4) of the FVM is interposed between the amplifier's differential input-pair (M_{N+} & M_{N-}). As a result, a voltage translinear loop (TL) is formed from V_H to V_L via the final output V_0. The negative feedback inherent in the buffer amplifier forces the input-pair (M_{N+} & M_{N-}) currents to equalize ($I_{N+} = I_{N-}$), thereby guaranteeing $V_{gs,N+} = V_{gs,N-}$. This equilibrium yields the TL equation: $V_H - (V_0 - V_{gs,3}) = V_0 - (V_L - V_{gs,4})$, as shown at the bottom-right of Fig. 24.5.2. Using $V_{gs,3} - V_{gs,4} = \Delta V_F$, the equation is re-expressed clearly as: $V_0 = (V_H + V_L)/2 + \Delta V_F/2$. Thus, when ΔV_F is positive (negative), V_0 shifts above (below) V_{MID} by the fine-voltage magnitude $|\Delta V_F/2|$. Because this TL behavior is established entirely in the true-DC domain, its linearity is immune to the magnitude of ΔV_{HL}; consequently, the coarse resolution M (related to ΔV_{HL}) can be safely reduced toward the fine resolution N—where area efficiency peaks.

The top of Fig. 24.5.3 depicts how the buffer amplifier's input stage is designed for a rail-to-rail voltage range. In addition to the N-type differential input-pair (DIP) and N-type FVM of Fig. 24.5.2, a complementary P-type DIP and its corresponding P-type FVM—biased by a tail current of $I_{BIAS,P}$—are added. When the selected V_H and V_L approach the lower rail (V_{REFL}), the P-type DIP activates while the N-type DIP turns off, and vice versa near the upper rail (V_{REFH}), which is typical behavior of a rail-to-rail N/P input stage. In contrast, the N- and P-type FVMs remain continuously active regardless of V_H and V_L, since their fine-voltage pair (V_{F+} and V_{F-}) has a fixed common-mode level. This mismatch in active regions between FVMs and DIPs results in differential-nonlinearity (DNL) errors, as shown in the "No Control" case. A straightforward remedy is the "N/P Separation" scheme, which aligns the active regions by preemptively deactivating the same-type DIP/FVM pair well before its DIP would turn off near the rails: $I_{BIAS,P}$ is enabled when V_L is below mid-supply ($V_{DD}/2$), and $I_{BIAS,N}$ when V_L is above it. Unfortunately, the abrupt handoff at $V_{DD}/2$ exposes the offset (V_{OS}) mismatch between the N- and P-type DIPs, creating a sharp DNL spike at the transition point. To eliminate this discontinuity, the design replaces the hard switch with a gradual handover. Around mid-supply, $I_{BIAS,N}$ is ramped down while $I_{BIAS,P}$ is ramped up (or vice versa), so the relative contribution of the N- and P-type DIP/FVM paths shifts smoothly as the input V_L moves through $V_{DD}/2$. Because this transition is gradual, the composite transfer characteristic remains continuous, and the DNL stays well within specification across the full rail-to-rail range.

The bottom-left of Fig. 24.5.3 presents the simulated DNL versus the main R-string's unit-segment voltage, $\Delta V_{HL} = V_H - V_L$. The PTL-DAC-in-buffer—whose conversion occurs entirely in the true-DC domain—holds DNL to a consistently low level, regardless of how wide ΔV_{HL} becomes. By contrast, conventional g_m-based DAC-embedded amplifiers [2–4] suffer a steep rise in DNL as ΔV_{HL} widens. This immunity allows the coarse M-bit decoder—the dominant contributor to channel area—to be safely downsized, breaking the trade-off between linearity and silicon area. The bottom-right of Fig. 24.5.3 compares the channel area and 1-H scan speed of the proposed design with those of prior ultra-compact DDI chips. Unlike switched-capacitor-based [1, 12, 13] or PWM-based [15] architectures, the PTL-DAC-in-buffer operates without any clocked phases. This achieves the shortest intrinsic D/A-conversion time (T_{DAC}), thereby providing a larger timing margin for the output V_0 to settle.

The top of Fig. 24.5.4 shows the complete schematic of the PTL-DAC-in-buffer. The input section that carries out the translinear-loop interpolation also functions as a built-in class-AB stage. This class-AB operation greatly boosts slew rate when the buffer amplifier must drive the highly capacitive load at V_0. During a large rising step on V_L (and V_H), the internal node V_X can swing all the way to V_{DD}, sourcing a boosted current (I_{BST}) that rapidly charges the load and shortens the V_0-settling interval. This slew-rate enhancement is inherent to the PTL-DAC-in-buffer, requiring no extra assist circuitry unlike prior works [7,11,13,14].

418 • 2026 IEEE International Solid-State Circuits Conference

979-8-3315-8937-0/26 $31.00 © 2026 IEEE

The 390-channel, 10b DDI was fabricated in a 180nm CMOS process (Fig. 24.5.7). The bottom-right of Fig. 24.5.4 shows the measured transient V_O waveforms obtained by sweeping input codes, D<9:0>, under a panel load of 10kΩ and 70pF. A slew rate of 10V/μs was achieved, and the available minimum 1-H scan time was measured to be 1μs while maintaining 10b accuracy. Figure 24.5.5 plots the measured static performance. The integral-nonlinearity (INL) and DNL peak at 0.63LSB and 0.21LSB, respectively; the dramatic DNL improvement confirms the effectiveness of the gradual tail-current hand-over scheme. The inter-channel deviation of voltage outputs (DVO) was evaluated across 40 randomly selected channels, yielding a worst-case spread of 14.8mV without any offset-cancellation techniques. This DVO value can be reduced via standard chopping [16]. Figure 24.5.6 summarizes the overall performance and shows the 1-channel layout. This work achieves a balanced improvement in both speed (1-H time) and silicon area, without harming linearity (DNL error).

Acknowledgement:
This work was supported by the IC Design Education Center (IDEC), Korea.

References:
[1] S. Shin *et al.*, "A Display Source-Driver IC Featuring Multistage-Cascaded 10-Bit DAC and True-DC-Interpolative Super-OTA Buffer," *IEEE JSSC*, vol. 59, no. 4, pp. 1050–1066, April 2024. http://doi.org/10.1109/JSSC.2024.3350240
[2] J.-S. Kang *et al.*, "10-bit Driver IC Using 3-bit DAC Embedded Operational Amplifier for Spatial Optical Modulators (SOMs)," *IEEE JSSC*, vol. 42, no. 12, pp. 2913–2922, Dec. 2007. http://doi.org/10.1109/JSSC.2007.908690
[3] Y.-J. Jeon *et al.*, "A Piecewise Linear 10 Bit DAC Architecture with Drain Current Modulation for Compact LCD Driver ICs," *IEEE JSSC*, vol. 44, no. 12, pp. 3659–3675, Dec. 2009. http://doi.org/10.1109/JSSC.2009.2035547
[4] H.-M. Lee *et al.*, "An Area and Power Efficient Interpolation Scheme Using Variable Current Control for 10-Bit Data Drivers in Mobile Active-Matrix LCDs," *IEEE Trans. Consumer Electronics*, vol. 65, no. 2, pp. 253–262, May 2019. http://doi.org/10.1109/TCE.2019.2900512
[5] C.-W. Lu *et al.*, "A 10-bit 1026-Channel Column Driver IC with Partially Segmented Piecewise Linear Digital-to-Analog Converters for UHD TFT-LCDs With One Billion Color Display," *IEEE JSSC*, vol. 54, no. 10, pp. 2703–2716, Oct. 2019. http://doi.org/10.1109/JSSC.2019.2927444
[6] C.-W. Lu, "A 10-b Two-Stage DAC with an Area-Efficient Multiple-Output Voltage Selector and a Linearity-Enhanced DAC-Embedded Op-Amp for LCD Column Driver ICs," *IEEE JSSC*, vol. 48, no. 6, pp. 1475–1486, June 2013. http://doi.org/10.1109/JSSC.2013.2252525
[7] K. Ryu *et al.*, "A Source-Driver IC Including Power-Switching Fast-Slew-Rate Buffer and 8Gb/s Effective 3-Tap DFE Receiver Achieving 4.9mV DVRMS and 17V/μs Slew Rate for 8K Displays and Beyond," *ISSCC*, Feb. 2023, pp. 382–384. http://doi.org/10.1109/ISSCC42615.2023.10067592
[8] Y.-C. Sung *et al.*, "10 Bit Source Driver with Resistor-Resistor-String Digital to Analog Converter," *SID Symp.*, vol. 36, May 2005, pp. 1099–1101. http://doi.org/10.1889/1.2196513
[9] C.-W. Lu *et al.*, "A 10-bit Resistor-Floating-Resistor-String DAC (RFR-DAC) for High Color-Depth LCD Driver ICs," *IEEE JSSC*, vol. 47, no. 10, pp. 2454–2466, Oct. 2012. http://doi.org/10.1109/JSSC.2012.2206684

[10] P.-Y. Yin *et al.*, "A 10-Bit Low-Power High-Color-Depth Column Driver with Two-Stage Multi-Channel RDACs for Small-Format TFT-LCD Driver ICs," *J. Display Technology*, vol. 11, no. 12, pp. 1061–1068, Dec. 2015. http://doi.org/10.1109/JDT.2015.2434414
[11] G.-G. Kang *et al.*, "A 12-Bit Mobile OLED/μLED Display Driver IC with Cascaded Loading-Free Capacitive Interpolation DAC and 6.24V/μs-Slew-Rate Buffer Amplifier," *IEEE Symp. VLSI Circuits*, June 2021, pp. 1–2. http://doi.org/10.23919/VLSICircuits52068.2021.9492490
[12] G.-W. Lim *et al.*, "An Area-Efficient 10-Bit Source-Driver IC With LSB-Stacked LV-to-HV-Amplify DAC for Mobile OLED Displays," *IEEE JSSC*, vol. 58, no. 11, pp. 3164–3178, Nov. 2023. http://doi.org/10.1109/JSSC.2023.3289503
[13] Y. Park *et al.*, "A 10-bit Source-Driver IC with Charge-Modulation DAC for Enhanced Frame-Rate Mobile OLED Displays," *IEEE JSSC*, vol. 59, no. 11, pp. 3511–3524, Nov. 2024. http://doi.org/10.1109/JSSC.2024.3442248
[14] J. Ahn *et al.*, "A Fully Nonlinear Compact 10b Source Driver with Low-Voltage Gamma Slope DAC and Data/Phase Dependent Current Modulation Achieving 2411μm²/Channel for Mobile OLED Displays," *ISSCC*, Feb. 2024, pp. 434–435. http://doi.org/10.1109/ISSCC49657.2024.10454504
[15] J. Y. An *et al.*, "A Compact 10b Source Driver IC with Delta-Sigma Pulse Width Modulation for Low-Voltage Digital Interpolation Achieving 1884μm²/Channel," *ISSCC*, Feb. 2025, pp. 130–132. http://doi.org/10.1109/ISSCC49661.2025.10904796
[16] A. Milanesi and P. Buchschacher, "A Novel Offset Cancellation Circuit for TFT-LCD Driver," *SID Symp.*, May 2004, pp. 1568–1571. http://doi.org/10.1889/1.1825786

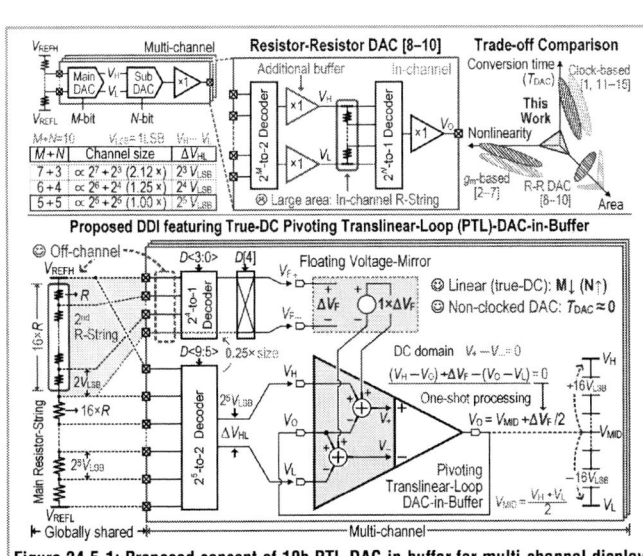

Figure 24.5.1: Proposed concept of 10b PTL-DAC-in-buffer for multi-channel display-driver IC (DDI).

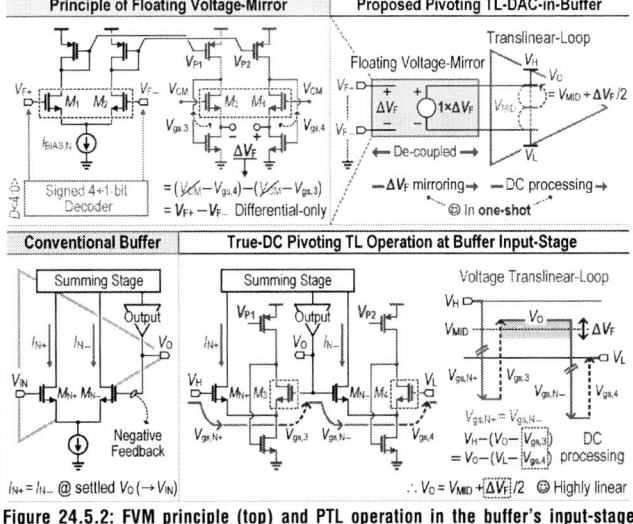

Figure 24.5.2: FVM principle (top) and PTL operation in the buffer's input-stage (bottom).

ISSCC 2026 / SESSION 24 / DISPLAYS / 24.5

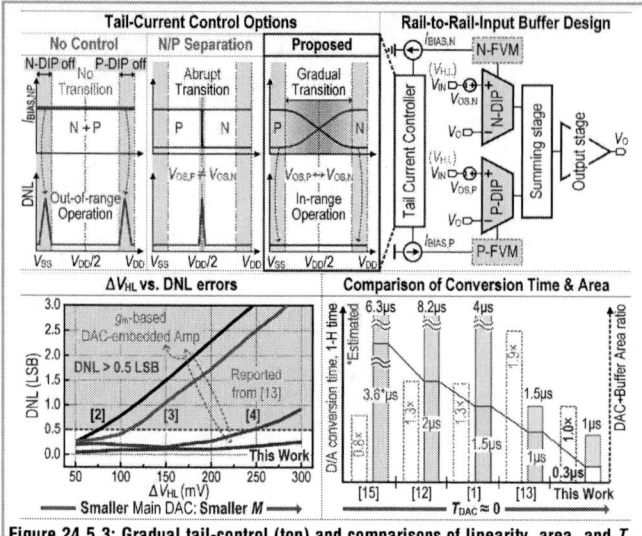

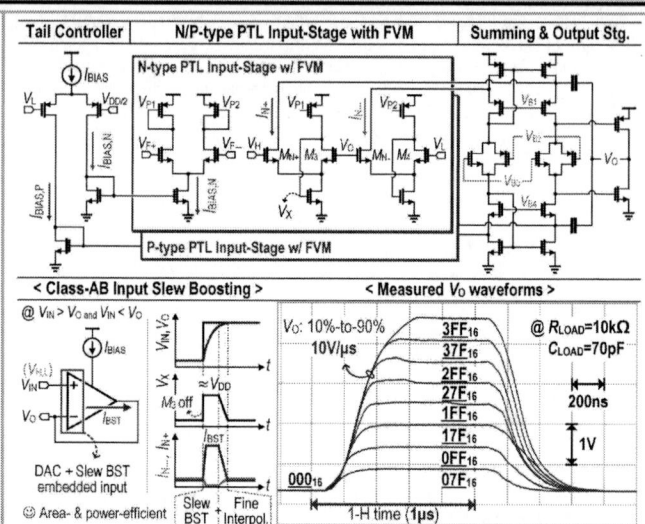

Figure 24.5.3: Gradual tail-control (top) and comparisons of linearity, area, and T_{DAC} (bottom).

Figure 24.5.4: Schematic of the PTL-DAC-in-buffer (top) and measured V_O waveforms (bottom).

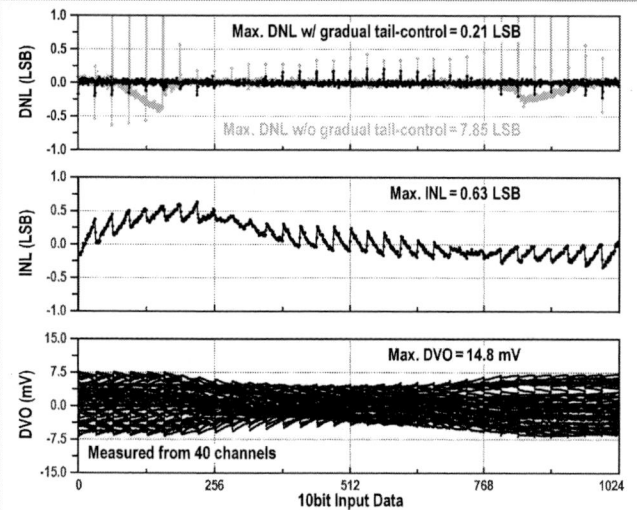

Figure 24.5.5: Measured DNL (w/ and w/o gradual tail-control), INL, and inter-channel deviation of voltage outputs (DVO) in 10b resolution.

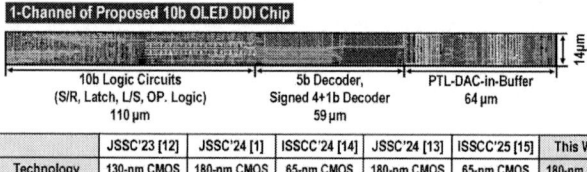

	JSSC'23 [12]	JSSC'24 [1]	ISSCC'24 [14]	JSSC'24 [13]	ISSCC'25 [15]	This Work
Technology	130-nm CMOS	180-nm CMOS	65-nm CMOS	180-nm CMOS	65-nm CMOS	180-nm CMOS
Resolution	10-bit (8b + 2b)	10-bit (5b + 4b + 1b)	10-bit (2b + 8b)	10-bit (6b + 4b)	10-bit (6b + 4b)	10-bit (5b + 5b)
DAC Architecture	SC-multiplier + LSB stack-up	SC-DAC+ DAC-embedded Amp.	Global Slope DAC	Charge Modulation DAC	Sigma-Delta PWM DAC	PTL-DAC-in-Buffer
Output Range	0.3 to 4.5 V	0.1 to 4.9 V	0.3 to 4.7 V	0.25 to 4.85 V	0.2 to 4.8 V	0.4 to 4.6 V
DNL / INL	0.39 / 0.9 LSB	0.37 / 1.17 LSB	0.49 / 0.85 LSB	0.21 / 1.21 LSB	0.42 / 1.44 LSB	0.21 / 0.63 LSB
Max. DVO	4.82 mV	15.5 mV	10 mV	11.5 mV	14.3 mV	14.8 mV
Static Current (I_Q)	1.8 µA/ch	1.7 µA/ch	1.5 µA/ch	2.4 µA/ch	1.5 µA/ch	2.6 µA/ch
Slew Rate @ R_{LOAD} / C_{LOAD}	8.1*** V/µs @ 30kΩ / 30pF	44 V/µs @ 3kΩ / 100pF	4.7*** V/µs @ 30kΩ / 30pF	22 V/µs @ 3kΩ / 100pF	4.6 V/µs @ unknown	10 V/µs @ 10kΩ / 70pF
DAC + Buffer Area per Channel (Area Ratio*)	144 × 16 µm² (1.33×)	132.4 × 16.7 µm² (1.28×)	135.7 × 16 µm² (1.26×)	221 × 15 µm² (1.93×)	90 × 16.5 µm² (0.86×)	123 × 14 µm² (1×)
1-Horizontal Time (Time Ratio**)	8.2 µs (8.2×)	4 µs (4×)	8.2 µs (8.2×)	1.5 µs (1.5×)	6.3 µs (6.3×)	1 µs (1×)

* 1-channel (DAC + Buffer) area of each reference paper / 1-channel (DAC + Buffer) area of this work
** 1-horizontal time of each reference paper / 1-horizontal time of this work *** Estimation from measured waveforms

Figure 24.5.6: Performance summary with 1-channel layout of the proposed DDI chip.

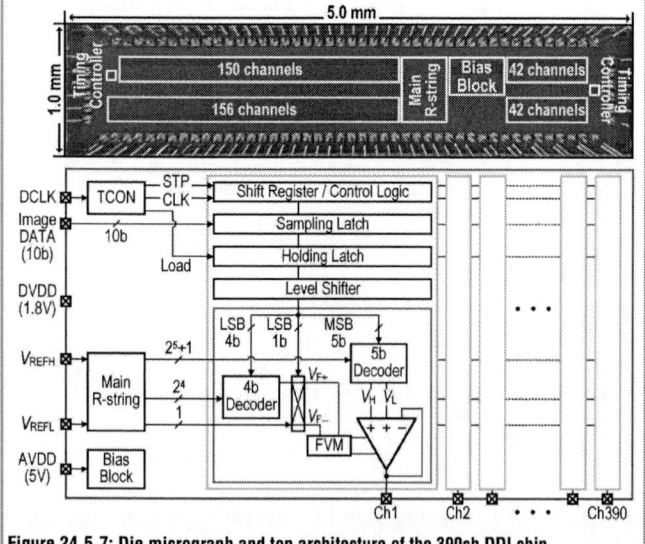

Figure 24.5.7: Die micrograph and top architecture of the 390ch DDI chip.

• 2026 IEEE International Solid-State Circuits Conference

ISSCC 2026 / SESSION 24 / DISPLAYS / 24.6

24.6 A Touch-Embedded OLED Display-Driver IC with Display Noise Referencing and Display Coupling Noise Prediction Based on Dedicated Neural Networks for Mobile Applications with CoE Display

Yun-Rae Jo, Sewhan Na, Sanho Byun, Junchul Park, Gyeongmin Ha, Jiheon Ok, Tae-Gyun Song, Woo-Nyoung Lee, Jihyun Lee, Hyun-Wook Lim, Siwoo Kim, Jae-Youl Lee

Samsung Electronics, Hwaseong, Korea

Abstract

This paper proposes a TDDI, integrating display driver and touch controller to overcome the interference issue between touch and display electrodes. In addition, an analog front-end (AFE) circuit is introduced to suppress coupling noise from display, and a dedicated neural network engine is used to improve interference based on displayed image and display driving condition. The TDDI can achieve SNRs of 43.4 and 41.8dB with white and 1H zebra images, respectively.

Recently, organic light emitting diode (OLED) displays adopting a color filter on encapsulation (CoE) structure have been introduced, which eliminate the polarizer of the display panel stack to improve optical efficiency. This leads to thinner encapsulation layers, causing significant coupling noise issues between display and touch electrodes. This noise degrades touch accuracy and overall system reliability, making effective noise-suppression methods essential. Furthermore, the demand for slimmer mobile devices and long battery time drives the need to reduce the form factor of a device. In this paper, we propose a touch-embedded display-driver IC (TDDI) that integrates the display driver and touch controller while leveraging display image data in the touch signal processing stage and minimizing module form factor by eliminating a touch controller IC (T-IC), routing, and power sources. The conventional OLED display has two ICs, namely a display driver IC (DDI) and a T-IC, which operate with four power rails (AVDD, VCI, VDDI, and DVDD) and two power rails (AVDD_T and VDDI_T), respectively, as shown in Fig. 24.6.1(a). Additionally, the channel routing between the T-IC and panel occupies a broad area of the display system. In contrast, the TDDI can reduce the number of external power sources and the area by integrating the DDI and T-IC. Figure 24.6.1(b) shows the TDDI operates with only 4 power rails by sharing the power source for the display and touch blocks, and also can remove the T-IC and touch channel routing, so that the system area of the display system can be optimized. In addition, we introduce an analog front-end (AFE) circuit to suppress the coupling noise from display and a dedicated neural network engine to improve interference between displays and touch electrodes.

As aforementioned, in the OLED panels adopting the CoE as shown in Fig. 24.6.2(a), the encapsulation layer thickness is reduced compared to conventional designs to lower optical interference between RGB pixels. Since the encapsulation layer serves as a dielectric between the display and the touch electrodes, reducing its thickness increases the parasitic capacitance between them. As a result, display-driving signals are more strongly coupled into the touch circuitry, intensifying display-to-touch interference. By using the circuit-level modeling for the display structure as shown in Fig. 24.6.2(b), two major interference mechanisms are observed: (1) that induced by source line toggling, and (2) that resulting from the display image status changing the impedance of ELVSS, as shown in Fig. 24.6.2(c).

The first mechanism is the touch signal fluctuation by the coupling noise from the source lines toggling rapidly during pixel charging for each horizontal time. Because of the parasitic capacitor (C_{DATA}) between the source lines and the ELVSS, fluctuations occur in the ELVSS voltage. ELVSS fluctuations ($\Delta ELVSS$) directly translate into coupling noise on the touch sensing signals through the parasitics between ELVSS and touch channels (C_{RX} and C_{TX}) [1]. As shown in top of Fig. 24.6.2(c), $\Delta ELVSS$ affects to the output ($V_{I/V}$) of current-to-voltage converter (I/V), so the measured touch profile randomly fluctuates according to the TX direction that means time. Since this coupling noise is the time-variant random noise having a frequency associated with display horizontal time, it is necessary for the AFE to suppress it. Another interference mechanism is the variation of the touch data offset induced from the impedance change of ELVSS resulting from a change in the displayed image status, such as pattern and brightness. When the image status changes, the OLED current (I_{OLED}) changes, and then according to the small-signal model of the diode, the impedance of ELVSS (Z_{ELVSS}) changes in inverse proportion. This impedance change causes a variation in settling time of ELVSS, which causes a signal offset on the AFE circuit of touch controller. As shown in the bottom of Fig. 24.6.2(c), in the white image with high OLED current, the ELVSS settles faster than in the black image, and the measured touch data is shifted by an image transition from white to black. Because this interference shows similar characteristics according to the display pattern and brightness, it can be improved by estimating the touch data change from the image status.

Figure 24.6.3 shows the AFE structure and driving method to suppress the coupling noise by source line toggle. The AFE consists of a current conveyor (CC), a current-to-voltage converter (I/V), a low pass filter (LPF), a programmable gain amplifier (PGA), and an analog-to-digital converter (ADC), as shown in Fig. 24.6.3(a). The negative input of CC is connected to the RX channel that is fed a signal current (I_{SIG}) by toggling TX channel and the positive input of CC is connected to the unused TX channel that is operated as the display coupling noise (D-noise) referencing sensor. In Fig. 24.6.3(b), an example case with 8 TX channels with 4-channel coded driving is described. TX buffers drive the first 4 channels, but the bottom 4 channels are inactive and stay floating, so we can use the unused TX channels as a D-noise referencing sensor. The coupling voltages on the RX ($V_{DNOISE,RX}$) and TX as a D-noise referencing sensor ($V_{DNOISE,TX}$) are slightly different since the parasitic capacitances between ELVSS-RX (C_{RX}) and ELVSS-TX (C_{TX}) are different. But, these parasitic capacitances from the panel are much larger than the parasitic capacitances of the RX and TX nodes ($C_{P,RX}$ and $C_{P,TX}$) in the IC, so the $V_{DNOISE,RX}$ and $V_{DNOISE,TX}$ are represented the $\Delta ELVSS$. The output current of the CC (I_{OUT}) does not include the $\Delta ELVSS$ because the CC operates as a unity-gain amplifier and generates the I_{OUT} by the difference between the positive and negative input voltages. Finally, the I/V generates a signal voltage ($V_{I/V}$) that is not related to D-noise, and it is amplified by a PGA and converted to digital data by an ADC. The TDDI can obtain the noise-free touch data through the AFE with D-noise referencing as described in Fig. 24.6.3(c) that shows the SNR is increased by 13dB for the worst image pattern (1-H zebra image).

To improve the interference between touch and display electrodes due to the impedance change of ELVSS, a coupling noise predictor uses neural networks, as described in Fig. 24.6.4(a). The coupling noise predictor, which is composed of pre-processing and neural networks, estimates the induced noise level on each touch sensor node based on input image and the display driving condition such as the display brightness value (DBV). The final touch data is then obtained by subtracting the predicted noise data from the touch raw data. To reduce the channel dimension of the neural networks, the pre-processing block converts image into a one-channel current map data combining display condition channel and the gray level of each RGB channel by using transfer LUTs as described in Fig. 24.6.4(b). In addition, the image data is down-sampled with the same resolution as the touch sensor node from 1,280×2,800 of display resolution to 20×40 of touch resolution. The down-scaling uses a nearest-neighbor algorithm, and the values for transfer LUTs are determined by a tuning procedure to find the mapping curve of RGB pixel value to the amount of current driven.

To ensure that the predictor can be implemented with sufficient accuracy within optimized hardware size and power consumption, a simple custom neural-network unit is designed through a single multiply-accumulate (MAC) hardware. The task of predicting display coupling noise using a display image should be extended to multi-tasks due to various usage scenarios of display and touch sensing. To solve this issue, the neural network applies transfer learning to re-tune only some of the fine-tuning layers for each task based on a pre-trained base layer as shown in Fig. 24.6.4(c). The networks consist of six layers, of which the four layers at the front are base layers and the remaining layers are used for fine-tuning. Each usage scenario shares the weight and bias parameters of the base layers and has its own-trained parameters for scenarios derived only from the fine-tuning layers.

In addition, for further optimized hardware implementation, we designed a 1D-locally connected neural network layer as a unit layer for a predictor neural-network block according to a display scanning and touch readout scheme in the vertical or horizontal direction. The unit layer calculates one feature map using a 1D-kernel selected from one of horizontal or vertical direction, as shown in Fig. 24.6.5(a). The dedicated neural-network block can be implemented using one simple MAC, as in Fig. 24.6.5(b). There is one MAC unit to calculate the feature map with one directional iterative operation. The three SRAMs configure one SRAM for storing trained neural-network parameter values, and two SRAMs for reading the feature map of the previous layer and writing the output in current layer calculations. The operation step is divided into four stages: data fetch (DF), multiplication (MULT), accumulation (ACC), and write memory (WM). At the DF stage, the trained network parameter values for the current layer stored in parameter-SRAM are loaded and the feature map values of the previous layer are loaded from SRAM1 to the data fetch registers sequentially addressed by the 1D-kernel direction. These two input values are multiplied in the MULT stage, and then the result is then sent to the ACC stage. The ACC stage accumulates the output values over one layer duration. Finally, the accumulated data are written to SRAM2 in WM stage. These operations are pipelined repeatedly for column and row layers, as shown in Fig. 24.6.5(c).

The TDDI was fabricated with 28nm process with high voltage device with 8V/25V transistors, as described in Fig. 24.6.7, which has a display resolution of 1,280×2,800 and touch channel TX20×RX40. The AFE with D-noise referencing sensor can increase the touch SNR by 13dB in the worst displayed image pattern, as shown in Fig. 24.6.3(c). Figure 24.6.6(a) shows the noise profiles induced from the low, middle, and high levels of gray patterns at 60Hz and 120Hz. The mean absolute error (MAE) is a noise-estimation error calculated by a noise level difference between noise raw data and estimated noise data. The resulting values show that noise levels can be compensated by from 84% to 95% for various displays. The performance summary and comparison table are summarized in Fig. 24.6.6(b) [2-5]. The proposed TDDI compensates for the interference between display and touch electrodes by using the display image data and conditions, because TDDI can access to DDI information using the touch MCU. In addition, the AFE with D-noise referencing can suppress noise induced by source line toggle, so the SNR can achieve 43.4dB and 41.8dB with white and 1H zebra images, respectively. Therefore, the proposed TDDI is a promising architecture for beyond-OLED display systems such as CoE display for ultra-thin mobile set applications.

References:

[1] S.-H. Choi et al., "Implementation of Full-Panel Circuit Models for Interference Estimation Between Touch and Display Operation in On-Cell Touch AMOLED," *SID International Symposium*, pp. 24-27, June 2022. https://doi.org/10.1002/sdtp.15406
[2] S. Byun et al., "A 45.8dB-SNR 120fps 100pF-Load Self-Capacitance Touch-Screen Controller with Enhanced In-Band Common Noise Immunity Using Noise Antenna Reference," *ISSCC*, pp. 386–387, Feb. 2023.
https://doi.org/10.1109/ISSCC42615.2023.10067374
[3] J. Lee et al., "A 620pF-Compensated Dual-Mode Capacitance Readout IC for Sub-Display TSP with VRR Scan," *ISSCC*, pp. 438–439, Feb. 2024.
https://doi.org/10.1109/ISSCC49657.2024.10454341
[4] J.Y. An et al., "Noise Immunity in Capacitive Sensing: Single-Ended AFE Design with Common-Current Subtraction for Mutual- and Self-Capacitance Sensing in 390pF Load," *ISSCC*, pp. 436–437, Feb. 2024. https://doi.org/10.1109/ISSCC49657.2024.10454465
[5] S.H. Choi et al., "A Hybrid Touch Sensing AFE with Common-CVQ (Currents, Voltages, and Charges) Subtraction to Improve Display Noise Immunity for Large Sensing Load up to 820pF," *Symposium on VLSI Circuits*, June 2025.
https://doi.org/10.23919/VLSITechnologyandCir65189.2025.11075171

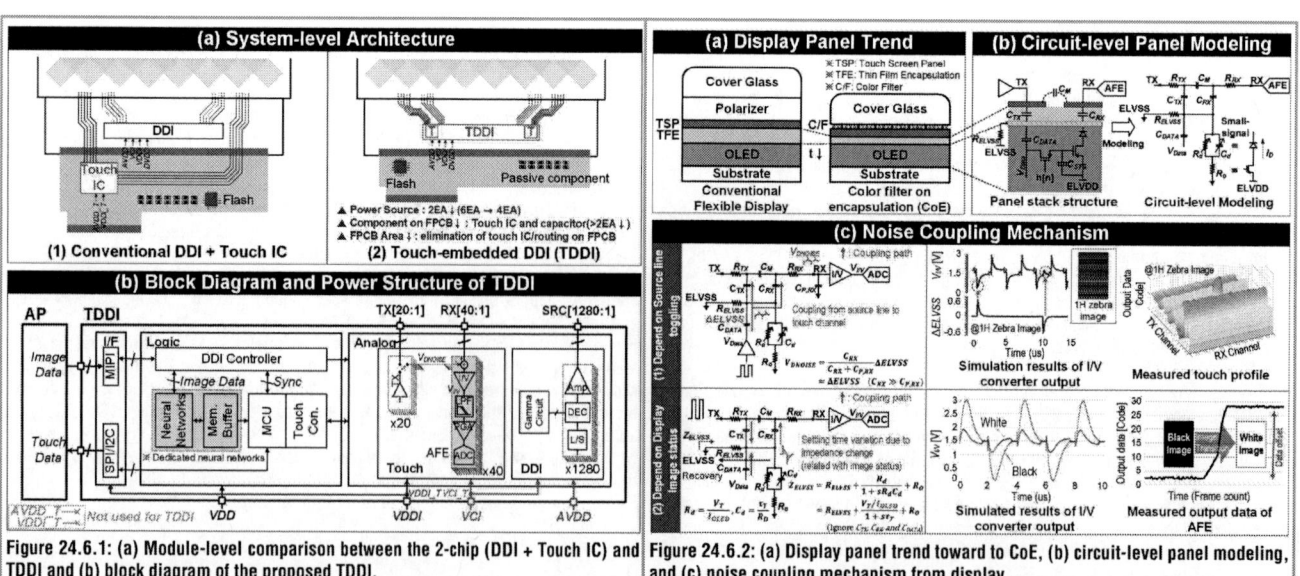

Figure 24.6.1: (a) Module-level comparison between the 2-chip (DDI + Touch IC) and TDDI and (b) block diagram of the proposed TDDI.

Figure 24.6.2: (a) Display panel trend toward to CoE, (b) circuit-level panel modeling, and (c) noise coupling mechanism from display.

ISSCC 2026 / SESSION 24 / DISPLAYS / 24.6

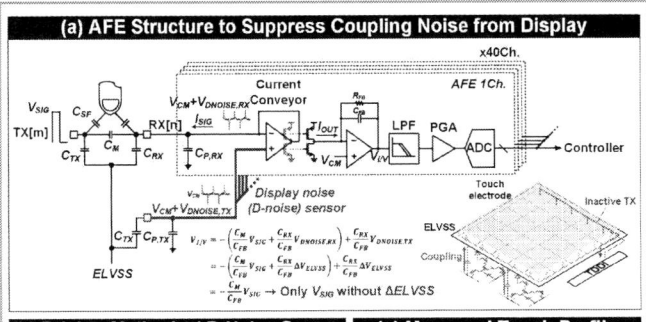

(a) AFE Structure to Suppress Coupling Noise from Display

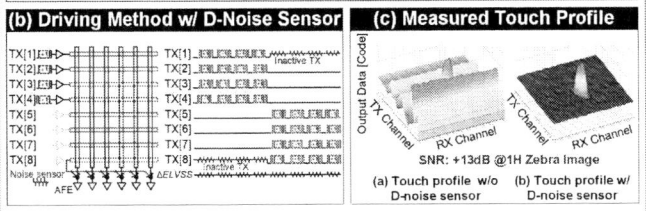

(b) Driving Method w/ D-Noise Sensor
(c) Measured Touch Profile

SNR: +13dB @1H Zebra Image

(a) Touch profile w/o D-noise sensor
(b) Touch profile w/ D-noise sensor

Figure 24.6.3: (a) Schematic diagram, (b) driving method, and (c) measured results of the AFE with D-noise referencing sensor.

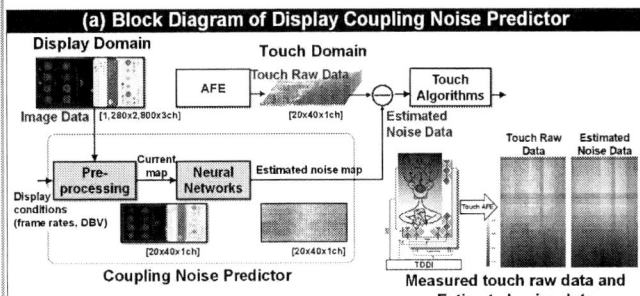

(a) Block Diagram of Display Coupling Noise Predictor

Measured touch raw data and Estimated noise data

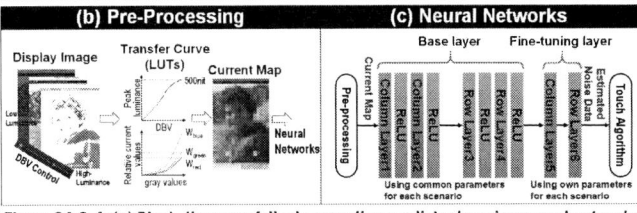

(b) Pre-Processing
(c) Neural Networks

Figure 24.6.4: (a) Block diagram of display coupling predictor by using neural networks and the detailed diagram for (b) pre-processing and (c) neural networks.

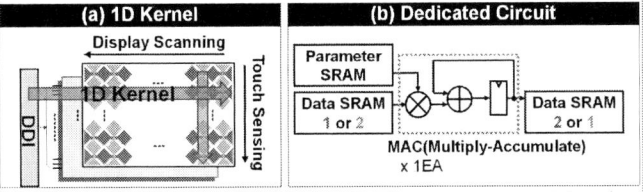

(a) 1D Kernel
(b) Dedicated Circuit

MAC(Multiply-Accumulate) x 1EA

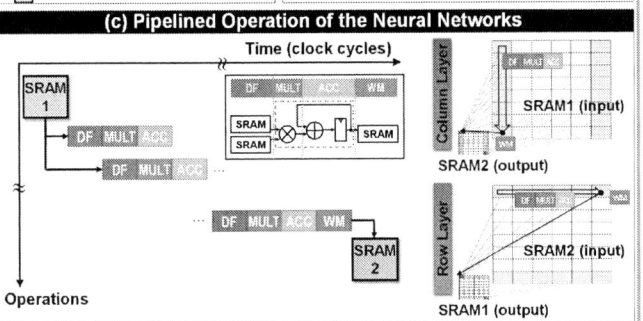

(c) Pipelined Operation of the Neural Networks

Figure 24.6.5: (a) Concept of 1D kernel operation, and (b) schematic diagram and (c) operating principles of the dedicated circuit for neural networks.

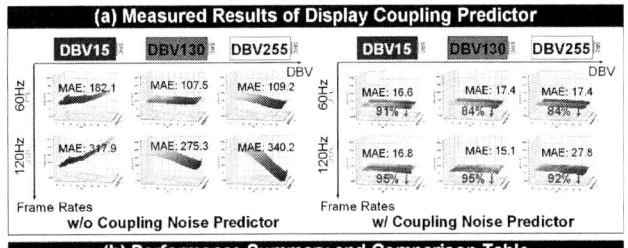

(a) Measured Results of Display Coupling Predictor

DBV15	DBV130	DBV255	DBV15	DBV130	DBV255
MAE: 182.1	MAE: 107.5	MAE: 109.2	MAE: 16.6 (91%↓)	MAE: 17.4 (84%↓)	MAE: 17.4 (84%↓)
MAE: 317.9	MAE: 275.3	MAE: 340.2	MAE: 16.8 (95%↓)	MAE: 15.1 (95%↓)	MAE: 27.8 (92%↓)

w/o Coupling Noise Predictor | w/ Coupling Noise Predictor

(b) Performance Summary and Comparison Table

	ISSCC '23 [2]	ISSCC '24 [3]	ISSCC '24 [4]	VLSI '25 [5]	This Work
Technology	45nm	130nm/350nm	65nm	65nm	28nm
TDDI	X	X	X	X	O (1,280×2,800)
Sensor type	Self-Cap	Hybrid(Mutual+Self)	Hybrid(Mutual+Self)	Hybrid(Mutual+Self)	Hybrid(Mutual+Self)
D-noise Mitigation induced by source-line	NARS Differential	Differential	Common Signal Rejection	Common Signal Rejection	D-noise sensor
D-noise Mitigation induced by image pattern	None	None	None	None	Predictor w/ Neural networks
SNR [dB] / Signal	45.8 / 500fF	47.2 / Finger	39.3 / 100fF	39.1 / 100fF	43.4, 41.8 / 70fF (White, 1H Zebra)
Reporting Rate [Hz]	120	184	240	240	360

Figure 24.6.6: (a) Measured results of touch data without and with display coupling predictor according to display conditions, and (b) performance summary and comparison table.

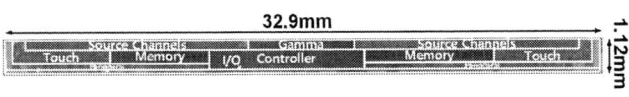

Figure 24.6.7: Die micrograph of the OLED TDDI with AFE with d-noise referencing sensor and display coupling predictor by using neural networks.

• 2026 IEEE International Solid-State Circuits Conference

ISSCC 2026 / SESSION 25 / HARDWARE SECURITY / OVERVIEW

Session 25 Overview: *Hardware Security*

SECURITY SUBCOMMITTEE

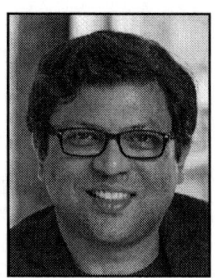

Session Chair: Shreyas Sen
Purdue University, West Lafayette, IN

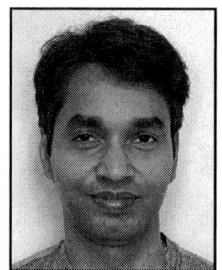

Session Co-Chair: Santosh Ghosh
NVIDIA, Hillsboro, OR

This session presents hardware security solutions advancing the state-of-the-art across design hierarchies, from cryptographic processors to circuit-level primitives. The first five papers showcase accelerators for fully homomorphic encryption (FHE) and post-quantum cryptography (PQC) with various architectural enhancements for high performance and energy efficiency. The next two papers introduce countermeasures against power and electromagnetic (EM) side-channel attacks and chip-to-chip interface probing attacks, respectively. These are followed by a reconfigurable physically unclonable-function (PUF) circuit design. The final two papers present true-random-number-generator (TRNG) circuits with physical attack tolerance.

8:00 AM

25.1 HERACLES: 8192-Way SIMD Programmable Scalable Fully-Homomorphic Encryption SoC for Privacy-Preserving Cloud Computing in Intel 3 CMOS

Anupam Golder, Intel, Hillsboro, OR

In Paper 25.1, Intel presents HERACLES, an FHE accelerator SoC in Intel 3 CMOS with single-cycle 8k polynomial arithmetic throughput using a massively-parallel 8192-way SIMD vector compute engine. The programmable SoC achieves a throughput of 29.49TOPs at 0.75V and 1.2GHz, delivering up to 5547× higher FHE performance than general-purpose CPUs.

8:25 AM

25.2 A 28nm 0.48mJ/boot Torus FHE Processor for Arbitrary Computation on Encrypted Data

Xinglong Yu, Fudan University, Shanghai, China

In Paper 25.2, Fudan University presents a 28nm Torus FHE (TFHE) processor for arbitrary computation on encrypted data. With a security-preserving low-bit-width torus quantization and a divide-and-conquer polynomial multiplication engine, the proposed accelerator demonstrates 0.48mJ/boot energy consumption and 466boots/s throughput at 128b security level.

8:50 AM

25.3 A 16nm 0.042mm² 0.66µJ/Ops Lightweight MLWE PQC KEM with Cryptanalysis-ASIC Co-Optimization

Archisman Ghosh, Purdue University, West Lafayette, IN

In Paper 25.3, Purdue University, KU Leuven and IIT Kanpur jointly present a co-optimized algorithm-ASIC design for quantum safe key establishment in resource-constrained devices. The 0.042mm² design in 16nm FinFET is 8× smaller than the NIST standard ML-KEM and provides a 509× improvement in energy-area product over generic PQC processors.

9:15 AM

25.4 OmniCrypt: A 435.86M-GOPS/W Bootstrappable Multi-Scheme FHE Accelerator with On-Chip Data Generation for Privacy-Preserving Computation

Adiwena Putra, KAIST, Daejeon, Korea

In Paper 25.4, KAIST presents OmniCrypt, a 435.86M-GOPS/W bootstrappable multi-scheme FHE accelerator with on-chip data generation for privacy-preserving computation. The 28nm accelerator achieves 252.8M-GOPS, 19.5M-GOPS/mm² and 435.9M-GOPS/W, delivering the highest CKKS bootstrapping throughput and 107× CPU speedup in application benchmarks.

422 • 2026 IEEE International Solid-State Circuits Conference

979-8-3315-8937-0/26 $31.00 © 2026 IEEE

9:30 AM

25.5 A 0.05mm² 1.19-to-7.34mW SQIsign-1D Isogeny-Based Post-Quantum Signature Verification Accelerator for IoT

Krishna Sai Tarun Ramapragada, Indian Institute of Science, Bengaluru, India

In Paper 25.5, Indian Institute of Science presents a 0.05mm² and 1.19-to-7.34mW isogeny-based digital signature verification accelerator supporting the NIST post-quantum candidate SQIsign for resource-constrained IoT applications. The 28nm design achieves up to 39× energy savings in SQIsign-1D signature verification compared to state-of-the-art in a small form factor.

10:05 AM

25.6 A 17%/27% Area-/Energy-Overhead Glitch-Transition Secure SHA-3 Engine Fusing Dual-Rail Precharge Logic and Asymmetric Masking

Cankun Zhao, School of Integrated Circuits, Tsinghua University, Beijing, China,
      State Key Laboratory of Cryptography and Digital Economy Security, Tsinghua University, Beijing, China

In Paper 25.6, Tsinghua University and Micro Innovation Integrated Circuit Design jointly present a 0.046mm² robust SHA-3 engine in 65nm CMOS that combines dual-rail precharge logic with asymmetric masking to prevent power and EM side-channel attacks. The protected SHA-3 engine achieves >100M MTD with 0% latency, 17% area and 27% energy overheads which are significantly lower than state-of-the-art.

10:30 AM

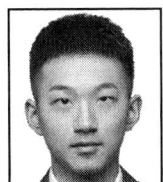

25.7 TinyPAD: A 166μm²/lane Variation-Tolerant Probing-Attack Detector for an 8Gb/s/lane Chip-to-Chip Interface in 16nm FinFET

Mao Li, Columbia University, New York, NY

In Paper 25.7, Columbia University and Intel jointly present TinyPAD, a 166μm²/lane variation-tolerant probing-attack detector for an 8Gbps/lane chip-to-chip interface. Compared to state-of-the-art, the 16nm FinFET design achieves 3.3× maximum supported loading and 2.5× smaller minimum detectable capacitance over wide VT conditions at 11× smaller area overhead.

10:55 AM

25.8 A Sub-Threshold All-NMOS Reconfigurable PUF with Secure Configuration Selection for Stable 6b/Cell

Shufan Xu, Kyoto University, Kyoto, Japan

In Paper 25.8, Kyoto University presents a sub-threshold all-NMOS reconfigurable PUF with secure configuration selection that achieves 6 stable bits per cell and "zero" bit error (<6.5E–8 BER) across 0°C to 70°C and 0.6V to 0.8V. The 65nm design consumes only 253F² feature area per bit, low energy of 16.8fJ/b and mitigates long-term NBTI aging.

11:20 AM

25.9 A PVT Variation- and Attack-Tolerant Metastability-Based TRNG Using Binary Search in 2nm

Yelim Youn, Samsung Electronics, Hwaseong, Korea

In Paper 25.9, Samsung Electronics presents a PVT variation and attack tolerant metastability-based TRNG using binary search with offset-tunable comparator. The 2nm TRNG occupies 320 μm² and operates without warm-up time at 20Mb/s while consuming 1.8pJ/b. Measurement results show NIST- and AIS-compliant randomness across all PVT corners.

11:45 AM

25.10 A 65nm 0.066pJ/b Floating-Latch-Based True Random Number Generator Resilient to Power-Noise Injection Attacks

Kai Cheng, Institute of Microelectronics of the Chinese Academy of Sciences, Beijing, China

In Paper 25.10, Institute of Microelectronics of the Chinese Academy of Sciences, University of Macau and Tsinghua University jointly present a 0.066pJ/b floating-latch-based TRNG in 65nm. A floating reservoir capacitor scheme provides resilience against power-supply noise injection and current-starved inverters perform mismatch compensation to improve PVT tolerance.

ISSCC 2026 / SESSION 25 / HARDWARE SECURITY / 25.1

25.1 HERACLES: 8192-Way SIMD Programmable Scalable Fully-Homomorphic Encryption SoC for Privacy-Preserving Cloud Computing in Intel 3 CMOS

Anupam Golder[1], Raghavan Kumar[1], Sachin Taneja[1], Kylan Race[2], Paolo Aseron[1], James Greensky[1], Wen Wang[1], Huijing Gong[1], Lalith Kethareswaran[2], Vikram Suresh[1], Adish Vartak[1], AppaRao Challagundla[2], Jeremy Casas[1], Poornima Lalwaney[1], Duhyeong Kim[1], Christopher N. Gutierrez[1], Ernesto Zamora Ramos[1], Wonhee Cho[1], Jose M. Rojas Chaves[1], Michael Steiner[1], Dan Lake[1], Nataraj Yennampelli[3], Karthik Nivarthi[3], Kamalakanth Bijinapally[3], Bala Prasad Talamala[3], Sravanth Valluri[3], Vasantha Srirambhatla[3], Chris Wilkerson[1], Rosario Cammarota[1], Sanu Mathew[1]

[1]Intel, Hillsboro, OR, [2]Intel, Austin, TX, [3]Intel, Hyderabad, India

Abstract

A fully homomorphic encryption (FHE) accelerator SoC fabricated in Intel 3 CMOS with single-cycle 8k polynomial arithmetic throughput using a massively-parallel 8192-way SIMD vector compute engine is presented. The programmable SoC employs a 3-tiered memory subsystem, optimized Cooley-Tukey NTT/iNTT butterflies and a 2D mesh NoC to support multiple FHE schemes and parameters, with measured compute engine throughput of 29.49TOPs at 0.75V, 1.2GHz, delivering up to 5547× higher FHE performance than general-purpose CPUs.

Fully homomorphic encryption (FHE) is regarded as the holy-grail of secure private computing, enabling arbitrary computations on encrypted data without requiring users to share their secret keys with untrusted third parties. Unlike classical cryptographic schemes such as advanced encryption standard (AES), that protect data only during transmission and storage, FHE extends confidentiality to the computation phase itself, preserving data privacy on untrusted infrastructure, making FHE a foundational technology for privacy-preserving cloud computing. However, FHE poses significant challenges that make real-time interactive execution infeasible on present-day systems, including ~10^5× increase in ciphertext sizes, massively parallel modular arithmetic and complex wide-vector permutations (Fig. 25.1.1). FHE applications, such as encrypted artificial intelligence (AI) and private database analytics require large multiplicative depths, necessitating bootstrapping to refresh accumulated noise. Bootstrapping consumes 60 to 95% of overall run-time [1] and >740b of modulus space, requiring large ring dimensions for efficient computations with 128-bit security. While special-purpose FHE accelerators have been previously reported, they lack support for bootstrappable parameters [2], have limited on-die memory [2-4] leading to off-chip communication bottlenecks, or employ separate number-theoretic-transform (NTT), MAC, and automorphism modules resulting in complex data movement scenarios [2-5]. In this paper, we present **HERACLES**, a programmable, scalable FHE SoC fabricated in Intel 3 CMOS [6], featuring: (a) a polynomial instruction set architecture with native support for 8k modular arithmetic and 16k-point NTT/iNTT using an 8192-way SIMD vector compute engine (CE); (b) seamless transfer of 32kB instruction and data bundles along independent 512B polynomial coefficient lanes in a 3-tiered memory sub-system composed of 48GB HBM, 64MB scratchpad memory and a distributed 9MB register file (RF), delivering bandwidths of 1.6TB/s, 9.6TB/s and 307TB/s, respectively; (c) a 32b NTT/iNTT-optimized carry-save Cooley-Tukey (CT) butterfly datapath with single-cycle throughput and 30% area reduction compared to reconfigurable CT/Gentleman-Sande (GS) implementations [7]; (d) a 2D mesh-based NoC programmed apriori with contention-free constant-geometry routes achieving 5-cycle throughput for 16k-point NTT/iNTT data shuffles; (e) an on-the-fly twiddle generator supporting 1k-to-64k-point NTT/iNTT operations, reducing on-die storage by up to 512×; (f) on-the-fly polynomial sampling for key-switch hint and fresh ciphertext generation, (g) automorphisms implemented via scaled primitive-root iNTTs; (h) a measured CE throughput of 29.49TOPS at 0.75V, 1.2GHz, 25°C, delivering up to 5547× higher performance than CPUs across a range of FHE schemes and parameters.

The HERACLES SoC architecture is designed to efficiently transport large instruction and data bundles from an external host through a 3-tiered memory subsystem, for execution on a massively-parallel 8192-way vector SIMD CE structured as an 8x8 array of tile-pairs $TP_{63:0}$ (Fig. 25.1.2). The HERACLES Encrypted Computing SDK decomposes FHE applications, written using OpenFHE/SEAL libraries into pre-assembled optimized binaries of operators like HEAdd, HEMul, HERotate, etc. Executable kernel bundles are transferred over a 16-lane PCIe link to a pair of on-package 24GB HBM3 dies. Two HBM sub-systems manage connectivity to HBM dies through 819GB/s EMIB interconnect links, while also directing bundle traffic to 16 scratchpad banks $SPAD_{15:0}$. Each scratchpad bank connects to four CE tile-pairs through daisy-chained 512B load/store buses, delivering aggregate SPAD→CE bandwidth of 9.6TB/s. A 2D mesh NoC connects all 64 tile-pairs, using pre-programmed routes for NTT/iNTT coefficient shuffles. To decouple memory and compute latencies and minimize control overheads across the 197mm² die, HERACLES programs are partitioned into 3 execution streams: *Minstructions*, *Cinstructions*, and *Xinstructions*, which are stored in separate queues within the centrally located CE control block (CCB). *MinstQ* handles HBM↔scratchpad load/stores, *CinstQ* manages scratchpad↔CE transfers, and *XinstQ* executes vectored polynomial math instructions and CE control/dataflow operations.

Efficient movement of large payloads is a critical challenge in FHE workloads. The HERACLES memory subsystem addresses this with a streamlined dataflow architecture organized along memory lanes, leveraging the SIMD nature of double-Chinese remainder theorem (CRT) polynomial operations, where coefficients do not interact across lanes except during transformations such as NTT, iNTT, and automorphisms (Fig. 25.1.3). Each 32kB data word (8k polynomial) is distributed across 16 parallel lanes, with each lane comprising 4 HBM pseudo-channels, a 4MB SPAD bank, and 4 CE tile-pairs. A 32kB load requires each of the 16 lanes to concurrently move 2kB data words from HBM to SPAD along a 128B bus (*mload*) followed by 4 hops (*cload*) along the daisy-chained CE tile-pair data bus. *cload/cstore* instructions can be overlapped with *XinstQ* polynomial math operations to mitigate SPAD access latencies. The 3-tiered memory sub-system allows concurrent, decoupled data movement isolating the intrinsic variability of HBM latencies from X/Cinstruction scheduling. 32kB instruction bundles use a streaming dataflow using *Xinst/Cinst/Minst* finite-state-machines (FSMs) to fetch, schedule and dispatch X/C/Minstructions from the two HBM sub-systems to the corresponding left/right instruction queues. A global scheduler populates global queues from left/right buffer queues in a ping-ponged manner followed by queue synchronizations and resolution of *cnops* prior to instruction dispatch. While *M/Cinstructions* are dispatched by the FSM to the 16 SPAD banks for execution, *Xinstructions* are further daisy-chained into their 4 tile-pair rows using *ifetch* instructions. Dependencies between *Minsts* and *Cinsts* are enforced through bulk synchronization points specified by *csyncm* and *msyncc*. A-priori knowledge of *C/Xinst* latencies and dependencies eliminates the need for dynamic scoreboarding, enabling assembler-constructed static trace execution.

HERACLES leverages coefficient residue-level parallelism of FHE operands to execute native 8k polynomial arithmetic with single-cycle throughput using an 8192-way SIMD vector compute engine. The left and right tiles within the CE tile-pairs (Fig. 25.1.4) are each composed of 64 butterfly (BF) circuits that are reconfigured by *Xinstructions* to perform 32b modular add/subtract/multiply/multiply-accumulate/NTT/iNTT operations. With CE dominating timing and area budgets, three butterfly datapath optimizations were critical to meeting HERACLES design targets: (a) CT butterfly reuse: In contrast to using reconfigurable CT/GS butterflies in conjunction with Decimation-in-Time/Frequency(DiT/DiF) inter-stage shuffles for NTT/iNTT, respectively [7], the HERACLES CE avoids explicit coefficient reordering [8] by employing CT butterflies for both NTT and iNTT compute while switching between DiT/DiF for inter-stage NTT/iNTT shuffles; (b) Montgomery reduction optimization: The BF datapath uses Montgomery reduction with moduli constrained to NTT-optimal values [8] $q=q_H 2^{16}+1$, which converts 16x32b multiplications in the 2 reduction stages to 16b multiplications; (c) Carry-save representation: Multiplier and reduction datapaths are implemented in redundant carry-save representation, postponing expensive carry-propagations to the final add/subtract. The CT NTT-optimal carry-save BF delivers 30% area reduction at iso-delay compared to a CT/GS non-redundant Montgomery BF [7]. Four-way banked 1R1W RF macros feed input operands (a, b, ω) to the butterfly datapath, with results written back to the destination RF address, providing 9MB tightly-coupled local storage with single-cycle throughput of 307TB/s distributed across 64 CE tile-pairs. NTT/iNTT computations require twiddle factors that occupy 168MB of storage for 42-residue 64k rings. To mitigate area and memory bandwidth constraints, HERACLES only stores successive squares of primitive roots ω and ω^{-1} as compact 336kB metadata (512× reduction). An on-the-fly twiddle factor generator leverages CE butterflies and *twntt/twintt* Xinstructions to compute next-stage factors in real-time prior to *pntt/pintt* computations.

Inter-stage NTT/iNTT permutations are orchestrated by the *rshuffle* instruction in 2 steps – within-tile-pair 256-point DiT/DiF permutations using dedicated point-to-point wires, followed by tilepair-to-tilepair shuffles (Fig. 25.1.5). HERACLES implements a modular 2D-mesh NoC to map inter-tile-pair shuffles as pre-programmed multi-hop routes over 512B links that connect adjacent tile-pairs in each direction. An NTT/iNTT router embedded within each tile-pair directs 512B data packets originating at the local RF or in-flight packets towards non-contending output ports. A 6b source address TP_{addr} indexes the *nload*-programmed contention-free-by-construction NTT/iNTT routing table to obtain a 3b output address for each packet. To maximize throughput, the router implements packet stalls that prioritize long-hop packets without impacting worst-case hop latency. Short-hop stalled packets are held in a local register for a time-period indicated by the stall count field in the 1B routing table entry. At the destination, packets are written into a 4-deep FIFO to

424 • 2026 IEEE International Solid-State Circuits Conference

979-8-3315-8937-0/26 $31.00 © 2026 IEEE

synchronize RF writeback across all tiles. The 2D-mesh NoC delivers contention-free 16k shuffle throughput of 5 cycles, while reducing worst-case link bundle width by 3.5× enabling a modular layout compared to dedicated point-to-point links. The NoC also computes k-rotation automorphisms by executing iNTTs with a scaled primitive root of $\omega^{-1/k}$. HERACLES's HERotate kernel further optimizes automorphisms by merging them with iNTTs that typically precede key-switching in conventional HERotate implementations.

The HERACLES SoC die is fabricated in Intel 3 CMOS process, occupying 197mm^2 (Fig. 25.1.7) and housed in a 56x45mm^2 FCBGA 3184-pin package. The SiP is mounted on a PCIe board supporting SPI, JTAG/UART side-bands and attached to a 2.5GHz, 128GB RAM, Intel Xeon-w7 3455 host. At nominal supply voltage of 750mV, 25°C, the HERACLES operates at maximum frequency of 1.2GHz with total power consumption of 176W (Fig. 25.1.6). Performance measurements of NTT, HEAdd, HEMul, HEModSwitch, Relin, HERotate, and HERescale kernels for multiple FHE schemes (BGV, BFV, CKKS) across a wide range of ciphertext polynomial parameters (ring dimensions: 2^{14} to 2^{16}, 10 to 42 residues) shows HERACLES speedup of 1000 to 5547× over a general-purpose CPU. HERACLES delivers 64k-point 42-residue negacyclic ciphertext NTT latency of 39.34μs (2355× speedup over CPU) with bit throughput of 4.17Tb/s (20.5× higher than [2]). Latency of 306ms was obtained for BGV bootstrapping+scaling with ring-dimension of 2^{16}, 62 residues, 64-slots, representing 2165× speedup over CPU. HERACLES delivers bootstrapping+scaling throughput of 3.27 boots/s, an 8.17× speedup over prior silicon-proven FHE implementations [4].

References:
[1] S. Kim et al., "BTS: An Accelerator for Bootstrappable Fully Homomorphic Encryption," *ISCA*, pp. 711-725, June 2022. https://doi.org/10.1145/3470496.3527415
[2] S. Lu et al., "A 28nm 4.05μJ/Encryption 8.72kHMul/s Reconfigurable Multi-Scheme Fully Homomorphic Encryption Processor for Encrypted Client-Server Computing," *ISSCC*, vol. 68, pp. 294-295, Feb. 2025. https://doi.org/10.1109/ISSCC49661.2025.10904812
[3] L. Lin et al., "A 30.4GOPS/mW MK-CKKS Processor for Secure Multi-Party Computation," *ISSCC*, vol. 68, pp. 296-297, Feb. 2025. https://doi.org/10.1109/ISSCC49661.2025.10904776
[4] H. Lee et al., "A 2.7-to-13.3μJ/boot/slot Flexible RNS-CKKS Processor in 28nm CMOS Technology for FHE-Based Privacy-Preserving Computing," *ISSCC*, vol. 67, pp. 296-297, Feb. 2024. https://doi.org/10.1109/ISSCC49657.2024.10454420
[5] R. Geelen et al., "BASALISC: Programmable Hardware Accelerator for BGV Fully Homomorphic Encryption," *Transactions on Cryptographic Hardware and Embedded Systems* (*TCHES*), vol. 2023 No. 4, pp. 32-57, Aug. 2023. https://doi.org/10.46586/tches.v2023.i4.32-57
[6] W. Hafez et al., "An Intel 3 Advanced FinFET Platform Technology for High Performance Computing and SOC Product Applications," *IEEE Symp. VLSI Circuits*, pp. 1-2, Jun. 2024. https://doi.org/10.1109/VLSITechnologyandCir46783.2024.10631513
[7] U. Banerjee et al., "An Energy-Efficient Configurable Lattice Cryptography Processor for the Quantum-Secure Internet of Things," *ISSCC*, pp. 46-47, Feb. 2019. https://doi.org/10.1109/ISSCC.2019.8662528
[8] A. C. Mert et al., "Design and Implementation of Encryption/Decryption Architectures for BFV Homomorphic Encryption Scheme," *TVLSI*, vol. 28, no. 2, pp. 353-362, Feb. 2020. https://doi.org/10.1109/TVLSI.2019.2943127

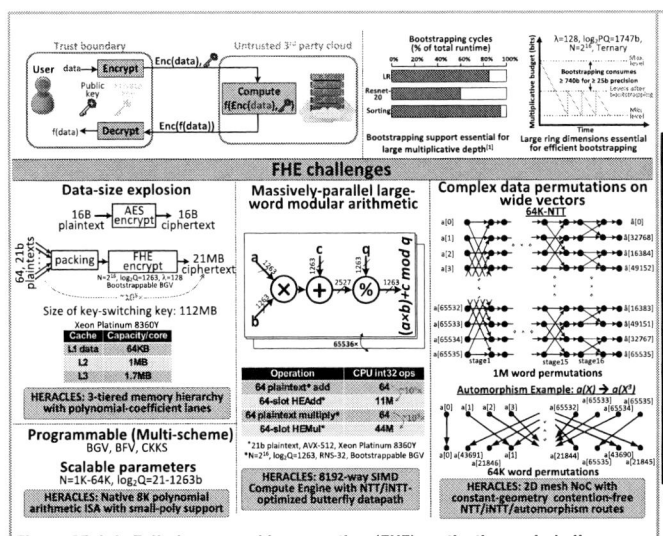

Figure 25.1.1: Fully homomorphic encryption (FHE) motivation and challenges on current computing systems.

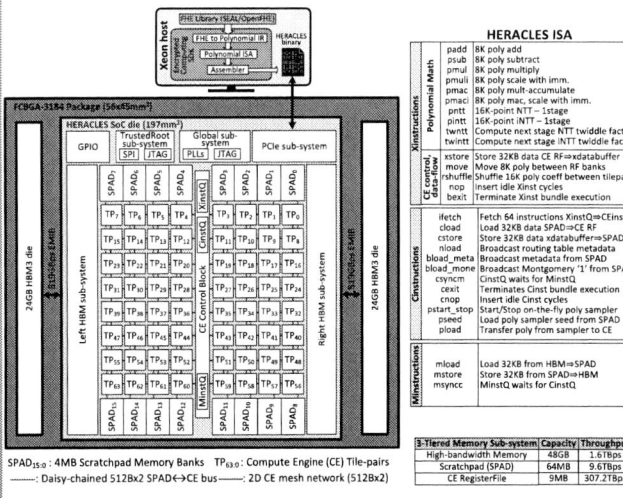

Figure 25.1.2: The HERACLES SoC architecture and instruction-set.

ISSCC 2026 / SESSION 25 / HARDWARE SECURITY / 25.1

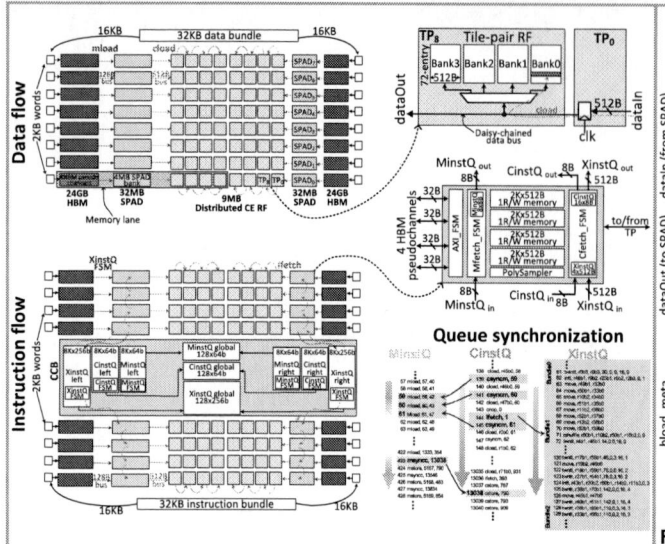

Figure 25.1.3: 3-tiered memory sub-system and data/instruction-flow.

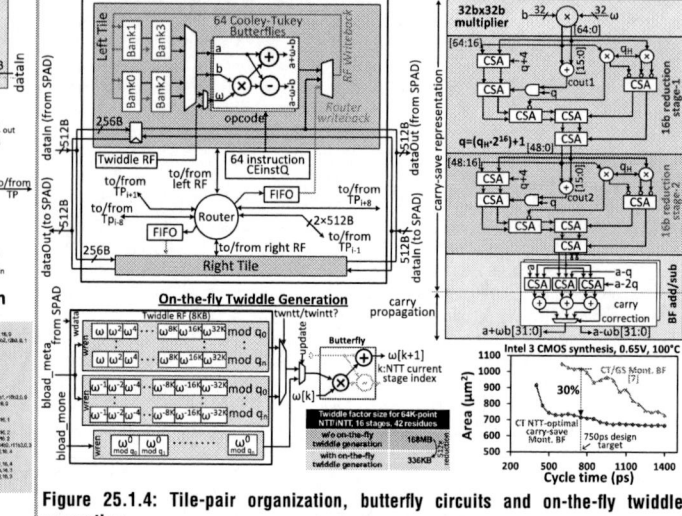

Figure 25.1.4: Tile-pair organization, butterfly circuits and on-the-fly twiddle generation.

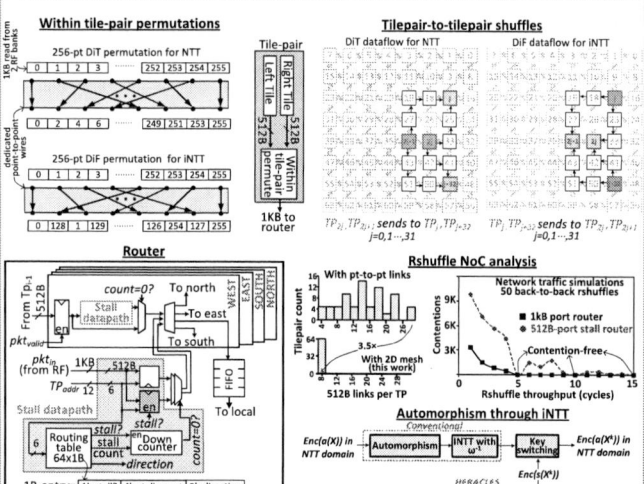

Figure 25.1.5: NTT/iNTT/Automorphism data permutations.

HERACLES measured FHE Performance (1.2GHz, 0.75V, 25°C) and comparisons with CPU

Scheme	Operation	Ring dimension N	Number of 32b residues in each ciphertext ring polynomial n_{res}	This work (µs)	CPU latency[1] (ms)	Speedup over CPU
BGV, BFV, CKKS	Ciphertext negacyclic NTT[4]	2^{14}	10	2.13	5	2347×
		2^{15}	21	8.94	21.95	2455×
		2^{16}	42	39.34	92.66	2355×
BGV, BFV, CKKS	HEadd	2^{14}	10	0.6	0.64	1067×
		2^{15}	21	2.61	2.61	1000×
		2^{16}	42	9.95	10.69	1074×
BGV, BFV, CKKS	HEMul (without key switching)	2^{14}	10	0.7	2.27	3243×
		2^{15}	21	3.1	9.48	3058×
		2^{16}	42	13	38.48	2960×
BGV	HEModSwitch	2^{14}	10	2.56	6.49	2535×
		2^{15}	21	10.99	28.31	2576×
		2^{16}	42	39.64	120.94	3051×
BGV	Relin	2^{14}	10	10.22[3]	32.05[6]	3136×
		2^{15}	21	62.48[3]	252.51[6]	4041×
		2^{16}	42	349.18[3]	1936.81[6]	5547×
BGV	HERotate (with key switching)	2^{14}	10	12.28[3]	32.29[6]	2629×
		2^{15}	21	70.84[3]	254.72[6]	3596×
		2^{16}	42	386.11[3]	1944.67[6]	5037×
CKKS	HERescale	2^{14}	10	2.64	4.68	1773×
		2^{15}	21	11.5	20.74	1803×
		2^{16}	42	57.36	89.65	1563×
BGV	Bootstrapping + scaling[2]	2^{16}	62[5]	305684.39[6]	661903[6]	2165×

[1] FHE Library : SEALv4.1.1 Default cmake options, CPU configuration: Intel Xeon 8360Y @ 3.5GHz, single-core/single-thread.
[2] Implements programmable bootstrapping (with scaling by p = 127). Plaintext modulus $p^3 = 127^3$. Number of slots = 64.
[3] Implements hybrid key switching with decomposition number, dnum = 4.
[4] Implements NTT of ciphertext $\in (R_Q)^2$ evaluated at the primitive 2Nth roots of unity, includes pre-multiplication.
[5] Maximum number of residues. The difference of levels between fresh ciphertext vs freshly bootstrapped ciphertext is 54 due to scaling functionality and readjust for 32 bit RNS.
[6] Algorithm uses full digit decomposition

Figure 25.1.6: Intel 3 silicon measurements and performance data.

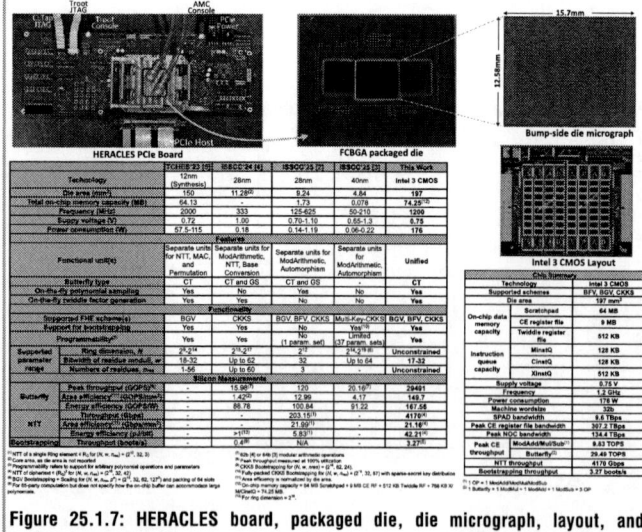

Figure 25.1.7: HERACLES board, packaged die, die micrograph, layout, and performance summary.

• 2026 IEEE International Solid-State Circuits Conference

979-8-3315-8937-0/26 $31.00 © 2026 IEEE

ISSCC 2026 / SESSION 25 / HARDWARE SECURITY / 25.2

25.2 A 28nm 0.48mJ/boot Torus FHE Processor for Arbitrary Computation on Encrypted Data

Xinglong Yu, Yi Sun, Yifan Zhao, Honglin Kuang, Song Wang, Zhen Yang, Jun Han

Fudan University, Shanghai, China

Abstract

This work presents a 3.24mm² torus fully homomorphic encryption (FHE) processor fabricated in 28nm CMOS, featuring: 1) a security-preserving low-bit-width torus quantization with keys hints reducing latency by 62%, 2) a divide-and-conquer innovative polynomial multiplication engine improving energy efficiency by 29%, and 3) a conflict-free atomic swizzling with full-depth pipelining reducing memory size by 96%. It delivers 196× higher energy efficiency in bootstrapping compared to a CPU.

Fully homomorphic encryption (FHE) has recently emerged as a promising technology for privacy-preserving applications, as it enables computations directly on encrypted data without requiring decryption. State-of-the-art FHE schemes can be broadly classified into two types: arithmetic FHE, exemplified by CKKS, which enables homomorphic arithmetic operations [2-4], and logic FHE, represented by TFHE, which supports homomorphic Boolean operations [5,6]. Notably, in a CKKS-based large language model (LLM) developed for privacy-preserving inference [16], non-linear operations can introduce relative errors of up to 297%, substantially compromising model accuracy [17]. In contrast, TFHE [18] facilitates bit-level exact arbitrary computation on encrypted data and features fast programmable bootstrapping (PBS) in unbounded computational depth. However, homomorphic evaluations in FHE entail substantial computational overhead [2-6], highlighting the need for ASIC-based processors that reduce user-perceived latency and thermal footprint while maintaining manufacturing cost-effectiveness. While dedicated processors have been developed for arithmetic FHE [2-4], silicon-proven logic FHE processors remain underexplored. As illustrated in Fig. 25.2.1, designing a high-performance TFHE processor must address three main challenges. First, the extensive off-chip movement of bootstrapping keys (BSK) imposes significant bandwidth demands, severely limiting computational latency. Second, the computation-intensive torus polynomial multiplication dominates TFHE workloads, forming a critical bottleneck in achieving high energy efficiency. Third, the large ciphertext storage requirements, combined with irregular data access patterns in transformations for efficient polynomial multiplication, considerably exacerbate on-chip memory usage, resulting in increased silicon area overhead.

To address the challenges above, this paper presents a silicon-proven TFHE processor that achieves a high energy efficiency of 0.48mJ/boot at the 128b security level. The architecture incorporates three key innovations: 1) A security-preserving low-bit-width torus quantization (TQ) technique combined with BSK hints, reducing PBS latency by 62.4%. 2) A divide-and-conquer (D&C) polynomial multiplication engine (DCPME) with bit-width-reduced multipliers, improving energy efficiency for PBS by 75.5% compared to prior design [5]. 3) A conflict-free atomic swizzling strategy integrated with full-depth PBS pipelining, reducing on-chip memory size by 96.5% relative to the architecture in [6].

Figure 25.2.2 depicts the overall architecture of the processor, which comprises six primary components: an instruction scheduling unit (ISU) for top-level control, a load-store unit (LSU) for ciphertext transfer, an affine unit (AFU) for ciphertext addition, a 114KB ciphertext file (CF) for ciphertext storage, a key-switching unit (KSU), and a bootstrapping unit (BSU). The BSU is designed to execute four-lane parallel PBS and integrates several key submodules: a BSK hint generator (BSKHG), a BSK unfolding unit (BSKU), a fused test polynomial generator (FTPG), a dual sample extraction unit (DSEU), and a divide-and-conquer external product unit (DCEXPU). The BSKHG expands a 128b PRNG seed [15] to generate ciphertext components on the fly, reducing BSK data movement. The FTPG and DSEU collectively enable concurrent evaluation of multiple homomorphic look-up table (HLUT) operations [20]. The DCEXPU serves as an innovative engine to improve polynomial multiplication efficiency. Its architecture consists of a blind rotating polynomial controller (BRPC), four gadget decomposition units (GDU), four polynomial multiplication dividing units (PMDU), four polynomial multiplication kernel units (PMKU), and two polynomial multiplication conquering units (PMCU). As also demonstrated in Fig. 25.2.2, the full-depth PBS pipelining technique achieves 94% DCEXPU utilization and near-100% bandwidth utilization, while fully concealing KSU latency within the blind rotation.

Figure 25.2.3 illustrates the BSKU engine, which employs an atomic swizzling strategy to facilitate bit-reversal permutation. This approach is inspired by the swizzling technique used in NVIDIA shared memory to mitigate bank conflicts in general matrix multiplication. However, the conventional swizzle mode operates on fixed 16-byte chunks within a depth of 8 as the basic repeating unit. As a result, when memory accesses span chunks with a depth stride exceeding 8, the fragment of 16-byte bundles across multiple banks leads to bank conflicts. To overcome this limitation, this work introduces atomic swizzling, which employs element-wise granularity and depth-aware arrangement to enable fully conflict-free polynomial multiplication and bit-reversal permutation. Polynomial coefficients are seamlessly distributed across rotating buffers of varying depths, while the dual-pointer unit incrementally generates unfolded polynomials at successive depths. The BSKU reduces

external product operations by 50% [19] and improves PBS throughput by 1.9×. As also illustrated in Fig. 25.2.3, the PMDU recursively decomposes an N-degree polynomial into three $N/2$-degree sub-polynomials, reducing the required number of multiplications from N^2 to $3N^2/4$. The PMDU adopts a ternary-tree streaming pattern to orchestrate hierarchical accumulation of sub-polynomials. Using an order counter, stack counter, and stack pointer generated by level counters S_i, the PMDU continuously fetches and accumulates degree-16 sub-polynomials from the origin and stack memory banks. This stack-based approach, combined with the atomic swizzling's conflict-free memory layout, eliminates the need for coefficient temporary storage across all recursion levels employed in the basic multi-level banking method. Consequently, the PMDU improves area efficiency by 46% and reduces memory bank usage by 90% compared to the baseline.

Figure 25.2.4 illustrates the intrinsic advantages of the DCPME, attained through security-preserving low-bit-width torus quantization. The energy consumption of polynomial multiplication is primarily determined by the input/output bit-width of multipliers and the total number of multiplication operations. To balance computational efficiency and cryptographic security, we reduce torus precision from conventional 32b [5,6,10] to 24b, while maintaining security guarantees as verified by the lattice estimator [21]. However, transformations for efficient polynomial multiplication typically require a higher intermediate bit-width to preserve numerical accuracy. Under a 32b torus setting, fast Fourier transform (FFT) relies on double floating-point precision, and number theoretic transform (NTT) requires double-word integers. A conventional finite impulse response (FIR) filter architecture [11] minimizes multiplier input/output bit-widths but retains the maximum multiplier invocation count, resulting in suboptimal performance. The DCPMU without TQ improves area efficiency by 51% and 49%, and energy efficiency by 62% and 61% compared to FFT and NTT hardware, respectively. With TQ, the DCPMU achieves an additional 1.8× improvement. To maximize efficiency, the PMKU employs 81 multipliers for degree-16 polynomial multiplication and leverages a ternary tree to expand the PMDU's temporal structure. A key challenge is sub-polynomial rotation, which necessitates coefficient reordering and multi-operand accumulation. Compared to a TQ-based FIR structure [11], the PMKU reduces area by 43% and power consumption by 47%. Unlike prior Karatsuba-based low-high split divide-and-conquer architectures [12,13], this work introduces the full-recursion odd-even split DCPME, which halves both the storage of conquering coefficients and the number of conquering adders. Although prior work [11] introduces the odd-even split approach, the lack of a fully recursive conquering framework prevents achieving minimal overhead. Based on a derivation for streaming reordered sub-polynomials, our PMCU integrates a multi-level buffer for temporary storage at each recursion level, unlocking the advantages of full recursion. To optimize memory usage, an intra-lane interleaving technique schedules one coefficient output every two cycles at level 2 and distributes a two-read and one-write operation across two cycles at level 3. Additionally, an inter-lane interleaving strategy enables 2-lane multiplexing to handle three concurrent memory accesses, reducing memory bank requirements. Collectively, these optimizations reduce baseline memory bank usage by 50% and decrease required area by 21%.

Figure 25.2.5 presents the measured performance of the TFHE processor in comparison with related works. To enable a fair comparison of polynomial multiplication efficiency, the performance of the DCPME is normalized to that of NTT hardware, demonstrating a 1.5× improvement in area efficiency and a 1.8× enhancement in energy efficiency over [8] and [3], respectively. Compared to the CPU platform (1-core 1-thread Intel(R) Xeon(R) Gold 5218 CPU @ 2.30GHz) running TFHE-rs 1.3.3 [22], this design achieves a throughput improvement of 11.4-25.5× and an energy efficiency gain of up to 196×. For various homomorphic non-linear operations, this work attains 96-148× higher energy efficiency and a speedup of 5.5-8.4×. Figure 25.2.6 provides a comparison of design features with prior works [1-6]. The TFHE processor, fabricated in 28nm CMOS technology, operates at 170-390MHz with a supply voltage range of 0.7-1.0V. Compared to recent ASIC-based designs supporting homomorphic arithmetic evaluation, this work presents the silicon-proven logic FHE processor capable of bootstrapping with a 2672× improvement in throughput. With a compact core area of 2.25mm², the design achieves an area efficiency of 475boots/s/mm² at 390MHz and 1.0V, representing a 1.7× gain over [5]. Moreover, it demonstrates an energy efficiency of 0.48mJ/boot at 170MHz and 0.7V, even outperforming the architecture-level studies [5] and [6] by 4.0× and 1.4×, respectively. Figure 25.2.7 shows the die micrograph and summarizes the chip specifications, including area and power breakdown, as well as voltage-frequency scaling characteristics.

Acknowledgement:
This work was supported by the National Natural Science Foundation of China (61934002, 62234008). The corresponding author of this paper is Jun Han (junhan@fudan.edu.cn).

References:
[1] G. Shi et al., "A 28nm 68MOPS 0.18μJ/Op Paillier Homomorphic Encryption Processor with Bit-Serial Sparse Ciphertext Computing," *ISSCC*, pp. 242-243, Feb. 2023. http://doi.org/10.1109/ISSCC42615.2023.10067522

[2] H. Lee, H. Kwon, and Y. Lee, "A 2.7-to-13.3μJ/boot/slot Flexible RNS-CKKS Processor in 28nm CMOS Technology for FHE-Based Privacy-Preserving Computing," *ISSCC*, pp. 296-297, Feb. 2024. http://doi.org/10.1109/ISSCC49657.2024.10454420

[3] S. Lu et al., "A 28nm 4.05μJ/Encryption 8.72kHMul/s Reconfigurable Multi-Scheme Fully Homomorphic Encryption Processor for Encrypted Client-Server Computing," *ISSCC*, pp. 294-295, Feb. 2025. http://doi.org/10.1109/ISSCC49661.2025.10904812

[4] L.-H. Lin, Y.-K. Yang, and C.-H. Yang, "A 30.4GOPS/mW MK-CKKS Processor for Secure Multi-Party Computation," *ISSCC*, pp. 296-297, Feb. 2025. http://doi.org/10.1109/ISSCC49661.2025.10904776

[5] A. Putra, P. Prasetiyo, Y. Chen, J. Kim, and J.-Y. Kim, "Strix: An End-to-End Streaming Architecture with Two-Level Ciphertext Batching for Fully Homomorphic Encryption with Programmable Bootstrapping," *IEEE/ACM MICRO*, pp. 1319-1331, Oct. 2023. http://doi.org/10.1145/3613424.3614264

[6] Prasetiyo, A. Putra, and J.-Y. Kim, "Morphling: A Throughput-Maximized TFHE-based Accelerator using Transform-domain Reuse," *IEEE HPCA*, pp. 249-262, Mar. 2024. http://doi.org/10.1109/HPCA57654.2024.00028

[7] U. Banerjee, A. Pathak, and A. P. Chandrakasan, "An Energy-Efficient Configurable Lattice Cryptography Processor for the Quantum-Secure Internet of Things," *ISSCC*, pp. 46-47, Feb. 2019. http://doi.org/10.1109/ISSCC.2019.8662528

[8] B. Kim, J. Park, S. Moon, K. Kang, and J.-Y. Sim, "Configurable Energy-Efficient Lattice-Based Post-Quantum Cryptography Processor for IoT Devices," *IEEE ESSCIRC*, pp. 525-528, Sept. 2022. http://doi.org/10.1109/ESSCIRC55480.2022.9911531

[9] S. Das et al., "A 10.33 μJ/encryption Homomorphic Encryption Engine in 28nm CMOS with 4096-degree 109-bit Polynomials for Resource-Constrained IoT Clients," *IEEE ESSCIRC*, pp. 193-196, Sept. 2023. http://doi.org/10.1109/ESSCIRC59616.2023.10268762

[10] T. Kong and S. Li, "Hardware Acceleration and Implementation of Fully Homomorphic Encryption Over the Torus," *IEEE TCAS-I*, vol. 71, no. 3, pp. 1116-1129, Mar. 2024. http://doi.org/10.1109/TCSI.2023.3338953

[11] W. Tan, A. Wang, X. Zhang, Y. Lao, and K.K. Parhi, "High-Speed VLSI Architectures for Modular Polynomial Multiplication via Fast Filtering and Applications to Lattice-Based Cryptography," *IEEE TC*, vol. 72, no. 9, pp. 2454-2466, Sept. 2023. http://doi.org/10.1109/TC.2023.3251847

[12] Y. Zhu et al., "LWRpro: An Energy-Efficient Configurable Crypto-Processor for Module-LWR," *IEEE TCAS-I*, vol. 68, no. 3, pp. 1146-1159, Mar. 2021. http://doi.org/10.1109/TCSI.2020.3048395

[13] Z.-Y. Wong et al., "KaratSaber: New Speed Records for Saber Polynomial Multiplication Using Efficient Karatsuba FPGA Architecture," *IEEE TC*, vol. 72, no. 7, pp. 1830-1842, July 2023. http://doi.org/10.1109/TC.2023.3238129

[14] S.G. Bhaskaracharya, J. Demouth, and V. Grover, "Automatic Kernel Generation for Volta Tensor Cores," *arXiv*:2006.12645, pp. 1-13, Aug. 2020. http://doi.org/10.48550/arXiv.2006.12645

[15] N. Samardzic et al., "CraterLake: A Hardware Accelerator for Efficient Unbounded Computation on Encrypted Data," *IEEE/ACM ISCA*, pp. 173-187, Jun. 2022. http://doi.org/10.1145/3470496.3527393

[16] J. Zhang et al., "Secure Transformer Inference Made Non-Interactive," *NDSS*, pp. 1-17, Feb. 2025. http://doi.org/10.14722/ndss.2025.230868

[17] T. Xu et al., "Breaking the Layer Barrier: Remodeling Private Transformer Inference with Hybrid CKKS and MPC," *USENIX Security Symposium*, pp. 2653-2672, Aug. 2025. http://doi.org/10.5281/zenodo.15590214

[18] I. Chillotti, N. Gama, M. Georgieva, and M. Izabachène, "TFHE: Fast Fully Homomorphic Encryption Over the Torus," *Journal of Cryptology*, vol. 33, no. 1, pp. 34-91, Jan. 2020. http://doi.org/10.1007/s00145-019-09319-x

[19] F. Bourse, M. Minelli, M. Minihold, and P. Paillier, "Fast Homomorphic Evaluation of Deep Discretized Neural Networks," *CRYPTO*, pp. 483-512, Aug. 2018. http://doi.org/10.1007/978-3-319-96878-0_17

[20] I. Chillotti, D. Ligier, J.-B. Orfila, and S. Tap, "Improved Programmable Bootstrapping with Larger Precision and Efficient Arithmetic Circuits for TFHE," *ASIACRYPT*, pp. 670-699, Dec. 2021. http://doi.org/10.1007/978-3-030-92078-4_23

[21] M.R. Albrecht, R. Player, and S. Scott, "On the Concrete Hardness of Learning with Errors," *Journal of Mathematical Cryptology*, vol. 9, no. 3, pp. 169-203, Oct. 2015. http://doi.org/10.1515/jmc-2015-0016

[22] Zama, "TFHE-rs: A Pure Rust Implementation of the TFHE Scheme for Boolean and Integer Arithmetics Over Encrypted Data," 2022. Accessed on Dec. 14, 2025. <https://github.com/zama-ai/tfhe-rs>

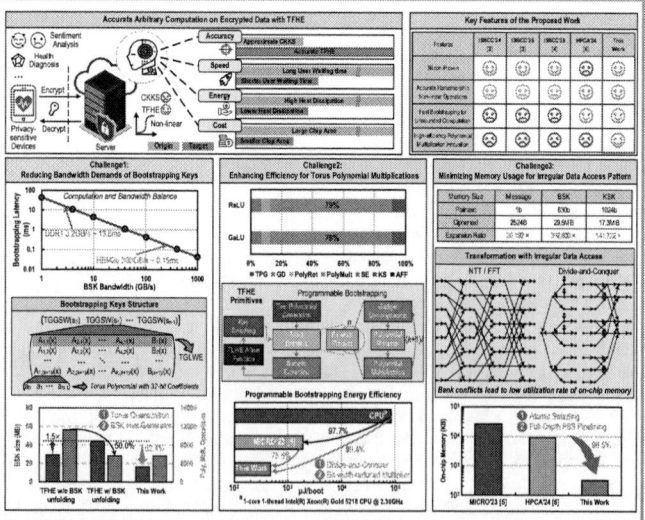

Figure 25.2.1: Challenges of designing a torus FHE processor for arbitrary computation on encrypted data.

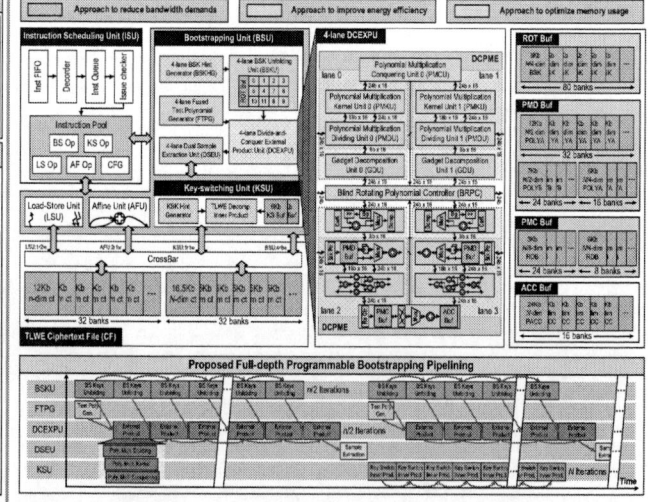

Figure 25.2.2: System architecture with full-depth programmable bootstrapping pipelining technique.

ISSCC 2026 / SESSION 25 / HARDWARE SECURITY / 25.2

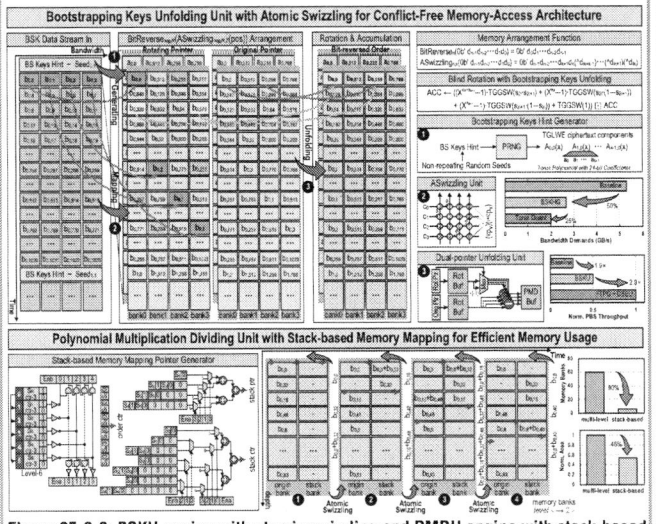

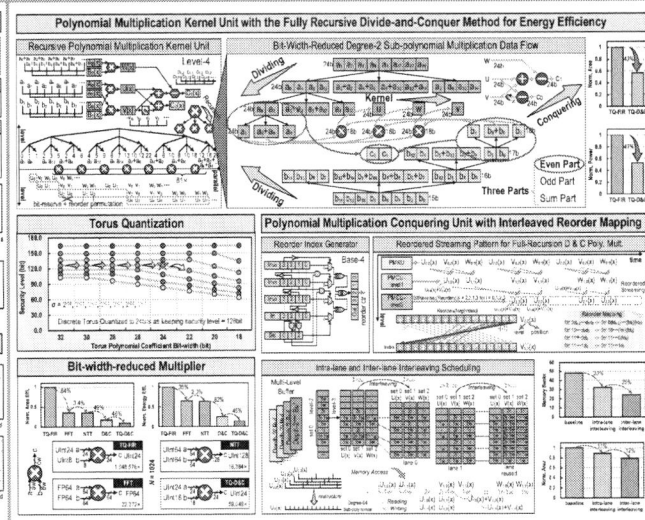

Figure 25.2.3: BSKU engine with atomic swizzling and PMDU engine with stack-based memory mapping.

Figure 25.2.4: PMKU engine with torus quantization and PMCU engine with interleaved reorder mapping.

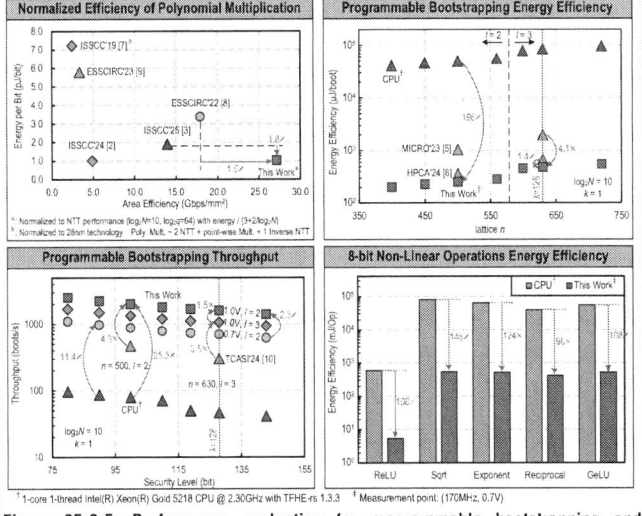

Figure 25.2.5: Performance evaluation for programmable bootstrapping and demonstration for design features.

	ISSCC'23 [1]	ISSCC'24 [2]	ISSCC'25 [3]	ISSCC'25 [4]	MICRO'23 [5]	HPCA'24 [6]	This Work
Silicon Proof	Yes	Yes	Yes	Yes	No	No	Yes
Technology	28nm	28nm	28nm	40nm	28nm (Synthesis)	28nm (Synthesis)	28nm
Frequency	500MHz	333MHz	125-625MHz	50-210MHz	1.2GHz*	1.2GHz*	170-390MHz
Voltage	0.90V	1.00V	0.70-1.10V	0.65-1.30V	-	-	0.70-1.00V
Power	4W/12W	160mW	138-1185mW	55-221mW	77W*	53W*	222-1097mW
Core Area	42.96mm²	11.25mm²	5.4mm²	2.85mm²	141.37mm²*	74.79mm²*	2.25mm²
HE Scheme	Paillier†	CKKS	BGV, BFV, CKKS	MK-CKKS	TFHE	TFHE	TFHE
HE Type	Partially	Fully	Leveled Fully^	Fully	Fully	Fully	Fully
HE Operation	HAdd, PMul	HAdd, HPMul, HRot	HAdd, HPMul, HRot	HAdd, HPMul, HRot	HAdd, HLUT	HAdd, HLUT	HAdd, HLUT
Poly. Mult. Algorithm	-	NTT	NTT	NTT	FFT	FFT	Divide-and-Conquer
Datapath Bit-width	-	62bit	32bit	64bit	64bit	64bit	24bit
Security Level	-	128bit	-	128bit	128bit	128bit	128bit
(N, n, l, k)	-	(65536, -, 24, 1)*	-	(65536, -, 34, 1)*	(1024, 630, 3, 1)	(1024, 630, 3, 1)	(1024, 630, 3, 1)
Throughput	-	0.4boots/s	-	0.1boots/s*	39600boots/s	78692boots/s	1069boots/s*
Area Efficiency	-	0.035boots/s/mm²	-	0.035boots/s/mm²*	280boots/s/mm²	1052boots/s/mm²	475boots/s/mm²*
Energy Efficiency	-	179.2mJ/boot	-	110.4mJ/boot*	1.94J/boot	0.67mJ/boot	0.48mJ/boot*

*: Architecture research with synthesized frequency, power, and area, ignoring the impact of interconnect RC
†: Paillier relies on the problem of computing n-th residue classes; therefore, it is represented as a large integer form, but not a polynomial form
^: Derived from the indirect results in the paper and the lattigo library
‡: Bootstrapping performance is not reported due to focusing on leveled FHE operations for client-server computing
#: The RNS-CKKS scheme lacks the parameter n without the LWE samples
Measurement points: a(390MHz, 1.0V); b(170MHz, 0.7V)

Figure 25.2.6: Performance comparison with the state-of-the-art works.

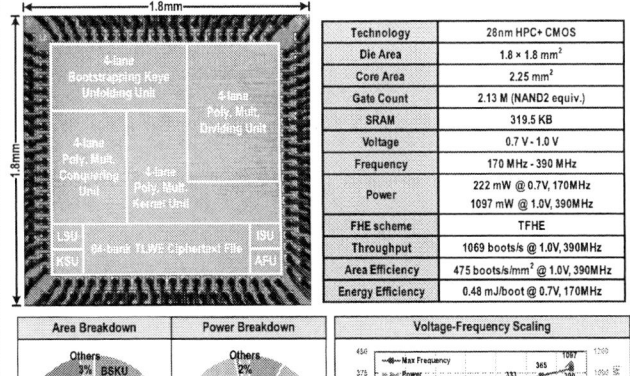

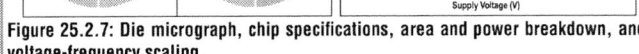

Figure 25.2.7: Die micrograph, chip specifications, area and power breakdown, and voltage-frequency scaling.

Technology	28nm HPC+ CMOS
Die Area	1.8 × 1.8 mm²
Core Area	2.25 mm²
Gate Count	2.13 M (NAND2 equiv.)
SRAM	319.5 KB
Voltage	0.7 V - 1.0 V
Frequency	170 MHz - 390 MHz
Power	222 mW @ 0.7V, 170MHz 1097 mW @ 1.0V, 390MHz
FHE scheme	TFHE
Throughput	1069 boots/s @ 1.0V, 390MHz
Area Efficiency	475 boots/s/mm² @ 1.0V, 390MHz
Energy Efficiency	0.48 mJ/boot @ 0.7V, 170MHz

• 2026 IEEE International Solid-State Circuits Conference

ISSCC 2026 / SESSION 25 / HARDWARE SECURITY / 25.3

25.3 A 16nm 0.042mm² 0.66µJ/Ops Lightweight MLWE PQC KEM with Cryptanalysis-ASIC Co-Optimization

Archisman Ghosh*[1], Ming-Che Li*[1], Lingke Ding[1], Suparna Kundu[2], Abdullah Sayeed[1], Angshuman Karmakar[3], Harshit Naman[1], Sarthak Antal[1], Ingrid Verbauwhede[2], Shreyas Sen[1]

[1]Purdue University, West Lafayette, IN, [2]KU Leuven ESAT, Leuven, Belgium, [3]Indian Institute of Technology, Kanpur, India
*Equally Credited Authors (ECAs)

Abstract

Resource-limited devices demand post-quantum cryptography (PQC) within stringent power, area, and memory limits. To our knowledge this work introduces the first cryptanalysis-ASIC co-optimized PQC IC, achieving minimal area, power, and memory footprint. This approach achieves an area reduction of 8× relative to the state-of-the-art

M-LWE accelerator and up to 509× compared to a generic PQC processor. Fabricated in Intel16, this IC operates over a frequency range of 0.1MHz to 1.03GHz and achieves an 8.5× reduction in energy-area product.

Resource-limited devices such as IoT edge devices, medical implants, wearables, etc., are increasingly intertwined with sensitive personal data, ranging from audio and video streams to physiological data and sensitive payment credentials. These devices form the backbone of emerging digital infrastructure, but their adoption remains hindered by the absence of advanced cryptographic primitives that can simultaneously ensure post-quantum security and meet stringent hardware resource budgets. Existing post-quantum cryptographic (PQC) algorithms [1, 2], designed for general-purpose platforms usually require large logic instances (>M gate instances), extensive memory (>100 kB), and significant energy, which is not suitable for resource-limited nodes (e.g. < hundreds of microwatts, <100k gate instances). This challenge is compounded by the siloed development of algorithm and hardware design. Most circuit designers primarily implement algorithms designed by cryptographers, the National Institute of Standard and Technology (NIST), or other standard bodies and try to optimize them from implementation perspective. However, algorithmic parameter choices disproportionally impact power, area, and memory footprint. Therefore, without Cryptanalysis-ASIC co-optimization, integrating PQC into the resource-limited device efficiently is highly impractical.

In 2025, NIST standardized ASCON [3] for lightweight devices as an alternate to AES for symmetric key encryption as well as to Keccak for use as a hash function, extended output function (XOF) or pseudo-random number generator (PRNG). A similar evolution for PQC Key Encapsulation Mechanism (KEM) for lightweight devices, minimizing area, power and memory is anticipated (Fig. 25.3.1).

In the state-of-the-art (SOTA) PQC core Kyber [4], the smallest area was achieved through digital circuit optimization. However, it still relied on the resource-intensive Keccak for PRNG, XOF, and hashing due to strict adherence to NIST standard parameters, leading to significant area and power overhead, with 0.34mm² area [4]. While massively parallel datapaths lead to agile PQC processors [5, 6], the associated logic instances exceed 2 MGE and energy above 3µJ/Op, making them prohibitively unsuitable for resource-limited devices. In contrast, the presented IC, fabricated in Intel16, introduces an algorithm-IC co-optimization flow for minimum area implementation that replaces Keccak with ASCON (Fig. 25.3.1), explores parameters of module learning with Error (MLWE) beyond the standard sets while being quantum secure, employs a single-butterfly datapath with efficient modulo arithmetic, and minimizes memory via tiled data processing. These innovations collectively yield active area of 0.042mm², 8× better than the current SOTA Kyber accelerator [4], and 509× better than generic SOTA PQC processor [5] while achieving 0.028µJ/Op/mm² energy-area-product (EAP), an 8.5× improvement compared to [4], with a wide operation frequency range of 100kHz to 1.03GHz. To the best of our knowledge, this work is the smallest PQC IC reported to date and the first to explicitly target resource-limited nodes via cryptanalysis-ASIC co-optimization strategy.

Three tightly integrated innovations (Fig. 25.3.2) drive the compactness of this architecture. First, an ASCON-based PRNG and hash replaces the area-intensive Keccak. Theoretical study (NIST) [7] shows a potential 12× improvement in area through replacing Keccak with ASCON. This substitution not only saves area but also reduces power by ~9.5× compared to Keccak-based implementations (estimated from [5]) in iso-frequency, iso-VDD. Second, a minimum computation datapath centered on a single butterfly engine for number theoretic transform (NTT) and inverse-NTT (INTT) is designed. Custom shift-and-add-based modulo reduction achieves up to 3× smaller area than the standard Montgomery reduction. With shift-and-add modulo division (MD), 9x area reduction is realized with respect to multiplier-based design [8]. Third, we optimize memory with concurrent data generation and computation with tiled data processing by two independent datapaths and in-place memory reuse with access scheduling for NTT/INTT, reducing total memory down to only 6.7kB. Together, these innovations enable the first PQC KEM IC that achieves both low area and low energy, paving the way for mass adoption in ultra-low-power IoT nodes and wearables.

The overall architecture (Fig. 25.3.2) integrates these innovations together into a compact system. The computation datapath and ASCON permutation datapath are coordinated by a shared scheduler, enabling parallel operation without idle cycles. A test FSM provides high-level control and is used for validation. 4 SRAM banks totaling 6.7kB support computation, with two banks active during NTT/INTT, one used for secret or matrix generation, and one

for output and test storage. The datapath is designed to sustain concurrent PRNG and arithmetic computation without memory access clashing, as the SRAM follows a 1-write/1-read (1W1R) scheme for minimum area and energy, and simultaneous multi-access would otherwise introduce additional stall cycles.

To achieve compactness while ensuring security, an exhaustive Pareto parameter search (Fig. 25.3.3) of LWE (NIST-preferred hardware-friendly quantum hard problem) across the algorithm-ASIC design space is conducted. Although a similar approach has been reported on FPGA [9], a thorough IC-level exploration is required to capture the constraints of resource-limited devices. The chosen Pareto frontier point of matrix rank l=9, polynomial size n=64, underlying lattice dimension n'=576, modulus q=7681, and standard deviation η=1 provides 100b core SVP security (NIST level 1) with a failure probability of 2^{-128}. Security is rigorously evaluated against both the classical and quantum adversaries using SOTA cryptanalysis techniques [10], including the Block Korkine-Zolotarev (BKZ) lattice reduction algorithm [11] and the Core Shortest Vector Problem (Core-SVP) estimator [10]. This exploration revealed that standard parameter sets used by NIST candidates could be optimized further for resource-limited hardware. By systematically navigating this design space, we identified a Pareto frontier point where area, memory, and energy are minimized while security is preserved. Any deviation from this operating point either increases area or reduces effective bit security below the required NIST level 1.

Figure 25.3.3 also demonstrates key architectural components. ASCON permutation block is reused across PRNG/XOF/Hash operations, yielding a unified lightweight primitive aligned with the standardized lightweight security requirements. The second innovation is a minimum computation datapath designed around a single butterfly unit (BU) reused for NTT, INTT and point-wise multiplication (PWM). Conventional PQC accelerators often instantiate multiple BUs in highly parallel architectures, which significantly inflate area and power. This design pipelines NTT into five stages and INTT into seven stages, the latter requiring additional modulo division due to the Gentleman–Sande algorithm. To keep the datapath compact, we introduce a shift-and-add modulo reduction (MR) unit which reduces area by 3× compared to the standard Montgomery reduction. A long-division-inspired modular division only needs one shift, one mux and one adder. Unlike prior massively parallel designs [5], the datapath achieves the required throughput in a fraction of the area, delivering efficiency tailored to resource-limited devices.

Figure 25.3.4 illustrates the memory footprint reduction techniques including the concurrent tiled generate-compute pipelined architecture and in-place memory reuse with access scheduling. Conventional PQC implementations treat secret and matrix generation as stages separate from NTT/INTT computation, which leads to idle datapaths and significantly larger memory requirements. In this design, secret and public matrix generation are decomposed into tiles, enabling memory reuse and minimizing buffer size. A concurrent generate-compute pipelined architecture further reduces idle time. This optimization is a key factor contributing to a reduction of 5.14× in memory footprint.

While concurrency reduces idle time, it is important to implement a minimum-area N(=64)-point NTT. Thus, three key constraints are imposed. First, only two 1W1R SRAMs are used to minimize area and energy. Second, only a single BU is employed. Third, the operation must be completed within the theoretical minimum latency $((N/2)\times logN+\Delta)$ cycles to fill the pipeline). Conventional techniques either require additional memory/buffer or additional BU. To demonstrate, in a simple case of three-stage 8-point NTT (real implementation is a 64-point one) with input data stored in MEM0 (a0 to a3) and MEM1 (a4 to a7), stage 0 output data is computed and stored back in-place (b0 to b7). However, this leads to memory access clashing as stage 1 computation requires data stored in the same 1W1R SRAM (Fig. 25.3.4), necessitating additional memory units or stalling cycles. This work adopts a unique memory swapping technique to detect the clash preemptively and swaps the required memory when writing back at each stage (Fig. 25.3.4, bottom). Thus, the computation data for stages 1 and 2 are stored separately in MEM0 and MEM1 (rather than sharing a single memory), thereby preventing memory access clashes. The same technique is also applied to the INTT operation.

428 • 2026 IEEE International Solid-State Circuits Conference

979-8-3315-8937-0/26 $31.00 © 2026 IEEE

Figure 23.3.5 shows the performance evaluation of the implemented IC. It is fabricated in Intel16 FinFET and characterized across 0.77-to-1.4V supply. The core consumes a power of 50µW from a 0.77V supply while operating at 100kHz and 30.2mW from a 1.4V supply while operating at 1.03GHz. A highly pipelined and optimized critical path helps to achieve this maximum frequency, which is the highest amongst the existing SOTA PQC KEM IC (Fig. 25.3.6). This IC occupies only 0.042mm² active area including the hash functions and memory, as well as computation circuitry which to the best of our knowledge makes it the smallest PQC IC reported to date. We introduce EAP as a figure-of-merit for compactness of PQC IC. We achieved 8.5× better EAP with respect to the smallest published Kyber core [4]. Figure 25.3.6 presents a comparative analysis with SOTA implementations, alongside a Shmoo plot illustrating the relationship between supply voltage and operating frequency, and a detailed breakdown of area distribution across architectural components. The die micrograph with specification and the test setup is shown in Fig. 25.3.7. With these techniques, as far as we are aware, this work establishes the smallest and most energy-efficient PQC IC reported to date, paving the way for adoption in resource-limited ultra-low-power IoT nodes.

References:

[1] NIST, "Module-Lattice-Based Key-Encapsulation Mechanism Standard," Federal Information Processing Standards (FIPS), FIPS 203, Aug. 2024 http://doi.org/10.6028/NIST.FIPS.203

[2] NIST, "Module-Lattice-Based Digital Signature Standard," Federal Information Processing Standards (FIPS), FIPS 204, Aug. 2024, http://doi.org/10.6028/NIST.FIPS.204

[3] M. Turan et al., "Ascon-Based Lightweight Cryptography Standards for Constrained Devices Authenticated Encryption, Hash, and Extendable Output Functions," NIST Special Publication 800-232, Aug. 2025. http://doi.org/10.6028/NIST.SP.800-232

[4] A. Li et al., "A 273µW 0.34mm² Efficient CRYSTALS-KYBER Processor for PQC Towards Edge Computing," *ESSERC*, pp. 472-475, Sep. 2024. http://doi.org/10.1109/ESSERC62670.2024.10719541

[5] Y. Zhu et al., "A 28nm 69.4kOPS 4.4µJ/Op Versatile Post-Quantum Crypto-Processor Across Multiple Mathematical Problems," *ISSCC*, pp. 298-299, Feb. 2024. http://doi.org/10.1109/ISSCC49657.2024.10454332

[6] Y. Zhu et al., "A 28nm 48KOPS 3.4µJ/Op Agile Crypto-Processor for Post-Quantum Cryptography on Multi-Mathematical Problems," *ISSCC*, pp. 514-515, Feb. 2022. http://doi.org/10.1109/ISSCC42614.2022.9731783

[7] T. Yalcin and S. Ghandali, "Need for Low-latency Ciphers: A Comparative Study of NIST LWC Finalists," NIST Lightweight Cryptography Workshop, May 2022. Accessed on Dec. 8, 2025. <https://csrc.nist.gov/csrc/media/Presentations/2022/need-for-low-latency-ciphers-a-comparative-study-o/images-media/session-1-yalcin-need-for-low-latency-ciphers.pdf>

[8] A. Ghosh et al., "A 334µW 0.158mm² Saber Learning with Rounding based Post-Quantum Crypto Accelerator," *IEEE CICC*, pp. 1-2, Apr. 2022. http://doi.org/10.1109/CICC53496.2022.9772859

[9] S. Kundu, A. Ghosh, A. Karmakar, S. Sen, and I. Verbauwhede, "Rudraksh: A Compact and Lightweight Post-Quantum Key-Encapsulation Mechanism," *IACR Transactions on Cryptographic Hardware and Embedded Systems*, vol. 2025, no. 2, pp. 647-680, Mar. 2025, 647–680. http://doi.org/10.46586/tches.v2025.i2.647-680

[10] D. Dachman-Soled, L. Ducas, H. Gong, and M. Rossi, "LWE with Side Information: Attacks and Concrete Security Estimation," *Advances in Cryptology*, *Lecture Notes in Computer Sceince*, vol. 12171, pp, 329-358, Aug. 2020. https://doi.org/10.1007/978-3-030-56880-1_12

[11] D. Hofheinz, K. Hövelmanns, and E. Kiltz, "A Modular Analysis of the Fujisaki-Okamoto Transformation," *Theory of Cryptography, Lecture Notes in Computer Science*, vol. 10677, pp. 341-371, Nov. 2017. http://doi.org/10.1007/978-3-319-70500-2_12

[12] A. Li et al., "A 40nm 2.76µJ/Op Energy-Efficient Secure Post-Quantum Crypto-Processor for Crystals-Kyber on Module LWE,' *A-SSCC*, pp. 1-3, Nov. 2023. http://doi.org/10.1109/A-SSCC58667.2023.10347915

[13] B. Kim et al., "Configurable Energy-Efficient Lattice-Based Post-Quantum Cryptography Processor for IoT Devices," *ESSCIRC*, pp. 525-528, Sep. 2022. http://doi.org/10.1109/ESSCIRC55480.2022.9911531

[14] U. Banerjee, A. Pathak, and A.P. Chandrakasan, "An Energy-Efficient Configurable Lattice Cryptography Processor for the Quantum-Secure Internet of Things," *ISSCC*, pp. 46-47, Feb. 2019. http://doi.org/10.1109/ISSCC.2019.8662528

[15] G. Bertoni, J. Daemen, M. Peeters, and G. Van Assche, "Keccak," Advances in Cryptology – EUROCRYPT, *Lecture Notes in Computer Science*, vol. 7881, pp. 313-314, May 2013. http://doi.org/10.1007/978-3-642-38348-9_19

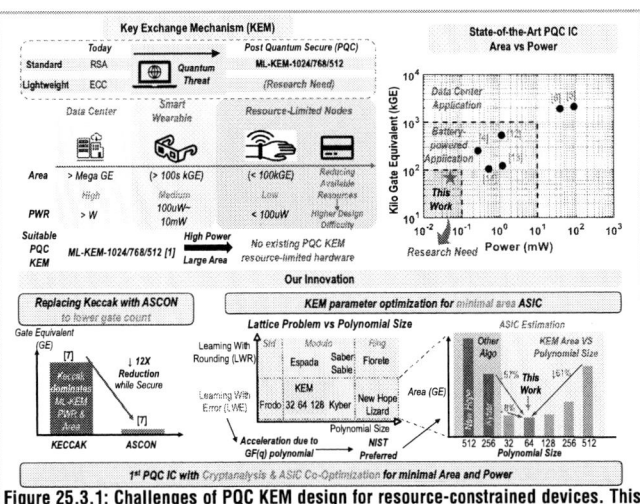

Figure 25.3.1: Challenges of PQC KEM design for resource-constrained devices. This work designs a lightweight PQC KEM core by using lightweight crypto standard ASCON based PRNG/XOF/Hash function, KEM parameter optimization for minimum area and is the 1st PQC IC with cryptanalysis and ASIC co-optimization for low area and power.

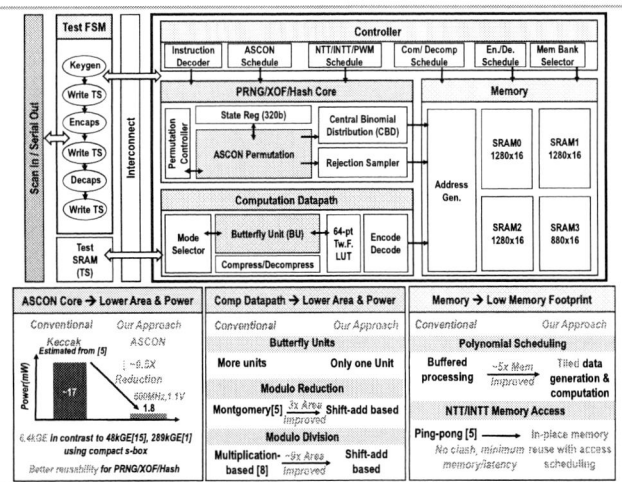

Figure 25.3.2: Full system architecture; Benefit of ASCON in reducing area and power; Improvements in computation data path in terms of butterfly unit, modulo reduction technique and modulo division technique; Memory reduction technique through in-place memory reuse with access scheduling.

ISSCC 2026 / SESSION 25 / HARDWARE SECURITY / 25.3

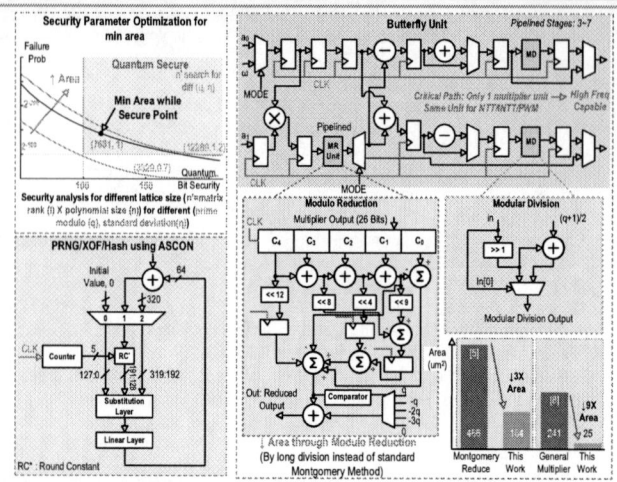

Figure 25.3.3: Security parameter optimization for minimum area; ASCON implementation; Butterfly unit implementation for NTT/INTT/PWM and critical path reduction for high frequency capability with modulo reduction and modular division techniques highlighted.

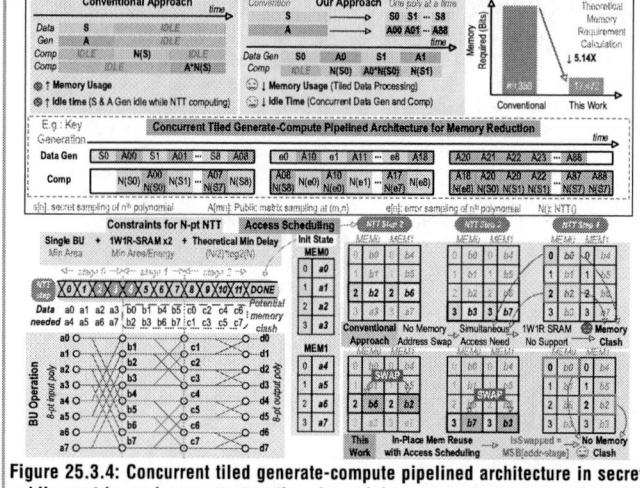

Figure 25.3.4: Concurrent tiled generate-compute pipelined architecture for minimum memory usage in secret, public matrix and error generation for minimum memory usage; N-pt NTT implementation with in-place memory reuse with access scheduling to fit resource-limited design.

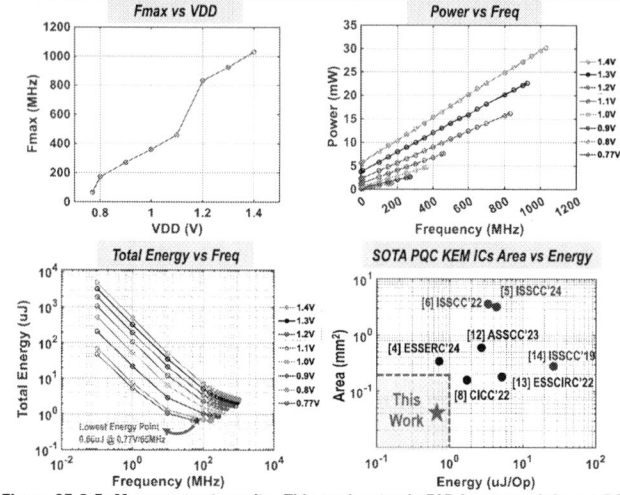

Figure 25.3.5: Measurement results. This work extends EAP for state-of-the-art PQC KEM IC by 8.5× and achieves ~509× better EAP than state-of-the art a generic PQC processor.

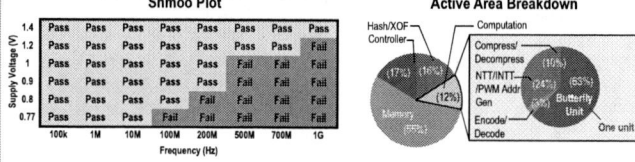

	This work	ESSERC'24 [4]	ISSCC'24 [5]	ASSCC'23 [12]	ESSCIRC'22 [13]	ISSCC'22 [6]	CICC'22 [8]	ISSCC'19 [14]
Technology	Intel16	40nm	28nm	40nm	28nm	28nm	65nm	40nm
Voltage (V)	0.77-1.4	0.62-1.2	0.7-1.1	0.8-1.2	0.65-1.35	0.9	0.7-1.1	0.68-1.1
Freq. (MHz)	0.1-1030	10-270	275-750	80-115	35-190	500	40-160	12-72
Active Area (mm²)	0.042	0.34	3.2	0.6	0.18	3.6	0.158f	0.28g
Logic Gates (kGE) / Instance (k)	155/15	253d	2,100	532	123e	1,900	NA	106g
Memory (KB)	6.7	7	228.5	14	31	448	10.19	40.25
Power (mW)	0.05-30.2	0.27-35.0	91-420	1.14-27.57	1.21	39-368	0.33	0.52
Cycles	87,598	21,143	7,197	12,245	83,148	10,433	57,014	347,826
Energy (uJ/Op)	0.66a	0.72	4.4	2.76	5.22	3.4	1.75	26.6
Throughput (Ops/s)	11,758b / 9,475c	12,821	69,473	9,434	1,924	47,925	2,841	207
EAP (uJ/ops*mm²)	0.028	0.24	14.08	1.7	0.94	12.24	0.28	7.45

a: Operating condition 0.77V/65MHz b: Operating condition 1.4V/1.04GHz c: Operating condition 1.2V/830MHz d: All circuit included SHA-3, on-chip SRAM, ROM, comm interface, and test circuit e: Cache and poly bank memory are not included f: Only computing circuit, excluding PRNG (Keccak) and SRAM g: Only computing circuit, w/o PRNG (Keccak) and SRAM

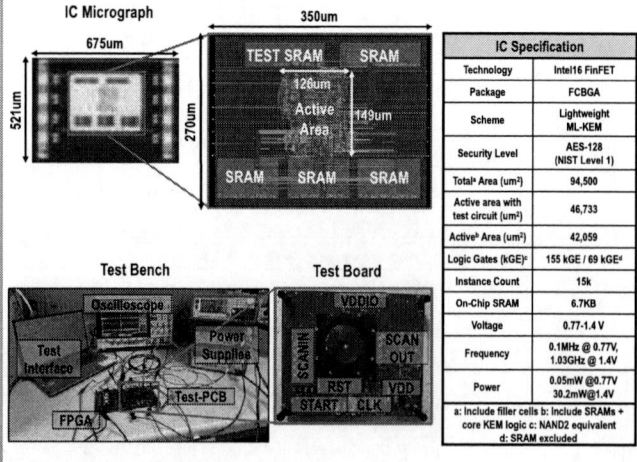

Figure 25.3.6: Comparison with state-of-the-art designs, measured Shmoo plot at different voltage and frequency, and area division of different architectural components.

IC Micrograph

IC Specification	
Technology	Intel16 FinFET
Package	FCBGA
Scheme	Lightweight ML-KEM
Security Level	AES-128 (NIST Level 1)
Totala Area (um²)	94,500
Active area with test circuit (um²)	46,733
Activeb Area (um²)	42,059
Logic Gates (kGE)c	155 kGE / 69 kGEd
Instance Count	15k
On-Chip SRAM	6.7KB
Voltage	0.77-1.4 V
Frequency	0.1MHz @ 0.77V, 1.03GHz @ 1.4V
Power	0.05mW @0.77V 30.2mW@1.4V

a: Include filler cells b: Include SRAMs + core KEM logic c: NAND2 equivalent d: SRAM excluded

Figure 25.3.7: IC micrograph; IC specification and test setup.

• 2026 IEEE International Solid-State Circuits Conference

979-8-3315-8937-0/26 $31.00 © 2026 IEEE

ISSCC 2026 / SESSION 25 / HARDWARE SECURITY / 25.4

25.4 OmniCrypt: A 435.86M-GOPS/W Bootstrappable Multi-Scheme FHE Accelerator with On-Chip Data Generation for Privacy-Preserving Computation

Adiwena Putra*, Hyunjun Cho*, Sungwoong Yune, Cuong Manh Duong, Jaeho Jeon, Joo-Young Kim

KAIST, Daejeon, Korea
*Equally Credited Authors (ECAs)

Abstract

OmniCrypt is a silicon-proven multi-scheme FHE accelerator that supports both bootstrapping and scheme switching. Fabricated in 28nm CMOS (12.96mm²), it integrates three specialized engines for efficient external product computation, on-chip data generation, and the SAM3 modular multiplier to minimize memory and compute overhead. OmniCrypt achieves 252.8M-GOPS, 19.5M-GOPS/mm², and 435.9M-GOPS/W, delivering the highest CKKS bootstrapping throughput and 107× CPU speedup in application benchmarks.

Fully homomorphic encryption (FHE) is cryptographic technique that enables direct computation on encrypted data, preserving privacy throughout the process. Security in FHE relies on adding noise to ciphertexts [9], but this noise grows with each operation, eventually limiting the number of computations that can be performed. To enable deeper computations, a process called bootstrapping (BTS) is used to refresh the ciphertext and reset its noise level [7]. Beyond depth, another key challenge is functionality, since different FHE schemes provide complementary strengths. For example, the CKKS scheme [10] is suitable for basic vector computations because it processes data in a SIMD manner, but it struggles with nonlinear functions, which need to be approximated by high-degree polynomials. In contrast, TFHE scheme [11] excels at nonlinear tasks by evaluating look-up tables (LUTs) on ciphertexts. To combine the advantages of both schemes, the scheme-switching (SS) technique has been introduced, enabling runtime conversion between different FHE schemes [6].

Both BTS and SS are essential in real-world applications. BTS enables long chains of operations, while SS expands the range of applications by supporting both vector processing and nonlinear functions (Fig. 25.4.1). Previous accelerators have primarily focused on a single scheme and targeted BTS operations [1-4, 12]. However, in multi-scheme scenarios, SS can introduce overhead comparable to or even greater than that of BTS, due to the sequential iterations required in blind rotation (BR) and bandwidth-intensive ring-packing (RP). This motivates our work, OmniCrypt, which to the best of our knowledge is the first silicon-proven accelerator that supports both BTS and SS in multi-scheme FHE setting.

To this end, we face two key challenges. The first is the area overhead, which comes from both the large memory footprint and the heavy computation. The memory footprint grows with ciphertext and key sizes at higher security levels, while the computation is further compounded by the diverse primitives required across multiple FHE schemes, each incurring significant area cost due to modular reduction. The second is the high latency caused by inherently sequential operations and bandwidth bottlenecks, particularly in BR, which involves thousands of iterations with expensive external products, and in RP, which demands significant bandwidth to combine two RLWE ciphertexts using homomorphic rotation (HRot).

To address the first challenge, OmniCrypt integrates several optimizations. 1) The general bootstrapping (G-BTS) [5] algorithm reduces memory capacity by enabling smaller polynomial degrees and bitwidths, cutting footprint by 73%. 2) On-chip data generation (ODG) produces twiddle factors, binomials, and keys on the fly, further reducing memory by 59%. 3) The shift-and-add Montgomery modular multiplier (SAM3), optimized with hardware-friendly primes, lowers compute area by 67%, while 4) reconfigurable function engines enable multiple tasks to share hardware, saving an additional 32%.

To address the second challenge, OmniCrypt applies 1) LWE key switching (L-KS) [8] before BR, shortening LWE ciphertexts and reducing BR latency by 93%. 2) Streamlined compute dataflow enables fully pipelined execution, achieving another 58% latency reduction. Finally, two new scheduling techniques further improve efficiency: 3) Extend-Limb-Decomp-Depth (ELDD), which overlaps the ModDown process to cut latency by 44%, and 4) BR-RP interleaving (BR2P), which enhances key and ciphertext reuse while reducing external memory accesses by half, lowering latency by an additional 50%. Putting all features together, OmniCrypt reduces total memory footprint by 89%, compute area by 78%, and overall latency by 89%.

Figure 25.4.2 shows the overall architecture of OmniCrypt, which consists of a DMA, an ADPLL, a top controller with scheduler, and three main compute engines. The first is the Gadget Product Engine (GPE), which integrates the twiddle generator (TGU), binomial generator (BGU), and evaluation key generator (KGU), along with two external product cores (EPCs). The second engine is the Reconfigurable Butterfly Engine (RBE), which performs polynomial transforms using four butterfly clusters that can be configured as either an INTT or an automorphism engine. The third is the Modular Vector Processor (MVP), which provides vectorized modular arithmetic through a polynomial rotator and eight modular lanes acting as a general-purpose modular processor. Figure 25.4.2 also illustrates the architectural mapping of the main workload, the external product (EP), which supports two operating modes: (1) a low-latency (LL) mode that splits the decomposition polynomial across two EPCs, and (2) a double-throughput (DT) mode that runs two EPs in parallel. These modes play a crucial role in enabling efficient BR2P task scheduling.

Figure 25.4.3 shows the GPE, which performs ModUp and polynomial multiply–accumulate (PMAC) operations for EP and L-KS. It contains two EPCs, where each EPC integrates a pipelined NTT (P-NTT), ModLogics, and two PMAC units. The EPC supports three datapaths by sharing the sub-blocks, as illustrated in the figure. The first two are the LL and DT modes introduced in Fig. 25.4.2, while the third datapath enables L-KS. Since the number of BR iterations equals the LWE ciphertext length, applying L-KS shortens the ciphertext and significantly reduces the iteration count. Although this introduces additional vector–matrix computations using ModLogics to generate the new ciphertext, the overhead is hidden by distributing the workload across BR iterations. Specifically, at iteration t, the LWE element for $t+1$ is computed concurrently during the NTT transform stage, when the ModLogics units would otherwise be idle. While L-KS overhead is effectively hidden, overall performance is still dominated by the NTT, the most compute-intensive operation. We implement a pipelined NTT (P-NTT) based on the FFT-MDC architecture [13]. The design has 13 stages and supports polynomials of degree up to $N=2^{13}-1$ in a single pass. While P-NTT provides high throughput, the on-the-fly data shuffling at every stage increases area overhead. To mitigate this, we introduce the area-optimized modular multiplier SAM3. By selecting hardware-friendly primes expressed as powers of two terms, SAM3 reduces multiplications from three to one, cutting Montgomery multiplier area by 67% and enabling P-NTT to achieve the highest area efficiency compared with prior works.

Figure 25.4.4 shows the on-chip data generators, which reduce memory footprint and provide essential auxiliary data for BTS and SS. These include the TGU, KGU, and BGU. The TGU is used in both the P-NTT (in GPE) and the INTT (in RBE). By exploiting data periodicity in modular arithmetic, TGU stores only a compact twiddle-factor seed at each stage, reducing twiddle memory by over 99%. The KGU adopts the Trivium CSPRNG [14] to ensure data security. It generates BTS, R-KS, and L-KS keys. For BTS and R-KS, it reduces memory storage by 50%, while for L-KS it generates keys fully on-chip, eliminating storage entirely. The BGU produces binomial values needed in the key preparation phase of each BR iteration. Using the binomial index derived from L-KS results, the BGU generates binomials on demand and reduces binomial storage by over 99%. Once the main EP or R-KS computation in the GPE finishes, data is transferred to the RBE to begin the ModDown process starting with INTT. The RBE consists of four butterfly clusters, each with 16 lanes, that can be configured as either an INTT or an automorphism engine. To eliminate complex data shuffling in both INTT and automorphism operations, we adopt a constant-geometry butterfly [15] combined with a ping–pong buffer that switches input and output memories at every stage. When configured as an automorphism engine, the RBE functions as a butterfly network capable of arbitrary data permutation. This reconfigurability allows hardware sharing between INTT and automorphism, saving area and achieving the highest area efficiency compared with prior works.

Figure 25.4.5 shows the detailed architecture of the MVP. The MVP is a general-purpose vector processor for modular arithmetic, with each lane equipped with a specialized pipeline to perform quick base conversion (BConv) given the cryptographic parameters use a single extended prime (single-P). Selecting single-P parameters offers both advantages and drawbacks: it simplifies BConv into outer-product operations and increases ciphertext levels, but also enlarges key sizes and raises the computation cost for EP and R-KS. Fortunately, given OmniCrypt features on-chip data generation and streamlined processing, these drawbacks can be effectively mitigated. The MVP also includes a rotator (monomial multiplier) optimized for bit-reverse–order polynomials. Bit-reverse rotation ensures that outputs remain within the same memory row, requiring only intra-row data permutations. Leveraging this property reduces the latency of polynomial rotation by 66%. Finally, workload execution is coordinated by OmniCrypt's scheduler, which combines Extend-Limb Decomp-Depth (ELDD) scheduling with BR–RP (BR2P) task interleaving. ELDD is an intra-task strategy that executes extended limbs first, then proceeds to the next decomposition polynomial within the same ring. This ordering hides ModDown latency and reduces overall runtime by 44%. At the inter-task level, BR2P scheduling maps two BR tasks to run in parallel on the GPC using DT mode. Although each BR executes more slowly, their operations are interleaved with RP in LL mode, which directly merges the outputs. This parallelism enables BTS keys to be shared across two BR tasks, reducing EMA pressure by 50% and cutting overall latency by half.

Figure 25.4,7 shows the OmniCrypt chip micrograph, fabricated in 28nm technology with a core area of 12.96mm². It operates at 100–400MHz with a supply voltage of 0.7–1.1V and supports CKKS, TFHE, and other representative algorithms, as listed in the figure. The measurement results of OmniCrypt and comparisons with prior works are presented in Figure 25.4.6. OmniCrypt achieves a peak throughput of 252.80Modular-GOPS (M-GOPS), with an area efficiency of 19.51M-GOPS/mm² and a power efficiency of 435.86M-GOPS/W. Compared with prior accelerators, it delivers the highest bootstrapping throughput for CKKS and to the best of our knowledge is the first fabricated chip to support TFHE bootstrapping. The accelerator also meets the baseline requirement for constant execution time, with runtime determined only by public parameters and independent of secret inputs. For G-BTS and scheme switching, it outperforms a server-class CPU [16] by 88× to 108×, and in hybrid application benchmarks achieves a speedup of 75× to 107× over CPU execution. As far as we are aware, these results establish OmniCrypt as the first silicon-proven accelerator to support multi-scheme bootstrapping and scheme switching, setting a new benchmark for area-efficient and high-performance FHE hardware.

Acknowledgement:
This work was supported by the Ministry of Science and ICT (MSIT), South Korea, under the Institute of Information & Communications Technology Planning & Evaluation (IITP) grants (No. 2022-0-01037, Development of High Performance Processing-in-Memory Technology based on DRAM and IITP-2025-RS-2020-II201847, Development of Semiconductor/Systems-Integrated Technologies for Remote/AI based Society) and the Graduate School of AI Semiconductor program (IITP-2025-RS-2023-00256472). Chip fabrication and EDA tool support were provided by IDEC.

References:
[1] S. Lu et al., "A 28nm 4.05μJ/Encryption 8.72kHMul/s Reconfigurable Multi-Scheme Fully Homomorphic Encryption Processor for Encrypted Client-Server Computing," *ISSCC*, pp. 294-295, Feb. 2025. http://doi.org/10.1109/ISSCC49661.2025.10904812
[2] L.-H. Lin, Y.-K. Yang, and C.-H. Yang, "A 30.4GOPS/mW Mk-CKKS Processor for Secure Multi-Party Computation," *ISSCC*, pp. 296-297, Feb. 2025. http://doi.org/10.1109/ISSCC49661.2025.10904776
[3] H. Lee, H. Kwon, and Y. Lee, "A 2.7-to-13.3μJ/boot/slot Flexible RNS-CKKS Processor in 28nm CMOS Technology for FHE-Based Privacy-Preserving Computing," *ISSCC*, pp. 296-297, Feb. 2024. http://doi.org/10.1109/ISSCC49657.2024.10454420
[4] G. Shi et al., "A 28nm 68MOPS 0.18μJ/Op Paillier Homomorphic Encryption Processor with Bit-Serial Sparse Ciphertext Computing," *ISSCC*, pp. 242-243, Feb. 2023. http://doi.org/10.1109/ISSCC42615.2023.10067522
[5] A. Kim et al., "General Bootstrapping Approach for RLWE-Based Homomorphic Encryption," *IEEE Transactions on Computers*, vol. 73, no. 1, pp. 86-96, Jan. 2024. http://doi.org/10.1109/TC.2023.3318405
[6] W. Lu, Z. Huang, C. Hong, Y. Ma, and H. Qu, "Pegasus: Bridging Polynomial and Non-Polynomial Evaluations in Homomorphic Encryption," *IEEE Symposium on Security and Privacy (SP)*, pp. 1057-1073, May 2021. http://doi.org/10.1109/SP40001.2021.00043

[7] K. Han and D. Ki, "Better Bootstrapping for Approximate Homomorphic Encryption," *Topics in Cryptology – CT-RSA 2020*, pp. 364-390, Feb. 2020. https://doi.org/10.1007/978-3-030-40186-3_16
[8] H. Chen, W. Dai, M. Kim, and Y. Song, "Efficient Homomorphic Conversion Between (Ring) LWE Ciphertexts," *Applied Cryptography and Network Security*, pp. 460-479, June 2021. https://doi.org/10.1007/978-3-030-78372-3_18
[9] C. Gentry, "Fully Homomorphic Encryption Using Ideal Lattices," *ACM Symposium on Theory of Computing*, pp. 169-178, May 2009. https://doi.org/10.1145/1536414.153644
[10] J.H. Cheon, A. Kim, M. Kim, and Y. Song, "Homomorphic Encryption for Arithmetic of Approximate Numbers," Advances in Cryptology, *Lecture Notes in Computer Science*, vol. 10624, pp. 409-437, Nov. 2017. http://doi.org/10.1007/978-3-319-70694-8_15
[11] I. Chillotti, N. Gama, M. Georgieva, and M. Izabachène, "TFHE: Fast Fully Homomorphic Encryption Over the Torus," *Journal of Cryptology*, vol. 33, no. 1, pp. 34-91, Apr. 2019. http://doi.org/10.1007/s00145-019-09319-x
[12] A. Putra, Prasetiyo, Y. Chen, J. Kim, and J.-Y. Kim, "Strix: An End-to-End Streaming Architecture with Two-Level Ciphertext Batching for Fully Homomorphic Encryption with Programmable Bootstrapping," *IEEE/ACM International Symposium on Microarchitecture*, pp. 1319-1331, Oct. 2023. http://doi.org/10.1145/3613424.3614264
[13] M. Garrido, "A Survey on Pipelined FFT Hardware Architectures," *Journal of Signal Processing Systems*, vol. 94, no. 11, pp. 1345-1364, Jul. 2021. http://doi.org/10.1007/s11265-021-01655-1
[14] C. De Cannière and B. Preneel, "Trivium," *New Stream Cipher Designs, Lecture Notes in Computer Science*, vol. 4986, pp. 244-266, 2008. http://doi.org/10.1007/978-3-540-68351-3_18
[15] G. Miel, "Constant Geometry Fast Fourier Transforms on Array Processors," *IEEE Transactions on Computers*, vol. 42, no. 3, pp. 371-375, Mar. 1993. http://doi.org/10.1109/12.210180
[16] Intel Corp., "Intel® Xeon® Platinum 8452Y Processor (67.5m cache, 2.00 GHz) - Product Specifications," Intel, 2023. Accessed Dec. 8, 2025. <https://www.intel.com/content/www/us/en/products/sku/231761/intel-xeon-platinum-8452y-processor-67-5m-cache-2-00-ghz/specifications.html>
[17] Tuneinsight, "Tuneinsight/lattigo: A Library for Lattice-Based Multiparty Homomorphic Encryption in Go," GitHub, Mar. 2025. Accessed on Dec. 8, 2025. <https://github.com/tuneinsight/lattigo>
[18] J.-P. Bossuat et al., "Security guidelines for implementing Homomorphic encryption," IACR Communications in Cryptology, vol. 1, no. 4, Jan. 2025. doi:10.62056/anxra69p1

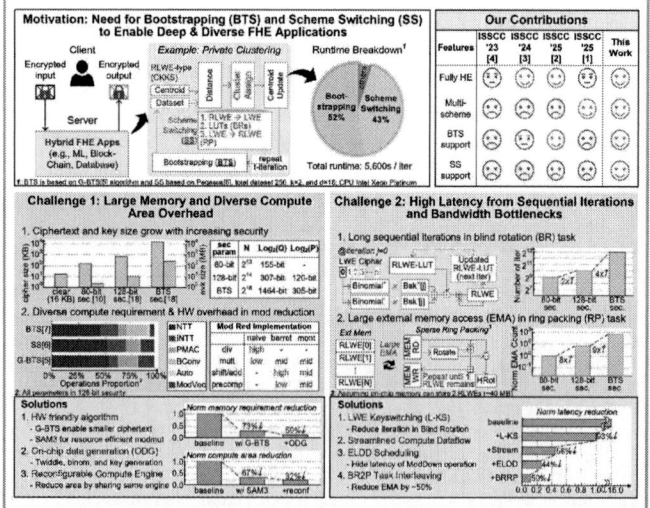

Figure 25.4.1: FHE accelerator design motivation & challenges for bootstrapping (BTS) and scheme switching (SS).

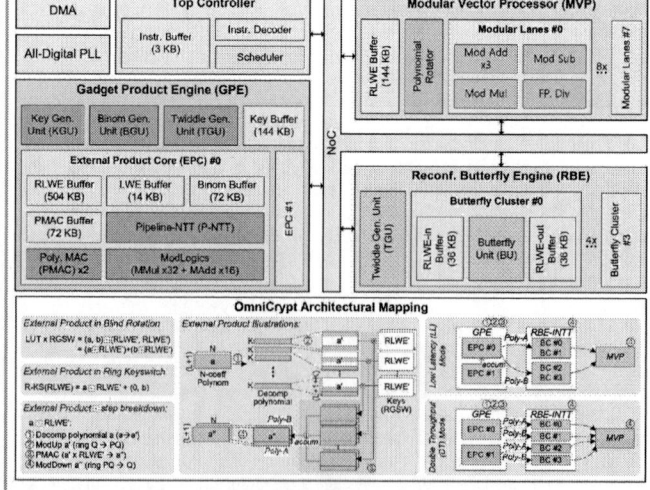

Figure 25.4.2: Overview of OmniCrypt architecture and its mapping for external product (EP).

ISSCC 2026 / SESSION 25 / HARDWARE SECURITY / 25.4

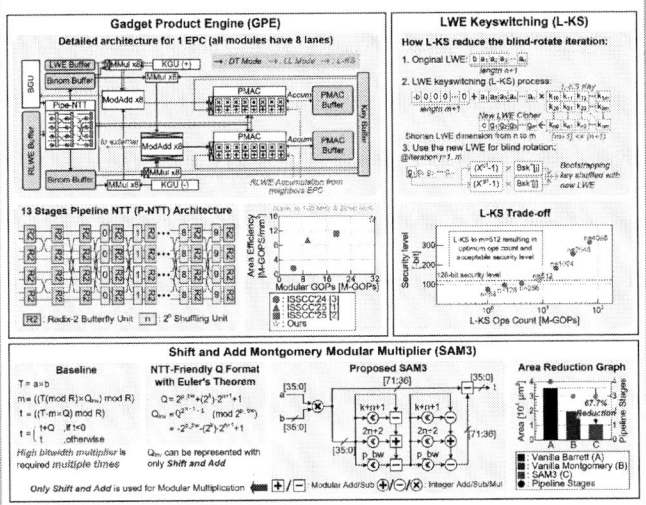

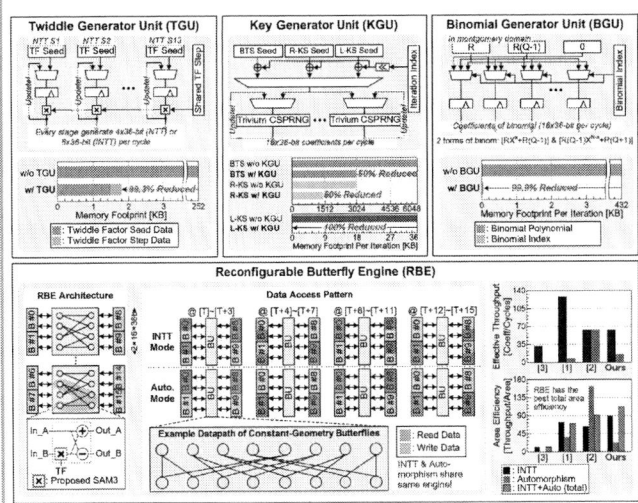

Figure 25.4.3: Gadget Product Engine (GPE) for EP and L-KS with area-efficient SAM3 modular multiplier.

Figure 25.4.4: On-chip data generators (TGU, KGU, BGU) and RBE for reconfigurable INTT/automorphism.

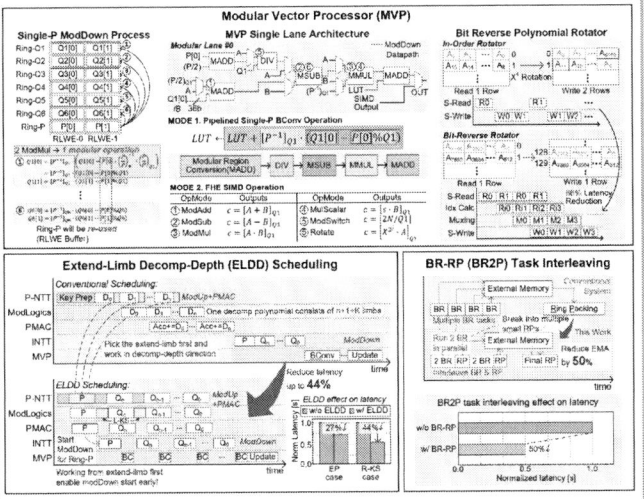

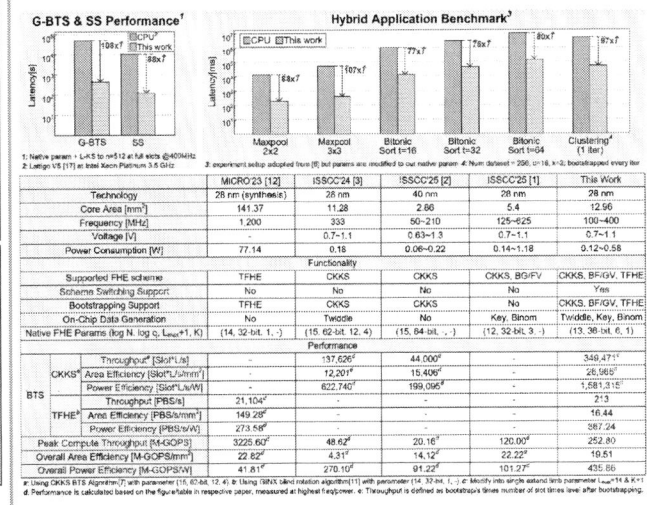

Figure 25.4.5: MVP for BConv and modular arithmetic with ELDD & BR2P scheduling optimization.

Figure 25.4.6: Performance comparison with prior works.

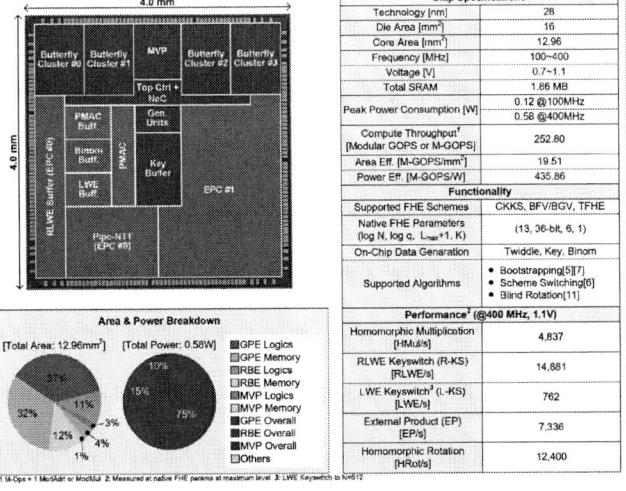

Figure 25.4.7: Die micrograph, chip specification, and overall area/power breakdown.

• 2026 IEEE International Solid-State Circuits Conference

ISSCC 2026 / SESSION 25 / HARDWARE SECURITY / 25.5

25.5 A 0.05mm² 1.19-to-7.34mW SQIsign-1D Isogeny-Based Post-Quantum Signature Verification Accelerator for IoT

Krishna Sai Tarun Ramapragada, Utsav Banerjee

Indian Institute of Science, Bengaluru, India

Abstract

This work presents a 0.05mm² and 1.19-to-7.34mW hardware accelerator for the isogeny-based NIST post-quantum digital signature candidate SQIsign. Our Montgomery multiplier design enables 2× speedup and energy savings, and efficient memory organization provides 31% area savings. We demonstrate SQIsign-1D uncompressed signature verification in a small form factor for IoT applications with order of magnitude improvement in performance and energy-efficiency compared to assembly-optimized software.

Rapid advances in quantum computing technologies have motivated the theoretical construction and practical implementation of post-quantum cryptography (PQC) [1]. After several years of rigorous analysis, NIST has recently announced its first set of PQC standards [2]. The primary recommendations for digital signatures are Dilithium (ML-DSA) and Falcon (FN-DSA) [3]. Both are lattice-based schemes known for their strong security and computational efficiency, as evident from their state-of-the-art optimized software [4, 5], hardware-software co-design [6] and dedicated hardware [7, 8, 9, 10] implementations. However, their large memory footprint and communication overhead make them prohibitively expensive for resource-constrained embedded systems [11, 12]. In case of typical IoT applications such as secure boot, firmware update, certificate validation and server authentication, the expensive signature generation is done by a powerful server while the constrained client devices only need to perform signature verification. Therefore, IoT devices often demand verification-friendly signature algorithms with small public key and signature sizes. This has led to the development of SQIsign, an isogeny-based quantum-safe signature scheme [13]. To diversify the portfolio of PQC digital signature standards beyond lattice-based, NIST has recently started an additional phase of standardization [14]. SQIsign has been shortlisted by NIST as a Round 2 candidate and has emerged as a strong contender with the smallest public key and signature size. However, this comes at the cost of significantly increased computation cost and SQIsign is orders of magnitude more computationally expensive compared to lattice-based signature schemes [15]. To address this challenge, we present a compact and low-power hardware accelerator for SQIsign-1D-Uncompressed signature verification which implements algorithm-architecture co-optimizations to achieve order of magnitude speedup and energy savings compared to state-of-the-art assembly-optimized software [16].

Figure 25.5.1 compares post-quantum isogeny-based cryptography (IBC) with pre-quantum elliptic curve cryptography (ECC). While the security of ECC is based on finding secret maps between points on a curve, IBC relies on the hardness of finding secret maps between curves. In particular, security of SQIsign is based on the hardness of computing the endomorphism ring of a supersingular elliptic curve, which is considered quantum-safe [13]. Different variants of SQIsign are under active development, and SQIsign-1D-Uncompressed currently provides the fastest signature verification [15, 16]. At the same security level, its combined public key and signature size is 10× and 4× smaller compared to Dilithium and Falcon respectively. This makes SQIsign ideal for embedded devices requiring signature verification with low communication cost.

Figure 25.5.2 shows the architecture of our hardware accelerator along with a high-level overview of the required computations. Uncompressed SQIsign-1D signature verification primarily involves computing two isogenies, which form the most expensive component, along with kernel point sampling and j-invariant calculations. Montgomery curves defined over the finite field Fp^2 with the 251b prime $p=5×2^{248}-1$ are used at the NIST-I security level. Operations in Fp^2 are further decomposed into modular arithmetic in Fp, and software profiling indicates that Fp additions, subtractions and multiplications together account for ~87% of the computation cost. Therefore, our design consists of an efficient Fp/Fp^2 arithmetic unit supporting 28 different operations. A 2.75kB top-level memory is used for storing inputs, outputs and intermediate computation results. Curve functions and isogeny strategies are stored in lookup tables implemented as combinational logic for area savings. We implement the SHAKE256-based smart hash technique [15] for fast kernel point sampling using a compact serialized core for plane-per-plane processing [17] of the underlying Keccak permutation [18]. Aggressive clock gating is used throughout the design for power savings.

Figure 25.5.3 presents our optimized algorithm and architecture for Fp multiplication exploiting the special structure of the SQIsign prime, also referred to as p_{5248}. Montgomery multiplication is well suited for large prime moduli, and the coarsely integrated operand scanning (CIOS) algorithm is the most efficient [19]. Large CIOS word sizes are known to provide the best energy-efficiency and performance [20]. The baseline implementation, with the Montgomery constant $R=2^{256}$ and each of the zero-padded 256b inputs a and b decomposed into four 64b words a[0], a[1], a[2], a[3] and b[0], b[1], b[2], b[3] respectively, requires 48 cycles per modular multiplication. However, in case of p_{5248}, we observe that p'[0]=1 and words of the prime modulus are p[0]=p[1]=p[2]=$2^{64}-1$ and p[3]=$5×2^{56}-1$, thus significantly simplifying the algorithm. This allows us to complete the CIOS second inner loop in a single cycle and eliminates the need for register t[5]. As a result, the overall cycle count is reduced to only 20 cycles, and we achieve a 2.43× energy savings with similar area footprint. The core components of our CIOS design are a 64b×64b multiplier and a 128b+64b+64b adder, both utilizing carry save logic for critical path optimization.

Figure 25.5.4 shows the architecture of our Fp/Fp^2 arithmetic unit which consists of a small 8×256b register file representing a memory hierarchy for close integration with our optimized CIOS Montgomery modular multiplier. This register file, having one write port and two read ports to allow two-input arithmetic computations, is implemented entirely using flip-flops and multiplexers for faster operation. Single cycle modular addition and subtraction are realized using two cascaded 256b adder-subtractors [20]. Compared to state-of-the-art software [16], our Fp addition / subtraction and Fp multiplication achieve 2 orders and 1 order of magnitude energy savings respectively. The Fp / Fp^2 arithmetic unit works in tandem with the function LUT, strategy LUT, instruction decoder and the larger top-level memory. The curve function LUT encodes various 32b instructions to implement point arithmetic, isogeny, j-invariant and other essential computations using sequences of Fp^2 operations. This LUT is realized using logic gates to achieve 13× and 1.7× area savings compared to SRAM-based and ROM-based implementations respectively. Through intelligent sharing of scratchpad memory locations among intermediate computation results and by observing that the number of splitting points can be reduced from 14 to 9, we are also able to compress the top-level memory by ~31% compared to the data memory organization from [16].

Figure 25.5.5 shows the breakdown of area and average power consumption of our accelerator, its voltage scaling performance and comparison with state-of-the-art software. Clearly, the Fp/Fp^2 arithmetic unit is the most important component accounting for 48% of total area and 94% of total power, thus justifying our design optimizations. Based on experimental characterization of our test chip, the accelerator can operate across a wide range of supply voltages while providing power-performance trade-offs. At nominal 0.9V, it completes a signature verification in 28.33ms while consuming 208µJ. At scaled 0.6V, it takes 85ms to complete a signature verification and consumes 101µJ. Compared to state-of-the-art software implementation on an ARM Cortex-M4 [16], our hardware accelerator achieves 12.8× (4.3×) speedup and 18.8× (38.8×) energy savings at 0.9V (0.6V) and 120MHz (40MHz).

Figure 25.5.6 compares our accelerator with previous work on hardware implementations supporting the NIST PQC standard digital signature schemes Dilithium and Falcon at the same security level. While state-of-the-art lattice-based cryptography accelerators [7, 8, 9, 10] achieve very high performance and energy-efficiency, they also have significantly high area, power and memory overheads. In comparison, our design has at least an order of magnitude lower area (5 to 72×) and power (8 to 50×) while still achieving reasonable performance suitable for IoT applications requiring only signature verification. Compared to the resource-constrained hardware-software co-design of lattice cryptography from [6], our accelerator achieves 6× smaller area, requires 12× lower memory and has similar power consumption. Since SQIsign is a very recent post-quantum signature scheme, existing literature has not yet explored its hardware implementation. Compared to assembly-optimized software implementations of SQIsign-1D-Uncompressed signature verification [15, 16], we demonstrate an order of magnitude improvement in both performance and energy-efficiency.

Figure 25.5.7 shows the chip micrograph and summarizes its performance. The test chip was fabricated in a 28nm HPC+ CMOS process and supports supply voltage scaling from 0.9V down to 0.6V. Our SQIsign-1D-Uncompressed signature verification accelerator occupies only 0.05mm² comprising 118k logic gates (NAND2 gate equivalent) and 3.25kB of SRAM. It consumes an average power of 1.19mW and 7.34mW at 0.6V/40 MHz and 0.9V/120MHz, respectively, during signature verification, which requires ~3.4M clock cycles. Signature verification involves computations using only public information. Therefore, our implementation is inherently isochronous, that is, its execution time is independent of any secret data. All measured performance metrics for our hardware accelerator have been obtained through experimental characterization of our test chip at room temperature. The functionality of the accelerator has been validated against test vectors from [15, 16] using a custom serial interface in the test chip.

In summary, this work demonstrates hardware-accelerated quantum-safe SQIsign-1D-Uncompressed signature verification with low communication overhead, compact area footprint and low power consumption, thus making it affordable for resource-constrained embedded systems. This is achieved through algorithm-architecture co-optimization of modular arithmetic, efficient memory organization and careful circuit design of all core building blocks. Our accelerator has area and performance very similar to pre-quantum elliptic curve cryptography accelerators [20]. Our hardware architecture, especially its finite field arithmetic implementation, can also be easily extended to other emerging variants of isogeny-based digital signature schemes such as SQIsign-2D and SQIsign-HD [16], thus enriching the way towards PQC migration.

Acknowledgement:
This work was supported by the Prime Minister's Research Fellowship (PMRF), Ministry of Education, Government of India and by the POWERGRID Center of Excellence in Cyber Security (PGCoE), Indian Institute of Science, Bengaluru.

References:
[1] I. Verbauwhede et al., "Circuit Challenges from Cryptography," *ISSCC*, pp. 428-429, Feb. 2015. https://doi.org/10.1109/ISSCC.2015.7063109
[2] L. Chen et al., "Report on Post-Quantum Cryptography," *NIST IR 8105*, April 2016. https://doi.org/10.6028/NIST.IR.8105
[3] G. Alagic et al., "Status Report on the Third Round of the NIST Post-Quantum Cryptography Standardization Process," *NIST IR 8413*, July 2022. https://doi.org/10.6028/NIST.IR.8413-upd1
[4] J. Zheng et al., "ESPM-D: Efficient Sparse Polynomial Multiplication for Dilithium on ARM Cortex-M4 and Apple M2," *arXiv preprint*, pp. 1-19, Apr. 2024. https://doi.org/10.48550/arXiv.2404.12675
[5] T. Pornin, "Falcon on ARM Cortex-M4: An Update," *IACR Cryptology ePrint Archive*, Paper 2025/123, pp. 1-18, Jan. 2025. Accessed on Dec. 8, 2025. <https://eprint.iacr.org/2025/123>
[6] U. Banerjee et al., "Sapphire: A Configurable Crypto-Processor for Post-Quantum Lattice-based Protocols," *IACR TCHES*, vol. 2019, no. 4, pp. 17-61, Aug. 2019. https://doi.org/10.13154/tches.v2019.i4.17-61
[7] A. Aikata et al., "KaLi: A Crystal for Post-Quantum Security Using Kyber and Dilithium," *IEEE TCAS-I*, vol. 70, no. 2, pp. 747-758, Feb. 2023. https://doi.org/10.1109/TCSI.2022.3219555
[8] Y. Zhu et al., "RePQC: A 3.4-uJ/Op 48-kOPS Post-Quantum Crypto-Processor for Multiple-Mathematical Problems," *IEEE JSSC*, vol. 58, no. 1, pp. 124-140, Jan. 2023. https://doi.org/10.1109/JSSC.2022.3216758
[9] Y. Ouyang et al., "FalconSign: An Efficient and High-Throughput Hardware Architecture for Falcon Signature Generation," *IACR TCHES*, vol. 2025, no. 1, pp. 203-226, 2024. https://doi.org/10.46586/tches.v2025.i1.203-226
[10] Y. Zhu et al., "PQPU: A 4.4-µJ/Op 69.4-kOPS Agile Post-Quantum Crypto-Processor Across Multiple Mathematical Problems," *IEEE JSSC*, vol. 60, no. 6, pp. 2261-2275, Jun. 2025. https://doi.org/10.1109/JSSC.2024.3476949

[11] M.J. Kannwischer et al., "pqm4: Testing and Benchmarking NIST PQC on ARM Cortex-M4," *IACR Cryptology ePrint Archive*, Paper 2019/844, pp. 1-22, July 2019. Accessed on Dec. 8, 2025. <https://eprint.iacr.org/2019/844>
[12] M.-J.O. Saarinen, "Mobile Energy Requirements of the Upcoming NIST Post-Quantum Cryptography Standards," *IEEE International Conference on Mobile Cloud Computing, Services, and Engineering* (*MobileCloud*), pp. 23-30, Aug. 2020. https://doi.org/10.1109/MobileCloud48802.2020.00012
[13] L. De Feo et al., "SQISign: Compact Post-Quantum Signatures from Quaternions and Isogenies," *IACR ASIACRYPT*, pp. 64-93, Dec. 2020. https://doi.org/10.1007/978-3-030-64837-4_3
[14] G. Alagic et al., "Status Report on the First Round of the Additional Digital Signature Schemes for the NIST Post-Quantum Cryptography Standardization Process," *NIST IR 8528*, Oct. 2024. https://doi.org/10.6028/NIST.IR.8528
[15] M. A. Aardal et al., "Optimized One-Dimensional SQIsign Verification on Intel and Cortex-M4," *IACR TCHES*, vol. 2025, no. 1, pp. 497-522, Dec. 2025. https://doi.org/10.46586/tches.v2025.i1.497-522
[16] F. C. Rodrigues et al., "Generation of Fast Finite Field Arithmetic for Cortex-M4 with ECDH and SQIsign Applications," *IACR TCHES*, vol. 2025, no. 4, pp. 588-620, Sep. 2025. https://doi.org/10.46586/tches.v2025.i4.588-620
[17] A. Dolmeta et al., "Comparative Study of Keccak SHA-3 Implementations," *Cryptography*, vol. 7, no. 4, pp. 1-16, Nov. 2023. https://doi.org/10.3390/cryptography7040060
[18] NIST, "SHA-3 Standard: Permutation-Based Hash and Extendable-Output Functions," *NIST FIPS 202*, Aug. 2015. https://doi.org/10.6028/NIST.FIPS.202
[19] C.K. Koc et al., "Analyzing and Comparing Montgomery Multiplication Algorithms," *IEEE Micro*, vol. 16, no. 3, pp. 26-33, Jun. 1996. https://doi.org/10.1109/40.502403
[20] U. Banerjee et al., "A Low-Power BLS12-381 Pairing Cryptoprocessor for Internet-of-Things Security Applications," *IEEE SSCL*, vol. 4, pp. 190-193, Oct. 2021. https://doi.org/10.1109/LSSC.2021.3124074

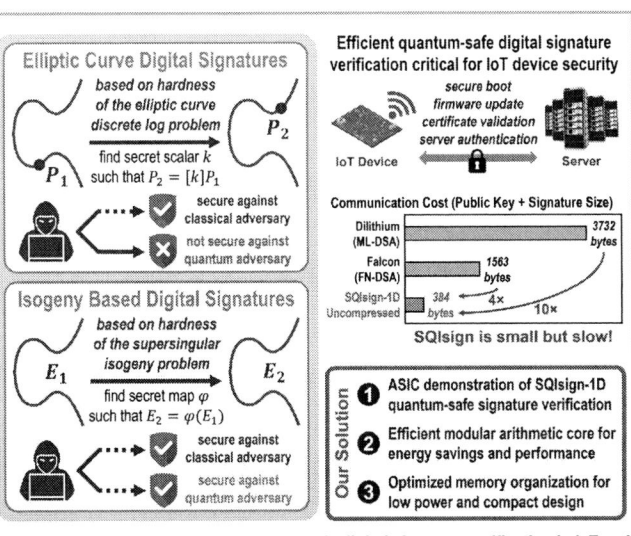

Figure 25.5.1: Need for efficient quantum-safe digital signature verification in IoT and our solution for SQIsign.

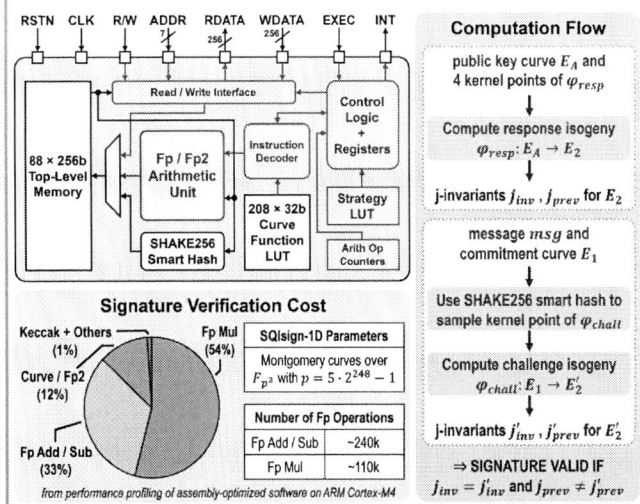

Figure 25.5.2: Our accelerator for SQIsign-1D uncompressed signature verification along with computation details.

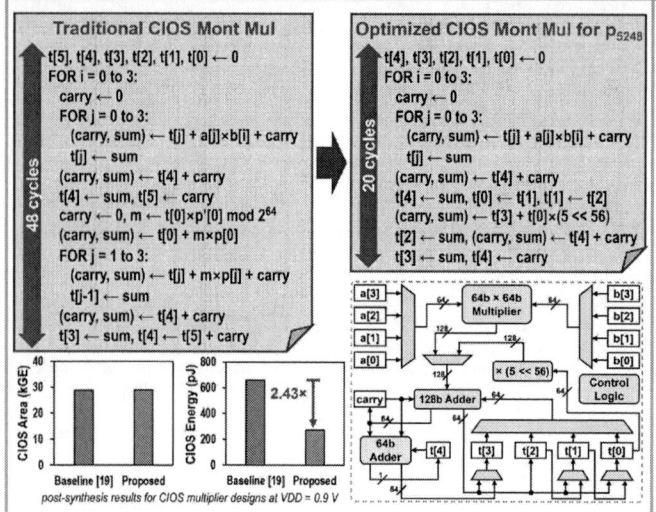

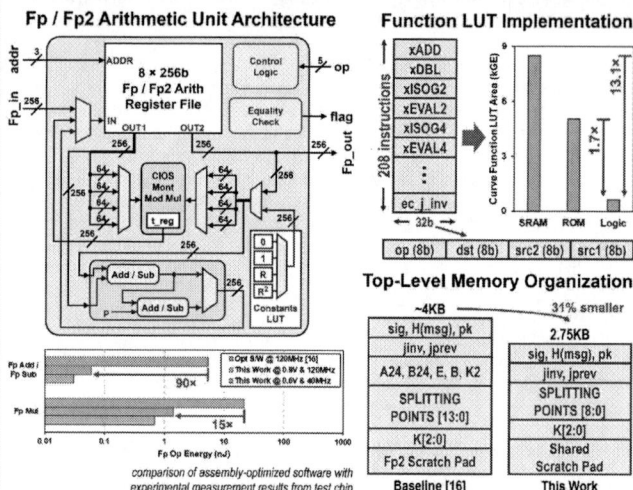

Figure 25.5.3: Optimized CIOS Montgomery modular multiplier for SQIsign-1D and comparison with traditional CIOS.

Figure 25.5.4: Architecture of Fp / Fp2 arithmetic unit, curve function LUT and top-level memory organization.

Figure 25.5.5: Accelerator area and power distribution, voltage scaling and comparison with optimized software.

Figure 25.5.6: Comparison with state-of-the-art SQIsign-1D software and lattice-based PQC signature hardware.

	TCHES 2019 [6]	TCAS-I 2023 [7]	JSSC 2023 [8]	TCHES 2024 [9]	JSSC 2025 [10]	TCHES 2025 [15]	TCHES 2025 [16]	This Work	
Design	RISC-V+H/W	H/W Accel	H/W Accel	H/W Accel	H/W Accel	Cortex-M4 S/W	Cortex-M4 S/W	H/W Accel	
Platform	ASIC [a]	ASIC [a, b]	ASIC [a]	ASIC [b]	ASIC [a]	MCU	MCU	ASIC	
Technology	40nm	28nm	28nm	28nm	28nm	.	.	28nm	
VDD (V)	1.1	0.9	0.9	0.9	1.1	1.2	1.2	0.6	0.9
Freq (MHz)	72	1000 / 2000 [c]	500	530	450 / 550	120	120	40	120
Total Area	0.28 mm² [d]	0.263 mm²	3.6 mm²	0.71 mm²	3.2 mm²	.	.	0.05 mm²	
Logic (GE)	106k [d]	747k	1900k	759k	2100k	.	.	118k	
Memory (KB)	40.25 [d]	34.82	484	136	228.5	67.69 [e]	68.56 [f]	3.25	
Power (mW)	7.49	367	237	62	282 / 392	10.8 [g]	10.8 [g]	1.19	7.34
PQC Sign Primitive	Lattice-Based	Lattice-Based	Lattice-Based	Lattice-Based	Lattice-Based	Isogeny-Based	Isogeny-Based	Isogeny-Based	
PQC Sign Scheme	Dilithium-II	Dilithium-II	Dilithium-II	Falcon-512	Dilithium-II / Falcon-512	SQIsign-1D Uncompressed	SQIsign-1D Uncompressed	SQIsign-1D Uncompressed	
Pub Key + Sig Size	3732 bytes	3732 bytes	3732 bytes	1563 bytes	3732 bytes / 1563 bytes	384 bytes	384 bytes	384 bytes	
Sign / Verify	Sign + Verify	Sign + Verify	Sign + Verify	Sign Only	Sign + Verify	Verify Only	Verify Only	Verify Only	
SQIsign-1D Uncompressed Isogeny-Based Digital Signature Verification Performance									
Clock Cycles	-	-	-	-	-	82.3M	43.6M	3.4M	
Latency (ms)	-	-	-	-	-	686	363	85	28.33
Energy (µJ)	-	-	-	-	-	7408.8	3920.4	101	208

[a] hardware support for multiple PQC algorithms [b] post-synthesis results reported [c] separate clocks used for logic and memory
[d] only area of cryptographic hardware accelerator reported (excluding RISC-V) [e], [f] total memory required for code and data
[g] average power obtained from datasheet of STM32L4R5xx (MCU in NUCLEO-L4R5ZI dev board) with external SMPS

Chip Specifications

Technology	28nm HPC+ CMOS 1P10M
Supply Voltage	0.6V – 0.9V
Package	24-pin QFN
Die Size	1mm × 1mm

Cryptographic Accelerator Core

Total Area	0.05 mm² (including both logic gates and memory)
Logic Gates	118k (NAND2 equivalent)
Memory	3.25KB (SRAM)
Maximum Frequency	40MHz at 0.6V 120MHz at 0.9V
Post-Quantum Crypto Scheme	SQIsign-1D-Uncompressed Digital Signature Verification
Base Field	F_{p^2} with $p = 5 \cdot 2^{248} - 1$
Hash Function	SHAKE256
Verify Latency	~3.4M clock cycles
Average Power	1.19mW at 0.6V & 40MHz 7.34mW at 0.9V & 120MHz

Important submodules of SQIsign-1D-Verify hardware cryptographic core shown in red to indicate their relative positions on test chip (not to scale)

SPI-like custom serial interface used for testing and functional validation of hardware accelerator core

Figure 25.5.7: Chip micrograph and performance summary.

ISSCC 2026 / SESSION 25 / HARDWARE SECURITY / 25.6

25.6 A 17%/27% Area-/Energy-Overhead Glitch-Transition Secure SHA-3 Engine Fusing Dual-Rail Precharge Logic and Asymmetric Masking

Cankun Zhao[1,2], Hanyue Shui[1,2], Bohan Yang[1,2], Wenping Zhu[1,2], Yuluan Cao[1,2], Zhang Hou[1,2], Yuqi Liu[1,2], Xiangdong Han[1,2], Shuying Yin[1,2], Weinan Chen[1,2], Hanning Wang[1,2], Jinjiang Yang[3], Min Zhu[4], Aoyang Zhang[1,2], Leibo Liu[1,2]

[1]School of Integrated Circuits, Tsinghua University, Beijing, China, [2]State Key Laboratory of Cryptography and Digital Economy Security, Tsinghua University, Beijing, China
[3]Wuxi Research Institute of Applied Technologies, Tsinghua University, Beijing, China, [4]Micro Innovation Integrated Circuit Design, Wuxi, China

Abstract

This 0.046mm^2 SHA-3 engine demonstrates overheads of only 17% area, 27% energy, and 0% latency while achieving provable glitch- and transition-robust security (power/EM TVLA MTD>100M). Glitch-free DPL reduces masked states by 3×, resulting in a 72,950μm^2 register area reduction and lower logic energy. LMDPL refresh preserves secure asymmetric-share mapping. Masked transitions enable partial replacement of DPL with SRL, saving 50,852μm^2 in registers, precharge gates, and multiplexers.

Cryptographic hash functions are fundamental components in modern security systems. Among them, Secure Hash Algorithm 3 (SHA-3) is the latest NIST standard [1], supporting diverse applications such as post-quantum cryptography (PQC) [2,3] and authentication [4,5]. In these settings, SHA-3 routinely processes sensitive data [6] and thus becomes a prime target for side-channel attacks (SCAs) [6-9]. A range of countermeasures [10-18] have been explored against such threats; among them, masking is indispensable for achieving provable security under well-defined leakage models [10]. However, conventional masked SHA-3 engines [11-15] incur prohibitive overheads—over 2.5× in area and 5.9× in energy—making them impractical for resource-constrained platforms such as IoT devices, which are both SCA-prone and limited by area and energy budgets. This work presents a SHA-3 engine with dual-rail precharge asymmetric masking (DPAM) that combines asymmetric masking (AM) with dual-rail precharge logic (DPL) to retain provable security at only 17% area and 27% energy overhead. Fabricated in 65nm CMOS and evaluated using both power and electromagnetic (EM) test-vector leakage assessments (TVLAs) [26,27], the design shows no detectable leakage over 100 million traces.

As shown in Fig. 25.6.1, unprotected SHA-3 modules already occupy 47% of the datapath area [2] and 38% of the datapath power [3] in PQC processors, which may rise by over 2.5× in area [11-15] and 5.9× in energy [11] when protected with conventional masking, such as threshold implementation (TI) [11,12]. Attempts to reduce area overhead are fundamentally constrained by the register footprint required to store three shares of SHA-3's 1600b state array under TI, which alone accounts for 94% of the masked core area [12]. Similarly, energy overhead is difficult to mitigate, as TI increases computational complexity by more than 3× [12]. This work addresses these challenges using AM, which splits the 1600b state into a 1600b main share and a 10b round share (Fig. 25.6.2), sharply reducing both the register count and computational complexity. Although AM has been explored in software [20,21], its adoption in hardware remains challenging [20] due to glitch- and transition-induced leakage [22]. To overcome this, we implement glitch-free hardware AM using DPL, and further introduce an AM share refresh scheme based on LUT-masked dual-rail with precharge logic (LMDPL) [23,24] to securely manage the round share in AM [20]. Additionally, a hybrid single-rail/dual-rail (SR/DR) masked logic style is introduced to mitigate the area and energy overheads typically associated with DPL [23,24].

Figure 25.6.2 illustrates the system architecture, the AM scheme, and the round-level timing behavior. The 1600b secret state is divided into a 1600b main share and a 10b round share, which is refreshed each round via the share updater (SU). The round share is split into two 5b parts that protect odd and even state slices, respectively; each bit corresponds to a row of slices, as color-coded in Fig. 25.6.2. The main share slices are processed sequentially through SHA-3's χ, ι, π, and θ operations using DPL to prevent glitch propagation in multi-stage logic. Every 64 cycles, the full 1600b main share is updated in parallel through the ρ operation. SRL is adopted for the ρ operation and multiplexers, yielding significant area and energy savings compared to full DPL.

Figure 25.6.3 illustrates how DPL addresses two key challenges in implementing AM in hardware. In AM, multiple main-share bits may share a single round-share bit—marked with the same color in Fig. 25.6.3—which becomes unsafe due to glitch propagation: switching activity in combinational logic can leak information from earlier-stage inputs [22]. This vulnerability is exemplified by the θ operation in Fig. 25.6.3 and such leakage has been observed in a previously published masked SHA-3 design lacking glitch-robust security [16]. To address this, we adopt glitch-free DPL [23], which confines each gate's side-channel signature to its immediate inputs. This eliminates the need to track multi-stage dependencies and enables a more compact AM design, achieving probing security under the glitch-robust model [25] by protecting each gate's inputs with independent randomness. As a result, the round share size is reduced from 1600b in conventional schemes to just 10b. The second challenge is maintaining the structured mapping between round and main shares across SHA-3 rounds. After one round of diffusion operations, the original one-bit-per-row mapping becomes disordered, potentially compromising security in the next round. To restore this structure, we introduce an LMDPL-based refresh mechanism that re-masks the main share output during the χ operation using a new 10b round share with the same pattern. This enables secure multi-round AM execution with minimal randomness overhead. Together, these techniques reduce register usage by 2.9× and randomness consumption by 14.4× compared to TI [12] while retaining first-order glitch-robust security.

Figure 25.6.4 illustrates the hybrid SR/DR architecture that reduces the area cost of DPL [23,24]. In conventional DPL, registers require inserting a precharge state between valid transitions to eliminate Hamming distance leakage, leading to two-stage pipelined registers [23,24]. In our design, the AM scheme and datapath are co-designed so that every transition is protected by independent randomness, allowing single-stage single-rail registers. The 1600b main share input is selected by a single-stage multiplexer (MUX) between a shift and a ρ operation. During the shift, odd and even slices of the main share are masked with different round-share bits, while in ρ, different rows are masked with different round-share bits. The y-axis rotation ensures that pre- and post-transition values are from different rows. Together, the AM mapping scheme and y-axis rotation provide transition-robust security [22] for registers and glitch-robust security [22] for MUX logic, eliminating the need for precharging MUX inputs and enabling single-rail implementation. The hybrid SR/DR design saves 7.7 kGE in registers, 3.9 kGE in precharge gates, and 3.6 kGE in MUXes, thereby improving area and energy efficiency.

Figure 25.6.5 shows the power and EM TVLA results of this engine compared to the unprotected core based on [12]. Measurements were performed on the fabricated 65nm ASIC at 100MHz, using a 500MHz bandwidth oscilloscope sampled at 625MS/s. Power traces were captured through a 1Ω shunt resistor with 20dB on-board amplification, while EM traces were acquired using a near-field probe with a 20dB external amplifier. Following [28], reference measurements on the unprotected engine and on the DPAM engine with randomness disabled revealed clear leakage within 10k traces. In contrast, the DPAM engine with active randomness passed 100 million traces of TVLA in both power and EM channels without detectable leakage, demonstrating >73,000× improvement in power minimum traces to disclosure (MTD) and >18,000× in EM MTD.

Compared to prior masked SHA-3 designs (Fig. 25.6.6), this work reduces area and energy overheads to only 17% and 27%, which are >8.9× and >9.2× lower than previous implementations. The engine satisfies a wide set of provable side-channel security models (glitch- and transition-robust probing [22,25]) and demonstrates empirical resistance with an MTD exceeding 100 million in TVLA. It further achieves low latency overhead and reduces external randomness usage to 250 bits (>6.6× reduction), whose generation and storage may incur significant area and energy costs [29].

Relative to the prior ASIC-based masked SHA-3 [11], energy efficiency improves by 16.7×. Compared to a recent masked SHA-256 [17], this engine achieves 25× better energy efficiency, >4.9× lower overheads, while retaining security under stronger leakage assumptions and achieving significantly larger TVLA MTD improvements. DPAM reduces the cost of provable security to levels comparable with hiding-based techniques [18,19], making practical deployment significantly more feasible. Figure 25.6.7 shows the die micrograph and chip characteristics.

Acknowledgement:
This work was supported in part by the National Key R&D Program of China (Grant No. 2023YFB4403500, No. 2024YFB3108103) and in part by the National Natural Science Foundation of China (Grant No. 62274102). The authors thank Shaonan Wu, Zhiqi Yang, and Guanxin Huang for their support in chip testing. Corresponding authors are Bohan Yang (email: bohanyang@tsinghua.edu.cn), Aoyang Zhang (email: aoyang@tsinghua.edu.cn), and Leibo Liu (email: liulb@tsinghua.edu.cn).

References:
[1] Federal Information Processing Standards Publication, "SHA-3 Standard: Permutation-Based Hash and Extendable-Output Functions," *FIPS 202*, Aug. 2015. https://doi.org/10.6028/NIST.FIPS.202
[2] U. Banerjee et al., "An Energy-Efficient Configurable Lattice Cryptography Processor for the Quantum-Secure Internet of Things," *ISSCC*, pp. 46-47, Feb. 2019. https://doi.org/10.1109/ISSCC.2019.8662528
[3] Y. Zhu et al., "A 28nm 48KOPS 3.4μJ/Op Agile Crypto-Processor for Post-Quantum Cryptography on Multi-Mathematical Problems," *ISSCC*, pp. 514-515, Feb. 2022. https://doi.org/10.1109/ISSCC42614.2022.9731783

434 • 2026 IEEE International Solid-State Circuits Conference

979-8-3315-8937-0/26 $31.00 © 2026 IEEE

[4] J. Kelsey et al., "SHA-3 Derived Functions: cSHAKE, KMAC, TupleHash, and ParallelHash," NIST SP 800-185, Dec. 2016. https://doi.org/10.6028/NIST.SP.800-185

[5] C. S. Juvekar et al., "A Keccak-Based Wireless Authentication Tag with per-Query Key Update and Power-Glitch Attack Countermeasures," ISSCC, pp. 290-291, Feb. 2016. https://doi.org/10.1109/ISSCC.2016.7418021

[6] M. J. Kannwischer et al., "Single-Trace Attacks on Keccak," IACR Transactions on Cryptographic Hardware and Embedded Systems (TCHES), vol. 2020, no. 3, pp. 243-268, Jun. 2020. https://doi.org/10.13154/tches.v2020.i3.243-268

[7] M. Taha et al., "Differential Power Analysis of MAC-Keccak at Any Key-Length," International Workshop on Security (IWSEC), pp. 68-82, Nov. 2013. https://doi.org/10.1007/978-3-642-41383-4_5

[8] P. Luo et al., "Power Analysis Attack on Hardware Implementation of MAC-Keccak on FPGAs," International Conference on ReConFigurable Computing and FPGAs (ReConFig), pp. 1-7, Dec. 2014. https://doi.org/10.1109/ReConFig.2014.7032549

[9] S. You et al., "Single-Trace Fragment Template Attack on a 32-Bit Implementation of Keccak," Smart Card Research and Advanced Applications (CARDIS), pp. 3-23, Nov. 2021. https://doi.org/10.1007/978-3-030-97348-3_1

[10] E. Prouff et al., "Masking against Side-Channel Attacks: A Formal Security Proof," EUROCRYPT, pp. 142-159, May 2013. https://doi.org/10.1007/978-3-642-38348-9_9

[11] M. Muehlberghuber et al., "Towards Evaluating DPA Countermeasures for KECCAK on a Real ASIC," Constructive Side-Channel Analysis and Secure Design (COSADE), pp. 222-236, Apr. 2015. https://doi.org/10.1007/978-3-319-21476-4_15

[12] B. Bilgin et al., "Efficient and First-Order DPA Resistant Implementations of KECCAK," Smart Card Research and Advanced Applications (CARDIS), pp. 187-199, Nov. 2013. https://doi.org/10.1007/978-3-319-08302-5_13

[13] C. Zhao et al., "Breaking Ground: A New Area Record for Low-Latency First-Order Masked SHA-3," IACR Transactions on Cryptographic Hardware and Embedded Systems (TCHES), vol. 2024, no. 4, pp. 231-257, Sep. 2024. https://doi.org/10.46586/tches.v2024.i4.231-257

[14] S. Zarei et al., "Low-Latency Keccak at any Arbitrary Order," IACR Transactions on Cryptographic Hardware and Embedded Systems (TCHES), vol. 2021, no. 4, pp. 388-411, Aug. 2021. https://doi.org/10.46586/tches.v2021.i4.388-411

[15] A. R. Shahmirzadi et al., "Second-Order SCA Security with almost no Fresh Randomness," IACR Transactions on Cryptographic Hardware and Embedded Systems (TCHES), vol. 2021, no. 3, pp. 708-755, Jul. 2021. https://doi.org/10.46586/tches.v2021.i3.708-755

[16] H. Gross et al., "Higher-Order Side-Channel Protected Implementations of KECCAK," Euromicro Conference on Digital System Design (DSD), pp. 205-212, Aug. 2017. https://doi.org/10.1109/DSD.2017.21

[17] S. Taneja et al., "A 2455µm² 1.7Gbps Side-Channel Attack-Resistant Masked HMAC-SHA256 Accelerator in Intel 4 CMOS," IEEE CICC, pp. 1-3, Apr. 2025. https://doi.org/10.1109/CICC63670.2025.10983609

[18] S. Oruganti et al., "Power and EM Side-Channel-Attack-Resilient AES-128 Core with Round-Aligned Globally-Synchronous-Locally-Asynchronous Operation Based on Tunable Replica Circuits," ISSCC, pp. 308-310, Feb. 2024. https://doi.org/10.1109/ISSCC49657.2024.10454574

[19] A. Ghosh et al., "An EM/Power SCA-Resilient AES-256 with Synthesizable Signature Attenuation Using Digital-Friendly Current Source and RO-Bleed-Based Integrated Local Feedback and Global Switched-Mode Control," ISSCC, pp. 499-501, Feb. 2021. https://doi.org/10.1109/ISSCC42613.2021.9365978

[20] H. Gross et al., "First-Order Masking with Only Two Random Bits," Theory of Implementation Security Workshop (TIS), pp. 10-23, Nov. 2019. https://doi.org/10.1145/3338467.3358950

[21] W. Wang et al., "Side-Channel Masking with Common Shares," IACR Transactions on Cryptographic Hardware and Embedded Systems (TCHES), vol. 2022, no. 3, pp. 290-329, Jun. 2022. https://doi.org/10.46586/tches.v2022.i3.290-329

[22] S. Faust et al., "Composable Masking Schemes in the Presence of Physical Defaults & the Robust Probing Model," IACR Transactions on Cryptographic Hardware and Embedded Systems (TCHES), vol. 2018, no. 3, pp. 89-120, Aug. 2018. https://doi.org/10.13154/tches.v2018.i3.89-120

[23] A. J. Leiserson et al., "Gate-Level Masking under a Path-Based Leakage Metric," Cryptographic Hardware and Embedded Systems (CHES), pp. 580-597, Sep. 2014. https://doi.org/10.1007/978-3-662-44709-3_32

[24] P. Sasdrich et al., "Low-Latency Hardware Masking with Application to AES," IACR Transactions on Cryptographic Hardware and Embedded Systems (TCHES), vol. 2020, no. 2, pp. 300-326, Mar. 2020. https://doi.org/10.13154/tches.v2020.i2.300-326

[25] N. Müller et al., "Robust but Relaxed Probing Model," IACR Transactions on Cryptographic Hardware and Embedded Systems (TCHES), vol. 2024, no. 4, pp. 451-482, Sep. 2024. https://doi.org/10.46586/tches.v2024.i4.451-482

[26] G. Goodwill et al., "A Testing Methodology for Side Channel Resistance Validation," NIST Non-Invasive Attack Testing Workshop, Sep. 2011. Avaiable online, Accessed on Nov. 19, 2025, <https://csrc.nist.gov/csrc/media/events/non-invasive-attack-testing-workshop/documents/08_goodwill.pdf>

[27] T. Schneider et al., "Leakage Assessment Methodology: A Clear Roadmap for Side-Channel Evaluations," Cryptographic Hardware and Embedded Systems (CHES), pp. 495-513, Sep. 2015. https://doi.org/10.1007/978-3-662-48324-4_25

[28] U. Banerjee et al., "Writing a Good Security Paper for ISSCC (2025)," May 2025, arXiv:2505.12700. https://doi.org/10.48550/arXiv.2505.12700

[29] T. Sugawara, "3-Share Threshold Implementation of AES S-box without Fresh Randomness," IACR Transactions on Cryptographic Hardware and Embedded Systems (TCHES), vol. 2019, no. 1, pp. 123-145, Nov. 2018. https://doi.org/10.46586/tches.v2019.i1.123-145

[30] P. Pessl et al., "Pushing the Limits of SHA-3 Hardware Implementations to Fit on RFID," Cryptographic Hardware and Embedded Systems (CHES), pp. 126-141, Aug. 2013. https://doi.org/10.1007/978-3-642-40349-1_8

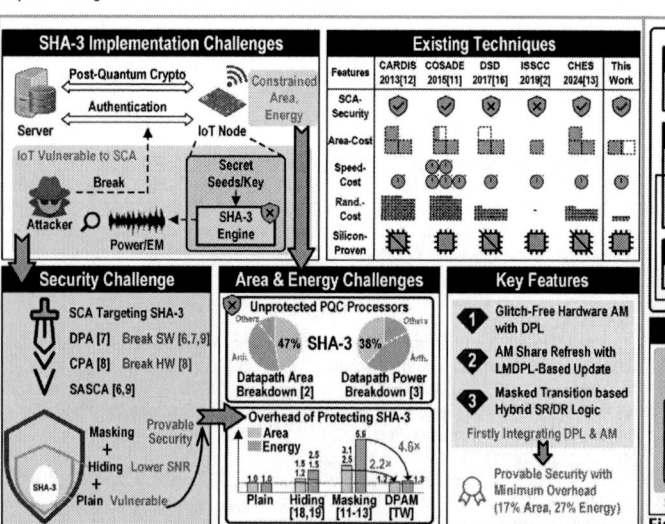
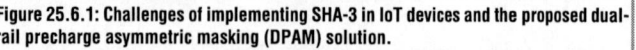

Figure 25.6.1: Challenges of implementing SHA-3 in IoT devices and the proposed dual-rail precharge asymmetric masking (DPAM) solution.

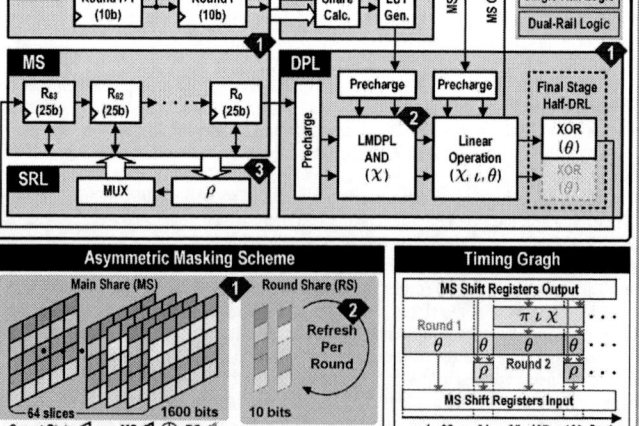

Figure 25.6.2: Architecture of the proposed DPAM SHA-3 engine, including MS-RS mapping in the asymmetric masking scheme and timing diagram of the SHA-3 step functions.

ISSCC 2026 / SESSION 25 / HARDWARE SECURITY / 25.6

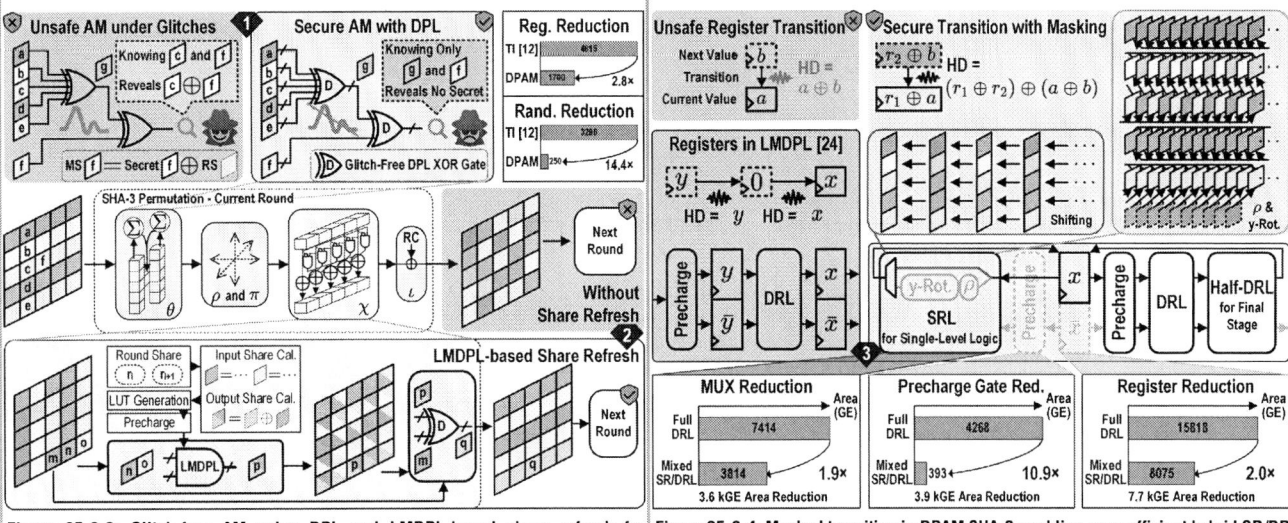

Figure 25.6.3: Glitch-free AM using DPL and LMDPL-based share refresh for reconstructing the MS-RS mapping.

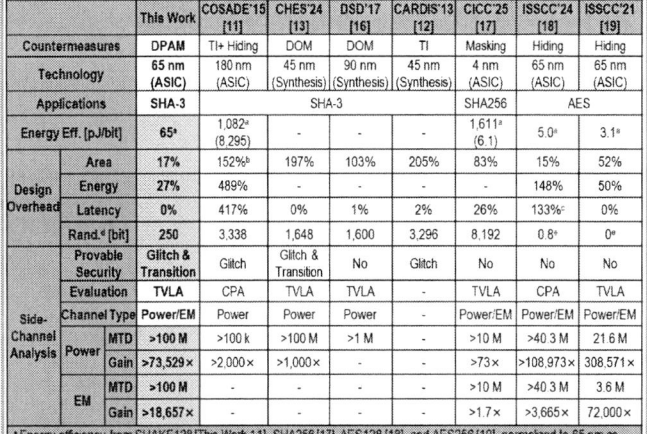

Figure 25.6.4: Masked transition in DPAM SHA-3 enabling area-efficient hybrid SR/DR logic.

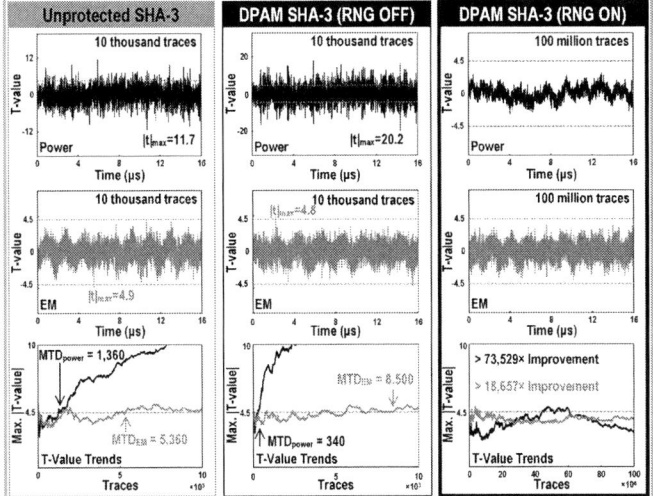

Figure 25.6.5: Power and EM TVLA results of DPAM SHA-3 with RNG on/off and unprotected baseline.

	This Work	COSADE'15 [11]	CHES'24 [13]	DSD'17 [16]	CARDIS'13 [12]	CICC'25 [17]	ISSCC'24 [18]	ISSCC'21 [19]
Countermeasures	DPAM	TI+ Hiding	DOM	DOM	TI	Masking	Hiding	Hiding
Technology	65 nm (ASIC)	180 nm (ASIC)	45 nm (Synthesis)	90 nm (Synthesis)	45 nm (Synthesis)	4 nm (ASIC)	65 nm (ASIC)	65 nm (ASIC)
Applications	SHA-3	SHA-3				SHA256	AES	
Energy Eff. [pJ/bit]	65[a]	1,082[a] (8,295)	-	-	-	1,611[a] (6.1)	5.0[a]	3.1[a]
Design Overhead — Area	17%	152%[b]	197%	103%	205%	83%	15%	52%
Design Overhead — Energy	27%	489%	-	-	-	-	148%	50%
Design Overhead — Latency	0%	417%	0%	1%	2%	26%	133%[c]	0%
Design Overhead — Rand.[d] [bit]	250	3,338	1,648	1,600	3,296	8,192	0.8[e]	0[e]
Provable Security	Glitch & Transition	Glitch	Glitch & Transition	No	Glitch	No	No	No
Side-Channel Analysis — Evaluation	TVLA	CPA	TVLA	TVLA	-	TVLA	CPA	TVLA
Side-Channel Analysis — Channel Type	Power/EM	Power	Power	Power	-	Power/EM	Power/EM	Power/EM
Side-Channel Analysis — Power MTD	>100 M	>100 k	>100 M	>1 M	-	>10 M	>40.3 M	21.6 M
Side-Channel Analysis — Power Gain	>73,529×	>2,000×	>1,000×	-	-	>73×	>108,973×	308,571×
Side-Channel Analysis — EM MTD	>100 M	-	-	-	-	>10 M	>40.3 M	3.6 M
Side-Channel Analysis — EM Gain	>18,657×	-	-	-	-	>1.7×	>3,665×	72,000×

[a] Energy efficiency from SHAKE128 [This Work,11], SHA256 [17], AES128 [18], and AES256 [19], normalized to 65 nm as × $(65/process)^2$; raw values in parentheses. [b] Compared with an unprotected design [30] of similar architecture. [c] Latency overhead considering the significant frequency drop. [d] Total randomness = randomness for mask initialization + fresh randomness during computation. [e] Designs integrated a PRNG; only external randomness requirements reported here.

Figure 25.6.6: Comparison of design overhead and SCA security with related works.

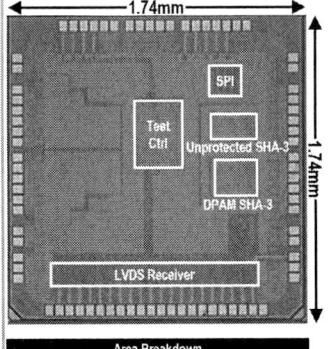

Technology	65 nm	
Die Area	3.03 mm²	
Supply Voltage	1.0 V	
SHA-3 Core	Unprotected	DPAM
Gate Counts	11.9 kGE	13.6 kGE
Layout Area	39,042 µm²	45,530 µm²
Fmax	650 MHz	650 MHz
Power	42.7 µW/MHz	54.4 µW/MHz
Latency	1600 cycles	1600 cycles

Figure 25.6.7: Die micrograph, chip specifications, and area/power breakdown.

ISSCC 2026 / SESSION 25 / HARDWARE SECURITY / 25.7

25.7 TinyPAD: A 166μm²/lane Variation-Tolerant Probing-Attack Detector for an 8Gb/s/lane Chip-to-Chip Interface in 16nm FinFET

Mao Li[1], Rentao Wan[1], Sanu K Mathew[2], Vivek De[2], Mingoo Seok[1]

[1]Columbia University, New York, NY, [2]Intel, Hillsboro, OR

Abstract

We present TinyPAD, a 166μm²/lane probing-attack detector in 16nm FinFET with 82pF maximum supported loading as well as 0.2pF minimum detectable capacitance (MDC) that operates robustly over -40 to125°C and from a 0.65-to-1.2V supply. We integrate TinyPAD in a 5-lane 8Gb/s/lane chip-to-chip interface. Compared to the prior arts, TinyPAD achieves 3.3× maximum supported loading, 2.5× smaller MDC over all voltage and temperature conditions, 1.3× wider range for temperature, 1.2× for voltage, at 11× smaller area overhead.

High-speed interconnect has been unprecedentedly important with the explosive growth of computing systems, as vast volumes of data need to be transmitted fast and efficiently between chips. However, their associated security vulnerabilities [1-4] have received comparatively little attention. Physical probing of signal traces between chips on the PCB represents a significant security vulnerability, enabling attackers to eavesdrop on data transmitted between chips, or even to take over a server memory entirely by probing the host-memory interconnect as in [5]. Tapping even faster signal traces without degrading the signal integrity is feasible thanks to commercially available multi-GHz bandwidth probes with < 0.3pF loading [6] (Fig. 25.7.1). Although standard cryptographic protocols such as AES could be utilized to encrypt the data [7-13], they introduce significant latency and power overheads. Given the low probability of a physical attack during a system's operational lifespan, always-on encryption is often inefficient and unnecessary. Therefore, a more practical security paradigm is to selectively activate protective measures only upon the detection of an active probing attack [14-25]. Recently, Li et al. have proposed a probing attack detector based on an SAR capacitance-to-digital converter (CDC), demonstrating excellent probe detection capability [26] (Fig. 25.7.1, bottom). However, it employs a large on-chip reference capacitor array, which impedes the miniaturization of the detector and thereby makes it unsuitable for modern chips that feature hundreds of IOs. Moreover, it cannot support long traces or large loadings as it would require proportionally increasing the reference capacitor array. Lastly, it requires an additional temperature sensor to compensate for temperature variations, which introduces area and calibration overhead.

To address these issues, we present TinyPAD, a 166μm²/lane probing-attack detector in 16nm FinFET with ~82pF maximum supported loading as well as 0.2pF minimum detectable capacitance (MDC) robustly over -40 to 125°C and from a 0.65-to-1.2V supply. We integrate TinyPAD in a 5-lane 8Gb/s/lane chip-to-chip interface [27]. The detector core drives the external PCB trace using an on-chip high-R resistor and measures the time to charge it to a certain level. It converts the time into a digital code using a counter. If the code goes out of a preset range, TinyPAD triggers an alarm to indicate a possible attack or tampering with the signal trace. We also propose a one-transistor leakage-based temperature compensation and a one-temperature-point calibration scheme, enabling TinyPAD to robustly operate over the industrial temperature range and against process variations. Compared to [26], TinyPAD achieves 3.3× maximum supported loading, 2.5× smaller MDC across all voltage and temperature (VT) conditions, 1.3× wider range for temperature, 1.2× for voltage, with 11× smaller area overhead.

Figure 25.7.2 shows the block diagram of TinyPAD. The chip contains five transmitters (TX0 to TX4) driving five lanes (Lane0 to Lane4), each equipped with an individual pattern generator and a separate detector core. A PLL generates a 2GHz clock (CLK2G) for the shared FSM, which controls all five detector cores. The 2GHz clock is then doubled by a DLL, feeding a 4GHz clock (CLK4G) to pattern generators to produce data for TXs. The shared FSM activates the detection only when the TX is not being used to avoid performance degradation. Also, it only enables a detector core one lane at a time to save power and silicon area.

Figure 25.7.3 (top) depicts the detailed schematics of one detector core controlled by the shared FSM. It consists mostly of digital standard cells without any precise analog circuits. Figure 25.7.3 (bottom left) shows the waveforms of the detection operation. To initiate the detection, the FSM first asserts *en*, connecting the detector core to Lane0. In the meantime, TX0 and the corresponding RX (in the other chip) are turned off to avoid impact from termination. The detector core also pulls down Lane0 to ground through a single NMOS. Subsequently, the FSM sets *start* high. This signal goes through a NAND gate, an inverter, and a high-R resistor to charge Lane0. We purposely employ this high-R resistor to slow down the charging process, making even a subtle loading change on Lane0 evident to the detector. The detector core then sets *cnten* high to start counting. As Lane0 slowly ramps up, the counter inside the shared FSM keeps counting the cycles it takes to charge. Once Lane0 reaches the threshold of *invsense*, its output flips, driving *cnten* low to stop the counter. Lane0 is also pulled down to the ground and stops charging to save power. At the same time, the detector core updates the flip-flop to set *done* to high, informing the shared FSM of the completion of the counting process. If the counter output is outside two preset thresholds (thH and thL), TinyPAD will trigger an alarm. To mitigate the impact of random noise, as shown in Fig. 25.7.3 (bottom right), the shared FSM can execute this detection up to 128 times for a single channel and average the output code before making a final comparison against the thresholds.

Robustness over process, voltage, and temperature (PVT) variations is critical for TinyPAD. We propose a set of solutions to compensate for those variations. In the design time, we create a novel leakage-based temperature compensation scheme, realized in only a single PMOS transistor (highlighted in red in Fig. 25.7.3). This transistor is designed to be in the subthreshold region. As the ambient temperature rises, the transistor's leakage current increases exponentially, which helps charge the trace capacitance faster, actively counteracting the inherent proportional-to-absolute-temperature (PTAT) characteristics of the high-R resistor and longer circuit delay at higher temperatures. This simple yet effective technique stabilizes the detection code across the entire operating temperature range without requiring a dedicated temperature sensor, as [26] requires. On the other hand, TinyPAD is inherently insensitive to voltage variations, as it leverages the RC time constant of each lane. As long as the loading does not change, the time it takes to charge should remain relatively constant over the operating voltage range.

To ensure the robustness to temperature and voltage across process variations, we perform low-cost, one-point calibration as illustrated in Fig. 25.7.5 (top left). The procedure is performed once per chip after the chip is mounted on the board. At a nominal condition (e.g., 0.8V, 25°C), the detector code from a known-secure trace is captured and stored as a digital offset. The alarm thresholds (Threshold high, Threshold low) are then established relative to this offset, defining the valid detection window. This single calibration step is sufficient for reliable operation across the full voltage and temperature range.

To demonstrate the detector's compatibility with a high-speed interface, we integrate TinyPAD with a 5-lane 8-Gb/s/lane TX. Figure 25.7.4 shows the schematics of one lane. The pattern generator clocked by CLK4G employs a double data rate (DDR) scheme, producing an 8Gbps data stream leveraging both edges of the clock. The TX features a 4b binary-weighted strength control, allowing for fine-grained impedance matching to the transmission line. The detector core is connected in parallel with the main TX driver and the ESD protection circuitry. During high-speed data transmission, the detector circuitry is completely disabled and presents a negligible capacitive load to the channel. This non-invasive design ensures that signal integrity is not compromised. The measured 8Gb/s eye diagrams (Fig. 25.7.4, bottom) before and after a detection are nearly identical, with the eye height changing by less than 1% (200.4mV vs 198.8mV), and the eye width remains constant, demonstrating minimal impact on the high-speed data link.

It is worth noting that to cheat the detector, some hackers may have the ability to change the physical characteristics of the PCB to reduce the lane capacitance before attaching a probe to it. For example, they may heat up a specific part of the board to modulate the thickness of the board slightly and thus reduce the trace capacitance. Then, they can place a small loading probe such that the total capacitance seen by TinyPAD remains almost the same. However, it is extremely difficult to precisely modulate the trace's capacitance with sub-pF precision in a single attempt. Even if they pull off this difficult task through several iterations, TinyPAD should be able to detect such tampering attempts since it has the fine detection precision as well as the ability to detect both an increase and a decrease in the loading capacitance.

TinyPAD was prototyped in a 16nm FinFET process. Figure 25.7.7 shows the chip micrograph and layout details. The detector core achieves a highly compact footprint of just 36μm² (6μm×6μm). The total detector area per lane, including the area of the shared controller and counter, amortized over the five channels, is only 166μm². At the typical condition, the detector's response is measured against various discrete capacitances added to the PCB signal trace (Fig. 25.7.5, top middle) and is also validated with attached commercial probes (Fig. 25.7.5, top right). It exhibits high linearity with a resolution of 0.02pF per code, enabling high-precision detection. We also measure the detection robustness across variations. The bottom-left graph of Fig. 25.7.5 shows that without compensation, the detector code exhibits significant thermal drift, which would lead to false alarms. In contrast, the bottom-middle graph shows that with the leakage-based compensation enabled, the code remains stable and well within the calibrated thresholds across the entire temperature range of -40 to 125°C. The detector also demonstrates robust operation across a wide supply voltage range of 0.65 to 1.2V, as shown in the bottom-right graph Fig. 25.7.5. The same high and low thresholds can be applied to the entire voltage range, ensuring robust detection over VT variations.

436 • 2026 IEEE International Solid-State Circuits Conference

979-8-3315-8937-0/26 $31.00 © 2026 IEEE

The overall performance of TinyPAD is compared against the prior works in Fig. 25.7.6. TinyPAD achieves the smallest typical minimum detectable capacitance (MDC) of 0.02pF. It also achieves the smallest MDC of 0.2pF across even wider temperature and voltage operating ranges, while supporting a maximum load of ~82pF, 3.3× larger than [26]. It requires only one-point calibration, which simplifies the chip testing. The energy consumption is a modest 151pJ per detection. It is higher than [28]; however, [28] cannot support a high-speed interface. Last but not least, TinyPAD occupies only 166µm²/lane, 11× smaller than [26]. This combination of a minimal area, high sensitivity, robust PVT-tolerant operation, and non-invasive integration establishes TinyPAD as one of the highly effective solutions for securing modern high-speed interfaces against probing attacks.

Acknowledgement:
This project is supported by SRC TxACE (Task: 3160.057) and by Intel University Silicon Fabrication.

References:
[1] I. Verbauwhede et al., "Circuit challenges from cryptography," *ISSCC*, pp. 428-429, Feb. 2015. http://doi.org/10.1109/ISSCC.2015.7063109
[2] M. Nagata et al., "Physical Attack Protection Techniques for IC Chip Level Hardware Security," *IEEE Transactions on Very Large Scale Integration (VLSI) Systems*, vol. 30, no. 1, pp. 5-14, Jan. 2022. http://doi.org/10.1109/TVLSI.2021.3073946
[3] T. Sugawara, "Performance Evaluation in Hardware Security: Upper-bounds by Attacks and Lower-bounds by Models," *International VLSI Symposium on Technology, Systems and Applications (VLSI TSA)*, pp.1-4, Apr. 2025. http://doi.org/10.1109/VLSITSA64674.2025.11046986
[4] R. T. Yazicigil et al., "Beyond Crypto: Physical-Layer Security for Internet of Things Devices," *IEEE Solid-State Circuits Magazine*, vol. 12, no. 4, pp. 66-78, Fall 2020. http://doi.org/10.1109/MSSC.2020.3021842
[5] A. Trikalinou et al., "Taking DMA Attacks to the Next Level: How to Do Arbitrary Reads/Writes in a Live and Unmodified System," Black Hat, July 2017. Accessed on Dec. 8, 2025. <https://www.blackhat.com/docs/us-17/wednesday/us-17-Trikalinou-Taking-DMA-Attacks-To-The-Next-Level-How-To-Do-Arbitrary-Memory-Reads-Writes-In-A-Live-And-Unmodified-System-Using-A-Rogue-Memory-Controller.pdf>
[6] Tektronix Low Voltage Differential Probes Datasheet, date accessed: Aug. 28, 2025. Accessed on Dec. 8, 2025. <https://www.tek.com/en/products/oscilloscopes/oscilloscope-probes/low-voltage-differential-probes>
[7] Guido Bertoni et al., "Power-efficient ASIC synthesis of cryptographic sboxes," *ACM Great Lakes Symposium on VLSI (GLSVLSI)*, pp.277-281, Apr. 2004. https://doi.org/10.1145/988952.989019
[8] I. Verbauwhede et al., "Design and Performance Testing of a 2.29-GB/s Rijndael Processor," *IEEE JSSC*, vol. 38, no. 3, pp. 569-572, Mar. 2003. http://doi.org/10.1109/JSSC.2002.808300
[9] A. Ghosh et al., "R-STELLAR: A Resilient Synthesizable Signature Attenuation SCA Protection on AES-256 With Built-In Attack-on-Countermeasure Detection," *IEEE Open Journal of the Solid-State Circuits Society*, vol. 5, pp. 167-179, May 2025. http://doi.org/10.1109/OJSSCS.2025.3571334
[10] U. Banerjee et al., "An Energy-Efficient Reconfigurable DTLS Cryptographic Engine for End-to-End Security in IoT Applications," *ISSCC*, pp. 42-43, Feb. 2018. http://doi.org/10.1109/ISSCC.2018.8310174
[11] A. Ghosh et al., "Power and EM SCA Resilience in 65nm AES-256 Exploiting Clock Slew Dependent Variability in CMOS Digital Circuits," *IEEE CICC*, pp. 1-2, Apr. 2023. http://doi.org/10.1109/CICC57935.2023.10121240

[12] S. Lu et al., "A 28nm 4.05µJ/Encryption 8.72kHMul/s Reconfigurable Multi-Scheme Fully Homomorphic Encryption Processor for Encrypted Client-Server Computing," *ISSCC*, pp. 294-295, Feb. 2025. http://doi.org/10.1109/ISSCC49661.2025.10904812
[13] J. Buchmann et al., "High-Performance and Lightweight Lattice-Based Public-Key Encryption," *ACM International Workshop on IoT Privacy, Trust, and Security (IoTPTS)*, pp. 2-9, May 2016. http://doi.org/10.1145/2899007.2899011
[14] V.K. Rajanna et al., "Fully-Digital Broadband Calibration-Less Impedance Monitor," *IEEE Symp. VLSI Circuits*, pp. 144-145, June 2022. http://doi.org/10.1109/VLSITechnologyandCir46769.2022.9830158
[15] S. Konno et al., "A 65-nm Delta-Sigma ADC-Based VDD-Variation-Tolerant Power-Side-Channel-Attack Sensor," *IEEE Solid-State Circuits Letters*, vol. 8, pp. 57-60, Jan. 2025. http://doi.org/10.1109/LSSC.2025.3527153
[16] M. Li et al., "EQZ-LDO: A Secure Digital Low Dropout Regulator Armed With Detection-Driven Protection Against Correlation Power Analysis," *IEEE Journal of Solid-State Circuits*, vol. 59, no. 11, pp. 3806-3815, Nov. 2024. http://doi.org/10.1109/JSSC.2024.3400621
[17] M. Li et al., "GUARD: A Fully-Digital TDC-Based Clock and Voltage Glitch Detector with On-Demand Protection in a 28nm CMOS," *IEEE Symp. VLSI Circuits*, pp. 1-3, Jun. 2025. http://doi.org/10.23919/VLSITechnologyandCir65189.2025.11075120
[18] N. Mehta et al., "A 100MHz Self-Calibrating RC Oscillator Capable of Clock-Glitch Detection for Hardware Security in a 3nm FinFET Process," *ISSCC*, pp. 302-303, Feb. 2025. http://doi.org/10.1109/ISSCC49661.2025.10904687
[19] Y. He et al., "A Synthesizable Design-Agnostic Timing Fault Injection Monitor Covering 2MHz to 1.26GHz Clocks in 65nm CMOS," *ISSCC*, pp. 304-305, Feb. 2024. http://doi.org/10.1109/ISSCC49657.2024.10454280
[20] S. Song, et al., "An FLL-Based Clock Glitch Detector for Security Circuits in a 5nm FINFET Process," *IEEE Symp. VLSI Circuits*, pp. 146-147, Jun. 2022. http://doi.org/10.1109/VLSITechnologyandCir46769.2022.9830157
[21] H. Zhang et al., "Sensor-Less Laser Voltage-Probing Attack Detection via Run-Time-Leakage-Shift Monitoring with 4.35% Area Overhead," *ISSCC*, pp. 292-293, Feb. 2025. http://doi.org/10.1109/ISSCC49661.2025.10904688
[22] B. Karpinskyy et al., "An Efficient Vth-Tilting PUF Design in 3nm GAA and 8nm FinFET Technologies," *ISSCC*, pp.298-299, Feb. 2025. http://doi.org/10.1109/ISSCC49661.2025.10904512
[23] D. Lee et al., "An Off-Chip Attack on Hardware Enclaves via the Memory Bus," *USENIX Security Symp.*, pp. 487-504, Aug. 2020. Accessed on Dec. 8, 2025. <https://www.usenix.org/conference/usenixsecurity20/presentation/lee-dayeol>
[24] Z. Xu et al., "Runtime Detection of Probing/Tampering on Interconnecting Buses," *International Symposium on Field-Programmable Custom Computing Machines (FCCM)*, pp.247-251, May 2021. http://doi.org/10.1109/FCCM51124.2021.00038
[25] A. Oksman, "A Method for Detecting DRAM Bus", Aalto University, Jan. 2020. Accessed on Dec. 8th, 2025. <https://aaltodoc.aalto.fi/server/api/core/bitstreams/f277b7ec-a6cc-48a2-9cf5-af4f11573e17/content>
[26] M. Li et al., "PACTOR: A Variation-Tolerant Probing-Attack Detector for a 2.5Gb/s×4-Channel Chip-to-Chip Interface in 28nm CMOS," *ISSCC*, pp. 306-307, Feb. 2024. http://doi.org/10.1109/ISSCC49657.2024.10454309
[27] Y. Nishi et al., "A 0.297-pJ/Bit 50.4-Gb/s/Wire Inverter-Based Short-Reach Simultaneous Bi-Directional Transceiver for Die-to-Die Interface in 5-nm CMOS," *IEEE Journal of Solid-State Circuits*, vol. 58, no. 4, pp. 1062-1073, April 2023. http://doi.org/10.1109/JSSC.2022.3232024
[28] M. Li et al., "A Fully-Digital Variation-Tolerant Runtime Detector for PCB-Level Probing Attack in a 28-nm CMOS," *IEEE Solid-State Circuits Letters*, vol. 6, pp. 245-248, Aug. 2023. http://doi.org/10.1109/LSSC.2023.3310266

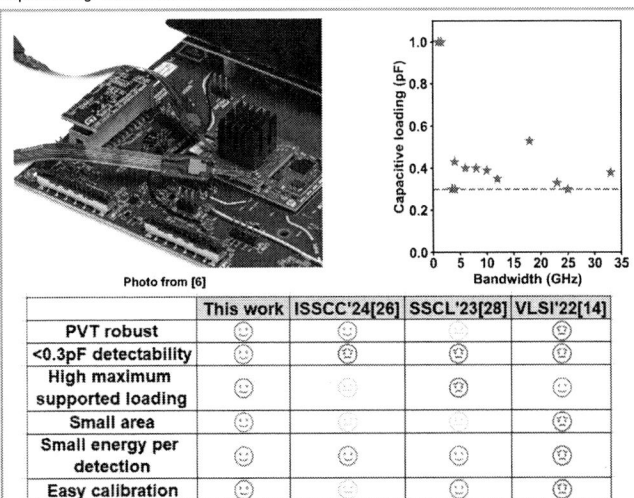

	This work	ISSCC'24[26]	SSCL'23[28]	VLSI'22[14]
PVT robust	☺	☺	☹	☺
<0.3pF detectability	☺	☹	☺	☹
High maximum supported loading	☺	☹	☺	☺
Small area	☺	☹	☹	☹
Small energy per detection	☺	☺	☺	☹
Easy calibration	☺	☺	☺	☹

Figure 25.7.1: Probing a high-speed link is possible thanks to the probe technology advances [6] (top left), the survey of high-bandwidth probes (top right), and the innovation of TinyPAD compared to the prior works (bottom).

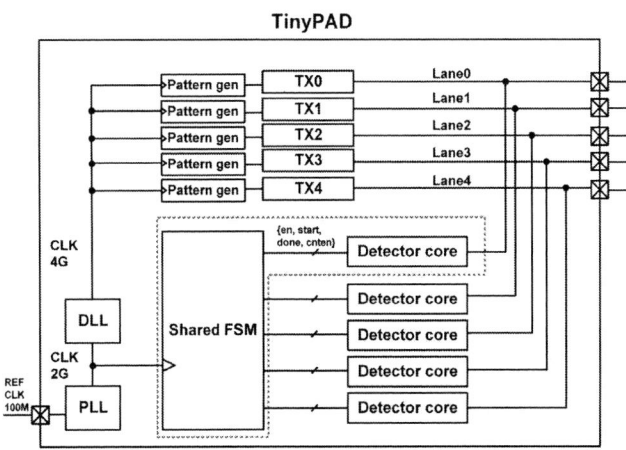

Figure 25.7.2: Block diagram of TinyPAD, including the clock generation (DLL & PLL), five TXs, shared FSM, and five detector cores, with detailed schematics in the red box shown in Fig. 25.7.3.

ISSCC 2026 / SESSION 25 / HARDWARE SECURITY / 25.7

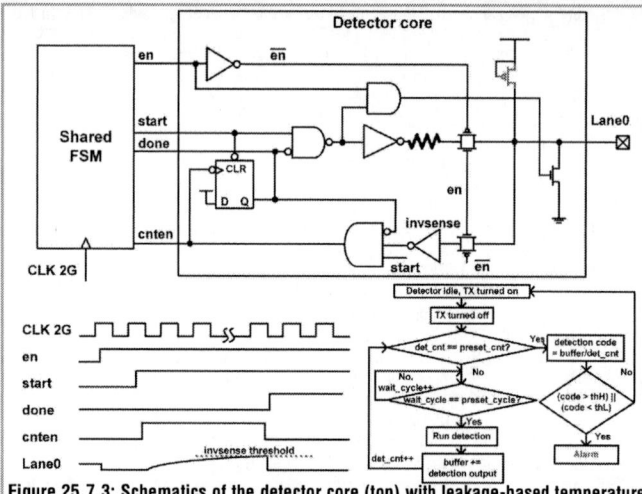

Figure 25.7.3: Schematics of the detector core (top) with leakage-based temperature compensation (in red), the timing diagram of the detection operation (bottom left), and the state diagram of a detection with multiple iterations to suppress noise (bottom right).

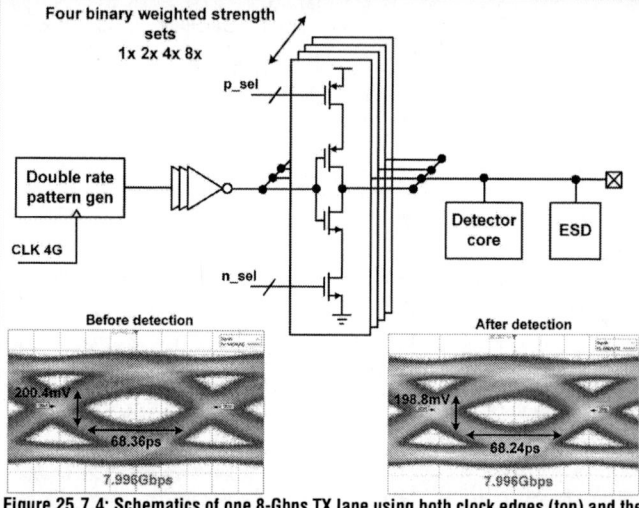

Figure 25.7.4: Schematics of one 8-Gbps TX lane using both clock edges (top) and the measured eye diagrams before (bottom left) and after (bottom right) running a detection operation.

Figure 25.7.5: Calibration steps to set the thresholds (top left); measurement results of detection at a typical condition (top middle) with real probes (top right), across different temperatures without (bottom left) and with (bottom middle) leakage-based temperature compensation, and across different supply voltages (bottom right), demonstrating the VT robustness.

	This work	ISSCC'24[26]	SSCL'23[28]	VLSI'22[14]
Technology	16nm	28nm	28nm	28nm
Fully digital?	Yes	No	Yes	Yes
Minimum detectable capacitance (MDC) at a typical condition (pF)	0.02	0.07	1	10^5
Working temperature (°C)	-40-125	-20-105	30-90	0-50
Supply voltage (V)	0.65-1.2	0.65-1.1	0.8-0.9	0.8-1.1
Detectable capacitance change polarity	Positive and negative	Positive	Positive	Positive
Calibration	One-point	Simulation assisted one-point	One-point	N/R
MDC across all VT conditions with the same threshold(s) (pF)	0.2	0.5	2	100
Maximum loading supported (pF)	~82[a]	~25[b]	~16[b]	>10[6]
Power consumption (mW)	0.36[c]	0.145	0.0368[d]	2.9[e]
Energy per detection per lane (pJ)	151[f]	232[b]	0.92[b]	N/R
IO data rate (Mbps)	8000	2500	160	N/A
Detection core area (µm²)	36	476	3910	6300
Controller and other area (µm²)	650	6665	N/A	N/A
Area per lane (µm²)	166	1785	3910	6300

a: Estimated based on a 12-bit counter, could be extended by simply adding more flip flops; b: estimated based on reported data; c: measured at 0.8V, 2GHz; d: measured at 0.9V, 40MHz; e: condition not reported; f: measured with a 10cm 50-Ω Rogers trace. N/R: not reported. N/A: not available

Figure 25.7.6: Comparisons with recent works.

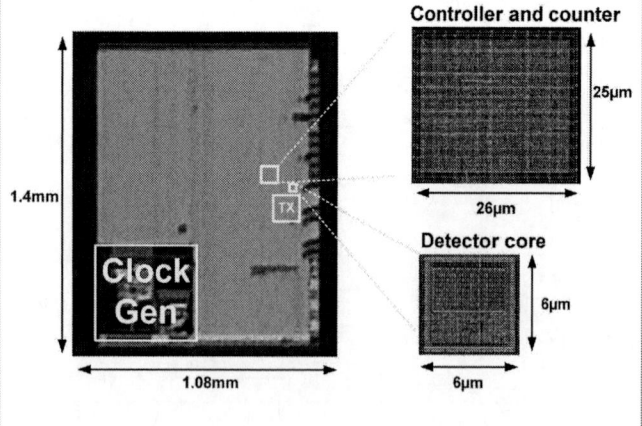

Figure 25.7.7: Die micrograph and zoom-in layout of the detector core and the shared controller.

- 2026 IEEE International Solid-State Circuits Conference

979-8-3315-8937-0/26 $31.00 © 2026 IEEE

ISSCC 2026 / SESSION 25 / HARDWARE SECURITY / 25.8

25.8 A Sub-Threshold All-NMOS Reconfigurable PUF with Secure Configuration Selection for Stable 6b/Cell

Shufan Xu*, Kunyang Liu*, Lando Chan, Hironori Tagawa, Hirofumi Shinohara, Kiichi Niitsu

Kyoto University, Kyoto, Japan
*Equally Credited Authors (ECAs)

Abstract

In this work, a reconfigurable PUF featuring secure configuration selection is presented. Without exposing secret PUF data, it achieves 6 stable bits per cell (b/cell) and "zero" bit error (<6.5E–8 BER) across 0 to 70°C and 0.6 to 0.8V. The area per effective bit is only 253F². Operating in the sub-threshold region, the PUF achieves low energy consumption of 16.8fJ/b. The all-NMOS bitcell structure mitigates long-term NBTI aging, which is verified through a 45-hour accelerated stress test.

Physically unclonable functions (PUFs) exploit inherent process variations to generate device-specific responses, serving as security primitives for cryptographic key generation [1]. However, these subtle variations are susceptible to noise and environmental fluctuations, particularly changes in supply voltage and temperature (VT), which can result in bit errors. In general, PUF cells with larger intrinsic mismatch and lower sensitivity to VT variations exhibit higher stability. To achieve cryptographic-level stability, prior works [2-6] detected dark bits with small mismatch and discarded (masked) them, but this over-eliminated otherwise stable cells and led to substantial bit-cell loss. Previous reconfigurable PUF designs [7-9] attempted to select configurations with larger mismatch among multiple candidates, but their efficiency was limited, and additional masking or error-correcting codes (ECCs) were still required. A 36-way reconfigurable PUF [10] achieved 4 stable b/cell, but similar to other approaches, it required access to secret PUF data to determine the optimal configuration. Only the in-cell burn-in design [11] and the eye-opening arbiter PUF [5] stabilized bit-cells without requiring secret information, but the former required long-time high-voltage stress, and the latter required a large cell footprint and >50% cell loss.

This work presents a reconfigurable PUF that achieves "zero" bit error rate (BER) across the temperature range of 0 to 70°C and the voltage range of 0.6 to 0.8V. The stability is achieved through precise configuration selection based on sense-delay time information, without exposing any secret PUF data. Unlike dark-bit masking, where the masking ratio often differs chip-by-chip, the developed reconfiguration generates fixed-length PUF data for cryptographic keys without bit-cell loss. It features: 1) Each cell supports 64 configurations, enabling generation of 6 stable and mutually uncorrelated response bits per cell. 2) A compact NMOS-only structure achieves a small effective area of 253F² per bit and low energy consumption of 16.8fJ/b. 3) A sense-delay detection circuit is introduced to identify unstable cells/configurations while ensuring high security, as PUF data are never exposed, thereby preserving the confidentiality of security-critical information. After reconfiguration based on the sense delay information, a stability of <6.5E–8 BER is achieved.

Figure 25.8.1 illustrates the bit-cell structure. Sixteen identical NMOS stacks are symmetrically arranged, with eight on each side. Reconfiguration is realized by selectively activating one stack on each side using gate selection signals GL_j/GR_k (j, k=0 to 7). Each bit-cell thus has a total of 64 configurations of NMOS stack pairs, enabling up to 8 non-overlapping bits to be generated. The process-induced mismatch between the selected stacks produces a differential voltage between VML and VMR, which is transferred to BLL and BLR. The BLL/BLR voltage difference is then sensed by a StrongARM sense latch. To raise BLL/BLR close to half of VDD for fast sense-latch operation and to minimize the stack-pair shoot-through current, a shared bottom NMOS transistor is inserted, biasing the cell in the sub-threshold region. As shown in Fig. 25.8.1, the footer NMOS raises the BLL/BLR levels by 4× (to ~0.4V) and accelerates the sensing speed by 480× in simulation, while simultaneously reducing shoot-through current by 342×. By eliminating PMOS devices, the all-NMOS structure mitigates NBTI-induced V_{TH} shifts and improves long-term stability. To further reduce power consumption, VD and WL are activated only when the corresponding row is selected.

For each configuration, the mismatch between BLL and BLR, denoted as $\Delta BL = |BLL–BLR|$, is quantified using a sense-delay detection circuit composed of a StrongARM latch, a voltage-to-time converter, and a time-to-digital converter (TDC). As shown in Fig. 25.8.1, a small ΔBL (i.e., a small mismatch for the configuration) results in a longer sense delay time (TD). This TD information, which is then digitized by the TDC, can therefore be used to identify configurations with small mismatch. Figure 25.8.2 shows the dependence of μ_{TD} and σ_{TD} on ΔBL, obtained from 100-run Monte-Carlo noise simulations. TD is shown in boxplots, where small ΔBL values exhibit not only a larger TD (i.e., larger μ_{TD}) but also a wider distribution and higher variability (i.e., larger σ_{TD}).

The TD variation σ_{TD} has a wider dynamic range than μ_{TD}. Therefore, stable configurations characterized by large ΔBL can be identified through their low σ_{TD} and μ_{TD}. The σ_{TD} and μ_{TD} metrics also enable the detection of VDD-induced data flipping without exposing secret PUF data. Figure 25.8.2 also depicts the VDD dependence of BLL/BLR and the associated σ_{TD} (normalized by μ_{TD}), demonstrating that peaks in σ_{TD} coincide with the BLL and BLR crossover points, where polarity reversal leads to data flipping and errors. Conversely, a reliable PUF bit is resolved when BLL/BLR preserves the polarity of (BLL–BLR), exhibiting no sharp σ_{TD} peak. Furthermore, measuring σ_{TD} and μ_{TD} across different VDD values (i.e., VDD sweep) captures temperature-induced instabilities with higher probability than observing TD only at the nominal VDD of 0.7V. This is due to the strong correlation between bit flips observed under low (high) VDD and those occurring at low (high) temperature, as shown in Fig. 25.8.3.

The bottom plot of Fig. 25.8.2 shows examples of VDD-sweep measurement results for four cells in which data flipped between 0.7V and 0.8V. In each case, peaks of μ_{TD} and σ_{TD} appear simultaneously with the increase in error rate that indicates the data flip. This suggests that by taking the maximum values of μ_{TD} and σ_{TD} across the VDD sweep, we can detect unstable cells/configurations in the corresponding VDD range and also some temperature-induced bit errors. Figure 25.8.3 shows the measured relation between the unstable cell/configuration ratio evaluated over VDD=0.6 to 0.8 V and the maximum μ_{TD} and σ_{TD} across VDD=0.58 to 0.82V at room temperature. The color map and the CDF plots of stable and unstable cells/configurations show clear separation based on max μ_{TD} and max σ_{TD}. With a threshold of [max μ_{TD} + max σ_{TD}]=2.9, all unstable cells/configurations are rejected, while 84% of the stable cells/configurations are retained in this case. Thus, for accurate and efficient best-configuration selection, the metric [max μ_{TD} + max σ_{TD}] is applied hereafter (μ_{TD} and σ_{TD} are normalized by the average value in each test condition). A combinatorial optimization algorithm is used to select the non-overlapping (i.e., do not use the same stack more than once) combination of configurations with the smallest [max μ_{TD} + max σ_{TD}] measured at 0, 25, and 70°C.

Test chips were fabricated in a 65nm LP CMOS process. Each die contains 64-word by 4b PUF cells, as shown in Fig. 25.8.4. With each bit-cell generating 6b, each chip provides 1536b of PUF data. A total of 5 chips were measured across VDD variations from 0.6 to 0.8V and temperature variations from –40 to 120°C, with 2000 evaluations performed at each test point. Figure 25.8.4 shows the measured BERs across VT variations before and after configuration selection. BERs are calculated based on golden data under the nominal VT of 0.7V/25°C. The native BER (i.e., without configuration selection) of all configurations under the nominal condition is 0.52%. After applying the secure configuration selection, no errors are observed across 0 to 70 °C at 0.7V. Bit errors across the VDD range of 0.6 to 0.8V at 25°C are also eliminated. The corresponding BER is <6.5E–8, calculated conservatively by assuming one error in the next evaluation. This demonstrates the effectiveness of the secure configuration selection technique.

To evaluate long-term reliability, an accelerated aging test was performed, as shown in Fig. 25.8.4. One chip was baked under an elevated VT condition of 1.5V and 120°C. After 45 hours of aging, the measured native BER increased only slightly from 0.472% to 0.579%. After configuration selection, no bit errors were observed, demonstrating the resilience of the NMOS-only PUF cell against long-term aging. Figure 25.8.4 also shows the measured core energy and throughput. The sub-threshold operation of the PUF cell achieves low core energy of 16.8 fJ/bit at 0.7 V/25°C. Including peripheral circuits such as the sense latch, the total energy is 61.5fJ/b. The corresponding 72Mb/s throughput is also achieved, enabled by the raised BLL/BLR design.

Figure 25.8.5 shows the statistical quality metrics. The top-left plot shows the autocorrelation of raw data under the same configuration across bit-cells before best-configuration selection, and the middle-left plot shows the autocorrelation after selection. The measured 95% confidence interval (CI) bounds of 0.05287 and 0.0215 are close to the ideal values of 0.05478 and 0.0224 under the i.i.d. assumption, respectively, indicating high randomness and minimal correlation across bit-cells and across the non-overlapping configurations even selected from the same cell. The top-right plot shows the distribution of normalized hamming distance (HD) among 256b PUF bitstreams after configuration selection, where the measured μ=0.4891 and σ=0.0259 are also close to the ideal values of 0.5000 and 0.03125. Furthermore, the PUF data from the five measured chips passed all applicable NIST SP800-22 and SP800-90B randomness tests, demonstrating high-quality randomness.

ISSCC 2026 / February 18, 2026 / 10:55 AM

Figure 25.8.6 shows the comparison with state-of-the-art works. This work achieves "zero" bit error without requiring access to secret PUF data or incurring bit-cell loss for stabilization. The effective area per bit, as well as the core energy, are highly competitive compared with prior works. Figure 25.8.7 presents the chip micrograph and the bit-cell layout.

Acknowledgement:

This work is supported in part by the Japan Science and Technology Agency (Grant No. JPMJCS24K9, Grant No. JPMJMS2214-5, Grant No. JPMJCR24U1), by the Japan Society for the Promotion of Science (Grant No. 24K07596, Grant No. 25K21173), and by the New Energy and Industrial Technology Development Organization (Grant No. JPNP14004). The authors also acknowledge Waseda University for instrument support for chip measurement, and VDEC, the University of Tokyo, in collaboration with Cadence Design Systems Inc., and Mentor Graphics Inc.

References:

[1] S.K. Mathew et al., "A 0.19pJ/b PVT-variation-tolerant hybrid physically unclonable function circuit for 100% stable secure key generation in 22nm CMOS," *ISSCC*, pp. 278-279, Feb. 2014. https://doi.org/10.1109/ISSCC.2014.6757433

[2] K. Liu et al., "A 373-F^2 0.21%-Native-BER EE SRAM Physically Unclonable Function With 2-D Power-Gated Bit Cells and VSS Bias-Based Dark-Bit Detection," *IEEE JSSC*, vol. 55, no. 6, pp. 1719-1732, June 2020. https://doi.org/10.1109/JSSC.2019.2963002

[3] S. S. Kudva et al., "High-density and low-power PUF designs in 5nm achieving 23× and 39× BER reduction after unstable bit detection and masking," *ISSCC*, pp. 302–303, Feb. 2024. https://doi.org/10.1109/ISSCC49657.2024.10454365

[4] B. Karpinskyy et al., "An Efficient Vth-Tilting PUF Design in 3nm GAA and 8nm FinFET Technologies," *ISSCC*, pp. 298-299, Feb. 2025. https://doi.org/10.1109/ISSCC49661.2025.10904512

[5] B. Driemeyer et al., "An eye-opening arbiter PUF for fingerprint generation using auto-error detection for PVT-robust masking and bit stabilization achieving a BER of 2e-8 in 28nm CMOS," *ISSCC*, pp. 300-301, Feb. 2025. https://doi.org/10.1109/ISSCC49661.2025.10904785

[6] T. Wang et al., "SRAM Physically Unclonable Function Extracting Static Entropy from Every Bitcell Transistor for 6 bit/bitcell and Data Fingerprinting Capability for Provenance Assurance," *A-SSCC*, pp. 1-3, Nov. 2024. https://doi.org/10.1109/A-SSCC60305.2024.10848665

[7] Y. He et al., "An Automatic Self-Checking and Healing Physically Unclonable Function (PUF) with <3×10-8 Bit Error Rate," *ISSCC*, pp. 506-507, Feb. 2021. https://doi.org/10.1109/ISSCC42613.2021.9365741

[8] J. Park et al., "A physically unclonable function combining a process mismatch amplifier in an oscillator collapse topology," *ISSCC*, pp. 504-505, Feb. 2021. https://doi.org/10.1109/ISSCC42613.2021.9365829

[9] J. Lee et al., "A 354F^2 leakage-based physically unclonable function with lossless stabilization through remapping for low-cost IoT security," *IEEE JSSC*, vol. 56, no. 2, pp. 648-657, Feb. 2021. https://doi.org/10.1109/JSSC.2020.3014386

[10] S. Xu et al., "A less than 6.5E–8 BER 36-way reconfigurable PUF with 4-bit stable responses per cell featuring machine learning-based best-configuration selection," accepted for presentation in *A-SSCC*, Nov. 2025.

[11] K. Liu et al., "A modeling attack resilient Strong PUF with feedback-SPN structure having <0.73% bit error rate through in-cell hot-carrier injection burn-in," *ISSCC*, pp. 502-503, Feb. 2021. https://doi.org/10.1109/ISSCC42613.2021.9365942

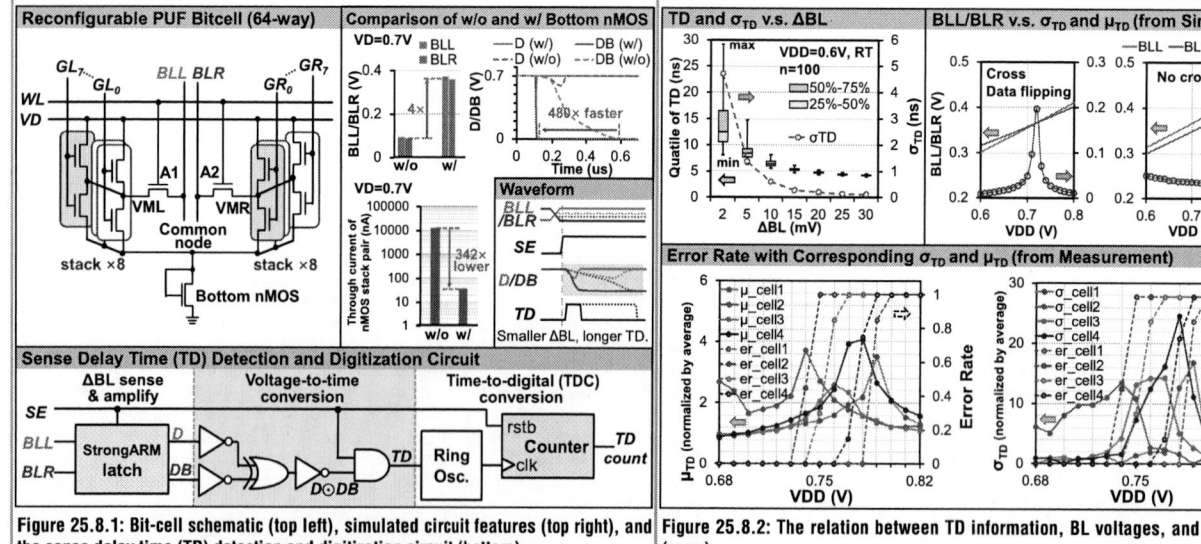

Figure 25.8.1: Bit-cell schematic (top left), simulated circuit features (top right), and the sense delay time (TD) detection and digitization circuit (bottom).

Figure 25.8.2: The relation between TD information, BL voltages, and data flipping (error).

ISSCC 2026 / SESSION 25 / HARDWARE SECURITY / 25.8

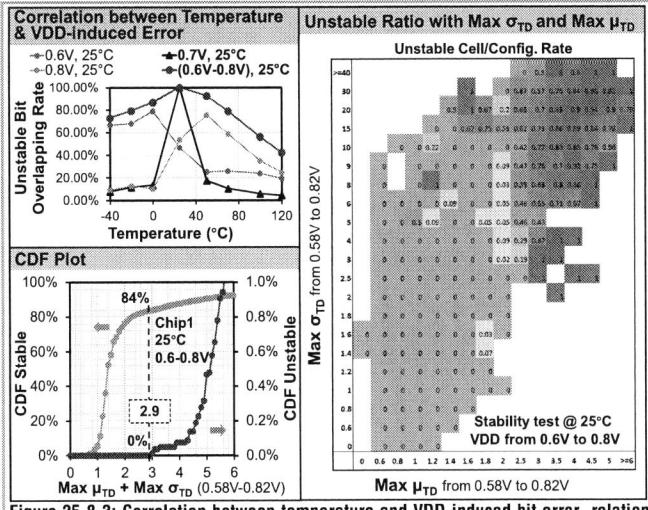

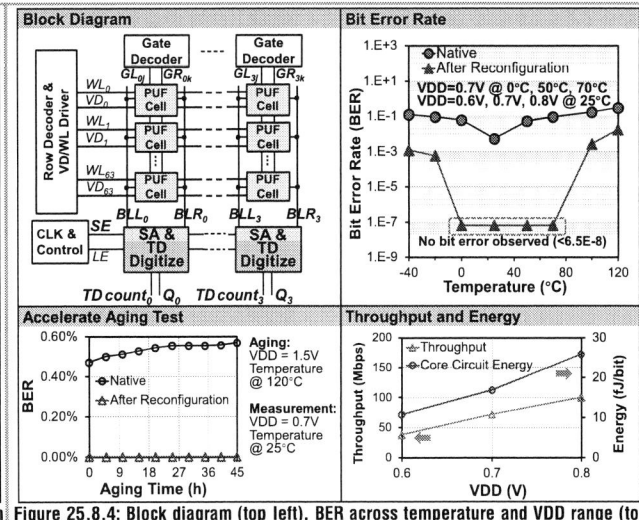

Figure 25.8.3: Correlation between temperature and VDD-induced bit error, relation map between max μ_{TD}, max σ_{TD}, and unstable config ratio, and the corresponding CDF plot.

Figure 25.8.4: Block diagram (top left), BER across temperature and VDD range (top right), BER against accelerated aging time (bottom left), throughput and energy (bottom right).

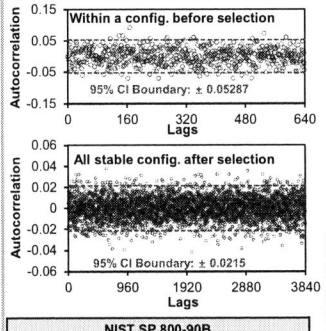

NIST SP 800-90B
(Selected configurations)

Test Name	PASS
IID Permutation	YES
χ^2 Independence	YES (Score=130.8053, dof=126)
χ^2 Goodness of Fit	YES (Score=20.1764, dof=9)
LRS Test	YES (Pr=0.97026)
Min-Entropy	0.93612

NIST SP 800-22
(Selected configs, 5chip × 6config = 30 groups)

Test Name	P-Value	Pass Rate	Pass?
Frequency	0.59333	30/30	YES
Block Frequency (M=32)	0.37701	30/30	YES
Runs	0.48643	30/30	YES
Longest Run of Ones	0.43879	30/30	YES
Cumulative Sums	0.58306	60/60	YES
FFT	0.37235	30/30	YES
Non-Overlapping Template Matching (m=6)	0.61219	591/600	YES
Serial (m=6)	0.52388	59/60	YES
Approximate Entropy (m=2)	0.60682	30/30	YES

Figure 25.8.5: Autocorrelation of raw data (top left), autocorrelation after selection (middle left), normalized hamming distances among 256b PUF data (top right), results of NIST SP800-90B tests and NIST SP800-22 tests (bottom).

Comparison table

	This Work	A-SSCC'25 [10]	ISSCC'25 [4]	ISSCC'25 [5]	A-SSCC'24 [6]	ISSCC'24 [3]	JSSC'21 [9]	ISSCC'21 [8]	ISSCC'21 [7]	JSSC'20 [2]
Technology [nm]	65	65	3	28	28	5	180	40	65	130
Native BER @ Nom. VT	**0.52%**	0.44%	N/A	3.49%	1.99%	1.94%	1.95%	0.43%	~10% [a]	0.29%	0.21%
Stabilization Method	Secure Reconfig (64-way)	Reconfig (36-way) + ML Prediction	Masking	Masking	TMV, Masking	Masking	Reconfig (2-way) / Masking	Reconfig (4-way), TMV, Masking	Reconfig (2-way), Masking	Masking
Without Secret Data Leakage?	Yes	No	No	Yes	No	No	No	No	No	No
Without Bitcell Loss?	Yes	Yes	No	No	No	No	No	No	No	No
Making Ratio	0	0	~75%	53%	50.6% [b]	23.2% 27.7%	15% [c]	3.64%	27%	67.4%
Aver. Stable Bits/Cell	6	4	0.25	0.47	2.96	0.768 0.723	0.85	0.963	0.73	0.326
Effective Cell Area/Bit (F²) *	253	674	145312	83894	271	758 5422	354	21675	814	1144
V_{DD} Range (V)	0.6-0.8	0.6-0.8	0.56-0.84	0.81-0.99	0.55-0.75	0.7-1.0	1.0-1.8	0.7-1.4	0.7-1.4	0.8-1.4
Temperature Range (°C)	0~70	-40~120	-40~125	-40~125	0~70	0~90	-80	-40~125	-40~125	-40~125
BER before Masking	0 (<6.5E-8)	0 (<6.51E-8)	N/A	0.0278	0.014 [d]	0.036 [d] 0.043 [d]	4.7E-3(R) [c]	1.1E-5	0.04	0.059 [d]
Stabilized BER Across Temp.	0 (<6.5E-8)	0 (<6.51E-8)	2.23E-3 [e]	2E-6	<2.22E-7	0.0118 5.5E-5	5.4E-3(M) [c]	9E-7	0 (<3.34E-8)	0 (<5.99E-7)
Core Energy (fJ/bit)	16.8 (0.7V)	1200 (0.6V) 2010 (0.7V)	N/A	259	226	84 8	465000	101	0.057	258

* Considering the bitcell loss ratio (post-masking).
[a]. Estimate from the BER after reconfiguration.
[b]. Initially 6 PUF bits/cell, 2.96 bits/cell after masking.
[c]. Masking and reconfiguration perform separately.
[d]. Estimate from the figures.
[e]. Including process and V_{DD} variations.

M = Masking; R = Reconfiguration.

Figure 25.8.6: Comparison table.

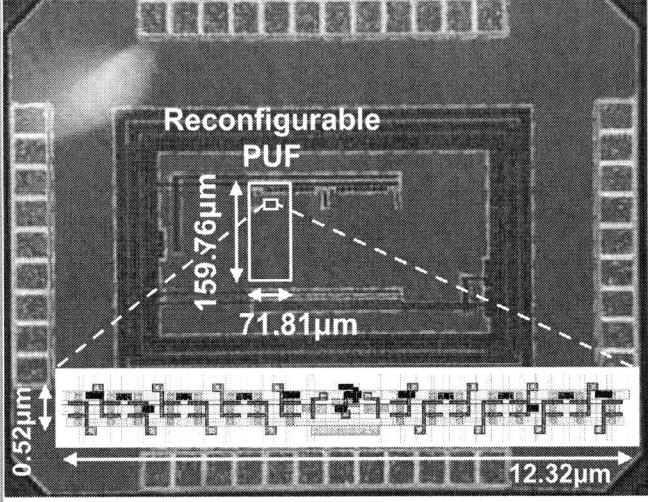

Figure 25.8.7: Chip micrograph and bit-cell layout.

• 2026 IEEE International Solid-State Circuits Conference

ISSCC 2026 / SESSION 25 / HARDWARE SECURITY / 25.9

25.9 A PVT Variation- and Attack-Tolerant Metastability-Based TRNG Using Binary Search in 2nm

Yelim Youn[1], Yong Lim[1,2], Jongmi Lee[1], Dongyeon Hong[1], Wan Kim[1], Yong-Sik Kwak[1], Kyoung-Jun Moon[1], Bogyeong Kang[1], Sangmin Yoo[1]

[1]Samsung Electronics, Hwaseong, Korea, [2]now with Yonsei University, Seoul, Korea

Abstract

This work presents a single source, metastability-based TRNG using binary search with offset-tunable comparator. The proposed TRNG operates without warm-up time and is robust against low-frequency noise, PVT variations and environmental attacks, while its stochastic model is analytically verified. Fabricated in 2nm MBCFET, the prototype occupies 320µm², operates at 20Mb/s with 1.8pJ/b. The measurement result of 320 chips shows NIST- and AIS-compliant randomness across all PVT corners.

It has become difficult to imagine daily life without smart devices, which store our valuable information from personal memories to sensitive financial data. To keep this information secure, these devices heavily rely on cryptographic protocols, for which a true-random-number generator (TRNG) is an essential building block. Since the quality and robustness of random numbers dedicate overall system security, reliable TRNG design has become an essential requirement for advanced SoC integration.

Metastability has been widely studied as a reliable entropy source for TRNGs. In [1], a back-to-back inverter structure with an auto-zeroing scheme is adopted to improve PVT tolerance. However, the large auto-zeroing capacitor, sized to dominate parasitic capacitances, imposes a fundamental speed limitation. In addition, since residual offsets remain even after auto-zeroing, it requires multiple entropy sources combined with XOR to achieve sufficient entropy. In [2], the SAR residue is used as the entropy source, however, it requires an external analog input to generate sufficient entropy, and relies on a bulky CDAC for the offset control, leading to increased area and reduced operation speed. Furthermore, stochastic analysis of the entropy source is not provided. In [3] and [4], TRNGs exploiting the comparator's inherent noise as the entropy source have been reported. Such approach requires iterative offset adjustment by observing long output bitstream, which results in start-up latency and the risk of increased autocorrelation. Moreover, it is vulnerable to voltage or temperature (VT) variations, necessitating additional regulation or filtering blocks at the system level. To overcome these limitations, this work introduces a single source, metastability-based TRNG using binary search that analytically validates its entropy source. The proposed TRNG operates without latency and achieves robustness under PVT variations with compact, digital-friendly implementation. It is practical solution for smart device applications, supporting reliable randomness in both continuous and intermittently invoked security functions.

Figure 25.9.1 provides a comparison between the prior comparator-based TRNG [3] and the proposed binary search based TRNG. The prior art relies on the comparator's inherent noise, where the offset is locked through a foreground calibration. To ensure statistical validity of outputs, the calibration is performed by monitoring long output bitstream with a counter and iteratively tuning the offset based on a look-up table (LUT). Thus, it inherently requires warm-up time and increases risk of autocorrelation since the output affects subsequent bit generation. Also, this architecture is vulnerable to offset shifts from VT variations or attacks, so faulty outputs can be generated until the calibration logic detects the error and re-calibrates. To mitigate low frequency noise such as flicker noise, the prior work employed multi-stage, computation-heavy post processing circuit such as 2-stage middle square method. In our proposed TRNG, the cumulative statistical property of the comparator revealed by the binary search is exploited. In every cycle, the random bit is produced following the binary search with an offset-tunable comparator. This approach eliminates not only warm-up delay but also risk of autocorrelation. Since binary search tracks offset variation in every cycle, any abrupt offset disturbance affects only a single cycle and the design exhibits negligible sensitivity to low frequency noise, ensuring reliable randomness. This advantage becomes more pronounced under harsh operating environments. Thanks to its robustness against PVT variations, the proposed TRNG achieves NIST- and AIS-compliant randomness with only a simple corrector, Dichtl's H3 [5].

In Fig. 25.9.2, both prior and proposed TRNGs are formulated within the same statistical framework. The analysis shows that the proposed entropy source delivers statistically valid randomness without requiring long-term calibration. Fig. 25.9.2 (top right) plots the probability density function (PDF) of the comparator decision noise as a function of the input-referred noise (v_{os}). The noise is modeled as a Gaussian distribution and the probability of obtaining '1' (P_1) can be expressed as $p_{comp}(1|v_{os})$ [6], where σ is input-referred noise. In the prior work [3, 4], the offset is locked by the foreground calibration, so the subsequent output directly reflects $p_{comp}(1|v_{os})$. To fulfill the NIST randomness requirement with this approach, the comparator offset must be remained within approximately ±0.24σ [4], corresponding P_1 in the range of 0.4 to 0.6. Since the required step size for offset tuning (1 LSB) is smaller than 1σ, the prior work observes long-term output bitstream and iteratively trims the comparator offset until the operating point falls into the NIST zone. Only after such trimming, can a statistically valid random sequence be obtained. However, the

comparator offset tuning must be repeated whenever the operating point drifts outside the NIST zone. This necessity for offset fine tuning results in long warm-up time and increased system complexity.

The proposed entropy source is characterized by the comparator's cumulative probability distribution, which can be derived in two steps (Fig. 25.9.2, Eq. 3). First, the probability of each DAC code (P_{DAC}) resulting of the binary search is expressed as a function of v_{os} and p_{comp}. Second, the overall bit probability of the proposed TRNG is obtained by weighting the comparator probability by the corresponding DAC code probabilities and summing over all DAC codes. A fundamental property of the binary search is that it produces a code distribution nearly symmetric around the zero-bias point (Fig. 25.9.2 bottom right). When this mirrored probability property is combined with the comparator decision model, the complementary relation, $p_{comp}(1|\Delta) + p_{comp}(1|-\Delta) = 1$, ensures that contribution from symmetric DAC codes naturally pulls the final probability $p_{proposed}$ toward 0.5. The calculated $p_{proposed}$ remains within the range of 0.48 to 0.52, while it varies slightly with target codes due to asymmetry of DAC code distribution. This ensures validity of the proposed entropy source.

In the proposed TRNG, a Strong-ARM comparator is employed as the entropy source, while the offset-tuning DAC is implemented by controlling the number of enabled input transistors, with both the negative and positive inputs tied to the supplies. The DAC consists of a 4b thermometer for MSB, and a 2b thermometer and a 2b binary for LSB array. Since the DAC linearity around the target code directly affects the symmetry of the binary search code probability distribution, a dynamic element matching technique is applied to suppress non-linearity. In parallel with the DAC, dummy input transistors are always enabled to ensure sufficient regeneration speed, even at minimum DAC code. The comparator noise level is designed to be around 10 LSB of the offset-tuning DAC, where the reset switches at the output nodes (VOLN, VOLP) and latch NMOS transistors are the dominant noise contributors. These devices are intentionally downsized so that the kT/C noise, left on the output nodes at reset release, becomes the primary noise source, while the reduced g_m of the latch NMOS further increases noise during the early stage of latch regeneration. Meanwhile, the reset switches, though minimized, remain sufficiently large to guarantee complete resetting at the asynchronous clock speed. This design ensures that the overall decision noise is dominated by well-defined thermal noise, and reduces sensitive to variations in input transistor sizing throughout the binary search. Unlike the prior work [3], the comparator incorporates memory-clearing reset switches to eliminate the dependency on previous decisions and prevent the history-induced bias.

Residual asymmetry arising from device mismatch, DAC non-idealities, or state-dependent noise property during the search process is effectively mitigated using Dichtl's H3 post processing. By leveraging the mirrored probability property of the binary search, a simple corrector is sufficient to suppress the remaining bias and satisfy statistical requirements with minimum overhead.

As shown in Fig. 25.9.3 (bottom), the proposed TRNG operates in three phases: (1) binary search, (2) random bit generation, and (3) reset. During the binary search, the comparator repeatedly triggered by the CLK_COMP signal, which is asynchronously generated clock using comparator outputs [7]. After each comparator decision, the search logic updates the DAC code prior to the next comparison begins. Once the binary search is completed, the comparator is triggered once more to generate the random bit output. As the final step, the DAC code is reset to center ensuring consistent statistical behavior from same starting point. Since the binary search is performed in every cycle, the proposed TRNG operates without warm-up time and the low frequency noise have negligible impact on the randomness quality. Also, sudden offset shifts induced by VT attacks affect only a single cycle, after which the search process re-centers automatically. Finally, it does not require a strict jitter specification and can function even with a simple digital enable signal instead of a dedicated clock. These advantages make this architecture well suited for SoC integration.

Figure 25.9.7 shows the die micrograph of the prototype, fabricated in 2nm MBCFET process, occupying an active area of 320µm², achieving 20Mb/s throughput with 51µA at 0.7V. A total of 320 chips with 5 corners were measured for randomness tests under 3

440 • 2026 IEEE International Solid-State Circuits Conference

979-8-3315-8937-0/26 $31.00 © 2026 IEEE

voltages and 3 temperatures.

In line with the AIS-31 version 3 guidelines, the evaluation is performed on the post-processed out with Dichtl's H3. Fig. 25.9.4 (top left) shows that the evaluated P_1 meets the strict AIS-31 monobit requirement ($P_1 \in [0.493, 0.507]$). The autocorrelation of the output sequence shows no significant correlation across lags, confirming the independence between bits (Fig. 25.9.4 top right). In all PVT corners, the averaged min-entropy remains above 0.998, indicating that entropy quality is maintained without noticeable sensitivity (Fig. 25.9.4 bottom left). The proposed TRNG achieved high energy efficiency, while maintaining min-entropy above 0.998 across the supply range from 0.55V to 1V (Fig. 4 bottom right). Figure 25.9.5 summarizes the pass rates of NIST SP 800-22 tests and averaged min-entropy of NIST SP800-90B non-IID tests. These results further demonstrate that the proposed TRNG satisfies stringent statistical requirements under all evaluated conditions.

Figure 25.9.6 summarizes a comparison with state-of-the-art TRNGs. Compared to prior works that do not specify explicit system-level requirements such as warm-up time and supply regulation, the proposed design achieves a moderate throughput and a compact area. Moreover, it supports operation across a wide temperature range, making it highly suitable for SoC integration and mass production.

References:

[1] J. Lee et al., "A 1.7 pJ/bit 10 MHz Calibration-Free PVT Variation and Mismatch Tolerant Latch-Based True Random Number Generator in 4 nm FinFET," *IEEE Symp. VLSI Circuits*, pp. 1-3, June 2025.
https://doi.org/10.23919/VLSITechnologyandCir65189.2025.11075211
[2] A. Jayaraj, N. Nitin Gujarathi, I. Venkatesh, and A. Sanyal, "0.6–1.2 V, 0.22 pJ/bit True Random Number Generator Based on SAR ADC," *IEEE Transactions on Circuits and Systems II: Express Briefs*, vol. 67, no. 10, pp. 1765-1769, Oct. 2020.
https://doi.org/10.1109/TCSII.2019.2949775
[3] J. Kim and H. Chae, "A 10-Gb/s True Random Number Generator Using ML-Resistant Middle Square Method," *IEEE Journal of Solid-State Circuits*, vol. 59, no. 7, pp. 2321-2329, July 2024. https://doi.org/10.1109/JSSC.2023.3346428
[4] S.K. Mathew et al., "2.4 Gbps, 7 mW All-Digital PVT-Variation Tolerant True Random Number Generator for 45 nm CMOS High-Performance Microprocessors," *IEEE Journal of Solid-State Circuits*, vol. 47, no. 11, pp. 2807-2821, Nov. 2012.
https://doi.org/10.1109/JSSC.2012.2217631
[5] M. Dichtl, "Bad and Good Ways of Post-processing Biased Physical Random Numbers," *International Workshop on Fast Software Encryption*, *Lecture Notes in Computer Science*, Springer, vol 4593, pp. 137-152, March 2007.
https://doi.org/10.1007/978-3-540-74619-5_9
[6] P. Nuzzo, F. De Bernardinis, P. Terreni, and G. Van der Plas, "Noise Analysis of Regenerative Comparators for Reconfigurable ADC Architectures," *IEEE Transactions on Circuits and Systems I: Regular Papers*, vol. 55, no. 6, pp. 1441-1454, July 2008.
https://doi.org/10.1109/TCSI.2008.917991
[7] Shuo-Wei Mike Chen and R. W. Brodersen, "A 6b 600MS/s 5.3mW Asynchronous ADC in 0.13/spl mu/m CMOS," *ISSCC*, pp. 1-2, Feb, 2006.
https://doi.org/10.1109/ISSCC.2006.1696298
[8] J. Park et al., "A 60Mb/s TRNG with PVT-Variation-Tolerant Design Based on STR in 4nm," *ISSCC*, pp. 310-311, Feb, 2024.
https://doi.org/10.1109/ISSCC49657.2024.10454373

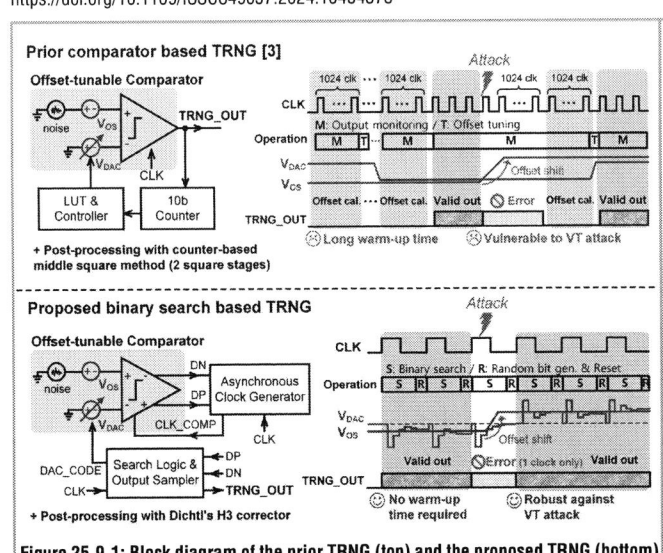

Figure 25.9.1: Block diagram of the prior TRNG (top) and the proposed TRNG (bottom) with system operation diagram at startup and under VT attack.

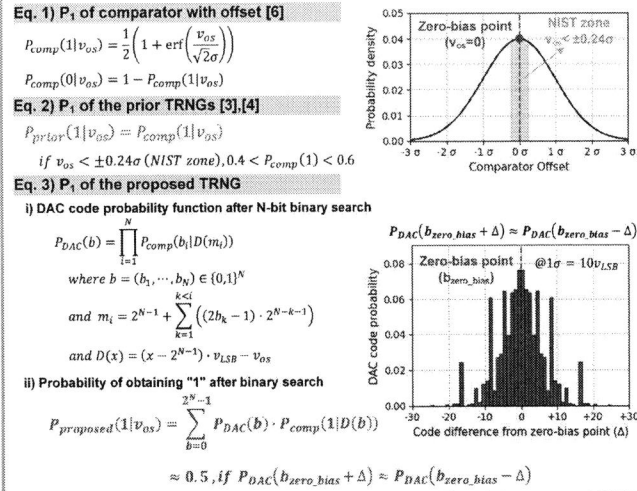

Figure 25.9.2: Probability characteristics of the prior TRNG and the proposed TRNG with probability density function of comparator decision with respect to offset (top) and DAC code probability distribution resulting of the binary search (bottom).

ISSCC 2026 / SESSION 25 / HARDWARE SECURITY / 25.9

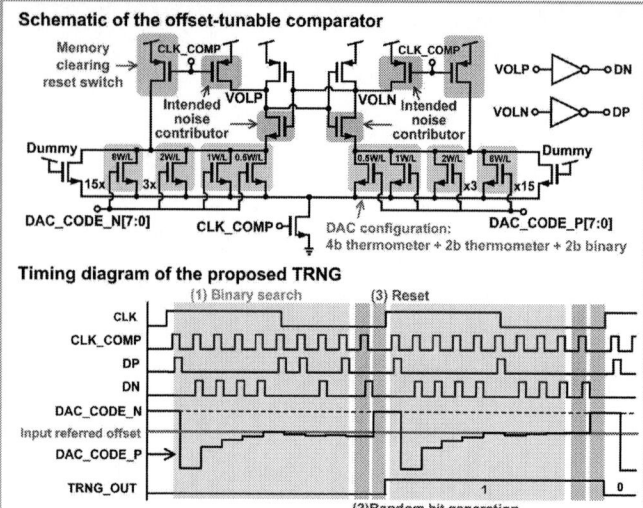

Schematic of the offset-tunable comparator

Timing diagram of the proposed TRNG

Figure 25.9.3: Schematic of the offset-tunable comparator (top) and timing diagram of the proposed TRNG (bottom).

Figure 25.9.4: Measured probability of obtaining "1" (P_1) of 320 chips under 3 temperatures and 3 voltages, autocorrelation function, averaged min-entropy across all PVT conditions, and energy efficiency and min-entropy with various supply voltages.

NIST SP 800-22 tests	Avg. p-value	Pass rate		NIST SP 800-90B non-IID tests	Avg. min entropy
Frequency Test (Monobit)	0.499	98.82 %		MCV	0.9966
Frequency Test within a Block	0.498	99.06 %		Collison	0.9274
Run Test	0.507	99.37 %		Markov	0.9987
Longest Run of Ones in a Block	0.508	98.78 %		Compression	0.8751
Binary Matrix Rank Test	0.506	99.23 %		T-Tuple	0.9275
Discrete Fourier Transform Test	0.493	98.85 %		LRS	0.9808
Non-Overlapping Template Matching Test	0.501	98.71 %		MultiMCW	0.9967
Overlapping Template Matching Test	0.496	98.57 %		Lag	0.9965
Maurer's Universal Statistical test	0.496	99.03 %		MultiMMC	0.9968
Linear Complexity Test	0.504	98.78 %		LZ78Y	0.9967
Serial Test ($\nabla\psi^2$)	0.491	99.30 %			
Serial Test ($\nabla^2\psi^2$)	0.494	98.85 %		**2Mbit data**	
Approximate Entropy Test	0.492	98.71 %		with Dichtl's H3 post processing	
Cummulative Sums (Forward) Test	0.500	98.82 %		Total 320 chips	
Cummulative Sums (Reverse) Test	0.499	98.92 %		5 corners (TT,SS,FF,SF,FS)	
Random Excursions Test	0.500	98.55 %		3 temp. (-40°C,25°C,125°C)	
Random Excursions Variant Test	0.502	98.94 %		3 voltages (0.63V, 0.7V, 0.77V)	

Figure 25.9.5: NIST SP 800-22 tests result table (left) and NIST SP 800-90B non-IID tests result table (right) of 320 chips with 5 corners (TT, SS, FF, SF, FS), 3 temperatures (-40°C, 25°C, 125°C), 3 voltages (0.63V, 0.7V, 0.77V).

	This Work	VLSI 2025 [1]	TCASII 2020 [2]	JSSC 2024 [3]	JSSC 2012 [4]	ISSCC 2024 [8]
Process [nm]	2	4	65	28	45	4
Entropy source	Metastability (Comparator)	Metastability (Latch)	SAR residue	Metastability (Comparator)	Metastability (Comparator)	Jitter (STR)
Supply voltage [V]	0.63 ~ 0.77	0.75 ~ 0.95	0.6 ~ 1.2	0.7 ~ 1.3	0.28 ~ 1.35	0.675 ~ 0.935
Temperature [°C]	-40 ~ 125	-40 ~ 85	-5 ~ 50	-40 ~ 120	50	-40 ~ 125
Power [µW]	36 @ 0.7V	N/A	0.27 @0.8V	1210 @0.7V	7000	N/A
Throughput [Mbps]	20	10	1.25	10000	2400 @1.1V	60
Area [µm²]	320	1679	90000	2254	4004	1289.75
Energy efficiency [pJ/bit]	1.8	1.7	0.22	0.121	2.9	N/A
Number of entropy sources	1	4	1	1	1	45
Warm-up time requirement	No	No	No	Yes	Yes	No
Supply regulation requirement	No	No	No	Yes	Yes	Yes
Evaluated process corner	5 corners TT,SS,FF,SF,FS	3 corners TT,SS,FF	N/A	N/A	N/A	5 corners TT,SS,FF,SF,FS

Figure 25.9.6: Comparison table of the proposed TRNG and prior works.

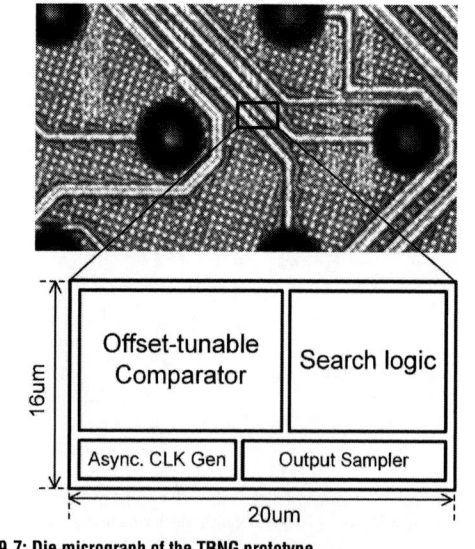

Figure 25.9.7: Die micrograph of the TRNG prototype.

- 2026 IEEE International Solid-State Circuits Conference

ISSCC 2026 / SESSION 25 / HARDWARE SECURITY / 25.10

25.10 A 65nm 0.066pJ/b Floating-Latch-Based True Random Number Generator Resilient to Power-Noise Injection Attacks

Kai Cheng[1], Yunbo Huang[1], Zunsong Yang[1], Li Wang[1], Hongyu Ren[1], Xiaoyu Shan[1], Pui-In Mak[2], Yong Chen[3], Bo Li[1]

[1]Institute of Microelectronics of the Chinese Academy of Sciences, Beijing, China, [2]University of Macau, Macau, China, [3]Tsinghua University, Beijing, China

Abstract

In Paper 25.10 a floating-latch-based true random number generator (FL-TRNG) is presented that achieves an energy efficiency of 0.066pJ/b by leveraging a floating reservoir capacitor as its power supply. This architecture inherently offers strong resilience against power-supply noise injection attacks. Furthermore, current-starved inverters are employed within the latch to perform mismatch compensation and enhance PVT tolerance.

True random number generators (TRNGs) play a crucial role in ensuring security across various Internet of Things (IoT) applications, including device authentication, secure boot, encrypted sensor data transmission, and dynamic key generation [1,9-10]. These applications demand not only robust cryptographic guarantees but also strict adherence to resource constraints, imposing simultaneous requirements for high energy efficiency, moderate data rates, and resilience against environmental disturbances. While ring oscillator (RO)-based TRNGs are capable of higher data rates via jitter accumulation [1-4], they suffer from high power consumption and supply noise sensitivity, limiting the applications in energy-sensitive and noisy IoT scenarios. Metastability-based approaches, such as latch-based TRNGs, are attractive due to their simplicity and high energy efficiency. For example, the prior work in [5] achieves an energy efficiency of 0.186pJ/b. Yet, its limited throughput (7.87kb/s at 0.3V) falls short for medium- to high-speed applications.

To overcome these challenges, we devise a floating-latch-based TRNG (FL-TRNG) architecture, where the kernel latch is powered by a floating reservoir capacitor, achieving both low power consumption and resilience to power injection attacks. Current-starved inverters are employed in the latch, and a successive approximation register (SAR) logic-based calibration loop is introduced to adaptively adjust the bias voltage of the current-starved inverters, thereby compensating for process-induced mismatch and enhancing the robustness of the latch entropy source.

In conventional latch-based TRNGs, the latch is first biased in a metastable state and then switches state under the influence of noise. Before the transition, the inverters within the latch are shorted, causing both PMOS and NMOS transistors to conduct simultaneously. This induces a static current (I_{short}) path from power supply to ground, the magnitude of which depends on the supply voltage (V_{DD}) and the transistor dimensions. The resulting power consumption is also proportional to the duration of the metastable state ($P_{short}=I_{short}V_{DD}t_{short}$), which dominates the overall power dissipation of the latch. Downscaling V_{DD} is a common approach to reduce power [5]. However, this significantly slows down the latch operation, thereby increasing the t_{short}. As a result, the static current remains the dominant contributor to power consumption. The FL-TRNG addresses this issue by periodically powering the latch using a capacitor C_{RES}. In the equalization phase, C_{RES} supplies the voltage to maintain latch metastability. Importantly, no short-circuit current flows through the inverter during this phase. During the amplification phase, V_{DD} recharges C_{RES}. Furthermore, compared to the conventional latch structures, the capacitor-powered scheme exhibits lower dynamic power and eliminates leakage current. As shown in Fig. 25.10.1, for identical transistor sizes and operating frequency, the average current of the floating latch architecture is significantly reduced by 98.7% at 1V V_{DD}, and 77.6% at 0.3V, respectively. In addition to its power efficiency, the floating-latch architecture demonstrates robust resilience against power supply noise attacks.

Figure 25.10.2 outlines the top architecture of the FL-TRNG (top-left), consisting of a floating latch entropy source, a SAR-Logic-based calibration loop, and the clock generator (top right). By monitoring the proportion between zeros and ones in the finite-length output sequence, the calibration circuit compensates for data offset caused by latch mismatch and (process, voltage, and temperature) PVT variations. Specifically, the integrator accumulates the output of the comparator for 128 cycles and decides whether the sum falls within the range [-2, 2]. If so, the enable signal transitions from "1" to "0", the SAR logic halts and retains the current code in its register, the calibration loop is shut down, and V_{CAL} remains constant. Otherwise, the Up or Down signal remains active "1", and the calibration process continues until the condition is satisfied. To mitigate the kick-back noise introduced by the switches connected to the reservoir capacitor C_{RES}, which causes significant fluctuation in V_{CAL}, the size ratio of M_{N2}:M_{N1} is set to 3:1. The V_{CAL} is connected to transistor M_{N1}, which reduces the kick-back noise without adjusting the transistor size, thereby avoiding excessive flicker noise by bias-transistor. The entropy generation in the FL-TRNG operates in three primary phases, as illustrated in the timing diagram in Fig. 25.10.2 (middle right). In the equalization phase, the latch is biased at its metastable point and powered by C_{RES}. The second phase is the amplification phase, since the charge-powering operation involves charge equalization between the upper and lower plates of C_{RES} through the inverters, the voltage on the capacitor settles to a level between V_{DD} and $V_{DD}/2$. The exact value depends on the capacitance and the duration of the metastable phase. As a result, when the latch is perturbed by noise and begins to amplify, the output voltage swing does not reach the V_{DD}.

If the capacitor were charged simultaneously during amplification, the switching activity between the supply and the capacitor would introduce clock feedthrough, adversely affecting the latch's transition. To avoid this issue while ensuring full-swing output, the design splits the amplification into two sub-phases: The first amplification uses the voltage present on the capacitor plates. In the second step, the capacitor is recharged while the latch is directly powered by V_{DD}, enabling a second amplification stage. This approach prevents supply noise from interfering during the first entropy amplification, achieving a full-swing output before comparison and thereby relaxing the matching requirements of the quantization comparator for the next stage. Additionally, this structure eliminates the need for a large capacitor C_{RES}.

As previously mentioned, the FL-TRNG utilizes a charge-powering technique to reduce the power consumption of the latch entropy source while significantly enhancing resilience against power supply noise attacks. As observed in Fig. 25.10.3, both power efficiency and noise immunity are closely related to the value of C_{RES}. Transient simulation results show that at a supply voltage of 1V and an operating frequency of 1MHz, when the C_{RES}=0.2pF, the limited voltage cannot drive the latch into positive feedback operation. In contrast, with $C_{RES} \geq 0.4pF$, the latch functions correctly. Moreover, the size of C_{RES} exhibits a linear relationship with the average current; larger C_{RES} leads to higher power consumption. Beyond its impact on latch gain and power, the architecture also offers improved noise immunity. As illustrated in Fig. 25.10.3 (bottom), since the latch is powered by C_{RES} during the metastable phase, supply noise injection attacks are ideally isolated. In practice, due to parasitic capacitance (C_{DS}) in the switches between the V_{DD} and C_{RES}, residual noise influence remains. The extent of this noise influence is determined by the ratio between C_{DS} and C_{RES}. Thus, a larger C_{RES} results in weaker power supply noise coupling. With a 200mV,100MHz supply noise injection, simulations show that the input-coupling noise of the FL-latch is reduced by >87% compared with the conventional latch structure, demonstrating substantially enhanced resilience to power supply noise attacks. In this design, C_{RES} is realized by a 0.5pF metal-oxide-metal (MOM) capacitor.

Fabricated in 65nm CMOS, the TRNG prototype occupies a core area of 2171µm^2 (die micrograph shown in Fig. 25.10.7). Figure 25.10.4 summarizes the randomness measurement results of the TRNG. The autocorrelation measurement performed on 1M consecutive bits, with lags ranging from 1 to 5000, shows that all values reside within the 95% confidence interval of ±0.00195. The fast Fourier transform (FFT) analysis applied to the same 1Mb stream using a Hanning window further confirms the randomness, exhibiting a flat amplitude spectrum. A speckle pattern visualization of the bit stream is also provided, showing no discernible color bias. Additionally, the figure plots the energy efficiency of the FL-TRNG across supply voltages from 0.3V to 1.3V at a fixed data rate of 1Mb/s. The corresponding energy efficiency ranges from 0.066 to 3.35pJ/b, demonstrating outstanding efficiency especially at near-threshold voltages.

Figure 25.10.5 (top) presents the measured Shannon entropy of the FL-latch TRNG under different voltage and temperature (VT) conditions, along with its performance under frequency injection attacks (bottom). As shown in the top subfigure, the TRNG achieves excellent entropy output (Shannon entropy > 0.95) across a wide supply voltage range from 0.3V to 1.3V and a temperature range from –40 to 120°C, while operating at a clock frequency of 1MHz. The bottom subfigure demonstrates the entropy performance under frequency injection attacks, with noise frequency swept from 1 to 100MHz and amplitude from 0 to 500mV at a 1V supply. Remarkably, the Shannon entropy remains consistently above 0.95 under all tested noise conditions, confirming the robustness of the FL-TRNG architecture against supply noise attacks.

In Fig. 25.10.6, the randomness of the generated bitstream was further validated using the NIST SP 800-20 (top-left) and NIST SP 800-90B (bottom-left) test suites with a 10Mb sequence. The results confirm that the output sequences exhibit strong statistical randomness. Furthermore, as summarized in Fig. 25.10.6 (right), comparing the performance with prior TRNGs demonstrates that the design achieves significantly superior energy efficiency. The performance of the FL-TRNG is summarized and compared with prior works in Fig. 25.10.7 (bottom right). Thanks to the floating latch technique, our proposed TRNG achieves the best energy efficiency (0.066pJ/b) among the state-of-the-art designs in the benchmark plot, while also achieving one of the smallest area footprints.

Acknowledgement:
This work was supported by the National Natural Science Foundation of China under Grant 62574222. Corresponding authors: Yunbo Huang, Zunsong Yang, Yong Chen.

References:
[1] J. Hao et al., "A 98fJ/bit Current-Starved-Ring-Oscillator-based TRNG with High PVT Tolerance and Resilience to Frequency Injection Attack to 1V," *IEEE CICC*, pp. 1-2, Apr. 2024. https://doi.org/10.1109/CICC60959.2024.10528979
[2] J. Park et al., "A 60Mb/s TRNG with PVT-Variation-Tolerant Design Based on STR in 4nm," *ISSCC*, 2024, vol. 67, pp. 310-311, Feb.2024. https://doi.org/10.1109/ISSCC49657.2024.10454373
[3] Y. He et al., "A Fully Synthesizable 100Mbps Edge-Chasing True Random Number Generator," *IEEE Symp. VLSI Circuits*, pp. 1-2, June 2023. https://doi.org/10.23919/VLSITechnologyandCir57934.2023.10185323
[4] Y. Cao et al., "A New Energy-Efficient and High Throughput Two-Phase Multi-Bit per Cycle Ring Oscillator-Based True Random Number Generator," *IEEE TCAS-I*, vol. 69, no. 1, pp. 272-283, Jun. 2022. https://doi.org/10.1109/TCSI.2021.3087512
[5] R. Zhang et al., "A 0.186-pJ per Bit Latch-Based True Random Number Generator Featuring Mismatch Compensation and Random Noise Enhancement," *IEEE JSSC*, vol. 57, no. 8, pp. 2498-2508, Mar. 2022. https://doi.org/10.1109/JSSC.2021.3137312
[6] J. Lee et al., "A 1.7 pJ/bit 10 MHz Calibration-Free PVT Variation and Mismatch Tolerant Latch-Based True Random Number Generator in 4 nm FinFET," *IEEE Symp. VLSI Circuits*, pp. 1-3, June 2025. https://doi.org/10.23919/VLSITechnologyandCir65189.2025.11075211
[7] J. Kim et al., "A 10-Gb/s True Random Number Generator Using ML-Resistant Middle Square Method," *IEEE JSSC*, vol. 59, no. 7, pp. 2321-2329, Jul. 2024. https://doi.org/10.1109/JSSC.2023.3346428
[8] V. R. Pamula et al., "An All-Digital True-Random-Number Generator with Integrated De-correlation and Bias Correction at 3.2-to-86 MB/S, 2.58 PJ/Bit in 65-NM CMOS," *IEEE Symp. VLSI Circuits*, pp. 1-2. Oct. 2018. https://doi.org/10.1109/VLSIC.2018.8502375
[9] M. Kim et al., "A 82-nW Chaotic Map True Random Number Generator Based on a Sub-Ranging SAR ADC," *IEEE JSSC*, vol. 52, no. 7, pp. 1953-1965, Jul. 2017. https://doi.org/10.1109/JSSC.2017.2694833
[10] S. K. Satpathy et al., "An All-Digital Unified Physically Unclonable Function and True Random Number Generator Featuring Self-Calibrating Hierarchical Von Neumann Extraction in 14-nm Tri-gate CMOS," *IEEE JSSC*, vol. 54, no. 4, pp. 1074-1085, Jan. 2019. https://doi.org/10.1109/JSSC.2018.2886350

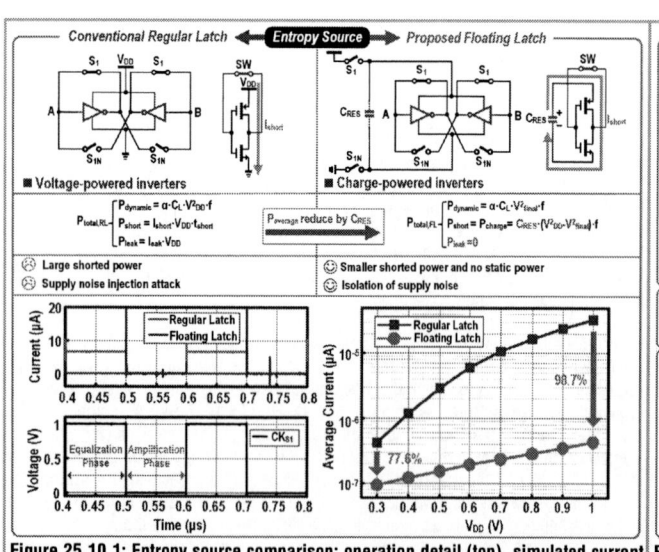

Figure 25.10.1: Entropy source comparison: operation detail (top), simulated current reduction (bottom).

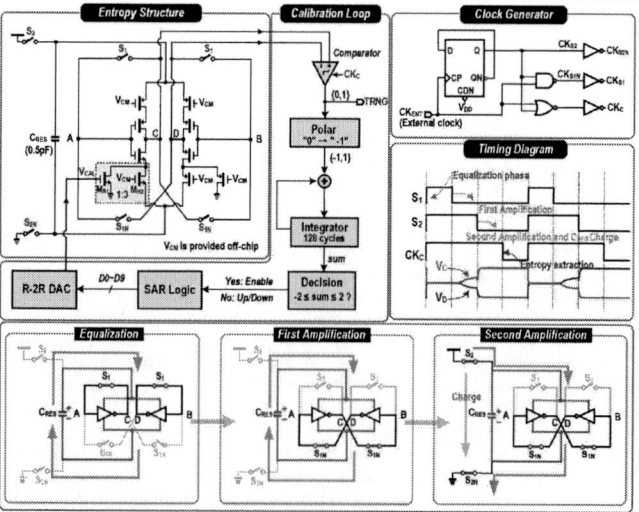

Figure 25.10.2: Architecture of the FL-TRNG (top-left), timing diagram (top-right), and operation details (bottom).

ISSCC 2026 / SESSION 25 / HARDWARE SECURITY / 25.10

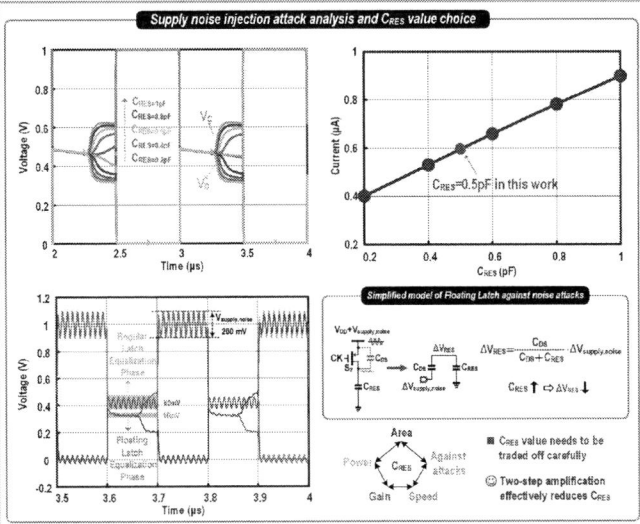

Figure 25.10.3: Simulation results of the supply noise injection attack analysis and optimization of C_{RES} value.

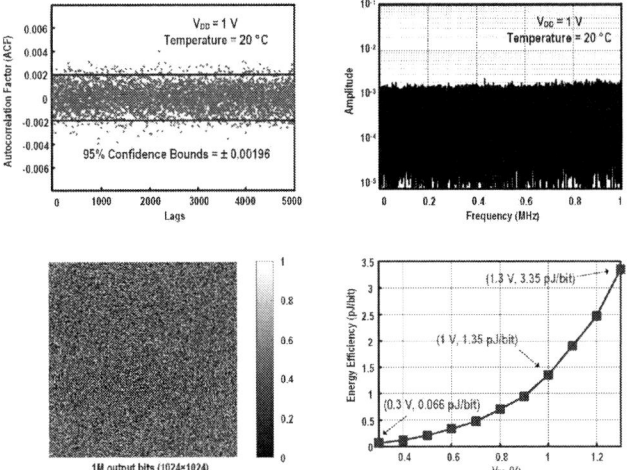

Figure 25.10.4: Measured results based on ACF, FFT, speckle diagram, and energy efficiency versus supply voltage.

Figure 25.10.5 plots (Shannon Entropy vs V_{DD}, Temperature, Noise Amplitude, Noise Frequency)

Figure 25.10.5: Measured Shannon entropy under different VT conditions, and frequency injection attack results.

NIST SP 800-22	P-value	Pass Rate
Frequency	0.350485	10/10
Block Frequency	0.534146	10/10
Cumulative Sums	0.911413	10/10
Runs	0.035174	10/10
Longest Run of Ones	0.911413	10/10
Rank	0.911413	10/10
Discrete Fourier Transform	0.122325	10/10
Non-Overlapping Template	PASS*	PASS*
Overlapping Template	0.534146	10/10
Universal Statistical	0.328236	PASS
Approximate Entropy	0.911413	10/10
Random Excursions	PASS*	PASS*
Random Excursions Variant	PASS*	PASS*
Serial	0.122325	10/10
Linear Complexity	0.350485	10/10
NIST Pub 800-90B	1 V, 20 °C	
Chi-Square Independence	P-value =0.8836	
Chi-Square Goodness-of-fit	P-value =0.66771	
LRS Test	Pr (X ≥1) = 0.9993	
IID Permutation Tests	PASS*	
Minimum Entropy	0.99488	

* With all sub tests passed.

	This Work	VLSI'25 [6]	CICC'24 [1]	JSSC'22 [5]	JSSC'24 [7]	VLSI'18 [8]
CMOS (nm)	65	4	40	130	28	65
Entropy Source	Meta-Stability	Meta-Stability	Jitter	Meta-Stability	Meta-Stability	Meta-Stability
Key Structure	Floating Latch	Latch	CSRO	Latch	StrongARM Latch	StrongARM Latch
V_{DD} Range (V)	0.3-1.3	0.65-0.95	0.6-1.3	0.3-1	0.7-1.3	0.53-1
Temperature Range (°C)	-40-120	-40-85	-65-140	-40-120	-40-120	-20-100
Throughput (Mb/s)	1	10	40	0.00787	10000	3.2-86
Power Consumption (μW)	0.066 @0.3 V	17 @0.75 V	3.916 @0.6 V	0.00147	1210 @0.7 V	8.33-523
Energy Efficiency (pJ/bit)	0.066 @0.3 V	1.7 @0.75 V	0.098 @0.6 V	0.186	0.121 @0.7 V	2.58 @0.53 V
Core Area ($\mu m^2/10^5 F^2$)	2171 /0.514	1679 /105	331.5 /0.207	5561 /0.329	2254 /2.875	10000 /2.37
Calibration or Tuning?	Yes	No	No	No	Yes	Yes
Post-Processing	No	No	No	Yes	Yes	Yes
Power Attack Tolerant	Yes	Yes	Yes	Yes	Yes	Yes

Figure 25.10.6: Performance summary and comparison with the state-of-the-art.

Area Breakdown

Main Circuits		Area (μm²)
A	Floating Latch	324
B	Comparator	36
C	Clock Generator	24
D	Digital Calibration	900
E	RDAC	887
Total Area		2171
[Normalized by F (×10⁶ F²)]		[0.514]

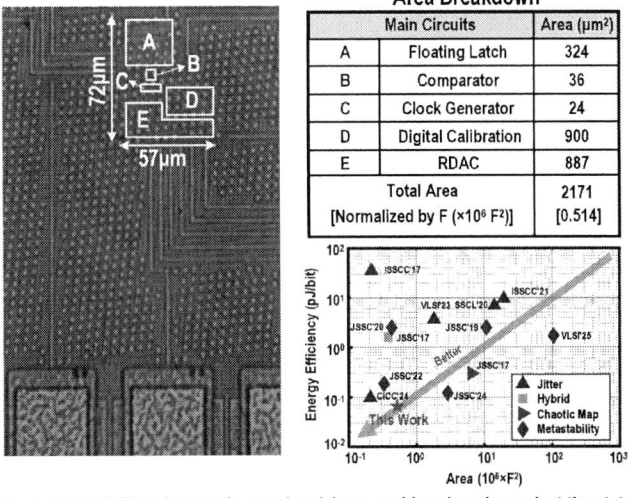

Figure 25.10.7: Die micrograph, area breakdown, and benchmarks against the state-of-the-art TRNG.

• 2026 IEEE International Solid-State Circuits Conference

ISSCC 2026 / SESSION 26 / COMPUTE POWER AND SUPPLY MODULATORS / OVERVIEW

Session 26 Overview: *Compute Power and Supply Modulators*
POWER MANAGEMENT SUBCOMMITTEE

Session Chair: Nicolas Butzen
Intel, Hillsboro, OR

Session Co-Chair: Dongsu Kim
Samsung, Hwaseong, Korea

As the density of compute is rapidly increasing at the CPU/XPU, rack, and datacenter levels, there is a tremendous challenge in providing power to these systems. Especially for the final point-of-load DC-DC converter, there is a need for scalable solutions that can provide good efficiency in ever-decreasing footprints and volumes. At the same time, for mobile systems, supply modulators are crucial components to limit the power consumption of 5G power amplifiers and improve overall battery life.

8:00 AM

26.1 Coupled-OSC-Based Converters Achieving 89.5% Peak Efficiency at 61MHz, 1.82W/mm² Power Density at 360MHz, and Inherent Even Phase Interleaving

Jianxin Yang, University of Macau, Macau, China

In Paper 26.1, University of Macau presents a high-frequency coupled-oscillator converter for GPU power delivery. The proposed converter achieves gate-charge recycling, inherent phase-interleaving, per-phase current balancing, and improved transient performance with a fast start-up schema. The paper demonstrates 89.5% peak-efficiency and up to 1.82W/mm² power density.

8:25 AM

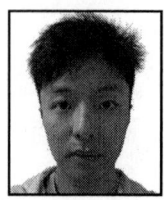

26.2 A Compact 4Vin 93.4%-Peak-Efficiency 12A Load and 20mV Undershoot Resonant Sigma Converter with PCB-Embedded Converter-on-Substrate Packaging

Yukan Du, Zhejiang University, Hangzhou, China

In Paper 26.2, Zhejiang University presents a 4V Vin Resonant Sigma Converter consisting of a high-side Switched-Capacitor Buck (SCB) and a low-side Resonant Switched-Capacitor (ReSC) converter. It achieves 93.4% peak-efficiency and 20mV undershoot for a 6A/10ns load step.

8:50 AM

26.3 A Multi-Phase Hybrid Converter with Q Samplers Enabling Simultaneous IL Auto-Balance and Arbitrary Phase Count

Jiacheng Yang, University of Macau, Macau, China

In Paper 26.3, University of Macau introduces a Multi-Phase Hybrid converter that enables an arbitrary number of phases while boasting both capacitor voltage and inductor current balancing using Q samplers. The 12V Vin converter achieves 90.5% peak-efficiency at 1V output voltage.

9:15 AM

26.4 A 12-to-1V 90.5%-Peak-Efficiency 721A/cm³-Current-Density Quad-Output Converter with One Shared DC Capacitor

Qingqing Min, University of Science and Technology of China, Hefei, China

In Paper 26.4, University of Science and Technology of China presents a 12V Vin Quad-Output Hybrid Converter for datacenter platforms. With a single shared DC capacitor, the system demonstrates zero cross-regulation, in addition to >90% peak-efficiency and 721A/cm³ power density.

9:30 AM

26.5 An Inductor-at-Middle Hybrid Buck Converter with Shared Power-Signal Path for Distributed Vertical Power Delivery

Zhongyao Zhu, Tsinghua University, Beijing, China, University of Macau, Macau, China

In Paper 26.5, Tsinghua University introduces a signal-over-power-path technique in a 12V Vin Inductor-at-Middle Hybrid converter to reduce signal routing in distributed vertical power delivery architectures. The converter demonstrates 85.3% peak-efficiency.

10:05 AM

26.6 A 92.4%-Peak-Efficiency 48V to 0.8-to-1.6V Hybrid Converter with Inductor-Interleaved Fibonacci Switched-Capacitor

Hyun-Woo Jeong, Sogang University, Seoul, Korea

In Paper 26.6, Sogang University presents a 48V to 0.8V-to-1.6V hybrid converter suitable for ultra-low voltage conversion ratio (VCR) applications. The proposed inductor-interleaved Fibonacci switched-capacitor structure enables high duty, low ACR and DCR losses under ultra-low VCR conditions. It achieves 91.3/92.4% peak efficiencies for 1.0V/1.6V output voltages.

10:30 AM

26.7 A 100A 93.4%-Peak-Efficiency LLC Resonant Converter with an Embedded Primary-Current-Extracted Regulator

Zeguo Liu, University of Science and Technology of China, Hefei, China

In Paper 26.7, University of Science and Technology of China presents a 100A LLC resonant converter with an embedded primary-current-extracted regulator. It eliminates the need for an external auxiliary buck converter or additional auxiliary windings and achieves 93.4% peak efficiency for 60V-to-2V conversion and 100A output current.

10:55 AM

26.8 A 1.2μs 1-to-12V Symbol Power Tracking Supply Modulator with Two-Step Subranging DVS Scheme and Boosted IL Slew Rate for 5G Mobile Devices

Changjin Chen, University of Science and Technology of China, Hefei, China

In Paper 26.8, University of Science and Technology of China presents a symbol power tracking supply modulator with a two-step subranging DVS scheme and an inductor current slew-rate enhancement technique. The proposed supply modulator achieves a fast transition time of 1.2μs for a wide voltage transition of 1V-to-12V to meet 5G requirements.

11:20 AM

26.9 A Compact Dual-Capacitor Relay SPT Supply Modulator with Overshoot-Free Adaptive On-Time Control for 5G FR2 CMOS PA

Zhewen Yu, Tsinghua University, Beijing, China, University of Macau, Macau, China

In Paper 26.9, Tsinghua University presents a symbol power tracking supply modulator for 5G FR2 CMOS power amplifiers. With a dual-capacitor relay and an overshoot/undershoot-free adaptive on-time control scheme, it ensures a worst-case settling time less than 290ns and an output voltage ripple below 65mV for a 0.8V-to-1.8V output voltage range.

ISSCC 2026 / SESSION 26 / COMPUTE POWER AND SUPPLY MODULATORS / 26.1

26.1 Coupled-OSC-Based Converters Achieving 89.5% Peak Efficiency at 61MHz, 1.82W/mm² Power Density at 360MHz, and Inherent Even Phase Interleaving

Jianxin Yang, Fuyao Zhang, Rui P. Martins, Mo Huang

University of Macau, Macau, China

Abstract

This work proposes coupled-OSC-based high-frequency converters. Retaining gate-charge recycling from prior single-phase design, this work achieves inherent even phase interleaving along with per-phase and per-inductor current balancing. It reduces the inductor count per phase, while enabling high-speed ON-OFF control and improving load-transient performance using a fast startup scheme. It achieves 89.5% peak PCE at 61MHz with air-core inductors, and 80% peak PCE and 1.82W/mm² density at 360MHz using bondwire inductors.

Power delivery for GPUs has become increasingly challenging with the rapid growth of AI computing capabilities. In many HPC accelerators, power modules already occupy more than 70% of the system area, becoming a bottleneck for compute scalability. This thereby places stringent demands on power density. In a two-stage 48V/24V-to-load architecture, careful selection of the intermediate rail voltage for the secondary voltage regulators (VRs) is critical. A growing trend is to lower this rail, e.g., to 3V (Fig. 26.1.1), enabling the use of low-V_{DS} devices with a better FoM in the secondary VR stage. Meanwhile, high-frequency (HF) conversion [1] is key to shrinking passive components and improving power density, e.g., inductor size can be reduced by 20× when moving from magnetic core to PCB-trace implementations.

However, even with low-V_{DS} devices, conventional HF buck converters (Fig. 26.1.1) still suffer from high power loss, dominated by gate-drive power (P_{gate}) [2]. Furthermore, multiphase (MP) topologies are required to support the large GPU load currents, but implementing HF MP controllers is highly challenging. In conventional active controls, propagation-delay mismatches in PWM signals, which are negligible at low frequencies, cause uneven phase interleaving and increased steady-state output ripples. Precise phase-interleave calibration requires a high overhead at HF. Meanwhile, accurate HF inductor current sensing is also difficult, leaving per-phase current imbalance almost unsolvable.

Prior work [3] proposed an oscillator-based converter (EMLC, Fig. 26.1.1) that recycles gate charge (Q_{gate}), mitigating the f_{SW}-P_{gate} trade-off in conventional HF bucks. The design uses a class-D power oscillator with two inductors, followed by a rectifier with cross-connected transistors (as diodes) and two inductors. On-chip coupled inductors are employed, and output voltage regulation is achieved via ON-OFF control. However, the four-inductor single-phase structure increases complexity and lacks per-inductor current balance, as discussed in Fig. 26.1.2. Moreover, parallelizing such single-phase units into an M-phase system still inherits the drawbacks of conventional HF bucks, together with uneven f_{SW} from LC mismatches across oscillators.

This work draws inspiration from quadrature LC oscillators (OSC) [4], commonly used in RF transceivers for generating precise-shift local oscillator signals. By coupling two class-D oscillators via auxiliary transistors paralleled to the main transistors, this topology achieves mutual injection locking that synchronizes f_{SW} (Fig. 26.1.1). Building on this concept, we propose coupled-OSC-based converters that couple multiple class-D power oscillator stages. Differential outputs (*Out*) of each stage are connected to the inputs (*In*) of the next. Like its RF counterpart, this topology inherently provides evenly interleaved phases and a uniform synchronized f_{SW}. While retaining the gate-charge recycling benefit from [3], we simplify the rectifier stage by replacing two inductors with two switches. An output inductor is introduced to ensure auto current balance across all inductors. These changes reduce the total inductor count to three per phase.

The class-D oscillator in the proposed converter (Fig. 26.1.2) operates between the input and output voltages (V_{IN} and V_0). It is comprised of two inductors (L_1 and L_2) and a cross-connected transistor pair (M_1 and M_2). The differential outputs (V_{TX+} and V_{TX-}) connect the rectifier inputs (V_{RX+} and V_{RX-}) through flying capacitors (C_{F1} and C_{F2}). The rectifier includes two cross-connected switch pairs (M_3-M_6) and an output inductor L_3.

The converter operates in two complementary states. During State$_1$, M_2 is off and L_2 resonates with M_2's drain-source parasitic capacitance (C_{P2}). The resulting resonant voltage V_{TX-} turns M_1 on, settling V_{TX+} to V_0 and energizing L_1. At the rectifier, V_{RX-} and V_{RX+} turn M_3 and M_6 on, allowing L_3 (acting as a current source) to draw current (I_{L3}) from the L_2-C_{P2} resonance through C_{F2}. Inductor charge balance is maintained during this state: the charge delivered by L_2 (Q_2) equals that received by L_3 (Q_3), as illustrated by the shaded waveforms. Figure 26.1.2 presents a simplified behavioral model of this state, where C_{F2} holds V_0 and serves as a DC voltage source. When V_{TX-} falls below the threshold voltage and turns off M_1, the oscillator transitions to State$_2$: L_1 resonates with M_2's capacitance C_{P1}, transferring charge Q_1 to L_3 (as Q_3'). Over a full cycle, charge balance in L_3 enforces $Q_3 = Q_3'$, leading to inherent balanced DC currents through L_1-L_3.

An alternative possible implementation of the proposed topology removes L_3 (Fig. 26.1.2), where C_{F2} then participates in the resonance rather than serving as a voltage source, as presented in the model. Since $C_{F2} >> C_{P2}$, the achievable resonant frequency is significantly lowered, leading to higher RMS current and conduction loss. For comparison, EMLC employs two rectifier inductors (L_4 and L_5), which act as independent current sources as presented in the simplified State$_1$ and State$_2$ models. Unlike the proposed design, where charge-balanced L_3 ensures equal inductor currents, EMLC provides no such mechanism. Figure 26.1.2 displays the simulated I_{L1} and I_{L2} imbalance due to L_1/L_2 variations, where the proposed scheme reduces the imbalance by around five times relative to EMLC.

Figure 26.1.3 presents the schematic of the prototype four-phase converter. Multiplexers are inserted at the input signals of each power stage, enabling flexible scaling of the active phase count. The outputs can be either direct- or cross-coupled to the inputs, following [5] and the Barkhausen criterion. Global output-voltage regulation can be achieved with input voltage regulation [6], while local regulation is implemented via ON-OFF control. An error amplifier (EA) with a type-II compensator integrates the error between V_{OUT} and reference voltage V_{REF} to generate control signal V_{EA}. This signal is compared with four low-frequency ramping signals ($V_{RAMP,i}$, i = 1 to 4). The intersection between V_{EA} and $V_{RAMP,j}$ produces an ON-OFF duty-cycle signal for the jth stage (EN_j), while phases 1 to (j–1) remain active and phases beyond j are disabled. Furthermore, in conventional MP bucks, per-phase current balance is often limited by duty-cycle mismatches [7]. In contrast, the proposed OSC-based topology inherently guarantees a 50% duty cycle per phase, thereby enabling automatic current balancing across all outputs. During load transients, a transient detector (TR det) senses the V_0 undershoot and instantaneously pulls up V_{EA} for maximum output current delivery.

Each power stage uses N-type devices for M_{1-4} and P-type for M_{5-6} (Fig. 26.1.3). All gates are AC-coupled to resonant nodes, with the coupling capacitors and intrinsic gate capacitances forming voltage dividers that prevent gate overvoltage. Their DC bias levels (V_{BIAS}, V_{BIASn} or V_{BIASp}) are selected to minimize loss in the power switches. To disable a power stage at light load, additional switches (controlled by EN) pull each gate to its source.

To address the slow oscillation build-up in oscillator designs, a fast-startup circuit reuses the phase-coupling devices M_{in1} and M_{in2} (Fig. 26.1.3). By momentarily forcing them on through pulled-up transistors, L_1 and L_2 are rapidly energized, significantly shortening the build-up time. Without this scheme, the effective ON duration becomes much shorter than $EN = 1$, severely limiting ON-OFF operation frequency. Furthermore, the fast-startup scheme reduces output undershoot during load transient events.

Two prototype chips were fabricated using a 65-nm CMOS process, occupying silicon areas of 3.8 and 0.64mm², respectively (Fig. 26.1.7). Chip1 features enlarged M_{1-2} devices and has 0402HP-18N inductors soldered on the chip (Design1), targeting high PCE. Chip2 targets HF and high power density, employing 2.2-nH PCB-trace (Design2) and 0.64-nH bondwire inductors (Design3). The total areas for their four-phase implementations are 20.2, 5.94 and 5.46mm², respectively, with single-phase Design3 at 1.375mm². Figure 26.1.4 presents the measured V_0 spectra of Designs 2 and 3, obtained through FFT analysis. Design2 oscillates at 230MHz, while Design3 achieves a higher oscillation frequency of 360MHz. Second-harmonic components, originating from the V_X nodes, are also observed. Measured PCEs versus output power (P_{OUT}) reveal peak efficiencies of 89.5%, 85% and 80% for Designs 1-3, respectively. Simulation results show that inductor losses are the dominant contributor to total loss. The four-phase implementations extend the maximum P_{OUT} to around 7W. Figure 26.1.4 further displays the PCE versus power density, where Design3 demonstrates the highest power density of 1.82W/mm².

Figure 26.1.5 compares the measured waveforms of a single-phase V_{TX}s under conditions with and without the proposed fast-startup scheme. With the scheme, oscillations are established within around 90ns, compared to >500ns without it, thereby facilitating fast ON-OFF operation. In addition, the measured per-phase V_X waveforms indicate a maximum phase-interleave non-uniformity of 4° in four-phase operation, consistent with the inherent advantage of coupled oscillators. These measurements require probing high-frequency nodes and thus are demonstrated using Design1, while the same benefits are expected to

446 • 2026 IEEE International Solid-State Circuits Conference

979-8-3315-8937-0/26 $31.00 © 2026 IEEE

hold in Design2 and 3. Measured load-transient waveforms display 35-mV undershoot and 69-mV overshoot in V_O under I_{LOAD} transitions between 0 and 1.2A with edge times of <10ns (Fig. 26.1.5). As shown, the proposed fast startup scheme improves the undershoot and shortens recovery time.

Figure 26.1.6 compares the proposed designs with state-of-the-art HF converters. Compared with conventional buck [8,9] and hybrid [10] topologies, this work recycles the gate charge of the power switches, achieving a competitive peak PCE. Within the VCR ranges of 0.5-0.6, this advantage is more evident, as reflected by the highest PCE improvement over baseline LDOs in Fig. 26.1.6 at the same VCRs. Operating at 360MHz enables the use of 0.6-nH bondwire inductors, supporting a maximum P_{OUT} of 6.08W and a 1.82W/mm² power density. The density could be further improved with advanced packages and inductors. Compared to prior OSC-based designs [3,11], this work reduces the number of inductors per phase and ensures inherent inductor current balancing. In addition, this work extends the OSC-based approach to multi-phase, with inherently even phase interleaving. In contrast, conventional MP bucks and hybrids require complex active control of interleaving, which is challenging at HF. Furthermore, the proposed fast startup scheme enables a 35mV output undershoot under a 1.2A load current transition.

Acknowledgement:
This work was supported in part by the Science and Technology Development Fund, Macau SAR (0029/2025/AMJ, 0042/2025/RIB1 and 004/2023/SKL), in part by the Guangdong-HongKong-Macao Joint Laboratories (2025B1212150003), in part by the Guangdong Basic and Applied Basic Research Foundation (2023B1515130001), in part by the Research Committee of University of Macau (MYRG-GRG2024-00041-IME).

References:
[1] C. Schaef, K. Kesarwani and J. T. Stauth, "A variable-conversion-ratio 3-phase resonant switched capacitor converter with 85% efficiency at 0.91W/mm² using 1.1nH PCB-trace inductors," *2015 IEEE International Solid-State Circuits Conference - (ISSCC) Digest of Technical Papers*, San Francisco, CA, USA, 2015, pp. 1-3. https://doi.org/10.1109/ISSCC.2015.7063075
[2] M. Lee, Y. Choi and J. Kim, "A 500-MHz, 0.76-W/mm Power Density and 76.2% Power Efficiency, Fully Integrated Digital Buck Converter in 65-nm CMOS," in *IEEE Transactions on Industry Applications*, vol. 52, no. 4, pp. 3315-3323, July-Aug. 2016. https://doi.org/10.1109/TIA.2016.2541079
[3] A. Novello, G. Atzeni, G. Cristiano, M. Coustans and T. Jang, "A 1.25GHz Fully Integrated DC-DC Converter Using Electromagnetically Coupled Class-D LC Oscillators," *2021 IEEE International Solid-State Circuits Conference (ISSCC)*, San Francisco, CA, USA, 2021, pp. 260-262. https://doi.org/10.1109/ISSCC42613.2021.9366037
[4] A. Mirzaei, M. E. Heidari, R. Bagheri, S. Chehrazi and A. A. Abidi, "The Quadrature LC Oscillator: A Complete Portrait Based on Injection Locking," in *IEEE Journal of Solid-State Circuits*, vol. 42, no. 9, pp. 1916-1932, Sept. 2007. https://doi.org/10.1109/JSSC.2007.903047
[5] Jae Joon Kim and Beomsup Kim, "A low-phase-noise CMOS LC oscillator with a ring structure," *2000 IEEE International Solid-State Circuits Conference. Digest of Technical Papers (ISSCC)*, San Francisco, CA, USA, 2000, pp.1-3. https://doi.org/10.1109/ISSCC.2000.839846

[6] S. Ren et al., "A 12A 89.3% Peak Efficiency and 26mV Undershoot 12-to-1V Two-Stage Converter with Regulated Resonant Switched-Capacitor Regulators," *2025 IEEE International Solid-State Circuits Conference (ISSCC)*, San Francisco, CA, USA, 2025, pp. 1-3. https://doi.org/10.1109/ISSCC49661.2025.10904628
[7] J. Yang, R. P. Martins and M. Huang, "A Segmented-Interlacing Multi-Phase Hybrid Converter with Inherently Auto-Balanced ILs and Boosted IL Slew Rate During Load Transients," *2025 IEEE International Solid-State Circuits Conference (ISSCC)*, San Francisco, CA, USA, 2025, pp.1-3. https://doi.org/10.1109/ISSCC49661.2025.10904561
[8] J. -H. Cho et al., "A 1.23W/mm² 83.7%-Efficiency 400MHz 6-Phase Fully Integrated Buck Converter in 28nm CMOS with On-Chip Capacitor Dynamic Re-Allocation for Inter-Inductor Current Balancing and Fast DVS of 75mV/ns," *2022 IEEE International Solid-State Circuits Conference (ISSCC)*, San Francisco, CA, USA, 2022, pp.1-3. https://doi.org/10.1109/ISSCC42614.2022.9731726
[9] S. Kim et al., "A Monolithic 10.5W/mm² 600 MHz Top-Metal and C4 Planar Spiral Inductor-Based Integrated Buck Voltage Regulator on 16nm-Class CMOS," *2024 IEEE International Solid-State Circuits Conference (ISSCC)*, San Francisco, CA, USA, 2024. pp1-3. https://doi.org/10.1109/ISSCC49657.2024.10454473
[10] J. -H. Cho, H. -H. Bae, G. -W. Lim, T. -H. Kong, J. -H. Yang and H. -S. Kim, "A Fully Integrated Multi-Phase Voltage Regulator With Flying-Capacitor-Based Inter-Inductor Current Self-Balancing Scheme and Charge-Recycling Gate Driver," in *IEEE Journal of Solid-State Circuits*, vol. 59, no. 8, pp. 2529-2544, Aug. 2024. https://doi.org/10.1109/JSSC.2024.3360708
[11] A. Novello, G. Atzeni, G. Cristiano, M. Coustans and T. Jang, "A 2.3GHz Fully Integrated DC-DC Converter based on Electromagnetically Coupled Class-D LC Oscillators achieving 78.1% Efficiency in 22nm FDSOI CMOS," *2021 Symposium on VLSI Circuits*, Kyoto, Japan, 2021, pp.1-3. https://doi.org/10.23919/VLSICircuits52068.2021.9492491

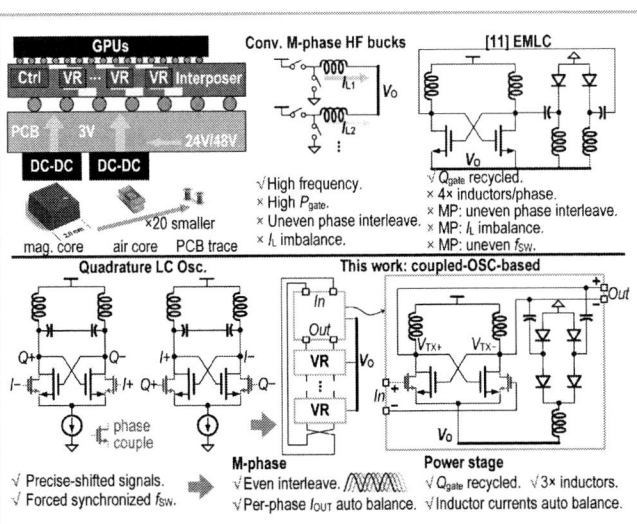

Figure 26.1.1: Motivations (top left), EMLC converter (top right), quadrature LC oscillators (bottom left), and proposed coupled-OSC-based converter (bottom right).

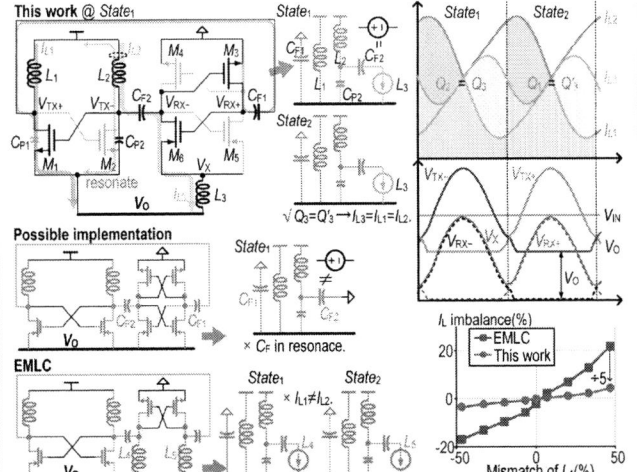

Figure 26.1.2: Working principles of proposed converter (top), false resonance if without L_3 (middle), L_1 and L_2 current imbalance in EMLC (bottom left), imbalance vs. L mismatch (bottom right).

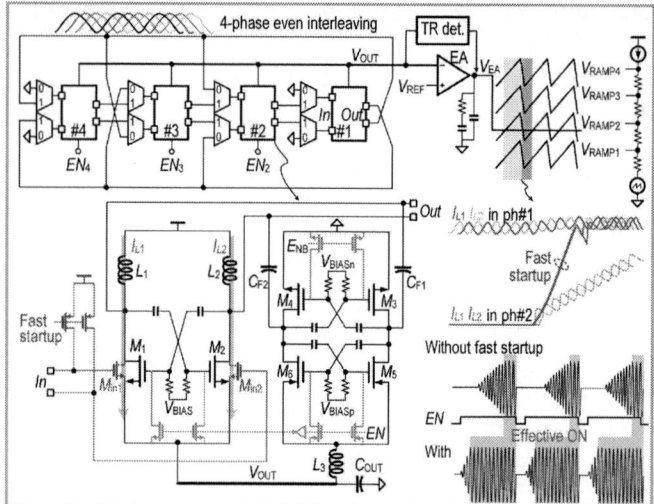

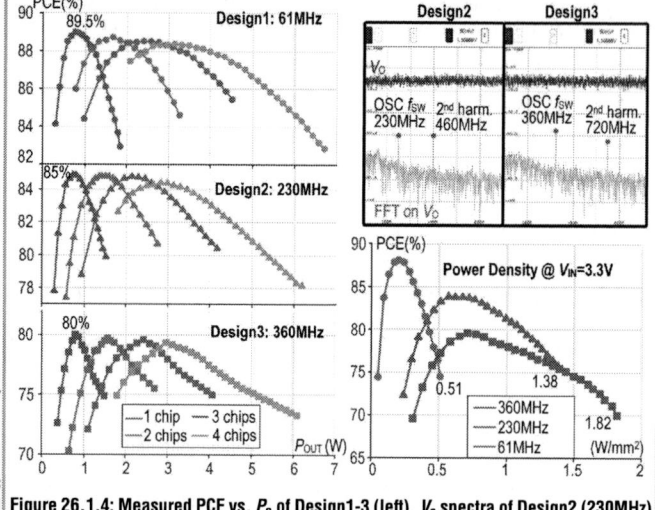

Figure 26.1.3: Proposed coupled-OSC-based converter (top), per-phase schematic (bottom left), fast startup of deactivated stages (middle right), enabling high-speed ON-OFF (bottom right).

Figure 26.1.4: Measured PCE vs. P_0 of Design1-3 (left), V_0 spectra of Design2 (230MHz) and Design3 (360MHz) (right top), PCE vs. power density (right bottom).

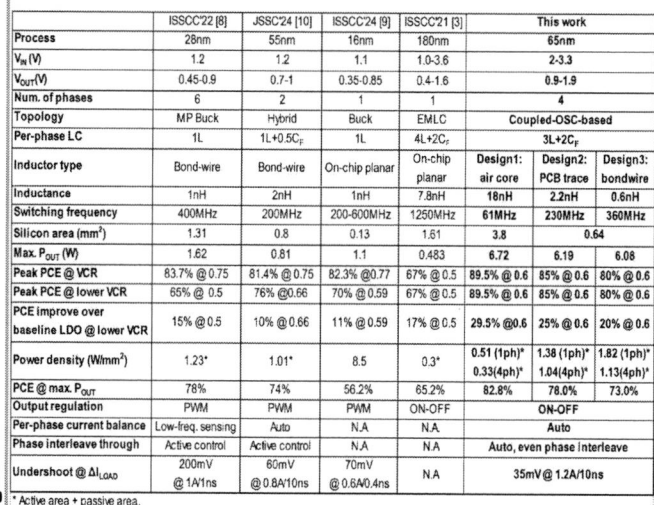

Figure 26.1.5: Measured single-phase waveforms with and without fast startup (top left), non-uniformity of phase interleave (top right), load transient response with 0-1.2A/10ns I_{LOAD} transition (bottom).

	ISSCC'22 [8]	JSSC'24 [10]	ISSCC'24 [9]	ISSCC'21 [3]	This work		
Process	28nm	55nm	16nm	180nm	65nm		
V_{IN} (V)	1.2	1.2	1.1	1.0-3.6	2-3.3		
V_{OUT}(V)	0.45-0.9	0.7-1	0.35-0.85	0.4-1.6	0.9-1.9		
Num. of phases	6	2	1	1	4		
Topology	MP Buck	Hybrid	Buck	EMLC	Coupled-OSC-based		
Per-phase LC	1L	1L+0.5C_F	1L	4L+2C_f	3L+2C_F		
					Design1	Design2	Design3
Inductor type	Bond-wire	Bond-wire	On-chip planar	On-chip planar	air core	PCB trace	bondwire
Inductance	1nH	2nH	1nH	7.8nH	18nH	2.2nH	0.6nH
Switching frequency	400MHz	200MHz	200-600MHz	1250MHz	61MHz	230MHz	360MHz
Silicon area (mm²)	1.31	0.8	0.13	1.61	3.8	0.64	
Max. P_{OUT} (W)	1.62	0.81	1.1	0.483	6.72	6.19	6.08
Peak PCE @ VCR	83.7% @ 0.75	81.4% @ 0.75	82.3% @ 0.77	67% @ 0.5	89.5% @ 0.6	85% @ 0.6	80% @ 0.6
Peak PCE @ lower VCR	65% @ 0.5	76% @0.66	70% @ 0.59	67% @ 0.5	89.5% @ 0.6	85% @ 0.6	80% @ 0.6
PCE improve over baseline LDO @ lower VCR	15% @0.5	10% @0.66	11% @0.59	17% @0.5	29.5% @0.6	25% @0.6	20% @0.6
Power density (W/mm²)	1.23*	1.01*	8.5	0.3*	0.51 (1ph)* 0.33(4ph)*	1.38 (1ph)* 1.04(4ph)*	1.82 (1ph)* 1.13(4ph)*
PCE @ max. P_{OUT}	78%	74%	56.2%	65.2%	82.8%	78.0%	73.0%
Output regulation	PWM	PWM	PWM	ON-OFF	ON-OFF		
Per-phase current balance	Low-freq. sensing	Auto	N.A	N.A	Auto		
Phase interleave through	Active control	Active control	N.A	N.A	Auto, even phase interleave		
Undershoot @ ΔI_{LOAD}	200mV @ 1A/1ns	60mV @ 0.8A/10ns	70mV @ 0.6A/0.4ns	N.A	35mV @ 1.2A/10ns		

* Active area + passive area.

Figure 26.1.6: Performance comparison with state-of-the-art works.

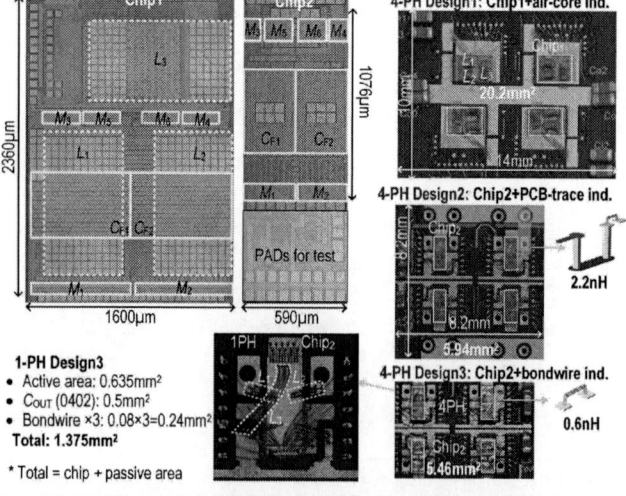

1-PH Design3
- Active area: 0.635mm²
- C_{OUT} (0402): 0.5mm²
- Bondwire ×3: 0.08×3=0.24mm²
Total: 1.375mm²

* Total = chip + passive area

Figure 26.1.7: Micrographs of Chip1 and Chip2 (top left), PCB of four-phase Design1-3 (right), PCB of single-phase Design3 with its total area calculation (bottom left).

ISSCC 2026 / SESSION 26 / COMPUTE POWER AND SUPPLY MODULATORS / 26.2

26.2 A Compact 4Vin 93.4%-Peak-Efficiency 12A Load and 20mV Undershoot Resonant Sigma Converter with PCB-Embedded Converter-on-Substrate Packaging

Yukan Du, Shengdao Ren, Zheyuan Fei, Ziteng Chen, Menglian Zhao, Chushan Li, Yong Ding, Wuhua Li, Zhichao Tan, Wanyuan Qu

Zhejiang University, Hangzhou, China

Abstract

This work presents a 4V input 12A load resonant sigma converter which consists of a 10MHz high-side switched-capacitor Buck (SCB) and a 5MHz low-side direct resonant switched-capacitor (ReSC) converter. The proposed converter features a low V·A metric, a compact volume, and more importantly, a fast response by providing an ~0.5A/ns current slew rate during the transient. The measurements show 93.4% peak efficiency, 20mV undershoot under a 6A load step and 0.38A/mm² system current density.

With the great advances in artificial intelligence, the power demand of xPUs has roared rapidly from 100W to 1000W levels, imposing enormous stresses on the power converters for ever higher efficiency, speed, and density (E·S·D). The classic solutions, however, face severe challenges in meeting all the three E·S·D targets at the same time. For example, for the classic Buck converters such as integrated-voltage-regulators (IVR), DrMOS, and 3-level converters, wide output range and excellent inductor current slew rate is preferable [1-4]. Nevertheless, due to their large voltage stress on both active and passive devices, Buck converters usually exhibit inferior efficiency and density under the same conditions. For hybrid topologies that mostly alleviate active and passive stresses, the efficiency is generally decent and density is improved [6-13]. However, hybrid converters often have limited duty cycles which can limit the output range and transient response. The reduced voltage stress also limits the maximum inductor slew rate, leading to inferior transient performance. Here, in order to assess these overall E·S·D performance of converters to some extent, an E·S·D metric is proposed which evaluates the efficiency using a normalized V·A metric [14], the speed using the maximum inductor current slew rate, and the density using the total passive volume [6-8] (using a typical energy density of 1μJ/mm³ for inductor and 1mJ/mm³ for capacitor), i.e., E·S·D Metric = V·A * Passive volume / (maximum inductor di/dt).

Figure 26.2.1 shows the proposed resonant sigma converter, which consists of a high-side switched-capacitor Buck (SCB) and a low-side direct resonant switched-capacitor (ReSC) converter. From the efficiency perspective, owing to the input series and output parallel sigma structure [9,16], the voltage stress of both high-side and low-side converters is greatly reduced with the high-side only requiring 2-transistor cascode devices and the low-side using 1.2V core devices. In addition, since the ReSC operates in ideal non-regulation mode during steady-state, excellent efficiency can be achieved under a resonant frequency of 5MHz. According to the normalized V·A metric with respect to $V_0 \cdot I_0$ for V_{IN} = 4V and V_0 = 0.4~1.2V shown in Fig. 26.2.1 (right), the proposed converter shows an excellent V·A metric over a wide output range. Besides, from the speed point of view, considering the resonant operation of the low-side ReSC, a tiny PCB trace inductor L_r of 2nH is designed. During the transients, with the M_{r1} and M_{r2} both turned on similar to 3-level converter operation, a large voltage is imposed on L_r, leading to an ampere-per-second (A/ns) inductor current slew rate, which is comparable to typical IVRs operating around 100MHz [17-18]. Therefore, the proposed converter achieves an A/ns slew rate during the transients with only 5MHz ReSC operation during steady-state. Figure 26.2.1 (right) shows the maximum inductor current di/dt of the proposed converter and other popular designs, all under the same V_{IN} (4V), inductor switching frequency, and ΔI_L ripples. From the density perspective, thanks to the low voltage stress on both high-side and low-side inductors, the high-side adopts only a 0505-sized 110nH inductor and the low-side 2nH PCB trace inductor. Moreover, a PCB-embedded converter-on-substrate packaging is proposed to further enhance the density and reduce the parasitics. As is shown in Fig. 26.2.1, all of the input and flying capacitors are vertically attached to the bottom side of the substrate to form a converter-on-substrate, with the converter chip partially embedded inside the PCB. Therefore, the proposed converter shows excellent density compared to other state-of-the art designs. Finally, according to the overall E·S·D metric depicted in Fig. 26.2.1 (bottom right), the proposed resonant sigma converter shows an improved overall performance.

Figure 26.2.2 illustrates the operation principles of the proposed converter. The high-side SCB involves two switching phases ΦH₁ and ΦH₂. During ΦH₁, transistor M₁ turns on with the flying capacitor C_F in charging and the inductor L in magnetizing mode. During ΦH₂, transistors M₂ and M₃ turn on with C_F discharging and L demagnetizing. Since C_F is in shunt with the input of the low-side converter during ΦH₂, the charge of C_F is reused as the input for the low-side converter and the voltage V_{CF} equals V_L. The key waveforms of the SCB are shown in Fig. 26.2.2 (top right). For the low-side ReSC, the converter works the same as a 3-level converter does, with the only difference being the choice of the tiny inductor and the resulting resonant inductor current. During the non-regulation operation, the direct ReSC works at a 2:1 ratio in full-resonance operation with the ΦL₂ and ΦL₄ phases. Step-down operation adds the ΦL₃ phase and step-up operation adds the ΦL₁ phase. Since the ReSC operates in non-regulation mode during steady-state, an average inductor current slew rate of 0.03A/ns is observed for a 2V-to-1V conversion with a 5A load. During the transients, as is shown in Fig. 26.2.2 (bottom), the converter operates in step-up mode with

an approximately 0.5A/ns slew rate. Therefore, a current slew rate enhancement of around 16.7× is obtained during the transients.

Figure 26.2.3 shows the overall circuits of the proposed converter. The power stage consists of two pieces of high-side SCB and four pieces of low-side ReSC, while the controller includes a local regulation loop for the ReSC and a global regulation loop for the SCB. The local ReSC loop adopts a simple four-phase voltage mode control with a 90-degree difference for each phase, and the global loop is closed by inspecting the gate voltages VG_{r1} and VG_{r3} for M_{r1} and M_{r3}, as reported in [19]. If the ReSC deviates from the ideal non-regulation mode, VG_{r1} and VG_{r3} will show phase differences. Therefore, a PLL-based reference generator with VG_{r1} and VG_{r3} as the input generates an optimum reference voltage V_{REF_HS} which regulates the SCB stage to dictate an optimum V_L for the low-side ReSC to settle in ideal non-regulation mode. The global SCB controller also adopts a simple two-phase voltage mode control with current balances.

The proposed prototype was fabricated using a 65nm CMOS process and verified with V_{IN} = 3.6-to-4.8V, V_0 = 0.6-to-1.2V, and the load current I_{Load} up to 12A. The switching frequencies of the individual high-side SCB and low-side ReSC converter pieces are 10MHz and 5MHz, respectively.

Figure 26.2.4 shows the measured steady-state waveforms for both 0A and 12A load, respectively, under a 4V-to-0.85V conversion. For the high-side SCB, the 10MHz switching frequency and balanced inductor current is observed. During 0A loading, V_L exactly equals $2V_0$ of 1.7V and the SCB shows a duty ratio of less than 0.5. With a 12A load, V_L increases to 2V, indicating that the low-side is operating with optimum non-regulation. The duty ratio of the SCB is also increased accordingly and the inductor current increases to 3A. For the low-side ReSC, under the 0A load condition, no significant resonance is observed on the resonance capacitor voltage V_{Cr} or on the capacitor top and bottom voltages V_{C+} and V_{C-}. Contrastingly, for the 12A load, the V_{Cr} shows a 0.5V peak-to-peak resonance with both V_{C+} and V_{C-} in visible resonance. Additionally, the 180-degree phase shift between the first and the third pieces of the ReSC is also verified.

Figure 26.2.5 gives the measured transient responses and efficiency curves of the proposed converter. Under a 4V-to-0.85V voltage conversion, a 0A to 6A load transient is performed with a rise time of 10ns and an output capacitance of 18.8uF. V_0 shows an under- and over-shoot voltages of 20mV and 28mV, respectively. From the zoomed-in view, V_0 settles in less than 400ns during the step-up transient, by which time V_0 is regulated by the low-side ReSC and the high-side SCB current has yet to have finished ramping up to the desired target. Besides, during the transient, the ReSC switching node V_{LX} exhibits a large pulsating voltage, which indicates a large voltage across the 2nH L_r that generates a large current slew rate. As is shown, after the high-side SCB current I_L reaches the desired target, the ReSC settles back into ideal non-regulation mode with the desired increased V_L voltage. Similarly, during the step-down transient, the ReSC firstly regulates V_0 with its switching node V_{LX} imposing a large negative voltage on L_r and the high-side SCB current gradually ramps down. Figure 26.2.5 (bottom) also shows that, with the ReSC in open-loop non-regulation and SCB in close-loop operation, the converter presents under- and over-shoot voltages of 31mV and 36mV, respectively. Finally, the measurements show that the ReSC gives a peak efficiency of 96.2% under steady-state, and the overall converter gives a peak efficiency of 93.4% and a full-load efficiency of 82.0% under 12A.

Figure 26.2.6 gives the performance summary of the proposed work and other state-of-the-art. As can be seen, among the converters presented in Fig. 26.2.6 with similar input voltage, output voltage and load current, the proposed design shows the best undershoot voltage during transients, the best system-level current density, and one of the best peak efficiencies. Figure 26.2.7 depicts the PCB-embedded converter-on-substrate packaging as mentioned above. The die with an area of 2.1mm × 3mm is attached on the top side of the substrate with an area of 6.5mm × 6.5mm. The input and flying capacitors are vertically attached on the bottom side of substrate to further enhance the density and reduce the parasitics. The other passive components are mounted within the substrate footprint, with a total passive volume of 4.854mm³.

Acknowledgement:

This work was supported in part by the National Natural Science Foundation of China (NSFC) under Grant 92473106, and in part by the NSFC under Grant 62274147.

References:

[1] N. Desai et al., "A 32-A, 5-V-Input, 94.2% Peak Efficiency High-Frequency Power Converter Module Featuring Package-Integrated Low-Voltage GaN nMOS Power Transistors," *IEEE JSSC*, vol. 57, no. 4, pp. 1090-1099, April 2022. http://doi.org/10.1109/JSSC.2022.3141779

[2] C. Schaef et al., "A 93.8% Peak Efficiency, 5V-Input, 10A Max ILOAD Flying Capacitor Multilevel Converter in 22nm CMOS Featuring Wide Output Voltage Range and Flying Capacitor Precharging," *ISSCC*, pp. 146-148, Feb. 2019. http://doi.org/10.1109/ISSCC.2019.8662475

[3] X. Sun et al., "A 10W 3.8-5V Input IVR Chiplet with 93%-Peak-Efficiency and 3.2A/mm2 Density Featuring Wide Load Range and Adaptive Ganging for 2.5D/3D Vertical Power Delivery," *IEEE Symp. VLSI Circuits*, pp. 1-3, June 2025. http://doi.org/10.23919/VLSITechnologyandCir65189.2025.11074928

[4] J. -I. Seo, B. -M. Lim, W. -J. Choi, Y. -S. Noh and S. -G. Lee, "A 95.1% Efficiency Hybrid Hysteretic Reconfigurable 3-Level Buck Converter With Improved Load Transient Response," *IEEE Trans. Power Electronics*, vol. 37, no. 12, pp. 14916-14925, Dec. 2022. http://doi.org/10.1109/TPEL.2022.3189187

[5] Y. Jingshu et al., "A Monolithic 5.7A/mm2 91% Peak Efficiency Scalable Multi-Stage Modular Switched Capacitor Voltage Regulator with Self-Timed Deadtime and Safe Startup for 3D-ICs," *IEEE Symp. VLSI Circuits*, pp. 1-2, June 2024. http://doi.org/10.1109/VLSITechnologyandCir46783.2024.10631399

[6] G. Cai, Y. Lu and R. Martins, "A Battery-Input Sub-1V Output 92.9% Peak Efficiency 0.3A/mm2 Current Density Hybrid SC-Parallel-Inductor Buck Converter with Reduced Inductor Current in 65nm CMOS," *ISSCC*, pp. 312-314, Feb. 2022. http://doi.org/10.1109/ISSCC42614.2022.9731576

[7] Z. Xia et al., "A Two-Stage Cascaded Hybrid Switched-Capacitor DC-DC Converter with 96.9% Peak Efficiency Tolerating 0.6V/μs Input Slew Rate During Startup," *ISSCC*, pp. 256-258, Feb. 2021. http://doi.org/10.1109/ISSCC42613.2021.9365763

[8] H. Han, J. -H. Cho, W. Jang, Y. Park, J. Lee and H. -S. Kim, "A 94.1%-Efficiency Parallel-SC Hybrid Buck Converter Designed Using VCR-Aware Topology Optimizer for a 4.2A/mm² Current-Density FoM," *ISSCC*, pp. 464-466, Feb. 2024. http://doi.org/10.1109/ISSCC49657.2024.10454535

[9] X. Yang et al., "A 5V Input 98.4% Peak Efficiency Reconfigurable Capacitive-Sigma Converter With Greater than 90% Peak Efficiency for the Entire 0.4~1.2V Output Range," *ISSCC*, pp. 108-110, Feb. 2022. http://doi.org/10.1109/ISSCC42614.2022.9731550

[10] C. Wang, Y. Lu and R. P. Martins, "A Highly Integrated Tri-Path Hybrid Buck Converter With Reduced Inductor Current and Self-Balanced Flying Capacitor Voltage," *IEEE TCAS-I*, vol. 69, no. 9, pp. 3841-3850, Sept. 2022. http://doi.org/10.1109/TCSI.2022.3182145

[11] C. Wang, Y. Lu, X. Li and R. P. Martins, "A Dual-Branch Series-Parallel Hybrid Buck DC-DC Converter With Flying Capacitor Voltage Auto-Balancing," *IEEE TCAS-I*, vol. 69, no. 12, pp. 4741-4750, Dec. 2022. http://doi.org/10.1109/TCSI.2022.3201166

[12] C. Hardy et al., "A Scalable Heterogeneous Integrated Two-Stage Vertical Power-Delivery Architecture for High-Performance Computing," *ISSCC*, pp. 182-184, Feb. 2023. http://doi.org/10.1109/ISSCC42615.2023.10067315

[13] G. Cai, Y. Lu and R. P. Martins, "A Compact 12V-to-1V 91.8% Peak Efficiency Hybrid Resonant Switched-Capacitor Parallel Inductor (ReSC-PL) Buck Converter," *ISSCC*, pp. 198-200, Feb. 2023. http://doi.org/10.1109/ISSCC42615.2023.10067303

[14] G. Cai, Y. Lu and R. P. Martins, "A Compact 12V-to-1V 91.8% Peak Efficiency Hybrid Resonant Switched-Capacitor Parallel Inductor (ReSC-PL) Buck Converter," *ISSCC*, pp. 198-200, Feb. 2023. http://doi.org/10.1109/ISSCC42615.2023.10067802

[15] Y. Ji et al., "A 12V-Input 1V-1.8V-Output 94.7%-Peak-Efficiency 685A/cm3-Current-Density Hybrid DC-DC Converter with a Charge Converging Phase," *ISSCC*, pp. 458-460, Feb. 2024. http://doi.org/10.1109/ISSCC49657.2024.10454573

[16] M. Ahmed, C. Fei, F. C. Lee and Q. Li, "High-efficiency high-power-density 48/1V sigma converter voltage regulator module," *IEEE APEC*, pp. 2207-2212, Mar. 2017. http://doi.org/10.1109/APEC.2017.7931005

[17] E. A. Burton et al., "FIVR — Fully integrated voltage regulators on 4th generation Intel® Core™ SoCs," *IEEE APEC*, pp. 432–439, Mar. 2014. http://doi.org/10.1109/APEC.2014.6803344

[18] C. Schaef et al., "A IMax Fully Integrated Multi-Phase Voltage Regulator with 91.5% Peak Efficiency at 1.8 to 1V, Operating at 50MHz and Featuring a Digitally Assisted Controller with Automatic Phase Shedding and Soft Switching in 4nm Class FinFET CMOS," *ISSCC*, pp. 1-3, Feb. 2022. http://doi.org/10.1109/ISSCC42614.2022.9731658

[19] S. Ren et al., "A 12A 89.3% Peak Efficiency and 26mV Undershoot 12-to-1V Two-Stage Converter with Regulated Resonant Switched-Capacitor Regulators," *ISSCC*, pp. 374-376, Feb. 2025. http://doi.org/10.1109/ISSCC49661.2025.10904628

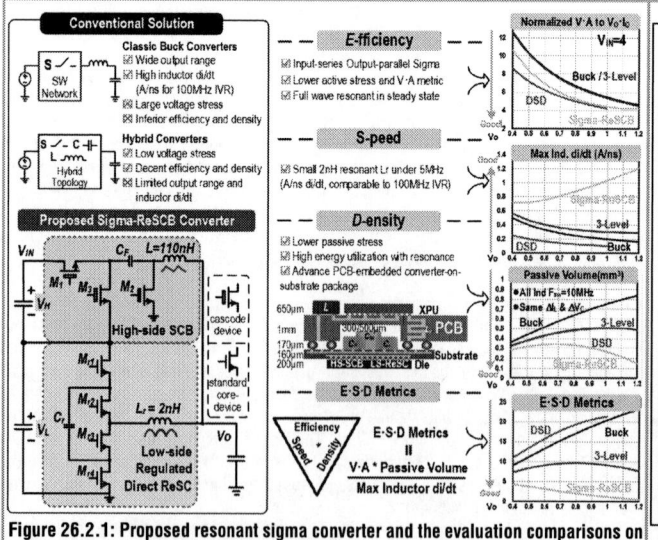

Figure 26.2.1: Proposed resonant sigma converter and the evaluation comparisons on the efficiency, speed, and density aspects.

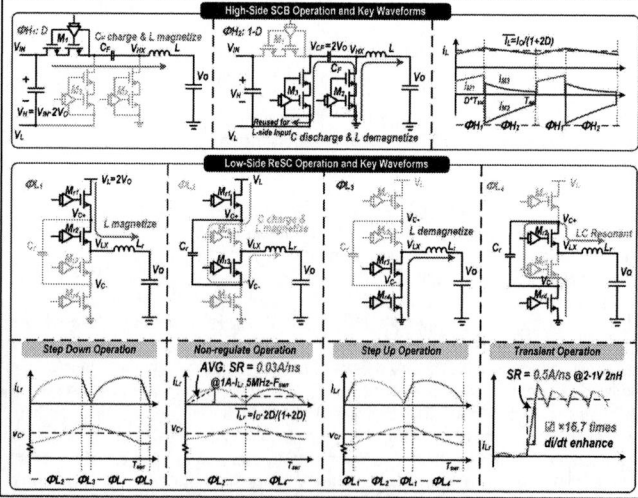

Figure 26.2.2: The operation principles and key waveforms of the proposed converter.

ISSCC 2026 / SESSION 26 / COMPUTE POWER AND SUPPLY MODULATORS / 26.2

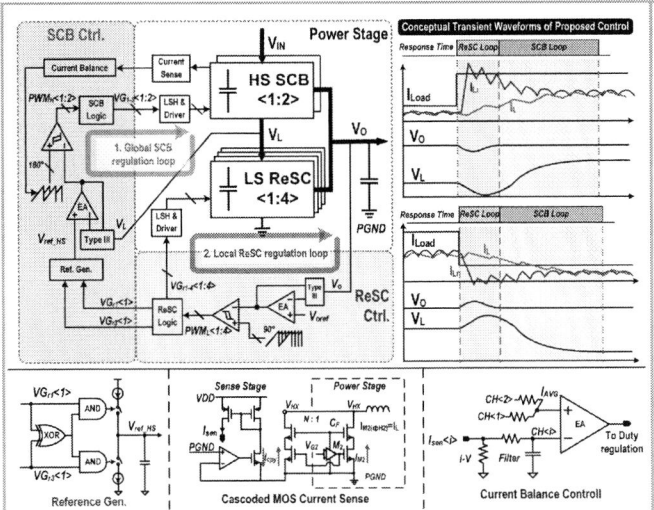

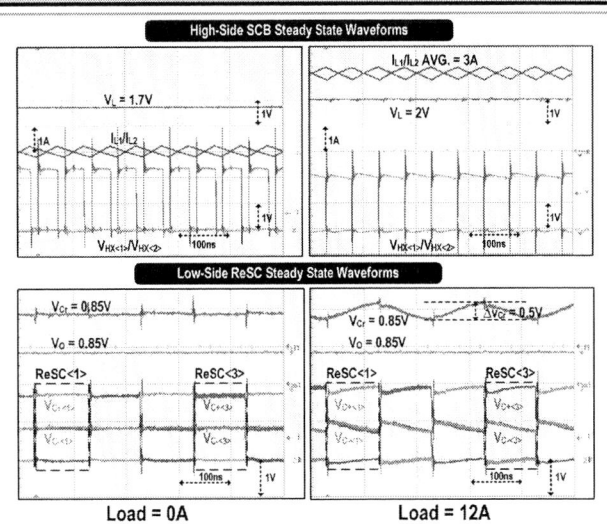

Figure 26.2.3: The overall circuit diagrams of the proposed converter, the circuit implementations, and the conceptual transient waveforms.

Figure 26.2.4: The measured steady-state waveforms with 0A and 12A load under a 4V-to-0.85V conversion.

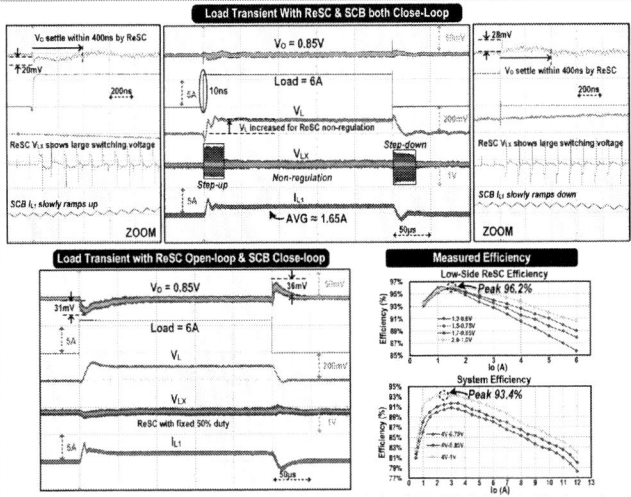

Figure 26.2.5: The measured load transients with (top) ReSC & SCB both close-loop, with (bottom left) ReSC open-loop & SCB close-loop, and (bottom right) the efficiency curves.

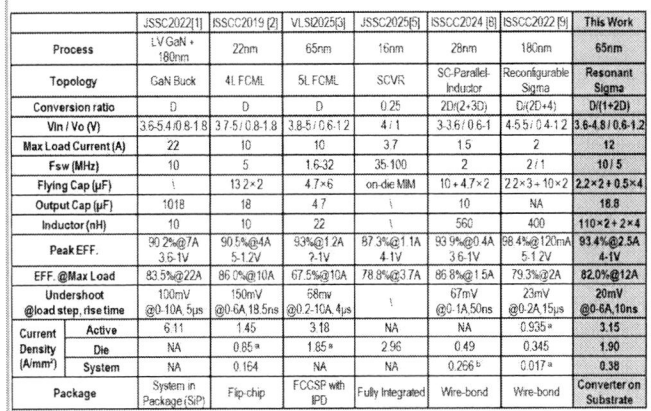

Figure 26.2.6: Performance comparison with the state-of-the-art works.

		JSSC2022[1]	ISSCC2019 [2]	VLSI2025[3]	JSSC2025[5]	ISSCC2024 [6]	ISSCC2022 [9]	This Work
Process		LV GaN + 180nm	22nm	65nm	16nm	28nm	180nm	65nm
Topology		GaN Buck	4L FCML	5L FCML	SCVR	SC-Parallel-Inductor	Reconfigurable Sigma	Resonant Sigma
Conversion ratio		D	D	D	0.25	2D(2+3D)	D/(2D+4)	D/(1+2D)
Vin / Vo (V)		3.6-5.4 / 0.8-1.8	3.7-5 / 0.8-1.8	3.8-5 / 0.6-1.2	4 / 1	3.3-6 / 0.6-1	4.5-5 / 0.4-1.2	3.6-4.8 / 0.6-1.2
Max Load Current (A)		22	10	10	3.7	1.5	2	12
Fsw (MHz)		10	5	1.6-32	35-100	2	2/1	10 / 5
Flying Cap (μF)		\	13.2×2	4.7×6	on-die MIM	10 + 4.7×2	2.2×3 + 10×2	2.2×2 + 0.5×4
Output Cap (μF)		1018	18	4.7	\	10	NA	18.8
Inductor (nH)		10	10	22	\	560	400	110×2 + 2×4
Peak EFF.		90.2%@7A 3.6-1V	90.5%@4A 5-1.2V	93%@1.2A ?-1V	87.3%@1.1A 4-1V	93.9%@0.4A 3.6-1V	98.4%@120mA 5-1.2V	93.4%@2.5A 4-1V
EFF. @Max Load		83.5%@22A	86.0%@10A	67.5%@10A	78.8%@3.7A	86.8%@1.5A	79.3%@2A	82.0%@12A
Undershoot @load step, rise time		100mV @0-10A, 5μs	150mV @0-6A,18.5ns	68mv @0-10A, 4μs	\	67mV @0-1A,50ns	23mV @0-2A,15μs	20mV @0-6A,10ns
Current Density (A/mm²)	Active	6.11	1.45	3.18	NA	NA	0.935 [a]	3.15
	Die	NA	0.85 [a]	1.85 [a]	2.96	0.49	0.345	1.90
	System	NA	0.164	NA	NA	0.266 [b]	0.017 [a]	0.38
Package		System in Package (SiP)	Flip-chip	FCCSP with IPD	Fully Integrated	Wire-bond	Wire-bond	Converter on Substrate

NA: Not Available; [a] Estimated from figure; [b] Passive Components Area Only

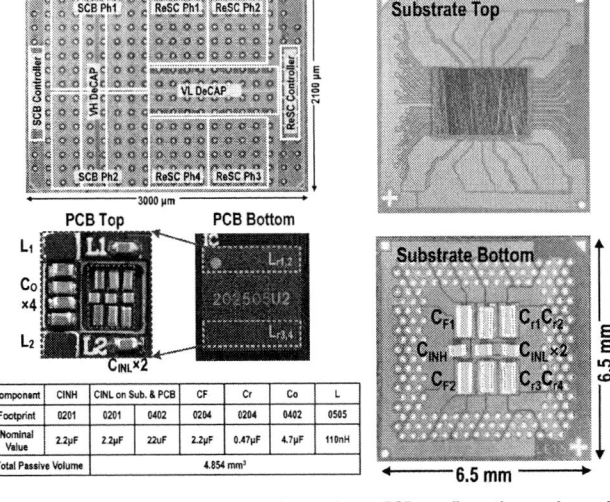

Figure 26.2.7: Die micrograph, package photo, PCB configuration and passive components summary.

Component	CINH	CINL on Sub. & PCB	CF	Cr	Co	L
Footprint	0201	0201	0402	0204	0402	0505
Nominal Value	2.2μF	2.2μF	22uF	2.2μF	0.47μF	4.7μF	110nH
Total Passive Volume			4.854 mm³			

- 2026 IEEE International Solid-State Circuits Conference

ISSCC 2026 / SESSION 26 / COMPUTE POWER AND SUPPLY MODULATORS / 26.3

26.3 A Multi-Phase Hybrid Converter with Q Samplers Enabling Simultaneous IL Auto-Balance and Arbitrary Phase Count

Jiacheng Yang*, Zihao Tang*, Rui P. Martins, Mo Huang

University of Macau, Macau, China
*Equally Credited Authors (ECAs)

Abstract

The paper presents a multiple-phase hybrid converter with Q samplers. Split-T_{ON} control enables V_{CF} auto-balance and arbitrary phase counts under all-phase interleaving, while 0-V Q samplers support I_L auto-balance with minimal reduction in power density. It achieves peak efficiencies of 90.5% and 91.5% at 1V and 1.8V V_0. With 6-phase operation and ALL-ON control, the converter attains the lowest normalized undershoot. This method is extendable to other hybrid converters.

Modern computing chips require high-current delivery from efficient point-of-load (PoL) converters. Multi-phase (MP) bucks [1], each phase comprising a half-bridge (HB) and an inductor (Fig. 26.3.1), are widely used. The phase count N should be flexibly scalable to meet the maximum load capability in the design specifications. During load transients, charging all inductors ("ALL-ON") mitigates output voltage (V_0) droop, while steady-state N-phase interleaving ($2\pi/N$ phase shift) suppresses V_0 ripple. However, at increased input voltage (V_{IN}) $\geq$12V commonly chosen to reduce input PDN loss, converters' power conversion efficiency (PCE) drops significantly [2]. Moreover, conventional MP bucks suffer from intrinsic inductor current (I_L) imbalance among phases [3], analogous to unequal water levels in different glasses, requiring per-phase current sensing. This increases circuit complexity and becomes challenging at the higher switching frequency (f_{SW}) that is desirable to minimize passive volume.

Hybrid converters, merging a pre-stage switched-capacitor (SC) network with cascaded buck stages, use low-V_{DS} power switches with a better figure-of-merit (FoM) [4], improving PCE in 12-to-PoL applications. Multi-phase hybrids inherit this benefit, and potentially support I_L auto-balance, such as in 3P4S [5] and MIH [6] topologies (Fig. 26.3.1). In these converters, the ith high-level switching node ($V_{Xi} = V_H$) during T_{ONi} reflects the charge (chg) in the flying capacitor (C_{Fi}), which then discharges (dchg) during $T_{ON(i+1)}$. Although C_F charge balance ensures I_L auto-balance [3], it inherently forbids overlap between high-level V_{Xi} and $V_{X(i+1)}$, which also indicates overlap of C_{Fi} charging and discharging [7]. This limits N to $T_{CYCLE}/T_{ON} = 1/D$, where D is the duty cycle. Since the voltage conversion ratio (VCR) is $V_H \times D = V_{IN}/N$ in [5,6], the boundary for the maximum achievable VCR is $1/N^2$, constraining N to three for 12-to-PoL applications. Meanwhile, [5,6] disallow ALL-ON operation, degrading transient performance [8,9]. Mesh-connected 3P4S stages [3] overcome both the VCR and ALL-ON limitations, but extending N remains difficult and extensive SC mesh routing raises complexity.

The HOOP converter [10] also achieves I_L auto-balance. With each SC stage directly supplied by V_{IN}, it makes $V_H = V_{IN}/2$ independent of N. The energizing of the ith phase inductor occurs during two T_{ON} intervals, corresponding to C_{Fi}'s charging and $C_{F(i-1)}$'s discharging (illustrated in Fig. 26.3.1 for $N = 4$). Compared with [5] and [6], HOOP enables ALL-ON. However, it only equalizes C_F voltages (V_{CF}) without setting them to the desired value [11,12]. In addition, reducing the phase count requires forming a smaller ring via auxiliary switches (SWs) and rerouted power paths, adding conduction loss and layout complexity. Furthermore, it only implements 2-phase interleaving (Fig. 26.3.1), even when a higher phase count N is used. Although an extension to N-phase interleaving is theoretically possible, N is thereby limited by the non-overlap requirement between charging and discharging of C_{F4}. Thus, the maximum N is restricted to 2/VCR, i.e., $\leq$ 6 for 12-to-1. MP CCC [13], with all phases coupled via a bus connection at their intermediate DC nodes, inherits most advantages and drawbacks of HOOP under the same control scheme, but enables phase scaling without auxiliary circuits (Fig. 26.3.1).

An alternate scheme, termed split-T_{ON} control, scales N freely while preserving VCR and N-phase interleaving. This is achieved by building a high-level V_X within a single T_{ON} from multiple C_Fs' operations (e.g., high-level V_{X1} formed by C_{F1} charging and C_{F4} discharging, Fig. 26.3.1). This prevents C_F from being simultaneously charged and discharged even under overlap. However, this scheme sacrifices I_L auto-balance. To address this trade-off, we link adjacent phases through a sampling capacitor C_S, referred to as a Q sampler, while retaining the split-T_{ON} control (Fig. 26.3.1). These sampling capacitors act like "communicating-vessels" in a water analogy, naturally equalizing "water levels". With zero DC voltage across each C_S, compact components can be used, reducing size and cost. Most importantly, this topology guarantees arbitrary phase count.

Figure 26.3.2 presents the four working states of each phase: State$_0$ (S_0): switch M_4 turns on, de-energizing the inductor. State$_1$ (S_1): M_1 and M_3 turn on, energizing the inductor through two current paths: (1) charges C_F and (2) discharges other C_Fs, enabling split-T_{ON} control. State2 (S_2): M_2 and M_4 turn on, de-energizing the inductor while discharging C_F. State3 (S_3): M_1 turns on, charging C_F and energizing the inductor.

The working principle is illustrated for $N = 4$ under non-overlapping T_{ON} ($D < 0.25$). During T_{ON1}, inductor L_1 energizes while L_{2-4} de-energizes. Phase$_{1-4}$, configured in S_1, S_0, S_0 and S_2, respectively, allow C_{F1} to receive charge Q_1 while C_{F4} and C_{S1-3} are allowed to release Q_4. Thus, the total charge received by L_1 during T_{ON1} is $Q_{ON1}=Q_1+Q_4$. During T_{ON2}, Phase$_{1-4}$ operate in S_2, S_1, S_0 and S_0, charging C_{F2} and C_{S1} while discharging C_{F1}. The charge received by L_2 is $Q_{ON2}=Q_1+Q_2$. Similarly, L_3 and L_4 receive $Q_{ON3}=Q_2+Q_3$ and $Q_{ON4}=Q_3+Q_4$ during T_{ON3} and T_{ON4}, respectively. In this process, C_{S1-3} sample $Q_4=Q_S$ during T_{ON1} and subsequently release it during T_{ON2-4}. The charge balance across C_{S1-3} enforces $Q_1=Q_2=Q_3=Q_4=Q_S$, resulting in equal charge received by each inductor ($Q_{ON}=2Q_S$) and hence inherent I_L balance [3]. Capacitor voltage equations in each T_{ON} yield steady-state $V_{CF}=V_{IN}/2$ and $V_{CS}=0$.

Figure 26.3.2 also illustrates the working principle when overlap occurs (D>0.25), exemplified by the scenario where the full high-level of V_{X1} splits into three parts, overlapping with V_{X4} and V_{X2} during T_{ON41} and T_{ON12}, while no overlap occurs during T_{ON1}. During T_{ON41}, C_{F1} charging establishes the high-level of V_{X1}, whereas C_{F4} charging and C_{F3} discharging establish the high level of V_{X4}, ensuring non-overlap of C_{F4} charging and discharging. Furthermore, we can derive similar combinations in other V_X overlaps. Charge balance of each C_S is preserved, maintaining I_L balance as in the D<0.25 case.

Figure 26.3.3 illustrates the schematic of the proposed 6-phase converter. The chip integrates the power MOSFETs, bootstrap (BST) gate drive circuits and controller. We employ synchronized hysteretic current-mode control [9] for fast transients. I_{L1} is sensed by an RC filter (R_1 and C_1) and amplified by an operational amplifier (AMP). A sawtooth slope compensation signal, synchronized to a clock signal CLK$_1$, is added to the AMP output to generate the ramping signal V_{RAMP}. An error amplifier (EA) with a type-II compensation compares V_0 with reference voltage V_{REF}, producing V_{EA}. A hysteretic comparator then compares V_{RAMP} and V_{EA}, with the output fed into D-copy circuit to generate PWM signals D_{1-6}. These signals are fed to the non-overlap block and multiplexed with the start-up control signals to produce the gate control signals G_{1-4} of each phase. Under a load current step, D_1 is set high when V_{RAMP} intersects with the lower boundary of the V_{EA} window. When V_0 drops below a threshold voltage V_{REFL}, D_{2-6} is pulled high immediately, maximizing the I_L slew rate and achieving ALL-ON control (Fig. 26.3.3).

Figure 26.3.3 presents capacitance degradation as a function of bias voltage from TDK C series 2.2-μF capacitors [14]. While they maintain nominal 2.2-μF at 0V, capacitance drops rapidly with increasing bias, particularly for small packages such as 0402 and 0603. Therefore, using 0V C_S in this work enables compact implementation while minimally reducing power density. This method extends to high-order hybrids with multiple DC nodes [15]. In such cases, C_S can be inserted into one of these DC nodes, while others remain directly connected (Fig. 26.3.3). Inherent inductor current balance is ensured using the same rationale.

The prototype 6-phase chip, fabricated in a 0.18-μm BCD process, occupies a silicon area of 3.2×3.6mm². We use 0603 C_F and 0402 C_S, with a PCB area of 8.6×10.7mm² (Fig. 26.3.7). The converter operates at 2MHz under a nominal 12-V V_{IN}, delivering a maximum current (I_{LOAD}) of 8A. Figure 26.3.4 displays the measured steady-state waveforms at 1-V and 1.8-V V_0, under N=6, I_{LOAD}=4A. We observe auto-balance in the inductor currents with each phase interleaved by $2\pi/N$, in contrast to the fixed $2\pi/2$ in [10]. The maximum current imbalances (calculated with Eq(1)) are 1.5% and 1.8% at V_0 = 1V and 1.8V, respectively. Meanwhile, D overlaps in 1.8-V V_0 enabled by the split-T_{ON} control. All switching-node swings are around 6V, indicating V_{CF} auto-balance. Figure 26.3.4 also plots the measured PCE versus I_{LOAD}. The peak PCE of 1-V and 1.8-V V_0 are 90.5% and 91.5%, respectively, while the MP operation extends the high-PCE load range.

Figure 26.3.5 exhibits the measured load transient waveforms at 12-V V_{IN} and 1-V V_0 under a 0-to-6A load step within 20ns. Waveforms with ALL-ON disabled are displayed for comparison, where the instantaneous inductor current slew rate (SR) is only ($V_{IN}-6V_0$)/L and the resultant undershoot becomes 182mV. With ALL-ON control, simultaneous charging of all inductors is enabled after the detect of transient events, maximizing the current SR to ($3V_{IN}-6V_0$)/L, reducing the undershoot to 65mV. Only I_{L1}, I_{L3} and I_{L5} are displayed due to the limited number of current probes. Figure 26.3.5 presents the start-up waveforms, showcasing C_Fs pre-charged to the desired value before normal operation. Figure 26.3.5 also illustrates the reference tracking waveforms when the reference voltage varies from 1V to 1.8V.

Figure 26.3.6 presents a performance summary and comparison with the state-of-the-art works. This work achieves a peak PCE of 90.5% in 12-to-1 conversion, which is comparable or better than prior work. Owing to the use of 0-V sampling capacitors and a compact PCB design, this work obtains a current density of 81.5mA/mm² at 85% efficiency. This topology achieves simultaneous I_L balance, V_{CF} balance and arbitrary phase count. In contrast, [13] does not ensure I_L balance, while [3], [10], and [16] restrict the phase count under all-phase interleaving (where the number of interleaving equals the number of inductors). These advantages imply that our design is suitable for targeting high-current applications. With six phases implemented and "ALL-ON" operation enabled during transient, the proposed converter achieves a normalized undershoot (US) of 0.41, calculated as the ratio of the measured US to the theoretical minimum value of single-phase operation ($US_{MIN,1PH}$, from Eq(2) [9]), which is the lowest among the works compared in Fig. 26.3.6. We can further extend this method to more hybrids.

Acknowledgement:
This work was supported in part by the Science and Technology Development Fund, Macau SAR (0029/2025/AMJ, 0042/2025/RIB1, 004/2023/SKL and 001/2024/COP), in part by the Guangdong-HongKong-Macao Joint Laboratories (2025B1212150003), in part by the Guangdong Basic and Applied Basic Research Foundation (2023B1515130001), in part by the Research Committee of University of Macau (MYRG-GRG2024-00041-IME).

References:
[1] J. -H. Cho *et al.*, "A 1.23W/mm2 83.7%-Efficiency 400MHz 6-Phase Fully Integrated Buck Converter in 28nm CMOS with On-Chip Capacitor Dynamic Re-Allocation for Inter-Inductor Current Balancing and Fast DVS of 75mV/ns," *2022 IEEE International Solid-State Circuits Conference (ISSCC)*, San Francisco, CA, USA, 2022, pp. 1-3. https://doi.org/10.1109/ISSCC42614.2022.9731726
[2] H. -J. Choi et al., "A 92.7% Peak Efficiency 12V-to-60V Input to 1.2V Output Hybrid DC-DC Converter Based on a Series-Parallel-Connected Switched Capacitor," *2024 IEEE International Solid-State Circuits Conference (ISSCC)*, San Francisco, CA, USA, 2024, pp. 156-158. https://doi.org/10.1109/ISSCC49657.2024.10454344
[3] J. Yang, R. P. Martins and M. Huang, "A Segmented-Interlacing Multi-Phase Hybrid Converter with Inherently Auto-Balanced ILs and Boosted IL Slew Rate During Load Transients," *2025 IEEE International Solid-State Circuits Conference (ISSCC)*, San Francisco, CA, USA, 2025, pp. 378-380. https://doi.org/10.1109/ISSCC49661.2025.10904561
[4] X. Yang et al., "An 8A 998A/inch3 90.2% Peak Efficiency 48V-to-1V DC-DC Converter Adopting On-Chip Switch and GaN Hybrid Power Conversion," *2021 IEEE International Solid-State Circuits Conference (ISSCC)*, San Francisco, CA, USA, 2021, pp. 466-468. https://doi.org/10.1109/ISSCC42613.2021.9366005
[5] Y. Huang, Y. Ramadass and D. B. Ma, "A 90.7% 4-W 3P4S Hybrid Switching Converter Using Adaptive VCF Rebalancing Technique and Switching Node Dual-Edge tdead Modulation for Extreme 48V/1V Direct DC-DC Conversion," *2022 IEEE Symposium on VLSI Technology and Circuits (VLSI Technology and Circuits)*, Honolulu, HI, USA, 2022, pp. 178-179. https://doi.org/10.1109/VLSITechnologyandCir46769.2022.9830419

[6] R. Das, G. -S. Seo, D. Maksimovic and H. -P. Le, "An 80-W 94.6%-Efficient Multi-Phase Multi-Inductor Hybrid Converter," *2019 IEEE Applied Power Electronics Conference and Exposition (APEC)*, Anaheim, CA, USA, 2019, pp. 25-29. https://doi.org/10.1109/APEC.2019.8721952
[7] M. Gong, H. Chen, X. Zhang, R. Jain and A. Raychowdhury, "A 90.4% Peak Efficiency 48-to-1-V GaN/Si Hybrid Converter With Three-Level Hybrid Dickson Topology and Gradient Descent Run-Time Optimizer," in *IEEE Journal of Solid-State Circuits*, vol. 58, no. 4, pp. 1002-1014, April 2023. https://doi.org/10.1109/JSSC.2022.3228233
[8] J. Yuan, Z. Liu, F. Wu and L. Cheng, "A 12V/24V-to-1V DSD Power Converter with 56mV Droop and 0.9μS Settling Time for a 3A/20ns Load Transient," *2022 IEEE International Solid-State Circuits Conference (ISSCC)*, San Francisco, CA, USA, 2022, pp. 1-3. https://doi.org/10.1109/ISSCC42614.2022.9731701
[9] T. Hu, M. Huang, R. P. Martins and Y. Lu, "A 12-to-1 Flying Capacitor Cross-Connected Buck Converter With Inserted D > 0.5 Control for Fast Transient Response," in *IEEE Journal of Solid-State Circuits*, vol. 58, no. 11, pp. 3207-3218, Nov. 2023. https://doi.org/10.1109/JSSC.2023.3297111
[10] Z. Tong et al., "HOOP: A Scalable Hybrid DC-DC Converter Ring for High-Performance Computing," *2025 IEEE International Solid-State Circuits Conference (ISSCC)*, San Francisco, CA, USA, 2025, pp. 1-3. https://doi.org/10.1109/ISSCC49661.2025.10904695
[11] X. Liu, C. Huang and P. K. T. Mok, "A High-Frequency Three-Level Buck Converter With Real-Time Calibration and Wide Output Range for Fast-DVS," in *IEEE Journal of Solid-State Circuits*, vol. 53, no. 2, pp. 582-595, Feb. 2018. https://doi.org/10.1109/JSSC.2017.2755683
[12] S. S. Amin and P. P. Mercier, "A Fully Integrated Li-Ion-Compatible Hybrid Four-Level DC–DC Converter in 28-nm FDSOI," in *IEEE Journal of Solid-State Circuits*, vol. 54, no. 3, pp. 720-732, March 2019. https://doi.org/10.1109/JSSC.2018.2880183
[13] J. Yang, T. Hu, Y. Lu, X. Ming, R. P. Martins and M. Huang, "Phase-Scalable CF-Cross-Connected-Based Hybrid DC–DC Converter With Auto VCF Balancing and Inactive CF Charging," in *IEEE Journal of Solid-State Circuits*. https://doi.org/10.1109/JSSC.2025.3557060
[14] TDK corporation, Multilayer Ceramic Chip Capacitors, C series. [Online]. Available:. https://product.tdk.com/en/search/capacitor/ceramic/mlcc/list#_l=20&_p=1&_c=pure_status&_d=0
[15] T. Hu, M. Huang, R. P. Martins and Y. Lu, "A 12-to-1 V Quad-Output Switched-Capacitor Buck Converter With Shared DC Capacitors," in *IEEE Journal of Solid-State Circuits*, vol. 58, no. 12, pp. 3492-3502, Dec. 2023. https://doi.org/10.1109/JSSC.2023.3301068
[16] M. H. K. Hmada, W. -C. B. Liu, G. Pillonnet and P. Mercier, "Merging Hybrid and Multi-Phase Topologies: A 6-Phase Triple-Step-Down DC-DC Converter Achieving up to a 60:1 Voltage Conversion Ratio and 868A/cm3 Current Density," *2025 IEEE International Solid-State Circuits Conference (ISSCC)*, San Francisco, CA, USA, 2025, pp. 386-388. https://doi.org/10.1109/ISSCC49661.2025.10904637

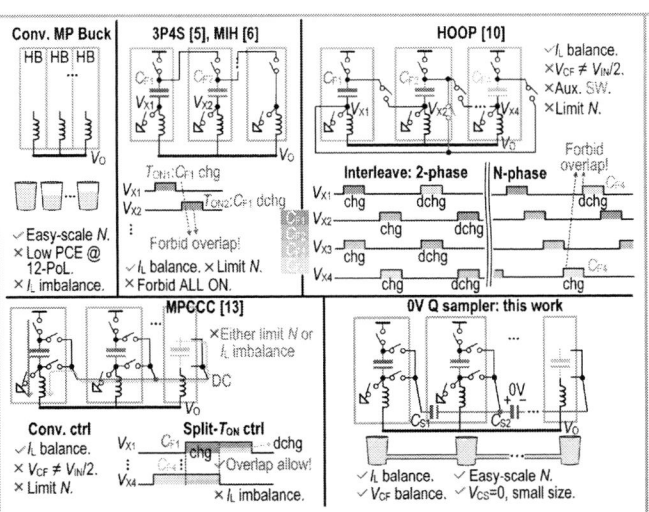

Figure 26.3.1: Advantages and drawbacks of conventional MP buck (top left), prior hybrid MP hybrids (top right and bottom left), and proposed MP hybrid with Q sampler (bottom right).

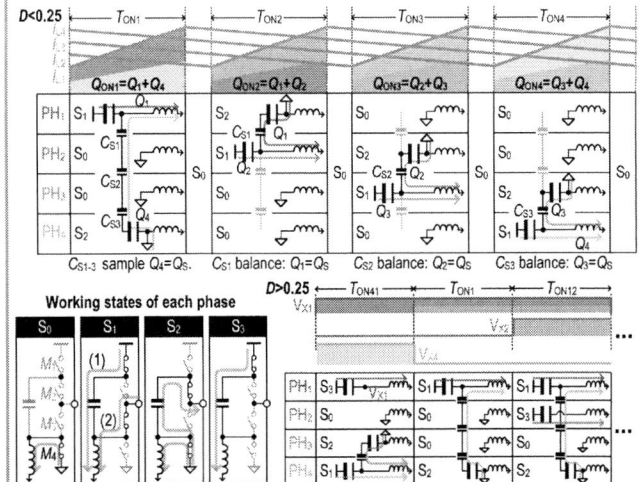

Figure 26.3.2: Working states of each phase (bottom left), working principles for $N = 4$ under $D < 0.25$ (top) and $D > 0.25$ (bottom).

ISSCC 2026 / SESSION 26 / COMPUTE POWER AND SUPPLY MODULATORS / 26.3

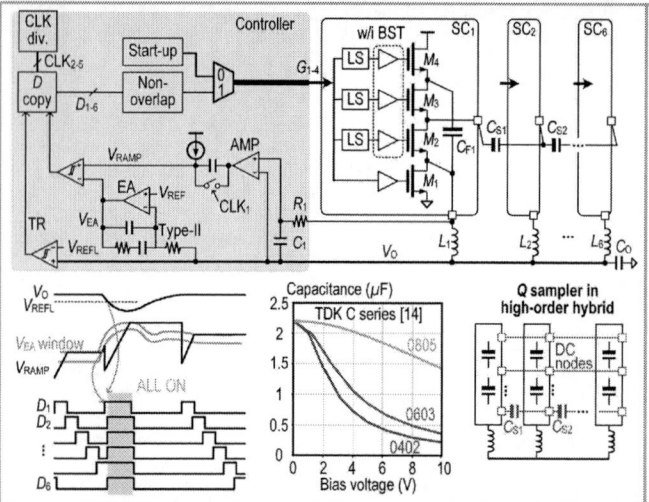

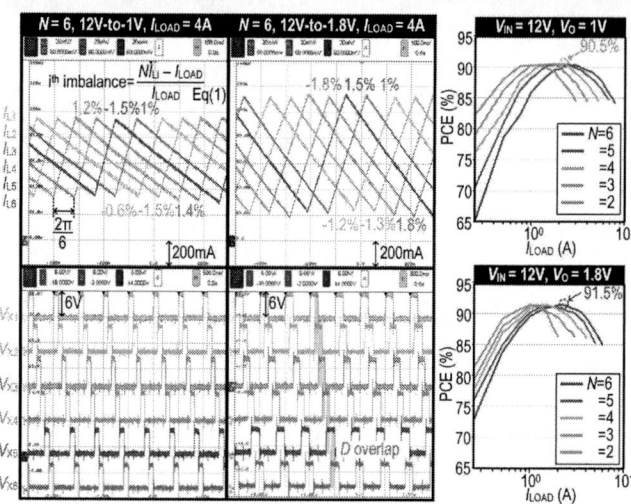

Figure 26.3.3: Schematic of proposed 6-phase converter (top), load-transient waveforms (bottom left), capacitance degradation vs. bias voltage (bottom middle) and Q sampler in high-order hybrid (bottom right).

Figure 26.3.4: Measured steady-state waveforms at 1-V and 1.8-V V_O, where D overlap in 1.8V-V_O, and I_Ls are auto-balanced in both cases (left), measured PCE vs. I_{LOAD} (right).

Figure 26.3.5: Measured waveforms during load transient with ALL-ON disabled (top left) and enabled (bottom left), start-up (top right) and reference tracking (bottom right).

	JSCC' 25 [13]	ISSCC' 25 [16]	ISSCC' 25 [3]	ISSCC' 25 [10]	This work
Process	180nm BCD	180nm BCD	180nm BCD	180nm BCD	180nm BCD
Topology	MP CCC	TSD	SI	HOOP	Bus+Q sampler
V_{IN}, V_O (V)	12, 1.2 – 1.8	12, 0.3 – 1.1	12, 1 – 1.8	12, 1 – 1.2	12, 1 – 1.8
f_{SW} (MHz)	2	1	2	1	2
Max. I_{OUT} (A)	4	7	6	16	8
C_{OUT} (μF)	4 × 4.7	10	10	8 × 10	20
Inductors (μH) @DCR (mΩ)	4×1 @30	3×0.68 @17	3×0.6* @29	8×1 @48	6×0.9* @26
# phases interleave	= #inductors	= #inductors	= #inductors	2	= #inductors
Peak PCE @ V_{IN}-to-V_O	90.3% @12-to-1.2V	89.7% ** @12-to-1V	91.7% @12-to-1V	89% @12-to-1V	90.5% @12-to-1V
Area of PCB (mm²)	247**	238	70.4	256	92
Current density @85% eff. (mA/mm²)	16.2	17.9 **	65.3 **	32.8 **	81.5
Arbitrary # phases?	Yes	No, fixed 3	No, fixed 3	No, max. 6 @ 12-to-1	Yes
Inherent I_L balance?	No	Yes	Yes	Yes	Yes
V_{CF} auto balance?	Yes	Yes	Yes	No	Yes
"ALL-ON"?	Yes	No	Yes	Yes	Yes
$\Delta I_{LOAD}/T_{EDGE}$	4A / 20ns	1A / 10ns	3.5A / 20ns	2.7A / 80ns	6A / 20ns
US (mV)	93	69.5	92	85	65
US+US$_{MIN,1PH}$***	1.08	6.4	0.77	2.73	0.41

* Effective value ** Estimated from figure ***$US_{MIN,1PH} = \frac{C_{OUT}}{2}\left(\frac{L \times I_{LOAD}^2}{V_H} - I_{LOAD} \times T_{EDGE}\right)$ Eq(2)

Figure 26.3.6: Performance summary and comparison with state-of-the-art works.

* Limited by the number of current probes, only I_{L1}, I_{L3} and I_{L5} are measured.

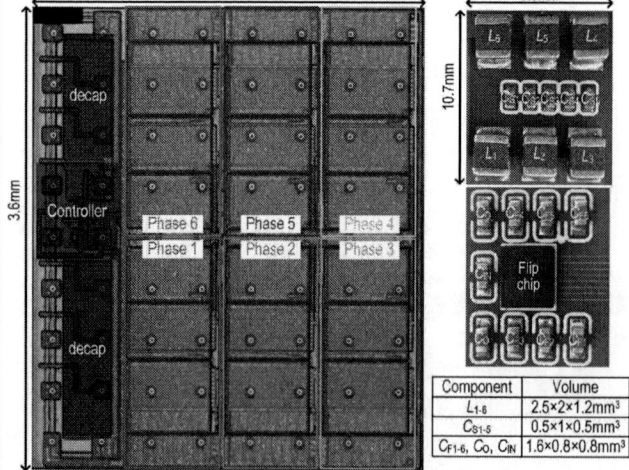

Component	Volume
L_{1-6}	2.5×2×1.2mm³
C_{S1-5}	0.5×1×0.5mm³
C_{F1-6}, C_O, C_{IN}	1.6×0.8×0.8mm³

Figure 26.3.7: Chip micrograph (left), PCB photo (top right) and passive components of power stage (bottom right).

• 2026 IEEE International Solid-State Circuits Conference

ISSCC 2026 / SESSION 26 / COMPUTE POWER AND SUPPLY MODULATORS / 26.4

26.4 A 12-to-1V 90.5%-Peak-Efficiency 721A/cm³-Current-Density Quad-Output Converter with One Shared DC Capacitor

Qingqing Min[1], Jinyi Yuan[1], Ji Jin[1], Yichao Ji[1], Weiwei Xu[2], Lin Cheng[1,2]

[1]University of Science and Technology of China, Hefei, China, [2]Hefei CLT Microelectronics, Hefei, China

Abstract

This paper presents a 12-to-1V 4-output hybrid DC-DC converter that shares a single DC capacitor among all outputs. This shared capacitor stores and redistributes energy, ensuring independent regulation at each output. The proposed design achieves a 721A/cm³ current density and 90.5% peak efficiency at 12-to-1.1V at 2MHz, delivering up to 16A of total load current. The measured results during load transients shows there is no cross-regulation issue.

12-to-1V DC-DC converters are ubiquitous in data-center platforms, many of which must generate several supply rails on the same board. In DDR5 DIMMs, for example (Fig. 26.4.1, top left), a PMIC typically supplies 1.1V and 1.8V for core and I/O, while ever-increasing memory bandwidth pushes requirements for higher efficiency and current density. To improve efficiency and current density, several hybrid topologies have been proposed. By utilizing flying capacitors to reduce the voltage stress on power transistors, these designs achieve improved performance [1-5]. Recent research has increasingly focused on multi-output hybrid topologies, which aim to enhance current density by extending conventional hybrid structures to support multiple outputs within a compact footprint. For instance, by sharing the DC capacitors of a 3:1 ladder converter [6], a quad-output architecture can reduce both the number and volume of DC capacitors. However, this topology still relies on multiple flying capacitors to generate low-voltage nodes. In [7], the multiphase series capacitor converter achieves multiple outputs by decoupling inductors. While this approach reduces component count, the output power is constrained by the need to maintain charge balance across the flying capacitors, and cross-regulation issues may occur. In [8], DC capacitors are employed to suppress cross-regulation effects among outputs. However, this approach introduces additional DC capacitor components, increasing the overall component count.

As shown in Fig. 26.4.1 (bottom left), this work proposes an N-output hybrid topology that shares a single DC capacitor among all outputs. Compared with a conventional multiphase series-capacitor buck converter, the proposed converter introduces N switches and one shared DC capacitor, where each switch is connected between the switching node SW_i and the DC capacitor. This shared capacitor stores and redistributes energy, ensuring independent regulation at each output and improving overall current density. In this design, the voltage across the flying capacitor C_{Fi} and the DC capacitor C_{DC} are $V_{IN}(N-i)/N$ and V_{IN}/N, respectively. To reduce power loss and simplify control, the switching sequences of the N outputs are phase-shifted by 360°/N. For each output V_{Oi}, when switch S_{Bi} is turned on with a duty cycle of D_{Bi} ($D_{Bi} = D_i$), the inductor L_i is magnetized, achieving a voltage conversion ratio of D_i/N. If $D_{i-1}<1/N$, then S_{Ai} and S_{Bi} turn on simultaneously; if $D_{i-1}>1/N$, then S_{Ai} is turned on with a delay to avoid overlap with $S_{A(i-1)}$. When switch S_{Ci} is turned on ($D_{Ci} = 1-D_i$), the inductor L_i is demagnetized. Due to the phase-shifting between outputs, the value of the DC capacitor C_{DC} does not scale with the number of outputs. Figure 26.4.1 (bottom right) illustrates a two-phase example to explain the charge transfer between the capacitors. During phase Φ_1, the flying capacitor C_{F1} and the DC capacitor C_{DC} are connected in series between V_{IN} and ground. At the beginning of this phase, they are rapidly charged in a hard-charging mode until $V_{CF} + V_{DC} = V_{IN}$. Subsequently, both capacitors discharge softly to the output V_{O1}. During phase Φ_2, the flying capacitor C_{F1} and the DC capacitor C_{DC} are connected in parallel, and therefore $V_{CF} = V_{DC}$. During phase Φ_3, the two capacitors are connected to the inductor L_2 and discharge softly to the output V_{O2}. During phase Φ_4, no charge transfer occurs between the capacitors, since $V_{CF} = V_{DC}$ is maintained in the previous phase. Thanks to the shared DC capacitor C_{DC}, the unbalanced charge is stored in it through hard charging and can be redistributed to the required output. As the converter scales to more outputs, the same operating principle applies, achieving independent output regulation. This topology shares a single DC capacitor and N−1 flying capacitors, effectively improving current density. The flying capacitors and DC capacitor provide charge to each output independently, ensuring the independence of each output and eliminating cross-regulation.

Figure 26.4.2 illustrates the proposed hybrid topology applied to a DDR5 DIMM PMIC. A dual-output converter with N = 2 is employed to support a wide input voltage range of 4.5-to-15V. This converter consists of six switches, one flying capacitor, and one DC capacitor. Two such dual-output converters are connected in parallel and share the DC capacitor C_{DC}, forming a quad-output converter that delivers multiple regulated voltages to the DIMM. Compared to other works, the proposed topology achieves both high density and independent outputs. The two sub-converters, A and B, operate with independent timing. To reduce power loss and simplify control, the switching sequences of the two outputs are phase-shifted by 180°. Figure 26.4.2 illustrates the operation principle of the proposed converter that encompasses four phases: Φ_1, Φ_2, Φ_3 and Φ_4. During phase Φ_1, switches S_1 and S_3 are turned on, while S_2 and S_4 are turned off, allowing inductor L_1 to magnetize and flying capacitor C_F to charge. Simultaneously, switch S_6 is turned on, while S_5 is turned off, allowing inductor L_2 to demagnetize. During phase Φ_3, switches S_2 and S_4 are turned on, while S_1 and S_3 are turned off, leading to the demagnetization of L_1 and the discharge of C_F;

switch S_5 is also turned on, while S_6 is turned off, allowing inductor L_2 to magnetize. During phase $\Phi_{2/4}$, both inductors L_1 and L_2 undergo demagnetization.

Figure 26.4.3 shows the system architecture of the proposed DC-DC converter. The power stage consists of two dual-output converters sharing a single C_{DC}, achieving four independent outputs. All power switches are implemented with 6V-LDMOS devices by utilizing flying capacitor C_F. Since the source of switch S_2 is connected to V_{DC}, it can be naturally driven using V_{IN}-V_D (V_D is the diode forward voltage) as the bootstrap voltage. Simple diode-based bootstrap circuits are used for switches S_1, S_3, and S_5, which is similar to the high-side switch in a buck converter. In contrast, S_4 and S_6 are directly driven by the 5V supply. The power stage transfer function of the proposed converter is similar to that of a two-phase buck converter with an input voltage of half of V_{IN}. Therefore, two Type-III compensators are adopted to generate $PWM_{1,2}$ and regulate the output voltages V_{O1} and V_{O2}, respectively. Two clock signals CLK_1 and CLK_2 with a 180° phase shift are used to maintain interleaving between outputs. During startup, the bottom plate of C_F is connected to ground by S_4 and it is paralleled with C_{DC}. Both capacitors are pre-charged from V_{IN} via a small on-chip power switch. Once their voltage reaches half of V_{IN}, the PWM controller is enabled to initiate soft-start operation.

The proposed converter is implemented in a 180nm BCD process, with a die area of 5.31mm². Each dual-output converter uses two 330nH inductors, one 10µF flying capacitor, and two 10µF output capacitors, while also tapping into the single shared 10µF DC capacitor. The converter operates at a switching frequency of 2MHz, and each output supports a peak current of 4A. Figure 26.4.4 shows the measured startup and steady-state waveforms of the dual-output converter. Before the switch is enabled, V_{DC} is pre-charged to $V_{IN}/2$ within 90µs. Under the input condition of 12V and at output voltages of 1.1V, the voltage swing of V_{SW1} and V_{SW2} ranges from 0V to 6V, and the output ripple voltage is approximately 10mV. Figure 26.4.4 also shows the measured load-transient responses under different output load step conditions. With an I_{O1} step of 2A at V_{IN}/V_{OUT} = 12V/1.1V, the measured undershoot and 1% settling time is 201mV and 4.18µs, respectively, while cross-regulation is negligible. A similar waveform can be observed during the load transient with I_{O2} step of 2A at V_{IN}/V_{OUT} = 12V/1.1V.

Figure 26.4.5 shows the efficiency of the proposed converter under different output voltage and load current conditions. With a 12V input, the dual-output converter achieves a peak efficiency of 92.3% at V_{O1} = V_{O2} = 1.8V and I_{O1} = I_{O2} = 0.8A (equal load), and 90.5% at V_{O1} = V_{O2} = 1.1V and I_{O1} = I_{O2} = 1A (equal load). Under unequal load conditions ($I_{O1} = I_{LOAD}$, I_{O2} = 0), the peak efficiency is 91.2% when V_{O1} = V_{O2} = 1.8V, I_{O1} = 0.8A, and I_{O2} = 0A. For the proposed dual-output converter, where two dual-output converters are paralleled and they share a DC capacitor, the peak efficiency reaches 91.4% at V_{O1} = 1.8V and V_{O2} = V_{O3} = V_{O4} = 1.1V. Compared to previous work, the proposed converter achieves both high efficiency and current density with reduced off-chip capacitor.

Figure 26.4.6 summarizes the performance and compares it with state-of-the-art designs. By sharing flying capacitors and a single DC capacitor, the proposed design achieves a peak efficiency of 90.5% and a current density of 721A/cm³, which is the highest current density among the state-of-the-art multi-output hybrid topologies listed in Fig. 26.4.6. Benefiting from the charge storage and redistribution principle, the proposed topology maintains independent outputs with no cross-regulation. Figure 26.4.7 shows the die micrograph of the converter alongside a photo of its PCB solution.

Acknowledgement:
This work was supported in part by the Anhui Provincial Key Research and Development Project under Grant No. 202423k09020002. Corresponding author: Lin Cheng (eecheng@ustc.edu.cn).

References:
[1] K. Wei, Y. Ramadass and D. B. Ma, "A Direct 12V/24V-to-1V 3W 91.2%-Efficiency Tri-State DSD Power Converter with Online VCF Rebalancing and In-Situ Precharge Rate Regulation," *2020 IEEE International Solid-State Circuits Conference - (ISSCC)*, San Francisco, CA, USA, 2020, pp. 190-192. http://doi.org/10.1109/ISSCC19947.2020.9063087

[2] X. Yang et al., "A 5A 94.5% Peak Efficiency 9~16V-to-1V Dual-Path Series-Capacitor Converter with Full Duty Range and Low V.A Metric," *2023 IEEE International Solid-State Circuits Conference (ISSCC)*, San Francisco, CA, USA, 2023, pp. 196-198. http://doi.org/10.1109/ISSCC42615.2023.10067802

[3] W. -L. Zeng *et al.*, "A 12V-Input 1V-1.8V-Output 93.7% Peak Efficiency Dual-Inductor Quad-Path Hybrid DC-DC Converter," *2023 IEEE International Solid-State Circuits Conference (ISSCC)*, San Francisco, CA, USA, 2023, pp. 10-12. http://doi.org/10.1109/ISSCC42615.2023.10067710

[4] Y. Ji, J. Jin and L. Cheng, "A 12V-Input 1V-1.8V-Output 94.7%-Peak-Efficiency 685A/cm3-Current-Density Hybrid DC-DC Converter with a Charge Converging Phase," *2024 IEEE International Solid-State Circuits Conference (ISSCC)*, San Francisco, CA, USA, 2024, pp. 458-460. http://doi.org/10.1109/ISSCC49657.2024.10454573

[5] J. Yuan, Z. Liu, F. Wu and L. Cheng, "A 12V/24V-to-1V DSD Power Converter with 56mV Droop and 0.9$\mu \mathrm{S}$ 1% Settling Time for a 3A/20ns Load Transient," *2022 IEEE International Solid-State Circuits Conference (ISSCC)*, San Francisco, CA, USA, 2022, pp. 1-3. http://doi.org/10.1109/ISSCC42614.2022.9731701

[6] T. Hu, M. Huang, Y. Lu and R. P. Martins, "A 12V-to-1V Quad-Output Switched-Capacitor Buck Converter with Shared DC Capacitors Achieving 90.4% Peak Efficiency and 48mA/mm3 Power Density at 85% Efficiency," *2023 IEEE International Solid-State Circuits Conference (ISSCC)*, San Francisco, CA, USA, 2023, pp. 184-186. http://doi.org/10.1109/ISSCC42615.2023.10067463

[7] R. Das and H. -P. Le, "Multi-Inductor Multi-Output Hybrid (MiMoH) Converter for Large Conversion Ratio and Multiple Outputs," *2023 IEEE Applied Power Electronics Conference and Exposition (APEC)*, Orlando, FL, USA, 2023, pp. 943-947. http://doi.org/10.1109/APEC43580.2023.10131378

[8] Y. Xie and J. Guo, "A 12V/24V-to-1V Shared Switched-Capacitor Multi-Inductor Multi-Output Converter with 90.9%/89.5%Peak Efficiency and Negligible Cross Regulation," *2025 IEEE Custom Integrated Circuits Conference (CICC)*, Boston, MA, USA, 2025, pp. 1-3. http://doi.org/10.1109/CICC63670.2025.10983344

[9] W. -C. Hung et al., "A Double Step-Down Dual-Output Converter with Cross Regulation of 0.025mV/mA and Improved Current Balance," *2023 IEEE International Solid-State Circuits Conference (ISSCC)*, San Francisco, CA, USA, 2023, pp. 188-190. http://doi.org/10.1109/ISSCC42615.2023.10067843

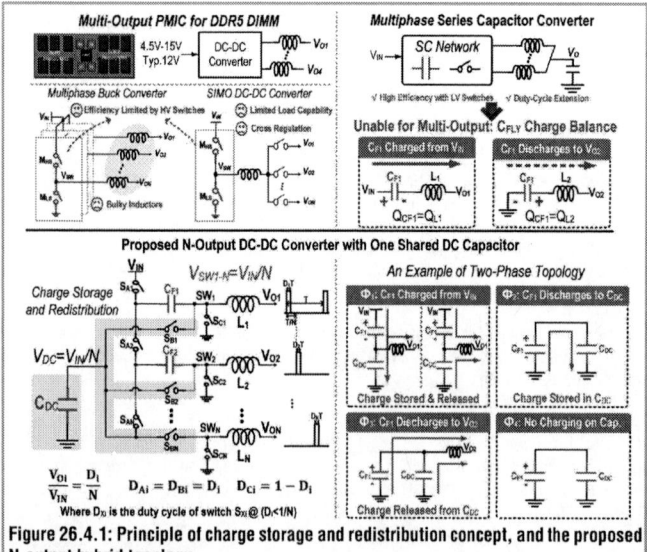

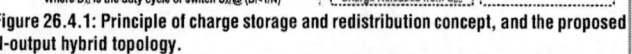

Figure 26.4.1: Principle of charge storage and redistribution concept, and the proposed N-output hybrid topology.

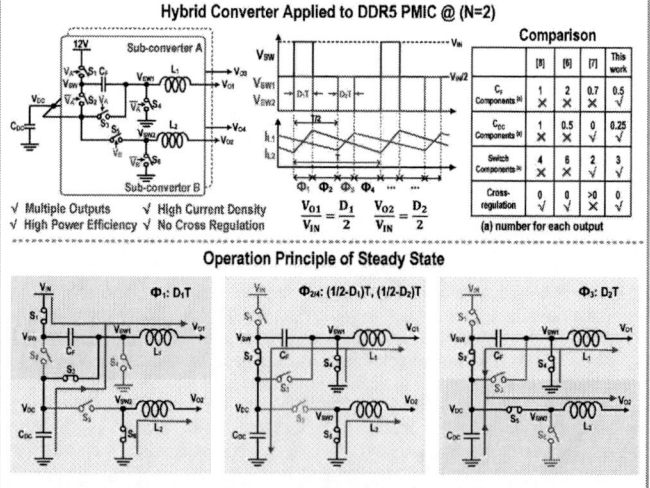

Figure 26.4.2: Working principle of the proposed hybrid converter applied to DDR5 DIMM PMIC.

ISSCC 2026 / SESSION 26 / COMPUTE POWER AND SUPPLY MODULATORS / 26.4

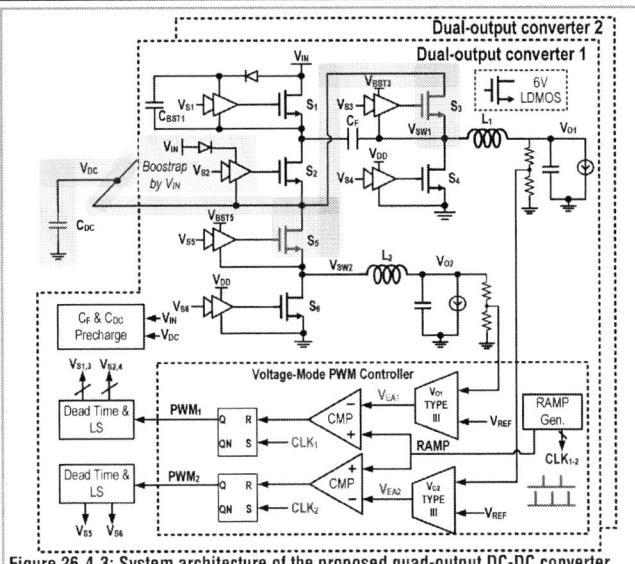

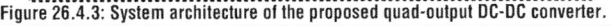

Figure 26.4.3: System architecture of the proposed quad-output DC-DC converter.

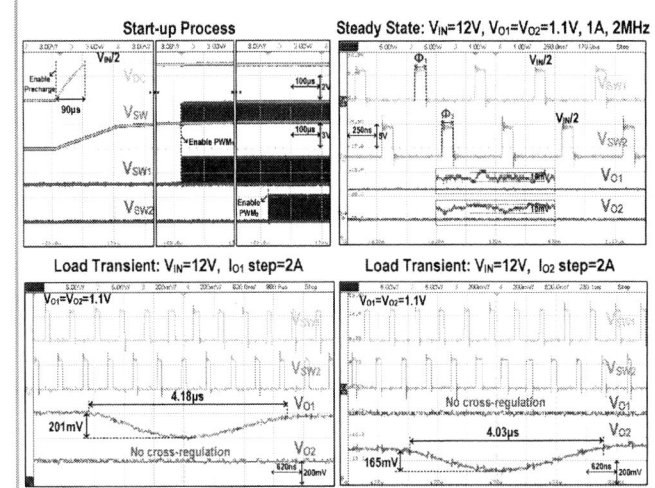

Figure 26.4.4: Measured start-up process, steady-state waveforms and load responses of the proposed converter.

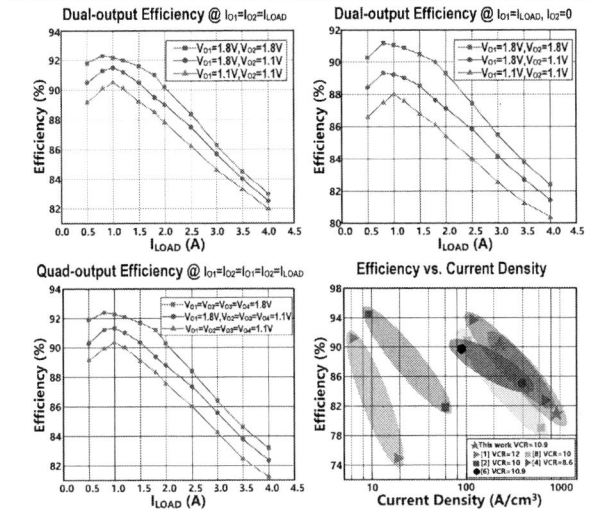

Figure 26.4.5: Measured efficiency of the proposed converter and performance comparison with the prior works.

	This work	This work	CICC'25 [8]	ISSCC'23 [9]	ISSCC'23 [6]	APEC'23 [7]
Technology	180nm BCD	180nm BCD	180nm BCD	150nm BCD	180nm BCD	Discrete
Norminal V_{IN}	12V	12V	12V/24V	48V	12V	24V/48V
Norminal V_O	1.1/1.8V	1.1/1.8V	0.8-1.8V	1.8/1.3V	1.1V	1.2-2.2V
Switching Frequency	2MHz	2MHz	2MHz	2MHz	2MHz	0.3MHz
Chip Area	5.31mm²	5.31mm²	2×5.31mm²	2×5.29mm²	7.29mm²	4×2.5mm²
Number of Outputs	2	4	4	2	4	3
Components of Each Output[a]	1L 1C 3S	1L 0.75C 3S	1L 2C 4S	1L 0.5C 2S	1L 2.5C 6S	1L 0.7C 2S
Max Load Current	2×4A	4×4A	4×4A	2×2A	4×4A	N.A.
Inductors	2×330nH	4×330nH	4×470nH	2×1μH	4×220nH	3×400nH
Output Capacitor	2×10μF	4×10μF	4×10μF	2×10μF	4×22μF	3×20μF
Cross-Regulation	0	0	0	25mV/A	0	N.A.
Peak Efficiency	90.5%(@12V-1.1V)	90.3%(@12V-1.1V)	90.9%(@12V-1.2V)	77%(@48V-1.8/1.3V)	89.8%(@12V-1.1V)	91.8%(@24V-1.2/1.5/1.8V)
Total Passive Volume[b]	12.65mm³	22.175mm³	37.8mm³	N.A.	69.65mm³	N.A.
Current Density	632A/cm³	721A/cm³	423A/cm³	N.A.	230A/cm³	N.A.

(a) Number of inductors (L), flying and DC capacitors (C), switches (S) adopted in the topology
(b) Counts flying capacitors, DC capacitors and inductors   (c) Estimated from graph

Figure 26.4.6: Performance summary and comparison with prior works.

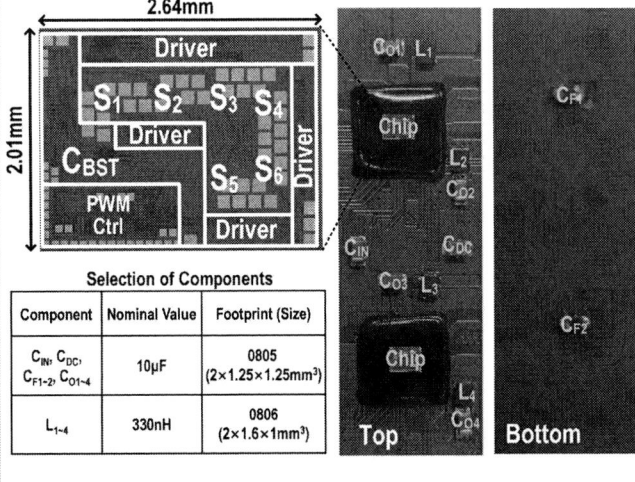

Selection of Components

Component	Nominal Value	Footprint (Size)
C_{IN}, C_{DC}, C_{F1-2}, C_{O1-4}	10μF	0805 (2×1.25×1.25mm³)
L_{1-4}	330nH	0806 (2×1.6×1mm³)

Figure 26.4.7: Die micrograph and a photo of the PCB of the proposed converter.

ISSCC 2026 / SESSION 26 / COMPUTE POWER AND SUPPLY MODULATORS / 26.5

26.5 An Inductor-at-Middle Hybrid Buck Converter with Shared Power-Signal Path for Distributed Vertical Power Delivery

Zhongyao Zhu[1,2], Junwei Huang[2], Zhibang Song[1,2], Zhewen Yu[1,2], Shousheng Han[2], Sai-Weng Sin[2], Yan Lu[1]

[1]Tsinghua University, Beijing, China, [2]University of Macau, Macau, China

Abstract

This work presents an inductor-at-middle hybrid buck converter for high-current density vertical power delivery, reusing the power inductors as signal feedback paths. The proposed switching-bus-multiplexing control reuses the power path to carry a feedback signal, mitigating the challenges in sophisticated signal and power network co-design for pin-intensive distributed power delivery networks. The prototype achieves a vertical current density of 3.17A/mm².

State-of-the-art XPU platforms, including CPUs, GPUs, and AI accelerators, demand ever-increasing supply currents that can exceed hundreds of Amperes. Meanwhile, conventional 2D power delivery from PCB to processor suffers from severe IR drop, routing congestion, and efficiency loss [1]. To alleviate these issues, vertical power delivery (VPD) has emerged, where voltage regulators are placed directly beneath the processor die, minimizing distribution path length. In a single-stage vertical DC-DC, however, the output current still needs to go through the substrate or package layers from the PCB backside to the XPUs, resulting in considerable conduction losses [2,3]. Two-stage vertical converters address this by using pre-stage voltage regulation modules (VRMs) on the backside of the PCB, and high-current post-stage integrated voltage regulators (IVRs) collocated with the processors, to reduce the currents through the board and the package. This two-stage approach also allows the second stage to exploit advanced low-voltage CMOS processes for an improved figure-of-merit (FoM) and for higher switching frequencies [4-6].

However, two-stage VPD has its limitations as shown in Fig. 26.5.1. First, for good power integrity of the power delivery network (PDN), both stages require dedicated decoupling capacitors: the first-stage output capacitor and the second-stage input capacitor are physically separated, occupying valuable footprint on both sides of the PCB. Second, routing complexity grows dramatically in a distributed point-of-load (PoL) systems, where multiple parallel converters serve different processor regions or chiplets. Each converter requires feedback paths to the load, often consuming scarce I/O resources and crowding the already congested silicon-package interface. The need to reserve vias and routing channels for feedback signals competes directly with power vias and output decoupling capacitors, degrading both electrical performance and integration density. These challenges highlight the need for a new power conversion architecture and a "via-less" feedback control scheme.

To address these challenges, this work introduces an inductor-at-middle hybrid buck converter that merges the benefits of both single- and two-stage architectures. As shown in Fig. 26.5.2, the inductor is placed between the two hybrid converter networks: a high-voltage double step-down (DSD) front stage implemented in 180nm BCD technology on the PCB backside, and a low-voltage inductor-first (L-first) stage implemented in 65nm CMOS collocated with the processor. The shared inductors provide electrical isolation between voltage domains, eliminate redundant decoupling capacitors, and allow each stage to be optimized for its voltage and current range. The DSD stage operates at low frequency (1MHz) to efficiently handle high input voltage and large voltage swing, while the L-first stage operates at higher frequency (20MHz) with a small voltage conversion ratio and low device voltage stress, enabling compact design and high total system efficiency.

Beyond topology, the most significant contribution of this work is a multiplexed power-signal architecture that uses the power switching node for both energy transfer and feedback signaling. In conventional distributed VPD systems, feedback signals must be routed from the processor side to the PCB backside, consuming I/O and introducing noise susceptibility. Here, by configuring the 65nm stage at a fixed 50% duty ratio, the switching node naturally generates a periodic 20MHz square wave. This stable waveform is leveraged as an embedded clock for Manchester coding. The locally generated duty signal, derived from V_{OUT} feedback at the processor side, is Manchester-encoded with this clock and directly modulates the switching node waveform. As a result, the DSD stage can recover the duty information locally by observing the inductor switching node, requiring no dedicated feedback vias from the output.

This multiplexing achieves several critical advantages. First, it completely eliminates the need for routing feedback signals across the package, thereby preserving I/O resources and reducing routing congestion under the XPUs. Second, by embedding control information within the power path itself, the system ensures that the distributed PoL converters can scale without additional wiring complexity. Third, the approach leverages the high-frequency, low-voltage capabilities of advanced CMOS process to generate encoded signals, while the BCD stage only needs to decode a low-frequency duty waveform, ensuring compatibility across heterogeneous integration. Taken together, this hybrid converter not only reduces power path losses and improves efficiency, but also fundamentally redefines how control information is delivered in distributed vertical PDNs. The detailed circuit implementation is illustrated in Fig. 26.5.2. The 180nm BCD section implements a DSD topology operating at 1MHz, where high-side switches S_1 and S_3 are realized with 12V LDMOS devices and low-side switches S_2 and S_4 with 6V LDMOS. This stage is optimized for robustness with high input voltage [7]. The 65nm technology section operates at 20MHz using 1V transistors in the inductor-first topology. A fixed 50% duty ratio is chosen to balance the current distribution and to minimize hard-charging loss between flying and output capacitors, which becomes significant if the duty cycle deviates too far from 50% [8]. Importantly, this fixed-duty operation transforms the 65nm switching node into a reliable 20MHz clock reference, naturally enabling Manchester coding without additional circuitry.

Local feedback from V_{OUT} is processed by a PID controller in the 65nm stage to generate the duty control signal D_{CTRL}. This signal performs an XOR with the embedded 20MHz clock to produce the Manchester-encoded signal D_{TX}. Depending on the voltage level of D_{CTRL}, the encoder generates two different patterns, i.e. Code=0 for $D_{CTRL}=0$, and Code=1 for $D_{CTRL}=1$. As a result, the encoded modulation is directly reflected on the switching node, whose voltage alternates between V_{OUT} and $2V_{OUT}$. On the 180nm converter side, the modulated waveform is decoded into the recovered duty signal D_{RX}, which controls the DSD power stage. This feedback mechanism allows the two technology nodes to communicate without explicit feedback routing.

The encoder-decoder signal chain is detailed in the bottom-left diagram of Fig. 26.5.2. At the decoding side, the switch-node voltage V_{SW1} varies between V_{OUT} and $2V_{OUT}$, exhibiting high frequency ringing due to switching transitions. To convert the noisy switch node voltage into a clear digital signal, the voltage range translator applies a two-stage filtering. A high bandwidth low-pass filter first attenuates high-frequency noise and reshapes the waveform, then a low bandwidth low-pass filter generates an adaptive threshold V_{REF} from the average V_{SW1}. Next, a comparator compares the filtered V_{SW1} waveform against the adaptive threshold V_{REF}, generating the digital signal for decoding. Moreover, Fig. 26.5.3 presents the schematic of the edge-aligned encoder and dual-sampling decoder, along with their waveforms. At the encoding side, a phase alignment circuit ensures the duty waveform is edge-aligned to the 20MHz carrier clock for avoiding glitches. Besides, in conventional Manchester decoding, each bit has a fixed duration, so the signal can be reliably sampled at a predefined moment, e.g. only at 3/4 bit time. However, for a PWM signal, as the duty transition can coincide with either the rising or falling edge of the clock, a dual-sampling strategy should be used. Finally, the duty slicer converts the recovered duty signal into two 180° out-of-phase control signals for the DSD stage. Overall, the encoder-decoder design combines phase-safe encoding, adaptive thresholding, and dual-sampling to achieve reliable communication.

Steady-state operation is demonstrated in Fig. 26.5.4, which presents measured waveforms under 12V-to-0.8V at no load and 12V-to-1.2V at 3A. The measured V_{SW1} waveform clearly illustrates the duty modulation encoded by the 65nm stage. On the receiving side, the BCD-based DSD stage accurately decodes the modulation, as shown by the waveforms at nodes V_X and V_Y. As the output voltage and load current increase, the PWM duty ratio correspondingly expands. The bottom plots further demonstrate that V_X and V_Y continue to follow the modulated signal. These results support that stable operation is maintained, confirming the feasibility of using the power switching node as a dual-purpose path for energy transfer and feedback signaling.

Dynamic performance is evaluated in Fig. 26.5.5 with a load step from 0A to 3A in 5ns. The output filter comprises six 3.3μF 0201 capacitors in parallel to minimize ESL and ESR. The measured voltage drop is 250mV with a recovery time of 24μs. During the transient, the DSD stage promptly adjusts its duty ratio in response to the decoded signal as shown in the bottom-left plot of Fig. 26.5.5. Even though V_{SW1} suffers from voltage variation under fast load changes, the adaptive V_{REF} dynamically tracks its average value, providing a stable reference for accurate duty decoding. The efficiency characteristics of the converter are illustrated in the bottom-right plot of Fig. 26.5.5, with a peak efficiency of 85.3%. Although the VA metric suggests the efficiency could be theoretically higher, the parasitics from wire-bond packaging in the 20MHz power stage considerably limit the performances. In practical processor power-delivery scenarios, flip-chip and advanced packaging techniques would significantly alleviate these parasitic effects.

454 • 2026 IEEE International Solid-State Circuits Conference

979-8-3315-8937-0/26 $31.00 © 2026 IEEE

Figure 26.5.6 compares this work against prior arts. By reusing the power path as a signaling channel, the architecture simplifies the co-design of power and signal networks in distributed vertical power delivery systems. To further evaluate the current density performance in VPD, a VPD current density for the second-stage DC-DC converter is defined, reflecting the current delivery capability within the limited footprint beneath the processors. The architecture achieves a VPD current density of 3.17A/mm².

As shown in Fig. 26.5.7, the vertical power delivery demo includes a DSD stage occupying 3.03mm², with a 0402 22μF flying capacitor mounted on die. The inductor-first stage occupies 1.89mm² with two 0201 1μF flying capacitors also placed on die.

Acknowledgment:
This work was supported in part by the National Natural Science Foundation of China (92573201), the Macau Science and Technology Development Fund (004/2023/AKP and 004/2023/SKL). Corresponding Author: Yan Lu.

References:
[1] K. Radhakrishnan, M. Swaminathan and B. K. Bhattacharyya, "Power Delivery for High-Performance Microprocessors—Challenges, Solutions, and Future Trends," in *IEEE Transactions on Components, Packaging and Manufacturing Technology*, vol. 11, no. 4, pp. 655-671, April 2021. https://doi.org/10.1109/TCPMT.2021.3065690
[2] Z. Tong et al., "HOOP: A Scalable Hybrid DC-DC Converter Ring for High-Performance Computing," *2025 IEEE International Solid-State Circuits Conference (ISSCC)*, San Francisco, CA, USA, 2025, pp. 1-3. https://doi.org/10.1109/ISSCC49661.2025.10904695
[3] G. Cai, Y. Lu and R. P. Martins, "A Compact 12V-to-1V 91.8% Peak Efficiency Hybrid Resonant Switched-Capacitor Parallel Inductor (ReSC-PL) Buck Converter," *2023 IEEE International Solid-State Circuits Conference (ISSCC)*, San Francisco, CA, USA, 2023, pp. 198-200. https://doi.org/10.1109/ISSCC42615.2023.10067303
[4] C. Hardy et al., "A Scalable Heterogeneous Integrated Two-Stage Vertical Power-Delivery Architecture for High-Performance Computing," *2023 IEEE International Solid-State Circuits Conference (ISSCC)*, San Francisco, CA, USA, 2023, pp. 182-184. https://doi.org/10.1109/ISSCC42615.2023.10067315
[5] S. Ren et al., "A 12A 89.3% Peak Efficiency and 26mV Undershoot 12-to-1V Two-Stage Converter with Regulated Resonant Switched-Capacitor Regulators," *2025 IEEE International Solid-State Circuits Conference (ISSCC)*, San Francisco, CA, USA, 2025, pp. 374-376. https://doi.org/10.1109/ISSCC49661.2025.10904628
[6] E. A. Burton et al., "FIVR — Fully integrated voltage regulators on 4th generation Intel® Core™ SoCs," *2014 IEEE Applied Power Electronics Conference and Exposition - APEC 2014*, Fort Worth, TX, USA, 2014, pp. 432-439. https://doi.org/10.1109/APEC.2014.6803344
[7] J. Yuan, Z. Liu, F. Wu and L. Cheng, "A 12V/24V-to-1V DSD Power Converter with 56mV Droop and 0.9μs 1% Settling Time for a 3A/20ns Load Transient," *2022 IEEE International Solid-State Circuits Conference (ISSCC)*, San Francisco, CA, USA, 2022, pp. 1-3. https://doi.org/10.1109/ISSCC42614.2022.9731701
[8] H. Han, J.-H. Cho, W. Jang, Y. Park, J. Lee and H.-S. Kim, "A 94.1%-Efficiency Parallel-SC Hybrid Buck Converter Designed Using VCR-Aware Topology Optimizer for a 4.2A/mm² Current-Density FoM," *2024 IEEE International Solid-State Circuits Conference (ISSCC)*, San Francisco, CA, USA, 2024, pp. 464-466. https://doi.org/10.1109/ISSCC49657.2024.10454535

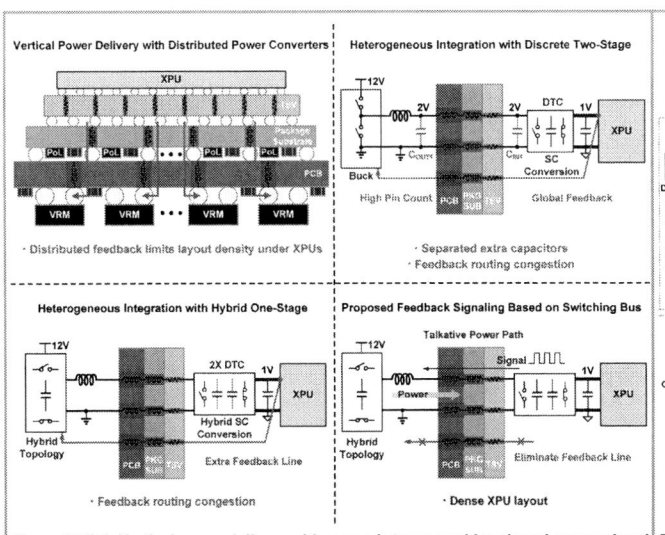

Figure 26.5.1: Vertical power delivery with merged stages enables shared power-signal paths for dense XPU layout.

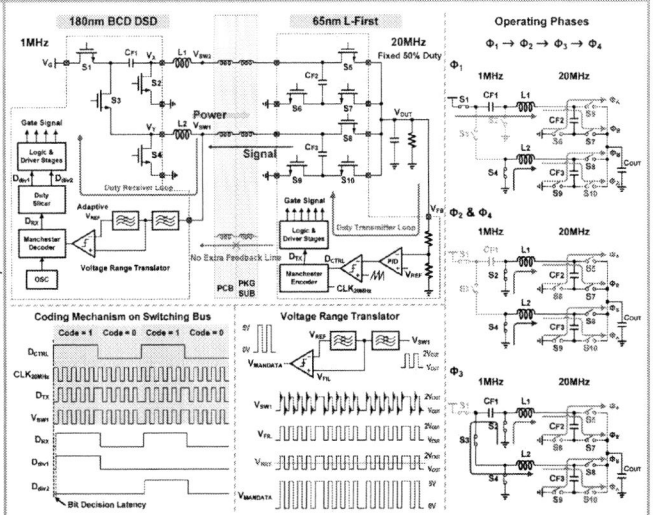

Figure 26.5.2: System architecture of the inductor-at-middle hybrid buck converter with shared power-signal path.

ISSCC 2026 / SESSION 26 / COMPUTE POWER AND SUPPLY MODULATORS / 26.5

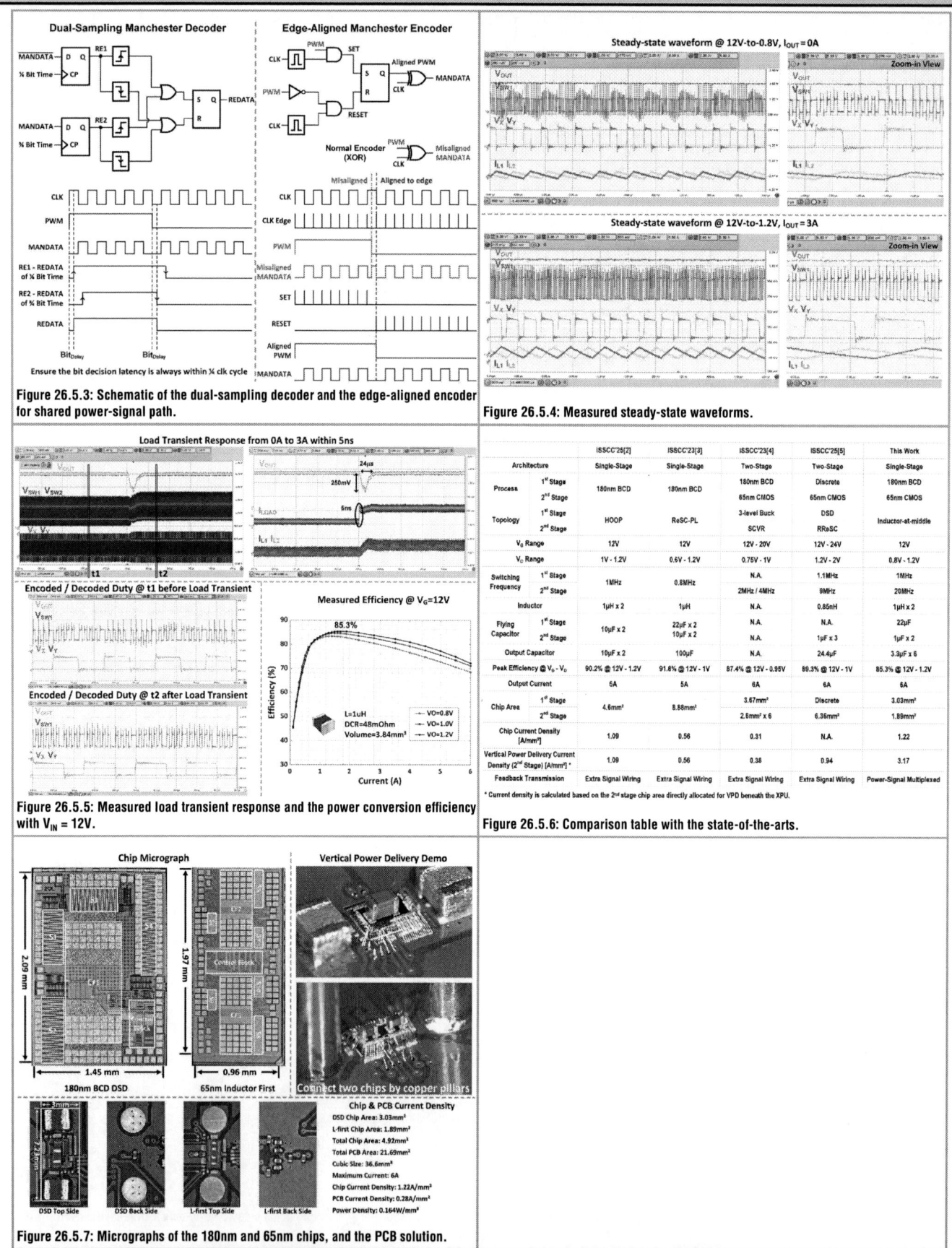

Figure 26.5.3: Schematic of the dual-sampling decoder and the edge-aligned encoder for shared power-signal path.

Figure 26.5.4: Measured steady-state waveforms.

Figure 26.5.5: Measured load transient response and the power conversion efficiency with V_{IN} = 12V.

		ISSCC'25[2]	ISSCC'23[3]	ISSCC'23[4]	ISSCC'25[5]	This Work
Architecture		Single-Stage	Single-Stage	Two-Stage	Two-Stage	Single-Stage
Process	1st Stage	180nm BCD	180nm BCD	180nm BCD	Discrete	180nm BCD
	2nd Stage			65nm CMOS	65nm CMOS	65nm CMOS
Topology	1st Stage	HOOP	ReSC-PL	3-level Buck	DSD	Inductor-at-middle
	2nd Stage			SCVR	RReSC	
V_G Range		12V	12V	12V - 20V	12V - 24V	12V
V_O Range		1V - 1.2V	0.6V - 1.2V	0.75V - 1V	1.2V - 2V	0.8V - 1.2V
Switching Frequency	1st Stage	1MHz	0.8MHz	N.A.	1.1MHz	1MHz
	2nd Stage			2MHz / 4MHz	9MHz	20MHz
Inductor		1µH x 2	1µH	N.A.	0.85nH	1µH x 2
Flying Capacitor	1st Stage	10µF x 2	22µF x 2	N.A.	N.A.	22µF
	2nd Stage		10µF x 2	N.A.	1µF x 3	1µF x 2
Output Capacitor		10µF x 2	100µF	N.A.	24.4µF	3.3µF x 6
Peak Efficiency @ V_G - V_O		90.2% @ 12V - 1.2V	91.8% @ 12V - 1V	87.4% @ 12V - 0.95V	89.3% @ 12V - 1V	85.3% @ 12V - 1.2V
Output Current		5A	5A	6A	6A	6A
Chip Area	1st Stage	4.6mm²	8.88mm²	3.67mm²	Discrete	3.03mm²
	2nd Stage			2.6mm² x 6	6.36mm²	1.89mm²
Chip Current Density [A/mm²]		1.09	0.56	0.31	N.A.	1.22
Vertical Power Delivery Current Density (2nd Stage) [A/mm²] *		1.09	0.56	0.38	0.94	3.17
Feedback Transmission		Extra Signal Wiring	Extra Signal Wiring	Extra Signal Wiring	Extra Signal Wiring	Power-Signal Multiplexed

* Current density is calculated based on the 2nd stage chip area directly allocated for VPD beneath the XPU.

Figure 26.5.6: Comparison table with the state-of-the-arts.

Figure 26.5.7: Micrographs of the 180nm and 65nm chips, and the PCB solution.

ISSCC 2026 / SESSION 26 / COMPUTE POWER AND SUPPLY MODULATORS / 26.6

26.6 A 92.4%-Peak-Efficiency 48V to 0.8-to-1.6V Hybrid Converter with Inductor-Interleaved Fibonacci Switched-Capacitor

Hyun-Woo Jeong, Chan-Ho Lee, Hyeon-Ji Choi, Sung-Wan Hong

Sogang University, Seoul, Korea

Abstract

This paper presents a 92.4%-peak-efficiency 48V to 0.8-to-1.6V hybrid converter based on an inductor-interleaved Fibonacci switched-capacitor (I^2-FSC). The proposed converter reduces ΔV_L with few flying capacitors, thereby achieving a high duty cycle under ultra-low VCRs and reducing ACR loss. It also minimizes switching and conduction losses, while current distribution lowers DCR loss and a fixed 0.5 duty in the capacitive path reduces V_0 ripple. The chip is fabricated in a 180nm BCD process.

As data centers become more advanced, point-of-load (PoL) converters that generate low output voltages (0.8-to-1.6V) directly from a 48V input are becoming increasingly important. To enable such direct conversion, converters must achieve high efficiency at ultra-low voltage conversion ratios (VCRs). Conventional half-bridge converters [1] support low VCR operation under ideal conditions, but they suffer from an extremely narrow duty cycle (D), making them highly sensitive to noise. Moreover, since the entire load current (I_0) flows through a single inductor, the DC resistance (DCR) loss is large. Furthermore, a large voltage swing across the inductor (ΔV_L), especially in 48V-to-1V conversion, increases the inductor current (I_L) ripple, causing significant AC resistance (ACR) loss and a large output voltage (V_0) ripple. Although a bulky inductor can mitigate these issues, it is unsuitable for compact systems.

To overcome the previous problems, several types of direct PoL converters have been proposed, as shown in Fig. 26.6.1 (bottom). Inductor-first converters place an N-to-1 step-down switched-capacitor (SC) stage after the inductor [2,3]. The SC stage lowers the DC level of I_L by a factor of N, thereby reducing DCR loss. However, this structure requires a large N under low VCR conditions, which limits the VCR, and still suffers from a large ACR loss and significant V_0 ripple (ΔV_0) due to the large ΔV_L. Furthermore, the capacitor-provided current (I_C) is non-uniform, which further worsens the ΔV_0.

Inductor-last structures place the SC stage before the inductor [4-9], increasing the D and reducing the ΔV_L, which lowers I_L ripple (ΔI_L) and ACR loss. When I_0 is supplied solely by the inductor [4-8], ΔV_0 is reduced, but DCR loss remains significant. In [9], I_0 is delivered to the output through both the inductor and the capacitor, reducing DCR loss but causing a large ΔV_0 due to the non-uniform I_C.

Inductor-middle structures combine features of both inductor-first and last structures [10,11] with moderate characteristics. However, they still experience a relatively large ACR loss and large ΔV_0.

To overcome these limitations, this work proposes the inductor-interleaved Fibonacci switched-capacitor (I^2-FSC) converter, which extends the Fibonacci-based SC configuration by adding two inductors, which can be regarded as a modified inductor-last structure, as shown in Fig. 26.6.1 (middle-right).

The Fibonacci-based SC design allows the converter to achieve a high D under ultra-low VCR conditions while using a practical number of flying capacitors (C_Fs). In addition, the converter delivers I_0 through two interleaved inductor paths and one minor capacitor path, distributing the current and reducing DCR loss. Also, the I^2-FSC converter further reduces ΔV_L, thereby reducing ACR loss. Furthermore, the capacitor path in the Fibonacci SC stage operates with a fixed D of 0.5, ensuring uniform I_C and minimizing ΔV_0.

Additionally, the I^2-FSC converter minimizes both conduction and switching losses at the switches. For high-voltage (HV) switches with large parasitic components, the conduction loss is reduced by fixing their D at 0.5, which theoretically minimizes the peak level of I_C. In addition, switching nodes in the HV region experience one complete voltage swing per switching cycle, which consists of four operational phases. Although the Fibonacci SC (FSC) structure has larger voltage swings at the switching nodes compared to other SC structures due to fewer C_Fs, the frequency of these swings can be reduced by half. As a result, the switching loss becomes comparable to, or even smaller than, that of previous structures.

For low-voltage (LV) switches that conduct large currents, I_L flows through only one LV switch or through multiple parallel paths that distribute the current. In addition, the switching nodes associated with the LV switches also swing once per switching cycle. Consequently, the I^2-FSC converter reduces both conduction and switching losses at the switches, while also minimizing the losses at the inductors.

Figure 26.6.2 shows the overall design flow and topology of the I^2-FSC converter. The converter is designed based on a five-stage FSC structure, where five flying capacitors (C_{F1}–C_{F5}) and their associated switches divide the input voltage (V_{IN}) according to the Fibonacci sequence. The baseline FSC achieves a VCR of 1/13 and operates with a fixed D of 0.5 for

the conduction loss optimization. To further reduce VCR and enable V_0 regulation, the proposed converter modifies the baseline FSC through a three-step design strategy, as shown in Fig. 26.6.2 (right).

First, switch S_{16} is replaced by a primary inductor (L_P). With this replacement, switches S_{13} and S_{14} no longer operate with a fixed D of 0.5 like the other switches but instead they use an adjustable duty cycle $D_{(1)}$, which is controlled to regulate V_0. Under this single-inductor configuration, the VCR is given by $V_0/V_{IN}=D_{(1)}/5+8D_{(1)}$.

Second, an auxiliary inductor (L_A) replaces switch S_{15} to further decrease VCR and distribute I_0. Similarly, switches S_{11} and S_{12} operate with an adjustable duty cycle $D_{(2)}$ instead of the fixed D of 0.5. This dual-inductor configuration relieves current stress on L_P and doubles the effective switching frequency owing to multi-phase operation, thereby reducing ΔV_0. In this configuration, ΔV_L is further reduced, allowing $D_{(2)}$ to be higher than $D_{(1)}$. The resulting VCR, given by $V_0/V_{IN}=D_{(2)}/8+5D_{(2)}$, becomes even lower.

Finally, to further enhance performance, the lower sides of switches S_7 and S_{10} are reconnected from ground to the upper nodes of inductors L_A and L_P, which correspond to switching nodes V_{x8} and V_{x10}, respectively. This modification further reduces ΔV_L, achieving an even lower VCR.

Through this design sequence, the I^2-FSC converter achieves a sufficiently high D while achieving ultra-low VCR, which makes the converter more robust against noise. In addition, ΔV_L is significantly reduced, even with the use of a small number of C_Fs, enabling the transistors that mainly conduct large inductor currents to be implemented as 5V CMOS devices.

As illustrated in Fig. 26.6.2 (bottom-left), the proposed I^2-FSC converter comprises two inductors, five C_Fs, nine LDMOS switches, and five 5V CMOS switches. Notably, L_A delivers less current than L_P, allowing a physically smaller inductor with higher DCR, reducing overall system volume while maintaining high efficiency.

Figure 26.6.3 illustrates the operational phases of the I^2-FSC converter. The entire operation consists of four phases (Φ_1–Φ_4), which are grouped as '$\Phi_a=\Phi_1+\Phi_2$' and '$\Phi_b=\Phi_3+\Phi_4$', each with a fixed D of 0.5.

During Φ_1, L_P is de-energized while L_A is energized. Simultaneously, C_{F1}, C_{F3}, and C_{F5} are charged, and C_{F2} and C_{F4} are discharged. The I_L of L_P (I_{LP}) flows through a single 5V CMOS switch, thereby minimizing the conduction loss. On the other hand, the I_L of L_A (I_{LA}) flows through multiple switches. However, it is distributed into I_{LA1} and I_{LA2} through C_{F2} and C_{F4} in a 1:3 ratio, which reduces the conduction loss even though it passes through several switches. This distribution ratio is determined by the current-second balance of C_{F2} and C_{F4} across other operational phases.

During Φ_2, both L_P and L_A are de-energized, and only C_{F1}, C_{F3}, and C_{F5} are charged. In this phase, I_{LP} and I_{LA} each flow through a single transistor (S_{13} and S_{12}), which reduces the conduction loss.

During Φ_3, L_P is energized while L_A is de-energized. At the same time, C_{F2} and C_{F4} are charged, while C_{F1}, C_{F3}, and C_{F5} are discharged. I_{LP} is distributed into I_{LP1} and I_{LP2} through C_{F3} and C_{F5} in a 2:5 ratio. Although I_{LP} can be delivered through a single CMOS switch, S_{14}, this distributed delivery reduces the conduction loss even further.

During Φ_4, both L_A and L_P are de-energized, and C_{F2} and C_{F4} are charged, while C_{F1} is discharged. Similar to the operation in Φ_2, I_{LP} and I_{LA} flow through a single transistor each (S_{13} and S_{12}).

In addition, it is observed that all switching nodes (V_{x1}–V_{x10}) swing once per cycle, reducing switching loss at these nodes.

456 • 2026 IEEE International Solid-State Circuits Conference

979-8-3315-8937-0/26 $31.00 © 2026 IEEE

ISSCC 2026 / February 18, 2026 / 10:05 AM

Figure 26.6.3 (bottom-left) shows several key parameters, including VCR and the DC levels of I_{LP}, I_{LA}, and I_C, which are calculated based on the voltage-second and current-second balances. Figure 26.6.3 (bottom-right) shows the charging and discharging of each C_F. According to the current-second balance, the amount of charge stored and released by each C_F must be equal. Based on this relationship, the ratio between the DC levels of I_{LP} and I_{LA} is determined to be 7:4, which is independent of the D.

Figure 26.6.4 (top-left) shows the D as a function of the VCR. For ultra-low VCRs, such as below 1/48, the D in most prior works drops to 0.1 or even lower. In contrast, the proposed converter achieves a D of 0.239, which is higher than all of the compared state-of-the-art works in Fig. 26.6.4 (top-left) except for [5], which uses eleven C_Fs. Figure 26.6.4 (top-right) analyzes and compares inductor losses, including DCR loss and ACR loss. In general, DCR loss is more dominant than ACR loss in previous designs. However, when V_{IN} is significantly high, such as 48V or above, ACR loss becomes large enough that it cannot be ignored due to the increased ΔI_L. Since the I²-FSC converter has comparable or even smaller ΔI_L than other designs, despite operating at a lower switching frequency (f_{SW}), it incurs significantly smaller ACR loss. As a result, it achieves the lowest overall inductor loss among the compared 48V input converters, even though its DCR loss is not the smallest. Figure 26.6.4 (bottom) shows the measured efficiency, with a peak efficiency of 92.4% at V_0=1.6V.

Figure 26.6.5 presents measured waveforms at V_{IN}=48V and V_0=1V. Figure 26.6.5 (top-left) shows the switching node voltages in steady-state, while Fig. 26.6.5 (top-right) illustrates the load transient response when the I_0 steps from 0.1A to 3A. Figure 26.6.5 (bottom) shows the I_Ls, V_0, and switching behaviors under various load conditions. As previously analyzed, the ratio between I_{LP} and I_{LA} is 7:4, regardless of I_0. The proposed converter maintains a small ΔV_0, measured as 10mV at I_0=1A and 13mV at I_0=4A.

Figure 26.6.6 shows a performance comparison with prior works. Since the DC levels of I_{LP} and I_{LA} are different, while their ripples are equal, the I²-FSC converter uses inductors of different sizes to minimize the overall system volume. In addition, it achieves the highest D and lowest ΔV_L of the compared works except for [5], which uses a significant number of C_Fs. This enables the proposed converter to maintain a lower ΔV_0 even at a lower f_{SW}, thereby further reducing overall losses and ensuring reliable operation of both the converter and the load. The proposed converter achieves 91.3% peak efficiency at 48V-to-1V, which is the highest among the compared state-of-the-art designs in Fig. 26.6.6 with similar VCRs.

Figure 26.6.7 shows the chip micrograph.

Acknowledgement:
This work was supported by National Research Foundation of Korea (NRF) Granted funded by the Korea government (MSIT) under Grant RS-2023-00207919 and IITP-2025-RS-2023-00260091.

References:
[1] S. -Y. Li *et al.*, "48-to-1 V Direct Conversion Using High-Voltage Storage and Low-Voltage Boost Bootstrap Technique and Early Comparison On-Time Generator for Precise Nanosecond Pulses and 90.3% Efficiency in Automotive Applications," *IEEE JSSC*, vol. 57, no. 11, pp. 3396-3406, Nov. 2022. https://doi.org/10.1109/JSSC.2022.3170882

[2] Z. Tong *et al.*, "A 42W Reconfigurable Bidirectional Power Delivery Voltage-Regulating Cable," *ISSCC*, pp. 192-194, Feb. 2023. https://doi.org/10.1109/ISSCC42615.2023.10067491
[3] X. Zhang *et al.*, "An Outphase-Interleaved Switched-Capacitor Hybrid Buck Converter With Relieved Capacitor Inrush Current and COUT-Free Operations," *IEEE JSSC*, vol. 59, no. 4, pp. 1078-1092, April 2024. https://doi.org/10.1109/JSSC.2023.3346292
[4] D. Yan *et al.*, "Direct 48-/1-V GaN-Based DC–DC Power Converter With Double Step-Down Architecture and Master–Slave AO2T Control," *IEEE JSSC*, vol. 55, no. 4, pp. 988-998, April 2020. https://doi.org/10.1109/JSSC.2019.2957237
[5] H. Cao *et al.*, "A 12-Level Series-Capacitor 48-1V DC–DC Converter With On-Chip Switch and GaN Hybrid Power Conversion," *IEEE JSSC*, vol. 56, no. 12, pp. 3628-3638, Dec. 2021. https://doi.org/10.1109/JSSC.2021.3104328
[6] Y. Huang *et al.*, "Design of Direct 48-V/1-V Three-Path Four-State Switching Power Converter With Adaptive VCF Rebalancing and Dual-Edge t_{dead} Modulation," *IEEE JSSC*, vol. 59, no. 5, pp. 1592-1602, May 2024. https://doi.org/10.1109/JSSC.2023.3311731
[7] C. Chen *et al.*, "A 2.5-5MHz 87% Peak Efficiency 48V-to-1V Integrated Hybrid DC-DC Converter with Capacitor-Assisted Dual-Inductor Filtering," *ISSCC*, pp. 234-236, Feb. 2022. https://doi.org/10.1109/ISSCC42614.2022.9731764
[8] M. Gong *et al.*, "A 90.4% Peak Efficiency 48-to-1V GaN/Si Hybrid Converter With Three-Level Hybrid Dickson Topology and Gradient Descent Run-Time Optimizer," *IEEE JSSC*, vol. 58, no. 4, pp. 1002-1014, April 2023. https://doi.org/10.1109/JSSC.2022.3228233
[9] X. Yang *et al.*, "A 5A 94.5% Peak Efficiency 9~16V-to-1V Dual-Path Series-Capacitor Converter with Full Duty Range and Low V.A Metric," *ISSCC*, pp. 196-198, Feb. 2023. https://doi.org/10.1109/ISSCC42615.2023.10067802
[10] H. -J. Choi *et al.*, "A 92.7% Peak Efficiency 12V-to-60V Input to 1.2V Output Hybrid DC-DC Converter Based on a Series-Parallel-Connected Switched Capacitor," *ISSCC*, pp. 156-158, Feb. 2024. https://doi.org/10.1109/ISSCC49657.2024.10454344
[11] Y. Ji *et al.*, "A 12V-Input 1V-1.8V-Output 94.7%-Peak-Efficiency 685A/cm3-Current-Density Hybrid DC-DC Converter with a Charge Converging Phase," *ISSCC*, pp. 458-460, Feb. 2024. https://doi.org/10.1109/ISSCC49657.2024.10454573

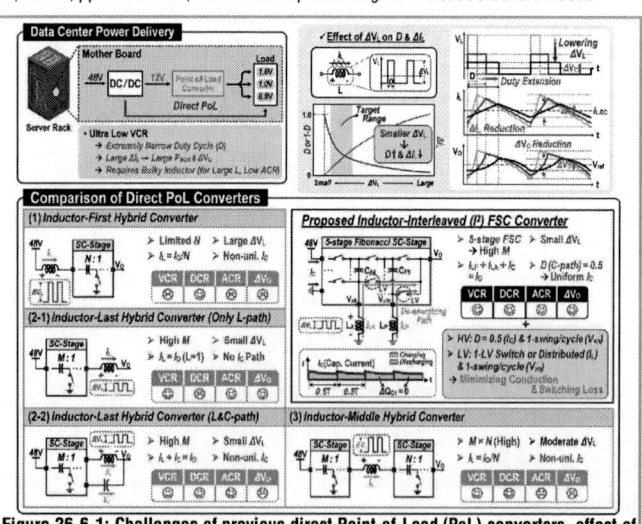

Figure 26.6.1: Challenges of previous direct Point-of-Load (PoL) converters, effect of inductor voltage swing (ΔV_L) on duty cycle and ripple, and comparison of direct PoL converters.

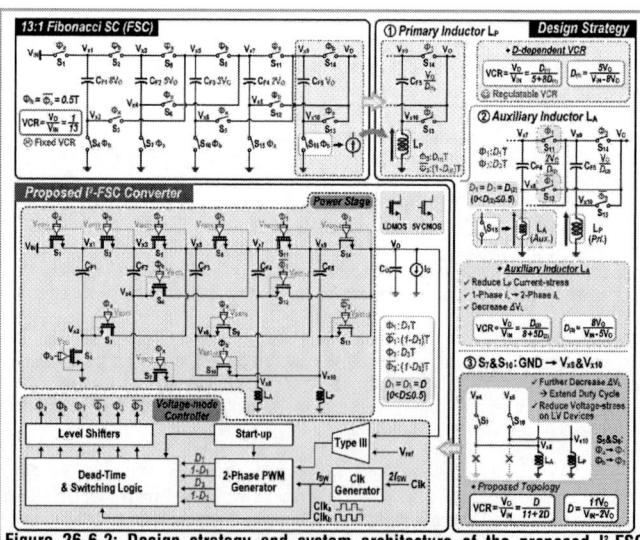

Figure 26.6.2: Design strategy and system architecture of the proposed I²-FSC converter.

DIGEST OF TECHNICAL PAPERS • 457

979-8-3315-8937-0/26 $31.00 © 2026 IEEE

ISSCC 2026 / SESSION 26 / COMPUTE POWER AND SUPPLY MODULATORS / 26.6

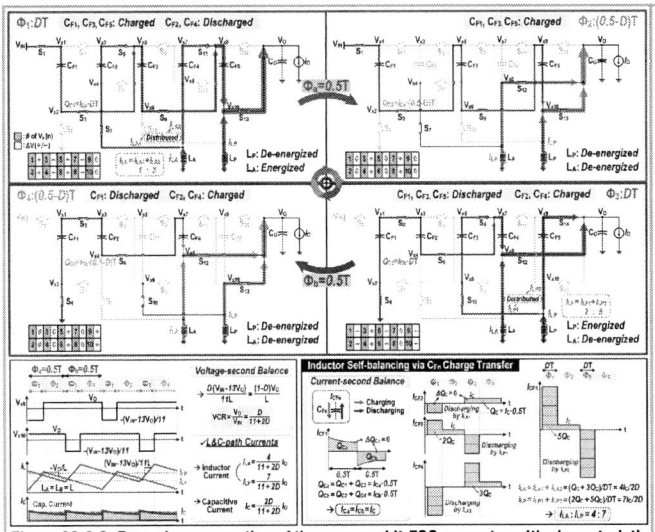

Figure 26.6.3: Four-phase operation of the proposed I²-FSC converter with characteristic analysis and principle of inductor current balancing.

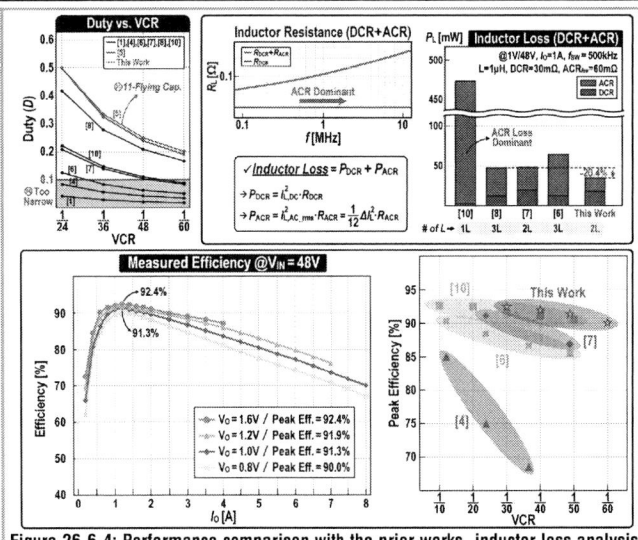

Figure 26.6.4: Performance comparison with the prior works, inductor loss analysis, and measured efficiency of proposed converter.

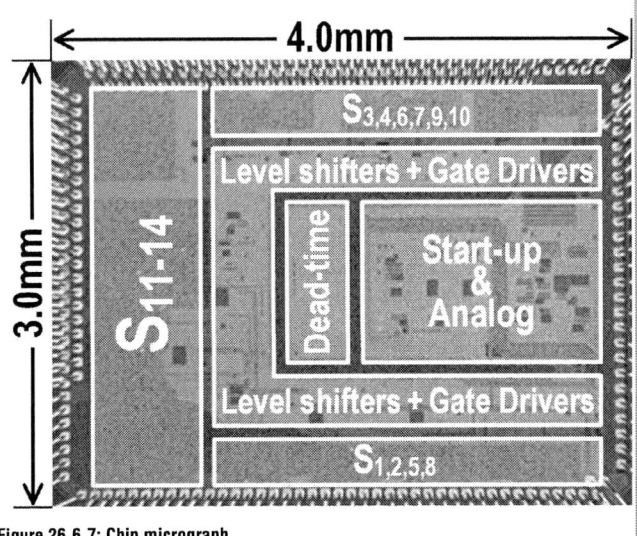

Figure 26.6.5: Measured switching nodes (V_{xn}) waveforms, load transient response, and inductor current waveforms w/ switching nodes (V_{x8}, V_{x10}) and ΔV_O at various I_O values.

	[1]	[4]	[5]	[6]	[7]	[8]	[10]	This Work
Topology	Buck	DSD	12-level	3P4S	3:1 Ladder-based CADI	3-level Hybrid Dickson	Buck + SPCSC	I²-FSC
V_{IN} [V]	12 - 48	48	36 - 60	12 - 48	36 - 55	48	12 - 60	48
V_O [V]	1	1	0.5 - 1	1	0.8 - 2	0.7 - 1	1.2	0.8 - 1.6
Maximum I_O	3A	1.5A	8A	4A	3A	12A	2A	8A
Max. Effective SW. Frequency	2.3MHz	2 × 2.0MHz	2 × 2.5MHz	3 × 1.5MHz	2 × 2.5MHz	3 × 2.0MHz	0.5MHz	2 × 0.5MHz
Power Transistors	MOSFET (On-chip)	GaN (Discrete)	GaN (Discrete) & MOSFET (On-chip)	MOSFET (On-chip)	MOSFET (On-chip)	GaN (Discrete) & MOSFET (On-chip)	MOSFET (On-chip)	MOSFET (On-chip)
Inductor	4.7µH	0.9µH × 2	0.11µH × 2	0.33µH × 3	0.82µH × 2	0.62µH × 3	10µH	1µH × 2
Inductor Size (DCR)	N/A	4.0×4.0×N/A mm³ (N/A)* (× 2)	4.0×4.0×N/A mm³ (3mΩ)* (× 2)	N/A	N/A	4.0×4.0×N/A mm³ (N/A)* (× 3)	N/A	2.5×2.0×1.0 mm³ (30mΩ, L_A) 4.0×4.0×2.1 mm³ (13.5mΩ, L_P)
Flying Capacitors	2.2nF	1µF	1µF × 11	1µF × 2	0.22µF × 4	1µF × 5	22µF × 4	10µF × 5
Output Capacitor	22µF	22µF	47µF × 4	4.7µF	15µF	10µF	N/A	22µF
Duty @VCR=1/48	1/48 = 0.021	2/48 = 0.042	12/48 = 0.25	3/48 = 0.0625	5/48 = 0.104	10/48 = 0.208	5.3/48 = 0.110	11/46 = 0.239
ΔV_O @I_O, f_{SW} (V_{IN}=48V)	N/A	50mV* @0.2A & 2.0MHz	12mV* @5A & 2.5MHz	N/A	10mV @1A & 1.25MHz	N/A	N/A	10mV @1A & 0.5MHz
Peak Eff. @1/48,V_{IN},f_{SW}	90.3% @1/48 & 0.417MHz	85.4% @1/48 & 100kHz	90.2% @1/48 & 2.5MHz	85.6% @1/48 & 1.5MHz	87.0% @1/48 & 1.25MHz	90.4% @1/48 & 1MHz	91.3% @1.2/48 & 0.5MHz	92.4% @1.6/48 & 0.5MHz 91.3% @1/48 & 0.5MHz

*Estimated from figure

Figure 26.6.6: Performance summary and comparison with state-of-the-art direct PoL converters.

Figure 26.6.7: Chip micrograph.

• 2026 IEEE International Solid-State Circuits Conference

ISSCC 2026 / SESSION 26 / COMPUTE POWER AND SUPPLY MODULATORS / 26.7

26.7 A 100A 93.4%-Peak-Efficiency LLC Resonant Converter with an Embedded Primary-Current-Extracted Regulator

Zeguo Liu[1], Yichao Ji[1], Zhiren Luo[1], Yu Ge[1], Weiwei Xu[2], Weiwei Huang[2], Lin Cheng[1,2]

[1]University of Science and Technology of China, Hefei, China, [2]Hefei CLT Microelectronics, Hefei, China

Abstract

This paper presents a 100A 93.4%-peak-efficiency LLC resonant converter with an embedded primary-current-extracted regulator. The proposed regulator powers the gate drivers of the LLC converter, eliminating the need for an external auxiliary buck converter or additional auxiliary windings. Fabricated in a 0.18µm BCD process, the proposed regulator occupies a die are of 8.3mm², and achieves a 60V-to-2V conversion at 1MHz with a peak efficiency of 93.4%.

LLC resonant converters are widely used for 48V-to-~1V xPU power delivery because primary-side zero-voltage switching (ZVS) and secondary-side zero-current switching (ZCS) enable high efficiency and compact magnetics. To meet the hundred-amperes load, multiple paralleled discrete GaN/Si power devices are employed, whose gate drivers require a dedicated low-voltage rail (e.g., 5V). In practice, this rail is often produced by a separate 48V auxiliary buck converter, which is hard-switched and lightly loaded much of the time, yielding poor efficiency and adding BOM, board area, and design complexity. Using an auxiliary winding on the LLC transformer to derive the rail is feasible [1-3], but the resulting voltage is unregulated, and it varies with line voltage due to the fixed turns ratio. The extra winding also alters coupling and leakage that increases AC copper loss.

In this paper, we propose an embedded primary-current-extracted regulator to supply the gate drivers of an LLC converter, eliminating the need for an external auxiliary buck converter or any auxiliary winding. Figure 26.7.1 shows the block diagram of the proposed regulator alongside the main LLC converter. The main LLC power stage employs four off-chip 80V GaN FETs (S_{P1}~S_{P4}) on the primary side and eighty 5V Si MOSFETs on the secondary side, supporting an output current (I_O) up to 100A while requiring the proposed regulator to deliver approximately 0.6A for the gate drivers when switching at 1MHz. The regulator is fully integrated, and its power stage comprises three on-chip power devices: two 65V nLDMOS transistors (S_{A1} and S_{A2}) and a 5V NMOS transistor S_{A3}. By alternately turning on S_{A1} and S_{A2} each half-cycle, the primary current I_{Lr} is redirected to a reservoir capacitor C_{DRV} instead of ground, thereby generating the driver rail V_{DRV}. S_{A3} is connected in series with S_{A1} (or S_{A2}) to form a bidirectional switch that disconnects V_{DRV} from V_{P1} (V_{P2}) when V_{P1} is grounded, and a small Schottky diode D_A is used to clamp node V_A and prevent it from floating. Since V_{IN} is much higher than V_{DRV}, the insertion of C_{DRV} has a negligible effect on the sinusoidal current slope of the resonant tank, and it does not impact the secondary-side output V_O. A dedicated controller adjusts the on-time of S_{A3} in each half-cycle to regulate V_{DRV}, as will be discussed later. By leveraging the primary-side ZVS feature of the LLC converter, the regulator achieves an even higher efficiency than an external buck converter. Moreover, it adds essentially no BOM, since C_{DRV} also serves as the decoupling capacitor of the gate drivers, which is already required.

Figure 26.7.2 illustrates the operating principle of the proposed converter with four phases, Φ_1 to Φ_4. Φ_1 and Φ_3 operate similarly to a conventional LLC stage but with duty cycles less than 50% because of the insertion of Φ_2 and Φ_4 for charging V_{DRV}. The gate signals of S_{A1} and S_{A2} are synchronized with those of S_{P2} and S_{P4}, respectively, and the on-time of S_{A3} defines the duty cycle of Φ_2 and Φ_4. During the dead-time from Φ_1 to Φ_2, S_{P4} is turned off first, and the primary current charges V_A up to $V_{DRV}+V_D$, where V_D is the body-diode drop of S_{A3}. After this charging S_{A3} is turned on with ZVS, which also occurs during the dead-time from Φ_3 to Φ_4. During the dead-time from Φ_2 to Φ_3, S_{A2} and S_{A3} are turned off first, and V_A is discharged through the body diode of S_{A1} to ground, after which S_{A1} is turned on with ZVS. A similar process occurs for S_{A2} during the dead-time from Φ_4 to Φ_1. As a result, ZVS is achieved for S_{A1}~S_{A3}. ZVS detectors are implemented to sense the source-drain voltages of S_{A1}~S_{A3}, ensuring their timely turn-on to minimize body-diode conduction at various current levels of I_{Lr}.

Figure 26.7.3 shows the V_{DRV} regulation circuit that generates the duty cycles D_2 and D_4 for Φ_2 and Φ_4. Because I_{Lr} represents as a current-source to the regulator, a Type-II compensator is sufficient for stable voltage regulation. To prevent secondary-current imbalance between the two half cycles, D_2 and D_4 need to be matched; otherwise, RMS current increases and efficiency degrades [4]. Accordingly, the compensator output V_{EA} is sampled once per switching cycle to obtain $V_{EA,s}$, which is compared with a ramp signal from the ramp generator to produce PWM_{AUX} that ensures D_2 and D_4 are identical. Therefore, PWM_{AUX} runs at twice the frequency of CLK_{LLC}, the 50%-duty clock that drives the main LLC converter. When I_{Lr} varies with I_O (e.g., from light load to heavy load), the increase in I_{Lr} lowers down $V_{EA,s}$ and thus adaptively shortens D_2 and D_4. PWM_{AUX} is designed to have a minimum on-time of 80ns for robust operation, and the regulator automatically enters pulse-skip mode (PSM) when $V_{EA,s}$ falls below the low threshold of the ramp.

Figure 26.7.3 (bottom left) also shows the delay-alignment circuit for gate-signal synchronization. Because the on-chip drivers for S_{A1}~S_{A3} have ~5ns of propagation delay while the off-chip drivers for S_{P1}~S_{P4} introduce ~40ns of external delay, their edges would otherwise be skewed. To measure this external delay directly, a phase-frequency detector (PFD) compares the pre-driver node feeding the low-side off-chip driver (S_{P2_BF}/S_{P4_BF}) with the resulting gate waveform (S_{P2}/S_{P4}), converting the edge time error t_d into a control voltage V_d. This V_d controls on-chip voltage-controlled delay lines (VCDLs) inserted in the S_{A1}~S_{A3} driving paths, adding a matching programmable delay and aligning the gate signals.

Figure 26.7.3 (bottom right) shows the start-up circuit for the proposed regulator. Upon V_{IN} ramp-up, a pre-driver generates PRE_VDD to bias the controller, while a dedicated start-up LDO charges C_{DRV} through a diode, establishing V_{DRV_LDO}. When V_{DRV_LDO} reaches 5V, LDO_OK is asserted, which enables the LLC stage and initiates a soft-start of V_O. As the primary current builds, the embedded regulator takes over and charges C_{DRV}, regulating V_{DRV}. A diode-OR between the start-up LDO and the primary-derived rail ensures automatic, bumpless handover and blocks reverse-current.

The proposed primary-current-extracted regulator was fabricated in a 0.18µm BCD process, occupying a die are of 8.3mm². The main LLC resonant converter operates at a switching frequency of 1MHz with a 48~60V input and a 1.6~2V output, utilizing eighty parallel secondary-side rectifiers, eighty 47µF output capacitors, and an LLC transformer implemented by the PCB planar transformer with a 30:1:1 turn ratio to transfer high power. Figure 26.7.4 presents the measured steady-state waveforms of switching nodes V_{P1}, V_{P2}, V_A, and V_{DRV} under different voltage conversion ratios and output currents. When V_{P1} or V_{P2} switches to V_{DRV}, Φ_2 or Φ_4 is inserted to charge C_{DRV}. The voltage levels after each dead-time confirm ZVS operation for all the primary-side power switches, which aligns with the analysis presented in Fig. 26.7.2. The waveform of V_A shows a V_{DRV} (5V) swing and operates at twice the frequency (2MHz) of V_{P1} and V_{P2} with a symmetric duty. At I_O=5A, to achieve V_{DRV}=5V, the duty cycles are 0.152 at V_{IN}/V_O=48V/1.6V and 0.105 at V_{IN}/V_O=60V/2V, respectively. At the full load of 100A, the duty cycles are reduced to 0.118 and 0.086, respectively, due to the higher primary-side current.

Figure 26.7.5 (top) shows the measured load-transient responses at 54V-to-1.8V conversion. With an I_O step of 45A, the undershoot of V_{DRV} is unnoticeable and the overshoot is 85mV, demonstrating tight regulation and robustness of the proposed regulator. Figure 26.7.5 (bottom) illustrates the measured efficiency under various V_{IN}, V_O and I_O conditions. At I_O=100A, the system achieves peak efficiencies of 92.6%, 93.09%, and 93.4% for 48V-to-1.6V, 54V-to-1.8V, and 60V-to-2V conversions, respectively. Compared with an external auxiliary buck converter and assuming 90% at 0.6A load, the proposed regulator improves system efficiency by up to 0.9% at light-to-medium loads. At a heavy load (>60A), these two schemes exhibit comparable efficiency, as gate driver loss is no longer dominant. It demonstrates the proposed embedded regulator achieves high efficiency without adding to the BOM.

Figure 26.7.6 shows the simulated power loss breakdown and the thermal image at the maximum output power. The integrated power switches S_{A1}~S_{A3} contribute only a small portion of the total power loss. Thanks to ZVS operation of S_{A1}~S_{A3}, the maximum temperature of the chip reaches only 48.3°C, indicating the potential for a higher power capability. Figure 26.7.6 also compares the performance of the proposed converter with state-of-the-art designs. By extracting current directly from the primary side and leveraging ZVS, the proposed regulator minimizes auxiliary-rail loss and improves overall efficiency. More importantly, it requires no additional BOM components and eliminates the need for an auxiliary buck converter or transformer winding, which improves power density. The proposed primary-current-extracted regulator can also be extended to other transformer-based full-bridge DC-DC converters such as current doublers and dual active bridge DC-DC converters, offering a simple yet effective method to generate an embedded low-voltage rail for gate drivers. Figure 26.7.7 shows the die micrograph of the proposed regulator and the controller, along with the PCB prototype of the implemented LLC converter and the parameters of the power stage.

Acknowledgement:
This work was supported in part by the National Natural Science Foundation of China under Grant 92373203. Corresponding author: Lin Cheng (eecheng@ustc.edu.cn).

458 • 2026 IEEE International Solid-State Circuits Conference

979-8-3315-8937-0/26 $31.00 © 2026 IEEE

ISSCC 2026 / February 18, 2026 / 10:30 AM

References:
[1] D. Pan, W. Xu, X. Wu, A. Li and L. Cheng, "A 24V-to-20V 6W 73.2%-Peak-Efficiency Isolated DC-DC Converter Using a Transformer-Based Supply-Generating Technique," *2024 IEEE Custom Integrated Circuits Conference (CICC)*, Denver, CO, USA, 2024, pp. 1-2. http://doi.org/10.1109/CICC60959.2024.10529013
[2] Texas Instruments, "UCC25640x LLC Resonant Controller with Ultra-Low Audible Noise and Standby Power," Accessed on Aug. 29, 2025, <https://www.ti.com/lit/ds/symlink/ucc256404.pdf>.
[3] Monolithic Power Systems, "HR1211 Multi-Mode PFC and Current Mode LLC Controller with Configurable Audible Noise Reduction Control," Accessed on Aug. 29, 2025, < https://www.monolithicpower.com/en/documentview/productdocument/index/version/2/d ocument_type/Datasheet/lang/en/sku/HR1211GM/document_id/9080/>.
[4] H. Shi et al., "A 2.4-to-240W, 95.04% Peak Efficiency LLC Isolate Converter Controller with Symmetric Pulse-Width Balancing and Fixed-Period Hysteresis Burst Control," *2024 IEEE Custom Integrated Circuits Conference (CICC)*, Denver, CO, USA, 2024, pp. 1-2. http://doi.org/10.1109/CICC60959.2024.10529004
[5] P. R. Prakash et al., "A 2400 W/in³ 1.8 V Bus Converter Enabling Vertical Power Delivery for Next-Generation Processors," *2024 IEEE Applied Power Electronics Conference and Exposition (APEC)*, Long Beach, CA, USA, 2024, pp. 910-917. http://doi.org/10.1109/APEC48139.2024.10509453
[6] X. Lou and Q. Li, "Single-Stage 48 V/1.8 V Converter With a Novel Integrated Magnetics and 1000 W/in3 Power Density," in *IEEE Transactions on Industrial Electronics*, vol. 71, no. 7, pp. 6601-6611, July 2024. http://doi.org/10.1109/TIE.2023.3310020
[7] M. Choi and D. -K. Jeong, "A 92.8%-Peak-Efficiency 60A 48V-to-1V 3-Level Half-Bridge DC-DC Converter with Balanced Voltage on a Flying Capacitor," *2020 IEEE International Solid-State Circuits Conference - (ISSCC)*, San Francisco, CA, USA, 2020, pp. 296-298. http://doi.org/10.1109/ISSCC19947.2020.9063061
[8] Z. Liu et al., "A 100A 48-60V to 1V Hybrid LLC Resonant Converter with 51mV Droop for a 70A/20ns Load Transient," *2025 IEEE Custom Integrated Circuits Conference (CICC)*, Boston, MA, USA, 2025, pp. 1-3. http://doi.org/10.1109/CICC63670.2025.10982863

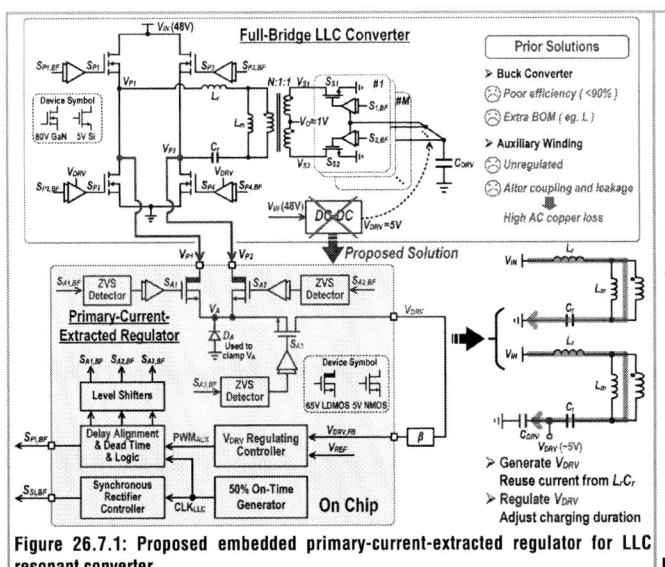

Figure 26.7.1: Proposed embedded primary-current-extracted regulator for LLC resonant converter.

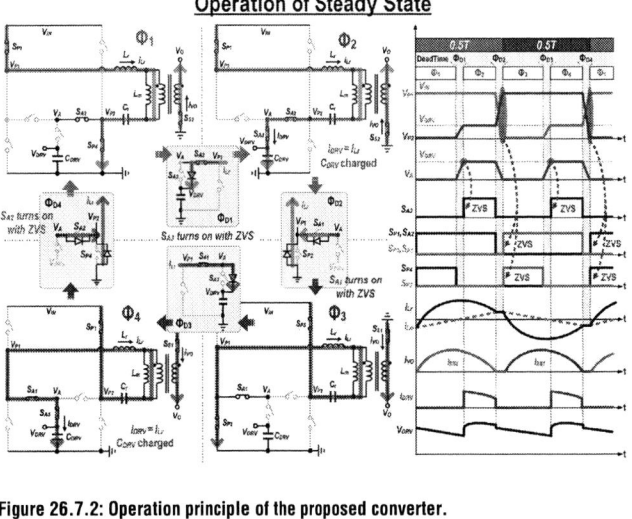

Figure 26.7.2: Operation principle of the proposed converter.

DIGEST OF TECHNICAL PAPERS • 459

979-8-3315-8937-0/26 $31.00 © 2026 IEEE

ISSCC 2026 / SESSION 26 / COMPUTE POWER AND SUPPLY MODULATORS / 26.7

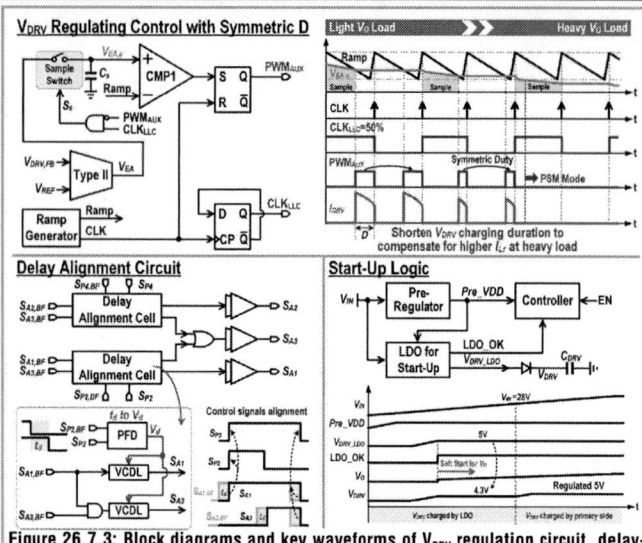

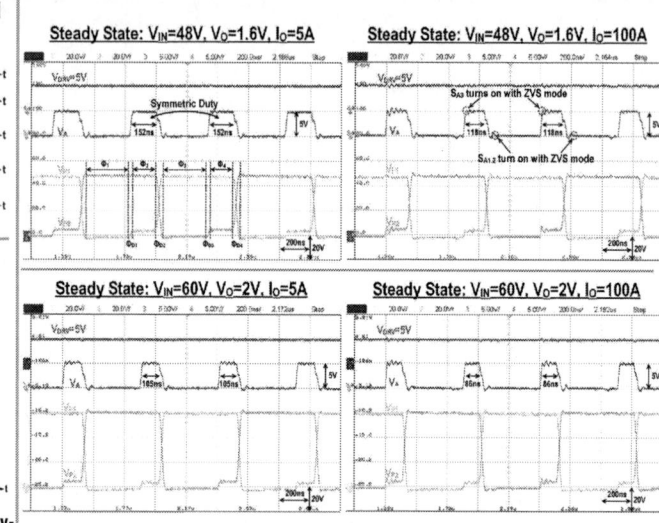

Figure 26.7.3: Block diagrams and key waveforms of V_{DRV} regulation circuit, delay-alignment circuit and start-up logic.

Figure 26.7.4: Measured steady-state waveforms of the proposed converter.

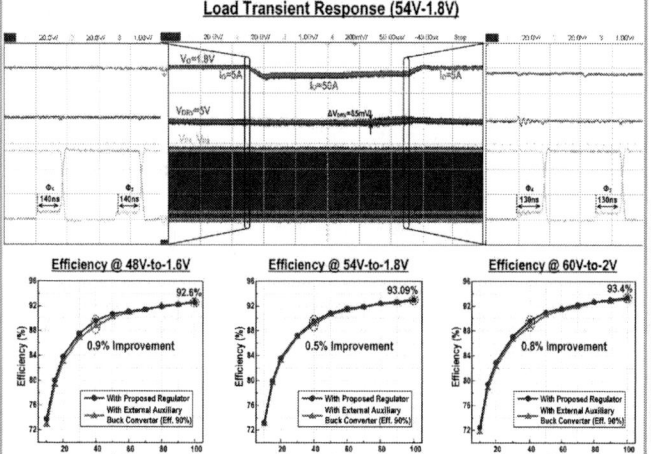

Figure 26.7.5: Measured load-transient responses and efficiency of the proposed converter.

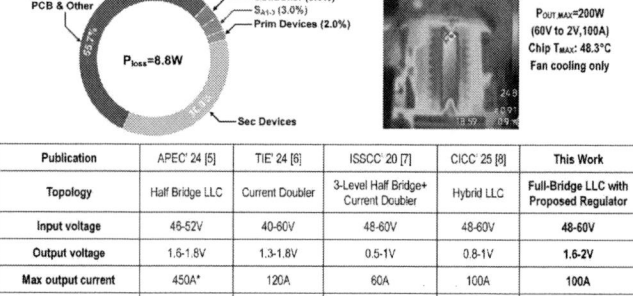

Publication	APEC' 24 [5]	TIE' 24 [6]	ISSCC' 20 [7]	CICC' 25 [8]	This Work
Topology	Half Bridge LLC	Current Doubler	3-Level Half Bridge + Current Doubler	Hybrid LLC	Full-Bridge LLC with Proposed Regulator
Input voltage	46-52V	40-60V	48-60V	48-60V	48-60V
Output voltage	1.6-1.8V	1.3-1.8V	0.5-1V	0.8-1V	1.6-2V
Max output current	450A*	120A	60A	100A	100A
Switching frequency	600kHz	600kHz	333kHz	1MHz	1MHz
Peak efficiency at V_O=1.8V	95.1%	93.1%	92.8% (V_O=1V)	89.0%(V_O=1V)	93.09%
External driver power supply required	YES	YES	YES	YES	NO

* Estimated from graph.

Figure 26.7.6: Performance comparison with prior works, simulated power loss breakdown and thermal image.

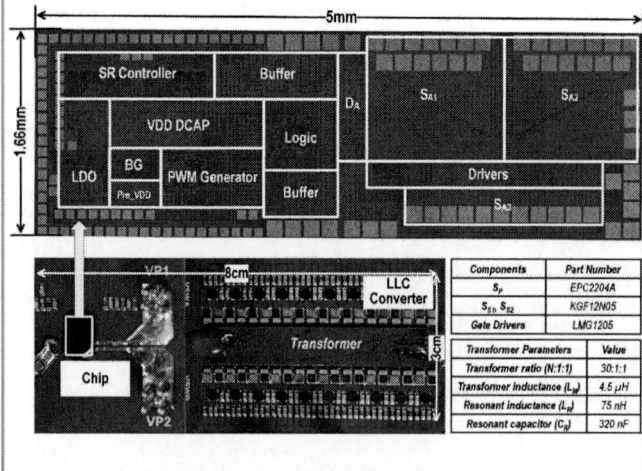

Figure 26.7.7: Die micrograph (top), PCB photo (bottom left), parameters of the LLC converter (bottom right).

• 2026 IEEE International Solid-State Circuits Conference

ISSCC 2026 / SESSION 26 / COMPUTE POWER AND SUPPLY MODULATORS / 26.8

26.8 A 1.2µs 1-to-12V Symbol Power Tracking Supply Modulator with Two-Step Subranging DVS Scheme and Boosted IL Slew Rate for 5G Mobile Devices

Changjin Chen, Minghao Shang, Yichao Ji, Lin Cheng

University of Science and Technology of China, Hefei, China

Abstract

This paper presents a symbol power tracking supply modulator that delivers a high output voltage and a wide output range for 5G high-voltage GaN PAs. To accelerate voltage transitions, a two-step subranging DVS scheme reduces the charge required for large voltage steps, while an inductor current slew-rate enhancement technique boosts the charging current. Fabricated in a 180-nm BCD process, the proposed modulator achieves a 1.2µs transition across a wide 1-to-12V range.

In 5G mobile devices, the power amplifier (PA) is a dominant source of power consumption, directly affecting battery life and thermal performance. To improve efficiency, average power tracking (APT) and envelope tracking (ET) have been widely used. However, extending the bandwidth of analog envelope tracking (AET) beyond 200MHz is technically challenging [1,2]. Therefore, digital envelope tracking (DET) [3-5] and symbol power tracking (SPT) [6] have been proposed to overcome bandwidth limitations. Digital ET extends the tracking bandwidth but suffers from high digital pre-distortion complexity and degraded linearity. The SPT approach adjusts the supply voltage at the symbol level, and PA's supply voltage transition must be completed within the cyclic prefix (CP) period. For 5G NR frequency-range 1 (FR1) with 60kHz subcarrier spacing (SCS), the maximum allowable transition time is only 1.2µs. At the same time, future PAs are trending toward higher supply voltages to reduce current stress, enhance efficiency, increase output power, and simplify output matching [7], with advances in GaN processes further accelerating this development [8]. Although these high-voltage PAs provide significant benefits, they also present substantial challenges for the SPT supply modulator (SM). Specifically, the 12V SM must not only achieve a high output voltage with a wide voltage conversion ratio (VCR), but also deliver nearly three times the charge to the load capacitor within the short CP duration than in a 1-to-5V transition, which greatly increases the implementation difficulty.

Previous works have attempted to shorten the transition time by employing auxiliary charging current for the output capacitor [9], extending charging time through pre- [6] or post- [10,11] charging or reducing the output capacitance [12]. However, they either require an auxiliary converter with an additional inductor or suffer from LDO charging losses. The work in [13] achieves fast dynamic voltage scaling (DVS) but offers only six discrete output voltage levels, which results in higher power loss in the PA. Moreover, the aforementioned works support a PA supply voltage of 5.5V at most, whereas the design of SPT SM capable of operating above 10V with fast DVS has not yet been explored and remains a significant challenge.

In this paper, we propose a new SPT structure that employs a two-step subranging DVS scheme and a boosted I_L slew rate to address high-voltage SM design challenges. As shown in Fig. 26.8.1, the proposed architecture consists of a modified reconfigurable ladder SC converter and a subranging DVS module. The modified ladder converter, with an inductor at the last stage, achieves a high voltage conversion ratio (VCR) and supports full duty range. Leveraging the linear characteristic of the ladder structure, the SC converter not only provides a maximum VCR of 4× but also generates intermediate nodes V_{X1}-V_{X3}, corresponding to 1×, 2×, and 3× VCRs. By selecting bus switches S_{B1}-S_{B4}, V_{X1}-V_{X4} can be connected to V_{SW}. In steady-state operation, one of the bus switches remains on, and the output voltage is regulated by modulating the duty cycle of V_{X1}-V_{X4} within the ladder converter. This operating principle differs from [3-5], where the SC converter's DC nodes are directly tied to the output. The inductor-last structure produces a "buck-like" continuous output current, eliminates right-half-plane (RHP) zero, and improves transient response. In addition, because the swing of V_{SW} is limited to V_{IN}, the converter inherently achieves low output ripple. For reference tracking operation, V_{X4} is fixed at $4V_{IN}$ while V_{SW} switches between V_{X4} and GND through S_{B4} and S_{B0}. With a fourfold expansion of the V_{SW} swing, the I_L slew rate is enhanced, reducing the transition time.

The two-step subranging DVS scheme transforms the wide-range voltage transition challenge of high-voltage PAs into a conventional small-range transition problem. As shown in Fig. 26.8.1, the subranging DVS module employs two output capacitors, C_{OL} and C_{OH}, each connected to the output through a switch. They divide the entire output voltage range into two equal parts, operating in: 0-to-$V_{O,MAX}$/2 and $V_{O,MAX}$/2-to-$V_{O,MAX}$, respectively, where $V_{O,MAX}$ is the maximum output voltage. The DVS operation proceeds in two steps. First, the appropriate output capacitor is selected through switches S_{OH} and S_{OL} according to the target output voltage. Then, the output voltage is precisely regulated to the target value by controlling switches S_{B0} and S_{B4} through the feedback loop. This coarse-to-fine procedure halves the voltage swing of the output capacitor during the maximum DVS transition, thereby reducing the required charging charge by 50%. Consequently, the transition time is significantly shortened, and the inductor's saturation current requirement is alleviated.

Figure 26.8.2 shows the functional block diagram of the proposed two-step subranging DVS scheme. At the beginning of the transition, the subranging comparator compares V_{REF} with $V_{REF, MAX}$/2 (half of the maximum reference voltage) and, based on the result, turns on

either S_{OH} or S_{OL}. Subsequently, the converter generates its own duty cycle by comparing V_{FB} with V_{REF}, thereby controlling switches S_{B0} and S_{B4} to regulate the output voltage. The designed SM uses fixed frequency hysteretic control [14,15] with a ripple injection feedback network [6,11,16] for fast response and noise attenuation. A 1/6 voltage divider is incorporated to ensure that V_{FB} remains in the sub-5V range of the control stage. Two identical feedback networks are connected to V_{CAPL} and V_{CAPH}, respectively, and switch along with the output capacitors, ensuring that the feedback voltages of C_{FBL1} and C_{FBL2}, whose DC levels depend on the output voltage, track both the inductor current and the capacitor voltages. Thus, the initial coarse step will not deteriorate the subsequent fine step. Figure 26.8.2 (bottom) illustrates the waveform diagram of the two-step subranging DVS scheme. Unlike pre-charging [6] or post-charging [10,11] approaches, all charging processes in this scheme are completed within the CP interval, eliminating the need for additional inductors or LDOs to charge the output capacitor. When switch S_{OH} or S_{OL} is turned off, the voltage of the corresponding capacitor is held at that instant. Upon the next turn-on, this stored voltage is used as the initial state for voltage adjustment. For a floating capacitor, maintaining its voltage within the normal range is sufficient; a protection circuit acts only if small control-stage currents drive it outside this range. Unlike [13], which provide only discrete output levels, the output voltage in this work is continuously tunable.

Figure 26.8.3 illustrates the operating principle of the modified ladder SC converter. As is well known, SC converters face challenges at extremely narrow or wide duty cycles, where large inrush currents can increase the RMS current value and even damage the power switches. To address this, the design targets an extended duty-cycle range to cover the full span of 0%-100%. The key in this work is to constrain the charging time of each capacitor to at least 30% of the period T. The voltages of C_{FL1}-C_{FL3} are all V_{IN}, whereas the voltages of C_{FR1}-C_{FR3} are V_{IN}, $2V_{IN}$, and $3V_{IN}$, respectively. In steady state, the SC operates in a three-phase mode. Taking S_{B3} as an example, it corresponds to the $2V_{IN}$-$3V_{IN}$ output-voltage range, with the duty cycle determining the voltage at node V_{X3}, which also serves as the inductor switching node (V_{SW}). For D<40%, the SC operates in Φ_1, Φ_2 and Φ_3 with duty cycles D_1-D_3. Φ_1 and Φ_3 correspond to the conventional two-phase operation. Since D_1 equals D, C_{FL2} has a short charging time of DT under narrow D. By introducing Φ_2 and setting D_1+D_2=40%, the charging time for C_{FL2} is extended to 0.4·T. For D>70%, the SC operates in Φ_4, Φ_5 and Φ_6 with duty cycles D_4-D_6. Φ_5 and Φ_6 correspond to the conventional two-phase operation. Since D_6=1−D, under high D, C_{FR1} and C_{FR2} experience a short charging time of (1−D)·T. By inserting Φ_4 and setting D_4 to be 40%, the minimum charging times for C_{FR1} and C_{FR2} are extended to 0.4·T, while C_{FL2} maintains a minimum charging time of 0.3·T. Thus, under both high and low duty cycles, a minimum charging time of 0.3·T is ensured for each capacitor, preventing inrush current and alleviating RMS losses at extreme duty conditions. In steady state, the 16-V high-voltage bus switches remain always on or off, thereby eliminating switching losses. The 5-V switches that do not conduct current switch at a reduced frequency to minimize power consumption. For the S_{B3} case, S_{R6}, S_{R7}, S_{R10} and S_{R11} can switch at a reduced frequency. During the transition, the SC operates in Φ_7 and Φ_8 in open-loop with a fixed duty cycle; V_{X4} remains at $4V_{IN}$, while the feedback network controls S_{B0} and S_{B4} to regulate the output voltage.

The proposed SPT SM is fabricated in a 0.18µm BCD process. The modified SC operates at 4MHz with a 1µH inductor and 470nF output capacitors for both C_{OL} and C_{OH}. Figure 26.8.4 (top) presents steady-state waveforms at V_{IN}=4V with S_{B3} enabled. At narrow D, V_{X3} corresponds to the designed minimum duty cycle, V_{X2} operates at 40%, and V_{X6} (V_{X4}) corresponds to the post-divide-by-8 minimum duty. This minimum duty ensures that the flying capacitors maintain their DC voltage, preventing excessive discharge during extended intervals when they are not delivering power. At high duty, V_{X3} and V_{X2} switch with the applied duty, while V_V operates at 40%. Figure 26.8.4 (bottom left) illustrates the mid-duty condition for a 12V output and 400mA load current, where V_{X4} operates at the appropriate duty cycle and V_{X3} at operates at 0.4·T. The voltage ripple at the output capacitor is only 30mV, while the ripple at VDD_PA is larger due to the S_{OH} on-resistance and can be reduced by decreasing it. Figure 26.8.4 (bottom right) shows the V_{REF} sweep, demonstrating an output range of 1-to-12V. Correspondingly, bus switches S_{B1}-S_{B4} define four distinct V_{SW} operating regions. Figure 26.8.5 presents the measured SPT transitions of the SM. For a 1-to-6V transition, the process is completed using only the C_{OL} region, with V_{SW} swinging between ground and $4V_{IN}$ to enable fast response. The switching threshold is set at 6.5V, so both C_{OL} and C_{OH} participate in the 6-to-12V transition. In the case of a 1-to-12V up transition, the two-step subranging DVS scheme shortens the transition time to nearly that of the 6-to-12V half transition. All up and down transitions are completed within 1.2µs, satisfying the 5G FR1 requirement.

Figure 26.8.6 (top left) shows the measured efficiency at V_{IN}=4V. The converter delivers up to 600mA load current at a 12V output. A peak efficiency of 92.3% is achieved at VDD_{PA}=6V and 350mA load. Figure 26.8.6 (bottom) compares the performance with state-of-the-art designs, where the proposed SM attains the highest output voltage, widest output range, and fastest normalized transition speed among the compared SMs with continuous output. Figure 28.6.7 presents the die micrograph along with the values of the off-chip components.

Acknowledgement:
This work was supported in part by the National Natural Science Foundation of China under Grant No. 62261160647. Corresponding author: Lin Cheng (eecheng@ustc.edu.cn).

References:
[1] C. Chen, X. Li, R. Hu and L. Cheng, "An 83.4%-Peak-Efficiency Envelope-Tracking Supply Modulator Using a Class-G Linear Amplifier and a Single-Inductor Dual-Input-Dual-Output Converter for 200MHz Bandwidth 5G New Radio RF Applications," *2024 IEEE International Solid-State Circuits Conference (ISSCC)*, San Francisco, CA, USA, 2024, pp. 496-498. http://doi.org/10.1109/ISSCC49657.2024.10454328
[2] C. -Y. Chen et al., "An 85.6%-Efficiency Supply Modulator with Auxiliary Bidirectional Power for 200MHz 5G NR Applications," *2025 Symposium on VLSI Technology and Circuits (VLSI Technology and Circuits)*, Kyoto, Japan, 2025, pp. 1-3. http://doi.org/10.23919/VLSITechnologyandCir65189.2025.11074847
[3] J. -S. Bang et al., "2-Tx Digital Envelope-Tracking Supply Modulator Achieving 200MHz Channel Bandwidth and 93.6% Efficiency for 2G/3G/LTE/NR RF Power Amplifiers," *ISSCC*, pp. 1-3, Feb. 2022. http://doi.org/10.1109/ISSCC42614.2022.9731655
[4] J. -H. Bae et al., "A Digital Envelope Tracking RF Power Amplifier Achieving 400MHz Channel Bandwidth and 91.9% Efficiency for Upper-Mid Band Extreme Massive MIMO 6G Communications," *2025 Symposium on VLSI Technology and Circuits (VLSI Technology and Circuits)*, Kyoto, Japan, 2025, pp. 1-3. http://doi.org/10.23919/VLSITechnologyandCir65189.2025.11075165
[5] H. M. Pham et al., "A 74W/48V Monolithic-GaN Integrated Adjustable Multilevel Supply Modulator for 5G Base-Station Massive-MIMO Arrays," *2025 IEEE International Solid-State Circuits Conference (ISSCC)*, San Francisco, CA, USA, 2025, pp. 1-3. http://doi.org/10.1109/ISSCC49661.2025.10904518
[6] J. -S. Paek, T. Nomiyama, J. Han, I. -H. Kim, Y. Lee, D. Kim, E. Park, S. Lee, J. Lee, T. B. Cho, and I. Kang, "A 90ns/V Fast-Transition Symbol-Power-Tracking Buck Converter for 5G mm-Wave Phased-Array Transceiver," in *2019 IEEE International Solid- State Circuits Conference - (ISSCC)*, San Francisco, CA, USA, Feb. 2019, pp. 240–242. http://doi.org/10.1109/ISSCC.2019.8662420
[7] J. -S. Paek, D. Kim, Y. Choo, Y. -S. Youn, J. Lee, and T. B. -H. Cho, "Design of Boosted Supply Modulator With Reverse Current Protection for Wide Battery Range in Envelope Tracking Operation," *IEEE Trans. Microwave Theory Techn.*, vol. 67, no. 1, pp. 183–194, Jan. 2019. http://doi.org/10.1109/TMTT.2018.2879323
[8] H. Wang. PA Survey Version 10: Saturated Output Power vs. Frequency (All Technologies). https://ideas.ethz.ch/Surveys/pa-survey.html.

[9] P. Xu, J. Kang, Z. Tong, P. Cao, Y. Wang, H. Shi, J. Xu, and Z. Hong, "A 0.15-µs/V Buck-Boost Symbol-Power-Tracking Supply Modulator With Dual Auxiliary Current Paths and EPP Scheme for 5G NR Power Amplifiers," *IEEE Trans. Circuits Syst. I*, vol. 70, no. 11, pp. 4660–4670, Nov. 2023. http://doi.org/10.1109/TCSI.2023.3310240
[10] J. -S. Bang, D. Kim, Y. Choo, I. -H. Kim, S. Park, J. Lee, S. -H. Lee, Y. -H. Jung, J. -Y. Ko, S. Jung, J. Han, W. Kim, J. -S. Paek, and J. Lee, "5G NR RF PA Supply Modulator Supporting 179ns 0.5-to-5.5V Symbol Power Tracking and Envelope Tracking," in *2023 IEEE Symposium on VLSI Technology and Circuits (VLSI Technology and Circuits)*, Kyoto, Japan, Jun. 2023, pp. 1–2. http://doi.org/10.23919/VLSITechnologyandCir57934.2023.10185384
[11] I. -H. Kim, J. -I. Seo, Y. Choo, S. Park, J. Han, W. Kim, S. Jung, T. Ko, D. Kim, J. Lee, and S. Kwak, "A 950ns 0.5-to-5.5V 5G NR RF PA Supply Modulator with Floating Capacitor Control for Symbol Power Tracking," in *2024 IEEE International Solid-State Circuits Conference (ISSCC)*, San Francisco, CA, USA, Feb. 2024, pp. 500–502. http://doi.org/10.1109/ISSCC49657.2024.10454517
[12] M. Shang, B. Wang, C. Chen, J. Jin, and L. Cheng, "A 102ns/V 94.3%-Peak-Efficiency Symbol-Power-Tracking Supply Modulator for 5G NR Power Amplifiers," in *2025 IEEE International Solid-State Circuits Conference (ISSCC)*, Feb. 2025, vol. 68, pp. 01–03. http://doi.org/10.1109/ISSCC49661.2025.10904803
[13] J. Baek and A. Niknejad, "A 400MHz Symbol-Power-Tracking (SPT) Supply Modulator with SPT-Adaptive-Biasing Network Supporting 5G FR2 CMOS PA," in *2025 IEEE International Solid-State Circuits Conference (ISSCC)*, Feb. 2025, vol. 68, pp. 1–3. http://doi.org/10.1109/ISSCC49661.2025.10904614
[14] M. K. Song, J. Sankman, and D. Ma, "A 6A 40MHz four-phase ZDS hysteretic DC-DC converter with 118mV droop and 230ns response time for a 5A/5ns load transient," in *2014 IEEE International Solid-State Circuits Conference Digest of Technical Papers (ISSCC)*, San Francisco, CA, USA, Feb. 2014, pp. 80–81. http://doi.org/10.1109/ISSCC.2014.6757346
[15] M. K. Song, D. Yan, and D. B. Ma, "A 8.9W/mm2 , 95.4%-Efficiency, CP-Length Tracking Switching Supply Modulator for 5G New Radio mmWave PA Arrays," in *ESSCIRC 2021 - IEEE 47th European Solid State Circuits Conference (ESSCIRC)*, Grenoble, France, Sep. 2021, pp. 319–322. http://doi.org/10.1109/ESSCIRC53450.2021.9567773
[16] J. -S. Paek, S. -C. Lee, Y. -S. Youn, D. Kim, J. -H. Choi, J. Jung, Y. -H. Choo, S. -J. Lee, J. -Y. Han, and T. B. Cho, "A –137 dBm/Hz Noise, 82% Efficiency AC-Coupled Hybrid Supply Modulator With Integrated Buck-Boost Converter," *IEEE Journal of Solid-State Circuits*, vol. 51, no. 11, pp. 2757–2768, Nov. 2016. http://doi.org/10.1109/JSSC.2016.2604296
[17] J. Baek, T. Nomiyama, S. Park, Y. -H. Jung, D. Kim, J. Han, J. -S. Bang, Y. Lee, I. -H. Kim, J. -S. Paek, J. Lee, and T. B. Cho, "A Voltage-Tolerant Three-Level Buck-Boost DC-DC Converter with Continuous Transfer Current and Flying Capacitor Soft Charger Achieving 96.8% Power Efficiency and 0.87µs/V DVS Rate," in *2020 IEEE International Solid- State Circuits Conference - (ISSCC)*, San Francisco, CA, USA, Feb. 2020, pp. 202–204. http://doi.org/10.1109/ISSCC19947.2020.9063105
[18] J. Ruan, J. Jiang, C. Ding, K. Yuan, K. N. Leung, and X. Liu, "A Li-ion-Battery-Input 1-to-6V-Output Bootstrap-Free Hybrid Buck-or-Boost Converter Without RHP Zero Achieving 97.3% Peak Efficiency 6µs Recovery Time and 1.13µs/V DVS Rate," in *2024 IEEE International Solid-State Circuits Conference (ISSCC)*, Feb. 2024, vol. 67, pp. 148–150. http://doi.org/10.1109/ISSCC49657.2024.10454342

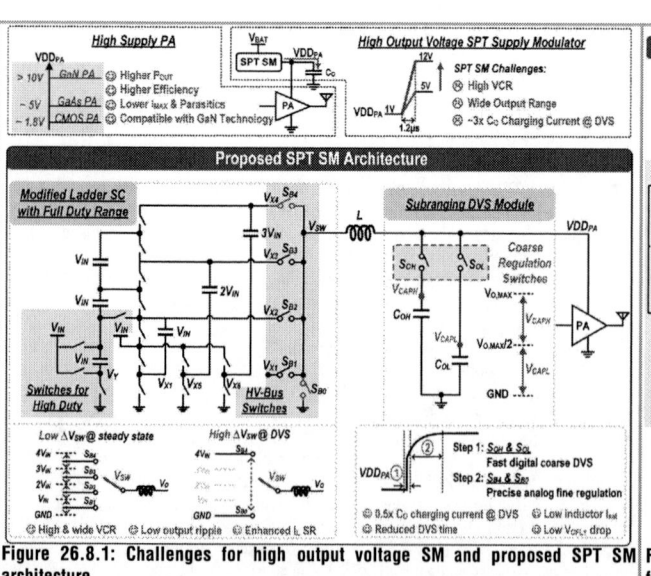

Figure 26.8.1: Challenges for high output voltage SM and proposed SPT SM architecture.

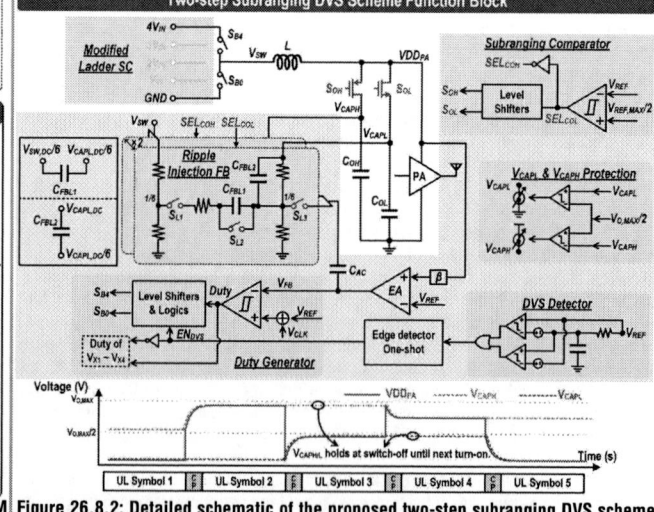

Figure 26.8.2: Detailed schematic of the proposed two-step subranging DVS scheme function block.

ISSCC 2026 / SESSION 26 / COMPUTE POWER AND SUPPLY MODULATORS / 26.8

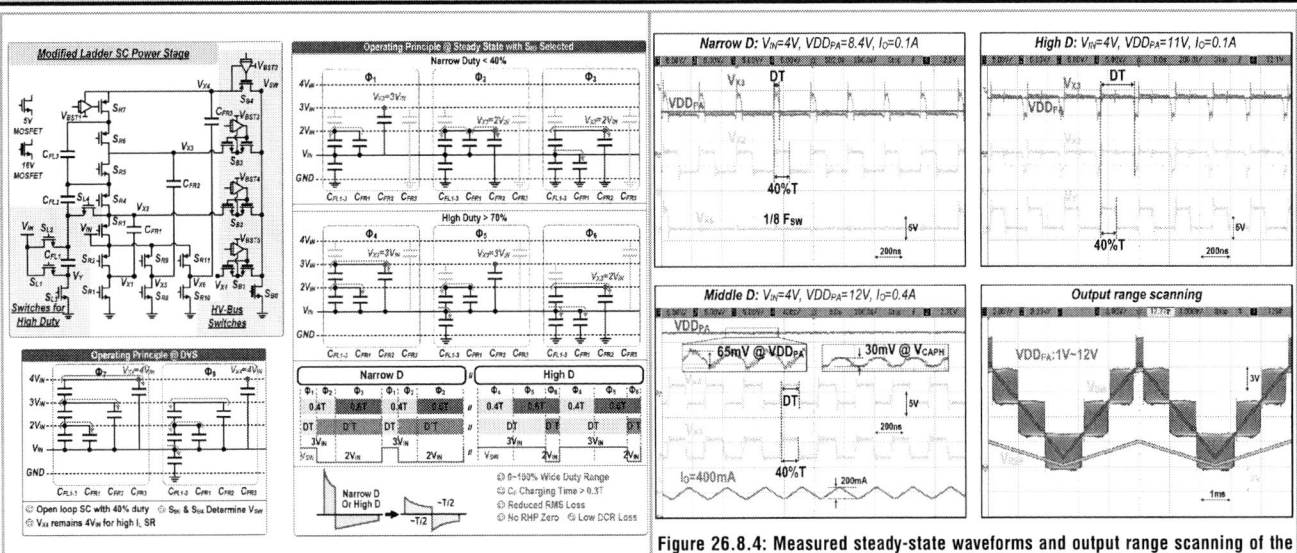

Figure 26.8.3: Operating principles of the modified Ladder SC converter.

Figure 26.8.4: Measured steady-state waveforms and output range scanning of the proposed SPT SM.

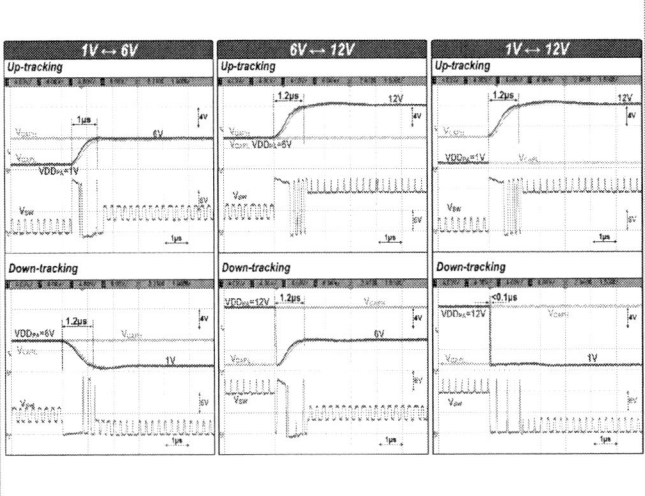

Figure 26.8.5: Measured voltage waveforms of SPT transitions.

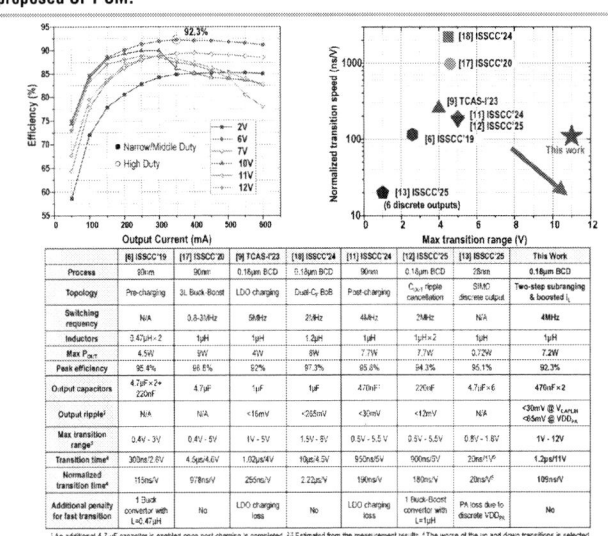

Figure 26.8.6: Measured SPT SM performances and comparison table.

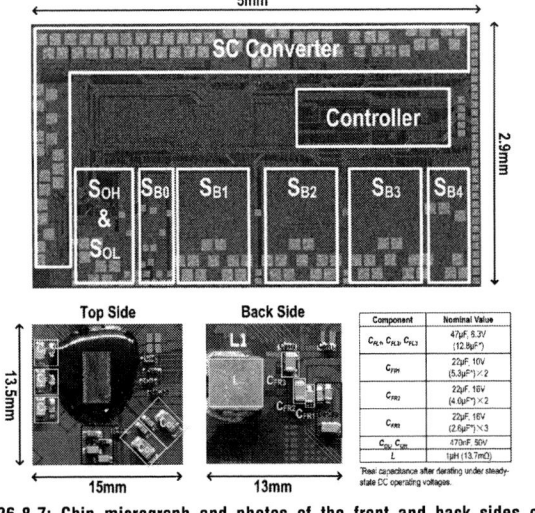

Figure 26.8.7: Chip micrograph and photos of the front and back sides of the measurement PCB.

• 2026 IEEE International Solid-State Circuits Conference

979-8-3315-8937-0/26 $31.00 © 2026 IEEE

ISSCC 2026 / SESSION 26 / COMPUTE POWER AND SUPPLY MODULATORS / 26.9

26.9 A Compact Dual-Capacitor Relay SPT Supply Modulator with Overshoot-Free Adaptive On-Time Control for 5G FR2 CMOS PA

Zhewen Yu[1,2], Zhiguo Tong[2], Junwei Huang[2], Jun Yin[2], Yan Lu[1]

[1]Tsinghua University, Beijing, China, [2]University of Macau, Macau, China

Abstract

5G new-radio frequency-range-2 CMOS PAs with high peak-to-average-power ratios suffer from low efficiency, calling for compact and fast symbol-power-tracking (SPT) supply modulators (SMs). Inductor-based SPT shows limited tracking speed, while multilevel switched-capacitor approaches need redundant capacitors and switches. The proposed prototype dual-capacitor relay SPT SM with overshoot-free adaptive constant on-time control achieves <290ns tracking using a 0402 inductor, suitable for compact PA arrays.

The 5G new-radio (NR) frequency-range-2 (FR2) standard supports up to 400MHz bandwidth and uses orthogonal frequency-division multiplexing (OFDM) to achieve high data rates with complex modulation, such as quadrature amplitude modulation (QAM). CMOS PAs consume high power at average output levels due to high peak-to-average-power ratios (PAPRs), which shortens battery life in mobile devices. Envelope tracking (ET) enhances PA efficiency at deep power back-off (PBO) by adjusting the PA supply voltage V_{DDPA} but this becomes challenging for a high bandwidth of 400MHz. Symbol power tracking (SPT) offers an alternative. In FR2 with 240kHz sub-carrier spacing (SCS), the SPT modulator must complete V_{DDPA} transitions within the 290ns cyclic prefix (CP), imposing stringent speed requirements on the supply modulator (SM) design. Hence, a compact SPT SM with a fast V_{DDPA} transition capability is required for mobile devices.

Figure 26.9.1 (bottom-left) illustrates a high-frequency inductive DC-DC converter with small LC values for fast V_{DDPA} transitions but it cannot meet both SPT speed and ripple requirements. State-of-the-art SPT modulators employ fast large-signal transition control, followed by ripple-cancellation techniques, to overcome the LC frequency limit [1-4]. These approaches require complex control schemes and an auxiliary linear regulator to suppress ripple and accelerate transient recovery. However, these control techniques are difficult to extend to multi-level supply generation, limiting further efficiency improvement. Recent works [5-7] employ multi-level switched-capacitor (MLSC) converters to generate multi-level supply voltages and realize fast V_{DDPA} transitions through the output switches. However, the output levels are interdependent, and the overall V_{DDPA} range is constrained by the MLSC topology. 5G FR2 CMOS PAs typically operate with a 0.8-to-1.8V supply range, where MLSC converters cannot provide this narrow supply range with high utilization and efficiency. To address this issue, single-inductor multiple-output (SIMO) SMs [8] have been proposed to generate arbitrary supply voltages (Fig. 26.9.1 bottom-middle). However, this approach still suffers from redundant output capacitors and output switches, especially when the number of output levels increases. In fact, MLSC solutions achieve sufficiently fast V_{DDPA} transitions; however, the transient capability of the SIMO converter is wasted, as it only provides fixed supplies. To reduce the redundancy of output capacitors and to minimize the number of output switches, this work proposes a dual-capacitor-relay SPT SM architecture (Fig. 26.9.1 bottom-right). It splits the SPT supply voltage requirement into two halved voltage domains and is implemented using a fast-dynamic DC-DC converter. The number of output capacitors is reduced to two, and arbitrary supply voltage levels can be generated within the 0.8-to-1.8V range.

Figure 26.9.2 illustrates the operation of the proposed dual-capacitor-relay SPT. Two output capacitors C_{O1} and C_{O2} are alternately connected to the V_{DDPA} according to the SPT ranges of 0.8-to-1.3V and 1.3-to-1.8V, respectively. When V_{DDPA} transitions between the two ranges, the channel-switching signal (CH_S) is enabled, and then V_{DDPA} switches between V_{O1} and V_{O2}. Prior work [8] regulates V_{O1} and V_{O2} simultaneously using SIMO control. The energy generation loop compares the total error of output voltages with V_{REF}, where multi-output coupling limits the loop bandwidth. In addition, the output switches are divided into the SIMO switches for voltage regulation and the SPT switches for output selection, leading to discontinuous current and large output ripple. Furthermore, the SIMO switches operate at a higher frequency than the SPT switches, resulting in high switching loss. To solve these problems, this work proposes a dual-output overshoot/undershoot-free adaptive constant on-time (AOT) control, where the control loop is shared between the two outputs and the feedback signal is switched according to the output selection. Since the control is ripple-based, unlike PWM control, no extra loop settling time is required during the output relay. And, the two outputs do not need simultaneous regulation, so the output switches can be merged into a single back-to-back SPT switch operating at the output switching frequency, resulting in a smaller output ripple and a lower switching frequency.

With only two output capacitors, it is essential to maximize the SPT range of each supply voltage. Conventional PWM control suffers from long settling time for SPT operation due to limited loop bandwidth (BW). Although AOT control enables fast response, the absence of derivative (D) control causes large voltage overshoots and undershoots, which degrades the PA error vector magnitude (EVM). By programming the V_{REF} signal with a fixed edge time and combining it with ripple-based control, the output voltage tracks V_{REF} in real time during voltage transitions, preventing excessive inductor current response. This control

achieves both fast response and overshoot/undershoot suppression. For SPT operation, this approach improves EVM performance while significantly reducing the settling time. As shown in Fig. 26.9.2 (bottom-right), the proposed AOT control demonstrates better consistency and faster settling time. At low frequency AOT control, a long on time leads to output overshoot, while at high frequencies, the fixed minimum off time limits the maximum duty ratio, thereby slowing down the transient response. Considering both output ripple and dynamic performance, an LC product of 33nH·μF is adopted and the AOT switching frequency is set to 17MHz.

Figure 26.9.3 shows the system block diagram. The power stage adopts a stacked half-bridge structure composed of M_1-M_4. For the low supply voltage V_{O1} (0.8-to-1.3V), back-to-back NMOS devices (M_5 and M_6) serve as the unified SPT switch. For the high supply voltage V_{O2} (1.3-to-1.8V), back-to-back PMOS devices (M_7 and M_8) are adopted. Both supply voltages are clamped to maintain their levels within the designated SPT range, ensuring fast and stable reactivation during long unselected intervals. M_1-M_4 operate at 17MHz, and during the DVS transition, the AOT control rapidly increases the switching frequency up to 50MHz. To prevent instability of the on-chip bootstrap capacitor's voltage V_{BT1} caused by ringing, a high-frequency bootstrap circuit is employed (Fig. 26.9.3 bottom-left). A back-to-back 1.8V PMOS is used to control the charging of C_{BT1}. After a short dead time, the DBN signal enables V_{BT1} charging after the ringing decays. For the double-CMOS stacked buck stage, the high-side bootstrap capacitor C_{BT2} is charged through a double-stacked 1.8V PMOS controlled by the DB signal.

To ensure consistent output ripples while keeping the switching frequency below the self-resonant frequency of the output capacitors, a feed forward AOT control is employed. In addition, to prevent subharmonic oscillation across the entire output voltage range, both outputs adopt on-chip ripple injection (Fig. 26.9.3 bottom-middle). The current ripple information is extracted between the stable voltage nodes (V_{O1}, V_{O2}) and the switching node V_{SW} through the RC network R_R and C_R, and then capacitively coupled into the feedback nodes V_{FB1} and V_{FB2} via C_{AC}.

Figure 26.9.3 (top-right) shows the control block diagram. For verification, an off-chip FPGA and DAC generate a V_{REF} signal with a fixed edge time and a synchronized channel-switching signal CH_S. The on-chip relay protection logic processes CH_S and completes the output relay within 8ns. Then, the feedback voltage is connected into the AOT loop. The corresponding output capacitor then takes over and performs the supply voltage transition within its SPT range. The detailed channel-switching logic is illustrated in Fig. 26.9.3 (bottom-right). When CH_S arrives, the buck stage is first disabled, avoiding erroneous on-time activation. Then, the output switch, feedback voltage, and reference voltage are toggled. Finally, the buck stage resumes operation once the signal switch completes.

Figure 26.9.4 shows the measured results at V_{IN} = 3.6V and R_L = 4Ω. The steady-state waveforms of V_{O1} = 1V and V_{O2} = 1.8V have V_{DDPA} output ripples of 65mV and 54mV, respectively. With AOT control operating at about 17MHz, the on-times are 15.5ns for V_{O1} and 31ns for V_{O2}. The lower plots show the worst case of single-channel SPT transitions. V_{O1} switches between 0.8-to-1.3V with rise/fall times of 205ns/220ns, while V_{O2} switches between 1.3-to-1.8V with rise/fall times of 230ns/220ns.

Figure 26.9.5 (top) demonstrates dual-capacitor relay SPT, enabling multi-level V_{DDPA} between 0.8-to-1.8V with <290ns of settling time for any transition. For a 0.8-to-1.8V step, C_{O1} hands over to C_{O2} within 15ns, followed by the inductor charging C_{O2} until V_{O2} reaches 1.8V in 200ns. When the output voltage steps down from 1.4-to-0.8V, the relay action is completed within 15ns. Figure 26.9.5 (bottom-left) illustrates the supply-voltage clamp function: V_{O2} is held at 1.55V when unselected, ensuring it remains within the SPT range, and it recovers to 1.7V within 200ns once reactivated. Figure 26.9.5 (bottom-right) shows the measured SM efficiency at V_{OUT} = 1.0V, 1.3V, and 1.8V. For space-constrained applications, we adopt a 1.0×0.6×0.5mm³ inductor, while for efficiency-first cases, we adopt an air-core inductor to reduce the AC loss.

Figure 26.9.6 compares this work with state-of-the-art SMs. Operating with a 3.6V input, it generates 0.8-to-1.8V supply voltages at 17MHz with a peak efficiency of 91.5%. With an

462 • 2026 IEEE International Solid-State Circuits Conference

979-8-3315-8937-0/26 $31.00 © 2026 IEEE

ultra-compact 0402 inductor and only two output capacitors, the total passive volume is reduced to 0.81mm³, enabling flexible multi-level V_{DDPA} SPT. This design achieves a worst-case settling time of <290ns and V_{DDPA} ripple <65mV.

Figure 26.9.7 shows the die micrograph of the prototype. The proposed SPT SM is fabricated in a 0.18µm BCD process, occupying a die area of 5.03mm². The top-right image shows the test PCB layout, which consists of the on-chip solder-mounted 0402 ferrite-core wire-wound 33nH inductor, two 0204 1µF low-ESL output capacitors with a 17MHz SRF, and one 0201 2nF capacitor C_0. For efficiency-oriented scenarios, a 3.5×3.1×2.9mm³ air-core RF inductor is mounted on the backside of the PCB, offering reduced AC loss and high saturation current, making it well suited for low-current steady-state operation as well as large transient current during SPT.

Acknowledgement:
This work was supported in part by the National Natural Science Foundation of China (92573201), the Macau Science and Technology Development Fund (FDCT/0103/2022/AFJ and 004/2023/SKL). Corresponding Author: Yan Lu.

References:
[1] M. Shang, B. Wang, C. Chen, J. Jin and L. Cheng, "A 102ns/V 94.3%-Peak-Efficiency Symbol-Power-Tracking Supply Modulator for 5G NR Power Amplifiers," in *IEEE International Solid-State Circuits Conference (ISSCC)*, Feb. 2025, pp. 01-03. https://doi.org/10.1109/ISSCC49661.2025.10904803
[2] I.-H. Kim et al., "A 950ns 0.5-to-5.5V 5G NR RF PA Supply Modulator with Floating Capacitor Control for Symbol Power Tracking," in *IEEE International Solid-State Circuits Conference (ISSCC)*, Feb. 2024, pp. 500-502. https://doi.org/10.1109/ISSCC49657.2024.10454517
[3] H.-H. Bae, J.-H. Cho, G.-G. Kang, Y. Park and H.-S. Kim, "A Power-Efficient 1–2-MHz Hysteretic Buck Converter With High DVS Rate Enabled by Isosceles-Triangular Shunt Current Push–Pull Technique," in *IEEE Solid-State Circuits Letters*, vol. 5, pp. 280-283, 2022. https://doi.org/10.1109/LSSC.2022.3225301
[4] J. Ruan, J. Jiang, C. Ding, K. Yuan, K. N. Leung and X. Liu, "A Li-ion-Battery-Input 1-to-6V-Output Bootstrap-Free Hybrid Buck-or-Boost Converter Without RHP Zero Achieving 97.3% Peak Efficiency 6µs Recovery Time and 1.13µs/V DVS Rate," in *IEEE International Solid-State Circuits Conference (ISSCC)*, Feb. 2024, pp. 148-150. https://doi.org/10.1109/ISSCC49657.2024.10454342
[5] A. Sepahvand, P. Momenroodaki, Y. Zhang, Z. Popović and D. Maksimović, "Monolithic multilevel GaN converter for envelope tracking in RF power amplifiers," in *IEEE Energy Conversion Congress and Exposition (ECCE)*, 2016, pp. 1-7. https://doi.org/10.1109/ECCE.2016.7855275
[6] H. Pham, R. Das, C. Hardy, D. Kimball, P. Asbeck and H.-P. Le, "Adjustable 4-Level Hybrid Converter for Symbol Power Tracking in 5G New Radio," in *IEEE Applied Power Electronics Conference and Exposition (APEC)*, Feb. 2023, pp. 1858-1861. https://doi.org/10.1109/APEC43580.2023.10131338
[7] H. M. Pham et al., "A 74W/48V Monolithic-GaN Integrated Adjustable Multilevel Supply Modulator for 5G Base-Station Massive-MIMO Arrays," in *IEEE International Solid-State Circuits Conference (ISSCC)*, Feb. 2025, pp. 1-3. https://doi.org/10.1109/ISSCC49661.2025.10904518

[8] J. Baek and A. Niknejad, "A 400MHz Symbol-Power-Tracking (SPT) Supply Modulator with SPT-Adaptive-Biasing Network Supporting 5G FR2 CMOS PA," in *IEEE International Solid-State Circuits Conference (ISSCC)*, Feb. 2025, pp. 1-3. https://doi.org/10.1109/ISSCC49661.2025.10904614
[9] C. Chen, X. Li, R. Hu and L. Cheng, "An 83.4%-Peak-Efficiency Envelope-Tracking Supply Modulator Using a Class-G Linear Amplifier and a Single-Inductor Dual-Input-Dual-Output Converter for 200MHz Bandwidth 5G New Radio RF Applications," in *IEEE International Solid-State Circuits Conference (ISSCC)*, Feb. 2024, pp. 496-498. https://doi.org/10.1109/ISSCC49657.2024.10454328

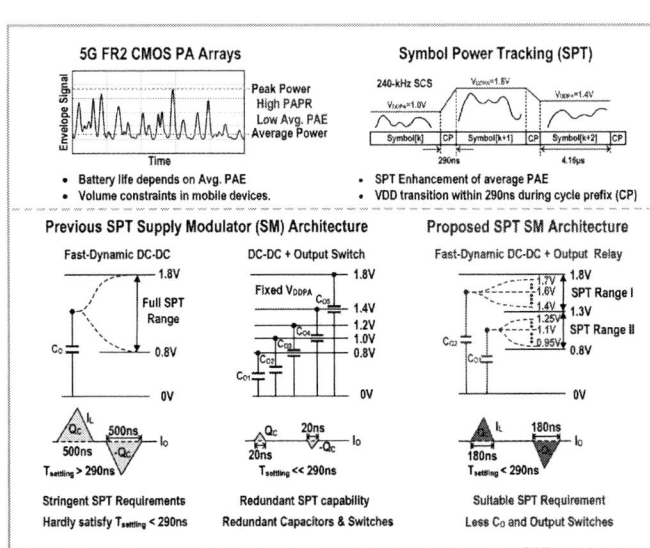

Figure 26.9.1: The conventional and proposed dual-capacitor relay SPT architectures for 5G FR2 CMOS PA.

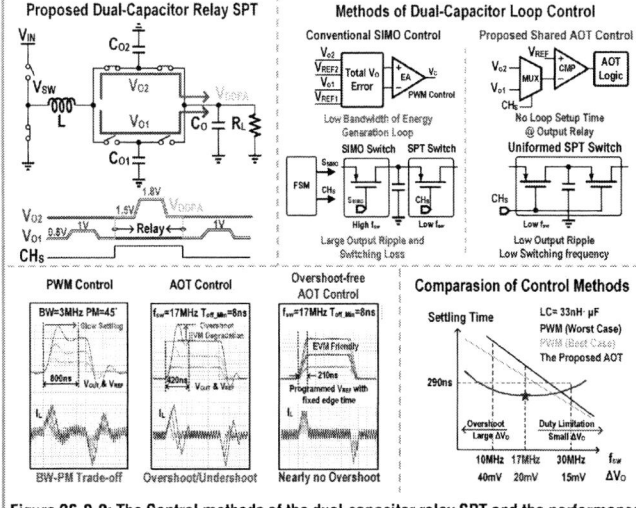

Figure 26.9.2: The Control methods of the dual-capacitor relay SPT and the performance comparison of different control methods.

ISSCC 2026 / SESSION 26 / COMPUTE POWER AND SUPPLY MODULATORS / 26.9

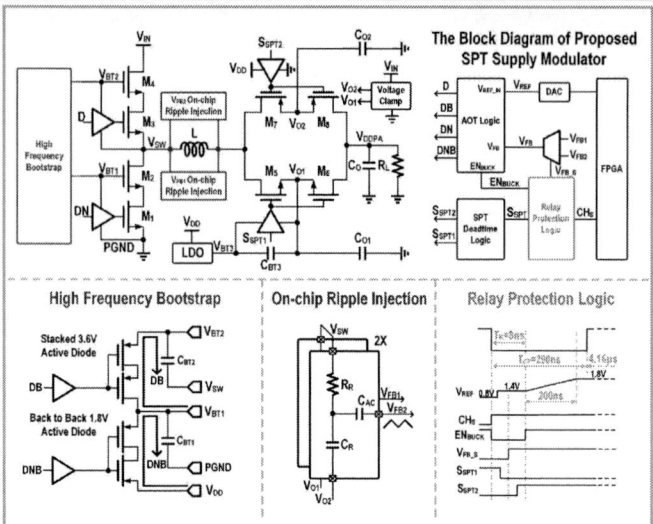

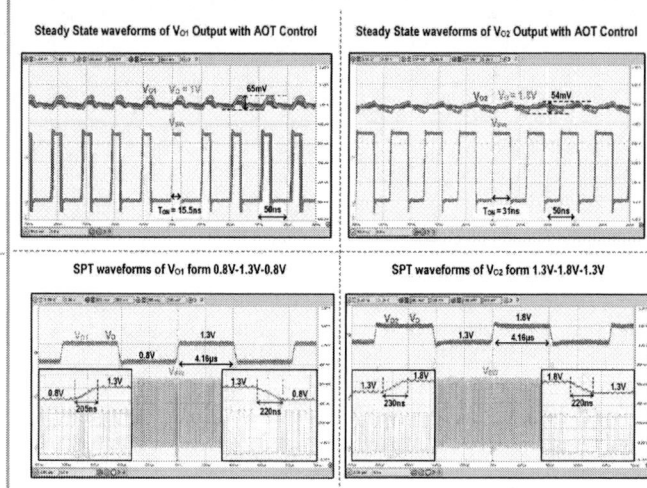

Figure 26.9.3: The block diagram of the proposed SPT SM and the circuit implementation details.

Figure 26.9.4: The measured steady state, and SPT waveforms.

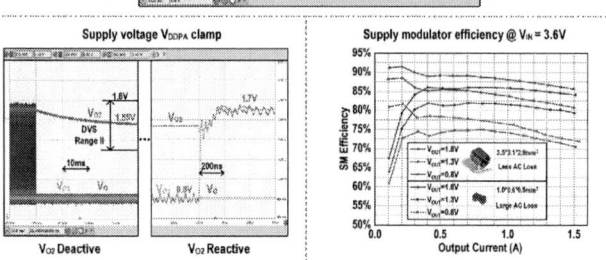

Figure 26.9.5: The measured multiple V_{DDPA} levels SPT, supply clamp, and efficiency waveforms.

	ISSCC'24 [2]	ISSCC'24 [9]	ISSCC'25 [1]	ISSCC'25 [6]	This Work	
Process	90nm	65nm	180nm	28nm	180nm	
Topology	Buck-Boost	Buck-Boost	3L Buck-Boost	SIMO BUCK	SIMO BUCK	
Control Methods	Floating Capacitor + LDO	SIDIDO + LA	Ripple Cancellation + LDO	Output Switch	AOT + Output Relay	
Nominal Input Voltage	4V	4V	4V	3.6V	3.6V	
Output Voltage Range	0.5V-5.5V	1V-5V	0.5V-5.5V	0.8V-1.8V	0.8V-1.8V	
Switching Frequency	4-7MHz	2MHz	2MHz	N. A.	17MHz	
Flying Capacitor	N. A.	N. A.	470nF	N. A.	N. A.	
Output Capacitor	4.7µF/470nF	4.7µF/2.2µF	220nF	N. A.	1µF	
Number of Output Capacitors	2	3	1	6	2	
Output Ripple	N. A.	<50mV*	<12mV	<100mV*	<65mV	
Chip/Package Area	3.96mm²	2.175mm²	6.6mm²	5.88mm²	5.03mm²	
Inductor	1µH	4.7µH	2*1µH	N. A.	33nH	34nH
Inductor Volume	N. A.	N. A.	74mm³	N. A.	0.3mm³	31.4mm³
Total Passive Volume	N. A.	N. A.	>74mm³*	N. A.	0.8mm³	31.9mm³
SM Peak Efficiency V_{IN}/V_{OUT}	95.8% 4V/3V	83.4%	94.3% 4V/5.5V	95.1% 3.6V/1.8V	86.1% 3.6V/1.8V	91.5% 3.6V/1.8V
Reference Tracking — Up Transient Speed	164ns/V	N. A.	102ns/V	20ns/V	230ns/V	
Up Settling Time	820ns	N. A.	510ns	20ns	230ns	
Down Transient Speed	190ns/V	N. A.	180ns/V	20ns/V	220ns/V	
Down Settling Time	950ns	N. A.	900ns	20ns	220ns	

* Estimated from the measurement results

Figure 26.9.6: Comparison with state-of-the-art works.

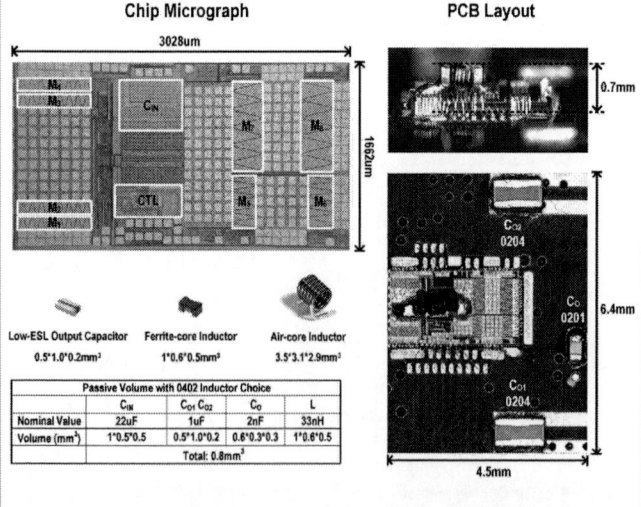

Figure 26.9.7: The chip micrograph, PCB implementation and components list.

• 2026 IEEE International Solid-State Circuits Conference

ISSCC 2026 / SESSION 27 / FREQUENCY GENERATORS, MULTIPLIERS AND MODULATORS / OVERVIEW

Session 27 Overview: *Frequency Generators, Multipliers and Modulators*

RF SUBCOMMITTEE

Session Chair: Danilo Manstretta
University of Pavia, Pavia, Italy

Session Co-Chair: Teerachot Siriburanon
University College Dublin, Dublin, Ireland

Advances in frequency synthesis and signal generation are driving higher precision, wider bandwidth, and improved efficiency across RF and mm-wave systems. The first three papers introduce innovative PLL and cascaded-PLL architectures achieving low jitter and spur levels. These are followed by wide-tuning, high-FoM VCOs, and broadband FMCW synthesizers with enhanced linearity. The session concludes with compact and efficient mm-wave building blocks, including a precision vector-summing phase shifter (VSPS), a low-voltage digitally controlled oscillating power amplifier (DCO-PA), and a high-efficiency frequency doubler.

8:00 AM

27.1 A 20GHz Frequency Synthesizer with Spur-Shaping Modulator Achieving 46.2fs Jitter and -76.5dBc Worst-Case Fractional Spur

Zonglin Ye, University of Electronic Science and Technology of China, Chengdu, China

In Paper 27.1, UESTC introduces a spur-shaping modulator (SSM) that suppresses inter-slice mismatch nonlinearity in the multi-path-based PLL, achieving ultra-low fractional spurs in near-integer channels. With the SSM, the prototype synthesizer attains 46.2fs jitter, -84.1dBc near-integer spur, and FoMs of –251.4dB and –102.6dB in 28nm CMOS.

8:25 AM

27.2 A 157fs$_{rms}$-Jitter, –73dBc-Fractional-Spur, Calibration-Free Cascaded SPLL Employing Robust Feedforward Noise Cancellation and MMD-Based Quantization-Error Cancellation with a 60MHz Reference

Haoming Zhang, University of Tokyo, Tokyo, Japan

In Paper 27.2, the University of Tokyo presents a 6.2-to-6.8GHz calibration-free fractional-N sampling PLL with cascaded architecture, achieving 157fs rms jitter and –73dBc worst-case fractional spur using feed-forward noise cancellation (FFNC) and MMD-based quantization-error cancellation (QEC).

8:50 AM

27.3 A 14GHz Ring-Based 3rd-Order Fractional-N PLL with 164fs$_{rms}$ Jitter and a 100MHz Reference

Zhaochen Zhu, Hong Kong University of Science and Technology, Guangzhou, China

In Paper 27.3, HKUST (GZ) shows a cascaded sub-sampling DLL with a narrow-pulsed integrator, delay calibration, and tri-state polarity-reversible SSPD, which reduces integrated jitter and supply noise, achieving higher-order VCO noise suppression and a 10dB better FoM$_N$ than prior ring-based fractional-N PLLs.

ISSCC 2026 / February 18, 2026 / 8:00 AM

9:15 AM

27.4 A 0.068mm² 8.5-to-12.7GHz Complementary Dual-Core VCO with Auto-2nd-Harmonic-Tracking Technique Achieving 202.7dBc/Hz Peak FoM_T and 0.9dB-FoM Variation at a 1MHz Offset in a 39.6% Tuning Range

Xincheng Du, University of Macau, Macau, China, UM Hetao IC Research Institute, Shenzhen, China

In Paper 27.4, the University of Macau presents a complementary dual-core VCO with auto-second-harmonic-tracking, which achieves consistently low phase noise and high FoM across a wide 8.5-to-12.7GHz (39.6%) tuning range. Fabricated in 65nm CMOS, it attains peak FoMs of 190.7/193.5dBc/Hz with only 0.9/1.6dB variation and occupies just 0.068mm².

9:30 AM

27.5 A 17.9-to-22.4GHz 195.6±1.3dBc/Hz FoM Quad-Core Class-F⁻¹ VCO Featuring Improved Synchronization Using a Circular Pentafilar Transformer-Based Tank

Yatao Peng, University of Macau, Macau, China, UM Hetao IC Research Institute, Shenzhen, China

In Paper 27.5, the University of Macau introduces a 17.9-to-22.4GHz quad-core inverse-Class-F VCO achieving –145.6 to –141dBc/Hz phase noise at a 10MHz and a 194.3-to-196.9dBc/Hz FoM. A compact (~0.15mm²) circular pentafilar transformer tank provides high Q, while multi-path synchronization equalizes impedances and co-locks $f_0/2f_0$.

10:05 AM

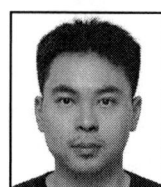

27.6 A 9.7GHz Self-Linearized-VCO-Based FMCW Chirp Generator Achieving 1.56GHz/µs Slope and 0.57µs Duration with 0.094% rms Frequency Error

Daxu Zhang, Institute of Science Tokyo, Tokyo, Japan

In Paper 27.6, Institute of Science Tokyo presents a 9.7GHz self-linearized VCO–based type-III FMCW synthesizer using a C_{Cal} bank and dual LMS calibrations for linearization and fast convergence. Fabricated in 65nm CMOS, it achieves a 0.57µs chirp, 40ns retrace, 1562.8MHz/µs rate, and 1GHz bandwidth with 0.094% rms FM error.

10:30 AM

27.7 A 40.5-to-58.5GHz 36%-Fractional-Chirp-Bandwidth 18GHz-Absolute-Chirp-Bandwidth 2.2GHz/µs-Chirp-Rate 0.02%-Chirp-Error Post-Mixing Bandwidth-Extending Sawtooth-FMCW Frequency Synthesizer Employing a Chirp-Tracking ILFT and a Fractional-Bandwidth Doubler

Yi Liu, HKUST, Hong Kong, China

In Paper 27.7, HKUST reveals an 18GHz-bandwidth sawtooth-FMCW synthesizer employing a chirp-tracking ILFT with feed-forward and frequency-tracking loops for robust fast-chirp tracking and an FBW doubler for sawtooth generation, bandwidth extension, and halved chirp error. It achieves a 36% FBW chirp from 40.5 to 58.5GHz with 0.02% error at 2.2GHz/µs.

10:55 AM

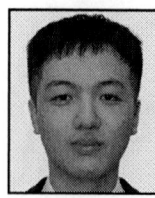

27.8 A 77GHz 8-bit CMOS Phase Shifter Adopting a Nested-Vector-Based Error Correction with 0.33°/0.07dB RMS-Error for MIMO Radar Applications

Geonho Park, Samsung Electronics, Hwaseong, Korea

In Paper 27.8, Samsung Electronics introduces an 8-bit nested VSPS operating from 71 to 80GHz, achieving 0.33° rms phase and 0.07dB rms gain errors using peak error smoothing (PES). With two sub-VSPSs each containing an I/Q VGA and buffer, the design doubles phase resolution over prior work while occupying only 0.17mm².

11:20 AM

27.9 A Dual-Mode DCO-PA with a Twisted 8-Shape Inductor for BLE Achieving 42% TX Efficiency at 1.6dBm and 0.29mW RX Clock

Jiawen Chen, University College Dublin, Dublin, Ireland, King's College London, London, United Kingdom

In Paper 27.9, University College Dublin presents a 0.3V dual-mode (2.4/4.8GHz) DCO-PA that achieves 1.6dBm output with 41% efficiency in TX mode and 0.29mW in RX clock mode. A twisted 8-shaped inductor enables octave switching, Class-D-like operation boosts efficiency to 75%, and a tail transistor lowers power and flicker noise (~40 kHz).

11:35 AM

27.10 A 48-to-82.5GHz CMOS Split-Tail Gilbert-Cell Frequency Doubler Achieving 11% PAE at 8.5dBm Output Power

Jismal Jamal, University of Pavia, Pavia, Italy

In Paper 27.10, the University of Pavia, Fondazione Chips-IT, and STMicroelectronics reveal a 48-to-82.5GHz frequency doubler in 28nm FDSOI CMOS that uses a modified Gilbert cell with split tail transconductors and capacitor-shorted switching quads. The resulting DC offset reduces duty cycle, boosting power conversion gain and efficiency. It delivers 8.5dBm output at 11% PAE.

27

DIGEST OF TECHNICAL PAPERS • 465

979-8-3315-8937-0/26 $31.00 © 2026 IEEE

ISSCC 2026 / SESSION 27 / FREQUENCY GENERATORS, MULTIPLIERS AND MODULATORS / 27.1

27.1 A 20GHz Frequency Synthesizer with Spur-Shaping Modulator Achieving 46.2fs Jitter and -76.5dBc Worst-Case Fractional Spur

Zonglin Ye, Yuxuan Sun, Longxiang Hou, Yuhan Ding, Yixuan Wen, Hongyang Zhang, Xinlin Geng, Qian Xie, Shiheng Yang, Zheng Wang

University of Electronic Science and Technology of China, Chengdu, China

Abstract

In this paper, a frequency synthesizer is proposed with low fractional spur and low jitter, featuring a spur-shaping-modulator (SSM) technique that suppresses the nonlinearity arising from inter-slice mismatch. Implemented in 28nm CMOS, the prototype synthesizer is measured to achieve a 46.2fs integrated jitter, a –84.1dBc near-integer fractional spur, and a -76.5dBc worst-case fractional spur with a –251.4dB FoM and –102.6dB FoM_{IBS}.

A low-jitter low-spur fractional-N frequency synthesizer is crucial in high-speed transceivers. Since the jitter contribution of the quantization noise from the delta-sigma modulator (DSM) has become significant in low-jitter frequency synthesizers, numerous quantization-noise reduction techniques have recently been investigated to mitigate this issue [1–11]. Using a digital-to-time converter (DTC) is a popular solution as it could generate a fine-resolution delay to cancel the quantization error [5–11]. Nevertheless, accurate operation of the DTC necessitates a calibration. A multi-path feedback approach has been proposed to cancel the quantization noise using current summation, eliminating calibration [2]. However, the additional nonlinearity induced by these quantization-noise-reduction techniques introduce unwanted fractional spur [12,13], making it a key design challenge in low-jitter fractional-N frequency synthesizers.

This work introduces a spur-shaping-modulator (SSM) technique that suppresses the nonlinearity arising from inter-slice mismatch in the multi-path-based PLL, thereby realizing ultra-low fractional spurs in near-integer channels. With the proposed SSM, the prototype synthesizer is measured to achieve a 46.2fs integrated jitter, a –84.1dBc near-integer fractional spur, and a -76.5dBc worst-case fractional spur with –251.4dB FoM and –102.6-dB FoM_{IBS} in 28nm CMOS.

A comparison between the nonlinearity generation mechanisms in conventional and multi-path CPs is depicted in Fig. 27.1.1. In conventional CPs, as all CP slices share the same control signal, the parallel slices behave collectively as a single slice with proportionally larger current, so inter-slice mismatch does not introduce additional nonlinearity. As a result, due to the large quantization-error range, the intrinsic nonlinearity of the CP dominates the spur performance. In a multi-path CP, each slice receives an independent control signal, and inter-slice mismatch is modulated by the instantaneous phase error of each slice, thereby translating into nonlinearity. Because the quantization-error range is narrower, the nonlinearity induced by this mismatch, denoted as B in (E 1.1), dominates the overall nonlinearity. To solve the problem of the inter-slice mismatch in a multi-path CP, the spur-shaping modulator is introduced, and its principle is illustrated in the bottom of Fig. 27.1.1. The core idea of the SSM is in using a control strategy that can shape the spectrum of B to be low-stop, so that the low-frequency fractional spur could be suppressed. As the integral operation in time domain corresponds to a 1/s transfer function, if we can make the integral of B, denoted as Q, bounded, the spectrum of B will be low-stop shaped. Therefore, the key to the spectrum shaping rests with the way to keep Q bounded.

As Q is tightly related to the specific mismatch of all slices, controlling it is difficult. From another perspective, Q can be decomposed into a sum of the products of A_n and $I_{mis,n}$, where $I_{mis,n}$ is the CP mismatch, and A_n is a coefficient defined in (E 1.4). Therefore, if A_n can be kept bounded, so can Q. Fortunately, A_n is decoupled from the specific mismatch of the slice and only relates to the modulation pattern. Therefore, controlling A_n is technically easier than controlling Q. The strategy for the proposed SSM that ensures bounded A_n is depicted in the bottom-right of Fig. 27.1.1. The key insight is that the additional degrees of freedom exist in selecting which slices receive the modulation, allowing A_n to be adjusted without compromising quantization-noise cancellation. For example, if one slice should be modulated by +1 in a 3-slice configuration, then the choice of the slice could be slice 1, slice 2, or slice 3. Modulating slice x increases A_x by 2/3 and decreases the others by 1/3. The SSM, therefore, selects the slice with the smallest A_x as the "modulated" slice to increase the corresponding A_x, thereby balancing A_n. Generally, the SSM chooses the L slices with the smallest A_x to be modulated, in the cases where L slices should be modulated. A time-domain simulation confirms that the SSM effectively bounds A_n as shown in the bottom of Fig. 27.1.1.

The implementation of the digital subsystem, consisting of the DSM and the SSM in this work, is illustrated in Fig. 27.1.2. The DSM with first-order noise shaping is employed to control the divider. The residue for the DSM is quantized by a multi-stage noise-shaping (MASH) 1-1 modulator to generate a number of PFD slices with an additional one-VCO-period delay modulation, denoted as CP_{bin}. The SSM then reads CP_{bin} and determines which specific slice is driven by an additional delay. In the SSM, a counter array tracks the number of times that each PFD input is delayed. Whenever a PFD input is given an extra delay, its corresponding counter is incremented. To prevent overflow, a global decrement-by-one is applied once all counters exceed zero. After the SSM reads CP_{bin}, it sorts the counter array

and enables the delay switches of PFDs with the CP_{bin} smallest counter value. To verify the effectiveness of the SSM, a simulation is conducted using System Verilog. The 8-slice CP is modeled with 1% mismatch and 1% intrinsic nonlinearity. The simulated spectrum is shown in the bottom of Fig. 27.1.2. In the cases when SSM is disabled, the nonlinearity is dominated by the inter-slice mismatch, therefore, resulting in a spur level of -60dBc. When the SSM is enabled, it modulates the mismatch and spurs are reduced to -86dBc. In addition, the mismatch-free case with the SSM disabled is also simulated, and the result shows the same spur level with the case that SSM is enabled, and mismatch exists. It indicates that a -86dBc spur is contributed by the intrinsic CP nonlinearity and, therefore, proves that the spur contribution from the mismatch is effectively suppressed by the SSM.

Figure 27.1.3 shows the overall architecture of the SSM-based harmonic-mixing (HM) synthesizer. The synthesizer contains two stages of the PLL where the stage I is a sampling PLL and the stage II is an HM PLL with multi-path PFDs. The gain of first stage sampling phase detector is 4-times configurable to support a flexible bandwidth. Moreover, the VCO in stage I covers a frequency band from 6.5 to 7GHz and is implemented in a Class-D topology for its low-power consumption. The output of stage I is divided into two paths. The first path is connected to a divider controlled by the DSM for the fractional-N operation. The output of the divider is two-stage retimed for the quantization-error cancellation using multipath PFDs. The second path subsamples the output of stage-II VCO, whose tuning range is 1.1GHz at a 20GHz center frequency to perform down-conversion. The intermediate frequency (IF) is extracted using an inverter-based band-pass filter. In the middle stages of the filter, a series capacitance, which removes the DC offset caused by the sub-sampler and first inverter, is added to improve robustness against process, voltage, and temperature (PVT) variations. The loop filter is implemented on-chip with a 6-times reconfigurable bandwidth.

The measurement results for near-integer fractional spurs with a frequency-control word (FCW) = 2^{-16} is depicted in Fig. 27.1.4. When the SSM is disabled, the spur is measured to be -71.02dBc, and when the SSM is enabled, the spur is reduced to be below the noise floor due to the 1Hz RBW limitation of the instrument. Figure 27.1.5 shows the measurement results for phase noise for stage I and stage II, where the dividing ratio is N_1=27 for stage I and N_2=67 for stage II. The jitter of stage I is measured to be 41.58fs. The jitter of the final output is measured to be 46.27fs with FCW = 2^{-16}. The maximum near-integer spur is marked in the bottom of Fig. 27.1.5 at 3.074kHz, which is 6.75GHz / 67 / $2^{16} \times 2$, at a level of -84.07dBc. Comparing with the configuration without the SSM, the maximum near-integer spur is reduced by 13.05dB (from -71.02 to -84.07dBc). The fractional spurs are also measured across different offsets, where the worst spur is -76.5dBc.

Figure 27.1.6 gives a comparison table between this work and prior-art low-jitter low-near-integer-spur synthesizers. This work achieves lower FoM_{IBS}, which is the near-integer spur level normalized to a 1GHz carrier [1], in the comparison table with sub-50fs jitter. The proposed synthesizer helps to move the design towards a more favorable FoM_{IBS}-jitter frontier (Fig. 27.1.7). As shown in Fig. 27.1.7, the 28nm-CMOS prototype consumes 7.01mW in stage I and 27.02mW in stage II, with their respective areas being 0.089mm^2 and 0.118mm^2.

Acknowledgement:

This work was supported in part by the National Key Research and Development Program of China under Grant 2023YFB4403900 and in part by the National Natural Science Foundation of China under Grant 62525404, 62034002 and 62374026. Corresponding authors: Qian Xie and Zheng Wang.

References:

[1] M. Kennedy et al., "A 45.5fs-Integrated-Random-Jitter and -75dBc-Integer-Boundary-Spur BiCMOS Fractional-N PLL with Suppression of Fractional, Horn, and Wandering Spurs," *ISSCC*, pp. 164-166, Feb. 2024.
https://doi.org/10.1109/ISSCC49657.2024.10454462

[2] C. Hung et al., "A Fractional-N PLL with 34fsrms Jitter and -255.5dB FoM Based on a Multipath Feedback Technique,"' *ISSCC*, pp. 328-329, Feb. 2025.
https://doi.org/10.1109/ISSCC49661.2025.10904550

[3] D. Yang et al., "A Harmonic-Mixing PLL Architecture for Millimeter-Wave Application," *IEEE JSSC*, vol. 57, no. 12, pp. 3552-3566, Dec. 2022. https://doi.org/10.1109/JSSC.2022.3209614

[4] H. Zhang et al., "A 96fs$_{rms}$-Jitter, -70.6dBc-Fractional-Spur Cascaded PLL Employing Two MMDs with Shared DSM for Quantization Noise Cancellation," *ISSCC*, pp. 326-329, Feb. 2025. https://doi.org/10.1109/ISSCC49661.2025.10904516

[5] H. Li et al., "A 27GHz Fractional-N Sub-Sampling PLL Achieving 57.9fs$_{rms}$ Jitter, -249.7dB FoM, and 1.98μs Locking Time Using a Polarity-Reversible SSPD," *ISSCC*, pp. 336-339, Feb. 2025. https://doi.org/10.1109/ISSCC49661.2025.10904556

[6] S. M. Dartizio et al., "A Low-Spur and Low-Jitter Fractional- N Digital PLL Based on an Inverse-Constant-Slope DTC and FCW Subtractive Dithering," *IEEE JSSC*, vol. 58, no. 12, pp. 3320–3337, Dec. 2023. https://doi.org/10.1109/JSSC.2023.3311681

[7] H. Liu et al., "A 0.18-μs-Locking-Time Fractional-N PLL with Stochastic Gradient Descent Tuning Curve Fitting, Initial Phase Error Zeroing, and Random DSM Achieving 44.4-fs Jitter at Near-Integer Channel," *IEEE CICC*, pp. 1-3, Apr. 2025. https://doi.org/10.1109/CICC63670.2025.10982962

[8] Y. Shin et al., "A 76 fs$_{rms}$-Jitter and -65dBc- Fractional-Spur Fractional-N Sampling PLL Using a Nonlinearity-Replication Technique," *ISSCC*, pp. 196-198, Feb. 2024. https://doi.org/10.1109/ISSCC49657.2024.10454557

[9] G. Castoro et al., "A 9.25GHz Digital PLL with Fractional-Spur Cancellation Based on a Multi-DTC Topology," *ISSCC*, pp. 82-84, Feb. 2023. https://doi.org/10.1109/ISSCC42615.2023.1006735

[10] M. Rossoni et al., "An 8.75GHz Fractional-N Digital PLL with a Reverse-Concavity Variable-Slope DTC Achieving 57.3fs$_{rms}$ Integrated Jitter and –252.4dB FoM," *ISSCC*, pp. 188-190, Feb. 2024. https://doi.org/10.1109/ISSCC49657.2024.10454388

[11] Y. Liu et al., "A 37.5fs-rms Jitter and –254.1dB FoM Fractional-N Sampling PLL with Reference-Phase-Selection and Complementary-DTC Achieving 8× DTC Range Reduction and Zero DTC Delay Offset," *IEEE CICC*, pp. 1-3, Apr. 2025. https://doi.org/10.1109/CICC63670.2025.10983724

[12] Y. Donnelly et al., "Prediction of Phase Noise and Spurs in a Nonlinear Fractional-N Frequency Synthesizer," *IEEE TCAS-I*, vol. 66, no. 11, pp. 4108–4121, Nov. 2019. https://doi.org/10.1109/TCSI.2019.2925181

[13] X. Wang and M. Peter Kennedy, "Spurs in Fractional-N Frequency Synthesizers Resulting from Resolution Mismatch Between the Divider Controller and the DTC: Manifestations, Analysis, and Mitigation," *IEEE TCAS-I*, vol. 72, no. 7, pp. 2998-3011, July 2025. https://doi.org/10.1109/TCSI.2025.3557258

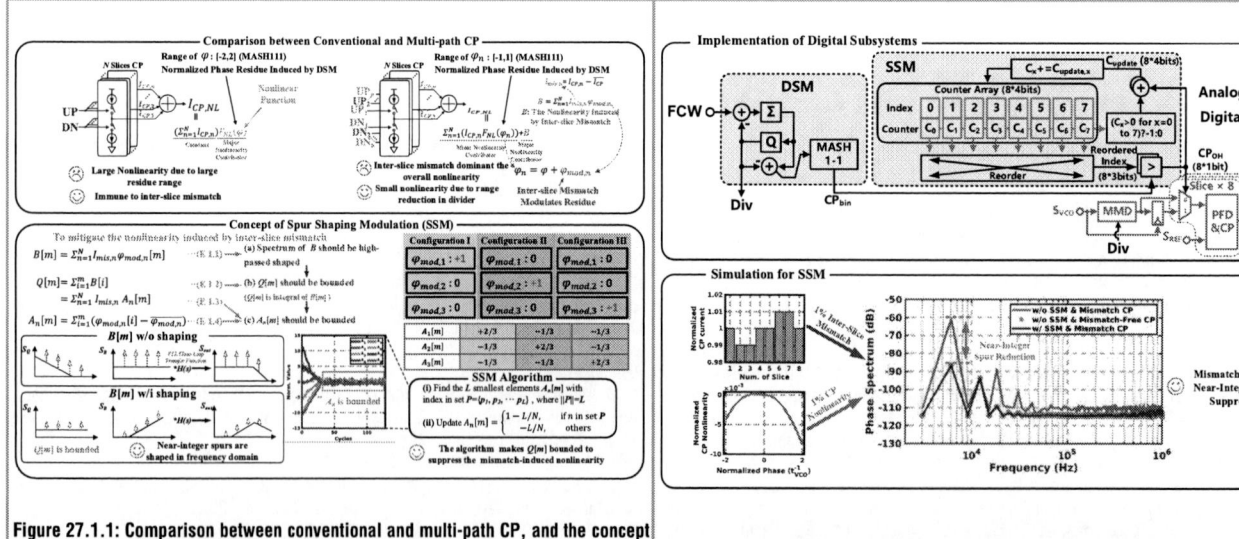

Figure 27.1.1: Comparison between conventional and multi-path CP, and the concept of spur-shaping modulator (SSM).

Figure 27.1.2: Implementation of digital subsystems and simulation of the SSM.

ISSCC 2026 / SESSION 27 / FREQUENCY GENERATORS, MULTIPLIERS AND MODULATORS / 27.1

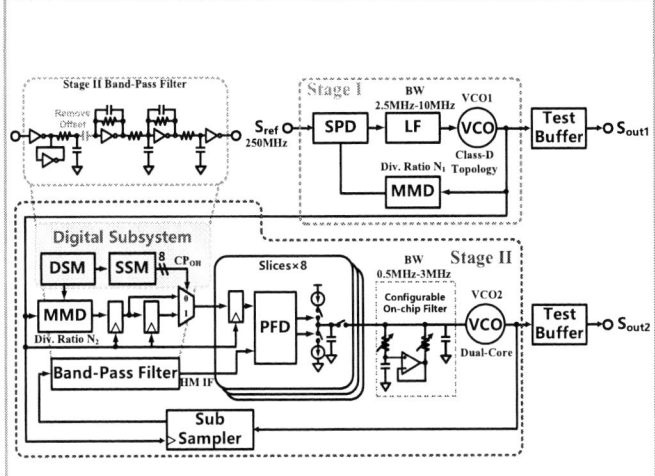

Figure 27.1.3: Overall architecture of the proposed SSM-based harmonic-mixing (HM) synthesizer.

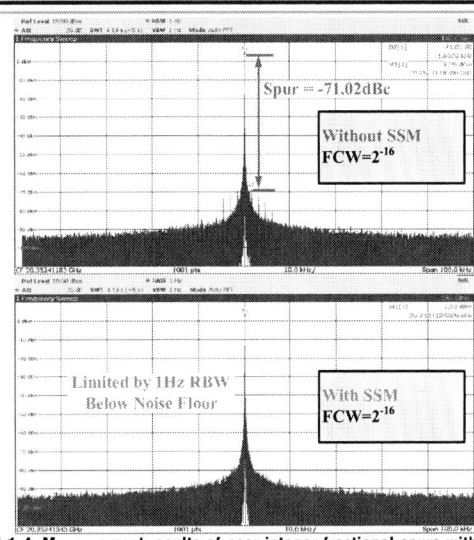

Figure 27.1.4: Measurement results of near-integer fractional spurs with frequency-control word (FCW) = 2^{-16}.

Figure 27.1.5: Measurement results of phase noise and near-integer fractional spur for stage I and stage II.

Figure 27.1.6: Comparison table of this work and prior-art low-jitter low-near-integer-spur synthesizers.

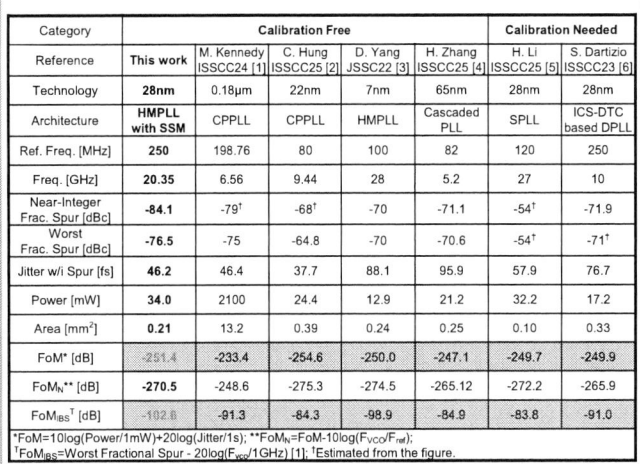

Category	Calibration Free					Calibration Needed	
Reference	This work	M. Kennedy ISSCC24 [1]	C. Hung ISSCC25 [2]	D. Yang JSSC22 [3]	H. Zhang ISSCC25 [4]	H. Li ISSCC25 [5]	S. Dartizio ISSCC23 [6]
Technology	28nm	0.18µm	22nm	7nm	65nm	28nm	28nm
Architecture	HMPLL with SSM	CPPLL	CPPLL	HMPLL	Cascaded PLL	SPLL	ICS-DTC based DPLL
Ref. Freq. [MHz]	250	198.76	80	100	82	120	250
Freq. [GHz]	20.35	6.56	9.44	28	5.2	27	10
Near-Integer Frac. Spur [dBc]	-84.1	-79[†]	-68[†]	-70	-71.1	-54[†]	-71.9
Worst Frac. Spur [dBc]	-76.5	-75	-64.8	-70	-70.6	-54[†]	-71[†]
Jitter w/i Spur [fs]	46.2	46.4	37.7	88.1	95.9	57.9	76.7
Power [mW]	34.0	2100	24.4	12.9	21.2	32.2	17.2
Area [mm²]	0.21	13.2	0.39	0.24	0.25	0.10	0.33
FoM* [dB]	-251.4	-233.4	-254.6	-250.0	-247.1	-249.7	-249.9
FoMₙ,** [dB]	-270.5	-248.6	-275.3	-274.5	-265.12	-272.2	-265.9
FoMᵢᵦₛ[†] [dB]	-102.8	-91.3	-84.3	-98.9	-84.9	-83.8	-91.0

*FoM=10log(Power/1mW)+20log(Jitter/1s); **FoMₙ=FoM-10log(F_{VCO}/F_{ref});
[†]FoMᵢᵦₛ=Worst Fractional Spur - 20log(F_{VCO}/1GHz) [1]; [†]Estimated from the figure.

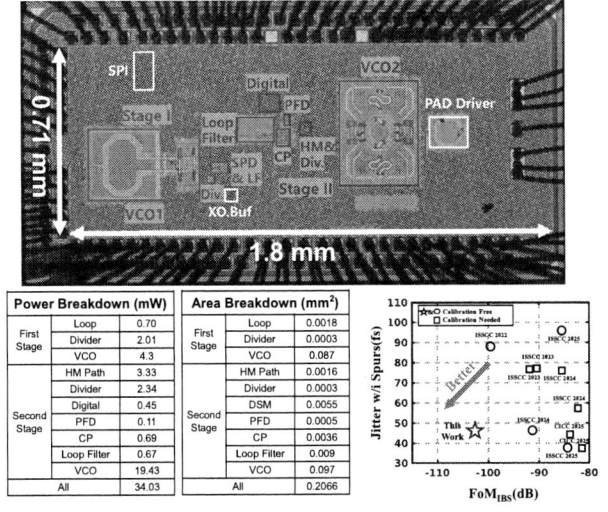

Figure 27.1.7: Die micrograph, power breakdown table, area breakdown table, and performance benchmarking against prior art.

• 2026 IEEE International Solid-State Circuits Conference

ISSCC 2026 / SESSION 27 / FREQUENCY GENERATORS, MULTIPLIERS AND MODULATORS / 27.2

27.2 A 157fs$_{rms}$-Jitter, −73dBc-Fractional-Spur, Calibration-Free Cascaded SPLL Employing Robust Feedforward Noise Cancellation and MMD-Based Quantization-Error Cancellation with a 60MHz Reference

Haoming Zhang, Yongjun He, Yuyang Zhu, Huanyu Ren, Tetsuya Iizuka

University of Tokyo, Tokyo, Japan

Abstract

This work presents a 6.2-to-6.8GHz calibration-free fractional-N sampling PLL with cascaded architecture, achieving 157fs$_{rms}$ jitter and a −73dBc fractional spur using feedforward noise cancellation (FFNC) and MMD-based quantization-error cancellation (QEC). The QEC reduces Q-error by N, enabling a low-noise SPD for fractional-N operation. With both stages as SPLLs, the FFNC-path gain and the original-path gain inherently match regardless of division-ratio changes, ensuring robust cancellation of the first-stage VCO noise.

Sampling PLLs (SPLLs) achieve low-jitter frequency synthesis thanks to the high gain of the sampling phase detector (SPD), which enables low in-band phase noise (PN). However, the narrow linear range of the SPD—typically only tens of ps—prevents direct fractional-N operation, since the delta-sigma modulator (DSM) induces quantization error (Q-error) of several hundred ps at the SPD input, far exceeding its linear range. To overcome this, a digital-to-time converter (DTC) is usually employed for Q-error cancellation (QEC) in front of the SPD [1,2] as shown in Fig. 27.2.1(a). Yet, a DTC requires complex calibration and suffers from its intrinsic nonlinearity, potentially leading to degraded in-band PN and fractional spurs. A DTC-free technique, named "MMD-based QEC," was proposed in [3,4]. As shown in Fig. 27.2.1(b), by using two MMDs driven by a shared DSM, the Q-error is reduced by a factor of the main-PLL division ratio N, immune to PVT variations. For example, with a 3rd-order DSM, $N = 20$, and $f_{out} = 5$GHz, the Q-error shrinks from 800 ps ($4T_{vco}$) to 40 ps, falling within the linear range of an SPD. Therefore, if the SPD is employed in the main loop in place of the PFD/CP, which has dominated the final in-band PN in [3], the structure becomes a strong candidate to leverage the advantages of the SPD without using a DTC. However, this method requires an auxiliary (AUX) PLL to generate a high frequency to perform the QEC. Although it has been demonstrated in [4] that, implemented as a wide-bandwidth (BW) SPLL using an LC oscillator (LCO), the jitter–power product of the extra AUX stage plus one MMD is comparable to that of a DTC, the extra LCO still doubles the chip area and raises concerns about magnetic coupling. Using a ring oscillator (RO) could mitigate these issues thanks to its compact, inductorless feature, but prior-art ROs exhibit ~160 dB FoM [5,6]—about 30dB worse than LCOs [7,8]—necessitating either higher f_{ref} to extend the AUX PLL BW or higher RO power to compensate for this gap, undermining the benefit of calibration-free QEC. As a potential solution, a technique called "feedforward noise cancellation (FFNC)" has recently been proposed to suppress the AUX VCO PN in cascaded PLL structures [9,10]. By adding the first-stage PD output to that of the second stage, the FFNC creates another path that cancels the AUX VCO noise at the main PD output when its gain matches the inter-stage gain, as illustrated in Fig. 27.2.1(c). As analyzed in [9], the FFNC provides an additional high-pass filtering effect with a cut-off frequency of $f_{ref}/2$ for the AUX VCO PN, without affecting the characteristics of the loops or increasing the noise contribution of other components. Therefore, as illustrated in Fig. 27.2.1(c), this work proposes a fractional-N SPLL free from calibration-intensive DTC schemes by employing FFNC with MMD-based QEC. With a low f_{ref}, a balanced jitter, power, and chip area can be achieved using only one LCO despite the cascaded structure.

Although proven effective, the FFNC architecture proposed in [9] has potential issues. As illustrated in Fig. 27.2.2(a), a gain mismatch between the two paths leads to degraded low-frequency noise filtering and compromises the FFNC, making it essential to minimize the gain error. In the structure presented in [9] (Fig. 27.2.2(b)), the gains of the two paths—shown in blue (harmonic mixer (HM), XOR-PD) and red (N_{int}, SPD, feedforward amplifier (FFAMP))—should be matched in the ideal case, but two issues arise in practice: first, the SPD gain—i.e., the slew rate SR—is PVT sensitive, which may dominate the gain error and degrade the FFNC [9]; second, although PVT unrelated, N_{int} in the FFNC path is not constant across different frequency setting, appearing as another source of gain error. Such an error could in principle be compensated either by replacing the fixed-gain FFAMP with a programmable variable-gain amplifier, which increases design complexity in both the overall structure and the amplifier itself, or by adopting a digital-domain feedforward approach [10], where the A/D conversion may add further PVT sensitivity. Interestingly, the above two issues are naturally solved in the proposed PLL illustrated in Fig. 27.2.2(c), where both stages are implemented as an SPLL. In this work, the two SPDs share an identical structure, so their SR can be assumed to be equal, reducing the PVT-induced gain error. More importantly, the gains of the red and blue paths both equal SR/f_{aux} regardless of N_{int} or N_{aux}, thereby eliminating the division-ratio-related gain error. The key difference lies in the phase-detection mechanism: the SPD detects the phase error in the time domain and converts it into a voltage ramp, giving a constant gain of SR in the time domain, and thus a phase-domain gain of $SR/2\pi f_{spd}$, where f_{spd} is the SPD input frequency. As a result, the gains of the AUX and main SPDs include f_{ref} and f_{main} in the denominator, compensating for the changes in N_{int} and N_{aux}. In contrast, the XOR-PD and PFD/CP do not have this characteristic as they detect the phase error directly in the phase domain, yielding a constant gain in the phase domain (e.g., VDD/π or $I_{cp}/2\pi$). This indicates that the magnitude of the VCO PN passing through a divider/SPD cascade is independent of the division ratio, whereas in a

divider/XOR-PD case it depends on the division ratio. To verify this, a circuit-level simulation is performed with the circuit in Fig. 27.2.2(d), where the PN of only the AUX VCO is considered, and N_{aux} on the inter-stage path is set to different values. Both f_{out} and the main PLL BW are kept the same by simultaneously varying N_{main} in the main loop along with N_{aux}. The results show that the proposed FFNC remains effective regardless of the value of N_{aux} when both PLLs employ the SPD. In contrast, Fig. 27.2.2(e) employs an XOR-PD for the second stage, where the FFNC—effective in the black case with matched path gains—becomes invalid when N_{aux} is 1.5×(blue) or 2× higher(red). Consequently, in the proposed PLL, the gains of the two paths inherently match regardless of the division ratios, eliminating the need for a tunable gain element in the FFNC path. Instead, the FFNC path can be implemented with a unity-gain buffer (UGB) in the proposed PLL, yielding a simple cascaded PLL structure.

The circuit implementation of the whole PLL is shown in Fig. 27.2.3. Both the AUX and main PLLs employ a type-II SPLL structure. Implemented with a three-stage RO, the AUX VCO has its PN suppressed by the FFNC realized through feeding the AUX SPD output to the main SPD output via a UGB and resistive summation. The UGB is implemented using a two-stage Miller-compensated operational transconductance amplifier (OTA). The OTA achieves over 80° phase margin and a unity-gain bandwidth of 400MHz, which is much wider than the FFNC filtering bandwidth, so the UGB gain can be treated as a constant in the noise analysis. The AUX and main SPDs, together with the subsequent G_M cells, adopt identical designs and share the same bias voltages to alleviate the PVT-related impact. An MMD sharing the same DSM with that in the main loop is inserted at the input of the main SPD, so that the phase fluctuation is reduced by a factor of N_{aux} when N_{main} differs from N_{aux} by one [4]. A divide-by-2 is inserted in the main feedback path to halve the main MMD input frequency and thus the AUX MMD input frequency—since the two are nearly identical when MMD-QEC is applied [4]—so as to reduce the RO frequency to ~3.5GHz, alleviating its design complexity as higher-frequency RO are relatively difficult to implement and generally exhibit poorer FoM. The ~7GHz main VCO is realized using a Class-B explicit second-order common-mode resonance topology [7].

The proposed PLL was implemented in a 65nm CMOS process, occupying an active area of 0.16mm², as shown in Fig. 27.2.7(left). In contrast to prior cascaded PLLs [3,9,11], where the AUX LCO nearly doubled the chip area, the RO in this work accounts for only 4% of the area while consuming 34% of the total power. As the BW of the FFNC filtering equals $f_{ref}/2$, the RO power scales directly with f_{ref} and could be further reduced if f_{ref} were increased. Despite this, a lower f_{ref} of 60MHz is adopted, accepting the modest RO power overhead for practical considerations. The output of the PLL is measured after the on-chip divide-by-4. Figure 27.2.4(a) shows the measured PN in the integer-N mode with the FFNC on or off near 6.4GHz. When the FFNC is off, the RO PN dominates the overall PN, resulting in 358fs jitter. When the FFNC is on, the RO PN is filtered, and the jitter decreases to 145fs. Figure 27.2.4(b) shows the PN in the near-integer fractional-N mode with the QEC enabled and the FFNC either on or off, demonstrating that the FFNC remains effective in the fractional-N operation. With both the QEC and FFNC on, the jitter is 157fs with in-band fractional spurs. Figure 27.2.4(c) shows the measured PN at another fractional-N mode near 6.7GHz. Note that, in this configuration, N_{aux} and N_{main} are changed from ~30 to ~20, while no gain error arises from this change in the proposed architecture; thus, the FFNC remains effective, improving the jitter from 442 to 160fs. To demonstrate the effectiveness of the QEC, it is disabled in Fig. 27.2.4(d), where the PN is significantly degraded by the Q-noise folding caused by the SPD nonlinearity. To strictly prove the robustness of the FFNC, the same verification performed in Fig. 27.2.2(d) is demonstrated with measurement in the integer-N mode, with measured PN profiles shown in Fig. 27.2.5(a). As the value of N_{aux} is increased by a factor of 1.5 or 2, which yields a 30% or 50% variation in the divider gain, the measured PN remains nearly unchanged, consistent with simulation results in Fig. 27.2.2(d), thus confirming that the FFNC gain is independent of the division ratio and robust in all frequency settings. The blue line in Fig. 27.2.5(b) shows the measured in-band fractional-spur levels across near-integer channels around 6.4GHz. Among them, the two worst-case spectra are shown in Figs. 27.2.5(c) and (d), with the spur levels of −72.7dBc and −72.8dBc, respectively, normalized to the output frequency. The red line in Fig. 27.2.5(b) presents the measured jitter at these channels, indicating that the fractional spurs are sufficiently low such that the jitter remains almost unchanged, regardless of whether the

---

468 • 2026 IEEE International Solid-State Circuits Conference

979-8-3315-8937-0/26 $31.00 © 2026 IEEE

spurs fall in-band or out-of-band, and are thus negligible even with the narrow-range SPD. Figure 27.2.6 compares this work with recent low-jitter fractional-N PLLs. Despite its cascaded architecture, the proposed PLL employs only one LCO while maintaining power and area comparable to single-loop architectures. At the same time, it achieves competitively low spur and jitter by leveraging robust, calibration-free cancellation techniques of the RO noise and the quantization noise, using a practical 60MHz reference.

Acknowledgement:
This work was supported in part by Japan Science and Technology Agency (JST)-CRONOS Grant Number JPMJCS24K1 and in part by Murata Science and Education Foundation Grant Number M24AN001. The chip design in this study was supported through the activities of the VLSI Design and Education Center (VDEC), The University of Tokyo, in collaboration with Cadence Design Systems, Inc., Nihon Synopsys G.K., and Siemens Electronic Design Automation Japan K.K.

References:
[1] W. Wu et al., "A 14-nm Ultra-Low Jitter Fractional-N PLL Using a DTC Range Reduction Technique and a Reconfigurable Dual-Core VCO," *IEEE JSSC*, vol. 56, no. 12, pp. 3756-3767, Dec. 2021. http://doi.org/10.1109/JSSC.2021.3111134
[2] G. Jin et al., "A Fractional-N Sampling PLL with a Merged Constant-Slope DTC and Sampling PD," *IEEE JSSC*, vol. 59, no. 8, pp. 2407-2417, Aug. 2024. http://doi.org/10.1109/JSSC.2024.3358564
[3] H. Zhang et al., "A 96fs$_{rms}$-Jitter, -70.6dBc-Fractional-Spur Cascaded PLL Employing Two MMDs with Shared DSM for Quantization Noise Cancellation," *ISSCC*, pp. 326-327, Feb. 2025. http://doi.org/10.1109/ISSCC49661.2025.10904516
[4] H. Zhang et al., "A Fractional-N Cascaded PLL With MMD-Based Quantization-Error Cancellation," *IEEE JSSC*, vol. 60, no.12, pp. 4543-4556, Dec. 2025. http://doi.org/10.1109/JSSC.2025.3595657
[5] C. Hwang et al., "A Low-Jitter and Low-Fractional-Spur Ring-DCO-Based Fractional-N Digital PLL Using a DTC's Second-/Third-Order Nonlinearity Cancellation and a Probability-Density-Shaping ΔΣM," *IEEE JSSC*, vol. 57, no. 9, pp. 2841-2855, Sep. 2022. http://doi.org/10.1109/JSSC.2022.3141782
[6] H. Park et al., "A 365fs$_{rms}$-Jitter and -63dBc-Fractional Spur 5.3GHz-Ring-DCO-Based Fractional-N DPLL Using a DTC Second/Third- Order Nonlinearity Cancelation and a Probability-Density-Shaping ΔΣM," *ISSCC*, pp. 442-444, Feb. 2021. http://doi.org/10.1109/ISSCC42613.2021.9365798
[7] D. Murphy et al., "Implicit Common-Mode Resonance in LC Oscillators," *IEEE JSSC*, vol. 52, no. 3, pp. 812-821, Mar. 2017. http://doi.org/10.1109/JSSC.2016.2642207
[8] Y. Hu et al., "A Low-Flicker-Noise 30-GHz Class-F$_{23}$ Oscillator in 28-nm CMOS Using Implicit Resonance and Explicit Common-Mode Return Path," *IEEE JSSC*, vol. 53, no. 7, pp. 1977-1987, July 2018. http://doi.org/10.1109/JSSC.2018.2818681
[9] H. Zhang et al., "A Harmonic-Mixer-Based Fractional-N PLL Employing Voltage-Domain Feed-Forward Noise Cancellation," *IEEE JSSC*, vol. 60, no. 8, pp. 2820-2831, Aug. 2025. http://doi.org/10.1109/JSSC.2024.3516139
[10] Y. Duan et al., "A PVT-Robust 5.5GHz Fractional-N Cascaded RO-Based Digital PLL with Voltage-Domain Feedforward Noise Cancellation," *ISSCC*, pp. 324-326, Feb. 2025. http://doi.org/10.1109/ISSCC49661.2025.10904636
[11] D. Yang et al., "A Harmonic-Mixing PLL Architecture for Millimeter-Wave Application," *IEEE JSSC*, vol. 57, no. 12, pp. 3552-3566, Dec. 2022. h ttp://doi.org/10.1109/JSSC.2022.3209614

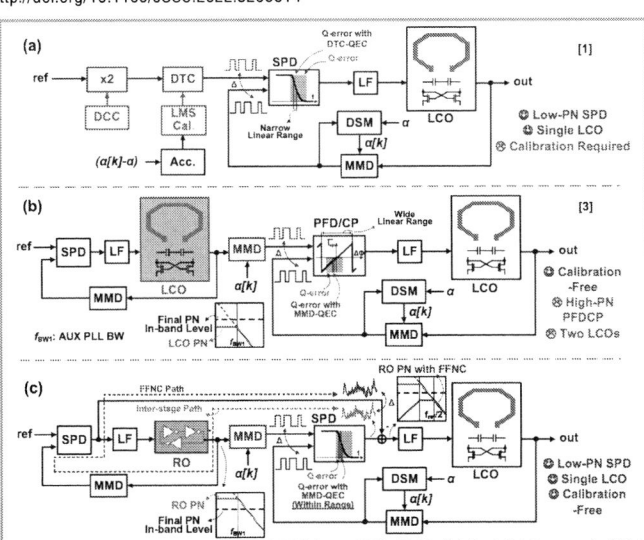

Figure 27.2.1: Block diagrams of (a) DTC-based PLL [1,2], (b) the LC-LC cascaded PLL with MMD-QEC [3], and (c) proposed ring-LC cascaded PLL with MMD-QEC and FFNC.

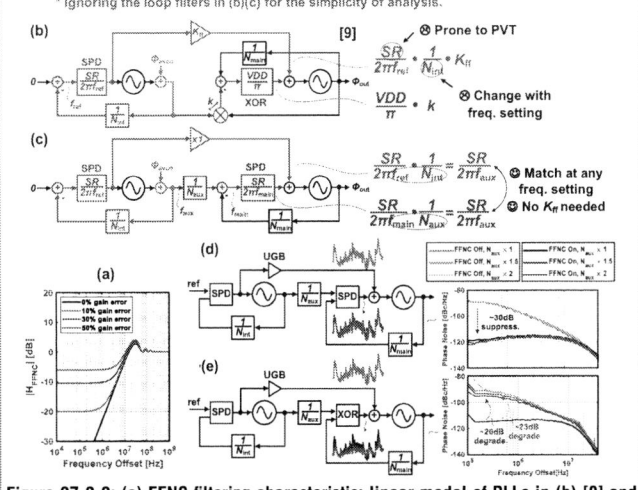

Figure 27.2.2: (a) FFNC filtering characteristic; linear model of PLLs in (b) [9] and (c) this work; a comparison between the FFNC effect using (d) SPD+SPD or (e) SPD+XOR-PD.

ISSCC 2026 / SESSION 27 / FREQUENCY GENERATORS, MULTIPLIERS AND MODULATORS / 27.2

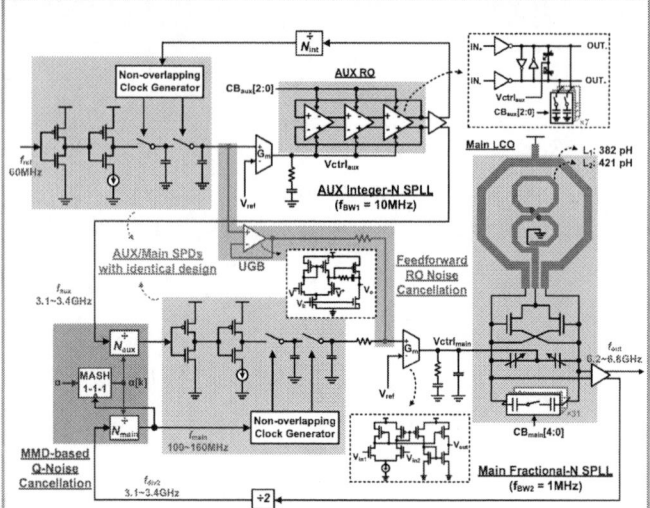

Figure 27.2.3: Circuit implementation of the proposed PLL.

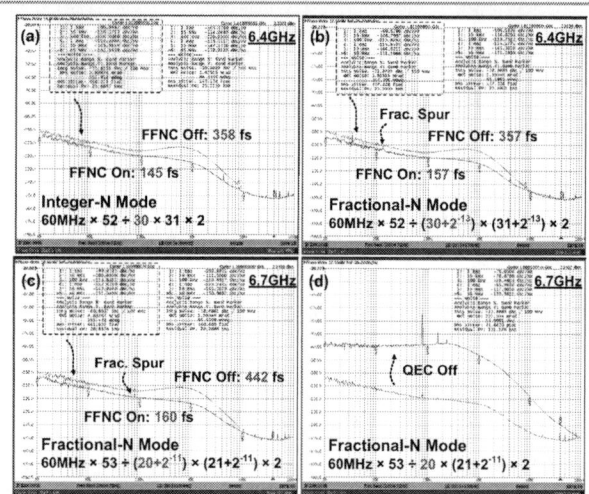

Figure 27.2.4: Measured PN near 6.4GHz in (a) integer-N mode and (b) fractional-N mode with QEC on, FFNC on or off; measured PN near 6.7GHz in (c) fractional-N mode with QEC on, FFNC on or off and (d) fractional-N mode with FFNC on, QEC on or off.

Figure 27.2.5: (a) FFNC robustness verification based on measurement results; (b) measured fractional spur levels at near-integer channels; (c)(d) measured output spectra with fractional spurs.

	This Work	H. Zhang ISSCC'25	D. Yang ISSCC'22	D. Xu JSSC'25	G. Jin JSSC'24	M. Mercandelli JSSC'22	W. Wu JSSC'21
Phase Detector	SPD/SPD	SPD/PFDCP	SPD/XORPD	Pulse Gen. +BBPD	SPD	SPD	SPD
Noise Cancell./ Filter. Technique	MMD-QEC +FFNC	MMD-QEC	Harmonic-mixing	DTC-QEC	DTC-QEC	DTC-QEC	DTC-QEC
Calibration required?	No	No	No	Yes	Yes	Yes	Yes
VCO Type	Ring+LC	LC+LC	LC+LC	Ring+LC	LC	LC	LC
Reference Frequency [MHz]	60	82	74	50	100	500	76.8x2
Output Frequency [GHz]	6.2 to 6.8	4.7 to 5.7	25 to 28	6.5 to 8	3.3 to 4.5	11.9 to 14.1	5 to 7
Integrated Jitter [fs]	157 (1k to 100M)	95.9 (1k to 100M)	88 (10k to 40M)	191 (10k to 10M)	203 (10k-100M)	58.2 (1k-100M)	80~95 (10k to 100M)
In-band Fractional Spur [dBc]	−72.7	−70.6	−70	−52.7	−57	−63.2	<−72
Power [mW]	17.2	21.2	12.9	14.2	2.4	18	14.2
FoM* [dB]	−243.7	−247.1	−250	−242.9	−250.0	−252.1	−250.4 ~ −249.0
FoM$_{REF}$** [dB]	−235.9	−238.0	−241.3	−235.9	−240.0	−235.1	−241.5 ~ −240.1
CMOS Process [nm]	65	65	7	65	40	28	14
Active Area [mm²]	0.16	0.25	0.24	0.48	0.36	0.16	0.31

FoM = 10log((Power/1mW)(Jitter/1s)²)   **FoM$_{ref}$ = 10log((Power/1mW)*(Jitter/1s)²)+10log(f$_{ref}$/10MHz)

Figure 27.2.6: Performance comparison of the proposed PLL with prior-art designs.

Power (mW)		
AUX Loop	0.9	5%
RO	5.8	34%
Main Loop	1.7	10%
DSM	0.4	2%
LCO	8.4	49%
Total	17.2	100%

Area (mm²)		
AUX Loop	0.010	6%
RO	0.006	4%
Main Loop	0.023	14%
DSM	0.003	2%
LCO	0.119	74%
Total	0.162	100%

Figure 27.2.7: Die micrograph (left); measured power consumption and area (right).

• 2026 IEEE International Solid-State Circuits Conference

ISSCC 2026 / SESSION 27 / FREQUENCY GENERATORS, MULTIPLIERS AND MODULATORS / 27.3

27.3 A 14GHz Ring-Based 3rd-Order Fractional-N PLL with 164fs$_{rms}$ Jitter and a 100MHz Reference

Zhaochen Zhu, Qingxuan Lin, Zhiqiang Huang

Hong Kong University of Science and Technology, Guangzhou, China

Abstract

This work presents a 14GHz ring-based 3rd-order fractional-N PLL with 164fs$_{rms}$ jitter and a 100MHz reference. A sub-sampling DLL is cascaded in the type-II PLL output for extra phase-noise and supply-noise suppression. A narrow-pulsed integrator is used in the sub-sampling DLL to reduce integrated jitter. A tri-state polarity-reversible SSPD is employed to reduce the DTC range and bandwidth degradation. Delay calibration is applied to reduce the VCDL delay range and implement the VCDL duty-cycle calibration.

A low-jitter phase-locked loop (PLL) is a critical building block in high-speed wireline transceivers, such as serializer/deserializer (SerDes) systems. Fractional-N operation with fine frequency resolution is required to accommodate different crystal frequencies. Ring-oscillator (RO)-based PLLs are particularly attractive for such applications due to their compact area and ease of integration. However, RO-based PLLs suffer from poor phase noise and supply sensitivity, limiting their use in low-jitter applications. A wide loop bandwidth is needed to suppress voltage-controlled-oscillator (VCO) phase noise up to high offset frequencies. Using a high reference frequency [1,3] increases the loop bandwidth but raises system cost. Injection-locked PLLs [2] provide wide VCO-phase-noise suppression bandwidth. However, direct injection into the VCO significantly degrades the reference spurs due to the rich harmonics in the narrow injection pulse and injection-timing misalignment. Besides, injection locking and the feedback loop attempt to lock the output at different phases, resulting in competition and a slower settling time. To greatly suppress VCO supply sensitivity, higher-order loops such as a type-III PLL [3] can be used to provide a large low-frequency loop gain. However, this approach faces stability issues and needs a smaller loop bandwidth and a low-frequency compensation zero with a large capacitor area to stabilize the loop. Cascading a sub-sampling delay-locked loop (SS-DLL) [4,5] achieves stable and wideband 3rd-order VCO phase-noise suppression, but it is limited to integer-N mode. Besides, the delay range of the voltage-controlled delay line (VCDL) in an SS-DLL needs to be wide enough to cover the time shift from the reference to the PLL output. Furthermore, the SS-DLL employs wide-pulsed integrators, limiting the phase-error-correction speed. Isolating the RO from the supply using a regulator can reduce supply-noise sensitivity but introduces additional voltage drop and nearly doubles the power consumption of the VCO [3]. To reduce jitter and supply sensitivity, this work presents a 14GHz ring-based 3rd-order fractional-N PLL with a cascaded sub-sampling DLL (SS-DLL) achieving 164fs$_{rms}$ jitter integrated from 1kHz to 30MHz while consuming 11.27mW with a 100MHz reference by employing a tri-state polarity-reversible sub-sampling phase detector (TPR-SSPD), a narrow-pulsed integrator, delay calibration, and digital-to-time-converter (DTC) range reduction.

Figure 27.3.1 illustrates the overall architecture of the presented fractional-N PLL with a cascaded SS-DLL. A type-II sampling-PLL architecture is adopted as the 1st stage for its zero steady-state input phase error. The 1st-stage PLL is then cascaded with an SS-DLL to achieve an additional order of VCO phase-noise and supply-noise suppression. Thanks to the 1st-order frequency response of the SS-DLL, a wide loop bandwidth can be achieved without stability concerns. To achieve fraction-N operation, the division ratio of the multi-modulus divider (MMD) in the 1st-stage PLL is dithered by a 1st-order delta-sigma modulator (DSM). A DTC (DTC1) is used to cancel out the DSM quantization noise. The SS-DLL comprises a TPR-SSPD, a narrow-pulsed integrator, and a VCDL. The TPR-SSPD samples and converts the phase error between the DTC and the VCDL output into voltage. The converted result is then fed into the integrator, which adjusts the VCDL delay to compensate the phase error and realign the DLL output with the DTC1. To reduce the DTC phase-noise contribution, a DTC range-reduction technique is adopted by switching the rising and falling edges of the VCO to halve the delay range of the DTC1 from Tvco to Tvco/2 [6]. Both the PLL and the SS-DLL must support such edge switching. In the PLL, the MMD output is split into two paths, re-timed by DFFs driven by VCO$_P$ and VCO$_N$, with one path delayed by Tvco/2 to meet setup and hold time requirements. The sign bit of the DSM accumulated quantization (Q$_E$) is used as a selection signal (SEL) to select different retiming edges, while the Q$_E$ residue (φ_e) after correction by a calibration engine is used to control the DTC1. During a steady state of the SS-DLL, the VCDL delay Δt_{vcdl} equals to the PLL delay Δt_{pll}. To reduce the VCDL delay-range requirement, another DTC (DTC2) is inserted before the sampling phase detector (SPD), introducing an additional delay Δt_{DTC2}. The DTC2 consists of a tree-like coarse DTC [7] for wide tuning range and a fine DTC for high resolution tuning. Since the noise of the DTC2 is high-pass filtered by the SS-DLL, it can be sized down to save power. To correct the DTC gain error and compensate the VCO and VCDL duty-cycle errors, the DTC gain [8], VCO and VCDL duty-cycle calibration engine is used. To ensure correct operation of the VCDL duty-cycle calibration (DCC), a delay calibration is used to adjust the DTC2 delay to track the VCDL delay under PVT variation. As such, the VCDL delay range does not need to cover the PLL delay variation and can therefore be minimized to reduce jitter and power consumption.

Figure 27.3.2 shows the details of the differential SS-DLL. As the DTC1 delay is toggled by half the VCO period after SEL transitions, the TRP-SSPD samples different VCDL output edges and reverses the detection-gain polarity. The polarity-reversible SSPD (PR-SSPD) [9] is needed to maintain the correct polarity for stable SS-DLL locking. However, the conventional PR-SSPD, comprised of a two-stage sample-and-hold (S&H) circuit, forms a low-pass filter due to charge sharing between a sample capacitor and a hold capacitor, reducing the loop gain and limiting the DLL bandwidth. To overcome this limitation, a TPR-SSPD is presented. A pseudo-differential tri-state inverter driven by the VCDL output is enabled for a short time after the rising edge of CLK$_{DTC1}$ to charge or discharge a sample capacitor Cs. Then the sampled voltage, V_{SP} and V_{SN}, is held on Cs until the arrival of the next falling edge, which resets V_{SP} and V_{SN} to V_{CM}. The polarity reverse operation is performed by using a passive mixer controlled by the SEL. The integrator after the TPR-SSPD is turned on only after sampling and before reset to avoid voltage disturbance during the sampling-and-reset process. Since there is no charge sharing forming an additional pole in the TPR-SSPD, a wide loop bandwidth can be guaranteed. Furthermore, compared to an isolated SSPD [10], the TPR-SSPD adopted both NMOS and PMOS transistors in a tri-state gate, which can achieve high detection gain to reduce input-referred noise and extend the loop bandwidth. Besides, the sample capacitor is isolated, which can mitigate the phase modulation of the VCDL induced by load variations caused by switching during the sampling process. As shown in the bottom-right of Fig. 27.3.2, a conventional always-on integrator accumulates phase error throughout the entire sampling interval, requiring a full reference cycle to correct the phase deviation. To further reduce the output phase error after the SS-DLL, a narrow-pulsed integrator is utilized to achieve a short integration window. Similar to the fast phase-error correction technique (FPEC) [11], in which the pulse is directly applied to the VCO, the narrow-pulsed integrator applied to the VCDL shortens the correction time and reduces error accumulation. The shaded regions in the phase-domain plots represent the accumulated phase error, clearly illustrating a smaller output phase error with the narrow-pulsed integrator.

The top-left of Fig. 27.3.3 shows the detailed circuit of the narrow-pulsed integrator, which employs a differential-to-single-ended conversion realized by a classical transconductance amplifier followed by a common-source stage to restore the differential signal, eliminating the need for a common-mode feedback loop. The top-right of Fig. 27.3.3 shows the ring VCO schematic. Each inverter cell includes coarse and fine-tuning circuits. The coarse bank comprises parallel inverters with binary weighted sizes to adjust effective transistor sizes, while fine bank uses parallel switches forming a tunable resistor to control the small inverter current. The bottom of Fig. 27.3.3 presents the DSM-driven background calibration circuits. The implemented delay calibration serves two purposes. First, it compensates for the phase shift between the PLL and DLL loops while simultaneously reducing the delay range required of the VCDL. Its operation involves dynamically adjusting the delay of the DTC2 based on the error signal $d[k]$, generated by comparing control voltages V_{CP} and V_{CN}. It forces the VCDL control voltage to the point of $V_{CP} = V_{CN} = VDD/2$ over PVT variations. Second, delay calibration is essential for the DTC gain calibration and the VCDL DCC to function correctly. The phase errors due to the DTC gain mismatch and the VCDL duty-cycle error become observable in the $d[k]$ signal only after the delay-calibration converges to make $d[k]$ dithers between 0 and 1. To eliminate these phase errors, the DTC gain calibration and the VCDL DCC adjust the delay of the DTC1, ensuring that the DSM noise is canceled out in the SSDLL. Simultaneously, the VCO duty-cycle error is compensated by push-pulling the reference edge within the DTC2.

This FN-SSDLL was implemented in 28nm CMOS, occupying 0.036mm², and consuming 11.27mW in the fractional-N mode (Fig. 27.3.7). Figure 27.3.4 shows the measured phase noise and jitter at a near-integer fractional-N frequency setting with a frequency control word (FCW) = 140+2^{-14}. With a wide pulse of about 5ns, the rms jitter integrated from 1kHz to 30MHz and 1kHz to 100MHz with the optimal loop bandwidth is 300fs and 354fs, respectively. With a narrow pulse of around 300ps, the optimal rms jitter for 1kHz-to-30MHz and 1kHz-to-100MHz integration is reduced to 164fs and 228fs. The worst-case fractional spur is −57.8dBc at a 6.1kHz offset. Figure 27.3.5 demonstrates the supply-noise rejection of the PLL. Since the VCDL is much less sensitive to supply noise than the VCO, the supply noise is only injected into the VCO, and the noise frequency is swept from 200kHz to 25.6MHz to evaluate noise rejection. The injection power is adjusted to maintain a −35dBc spur without the SS-DLL. The measured results demonstrate that the SS-DLL provides more than 20dB suppression of supply noise within 6.4MHz. Comparable rejection is also observed in the integer-N mode. Deviations from the ideal 20dB/dec trend are observed, which are attributed to substrate and ground coupling as well as AM-to-PM conversion. In the fractional-N mode, sweeping the fractional FCW (FCW_F), the measured worst-case fractional spurs are all below -50dBc and rms jitter integrated from 1kHz to 30MHz with the spur is lower than 200fs. Figure 27.3.6 summarizes the measured performance of the implemented fractional-N PLL and compares it with prior-art ring-based fractional-N PLLs. The PLL achieves an FoM of -245.2dB and -242.4dB for 1 kHz-to-30MHz and 1 kHz-to-100MHz integration ranges, respectively, along with an FoM_N of -266.6dB and -263.8dB demonstrating a 10dB-better FoM_N among prior ring-based fractional-N PLLs in Fig. 27.3.6.

Acknowledgement:
This work is partially supported by GDIC. Corresponding author: Zhiqiang Huang (zqhuang@hkust-gz.edu.cn).

References:
[1] Y. Jo et al., "A 135fs$_{rms}$-Jitter 0.6-to-7.7GHz LO Generator Using a Single LC-VCO-Based Subsampling PLL and a Ring-Oscillator-Based Sub-Integer-N Frequency Multiplier," *ISSCC*, pp. 76-78, Feb. 2023. http://doi.org/10.1109/ISSCC42615.2023.10067748
[2] T.-H. Tsai et al., "A Cascaded PLL (LC-PLL + RO-PLL) with a Programmable Double Realignment Achieving 204fs Integrated Jitter (100kHz to 100MHz) and −72dB Reference Spur," *ISSCC*, pp. 376-378, Feb. 2022. http://doi.org/10.1109/ISSCC42614.2022.9731676
[3] M. A. Khalil et al., "A 69.3fs Ring-Based Sampling-PLL Achieving 6.8GHz-14GHz and −54.4dBc Spurs Under 50mV Supply Noise," *ISSCC*, pp. 138-140, Feb. 2024. http://doi.org/10.1109/ISSCC49657.2024.10454445
[4] Z. Huang et al., "A 4.2µs-Settling-Time 3rd-Order 2.1GHz Phase-Noise-Rejection PLL Using a Cascaded Time-Amplified Clock-Skew Sub-Sampling DLL," *ISSCC*, pp. 40–41, Jan. 2016. http://doi.org/10.1109/ISSCC.2016.7417896
[5] Z. Huang et al., "A 2.1-GHz Third-Order Cascaded PLL with Sub-Sampling DLL and Clock-Skew-Sampling Phase Detector," *IEEE TCAS-I*, vol. 65, no. 7, pp. 2118-2126, July 2018. http://doi.org/10.1109/TCSI.2017.2779514
[6] W. Wu *et al.,* "A 14-nm Ultra-Low Jitter Fractional-N PLL Using a DTC Range Reduction Technique and a Reconfigurable Dual-Core VCO," *IEEE JSSC*, vol. 56, no. 12, pp. 3756-3767, Dec. 2021. http://doi.org/10.1109/JSSC.2021.3111134
[7] Z. Huang et al., "A 5GHz Fractional-N PLL with 97fs$_{rms}$ Jitter and -255.3dB FoM," *IEEE VLSI Tech. & Circuits* pp. 1–2, June 2024. http://doi.org/10.1109/VLSITechnologyandCir46783.2024.10631416
[8] Z. Huang et al., "A 6.4GHz Fractional-N PLL with 96.6fs$_{rms}$ Jitter and -257.4dB FoM," *IEEE VLSI Tech. & Circuits*, pp. 1–3, June 2025. http://doi.org/10.23919/VLSITechnologyandCir65189.2025.11075194
[9] H. Li et al., "A 27GHz Fractional-N Sub-Sampling PLL Achieving 57.9fs$_{rms}$ Jitter, −249.7dB FoM, and 1.98µS Locking Time Using Polarity-Reversible SSPD," *ISSCC*, pp. 336-338, Feb. 2025. http://doi.org/10.1109/ISSCC49661.2025.10904556
[10] Z. Yang et al., "A 25.4-to-29.5GHz 10.2mW Isolated Sub-Sampling PLL Achieving -252.9dB Jitter-Power FoM and -63dBc Reference Spur," *ISSCC*, pp. 270-272, Feb. 2019. http://doi.org/10.1109/ISSCC.2019.8662364

[11] Y. Lee et al., "A Low-Jitter and Low-Reference-Spur Ring-VCO-Based Switched-Loop Filter PLL Using a Fast Phase-Error Correction Technique," *IEEE JSSC*, vol. 53, no. 4, pp. 1192-1202, Apr. 2018. http://doi.org/10.1109/JSSC.2017.2768411
[12] Y. Duan et al., "A PVT-Robust 5.5GHz Fractional-N Cascaded RO-Based Digital PLL with Voltage-Domain Feedforward Noise Cancellation," *ISSCC*, pp. 324-326, Feb. 2025. http://doi.org/10.1109/ISSCC49661.2025.10904636
[13] C. Hwang et al., "A 188fs$_{rms}$-Jitter and −243d8-FoM$_{jitter}$ 5.2GHz-Ring-DCO-Based Fractional-N Digital PLL with a 1/8 DTC-Range-Reduction Technique Using a Quadruple-Timing-Margin Phase Selector," *ISSCC*, pp. 378-380, Feb. 2022. http://doi.org/10.1109/ISSCC42614.2022.9731646
[14] H. Park et al., "A 365fs$_{rms}$-Jitter and -63dBc-Fractional Spur 5.3GHz-Ring-DCO-Based Fractional-N DPLL Using a DTC Second/Third-Order Nonlinearity Cancelation and a Probability-Density-Shaping ΔΣM," *ISSCC*, pp. 442-444, Feb. 2021. http://doi.org/10.1109/ISSCC42613.2021.9365798
[15] A. Santiccioli et al., "A 1.6-to-3.0-GHz Fractional-N MDLL with a Digital-to-Time Converter Range-Reduction Technique Achieving 397fs Jitter at 2.5-mW Power," *IEEE CICC*, pp. 1-4, Apr. 2019. http://doi.org/10.1109/CICC.2019.8780235
[16] A. Elmallah et al., "A 3.2-GHz 405 fs$_{rms}$ Jitter −237.2 dB FoM$_{JIT}$ Ring-Based Fractional-N Synthesizer," *IEEE JSSC*, vol. 57, no. 3, pp. 698-708, Mar. 2022. http://doi.org/10.1109/JSSC.2022.3143468
[17] T. Seong et al., "A −58dBc-Worst-Fractional-Spur and −234dB-FoM$_{jitter}$, 5.5GHz Ring-DCO-Based Fractional-N DPLL Using a Time-Invariant-Probability Modulator, Generating a Nonlinearity-Robust DTC-Control Word," *ISSCC*, pp. 270-272, Feb. 2020. http://doi.org/10.1109/ISSCC19947.2020.9062948

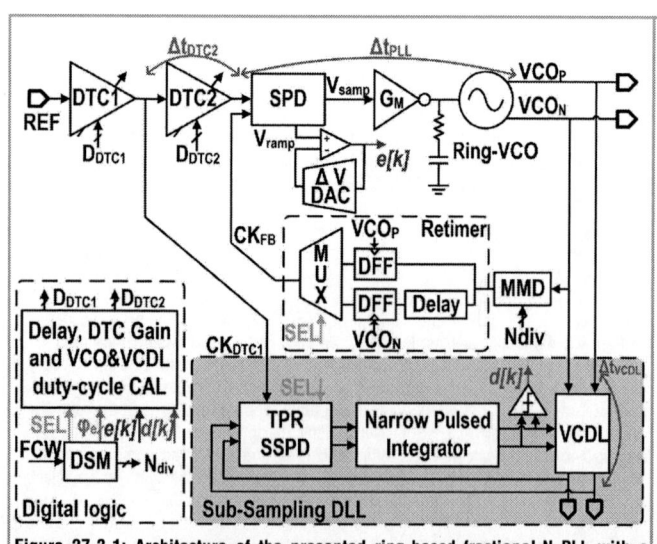

Figure 27.3.1: Architecture of the presented ring-based fractional-N PLL with a cascaded sub-sampling DLL.

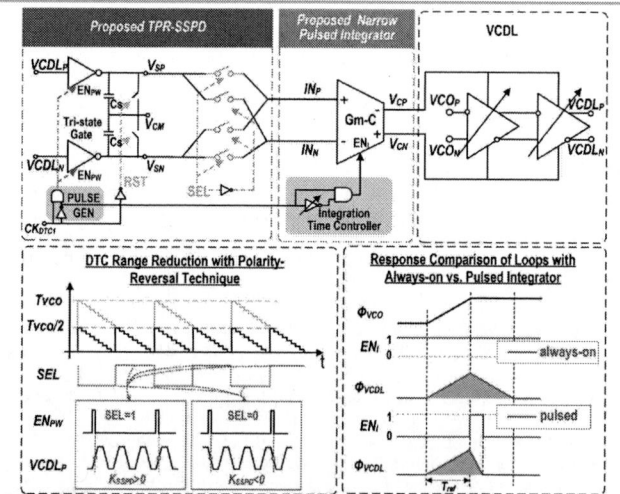

Figure 27.3.2: Schematic of the SS-DLL (top); timing diagram of the DTC range reduction with polarity reversal (bottom left); loop response with always-on/pulsed integrator (bottom right).

ISSCC 2026 / SESSION 27 / FREQUENCY GENERATORS, MULTIPLIERS AND MODULATORS / 27.3

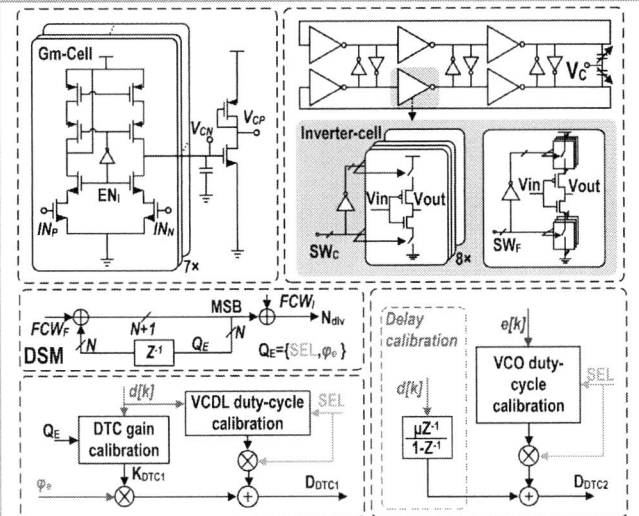

Figure 27.3.3: Schematic of the differential integrator and ring VCO (top); block diagram of the DSM, DTC gain, delay, VCO and VCDL duty-cycle calibration (bottom).

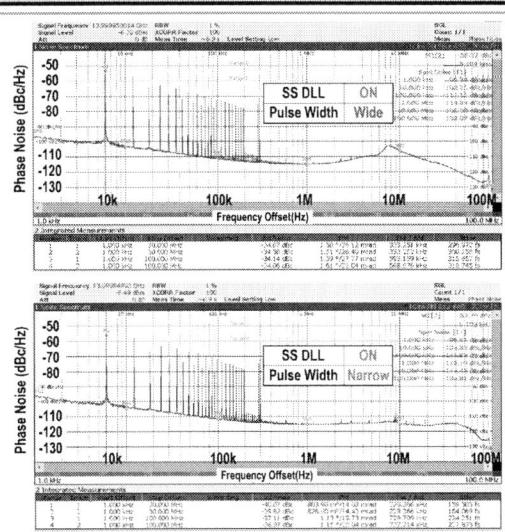

Figure 27.3.4: Measured phase noise of the DLL at a near-integer fractional-N frequency (FCW=140+2⁻¹⁴) with a wide-pulsed integrator and narrow-pulsed integrator.

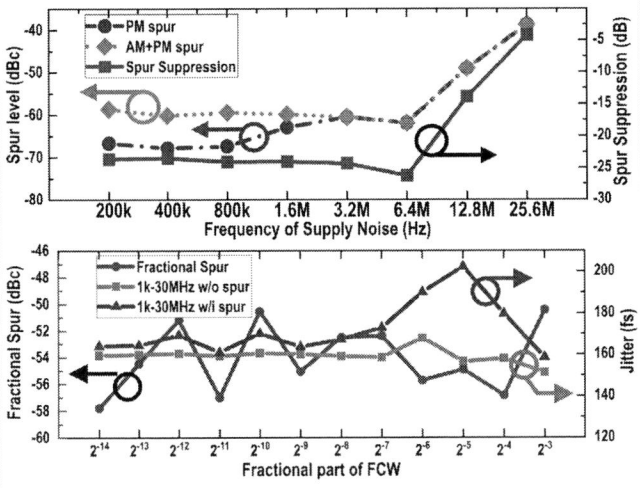

Figure 27.3.5: Measured spur level under different supply-noise frequencies (top); measured a near-14GHz worst fractional spur and jitter of the DLL versus different FCW_F (bottom).

	This Work		ISSCC25 Y. Duan [12]	ISSCC 22 C. Hwang [13]	ISSCC21 H. Park [14]	JSSC 19 A. Santiccioli [15]	JSSC 22 A. Elmallah [16]	ISSCC20 T. Seong [17]
Technology[nm]	28		28	65	65	65	65	65
Architecture	PLL+Cascade DLL		DPLL	DPLL	DPLL	MDLL	DPLL	DPLL
f_{REF} [MHz]	100		200	100×2	100	100	96	100
f_{OUT} [GHz]	14		5.5	5.2	5.3	1.65	3.26	5.5
Ref. Spur [dBc]	-52		-67	-64	-77	-56	-52	N/A
Worst frac. Spur [dBc]	-50		-62	-60	-63	-52	-51	-58
Power [mW]	11.27		20.6	15.7	9.3	2.5	11.7	9.9
RMS jitter [fs]	164 (1k to 30M)	228 (1k to 100M)	291 (1k to 40M)	188 (1k to 30M)	365 (10k to 30M)	397 (30k to 30M)	405 (1k to 30M)	648 (1k to 30M)
FoM* [dB]	-245.2	-242.3	-237.6	-242.6	-239.1	-244	-237.2	-233.8
FoMₙ** [dB]	-266.6	-263.8	-252	-256.7	-256.3	-256.2	-252.5	-251.2
VCO phase noise suppression order	3		2	2	2	1	2	2
Core Area [mm²]	0.036		0.099	0.139	0.146	0.0275	0.134	0.108

*FoM=10log10[(Jitter/1s)²·(Power/1mW)] **FoMₙ=FoM+10log10(f_{REF}/f_{OUT})

Figure 27.3.6: Performance summary and comparison with prior-art ring-based fractional-N PLLs.

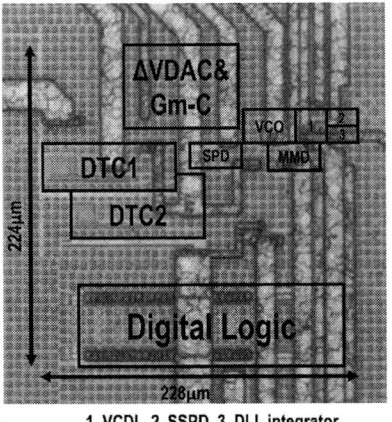

Power Consumption [mW]	
VCO	8.75
VCDL	1.19
MMD	0.36
SPD&Gm-C &ΔVDAC&DTC2	0.42
SSPD	0.02
DLL Integrator	0.03
DTC1	0.36
Digital Logic	0.14
Sum	11.27

1. VCDL 2. SSPD 3. DLL integrator

Figure 27.3.7: Die micrograph and power consumption.

• 2026 IEEE International Solid-State Circuits Conference

979-8-3315-8937-0/26 $31.00 © 2026 IEEE

ISSCC 2026 / SESSION 27 / FREQUENCY GENERATORS, MULTIPLIERS AND MODULATORS / 27.4

27.4 A 0.068mm² 8.5-to-12.7GHz Complementary Dual-Core VCO with Auto-2nd-Harmonic-Tracking Technique Achieving 202.7dBc/Hz Peak FoMT and 0.9dB-FoM Variation at a 1MHz Offset in a 39.6% Tuning Range

Xincheng Du[1,2], Xiangxun Zhan[1], Tincheng Ou[1], Zaize Chen[1], Jinge Li[1], Haoran Li[1], Zhizhan Yang[3], Zhuo Xu[1], Pui-In Mak[1], Rui P. Martins[1], Jun Yin[1]

[1]University of Macau, Macau, China, [2]UM Hetao IC Research Institute, Shenzhen, China, [3]Southwest Jiaotong University, Chengdu, China

Abstract

A complementary dual-core VCO with auto f_{2nd} tracking is proposed to achieve consistently low PN and a high FoM over a wide tuning range (TR). Prototyped in 65nm CMOS, the proposed VCO achieves a peak $FoM_{@1/10MHz}$ of 190.7/193.5dBc/Hz with only 0.9/1.6dB

variation across the TR from 8.5 to 12.7GHz (39.6%). The VCO occupies a compact area of 0.068mm² including a small decoupling capacitor of 18pF.

To satisfy the requirements of emerging multi-standard communications, a wideband voltage-controlled oscillator (VCO) with low phase noise (PN) and a high figure of merit (FoM) is required. Utilizing a tail resonator to provide a common-mode (CM) resonance (i.e., 2nd-harmonic resonance) is an effective method for improving the PN and power efficiency [1]. Although a high-order head resonator is employed to provide a wideband CM resonance [2] (Fig. 27.4.1), the tuning range (TR) for maintaining low PN and high FoM is limited to 24%. Furthermore, the tail or head resonator occupies extra chip area. The implicit CM resonance [3] merges the function of the tail/header resonator into the main tank to save area. However, to ensure that the 2nd-harmonic resonance frequency (f_{2nd}) aligns to twice the oscillation frequency ($2f_0$), two types of switched-capacitor arrays (SCAs), i.e., differential and single-ended SCAs, must be independently tuned. If only the differential SCA is tuned [4], f_{2nd} to $2f_0$ can only be aligned within a narrow bandwidth, resulting in a 4.1dB FoM variation in a TR of 21.7%. Although a calibration loop can automatically set the differential and single-ended SCAs to align f_{2nd} to $2f_0$ [5], it requires extra hardware and calibration time. With an F_2 inductor, only tuning the single-ended SCA ensures f_{2nd} tracking $2f_0$ automatically [6], achieving a low $1/f^3$ PN corner in a large TR of 31%. Nevertheless, single-ended SCA inherently has a lower Q than the differential one, limiting the PN and FoM performance, especially for high-frequency VCOs.

To overcome these limitations, this paper proposes a complementary dual-core VCO with consistently low PN and a high FoM over a wide TR. In this work, only the differential SCA is utilized to tune the 1st- and 2nd-harmonic resonances simultaneously. Besides, compensated capacitors are added to preserve auto f_{2nd} tracking from parasitic capacitors. Furthermore, the two inductors between the VCO cores are tightly coupled to reduce area. Prototyped in 65nm CMOS, the proposed wideband VCO achieves a peak $FoM_{@1/10MHz}$ of 190.7/193.5dBc/Hz with only 0.9/1.6dB variation across the TR from 8.5 to 12.7GHz (39.6%). The VCO occupies a compact area of 0.068mm² including a small decoupling capacitor of 18pF.

In the conventional complementary implicit-CM-resonance VCO [3] (Fig. 27.4.2, top left), f_0 can be tuned by both the differential SCA (C_D) and the single-ended SCA (C_{SE}), while the f_{2nd} of the CM resonance can only be tuned by C_{SE}. Due to the parasitic capacitor from switch transistors, it is difficult to maintain a constant ratio between the tank capacitance seen by 1st and 2nd harmonics across the TR. Therefore, for f_{2nd}-to-$2f_0$ alignment, manual tuning is necessary to uphold an optimal performance through independently controlling C_D and C_{SE}. In the proposed VCO, C_D is removed and only one set of the differential SCA (C_T) is reserved for frequency tuning (Fig. 27.4.2, top right). Since both the 1st- and 2nd-harmonic voltages in V_{P1} and V_{N1} (V_{P2} and V_{N2}) are out of phase, C_T contributes to both the 1st- and 2nd-harmonic tank capacitance ($C_{1st} = C_{2nd} = C_T$). The fundamental and 2nd-harmonic frequencies can be expressed as $\omega_0 = 1/\sqrt{(1+k)LC_{1st}}$ and $\omega_{2nd} = 1/\sqrt{(1-k)LC_{2nd}}$, where k is the coupling coefficient between the two tank inductors. By choosing k = 0.6, f_{2nd} can automatically track $2f_0$ across the TR regardless of C_T. However, f_{2nd} can deviate from $2f_0$ due to the parasitic capacitor from the cross-coupled pairs. The parasitic capacitor contains two portions: 1) the single-ended capacitor ($C_{GS}+C_{DB}$), which contributes to both the C_{1st} and C_{2nd}; 2) the differential capacitor ($2C_{GD}$), which only contributes to C_{1st}. To re-align f_{2nd} to $2f_0$, two compensated capacitors (C_C) are added between V_{P1} and V_{N2}, and between V_{P2} and V_{N1}, where C_C can be seen by only the 2nd harmonic for C_{2nd} compensation. Thus, $C_{1st} = C_T + (C_{GS}+C_{DB})/2 + 2C_{GD}$ and $C_{2nd} = C_T + (C_{GS}+C_{DB})/2 + C_C$. By choosing $C_C = 2C_{GD}$, $C_{1st} = C_{2nd}$ is guaranteed, and f_{2nd} is able to automatically track $2f_0$ again across a wide TR.

In this design, the NMOS and PMOS transistors are sized with the same dimensions of 34μm/60nm to maintain similar parasitic capacitance. As illustrated by simulation results (Fig. 27.4.2, bottom left), the minimum $1/f^3$ PN corner is achieved when $C_C = 2C_{GD} = 26fF$. When C_C varies by ±20%, the $FoM_{@1/10MHz}$ slightly changes by 0.4dB. The VCO may oscillate in an undesired mode where the fundamental voltages in V_{P1} and V_{N1} (V_{P2} and V_{N2}) are in phase (Fig. 27.4.2, bottom right). By connecting a small resistor R_{rej} of 10Ω between the center taps (T_1 and T_2) of two tank inductors, the tank impedance of the undesired mode is suppressed by >20×, avoiding the mode ambiguity issue. Since T_1 and T_2 are virtual grounds in the desired mode, adding R_{rej} does not impair inductor Q.

Figure 27.4.3 (left) details the layout design. Two tank inductors are tightly coupled to realize k = 0.6. To facilitate the layout of C_T and C_C, the output nodes (V_{P1}, V_{P2}, V_{N1}, and V_{N2}) are all

arranged on the same side of the inductors. This floorplan also enables NMOS and PMOS cross-coupled pairs to be close in the layout, as well as V_{DD} and ground nodes, to be placed close together, minimizing the decoupling capacitor required to provide a low-impedance current return path. The proposed VCO can be viewed as a dual-core VCO for PN reduction. Separated by Y-axis, core #1 (core #2) consists of complementary negative transconductors M_{P1} and M_{N1} (M_{P2} and M_{N2}) connected with an LC resonator between V_{P1} and V_{N1} (V_{P2} and V_{N2}). Compared to the conventional dual-core VCO using two separate inductors or transformers [4], the proposed tightly coupled inductors help reduce the chip area and strengthen the synchronization between the two cores. The tank capacitor C_T is implemented by a 7-bit binary-sized SCA for coarse tuning and varactors for fine tuning. R_{rej} is implemented with vias and thin metal traces. To determine the virtual ground nodes (T_1 and T_2) of the tank inductors, a number of pins are added in the region near the midpoints of the inductors when obtaining the inductor model from the electromagnetic (EM) simulation. In the VCO simulation, two pins with minimal voltage amplitude are chosen as the virtual ground nodes (T_1 and T_2).

The variation of k can deviate f_{2nd} from $2f_0$, which affects the VCO performance. To study this effect, k is altered by changing the distance between the two tank inductors. The post-layout simulations (Fig. 27.4.3, right) reveal that the $FoM_{@1/10MHz}$ is also not sensitive to the k variation. When k < 0.6, more $1/f$ noise upconversion occurs. However, since the Q of the equivalent tank inductor increases when k is reduced, a high $FoM_{@1MHz}$ is still maintained. As a result, simulated by 10% variation when k < 0.6, $FoM_{@1MHz}$ only degrades by at most 0.5dB and 1dB under the minimum and maximum oscillation frequency (f_{min} and f_{max}).

The proposed complementary VCO is prototyped in 65nm LP CMOS (standard V_{DD} = 1.2V), occupying a core area of 0.068mm² including the decoupling capacitor (Fig. 27.4.7). The measured frequency is from 8.5 to 12.7GHz (39.6%), and the power dissipation is from 4.22 to 3.39mW under 1.1V V_{DD}. According to Fig. 27.4.4 (left), the measured $PN_{@1/10MHz}$ is −118.3/−139.5dBc/Hz at 8.53GHz (f_{min}) and −113.9/−136.7dBc/Hz at 12.73GHz (f_{max}), corresponding to the $FoM_{@1/10MHz}$ of 190.7/191.9dBc/Hz (f_{min}) and 190.7/193.5dBc/Hz (f_{max}). Figure 27.4.4 (right) depicts the measured PN and FoM across the TR. The $FoM_{@1/10MHz}$ exhibits small variations of 0.9dB (189.8 to 190.7dBc/Hz) and 1.6dB (191.9 to 193.5dBc/Hz) thanks to the auto-2nd-harmonic-tracking technique. The above measurements are all performed by using the same control word for SCA_1 and SCA_2, i.e, $B_1[6:0] = B_2[6:0]$.

The PN and FoM are also measured by setting $B_1[6:0]$ different from $B_2[6:0]$, emulating the situation when capacitor mismatch exists between SCA_1 and SCA_2, i.e., $C_{SCA1} \neq C_{SCA2}$, which leads to a frequency mismatch between two cores. As shown in Fig. 27.4.5 (top), the $PN_{@1/10MHz}$ and $FoM_{@1/10MHz}$ only degrade by at most 0.34/0.56dB and 0.42/0.64dB within a 20% capacitor mismatch between SCA_1 and SCA_2. Since C_{SCA1} and C_{SCA2} can be seen by both 1st- and 2nd-harmonic voltages, $C_{1st} = C_{2nd}$ still holds for each core even when $C_{SCA1} \neq C_{SCA2}$, ensuring $f_{2nd} = 2f_0$. The precise f_{2nd}-to-$2f_0$ tracking together with the strong magnetic coupling between the two cores guarantees that the PN and FoM are insensitive to the capacitor mismatch between SCA_1 and SCA_2.

Compared with prior-art single-mode VCOs operating at similar frequency bands (Fig. 27.4.6), our design achieves maximum TR (39.6%) with competitive $FoM_{@1/10MHz}$, resulting in the highest $FoM_{T@1/10MHz}$. Relative to other designs in Fig. 27.4.6, our design also achieves a best-in-class $FoM_{@1MHz}$ variation of 0.9dB. Compared with the dual-core VCO using two separate transformers [4], our design with two tightly coupled inductors reduces the core area (including the decoupling capacitor) by 1.8×. Figure Fig. 27.4.6.5 (bottom) provides comprehensive benchmarking with prior-art wideband single-mode oscillators operating between 2 and 30GHz. This design demonstrates a compact, wideband, robust, low-PN, and low-power VCO solution for multi-standard communication terminals.

Acknowledgement:
The work is funded by the University of Macau Research Fund -MYRG-GRG2024-00298-IME, the Macau Science and Technology Development Fund (FDCT) -0103/2022/AFJ, -001/2024/COP, -004/2023/SKL, and Hetao SZ-HK S&T Innovation Cooperation Zone Project (HTHZQSWS-KCCYB-2023030). Corresponding authors: Xiangxun Zhan and Jun Yin.

472 • 2026 IEEE International Solid-State Circuits Conference

979-8-3315-8937-0/26 $31.00 © 2026 IEEE

References:

[1] E. Hegazi et al., "A Filtering Technique to Lower LC Oscillator Phase Noise," *IEEE JSSC*, vol. 36, no. 12, pp. 1921–1930, Dec. 2001. https://doi.org/10.1109/4.972142

[2] H. Guo et al., "A 5.0-to-6.36GHz Wideband-Harmonic-Shaping VCO Achieving 196.9dBc/Hz Peak FoM and 90-to-180kHz 1/f³ PN Corner Without Harmonic Tuning," *ISSCC*, pp. 294–296, Feb. 2021. https://doi.org/10.1109/ISSCC42613.2021.9365761

[3] D. Murphy et al., "Implicit Common-Mode Resonance in LC Oscillators," *IEEE JSSC*, vol. 52, no. 3, pp. 812–821, Mar. 2017. https://doi.org/10.1109/JSSC.2016.2642207

[4] Q. Wu et al., "An Enhanced Class-F Dual-Core VCO with Common-Mode-Noise Self-Cancellation and Isolation Technique," *IEEE JSSC*, pp. 2441-2454, 2024. https://doi.org/10.1109/JSSC.2024.3367351

[5] J. Gong et al., "A 200dB FoM 4-to-5GHz Cryogenic Oscillator with an Automatic Common-Mode Resonance Calibration for Quantum Computing Applications," *ISSCC*, pp. 308–310, Feb. 2020. https://doi.org/10.1109/ISSCC19947.2020.9062913

[6] M. Shahmohammadi et al., "A 1/f Noise Upconversion Reduction Technique for Voltage-Biased RF CMOS Oscillators," *IEEE JSSC*, vol. 51, no. 11, pp. 2610–2624, Nov. 2016. https://doi.org/10.1109/JSSC.2016.2602214

[7] Z. Lin et al., "A Low-Phase-Noise VCO with Common-Mode Resonance Expansion and Intrinsic Differential 2nd-Harmonic Output Based on a Single Three-Coil Transformer," *IEEE JSSC*, pp. 253-267, 2023. https://doi.org/10.1109/JSSC.2023.3274178

[8] G. Zhang et al., "A Calibration-Free 12.8-16.5GHz Cryogenic CMOS VCO with 202dBc/Hz FoM for Classic-Quantum Interface," *ISSCC*, pp. 512–514, Feb. 2023. https://doi.org/10.1109/ISSCC42615.2023.10067803

[9] J. Guo et al., "A Differential Series-Resonance CMOS VCO with Pole-Convergence Technique Achieving 202.1dBc/Hz FoM$_{TA}$ at 10MHz Offset," *ISSCC*, pp. 332–334, Feb. 2025. https://doi.org/10.1109/ISSCC49661.2025.10904629

[10] H. Ge et al., "A 13.7-to-41.5GHz 214.1dBc/Hz FoM$_T$ Quad-Core Quad-Mode VCO Using an Oscillation-Mode-Splitting Technique," *ISSCC*, pp. 356–358, Feb. 2024. https://doi.org/10.1109/ISSCC49657.2024.10454356

[11] H. Guo et al., "A 9.05-to-37.0GHz LO Generator with Magnetic Mode Switching and Tuning-Free Octave-Bandwidth Common-Mode Resonator Achieving >190.7dBc/Hz FoM", *ISSCC*, pp. 560–562, Feb. 2025. https://doi.org/10.1109/ISSCC49661.2025.10904644

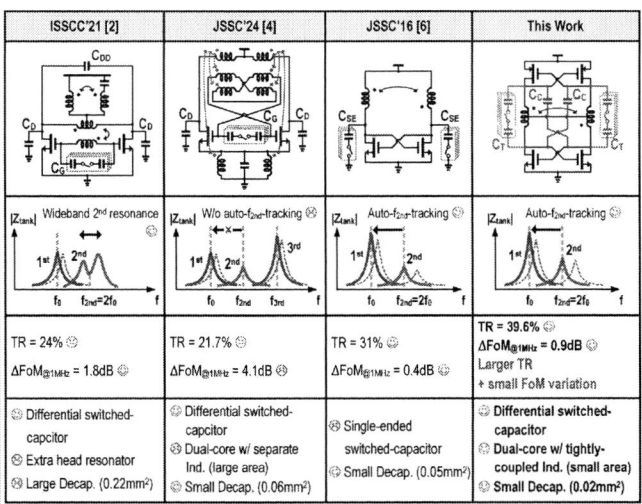

Figure 27.4.1: Comparison and key features of single-mode VCOs utilizing 2ⁿᵈ-harmonic resonance w/o manual harmonic tuning.

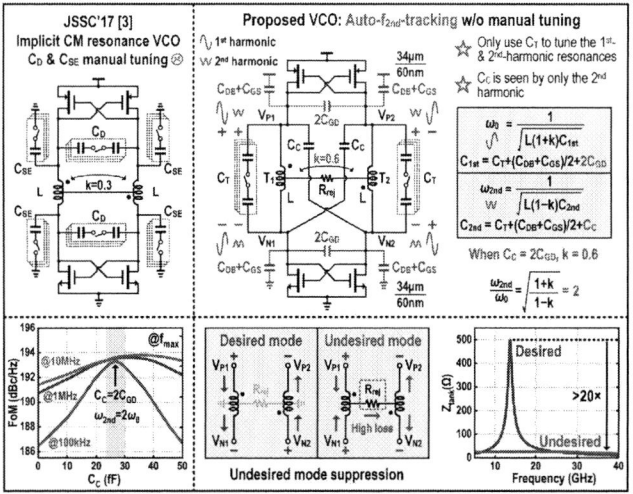

Figure 27.4.2: (Top left) Implicit CM resonance VCO; (Top right) proposed VCO with auto f$_{2nd}$ tracking; (Bottom left) simulated FoM versus C$_C$ of the proposed VCO; (Bottom right) undesired mode suppression.

ISSCC 2026 / SESSION 27 / FREQUENCY GENERATORS, MULTIPLIERS AND MODULATORS / 27.4

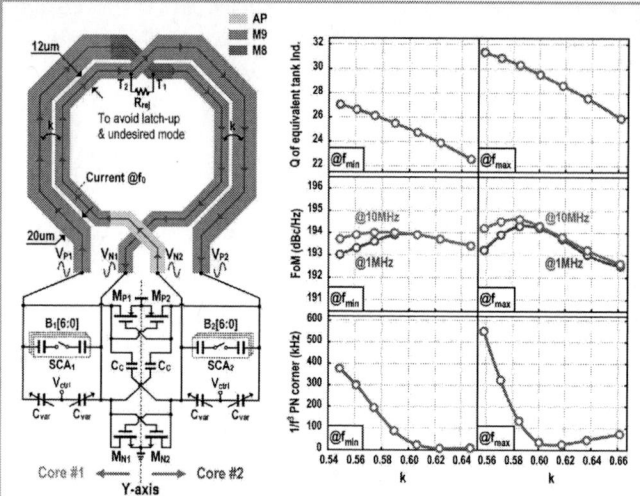

Figure 27.4.3: (Left) Layout floorplan and detailed implementation of the proposed VCO; (Right) simulated Q of the equivalent tank inductor, FoM$_{@1/10MHz}$, and 1/f^3 PN corner versus k variation.

Figure 27.4.4: (Left) Measured PN profiles at lowest and highest frequencies; (Right) measured PN and FoM across the TR using the same control word for SCA$_1$ and SCA$_2$ (B$_1$[6:0]=B$_2$[6:0]) under 1.1V V$_{DD}$.

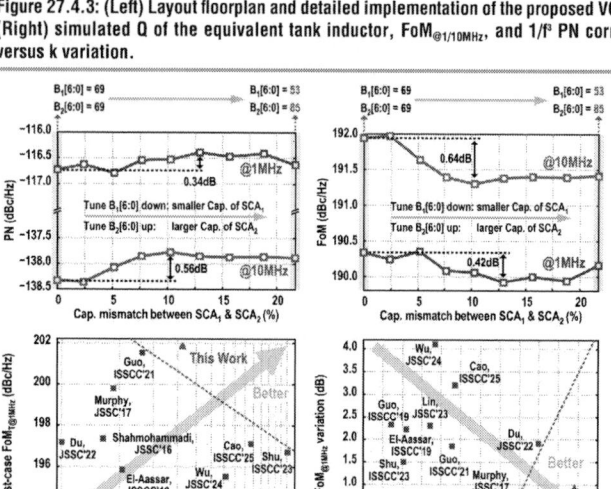

Figure 27.4.5: (Top) Measured PN and FoM versus capacitor mismatch between SCA$_1$ and SCA$_2$. (Bottom) Benchmarks of worst-case FoM$_{T@1MHz}$ and FoM$_{@1MHz}$ variation with wideband single-mode VCOs operating from 2 to 30GHz.

Figure 27.4.6: Performance summary and comparison with prior-art VCOs above 7GHz.

	This Work	Q. Wu JSSC'24 [4]	Z. Lin JSSC'23 [7]	G. Zhang ISSCC'23 [8]	J. Guo ISSCC'25 [9]	H. Ge ISSCC'24 [10]	H. Guo ISSCC'25 [11]
CMOS Technology	65nm	65nm	65nm	65nm	65nm	65nm	65nm
Number of Modes	Single-mode					Quad-mode	Dual-mode
Number of Cores	Dual-core	Dual-core	Single-core	Single-core	Single-core	Quad-core	Dual-core
Resonance @2f$_0$ Bandwidth	Narrowband	Narrowband	Wideband	Wideband	N/A	Narrowband	Wideband
Resonance @2f$_0$ Tracking	Auto	None	None	None		None	None
V$_{DD}$ (V)	1.1	N/A	0.6	1.2	1.2	0.4/0.55	0.6
Frequency (GHz)	8.5 to 12.7	11.5 to 14.3	12.3 to 15.2	12.3 to 15.8	7.7 to 9.1	13.7 to 41.5	9.05 to 18.5§
TR	38.6%	21.7%	21.1%	24.6%	17.7%	101%	68.6%
PN (dBc/Hz) @1MHz	−116.3 to −113.7	−119.2 to −115.6*	−118.0 to −113.6	−117.0 to −114.0	−131.9 to −130.6*	−111.2 to −98.2*	−126.1 to −115.8*
PN (dBc/Hz) @10MHz	−139.5 to −136.4	−138.6 to −135.5*	−141.4 to −138.8	−138.0 to −135.2	−153.1 to −150.6*	−133.2 to −124.4	−147.6 to −139.6
FoM (dBc/Hz) @1MHz	189.8 to 190.7	188.5 to 192.9*	188.1 to 190.4	185.5 to 186.4	186.3 to 188.8*	180.3 to 192.0*	189.1 to 192.8*
FoM (dBc/Hz) @10MHz	191.9 to 193.5	190.2 to 192.3*	193.1 to 193.7	186.3 to 187.7	186.4 to 190.0*	184.4 to 194.0	192.3 to 194.4
FoM$_T$ (dBc/Hz) @1MHz	201.8 to 202.7	195.5 to 199.6	194.6 to 196.9	193.4 to 194.3	191.3 to 193.6	200.4 to 212.1	205.9 to 209.6
FoM$_T$ (dBc/Hz) @10MHz	203.9 to 205.5	196.9 to 199.0	199.6 to 200.2	194.2 to 195.6	191.4 to 195.0	204.5 to 214.1	209.1 to 211.2
FoM$_{@1MHz}$ variation (dB)	0.9	4.1	2.3	0.9	2.5	11.7	3.7
1/f^3 PN Corner (kHz)	360 to 830	360 to 600	450 to 700	500 to 900	100 to 700	140 to 1300*	170 to 720
Power (mW)	3.39 to 4.22	5.6	8.4 to 9.6	17	159 to 164#	2.0 to 8.8	9.6 to 24.9
Core Area (mm^2)	0.048	0.065	0.06	0.093	0.192	0.14	0.14
Area w/ Decap. (mm^2)	0.068	0.124*	N/A		0.233*	0.234*	0.204*

FoM=|PN|(Δf)+20log10(f$_0$/Δf)−10log10(P$_{DC}$/1mW)   FoM$_T$=FoM+20log10(TR(%)/10)   *Estimated from figure #Series-resonance VCO §Only report VCO

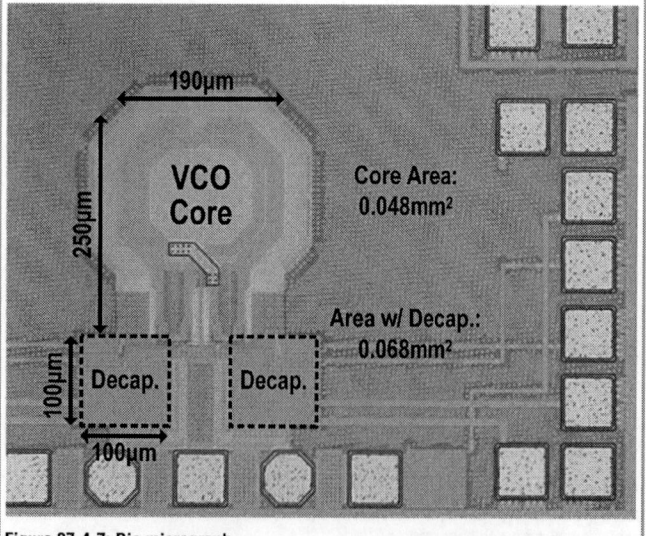

Figure 27.4.7: Die micrograph.

• 2026 IEEE International Solid-State Circuits Conference

ISSCC 2026 / SESSION 27 / FREQUENCY GENERATORS, MULTIPLIERS AND MODULATORS / 27.5

27.5 A 17.9-to-22.4GHz 195.6±1.3dBc/Hz FoM Quad-Core Class-F^{-1} VCO Featuring Improved Synchronization Using a Circular Pentafilar Transformer-Based Tank

Yue Wu[1], Yatao Peng[1,2], Jiawei Li[1], Fengen Yuan[1], Jinge Li[1], Jun Yin[1], Rui P. Martins[1], Pui-In Mak[1]

[1]University of Macau, Macau, China, [2]UM Hetao IC Research Institute, Shenzhen, China

Abstract

A 17.9-to-22.4GHz quad-core inverse-Class-F VCO achieving PN of –145.6 to –141dBc/Hz and an FoM of 194.3 to 196.9dBc/Hz is reported. The VCO features a circular pentafilar transformer tank that provides high Q1/Q2 without requiring extra CM resonators. Multi-path synchronizations—in-phase at gates, out-of-phase at drains, and an auxiliary dual-path coil—equalize impedances at both f_0 and $2f_0$. A segmented alignment technique is proposed to preserve 1D tuning with improved robustness against PVT variations.

Harmonic-shaping (HS) and multi-core (MC) techniques represent two critical approaches to lower PN of mm-wave VCOs [1]. In MC-VCOs, sufficient synchronization paths must be introduced to reduce the PN penalty caused by the mismatch of the cores [2,3]. In HS-VCOs, optimal performance requires both precise harmonic-resonance alignment and sufficiently high harmonic impedance to enhance the harmonic components. The Class-F$_{23}$ VCO presented in [1] integrates the MC and HS techniques into a multi-core harmonic-shaping (MC-HS) VCO, yielding low PN; however, such differential VCOs need additional resonators to ensure wideband high impedances at the common-mode (CM) 2nd harmonic as shown in Fig. 27.5.1 (top-left), thus increasing area. Despite the Q factors > 40 for both CM and differential-mode (DM) inductors (at the expense of increased area), the PN at a 10MHz offset is only ~-138dBc/Hz given equal quad-core coupling. Compared to Class-F$_{23}$ oscillators, the transformer-based class-F^{-1} VCO [4] is more power and area efficient. Its inherently high Q_1 and Q_2 enable excellent PN even without 3rd-harmonic shaping [5]. The work in [6] presents a dual-core-coupled Class-F^{-1} VCO (Fig. 27.5.1 top-middle), operating at ~25GHz with high-Q circular inductors. Its average PN at 10MHz is ~-134.5dBc/Hz, not fully leveraging the structure merits. Moreover, it overlooks the synchronization in MC-VCOs and requires 2D tuning for harmonic alignment.

The Q-factor of a circular LC tank (Q_T) in mm-wave VCOs at 20 to 30GHz is comparable to, or even better than, that of sub-6GHz designs [7]. Thus, a well-designed ~20GHz HS-VCO should achieve PN close to sub-6GHz (<-140dBc/Hz), yet this has not been consistently observed [1,8,9]. Figure 27.5.1 (top-right) present the proposed quad-core Class-F^{-1} VCO with a circular pentafilar transformer-based tank achieving -145.6dBc/Hz PN and a 194.3-to-196.9dBc/Hz FoM from 17.9 to 22.4GHz, comparable to sub-6GHz levels [10]. Beyond the quad-core circular Class-F^{-1} topology, two techniques are proposed to further enhance the PN and FoM: 1) improved both drain and gate synchronization across the cores, reducing the PN penalty from frequency mismatches in the MC-HS-VCOs [11,12]; and 2) segmented harmonic alignment technique to ensure 1D frequency tuning, minimizing 2nd-harmonic impedance and phase variations in the FTR.

The transformer-based Class-F^{-1} VCO core utilizes a well-aligned 2nd-harmonic component (V_{P2}) to shape the drain waveforms with a wide ISF flat region. Three key design criteria minimize PN in the oscillating core design: 1) sufficiently high R_{P2} to boost voltage ratio $V_{P2}/V_{P1}\approx0.33$; 2) tight phase alignment between V_{P2} and V_{P1} (<2.5°); and 3) the suppression of misaligned 3rd-harmonic voltage V_{P3} by lowering 3rd-harmonic impedance $|Z_T(3\omega_0)|$, enabling Class-F^{-1} waveforms with higher power efficiency [5,13].

In a dual-core HS VCO with core mismatches, i.e., $f_1 \neq f_2$, (Fig. 27.5.1 bottom-left), unlike conventional MC VCOs [2,7,8,14], synchronization in MC-HS VCOs must account for not only the fundamental but also the harmonic components. Moreover, in transformer-based Class-F^{-1} VCOs, the inherent weak drain-gate coupling ($k_m<0.4$) requires careful design of both drain and gate synchronization. As shown in Fig. 27.5.1 (bottom-middle), two harmonic well-aligned Class-F^{-1} oscillating cores with 10% mismatch are in-phase coupled (IPC) via a gate-to-gate connection. Gate-port impedances Z_{G1} and Z_{G2} are consistent at both ω_0 and $2\omega_0$, indicating good synchronization. However, the drain-port impedances Z_{D1} and Z_{D2} match only at ω_0 and remain mismatched at harmonics, causing 2nd-harmonic phase misalignment and worsening the PN. When two cores are out-of-phase coupling (OPC) in a circular configuration (Fig. 27.5.1, bottom-right), frequency mismatch shifts virtual ground to equalize the effective resonant frequencies of the cores, achieving synchronization. Yet, this leads to asymmetric impedance magnitudes at ω_0 and $2\omega_0$, i.e., $|Z_{G1}|\neq|Z_{G2}|$ and $|Z_{D1}|\neq|Z_{D2}|$, disrupting optimal Class-F^{-1} oscillation conditions. This multi-harmonic, multi-port synchronization issue has been overlooked in prior MC-HS VCOs design [1, 6,11,12].

Motivated by the above observations, we propose an enhanced synchronization scheme for quad-core VCOs, which ensures that, despite frequency mismatch among the cores, the resonant frequencies (both 1st and 2nd harmonics) and impedance amplitudes at the drain and gate ports remain identical after synchronization. Figure 27.5.2 illustrates the gate dual-path synchronization scheme [3], which implements the IPC for excellent gate synchronization. However, the drains of core #1 and core #3 have neither IPC nor OPC, causing frequency misalignment at 2ω_0 (see Fig. 27.5.1). Alternatively, applying dual-path synchronization to drains ensure a robust drain synchronization, but the gate port

impedances of non-adjacent mismatched cores cannot be maintained consistently. To leverage the benefits of both approaches, we combined the two structures into a 5-coil transformer, reaching the proposed VCO. The proposed structure provides a triple synchronization mechanism accommodating the frequency mismatches between the four cores. Sync1: the OPC between adjacent cores aligns drain-impedance resonances with ω_0 and $2\omega_0$, but the impedance magnitude may still differ. Sync2: the dual-path IPC between both adjacent and non-adjacent cores ensures good matching of gate port impedances. Meanwhile, the drain-port impedance magnitude can essentially remain identical: $|Z_{D1}(\omega_0)|=|Z_{D2}(\omega_0)|=|Z_{D3}(\omega_0)|=|Z_{D4}(\omega_0)|$. Sync 3: Auxiliary L_T-C_T resonances introduce an additional resonance near $2\omega_0$, forming a high-impedance band that ensures consistent drain-port-impedance resonances, addressing the residual inconsistencies in the drain-port resonance after Sync1 and Sync2. With 10% mismatches applied to both C_{DS} and C_{GS}, the synchronized gate and drain impedances of the four cores show no significant inconsistency in resonance frequencies and magnitudes compared to the mismatch-free case (Fig. 27.5.1 left), which is also confirmed with the closely matched waveforms of oscillating signals across the cores, indicating a better alignment of both 1st and 2nd harmonic impedances. Consequently, the PN penalty is ~3.1dB under a 10% mismatch, about ~10.1dB reduction compared to the quad-core Class-F^{-1} VCO (extended from the dual-core VCO [12]) with solely OPC synchronization. Compared with a counterpart with only synchronization at 1st harmonic via IPC at the gates (refer to [8]), a 3.3dB PN penalty reduction can still be achieved by the proposed structure.

Figure 27.5.3 shows the schematic of the proposed quad-core Class-F^{-1} VCO employing the circular pentafilar-transformer-based tank, critical for enhanced synchronization. $R_D=32\Omega$ and $R_G=1.2\Omega$ are inserted between the CM nodes A1 to A4, B1 to B4 to resolve mode ambiguity [14]. The five inductors exhibit 10 mutual coupling pairs. However, strong coupling (k_{gg} and $k_{tt}\approx0.7$) between L_{G1}-L_{G2} and L_{T1}-L_{T2} makes each pair equivalent to a single inductor (L_G and L_T), degenerating the five-tank system into three-tank coupled configuration (Fig. 27.5.3 right), featuring three resonances ($\omega_1/\omega_2/\omega_3$) after the ambiguous modes are damped by R_D and R_G. The equivalent inductances L_{TE}, L_{GE}, L_{DE} and coupling coefficients k_{GDE}, k_{GTE}, k_{DTE} are determined by the inductors self-inductances and intrinsic coupling. When $\omega_1=\omega_0$, a straightforward approach to realizing a 1D frequency-tuning VCO is to design ω_2 and ω_3 to be close to each other, thereby forming a wideband high-impedance region covering FTR of $2\omega_0$. Thus, when tuning ω_0 via C_G, $Z_T(2\omega_0)$ remains relatively high, without adjusting C_D and C_T for harmonic alignment. This approach is widely adopted in explicit CM VCOs [1,10]. However, the in-band $Z_T(2\omega_0)$ variation remain at 90 to 180Ω for magnitude and ~±25° for phase. Also, a wideband $Z_T(2\omega_0)$ inevitably raises $|Z_T(3\omega_0)|$. These traits render this strategy unsuitable to lower PN for Class-F^{-1} VCOs. To preserve the harmonic tuning-free operation while achieving low PN, we propose a segmented high-impedance tuning scheme, as shown in Fig. 27.5.3 (right-bottom). The $2\omega_0$ tuning range (18 to 23GHz) is divided into 8 overlapping segments. Within each segment, the phase variation of $Z_T(2\omega_0)$ is kept within ±2.5° with $|Z_T(2\omega_0)|>190\Omega$, ensuring a good phase alignment and $V_{P2}/V_{P1}\approx0.33$. The MSBs (B4-B2) of the C_G SCB also control the 3-bit SCBs of C_D and C_T ($C_D=C_T$). The sliding adjustment of B4:B2 tunes C_G, C_D, and C_T for segment switching, while LSBs (B1:B0) and a varactor handle intra-segment frequency adjustment. This segment alignment with a ~3% bandwidth high impedance, offers superior robustness against process variations compared to the harmonic tuning-free structure with k=0.6 and $C_S=C_D$ [15], while ensuring sufficient impedance to boost V_{P2}/V_{P1}. Moreover, its narrowband nature reduces $|Z_T(3\omega_0)|$, lowering $|Z_T(3\omega_0)|/R_{P1}$ by 2.8× versus the wideband case, as shown in Fig. 27.5.4 (top-left). The PN simulation in Fig. 27.5.4 (top-right) reveals that the proposed segment alignment achieves lower PN with reduced in-band PN fluctuation.

The proposed VCO was fabricated in 65nm CMOS, occupying a 0.15mm^2 core area (Fig. 27.5.7). The VCO was measured by an RS FSWP PN analyzer with a D-S balun. Figure 27.5.4 (bottom) shows the measured PN of –121.1/–145.6dBc/Hz at a 1MHz/10MHz offset at 17.9GHz, yielding an FoM of 192.4/196.9dBc/Hz. At 20.1GHz (Fig. 27.5.5 top), the PN is –118.3/–141.6dBc/Hz and the FoM is 192.5/195.8dBc/Hz. Figure 27.5.5 (bottom) summarizes the performance across the FTR: the PN at a 10MHz offset ranges from -145.6 to -141.0dBc/Hz, with the FoM from 194.3 to 196.9dBc/Hz. With a 22.4% FTR, the FoM$_T$ peaks at 203.9dBc/Hz. The power consumption varies from 18.5 to 24mW within the FTR. The 1/f^3 PN corner exceeds simulation (110 to 430kHz) due to unbalanced PMOS/NMOS

ISSCC 2026 / February 18, 2026 / 9:30 AM

transconductance. We measured a 1-to-1.4dBc power offset between the in- and out-phase paths, which degrades the $1/f^3$ PN corner.

Figure 27.5.6 benchmarks the prototype against recent low-PN MC and HS oscillators. Among the MC oscillators, our chip achieves the lowest 10MHz-offset PN, indicating a minimal PN penalty due to the proposed improved synchronization. In addition, by using segment harmonic auto-alignment technique, the prototype achieves superior PN/FoM. It further leads CMOS or BiCMOS multi-core peers in FoMs and FoM$_T$s at 1MHz and 10 MHz offsets. The extended survey in Fig. 27.5.7 (right), spanning 10 to 60GHz, confirms best-in-class 10MHz-offset PN and FoM.

Acknowledgement:
This work is funded by the Macau Science and Technology Development Fund-0151/2022/A3 and 004/2023/SKL, and by the University of Macau-MYRG-GRG2024-00304-IME, and by Hetao SZ-HK S&T Innovation Cooperation Zone Project (HTHZQSWS-KCCYB-2023030). Corresponding author: Yatao Peng.

References:
[1] H. Cao et al., "An 18.5-to-23.6GHz Quad-Core Class-F$_{23}$ Oscillator Without 2nd/3rd Harmonic Tuning Achieving 193dBc/Hz Peak FoM and 140-to-250kHz $1/f^3$ PN Corner in 65nm CMOS," *ISSCC*, pp. 562–563, Feb. 2025.
https://doi.org/10.1109/ISSCC49661.2025.10904566
[2] L. Iotti et al., "Insights Into Phase-Noise Scaling in Switch-Coupled Multi-Core LC VCOs for E-Band Adaptive Modulation Links," *IEEE JSSC*, vol. 52, no. 7, pp. 1703–1718, June 2017. https://doi.org/10.1109/JSSC.2017.2697442
[3] S. A.-R. Ahmadi-Mehr et al., "Analysis and Design of a Multi-Core Oscillator for Ultra-Low Phase Noise," *IEEE TCAS-I*, vol. 63, no. 4, pp. 529–539, Apr. 2016.
https://doi.org/10.1109/TCSI.2016.2529218
[4] C. C. Lim et al., "An Inverse-Class-F CMOS Oscillator with Intrinsic-High-Q First Harmonic and Second Harmonic Resonances," *IEEE JSSC*, vol. 53, no. 12, pp. 3528–3539, Dec. 2018. https://doi.org/10.1109/JSSC.2018.2875099
[5] Y. Wu et al., "An Inverse Class-F VCO with Reduced Third Harmonic Detriment Using a High Fundamental and Second Harmonic Q-Factor Resonator Achieving a 198.9 dBc/Hz Peak FoM," *IEEE RFIC*, pp. 239–242, June 2025.
https://doi.org/10.1109/RFIC61188.2025.11082848
[6] H. Ge et al., "A 22.0-to-28.4GHz 192.2dBc/Hz FoM and 206.2dBc/Hz FoM$_A$ Dual-Core VCO Using Circular-Inverse-Class-F Topology Under Standard Supply Voltage in 65nm CMOS Process," *IEEE CICC*, pp. 1–3 Apr. 2025.
https://doi.org/10.1109/CICC63670.2025.10982955
[7] D. Murphy et al., "A 27-GHz Quad-Core CMOS Oscillator with No Mode Ambiguity," *IEEE JSSC*, vol. 53, no. 11, pp. 3208–3216, Nov. 2018.
https://doi.org/10.1109/JSSC.2018.2865460
[8] X. Zhan et al., "A 22.4-to-26.8GHz Dual-Path-Synchronized Quad-Core Oscillator Achieving -138dBc/Hz PN and 193.3dBc/Hz FoM at 10MHz Offset from 25.8GHz," *ISSCC*, pp. 148-150, Feb. 2023. https://doi.org/10.1109/ISSCC42615.2023.10067277
[9] Y. Shu et al., "A 28GHz Scalable Inter-Core-Shaping Multi-Core Oscillator with DM/CM-Configured Coupling Achieving 193.3dBc/Hz FoM and 205.5dBc/Hz FoM$_A$ at 1MHz Offset," *ISSCC*, pp. 150–152, Feb. 2023.
https://doi.org/10.1109/ISSCC42615.2023.10067826

[10] H. Guo et al., "A 5.0-to-6.36GHz Wideband-Harmonic-Shaping VCO Achieving 196.9dBc/Hz Peak FoM and 90-to-180kHz $1/f^3$ PN Corner Without Harmonic Tuning," *ISSCC*, pp. 294–296, Feb. 2021. https://doi.org/10.1109/ISSCC42613.2021.9365761
[11] X. Meng et al., "Analysis and Design of a 15.2-to-18.2-GHz Inverse-Class-F VCO with a Balanced Dual-Core Topology Suppressing the Flicker Noise Upconversion," *IEEE TCAS-I*, vol. 70, no. 12, pp. 5110–5123, Dec. 2023.
https://doi.org/10.1109/TCSI.2023.3312817
[12] C.-C. Hung et al., "A Fractional-N PLL with 34fs$_{rms}$ Jitter and -255.5dB FoM Based on a Multipath Feedback Technique," *ISSCC*, pp. 328–329, Feb. 2025.
https://doi.org/10.1109/ISSCC49661.2025.10904550
[13] A. Grebennikov et al., "Switchmode RF Power Amplifiers." Communications engineering series. Amsterdam: Elsevier/Newnes, 2007. https://doi.org/10.1016/C2011-0-04475-7
[14] H. Jia et al., "A Low-Phase-Noise Quad-Core Millimeter-Wave Fundamental VCO Using Circular Triple-Coupled Transformer in 65-nm CMOS," *IEEE JSSC*, vol. 58, no. 2, pp. 371–385, Feb. 2023. https://doi.org/10.1109/JSSC.2022.3196181
[15] P. Chen et al., "A 529-µW Fractional-N All-Digital PLL Using TDC Gain Auto-Calibration and an Inverse-Class-F DCO in 65-nm CMOS," *IEEE TCAS-I*, vol. 69, no. 1, pp. 51-63, Jan. 2022. https://doi.org/10.1109/TCSI.2021.3094094

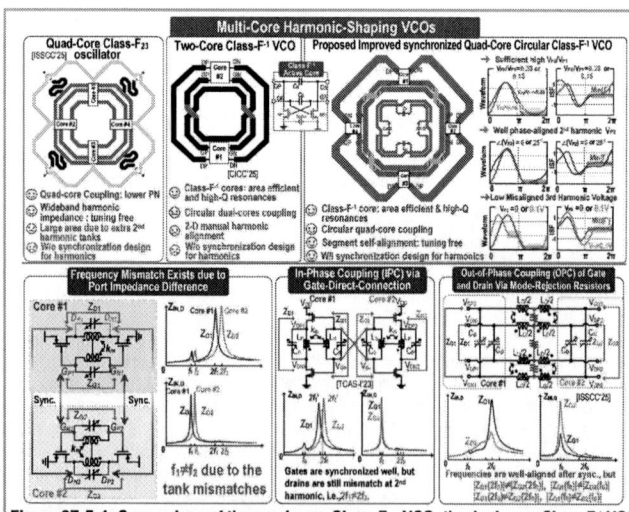

Figure 27.5.1: Comparison of the quad-core Class-F$_{23}$ VCO, the dual-core Class-F^{-1} VCO, and the proposed VCO (upper); Synchronization challenges in MC-HS VCOs with mismatches (bottom), exemplified by the Class-F^{-1} VCO.

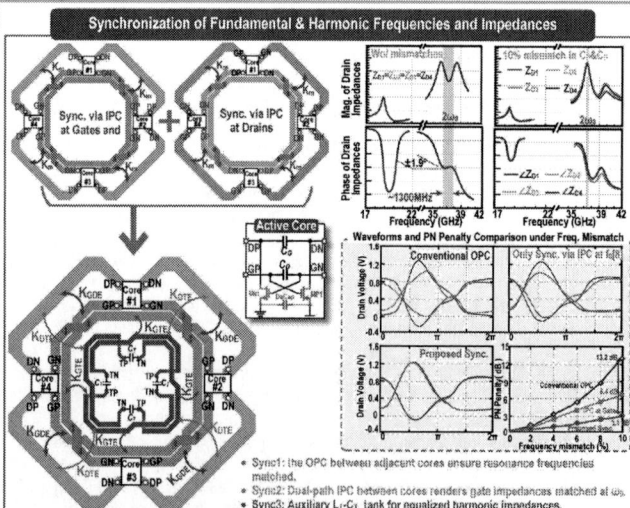

Figure 27.5.2: Evolution of the proposed quad-core VCO with improved synchronization (left); the effect of the synchronization of the proposed VCO (right).

DIGEST OF TECHNICAL PAPERS • 475

ISSCC 2026 / SESSION 27 / FREQUENCY GENERATORS, MULTIPLIERS AND MODULATORS / 27.5

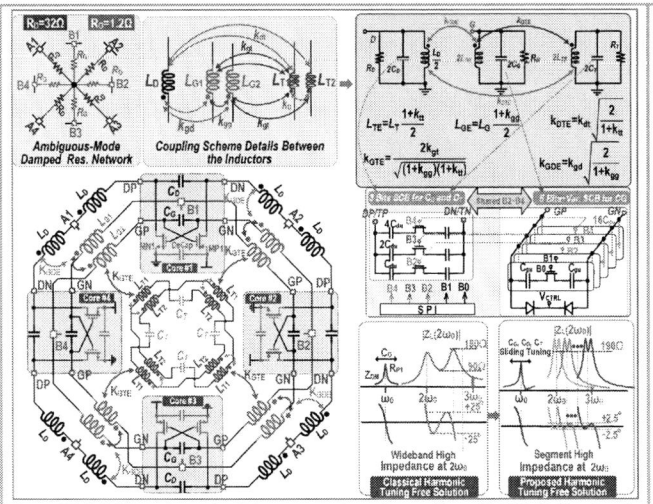

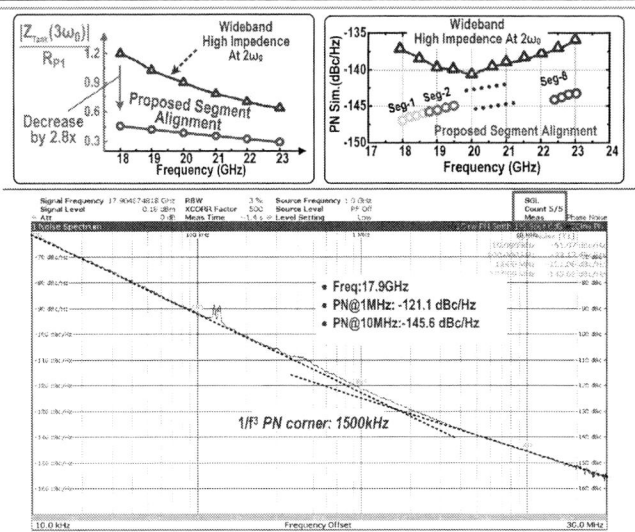

Figure 27.5.3: The schematic of the proposed quad-core Class-F⁻¹ VCO (left) using the circular pentafilar-transformer-based tank and the theory of segmented harmonic alignment (bottom-right).

Figure 27.5.4: The comparison of impedance at 3ω₀ and PN in the FTR (top); measured phase noise at 17.9GHz (bottom).

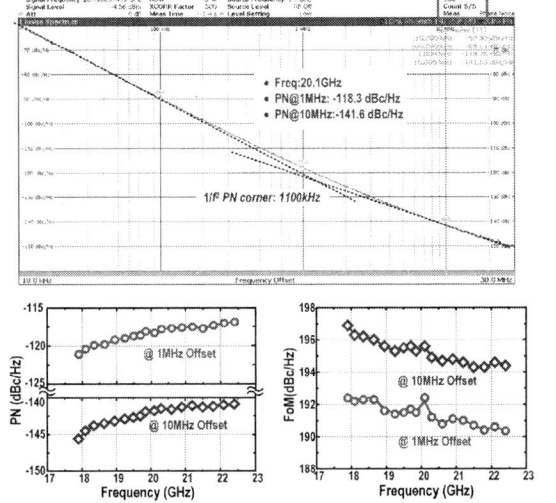

Figure 27.5.5: Measured phase noise at 20.1GHz (top). Measured PN and FoM across the frequency-tuning range (bottom).

Figure 27.5.6: Comparison of the prior-art low-PN high-FoM multi-core and harmonic-shaping oscillators at >10GHz.

Parameters		This Work	ISSCC'25 [1]	CICC'25 [6]	ISSCC'23 [8]	ISSCC'23 [9]	ISSCC'18	JSSC'18 [7]	JSSC'17 [2]
Technology		Sync. Circular class-F⁻¹ with Segment Harmonic Alignment	Wideband Class-F₂₁	Circular class-F⁻¹	Dual-path Sync. Circular	Inter-Core-Shaping	Concentric	Circular	Concentric
Num. of Cores		4	4	2	4	20	4	4	4
Frequency (GHz)		17.9 to 22.4	18.5 to 23.6	22 to 28.4	22.4 to 26.8	25.3 to 30.1	13.1 to 15.4	23 to 29.9	17.4 to 20.3
Tuning Range (%)		22.3	24.3	25.3	18.2	17.3	16	26	15
Power (mW)		18.5 to 24	17.3 to 21	11.8	17.6 to 25	56 to 62	72	16.4	50
PN (dBc/Hz)	@ 1MHz	-121.1 to 116.9	-118.3 to -116.8	-110.8 to -107.3	-115.3	-121.7 to -119.8	-124	-110	-118.5
	@ 10MHz	-145.6 to -141	-139.3 to -138	-135.6 to -133.5	-138	-142.8 to -140.5	-144ˢ	-130ᴬ	-140
FoM (dBc/Hz)	@ 1MHz	190.4 to 192.5	189.4 to 192.6	NA	190.6	191.3 to 192.7	189	187	187.5
	@ 10MHz	(195.6) 194.3 to 196.9	190.1 to 193	191.5 to 192.2	193.3	192.2 to 193.4	189	187ᴬ	189
FoMᴛ (dBc/Hz)	@ 1MHz	197.4 to 199.4	197.1 to 200.3	NA	195.8	195.8	193	195.3	191
	@ 10MHz	201.3 to 203.9	197.8 to 200.8	199.5 to 200.2	198.5	198.5	193	195.3	192.5
Technology		65nm CMOS	65nm CMOS	65nm CMOS	65 nm CMOS	40 nm CMOS	0.13 µm BiCMOS	40 nm CMOS	55 nm CMOS
Core Area (mm²)		0.15	0.28	0.04	0.16	0.33	1	0.1	0.6

FoM = |PN| + 20log₁₀(f₀/Δf)-10log₁₀(P_DC/1mW) FoMᴛ = FoM + 20log₁₀(TR/10) ˢ Estimated from figures

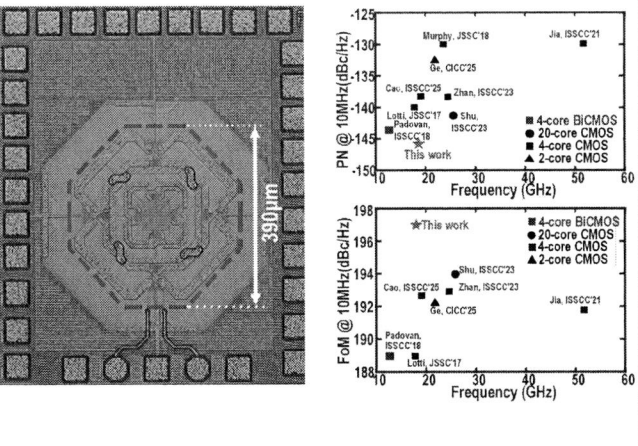

Figure 27.5.7: Die micrograph (left) and comparison of the PN and FoM at a 10MHz offset with the prior-art oscillators operating between 10 and 60GHz (right).

• 2026 IEEE International Solid-State Circuits Conference

ISSCC 2026 / SESSION 27 / FREQUENCY GENERATORS, MULTIPLIERS AND MODULATORS / 27.6

27.6 A 9.7GHz Self-Linearized-VCO-Based FMCW Chirp Generator Achieving 1.56GHz/µs Slope and 0.57µs Duration with 0.094% rms Frequency Error

Daxu Zhang, Yuncheng Zhang, Zezheng Liu, Yuang Xiong, Michele Rossoni, Wenqian Wang, Ashbir Aviat Fadila, Duo Li, Minzhe Tang, Dongfan Xu, Carrel de Gomez, Dingxin Xu, Kazuaki Kunihiro, Hiroyuki Sakai, Kenichi Okada

Institute of Science Tokyo, Tokyo, Japan

Abstract

A 9.7GHz self-linearized-VCO-based Type-III FMCW synthesizer is presented. A C_{Cal} bank calibrates the nonlinearity of the C_{Mod} bank, and two LMS-based calibrations (gain and offset) reduce digital overhead and ensure stable convergence under fast chirp rates. Fabricated in 65nm CMOS, it achieves the shortest chirp duration of 0.57µs, a 40ns retrace, and a 1562.8MHz/µs chirp rate, delivering a 1GHz bandwidth with a 0.094% rms FM error.

FMCW radars play a pivotal role in automotive sensing systems, where high-precision ranging and rapid velocity estimation are essential. With the recent allocation of the FR3 band for high-resolution sensing and communication, achieving both a wide modulation bandwidth and fast chirp generation has become increasingly important to meet stringent performance requirements. To maintain high DCO linearity in such fast chirp generation, the DPD-based TPM technique is commonly employed in phase-locked loops (PLLs) [1,2], enabling the use of narrowband loop filters to preserve excellent phase-noise performance even at high chirp rates [3-6]. However, in the DPD-based TPM technique, the coefficients are iteratively updated using the phase difference measured by the TDC, where the phase is sampled once per reference cycle, forming a sequence of discrete-time samples. The sawtooth or triangular chirp is divided into 32 or 64 segments, and as the chirp rate increases, the number of samples per segment may become insufficient (e.g., fewer than two), leading to instability in the LUT coefficient update and consequently degrading VCO linearity, particularly at lower reference frequencies. To address this issue, this work introduces a self-linearized-VCO-based Type-III architecture that realizes a reduced chirp duration and an increased chirp rate, while preserving competitive frequency-error performance. The fabricated PLL, centered at 9.7GHz, leverages the self-linearized VCO architecture to achieve a chirp rate of 1.56GHz/µs and a duration time of 0.57µs. Under this condition, the generated chirp attains a 1GHz bandwidth with an rms FM error of 0.094% and a retrace time of 40ns ($4 \times T_{ref}$).

Figure 27.6.1 (left) depicts the architecture of a conventional PWL-based FMCW synthesizer, in which a 6-bit LUT (64 segments) is employed. At higher chirp rates, the duration time (T_C) is significantly reduced (e.g., to less than 1µs), thereby reducing the number of samples per segment [7-9]. Insufficient samples per segment (e.g., fewer than two) lead to instability in the calibration coefficients, resulting in the degradation of the DCO linearity. Reducing the LUT depth, for example, employing a 4-bit LUT (16 segments), enables convergence at higher chirp rates; however, the residual frequency error after linearization increases accordingly. Therefore, a fundamental trade-off exists between linearity and duration time.

Figure 27.6.1 (right) illustrates the architecture of the self-linearized-VCO-based Type-III based FMCW synthesizer. The architecture consists of a ramp tracker (RT) and a VCO-nonlinearity calibration block (NL$_{VCO}$ Cal.). The RT generates frequency chirps while mitigating the frequency gain error in the C_{Mod} bank. Subsequently, the NL$_{VCO}$ Cal. employs a C_{Cal} bank to compensate for the chirp nonlinearity introduced by the C_{Mod} bank. Since the C_{Mod} bank is implemented with NMOS varactors whereas the C_{Cal} bank uses PMOS varactors, their nonlinearities exhibit opposite characteristics, allowing the C_{Cal} bank to effectively cancel the nonlinearity of the C_{Mod} bank. This allows for the realization of high chirp rates while maintaining suppressed FM error. The NL$_{VCO}$ Cal. linearizes the nonlinearities of the VCO by tuning the gain and offset of the C_{Cal} bank, so that only two LMS-based LUTs are required, thereby significantly reducing the digital overhead. Moreover, sufficient samples can be utilized with the two LUTs even under fast chirp-rate conditions to accurately update the gain and offset coefficients of the C_{Cal} bank, effectively breaking the trade-off between nonlinearity and chirp rate.

Figure 27.6.2 (top left) shows the implementation details of the self-linearized VCO, which consists of a 256-NMOS-varactor capacitor array (C_{Mod} bank) generating a chirp bandwidth of approximately 1.2GHz. This bank employs 128 RVT NMOS and 128 LVT NMOS devices, and the multi-Vth structure effectively suppresses the intrinsic nonlinearity of C_{Mod} bank, as depicted in Fig. 27.6.2 (bottom left). Compared with a conventional 256-RVT-NMOS design [10-12], the proposed multi-Vth structure reduces the VCO sensitivity K_{VCO} variation by 30%. As shown in Fig. 27.6.2 (bottom right), compared with the RVT-only design, the multi-Vth structure reduces the nonlinearity from ±12 to ±7.5MHz. In addition, a 32-PMOS-varactor capacitor array (C_{Cal} bank) covers 0.2GHz and introduces significant nonlinearity, which is exploited to cancel the nonlinearity of C_{Mod} bank. The operating principle of C_{Cal} bank is illustrated in Fig. 27.6.2 (top right). The offset parameter λ_0 is employed to calibrate the quasi second-order nonlinearity, where λ_{0B} denotes the optimum value. Deviation of λ_0 from λ_{0B} results in residual second-order nonlinearity. $\lambda_{0A} < \lambda_{0B}$ results in a downward-opening quadratic response, whereas $\lambda_{0C} > \lambda_{0B}$ leads to an upward-opening one. At the optimum point $\lambda_0 = \lambda_{0B}$, the second-order nonlinearity is canceled, leaving a third-order characteristic. The gain parameter λ_1 is employed to calibrate the quasi third-order

nonlinearity, where λ_{1B} denotes the optimum value. For $\lambda_1 > \lambda_{1B}$ or $\lambda_1 < \lambda_{1B}$, the residual third-order nonlinearity characteristic exhibits a polarity determined by whether the PMOS-based C_{Cal} bank or the NMOS-based C_{Mod} bank dominates. At the optimum point $\lambda_1 = \lambda_{1B}$, a negligible residual nonlinearity remains. As shown in Fig. 27.6.2 (bottom right), the proposed self-linearized VCO suppresses the uncalibrated multi-Vth nonlinearity from ±7.5 to ±0.3MHz, achieving an improvement by a factor of 25. Furthermore, it remains robust under process variations, maintaining the VCO nonlinearity within ±0.3MHz across the three process corners.

Figure 27.6.3 shows the block diagram of the self-linearized-VCO-based Type-III FMCW synthesizer. The ramp tracker consists of two cascaded accumulators. The first accumulator outputs the constant frequency-ramp slope µ in the steady state, which determines the step size of the second accumulator output. After quantization in the Q$_{mod}$ module, the upper 10 bits are delivered to tw$_{mod}$ to drive the C_{Mod} bank, while the quantization residue is forwarded to tw$_{track}$ and added for quantization-error compensation. The gain and offset of the C_{Cal} bank are tuned to compensate the VCO nonlinearity caused by the C_{Mod} bank. The offset calibration (Offset Cal.) module computes λ_0 by accumulating the frequency error obtained from the differentiated TDC output. As illustrated in the error analysis of the C_{Cal} bank (bottom left of Fig. 27.6.3), when λ_0 is above or below its optimum value λ_{0_opt}, the accumulated value of E_K (ΣE_K) becomes negative or positive, respectively. When λ_0 equals λ_{0_opt}, the accumulator output converges toward zero, and λ_0 stabilizes at its optimum value. The gain calibration (Gain Cal.) module compensates for quasi third-order nonlinearity. It determines the iterative update of λ_1 by computing the accumulated frequency errors in the first half (ΣE_{K1}) and the second half (ΣE_{K2}) of the chirp. If λ_1 is above its optimum value λ_{1_opt}, ΣE_{K1} is positive and ΣE_{K2} is negative, yielding $\Sigma E_{K1} - \Sigma E_{K2} > 0$. Conversely, if λ_1 is below λ_{1_opt}, ΣE_{K1} is negative and ΣE_{K2} is positive, yielding $\Sigma E_{K1} - \Sigma E_{K2} < 0$. At the optimum point $\lambda_1 = \lambda_{1\_opt}$, both ΣE_{K1} and ΣE_{K2} converge to nearly zero, and thus $\Sigma E_{K1} - \Sigma E_{K2} \approx 0$, driving λ_1 toward a stable optimum. As shown in Fig. 27.6.3 (bottom right), the convergence behavior demonstrates that the C_{Mod} bank gain, the C_{Cal} bank gain, and offset settle to stable values within a short period (<100µs).

The proposed FMCW synthesizer was fabricated in 65nm CMOS, and the die micrograph is shown in Fig. 27.6.7. The core area is 0.24mm², excluding decoupling capacitors and test circuits. The prototype consumes approximately 9.8mW during operation. The PLL covers a frequency range from 8.6 to 10.3GHz with a 100MHz reference clock. As shown in Fig. 27.6.4, the chirp is centered at 9.7GHz with a bandwidth of approximately 1GHz, as measured at the divider-by-2 output. The rms FM errors are 20.5kHz, 158.7kHz, and 850.8kHz when the chirp rates are configured to 3.05MHz/µs, 9.06MHz/µs, and 1562.8MHz/µs, respectively. At the maximum chirp rate, the retrace time is 40ns. As illustrated in Fig. 27.6.4 (bottom right), the proposed design achieves consistently lower FM error than the PWL-based DPD approach across chirp rates from 3.05 to 1562.8MHz/µs, with significant advantages at high chirp rates beyond 781.5 and 1562.8MHz/µs. Figure 27.6.5 (top left) shows the output frequency spectrum of the FMCW synthesizer in the steady-state operation. When tested in the PLL mode with an FCW setting of $97+10^{-12}$, the worst fractional spur is –52.8dBc at 24.4kHz, and the integrated jitter is 278fs (10kHz to 100MHz). The measured phase noise is –103.8dBc/Hz at a 1MHz offset, corresponding to –103.5dBc/Hz when normalized to 10GHz.

Figure 27.6.6 summarizes the performance comparison with prior works. Due to the self-linearized-VCO-based Type-III architecture, the number of LMS units in the NL$_{VCO}$ calibration is reduced to two, enabling more stable coefficient convergence under fast-chirp conditions. As a result, among designs in Fig. 27.6.6, the proposed design achieves the shortest duration time (0.57µs) and a retrace time as low as four reference cycles. Furthermore, at a chirp rate of 1562.8MHz/µs, this work achieves the best rms FM error among prior work operating beyond 1GHz/µs in Fig. 27.6.6.

Acknowledgement:
This work is partially supported by NICT (JPJ012368C00801), STAR, and VDEC in collaboration with Cadence Design Systems, Inc., Mentor Graphics, Inc., and Keysight Technologies Japan, Ltd. Corresponding Author: Yuncheng Zhang.

476 • 2026 IEEE International Solid-State Circuits Conference

979-8-3315-8937-0/26 $31.00 © 2026 IEEE

ISSCC 2026 / February 18, 2026 / 10:05 AM

References:

[1] H. Li et al., "A 27-GHz Fractional-N Sub-Sampling PLL Achieving 57.9fs RMS Jitter, –249.7dB FoM, and 1.98µs Locking Time Using Polarity-Reversible SSPD," *ISSCC*, pp. 336-337, Feb. 2025. https://doi.org/10.1109/ISSCC49661.2025.10904556

[2] G. Jin et al., "A Fractional-N Sampling PLL With a Merged Constant-Slope DTC and Sampling PD," *IEEE JSSC*, vol. 59, no. 8, pp. 2407-2417, Aug. 2024. https://doi.org/10.1109/JSSC.2024.3358564

[3] F. Tesolin et al., "A 10GHz FMCW Modulator Achieving 680MHz/µs Chirp Slope and 150kHz RMS Frequency Error Based on a Digital-PLL with a Non-Uniform Piecewise-Parabolic Digital Predistortion," *ISSCC*, pp. 198-200, Feb. 2024. https://doi.org/10.1109/ISSCC49657.2024.10454289

[4] X. Wang et al., "An 11GHz 2nd-Order DPD FMCW Chirp Generator With 0.051% RMS Frequency Error Under a 2.3GHz Chirp Bandwidth, 2.3GHz/µs Slope, and 50ns Idle Time in 65nm CMOS," *ISSCC*, pp. 200-202, Feb. 2024. https://doi.org/10.1109/ISSCC49657.2024.10454283

[5] P. T. Renukaswamy et al. "A 16-GHz Background-Calibrated Duty-Cycled FMCW Charge-Pump PLL," *IEEE JSSC*, vol. 59, no. 6, pp. 1684-1696, June 2024. https://doi.org/10.1109/JSSC.2023.3331519

[6] A. Yan et al., "An 11-to-16.4GHz, 3.4GHz/µs-Slope, 5.32GHz-Chirp-Bandwidth, 0.043%-RMS-Frequency-Error FMCW Digital PLL with Posterior-Segment DPD Featuring 5-Chirp-Cycle Convergence Time," *ISSCC*, pp. 340-341, Feb. 2025. https://doi.org/10.1109/ISSCC49661.2025.10904547

[7] Z. Shen et al., "A 24-GHz Self-Calibrated All-Digital FMCW Synthesizer With 0.01% RMS Frequency Error Under 3.2-GHz Chirp Bandwidth and 320-MHz/µs Chirp Slope," *IEEE JSSC*, vol. 57, no. 7, pp. 2167-2180, July 2022. https://doi.org/10.1109/JSSC.2021.3123693

[8] Q. Shi et al., "A Self-Calibrated 16GHz Subsampling-PLL-Based 30s Fast Chirp FMCW Modulator With 1.5GHz Bandwidth and 100kHz RMS Error," *ISSCC*, pp. 408-410, Feb. 2019. https://doi.org/10.1109/ISSCC.2019.8662348

[9] D. Cherniak et al., "A 23GHz Low-Phase-Noise Digital Bang-Bang PLL for Fast Triangular and Sawtooth Chirp Modulation," *ISSCC*, pp. 248-250, Feb. 2018. https://doi.org/10.1109/ISSCC.2018.8310277

[10] Y. Wu et al., "A 3.5–6.8-GHz Wide-Bandwidth DTC-Assisted Fractional-N All-Digital PLL with a MASH ΔΣ-TDC for Low In-Band Phase Noise," *IEEE JSSC*, vol. 52, no. 7, pp. 1885-1903, July 2017. https://doi.org/10.1109/JSSC.2017.2682841

[11] M. Shahmohammadi et al., "A 1/f Noise Upconversion Reduction Technique Applied to Class-D and Class-F Oscillators," *ISSCC*, pp. 444-445, Feb. 2015. https://doi.org/10.1109/ISSCC.2015.7063117

[12] I. Apostolina and D. Manstretta, "A 14.5–17.9-GHz Harmonically-Coupled Quad-Core P-N Class-B DCO with –117.3-dBc/Hz Phase Noise at 1-MHz Offset in 28-nm CMOS," *IEEE RFIC*, pp. 211-214, June 2022. https://doi.org/10.1109/RFIC54546.2022.9863180

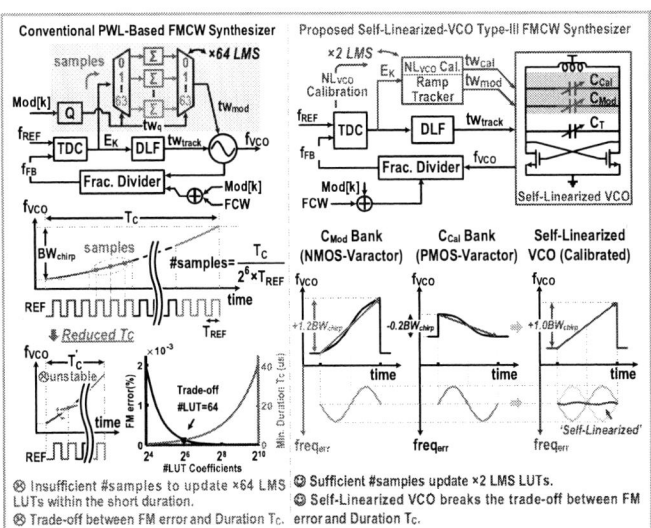

Figure 27.6.1: Limitations of traditional PWL-based FMCW synthesizer (left), proposed self-linearized-VCO-based Type-III FMCW synthesizer(right).

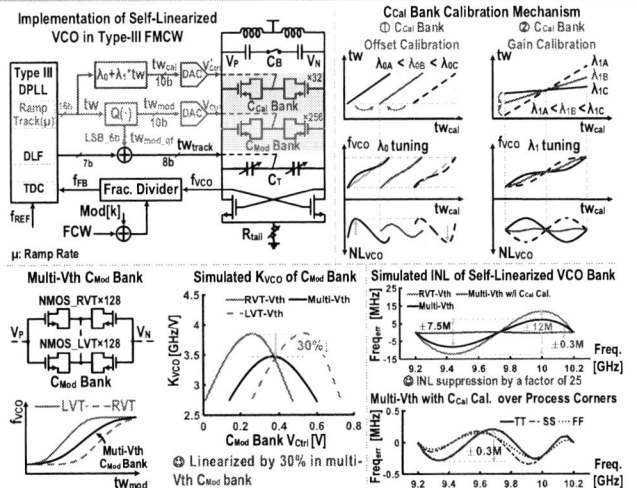

Figure 27.6.2: Self-linearized VCO architecture and operation (top left), C_{Cal} bank calibration mechanism (top right), multi-Vth C_{Mod} bank (bottom left), and simulated VCO INL (bottom right).

DIGEST OF TECHNICAL PAPERS • 477

979-8-3315-8937-0/26 $31.00 © 2026 IEEE

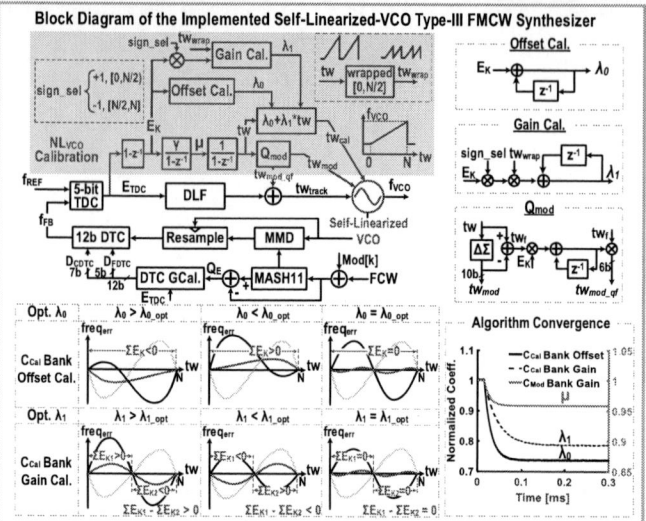

Block Diagram of the Implemented Self-Linearized-VCO Type-III FMCW Synthesizer

Figure 27.6.3: Block diagram of the implemented FMCW synthesizer (top), background calibration of the C_{Cal} bank (bottom left), and algorithm convergence (bottom right).

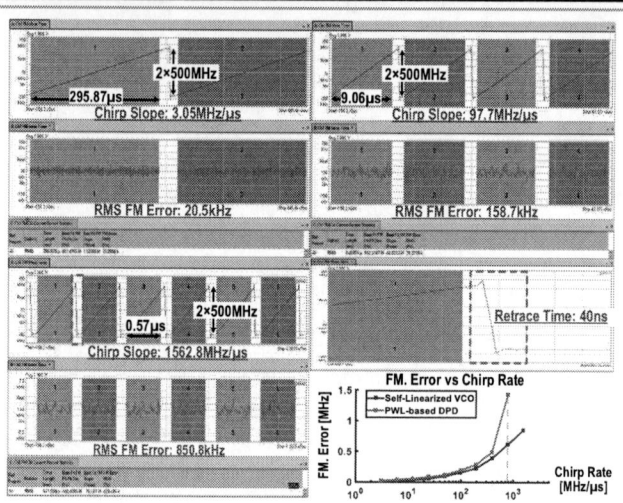

Figure 27.6.4: Measured divide-by-2 output: chirp rate and FM error for chirp durations of 295.87µs (top left), 9.06µs (top right), and 0.57µs (bottom left); retrace time at the fastest chirp rate and the FM error versus chirp rate (bottom right).

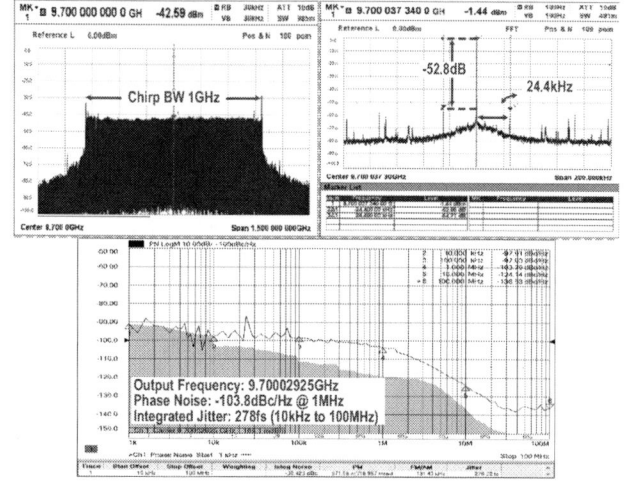

Figure 27.6.5: Measured chirp spectrum (top left) and phase-noise and fractional-spur spectra around a 9.7GHz near-integer channel with a 24.4kHz offset (bottom and top right).

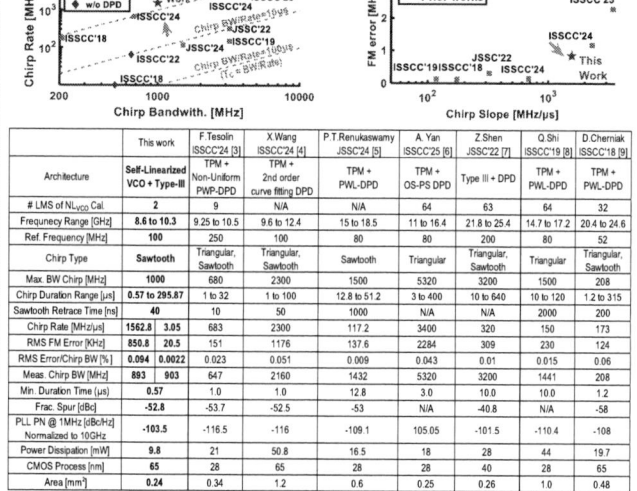

Figure 27.6.6: Performance comparison table with prior works.

	This work	F.Tesolin ISSCC'24 [3]	X.Wang ISSCC'24 [4]	P.T.Renukaswamy JSSC'24 [5]	A.Yan ISSCC'25 [6]	Z.Shen JSSC'22 [7]	Q.Shi ISSCC'19 [8]	D.Cherniak ISSCC'18 [9]
Architecture	Self-Linearized VCO + Type-III	TPM + Non-Uniform PWP-DPD	TPM + 2nd order curve fitting DPD	TPM + PWL-DPD	TPM + OS-PS DPD	Type III + DPD	TPM + PWL-DPD	TPM + PWL-DPD
# LMS of NL$_{VCO}$ Cal.	2	9	N/A	N/A	64	63	64	32
Frequency Range [GHz]	8.6 to 10.3	9.25 to 10.5	9.6 to 12.4	15 to 18.5	11 to 16.4	21.8 to 25.4	14.7 to 17.2	20.4 to 24.6
Ref. Frequency [MHz]	100	250	100	80	80	200	80	52
Chirp Type	Sawtooth	Triangular, Sawtooth	Triangular, Sawtooth	Sawtooth	Triangular	Triangular, Sawtooth	Triangular	Triangular, Sawtooth
Max. BW Chirp [MHz]	1000	680	2300	1500	5320	3200	1500	208
Chirp Duration Range [µs]	0.57 to 295.87	1 to 32	1 to 100	12.8 to 51.2	3 to 400	10 to 640	10 to 120	1.2 to 315
Sawtooth Retrace Time [ns]	40	10	50	1000	N/A	N/A	2000	200
Chirp Rate [MHz/µs]	1562.8 3.05	683	2300	117.2	3400	320	150	173
RMS FM Error [KHz]	850.8 20.5	151	1176	137.6	2284	309	230	124
RMS Error/Chirp BW [%]	0.094 0.0022	0.023	0.051	0.009	0.043	0.01	0.015	0.06
Meas. Chirp BW [MHz]	893 903	647	2160	1432	5320	3200	1441	208
Min. Duration Time [µs]	0.57	1.0	1.0	12.8	3.0	10.0	10.0	1.2
Frac. Spur [dBc]	-52.8	-53.7	-52.5	-53	N/A	-40.8	N/A	-58
PLL PN @ 1MHz [dBc/Hz] Normalized to 10GHz	-103.5	-116.5	-116	-109.1	105.05	-101.5	-110.4	-108
Power Dissipation [mW]	9.8	21	50.8	16.5	18	28	44	19.7
CMOS Process [nm]	65	28	65	28	28	40	28	65
Area [mm²]	0.24	0.34	1.2	0.6	0.25	0.26	1.0	0.48

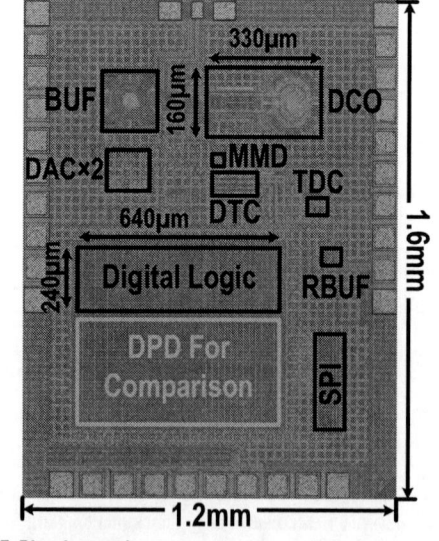

Figure 27.6.7: Die micrograph.

• 2026 IEEE International Solid-State Circuits Conference

979-8-3315-8937-0/26 $31.00 © 2026 IEEE

ISSCC 2026 / SESSION 27 / FREQUENCY GENERATORS, MULTIPLIERS AND MODULATORS / 27.7

27.7 A 40.5-to-58.5GHz 36%-Fractional-Chirp-Bandwidth 18GHz-Absolute-Chirp-Bandwidth 2.2GHz/µs-Chirp-Rate 0.02%-Chirp-Error Post-Mixing Bandwidth-Extending Sawtooth-FMCW Frequency Synthesizer Employing a Chirp-Tracking ILFT and a Fractional-Bandwidth Doubler

Yi Liu*, Zixi Jing*, Wen Yang, Hanlin Yang, Bodong Zhang, Howard Cam Luong

HKUST, Hong Kong, China
*Equally Credited Authors (ECAs)

Abstract

An 18GHz-bandwidth (BW) sawtooth-FMCW synthesizer is presented. First, a chirp-tracking ILFT with feed-forward and frequency-tracking loop ensures robust and continuous tracking of fast chirps. Second, a fractional-BW (FBW) doubler enables sawtooth generation and extends both the BW and FBW while halving the chirp error. The synthesizer measures a 36% FBW sawtooth FMCW chirp from 40.5 to 58.5GHz with a 0.02% chirp error at a chirp rate of 2.2GHz/µs without DCO-nonlinearity digital pre-distortion.

Mm-Wave (mmW) FMCW radar systems require frequency synthesizers (FSs) with wide chirp absolute bandwidths (ABWs) and chirp fractional bandwidths (FBWs), as well as fast chirp rates and high chirp linearity. Sawtooth FMCW chirps are preferred over triangular ones for multi-object detection [1-6] but are more challenging to generate. Conventional sawtooth FMCW FSs rely on complex DCO-nonlinearity (NL) digital pre-distortion (DPD) to improve linearity under fast chirps [1-3], but suffer from large DPD overhead and convergence issues, limited chirp BW, and increased phase noise (PN). [7] employs posterior-segment DPD in a triangular chirp to reduce convergence time, but this approach is not suitable for sawtooth generation. Using frequency multipliers (FMs) can extend the ABW [10-13], but it does not improve the FBW nor chirp linearity. A pre-mixing FBW-boosting method [14] down-converts the center frequency of a main PLL (MPLL) to boost the FBW while preserving the ABW and is followed by an injection-locked frequency multiplier (ILFM) to recover the original frequency as shown in Fig. 27.7.1 (top left). Such an ILFM was not demonstrated in [14] as shown in Fig. 27.7.1 (bottom left), but its implementation would face two key design challenges. First, because mixers typically provide low output power, the ILFM may exhibit a narrow locking range (LR) given the limited input loading and power budget. Moreover, the center frequency is not increased compared to the MPLL even with the ILFM. Second, harmonic mixing spurs may fall into the mixer and ILFM frequency band. Finally, [14] requires a complicated fractional frequency divider (FFD) to correlate and cancel the MPLL PN, which inevitably adds extra spurs and PN.

This work proposes a post-mixing BW-extending sawtooth FMCW FS, as shown in Fig. 27.7.1 (top right). The system introduces four features. First, a chirp-tracking injection-locked frequency tripler (CT-ILFT) is placed right after an MPLL and before a mixer. As such, the CT-ILFT is driven by a large-swing and clean output from the MPLL and can thus achieve improved input injection strength, LR, and spectrum purity. Second, an adaptive tracking scheme of the CT-ILFT self-oscillation frequency f_{osc} is proposed to further extend the CT-ILFT LR with high spectrum purity and lower power. Third, an FBW doubler with frequency and phase alignment and a slope-polarity reverser (SPR) are proposed to generate sawtooth chirps and simultaneously boost the ABW and FBW without lowering the center frequency. Fourth, the FBW doubler halves the chirp error without the need of a complicated DCO NL DPD.

Figure 27.7.1 (bottom) shows the operation of this FMCW FS. A triangular chirp f_{MPLL} from the MPLL is generated and fed to the CT-ILFT to multiply the MPLL ABW_0, center frequency and chirp rate by 3. The CT-ILFT up-chirp $f_{CT-ILFT}$ forms the low-band (LB) output while its down-chirp is folded to a high-band (HB) up-chirp by the FBW doubler. By aligning their frequencies and phases at the band-switching (BS) point f_{BS}, the LB and HB chirps are combined into a seamless sawtooth chirp. Since the absolute chirp error Δf_{rms} remains constant across the LB and HB, the normalized chirp error, $\Delta f_{rms}/ABW$ (%), is halved compared to that of the MPLL and CT-ILFT. In this post-mixing scheme, since the FBW at the CT-ILFT input is not boosted, its LR (%) requirement is significantly relaxed as compared to the pre-mixing FBW-boosting scheme in [14]. Crucially, the CT-ILFT fully covers $ABW=3\times ABW_0$ by adaptively aligning its f_{osc} with $3\times f_{MPLL}$ as will be explained later in Fig. 27.7.3. The chirp profiles of each building block are summarized in Fig. 27.7.1 (bottom right). Clearly, this FS simultaneously boosts up the MPLL ABW, center frequency, FBW, and chirp rate while improving the chirp linearity.

Figure 27.7.2 (top) shows the block diagram of this FMCW FS. The MPLL produces a triangular chirp that is controlled by $FCW_{MPLL}(t)=FCW_{chirp}(t)+FCW_0$ to ramp up and down alternately within 13.5 to 16.5GHz, which is then multiplied by the CT-ILFT to ramp down from 40.5 to 49.5GHz. The up-chirp forms the LB output as $f_{LB}(t)=3f_{MPLL}=3f_{REF}\times FCW_{MPLL}(t)$, where the f_{REF} is the reference frequency. The down-chirp is reserved to another HB up-chirp through a down-conversion mixer with a constant auxiliary frequency $f_{aux}=2\times f_{LB,max}=2\times f_{BS}=99GHz$ to realize the HB chirp $f_{HB}=f_{aux}-3f_{MPLL}(t)$ from 49.5 to 58.5GHz. Both LB and HB signals are input to a multiplexer (MUX) and conditionally selected as the FS output by HB_EN. To generate f_{aux}, a low-jitter integer-N auxiliary PLL (APLL) employing a Class-F VCO operating at 11GHz but with a strong 3rd harmonic is cascaded by an injection-locked oscillator (ILO_{aux1}) as a buffer at 33GHz, an $ILFM_{aux}$ ×3, and finally another buffer ILO_{aux2} at 99GHz. As a result, f_{aux} is controlled by the FCW of the APLL as $9f_{REF}\times FCW_{APLL}$ at

a nominal value of 99GHz. To avoid the frequency discontinuity at the BS point f_{BS}, the LB and HB edge frequencies are precisely aligned as $f_{LB,max}=f_{HB,min}=f_{BS}=49.5GHz$ by controlling $FCW_{MPLL,max}=1.5FCW_{APLL}$ in the digital domain. The phase continuity is ensured by a phase calibration (PC) scheme as shown in Fig. 27.7.2 (bottom right). The output phases $\Phi_{out,LB/HB}$ are sequentially compared with Φ_{ILFT} by a phase detector (PD) and digitized by an off-chip ADC as $K_{PD}\times(\Phi_{out}-\Phi_{ILFT})$. Their difference is then accumulated to generate a feedback signal, *phase_cal*, which drives a narrow-range on-chip DTC operating at f_{REF} as the reference of the APLL (10ps range, equivalent to 2π at f_{aux}) for phase alignment between the LB and HB. In this work, the PC is performed in foreground during power-on, but it can also work in background with an on-chip synchronous ADC if necessary. Notably, there is no LR concern in the f_{aux} generation thanks to the single frequency point operation and that the f_{aux} generation is only enabled in HB with ~50% duty cycle to save power. Furthermore, since only half of the FBW is present at the mixer output, negligible in-band harmonic mixing spur exists within the MUX passband. Finally, the f_{REF} PN is correlatedly maintained at the mixer inputs and cancelled at the mixer output at HB.

As shown in Fig. 27.7.2 (bottom left), the MPLL incorporates a digital loop filter (DLF) with 3 paths: proportional (P), integral (I), and feedback (F), similar to [7-10,12]. The F-path cancels the TDC offset proportional to chirp rate [12]. The P and I signals are combined into the output, *slope_out*, which is fed to an integrator to generate a ramp control to the DCO and needs to be reversed at the BS point from SL(@f_{BS}) to -SL(@f_{BS}). The seamless polarity reversal is achieved by an SPR to explicitly add a -2SL(@f_{BS}) offset to the I-path instead of relying on the MPLL loop dynamics. Consequently, a triangular chirp from the MPLL with a negligible chirp error at f_{BS} is generated for seamless combination between LB and HB. Importantly, this work reduces the chirp error from the system level using the FBW doubler without DCO-NL DPD in the MPLL, avoiding the potential convergence issue [1-3,7].

As shown in Fig. 27.7.3 (top left), to extend the ILFT LR for fast CT, the f_{osc} must continuously track the 3rd harmonic of the f_{MPLL}, minimizing the frequency deviation $f_{dev}=3f_{MPLL}-f_{osc}$. This is commonly achieved with a frequency-tracking loop (FTL) [15-17], see Fig. 27.7.3 (top middle). An FTL similar to [17] is adopted in this work (Fig. 27.7.3, bottom left). However, to track a fast chirp, its loop BW needs to be much larger than $1/T_{chirp}$, which would cause instability and chirp linearity degradation [17]. To overcome this limitation, a feed-forward (FF) CT scheme is first considered (Fig. 27.7.3, top right). The FF path predicts the desired f_{osc} based on FCW_{chirp}, $f_{osc}=3f_{MPLL}=3(FCW_{chirp}+FCW_0)f_{ref}$. However, unlike FTL, this open-loop FF path must address the nonlinear f_{osc}-versus-DCW_{FF} curve, where DCW_{FF} is the digital control word of the ILFT switched-capacitor array (SCA). This nonlinearity is pre-distorted using a piecewise linear function with initialized gain coefficients $g_{0,1}$ based on simulation under the typical PVT conditions. The coefficients $g_{0,1}$ are proportional to $1/k_{ILFT}$, where k_{ILFT} is the gain from DCW_{FF} to f_{osc}. Although this FF scheme removes the chirp-rate limit, it is highly sensitive to PVT variations as k_{ILFT} drifts while $g_{0,1}$ stay constant. Severe PVT variations would increase f_{dev} and even cause ILFT to fail to lock. As a solution, this work further combines the FF path with an FTL to compensate for the residual f_{dev} caused by the PVT-induced k_{ILFT} and f_{osc} variations. As shown in Fig. 27.7.3 (bottom), FTL senses the residual f_{dev} and generates a correction code DCW_{FTL} that is summed with DCW_{FF}. As such, the FTL only needs to minimize the small residual error from the FF path, thereby maintaining robust CT. Dynamically controlling f_{osc} does not degrade the normalized chirp error or PN under locking. Additionally, this decoupling of f_{osc} from f_{ILFT} relaxes the accuracy and resolution requirements of the CT scheme.

Figure 27.7.4 shows the demodulated chirp profile at the output divided by 16 with (i) LB-only (top left); (ii) both HB and LB on + MPLL SPR and phase calibration (PC) OFF (top right); and (iii) both HB and LB on + MPLL SPR and PC ON (bottom left). Obviously, the FBW doubler doubles the FBW and ABW compared to the LB output from the CT-ILFT. With the MPLL SPR and the PC turned OFF, the rms chirp error degrades from 0.04% at the MPLL and ILFT outputs to 0.2% at the FS output at a chirp rate of 2.2GHz/ms due to the frequency and phase discontinuity at f_{BS}. In contrast, with the MPLL SPR and PC ON, the rms chirp error at the FS output is reduced by 10 times to only 0.02% at the same chirp rate of 2.2GHz/ms, which is half of that at the CT-ILFT output as expected. The sawtooth retrace time is 16ns. Figure 27.7.4 (bottom right) shows the chirp spectrum at the FS output

478 • 2026 IEEE International Solid-State Circuits Conference

979-8-3315-8937-0/26 $31.00 © 2026 IEEE

divided by 2 with the equivalent ABW of 18GHz and the FBW of 36% at the FS output. Figure 27.7.5 (top left) shows the chirp error at a chirp rate up to 2.2GHz/ms with estimation errors in the FF gain coefficients $g_{0,1}$ intentionally introduced to mimic the PVT variations of the CT-ILFT f_{osc} tuning curve. With FF CT scheme only, the CT-ILFT can tolerate only ±8% estimation errors. However, with both FF and FTL ON, the tolerance is increased to ±35%, which is more than enough to cover the worst-case estimation errors. Figure 27.7.5 (top right) shows the measured chirp errors at different chirp rates and the measured PN at the near-integer channel at 45GHz (bottom).

Figure 27.7.6 compares this work with the prior art. Among the sawtooth FMCW FSs, this FS not only achieves wide FBW and ABW but also significantly improves the ratio of the chirp rate to the chirp error, even without DCO NL DPD. Figure 27.7.7 shows the die micrograph in a 28nm CMOS process.

Acknowledgement:
This research was supported by Hong Kong General Research Fund 16207223. Corresponding authors: Yi Liu and Zixi Jing.

References:
[1] F. Tesolin et al., "A 10GHz FMCW Modulator Achieving 680MHz/μs Chirp Slope and 150kHz rms Frequency Error Based on a Digital-PLL with a Non-Uniform Piecewise-Parabolic Digital Predistortion," *ISSCC*, pp. 198-200, Feb. 2024. http://doi.org/10.1109/ISSCC49657.2024.10454289
[2] X. Wang et al., "An 11GHz 2nd-order DPD FMCW Chirp Generator with 0.051% rms Frequency Error under a 2.3GHz Chirp Bandwidth, 2.3GHz/μs Slope, and 50ns Idle Time in 65nm CMOS," *ISSCC*, pp. 200-202, Feb. 2024. http://doi.org/10.1109/ISSCC49657.2024.10454283
[3] P. T. Renukaswamy et al., "A 16-GHz Background-Calibrated Duty-Cycled FMCW Charge-Pump PLL," *IEEE JSSC*, vol. 59, no. 6, pp. 1684-1696, June 2024. http://doi.org/10.1109/JSSC.2023.3331519
[4] Z. Shen et al., "A 24 GHz Self-Calibrated All-Digital FMCW Synthesizer with 0.01% RMS Frequency Error Under 3.2 GHz Chirp Bandwidth and 320 MHz/μs Chirp Slope," *IEEE JSSC*, vol. 57, no. 7, pp. 2167-2180, July 2022. http://doi.org/10.1109/JSSC.2021.3123693
[5] H. Shanan et al., "A 9-to-12GHz Coupled-RTWO FMCW ADPLL with 97fs RMS Jitter, -120dBc/Hz PN at 1MHz Offset, and With Retrace Time of 12.5ns and 2μs Chirp Settling Time," *ISSCC*, pp. 146-148, Feb. 2022. http://doi.org/10.1109/ISSCC42614.2022.9731575
[6] D. Cherniak et al., "A 23-GHz Low-Phase-Noise Digital Bang–Bang PLL for Fast Triangular and Sawtooth Chirp Modulation," *IEEE JSSC*, vol. 53, no. 12, pp. 3565-3575, Dec. 2018. http://doi.org/10.1109/JSSC.2018.2869097
[7] A. Yan et al., "An 11-to-16.4GHz, 3.4GHz/μs-Slope, 5.32GHz-Chirp-Bandwidth, 0.043%-RMS-Frequency-Error FMCW Digital PLL with Posterior-Segment DPD Featuring 5-Chirp-Cycle Convergence Time," *ISSCC*, pp. 340-341, Feb. 2025. http://doi.org/10.1109/ISSCC49661.2025.10904547
[8] A. Yan et al., "A Cycle-Slip Compensated FMCW Digital PLL with Background Back-Tracking DPD," *IEEE JSSC,* Early Access, 2025. http://doi.org/10.1109/JSSC.2025.3590136
[9] W. Deng et al., "A Self-Adapted Two-Point Modulation Type-II Digital PLL for Fast Chirp Rate and Wide Chirp-Bandwidth FMCW Signal Generation," *IEEE JSSC*, vol. 57, no. 4, pp. 1162-1174, Apr. 2021. http://doi.org/10.1109/JSSC.2021.3129900

[10] W. Deng et al., "A D-Band Joint Radar-Communication CMOS Transceiver," *IEEE JSSC*, vol. 58, no. 2, pp. 411-427, Feb. 2023. http://doi.org/10.1109/JSSC.2022.3185160
[11] B. P. Ginsburg et al., "A Multimode 76-to-81GHz Automotive Radar Transceiver with Autonomous Monitoring," *ISSCC* pp. 158-160, Feb. 2018. http://doi.org/10.1109/ISSCC.2018.8310232
[12] Y. Liu et al., "A 74GHz-80GHz 1.2GHz/μs-Slope 20.9mW FMCW Synthesizer with TDC-Gain-Independent Loop-Bandwidth Employing a TDC-Offset-Free Type-II Digital PLL and a Linearized Hybrid-Tuning DCO," *IEEE RFIC*, pp. 207-210, June 2024. http://doi.org/10.1109/RFIC61187.2024.10600027
[13] C. Xu et al., "A Packaged 54-to-69-GHz Wideband 2T2R FMCW Radar Transceiver Employing Cascaded-PLL Topology and PTAT-Enhanced Temperature Compensation in 40-nm CMOS," *IEEE JSSC*, vol. 59, no. 10, pp. 3156-3171, Oct. 2024. http://doi.org/10.1109/JSSC.2024.3404263
[14] Y. Liu et al., "A 4.25GHz-8.45GHz 67%-Chirp-Fractional-Bandwidth -121.5dBc/Hz-PN@1MHz 88fs-Jitter FMCW Synthesizer with Bandwidth-Boosting and Phase-Noise-Cancellation Techniques," *IEEE RFIC*, pp. 211-214, June 2024. http://doi.org/10.1109/RFIC61187.2024.10599999
[15] D. Shin and K. -J. Koh, "An Injection Frequency-Locked Loop—Autonomous Injection Frequency Tracking Loop with Phase Noise Self-Calibration for Power-Efficient mm-Wave Signal Sources," *IEEE JSSC*, vol. 53, no. 3, pp. 825-838, Mar. 2018. http://doi.org/10.1109/JSSC.2017.2782762
[16] H. Yoon et al., "A Low-Jitter Injection-Locked Multi-Frequency Generator Using Digitally Controlled Oscillators and Time-Interleaved Calibration," *IEEE JSSC*, vol. 54, no. 6, pp. 1564-1574, June 2019. http://doi.org/10.1109/JSSC.2019.2893513
[17] Z. Jing et al., "A 35.2-51.4GHz Frequency-Tracking Injection-Locked Frequency Tripler Achieving >28.5dBc Harmonic Rejection Ratios, -7.3dBm Output Power, and 4.3dB Output Power Variation," *IEEE RFIC*, pp. 347-350, Feb. 2025. http://doi.org/10.1109/RFIC61188.2025.11082938

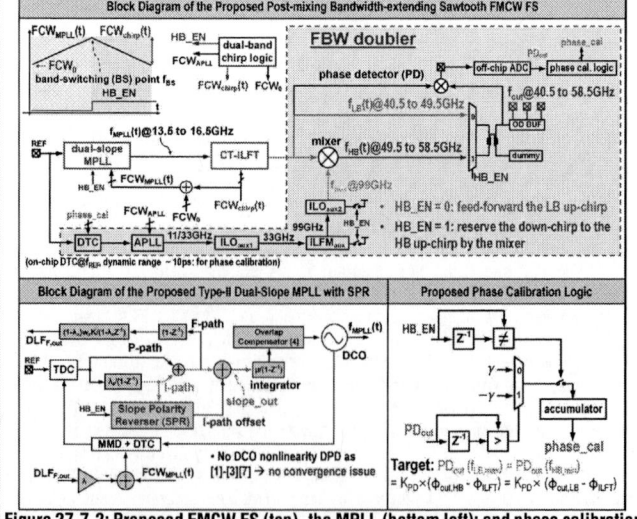

Figure 27.7.1: Existing BW-boosting FMCW FS (top left), proposed FMCW FS (top right); and the comparison (bottom).

Figure 27.7.2: Proposed FMCW FS (top), the MPLL (bottom left); and phase calibration logic (bottom right).

ISSCC 2026 / SESSION 27 / FREQUENCY GENERATORS, MULTIPLIERS AND MODULATORS / 27.7

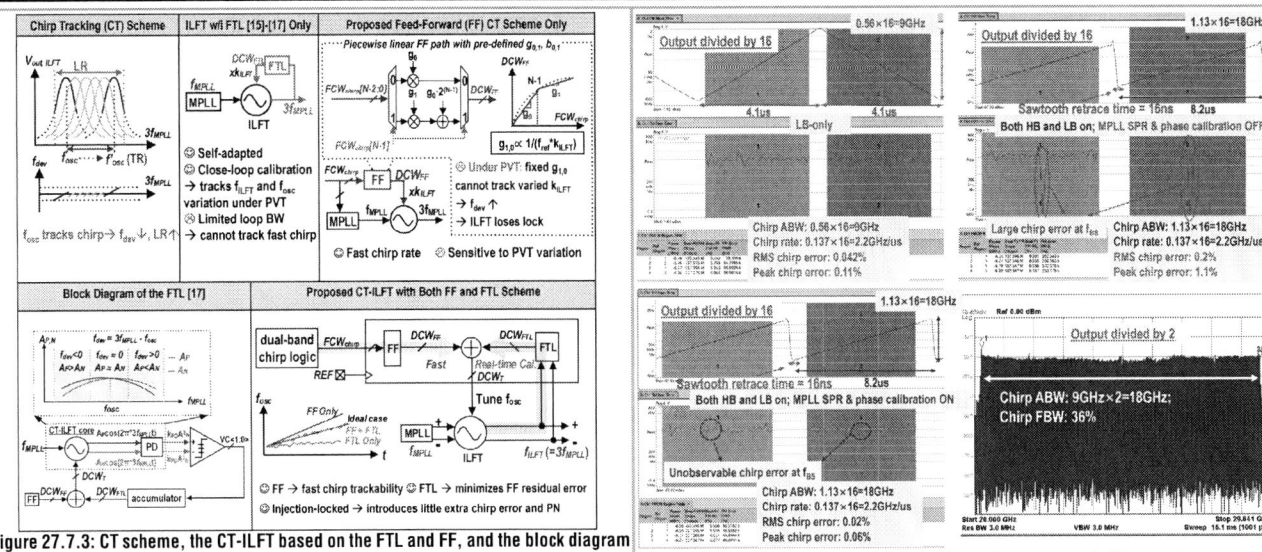

Figure 27.7.3: CT scheme, the CT-ILFT based on the FTL and FF, and the block diagram of the FTL.

Figure 27.7.4: Measured demodulated chirp performance and chirp spectrum.

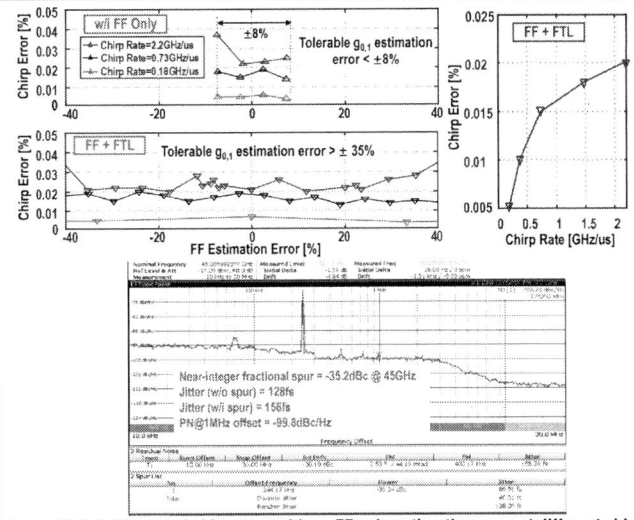

Figure 27.7.5: Measured chirp error with an FF gain estimation error at different chirp rates and measured PN.

Figure 27.7.6: Performance comparison with prior-art FMCW frequency synthesizers.

	This Work	Sawtooth + Triangular FMCW FS					Triangular-only FMCW FS			
		ISSCC'24 [1]	ISSCC'24 [2]	JSSC'24 [12]	JSSC'22 [4]	ISSCC'22 [5]	ISSCC'25 [7]	JSSC'25 [8]	JSSC'22 [9]	JSSC'22 [10]
Architecture	CT-ILFT + FBW doubler + DPLL	DPLL + DPD	SPLL + DPD	cascaded DPLL	Type-III DPLL	DPLL + RTWO	DPLL	DPLL	DPLL	DPLL + doubler
PLL DCO DPD method	Overlap only	Overlap + Nonlinearity	Overlap + Nonlinearity	No	No	Nonlinearity	Overlap + Nonlinearity	Overlap only	No	No
Center freq. [GHz]	50	10	11	61	23.6	10.4	13.7	13.9	12.5	150
Chirp fractional BW (FBW) [%]	36	6.8	20.6	11.8	13.3	6.5	39	24.5	19.5	20
Chirp absolute BW (ABW) [GHz]	18	0.68	2.16	7.2	3.2	0.5	5.35	4.75	2.4	30
Max. Chirp rate [MHz/us]	2197	683	2300	468.5	128.8	65	3400	960	132	1500
RMS Chirp error [%]	0.02	0.072	0.051	0.019	0.012	NA	0.043	0.034	0.1	0.21
Peak Chirp error [%]	0.06	0.22	0.14	NA	NA	NA	0.15	0.71	0.3	NA
23log(Chirp rate /Chirp error) [dB]	100.8	79.5	93.1	87.8	80.6	79.4	98	89	62.4	77.1
Sawtooth retrace time [ns]	16	10	50	NA	NA	12.5	NA	NA	NA	NA
Frequency Tuning Range [GHz / %]*	38.4 to 71.4 (60%)	9.25 to 10.5 (12.7%)	9.6 to 12.4 (25.4%)	54 to 69 (24.5%)	21.8 to 25.4 (15.2%)	8.8 to 12 (30.1%)	11 to 16.4 (39.4%)	11.4 to 16.4 (36%)	11.1 to 14.2 (24.5%)	135 to 165 (20%)
Power [mW]	70.3	21	51	386##	28	187	18	10	23	301##
PN@(1MHz) [dBc/Hz]*	-94.9	-98.5	-98	-90.9	-83.6	-102.6	-87	-84.4	-95.7	-95.3
FoM$_N$@1MHz [dB]**	-190	-185.2	-187.0	-170.8	-170.7	-187.4	-184.2	-183.5	-187.9	-174.5
Worst fractional spur [dBc]	-30.3	-35.7	-34.5	NA	-30.1	NA	NA	NA	NA	NA
Worst mixer spur [dBc]	-48.9	NA	NA	NA	NA	NA	NA	NA	NA	NA
Jitter w/o spur [fs]	128	NA	NA	NA	NA	97	NA	572.4	NA	NA
Jitter w/ spur [fs]	156	87.1	148.7	NA	NA	NA	299.2	NA	NA	NA
Reference frequency [MHz]	250	250	100	NA	200	200	80	76.8	100	100
FoM$_{SL}$ (w/o spur) [dBc]***	-261.9	NA	NA	NA	NA	-254.7	NA	-257.2	NA	NA
FoM$_{SL}$ (w/ spur) [dBc]***	-260.2	-263.9	-259.9	NA	NA	NA	-260.3	NA	NA	NA
Core Area [mm²]	0.52	0.34	1.2	NA	0.26	2	0.34	0.19	0.31	NA
CMOS Technology	28nm	28nm	65nm	40nm	40nm	28nm	28nm	28nm	28nm	28nm

*Normalized to 79GHz
#Both f_{LB} and f_{HB} can be tuned for non-FMCW mode
##Only LO chain for TX is included
**FoM$_N$=PN-20*log(f$_o$/Δf*FTR/10%)+10*log(power/1mW)
###FoM$_{SL}$=20*log10(jitter/1s) + 10*log10(power/1mW) − 10*log10(f$_{out}$/f$_{ref}$)

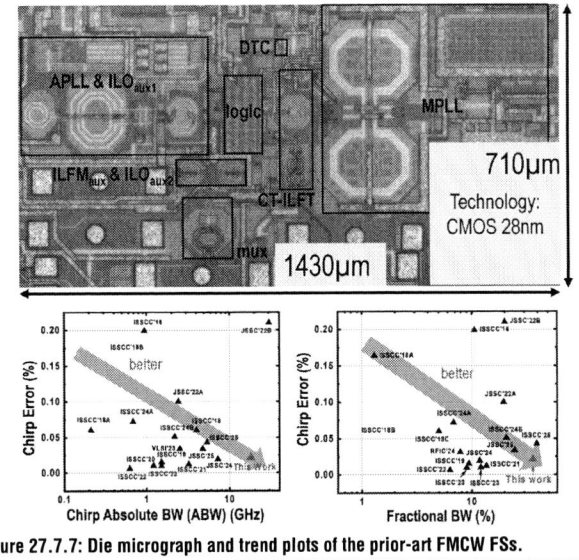

Figure 27.7.7: Die micrograph and trend plots of the prior-art FMCW FSs.

• 2026 IEEE International Solid-State Circuits Conference

ISSCC 2026 / SESSION 27 / FREQUENCY GENERATORS, MULTIPLIERS AND MODULATORS / 27.8

27.8 A 77GHz 8-bit CMOS Phase Shifter Adopting a Nested-Vector-Based Error Correction with 0.33°/0.07dB RMS-Error for MIMO Radar Applications

Geonho Park, Byeong-Taek Moon, Kyunghwan Kim, Doyoon Kim, Goeun Baek, Byungho Yook, Hyun-Chul Park, Chan-Hong Park

Samsung Electronics, Hwaseong, Korea

Abstract

In this paper, we present a 77GHz 8-bit nested vector-sum phase shifter in 28nm CMOS. The architecture performs vector modulation by splitting a single outer vector into two inner vectors generated by a two-stage I/Q coupler. Owing to its insensitivity to I/Q mismatch and the peak error smoothing technique, the phase shifter achieves 0.33° and 0.065dB rms phase and gain errors while occupying a core area of 0.17mm².

With the recent advancements in advanced driver-assistance systems (ADAS), millimeter-wave (mm-wave) phased-array RFICs have attracted attention as key technologies for radar systems in autonomous driving [1] and are expected to expand into various fields such as robotics and mobile applications. In multiple-input-multiple-output (MIMO) radar systems, phased-array architecture is required to increase the detection range and achieve high angular resolution. A vector-sum phase shifter (VSPS) achieves 360° phase-shifting range by controlling and combining two quadrature signals (I and Q) over a wide bandwidth while occupying a small die area [2-6]. However, it is sensitive to both I/Q mismatch (IQMM) and an I/Q gain-weighting error, which limit the achievable phase resolution; consequently, most implementations are limited to a phase resolution of 7-bit or fewer in the mm-wave band. This work presents a nested VSPS architecture for an 8-bit operation from 71 to 80GHz, achieving a 0.33° rms (root-mean-square) phase error and a 0.07dB rms gain error by adopting a peak-error-smoothing (PES) technique. The nested architecture divides a single vector-modulated outer vector into two vector-modulated inner vectors for more precise control, leading two advantages. First, the presented architecture generates an intentional phase difference through independent control of the two inner vectors, thereby reducing the phase and gain errors of the outer vector. Second, by leveraging a multi-stage I/Q-generation technique [7,8] to synthesize the two inner vectors, the outer vector becomes insensitive to IQMM.

Figure 27.8.1 (top-left) shows the block diagram of a conventional VSPS together with its error sources. Because the phase shift is generated from the I and Q signals, the VSPS is highly sensitive to IQMM. Moreover, the phase shift is determined by the ratio of the I-path gain (G_I) to the Q-path gain (G_Q); consequently, any gain-weighting error (G_Q/G_I) due to the non-ideality of variable-gain amplifiers (VGAs) in these paths or any process, voltage, and temperature (PVT) variations leads to increased phase and gain errors in the output signal. To overcome these limitations, we present a nested VSPS, as illustrated in Fig. 27.8.1 (top-right). The basic idea of the presented architecture is the decomposition of the outer vector into two inner vectors, each having roughly half the magnitude of the outer vector. These inner vectors are individually vector modulated by Sub-VSPS1 and Sub-VSPS2, resulting in a nested VSPS configuration. As described earlier, the combination of the PES technique, which imposes an intentional phase offset between the two inner vectors and the self IQMM calibration using a multi-stage I/Q generation approach enables the outer vector to achieve a precise phase shifting. The circuit realization and design details of the implemented nested VSPS are shown in Fig. 27.8.1 (bottom). A transformer-based 90° hybrid coupler is employed to generate I/Q signals compactly. VGAs, controlled by complementary switching of sliced tail-current sources, are placed in the I1, Q1, Q2, and I2 paths, and they are programmed with a 7-bit gain code. In each Sub-VSPS, the first summation uses a transformer (TF6 and TF7) whose three windings are magnetically coupled to minimize occupied area, and the combined signal is amplified in a buffer. The buffer is inserted to improve isolation and compensate for loss, and it is implemented as a neutralized common-source stage using cross-coupled capacitors. The amplified signal is then current combined in front of the output-matching transformer (TF8), realizing the second summation.

Figure 27.8.2 (top-left) shows the self IQMM calibration process of the nested VSPS. In a conventional architecture, the phase and gain errors caused by IQMM appear directly at the outer vector, while in the nested VSPS, the IQMM of Sub-VSPS 1 and Sub-VSPS 2 overlap and appear at the outer vector. Consequently, as shown in the vector diagram, the relative phase and amplitude with respect to the reference state are preserved even in the presence of IQMM. This can also be verified quantitatively as shown in Fig. 27.8.2 (bottom). The factor that causes the phase error of the outer vector, which is the outer vector G_Q/G_I ratio is defined as the error factor (EF). The plots of EF for amplitude-only mismatch ($\theta = 0$) and phase-only mismatch ($\varepsilon = 1$) show that in the presented architecture, EF stays close to the ideal value of 1, unlike for the conventional architecture. Moreover, in the conventional VSPS, the EF associated with phase mismatch depends on the phase state, whereas in the nested VSPS, it does not. The rms phase error and rms gain error contours, displayed in Fig. 27.8.2 (top right), clearly illustrate this phenomenon in greater detail. To achieve an rms phase error ≤0.7°, the conventional structure requires essentially zero I/Q amplitude/phase mismatch, while the presented structure meets this requirement in a considerably relaxed IQMM condition; the same holds for the rms gain error. Consequently, the nested VSPS is far less sensitive to IQMM. Under identical IQMM conditions, the nested VSPS attains an rms phase error ≤0.7° and an rms gain error ≤ 0.1dB, whereas the conventional VSPS exhibits an rms phase error ≥5° and an rms gain error ≥ 0.7dB.

Figure 27.8.3 (top-left) illustrates the concept of the peak-error-smoothing (PES) technique. In a conventional VSPS, the I-gain (G_I) and Q-gain (G_Q) change for each phase state (or code), which results in an almost random distribution of gain- and phase-error trends across the phase states. Consequently, certain phase states exhibit large errors while others show small errors, and the errors can be either positive or negative. Therefore, if the VSPS error plot is shifted and then averaged with the original error, the peak error can be reduced and the errors become concentrated around the mean value. However, because a large phase difference between the two summed vectors can increase loss, this work limits the offset to at most one or two phase-code steps. Figure 27.8.3 (top-right) shows how the PES technique is applied in the nested VSPS. The nested VSPS can be regarded as splitting a large outer vector into two inner vectors, each with roughly half the magnitude. Consequently, if the phase codes of Sub-VSPS1 and Sub-VSPS2 are not identical but one of them is shifted by one or two codes, the resulting outer vector is the sum of two vectors—one following the original error profile and the other following the shifted error profile—thereby enabling the PES technique. Moreover, because only the I/Q gain weighting changes in one of the Sub-VSPSs, the inherent self IQMM compensation effect remains effective. Figure 27.8.3 (bottom) presents the rms phase error and rms gain error curves for the conventional VSPS and the nested VSPS with the PES technique, evaluated under IQMM. Random variations of up to ±10% were applied to both G_I and G_Q, and the results were averaged over 1000 runs. In the nested VSPS, the phase codes of Sub-VSPS1 and Sub-VSPS2 were set with a one-code offset. The graphs demonstrate that the nested VSPS with the PES technique attains lower phase and gain errors. Compared with the conventional VSPS, the minimum rms gain error is reduced from 0.43 to 0.31dB (a 28% reduction), and the minimum rms phase error drops from 1.65 to 1.17° (a 29% reduction). In addition, the presented design continues to exhibit insensitivity to IQMM.

The nested VSPS was fabricated in a 28nm bulk CMOS process, and the core die measures 693μm × 252μm. The fabricated chips were wire-bonded onto a PCB for measurement. DC supply voltages were generated by a DC power supply (E36312A), and digital control signals were applied via an external SPI host board. S-parameter and large-signal measurements were performed with a vector network analyzer (R&S ZNA43) equipped with a frequency-extension module (R&S ZC90) and a power sensor (R&S NRP110T). On-chip probing was carried out with a 100μm-pitch GSG probe. Figure 27.8.4 shows the measured S-parameter results for the 256 phase states and the large signal performance result for the 0° phase state. The peak average S21 is −2.2dB at 80GHz, and the variation of S21 over the 256 phase states is only 0.45dB, indicating virtually no gain change during phase shifting. The relative phase for the 0° state demonstrates that the fabricated VSPS covers the full 360° range with a phase step of approximately 1.4°. In addition, the measured input-referred 1dB compression point (IP1dB) is 1dBm. Figure 27.8.5 shows the phase error and S21 versus phase state at 79GHz, and the rms phase error and rms gain error versus frequency with and without the PES technique. The graphs show that, owing to the PES technique, the phase and gain errors are reduced while the gain remains almost unchanged. The target rms phase error of less than 0.7° is achieved from 71 to 80GHz, and an rms gain error of less than 0.12dB is achieved from 70 to 85GHz. The minimum rms phase and gain errors are 0.33° and 0.065dB, respectively. The performance of the nested VSPS and previously reported VSPSs is summarized and compared in the table of Figure 27.8.6. Thanks to the nested structure and the PES technique, the presented VSPS achieves the highest phase resolution (8-bit) and the lowest rms gain and phase errors among designs in Fig. 27.8.6 while occupying a core die area of only 0.17mm². Figure 27.8.7 shows a die micrograph and a detailed breakdown of DC power consumption.

References:
[1] K. Dandu et al., "High-Performance and Small Form-Factor mm-Wave CMOS Radars for Automotive and Industrial Sensing in 76-to-81GHz and 57-to-64GHz Bands," *ISSCC*, pp. 39-41, Feb. 2022. http://doi.org/10.1109/ISSCC42613.2021.9365838
[2] D. Pepe et al., "Two mm-Wave Vector Modulator Active Phase Shifters with Novel IQ Generator in 28 nm FDSOI CMOS," *IEEE JSSC*, vol. 52, no. 2, pp. 344-356, Feb. 2017. http://doi.org/10.1109/JSSC.2016.2605659

[3] Q. Lu et al., "A 94-GHz 6-bit High-Precision Phase Shifter Based on Dual-Injection Gilbert Vector Modulator," *IEEE TMTT*, vol. 73, no. 8, pp. 5496-5506, Aug. 2025. http://doi.org/10.1109/TMTT.2025.3545028

[4] S. Kwon et al., "A Miniaturized Marchand Balun-Based Broadband Vector Sum Phase Shifter with 0.49° RMS Phase Error," *IMS*, pp. 790-793, June 2025. http://doi.org/10.1109/IMS40360.2025.11104054

[5] P. Gu et al., "Analysis and Design of a CMOS Bidirectional Passive Vector-Modulated Phase Shifter," *IEEE TCAS-I*, vol. 68, no. 4, pp. 1398-1408, Apr. 2021. http://doi.org/10.1109/TCSI.2020.3048816

[6] L. Piotto et al, "A 125-to-170GHz Power-Efficient Phase Shifter in SiGe BiCMOS with Outphasing Gain and Phase Corrections," *ISSCC*, pp. 546-548, Feb. 2025. http://doi.org/10.1109/ISSCC49661.2025.10904520

[7] F. Behbahani et al., "CMOS mixers and polyphase filters for large image rejection," *IEEE JSSC*, vol. 36, no. 6, pp. 873-887, June 2001. http://doi.org/10.1109/4.924850

[8] J. S. Park et al., "A Transformer-Based Poly-Phase Network for Ultra-Broadband Quadrature Signal Generation," *IEEE TMTT*, vol. 63, no. 12, pp. 4444-4457, Dec. 2015. http://doi.org/10.1109/TMTT.2015.2496187

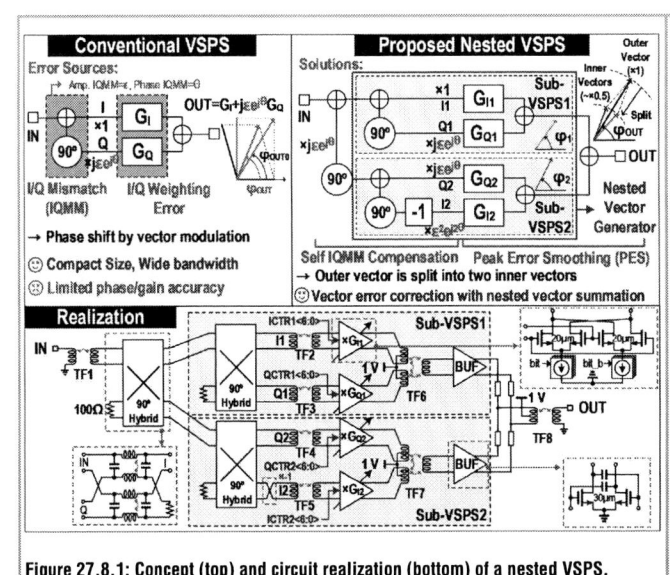

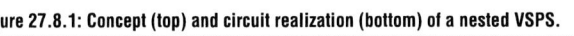

Figure 27.8.1: Concept (top) and circuit realization (bottom) of a nested VSPS.

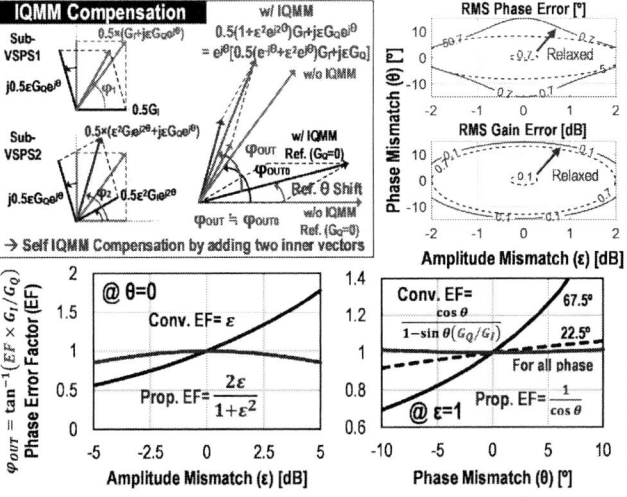

Figure 27.8.2: Self IQMM compensation in a nested VSPS (top) and calculated EF (bottom).

ISSCC 2026 / SESSION 27 / FREQUENCY GENERATORS, MULTIPLIERS AND MODULATORS / 27.8

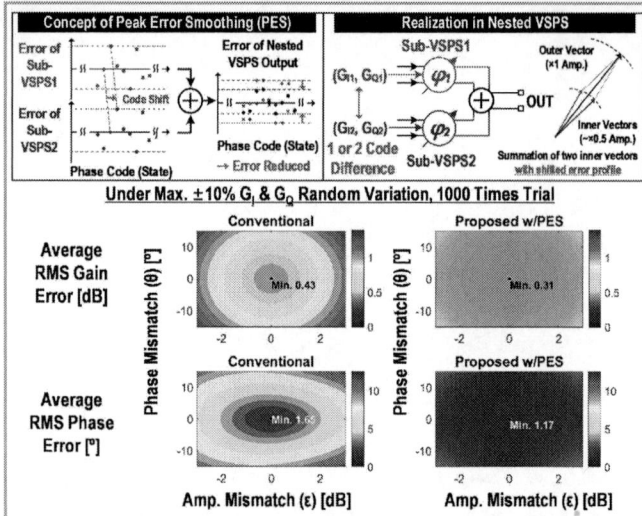

Figure 27.8.3: Concept and realization of the PES technique (top) and calculated rms gain and phase error (bottom).

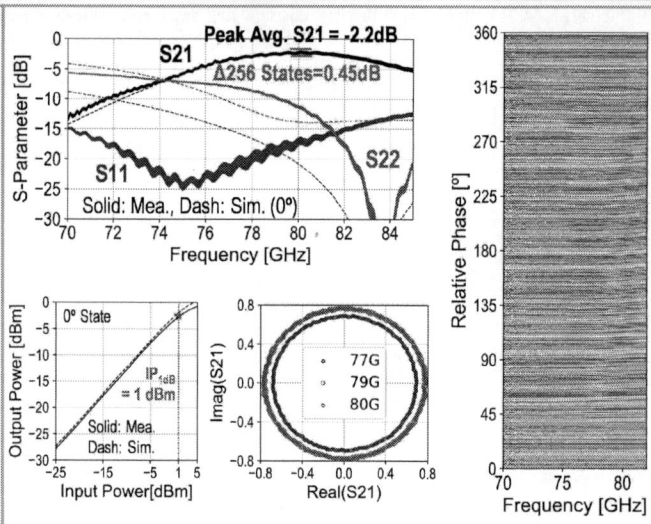

Figure 27.8.4: Measured S-parameters, relative phase for 256 phase states, and P1dB at 0° state.

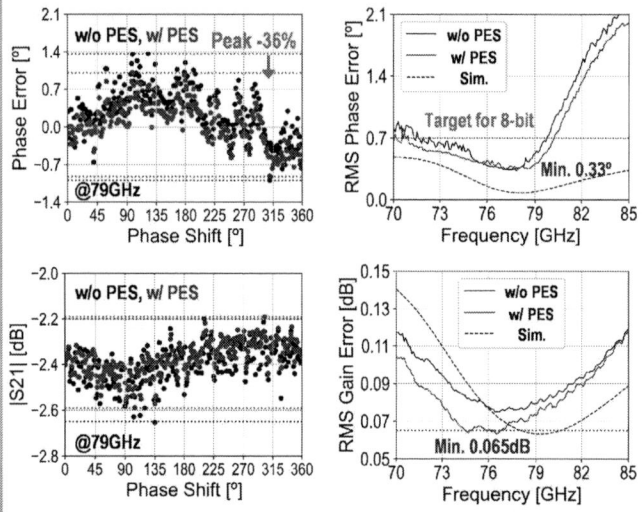

Figure 27.8.5: Measured phase and gain errors with and without the PES technique.

Figure 27.8.6: Performance comparison with prior works.

	This Work	[2] JSSC 2017	[3] TMTT 2025	[6] ISSCC 2025	[4] IMS 2025	[5] TCAS-I 2021
Type	Active VSPS				Passive VSPS	
Architecture	Nested Structure	Conventional	Conventional	Hybrid Structure	Conventional	Conventional
Process	28nm CMOS	28nm CMOS FDSOI	0.13μm SiGe BiCMOS	55nm SiGe BiCMOS	28nm CMOS	40nm CMOS
Frequency [GHz]	71 to 81	78.7 to 92.8	91.5 to 98	125 to 170	64.7 to 93.3	70 to 90
Peak Average Gain [dB]	−2.2	2.3	14.4	−2.3	−11.8	−15.1
Phase Resolution [bit]	8 (1.41°)	4 (22.5°)	6 (5.62°)	5.3 (9°)	7 (2.81°)	6 (5.62°)
Gain Variation [dB]	< ±0.23	±2.5*	±1*	±2*	±0.25**	< ±0.6
RMS Gain Error [dB]	0.07 to 0.09	< 2	0.58 to 0.85	0.5 to 0.8	0.12 to 0.33	N/A
RMS Phase Error [°]	0.33 to 0.7	< 11.9	1.2 to 2.9	1 to 5	0.49 to 2	1.1 to 3.7
Error Reduction Technique?	YES (PES)	NO	NO	NO	NO	NO
IP1dB [dBm]	1	−7	−14	5.2 to 6.8***	N/A	N/A
DC Power [mW]	30	21.6	42	31	0	0
Core Size [mm²]	0.17	0.12	0.33	0.2	0.15	0.15

*Graphically estimated **at 75GHz ***Calculated from OP1dB

DC Power Consumption				
	Sub-VSPS1	I1-VGA	4mW	13.3%
		Q1-VGA	4mW	13.3%
		CS Buffer	7mW	23.4%
	Sub-VSPS2	Q2-VGA	4mW	13.3%
		I2-VGA	4mW	13.3%
		CS Buffer	7mW	23.4%
	Total		30mW	100%

Figure 27.8.7: Die micrograph and DC power breakdown.

• 2026 IEEE International Solid-State Circuits Conference

979-8-3315-8937-0/26 $31.00 © 2026 IEEE

ISSCC 2026 / SESSION 27 / FREQUENCY GENERATORS, MULTIPLIERS AND MODULATORS / 27.9

27.9 A Dual-Mode DCO-PA with a Twisted 8-Shape Inductor for BLE Achieving 42% TX Efficiency at 1.6dBm and 0.29mW RX Clock

Jiawen Chen[1,2], Kai Xu[2], Luyi Guo[1], Teerachot Siriburanon[1], Jun Yin[3], Bashir M. Al-Hashimi[2], Robert Bogdan Staszewski[1]

[1]University College Dublin, Dublin, Ireland, [2]King's College London, London, United Kingdom, [3]University of Macau, Macau, China

Abstract

This paper presents a 0.3V dual-mode (2.4/4.8GHz) DCO-PA for BLE, achieving 42% efficiency in the TX mode and 0.29mW power consumption in the RX clock mode. It features: 1) a twisted 8-shape inductor enabling octave tuning, with Class-D-like operation boosting transistor efficiency to 75%; 2) a tail transistor in the RX mode that minimizes power consumption and lowers the 1/f phase-noise corner to ~40kHz.

With the rapid growth of IoT, demand for ultra-low-power (ULP) wireless devices has surged. Among short-range standards, Bluetooth low energy (BLE) dominates ULP transceiver (TRX) design. In conventional architectures (Fig. 27.9.1, top left), multiple blocks (oscillator, driver, PA) consume significant power, limiting efficiency even at 0dBm output [1-7]. To improve efficiency, function-reuse oscillating power amplifiers (Osc-PAs) merge oscillation and power amplification into a single block. While effective in standalone transmitters (TX) [8-10], direct TRX integration is fundamentally challenging due to continuous antenna power delivery. In a full TRX (Fig. 27.9.1, top right), the Osc-PA must also generate a $2f_0$ clock for RX I/Q, while minimizing antenna leakage and power consumption. A recent dual-mode DCO-PA addresses this by leveraging differential/common-mode (DM/CM) switching across a multi-coil inductor to enable mode-switching between 2.4 and 4.8GHz [10]. However, its Colpitts-based operation constrains efficiency to approximately 30%, and the CM Q-degradation (i.e. $Q_{eq}=(1-k_m)Q_0$, $k_m=0.6$) incurs high RX clock power (~3mW). Additionally, while the output transformer suppresses any CM leakage to an antenna, it fails to preserve differential signaling in this mode.

This paper presents a dual-core even/odd-mode switching DCO-PA tailored for TRX integration. The key enabler is a twisted 8-shape inductor (Fig. 27.9.1, bottom right) that provides a large inductance difference (L) between even (2.4GHz) and odd (4.8GHz) modes without Q-degradation (i.e. $Q_{eq}=Q_0$). In the TX (even) mode, it delivers inherent electromagnetic (EM) boost for high efficiency, while in the RX (odd) mode it provides EM cancellation to suppress antenna leakage. The prototype achieves 42% efficiency at 1.6dBm with Class-D-like operation in the TX mode, and only 0.29mW consumption with a sub-100kHz flicker phase-noise (PN) corner and the antenna leakage of merely -67dBm in the RX clock mode. Both modes meet the BLE PN requirements.

Figure 27.9.2 (left) depicts the evolution of our twisted 8-shape inductor from the conventional parallel 8-shape inductor. In the prior design [11], the common path inductance (L_c) supports dual-mode oscillation but restricts operation to moderately spaced frequencies, preventing octave switching (e.g., 2.4 to 4.8GHz). By twisting L_c around the two main inductors (L_0), controlled magnetic coupling is introduced, enlarging effective inductance difference L (from small L_c to large L_c+M) and enabling wideband mode switching. At the same time, this floorplan shortens the power delivery path, which is favorable for low flicker PN.

The full dual-mode DCO-PA (Fig. 27.9.2, right) adds an output coil (L_s), magnetically coupled with L_c and L_0 in a transformer configuration to directly drive the antenna. Vertically stacked capacitor banks and cross-coupled pair cores are placed inside the innermost L_0. Sized tail transistors reduce the power consumption and flicker PN in the RX clock mode. Four switches configure the two cores for even/odd-mode operation, while a small 0.5pF off-chip capacitor suppresses third-harmonic distortion (HD3) [8]. Magnetic boost enhances the primary coil equivalent Q-factor ($Q_{p,eq}$) in the TX (even) mode, while EM cancellation prevents Q-degradation in the RX (odd) mode.

Figure 27.9.3 illustrates the operation of the twisted 8-shape inductor by showing the magnetic flux, current directions, and simplified equivalent circuits for the two modes. In the even mode, the currents in L_c and L_0 flow in the same direction, producing constructive coupling and magnetic boost. Under this condition, L_c and L_s are actively engaged, forming a transformer-based resonant tank that directly drives the antenna. EM simulation indicates that the equivalent primary inductance ($L_{p,eq}$) increases to ~2.8nH with a Q-factor of 11, while L_s is 1.8nH with a coupling factor k_m (between $L_{p,eq}$ and L_s) of 0.48. All tail transistors are activated as closed switches (512μm/20nm) in this mode to maximize the oscillation current. The small on-resistance results in a slight ~2% efficiency degradation. A hybrid (current and voltage) power combination is deployed to achieve ~0dBm at a supply of 0.3V. Two L_0 are weakly coupled with L_s creating voltage-mode combination, while the current of the two L_0 flows into the shared L_c, which strongly couples with L_s, resulting in a current-mode combination. The output efficiency is determined by passive efficiency (η_M) and transistor efficiency (η_T). As shown in the equivalent model in Fig. 27.9.3 (left), the oscillation current splits between the transformer parasitic resistance and the load. Increasing the effective parallel resistance of the transformer directs more current into the load, thereby improving η_M to ~65%, even at $k_m=0.48$. Meanwhile, the Class-D-like switching

operation [12] enhances η_T by keeping the drain voltage near zero during conduction, minimizing power loss, while stored inductor energy drives the voltage swing up to 2.4× VDD, boosting the output power. This raises η_T to ~75% at 1dBm (compared to ~72% at 7dBm in Class-F [8] and ~42% at 0dBm in Colpitts [10]), yielding an overall efficiency of $\eta_M\eta_T$=48.8%. Fine power tuning is achieved via back-gate voltage control rather than supply scaling. As shown in the lower left of Fig. 27.9.3, the FD-SOI technology with flip-well option ties the NMOS back-gate to an n-well, allowing control voltages up to 2V without triggering parasitic PN diode conduction. Output power can also be adjusted from -1.6 to 4.4dBm with a supply voltage from 0.25 to 0.4V. These features collectively enable efficient even-mode operation for the 2.4GHz TX mode.

In the odd mode (Fig. 27.9.3, right), the current in L_0 bypasses L_c, causing the magnetic fluxes in the two cores flow to opposite each other. Consequently, both L_c and L_s are effectively disabled, leaving the resonant tank dominated by L_0. In this case, $L_{p,eq}$ equals L_0 (~0.8nH) with $Q_{p,eq}=Q_0=11$ (compared to $Q_{eq}=5$ in [10]), while the coupling k_m approaches zero due to flux cancellation. The tank impedance remains high (~250), which is favorable for low-power operation. Since a robust startup is ensured by the large core transistors originally designed for the Class-D-like operation, the tail transistors are sized at 16μm/200nm and biased at 0.8V. Operating in the triode region, the tail transistor behaves as a resistive element, simultaneously reducing power consumption and suppressing flicker PN. Simulations confirm this effect: without the tail transistor, the oscillator consumes 1.9mW with a flicker PN corner of ~550kHz, whereas enabling it lowers power consumption to 0.3mW and shifts the flicker PN corner down to ~10kHz. These properties allow the odd-mode operation to serve as a highly energy-efficient differential 4.8GHz RX clock source, achieving both low flicker PN and minimal power overhead.

The fabricated DCO-PA was characterized in the TX mode using an R&S FSW85 spectrum analyzer. Figure 27.9.4 summarizes the measured PN, output power and efficiency, harmonic distortion (HD2 and HD3), fine power tuning, and a Gaussian frequency-shift-keying (GFSK) modulation spectrum. The frequency tuning range (FTR) spans 2.14 to 2.63GHz, fully covering the BLE band. With a 0.3V supply, the DCO-PA delivers 1.3-to-1.7dBm output power with 39-to-45% efficiency over the FTR across three samples, demonstrating strong chip-to-chip consistency. The measured PN at 2.5MHz offset ranges from -130 to -123dBc/Hz, well below the BLE requirement. At 2.42GHz, the flicker PN corner is ~80kHz, ensuring stable open-loop GFSK modulation [1]. The bondwire at the RF output is modeled as a 2nH inductor. Together with a 0.5pF off-chip capacitor, it suppresses the HD3 to -41dBm in simulation. Both HD2 and HD3 remain below the BLE limit of -41dBm. Fine power tuning via back-gate voltage control (0 to 1.5V) enables ~1dB power adjustment with <2% efficiency variation. The design also supports 1Mbps GFSK modulation in open loop using a 7-bit fine-tuning capacitor bank. The measured modulation spectrum (BT = 0.5 and modulation index m = 0.5) complies with the BLE spectral mask, with a 3.6% FSK error.

In the RX clock mode, the oscillator was characterized using Keysight E5052B. Figure 27.9.5 summarizes the performance in the RX clock mode. Operating in the $2f_0$ band around 4.8GHz, it supports I/Q generation when cascaded with a divider. Benefiting from the tail transistor, it consumes only 0.29mW from a 0.3V supply, eliminating the need for supply regulation across modes. The measured FTR spans 4.2 to 5.3GHz, fully covering twice the 2.4-to-2.48GHz BLE band. At 2.5MHz offset, the measured PN ranges from -116 to -109dBc/Hz, providing ~10dB margin over the BLE specification (normalized to 4.8GHz). The flicker PN corner is effectively suppressed below 100kHz, varying from 30 to 85kHz across frequencies. Additionally, the undesired antenna leakage in the RX clock mode is measured at -67dBm at 4.84GHz.

Figure 27.9.6 summarizes the results, highlighting the superior efficiency-power trade-off of this work compared with prior-art energy-efficient BLE Osc-PAs and transmitters. This design employs an 8-shape inductor that enables a dual mode operation: a Class-D-like DCO-PA mode for high-efficiency transmission and a sub-mW RX clock mode via octave frequency switching, thereby facilitating seamless Osc-PA integration into a full TRX. Specifically, among designs in Fig. 27.9.6, the DCO-PA achieves the highest reported open-loop efficiency of 42% at 1.6dBm, surpassing prior DCO-PAs [10] (31.92% at 2dBm) and transmitters with separate oscillator-PA architectures [6] (41.1% at 3dBm with external matching). In the RX clock mode, the power consumption is reduced to 0.29mW, over an order of magnitude lower than the DCO-PA in [10] (3.69mW), while other Osc-PA architectures inherently lack the RX clock capability.

The DCO-PA is fabricated in GlobalFoundries 22nm FD-SOI technology. Figure 27.9.7 presents the die micrograph.

Acknowledgement:
This work was supported in part by Science Foundation Ireland (SFI) Grants 21/RP-2TF/10019 and 20/FFP-P/8437, and in part by King's College London NMES start-up package. The authors thank GlobalFoundries for chip fabrication.

References:
[1] M. Babaie et al., "A Fully Integrated Bluetooth Low-Energy Transmitter in 28 nm CMOS With 36% System Efficiency at 3 dBm," *IEEE JSSC*, vol. 51, no. 7, pp. 1547-1565, July 2016. http://doi.org/10.1109/JSSC.2016.2551738
[2] J. Yin et al., "A 0.2V Energy-Harvesting BLE Transmitter with a Micropower Manager Achieving 25% System Efficiency at 0dBm Output and 5.2nW Sleep Power in 28nm CMOS," *ISSCC*, pp. 450-452, Feb. 2018. http://doi.org/10.1109/ISSCC.2018.8310378
[3] M. Ding et al., "A Bluetooth 5 Transceiver with a Phase-Tracking RX and Its Corresponding Digital Baseband in 40-nm CMOS," *IEEE JSSC*, vol. 56, no. 1, pp. 254-266, Jan. 2021. http://doi.org/10.1109/JSSC.2020.3005788
[4] K. Shibata et al., "A 22nm 0.84mm² BLE Transceiver with Self IQ-Phase Correction Achieving 39dB Image Rejection and on-Chip Antenna Impedance Tuning," *ISSCC*, pp. 398-400, Feb. 2022. http://doi.org/10.1109/ISSCC42614.2022.9731558
[5] N. Scolari et al., "A 1mm² Software-Defined Dual-Mode Bluetooth Transceiver with 10dBm Maximum TX Power and -98.2dBm Sensitivity 2.96mW RX Power at 1Mb/s," *ISSCC*, pp. 402-404, Feb. 2024. http://doi.org/10.1109/ISSCC49657.2024.10454304
[6] Y. Shen et al., "A Crystal-Less Frequency-Modulation Transmitter IC with Joint Neural-Network-Driven Modulation and Coding for Low-Power Connectivity," *ISSCC*, pp. 462-463, Feb. 2025. http://doi.org/10.1109/ISSCC49661.2025.10904502
[7] L. Feng et al., "A 0.6-V High-Efficiency BLE Transmitter Using a Hybrid PLL With RDAC-Based Common-Mode Ripple Cancellation in 65-nm CMOS," *IEEE JSSC*, Early Access, 2025. http://doi.org/10.1109/JSSC.2025.3569172
[8] X. Peng et al., "A 2.4-GHz ZigBee Transmitter Using a Function-Reuse Class-F DCO-PA and an ADPLL Achieving 22.6% (14.5%) System Efficiency at 6-dBm (0-dBm) P_{out}," *IEEE JSSC*, vol. 52, no. 6, pp. 1495-1508, June 2017. http://doi.org/10.1109/JSSC.2017.2672990
[9] K. Xu et al., "A Single-Pin Antenna Interface RF Front End Using a Single-MOS DCO-PA and a Push–Pull LNA," *IEEE JSSC*, vol. 55, no. 8, pp. 2055-2068, Aug. 2020. http://doi.org/10.1109/JSSC.2020.2991520
[10] B. Yu et al., "A 31.92% Efficiency Colpitts DCO-PA for BLE TX with Implicit RX Clock Generation Using DM-to-CM Mode-switching," *IEEE ASSCC*, pp. 1-3, Nov. 2024. http://doi.org/10.1109/A-SSCC60305.2024.10848745
[11] W. Deng et al., "An 8.2-to-21.5 GHz Dual-Core Quad-Mode Orthogonal-Coupled VCO with Concurrently Dual-Output using Parallel 8-Shaped Resonator," *IEEE CICC*, pp. 1-2, Apr. 2021. http://doi.org/10.1109/CICC51472.2021.9431447
[12] L. Fanori and P. Andreani, "Class-D CMOS Oscillators," *IEEE JSSC*, vol. 48, no. 12, pp. 3105-3119, Dec. 2013. http://doi.org/10.1109/JSSC.2013.2271531

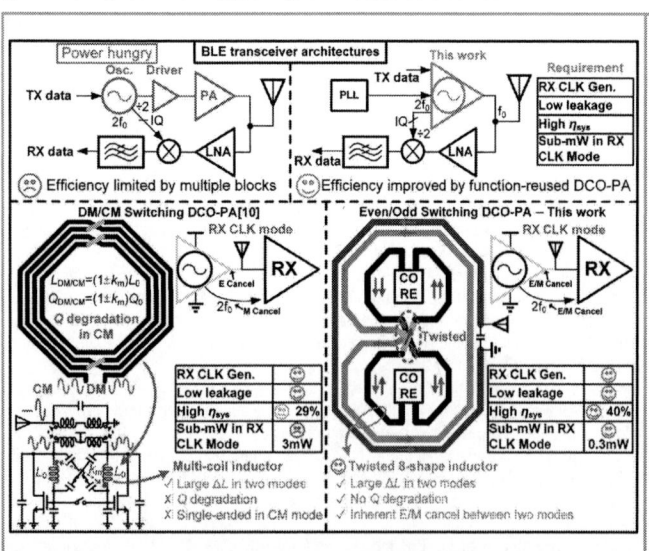

Figure 27.9.1: Criteria for Osc-PA in BLE TRX and comparison with prior DCO-PA.

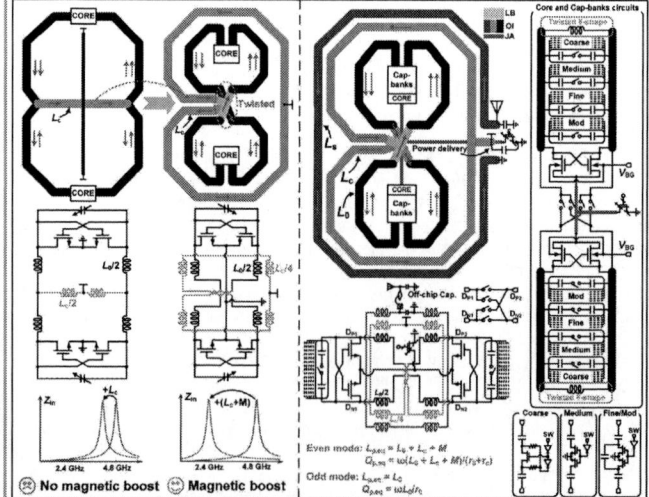

Figure 27.9.2: Octave frequency switching with twisted 8-shape inductor and the detailed implementation of the DCO-PA.

ISSCC 2026 / SESSION 27 / FREQUENCY GENERATORS, MULTIPLIERS AND MODULATORS / 27.9

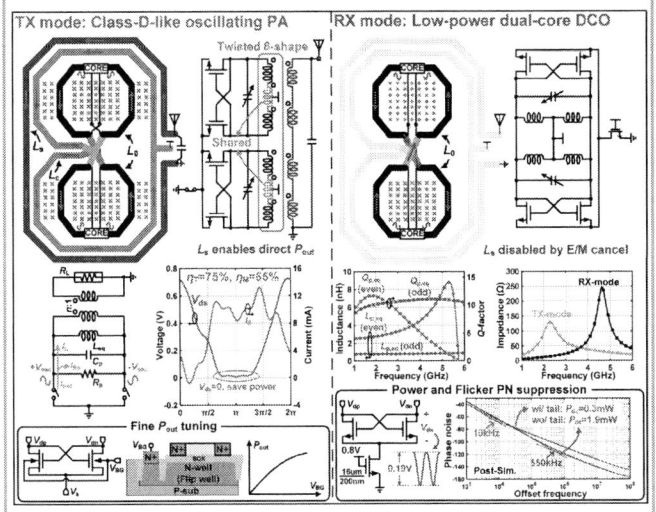

Figure 27.9.3: Operational principle of the two modes; EM simulations of the inductors and tank impedance.

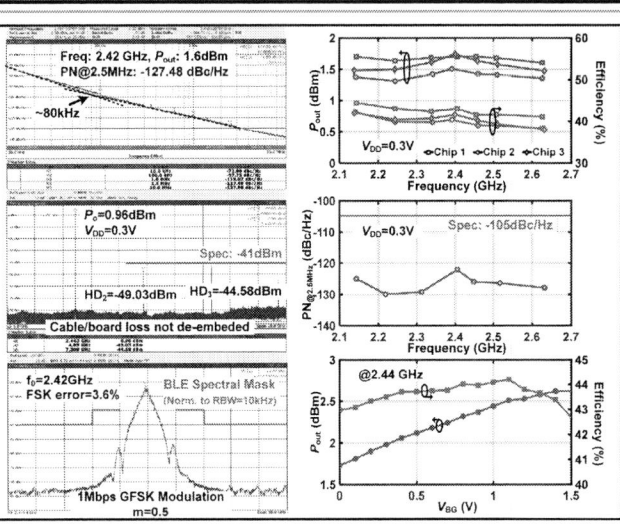

Figure 27.9.4: Measured output power, efficiency, HD, PN, GFSK spectrum, and fine power tuning in the TX mode.

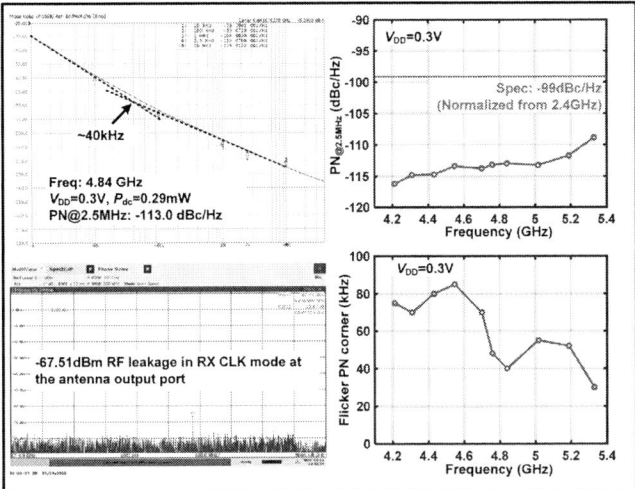

Figure 27.9.5: Measured PN, flicker PN corner, and antenna leakage in the RX clock mode.

	This work	ASSCC24[10]	JSSC20[9]	JSSC17[8]	ISSCC18[2]	JSSC25[7]	ISSCC25[6]	JSSC21[3]
Topology	Twisted 8-shape Mode-switching DCO-PA	Multi-Coils Mode-switching DCO-PA	Single-MOS DCO-PA	Class-F DCO-PA	uPM/PA/PLL TX	PA/PLL TX	Neural-FM TX	PA/PLL TRX
Technology	22nm FDSOI	28nm CMOS	65nm CMOS	65nm CMOS	28nm CMOS	65nm CMOS	28nm CMOS	40nm CMOS
Core Area [mm²]	0.28	0.35	0.17	0.39	0.32 (Excl. uPM)	0.46	2.16 (Incl. PADs)	0.8
Supply [V]	0.3 (0.25 to 0.4)	~0.3	0.7 (0.2 to 0.7)	0.7 (0.3 to 0.7)	0.2	0.6/0.5	-	0.8
P_{out} [dBm]	1.6 (-1.6 to 4.4)	-1 to 2	1.6 (-20 to 0)	6 (-5 to 6)	0	0	3	1.8
Open-loop Efficiency [%]	42@1.6dBm 38.4@0dBm	31.92@2dBm 29.37@0dBm	20.8@0dBm	22.6@0dBm 26.1@6dBm	25@0dBm	25.1@0dBm (close-loop)	41.1ᴬ (w/ DBB)	27@1.8dBm
RX Clock Generation	Yes	Yes	No	No	Yes*	Yes*	Yes*	Yes*
RX clock P_{dc} [mW]	0.29	3.69	-	-	0.67	0.4	0.37***	0.35
Frequency [GHz]	TX mode: 2.1 to 2.6 RX mode: 4.2 to 5.3	TX mode: 1.9 to 2.6 RX mode: 4.0 to 5.6	2.2 to 2.5	2.3 to 2.4	2.2 to 2.6	2.23 to 2.52	2.4 to 2.48	2.1 to 2.7
PN@2.5MHz (Spec: -105) [dBc/Hz]	TX mode: -127@1.6dBm RX mode: -113 (2f_0)	TX mode: -125@0dBm RX mode: -120 (2f_0)	-108@0dBm	-126ᴬ@6dBm	-128ᴱ	-115.3**	-109***	-115**

ᴬ External filtering/matching required
* Using individual VCO
** At 1MHz offset
*** Including divider for digital baseband
ᴱ Graphically estimated

Figure 27.9.6: Performance summary and comparison with prior-art Osc-PAs and ULP TXs.

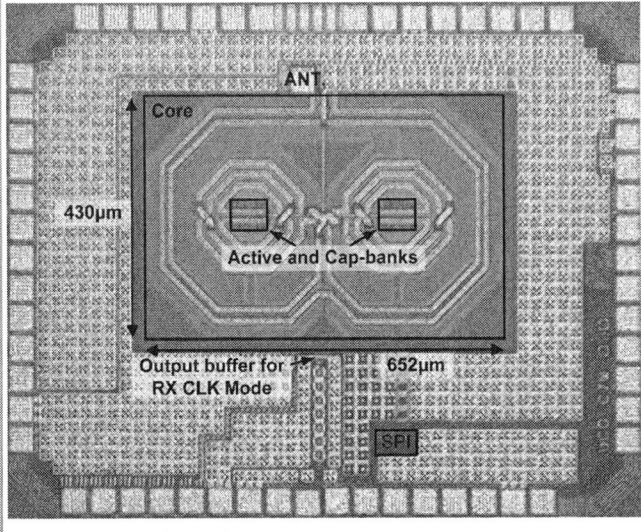

Figure 27.9.7: Die micrograph.

• 2026 IEEE International Solid-State Circuits Conference

979-8-3315-8937-0/26 $31.00 © 2026 IEEE

ISSCC 2026 / SESSION 27 / FREQUENCY GENERATORS, MULTIPLIERS AND MODULATORS / 27.10

27.10 A 48-to-82.5GHz CMOS Split-Tail Gilbert-Cell Frequency Doubler Achieving 11% PAE at 8.5dBm Output Power

Jismal Jamal[1], Lorenzo Piotto[2], Federico Vecchi[3], Mahmoud M. Pirbazari[3], Andrea Mazzanti[1]

[1]University of Pavia, Pavia, Italy, [2]Fondazione Chips-IT, Pavia, Italy, [3]STMicroelectronics, Pavia, Italy

Abstract

A 48-to-82.5GHz frequency doubler in 28nm FDSOI CMOS using a modified Gilbert cell is presented. The tail transconductors are split and the switching-quad transistors AC-shorted by capacitors. DC offset builds up across the capacitors, reducing the switching-quad duty cycle in steady state. This mechanism enhances power-conversion gain and efficiency. Measurements show 8.5dBm output power at 11% PAE.

Despite sustained research, mm-wave CMOS VCOs still struggle to meet the phase-noise and tuning-range requirements of emerging wireless communication systems and radar sensors. As a result, frequency multiplication from a lower-frequency VCO remains the only viable solution for high-performance mm-wave frequency synthesis. In this context, CMOS frequency doublers continue to attract research attention, with several recent works reporting notable performance improvements.

Among various architectures, the push-push pair is commonly preferred for its simplicity and solid baseline performance [1-9]. In [1,2], the standard common-source structure is modified into a common-source/common-gate topology, achieving broader bandwidth and higher second-harmonic output. To enable differential input/output, [3,4] use lumped hybrid couplers to drive push-push pairs in quadrature, yielding anti-phase outputs. A more compact and efficient solution, introduced in [5,6], is the complementary (NMOS–PMOS) push-push with Class-C biased transistors, offering differential output and improved efficiency. Second-harmonic input traps or positive feedback are frequently incorporated to boost output power [1,2,4,5,7], though at the cost of complexity and bandwidth penalty.

Recent works have improved considerably the drain power efficiency ($\eta_D = P_{o,2f0}/P_{DC}$, with $P_{o,2f0}$ the 2nd-harmonic output power and P_{DC} power consumption) but often at the cost of low conversion gain ($CG = P_{o,2f0}/P_{in}$ with P_{in} the input power at f_0). As a result, most CMOS doublers operating in the 50-to-100GHz range exhibit negative power-added efficiency ($PAE = (P_{o,2f0} - P_{in})/P_{DC}$), with P_{in} exceeding $P_{o,2f0}$.

A promising yet much less explored alternative to the push-push pair is the use of a Gilbert cell to multiply the input signal by itself, thereby generating an output at twice the frequency [10,11]. In [11], it has been shown that both η_D and CG are remarkably improved by operating the switching quad of the cell at a reduced duty cycle. However, the circuit realization, in a SiGe BiCMOS process, is hardly compatible with a low-voltage CMOS process, limiting the adoption in mainstream integration platforms.

This paper introduces a simple and elegant implementation of the concept in CMOS. The presented Gilbert-cell doubler inherently provides a differential output and does not need complex passive networks (except input and output baluns, required for measurements), resulting in a compact and wideband design. Fabricated in 28nm FDSOI CMOS and operating across the 48-to-82.5GHz range, the doubler achieves wide bandwidth and high output power. Owing to its high conversion gain, it reaches 11% PAE at the center frequency, an order of magnitude greater than previously reported for CMOS doublers in this band in Fig. 27.10.6.

Figure 27.10.1 shows the idealized schematic of a Gilbert-cell doubler with relevant waveforms. The tail current sources model a differential Class-A transconductor delivering sinusoidal currents ($i_{IN,P}$, $i_{IN,N}$) of amplitude I_{IN}, at frequency f_0, with a DC component I_{DC}. The switching quad is driven by square-waves (s_{PN}, s'_{PN}) of amplitude ±1 at f_0 with duty cycle δ. The differential output current, i_0, is given by the equation in the figure. For δ =50%, i_0 is a rectified sinewave, the same of a push-push pair of Class-B transistors. The spectrum (on the bottom of the figure) shows a large DC offset and a relatively small 2nd-harmonic component, $I_{0,2f0}$. With the transconductors at the edge of Class-A (i.e., $I_{IN} = I_{DC}$) the current conversion efficiency is $\eta_I = I_{0,2f0}/I_{DC} = 0.42$ only. If δ is reduced to 25%, i_0 becomes nearly square wave at $2f_0$, thus raising $I_{0,2f0}$ by ideally three times, to 1.23. Notably, with $I_{IN} = I_{DC}$, the transconductance conversion gain of the stage is $g_{m,2f0} = g_m \eta_I$ (where g_m is the transconductance of the tail current sources). Thus, compared to the push-push-like waveform at δ=50%, the threefold increase in η_I with δ=25% also yields a remarkable 9.5dB higher power conversion gain. Still, with $g_{m,2f0} > g_m$ (with $\eta_I > 1$) the doubler conversion gain ideally surpasses that of a simple amplifier using the same transconductors.

The bottom-right plot in Fig. 27.10.1 shows that at the duty cycle where η_I peaks (blue curve), $\bar{I}_0$, the steady-state average of i_0 (orange curve), is nearly zero, i.e., the two single-ended outputs carry the same average current, equal to I_{DC}. In [11], the switching-quad duty cycle is controlled by superimposing a differential DC offset voltage onto the sinusoidal driving signals. $\bar{I}_0$ is sensed via two resistors in series with the supply, and an op-amp uses the resulting differential IR drop to generate the DC offset, thereby driving $\bar{I}_0$ toward zero and thus to δ≈25%. In SiGe BiCMOS, the voltage swing required for current commutation in the switching quad is relatively low (about 100mV), making it easier to adjust δ by adding a small offset to the driving signals. However, a feedback loop around the doubler is inherently slow and may compromise noise and stability if not carefully designed. In addition, the resistors used to sense the differential average current need a non-negligible voltage drop, which can significantly impact power efficiency at low supply voltage.

A much simpler and more efficient way to enforce operation at δ≈25% is found by observing that, when the average of the two single-ended output currents equals I_{DC}, KCL dictates equal DC current, $I_{DC}/2$, in each of the four switches. A straightforward modification of the Gilbert cell to impose this condition is shown in Fig. 27.10.2. Each tail transconductor is split into two equal-size devices, and the source terminals of each pair in the switching quad are AC shorted through capacitors C_S, ensuring equal average current in all four switching transistors at steady state. The top and bottom transistors are driven by differential signals, and their dimensions are optimized according to the simulations shown on the right. The gate-voltage amplitude on each transconductor is set to 350mV$_{0-pk}$, placing the devices at the edge of Class-A. The top-right plot illustrates the dependence of η_I on the amplitude at each gate of the switching quad, V_2, and the size ratio between top and bottom devices (W_2/W_1). Sharper switching transitions, achieved by increasing the drive amplitude and device sizes, improve performance. With $W_2/W_1 \approx 1.5$ and $V_2 = 750mV_{0-pk}$, $\eta_I \approx 1$ is obtained. The bottom-right plot shows η_I versus frequency for different values of the AC-shorting capacitors. With large C_S, $\eta_I \approx 1$ is maintained from low frequencies up to 100GHz. A practical choice of C_S=500fF ensures $\eta_I > 0.9$ for output frequencies above 40GHz. Transient simulations in the bottom-left plots provide insight into the circuit operation. The input signal is at 35GHz, and ideal switches initially short the capacitors C_S. Under this condition, the circuit operates as a conventional Gilbert-cell doubler. The differential output current i_0 resembles that of a push–push pair, exhibiting a large DC offset and a limited $I_{0,2f0}$. At t = 200ps, the switches open and C_S establish equal DC currents in all transistors in steady state, as evidenced by the final output current waveform with a zero average. In this situation, a DC offset voltage builds up across C_S, shifting the gate–source voltage of each switching-quad transistor, thereby modifying their on/off times, i.e., the duty cycle. The C_S charging transient lasts only ~200ps. Analysis on the idealized circuit yields a time constant $\tau = (\pi/\sqrt{2})(C_S V_2/I_{DC}) = 45$ps, in agreement with the simulation. The operation at the reduced duty cycle ultimately enhances the second harmonic, as evidenced by a peak-to-peak i_0 amplitude nearly twice that of the initial transient, when C_S were shorted.

The schematic of the doubler, implemented in a 28nm FDSOI CMOS technology, is reported in Fig. 27.10.3 with component values. The bias current (10mA per branch) and transistor sizes are chosen to reach a saturated output power of 10dBm, in line with the highest reported for CMOS doublers in [3,4]. If a lower $P_{o,2f0}$ is required, the circuit can be scaled down to reduce power consumption. Input and output transformers (layout on the right) provide single-ended-to-differential conversion. The input balun additionally performs impedance matching and sets the switching-quad bias voltage through a center tap on the secondary coil. R_1-C_3 set the bias and attenuate the signal amplitude at the gate of the transconductors. Small degeneration inductors (L_s=40pH) are included to increase the real part of the impedance at gates, simplifying input matching. According to simulations, L_s also provide a slight performance improvement by introducing a small phase shift in the injected currents and by allowing the voltage at the source of the transconductors to swing partially negative.

Figure 27.10.4 shows the measured input reflection coefficient (S_{11}, bottom-left), with a notch at the 33GHz input frequency. The top plot reports the measured $P_{o,2f0}$, CG, and PAE at the 65GHz output frequency versus P_{in}. The saturated output power at a 1.1V supply is $P_{o,2f0-sat}$ = 9.5dBm, while the maximum CG is 7.3dB. The PAE peaks at 11% for $P_{o,2f0}$ = 8.5dBm, only 1dB below $P_{o,2f0-sat}$, with a corresponding CG of 5.5dB. η_D (not shown) is 16%. The phase noise at the output is 6dB higher than at the input (bottom-right), confirming negligible deterioration introduced by the doubler.

484 • 2026 IEEE International Solid-State Circuits Conference

979-8-3315-8937-0/26 $31.00 © 2026 IEEE

Figure 27.10.5 shows the measured performance versus the output frequency. The bandwidth where $P_{o,2f0\text{-sat}}$ is within 3dB of its maximum extends from 48 to 82.5GHz. Across the entire band, the output power at peak PAE remains roughly 1 to 2dB below $P_{o,2f0\text{-sat}}$. The rejection of the fundamental component (FR), impaired by the finite CMRR of the input and output baluns as confirmed by simulations, is higher than 30dB. The bottom plot shows PAE, the maximum CG, and the CG at the peak of PAE. PAE is above 10% with CG greater than 5dB across most of the operating bandwidth.

Experiments are summarized in Fig. 27.10.6 and compared with CMOS doublers with output frequency above 40GHz. $P_{o,2f0\text{-sat}}$ and fractional bandwidth (BW) are comparable to the best values reported in the figure. Although higher η_D has been achieved, only a few works demonstrated positive, though limited, CG, with PAE not exceeding 1.5%. In contrast, the implemented doubler outperforms prior works in this figure with 11% PAE, an order of magnitude higher than previously reported, while keeping $P_{o,2f0}$ at the peak PAE of ≈1dB below $P_{o,2f0\text{-sat}}$. The die micrograph is shown in Fig. 27.10.7.

References:
[1] S. Li et al., "A Buffer-Less Wideband Frequency Doubler in 45-nm CMOS-SOI With Transistor Multiport Waveform Shaping Achieving 25% Drain Efficiency and 46–89 GHz Instantaneous Bandwidth," IEEE SSCL, vol. 2, no. 4, pp. 25-28, Apr. 2019. http://doi.org/10.1109/LSSC.2019.2918943
[2] A. Aghighi et al., "A Frequency Doubler with Second Harmonic Feedback for Wideband, Efficient Frequency Multiplication at Millimeter-Wave," IEEE TMTT, vol. 72, no. 5, pp. 2704-2715, May 2024. http://doi.org/10.1109/TMTT.2024.3367882
[3] S. Vehring et al., "A 3.1-dBm E-Band Truly Balanced Frequency Quadrupler in 22-nm FDSOI CMOS," IEEE MWCL, vol. 30, no. 12, pp. 1165-1168, Dec. 2020. http://doi.org/10.1109/LMWC.2020.3028053
[4] D. Yoo and B.-W. Min, "A High-Conversion-Gain Compact W-Band Distributed Doubler with Second Harmonic Positive Feedback Using Cross-Coupled Capacitor," IEEE RFIC, pp. 351-354, July 2025. http://doi.org/10.1109/RFIC61188.2025.11082906
[5] J. Moody, "A Double Balanced Frequency Doubler Achieving 70% Drain Efficiency and 25% Total Efficiency," IEEE RFIC, pp. 157-160, July 2023. http://doi.org/10.1109/RFIC54547.2023.10186147
[6] J. Yoo and S. Hong, "Highly Efficient Differential Frequency Doubler with Output Resistance Boosting Feedback," IEEE JSSC, vol. 59, no. 2, pp. 414-423, Feb. 2024. http://doi.org/10.1109/JSSC.2023.3289512
[7] H. Fu et al., "A High Conversion Gain Frequency Doubler Using Transformer-Based Second Harmonic Feedback Technique in 28-nm CMOS," IEEE MWCL, vol. 32, no. 9, pp. 1071-1074, Sep. 2022. http://doi.org/10.1109/LMWC.2022.3168580
[8] Y. Ye et al., "A High Efficiency E-Band CMOS Frequency Doubler with a Compensated Transformer-Based Balun for Matching Enhancement," IEEE MWCL, vol. 26, no. 1, pp. 40-42, Jan. 2016. http://doi.org/10.1109/LMWC.2015.2505617
[9] J. Oh et al., "A W-Band High-Efficiency CMOS Differential Current-Reused Frequency Doubler," IEEE MWCL, vol. 25, no. 5, pp. 307-309, May 2015. http://doi.org/10.1109/LMWC.2015.2409773
[10] J. Wan et al., "A Truly Balanced Q-Band CMOS Frequency Doubler Based on Hybrid Quadrature Coupler," IEEE MWCL, vol. 27, no. 2, pp. 165-167, Feb. 2017. http://doi.org/10.1109/LMWC.2016.2646909

[11] L. Piotto et al., "A 14–32 GHz SiGe-BiCMOS Gilbert-Cell Frequency Doubler with Self-Adjusted Reduced Duty-Cycle Performance Enhancement," IEEE JSSC, vol. 59, no. 3, pp. 878-888, Mar. 2024. http://doi.org/10.1109/JSSC.2023.3313501

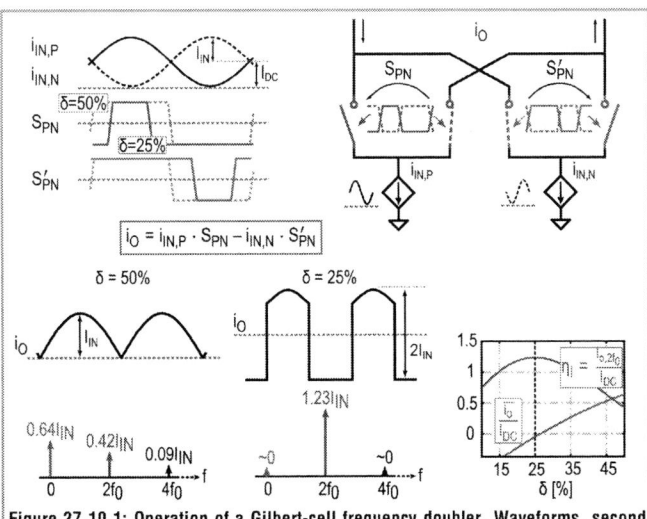

Figure 27.10.1: Operation of a Gilbert-cell frequency doubler. Waveforms, second harmonic, and the average value of the output current changing the switching-quad duty cycle.

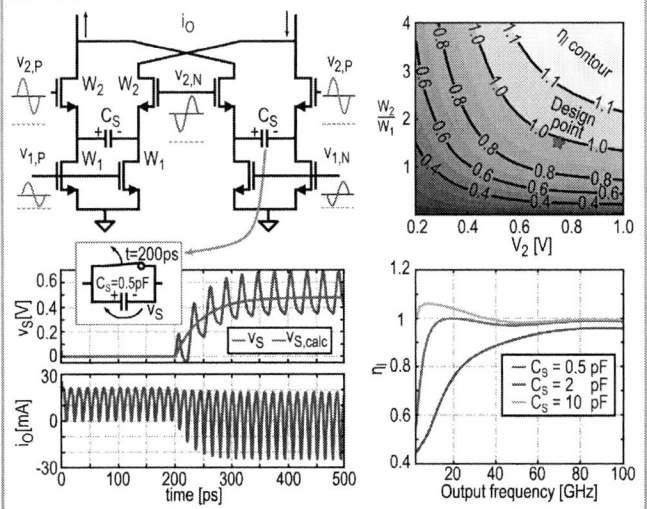

Figure 27.10.2: Split-tail Gilbert-cell doubler with reduced duty cycle, simulated current conversion efficiency, and transient waveforms.

ISSCC 2026 / SESSION 27 / FREQUENCY GENERATORS, MULTIPLIERS AND MODULATORS / 27.10

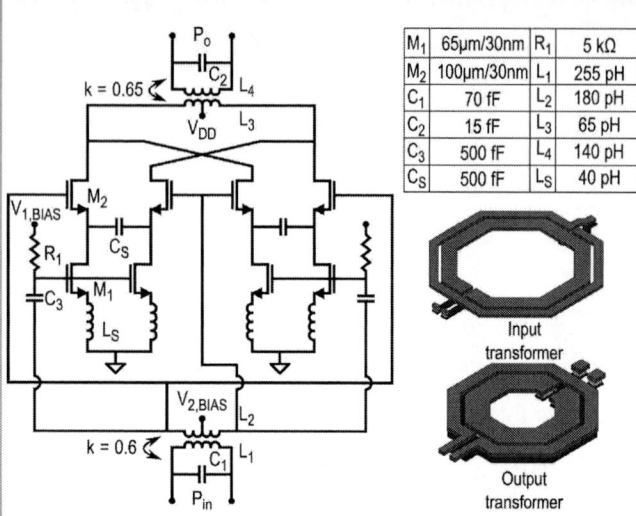

Figure 27.10.3: Schematic and component values of the realized frequency doubler.

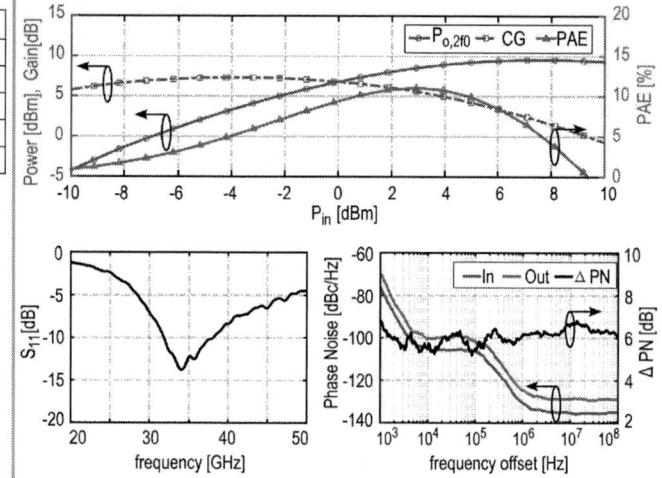

Figure 27.10.4: Output power, gain, and PAE at 65GHz output frequency (top). Input reflection coefficient and phase noise (bottom).

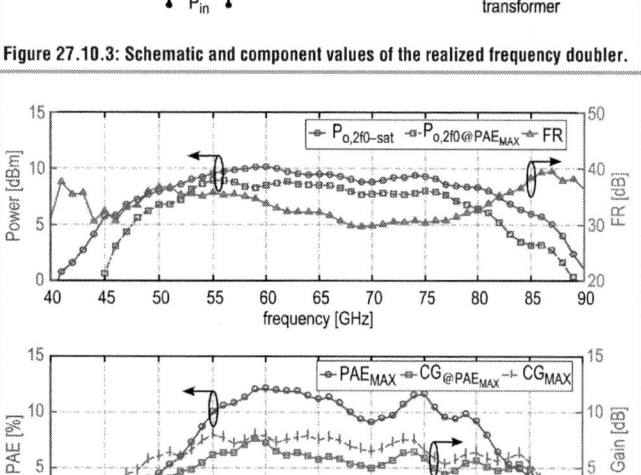

Figure 27.10.5: Output power, fundamental rejection, PAE, and conversion gain versus frequency.

Publication Year	This Work	[4] RFIC 2025	[6] JSSCC 2024	[2] TMTT 2024	[5] RFIC 2023	[7] MWCL 2022	[3] MWCL 2020	[1] SSCL 2019	[10] MWCL 2017	[8] MWCL 2016	[9] MWCL 2015
Technology	28nm FDSOI	28nm FDSOI	40nm CMOS	65nm CMOS	45nm SOI	28nm CMOS	22nm FDSOI	45nm SOI	90nm CMOS	65nm CMOS	65nm CMOS
V_{SUPPLY} [V]	1.1	1.0	-	1.2	1.0	0.9	0.8	1.1	1.2	1.0	1.5
Topology	Gilbert	Push Push	Push Push	Push Push	Push Push	Push Push	Push Push	Push Push	Gilbert	Push Push	Push Push
f_{out} [GHz]	48 to 82.5	75 to 89	50 to 57	47 to 70.5	53 to 74	48 to 64	69 to 80	46 to 89	40 to 49	62 to 90	73 to 88
BW [%]	53	17	13	39#	33	29	15	64	20	37	19
$P_{o,2f0-sat}$ [dBm]	9.5	6.7	5.3	10.1	9.8	5	2.4	7.4	5#	2	0#
FR [dB]	30	36	19#	33	32	24	36	38	25	20	19
η_D [%]	16	7#	18	33#	74	12.3#	4.9	25	15#	13#	7.14#
CG_{MAX} [dB]	7.3	0.1	0	-1.8	-4.1	1.2	-0.6#	-4.3	-1.6	-2.5	0.8
PAE_{MAX} [%]	11	0.13#	0#	<0	<0	1.5#	<0	<0	<0	<0	0.6
$CG @ PAE_{MAX}$ [dB]	5.5	0.1	0#	-	-	1.0#	-	-	-	-	0.8#
$P_{o,2f0} @ PAE_{MAX}$ [dBm]	8.5	3.1	0#	-	-	2.9#	-	-	-	-	-3
Area [mm²]	0.1	0.18*	0.12	0.18	0.058	0.17	0.17	0.088	0.39*	0.12*	0.096*

Estimated from the graphs and tables.
* Estimated from the chip micrograph, area including core, input and output passives.

Figure 27.10.6: Summary and comparison table.

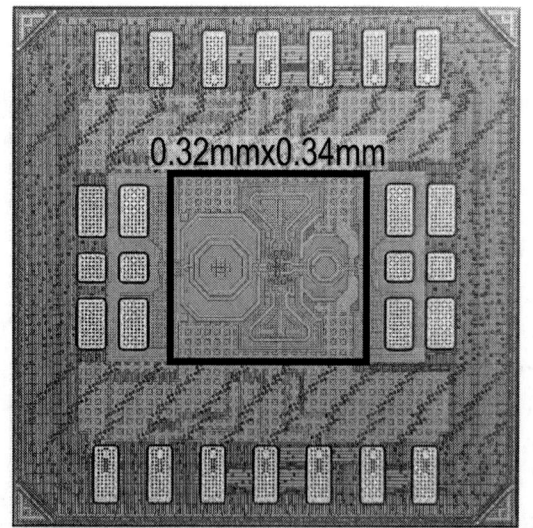

0.32mmx0.34mm

Figure 27.10.7: Die micrograph.

• 2026 IEEE International Solid-State Circuits Conference

ISSCC 2026 / SESSION 28 / INNOVATIONS FROM OUTSIDE THE (ISSCC) BOX / OVERVIEW

Session 28 Overview: *Innovations from Outside the (ISSCC) Box*

TECHNOLOGY DIRECTIONS SUBCOMMITTEE

Session Chair: Kaushik Sengupta
Princeton University
Princeton, NJ

Session Co-Chair: Firooz Aflatouni
University of Pennsylvania
Philadelphia, PA

Exciting advances are unfolding beyond the traditional circuit community, with the potential to profoundly shape the future of the Solid-State Circuits Society—opening new applications and creating fresh platforms for processing, communication, and sensing. This invited session introduces a few of these groundbreaking developments to the ISSCC audience, aiming to spark new ideas in circuit design and inspire collaborations across diverse fields. This year, we feature leading experts in materials science and device technology who will illustrate how recent breakthroughs are poised to transform the future of radio-frequency and wireless systems. We also highlight emerging innovations in flexible electronics and computer vision/AI that promise to revolutionize personal healthcare and mental health.

8:00 AM

28.1 Importance of GaN for 5G and Solid-State mm-Wave Circuits of the Future

Umesh K. Mishra, University of California, Santa Barbara, CA

In Paper 28.1, UC Santa Barbara presents the latest advances in Gallium nitride (GaN), a wide-bandgap semiconductor, that has revolutionized high-frequency and high-power applications such as 5G communications and mm-wave systems due to its exceptional electron mobility and breakdown field. Recent advances in N-polar GaN technology offer even greater efficiency, power density, and thermal performance than traditional Ga-polar GaN, promising major improvements in compact, energy-efficient RF and satellite communication systems.

8:25 AM

28.2 High-Selectivity, Tunable Filters with Zero Static Power Consumption Formed Using Micromachined Magnetostatic Wave Cavities

Roy H. Olsson III, University of Pennsylvania, Philadelphia, PA

In Paper 28.2, U Penn presents a new paradigm of miniaturized and high-efficiency tunable RF filters with micromachined YIG magnetostatic wave filters that achieve wide tunability, high selectivity, and low loss while operating with zero static power. The technology significantly improves insertion loss, rejection, and power handling compared to prior tunable filter technologies, making it a scalable solution for compact, low-power 6G and beyond communication front-ends.

8:50 AM

28.3 Body-Interfaced Biosensors

Wei Gao, California Institute of Technology, Pasadena, CA

In Paper 28.3, Caltech reviews advances in wearable, implantable, and ingestible biosensors that enable continuous molecular monitoring across diverse body fluids such as sweat, breath, wounds, and the gut. By integrating nanomaterials, microfluidics, wireless communication, and self-powered systems, these platforms pave the way for predictive, personalized, and sustainable digital healthcare.

9:15 AM

28.4 The Sensing, Computing, and Devices Opportunities of Computational Behavioral Phenotype: An Autism Spectrum Disorder Case Study

Guillermo Sapiro, Princeton University, Princeton, NJ

In Paper 28.4, Princeton presents a new domain for computer vision at the intersection of AI and mental health. The paper demonstrates how combining computer vision and machine learning, one can allow early detection of autism-related behaviors through short, interactive videos on smartphones or tablets. By quantifying multiple digital behavioral markers such as gaze, facial movement, and response to name, the app achieves highly accurate, equitable, and scalable early autism screening, highlighting opportunities for innovation in sensors, edge computing, and AI-driven healthcare.

ISSCC 2026 / SESSION 28 / INNOVATIONS FROM OUTSIDE THE (ISSCC) BOX / 28.1

28.1 Importance of GaN for 5G and Solid-State mm-Wave Circuits of the Future

Umesh K. Mishra, Christopher J. Clymore, Emre Akso, Matthew Guidry

University of California, Santa Barbara, CA

Abstract

GaN has played an important role in both RF and power devices. GaN made the rapid development and deployment of 5G base stations possible with use of its Ga-polar face. Meanwhile, the N-polar face of GaN has shown great promise through current research, with record power and efficiency at W-band. The benefits of N-polar GaN for systems include a smaller size with fewer power amplifier chips combined due to its higher power density, and lower cost of operation with its improved efficiency.

Wide-bandgap semiconductors play an important role in solid-state systems and circuits. Due to their large bandgap, these semiconductors, such as GaN and SiC, can handle large breakdown fields enabling power devices that can handle kilovolts[1]–[4]. GaN is a polarized material, which can be used with other III-Nitrides for use in high-electron-mobility transistors (HEMTs) with a formation of a polarization discontinuity-induced two-dimensional electron gas (2DEG). These properties combined with its high mobility and saturated electron velocity, enable high current densities and fast switching which allows GaN to be an important semiconductor material for RF and mm-Wave technologies, with high power and efficiency. Due to the polarization, there are two faces of GaN used in HEMTs: Ga-polar and N-polar. Ga-polar is the negatively charged surface, while N-polar is positively charged, illustrated in Fig. 28.1.1 [5]. Initial research in GaN started with Ga-polar, with the first HEMT produced by APA Optics Inc in 1993 [6], and the first RF power GaN HEMT in 1996 at UCSB [7].Since then, further work based on several breakthroughs at UCSB and across academia and industry, has pushed GaN to remarkable improved efficiency and output power. Commercially, Ga-polar GaN HEMTs, such as the one shown in Fig. 28.1.1(a) [5], are heavily used up to 30GHz for electronic warfare, defense applications, and communication. GaN has made massive commercial rollout of 5G communication possible due to its RF capabilities including high output power density, improved efficiency at saturated and linear power [8]–[10], and wide bandwidth, which can be traced to the extensive research on GaN done at UCSB.

While N-polar GaN has been researched far less, the reverse polarization fields within it, the fundamental difference between N- and Ga-polar GaN, unlocks a myriad of ways to improve transistor performance. The charge inducing barrier layer places a critical role in device design. In a Ga-polar GaN HEMT, the barrier layer is on top of the channel. Therefore, scaling the gate-to-channel distance reduces the charge density and current[11]; thus, designs for high gain and high-power operations are at odds. In contrast, the N-polar orientation has the barrier layer underneath the channel, as shown in Fig. 28.1.2 [12]. This allows for independent tuning of charge density and gate-channel spacing, thereby enabling high current, power, gain, and efficiency simultaneously. Also, N-polar GaN technology uses a GaN cap in the source and drain access regions, which improves the conductivity in these access regions and eliminates DC-RF dispersion caused by traps [13]. Ga-polar GaN HEMTs suffer from such dispersion contributing to lower power density and efficiency than N-polar GaN especially at mm-wave frequencies. N-polar GaN enables near-zero dispersion operation [14] without current collapse, knee walkout, and reduced memory effects. Thus, the ideal RF power transistor behavior with high power, high efficiency, and excellent linearizability can be achieved. So far, N-polar GaN HEMTs have been optimized for W-band frequencies (75-to-110GHz), where it has shown improved performance compared with Ga-polar GaN, Fig. 28.1.3.

There are multiple substrates being used in research of N-polar GaN, such as silicon, GaN, SiC, and sapphire. Each has their own positives and negatives such as cost, thermal conductivity, substrate and epi defect density, available wafer sizes, and lattice matching. Overall, an ideal substrate would be low cost with high thermal conductivity, available in large diameter with a low defect density, and a well-matched lattice constant of that to GaN. A closely matched lattice constant will result in fewer dislocations within the epitaxially grown layers, enabling improved device performance with fewer trap-induced effects and possible elimination or reduction buffer layer thickness, which would reduce the thermal resistance at the transistor level. In this regard, GaN substrates could be the best[15], [16], but due to their current high cost they are not used in large-scale manufacturing at the present but is attractive in the future. The two main substrates that are used for RF GaN and N-polar GaN research have been sapphire and SiC. Sapphire is an inexpensive substrate relative to SiC available in large diameter at low defect density, enabling more cost-effective transistors and systems. SiC has a higher thermal conductivity, which can simplify cooling for very high power applications. The high efficiency N-polar GaN on sapphire has exhibited at 94GHz increases the viability of sapphire substrates due to the reduction in dissipated power and therefore reduction in heat. New technologies such as integrated diamond on GaN proximal to the device [17], [18] can additionally reduce thermal resistances making the substrate material a less critical component in the thermal environment.

Today, Ga-polar GaN plays an important role in the RF front-end of communication systems including 5G base station, satcom, and backhaul. New N-polar GaN technology can play an important role in improving RF and mm-Wave power amplifiers (PAs) in such systems by reducing their size and operating costs. It can also improve satellite-to-satellite communication, for which photonics technology is also a contender. With the use of high efficiency mm-Wave N-polar HEMTs, this communication can be handled by compact solid-state phased arrays with a reduced size and weight, while keeping power consumption and heat dissipation small. The next generations of backhaul are also expanding into E-band, and D-band frequencies, in addition to lower bands. Phased arrays are important for many of these applications, especially at mm-Wave frequencies due to the low power available from a single amplifier and the need for steerable high directivity antennas to overcome high path losses. A high power per antenna element in a phased array is still desired due to antenna aperture size (and thus directivity) constraints, but this becomes a challenge in many technology due to the reduction in power density as frequency is increased into the mm-Wave, compounded by the area available for each element in a 2D array, $\lambda/2^2$ with element spacing of $\lambda/2$, dropping by $4\times$ with every doubling of frequency. Therefore, technologies with a high transistor-level power density and per-transistor power with high PAE are highly desired for such applications. N-polar GaN has significantly expanded the state-of-the-art with the first demonstration of constant power density of 8 W/mm from 10-to-94GHz [19], a ~$3\times$ improvement of transistor power density to 8.8W/mm [20], and a record 1W of output power from a single transistor at 94GHz [21]. These results prove N-polar's capability of reducing the circuit size with high output power from single transistors, enabling high-power long-range wireless links for mm-Wave applications.

With power also comes heat, and a dense array of heat sources becomes difficult to cool, especially for space-based applications where all heat must be radiated into space as air is not available as a heatsink medium. N-polar GaN has also distinguished itself here with record efficiency at 94GHz, Fig. 28.1.3. The use of Schottky gates has also enabled improved efficiency, with record power added efficiency, or PAE, of over 50% at W-band [22], [23]. This high efficiency reduces the power which needs to be dissipated as heat through the system's heatsinks, making new and more compact system architectures possible. Although the superior performance has so far been demonstrated mainly at >75GHz, N-polar GaN has a tremendous potential at lower frequency bands due to its inherent high breakdown field and dispersion-free high current density per unit output capacitance, which means not only high power but also higher Bode-Fano bandwidth compared to Ga-polar GaN.

The first circuit demonstration of N-polar GaN at >100GHz has been reported recently [24]. Because of limited availability of loadpull capability and limited loadpull drive powers at D-band, the pre-matched N-polar GaN HEMT circuit was designed using 94GHz loadpull measurements, small signal parameters, and pulsed-IV data to obtain the target load and source reflection coefficients at 130GHz. Co-planar waveguide matching networks were designed with metal-air-metal airbridge capacitors on the shunt lines to achieve compact circuit size, as shown in Fig. 28.1.6. Although it is a one-stage circuit with no driver stage, the circuit outperformed the previously reported Ga-polar GaN based circuits in power density in 120-to-150GHz range, as illustrated in Fig. 28.1.7. This result further proved the feasibility of this emerging technology in circuits and systems beyond the transistor level performance.

Additionally, there has been growing research in ultra-wide bandgap materials such as AlN and Ga_2O_3 that can handle extreme breakdown fields due to their larger bandgaps. Each material has its own challenges and advantages, but both are still in the early stages of research and are many years away from commercial impact. The next step in the evolution of GaN as the preferred RF semiconductor alongside Si is N-polar GaN.

References:

[1] C. E. Weitzel *et al.*, "Silicon carbide high-power devices," *IEEE Trans. on Electron Devices*, vol. 43, no. 10, pp. 1732-1741, Oct. 1996, doi: 10.1109/16.536819. https://doi.org/10.1109/16.536819

[2] F. Nouketcha *et al.*, "Detailed study of breakdown voltage and critical field in wide bandgap semiconductors," *2019 IEEE 7th Workshop on Wide Bandgap Power Devices and Applications, (WiPDA)*, pp. 200–207, Oct. 2019, doi: 10.1109/WIPDA46397.2019.8998828. https://doi.org/10.1109/WIPDA46397.2019.8998828

[3] B. Jayant Baliga, "Silicon Carbide Power Devices: Progress and Future Outlook," *IEEE J. of Emerging and Selected Topics in Power Electronics,* vol. 11, no. 3, 2023, doi: 10.1109/JESTPE.2023.3258344. https://doi.org/10.1109/JESTPE.2023.3258344

[4] W. Liu, "GaN Power Device Technology and monolithic integrated GaN power systems," *Int. Conf. on IC Design and Technology,* pp. xxxix, Dec. 2023, doi: 10.1109/ICICDT59917.2023.10332321. https://doi.org/10.1109/ICICDT59917.2023.10332321

[5] S. Keller *et al.*, "Recent progress in metal-organic chemical vapor deposition of (000) N-polar group-III nitrides," *Semiconductor Science and Technology,* vol. 29, no. 11, p.113001, Aug. 2014, doi: 10.1088/0268-1242/29/11/113001. https://doi.org/10.1088/0268-1242/29/11/113001

[6] M. Asif Khan, A. Bhattarai, J. N. Kuznia, and D. T. Olson, "High electron mobility transistor based on a GaN-AlxGa1−xN heterojunction," *Applied Physics Letters,* vol. 63, no. 9, pp. 1214–1215, Aug. 1993, doi: 10.1063/1.109775. https://doi.org/10.1063/1.109775

[7] Y.-F. Wu *et al.*, "Measured microwave power performance of AlGaN/GaN MODFET," *IEEE Electron Device Letters,* vol. 17, no.9, pp. 455–457, 1996, doi: 10.1109/55.536291. https://doi.org/10.1109/55.536291

[8] "Why GaN is 5G's Super 'Power' - Qorvo." https://www.qorvo.com/design-hub/blog/why-gan-is-5gsuper-power (accessed Sep. 01, 2025). https://www.qorvo.com/design-hub/blog/why-gan-is-5g-super-power

[9] H. Lu *et al.*, "A review of GaN RF devices and power amplifiers for 5G communication applications," *Fundamental Research,* vol. 5, no. 1, pp. 315–331, Jan. 2023, doi: 10.1016/J.FMRE.2023.11.005. https://doi.org/10.1016/J.FMRE.2023.11.005

[10] M. B. Yaseen, F. Wan, F. Siddique, and A. Thakur, "GaN radiofrequency components and power amplifiers for next-generation 5G communications," *Microelectronic Engineering,* vol. 297, p.112305, Mar. 2025, doi: 10.1016/J.MEE.2024.112305. https://doi.org/10.1016/J.MEE.2024.112305

[11] M. H. Wong *et al.*, "N-polar GaN epitaxy and high electron mobility transistors," *Semiconductor Science and Technology,* vol. 28, no. 7, Jul. 2013, doi: 10.1088/0268-1242/28/7/074009. https://doi.org/10.1088/0268-1242/28/7/074009

[12] W. Li *et al.*, "Record RF Power Performance at 94 GHz From Millimeter-Wave N-Polar GaN-on-Sapphire Deep-Recess HEMTs," *IEEE Trans Electron Devices,* pp. 1–6, Feb. 2023, doi: 10.1109/ted.2023.3240683. https://doi.org/10.1109/ted.2023.3240683

[13] S. Wienecke *et al.*, "N-polar GaN cap MISHEMT with Record Power Density Exceeding 6.5 W/mm at 94 GHz," *IEEE Electron Device Letters,* vol. 38, no. 3, pp. 359–362, Mar. 2017, doi: 10.1109/LED.2017.2653192. https://doi.org/10.1109/LED.2017.2653192

[14] S. Diez, S. Mohanty, C. Kurdak, and E. Ahmadi, "Record high electron mobility and low sheet resistance on scaled-channel N-polar GaN/AlN heterostructures grown on on-axis N-polar GaN substrates by plasma-assisted molecular beam epitaxy," *Appl Phys Lett,* vol. 117, no. 4, p. 42102, 2020, doi: 10.1063/5.0014460. https://doi.org/10.1063/5.0014460

[15] M. Saro *et al.*, "Deep Level Effects in N-Polar AlGaN/GaN High Electron Mobility Transistors: Toward Zero Dispersion Effects," *IEEE International Reliability Physics Symposium Proceedings,* 2024, doi: 10.1109/IRPS48228.2024.10529479. https://doi.org/10.1109/IRPS48228.2024.10529479

[16] O. Odabasi *et al.*, "Record-high electron mobility at near pinch-off in N-polar GaN HEMT structures grown on on-axis N-polar GaN substrates by plasma-assisted molecular beam epitaxy," *Appl Phys Lett,* vol. 127, no. 13, p. 132101, 2025, doi: 10.1063/5.0258504. https://doi.org/10.1063/5.0258504

[17] M. Malakoutian *et al.*, "Record-Low Thermal Boundary Resistance between Diamond and GaN-on-SiC for Enabling Radiofrequency Device Cooling," *ACS Appl Mater Interfaces,* vol. 13, no. 50, pp. 60553–60560, Dec. 2021, doi: 10.1021/ACSAMI.1C13833. https://doi.org/10.1021/ACSAMI.1C13833

[18] R. Soman *et al.*, "Integration of 150 nm gate length N-polar GaN MIS-HEMT devices with all-around diamond for device-level cooling," *Applied Physics Express,* vol. 18, no. 4, p. 46503, Apr. 2025, doi: 10.35848/1882-0786/adcb87. https://doi.org/10.35848/1882-0786/adcb87

[19] B. Romanczyk *et al.*, "Demonstration of constant 8 W/mm power density at 10, 30, and 94 GHz in state-of-the-art millimeter-wave N-polar GaN MISHEMTs," *IEEE Transactions on Electron Devices,* vol. 65, no. 1, pp. 45–50, Jan. 2018, doi: 10.1109/TED.2017.2770087. https://doi.org/10.1109/TED.2017.2770087

[20] B. Romanczyk *et al.*, "W-Band Power Performance of SiN-Passivated N-Polar GaN Deep Recess HEMTs," *IEEE Electron Device Letters,* vol. 41, no. 3, pp. 349–352, 2020, doi: 10.1109/LED.2020.2967034. https://doi.org/10.1109/LED.2020.2967034

[21] E. Akso *et al.*, "Record 1 W output power from a single N-Polar GaN MISHEMT at 94 GHz," *2023 Device Research Conference (DRC),* Jun. 2023, vol. 2023-June, pp. 1–2, doi: 10.1109/DRC58590.2023.10187008. https://doi.org/10.1109/DRC58590.2023.10187008

[22] E. Akso *et al.*, "Schottky Barrier Gate N-Polar GaN-on-Sapphire Deep Recess HEMT With Record 10.5 dB Linear Gain and 50.2% PAE at 94 GHz," *IEEE Microwave and Wireless Technology Letters,* vol. 34, no. 2, pp. 183–186, Feb. 2024, doi: 10.1109/LMWT.2023.3345531. https://doi.org/10.1109/LMWT.2023.3345531

[23] H. Collins *et al.*, "N-Polar Deep Recess GaN HEMT With a TiN Schottky Gate Contact Demonstrating 53.4% PAE and 3.7 W/mm Associated Pout at 94 GHz," *IEEE Microwave and Wireless Technology Letters,* vol. 34, no. 7, pp. 907–910, 2024, doi: 10.1109/LMWT.2024.3402558. https://doi.org/10.1109/LMWT.2024.3402558

[24] E. Akso *et al.*, "Record D-Band Performance From Prematched N-Polar GaN-on-Sapphire Transistor With 2 W/mm and 10.6% PAE at 132 GHz," *IEEE Microwave and Wireless Technology Letters,* vol. 34, no. 4, pp. 395–398, Apr. 2024, doi: 10.1109/LMWT.2024.3365145. https://doi.org/10.1109/LMWT.2024.3365145

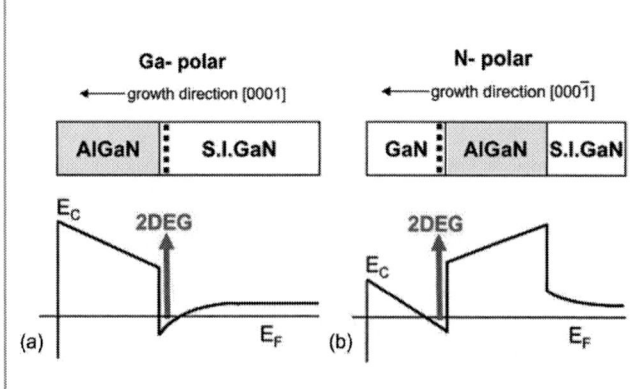

Figure 28.1.1: (a) Basic material stack and band diagram for a Ga-polar HEMT. Showcases the 2DEG behind the charge inducing barrier. (b) Basic material stack and band diagram for an N-polar HEMT. Showcasing the charge inducing barrier is behind the 2DEG. [5]

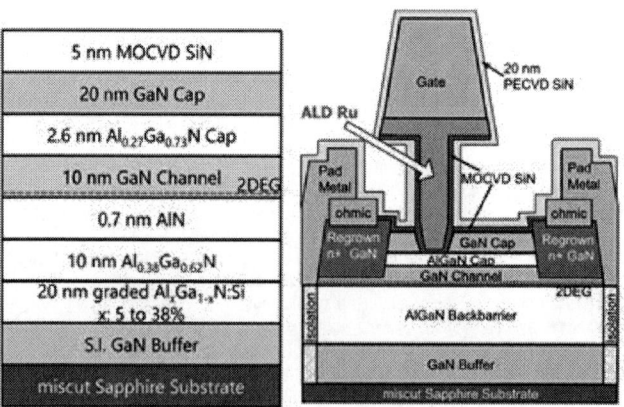

Figure 28.1.2: Material stack of a deep recess N-polar HEMT next to the device cross section of a mm-Wave deep recess N-polar HEMT. [12]

ISSCC 2026 / SESSION 28 / INNOVATIONS FROM OUTSIDE THE (ISSCC) BOX / 28.1

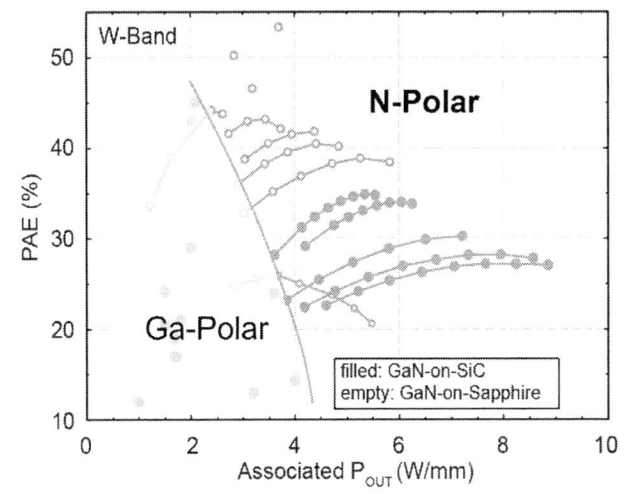

Figure 28.1.3: Comparison of peak PAE and its associated Output power in W//mm for N-polar and Ga-polar at W-band frequencies. N-polar shows a higher output power and PAE, illustrating its ability of obtaining larger breakdown fields than Ga-polar.

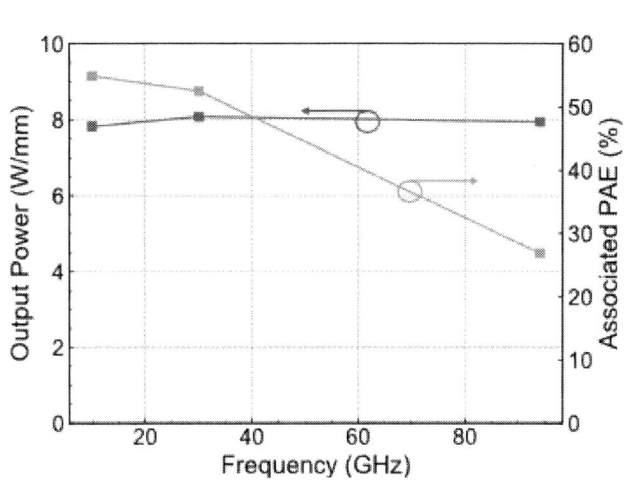

Figure 28.1.4: Demonstration of N-polar deep recess HEMTs having 8W/mm from 10-to-94GHz with a drain bias of 20V. [15]

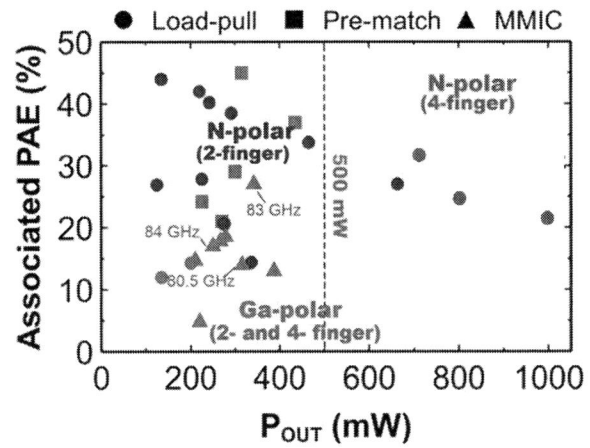

Figure 28.1.5: Demonstration of the first 1 Watt transistor at 94GHz with use of 4-finger N-polar deep recess HEMT technology. Proves the ability for N-polar GaN to have large output powers per device. [17]

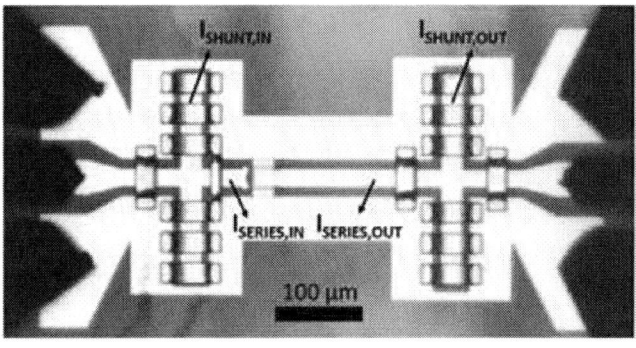

Figure 28.1.6: Prematched N-Polar GaN HEMT circuit with 2 × 25µm gate periphery designed and fabricated for operation at 130GHz.

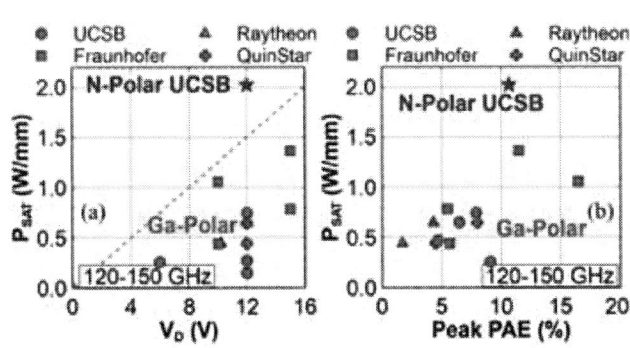

Figure 28.1.7: (a) saturated output power in W/mm versus drain bias and (b) saturated output power in W/mm versus peak PAE at 132GHz. N-polar demonstrates improved performance despite not fully pushing the device to maximum power and PAE due to limited drive power. [18]

• 2026 IEEE International Solid-State Circuits Conference

979-8-3315-8937-0/26 $31.00 © 2026 IEEE

ISSCC 2026 / SESSION 28 / INNOVATIONS FROM OUTSIDE THE (ISSCC) BOX / 28.2

28.2 High-Selectivity, Tunable Filters with Zero Static Power Consumption Formed Using Micromachined Magnetostatic Wave Cavities

Shun Yao, Xingyu Du, Yixiao Ding, Shuxian Wu, Chin-Yu Chang, Mark Allen, Roy H. Olsson III

University of Pennsylvania, Philadelphia, PA

Abstract

This work reports zero-static-power, high-Q magnetostatic wave filters in YIG films with wide frequency tuning range and high selectivity. A compact 1.2-cc magnetic bias circuit enables continuous operation without DC power. This supports bandpass filter tuning from 3.3-9.7GHz with 2-3dB insertion loss and >35dB rejection, and notch filter tuning from 3.5-to-16.8GHz with <0.9dB loss and >38dB notch. Thickness-scaled YIG extends operation to 32GHz with fewer spurs and higher power handling.

In comparison to switched filter banks, which implement frequency control in 4G and 5G front-ends, a single tunable filter shows great potential to reduce front-end module size, cost and complexity. When scaling to emerging bands, such as those targeted for 6G, low-loss tunable filters eliminate the high losses that are introduced by RF switch matrices at high frequency [2]. Among candidate technologies, tunable filters based on magnetostatic waves (MSW) propagating in yttrium iron garnet (YIG) films are especially promising due to their miniature size, extremely low-loss and large frequency tuning range [3,4]. The low phase velocity of MSW in YIG [3] compared with the velocity of electromagnetic waves enables miniature resonant structures that are compatible with microwave assemblies. Also, the intrinsic low damping of MSW in YIG yields devices with high quality factors, which are usually larger than 1000 [3]. MSW phase velocity is tunable by adjusting the external magnetic bias field, allowing for filter frequency tuning. This allows much larger tuning range compared with other tunable filter technologies like varactor based or MEMS based tunable filters. However, commercial YIG filters [8] and some recently published research [9-10] rely on bulky electromagnets to realize the magnetic field, which have a large volume >20 cubic cm (cc) and a power consumption of several Watts, limiting their adoption in compact wireless systems.

In this paper, we first introduce a zero-static power consumption magnetic bias circuit, with a size less than 1.2cc. Then a tunable 3-to-10GHz bandpass filter design with an insertion loss of 2-to-3dB and over 35dB out-of-band (OOB) rejection is reported. Next, we report a tunable 3.5-to-17GHz bandstop filter with less than 1dB of insertion loss and >38dB of notch rejection. Both filters are fabricated using 3μm thick YIG films and are measured inside the zero-static power magnetic bias circuit. Finally, we present a tunable bandpass filter operating over an 8-to-32 GHz frequency range with 2.5-to-5.4dB insertion loss and >28dB of OOB rejection based on a 15μm thick YIG film.

Figure 28.2.1 shows the overall tunable filter concept including the magnetic bias circuit design. YIG thin films on gadolinium gallium garnet (GGG) are micromachined into resonant cavities using the processes reported in [1,7]. Al is deposited and patterned on top of the YIG cavities which serves as transducers to launch and detect magnetostatic surface waves (MSSW). For 3μm thick YIG devices, the Al is directly deposited on the YIG/GGG surface, whereas a planarization layer of benzocyclobutene (BCB) is required for 15μm-thick films to ensure reliable transducer integration. In terms of the magnetic bias circuit, two NdFeB permanent magnets (blue) and two non-volatile AlNiCo programmable magnets are used to provide the external bias magnetic field. Two coils (red) are wound around the AlNiCo non-volatile magnets. When a short current pulse is generated by the discharging capacitor and applied to the coils, the magnetic remanence of the AlNiCo magnets will be altered due to its moderate coercivity. The magnetic field is directed by two yoke pieces to the center region to provide a uniform bias field for the YIG cavities which determines the resonant frequency of the filter. As reported in [11], the maximum switching duration and energy consumption is approximately 100μs and 116mJ, respectively, when tuning from the lowest field (0 Gauss) to the highest field (5700 Gauss) where the filter operates at 18GHz. For smaller tuning spans the time and energy is significantly reduced. For example, tuning from 0-to-1800 Gauss (0-to-7 GHz) requires only 14μs and 13mJ. Because the AlNiCo remanence is non-volatile, energy is only required during reprogramming, enabling operation at zero DC power.

Figure 28.2.2 is an image of a magnetic bias circuit with a field tuning range from -128 Gauss to 5157 Gauss, ensuring a frequency tunability up to 16.75GHz. Figure 28.2.3 demonstrates a 9-stage YIG notch filter [5] placed inside the magnetic bias circuit.

Figure 28.2.4(a) presents a scanning electron microscope (SEM) image of a two-cavity, 6-pole bandpass filter using a 3-stage meander line transducer at each port. Unlike the straight line transducers reported in [1], meander lines are used to enhance the cavity quality factor and Figure of Merit (FoM = K²Q) [6]. This meander geometry, however, introduces higher series and magnetostatic resistance, leading to over coupling to the 50Ω filter terminations. In [6], filters with various stages of meander line are fabricated, showing that although additional stages further raise FoM, they also increase impedance mismatch and degrade the overall insertion loss. To mitigate the resulting mismatch loss, the filter is implemented in a parallel two-cavity configuration. Figure 28.2.4(b) shows the measured frequency

response as the filter is tuned from 3.3GHz to 9.7GHz via the magnetic bias circuit. In the measurement, tuning is achieved by charging and discharging a capacitor to generate the short current pulses in the coils that magnetize the AlNiCo magnets. The voltage in the legend corresponds to the voltage used to charge the capacitor which determines the magnitude of the current pulse and the resulting frequency. The AlNiCo magnets can be demagnetized by altering the direction of the applied voltage and current, allowing bidirectional frequency tuning. The measured insertion loss is between 2.1dB to 3dB, while the OOB rejection is approximately 45dB at low frequency and more than 35dB at 10GHz. The out-of-band rejection is primarily determined by the electromagnetic coupling between the input and output transducers. As reported in [11], where a similar 6-pole meander-line structure was employed on 18μm thick YIG, increasing the transducer spacing enhanced the rejection to 53dB at 9.5GHz, with less than 0.5dB increase in insertion loss, owing to the intrinsically low propagation loss of MSSW in YIG.

Bandstop filters are also implemented and characterized in the magnetic bias circuit. Figure 28.2.5(a) shows an SEM image of 5 stages of a 9-stage bandstop filter. For the notch filters, a wideband impedance matching technique is employed to minimize the insertion loss. Shunt capacitors are added to absorb the inductance of the MSW transducers, which creates a 50Ω LC transmission line that ensures good matching and extremely low loss over a wide frequency range [5]. Figure 28.2.5(c) shows the measured S_{21} response of the notch filter when tuning the notch from 3.5GHz to 16.76GHz. The rejection increases from 24.5dB at 3.5GHz to more than 55dB at 16.76GHz. Similar to the bandpass filter, the magnetostatic resistance of the transducer increases with frequency, which generates a stronger reflection and thus a deeper notch at higher frequencies. Also, as shown in Fig. 28.2.5(b), due to the wideband matching technique, the insertion loss in the passband is less than 0.9dB from 2-to-18 GHz. Compared to the notch filter reported in [5], the maximum frequency increased from 10.3GHz to 16.76GHz, owing to a revised design of the magnetic bias circuit. The improved field uniformity of the new bias circuit also suppresses spurious modes and maintains sharp notches across the extended tuning range.

Recent studies indicate that increasing the YIG film thickness enhances MSW coupling, suppresses spurious responses and improves power handling [7]. To enable reliable integration of transducers on patterned 15μm thick YIG films, a planarization process using BCB was developed. This mitigates transducer breakage at steep sidewalls and significantly improves fabrication yield [7]. Figure 28.2.6(a) shows optical microscope images of the fabricated filters. The yellow rectangular region indicates the BCB openings. The remaining areas are covered with BCB, which introduces no measurable performance difference compared to devices without BCB, owing to the intrinsically low dielectric loss of BCB [11]. Then straight-line bandpass filters similar to those reported in [1] are fabricated and measured. Figure 28.2.6(d) shows the frequency response of the filter on thick YIG with a frequency tuning range of 8-to-32GHz. Compared to the filter response of a 3μm thick YIG cavity [1], devices on thick YIG show higher selectivity with fewer spurs and a steeper filter skirt. The spurs originally appearing above the passband are identified as higher-order length modes, which are effectively suppressed by the thick YIG due to the increased radiation resistance and shaper dispersion relation [7]. The remaining spurs below the passband are attributed to higher-order width modes, which can be mitigated by tailoring the transducer geometry to shape the current density and spatial excitation profile, as reported in [12]. As in Fig. 28.2.6(a), a two-cavity, parallel design is utilized to improve impedance matching and reduce insertion loss across the frequency tuning range. As shown in Fig. 28.2.6(b), the filter is well-matched to 50 Ohms at around 10GHz and is overcoupled at higher frequency. Figure 28.2.6(c) summarizes the insertion loss across the frequency tuning range. The filter shows an insertion loss ranging from 2.5dB to 5.4dB and more than 28dB of OOB rejection over the entire range. In addition, the thicker YIG raises the 1-dB compression point (P1dB) to +13dBm [7], compared to −14dBm in 3μm thick YIG devices [1], demonstrating a substantial improvement in power handling.

In summary, the high-Q YIG thin-film tunable filters with zero-static-power magnetic bias circuits provide a compact (<1.2cc), low-cost solution for integration into microwave front-ends. Compared to switched filter banks and other tunable technologies, these designs eliminate switching loss, require no static power, and maintain low insertion loss across a wide tuning span. The updated magnetic bias circuit achieves a field range of −128 Gauss to 5157 Gauss, enabling a notch tuning range from 3.5-to-16.76GHz, extending well beyond the limits of prior work [5]. The notch filter maintains <0.9dB insertion loss across 2-to-18GHz, with notch depth exceeding 55dB at 16.76GHz due to better field uniformity. Leveraging thickness-scaled YIG films, we further demonstrate filters with extended operating range up to 32GHz, with higher selectivity and fewer spurs, achieving 2.5-to-5.4dB insertion loss, and >28dB out-of-band rejection. Additionally, the thicker YIG enhances power handling, with P1dB increased from −14dBm in thin YIG to +13dBm in thick YIG devices. Figure 28.2.7 presents a comparison between the YIG tunable filters reported in this work and other tunable filter technologies. These results highlight the potential of YIG-based magnetostatic wave filters as a scalable technology for wideband, size-weight-and-power constrained applications in future 6G and beyond.

Acknowledgement:
The authors would like to thank Dr. Todd Bauer, Dr. David Abe, and Dr. Tim Hancock of the Defense Advanced Research Projects Agency (DARPA) and Dr. Michael Page of the Air Force Research Laboratory for their guidance and support of this work under the DARPA Wideband Adaptive RF Protection (WARP) program, contract FA8650-21-1-7010. The fabrication of devices was performed at the Singh Center for Nanotechnology, supported by the NSF National Nanotechnology Coordinated Infrastructure Program (No. NNCI-1542153).

References:
[1] X. Du *et al.*, "Frequency Tunable Magnetostatic Wave Filters with Zero Static Power Magnetic Biasing Circuitry," *Nature Communications,* 15, 3582 (2024). https://doi.org/10.1038/s41467-024-47822-3
[2] G. Slovin *et al.*, "SPNT Circuits using Phase-Change Material RF Switches For 5G And Millimeter Wave Applications". *IEEE MTT-S International Microwave Symposium (IMS) (2021).* http://doi.org/10.1109/IMS19712.2021.9574818
[3] W. S. Ishak, "Magnetostatic Wave Technology: A Review," *Proceedings of the IEEE,* vol. 76, no. 2, pp. 171-187, Feb. 1988. http://doi.org/10.1109/5.4393
[4] Khrystyna O. Levchenko *et al.*, "Review on spin-wave RF applications." arXiv preprint arXiv:2411.19212 (2024). https://arxiv.org/abs/2411.19212
[5] X. Du *et al.*, "Magnetostatic Wave Notch Filters Frequency Tuned Via a Zero DC Power Magnetic Bias Circuit," *IEEE International Microwave Filter Workshop (IMFW), 2024,* pp. 176-179. http://doi.org/10.1109/IMFW59690.2024.10477145
[6] X. Du *et al.*, "Meander Line Transducer Empowered Low-Loss Tunable Magnetostatic Wave Filters with Zero Static Power Consumption," *IEEE/MTT-S International Microwave Symposium - IMS 2024,* pp. 42-45. http://doi.org/10.1109/IMS40175.2024.10600197
[7] X. Du *et al.*, "A Magnetostatic Surface Wave Filter Tunable Over 8-32 GHz Realized in Thickness Scaled Yttrium Iron Garnet," *IEEE/MTT-S International Microwave Symposium - IMS 2025, San Francisco, CA, USA, 2025,* pp. 890-893. http://doi.org/10.1109/IMS40360.2025.11103895
[8] Micro Lambda Wireless Inc. MLFP 4 Stage Filter Data Sheet. MLFP-42018. https://www.microlambdawireless.com/uploads/pdfs/MLFP%204%20Stage%20Filter%2 0Data%20Sheet%20-%20Copy%201.pdf
[9] Y. Feng *et al.*, "Micromachined Tunable Magnetostatic Forward Volume Wave Bandstop Filter," *IEEE Microwave and Wireless Technology Letters,* 2023. http://doi.org/10.1109/LMWT.2023.3267449

[10] C. Devitt *et al,* "An Edge-Coupled Magnetostatic Bandpass Filter, vol. 15, no. 1, p. 7764, 2024. https://doi.org/10.1038/s41467-024-51735-6
[11] X. Du *et al.*, "A Wideband Tunable, Nonreciprocal Bandpass Filter Using Magnetostatic Surface Waves with Zero Static Power Consumption." arXiv preprint arXiv:2505.09845 (2025). https://doi.org/10.48550/arXiv.2505.09845
[12] S. Wu *et al.*, "Spatially Tailored Spin Wave Excitation for Spurious-Free, Low-Loss Magnetostatic Wave Filters with Ultra-Wide Frequency Tunability." arXiv preprint arXiv:2507.14469 (2025). http://doi.org/10.48550/arXiv.2507.14469
[13] C. Devitt *et al.*, "Spinwave Bandpass Filters for 6G Communication." arXiv preprint arXiv: 2507.18931 (2025). https://doi.org/10.48550/arXiv.2507.18931
[14] Analog Devices, ADMV8432: Tunable band-pass filter, 15–30 GHz, datasheet. https://www.analog.com/media/en/technical-documentation/data-sheets/admv8432.pdf
[15] S. Hari *et al.*, "A Reflection-Mode N-Path Filter Tunable From 6 to 31 GHz," *IEEE Journal of Solid-State Circuits,* vol. 58, no. 7, pp. 1973-1986, July 2023. http://doi.org/10.1109/JSSC.2023.3235976

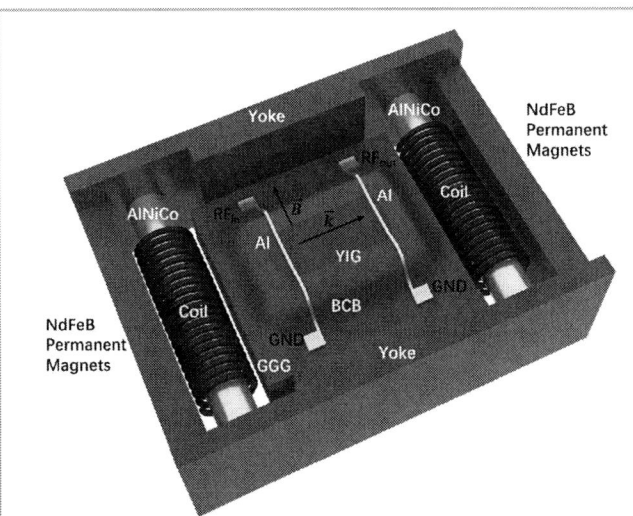

Figure 28.2.1: A 3D schematic of the MSSW filter and the magnetic bias circuit assembly.

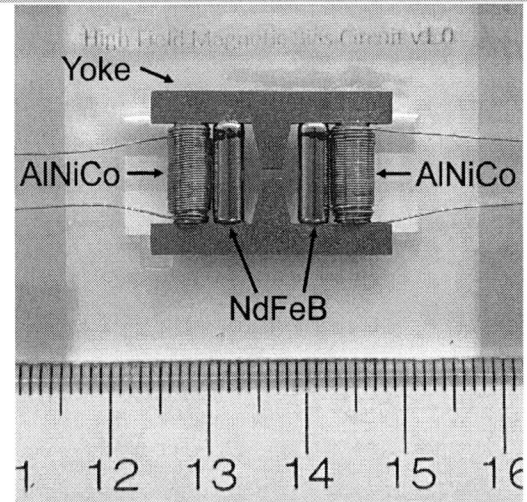

Figure 28.2.2: Zero power magnetic bias circuit with a field tunable range of -128 to 5157 Gauss and a small volume of 1.2 cc.

ISSCC 2026 / SESSION 28 / INNOVATIONS FROM OUTSIDE THE (ISSCC) BOX / 28.2

Figure 28.2.3: Zoomed in image showing a 9-stage notch filter [5] placed inside the magnetic bias circuit.

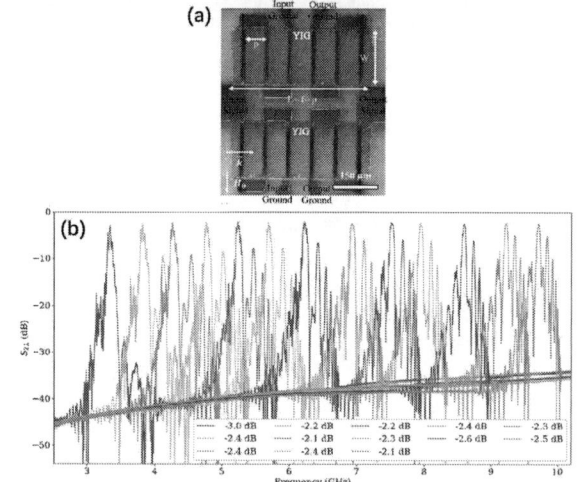

Figure 28.2.4: (a) SEM image of a two-cavity, 6-pole bandpass filter. (b) Measured 6-pole bandpass filter response in the magnetic bias circuit with frequency tuning from 3.3 to 9.7GHz. Adapted from [6].

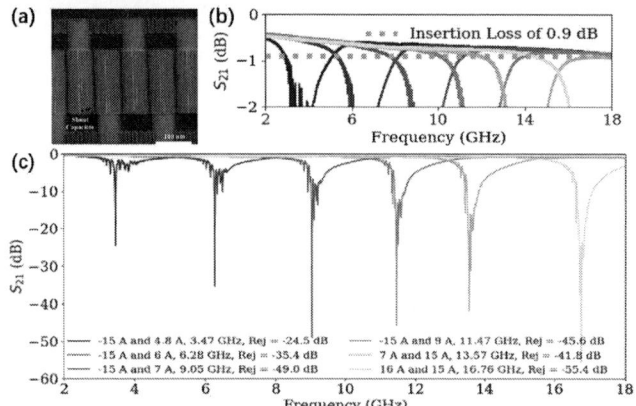

Figure 28.2.5: (a) SEM image of a multi-stage bandstop filter. (b) Zoomed in frequency responses showing insertion loss <0.9dB. (c) Measured 9-stage bandstop filter response measured in the magnetic bias circuit when tuning frequency from 3.5GHz to 16.76GHz. The two current values indicate the peak of the programming current pulses separately applied to the two AlNiCo magnet coils, which is measured using a current probe.

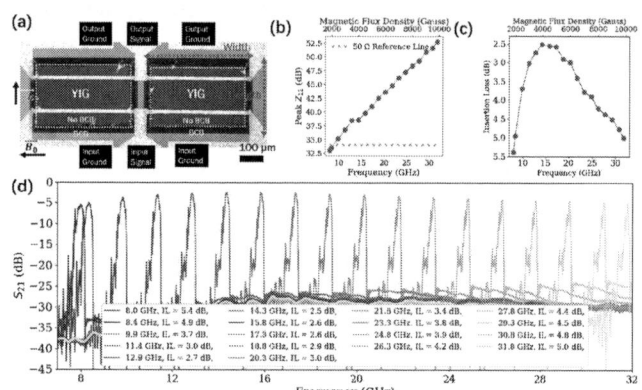

Figure 28.2.6: (a) Optical microscope image of a two-cavity straight-line bandpass filter implemented in a 15μm thick YIG film. The yellow dashed line shows areas where the BCB was removed, exposing the YIG surface for transducer integration. (b) Peak Z_{11} vs. frequency tuning. (c) Insertion loss vs. frequency tuning. (d) Measured 8-to-32 GHz frequency response of the two-cavity bandpass straight-line filter realized in a 15μm thick YIG film. Adapted from [7].

Tuning Type	Frequency range (GHz)	Insertion Loss (dB)	OOB rejection (dB)	Size (cc)	DC power (mW)	In-band IIP3 (dBm)	OOB IIP3 (dBm)
6-pole YIG [6]	3.3-9.7	2.1-3	>35	1.2	0	-0.7	>60[1]
Thickness-scaled YIG [7]	8-32	2.5-5.4	>28	NA[2]	NA[2]	26	>60[1]
YIG sphere [8]	2-18	<6	>80	~22	~2000	Not mentioned	Not mentioned
YIG [10]	4.5-10.1	<6	>25	NA[2]	NA[2]	-4.85	>25
YIG [13]	7.1-21.6	2.5-5.8	3.6-13.8	NA[2]	NA[2]	>11	Not mentioned
Varactor [14]	25.7-30.1	9	>30	0.027	8.5	37	Not mentioned
Varactor [14]	16.6-22.5	8	>30			34	
N-path [15]	6-31	4.5-6.6	15-50	<0.001	146-410	1.4-6.3	14-20

1. Out-of-Band IIP3 is limited by the test equipment.
2. NA indicates that the YIG filter is not paired with a DC magnetic bias circuit.

Figure 28.2.7: Performance comparison with other tunable bandpass filters.

ISSCC 2026 / SESSION 28 / INNOVATIONS FROM OUTSIDE THE (ISSCC) BOX / 28.3

28.3 Body-Interfaced Biosensors

Wei Gao

California Institute of Technology, Pasadena, CA

Abstract

The rise of personalized medicine is transforming healthcare through predictive and tailored strategies. I will present our progress on wearable, implantable, and ingestible biosensors for real-time molecular analysis across diverse body fluids. These scalable, nanomaterial-based systems enable continuous monitoring of metabolites, hormones, and drugs, with applications in stress, nutrition, chronic disease, and personalized therapy.

Body-Interfaced Biosensors for Digital and Precision Medicine

The human body relies on a complex network of physiological and biochemical processes that continuously generate signals reflective of health and disease. Traditional approaches to health monitoring have centered on discrete measurements performed in clinical settings or on wearable devices that track physical activity and vital signs. While valuable, these measurements provide only limited insight into the underlying molecular events that govern physiology. The ability to continuously monitor biochemical markers from the body in real time offers an unprecedented opportunity to shift medicine toward predictive, preventive, and personalized models of care. Advances in nanomaterials, flexible electronics, and bioelectronic systems are now converging to make this vision achievable, through a new generation of body-interfaced biosensors that can seamlessly access diverse body fluids, analyze multiple analytes, and wirelessly transmit clinically relevant information.

Recent progress in our laboratory has focused on developing wearable, implantable, and ingestible biosensors that autonomously sample sweat, interstitial fluid, wound exudate, gastrointestinal fluids, and exhaled breath condensate. These bioelectronic systems combine high-performance electrochemical transducers with microfluidics, wireless modules, and scalable fabrication techniques such as laser engraving, inkjet printing, and 3D direct ink writing. Through the use of advanced nanomaterials—including laser-engraved graphene, printed MXene-gold nanocomposites, and molecularly imprinted polymer nanoparticles—these sensors achieve high sensitivity, broad chemical selectivity, and mechanical robustness. Collectively, they establish a versatile platform for tracking metabolites, electrolytes, hormones, proteins, nutrients, and therapeutic drugs across multiple compartments of the body, enabling molecular-level precision in health monitoring.

Sweat-based biosensing has served as a foundational technology in this field. Our early work demonstrated the fully integrated wearable sensor array for multiplexed sweat analysis, which laid the groundwork for real-time biochemical monitoring outside the laboratory. Building on this foundation, we created scalable, low-cost fabrication strategies for sweat patches capable of simultaneously measuring metabolites (e.g., glucose, lactate, and uric acid), electrolytes, nutrients (e.g., amino acids), drugs, hormones (e.g., female hormones and stress hormones), and proteins [1], [2], [3], [4]. To overcome the reliance on exercise-induced perspiration, we integrated iontophoresis modules for on-demand sweat induction, allowing continuous measurement even during sleep (Fig. 28.3.1) [1]. These patches leverage microfluidic sampling and nanomaterial-based electrodes to achieve robust operation in daily life, and have been applied in clinical validation studies ranging from gout management to cardiometabolic monitoring.

Integration of multimodal sensing modalities further elevates diagnostic power. By combining chemical and physical signals with machine learning analytics, we created soft electronic skins that capture complex physiological states. In one example, our system monitored multiple biochemical markers in sweat together with galvanic skin response, skin temperature, and sweat dynamics continuously over a 24-hour period and during different daily activities (Fig. 28.3.2). Feeding these multimodal datasets into supervised learning models accurately assessed stress levels and predicted neurocognitive impairments (Fig. 28.3.3) [5]. These capabilities represent a move from single-analyte sensing toward system-level physiological modeling, enabling prediction of behavioral and cognitive performance in both everyday and mission-critical environments.

A critical barrier to widespread adoption of body-interfaced biosensors has been their reliance on external batteries. To overcome this limitation, we pioneered energy-autonomous systems that harvest energy from biochemical, mechanical, and optical sources. Sweat lactate biofuel cells integrated into wearable patches achieved record power densities sufficient to sustain multiplexed biosensing and wireless communication [6]. Most recently, we developed flexible perovskite solar modules with exceptional indoor efficiency, enabling light-powered wearable sensors that operate continuously across daily environments (Fig. 28.3.4) [7]. These self-powered devices eliminate the need for battery replacement or recharging, paving the way for sustainable, long-term biosensing in real-world use.

Clinical validation is essential for translation. Through collaborations with major medical centers, we have demonstrated that sweat uric acid levels track serum dynamics in gout patients, that sweat C-reactive protein correlates with systemic inflammation in COPD and heart failure (Fig. 28.3.5) [2], and that wearable sensors can monitor cancer drugs such as busulfan and cyclophosphamide in patients [4]. Smart bandages have been evaluated in wound clinics, predicting healing outcomes and guiding therapy [8]. Smart capsules, with miniaturized form factor and low-power electronics, could continuously profile biomarkers inside gastrointestinal tract for metabolic monitoring (Fig. 28.3.6) [9]. These examples underscore the translational readiness of body-interfaced biosensors and their potential to impact clinical decision-making across a spectrum of diseases.

Despite rapid progress, significant challenges remain before body-interfaced biosensors can achieve widespread clinical and consumer deployment. From a technical standpoint, improving calibration stability, biofouling resistance, and molecular selectivity over multi-day use continues to be a key materials challenge. Current electrochemical limits of detection for small molecules are typically in the low μM–nM range, sufficient for many metabolites and hormones but still above physiological levels for certain cytokines and low-abundance proteins. On the systems level, integrating reliable wireless communication, on-sensor data processing, and adaptive power management within sub-$100\mu W$ power budgets remains a bottleneck for long-term operation. Translationally, inter-subject variability in sweat rate, gland density, and skin microenvironment can introduce more than 20-to-30% signal variation across populations, underscoring the need for adaptive calibration and personalized algorithms.

Looking ahead, close coupling between circuit design and biosensor development—such as integrated potentiostat ASICs, ultra-low-noise amplifiers, and edge-AI co-processors—will be essential for scaling these systems. Further opportunities lie in closed-loop therapeutic feedback, multiplexed molecular panels for early disease detection, and standardized clinical validation protocols that bridge regulatory and consumer health domains. Addressing these quantitative and qualitative gaps will accelerate the translation of body-interfaced biosensors from laboratory prototypes to ubiquitous healthcare tools.

In summary, body-interfaced biosensors represent a frontier in digital and precision medicine. By combining advances in nanomaterials, scalable manufacturing, multimodal sensing, AI-driven analytics, and energy harvesting, these systems extend molecular analysis beyond the laboratory into continuous, real-world monitoring. They establish the foundation for closed-loop diagnostics and therapeutics that can improve outcomes in chronic disease, enhance resilience in extreme environments, and democratize access to personalized healthcare. As bioelectronics evolves into an integral component of medicine, these technologies offer a glimpse into a future where real-time molecular sensing and intelligent health management are seamlessly embedded into daily life.

References:

[1] M. Wang et al., "A wearable electrochemical biosensor for the monitoring of metabolites and nutrients," *Nat. Biomed. Eng*, vol. 6, no. 11, pp. 1225–1235, Aug. 2022, doi: 10.1038/s41551-022-00916-z. https://doi.org/10.1038/s41551-022-00916-z

[2] J. Tu et al., "A wireless patch for the monitoring of C-reactive protein in sweat," *Nat. Biomed. Eng*, vol. 7, no. 10, pp. 1293–1306, June 2023, doi: 10.1038/s41551-023-01059-5. https://doi.org/10.1038/s41551-023-01059-5

[3] C. Ye et al., "A wearable aptamer nanobiosensor for non-invasive female hormone monitoring," *Nat. Nanotechnol.*, vol. 19, pp. 330–337, 2024, doi: 10.1038/s41565-023-01513-0. https://doi.org/10.1038/s41565-023-01513-0

[4] M. Wang et al., "Printable molecule-selective core–shell nanoparticles for wearable and implantable sensing," *Nat. Mater.*, vol. 24, no. 4, pp. 589–598, Apr. 2025, doi: 10.1038/s41563-024-02096-4. https://doi.org/10.1038/s41563-024-02096-4

[5] C. Xu et al., "A physicochemical-sensing electronic skin for stress response monitoring," *Nat Electron*, vol. 7, no. 2, pp. 168–179, Jan. 2024, doi: 10.1038/s41928-023-01116-6. https://doi.org/10.1038/s41928-023-01116-6

[6] Y. Yu et al., "Biofuel-powered soft electronic skin with multiplexed and wireless sensing for human-machine interfaces," *Science Robotics*, vol. 5, no. 41, p. eaaz7946, 2020, doi: 10.1126/scirobotics.aaz7946. https://doi.org/10.1126/scirobotics.aaz7946

[7] J. Min et al., "An autonomous wearable biosensor powered by a perovskite solar cell," *Nat Electron*, vol. 6, no. 8, pp. 630–641, July 2023, doi: 10.1038/s41928-023-00996-y. https://doi.org/10.1038/s41928-023-00996-y

[8] C. Wang et al., "A microfluidic wearable device for wound exudate management and analysis in human chronic wounds," Sci. Transl. Med., vol. 17, no. 795, p. eadt0882, Apr. 2025. https://doi.org/10.1126/scitranslmed.adt0882

[9] J. Min et al., "Continuous biochemical profiling of the gastrointestinal tract using an integrated smart capsule," Nature Electronics, vol. 8, pp. 844–855, 2025. https://doi.org/NODOI

ISSCC 2026 / February 18, 2026 / 8:00 AM

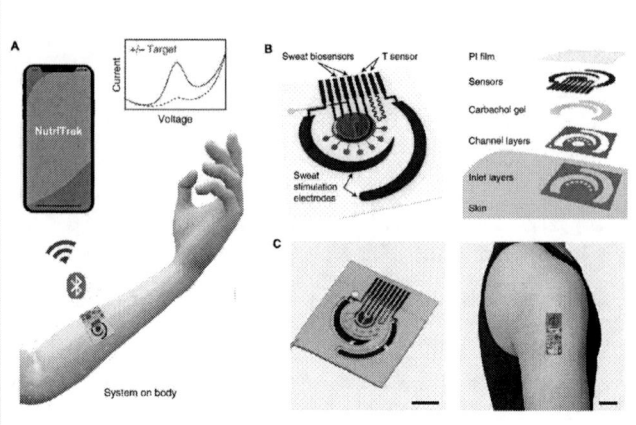

Figure 28.3.1: Wearable sweat biosensing platforms integrating nanomaterial-based electrodes, microfluidics, and iontophoresis for multiplexed monitoring of metabolites, electrolytes, nutrients, hormones, drugs, and proteins. Adapted from Ref [1].

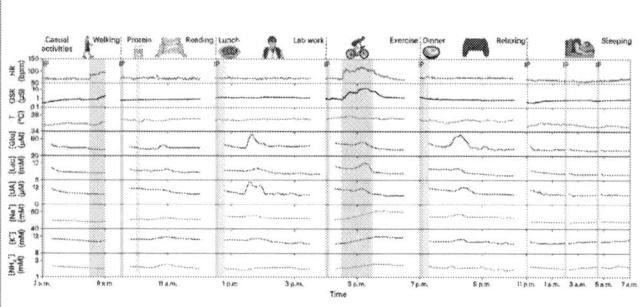

Figure 28.3.2: Continuous 24 h multimodal monitoring during a subject's daily activities using wearable biosensors. Adapted from Ref [5].

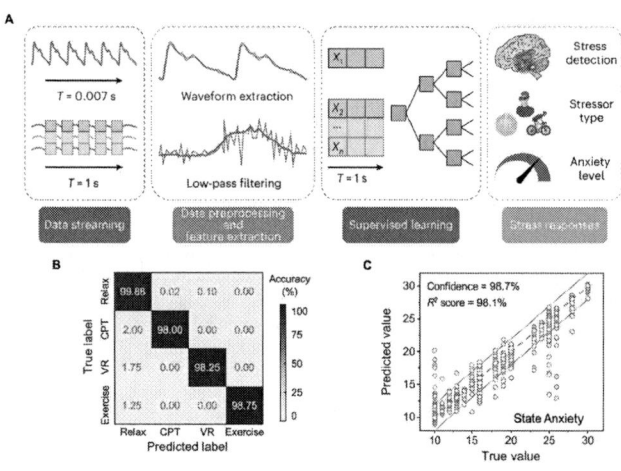

Figure 28.3.3: CARES, a multimodal sensing patch that continuously monitors physiological and biochemical responses from skin and performs artificial-intelligence-powered stress assessment. Adapted from Ref [5].

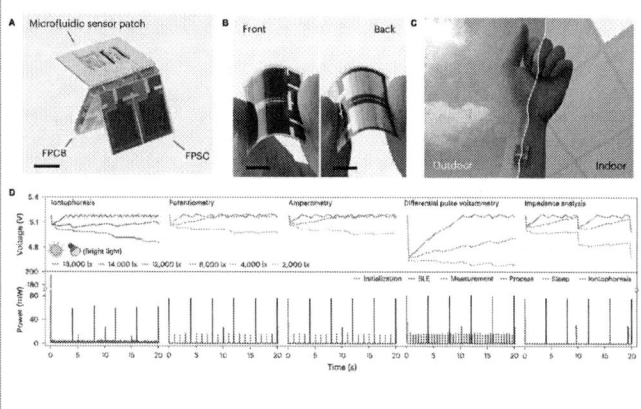

Figure 28.3.4: An autonomous wearable sensor uses a perovskite solar cell to harvest energy from ambient light, enabling battery-free multimodal biosensing across the daily activities. Adapted from Ref [7].

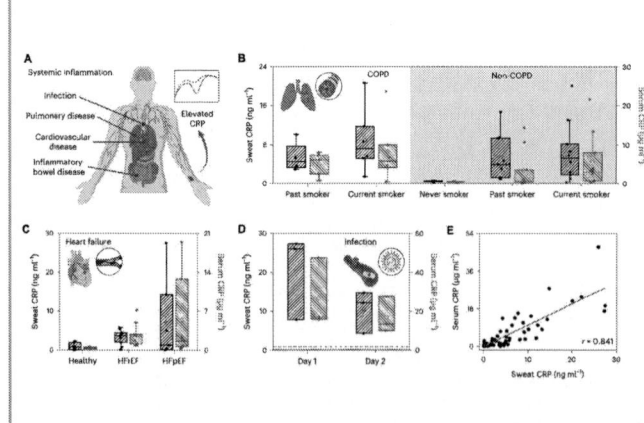

Figure 28.3.5: Evaluation of sweat c-reactive protein for the non-invasive monitoring of systemic inflammation in healthy and patients with COPD, heart failure, and infection. Adapted from Ref [2].

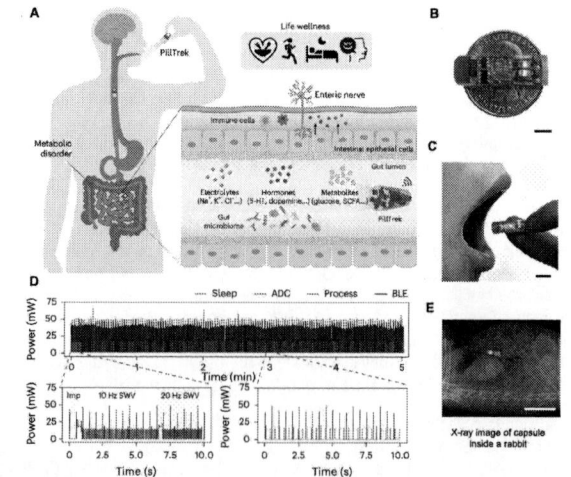

Figure 28.3.6: Minimally invasive ingestible capsule combining multiplexed biochemical sensors, wireless communication electronics, and biocompatible packaging for gastrointestinal metabolic profiling. Adapted from Ref [9].

DIGEST OF TECHNICAL PAPERS • 493

28

979-8-3315-8937-0/26 $31.00 © 2026 IEEE

ISSCC 2026 / SESSION 28 / OUTSIDE THE ISSCC BOX / 28.4

28.4 The Sensing, Computing, and Devices Opportunities of Computational Behavioral Phenotype: An Autism Spectrum Disorder Case Study

Guillermo Sapiro[1], Matias Di Martino[2], Geraldine Dawson[2]

[1]Princeton University, Princeton, NJ, [2]Duke University, Durham, NC

Abstract

The AI revolution will come when high-quality healthcare is accessible to all. We describe steps towards this in autism screening. We developed an app displaying stimuli eliciting behaviors that allow to detect signs of autism. These are automatically detected and quantified using machine learning, enabling an early, scalable, and accurate detection. This exemplifies the potential for mobile AI to provide an accessible approach to health and the technology challenges that come with it.

While we are already experiencing some impacts of AI and machine learning, the revolution will come with applications in healthcare. One such opportunity is to bring access to state-of-the-art knowledge and care to all. In this invited paper and presentation, we describe some of our efforts towards this goal in early autism screening. This will illustrate how we need an integrative approach, with challenges and opportunities ranging from regulatory aspects to the algorithm development and the design of new and efficient sensors and processors.

Autism Spectrum Disorder (ASD) is a neurodevelopmental condition characterized by challenges in social communication and the presence of restricted, repetitive behaviors. Clinical signs can emerge as early as 9-to-18 months, being early detection crucial as it allows for timely access to interventions that can significantly improve long-term outcomes. The standard screening tool used in pediatric primary care is the Modified Checklist for Autism in Toddlers-Revised with Follow-Up, a parent-completed questionnaire. However, its effectiveness has been shown to be considerably lower in real-world clinical settings compared to research environments. Key limitations include low sensitivity, missing many autistic children; low positive predictive value, indicating that most children who screened positive did not actually have autism; and disparities in accuracy. The questionnaire's accuracy is particularly compromised for girls and children of color. Cultural and linguistic factors contribute to the accuracy gap, influencing parental reporting and low rates of pediatricians completing the necessary follow-up interview. Objective methods like eye-tracking, which measures a child's visual preferences, have shown promise. However, when used alone, it has also demonstrated poor sensitivity. Autism is a diverse and heterogeneous condition, a screening tool that relies on a single behavioral marker is often insufficient. The need to detect multiple behaviors already indicates the need for diverse sensors and devices. To address these gaps, we developed **SenseToKnow** [1,2], a digital phenotyping app designed to capture and quantify a wide array of autism-related behaviors simultaneously. By using computer vision and machine learning, the app provides objective measure of a child's behavior, moving beyond reliance on subjective caregiver reports. The app can be administered on a standard iPhone or iPad. The child sits on a parent's lap and watches a series of short, engaging movies while the device's front-facing camera records their responses (Figs. 28.4.1-28.4.3 and 28.4.6). The session takes <10 minutes and includes a bubble-popping game. The app extracts and quantifies 23 distinct "digital phenotypes" from the video and touch data (Figs. 28.4.1 and 28.4.2): Social Attention & Gaze; Attention to Speech; Head and Facial Dynamics; Response to Name; Attentional Engagement; and Visual-Motor Skills. XGBoost combines these features to classify children as either autistic or nonautistic. A fully explainable result is then presented (Figs. 28.4.4).

Example of Results: High Accuracy and Equitable Performance. The app was deployed in pediatric clinics as part of the regular well-child visit [1]. The study included 475 (17-to-36 months old) children (269 boys and 206 girls), of which 49 were diagnosed with autism and 98 were diagnosed with developmental delay without autism. The results included AUC=0.90, sensitivity=87.8%, specificity=80.8%, NPV=97.8%, and PPV=40.6%. The algorithm had similar sensitivity performance across demographics. In [2] it was investigated how the app performs at home, showing comparable performance. The SenseToKnow app was administered by caregivers of 620 toddlers 16-to-40 months of age, 188 of whom were subsequently diagnosed with autism. SenseToKnow demonstrated again a high level of accuracy with AUC=0.92, sensitivity=83.0%, specificity=93.3%, PPV=84.3%, and NPV= 92.6%. Accuracy of the app for detecting autism was similar across demographics. Performance was also studied performance for older children [3]. Results revealed significant group differences between autistic and neurotypical children in terms of several behavioral features. Examining correlations between the Vineland Adaptive Behavior Scales and individual digital phenotypes, autistic children with higher levels of communication, daily living, socialization, motor, and adaptive skills exhibited higher levels of social attention and coordinated gaze with speech, less frequent head movements, higher complexity of facial movements, higher overall attention, lower blink rates, and higher visual motor skills, demonstrating convergent validity between app features and clinical measures. App features were also significantly correlated with ratings on the Social Responsiveness Scale.

Discussion. This study highlights the positive impact of long-term investment in AI on healthcare, particularly in expanding access to high-quality developmental and mental health support. Investments in AI and healthcare offer the potential for universal screening, leading to significant, though often immeasurable, human and monetary savings. This example demonstrates how interdisciplinary teams, by combining advanced yet accessible devices with cutting-edge machine learning, can effectively leverage AI. We are rapidly observing similar promising results, reinforcing our optimism for the future of AI in health. While current analyses are performed on protected servers, the ideal scenario involves on-device processing. Though current technology allows for this, scaling mobile health will make edge computations a bottleneck. Therefore, future advancements in chip technology, energy-efficient processors, and body sensors—all working in tandem—will be crucial for truly scalable health solutions. For instance, body sensors that measure fine and gross motor skills are essential for nearly all developmental and neurodegenerative conditions. These sensors must have long battery life and be virtually unnoticeable to the user. Another significant challenge is the coordinated use of all available sensors and devices. Imagine home devices working in full privacy and coordination with wearables, with data processed on edge devices. This approach necessitates the development of novel technologies and algorithms, as device and sensor collaboration is still in its early stages. This also includes the need for new, low-cost, low-power, and small-form-factor sensors to passively measure biomarkers ranging from stress to glucose and brain signals. Some of these technologies still require development at the necessary scale. Finally, we must address the ethical considerations of delivering sensitive health information without the professional support typically found in clinical settings. While the potential benefits likely outweigh the risks, these aspects require careful consideration and study. Privacy must be central to this development, and the more information we can keep on-device, the greater our ability to guarantee it. As mentioned, significant challenges, from sensor design to mobile computing, still need to be overcome to achieve this.

Acknowledgments and Disclosures:
We thank the families that participated in this important research. Full disclosure and conflict of interest are reported in [1,2,3].

References:
[1] S. Perochon et al., "Early detection of autism using digital behavioral phenotyping", *Nature Medicine*, Oct. 2023. https://doi.org/10.1038/s41591-023-02574-3
[2] P. Raj Krishnappa Babu et al., "Validation of a mobile app for remote autism screening in toddlers", *NEJM AI*, Sept. 2024. https://ai.nejm.org/doi/10.1056/AIcs2400510
[3] V. Aikat et al., "Autism digital phenotyping in preschool- and school-age children", *Autism Research*, April 2025. https://pubmed.ncbi.nlm.nih.gov/40176346/

ISSCC 2026 / February 18, 2026 / 8:00 AM

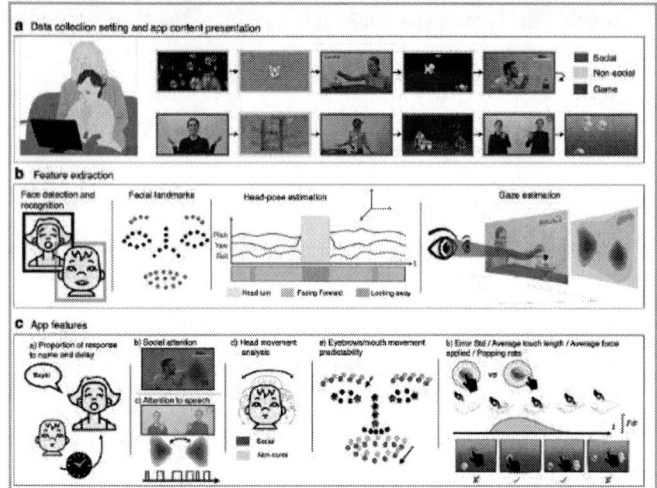

Figure 28.4.1: App variables and workflow. a, Administration and stimuli. b, Faces and landmarks automatically detected c, Automatic computation of digital phenotypes. (Reproduced from [1].)

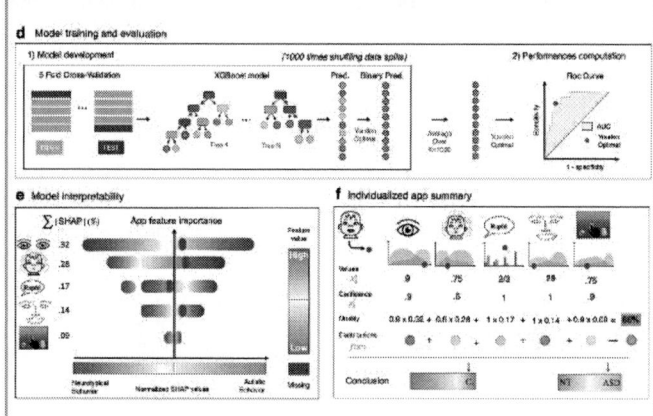

Figure 28.4.2: (cont. from Fig. 28.4.1) App variables and workflow. d, Training of XGBoost classifier from multiple phenotypes. e, Model interpretability using SHAP values analysis. f, Report of an individualized app administration showing the participant's unique profile. (Reproduced from [1].)

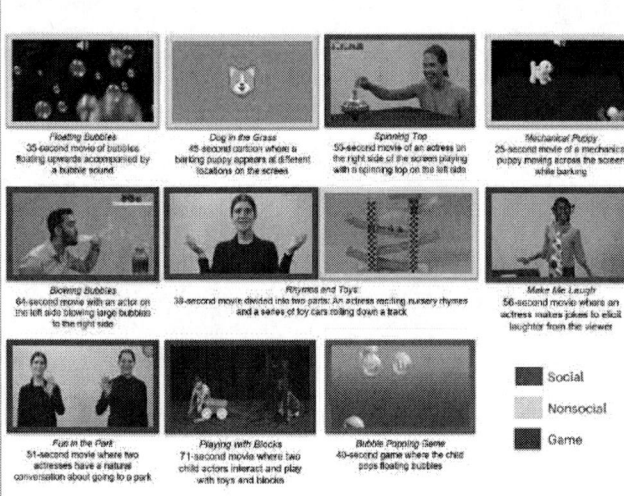

Figure 28.4.3: Frames of the app movies and game. (Reproduced from [3].)

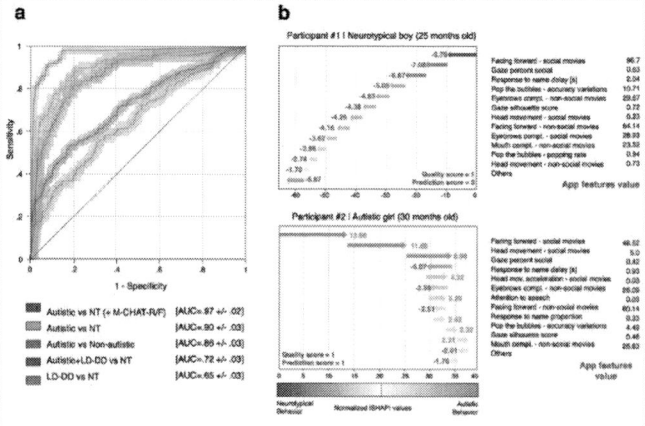

Figure 28.4.4: Accuracy metrics and normalized SHAP value analysis. a, ROC curve. The M-CHAT-R/F screening score was used when available. Error bands correspond to 95% CI computed by the Hanley McNeil method. b, Examples of app administration reports, including each child's app quality score, confidence score, and the contributions of each app variable. (Reproduced from [1].)

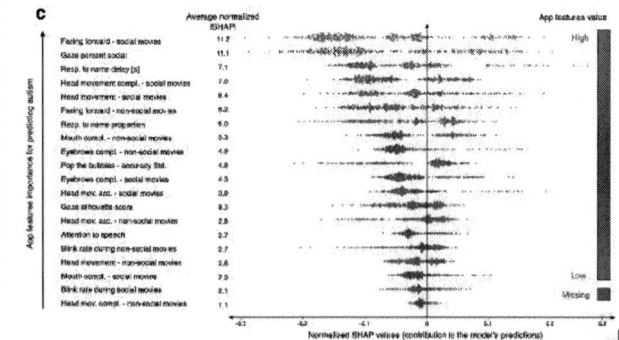

Figure 28.4.5: (cont. from Fig. 28.4.4) Accuracy metrics and normalized SHAP value analysis. c, Normalized SHAP value analysis showing the app variables importance. The blue–red color gradient indicates the relevance of each variable to the score; gray indicates missing variables. For each app variable, a point represents the normalized SHAP value of an individual participant. NT, neurotypical. (Reproduced from [1].)

Figure 28.4.6: Participant interacting with the app.

DIGEST OF TECHNICAL PAPERS • 495

979-8-3315-8937-0/26 $31.00 © 2026 IEEE

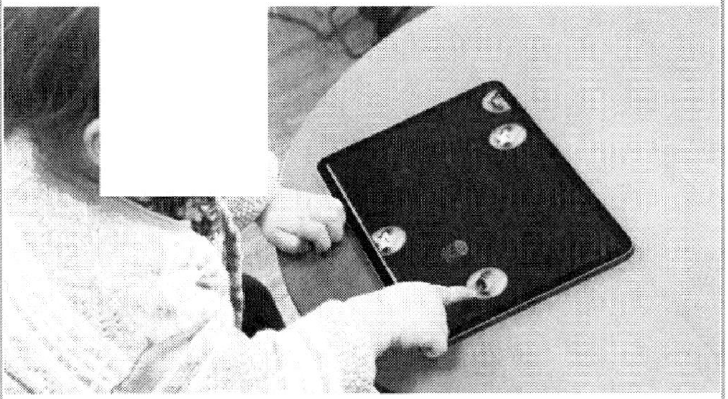

Figure 28.4.7: Participant playing with the app game.

979-8-3315-8937-0/26 $31.00 © 2026 IEEE

ISSCC 2026 / SESSION 29 / BIOCHEMICAL SENSORS FOR LIFE SCIENCES AND AGRICULTURE / OVERVIEW

Session 29 Overview: *Biochemical Sensors for Life Sciences and Agriculture*

TECHNOLOGY DIRECTIONS SUBCOMMITTEE

Session Chair: Kaiyuan Yang
Rice University, Houston, TX

Session Co-Chair: Albert Wang
Apple, Cupertino, CA

Innovations in biochemical sensors integrated with CMOS platforms are revolutionizing a wide range of applications in life sciences and agriculture. The first paper describes an epidermal pasting bioelectronic device for the detection and treatment of type-1 diabetes, followed by an ultra-high scan rate EPR-on-a-chip and a 256-channel solid-state nanopore readout chip. A CMOS-integrated shape-memory multi-size micro-cage array for biosample manipulation and sensing comes next, followed by two agricultural sensing platforms for pH-based plant stress monitoring and hydrogel-based dual-fluorescence bioavailable phosphorus (P) detection.

ISSCC 2026 / February 18, 2026 / 10:05 AM

10:05 AM

29.1 A 48-Day-Duration Pasting Bioelectronic Device Realizing Closed-Loop Realtime Detection and Precision Treatment for Type-1 Diabetes

Xiaoyuan Wu, East China Normal University, Shanghai, China

In Paper 29.1, East China Normal University presents an epidermal pasting bioelectronic device for closed-loop real-time detection and precision treatment of type-1 diabetes. It integrates a 19fA-resolution bioluminescence analog front-end with TDM-SAR calibration and a hybrid-mode LED driver.

10:30 AM

29.2 A 1400THz/s Ultra-Fast-Scan 14GHz EPR-on-a-Chip Based on Injection-Locked Phase Detection Featuring 120µM Concentration Sensitivity

Jean-Baptiste David, CEA-LETI-MINATEC, Grenoble, France

In Paper 29.2, CEA-LETI-MINATEC describes a battery-operated CMOS 22nm-FDSOI EPR-on-a-Chip based on injection-locked phase detection, achieving 96mW power, 120µM sensitivity, 100Gauss span, and 1400THz/s scan rate.

10:55 AM

29.3 A 256-Channel Event-Driven Readout for Solid-State Nanopore Single-Molecule Sensing with 193pArms Noise in a 1MHz Bandwidth

Sander Crols, KU Leuven - MICAS, Leuven, Belgium, imec, Leuven, Belgium

In Paper 29.3, imec and KU Leuven report a 256-channel solid-state nanopore readout IC in 65nm CMOS, featuring an integrate-and-hold transimpedance amplifier architecture and dynamic-allocation event-detection to achieve 1mW power and 0.019mm² area per pore.

11:20 AM

29.4 Shape-Memory Multi-Size Micro-Cage Array on CMOS with Integrated Electrochemical Sensors for Joint Bio-Sample Manipulation and Sensing

Zhikai Huang, ETH Zurich, Zurich, Switzerland

In Paper 29.4, ETH Zurich presents a CMOS-integrated shape-memory multi-size micro-cage array with multi-material electrochemical potentiostats and power-free immobilization (30-to-400µm), enabling joint on-chip biosample manipulation and sensing.

11:35 AM

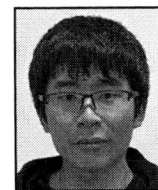

29.5 A Sub-Gram Individual Plant Stress Sensor Tag for Smart Farming

Daiki Nitto, University of Osaka, Suita, Japan

In Paper 29.5, University of Osaka introduces a 0.62g light-weight battery-less sensor tag for continuous monitoring of pH in the plant as a key stress indicator. A single miniature antenna establishes a simultaneous wireless power and data link over an 80cm distance.

29

11:50 AM

29.6 A 65nm CMOS Hydrogel-Based Dual Fluorescence Sensor for Bioavailable Phosphorus Detection

Ting-Yu Cheng, California Institute of Technology, Pasadena, CA

In Paper 29.6, California Institute of Technology shows a hydrogel-based single-culture dual-fluorescence sensor for bioavailable phosphorus (P) detection, achieving a 6.6fA sensitivity with a 15.5dB SNR and an output noise of 1.97mV at 182ms integration time. The sensor was validated using hydrogels embedded with a bioreporter strain and successfully detected both P limitation and cell viability.

DIGEST OF TECHNICAL PAPERS • 497

979-8-3315-8937-0/26 $31.00 © 2026 IEEE

ISSCC 2026 / SESSION 29 / BIOCHEMICAL SENSORS FOR LIFE SCIENCES AND AGRICULTURE / 29.1

29.1 A 48-Day-Duration Pasting Bioelectronic Device Realizing Closed-Loop Realtime Detection and Precision Treatment for Type-1 Diabetes

Xiaoyuan Wu*[1], Kangjie Zhao*[1], Mengyao Liu*[1], Shixiang Ding[1], Xuyang Li[1], Yuri Lu[1], Ruohan Wang[1], Ziyan Zhao[1], Liwen Qian[1], Yang Yang[1], Jing Wu[1], Leilei Huang[1], Chunqi Shi[1], Yuhong Wei[1], Yue Lu[1], Ningzi Guan[1], Haifeng Ye[1], Jinghong Chen[2], Runxi Zhang[1]

[1]East China Normal University, Shanghai, China, [2]University of Houston, Houston, TX
*Equally Credited Authors (ECAs)

Abstract

Conventional management of type-1 diabetes (T1D) relies on finger-prick blood sampling or electrochemical methods, which prevent continuous monitoring and treatment. This paper presents a novel epidermal pasting bioelectronic device (PBD). The device integrates a 19fA-resolution bioluminescence analog front-end (BAFE) with TDM-SAR calibration and a hybrid-mode LED driver. It enables closed-loop realtime detection and precision treatment for T1D. The battery life reaches 48 days.

Traditional type-1 diabetes (T1D) management primarily relies on finger-prick blood sampling or electrochemical method [1, 2], followed by manual administration of oral medications or injections, which limits realtime detection and precision treatment of T1D. Synthetic bioelectronic technology, known for its high sensitivity and long-term monitoring capabilities, has recently been applied to disease detection and treatment. Though references [3, 4, 5] (Fig. 29.1.1 top-left) demonstrate its use in realtime detection of light intensity from engineered bacterial cells for diagnosing gastric bleeding, they lack therapeutic functionality. Reference [6] (Fig. 29.1.1 top-middle) employs 660nm red light to irradiate REDMAP-functional (photo-responsive) cells, triggering insulin release for T1D treatment. However, it lacks glucose monitoring capability. This paper proposes a novel epidermal pasting bioelectronic device (PBD), integrating realtime detection and precision treatment modules to establish a closed-loop response system with 48-day battery life, offering a new intelligent solution for T1D.

Figure 29.1.1 (top-right) presents the proposed ultra-low-power PBD, which integrates a PBD chip, a detection chamber containing engineered bacterial cells, a treatment chamber equipped with REDMAP cells [6], four photosensitive devices, a CR1612 battery, and two 660nm LEDs, realizing closed-loop detection and treatment functions. Figure 29.1.1 (bottom) illustrates five working steps of the PBD system. Step 1, the engineered bacterial cells interact with glucose molecules in the blood to produce 460nm bioluminescence signal. Step 2, the PBD periodically quantifies the bioluminescence signal into duration data (e.g., N1, N2) and transmits it to an external terminal, enabling realtime monitoring. Step 3, the external intelligent terminal provides auto-analysis, deciding whether treatment is needed and its intensity. Step 4, the PBD receives treatment information and controls the emitting intensity and duration of 660nm luminescence, ensuring precision therapy. Step 5, the REDMAP cells respond to the luminescence, releasing insulin to reduce blood glucose.

Figure 29.1.2 (top) presents system architecture of the PBD chip, which consists of five subsystems: an ultra-low-power realtime detection path (DP), a precise insulin-release treatment path (TP), a 0.4V 2Hz relaxation-oscillator-driven system controller (SC), a sub-100nW power management unit (PMU), and a clock management unit (CMU) based on a fast-start chirp-injection 26MHz crystal oscillator. The DP includes a 4-channel 19fA-resolution bioluminescence analog front-end (BAFE) and a 5-bit voltage-mode digital TX. The TP comprises a wake-up RX and a hybrid-mode LED driver. Both paths are driven by the PMU and CMU and controlled by the SC. Figure 29.1.2 (bottom) shows the closed-loop timing and flow chart. After initialization, the PBD enters a configurable sleep period ranging from 0 to 3.2 hours. Upon waking, the BAFE reads the bioluminescence signal and performs a preliminary judgment. If the transmission condition is not met, the system returns to sleep. Otherwise, the TX digital baseband (TX DBB) drives the TX to complete transmission and enters the analysis step. During this step, a smartphone receives the detection data transmitted by the PBD's TX and synchronizes with it. The data is analyzed by an application (APP), and then therapeutic information is returned to the PBD's RX. The PBD initiates therapy if treatment condition is met, otherwise, the system returns to sleep.

Figure 29.1.3 (top) illustrates implementation of the realtime DP. Bioluminescence signals generated by engineered bacterial cells are detected by four NPN phototransistors (SFH3710). The bioluminescence current (I_{PD}) is injected into a 20nF capacitor to obtain an integrated voltage, which is then sampled by the BAFE. A comparator controls a 20-bit counter to convert the I_{PD} intensity into a digital code. To ensure realtime detection under prolonged operation, a pre-decision algorithm logic is applied to avoid additional power consumption caused by frequent TX communication [3, 4]. Pre-decision requires cell-free inter-channel I_{PD} mismatch < 2pA [5], necessitating solutions for comparator offset and inherent dark current mismatch calibration among phototransistors. The former is compensated via 4-channel TDM comparator [3, 4], while the latter is addressed by the proposed TDM-SAR calibration technique. A 7-bit RDAC replaces the fixed-voltage reference circuit, and the four channels (CH_1-CH_4) share the SAR logic module through time-division multiplexing. Initially, the most significant bit (MSB) of the 7-bit RDAC is set to 1, with the remaining bits set to 0. The code differences ΔCH_{N1} (CH_N-CH_1) are fed into SAR logic for decision: if $|\Delta CH_{N1}| < 30$ (corresponding to 0.5pA), calibration completes and V_{REF} for each channel is determined. Figure 29.1.3 (bottom-right) shows < 0.1min calibration response

time and < 0.5pA calibrated I_{PD} mismatch with 116fA$_{rms}$ random noise floor across four channels under 18-hour continuous testing at 37°C. The proposed BAFE consumes 39nW with 1.9×10^{-6}J per detection. The digital TX employs a switched-capacitor power amplifier (SCPA) to improve energy efficiency. A 7-phase injection-locked ring oscillator (ILRO) combined with edge-combing (EC) provides a 910MHz carrier signal. The TX achieves 30dB output power range for 0-to-10m transmission. Compared to conventional low-power TXs [4], energy efficiency improves to 32pJ/bit with chip area reduced to 0.07mm².

Figure 29.1.4 (top) presents the detailed TP, which includes the wake-up RX, RX digital baseband (RX DBB), hybrid-mode LED driver, and READMAP cells. The noise-canceling RX LNA amplifies weak signals transmitted by smartphones, while the envelope detector (ED) and comparator perform envelope detection to extract 125kbps data. The RX DBB identifies packet headers, extracts therapeutic information, and performs verification. The hybrid-mode LED driver controls intensity and duration of the corresponding driving current (I_{LED}) based on the therapeutic information. The RX employs a 36dB variable gain, balancing low power consumption and high sensitivity. The RX DBB operates at 13MHz for precise data parsing while switching to 2Hz during idle periods to save power. Figure 29.1.4 (mid-left) shows the secreted embryonic alkaline phosphatase (SEAP) characteristic curve of REDMAP cells, reflecting insulin production under controlled illumination duration and intensity [6]. The SEAP production increases rapidly within 4 seconds and then saturates. Due to a 10% frequency deviation in the 2Hz RXO clock, the illumination time is fixed at 10s and the release amount of REDMAP cells is regulated by adjusting I_{LED}. Comparing with the traditional 2T VREF-only and IDAC-only driver circuits, the proposed hybrid-mode LED driver consisting of a 3-bit IDAC and a 3-bit VREF, as shown in Fig. 29.1.4 (bottom-right), can achieve wide range and high precision I_{LED}, simultaneously. Additionally, the 80dB loop gain in the current source further ensures current accuracy. Figure 29.1.4 (bottom-left) displays the measurement results of the LED driver, demonstrating its ability to control I_{LED} within a range of 0-to-49.98mA with a minimum step of 0.89mA, thereby precisely regulating the release of REDMAP cells. When I_{LED} is up to 49.98mA, the LED driver achieves an efficiency of 87.5%. V_{stable} can be set to 1.7-to-2.7V using R_{CS}, enabling 580-to-660nm light irradiation for different therapy purposes.

Figure 29.1.5 (top-left) displays the PBD structure. The flexible printed circuit board (FPCB) is mounted with four SFH3710 and a 26MHz XTAL on the front, a 910MHz patch antenna on the side, and a PBD chip with customized 660nm LEDs for treatment on the back. The FPCB is sealed with PDMS and placed in a PLA-material cavity. In the detection chamber, the blank CH_1 serves as reference, while CH_2-CH_4 are all filled with 5×10^6 engineered bacterial cells (30µL Luria-Bertani medium). The treatment cavity contains 1×10^7 REDMAP cells (500µL Dulbecco's-modified-Eagle's medium) for therapy. Figure 29.1.5 (mid-left) shows the experimental setup of the bioelectronic system and the analysis APP GUI. A 37°C incubator simulates mouse body temperature, and the beaker contains 50mL phosphate buffered saline. Figure 29.1.5 (bottom-left) presents measurement results of the DP and TP. For detection, 0, 7.5×10^{-5}, and 4.5×10^{-4}mMol glucose are introduced into CH_2-CH_4, respectively. A plate reader (PR) serves as the benchmark, showing that the photocurrent measured by PBD aligns with the luminescence intensity measured by the PR (Envision, 2015). For treatment, SEAP release in the beaker is measured after 10s of LED irradiation under 1.5, 2 and 2.5mA I_{LED}. The standard deviation (STD) of triplicate experiments is below 3.60. Figure 29.1.5 (top-right) shows the PBD diagnostic and therapeutic measurement results in three mice (blood glucose measured manually with an electronic glucometer). 1# and 2# demonstrate the therapeutic effect of PBD on T1D mice when activated versus deactivated, while 3# confirms PBD has no impact on healthy mouse. Figure 29.1.5 (mid-right) presents the power consumption of each step, here each step is configured for 1s operation. Thanks to the proposed pre-decision mechanism, the PBD can operate continuously for 48 days on a 30mAh battery, performing one detection per minute, while one communication every 30 minutes and one treatment every hour on average.

Figure 29.1.6 summarizes performance of the proposed PBD and compares it with previously reported synthetic bioelectronic chips and systems. The PBD achieves a 1-minute detection cycle using on-chip TDM-SAR calibration, maintaining a 4-channel mismatch of less than 0.5pA. It employs a hybrid-mode LED driver to cover a linear range of 0-to-49.98mA with an I_{LED} step of 0.89mA for precise therapeutic intensity control. This work demonstrates, for the first time, a synthetic bioelectronic technology that enables an integrated solution for realtime detection and precision treatment of T1D, forming a closed-loop responsive intelligent system with a 48-day battery life. Figure 29.1.7 displays a microphotograph of the PBD chip, which is fabricated using a 40nm CMOS process and occupies an area of 3.3mm².

Acknowledgement:
This work was supported in part by the National Key Research and Development Program of China under grant 2019YFA0904502, in part by the Fundamental Research Funds for the Central Universities under Grant 40500-20103-222178, in part by the Eastern Talent Program Youth Project under grant 15904-412214-24013, and in part by the Science and Technology Commission of Shanghai Municipality under Grant 22DZ2229004. Runxi Zhang, Haifeng Ye, Chunqi Shi and Leilei Huang are co-corresponding authors.

References:
[1] Abbott Inc. FreeStyle Libre 3 System. [Online] Available: http://www.freestylelibre.com
[2] S. Alva, A. Bhargava, B. Bode *et al.*, "Accuracy of a 15-day Factory-Calibrated Continuous Glucose Monitoring System with Improved Sensor Design," *Journal of Diabetes Science and Technology*. 2025;0(0). http://doi.org/10.1177/19322968251329364
[3] P. Nadeau *et al.*, "Nanowatt Circuit Interface to Whole-Cell Bacterial Sensors," *ISSCC*, pp. 352-353, Feb. 2017. http://doi.org/10.1109/ISSCC.2017.7870406
[4] Q. Liu *et al.*, "A Threshold-Based Bioluminescence Detector with a CMOS-Integrated Photodiode Array in 65 nm for a Multi-Diagnostic Ingestible Capsule," *IEEE JSSC*, vol. 58, no. 3, pp. 838-851, Mar. 2023. http://doi.org/10.1109/JSSC.2022.3197465
[5] Mark Mimee *et al.*, "An Ingestible Bacterial-Electronic System to Monitor Gastrointestinal Health," *Science*, vol. 360, no. 6391, pp. 915-918, May 2018. http://doi.org/10.1126/science.aas9315
[6] Y. Zhou *et al.*, "A Small and Highly Sensitive Red/Far-Red Optogenetic Switch for Applications in Mammals," *Nature Biotechnology*, vol. 40, no. 2, pp. 262-272, Feb. 2022. http://doi.org/10.1038/s41587-021-01036-w
[7] M. Seok *et al.*, "A Portable 2-Transistor Picowatt Temperature-Compensated Voltage Reference Operating at 0.5 V," *IEEE JSSC*, vol. 47, no. 10, pp. 2534-2545, Oct. 2012. http://doi.org/10.1109/JSSC.2012.2206683

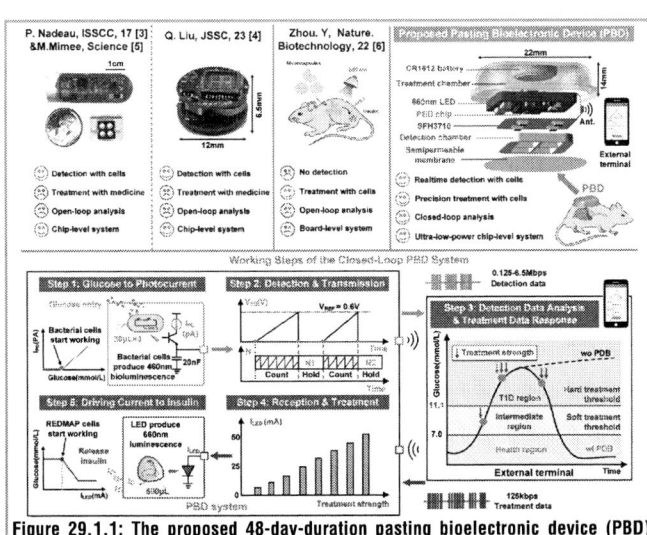

Figure 29.1.1: The proposed 48-day-duration pasting bioelectronic device (PBD) realizing realtime detection and precision treatment with closed-loop response for type-1 diabetes.

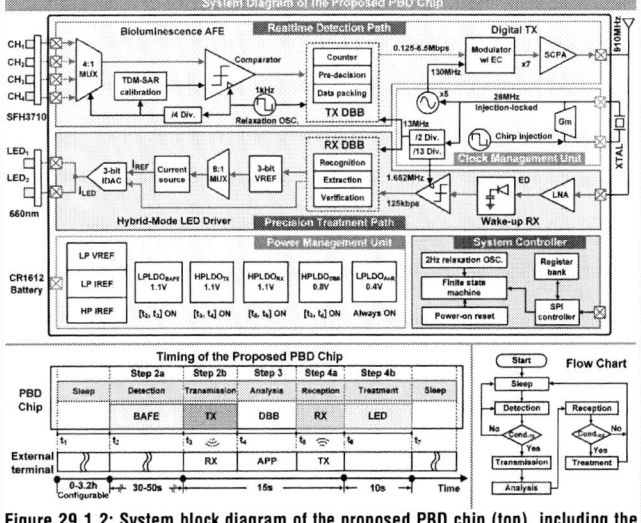

Figure 29.1.2: System block diagram of the proposed PBD chip (top), including the closed-loop system timing (bottom-left) and flow chart (bottom-right).

ISSCC 2026 / SESSION 29 / BIOCHEMICAL SENSORS FOR LIFE SCIENCES AND AGRICULTURE / 29.1

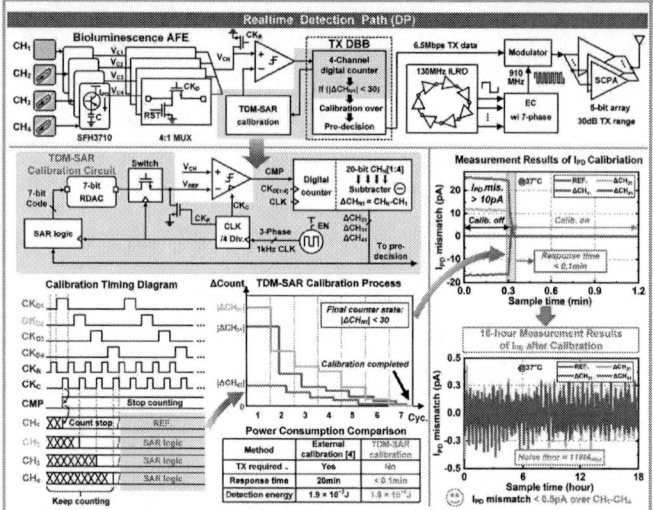

Figure 29.1.3: The proposed realtime detection path based on TDM-SAR calibration to achieve < 0.5pA mismatch and 116fA$_{rms}$ noise floor over 4 channels.

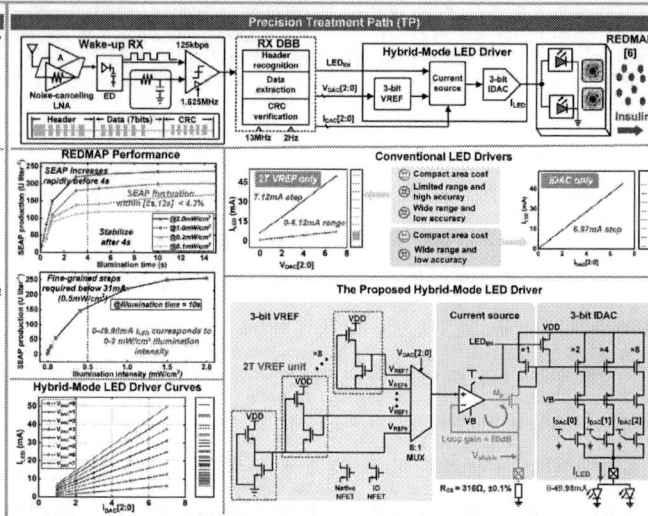

Figure 29.1.4: The proposed insulin-release treatment path with 0-to-49.98mA I$_{LED}$ driving current range and a minimum step of 0.89mA.

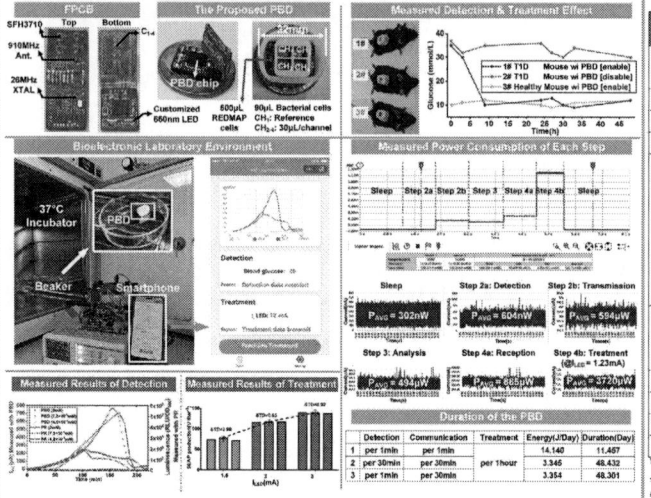

Figure 29.1.5: The proposed PBD, measurement environment and results of the bioelectronic system.

Figure 29.1.6: Comparison with the state-of-the-art bioelectronic systems.

Specifications		This Work	ISSCC2017 [2]	JSSC2023 [3]	Nature Biotechnology 2022 [5]
Application	Detection	✓	✓	✓	✗
	Treatment	✓	✗	✗	✓
Technology (nm)		40	65	65	Board-level
VDD (V)		3	2.5	2.5	12
Die area (mm²)		3.3	0.67	3.91	
Detection path	BAFE on-chip calibration	✓	✗	✗	
	Mismatch of channels (pA)	< 0.5		-	
	Resolution (fA/count)	19	53	25	
	Noise floor (fA$_{rms}$)	116	380	59/71	
	Power consumption (mW)	0.2-1.32		16.5	
	Energy efficiency (pJ/bit)	32-211†			
	Detection period (min)	1.0			
Treatment path	RX sensitivity(dBm)	-62	-	-	
	I$_{LED}$ range (mA)	0-49.98			
	I$_{LED}$ step (mA)	0.89			
	Wireless distance (m)	0-10			
Average power wo treatment (µW)		0.8*	12.7	15.55	
Average power wi treatment (µW)		1040§			Watt-level
Working duration (days)		48			

† The 32pJ/bit energy efficiency is achieved under -24dBm TX P$_{out}$, while 211pJ/bit under 3dBm TX P$_{out}$.
* Achieved under the assumption of one detection every 1minute, one communication every 30 minutes and no treatment is applied.
§ Achieved under the assumption of one detection every 1minute, one communication every 30 minutes and one treatment every hour.

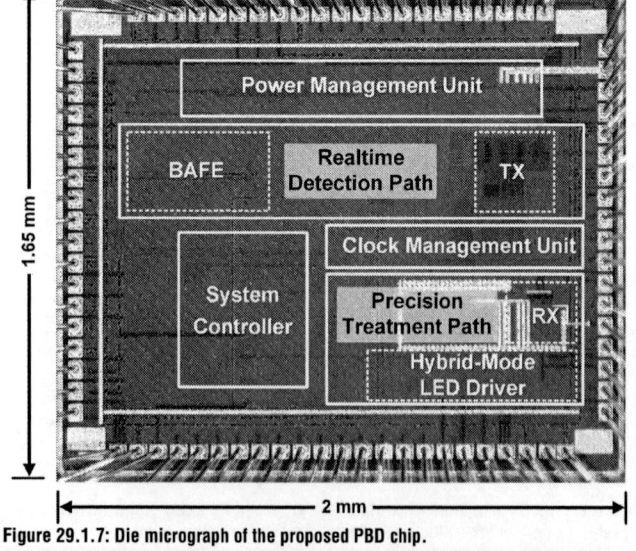

Figure 29.1.7: Die micrograph of the proposed PBD chip.

• 2026 IEEE International Solid-State Circuits Conference

ISSCC 2026 / SESSION 29 / BIOCHEMICAL SENSORS FOR LIFE SCIENCES AND AGRICULTURE / 29.2

29.2 A 1400THz/s Ultra-Fast-Scan 14GHz EPR-on-a-Chip Based on Injection-Locked Phase Detection Featuring 120µM Concentration Sensitivity

Jean-Baptiste David[1], Alexandre Siligaris[1], Cédric Dehos[1], José Luis Gonzalez Jimenez[1], Jean-François Jacquot[2], Christian Lombard[2], Kevin Chighine[2], Vincent Maurel[2], Serge Gambarelli[2]

[1]CEA-LETI-MINATEC, Grenoble, France, [2]CEA-IRIG, Grenoble, France

Abstract

A battery-operated low power (96mW), high sensitivity (120µM), wide span (100Gauss) Ultra-fast scan (1400THz/s) EPR-on-a-Chip based on injection-locked phase detection, implemented in CMOS 22nm-FDSOI technology is presented. 14GHz coherent frequency generation and sensing signal down-conversion is used to implement a stable and low noise EPR readout at around 500MHz without requiring lock-in based demodulation. A single inductor is used as paramagnetic sensor.

The detection and direct analysis of paramagnetic species using Electron Paramagnetic Resonance (EPR) spectrometry is becoming a rapidly growing activity. Applications range from chemistry to biomedical and a wide range of materials characterization for energy production, quantum computing, semiconductors, etc. In EPR, two fields are applied to the sample. A strong static magnetic field B_0 splits electron spin levels (Zeeman effect). A weaker oscillating electromagnetic field B_1 of frequency f_{B1} provides the energy for spin transition. EPR is implemented either by sweeping B_0 (requiring a bulky electromagnet) with a fixed f_{B1}, or by fixing B_0 and varying f_{B1}. The sweep span determines spectral window and its duration determines the scan speed. Conventional approaches suffer from slow electromagnet B_0 scan, which leads to sensitivity loss for long relaxation time paramagnetic species, because of saturation effects under long time B_1 exposure [4]. Furthermore, they preclude the observation of rapid spectral changes, such as those occurring during fast chemical reactions. A fixed B_0 of 0.5T can be easily realized with compact magnets. For that field value, the resonance is observed for B_1 continuous wave (CW) frequencies around 14GHz, set by the electron gyromagnetic ratio ($|\gamma|$= 28GHz/T). EPR-on-a-Chip has been recently demonstrated, enabling miniaturized and low-cost spectrometers [1-5]. The sample is deposited onto a micro-chip inductor being part of a Voltage Controlled Oscillator (VCO) that drives B_1 and is exposed to the fixed B_0. The reported f_{B1} scan speeds open new opportunities both for measurement of fast kinetics of paramagnetic species and for application of alternative measurement modes like in-situ/in-operando Free Induction Decay and Rapid-Scan (RS) [5]. For 14GHz frequency range, this inductor is typically ~200-to-300µm in diameter, which limits the sensitive volume. The challenge is to keep an equivalent concentration sensitivity (CS) to that of bulkier electromagnet-based equipment, despite the system integration. All the reported EPR-on-a-Chip are based on the conversion of a spin transition into a VCO self-oscillation frequency (f_0) variation. The CW f_{B1} sweep is implemented by tuning f_0. The EPR is obtained from the oscillator output measured using lock-in detection with a superposed frequency modulation in order to greatly increase Signal Noise Ratio (SNR) and remove baseline [2-5]. However, lock-in detection requires a Phase Lock Loop (PLL) with tens of MHz bandwidth that limits scan speed. In [4] an open-loop experiment reached 402THz/s on a small spectral window equivalent to 22.8 Gauss in conventional approach, but Amplitude Modulation (AM) detection degrades significantly the sensor sensitivity. In this work we present an innovative compact and portable EPR-on-a-chip spectrometer that allows to attain unprecedent scan speeds, three times faster than the state-of-the-art (1400THz/sec) over large f_{B1} spans (equ. to 100 Gauss) with an outstanding CS of 0.12mM.

The proposed innovation, fabricated in a CMOS 22nm-FDSOI technology, is based on the phase response of Injection Locked Oscillators (ILO), as in Fig. 29.2.1. In free running mode, the ILO oscillation frequency is f_0. When locked, the output signal oscillates at the input frequency f_{in} with a phase shift (Θ) that depends on $f_{in}-f_0$, the ILO tank quality factor (Q), and the free running to injection signal level ratio E_0/E_{inj}. According to the Adler's law, Θ follows an arcsin() function. In this work the sample is deposited on the inductor of an ILO that generates a CW B_1 field. When locked, f_{B1} is equal to f_{in}. The spin transition that pulls de ILO from its self-oscillation frequency f_0 is detected by measuring Θ as f_{in} is swept, plainly exploiting the Adler's law. As shown in the figure, when $B_0=0$, the ILO output signal phase varies as a function of $f_{in}-f_0$ with fixed f_0. When the sample is introduced in a constant B_0 field, f_0 gets perturbed by the EPR phenomena around the electron Larmor frequency (f_{EPR}), and this is reflected in the output phase vs f_{in} sweep. The EPR spectrogram can be retrieved by subtracting the baseline (Θ for B_0 = 0) from the phase response with constant B_0 field.

The proposed EPR sensing ILO is embedded into a more complex architecture to facilitate the output phase reading and to stabilize the measurement, as shown in Fig. 29.2.2. The sensing ILO is part of an integer-N frequency multiplier chain (DET CHAIN in Fig. 29.2.2) inspired in [6]. The ILO is locked to a large integer multiple (N=28) of the reference input (REF IN), whose frequency is selected to be around 500MHz, so that the ILO oscillates around 14GHz. The frequency multiplier is composed of a REF IN signal controller that generates the signals required to turn on and off periodically a pulsed oscillator (P-OSC in Fig. 29.2.2 and Fig. 29.2.3). The P-OSC output is coherently synchronized with the REF IN and consists in a set of harmonics at integer multiples of REF IN frequency. The two consecutive ILOs (ILO1 and ILO2 DET in Fig. 29.2.2 and Fig. 29.2.3) lock on one of these

harmonics (the Nth) and filter out the rest, keeping the phase coherence with the REF IN signal at $f_{ILO\_DET} = Nxf_{REF}$. Buffers are inserted in between the ILOs using transformer coupling (except for ILO2 DET), as in Fig. 29.2.3. They allow to modify E_{inj} for each ILO. And especially for ILO2 DET, it provides an additional tuning knob to adjust the frequency-to-phase conversion gain and hence adjust the sensitivity to improve the dynamic range.

ILO2 DET is the only EPR sensing oscillator in the architecture, thanks to its exposed inductor, where the sample is placed. The other ILOs inductors use stacked transformers (see Fig. 29.2.3) widely immune to their external environment. All ILOs are actually injected VCOs. Their tuning curves are shown in Fig. 29.2.4. The ILOs1 and ILO2 REF cover a large tuning range to fit with the needs of the two chains of the architecture, and to compensate process, temperature and voltage (PVT) variations. The ILO2 DET tuning range, from 13.5-to-14.8GHz, is optimized using a tradeoff to allow for a reasonably large inductor and to accommodate some tuning for PVT variations as well as for the permanent magnet B_0 variability compensation. For typical E_{inj} values, the ILOs locking range (LR) cover a few hundreds of MHz around their f_0. This LR can be centered anywhere across their tuning curves. The ILOs phase response is also centered around f_0, i.e., centered in their LR. Therefore, the tuning mechanism is used to center the EPR sensor phase response in the middle of the desired f_{B1} sweep range. The f_{B1} is swept by applying a frequency modulated continuous wave (FMCW) triangular signal to REF IN. Note that the frequency span of the REF IN FMCW is multiplied by N=28 to give the $f_{in}=f_{ILO\_DET}$ FMCW span shown in Fig. 29.2.1. The E_{inj} is set for all ILOs so that they remain locked across the entire f_{B1} sweep with some extra margins to accommodate frequency drifts during the averaging process. For example, when REF IN FMCW covers 502MHz ± 4MHz, the ILOs in the DET CHAIN are locked to a FMCW signal covering 14.056GHz ± 112MHz (i.e., equ. to 80 Gauss).

A second similar frequency multiplier chain sharing the same REF IN signal is set to provide a frequency multiplied signal around 13.5GHz (N = 27). Hence, with the previous example of REF IN FMCW signal, the CHAIN REF is locked at 13.554GHz ± 108MHz. In this REF CHAIN, the circuits are the same as in the DET CHAIN, excepted for the ILO2, which is identical to ILO1. This approach has three advantages: i) By mixing the output signals of the DET CHAIN and the REF CHAIN, the IC output can be measured at around 500MHz and compared with the REF IN signal. This reduces the power consumption of the digital processing and facilitates the sampling with commercial FPGA platforms (sampling frequencies in the order of 6 GS/s). ii) Since both chains are phase coherent, the subtraction of the REF CHAIN signal from the DET CHAIN signal implemented by the mixer allows compensating the VCOs self-oscillation frequency drifts that are common to both chains. iii) Compared to a frequency divider that would add noise to the output signal and divide as well the EPR phase response amplitude, the phase noise originated from coherent sources in the two chains is mitigated by the mixer operation, as confirmed by the phase noise (PN) plot shown in Fig. 29.2.4. A PN reduction of more than 10dB is achieved.

The intrinsic performances of the proposed EPR-spectrometer are evaluated through a set of ultra-fast scan analysis over wide f_{B1} spans, as shown in Fig. 29.2.5. The top-left graph exhibits the response of a DPPH sample under non-adiabatic rapid scan regime. The FMCW ramp duration is T=200ns and a f_{B1} speed of 1400THz/sec is achieved on a span of 14.056GHz ± 140MHz (equ. to 100 Gauss). Only 1000 averaged sweeps are used (0.2ms of integration time) with 50GS/s sampling rate. The curves at the top-right are obtained using a more reasonable sampling rate of 6.25GS/s, compatible with commercial FPGAs, and allow to compare the DPPH response for different scan rates (from 1/30µs to 1/0.25µs). Since the integration time is the same in all cases, the number of averaged sweeps is smaller for slower sweeps, hence the remaining noise is higher for slower scans. The curves at the bottom are obtained with a solution of TEMPOL in glycerol 90%. The bottom-left graph is for a 20mM sample over a f_{B1} sweep equivalent to 80 Gauss with T=10µs and 800 averages (8ms integration time) that is the limit of the scope memory at 6.25GS/s. The expected triple line spectrum of TEMPOL radical is observed. The rms noise is 25.8 mdeg resulting in an SNR of 9dB. The SNR can be improved by increasing the number of averaged sweeps, but due to the scope memory limit, this implies to reduce the scan time. The bottom-right curves are obtained by averaging 16000 sweeps of 500 ns, for various concentrations of TEMPOL. The rms noise is reduced to 1.3 mdeg. According to the methodology in [2-5], the estimated concentration sensitivity is extracted by the intercept at SNR = 5dB of the

ISSCC 2026 / February 18, 2026 / 10:30 AM

measured SNR *vs* concentration trend, extrapolated to 1s of integration time. It results in CS = 0.12mM, an outstanding sensitivity considering that the sensor is made with only one inductor.

The system is shown in Fig. 29.2.6. The 4.4mm² 22nm-FDSOI IC (Fig. 29.2.7) power consumption is 96mW, which enables battery operation and portability. Compared with previous works, the proposed injection-locked based phase demodulation EPR measurement technique achieves an estimated CS better than the state-of-art, even with a single sensing inductor, and without lock-in detection. It also features the fastest reported scan speed (1400THz/sec) and a wide f_{B1} scan range equivalent to 100 Gauss (Fig. 29.2.5). The extremely fast scan rate and high accuracy open the door to novel observations in the domain of paramagnetic species at a time scale never attainable before.

References:
[1] M. Kern, A. Chu and J. Anders, "Current Trends in VCO-Based EPR". *Appl Magn Reson* 55, 1065–1089 (2024). https://doi.org/10.1007/s00723-024-01698-0
[2] A. Chu *et al.*, "An 8-channel 13GHz ESR-on-a-Chip injection-locked vco-array achieving 200µM-concentration sensitivity," *IEEE International Solid-State Circuits Conference (ISSCC)*, Feb. 2018, pp. 354-356. http://doi.org/10.1109/ISSCC.2018.8310330
[3] B. Schlecker *et al.*, "Towards Low-Cost, High-Sensitivity Point-of-Care Diagnostics Using VCO-Based ESR-on-a-Chip Detectors," *IEEE Sensors Journal*, vol. 19, no. 20, pp. 8995-9003, 15 Oct.15, 2019. http://doi.org/10.1109/JSEN.2018.2875767
[4] S. Künstner *et al.*, "Rapid-scan electron paramagnetic resonance using an EPR-on-a-Chip sensor", *Magn. Reson.*, 2, 673–687, 2021. https://doi.org/10.5194/mr-2-673-2021
[5] J. -H. Sun *et al.*, "A Portable 14GHz Dual-Mode Pulse and Continuous-Wave Electron Paramagnetic Resonance Spectrometer Using a Subharmonic Direct Conversion Receiver," *IEEE International Solid-State Circuits Conference (ISSCC)*, Feb. 2024, pp. 478-480. http://doi.org/10.1109/ISSCC49657.2024.10454384
[6] J. -B. David *et al.*, "A 22-nm FDSOI 35-41 GHz Frequency Synthesizer," *2024 19th European Microwave Integrated Circuits Conference (EuMIC)*, 2024, pp. 18-21. http://doi.org/10.23919/EuMIC61603.2024.10732855

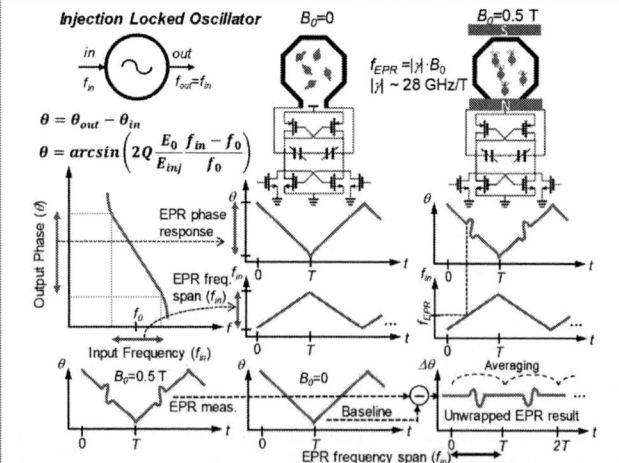

Figure 29.2.1: Left, injection locked oscillator (ILO) used as a frequency to phase transducer. Center and right: phase response of the ILO for a triangular FMCW input signal of slope duration T with EPR sample, without and with constant magnetic field (B0), respectively.

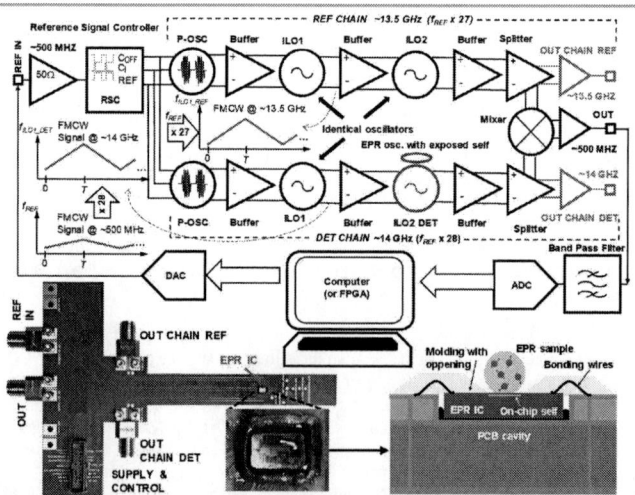

Figure 29.2.2: Top, block diagram of the EPR IC and other components of the measurement set-up. Bottom left, portable EPR board. Bottom right, cross-section of the PCB and IC assembly and sample deposition.

ISSCC 2026 / SESSION 29 / BIOCHEMICAL SENSORS FOR LIFE SCIENCES AND AGRICULTURE / 29.2

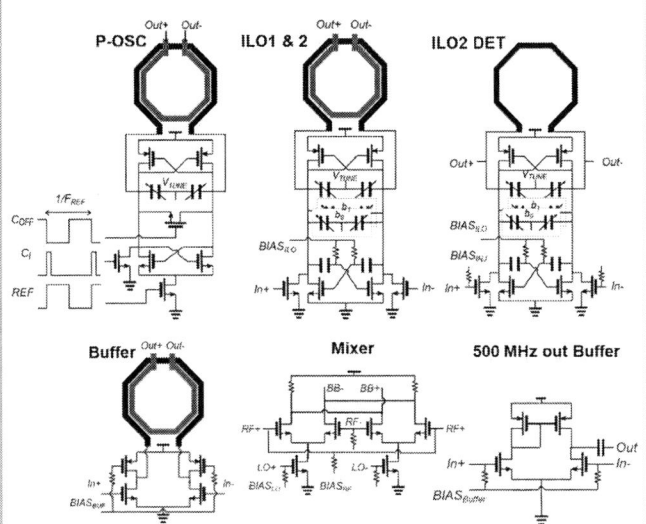

Figure 29.2.3: Circuit schematics of main blocks.

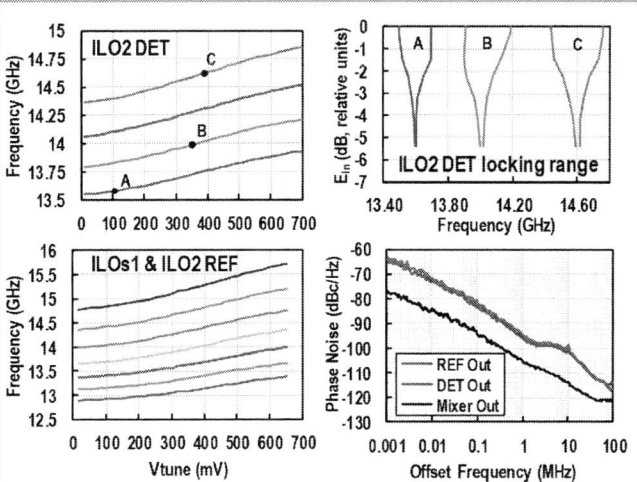

Figure 29.2.4: Measured tuning curves of the different ILOs, some examples of ILO2 DET locking range for different tuning values (A, B, C), and phase noise of the two chains outputs (at 14GHz and 13.5GHz), and of the mixer output (at 500MHz).

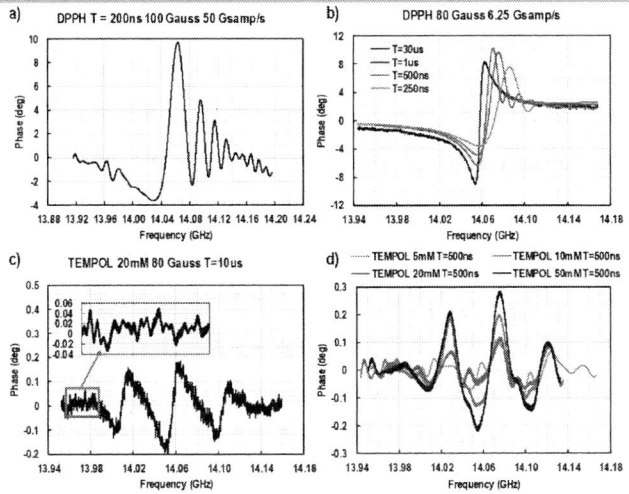

Figure 29.2.5: EPR experiments. a) DPPH with ultra-fast scan of 1400THz/s on 280MHz. b) Various scan speeds from 8THz/s up to 960THz/s for the same DPPH sample. c) 20mM TEMPOL at 24THz/s on 240MHz. d) Various concentrations of TEMPOL at 480 THz/s.

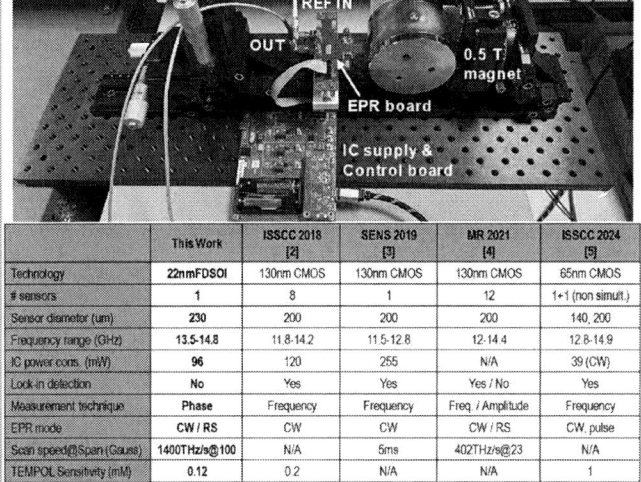

Figure 29.2.6: Picture of the portable ILO based EPR spectrometer. The table shows its main performances and a comparison with the state-of-the art.

	This Work	ISSCC 2018 [2]	SENS 2019 [3]	MR 2021 [4]	ISSCC 2024 [5]
Technology	22nmFDSOI	130nm CMOS	130nm CMOS	130nm CMOS	65nm CMOS
# sensors	1	8	1	12	1+1 (non simult.)
Sensor diameter (um)	230	200	200	200	140, 200
Frequency range (GHz)	13.5-14.8	11.8-14.2	11.5-12.8	12-14.4	12.8-14.9
IC power cons. (mW)	96	120	255	N/A	39 (CW)
Lock-in detection	No	Yes	Yes	Yes / No	Yes
Measurement technique	Phase	Frequency	Frequency	Freq. / Amplitude	Frequency
EPR mode	CW / RS	CW	CW	CW / RS	CW, pulse
Scan speed@Span (Gauss)	1400THz/s@100	N/A	5ms	402THz/s@23	N/A
TEMPOL Sensitivity (mM)	0.12	0.2	N/A	N/A	1

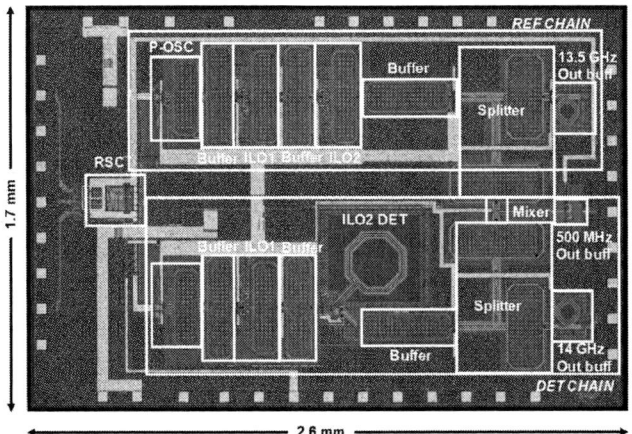

Figure 29.2.7: On-chip EPR spectrometer CMOS 22nm FD-SOI Integrated Circuit micrograph. Inductor diameter is 230µm.

ISSCC 2026 / SESSION 29 / BIOCHEMICAL SENSORS FOR LIFE SCIENCES AND AGRICULTURE / 29.3

29.3 A 256-Channel Event-Driven Readout for Solid-State Nanopore Single-Molecule Sensing with 193pArms Noise in a 1MHz Bandwidth

Sander Crols[1,2], Carolina Mora Lopez[2], Filip Tavernier[1], Marian Verhelst[1,2], Nick Van Helleputte[2]

[1]KU Leuven - MICAS, Leuven, Belgium, [2]imec, Leuven, Belgium

Abstract

The paper presents a 256-channel solid-state nanopore readout IC in 65nm CMOS. A new integrate-and-hold TIA cuts per-pore power and area >6× while achieving 1MHz BW and 193pA$_{rms}$ at 10nA. Event detection with dynamic slot allocation exploits sub-1 percent activity to reduce ADC data 32×. The chip delivers 1mW and 0.019mm^2 per pore and records labeled DNA translocations, enabling scalable, low-cost single-molecule diagnostics.

Nanopore single-molecule sensing holds great promise for enabling affordable personalized healthcare and medical diagnostics [1]. Reducing the cost of such diagnostic tools opens new application domains, including preventive cancer screening and the development of patient-specific therapies. While large biological nanopore arrays have already been deployed in commercial platforms, solid-state nanopores can provide larger signal amplitudes, enabling faster readouts per pore. This capability is critical for improving the sensitivity, throughput, and cost-efficiency of molecule sensing. Although several integrated circuits for the readout of solid-state nanopores have been presented in literature [2]-[7], limitations in area, power and data rate restrict their scalability to only a few dozen pores while maintaining adequate SNR. As a result, the molecular throughput remains far below that of biological nanopore arrays, which operate with millions of pores at kHz speeds. The primary circuit challenge in scaling solid-state nanopore readouts arises from the power and area required by the transimpedance amplifiers (TIAs) to meet stringent low-noise requirements, as well as the enormous data rates generated by large, high-speed pore arrays. This work addresses these challenges by (a) introducing a TIA architecture that reduces power and area by more than 6× compared to [7] and (b) integrating molecule-event detection to exploit the time sparsity of nanopore signals, thereby enabling drastic channel-count scaling. The presented chip implements a complete 256-channel frontend for event-driven single-molecule sensing using solid-state nanopores. Fabricated in 65nm CMOS, this prototype achieves a bandwidth (BW) of 1MHz and an input-referred noise of 193pA$_{rms}$ at 10nA DC, while consuming only 1mW and 0.019mm^2 per pore.

State-of-the-art nanopore readouts generally rely on low-noise current sensing as shown in Fig. 29.3.1 [8]. In a typical setup, a solid-state nanopore is immersed in an ionic solution, and a DC voltage bias is applied across the membrane using two Ag/AgCl electrodes. When a charged molecule such as DNA translocates through the pore, it partially blocks the ion flow, producing current modulations that depend on the molecule's geometry. The current amplitude and event duration (dwell time) are determined by factors including pore size, ionic strength, molecule charge, geometry, and applied bias. Typical currents are on the order of a few tens of nA, with dwell times ranging from ~100µs to several ms. Accurate molecular identification needs high temporal and amplitude resolution, which in turn requires a readout with MHz bandwidth and input-referred noise below 500pA$_{rms}$ [9]. Most state-of-the-art readout circuits employ feedback TIAs (R-TIA/C-TIA) [2]-[6] due to their favorable noise performance and architectural simplicity [10]. However, their noise performance degrades with increasing DC current and input capacitance [10]. As a result, low-noise TIAs for solid-state nanopores typically require >30mW and 0.1mm^2 per pore to operate under conditions of tens of nA DC current and 5-to-10pF input capacitance [11]. Such power and area requirements severely constrain scalability.

To overcome the limitations of feedback TIA's, this work introduces a new TIA architecture, illustrated in Fig. 29.3.2. The architecture consists of an integration capacitor (C_{INT}), a reset switch, an output buffer, and an input stage. The input stage employs a cascode stack of NMOS transistors (T_{in} and T_{casc}), enabling the signal current to flow from C_{INT} into the nanopore. To lower the input impedance and provide precise DC bias control across the nanopore, a gain-boosting opamp with capacitive compensation (C_{ext}) [12] is incorporated, extending the BW. During operation, the current flowing into the nanopore discharges C_{INT} over an integration period T_{INT}, after which the reset switch is activated. C_{INT} is sized as small as 8fF to maximize the transimpedance gain ($1/d_{INT}C_{INT}f_{RST}$) and minimize the input-referred kT/C noise. A compact source-follower drives subsequent stages. The presented architecture incorporates three key concepts to improve the noise performance while reducing power and area. First, it leverages the high output resistance R_{pore} of the nanopore device, typically tens of MΩ's [11]. By stacking the input transistor T_{in} directly on top of the nanopore, the pore effectively acts as source degeneration for T_{in}, reducing its transconductance at frequencies below $1/R_{pore}C_{IN}$ and thereby suppressing the noise contributions from both T_{in} and the opamp. A second innovation is introduced at the cascode transistor T_{casc}, where an additional node in the signal path allows the configurable capacitor C_{BW}, together with $1/gm_{casc}$, to limit the BW of the input stage without adding significant noise. This mechanism provides consistent BW and noise performance for varying nanopore characteristics and signal levels. However, at higher frequencies, C_{BW} reduces the source degeneration of T_{casc}, creating a trade-off between noise suppression by BW limitation and the additional noise introduced. Finally, the integrate-and-hold operation of the TIA averages high-frequency noise over the integration period, thereby minimizing its impact [10].

While recording precise, low-noise, high-BW data from a single pore is essential, scaling the presented architecture to larger pore arrays would quickly lead to data bottlenecks and high power consumption from analog-to-digital conversion and I/O transmission. To address this, we introduce a dynamic-allocation event-detection circuit for each nanopore, as shown in Fig. 29.3.2, exploiting the measured sub-1% time sparsity of useful nanopore data by monitoring each pore's input signal for molecular activity. With this approach, only data from active pores is forwarded to the ADCs, significantly reducing the data rate. The event-detection circuit consists of a switched-capacitor low-pass filter and a comparator that detects sudden deviations in current from the baseline, indicating molecular translocation through the pore. A 1kHz low-pass filter extracts the baseline current, suppressing <1ms translocations while still tracking slow baseline drift. On the comparator's signal side, a resistor and a current source generate a configurable DC offset that serves as a threshold. Upon detecting an event, the circuit communicates with a global register to identify the next available S&H slot and issues a claim request, after which the corresponding TIA is routed to the assigned slot.

Figure 29.3.3 (left) shows the block diagram of the complete ASIC, which integrates 256 TIAs with embedded event-detection, eight sample-and-hold (S&H) slots and an ADC. Each of the 256 TIAs can dynamically connect to any of the eight available S&H slots. Compared to static allocation, dynamic allocation reduces the number of missed events for the same slot count (Fig. 29.3.3, right). This lowers the overall ADC data rate by 32×, enabling slower ADCs, smaller die area, reduced power consumption, and simplified offline signal processing. The signaling between the event-detection circuit and the global register is shown in Fig. 29.3.3 (bottom). Because this architecture can only record from a limited number of pores simultaneously, data loss - defined as the probability that more pores are active than S&H slots are available - was simulated using binomial distribution models. As shown in Fig. 29.3.3 (right), eight S&H slots provide a good trade-off, limiting data loss to 0.1% while providing a 32× reduced data rate and the associated savings in area and power. Once sampled, the signal is digitized on chip by a 12-bit pipelined SAR ADC operating at 16MHz [13]. The ADC uses a semi time-interleaved architecture where two SAR stages alternate but the residue amplifier is shared.

The prototype readout system is manufactured in a 65nm bulk CMOS process and mounted on a PCB together with nanopore dies for verification. The die micrograph and measurement setup are shown in Fig. 29.3.7. The chip was evaluated both electrically and biologically. Digital data were captured using an Opal Kelly XEM8310 FPGA with LVDS transceivers. Figure 29.3.4 presents the measured input-referred noise and transfer curves for different gain settings of a single TIA. For a typical 10nA nanopore baseline current and 7.1pF input capacitance, the TIA achieves an integrated input-referred noise of 193pA$_{rms}$ over a 1MHz bandwidth. Due to noise folding from an overdesigned ADC driver, the full system noise increases to 258pA$_{rms}$, which can be reduced by optimizing the driver BW. The gain-boosting opamp extends the TIA BW from 200kHz to 1MHz. Thanks to the improved TIA architecture and on-chip event-detection, the chip achieves a core power consumption of only 1mW/pore and an area of 0.019mm^2/pore, while reducing the total data rate by 32× to just 288Mbit/s.

The prototype was biologically tested using SiN solid-state nanopores with diameters between 10 and 30nm, estimated from the open-pore conductance. Measurements were performed in 4M LiCl solution buffered to a pH of 8.0 with Ag/AgCl electrodes. DNA molecules were modified following the DNA-origami protocol of [14] to include three 17-dumbbell labels, as illustrated in Fig. 29.3.5. The molecules were introduced on the low-potential side of the membrane (cis). Driven by a 200mV DC cross-membrane bias, the negatively charged DNA molecules migrated through the nanopore towards the positive electrode (trans), producing current dips of approximately 0.5nA. Figure 29.3.5 shows representative translocation events recorded by the chip, where the expected molecular signature with 3 distinct spikes corresponding to the dumbbells is clearly visible. The initial spike at the start of a translocation is larger than the subsequent ones, which is attributed to molecule folding.

The system performance is summarized in Fig. 29.3.6 and benchmarked against state-of-the-art nanopore readouts. The presented TIA achieves the highest channel count, the smallest area per channel, and low power per channel, while maintaining competitive noise performance and integrating on-chip event detection for data-rate reduction. It should be noted that the competitive noise results in [2] were obtained over a larger bandwidth but without including a representative capacitive input load, making the reported performance unrepresentative for solid-state nanopore devices.

To conclude, this work presented a scalable readout system for solid-state nanopores, enabled by the presented TIA architecture and dynamic-allocation event detection. These innovations achieved a 10× improvement in scalability, culminating in the first 256-channel solid-state nanopore readout interface with significantly improved power and area efficiency per pore. The resulting increase in molecular throughput has the potential to shorten diagnostic lead times and lower the costs in point-of-care applications.

Acknowledgement:
The authors would like to thank Donald Raddoux, and Rudi Vanlaer for their technical support. This work was supported by Research Foundation-Flanders (FWO) under grant 1S17325N.

References:
[1] M. Jain *et al.*, "Nanopore sequencing and assembly of a human genome with ultra-long reads," *Nature Biotechnology*, vol. 36, no. 4, pp. 338–345, 2018. [Online]. Available: https://doi.org/10.1038/nbt.4060. https://doi.org/10.1038/nbt.4060
[2] G. Ferrari, F. Gozzini, A. Molari, and M. Sampietro, "Transimpedance amplifier for high sensitivity current measurements on nanodevices," *IEEE Journal of Solid-State Circuits*, vol. 44, no. 5, pp. 1609–1616, 2009. https://doi.org/10.1109/JSSC.2009.2016998
[3] S. Shekar *et al.*, "Measurement of DNA translocation dynamics in a solid-state nanopore at 100 ns temporal resolution," *Nano Letters*, vol. 16, no. 7, pp. 4483–4489, 2016, pMID: 27332998. [Online]. Available: https://doi.org/10.1021/acs.nanolett.6b01661. https://doi.org/10.1021/acs.nanolett.6b01661
[4] S. Dai, R. T. Perera, Z. Yang, and J. K. Rosenstein, "A 155-db dynamic range current measurement front end for electrochemical biosensing," *IEEE Transactions on Biomedical Circuits and Systems*, vol. 10, no. 5, pp. 935– 944, 2016. https://doi.org/10.1109/TBCAS.2016.2612581
[5] M. Taherzadeh-Sani *et al.*, "A 170-db ω cmos tia with 52-pa input-referred noise and 1-mhz bandwidth for very low current sensing," *IEEE Transactions on Very Large Scale Integration (VLSI) Systems*, vol. 25, no. 5, pp. 1756–1766, 2017. https://doi.org/10.1109/TVLSI.2017.2654452
[6] A. Das *et al.*, "A 376w per-channel, drift-tolerant translocation recording frontend with event detection for nanopore sensor arrays," *IEEE European Solid-State Electronics Research Conference (ESSERC)*, 2024, pp. 420–423. https://doi.org/10.1109/ESSERC62670.2024.10719545
[7] D.-P. Wiens *et al.*, "A direct digitizing, 1mhz bandwidth, 28fa/p (hz) current sensing front-end based on a mixed-signal integrator-differentiator tia in 28nm cmos," *IEEE Custom Integrated Circuits Conference (CICC)*, 2025, pp. 1–3. https://doi.org/10.1109/CICC63670.2025.10983397
[8] Q. Lin *et al.*, "Advances and challenges in integrated circuits for electrochemical sensing: Enabling next-generation biomedical and molecular applications," *IEEE Transactions on Biomedical Circuits and Systems*, pp. 1–21, 2025. https://doi.org/10.1109/TBCAS.2025.3589027

[9] C.-C. Chien *et al.*, "Single-stranded dna translocation recordings through solid-state nanopores on glass chips at 10 mhz measurement bandwidth," *ACS Nano*, vol. 13, no. 9, pp. 10 545–10 554, 2019, pMID: 31449393. [Online]. Available: https://doi.org/10.1021/acsnano.9b04626. https://doi.org/10.1021/acsnano.9b04626
[10] M. Crescentini, M. Bennati, M. Carminati, and M. Tartagni, "Noise limits of cmos current interfaces for biosensors: A review," *IEEE Transactions on Biomedical Circuits and Systems*, vol. 8, no. 2, pp. 278–292, 2014. https://doi.org/10.1109/TBCAS.2013.2262998
[11] J. K. Rosenstein and K. L. Shepard, "Temporal resolution of nanopore sensor recordings," *35th Annual International Conference of the IEEE Engineering in Medicine and Biology Society (EMBC)*, 2013, pp. 4110–4113. https://doi.org/10.1109/EMBC.2013.6610449
[12] F. Tavernier and M. Steyaert, "A bandwidth enhanced transimpedance amplifier with improved noise performance," *Analog Integrated Circuits and Signal Processing*, vol. 66, no. 2, pp. 277–283, 2011. Available: https://doi.org/10.1007/s10470-010-9545-x. https://doi.org/10.1007/s10470-010-9545-x
[13] Z. Li *et al.*, "An 80ms/s 70.79db-sndr 60.7fj/conv-step radiation-tolerant semi-timeinterleaved pipelined-sar adc," *IEEE Custom Integrated Circuits Conference (CICC)*, 2024, pp. 1–2. https://doi.org/10.1109/CICC60959.2024.10529093
[14] N. A. W. Bell and U. F. Keyser, "Digitally encoded DNA nanostructures for multiplexed, single-molecule protein sensing with nanopores," *Nature Nanotechnology*, vol. 11, no. 7, pp. 645–651, jul 2016. [Online]. Available: https://doi.org/10.1038/nnano.2016.50. https://doi.org/10.1038/nnano.2016.50

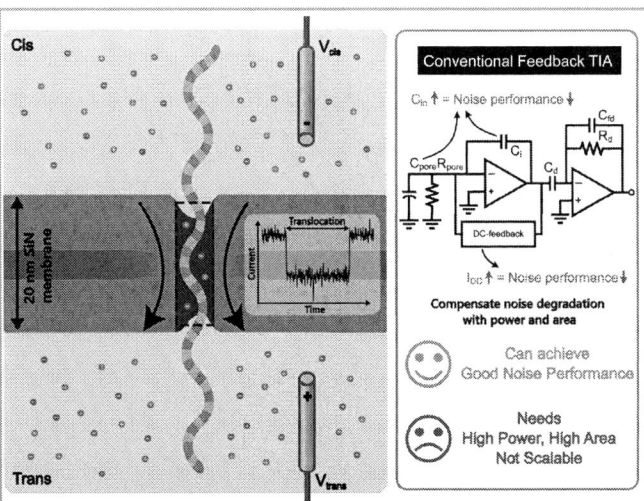

Figure 29.3.1: (Left) DNA molecule translocation through a solid-state nanopore in a SiN membrane surrounded by 4M LiCl solution resulting in a measurable electrode current dip. (Right) Discussion of conventional architecture challenges.

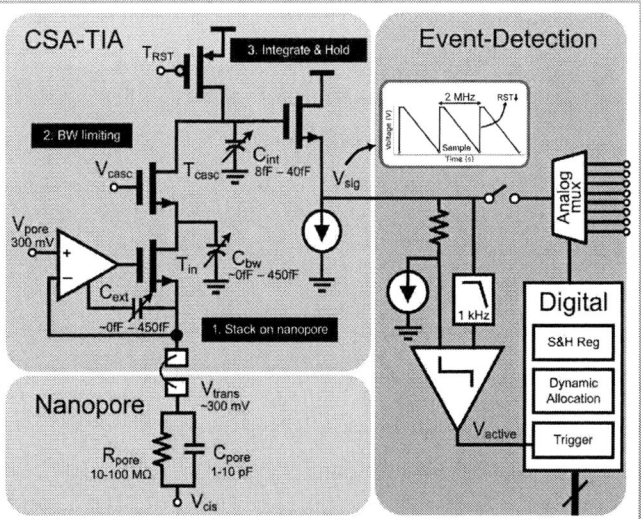

Figure 29.3.2: Channel readout architecture containing the proposed Charge Sensitive TIA and event-detection topology together with a first-order nanopore RC model.

ISSCC 2026 / SESSION 29 / BIOCHEMICAL SENSORS FOR LIFE SCIENCES AND AGRICULTURE / 29.3

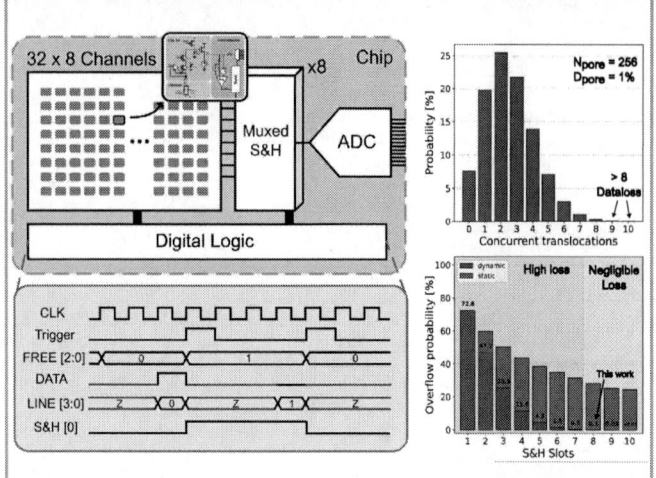

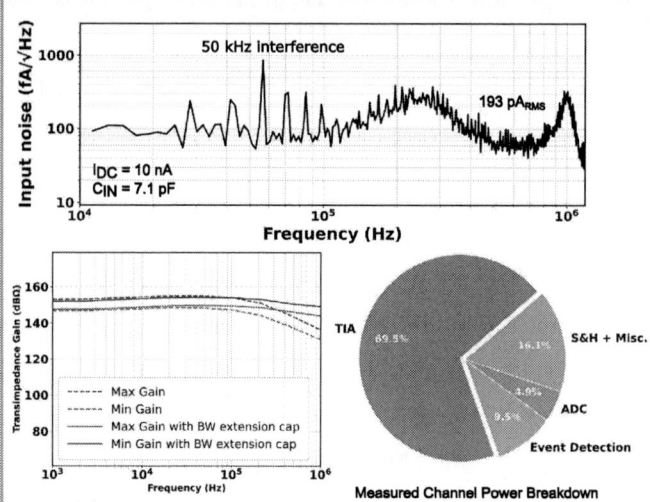

Figure 29.3.3: (Left) System architecture with Timing diagram for S&H slot negotiation. (Right) Simulation results showing probability of S&H occupation and estimated data loss for different system configurations when using static or dynamic allocation.

Figure 29.3.4: TIA noise PSD measurement, Transfer function for min and max gain settings showing the need for bandwidth compensation capacitors, power breakdown.

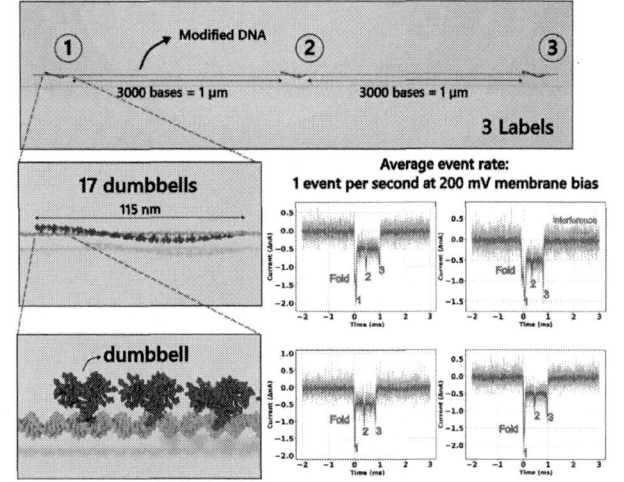

Figure 29.3.5: Modified DNA molecule with 3× equidistant 17-dumbbell label attachment, Translocation time traces showing proper label detection (raw and 100 kHz filtered).

	This work	[2] JSSC 2009	[3] Nano 2016	[4] TBCAS 2016	[5] VLSI 2017	[6] ESSERC 2024	[7] CICC 2025
Architecture	CSA	C-TIA	R-TIA	DT-CTIA	C-TIA	R-TIA	MS-TIA
Channels	256	1	25	1	1	16	1
Technology [nm]	65	350	180	180	130	180	28
Bandwidth [MHz]	1	4	5	1.4	1	0.92	1
Area per pore [mm²]	0.019	0.34	0.16	0.091	0.2	0.024	0.13
Power per pore [mW]	1	45	/	5.22	30	0.376	8.9
TIA noise [pA RMS][1]	193	>110[2]	1430	~1000	52	/	185
System noise [pA RMS][1]	258	>110[2]	1430	~1000	52	646	185
TIA Gain [Ohm]	60M	60M	7.5M	9.5M	330M	5M	2M
Input Capacitance [pF]	7.1	0.6	4	/	6	10	2.1
Event Detection	YES	NO	NO	NO	NO	YES	NO
Digital Output	YES	NO	NO	NO	NO	YES	YES

[1] At 10 nA DC input current
[2] Estimated from paper

Figure 29.3.6: Performance summary and comparison to prior works.

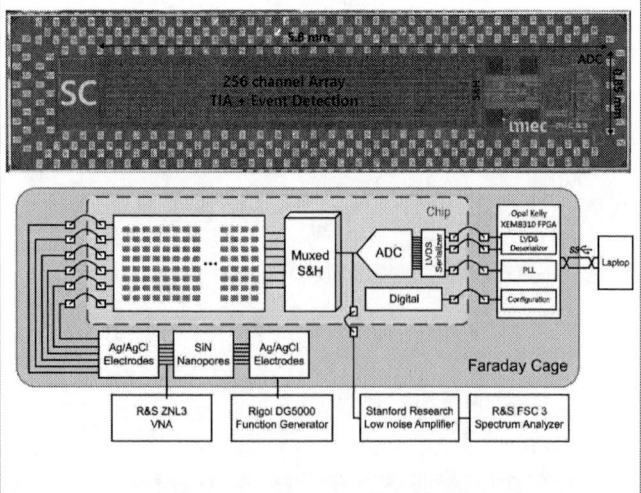

Figure 29.3.7: Annotated die micrograph and measurement setup illustration.

ISSCC 2026 / SESSION 29 / BIOCHEMICAL SENSORS FOR LIFE SCIENCES AND AGRICULTURE / 29.4

29.4 Shape-Memory Multi-Size Micro-Cage Array on CMOS with Integrated Electrochemical Sensors for Joint Bio-Sample Manipulation and Sensing

Zhikai Huang, Fuze Jiang, Hangxing Liu, Adam Wang, Yuguo Sheng, Marco Saif, Hua Wang

ETH Zurich, Zurich, Switzerland

Abstract

Lab-on-CMOS platforms require joint sensing–manipulation capabilities for biosamples from single cells to organoids. Existing methods such as DEP, optical, and acoustic tweezers need external setups, continuous power, and face limitations in physiological media. We present a CMOS-integrated shape-memory multi-size micro-cage array with multi-material electrochemical potentiostats, enabling power-free immobilization and multiplexed sensing, advancing multiscale biosample culture and monitoring.

Lab-on-CMOS systems increasingly need on-chip multifunctional joint sensing-manipulation capabilities to address complex biological samples from single cells to multicellular structures. For instance, organoids as miniaturized *in vitro* 3D cellular models widely used in drug discovery, disease modeling, and regenerative medicine, often require reliable on-chip immobilization to ensure stable growth, reproducible measurements, and optimal spacing that prevents unwanted organoid fusion and hypoxia-induced necrosis [1]. Most existing lab-on-CMOS platforms rely on manipulation techniques such as dielectrophoresis (DEP) [2], optical tweezers [3], or acoustic tweezers [4], which often require external setups and continuous power for operations. Moreover, these manipulation methods exhibit limitations in long-term culture applications with physiologically relevant conditions. For example, DEP fields are attenuated in high-ionic-strength electrolytes, while optical and acoustic tweezers suffer from heating and stability issues, making them unsuitable for long-term culture. In addition, they are generally optimized for narrow ranges of sample sizes, whereas practical applications often demand robust manipulation of diverse samples, from single cells to large organoids.

To address these challenges, we demonstrate an on-CMOS shape-memory multi-size micro-cage array integrated with in-pixel, multi-material electrochemical potentiostats for joint on-chip biosample manipulation and sensing. A conceptual illustration of the platform is shown in Fig. 29.4.1. Both the manipulation device (multi-size micro-cage) and the sensing device (multi-material electrochemical sensors with three metal materials on the working electrode, WE) are monolithically post-fabricated on a CMOS chip. The multi-size micro-cage is constructed from surface-electrochemical actuators composed of a Ti/Pt bilayer, where tuning the electrochemical state of Pt/PtO_x dictates bending or flattening of the arms of the micro-cage [5]. Owing to its electrochemical actuation mechanism and shape-memory property, the device enables highly localized and individually addressable control, operates in any physiological electrolyte including standard culture media, and retains its shape without continuous power supply. Three different sizes of cages are grouped to form the multi-size micro-cage, with each cage individually driven by an on-CMOS actuation electrode. It allows selective trapping and immobilization of biological samples ranging from single cells (~30µm) to large organoids (300–400µm). For biochemical monitoring, the integrated in-pixel potentiostats employ three distinct electrode materials (Au, Pt, and Pd) for the working electrodes, which are patterned so that all three electrode materials are exposed to the medium to enhance selectivity, sensitivity, and robustness over different analytes compared to single-material potentiostats.

Figure 29.4.2 shows the circuit schematics and system architecture of a proof-of-concept chip for the joint manipulation and sensing micro-cage array platform. The array consists of 3×3 multi-function pixels with each pixel measuring approximately 500µm × 520µm. Each multi-function pixel integrates three differently sized micro-cages (small, medium, and large) stacked together and three electrochemical potentiostats. In the flat state, the distances from the arm tip to the pixel center of the small, medium, and large micro-cages are approximately 100µm, 150µm, and 280µm, respectively, enabling the capture of biosamples ranging from ~30µm (single cells) to ~400µm (large organoids). Each micro-cage is connected to a dedicated actuation electrode, and by controlling the voltage between the cage and the actuator counter electrode, individual cages can be selectively programmed into flat or bent states via on-chip digital serial-to-parallel-interface (SPI) control. The in-pixel potentiostats are implemented using capacitive transimpedance amplifiers (CTIAs) with low-noise, low-mismatch amplifiers (standard deviation σ = 4.7mV) with adjustable gain–bandwidth product to enable high-precision electrochemical detection for various assays. To further enhance sensitivity and selectivity for different electrolytes, three working electrode materials (Au, Pt, and Pd) are incorporated, one for each potentiostat.

Figure 29.4.3 shows the post-CMOS fabrication process. CMOS chip power delivery, control, and electrochemical sensing rely on the top metal electrodes, which are passivated by the native SiO_2 and Si_3N_4 of the CMOS chip. The in-house process begins with opening the passivation layer by reactive ion etching (RIE), followed by Ti/Au deposition as a base electrode to cover the native Al electrode. An Al sacrificial layer is then patterned to define the release layer of the multi-size micro-cage. Next, the actuator layer of Ti/Pt is deposited on the chip and patterned by ion milling to form the micro-cage geometry. A Ti/Au panel layer is subsequently deposited atop the actuator to regulate the desired bending profile

after release. Finally, each micro-cage actuator is connected to three individually addressable electrodes, and two additional materials (Pt and Pd) are deposited onto the Au working electrodes of the potentiostats to expand sensing capabilities.

Figure 29.4.4 presents the fabrication results and material characterization of the 3×3 array comprising multi-size micro-cages and integrated electrochemical potentiostats. Energy-dispersive X-ray spectroscopy (EDS) imaging confirms the patterned deposition of Pt and Au, corresponding to the actuator bilayer composition and the panel layer above the Al release layer, as well as the three distinct electrode materials (Au, Pt, and Pd) used in the working electrode of the multi-material potentiostats. The surface roughness of the fabricated micro-cage actuators was characterized using atomic force microscopy (AFM) and compared with that of the native CMOS substrate and a reference Si wafer. The results indicate that underlying metal fillings significantly influence surface topography, an important factor for device design and optimization. Scanning transmission electron microscopy (STEM) combined with EDS on actuator cross-sections further confirm accurate deposition thicknesses. Finally, the on-chip electrochemical potentiostats with three different electrode materials were characterized and compared by cyclic voltammetry in $Fe(CN)_6^{3-/4-}$ electrolytes at different concentrations. The results exhibit characteristic sigmoid responses, with Pd showing the lowest sensitivity compared to Au and Pt, verifying different behaviors of the multi-material electrodes.

Figure 29.4.5 shows the packaged chip before and after the release process. The release is conducted at Al etchant type A. Once the device is fully bent, the Al etchant is gently exchanged with PBS 1× for all the following device electrical testing. Subcages of different sizes within the multi-size micro-cage can all bend to the desired shape, as verified by optical and SEM images. To verify the sample capture capability, we placed 300µm glass beads during the release process and observed the capturing after cage bending. The cage can be actuated on-chip reversibly, as shown in the time-lapse images. Further actuator characterization shows the desired fast response, robust durability, and shape-memory behavior.

Figure 29.4.6 (top) shows the multi-material electrochemical potentiostat sensing result. For L-ascorbic acid and $Fe(CN)_6^{3-/4-}$ oxidation peaks overlap with each other for all three working-electrode materials. High sensitivity of Au to $Fe(CN)_6^{3-/4-}$ blurs the peaks generated by L-ascorbic acid, making it impossible to distinguish the origin of the peak. However, Pd is less sensitive to $Fe(CN)_6^{3-/4-}$ and has decent sensitivity to L-ascorbic acid, allowing discrimination against the addition of L-ascorbic acid.

Figure 29.4.6 (bottom) compares reported CMOS manipulation platforms. Our design offers multi-size biosample manipulation and multi-material electrochemical sensing, greatly advancing the lab-on-CMOS technologies.

Acknowledgement:
The authors would like to thank the GlobalFoundries University Program for chip fabrication, the technical staff at the ETH FIRST and IBM BRNC cleanrooms, and ScopeM for their assistance. This work was in part sponsored by the Swiss State Secretariat for Education, Research, and Innovation (SERI) under the SwissChips initiative, an ETH Zurich grant, and the Swiss National Science Foundation "MICA" project under Project 207914.

References:
[1] D. Kim *et al.*, "Scalable production of uniform and mature organoids in a 3D geometrically-engineered permeable membrane," *Nature Communications*, vol. 15, no. 1, p. 9420, 2024. https://doi.org/10.1038/s41467-024-53073-z
[2] K. Park *et al.*, "Dielectrophoretic lab-on-CMOS platform for trapping and manipulation of cells," *Biomedical Microdevices*, vol. 18, no. 1, p. 6, 2016. https://doi.org/10.1007/s10544-016-0030-x
[3] T. Sneh *et al.*, "Optical tweezing of microparticles and cells using silicon-photonics-based optical phased arrays," *Nature Communications*, vol. 15, no. 1, p. 8493, 2024. https://doi.org/10.1038/s41467-024-52273-x

[4] F. Guo *et al.*, "Three-dimensional manipulation of single cells using surface acoustic waves," *Proceedings of the National Academy of Sciences*, vol. 113, no. 6, pp. 1522–1527, 2016. https://doi.org/10.1073/pnas.1524813113

[5] M. Z. Miskin *et al.*, "Electronically integrated, mass-manufactured, microscopic robots," *Nature*, vol. 584, no. 7822, pp. 557–561, 2020. https://doi.org/10.1038/s41586-020-2626-9

[6] Y. Liu *et al.*, "Cell-lab on a chip: a CMOS-based microsystem for culturing and monitoring cells," *IEEE EMBC*, pp. 2534-2537, 2004. https://doi.org/10.1109/iembs.2004.1403729

[7] D. Lee *et al.*, "A CMOS Multi-Functional Biosensor Array for Rapid Low-Concentration Analyte Detection with On-Chip DEP-Assisted Active Enrichment and Manipulation with No External Electrodes," *ISSCC*, pp. 316-318, Feb. 2023. https://doi.org/10.1109/ISSCC42615.2023.10067525

[8] D. Lee *et al.*, "Fully Integrated CMOS Ferrofluidic Biomolecular Processing Platform with On-Chip Droplet-Based Manipulation, Multiplexing and Sensing," *ISSCC*, pp. 324-326, Feb. 2024. https://doi.org/10.1109/ISSCC49657.2024.10454430

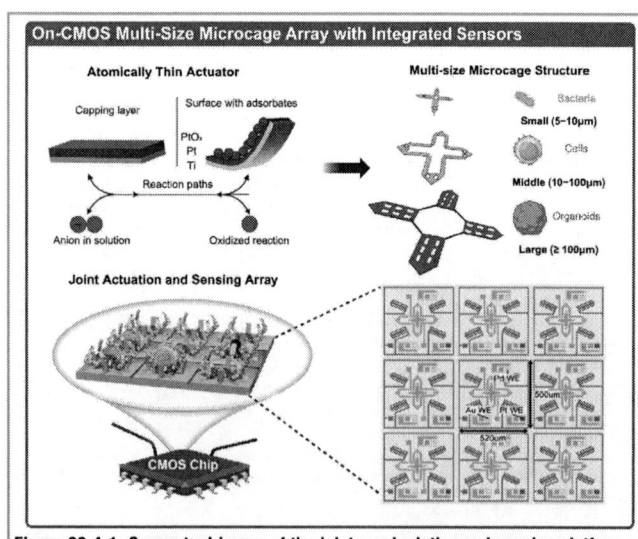

Figure 29.4.1: Conceptual image of the joint manipulation and sensing platform on CMOS chip.

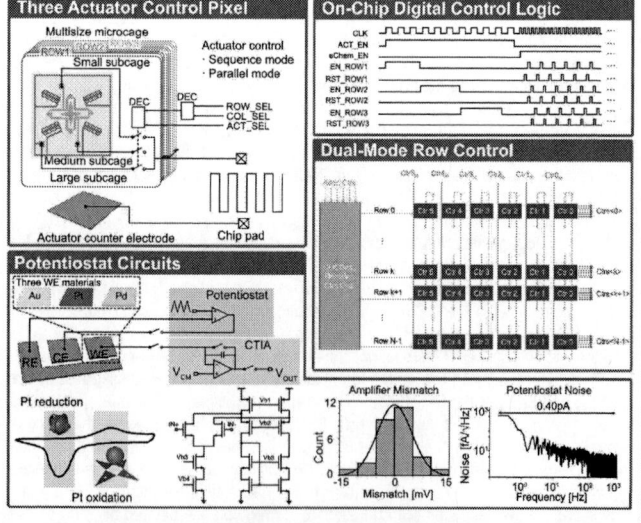

Figure 29.4.2: Circuit schematics and system architecture of the demonstrated platform.

ISSCC 2026 / SESSION 29 / BIOCHEMICAL SENSORS FOR LIFE SCIENCES AND AGRICULTURE / 29.4

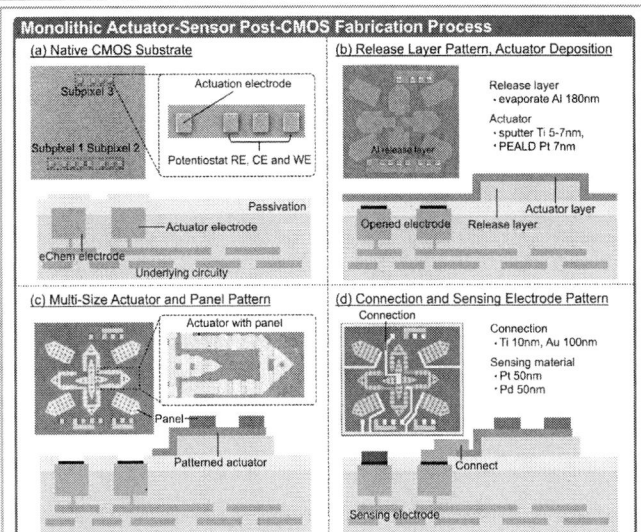

Figure 29.4.3: Device fabrication process.

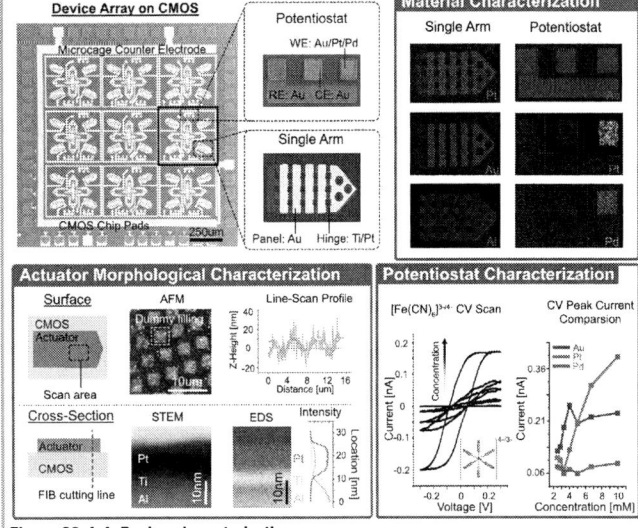

Figure 29.4.4: Device characterization.

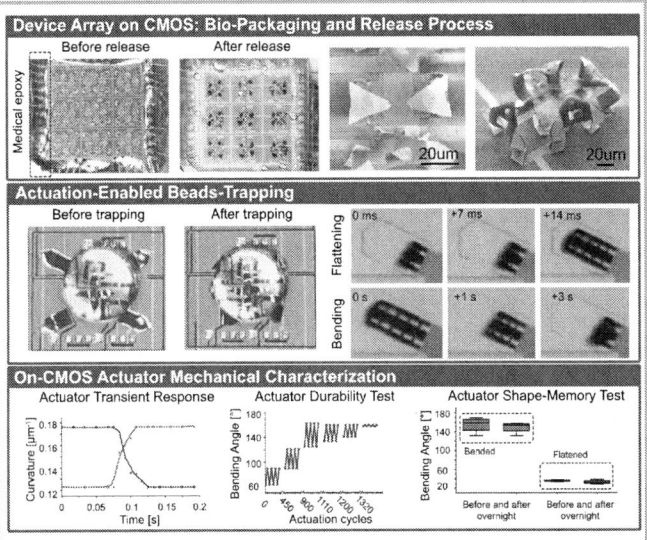

Figure 29.4.5: On-CMOS actuation experiment results.

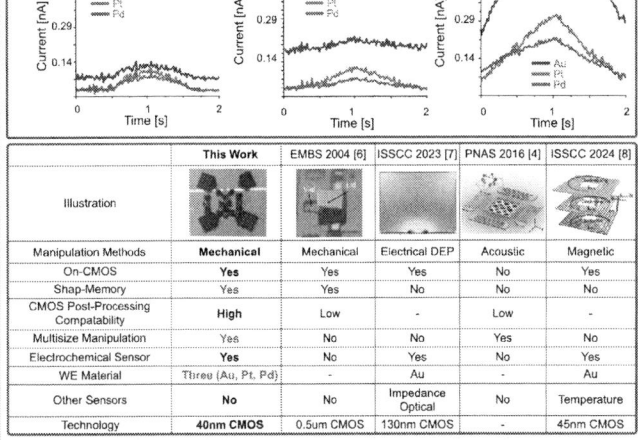

	This Work	EMBS 2004 [6]	ISSCC 2023 [7]	PNAS 2016 [4]	ISSCC 2024 [8]
Illustration					
Manipulation Methods	**Mechanical**	Mechanical	Electrical DEP	Acoustic	Magnetic
On-CMOS	**Yes**	Yes	Yes	No	Yes
Shap-Memory	Yes	Yes	No	No	No
CMOS Post-Processing Compatability	**High**	Low	-	Low	-
Multisize Manipulation	Yes	No	No	Yes	No
Electrochemical Sensor	**Yes**	No	Yes	No	Yes
WE Material	Three (Au, Pt, Pd)	-	Au	-	Au
Other Sensors	**No**	No	Impedance Optical	No	Temperature
Technology	**40nm CMOS**	0.5um CMOS	130nm CMOS	-	45nm CMOS

Figure 29.4.6: On-CMOS sensing experiment results and comparison table.

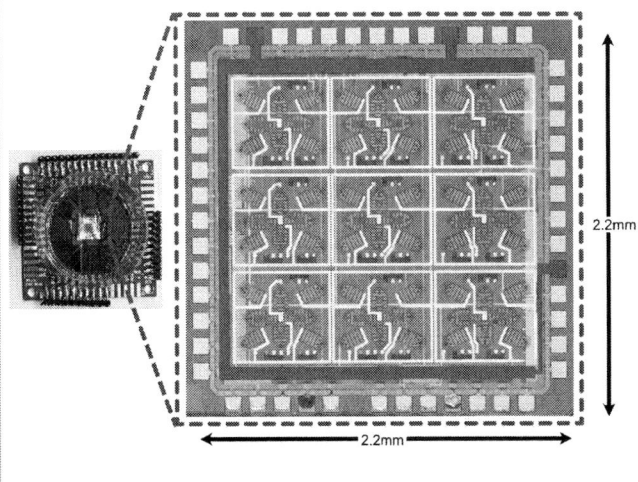

Figure 29.4.7: Chip micrograph.

ISSCC 2026 / SESSION 29 / BIOCHEMICAL SENSORS FOR LIFE SCIENCES AND AGRICULTURE / 29.5

29.5 A Sub-Gram Individual Plant Stress Sensor Tag for Smart Farming

Daiki Nitto[1], Keishi Yanase[1], Yui Fujisawa[1], Jun Shiomi[1], Yoshihiro Midoh[1], Takuya Wadatsumi[2], Makoto Nagata[2], Ahmet Oncu[3], Constantine Sideris[4], Chetphilin Suriyasak[5], Yushi Ishibashi[5], Noriyuki Miura[1]

[1]University of Osaka, Suita, Japan, [2]Kobe University, Kobe, Japan, [3]Bogazici University, Istanbul, Turkey, [4]University of Southern California, Los Angeles, CA, [5]Kyushu University, Fukuoka, Japan

Abstract

This paper presents a 0.62g light-weight battery-less sensor tag for continuous monitoring of plant stress. A wireless link at 920MHz powers the tag for in-vivo measuring pH in the plant as a key stress indicator. The pH data are backscattered to the link through an XOR-based multiplexer, enabling a single miniature antenna to establish simultaneous power and data link over 80cm distance. More than 30-days in-vivo plant stress monitoring was successfully demonstrated for future smart farming.

This paper presents a 0.62g ultra-light-weight battery-less sensor tag for continuous monitoring of plant stress conditions. A wireless link at 920MHz powers the tag for measuring pH in the plant in-vivo as a key stress indicator. The pH data are then backscattered to the link through an XOR-based RF signal multiplexer, enabling a single miniature antenna to establish simultaneous power and data links over 80cm distance. More than 30-days of continuous in-vivo plant stress monitoring was successfully demonstrated, paving the way for future smart farming.

Smart farming is a promising agricultural direction not only for economic profitability but also for environmental sustainability. Environmental field condition monitoring and remote sensing such as temperature, sunlight, rainfall, humidity, and soil-pH are actively studied [1-3]. One critical metric that remains largely unmonitored in such environments is the condition of the individual plant itself. Only a few studies [4-7] have attempted to monitor an individual plant's condition; however, none has addressed noninvasive, continuous monitoring in an economically viable manner. Although the sensor frontends from prior work are relatively compact and lightweight, the overall sensing system requires bulky external measurement instruments (i.e., readout electronics and power supply) and mechanical supporting devices (jig or fixture) to attach or insert the frontend into the plant. In this paper, a 0.62g ultra-lightweight, battery-less sensor tag is proposed and demonstrated, enabling long-term and non-invasive monitoring of stress conditions in individual plants (Fig. 29.5.1). The tag measures the internal pH (H+ concentration) of individual plants, which is a well-known stress indicator in response to drought and disease [8-10]. The measured pH data are reported back to a host reader (assumed to be a drone) through 920MHz wireless backscattered power and data links. No battery is required and a compact light-weight single antenna on an ultra-thin PCB significantly reduces the total tag weight down to 0.62g, obviating the need for mechanical support due to the sub-gram weight. Additionally, mechanically and electrically stable micro needles with only 0.1mm-ϕ diameter are developed and used as the pH sensing electrodes. This ultra-light-weight design and the non-invasive micro needles insertion enables low-cost, non-invasive, and hence long-term individual plant stress monitoring for future applications in large-scale autonomous smart farming.

Figure 29.5.2 depicts the block and the circuit diagram of the proposed individual plant stress sensor system. The host reader is a self-mixing type backscattering transceiver operating at 920MHz RF for simultaneous power and data links wirelessly. The reader board is also designed to be lightweight to minimize excessive load on the drone. A minimum set of compact discrete components (oscillator, PA, directional coupler, attenuator, and mixer) are assembled on a PCB. On the tag, a compact directional antenna, matching network, Power Management Unit (PMU), reference generators, pH sensing needles, sensor Analog Front-End (AFE), Digital Back-End (DBE), and load modulator are fully integrated. The PMU consisting of a 5-stage rectifier [11] and LDO [12] generates a stable 1.7V DC supply V_{DD} for all the on-chip circuitry. A low-power subthreshold BGR (sBGR) [13] provides a stable reference voltage V_{REF} for LDO and biases the plant potential through a Reference Electrode (RE). The proposed pH sensor core is based on an extended ISFET architecture. The gate terminal of the ISFET M_{IS} is extended by an Ion Electrode (IE) coated by an ion sensitive membrane (Ta$_2$O$_5$ in this work) used as a pH transducer. The IE potential V_{ION} relative to RE is proportional to the internal pH of the plant. This V_{ION} input to the M_{IS} gate is buffered by Ion Potential Buffer (IPB) [14] to the sensor AFE output V_{PH}. The sensor DBE is composed of two racing Relaxation Oscillators (ROs), RO1 for generating a clock frequency f_{PH} proportional to V_{PH}, and RO2 for generating f_{REF} proportional to V_{REF} stably generated by sBGR. By taking the ratio of f_{PH} and f_{REF}, the pH can be determined in a manner robust to static and dynamic variations in PVT by canceling out common disturbances in both f_{PH} and f_{REF}. These two tones need to be backscattered. In this work, an XOR-based RF signal multiplexer is proposed to reflect back multiple tones with only a single backscattering antenna. This scheme significantly simplifies the circuit components required in the tag to realize sub-gram ultra-light-weight solution. Finally, the reader extracts the two tones through simple BPF and FFT.

The pH sensor tag IC was designed and fabricated in 0.18µm CMOS (Fig. 29.5.3 bottom right). The total silicon layout area of the circuits is only 905µm x 460µm including IO pads. The IC chip was mounted on a tag board fabricated by using a 0.2mm-ultra-thin FR4

substrate. The thickness was carefully designed to guarantee mechanical stability for long-term operation while minimizing the overall tag weight. A gold-plated 0.1mm-ϕ micro needle made of High-Speed Steel (HSS) alloy is selected for RE and IE for non-invasive needle insertion into the plant while maintaining mechanical toughness. Ta$_2$O$_5$ was RF sputtered as a hydrogen ion (H+) sensitive membrane on the IE needle surface. Both the IE and RE needles were vertically soldered on the backside of the tag board (Fig. 29.5.3 bottom). A 0.1g ultra-light-weight antenna at 920MHz and its matching network were mounted on the frontside. The total tag weight was measured to be only 0.62g. The host reader board was fabricated on a common 1.57mm-thick FR4 PCB substrate (Fig. 29.5.3 bottom left). Careful selection of compact discrete components reduced the total reader board weight down to 46g. The electrical operation of the proposed pH sensor tag system was tested in an anechoic chamber (Fig. 29.5.3 top). A Yagi antenna was used for evaluating the functionality and performance of wireless backscattered power and data links. The antenna weighed only 320g. The combined weight of the host board and the antenna is light enough to be carried by the drone. The pH sensor performance was also evaluated by using standard pH solutions.

The electrical operation was first verified by probing the tag internal nodes. Figure 29.5.4 presents the measurement results of the proposed pH sensor AFE and DBE. The AFE successfully captures the standard solution pH (Fig. 29.5.4 top left) and outputs corresponding V_{PH} proportional to the pH values (4.01, 7.01, and 9.18). The conversion gain was measured to be 12.9mV/pH, which was close to our design target. In addition, the successful DBE operation was also confirmed by the measured waveform snapshots and their frequency spectra of the XOR-based multiplexer output V_{XOR} (Fig. 29.5.4 bottom). The sensor readout operation was also successfully demonstrated as being robust against temperature variations (Fig. 29.5.4 top right). The frequency drifts due to the temperature variations from 10-to-50°C were successfully canceled by taking the ratio f_{REF}/f_{PH} and a linear response against V_{ION} ($\sim V_{PH}$) was obtained.

Figure 29.5.5 presents the measurement results of the proposed backscattered power and data links. A stable 1.7V DC supply V_{DD} was successfully delivered to the tag over 80cm, the target range, through the 920MHz wireless power link operating at >+23dBm reader power setting. The maximum power link range can be extended up to 220cm by +30dBm reader power (Fig. 29.5.5 top left). The backscattered data link correctly recovered the signal load modulated by the XOR-based multiplexer V_{XOR} at the mixer output V_{MIX} (Fig. 29.5.5 top right). The FFT results clearly exhibit the spectral peaks of the two tones at $|f_{PH} - f_{REF}|$ and $f_{PH}+f_{REF}$ and the pH was successfully measured (Fig. 29.5.5 bottom right). The V_{ION} input range was measured to be 0.4-to-1.2V, which is wide enough to cover the full range of possible pH values in the plant. The measured maximum communication distance of the backscattered data link was >80cm at +23dBm reader power (Fig. 29.5.5 bottom left). This communication range is long enough to make a drone-based pH data collection system practical in an open field.

Finally, in-vivo plant experiments were conducted. Figure 29.5.6 presents the measurement results. In our experiments, soy and tomato were chosen as the target test plants. Our ultra-light-weight tag does not physically affect the plant (soy) posture at all, while an intentionally weighted tag of only 3g strongly tilts the plant and disturbs its growth (Fig. 29.5.6 top left). Such weight impact would become significantly more apparent for more flexible plants such as wheat and rice. The long-term comparative test with and without our tag also clearly exhibits its non-invasive nature throughout 30-days observation (Fig. 29.5.6 top right). Also, microscopic observation was conducted on the sensor needle insertion sites after the 30-day long-term experiment (Fig. 29.5.6 bottom left). Corken cells periderms were observed at the plant surface only while the internal organs were fresh and no clear rejection reactions were observed. During the 30-days experiment, our sensor tag measured the pH of soy and tomato and a host terminal stored the pH data continuously. The measured results clearly demonstrate the long-term stable and reliable operation of the pH sensor circuit and needle electrodes (Fig. 29.5.6 bottom right). Additionally, the proposed stress monitoring system precisely captured a stressful event on the plants (intentional termination of watering on the 14th day). A clear response was observed in both plants. Approximately 1.5 days after the termination, the pH potential shifted toward alkaline and its gradual recovery toward acidic (a well-known plant reaction) was captured. Interestingly, our system successfully

506 • 2026 IEEE International Solid-State Circuits Conference

979-8-3315-8937-0/26 $31.00 © 2026 IEEE

captured acidity over-drift in the in-vivo soy experiment, which is consistent with measurement results reported in a recent literature [15].

The system specifications are summarized in Fig. 29.5.7 and compared to prior works [4, 5]. Our proposed 0.62g ultra-light-weight individual plant stress sensor tag can be attached standalone onto the plant without any mechanical fixtures or supporting devices. No battery and no wired connections are required for both the power and data links, enabling low-cost, maintenance-free operation. Our non-invasive >30-days long-term stress monitoring experiment on in-vivo plants successfully demonstrated the capability and potential for using the tag to enable future low-cost autonomous smart farming. We anticipate that this system may provide a promising new technology direction that opens up a new frontier in agriculture and plant science.

Acknowledgement:
This work is supported by JST AdCORP Grant Number JPMJKB2307. Th authors are grateful to Shinko Co., Ltd. for providing micro needles, and to MEXT ARIM for RF sputtering processing.

References:
[1] H. Yin *et al.*, "Soil Sensors and Plant Wearables for Smart and Precision Agriculture," *Adv. Mater*, vol. 33, Art. no. 2007764, Apr. 2021.
https://doi.org/10.1002/adma.202007764
[2] M. Srbinovska *et al.*, "Environmental Parameters Monitoring in Precision Agriculture Using Wireless Sensor Networks," *Journal of Cleaner Production*, vol. 88, pp. 297-307, Feb. 2015. https://doi.org/10.1016/j.jclepro.2014.04.036
[3] Y. Inoue, "Satellite- and Drone-Based Remote Sensing of Crops and Soils for Smart Farming – a Review," *Soil Science and Plant Nutrition*, vol. 66, pp. 798–810, Sept. 2019. https://doi.org/10.1080/00380768.2020.1738899
[4] T. Yoshida *et al.*, "Development of Plant Growth Monitoring Sensor to Visualize Ion Dynamics in Plants and its Functional Validation in Long-Term Measurements," *Transducers*, pp. 725-728, Jun. 2023.
[5] A. Ruiz-Gonzalez *et al.*, "In Vivo Sensing of pH in Tomato Plants Using a Low-Cost and Open-Source Device for Precision Agriculture," *Biosensors*, vol. 12, no. 7, Art. no. 447, Feb. 2022. https://doi.org/10.3390/bios12070447
[6] S. Yin *et al.*, "Plant Tattoo Sensor Array for Leaf Relative Water Content, Surface Temperature, and Bioelectric Potential Monitoring," *Adv. Mater. Technol.*, vol. 9, Art. no. 2302073, Apr. 2024. https://doi.org/10.1002/admt.202302073
[7] F. Gentile *et al.*, "A Biomimetic, Biocompatible OECT Sensor for the Real-Time Measurement of Concentration and Saturation of Ions in Plant Sap," *Adv. Electron. Mater.*, vol. 8, Art. no. 2200092, June 2022. https://doi.org/10.1002/aelm.202200092
[8] S. Wilkinson *et al.*, "Effects of Xylem pH on Transpiration from Wild-Type andflacca Tomato Leaves: A Vital Role for Abscisic Acid in Preventing Excessive Water Loss Even from Well-Watered Plants," *Plant Physiology*, Vol. 117, no. 2, pp. 703–709, June 1998. https://doi.org/10.1104/pp.117.2.703
[9] Kesten, Christopher *et al.*, "Pathogen-Induced pH Changes Regulate The Growth-Defense Balance in Plants," *The EMBO Journal*, vol. 38, no. 24, Art. no. e101822, Nov. 2019. https://doi.org/10.15252/embj.2019101822

[10] V.M. Gallegos-Cedillo *et al.*, "Influence of Salinity on Transport of Nitrates and Potassium by Means of The Xylem Sap Content Between Roots and Shoots in Young Tomato Plants," *J. Soil Sci. Plant Nutr.*, vol. 16, no. 4, pp. 991-998, Dec. 2016. https://doi.org/10.4067/S0718-95162016005000072
[11] H. Lyu and A. Babakhani, "A 13.56-MHz –25-dBm-Sensitivity Inductive Power Receiver System-on-a-Chip with a Self-Adaptive Successive Approximation Resonance Compensation Front-End for Ultra-Low-Power Medical Implants," *IEEE TBioCAS*, vol. 15, no. 1, pp. 80-90, Feb. 2021. https://doi.org/10.1109/TBCAS.2020.3047827
[12] J. Pérez-Bailón *et al.*, "Transient-Enhanced Output-Capacitorless CMOS LDO Regulator for Battery-Operated Systems," *ISCAS*, pp. 1-4, May 2017. https://doi.org/10.1109/ISCAS.2017.8050961
[13] Y. Osaki *et al.*, "1.2-V Supply, 100-nW, 1.09-V Bandgap and 0.7-V Supply, 52.5-nW, 0.55-V Subbandgap Reference Circuits for Nanowatt CMOS LSIs," *IEEE JSSC*, vol. 48, no. 6, pp. 1530-1538, June 2013. https://doi.org/10.1109/JSSC.2013.2252523
[14] K. Nakazato *et al.*, "CMOS Cascode Source-Drain Follower for Monolithically Integrated Biosensor Array," *IEICE Trans. on Electronics*, vol. E91-C, no. 9, pp. 1505-1515, Sept. 2008. https://doi.org/10.1093/ietele/e91-c.9.1505
[15] C. M. Geilfus, "The pH of The Apoplast: Dynamic Factor with Functional Impact Under Stress," *Molecular Plant*, vol. 10, no. 11, pp. 1371-1386, Nov. 2017. https://doi.org/10.1016/j.molp.2017.09.018

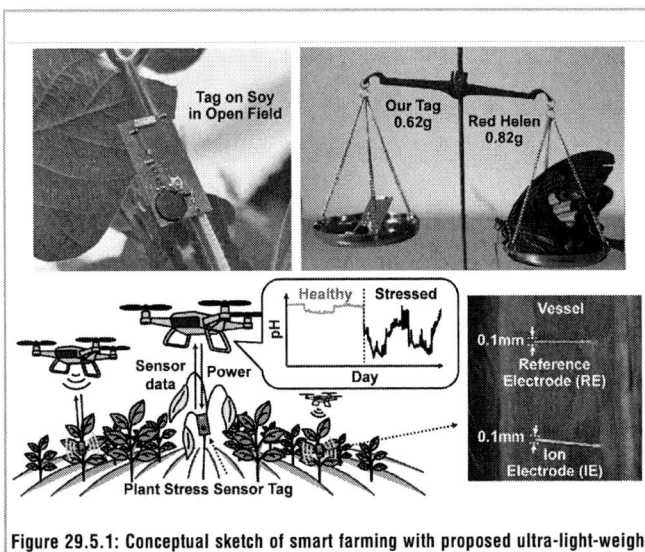

Figure 29.5.1: Conceptual sketch of smart farming with proposed ultra-light-weight individual plant stress sensor tag.

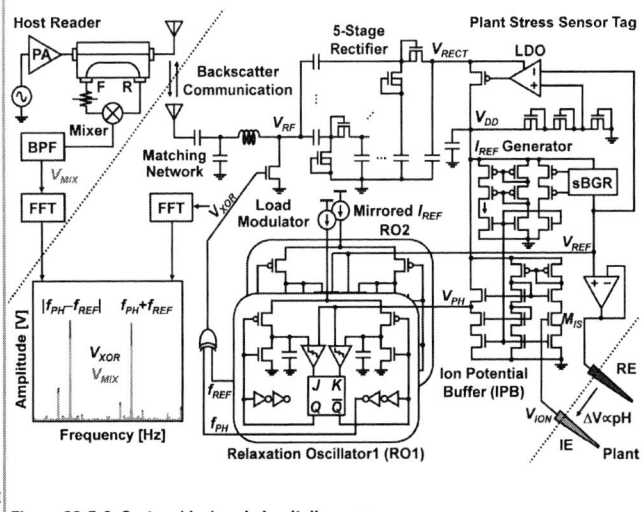

Figure 29.5.2: System block and circuit diagrams.

ISSCC 2026 / SESSION 29 / BIOCHEMICAL SENSORS FOR LIFE SCIENCES AND AGRICULTURE / 29.5

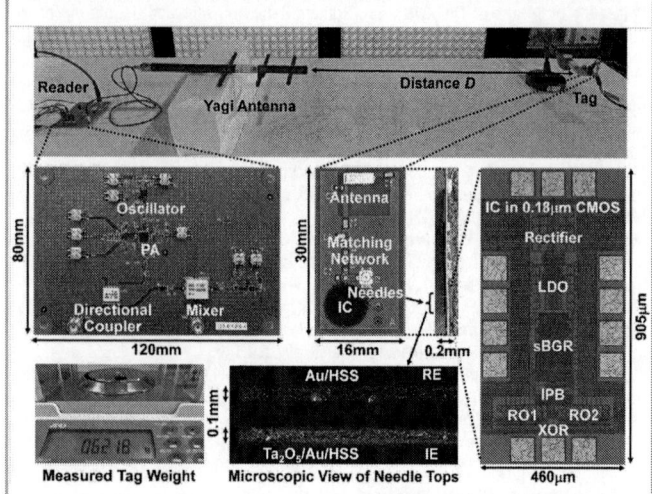

Figure 29.5.3: Test setup, host reader board, tag, needles, and die photos.

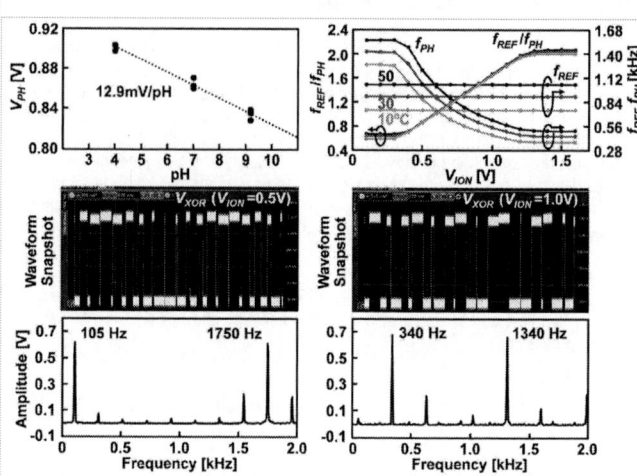

Figure 29.5.4: Measurement results of pH sensor AFE and DBE.

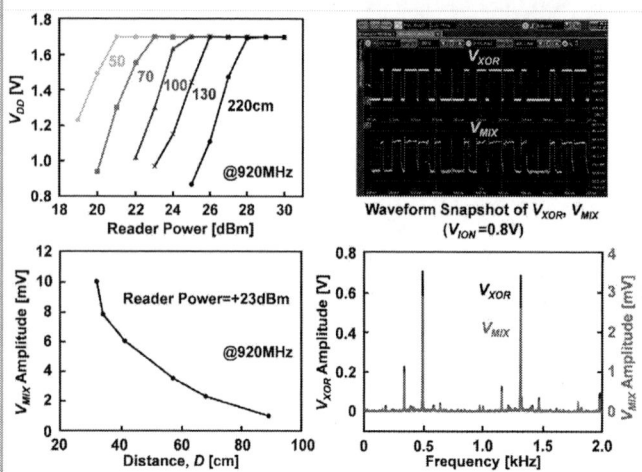

Figure 29.5.5: Measurement results of backscattered power and data links.

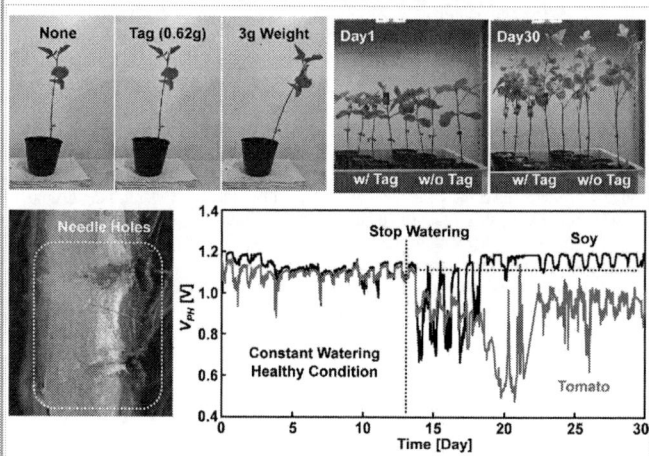

Figure 29.5.6: Measurement results of invasive test on plant and long-term in-vivo plant stress monitoring by proposed sensor tag.

	Transducers'23 [4]	Biosensors'22 [5]	This Work
Sensor Type	ISFET-Based pH Sensor	ISFET-Based pH Sensor	ISFET-Based pH Sensor
Sensor Size	25mm x 42mm	70mm x 57mm	16mm x 30mm
Sensor Weight	>7.5g* (Fixture Required)	50g**	0.62g
Plant Insertion Part	1.76mm x 3.2mm 0.1mm-Thin Chip	0.1mm-Width Thin Film	0.1mm-ϕ Needle
Sensitivity	56.5mV/pH	53.7mV/pH	12.9mV/pH
Power Supply	DC Supply	Consumer Charger	Wireless (No Battery)
Supply Voltage	N/A	5V**	1.7V
Communication	USB Connection to Laptop	WiFi	Wireless (Backscatter)
Long-Term Invasive Testing	No	Yes	Yes
Test Location	Lab. Only	Lab. Only	Lab. and Open Field

*Estimated Weight of Readout Board with Common 1.57mm-Thick FR4 Substrate.
**Catalog Specification of Wio Terminal, Main Readout Electronics Required in This System.

Figure 29.5.7: System specification summary and comparison.

ISSCC 2026 / SESSION 29 / BIOCHEMICAL SENSORS FOR LIFE SCIENCES AND AGRICULTURE / 29.6

29.6 A 65nm CMOS Hydrogel-Based Dual Fluorescence Sensor for Bioavailable Phosphorus Detection

Ting-Yu Cheng*, Raj S. Mukkamala*, Reinaldo E. Alcalde, Mohamadamin Panahandeh, Akshit Agarwal, Julia A. Kornfield, Dianne K. Newman, Azita Emami

California Institute of Technology, Pasadena, CA
*Equally Credited Authors (ECAs)

Abstract

A hydrogel-based single-culture dual-fluorescence sensor is presented for bioavailable phosphorus (P) detection. The sensor achieves 6.6fA sensitivity with a 15.5dB SNR and an output noise of 1.97mV at 182ms integration time. A large dynamic range is enabled through high-precision automatic background calibration with a coarse and fine control loop. The sensor was validated using hydrogels embedded with a bioreporter strain and successfully detected both P limitation and cell viability.

Dynamically sensing bioavailable analytes in the environment is currently a very challenging task. For example, phosphorus (P), an essential nutrient for ecosystem health, is typically monitored on active farms at the frequencies of once every several years. Elevated P levels from agricultural runoff can lead to eutrophication in water bodies, while P deficiency in soil limits crop growth and results in economic losses. Traditional methods for water quality tests and soil tests such as Mehlich-3, Olsen, and Bray-P1 procedures [1,2,3] are often expensive and time-consuming. A miniaturized platform based on CMOS technology offers a promising alternative, enabling real-time, low-cost, and high-sensitivity monitoring of environmental chemical levels (Fig. 29.6.1). Fluorescence (FL)-based sensing has emerged as a reliable method due to its selectivity and the ability to report bioavailable responses. Using FL proteins (FPs) expressed in bacteria is more biocompatible and self-sustaining compared to using strong FL sources such as QDots and FL dyes [4,5] for long-term in situ biosensing applications. Prior work [6] monitored FP production using a triple-well microfluidic chamber that requires a separate reference culture to be placed in the center well. Such design needs precise alignment above the sensor and can lead to cell leakage if deployed in soil or water systems. In this work, we use bioengineered bacteria encapsulated in a hydrogel to achieve dual-FL sensing in single culture for simultaneously detecting cell viability and reporting bioavailable P. The proposed integrated CMOS chip enables single-culture single-ended dual-FL measurements with large dynamic range through a high-precision on-chip automatic background calibration. The detection of weak fluorescence signals in the presence of strong scattered excitation background light, exacerbated by the thin geometry of hydrogels, is a major challenge. A feedback loop with coarse and fine control is designed to eliminate the scattered excitation signals with high resolution, thereby enhancing the sensor's dynamic range. Programmable integration time that spans over three orders of magnitude allows the optimum use of the dynamic range of the sensor as higher signal-to-noise ratio (SNR) can be achieved with longer integration time.

Figure 29.6.2 shows the block diagram of the fluorescence sensor. The system integrates on-chip photodiodes (PDs), capacitive transimpedance amplifiers (CTIA), correlated double sampling (CDS) circuits, automatic background calibration, a SAR ADC, a power management unit (PMU), clock generation, LED drivers, and I²C logic for communication with a nRF BLE module enabling future wireless deployment of the sensor in the soil. The targeted bioreporter strain, *Pseudomonas synxantha 2-79*, is encapsulated in a PEGDA (700 Da)/PEGSH (5,000 Da) hydrogel and engineered to express dual FPs: mScarlet-I (RFP) as a constitutive reporter to serve as a proxy for cell viability, and mNeonGreen (GFP) for reporting on the lack of bioavailable P [7]. Dual-band optical filters are used to spectrally separate excitation and emission wavelengths. The emission filter sits atop the CMOS sensor. Excitation LEDs, powered by on-chip LED drivers, illuminate the sample laterally and the dual FL signals are separated via time-multiplexed LED excitation.

Four N+/PW/DNW/P-sub triple junction PDs, 800um x 600um each, are employed as the sensing elements. These PDs are connected to two sets of CTIA and CDS readout circuits. Low-leakage switches are used to interface the PDs with the CTIA inputs. Additionally, dark PDs are connected to the opposite side of the CTIA differential input to partially cancel the dark photocurrent. The CTIA incorporates a 7-bit binary-weighted capacitive feedback network, with the MSB corresponding to a feedback capacitance of 138.56fF. The sensor exhibits linear response across varying input light power, as well as at different integration times under a fixed input power, as shown in Fig. 29.6.3. The readout circuits achieve a minimum detectable front-end current of 6.6fA (Fig. 29.6.4.). The responsivity of the on-chip PDs is measured to be 9.97mA/W at a wavelength of 600nm. The root-mean-square output noise of the sensor under dark conditions is 1.97mV at an integration time of 182ms. A SAR ADC [8] is used to digitize the CDS output, and the resulting data is transmitted over I²C for logging and further analysis.

The emission intensity of fluorescence proteins is governed by equation (1): $I_{FL} = QY \times I_{EX} \times (1 - 10^{-EC \cdot C \cdot L})$, where QY is the quantum yield of the target FP, EC is the extinction coefficient, C is the concentration of the fluorophores, and L is the optical path length. For a 10pM mScarlet-I sample with a 1mm optical path length for miniaturized environmental sensors, the excitation-to-emission intensity ratio is greater than 10^6. In addition, since the excitation background varies depending on environmental conditions, concentration of cells, and material being used in the encapsulates, the background signal is difficult to precisely predict. The background mainly arises from stray and reflected LED excitation light, as well as scattered light from interaction with the samples, which can easily saturate the FL sensor. To address this, two 9-bit current DACs were implemented for coarse and fine tuning, with maximum output currents of 93pA and 1.65pA, respectively. The outputs of both DACs are summed to generate the VCCS compensation current. The output current range can be further adjusted by selecting one of four available gate voltages (V_{G1}), ranging from 1.45-to-1.6V for coarse control and 1.5-to-1.65V for fine control, to account for process variations. As characterized in Fig. 29.6.3, the measured waveforms using the FL sensor show good agreement with simulation results. Two operating modes are designed for VCCS control: (1) automatic mode and (2) manual mode. In automatic mode, a comparator is used to detect the polarity of the CDS differential output. When the output is positive, the 9-bit counter steps up until the polarity flips. The control logic next reverts the previous count, holds the value for the coarse-tuning current DAC, and starts the next counter for fine-tuning current DAC until the polarity flips again. Figure 29.6.4 demonstrates the calibration process conducted prior to FL measurements using a hydrogel embedded with a dual reporter strain. Calibration was activated at the beginning of the measurement before the hydrogel started to produce significant FL signal. Calibration was performed sequentially for each set of readout circuits with the corresponding excitation LED illuminated. The noticeable excitation background occupying most of the dynamic range was reduced gradually as the automatic closed-loop circuitry began the calibration process, until the effect of background was minimized. The dual reporter FL responses were recorded next. The simulated worst-case VCCS resolutions, which occur at the steepest regions of the characteristic curves, are 500fA for coarse tuning and 22fA for fine tuning. The manual mode is done by sending external commands via the nRF BLE module to overwrite the 9-bit control registers through I²C communication. To further suppress stray light, the chip employs metal shielding around all photodiodes using metal layers M1–M7. This prevents sidewall light leakage through the 0.3mm thick die.

Band-gap reference circuits are used to generate stable reference voltages and currents across the chip. Multiple LDOs provide regulated supply voltages of 1.8V, 1.2V, 1V and 0.5V to power the core FL sensing circuits, readout chain, and all the other on-chip circuitry. Because the absolute FL intensity is primarily estimated through theoretical calculations using equation (1), and can vary due to sample geometry, optical coupling efficiency and biological factors such as cell growth, the integration time is highly programmable. The clock generation circuits support integration times from 6ms to 24.6s, enabling signal detection across a wide dynamic range. A current-starved ring oscillator generates a base frequency of 340kHz. This clock is used for generating CLK1-CLK4 through a clock phase generator with unit delay elements to create timing difference in controlling the readout circuits. CLK5 with a 90-degree phase shift is used for digital calibration logic, ensuring it queries the CDS output only after charge redistribution is fully completed.

The sensor was tested with bacterial cells contained within hydrogels. The hydrogels were prepared by encapsulating the dual reporter strain engineered to express GFP under P-limited conditions and RFP constitutively to indicate cell viability and growth. The hydrogels were transferred onto 25mm × 25mm No. 1 coverslips for sensor measurements. Two sample groups were tested 24 hours after gel preparation: (1) hydrogels preloaded with 0.5mM phosphate (P-limited) and (2) those preloaded with 5mM phosphate (P-replete). The sensor successfully detected P limitation by distinguishing between the two conditions (Fig. 29.6.5. (top)). In order to verify the results, wild-type (WT) samples, which contain bacteria without FPs, were also measured, and net FL signals were calculated by subtracting the signals of WT samples from the bioreporter samples. A clear signal difference was seen in the green channel, with P-limited samples exhibiting stronger signals than the P-replete samples. In contrast, the red channel showed similar signal levels for both groups, indicating comparable cell loading and growth. The sensor can also monitor cell growth and nutrient response over time under aqueous environments. For these experiments, a well filled with phosphate solution, matching the gel's phosphate concentration, was placed on each hydrogel to maintain a controlled nutrient environment. According to prior characterization [7], the bioreporter activates GFP expression when the phosphate concentration is consumed below 50uM by cells. The time resolved response (Fig. 29.6.5. (bottom)) showed a GFP signal induction in the P-limited sample after 10 hours of experiment, indicating phosphate depletion, while the P-replete counterpart showed no induction. RFP signals were constitutively expressed in both samples, suggesting active cell viability.

508 • 2026 IEEE International Solid-State Circuits Conference

979-8-3315-8937-0/26 $31.00 © 2026 IEEE

The comparison table and the die micrograph implemented in 65nm CMOS are shown in Fig. 29.6.6 and Fig. 29.6.7, respectively. To the best of our knowledge, this is the first FL chip that detects dual-FL signals from a single-culture bioreporter for monitoring bioavailable P using hydrogel encapsulates. The small size, large dynamic range, low power, and hydrogel-based design enables miniaturized wireless sensors for applications in environmental sensing and precision agriculture through real-time monitoring of analyte bioavailability.

Acknowledgement:
This work was supported by the Resnick Sustainability Institute at Caltech and the Institute for Collaborative Biotechnologies. The authors would like to thank them for their generous funding that made this research possible.

References:
[1] H. Zhang *et al.*, "Interlaboratory validation of the Mehlich 3 method for extraction of plant-available phosphorus". *J AOAC Int.*, 2009 Jan-Feb;92(1):91-102. https://doi.org/10.1093/jaoac/92.1.91
[2] M.d.C. Horta, J. Torrent, "The Olsen P method as an agronomic and environmental test for predicting phosphate release from acid soils". *Nutr Cycl Agroecosyst* 77, 283–292 (2007). https://doi.org/10.1007/s10705-006-9066-2
[3] A.M. Ebeling, L.G. Bundy, A.W. Kittell and D.D. Ebeling, "Evaluating the Bray P1 Test on Alkaline, Calcareous Soils". *Soil Sci. Soc. Am. J.*, 72: 985-991, 2008. https://doi.org/10.2136/sssaj2006.0347
[4] C. Zhu *et al.*, "An Ingestible Pill with CMOS Fluorescence Sensor Array, Bi-Directional Wireless Interface and Packaged Optics for in-Vivo Bio-Molecular Sensing," *IEEE Transactions on Biomedical Circuits and Systems*, vol. 17, no. 2, pp. 257-272, April 2023. http://doi.org/10.1109/TBCAS.2023.3244570
[5] M. Roschelle *et al.*, "A Wireless, Multicolor Fluorescence Image Sensor Implant for Real-Time Monitoring in Cancer Therapy," *IEEE Journal of Solid-State Circuits*, vol. 59, no. 11, pp. 3580-3598, Nov. 2024. http://doi.org/10.1109/JSSC.2024.3435736
[6] F. Aghlmand *et al.*, "A 65-nm CMOS Fluorescence Sensor for Dynamic Monitoring of Living Cells," *IEEE Journal of Solid-State Circuits*, vol. 58, no. 11, pp. 3003-3019, Nov. 2023. http://doi.org/10.1109/JSSC.2023.3308853
[7] E.M. Larsson, R.M. Murray, D.K. Newman, "Engineering the Soil Bacterium Pseudomonas synxantha 2–79 into a Ratiometric Bioreporter for Phosphorus Limitation". *ACS Synth. Biol.* 2024, 13, 1, 384–393. https://doi.org/10.1101/2023.10.20.563366
[8] S. Sharma *et al.*, "A Monolithic 3D Magnetic Sensor in 65nm CMOS with <10μTrms Noise and 14.8μW Power," *IEEE Custom Integrated Circuits Conference (CICC)*, 2023, pp. 1-2. http://doi.org/10.1109/CICC57935.2023.10121313
[9] A. Manickam *et al.*, "A Fully Integrated CMOS Fluorescence Biochip for DNA and RNA Testing," *IEEE Journal of Solid-State Circuits*, vol. 52, no. 11, pp. 2857-2870, Nov. 2017. http://doi.org/10.1109/JSSC.2017.2754363
[10] Q. Liu *et al.*, "A Threshold-Based Bioluminescence Detector With a CMOS-Integrated Photodiode Array in 65 nm for a Multi-Diagnostic Ingestible Capsule," *IEEE Journal of Solid-State Circuits*, vol. 58, no. 3, pp. 838-851, March 2023. http://doi.org/10.1109/JSSC.2022.3197465

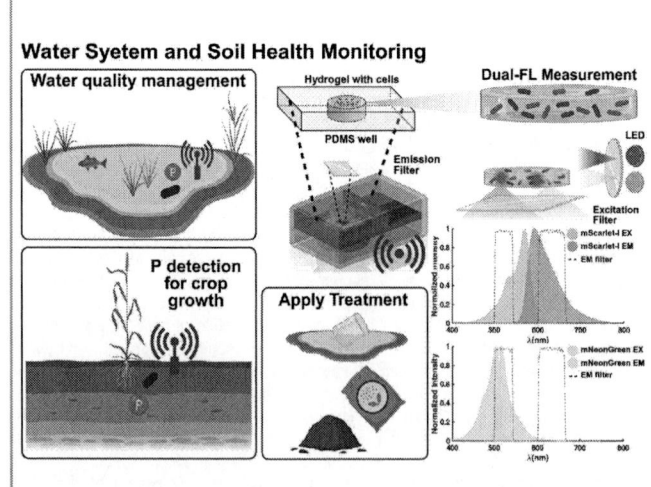

Figure 29.6.1: Integrated dual fluorescence sensor with bioreporter encapsulated in hydrogels and its applications.

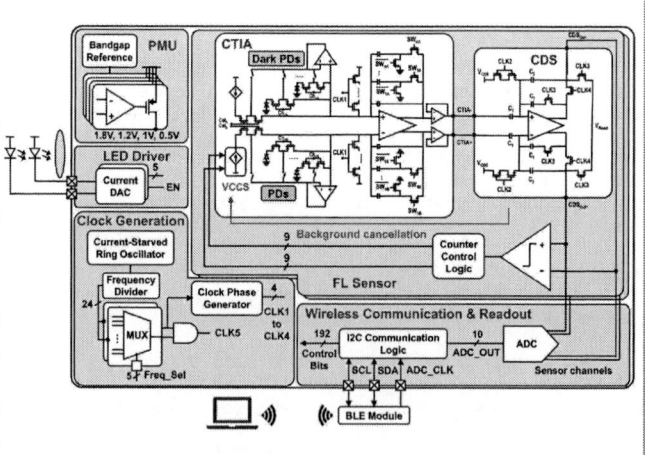

Figure 29.6.2: Block diagram of the wireless dual fluorescence sensor.

ISSCC 2026 / SESSION 29 / BIOCHEMICAL SENSORS FOR LIFE SCIENCES AND AGRICULTURE / 29.6

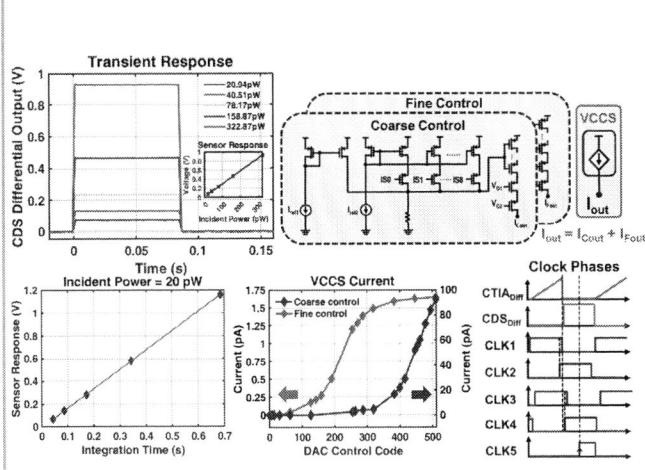

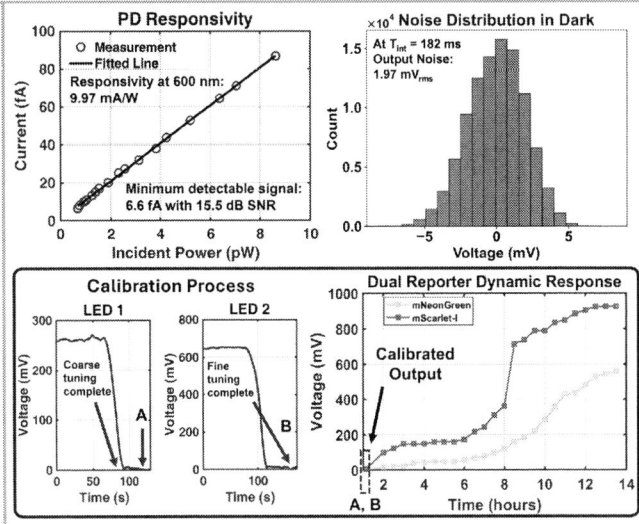

Figure 29.6.3: Sensor responses, VCCS circuit and current, and clock phases for sensor readout.

Figure 29.6.4: Photodiode responsivity, noise under dark condition, and background calibration process.

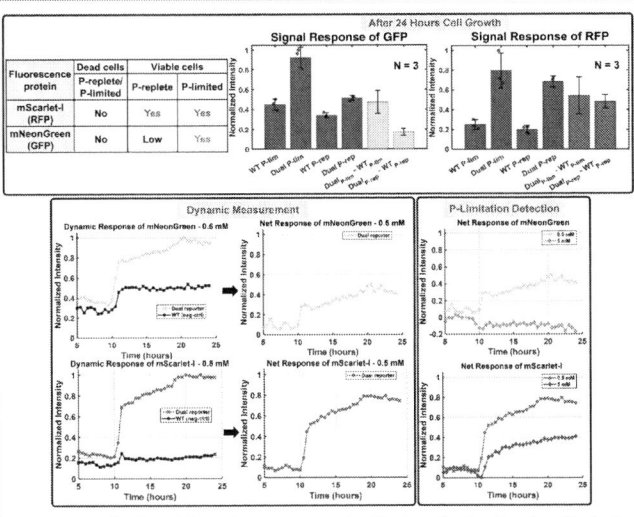

Figure 29.6.5: Static and dynamic measurements of hydrogels showing signal induction under P-limited condition.

Metric	JSSC 2017 [9]	JSSC 2022 [10]	TBioCAS 2023 [4]	JSSC 2023 [6]	JSSC 2024 [5]	This Work
Technology	250 nm CMOS	65 nm CMOS	65 nm CMOS	65 nm CMOS	180 nm CMOS	65 nm CMOS
Application	DNA/RNA testing	Biochemical detection	Bio-molecular sensing	Dynamic cell monitoring	Imaging of treatment response	Bioavailable P detection
Detector Type	Fluorescence	Bioluminescence	Fluorescence	Fluorescence	Fluorescence	Fluorescence
Reporter Type	DNA/RNA	In vitro bioengineered bacterial sensors	In vitro DNA assay	Fluorescence proteins in E. coli bacteria	Ex vivo immune cells	Rhizosphere bacteria encapsulated in hydrogels
Excitation/Emission Peak (nm)	490/~580	N/A	405/800 (QDot800)	387/703 (miRFP), 437/573 (LSSmOrange), 586/610 (mCherry), ~515/~530 (sfYFP)	455/500 (FAM), 650/670 (Cy5), 785/800 (beads)	505/517 (mNeonGreen), 570/594 (mScarlet-I)
Measurement Type	Static	Static / Dynamic	Static	Static/Dynamic	Static	Static / Dynamic
Excitation Source	LED	N/A	UV LED	SMD LED	Laser diode	SMD LED
Supply Voltage	2.5 V	2.5 V	N/R	3.3 V	5.5 V#	2 V
Chip Size	63 mm²	14.9 mm²	4 mm²	3 mm²	12.5 mm²	5 mm²
Output Noise*	N/R	N/R	1 - 2 mV (Photon shot noise), 0.4 mV (CTIA)	3.846 mVrms	5.4 mV	1.97 mVrms
SNR	> 20 dB	N/R	1	18.3 dB	20 dB**	15.5 dB
Sensor Sensitivity	~10fA	59 fA	40 dots/μm²	1.05 fA	N/R	6.6 fA
On-Chip Excitation Background Removal	No	N/A	No	Yes**	No	Yes
Integration Time	N/R	25.8 s	N/R	1 s	0 ms – 248 ms	6 ms – 24.6 s
Power	118 mW	16.3 uW	N/R	7.05 mW$	2.09 mW	1.66 mW

N/A: Not applicable, N/R: Not reported, * Under dark condition, ** T_{int} = 98 ms
Remains at 5.5 V throughout imaging and drops to 3.5 V during readout. ## Manual control
$ One row: include CTIA + 4 buffers + 1/3 of a shared CDS for 4 pixels

Figure 29.6.6: Comparison table of state-of-the-art fluorescence sensors.

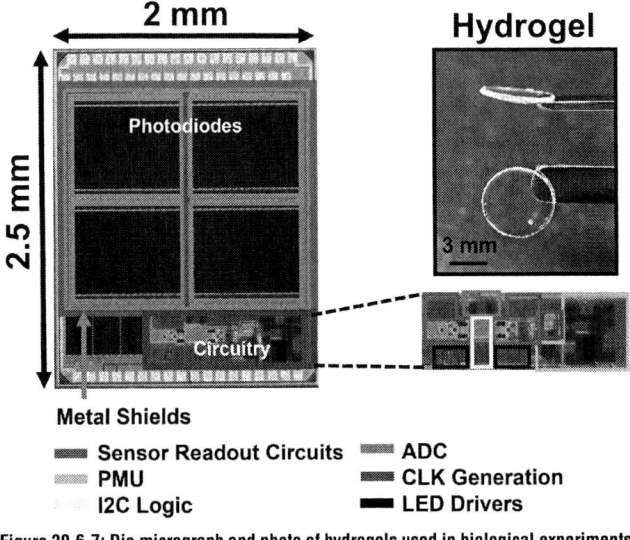

Figure 29.6.7: Die micrograph and photo of hydrogels used in biological experiments.

• 2026 IEEE International Solid-State Circuits Conference

Session 30 Overview: *Compute-in-Memory*
MEMORY SUBCOMMITTEE

Session Chair: Vita Pi-Ho Hu
National Taiwan University
Taipei, Taiwan

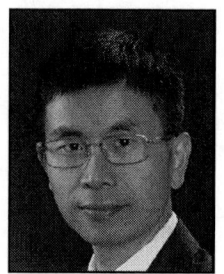

Session Co-Chair: Xueqing Li
Tsinghua University
Beijing, China

Compute-in-memory (CIM) continues to advance as a critical technology to overcome the memory–compute bottleneck in AI edge systems, enabling massive data throughput with ultra-low energy consumption. This session presents nine papers that demonstrate innovations across SRAM, gain-cell, ReRAM, charge-trap, and ferroelectric NAND technologies, achieving record-breaking energy efficiency and area efficiency. They collectively showcase breakthroughs in precision adaptability, hybrid analog-digital computing, 3D near-memory integration, and versatile data representations, driving CIM from device-level design to AI-edge deployment.

8:00 AM

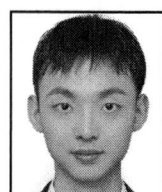

30.1 A 28nm 127.54TFLOPS/W MXFP6 and 117.42TFLOPS/W MXFP8 Compute-in-Memory Macro with Adaptive-Preserved-Bit-Width and Serial-Dual-Bit-Sliding Schemes

Xing Wang, Southeast University, Nanjing, China, National Center of Technology Innovation for EDA, Nanjing, China

In Paper 30.1, Southeast University introduces the first adaptive preserved-bit-width MXFP6/8 CIM macro, achieving 127.54TFLOPS/W and 4.44TFLOPS/mm² in MXFP6/6 mode through dual-bit-sliding and twin-stage precision-allocation schemes.

8:25 AM

30.2 A 12nm 4Mb 104.56-to-137.75TFLOPS/W Charge-Trap Transistor-Based Computing-in-Memory Macro Using Analog-Predict-Digital-Compute for AI Edge Devices

Junzhe Shen, Institute of Microelectronics of the Chinese Academy of Sciences, Beijing, China
University of Chinese Academy of Sciences, Beijing, China

In Paper 30.2, Chinese Academy of Sciences demonstrates a 12nm 4Mb charge-trap transistor-based CIM macro, supporting INT/FP4 MACs with an analog-predict/digital-compute scheme, achieving an energy efficiency of 137.75TFLOPS/W.

8:50 AM

30.3 A 22nm 96Mb 50.6-to-90.2TFLOPS/W Non-Linear MLC ReRAM CIM Macro with High-Retention for Mamba/Transformer/CNN

Hung-Hsi Hsu, National Tsing Hua University, Hsinchu, Taiwan

In Paper 30.3, National Tsing Hua University and TSMC present a 22nm 96Mb non-linear MLC ReRAM CIM macro featuring reconfigurable compute modes and high-retention capability, reaching 50.6–90.2TFLOPS/W (BF16) with a 79.17% lower accuracy loss at 10 years.

ISSCC 2026 / February 18, 2026 / 8:00 AM

9:15 AM

30.4 A 28nm 106.85TOPS/W and 77.68TFLOPS/W CIM Macro with Stage-Wise-Enabled Lossless Compressors Based on Sign-Bit-Embedded Transition-Counting-Lines for Edge-AI Devices

Yunlong Liu, Xidian University, Xi'an, China

In Paper 30.4, Xidian University presents a 28nm bit-parallel digital CIM macro with stage-wise-enabled lossless compressors based on sign-bit-embedded transition-counting lines, achieving 106.85TOPS/W (INT8) and 77.68TFLOPS/W (BF16).

10:05 AM

30.5 A 16nm 72kb 120.5TFLOPS/W Versatile-Format Dual-Representation Gain-Cell CIM Macro for General Purpose AI Tasks

Jen-Chun Tien, National Tsing Hua University, Hsinchu, Taiwan

In Paper 30.5, National Tsing Hua University and TSMC present a 16nm versatile-format gain-cell CIM macro supporting MX, FP, LNS, and INT modes with dual 2's-complement/sign-magnitude representation, achieving 120.5TFLOPS/W and 3.18TOPS/mm² with <0.01% accuracy loss.

10:30 AM

30.6 A 16Mb 166.8TOPS/W Near-Memory Phase-Domain-Computing Ferroelectric NAND Flash for Approximate Nearest Neighbor Search on Edge Devices

Weizeng Li, Institute of Microelectronics of the Chinese Academy of Sciences, Beijing, China,
University of Chinese Academy of Sciences, Beijing, China

In Paper 30.6, Chinese Academy of Sciences reports the first 16Mb near-memory phase-domain computing FeNAND, performing 512 cosine-similarity searches between 256-dimensional vectors in one operation, achieving 166.8TOPS/W and >12.8× latency reduction.

10:55 AM

30.7 A 1.2GHz 12.77GB/s/mm² 3D Two-DRAM-One-Logic Process-Near-Memory Chip for Edge LLM Applications

Yue Cao, Fudan University, Shanghai, China, Institute of Microelectronics of the Chinese Academy of Sciences, Beijing, China
Zhangjiang Laboratory, Shanghai, China

In Paper 30.7, Fudan University introduces a 3D two-DRAM one-logic process-near-memory chip, achieving 1.2GHz, 12.77GB/s/mm² bandwidth, and 0.67pJ/b efficiency, reducing memory access latency by 93% and GEMM execution time by 98%.

11:20 AM

30.8 A 16nm, 1Mb, 1-to-8b-Configurable 444.21TOPS/W Fully Digital SRAM Compute-In-Memory Macro for Hybrid SNN-CNN Edge Computing

Yao-Kai Yeh, National Tsing Hua University, Hsinchu, Taiwan

In Paper 30.8, National Tsing Hua University and ITRI present the first fully-digital 16nm SRAM CIM macro for hybrid SNN–CNN computing, achieving 444.21TOPS/W (1bIN–8bW–14bOUT, SNN) and 62.84TOPS/W (8bIN–8bW–22bOUT, CNN).

30

11:45 AM

30.9 A 147TOPS/W, 250TOPS/mm², Fully Synthesizable, Digital Compute-in-Memory Accelerator Supporting INT8×INT8 with Zero-Point Quantization in Intel 18A Technology

Amit Agarwal, Intel, Hillsboro, OR

In Paper 30.9, Intel showcases a fully-synthesizable digital-CiM accelerator in 18A technology supporting INT8xINT8 zero-point quantization, operating at 2.62GHz, delivering 147TOPS/W and 250TOPS/mm² area efficiency at 25°C.

DIGEST OF TECHNICAL PAPERS • 511

979-8-3315-8937-0/26 $31.00 © 2026 IEEE

ISSCC 2026 / SESSION 30 / COMPUTE-IN-MEMORY / 30.1

30.1 A 28nm 127.54TFLOPS/W MXFP6 and 117.42TFLOPS/W MXFP8 Compute-in-Memory Macro with Adaptive-Preserved-Bit-Width and Serial-Dual-Bit-Sliding Schemes

Xing Wang[1,2], Yucheng Du[1], Tianhui Jiao[1], Defa Wu[1], Xi Chen[1], Miaoyu Tang[1], Yi Yang[1], Zhichao Liu[1], An Guo[1], Gaoming Fu[3], Peng Li[3], Jun Dong[3], Bo Liu[1], Xinning Liu[1], Weiwei Shan[1], Hao Cai[1], Guangyu Sun[4], Lin Tong[3], Jun Yang[1,2], Xin Si[1]

[1]Southeast University, Nanjing, China, [2]National Center of Technology Innovation for EDA, Nanjing, China, [3]Xiaomi, Beijing, China, [4]Peking University, Beijing, China

Abstract

Conventional FP-CIMs suffer from fixed preserved bit-width (PBW), limiting their adaptability and efficiency. This work proposes the first MXFP-CIM macro enabling wide-range adaptive PBW, featuring: (1) A serial dual-bit-sliding scheme; (2) A harmless data mapping scheme with a hierarchical hidden-bit decoder; (3) An adjustable-PBW MXFP-MAC circuit via twin-stage allocation. The 28nm MXFP-CIM macro achieves a peak energy efficiency of 127.54TFLOPS/W in the MXFP6/6 mode.

With the rapid advancement of artificial intelligence (AI), neural network model sizes have grown exponentially, placing increasing demands on computational bandwidth, memory capacity, and energy efficiency. To alleviate these challenges, low-bit floating-point (FP) formats such as MXFP6 and MXFP8 have emerged as promising alternatives [1,2]. Notably, Deepseek-V3 [3] demonstrated successful FP8-based training and large-scale commercial deployment, highlighting the strong potential of low-bit FP representations. However, conventional FP computing architectures rely on fixed preserved bit-width (PBW), constraining both flexibility and efficiency when adapting to diverse model and layer requirements. This necessitates floating-point compute-in-memory (FP-CIM) solutions that: (1) Support low-bit FP formats, and (2) Offer adjustable PBW.

While existing FP-CIM designs [4-12] have demonstrated impressive performance and energy efficiency, they suffer from three critical challenges (Fig. 30.1.1): (1) Suboptimal arithmetic efficiency in input-alignment schemes [6,7,9,11] due to non-vertical product truncation, which wastes compute resources; (2) Memory capacity underutilization during reconfiguration from variable mantissa/exponent bit-widths, as fixed-precision architectures cannot adapt to dynamic data requirements; (3) Limited variable PBW support due to hardware/software constraints, restricting flexibility for diverse workloads. This work proposes the first MXFP-CIM macro enabling wide-range adaptive PBW, featuring: (1) A serial dual-bit-sliding (SDBS) scheme that enhances FP multiply-accumulate (MAC) computation efficiency by eliminating redundant operations; (2) A harmless data mapping (HDM) scheme with a hierarchical hidden-bit decoder for MXFP6/8, which fully utilizes memory capacity by dynamically adjusting to variable bit-widths; (3) An adjustable-PBW MXFP-MAC circuit via twin-stage allocation, which balances accuracy and efficiency by adaptively scaling precision for different layers/operations. Fabricated in 28nm CMOS technology, a 54kb (24×18×128b) adaptive-PBW MXFP-CIM macro supports MXFP6 and MXFP8 MAC operations for the first time. In the MXFP6/6 mode, it achieves a peak energy efficiency of 127.54TFLOPS/W and a peak area efficiency of 4.44TFLOPS/mm², setting a new benchmark for adaptive-precision FP-CIM designs.

Figure 30.1.2 illustrates the overall architecture, configuration table, and computation flow of the proposed MXFP-CIM macro. The design supports three MXFP6/8 data-type combinations with 3b to 17/23/27b wide-range adaptive preserved bit-width, enabling optimization for workloads spanning efficient inference (low PBW) to high-precision training (high PBW). The architecture comprises a controller, global FP-decoder, bit-width allocator, shared-scale buffer, accumulator, normalizer, input buffer, 24 FP-CIM banks, and other peripherals. Each bank incorporates an 18-row×128-column split-WL 6T-SRAM array alongside local decoders, an exponent-processing unit (EPU), and a mantissa MAC unit. The proposed SDBS FP-MAC flow slides the mantissa input (M_{IN}) 2 bits/cycle (with sign extension) relative to the mantissa weight (M_W), decomposing the multiplication operation across cycles while maintaining vertical truncation. Parallel accumulation aligns products based on exponent sums (E_{SUM}): a higher E_{SUM} triggers earlier M_{IN} sliding to prioritize critical bits. Critically, PBW scales with computation cycles (2-to-14 cycles), enabling dynamic precision via the bit-width allocator.

Figure 30.1.3 illustrates the SDBS FP-MAC flow and implementations of the DBSU, product-compressor (PC), and adder-tree. In the SDBS FP-MAC flow, the exponent adders (EAs) compute $E_{SUM}[4:0]$, the maximum finder (MF) selects the maximum $E_{SUM}[4:1]$, and the sliding destination tracer (SDT) generates sliding controls (SCs). Guided by $E_{SUM}[0]$ and SC, the dual-bit-sliding unit (DBSU) shifts M_{IN}. PCs then compute $M_{IN} \times M_W$ bit-products and compress them into 4-bit values (CVs), which are accumulated by a channel-wise adder tree and normalized in the accumulator & normalizer (A&N) to yield the final mantissa results. The DBSU employs clock gating, 1b shifter, M_{IN}-storage module, and sliding-result (SR) module (8T-C²MOS dynamic registers for area/energy efficiency). At the initial stage, M_{IN}-storage module and SR module are set to $M_{IN}[4:0]$ and the sign bit ($M_{IN}[4]$), respectively. When SC switches from 0 to 1, CLK is enabled, and values in the registers are shifted by 2b per cycle. The 1b shifter determines the connection between the M_{IN}-storage module and SR module based on the LSB value of E_{SUM} (odd/even parity). For the product-compressor, the design uses only six full-adders (FAs) to compress the bit-products, reducing adder tree input bit-width by 42.9% compared to state-of-the-art input-alignment schemes [6,7,9,11]. This reduction in hardware complexity translates to 35.9% less area

overhead and 40.0% lower power consumption. A waveform is presented to show an example of the SDBS FP-MAC, where its adjustable computation cycle provides circuit-level support for adaptive PBW.

Figure 30.1.4 (top) illustrates the harmless data mapping scheme and the hierarchical hidden-bit decoder. In MXFP8 mode, nine MXFP8 (E4M3) numbers are mapped to each 18-row×4-column sub-array, where even rows store sign bits and 3-bit mantissas, while odd rows store 4b exponents. In MXFP6 (E2M3) mode, twelve MXFP6 numbers are mapped per sub-array, with each odd row storing two MXFP6 exponents to avoid memory waste. To support both normal and subnormal numbers in MXFP formats, a hidden-bit decoder is required. In contrast to existing direct hidden-bit decoders for FP-weights [7], this work proposes a hierarchical decoder comprising a global pre-decoder and 24 local decoders. During weight writing period, the global decoder first pre-decodes the hidden-bit and converts weights from sign-magnitude (SM) format to 2's complement (2C) format ($\{S_W, M_W[3:0]\}$); subsequently, $M_W[3]$ is discarded and $\{S_W, M_W[2:0]\}$ is written to the SRAM array. During computation, each local decoder restores $M_W[3]$ from $\{S_W, M_W[2:0]\}$ and E_W. This approach eliminates the SRAM capacity overhead for storing $M_W[3]$ and reduces local decoder area. Compared to direct decoders [7], the proposed hierarchical decoder achieves 63.1% area reduction.

Figure 30.1.4 (bottom) illustrates the proposed twin-stage allocation scheme, aiming to efficiently leverage adjustable PBW FP-MAC circuits. This scheme searches for the optimal preserved bit-width at both layer and group stages to maximize PPA benefits without sacrificing accuracy. In the layer-wise coarse-grained allocation (CGA), a lookup table construction flow iterates through each layer to build a small-size layer-wise allocation table offline, which can be stored in the bit-width allocator. The layer-wise CGA ultimately achieves 14.8% average PBW reduction without accuracy loss on LLaMA-3.1 8B. To further enhance efficiency, group-wise fine-grained allocation (FGA) is introduced. Under the MXFP format, each group contains 32 numbers with an E8M0 shared scale (SS). Since an accumulation vector comprises multiple groups, larger SS values indicate higher contribution to accumulation results and thus require more CIM computation cycles, whereas smaller SS values permit cycle reduction due to lower significance. Consequently, the FGA dynamically allocates group-wise PBW based on ΔSS, yielding 29.3% average PBW reduction. The proposed MXFP-CIM supports three configurations: (1) fixed preserved bit-width; (2) layer-wise CGA; or (3) layer-wise CGA + group-wise FGA. A bit-width allocator and a configurable A&N are designed to support wide-range variable PBW.

Figure 30.1.5 presents the simulation results of the proposed MXFP-CIM macro. Compared to bit-parallel [8,10] and bit-serial [11] alignment schemes, the proposed bit-sliding alignment scheme with C²MOS registers achieves area reductions of 3.85× and 1.53×, and energy reductions of 1.44× and 1.43×, respectively. Owing to the twin-stage allocation scheme, the proposed MXFP-CIM achieves the sweet point of accuracy-performance trade-off: it delivers 1.20× to 1.43× higher throughput than global fixed 19b PBW with negligible accuracy loss on LLaMA-3.1 8B; compared to global fixed 11b PBW, it achieves 1.49× accuracy improvement at comparable throughput. As preserved bit-width increases, mean relative error (MRE) monotonically decreases while PPA overhead grows. This work supports a wide-range adjustable PBW, covering simple/complex inference and training applications. For BERT-base, only 3b-to-9b PBW maintains < 1% accuracy loss; for LLaMA-3.1 8B, 11b-to-17b PBW achieves near-lossless accuracy; 19b-to-27b high-precision range optimally supports training applications, and it slightly outperforms BF16 format when training ResNet-18 on CIFAR-100. Compared to existing FP-MAC schemes, the superiority of this work lies in two aspects: (1) higher energy-area efficiency under identical MRE; (2) wide-range adaptive PBW that efficiently adapts to diverse application tasks.

Figure 30.1.6 presents measured results of the adaptive-PBW MXFP-CIM macro, fabricated using 28nm CMOS technology. In MXFP8/8 mode, the peak energy efficiency and area efficiency are 117.42TFLOPS/W at 0.55V and 4.44TFLOPS/mm² at 0.9V, respectively. The shmoo plot confirms that the MXFP-CIM can operate under a 1.2ns clock cycle at 0.9V. Compared to prior SRAM-based FP-CIM designs [5-11] targeting CNN and Transformer inference, this work achieves a 2.8× to 123× improvement in the figure-of-merit (FoM=average energy efficiency×normalized area efficiency×memory density), while

512 • 2026 IEEE International Solid-State Circuits Conference

979-8-3315-8937-0/26 $31.00 © 2026 IEEE

preserving near-lossless end-to-end model accuracy. In addition, by leveraging the proposed wide-range adaptive PBW scheme, this architecture supports not only complex inference workloads (e.g., LLaMA-3.1 8B) but also training tasks (e.g., ResNet-18), delivering of 49.53TFLOPS/W for inference and 19.11TFLOPS/W for training. Figure 30.1.7 presents the die micrograph and performance summary table.

Acknowledgement:
This work was supported in part by the National Natural Science Foundation of China under Grant 62522403, 92264203, 92464202, 92464302, 6232B2022, and 62204036; in part by the National Science and Technology Major Project under Grant 2022ZD0118902, 2023YFB4405100, 2023YFB4405102; in part by the Key Research and Development Program of Jiangsu Province under Grant BE2023020-1; and in part by the Fundamental Research Funds for the Central Universities.

References:
[1] B.D. Rouhani et al., "Microscaling Data Formats for Deep Learning," *arXiv preprint*, pp. 1-9, Oct. 2023. https://doi.org/10.48550/arXiv.2310.10537
[2] S.-Y. Liu., "LLM-FP4: 4-Bit Floating-Point Quantized Transformers," *arXiv preprint*, pp. 1-14, Oct. 2023. https://doi.org/10.48550/arXiv.2310.16836
[3] DeepSeek-AI et al., "DeepSeek-V3 Technical Report," *arXiv preprint*, pp. 1-53, Feb. 2025. https://doi.org/10.48550/arXiv.2412.19437
[4] S. Lee et al., "A 1ynm 1.25V 8Gb, 16Gb/s/pin GDDR6-based Accelerator-in-Memory supporting 1TFLOPS MAC Operation and Various Activation Functions for Deep-Learning Applications," *ISSCC*, pp. 176-77, Feb. 2022. https://doi.org/10.1109/ISSCC42614.2022.9731711
[5] A. Guo et al., "A 28nm 64-kb 31.6-TFLOPS/W Digital-Domain Floating-Point-Computing-Unit and Double-Bit 6T-SRAM Computing-in-Memory Macro for Floating-Point CNNs," *ISSCC*, pp. 128-129, Feb. 2023. https://doi.org/10.1109/ISSCC42615.2023.10067260
[6] P.-C. Wu et al., "A 22nm 832Kb Hybrid-Domain Floating-Point SRAM In-Memory-Compute Macro with 16.2-70.2TFLOPS/W for High-Accuracy AI-Edge Devices," *ISSCC*, pp. 126-127, Feb. 2023. https://doi.org/10.1109/ISSCC42615.2023.10067527
[7] W.-S. Khwa et al., "A 16nm 96Kb Integer/Floating-Point Dual-Mode-Gain-Cell-Computing-in-Memory Macro Achieving 73.3-163.3TOPS/W and 33.2-91.2TFLOPS/W for AI-Edge Devices," *ISSCC*, pp. 568-569, Feb. 2024. https://doi.org/10.1109/ISSCC49657.2024.10454447
[8] Y. Yuan et al., "A 28nm 72.12TFLOPS/W Hybrid-Domain Outer-Product Based Floating-Point SRAM Computing-in-Memory Macro with Logarithm Bit-Width Residual ADC," *ISSCC*, pp. 576-577, Feb. 2024. https://doi.org/10.1109/ISSCC49657.2024.10454313
[9] W.-S. Khwa et al., "A 16nm 216kb, 188.4TOPS/W and 133.5TFLOPS/W Microscaling Multi-Mode Gain-Cell CIM Macro Edge-AI Devices," *ISSCC*, pp. 252-253, Feb. 2025. https://doi.org/10.1109/ISSCC49661.2025.10904606
[10] Z. Yue et al., "A 51.6TFLOPs/W Full-Datapath CIM Macro Approaching Sparsity Bound and <2-30 Loss for Compound AI," *ISSCC*, pp. 256-257, Feb. 2025. https://doi.org/10.1109/ISSCC49661.2025.10904702
[11] X. Wang et al., "A 28nm 17.83-to-62.84TFLOPS/W Broadcast-Alignment Floating-Point CIM Macro with Non-Two's-Complement MAC for CNNs and Transformers," *ISSCC*, pp. 254-255, Feb. 2025. https://doi.org/10.1109/ISSCC49661.2025.10904738

[12] Y. Wang et al., "A 28nm 83.23TFLOPS/W POSIT-Based Compute-in-Memory Macro for High-Accuracy AI Applications," *ISSCC*, pp. 566-567, Feb. 2024. https://doi.org/10.1109/ISSCC49657.2024.10454567

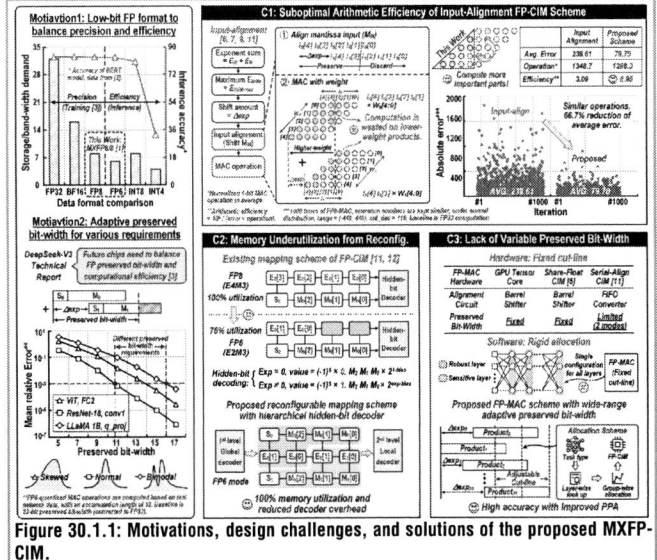

Figure 30.1.1: Motivations, design challenges, and solutions of the proposed MXFP-CIM.

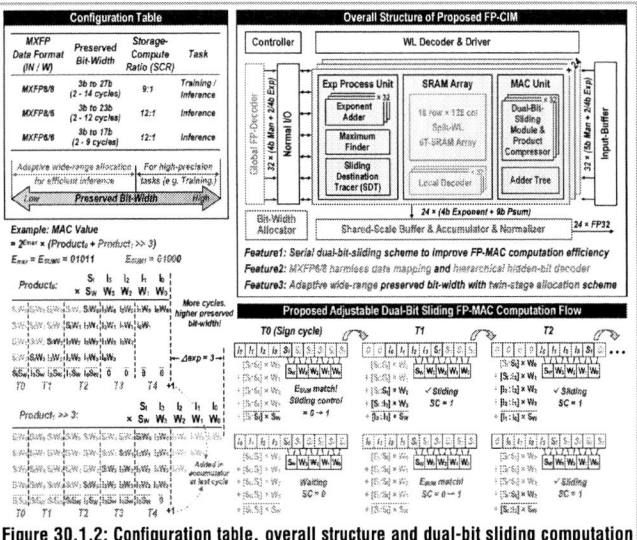

Figure 30.1.2: Configuration table, overall structure and dual-bit sliding computation flow.

ISSCC 2026 / SESSION 30 / COMPUTE-IN-MEMORY / 30.1

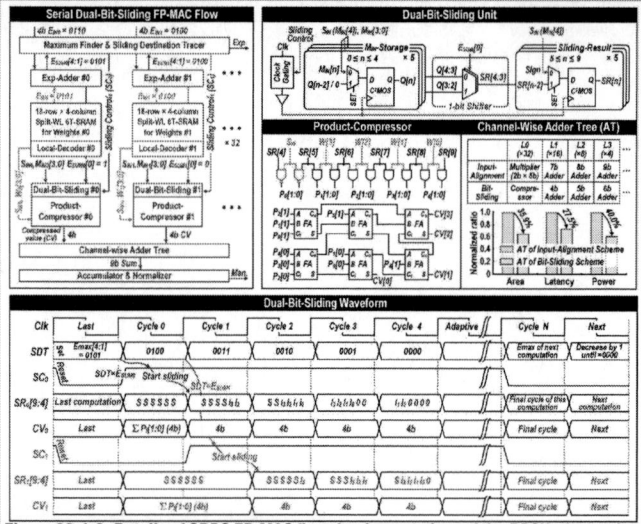

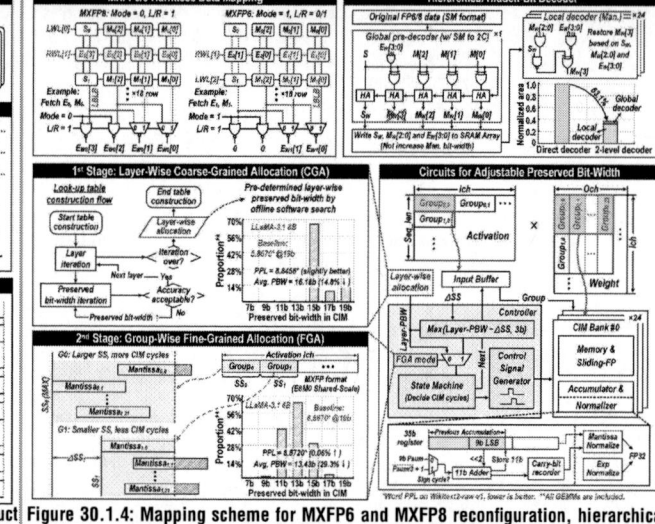

Figure 30.1.3: Details of SDBS FP-MAC flow, implementations of the DBSU, the product compressor, and the channel-wise adder tree.

Figure 30.1.4: Mapping scheme for MXFP6 and MXFP8 reconfiguration, hierarchical hidden-bit decoder and layer-group twin-stage allocation scheme.

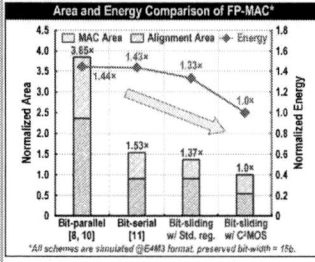

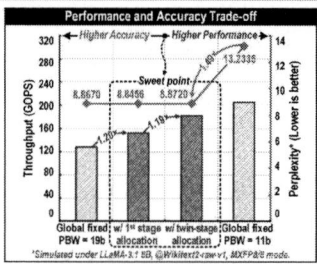

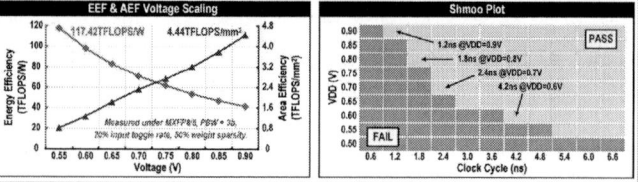

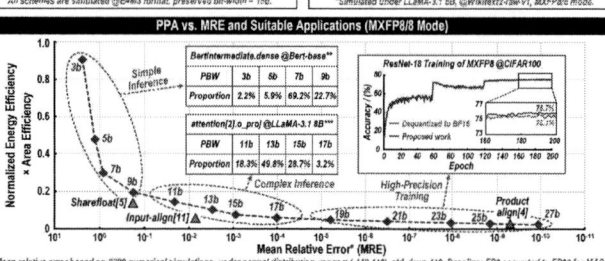

Figure 30.1.5: Simulated performance of the proposed work.

Figure 30.1.6: Measurement results and comparison table.

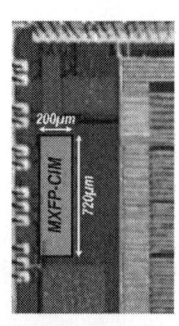

CHIP SUMMARY			
Technology	28nm CMOS		
Alignment Scheme	Bit-sliding, product alignment		
Macro Size	54Kb		
Macro Area (mm²)	0.144		
Input Precision	MXFP6/8		
Weight Precision	MXFP6/8		
Output Precision	FP32		
Supply Voltage (V)	0.55 - 0.9		
Frequency (MHz)	154 - 833		
Memory Density (Kb/mm²)	375		
Applicable Task	CNN/Transformer inference (Simple)	LLM inference (Complex)	Training
Peak Energy Efficiency (TFLOPS/W)	117.42 / 121.15 / 127.54[*1]		
Average Energy Efficiency (TFLOPS/W)	68.73 / 70.85 / 74.58[*2]	49.53[*2]	19.11[*2]
Area Efficiency (TFLOPS/mm²)	2.06[*2]	1.24[*2]	0.741[*2]
Network Accuracy	-0.81% / -0.32% / -0.38%[*3] (Bert-base @THUCNews)	+0.1466 PPL[*4] (LLaMA-3.1 8B @Wikitext2-raw-v1)	75.7%[*5] (ResNet-18 @CIFAR100)

*1: Measured under MXFP8/8, MXFP8/6 and MXFP6/6 @0.55V, PBW = 3b, 20% input toggle rate, 50% weight sparsity;
*2: Measured under real data distribution from networks, with twin-stage adaptive PBW, EEF@0.55V, AEF@0.9V;
*3: Under MXFP8/8, MXFP8/6, and MXFP6/6 with twin-stage adaptive PBW, all GEMMs are included, software baseline = 92.16% / 91.87% of 10%;
*4: Under MXFP8/8 with twin-stage adaptive PBW, all GEMMs are included, software baseline = 8.7254 (word PPL, lower is better);
*5: Under MXFP8/8, software baseline = 78.1%.

Figure 30.1.7: Die micrograph and chip summary table.

• 2026 IEEE International Solid-State Circuits Conference

979-8-3315-8937-0/26 $31.00 © 2026 IEEE

ISSCC 2026 / SESSION 30 / COMPUTE-IN-MEMORY / 30.2

30.2 A 12nm 4Mb 104.56-to-137.75TFLOPS/W Charge-Trap Transistor-Based Computing-in-Memory Macro Using Analog-Predict-Digital-Compute for AI Edge Devices

Junzhe Shen[*1,2], Zhidao Zhou[*1,2], Wenfeng Zha[1,2], Zhi Li[1,2], Weizeng Li[1,2], Bohan Wang[1,2], Junyu Zhu[1,2], Hanghang Gao[1,2], Zhongze Han[1,2], Yiman Wang[1,2], Linfang Wang[3], Hongyang Hu[1,2], Qing Luo[1,2], Chunmeng Dou[1,2], Ming Liu[1]

[1]Institute of Microelectronics of the Chinese Academy of Sciences, Beijing, China, [2]University of Chinese Academy of Sciences, Beijing, China
[3]Columbia University, New York, NY
*Equally Credited Authors (ECAs)

Abstract

Previous non-volatile CIM (nvCIM) macros suffer from low storage density, unnecessary multiply-and-accumulate (MAC) operations, and large hardware cost for floating point computations. A 4Mb CTT nvCIM macro, fabricated in 12nm CMOS, supports INT/FP4 MAC operations with the analog-predict-digital-compute scheme for power saving, achieving an energy-efficiency of 137.75TFLOPS/W and >40 times improved density FoM (storage density×computing density).

Recent AI workloads increasingly consist of low-computation-intensity operations, where memory access dominates power and latency [1]. Non-volatile computing-in-memory (nvCIM) addresses this by enabling on-chip data storage and in-place computation [2]. Prior works have mainly focused on advancing the computing performance of nvCIM [3-6], including energy efficiency (EF), computing density (CD), and precision. The storage density (SD) has received relatively less attention, despite being critical to maximizing the benefits of nvCIM. Furthermore, digital or hybrid-domain nvCIM solutions are rarely discussed.

As shown in Fig. 30.2.1, charge-trap transistors (CTTs), which have tunable threshold voltage (V_{TH}) by applying proper biasing to trap and de-trap electrons in the high-k dielectrics, offer a high-density embedded nonvolatile memory (eNVM) solution in advanced technology nodes without increasing process complexity [8]. They provide a potential technology path to boost on-chip storage density of nvCIM [9]. However, there are several critical challenges: (1) Despite their compact bit-cell sizes (0.017μm² for 1cell/b and 0.034μm² for 2cell/b), CTTs usually exhibit a small V_{TH} shift (70 to 300mV) [10,11], which makes it difficult to implement digital or hybrid-domain nvCIM that usually requires large signal ratios; (2) After going through activation (e.g., ReLU) and quantization, many negative or very small multiply-and-accumulate (MAC) values become zero or negligible, and therefore do not affect the final model output and become "unnecessary" to compute; (3) To process high-precision floating-point (FP) format data in nvCIM [5,12,13], considerable hardware costs are incurred, including dedicated circuits for pre-alignment (PA) and additional storage room for aligned mantissa.

We address these issues as follows: (1) a high-density differential-gain CTT computing array (DG-CCA) capable of supporting both analog and digital MAC with small area overhead; (2) a highly energy-efficient sign-and-magnitude predictable analog data-path (SMP-AD) to skip unnecessary partial MAC (pMAC) operations; and (3) a low PA- and storage-cost INT4/8 and FP4 digital data-path (INT/FP4-DD) for common AI workloads. The presented 12nm 4Mb CTT nvCIM macro, consisting of 16 256kb-banks (Bank$_{0-15}$), supports INT/FP4 MAC operations with analog-predict-digital-compute (APDC) for power saving. It achieves competent computing efficiency (EF=104.56-137.75 TFLOPS/W, CD=2.61-3.64 TFLOPS/mm²) and storage density (SD=1.33 Mb/mm²) at the same time.

Figure 30.2.2 illustrates the CTT MAC array of the nvCIM macro along with its operational waveforms for the APDC flow. Each bank performs MAC operations with 256 input channels and 32 output channels, supporting INT4/8 and FP4 (INT/FP4) precision for input (IN) or weight (W), and INT16-to-24 precision for output (OUT). The architecture includes: 1) Lookup table (LUT)-FP4 pre-processing for on-chip pre-alignment; 2) Dual-mode input drivers to drive the bit-lines (BLs) and the complementary bit-lines (BLBs) for different logic operations (XOR or AND); 3) Word-line (WL) drivers for weight data addressing; and 4) 32 CTT MAC arrays (CTT MAC Array$_{0-31}$) in a column along with SMP-ADs and INT/FP4-DDs for APDC. Within each CTT MAC array there are 256 differential-gain (DG)-CTT subarrays in a row, paired with local gain and pre-charge units (LGPU). Each DG-CTT subarray consists of 32×2 CTTs arranged between BL and BLB with a local source-line (LSL) and a globe source-line (GSL). Weight data are encoded using a differential pair of CTTs with V_{TH} in the high state (VT_H) and low state (VT_L). Weight-0 is encoded as a VT_H cell at the BL side and a VT_L cell at the BLB side. Weight-1 is encoded with the opposite state. To perform memory read and write, LSL is connected to GSL. To perform in-memory computing, the LSL is disconnected from the GSL, the input voltages are applied via BL and BLB and the LSL voltage (V_{LSL}) is determined by the voltage division across the selected CTT pair. The LGPU is designed to amplify the V_{LSL} changes and generate a large signal ratio for the analog or digital data-lines (ADL and DDL). It consists of two low V_{TH} (LVT) logic transistors, in which one NMOS (NM1) serves as the local gain cell (LGC) and one PMOS (PM1) serves as the local pre-charge (LPR) cell. Notice that the LVT LGC is used to reduce the driving voltage of the BL and BLB, and the LVT LPR is used to increase the ADL swing range for analog-domain computing. In the APDC flow, it takes 1+n² clock (CLK) cycles for MAC with n-bit IN and n-bit W. In the first CLK, the SMP-AD is triggered (AP_EN=1) to predict whether the MAC operations in a CTT MAC array are necessary based on the most-significant bits (MSBs) of IN and W (IN[n-1] and W[n-1]). If so, the INT/FP4-DD is triggered (DC_EN=1).

The n-bit inputs are sent into the array bitwise in n iterations. The n-bit weights stored in the same CTT MAC array are multiplied by the input data in n sub-iterations. Otherwise, the INT/FP4-DD is disabled (DC_EN=0) to avoid energy consumption from unnecessary computations.

Figure 30.2.3 shows the structure of SMP-AD and its operational flow for analog prediction (AP). The SMP-AD consists of one voltage-type differential input sense amplifier (VSA), one MUX for different AP modes, one PMOS for global pre-charge (GPR), one PMOS as a weak voltage keeper (WKP), one switch controlled by the bank grouping signal (BG), and one switch controlled by its opposite signal (BGB). The AP is configured by three steps. (1) Input mode selection: XOR mode is used for sign prediction based on the MSBs of signed IN and signed W, where BL is high and BLB is low for input-1 and the opposite for input-0. AND mode is used for magnitude prediction based on the MSBs of unsigned IN and signed W, where BL is high and BLB is low for input-1 and both BL and BLB are low for input-0. (2) Bank grouping selection: To fulfill different requirements on the accumulation lengths (ALs) in different models, different banks can be grouped. When BG=1 and BGB=0, the bank is grouped with the next one. (3) References (REF$_{0-3}$) are selected based on a pre-defined threshold value (PTV) derived from the model using the training dataset. For example, the relationship between MACV with the number of the positive product (N_{PP}) of sign-bits of IN and W in a MAC channel in layer #15 of ResNet-18 is analyzed. By setting a PTV of 710, 45% of the MACVs can be skipped. The REF corresponding to the largest N_{PP} smaller than PTV is selected to maximize the skip ratio (SR). In a typical AP process, the LPR is kept low and the WKP is weakly activated. The ADL is first pre-charged to VDD by GPR. After the BL and BLB inputs, the WL is activated for ADL development. VSA is enabled by SA_EN to output DC_EN. Notice that the WKP is important to improve ADL discharging linearity. With WKP, the ADL voltage shows good linearity with respect to N_{PP}.

Figure 30.2.4 shows the structure of INT/FP4-DD and its computing flow. Each INT/FP4-DD consists of 256 D-type flip-flops (DFF$_{0-255}$), one 8-stage adder tree, and a sign-aware digital accumulator (DSA). In a DC cycle, the ADL is kept at VDD (GPR = 0, WKP = 0). The input driver is set in the AND mode for BL/BLB input. The DDL is first pre-charged to VDD (LPR = 0). Then, the WL of the selected cells is activated for multiplication, and the DDL develops a voltage that depends on the output results. The DFFs latch the DDL data. The data is then summed up by the adder tree to produce 8b output data (DOUT [7:0]). The accumulator combines the DOUTs over multiple CLK cycles to output 16-to-24b ACC_OUT depending on IN/W precision. To reduce the hardware cost for FP computing, we introduce a pre-processing scheme based on a LUT-FP4 [14], which is used to maximize the numerical representation within a limited bit-width and thereby reduce quantization loss. The LUT-FP4 input data is pre-processed on-chip in three steps. (1) The mantissa (M) is expanded based on exponent bits (E2 and E1) using the LUT. For normalized values (E2E1 ≠ (00)₂), it is set to (1.M)₂; for subnormal values (E2E1 = (00)₂), it is set to (1.0)₂ if M=1 or (0.0)₂ if M=0. (2) The current exponent value (E_i=(E2E1)₂) is subtracted from a fixed maximum (E_{max}=(11)₂) to compute the shift length (SHL=(11)₂–E_i), reducing the overhead of finding a local E_{max} as in conventional FP CIM. (3) The mantissa is shifted and aligned according to SHL. Although the full width is 6b, the LSB is always 0 in LUT-FP4, allowing the aligned mantissa to be truncated to 5b without accuracy loss. Notice that the LUT-FP4 weight data is pre-processed off-line using similar steps and pre-loaded into the CTT MAC arrays.

Figure 30.2.5 shows the characterization results of the proposed 4Mb CTT nvCIM macro that is fabricated in a 12nm CMOS technology. The typical I-V curves of the CTT cells shows a clear memory window about 220mV between the programed (VT_H) and erased (VT_L) states. The read polarity between BL and GSL has rarely impact on the cell properties, supporting the XOR and AND input modes. Typical DFF-based path-delay excluding scheme is used to test the access time (T_{AC}). The captured waveforms confirm the macro achieved T_{AC}=1.25ns for AP and T_{AC}=1.34ns for one DC cycle. The shmoo test for FP4IN-FP4W-18bOUT confirms that T_{AC}=36ns at VDD=0.8V. The system-level measurement results show the macro achieves an average EF=104.56TFLOPS/W and a peak EF=137.75TFLOPS/W for ResNet-18 processing.

Compared to previous nvCIM works (Fig. 30.2.6), this work demonstrates a 12nm CTT nvCIM macro. By adopting the high-density CTT cells and the APDC hybrid-domain design, the implemented nvCIM achieves a 44.49× improvement in FoM$_1$ (Storage Density×Compute Density) and a 14.62× improvement in FoM$_2$ (Storage Density×Energy Efficiency×Compute Density) over prior INT8 nvCIM designs [3,4]. While SD and CD are often traded off against each other in conventional designs, this work achieves a more balanced optimization of both. The die micrograph, chip performance summary and its position chart are shown in Fig. 30.2.7.

Acknowledgement:
This work was supported by the National Natural Science Foundation of China (NSFC) under Grant No. 92364202, 62488101, 62404249. Corresponding Author: Chunmeng Dou.

References:
[1] A. Gholam et al., "AI and Memory Wall," *IEEE Micro*, vol. 44, no. 3, pp. 33-39, May-June 2024. https://doi.org/10.1109/MM.2024.3373763
[2] W.-H. Chen et al., "CMOS-integrated Memristive Non-Volatile Computing-in-Memory for AI Edge Processors," *Nature Electronics*, vol. 2, pp. 420-428, Aug. 2019. https://doi.org/10.1038/s41928-019-0288-0
[3] D.-Q. You et al., "A 22nm 104.5TOPS/W μ-NMC-Δ-IMC Heterogeneous STT-MRAM CIM Macro for Noise-Tolerant Bayesian Neural Networks," *ISSCC*, pp. 250-251, Feb. 2025. https://doi.org/10.1109/ISSCC49661.2025.10904540
[4] L. Wang et al., "A Flash-SRAM-ADC-Fused Plastic Computing-in-Memory Macro for Learning in Neural Networks in a Standard 14nm FinFET Process," *ISSCC*, pp. 582-583, Feb. 2024. https://doi.org/10.1109/ISSCC49657.2024.10454372
[5] T.-H. Wen et al., "A 22nm 16Mb Floating-Point ReRAM Compute-in-Memory Macro with 31.2TFLOPS/W for AI Edge Devices," *ISSCC*, pp. 580-581, Feb. 2024. https://doi.org/10.1109/ISSCC49657.2024.10454468
[6] W.-H. Huang et al., "A Nonvolatile AI-Edge Processor with 4MB SLC-MLC Hybrid-Mode ReRAM Compute-in-Memory Macro and 51.4-251TOPS/W," *ISSCC*, pp. 15-16, Feb. 2023. https://doi.org/10.1109/ISSCC42615.2023.10067610
[7] H. Cai et al., "A 28nm 2Mb STT-MRAM Computing-in-Memory Macro with a Refined Bit-Cell and 22.4 - 41.5TOPS/W for AI Inference," *ISSCC*, pp. 500-501, Feb. 2023. https://doi.org/10.1109/ISSCC42615.2023.10067339
[8] B. Jayaraman et al., "80-kb Logic Embedded High-K Charge Trap Transistor-Based Multi-Time-Programmable Memory With No Added Process Complexity," *IEEE JSSC*, vol. 58, no. 3, pp. 949-960, Mar. 2018. https://doi.org/10.1109/JSSC.2017.2784760
[9] S. Qiao et al., "Demonstration of Analog Compute-In-Memory Using the Charge-Trap Transistor in 22 FDX Technology," *IEEE IEDM*, pp. 2.5.1-2.5.4, Dec. 2022. https://doi.org/10.1109/IEDM45625.2022.10019527
[10] E. Hunt-Schroeder et al., "14nm FinFET 1.5MB Embedded High-K Charge Trap Transistor One Time Programmable Memory Using Dynamic Adaptive Programming," *IEEE Symp. VLSI Circuits*, pp. 87-88, June 2018. https://doi.org/10.1109/VLSIC.2018.8502415
[11] F. Khan et al., "Charge Trap Transistor (CTT): An Embedded Fully Logic-Compatible Multiple-Time Programmable Non-Volatile Memory Element for High-*k*-Metal-Gate CMOS Technologies," *IEEE LED*, vol. 38, no. 1, pp. 44-47, Jan. 2017. https://doi.org/10.1109/LED.2016.2633490

[12] W.-S. Khwa et al., "A Mixed-Precision Memristor and SRAM Compute-in-Memory AI Processor," *Nature*, vol. 639, pp. 617-623, Mar. 2025. https://doi.org/10.1038/s41586-025-08639-2
[13] X. Wang et al., "A 28nm 17.83-to-62.84TFLOPS/W Broadcast-Alignment Floating-Point CIM Macro with Non-Two's-Complement MAC for CNNs and Transformers," *ISSCC*, pp. 254-255, Feb. 2025. https://doi.org/10.1109/ISSCC49661.2025.10904738
[14] R. Wang et al., "Optimizing Large Language Model Training Using FP4 Quantization," *arXiv preprint*, pp. 1-21, May 2025. https://doi.org/10.48550/arXiv.2501.17116
[15] C.-C. Chou et al., "An N40 256K×44 embedded RRAM macro with SL-precharge SA and low-voltage current limiter to improve read and write performance," *ISSCC*, pp. 478-479, Feb. 2018. https://doi.org/10.1109/ISSCC.2018.8310392
[16] Y.-C. Shih et al., "Logic Process Compatible 40-nm 16-Mb, Embedded Perpendicular-MRAM With Hybrid-Resistance Reference, Sub-μA Sensing Resolution, and 17.5-nS Read Access Time," *IEEE JSSC*, vol. 54, no. 4, pp. 1029-1038, Apr. 2019. https://doi.org/(16)10.1109/JSSC.2018.2889106
[17] J. Yang et al., "A 28nm 1.5Mb Embedded 1T2R RRAM with 14.8 Mb/mm² using Sneaking Current Suppression and Compensation Techniques," *IEEE Symp. VLSI Circuits*, pp. 1-2, June 2020. https://doi.org/10.1109/VLSICircuits18222.2020.9163035
[18] Y.-D. Chih et al., "A 22nm 32Mb Embedded STT-MRAM with 10ns Read Speed, 1M Cycle Write Endurance, 10 Years Retention at 150°C and High Immunity to Magnetic Field Interference," *ISSCC*, pp. 222-223, Feb. 2020. http://doi.org/10.1109/ISSCC19947.2020.9062955
[19] P. Jain et al., "A 3.6Mb 10.1Mb/mm² Embedded Non-Volatile ReRAM Macro in 22nm FinFET Technology with Adaptive Forming/Set/Reset Schemes Yielding Down to 0.5V with Sensing Time of 5ns at 0.7V," *ISSCC*, pp. 212-213, Feb. 2019. https://doi.org/10.1109/ISSCC.2019.8662393
[20] P.-H. Lee et al., "16nm 32Mb Embedded STT-MRAM with a 6ns Read-Access Time, a 1M-Cycle Write Endurance, 20-Year Retention at 150°C and MTJ-OTP Solutions for Magnetic Immunity," *ISSCC*, pp. 494-495, Feb. 2023. https://doi.org/10.1109/ISSCC42615.2023.10067837
[21] J. Yang et al., "A 14nm-FinFET 1Mb Embedded 1T1R RRAM with a 0.022μm² Cell Size Using Self-Adaptive Delayed Termination and Multi-Cell Reference," *ISSCC*, pp. 336-337, Feb. 2021. https://doi.org/10.1109/ISSCC42613.2021.9365945
[22] Y.-C. Huang et al., "A 32Mb RRAM in a 12nm FinFet Technology with a 0.0249μm² Bit-Cell, a 3.2GB/S Read Throughput, a 10KCycle Write Endurance and a 10-Year Retention at 105°C," *ISSCC*, pp. 288-289, Feb. 2024. http://doi.org/10.1109/ISSCC49657.2024.10454367
[23] X. Chen et al., "A 28nm 64kb Bit-Rotated Hybrid-CIM Macro with an Embedded Sign-Bit-Processing Array and a Multi-Bit-Fusion Dual-Granularity Cooperative Quantizer," *ISSCC*, pp. 260-261, Feb. 2025. https://doi.org/10.1109/ISSCC49661.2025.10904646
[24] A. Guo et al., "A 22nm 64kb Lightning-Like Hybrid Computing-in-Memory Macro with a Compressed Adder Tree and Analog-Storage Quantizers for Transformer and CNNs," *ISSCC*, pp. 570-571, Feb. 2024. https://doi.org/10.1109/ISSCC49657.2024.10454278

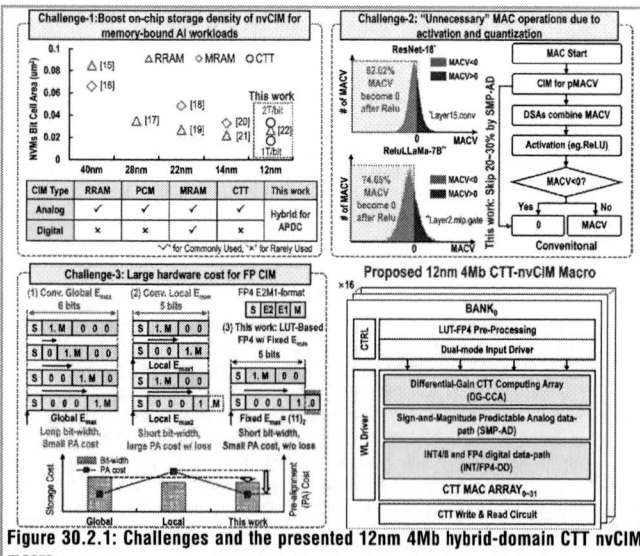

Figure 30.2.1: Challenges and the presented 12nm 4Mb hybrid-domain CTT nvCIM macro.

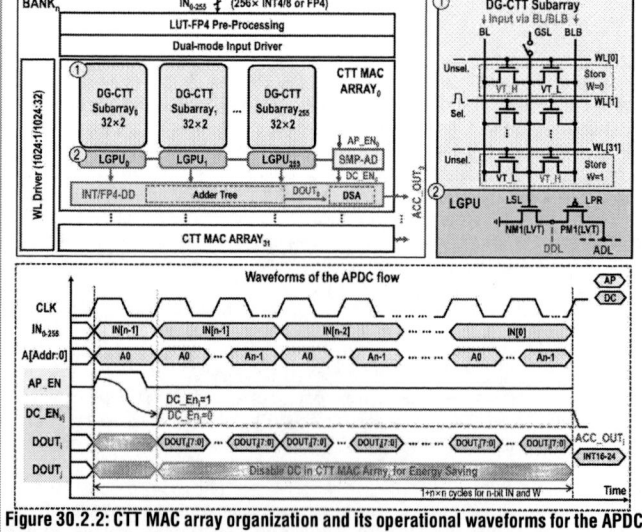

Figure 30.2.2: CTT MAC array organization and its operational waveforms for the APDC flow.

ISSCC 2026 / SESSION 30 / COMPUTE-IN-MEMORY / 30.2

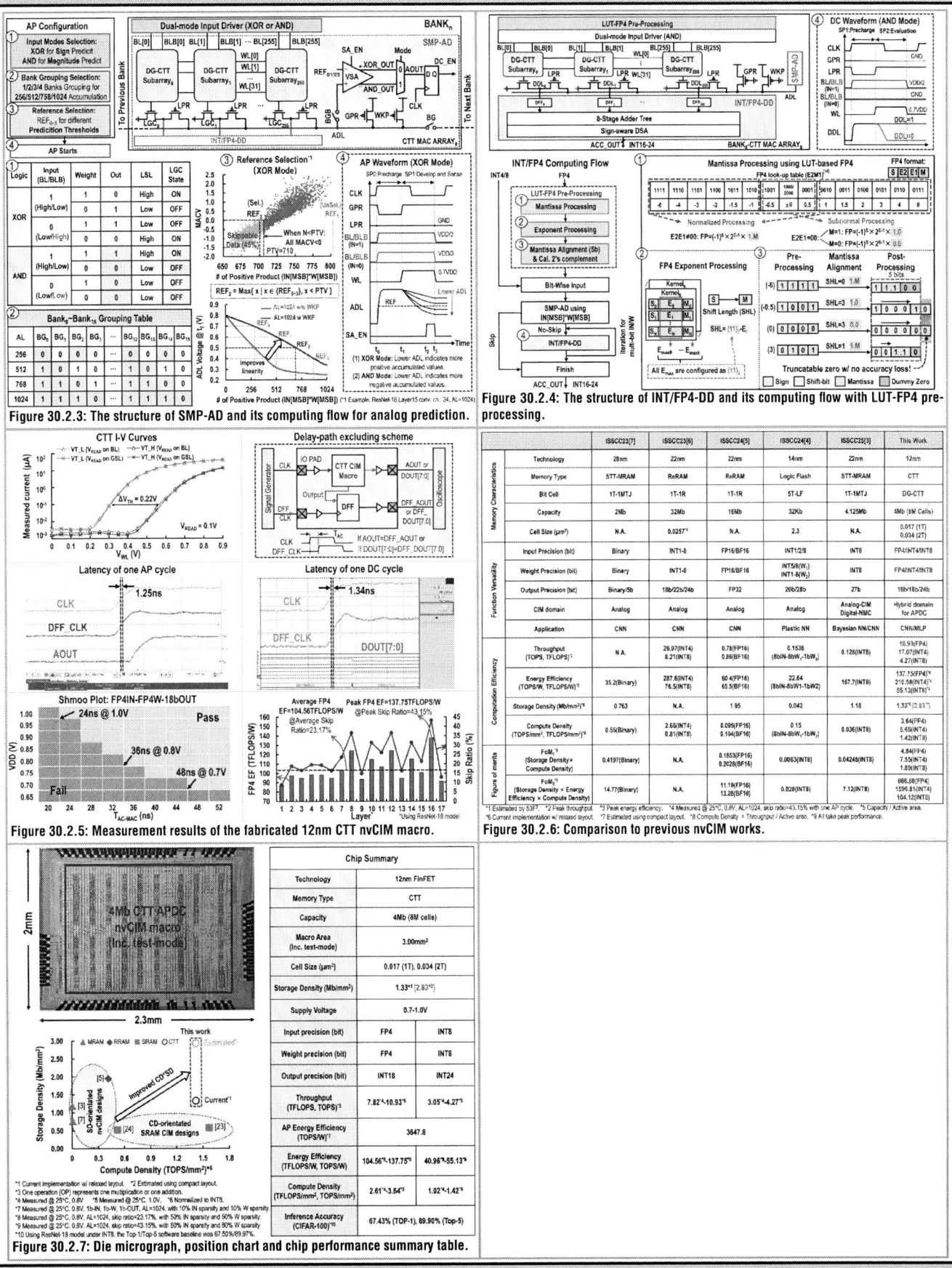

Figure 30.2.3: The structure of SMP-AD and its computing flow for analog prediction.

Figure 30.2.4: The structure of INT/FP4-DD and its computing flow with LUT-FP4 pre-processing.

Figure 30.2.5: Measurement results of the fabricated 12nm CTT nvCIM macro.

Figure 30.2.6: Comparison to previous nvCIM works.

Figure 30.2.7: Die micrograph, position chart and chip performance summary table.

• 2026 IEEE International Solid-State Circuits Conference

979-8-3315-8937-0/26 $31.00 © 2026 IEEE

ISSCC 2026 / SESSION 30 / COMPUTE-IN-MEMORY / 30.3

30.3 A 22nm 96Mb 50.6-to-90.2TFLOPS/W Non-Linear MLC ReRAM CIM Macro with High-Retention for Mamba/Transformer/CNN

Hung-Hsi Hsu[*1], Win-San Khwa[*2], Yao-Kai Yeh[1], Chih-Ling Wu[1], Chang-Yuan Chen[1], Yen-Che Huang[1], Yen-Hua Lin[1], Cheng-Feng Chang[1], Yen-Tung Shao[1], Jen-Chun Tien[1], De-Qi You[1], Ping-Sheng Wu[2], Bo Zhang[3], Ren-Shuo Liu[1], Chih-Cheng Hsieh[1], Kea-Tiong Tang[1], Meng-Fan Chang[1,2]

[1]National Tsing Hua University, Hsinchu, Taiwan, [2]TSMC Corporate Research, Hsinchu, Taiwan, [3]TSMC Corporate Research, San Jose, CA
[*]Equally Credited Authors (ECAs)

Abstract

We present a DTCO-designed high-retention multi-mode ReRAM nvCIM macro supporting linear MLC (L-MLC), non-linear MLC (NL-MLC), and SLC modes for Mamba, Transformer, and CNN workloads, addressing key challenges in inference accuracy, energy, and readout robustness. The proposed macro incorporates: 1) a reconfigurable compute mode for accuracy/efficiency tradeoff, 2) ISE-HF encoding to enhance input sparsity and save energy, 3) an S²R-Db ADC that enables high-yield, low-power NL-MLC readout. The design achieves 50.6-to-90.2TFLOPS/W (BF16) with ImageNet accuracy loss that is 79.17% lower than L-MLC after 10 years of retention.

Advanced non-volatile AI-edge processors [1-8] require non-volatile compute-in-memory (nvCIM) macros [8-16] that can retain complete neural-network parameters during power-down without losing weight data, while simultaneously supporting high-precision high energy-efficiency (EF) multiply-and-accumulate (MAC) operations. Non-linear multi-level-cell (NL-MLC) ReRAM [17] has emerged as a compelling candidate for such macros, offering longer retention times than conventional linear-MLC (L-MLC) devices, while achieving higher EF (TOPS/W) and computational density (TOPS/mm²) compared with single-level-cell (SLC) ReRAM. These advantages make NL-MLC ReRAM particularly suitable for modern architectures such as Mamba [18-21], which deliver high performance on edge workloads involving high-resolution images or long-sequence-length text and thus demand larger on-chip memory capacity.

However, implementing an NL-MLC ReRAM CIM macro introduces several challenges (Fig. 30.3.1), including: (1) navigating the trade-offs among SLC, NL-MLC, L-MLC modes with respect to inference accuracy, energy consumption, and area efficiency over device lifespan, (2) mitigating the high in-memory-compute (IMC) energy consumption associated with low input sparsity in conventional activation formats, an issue exacerbated by the increasingly adoption of activation functions such as GeLU and SiLU in the modern models, and (3) achieving accurate and energy-efficient readout, which is particularly difficult for NL-MLC devices.

This work presents a design technology co-optimization (DTCO) high-retention multi-mode ReRAM CIM Macro for Mamba, Transformer, and CNN architectures. It addresses the aforementioned challenges with three key schemes: (1) A DTCO reconfigurable compute mode that balances inference accuracy and energy efficiency by flexibly combining NL-MLC, L-MLC, and SLC configurations within the macro, supporting both near-memory computing (NMC) and IMC operations. The proposed Twin Non-Linear Cell Mapping (TNLCM) enables IMC with NL-MLCs, while the integrated Dual-Mode Data Converter (DMDC) reduces compute energy during NL-MLC operation. (2) An Input-Sparsity-Enhanced Hybrid Format (ISE-HF) that combines conventional 2C with the proposed Sign-Reversed (SR) representation to increase input sparsity and reduce the CIM macro energy consumption. (3) A Single-Self-Reference Dual-bit (S²R-Db) ADC that enables accurate two-bit readout while optimizing energy and area overhead for NL-MLC readout.

Figure 30.3.2 shows the breakdown of accumulation numbers (N_{ACCU}) for AI-edge models and the proposed DTCO reconfigurable compute mode. NMC offers better energy efficiency when N_{ACCU} is small, while IMC becomes advantageous as N_{ACCU} grows. Mamba and CNN models exhibit particularly low N_{ACCU} operation, especially in Element-Wise Multiplication (EWM) where $N_{ACCU}=1$ and the operation is IN × W without accumulation—while Transformers typically require higher N_{ACCU} values. To accommodate diverse workloads, we developed a DTCO multi-mode macro capable of adapting to a broad range of N_{ACCU} values, thereby maintaining energy efficiency across model types. Note that the N_{ACCU} threshold (TH=16 in this work) at which IMC outperforms NMC depends on the hardware design.

Conventional L-MLC maps weights via current ratios but suffers limited retention. Directly applying NL-MLC to IMC introduces challenges. Prior work [17] imitating L-MLC mapping works for NMC but fails for IMC due to inconsistent I_{MAC} values; e.g., MAC=3=S1+S4=S2+S3 can yield I_{MAC} currents of 4 or 5. Moreover, naïve current-ratio mapping cannot represent a weight of 1. To resolve this, we propose a TNLCM scheme that utilizes twin NL-MLC cells—$C_K[i]$ and $C_K[i+1]$—where $C_K[i]$ encodes positive weights via current magnitude and $C_K[i+1]$ encodes negative weights, enabling representation of weights from −4 to +4 and overcoming the limitations of the aforementioned methods. The increase in energy consumption when using multiple cells is alleviated by the proposed DMDC, which can be used in combination with the TNLCM strategy to reduce readout energy while retaining flexibility. Our DTCO reconfigurable compute mode integrates SLC, L-MLC, and NL-MLC to balance energy efficiency, inference accuracy, and retention characteristics. L-MLC offers higher storage density with lower area and energy consumption, but its retention is inferior to NL-MLC, making it better suited for storing less critical least-significant-bit (LSB) data. The proposed DMDC supports two operating modes. In NL-MLC Mode (NL-mode), designed for IMC with NL-MLC, the complementary NL-MLC weight mapping allows two bitline (BL) currents to be merged and subtracted before ADC conversion. This process generates a sign bit indicating which BL current is larger, enabling one ADC channel to be disabled for energy savings. In SLC/L-MLC Mode (SL-mode), the converter functions in a conventional manner for general usage.

Figure 30.3.3 illustrates the proposed ISE-HF scheme, designed to increase input bitwise sparsity during CIM operations and reduce compute energy. After input pre-alignment, values typically exhibit near-normal distribution. Conventional representations suffer from low bitwise sparsity. In Two's Complement (2C) format, near-zero negative values produce many sign-extension bits set to "1," while both 2C and sign-magnitude (SM) formats generate numerous "1" bits for large-magnitude values. These patterns reduce sparsity and limit CIM energy efficiency. The proposed ISE-HF adaptively selects either the 2C format or the proposed Sign-Reversed (SR) format, ensuring each input achieves at least 50% bitwise sparsity while preserving the numeric value.

The encoding process begins with format unification, wherein SM inputs are converted to 2C, while existing 2C inputs proceed directly to sparsity evaluation. For each input $IN_K<Q-1:0>$, the circuit calculates the ratio of zero bits to total bit width. If sparsity ≥ 50%, the input remains 2C and is flagged "0." If sparsity < 50%, the input is converted to SR by bitwise inversion with bit-index polarity reversal, subtracting one (equivalent to adding one to the absolute coefficient), and flagged "1." This transformation produces complementary sparsity relative to 2C, while reusing SM-to-2C hardware to improve implementation efficiency.

During computation, ISE-HF processes inputs bit-serially with zero-skipping [8], activating a fixed number of wordlines (WLs), which corresponds to the number of "1" bits being processed. Since 2C and SR have different sign interpretations, computation is sequential: bits with value "1" and flag "0" (2C) generate partial MAC values ($pMACV_{2C}$), followed by bits with "1" and flag "1" (SR) generating $pMACV_{SR}$. Partial results are combined according to bit position. For MSBs, the combined result is $pMACV_{SR} - pMACV_{2C}$, while for other bits, it is $pMACV_{2C} - pMACV_{SR}$. In each cycle, pMACV is accumulated by shift-and-add circuits with suitable bit-index scaling. Although sequential, this approach reduces overall latency compared to conventional methods because higher sparsity allows aggressive zero-skipping, enhancing macro-level EF.

Figure 30.3.4 illustrates the proposed S²R-Db ADC, which outputs dual-bit results in a single sensing operation using one self-reference current (I_{SR}). This enables accurate sensing while reducing energy consumption and area. Conventional schemes, such as SAR, Flash, and dual-bit small-offset (DbSO) ADC [14], require multiple midpoint reference currents (I_{MRs}), with each MR representing the midpoint between two data pattern distributions. Generating these MRs incurs significant power and area overhead due to numerous dummy array columns and dedicated midpoint-generation circuits. In contrast, the S²R-Db ADC needs only a single self-reference, minimizing overhead.

The S²R-Db ADC operates in five phases (Ph1 to Ph5). In Ph1, PREB = 0, and nodes G0, G1, S0, and S1 are pre-charged to VDD. In Ph2 (1st stage P0 & P1 offset sampling), BLPRE=PREB=1, and both the self-reference (SR) and data line (DL) are pre-charged to V_R (read-disturb-free voltage). Capacitors C0 and C1 sample the threshold voltages of P0 and P1, resulting in $V_{S0}=V_R+V_{THP0}$ and $V_{S1}=V_R+V_{THP1}$. In Ph3 (BL development and 2nd stage sampling), the WL is activated, discharging SR (10) and DL according to stored data. DL is discharged by ΔV_{DL} to G1 ($V_R-\Delta V_{DL}$) and coupled to S_0 ($V_R+V_{THP0}-\Delta V_{DL}$) via C0, while SR is discharged by ΔV_{SR} to G_0 ($V_R-\Delta V_{SR}$) and coupled to S_1 ($V_R+V_{THP1}-\Delta V_{SR}$) via C1. This generates differential overdrive voltages $V_{OVP0}=\Delta V_{SR}-\Delta V_{DL}$ and $V_{OVP1}=\Delta V_{DL}-\Delta V_{SR}$, independent of threshold voltages and induces voltage swings on Q2/QB2 and Q3/QB3. For read 00/01, $\Delta V_{DL}<\Delta V_{SR}$ yields $V_{OVP0}>0$ and $V_{OVP1}<0$, turning P0 on strongly and charging Q ($\Delta Q \propto V_{OVP0}$), while P1 remains off (QB≈0, Far+ region). For read 10, $\Delta V_{DL} \approx \Delta V_{SR}$, both $V_{OVP} \approx 0$, placing P0 and P1 in subthreshold, slightly charging Q and QB (Near region). For read 11, $\Delta V_{DL} > \Delta V_{SR}$, $V_{OVP0} < 0$, $V_{OVP1} > 0$, P1 turns on strongly and charges QB ($\Delta QB \propto V_{OVP1}$), P0 remains off (Far− region). In Ph4, small charge differences (ΔQ, ΔQB) are amplified to near full-swing signals Q3 and QB3, and the P5–N0 amplifier detects Q2 vs. QB2 to generate decision output D, distinguishing Very Far+ and Far+. Finally, in Ph5, Schmitt trigger inverters convert Q3, QB3, and D into digital outputs SA<2:0>, which are encoded into two-bit output OUT<1:0>, effectively digitalizing all four levels.

516 • 2026 IEEE International Solid-State Circuits Conference

979-8-3315-8937-0/26 $31.00 © 2026 IEEE

Figure 30.3.5 presents the simulated performance results of the proposed schemes. The TNLCM with DMDC method improved FoM1 (defined as energy efficiency×area efficiency×inference accuracy @ 10 years) by 1.6× compared to the all-SLC method, while the all-L-MLC approach exhibited an inference accuracy <1% @ 10 years. The ISE-HF enhanced macro-level EF by 1.23 to 1.45×, with only a slight area overhead. The S²R-Db ADC increased FoM2 (defined as 1/(Readout Energy×Area×Worst Case Readout Yield Loss)) by at least 5.3×. Overall, the macro-level EF improved by 1.86× compared to a SLC macro without the proposed schemes.

Figure 30.3.6 shows the simulated inference accuracy and measurement results of the 22nm 96Mb DTCO high-retention multi-mode ReRAM CIM macro, along with a comparison table of previous silicon-verified ReRAM CIM macros. The proposed macro can be adapted to diverse target applications. For example, Hybrid-II mapping is well suited for AI-edge workloads, where 3 to 5 years of retention is sufficient and energy efficiency is the primary concern, while Hybrid-I is more appropriate for scenarios in which device updates are infrequent and long data retention is essential. The simulated inference accuracy in hybrid mode is 79.17% higher than that of all L-MLC at 10 years using TinyViM-S (Mamba) under BF16. The measured shmoo plot confirms that the macro supports a 5.0ns cycle time @0.8V. Figure 30.3.7 shows the die micrograph alomg with a summary of chip performance.

Acknowledgement:
The authors would like to thank NTHU-TSMC major league, TSMC JDP, and NSTC for financial and manufacturing support.

References:
[1] W.-S. Khwa et al., "A Mixed-Precision Memristor and SRAM Compute-in-Memory AI Processor," *Nature*, vol. 639, pp. 617–623, Mar. 2025. https://doi.org/10.1038/s41586-025-08639-2
[2] H.-H. Hsu et al., "A Nonvolatile AI-Edge Processor With SLC–MLC Hybrid ReRAM Compute-in-Memory Macro Using Current–Voltage-Hybrid Readout Scheme," *JSSC*, vol. 59, no. 1, pp. 116-127, Jan. 2024. https://doi.org/10.1109/JSSC.2023.3314433
[3] D.-Q. You et al., "A 22nm Nonvolatile AI-Edge Processor with 21.4TFLOPS/W using 47.25Mb Lossless-Compressed-Computing STT-MRAM Near-Memory-Compute Macro," *IEEE Symp. VLSI Circuits*, pp. 1-2, June 2024. https://doi.org/10.1109/VLSITechnologyandCir46783.2024.10631408
[4] M. Chang et al., "A 73.53TOPS/W 14.74TOPS Heterogeneous RRAM In-Memory and SRAM Near-Memory SoC for Hybrid Frame and Event-Based Target Tracking," *ISSCC*, pp. 426-427, Feb. 2023. https://doi.org/10.1109/ISSCC42615.2023.10067544
[5] K. Prabhu et al., "CHIMERA: A 0.92-TOPS, 2.2-TOPS/W Edge AI Accelerator With 2-MByte On-Chip Foundry Resistive RAM for Efficient Training and Inference," *JSSC*, vol. 57, no. 4, pp. 1013-1026, Apr. 2022. https://doi.org/10.1109/JSSC.2022.3140753
[6] T.-H. Wen et al., "Fusion of Memristor and Digital Compute-in-Memory Processing for Energy-Efficient Edge Computing," *Science*, vol. 384, n. 6693, pp. 325-332, Apr. 2024. https://doi.org/10.1126/science.adf5538
[7] H.-H. Hsu et al., "A 22nm 41.8TFLOPS/W AI-Edge Transformer/CNN Nonvolatile-Processor Using QKV-Softmax-Layer-Fused Hybrid ReRAM-CIM and Concurrent-Transpose/Non-Transpose SRAM-CIM," *IEEE Symp. VLSI Circuits*, pp. 1-3, June 2025. https://doi.org/10.23919/VLSITechnologyandCir65189.2025.11074985

[8] W.-H. Huang et al., "A Nonvolatile AI-Edge Processor with 4MB SLC-MLC Hybrid-Mode ReRAM Compute-in-Memory Macro and 51.4-251TOPS/W," *ISSCC*, pp. 15-16, Feb. 2023. https://doi.org/10.1109/ISSCC42615.2023.10067610
[9] T.-H. Wen et al., "A 22nm 16Mb Floating-Point ReRAM Compute-in-Memory Macro with 31.2TFLOPS/W for AI Edge Devices," *ISSCC*, pp. 580-581, Feb. 2024. https://doi.org/10.1109/ISSCC49657.2024.10454468
[10] W. Wan et al., "A compute-in-memory chip based on resistive random-access memory," *Nature*, vol. 608, pp. 504–512, Aug. 2022. https://doi.org/10.1038/s41586-022-04992-8
[11] T.-H. Wen et al., "A 22nm 16Mb Floating-Point ReRAM Compute-in-Memory Macro with 31.2TFLOPS/W for AI Edge Devices," *ISSCC*, pp. 580-581, Feb. 2024. https://doi.org/10.1109/ISSCC49657.2024.10454468
[12] H.-H. Hsu et al., "A 22 nm Floating-Point ReRAM Compute-in-Memory Macro Using Residue-Shared ADC for AI Edge Device," *JSSC*, vol. 60, no. 1, pp. 171-183, Jan. 2025. https://doi.org/10.1109/JSSC.2024.3470211
[13] J. M. Correll et al., "An 8-bit 20.7 TOPS/W Multilevel Cell ReRAM Macro With ADC-Assisted Bit-Serial Processing," *JSSC*, vol. 60, no. 8, pp. 2995-3008, Aug. 2025. https://doi.org/10.1109/JSSC.2025.3540114
[14] C.-X. Xue et al., "A 22nm 2Mb ReRAM Compute-in-Memory Macro with 121-28TOPS/W for Multibit MAC Computing for Tiny AI Edge Devices," *ISSCC*, pp. 244-245, Feb. 2020. https://doi.org/10.1109/ISSCC19947.2020.9063078
[15] D.-Q. You et al., "A 22nm 104.5TOPS/W μ-NMC-Δ-IMC Heterogeneous STT-MRAM CIM Macro for Noise-Tolerant Bayesian Neural Networks," *ISSCC*, pp. 250-251, Feb. 2025. https://doi.org/10.1109/ISSCC49661.2025.10904540
[16] Y.-C. Chiu et al., "A 22nm 8Mb STT-MRAM Near-Memory-Computing Macro with 8b-Precision and 46.4-160.1TOPS/W for Edge-AI Devices," *ISSCC*, pp. 496-497, Feb. 2023. https://doi.org/10.1109/ISSCC42615.2023.10067563
[17] C.Y. Tsai et al., "A CMOS-Compatible 12nm 8Mb MLC RRAM Enabling Producible 2-Bit Per Cell for High Energy Efficiency Compute-In-Memory in Edge AI Applications," *IEEE Symp. VLSI Circuits*, pp. 1-3, June 2025. https://doi.org/10.23919/VLSITechnologyandCir65189.2025.11074924
[18] A. Gu et al., "Mamba: Linear-time sequence modeling with selective state spaces," *arXiv preprint*, pp. 1-36, May 2024. https://doi.org/10.48550/arXiv.2312.00752
[19] L. Zhu et al., "Vision Mamba: Efficient Visual Representation Learning with Bidirectional State Space Model," *arXiv preprint*, pp.1-14, Nov. 2024. https://doi.org/10.48550/arXiv.2401.09417
[20] Z. Wang et al., "Mamba YOLO: A Simple Baseline for Object Detection with State Space Model," *AAAI*, vol. 39, no. 8, pp. 8205-8213, Apr. 2025. https://doi.org/10.1609/aaai.v39i8.32885
[21] X. Ma et al., "Tinyvim: Frequency Decoupling for Tiny Hybrid Vision Mamba," *arXiv preprint*, pp. 1-14, Dec. 2025. https://doi.org/10.48550/arXiv.2411.17473
[22] S. Fukuyama et al., "Comprehensive Analysis of Data-Retention and Endurance Trade-Off of 40nm TaOx-based ReRAM," *IEEE International Reliability Physics Symposium (IRPS)*, pp. 1-6, Mar.-Apr. 2019. https://doi.org/10.1109/IRPS.2019.8720436

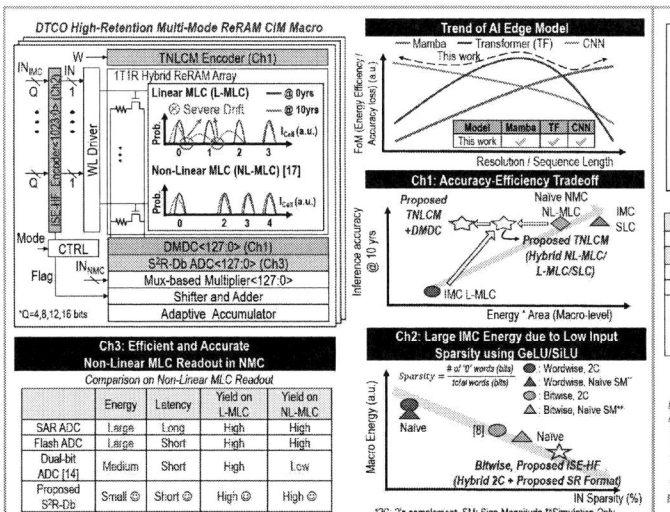

Figure 30.3.1: Advantages of Non-Linear MLC, challenges and proposed High-Retention Multi-Mode ReRAM CIM macro.

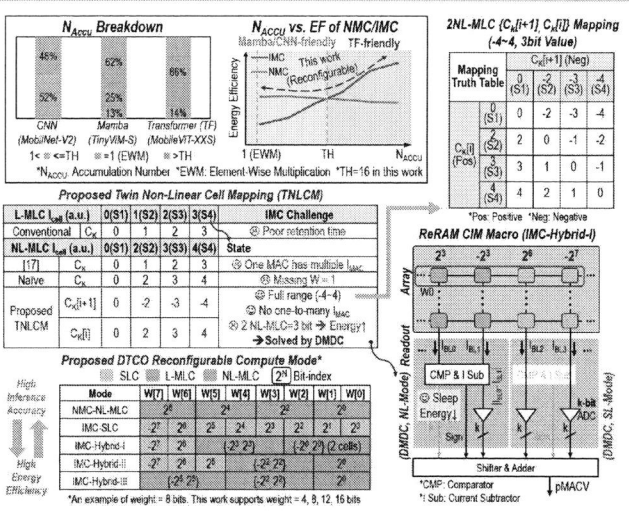

Figure 30.3.2: Accumulation number breakdown of recent models and proposed DTCO reconfigurable compute mode with TNLCM and DMDC.

ISSCC 2026 / SESSION 30 / COMPUTE-IN-MEMORY / 30.3

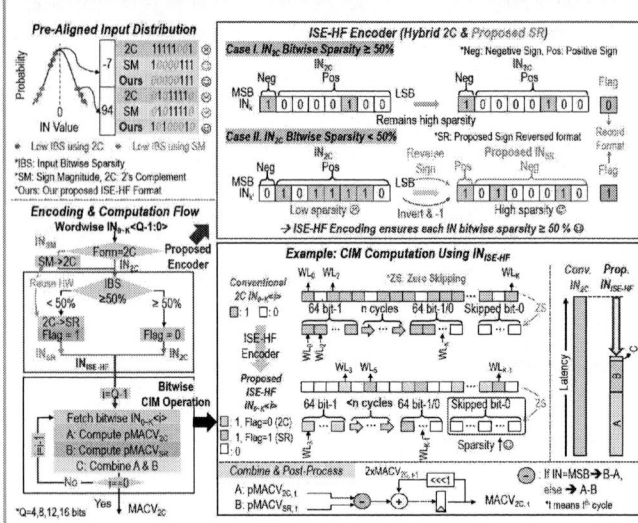

Figure 30.3.3: Flow chart and illustrations of proposed ISE-HF.

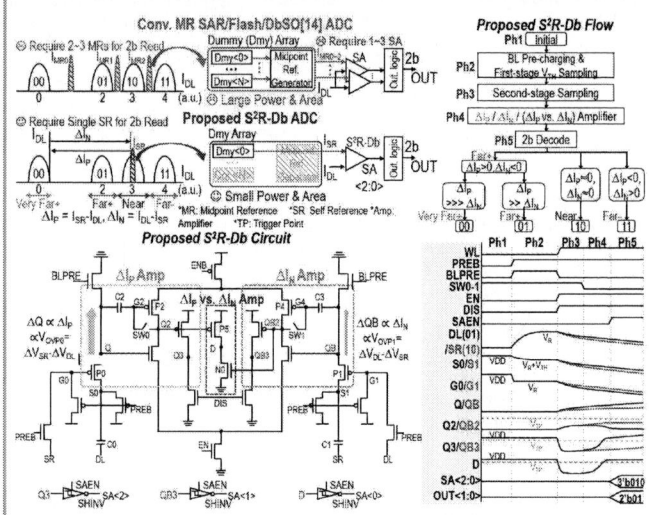

Figure 30.3.4: Illustration and waveform of proposed S²R-Db ADC.

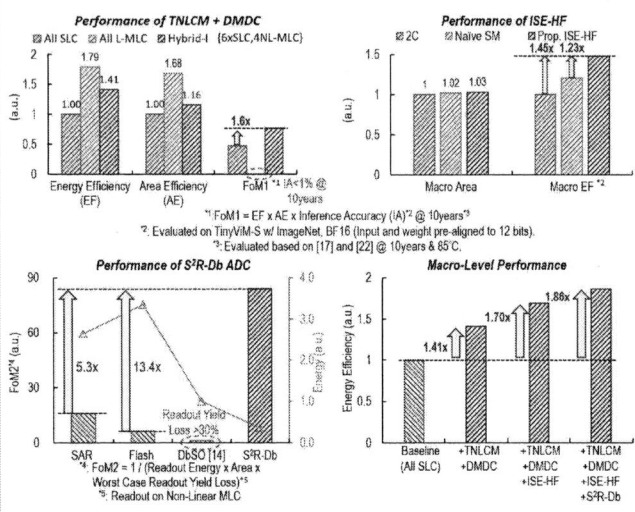

Figure 30.3.5: Simulated performance of proposed schemes.

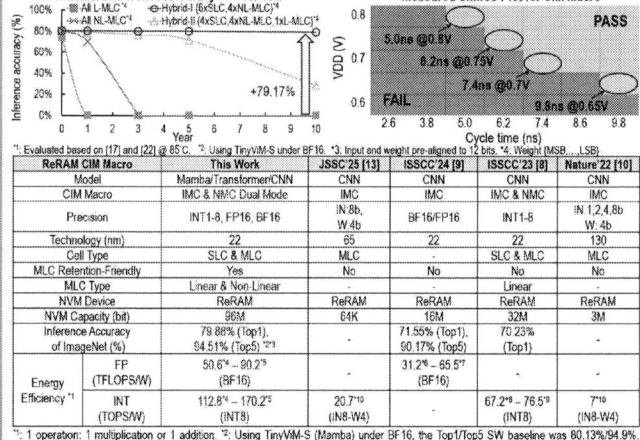

Figure 30.3.6: Simulated inference accuracy over years, measured shmoo plot and comparison table of recent ReRAM CIM macros.

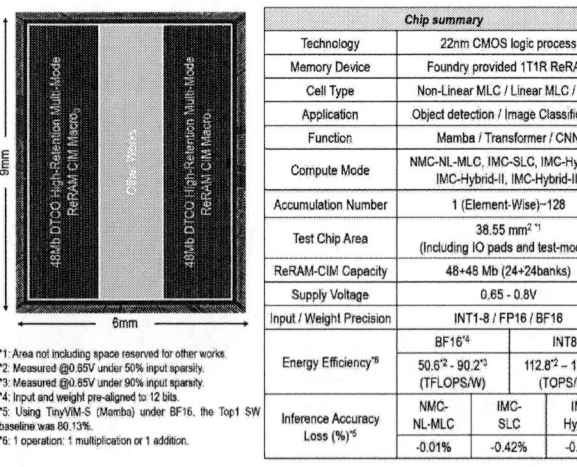

Figure 30.3.7: Die micrograph and chip performance summary.

- 2026 IEEE International Solid-State Circuits Conference

ISSCC 2026 / SESSION 30 / COMPUTE-IN-MEMORY / 30.4

30.4 A 28nm 106.85TOPS/W and 77.68TFLOPS/W CIM Macro with Stage-Wise-Enabled Lossless Compressors Based on Sign-Bit-Embedded Transition-Counting-Lines for Edge-AI Devices

Lichen Feng, Yunlong Liu, Lin Wu, Dong Li, Zhangming Zhu

Xidian University, Xi'an, China

Abstract

This paper proposes a bit-parallel digital CIM macro featuring a lossless compressor based on transition-counting-lines (TCLs) for bit-column addition. The bus TCL incorporates sign-bit extension to support signed 2's-complement MAC operations with minimal area overhead. A stage-wise enabling circuit prevents unnecessary toggling of adders, improving

energy efficiency by 42%. Leveraging these techniques, the proposed CIM macro, which is fabricated in 28nm CMOS, achieves 106.85TOPS/W for INT8 operations and 77.68TFLOPS/W for BF16 operations.

SRAM CIM macros improve energy efficiency for edge-AI devices. In contrast to "bit-serial" CIM macros [1], "bit-parallel" CIM directly matches the data format of computing systems, eliminating the parallel-to-serial conversion and specialized weight mapping strategies [2,3]. It can also reduce the MSB toggle rate [4], thereby lowering power consumption. Current bit-parallel CIM macros have achieved notable progress [5-7], however, as illustrated in Fig. 30.4.1, several challenges remain: (1) For 8b MAC, "bit-parallel" CIM suffers from routing congestion [7]. Some approaches concatenate multi-cycle computation results from low-precision MAC units (e.g., INT4) [6][8], while others employ ADCs as local compressors [5][9] to perform bit-column addition. A tradeoff between throughput and area overhead is required. Approximate compressors [6,10] can reduce resource consumption, but with accuracy loss. (2) Handling sign bits for 2's-complementary MAC in these macros often demands large extra area [6][8] or additional cycles [11]. (3) During multiplication and local compression before generating the final outer-product results, intermediates induce unnecessary toggling in high-fan-in adders, resulting in power inefficiency [12].

To address these challenges, this paper presents a fully-digital bit-parallel CIM macro with three key features: (1) Lossless digital local compressors based on transition-counting-lines (TCLs), realizing sparsity-adaptive and area-efficient (minimized transistor count) bit-column addition for outer-product matrix; (2) A bus TCL embedding sign-bit extension for signed 2's-complementary MAC, incurring only 2.95% extra area without additional delay; (3) A shared local 10-phase generator, enabling TCLs and adder trees to operate in a stage-wise manner within a computation cycle, thereby avoiding unnecessary toggling of adders and improving energy efficiency by 42%. A 28nm 32kb CIM macro based on TCLs is fabricated to support INT8 and BF16 MAC operations, achieving energy efficiencies of 106.85TOPS/W for INT8 and 77.68TFLOPS/W for BF16.

Figure 30.4.2 illustrates the architecture of the presented TCL-based CIM macro, which includes a global controller (Ctrl.), an SRAM read/write controller, four CIM banks, a global input buffering unit, and an exponent post-processing unit for BF16. Each CIM bank consists of a local 10-phase generator, an Out-Block Adder Tree, and four CIM blocks (each containing 16 CIM cells and an in-block adder tree). The in-block adder tree adds the 16 16b products from the 16 CIM cells, and the out-block adder tree adds the four 20b sums from the four In-Block Adder Trees, generating the 22b MAC Value (MACV). A CIM cell integrates a 16×8b 6T-SRAM for storing unsliced weights and a computing cell that performs 8b×8b bit-parallel multiplication. The computing cell comprises 64 multiply-multiplex units (MMUs), 15 counters (CNTs), and a combine adder, reading an 8b weight per cycle from the SRAM. The MMUs at the same bit-position are connected hierarchically, with a CNT at the end to form a TCL, performing bit-column additions. Following the outer-product rule, 15 TCLs are used for 8b×8b multiplication, and the 15 bit-column sums are accumulated by the Combine Adder to obtain a 16b product. The MMUs for sign-bits in the bus TCL (TCL-14) are reused across TCL-7~14 as embedded sign-bit extension for 2's-complementary multiplication. A clock cycle includes 10 phases: phases 0~7 (Ph0~7) sequentially enable MMUs in each TCL, Ph8 enables the Combine Adder, and Ph9 enables the In-Block Adder Tree. The CIM macro supports both INT8 and BF16 MAC operations. In INT8 mode, 64 features (XIN, 8b) are applied to four CIM banks, which hold 64×4 pre-read 8b weights, yielding four 22b MACVs per clock cycle. In BF16 mode, the exponents of the input features are pre-aligned [1], and the CIM banks store mantissas of the pre-aligned weights. The four mantissa MACVs obtained in the four CIM banks are post-processed to generate the FP32 output. These three stages are pipelined to generate the FP32 output per clock cycle.

Figure 30.4.3 illustrates the circuit structures and operations of the 10-phase generator and the TCL. The 10-phase generator comprises ten identical Rising-edge Delay Cells (RDCs) connected in cascade. V_{rst} controls each RDC's output state, while V_{delay} regulates its input-output delay. V_{rst} is synchronously asserted high (logic 1) with the rising edge of the main clock (Clk) and then quickly deasserted low (logic 0). When $V_{rst}=1$, all RDC outputs (Ph0~9) are set to 0. Ph0 is driven high when $V_{rst}=0$. After the delay of RDC #0 (Δt_{r0}), Ph1 is driven high as well. This process propagates through the 10 RDCs, driving Ph0~9 to high sequentially, generating the 10 phases as shown in Fig. 30.4.3. The TCL is composed of MMUs and a CNT. Each MMU comprises a Multiply section and a Multiplex section. The

Multiplex sections of the MMUs in a TCL are cascaded: the output (MOut) of each MMU serves as the input (MIn) of the next MMU. The MIn of the first MMU is tied to V_{DD} and the MOut of the last MMU is connected to the input of the CNT. The Multiply section determines whether Path1 or Path2 is activated in the Multiplex section, depending on the values of Phk, XIN[m], and W[n] (k, m, n=0~7). Only when phase Phk=1 and P_{mn}=XIN[m]×W[n]=1 is Path1 activated. In that case, MOut is the inversion of MIn, producing a transition (edge) that CNT counts. Taking TCL-9 with eight MMUs as an example, where the bit-products are {1, 1, 0, 0, 1, 0, 0, 1}, respectively: During Δt_{r0} (Ph0=1 and Ph1~9=0), only the first MMU (corresponding to P_{07}) activates Path1; since the MIn of the first MMU is tied to V_{DD}, Path1 yields a falling-edge-transition at MOut. The remaining MMUs activate Path2 and pass the signal unmodified, allowing the falling edge to propagate to the CNT and increment the count by 1. Similarly, transitions occur during $\Delta t_{r1,4,7}$. In total, the CNT detects four transition edges as the bit-column sum. The eight bit-products are encoded as the 4b count, forming an 8-to-4 lossless compression. Ph8 and Ph9 remain low during Ph0~7, so the Combine Adder and Adder Tree remain inactive, reducing dynamic power consumption.

Figure 30.4.4 illustrates how the bus TCL implements embedded sign-bit extension, and the enabling-circuit structures for Combine Adder and In-Block Adder Tree. For 2's-complementary signed multiplication, TCL-14 forms the bus TCL by connecting P_{m7}, m=0~7. P_{07} is used by each of TCL-7~14 (thus reused 8×), P_{17} is used by each of TCL-8~14 (reused 7×), and so forth, thereby realizing sign-bit extension. To implement the complete 2's-complementary multiplication rule, the inverted W_n's (n=0~7) are applied to the invert-bit MMUs, respectively, and the CNT in TCL-7 is initialized to XIN[7]. For unsigned computation, we only need to disconnect P_{m7} (m=0~7) in TCL-14, multiplex W_n's to the invert-bit MMUs, and initialize the CNT in TCL-7 to 0. Switching between the two modes requires only seven transmission gates (TGs) and eight 2-to-1 multiplexers, with an additional area overhead of 2.95%. Before Ph8, the inputs of the combine adder and the in-block adder tree are forced to ground using single-pole double-throw switches, preventing the propagation of spurious toggling. As shown in Fig. 30.4.2, the In-Block Adder Tree is of four stages. Each stage can include phase-controlled switches to improve energy efficiency, while they increase area overhead. We simulated the area and efficiency, versus the number of enabling stages. As a tradeoff, one enabling stage is selected for Combine Adder and one for In-Block Adder Tree, achieving a 42% energy efficiency improvement with only 6% area efficiency degradation.

Figure 30.4.5 illustrates the performance of the presented design. The 8b bit-parallel MAC computing cells are compared in terms of MAC cell FoM (Energy Efficiency/Transistor Count). In comparison with standard digital bit-multiplication with standard digital compressors, analog bit-multiplication with two 3b ADCs as compressors [5], and the design [6] employing two 4b bit-parallel MAC cells with ripple carry adders, the presented MAC cell with TCLs achieves the FoM improvement by 1.66 to 1.87×. To better demonstrate the impacts of sign-bit extension and stage-enabling, two additional implementations are evaluated: (1) a baseline unsigned MAC cell without bus TCLs or the adder-enabling circuit, and (2) a signed MAC cell without the enabling circuit. Relative to the baseline design, connecting P_{m7}'s (m=0~7) for bus TCL and increasing CNT bit-widths in signed MAC cell consume 1.03× area and more power. With the introduction of the two-stage adder-enabling circuit, the final fabricated design achieves a 1.35× energy efficiency with only a 1.09× area overhead compared to the baseline unsigned design, and a 1.42× improvement with a 1.06× area overhead compared to the signed design without adder enabling. The 10k Monte Carlo simulation results of the transition delay of MMUs (Δt_{MMU}) and the transition delay of RDCs ($\Delta t_{r0~7}$, annotated as Δt_{RDC}) indicate that, Δt_{MMU} is consistently smaller than Δt_{RDC}, ensuring correct TCL operation under process and supply voltage variations. The energy efficiency of the presented signed MAC macro is measured across various input toggle rates and input sparsity levels, with the 50% weight sparsity. A peak performance of 106.85TOPS/W is achieved.

Figure 30.4.6 shows the measurement results of the TCL-based CIM chip, which is fabricated in a 28nm CMOS technology. The measured access time is 4.1ns at 0.9V for BF16 inputs and weights, and FP32 outputs. The measured maximum energy and area efficiency are 77.68TFLOPS/W and 1.24TFLOPS/mm² for BF16 MAC operation under 0.55V

518 • 2026 IEEE International Solid-State Circuits Conference

979-8-3315-8937-0/26 $31.00 © 2026 IEEE

ISSCC 2026 / February 18, 2026 / 9:15 AM

and 0.9V supply voltages, respectively, with 90% input sparsity and 10% toggle rate, while the weight sparsity is 50%. By adopting the sign-bit-embedded TCL and stage-wise-enabling, the presented bit-parallel CIM macro improves the output precision×TOPS/W×TOPS/mm² (FoM1) by 1.25 to 2.88× and the TFLOPS/W×TFLOPS/mm² (FoM2) by 2.20 to19.9×, in comparison to prior signed CIM macros [2,3,6,7]. In BF16 mode, this work achieves an 81.05% inference accuracy when used on ViT-B @ImageNetV2 and an 81.19% inference accuracy when used on ResNet50 @ImageNetV2. Figure 30.4.7 presents a die micrograph and a summary of chip performance.

Acknowledgement:

This work is supported in part by the National Key R&D Program of China (2022ZD0118903), in part by the National Natural Science Foundation of China (62474128, U22A2013, U24A20291). (Corresponding authors are Zhangming Zhu, zmyh@263.net; Yunlong Liu, liuyunlongahu@163.com)

References:

[1] A. Guo et al., "A 28nm 64-kb 31.6-TFLOPS/W Digital-Domain Floating-Point-Computing-Unit and Double-Bit 6T-SRAM Computing-in-Memory Macro for Floating-Point CNNs," *ISSCC*, pp. 128-129, Feb, 2023.
https://doi.org/10.1109/ISSCC42615.2023.10067260
[2] X. Wang et al., "A 28nm 17.83-to-62.84TFLOPS/W Broadcast-Alignment Floating-Point CIM Macro with Non-Two's-Complement MAC for CNNs and Transformers," *ISSCC*, pp. 254-255, Feb. 2025. https://doi.org/10.1109/ISSCC49661.2025.10904738
[3] X. Chen et al., "A 28nm 64kb Bit-Rotated Hybrid-CIM Macro with an Embedded Sign-Bit-Processing Array and a Multi-Bit-Fusion Dual-Granularity Cooperative Quantizer," *ISSCC*, pp. 260-261, Feb. 2025. https://doi.org/10.1109/ISSCC49661.2025.10904646
[4] H. Fujiwara et al., "A 3nm, 32.5TOPS/W, 55.0TOPS/mm2 and 3.78Mb/mm2 Fully-Digital Compute-in-Memory Macro Supporting INT12 × INT12 with a Parallel-MAC Architecture and Foundry 6T-SRAM Bit Cell," *ISSCC*, pp. 572-573, Feb. 2024. https://doi.org/10.1109/ISSCC49657.2024.10454556
[5] Y. Yuan et al., "A 28nm 72.12TFLOPS/W Hybrid-Domain Outer-Product Based Floating-Point SRAM Computing-in-Memory Macro with Logarithm Bit-Width Residual ADC," *ISSCC*, pp. 576-577, Feb. 2024.
https://doi.org/10.1109/ISSCC49657.2024.10454313
[6] Y. Yuan et al., "A 28nm 192.3TFLOPS/W Accurate/Approximate Dual-Mode-Transpose Digital 6T-SRAM CIM Macro for Floating-Point Edge Training and Inference," *ISSCC*, pp. 258-259, Feb. 2025. https://doi.org/10.1109/ISSCC49661.2025.10904659
[7] P.-C. Wu et al., "A 22nm 832Kb Hybrid-Domain Floating-Point SRAM In-Memory-Compute Macro with 16.2-70.2TFLOPS/W for High-Accuracy AI-Edge Devices," *ISSCC*, pp. 126-127, Feb. 2023. https://doi.org/10.1109/ISSCC42615.2023.10067527
[8] H. Mori et al., "A 4nm 6163-TOPS/W/b 4790–TOPS/mm2/b SRAM Based Digital-Computing-in-Memory Macro Supporting Bit-Width Flexibility and Simultaneous MAC and Weight Update," *ISSCC*, pp. 132-133, Feb. 2023. https://doi.org/10.1109/ISSCC42615.2023.10067555
[9] S. Hong et al., "Dyamond: Compact and Efficient 1T1C DRAM IMC Accelerator With Bit Column Addition for Memory-Intensive AI," *IEEE Journal of Solid-State Circuits*, vol. 60, no. 4, pp. 1299-1310, April 2025. https://doi.org/10.1109/JSSC.2025.3538899
[10] D. Wang et al., "DIMC: 2219TOPS/W 2569F2/b Digital In-Memory Computing Macro in 28nm Based on Approximate Arithmetic Hardware," *ISSCC*, pp. 266-267, Feb. 2022. https://doi.org/10.1109/ISSCC42614.2022.9731659

[11] F. Tu et al., "A 28nm 29.2TFLOPS/W BF16 and 36.5TOPS/W INT8 Reconfigurable Digital CIM Processor with Unified FP/INT Pipeline and Bitwise In-Memory Booth Multiplication for Cloud Deep Learning Acceleration," *ISSCC*, pp. 254-255, Feb. 2022. https://doi.org/10.1109/ISSCC42614.2022.9731762
[12] C.-F. Lee et al., "A 12nm 121-TOPS/W 41.6-TOPS/mm2 All Digital Full Precision SRAM-based Compute-in-Memory with Configurable Bit-width For AI Edge Applications," *IEEE Symp. VLSI Circuits*, pp. 24-25, June 2022. https://doi.org/10.1109/VLSITechnologyandCir46769.2022.9830438

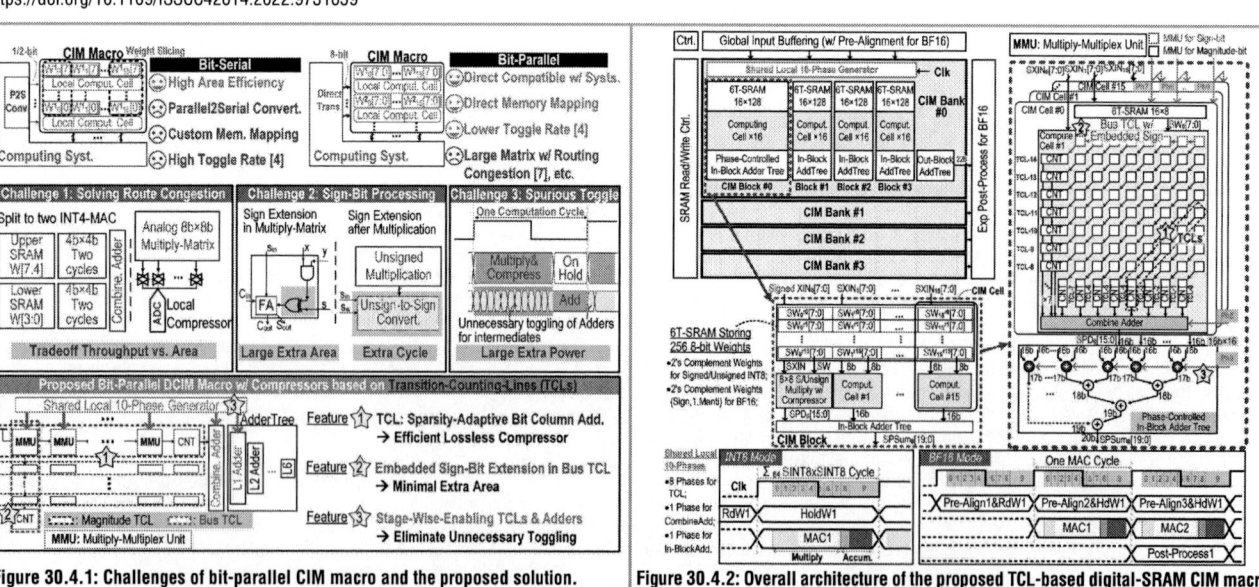

Figure 30.4.1: Challenges of bit-parallel CIM macro and the proposed solution.

Figure 30.4.2: Overall architecture of the proposed TCL-based digital-SRAM CIM macro.

DIGEST OF TECHNICAL PAPERS • 519

979-8-3315-8937-0/26 $31.00 © 2026 IEEE

ISSCC 2026 / SESSION 30 / COMPUTE-IN-MEMORY / 30.4

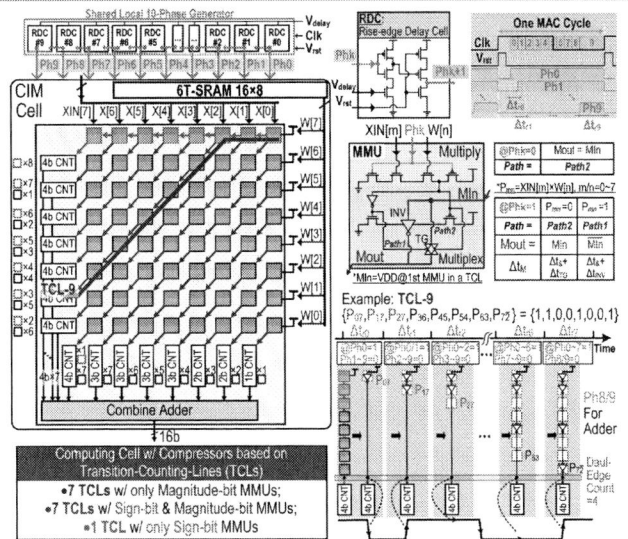

Figure 30.4.3: The structures and workflows of the TCL and the 10-phase generator.

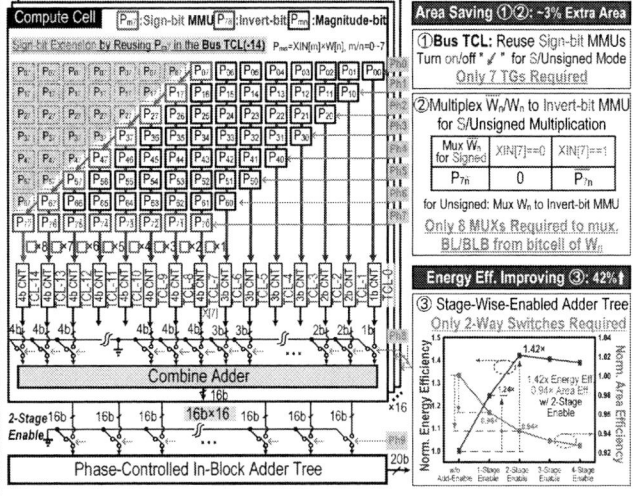

Figure 30.4.4: Embedded sign-bit extension in bus TCL-14 and the enabling-circuit structure for Combine Adder and In-Block Adder Tree.

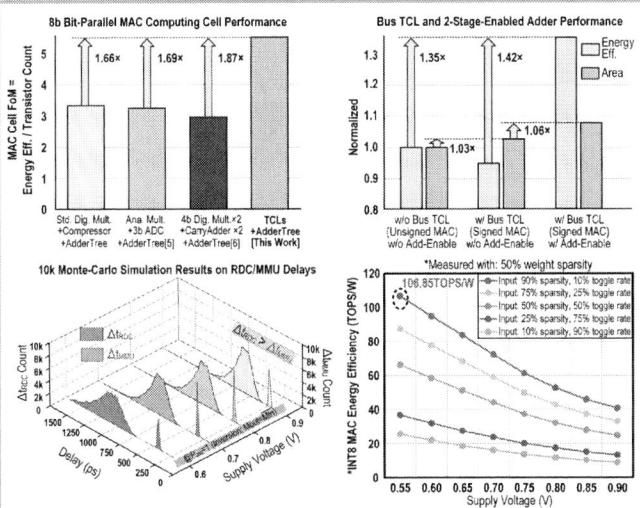

Figure 30.4.5: Simulated MAC computing cell performance and measured energy efficiencies of the proposed CIM macro.

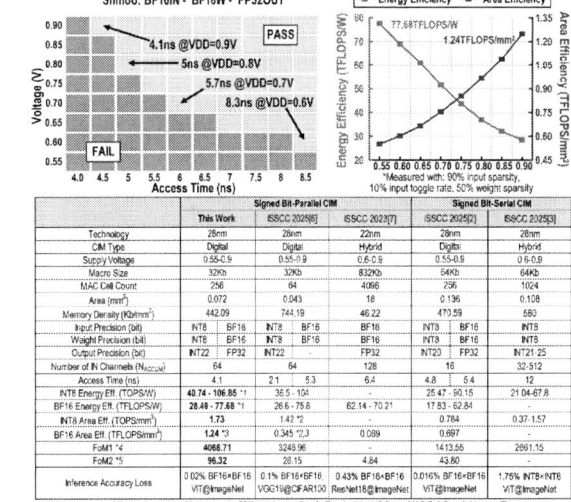

Figure 30.4.6: Measurement results and comparison table.

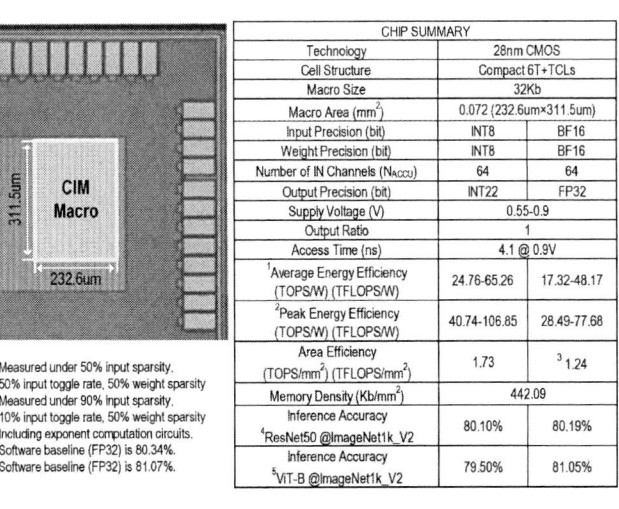

CHIP SUMMARY		
Technology	28nm CMOS	
Cell Structure	Compact 6T+TCLs	
Macro Size	32Kb	
Macro Area (mm²)	0.072 (232.6um×311.5um)	
Input Precision (bit)	INT8	BF16
Weight Precision (bit)	INT8	BF16
Number of IN Channels (N_ACCU)	64	64
Output Precision (bit)	INT22	FP32
Supply Voltage (V)	0.55-0.9	
Output Ratio	1	
Access Time (ns)	4.1 @ 0.9V	
[1]Average Energy Efficiency (TOPS/W) (TFLOPS/W)	24.76-65.26	17.32-48.17
[2]Peak Energy Efficiency (TOPS/W) (TFLOPS/W)	40.74-106.85	28.49-77.68
Area Efficiency (TOPS/mm²) (TFLOPS/mm²)	1.73	[3]1.24
Memory Density (Kb/mm²)	442.09	
Inference Accuracy [4]ResNet50 @ImageNet1k_V2	80.10%	80.19%
Inference Accuracy [5]ViT-B @ImageNet1k_V2	79.50%	81.05%

[1] Measured under 50% input sparsity, 50% input toggle rate, 50% weight sparsity.
[2] Measured under 90% input sparsity, 10% input toggle rate, 50% weight sparsity.
[3] Including exponent computation circuits.
[4] Software baseline (FP32) is 80.34%.
[5] Software baseline (FP32) is 81.07%.

Figure 30.4.7: Die micrograph and summary of key performance metrics.

• 2026 IEEE International Solid-State Circuits Conference

ISSCC 2026 / SESSION 30 / COMPUTE-IN-MEMORY / 30.5

30.5 A 16nm 72kb 120.5TFLOPS/W Versatile-Format Dual-Representation Gain-Cell CIM Macro for General Purpose AI Tasks

Jen-Chun Tien*[1], Win-San Khwa*[2], Le-Jung Hsieh[1], Tsung-Han Lou[1], Jyun-Cheng Bai[1], Yu-Sheng Kao[1], Ting-Hao Hsu[1], Mai Tseng[1], Hung-Hsi Hsu[1],
Yao-Kai Yeh[1], De-Qi You[1], Ashwin Sanjay Lele[3], Brian Crafton[3], Bo Zhang[3], Ping-Sheng Wu[2], Ya-Tang Yang[1], Chung-Chuan Lo[1], Ren-Shuo Liu[1],
Chih-Cheng Hsieh[1], Kea-Tiong Tang[1], Meng-Fan Chang[1,2]

[1]National Tsing Hua University, Hsinchu, Taiwan, [2]TSMC Corporate Research, Hsinchu, Taiwan, [3]TSMC Corporate Research, San Jose, CA
*Equally Credited Authors (ECAs)

Abstract

Diverse AI workloads require flexible data representations and numerical formats. In this work, we present a reconfigurable 2's-complement and sign-magnitude scheme integrated within a versatile-format CIM macro supporting MX, LNS, FP, and INT for MAC operations.

Implemented in a 16nm 72kb gain-cell array, the macro achieves record energy efficiency (120.5TFLOPS/W) and throughput density (3.18 TOPS/mm2) in MXINT8 mode.

Diverse AI workloads require flexible compute-in-memory (CIM) architectures that support multiple numerical formats (e.g., floating-point (FP) and integer (INT)) and data representations (e.g., 2's complement (2C) and sign magnitude (SM)) to balance accuracy, area, and energy efficiency. Commercial accelerators have begun adopting micro-scaling (MX) to train low-precision large language models (LLMs) [1,4], while the logarithmic number system (LNS) has shown promotion for spectral-based AI kernels (e.g., Fourier neural operator) [24-29] due to its multiplication efficiency. However, most existing CIM architectures are tailored to a specific representation or format [5-22], limiting adaptability across diverse AI tasks and deployment scenarios. Figure 30.5.1 illustrates the key challenges in developing CIM for general-purpose AI workloads: (1) 2C versus SM area-power trade-off: 2C representation suffers from high toggle rates due to low bit-wise sparsity near negative zero, whereas SM increases hardware area overhead. (2) Fixed MX block size: Using a single MX block-size (k) constrains the energy-accuracy trade-off as it cannot adapt to different input distributions. (3) LNS area overhead: Supporting LNS in CIM requires replacing MAC operations with additions and look-up table (LUT) accesses, increasing area overhead and reducing area utilization since hardware cannot be shared across formats.

To address these challenges, this work presents a versatile-format (MX-LNS-FP-INT) dual-representation (2C-SM) gain-cell CIM (VF-DR GC-CIM) for general-purpose AI tasks, featuring: (1) Compact in-situ weight (W) sparsity booster (CiWSB) with polarity-shift W control (PSWC): Transfers weight polarity to inputs (INs) to overcome the area-energy trade-off between 2C and SM and enables reconfigurable 2CSM computation in CIM macros with minimal area overhead; (2) Input distribution-aware MX quantizer (IDA-MXQ): Adaptively selects the optimal MX block-size k based on IN distributions, improving both energy efficiency and accuracy; (3) Multi-format adaptive computing cell (MFA-CC) with delta exponent subtractor (ΔES) and LUT support: Shares hardware across LNS, FP, and INT modes. The ΔES increases bit-wise sparsity in LNS mode, enabling high area utilization and energy efficient computation.

We fabricated a 16nm 72kb VF-DR GC-CIM supporting versatile-format MAC operations. Leveraging gain-cell memory for high storage density, the macro achieves energy efficiency of 120.5TFLOPS/W with area efficiency of 3.18TOPS/mm² in MXINT8 mode, 98.1TFLOPS/W with 3.18TOPS/mm² in LNS8 mode, 50.4TFLOPS/W with 1.59TOPS/mm² in BF16 mode, and 138.5TOPS/W with 3.18TOPS/mm² in INT8 mode.

Figure 30.5.1 illustrates the structure of the proposed VF-DR GC-CIM, comprising four tile-based matrix-matrix CIM clusters. Each cluster contains an IDA-MXQ and four CIM banks. Each CIM bank includes 32 sub-banks, an alignment and adder tree, and a FP converter. Each sub-bank consists of a gain-cell array (8 rows×18 columns), an IN processor, a CiWSB, and four MFA-CC units. Each CIM bank provides four output channels, broadcasting the same weight to four different inputs across MFA-CC units. The outputs from the four CIM banks are combined to produce the final matrix–matrix result.

Figure 30.5.2 shows the proposed CiWSB. By shifting the polarity of Ws to INs and retaining Ws in positive 2C (P2C) representation, the average MFA-CC power consumption is reduced by 31%. This reduction arises because the MUX-based multiplier [12] in MFA-CC is more sensitive to W sparsity than IN sparsity. Conventional digital multipliers also benefit from this approach, as cases where (IN, W)=(−, −) can be converted to (+, +), effectively increasing sparsity. CiWSB increases W bit-wise sparsity by 20.1%, improving power efficiency in weight-stationary CIM flows. Moreover, the positive SM representation aligns with P2C, enabling direct execution of SM computations on 2C MAC units. This eliminates the for negative adders or subtractors [23] improving area efficiency for SM MAC operations.

The CiWSB comprises 18 stationary gated units (SGUs) and two PSWCs. During computation, the SGUs retain the weights read from the gaincell array, while the PSWCs and SGUs together convert them into a P2C representation. The encoding flow operates as follows: In 2C mode, when a W is negative, the PSWC converts it to positive by locating the first "1" from the LSB and setting all CTRL signals from the MSB down to this bit at 1. This reconfigures the SGU readout path, after which the SGU inverts the readout value. Unlike the conventional invert-plus-one method [30], which relies on numerous power-hungry halfadders, PSWC achieves the same function while reducing power consumption by 91.24% and area overhead by 76.4%. In SM mode, the PSWC sets the MSB of CTRL to 1 for negative W, enabling direct computation of SM data on 2C MAC units to enhance energy

efficiency. Finally, IN processor converts INs (2C or SM) into corresponding 2C, while adjusting their polarity according to the polarity of W.

Figure 30.5.3 illustrates the proposed IDA-MXQ, which dynamically adjusts the block-size (k) for each IN data patch within a single cycle, thereby enhancing computational efficiency without compromising accuracy. Our analysis shows that the k of each MX group should be adaptively determined according to the variance of the data distribution, rather than adopting a uniform k across all neural network models [24]. For high-variance data, a smaller k preserves the data range during quantization, though at the cost of higher computational overhead. For low-variance data, a larger k improves computational efficiency by grouping more inputs into an MX block, with only a minor loss due to quantization distortion, particularly for outliers. IDA-MXQ adaptively quantizes FP32 data patches to MX-INT8/FP8/LNS8 with k ranging from 4 to 64, depending on the distribution variance. Unlike conventional approaches that combine computation across multiple MX blocks and thereby accumulate both partial results and quantization errors [6], this design completes MAC operations entirely within each MX block and outputs FP32 results directly, ensuring that MX acceleration does not significantly degrade accuracy.

The IDA-MXQ consists of two main components: (1) an input distribution-aware k detector (IDA-KD), which selects the optimal k for each patch of 64 INs, and (2) an FP to MX quantizer (FP2MX), which quantizes FP32 data into MX format to accelerate MAC operations. IDA-KD and FP2MX both complete their operations within one cycle.

IDA-KD operates in three phases. In phase 1, an exponent frequency counter (EFC) records the frequency of each input exponent (IN_E) within the high-probability region (HPR). In most neural network models, the IN_E distribution is concentrated within a narrow HPR, enabling IDA-KD to significantly reduce area and latency overhead in variance detection. For ImageNet-1K inference with MobileViT, HPR is set between −15 and 2, predetermined during training. In phase 2, a region aggregator (RA) aggregates frequencies within regions A, B, and C (F_A, F_B, F_C) and extracts the most frequent value (MFV). In Phase 3, a k finder (KF) determines the optimal k based on F_A, F_B, F_C, and MFV. In the following cycle, FP2MX quantizes the FP32 data patch to MX with the optimal k. IDA-MXQ adaptively selects the optimal k by balancing the trade-off between precision and efficiency. Moreover, by limiting detection to HPR, the IDA-MXQ significantly reduces latency and increases area efficiency compared to full-range detection methods.

Figure 30.5.4 illustrates the proposed MFA-CC. Each sub-bank integrates 4 MFA-CCs, which share the same Ws ($M_0[8:0]$, $M_1[8:0]$) broadcasted from the array, while receiving four independent inputs from IDA-MXQ for matrix-matrix MAC operations. An MFA-CC consists of 18 bit-wise reconfigurable units (BRUs), two ΔES, and two LUTs. Each BRU can be configured as either a bit-wise full adder or a multiplier [12], enabling support for LNS, FP, and INT MAC operations within the same hardware and improving area utilization across formats. In 8b reconfigurable mode, BRU is divided into two parts to support reconfigurable MAC operations.

Unlike FP or INT multiplication, which requires costly multipliers, LNS multiplication uses simple adders followed by a LUT [27-28]. The proposed ΔES with LUT improves energy efficiency by enhancing sparsity in LNS-based LUT results. During alignment, conventional LNS LUTs read out the entire LUT length, including truncated bits [27-28], causing unnecessary bit toggles in both the LUT and alignment logic. In contrast, the ΔES generates an access mask based on the delta product exponent (ΔPD_E). The access mask bit-width (AMB) is set as 8−ΔPD_E, where 8 is the LUT length. By masking out truncated bits, the ΔES avoids unnecessary readouts without a loss of accuracy, increases LSB sparsity during alignment, and reduces alignment energy. In FP mode, the sign and leading-1 are pre-stored in memory and read out when combined with multiplication. ΔES is reconfigured to compute the number of shifting bits (NS) of products, further improving area utilization across formats beyond LNS.

Figure 30.5.5 summarizes the performance of the proposed schemes. CiWSB improved FoM1 by 1.37 to 6.81× compared with fixed-representation CIM designs by increasing weight bit-wise sparsity to improve energy efficiency in 2C mode, while reusing the same mechanism to increase area efficiency in SM mode. IDA-MXQ achieved a 1.29× FoM2 gain over MX-CIM with a fixed block-size by adaptively selecting the optimal block-size according to the input data distribution. MFACC improved FoM2 by 1.49 to 1.87× over conventional versatile-format CIM implementations, by enabling hardware sharing across

520 • 2026 IEEE International Solid-State Circuits Conference

979-8-3315-8937-0/26 $31.00 © 2026 IEEE

MX-LNS-FP-INT formats and providing support for LNS. Finally, VF-DR GC-CIM achieved a 1.7 to 2.3× improvement in FoM1 across representative AI workloads, demonstrating that supporting MXLNSFPINT format provides a range of benefits in general-purpose CIM.

Figure 30.5.6 presents measured results and a comparison table of the proposed VFDR GCCIM performing versatile-format MAC operations. Shmoo plots indicated a 1.2ns access time for MXINT8-MAC operation using a 0.8V power supply. Figure 30.5.7 presents a die micrograph and chip performacne summary table. Product alignment and full-output precision preserved inference accuracy, with only a <0.01% loss across all formats, when applying ResNet-20 to CIFAR-100 and MobileViT to ImageNet-1K.

Acknowledgement:
The authors would like to thank TSMC and NSTC for manufacturing and financial support.

References:
[1] H. Yang et al., "An Empirical Study of Microscaling Formats for Low-Precision LLM Training," *IEEE Symp. on Computer Arithmetic*, pp. 1-8, May 2025. https://doi.org/10.1109/ARITH64983.2025.00011
[2] C. Xiao et al., "Microscaling Vision Transformers on FPGAs," *IEEE Annu. Int. Symp. on Field-Programmable Custom Computing Machines*, p. 275, May 2025. https://doi.org/10.1109/FCCM62733.2025.00035
[3] H. Mun et al., "ASAP: A 28nm Transformer Training Accelerator with Alternating Sparsity and Asymmetrical Microscaling Floating-Point Precision," *IEEE Symp. VLSI Circuits*, pp. 1-3, June 2025. https://doi.org/10.23919/VLSITechnologyandCir65189.2025.11075114
[4] Y. Jang et al., "Redefining PIM Architecture with Compact and Power-Efficient Microscaling," i*Int. Conf. on Electron., Inf., and Communication*, pp. 1-4, Jan. 2025. https://doi.org/10.1109/ICEIC64972.2025.10879742
[5] D.-Q. You et al., "A 22nm 104.5TOPS/W µ-NMC-Δ-IMC Heterogeneous STT-MRAM CIM Macro for Noise-Tolerant Bayesian Neural Networks," *ISSCC*, pp. 250-251, Feb. 2025. https://doi.org/10.1109/ISSCC49661.2025.10904540
[6] W.-S. Khwa et al., "A 16nm 216kb, 188.4TOPS/W and 133.5TFLOPS/W Microscaling Multi-Mode Gain-Cell CIM Macro Edge-AI Devices," *ISSCC*, pp. 252-253, Feb. 2025. https://doi.org/10.1109/ISSCC49661.2025.10904606
[7] X. Wang et al., "A 28nm 17.83-to-62.84TFLOPS/W Broadcast-Alignment Floating-Point CIM Macro with Non-Two's-Complement MAC for CNNs and Transformers," *ISSCC*, pp. 254-255, Feb. 2025. https://doi.org/10.1109/ISSCC49661.2025.10904738
[8] Z. Yue et al., "A 51.6TFLOPs/W Full-Datapath CIM Macro Approaching Sparsity Bound and <2-30 Loss for Compound AI," *ISSCC*, pp. 256-257, Feb. 2025. https://doi.org/10.1109/ISSCC49661.2025.10904702
[9] Y. Yuan et al., "A 28nm 192.3TFLOPS/W Accurate/Approximate Dual-Mode-Transpose Digital 6T-SRAM CIM Macro for Floating-Point Edge Training and Inference," *ISSCC*, pp. 258-259, Feb. 2025. https://doi.org/10.1109/ISSCC49661.2025.10904659
[10] X. Chen et al., "A 28nm 64kb Bit-Rotated Hybrid-CIM Macro with an Embedded Sign-Bit-Processing Array and a Multi-Bit-Fusion Dual-Granularity Cooperative Quantizer," *ISSCC*, pp. 260-261, Feb. 2025. https://doi.org/10.1109/ISSCC49661.2025.10904646
[11] Y. Wang et al., "A 28nm 83.23TFLOPS/W POSIT-Based Compute-in-Memory Macro for High-Accuracy AI Applications," *ISSCC*, pp. 566-567, Feb. 2024. https://doi.org/10.1109/ISSCC49657.2024.10454489
[12] W.-S. Khwa et al., "A 16nm 96Kb Integer/Floating-Point Dual-Mode-Gain-Cell-Computing-in-Memory Macro Achieving 73.3-163.3TOPS/W and 33.2-91.2TFLOPS/W for AI-Edge Devices," *ISSCC*, pp. 568-569, Feb. 2024. https://doi.org/10.1109/ISSCC49657.2024.10454447
[13] A. Guo et al., "A 22nm 64kb Lightning-Like Hybrid Computing-in-Memory Macro with a Compressed Adder Tree and Analog-Storage Quantizers for Transformer and CNNs," *ISSCC*, pp. 570-571, Feb. 2024. https://doi.org/10.1109/ISSCC49657.2024.10454278

[14] H. Fujiwara et al., "A 3nm, 32.5TOPS/W, 55.0TOPS/mm² and 3.78Mb/mm² Fully-Digital Compute-in-Memory Macro Supporting INT12 × INT12 with a Parallel-MAC Architecture and Foundry 6T-SRAM Bit Cell," *ISSCC*, pp. 572-573, Feb. 2024. https://doi.org/10.1109/ISSCC49657.2024.10454556
[15] K. Yoshioka, "A 818-4094TOPS/W Capacitor-Reconfigured CIM Macro for Unified Acceleration of CNNs and Transformers," *ISSCC*, pp. 574-575, Feb. 2024. https://doi.org/10.1109/ISSCC49657.2024.10454489
[16] Y. Yuan et al., "A 28nm 72.12TFLOPS/W Hybrid-Domain Outer-Product Based Floating-Point SRAM Computing-in-Memory Macro with Logarithm Bit-Width Residual ADC," *ISSCC*, pp. 576-576, Feb. 2024. https://doi.org/10.1109/ISSCC49657.2024.10454313
[17] Y. He et al., "A 28nm 2.4Mb/mm² 6.9-16.3TOPS/mm² eDRAM-LUT-Based Digital-Computing-in-Memory Macro with In-Memory Encoding and Refreshing," *ISSCC*, pp. 578-579, Feb. 2024. https://doi.org/10.1109/ISSCC49657.2024.10454323
[18] T.-H. Wen et al., "A 22nm 16Mb Floating-Point ReRAM Compute-in-Memory Macro with 31.2TFLOPS/W for AI Edge Devices," *ISSCC*, pp. 580-581, Feb. 2024. https://doi.org/10.1109/ISSCC49657.2024.10454468
[19] S. Um et al., "Dial: An Energy-Efficient DRAM In-Memory Computing Accelerator with Compact Partial Product LUT and Twisted Differential ADC," *IEEE Symp. VLSI Circuits*, pp. 1-3, June 2025. https://doi.org/10.23919/VLSITechnologyandCir65189.2025.11075215
[20] H.-H. Hsu et al., "A 22nm 41.8TFLOPS/W AI-Edge Transformer/CNN Nonvolatile-Processor Using QKV-Softmax-Layer-Fused Hybrid ReRAM-CIM and Concurrent-Transpose/Non-Transpose SRAM-CIM," *IEEE Symp. VLSI Circuits*, pp. 1-3, June 2025. https://doi.org/10.23919/VLSITechnologyandCir65189.2025.11074985
[21] Z. Wu et al., "CELLA: A 28nm Compute-Memory Co-Optimized Real-Time Digital CIM-Based Edge LLM Accelerator with 1.78ms-Response in Prefill and 31.32 token/s in Decoding," *IEEE Symp. VLSI Circuits*, pp. 1-3, June 2025. https://doi.org/10.23919/VLSITechnologyandCir65189.2025.11075101
[22] M.-E. Shih et al., "NVE: A 3nm 23.2TOPS/W 12b-Digital-CIM-Based Neural Engine for High-Resolution Visual-Quality Enhancement on Smart Devices," *ISSCC*, pp. 360-361, Feb. 2024. https://doi.org/10.1109/ISSCC49657.2024.10454482
[23] H. An et al., "An 8.09TOPS/W Neural Engine Leveraging Bit-Sparsified Sign-Magnitude Multiplications and Dual Adder Trees," *ISSCC*, pp. 422-423, Feb. 2023. https://doi.org/10.1109/ISSCC42615.2023.10067269
[24] M. Li et al., "A 19.7 TFLOPS/W Multiply-Less Logarithmic Floating-Point CIM Architecture with Error-Reduced Compensated Approximate Adder," *ISCAS*, pp. 1-5, May 2024. https://doi.org/10.1109/ISCAS58744.2024.10558433
[25] Z. Li et al., "Fourier Neural Operator for Parametric Partial Differential Equations," *arXiv preprint*, pp. 1-16, May 2021. https://doi.org/10.48550/arXiv.2010.08895
[26] J. Zhao *et al.*, "LNS-Madam: Low-Precision Training in Logarithmic Number System Using Multiplicative Weight Update," *IEEE Transactions on Computers*, vol. 71, no. 12, pp. 3179-3190, Dec. 2022. https://doi.org/10.1109/TC.2022.3202747
[27] S. Vogel et al., "Efficient Hardware Acceleration of CNNs Using Logarithmic Data Representation with Arbitrary Log-Base," *ICCAD*, pp. 1-8, Nov. 2018. https://doi.org/10.1145/3240765.3240803
[28] P. Haghi et al., "Bridging the Gap Between LLMs and LNS with Dynamic Data Format and Architecture Codesign," *MICRO*, pp. 1617-1631, Nov. 2024. https://doi.org/10.1109/MICRO61859.2024.00118
[29] B. Crafton et al., "Finding the Pareto Frontier of Low-Precision Data Formats and MAC Architecture for LLM Inference," *DAC*, pp. 1-7, June 2025. https://doi.org/10.1109/DAC63849.2025.11132989
[30] M.M. Mano and M.D. Ciletti, *Digital Design: With an Introduction to the Verilog HDL, VHDL, and SystemVerilog*, 6th ed. Pearson, p. 51, 2017. Accessed on Dece. 8, 2025. <https://www.pearson.com/en-us/subject-catalog/p/digital-design-with-an-introduction-to-the-verilog-hdl-vhdl-and-systemverilog/P200000003241/9780137501984>

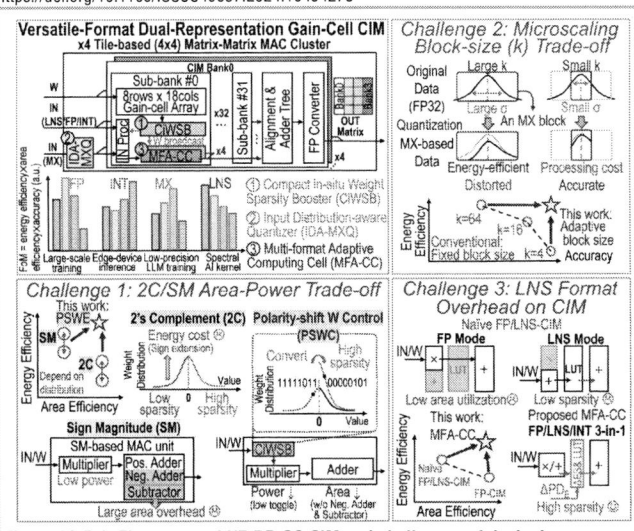

Figure 30.5.1: The proposed VF-DR GC-CIM and challenges of designing a general purposed CIM macro.

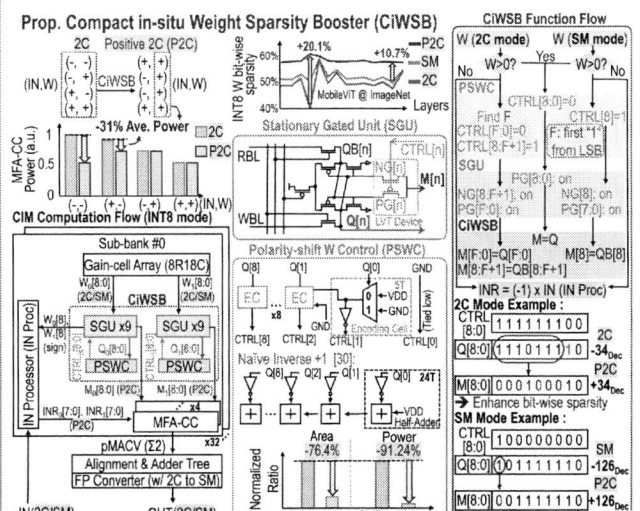

Figure 30.5.2: Proposed CiWSB with SGU, PSWC, and CiWSB function flow.

ISSCC 2026 / SESSION 30 / COMPUTE-IN-MEMORY / 30.5

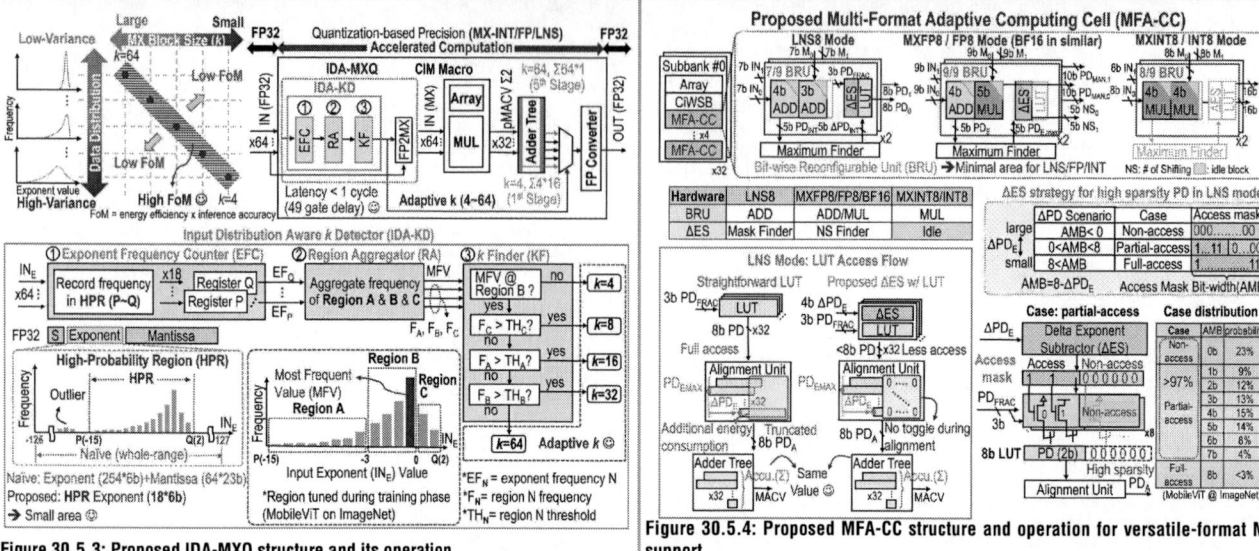

Figure 30.5.3: Proposed IDA-MXQ structure and its operation.

Figure 30.5.4: Proposed MFA-CC structure and operation for versatile-format MAC support.

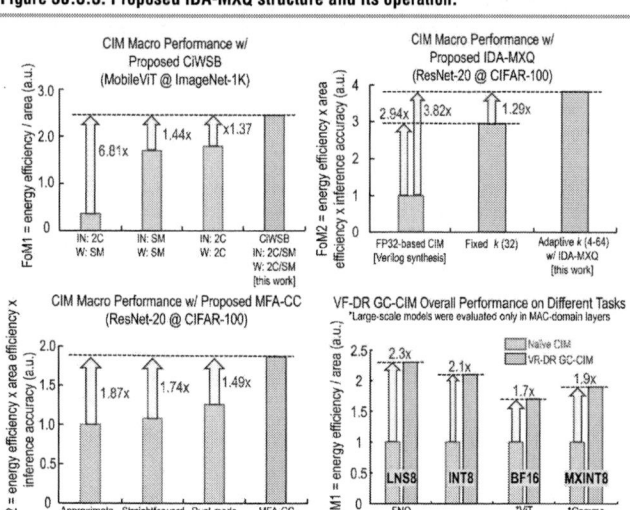

Figure 30.5.5: Simulated performance of proposed schemes.

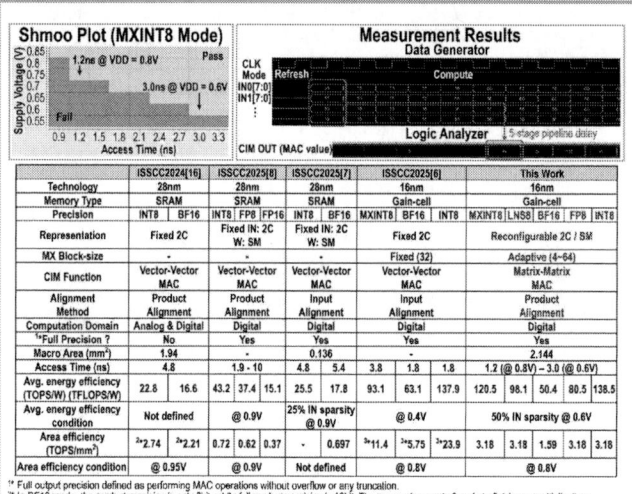

Figure 30.5.6: Measurement results and comparison table.

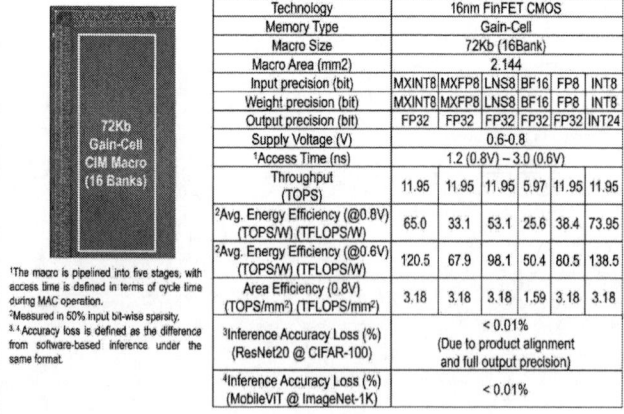

Figure 30.5.7: Die micrograph and chip performance summary table.

• 2026 IEEE International Solid-State Circuits Conference

979-8-3315-8937-0/26 $31.00 © 2026 IEEE

ISSCC 2026 / SESSION 30 / COMPUTE-IN-MEMORY / 30.6

30.6 A 16Mb 166.8TOPS/W Near-Memory Phase-Domain-Computing Ferroelectric NAND Flash for Approximate Nearest Neighbor Search on Edge Devices

Weizeng Li*[1,2], Bohan Wang*[1,2], Zhidao Zhou[1,2], Junyu Zhu[1,2], Zhi Li[1,2], Junzhe Shen[1,2], Wenfeng Zha[1,2], Zhongze Han[1,2], Yiman Wang[1,2], Linfang Wang[3], Hongyang Hu[1,2], Qing Luo[1,2], Chunmeng Dou[1,2], Ming Liu[1]

[1]Institute of Microelectronics of the Chinese Academy of Sciences, Beijing, China, [2]University of Chinese Academy of Sciences, Beijing, China
[3]Columbia University, New York, NY
*Equally Credited Authors (ECAs)

Abstract

Previous near-memory computing (NMC) or in-memory-computing (IMC) NANDs suffers from limited IO width, large energy-delay-product, and an inability to support diverse vector formats. This work presents a fabricated 16Mb near-memory phase-domain-computing (NM-PDC) FeNAND chip can compute the 512 similarity distances between 256-dimensional 4b vectors in a single search operation, achieving 166.8TOPS/W energy efficiency and a 12.8× reduction in end-to-end search latency.

Approximate nearest neighbor search (ANNS) for sparse, high-dimensional data, such as text embeddings, is widely used in applications like natural language processing, recommendation systems, and vector search [1-3]. Deploying ANNS on edge devices enhances real-time responsiveness, enables offline operation, and improves data privacy and security. Near-memory computing (NMC) with NAND flash offers a viable approach to accelerating on-chip ANNS by integrating compute units near the high-density storage. This design reduces latency and energy costs associated with data movement. However, due to high fabrication costs, NMC NAND prototypes are seldom realized or tested in silicon.

Ferroelectric NAND (FeNAND) memory has been under intensive research and development in recent years due to its foundry-compatible process, low write power, high endurance, and scalability [4-8]. It therefore provides a promising platform for developing next-generation NMC NAND. Yet, several circuit-level challenges persist (Fig. 30.6.1): (1) The small cell pitch and highly parallel readout scheme in NAND impose tight constraints on the area and power budgets for the NMC units; (2) The large bit-line capacitance results in a large energy-delay product (EDP) for readout, hindering performance in frequently accessed scenarios; (3) ANNS workloads for different applications incorporate diverse vector types with various bit-widths, sign formats, and sparsity, whereas prior efforts have focused mainly on binary, unsigned, and dense vector search operations [9-11].

To address these challenges, we developed: (1) A near-memory phase-domain-computing (NM-PDC) FeNAND with that integrates highly energy- and area-efficiency PDC units near the BEOL FeNAND array for graph-based ANNS; (2) A charge-discharge interleaving (CDI) sensing scheme with an auto-polarity-switching sense amplifier (APS-SA) that reduces the EDP up to 80.5% during continuous readout during the search operation; (3) Multi-bit and sign-aware phase-domain computing units (MS-PDCUs) and sparsity-aware input vector processing unit (SA-IVPU) that for highly sparse multi-bit vectors searching. On this basis, we present a 16Mb NM-PDC FeNAND chip fabricated using a 180nm logic process with BEOL ferroelectric capacitors. In a search operation, the chip computes the 512 cosine similarities between a 256-dimensional, 4b input vector ($IV_{0-255}[3:0]$) and 512 256-dimensional, 4b stored vectors ($SV_{0-255,0-511}[3:0]$) through highly parallel vector-matrix multiplication (VMM). It achieves an energy-efficiency (EF) of 166.8 TOPS/W for VMM operations, and 12.8 times reduced overall search latency in a typical ANNS task.

Figure 30.6.2 illustrates the overall structure of the NM-PDC FeNAND chip for the graph-based ANNS. It consists of a 16Mb FeNAND cell array, row decoder and drivers, write and read (W/R) circuitry with 2048 APS-SAs, 2048 MS-PDCUs, 512 similarity distance computing units (SDCUs), one SA-IVPU, and the main control (Ctrl). The FeNAND cell array comprises 128×2048 FeNAND chains (FNCs) with parallel bit-lines (BLs) and source-lines (SLs). Each FNC contains 64 stacked FeNAND cells, which can be addressed by word-lines (WLs), the source-select-gate (SGS), and the drain-select-gate (SGD). The FNC is fabricated by integrating $Hf_xZr_{1-x}O_2$ (HZO)-based metal-ferroelectric-metal (MFM) capacitors in the contact (CT) layer on top of the gate of standard logic transistors. The graph-based ANNS process [2-3] implemented on the NMC-FeNAND chip is described below. First, we use truncated singular value decomposition (TSVD) to reduce the dimensions (DIMs) of the input query and the dataset to 256. The IVs and SVs are used to construct a graph-based ANNS index, which generates the address bias (ADD_BIAS) to locate different SVs. The IV and ADD_BIAS are then sent to the SA-IVPU for similarity computing. The chip performs VMM between $IV_{0-255}[3:0]$ and $SV_{0-255,0-511}[3:0]$ in a search operation within 0 to 256 CLK cycles, depending on the IV sparsity, and generates the similarity outputs ($SO_{0-511}[14:0]$). The SO is then sent back to the ANNS index to update the ADD_BIAS. This process iterates until no SV with higher similarity can be identified, and then outputs the top-k most similar vectors. The TSVD can reduce the DIM of vectors to reduce data transfer, yet may cause increased recall loss if the DIM becomes excessively small. By truncating the DIM to 256, the data transfer is reduced by 664 times with only a 0.1% increase in recall loss for the ANNS task using the 20NewsGroup dataset [12].

Figure 30.6.3 illustrates the APS-SA using CDI sensing scheme for continuous read operations. The APS-SA consists of two multiplexers (MUX1 and MUX2) for BL/SL pre-charge control, a latch-type differential input sense amplifier (SA) to compare the BL voltage (V_{BL}) with the reference voltage (V_{REF}), a DFF to store the previous SA output (SAO), an XOR gate to output SV based on the polarity switching logic, and two switches for read enable (RD_EN) and pre-charge (PRE). The APS-SA is designed to output correct data in the CDI sensing scheme. The polarity (POL) of the APS-SA is set to zero for initialization. A typical read operation is divided into two sub-phases (SP0 and SP1) and activated by RD_EN. In SP0 (PRE=ON), if POL=0, the BL is grounded and the SL is charged to V_{PR}; if POL=1, the BL is charged to V_{PR} and the SL is grounded. In SP1 (PRE=OFF), the SGD of the selected FNC is activated to allow the BL to develop. The BL develops toward V_{PR} if POL=0, or toward GND if POL=1. The SA is then triggered (SA_EN=1) to output the SAO. If the current SAO is the same as the current POL, the POL is kept the same; otherwise, it is switched. By performing an XOR operation between the SAO and POL signals, the APS-SA generates the correct logic value for SV output. The APS-SA with the CDI scheme reduces pre-charge energy by eliminating the need for pre-charge and discharge of the selected BL in each CLK cycle, which is required in the conventional scheme. In addition, the CDI scheme employs a low V_{PR} of 400 mV instead of V_{PR}=VDD in the conventional scheme, further reducing the energy and delay during BL development. Compared to the conventional scheme, the scheme achieves 49.8% lower energy, 59.2% lower delay, and 80.5% lower EDP.

Figure 30.6.4 illustrates the circuit blocks to perform highly efficient VMM in the phase-domain, including the SA-IVPU, MS-PDCUs, and SDCU. The SA-IVPU consists of a sparsity-aware vector aligner (SAVA) and an inverter-based digital-to-time converter with a unit time of τ (τ-DTC). While the SAVA is used to skip over the sparsity in IV and align the IV with the SV, the τ-DTC is used to convert the IV data into the time-domain signal (DTC_OUT). The SAVA is composed of a scan I/O for data input, an 8b counter to count the number of zero-valued entries (N_Z), an 8b WL counter to generate the number of WLs (N_{WL}) for WLs scanning, an adder to calculate the WL address (WADD), and a finite state machine (FSM) for control. To start the vector searching process, N_Z and N_{WL} are reset to 0. $IV_{0-255}[3:0]$ is input to SAVA serially. In the ith cycle, if $IV_i[3:0]$ is zero, N_Z and i are incremented ($N_Z=N_Z+1$, i=i+1), and the scan I/O continues scanning in new data until a non-zero value appears or i reaches 256. If $IV_i[3:0]$ is non-zero, WADD is updated as WADD=ADD_BIAS+N_Z+N_{WL} to fetch the corresponding $SV_{i,0-511}[3:0]$. Then, N_{WL} and i are incremented ($N_{WL}=N_{WL}+1$, i=i+1), N_Z is reset to 0, and the next data is scanned in. To perform VMM with IV[3:0] and SV[3:0] in the sign magnitude format, three MS-PDCUs and one SDCU are grouped. Each MS-PDCU consists of one XOR gate for multiplying between the sign bit (IV[3] and SV[3]) to generate SIGN signal, one AND gate for multiplying between the unsigned ones (IV[2:0] and SV[2:0]), two five-stage voltage-controlled-oscillators (VCOs) for phase accumulation in the positive (SIGN=0) or negative (SIGN=1) directions, a MUX for direction selection, an 8b up and down counter (UDC) and D-type flip-flops (DFFs) to sample the cycles and phase information in the VCO. The SDCU converts the phase data (PO[4:0]) to a digital value, multiplies the counter value (CO[7:0]) by ten, and combines these values at their place values using a digital shifter-and-adder (DSA) to produce the SO[14:0].The NM-PDC FeNAND operates in two phases across separate CLK cycles. In the readout phase, the SV is readout from FeNAND according to WADD. In the computing phase, phase-domain VMM is performed in the MS-PDCU. The readout and computing phases operate in a pipelined manner to reduce overall latency. Notably, the unit delay cells of the inverter-based DTCs are designed to be the same as those in the VCO to match their unit time, ensuring robustness against process variation.

Figure 30.6.5 shows the measurement results of the fabricated NM-PDC FeNAND. The FeNAND cells in FNC can be programed and erased (P/E) using fixed-amplitude pulses (V_{PRG}=4.0V, V_{ERS}=3.5V) with a duration of 100ns. A pass voltage (V_{PASS}) of 2.5V is used for the unselected cells. A clear memory window (MW) > 0.45V can be maintained after 10^9 P/E cycles. Typical ANNS tasks of searching k similar vectors out of 18k vectors in the 20NewsGroup dataset are demonstrated on the fabricated chip. Captured functional waveforms show the results for one mismatched and matched sample. A shmoo test confirms that the chip achieves a peak EF of 166.8TOPS/W at 27MHz and a peak throughput (TP) of 471.0GOPS at 36MHz. With the SA-IVPU, the chip achieves a 12.8× reduction in search time and a 2.32× improvement in EF. It achieves recall rates of 92.1% (k=100) and 91.9% (k=10). Compared with previous works on phase-domain computing units and near- or in-memory computing (NMC/IMC) NAND (Fig. 30.6.6), this work presents a NMC FeNAND chip featuring competitive performance metrics, including cell endurance (>10⁹ P/E cycles), storage capacity (16Mb), and EF (166.8TOPS/W). It achieves a 1.29× improvement in EF [16] and a 2.56× improvement in FoM [15] (Capacity×IV-precision×SV-precision×Signal Margin×EF) over prior works. The die micrograph and chip performance summary are shown in Fig. 30.6.7.

Acknowledgement:
This work was supported by the National Natural Science Foundation of China (NSFC) under Grant No. 62488101, 92364202, 62425407. Corresponding Author: Chunmeng Dou.

References:
[1] I. Azizi et al., "Graph-Based Vector Search: An Experimental Evaluation of the State-of-the-Art," *ACM Manag. Data*, vol. 3, no. 1, pp. 1-31, Feb. 2025. https://doi.org/10.1145/3709693

[2] M. Wang et al., "A Comprehensive Survey and Experimental Comparison of Graph-Based Approximate Nearest Neighbor Search," *VLDB Endowment*, vol. 14, no. 11, pp. 1964-1978, Jul. 2021. https://doi.org/10.14778/3476249.3476255

[3] Y.A. Malkov et al., "Efficient and Robust Approximate Nearest Neighbor Search Using Hierarchical Navigable Small World Graphs," *IEEE TPAMI*, vol. 42, no. 4, pp. 824-836, 1 Apr. 2018. http://doi.org/10.1109/TPAMI.2018.2889473

[4] T. Hatanaka et al., "Ferroelectric (Fe)-NAND Flash Memory With Batch Write Algorithm and Smart Data Store to the Nonvolatile Page Buffer for Data Center Application High-Speed and Highly Reliable Enterprise Solid-State Drives," *IEEE JSSC*, vol. 45, no. 10, pp. 2156-2164, Oct. 2010. http://doi.org/10.1109/JSSC.2010.2061650

[5] M. Song et al., "Ferroelectric NAND for Efficient Hardware Bayesian Neural Networks," *Nat. Commun.*, vol. 16, pp. 1-14, July 2025. https://doi.org/10.1038/s41467-025-61980-y

[6] S. Yoon et al., "QLC Programmable 3D Ferroelectric NAND Flash Memory by Memory Window Expansion using Cell Stack Engineering," *IEEE Symp. VLSI Circuits*, pp. 1-2, June 2023. http://doi.org/10.23919/VLSITechnologyandCir57934.2023.10185294

[7] H. Joh et al., "Oxide Channel Ferroelectric NAND Device with Source-Tied Covering Metal Structure: Wide Memory Window (14.3 V), Reliable Retention (> 10 Years) and Disturbance Immunity ($\Delta V_{th} \leq 0.1V$) for QLC Operation," *IEEE IEDM*, pp. 1-4, Dec. 2024. http://doi.org/10.1109/IEDM50854.2024.10873376

[8] M.-K. Kim et al., "CMOS-Compatible Ferroelectric NAND Flash Memory for High-Density, Low-Power, and High-Speed Three-Dimensional Memory," *Sci. Adv.*, vol. 7, no. 3, pp. 1-10. Jan. 2021. http://DOI.org/10.1126/sciadv.abe1341

[9] H.-W. Hu et al., "A 512Gb In-Memory-Computing 3D-NAND Flash Supporting Similar-Vector-Matching Operations on Edge-AI Devices," *ISSCC*, pp. 138-139, Feb. 2022. http://doi.org/10.1109/ISSCC42614.2022.9731775

[10] C.-K. Liu et al., "COSIME: FeFET Based Associative Memory for In-Memory Cosine Similarity Search," *IEEE/ACM ICCAD*, pp. 1-9, Dec. 2022. https://doi.org/10.1145/3508352.3549412

[11] A. Kazemi et al., "In-Memory Nearest Neighbor Search with FeFET Multi-Bit Content-Addressable Memories," *DATE*, pp. 1084-1089, Feb. 2021. http://doi.org/10.23919/DATE51398.2021.9474025

[12] K. Lang, "NewsWeeder: Learning to Filter Netnews," *ICML*, pp. 331-339, July 1995. https://doi.org/10.1016/B978-1-55860-377-6.50048-7

[13] Y. Toyama et al., "An 8 Bit 12.4 TOPS/W Phase-Domain MAC Circuit for Energy-Constrained Deep Learning Accelerators," *IEEE JSSC*, vol. 54, no. 10, pp. 2730-2742, Oct. 2019. http://doi.org/10.1109/JSSC.2019.2926649

[14] S. Gweon et al., "FlashMAC: A Time-Frequency Hybrid MAC Architecture With Variable Latency-Aware Scheduling for TinyML Systems," *IEEE JSSC*, vol. 57, no. 10, pp. 2944-2956, Oct. 2022. http://doi.org/10.1109/JSSC.2022.3182699

[15] H.-T. Lue et al., "Optimal Design Methods to Transform 3D NAND Flash into a High-Density, High-Bandwidth and Low-Power Nonvolatile Computing in Memory (nvCIM) Accelerator for Deep-Learning Neural Networks (DNN)," *IEEE IEDM*, pp. 38.1.1-38.1.4, Dec. 2019. http://doi.org/10.1109/IEDM19573.2019.8993652

[16] M. Kim et al., "An Embedded NAND Flash-Based Compute-In-Memory Array Demonstrated in a Standard Logic Process," *IEEE JSSC*, vol. 57, no. 2, pp. 625-638, Feb. 2021. http://doi.org/10.1109/JSSC.2021.3098671

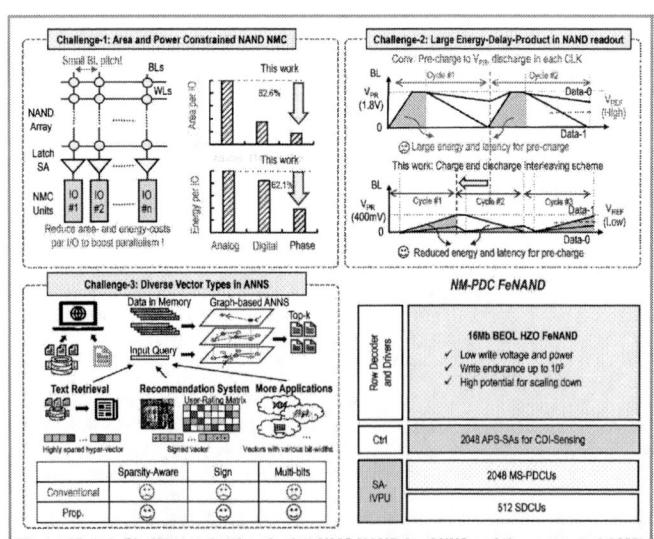

Figure 30.6.1: Challenges for developing NMC NAND for ANNS and the presented 16Mb HZO-based NM-PDC FeNAND.

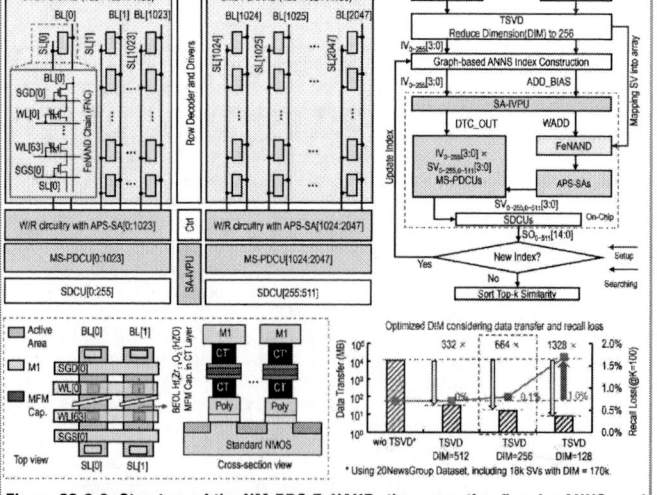

Figure 30.6.2: Structure of the NM-PDC FeNAND, the computing flow for ANNS, and vector dimension optimization.

ISSCC 2026 / SESSION 30 / COMPUTE-IN-MEMORY / 30.6

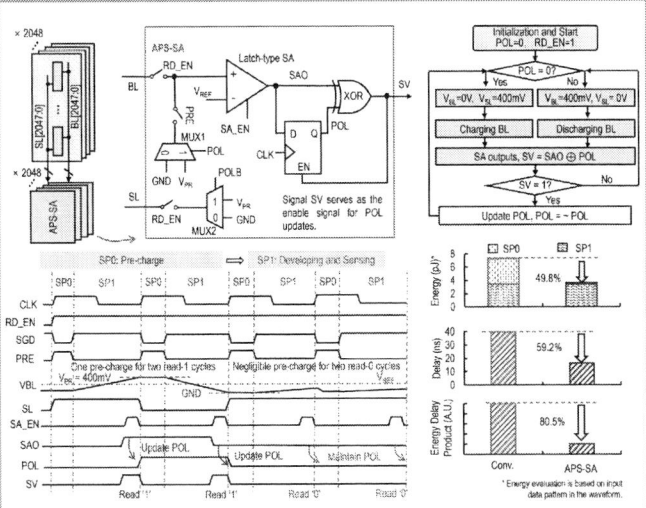

Figure 30.6.3: Structure and operation of the APS-SA with CDI-sensing scheme for FeNAND readout.

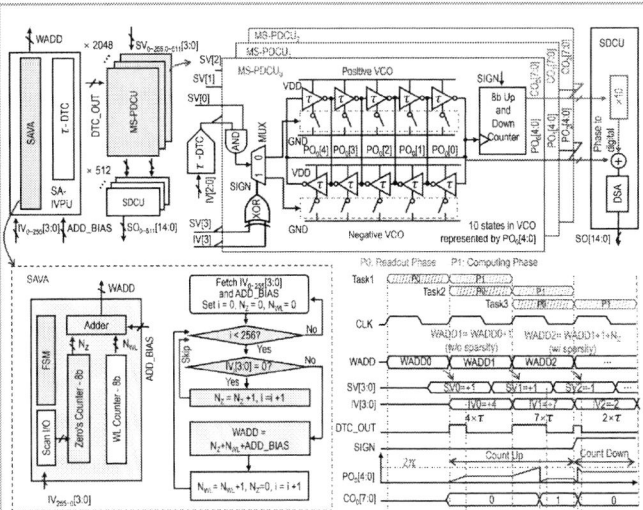

Figure 30.6.4: Structure and operation of the NM-PDC Units, including SA-IPVU, MS-PDCU, and the SDCU.

Figure 30.6.5: Measurement results of the fabricated NM-PDC FeNAND chip.

	JSSC19[13]	JSSC22[14]	IEDM19[15]	JSSC21[16]	ISSCC22[9]	NC25[5]	This work
On-Chip IMC/NMC	N.A.	N.A.	IMC	IMC	IMC	N.A.	NMC
On-chip Memory	N.A.	N.A.	NAND	Logic FLASH NAND	3D-NAND	FeNAND	FeNAND
Capacity	N.A.	N.A.	6.5 Gb	20 kb	512 Gb	0.28 kb	16 Mb
Endurance	N.A.	N.A.	N.A.	N.A.	N.A.	10^7	10^8
Computing Domain	Phase	Phase	Current	Current	Current	Current	Phase
IN/IV Precision (bit)	8	4,7	4	8	1	1	4
W/SV Precision (bit)	8	4,7	4	8	3	1	4
Output Precision (bit)	8	16	8	N.A.	8	N.A.	15 (SDCU)
Signal Margin*1	1	1	1/256	1/7	1/256	N.A.	1
Sparsity Aware	No	No	No	No	No	No	Yes
Sign Aware	IN and W	IN	No	IN and W	No	No	IV and SV
Energy Efficiency (TOPS/W)	12.4	47.19	40*2	129*7	N.A.	N.A.	72.0*3,4,5 - 166.8*3,4,6 (1.29×)
Application	Image Recognition /Anomaly Detection	Image Recognition	Image Recognition	Image Recognition	Similar Vector Matching	Bayesian Neural Networks	Graph-based ANNS
FoM*7	N.A.	N.A.	16640.0	23.0	N.A.	N.A.	42575.2 (2.56×)

*1. Signal margin is defined as 1/# of the analog states needs to be differentiated in the memory array *2. Simulated.
*3. Measured at 25°C, 1.45V and 27 MHz *4. 92.18% sparsity on 20NewsGroup Dataset *5. w/o SAVA *6. w/ SAVA
*7. FoM = Capacity × IN/IV-precision × W/SV-precision × Signal Margin × EF

Figure 30.6.6: Comparison with previous works on phase domain MAC units and IMC

Technology	180nm CMOS 1P6M
Memory	HZO MFM FeNAND
Chip Mode	NMC/Storage
Capacity	16 Mb
Array Area	50.63 mm²
Memory Density	0.316 Mb/mm² (9.42 Mb/mm²)*1
Endurance	10^8
Frequency	4 – 36 MHz
Supply Voltage	Core: 1.2 - 2.2 V; I/O: 3.3 V
IV Precision	4 bit
SV Precision	4 bit
Output Precision	15 bit (SDCU)
Search Time*2	19.5 μs
Measured TP	471.0 *3 GOPS
Measured EF	72.0*2,4,5 - 166.8*2,4,6 TOPS/W
Application	Graph-based ANNS
Hardware Recall Rate (%) 20NewsGroup Dataset (DIM=256) *6,7	92.1% (@k=100) 91.9% (@k=10)

*1. Estimated using a standard 28nm logic process
*2. Using 20NewsGroup Dataset with 92.18% sparsity, matching 100 vectors in 18k SVs.
*3. Measured at 25°C, 2.2V and 36MHz, 4b-IV, 4b-SV, 256 accumulation.
*4. Measured at 25°C, 1.45V and 27 MHz, 4b-IV, 4b-SV, 256 accumulation.
*5. w/o SAVA *6. w/ SAVA
*7. Software baseline: 92.9% (@k=100) 96.4% (@k=10).

Figure 30.6.7: Die micrograph and chip performance summary.

• 2026 IEEE International Solid-State Circuits Conference

ISSCC 2026 / SESSION 30 / COMPUTE-IN-MEMORY / 30.7

30.7 A 1.2GHz 12.77GB/s/mm² 3D Two-DRAM-One-Logic Process-Near-Memory Chip for Edge LLM Applications

Yue Cao[1,2,3], Jinghao Jiang[1], Haijun Jiang[3], Qian Zhang[3], Xuanzhi Liu[2], Jinhui Cheng[2], Zhongze Han[2], Xiping Jiang[4], Fengguo Zuo[4], Song Wang[4], Fujun Bai[4], Yixin Guo[4], Chunmeng Dou[2], Jianguo Yang[1,2,3], Hangbing Lv[2], Qi Liu[1], Ming Liu[1,2]

[1]Fudan University, Shanghai, China, [2]Institute of Microelectronics of the Chinese Academy of Sciences, Beijing, China, [3]Zhangjiang Laboratory, Shanghai, China
[4]Xi'an UniIC Semiconductors, Xi'an, China

Abstract

A high-bandwidth-density (12.77GB/s/mm²) high-memory-density (99.4Mb/mm²) low-energy-consumption (0.67pJ/b) 3D PNM design that operates at 1.2GHz is presented. The design adopts a two-DRAM-one-logic architecture that enables near-DRAM computing through a high-density 3D integration path, reducing memory-access latency by up to 93% and GEMM execution time by up to 98%, demonstrating strong potential for edge-LLM workloads.

Upon the rapid development of generative AI, there is an urgent need to embed LLM inference support into client devices (PCs, mobiles) to improve response time, data privacy, and customization. However, as shown in Fig. 30.7.1, challenges remain for edge-LLM applications. Compared to cloud deployments, edge tasks typically use much smaller batch sizes, which significantly reduces data reuse during the generation stage of LLM inference and makes the task more memory-intensive [1]. Meanwhile, existing client products are limited by either computing power (CPU) or memory bandwidth (GPU), and GPU energy consumption is prohibitive for battery-powered devices such as phones. Therefore, supporting offline edge LLM inference requires an LLM-optimized design that delivers sufficient computing power and memory bandwidth, high energy efficiency, and easy integration with existing edge platforms.

To mitigate the memory bottleneck, several processing-near-memory (PNM) architectures have been proposed [2-5]. However, most prior works integrate compute directly within the DRAM die, constraining compute capability due to DRAM process limitations. In parallel, recent advances in 3D HB + mini-TSV integration enable high-bandwidth low-energy memory access for 3D near-memory designs, motivating this work. Our design stacks two DRAM dies with one logic die, providing 1GB on-chip DRAM capacity and a memory density of 99.4Mb/mm². Decoupling logic from DRAM allows implementing compute logic in standard CMOS and seamlessly leveraging future improvements via advances in process nodes. The 3D-stacked DRAM delivers GB-scale on-chip memory for LLM deployment, greatly reducing off-chip traffic, while the high-density HB + mini-TSV interconnect enables near-in-die interconnects that offer low-latency, low-energy memory access, achieving a bandwidth density of 12.77GB/s/mm² to alleviate the LLM memory bottleneck. The design also supports a DDR4-protocol-compatible external interface for flexible system integration; when PNM is disabled, the chip operates as conventional host DRAM.

Figure 30.7.2 illustrates the overall architecture of the presented 3D NMC design. The presented design comprises eight 1Gb compute blocks, denoted as bankgroup (BG) blocks. Each BG block consists of two 512Mb DRAM blocks on the DRAM_N and DRAM_F dies, and a compute module on the logic die. Each DRAM block contains 8 banks and each bank contains 8 segments. Within each segment, 16 subarrays are arranged such that two adjacent subarrays share a 512-SA stage to generate local DQ (LDQ) signals. By controlling MDQ_sel, target column data are selected onto MDQ lines then driven into a second-stage SA for faster reads. Each segment produces a 16b RDL output, so one BG block provides a 2048b bus (=16×8×8×2). These RDLs connect to the logic die via 3D paths: signals from DRAM_N use direct HB connections, while those from DRAM_F traverse HB-TSV-HB chains to pass through DRAM_N. For TSV reliability, two HB connections accompany each TSV. The RDLs from the DRAM dies connect to a custom 3D DRAM PHY. Because the 3D stack yields a much wider bus, the pin data rate required for equivalent bandwidth is substantially lower; consequently, the 3D DRAM PHY is simpler than DDR/HBM PHYs and primarily serves as a data synchronizer and level shifter, reducing design complexity and power consumption.

Each compute module integrates two accelerator (ACC) units that perform near-memory computation and access DRAM through the memory controller. The general-purpose system host accesses DRAM and manages the ACCs via a DDR-compatible interface. Two additional DDR4-formatted commands, T-ACC and Act-Reg, are introduced to trigger target ACCs and set configuration registers. When the host issues a command, the DDR interface and decoder accept and decode it, and then assert the corresponding control signals to the target module (ACC or memory controller). This scheme maintains full host compatibility while providing a lightweight command extension mechanism for near-memory control.

Figure 30.7.3 shows the 3D DRAM datapath, including ACC-DRAM and system-memory-DRAM paths. Our RISC-V-based ACC includes a core that supports common RISC-V instructions. L1 instruction and data caches reduce access latency. An SRAM-based system memory is connected to the ACC via the memory bus for local data storage. We designed a custom FPU to accelerate critical floating-point operations for LLM workloads (fadd, fsub, fmul, fdiv); other floating-point instructions, such as fsqrt, are also supported by the core. As LLM workloads involve extensive consecutive loads and stores, a DMA module manages bulk transfers to relieve the core of data-movement tasks. The ACC can issue direct DRAM access commands or trigger DMA transfers between system memory and 3D DRAM space. The memory controller decodes access requests from the system bus and forwards them to the command scheduler, which generates DRAM commands. The scheduler FSM is also shown in Fig. 30.7.3. We employ an open-page policy and add a 2048b buffer to hold the most recently accessed data. On a new request, if buf_hit is asserted, the scheduler serves data from the buffer; otherwise, it starts a DRAM request. For a row hit, the read/write command is issued directly; on a row miss, precharge and activate commands are scheduled before read/write. On DRAM reads, the returned data updates the buffer. Scheduled commands flow to a signal generator to form actual DRAM control signals, and the corresponding timing waveform is presented in Fig. 30.7.3. A host programmable config module enforces timing and refresh scheduling. A refresh control issues refresh requests per the configuration and forwards them to the scheduler. Generated DRAM signals and external host signals are selected by a switch according to the current mode and remapped to DRAM addresses. The 3D DRAM PHY comprises four 256Mb PHY submodules to service the 1Gb DRAM space.

Figure 30.7.3 also illustrates the GEMM dataflow. GEMM is the dominant kernel in LLM computation, thus our evaluation focuses on this task. Cross-BG or cross-chip data accesses require host intervention and should be avoided whenever possible. For example, in a full-chip assignment for an M×N×K GEMM (M×N input multiplied by N×K weight), M and K are each partitioned by four, and the resulting submatrices are assigned to the 16 ACCs across the 8 BGs (e.g., IOW0→BG0/ACC0, IOW1→BG0/ACC1, IOW2→BG1/ACC0,...). Each accelerator computes a submatrix of the final output, which the host can directly concatenate without additional accumulation. When scaling to multiple NMC chips, M and K can be further partitioned. Within each accelerator, all weight data are stored consecutively and read via DMA, while each input row is evenly distributed across all banks accessed sequentially, enabling complete input reads through all-bank parallelism. When an input row is ready, it will be sent to the digital logic with the corresponding weight data for MAC and potentially activation function (AF) computation. Normalized GEMM performance results show that adding DMA and the 2048b buffer yields substantial improvement: on average, throughput increases by 53.8% relative to a baseline without DMA or buffer, demonstrating the benefits of local buffering plus DMA-assisted data movement. Furthermore, in some specific scenarios, parameters like weights can be uploaded to 3D DRAM only once after powering on, which could greatly reduce the off-chip access amount.

Figure 30.7.4 presents the measurement system and prototype board. The setup uses one host FPGA and two test PNM chips to verify scalability. A host CPU in an FPGA (serves as the general-purpose CPU system in application scenarios) accesses the test chips via the host memory controller and a DDR4-compatible interface. Control signals (commands and addresses) are broadcast to both chips. The 16b DDR4 data bus is evenly split and connected to the two test chips. Inside each test chip, control and data signals are broadcast to the 8 BGs, and the target BGs activate according to decoded information. A high-speed IO chip is used for ACC program loading and debug transfer, providing program flexibility.

Figure 30.7.4 also shows DRAM measurement results. The DRAM shmoo indicates operation at 526Mbps at 1.2V, corresponding to 12.77GB/s/mm² bandwidth density, which exceeds GDDR6 by 8.63× and HBM3E by 1.22×. DRAM retention testing (96ms at 98°C) failed on <0.05% of wafer chips, indicating good DRAM quality. The results of 3D integration yield verification, and the corresponding testing circuit is also presented in Figure 30.7.4. Only 0.58% additional failure rate is detected after 3D integration, proving that the integration technology maintains a good yield rate. The 3D integration architecture enables 0.67pJ/b memory energy efficiency, which is 9% of GDDR6 and 28% of HBM3E.

As shown in Fig. 30.7.5, the 3D-stacked chip integrates high-density HB and mini-TSV interconnects with HB pitch=3µm and TSV pitch=5µm. Figure 30.7.5 reports voltage-frequency scaling, logic-die area breakdown, and chip power breakdown. It also presents memory-access latency and GEMM execution results. In an 8192-access consecutive test, we measure 3.21ns write and 3.24ns read latency, reductions of 90% and 93% versus a scaled LPDDR4 system (system frequency scaled to 1.2GHz); for random access, we measure 45.12ns write and 50.01ns read latency, reductions of 37% and 44%. These results

524 • 2026 IEEE International Solid-State Circuits Conference

979-8-3315-8937-0/26 $31.00 © 2026 IEEE

show that the 3D near-memory architecture substantially reduces memory latency, benefiting latency-sensitive edge LLM workloads. Compared to a 2.1GHz Intel i7-13700F, our design reduces execution time for a 128×128×128 GEMM by 86%. When scaling our work to an 8-chip DIMM system with 8GB DRAM capacity and 128 ACCs, the reduction can be further improved to 98%, illustrating practical gains for common inference kernels.

Figure 30.7.6 compares our design with SOTA PNM designs. Unlike prior works, our design supports a DDR4-compatible interface for general-purpose integration. The two-die stacked DRAM with 3D integration achieves 12.77GB/s/mm² bandwidth density, 99.4Mb/mm² memory density, and 0.67pJ/b energy efficiency, while a separate 28nm CMOS logic die enables 1.2GHz compute frequency, which provides suitable computing power for client LLMs. The design FoM is 7.33× higher than a prior 3D PNM design and 5.13× higher than an active-interposer design [7]. By consolidating the logic and DRAM dies through 3D integration, this work leverages PNM-DRAM advancements without compromising the compute capability provided by standard CMOS logic, achieving outstanding bandwidth density and logic frequency compared with existing PNM designs. Moreover, this industrial-standard fabricated prototype demonstrates the feasibility and commercial potential of 3D PNM solutions for edge LLM deployment. The die micrographs and key specifications are shown in Fig. 30.7.7.

Acknowledgement:
This research was supported by the National Natural Science Foundation of China (NSFC) under grants 62488101, 62222119, 62025406. Corresponding author: Jianguo Yang.

References:
[1] C. Li et al., "H2-LLM: Hardware-Dataflow Co-Exploration for Heterogeneous Hybrid-Bonding-based Low-Batch LLM Inference", *ISCA*, pp. 194-210, June 2025.
https://doi.org/10.1145/3695053.3731008
[2] H. Shin et al., "McDRAM: Low Latency and Energy-Efficient Matrix Computations in DRAM," *IEEE Transactions on Computer-Aided Design of Integrated Circuits and Systems*, vol. 37, no. 11, pp. 2613-2622, Nov. 2018. https://doi.org/10.1109/TCAD.2018.2857044
[3] F. Devaux, "The True Processing In Memory Accelerator," *IEEE Hot Chips*, pp. 1-24, Aug. 2019. https://doi.org/10.1109/HOTCHIPS.2019.8875680
[4] Y.-C. Kwon et al., "A 20nm 6GB Function-In-Memory DRAM, Based on HBM2 with a 1.2TFLOPS Programmable Computing Unit Using Bank-Level Parallelism, for Machine Learning Applications," *ISSCC*, pp. 350-351, Feb. 2021.
https://doi.org/10.1109/ISSCC42613.2021.9365862
[5] S. Lee et al., "A 1ynm 1.25V 8Gb, 16Gb/s/pin GDDR6-based Accelerator-in-Memory supporting 1TFLOPS MAC Operation and Various Activation Functions for Deep-Learning Applications," *ISSCC*, pp. 176-177, Feb. 2022.
https://doi.org/10.1109/ISSCC42614.2022.9731711
[6] D. Niu et al., "184QPS/W 64Mb/mm² 3D Logic-to-DRAM Hybrid Bonding with Process-Near-Memory Engine for Recommendation System," *ISSCC*, pp. 462-463, Feb. 2022.
https://doi.org/10.1109/ISSCC42614.2022.9731694
[7] B. Jiao et al., "SHINSAI: A 586mm2 Reusable Active TSV Interposer with Programmable Interconnect Fabric and 512Mb 3D Underdeck Memory," *ISSCC*, pp. 612-613, Feb. 2025. https://doi.org/10.1109/ISSCC49661.2025.10904819

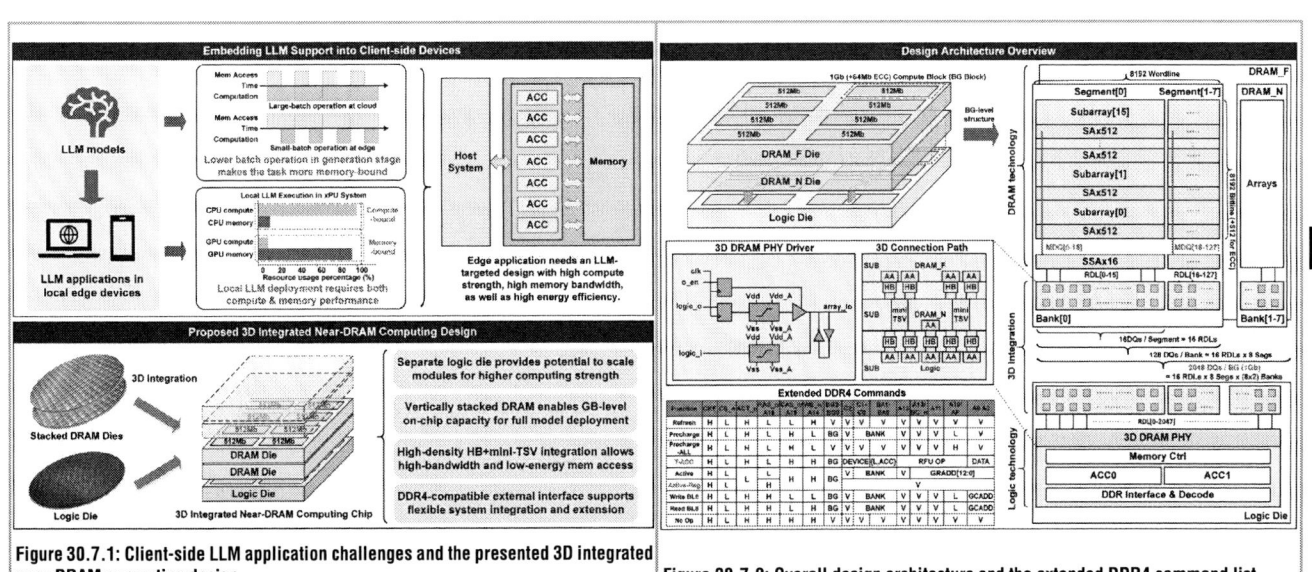

Figure 30.7.1: Client-side LLM application challenges and the presented 3D integrated near-DRAM computing design.

Figure 30.7.2: Overall design architecture and the extended DDR4 command list.

ISSCC 2026 / SESSION 30 / COMPUTE-IN-MEMORY / 30.7

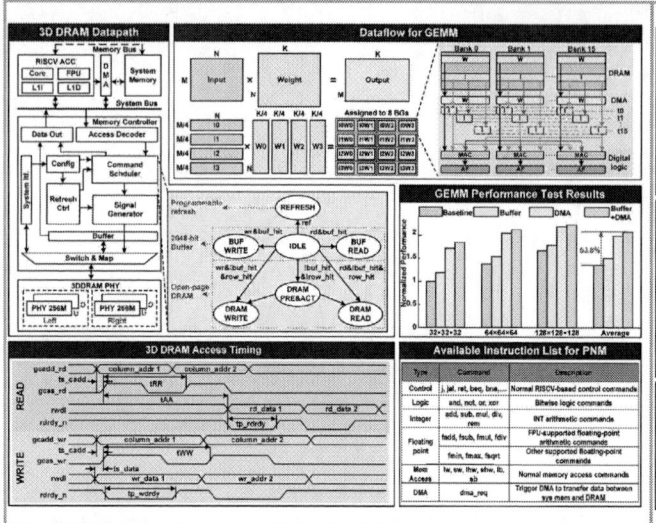

Figure 30.7.3: Datapath and access timing for 3D DRAM, and dataflow and performance test for GEMM.

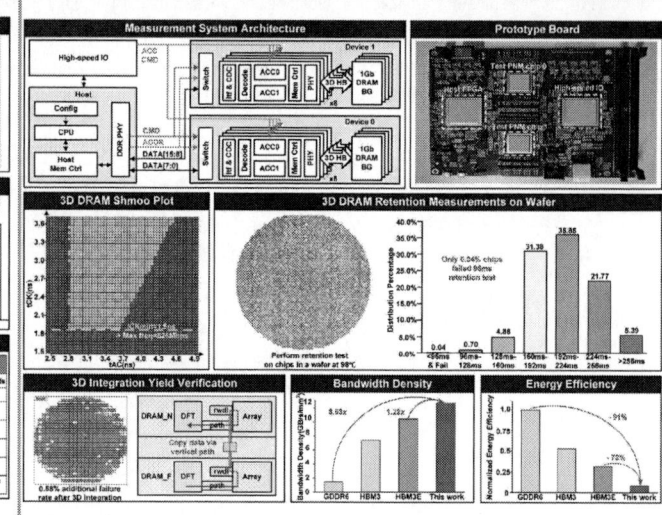

Figure 30.7.4: Measurement system architecture and 3D DRAM measurement results.

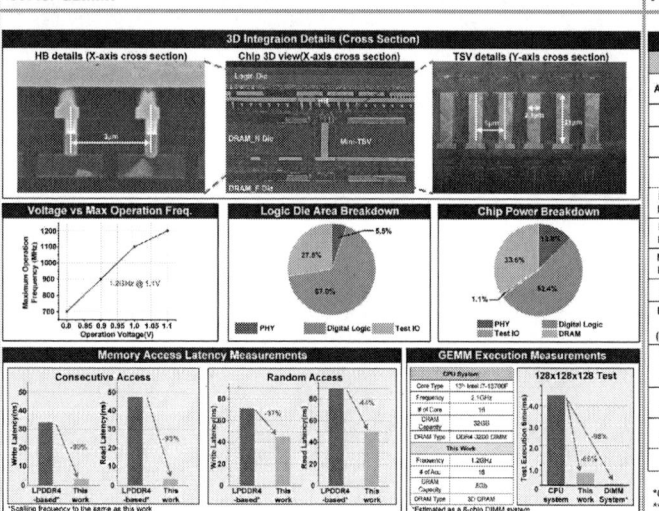

Figure 30.7.5: 3D integration details, area and power breakdown, and design measurements.

Comparison Table

	TCAD 2018 [2]	HCS 2019 [3]	ISSCC 2021 [4]	ISSCC 2022 [6]	ISSCC 2022 [5]	ISSCC 2025 [7]	This work
Architecture	DRAM +in-die logic	DRAM +in-die logic	HBM +in-die logic	1 DRAM die +1 logic die	DRAM +in-die logic	Logic chiplets +active interposer	2 DRAM die +1 logic die
Interface	LPDDR4	DDR4	HBM2	Custom	GDDR6	Custom	DDR4-compatible
Process	20nm DRAM	2xnm DRAM	20nm DRAM	25nm DRAM 55nm Logic	1ynm DRAM	28nm Logic	25nm DRAM 28nm Logic
Mem Capacity	8GB per chip	8GB per DIMM	6GB per cube	4.5GB per chip	8GB per chip	512Mb per interposer	8GB per chip
Near-mem Bandwidth	25.6GB/s per chip	19.2GB/s per DIMM	307GB/s per chip	1382GB/s per chip	64GB/s per chip	307.2GB/s per interposer	1052GB/s per chip
Max Logic Frequency	250MHz	500MHz	300MHz	300MHz	1GHz	400MHz	1.2GHz
Max DRAM Frequency	250MHz	500MHz	300MHz	150MHz	1GHz	400MHz	526MHz
Chip Area	–	–	*~96mm²	602.22mm²	*~40mm²	586mm²	82.4mm²
Bandwidth density (GB/s/mm²)	–	–	3.20	2.29	1.6	0.52	12.77
Memory density (Mb/mm²)	–	–	512	64	204.8	0.873	99.4
Peak Throuput	0.5TOPS (250M x 8B x 256)	0.5TOPS (500Mx8Bx128)	1.2TFLOPS (300Mx32Bx128)	0.68TOPS (150Mx128Bx36)	1TFLOPS (1Gx32Bx32)	0.15TOPS (400Mx24Bx16)	1.05TFLOPS (526Mx256Bx8)
Energy Efficiency	–	~25pJ/b	~2.75pJ/b	0.88pJ/b	–	0.14pJ/b	0.67pJ/b
**FoM	–	1.16	2.60	–	3.71	19.06

*Chip area is estimated by reported HBM2 and GDDR6 chip size.

**FoM=Bandwidth density / Energy efficiency

Figure 30.7.6: Comparison with prior works.

DRAM Die		Logic Die	
Technology	25nm	Technology	28nm
Area	82.4mm²	Area	82.4mm²
Voltage	1.2V	Voltage	0.8~1.1V
Max Frequency	526Mbps/pin	Frequency	700M~1.2GHz
Bandwidth*	1052GB/s per chip	ACC number	16
Energy Efficiency	0.67pJ/b	Power	3.72W@1.1V

Figure 30.7.7: Die micrographs and design specifications.

- 2026 IEEE International Solid-State Circuits Conference

979-8-3315-8937-0/26 $31.00 © 2026 IEEE

ISSCC 2026 / SESSION 30 / COMPUTE-IN-MEMORY / 30.8

30.8 A 16nm, 1Mb, 1-to-8b-Configurable 444.21TOPS/W Fully Digital SRAM Compute-In-Memory Macro for Hybrid SNN-CNN Edge Computing

Yao-Kai Yeh*[1], Jian-Wei Su*[2], Ting-Hao Hsu[1], Mai Tseng[1], Jen-Chun Tien[1], Ko-Chi Chen[1], Chih-Yen Yue[1], Yu-En Lin[1], Yu-Jia Hu[1], Le-Jung Hsieh[1], Jyun-Cheng Bai[1], Yu-Sheng Kao[1], Tsung-Han Lou[1], Hung-Hsi Hsu[1], De-Qi You[1], Sheu Shyh-Shyuan[2], Wei-Chung Lo[2], Shih-Chieh Chang[1,2], Ya-Tang Yang[1], Chung-Chuan Lo[1], Ren-Shuo Liu[1], Chih-Cheng Hsieh[1], Kea-Tiong Tang[1], Meng-Fan Chang[1]

[1]National Tsing Hua University, Hsinchu, Taiwan, [2]Industrial Technology Research Institute, Hsinchu, Taiwan
*Equally Credited Authors (ECAs)

Abstract

We present a silicon-verified fully digital SRAM-CIM macro that supports multi-bit hybrid SNN-CNN processing. Key innovations include: (1) a compact CFD-SC-RC architecture that shares SNN-CNN MAC hardware to minimize area and energy overhead while enabling adaptive model and precision selection; and (2) a PS-WR-ODM scheme that balances partial-sum storage and weight-update energy. Fabricated in 16nm FinFET CMOS, the test chip achieves 444.21TOPS/W (1bIN-8bW-14bOUT, SNN) and 62.84TOPS/W (8bIN-8bW-22bOUT, CNN).

Spiking neural networks (SNNs) [1] have demonstrated superior energy efficiency in event-driven applications such as event-based vision and audio processing, while convolutional neural networks (CNNs) [2-3] provide high accuracy in image and voice recognition tasks at the edge. Hybrid SNN-CNN models [4-6] have shown promise in balancing accuracy and compute energy, enabling robust edge inference for various applications. Meanwhile, SRAM-based digital compute-in-memory (SRAM-DCIM) [7] has been widely explored to reduce data movement and improve energy efficiency, while outperforming analog SRAM-CIM (SRAM-ACIM) [8] in MAC accuracy. Combining SRAM-DCIM with hybrid SNN-CNN models is a promising solution for diverse edge-AI tasks.

However, as shown in Fig. 30.8.1, designing SRAM-DCIM macros supporting hybrid SNN-CNN introduces two key challenges: (1) extra hardware overhead for supporting both SNN and CNN, and excessive compute energy in the digital domain; and (2) area-energy trade-off between partial sum (PSUM) storage and weight (W) update during input (IN) and W data mapping.

This work addresses these challenges with two key techniques: (1) a compact fully-digital SNN-CNN reconfigurable computing (CFD-SC-RC) architecture that shares SNN-CNN MAC circuit to reduce area and energy overhead, and supports adaptive model and precision selection across layers; and (2) a PSUM storage–weight reuse optimized data mapping (PS-WR-ODM) scheme that balances PSUM-storage minimization and weight-reuse maximization, achieving optimized register usage and weight-update energy. The test chip, fabricated in a foundry 16nm FinFET CMOS process, demonstrates 444.21 TOPS/W with a 1bIN-8bW-14bOUT configuration in SNN mode, and 62.84TOPS/W with an 8bIN-8bW-22bOUT configuration in CNN mode.

Figure 30.8.2 illustrates the proposed 1Mb CFD-SC-RC SRAM-DCIM macro, consisting of 16 64kb banks. Each bank integrates a control block, I/O block, IN/W scheduler, WL driver, 32 reconfigurable inference blocks (RIBs), an adder tree, and a dual multi-membrane potential dynamics behavior (MPDB) unit (DM-MPDBU). Each RIB contains a 2kb SRAM array, a weight stationary cell (WSC), eight cross-model dual-mode local computing cells (CMDM-LCCs), and a shift-and-add (SaA) circuit. Naïve SRAM-ACIM based hybrid SNN-CNN macros are vulnerable to process variations in MPDB functions, while prior SRAM-DCIM based hybrid SNN-CNN macros [4] support only a single MPDB type, limiting flexibility for diverse model requirements. Both architectures lack hardware sharing, incurring large area overhead. The proposed CFD-SC-RC architecture shares SNN-CNN MAC circuits to reduce area overhead, provides high tolerance to process variation, and, via the DM-MPDBU, additionally supports multiple MPDB functions beyond prior works [4], including integrate-and-fire (IF), leaky integrate-and-fire (LIF), and integer quadratic integrate-and-fire (IQIF) [9]. The IF mode provides simple spike generation; the LIF mode incorporates leakage dynamics for temporal noise filtering; and the IQIF mode introduces nonlinear integration for improved accuracy, enabling general-purpose SNN operations at the edge.

Figure 30.8.3 illustrates the CFD-SC-RC computation flow using the proposed CMDM-LCC and DM-MPDBU. Conventional bit-wise (BW)-IN multipliers (BWMs) for SNN cannot support word-wise (WW)-IN multiplication for CNN. To enable both BW-IN MAC for SNN and configurable WW-IN MAC for CNN, the proposed CMDM-LCC integrates two MUX-based multipliers (MULs), an INT8 toggle reducer (TR), dual-mode weight mapping (INT4/INT8) in the array, and an SaA circuit. Within one bank, each of eight CMDM-LCCs performs MAC operations with INs and Ws (W_{0-3}). In INT4 mode, the TR is disabled and four 4b MACs are executed per cycle: W_{0-1} map to two 4b weights (A[3:0], B[3:0]) and W_{2-3} to another two (C[3:0], D[3:0]). In INT8 mode, the TR eliminates redundant toggles, enabling two 8b MACs per cycle: W_{0-1} correspond to the four LSBs of A[7:0] and B[7:0], while W_{2-3} correspond to their four MSBs. The partial MACs (pMAC$_{0-7}$) are sent directly to the adder tree in SNN mode, while in CNN mode pMAC$_{0-7}$ are passed to the SaA circuit for WW-accumulation before entering the adder tree. Meanwhile, conventional MPDB blocks only support a single MPDB function, limiting flexibility for diverse SNN models. The proposed DM-MPDBU supports three functions: (1) IF mode—membrane potential (MP) accumulates until reaching threshold (peak), then fires and resets; (2) LIF mode—MP accumulates with leakage subtraction before proceeding as IF; and (3) IQIF mode—MP accumulates with an integer quadratic (IQ) term dynamically defined by parameters (pdeth, rest, thresh, α, β): if MP[t] < pdeth, IQ=($\alpha\times$rest–MP[t])) >> 3; otherwise, IQ=($\beta\times$(MP[t]–thresh)) >> 3.

Figure 30.8.4 illustrates the proposed PS-WR-ODM scheme. In conventional timestep (TS)-first data mapping, all input neurons (IN$_{0-K}$) are accessed sequentially within each TS, minimizing PSUM storage but incurring high weight-update energy and long latency because weight reuse cannot be leveraged for early MPDB operations. In contrast, prior weight-reuse data mapping [10] accesses same-index INs (e.g., IN$_0$) across all TSs (TS$_{0-N}$), maximizing weight reuse and throughput but requiring large PSUM storage. The proposed PS-WR-ODM adopts a hybrid strategy in which TSs are grouped into blocks with user-defined size (e.g., pairs of adjacent TSs), and weight-reuse mapping is applied within each block. This approach achieves balanced performance in weight-update energy and latency with only marginal PSUM buffer overhead. For example, as shown in the timing diagrams, the conventional TS-first data mapping incurs long weight-load cycles, while the previous weight-reuse mapping demands excessive PSUM buffering. The proposed PS-WR-ODM achieves comparable latency reduction and energy savings without significant PSUM overhead.

Figure 30.8.5 shows the simulated performance of the proposed schemes. The CFD-SC-RC reduces macro area by 33% and improves energy efficiency by 2.31× over prior hybrid SNN-CNN CIM design [4]. The PS-WR-ODM achieves >1.22× improvement in FoM1, defined as: energy efficiency/(area×latency), compared to conventional timestep-first and weight-reuse-first [10] mappings. The combined implementation of CFD-SC-RC, CMDM-LCC, and PS-WR-ODM further delivers an FoM2, defined as: inference accuracy×energy efficiency/(area×latency)], of 5.48× over baseline designs [4,10].

Figure 30.8.6 presents the measurement results of the 16nm SRAM-DCIM macro and the position chart versus prior silicon-verified works. The Shmoo plot confirms 1.3ns computing latency (T_{MAC}) at 0.8V with an 8bIN-8bW-22bOUT configuration. The proposed macro achieves up to 8.2× improvement in FoM3, defined as: IN precision×W precision×energy efficiency, over prior works. Figure 30.8.7 presents the die micrograph and chip performance summary. To the best of our knowledge, this is this chip is the first silicon-verified fully digital SRAM-CIM macro that supports multi-bit hybrid SNN–CNN processing.

Acknowledgement:
The authors would like to thank NTHU-ITRI collaboration and NSTC for their financial and manufacturing support.

References:
[1] D. Huh and T. J. Sejnowski, "Gradient Descent for Spiking Neural Networks," NeurIPS, vol. 31, pp. 1–11, Dec. 2018. Accessed on Dec. 9, 2025.
<https://proceedings.neurips.cc/paper_files/paper/2018/file/185e65bc40581880c4f2c82958 de8cfe-Paper.pdf>
[2] K. He et al., "Deep Residual Learning for Image Recognition," *IEEE Conf. Comp. Vision and Pattern Recognition* (*CVPR*), pp. 770-778, Dec. 2016. https://doi.org/10.1109/CVPR.2016.90
[3] M. Sandler et al., "MobileNetV2: Inverted Residuals and Linear Bottlenecks," *IEEE/CVF CVPR*, pp. 4510-4520, Dec. 2018. https://doi.org/10.1109/CVPR.2018.00474
[4] C. Zha et al., "A Reconfigurable Digital Compute-In-Memory Heterogeneous Macro for Differential Frame Convolution and Spiking Neural Network," *ISCAS*, pp. 1-5, May 2025. https://doi.org/10.1109/ISCAS56072.2025.11043929
[5] S. Kim et al., "C-DNN: A 24.5-85.8TOPS/W Complementary-Deep-Neural-Network Processor with Heterogeneous CNN/SNN Core Architecture and Forward-Gradient-Based Sparsity Generation," *ISSCC*, pp. 334-335, Feb. 2023. https://doi.org/10.1109/ISSCC42615.2023.10067497
[6] S. Kim, S. Park, B. Na, and S. Yoon, "Spiking-YOLO: Spiking Neural Network for Energy-efficient Object Detection," *AAAI Conference on Artificial Intelligence*, vol. 34, no. 7, pp. 11270–11277, Feb. 2020. https://doi.org/10.1609/aaai.v34i07.6787

[7] H. Fujiwara et al., "A 3nm, 32.5TOPS/W, 55.0TOPS/mm2 and 3.78Mb/mm2 Fully-Digital Compute-in-Memory Macro Supporting INT12 × INT12 with a Parallel-MAC Architecture and Foundry 6T-SRAM Bit Cell," *ISSCC*, pp. 572-573, Feb. 2024. https://doi.org/10.1109/ISSCC49657.2024.10454556

[8] P. Chen et al., "A 22nm Delta-Sigma Computing-In-Memory (ΔΣCIM) SRAM Macro with Near-Zero-Mean Outputs and LSB-First ADCs Achieving 21.38TOPS/W for 8b-MAC Edge AI Processing," *ISSCC*, pp. 140-141, Feb. 2023. https://doi.org/10.1109/ISSCC42615.2023.10067289

[9] W.-C. Wu et al., "Integer Quadratic Integrate-and-Fire (IQIF): A Neuron Model for Digital Neuromorphic Systems," *AICAS*, pp. 1-4, June 2021. https://doi.org/10.1109/AICAS51828.2021.9458572

[10] S. Kim et al., "SNPU: An Energy-Efficient Spike Domain Deep-Neural-Network Processor With Two-Step Spike Encoding and Shift-and-Accumulation Unit," *JSSC*, vol. 58, no. 10, pp. 2812-2825, Oct. 2023. https://doi.org/10.1109/JSSC.2023.3270442

[11] K. Park et al., "A 701.7 TOPS/W Compute-in-Memory Processor With Time-Domain Computing for Spiking Neural Network," *IEEE Transactions on Circuits and Systems I: Regular Papers*, vol. 72, no. 1, pp. 25-35, Jan. 2025. https://doi.org/10.1109/TCSI.2024.3480350

[12] H. Fu et al., "NeuC-CIM: A 1.3pJ/SOP Neuromorphic Charge-Domain Compute-in-Memory Macro for Spiking Neural Network," *IEEE Symp. VLSI Circuits*, pp, 1-3, June 2025. https://doi.org/10.23919/VLSITechnologyandCir65189.2025.11074927

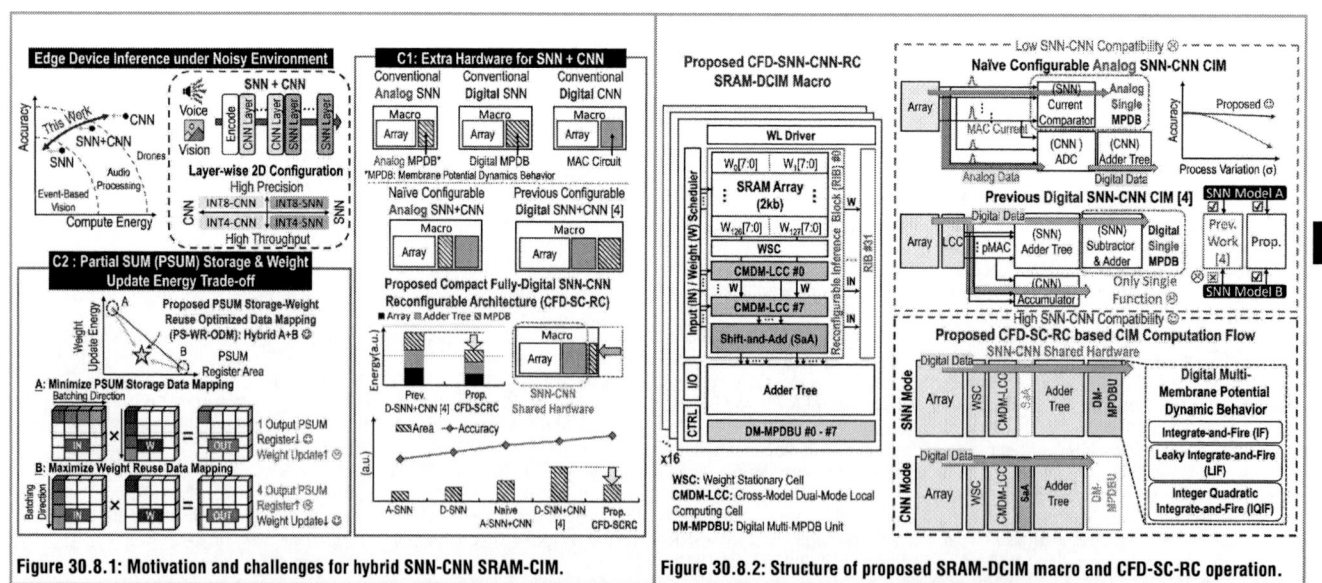

Figure 30.8.1: Motivation and challenges for hybrid SNN-CNN SRAM-CIM.

Figure 30.8.2: Structure of proposed SRAM-DCIM macro and CFD-SC-RC operation.

ISSCC 2026 / SESSION 30 / COMPUTE-IN-MEMORY / 30.8

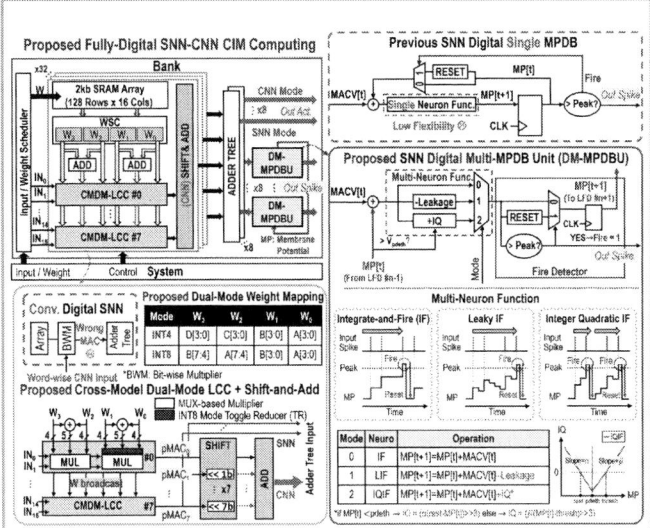

Figure 30.8.3: CMDM-LCC and DM-MPDBU structure and operation.

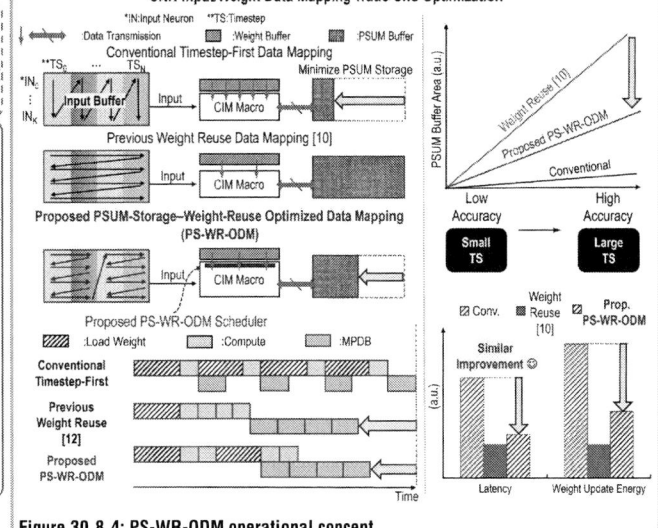

Figure 30.8.4: PS-WR-ODM operational concept.

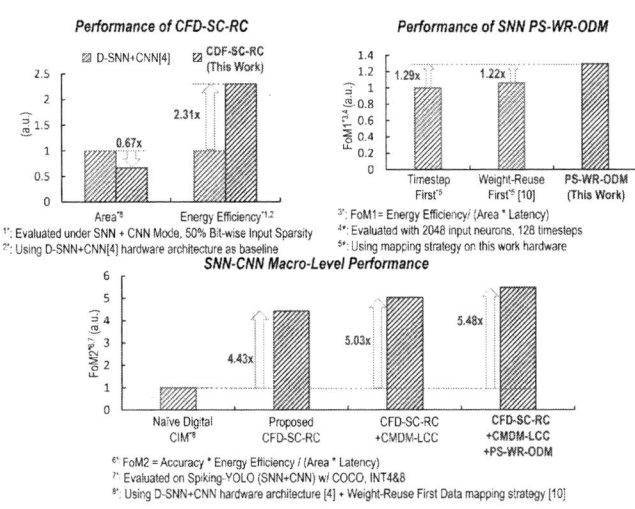

Figure 30.8.5: Simulated performance of proposed schemes.

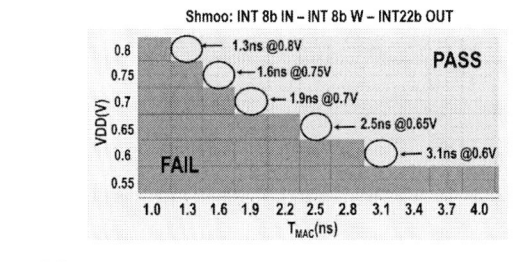

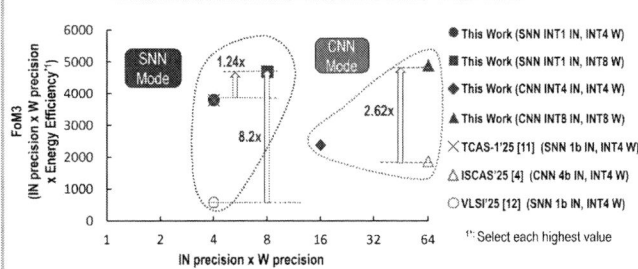

Figure 30.8.6: Measurement results and position chart.

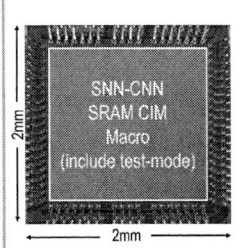

CHIP SUMMARY				
Technology	16nm FinFET CMOS			
Domain	Digital			
Cell Type	Compact 6T-SRAM			
Macro size	1Mb			
Mode	SNN		CNN	
Input precision (bit)	1		4	8
Weight precision (bit)	4	8	4	8
Number of input channels	128	64	128	64
Number of output channels	128		16	
Output precision (bit)	11	14	15	22
Supply voltage (V)	0.6 - 0.8			
Access time[1] (ns)	1.3			
Throughput[1,2] (TOPS)	25.21	12.60	3.15	1.58
Energy Efficiency[2] (TOPS/W) (Average Performance[4])	870.28	444.21	123.62	62.84
Energy Efficiency[2] (TOPS/W) (Peak Performance[3])	949.41	586.36	148.35	76.35
Area Efficiency[1,2] (TOPS/mm²)	1.69 - 26.99			
SNN MPDB Type	LIF / IF / IQIF			
Accuracy Loss	No			
Silicon-Verified	Yes			

[1]: MAC computing @VDD = 0.8V
[2]: One operation (OP) represents one multiplication or one addition
[3]: Measured @VDD = 0.8V, 50% input sparsity
[4]: Measured @VDD = 0.6V, 90% input sparsity

Figure 30.8.7: Die micrograph and performance summary table.

• 2026 IEEE International Solid-State Circuits Conference

979-8-3315-8937-0/26 $31.00 © 2026 IEEE

ISSCC 2026 / SESSION 30 / COMPUTE-IN-MEMORY / 30.9

30.9 A 147TOPS/W, 250TOPS/mm², Fully Synthesizable, Digital Compute-in-Memory Accelerator Supporting INT8×INT8 with Zero-Point Quantization in Intel 18A Technology

Amit Agarwal[1], Steven K. Hsu[1], Mark A. Anders[1], Arnab Raha[2], Deepak A. Mathaikutty[3], Ram Krishnamurthy[1]

[1]Intel, Hillsboro, OR, [2]Intel, Santa Clara, CA, [3]Intel, Chandler, AZ

Abstract

A fully synthesizable DCiM on-die accelerator with 128 input channels, 32 output channels, and 2 weight sets, supporting zero-point-quantized INT8×INT8 computation, is fabricated in Intel 18A technology and occupies 0.0856mm². The DCiM accelerator operates at 2.62GHz

for a supply voltage of 1.1V, and 25°C, achieving a peak area-efficiency of 250TOPS/mm². Robust operation of DCiM is maintained down to 400mV, 25°C, where it delivers a peak energy-efficiency of 147TOPS/W at 25% input activity, with a power consumption of 6.1mW.

Artificial Intelligence (AI) hardware accelerators continue to drive advances in computer vision, speech recognition, large language models (LLMs), and image and video processing. Compute-in-memory (CiM) architectures reduce data-movement overhead by integrating distributed memory and compute with wide adder-trees, enabling highly parallel computation. Digital compute-in-memory (DCiM) provides deterministic accuracy, with process and voltage scalability along with immunity to parametric variation, while a fully synthesizable design enables portability across technology nodes, and simplifies integration using standard design tools and cell libraries, promoting broader adoption. This work presents a fully synthesizable DCiM on-die accelerator featuring 128 input channels, 32 output channels, 2 weight sets, high-density multibit interruptible 12T-latch-based memory storage, radix-4 stationary weight encoded Booth computation, a unique unsigned-to-signed mapping scheme, and support for zero-point-quantized INT8×INT8 data format. Fabricated in Intel 18A technology [1] (Fig. 30.9.7), the DCiM accelerator operates at 2.62GHz measured at maximum supply voltage of 1.1V and 25°C, occupying a compact area of 0.0856mm² while achieving: (i) peak area efficiency of 250TOPS/mm² and throughput of 21.5TOPS, at 1.1V and 25°C, (ii) robust functionality of DCiM measured down to 400mV at 25°C with peak energy efficiency of 147TOPS/W and power consumption of 6.1mW, at 25% input activity, (iii) energy efficiency scalable to 254TOPS/W at 3.5mW for 10% input activity for supply voltage of 400mV and 25°C, (iv) unique unsigned-to-signed mapping and zero-point expansion with pre-computing saving 20% power/17% area, (v) Booth encoding on stationary weights with 28% reduction in compute power, and (vi) use of high-density multibit latch bitcell resulting in 20% reduction in weight-storage area.

The DCiM accelerator organization consists of 128 input channels and 32 output channels, supporting zero-point quantized 8b unsigned/signed activations and weights (Fig. 30.9.1). Zero-point quantization improves computation accuracy by quantizing more precisely for data not symmetrically centered around zero. However, this requires addition of an origin shift constant from activations and weights, resulting in an effective signed 9b computation. Pre-computing both an unsigned-to-signed bias and zero-point expansion converts the datapath to signed 8b×8b. The converted signed 8b weights are Booth encoded as four 3b Booth selects (single, neg, pos). These Booth selects for a compute column of 128 weights are written every cycle, completing the entire weight-set write in 32 cycles. A double buffered scheme stores two weight sets per compute column. Computations are performed on the selected weight set, which is read through a static 2:1 multiplexor only at the beginning of each compute operation for reduced read power. The second weight location can be updated in the background, eliminating any update stalls when the latency of compute operation is longer than writing a new set of weights. Balanced optimization of throughput and power results in four pipeline stages to complete 128-way MAC. The unsigned-to-signed conversion with zero-point offset is performed in the first pipeline stage while the next two stages complete the 32-way bank results. The converted signed 8b activations from the first stage are broadcast across all compute columns, generating four partial products per multiplication. Partial products are added using 32-way carry-save adder trees, first with addition across multipliers, followed by merging the shifted partial-products sum. Finally, the outputs from four banks are summed to complete 128-way MAC in the last pipeline stage.

When optimizing for power and area, Booth multipliers offer unique benefits when applied to DCiM datapaths. Radix-4 Booth encoding reduces partial products by half but requires one multiplier input to be Booth encoded. Therefore, this encoder overhead can limit the power and area savings of Booth multipliers, especially for low data bit widths. The DCiM dataflow is unique, because weight inputs are stored and remain stationary for multiple compute cycles, while activation inputs are broadcast across compute columns. Booth encoding can be applied to either the broadcast activation or the stored weights. The 8b activation can be encoded once in the periphery, with 12b Booth selects broadcast across the compute columns. This amortizes the encoder area and power across all 32 compute columns but increases the number of input bits switching every cycle from 8 to 12, thereby increasing compute power. Alternatively, pre-encoding 8b weights and storing 12b Booth selects eliminates encoder power during compute but increases storage area (Fig. 30.9.2). Since weights are stored and written per column, moving the weight encoder to the memory write path enables reuse and reduces encoder area by 32× due to lower throughput requirements. Booth S-bits, which only depend on multipliers (weights), are pre-computed

and stored in the weight memory for later addition with the final sum. Furthermore, encoding of the weights preserves typically low MSB activity of activation inputs into the adder-tree. Overall, the implemented weight encoding design results in only 7% increase in area, while enabling 28% lower compute power with stationary weights, compared to activation encoding. Double buffered weights are stored using interruptible 12T multibit latch standard cells. This enables robust low supply voltage write operation and memory read through static 2:1 multiplexor followed by Booth selectors. A modified Booth encoding is designed with single/neg/pos instead of single/double/neg select signals, which encodes ±0 by forcing both neg/pos signal to either 0 or 1. This modified Booth encoding results in up to 12% lower Booth selector area.

Radix-4 Booth multiplication with 8b unsigned/signed activations and weights results in five partial products. Five carry-save adder trees sum these partial products in groups with the same alignment across multipliers. Zero-point quantization requires addition of a constant, converting these inputs to 9b signed integers. The radix-4 Booth with 9b signed multiplication would not only generate 5 partial products but would also increase their width from 9b to 10b (Fig. 30.9.3). Instead, a unique unsigned-to-signed mapping converts the 8b unsigned integer to 8b signed integer by subtracting a bias of 128. This conversion can be done by conditionally inverting the MSB based on input format. Likewise, the zero-point constant must also be added from every activation and weight. The bias and zero-point constant are combined and added to the mapped 8b signed integer, representing final activation and weights with zero-point quantization. The DCiM sum of product computation is represented by the equation ($\Sigma_N A_Z W_Z$) shown in Fig. 30.9.2. Expanding this equation splits each 9b×9b signed multiplication ($\Sigma_N A_Z W_Z$) into an 8b×8b signed multiplication ($\Sigma_N A_S W_S$) plus three extra terms. In the DCiM dataflow the activation inputs are shared across multiple compute columns. The weight inputs are stationary (stored in memory) and constant for multiple cycles. Therefore, the compute cost of the three extra terms is amortized to reduce area and power. The sum of the activations ($\Sigma_N A_S$) multiplied with the constant ($C + Z_W$) is computed once every cycle and shared across multiple compute columns. The sum of weights ($\Sigma_N W_S$) multiplied with the constant ($C + Z_A$) is precomputed and stored in memory during weight writes, and therefore does not consume power during compute operations. Finally, the third term $N(C + Z_W)(C + Z_A)$ is a fixed constant. All three terms are added at the end of the compute datapath. This intrinsic zero-point compute with unsigned-signed mapping within the DCiM dataflow reduces the 9b×9b signed multiplication to an 8b×8b signed multiplication. This reduces the partial product width by 10% and decreases the number of partial products from 5 to 4, resulting in only 4 adder trees to complete the MAC operation. This technique lowers overall DCiM power by 17% and area by 21%.

DCiM area and power breakdown are shown in Fig. 30.9.4. The storage latch and read multiplexor circuits contribute 21% of the area, even though they consume only 11% of total write power. However, the write data driver and decoder contribute 56% of write power but occupy only 3% of area. The Booth encoder along with S-bit and zero-point pre-computation contribute only 16% of write power and 2% of total area, because of sharing across 32 columns. Double buffering allows concurrent update and compute. When compute latency is higher than writing a new set of weights, the DCiM becomes compute power bound, leading to maximum energy-efficiency. Booth selectors and adder trees, which contribute 74% of the DCiM area, reach 98% of the total compute power. The design is implemented with five different latch types ranging from single-bit to high-density multibit with two cycle compute-after-write timing constraints. This optimized timing achieves 95.5% high-density multibit insertion ratio, resulting in 20% weight storage area reduction.

The DCiM is synthesized using Intel 18A standard cell library, occupying total accelerator area of 0.0856mm². The DCiM implements 128 input channels, 32 output channels and 2 weight sets supporting 8b unsigned/signed zero-point quantized activations and weights. Weights are stored using interruptible 12T latch standard cells for robust low supply voltage write operation. The fabricated test-chip demonstrates a maximum throughput of 21.5TOPS at 2.62GHz, consuming 1.23W, with peak area efficiency of 250TOPS/mm² (measured at 1.1V and 25°C) (Fig. 30.9.5). The DCiM achieves peak energy efficiency of 147TOPS/W for input activity of 25%, weight 1=50%, consuming 6.1mW (measured at 400mV and 25°C). Energy efficiency scales to 254TOPS/W at 10% input activity, consuming 3.5mW. The double buffered scheme enables MAC computations on the selected weight set, while the

528 • 2026 IEEE International Solid-State Circuits Conference

979-8-3315-8937-0/26 $31.00 © 2026 IEEE

second weight set is updated in the background for 32 cycles. As the number of total compute cycles increases, the write power contribution reduces to 1% and the DCiM becomes compute power bound reaching peak energy efficiency. Figure 30.9.6 compares the proposed work with recent state-of-the-art digital CiMs, showing best-in-class area efficiency (TOPS/mm²) and energy efficiency (TOPS/W). Figure 30.9.7 presents the photograph of the test-chip, its die dimensions, and a performance summary table.

Acknowledgment:
The authors thank M. Langan, V. Ilderem, V. De, and J. Tschanz, for encouragement and discussions.

References:
[1] K. Fischer et al., "Intel 18A Platform Technology Featuring RibbonFET (GAA) and PowerVia for Advanced High-Performance Computing," *IEEE Symp. VLSI Circuits*, pp.1-3, June 2025. https://doi.org/10.23919/VLSITechnologyandCir65189.2025.11075006
[2] H. Mori et al., "A 4nm 6163-TOPS/W/b 4790–TOPS/mm²/b SRAM Based Digital-Computing-in-Memory Macro Supporting Bit-Width Flexibility and Simultaneous MAC and Weight Update," *ISSCC*, pp. 132-133, Feb. 2023. https://doi.org/10.1109/ISSCC42615.2023.10067555
[3] G. Desoli et al., "A 40-310TOPS/W SRAM-Based All-Digital Up to 4b In-Memory Computing Multi-Tiled NN Accelerator in FD-SOI 18nm for Deep-Learning Edge Applications," *ISSCC*, pp. 260-261, Feb. 2023. https://doi.org/10.1109/ISSCC42615.2023.10067422
[4] H. Fujiwara et al., "A 3nm, 32.5TOPS/W, 55.0TOPS/mm² and 3.78Mb/mm² Fully-Digital Compute-in-Memory Macro Supporting INT12 × INT12 with a Parallel-MAC Architecture and Foundry 6T-SRAM Bit Cell," *ISSCC*, pp. 572-574, 2024. https://doi.org/10.1109/ISSCC49657.2024.10454556
[5] H. Mori et al., "A 3nm 125 Tops/W-29 TFLOPS/W, 90 TOPS/mm²-17 TFLOPS/mm2 SRAM-Based INT8 and FP16 Digital-CIM Compiler with Multi-Weight Update/Cycle," *IEEE Symp. VLSI Circuits*, pp. 1-3, June 2025. https://doi.org/10.23919/VLSITechnologyandCir65189.2025.11074897

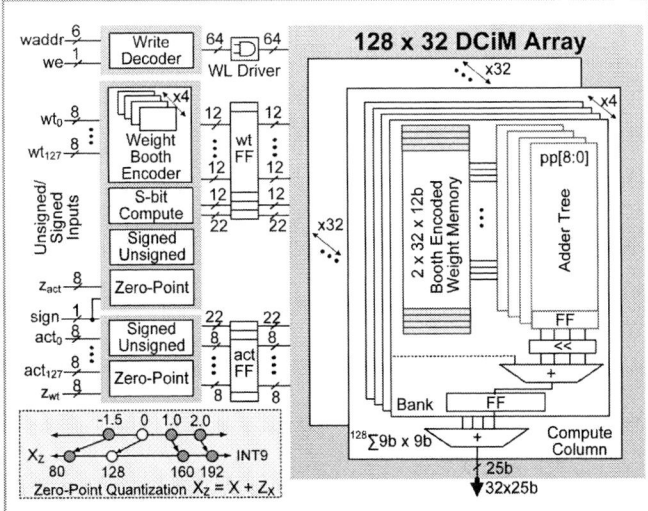

Figure 30.9.1: DCiM accelerator organization supporting zero-point quantized 8b unsigned/ signed activation and weights.

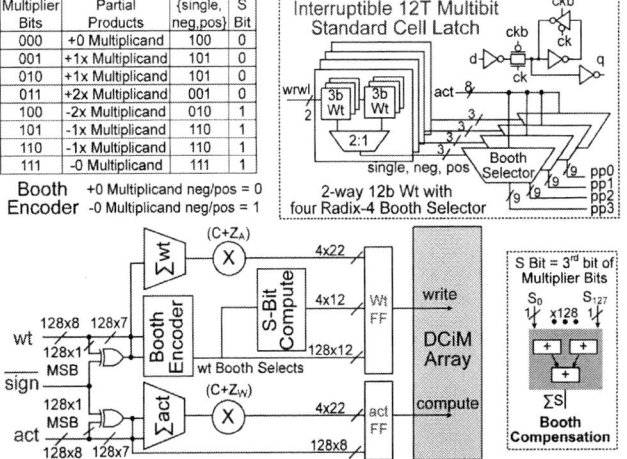

Figure 30.9.2: Radix-4 Booth encoded weights, interruptible 12T multibit latch storage with static 2:1 multiplexor read, S-bit compute, and modified Booth encoder.

ISSCC 2026 / SESSION 30 / COMPUTE-IN-MEMORY / 30.9

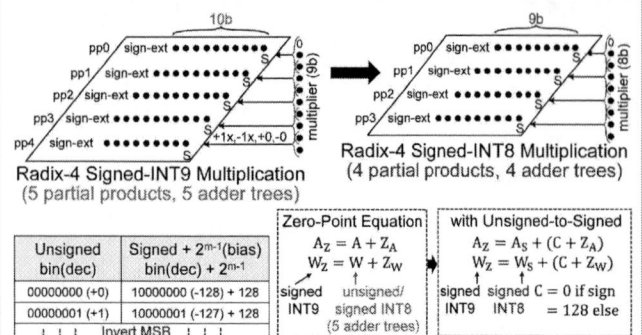

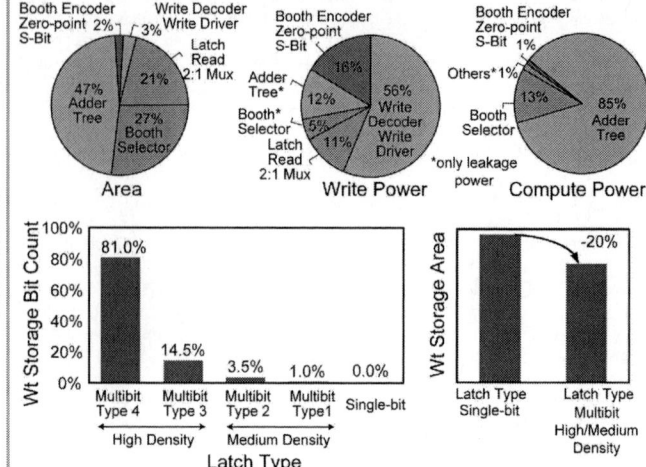

Figure 30.9.3: Intrinsic zero-point compute with unsigned-signed mapping within the DCiM dataflow.

Figure 30.9.4: Area/power breakdown and weight storage area savings.

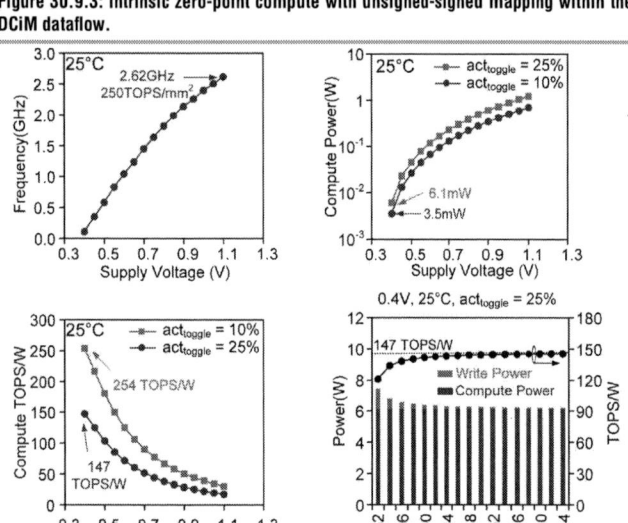

Figure 30.9.5: Frequency, power, and energy efficiency measurements in Intel 18A technology.

	ISSCC'23[2]	ISSCC'23 [3]	ISSCC'24 [4]	VLSI'25 [5]	This work
Technology	4nm	18nm	3nm	3nm	Intel 18A
MAC Operation	Digital	Digital	Digital	Digital	Digital
BitCell Type	8T	8T	6T	12T	12T
Macro Area	0.0172mm²	0.0656mm²	0.0167mm²	0.0175mm²	0.0856mm²
Voltage	0.32-1.1V	0.525-1.0V	0.36-1.1V	0.37-1.1V	0.4-1.1V
Input Channel	144	32	72	64	128
Output Channel	16	64	4	8	32
Weight Set	2	32	18	18	2
Format	signed INT8/INT12 bit-serial	unsigned/ signed INT4	signed INT12	unsigned/ signed INT8	unsigned/signed INT8 zero-point INT9
Simultaneous Mac + Write	Yes	Yes	Yes	Yes	Yes
TOPS/mm²	*49.9@0.9V	*3.4@1.0V	*124@0.9V	90.2@1.1V	250@1.1V
TOPS/W act$_{toggle}$=10%	--	--	*121@0.5V	124.6@0.5V	254@0.4V 180@0.5V
TOPS/W act$_{toggle}$=25%	*59.4@0.5V	*14.5@0.525V	*73.2@0.5V	70.4@0.5V	147@0.4V 104@0.5V

(*) Normalized to INT8

Figure 30.9.6: Power, performance, area summary, and comparison to prior work.

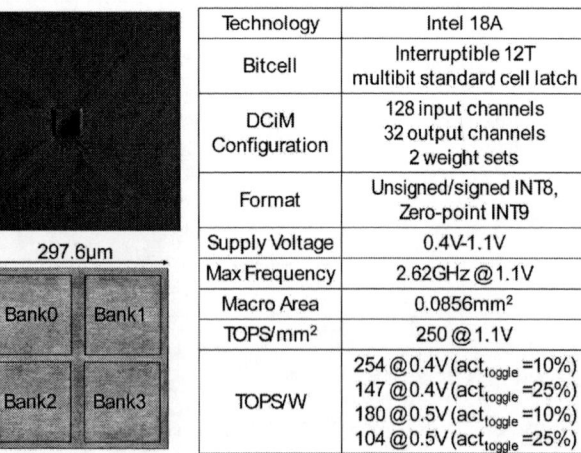

Technology	Intel 18A
Bitcell	Interruptible 12T multibit standard cell latch
DCiM Configuration	128 input channels 32 output channels 2 weight sets
Format	Unsigned/signed INT8, Zero-point INT9
Supply Voltage	0.4V-1.1V
Max Frequency	2.62GHz @ 1.1V
Macro Area	0.0856mm²
TOPS/mm²	250 @ 1.1V
TOPS/W	254 @ 0.4V (act$_{toggle}$=10%) 147 @ 0.4V (act$_{toggle}$=25%) 180 @ 0.5V (act$_{toggle}$=10%) 104 @ 0.5V (act$_{toggle}$=25%)

Figure 30.9.7: Packaged Intel 18A test-chip photograph, die dimensions, and performance summary table.

• 2026 IEEE International Solid-State Circuits Conference

ISSCC 2026 / SESSION 31 / AI ACCELERATORS / OVERVIEW

Session 31 Overview: *AI Accelerators*
DIGITAL ARCHITECTURES AND SYSTEMS SUBCOMMITTEE

Session Chair: Giuseppe Desoli
STMicroelectronics, Cornaredo, Italy

Session Co-Chair: Zhengya Zhang
University of Michigan, Ann Arbor, MI

This session presents a series of processors targeting large language models, diffusion networks, and multimodal generative workloads. The papers cover innovations in memory integration, low-bit quantization, and hybrid computing for efficient LLM and diffusion processing. Architectures range from high-performance and scalable inference engines to fully on-device systems that support personalization, fine-tuning and real-time interaction.

1:30 PM

31.1   A 14.08-to-135.69Token/s ReRAM-on-Logic Stacked Outlier-Free Large-Language-Model Accelerator with Block-Clustered Weight-Compression and Adaptive Parallel-Speculative-Decoding

Pingcheng Dong, Hong Kong University of Science and Technology, Hong Kong, China
AI Chip Center for Emerging Smart System, Hong Kong, China

In Paper 31.1, Hong Kong University of Science and Technology proposes the use of ReRAM-on-logic wafer stacking to improve memory bottlenecks for a 55nm LLM inference accelerator. Speculative decoding in both prefiling and decoding results in more than 7× higher performance and 4.8× energy savings.

1:55 PM

31.2   Revolver: Low-Bit GenAI Accelerator for Distilled-Model and CoT with Phase-Aware-Quantization and Rotation-Based Integer-Scaled Group Quantization

Sangjin Kim, KAIST, Daejeon, Korea

In Paper 31.2, KAIST introduces a 28nm 20.25mm² low-bitwidth generative AI accelerator with phase-aware-quantization and rotation-based integer-scaled group-quantization. The works achieves up to 2.36× lower energy on LLM prefill and DM workloads, and up to 2.05× faster LLM decoding compared to prior GenAI processors using simple quantization.

2:20 PM

31.3   A 51.6µJ/Token Subspace-Rotation-Based Dual-Quantized Large-Language-Model Accelerator with Fused Scale-Activation INT Datapath and Rearranged Bit-Slice LUT Computation

Xilong Kang, Southeast University, Nanjing, China

In Paper 31.3, Southeast University introduces a 51.6µJ/Token accelerator for rotation-based dual-quantized LLMs. The 1.37mm² chip, fabricated in 28nm CMOS, achieves an energy efficiency of 267.1 to 51.6µJ/token and latency of 621 to 2628ms, when generating 1024 tokens. A subspace-rotation method demonstrates 32.6% lower per-token energy than state-of-the- art accelerators under an equal-accuracy constraint.

530 • 2026 IEEE International Solid-State Circuits Conference

979-8-3315-8937-0/26 $31.00 © 2026 IEEE

ISSCC 2026 / February 18, 2026 / 1:30 PM

2:45 PM

31.4   VARSA: A Visual Autoregressive Generation Accelerator Using Performance-Scalable Multi-Precision PE-LUT and Grid-Similarity Attention Compression

Jiaqi Zhou, Peking University, Beijing, China

In Paper 31.4, Peking University presents VARSA, a 22nm 4.94mm^2 visual autoregressive generation accelerator featuring a scalable PE-LUT engine, multi-precision processing, and attention-map compression. It achieves 18.38TOPS/W and 503mJ/inference for 512×512 image generation, offering 2.7-to-8.9× better efficiency than prior diffusion accelerators.

3:35 PM

31.5   SoulMate: A 9.8mW Mobile Intelligence System-on-Chip with Mixed-Rank Architecture for On-Device LLM Personalization

Seongyon Hong, KAIST, Daejeon, Korea

In Paper 31.5, KAIST represents a fully on-device mobile intelligence system-on-chip, integrating retrieval-augmented generation (RAG) and fine tuning of a personal LLM. The 20.25mm^2 SoC, fabricated in 28nm CMOS, demonstrates real-time user interaction with high energy efficiency, such as 26.3μJ/token for inference and 56.8μJ/token for fine tuning.

4:00 PM

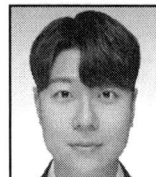

31.6   Tri-Oracle: A 17.78μJ/Token Vision-Language Model Accelerator with Token-Attention-Weight Redundancy Prediction

Seungjae Yoo, KAIST, Daejeon, Korea

In Paper 31.6, KAIST presents Tri-Oracle, a VLM accelerator that removes token, attention, and weight redundancies without fine tuning. It incorporates functionality to reduce tokens, predict key attention heads, and skip near-zero weights, cutting latency and memory use. Fabricated in 28nm CMOS, the 12.96mm^2 Tri-Oracle achieves 17.78μJ/token and 1.19 to 3.99TOPS/mm^2 without fine tuning.

4:25 PM

31.7   LUT-SSM: A 99.3TFLOPS/W LUT-Based State-Space Model Accelerator Using Energy-Efficient Element-Wise Layer Fusion and LUT-Friendly Weight-Only Quantization

Sunwoo Yoo, Korea Advanced Institute of Science and Technology, Daejeon, Korea,
        Pohang University of Science and Technology, Pohang, Korea,

In Paper 31.7, KAIST presents LUT-SSM, a 28nm 16.0mm^2 accelerator for state-space models (SSMs) with weight-only quantization, many-to-many LUT-based compute and element-wise layer fusion, demonstrating 99.3TFLOPS/W on the Mamba-1.4B model.

4:50 PM

31.8   A 28nm Speculative-Decoding LLM Processor Achieving 105-to-685μs/Token Latency for Billion-Parameter Models

Huanyu Wang, Tsinghua University, Beijing, China

In Paper 31.8, Tsinghua University presents an LLM processor that accelerates speculative decoding by addressing hardware inefficiencies. It introduces three key features: an exponent dual-reuse MAC, using draft model back-propagation for target model pruning and quantization, overlapping draft and target model execution for raising throughput. Fabricated in 28nm CMOS, the 56.82mm^2 LLM processor achieves up to 109.7TFLOPS/W and ~10× lower latency vs. the baseline FP16 model.

5:15 PM

31.9   ALPhA-Vision: A Real-Time Always-On Vision Processor with 787μs Face Detection Latency in <5mW

Ben Keller, Nvidia, Santa Clara, CA

In Paper 31.9, Nvidia presents a 16nm 4.20mm^2 real-time always-on vision processor for DNN-inference-based vision tasks in edge SoCs. The processor can perform face detection with 787μs latency and 99.3% detection accuracy, while consuming only 4.6mW average power at 60fps.

ISSCC 2026 / SESSION 31 / AI ACCELERATORS / 31.1

31.1 A 14.08-to-135.69Token/s ReRAM-on-Logic Stacked Outlier-Free Large-Language-Model Accelerator with Block-Clustered Weight-Compression and Adaptive Parallel-Speculative-Decoding

Pingcheng Dong[1,2], Yonghao Tan[1,2], Xuejiao Liu[2], Peng Luo[2], Yu Liu[2], Di Pang[2], Songchen Ma[1,2], Xijie Huang[1], Shih-Yang Liu[1], Dong Zhang[1,2], Zhichao Lu[3], Luhong Liang[2], Chi-Ying Tsui[1,2], Fengbin Tu[1,2], Liang Zhao[4], Kwang-Ting Cheng[1,2]

[1]Hong Kong University of Science and Technology, Hong Kong, China, [2]AI Chip Center for Emerging Smart System, Hong Kong, China, [3]Hefei Reliance Memory, Hefei, China, [4]Zhejiang University, Hangzhou, China

Abstract

This work presents a 55nm speculative decoding-based LLM accelerator with bumping-based face-to-face ReRAM-on-logic stacking technology. It features a local rotation unit for outlier-free low-bit quantization, a stacking-aware PNM architecture co-designed with blockwise vector quantization to reduce weight EMA overheads, and an adaptive parallel speculative decoding scheme with out-of-order scheduler for high resource and bandwidth utilization. Our chip achieves 14.08-to-135.69token/s and 4.46-to-7.17× speedup over vanilla speculative decoding.

Large Language Models (LLMs) [1-2] have achieved exceptional performance in natural language processing tasks, but their token-by-token autoregressive decoding (AD) paradigm incurs severe latency overhead due to extensive weight external memory access (EMA) [3-7]. Recently, speculative decoding (SD) [8-9] shown in Fig. 31.1.1 has been proposed and widely adopted on GPUs to resolve this issue by using a small-scale draft LLM (DLM) to decode multiple tokens in advance, which are then verified in parallel by a large-scale target LLM (TLM). However, on resource-constrained edge accelerators, LLM latency under SD is still dominated by weight EMA, with over 60% stemming from TLM. Although longer draft length (DL) can reduce TLM weight EMA by yielding more accepted tokens, increasing DLM latency diminishes this potential latency improvement. Thus, the TLM and DLM synergistically become the bottlenecks of SD, raising three challenges: 1) Low-bit post-training quantization [10] is frequently employed to reduce TLM EMA, but it suffers from severe accuracy degradation due to activation outliers [11]. Recent post-training quantization (PTQ) methods based on Fast Walsh-Hadamard Transform (FWHT) [11-13] can eliminate outliers, while preserving computational invariance due to the orthogonal property of the Hadamard matrix, but the deep FWHT array required by varied dimensions in TLM occupies nearly 4.37× the area of a 4K INT8 multiply-accumulate (MAC) array, incurring heavy area overheads. 2) Although DLM is much smaller and easily quantized to low bit precision via quantization-aware training (QAT), limited on-chip memory capacity in edge accelerators still fails to buffer all DLM weights, forcing frequent EMA where constrained external bandwidth further exacerbates latency overheads. 3) At long DL, over 90% of draft tokens decoded from DLM are rejected by TLM, whose latency overhead outweighs the latency savings from reduced TLM EMA.

To overcome these challenges, we develop an SD-based LLM accelerator with three key features: 1) We design a local rotation unit (LRU) that approximates global rotation by decomposing the deep FWHT into overlapped upper and lower low-cost 6-depth FWHTs. By rotating token features in two stages, the LRU removes activation outliers with little area burden and attains a 3.82-to-3.93× speedup over vanilla SD. 2) To extend on-chip memory capacity and avoid DLM EMA in a low-power yet cost-effective way, we utilize bumping-based ReRAM-on-logic stacking technology to design a ReRAM-stacked process-near-memory (RS-PNM) architecture with blockwise vector quantization (BVQ) algorithm. BVQ clusters DLM weights into block-level codebooks (CBs) stored in the high-density ReRAM, while RS-PNM reconstructs weights by retrieving CBs via a high-bandwidth stacking interface, which achieves 1.1-to-1.46× speedup over W4A8 SD with a LRU. 3) Inspired by recent parallel SD [14], we propose an adaptive parallel SD (APSD) scheme that combines the low rejection ratio of short DL and the high accepted token yield of long DL. APSD begins with non-parallel short DL drafting, followed by parallel draft-and-verify. Then, it dynamically adapts the drafting strategy based on feedback from TLM verification. If TLM accepts all previous draft tokens and its newest generated token matches the first draft token from concurrent DLM drafting, the parallel draft-and-verify continues. Otherwise, the draft tokens are discarded and APSD reverts to non-parallel DLM drafting. We further design an out-of-order scheduler with 4 parallel instruction queues decoupled from APSD workloads to avoid resource competition in parallel draft-and-verify as mentioned by [14], which attains 1.1-to-1.29× speedup and 10-to-14% rejected token ratio reduction.

Figure 31.1.2 shows the overall architecture of the proposed LLM accelerator with 4 ReRAM dies stacked on the logic die via 2048 face-to-face bumps for parallel read, delivering 25.6GB/s at 100MHz and 8MB memory capacity. The logic die mainly consists of a top controller, an MCU, a 64KB ISA buffer, 4 PLLs, an interconnect bus, a 1MB weight buffer, a 2MB global token buffer, an EMA controller (EMAC), an inter-chip transceiver, an LRU, a workload-decoupled out-of-order scheduler (WDOS), and an RS-PNM that includes a codebook fetcher unit (CFU), a ReRAM load interface (RLI), a tile-fused tensor engine (TFTE), and a non-linear processing unit (NLPU). During SD, the LRU performs local rotation for TLM, where the token allocator (TAU) sends upper/lower features to the reconfigurable FWHT array (RFA) or Hadamard accumulator unit (HAU) for decomposed FWHT. The rotated token is dynamically quantized with scaling factors bypassed to the TFTE for subsequent layer quantization. The RS-PNM utilizes the high bandwidth to load DLM CBs in ReRAM via RLI and fuses the token tiles that share the same CB entry by TFTE to avoid redundant CB loading. WDOS enables APSD by decoupling workloads into 4 parallel instruction queues, which are scheduled in an out-of-order manner with dependency-aware synchronization to maximize resource and bandwidth utilization.

Figure 31.1.3 illustrates the LRU with decomposed FWHT for low-bit outlier-free TLM quantization. The Hadamard matrix in the FWHT is built via a Kronecker product for power-of-two dimensions. However, a TLM often contains non-power-of-two (npot) channels. To handle them, the npot dimension n is typically factorized into 2^k×m dimensions (e.g., 14336=2^9×28 for the LLaMA3-8B down_proj layer), where m is the size of a pre-computed npot Hadamard matrix [15]. Such factorization results in a cascaded FWHT-GEMM array, which requires numerous high-precision operators, leading to significant area overhead. To address this, we utilize npot Hadamard construction to limit the FWHT depth from 9 to a low-cost 6, and approximate the global rotation using overlapped upper and lower rotations with searched with (m, k) pairs such that their combined coverage spans the original dimension n. Then, LRU starts two-stage local rotation with (m, k). In each stage, the TAU assigns the token tiles to RFAs first, which are reconfigurable to support 2^1-2^6 FWHT. To reduce the need for high-precision adders, adjacent FWHTs at early stages are merged to share inputs, and a lightweight router network further dispatches desired data based on the selected mode. Subsequently, the TAU allocates RFA outputs and the binary part of npot Hadamard tiles to the HAU for MAC-free accumulation by fusing FP16 Hadamard and the dynamic quantizers' scales. The LRU enables accurate W4A8 TLM quantization, achieving 3.82-to-3.93× speedup over BF16 SD while saving 92.7% area compared to global rotation.

Figure 31.1.4 depicts the RS-PNM architecture with the proposed BVQ algorithm. Unlike traditional vector quantization (VQ) methods [16-17] that incur heavy area overheads from index buffers and multi-port decoders [18]. BVQ performs block-level clustering by jointly learning blockwise CBs with INT4 QAT and block indices with Gumbel softmax reparameterization inspired by [19-20], which only requires a lightweight ISA decoder to retrieve the CBs. In RS-PNM, the MCU stores the DLM CBs in stacked ReRAM dies via SPI interface, after which the CFU triggers the ReRAM controller and RLI to load CBs from ReRAM to weight buffer. The RLI uses a double clock rate (200MHz) to stabilize the read data before transferring it to asynchronous FIFO groups for reliable clock domain crossing. To avoid data congestion caused by horizontal CB mapping at limited clock frequency, vertical CB mapping is employed. In addition, block dimensions within each CB are constrained by the per-die ReRAM bank width to maximize bandwidth utilization. However, vertical mapping results in redundant CB access issues when the complete weight is reconstructed on-chip, leading to extra latency overhead. To mitigate this, the tile fusion unit (TFU) fuses token tiles that share the same CB entry, ensuring each CB is fetched only once and thereby halving CB read latency. Moreover, TFUs facilitate both intra- and inter-layer parallelism by performing token fusion independently. Compared to W4A8 SD with LRU, the RS-PNM with INT4 BVQ achieves 1.1-to-1.46× speedup.

Figure 31.1.5 shows the APSD scheme with WDOS. A recent parallel SD method [14] enhances vanilla SD by inter-chip parallel draft-and-verify, which isolates DLM drafting and TLM verification workloads to avoid resource competition issues in intra-chip parallelism. However, the slow TLM verification leads to severe DLM idle time and wastes ReRAM bandwidth. The inter-chip parallelism further lowers both the ReRAM and DRAM bandwidth utilization as the memory interface in each chip cannot be shared across workloads. In contrast, APSD adaptively switches between short DL drafting and long DL parallel draft-and-verify based on whether previous draft tokens are all accepted and the first draft token matches TLM output, which alleviates DLM idleness. Moreover, APSD employs intra-chip parallelism enabled by WDOS with CB-interleaved intra-layer mapping (CILM). The CILM evenly allocates intra-layer CBs across chips and interleaves them within each DLM block to ensure full ReRAM bandwidth utilization at loading. Moreover, APSD workloads are simplified and decoupled into four instruction queues: inter-chip transceiver, compute, ReRAM load, and EMAC. Then, the intra-queue decoders extract dependency markers and send them to the inter-queue synchronizers, which jointly maintain a synchronous counter matrix for tracking readiness. An instruction is issued when its parent queues are ready, and its daughter queues are then notified. This dependency-aware scheduling efficiently enables intra-chip parallel draft-and-verify with high resource and bandwidth utilization, achieving 1.1-to-1.29× speed up over RS-PNM with 10-to-14% rejected DLM latency reduction.

Figure 31.1.6 shows the measurement results of the LLM accelerator, fabricated in 55nm using bumping-based face-to-face ReRAM-on-logic stacking technology. Die photos, specifications, 4-chip system, SEM/TEM images, and a ReRAM resistance distribution curve are shown in Fig. 31.1.7. The logic die operates at 63.5 to 285MHz at 0.89 to 1.40V, achieving 2.33TOPS peak performance. Each ReRAM die runs at 100MHz at 1.1V and consumes 49.54mW. The chip achieves 4.46-to-7.17× speedup and 3.74-to-4.85× energy savings over the BF16 SD baseline across various TLM/DLM pairs. Compared to the latest works [3-7], our chip integrates 3.43MB SRAM and 8MB stacked ReRAM with 25.6GB/s bandwidth. In a 4-chip system, ReRAM further scales to 32MB with 102.4GB/s bandwidth, sufficient to store all DLM CBs. The LRU supports reliable W4A8 quantization, achieving perplexity comparable to a SOTA W8A16 LLM accelerator [4] and better than works [6-7] that leverage 4b or sub-4b weight quantization. While most prior works [3–6] focus on compute-bound LLM prefilling optimization, the critical bottleneck lies in decoding. For a fair system-level comparison, we enhance each work with an LPDDR3 interface [21] to include EMA and 4-chip parallelism. As for LLaMA2-7B on the MT-Bench dataset that contains both prefilling and decoding, our chip achieves high decoding throughput of 17.82tokens/s and low energy consumption of 123.41mJ/token.

Acknowledgement:
This research was supported by ACCESS – AI Chip Center for Emerging Smart Systems, sponsored by InnoHK funding, Hong Kong SAR. The corresponding authors of this paper are Kwang-Ting Cheng (timcheng@ust.hk) and Liang Zhao (lzhao2020@zju.edu.cn).

References:
[1] A. Dubey et al., "The Llama 3 Herd of Models," arXiv: 2407.21783, 2024. https://arxiv.org/abs/2407.21783
[2] H. Touvron et al., "Llama 2: Open Foundation and Fine-Tuned Chat Models," arXiv: 2307.09288, 2023. https://arxiv.org/abs/2307.09288
[3] S. Kim et al., "C-Transformer: A 2.6-18.1µJ/token Homogeneous DNN-Transformer/Spiking-Transformer Processor with Big-Little Network and Implicit Weight Generation for Large Language Models," *ISSCC*, pp. 368-370, 2024. https://doi.org/10.1109/ISSCC49657.2024.10454330
[4] Y. Qin et al., "An 88.36TOPS/W Bit-Level-Weight-Compressed Large-Language-Model Accelerator with Cluster-Aligned INT-FP-GEMM and Bi-Dimensional Workflow Reformulation," *ISSCC*, pp. 420-422, 2025. https://doi.org/10.1109/ISSCC49661.2025.10904774
[5] S. Kim et al., "Slim-Llama: A 4.69mW Large-Language-Model Processor with Binary/Ternary Weights for Billion-Parameter Llama Model," *ISSCC*, pp. 421-423, 2025. https://doi.org/10.1109/ISSCC49661.2025.10904761
[6] Y. Wang et al., "LLM-CIM: A 28nm 126.7 TOPS/W Input-LUT-Based Digital CIM Macro with Reconfigurable Matrix Multiplication and Nonlinear Operation Modes for LLMs," *IEEE Symp. VLSI Circuits*, 2025. https://doi.org/10.23919/VLSITechnologyandCir65189.2025.11074939

[7] Z. Wu et al., "CELLA: A 28nm Compute-Memory Co-Optimized Real-Time Digital CIM-Based Edge LLM Accelerator with 1.78 ms-Response in Prefill and 31.32 Token/s in Decoding," *IEEE Symp. VLSI Circuits*, 2025. https://doi.org/10.23919/VLSITechnologyandCir65189.2025.11075101
[8] Y. Leviathan et al., "Fast Inference from Transformers via Speculative Decoding," *ICML*, pp. 19274-19286, 2023. https://arxiv.org/abs/2211.17192
[9] T. Li et al., "EAGLE: Speculative Sampling Requires Rethinking Feature Uncertainty," *ICML*, pp. 28935-28948, 2024. https://arxiv.org/abs/2401.15077
[10] E. Frantar et al., "GPTQ: Accurate Post-Training Quantization for Generative Pre-trained Transformers," *ICLR*, 2023. https://arxiv.org/abs/2210.17323
[11] S. Ashkboos et al., "QuaRot: Outlier-Free 4-bit Inference in Rotated LLMs," *NeurIPS*, pp.100213-100240, 2024. https://arxiv.org/abs/2404.00456
[12] Z. Liu et al., "SpinQuant: LLM Quantization with Learned Rotations," *ICLR*, 2025. https://arxiv.org/abs/2405.16406
[13] X. Huang et al., "RoLoRA: Fine-tuning Rotated Outlier-free LLMs for Effective Weight-Activation Quantization," *Empirical Methods in Natural Language Proc.*, pp. 7563-7576, 2024. https://doi.org/10.18653/v1/2024.findings-emnlp.444
[14] T. Liu et al., "PEARL: Parallel Speculative Decoding with Adaptive Draft Length," *ICLR*, 2025. https://arxiv.org/pdf/2408.11850
[15] N. Sloane, "A Library of Hadamard Matrices," 2024. http://neilsloane.com/hadamard/
[16] V. B. Mart et al., "GPTVQ: The Blessing of Dimensionality for LLM Quantization," arXiv: 2402.15319, 2024. https://arxiv.org/abs/2402.15319
[17] Y. Liu et al., "VPTQ: Extreme Low-bit Vector Post-Training Quantization for Large Language Models," *ACL*, pp. 8181-8196, 2024. https://doi.org/10.18653/v1/2024.emnlp-main.467
[18] S. Li et al., "MVQ: Towards Efficient DNN Compression and Acceleration with Masked Vector Quantization," *ACM ASPLOS*, pp. 731-745, 2025. https://arxiv.org/abs/2412.10261
[19] F. Gong et al., "MaskLLM: Learnable Semi-Structured Sparsity for Large Language Models," *NeurIPS*, pp.7736-7758, 2024. https://arxiv.org/abs/2409.17481
[20] P. Dong et al., "A 28nm 0.22µJ/Token Memory-Compute-Intensity-Aware CNN-Transformer Accelerator with Hybrid-Attention-Based Layer-Fusion and Cascaded Pruning for Semantic-Segmentation," *ISSCC*, pp. 408-409, 2025. https://doi.org/10.1109/ISSCC49661.2025.10904499
[21] M. Gao et al., "Tetris: Scalable and Efficient Neural Network Acceleration with 3D Memory," *ACM ASPLOS*, pp. 751-764, 2017. https://doi.org/10.1145/3093337.3037702

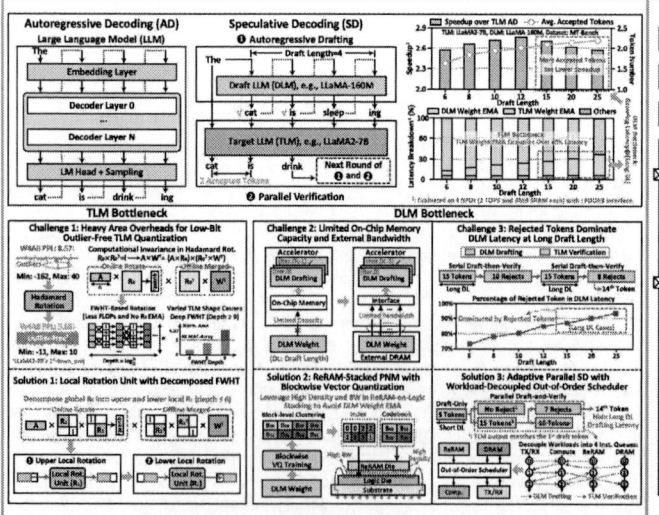

Figure 31.1.1: Challenges raised by target and draft large language model (LLM) in speculative decoding (SD) and proposed solutions.

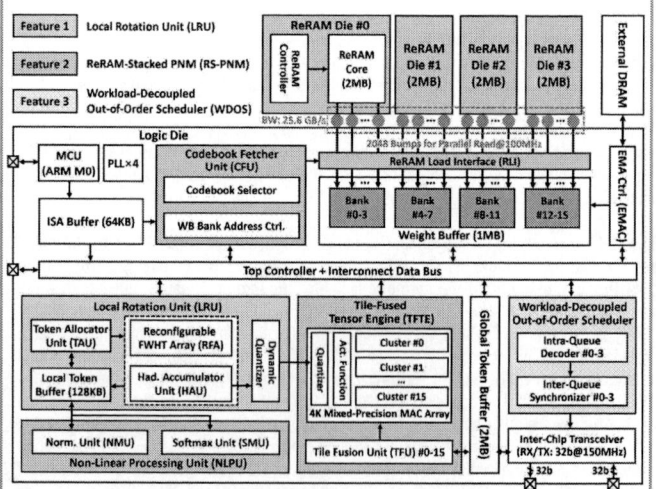

Figure 31.1.2: Overall architecture and three main features of the LLM accelerator with bumping-based ReRAM die on logic wafer face-to-face stacking technology.

ISSCC 2026 / SESSION 31 / AI ACCELERATORS / 31.1

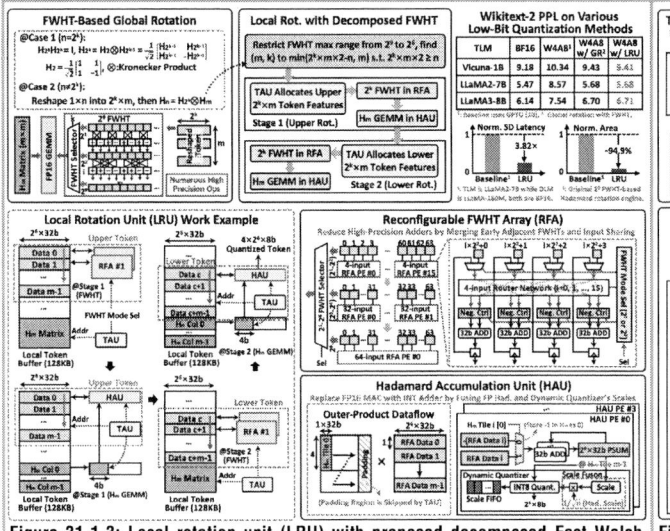

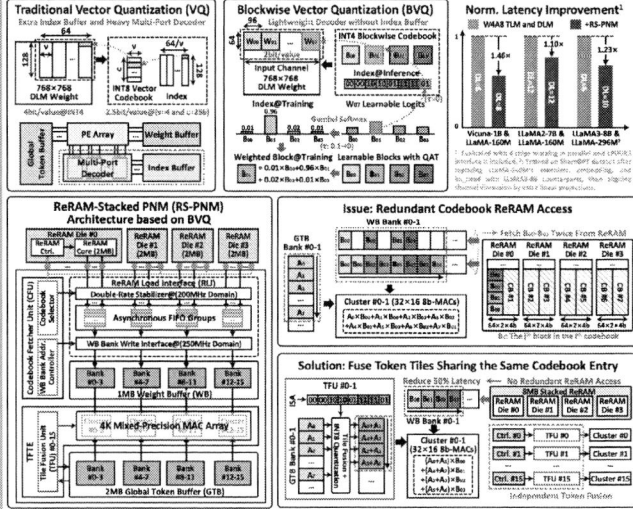

Figure 31.1.3: Local rotation unit (LRU) with proposed decomposed Fast Walsh-Hadamard Transform (FWHT) for outlier-free low-bit target LLM quantization.

Figure 31.1.4: ReRAM-stacked processing-near-memory (RS-PNM) architecture with blockwise vector quantization (BVQ) to avoid draft LLM external memory access (EMA).

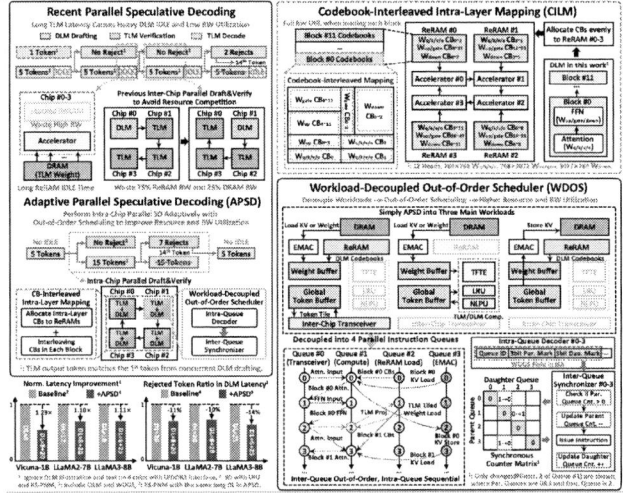

Figure 31.1.5: Adaptive parallel speculative decoding (APSD) with workload-decoupled out-of-order scheduler (WDOS) to achieve intra-chip parallel draft-and-verify with high resource utilization.

Figure 31.1.6: Measurement results and comparison with state-of-the-art LLM accelerators.

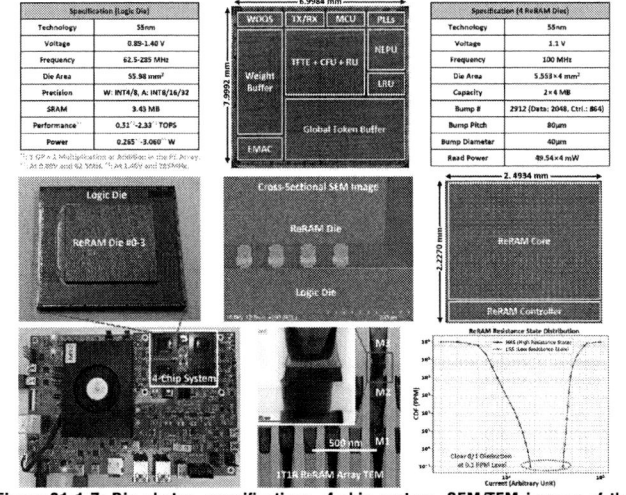

Figure 31.1.7: Die photos, specifications, 4-chip system, SEM/TEM images of the ReRAM-on-logic stacking interface/ReRAM chip, and ReRAM resistance distribution curve.

• 2026 IEEE International Solid-State Circuits Conference

ISSCC 2026 / SESSION 31 / AI ACCELERATORS / 31.2

31.2 Revolver: Low-Bit GenAI Accelerator for Distilled-Model and CoT with Phase-Aware-Quantization and Rotation-Based Integer-Scaled Group Quantization

Sangjin Kim, Jungjun Oh, Byeongcheol Kim, Yuseon Choi, Gwangtae Park, Hoi-Jun Yoo

KAIST, Daejeon, Korea

Abstract

Revolver is a low-bit GenAI accelerator that enables reasoning and multi-turn chat on edge devices under tight memory and power budgets. It introduces Phase-Aware Precision Selection (PAPS) with Multi-Precision Residual Encoding (MPRE) for memory-efficient multi-phase execution, Local Rotation with Harmonic-Aligned Permutation (LR-HAP) for low-cost rotation, and a Sliced Integer-based Dequantization Unit (SIDU) for efficient dequantization, achieving 3.99× energy savings and 2.10× speedup.

Multi-step reasoning [1, 2] and multi-turn chatting [3] are becoming dominant edge-AI tasks, requiring larger models with higher computation and memory capacity than conventional single-turn AI assistants [4–6]. However, on-device deployment is almost impossible due to limited memory capacity (≤~12GB [7]) and computing capability. To realize these tasks under limited model size, two trends have emerged: (i) knowledge distillation [8, 9] or fine-tuning [10, 11] that transfer capability from a larger teacher model or task-aligned supervision, and (ii) chain-of-thought (CoT), which generates additional "thought" tokens prior to final answers to improve reasoning performance [12, 13] (Fig. 31.2.1). However, for these harder tasks and distilled models, prior low-bit GenAI accelerators [14-16] that adopt simple quantization with perplexity (PPL) evaluation are not effective anymore. For example, although the PPL loss is <1 with an INT4 configuration, reasoning accuracy drops by ~6 percentage points; with distilled models and CoT the drop widens to ~15 percentage points. In addition, previous accelerators use fixed precision across all phases, limiting the hardware performance and efficiency. In an LLM, the prefill phase is precision critical, while the thought/answer decoding phase is more tolerant [17,18]. In diffusion models (DMs), early iterations are robust, whereas late iterations are sensitive [19]. Provisioning precision to cover the sensitive phases forces end-to-end high-bit execution, wasting resources on robust phases.

In this paper, Revolver is proposed to enable compute- and memory-efficient low-bit inference on edge with distilled and CoT-enabled models for harder tasks by: (1) phase-aware quantization that concentrates higher precision only on sensitive phases, and an advanced quantization algorithm that combines (2) rotation [20, 21], and (3) asymmetric group quantization, to preserve accuracy at low bit. However, implementing phase-aware quantization and rotation-based asymmetric group quantization on edge introduces three hardware challenges (Fig. 32.2.1). First, using different precisions across phases requires storing both high precision (HP) and low precision (LP) weights since calibration-based quantizers [22] prevent direct conversion from W_{HP} to W_{LP}, creating capacity overhead for duplicated weights. Second, non-Power-of-Two (PoT) channel sizes further preclude direct fast-Hadamard-transform (FHT) and inflate compute/energy for separate rotation units with FHT and matrix multiplication (MM). Third, asymmetric + group quantization demands per-group FP logic to handle the scale factor (SF) and zero point (ZP), where its utilization in a bit-scalable MAC architecture drops while partial-sum precision varies. To address these three challenges, Revolver introduces three key features: 1) Phase-aware Precision Selection (PAPS) with the Multi-precision Residual Encoding (MPRE) for memory-efficient multi-precision support. 2) Local Rotation with Harmonic Aligned Permutation (LR-HAP) and a Paired Unit-FHT Block (PUB) for a low-cost rotation unit design and SF precision reduction, and 3) a Sliced Integer-based Dequantization Unit (SIDU) for efficient integer-based dequantization.

Figure 31.2.2 shows the overall chip architecture of Revolver. It comprises the 4 Asymmetric Group Quantization Cores (AGC), Multi-Precision Loader (MP-Loader), a vector unit with a 1D SIMD core and a Rotation Unit (RU), 1.28MB global memory, and a top controller. MP-Loader supports multi-precision weight (W_{HP}/W_{LP}) fetch from external DRAM that stores only W_{LP} and W_R with MPRE. It consists of a residual decoder (RD) and a high-precision decoder (HPD) with 52KB for a weight buffer. The rotation unit consists of a 32-way FHT unit implemented with PUB and a 2KB intermediate Transposable Register File (TRF). TRF allows selective row/column-wise access for 2-stage hierarchical FHT operation, therefore it supports up to 1024-way FHT. The AGC consists of a 16×16 Asymmetric GQ-PE (AGPE) array and Input, Weight, Output, Scale Factors, and Zero Point memories. Also, a single row of weight sum units is integrated to compute the weight sum (W_{SUM}) of each group for asymmetric quantization support. Single AGPE consists of 32-way 4b-4b unit slice MACs and SIDU for dequantization operation. Each 2×2 neighboring AGPE shares a slicing unit to combine multiple unit-slice MAC and SIDUs in higher-bit precision operations.

Figure 31.2.3 illustrates the concept of PAPS and MPRE with its MP-Loader implementation. PAPS utilizes different precision according to phase-specific sensitivity and bottlenecks in an LLM and DM. In LLMs, the prefill processes all input tokens in parallel, therefore, becomes sensitive to weight precision; thus, LP is used for IA, and HP for W. During decoding which generates tokens one-by-one, weight EMA dominates energy and latency but the computational work is more robust to weight precision, while

activations have little impact on system efficiency. Thus, HP is used for IA, and LP for W. Similarly, in DM, early iterations for naïve outlining are robust and run with LP, whereas late iterations for image detailing demand HP for both W & IA.

To realize PAPS, high and low precision W (W_{HP} & W_{LP}) must be stored separately in external DRAM, since W_{LP} and W_{HP} are not directly overlaid. MPRE reduces weight storage by replacing W_{HP} with a weight residual (W_R) that uses fewer bits. W_R is encoded as the difference between W_{HP} and W_{LP} scaled by the ratio of scale factors (S_{LP}/S_{HP}). Although W_{HP} and W_{LP} are not overlaid, the distribution of W_R becomes much narrower than that of W_{HP} because W_{HP} and W_{LP} are approximately proportional. Therefore, W_R can be further compressed by splitting into the $(b_{LP}-1)$-bit sparse upper parts ($W_{R,S}$) encoded in COO format and the $(b_{HP}-b_{LP}+1)$-bit dense lower parts ($W_{R,D}$), saving up to 17.9% capacity compared to separate dual-copy storage. During the LP phase, only W_{LP} and S_{LP} are fetched, whereas the HP phase additionally retrieves W_R and S_{HP}, and then passes them to the MP-Loader, which reconstructs W_{HP} as follows: First, the DMA controller loads data for a 256×256 weight tile based on the weight-address table. Next, the RD decompresses the sparse residual ($W_{R,S}$) and concatenates it with the dense residual ($W_{R,D}$) to recover W_R. Finally, the HPD streams W_R, multiplies by S_{LP}/S_{HP}, and adds W_R to reconstruct W_{HP}. PAPS achieves 1.72-to-2.10× speedup on reasoning tasks with CoT. Combined with MPRE, weight EMA is reduced by up to 52.4%, and overall system energy by 50.4%.

Figure 31.2.4 illustrates the concept of LR-HAP and the implementation of PUB for a low-cost rotation unit. Unlike global rotation (GR) with a Ch-sized Hadamard Matrix (H_{Ch}), Local rotation (LR) forces the rotation sizes to PoT by using a small Hadamard Matrix (H_R, $R = 2^n$) and applying H_R to each set of R channels. Therefore, even with non-PoT Ch, LR avoids MM-based rotation, enabling an FHT-only rotation with 1.58-to-4.97× fewer operations. Additionally, HAP reshapes the rotated IA distribution to lower SF requirement with simple channel permutation. For asymmetric group quantization with group size G (=32), HAP permutes large-magnitude outlier channels to every G^{th} rows (0th, G^{th}, $2G^{th}$, …) which correspond to the coarse ±1 patterns (G-sized all +1 or all -1) in the Hadamard matrix. This permutation is applied in an offline manner by permuting output channels of the previous layer's weight. During rotation, outliers are multiplied with a consistent sign vector within each group, thus, the group-wise range of the rotated IA becomes narrower. Therefore, under asymmetric group quantization, the narrowed range drops the SF requirement from FP16 to INT8.

The PUB implements the FHT unit with paired logic for FP-ADD/SUB operations. In the FHT unit, each stage forms a pair with the same operands; therefore, PUB shares pre-alignment logic within a pair. Because one path is always a same-sign addition ±(|A|+|B|), and the other a different-sign addition ±(|A|-|B|), PUB first performs unsigned add/sub with aligned mantissa without sign handling. The unsigned subtraction is followed by normalization, whereas normalization for unsigned addition is omitted because no MSB-side zeros occur. A post sign-processing unit determines the output sign from the input sign (S_A & S_B) and the magnitude comparison results before subtraction (Swap). Such logic optimizations reduce FP ADD/SUB power by 56.3% and area by 50.0%. Finally, LR-HAP + PUB saves 59.4% of power and 60.8% of area of the rotation unit compared to the GR baseline.

Figure 31.2.5 details quantization/dequantization flow and the implementation of the SIDU. To handle per-group bias with LR-HAP, IA adopts asymmetric group quantization (A_Q) with per-group SF (S_a) and ZP (Z_a), while weights use symmetric group quantization (W_Q) with a per-group SF (S_w). Consequently, the dequantization operation consists of two terms based on the group-wise inner product of quantized W_Q/A_Q (PSUM) and the group-wise sum of W_Q (WSUM). With LR-HAP, INT8-SF enables a fully integer dequantization datapath (INT mult/sub, INT accumulator), removing FP logic. However, the PSUM/WSUM widths vary by the bit-mode of unit slice MAC (e.g., W4A4 uses 13/9b PSUM/WSUM, whereas W8A8 requires 21/13b), so W8A8 dequantization unit incurs an overhead of ~45% in power and ~46% in area compared to W4A4. The SIDU removes this overhead by fixing the dequantization unit with a minimal mode configuration (W4A4) and slicing wider PSUM/WSUM values with slicing unit for the higher-bit modes. The sliced mapping aligns with the 4 MAC granularity (W4A4, W4A8, W8A4, W8A8) and sustains high utilization

534 • 2026 IEEE International Solid-State Circuits Conference

979-8-3315-8937-0/26 $31.00 © 2026 IEEE

without over-width datapaths. Measured at W4A4, INT-SF using LR-HAP and SIDU lowers PE logic by 45.1% in power and 43.1% in area compared to an FP-SF baseline, and improves the energy efficiency of AGC by up to 66.7%.

Figure 31.2.6 summarizes measured results and presents a comparison table. Revolver is evaluated on LLMs for multi-step reasoning [1] and multi-turn chat [3] and on DM for text-to-image [23], using mobile-targeted models [8,10,11] with distillation and fine-tuning. Combining PAPS with LR-HAP yields 2.46-to-3.99× energy savings and 1.72-to-2.10× speedup on LLMs, and 2.48-to-3.47× energy savings and 1.94-to-2.83× speedup on DM. With LR-HAP and the SIDU, it achieves 13.73TOPS/W and outperforms previous transformer accelerators [25, 26] by 1.88 to 2.07×. Also, PAPS allows 3b weights in the decode stage, reducing weight EMA by 4.10× versus FP16. Under the same 32Gb/s external DDR4 SDRAM bandwidth [24], the proposed processor provides hardware-friendly rotation-based quantization and phase-aware optimization, achieving >1.76-to-2.36× lower energy on LLM prefill and DM workloads, and 1.68-to-2.05× faster LLM decoding compared to prior GenAI processors using simple quantization [15, 16]. Revolver is fabricated in 28nm CMOS technology and occupies 20.25mm² die area, as shown in Fig. 31.2.7. In summary, Revolver is a low-bit GenAI accelerator with phase-aware quantization and rotation-based asymmetric group quantization suitable for memory-limited edge devices.

References:
[1] K. Cobbe et al., "Training Verifiers to Solve Math Word Problems," *arXiv preprint arXiv:2110.14168,* 2021. https://doi.org/10.48550/arXiv.2110.14168
[2] S. Quan et al. , "CodeElo: Benchmarking Competition-Level Code Generation of LLMs with Human-Comparable Elo Ratings," *arXiv preprint* arXiv:2501.01257, 2025. https://doi.org/10.48550/arXiv.2501.01257
[3] L. Zheng et al., "Judging LLM-as-a-Judge with MT-Bench and Chatbot Arena," *NeurIPS,* vol. 36, pp. 46595–46623, 2023. https://doi.org/10.48550/arXiv.2306.05685
[4] S. Merity et al., "Pointer sentinel mixture models," *ICLR,* San Juan, Puerto Rico, May 2016. https://doi.org/10.48550/arXiv.1609.07843
[5] R. Zellers et al., "HellaSwag: Can a Machine Really Finish Your Sentence?" *ACL,* pp. 4791-4800, 2019. https://doi.org/10.48550/arXiv.1905.07830
[6] K. Sakaguchi et al., "WinoGrande: An Adversarial Winograd Schema Challenge at Scale," *arXiv preprint* arXiv:1907.10641, 2019. https://doi.org/10.48550/arXiv.1907.10641
[7] Samsung, "Galaxy Z Fold7," https://www.samsung.com/us/smartphones/galaxy-z-fold7/, accessed Sep. 2025. https://www.samsung.com/us/smartphones/galaxy-z-fold7/
[8] DeepSeek AI, "DeepSeek-R1-Distilled LLaMA-8B," https://huggingface.co/deepseek-ai/DeepSeek-R1-Distill-Llama-8B, accessed Sep. 2025. https://doi.org/10.48550/arXiv.2501.12948
[9] A. Shirgaonkar et al., "Knowledge distillation using frontier open-source LLMs: Generalizability and the Role of Synthetic Data," *arXiv preprint* arXiv:2410.18588, 2024. https://doi.org/10.48550/arXiv.2410.18588
[10] Meta AI, "Meta-Llama-3-8B-Instruct," https://huggingface.co/meta-llama/Meta-Llama-3-8B-Instruct, accessed Sep. 2025. https://doi.org/10.48550/arXiv.2407.21783
[11] J. Chen et al., "PixArt-Σ: Weak-to-Strong Training of Diffusion Transformer for 4K Text-To-Image Generation," *arXiv preprint* arXiv:2403.04692, 2024. https://doi.org/10.48550/arXiv.2403.04692
[12] J. Wei et al., "Chain-of-Thought Prompting Elicits Reasoning in Large Language Models," *NeurIPS,* vol. 35, pp. 24824–24837, 2022.

https://doi.org/10.48550/arXiv.2201.11903
[13] T. Kojima et al., "Large Language Models are Zero-Shot Reasoners," *NeurIPS,* vol. 35, pp. 22199–22213, 2022. https://doi.org/10.48550/arXiv.2205.11916
[14] S. Kim et al., "Slim-Llama: A 4.69mW Large-Language-Model Processor with Binary/Ternary Weights for Billion-Parameter Llama Model," *ISSCC,* pp. 421-423, 2025. https://doi.org/10.1109/ISSCC49661.2025.10904761
[15] S. Kim et al., "EdgeDiff: 418.4mJ/inference Multi-Modal Few-Step Diffusion Model Accelerator with Mixed-Precision and Reordered Group Quantization," *ISSCC,* pp. 410-411, 2025. https://doi.org/10.1109/ISSCC49661.2025.10904594
[16] D. Han and A. P. Chandrakasan, "MEGA.mini: A Universal Generative AI Processor with a New Big/Little Core Architecture for NPU," *ISSCC,* pp. 416-417, 2025. https://doi.org/10.1109/ISSCC49661.2025.10904514
[17] H. M. Chen et al., "Progressive Mixed-Precision Decoding for Efficient LLM Inference," *arXiv preprint* arXiv:2410.13461, 2024. https://doi.org/10.48550/arXiv.2410.13461
[18] J. Zhao et al., "LLM-PQ: Serving LLM on Heterogeneous Clusters with Phase-Aware Partition and Adaptive Quantization," *arXiv preprint* arXiv:2403.01136, 2024. https://doi.org/10.48550/arXiv.2403.01136
[19] Y. Kim et al., "Leveraging Early-Stage Robustness in Diffusion Models for Efficient and High-Quality Image Synthesis," *NeurIPS,* vol. 36, pp. 1229–1244, 2023. https://dl.acm.org/doi/abs/10.5555/3666122.3666181
[20] S. Ashkboos et al., "QuaRot: Outlier-Free 4-Bit Inference in Rotated LLMs," *arXiv preprint* arXiv:2404.00456, 2024. https://doi.org/10.48550/arXiv.2404.00456
[21] Z. Liu et al., "SpinQuant: LLM Quantization with Learned Rotations," *arXiv preprint* arXiv:2405.16406, 2024. https://doi.org/10.48550/arXiv.2405.16406
[22] E. Frantar et al., "GPTQ: Accurate Post-Training Quantization for Generative Pre-trained Transformers," *arXiv preprint* arXiv:2210.17323, 2022. https://doi.org/10.48550/arXiv.2210.17323
[23] X. Wu et al., "Human Preference Score v2: A Solid Benchmark for Evaluating Human Preferences of Text-To-Image Synthesis," *arXiv preprint* arXiv:2306.09341, 2023. https://doi.org/10.48550/arXiv.2306.09341
[24] Micron Technology, "DRAM Power Calculator," https://www.micron.com/sales-support/design-tools/dram-power-calculator, accessed Sep. 2025. https://www.micron.com/sales-support/design-tools/dram-power-calculator
[25] T. Tambe et al., "A 12nm 18.1TFLOPs/W Sparse Transformer Processor with Entropy-Based Early Exit, Mixed-Precision Predication and Fine-Grained Power Management," *ISSCC,* pp. 342–344, 2023. https://doi.org/10.1109/ISSCC42615.2023.10067817
[26] S. Kim et al., "C-Transformer: A 2.6–18.1µJ/token Homogeneous DNN-Transformer/Spiking-Transformer Processor with Big-Little Network and Implicit Weight Generation For Large Language Models," *ISSCC,* pp. 368–370, 2024. https://doi.org/10.1109/ISSCC49657.2024.10454330

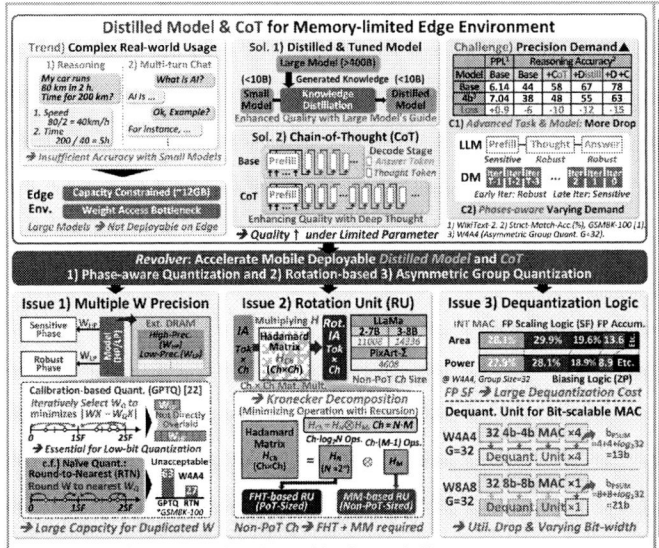

Figure 31.2.1: GenAI model for edge and hardware challenges.

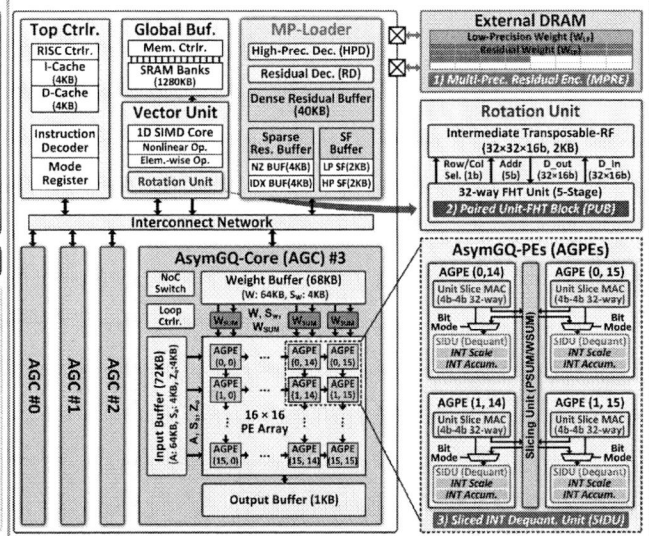

Figure 31.2.2: Overall chip architecture of proposed Revolver.

ISSCC 2026 / SESSION 31 / AI ACCELERATORS / 31.2

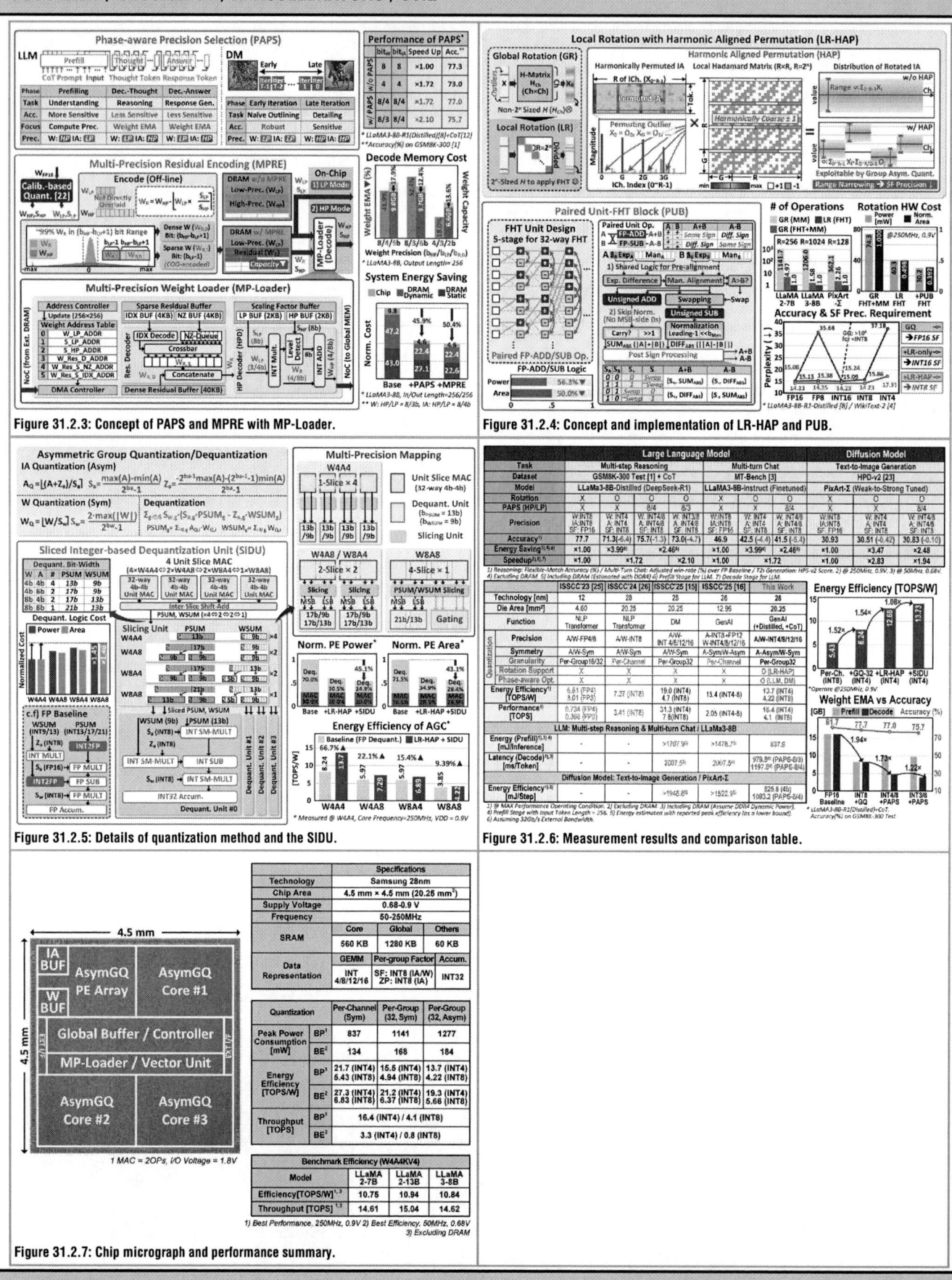

Figure 31.2.3: Concept of PAPS and MPRE with MP-Loader.

Figure 31.2.4: Concept and implementation of LR-HAP and PUB.

Figure 31.2.5: Details of quantization method and the SIDU.

Figure 31.2.6: Measurement results and comparison table.

Figure 31.2.7: Chip micrograph and performance summary.

• 2026 IEEE International Solid-State Circuits Conference

979-8-3315-8937-0/26 $31.00 © 2026 IEEE

ISSCC 2026 / SESSION 31 / AI ACCELERATORS / 31.3

31.3 A 51.6µJ/Token Subspace-Rotation-Based Dual-Quantized Large-Language-Model Accelerator with Fused Scale-Activation INT Datapath and Rearranged Bit-Slice LUT Computation

Bo Liu, Zihan Zou, Xinming Yan, Xilong Kang, Xinyang Chen, Bo Hu, Jiaming Lin, Haoran Du, Jun Yang, Xin Si, Hao Cai

Southeast University, Nanjing, China

Abstract

A 51.6µJ/token accelerator for rotation-based dual-quantized LLMs is presented. A subspace-rotation method with parallel Hadamard transposer reduces on-chip rotation power by 62.3% and area by 59.7%. A fused scale-activation unit lowers energy by 61.5% vs. the naive FP design. Rearranged bit-slice LUT computation achieves 2.28× better energy efficiency compared to a direct bit-parallel MAC implementation, while supporting flexible bit-width. The chip reduces per-token energy by 32.6% over SOTA under an equal accuracy constraint.

Large language models (LLMs) have demonstrated outstanding performance in natural language processing (NLP) tasks, pushing the boundaries of artificial intelligence (AI) applications [1-4]. However, this performance is achieved by scaling parameter counts into the billions, which greatly limits their deployment on cloud and edge devices [5]. As shown in Fig. 31.3.1, recent advanced quantization algorithms, such as group-wise and rotation-based quantization, enable low-bit-width deployment of LLMs with negligible accuracy degradation, even under dual-quantized settings [6-10]. While numerous accelerators exploit quantization, hardware support for rotation-based quantization remains limited [11-16]. Recent efforts such as LightRot [17] explored co-designed rotation with hardware, but their datapaths still do not maintain an end-to-end INT flow, limiting the potential of rotation-based dual quantization. These limitations are attributed to three major challenges. First, the rotation operator, which smooths activation distributions, is not efficiently supported by current accelerators. To preserve the dual-quantized inference flow, PE arrays must execute rotation using naïve general matrix multiplication (GEMM), leading to 40.1% performance degradation compared to the ideal case. Second, the INT computation path is frequently disrupted by group-wise scaling and embedded activation functions (EAFs). Although INT-INT MAC operations can be applied within a single quantization group, channel-wise accumulations require frequent FP scaling. Moreover, the element-wise nature of EAFs (e.g., Swish, RoPE) interrupts the INT pipeline between consecutive linear layers. As a result, PEs must integrate group-scaling and INT2FP units, incurring 4.56× area and 3.82× power overhead compared to conventional INT-based execution. Third, dual-quantized LLMs suffer from low efficiency when supporting multiple precisions and bit-level sparsity. Bit-serial MAC units inherently allow precision scalability and exploit bit sparsity, but their efficiency is limited by varying channel-wise zeros and low integration density [11]. In practice, only 40.3% PE utilization and 0.91× energy efficiency are achieved compared to conventional MAC units.

This paper presents a dual-quantized LLM accelerator with three key features: 1) A Subspace Hadamard-Quantization (SHQ) transformation that realizes the rotation operator outside the PE array. It is fully pipelined with PE execution and reduces area by 59.7% and power by 62.3% compared to MAC-based implementations. 2) A Fused Scale-Activation (FSA) unit that unifies group-scaling and element-wise activation in the INT domain. RoPE is implemented with LUT-based stepwise trigonometric approximation, while Swish is computed by magnitude-aware polynomial fitting. The FSA array reduces energy by 59.9% for scaling and 61.5% for activation functions compared to a naïve FP design. 3) A Rearranged Bit-Slice LUT (RBLUT) engine that parallelizes bit-slice computation across weight tensors, avoiding utilization loss inherent in bit-serial MACs. By rearranging weight patterns and precomputing activation partial sums, RBLUT achieves theoretical precision scalability and improves energy efficiency by up to 2.28× over conventional MAC units.

Figure 31.3.2 shows the overall architecture of the accelerator. It consists of 4 Subspace-Rotation RBLUT (SR-RBLUT) cores, a 64KB global buffer, a network-on-chip (NoC), and a top controller. Each SR-RBLUT core integrates two PE lines, one SHQ unit, and two FSA arrays. The SHQ unit implements rotation with a butterfly network and adders, together with a quantization-parameter generator and a division-free quantizer for on-the-fly scaling. Each FSA array is tightly coupled with a PE line and supported by the SR-RBLUT 8KB local buffer that stores partial sums, nonlinear coefficients, and stepwise trigonometric tables. The FSA array reuses the same hardware resources in a reconfigurable manner to execute both activation functions and group-scaling operations. Each PE line contains one LUT generator and 16 PEs. The LUT generator produces all possible partial sums for the current activation set. Each PE integrates a local LUT and 32 lookup-accumulation (LAC) units. The local LUT stores partial sums delivered by the generator and distributes them to the 32 LACs for parallel processing.

Figure 31.3.3 presents the details of the SHQ unit. Direct implementation of Hadamard transformation (HT) on high-dimensional LLM activations, such as LLaMA2-7B with 11008 FFN channels, introduces significant challenges. Traditional PE-based approaches rely on GEMM operations with $O(N^2)$ complexity, leading to excessive computational overhead and data dependencies. To mitigate this, SHQ employs Kronecker decomposition to break down the large-scale HT into vector-based operations, with each vector tile consisting of 128 elements. This decomposition allows for efficient pipelined processing, followed by quantization and concatenation. For each vector tile, SHQ operates on 8×16 sub-blocks, applying intra-vector Kronecker decomposition. First, a linear combination of the 8 sub-blocks is computed, followed by a 16-point Fast Walsh-Hadamard Transformation (FWHT) applied to each reconstructed sub-block. This approach reduces computational complexity to $O(Nlog_2N)$, enabling efficient parallel processing and reducing the number of operations. Compared to traditional MAC-based methods, SHQ achieves up to 33.14% operation reduction in rotation-based MHA, 47.06% operation reduction in rotation-based FFN, and a 1.74× overall speedup. The group-wise transposer uses a log_2-based group-adder tree to efficiently handle the required linear combination. The output of the group-wise transposer is subsequently processed by a parallel HT tree, which employs a butterfly network and adders to perform FWHT on the generated results. This entire HT unit reduces area by 59.7% and power by 62.3% compared to the MAC-based implementation. Finally, the output of the HT is processed by a quantizer, which avoids division operations by utilizing a log_2 MinMax selector and LUT-based gradient approximation. The first 8 most significant bits (MSB) of $X_{max}-X_{min}$ are precomputed using $y = 1/x$ and stored in an $M/2^N$ format. Linear gradients between adjacent 1/x MSB values are also stored for on-chip decoding of the remaining bits. In this design, division is replaced by LUT lookups and MAC operations, aligning throughput between HT and on-the-fly quantization.

Figure 31.3.4 details the FSA. In LLMs, while the magnitudes of activations vary significantly, per-channel scale factors exhibit very small exponent spread. For instance, 97.26% of channels in LLaMA2-13B have an exponent difference ≤ 2. Hence, FSA pre-aligns scale factors within each channel without significant expansion of mantissa bitwidth. Group partial sums are then multiplied and accumulated in INT, guaranteeing software-hardware numerical equivalence, while avoiding INT2FP conversions. FSA also implements EAFs on the same MAC pipeline via two reconfigurable modes. In RoPE activation mode, a step-angle LUT supplies sin/cos values for an optimized MAC-based RoPE operation. The current token uses the lookup values directly, and an angle accumulation datapath updates the trigonometric values for the next token. This eliminates the need to compute or store full trigonometric tables. In Swish activation mode, a magnitude-aware piecewise polynomial is used, with higher order near zero (high curvature) and lower order otherwise. Polynomials are processed in INT using a Horner-style sequence, so powers are produced iteratively by the same MAC lanes without extra hardware. By fusing group scaling and activations into one INT engine, FSA sustains PE throughput and removes format switching. Compared with a naïve FP design, FSA achieves 59.9% and 61.5% energy reductions for group-scaling and activation functions, respectively.

Figure 31.3.5 details the RBLUT. Bit-slice GEMM expands the tensor weights into binary planes and executes inner products via shift-and-add. A LUT-based variant precomputes all partial-sum patterns of a n-element group and indexes them by the n weight bits, but the LUT cardinality grows with the number of nonzero patterns (up to $2^4-1=15$ when n=4), incurring significant area and energy overheads. RBLUT exploits inherent bit sparsity and Hamming-weight awareness to shrink the table. Specifically, weight bits are rearranged across neighboring quartets (n=4) so that each quartet belongs to a small set of canonical patterns with Hamming weight $h_i\in\{1,3,4\}$. Pairs of h_2 quartets are cross paired to become h_1/h_3 by swapping a single '1' between them. The LUT then stores only five entries with the four "one-hot" sums (h_1) and the all-ones sum (h_4). The h_3 sums are obtained by complementation ($h_3=h_4-h_1$), while h_0 is zero and needs no entry. This method keeps the inner-product expectation unchanged and reduces table size by 66.7%. Moreover, a fixed quartet width is processed at every step to avoid conventional bit-serial under-utilization. The PE array is organized in a systolic manner. Each PE integrates a local LUT that is forwarded downstream for reuse, and 32 reconfigurable LAC units that share the LUT. Lookup is realized by a lightweight gate-level decoder with a multiplexer path, while accumulation proceeds in parallel across the 32 LACs. Measured on W2 configurations, it improves energy efficiency by 1.75× and 2.28× over bit-serial/parallel PE arrays, respectively.

Figure 31.3.6 summarizes the measurement results. The chip operates from 0.6-1.0V up to 450MHz. Measured at the best performance/efficiency points and the generation of 1024 tokens, the latency is 2628 to 621ms and the energy is 267.1 to 51.6µJ/token (including external memory access) for the three LLM benchmarks, respectively. While prior state-of-the-art (SOTA) works focus on conventional transformer models or quantization deployment [11-14], our chip accelerates rotation-based dual-quantized LLMs with negligible accuracy loss (e.g., 0.56 at LLaMA2-7B), achieving 32.6% system-level energy savings compared to the SOTA on LLaMA2-7B [11]. Considering on-chip performance, the staged deployment of SHQ, FSA, and RBLUT improve energy efficiency by 1.41×, 1.92×, and 3.17× compared to the full 4b baseline, respectively, achieving 118.6TOPS/W (A4W4KV4, 1V) at LLaMA2-13B benchmark. Figure 31.3.7 shows the chip summary and detailed performance sheet.

Acknowledgement:
This work was supported by the National Key Research and Development Program of China under Grant 2023YFB4403103. (Corresponding author: Xin Si, xinsi@seu.edu.cn; Hao Cai, hao.cai@seu.edu.cn)

References:
[1] D. Guo et al., "Deepseek-r1: Incentivizing Reasoning Capability in LLMs via Reinforcement Learning," *arXiv preprint* arXiv:2501.12948, 2025. https://arxiv.org/abs/2501.12948
[2] A. Grattafiori et al., "The Llama 3 Herd of Models," arXiv preprint arXiv:2407.21783, 2024. https://doi.org/10.48550/arXiv.2407.21783
[3] Gemini Team, "Gemini: A Family of Highly Capable Multimodal Models," *arXiv preprint* arXiv:2312.11805, 2023. https://doi.org/10.48550/arXiv.2312.11805
[4] OpenAI, "GPT-4 Technical Report," *arXiv preprint* arXiv:2303.08774, 2023. https://doi.org/10.48550/arXiv.2303.08774
[5] W. Zhao et al., "A Survey of Large Language Models," *arXiv preprint* arXiv:2303.18223 1.2, 2023. https://doi.org/10.48550/arXiv.2303.18223
[6] G. Xiao et al., "SmoothQuant: Accurate and Efficient Post-Training Quantization for Large Language Models," *ICML*, pp. 38087-38099, 2023. https://doi.org/10.48550/arXiv.2211.10438
[7] H. You et al., "ShiftAddLLM: Accelerating Pretrained LLMs via Post-Training Multiplication-Less Reparameterization," *NeurIPS*, pp. 24822-24848, 2024. https://doi.org/10.48550/arXiv.2406.05981
[8] M. Chen et al., "EfficientQAT: Efficient Quantization-Aware Training for Large Language Models," *arXiv preprint* arXiv:2407.11062, 2024. https://doi.org/10.48550/arXiv.2407.11062
[9] Z. Liu et al., "Spinquant: LLM Quantization with Learned Rotations," *arXiv preprint* arXiv:2405.16406, 2024. https://doi.org/10.48550/arXiv.2405.16406
[10] L. He et al., "BASE-Q: Bias and Asymmetric Scaling Enhanced Rotational Quantization for Large Language Models," *arXiv preprint* arXiv:2506.15689, 2025. https://doi.org/10.48550/arXiv.2506.15689
[11] Y. Qin et al., "An 88.36 TOPS/W Bit-Level-Weight-Compressed Large-Language-Model Accelerator with Cluster-Aligned INT-FP-GEMM and Bi-Dimensional Workflow Reformulation," *ISSCC*, pp. 420-422, 2025. https://doi.org/10.1109/ISSCC49661.2025.10904774
[12] S. Kim et al., "C-Transformer: A 2.6-18.1μJ/Token Homogeneous DNN-Transformer/Spiking-Transformer Processor with Big-Little Network and Implicit Weight Generation for Large Language Models," *ISSCC*, pp. 368-370, 2024. https://doi.org/10.1109/ISSCC49657.2024.10454330

[13] R. Guo et al., "A 28nm 74.34TFLOPS/W BF16 Heterogenous CIM-Based Accelerator Exploiting Denoising-Similarity for Diffusion Models," *ISSCC*, pp. 362-364, 2024. https://doi.org/10.1109/ISSCC49657.2024.10454308
[14] S. Liu et al., "A 28nm 53.8TOPS/W 8b Sparse Transformer Accelerator with In-Memory Butterfly Zero Skipper for Unstructured-Pruned NN and CIM-Based Local-Attention-Reusable Engine," *ISSCC*, pp. 250-252, 2023. https://doi.org/10.1109/ISSCC42615.2023.10067360
[15] F. Tu et al., "MuITCIM: A 28nm 2.24μJ/Token Attention-Token-Bit Hybrid Sparse Digital CIM-Based Accelerator for Multimodal Transformers," ISSCC, pp. 248-249, 2023. https://doi.org/10.1109/ISSCC42615.2023.10067842
[16] Z. Mo et al., "LUT Tensor Core: A Software-Hardware Co-Design for LUT-Based Low-Bit LLM Inference," *IEEE/ACM ISCA*, pp. 514-528, 2025. https://doi.org/10.1145/3695053.3731057
[17] S. Kim et al., "LightRot: A Light-Weighted Rotation Scheme and Architecture for Accurate Low-Bit Large Language Model Inference," *IEEE Journal on Emerging and Selected Topics in Circuits and Systems*, vol. 15, no. 2, pp. 231-243, 2025. https://doi.org/10.1109/JETCAS.2025.3558300

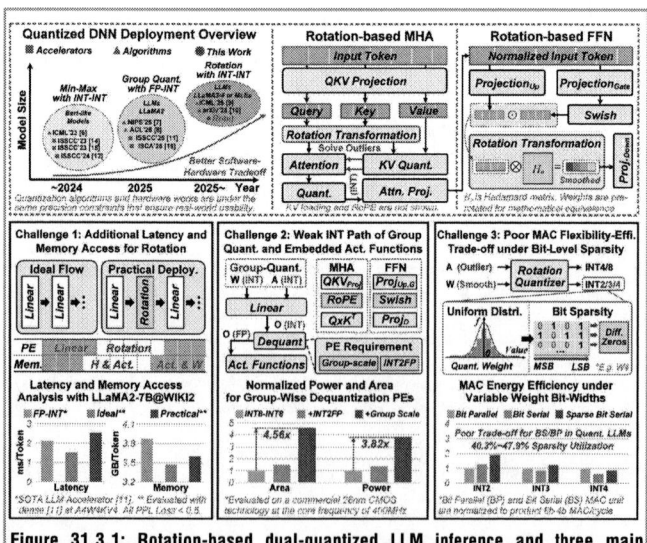

Figure 31.3.1: Rotation-based dual-quantized LLM inference and three main challenges.

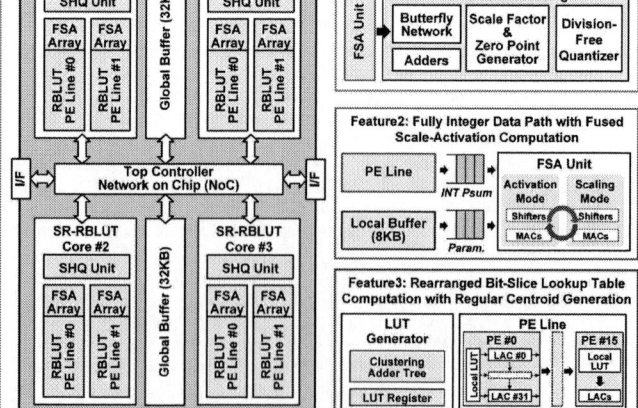

Figure 31.3.2: Overall architecture of the LLM accelerator.

ISSCC 2026 / SESSION 31 / AI ACCELERATORS / 31.3

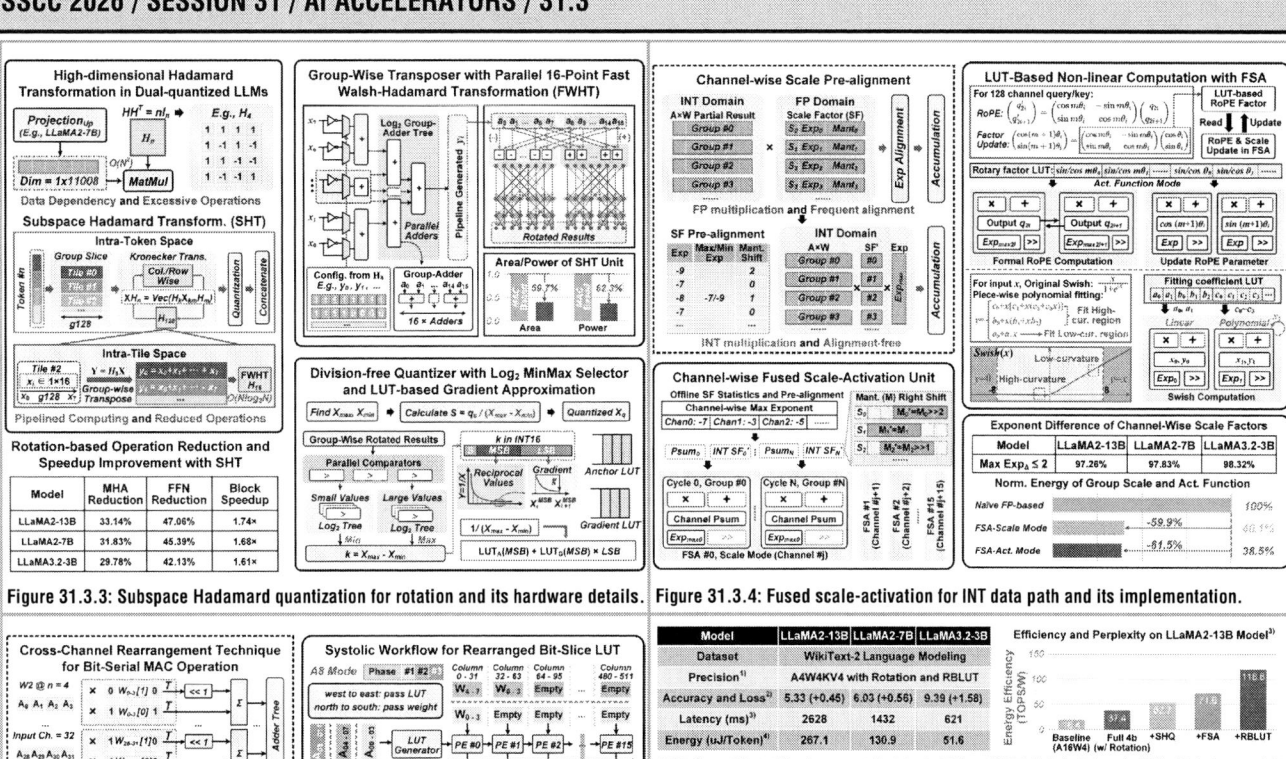

Figure 31.3.3: Subspace Hadamard quantization for rotation and its hardware details.

Figure 31.3.4: Fused scale-activation for INT data path and its implementation.

Model	LLaMA2-13B	LLaMA2-7B	LLaMA3.2-3B
Dataset	WikiText-2 Language Modeling		
Precision[1]	A4W4KV4 with Rotation and RBLUT		
Accuracy and Loss[2]	5.33 (+0.45)	6.03 (+0.56)	9.39 (+1.58)
Latency (ms)[3]	2628	1432	621
Energy (uJ/Token)[4]	267.1	130.9	51.6

Workloads	ISSCC'23 [14]	ISSCC'24 [13]	ISSCC'24 [12]	ISSCC'25 [11]	This work
	Transformers	Diffusion Models	Language Models	General LLMs	General LLMs
Quantization	General INT	General FP	General INT	Uniform Weight Only	Rotation-based Uniform Act. & W
Technique (nm)	28	28	28	28	28
Die Area (mm²)	3.93	3.67	20.25	3.52	1.37
Supply Voltage (V)	0.64-1.03	0.6-1.0	0.7-1.1	0.63-1.0	0.6-1.0
Frequency (MHz)	20-320	50-540	50-200	50-460	50-450
Precision	INT8	BF16/FP16	INT8/INT16	BF16 (A) INT8 (W)	INT8/4 (A) INT2/3/4 (W)
Power (mW)	8.27-250.65	8.268-170.0	47.5-469.2	16.95-105.0	11.6-68.4
Performance (TOPS)	0.49-3.33	6.636 (BF16) 4.424 (FP16)	3.41 (INT8)	0.70-3.58	3.87-15.79
Energy Efficiency (TOPS/W)	1.96-53.83	74.34 (BF16) 67.89 (FP16)	22.9-47.8 (INT8)	10.07-88.36	82.08-318.21

Figure 31.3.6: Measurement results and comparison with the state-of-the-art accelerators.

Figure 31.3.5: Rearranged bit-slice LUT computation for precision flexible PE.

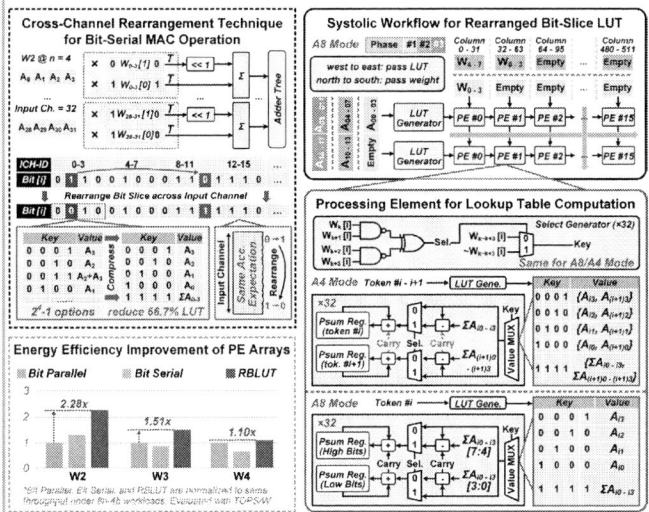

Specification	
Technique (nm)	28
Die Area (mm²)	1.37
Supply Voltage (V)	0.6-1.0
SRAM (KB)	96
Frequency (MHz)	50-450
Data Precision	INT8/4 (A) INT2/3/4 (W)
Power (mW)	11.6-68.4
Peak Performance (TOPS)[1]	3.87[2]-15.79[3]
Area Efficiency (TOPS/mm²)[1]	2.82[3]-11.51[3]
Energy Efficiency (TOPS/W)[4] 0.69V 188MHz[5]	82.08[3]-318.21[3]
Energy Efficiency (TOPS/W)[4] 1.0V 450MHz	57.72[3]-236.20[3]

1) Evaluated at 1.0V, 450MHz core frequency, 1 MAC equivalent to 2 operations 2) Configured with W4A8KV8 quantization 3) Configured with W2A4KV4 quantization. 4) External memory access is excluded. 5) Results at highest energy efficiency point.

Figure 31.3.7: Chip summary and detailed performance sheet.

• 2026 IEEE International Solid-State Circuits Conference

ISSCC 2026 / SESSION 31 / AI ACCELERATORS / 31.4

31.4 VARSA: A Visual Autoregressive Generation Accelerator Using Performance-Scalable Multi-Precision PE-LUT and Grid-Similarity Attention Compression

Jiaqi Zhou, Hongou Li, Keyao Jiang, Yiyang Sun, Tianyu Jia

Peking University, Beijing, China

Abstract

This paper presents VARSA, a 22nm visual autoregressive accelerator for efficient text-to-image generation, featuring: 1) a performance-scalable hybrid PE-LUT core; 2) multi-precision parallel processing with runtime precision management; 3) attention map compression leveraging inter-grid similarity. The innovations enable VARSA to achieve 503mJ/inference for 512×512 image generation, which is 2.7-to-8.9× better than prior diffusion-based SOTA accelerators.

AI-generated content (AIGC) has become an essential technique to support various content generation, e.g., images, videos. In recent years, the model architecture of visual AIGC has rapidly evolved from the Diffusion Transformer (DiT) [1-3], to the autoregressive (AR) Transformer [4-6] to the visual autoregressive model (VAR) [7-10]. As shown in Fig. 31.4.1, DiT or AR models require either extensive denoising timesteps or inefficient serial raster-scan, leading to high inference latency. For example, generating a 512×512 image using a DiT model could take >10s on GPUs due to the large number of denoising steps. A VAR revolutionizes image generation by modifying an autoregressive transformer with a coarse-to-fine scheme, which generates content with progressively increasing resolution. For example, a VAR generates an image represented by a token map, which initially only consists of a single patch of 16×16 pixels. The number of patches in the token map increases by interpolating at each generation step to achieve a larger scale, e.g., having 16×16 patches represent a resolution of 256×256. Furthermore, since VAR works in an autoregressive manner, which is used for large language models (LLMs), it has the natural advantage of generating multimodal content, including both image and text, thereby targeting a broader range of applications [10-12].

Although a VAR has its model advantages, it also introduces unique challenges: increased computational operations, wide-range data precision, and heavy memory access from the attention map. First, the interpolation block in a VAR transformer expands the patch number of the token map at each generation step, leading to increasing requirements on computational performance, e.g., an 171× operation-count increase for growing patches by 256×. Second, the activation data in a VAR requires a wide value range, e.g., 190× spatial and 238× temporal variations due to context differences. This leads to a wide data-precision requirement. Third, as the KV cache stores all previous patches for the growing token map, the attention layers dominate the memory cost by >60%. In fact, there is significant similarity between subregions, i.e. grids of a given size, of the attention map, e.g., >0.92 cosine similarity. Prior accelerators [13-18] only explored diffusion-based image generation, and consequently, prior work cannot be deployed effectively for VAR model. In this work, we present an accelerator, VARSA, for edge VAR with the following features: 1) a performance-scalable hybrid PE-LUT core with configurable performance to accommodate the increased performance demand, 2) multi-precision parallel processing with runtime distribution-aware precision management for the wide range of activation data, 3) attention map compression leveraging inter-grid similarity to reduce KV cache access and nonlinear operation overhead. Our VARSA achieves a content generation efficiency of 503mJ/Inference, which is 2.7-to-8.9× better than prior SOTA accelerators [16-17].

Figure 31.4.2 shows the overall architecture of VARSA. It comprises 8 PE-LUT clusters, each containing two PE-LUT cores with memory cross-interconnection, a precision-aware vector unit (PAVU), an attention compression unit (ACU), 128KB on-chip global buffer, a host RISC-V core, and peripheral circuits. The PE-LUT core is designed with a hybrid array structure comprising one PE column (PEC) and 8 LUT columns (LUTC), which can be configured as either compute units or registers. The PE-LUT array can flexibly fetch data from either 8KB local SRAM or a global buffer. To accommodate multi-precision operations, the RISC-V core profiles the activation data distribution and configures the computation precision of each PE-LUT core for multi-precision parallel processing. For the nonlinear and quantization operations, the PAVU consists of 32 high-precision FP units and 32 low-precision LUT units. The ACU integrates a shared 3×3 PE array to support grid convolution, similarity calculation, and equation-solving operations for attention map compression. Our chip also incorporates an Infineon HyperBus DRAM interface for off-chip data access.

Figure 31.4.3 illustrates the details of our performance-scalable PE-LUT core. In our design, each LUTC is equipped with 2×16 efficient dual-function LUTs (DF-LUT), which can either perform an INT4×INT4 multiplication or store 128b data. For compute-intensive workloads, the DF-LUT stores preloaded INT8 products for INT4×INT4 multiplication. In INT4×INT4 multiplication, weight data are used as the address to retrieve the corresponding product values, which are then sent to the adder tree for accumulation. Furthermore, two LUTs can be grouped to perform higher-precision multiplications, such as INT4×INT8 or INT4×INT12. For memory-intensive workloads, the LUTC is reconfigured as a supplement of the local SRAM with the address ports connected to the PE-LUT array controller. To reduce power, the corresponding adder trees are disabled by gating their inputs, and the local SRAM is

bypassed. The DF-LUT is also designed with two low-power techniques, i.e., address-aware masking and hierarchical clock gating. The LUTs in the DF-LUT are divided into three regions based on the Gaussian distribution of the model weight values, where two regions with the address (i.e., weight) around zero, i.e., 0 to 3 and -4 to -1, are more likely to be selected. The address-aware masking prevents toggling in non-hit regions, and the hierarchical clock gating shared across 16 DF-LUTs further reduces 48.2% and 38.8%/33.7% power during computation or read/write operations. Our PE-LUT design enables a 16× computational performance scaling from 0.435TOPS to 6.96TOPS to adapt to VAR workload characteristics.

Figure 31.4.4 depicts multi-precision parallel processing for a VAR. Due to wide data distribution, activations are partitioned into 16×8-element tiles and processed using three precisions, i.e., INT4, INT8, and INT12. For each tile, the RISC-V core gathers the maximum absolute activation value and compares it with two predefined thresholds to determine the precision (0 for INT4, 1 for INT8, and 2 for INT12). A precision map is generated for tile scheduling, and two INT4 tiles are concatenated to 16×16 elements to balance computation cycles between different precisions. The computation of each tile is scheduled to one PE-LUT core to trigger multi-precision parallel processing. A reconfigurable accumulator comprises four 12-to-20b adder trees and three 14-to-23b adders to accumulate PE-LUT core results with different precisions. Large bitwidth output vectors are sent to the PAVU for efficient FP nonlinear activations and then restored back to INT, while low bit-width vectors are truncated and processed via the LUT. To support multi-precision operations, every two DF-LUTs in LUTC are grouped with the same address for INT4×INT12 (W4A12) or INT4×INT8 (W4A8) operations by concatenating the outputs of two LUTs. For W4A12 computation, the products of INT4×INT12 are generated by INT8×INT4 and INT4×INT4 multipliers in PE units and preloaded to the grouped DF-LUT for subsequent computation. For W4A8 computation, only a 12b result is required and the highest 4b of the first LUT are masked. The 16b adder tree is split into 4b/4b/8b to bypass the accumulation of the highest 4b to improve computation efficiency. The 8b weight computation can be executed by combining two 4b partial products in two cycles. Overall, the DF-LUT with multi-precision parallel processing can achieve 2.8× memory and 3.8× compute energy reduction compared to a baseline FP16 PE.

Figure 31.4.5 presents our attention map compression scheme. For attention map generation, patches in the token map are row-wise reordered in a raster-scan order to form the Query vector, while the Key vector concatenates cached Keys from previous generation steps with reordered patch rows. Due to high contextual similarity across adjacent rows, e.g., row 1 & 2, the generated attention map by Q×K^T shows a high inter-grid similarity in its data distribution, where grids are equal-sized submatrices within the attention map. After profiling for different attention heads, we observe >90% attention maps exhibit three distinct patterns, i.e., attention heads with mostly small values (Pattern A), large values (Pattern B), or a diagonal distribution (Pattern C). The attention maps with Patterns A and B can be easily optimized using traditional Top-k sparsity, while a specific compression is applied for Pattern C maps. First, two grids with maximum (Grid Basis 1) and minimum (Grid Basis 2) mean values in the first column are selected from the attention map to represent the diagonal pattern. Second, for each remaining grid, two sample points are selected for linear regression, with regression coefficients α and β being solved for, and then and utilized to represent and compress the grid. To support the compression scheme, an attention pattern classifier in the ACU first employs a 3×3 diagonal kernel and a 3×3 summation kernel to perform convolution on the grid basis, comparing results with predefined thresholds to classify attention patterns. Two grid bases with diagonal distribution patterns are then processed in the cosine similarity calculator to compute their similarity. If a high inter-grid similarity is identified, a parameter extraction unit is enabled to solve the linear equation for α and β. Our attention compression brings 40.5% memory and 20.4% compute energy savings in 512×512 image generation.

Figure 31.4.6 shows the measurement results and the comparison table. VARSA is fabricated in a 22nm process and evaluated using the FastVAR model [8] on the LAION dataset [20]. Compared to baseline FP16 image generation, a fixed quantization method using either W4A4 or W4A8 leads to significant FID quality loss, while our mixed-precision quantization and attention compression result in negligible degradation. Compared to a

538 • 2026 IEEE International Solid-State Circuits Conference

979-8-3315-8937-0/26 $31.00 © 2026 IEEE

baseline FP16 digital design, our performance-scalable PE-LUT, mixed-precision processing, and attention inter-grid compression achieve 1.4×, 2.5×, and 1.6× energy reduction and an overall 2.3× speedup. Compared to prior text-to-image accelerators, our work targets 8× higher resolution of 512×512 image generation than [14]. VARSA achieves 503mJ and 1.92s per inference including external memory access (EMA) latency, which is an 8.9× energy and 3.8× latency improvement relative to [17]. In contrast to [16], which did not consider EMA, our EMA-excluded efficiency is 290.5mJ/inference, which is 2.7× better than [16]. VARSA achieves a peak energy efficiency of 33.45TOPS/W at W4A4 and an average 18.38TOPS/W for mixed precision, which is 1.8× higher than a prior AR-based LLM accelerator [19]. The die photo and chip summary are shown in Fig. 31.4.7.

Acknowledgement:
This work was supported in part by NSFC Grant No. 92464202. Corresponding author: Tianyu Jia.

References:
[1] W. Peebles et al., "Scalable Diffusion Models with Transformers," *ICCV*, pp. 4195-4205, 2023. https://doi.org/10.1109/iccv51070.2023.00387
[2] Z. Wang et al., "LaVin-DiT: Large Vision Diffusion Transformer," *CVPR*, pp. 20060-20070, 2025. https://doi.org/10.1109/CVPR52734.2025.01868
[3] Z. Zhang et al., "Tora: Trajectory-oriented Diffusion Transformer for Video Generation," *CVPR*, pp. 2063-2073, 2025. https://doi.org/10.1109/CVPR52734.2025.00198
[4] P. Esser et al., "Taming Transformers for High-Resolution Image Synthesis," *CVPR*, pp. 12873-12883, 2021. https://doi.org/10.1109/CVPR46437.2021.01268
[5] A. Ramesh et al., "Zero-shot text-to-image generation," *PMLR*, pp. 8821-8831, 2021. https://doi.org/10.48550/arXiv.2102.12092
[6] D. Lee et al., "Autoregressive Image Generation Using Residual Quantization," *CVPR*, pp. 11523-11532, 2022. https://doi.org/10.1109/cvpr52688.2022.01123
[7] K. Tian et al., "Visual Autoregressive Modeling: Scalable Image Generation via Next-Scale Prediction," *NeurIPS*, pp. 84839-84865, 2024. https://doi.org/10.48550/arXiv.2404.02905
[8] H. Guo et al., "FastVAR: Linear Visual Autoregressive Modeling via Cached Token Pruning," *ICCV*, 2025. https://doi.org/10.48550/arXiv.2503.23367
[9] E. Liu et al., "Distilled Decoding 1: One-step Sampling of Image Auto-regressive Models with Flow Matching," *ICLR*, 2025. https://doi.org/10.48550/arXiv.2412.17153
[10] J. Han et al., "Infinity: Scaling Bitwise Autoregressive Modeling for High-Resolution Image Synthesis," *CVPR*, pp. 15733-15744, 2025. https://doi.org/10.1109/CVPR52734.2025.01467
[11] L. Qu et al., "Tokenflow: Unified Image Tokenizer for Multimodal Understanding And Generation," *CVPR*, pp. 2545-2555, 2025. https://doi.org/10.1109/cvpr52734.2025.00243
[12] Y. Chen et al., "SAR3D: Autoregressive 3D Object Generation and Understanding Via Multi-Scale 3D VQVAE," *CVPR*, pp. 28371-28382, 2025. https://doi.org/10.1109/CVPR52734.2025.02642
[13] W. Kong et al., "Cambricon-D: Full-Network Differential Acceleration for Diffusion Models," *ISCA*, pp, 903-914, 2024. https://doi.org/10.1109/isca59077.2024.00070
[14] Y. Qin et al., "A 52.01 TFLOPS/W Diffusion Model Processor with Inter-Time-Step Convolution-Attention-Redundancy Elimination and Bipolar Floating-Point Multiplication," *IEEE Symp. VLSI Circuits*, 2024. https://doi.org/10.1109/vlsitechnologyandcir46783.2024.10631322

[15] R. Guo et al., "A 28nm 74.34TFLOPS/W BF16 Heterogenous CIM-Based Accelerator Exploiting Denoising-Similarity for Diffusion Models," *ISSCC*, pp. 362-364, 2024. https://doi.org/10.1109/ISSCC49657.2024.10454308
[16] S. Kim et al., "EdgeDiff: 418.4 mJ/Inference Multi-Modal Few-Step Diffusion Model Accelerator with Mixed-Precision and Reordered Group Quantization," *ISSCC*, pp. 410-412, 2025. https://doi.org/10.1109/isscc49661.2025.10904594
[17] D. Han et al., "MEGA. mini: A Universal Generative AI Processor with a New Big/Little Core Architecture for NPU," *ISSCC*, pp. 416-418, 2025. https://doi.org/10.1109/isscc49661.2025.10904514
[18] Y. Jing et al., "A 22nm 60.81TFLOPS/W Diffusion Accelerator with Bandwidth-Aware Memory Partition and BL-Segmented Compute-in-Memory for Efficient Multi-Task Content Generation," *ISSCC*, pp. 616-618, 2025. https://doi.org/10.1109/isscc49661.2025.10904573
[19] Y. Qin et al., "An 88.36TOPS/W Bit-Level-Weight-Compressed Large-Language-Model Accelerator with Cluster-Aligned INT-FP-GEMM and Bi-Dimensional Workflow Reformulation," *ISSCC*, pp. 420-422, 2025. https://doi.org/10.1109/isscc49661.2025.10904774
[20] C. Schuhmann et al., "LAION-400M: Open Dataset of CLIP-Filtered 400 Million Image-Text Pairs," *NeurIPS*, 2021. https://doi.org/10.48550/arXiv.2111.02114

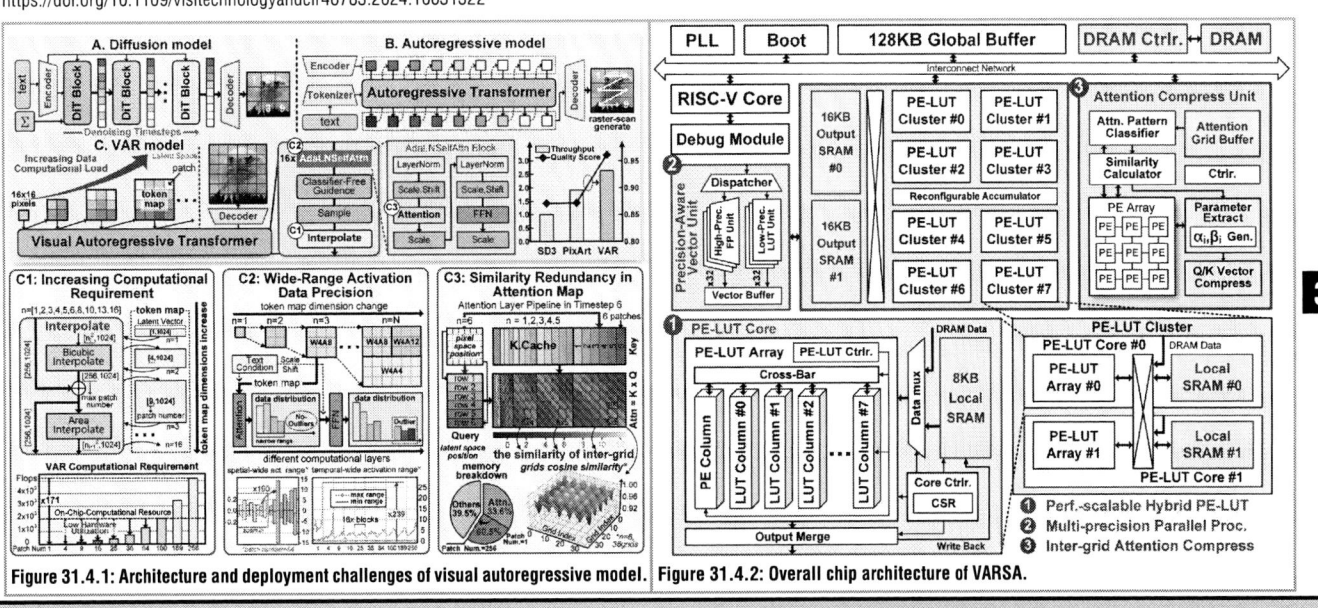

Figure 31.4.1: Architecture and deployment challenges of visual autoregressive model.

Figure 31.4.2: Overall chip architecture of VARSA.

ISSCC 2026 / SESSION 31 / AI ACCELERATORS / 31.4

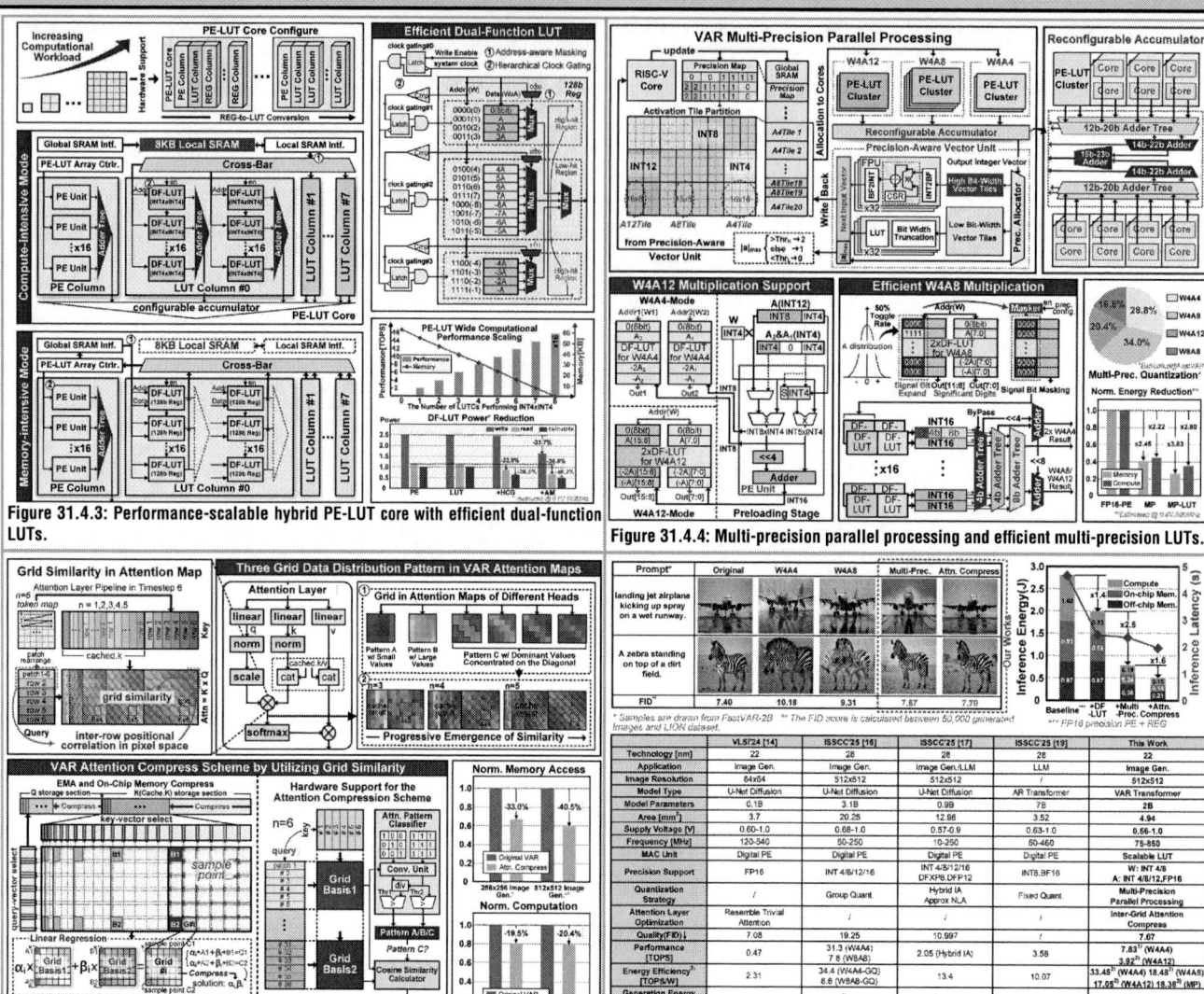

Figure 31.4.3: Performance-scalable hybrid PE-LUT core with efficient dual-function LUTs.

Figure 31.4.4: Multi-precision parallel processing and efficient multi-precision LUTs.

Figure 31.4.5: Inter-grid similarity in attention map and our compression scheme.

Figure 31.4.6: Measurement results and comparison table.

	VLSI'24 [14]	ISSCC'25 [16]	ISSCC'25 [17]	ISSCC'25 [19]	This Work
Technology [nm]	22	28	28	28	22
Application	Image Gen.	Image Gen.	Image Gen./LLM	LLM	Image Gen.
Image Resolution	64x64	512x512	512x512		512x512
Model Type	U-Net Diffusion	U-Net Diffusion	U-Net Diffusion	AR Transformer	VAR Transformer
Model Parameters	0.1B	3.1B	0.9B	7B	2B
Area [mm²]	3.7	20.25	12.96	3.52	4.94
Supply Voltage [V]	0.60-1.0	0.68-1.0	0.57-0.9	0.63-1.0	0.56-1.0
Frequency [MHz]	120-540	50-250	10-250	50-460	75-850
MAC Unit	Digital PE	Digital PE	Digital PE	Digital PE	Scalable LUT
Precision Support	FP16	INT 4/6/12/16	INT 4/8/12/16 DFXP8,DFP12	INT8,BF16	W: INT 4/8 A: INT 4/8/12,FP16
Quantization Strategy		Group Quant.	Hybrid IA Approx NLA	Fixed Quant.	Multi-Precision Parallel Processing
Attention Layer Optimization	Resemble Trivial Attention	✓	✓	✓	Inter-Grid Attention Compress
Quality(FID)↓	7.08	19.25	10.997		7.67
Performance [TOPS]	0.47	31.3 (W4A4) 7.8 (W8A8)	2.05 (Hybrid IA)	3.58	7.83¹⁾ (W4A4) 3.92²⁾ (W4A12)
Energy Efficiency³⁾ [TOPS/W]	2.31	34.4 (W4A4-GQ) 8.6 (W8A8-GQ)	13.4	10.07	33.45¹⁾(W4A4) 18.48¹⁾(W4A8) 17.05¹⁾(W4A12) 18.38¹⁾(MP)
Generation Energy [mJ/inf.]		786.3⁵⁾ (4 Step)	4480²⁾ (20 Step)		290.5⁴⁾⁵⁾ / 503¹⁾⁵⁾ (10 Step)
Execution Time [s/inf.]	2.21(1K Step)	7.33 (4 Step)	5.34 (20 Step)		1.92 (10 Step)⁴⁾⁵⁾

1) Measured @1.0V, 850MHz; 2) Measured @0.56v, 75MHz; 3) Peak hardware efficiency
4) Measured @ FastVAR-2B, Cfg=1, 512x512 Image Size; 5) EMA is sampled (32Gbit); 5) EMA is excluded

Figure 31.4.7: Chip photograph and more specifications.

	Specifications		
Technology	22nm 1P9M CMOS		
Chip Area	2.22 mm x 2.22mm (4.94 mm²)		
Supply Voltage	0.56 V - 1.0 V		
Frequency	75 MHz - 850 MHz		
On-Chip Memory	Global SRAM	Local SRAM	DF-LUT
	128KB	160KB	0-64KB
Data Type	Weight	Activation	Nonlinear
	INT 4/8	INT 4/8/12	FP16/LUT
# of MAC	PE Unit	LUT Unit	Vector Unit
	256	4096	64
Energy Efficiency¹⁾ [TOPS/W]	W4A4	W4A8	W4A12
	33.45	18.48	17.05

1) 5% Activation Sparsity and 37.5% Weight Sparsity, which is consistent with the FastVAR-2B.

ISSCC 2026 / SESSION 31 / AI ACCELERATORS / 31.5

31.5 SoulMate: A 9.8mW Mobile Intelligence System-on-Chip with Mixed-Rank Architecture for On-Device LLM Personalization

Seongyon Hong, Jiwon Choi, Jeonggyu So, Nayeong Lee, Wooyoung Jo, Zhamaliddin Kalzhan, Woojin Chin, Hoi-Jun Yoo

KAIST, Daejeon, Korea

Abstract

This work presents SoulMate, a fully on-device mobile intelligence system-on-chip, integrating retrieval-augmented generation (RAG) and fine tuning of a personal LLM. SoulMate is fabricated in 28nm CMOS with a novel mixed-rank token processing and similarity-aware sequence processing architecture. It demonstrates real-time user interaction consuming only 9.8-to-180.5mW power, and state-of-the-art energy efficiency, such as 26.3μJ/token for inference and 56.8μJ/token for fine-tuning.

Mobile intelligence systems with large language models (LLMs) provide personalized conversational assistance tailored to each user's characteristics [1-3]. They respond to user queries with awareness of individual preferences as well as factual knowledge, providing more personalized and relevant answers. However, the LLMs of existing systems require >10B parameters and >8GB RAM, while executing >1T operations per query, far exceeding the hardware capabilities of typical mobile devices. Therefore, most execution is offloaded to cloud servers, where all data, including private information, must be transmitted, resulting in privacy concerns. In addition, due to the network delay to the server, a time-to-first-token (TTFT) latency may exceed 400ms, disrupting user engagement and attention [4].

In this paper, Soulmate is presented as a fully on-device mobile intelligence system that enables personalization with a compact LLaMA3.2-1B model by combining retrieval-augmented generation (RAG) [5] from dialogue history and fine tuning from user feedback. As shown in Fig. 31.5.1, the system operates in two modes to adapt to evolving user preferences: user interaction (UI) for immediate personalization and user adaptation (UA) for progressive personalization. In UI, multimodal user inputs are processed and augmented with accumulated dialogue history and system prompts through RAG from a 32MB database, followed by LLM inference with ~1K context tokens. In UA, user feedback is collected and stored in a 4MB off-chip replay buffer, and LoRA-based fine tuning [6] updates the LLM using the stored data. The accepted and rejected responses from the user feedback are used as win sequences and lose sequences, respectively [7-8].

However, its realization on mobile platforms still entails three major hardware challenges. First, response latency during UI increases by over 10× because the augmented context length for personalization extends the input sequence, leading to a heavier prefill workload. As the prefill stage requires parallel processing of the entire input sequence, TTFT still exceeds 400ms, disrupting user engagement. Second, during UA, redundant model updates from user feedback consume excessive computation energy, accounting for 73% of total system energy. Since win and lose sequences in a feedback pair overlap with more than 70% similarity, their opposite signs of gradient updates cause unnecessary computation. Third, the computation power with the micro-scaling FP (MXFP) format [9], widely used for efficient LLMs, remains a bottleneck and contributes 82% of the chip power. The low bit sparsity of MXFP limits energy efficiency of computation because it adopts block-wise quantization to fully utilize the data range.

The proposed SoulMate SoC addresses these challenges with three key features: 1) Mixed-rank token processing with a token management unit (TMU) and a mixed-rank neural engine (MRNE) to reduce UI latency, 2) Similarity-aware sequence processing with a sequence management unit (SMU) to lower UA energy consumption, and 3) a Boolean-primitive MX (BPMX) tensor core to reduce peak power of MAC computations. As a result, SoulMate offers a mobile intelligence system within 180.5mW and achieves 216.4ms UI latency.

Figure 31.5.2 illustrates the overall architecture of SoulMate. The SoC consists of a RISC-V top controller, token management unit (TMU), sequence management unit (SMU), 4 mixed-rank neural engines (MRNEs), and 2 token buffers (TBUFs). They are interconnected through a 2D mesh network-on-chip (MNoC), while weights from off-chip are delivered to the MRNEs through a broadcast network-on-chip (BNoC). The RISC-V top controller generates kernel instructions and manages data movement across system I/O and cores. The TMU with attention core performs low-rank token evaluation and rank assignment required in mixed-rank token processing. The SMU tracks token similarity within feedback pairs to eliminate redundancy during fine tuning for similarity-aware sequence processing. Each MRNE realizes general matrix-matrix multiplication (GeMM) operations of linear layers and integrates 512KB IMEM, 128KB OMEM, 64KB WMEM, a local token router (LTR), a token and weight loader, a BPMX tensor core, and multi-function SIMD. It supports mixed-rank computation with MXFP6 input, MXFP4 weight, and MXFP8 gradient. The LTR routes tokens from the MNoC or OMEM to the IMEM for subsequent mixed-rank processing, and the token loader controls skipping and offloading while feeding them to BPMX core. The BPMX core factorizes the MXFP format and performs Boolean MAC to reduce computation power when operating on MXFP data.

Figure 31.5.3 presents the mixed-rank processing architecture with the TMU and MRNE to reduce latency and energy during UI. It applies weight ranks differently according to token importance, adopting weight truncation with singular value decomposition (SVD) [10] to reduce the computational workload. Since token importance varies across transformer blocks, dynamically assigning different weight ranks to each input token can eliminate redundant computation, resulting in 69.7-to-71.4% UI energy reduction. Each MRNE first executes low-rank GeMM for linear layers to quickly approximate token importance before the next attention layer within each decoder block. The token evaluation unit (TEU) then calculates the token importance, tracking the maximum attention score of each token across heads. The rank assignment unit (RAU) determines the token-wise weight ranks from the rank look-up table and stores them in the token management table (TMT). The TMT sorts the tokens by their weight ranks, recording their physical addresses for TBUF access. The global token router (GTR) in the TBUF distributes the token workloads of different ranks to the MRNEs, balancing them based on the sorted rank information. Within each MRNE, the token loader accelerates alternating U and V operations of SVD weights, where the U operation skips a group of similar small ranks and the V operation offloads operations of different sizes to feed processing element (PE) lines from large-rank to small-rank. After half of the PE line operations are complete, the token loader repeats offloading of the remaining half to the idle lines. Maximally, three times halving leads to 75.0-to-82.5% UI latency reduction.

Figure 31.5.4 shows the similarity-aware sequence processing architecture with the SMU to reduce energy and latency during UA. The SMU focuses gradient computation only on less-similar tokens, while reducing the computation for similar tokens to improve overall error propagation efficiency. User feedback of win/lose sequence pairs exhibits similarity within certain segments, including (1) lexical similarity from identical tokens and (2) semantic similarity from close embeddings in latent space. Such similarities enable skipping parts of gradient computation because the error gradients from positive (win) and negative (lose) losses may cancel each other. The top-down search controller (TSC) performs recursive segmentation of the sequence pairs, starting from the longest search length to squeeze the candidate space, improving search latency by 81.3%. The similarity search unit (SSU) comprises two stages: a lexical search unit that detects similar segments with matched IDs, and a semantic search unit that identifies latent semantic relationships with high cosine similarities. Detected similarities are recorded in the sequence management table (SMT), which stores the start position, segment length, and similarity length of each segment. Finally, the TMU reads this information from the SMT and dynamically determines per-token ranks according to the similarity types, reducing gradient computations. Fine tuning is performed with feedback data using LoRA and the Adam optimizer. As a result, UA energy and latency are reduced by 61.7 to 76.2% and 70.7%, respectively.

Figure 31.5.5 illustrates the BPMX core, which reduces the power consumption of MXFP computation. Operand sparsity boosting (OSB) and product sparsity boosting (PSB) are proposed to reduce the bit toggle count and implement FP MAC with a simple AND/OR combination. The MX factorizer performs prime factorization on the MXFP6/8 input and MXFP4 weight for OSB to enhance the operand sparsity. OSB extracts the power-of-two factor from the mantissa with a trailing zero counter to transfer a part of the mantissa to the exponent part, while the remaining value is encoded into a high-sparsity prime one-hot code. The BPMX multiplier selects product results by AND operations on the one-hot prime codes, then accumulates them across bit positions through an OR tree to generate the product, reducing multiplier power by 85.1%. Concurrently, PSB removes consecutive '1' bits and generates don't-care regions in the full adder truth table to enhance the product sparsity and simplify adder circuits, reducing adder tree power by 84.1%. Through the BPMX core with AND/OR-based FP MAC by OSB and PSB, the core peak power is reduced by 66.1% with only 3.4% area overhead.

Figure 31.5.6 presents the measurement results with a comparison table. SoulMate is evaluated on PersonaMem [11], a personalized dialogue dataset to test whether LLMs can respond to evolving user preferences. Compared to the baseline model, applying SoulMate's UI and UA consistently improves the benchmark accuracy across tasks such as tracking preference evolution, recalling facts, and acknowledging user preferences. The system-level evaluation demonstrates that the SoulMate UI with TMU, MRNE, and BPMX achieves short latency at a given power consumption compared with existing LLM inference processors

ISSCC 2026 / February 18, 2026 / 3:35 PM

[12-14]. Also, integrating SMU, MRNE, and BPMX reduces UA energy by 82.9%. The system with a 1B-parameter LLM achieves a real-time TTFT of 216.4ms, and consumes 9.8-to-180.5mW power. Unlike previous LLM or transformer processors [12-17], it supports not only inference but also on-chip training (fine-tuning) of an LLM with high energy efficiency, such as 26.3μJ/token inference energy and 56.8μJ/token fine tuning energy.

Figure 31.5.7 shows the chip photograph and performance summary of SoulMate. The chip is implemented in 28nm CMOS technology with a die area of 20.25mm². The design integrates a RISC-V core with the RV64-IMAFDC instruction set architecture and 3.9MB of on-chip SRAM. It achieves up to 2.2TFLOPS at 250MHz at 0.82V, and consumes 9.8mW at the peak energy efficiency point of 25MHz at 0.58V. It achieves UI and UA latencies of 216.4ms and 622.9ms, with energy efficiencies of 26.3μJ/token and 56.8μJ/token, respectively. In conclusion, the fully on-device mobile personalized intelligence system is successfully realized to demonstrate real-time user interaction with RAG and high energy-efficiency user adaptation with fine tuning. SoulMate enables on-device personalization with two schemes, fine tuning, as well as RAG and inference.

References:
[1] Gunter, Tom et al., "Apple Intelligence Foundation Language Models," arXiv preprint arXiv:2407.21075, 2024. https://arxiv.org/abs/2407.21075
[2] Team, Gemma et al., "Gemma: Open Models Based on Gemini Research and Technology," *arXiv preprint* arXiv:2403.08295, 2024. https://arxiv.org/abs/2403.08295
[3] Pham, Thang M. et al., "SlimLM: An Efficient Small Language Model for On-Device Document Assistance," *arXiv preprint* arXiv:2411.09944, 2024. https://arxiv.org/abs/2411.09944
[4] Yan, Xiao, and Yi Ding. "Are We There Yet? A Measurement Study of Efficiency for LLM Applications on Mobile Devices," *Proc. Int'l Workshop on Foundation Models for Cyber-Physical Systems & Internet of Things*, 2025. https://doi.org/10.1145/3722565.3727192
[5] Lewis, Patrick et al., "Retrieval-Augmented Generation for Knowledge-Intensive NLP Tasks," *NeurIPS*, pp. 9459-9474, 2020. https://arxiv.org/pdf/2005.11401
[6] Hu, Edward J. et al., "LoRA: Low-Rank Adaptation of Large Language Models," *ICLR*, 2022. https://arxiv.org/pdf/2106.09685v1/1000
[7] Yu, Tianshu et al., "Constructive Large Language Models Alignment with Diverse Feedback," *arXiv preprint* arXiv:2310.06450, 2023. https://arxiv.org/pdf/2310.06450
[8] Rafailov, Rafael et al., "Direct Preference Optimization: Your Language Model is Secretly a Reward Model," *NeurIPS*, pp. 53728-53741, 2023. https://arxiv.org/pdf/2305.18290
[9] Rouhani, Bita Darvish et al., "Microscaling Data Formats for Deep Learning," *arXiv preprint* arXiv:2310.10537, 2023. https://arxiv.org/pdf/2310.10537
[10] Qinsi, Wang et al., "Dobi-SVD: Differentiable SVD for LLM Compression and Some New Perspectives," *ICLR*, 2025. https://arxiv.org/abs/2502.02723
[11] Jiang, Bowen et al., "Know Me, Respond to Me: Benchmarking LLMs for Dynamic User Profiling and Personalized Responses At Scale," *arXiv preprint* arXiv:2504.14225, 2025. https://arxiv.org/pdf/2504.14225
[12] Kim, Sangyeob et al., "C-transformer: A 2.6-18.1μJ/token Homogeneous DNN-Transformer/Spiking-Transformer Processor with Big-Little Network and Implicit Weight Generation for Large Language Models," *ISSCC*, pp. 368-369, 2024. https://doi.org/10.1109/ISSCC49657.2024.10454330

[13] Jo, Wooyoung et al., "BROCA: A 52.4-to-559.2mW Mobile Social Agent System-on-Chip with Adaptive Bit-Truncate Unit and Acoustic-Cluster Bit Grouping," *ISSCC*, pp. 418-419, 2025. https://doi.org/10.1109/ISSCC49661.2025.10904658
[14] Kim, Sangyeob et al., "Slim-Llama: A 4.69 mW Large-Language-Model Processor with Binary/Ternary Weights for Billion-Parameter Llama Model," *ISSCC*, pp. 422-423, 2025. https://doi.org/10.1109/ISSCC49661.2025.10904761
[15] Qin, Yubin et al., "An 88.36 TOPS/W Bit-Level-Weight-Compressed Large-Language-Model Accelerator with Cluster-Aligned INT-FP-GEMM and Bi-Dimensional Workflow Reformulation," *ISSCC*, pp. 420-421, 2025. https://doi.org/10.1109/ISSCC49661.2025.10904774
[16] Wu, Ping-Sheng et al., "A 99.2 TOPS/W Transformer Learning Processor with Approximated Attention Score Gradient Computation and Ternary Vector-based Speculation," *IEEE Symp. VLSI Circuits*, 2024. https://doi.org/10.1109/VLSITechnologyandCir46783.2024.10631391
[17] Mun, HanGyeol et al., "ASAP: A 28nm Transformer Training Accelerator with Alternating Sparsity and Asymmetrical Microscaling Floating-Point Precision," *IEEE Symp. VLSI Circuits*, 2025. https://doi.org/10.23919/VLSITechnologyandCir65189.2025.11075114

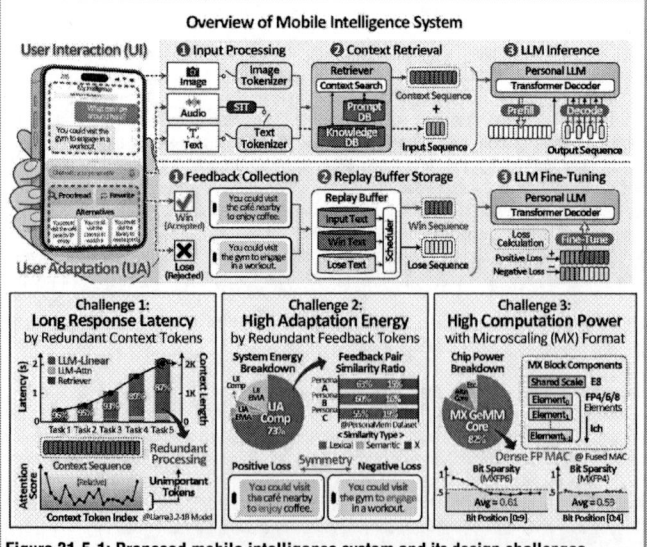

Figure 31.5.1: Proposed mobile intelligence system and its design challenges.

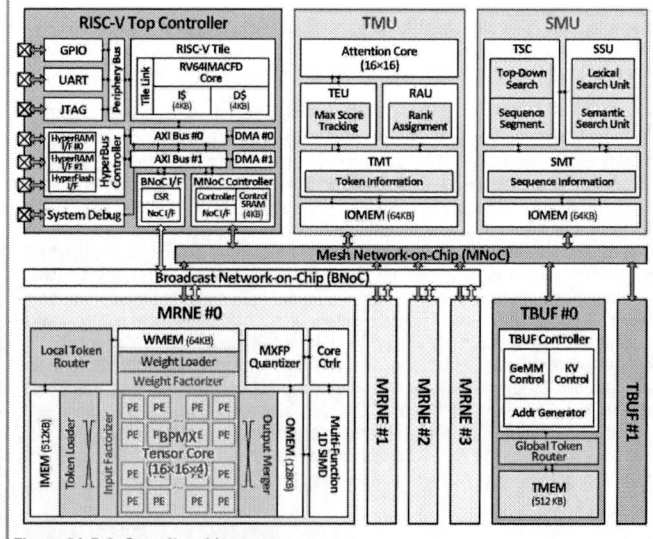

Figure 31.5.2: Overall architecture.

DIGEST OF TECHNICAL PAPERS • 541

979-8-3315-8937-0/26 $31.00 © 2026 IEEE

ISSCC 2026 / SESSION 31 / AI ACCELERATORS / 31.5

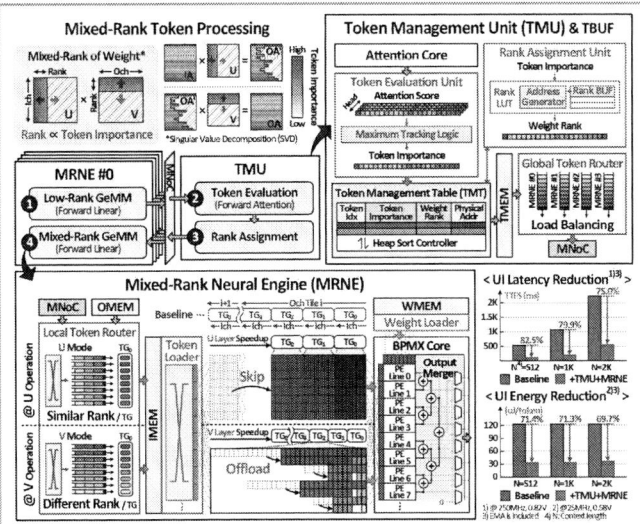

Figure 31.5.3: Mixed-rank token processing with TMU and MRNE for UI latency reduction.

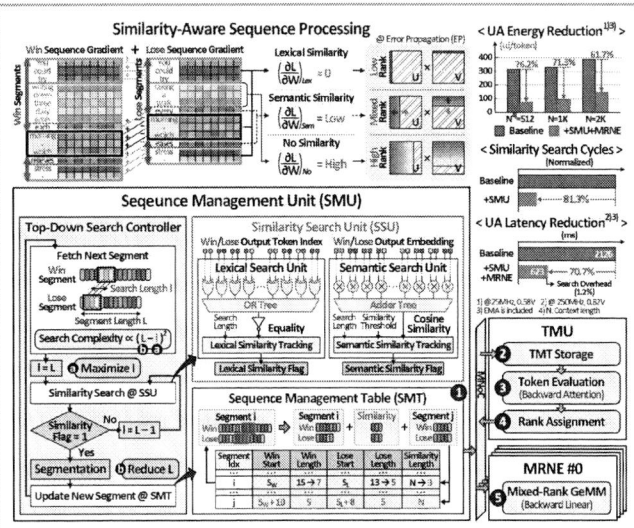

Figure 31.5.4: Similarity-aware sequence processing with SMU for UA energy reduction.

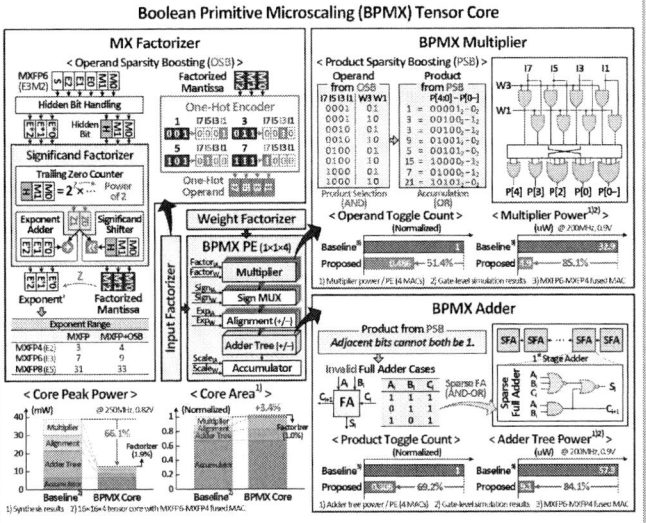

Figure 31.5.5: BPMX tensor core for peak power reduction.

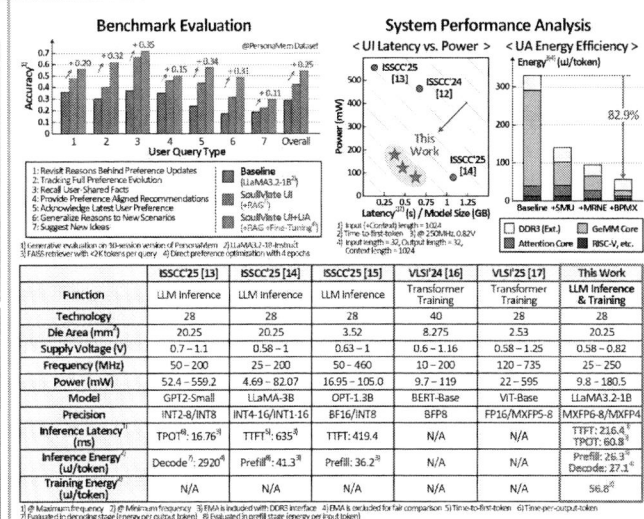

Figure 31.5.6: Measurement results and comparison table.

Function	ISSCC'25 [13]	ISSCC'25 [14]	ISSCC'25 [15]	VLSI'24 [16]	VLSI'25 [17]	This Work
Function	LLM Inference	LLM Inference	LLM Inference	Transformer Training	Transformer Training	LLM Inference & Training
Technology	28	28	28	40	28	28
Die Area (mm²)	20.25	20.25	3.52	8.275	2.53	20.25
Supply Voltage (V)	0.7 – 1.1	0.58 – 1	0.63 – 1	0.6 – 1.16	0.58 – 1.25	0.58 – 0.82
Frequency (MHz)	50 – 200	25 – 200	50 – 460	10 – 200	120 – 735	25 – 250
Power (mW)	52.4 – 559.2	4.69 – 82.07	16.95 – 105.0	9.7 – 119	22 – 595	9.8 – 180.5
Model	GPT2-Small	LLaMA-3B	OPT-1.3B	BERT-Base	ViT-Base	LLaMA3.2-1B
Precision	INT2-8/INT8	INT4-16/INT1-16	BF16/INT8	BFP8	FP16/MXFP5-8	MXFP6-8/MXFP4
Inference Latency¹⁾ (ms)	TPOT⁴⁾: 16.76¹⁾	TTFT⁷⁾: 635⁴⁾	TTFT: 419.4	N/A	N/A	TTFT: 216.4⁴⁾ TPOT: 60.8⁵⁾
Inference Energy¹⁾ (uJ/token)	Decode⁷⁾: 2920⁴⁾	Prefill⁸⁾: 41.3⁴⁾	Prefill: 36.2³⁾	N/A	N/A	Prefill: 26.3⁴⁾ Decode: 27.1⁴⁾
Training Energy (uJ/token)	N/A	N/A	N/A	N/A	N/A	56.8⁴⁾

1) @ Minimum frequency 2) @ Minimum frequency 3) EMA is included with DDR2 interface 4) EMA is excluded for fair comparison 5) Time-to-first-token 6) Time-per-output-token
7) Evaluated in decoding stage (energy per output token) 8) Evaluated in prefill stage (energy per input token)

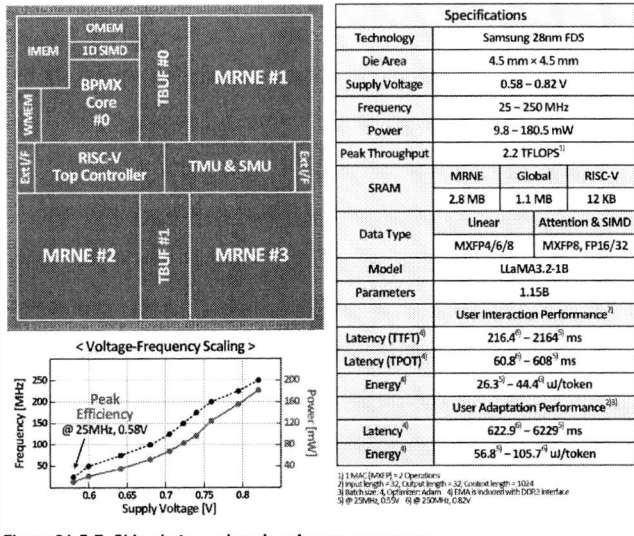

Figure 31.5.7: Chip photograph and performance summary.

Specifications	
Technology	Samsung 28nm FDS
Die Area	4.5 mm × 4.5 mm
Supply Voltage	0.58 – 0.82 V
Frequency	25 – 250 MHz
Power	9.8 – 180.5 mW
Peak Throughput	2.2 TFLOPS¹⁾

SRAM	MRNE	Global	RISC-V
	2.8 MB	1.1 MB	12 KB

Data Type	Linear	Attention & SIMD
	MXFP4/6/8	MXFP8, FP16/32

Model	LLaMA3.2-1B
Parameters	1.15B

User Interaction Performance²⁾	
Latency (TTFT)⁴⁾	216.4⁴⁾ – 2164⁵⁾ ms
Latency (TPOT)⁴⁾	60.8⁵⁾ – 608⁵⁾ ms
Energy⁴⁾	26.3⁵⁾ – 44.4⁴⁾ uJ/token

User Adaptation Performance²⁾³⁾	
Latency	622.9⁵⁾ – 6229⁵⁾ ms
Energy⁶⁾	56.8⁵⁾ – 105.7⁴⁾ uJ/token

1) 1 MAC (MXFP) = 2 Operations
2) Input length = 32, Output length = 32, Context length = 1024
3) Batch size: 4, Optimizer: Adam 4) EMA is included with DDR2 interface
5) @ 25MHz, 0.58V 6) @ 250MHz, 0.82V

ISSCC 2026 / SESSION 31 / AI ACCELERATORS / 31.6

31.6 Tri-Oracle: A 17.78µJ/Token Vision-Language Model Accelerator with Token-Attention-Weight Redundancy Prediction

Seungjae Yoo, Hangyeol Kim, Muyoung Son, Yi Chen, Suheon Jeong, Joo-Young Kim

KAIST, Daejeon, Korea

Abstract

This paper presents Tri-Oracle, a VLM accelerator exploiting token, attention, and weight redundancies. A Token Merging Unit (TMU) merges 68% of redundant tokens. An Attention Head Prediction Unit (AHPU) predicts streaming heads and skips them, reducing attention latency by 44% and balancing load. A Strip-wise SNZV Computation Unit (SSCU) predicts near-zero values, cutting FFN latency by 45% and EMA by 57%. Fabricated in 28nm CMOS, Tri-Oracle achieves 17.78µJ/token and 1.19 to 3.99TOPS/mm² without fine tuning.

Recently, the success of transformer-based large language models (LLMs) has expanded into multimodal domains, particularly vision-language models (VLMs) [1] that combine visual perception with natural language understanding. As shown in Fig. 31.6.1, VLMs are increasingly utilized in vision questioning and reasoning tasks by co-processing image and text tokens with an LLM. Its emerging applications include e-commerce and security cameras, making on-device deployment desirable. However, two significant trends hinder on-device deployment. First, VLM input size is rapidly growing due to the demand for high-resolution image quality, which exponentially increases the number of tokens and the overall computational load. Second, VLMs composed of a vision encoder (Vision Transformer, ViT) and an LLM tend to adopt ever-larger models (>1B parameters) to improve their answering and reasoning performance [2-4].

Given these trends, we observe three inherent sources of redundancy in VLMs. (1) Token-level redundancy: high-resolution images account for over 90% of input tokens, while 86% of them are highly similar due to spatial locality. (2) Attention-level redundancy: only a few attention heads are critical, as the attention-sink [5] phenomenon categorizes the heads into Retrieval Heads (RHs) and Streaming Heads (SHs), with SHs accounting for 50–75% of all heads. Since a substantial portion of SHs can be skipped compared to RHs, Grouped Query Attention (GQA) propagates this imbalance from the key level to the group level, further intensifying inherent key-wise variation that leads to load fluctuations. (3) Weight-level redundancy: gated FFN operations produce 70–90% near-zero or small near-zero values (NZVs/SNZVs), making many corresponding weights redundant, and given that FFN weights account for ~64% of the model, this is far from negligible. These observations call for a unified approach to predict and exploit such redundancies. We present Tri-Oracle, a VLM accelerator that aggressively addresses redundancy in token, attention, and weight to overcome the computational complexity. For tokens, runtime token merging (TM) is applied during LLM computation to reduce the number of tokens based on similarity patterns. For attention, head type prediction pre-computes attention-sink-related operations to skip unnecessary computations and balance the load. For FFN weights, SNZV-related weight prediction reduces both computation and external memory access (EMA). Unlike prior transformer processors [6,7] that require fine tuning or specialized training, Tri-Oracle adopts an off-the-shelf methodology that exploits the model's inherent characteristics, requiring only minimal hardware support and incurring negligible overhead.

Figure 31.6.2 shows the overall architecture of Tri-Oracle, highlighting three key blocks: 1) The Token Merging Unit (TMU), which performs runtime token merging to dynamically reduce redundant tokens using the Sign Magnitude-Similarity Estimation Unit (SMSE); 2) an Attention Head Prediction Unit (AHPU), which predicts the head type before full attention computations; and 3) a Strip-wise SNZV Computation Unit (SSCU), which predicts SNZV locations via the SMSE to skip related computations and avoid fetching corresponding weights, thereby reducing EMA. The architecture comprises five Weight-Attention-Token Efficient Clusters (WAT_ECs), a top controller, an all-digital PLL, and a Network-on-Chip (NoC). Each WAT_EC contains four Weight-Attention Cores (WACs) and one Aggregator. The Aggregator collects partial sums from the WACs and includes its own SIMD unit, IOMEM, and controller. The TMU resides inside the Aggregator to gather token information from the four WACs and apply token merging. Each WAC consists of memories, a controller, and a PE array that performs matrix multiplications, the dominant model computation. Additionally, it is equipped with an AHPU and an SSCU for redundancy handling of attention and weight.

Figure 31.6.3 illustrates the details of the TMU and its application of runtime token merging through its components. First, the SMSE performs approximate similarity computation while the Semantic Address Translator (SAT) identifies the token being processed. Using this information, the SMSE retrieves sign bits from the Survived Sign Bank (SSB). Second, the Token Merge Table (TMT) tracks the status of merging. Finally, the On-the-fly Token Size Unit (OTSU) refers to the TMT to compute the token size — the number of tokens merged into a given surviving token — which must be considered in a proportional softmax [8] computation. In the SMSE, similarity is computed using only the sign bits of each vector, as cosine similarity can be well-approximated by sign-bit comparison [9], previously applied to ViTs without causal dependencies. Specifically, the sign bits of the current token are extracted from the SIMD and compared with those of the left and upper neighboring tokens

via XNOR gates. A popcount module then counts the number of matching bits. This count is thresholded to decide merging, and the result is recorded in the TMT. The SSB stores the sign bits of surviving tokens (i.e., the tokens remaining after token merging) for subsequent merging. To limit storage overhead to the image-width level and avoid frequent data movement during dynamic merging, the SAT maps semantic token locations (left, up) to their physical addresses in the SSB. Compared with a naïve approach that stores all previous tokens' sign bits, this reduces SSB storage overhead by 91% and data migration by 51%. The OTSU monitors the TMT, filling its location table via iterative cursor movement to count merged tokens. When the cursor encounters a new origin (i.e., a surviving token) in the TMT, the table is updated, and the corresponding token size is recorded to allow causal-mask-aware softmax computation. The final token size is sent to the SIMD for the softmax stage. Overall, with a 5.14% area overhead in the WAT_EC, the TMU reduces the token count by 68% and cuts layer latency by 67%.

Figure 31.6.4 illustrates the head type prediction mechanism. To predict the head type before full attention computation, a small set of attention-sink-related tiles is selected and partially computed. These pre-computed tiles are used to find the row-wise maximum and calculate intensity. First, the row-wise maximum from the attention-sink tiles is obtained. Using this and the softmax property that values far smaller than the maximum produce negligible outputs, the number of elements exceeding a threshold is counted. Finally, the intensity of inner and outer tiles is compared to determine the head type (RH or SH). The AHPU, responsible for head type prediction, comprises the Head Max & Intensity Unit (HMIU) and the Type Decision Unit (TDU). Results from the pre-computed tiles, generated in the PE array, are sent to the HMIU, which updates the row-wise maximum (max) in a register. Once found, the results of tiles 1 and 2 are compared with max−α (see top-left of Fig. 31.6.4). A popcount module accumulates comparison results, and the intensity value is stored in the TDU. After all intensity computations, the TDU divides the intensity values of tiles 1 and tiles 2, applies a threshold, and predicts the head type. While predicting all head types can cause group-wise load imbalance under GQA, the early availability of prediction results enables workload balancing to mitigate it. With token merging (Fig. 31.6.3), the merging degree determines the token count, which in turn decides whether head-wise mapping occurs at the cluster or core level. If limited merging requires cluster-level mapping, top-bottom and middle-middle offloading strategies are applied based on cluster-wise variation. If merging sufficiently reduces tokens for core-level mapping, RH workloads are evenly distributed. Overall, while the AHPU occupies less than 1% of the WAC area, attention latency is reduced by 44% from RH/SH prediction alone. With workload balancing, cluster utilization and latency are further improved by 24.6% and 19.2%, on average, over the baseline.

Figure 31.6.5 illustrates the mechanism of SNZV-related weight prediction. The key idea is to approximate the dot product by leveraging only the sign bits of outlier elements. Before performing the full dot product, outlier elements in a vector are detected, and the outlier sign bits are used to estimate the magnitude of the final output. This estimation enables early prediction of whether the result will be near zero, allowing redundant computations and EMA to be skipped. Based on this concept, outlier detection is first applied to the activation vector, and the corresponding weight elements are aligned. The outlier sign bits from both activations and weights are fed to the SMSE for SNZV prediction, and the EMA overhead is minimal because only sign bits are fetched. By performing magnitude approximation for both W_gate and W_up, SNZV locations are determined by identifying where NZVs appear in the same position across both weight matrices. The SMSE, designed for token merging, is also instantiated for SNZV prediction, where it operates only on the outlier sign bits. It produces NZV location predictions in a bitmask format through thresholding, and the final SNZV locations are determined by the Skip Index Decision Unit (SIDU). Outlier detection, a prerequisite for SNZV prediction, is handled in the SIMD using RMSNorm. During RMSNorm computation, the standard deviation can be approximated to 1, allowing for the efficient application of the 3-sigma criterion for detecting outliers without significant overhead. Overall, the SSCU accounts for only 2.3% of the WAC area yet achieves a 45% reduction in gated FFN latency and a 57% reduction in EMA even when accounting for the additional cost of fetching weight sign bits.

ISSCC 2026 / February 18, 2026 / 4:00 PM

Figure 31.6.6 shows the measurement results of Tri-Oracle. The chip operates at a 0.75-to-1.1 V supply voltage with a maximum frequency of 580MHz, and supports vision questioning and reasoning tasks with TinyLLaVA-v1 and LLaVA-1.5. Compared to the baseline, Tri-Oracle achieves a 0.36-to-0.55× reduction in computation energy and a 0.53-to-0.67× reduction in EMA energy. Tri-Oracle is also compared with prior accelerators in the comparison table, indicating whether additional training or fine tuning is required. Under off-the-shelf conditions, Tri-Oracle achieves 17.78μJ/token and 1.19 to 3.99TOPS/mm², delivering state-of-the-art system performance and area efficiency, and remains comparable to prior works that rely on fine-tuned settings. Figure 31.6.7 shows the chip photograph of Tri-Oracle, fabricated in 28nm CMOS technology. By exploiting token-attention-weight redundancies, Tri-Oracle delivers state-of-the-art system performance and area efficiency with a single-digit area overhead, demonstrating the feasibility of efficient on-device VLM acceleration.

Acknowledgement:
This work was supported by the Institute of Information & Communications Technology Planning & Evaluation (IITP) through the Graduate School of AI Semiconductor (IITP-2025-RS-2023-00256472), the IITP-ITRC program (IITP-2025-RS-2020-II201847), and the IITP grant (No. RS-2025-02264029, Integration and Validation of an AI Semiconductor-Based Data Center Training and Inference System), all funded by the Korea government (MSIT).

References:
[1] Liu, Haotian et al., "Visual Instruction Tuning," *NeuIPS,* pp. 34892-34916, 2023. https://doi.org/10.48550/arXiv.2304.08485
[2] Zhou, Baichuan et al., "Tinyllava: A Framework of Small-Scale Large Multimodal Models," *arXiv preprint arXiv:2402.14289,* 2024. https://doi.org/10.48550/arXiv.2402.14289
[3] Liu, Haotian et al., "Improved Baselines with Visual Instruction Tuning," *IEEE CVPR,* 2024. https://doi.org/10.48550/arXiv.2310.03744
[4] Li, Feng et al., "Llava-Next-Interleave: Tackling Multi-Image, Video, and 3D in Large Multimodal Models," *arXiv preprint arXiv:2407.07895,* 2024. https://doi.org/10.48550/arXiv.2407.07895
[5] Xiao, Guangxuan et al., "DuoAttention: Efficient Long-Context LLM Inference with Retrieval and Streaming Heads," *arXiv preprint arXiv:2410.10819,* 2024. https://doi.org/10.48550/arXiv.2410.10819
[6] Kim, Sangyeob et al., "C-Transformer: A 2.6-18.1μJ/token Homogeneous DNN-Transformer/Spiking-Transformer Processor with Big-Little Network and Implicit Weight Generation For Large Language Models," *ISSCC,* pp. 368-369, 2024. https://doi.org/10.1109/ISSCC49657.2024.10454330
[7] Moon, S. et al., "T-REX: A 68-to-567μs/Token 0.41-to-3.95μJ/Token Transformer Accelerator with Reduced External Memory Access and Enhanced Hardware Utilization in 16nm FinFET," *ISSCC,* pp. 406-407, 2025. https://doi.org/10.1109/ISSCC49661.2025.10904793
[8] Bolya, Daniel et al., "Token Merging: Your ViT but Faster," *arXiv preprint arXiv:2210.09461,* 2022. https://doi.org/10.48550/arXiv.2210.09461
[9] Yoo, Seungjae et al., "AdapTiV: Sign-Similarity Based Image-Adaptive Token Merging for Vision Transformer Acceleration," *ACM/IEEE Micro,* pp. 64-77, 2024. https://doi.org/10.1109/MICRO61859.2024.00015

[10] Tu, Fengbin et al., "MuITCIM: A 28nm 2.24μJ/Token Attention-Token-Bit Hybrid Sparse Digital CIM-Based Accelerator for Multimodal Transformers," *ISSCC,* pp. 248-249, 2023. https://doi.org/10.1109/ISSCC42615.2023.10067842
[11] Liu, Shiwei et al., "A 28nm 53.8 TOPS/W 8b Sparse Transformer Accelerator with In-Memory Butterfly Zero Skipper for Unstructured-Pruned NN and CIM-Based Local-Attention-Reusable Engine," *ISSCC,* pp. 250-251, 2023. https://doi.org/10.1109/ISSCC42615.2023.10067360
[12] Tambe, Thierry et al., "A 12nm 18.1 TFLOPs/W Sparse Transformer Processor with Entropy-Based Early Exit, Mixed-Precision Predication and Fine-Grained Power Management," *ISSCC,* pp. 342-243, 2023. https://doi.org/10.1109/ISSCC42615.2023.10067817
[13] Wang, Yang et al., "A 28nm 77.35 tops/w Similar Vectors Traceable Transformer Processor with Principal-Component-Prior Speculating and Dynamic Bit-Wise Stationary Computing," *IEEE Symp. VLSI Circuits,* 2023. https://doi.org/10.23919/VLSITechnologyandCir57934.2023.10185403

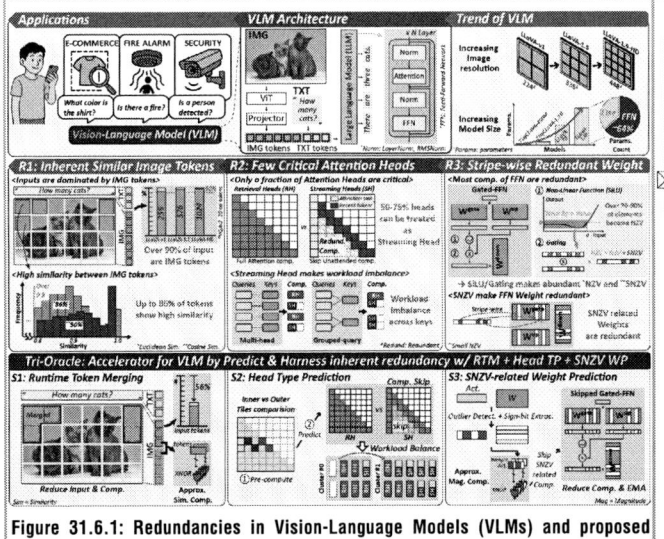

Figure 31.6.1: Redundancies in Vision-Language Models (VLMs) and proposed solutions for efficient on-device VLMs.

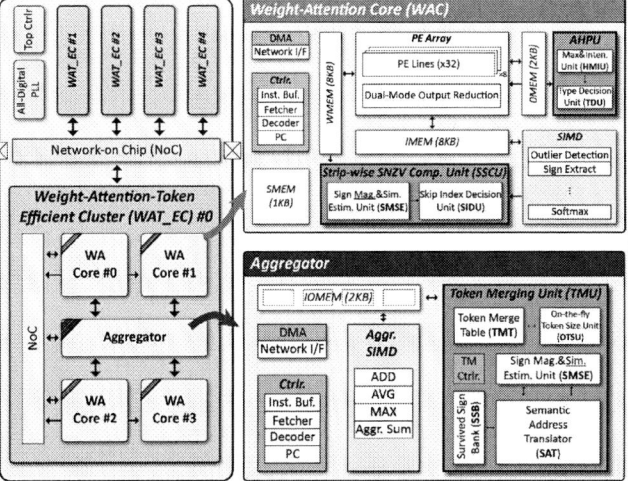

Figure 31.6.2: Overall architecture of Tri-Oracle.

DIGEST OF TECHNICAL PAPERS • 543

979-8-3315-8937-0/26 $31.00 © 2026 IEEE

ISSCC 2026 / SESSION 31 / AI ACCELERATORS / 31.6

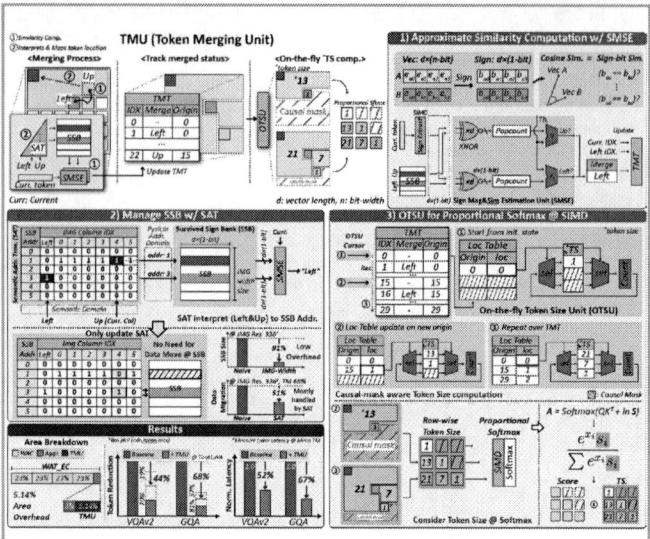

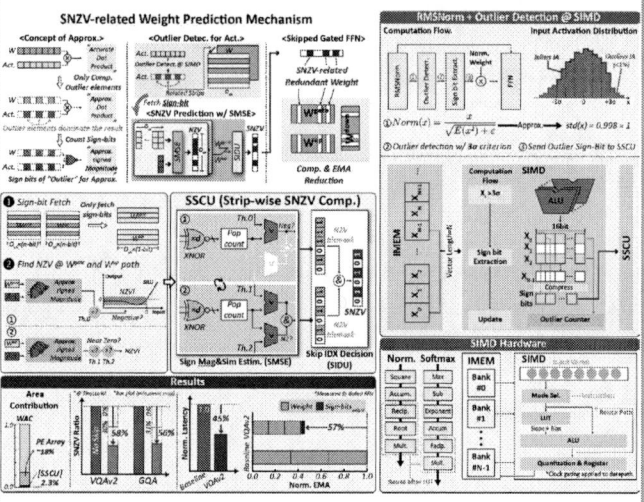

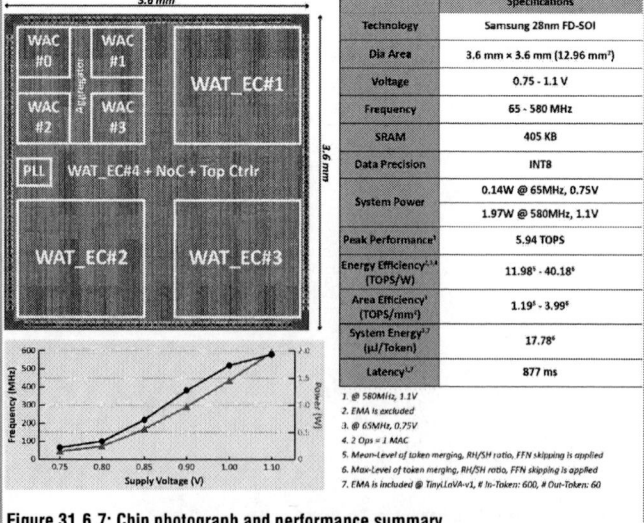

Figure 31.6.3: Details of the Token Merging Unit (TMU) for token-level redundancy.

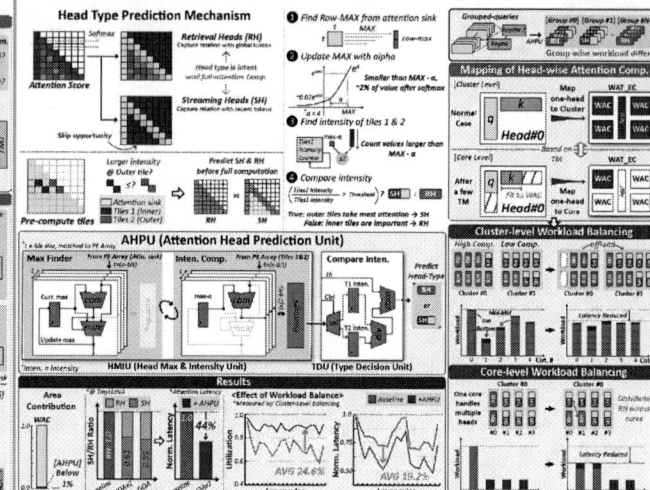

Figure 31.6.4: Details of the Attention Head Prediction Unit (AHPU) and workload balancing, for attention-level redundancy.

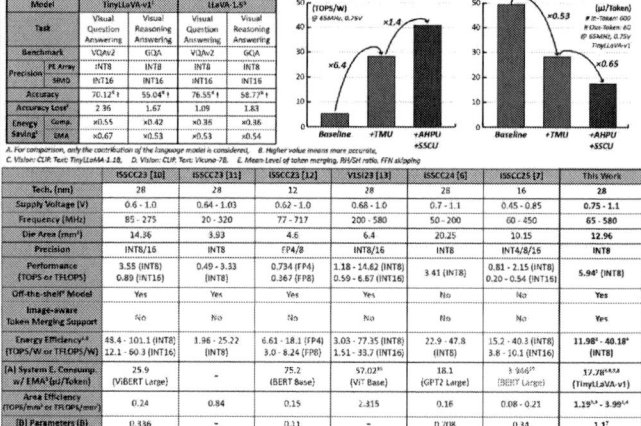

Figure 31.6.5: Details of the Strip-wise SNZV Computation Unit (SSCU) for weight-level redundancy.

Evaluation on Different Tasks[A]

Model	TinyLLaVA-v1[C]		LLaVA-1.5[D]	
Task	Visual Question Answering	Visual Reasoning Answering	Visual Question Answering	Visual Reasoning Answering
Benchmark	VQAv2	GQA	VQAv2	GQA
Precision PE Array	INT8	INT8	INT8	INT8
SIMD	INT16	INT16	INT16	INT16
Accuracy	70.12[†1]	55.04[†1]	76.55[†1]	58.77[†1]
Accuracy Loss[B]	2.36	1.67	1.09	1.83
Energy Saving[E]	×0.55	×0.36	×0.35	×0.36
EMA	×0.67	×0.53	×0.53	×0.54

A. For comparison, only the contribution of the language model is considered. B. Higher value means more accurate. C. Vision: CLIP. Text: TinyLLaMA-1.1B. D. Vision: CLIP, Text: Vicuna-7B. E. Mean-Level of token merging, RH/SH ratio, FFN skipping is applied.

	ISSCC23 [10]	ISSCC23 [11]	ISSCC23 [12]	VLSI23 [13]	ISSCC24 [6]	ISSCC25 [7]	This Work
Tech. (nm)	28	28	12	28	28	16	28
Supply Voltage (V)	0.6 - 1.0	0.64 - 1.03	0.62 - 1.0	0.68 - 1.0	0.7 - 1.1	0.45 - 0.85	0.75 - 1.1
Frequency (MHz)	85 - 275	20 - 320	77 - 717	200 - 580	50 - 200	60 - 450	65 - 580
Die Area (mm²)	14.36	3.93	4.6	6.4	20.25	10.15	12.96
Precision	INT8/16	INT8	FP4/8	INT8/16	INT8	INT4/8/16	INT8
Performance (TOPS or TFLOPS)	3.55 (INT8) 0.89 (INT16)	0.49 - 3.33 (INT8)	0.734 (FP4) 0.367 (FP8)	1.18 - 14.62 (INT8) 0.59 - 6.67 (INT16)	3.41 (INT8)	0.81 - 2.15 (INT8) 0.20 - 0.54 (INT16)	5.94[3] (INT8)
Off-the-shelf Model	Yes	Yes	Yes	Yes	No	No	Yes
Image-aware Token Merging Support	No	No	No	No	No	No	Yes
Energy Efficiency[4] (TOPS/W or TFLOPS/W)	48.4 - 101.1 (INT8) 12.1 - 60.3 (INT16)	1.96 - 25.22 (INT8)	6.61 - 18.1 (FP4) 3.0 - 8.24 (FP8)	3.03 - 77.35 (INT8) 1.51 - 33.7 (INT16)	22.9 - 47.8 (INT8)	15.2 - 40.3 (INT8) 3.8 - 10.1 (INT16)	11.98[5] - 40.18[6] (INT8)
[A] System E. Consuming w/ EMA[5] (μJ/Token)	25.9 (ViBERT Large)	-	75.2 (iBERT Base)	57.02[9] (ViT Base)	18.1 (GPT2 Large)	9.66[25] (iBERT Large)	17.78[5,A,6] (TinyLLaVA-v1)
Area Efficiency (TOPS/mm² or TFLOPS/mm²)	0.24	0.84	0.15	2.315	0.15	0.08 - 0.21	1.19[5,1] - 3.99[6,1]
[B] Parameters (B)	0.336	-	0.11	-	0.708	0.33	1.7
TOM ([A]/[B])	77	-	684	-	26	26	16

Figure 31.6.6: Measurement results and performance comparison table.

Figure 31.6.7: Chip photograph and performance summary.

Specifications	
Technology	Samsung 28nm FD-SOI
Die Area	3.6 mm × 3.6 mm (12.96 mm²)
Voltage	0.75 - 1.1 V
Frequency	65 - 580 MHz
SRAM	405 KB
Data Precision	INT8
System Power	0.14W @ 65MHz, 0.75V
	1.97W @ 580MHz, 1.1V
Peak Performance[3]	5.94 TOPS
Energy Efficiency[1,4] (TOPS/W)	11.98[5] - 40.18[6]
Area Efficiency[1] (TOPS/mm²)	1.19[5] - 3.99[6]
System Energy[3,7] (μJ/Token)	17.78[6]
Latency[1,7]	877 ms

1. @ 580MHz, 1.1V
2. EMA is excluded
3. @ 65MHz, 0.75V
4. 2 Ops = 1 MAC
5. Mean-Level of token merging, RH/SH ratio, FFN skipping is applied
6. Max-Level of token merging, RH/SH ratio, FFN skipping is applied
7. EMA is included @ TinyLLaVA-v1, # In-Token: 600, # Out-Token: 60

Figure 31.6.7: Chip photograph and performance summary.

• 2026 IEEE International Solid-State Circuits Conference

ISSCC 2026 / SESSION 31 / AI ACCELERATORS / 31.7

31.7 LUT-SSM: A 99.3TFLOPS/W LUT-Based State-Space Model Accelerator Using Energy-Efficient Element-Wise Layer Fusion and LUT-Friendly Weight-Only Quantization

Sunwoo Yoo[*1,2], Dongyun Kam[*3], Gunho Park[4], Soonhyun Kwon[1], Dongsoo Lee[4], Youngjoo Lee[1]

[1]Korea Advanced Institute of Science and Technology, Daejeon, Korea, [2]Pohang University of Science and Technology, Pohang, Korea,
[3]Ulsan National Institute of Science and Technology, Ulsan, Korea, [4]Naver Cloud, Seongnam, Korea
*Equally Credited Authors (ECAs)

Abstract

State-space models (SSMs) and weight-only quantization alleviate huge external memory access with a minimal accuracy degradation. To support both efficiently, we propose LUT-SSM, a LUT-based SSM accelerator with a many-to-many LUT-ACC and an element-wise (EW) layer fusion. LUT-SSM efficiently supports INT-FP GEMM and sequential EW operations in the SSM blocks. Fabricated in 28nm FD-SOI, LUT-SSM achieves 99.3TFLOPS/W peak efficiency and 96.1TFLOPS/W on Mamba 1.4B, improving energy efficiency by 1.28× over recent designs.

Recently, attention-based models [1] have been deployed in a variety of applications, e.g., AI agents, autonomous driving, robotics, and healthcare. However, they suffer from significant external memory access (EMA) for weights and intermediate caches, leading to long latency and high energy consumption. The recent weight-only quantization methods [2-4] significantly reduce weight EMA while preserving model quality. In addition, as illustrated in Fig. 31.7.1, the recent models, e.g., Mamba [5], Samba [6], and Hymba [7], have actively reduced cache EMA by replacing attention layers with state-space models (SSMs), which utilize only a fixed cache size regardless of the sequence length. Despite these advances, current AI processors [8-15] still face limitations in efficiently executing SSM-based models. To overcome the challenges, this work presents a Look-up Table (LUT)-based SSM accelerator and specialized optimizations that enhance energy efficiency and inference speed for SSM-based models.

Figure 31.7.1 shows the SSM-based Mamba model incorporating weight-only quantization, which requires matrix multiplications with integer (INT) weights and 16b floating point (FP) activations in linear layers. Unlike conventional GPU platforms, previous LUT-based processors [11, 13] efficiently perform integer-domain computations, with [13] extending the support to INT–FP operations without dequantization. However, three challenges remain for efficient inference of SSM-based models: 1) Redundant area and costly write operations in the previous many-to-one LUT-ACC architecture [11]. 2) Long latency caused by sequential FP-FP element-wise (EW) operations. 3) Extra scaling multiplications needed to support state-of-the-art column-wise quantization [16]. First, prior LUT-based operators use a shared accumulator (ACC) across multiple LUTs, leading to identical activation values being stored in different LUTs during general matrix–vector multiplication (GEMV). This design results in redundant area usage and repeated LUT writes, reducing both area and energy efficiency. Second, the hidden-state update loop in an SSM requires sequential FP–FP EW operations, which are memory bound. They mainly cause long latency in SSM-based models on GPU platforms (as shown in Fig. 31.7.1), while demanding additional FP-FP operators even in LUT-based INT-FP processors. Third, in some SSM blocks, linear layers exhibit column-wise weight distributions, where column values are highly similar. Thus, the recent column-wise weight-only quantization [16] successfully enables low bit-precision with minimal accuracy degradation by reducing the number of outliers. However, it requires an additional FP-FP scaling multiplication for each INT-FP multiplication, leading to increased area and energy consumption. To address these limitations, we design a LUT-based SSM accelerator (LUT-SSM) by introducing a many-to-many LUT-ACC architecture, a LUT-based EW layer fusion, and a scale-reflected LUT generator.

Figure 31.7.2 shows the overall architecture of LUT-SSM. It consists of 8 LUT-SSM Cores (LSCs), a RISC-V–based top controller, a network-on-chip (NoC), a core controller, a global memory, and an I/O unit. Each LSC includes 4 Many-to-Many LUT-based Processing Elements (MM-LPEs), a Scale-Reflected LUT (SR-LUT) generator, an accumulator, a partial-sum (PS) buffer, and an element-wise processing unit (EWU). An MM-LPE is built from 8 One-to-Many LUTs (OM-LUTs) and 8 Many-to-One accumulators (MO-ACCs). Each OM-LUT consists of a 16b × 8-entry LUT, a double-buffered 4b × 8 weight buffer, and eight multiplexers. Each multiplexer selects a LUT value by using a 4b weight as an index. The outputs are then accumulated along the input-channel dimension by the MO-ACC. In matrix multiplication (Matmul) mode, the accumulated results are stored in the local PS buffer. When computations along the bit-plane dimension are finished, the results are moved to the LSC PS buffer. After the output-stationary phase, the results are transferred to global memory. In SSM mode, the MO-ACC integrates an FP16 slicer to support FP–FP EW operations and to enable layer fusion with LUT-based INT-FP operators. The exact outputs of the EW operation are generated by the EW compensator. The EWU handles nonlinear functions and bias operations in Matmul mode, while in SSM mode, it collaborates with the MM-LPE to enable a loop-level layer fusion and reduce latency and energy consumption.

Figure 31.7.3 illustrates an example of GEMV computed by 8 many-to-one LUT-ACCs [11] or a MM-LPE. In the many-to-one design, eight LUTs share a single ACC. As a result, each LUT-ACC is required to redundantly store identical values, which prevents spatial LUT reuse. This leads to extra area cost and redundant energy consumption in GEMV computation. In contrast, the MM-LPE employs multiple ACCs, each connected to multiple LUTs. This enables spatial LUT reuse and optimizes LUT read/write operations, resulting in 72.3% power reduction and 62.8% area reduction for the same workloads. When evaluating energy consumption with the Mamba model, the proposed MM-LPE successfully reduces core energy by 28.5% compared to the many-to-one design. To efficiently support GEMM/GEMV computations with LUT-based PEs, we analyze the weight configurations in the SSM block and classify them into two types: SOLI (Short Output Long Input) and LOSI (Long Output Short Input). Then, we develop two LUT-stationary dataflows specialized for each weight type. In SOLI, the LSC simultaneously broadcasts four weight-row tiles and four activation-column tiles to each MM-LPE, maximizing spatial LUT reuse. In LOSI, the LSC's MM-LPEs process four input-channel tiles to accumulate results, thereby significantly reducing EMA along the input-channel dimension. Compared to the weight-stationary dataflow in prior LUT-based processors [13], the weight shape-aware LUT-stationary dataflows employ appropriate tiling for SOLI and LOSI, thereby improving energy efficiency by 42% and 18%, respectively, even under varying activation sizes caused by sequence length.

Figure 31.7.4 illustrates the sequential FP-FP EW layers in the SSM block and their fusion, which improves both energy efficiency and throughput by reducing EMA and on-chip data movement. We classify EW layers into two types: reusable EW (r-EW) and non-reusable EW (nr-EW). In r-EW layers, one multiplicand is reused across multiple multiplications. This makes the MM-LPE highly suitable, since it naturally supports LUT reuse. However, the MM-LPE natively supports only INT-FP multiplications. To address this, we propose a LUT-friendly FP slicing method that maps an FP-FP multiplication into four INT-FP multiplications. Specifically, the mantissa, effectively 11b including the implicit leading bit, is partitioned into four 4b slices, which are broadcast to four LUTs as indices and accumulated in a single cycle. To ensure exact results, an EW compensator feeds the 5b exponent and 1b sign into the ACC output using an FP adder and a sign inversion unit. Figure 31.7.4 details this r-EW computation in MM-LPEs. At the first cycle, each MM-LPE generates four ΔA values and four Bx values as EW results, computing a total of eight FP-FP EW operations in parallel. At the second cycle, $\Delta * Bx$ is computed by slicing the Bx output and feeding it into weight buffers after FP slicing. In this process, the pre-stored Δ LUT is reused to compute $\Delta * Bx$. After finishing r-EW operations, the pipelined EWU processes nr-EW layers (e.g., exponentials and the state-update loop) by updating hidden states and partial sums stored in local buffers. By minimizing intermediate data movement into memory, the proposed EW layer fusion reduces on-chip energy consumption by 40% and EMA by 73%, compared to a prior SIMD architecture [8].

Figure 31.7.5 shows an SR-LUT generator that efficiently supports column-wise group quantization. Column-wise group quantization achieves competitive model quality by leveraging linear layers with column-wise distributions [16]. However, its column-oriented scale factors require an additional FP-FP scale multiplication for each INT-FP operation, which increases energy consumption and latency. To minimize overhead, while leveraging column-wise quantization, we develop an SR-LUT with four FP multipliers to scale activations before generating possible combinations. Although adding four multipliers to the LUT generator increases power by 42%, the proposed SR-LUTs significantly reduce the number of scaling multiplications, thereby reducing overall energy consumption by 31% compared to a straightforward GEMM/GEMV implementation with column-wise quantization.

The LUT-SSM further supports layer-wise mixed-precision quantization [4] natively, as it processes weight bit planes serially without additional hardware. As shown in Fig. 31.7.5, we evaluate both energy efficiency and perplexity when running the Mamba-1.4B model [5] with mixed-precision quantization, ranging from an average of 3b to 4b weights. The results reveal a trade-off: lower precision improves energy efficiency while slightly degrading perplexity.

Figure 31.7.6 shows the measurement results of LUT-SSM fabricated in 28nm FD-SOI. LUT-SSM operates at 1.0V, 200MHz, consuming 54.46mW. It efficiently accelerates SSM-based models such as Mamba by supporting the INT-FP GEMM with MM-LPEs, LUT-stationary

544 • 2026 IEEE International Solid-State Circuits Conference

dataflows, optimized EW layer fusion, and column-wise weight-only quantization, achieving 0.55s per token processing for the Mamba model, which is faster than the A100 (as shown in Fig. 31.7.1). Based on SSM-aware optimizations, LUT-SSM achieves 99.3 TFLOPS/W peak efficiency and 96.1 TFLOPS/W for the 4-bit–quantized Mamba-1.4B, which is competitive performance with transformer accelerators, while maintaining robust perplexity under low-bit quantization. Compared to previous LLM accelerators [10, 11], LUT-SSM provides an integrated architecture that addresses both the limitations of prior LUT structures and the unique challenges of SSM computation. Although based on FP operations and large on-chip memory, which slightly impact area efficiency, the architecture achieves strong energy-performance characteristics. Specifically, LUT-SSM offers 3.92×, 1.28×, and 1.12× higher energy efficiency compared to recent designs [10-12] at the same operating frequency due to the optimized LUT operators and LUT-stationary dataflows. Figure 31.7.7 shows the 16.0 mm² LUT-SSM's chip micrograph and a summary of key performance metrics, including a comparison with previous Mamba accelerators [14, 15].

Acknowledgement:
This work was supported by the National Research Foundation of Korea (NRF) grant funded by the Korea government (MSIT) (No. RS-2025-02264052 and No. RS-2025-09322969). The chip fabrication and EDA tools were supported by the IC Design Education Center (IDEC), Korea.

References:
[1] A. Grattafiori et al., "The Llama 3 Herd of Models," *arXiv preprint* arXiv:2407.21783, 2024. https://doi.org/10.48550/arXiv.2407.21783
[2] E. Frantar et al., "GPTQ: Accurate Post-Training Quantization for Generative Pre-trained Transformers," *ICLR*, 2023. https://doi.org/10.48550/arXiv.2210.17323
[3] G. Park et al., "LUT-GEMM: Quantized Matrix Multiplication based on LUTs for Efficient Inference in Large-Scale Generative Language Models," *ICLR*, 2024. https://doi.org/10.48550/arXiv.2206.09557
[4] H. You et al., "ShiftAddLLM: Accelerating Pretrained LLMs via Post-Training Multiplication-Less Reparameterization," *NeurIPS*, 2024. https://doi.org/10.48550/arXiv.2406.05981
[5] A. Gu et al., "Mamba: Linear-Time Sequence Modeling with Selective State Spaces," *arXiv preprint* arXiv:2312.00752, 2024. https://doi.org/10.48550/arXiv.2312.00752
[6] L. Ren et al., "Samba: Simple Hybrid State Space Models for Efficient Unlimited Context Language Modeling," *ICLR*, 2025. https://doi.org/10.48550/arXiv.2406.07522
[7] X. Dong et al., "Hymba: A Hybrid-head Architecture for Small Language Models," *ICLR*, 2025. https://doi.org/10.48550/arXiv.2411.13676
[8] B. Keller et al., "A 95.6-TOPS/W Deep Learning Inference Accelerator With Per-Vector Scaled 4-bit Quantization in 5 nm," *IEEE JSSC*, vol. 58, no. 4, pp. 1129-1141, 2023. https://doi.org/10.1109/JSSC.2023.3234893
[9] S. Kim et al., "C-Transformer: A 2.6-18.1µJ/Token Homogeneous DNN-Transformer/Spiking-Transformer Processor with Big-Little Network and Implicit Weight Generation for Large Language Models," *ISSCC*, pp. 368-370, 2024. https://doi.org/10.1109/ISSCC49657.2024.10454330
[10] J. Kim et al., "Adelia: A 4nm LLM Accelerator with Streamlined Dataflow and Dual-Mode Parallelization for Efficient Generative AI Inference," *IEEE Symp. on VLSI Circuits*, 2025. https://doi.org/10.23919/VLSITechnologyandCir65189.2025.11075108

[11] S. Kim et al., "Slim-Llama: A 4.69mW Large-Language-Model Processor with Binary/Ternary Weights for Billion-Parameter Llama Model," *ISSCC*, pp. 422-424, 2025. https://doi.org/10.1109/ISSCC49661.2025.10904761
[12] Y. Qin et al., "An 88.36TOPS/W Bit-Level-Weight-Compressed Large-Language-Model Accelerator with Cluster-Aligned INT-FP-GEMM and Bi-Dimensional Workflow Reformulation," *ISSCC*, pp. 420-422, 2025. https://doi.org/10.1109/ISSCC49661.2025.10904774
[13] G. Park et al., "FIGLUT: An Energy-Efficient Accelerator Design for FP-INT GEMM Using Look-Up Tables," *IEEE HPCA*, pp. 1098-1111, 2025. https://doi.org/10.48550/arXiv.2503.06862
[14] J. Li et al., "MARCA: Mamba Accelerator with Reconfigurable Architecture," *IEEE/ACM ICCAD*, 2024. https://doi.org/10.1145/3676536.3676798
[15] J. Zhang et al., "Memristor-Based Circuit Implementation and Circuitry Optimized Algorithm for Mamba Language Network," *IEEE TCAS-I*, 2025. https://doi.org/10.1109/TCSI.2025.3584247
[16] D. Kam et al., "Panacea: Novel DNN Accelerator Using Accuracy-Preserving Asymmetric Quantization and Energy-Saving Bit-Slice Sparsity", *IEEE HPCA*, pp. 701-715, 2025. https://doi.org/10.48550/arXiv.2412.10059
[17] J. Heo et al., "Rethinking Channel Dimensions to Isolate Outliers for Low-bit Weight Quantization of Large Language Models," *ICLR*, 2024. https://doi.org/10.48550/arXiv.2309.15531
[18] W. Dong et al., "Fusion-Mamba for Cross-Modality Object Detection," *IEEE Trans. on Multimedia*, vol. 27, pp. 7392-7406, 2025. https://doi.org/10.1109/TMM.2025.3599020
[19] Z. Xing et al., "SegMamba: Long-range Sequential Modeling Mamba For 3D Medical Image Segmentation," *Medical Image Computing and Computer Assisted Intervention (MICCAI)*, pp. 578-588, 2024. https://doi.org/10.48550/arXiv.2401.13560

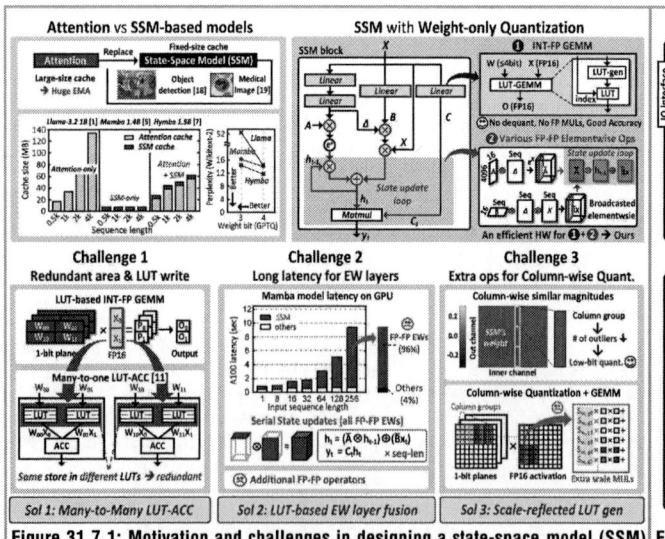

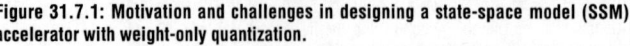

Figure 31.7.1: Motivation and challenges in designing a state-space model (SSM) accelerator with weight-only quantization.

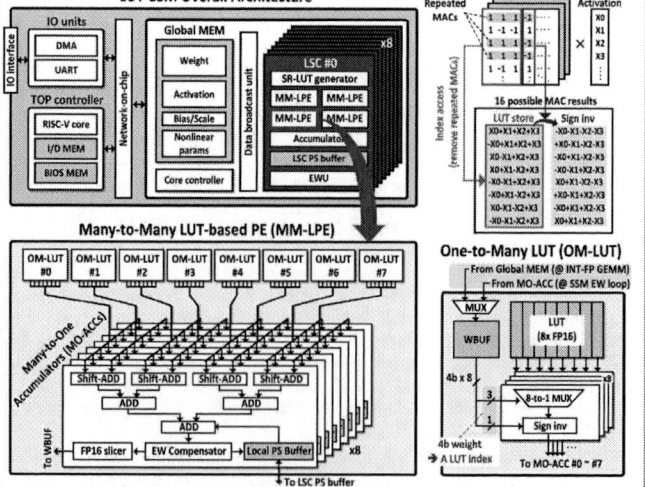

Figure 31.7.2: Overall architecture of the LUT-SSM with many-to-many LUT-based PEs (MM-LPEs) supporting INT–FP matrix multiplication.

ISSCC 2026 / SESSION 31 / AI ACCELERATORS / 31.7

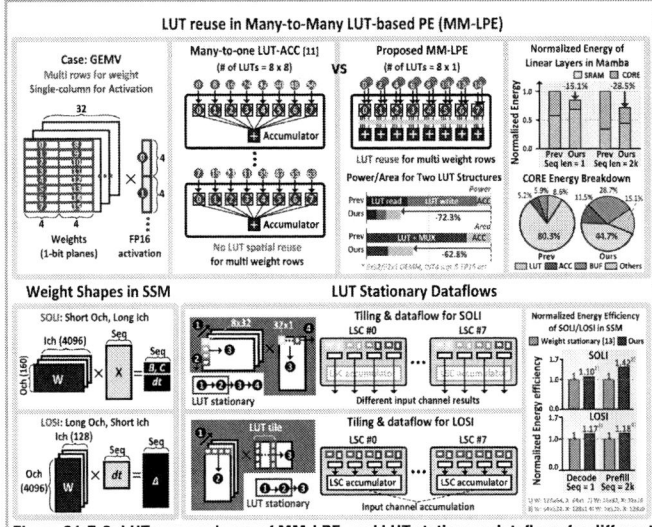

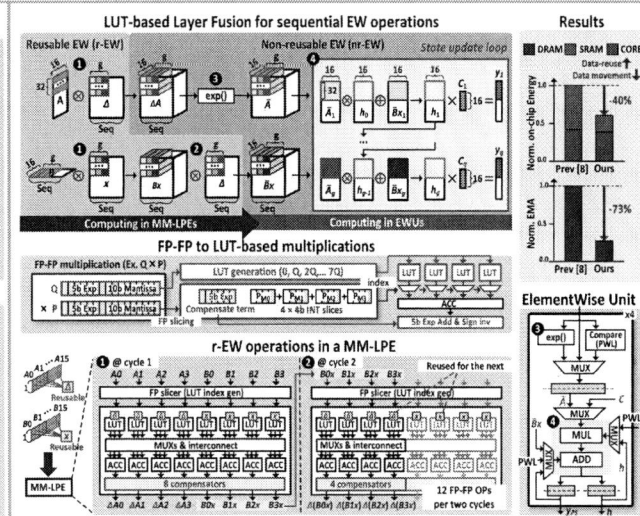

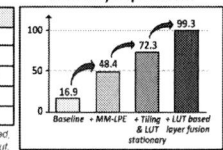

Figure 31.7.3: LUT reuse scheme of MM-LPE, and LUT stationary dataflows for different weight shapes in the SSM block.

Figure 31.7.4: LUT-based layer fusion and reusable elementwise multiplications for two FP–FP operation types in the MM-LPE.

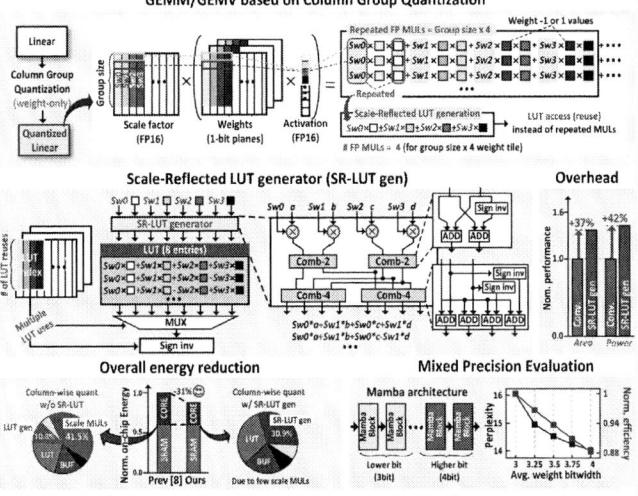

Figure 31.7.5: Implementation details of the scale-reflected LUT generator to support LUT-friendly column-wise quantization, and mixed-precision evaluation.

Evaluation on Mamba Model

Model	Mamba 1.4B		Llama-3.2 1B[3]
Sequence length	2048(Prefill)	1(Decode)	1(Decode)
Dataset	Wikitext-2 (Language Modeling)		
Perplexity (loss)[1]	14.02 (+2.45)		14.13 (+0.56)
Precision	W:INT4, A:FP16		
Latency(sec)[2]	74.75	0.55	0.54[4]

1) Perplexity (lower is better) compared to the FP16 baseline performance. 2) Sequence length is included.
3) 4-bit KV quantization applied. 4) Decoding with input length of 2048 and single-token output.

	JSSC'23 [8]	ISSCC'24 [9]	VLSI'25 [10]	ISSCC'25 [11]	ISSCC'25 [12]	This Work
Tech. (nm)	5	28	4	28	28	28
Supply Voltage (V)	0.46-1.05	0.7-1.1	0.65-0.9	0.58-1.0	0.63-1.0	1.0
Frequency (MHz)	152-1760	50-200	25-1000	25-200	50-460	200
Die Area (mm²)	0.153	20.25	5.28	20.25	3.52	16
Precision	INT4/8	INT8	FP16	INT4/8/16(A) INT1-16, 1.58(W)	BF16(A) INT8(W)	FP16(A) INT1-4/FP16(W)
Target Workloads	Transformer	Spiking NN & Transformer	Transformer	Transformer	Transformer	SSM & Transformer
Peak Performance (TOPS or TFLOPS)	3.6 (INT4) 1.8 (INT8)	3.41	8.19	4.92[5]-13.1[5] (A:8b,W:1/1.58b)	3.58	5.408
Power (mW)	-	-	11-735	4.69-82.70	16.95-105.0	54.469
Energy Efficiency (TOPS/W or TFLOPS/W)	91.1 (INT4) 39.1 (INT8)	22.9[1]-47.8[2]	25.3	77.6[6]-189.8[6] (A:8b,W:1/1.58b)	88.36	99.3
Benchmark Energy Efficiency[6] (TOPS/W or TFLOPS/W)	BERT-Base : 38.7 (INT4) 17.2 (INT8)	GPT-2 Large : 33.4		Llama 3B : 153.6[6] (1.58b)	-	Mamba 1.4B : 96.1

1) No spike-level zero skipping, 2) Spike-level zero skipping, 3) External memory access is excluded, 4) No sparsity condition.
5) 87.5 sparsity by output reuse. 6) INT4/8/16 mixed precision for activation, 67.4% sparsity and 53.3% transition reduction ratio.

Figure 31.7.6: Measurement results and comparison table.

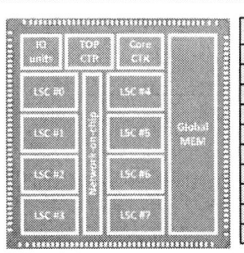

Specification

Technology (nm)	28nm FD-SOI
Die Area (mm²)	16
Supply Voltage (V)	1
Frequency (MHz)	200
SRAM	512KB
Data Precision	FP16 (A), INT1-4/FP16(W)
Power (mW)	54.469
Peak Performance (TFLOPS)	5.408
Area Efficiency (TFLOPS/mm²)	0.34
Energy Efficiency (TOPS/W or TFLOPS/W)	99.3

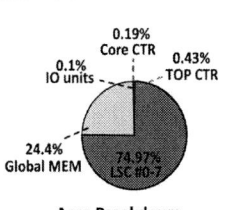

Comparison with the previous Mamba accelerators

	ICCAD [14][1]	TCAS-1 [15][2]	This Work
Tech. (nm)	28	-	28 FD-SOI
Implementation level	Synthesis	ASIC[2]	ASIC
INT-FP Operation	X	O	O
Reusable EW	X	X	O
Layer fusion	△[3]	X	O
Mixed precision	X	X	O
Frequency	1GHz	-	200MHz
Core/Die Area (mm²)	221.88	-	12.5/16
SRAM size	24MB	-	512KB
Power	10.44W	S85.32mW	54.469mW

1) Energy/area efficiency not reported, 2) Memristor-based analog ASIC design, 3) DRAM access evoided while SRAM remains accessed within the core.

Figure 31.7.7: Chip micrograph and performance summary of LUT-SSM, including a comparison with previous Mamba accelerators.

• 2026 IEEE International Solid-State Circuits Conference

ISSCC 2026 / SESSION 31 / AI ACCELERATORS / 31.8

31.8 A 28nm Speculative-Decoding LLM Processor Achieving 105-to-685µs/Token Latency for Billion-Parameter Models

Yang Wang[1], Huanyu Wang[1], Jiaxin Yang[1], Yutong Su[1], Ruiqi Guo[1], Zhiheng Yue[1], Jiangyuan Gu[1], Shaojun Wei[1], Yang Hu[1], Shouyi Yin[1,2,3]

[1]Tsinghua University, Beijing, China, [2]Shanghai Artificial Intelligence Laboratory, Shanghai, China, [3]International Innovation Center of Tsinghua University, Shanghai, China

Abstract

LLMs face decoding bottlenecks. Speculative Decoding (SD) reduces latency via a small draft model for serial decoding and a large target model to verify in parallel. Despite this advantage, it still suffers from exponent redundancy, inefficient weights/KV, and hardware underutilization. We propose an SD LLM processor with three innovations: an exponent dual-reuse MAC, draft back-propagation target mixed-precision dataflow, and draft early start parallel compute, achieving 10.29× speed increase.

Large language models (LLMs) have achieved remarkable advancements in various AI tasks [1-4]. However, the autoregressive sequential decoding introduces a performance bottleneck [5-6]. Speculative Decoding (SD) has emerged to mitigate the latency of autoregressive decoding via alternating rapid drafting and parallel verification [7-14]. Generally, SD contains a small draft model and a large target model, where the draft model is 10-to-20× smaller than the target model and is distilled from it [10-12]. During SD, the draft model performs N (typically N=4,5,6) autoregressive decodings, and each decoding selects the top-k (typically k=2,3) potential tokens as the candidates. It creates a multi-branch draft tree with $1+k+k^2+\cdots+k^{N-1}=(k^N-1)/(k-1)$ tokens. Each branch indicates a probable sequence with N tokens, as shown in Fig. 31.8.1. After that, the target model receives all $(k^N-1)/(k-1)$ tokens from the draft tree and calculates their probabilities. It compares these probabilities against those of the draft tokens, accepting M tokens in a specific branch if the comparison results exceed the acceptable margin. When employing a TinyLLaMA 1B as a draft model and a Llama 13B as a target model, SD reduces latency by 55.2% and improves energy efficiency by 2.56×.

Despite the theoretical superiority, designing an energy-efficient SD LLM processor has 3 challenges. (1) The draft model incurs implicit exponent redundancy. It randomly generates 39.5% of duplicate tokens with different positions. These semantically identical, but positionally distinct tokens result in similar vectors, where 95.7% of the corresponding elements have the duplicated exponent. (2) The target model contains inefficient weights and KV cache. The goal of a target model is to verify whether the candidate tokens in the draft tree are acceptable, where 78.3% of the head-wise weights and KV-cache entries are inefficient for verification. (3) The SD suffers from dual hardware underutilization. The autoregressive decoding of the draft model is memory bound, leading to 76.7% of MACs underutilized. The parallel verification with $(k^N-1)/(k-1)$ × more computation is computation bound, incurring 49.7% memory bandwidth underutilization.

This paper proposes an energy-efficient SD LLM processor, tackling the above challenges with 3 features. (1) An exponent dual-reuse MAC (EDRM) compute engine reuses the exponent addition in the multiplier and the exponent alignment logic in accumulators for duplicate tokens related vectors. It reduces the MAC energy by 30.1%. (2) A draft back-propagation target mixed-precision (DBTM) dataflow performs back-propagation on the draft model to obtain weight/KV gradients. According to the gradients, DBTM uses low precision or prunes the inefficient weight/KV in the target model, improving speed by 2.35×. (3) A draft early start parallel compute (DEPC) utilizes layer-wise verification results to start the draft model early, enabling draft-verify parallel computing. Thus, the draft and target models can mutually utilize the underutilized resources from each other, achieving 1.69× higher peak throughput.

Figure 31.8.2 shows the overall architecture of the SD LLM processor, consisting of 4 EDRM compute engines, a DBTM dataflow scheduler, a DEPC controller, a RISC-V core, and 4MB global SRAM. Each EDRM contains 4 16×16 processing elements (PEs). The PE supports BF16/FP16/FP8 data precision. Generally, the draft model uses FP8, while the target model uses FP16/BF16. EDRM has 2 work modes. For semantically identical tokens, EDRM works in an exponent reuse mode. It shares 1 PE's exponent addition and alignment results to multiple PEs for reuse, reducing MAC energy. For other tokens, it works in a standard FP MAC mode. The DBTM works in the draft stage. After N autoregressive decodings with the draft model, DBTM performs an extra back propagation to calculate gradients of the weight head and KV cache. In the following verification stage, based on the gradients, DBTM determines whether to assign FP16/FP8 precision to different weight heads and KV caches, or to prune them. It saves the cost of inefficient weight/KV. The DEPC works with the target model, comparing the output tokens of each target model layer with the draft tokens. If 4 successive layers have the same comparison result, indicating an identical token accept or decline decision, DEPC controls the next draft start early. With the parallel draft-verify computing, the draft and target models can utilize each other's wasted resources, improving system throughput.

Figure 31.8.3 depicts the EDRM compute engine, reusing the exponent processing logic in the multiplier and accumulator to save MAC energy. The EDRM contains 16×16 PEs, working with 2 modes. Before computing, EDRM collects similar vectors caused by the

semantically identical, but positionally distinct tokens, naturally generated in the draft tree due to the same context. After position embedding and attention, these tokens become similar vectors, where 95.7% of the corresponding elements have an identical exponent. EDRM handles these vectors with exponent reuse mode. Assuming A_0/B_0 and A_1/B_1 belong to 2 similar vectors, where $E_{A0}=E_{A1}$ and $E_{B0}=E_{B1}$. When performing MAC with W_0/W_1, it has outputs of $O_0=A_0\times W_0+B_0\times W_1$ and $O_1=A_1\times W_0+B_1\times W_1$. Since $E_{A0}=E_{A1}$ ($E_{B0}=E_{B1}$), the multiplications of $A_0\times W_0$ ($B_0\times W_1$) and $A_1\times W_0$ ($B_1\times W_1$) have the same exponent addition result. Additionally, as the exponent of $A_0\times W_0$ equals $A_1\times W_0$ ($B_0\times W_1$ equals $B_1\times W_1$), the accumulators of O_0 and O_1 also have the same exponent alignment results. Rather than computing O_0 and O_1 separately, EDRM reuses the exponent processing results of O_0 for O_1. Generally, M semantically identical tokens indicate M× exponent logic reuse. EDRM reduces 63.8% of the exponent processing, which consumes considerable power, especially for accumulators. Specifically, the exponent comparison, alignment shift, and normalization are more complicated than full adders in FP accumulators, comprising 69.8% of the power. For vectors from semantically distinct tokens, EDRM works in standard FP mode. The dual-mode EDRM saves 30.1% of MAC energy consumed in draft decoding and parallel verification.

Figure 31.8.4 shows the DBTM, using back propagation to get the draft model's weight/KV gradients that control the target model to use low precision or prune inefficient head-wise weights and KV-cache entries. DBTM works in 3 stages. First, it controls the draft model to perform N standard decodings, generating a draft tree. Second, it uses the generated draft tree to perform back propagation for the draft model, getting the average gradient of weight heads ($Ave_{\Sigma W}$) and KV ($Ave_{\Sigma K/V}$). Generally, 86.7% of gradients are near-zero values, indicating less importance for the draft tree. Third, based on the gradients, DBTM uses low precision for the inefficient weight heads and KV, or prunes them. Since the draft model is distilled from the target model, the $Ave_{\Sigma W}$ and $Ave_{\Sigma K/V}$ from the draft model can characterize the importance of the target model's weight/KV. Each time, a softmax unit fetches 8 $Ave_{\Sigma W}$ (32 $Ave_{\Sigma K/V}$) to calculate the probabilities of P_i, where $\Sigma P_i=1$. Then, an analyzer detects the distribution of P_i. For a 3-level distributed P_i, where the top-20% P_i satisfy $\Sigma P_{top-20\%}>0.7$ and top-40% P_i satisfy $\Sigma P_{top-40\%}>0.9$, DBTM maintains FP16 precision for the top-20% weight/KV in the target model, and it uses FP8 for the weight/KV belonging to top-20% to top-40%. For the rest of the weight/KV, DBTM directly prunes them. For a 2-level distributed P_i, defined as $\Sigma P_{top-20\%}>0.9$, DBTM maintains FP16 for the target model's top-20% weight/KV and prunes the remaining 80% of weight/KV. Otherwise, if all P_i are approximately equal, indicating a 1-level even distribution, DBTM uses FP16 for all weight/KV in the target model. DBTM reduces 74.9% of the inefficient verification costs and increases the throughput by 2.2-2.5×.

Figure 31.8.5 details the DEPC, starting the draft model early in parallel with the target model, where the two models can utilize the wasted resources from each other to improve hardware utilization. DEPC works in 2 stages. The first stage is early start control. Instead of comparing the target model's final layer output tokens with the draft tokens, DEPC compares each target layer's output tokens with the draft tokens by computing their cosine similarity. In 89.6% of cases, starting from the 6th layer, all subsequent layers maintain an identical cosine similarity (CS) comparison result with the final layer, such as accepting the former 3 tokens, but declining the last 1 token. Based on this early consistency property, DEPC schedules the start time of the draft model. Specifically, once 4 successive target layers have the same CS results, it invokes the draft model early for parallel computing with the target model. This way, the memory-bound draft decoding can utilize the wasted bandwidth from the computation-bound target verification, and vice versa. It increases the hardware utilization by an average of 1.7×. The second stage is the output check. DEPC double checks the former 4 successive layers' acceptance result by comparing them with the final layer's output. If the comparison results are consistent, DEPC uses the draft tree generated by the early start draft model for the subsequent target verification. Otherwise, for the inconsistent cases (less than 5% of cases), DEPC discards the current results from the early start draft model and restarts the draft decoding with the final layer's outputs. The restart mechanism ensures DEPC is lossless. DEPC improves the throughput by 1.49-1.89×.

Figure 31.8.6 shows the measurement results. The proposed SD LLM processor consumes 129.8-to-887.6mW of energy with a voltage of 0.58-1.0V, 180-550MHz. The baseline

546 • 2026 IEEE International Solid-State Circuits Conference

979-8-3315-8937-0/26 $31.00 © 2026 IEEE

throughput is 4.5TFLOPS@FP16, 550MHz, where 1 operation indicates 1 multiplication or 1 accumulation. The peak energy efficiency is 109.7TFLOPS/W@FP8, achieved at 0.68V, 300MHz. It is 2.29× higher than [15] and 1.24× higher than [6] for three reasons. First, EDRM reduces MAC energy by reusing the duplicated exponent in draft tokens. Second, DBTM saves the cost of inefficient weight and KV cache leveraging mixed precision. Third, DEPC improves the hardware utilization via draft-verify parallel computing. In addition, SD LLM achieves 3.04× speedup over [6] as DBTM reduces the memory access time for inefficient weight and KV cache, and DEPC increases bandwidth utilization with the draft early start mechanism. When evaluated on Gemma-3-1B, Phi-2-2.7B, and Llama-2-7B models, SD LLM reduces the token-to-token latency by 9.97×, 10.08×, and 10.29×, respectively. Figure 31.8.7 shows a summary and die photo of the SD LLM processor fabricated in 28nm CMOS technology with an area of 56.82mm².

Acknowledgement:
This work was supported in part by the NSFC under Grant 62125403, Grant 92464302, Grant 92164301 and Grant 62304121; in part by the National Key R&D Project under Grant 2023YFB4403100; in part by the National Key Research and Development Program under Grant 2021ZD0114400; in part by the National Science and Technology Major Project under Grant 2022ZD0115201; Beijing S&T Project Z221100007722023; in part by the Beijing National Research Center for Information Science and Technology; and in part by the Beijing Advanced Innovation Center for Integrated Circuits. (Corresponding author: Shouyi Yin.)

References:
[1] OpenAI et al., "GPT-4 Technical Report," arXiv, Mar. 04, 2024. http://arxiv.org/abs/2303.08774
[2] G. Team et al., "Gemma 3 Technical Report," arXiv, Mar. 25, 2025. http://arxiv.org/abs/2503.19786
[3] A. Grattafiori et al., "The Llama 3 Herd of Models," arXiv, Nov. 23, 2024. http://arxiv.org/abs/2407.21783
[4] G. Comanici et al., "Gemini 2.5: Pushing the Frontier with Advanced Reasoning, Multimodality, Long Context, and Next Generation Agentic Capabilities," arXiv, Jul. 22, 2025. http://arxiv.org/abs/2507.06261
[5] M. Davies, N. Crago, K. Sankaralingam, and C. Kozyrakis, "Efficient LLM Inference: Bandwidth, Compute, Synchronization, and Capacity are all you Need," arXiv, Jul. 18, 2025. http://arxiv.org/abs/2507.14397
[6] Y. Qin et al., "An 88.36TOPS/W Bit-Level-Weight-Compressed Large-Language-Model Accelerator with Cluster-Aligned INT-FP-GEMM and Bi-Dimensional Workflow Reformulation," *ISSCC*, pp. 420–422, 2025. http://doi.org/10.1109/ISSCC49661.2025.10904774
[7] Y. Leviathan, M. Kalman, and Y. Matias, "Fast Inference from Transformers via Speculative Decoding," *ICML*, pp. 19274–19286, 2023. https://arxiv.org/abs/2211.17192
[8] B. F. Spector and C. Re, "Accelerating LLM Inference with Staged Speculative Decoding," *Workshop on Efficient Systems for Foundation Models at ICML*, 2023. https://arxiv.org/abs/2308.04623
[9] T. Cai et al., "MEDUSA: Simple LLM Inference Acceleration Framework with Multiple Decoding Heads," *ICML*, pp. 5209–5235, 2024. https://arxiv.org/abs/2401.10774
[10] X. Miao et al., "SpecInfer: Accelerating Large Language Model Serving with Tree-based Speculative Inference and Verification," *ACM ASPLOS*, pp. 932–949, 2024. http://doi.org/10.1145/3620666.3651335

[11] Y. Zhou et al., "DistillSpec: Improving Speculative Decoding via Knowledge Distillation," *ICLR*, 2023. https://arxiv.org/abs/2310.08461
[12] M. Yan et al., "Decoding Speculative Decoding," *ACL*, pp. 6460–6473, 2025. http://doi.org/10.18653/v1/2025.naacl-long.328
[13] H. Zheng and X. Wang, "Faster Speculative Decoding via Effective Draft Decoder with Pruned Candidate Tree," *ACL*, pp. 9856–9868, 2025. http://doi.org/10.18653/v1/2025.acl-long.486
[14] J. Wang et al., "OPT-Tree: Speculative Decoding with Adaptive Draft Tree Structure," *Trans. of the Association for Computational Linguistics*, vol. 13, pp. 188–199, 2025. http://doi.org/10.1162/tacl_a_00735
[15] S. Kim et al., "C-Transformer: A 2.6-18.1μJ/Token Homogeneous DNN-Transformer/Spiking-Transformer Processor with Big-Little Network and Implicit Weight Generation for Large Language Models," *ISSCC*, pp. 368–370, 2024. http://doi.org/10.1109/ISSCC49657.2024.10454330
[16] T. Tambe et al., "A 12nm 18.1TFLOPs/W Sparse Transformer Processor with Entropy-Based Early Exit, Mixed-Precision Predication and Fine-Grained Power Management," *ISSCC*, pp. 342–344, 2023. http://doi.org/10.1109/ISSCC42615.2023.10067817
[17] S. Moon et al., "T-REX: A 68-to-567μs/Token 0.41-to-3.95μJ/Token Transformer Accelerator with Reduced External Memory Access and Enhanced Hardware Utilization in 16nm FinFET," *ISSCC*, pp. 406–408, 2025. http://doi.org/10.1109/ISSCC49661.2025.10904793
[18] S. Kim et al., "Slim-Llama: A 4.69mW Large-Language-Model Processor with Binary/Ternary Weights for Billion-Parameter Llama Model," *ISSCC*, pp. 421–423, 2025. http://doi.org/10.1109/ISSCC49661.2025.10904761

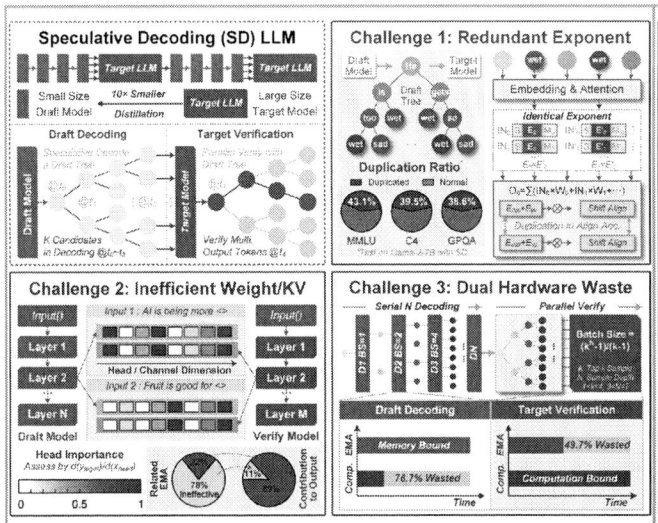

Figure 31.8.1: Speculative decoding LLM achieves performance advantages with rapid drafting and parallel verification, but suffers from three challenges.

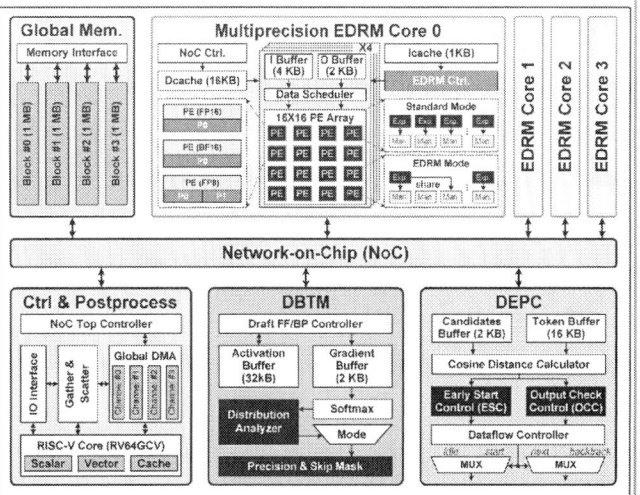

Figure 31.8.2: Overall architecture of the proposed energy-efficient speculative decoding LLM processor for billion-scale LLM applications with three key features.

ISSCC 2026 / SESSION 31 / AI ACCELERATORS / 31.8

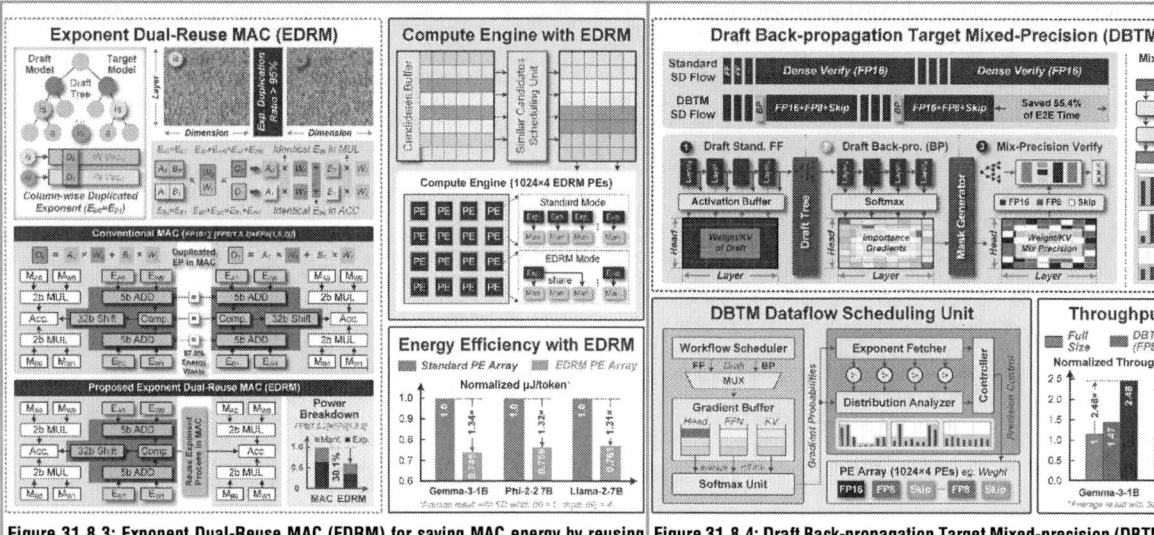

Figure 31.8.3: Exponent Dual-Reuse MAC (EDRM) for saving MAC energy by reusing exponent processing of semantically identical, but positionally distinct tokens.

Figure 31.8.4: Draft Back-propagation Target Mixed-precision (DBTM) for reducing cost of the inefficient weights and KV cache identified by gradients of draft model.

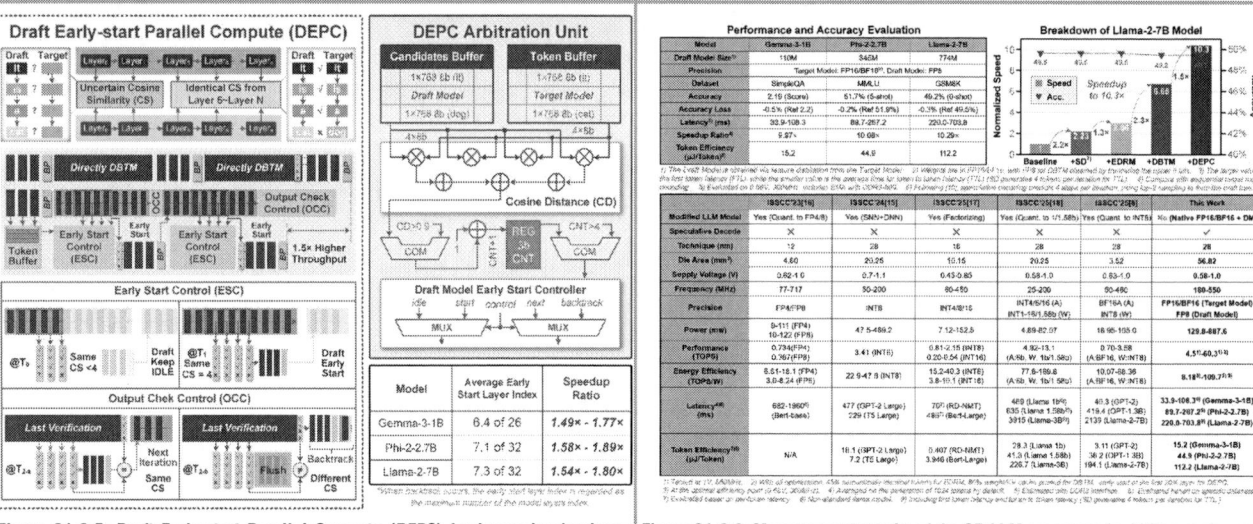

Figure 31.8.5: Draft Early-start Parallel Compute (DEPC) for improving hardware utilization by early start draft decoding in parallel with target verification.

Figure 31.8.6: Measurement results of the SD LLM processor for billion-scale models and performance comparison with state-of-the-art LLM processors.

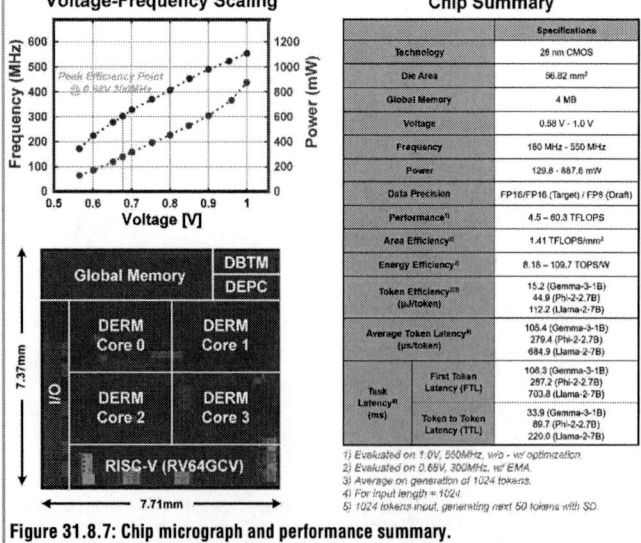

	Specifications
Technology	28 nm CMOS
Die Area	56.82 mm²
Global Memory	4 MB
Voltage	0.58 V - 1.0 V
Frequency	180 MHz - 550 MHz
Power	129.8 - 887.6 mW
Data Precision	FP16/FP16 (Target) / FP8 (Draft)
Performance[1]	4.5 – 60.3 TFLOPS
Area Efficiency[2]	1.41 TFLOPS/mm²
Energy Efficiency[2]	8.18 – 109.7 TOPS/W
Token Efficiency[2][3] (μJ/token)	15.2 (Gemma-3-1B) 44.9 (Phi-2-2.7B) 112.2 (Llama-2-7B)
Average Token Latency[3] (μs/token)	105.4 (Gemma-3-1B) 279.4 (Phi-2-2.7B) 684.9 (Llama-2-7B)
Task Latency[4] (ms)	First Token Latency (FTL)	106.3 (Gemma-3-1B) 287.2 (Phi-2-2.7B) 703.8 (Llama-2-7B)
	Token to Token Latency (TTL)	33.9 (Gemma-3-1B) 89.7 (Phi-2-2.7B) 220.0 (Llama-2-7B)

1) Evaluated on 1.0V, 550MHz, w/o - w/ optimization.
2) Evaluated on 0.58V, 300MHz, w/ EMA.
3) Average on generation of 1024 tokens.
4) For input length = 1024.
5) 1024 tokens input, generating next 50 tokens with SD.

Figure 31.8.7: Chip micrograph and performance summary.

• 2026 IEEE International Solid-State Circuits Conference

979-8-3315-8937-0/26 $31.00 © 2026 IEEE

ISSCC 2026 / SESSION 31 / AI ACCELERATORS / 31.9

31.9 ALPhA-Vision: A Real-Time Always-On Vision Processor with 787µs Face Detection Latency in <5mW

Ben Keller[1], Rangharajan Venkatesan[1], Steve Dai[1], Jason Clemons[2], Matthew Fojtik[3], Muya Chang[1], Thierry Tambe[4], Nathaniel Pinckney[2], Stephen G. Tell[3], Qijing Huang[1], Shalini De Mello[1], Brucek Khailany[2]

[1]Nvidia, Santa Clara, CA, [2]Nvidia, Austin, TX, [3]Nvidia, Durham, NC, [4]Stanford University, Stanford, CA

Abstract

ALPhA-Vision is an always-on low-power subsystem for DNN-inference-based vision tasks in edge SoCs. Flexible and programmable, the subsystem supports CNN and ViT inference and employs hardware/software co-design to enable fully end-to-end execution with no external memory accesses. Fine-grained power management features to mitigate leakage enable the subsystem to perform face detection with 787µs latency and 99.3% detection accuracy with 4.6mW average power at 60fps.

Always-on computer vision (CV) tasks are key workloads in applications such as user presence detection, safety monitoring, and gesture-based control, with always-on vision (AoV) poised to become even more prevalent as embodied artificial intelligence enters the mainstream [1]. In consumer devices, use cases such as face detection for opportunistic laptop display sleep can enable significant system-level power savings if a low-latency vision pipeline can be continuously executed in a low power footprint. Typical vision pipelines include both deep neural network (DNN) inference tasks and classical CV kernels, requiring a flexible and capable platform for end-to-end execution. GPU-based edge SoCs can execute these workloads, but their high power consumption (≥10W) limits their utility for AoV applications [2].

This work presents ALPhA-Vision, an Always-on Low-Power Accelerator for vision tasks that can serve as a dedicated subsystem in a larger high-performance SoC, enabling the rest of the SoC to sleep while the standalone subsystem runs continuously in a low power footprint (see Fig. 31.9.1). The programmable subsystem supports a variety of different DNN inference workloads, including both convolutional neural networks (CNNs) and vision transformers (ViTs), and specialized hardware accelerates not only matrix multiplication (GEMM) kernels, but many other layer types commonly found in CV DNNs, ensuring that non-GEMM operations do not bottleneck low-latency execution. Aggressive quantization and a capacity-maximizing hardware implementation enable end-to-end localized processing of image inputs without accessing external memory. Hardware-software codesign for fine-grained power management enables a sub-5mW power budget even at high framerates. A companion full-stack compiler flow lowers application source code to execution binaries after performing hardware-compliant quantization-aware training (QAT) to maintain task accuracy. Together, these components enable end-to-end execution of AoV pipelines without waking the high-power CPU, GPU, or main memory system of the full SoC.

Figure 31.9.2 shows the ALPhA-Vision architecture, which comprises a deep-learning accelerator (DLA) to execute GEMMs, a near-memory processor (NMP) dedicated to low-compute-intensity kernels, a 32b RISC-V Rocket processor (RV) [3], and a global memory system containing 2.125MB of SRAM, which fills much of the available chip area to accommodate fully resident weights and activations. Unlike prior works that target ultra-low-power edge devices using specialized low-leakage processes [4-10], ALPhA-Vision is designed as a subsystem for a larger SoC and therefore uses a standard high-performance process. Accordingly, SRAM leakage is a large component of overall power consumption, and minimizing this leakage power was a first-class consideration for system design. The most important factor in reducing static power consumption is for the system to "race to sleep" so it can spend as much time as possible in a low-power sleep mode. To accomplish this, compute units are optimized to target sub-ms end-to-end latency for typical workloads so that the system can sleep for most of each frame, even in high-framerate (60fps) scenarios.

The ALPhA-Vision DLA accelerates convolution and GEMM kernels that make up the bulk of DNN compute. The DLA, which supports both INT8 and INT4 with microscaling (INT8 scale factors with a block size of 32) [11,12] math operations, performs matrix-vector multiplication in 32 parallel lanes of 32-element vector MACs (16 for INT8) each cycle, for a total of 1024 (512 for INT8) MACs/cycle when fully utilized. Weights and activations are initially stored in global memory, with local buffers, collectors, and address generators orchestrating an efficient output-stationary, local-weight-stationary dataflow to maximize reuse [13], while affording high utilization on a range of different GEMM shapes. A post-processing unit (PPU) performs bias addition, scaling, and simple activation operations, such as ReLU and Tanh.

The compute acceleration afforded by the DLA is not sufficient to realize low latency on end-to-end DNN inference workloads, as a "long tail" of operations with low compute intensity would dominate the remaining runtime. The NMP, a programmable vector engine with high (128B/cycle) bandwidth to global memory, accelerates these remaining workloads. The NMP supports depthwise convolutions (DWConv) with a specialized engine performing 32 MACs/cycle, non-GEMM operations such as Groupnorm and SoftMax, elementwise operations, and memory-manipulation operations like shuffle and transpose, all of which are performed in INT8. A dedicated perspective warp engine offloads this common classical CV function from the CPU.

The global memories were co-designed with the accelerators to avoid memory-induced bottlenecks, while maximizing capacity. The memory system consists of a global weight memory (GWM, 16 banks, 68B/cycle peak read bandwidth) that stores most weights and a global scratchpad (GS, 64 banks, 136B/cycle peak read bandwidth, 65B/cycle peak write bandwidth) that stores the input image, intermediate activations, DWConv layer weights, and the RV program. The high GS bank count enables fine-grained memory management and supports multiple parallel accesses to different banks, with 14 physical banks statically grouped into virtual bank clusters for higher bandwidth. Virtual-to-physical address translation is hardware managed by accelerator address generators at runtime and can be reconfigured per DNN layer. The global memories minimize large, expensive crossbar structures by a compiler-enforced requirement that each bank be reserved for access by at most one requestor at a time.

The ALPhA-Vision design minimizes static power, the dominant component of energy per frame, both through direct optimization and by minimizing active-mode latency to maximize the time the subsystem is in sleep mode (see Fig. 31.9.3). Each of the GWM and GS banks can be independently placed into retention (power-gated periphery) or sleep (power-gated macros) modes, dramatically reducing leakage power, and these settings can be adjusted on-the-fly for fine-grained sleep state control during active-mode execution [14]. Outside of clock trees, HVT standard cells are favored to minimize leakage. In addition to hardware-managed leaf-level clock gating (CG), the system implements software-controlled block-level CG for the DLA and NMP, as well as a software-programmable root CG that halts the entire clock distribution network at its root, leaving only a single 31b counter active as a timer to restart the clock. To reduce control overhead and the associated latency, the DLA and NMP have deep configuration memories enabling up to 384 kernels to be fused together and executed back-to-back with no intervention from the RV. To further reduce control overhead and to minimize the latency of any workloads that lack hardware acceleration support, the RV supports a double-clock execution mode in which its clock runs at twice the frequency of the rest of the system with zero-cycle clock domain crossing latency between the 2× and 1× domains.

Figure 31.9.4 provides an overview of the application software flow used to execute end-to-end DNN inference workloads on ALPhA-Vision. Application software in PyTorch is quantized with a per-layer configuration, which accounts for both hardware constraints and maintaining accuracy, that is used to perform QAT, yielding quantized per-layer weights, biases, and scaling factors that can be executed in hardware. The software stack also allocates memory in the GWM and GS, tracking the liveness of activation data to maximize the number of activation memory banks that can be placed in retention or sleep mode during each layer execution while striding data across banks to accommodate high-bandwidth accesses. Timeloop [15] uses a high-level hardware model to map each GEMM onto the DLA, selecting loop nest dimensions to maximize utilization. Finally, the quantized layers are fused into one or more execution kernels. Using this flow, the weight and peak activation memory footprints of Yolov5n-0.5 [16] are reduced to 318KB and 461KB, respectively, while maintaining task accuracy, enabling fully on-chip execution with no external memory accesses.

A standalone testchip prototype of the ALPhA-Vision subsystem was taped out in 16nm with an active area of 4.20mm². In contrast to prior works that run at very low frequencies [4,7,17,18], the testchip operates at 359MHz at 0.6V, a moderate clock period target that minimized the need for LVT cell swaps and gate upsizing, while minimizing latency for race-to-sleep. Weights and input activations are loaded into the scratchpads before each measurement test, and the RV controller programs SRAM sleep states and block CG between kernel launches, as well as executing the non-maximum suppression (NMS) workload at the end of Yolov5.

Figure 31.9.5 summarizes key measurement results, which are measured at 0.6V and 359MHz (RV at 718MHz) unless otherwise noted. We evaluate the end-to-end performance of the system with face detection using Yolov5 trained on WiderFace [19] and tested on 192×192 grayscale input images. At 0.6V, the system achieves a face detection latency of 787ms and a peak throughput of 1270fps, but a more realistic use case operates at 60fps with 95.3% of each frame spent in sleep mode. The runtime breakdown demonstrates that the accelerators are effective at mitigating compute bottlenecks, with non-GEMM operations

548 • 2026 IEEE International Solid-State Circuits Conference

979-8-3315-8937-0/26 $31.00 © 2026 IEEE

using just 21% of active cycles. At 60fps, the system operates with 4.6mW average power, compared to 1.8mW of unavoidable leakage power with clocks halted and all SRAMs in sleep mode. Our power-saving techniques reduce active and sleep energy by 21% and 94% respectively, with 10% active energy savings from partitioning the end-to-end workload into multiple kernels to allow fine-grained reprogramming of SRAM sleep states. To showcase the versatility of the subsystem, we also evaluate performance on a wide range of kernels from MobileVitv2 [20] and Yolov5, as well as a perspective warp kernel. Across these diverse workloads, the DLA achieves an average compute utilization of 81% and the NMP achieves an average bandwidth utilization of 86%. Figure 31.9.6 shows the comparison to prior works, and Fig. 31.9.7 shows the die photo and kernel measurement results. Taken together, the measurement results demonstrate that the ALPhA-Vision subsystem can enable AoV applications in power-constrained edge devices.

Acknowledgement:
The authors would like to thank Miguel Rodriguez, Amey Kulkarni, Sirisha Jayanti, Oliver Li, Walter Li, Vijayan Rathnam, Mike Sekulic, Tom Gray, and Tezaswi Raja for tapeout, package, and PCB support.

References:
[1] J. Duan et al., "A Survey of Embodied AI: From Simulators to Research Tasks," *IEEE TETCI*, vol. 6, no. 2, pp. 230–244, 2022. https://doi.org/10.1109/TETCI.2022.3141105
[2] NVIDIA, *Jetson Orin NX Module*. https://developer.nvidia.com/downloads/jetson-orin-nx-series-data-sheet
[3] K. Asanović et al., "The Rocket Chip Generator," Tech. Rep., EECS Department, University of California, Berkeley, 2016.
https://www2.eecs.berkeley.edu/Pubs/TechRpts/2016/EECS-2016-17.html
[4] Z. Fan et al., "Audio and Image Cross-Modal Intelligence via a 10 TOPS/W 22 nm SoC with Back-Propagation and Dynamic Power Gating," *IEEE Symp. VLSI Circuits*, pp. 18–19, 2022. https://doi.org/10.1109/VLSITechnologyandCir46769.2022.9830226
[5] V. Jain et al., "TinyVers: A 0.8–17 TOPS/W, 1.7 µW–20 mW, Tiny Versatile System-on-Chip with State-Retentive eMRAM for Machine Learning Inference at the Extreme Edge," *IEEE Symp. VLSI Circuits*, pp. 20–21, Jun. 2022.
https://doi.org/10.1109/VLSITechnologyandCir46769.2022.9830409
[6] P. Jokic et al., "A Sub-mW Dual-Engine ML Inference System-on-Chip for Complete End-to-End Face-Analysis at the Edge," *IEEE Symp. VLSI Circuits*, 2021.
https://doi.org/10.23919/VLSICircuits52068.2021.9492401
[7] H. An et al., "A 170 µW Image Signal Processor Enabling Hierarchical Image Recognition for Intelligence at the Edge," *IEEE Symp. VLSI Circuits*, 2020.
https://doi.org/10.1109/VLSICircuits18222.2020.9162810
[8] I. Miro-Panades et al., "SamurAI: a 1.7MOPS–36GOPS Adaptive Versatile IoT Node with 15,000× Peak-to-Idle Power Reduction, 207 ns Wake-Up Time and 1.3 TOPS/W ML Efficiency," *IEEE Symp. VLSI Circuits*, 2020.
https://doi.org/10.1109/VLSICircuits18222.2020.9163000
[9] D. Garrett et al., "A 1mW Always-on Computer Vision Deep Learning Neural Decision Processor," *ISSCC*, pp. 158–159, 2023.
https://doi.org/10.1109/ISSCC42615.2023.10067588
[10] F. Conti et al., "A 12.4TOPS/W @ 136 GOPS AI-IoT System-on-Chip with 16 RISC-V, 2-to-8b Precision-Scalable DNN Acceleration and 30%-Boost Adaptive Body Biasing," *ISSCC*, pp. 21–23, 2023. https://doi.org/10.1109/ISSCC42615.2023.10067643

[11] S. Dai et al., "VS-Quant: Per-vector Scaled Quantization for Accurate Low-Precision Neural Network Inference," *MLSys*, pp. 873–884, 2021.
https://doi.org/10.48550/arXiv.2102.04503
[12] B. D. Rouhani et al., "Microscaling Data Formats for Deep Learning," *arXiv preprint*, arXiv:2310.10537, 2023. https://doi.org/10.48550/arXiv.2310.10537
[13] B. Keller et al., "A 95.6-TOPS/W Deep Learning Inference Accelerator With Per-Vector Scaled 4-bit Quantization in 5 nm," *IEEE JSSC*, vol. 58, no. 4, pp. 1129–1141, 2023.
https://doi.org/10.1109/JSSC.2023.3234893
[14] K. Prabhu et al., "MINOTAUR: A Posit-Based 0.42–0.50-TOPS/W Edge Transformer Inference and Training Accelerator," *IEEE JSSC*, vol. 60, no. 4, pp. 1311–1323, 2025.
https://doi.org/10.1109/JSSC.2025.3545731
[15] A. Parashar et al., "Timeloop: A Systematic Approach to DNN Accelerator Evaluation," *IEEE ISPASS*, pp. 304–315, 2019. https://doi.org/10.1109/ISPASS.2019.00042
[16] Q. Delong et al., "YOLO5Face: Why Reinventing a Face Detector," *arXiv preprint*, arXiv:2105.12931, 2021. https://doi.org/10.48550/arXiv.2105.12931
[17] Z. Yuan et al., "A 65nm 24.7µJ/Frame 12.3mW Activation-Similarity-Aware Convolutional Neural Network Video Processor Using Hybrid Precision, Inter-Frame Data Reuse and Mixed-Bit-Width Difference-Frame Data Codec," *ISSCC*, pp. 232–234, 2020.
https://doi.org/10.1109/ISSCC19947.2020.9063155
[18] S. Kwon et al., "Monolithic in-Memory Computing Microprocessor for End-to-End DNN Inferencing in MRAM-Embedded 28 nm CMOS Technology with 1.1 Mb Weight Storage," *ISSCC*, pp. 610–612, 2025.
https://doi.org/10.1109/ISSCC49661.2025.10904609
[19] S. Yang et al., "WIDER FACE: A Face Detection Benchmark," *CVPR*, pp. 5525–5533, 2016. https://doi.org/10.1109/CVPR.2016.596
[20] S. Mehta and M. Rastegari, "Separable Self-attention for Mobile Vision Transformers," *TMLR*, 2023. https://doi.org/10.48550/arXiv.2206.02680
[21] G. Fanelli et al., "Random Forests for Real Time 3D Face Analysis," *ICCV*, vol. 101, no. 3, pp. 437–458, 2013. https://doi.org/10.1007/s11263-012-0549-0
[22] V. Jain and E. Learned-Miller, "FDDB: A Benchmark for Face Detection in Unconstrained Settings," UMass Amherst Tech. Rep. UM-CS-2010-009.
https://people.cs.umass.edu/~elm/papers/fddb.pdf
[23] A. Gupta et al., "CogniVision: End-to-End SoC for Always-on Smart Vision with mW Power in 40 nm," *IEEE Symp. VLSI Circuits*, 2024.
https://doi.org/10.1109/VLSITechnologyandCir46783.2024.10631426
[24] E. Chang et al., "A 12-nm 0.62-1.61 mW Ultra-Low Power Digital CIM- based Deep-Learning System for End-to-End Always-on Vision," *IEEE Symp. VLSI Circuits*, 2023.
https://doi.org/10.23919/VLSITechnologyandCir57934.2023.10185296

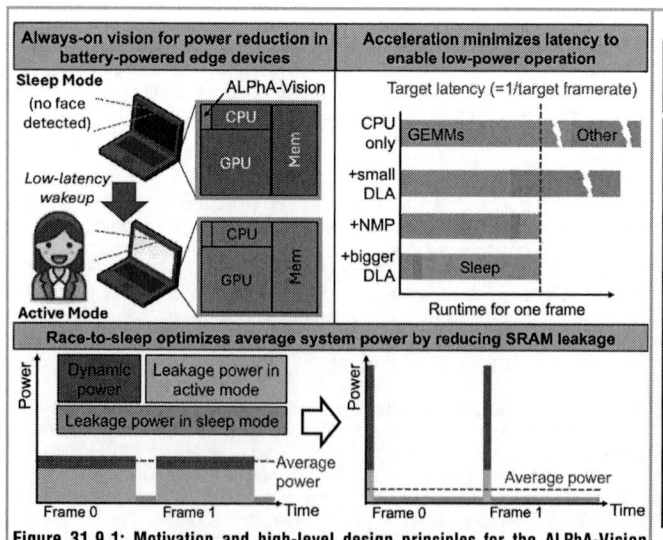

Figure 31.9.1: Motivation and high-level design principles for the ALPhA-Vision subsystem.

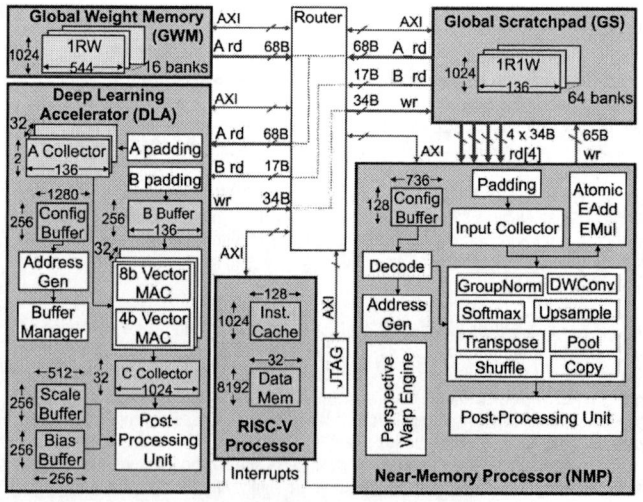

Figure 31.9.2: Block diagram of the ALPhA-Vision architecture.

ISSCC 2026 / SESSION 31 / AI ACCELERATORS / 31.9

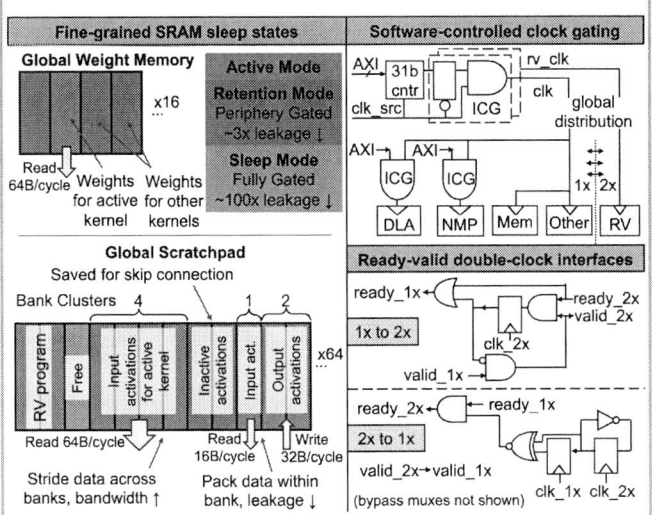

Figure 31.9.3: Overview of power-saving features in the ALPhA-Vision subsystem.

Figure 31.9.4: The ALPhA-Vision application software flow and impact of quantization of Yolov5 face detection.

Yolov5n-0.5 Evaluation	FP32	Hybrid INT8/INT4
Weight Memory (MB)	1.7	.377
Peak Activation Memory (MB)	2.3	.461
BIWI [21] Detection Rate	99.99	95.9
BIWI Recall (confidence>.3)	100	99.99
FDDB [22] Detection Rate	100	99.3
FDDB Recall (confidence>.3)	99.1	93

Global Scratchpad Bank Index

Layer Name	0	1	2	3	4	5	6
model.4.6.branch2.0							
model.4.6.branch2.3							
model.4.6.branch2.5							
model.4.6.shuffle							

Weights Active · Weights Retention · Activations Active · Activations Retention · Free

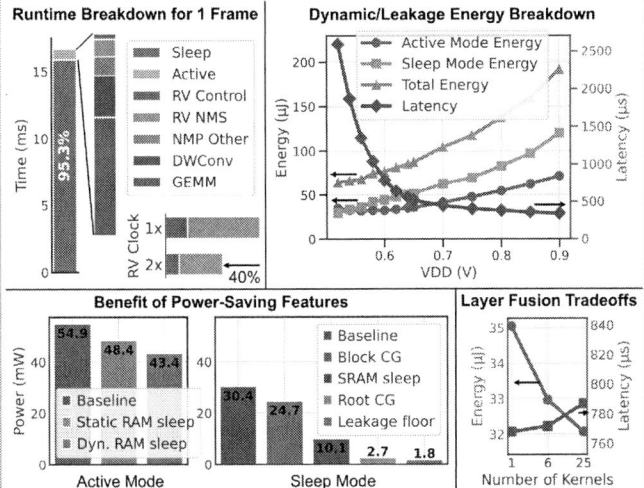

Figure 31.9.5: ALPhA-Vision measurement results on end-to-end Yolov5 face detection at 60fps.

	This Work	[23]	[24]	[18]	[6]
Application	Always-on vision	Always-on vision	Always-on vision	Low-power inference	Face analysis
Process (nm)	16	40	12	28	22
Area (mm²)	4.20	30	1.76	19.95	3.4
V_{DD} (V)	0.52-0.9	0.6-1.1	0.6	1	0.59-0.72
Frequency (MHz)	90-830	350	100-200	62.5	180-220
Power (mW)	3.9-11.5[1]	2.1[5]	0.62-1.61[7]	19-32[6]	0.16-12.77[9]
Memory (MB)	2.39	1.798	0.5	0.125	1.176
Throughput (GOPS)	1465[2]	2250	51	129	5
Energy Eff. (TOPS/W)	16.3[3] (4b)	29[6] (8b)	577[7,8] (8b/4b)	20[6] (1b)	1.1[9] (1b/16b)
Latency (ms)	0.787[1,4]	33[5]	67-500[7]	-	50-1000[9]
Energy/frame (µJ)	76.7[1,4]	91[5]	107-310[7]	-	409[9]
Energy/pixel (nJ)	2.1[1,4]	1.16[5]	6.5-18.9[7]	-	4.0[9]
FDDB Face Detection Accuracy	99.3%	-	-	91.3%	98%
Fully On-Chip End-to-End Execution	Yes	Yes	No	No	Yes

One MAC = 2 ops, other ops ignored. [1]Yolov5 face detection. [2]0.9V, 830MHz. [3]Benchmark GEMM kernel, 50% sparse, 0.54V, 146MHz. [4]0.6V, 359MHz. [5]SqueezeNet image classification. [6]Workload unspecified. [7]MobileNet human detection. [8]DLA only. [9]BDT face detection + CNN face recognition.

Figure 31.9.6: Comparison table. All measurements for this work report full system power, including leakage power.

Kernel	Active Units	Latency (µs)	Energy (µJ)	Util.
Conv (INT8)*	DLA	16.7	0.87	0.79
GEMM (INT8)†	DLA	8.43	0.38	0.87
Conv (INT4)*	DLA	8.44	0.35	0.78
DWConv*	NMP	14.94	0.66	1.00
GroupNorm†	NMP	1.91	0.09	0.96
GEMM+EAdd+Norm†	DLA,NMP	6.21	0.32	0.89
Transpose†	NMP	3.25	0.09	0.68
Shuffle*	NMP	3.43	0.16	0.48
MaxPool*	NMP	26.49	0.87	1.00
AveragePool†	NMP	0.34	0.01	0.8
GEMM+EAdd†	DLA,NMP	4.33	0.23	0.85
GEMM+Concat*	DLA	2.66	0.11	0.62
Split+GEMM*	DLA	4.31	0.22	0.77
Upsample*	NMP	0.89	0.02	0.92
Fold+GEMM†	DLA	8.43	0.39	0.87
GEMM+Unfold†	DLA	4.30	0.19	0.85
Perspective Warp‡	NMP	469.92	11.87	1.00

*Yolov5 kernel †MobileViTv2 kernel ‡640x480 to 128x320 transform

Figure 31.9.7: Annotated die plot and evaluation summary of DNN inference kernels (50% weight/activation sparsity).

• 2026 IEEE International Solid-State Circuits Conference

Session 32 Overview: *Low-Power Noise-Shaping ADCs*

DATA CONVERTERS SUBCOMMITTEE

Session Chair: Hajime Shibata
Analog Devices
Toronto, Canada

Session Co-Chair: Lucien Breems
NXP Semiconductors
Eindhoven, The Netherlands

Ultra-low-power high-dynamic range ADCs are vital for sensor interfaces in IoT, biomedical and industrial systems. In this session four highly energy-efficient sub-mW ADCs are presented exploiting noise-shaping techniques in different ways. The first part of the session shows that both a continuous-time zoom ADC and a fully dynamic discrete-time delta-sigma ADC architecture can achieve the same state-of-the-art power efficiency. The final 2 papers employ exponential noise coupling for efficient noise-shaping. The Schreier figure-of-merit (DR) for all ADCs exceeds 180dB.

ISSCC 2026 / February 18, 2026 / 1:30 PM

1:30 PM

32.1 A 98.5dB-SNDR 250kHz-BW 1V-Supply Continuous-Time Zoom ADC with Smart-Tracking and Floating-Tail-Resistor Linearized Gm-C Loop Filter

Chaoyang Xing, Tsinghua University, Beijing, China

In Paper 32.1, Tsinghua University presents a 28nm CMOS continuous-time incremental Zoom ADC achieving 98.5dB SNDR over 250kHz bandwidth at 1V supply, consuming 400µW. Smart-Tracking and Floating-Tail-Resistor techniques enable wideband input tracking and high linearity, delivering a 186dB FOM_S.

1:55 PM

32.2 A PVT-Robust Frequency-Scalable Fully Dynamic ΔΣ ADC with Bottom-Plate Level Shift

Tohru Kaneko, Asahi Kasei Microdevices, Yokohama, Japan

In Paper 32.2, Asahi Kasei Microdevices and Oregon State University reveal a fully dynamic, calibration-free ΔΣ ADC using capacitive-degeneration integrators and bottom-plate level shift (BPLS), achieving 95.6dB DR and 186dB $FOM_{S,DR}$ at 10kHz BW in 65nm CMOS. It maintains > 184dB $FOM_{S,DR}$ across 1 to 40kHz BW with < 2dB DR degradation under ±10% VDD or –40°C to 100°C, consuming only 10.1µW at 10kHz BW.

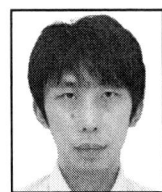

2:20 PM

32.3 An 85.1dB-SNDR 8MS/s Incremental Pipeline ADC with Dual-Residue-Assisted Exponential Quantization

Zongnan Wang, Peking University, Beijing, China

In Paper 32.3, Peking University discusses an 8MS/s incremental pipeline ADC with dual-residue-assisted exponential quantization, achieving 85.1dB SNDR and a 183dB FOM_S in 22nm CMOS. The design eliminates backend interpolators by leveraging an exponential ΔΣ loop, enabling fast, high-precision conversion while consuming only 618µW.

2:45 PM

32.4 A 103.9dB-SFDR 83.8dB-SNDR 3MHz-BW Multi-Bit Quadratic-Exponential Noise-Coupled IDSM with High Tolerance to DAC Non-Linearity

Zhensheng Li, University of Macau, Macau, China, UM Hetao IC Research Institute, Shenzhen, China

In Paper 32.4, University of Macau describes a 28nm CMOS multi-bit quadratic-exponential noise-coupled incremental DSM achieving 83.8dB SNDR and 103.9dB SFDR over 3MHz BW at 794µW. DAC and adder reuse enables high tolerance to DAC mismatch and delivers 179.6dB FOM_S.

32

DIGEST OF TECHNICAL PAPERS • 551

979-8-3315-8937-0/26 $31.00 © 2026 IEEE

ISSCC 2026 / SESSION 32 / LOW-POWER NOISE-SHAPING ADCs / 32.1

32.1 A 98.5dB-SNDR 250kHz-BW 1V-Supply Continuous-Time Zoom ADC with Smart-Tracking and Floating-Tail-Resistor Linearized Gm-C Loop Filter

Chaoyang Xing, Yilu Cui, Sining Pan, Yi Zhong, Nan Sun, Lu Jie

Tsinghua University, Beijing, China

Abstract

This paper presents a 98.5dB SNDR, 250kHz bandwidth CT incremental Zoom ADC. The design features a chopped capacitive front-end, Gm-C residue integrator, and 10b SAR ADC. A smart-tracking (ST) technique with dynamic step-sizing extends the input tracking bandwidth to the Nyquist frequency. A floating-tail-resistor (FTR) linearized Gm-C integrator achieves high linearity and efficiency. Fabricated in 28nm CMOS, the ADC consumes 400µW and achieves a superior 186.5dB FoMs.

High-resolution, energy-efficient ADCs with sampling rates exceeding 100kHz are in high demand for emerging IoT and wearable devices, posing a challenge to conventional ADC architectures. The Zoom ADC, which combines a front-end coarse quantizer with a back-end delta-sigma modulator (DSM), achieves high efficiency and linearity, making it a popular solution for these applications. Recently, continuous-time (CT) zoom ADCs with capacitive frontends have emerged as a promising solution due to their easy-driving input and absence of DAC thermal noise. However, a significant challenge for wideband CT zoom ADCs is maintaining a bounded coarse quantization residue. As input frequency increases, the coarse quantizer may lose track of the signal. Several techniques have been proposed to address this issue. Dynamic zoom ADCs [1], [2] utilize high-speed coarse quantization for rapid input tracking. However, this approach incurs high power consumption and increased driving effort (Fig. 32.1.1 top-left). A low-speed coarse quantizer can be employed [3] with a tracking mechanism that steps the feedback DAC based on the DSM output voting. This method necessitates a very high oversampling ratio (OSR), which increases design complexity and clock power (Fig. 32.1.1 top-middle). Tracking logic that monitors the DSM integrator's output to mitigate tracking delay can be used [4]. However, its tracking capability (i.e., input bandwidth) remains limited (Fig. 32.1.1 top-right).

To address this challenge, this work presents a CT incremental zoom ADC incorporating a smart-tracking (ST) technique (Fig. 32.1.1 bottom). The proposed architecture employs a chopped capacitive front-end, an efficient Gm-C residue integrator with 1b feedback (i.e., a 1b DSM), and a 10b SAR ADC for second-stage quantization. A smart-tracking logic controls the capacitor DAC (C_{DAC}) to generate a proper residue for the DSM. To get the initial DAC code, the C_{DAC} and the comparator (DM,I) are configured as an asynchronous 7b SAR ADC to perform an initial coarse quantization at the beginning of each conversion. Subsequently, the smart-tracking logic monitors the differential and common modes of the residue, as well as the integrator output, to steer the coarse DAC and enable wideband tracking. Another critical challenge in high-resolution ADC design is implementing highly linear and power-efficient amplifiers. This work incorporates a floating-tail-resistor (FTR) linearization technique to enable a transconductor circuit (Gm cell) with high power efficiency, linearity, CMRR, and PSRR. Thanks to the proposed techniques, the prototype design achieves 98dB SNDR over a 250kHz bandwidth with high power efficiency.

Figure 32.1.2 illustrates the smart-tracking technique, which includes differential-mode (DM) and common-mode (CM) tracking loops. The C_{DAC} codes $D_{DAC,P}$ and $D_{DAC,N}$ are constructed from a differential code D_{DM} and a common-mode code D_{CM}, which are generated by the two tracking loops, respectively. The primary goal of the DM tracking loop is to regulate the residue integrator output while avoiding unnecessary DAC toggling. To achieve this, the DM loop monitors the sign of both the integrator input and output, $D_{DM,I}$ and $D_{DM,O}$. If the signs differ (e.g., $D_{DM,I} = 0$ and $D_{DM,O} = 1$), the coarse DAC is left unchanged because the integrator output is already moving toward zero. Otherwise, the coarse DAC code is stepped up or down based on the direction of the signal change (DIR). In prior tracking techniques [3], [4], the coarse DAC step size is fixed at 1 LSB, which severely limits the maximum input signal frequency. This work proposes to use dynamic step size: If the coarse DAC is stepped in the same direction for two successive cycles (DIR[n] = DIR[n-1]), the step size increases to 2 LSB to track the input signal more rapidly. Otherwise, if the direction changes (DIR[n] ≠ DIR[n-1]), the step size resets to 1 LSB. A demonstration is shown in Fig. 32.1.2 bottom, the fixed-step tracking can only handle low-frequency inputs, and the integrator may overflow if the input frequency goes high, causing severe nonlinearity or functional failure. In contrast, the proposed dynamic-step tracking significantly improves the acceptable input bandwidth by 5 times, from $0.1f_{s,nyq}$ in [2] to $0.5f_{s,nyq}$, while maintaining a small integrator input swing for high linearity. Since the DM tracking loop only monitors differential signals, any common-mode variation in the input signal passes to the integrator without attenuation. Based on our simulation, a 20mV common-mode voltage shift can noticeably degrade the Gm cell's performance under low supply voltages, and a 50mV shift can render the Gm cell completely unusable. To address this, a CM tracking loop is introduced to monitor the integrator input common-mode. If the common-mode voltage shifts outside the predefined bounds ($V_{CM,L} = V_{CM,RST} - 10mV$ and $V_{CM,H} = V_{CM,RST} + 10mV$), the CM tracking loop adjusts the coarse DAC to re-center the input common-mode of the Gm cell. Thereby allowing the ADC to accept a large common-mode variation or even a single-ended signal.

An open-loop Gm-C integrator is chosen for high power efficiency. However, a simple differential pair exhibits poor linearity and is unsuitable for high-resolution zoom ADCs (Fig. 32.1.3 left). While source degeneration can improve linearity [2], it severely undermines the power efficiency by the degeneration factor ($1+g_mR$). Alternatively, the tail-resistor (TR) linearization technique [5], [6] mitigates the nonlinearity while maintaining the high efficiency of a differential pair by replacing the tail current source with a resistor. However, the TR technique is susceptible to input common-mode and supply voltage variations, resulting in unacceptably low CMRR (<6dB) and PSRR (<10dB). To overcome these limitations, this work proposes a floating-tail-resistor (FTR) technique, which utilizes a flipped-voltage-follower (FVF) to form a virtual ground that tracks the input common-mode voltage (Fig. 32.1.3). This mechanism ensures that the input transistors operate under consistent bias conditions, regardless of changes in the common-mode and supply voltages. Consequently, besides providing high linearity and high power efficiency, the FTR design achieves a CMRR of over 55dB and a PSRR of over 45dB, which represents a significant improvement over the original TR approach. Additionally, the CMFB current source in the FTR is connected to the common-source node of the differential pair, contributing substantially less differential noise than in the TR configuration.

Figure 32.1.3 (right) illustrates the proposed Gm cell. The deadband technique [2], [3] is introduced to prevent ISI errors from DAC settling. During the deadband phase, the integrator is temporarily disabled, which requires the Gm cell to switch and settle rapidly between its enabled and disabled states. This can be addressed with single-ended auxiliary paths [2], which ensure a rapid switching between phases. However, this approach significantly increases noise and power consumption. Incorporating a fully differential dummy branch [7], i.e., a replica of the main branch, can achieve low noise contribution, but extra nonlinearity is introduced during the transition phases. This work combines the advantages of both by managing operations across three phases: During the integration phase (1), only the main branch is active, integrating the residue signal while the dummy path is disabled to save power. In the deadband phase (2), the connections of cascode transistors in the main and dummy branches are swapped to null the output current. Additionally, since a direct transition between the integration and deadband phases can lead to a nonlinear and long settling process, a dummy enable phase (3) is introduced immediately before the deadband phase to solve this issue. During this phase, the dummy path is re-enabled, allowing its internal nodes to settle and facilitating a fast, clean transition. Furthermore, the dummy path is also chopped to suppress its offset and flicker noise. Consequently, this differential dummy-path Gm cell provides a nearly ideal deadband while contributing substantially less noise than the single-ended approach in [2].

The prototype ADC was fabricated in a 28nm CMOS process and occupies a core area of 0.022mm² (Fig. 32.1.7). A one-time, foreground, off-chip code weight calibration was applied to compensate for CDAC mismatch and integrator gain error. Measurement results are detailed in Fig. 32.1.4 and Fig. 32.1.5. With a 12.3kHz input signal, a peak SNDR of 98.5dB and an SFDR of 117.3dB were measured. To evaluate the proposed techniques, the prototype was also measured under high-frequency and single-ended inputs. With a 202.3kHz input, the smart-tracking technique achieves an SNDR of 93.5dB and an SFDR of 110.3dB. In contrast, the SNDR degrades to 90.9dB when switching to a conventional fixed-step tracking method, primarily due to the additional nonlinearity introduced by the larger output swing of the Gm cell. The CM tracking loop maintains a steady input common-mode voltage for the Gm cell, improving the SNDR from 81.8dB to 90.5dB for a full-scale single-ended (-6dBFS) input signal at 40.4kHz. With a high-frequency (>80kHz) full-scale single-ended input signal, the ADC fails if the proposed CM tracking loop is not enabled, further demonstrating the necessity of this feature. Additionally, the measured CMRR and PSRR exceed 70dB and 60dB, respectively, across the full signal bandwidth, validating the effectiveness of the CM tracking loop and the CM-insensitive Gm cell. A dynamic range of 99.7dB was measured by sweeping the input magnitude. The prototype performance varies by less than 3dB across three measured chips and ±10% supply voltage variation (without re-calibration). Operating at 500kS/s, the prototype consumes 400µW, with the power breakdown as follows: 63% for the Gm cell, 23% for the digital logic, and 12% for the reference. Figure 32.1.6 compares this work with prior art, highlighting its advantages. The capacitively-coupled input simplifies driving and reduces noise, and the efficient Gm-C integrator helps achieve a high Schreier figure-of-merit (FoMs) of 186.5dB under only 1V supply and $1V_{peak}$ input swing, which is the highest reported among compared ADCs with >90dB SNDR and >100kHz bandwidth.

References:

[1] B. Gönen, S. Karmakar, R. v. Veldhoven and K. Makinwa, "A Low Power Continuous-Time Zoom ADC for Audio Applications," 2019 Symposium on VLSI Circuits, Kyoto, Japan, 2019, pp. C224-C225. https://doi.org/10.23919/VLSIC.2019.8778021

[2] C. Xing, Y. Zhong, N. Sun and L. Jie, "A 94.4dB-SNDR 500kHz-BW Multi-Rate MASH 0-1-0 ADC with Easy-to-Drive Capacitive Input and Deadband-Embedded Gm-C Loop Filter," 2025 Symposium on VLSI Technology and Circuits (VLSI Technology and Circuits), Kyoto, Japan, 2025, pp. 1-3.
https://doi.org/10.23919/VLSITechnologyandCir65189.2025.11074988

[3] L. Jie, M. Zhan, X. Tang and N. Sun, "A 0.014mm2 10kHz-BW Zoom-Incremental-Counting ADC Achieving 103dB SNDR and 100dB Full-Scale CMRR," 2022 IEEE International Solid-State Circuits Conference (ISSCC), San Francisco, CA, USA, 2022, pp. 1-3. https://doi.org/10.1109/ISSCC42614.2022.9731742

[4] C. Yao et al., "A 44μW 140dB-DR Hybrid Light-to-Digital Converter with Current-Tracking Dynamic Zoom and Power-Scaling OTA," 2024 IEEE Custom Integrated Circuits Conference (CICC), Denver, CO, USA, 2024, pp. 1-2.
https://doi.org/10.1109/CICC60959.2024.10529087

[5] S. Li, S. Javvaji, V. Pecanins-Martinez, E. Aydin, R. van Veldhoven and K. A. A. Makinwa, "A Beyond-the-rail Audio CTΔΣM with a Passive Input Stage and 99.2dB SNDR," 2024 IEEE Symposium on VLSI Technology and Circuits (VLSI Technology and Circuits), Honolulu, HI, USA, 2024, pp. 1-2.
https://doi.org/10.1109/VLSITechnologyandCir46783.2024.10631367

[6] S. Pan and K. A. A. Makinwa, "A CMOS Resistor-Based Temperature Sensor with a 10fJ·K2 Resolution FoM and 0.4°C (30) Inaccuracy From –55°C to 125°C After a 1-point Trim," 2020 IEEE International Solid-State Circuits Conference - (ISSCC), San Francisco, CA, USA, 2020, pp. 68-70. https://doi.org/10.1109/ISSCC19947.2020.9063064

[7] G. Yun et al., "A 189.3dB-FoMs 14.5fJ/Conversion-Step Continuous-Time Noise-Shaping SAR Capacitance-to-Digital Converter," 2025 IEEE International Solid-State Circuits Conference (ISSCC), San Francisco, CA, USA, 2025, pp. 1-3.
https://doi.org/10.1109/ISSCC49661.2025.10904611

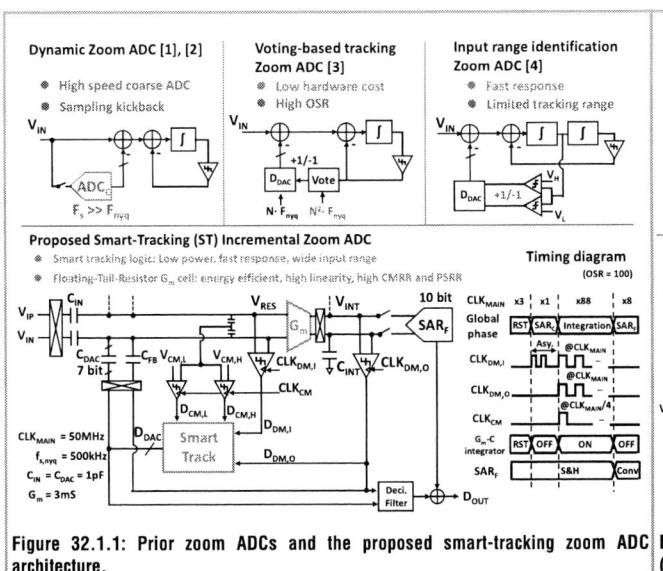

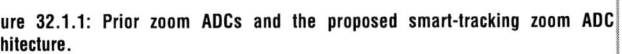

Figure 32.1.1: Prior zoom ADCs and the proposed smart-tracking zoom ADC architecture.

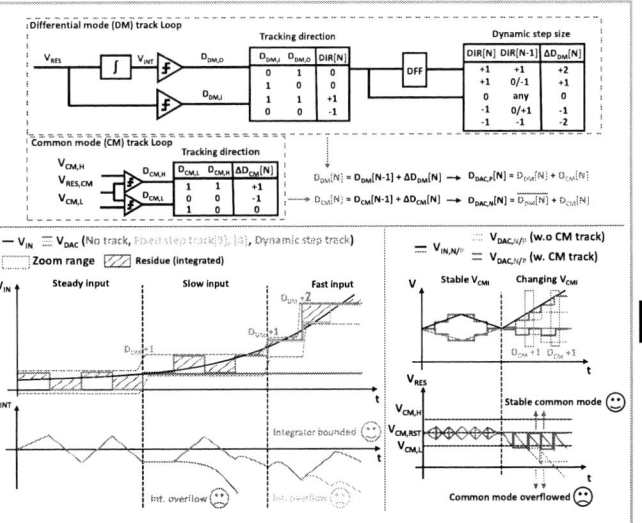

Figure 32.1.2: Smart-tracking logic implementation (top) and working waveforms (bottom).

ISSCC 2026 / SESSION 32 / LOW-POWER NOISE-SHAPING ADCs / 32.1

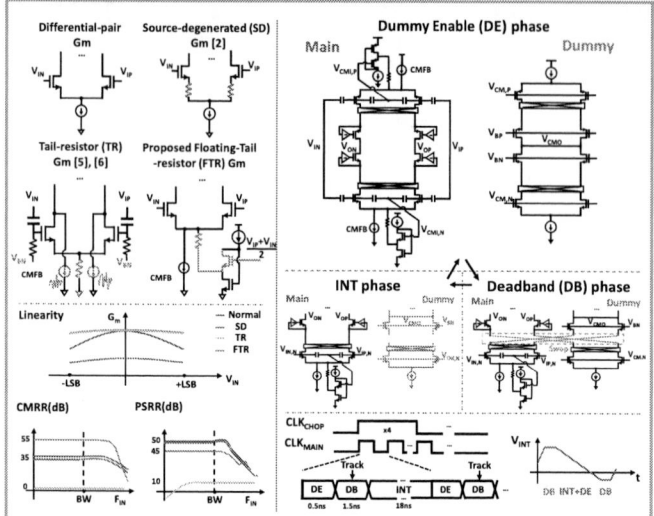

Figure 32.1.3: Comparison of different G_m cell structures and the details of the proposed FTR G_m cell.

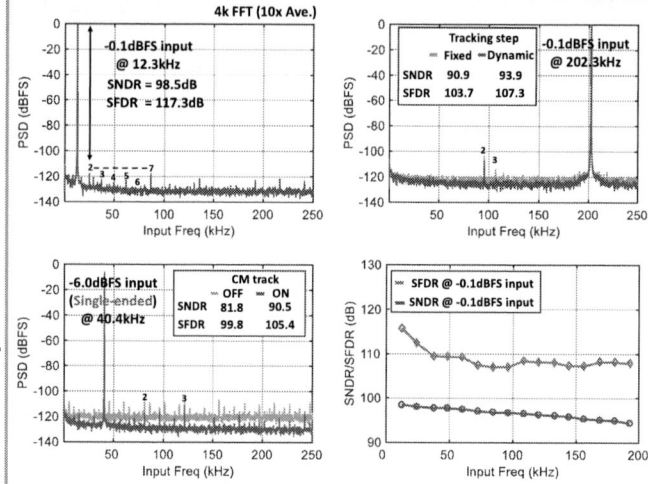

Figure 32.1.4: Measured spectra with low input frequency (top-left), high input frequency (top-right), single-ended input (bottom-left), and performance vs. input frequency.

Figure 32.1.5: Measured dynamic range, CMRR/PSRR vs. input frequency, performance variations of chips and analog supply voltage, and power breakdown.

	J. Steen ISSCC'22	Z. Wang ISSCC'25	B. Gönen VLSI'19	S. Li VLSI'24	L. Jie ISSCC'22	C. Xing VLSI'25	This work
Architecture	Pipe-SAR	NS-SAR (+kT/C cancel)	R-input CT Zoom (+Coarse SAR)	C-input CT Zoom (Voting)	C-input CT Zoom	C-input CT Zoom (+Coarse SAR)	C-input CT Zoom (Smart-Tracking)
Process (nm)	180	28	160	65	28	28	28
Easy-drive?	×	√	√	√	√	√	√
Support single-ended input?	×	×	×	Only Low-freq	Only Low-freq	×	√
Gm/OTA structure & linearization method	-	FIA	Telescopic	Telescopic +TR	Folded cascode	Telescopic +SD	Telescopic +FTR
Area (mm²)	0.78	0.034	0.27	0.28	0.014	0.023	0.022
Supply voltage (V)	1.8/5	0.9	1.8	1	1.2	1/1.2(G_m cell)	1
Power (mW)	8.5	0.47	0.62	0.08	0.62	0.79	0.40
F_S (MHz)	1	8	3	7.2	500	80	50
Bandwidth (kHz)	1000	800	20	24	10	500	250
F_{IN}(kHz)	1	100	5	200	1	1	2.3	23.4	192.4	12.3	202.3
SNDR (dB)	106	105.3	92.5	91.8	106.4	99.2	102.9	94.4	91.2	98.5	93.9
SFDR (dB)	>133	>114	112.2	113.1	114.0	112.0	113.1	108.6	103.4	117.3	107.3
DR (dB)	107	93.1	108.5	100.0	104.5	95.5	99.7
FoM₁* (dB)	186.0	185.3	184.8	184.1	181.5	184.0	176.2	182.3	179.1	186.5	181.9
FoM₂** (dB)	187.0	185.4	183.6	184.7	177.8	183.4	187.8

*FOM₁ = SNDR + 10log(BW/Power)   **FOM₂ = DR + 10log(BW/Power)

Figure 32.1.6: Performance summary and comparison with state-of-art ADCs.

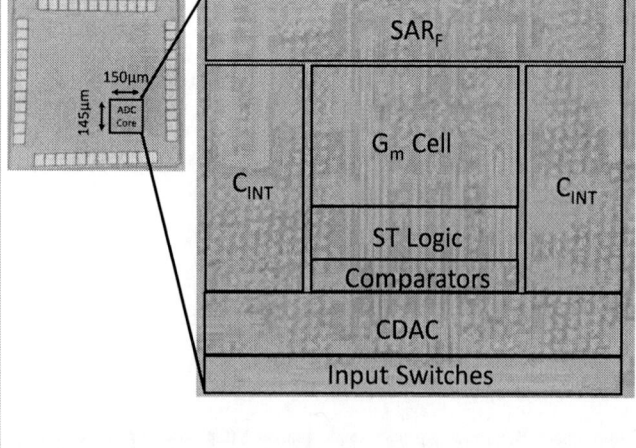

Figure 32.1.7: Chip micrograph.

- 2026 IEEE International Solid-State Circuits Conference

979-8-3315-8937-0/26 $31.00 © 2026 IEEE

ISSCC 2026 / SESSION 32 / LOW-POWER NOISE-SHAPING ADCs / 32.2

32.2 A PVT-Robust Frequency-Scalable Fully Dynamic ΔΣ ADC with Bottom-Plate Level Shift

Tohru Kaneko[1], Hideaki Ishida[1], Yuzuru Yamaki[1], Satoshi Takehara[1], Un-Ku Moon[2]

[1]Asahi Kasei Microdevices, Yokohama, Japan, [2]Oregon State University, Corvallis, OR

Abstract

A fully dynamic calibration-free ΔΣ ADC using capacitive-degeneration integrators with bottom-plate level shift (BPLS) is presented. BPLS samples negative charge to reduce threshold-voltage dependence, improve settling, and cancel kT/C noise, enabling 0.5pF sampling capacitance. The ADC achieves 91.7dB peak SNDR and 95.6dB DR with 10.1μW at 10kHz bandwidth (BW). Up to 40kHz BW, the worst-case DR degradation is ≤2dB under ±10%V_{DD} or -40 to 100°C. FoM$_{S,DR}$ is 185.6dB (10kHz BW) and >184dB (1-to-40kHz BW).

High-dynamic-range, frequency-scalable ADCs are useful across many sensing tasks and can reduce development costs. Noise-shaping architectures employing dynamic integrators are attractive for energy-efficient conversion and frequency scalability. The left side of Fig. 32.2.1 shows three established implementations. Passive integrators [1-3] are power-efficient, but their low loop gain makes it difficult to achieve high SQNR. Closed-loop integrators [3-6] improve SQNR, but negative feedback requires a high-gain, wide-bandwidth amplifier. With dynamic amplifiers, the effective gain and bandwidth are PVT- and time-dependent, so achieving accurate integration often needs multistage or complex designs and additional power consumption. Open-loop structures [7-10] are simpler, yet the nonlinearity and PVT sensitivity of dynamic open-loop amplifiers (or Gm-cells) remain challenges.

This paper presents a ΔΣ ADC with a calibration-free dynamic open-loop integrator. The integrator employs the capacitive-degeneration technique used in capacitive-degeneration amplifiers (CDAs) reported in [11,12]. Since the integration gain of the capacitive-degeneration integrator (CDI) is set by charge transfer, it enables dynamic operation and PVT-robust integration. We further propose a bottom-plate level shift (BPLS) that relaxes settling-speed limitations and improves PVT robustness. Moreover, BPLS cancels the kT/C noise of sampling capacitors by reusing degeneration capacitors. Fabricated in 65nm CMOS, the ADC achieves 95.6dB DR and 185.6dB FoM$_{S,DR}$ at 10kHz bandwidth (BW). Across 1-40kHz, FoM$_{S,DR}$ remains >184dB, with worst-case DR degradation ≤2dB under ±10%V_{DD} or -40 to 100°C.

As shown on the right side of Fig. 32.2.1, the CDI operation is based on charge transfer between C_1 and C_2. In each cycle, the sampled charge on C_1 is steered onto the integrating capacitor C_2. The transfer current naturally stops when the driving transistors run out of gate overdrive, so in the ideal case the integration step is set solely by the ratio of C_1 to C_2 and is independent of g_m and the settling-time budget. However, a key limitation is headroom-limited settling. The time constant is set by C_1 and the effective g_m of the transistors. Note that the effective g_m also scales with C_1 and headroom voltage $V_{headroom}$ because the drain current during the transfer is set by the amount of transferred charge. Thus, the time constant is ultimately determined by the headroom. This means that settling speed is strongly dependent on process variation because headroom voltage is obtained by the equation ($V_{com} - V_{TH}$). When headroom voltage is not enough, settling error causes SQNR deterioration in the ADCs, and the effect is severe in the SS process corner.

To alleviate this limitation, we focused on the C_1 charge at the start of integration. If C_1 samples a fixed negative charge before the integration, the transfer charge during integration is increased by the negative pre-charge amount. This is equal to the headroom extension/improvement. Furthermore, by applying an appropriate negative pre-charge, the process dependency can be eliminated at the same time. To realize it, we propose BPLS as shown in Fig. 32.2.2. BPLS has three phases. Firstly, C_1 is pre-charged to $-V_{com}$ by connecting the bottom plate of C_1 to V_{com} instead of V_{SS}. This establishes the charge needed for the subsequent negative sampling. Secondly, the top plate of C_1 is connected to the input-transistor source and samples $-V_{TH}$. Note that the transistor drain is not connected to C_2 to avoid affecting the integrator output during the negative charge sampling. Finally, the bottom plate of C_1 is returned from V_{com} to V_{SS}, initiating integration charge transfer from C_1 to C_2. Headroom for the integration is decided by the voltage difference between Φ_1 and Φ_2. Thanks to the negative-V_{TH} sampling, the headroom is decided solely by V_{com}. This means that settling speed is improved and process dependency is resolved. Simulation results in Fig. 32.2.2 show that the estimated SQNR regarding settling error and nonlinearity of CDI is improved by BPLS with short settling-time conditions. Although deterioration on the slow side is caused by subthreshold leakage, this can be ignored because a PVT-tracking timer, shown in Fig. 32.2.3, stops settling at the appropriate time. Since the final value of CDI is determined by capacitor ratios, and BPLS accelerates the initial settling while the long-time degradation is gradual, the timer accuracy requirement is very relaxed. Owing to this, temperature tolerance is improved.

BPLS also cancels kT/C noise of the CDAC. As shown in Fig. 32.2.2, after the determination of kT/C noise at Φ_0, the same noise remains until the end of Φ_2 because the gate node is floating. During Φ_1, the noise is sampled on C_1 with negative-V_{TH} sampling. After that, C_2 is

connected to the drain and the integration starts. However, the kT/C noise is not sent to C_2 since the gate and source have the same noise already. As a result, the CDAC kT/C noise does not appear in the output. This noise cancellation obtained by reusing C_1 also applies to the input transistor's low-frequency noise (and offset voltage). Therefore, BPLS offers four benefits: settling speed, process robustness, kT/C cancellation, and flicker noise suppression.

The bottom of Fig. 32.2.2 represents the final configuration of the CDI with BPLS. A CMFB amplifier uses a CDA architecture for full dynamic operation. The output of the CMFB amplifier, V_{CMFB}, is fed back to the bottom plate of C_1. Cascode transistors are inserted to prevent charge backflow from C_2 to C_1. For short-channel devices, the effective threshold depends on the drain voltage. As a result, the charge transferred from C_1 to C_2 in each integration cycle depends on the output level, and a small portion of the stored charge on C_2 is pulled back toward C_1 each cycle. Using a cascode topology keeps the V_{DS} of the transistors connected to C_1 nearly constant and suppresses the cumulative backflow of charge from C_2 to C_1.

Figure 32.2.3 illustrates the ADC implementation. The third-order ΔΣ loop uses two integrators and a first-order noise-shaping SAR (NS-SAR) quantizer. The oversampling ratio (OSR) is 32. A PVT-tracking timer uses a capacitor, diode-connected transistors, and a comparator to set the per-cycle settling-time budget that primarily follows V_{TH}. This budget self-adapts across PVT (earlier at hot/FF, later at cold/SS) without calibration. The timer clamps the phase window to prevent subthreshold-leakage–induced SQNR deterioration (right of Fig. 32.2.2). Both integrators are CDI, and the first CDI employs BPLS to cancel the CDAC kT/C noise. This enables a small 0.4pF CDAC, which would otherwise limit the ADC noise to -90.6dBFS, with no noise penalty. To relax the input-driver requirements, A$_{IN}$ is applied through a dual-branch passive RC network (coarse/fine) [13] implemented on-chip. The coarse branch absorbs kickback from the first integrator. The fine branch then connects and passes only the residual A$_{IN}$. With the smaller CDAC and the RC branches, the input is easy to drive. The reference path is likewise split. The coarse feed first pre-charges the CDAC and absorbs most kickback. The fine branch then connects and supplies only the residual reference charge. Because this residual is small, an RC filter is placed on the fine branch to suppress out-of-band reference ripple and ringing without affecting settling. As a result, no on-chip input or reference buffers are required. An attenuator ahead of the NS-SAR reduces the input swing of the first integrator to relax its linearity requirement. Partial feedback [6] is used between the 6b NS-SAR and the 4b CDAC to simplify DWA logic and the CDAC. The NS-SAR contains a 0.1pF sampling capacitor from A$_{IN}$ via an RC filter. Because a CDI cannot directly drive the subsequent circuits, a summation buffer is used to combine signals and to drive the NS-SAR. To retain the previous Q_2 term, the summation buffer uses a CDI with common-mode cancellation implemented by split switched capacitors. After sampling both differential and common-mode charge, one half of the divided sampling capacitance is connected to the opposite output node with reversed polarity. By charge sharing, the common-mode charge cancels, whereas the differential charge is preserved. With this switching, no CMFB is required. There is no static bias current anywhere in the ADC, enabling frequency scalability.

The prototype ADC is fabricated in 65nm CMOS, occupies 0.168mm², and operates from 1.2V analog, 1.2V reference (connected to analog V_{DD}), and 1.0V digital supplies. Common-mode reference V_{com} (0.6V) is generated on-chip by a dynamic V_{com} generator and requires no external supply. Figure 32.2.4 shows measured spectra exhibiting third-order noise shaping for 1kHz, 10kHz, and 40kHz bandwidths. The proposed BPLS suppresses low-frequency noise, and the CDAC kT/C component (-90.6dBFS) is canceled. The noise floor is set by the kT/C of C_1 (4pF). The FFT shape is essentially unchanged across clock settings. At 10kHz BW, peak SNDR is 91.7dB at -1.2dBFS input, and the DR is 95.6dB. Across five chips at 10kHz BW, SNDR and DR vary by only 0.31dB and 0.28dB, respectively. As summarized in Fig. 32.2.5, with fixed settings the bandwidth sweep is realized by changing only the clock and input frequency. The input frequency tracks the bandwidth at a constant 4% ratio over the sweep. Under nominal conditions (nominal V_{DD}, 27°C), SNDR and DR remain nearly constant from 0.125 to 64kHz BW. For BW >1kHz, power scales approximately linearly with BW (CLK/64) while maintaining performance. As a result, FoM$_{S,DR}$ rises above 184dB beginning at 1kHz BW and stays above that level across much of the subsequent sweep. At 0.125kHz BW, it is still 180.8dB. In addition, relative to the nominal conditions, the DR and SNDR degrade by <2.3dB under ±10%V_{DD} or −40 to 100°C up to 40kHz BW, confirming supply/temperature robustness. At −10%V_{DD} and BW≥40kHz, DR and SNDR gradually degrade because the reduced timing margin increasingly leads to incomplete SAR conversions. At 10kHz BW, the reference consumes only 0.6μW (6%) thanks to the small CDAC. Compared with recent NS-ADC works (Fig. 32.2.6), this work combines wide-range frequency scalability without retuning or recalibration, an easy-to-drive input (small 0.5pF total sampling capacitance via resistive interfaces), and calibration-free PVT robustness. With a single configuration (OSR=32), it maintains >184dB FoM$_{S,DR}$ over a wide range of bandwidths and reaches 185.6dB at 10kHz BW.

References:

[1] J. Liu et al., "A 90-dB-SNDR Calibration-Free Fully Passive Noise-Shaping SAR ADC With 4× Passive Gain and Second-Order DAC Mismatch Error Shaping," IEEE Journal of Solid-State Circuits, vol. 56, no. 11, pp. 3412–3423, Nov. 2021. https://doi.org/10.1109/JSSC.2021.3087661

[2] Z. Chen, M. Miyahara, and A. Matsuzawa, "A 2nd-order fully-passive noise-shaping SAR ADC with embedded passive gain," IEEE Asian Solid-State Circuits Conference (A-SSCC), Toyama, Japan, Nov. 2016, pp. 309–312. https://doi.org/10.1109/ASSCC.2016.7844197

[3] Y. Shen et al., "An 18-Bit 183.9 dB-FoMS,DR MES/Calibration-Free Scalable Zoom ADC Using Fully Passive Coarse Modulator and Gain-Linearity-Enhanced FIA with Sub-1-ppm THD at Full-Scale Input in 65-nm CMOS," IEEE Custom Integrated Circuits Conference (CICC), Boston, MA, USA, 2025, pp. 1–3. https://doi.org/10.1109/CICC63670.2025.10983665

[4] A. Matsuoka, T. Nezuka, and T. Iizuka, "Fully Dynamic Discrete-Time ΔΣ ADC Using Closed-Loop Two-Stage Cascoded Floating Inverter Amplifiers," IEEE Transactions on Circuits and Systems II: Express Briefs, vol. 69, no. 3, pp. 944–948, Mar. 2022. https://doi.org/10.1109/TCSII.2021.3134963

[5] S. Ma, L. Liu, T. Fang, J. Liu, and N. Wu, "A Discrete-Time Audio ΔΣ Modulator Using Dynamic Amplifier With Speed Enhancement and Flicker Noise Reduction Techniques," IEEE Journal of Solid-State Circuits, vol. 55, no. 2, pp. 333–343, Feb. 2020. https://doi.org/10.1109/JSSC.2019.2941540

[6] L. Meng, J. Chen, M. Zhao, and Z. Tan, "A Partially Feedback NSSAR Embedded Third-Order ΔΣ Modulator With Gain-Boosted Two-Stage FIAs," IEEE Journal of Solid-State Circuits, vol. 59, no. 9, pp. 2735–2746, Sept. 2024. https://doi.org/10.1109/JSSC.2024.3382007

[7] B. Zhang, R. Dou, L. Liu, and N. Wu, "A 91.2dB SNDR 66.2fJ/conv. dynamic amplifier based 24kHz ΔΣ modulator," IEEE Asian Solid-State Circuits Conference (A-SSCC), Toyama, Japan, Nov. 2016, pp. 317–320. https://doi.org/10.1109/ASSCC.2016.7844199

[8] R. Fan, W. Qiao, M. Lin, H. Feng, Z. Wang, and X. Xing, "A One-time Self Calibration Technique for Open-loop FIA in Noise Shaping SAR Achieving Ultra Gain PVT-robustness," IEEE International Symposium on Circuits and Systems (ISCAS), London, United Kingdom, 2025, pp. 1–5. https://doi.org/10.1109/ISCAS56072.2025.11044287

[9] M. Miyahara and A. Matsuzawa, "An 84 dB dynamic range 62.5–625 kHz bandwidth clock-scalable noise-shaping SAR ADC with open-loop integrator using dynamic amplifier," IEEE Custom Integrated Circuits Conference (CICC), Austin, TX, USA, 2017, pp. 1–4. https://doi.org/10.1109/CICC.2017.7993655

[10] H. Li, Y. Shen, H. Xin, E. Cantatore, and P. Harpe, "A 7.3-μW 13-ENOB 98-dB SFDR Noise-Shaping SAR ADC With Duty-Cycled Amplifier and Mismatch Error Shaping," IEEE Journal of Solid-State Circuits, vol. 57, no. 7, pp. 2078–2089, Jul. 2022. https://doi.org/10.1109/JSSC.2022.3168588

[11] A. Mao, S. Mo, and Z. Cai, "A Low-Power Cyclic ADC with Capacitive Degeneration Amplifier and Charge Mode Operation," IEEE International Symposium on Circuits and Systems (ISCAS), London, United Kingdom, 2025, pp. 1–5. https://doi.org/10.1109/ISCAS56072.2025.11043443

[12] H. Yoon, C. Lee, T. Kim, Y. Kwon, and Y. Chae, "A 65-dB-SNDR Pipelined SAR ADC Using PVT-Robust Capacitively Degenerated Dynamic Amplifier," IEEE Journal of Solid-State Circuits, vol. 58, no. 4, pp. 961–971, Apr. 2023. https://doi.org/10.1109/JSSC.2023.3235521

[13] D. A. Kerth et al., "Sampling circuit charge management," U.S. Patent 5,644,257, Jul. 1, 1997. https://patents.google.com/patent/US5644257

[14] Y. Luan et al., "A 12.2μW 99.6dB-SNDR 184.8dB-FoMs DT Zoom PPD ΔΣM with Gain-Embedded Bootstrapped Sampler," IEEE International Solid-State Circuits Conference (ISSCC), San Francisco, CA, USA, 2025, pp. 308–310. https://doi.org/10.1109/ISSCC49661.2025.10904732

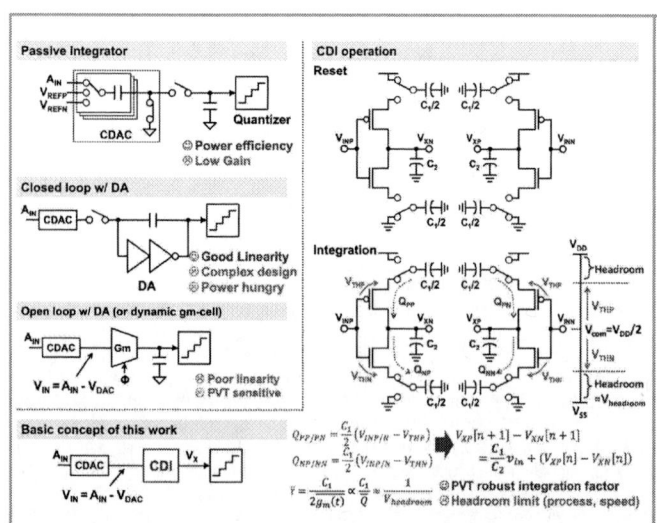

Figure 32.2.1: Dynamic integrator. Left: passive/closed loop/open loop/our approach. Right: CDI and headroom issue.

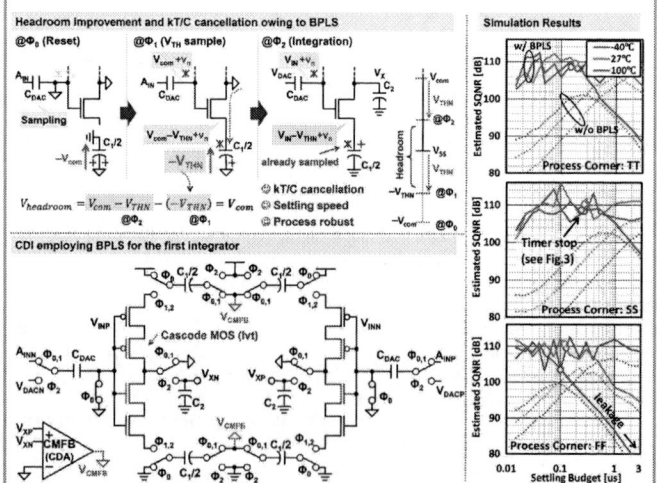

Figure 32.2.2: Proposed BPLS. Top: basic concept. Bottom: CDI with BPLS. Right: estimated SQNR in a ΔΣ ADC.

ISSCC 2026 / SESSION 32 / LOW-POWER NOISE-SHAPING ADCs / 32.2

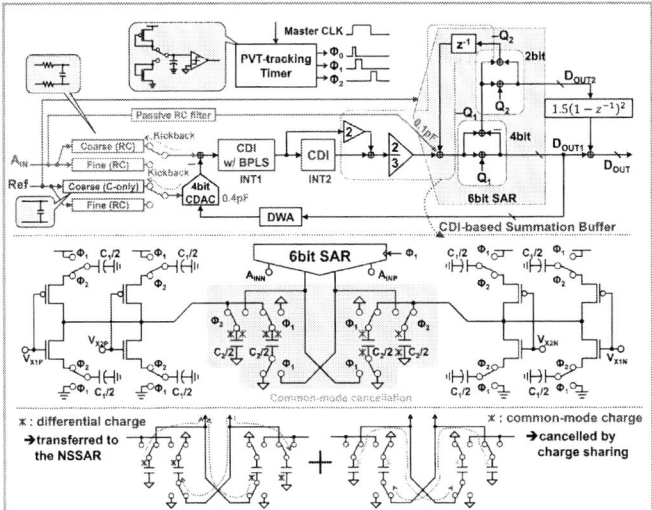

Figure 32.2.3: ADC implementation. Top: overview. Bottom: summation buffer and common-mode cancellation switching.

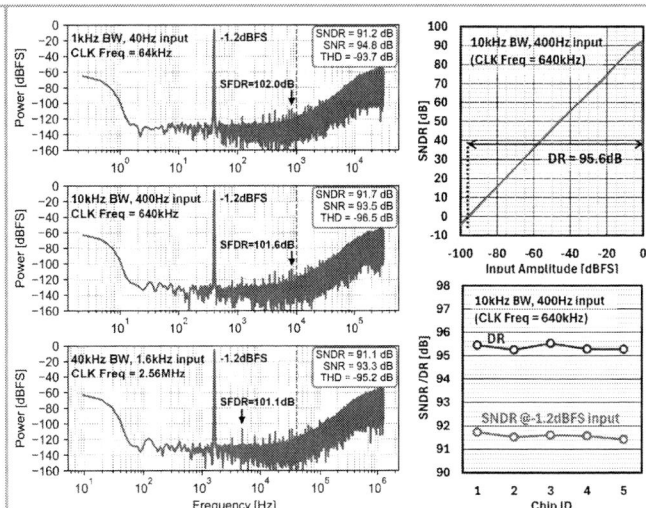

Figure 32.2.4: Measured FFT/SNDR/DR. Left: FFT. Top-right: SNDR vs input amplitude. Bottom-right: 5-chips results.

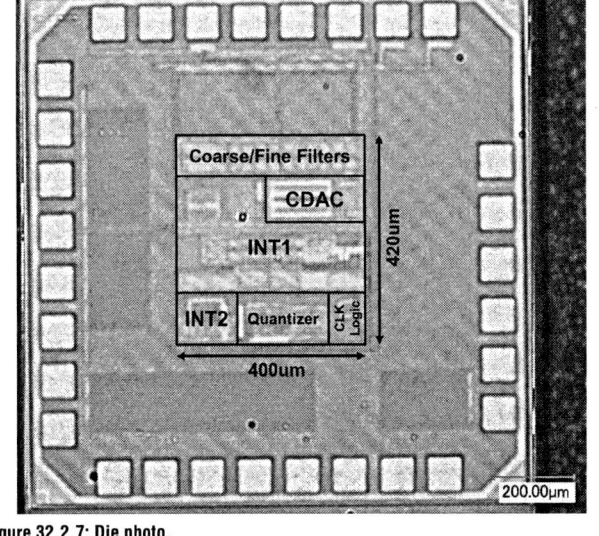

Figure 32.2.5: Bandwidth sweep results. Top: DR/SNDR. Bottom: FoMs/power at nominal conditions and power breakdown.

	This Work	Luan [14]	Shen [3]	Meng [6]	Matsuoka [4]	Liu [1]
		ISSCC '25	CICC '25	JSSC '24	TCAS-II '22	JSSC '21
Architecture	DT-Δ Σ	DT-PPD-Zoom	DT-Zoom	DT-Δ Σ	DT-Δ Σ	NS-SAR
Integrator type	CDI	Closed loop w/ OTA	Passive+Closed w/ FIA	Closed loop w/ FIA	Open loop w/ FIA	Passive
Process [nm]	65	55	65	55	65	40
Supply [V]	1.2/1.0	1.2/1.0	1.2	1.2	1	1.1
Area [mm²]	0.18	0.029	0.24	0.306	0.04	0.061
OSR	32 32 32	125	256	125 40	256	25
Fs [MSps]	0.064 0.64 2.56	1	5.12	0.4 0.8	10	2
BW [kHz]	1 10 40	4	10	1.6 10	19.5	40
Power [uW]	1.2 10.1 39.6	12.2	100.7	18.2 33.2	43.5	67.4
Peak SNDR [dB]	91.2 91.7 91.1	99.6	101.1	98.9 93.7	88.5	90.5
DR [dB]	94.8 95.6 95.2	102	104	101.1 94.7	91.7	94.3
FoMs,SNDR [dB] *1	180.5 181.7 181.2	184.8	181.0	178.3 178.5	175.0	178.2
FoMs,DR [dB] *1	184.1 185.6 185.3	187.2	183.9	180.5 179.5	178.2	182.0
Driving Friendly *2	Yes (Cs=0.5pF via R)	Yes (Cs~1pF via R)	- (Cs=7.4pF)	- (Cs=4pF)	*6	- (Cs=18pF)
Freq. Scalability *3	Yes (0.008-3.1MSps)	-	*5	(0.1-1.2MSps)	Yes (4-10MSps)	-
PVT robustness *4	Yes	-	-	Yes	Yes	*5

*1: FoMs = SNDR or DR + 10log(BW/Power)   Here, SNDR means peak SNDR.
*2: Indicates an input that does not require a large external driver/buffer (e.g., small effective Cs, R-input, etc.).
*3: Mark Yes when measurements at ≳2 distinct sampling frequencies are shown without retuning or recalibration.
*4: Mark Yes when TT/SS/FF, temperature, and VDD variation simulations or measurement results are shown.
*5: PVT robustness or frequency scalability are claimed without simulation or measurement data.
*6: Conventional sampling. Cs is not disclosed.

Figure 32.2.6: Performance summary and comparison with state-of-the-art works.

Figure 32.2.7: Die photo.

ISSCC 2026 / SESSION 32 / LOW-POWER NOISE-SHAPING ADCs / 32.3

32.3 An 85.1dB-SNDR 8MS/s Incremental Pipeline ADC with Dual-Residue-Assisted Exponential Quantization

Zongnan Wang, Bingrui Li, Haoyang Luo, Chenyuan Chu, Jiachang Yang, Yuan Wang, Xiyuan Tang

Peking University, Beijing, China

Abstract

This paper presents an incremental pipeline ADC with dual-residue architecture. An exponential $\Delta\Sigma$ loop is proposed to directly quantize the residue, replacing conventional interpolators that are complex and less accurate. It enables fast and high-precision backend conversion while maintaining insensitivity to inter-stage gain variations. The prototype achieves 85.1dB SNDR, the highest among comparable dual-residue ADCs, and 8MS/s conversion speed, the fastest among comparable high-precision incremental ADCs.

Incremental ADCs are gaining increasing attention for their high resolution, low latency, and easy system multiplexing [1]–[6]. Recent developments have leveraged pipelined incremental architectures for a better balance between power efficiency and conversion speed [7], [8]. However, ensuring accurate residue amplifier (RA) gain remains a critical challenge in high-resolution pipelined designs. Closed-loop amplifiers offer precise inter-stage gain but often incur high power consumption and stability concerns [9]–[11]. Techniques such as gain-error shaping (GES) [12]–[14] and dual-residue architecture [15]–[22] have been proposed to improve the inter-stage gain error tolerance, thus enabling the use of energy-efficient open-loop amplifiers without complex calibration. Nevertheless, the GES technique requires complicated auxiliary circuitry and is prone to error aliasing during output decimation in incremental ADCs. Alternatively, the dual-residue architecture alleviates the RA gain accuracy burden by performing backend interpolation on complementary residues. However, due to hardware cost, parasitic effects, and mismatch errors, interpolation resolution is typically limited to 5b, restricting its applicability in high-resolution systems.

This paper presents an incremental pipeline ADC with a dual-residue-assisted exponential quantizer. Unlike conventional dual-residue architectures that rely on interpolators, this work adopts an exponential $\Delta\Sigma$ loop for high-resolution backend quantization. The key insight is that the pre-generated complementary residues ($V_{RES} \pm V_R$) inherently represent the error of $\pm$1LSB quantization, allowing them to be directly employed for loop filter updating in single-bit $\Delta\Sigma$ modulation. To fully exploit its benefit, a capacitor-stacking-based error feedback (EF) loop filter is designed. It preserves the pre-stored residues during the entire conversion; thus, no additional sampling or residue generation is required. By designing the EF coefficient to be greater than 1, exponential quantization further boosts the conversion efficiency. Overall, the proposed exponential incremental quantizer contributes to over 50dB SQNR with only 16 single-bit conversions, realizing fast and high-precision backend quantization. It also remains insensitive to inter-stage gain variation, enabling the use of a low-power two-stage floating-inverter amplifier (FIA) as the residue amplifier [23]. A 3-step conversion scheme is adopted in the 1st stage to mitigate offset errors. Compared with the noise-shaping dual-residue ADC [22], the proposed design eliminates the backend interpolator, largely reducing design complexity, and requires only one-time sampling and residue generation, thereby saving considerable power and timing cost. Prototyped in 22nm CMOS, the proposed dual-residue pipeline ADC realizes 85.1dB SNDR while consuming only 618.1μW at 8MS/s, leading to a Schreier FoM of 183.2dB.

The architecture of the proposed incremental pipeline ADC is illustrated in Fig. 32.3.1. After the 1st-stage SAR conversion, the DAC input D_R is toggled between +1 and -1, effectively shifting the 1st-stage residue V_{RES} by $\pm V_R$. The shifted residues are amplified by the RA during Φ_{AU} and Φ_{AL} sequentially, producing two inherently gain- and offset-matched voltages, V_{IU} and V_{IL}. In conventional dual-residue architectures, the 2nd stage measures the ratio between V_{IU} and V_{IL} rather than their absolute values, decoupling conversion resolution from inter-stage gain but necessitating complex and error-prone interpolator designs. In contrast, this work proposes to utilize the pre-generated V_{IU} and V_{IL} to directly convert the 1st-stage residue in a high-resolution $\Delta\Sigma$ manner, as illustrated in Fig. 32.3.1 (right). The 2nd stage first performs a single-bit quantization of V_{RES} by merging V_{IU} and V_{IL}. Depending on the decision result, one of the dual residues, which essentially represents the single-bit quantization error, is chosen to update the loop filter. In the following cycle, the loop filter output is added back to the generated V_{IU} and V_{IL} for subsequent conversions. This conversion process can be regarded as a single-bit EF-based incremental $\Delta\Sigma$ modulation. Exponential quantization is achieved by setting the EF coefficient k greater than 1, which largely enhances the backend resolution [24]. In this work, k is designed as 1.4 to provide sufficient quantizer resolution while avoiding loop saturation. As a result, the proposed exponential quantizer achieves an effective resolution of 9b using only 16 single-bit conversions, outperforming the backend stage resolution in prior dual-residue designs. Meanwhile, it maintains insensitivity to inter-stage gain variation, facilitating high-resolution and robust ADC design.

Figure 32.3.2 depicts the schematic and timing diagram of the proposed incremental pipeline ADC. It consists of an 11b SAR quantizer as the 1st stage and the proposed exponential quantizer as the backend stage, with 2b inter-stage redundancy. A 2-stage FIA is designed to perform dual-residue amplification with a large inter-stage gain of ~100, providing sufficient backend noise attenuation. The noise cancellation capacitor C_{NC} is placed between two FIA stages for kT/C noise cancellation, allowing for a small input capacitance of 1.6pF and improving the drivability of the proposed ADC. During the residue amplification phase, the FIA produces two complementary differential outputs, V_{IU} and V_{IL}, which are stored onto $C_{UP/N}$ and $C_{LP/N}$, respectively. The loop filter results are stored on $C_{U/LA}$, and added to the quantization path through capacitor stacking. The generated differential signals ($V_{CU/L}$) are quantized by a 4-input comparator. Based on the decision result, a floating gm-ratio-based amplifier (FGA) amplifies either V_{CU} or V_{CL} onto the capacitors $C_{U/LB}$, which will be used to update the loop filter ($C_{U/LA}$) by charge sharing. The gm ratio of the FGA is designed to be 4:1, providing an open-loop gain of 2.8 given body effects. Considering the 0.5$\times$ signal attenuation during the loop filter update, the EF coefficient is designed to be 1.4, facilitating highly efficient exponential quantization. Thanks to the capacitor stacking-based EF loop filter design, the pre-stored dual residues can be preserved on $C_{UP/N}$ and $C_{LP/N}$ during the entire incremental conversion, and thus, no additional sampling and residue generation are required, yielding significant energy and time savings.

Although dual-residue architectures provide large inter-stage gain variation tolerance, they are susceptible to offset errors that can compromise the opposite-polarity requirement of V_{IU} and V_{IL}. Prior works necessitate backend overrange or offset calibration [16], [18]–[20], but at a cost of limited 1st-stage resolution and increased system complexity. While kT/C noise cancellation mitigates the amplifier offset [22], it relies on a power-hungry static amplifier. This work adopts a 3-step conversion scheme that mitigates the offset impact. As illustrated in Fig. 32.3.3 (top), during the sampling phase, the noise cancellation capacitor C_{NC} stores the amplified input difference between two sampling moments, frontend sampling noise, together with the offset of FIA1. The 1st-step conversion is performed before FIA1 with RA disabled. It is followed by a 2nd-step conversion at the FIA2 input, during which FIA1 is powered by a small reservoir capacitor for energy savings. It reduces the FIA2 input by quantizing the voltage stored on C_{NC}, together with previous errors. Finally, the 3rd-step conversion is performed at FIA2 output, compensating for the residual offset and conversion errors. The offset of the 3rd comparator is suppressed by A_1A_2 times. As a result, through the 3-step conversion, the offset issues are largely alleviated. Thus, up to $\pm$28mV offset errors can be tolerated with an 11b 1st-stage SAR quantizer and a fully-dynamic residue amplifier, ensuring correct opposite polarities of V_{IU} and V_{IL} without requiring any offset calibration.

To ensure accurate residue amplification, the two-stage FIA operates sequentially, rather than concurrently as in conventional multi-stage FIA designs [9], [11], [25]. It is necessitated by the voltage stored on C_{NC}, which causes the FIA2 input to start from an incorrect large value. With a large input, the initial amplification can rapidly drain the reservoir charge. Even though the input eventually returns to the correct value, the depleted C_{R2} cannot support FIA2 to generate the correct output. This challenge is generally applicable to multi-stage dynamic amplifiers with kT/C noise cancellation, and is further exacerbated at higher input frequencies, where the C_{NC} stores a larger voltage. In this work, the two FIAs operate sequentially, improving THD by 35dB compared to concurrent amplification under a 1MHz input. By matching the load capacitance C_L to C_{NC}, a constant FIA gain is maintained across kT/C noise extraction, 3rd-step conversion, and residue amplification phases. Thanks to the dual-residue architecture, the RA gain variation has a negligible impact on ADC performance. Figure 32.3.3 (bottom-right) shows that 0.75~1.25V FIA supply variation causes ~$\pm$40% gain variation. The measured SNDR only degrades less than 1.5dB, which is primarily due to the changes in RA noise, as SFDR remains above 105dB across the supply voltage range. This firmly confirms the robustness of the proposed incremental ADC with dual-residue-assisted quantizer.

The prototype ADC is fabricated in 22nm CMOS and occupies an active area of 0.038mm² (Fig. 32.3.7). Measurement results are shown in Fig. 32.3.4 and Fig. 32.3.5. One-time foreground calibration is used to address the capacitor mismatches in the 1st stage. With a 20kHz 1.8V$_{pp}$ input, the measured SNDR and SFDR are 85.5dB and 112.7dB. With a 1MHz 1.8V$_{pp}$ input, the measured SNDR and SFDR are 85.1dB and 100.3dB. The measured dynamic range is 86dB. The ADC core consumes 618.1μW at 8MS/s (11.9μW for reference, 285.6μW for FIA, 99μW for FGA, 90.4μW for other analog components, and 131.2μW for digital). Five chips are tested over a 0.8-1V analog supply voltage range and a -20~80°C temperature range, demonstrating an SNDR variation of less than 1.6dB, owing to the combination of dual-residue architecture and robust EF implementation. Figure 32.3.6 summarizes the performance of this work and compares it with prior publications with similar performance. Thanks to the proposed high-resolution exponential quantizer, the proposed incremental pipeline ADC realizes the best SNDR compared to prior dual-residue ADCs and the highest conversion speed compared to prior high-resolution incremental

556 • 2026 IEEE International Solid-State Circuits Conference

979-8-3315-8937-0/26 $31.00 © 2026 IEEE

ADCs (SNDR≥80dB). Furthermore, the proposed ADC is fully dynamic, resulting in significant power saving and leading to the best FoMs of 183.2dB among all compared ADCs operating above 5MS/s.

Acknowledgement:
This work is supported in part by National Key R&D Program of China (2022YFB4401900) and the NSFC (62274005). The corresponding author is Xiyuan Tang.

References:
[1] Y. Chae, K. Souri and K. A. A. Makinwa, "A 6.3 μW 20 bit Incremental Zoom-ADC with 6 ppm INL and 1 μV Offset," in IEEE Journal of Solid-State Circuits, vol. 48, no. 12, pp. 3019-3027, Dec. 2013. https://doi.org/10.1109/JSSC.2013.2278737
[2] B. Gönen, F. Sebastiano, R. van Veldhoven and K. A. A. Makinwa, "A 1.65mW 0.16mm² dynamic zoom-ADC with 107.5dB DR in 20kHz BW," 2016 IEEE International Solid-State Circuits Conference (ISSCC), San Francisco, CA, USA, 2016, pp. 282-283. https://doi.org/10.1109/ISSCC.2016.7418017
[3] Y. Liu et al., "A 4.96μW 15b Self-Timed Dynamic-Amplifier-Based Incremental Zoom ADC," 2022 IEEE International Solid-State Circuits Conference (ISSCC), San Francisco, CA, USA, 2022, pp. 170-172. https://doi.org/10.1109/ISSCC42614.2022.9731631
[4] L. Jie, M. Zhan, X. Tang and N. Sun, "A 0.014mm² 10kHz-BW Zoom-Incremental-Counting ADC Achieving 103dB SNDR and 100dB Full-Scale CMRR," 2022 IEEE International Solid-State Circuits Conference (ISSCC), San Francisco, CA, USA, 2022, pp. 1-3. https://doi.org/10.1109/ISSCC42614.2022.9731742
[5] Z. Wang et al., "A 150kHz-BW 15-ENOB Incremental Zoom ADC with Skipped Sampling and Single Buffer Embedded Noise-Shaping SAR Quantizer," 2023 IEEE International Solid-State Circuits Conference (ISSCC), San Francisco, CA, USA, 2023, pp. 9-11. https://doi.org/10.1109/ISSCC42615.2023.10067696
[6] T. -H. Wang et al., "A 50-kHz BW 92.1-dB SNDR Incremental ADC Using a Back-End Sampling Two-Step NS-SAR Architecture with Concurrent Gain-Error + Noise Suppression," 2025 IEEE Custom Integrated Circuits Conference (CICC), Boston, MA, USA, 2025, pp. 1-3. https://doi.org/10.1109/CICC63670.2025.10983053
[7] S. Baek et al., "A 12b 600MS/s Pipelined SAR and 2x-Interleaved Incremental Delta-Sigma ADC with Source-Follower-Based Residue-Transfer Scheme in 7nm FinFET," 2021 IEEE International Solid-State Circuits Conference (ISSCC), San Francisco, CA, USA, 2021, pp. 172-174. https://doi.org/10.1109/ISSCC42613.2021.9366051
[8] Z. Wang et al., "A 184.8dB-FoMs 1.6MS/s Incremental Noise-Shaping Pipeline ADC with Single-Amplification-Based kT/C Noise Cancellation Technique," 2025 IEEE International Solid-State Circuits Conference (ISSCC), San Francisco, CA, USA, 2025, pp. 1-3. https://doi.org/10.1109/ISSCC49661.2025.10904507
[9] X. Tang, X. Yang, J. Liu, W. Shi, D. Z. Pan and N. Sun, "A 0.4-to-40MS/s 75.7dB-SNDR Fully Dynamic Event-Driven Pipelined ADC with 3-Stage Cascoded Floating Inverter Amplifier," 2021 IEEE International Solid-State Circuits Conference (ISSCC), San Francisco, CA, USA, 2021, pp. 376-378. https://doi.org/10.1109/ISSCC42613.2021.9365753
[10] M. Li et al., "A Rail-to-Rail 12MS 91.3dB SNDR 94.1dB DR Two-Step SAR ADC with Integrated Input Buffer Using Predictive Level-Shifting," 2023 IEEE International Solid-State Circuits Conference (ISSCC), San Francisco, CA, USA, 2023, pp. 1-3. https://doi.org/10.1109/ISSCC42615.2023.10067703
[11] Z. Wang, H. Luo, R. Bao, L. Jie and X. Tang, "An 88.8dB-SNDR 6-MS/s Pipelined SAR ADC with A Closed-Loop Dynamic Amplifier Featuring Highly-Linear Full-Scale Output Swing," 2025 Symposium on VLSI Technology and Circuits (VLSI Technology and Circuits), Kyoto, Japan, 2025, pp. 1-3. https://doi.org/10.23919/VLSITechnologyandCir65189.2025.11075015

[12] C. -K. Hsu and N. Sun, "A 75.8dB-SNDR Pipeline SAR ADC with 2nd-order Interstage Gain Error Shaping," 2019 Symposium on VLSI Circuits, Kyoto, Japan, 2019, pp. C68-C69. https://doi.org/10.23919/VLSIC.2019.8778032
[13] H. Zhang, Y. Zhu, C. -H. Chan and R. P. Martins, "A 25MHz-BW 75dB-SNDR Inherent Gain Error Tolerance Noise-Shaping SAR-Assisted Pipeline ADC with Background Offset Calibration," 2021 IEEE International Solid-State Circuits Conference (ISSCC), San Francisco, CA, USA, 2021, pp. 380-382. https://doi.org/10.1109/ISSCC42613.2021.9365833
[14] H. Zhang, Y. Zhu, C. -H. Chan and R. P. Martins, "A 25MHz-BW 77.2dB-SNDR 2nd-Order Gain-Error-Shaping and NS Pipelined SAR ADC Based on a Quantization-Prediction-Unrolled Scheme," 2023 IEEE International Solid-State Circuits Conference (ISSCC), San Francisco, CA, USA, 2023, pp. 174-176. https://doi.org/10.1109/ISSCC42615.2023.10067438
[15] A. Matsuzawa, M. Kagawa, M. Kanoh, K. Tatehara, T. Yamaoka and K. Shimizu, "A 10 b 30 MHz two-step parallel BiCMOS ADC with internal S/H," 1990 37th IEEE International Conference on Solid-State Circuits, San Francisco, CA, USA, 1990, pp. 162-163. https://doi.org/10.1109/ISSCC.1990.110177
[16] C. Mangelsdorf, H. Malik, S. . -H. Lee, S. Hisano and M. Martin, "A two-residue architecture for multistage ADCs," 1993 IEEE International Solid-State Circuits Conference Digest of Technical Papers, San Francisco, CA, USA, 1993, pp. 64-65. https://doi.org/10.1109/ISSCC.1993.280082
[17] H. Miyahara, H. Lee, D. Paik and A. Matsuzawa, "A 10b 320 MS/s 40 mW open-loop interpolated pipeline ADC," 2011 Symposium on VLSI Circuits - Digest of Technical Papers, Kyoto, Japan, 2011, pp. 126-127.
[18] D. Vecchi et al., "An 800 MS/s Dual-Residue Pipeline ADC in 40 nm CMOS," in IEEE Journal of Solid-State Circuits, vol. 46, no. 12, pp. 2834-2844, Dec. 2011. https://doi.org/10.1109/JSSC.2011.2164301
[19] J. Lin, D. Paik, S. Lee, M. Miyahara and A. Matsuzawa, "An Ultra-Low-Voltage 160 MS/s 7 Bit Interpolated Pipeline ADC Using Dynamic Amplifiers," in IEEE Journal of Solid-State Circuits, vol. 50, no. 6, pp. 1399-1411, June 2015. https://doi.org/10.1109/JSSC.2015.2415472
[20] M. -J. Seo, Y. -D. Kim, J. -H. Chung and S. -T. Ryu, "A 40nm CMOS 12b 200MS/s Single-amplifier Dual-residue Pipelined-SAR ADC," 2019 Symposium on VLSI Circuits, Kyoto, Japan, 2019, pp. C72-C73. https://doi.org/10.23919/VLSIC.2019.8778005
[21] K. -I. Cho, Y. -S. Kwak, H. -J. Kim, J. -H. Boo, S. -H. Lee and G. -C. Ahn, "A 10-b 320-MS/s Dual-Residue Pipelined SAR ADC with Binary Search Current Interpolator," 2019 IEEE Custom Integrated Circuits Conference (CICC), Austin, TX, USA, 2019, pp. 1-4. https://doi.org/10.1109/CICC.2019.8780117
[22] J. -H. Chung et al., "A 1.5-MHz BW 81.2-dB SNDR Dual-Residue Pipeline ADC With a Fully Dynamic Noise-Shaping Interpolating-SAR ADC," in IEEE Journal of Solid-State Circuits, vol. 59, no. 8, pp. 2481-2491, Aug. 2024. https://doi.org/10.1109/JSSC.2024.3360944
[23] X. Tang et al., "An Energy-Efficient Comparator With Dynamic Floating Inverter Amplifier," in IEEE Journal of Solid-State Circuits, vol. 55, no. 4, pp. 1011-1022, April 2020. https://doi.org/10.1109/JSSC.2019.2960485
[24] B. Wang, S. -W. Sin, S. -P. U., F. Maloberti and R. P. Martins, "A 550-μW 20-kHz BW 100.8-dB SNDR Linear- Exponential Multi-Bit Incremental ΣΔ ADC With 256 Clock Cycles in 65-nm CMOS," in IEEE Journal of Solid-State Circuits, vol. 54, no. 4, pp. 1161-1172, April 2019. https://doi.org/10.1109/JSSC.2018.2888872
[25] X. Tang et al., "A 13.5b-ENOB Second-Order Noise-Shaping SAR with PVT-Robust Closed-Loop Dynamic Amplifier," 2020 IEEE International Solid-State Circuits Conference - (ISSCC), San Francisco, CA, USA, 2020, pp. 162-164. https://doi.org/10.1109/ISSCC19947.2020.9063058

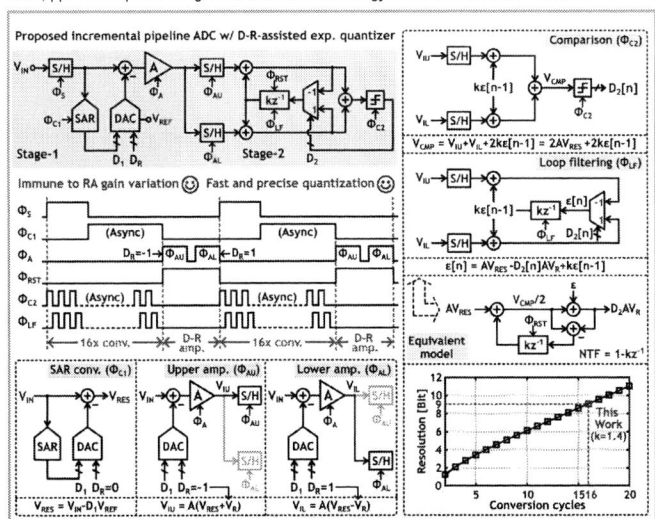

Figure 32.3.1: Simplified block and timing diagram of the proposed incremental pipeline ADC with dual-residue-assisted exponential quantizer.

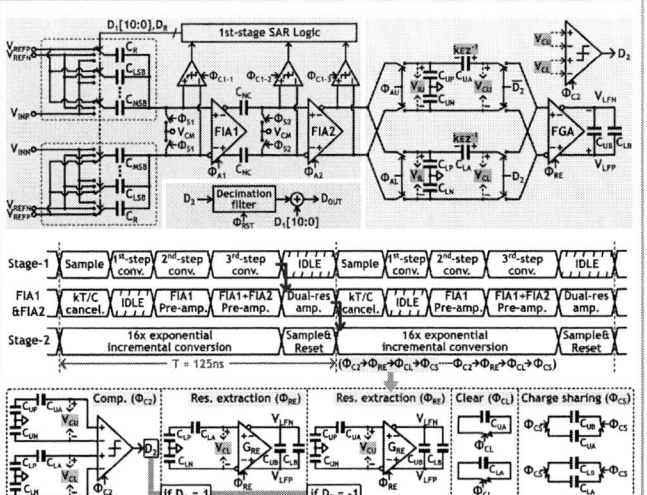

Figure 32.3.2: Simplified ADC schematic, timing diagram, and operating principle of the proposed exponential incremental quantizer.

ISSCC 2026 / SESSION 32 / LOW-POWER NOISE-SHAPING ADCs / 32.3

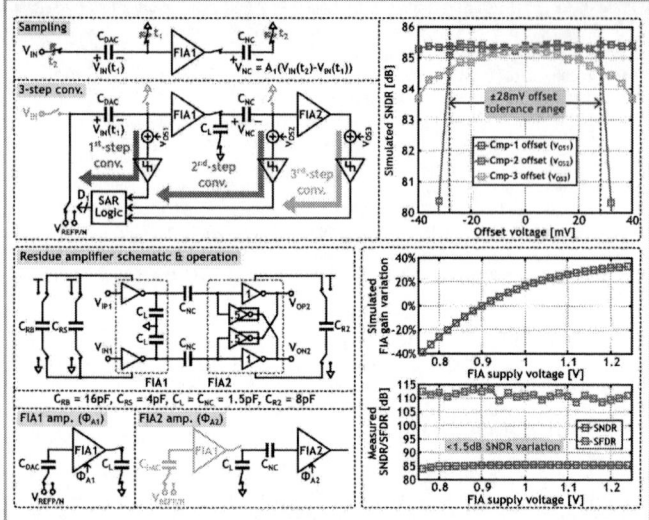

Figure 32.3.3: 3-step 1st-stage SAR conversion and fully dynamic residue amplifier operation.

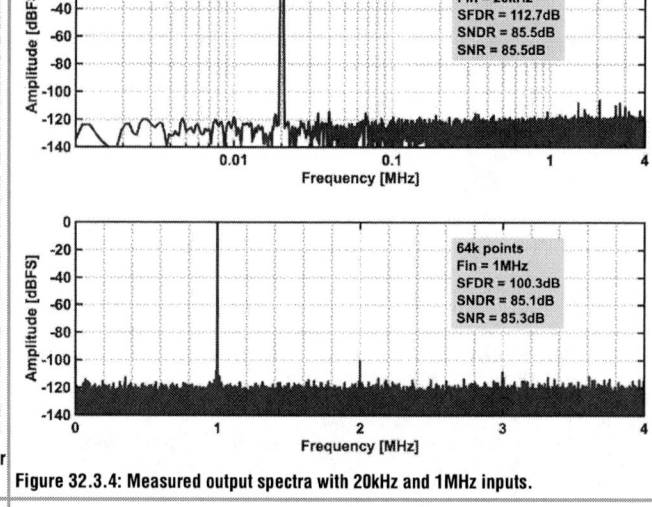

Figure 32.3.4: Measured output spectra with 20kHz and 1MHz inputs.

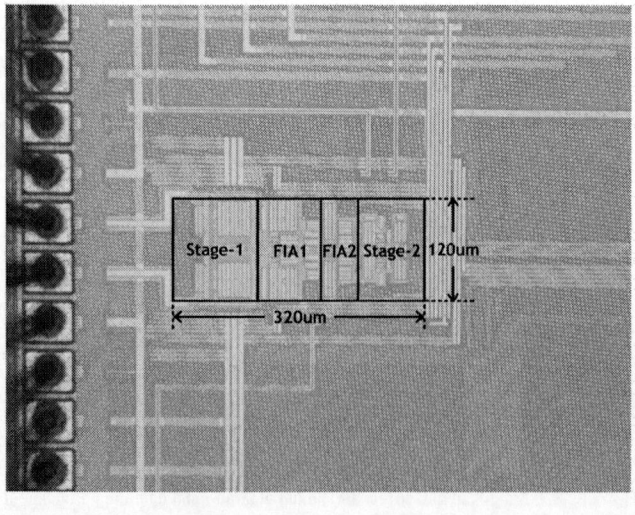

Figure 32.3.5: Measured dynamic range, power breakdown, and performance variations of 5 chips with analog supply voltage and temperature variations.

Figure 32.3.6: Performance summary and comparison with state-of-art ADCs.

	ISSCC-22 Y. Liu	CICC-25 T.-H. Wang	ISSCC-25 Z. Wang	ISSCC-21 X. Tang	ISSCC-23 M. Li	JSSC-24 J.-H. Chung	This Work[1]
Architecture	Incremental Zoom (SAR+ΔΣ)	Incremental Two-step (SAR+NS-SAR)	Incremental Pipelined (SAR+NS-SAR)	Pipelined SAR	Pipelined SAR	NS Pipelined (SAR+Interpolator)	Incremental Pipelined (SAR+Exp. ΔΣ)
Process [nm]	55	65	28	40	180	180	22
Multiplexable	✓	✓	✓	✓	✗	✗	✓
Fully dynamic	✗	✗	✓	✓	✗	✗	✓
Gain error suppression	N/A	Gain error shaping	Calibration	Closed-loop amplifier	Closed-loop amplifier	Dual-residue	Dual-residue
Conversion speed [MS/s]	0.0027	0.1	1.6	40	12	3	8
OSR	128	8	10	1	1	8	1
Sample cap [pF]	4	<0.1	0.8	4	17	0.52	1.6
Supply [V]	1	1.2	0.9	1.2	1.8/3.3	1.8	0.9
Area [mm²]	0.23	0.167	0.034	0.056	0.6	0.36	0.038
Power [uW]	4.96	44.8	467.3	820	9430	1800	618.1
Input freq. [MHz]	Quasi-static	0.047	0.2	18	4.04	0.5	1
SNDR [dB]	93	91	91.8	75.7	91.3	81.2	85.1
FoMw² [fJ/c-s]	50.2	15.4	9.2	4.1	26.2	63.9	5.3
FoMs³ [dB]	177.3	181.5	184.1	179.6	179.3	170.4	183.2

[1] One-time off-chip foreground calibration for capacitor mismatches
[2] $FoM_W = Power/(2^{ENOB} \cdot 2 \cdot BW)$
[3] $FoM_S = SNDR + 10 \cdot \log_{10}(BW/Power)$

Figure 32.3.7: Chip micrograph.

• 2026 IEEE International Solid-State Circuits Conference

ISSCC 2026 / SESSION 32 / LOW-POWER NOISE-SHAPING ADCs / 32.4

32.4 A 103.9dB-SFDR 83.8dB-SNDR 3MHz-BW Multi-Bit Quadratic-Exponential Noise-Coupled IDSM with High Tolerance to DAC Non-Linearity

Zhensheng Li[1,2], Biao Wang[3], Mingqiang Guo[1], Rui P. Martins[1], Sai-Weng Sin[1]

[1]University of Macau, Macau, China, [2]UM Hetao IC Research Institute, Shenzhen, China, [3]Formula Microelectronics, Shanghai, China

Abstract

This paper presents a quadratic-exponential noise-coupled (NC) IDSM to achieve a quantization noise shaping effect greater than 4th order with OSR 22, while having high tolerance to DAC non-linearity and a small noise penalty factor. The DAC2 and adder in NC are reused for a larger effective OSR and improve energy efficiency. Implemented in 28nm CMOS, the IDSM achieves an SNDR/SFDR of 83.8dB/103.9dB, with a 3MHz bandwidth, and 794µW, resulting in Schreier FoM$_{SNDR}$/FoM$_{DR}$ of 179.6dB/181.4dB.

Incremental ADCs are increasingly attracting interest for their Nyquist-like single-shot conversion, easy multiplexing, and simple decimation filter, making them ideal for high-resolution IoT devices, sensor interfaces, and instrumentation [1-4]. To boost resolution and bandwidth with high efficiency, high-order loop filters and multi-bit quantizers are frequently utilized in IDSMs. However, the non-uniform weighting of high-order IDSMs exacerbates multi-bit DAC mismatch, particularly at lower oversampling rates, as the "effective OSR" for averaging element mismatch errors decreases. Simultaneously, the effectiveness of the simple first-order DWA also degrades as the order of IDSM increases. In [5], a calibration-free 3-0 Incremental SMASH achieving 86dB SNDR with a 60× OSR was implemented by utilizing different weights of multi-bit and low-bit DACs. The output swing of the integrator was increased during low-bit DAC resolution. In [6] and [7], a first-order-then-exponential (linear-exp) algorithm was introduced, achieving the SNDR greater than 100dB with an OSR of 256. The exponential IDSM (Fig. 32.4.1 top) enables rapid SQNR growth, while the combination of linear and exponential cycles effectively mitigates the impact of non-uniform weighting. While the DAC1 mismatch can be mitigated by 246 1st-order cycles averaging, DAC2 will only participate in the noise-coupling operation in the last 10 cycles of the exponential phase. As a result, the DAC2 mismatch averaging effect is weak even if it is in the backend, compromising the adder stage capacitance, area, and power consumption.

In this paper, a multi-bit quadratic-exponential noise-coupled (NC) IDSM is proposed, achieving the following advantages:

1) Combining the quadratic and exponential phase, with a quantization noise shaping effect greater than 4th-order IDSM, fast accumulation is achieved under small OSR, enabling wider bandwidth while maintaining a small penalty factor of 1.64.

2) Reusing DAC2 in all the cycles with a maximized effective OSR to mitigate DAC mismatch, thereby enabling the adoption of a smaller DAC unit to reduce the power consumption of the noise coupling network and its adder.

3) Reusing the adder with closer feedback factors between the two phases to improve its energy efficiency. The reuse of the adder increases the linear phase to a quadratic phase, thus simultaneously reducing the number of cycles and increasing bandwidth.

4) The multi-path input feedforward enables noise and signal separation, allowing most of the switched-capacitor circuit to process only quantization noise with a low internal swing, thus relaxing linearity requirements.

Figure 32.4.1 illustrates a comparison between the linear-exp IDSM and the quadratic-exponential NC IDSM. In the linear-exp IDSM, DAC1 receives strong DWA averaging due to the large OSR provided by the 1st-order phase. However, for DAC2, the noise coupling path is only activated during the exponential phase, with a small effective OSR of 3 [6]. So, DAC2 cannot be too small, which increases the power consumption of the adder. Another problem in linear-exp IDSM is the utilization effectiveness of the adder. The linear-exp IDSM employs an active adder with feedback factors of 0.67 and 0.17 during the linear and exponential phases – a fourfold difference. The lower feedback factor imposes a higher GBW demand on the amplifier. The difference in the feedback factors also makes stability (and phase margin) consideration very different in the two phases. Consequently, to meet the settling requirements of both phases, the GBW of the adder needs to be significantly enlarged, which represents an over-design for that in the linear phase, resulting in wasted energy.

While in the quadratic-exponential NC IDSM, we inject the noise-coupled DAC2 for almost the entire cycle of conversion. The NC IDSM employs noise coupling to construct the NTFs of $(1 - z^{-1})\varepsilon_q$ and $(1 - (1 + k_e)z^{-1})\varepsilon_q$, achieving quadratic and exponential phase, respectively. It achieves an OSR of 20, with an effective OSR for DAC2 equal to 15/1.3 + 3 = 14.5, where 1.3 represents the penalty factor of a quadratic IDSM. The 5× effective OSR contributes to better suppression of DAC2 mismatch and enhanced DWA effectiveness. As a result, the whole adder and noise-coupling network can be scaled down for better power efficiency. The NC IDSM efficiently reuses the adder with feedback factors of 0.33 for the quadratic phase and 0.22 in the exponential phase, differing by 1.5× only, thereby ensuring optimal energy utilization for the whole conversion cycle. Although the lower overall feedback factor slows down the settling, the usage of quadratic accumulation speeds up the conversion cycles, and the IDSM bandwidth is improved.

The block diagram of the quadratic-exponential NC IDSM is presented in Fig. 32.4.2. Initially, during residual extraction preparation, STF is set to $\frac{1}{1-z^{-1}}$. The process then moves to the quadratic phase where STF changes to $(\frac{1}{1-z^{-1}})^2$. Finally, in the exponential phase, STF transforms to $\frac{1}{1-z^{-1}} \frac{1}{1-2z^{-1}}$, enabling exponential accumulation of samples. The Monte Carlo simulations with 0.2% DAC (including DAC2) element mismatch at a conversion cycle of 22 are shown in Fig. 32.4.2 top. The linear-exp IDSM with 10-exp cycles exhibits a mean SNDR of 90.8dB, while the NC IDSM achieves a mean SNDR of 99.5dB, representing an 8.7dB improvement. The NC IDSM maintains an SNDR of greater than 90dB with DAC element mismatch varying from 0 to 1%, demonstrating high tolerance to DAC mismatch. The comparison also shows that the 2nd-order, 3rd-order, and linear-exp with 5-exp cycles IDSMs are limited by low SQNR, while the 4th-order, exponential, and linear-exp with 10-exp cycles IDSMs are limited by DAC2 when DAC element mismatch exceeds 0.1%.

Figure 32.4.3 presents the circuit and timing diagrams. The two capacitors, C_e and C_o, operate in even and odd phases in a ping-pong manner: one captures Vy from the current cycle's adder output while the other feeds back the prior cycle's value to the adder input, enabling continuous noise coupling in each cycle. Utilizing the high mismatch tolerance of the NC IDSM, the DAC2 unit is implemented with 6fF capacitance to reduce the power consumption of the noise coupling network effectively. DAC2, an active adder, and ping-pong capacitors are reused in both the quadratic and exponential phases to get benefits from the averaging effect of the longer quadratic phase. The transition between these two phases is achieved by switching the capacitors at the output of the first-stage integrator and the feedback capacitor of the active adder (red switches in Fig. 32.4.3). These capacitors are connected in the quadratic phase and disconnected in the exponential phase. Signal and quantization noise separation is achieved in this work through two input feedforward paths. One path connects to the CDAC of the SAR and is active throughout the entire conversion cycle. The passive-adder embedded 4b SAR ADC performs passive summation and sampling at Φ_1, and comparison with reference voltage scaling at Φ_2. The other feedforward path is connected to the input of the active adder and remains active for 20 cycles (including both quadratic and exponential phases) when the noise coupling is enabled. In comparison to the implementation in [6], the signal components of the noise coupling adder and the first-stage integrator are both eliminated at the output of DACs in this work, meaning that only quantization noise needs to be processed in both phases for the active noise-coupling adder. This relaxes the linearity requirements of the circuit with a small internal swing. Taking advantage of the low internal swing, we adopted an energy-efficient two-stage FIA. The first stage is an inverter with reset bias voltages reduced to VDD–V$_{rst1}$ and V$_{rst2}$. The second stage is an adaptively biased cascode FIA with a low output swing, designed to enhance the amplifier gain. The NC IDSM achieves a thermal noise penalty factor of 1.64 and employs sampling in every cycle to maximize the effectiveness of OSR in mitigating sampled thermal noise. Consequently, we implemented a 1 pF sampling capacitor. Compared to [6], the increased penalty of 1.64 is moderate for DEM and noise, while the quadratic accumulation allows higher BW in this work. Dynamic range scaling is also applied in Fig. 32.4.3. The quantizer has a full-scale range of FS/3, while DAC1 and DAC2 have a full-scale range of FS, resulting in a 3× gain in the feedback DAC. A gain of 2/3 is applied through the passive adder. In the multi-path input feedforward, the feedforward factor is 0.5 for the second-order phase and 1 for the exponential phase. This approach enables the cancellation of the input signal for both phases.

For IDSMs, all Nyquist samples have equal weight. Therefore, any signal within the loop filter should not accumulate between two Nyquist conversions. However, in normal DWA, the pointer accumulates continuously, which implies that the pointer has a memory effect. It further worsens the non-uniform weighting feature. In this work, we align the pointer at the beginning of each conversion. With an aligned pointer, we can regard the DAC as having same pattern for two adjacent conversions. The pointer-aligned DWA is shown in Fig. 32.4.3. When the accumulated loop is reset, the pointer is aligned to the first DAC unit. In the following cycles, the pointer is updated based on the quantizer output. By implementing the pointer-aligned DWA, the memory effect of the pointer between two adjacent conversions is eliminated.

The prototype is fabricated in 28nm CMOS with an area of 0.049mm² (Fig. 32.4.7). The IDSM is clocked at 132MHz with an oversampling rate of 22, leading to a 3MHz bandwidth. Figure 32.4.4 shows that the measured SNDR/SFDR is 83.8dB/103.9dB, while the

ISSCC 2026 / February 18, 2026 / 2:45 PM

performance is limited to 64.4dB/88.5dB without exponential phase. The SNDR/SNR increases by 2 to 3dB per cycle within the quadratic phase. At the beginning of the exponential phase, there is a 6dB increase in each cycle. In the last two cycles, the upper limit of SNDR/SNR is reached due to thermal noise. The measured SNDR/SFDR sweep vs. fin is also shown. As shown in Fig. 32.4.5, the measured dynamic range is 85.6dB. The measured DNL and INL stay limited to +0.56/-0.53 LSB and +0.83/-0.71 LSB (14b level), respectively. Across five chips, the SNDR variation is less than 1dB. The IDSM also shows good stability under ±5% supply variation and -40°C~80°C temperature range.

Figure 32.4.6 summarizes the performance and compares with state-of-the-art high-SNDR MHz-band ADCs. It demonstrates that our work is in line with the state-of-the-art. Thanks to the proposed quadratic-exponential NC IDSM, this work achieves Schreier FoM$_{SNDR}$/FoM$_{DR}$ of 179.6dB/181.4dB and a high SFDR of 103.9dB.

Acknowledgement:
This work was supported by The Science and Technology Development Fund, Macao (File no. 0004/2023/AKP,0207/2024/AGJ,004/2023/SKL), Hetao SZ-HK S&T Innovation Cooperation Zone Project (HTHZQSWS-KCCYB-2023030).

References:
[1] Z. Wang et al., "A 150kHz-BW 15-ENOB Incremental Zoom ADC with Skipped Sampling and Single Buffer Embedded Noise-Shaping SAR Quantizer," 2023 IEEE International Solid-State Circuits Conference (ISSCC), San Francisco, CA, USA, 2023, pp. 9-11. http://doi.org/10.1109/ISSCC42615.2023.10067696
[2] T. -H. Wang et al., "A 50-kHz BW 92.1-dB SNDR Incremental ADC Using a Back-End Sampling Two-Step NS-SAR Architecture with Concurrent Gain-Error + Noise Suppression," 2025 IEEE Custom Integrated Circuits Conference (CICC), Boston, MA, USA, 2025, pp. 1-3. http://doi.org/10.1109/CICC63670.2025.10983053
[3] L. Jie, M. Zhan, X. Tang and N. Sun, "A 0.014mm2 10kHz-BW Zoom-Incremental-Counting ADC Achieving 103dB SNDR and 100dB Full-Scale CMRR," 2022 IEEE International Solid-State Circuits Conference (ISSCC), San Francisco, CA, USA, 2022, pp. 1-3. http://doi.org/10.1109/ISSCC42614.2022.9731742
[4] Y. Zhao, Y. Ye, S. Yuan and Y. Qin, "A 133.6-µW 1kHz-BW Multi-Bit 2nd-Order Incremental ADC Achieving 115.4-dB SNDR with Low-Cost Coarse-Sorting DEM and Zip-Extended-Counting," 2025 IEEE Custom Integrated Circuits Conference (CICC), Boston, MA, USA, 2025, pp. 1-3. http://doi.org/10.1109/CICC63670.2025.10982735
[5] M. A. Mokhtar, A. Abdelaal, M. Sporer, J. Becker, J. G. Kauffman and M. Ortmanns, "A 0.9-V DAC-Calibration-Free Continuous-Time Incremental Delta–Sigma Modulator Achieving 97-dB SFDR at 2 MS/s in 28-nm CMOS," in IEEE Journal of Solid-State Circuits, vol. 57, no. 11, pp. 3407-3417, Nov. 2022. http://doi.org/10.1109/JSSC.2022.3160325
[6] B. Wang, S. -W. Sin, U. Seng-Pan, F. Maloberti and R. P. Martins, "A 550µW 20kHz BW 100.8-dB SNDR Linear-Exponential Multi-Bit Incremental SD ADC with 256 Clock Cycles in 65-nm CMOS," in IEEE Journal of Solid-State Circuits, vol. 54, no. 4, pp. 1161-1172, Apr. 2019. http://doi.org/10.1109/JSSC.2018.2888872
[7] B. Wang, S. -W. Sin, S. -P. U., F. Maloberti and R. P. Martins, "A 1.2V 86dB SNDR 500kHz BW Linear-Exponential Multi-Bit Incremental ADC Using Positive Feedback in 65nm CMOS," 2019 IEEE Asian Solid-State Circuits Conference (A-SSCC), Macau, Macao, 2019, pp. 117-120. http://doi.org/10.1109/A-SSCC47793.2019.9056948

[8] S. Song, T. Kang, A. Knowlton, S. Lee and M. P. Flynn, "An NS-SAR Quantizer-Based Pipeline Incremental Delta-Sigma ADC Using a Current-Regulated Floating Ring Amplifier and Two-Phase Miller Negative-C, " 2025 IEEE Symposium on VLSI Technology and Circuits (VLSI Technology and Circuits), Kyoto, Japan, 2025, pp. 1-2. http://doi.org/10.23919/VLSITechnologyandCir65189.2025.11075190
[9] M. Fukazawa and T. Matsui, "A 24-OSR to Simplify Anti-Aliasing Filter 2MHz-BW 83dB-DR 3rd-order DT-DSM using FIA-Based Integrator and Noise-Shaping SAR Combined Digital Noise-Coupling Quantizer," 2023 IEEE Symposium on VLSI Technology and Circuits (VLSI Technology and Circuits), Kyoto, Japan, 2023, pp. 1-2. http://doi.org/10.23919/VLSITechnologyandCir57934.2023.10185310
[10] Y. Guo, J. Jin, X. Liu and J. Zhou, "A 60-MS/s 5-MHz BW Noise-Shaping SAR ADC With Integrated Input Buffer Achieving 84.2-dB SNDR and 97.3-dB SFDR Using Dynamic Level-Shifting and ISI-Error Correction," in IEEE Journal of Solid-State Circuits, vol. 58, no. 2, pp. 474-485, Feb. 2023. http://doi.org/10.1109/JSSC.2022.3185501
[11] S. Ye et al., "A Rail-to-Rail 3rd-Order Noise-Shaping SAR ADC Achieving 105.4dB SFDR with Integrated Input Buffer Using Continuous-Time Correlated Level Shifting," 2025 IEEE International Solid-State Circuits Conference (ISSCC), San Francisco, CA, USA, 2025, pp. 314-315, Feb. 2025. http://doi.org/10.1109/ISSCC49661.2025.10904773
[12] C. -C. Liu and M. -C. Huang, "A 0.46mW 5MHz-BW 79.7dB-SNDR noise-shaping SAR ADC with dynamic-amplifier-based FIR-IIR filter," 2017 IEEE International Solid-State Circuits Conference (ISSCC), San Francisco, CA, USA, 2017, pp. 466-467. Feb. 2017. http://doi.org/10.1109/ISSCC.2017.7870463

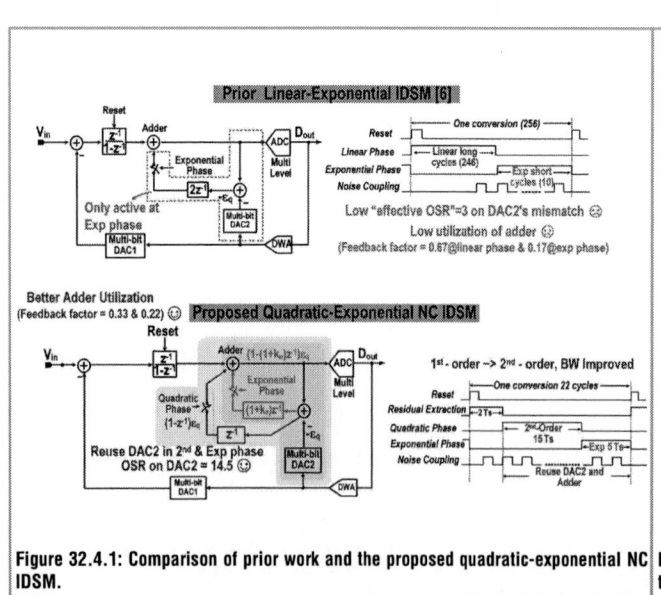

Figure 32.4.1: Comparison of prior work and the proposed quadratic-exponential NC IDSM.

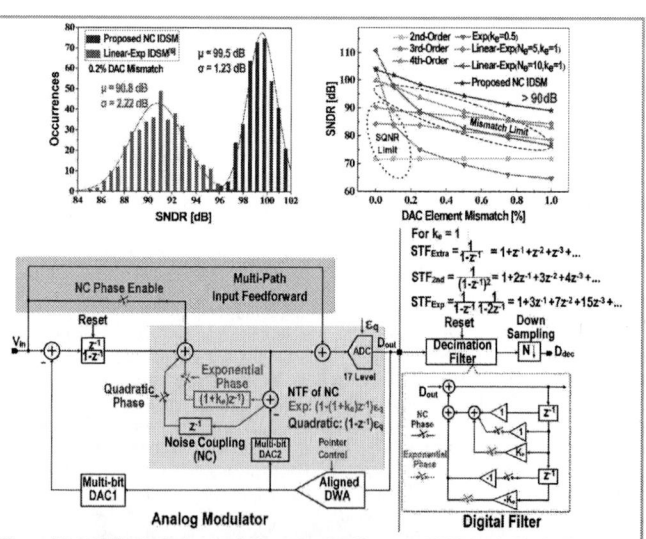

Figure 32.4.2: Monte Carlo simulation adopting the same DWA (top), block diagram of the NC IDSM (bottom).

DIGEST OF TECHNICAL PAPERS • 559

ISSCC 2026 / SESSION 32 / LOW-POWER NOISE-SHAPING ADCs / 32.4

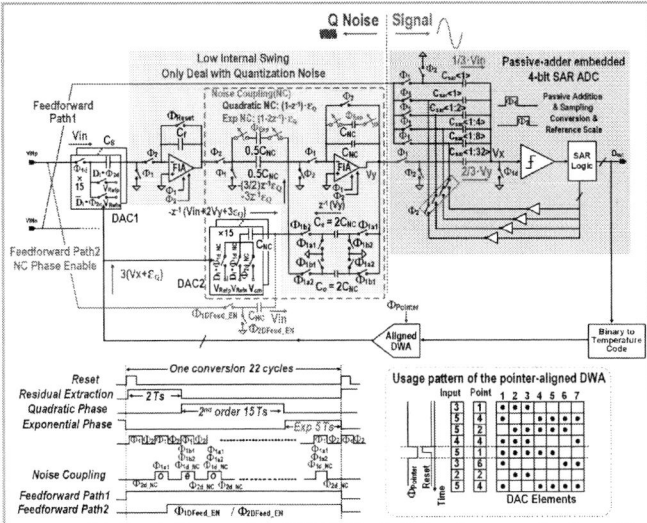

Figure 32.4.3: Circuit diagram of the NC IDSM (top), timing diagram and pattern of the aligned DWA (bottom).

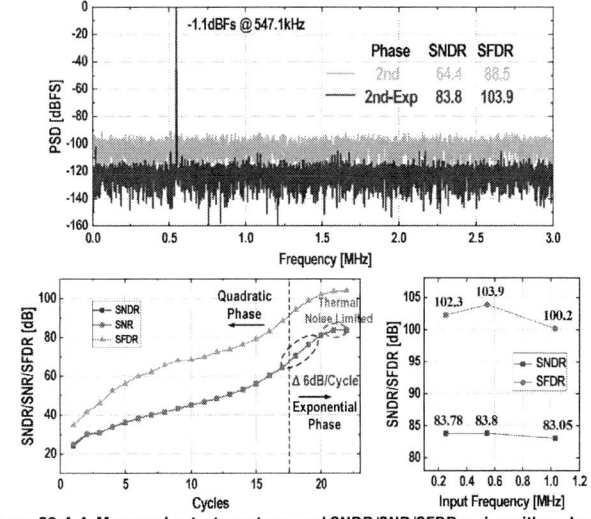

Figure 32.4.4: Measured output spectrum, and SNDR/SNR/SFDR varies with cycles and input frequency.

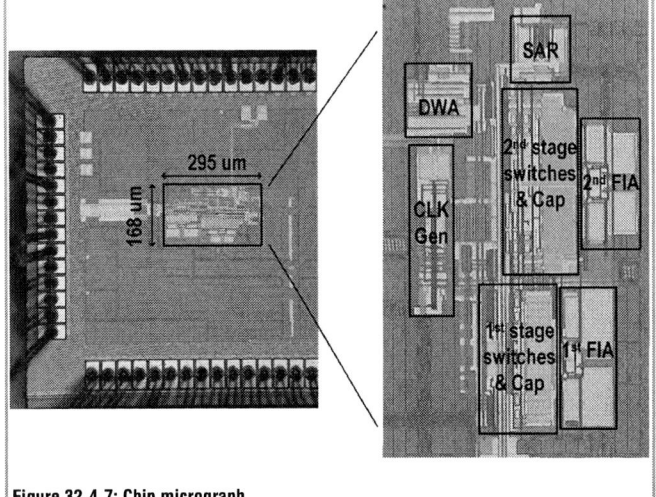

Figure 32.4.5: Measured dynamic range, DNL/INL, SNDR/SFDR varies with multiple chips, temperature and supply.

	This Work	VLSI-25 S. Song [8]	VLSI-23 M. Fukazawa [9]	ISSCC-25 S. Ye [11]	ISSCC-17 C.-C. Liu [12]	JSSC-23 Y. Guo [10]
Architecture	Quadratic-Exp NC IDSM	Pipeline IDSM	DSM	NS-SAR	NS-SAR	NS-SAR
NS Order	> 4	4	3	3	1	2
Process [nm]	28	28	22	55	28	40
Area [mm^2]	0.049	0.052	0.025	0.043	0.0049	0.36
Supply [V]	1	1.05	1	1.2	1	1.1/2.5
F$_S$ [MHz]	132	80	96	40	132	60
BW [MHz]	3	5	2	2.5	5	5
SNDR$_{Peak}$ [dB]	83.8	80.1	80.1	82.9	79.74	84.2
SFDR [dB]	103.9	97.2	\	105.4	92.6	97.3
DR [dB]	85.6	81.1	83.1	83.2	81.8	86.4
Power [μW]	794	1850	1040	470/1150*	460	1840/8060*
FoM$_{S, SNDR}$[1] [dB]	179.6	174.4	172.9	180.2/176.3*	180.1	178.6/172.1*
FoM$_{S,DR}$[2] [dB]	181.4	175.4	175.9	180.5/176.6*	182.16	180.8/174.3*

*With input buffer

[1]FOM$_{S,SNDR}$ = SNDR+10·log10(BW/Power)

[2]FOM$_{S,DR}$ = DR+10·log10(BW/Power)

Figure 32.4.6: Performance summary and comparison with state-of-art ADCs.

Figure 32.4.7: Chip micrograph.

ISSCC 2026 / SESSION 33 / TIME-VARYING CIRCUIT TECHNIQUES FROM RF TO mm-WAVE / OVERVIEW

Session 33 Overview: *Time-Varying Circuit Techniques from RF to mm-Wave*

RF SUBCOMMITTEE

Session Chair: Jeremy Dunworth
Qualcomm Technologies Inc.
San Diego, CA

Session Co-Chair: Wooyeol Choi
Seoul National University
Seoul, Korea

Switch-based linear time-varying circuits are employed to implement highly integrated digital transmitters, non-reciprocal phase shifters for non-magnetic isolators and circulators, and a tunable-bandwidth delay line. With the continued improvement of CMOS switch performance, time-varying techniques are being applied across a wide frequency range from sub-1GHz to mm-wave. The session features a 22-to-25GHz balanced circulator, an 11.1GHz power amplifier with an integrated isolator, and three switched capacitor power amplifiers and a delay line operating in the sub-6GHz range. These papers demonstrate the versatility of linear time-varying circuits in enabling robust, high-performance, high-frequency microsystems.

ISSCC 2026 / February 18, 2026 / 3:35 PM

3:35 PM

33.1   A 22-to-25GHz CMOS Non-Magnetic Balanced Circulator Achieving at Least 20dB TX-RX Isolation for an Antenna VSWR of 2

Heqi Deng, TU Delft, Delft, The Netherlands

In Paper 33.1, the Delft University of Technology presents a 22-to-25GHz non-magnetic balanced circulator achieving more than 20dB TX-RX isolation for an antenna VSWR of 2. The circulator exhibits less than 4.8dB loss while in a 1.42mm^2 area. It achieves a competitive noise figure of 4-to-6dB, an IIP3 of 23dBm, over 30dB isolation across a 1GHz bandwidth, and over 20dB isolation across a 4GHz bandwidth.

4:00 PM

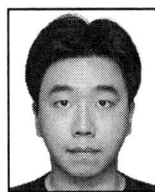

33.2   An Infinite-Loop CMOS-Compatible Isolator Enabled True VSWR-Resilient Power Amplifier for 6G FR3 in Massive MIMO and Phased-Array Systems

Mohsen Ghorbanpoor, CSEM, Neuchatel, Switzerland, ETH Zurich, Zurich, Switzerland

In Paper 33.2, CSEM and ETH Zurich demonstrate an 11.1GHz balanced power amplifier incorporating a CMOS-compatible isolator for 6G FR3. The "infinite-loop" isolator, composed of two quadrature couplers and non-reciprocal phase shifters (NRPS), breaks the tradeoff between linearity and insertion loss found in conventional NRPS designs. The PA exhibits less than ±0.75dB OP$_{1dB}$ variation over 3:1 and 2:1 VSWR, delivering 23.9dBm saturated output power at 50Ω.

4:25 PM

33.3   A 6GHz Quadrature Digital Transmitter Supporting a 1GHz Signal Bandwidth with <-40dB EVM Floor and >55dB Dynamic Range in 28nm CMOS

Feng Xie, Fudan University, Shanghai, China

In Paper 33.3, Fudan University showcases a 6GHz digital transmitter supporting a 1GHz signal bandwidth and delivering 23.1dBm peak output power in a 0.67mm^2 core area. The design achieves an EVM floor smaller than -40dB and a dynamic range greater than 55dB through careful mitigation of static and dynamic nonlinearities.

4:50 PM

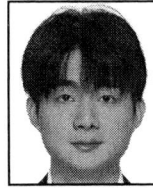

33.4   A 4.5-to-7.2GHz Beyond Rail-to-Rail Output SCPA with 27.9dBm P$_{out}$ and 46.2% DE at 5.1GHz Using Periodic Voltage-Pacing Network

Bingzheng Yang, Shenzhen University, Shenzhen, China, University of Electronic Science and Technology of China, Chengdu, China

In Paper 33.4, Shenzhen University and the University of Electronic Science and Technology of China demonstrate a switched capacitor power amplifier achieving 27.9dBm P$_{out}$ and 46.2% DE at 5.1GHz, with a 4.5-to-7.2GHz P$_{out}$ 1dB bandwidth. A "periodic voltage pacing network" enables output swing greater than VDD and below GND to enhance output power and efficiency.

5:05 PM

33.5   A 0.6-to-0.9GHz, 28.06dBm P$_{out}$, 43.73% SE, 6-Phase Switched-Capacitor Power Amplifier Using In-Cell Digital Waveform Synthesis for the 2nd-, 3rd-, and 4th-Harmonic Suppression

Xiqing Liang, University of Electronic Science and Technology of China

In Paper 33.5, the University of Electronic Science and Technology of China and Shenzhen University present a switched-capacitor power amplifier achieving 28dBm peak output power and 43.7% peak system efficiency operates across 0.6 to 0.9GHz. Employing unit cells that digitally synthesize a multi-level stepped waveform, the design suppresses 2nd, 3rd, and 4th harmonics.

5:20 PM

33.6   A Frequency-Translated, 1-to-5GHz Centred, All-Passive, Programmable-Bandwidth, Switched-Capacitor Delay Line with 6.5-to-8.2dB Delay-Independent Insertion Loss in 65nm CMOS

Mohmad Aasif Bhat, Indian Institute of Technology, Kanpur, India

In Paper 33.6, the Indian Institute of Technology Kanpur describes a switched-capacitor delay line tunable over a 1-to-5GHz center frequency. This all-passive CMOS design offers tunable center frequency and bandwidth by employing a switch-mixer-based frequency translation topology. The delay line supports a group delay range of 2.5 to 17.5ns and achieves delay-independent insertion loss between 6.5 and 8.2dB.

33

DIGEST OF TECHNICAL PAPERS • 561

979-8-3315-8937-0/26 $31.00 © 2026 IEEE

ISSCC 2026 / SESSION 33 / TIME-VARYING CIRCUIT TECHNIQUES FROM RF TO MM-WAVE / 33.1

33.1 A 22-to-25GHz CMOS Non-Magnetic Balanced Circulator Achieving at Least 20dB TX-RX Isolation for an Antenna VSWR of 2

Heqi Deng, Ehsan Shokrolahzade, Niels Fakkel, Marco Spirito, Fabio Sebastiano, Masoud Babaie

TU Delft, Delft, The Netherlands

Abstract

This paper presents a mm-wave non-magnetic balanced circulator that bridges the gap between the high insertion loss (IL) of electrical balance duplexers and the transmitter (TX)-to-receiver (RX) isolation degradation of conventional circulators under varying antenna (ANT) voltage standing wave ratio (VSWR). Using quadrature couplers, a balance network, and (non)-reciprocal branches, it achieves <5.2dB TX-to-ANT IL, <4.2dB RX-to-ANT IL, and >20dB TX-to-RX isolation over a >2.2GHz bandwidth at VSWR=2.

Full-duplex (FD) operation allows simultaneous transmission and reception at the same frequency, thereby enhancing spectral efficiency in communication systems [1] and enabling real-time sensing in radar applications [2]. However, a key challenge in FD systems lies in designing low-cost, compact, CMOS-compatible integrated-antenna (ANT) interfaces that do not only provide low insertion loss (IL), low noise, and large transmit-power-handling capability, but also maintain high transmitter-to-receiver (TX-to-RX) isolation under varying antenna voltage-standing-wave-ratio (VSWR) conditions caused by real-world user interaction. Such isolation is crucial, as it relaxes the dynamic range requirements of the self-interference cancellation blocks in the RF, analog baseband, and digital domains.

One widely studied solution is the electrical-balance duplexer (EBD) [3-5], which employs hybrid transformers to cancel TX leakage at the RX through impedance balancing between the antenna (Z_{ANT}) and a tunable balance network (Z_{BN}), as depicted in Fig. 33.1.1(a). Moreover, owing to its fully passive architecture, the EBD ensures zero DC power consumption and high linearity. However, EBDs inherently suffer from a theoretical 3dB intrinsic IL for both TX-to-ANT transmission and ANT-to-RX reception, since half of the signal is dissipated in the balance network. To address this limitation, non-magnetic circulators employing linear periodically time-varying circuits—such as N-path filters [6] or switch-based spatio-temporal conductivity modulation [7-11]—have been introduced. As shown in Fig. 33.1.1(b), these circuits leverage non-reciprocity to achieve low-loss unidirectional signal routing from TX to ANT, and from ANT to RX. While non-magnetic circulators can, in theory, achieve 0dB intrinsic IL, their linearity is constrained by switch non-idealities, and they require DC power to drive the switches. More importantly, their TX–RX isolation degrades proportionally with the antenna reflection coefficient (Γ_{ANT}), a limitation not present in EBDs.

To bridge the gap between EBDs and conventional circulators, this paper introduces a non-magnetic balanced circulator that achieves an intrinsic IL of 1.25dB, while maintaining 20dB TX-RX isolation for antenna VSWR up to 2. Relative to EBDs, it reduces the intrinsic IL by 1.75dB, albeit with a narrower VSWR tolerance. Compared to conventional circulators, it incurs slightly higher IL but uniquely preserves high isolation under antenna mismatch. This is enabled by introducing a balanced circulator architecture that 1) employs a compact quadrature hybrid coupler (QHC) to implement the required reciprocal branches with reduced area and loss, 2) repurposes the QHC isolation port to host a tunable balance network, and 3) redirects a small portion of the TX power to the balance network, so that by tuning its impedance, TX-to-RX leakage is effectively cancelled under antenna reflections.

Figure 33.1.2 illustrates the evolution and operating principles of the proposed non-magnetic balanced circulator. The topmost architecture shows the preliminary schematic, consisting of two identical QHCs, and a non-reciprocal phase shifter. The input (IN) and isolation (ISO) ports of the first QHC are connected to the antenna and a termination impedance Z_{BN}, respectively, while the IN and ISO ports of the second QHC are connected to the TX and RX. The non-reciprocal phase shifter is inserted between the coupled (CPL) ports of the QHCs, introducing a direction-dependent phase response of 0° in the forward path and 180° in the reverse path. During transmission, the TX signal splits: one path propagates through the through (THR) ports of the QHCs, while the other passes via the coupled ports and the non-reciprocal phase shifter. At the antenna port, the two paths combine constructively. Similarly, the received antenna signal is split into two branches and then constructively recombined at the RX. Ideally, this architecture functions as a circulator with 0dB intrinsic IL. Compared with prior-art circulators [6-11], it achieves a more compact implementation since the long transmission lines are replaced by QHCs. However, it retains the same inherent drawback as conventional circulators: under antenna mismatch, a portion of the TX signal is reflected and leaks into the RX, which reduces TX-to-RX isolation to $20\log(|\Gamma_{ANT}|)$, independent of Z_{BN}.

To mitigate TX-to-RX leakage when $|\Gamma_{ANT}|>0$, a reciprocal branch with a phase shift θ is added between the THR ports of the QHCs to direct a fraction of the TX signal to the ISO port of QHC1, where the balance network is located, as shown in Fig. 33.1.2(c-f). Due to circuit symmetry, the ANT-to-RX and TX-to-ANT ILs are identical; here, only the TX-to-ANT loss is discussed to avoid repeating similar calculations, although Fig. 33.1.2(e) illustrates both cases. During transmission, the antenna receives $V_{ANT-TX}^{+} = 0.5V_{TX}^{+}(1+\cos\theta -j\sin\theta)$,

where V_{TX}^{+} is the forward TX voltage wave, increasing the TX-to-ANT intrinsic IL from 0dB to $20\log|\cos(\theta/2)|$ (e.g., 1.25dB for θ=60°). Simultaneously, the balance network (BN) receives $V_{BN-TX}^{+} = 0.5V_{TX}^{+}(1-\cos\theta-j\sin\theta)$, resulting in a TX-to-BN leakage of $20\log|\sin(\theta/2)|$ (e.g., -6dB for θ=60°).

When the antenna port is mismatched, reflected TX signals from both the ANT and BN ports pass through the QHCs and the reciprocal/non-reciprocal phase shifters and appear at the RX port. As proven in Fig. 33.1.2, if the balance network reflection coefficient (Γ_{BN}) is adjusted to $e^{j2\theta}\Gamma_{ANT}\cot^2(\theta/2)$, then perfect TX cancellation at the RX occurs. Since $|\Gamma_{BN}|$ must be below 1, the perfect cancellation is possible only when $|\Gamma_{ANT}| \leq \tan^2(\theta/2)$ or VSWR ≤ $1/|\cos(\theta)|$. The reciprocal phase shift θ controls the trade-off between intrinsic IL and compensation range: a larger θ directs more power to the balance network, enabling compensation over a wider antenna VSWR range, but increases intrinsic IL. In this design, for antenna VSWR variations up to 2:1 at the ANT port, θ=60° is selected, resulting in an intrinsic loss of 1.25dB. On the other hand, as shown in Fig. 33.1.2(f), the input return loss of the proposed balanced circulator is slightly improved, from Γ_{RX} to $0.5\sin\theta(\Gamma_{TX}-\Gamma_{RX})$. While designing an automatic calibration loop to optimize Z_{BN} under unknown Z_{ANT} is beyond the scope of this work, two approaches are outlined. A conventional method tunes Z_{BN} by minimizing TX leakage at the RX port [12], achieving isolation iteratively but requiring multiple steps. Alternatively, Γ_{ANT} can be estimated using a joint true power detector and impedance sensor [13], and Z_{BN} is adjusted accordingly via $\Gamma_{BN}=e^{j2\theta}\Gamma_{ANT}\cot^2(\theta/2)$.

Figure 33.1.3 shows the schematic of the non-magnetic balanced circulator, designed to operate differentially at f_{in}=24GHz. The non-reciprocal phase shifter is based on the spatio-temporal conductance modulation concept [7-11], comprising a passive network providing a delay of $T_d=1/(4f_m)$, and two sets of I/Q differential passive mixers driven at the modulation frequency $f_m=f_{in}/(2n-1)$, where n is an integer. In this work, $f_m=f_{in}/7$ is chosen to significantly reduce the LO power consumption; however, this requires a larger T_d, which in turn demands higher inductance and area. To realize this, a 5-section transformer-based hybrid lattice–ladder LC network is employed. This network not only absorbs the parasitic capacitance of the passive mixers but also achieves the required delay with only half the number of LC sections compared to an artificial transmission line [14], thanks to its high Bragg frequency, thereby reducing overall loss.

The NMOS devices in the balanced Gilbert quads are sized 32×1.6µm/40nm, with each placed in a separate N-well so that their bodies are tied to GND through a 5.4kΩ resistor. This makes the bulk floating at high frequencies, improving switch linearity [15]. The required reciprocal 60° phase shift is implemented using a transformer-capacitor network.

The QHC has four differential ports and is realized by folding two identical single-ended transformer-based QHCs, thereby reducing area and providing positive magnetic coupling between the transformer coils, which lowers QHC IL [16]. The coils are implemented using the top two metal layers, which inherently provide capacitive coupling, thereby eliminating the need for explicit capacitors and further reducing loss. Electromagnetic (EM) simulations show a passive loss of 0.34dB, port-to-port isolation of >29dB, and magnitude and phase imbalance below 0.2dB and 0.6°, respectively, across 22 to 30GHz.

The balance network is implemented differentially, consisting of one transformer, two 4-bit differential tunable capacitors, and one 8-bit differential tunable resistor, enabling coverage of all points within a $|\Gamma_{BN}|$<0.6 circle on the Smith chart. As a result, ideal TX-to-RX cancellation is achieved for $|\Gamma_{ANT}|$<0.2, while isolation gradually decreases for larger $|\Gamma_{ANT}|$. Nevertheless, 20-to-30dB isolation is still maintained even at $|\Gamma_{ANT}|$=0.33.

The balanced circulator, fabricated in 40nm CMOS, occupies a core area of 0.63mm² (Fig. 33.1.7) and consumes 11.6mW from a 1.1V supply. All measurements are performed after de-embedding the on-chip baluns using test structures. In Fig. 33.1.4, TX-to-ANT IL, ANT-to-RX IL, and TX-to-RX isolation were measured with Z_{ANT}=50Ω while Z_{BN} was swept across 256 settings. As predicted, Z_{BN} negligibly affects TX-to-ANT and ANT-to-RX ILs, impacting only the TX-to-RX isolation depth and bandwidth (BW). TX-to-RX isolation exceeds 30dB (20dB) over 1GHz (4GHz) BW from 22.7 to 23.7GHz (21 to 25GHz). At the center of the TX-to-RX isolation notch, TX-to-ANT and ANT-to-RX ILs are 4.8dB and 4.2dB, respectively.

562 • 2026 IEEE International Solid-State Circuits Conference

979-8-3315-8937-0/26 $31.00 © 2026 IEEE

The return loss is <-15dB for all ports over 20 to 26GHz. The ANT-to-RX noise figure aligns with the IL (4 to 6dB) over 22 to 26GHz, showing negligible degradation from the LO phase noise. TX–ANT and ANT–RX IIP3 are ~+24dBm and ~+22dBm, with OP1dB exceeding +8.5dBm and +9.5dBm, limited by the measurement setup. The TX–RX isolation is >30dB for the TX power up to +7.4dBm. In-band LO leakage at the ANT port due to the 6th and 7th LO harmonics is –83 dBm and –72 dBm, respectively.

The experiment was repeated under different angles of $|\Gamma_{ANT}|$ of 0.2 (VSWR=1.5) and 0.33 (VSWR=2), with results shown in Fig. 33.1.5. Even at VSWR=2, tuning Z_{BN} restores >20dB isolation over 22.7 to 23.7GHz, while the ANT-to-RX and TX-to-ANT ILs remain 4.2dB and 4.8dB at the center of the isolation notch. The TX–RX isolation stays above 20dB over more than 3GHz of bandwidth for different $\angle\Gamma_{ANT}$. As shown in Fig. 33.1.6, compared to mm-wave circulators [7,8,11], this work—despite being implemented in bulk CMOS, which offers lower-quality switches and passives than SOI—incurs slightly higher ILs but maintains TX–RX isolation even under large VSWR, thus bridging the performance gap between conventional EBDs and non-magnetic circulators.

Acknowledgment:
EU grant HORIZON-EIC-2022-PATHFINDEROPEN-01-101099697, QUADRATURE, supported this work.

References:
[1] Z. Zhang et al., "Full-Duplex Wireless Communications: Challenges, Solutions, and Future Research Directions," *Proc. of the IEEE*, vol. 104, no. 7, pp. 1369-1409, July 2016. https://doi.org/10.1109/JPROC.2015.2497203
[2] S. A. Hassani et al., "In-Band Full-Duplex Radar-Communication System," *IEEE Systems Journal*, vol. 15, no. 1, pp. 1086-1097, Mar. 2021. https://doi.org/10.1109/JSYST.2020.2992689
[3] B. van Liempd et al., "A +70dBm IIP3 single-ended electrical-balance duplexer in 0.18µm SOI CMOS," *ISSCC*, pp. 32-33, Feb. 2015. http://doi.org/10.1109/ISSCC.2015.7062851
[4] B. Hershberg et al., "A dual-frequency 0.7-to-1GHz balance network for electrical balance duplexers," *ISSCC*, pp. 356-357, Feb. 2016. http://doi.org/10.1109/ISSCC.2016.7418054.
[5] M. Elkholy et al., "Low-Loss Integrated Passive CMOS Electrical Balance Duplexers with Single-Ended LNA," *IEEE TMTT*, vol. 64, no. 5, pp. 1544-1559, May 2016. http://doi.org/10.1109/TMTT.2016.2541118
[6] J. Zhou et al., "Receiver with Integrated Magnetic-Free N-Path-Filter-Based Non-Reciprocal Circulator and Baseband Self-Interference Cancellation for Full-Duplex Wireless," *ISSCC*, pp. 178–179, Feb. 2016. http://doi.org/10.1109/ISSCC.2016.7417965
[7] T. Dinc and H. Krishnaswamy, "A 28GHz Magnetic-Free Non-Reciprocal Passive CMOS Circulator Based on Spatio-Temporal Conductance Modulation," *ISSCC*, pp. 294–295, Feb. 2017. http://doi.org/10.1109/ISSCC.2017.7870377.
[8] T. Dinc et al., "A Millimeter-Wave Non-Magnetic Passive SOI CMOS Circulator Based on Spatio-Temporal Conductivity Modulation," *IEEE JSSC*, vol. 52, no. 12, pp. 3276-3292, Dec. 2017. http://doi.org/10.1109/JSSC.2017.2759422
[9] A. Ruffino et al., "A Wideband Low-Power Cryogenic CMOS Circulator for Quantum Applications," *IEEE JSSC*, vol. 55, no. 5, pp. 1224-1238, May 2020. http://doi.org/10.1109/JSSC.2020.2978020

[10] A. Nagulu and H. Krishnaswamy, "Non-Magnetic CMOS Switched-Transmission-Line Circulators with High Power Handling and Antenna Balancing: Theory and Implementation," *IEEE JSSC*, vol. 54, no. 5, pp. 1288-1303, May 2019. http://doi.org/10.1109/JSSC.2019.2905146
[11] A. Nagulu and H. Krishnaswamy, "Non-Magnetic 60GHz SOI CMOS Circulator Based on Loss/Dispersion-Engineered Switched Bandpass Filters," *ISSCC*, pp. 452–453, Feb. 2019. http://doi.org/10.1109/ISSCC.2019.8662467
[12] S. H. Abdelhalem et al., "Tunable CMOS Integrated Duplexer with Antenna Impedance Tracking and High Isolation in the Transmit and Receive Bands," *IEEE TMTT*, vol. 62, no. 9, pp. 2092-2104, Sep. 2014. https://doi.org/10.1109/TMTT.2014.2338271
[13] D. J. Munzer et al., "A Broadband Mm-Wave VSWR-Resilient Joint True-Power Detector and Impedance Sensor Supporting Single-Ended Antenna Interfaces," *ISSCC*, pp. 326-327, Feb. 2022. https://doi.org/10.1109/ISSCC42614.2022.9731769
[14] M. Koochakzadeh and A. Abbaspour-Tamijani, "Miniaturized Transmission Lines Based on Hybrid Lattice-Ladder Topology," *IEEE TMTT*, vol. 58, no. 4, pp. 949-955, Apr. 2010. http://doi.org/10.1109/TMTT.2010.2042847
[15] Q. Li and Y. P. Zhang, "CMOS T/R Switch Design: Towards Ultra-Wideband and Higher Frequency," *IEEE JSSC*, vol. 42, no. 3, pp. 563-570, Mar. 2007. http://doi.org/10.1109/JSSC.2006.891442
[16] J. S. Park and H. Wang, "A Transformer-Based Poly-Phase Network for Ultra-Broadband Quadrature Signal Generation," *IEEE TMTT*, vol. 63, no. 12, pp. 4444-4457, Dec. 2015. http://doi.org/10.1109/TMTT.2015.2496187

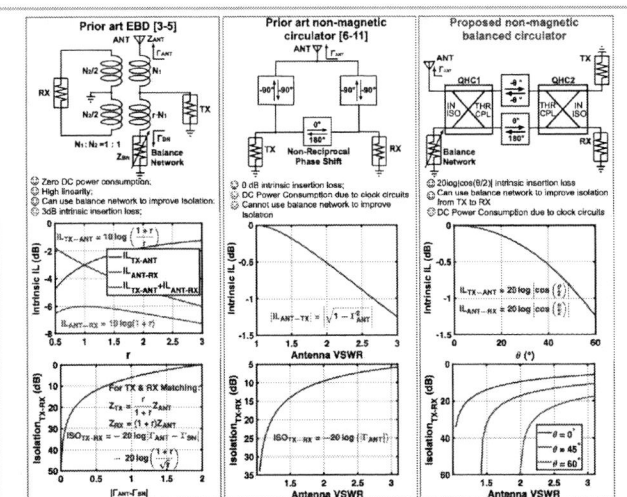

Figure 33.1.1: Comparison between the intrinsic insertion loss and the TX-to-RX isolation of the electrical balance duplexer, non-magnetic circulator, and proposed balanced circulator.

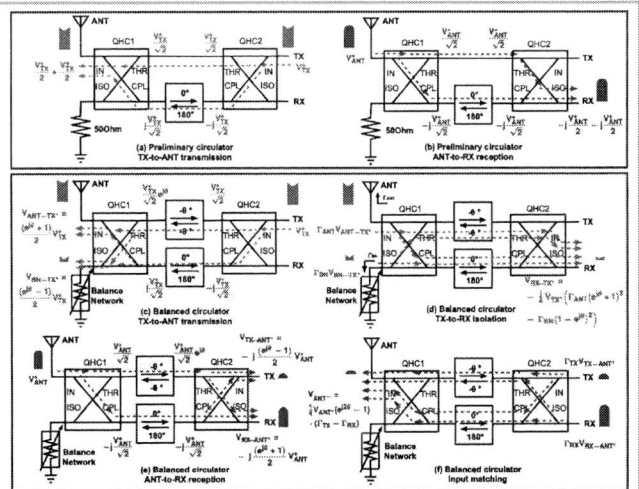

Figure 33.1.2: TX–ANT–RX transmission, reception, isolation, and input matching for (a, b) a preliminary non-magnetic circulator, and (c to f) the proposed non-magnetic balanced circulator.

ISSCC 2026 / SESSION 33 / TIME-VARYING CIRCUIT TECHNIQUES FROM RF TO MM-WAVE / 33.1

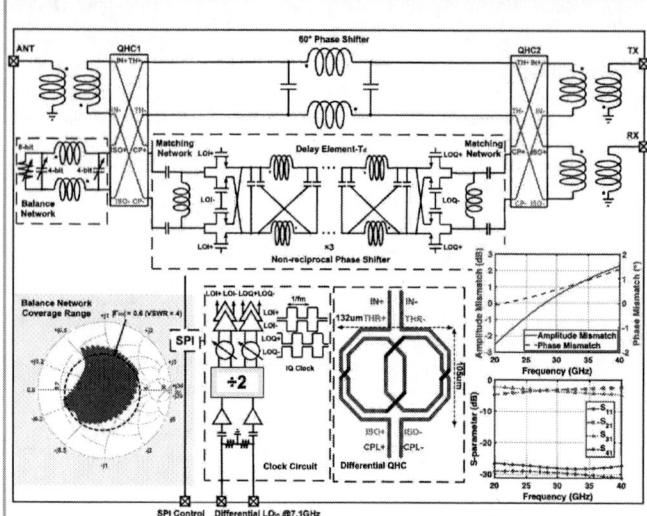

Figure 33.1.3: Schematic of the proposed non-magnetic balanced circulator.

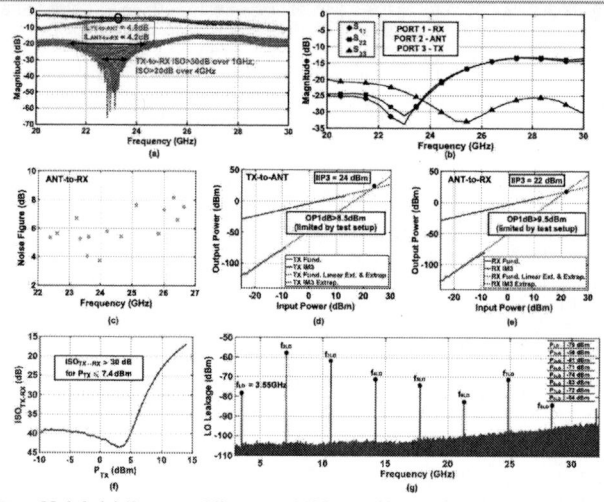

Figure 33.1.4: (a) IL_{TX-ANT} and IL_{ANT-RX} and ISO_{TX-RX}; (b) return loss; (c) NF_{ANT-RX}; (d, e) IIP3 & OP1dB; (f) TX–RX large-signal isolation; (g) LO leakage at the ANT when $Z_{ANT}=50\Omega$.

Figure 33.1.5: TX-to-ANT IL, ANT-to-RX IL, and TX-to-RX isolation vs. frequency for different ANT reflection-coefficient angles, with (a to c) magnitude = 0.2 and (d to f) magnitude = 0.33.

Architecture	This work	ISSCC 17 [7]	ISSCC 19 [11]	JSSC 20 [9]	JSSC 19 [10]	ISSCC 15 [3]
	Non-magnetic Balanced Circulator	mmWave Non-magnetic Circulator	mmWave Non-magnetic Circulator	Non-magnetic Circulator	Non-magnetic Circulator with TX-ANT and ANT-RX feed capacitors	EBD
VSWR Resilience	Yes	No	No	No	Yes	Yes
Technology	40nm CMOS	45nm SOI	45nm SOI	40nm CMOS	0.18μm SOI	0.18μm SOI
Frequency(GHz)	21 to 25	25	50 to 56.8	5.6 to 7.4	0.85 to 1.08	1.9 to 2.2
TX-ANT IL at isolation notch center(dB)	4.8	3.3	3.9	2.2	2.1	3.7
ANT-RX IL at isolation notch center(dB)	4.2	3.2	3.1	2.2	2.9	3.9
BW of TX-RX Isolation With VSWR=1	4%(>30dB) 17%(>20dB)	18%(>18.5dB)	2.5%(>40dB) 14%(>20dB)	30%(>18dB)	3.1%(>40dB) 17%(>25dB)	11%(>45dB)
Max VSWR Tolerance	2:1	N/A	N/A	N/A	2:1	1.5:1
BW of TX-RX Isolation With VSWR=1.5	4%(>30dB) 17%(>20dB)	N/A	N/A	N/A	12%(>30dB) 32%(>20dB)	N/A
BW of TX-RX Isolation With VSWR=2	4%(>25dB)	N/A	N/A	N/A	17%(>30dB) 30%(>20dB)	N/A
TX-ANT IIP3(dBm)	24	20.1	19.4	18.7	50.0	70
Power(mW)	11.6	78.4	24.1	12.4	170	0
Core Area(mm²)	0.63/0.7*	2.16	1.72	0.45	16.5	1.75

* Includes on-chip baluns for TX, RX, and ANT ports.

Figure 33.1.6: Performance summary and comparison table.

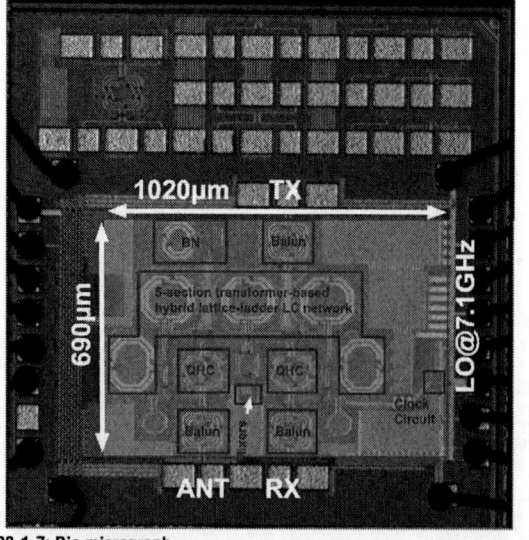

Figure 33.1.7: Die micrograph.

• 2026 IEEE International Solid-State Circuits Conference

ISSCC 2026 / SESSION 33 / TIME-VARYING CIRCUIT TECHNIQUES FROM RF TO mm-WAVE / 33.2

33.2 An Infinite-Loop CMOS-Compatible Isolator Enabled True VSWR-Resilient Power Amplifier for 6G FR3 in Massive MIMO and Phased-Array Systems

Mohsen Ghorbanpoor[1,2], Mohamed Eleraky[2], Konstantinos Manetakis[1], Pascal Nussbaum[1], Hua Wang[2]

[1]CSEM, Neuchatel, Switzerland, [2]ETH Zurich, Zurich, Switzerland

Abstract

This work presents a balanced power amplifier (PA) featuring a time-varying output matching network designed to suppress unwanted reflections and couplings in phased array and MIMO systems. By leveraging lossy non-reciprocal phase shifters in an indirect infinite loop, the architecture minimizes transmission loss while maintaining perfect isolation. Fabricated using a 22nm CMOS SOI process, the proposed PA achieves IP1dB variation within ±0.43dB across 3:1/2:1 VSWR contours and bandwidth (BW).

The rapid adoption of cm-wave and mm-wave wireless technologies, including satellite communications (SATCOM) and emerging 5G/6G systems, drives the demand for highly reliable, large-scale phased arrays and massive-MIMO systems. PAs play a crucial role in determining key system figures of merit (FoMs) such as output power, efficiency, signal quality, and spectral purity. Large arrays exhibit two inherent challenges: voltage standing wave ratio (VSWR) and reverse intermodulation (RIMD), which significantly degrade PA performance. Considerable mutual coupling exists among antenna elements in phased-array and MIMO systems. This results in significant antenna-active-impedance variations across wide scanning angles (Θ) and transmission BW introducing VSWR load mismatches and stress on the PA. On the other hand, the same coupling leads to RIMD in MIMO arrays, causing beam-dependent nonlinear distortions that are difficult to correct (Fig. 33.2.1 top). As most PA designs are optimized for 50Ω load operation, the output 1dB compression point (OP1dB), power-added efficiency (PAE), power gain (PG), and adjacent-channel power ratio (ACPR) are highly impacted by such load mismatches. Moreover, severe VSWR mismatches also accelerate device aging and even cause device breakdown. The dynamic behavior of these VSWR mismatches and their dependency on Θ, carrier frequency (f_c), and element position makes it exceedingly challenging to address using offline or look-up-table (LUT)-based calibrations.

One way to address this problem is with reconfigurable PAs [1] (Fig. 33.2.1 middle). While they partially restore PA performance, they rely on complex yet accurate impedance/power sensing at the PA-antenna interface [2] along with extensive digital processing. An alternative involves using off-chip isolators enabled by ferromagnetic materials (Fig. 33.2.1 middle). However, they are too bulky to fit in compact-size high-frequency array elements. The implementation of non-magnetic on-chip isolators requires non-reciprocal phase shifters (NRPS) realized with either time-varying conductivity modulation [3], or active transistors. The former exhibits high insertion loss (IL), usually >3dB, while the latter is limited by linearity due to the PA high output power. A conceptual on-chip isolator is shown in Fig. 33.2.1 middle. Its linearity and IL are directly affected by the NRPS performance. [4] realizes an on-chip isolator with a NRPS-based circulator. The good Tx-to-antenna IL in this work is attributed to the intentional imbalance between the NRPS and the direct-path impedances. Thus, the antenna-to-TX power flow is mainly through the direct-path and is not influenced by the NRPS, leading to limited antenna-to-TX isolation, unlike the good Tx-to-Rx isolation. Designing a PA with good S_{22}, such as the complex-cascode LC-neutralized [5] or the 8-way balanced PA [6], can slightly improve the performance, but the PAs are still subject to the load VSWR mismatch and susceptible to device breakdown (Fig. 33.2.1 bottom).

This work introduces a balanced PA featuring a modified output matching network that forms an infinite-loop trap isolator (Fig. 33.2.1 bottom). It uses NRPSs within an indirect loop to reduce the IL, achieving a 0.5dB IL when IL_{NRPS}=3dB, while simultaneously relaxing the linearity requirements on the NRPS and maintaining perfect reverse isolation. A further benefit is the complete cancellation of reflections at the outputs of the PAs, rendering them immune to performance degradation and potential breakdown.

Figure 33.2.2 explains the operation of the infinite-loop isolator. Ideal lossless 90° couplers are assumed, exhibiting gain imbalance between the through port (THR) and coupled port (CPL) ($\alpha_{THR}^2 + \alpha_{CPL}^2 = 1$), where α denotes voltage gain. In the forward direction (PA to antenna), the coupled signals from the in-phase PA (PA$_1$) and the quadrature-phase PA (PA$_2$) add up out-of-phase within the loop leading to heavily attenuated signals and at the inputs of the NRPSs. This has two advantages: 1) The loop power contributes negligibly to the isolator output power, resulting in minimal influence of the NRPS IL and P1dB on the isolator performance. 2) The decreased NRPS power level reduces the stress on their transistors/switches. The signal at the THR ports comes from two paths, one directly from the input port (IN) and the other from the isolation port (ISO) via the loop. These two signals are in-phase, improving the isolator forward IL. In the backward direction (antenna to PA), the signal rotation within the loop is opposite to that of the forward direction. With the coupler power combiner at the isolator output, the backward/reflection signals exhibit relative phases of 0° and –90° at the I and Q THR ports of the isolator. The backward signals thus add in-phase in the loop and get attenuated by the NRPSs. The out-of-phase combination of the signal from the THR to the IN ports with the loop signal coming from

the CPL ports enhances the suppression in the backward path and achieves ideal isolation provided that $\alpha_{THR}=\alpha_{NRPS}$. Determining the forward and backward gains involves evaluating an infinite series. Figure 33.2.2 (right) shows the resulting expression along with the forward and backward voltage gain plots. Although the α_{NRPS} is limited by the topology and technology, perfect isolation can be achieved through tuning of the gain imbalances within the coupler for any α_{NRPS}. Figure 33.2.2 (bottom) shows the NRPS input power and isolator linearity. With a balanced coupler and IL_{NRPS} = 3dB, the input power of the NRPS is 9.55dB lower than that of the isolator. Additionally, the P1dB of the isolator exceeds that of the NRPS by 16.5dB. Although the forward performance still depends on the NRPS IL, the linearity advantage of this topology allows the design effort to focus on minimizing loss, rather than simultaneously optimizing NRPS linearity and loss.

Figure 33.2.3 presents the complete schematic of the proposed PA. The core PA employs a three-stack differential topology, and transformer-based input and output matching networks as in our previous work [7]. All couplers are identical 90° balanced (hybrid) couplers, implemented using on-chip transformers [8]. The NRPS is realized using the time-varying conductivity-modulation method [3], complemented by additional step-down transformers to enhance linearity. These transformers employ parallel secondary windings to improve both the quality factor and the coupling coefficient. The artificial λ/4 transmission lines (TLs) are constructed with metal stacking to minimize IL. The NRPS uses a four-phase clock at the RF carrier third subharmonic and achieves 15dBm IP1dB, suitable for isolators requiring P1dB>31dBm. The NRPS IL depends on the phase alignment between the RF and clock signals. Therefore, the two NRPS clocks are 90° phase shifted to enable simultaneous synchronization of both NRPS clocks with the RF signal. Clock feedthrough appears as an even-harmonic common-mode signal, which is blocked by the transformers. If there are device mismatches, the third harmonic of the clock feedthrough falls within the BW of the coupler and transformer and pass through. Due to 90° phase shift of the NRPS clocks, this harmonic reaches output coupler with a 270° phase shift and is mostly absorbed by the termination resistor. The clock distribution network incorporates type-I polyphase filters (PPFs), followed by driver stages that amplify and reshape the clock into a rectangular waveform. Type-I PPFs are preferred due to wide BW quadrature accuracy [9], while driver corrects any amplitude mismatch. The PPF operates over a 3-to-5GHz BW with ±1° phase accuracy. At the driver last stage, a stacking technique is employed to achieve a 1.6V output swing, effectively suppressing unintended switch activation under high RF signal power. As the absolute phase shift over a loop cycle must be multiple of 2π, and transformer-based couplers inherently violate this, a loop phase adjustment is required. The co-design of the loop components ensures that the phase shift of the couplers and TLs counterbalance that of the step-down transformer.

Fabricated using a 22nm CMOS SOI process, the infinite-loop PA occupies a core/total area of 3/3.75mm² (Fig. 33.2.7). Figure 33.2.4 illustrates both small-signal and continuous-wave (CW) measurement results. The small-signal measurements show a 13.8dB peak S_{21} at 11.7GHz, with a 10.5-to-13GHz 3dB BW. Additionally, the S_{22} remains better than –15dB from 8 to 15GHz. Synchronization between the RF and clock signals during CW measurements is facilitated by a Keysight M8195A arbitrary waveform generator. Under a 50Ω load, the PA achieves 23.9dBm saturated output power (P$_{sat}$) and 23.3dBm OP1dB with a peak PAE (PAE$_{sat}$) of 30.6/28% excluding/including the clock driver consumption at 11.1GHz. A summary of CW measurement results across frequency under 50Ω is presented in Fig. 33.2.4 (left). CW measurements under 3/2:1 VSWR contours are performed using a Maury MT985AL impedance tuner to evaluate the PA robustness against impedance mismatches. To ensure measurement accuracy, all cables, probes, and connectors are calibrated across the Smith chart and at the relevant frequencies. The PA shows worst-case variations of less than ±0.75dB, ±0.71dB, ±0.43dB, ±0.2dB in PG, OP1dB, IP1dB, and PG peaking respectively across 10.2/10.8/11.1/11.4/12GHz frequencies and 3/2:1 VSWR contours (Fig. 33.2.4 right). Careful analysis shows that IP1dB and PG peaking exhibit minimal variations compared to PG and OP1dB, indicating that these variations are primarily affected by the IL of the output matching network rather than the PA itself. Unlike PG-peaking and IP1dB, both PG and OP1dB show the same trend with maxima/minima around 0/180°. This behavior is attributed to the increased series resistance loss in the matching network.

ISSCC 2026 / February 18, 2026 / 4:00 PM

The PA is further characterized with modulated signals, with the constellation plots and performance metrics presented in Fig. 33.2.5. With a 400MHz 64-QAM signal at 11.1GHz, it delivers an average power (P_{AVG}) of 18.69dBm and an average PAE (PAE_{AVG}) of 19.53%. Across 10.2/11.1/12GHz frequencies and under a 3:1 VSWR contour, the P_{AVG} variation remains within ±0.4dB, demonstrating strong resilience to VSWR mismatches. To evaluate the performance in a MIMO array, an RIMD measurement [10] is conducted at 10.2/11.1/12GHz with a 100MHz offset between the PA carrier and the reverse tone. The PA operates at OP1dB, and a Keysight E8267D signal generator injects the reverse tone through an off-chip circulator at -6dBc, equivalent to a 3:1 VSWR return loss. Figure 33.2.5 shows the PA output spectrum under the RIMD test. At 11.1GHz, the PA achieves third-order RIMD better than −35.5dBc, confirming both its immunity to reverse tones and excellent linearity under this condition.

Figure 33.2.6 presents a comparison between the infinite-loop balanced PA and existing designs, relative to which, our work demonstrates the lowest output power variation.

Acknowledgement:
The authors would like to thank the GlobalFoundries University Program for chip fabrication. This work was in part sponsored by the Swiss State Secretariat for Education, Research, and Innovation (SERI) under the SwissChips initiative, an ETH Zurich grant, and the HORIZON-JU-SNS-2023 "6G-REFERENCE" project under Project 101139155.

References:
[1] N. S. Mannem et al., "A Reconfigurable Series/Parallel Quadrature-Coupler-Based Doherty PA in CMOS SOI with VSWR Resilient Linearity and Back-Off PAE for 5G MIMO Arrays," *ISSCC*, pp. 364-366, Feb. 2020.
https://doi.org/10.1109/ISSCC19947.2020.9062944
[2] E. Liu et al., "A Compact Broadband VSWR-Resilient True-Power-and-Gain Sensor with Dynamic-Range Compensation for Phased-Array Applications," *ISSCC*, pp. 538-540, Feb. 2024. https://doi.org/10.1109/ISSCC49657.2024.10454458
[3] T. Dinc et al., "A Millimeter-Wave Non-Magnetic Passive SOI CMOS Circulator Based on Spatio-Temporal Conductivity Modulation," *IEEE JSSC*, vol. 52, no. 12, pp. 3276-3292, Dec. 2017. https://doi.org/10.1109/JSSC.2017.2759422
[4] M. Pashaeifar et al., "A Millimeter-Wave Power Amplifier with an Integrated CMOS Isolator/Circulator/Receiver," *IEEE JSSC*, Early Access, 2025.
https://doi.org/10.1109/JSSC.2025.3570656
[5] M. Eleraky et al., "An Ultra-Compact Wideband Load-Insensitive Complex-Cascode LC-Neutralized Power Amplifier for 4:1-VSWR-Resilient Operations in Large-Scale Phased Arrays," *ISSCC*, pp. 98-99, Feb. 2025.
https://doi.org/10.1109/ISSCC49661.2025.10904694
[6] M. Pashaeifar et al., "A 25.2dBm P_{SAT}, 35-to-43GHz VSWR-Resilient Chain-Weaver Eight-Way Balanced PA with an Embedded Impedance/Power Sensor," *ISSCC*, pp. 532-534, Feb. 2024. https://doi.org/10.1109/ISSCC49657.2024.10454427
[7] M. Ghorbanpoor et al., "12 GHz Stacked Power Amplifier with 22.9 dBm P_{sat} and 44.9% PAE_{sat} for 6G in 22nm FDSOI," *EuMIC*, pp. 62-65, Sep. 2024.10732645.
https://doi.org/10.23919/EuMIC61603.2024.10732645
[8] J. S. Park and H. Wang, "A Transformer-Based Poly-Phase Network for Ultra-Broadband Quadrature Signal Generation," *IEEE TMTT*, vol. 63, no. 12, pp. 4444-4457, Dec. 2015. https://doi.org/10.1109/TMTT.2015.2496187

[9] J. Kaukovuori et al., "Analysis and Design of Passive Polyphase Filters," *IEEE TCAS-I*, vol. 55, no. 10, pp. 3023-3037, Nov. 2008. https://doi.org/10.1109/TCSI.2008.917990
[10] A. N. Atanasov et al., "Reverse Intermodulation in Multi-Tone Array Transmitters," *IEEE BCICTS*, pp. 1-4, Nov. 2020. https://doi.org/10.1109/BCICTS48439.2020.9392972
[11] G. Diverrez et al., "A 22-44 GHz 28nm FD-SOI CMOS 5G Doherty Power Amplifier with Wideband PAE_{6dBPBO} Enhancement and 3:1 VSWR Resiliency," *IEEE RFIC*, pp. 131-134, June 2024. https://doi.org/10.1109/RFIC61187.2024.10600014
[12] J.-H. Kim et al., "An Efficient Ku-Band Two-Way Vertical-like Power-Combining Power Amplifier using Merged Inter-stage Transformers Achieving 23-23.4 dBm P_{sat} and 45.2-46.6% Peak PAE in 65nm CMOS," *IEEE RFIC*, pp. 299-302, June 2024. https://doi.org/10.1109/RFIC61187.2024.10599991
[13] J. Xu et al., "A Compact Doubly Neutralized Ku-Band Power Amplifier With 39% Peak PAE and 23-dBm Output Power in 22FDX+ EDMOS for 6G FR3," *IEEE MWTL*, vol. 35, no. 6, pp. 856-859, June 2025. https://doi.org/10.1109/LMWT.2025.3566279

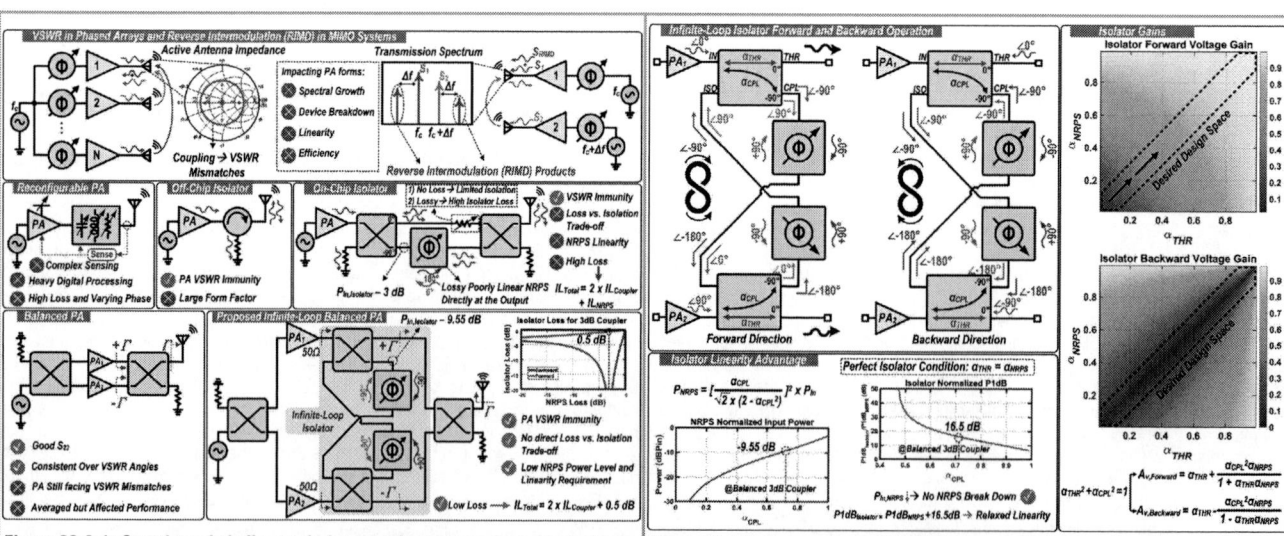

Figure 33.2.1: Overview of challenges in large-scale arrays and on-chip isolators, existing solutions, and the proposed infinite-loop balanced PA.

Figure 33.2.2: Proposed on-chip infinite-loop isolator operation, loss and isolation performance, and the linearity advantages.

DIGEST OF TECHNICAL PAPERS • 565

979-8-3315-8937-0/26 $31.00 © 2026 IEEE

ISSCC 2026 / SESSION 33 / TIME-VARYING CIRCUIT TECHNIQUES FROM RF TO mm-WAVE / 33.2

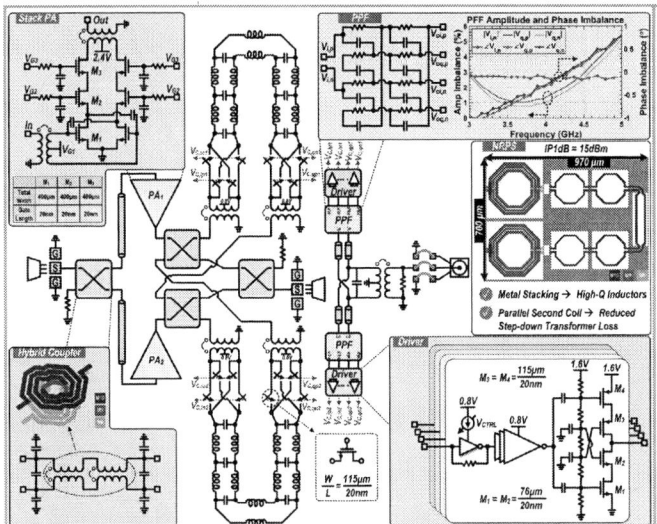

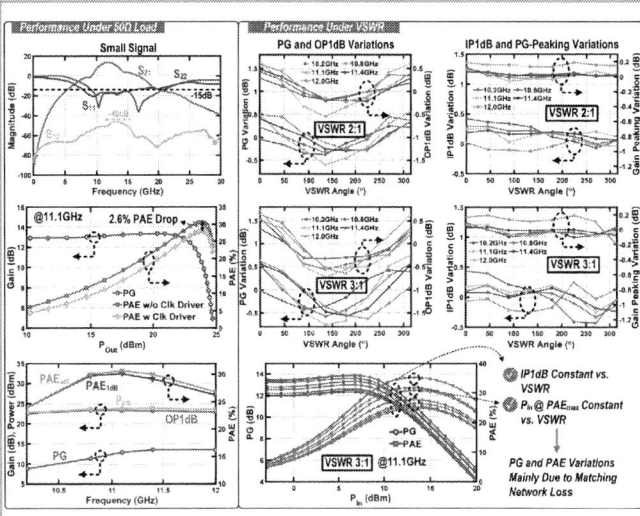

Figure 33.2.3: Complete schematic of the proposed VSWR-resilient balanced amplifier, including a clock distribution and driver network.

Figure 33.2.4: Small-signal and large-signal performance summary across the BW and under two VSWR circles of 3/2:1.

Figure 33.2.5: Modulation measurement results across 3:1 VSWR and reverse intermodulation with a 100MHz spacing at multiple frequencies.

Figure 33.2.6: Performance summary of the proposed balanced PA and comparison with the prior-art VSWR-resilient PAs.

	This Work	Eleraky ISSCC 2025 [5]	Pashaeifar ISSCC 2024 [6]	Diverrez RFIC 2025 [11]	Mannem ISSCC 2020 [1]	Kim RFIC 2024 [12]	Xu MWTL 2025 [13]
Technology	22nm SOI	22nm SOI	40nm CMOS	28nm SOI	45nm SOI	65nm CMOS	22nm SOI
Architecture	Infinite Loop Balanced Three-Stack PA	Complex-Cascode LC-Neutralized PA	Chain Weaver 8-Balanced PA	Inductive Doherty PA	Reconfigurable Doherty PA	Two-Stack PA	Two-Stack PA
Supply (V)	2.4	1.6	2	NA	2	2.2	2.4
Frequency (GHz)	11.1	39	37	28	39	13	12
Gain (dB)	12.9	22.7	29.9	22	12.4	27.3	25.17
P_{sat} (dBm)	23.9	16.4	25.19	20.3	20.8	23.4	23.55
OP1dB (dBm)	23.3	16.3	22.67	19.8	20.2	22.9	22.81
Peak PAE (%)	30.6[$\dagger$]/28[#]	34.1	16.18	34.4	33.3	46.6	39.27
PAE_{1dB} (%)	30.1[*]	34	NA	34[*]	32.2	45.5	37.92
Core Area (mm^2)	3	0.093	2.08	0.82	1.18	0.234	0.152
VSWR							
VWSR Ratio	(3:1) to (2:1)	(4:1) to (2:1)	(3:1) to (2:1)	(3:1)	(3:1)	NA	NA
Frequency Range (GHz)	10.2 to 12	37 to 43	37 to 40	24 to 30	39	NA	NA
PG Variation (dB)	0.75	1.5	0.7	NA	0.9	NA	NA
P1dB Variation (dB)	0.71[*]/0.43[#]	3[*]	0.79[*]	2.1[*]	0.85[*]	NA	NA
PG-Peaking Variation (dB)	0.2	NA	NA	NA	NA	NA	NA
Modulation							
Load	50Ω	3:1 VSWR	3:1 VSWR	50Ω	50Ω	3:1 VSWR	50Ω
Modulation Scheme	64-QAM	64-QAM	64-QAM	64-QAM	64-QAM	64-QAM	64-QAM
Frequency	11.1	41	37	26	39	13	12
Bandwidth (MHz)	400	100	2000	200	100	100	4250
EVM_{RMS} (dB)	-25	-25	-24.6	-25	-22	-25	-25
P_{AVG} (dBm)	18.69	18.2 ↔ 19	7 ↔ 9.8	16	10.6	>11.2	17.7	15.54
PAE_{AVG} (%)	19.53	17.2 ↔ 20.9	9.7 ↔ 17	4.1	14.9	>9.6	24.1	21.67
ACPR (dB)	-28	-26	-23.8	-30.7	-27.5	-22.8	-30	-29.2
RIMD3 @ OP1DB (dBc)	-35.5	NA	NA	NA	NA	NA	NA

NA: Not Available, [*]Excluding Clock Driver, [#]Including Clock Driver, [*]Estimated from The Figures, [^{0}]Output Referred, [^{i}]Input Referred, [$\dagger$]Reverse Tone 9dB Below Main Tone

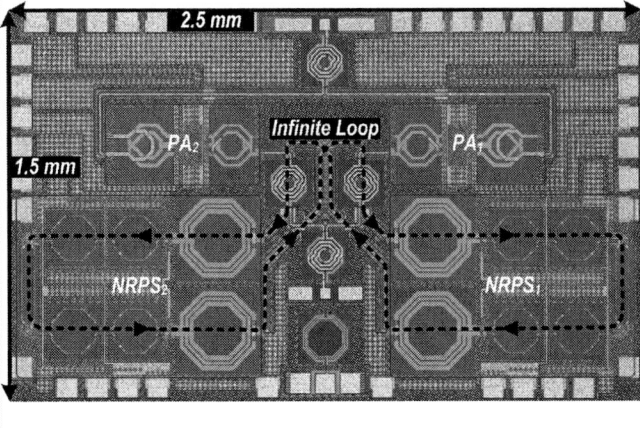

Figure 33.2.7: Die micrograph of the proposed balanced PA.

- 2026 IEEE International Solid-State Circuits Conference

ISSCC 2026 / SESSION 33 / TIME-VARYING CIRCUIT TECHNIQUES FROM RF TO MM-WAVE / 33.3

33.3 A 6GHz Quadrature Digital Transmitter Supporting a 1GHz Signal Bandwidth with <-40dB EVM Floor and >55dB Dynamic Range in 28nm CMOS

Yicheng Li*, Feng Xie*, Yun Yin, Lixuan Cao, Tianze Yang, Jiaxiang Li, Hongtao Xu

Fudan University, Shanghai, China
*Equally Credited Authors (ECAs)

Abstract

A 6GHz quadrature DTX supporting a 1GHz signal bandwidth is presented in 28nm CMOS, where static and dynamic nonlinearities are carefully optimized. Occupying a core area of 0.67mm² and packaged in a fanout SiP, the DTX delivers 23.1dBm peak output power with a 30.3% peak PAE at 6.0GHz, and the 1dB RF bandwidth is from 5.7 to 6.5GHz. For 1GHz 64-QAM signals, it achieves 18.0dBm P_{avg} and 16.2% PAE_{avg} with -25.5dB EVM. The dynamic power range is >55dB, and its EVM floor is -40.8dB.

Beyond the sub-6GHz band, the 6-to-7GHz frequency band has gained much attention and adoption for supporting signal bandwidths (BWs) up to 200 to 400MHz in applications such as 3GPP U6G and 802.11be. However, the high modulation order with stringent linearity requirements and low power efficiency at power backoff (PBO) are main design challenges for transmitters (TXs). Analog TXs [1-3] offer good linearity but normally consume much more power than digital TXs (DTXs) due to multiple modules, especially for wideband applications, and hardly benefit from CMOS process scaling. On the other hand, DTXs show good scalability as well as better system efficiency with direct interface to digital baseband and a compact structure [4-6]. Polar DTXs are good candidates when high power and efficiency are required, and recent studies have demonstrated their ability to support signal BWs of 160 to 320MHz with the aid of digital pre-distortion (DPD) [7-9]. However, its potential for further signal BW increase is limited by the inherent spectrum-expansion issue due to coordinate conversion, as depicted at the top of Fig. 33.3.1.

Compared to polar DTXs, quadrature DTXs use a more compact architecture without requiring a Cordic or phase modulator. Moreover, quadrature DTXs are theoretically better suited for wideband modulations because they do not need to handle signals with expanded spectra. Recent research on quadrature DTX/modulators has also realized signal BWs of 200 to 320MHz [10,11], but the potential upper limit on signal BWs enabled by the quadrature architecture should be much higher. Prior studies found that time-variant output resistance of the power stage and mismatch among unit cells are the two key nonlinearity sources in DTXs and switched-capacitor digital power amplifiers (DPAs) [12-14]. These issues can be mitigated by careful design of the LO overlap and optimized DPA array layout, respectively. Nevertheless, such measures are insufficient for ultra-wideband applications. In this work, the inherent sources that contribute to nonlinearities in a quadrature DTX are comprehensively discussed. A 6GHz quadrature DTX is presented that achieves a 1GHz signal BW with an EVM floor <-40dB and a dynamic range >55dB, while achieving good efficiency within a compact core size.

Figure 33.3.1 lists multiple nonlinearity sources in the quadrature DTX, including the aforementioned time-variant output resistance of the power stage and the unit cell mismatch, which are classified as static nonlinearity sources (SNS) because they are unrelated to the signal BW. Here, the nonlinearity of the DTX can be evaluated conveniently by the spurious-free dynamic range (SFDR) of a two-tone signal with a corresponding signal BW. As the signal BW increases, the nonlinearity of the whole DTX is gradually dominated by dynamic nonlinearity sources (DNS), which contain AM-PM asynchronization (DNS1), signal feedthrough in the mixer (DNS2), asymmetry in the power combiner (DNS3), and parasitic impedance in the supply network (DNS4). Specifically, DNS1 exists at the architecture level, while DNS2,3, and 4 are all caused by parasitics at the circuit implementation level.

Figure 33.3.2 demonstrates the block diagram of the proposed quadrature DTX. To support the signal BW of up to 1GHz, a sampling clock of at least 2GHz is required in this 14-bit DTX. The high-speed parallel data I/Q<13:0> is generated by an interface module from 4 pairs of differential serial signals (I1+/I1-/I2+/I2-/Q1+/Q1-/Q2+/Q2-). The interface contains an 8-phase clock generator to enable a multi-phase operation of a comparator array, whose outputs are synchronized by the following D flip-flops. The switched-capacitor structure with IQ-sharing and Doherty techniques [15] is adopted in the DPA for good linearity and efficiency. The sampling clock (CLK) for the DPA is also available after a suitable delay. I/Q<13:0> are converted to thermometer-binary hybrid codes by decoder modules and then sent into two sub-DPA arrays (DPA1/2). Here, I/Q<6:0> are set as logic signals for all binary cells, while I/Q<11:10> and I/Q<9:7> are decoded to thermometer codes of $G_{I/Q}$<2:0> and $T_{I/Q}$<6:0>, which manipulate the 4 hybrid groups (group0 to 3) and the unary cells inside, respectively. Meanwhile, I/Q<12> acts as the MSB to control the IQ-sharing with Doherty operation for 0-to-6dB PBO efficiency enhancement, and I/Q<13> is the sign bit that determines the quadrant on the complex plane. These decoded words ($G_{I/Q}$<2:0>, $T_{I/Q}$<6:0>, MSB) are eventually translated into the local EN signals for each unit cell.

To solve DNS1 caused by the AM-PM asynchronization, the unsigned quadrature LOs ($LO_{I/Q}$) are directly sent into each sub-DPA array without phase selection. The sign bit that contains PM information is sent into the sub-DPA array together with the AM code, and the synchronization between AM-PM is optimized by inserting the LO phase selection in each unit cell, as depicted in Fig. 33.3.2. Two D flip-flops are utilized to eliminate the delay mismatch caused by the extra logic operation of the AM code. Then, a dummy delay is added on the AM path as the sign bit completes the LO phase selection. Therefore, the signed LO (PM) and the AM code are synchronized before entering the mixer, which reduces the AM-PM delay mismatch to less than 2ps and leads to an SFDR of <-50dBc for the signal BW of 1GHz. Delay aligners are utilized in [11] to obtain about 5ps delay mismatch, which is limited by the resolution of the delay cell. As shown in the SFDR curves with various signal BWs, even such minor increasement of delay mismatch deteriorates the SFDR by more than 10dB. In addition, simple inverter instead of a cascode inverter is adopted to reduce the output resistance variance of the power stage, thus alleviating SNS1. The supply voltage of the driver and power stages is 1.1V, and a 0.9V supply is used for the preceding logic circuits.

Furthermore, the solution to DNS2 is shown in Fig. 33.3.3, targeting the signal feedthrough in the mixer. Normally, the conventional AND gate is used as the mixer in a DPA unit cell. The parasitic on-resistance (R_{on}) and capacitance between the gate and drain (C_{gd}) of the transistors cause feedthrough from input LO/EN signals towards drain node, adding certain disturbance on its waveform. Although the following rail-to-rail inverter eliminates the fluctuation on the magnitude, the significant glitches and edge timing delay still remain at the output of the AND gate, thus impairing the DPA linearity. The severity of signal feedthrough is related to the value of R_{on} and C_{gd}, which means that nonlinearity caused by DNS2 can be reduced directly in more advanced CMOS process. Evaluating 28nm CMOS, the SFDR becomes worse than -50dBc when the signal BW >400MHz, and DNS2 cannot be ignored. A structure based on a transmission gate (TG) is proposed to address this issue, in which M1 to M3 complete the function of an AND gate, and M4 is used to compensate the feedthrough from EN_P and EN_M. The LO signal is connected at the source node of the TG with low feedthrough, benefiting from a much smaller parasitic capacitance between the drain and source terminals (C_{ds}). Additionally, the differential nature of the EN signal cancels out the feedthrough at the TG drain node. The simulated SFDR can be improved by about 15dB with this mixer structure.

Figure 33.3.3 depicts the top floorplan and power combiner, which are related to SNS2 and DNS3, respectively. The number of groups in each sub-DPA is restricted to obtain compact array size (240×188μm²), thus reducing the unit-cell mismatch. The simulated magnitude and phase mismatches among unit cells correspond to an SFDR of better than -50dBc. Additionally, a symmetrical serial-combining-transformer (SCT) power combiner is adopted for good symmetry, which also guarantees the SFDR to be better than -50dBc according to the simulated results. The size of power combiner is 528×286μm², including the pick-up routings.

The proposed quadrature DTX was implemented in a 28nm CMOS process (Fig. 33.3.7), occupying a core area of 0.67mm². The die is set in a 4mm × 4mm fanout flip-chip SiP (System in Package) with embedded decoupling capacitors (decaps) to alleviate the nonlinearity caused by DNS4. The supply impedances at higher frequencies are lowered by decaps on-chip. Figure 33.3.4 demonstrates the measured frequency response, power-added efficiency (PAE) versus output power on the 0deg/45deg axis, and AM-AM/AM-PM distortions. The DTX delivers a 23.1dBm peak output power with a 30.3% peak PAE at 6.0GHz, while the 1dB RF bandwidth is from 5.7 to 6.5GHz. The PAE at 3/6dB PBO is 19.1%/20.2%, revealing the efficiency enhancement acquired by IQ sharing with Doherty. Moreover, AM-AM<0.9dB and AM-PM<14deg at 6.0GHz are obtained. Figure 33.3.5 presents the measured results of wideband modulated signals. For 400MHz 256-QAM signals, the DTX achieves an 18.2dBm average output power (P_{avg}) and a 17.2% average PAE (PAE_{avg}) with -30.2dB EVM. The dynamic power range is >55dB at a -30dB EVM limit, and its EVM floor is -47.1dB. For 1GHz 64-QAM signals, the DTX achieves a 18.0dBm P_{avg} and a 16.2% PAE_{avg} with -25.5dB EVM. The corresponding dynamic power range is also >55dB at the -25dB EVM limit, and the EVM floor is -40.8dB. Figure 33.3.6 summarizes the performance and compares our DTX to prior works. Owing to the effective solutions to multiple nonlinearity sources, this quadrature DTX successfully supports the widest signal bandwidth with the lowest EVM floor reported in the figure. Its power efficiency and dynamic power range are also among the best in the figure while occupying a compact core size.

566 • 2026 IEEE International Solid-State Circuits Conference

979-8-3315-8937-0/26 $31.00 © 2026 IEEE

ISSCC 2026 / February 18, 2026 / 4:25 PM

Acknowledgement:
This work was supported by the National Natural Science Foundation of China under Grant 62322105 and by the Project Fund of State Key Laboratory of Integrated Chips and Systems at Fudan University under Grant SKLICS-Z202506. Corresponding authors: Yun Yin, Hongtao Xu.

References:
[1] B. Jann et al., "A 5G Sub-6GHz Zero-IF and Wm-Wave IF Transceiver with MIMO and Carrier Aggregation," *ISSCC*, pp. 352-353, Feb. 2019.
https://doi.org/10.1109/ISSCC.2019.8662417
[2] E. Lu et al., "A 4×4 Dual-Band Dual-Concurrent WiFi 802.11ax Transceiver with Integrated LNA, PA and T/R Switch Achieving +20dBm 1024-QAM MCS11 Pout and −43dB EVM Floor in 55nm CMOS," *ISSCC*, pp. 178-179, Feb. 2020.
https://doi.org/10.1109/ISSCC19947.2020.9063127
[3] J. Lee et al., "A Tri-Band Dual-Concurrent Wi-Fi 802.11be Transceiver Achieving − 46 dB TX/RX EVM Floor at 7.1 GHz for a 4 K-QAM 320 MHz Signal," in *IEEE JSCC*, vol. 59, no. 12, pp. 3966-3979, Dec. 2024. https://doi.org/10.1109/JSSC.2024.3441858
[4] Z. Deng et al., "A dual-band digital-WiFi 802.11a/b/g/n Transmitter SoC with Digital I/Q Combining and Diamond Profile Mapping for Compact Die Area and Improved Efficiency in 40nm CMOS," *ISSCC*, pp. 172-173, Feb. 2016.
https://doi.org/10.1109/ISSCC.2016.7417962
[5] P. Madoglio et al., "A 2.4GHz WLAN Digital Polar Transmitter with Synthesized Digital-to-Time Converter in 14nm Trigate/FinFET Technology for IoT and Wearable Applications," *ISSCC*, pp. 226-227, Feb. 2017. https://doi.org/10.1109/ISSCC.2017.7870343
[6] M. Fulde et al., "A Digital Multimode Polar Transmitter Supporting 40MHz LTE Carrier Aggregation in 28nm CMOS," *ISSCC*, pp. 218-219, Feb. 2017.
https://doi.org/10.1109/ISSCC.2017.7870339
[7] A. Ben-Bassat et al., "A Fully Integrated 27-dBm Dual-Band All-Digital Polar Transmitter Supporting 160 MHz for Wi-Fi 6 Applications," *IEEE JSSC*, vol. 55, no. 12, pp. 3414-3425, Dec. 2020. https://doi.org/10.1109/JSSC.2020.3024973
[8] B. Khamaisi et al., "A 16nm, +28dBm Dual-Band All-Digital Polar Transmitter Based on 4-core Digital PA for Wi-Fi6E Applications," *ISSCC*, pp. 324-325, Feb. 2022.
https://doi.org/10.1109/ISSCC42614.2022.9731624
[9] N. R. Shay et al., "A Watt level, 5-7GHz all digital polar TX based on 3.3V switched capacitor digital PA in 16nm Fin-FET for Wi-Fi7 applications," *IEEE RFIC*, pp. 255-258, June 2024. https://doi.org/10.1109/RFIC61187.2024.10600000
[10] Y. Shen et al., "A Wideband IQ-Mapping Direct-Digital RF Modulator for 5G Transmitters," *IEEE JSSC*, vol. 57, no. 5, pp. 1446-1456, May 2022.
https://doi.org/10.1109/JSSC.2022.3144362
[11] M. Beikmirza et al., "A Wideband Energy-Efficient Multi-Mode CMOS Digital Transmitter," *IEEE JSSC*, vol. 58, no. 3, pp. 677-690, Mar. 2023.
https://doi.org/10.1109/JSSC.2022.3222028
[12] W. Luo et al., "Nonlinear Analytical Model for Switched-Capacitor Class-D RF Power Amplifiers," *IEEE TCAS-I*, vol. 66, no. 6, pp. 2309-2321, June 2019.
https://doi.org/10.1109/TCSI.2019.2892566
[13] S.-W. Yoo et al., "A 0.26mm² DPD-Less Quadrature Digital Transmitter With <−40dB EVM Over >30dB Pout Range in 65nm CMOS," *ISSCC*, pp. 184-185, Feb. 2020.
https://doi.org/10.1109/ISSCC19947.2020.9063070

[14] H. Wang et al., "A Highly-Efficient Multi-Band Multi-Mode All-Digital Quadrature Transmitter," *IEEE TCAS-I*, vol. 61, no. 5, pp. 1321-1330, May 2014.
https://doi.org/10.1109/TCSI.2014.2309811
[15] D. Zheng et al., "A 15b Quadrature Digital Power Amplifier with Transformer-Based Complex-Domain Power-Efficiency Enhancement," *ISSCC*, pp. 370-371, Feb. 2020.
https://doi.org/10.1109/ISSCC19947.2020.9062959
[16] C. Hu et al., "A Wideband Replicas-Rejection Digital Transmitter Using Joint-Digital-Analog Interpolation and Filtering in 28nm CMOS," *ISSCC*, pp. 460-461, Feb. 2025.
https://doi.org/10.1109/ISSCC49661.2025.10904539
[17] M. Beikmirza et al., "A 4-Way Doherty Digital Transmitter Featuring 50%-LO Signed IQ Interleave Upconversion with more than 27dBm Peak Power and 40% Drain Efficiency at 10dB Power Back-Off Operating in the 5GHz Band," *ISSCC*, pp. 92-93, Feb. 2021.
https://doi.org/10.1109/ISSCC42613.2021.9365831
[18] S.-C. Hung et al., "A Quadrature Class-G Complex-Domain Doherty Digital Power Amplifier," *IEEE JSSC*, vol. 56, no. 7, pp. 2029-2039, July 2021.
https://doi.org/10.1109/JSSC.2020.3040973
[19] B. Yang et al., "Watt-Level Triple-Mode Quadrature SFCPA with 56 Peaks for Ultra-Deep PBO Efficiency Enhancement Using IQ Intrinsic Interaction and Adaptive Phase Compensation," *IEEE CICC*, pp. 1-2, Apr. 2022.
https://doi.org/10.1109/CICC53496.2022.9772776
[20] H. Tang et al., "A Self-Calibration SCPA With Storage Capacitor Array Supporting 64-/256-/1024-QAM," *IEEE JSSC*, vol. 58, no. 5, pp. 1241-1255, May 2023.
https://doi.org/10.1109/JSSC.2023.3246634
[21] C. Hu et al., "A 0.7-to-2.5GHz Sliding Digital-IF Quadrature Digital Transmitter Achieving >40% System Efficiency for Multi-Mode NB-IoT/BLE Applications," *ISSCC*, pp. 472-473, Feb. 2023. https://doi.org/10.1109/ISSCC42615.2023.10067825
[22] T. Wang et al., "A Fully Integrated Digital Polar Transmitter with Single-Ended Doherty PA and DLL-Based Three-Segment Hybrid DTC in 28 nm CMOS," *IEEE JSSC*, vol. 59, no. 2, pp. 388-399, Feb. 2024. https://doi.org/10.1109/JSSC.2023.3282018
[23] Y. Li et al., "A Quadrature Digital Power Amplifier with Wide Efficiency Enhancement Coverage and High Dynamic Power Range," *IEEE JSSC*, vol. 59, no. 7, pp. 2133-2144, July 2024. https://doi.org/10.1109/JSSC.2023.3347309

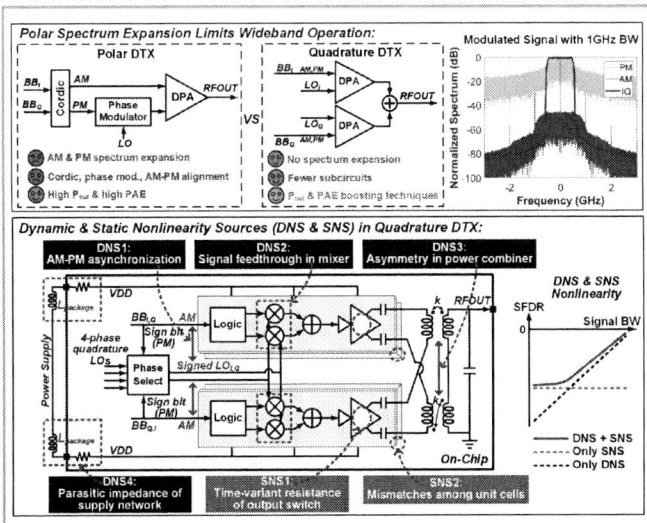

Figure 33.3.1: Spectrum expansion issue in the polar DTX, and nonlinearity contribution of dynamic and static sources in the quadrature DTX.

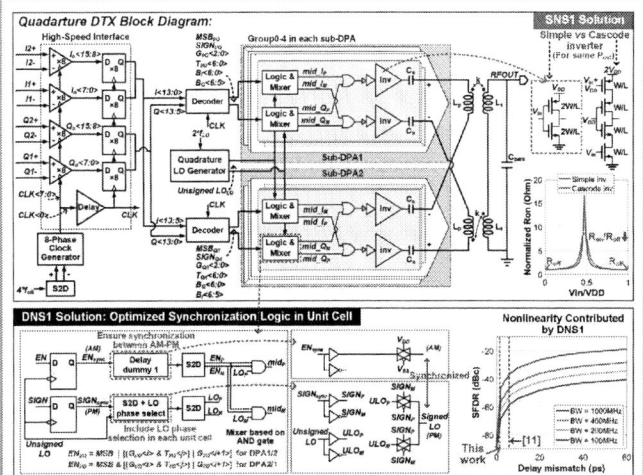

Figure 33.3.2: Block diagram of the proposed quadrature DTX, the structure of power stage for solution to SNS1, and the optimized AM-PM synchronization for solution to DNS1.

DIGEST OF TECHNICAL PAPERS • 567

979-8-3315-8937-0/26 $31.00 © 2026 IEEE

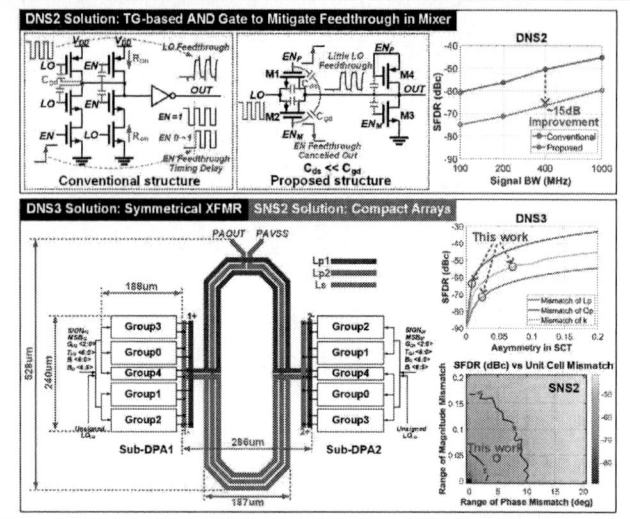

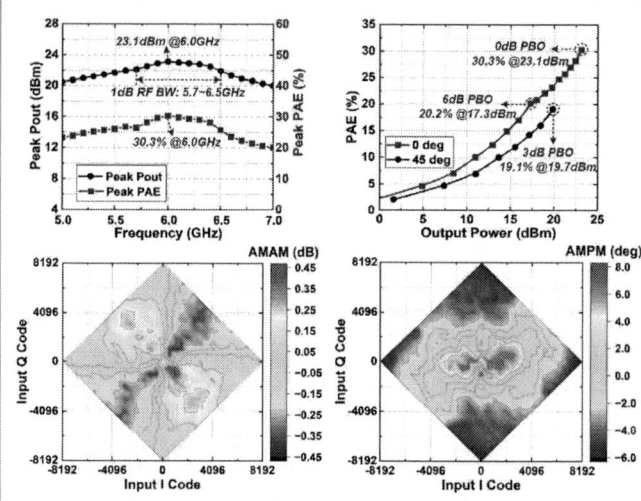

Figure 33.3.3: The TG-based AND gate for solution to DNS2, the SCT power combiner and floorplan for solutions to DNS3 and SNS2, respectively.

Figure 33.3.4: Measured frequency response, PAE versus output power over the 0/45deg axis, and AM-AM/AM-PM distortions at 6.0GHz.

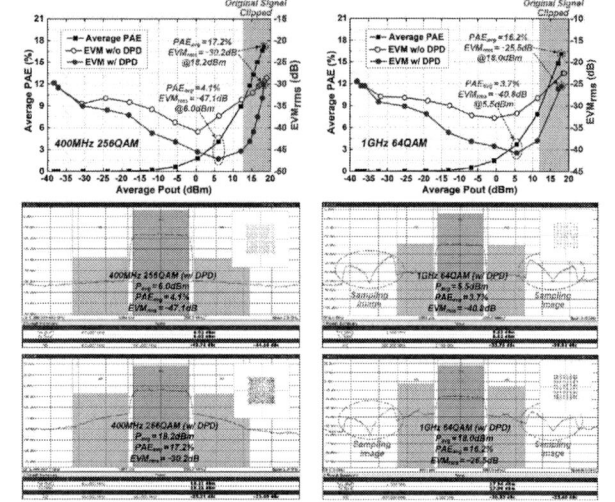

Figure 33.3.5: Measured dynamic power range, spectrum and constellation diagrams of 400MHz 256-QAM and 1GHz 64-QAM modulated signals.

	This work	JSSC22 [10]	ISSCC22 [8]	JSSC23 [11]	RFIC24 [9]	ISSCC25 [16]
Architecture	Quadrature DTX IQ-sharing Doherty DPA	Direct-digital RF modulator	Polar 4-core DPA	Multi-mode DTX	2-cores polar DTX 3-stack SC-DPA	Quadrature DTX
Balun	On-chip	Off-chip	On-chip	Off-chip	On-chip	On-chip
Frequency (GHz)	6.0	2	6	2.4	6.1	3.6
Supply (V)	0.9/1.1	1.1/1.7	1.2	0.95	1.1/3.3	1.1
Peak Pout (dBm)	23.1	14.1	27.8	23.18	29	27.3
Peak PAE (%)	30.3	7.4	25	52.59	31	30.5
Modulation Signal	400MHz 256QAM / 1GHz 64QAM	320MHz 256QAM (@2.4GHz)	160MHz 1024QAM WI-FI 6E (@6GHz)	200MHz 1024QAM OFDM	320MHz WI-FI 7 4096QAM	80MHz 256QAM 802.11ax (@3.84GHz)
P_{avg}(dBm)	18.2 / 18.0	5.15	18	12.23	20	17.5
PAE_{avg}(%)	17.2 / 16.2	-	10.3	19.34	12.0	10.3
EVM (dB)	-30.2 / -25.5	-32	-35	-33.99	-38	-29.5
EVM Floor (dB)	-47.1 / -40.8	-	-38*	-39*	-41*	-33*
Dynamic Range (dB)	>55 / >55	-	>5*	>10*	>9*	>33
DPD	Yes	No	Yes	Yes	Yes	Yes
Core Area (mm²)	0.67	0.96	-	0.72	0.5 (DPA)	0.88
CMOS Technology	28nm	40nm	16nm	40nm	16nm	28nm

* Estimated from figure.

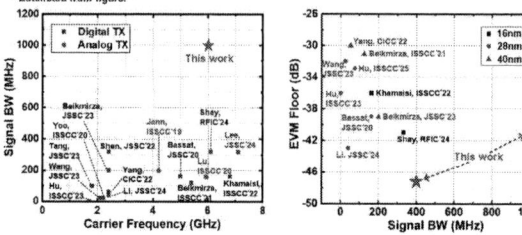

Figure 33.3.6: Performance comparison table and diagrams.

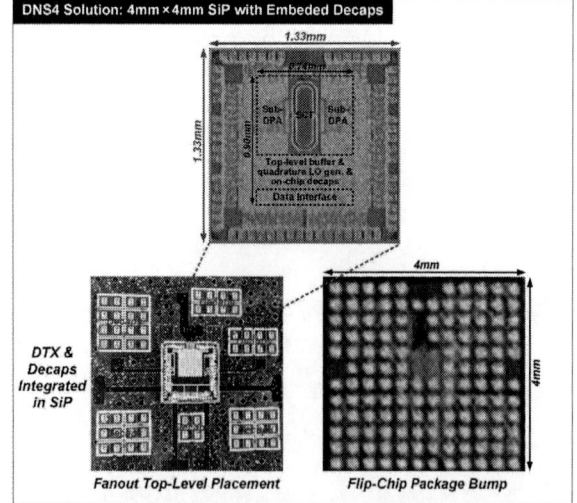

Figure 33.3.7: Die micrograph of the DTX packaged in a 4mm × 4mm SiP with embedded decaps (solution to DNS4).

ISSCC 2026 / SESSION 33 / TIME-VARYING CIRCUIT TECHNIQUES FROM RF TO MM-WAVE / 33.4

33.4 A 4.5-to-7.2GHz Beyond Rail-to-Rail Output SCPA with 27.9dBm P_{out} and 46.2% DE at 5.1GHz Using Periodic Voltage-Pacing Network

Bingzheng Yang[1,2], Jie Zhou[1], Junfa Mao[1], Xun Luo[1]

[1]Shenzhen University, Shenzhen, China, [2]University of Electronic Science and Technology of China, Chengdu, China

Abstract

This work presents a switched-capacitor power amplifier (SCPA) using periodic voltage-pacing network, capable of generating output voltage beyond conventional rail-to-rail limit, which enhances both the output power (P_{out}) and efficiency. Fabricated in a 40nm CMOS process, it shows a competitive 27.92dBm peak P_{out} and 46.2% peak drain efficiency covering a wide 1dB bandwidth from 4.5 to 7.2GHz. Furthermore, it supports 200MHz 64QAM and 80MHz 256QAM, confirming its suitability for 5G/6G and WiFi systems.

Digital power amplifiers (DPAs) are highly valued for their high efficiency, high output power (P_{out}), and high integration, making them essential for sub-6G wireless applications [1-5]. Besides, with the advancement of 5G/6G communication and WiFi systems, operating frequencies continue to increase, e.g., 5 to 7GHz. Current-mode DPAs have demonstrated high efficiency and good performance at high frequencies (e.g., >5GHz). However, they suffer from AM-AM nonlinearity that requires correction [6-8]. In contrast, voltage-mode DPAs—such as switching-capacitor power amplifiers (SCPAs)—inherently provide high linearity. Nevertheless, such topologies lead to increased dynamic power loss at higher operating frequencies, which limits their efficiency [9]. The low-loss asymmetrical power-combining transformer has been utilized in SCPAs to achieve both high P_{out} and efficiency at a cost of larger circuit size [10]. Consequently, it is still challenging to design the SCPA operating at high frequencies (e.g., >5GHz) with both high P_{out} and efficiency. To alleviate those drawbacks, this work presents an SCPA featuring beyond rail-to-rail output voltage using a periodic voltage-pacing network (PVPN) with increased P_{out} and efficiency, simultaneously. The proof-of-concept prototype shows a 1dB bandwidth of 4.5 to 7.2GHz, which achieves a competitive 27.92dBm peak output power at 5.0GHz and 46.2% peak drain efficiency at 5.1GHz, respectively.

As shown in Fig. 33.4.1, for a conventional differential SCPA, the output voltage Vout+ and Vout- switches between voltage VD and VS at the carrier frequency f_0, effectively producing a rail-to-rail square wave with a 50% duty cycle. The peak output voltage Vout+ at f_0 is limited to $2VDD/\pi$ [11], which constrains the maximum achievable P_{out}. Moreover, the dynamic power consumption increases at higher operating frequencies, further degrading the efficiency of the SCPA. To address this limitation, this work presents a beyond rail-to-rail output-voltage SCPA using PVPN, as depicted in Fig. 33.4.1 (bottom). A coupled line is introduced into the SCPA array to implement the PVPN. The charging and discharging currents of the SCPA induce the voltage swings of VD and VS at a frequency of $2f_0$. Note that the charging and discharging currents have a period of $1/2f_0$, considering both the positive and negative SCPA cells when switching. When switching from VS to VD, Vout+ rises above VDD. Conversely, when switching from VD to VS, Vout+ falls below GND. Vout- behaves similarly to Vout+. As a result, the peak Vout+ at the carrier frequency f_0 becomes $2VDD/\pi+|VD|+|VS|$, exceeding that of a conventional SCPA. Therefore, with the same load impedance Rload, the proposed SCPA using the PVPN achieves enhanced output power and improved efficiency.

The PVPN implemented with a symmetric coupled line is shown in Fig. 33.4.2 (top). Both transmission lines have identical widths and lengths. Seen from the VDD terminal, the input impedance Zin exhibits a band-pass filtering frequency response with a reduced impedance around $2f_0$. For a conventional SCPA (i.e., without the PVPN), the charging capacitor leads to the current pulse with a period of $1/2f_0$. A Fast Fourier Transform (FFT) analysis reveals that the current ID comprises a DC component and significant AC harmonics at integer multiples of $2f_0$. After introducing the PVPN, the transient current ID is filtered and exhibits a smoother waveform without pulses. The corresponding frequency spectrum indicates that AC components above $2f_0$ are suppressed, with minimal impact on the DC component. Similarly, the discharging current IS exhibits analogous characteristics to ID. The process leading to periodic boosting of the output voltage Vout+ is depicted in Fig. 33.4.2 (bottom). In state I, the charging current ID of the positive SCPA cells induces a current IS1 in the adjacent transmission line. Conversely, in state II, the discharging current IS of the positive SCPA cells induces a current ID1. The induced current IS1 flows in the same direction as IS, while the direction of ID1 is consistent with ID. The current operation for the negative part of the SCPA cell is similar with the positive part. Thus, when the differential SCPA switches alternately between the positive and negative cells, the induced currents IS1 and ID1 reinforce ID and IS, respectively. As shown in Fig. 33.4.2 (top-right), simulated spectra of ID with the PVPN confirm an increased current magnitude at the frequency of $2f_0$. The enhanced currents ID and IS lead to the raised voltage swings of VD and VS. Consequently, the output voltage swing can exceed VDD after using the PVPN.

The architecture of the proposed polar SCPA is shown in Fig. 33.4.3, which consists of two identical differential sub-SCPAs. Each sub-SCPA is composed of 6-bit MSB unit cells, controlled by thermometer code, and 3-bit LSB unit cells, controlled by binary code. The control signals BB1[65:0] and BB2[65:0] (considering both of MSB and LSB) are generated by the six parallel 1:3 deserializers and decoders with the input parallel 6-bit baseband signal D[5:0]. The metal-stacked voltage-power-combining transformer (MSVPCT) with a minimal insertion loss of 1.02dB is used for output impedance matching. Meanwhile, the Doherty operation is introduced to achieve the -6dB PBO efficiency enhancement. To ensure a symmetric layout floorplan and enough DC current capacity of the SCPA array, two identical PVPNs are integrated within each sub-SCPA. As shown in the 3D view, the coupled lines for the PVPN are realized using top three thick metal layers (i.e., AP, M9, and M8) in the stacked topology, which decreases the parasitic resistance of the coupled line. The electric length is adjusted by tuning the physical length of the coupled line. The coupling factor k is tuned by changing the gap between two transmission lines. By optimizing the length and gap, simulations of half of the differential sub-SCPA array demonstrate a DE of 65% and a P_{out} of 23.5dBm at 5GHz. Both the MSB and LSB unit cells use a cascode inverter with thin-oxide transistors for higher output power. The delay cell is utilized to compensate the time delay of the lower signal path without a level shifter. The drain-source voltage swing of each NMOS and PMOS thin-oxide transistors in unit cells is carefully designed to remain below 2.2V, ensuring robust reliability in accordance with process limits. Under a supply voltage of 2.5V, the simulated output voltage Vout+ exhibits a peak-to-peak voltage swing exceeding 2.5V, which demonstrates the proposed concept and implies higher P_{out} and efficiency. Note that the voltage swing of Vout+ is affected by the coupling factor k.

The proposed SCPA using the PVPN is fabricated in a 40nm CMOS technology (Fig. 33.4.7), occupying a compact area of 0.99mm², including all I/O pads. Figure 33.4.4 shows the CW measured results. The SCPA delivers a peak P_{out} of 27.92dBm at 5GHz, a peak DE of 46.2%, and a peak PAE of 38.6% at 5.1GHz with a 1dB bandwidth from 4.5 to 7.2GHz. Meanwhile, the P_{out} at 5.1GHz is 27.90dBm. The power dissipation of the core circuits, driving circuits, deserializers, and decoders are considered in the PAE calculation. Such an SCPA with the Doherty operation exhibits 38.3% DE at -6dB PBO, representing a $1.7\times$ improvement over the standard Class-B PA. The AM-AM and AM-PM linearity is measured at 5GHz. The output phase variation is 18°, which is mainly deteriorated by the limited switching speed of transistors. The measured DNL and INL at 5GHz are +0.3/-0.3 LSB and +1.8/-2.7 LSB, respectively, which demonstrates good AM-AM linearity.

As shown in Fig. 33.4.5, the dynamic performance of the SCPA is firstly evaluated by a 200MHz single-carrier 64-QAM signal at 5GHz with an average P_{out} (P_{avg}) of 21.98dBm, EVM of -25.1dB, ACLR of -29.58dBc, and an average DE of 35.4%. Besides, an 80MHz single-carrier 256-QAM signal at 5GHz is measured featuring a P_{avg} of 21.09dBm, EVM of -31.3dB, ACLR of -34.98dBc, and an average DE of 29.2%. In addition, 40/50/80/100/160/200MHz 64-QAM and 10/20/40/80MHz 256-QAM signals are tested with the EVM of -32.8/-32.1/-31.1/-28.9/-27.3/-25.1dB and EVM of -34.0/-33.8/-33.0/-31.3dB, respectively. The DPD is used to minimize the AM-AM and AM-PM nonlinearities. The periodic voltage-pacing SCPA achieves a peak P_{out} of 27.92dBm and a peak DE of 46.2% while occupying a compact overall circuit area of 0.99mm², compared to the prior art at similar carrier frequencies reported in recent years (Fig. 33.4.6). Besides, it operates over a wide bandwidth of 4.5 to 7.2GHz and supports high data-rates, which are attractive for wireless applications.

Acknowledgement:
The authors thank the support of National Natural Science Foundation of China under Grant 62574034. Corresponding author: Xun Luo (xun-luo@ieee.org).

References:
[1] Y. Zhang et al., "A Power-Efficient CORDIC-less Digital Polar Transmitter Using 1b DSM-Based PA Supporting 256-QAM," *ISSCC*, pp.100-101, Feb. 2025.
http://doi.org/10.1109/ISSCC49661.2025.10904639
[2] S.-W. Yoo et al., "A Watt-Level Multimode Multi-Efficiency-Peak Digital Polar Power Amplifier with Linear Single-Supply Class-G Technique," *ISSCC*, pp. 368-370, Feb. 2020.
http://doi.org/10.1109/ISSCC19947.2020.9063069
[3] C. Hu et al., "A Wideband Replicas-Rejection Digital Transmitter Using Joint-Digital-Analog Interpolation and Filtering in 28nm CMOS," *ISSCC*, pp. 460-461, Feb. 2025.
http://doi.org/10.1109/ISSCC49661.2025.10904539

[4] A. Zhang et al., "A 5-to-6GHz Current-Mode Subharmonic Switching Digital Power Amplifier for Enhancing Power Back-Off Efficiency," *ISSCC*, pp. 364-365, Feb. 2021. http://doi.org/10.1109/ISSCC42613.2021.9365998

[5] V. Vorapipat et al., "A Class-G Voltage-Mode Doherty Power Amplifier," ISSCC, pp. 46-47, Feb. 2017. http://doi.org/10.1109/ISSCC.2017.7870253

[6] J. S. Park et al, "A CMOS Wideband Current-Mode Digital Polar Power Amplifier with Built-In AM–PM Distortion Self-Compensation," *IEEE JSSC*, vol. 53, no. 2, pp. 340-356, Feb. 2018. http://doi.org/10.1109/JSSC.2017.2760898

[7] M. Hashemi et al., "An Intrinsically Linear Wideband Digital Polar PA Featuring AM-AM and AM-PM Corrections Through Nonlinear Sizing, Overdrive-Voltage Control, and Multiphase RF Clocking" *ISSCC*, pp. 300-301, Feb. 2017. http://doi.org/10.1109/ISSCC.2017.7870380

[8] B. Yang et al., "A CMOS Wideband Watt-Level 4096-QAM Digital Power Amplifier Using Reconfigurable Power-Combining Transformer," *IEEE JSSC*, vol. 58, no. 2, pp. 357-370, Feb. 2023. http://doi.org/10.1109/JSSC.2022.3191975

[9] A. Zhang et al., "A Subharmonic Switching Digital Power Amplifier for Power Back-Off Efficiency Enhancement," *IEEE JSSC*, vol. 54, no. 4, pp. 1017-1028, Apr. 2019. http://doi.org/10.1109/JSSC.2019.2893534

[10] H. Tang et al., "A 4.7 GHz, 27.7dBm P$_{out}$, 37.8% PAE, 5.8° AM-PM Distortion Polar SCPA Using In-Cell Fast Slope-to-Phase Self-Calibration and Asymmetrical 4-to-1 Differential Power-Combining Transformer," *ISSCC*, pp. 102-103, Feb. 2025. http://doi.org/10.1109/ISSCC49661.2025.10904563

[11] S.-M. Yoo et al., "A Switched-Capacitor RF Power Amplifier," *IEEE JSSC*, vol. 46, no. 12, pp. 2977-2987, Dec. 2011. http://doi.org/10.1109/JSSC.2011.2163469

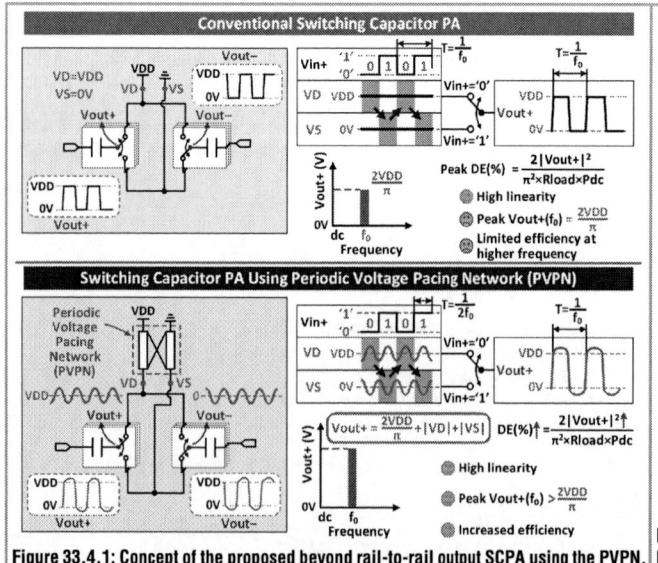

Figure 33.4.1: Concept of the proposed beyond rail-to-rail output SCPA using the PVPN.

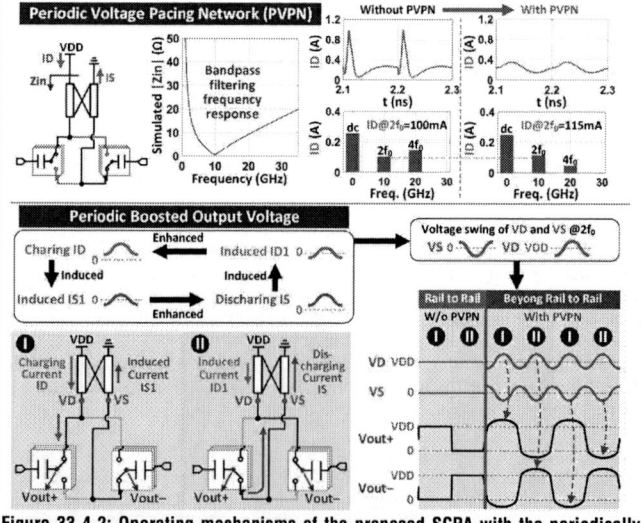

Figure 33.4.2: Operating mechanisms of the proposed SCPA with the periodically boosted output voltage.

ISSCC 2026 / SESSION 33 / TIME-VARYING CIRCUIT TECHNIQUES FROM RF TO mm-WAVE / 33.4

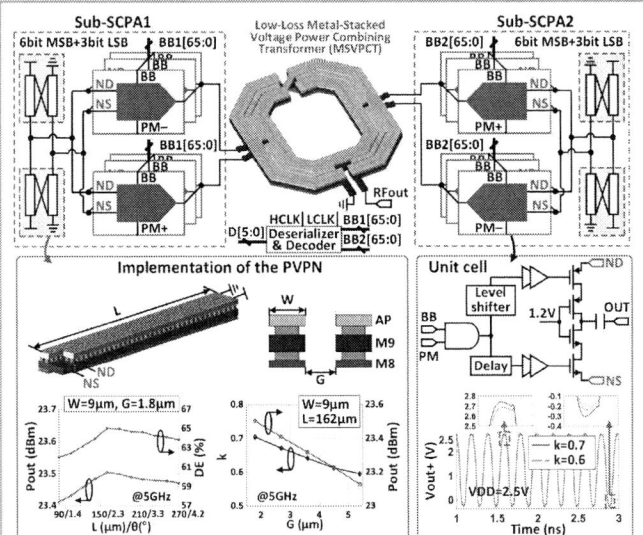

Figure 33.4.3: Simplified block diagram of the proposed SCPA using the PVPN.

Figure 33.4.4: The measurement results of the frequency response, efficiency at power back-off, AM-AM/AM-PM linearity, and DNL/INL.

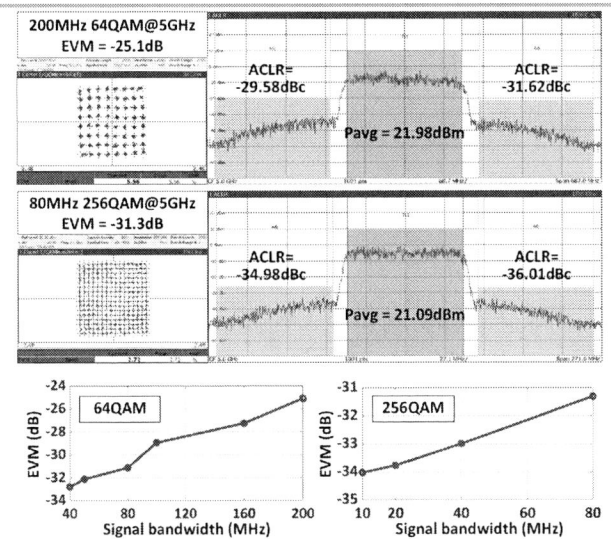

Figure 33.4.5: Measured results of modulated signals.

		This work	A. Zhang ISSCC'21	H. Tang ISSCC'25	C. Hu ISSCC'25	M. Hashemi ISSCC'17	J. S. Park JSSC'18
Architecture		Voltage pacing polar SCPA	Current-Mode Subharmonic Switching DPA	Self-Calibration Polar SCPA	Digital Polar Transmitter	Class-G Voltage-Mode DPA	Current-Mode Polar DPA
Technology		40nm CMOS	65nm CMOS	40nm CMOS	28nm CMOS	45nm CMOS	28nm CMOS
Supply (V)		2.5	N/A	1.2	1.1	1.2/2.4	1.4
Frequency (GHz)		4.5 to 7.2 (1dB)	5.2 to 6.2 (1dB)	4.4 to 5.2 (1dB)	2.6/3.6/5.1	2.9 to 4.3	2 to 4.3 (1dB)
Max P$_{out}$ (dBm)		27.92 (5GHz)	27 (5.7GHz)	27.7 (4.7GHz)	27.8/27.3/24.8	25.3(3.5GHz)	24.9(3.1GHz)
Peak PAE (%)		38.6 (5.1GHz)	N/A	37.8 (4.7GHz)	30.4/30.5/14.9#	30.4(3.5GHz)	N/A
Peak DE (%)		46.2 (5.1GHz)	40.1 (5.4GHz)	44.1 (4.7GHz)	N/A	N/A	42.7(2.5GHz)
-6dB PBO DE (%)		38.3 (5.1GHz)	26.3 (5.4GHz)	N/A	N/A	25.3*	N/A
Modulation (Frequency)		64 QAM (5GHz) / 256 QAM (5GHz)	256QAM (5.4GHz)	64QAM (4.7GHz)	WiFi 64QAM (2.56G)	32 Carriers 256QAM (3.5GHz)	64QAM (2.8GHz)
BW (MHz)		200 / 80	20	160	40	10	20
P$_{avg}$ (dBm)		21.98 / 21.09	22	21.63	21.1	17.1	18.2
Average η (%)		35.4(DE) / 29.2(DE)	27.4	N/A	21.6#	21.4*	23.7 (DE)
EVM (dB)		-25.1 / -31.3	-33.5	-25.19	-26	-40.1	-32.4
Chip size (mm²)		0.99/0.72**	7.1	3.59/1.12**	2.1/0.88**	1.2	3.24

*Power added efficiency (PAE) **Core size
#System efficiency

Figure 33.4.6: The comparison with prior-art DPAs and DTXs with similar operating frequencies.

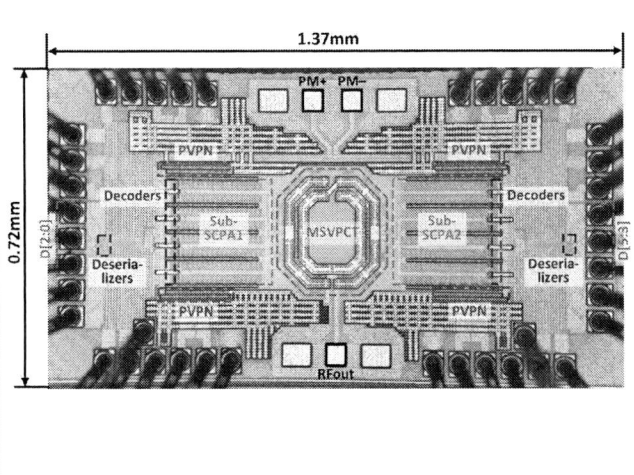

Figure 33.4.7: Die micrograph.

• 2026 IEEE International Solid-State Circuits Conference

979-8-3315-8937-0/26 $31.00 © 2026 IEEE

ISSCC 2026 / SESSION 33 / TIME-VARYING CIRCUIT TECHNIQUES FROM RF TO mm-WAVE / 33.5

33.5 A 0.6-to-0.9GHz, 28.06dBm P_out, 43.73% SE, 6-Phase Switched-Capacitor Power Amplifier Using In-Cell Digital Waveform Synthesis for the 2nd-, 3rd-, and 4th-Harmonic Suppression

Xiqing Liang[1], Bingzheng Yang[1,2], Hongxin Tang[1,2], Xun Luo[2]

[1]University of Electronic Science and Technology of China, Chengdu, China, [2]Shenzhen University, Shenzhen, China

Abstract

This work presents a 6-phase 0.6-to-0.9GHz switched-capacitor power amplifier with a peak output power of 28.06dBm and a peak system efficiency of 43.73%, using digital waveform synthesis for harmonic suppression. A reused differential-switch topology is utilized to generate multi-level stepped waveform with intrinsically canceled 2nd, 3rd, and 4th harmonics. Such an SCPA achieves 38.7/52.8/39.5dBc suppression level of the 2nd/3rd/4th harmonics at 850MHz and supports 40MHz 64-QAM and 20MHz 256-QAM signals.

The sub-1GHz frequency band has emerged as a potential candidate for Satellite Internet of Things applications, due to its superior propagation and penetration characteristics [1]. For power amplifiers operating in the sub-1GHz frequency band, high output power, high system efficiency, and clean signals with minimal harmonic distortion and spurious emissions are strongly demanded. Recently, digital power amplifiers (DPAs) have been widely investigated due to their high power efficiency with high integration levels [1-10]. However, the DPAs, operating in a switching mode, generally introduce significant unwanted harmonics (e.g., the 2nd and 3rd harmonics). These out-of-band harmonics of the DPA (especially when operating in a sub-1GHz range) interfere with other wireless systems operating at higher frequency bands [2].

A switched-capacitor power amplifier (SCPA) featuring high linearity is attractive for sub-1GHz applications. As shown in Fig. 33.5.1-top, the conventional SCPA with a 1/2 duty-cycle LO signal input leads to the intrinsic odd harmonics (e.g., 3rd harmonic). To suppress the 3rd harmonic, the SCPA with a 1/3 duty-cycle LO signal was presented in [2] as shown in Fig. 33.5.1-middle. However, the 1/3 duty-cycle LO results in inherent 2nd and 4th harmonics, which require a strictly symmetrical layout to avoid. Therefore, to achieve a pure frequency spectrum of the output signal without 2nd/3rd/4th harmonics, this work proposes the in-cell-digital-waveform-synthesis technique in Fig. 33.5.1-bottom. Such a digital-waveform-synthesis-technique SCPA cell converts the 1/2 duty-cycle LOs into a multi-level stepped waveform. A sine waveform operating at the 3rd harmonic is subtracted in the 1/2 duty-cycle LO waveform. Then, the synthesized waveform achieves high rejection of the 3rd harmonic with the inherent absence of 2nd and 4th harmonics. A Fast Fourier Transform (FFT) analysis shows that the synthesized multi-level stepped waveform cancels 2nd, 3rd, and 4th harmonics, as depicted in Fig. 33.5.1-bottom-right. Thus, the proposed in-cell-digital-waveform-synthesis technique features a multi-order harmonics rejection.

The proposed SCPA cell with in-cell-digital-waveform-synthesis technique and the diagram of the operation process are shown in Fig. 33.5.2. In Fig. 33.5.2-top-left, the proposed SCPA cell applies a reused differential-switch topology to generate the multi-level stepped waveform. Besides, the conventional power-charge and ground-discharge paths in an SCPA, the proposed reused differential-switch introduces a current reuse path between the positive and negative cells for the redistribution of electric charge. The current-reuse path (in green) is composed of two NMOS transistors (i.e., N_1, N_3) and two PMOS transistors (i.e., P_5, P_6). The VDD is connected between the drain terminals of P_5 and P_6 to ensure proper operation conditions. The enabled signals of the current-reuse path are LO_{EN1} and LO_{EN2} in Fig. 33.5.2-middle-left. When the LO_{EN1} is set to 1 and the LO_{EN2} is set to 0, the current-reuse path is enabled. When both the LO_{EN1} and LO_{EN2} are set to VDD, the current-reuse path is disabled. Therefore, within the operation process, the switch cell has four states: A) the power supply VDD2 charges the switched capacitor (i.e., C_P, C_N); B) the capacitor discharges to the ground; C) the positive capacitor C_P charges the negative capacitor C_N; D) the negative capacitor C_N charges the positive capacitor C_P. The diagram of the operation process is shown in the Fig. 33.5.2-right. In the initial state, the positive and negative output are assumed to 0 and VDD2, respectively, i.e., C_P is connected to GND and C_N is connected to VDD2. Therefore, the current-reuse path is enabled as state D. The C_P and C_N are electrically connected, and the C_N charges C_P until their voltages are same. Then, C_P is connected to VDD2, and C_N is connected to GND. Meanwhile, the current-reuse path is disabled for the variation of LO_{EN1} and LO_{EN2}. Next, the current-reuse path is re-opened as state C, and the charge direction is opposite to state D, i.e., C_P charges C_N. Finally, C_P is connected to GND, and C_N is connected to VDD2. Simultaneously, the current-reuse path is disabled. During this process, the generation of multi-level waveforms is realized through the current reuse between C_P and C_N. Then, the 2nd, 3rd, and 4th harmonics are intrinsically cancelled. It is worth noting that the positive and negative outputs exhibit excellent symmetry, which maintains the original voltage-symmetry characteristic of the square wave.

Figure 33.5.3 shows the configuration of the proposed in-cell-digital-waveform-synthesized SCPA using a six-phase topology. The six-phase LOs for digital waveform synthesis are generated by the frequency divider with a 3×LO input. Each sub-DPA array is composed of a 7-bit MSB and a 2-bit LSB. As shown in Fig. 33.5.3-bottom-left, the 7-bit MSB cells are controlled by thermometer codes, while the 2-bit LSB cells are controlled by binary codes. The waveform synthesis logic circuits are constructed by digital standard cells for low area

cost. As the digital control signal is generated without connecting the matching network, the digital waveform synthesis is independent of the matching network. The cell-reuse technique is applied to improve the average output power. Meanwhile, the sign-map module outputs three LO signals (i.e., $LO_\phi1$, $\phi2$, $\phi3$) with adjacent phases in each quadrant. In this work, the 4-to-1 power combining transformer in Fig. 33.5.3-bottom-middle is adopted for higher output power. Due to the enhancement of the magnetic fields of the two primary coils, the insertion loss of the power-combining transformer is <1.75dB over 0.6 to 2GHz with the minimal loss of 1dB. The phase difference and the rising/falling edges of the LO signals deteriorate the harmonic suppression effect. The simulation result shows that the phase difference of ±5° leads to about 20dB degradation of the third harmonic suppression. Meanwhile, the phase difference of ±30° introduces 5dB and 10dB variations of the second- and fourth-harmonic rejection levels, respectively.

The in-cell digital-waveform-synthesized SCPA chip was fabricated in a conventional 40nm bulk CMOS technology as shown in Fig. 33.5.7. The proposed SCPA occupies 0.994mm² of core area with the power supply of 1.2 and 2.4V. As shown in the Fig. 33.5.4-top-left, the measured chip achieves a 28.06dBm peak output power and a 43.73% peak system efficiency (SE) at 800MHz. The calculation of SE includes the power consumption of the core circuits with the digital waveform synthesis, decoders, dividers, driving buffers, and deserializers. The 2nd, 3rd, and 4th harmonic rejection are measured and shown in Fig. 33.5.4-top-right. This chip achieves >30dBc HR2, HR3, and HR4 over 600 to 900MHz, where the peak harmonic rejection reaches 38.7dBc HR2, 52.8dBc HR3, and 39.5dBc HR4 at 850MHz. The measured far-out spectrum of this chip operating at 850MHz is shown in Fig. 33.5.4-bottom.

Figure 33.5.5 shows the measured constellations and spectra of modulated signals at 850MHz. This chip supports 40MHz 64-QAM and 20MHz 256-QAM signals. For the 40MHz 64-QAM signal, it achieves an average output power (P_{avg}) of 20.69dBm, EVM of −27.25dB, and ACLR ≤ −30.11dBc. Meanwhile, for the 20MHz 256-QAM signal, it achieves a P_{avg} of 19.65dBm, EVM of −30.96dB, and ACLR ≤ −34.97dBc. In addition, the 20MHz 64QAM without power back-off is measured at 650MHz and 750MHz with the EVM of −27.47dB and −28.05dB, respectively. The comparison with the prior-art DPAs/DTXs is shown in Fig. 33.5.6. The proposed six-phase SCPA using the in-cell-digital-waveform-synthesis technique exhibits high output power and competitive system efficiency supporting high-order modulation signals including 40MHz 64QAM and 20MHz 256QAM. Meanwhile, such an SCPA simultaneously cancels the 2nd, 3rd, and 4th harmonics without any additional harmonic-rejection filters. With these advantages, the proposed SCPA is attractive for sub-1GHz wireless applications, offering improved signal quality and enhanced spurious suppression.

Acknowledgement:
The authors thank the support of National Natural Science Foundation of China under Grant 62304032 and 62574034, and Sichuan Science and Technology Program under Grant 2025ZNSFSC0525 and 2025ZNSFSC0519. Corresponding authors: Bingzheng Yang (bzyang1020@hotmail.com) and Xun Luo (xun-luo@ieee.org).

References:
[1] C. Hu et al., "A 0.7-to-2.5GHz Sliding Digital-IF Quadrature Digital Transmitter Achieving >40% System Efficiency for Multi-Mode NB-IoT/BLE Applications," *ISSCC*, pp. 472-474, Feb. 2023. http://doi.org/10.1109/ISSCC42615.2023.10067825
[2] J. Li et al., "A Highly-Integrated 6-Phase Cell-Reused Digital Transmitter Using 1/3 Duty-Cycle LO Signals for Harmonic Rejection," *ISSCC*, pp. 82-84, Feb. 2024. http://doi.org/10.1109/ISSCC49657.2024.10454514
[3] B. Yang et al., "A CMOS Wideband Watt-Level 4096-QAM Digital Power Amplifier Using Reconfigurable Power-Combining Transformer," *IEEE JSSC*, vol. 58, no. 2, pp. 357-370, Feb. 2023. http://doi.org/10.1109/JSSC.2022.3191975
[4] S.-W. Yoo et al., "A Multimode Multi-Efficiency-Peak Digital Power Amplifier," *IEEE JSSC*, vol. 55, no. 12, pp. 3322-3334, Dec. 2020. http://doi.org/10.1109/JSSC.2020.3022012
[5] S.-C. Hung et al., "A Quadrature Class-G Complex-Domain Doherty Digital Power Amplifier," *IEEE JSSC*, vol. 56, no. 7, pp. 2029-2039, July 2021. http://doi.org/10.1109/JSSC.2020.3040973

570 • 2026 IEEE International Solid-State Circuits Conference

979-8-3315-8937-0/26 $31.00 © 2026 IEEE

[6] A. Zhang et al., "A 5-to-6 GHz Current-Mode Subharmonic Switching Digital Power Amplifier for Enhancing Power Back-Off Efficiency," *ISSCC*, pp. 364-365, Feb. 2021. http://doi.org/10.1109/ISSCC42613.2021.9365998

[7] S.-W. Yoo et al., "A Watt-Level Quadrature Class-G Switched-Capacitor Power Amplifier with Linearization Techniques," *IEEE JSSC*, vol. 58, no. 5, pp. 1274-1287, May 2019. http://doi.org/10.1109/JSSC.2019.2904209

[8] S.-M. Yoo et al., "A Switched-Capacitor RF Power Amplifier," *IEEE JSSC*, vol. 46, no. 12, pp. 2977-2987, Dec. 2011. http://doi.org/10.1109/JSSC.2011.2163469

[9] V. Vorapipat et al., "A Class-G Voltage-Mode Doherty Power Amplifier," *IEEE JSSC*, vol. 52, no. 12, pp. 3348-3360, Dec. 2017. http://doi.org/10.1109/JSSC.2017.2748283

[10] M. Beikmirza et al., "A Wideband Four-Way Doherty Bits-In RF-Out CMOS Transmitter," *IEEE JSSC*, vol. 56, no. 12, pp. 3768-3783, Dec. 2021. http://doi.org/10.1109/JSSC.2021.3105542

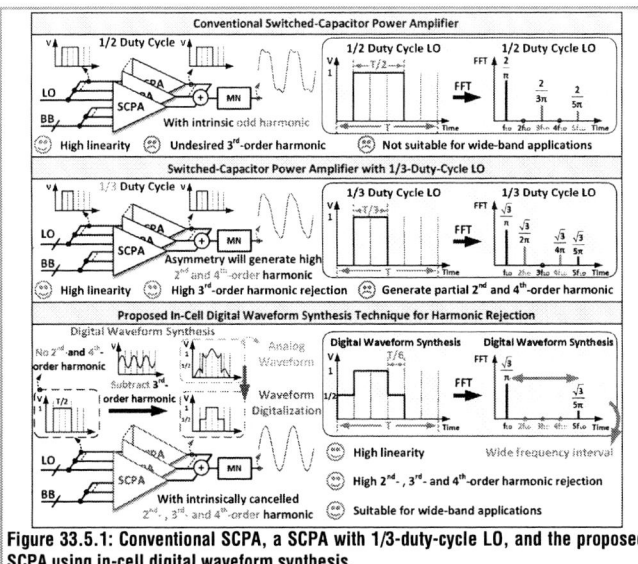

Figure 33.5.1: Conventional SCPA, a SCPA with 1/3-duty-cycle LO, and the proposed SCPA using in-cell digital waveform synthesis.

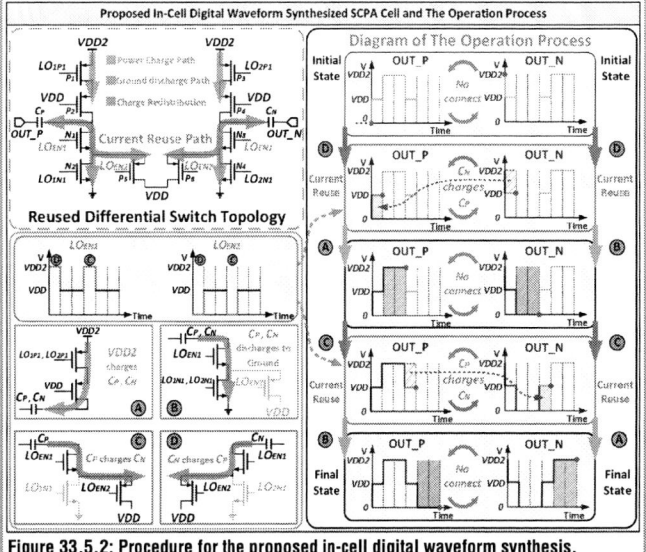

Figure 33.5.2: Procedure for the proposed in-cell digital waveform synthesis.

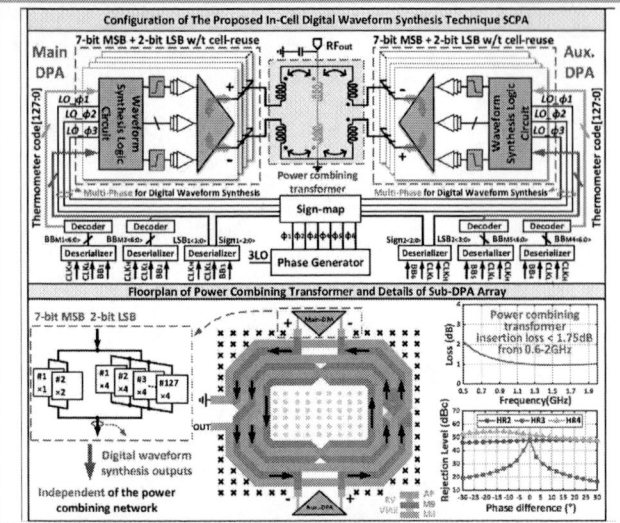

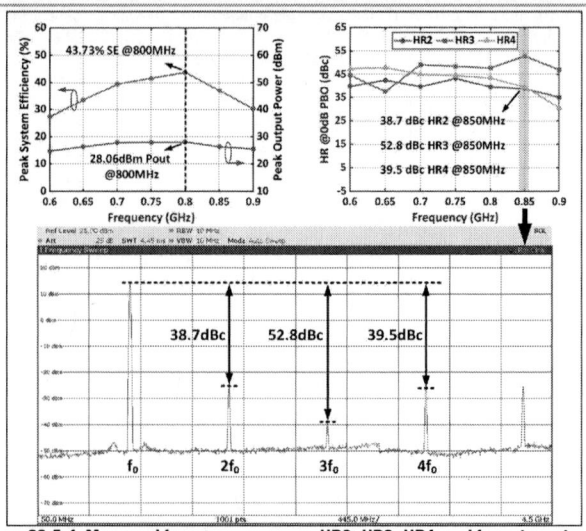

Figure 33.5.3: Block diagram of the proposed SCPA using digital waveform synthesis and the configuration of the power combining transformer.

Figure 33.5.4: Measured frequency response, HR2, HR3, HR4, and far-out spectrum at f_0=850MHz.

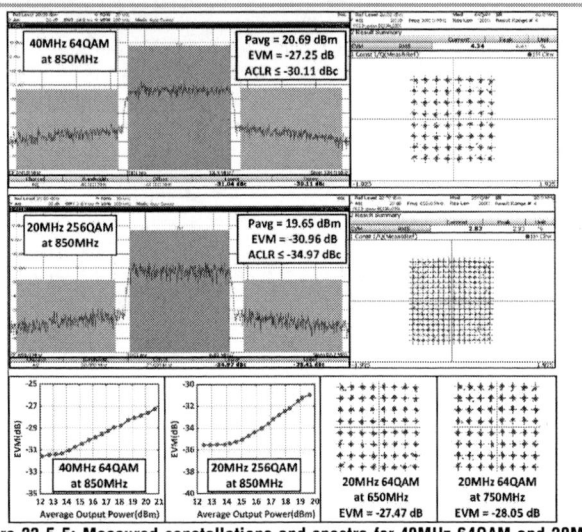

Figure 33.5.5: Measured constellations and spectra for 40MHz 64QAM and 20MHz 256QAM at 850MHz.

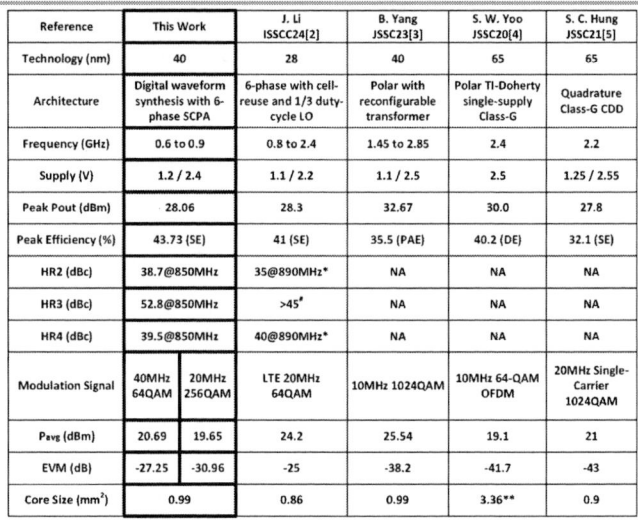

Reference	This Work		J. Li ISSCC24[2]	B. Yang JSSC23[3]	S. W. Yoo JSSC20[4]	S. C. Hung JSSC21[5]
Technology (nm)	40		28	40	65	65
Architecture	Digital waveform synthesis with 6-phase SCPA		6-phase with cell-reuse and 1/3 duty-cycle LO	Polar with reconfigurable transformer	Polar TI-Doherty single-supply Class-G	Quadrature Class-G CDD
Frequency (GHz)	0.6 to 0.9		0.8 to 2.4	1.45 to 2.85	2.4	2.2
Supply (V)	1.2 / 2.4		1.1 / 2.2	1.1 / 2.5	2.5	1.25 / 2.55
Peak Pout (dBm)	28.06		28.3	32.67	30.0	27.8
Peak Efficiency (%)	43.73 (SE)		41 (SE)	35.5 (PAE)	40.2 (DE)	32.1 (SE)
HR2 (dBc)	38.7@850MHz		35@890MHz*	NA	NA	NA
HR3 (dBc)	52.8@850MHz		>45#	NA	NA	NA
HR4 (dBc)	39.5@850MHz		40@890MHz*	NA	NA	NA
Modulation Signal	40MHz 64QAM	20MHz 256QAM	LTE 20MHz 64QAM	10MHz 1024QAM	10MHz 64-QAM OFDM	20MHz Single-Carrier 1024QAM
Pavg (dBm)	20.69	19.65	24.2	25.54	19.1	21
EVM (dB)	-27.25	-30.96	-25	-38.2	-41.7	-43
Core Size (mm²)	0.99		0.86	0.99	3.36**	0.9

* Estimate from figure   ** Chip size   # HR3 @6dB PBO

Figure 33.5.6: Performance summary and comparison with prior DPAs/DTXs.

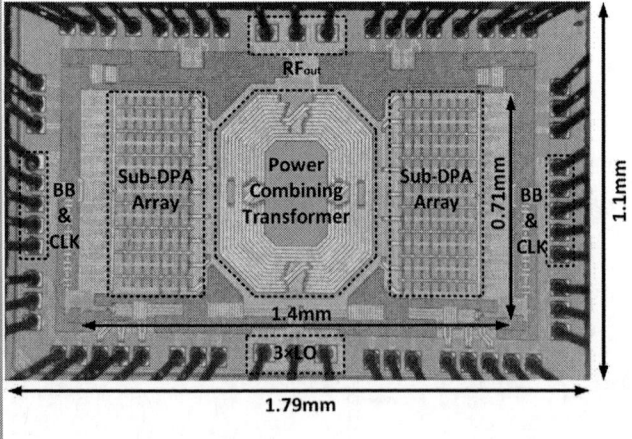

Figure 33.5.7: Die micrograph.

ISSCC 2026 / SESSION 33 / TIME-VARYING CIRCUIT TECHNIQUES FROM RF TO MM-WAVE / 33.6

33.6 A Frequency-Translated, 1-to-5GHz Centred, All-Passive, Programmable-Bandwidth, Switched-Capacitor Delay Line with 6.5-to-8.2dB Delay-Independent Insertion Loss in 65nm CMOS

Mohmad Aasif Bhat, Imon Mondal

Indian Institute of Technology, Kanpur, India

Abstract

We propose an all-passive, programmable-bandwidth, frequency-translated switched-capacitor delay line (DL) with a 6.5-to-8.2dB delay-independent insertion loss in 65nm CMOS. The DL offers tunable bandwidth (0.2 to 1GHz) and tunable centre frequency (1 to 5GHz) while supporting ns delay range. The results demonstrate a delay-range × BW of 3,

NF<11dB, and P1dB > +5.5dBm. In a narrow-bandwidth mode the DL has the OOB-IIP3 of +28dBm. The DL consumes an average power between 12 and 33mW across all bandwidth settings.

Linear periodically time-varying (LPTV) two-port true-time-delay (TTD) lines have gained importance because they can decouple the achievable delay from the physical (or emulated) length of the delay line (DL). They operate by catching the input from port-1 and storing it on a memory element in one phase and releasing it to port-2 in another phase in a time-interleaved manner (Fig. 33.6.1). The delay between the clock phases performing the catch-and-release operations constitute the delay of the analog signal between the ports. Thus, it uses the digitally implementable clock-phase delays to impart delay to an analog signal. This is particularly useful for the sub-6GHz frequency bands, as in this frequency range, the LC/transmission-line-based delay lines [1,2] tend to become increasingly bulky and lossy (to realize ns range delays), and the active/Gm-C based delay lines [3,4] suffer from frequency limitations (above 2GHz in bulk-CMOS) due their inherent parasitic capacitances. Increasing the bandwidth (BW) further, requires prohibitively large power. Additionally, an all-active implementation increases the noise figure (NF) and worsens distortion.

The TTD offered by the LPTV-based delay lines is only dependent on the clock phases, thereby de-linking the realizable delay from the chip area, resulting in a compact form factor with respect to all other passive linear time-invariant (LTI) DLs and low power and distortion in comparison to the active LTI-based delay lines. Leveraging these principles, an LPTV-based DL was proposed in [5], where the RC time constant was kept approximately equal to the on-time of the switches to prevent it from operating in the "sampling region," but small enough to not cause reduction in BW. This reduced the insertion loss (IL) with respect to a resistively loaded sampler and helped reach the BW approximately equal to N×f_{BW}, (f_{BW} = BW of one harmonic band of an N-path filter) by stitching together the harmonic bands of an N-path filter (Fig. 33.6.1). However, in this architecture, the overall group delay (GD) is centred around DC. In the work in [6,7], the memory element was replaced with an LC tank, thereby increasing the bandwidth of the DL, and moving the centre frequency to the resonant frequency of the tank (f_0). The stitching of the harmonic pass bands (by careful choice of the switching frequency) was used to increase the BW beyond that of a stand-alone RLC tank. The work in [8] improved upon the architecture by using inductors for bandwidth enhancement, but not for memory, thereby reducing the IL and also making it delay independent. While these improvements are significant, the realized IL and NF in the all-passive implementations in the sub-6GHz range are still >15dB (in 65nm CMOS), the primary contributor for which are the lossy inductors. Additionally, since the centre frequencies and BW are dependent on the inductors, they are not electronically programmable (without affecting the IL), thus limiting the efficacy of the DL for multi-band applications with different BWs.

In this work, we improve the LPTV delay lines by realizing the GD in baseband (around DC) using a harmonic-stitching N-path DL (similar to [5]), and then translating its driving-point impedance to f_{RF} using the frequency-translational property of an N-path filter (similar to [9,10]). This is shown in Fig. 33.6.2(top). The variable delay at RF is set by changing the phase difference between the baseband clocks operating at f_s (Fig. 33.6.2 right and bottom). Since the GD is now set by a baseband clock, $\Phi_i - \Phi_{id}$, its realizable value is also independent of the centre frequency (f_{RF}). This is particularly useful for beamforming applications with wide modulation bandwidths and also for large array sizes where the on-board signal routing adds to delays greater than one RF clock period.

The proposed architecture—henceforth referred to as frequency-translated switched-capacitor delay line (FTSCDL)—has the following advantages:
1. Since fixed inductors are no longer necessary, it removes the dependence of f_{RF} on the component values.
2. The losses associated with on-chip inductors are eliminated, which naturally improves both the insertion loss and NF significantly.
3. The translation of the driving-point impedance to f_{RF}, helps set the RF BW and GD independently of f_{RF} (tunable over frequency from 1 to 5 GHz).
4. The achievable BW is determined by the switched-capacitor (SC) delay cell operating at baseband and can be adjusted by trimming the capacitors and changing f_s, while keeping RC × f_s constant. Consequently, this architecture offers multi-bandwidth tuning flexibility that has not been available in conventional LPTV delay lines.
5. With an appropriate choice of f_s and the capacitor value (i.e., large RC and small f_s), the

DL can be configured as a narrow-band phase shifter. In this mode, thanks to the frequency translation of the driving-point impedance, the DL provides steep stop-band attenuation, enabling rejection of out-of-band (OOB) blockers. From a system-level perspective, the proposed architecture thus offers spectral filtering at RF in addition to spatial filtering when configured for phased-array applications, combining spatial and spectral filtering in a way that is not present in the literature on phase shifters.

Figure 33.6.3 shows the diagrammatic representation of the proposed chip. It uses 8-path SC time-interleaved delay cells at baseband to realize the desired baseband response, whose -3dB-magnitude bandwidth is controlled by the baseband clock frequency (f_s) and the RC product. The delay between clock phases, $\Phi_i - \Phi_{id}$, realizes the GD. The low-pass magnitude and the GD response thus realized are translated to RF by an extra set of switches (S_{1-4}) operating at switching frequency f_{RF}.

The implementation consists of differential four paths at RF, driven by 4-phase 25% duty-cycle clocks operating at f_{RF}. Note that the RF clocks in both ports operate in unison without any phase shift between them. The RF switches are designed using lvt NMOS transistors with a size of 20μm/60nm. They are optimized for IL due to the on-resistance and shunt capacitances through simulations. Two all-passive differential 8-path SC delay lines are added at baseband, sandwiched between the RF switches. Each path in the SC delay line has a grounded metal-insulator-metal (MIM) capacitor bank, with a three-bit digital control. The baseband switches are also implemented using the lvt NMOS transistors with a size of 64μm/60nm. To allow for the independent biasing of the delay line and switches, the gate terminal of each switch is AC coupled with a bias voltage $V_{sw} \approx 600mV$, and the source/drain terminals of the NMOS transistors are kept biased at $V_{cm} \approx 450mV$. An input sinusoidal signal at 4×f_s is applied directly to one of the clock-distribution blocks (to generate Φ_1 to Φ_8) and also through an off-chip digital delay network to another identical on-chip clock distributer (to generate Φ_{1d} to Φ_{8d}). Each of the clock distributers generates eight non-overlapping clock phases with a 12.5% duty cycle. Another on-chip clock distribution circuit with a sinusoidal input at 2×f_{RF} is used to generate four non-overlapping phases with 25% duty-cycle clocks to drive the RF switches (S_{1-4}). A digitally controlled on-chip duty-cycle-controlling circuit is added in each of the clock distribution blocks. An on-chip output buffer is used to de-embed the test board and other interface parasitics, following the procedure similar to [11]. There is also a provision to by-pass the buffer for linearity and noise-figure measurements. The overall chip is digitally controlled by an on-chip SPI module.

Figure 33.6.7 depicts the micrograph of the die implemented in a standard 65nm CMOS process and occupying an active area of 0.7mm×1mm. Most of the active area is dictated by the size of the capacitors, chosen to support the lower-bandwidth mode (large RC, small f_s). The chip was wire bonded on a 4-layer Rogers PCB. The RF clock and the input signals were routed at 90° to each other to minimize mutual inductive coupling (both on-board and on-chip), and each RF input to the chip was connected to two bond pads to reduce the inductance associated with the bond wires.

The chip was tested for multiple bandwidth modes, controlled by trimming the baseband capacitors and switching frequency, f_s, for varying f_{RF}. The effects of board and other parasitics were de-embedded from the measurement results using the output buffer, by following the same procedure as outlined in [4,11]. Figure 33.6.4 depicts the measured response for a 200MHz 3dB BW mode for f_{RF} between 1 and 5GHz. The measured results demonstrate frequency tunability of the magnitude response, with in-band loss of 6.5 to 8.2dB and S_{11} <-10dB for f_{RF} between 1 and 5GHz. The granular measured GD response around f_{RF}=2GHz and 4GHz are also shown, which demonstrates a clock-controlled flat GD range of 2.5 to 17.5ns around each f_{RF}. The measured small-signal results for a 1GHz instantaneous bandwidth mode are shown in Fig. 33.6.5. With f_s=250MHz, a 1GHz bandwidth is achieved over 2 to 5GHz f_{RF}, with a flat GD ranging from 0.5 to 3.5ns across the same BW. The measured NF for the 200MHz and 1GHz BW modes is also shown in Fig. 33.6.5, varying from 9 to 11dB across f_{RF}.

The measured large-signal performance of the chip is shown in Fig. 33.6.5, when the DL operates at f_s=50MHz (BW = 200MHz) around f_{RF}=3GHz. The delay line demonstrates the measured in-band IP1dB of +5.6dBm (for BW = 200MHz and 1GHz) and the in-band IIP3 of +18dBm with 3GHz and 3.01GHz in-band tones (for BW = 200MHz). To measure the out-of-band (OOB) performance of the architecture, a blocker was applied one-bandwidth away from f_{RF} and the measured in-band blocker P1dB was +6dBm. For OOB characterization (for RF BW=200MHz), two OOB tones at 3.2 and 3.4GHz were applied such that the OIM3 falls at 3GHz. The measured OOB blocker IIP3 is +28dBm. This demonstrates the strong OOB blocker rejection capability of the proposed DL in the narrow-band mode, while being centred around high f_{RF}. The measured average power consumption of the DL is 12 to 33mW with a 1.2V supply voltage, over multiple bandwidth modes across 1 to 5GHz RF. Figure 33.6.6 presents the performance comparison with prior-art LPTV-based delay lines. Unlike other LPTV DLs, the proposed architecture offers both the tunable bandwidth and tunable centre frequency across a sub-6GHz RF range. The results demonstrate a high delay-range-bandwidth product (DBW) of 3 and minimum IL and NF (per DBW) in comparison to reported LPTV delay lines operating beyond >2GHz RF. Also, unlike the works in Fig. 33.6.6, the proposed architecture achieves excellent blocker performance, a key feature of the FTSCDL.

Acknowledgement:
This work was supported in part by the Chip-to-Startup Program of Ministry of Electronics and Information Technology of the Government of India.
The authors thank Swami Sharan Donthagani for the layout design of the chip.

References:
[1] J. Roderick et al., "Silicon-Based Ultra-Wideband Beam-Forming," *IEEE JSSC*, vol. 41, no. 8, pp. 1726-1739, Aug. 2006. https://doi.org/10.1109/JSSC.2006.877257
[2] N. Rajesh and S. Pavan, "Design of Lumped-Component Programmable Delay Elements for Ultra-Wideband Beamforming," *IEEE JSSC*, vol. 49, no. 8, pp. 1800-1814, Aug. 2014. https://doi.org/10.1109/JSSC.2014.2317132
[3] S. K. Garakoui et al., "Compact Cascadable g_m -C All-Pass True Time Delay Cell With Reduced Delay Variation Over Frequency," *IEEE JSSC*, vol. 50, no. 3, pp. 693-703, Mar. 2015. https://doi.org/10.1109/JSSC.2015.2390214
[4] I. Mondal and N. Krishnapura, "A 2-GHz Bandwidth, 0.25–1.7 ns True-Time-Delay Element Using a Variable-Order All-Pass Filter Architecture in 0.13 µm CMOS," *IEEE JSSC*, vol. 52, no. 8, pp. 2180-2193, Aug. 2017. https://doi.org/10.1109/JSSC.2017.2693229
[5] A. Nagulu et al., "A Full-Duplex Receiver with True-Time-Delay Cancelers Based on Switched-Capacitor-Networks Operating Beyond the Delay–Bandwidth Limit," *IEEE JSSC*, vol. 56, no. 5, pp. 1398-1411, May 2021. https://doi.org/10.1109/JSSC.2021.3063658
[6] S. Ming et al., "A Commutated-LC RF Broadband Delay Circuit," *IEEE JSSC*, vol. 57, no. 11, pp. 3370-3383, Nov. 2022. https://doi.org/10.1109/JSSC.2022.3167431
[7] S. Ming et al., "A C-Band Commutated-LC-Negative-R Delay Circuit with Harmonic Power Recycling Achieving 1.5-ns Delay, 1.4-GHz BW, and 6-dB IL," *IEEE RFIC*, pp. 271-274, June 2022. https://doi.org/10.1109/RFIC54546.2022.9863138
[8] M. A. Bhat and I. Mondal, "A 3–5.5 GHz Compact, Power-Efficient, Switched LC Delay-Line With 0.2–1.1 ns Delay Range," *IEEE JSSC*, Early Access, 2025. https://doi.org/10.1109/JSSC.2025.3590509
[9] A. Nagulu et al., "A 1-to-5GHz All-Passive Frequency-Translational 4th-Order N-path Filter with Low-Power Clock Boosting for High Linearity and Relaxed Pdc-Frequency Trade-Off," *ISSCC*, pp. 378-380, Feb. 2023. https://doi.org/10.1109/ISSCC42615.2023.10067862

[10] M. A. Bhat and I. Mondal, "An LPTV Programmable Bandpass True-time-delay Line Without External Clock-phase Shifter," *IEEE ISCAS*, pp. 1-5, May 2025. https://doi.org/10.1109/ISCAS56072.2025.11042936
[11] S. Pavan and T. Laxminidhi, "Accurate Characterization of Integrated Continuous-Time Filters," *IEEE JSSC*, vol. 42, no. 8, pp. 1758-1766, Aug. 2007. https://doi.org/10.1109/JSSC.2007.900288

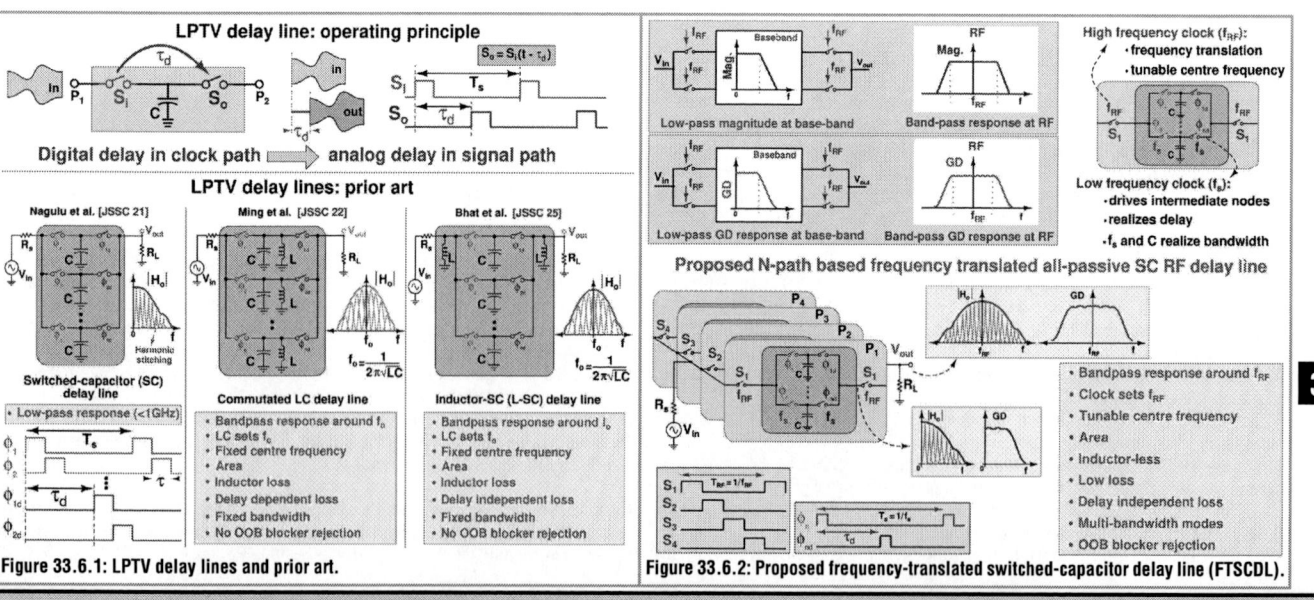

Figure 33.6.1: LPTV delay lines and prior art.

Figure 33.6.2: Proposed frequency-translated switched-capacitor delay line (FTSCDL).

ISSCC 2026 / SESSION 33 / TIME-VARYING CIRCUIT TECHNIQUES FROM RF TO mm-WAVE / 33.6

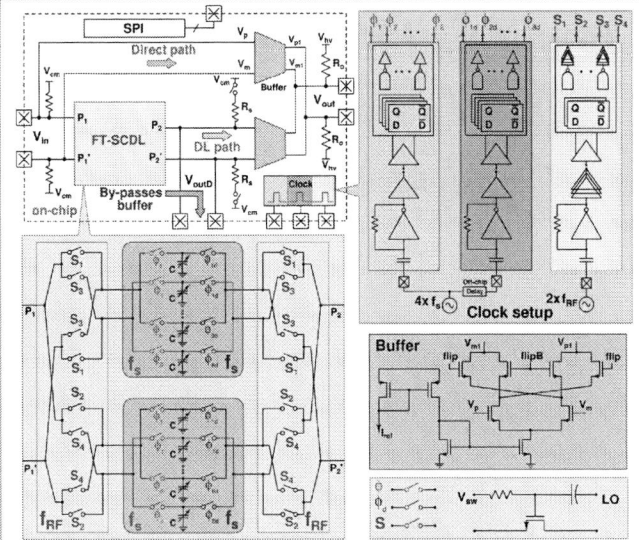

Figure 33.6.3: Top-level representation of the implemented design.

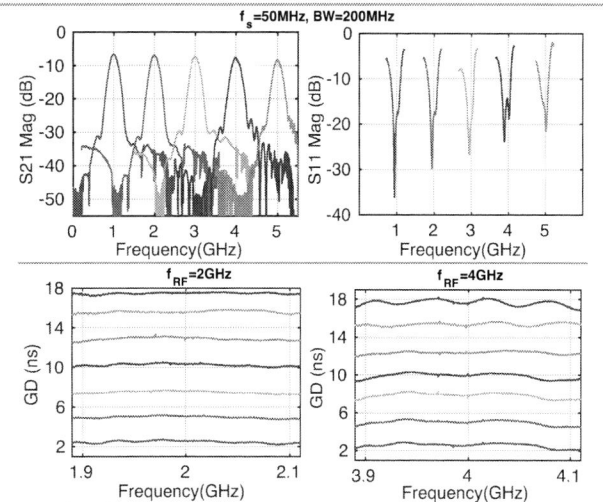

Figure 33.6.4: 200MHz BW mode: measured small-signal performance, S21 and S11 of the delay line at f_s =50MHz and across f_{RF}=1 to 5GHz (top), and the measured GD response around f_{RF} =2GHz and 4GHz (bottom).

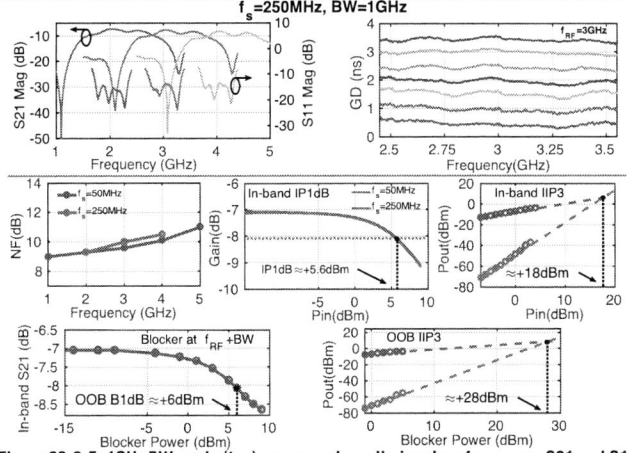

Figure 33.6.5: 1GHz BW mode (top): measured small-signal performance, S21 and S11 of the delay line at f_s =250MHz across f_{RF} =2 to 4GHz, and the measured GD response around f_{RF} =3GHz. Measured NF across f_{RF}. Measured IB IP1dB at f_s =50MHz and 250MHz, and IB-IIP3, out-of-band blocker B1dB and IIP3 at f_s =50MHz around f_{RF} =3GHz (bottom).

	This work				JSSC 2025 [8]	RFIC 2022 [7]	JSSC 2022 [6]	JSSC 2021 [5]
Delay-line architecture	FT-SCDL				L-SC	LC-nR RF	LC RF	SC RF
Technology (CMOS)	65nm				65nm	65nm	65nm	65nm
Active/Passive	Passive				Passive	Active	Passive	Passive
Frequency Range (GHz)	1 to 5				3 to 5.5	3.6 to 4.9	3.6 to 4.9	DC to 0.19
Tunable center frequency	Yes				No	No	No	No
Tunable bandwidth	Yes				No	No	No	No
Switching freq. (fs) (MHz)	50	100	150	250	800	250	250	50 to 75
3dB Instant. BW (GHz)	0.2	0.4	0.6	1	2.5	1.4	1.3	0.19
Max. Delay (ns)	17.5	8.7	5.8	3.5	1.15	1.5	1.4	11
Delay range × BW (DBW)	3	3	3	3	2.25	1.7	1.7	1.9
IB Insertion Loss (dB)	6.5 to 8.2				15	5.5 to 23	23	2.5
IP1dB (dBm)	+5.6				+2	-9 to +2	+1	N.R.
IB-IIP3(dBm)	+18				N.R.	N.R.	+16	N.R.
OOB-B1dB (dBm)	+6				N.A.	N.A.	N.A.	N.A.
OOB-IIP3 (dBm)	+28				N.A.	N.A.	N.A.	N.A.
Area (mm²)	0.7				0.325	1.2	1.2	2
Power, DC (mW)	12 to 28	14 to 29	16 to 31	26 to 33	16	26 to 36	26	N.R.
NF (dB)	9 to 11				15	11 to 23	23	N.R
NF/DxBW (dB)	3 to 3.6				6.6	5.23 to 10.95	13.52	N.A.
IL/DxBW (dB)	2.1 to 2.73				6.6	2.6 to 10.95	13.52	1.31

Note for IP1dB/IB-IIP3/OOB-B1dB/OOB-IIP3 rows: @ fs=50MHz (BW=200MHz)

Figure 33.6.6: Measured performance of the proposed delay line across different bandwidth modes and comparison with prior-art LPTV delay lines.

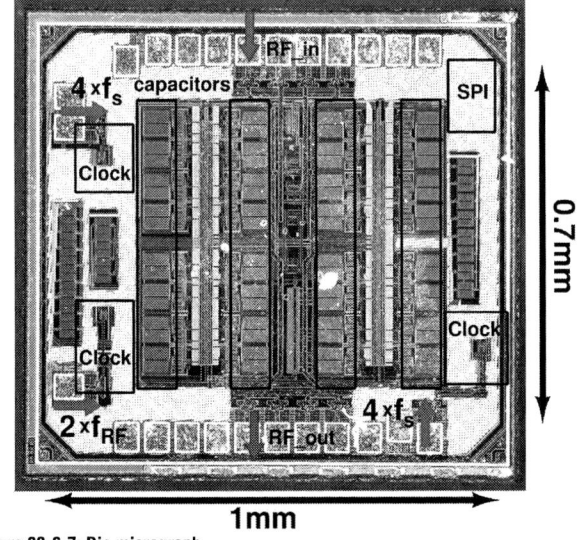

Figure 33.6.7: Die micrograph.

- 2026 IEEE International Solid-State Circuits Conference

ISSCC 2026 / SESSION 34 / INTEGRATED RADAR AND UWB TRANSCEIVERS FROM MICROWAVE TO SUB-THz / OVERVIEW

Session 34 Overview: *Integrated Radar and UWB Transceivers from Microwave to Sub-THz*

WIRELESS SUBCOMMITTEE

Session Chair: Zhiwei Xu
Zhejiang University, Zhejiang, China

Session Co-Chair: Wu-Hsin Chen
Qualcomm, San Diego, CA

This session highlights recent advances in radar SoC technologies. The first paper introduces a 20GHz radar-on-chip with respiration-synchronized PMCW, enhancing cardiac signal detection. The second paper presents 60/77GHz 4T/4R FMCW radar RFICs with 3cm range resolution and 200m detection range, followed by a 245GHz dual-polarized radar with AiP achieving 69% radiation efficiency. The fourth paper develops a 1.6-to-3.8GHz FMCW radar SoC for non-line-of-sight life detection through walls, and the fifth paper showcases an all-digital IR-UWB TX with 84% spectrum utilization and minimal sidelobes, meeting global compliance in a compact form.

ISSCC 2026 / February 18, 2026 / 1:30 PM

1:30 PM

34.1 A 128mW 2×4 Radar-on-Chip with Forward-ΔΣ DPLL-Locked Multi-Injection RTWO in 22nm CMOS Enabling ADC-Free Digitization and PS-Free Beamforming Demonstrated in In-Cabin Vital-Sign Monitoring

Liheng Lou, University of Science and Technology of China, Hefei, China

In Paper 34.1, USTC presents a radar-on-chip that leverages a 20GHz, forward-ΔΣ DPLL-locked RTWO as a (de)modulator, ±48° FoV beamformer, and ~12-ENOB data converter. By adopting respiration-synchronized PMCW, the SFDR of the cardiac signal is boosted by 9dB, enabling reliable DSP extraction.

1:55 PM

34.2 Compact, Low-Power 60-and-77GHz 4T/4R Multi-Mode FMCW Radar RFICs in 28nm CMOS

Hyun-Chul Park, Samsung Electronics, Hwaseong, Korea

In Paper 34.2, Samsung presents 60-and-77GHz 4T/4R multi-mode FMCW radar RFICs that achieve excellent range resolutions of 3cm, angle detection of 61.9° (2T/3R 60GHz radar), and detection ranges of around 200 meters (8T/8R 77GHz radar). The designs incorporate high-performance RF/analog blocks, featuring a TX P_{SAT} exceeding 14.4dBm, RX noise figures between 12 and 18dB, and an ABB/ADC with over 36MHz bandwidth and ENOB of 10.6. The proposed radars support multiple TX/RX modes and can be cascaded to larger system configurations.

2:20 PM

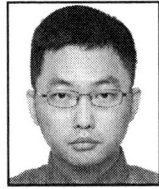

34.3 A 234-to-252GHz Dual-Polarized Transceiver Using Antenna-in-Package Technologies for Cross-Polarimetric Sensing

Xibi Chen, Massachusetts Institute of Technology, Cambridge, MA

In Paper 34.3, MIT and Intel introduce a 245GHz dual-polarized monostatic radar transceiver system designed for cross-polarimetric sensing. The system integrates AiP technology and features a modular assembly approach. Measurement results demonstrate a peak radiation efficiency of 69% at 250GHz, highlighting the effectiveness of the design for high-frequency sensing applications.

2:45 PM

34.4 A 1.6-to-3.8GHz Reconfigurable FMCW Radar SoC with 81.5% Relative-Bandwidth PLL for Real-Time Life Detection in Disaster Response

Fanxun Cai, Beihang University, Beijing, China

In Paper 34.4, Beihang University describes a 1.6-to-3.8GHz FMCW radar SoC for real-time non-line-of-sight life detection. The chip integrates a wideband PLL supporting chirp slope from 0.78 to 400MHz/μs, a PA delivering 13.43dBm output power, and an RX with 8.1dB NF. Additionally, a RISC-V core equipped with FPU and VPU enables efficient DSP. The system successfully detects human presence through six walls and vital sign behind two.

3:00 PM

34.5 A 0.0523mm² 11.4mW IEEE 802.15.4a/z/ab Compatible Aliasing-Suppressing All-Digital IR-UWB Transmitter Featuring Comb-Notched Maximally Flat Amplitude Spectral Shaping

Jihun Son, Daegu Gyeongbuk Institute of Science and Technology, Daegu, Korea

In Paper 34.5, DGIST shows an IEEE 802.15.4a/z/ab-compatible all-digital IR-UWB TX with a novel maximally flat amplitude spectral shaping while meeting global spectrum compliance. Consuming 11.4mW, it achieves the highest spectrum utilization (84%) and lowest near-in sidelobe (≤−35dBr at 489MHz) in a much smaller area than prior arts.

34

ISSCC 2026 / SESSION 34 / INTEGRATED RADAR AND UWB TRANSCEIVERS FROM MICROWAVE TO SUB-THz / 34.1

34.1 A 128mW 2×4 Radar-on-Chip with Forward-ΔΣ DPLL-Locked Multi-Injection RTWO in 22nm CMOS Enabling ADC-Free Digitization and PS-Free Beamforming Demonstrated in In-Cabin Vital-Sign Monitoring

Liheng Lou[1], Zeyu Zhou[1], Luyao Yuan[1], Yucheng Long[1], Gong Chen[2], Xiaoyu Li[3], Shaoqi Yang[1], Han Cui[1], Ji Peng[1], Weichen Tao[1], Juncheng Deng[1], Yizhe Hu[1]

[1]University of Science and Technology of China, Hefei, China, [2]National University of Singapore, Singapore, Singapore, [3]Shanghai Jiao Tong University, Shanghai, China

Abstract

A 2×4 phased-array SIL radar-on-chip in 22nm CMOS is demonstrated for in-cabin vital-sign monitoring. Built on a forward-ΔΣ DPLL-locked multi-injection RTWO at 20GHz, it concurrently serves as (de)modulator, beamformer, and data converter. The chip delivers 0dBm CW/PMCW for respiration cancellation, boosting SFDR by 9dB for cardiac extraction. RTWO-based beamforming covers ±48° FoV in ~15° AoA step. The 1.42mm² chip consumes 128mW and achieves RR/HR errors <±6bpm with 1.2mW DSP on-chip.

Vital-sign detection radars provide comprehensive, non-contact physiological monitoring, with applications spanning survivor detection, sleep analysis, clinical care, and in-cabin monitoring [1,2]. To meet these demands in compact and energy-efficient platforms, recent efforts have integrated radar functions onto a single chip, balancing sensitivity, efficiency, and scalability. Figure 34.1.1 illustrates a conceptual in-cabin monitoring (IM) system [3-5], including a driver monitoring system (DMS) and an occupancy monitoring system (OMS), based on a vital-sign radar-on-chip (RoC). While current DMS solutions predominantly rely on cameras, RoC offers a feasible alternative by extracting respiratory and heart rate (RR/HR) without imaging, enabling continuous, all-weather vital-sign monitoring with strong privacy [6]. For integrated IM, achieving balanced angular resolution is essential, making arrayed RoC configurations indispensable for robust occupant localization and monitoring. This paper presents a 20GHz phased-array RoC with a die area of 1.42mm², resolving RR/HR error <±6bpm at 128mW, demonstrated in a cabin-sized setup.

Conventional mixer-based vital-sign radar architectures, as in Fig. 34.1.1, typically achieve detection ranges of ~1.5m with 5dBm output power [7]. Although direct-conversion approaches leveraging range correlation can extract low-frequency cardiopulmonary signals, their RX NF (~7dB) necessitates >20dB LNAs, resulting in ~20mW per-channel power at mm-wave and >0.8mm² area. These systems also require ≥10-bit ADCs (~0.1mW, 0.1mm² each) and complex IQ DSP with offline data training to recover HR from RR–HR intermodulation harmonics [8,9]. Moreover, TRX arrays are needed to cover wide field-of-view (FoV) (e.g. >±45°), making fully integrated RoCs challenging in area, power, and real-time processing [8,10]. On the other hand, injection-locked-oscillator (ILO)-based radars inherently provide Doppler sensitivity and data efficiency. However, they still rely on mixers and time-delay elements for demodulation [11], followed by high-resolution ADCs for digitization, similar to closed-loop counterparts [12]. Furthermore, ILO-based architectures are difficult to scale into arrays, as only time-division schemes switching among paths with fixed phase offsets are feasible due to the narrow injection dynamic range [12,13], yielding degraded angular resolution even with complex channel hopping and synchronization. To overcome these limitations, we propose an ADC- and mixer-free RoC architecture (Fig. 34.1.1) built around a rotary traveling-wave oscillator (RTWO) locked by a digital PLL (DPLL). The received signals are injected into selected RTWO ports for beamforming without phase shifters (PSs), and the induced phase fluctuations are directly digitized through forward ΔΣ modulation in the DPLL, eliminating standalone ADCs. Occupying 1.42mm², the complete RoC, including an efficient respiration-cancelling DSP for RR/HR extraction, serves as a prototype SoC for intelligent in-cabin vital-sign monitoring.

This paper presents a phased-array RoC for hierarchical sensing to resolve RR and HR for in-cabin monitoring, as shown in Fig. 34.1.2. The RTWO is locked by a DPLL to generate a ~20GHz carrier from a 50MHz reference, which drives two TX channels. Echoes are injected through four LNAs into the RTWO, inducing slow periodic phase-pulling correlated with chest displacement and reflecting RR/HR. The DPLL compensates this deviation by adjusting the frequency control word (FCW), such that the FCW inherently digitizes the cardiopulmonary motion. A feedforward MASH 1-1-1 ΔΣ modulator enhances the RTWO's frequency resolution, achieving ~12ENOB at a 1.67GHz sampling rate. The two TXs employ phase interpolators (PIs) for independent absolute- and relative-phase control, enabling continuous phase trimming for respiration cancellation and, through 2-Tx beamsteering, forming an 8-channel virtual array with the four RXs. A spillover-cancellation path employing an attenuator and a PI preserves injection-locking (IL) sensitivity. Respiration-induced chest displacement is typically 5-to-20× larger than that from heartbeat, and the resulting RR–HR intermodulation produces harmonics that mask the cardiac spectrum, making HR extraction difficult even with powerful DSP. While RR can be readily obtained using IIR filtering, accurate HR recovery requires suppressing the source of intermodulation rather than relying on higher-order algorithms. In this work, this is achieved by transmitting a periodic phase-modulated continuous wave (PMCW) synchronized with RR, which cancels respiration at the modulation source and suppresses its harmonics, thereby unmasking the weaker cardiac signature. Benefiting from the high sensitivity of the self-injection-locked (SIL) RTWO-DPLL, only single-stage PAs and LNAs are required, with relaxed output power and NF.

Circuit-level implementation details are shown in Fig. 34.1.3. The RTWO combines the low PN of an LC oscillator with the wide locking range of a ring oscillator, making it well suited for SIL vital-sign radars. Gm is provided by current-starved inverters, enabling swing control to optimize locking range and hence detection sensitivity. Four out of eight injection ports are selected via 4-to-1 multiplexers along the Möbius ring, improving sensitivity by 12dB compared to single-point injection. A current phasor diagram in Fig. 34.1.3 illustrates optimal injection combinations for different angle-of-arrivals (AoAs), maximizing the locking range. RX-side beamforming is realized by signal re-routing to selected RTWO ports among 8 injection points, supporting a ±48° FoV with ~15° step. Three 7-bit integrating-mode PIs with DNL/INL of 0.4/2.0 LSB [14] and step size of 0.39ps (~2.8° at 20GHz) are employed with distinct functions: asynchronous tuning of PI.1/PI.2 enables beamsteering over ±60°; synchronous tuning of PI.1/PI.2 enables respiration cancellation for HR extraction, and also provides independent control for null-point avoidance; and PI.3 adjusts the phase of a loopback path for spillover suppression, with amplitude controlled by a cascaded 15-to-72dB attenuator. The DSP module then extracts RR and HR, where an IIR filter recovers RR and the respiration-cancellation loop suppresses intermodulation harmonics, revealing the cardiac component for simple peak detection. To hierarchically resolve RR and HR, the DSP acquires the ΔΣ code from the DPLL via a cascaded integrator comb (CIC) filter. RR is extracted using FFT and simultaneously mapped to a phase step (f_R–φ mapping in Fig. 34.1.3) for velocity control of phase modulation. In parallel, the filtered ΔΣ code is processed by a max–min detector, and the phase deviation range of RR is stored in registers. A scaling factor, k, is applied to this recorded range as a threshold. After respiration cancellation is engaged, the max–min detector continuously monitors the phase deviation range, and a comparator evaluates it against the stored threshold. When the current range (indicating the strength of f_R and associated harmonics) remains excessive, the comparator outputs a positive sign. An accumulator records these events and configures the CW mapping with updated phase-span information. In this system, $k = 0.88$ is adopted, ensuring the suppression of RR–HR intermodulation harmonics up to ~45dB theoretically, which enables reliable HR extraction without complex algorithms and allows full on-chip integration of all DSP functions, including FFT.

The phased array SIL RoC, fabricated in a 22nm CMOS process (Fig. 34.1.7), consumes 128mW from a 0.8V/3.3V supply (18.1mW RTWO, 20.2mW RXs, 27.8mW PAs, 41.3mW PIs, 19.8mW DPLL with ΔΣ, and 1.2mW DSP). The chip is mounted on a PCB for testing. Shown in Fig. 34.1.4, the IL-RTWO exhibits an injection-locking sensitivity of −90dBm under 4-point in-phase injection, ~11dB higher than single-point injection, and −75dBm under out-of-phase injection, providing ~15dB rejection for AoAs outside the main lobe in beamforming. The TX output spectrum is measured with PI modulated to provide continuous PMCW correlated with f_R for respiration cancellation, supporting ~0dB TX power. The sideband level, whose frequency offset indicates the phase modulation (PM) ratio, is −15dBc (limited by >1Hz VBW/RBW of the equipment, 5Hz/20Hz PM were tested), and remains compliant with ETSI EN regulations [15] since the PM is at the Hz-order, aligned with RR. With PI.3 and the attenuator optimally set, spillover cancellation achieves up to 63dB suppression by reducing the spillover-induced FCW offset to 2 LSBs. A TX–RX loopback test was performed by applying 0.08Hz/0.5Hz PM, covering RR from 0.083 to 0.42Hz. The control code was read out and reconstructed based on the ΔΣ frame, and FFT results confirmed that the demodulated signal matched the preset PM. A 5Hz PM is tested, producing the expected tone at 5Hz, which is also verified in TX output spectrum measurement.

The antenna array on PCB and its performance are shown in Fig. 34.1.5. Each parallel-fed 4-patch antenna element achieves 1.2GHz bandwidth and ~10dBi H/V gain. The RX antennas are spaced at λ/2, and radiation patterns of the synthesized 8 virtual channels demonstrate angular coverage up to ±48° with 12°-to-17° HPBW at ~15° step. Demodulated RR/HR spectra before and after respiration cancellation are reconstructed from the control code of a subject at 1m, 0° indoors. The most significant adjacent harmonics at both sides are suppressed by 14dB and 30dB, respectively, yielding ~9dB SFDR despite a 3dB drop at f_H. RR and HR are extracted across a FoV spanning 0.5-to-4m in range and ±48° in angle. Results show RR/HR errors ≤ ±2bpm/4bpm at 0° and ≤ ±4bpm/5bpm within ±48° AoA at 1m. A performance summary of the phased-array RoC prototype compared to prior arts [9,16-19] is presented in Fig. 34.1.6. While SIL radar modules have been widely studied, few IC-level implementations have been reported. By integrating hierarchical vital-sign sensing with high data efficiency and accuracy, enabled by PMCW-based respiration cancellation and harmonic suppression, the proposed fully integrated RoC provides a compact and high-performance solution for in-cabin vital-sign monitoring.

Acknowledgement:
This work was supported in part by the National Natural Science Foundation of China under Grant 62374156. Corresponding author: Yizhe Hu (huyz@ustc.edu.cn)

References:
[1] C. Li et al., "A Review on Recent Advances in Doppler Radar Sensors for Noncontact Healthcare Monitoring," *IEEE TMTT*, vol. 61, no. 5, pp. 2046-2060, May 2013. http://doi.org/10.1109/TMTT.2013.2256924

[2] G. Wang et al., "Application of Linear-Frequency-Modulated Continuous-Wave (LFMCW) Radars for Tracking of Vital Signs," *IEEE TMTT*, vol. 62, no. 6, pp. 1387-1399, June 2014. http://doi.org/10.1109/TMTT.2014.2320464

[3] American Prosecutors Research Institute, "Children and Cars – A Potentially Lethal Combination," Online. Available: https://www.ndaa-apri.org.

[4] Euro NCAP, "Safe Driving and Occupant Monitoring," Online. Available: https://www.euroncap.com.

[5] Euro NCAP, "Safe Driving and Driver Engagement," Online. Available: https://www.euroncap.com.

[6] Infineon, "60GHz In Cabin Monitoring System." Online. Available: https://www.infineon.com/assets/row/public/documents/cross-divisions/53/infineon-60Ghz-in-cabin-monitoring-system-2022-tc-applicationbriefen.pdf?fileId=8ac78c8c80f4d329018119612d3c4d59.

[7] Ping-Hsun Wu et al., "Vital-Sign Detection Doppler Radar Based on Phase Locked Self- Injection Oscillator," *IEEE MTT-S*, Montreal, QC, Canada, 2012, pp. 1-3. http://doi.org/10.1109/MWSYM.2012.6259524

[8] C. -C. Chou, W. -C. Lai, Y. -K. Hsiao and H. -R. Chuang, "60-GHz CMOS Doppler Radar Sensor with Integrated V-Band Power Detector for Clutter Monitoring and Automatic Clutter-Cancellation in Noncontact Vital-Signs Sensing," in *IEEE TMTT*, vol. 66, no. 3, pp. 1635-1643, March 2018. http://doi.org/10.1109/TMTT.2017.2777467

[9] N. Andersen et al., "A 118-mW Pulse-Based Radar SoC in 55-nm CMOS for Non-Contact Human Vital Signs Detection," in *IEEE JSSC*, vol. 52, no. 12, pp. 3421-3433, Dec. 2017. http://doi.org/10.1109/JSSC.2017.2764051

[10] M. Nosrati, S. Shahsavari, S. Lee, H. Wang and N. Tavassolian, "A Concurrent Dual-Beam Phased-Array Doppler Radar Using MIMO Beamforming Techniques for Short-Range Vital-Signs Monitoring," in *IEEE TAP*, vol. 67, no. 4, pp. 2390-2404, April 2019. http://doi.org/10.1109/TAP.2019.2893337

[11] W. -C. Su, M. -C. Tang, R. E. Arif, T. -S. Horng and F. -K. Wang, "Stepped-Frequency Continuous-Wave Radar with Self-Injection-Locking Technology for Monitoring Multiple Human Vital Signs," in *IEEE TMTT*, vol. 67, no. 12, pp. 5396-5405, Dec. 2019. http://doi.org/10.1109/TMTT.2019.2933199

[12] P. -H. Wu, J. -K. Jau, C. -J. Li, T. -S. Horng and P. Hsu, "Phase- and Self-Injection-Locked Radar for Detecting Vital Signs with Efficient Elimination of DC Offsets and Null Points," in *IEEE TMTT*, vol. 61, no. 1, pp. 685-695, Jan. 2013. http://doi.org/10.1109/TMTT.2012.2228222

[13] W. -C. Su et al., "2-D Self-Injection-Locked Doppler Radar for Locating Multiple People and Monitoring Their Vital Signs," *IEEE TMTT*, vol. 69, no. 1, pp. 1016-1026, Jan. 2021. http://doi.org/10.1109/TMTT.2020.3037519

[14] A. K. Mishra et al., "Improving Linearity in CMOS Phase Interpolators," *IEEE JSSC*, vol. 58, no. 6, pp. 1623-1635, June 2023. http://doi.org/10.1109/JSSC.2023.3243305

[15] ETSI, "Fixed Radio Systems; Characteristics and Requirements for Point-To-Point Equipment and Antennas; Part 2: Digital Systems Operating In Frequency Bands from 1 GHz to 86 GHz; Harmonised Standard for Access To Radio Spectrum", Online. Available: https://www.etsi.org.

[16] L. Lou et al., "A 253mW/channel 4TX/4RX Pulsed Chirping Phased-Array Radar TRX in 65nm CMOS for X-Band Synthetic-Aperture Radar Imaging," *ISSCC*, San Francisco, CA, USA, 2018, pp. 160-162. http://doi.org/10.1109/ISSCC.2018.8310233

[17] Y. -H. Liu et al., "A 680 µW Burst-Chirp UWB Radar Transceiver for Vital Signs and Occupancy Sensing up to 15m Distance," *ISSCC*, San Francisco, CA, USA, 2019, pp. 166-168. http://doi.org/10.1109/ISSCC.2019.8662536

[18] H. Reggad et al., "A Single-Chip Single-Antenna Radar for Remote Vital Sign Monitoring," *IEEE TMTT*, vol. 71, no. 10, pp. 4519-4532, Oct. 2023. http://doi.org/10.1109/TMTT.2023.3267554

[19] P. Luo et al., "A 28-nm 9-mm High-Resolution Multi-Mode IR-UWB Radar SoC with 16-GS/s Equivalent-Time Sampling for Non-Contact Detection of Human Vital Signs," *IEEE RFIC*, San Francisco, CA, USA, 2025, pp. 407-410. http://doi.org/10.1109/RFIC61188.2025.11082777

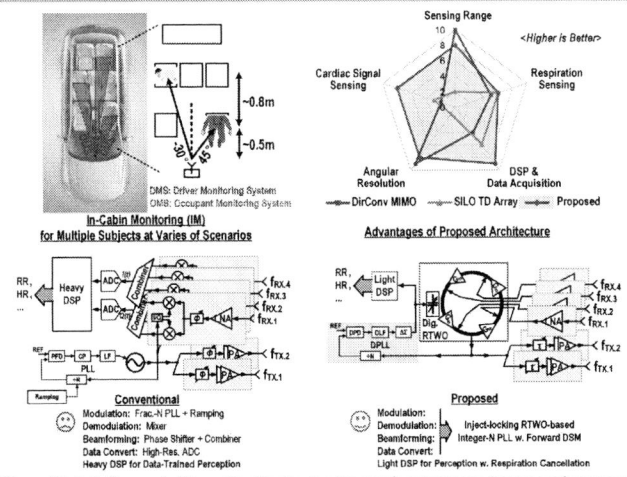

Figure 34.1.1: Concept of In-cabin Monitoring (top left), advantage features of proposed over the conventional direct-conversion TRX MIMO and Self-Injection-Locked (SIL) Time-Division (TD) Array for AoA resolution(top right), and implementational comparison between conventional and proposed radar-on-chip (bottom).

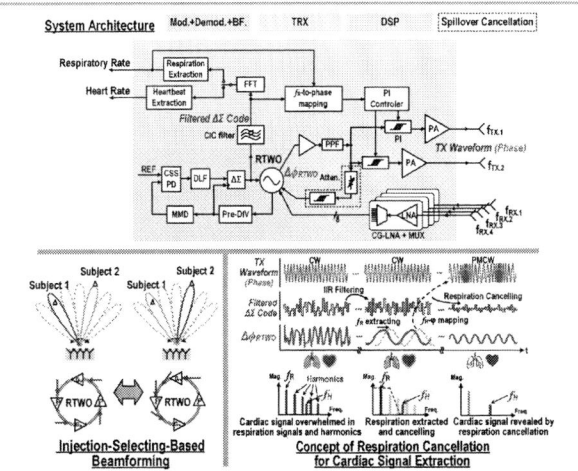

Figure 34.1.2: Architecture of the proposed self-injecting-locked phased-array RoC(top) for vital-sign sensing, conceptual illustration of the Injection-based beamforming for AoA estimation(bottom left), and principle of forward ΔΣ-based data acquisition and respiration cancellation(bottom right).

ISSCC 2026 / SESSION 34 / INTEGRATED RADAR AND UWB TRANSCEIVERS FROM MICROWAVE TO SUB-THz / 34.1

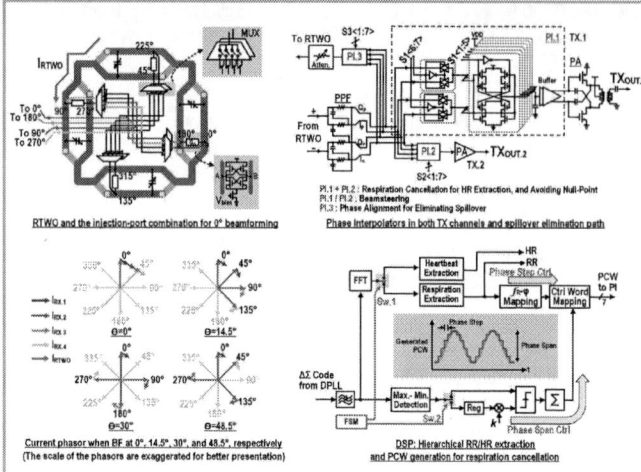

Figure 34.1.3: Circuit schematics of the RTWO with injecting-port MUX and injecting combination on phasor diagram (left), and Integrating-Mode PI and 1-stage PA (top right), and hierarchical RR/HR extraction and phase control word (PCW) generating implementation for respiration cancellation (bottom right).

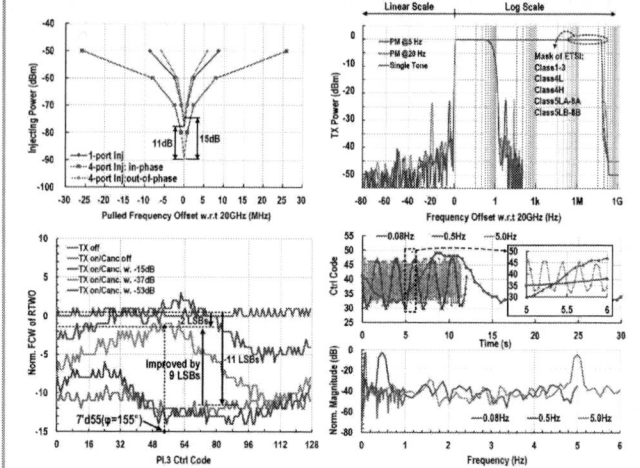

Figure 34.1.4: Measured results of the injecting sensitivity curves (top left), PMCW spectrum in linear scale and log scale with ETSI mask (top right), spillover cancellation (bottom left), and loopback FFT spectra of PMCW with 0.08Hz/0.5Hz/5Hz modulation rate (bottom right).

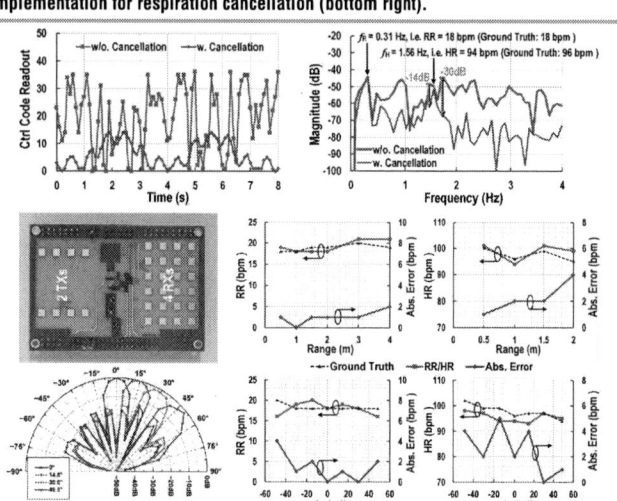

Figure 34.1.5: Measurement of the hierarchical RR/HR extraction with respiration cancellation (top), the 2×4 antenna array on the EVB and 8-virtual-channel beamforming pattern (bottom left), and RR/HR extracted and absolute error (bottom right).

	N. Anderson, JSSC'17 [9]	L. Lou, ISSCC'18 [16]	Y-H. Liu, ISSCC'19 [17]	H. Reggad, TMTT'23 [18]	P. Luo, RFIC'25 [19]	This Work
Transceiver Architectrue	TRX+PLL+ADC	TRX+PS+DLL+DDS+PLL	TRX+FLL+ADC	RO-PLL	TRX+ADC	TRX+PI+IL-RTWO-ΔΣDPLL
Channel Number	1TX+1RX	4TX+4RX	1TX+1RX	—	1TX+1RX	2TX+4RX
Modulation	Impulse	FMCW	Burst Chirp	CW	Impulse	CW+PMCW
Center Freq. (GHz)	7.2-8.5	10	6.8-8.2	1.5	3-5	20
Beamsteering	—	±60°	—	—	—	±60°
Beamforming	—	±60°	—	—	—	±48°
TX Pout (dBm)	0.7	10	0	-10	8.6	0
RX RFFE NF (dB)	6.3	5.6-6.5	12.5	—	—	3.9
Pwr per Chl. (mW) TX	4.86	179.3	6.6	—	—	27.3
Pwr per Chl. (mW) RX	76.5	73.7	9.7	—	—	5.3
Analog to Digital	ST ADC	—	SAR ADC	—	TI-ADC	Forward-ΔΣ DPLL
On-chip DSP	No	No	No	No	No	Yes
RR Range (m)	9	—	15	0 (Contacted)	2.5	>4
HR Range (m)	5	—	5	0 (Contacted)	—	>2
RR/HR Error (bpm)	—	—	—	—	—	≤ ±4/5
Peak DC Pwr (mW)	118	1012	19	7	154	128
Supply Voltage (V)	1.8	1.2	1/1.1	1.8	—	0.8/3.3
Die Area (mm²)	8.6	7.8	1.8	0.18	1.215	1.42
Technology	55nm CMOS	65nm CMOS	40nm CMOS	65nm CMOS	28nm CMOS	22nm CMOS

Figure 34.1.6: Performance summary and comparison.

Figure 34.1.7: Chip micrograph.

• 2026 IEEE International Solid-State Circuits Conference

979-8-3315-8937-0/26 $31.00 © 2026 IEEE

ISSCC 2026 / SESSION 34 / INTEGRATED RADAR AND UWB TRANSCEIVERS FROM MICROWAVE TO SUB-THz / 34.2

34.2 Compact, Low-Power 60-and-77GHz 4T/4R Multi-Mode FMCW Radar RFICs in 28nm CMOS

Byeong-Taek Moon[1]*, Mingyuan Li[2]*, Junseuk Suh[1]*, Minseob Lee[1], Doyoon Kim[1], Kyunghwan Kim[1], Geonho Park[1], Goeun Baek[1], Byungho Yook[1], Dooseok Choi[1], Kyungwoo Yoo[1], Junseong Kim[1], Taewoo Yu[1], Seonghyeon Kang[1], Hyeonsu Jo[1], Juhee Son[1], Sangsung Lee[1], Pak-Kim Lau[3], Siuchuang Ivan Lu[3], Gregory Eric Rogers[3], Ajaypat Jain[3], Jian Wang[2], Viduneth Ariyarathna[2], Wan Jong Kim[2], Oren Eliezer[2], Gennady Feygin[2], Wen Zhou[2], Kyoung-Jun Moon[1], Jaehyun Chung[1], Woncheol Lee[1], Seongjung Kim[1], Jonghyun Kim[1], Joonggeun Lee[1], Taeyeon Kim[1], Sungjoo Kim[1], Youngki Lee[1], Yonghwan Harold Jang[2], Sai Krishna Rayudu[3], Shihchieh Chien[3], Ying Chen[3], Hyungsun Lim[1], Kidong Kang[1], Sungjun Lee[1], Joonhee Lee[1], Jeongyeol Bae[1], Hyun-Gi Seok[1], Pranav Dayal[2], Wanghua Wu[3], Hyun-Chul Park[1], Joonhoi Hur[3], Sangmin Yoo[1], Chan-Hong Park[1], Joonsuk Kim[1]

[1]Samsung Electronics, Hwaseong, Korea, [2]Samsung Electronics, San Diego, CA, [3]Samsung Electronics, San Jose, CA
*Equally Credited Authors (ECAs)

Abstract

This paper presents 60-and-77GHz FMCW radar RFICs with 4T/4R MIMO and cascading capability in 28nm CMOS. The RFICs demonstrate excellent RF/analog/digital performance, including >14.8dBm TX P_{SAT}, 7-bit phase resolution, and >36MHz RX IF BW, with a compact 4.5x4.5mm^2 chip size and <2.1W power consumption. The RFICs support multiple radar modes and are validated through OTA tests, demonstrating their suitability for various radar applications.

Millimeter-wave (mm-Wave) frequency-modulated-continuous-wave (FMCW) radar has gained widespread adoption in home appliances, consumer devices, robotics, drones, industrial, and automotive applications due to its wide bandwidth (BW) and high-resolution sensing capabilities [1-4]. The 60GHz band, which provides license-exempt spectrum where high atmospheric absorption is experienced, is well-suited for short-range indoor and industrial applications [3], while the 77GHz band dominates automotive applications, supporting long-, mid-, and short-range detections for advanced driver-assistance systems (ADAS) and autonomous vehicles [1,4]. Recent trends toward multi-cascaded radio frequency integrated circuits (RFICs), such as 16T/16R systems built from four 4T/4R radar chips, offer cost-effective alternatives to LiDAR by enabling sub-1° angular resolution and >200-meter range when combined with centralized/imaging radar processing units [1,4]. Achieving this performance requires scalable, power-efficient RFICs with precise hardware (HW)-controlled timing, while reusing versatile analog and digital blocks across both 60-and-77GHz RFICs significantly reduces development complexity and cost [3,5]. This paper presents the system, design, implementation, and validation of separate 60-and-77GHz multi-purpose radar RFICs, each having fully integrated digital HW for chirp generation and system control and supporting 4T/4R multi-input multi-output (MIMO) operation.

Figure 34.2.1 shows the top-level schematic of the 4T/4R RFICs for 60 and 77GHz, highlighting shared functional blocks (white) and other blocks (gray) specifically parallelly designed/optimized for each band. The chips integrate mm-wave transceivers, two-stage cascaded phase-locked-loops (PLLs) with chirp generator, frequency-multiplier and divider networks, loopback-based built-in-self-test (BIST), high-speed camera-serial-interface-2 (CSI-2) data and serial-peripheral-interface (SPI) control IO interfaces, fully integrated digital front-end (DFE), LINK-PHY, microcontroller unit (MCU), and other digital functions. Each transceiver comprises four transmitter (TX) and four receiver (RX) paths. The TX chain incorporates high-resolution phase-shifters (PS), variable-gain-amplifiers (VGA), and driver/power amplifiers (DAs/PAs) to further achieve high angular resolution and extended detection ranges. Each RX path comprises a low-noise amplifier (LNA), highly linear mixer, widely tunable analog baseband (ABB) circuitry, and wide-BW/low-power analog-to-digital converter (ADC) for enhanced dynamic range and TX spillover/blocker immunity. The two-stage cascaded PLL architecture enables flexible chirp configurations, fast settling for steep chirp slopes, wide chirp BW, and improved phase noise (PN). Frequency multipliers (×3/×4) and divider networks ensure proper LO generation and balanced power distribution. TX-to-RX loopback-based BIST with on-chip power detectors (PDETs) supports accurate TX/RX calibrations. A DFE that is shared between all 4 paths delivers global timing references to all radar subsystems and performs time-domain pre-processing, including filtering and equalization. The LINK function consolidates and formats the data by combining DFE outputs, chirp parameters, and quality metrics. The CSI-2 and SPI interfaces allow external data/control access. An embedded MCU coordinates the data capture for BIST, monitors the radar system, and communicates with the radar host via inter-process communication (IPC) and GPIOs.

Figure 34.2.2 illustrates the overall operation of the FMCW radar RFICs with fully HW-controlled chirp generation. Single or cascaded RFIC operation is triggered by a 'Start' command, which generates the 'Sync_Out' and receives the 'Sync_In' signals. A precise clocking scheme enables counter-based chirp ($C_{1,1}, C_{1,2}, ...$), burst ($Burst_1, Burst_2, ...$), and frame ($Frame_1, Frame_2, ...$) timing, following predefined configurations. TX/RX settings for time-division multiplexing (TDM), bi-phase modulation (BPM), and Doppler-division multiplexing (DDM) modes are stored in on-chip SRAM lookup tables, allowing rapid reconfiguration. Each frame supports up to 4,096 chirps (up to 1,024 independent chirps per burst, 16 bursts per frame), maintaining accurate timing across both single-chip and cascaded operation. Chirp and mode variability enables dithering for interference mitigation, detecting/avoiding RX saturation, and supporting rapid mode switching for long-, mid-, and short-range detection. The two-stage cascaded PLL consists of a sampling analog PLL operating in integer-N mode to generate a low noise 3.84GHz reference, followed by a charge-pump PLL with a wide loop BW that enables fast lock/settling time and supports

chirp slopes up to 1,024MHz/μs. Chirp and modulation profiles can be flexibly configured through registers dedicated to 'Chirp_Profiles' and 'TX/RX Variables'.

Figure 34.2.3 shows the key RF/analog transceiver blocks for the 60-and-77GHz designs. While the RF front-end is independently designed and optimized for each frequency, the design of the ABB and ADC blocks is reused. These building blocks critically impact radar system performance including RX noise figure (NF), IP1dB, ABB filter corner frequency and gain control, ADC effective number-of-bit (ENOB) and sampling-rate (F_s), TX P_{SAT}, PS resolutions, and total power consumption (P_{DC}). Collectively, these factors define signal-to-noise ratio (SNR), detection range, and clutter suppression. The RX chain uses differential transformer (TF)-based amplifiers with neutralization capacitors and a current-mode passive mixer to enhance gain and linearity while minimizing NF at 60 and 77GHz. The two-stage ABB uses a TIA-based lowpass filter (LPF) and highpass filter (HPF) with feedback in the 1st-stage to suppress TX spillover interference and a widely tunable bandpass-filter-(BPF) response in the 2nd-stage, leading to flexible RX BW settings. The ADC adopts a 4th-order continuous-time (CT) delta-sigma modulator (DSM) using a current-steering digital-to-analog converter (DAC) with data-weighted averaging (DWA), for both high-speed operation and enhanced linearity. The ADC employs a 2nd-order CT loop filter and a 4-bit 2nd-order passive noise-shaping (NS) successive-approximation register (SAR), leading to a wide BW of 36MHz and significantly reduced P_{DC}. A dual-chain (77GHz TX only) vector-sum PS is proposed for the TX chain, as well as a four-way series-parallel power-combining PA and TR switches at each path to achieve up to 8-bit phase control resolution and high P_{SAT} (approximately 15dBm) with digitally controlled power adjustment.

Figure 34.2.4 presents the measured RX, TX, and BIST performance of the fabricated/packaged 60-and-77GHz radar RFICs in 28nm bulk CMOS and flip-chip ball-grid-array (FCBGA) package. The RX NF is 15 to 18dB (57 to 64GHz) and 12 to 15.5dB (74 to 81GHz), and OOB IP1dB is −11.6dBm and -8dBm, respectively. The ABB provides 20dB gain tuning in 2dB steps and adjustable LPF (10 to 40MHz) and HPF (0.3 to 4.2MHz) corner frequencies to support diverse applications. The ADC achieves 65.4dB SNDR (10.6-bit ENOB) and 74.3dB SFDR over 40MHz BW, operating at a sampling-rate of 960MS/s. TX P_{SAT} reaches 14.4dBm (60GHz) and 15.2dBm (77GHz), with phase-shifter resolutions of 7 bits. Most measurements are reproduced with on-chip loopback BIST with integrated PDETs for accurate calibration. For example, the internal loopback path can fully capture the ABB HPF frequency response (see Fig. 34.2.4), and detection/monitoring routines can be supported for RX gain/NF/IP1dB/IQ mismatch and TX P_{OUT}/gain/phase calibrations with the PDETs and DFE. TX P_{OUT} levels of P_0 and P_3 have also been measured accurately within ±1.0dB error. The package supports both 60-and-77GHz bands and its insertion loss of 1.5 to 2.0dB is included in the reported TX/RX measurement results.

Figure 34.2.5 demonstrates the over-the-air (OTA) measurements validating of the 60-and-77GHz RFICs with antenna arrays and field-programmable-gate-array (FPGA) boards (AMD K26 SOM). The evaluation boards and antenna arrays are fabricated using low-cost FR4 materials. Two configurations were tested: 2T/3R at 60GHz and 8T/8R at 77GHz. In the 60GHz cold-sky measurement for a single chirp, the noise floor reached −110dBm with observable spurs, clock mismatch, coupling, or ghost images. With chirp BW of 5.944GHz and slope of 117MHz/μs, the system achieved range of better than 3cm, angle detection of 61.9°, and field-of-view (FoV) of >70° (at -6dB). At 77GHz, measurements were performed using one 4T/4R chip with beamforming and a cascaded two-chip 8T/8R system with TDM MIMO configurations, detecting targets within ranges exceeding 100 meters, as verified using a radar-target-simulator from R&S. The 4T/4R MIMO radar detected a boresight range of 200/100 meters and at up to velocities of 78/40m/s with a BW of 0.24GHz, slope time of 26MHz/μs, and 1,024 chirps (radar-cross-section (RCS) = 25dBsm). The 8T/8R MIMO radar clearly detected a stationary target at 150 meters and a moving target with a velocity of 1m/s at 200 meters without requiring circuit or system-level calibration. The 3D range-Doppler map was extracted using non-coherent integration across the 64 virtual array elements. Estimated range and velocity information of each target was well matched to the

578 • 2026 IEEE International Solid-State Circuits Conference

979-8-3315-8937-0/26 $31.00 © 2026 IEEE

emulated conditions with sufficient SNR for radar post-processing. The observation that the outcome of the range-Doppler compression appears as distinct points serves as verification of proper timing synchronization between the two chips and chirp distribution from the primary to the secondary chip. With a BW of 0.33GHz, slope time of 10MHz/μs, and 512 chirps (RCS = T1: 30dBsm, T2: 50dBsm), the system demonstrated robust performance. Note that additional de-embedding was applied to compensate for the trace loss in the PCB and antenna degradation due to the high-loss FR4 substrate. While only a dual-RFIC system (8T/8R) has been demonstrated, the architecture can support up to four RFICs (16T/16R).

Figure 34.2.6 compares recently published radar RFICs at 60 and 77GHz. Both RFICs and their OTA platforms are independently implemented and demonstrate competitive RF/analog performance: TX P_{SAT} >14.8dBm, PS resolution ~7 bits, RX linearity (input P1dB) −11.6dBm, IF BW 36MHz, and compact 4.5×4.5mm² die size. P_{DC} is 1.9W (60GHz) and 2.1W (77GHz) at 4T/4R with 100% duty cycle. Note that duty cycle control and inter-burst chirp and TX/RX profile controls can further reduce P_{DC}. HW-based chirp control (up to 4,096 chirps) and versatile TX/RX modes (TDM/BPM/DDM) extend applicability across diverse radar applications. Figure 34.2.7 shows the die micrographs and FCBGA package. An additional major accomplishment is the reuse of analog/digital blocks (Fig. 34.2.1) and IC package design, resulting in a great reduction in the development effort and cycle time, both of which are critical in the industry. These fully implemented RFICs, offering robust FMCW radar operation and leading RF/analog/digital performance, are suitable for a wide range of radar applications.

References:
[1] F. G. Jansen et al., "Simultaneous Multi-Mode Automotive Imaging Radar Using Cascaded Transceivers," 2021 *18th European Radar Conference (EuRAD)*, London, United Kingdom, 2022, pp. 441-444. http://doi.org/10.23919/EuRAD50154.2022.9784489
[2] Z. Zhang et al., "A 77GHz Hybrid TDMA-MIMO Phased-Array Radar with 186m Detection Range and 3cm Range Resolution," *ISSCC*, San Francisco, CA, USA, 2025, pp. 01-03. http://doi.org/10.1109/ISSCC49661.2025.10904816
[3] K. Dandu et al., "High-Performance and Small Form-Factor mm-Wave CMOS Radars for Automotive and Industrial Sensing in 76-to-81GHz and 57-to-64GHz Bands," *ISSCC*, San Francisco, CA, USA, 2021, pp. 39-41.
http://doi.org/10.1109/ISSCC42613.2021.9365838
[4] B. P. Ginsburg et al., "A Multimode 76-to-81GHz Automotive Radar Transceiver with Autonomous Monitoring," *ISSCC*, San Francisco, CA, USA, 2018, pp. 158-160. http://doi.org/10.1109/ISSCC.2018.8310232
[5] K. Subburaj et al., "Digitally Assisted mm-Wave FMCW Radar for High Performance," *IEEE RFIC*, pp. 1-4, 2020. http://doi.org/10.1109/RFIC49505.2020.9218296

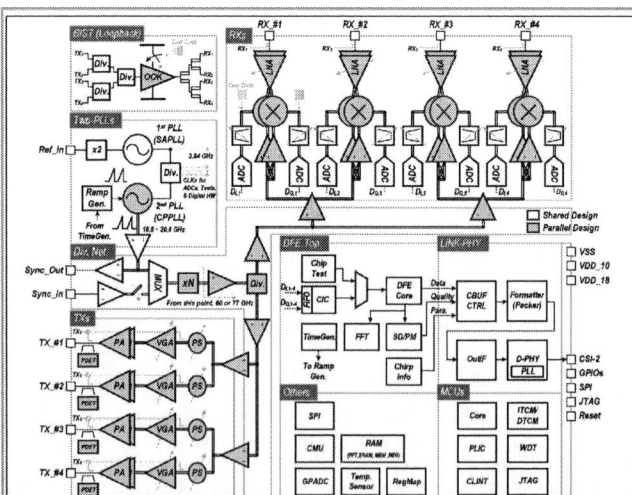

Figure 34.2.1: Top-level block diagram of 60-and-77GHz 4T/4R MIMO RFICs, including RF/analog, PLL with chirp generator, digital blocks, and BIST (blocks in gray were designed per system/band and those in white were shared/reused).

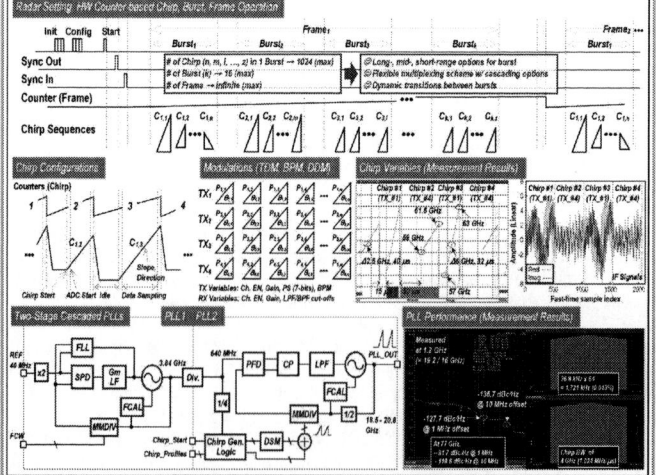

Figure 34.2.2: Overall operation of the FMCW radar RFICs with fully HW-controlled chirp generation and TX/RX modulations, a two-stage cascaded PLL block diagram, and measured chirp characteristic and PLL performance.

ISSCC 2026 / SESSION 34 / INTEGRATED RADAR AND UWB TRANSCEIVERS FROM MICROWAVE TO SUB-THz / 34.2

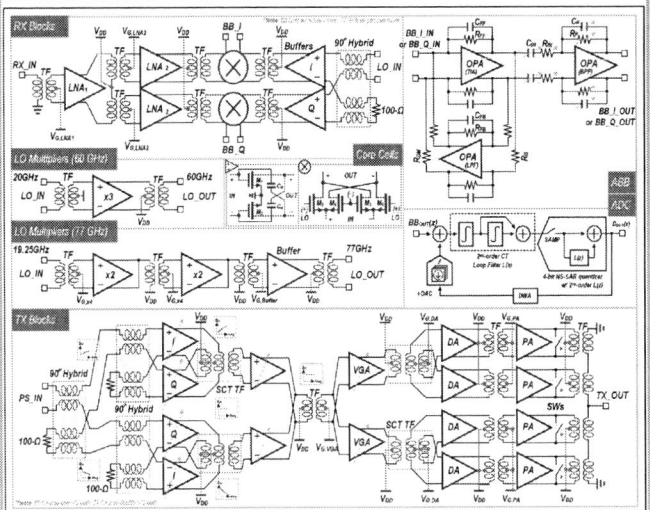

Figure 34.2.3: Key RF/analog transceiver blocks for the 60-and-77GHz designs, including RX chain, ABB/ADC, LO multipliers (×3, ×4), and TX chain.

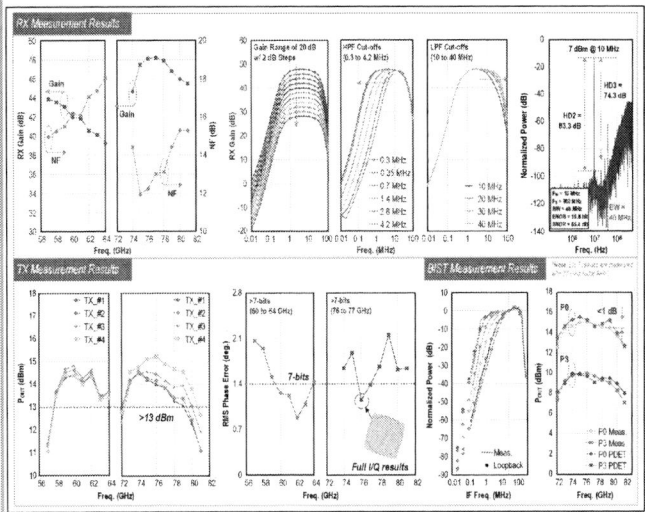

Figure 34.2.4: The measured RX, TX, and BIST performance of 60-and-77GHz radar RFICs, including the insertion loss of 1.5 to 2.0dB in TX/RX paths.

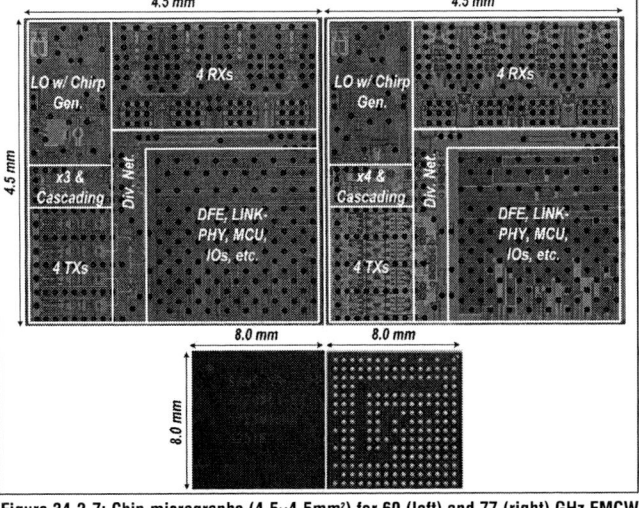

Figure 34.2.5: Empirical OTA validation of the 60GHz 2T/3R and 77GHz 8T/8R radar systems, including cold-sky tests and range/angle detection (60GHz), and long-range detection (77GHz).

	This Work		ISSCC25 [2]	ISSCC21 [3]	ISSCC21 [3]	ISSCC18 [4]
Technology	28-nm CMOS		55-nm CMOS	45-nm CMOS	45-nm CMOS	45-nm CMOS
Integration	FE, Chirp Gen, ABB, ADC, DFE, LINK-PHY, MCU, BIST		FE, Chirp Gen., ABB, ADC	FE, Chirp Gen, ABB, ADC, DFE, LINK-PHY, MCU, BIST	FE, Chirp Gen, ABB, ADC, DFE, LINK-PHY, MCU, BIST	FE, Chirp Gen, ABB, ADC, DFE, LINK-PHY, MCU, BIST
# of Channels (#T/#R)	4T/4R		8T/8R	3T/4R	3T/4R	3T/4R
Frequency (GHz)	57-64	76-81	77/79	76-81	57-64	76-81
Modulations	FMCW		FMCW	FMCW	FMCW	FMCW
Max Chirp Bandwidth (GHz)	7	5	8	5	4	4
Chirp Slope (MHz/μs)	24-200	32-1,024	545	266	250	100
Chirp Linearity (%)	0.01	0.001*	0.007**	-	-	-
Chirp Flexibilities	TDM, BPM, DDM		TDM	TDM, BPM, DDM	TDM, BPM, DDM	TDM, BPM, DDM
PN (dBc/Hz) at 1 MHz	-89.7	-91.7	-91	-93	-93	-91
Cascading	Yes (Up to 4 RFICs)		-	Yes	-	Yes
Max TX P_{OUT} (dBm)	14.8	15.2	-	13.1*	12.3*	13*
PS Res., RMS (Bit)	7 (4 GHz BW) 6 (8 GHz BW)	7 (2 GHz BW) 6 (8 GHz BW)	4	6	6	6
PS Gain Error, MAX (dB)	1.0	0.25	-	-	-	-
RX NF (dB)	15-18	12-15.5	16.8	12-12.8*	9.9-11.9*	14.5-17.2*
OOB IP1dB (dBm)	-11.6	-8	-	-10	-14	-7
RX Conversion Gain (dB)	39-44	44-48	26.5	55-60 (System)	-	-
RX Gain Range (dB)	20		-	-	-	-
IF BW (MHz)	36		0.4-20	20	10	15
Sampling Rate (MS/s)	40 (Real, 1x/2x Complex)		-	-	-	-
I/Q Support	Yes		Yes	Yes	Yes	Yes
ADC Res., ENOB (bits)	10.6		-	-	-	10.5
PDC (W), 100% Duty-Cycle	1.9	2.4	0.84	3.5	3.1	3.5
IOs for Data / Control	CSI-2 / SPI		GPIO / -	CSI-2 / SPI	-	CSI-2 / SPI
Chip Size (mm²)	4.5 x 4.5 (20.25)		8.2 x 8.2 (67.24)	-	-	22
PKG Type	FCBGA		-	FCBGA	FCCSP - AoP	FCBGA
PKG Size (mm²)	8.0 x 8.0		-	-	15.0 x 15.0	-

*Graphically estimated, **Measured at 128 MHz/μs, ***Measured at 16.8 MHz/μs.

Figure 34.2.6: Comparison table with state-of-the-art 60-and-77GHz FMCW MIMO radar RFICs.

Figure 34.2.7: Chip micrographs (4.5×4.5mm²) for 60 (left) and 77 (right) GHz FMCW radar RFICs, and FCBGA package (8.0×8.0mm²).

• 2026 IEEE International Solid-State Circuits Conference

ISSCC 2026 / SESSION 34 / INTEGRATED RADAR AND UWB TRANSCEIVERS FROM MICROWAVE TO SUB-THz / 34.3

34.3 A 234-to-252GHz Dual-Polarized Transceiver Using Antenna-in-Package Technologies for Cross-Polarimetric Sensing

Xibi Chen[1], Georgios C. Dogiamis[2,3], Ruonan Han[1]

[1]Massachusetts Institute of Technology, Cambridge, MA, [2]Intel, Chandler, AZ, [3]Deca Technologies, Tempe, AZ

Abstract

We present a 234-to-252GHz dual-polarized monostatic CMOS (Intel16) transceiver for cross-polarimetric sensing, with antenna-in-package (AiP) technology and a 10×10mm² modular system assembly. The AiP achieves 69% peak radiation efficiency estimated from measurements, which demonstrates the feasibility of >200GHz passive antenna/interconnects implemented in low-cost standard organic packages for enhanced performance and silicon area-saving, as opposed to the traditional all-on-chip scheme.

Sub-THz transceivers are favored in radar sensing applications for their low cost, small size, and high resolution. Typical automotive and security sensing requires the radar transceivers to receive echoed signal within the same polarization as the transmitted wave, since most radar objects have isotropic radar cross-section (RCS), and only generate tiny cross-polarized reflection. On the other hand, for low-power THz applications, such as THz IDs [1,2], it is common for the ID tag to have cross-polarized backscattering, so that the large background scattering does not affect the tag interrogation. Therefore, a dual-polarized transceiver with orthogonal TX/RX polarizations could be implemented as a tag reader, which effectively filters out the background scattering and only receives the tag response in cross-polarization (Fig. 34.3.1). Despite those advantages, sub-THz transceivers with on-chip antennas [3-7] typically suffer from the trade-offs between poor antenna efficiency and cumbersome assembly or post-processing. For example, to circumvent the inherently low efficiency from front-side radiation, backside radiation would require bulky silicon lens [3,6] or additional silicon substrate etching [7]. On-chip antennas, especially when forming large apertures, also occupy a large silicon chip area. In this paper, we present a 245GHz dual-polarized monostatic radar transceiver system for cross-polarimetric sensing. Through flip-chip C4 bumps, the silicon die is integrated with an off-chip step-horn antenna inside a standard organic chip package, and is housed in a low-cost modular assembly allowing for reliable thermal dissipation (Fig. 34.3.1). The system measurements indicate an estimated peak AiP radiation efficiency of 69% at 250GHz.

Figure 34.3.2 shows the transceiver AiP geometry, as well as the matching structures between the chip and package. At sub-THz frequencies, the chip-to-package interface dimension is electrically large, so that a dedicated matching circuit is needed to ensure a smooth THz signal transition. One big hurdle of the matching is the large parasitic shunt capacitance due to the large area of chip pads and the close distance between chip pads and chip ground. To resolve that issue, inductive matching structures are introduced on both the chip and the package sides. The AiP itself is designed based on a 6-layer organic package process, with a shared antenna interface for both TX and RX circuits on the chip. To achieve high transmission efficiency and to save metal layers in the package, M1 and M2 layers are used for substrate-integrated-waveguide (SIW) feedlines of the antenna. A cross-slot radiator is opened at the center of M2, so that the guided wave inside SIW can be efficiently converted for radiation with both 45° and 135° linear polarizations. A step-horn antenna is then built on the top with the remaining 4 metal layers, so that waves coming out of the slots are guided upwards to free space. Different opening sizes in each layer of the step-horn antenna, together with the slot radiators, enable multi-resonance of the AiP, which ensures >30GHz matching bandwidth (Fig. 34.3.2). To mitigate the cross-coupling between TX and RX, the AiP is differentially fed at the 4 corners of the chip. The two chip diagonals serve as differential pairs for the TX and RX, respectively. When the TX is fed differentially, the 135° slot in Fig. 34.3.2 is excited and generates a 45°-polarized radiation; while the odd symmetry rejects the TE_{10} mode in the RX SIW, minimizing the leakage towards RX. Similarly, when a 135°-polarized wave impinges on the AiP, only the 45° slot with RX SIW responds. At least 26dB of TX-to-RX isolation between 220 to 270GHz is obtained in the simulation. Such high isolation is also shown in the E-field simulation in Fig. 34.3.2.

The step-horn antenna with SIW feeds improves the tolerance of package manufacturing by using nearly all in-package vias – the primary bottleneck in fabrication resolution - as vertical ground walls, making the AiP resonance insensitive to the sizes or positions of vias. However, it occupies a relatively large in-package footprint compared to the transceiver chip, causing the AiP differential feeds to be placed near the chip corners rather than compactly grouped within a localized chip area. Thus, to avoid long-distance (i.e. lossy) THz transmission on chip, instead of routing the THz signals from a centralized circuit to the chip corners, the on-chip circuit system is divided into two phase-coherent slices (Fig. 34.3.2) near the chip corners. The amplifier-multiplier-chain (AMC) in Slice 1 provides a 0° phase directly to the TX+ port, while the AMC in Slice 2 provides a 180° phase to the TX- port. Fig. 34.3.3 illustrates the principle of this phase flip. Since the AMC contains a doubler, any differential excitation before AMC is converted to a common mode after the doubler. Therefore, a final-stage balun is implemented for the AMC. By flipping the terminal connections in Slice 1 and Slice 2, a 180° phase difference can be reliably achieved. An off-chip input at ~15GHz is first power-divided into the two slices (Fig. 34.3.1), the signal in each slice goes through a ×4 chain and multiplies to 60GHz. The 60GHz signal is further divided into two paths for TX and LO, where each path feeds to a doubler for 120GHz output, and a last-stage AMC for the final 240GHz output. Each final-stage AMC has a two-way power combining. Together with the differential feeds to the TX AiP, the complete TX performs an equivalent four-way power combining.

The transceiver chip was fabricated using an Intel16 CMOS FinFET process, and the photos of the die and its package are shown in Fig. 34.3.7. The transceiver assembly, shown in Fig. 34.3.3, follows the vertical stack illustrated in Fig. 34.3.1: the 2×2mm² chip is first flip-chip attached to the 7×7mm² package; they both are then flipped again and connected to a modular PCB. For better thermal dissipation of the transceiver chip, the modular PCB has a center opening (Fig. 34.3.3), so that the bare die is directly exposed from its backside, attached with thermal paste & heatsink. This thermal solution allows the 10×10mm² modular assembly to dissipate 1.7W of DC power with <50°C temperature, verified by a thermal camera. Compared to traditional all-on-chip backside-radiating systems, which require a bulky silicon lens while suffering from thermal challenges, our modular assembly provides both a compact system footprint and a favorable thermal condition. The modular PCB transfers all the transceiver I/Os to the castellated-holes (e.g. plated half-via) on its four edges, so that the whole assembly can be flexibly soldered to a motherboard.

The electrical characterizations of the transceiver are shown in Fig. 34.3.4. The TX chain is verified by using a VDI WR3.4 sub-harmonic mixer. The measured peak EIRP is 9.9dBm. Based on the simulated AiP directivity values, the total radiated power is estimated to be 0.3dBm. The RX chain is verified by using a VDI WR3.4 VNA extender (with its output power calibrated) and an off-chip IF amplifier. The measured minimum noise figure (NF) and maximum RX conversion gain (CG), including AiP radiation loss, are 22.8dB and 3.4dB, respectively. Radiation patterns in both the azimuth and elevation plane cuts are measured, as shown in Fig. 34.3.4. The measured radiation patterns indicate that the AiP is functioning as expected. A scanning electron microscope (SEM) image of the package cross-section is taken after the chip assembly (Fig. 34.3.4), based on which, the actual diameter (d) and height (h) of the chip-to-package solder balls, can be extracted. By updating those numbers in simulations, together with consideration of surface roughness of copper materials in the package, a modified simulated radiation efficiency curve can be obtained (Fig. 34.3.4), leading to an ~9% efficiency drop. With this modified efficiency, the chip-generated TX power without AiP loss is calculated from the measured radiated power (Fig. 34.3.4). Since the TX AMC and LO AMC are identical, this calculated TX power is also assumed to be the same as the LO power for the RX chain. With this updated LO power level in simulation, both the simulated NF and CG fit the measured results well. All these consistencies indicate that the above measurement-extracted radiation efficiency value (69% at 250GHz) is an accurate estimation.

Figure 34.3.5 shows the radar measurement for cross-polarimetric sensing. The transceiver module is paired with another 3D-printed dielectric lens to improve its link budget. To mimic the sensing target, a polarization plate is implemented on a PCB, where an array of anisotropic periodic rods flip the 45° linearly polarized incidence into 135° linearly polarized reflection. An FMCW signal is generated by an RF source (Keysight N5173B) with chirp modulation from a function generator (Tektronix AFG3102), and fed into the transceiver. After the ×16 multiplier chain on-chip, the transceiver gets 5.12GHz of FMCW bandwidth. Three different object scenarios are measured and compared. As the measured radar output spectra indicates (Fig. 34.3.5), the radar module only responds to the cross-polarized echo from the polarization plate, which proves the function. The measured range resolution (3.7cm) agrees well with the effective FMCW bandwidth (4.35GHz) after radar signal processing. Lastly, Fig. 34.3.6 gives a comparison with state-of-the-art sub-THz integrated transceivers. Compared with prior arts, this work demonstrates the feasibility of >200GHz passive antenna/interconnects implemented in low-cost standard organic packages, achieving 69% radiation efficiency, 10×10mm² system footprint, efficient thermal solution, and silicon area-saving, as opposed to the traditional all-on-chip scheme.

ISSCC 2026 / February 18, 2026 / 2:20 PM

Acknowledgement:
This work is funded by Intel Corporation through Semiconductor Research Corporation Member Specific Research Grants (CG72783377 & CG94330919) to MIT. The authors thank Intel University Shuttle Team for the chip fabrication, and Intel Technology Research Team, System & Advanced Packaging Technology Group for the assembly support. The authors also thank MIT T. J. Rodgers Laboratory and Dr. Steven F. Nagle, for providing prototyping facilities and helpful instructions

References:
[1] M. I. Ibrahim et al., "THzID: A 1.6mm² Package-Less Cryptographic Identification Tag with Backscattering and Beam-Steering at 260GHz," *ISSCC*, San Francisco, CA, USA, 2020, pp. 454-456. http://doi.org/10.1109/ISSCC19947.2020.9063068
[2] E. Lee, X. Chen, M. Ashok, J. Won, A. Chandrakasan and R. Han, "A Packageless Anti-Tampering Tag Utilizing Unclonable Sub-THz Wave Scattering at the Chip-Item Interface," *ISSCC*, San Francisco, CA, USA, 2024, pp. 226-228. http://doi.org/10.1109/ISSCC49657.2024.10454498
[3] E. Turkmen et al., "A 223–276-GHz Cascadable FMCW Transceiver in 130-nm SiGe BiCMOS for Scalable MIMO Radar Arrays," in *IEEE TMTT*, vol. 71, no. 12, pp. 5393-5412, Dec. 2023. http://doi.org/10.1109/TMTT.2023.3279872
[4] X. Chen et al., "A 140GHz Transceiver with Integrated Antenna, Inherent-Low-Loss Duplexing and Adaptive Self-Interference Cancellation for FMCW Monostatic Radar," *ISSCC*, San Francisco, CA, USA, 2022, pp. 80-82. http://doi.org/10.1109/ISSCC42614.2022.9731637
[5] X. Yi, C. Wang, X. Chen, J. Wang, J. Grajal and R. Han, "A 220-to-320-GHz FMCW Radar in 65-nm CMOS Using a Frequency-Comb Architecture," in *IEEE JSSC*, vol. 56, no. 2, pp. 327-339, Feb. 2021. http://doi.org/10.1109/JSSC.2020.3020291
[6] J. Grzyb, K. Statnikov, N. Sarmah, B. Heinemann and U. R. Pfeiffer, "A 210–270-GHz Circularly Polarized FMCW Radar with a Single-Lens-Coupled SiGe HBT Chip," in *IEEE Transactions on Terahertz Science and Technology*, vol. 6, no. 6, pp. 771-783, Nov. 2016. http://doi.org/10.1109/TTHZ.2016.2602539
[7] M. Kucharski, W. A. Ahmad, H. J. Ng and D. Kissinger, "Monostatic and Bistatic G-Band BiCMOS Radar Transceivers with On-Chip Antennas and Tunable TX-to-RX Leakage Cancellation," in *IEEE JSSC*, vol. 56, no. 3, pp. 899-913, March 2021. http://doi.org/10.1109/JSSC.2020.3041045
[8] B. Liu et al., "A 132-to-148GHz CMOS 4TX-4RX FMCW Radar Transceiver Array with Cavity-Backed Antenna-in-Package Achieving 28dBm EIRP," 2025 *IEEE International Solid-State Circuits Conference (ISSCC)*, San Francisco, CA, USA, 2025, pp. 1-3. http://doi.org/10.1109/ISSCC49661.2025.10904524
[9] Y. Chen, H. Beshary, E. Chou, M. Wei, N. Baniasadi and A. Niknejad, "A 140GHz FMCW Radar with 22dB Wideband RF-Domain Multipath Self-Interference Cancellation in 28nm CMOS," *IEEE RFIC*, San Francisco, CA, USA, 2025, pp. 307-310. http://doi.org/10.1109/RFIC61188.2025.11082831

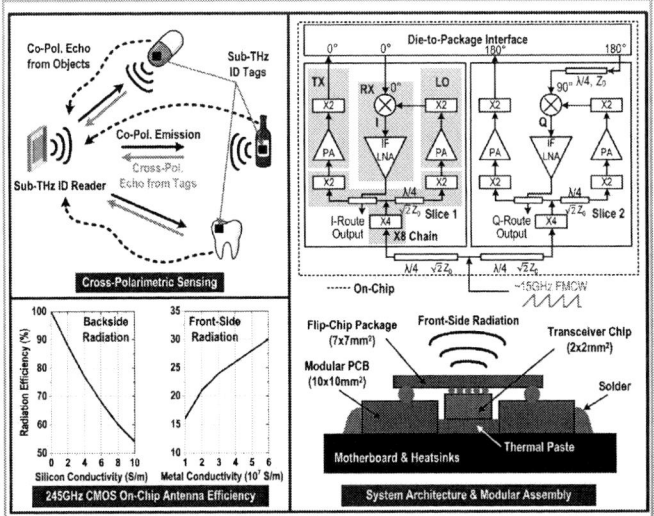

Figure 34.3.1: System architectures, cross-polarimetric sensing, and sub-THz on-chip antenna efficiencies.

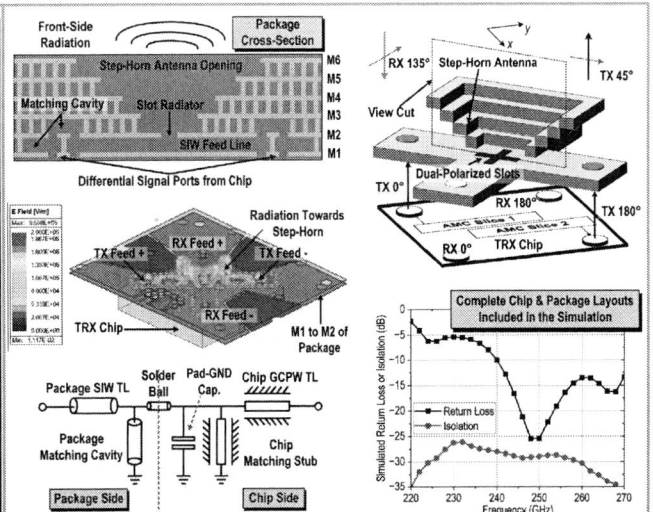

Figure 34.3.2: Dual-linear-polarized in-package antenna (AiP), chip-to-package matching circuit and simulations.

DIGEST OF TECHNICAL PAPERS • 581

979-8-3315-8937-0/26 $31.00 © 2026 IEEE

ISSCC 2026 / SESSION 34 / INTEGRATED RADAR AND UWB TRANSCEIVERS FROM MICROWAVE TO SUB-THz / 34.3

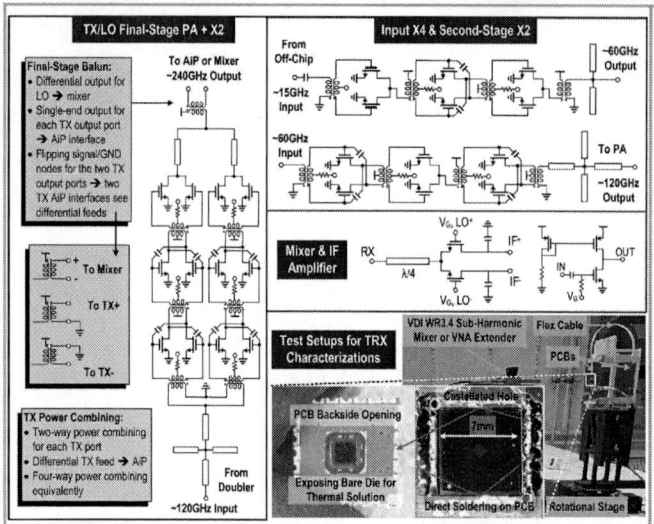

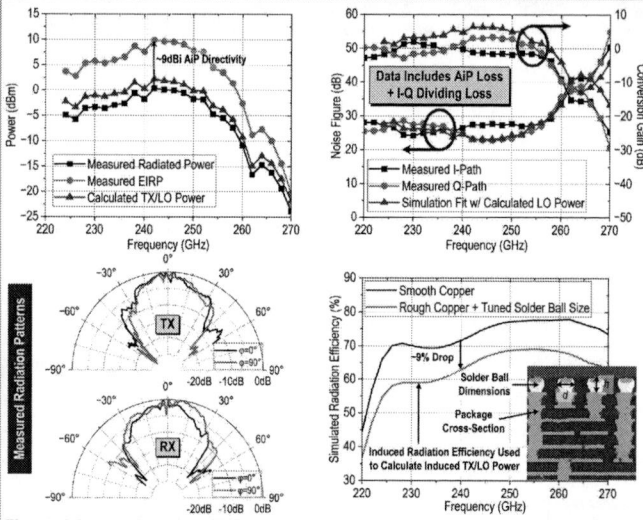

Figure 34.3.3: Schematics of the transceiver circuits. Test bench for TRX characterizations.

Figure 34.3.4: Measured TX/RX performance, AiP radiation patterns. 69% peak efficiency estimated from measurement.

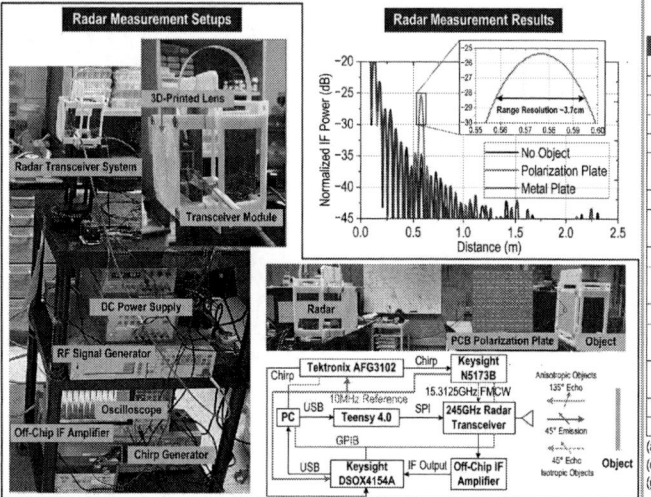

Figure 34.3.5: Measured radar detection results. Block diagram and photos of the radar test setups.

Figure 34.3.6: Comparison with state-of-the-art sub-THz integrated transceivers.

References	This Work	ISSCC 2025 [8]	RFIC 2025 [9]	T-MTT 2023 [3]	ISSCC 2022 [4]	JSSC 2021 [5]
Technology	Intel16 CMOS	28nm CMOS	28nm CMOS	130nm SiGe	65nm CMOS	65nm CMOS
Frequency (GHz)	234~252	132~148	130~140	223~276	134~148	220~320
Transceiver Type	Monostatic[a]	Bistatic	Bistatic	Monostatic	Monostatic	Multi-static
EIRP (dBm)	9.9	28.7	14.06	27	9.8, 25.2[g]	0.6, 20[i]
TX Power (dBm)	2.2	15.1	8.16	3.1	11.2[h]	N/A
Total Radiated Power (dBm)	0.3	N/A	N/A	N/A	6.2	N/A
RX NF$_{min}$ (dB)	22.8	7.8	12.78	15.7	12.9	22.2
Radiation Efficiency (%)	78[b], 69[c]	<56[b][d]	60[b]	74[b]	32[b]	25[b]
Isolation (dB)	26[b]	N/A	22[e], 44.7[e][f]	23.3[g]	33.3	N/A
Antenna Type	In-Package	In-Package	In-Package	On-Chip + Silicon Lens	On-Chip	On-Chip
Package Type	Organic	LTCC	N/A	N/A	N/A	N/A
Packaging Technique	Flip-Chip	Flip-Chip	Flip-Chip	Wire-Bonding	Wire-Bonding	Wire-Bonding
Package Footprint (mm²)	49	195	441	N/A	N/A	N/A
Die Area (mm²)	4	13.6	6.77	2.7	3.12	5
DC Power (mW)	1700	1890	255	770	405	840

(a) cross-polarimetric sensing (b) from simulation (c) estimated from measurement
(d) report 2.5dB feed line loss (e) self-interference cancellation level (f) achieved in narrowband
(g) with 3D-printed lens (h) assuming 32% radiation efficiency (i) with commercial lens

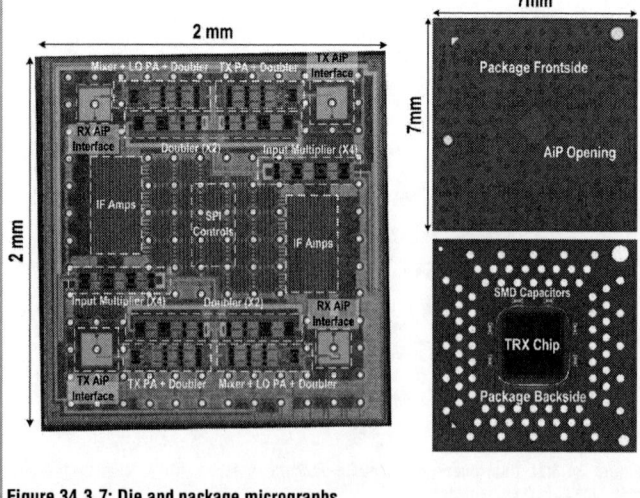

Figure 34.3.7: Die and package micrographs.

• 2026 IEEE International Solid-State Circuits Conference

ISSCC 2026 / SESSION 34 / INTEGRATED RADAR AND UWB TRANSCEIVERS FROM MICROWAVE TO SUB-THz / 34.4

34.4 A 1.6-to-3.8GHz Reconfigurable FMCW Radar SoC with 81.5% Relative-Bandwidth PLL for Real-Time Life Detection in Disaster Response

Renjie Fu*[1], Linxu Shi*[1], Fanxun Cai[1], Yunqi Yang[1], Hao Lei[1], Zhigang Fu[1], Yilin Xu[1], Pengcheng Wang[1], Ziyang Li[1], Shixuan Wang[1], Junkai Jiao[1], Dingqi Liang[1], Lianbo Wu[1,2], Guangzhong Zhang[3], Naike Du[3], Zijie Wang[1], Yucong Gu[1], Xiuzhu Ye[3], Xiao Fang[1], Weisheng Zhao[1,2], Hui Zhang[1,2]

[1]Beihang University, Beijing, China, [2]Tianmushan Laboratory, Hangzhou, China, [3]Beijing Institute of Technology, Beijing, China
*Equally Credited Authors (ECAs)

Abstract

A 1.6-to-3.8GHz FMCW radar SoC for real-time non-line-of-sight life detection is implemented in 40nm CMOS. An 81.5% BW ring-VCO PLL with DTC achieves 29.6kHz frequency error and 0.78-to-400MHz/μs chirp slopes. The TX integrates a 13.43dBm PA, while the RXRF features 8.1dB NF, baseband chain with 100dB-rejection SC-HPF at 1kHz and CT Δ-Σ ADC. A RISC-V core with FPU/VPU enables DSP, detecting people behind 6 walls and vital signs behind 2, occupying 7.65mm² and consuming 0.395W.

Disaster rescue operations require radar systems that can detect vital signs and motion of survivors behind solid obstructions such as collapsed concrete and rubble. This capability is essential in scenarios including earthquake-damaged structures and fire-compromised buildings with dense smoke. However, current solutions remain inadequate for these critical applications. Discrete low-frequency radars [1] exceed the size and power constraints of portable, battery-operated platforms. Meanwhile, existing integrated FMCW transceivers [2-20] lack the combination of sub-4GHz penetration and wide instantaneous bandwidth control necessary for high-resolution non-line-of-sight sensing. To address this gap, this work presents a 1.6-to-3.8GHz reconfigurable radar SoC implemented in 40nm CMOS for through-wall life detection. The chip features an 81.5% relative bandwidth ring-VCO-based PLL supporting frequency-agile chirp generation, a reconfigurable transceiver, and an integrated RISC-V processor that enables adaptive closed-loop control and on-chip digital signal processing. This enables autonomous adjustment of transceiver parameters based on environmental feedback, permitting real-time life-form detection through complex obstructions.

This work presents a radar SoC optimized for through-wall life detection applications. Through-wall detection is challenged by strong reflections from the nearest wall that can saturate the receiver, while desired behind-wall signals are significantly weaker. Effective operation requires suppression of wall returns and enhancement of target echoes, motivating extensive programmability across the transceiver chain for dynamic optimization. The SoC employs a software-defined architecture tailored for through-wall detection in disaster scenarios, as shown in Fig. 34.4.1. The transmitter operates from 1.6 to 3.8GHz, ensuring robust performance in cluttered environments. It includes a compact power amplifier (PA) with programmable output power of 5.67dBm, 10.88dBm, 13.43dBm. A wideband ring-VCO PLL achieves 81.5% relative bandwidth and provides chirps with dynamically configurable slopes ranging from 0.78MHz/μs to 400MHz/μs, occupying only 0.223mm² area (Fig. 34.4.7). The adoption of ring-VCO eliminates the need for external phase shifters or dividers by leveraging intrinsic multi-phase generation. The receiver integrates a noise-canceling RF front-end with I/Q baseband processing, offering a maximum tunable gain of 87dB and programmable filter cut-off frequency. A switched-capacitor highpass filter (HPF) provides 100dB of rejection at 1kHz to suppress low-frequency clutter, and a continuous-time Δ-Σ ADC (CT-DSM) achieves 69dB SNR for precision vital-sign detection. On-chip RC calibration ensures stable filter time constants, and an auxiliary ADC supports external signal acquisition. A RISC-V core with FPU/VPU acceleration executes real-time FFT/FIR algorithms, synchronized to the PLL start/stop command. System parameters including chirp slope, bandwidth, TX/RX gain, and filter corners are dynamically optimized under software control based on environmental feedback, enabling robust operation across diverse through-wall detection scenarios.

As shown in Fig. 34.4.2(a), the transmitter integrates a wideband PLL that drives a reconfigurable 1.6-to-3.8GHz PA through a matched buffer. The PA architecture in Fig. 34.4.2(c) employs three parallel differential cascode branches coupled via an on-chip high-k transformer (k=0.85) [21]. Figure 34.4.2(d) shows the 8-shaped transformer in a single metal layer, with a center tap providing a common-mode rejection path. This yields high impedance in differential mode, low impedance in common mode, and suppresses even-order harmonics. Amplitude flatness is improved, enabling direct 50Ω drive without external matching [22] in Fig. 34.4.4(b). The input stage uses core devices for matching and linearity, while the output stage employs thick-oxide transistors for voltage swing [23]. Each PA branch supports Class-A to Class-AB switching, enabling programmable power, linearity, and efficiency via digital control. The PA achieves 13.43dBm maximum power in a 0.29mm² core area, as shown in Fig. 34.4.4(b).

Compared to LC-VCO, the multi-phase outputs of ring VCO enable direct TX/RX phase distribution without dividers or shifters, and wide tuning range provides necessary bandwidth to achieve <10cm range resolution. To address target masking and ghost artifacts in wideband operation by non-linearity [24], a 1.6-to-3.8GHz PLL with 0.78-to-400MHz/μs programmable chirp generation is implemented in Fig. 34.4.2(a). The PLL employs a wide loop bandwidth to optimize chirp slope linearity while suppressing ring-VCO phase noise. As illustrated in Fig. 34.4.2(b), the 4-stage low-noise ring oscillator [25] provides eight phase outputs. The architecture of a Gm delay cell within this VCO is depicted in Fig. 34.4.2(b). During transitions where IN+ and OUT+ both switch from low to high, OUT– switches from high to low. Notably, since transistor M5 is disconnected 45° ahead of this transition due to the phase lead at its gate, both both M5 and Mc1 contribute zero noise during switching. As a result, the noise at OUT– stems solely from M1 and M3, enabling thermal noise cancellation and improved overall phase noise. Different VCO output buffers are used for TX and RX in Fig. 34.4.2(b). The TX buffer employs a π-type matching load to suppress FMCW harmonics, while the RX buffer uses a transmission line load. Both are implemented as AC-coupled self-biased inverters for area efficiency. Although wide bandwidth typically weakens noise filtering, a DPI-based DTC [26] effectively suppresses quantization noise from the Δ-Σ modulator. As shown in Fig. 34.4.4, measured results confirm a 1.6-to-3.8GHz chirp at 6.25MHz/μs slope, and the embedded DTC reduces RMS frequency error from 68.24kHz to 29.6kHz by mitigating Δ-Σ quantization noise. Additionally, a peak chirp slope of 400MHz/μs is achieved across 2.0 to 3.5GHz. In summary, this PLL achieves a 2.2GHz/81.5% chirp bandwidth and a maximum slope of 400MHz/μs within an active area of 0.223mm².

The receiver implements a software-defined RF-to-baseband chain for adaptive through-wall detection, providing 5.59 to 87dB variable gain for operation in complex environments in Fig. 34.4.3(a). A noise-canceling LNA [27], passive mixer, and TIA achieve 8.1dB NF at center frequency in Fig. 34.4.4(c). The LNA supports amplification, bypass, and attenuation modes, offering -15.08 to 11.3dB gain. ADC clip detection dynamically adjusts front-end gain to prevent saturation in Fig. 34.4.3(a). The RF front-end in Fig. 34.4.3(b) uses transistor M1 for real input impedance matching, with an off-chip inductor for biasing and reduced area. Bypass/attenuation modes employ parallel resistors for programmable attenuation. The mixer downconverts the input signal into I/Q baseband channels, while the TIA converts current to voltage, followed by a programmable bandpass filter with 40dB gain in 6dB steps. The filter highpass corner can be programmed from 100kHz to 300kHz, while the lowpass corner is tunable from 800kHz to 1.6MHz, measured by an on-chip BIST system, as shown in Fig. 34.4.4(d). To further save area, a switched-capacitor HPF is adopted in Fig. 34.4.3(c), and the highpass corner is tunable by adjusting the sampling frequency rather than RC time constant. The overall HPF provides 100dB rejection at 1kHz to nullify wall-reflected interference, occupying 33% less area than RC filters. A continuous-time Δ-Σ ADC follows the BPF to suppress the filter's out-of-band spurs through its intrinsic anti-aliasing characteristic. This ADC employs source-degenerated DACs and modulo-15 barrel-shifter DEM logic to eliminate the dummy cells in traditional DEMs as shown in Fig. 34.4.3(d), thus improving the overall ADC linearity, suppressing the second harmonic distortion (HD2) by 26dB in Fig. 34.4.4(d). The ADC provides an additional 3rd-order low pass property and achieves 69dB SNR and 76dB SFDR.

The system is implemented in a 40nm CMOS process, occupying 7.65mm², including 2.45mm² for analog and RF front-ends. The SoC is mounted on a PCB with the compact magneto-electric dipole antenna [28] positioned behind a metallic backplane in Fig. 34.4.5(a). Real-time data is transmitted via USART to an external display for immediate NLOS visualization. The system achieves 170ms latency per frame and consumes 0.395W during full operation. In motion-tracking mode, the system reconstructs human trajectories through multiple wall barriers; Fig. 34.4.5(b) shows bidirectional walking behind a 9.4cm-thick concrete wall over a 2-to-8m range. Figure 34.4.5(c) demonstrates penetration through six 9.4cm-thick walls, with accurate motion recovery from 13 to 17m. Wall outlines are partially superimposed on the visual output to improve spatial awareness. In static mode, the system concurrently monitors respiration of three stationary individuals at 5.8m, 7.4m, and 8.9m, fully occluded by two walls in Fig. 34.4.5(d). Despite significant attenuation, clear respiration signals using a Doppler radar-based respiration detection method [29] and spatial intensity maps are captured, confirming high sensitivity and deep NLOS penetration capability.

Figure 34.4.6 benchmarks this work against prior art, highlighting the highest level of integration and unique penetration capability through concrete barriers. This work presents a fully integrated SoC enabling real-time, non-contact lifeform detection and motion tracking through walls. The prototype achieves 7.81cm range resolution and 100m maximum detection range in free space, which decreases to 60m with a 9.4cm concrete wall. The system supports respiration sensing through two, and motion tracking through six, 9.4cm-thick concrete walls. With 0.395W power consumption at 3.3V under full algorithm execution, the design is optimized for portable deployment on handhelds, UAVs, and legged robots. Compared to previous systems constrained to free-space or glass-barrier scenarios [1], this SoC offers a scalable and power-efficient solution for through-wall sensing in complex, obstructed environments.

Acknowledgement:

The authors would like to thank Huiqun Huang, Haoyu Shen, Jingyi Peng, Xuefeng Ren, Yetao Zhuang, Yinghua Huang, and Yuchao Guo for their valuable support and contributions to this work. The corresponding authors are Hui Zhang and Lianbo Wu ({zhang_hui, lianbowu}@buaa.edu.cn).

References:

[1] Z. Peng et al., "A Portable FMCW Interferometry Radar with Programmable Low-IF Architecture for Localization, ISAR Imaging, and Vital Sign Tracking," *IEEE TMTT*, vol.65, no. 4, pp.1334–1344, Apr.2017. https://doi.org/10.1109/TMTT.2016.2633352

[2] Z. Zhang et al., "A 77GHz Hybrid TDMA-MIMO Phased-Array Radar with 186m Detection Range and 3cm Range Resolution," *ISSCC*, pp. 01–03, Feb. 2025. https://doi.org/10.1109/ISSCC49661.2025.10904816

[3] N. Andersen et al., "A 118-mW Pulse-Based Radar SoC in 55-nm CMOS for Non-Contact Human Vital Signs Detection," *IEEE JSSC*, vol. 52, no. 12, pp. 3421-3433, Dec. 2017. https://doi.org/10.1109/JSSC.2017.2764051

[4] B. Zhu et al., "A Digital-Intensive 1TX/2RX IEEE 802.15.4/4z-Compliant Joint-Radar-Communication-Location Transceiver SoC," *IEEE JSSC*, vol. 60, no. 3, pp. 1014-1029, March 2025. https://doi.org/10.1109/JSSC.2024.3451654

[5] H. Jia et al., "A 77 GHz Frequency Doubling Two-Path Phased-Array FMCW Transceiver for Automotive Radar," *IEEE JSSC*, vol. 51, no. 10, pp. 2299-2311, Oct. 2016. https://doi.org/10.1109/JSSC.2016.2580599

[6] A. Townley et al., "A 94-GHz 4TX–4RX Phased-Array FMCW Radar Transceiver with Antenna-in-Package," *IEEE JSSC*, vol. 52, no. 5, pp. 1245-1259, May 2017. https://doi.org/10.1109/JSSC.2017.2675907

[7] A. Mostajeran et al., "A 170-GHz Fully Integrated Single-Chip FMCW Imaging Radar with 3-D Imaging Capability," *IEEE JSSC*, vol. 52, no. 10, pp. 2721-2734, Oct. 2017. https://doi.org/10.1109/JSSC.2017.2725963

[8] Q. Shi et al., "A Self-Calibrated 16-GHz Subsampling-PLL-Based Fast-Chirp FMCW Modulator with 1.5-GHz Bandwidth," *IEEE JSSC*, vol. 54, no. 12, pp. 3503-3512, Dec. 2019. https://doi.org/10.1109/JSSC.2019.2941113

[9] T. Ma et al., "A CMOS 76–81-GHz 2-TX 3-RX FMCW Radar Transceiver Based on Mixed-Mode PLL Chirp Generator," *IEEE JSSC*, vol. 55, no. 2, pp. 233-248, Feb. 2020. https://doi.org/10.1109/JSSC.2019.2950184

[10] W. Lee, T. Dinc and A. Valdes-Garcia, "Multi-Mode 60-GHz Radar Transmitter SoC in 45-nm SOI CMOS," *IEEE JSSC*, vol. 55, no. 5, pp. 1187-1198, May 2020. https://doi.org/10.1109/JSSC.2020.2964150

[11] Y. -P. Su et al., "A 24-GHz Fully Integrated CMOS Transceiver for FMCW Radar Applications," *IEEE JSSC*, vol. 56, no. 11, pp. 3307-3317, Nov. 2021. https://doi.org/10.1109/JSSC.2021.3095137

[12] S. K. Sireesh et al., "A 4-bit RFDAC-Based FMCW Modulator for Automotive Radar," *IEEE JSSC*, vol. 60, no. 2, pp. 394-409, Feb. 2025. https://doi.org/10.1109/JSSC.2024.3441230

[13] Z. Chen et al., "A 122-168GHz Radar/Communication Fusion-Mode Transceiver with 30GHz Chirp Bandwidth, 13dBm Psat, and 8.3dBm OP1dB in 28nm CMOS," *IEEE Symp. VLSI Circuits*, pp. 1-2, June 2021. https://doi.org/10.23919/VLSICircuits52068.2021.9492460

[14] R. Wan et al., "A 132-to-163 GHz 4TX/4RX Distributed MIMO FMCW Radar Transceiver with Real-Time Reference-Clock Synchronization Enabling Cooperative Coherent Multistatic Imaging System," *IEEE Symp. VLSI Technology and Circuits*, pp. 1-2, June 2024. https://doi.org/10.1109/VLSITechnologyandCir46783.2024.10631434

[15] C. Xu et al., "A Packaged 54-to-69-GHz Wideband 2T2R FMCW Radar Transceiver Employing Cascaded-PLL Topology and PTAT-Enhanced Temperature Compensation in 40-nm CMOS," *IEEE JSSC*, vol. 59, no. 10, pp. 3156-3171, Oct. 2024. https://doi.org/10.1109/JSSC.2024.3404263

[16] J. Zhang et al., "A W-Band Transceiver Array with 2.4GHz LO Synchronization Enabling Full Scalability for FMCW Radar," *ISSCC*, pp.282–284, Feb. 2023. https://doi.org/10.1109/ISSCC42615.2023.10067317

[17] X. Liu et al., "A CMOS 49–63-GHz Phase-Locked Stepped-Chirp FMCW Radar Transceiver," *IEEE JSSC*, Early Access, pp. 1–15, Apr. 2025. https://doi.org/10.1109/JSSC.2025.3556649

[18] K. -I. Oh et al., "A 54–64-GHz 4TXs-4RXs CMOS Transceiver with 10-GHz Bandwidth Single Chirp for FMCW Radar Applications," *IEEE TMTT*, vol. 73, no. 3, pp. 1532-1544, Mar. 2025. https://doi.org/10.1109/TMTT.2024.3451482

[19] G. -H. Ko et al., "24-GHz 4TX–4RX Phased Array Transceiver with Automatic Beam Steering Mode for FMCW Radar Applications," *IEEE TMTT*, vol. 72, no. 5, pp. 3065-3075, May 2024. https://doi.org/10.1109/TMTT.2023.3320741

[20] S. M. H. Naghavi et al., "An Integrated 100-GHz FMCW Imaging Radar for Low-Cost Drywall Inspection," *IEEE TMTT*, vol. 72, no. 2, pp. 1070-1084, Feb. 2024. https://doi.org/10.1109/TMTT.2023.3305076

[21] H. Jia et al., "A Full Ka-Band Power Amplifier With 32.9% PAE and 15.3-dBm Power in 65-nm CMOS," *IEEE TCAS-I*, vol. 65, no. 9, pp. 2657-2668, Sep. 2018. https://doi.org/10.1109/TCSI.2018.2799983

[22] W. Ye et al., "A 2-to-6GHz Class-AB Power Amplifier with 28.4% PAE in 65nm CMOS Supporting 256QAM," *ISSCC*, pp. 1-3, Feb. 2015. https://doi.org/10.1109/ISSCC.2015.7062914

[23] H. Wang et al., "A 5.2-to-13GHz Class-AB CMOS Power Amplifier with a 25.2dBm Peak Output Power at 21.6% PAE," *ISSCC*, pp. 44-45, Feb. 2010. https://doi.org/10.1109/ISSCC.2010.5434059

[24] P. T. Renukaswamy et al., "Chirp Generation Techniques and Tradeoffs for Integrated FMCW Radar: A Tutorial Review," *IEEE Solid-State Circuits Magazine*, vol. 17, no. 1, pp. 86–96, 2025. https://doi.org/10.1109/MSSC.2024.3487144

[25] J.-M. Kim et al., "A Low-Noise Four-Stage Voltage-Controlled Ring Oscillator in Deep-Submicrometer CMOS Technology," *IEEE TCAS-II*, vol. 60, no. 2, pp. 71–75, Feb. 2013. https://doi.org/10.1109/TCSII.2012.2235734

[26] Z. Shen et al., "A 24 GHz Self-Calibrated All-Digital FMCW Synthesizer with 0.01% RMS Frequency Error Under 3.2 GHz Chirp Bandwidth and 320 MHz/μs Chirp Slope," *IEEE JSSC*, vol. 57, no. 7, pp. 2167–2180, Jul. 2021. https://doi.org/10.1109/JSSC.2021.3123693

[27] A. Bozorg et al., "A 20 MHz–2 GHz Inductorless Two-Fold Noise-Canceling Low-Noise Amplifier in 28-nm CMOS," *IEEE TCAS-I*, vol. 69, no. 1, pp. 42-50, Jan. 2022. https://doi.org/10.1109/TCSI.2021.3092960

[28] C. Sun, N. Du, Y. Guo, X. Zhang, H. Zhang and X. Ye, "A Compact Magneto-Electric Dipole Antenna for S-Band MIMO Through-Wall Radar," *IEEE Access*, vol. 12, pp. 67209-67218, 2024. https://doi.org/10.1109/ACCESS.2024.3399213

[29] A. D. Droitcour et al., "Range Correlation and I/Q Performance Benefits in Single-Chip Silicon Doppler Radars for Noncontact Cardiopulmonary Monitoring," *IEEE TMTT*, vol. 52, no. 3, pp. 838-848, Mar. 2004. https://doi.org/10.1109/TMTT.2004.823552

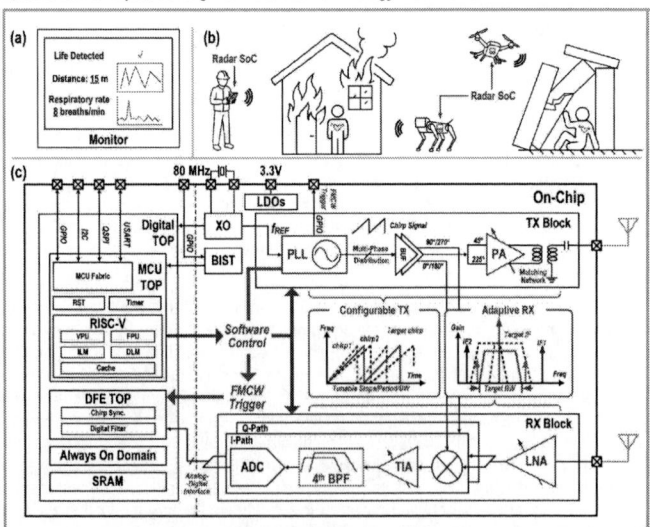

Figure 34.4.1: (a) Concept of life-form detection. (b) Disaster-rescue scenarios. (c) System-level block diagram.

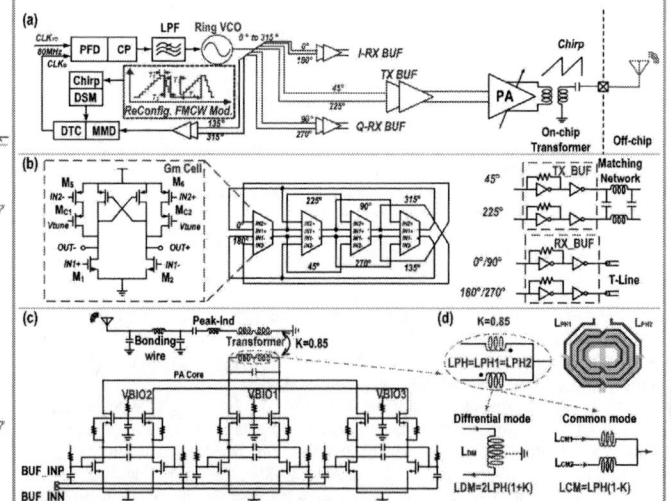

Figure 34.4.2: (a) TX Block diagram. (b) Ring VCO with output buffer. (c) PA with output stage. (d) On-chip Transformer.

ISSCC 2026 / SESSION 34 / INTEGRATED RADAR AND UWB TRANSCEIVERS FROM MICROWAVE TO SUB-THz / 34.4

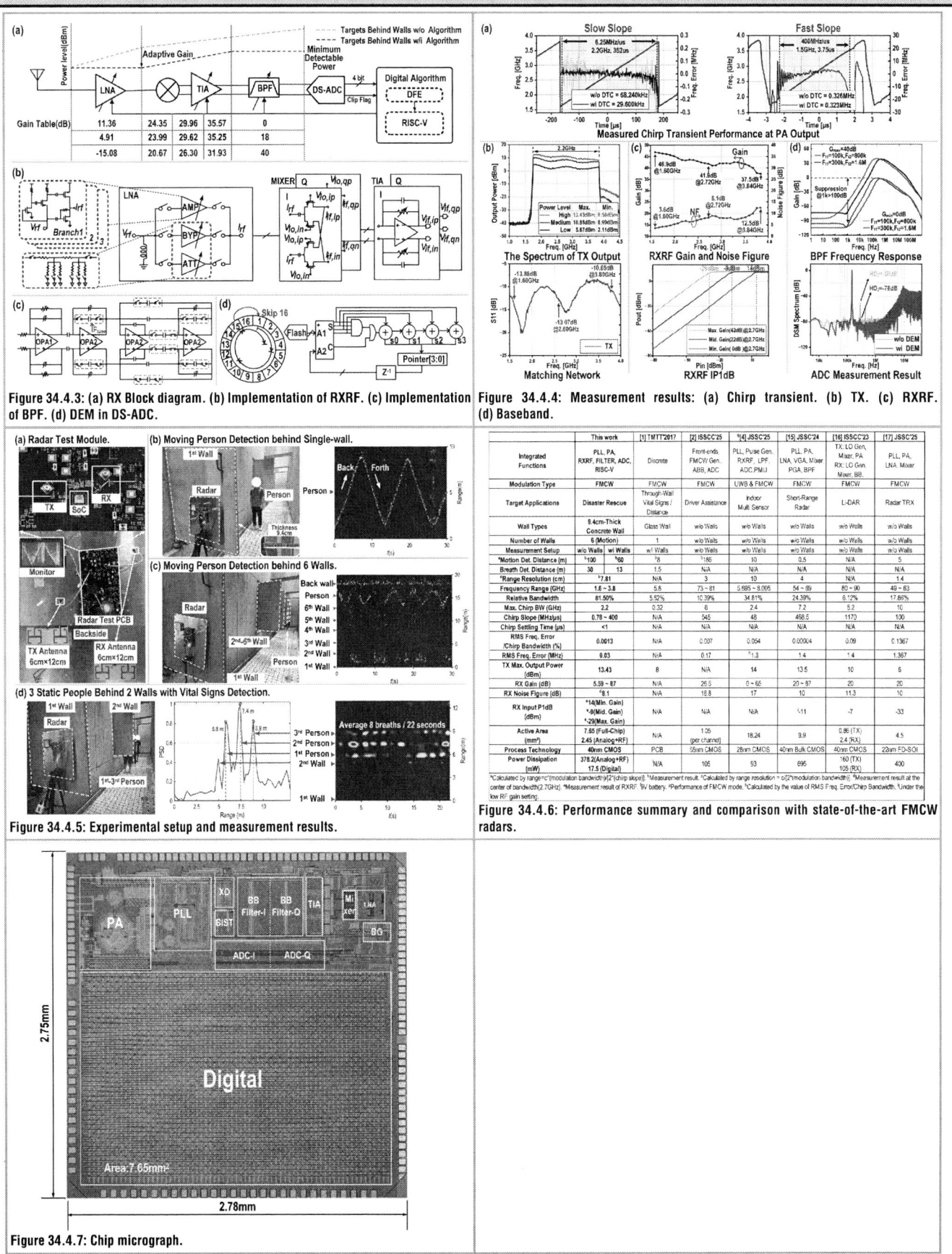

Figure 34.4.3: (a) RX Block diagram. (b) Implementation of RXRF. (c) Implementation of BPF. (d) DEM in DS-ADC.

Figure 34.4.4: Measurement results: (a) Chirp transient. (b) TX. (c) RXRF. (d) Baseband.

Figure 34.4.5: Experimental setup and measurement results.

Figure 34.4.6: Performance summary and comparison with state-of-the-art FMCW radars.

Figure 34.4.7: Chip micrograph.

• 2026 IEEE International Solid-State Circuits Conference

979-8-3315-8937-0/26 $31.00 © 2026 IEEE

ISSCC 2026 / SESSION 34 / INTEGRATED RADAR AND UWB TRANSCEIVERS FROM MICROWAVE TO SUB-THz / 34.5

34.5 A 0.0523mm² 11.4mW IEEE 802.15.4a/z/ab Compatible Aliasing-Suppressing All-Digital IR-UWB Transmitter Featuring Comb-Notched Maximally Flat Amplitude Spectral Shaping

Jihun Son, Minsu Park, Chansoo Park, Kyoungtae Lee, Jong-Hyeok Yoon, Junghyup Lee, Minyoung Song

Daegu Gyeongbuk Institute of Science and Technology, Daegu, Korea

Abstract

This work presents an IEEE 802.15.4a/z/ab compatible IR-UWB TX using an anti-phase delayed dual-Gaussian pulse with self-delay pulse comb-notch to maximize the spectrum utilization while meeting global masks. A slope-aware non-uniform interpolation suppresses sampling aliases by 11dB. A dual-level switched-capacitor PA cuts sidelobe emission up to 12dB. The TX draws 11.4mW, reaches 83.4% spectrum utilization and near-in sidelobe ($\leq$ -35dBr at 489MHz) with $\geq$9× smaller area than prior arts.

Impulse Radio Ultra-Wideband (IR-UWB) technology has rapidly gained popularity in wireless connectivity applications (e.g., IoT, smart homes, automotive, AR/VR), owing to its wide emission bandwidth ($\geq$ 500MHz) [1], thereby supporting high data throughput, and its fine time resolution enabling centimeter-level localization. Recently released IEEE 802.15.4z [2] strengthens the ranging security and introduces higher mean pulse repetition frequency (mPRF) modes at 124.8 and 249.6MHz, improving the localization accuracy. Building on these mPRFs, the next-generation IEEE 802.15.4ab [3] proposes higher data rates e. g., 62.4 and 124.8Mb/s, which tighten the link budget.

Since the TX power spectral density (PSD) is restricted to -41.3dBm/MHz EIRP [1], improving RX sensitivity can increase the link budget without increasing the TX power but this raises power consumption in the RX frontend and digital baseband. Instead, maximizing the TX spectrum utilization within the channel mask [2] through pulse shaping can be more power efficient to increase the link budget at constant average EIRP. For example, improving spectrum utilization from 65.5% to 83.4% yields a 1.05dB link budget gain (Fig. 34.5.1). Root-raised cosine [4] and Chebyshev type II [5] pulses can achieve high spectrum utilization by sharpening the frequency transition band, but their long temporal tails exacerbate truncation-induced spectral sidelobes (<-25dBr). This trade-off is the primary constraint for the global spectrum compliance. In particular, ETSI mandates -70dBm/MHz below 6GHz [6], which for channel 5 translates to 28.7dBr sidelobe suppression at roughly 490MHz offset from the channel center (6.489GHz). Refined apodized pulses (e.g., Gaussian) [7-9] can achieve acceptable sidelobe (<-35dBr) but yield only ~65% spectrum utilization [8].

Although analog RF upconversion TXs [10] offer precise spectral control, their high power consumption (>200mW) makes digital TXs [7-9,11,13,14] preferable for low-power IoT applications. However, digital TXs are constrained by finite quantization of digital PAs, which make apodized pulses for low sidelobe emission difficult and thus hinder achieving sufficient sidelobe suppression in practice. In addition, periodic sampling produces spectral alias at offset frequency of integer multiples of the sampling frequency, which can violate the spectral mask. Increasing sampling frequency such as doubling sampling frequency [9] and delay-line-based first-order hold [7] shifts the alias farther from the restricted bands and hybrid FIR filtering [11] filters out the alias directly in both baseband and RF domain. However, these approaches incur significant area and power overhead and require additional resources for calibration.

This paper presents an IEEE 802.15.4a/z/ab-compatible aliasing-suppressing all-digital IR-UWB TX that achieves the highest spectrum utilization (83.4%) while the lowest near-in sidelobe (<-35dBr at 489MHz) and in at least 9× smaller area compared to previously published works listed in Fig. 34.5.6. Three innovations are introduced: 1) anti-phase delayed dual-Gaussian (APDG) pulse with self-delay pulse comb-notch (SDPCN) for maximizing both spectrum utilization and near-in sidelobe suppression, 2) a slope-aware non-uniform interpolation (SANI) for reducing aliasing without significant area and power overhead, 3) a dual-level switched-capacitor power amplifier (DLSCPA) to reduce sidelobe emission due to AM-PM distortion.

The proposed APDG pulse for maximally flat-top amplitude spectrum is shown in Fig. 34.5.1. The pulse that consists of a main Gaussian and a delayed sub-pulse with anti-phase can be defined as $s(t) = g(t) - \alpha \cdot g(t-\tau)$, where $g(t) = 1/(\sigma\sqrt{(2\pi)}) \cdot exp(-t^2/(2\sigma^2))$; α and τ denote the scaling factor and the delay, respectively. Since $|S(f)|^2 = |G(f)|^2 \cdot (1 + \alpha^2 - 2\alpha \cdot cos(2\pi f\tau))$, where $S(f)$ and $G(f)$ are the Fourier transforms of $s(t)$ and $g(t)$, respectively, the Gaussian envelope is modulated by interference ripples, where notches appear with $\Delta f = 1/\tau$. Adjusting the notch depth ($10 \cdot log_{10}((1-\alpha)^2/(1+\alpha)^2)$) with α of 18.75%, $S(f)$ satisfy the maximally flat spectrum criterion introduced in [12]. A delay of τ is chosen as 2ns, for minimizing the inter-symbol interference (ISI). The proposed all-digital IR-UWB TX consists of two parts as shown in Fig. 34.5.1. Cascaded injection-locked digitally controlled oscillators (IL-DCOs) [14] are exploited to support 499.2MHz system clock, 6-9GHz LO frequency, and sampling clock for pulse shaper. The TX Analog Front-end (AFE) comprises main path and a delayed replica path that realizes self-delay pulse comb-notch filtering.

Figure 34.5.2 illustrates the block diagram of TX AFE. To perform the proposed APDG pulse, 2-bit ternary code representing amplitude (AM) and phase (PM) is input to each path. To reduce ISI in the pulse stream caused by the limited pulse rate (499.2MHz), the proposed polarized deserialization deserializes the AM signal into two parallel paths, denoted as AM_DES_P and AM_DES_N, according to the phase polarity (PM). Unlike the deserialization–serialization approach in [13], no additional phase deserialization is required. The simplified pipeline eases timing closure at the 4GHz sampling clock and minimizes power overhead. Main pulses (MAIN_P and MAIN_N) are first synthesized as 7-bit triangular (single pulse) or trapezoidal (pulse stream) waveforms with AM_DES_P and AM_DES_N and their synchronously delayed by 250ps (1/4GHz) ones by shift register. Similarly, 3-bit opposite-polarity sub-pulses (SUB_P and SUB_N) are also generated and delayed from the main pulses by 3.5ns (2/499.2MHz - 2/4GHz), which results in the 2ns delay (τ) of the APDG pulse (Fig. 34.5.1). The pulse shaper dissipates power only during pulse generation and has no static power consumption, and thus its average dissipation is duty-cycled and negligible.

Main and sub pulses are combined to reconstruct the pulse stream. Since all generated pulses are inherently polarized by the polarized deserialization, the composite envelope composed of pulses with opposite polarity is obtained by subtracting opposite-polarity contributions (i.e., a signed superposition). On top of the envelope, the phase of the pulse stream is derived by detecting phase transitions. Transitions are found at crossover instants where the magnitude dominance of each of the opposite-polarity pulses (MAIN_P+SUB_P and MAIN_N+SUB_N) flips. Rather than summing all bits, a ladder comparison from the midpoint code to the LSB and sub pulses on P and N paths finds the crossover, which minimizes hardware overhead. The reconstructed phase (PM_X) selects LO phase by phase multiplexer (PHMUX).

The reconstructed pulse stream is remapped via level duplication to form a 19-bit Gaussian envelope and fed into the proposed slope-aware non-uniform interpolation (SANI) to reduce the sampling aliasing. Two consecutive pulse amplitude bits (AM_X[2n] and AM_X[2n+1]) are simultaneously processed through both OR and AND gate, sampled by 0° and 180° phases, respectively. Therefore, the sampling phase is conditioned on the local pulse slope, effectively dithering the sampling interval and spreading alias energy, particularly for pulses with non-linear slope, e. g., Gaussian. The proposed SANI can be more effective to reduce the aliasing in the pulse stream, as the pulse slope also varies with the data-dependent pulse combination, which happens randomly (Fig. 34.5.2). As a result, sampling alias can be reduced without additional filters as in [7,11].

The proposed APDG pulse increases spectrum utilization and lowers sidelobe levels, yet the near-in sidelobe level is still marginal at about -30dBr (Fig. 34.5.2). To further suppress the near-in sidelobes, an SDPCN filtering is proposed. A 3-cycle (750ps) delayed replica path is combined with the main path at the PA output. It performs a 2-tap FIR filter ($H(z) = 1 + z^{-3}$) which creates notches at 667MHz and its odd multiples. Increasing the delay shifts the first notch closer to the mainlobe and deepens sidelobe suppression, but it also attenuates mainlobe energy and degrades spectrum utilization. A delay of 750ps was therefore chosen as the optimal trade-off between sidelobe suppression and mainlobe preservation, achieving -35dBr near-in sidelobe without additional filtering as in [10].

Switched-capacitor PAs (SCPAs) [14,15] are suitable for generating digitized impulse waveforms with high energy efficiency. However, finite channel resistance and transition time introduce AM–PM distortion [15], leading to increasing sidelobe emission. Upsizing the switch can mitigate this but coarsens the amplitude step under a fixed impulse amplitude, which in turn increases quantization-induced sidelobes. To minimize the AM-PM distortion while maintaining the quantization level, a dual-level SCPA (DLSCPA) is proposed in Fig. 34.5.3. Each unit cell supports two (full and half) amplitudes. In full-amplitude mode, the PA operates as a conventional SCPA. In half-amplitude mode, by shorting the differential PA output depending on the LO phase, both nodes swing at half amplitude (~0.5VDD). This reduces the effective quantization step without shrinking the cell, alleviating the AM–PM distortion. In simulation, the AM–PM distortion drops from 15° to 8.3°, a 45% reduction. Figure 34.5.4 shows the measurement results of the proposed DLSCPA. To quantify the impact of AM-PM distortion within the impulse, in-pulse phase deviation is measured. Compared with the conventional SCPA, the dual-level amplitude modulation reduces in-pulse phase deviation by 27.5% and suppresses the sidelobe emission by up to 12dB.

The proposed IR-UWB TX IC was fabricated in 28nm CMOS, occupying a core area of 0.0523mm². Total transient power consumption is 11.4mW and its breakdown is shown in Fig. 34.5.4. The measured output spectrum results are shown in Fig. 34.5.5. Using the proposed APDG pulse increases spectrum utilization by 17.9% over a Gaussian, owing to its maximally flat in-band amplitude spectrum (top-left), with slight increase in the near-in sidelobe (~–24dBr). Applying all proposed sidelobe suppression techniques (SDPCN and SANI) on channel 5 further cleans the spectrum (top-right): SDPCN forms a notch at 667MHz (and odd multiples) and reduces the 6GHz sidelobe by 11dB to –35dBr, meeting the ETSI mask, and SANI suppresses aliasing replicas by approximately 11dB. The measured output spectra across channels 5 (6.49GHz) to 9 (7.99GHz) satisfy FCC, ETSI, ARIB, KCC and China masks (bottom). Figure 34.5.6 summarizes the performance and benchmarks with state-of-the-art standard-compliant IR-UWB TXs. While achieving over 9× area efficiency thanks to all-digital TX topology, the presented IR-UWB TX achieves the highest spectrum utilization and the lowest near-in sidelobe (at 489MHz offset frequency) suppression compared to state-of-the-art shown in Fig. 34.5.6, leading to a competitive solution for next-generation high-data-rate IR-UWB standards while meeting global spectrum requirements.

Acknowledgement:
The chip fabrication and EDA tool were supported by the IC Design Education Center (IDEC), Korea. This work was supported by the National Research Foundation of Korea(NRF) grant funded by the Korea government(MSIT) (No. RS-2024-00416319, RS-2025-00519330, RS-2024-00439307), the Bio & Medical Technology Development Program of the National Research Foundation (NRF) funded by the Ministry of Science & ICT(RS-2025-02303581), and the Institute of Information & Communications Technology Planning & Evaluation(IITP) grant funded by the Korea government(MSIT) (No.RS-2025-02219277, AI Star Fellowship Support (DGIST)).

References:
[1] *Federal Communications Commission FCC 02–48. Published: Apr. 2002; Accessed on: Aug. 2025.*
https://transition.fcc.gov/Bureaus/Engineering_Technology/Orders/2002/fcc02048.pdf
[2] *IEEE Standard for Low-Rate Wireless Networks—Amendment 1: Enhanced Ultra-Wideband (UWB) Physical Layers (PHYs) and Associated Ranging Techniques*, IEEE Standard 802.15.4z-2020, (Amendment to IEEE Std 802.15.4-2020), Aug. 25, 2020, pp. 1–174. http://doi.org/10.1109/IEEESTD.2020.9179124
[3] *IEEE Draft Standard for Low-Rate Wireless Network Amendment 1: Enhanced Ultra-Wide Band (UWB) Physical Layers (PHYs) and Associated Medium Access and Control (MAC) sublayer Enhancements*, IEEE P802.15.4ab/D02, May 1, 2025, pp.1-256.
https://ieeexplore.ieee.org/document/10982429
[4] S. Kim, Y. Kim, X. Li, and J. Kang, "Orthogonal Pulse Design in Consideration of FCC and IEEE 802.15.4a Constraints," *IEEE Commun. Lett*, vol. 17, no. 5, pp. 896-899, May 2013. http://doi.org/10.1109/LCOMM.2013.040213.122936
[5] H. Chen, Y. Xiao, Z. Chen, R. Chen, Z. Wu, and B. Li, "A Digital IR-UWB Transmitter with High Spectrum Utilization and AM-PM Distortion Calibration," *IEEE TCAS-II: Express Br*, vol. 72, no. 1, pp. 53-57, Jan. 2025. http://doi.org/10.1109/TCSII.2024.3478774

[6] *Short Range Devices (SRD) Using Ultra Wide Band (UWB); Part 3: Worldwide UWB Regulations Between 3,1 and 10,6 GHz*, document ETSI TR 103 181-3-V2.1.1, 2019. https://www.etsi.org/deliver/etsi_tr/103100_103199/10318103/02.01.01_60/tr_1031810 3v020101p.pdf
[7] A. N. Bhat et al., "An IEEE802.15.4ab/a/z Compatible IR-UWB 2TRX with Dual-Antenna Full-Duplex 1x3 SIMO Radar Sensing and Aliasing Suppressing Semi-Synchronous TX," in *IEEE VLSI*, Jun. 2025, pp. 1-3.
http://doi.org/10.23919/VLSITechnologyandCir65189.2025.11074817
[8] H. Chen, Z. Chen, R. Ou, R. Chen, Z. Wu, and B. Li, "An IEEE 802.15.4z-Compliant Reconfigurable Pulse-Shaping UWB Digital Power Amplifier in 28-nm CMOS," in *IEEE Trans MTT*, vol. 71, no. 10, pp. 4366-4376, Oct. 2023.
http://doi.org/10.1109/TMTT.2023.3263898
[9] G. de Streel et al., "SleepTalker: A ULV 802.15.4a IR-UWB transmitter SoC in 28-nm FDSOI Achieving 14 pJ/b at 27 Mb/s with Channel Selection Based on Adaptive FBB and Digitally Programmable Pulse Shaping," *IEEE JSSC*, vol. 52, no. 4, pp. 1163-1177, Apr. 2017. http://doi.org/10.1109/JSSC.2016.2645607
[10] H. -G. Seok et al., "High-Sensitivity, Low-Power IR-UWB Radar Transceiver with Self-Interference Resistance for Child Presence Detection and Precision Positioning," *IEEE JSSC*, vol. 60, no. 4, pp. 1150-1161, Apr. 2025.
http://doi.org/10.1109/JSSC.2025.3541274
[11] Z. Huang, W. Deng, H. Jia, and B. Chi, "An 802.15.4/4z-Compliant UWB All-Digital Transmitter with Hybrid FIR Filtering Achieving 47dBr Sidelobe Suppression," in *IEEE RFIC*, Jun. 2025, pp. 11-14. http://doi.org/10.1109/RFIC61188.2025.11082915
[12] A. Milos, G. Molnar, and M. Vucic, "Spectrally Efficient UWB Pulse Shaping Based on Polynomially Weighted Gaussian Pulses with Maximally Flat Amplitude Spectra," *IEEE Commun. Lett*, vol. 27, no. 7, pp. 1869-1873, Jul. 2023.
http://doi.org/10.1109/LCOMM.2023.3271809
[13] M. Song et al., "An 8.7 mW/TX, 21 mW/RX 6-to-9GHz IEEE 802.15.4a/4z Compliant IR-UWB Transceiver with Pulse Pre-Emphasis Achieving 14mm Ranging Precision," in *IEEE VLSI*, Jun. 2023, pp. 1-2.
http://doi.org/10.23919/VLSITechnologyandCir57934.2023.10185245
[14] M. Song et al., "A 1.66Gb/s and 5.8pJ/b Transcutaneous IR-UWB Telemetry System with Hybrid Impulse Modulation for Intracortical Brain-Computer Interfaces," in *ISSCC*, Feb. 2022, pp. 394-396. http://doi.org/10.1109/ISSCC42614.2022.9731608.
[15] S. -M. Yoo, J. S. Walling, E. C. Woo, B. Jann and D. J. Allstot, "A Switched-Capacitor RF Power Amplifier," *IEEE JSSC*, vol. 46, no. 12, pp. 2977-2987, Dec. 2011. http://doi.org/10.1109/JSSC.2011.2163469

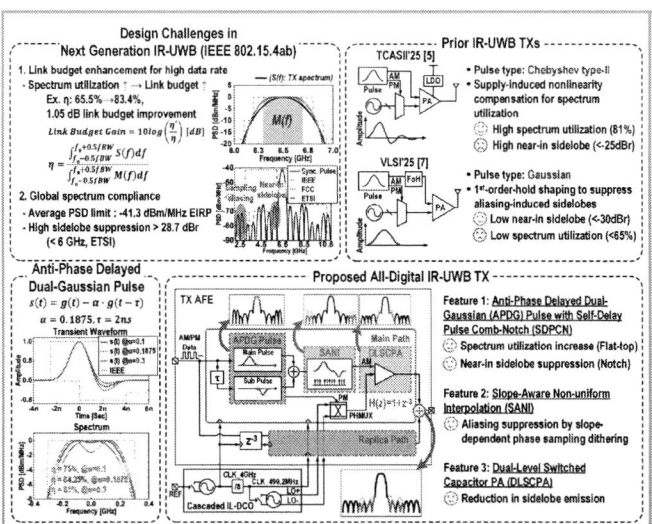

Figure 34.5.1: Design challenges in next generation IR-UWB, prior IR-UWB TXs, anti-phase delayed dual-Gaussian pulse, and overall architecture of proposed IR-UWB TX.

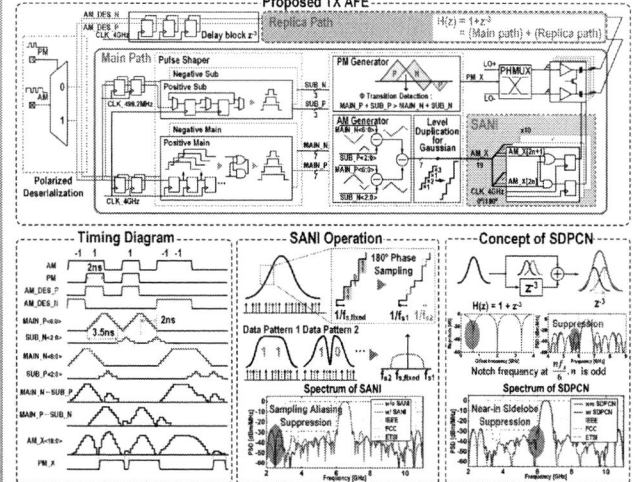

Figure 34.5.2: Detailed diagram of proposed TX AFE, timing diagram of TX AFE, operation of proposed slope-aware non-uniform interpolation, and self-delay pulse comb-notch.

ISSCC 2026 / SESSION 34 / INTEGRATED RADAR AND UWB TRANSCEIVERS FROM MICROWAVE TO SUB-THz / 34.5

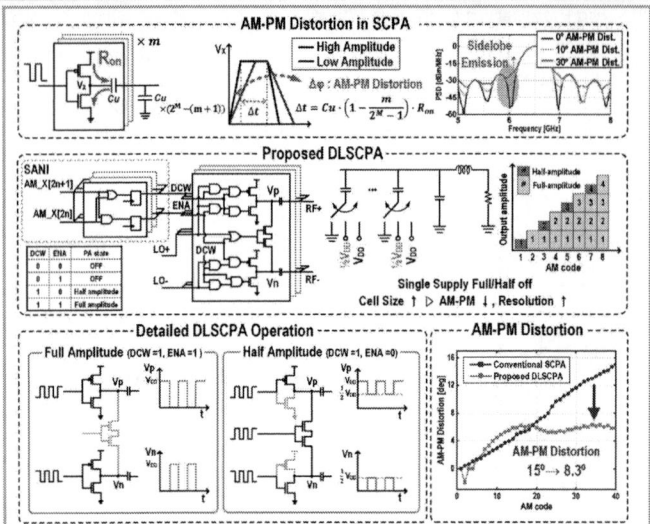

Figure 34.5.3: AM-PM distortion in SCPA, operation of proposed dual level switched capacitor PA (DLSCPA), and simulated AM-PM distortion results.

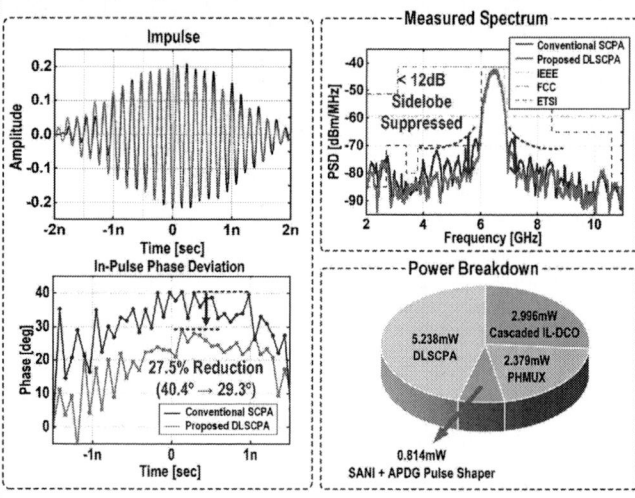

Figure 34.5.4: Measured in-pulse phase deviation, sidelobe suppression with DLSCPA, and power breakdown.

Figure 34.5.5: Measured results of spectrum utilization, sidelobe suppression with SANI and SDPCN, and output spectrum for channel 5/6/8/9.

	This work	RFIC 2025[11]	TCAS2 2025[5]	VLSI 2025[7]	JSSC 2025[10]
Technology [nm]	28	28	22 FDSOI	22	28
Supply [V]	1	0.9	1	0.8	3.6
Architecture	TX	TX	TX	TX	TRX
Standard Compliance	IEEE 802.15.4a/z/ab	IEEE 802.15.4a/z	IEEE 802.15.4z	IEEE 802.15.4a/z/ab	IEEE 802.15 4a/4z
UWB Regulation	FCC / ETSI / KCC / ARIB / CHI	FCC / ETSI / KCC / ARIB / CHI	FCC	FCC / ETSI	FCC / ETSI / KCC / ARIB
Frequency Range [GHz]	6 ~ 9	6.5 ~ 8	6.5 ~ 10	6 ~ 9	6.5 ~ 8
Modulation Bandwidth [MHz]	499.2	500 / 1000 / 1330	499.2	499.2	500
In-Band Spectrum Utilization [%]	83.4	64.83*	81	62.29*	68.19*
Sidelobe PSD [dBr] Δf = 489MHz **	-35 (Ch. 5)	-33.7* (Ch. 5)	-25* (Ch. 5)	-31.7* (Ch. 5)	-26.7* (Ch. 5)
Aliasing Suppression [dB]	11	20	N/A	12.1	N/A
Power Consumption [mW]	11.4	4.5***	33.4	28	254
Data Rate [Mb/s]	0.11 / 0.85 / 6.81 / 27.24 / 31.2 / 124.8	6.81	31.2	6.81 / 124.8	N/A
Supported PRF [MHz]	3.9 / 15.6 / 62.4 / 124.8 / 249.6	N/A	N/A	124.8	N/A
TX peak Power [dBm]	4.56	14.64	13	8	14
Energy Efficiency [pJ/bit]	1674 @ 6.81Mb/s	662 @ 6.81Mb/s ***	N/A	N/A	37400
Area [mm²]	0.0523	0.65	0.4896	10.2	3.97

*Estimate from figure **Δf is the offset frequency from center channel frequency (ex. 6.489GHz on Ch. 5)
*** Not including LO power consumption

Figure 34.5.6: Performance summary and comparison with state-of-art standard-compliant IR-UWB TXs.

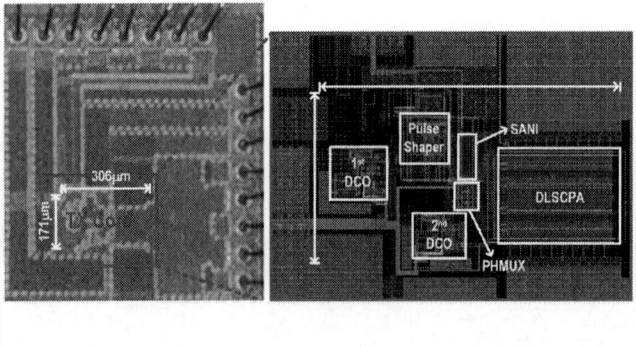

Figure 34.5.7: Die micrograph and layout details of implemented IR-UWB TX.

- 2026 IEEE International Solid-State Circuits Conference

979-8-3315-8937-0/26 $31.00 © 2026 IEEE

Session 35 Overview: *Low Power Wireless Transceivers for Localization and Communications*

WIRELESS SUBCOMMITTEE

Session Chair: Konstantinos Manetakis
CSEM, Neuchâtel, Switzerland

Session Co-Chair: Julian Tham
Infineon, San Jose, CA

This session focuses on low-power transceivers for communication and localization applications. It begins with a solution to the range–power trade-off in Ambient IoT backscatter systems by suppressing transmitter phase noise leakage in co-located access points. Next is a lightweight battery-free, crystal-less tag designed for small animal tracking. A blocker-tolerant receiver that employs non-uniform current sub-sampling techniques follows. Then, a dual-band RF transceiver compliant with both NearLink 2.0 and BT/BLE is introduced. Finally, the session presents a sub-milliwatt 802.11b backscatter WiFi transmitter with improved image and harmonic rejection.

ISSCC 2026 / February 18, 2026 / 3:35 PM

3:35 PM

35.1 CANCEL: A Cancellation-Aided Ambient IoT Nanopowered Communication System for Energy-Limited Tags

Hany Abolmagd, University of British Columbia, Vancouver, Canada, now with Nokia, Ottawa, Canada

In Paper 35.1, the University of British Columbia and the University of California present a backscatter system suppressing TX PN leakage in co-located TX/RX Ambient IoT access points. It enables low IF operation at the tag, reducing power consumption at low data rates. The system achieves 125dB self-interference rejection, supports tag operation at 92nW and achieves up to 32m range at 1kb/s.

4:00 PM

35.2 A 20mg Battery-Free Crystal-Less Miniaturized TX System for Flying Insects Localization with 1.45km Range

Yi Shen, University of Michigan, Ann Arbor, MI

In Paper 35.2, the University of Michigan presents a 20mg, battery-free, crystal-less localization tag. The system includes a custom antenna, photovoltaic cells, and SMD storage capacitor. It achieves localization accuracy of 0.9m over 1.45km. The tag was successfully deployed on a live wasp, which retained normal flight behavior, to demonstrate the system's unobtrusive design.

4:25 PM

35.3 A 0.052mm² Blocker-Tolerant Non-Uniform Current Sub-Sampling Receiver with a Discrete-Time FIR/IIR Filter Enabling 56dB Rejection in 28nm CMOS

Mostafa Ayesh, University of Southern California, Los Angeles, CA, now with Marvell, Irvine, CA

In Paper 35.3, the University of Southern California introduces a compact 18-to-22GHz receiver using non-uniform current sub-sampling and discrete-time FIR/IIR filtering. It achieves 41/56dB alias/non-alias blocker rejection, eightfold ADC speed relaxation, and -29.5dB EVM for a 64-QAM 150MS/s signal centered at 20GHz, and B1dB of 5.8dB while consuming 20mW.

4:50 PM

35.4 A NearLink 2.0 Compliant Dual-Band RF Transceiver for Smart Wireless Personal Audio Applications

Rui Yu, Huawei Technologies, Shenzhen, China

In Paper 35.4, Huawei Technologies presents a dual-band 2.4GHz and 5GHz RF transceiver compliant with both NearLink 2.0 and BT/BLE. It demonstrates 16Mb/s data-rate in 16-QAM Nearlink 2.0, -101.4dBm BLE 1Mb/s sensitivity, 30.9% 2.4GHz transmitter efficiency, and -55dBc carrier-to-interference rejection at +-3MHz offset.

5:15 PM

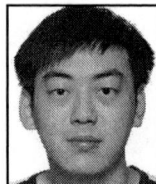

35.5 Fully Integrated Backscattered WiFi 802.11b Transmitter with Active Harmonics and Image Rejection for 30dB IRR and 36dB HRR at 0.88µW

Ruiyuan Yang, National University of Singapore, Singapore, Singapore

In Paper 35.5, the National University of Singapore presents a sub-µW 802.11b backscattered WiFi transmitter with 30dB image and 36dB harmonic rejection, using time-domain filtering and RF switch sequencing without digital filters. Temperature-driven calibration enhances performance, enabling reliable operation in dense wireless environments and better spectrum sharing.

35

DIGEST OF TECHNICAL PAPERS • 587

979-8-3315-8937-0/26 $31.00 © 2026 IEEE

ISSCC 2026 / SESSION 35 / LOW POWER WIRELESS TRANSCEIVERS FOR LOCALIZATION AND COMMUNICATIONS / 35.1

35.1 CANCEL: A Cancellation-Aided Ambient IoT Nanopowered Communication System for Energy-Limited Tags

Hany Abolmagd[1,2], Shih-Kai Kuo[3], Md Nazmul Hasan[1], Sreevatsank Kadaveru[3], Aboozar Ghorbani Nejad[1], Manideep Dunna[3], Ata Khorami[1], Yichen Yu[3], Dinesh Bharadia[3], Patrick Mercier[3], Sudip Shekhar[1]

[1]University of British Columbia, Vancouver, Canada, [2]now with Nokia, Ottawa, Canada, [3]University of California, San Diego, CA

Abstract

This paper presents a system that breaks the trade-off between range and power present in conventional wireless backscatter systems caused by phase noise leakage at low intermediate frequencies (IFs). By employing a self-interference cancellation chip together with a full-duplex antenna that achieves up to 125dB of cancellation, spectral leakage can be effectively eliminated, allowing for the developed backscatter tags to operate down to 92nW for communication at up to 32m at 1kb/s.

Ambient IoT envisions being able to power and interrogate smart sensor tags via a single access point (AP) such as a handheld device or router to enable convenient and low-cost data uplinks. Ambient IoT-style tags should: 1) operate with low-power - ideally ≤1μW - to meet the peak power restrictions of RF energy harvesters or small solid-state batteries; 2) utilize dynamic-frequency-scaling (DFS) techniques to further reduce data rate and power when possible, as many sensing tags are perfectly fine with even just 1kb/s; and 3) be able to operate with a single interrogation device (Fig. 35.1.1 top left). By eliminating high-power active RF circuits, backscatter communication techniques offer a promising pathway to meet the power needs of Ambient IoT-style tags. However, prior-art cannot yet meet these needs due to important, yet nuanced design trade-offs. Specifically, the power of a backscatter tag is dominated by its switching power, which in turn is dominated by the employed IF frequency (Δf from the incident signal); the only way to materially lower tag power is to reduce Δf. However, the AP's transmission of even just a tone (e.g., from a reversely-whitened BLE signal) has spectral leakage due to phase noise (PN) that, given the large transmit power (e.g., 30dBm) relative to the noise floor, results in a PN skirt that can easily overwhelm the weak received backscatter signal - ultimately placing a lower limit on Δf and/or an upper range limit (Fig. 35.1.1, bottom left). This is why the prior arts in [1-8] have used high IFs of 4 to 50MHz, and most have operated with two physically separated APs [1-4,6-7] (Fig. 35.1.1 top middle). This issue presents a trade-off between tag power and distance: reducing Δf reduces tag power, but degrades SNR due to the relatively larger PN skirt amplitude, which ultimately degrades the link budget. Decreasing the data rate doesn't help here, as the tag power is dominated by Δf, not data rate. The only way to break the power/distance trade-off under realistic co-located TX/RX conditions is to attempt to cancel the spectral leakage at the AP itself.

This paper presents CANCEL: a **C**ancellation-aided **A**mbient IoT **N**anopowered **C**ommunication system for **E**nergy-**L**imited tags that enables power scaling all the way down to 92nW while operating at up to 32m with a co-located TX/RX in a single AP (Fig. 35.1.1 top right) by: 1) implementing an AP utilizing a full-duplex (FD) antenna comprising two orthogonally-oriented patch elements that introduces 60dB of TX-RX-isolation over >20MHz; 2) deploying a passive RF self-interference cancellation (SIC) circuit that gives 65dB of additional isolation at a 1kHz offset, for a total isolation of 125dB, ultimately allowing for a Δf as low as 1kHz; 3) designing a backscatter tag that leverages DFS flexibly from 1Mb/s to 1kb/s, reducing power by 30×; and 4) maximizing range and spectral efficiency via a full-reflective 0.94dB-insertion-loss single-sideband backscatter modulator.

The top-level block diagram of the RF SIC, which can be added to PA/LNA combos in APs, is shown in Fig. 35.1.2 (top). The canceller consists of an input matching network, a balun connected to the PA's output, a passive polyphase filter (PPF) to generate differential I/Q signals, and a passive 18b RF Vector Modulator (VM). Two stages were selected for PPF as a compromise between the filter's sensitivity to its component values, bandwidth, and insertion loss.

The VM is designed using a hierarchical approach that boosts its resolution exponentially with low overhead in area and complexity. A moderate-resolution VM stage is repeated n times, with each handling the residual leakage from the preceding one [9]. Unlike the power-hungry (14.8mW/tap) implementation in [9], our VM comprises four 6b hierarchies that share the same passive VM core, resulting in a total theoretical resolution of 18 bits (after eliminating the redundancy) with no power consumption overhead. For each of the four VM hierarchies, both the incoming I & Q components are scaled independently using a MUX-based cap network, as shown in Fig. 35.1.2 (bottom). The passive MUX-based architecture avoids frequency translation or active gain stages [9,10], saving power, noise, and nonlinearity. The MUX stage switches the cap units between the output and a dummy load to maintain a constant load to the PPF and avoid code-dependent I/Q mismatches. Hierarchy is achieved by repeating the 6b core cap bank and adding attenuation to each subsequent hierarchy to bring its full scale to the LSB of the preceding stage (with 2b of overlap for margin). Input attenuation is implemented using a series input cap C_i, which performs a potential divider with the fixed input impedance of the cap bank. C_i also represents a high impedance to the input, reducing additional loading due to the subsequent hierarchical stages. Output attenuation is achieved using the current divider formed by the cap network C_{o1} & C_{o2}, which constitutes a low input impedance to reduce loading on the output of the current-based cap bank and minimize its linearity degradation. C_{o1} and C_{o2} also constitute an output impedance much higher than the LNA input impedance to avoid RX matching degradation. The cap bank MUX and the input and output attenuators are designed using the same 1fF unit MOM, C_u, to reduce process dependence and establish a well-defined full-scale relation between different hierarchy levels. SIC canceller optimization is performed using a simple CMOS-compatible search algorithm that iteratively achieves the optimum VM coefficients. This algorithm exploits the fact that each of the I & Q components can be optimized independently, and hence reduces the optimization problem to two consecutive 1D problems. Future on-chip integration of the algorithm can facilitate real-time tracking of fast environmental and antenna impedance variations.

The FD antenna (Fig. 35.1.3 (top)) comprises two orthogonally-oriented patch elements, acting as TX-RX elements fabricated on a 1.52mm-thick FR4 substrate with a ground plane below. The patch elements are separated by 4.125cm and fed by a microstrip line with a stub for impedance matching. The electric fields of the respective TX-RX patch elements are polarized orthogonally, thereby improving the isolation from 25dB to 46dB in simulation. However, coupled fields between the TX-RX patches generate coupled currents that travel as surface waves along the substrate and can worsen SI. To combat this, a decoupling structure consisting of a metal plate with cross-shaped slots is placed between the patches, and a degenerate ground slot is placed beneath it, which together impede the flow of surface waves between TX-RX patches and improve isolation by an additional 17dB in simulation. The measured antenna isolation at 2.47GHz is >60dB with <-10dB S11 and S22 (Fig. 35.1.3, bottom left).

A separate chip is designed for the backscatter tag, whose schematic is shown in Fig. 35.1.3 (bottom right). The tag features a fully-reflective switch matrix based on [2] that achieves an insertion loss of 0.94dB for maximizing backscatter range. The on-chip modulator supports BLE-compatible single-side-band reversely whitened-tone-to-FSK modulation, which can scale its data rate in a BLE-compatible manner through bit-repetition to as low as the employed Δf, enabling DFS and corresponding dynamic power savings by as much as 30×.

The SIC chip is fabricated in 65nm, and Fig. 35.1.4 (top left) shows the measured chip cancellation performance of a single tone versus the number of bits used for optimization, while also illustrating the contributions of each of the four hierarchies. Cancellation of up to 77dB is achieved at zero offset (i.e., the tone power itself). The measured VM constellation of the first two hierarchies is shown in Fig. 35.1.4 (bottom), plotted with coarse steps (8×8 points) for better visibility with the full scale of the first hierarchy (H1) normalized to unity. A comparison with state-of-the-art high-resolution cancellers is provided in Fig. 35.1.4 (top right). Our canceller achieves >2b higher resolution and >3× lower area. Importantly, unlike the prior art, our canceller consumes zero power due to its passive implementation, reducing its adoption overhead by existing PA/LNA pairs in conventional APs. Since the main purpose of cancellation in this work is to attenuate the PN skirt to enable low Δf for backscatter (see the PN illustration in Fig. 35.1.1, bottom), the SIC's cancellation performance vs. Δf is measured in Fig. 35.1.5 (top left), demonstrating 65dB of cancellation at 1kHz and 35dB cancellation at 1MHz. Lower cancellation at higher offset frequencies is typical for SIC circuits [11], though due to PN skirt roll-off at higher frequencies, this performance more than exceeds the requirement for robust backscatter operation at any Δf>1kHz in this work.

The backscatter chip is also fabricated in 65nm, and achieves a power consumption of 92 to 95nW, 116 to 145nW, 351 to 640nW and 2.66 to 4.69μW for 1, 10, 100, and 1000kb/s data rates, respectively, as shown in Fig. 35.1.5 (top right). The lower bound of power consumption at each data rate is limited by the required FSK frequency deviation, the employed Δf, and leakage. Wireless testing is performed at up to 32m, as shown in Fig. 35.1.5 (bottom left), where PTX=30dBm and the BER target is 10^{-3}. When using two conventional antennas for TX and RX spaced 10cm apart, less than 1m range is achieved for any data rate and any Δf≤10MHz due to severe self-interference of PN leakage, as shown by the gray bar at the bottom of Fig. 35.1.5 (bottom right). With our FD antenna, PN leakage is suppressed below the thermal noise floor for Δf>4MHz, resulting in successful 32m, 26m, 16m, and 10m operation for data rates of 1kb/s, 10kb/s, 100kb/s, 1Mb/s (solid lines), respectively, representing a 10-to-30× improvement in range over prior art that does not

588 • 2026 IEEE International Solid-State Circuits Conference

979-8-3315-8937-0/26 $31.00 © 2026 IEEE

employ an FD antenna. When the SIC chip is activated, thanks to the net 125dB of cancellation capabilities, the remaining PN skirt can be canceled to below the noise floor for Δf >1kHz, and operation can occur at any Δf (dashed lines), representing a >30× improvement in range, a >10⁴× reduction in IF, and a >30× reduction in tag power. Figure 35.1.6 compares results with prior art, showing that the antenna and RF SIC (that can be easily added to PA/LNA combos in APs with minimal power overhead) operating with the developed tag chip enables a system capable of achieving power reduction commensurate with DFS performance, all at a competitive range to prior art - all of which operate at significantly higher IFs, and almost all of which use two distally separated APs. Unlike RFID, which operates only down to 40kHz [12], CANCEL can operate to at least a 1kHz IF towards lower power and longer-range operation and, importantly, is operable with other standards such as BLE while supporting data rates compatible with Ambient IoT-style applications. Photos of the antenna, and SIC and backscatter chips are shown in Fig. 35.1.7.

Acknowledgement:
Authors would like to acknowledge NSERC, CMC, Schmidt Sciences and the UCSD Center for Wearable Sensors for their support to this work.

References:
[1] P. -H. Wang et al., "A 28μW IoT Tag that can Communicate with Commodity WiFi Transceivers via a Single-Side-Band QPSK Backscatter Communication Technique," *ISSCC*, San Francisco, CA, USA, 2020, pp. 312-314. http://doi.org/10.1109/ISSCC19947.2020.9063133
[2] S. -K. Kuo et al., "A WiFi and Bluetooth Backscattering Combo Chip Featuring Beam Steering via a Fully-Reflective Phased-Controlled Multi-Antenna Termination Technique Enabling Operation Over 56 Meters," *ISSCC*, San Francisco, CA, USA, 2022, pp. 1-3. http://doi.org/10.1109/ISSCC42614.2022.9731744
[3] L. Lin et al., "Battery-Less IoT Sensor Node with PLL-Less WiFi Backscattering Communications in a 2.5-μW Peak Power Envelope," *IEEE VLSI Circuits*, Kyoto, Japan, 2021, pp. 1-2. http://doi.org/10.23919/VLSICircuits52068.2021.9492358
[4] Z. Chang et al., "A Passive Bidirectional BLE Tag Demonstrating Battery-Free Communication in Tablet/Smartphone-to-Tag, Tag-to-Tablet/Smartphone, and Tag-to-Tag Modes," *ISSCC*, San Francisco, CA, USA, 2023, pp. 468-470. http://doi.org/10.1109/ISSCC42615.2023.10067538
[5] M. Meng et al., "Improving the Range of WiFi Backscatter Via a Passive Retro-Reflective Single-Side-Band-Modulating MIMO Array and Non-Absorbing Termination," *ISSCC*, San Francisco, CA, USA, 2021, pp. 202-204. http://doi.org/10.1109/ISSCC42613.2021.9366014
[6] J. Xiong et al., "A Wearable Backscatter System Featuring Concurrent RF Harvesting and Bidirectional Communication with Commodity BLE Transceivers," *CICC*, Boston, MA, USA, 2025, pp. 1-3. http://doi.org/10.1109/CICC63670.2025.10983441.
[7] K. A. Ahmed, R. Yang, P. Salamani, V. Rajanna and M. Alioto, "Single-Antenna Backscattered BLE5 Transmitter with up to 97m Range, 10.6 μW Peak Power for Purely-Harvested Green Systems," *ESSCIRC*, Lisbon, Portugal, 2023, pp. 49-52. http://doi.org/10.1109/ESSCIRC59616.2023.10268708
[8] S. -K. Kuo, M. Dunna, H. Lu, A. Agarwal, D. Bharadia and P. P. Mercier, "An LTE-Harvesting BLE-to-WiFi Backscattering Chip for Single-Device RFID-Like Interrogation," *ISSCC*, San Francisco, CA, USA, 2023, pp. 320-322. http://doi.org/10.1109/ISSCC42615.2023.10067815

[9] H. Abolmagd et al., "A Hierarchical Self-Interference Canceller for Full-Duplex LPWAN Applications Achieving 52–70-dB RF Cancellation," in *IEEE JSSC*, vol. 58, no. 5, pp. 1323-1336, May 2023. http://doi.org/10.1109/JSSC.2022.3200369
[10] K.-D. Chu, M. Katanbaf, T. Zhang, C. Su, and J. C. Rudell, "A Broadband and Deep-TX Self-Interference Cancellation Technique for Full-Duplex and Frequency-Domain-Duplex Transceiver Applications," in *ISSCC*, 2018, pp. 170-172. http://doi.org/10.1109/ISSCC.2018.8310238
[11] M. Katanbaf, K. -D. Chu, T. Zhang, C. Su and J. C. Rudell, "Two-Way Traffic Ahead: RF/Analog Self-Interference Cancellation Techniques and the Challenges for Future Integrated Full-Duplex Transceivers," in *IEEE Microwave Magazine*, vol. 20, no. 2, pp. 22-35, Feb. 2019. http://doi.org/10.1109/MMM.2018.2880489
[12] Q. Pan, Z. An, X. Zhao and L. Yang, "Revisiting Backscatter Frequency Drifts for Fingerprinting RFIDs: A Perspective of Frequency Resolution," *2023 20th Annual IEEE International Conference on Sensing, Communication, and Networking (SECON)*, Madrid, Spain, 2023, pp. 124-132. http://doi.org/10.1109/SECON58729.2023.10287428

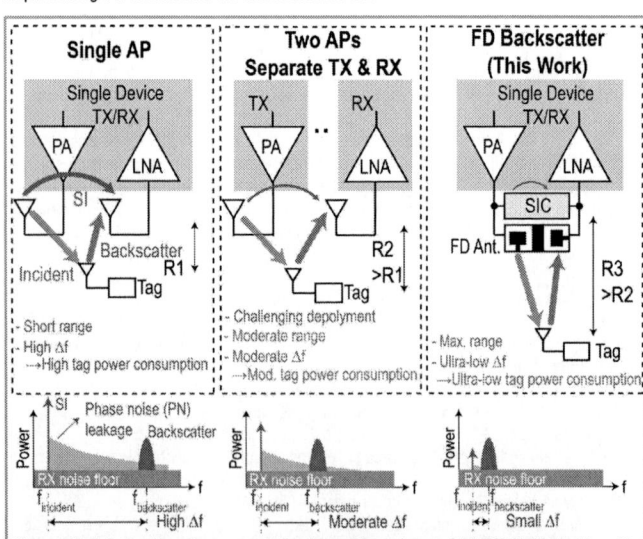

Figure 35.1.1: Backscatter communication with co-located TX/RX (left), backscatter communication TX & RX (middle), and the proposed FD backscatter (right).

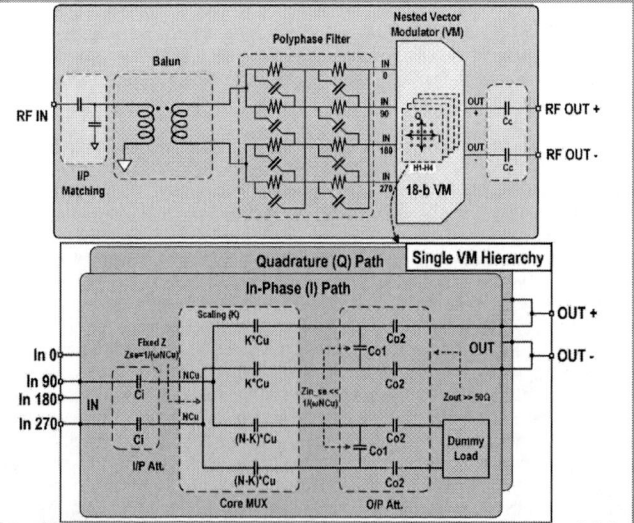

Figure 35.1.2: RF SIC block diagram (top) with circuit diagram of one vector modulator (VM) hierarchy shown (bottom).

ISSCC 2026 / SESSION 35 / LOW POWER WIRELESS TRANSCEIVERS FOR LOCALIZATION AND COMMUNICATIONS / 35.1

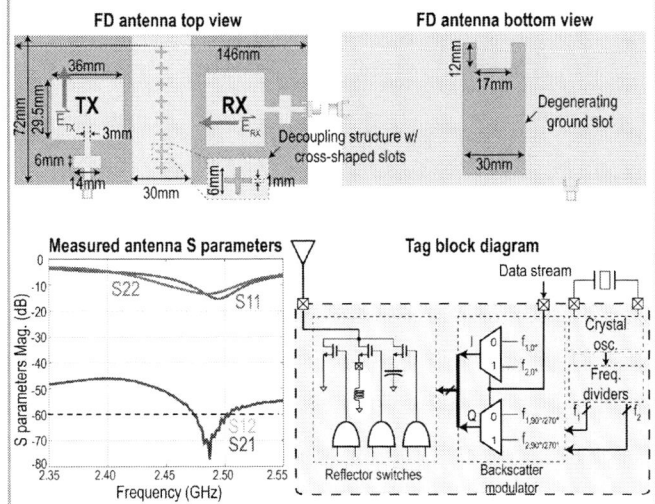

Figure 35.1.3: Designed FD antenna (top), measured antenna performance (bottom-left), and backscatter tag chip block diagram.

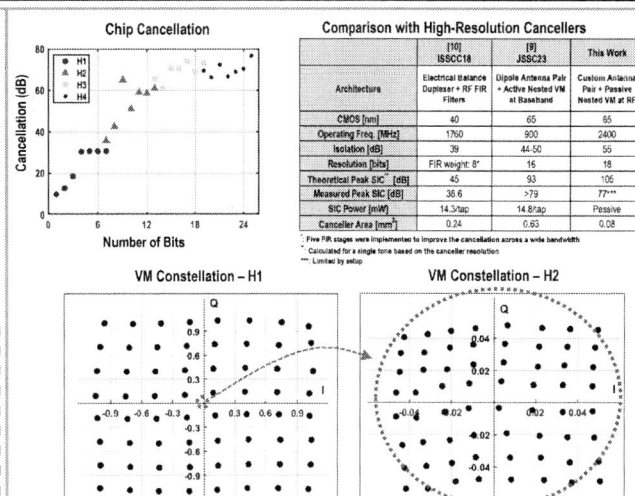

Figure 35.1.4: Measured canceller performance: Single-tone cancellation (top-left), VM constellation (bottom), and table of comparison with state-of-the-art high-resolution cancellers (top-right).

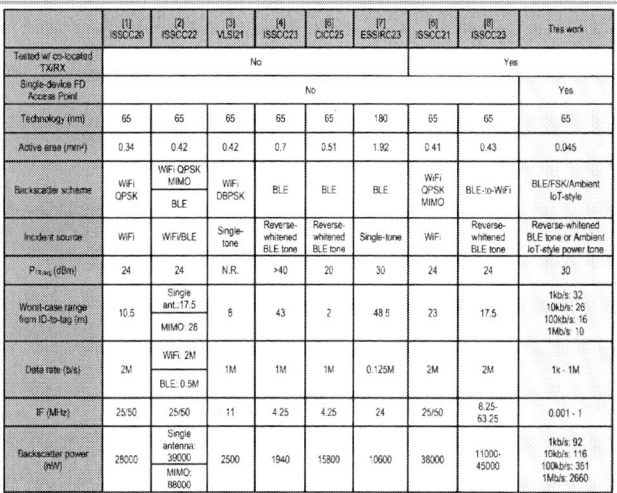

Figure 35.1.5: Canceller SIC vs. offset frequency (top-left), tag measured power consumption (top-right), distance vs. IF (bottom-right), and outdoor measurement setup (bottom-left).

Figure 35.1.6: Comparison table with state-of-the-art backscatter tags. Prior art cannot operate with a co-located ID and reduce the tag's IF and hence consumes significantly larger power.

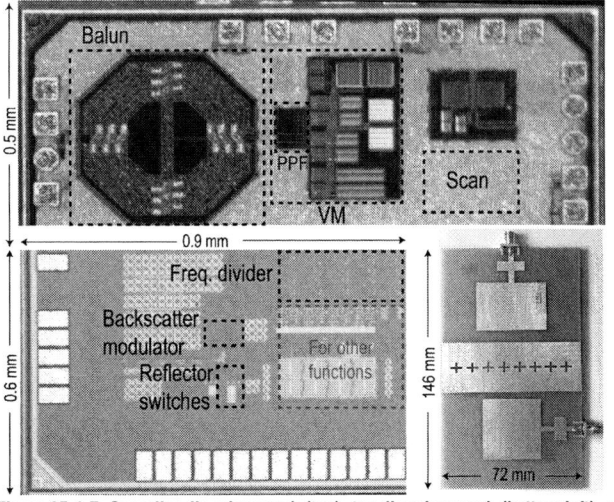

Figure 35.1.7: Canceller die micrograph (top), tag die micrograph (bottom-left), and fabricated FD antenna (bottom-right).

• 2026 IEEE International Solid-State Circuits Conference

ISSCC 2026 / SESSION 35 / LOW POWER WIRELESS TRANSCEIVERS FOR LOCALIZATION AND COMMUNICATIONS / 35.2

35.2 A 20mg Battery-Free Crystal-Less Miniaturized TX System for Flying Insects Localization with 1.45km Range

Yi Shen, Steve Young, Demba Komma, Ryan Strohman, Rahul Narasimha, Jeongtaek Chang, Andrea Bejarano-Carbo, Yunfan Wang, Guanren Tao, Hun-Seok Kim, David Blaauw

University of Michigan, Ann Arbor, MI

Abstract

A complete, 20mg, battery-free, crystal-less localization tag with a 28nm TX IC, custom antenna, PV cells, and storage cap achieves 0.9m accuracy over 1.45km. Designed for tracking wasps, this system also enables unobtrusive monitoring of other small insects/objects. We deployed a tag on a live wasp and demonstrated full system functionality while the wasp was able to fly normally. The system achieves the longest range, highest long-distance accuracy, and lowest weight, with battery-free operation, among the state-of-the-art entries in the comparison table in Figure 35.2.6.

In recent years, significant progress has been made in miniaturized tracking devices for new applications from equipment tracking to small animal studies. However, the ability to study the movements of small insects, such as honey bees and wasps, remains elusive despite their importance as key pollinators. Worker honey bees weigh ~80mg and most wasps are even lighter, placing a severe constraint on the tag weight, with a practical limit of 30mg—about half the weight of a waterdrop. At the same time, these insects can fly >1km on foraging trips and require meter-level localization accuracy for behavioral studies. The combination of these weight, size, distance, and accuracy constraints forms a formidable challenge.

Among standard approaches GNSS provides global range and reasonable accuracy [1], but is too heavy (> 2g [2]) because of its large power consumption (Fig. 35.2.1, top). BLE and Wi-Fi solutions have reduced weight, but lack the necessary range while still exceeding the weight constraint [3,4]. IR-UWB and back-scatter RF-ID approaches have the possibility for ultra-low weight but have a small range (< 10's of m) [5-7]. Recently, [8] introduced a Monarch butterfly tracking tag which collects light and temperature data along its 4000km flight. However, at 62mg, this tag remains too heavy while also lacking the needed accuracy. Further, [9] demonstrated a tag with 430m range and 1.6m accuracy but with a weight of 316mg, excluding antenna. Also, these last two approaches require unit recovery for data download which is not practical for most insect studies. Finally, [10] demonstrated a tag not requiring recovery weighing 100mg but with 80m range and poor accuracy (~10m). In sum, no localization method weighing < 60mg has been proposed and even those weighing 60-to-2000mg lack the needed range and/or accuracy.

To address this unmet need, this paper presents a complete, 20mg, battery-free and crystal-less localization tag that integrates a 28nm TX IC, a tiny custom antenna, PV cells, and SMD storage caps to achieve 0.9m accuracy at a measured distance of 1.45km. While our driving application is tracking paper-wasps, the system can be applied to unobtrusive tracking of other insects/objects.

In our approach, the PLL and crystal are removed and a free-running DCO is used to reduce power, weight, and start-up time while also improving frequency modulation (FM) speed to fulfill the bandwidth requirements for accurate localization. The subsequent loss in long-term frequency stability is addressed by increasing PA power and shortening the packet length, reducing intra-packet RF impairments, as well as advanced detection algorithms that address inter-packet drift.

To power the system, batteries add significant weight (~70mg [10]) and also cannot sustain the high current of our TX bursts. However, recent advances in small Multi-Layer Ceramic Capacitors (MLCCs) now provide capacities up to 10μF in 0201 SMD packages weighing just 0.83mg. Hence, the proposed system uses a battery-less energy architecture where PV cells harvest energy and 2 capacitors buffer it. This enables high instantaneous transmit power for short TX bursts, near-unlimited lifetime, and significant weight reduction.

However, this architecture also raises several challenges for RF signal generation and transmission for which we have the following solutions: 1) To address capacitor voltage drop during TX we use a specialized power management scheme and a pre-release RF template that captures tag-specific process-variation for packet detection/localization. 2) The high TX power and compact integration results in increased AM-to-FM frequency pulling and drift from chip self-heating which we address with EM-coupling mitigation and built-in temperature compensation. 3) To minimize the antenna size, we co-design a tiny loop antenna with on-chip matching networks.

The localization system consists of active tags (one shown), multiple RX anchors (three shown), and a TX beacon (Fig. 35.2.1. bottom right). Each tag embeds a unique 128-bit pseudo-random ID. In the digital baseband (DBB), symbols are binary modulated with chirp-spread spectrum (CSS), using a divided DCO clock and frequency-control words (FCWs). The DCO then directly synthesizes a 16MHz-wide CSS waveform via its switched-capacitor array (SCA) and feeds a PA for transmission. A beacon at a known location transmits a 40MHz QPSK Gold-code signal to synchronize the anchors.

Prior to field deployment, we collect a set of tag packets and apply principal component analysis to form a template that captures process-variation characteristics of the free-running DCO. During operation, we downconvert the RX signal with the conjugate of the template and then perform an FFT. If the signal lies within the search window, its energy collapses into a single FFT bin; the energy peak bin index reveals the CFO, which we use to drive sampling-rate correction, further refining the signal detection and timing recovery. All the timing peaks are refined with interpolation to achieve an accuracy higher than the sampling rate of 50MHz.

Figure 35.2.2 (top) shows the proposed 28nm TX IC organized into five power domains for burst-mode efficiency and signal quality. The always-on (PD1) domain is fed directly by the on-board PV cell, which provides 1.8 to 2.3V across lighting conditions. An ~1kHz leakage-based oscillator clocks the voltage detector and the always-on FSM. The detector compares the PV rail to a 2T-based low-power reference to choose charge vs. transmit. When the rail is below threshold, the FSM enters charge mode, closing two switches to replenish the storage capacitors of the TX-Top (PD2) and PA (PD3) domains. Once the voltage reaches the target, the FSM switches to TX mode and opens the charge switch to decouple the PA and DCO supplies prior to transmission. The BGR and LDO in TX-Top then turn ON to power the TX Core (PD4) from C_{PD2}, while isolating it from capacitor voltage droop. The TX-FSM wakes up and initializes the chain. In parallel, the DCO (PD5), with its own BGR and LDO, warms up and provides the baseband clock. The PA (PD3) draws current from C_{PD3}. The TX-FSM enables the PA only a few clock cycles before the packet transmission and shuts it OFF immediately after the last symbol, to save energy. Finally, the TX-FSM signals the always-on FSM to power down the TX domains, which closes the charge switches and waits for the next cycle.

For most of the cycle, the chip draws sub-uW—well below the PV supply capability—because all the other domains (except for the always-on) are power gated with low-leakage IO devices (Fig. 35.2.2, bottom right). When a packet is scheduled, the TX FSM and AFE start up in ~1 to 2 ms, preparing an ~250us high-power burst. During the TX burst, the PA dominates the energy budget, delivering high RF power, and consuming most of the harvested energy.

Figure 35.2.3 details the DCO and PA. To minimize intra-packet drift, the DCO uses a 6-bit fine SCA to support CSS modulation with various chirp bandwidths (8 to 80MHz) and adds dummy switch pairs, matched to the main switch to preserve phase continuity during symbol transitions. The co-designed LDO and BGR (Fig. 35.2.3, mid-right) employ dynamic bandwidth control to deliver a low-noise supply, avoiding phase-noise degradation with simultaneous low power overhead and fast start-up. The BGR also provides a CTAT reference that biases the varactor array for temperature compensation. Importantly, its fast V_{th}-based response enables intra-packet compensation necessary since the PA turns on just before transmission and then gradually heats the chip during the TX burst.

Through EM coupling, the free-running DCO experiences AM-to-FM frequency pulling. Figure 35.2.3 (top right) shows that with a conventional octagonal inductor and unoptimized bond-wire routing, the DCO can be pulled by >50MHz over the transmission window. Simply aligning the PA bond wire with the inductor center only reduces this to ~20MHz. To fully address the issue we instead use a figure-8 inductor and route the PA output bond through its center with a symmetric return reducing pull to 1.68MHz, 2.2 to 1.6V in silicon measurement.

Figure 35.2.3 (bottom) shows the high-efficiency Class-D PA implemented with I/O devices to reduce off-mode leakage. The PA output uses on-chip MOM capacitors in series and shunt configuration to match the customized electrically small loop antenna. With a low output match (10Ω), the PA delivers 10.2dBm at 2V with 34.2% drain efficiency (including the driver), and maintains relatively high output power as the capacitor voltage drops (8.1dBm @ 1.6V). Figure 35.2.3 (bottom-left) shows the antenna and its radiation pattern and impedance at different frequencies. The antenna weighs 9.55mg, exhibits 15.3nH inductance and 0.36Ω radiation resistance at 2.45GHz, and achieves −5.2dB radiation efficiency in EM simulation. The TX chain has an average EIRP of 0.37dBm during the transmission, including the cap voltage droop, matching, and parasitic loss.

We fabricated the TX chip in 28nm CMOS with a core area of 0.34mm² (Fig. 35.2.7). Figure 35.2.4 top left shows the BGR and LDO filtering reduces DCO phase noise at 1-to-100kHz offset, improving phase coherence for the 250us packet. The bottom-left panel shows the transmitted localization packet, instantaneous frequency (including a zoom-in of the 16MHz CSS symbols), and the packet phase angle showing the residual intra-packet frequency/phase drift remains within tolerance. The bottom-right panel displays the measured output spectrum of the tag.

We deployed a tag on a live wasp and demonstrated full system functionality (Fig. 35.2.5 right) while the wasp was able to fly normally. For controlled localization accuracy measurements, we used two RX anchors implemented with USRP B200s equipped with VERT2450 omnidirectional vertical antennas (3dBi gain), and a beacon implemented with a USRP X310. We performed time-difference-of-arrival (TDoA) ranging by varying the relative positions of the tag, beacon, and anchors outdoors under line-of-sight (LOS) conditions in a campus environment with moderate interference/multipath (up to 800m) and in a rural area with mild interference/multipath (up to 1.45km). The resulting TDoA measurements (left) yielded a maximum median accuracy of 3.0m in the 800m campus case and 1.7m in the 1.45km rural case. When the tag and anchors are aligned in a straight line, as in our testing configuration, TDoA is proportional to twice the actual range. As a result, the tag ranging error is half of the TDoA error (0.9m@1.45km) [11]. Figure 35.2.5 bottom right shows the relationship between the light intensity for energy harvesting and TX interval (<1s interval @ >2 klux). Figure 35.2.6 compares our results with prior low-power localization ICs/systems and mm-scale transmitters, showing longest range (up to 1.45 km, limited by testing setup), highest accuracy for long-distance systems, lowest weight (20mg), and battery-free operation.

Acknowledgment:
The authors would like to thank the TSMC University Shuttle Program for chip fabrication. The authors also thank Prof. Elizabeth Tibbetts and Joseph Caldwell from the Department of Ecology and Evolutionary Biology at the University of Michigan, Ann Arbor, for their assistance with the live wasp experiments. The support from Cellular Tracking Technologies (CTT) in providing the photovoltaic cells is also gratefully acknowledged.

References:
[1] C. G. Tan et al., "A Universal GNSS (GPS/Galileo/Glonass/Beidou) SoC with a 0.25mm² Radio in 40nm CMOS," *ISSCC*, San Francisco, CA, USA, 2013, pp. 334-335, doi: 10.1109/ISSCC.2013.6487758. http://doi.org/10.1109/ISSCC.2013.6487758
[2] "Mosaic-G5 P3 Ultra Compact Low Power GNSS / GPS Module." Accessed: Sep. 01, 2025. [Online]. Available: https://www.septentrio.com/en/products/gnss-receivers/gnss-receiver-modules/mosaic-G5-P3.
[3] "CTT BlüMorpho," Cellular Tracking Technologies. Accessed: Sep. 01, 2025. [Online]. Available: https://celltracktech.com/products/ctt-blumorpho.
[4] L. Lin et al., "A Battery-Free Crystal-Less BLE Transmitter Tag with Fully-Integrated RF Harvesting and Multitag TDD and FDD Broadcasting," *IEEE TMTT.*, vol. 73, no. 3, pp. 1837–1847, Mar. 2025. http://doi.org/10.1109/TMTT.2024.3458019.
[5] R. Nandakumar, V. Iyer, and S. Gollakota, "3D Localization for Sub-Centimeter Sized Devices," in *Proceedings of the 16th ACM Conference on Embedded Networked Sensor Systems*, in *SenSys*, New York, NY, USA: Association for Computing Machinery, 2018, pp. 108–119. http://doi.org/10.1145/3274783.3274851.

[6] A. Apsel, "A Simple Guide to Low-Power Wireless Technologies: Balancing the Tradeoffs for the Internet of Things and Medical Applications," *IEEE Solid-State Circuits Mag.*, vol. 10, no. 4, pp. 16–23, 2018. http://doi.org/10.1109/MSSC.2018.2867404.
[7] K.-K. Huang et al., "An Ultra-Low-Power 9.8 GHz Crystal-Less UWB Transceiver with Digital Baseband Integrated in 0.18 µm BiCMOS," *IEEE JSSC*, vol. 48, no. 12, pp. 3178–3189, Dec. 2013. http://doi.org/10.1109/JSSC.2013.2281523.
[8] I. Lee et al., "mSAIL: Milligram-Scale Multi-Modal Sensor Platform for Monarch Butterfly Migration Tracking," *Commun. ACM*, vol. 67, no. 6, pp. 93–101, 2024. http://doi.org/10.1145/3611105.
[9] A. Bejarano-Carbó et al., "An OFDMA Baseband Processor Enabling 165µW Long-Range IoT Localization," *IEEE VLSI Technology and Circuits*, Kyoto, Japan, 2025, pp. 1-3. http://doi.org/10.23919/VLSITechnologyandCir65189.2025.11075164
[10] V. Iyer, R. Nandakumar, A. Wang, S. B. Fuller, and S. Gollakota, "Living IoT: A Flying Wireless Platform on Live Insects," in *MobiCom* 2019, New York, NY, USA: Association for Computing Machinery, 2019, pp. 1–15. http://doi.org/10.1145/3300061.3300136
[11] F. Gustafsson and F. Gunnarsson, "Positioning Using Time-Difference of Arrival Measurements," in *IEEE International Conference on Acoustics, Speech, and Signal Processing (ICASSP '03).*, Apr. 2003, p. VI–553. http://doi.org/10.1109/ICASSP.2003.1201741
[12] M. Privitera, Y. Ruiyuan, K. Ali, A. Ballo, A. D. Grasso, and M. Alioto, "Sub-uW Battery- and Crystal-Free Tag Featuring 802.11ba/b-Compliant Wake-Up Receiver, Backscattered Transmitter and 3D Localization," in *IEEE VLSI Technology and Circuits*, Jun. 2025, pp. 1–3. http://doi.org/10.23919/VLSITechnologyandCir65189.2025.11074894
[13] E. Bechthum et al., "A Low-Power BLE Transceiver with Support for Phase-Based Ranging, Featuring 5µs PLL Locking Time and 5.3ms Ranging Time, Enabled by Staircase-Chirp PLL with Sticky-Lock Channel-Switching," in *ISSCC*, Feb. 2020, pp. 470–472. http://doi.org/10.1109/ISSCC19947.2020.9063073
[14] L.-X. Chuo, Z. Luo, D. Sylvester, D. Blaauw, and H.-S. Kim, "RF-Echo: A Non-Line-of-Sight Indoor Localization System Using a Low-Power Active RF Reflector ASIC Tag," in *MobiCom*, New York, NY, Association for Computing Machinery, 2017, pp. 222–234. http://doi.org/10.1145/3117811.3117840
[15] Z. Yang, J. Yin, W.-H. Yu, H. Zhang, P.-I. Mak, and R. P. Martins, "A ULP Long-Range Active-RF Tag with Automatic Antenna-Interface Calibration Achieving 20.5% TX Efficiency at -22dBm EIRP, and -60.4dBm Sensitivity at 17.8nW RX Power," in *ISSCC*, Feb. 2023, pp. 30–32. http://doi.org/10.1109/ISSCC42615.2023.10067628
[16] Y. Shi, X. Chen, H.-S. Kim, D. Blaauw, and D. Wentzloff, "A 606µW mm-Scale Bluetooth Low-Energy Transmitter Using Co-Designed 3.5×3.5mm² Loop Antenna and Transformer-Boost Power Oscillator," in *ISSCC*, Feb. 2019, pp. 442–444. http://doi.org/10.1109/ISSCC.2019.8662333

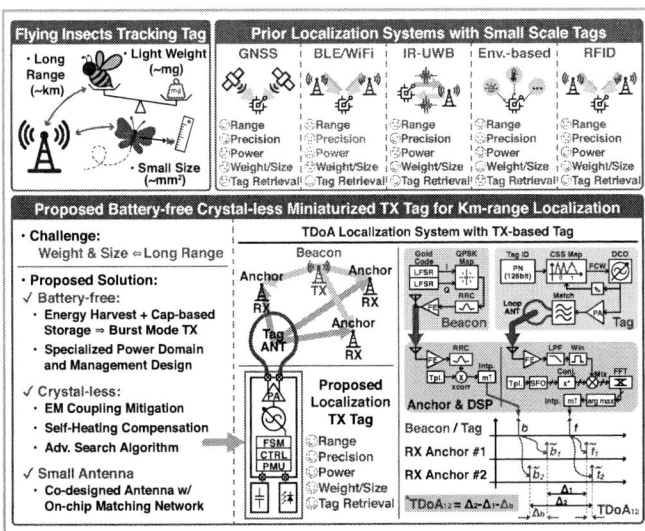

Figure 35.2.1: Overview of the small-scale localization systems (top) and the proposed work (bottom).

Figure 35.2.2: System diagram and working principles of the proposed system, including the power domain division and the voltage and power diagram of burst mode TX.

ISSCC 2026 / SESSION 35 / LOW POWER WIRELESS TRANSCEIVERS FOR LOCALIZATION AND COMMUNICATIONS / 35.2

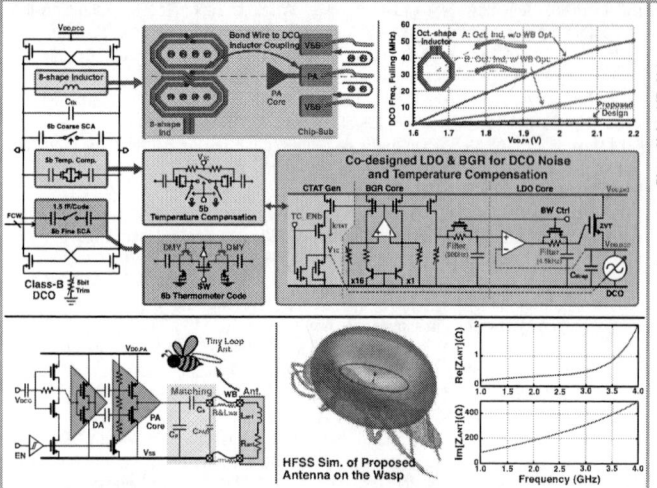

Figure 35.2.3: Simplified schematic of DCO and peripheral blocks (top), simulated PA-DCO pulling (top right). PA simplified schematic (bottom left) and EM simulation of custom loop antenna (bottom right).

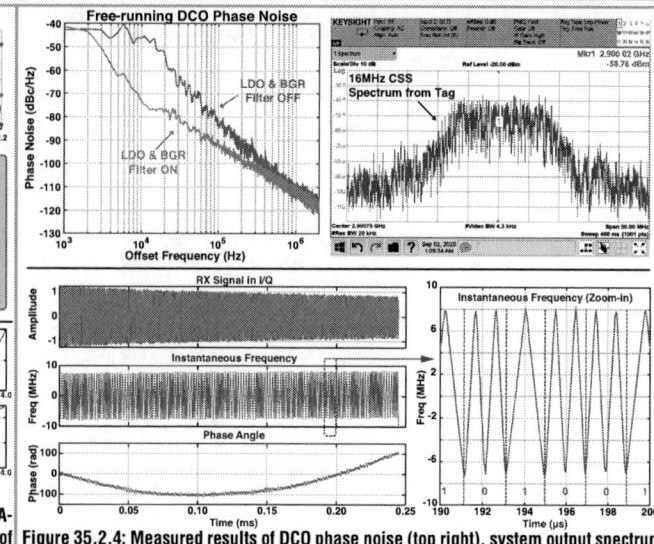

Figure 35.2.4: Measured results of DCO phase noise (top right), system output spectrum (top left). Measured chirp signal (bottom).

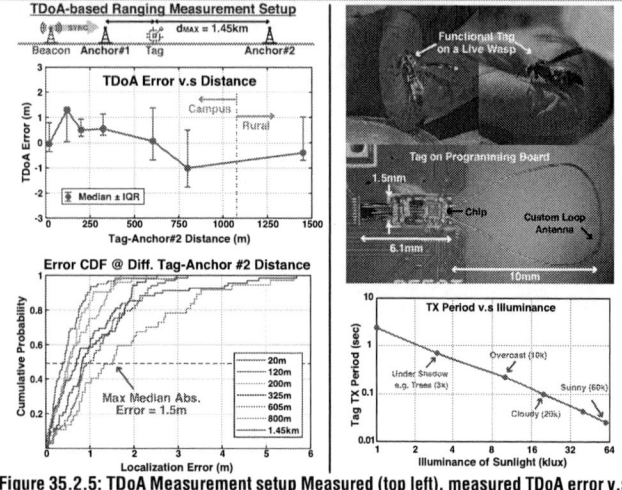

Figure 35.2.5: TDoA Measurement setup Measured (top left), measured TDoA error v.s. max tag-anchor distance (mid left), and measured location error CDF (bottom left). Live wasp and demonstrated full system functionality (top right), TX period vs sunlight (bottom right).

	Localization IC					Localization System		mm-scale TX w/ Small Antenna		
	This Work	[9] VLSI'25	[12] VLSI'25	[13] ISSCC'20	[14] MobiCom'17	[10] MobiCom'19	[5] SenSys'18	[4] TMTT'25	[15] ISSCC'23	[16] ISSCC'19
Technology	28nm	40+22nm	180nm	40nm	180nm	N/A	N/A	28nm	65nm	65nm
System	TX	RX+DBB	Backscatter	TRx	TRx	Envelope Detector	Backscatter	TX	TRX	TX
Max Range (m)	1450	430[a,b]	5	20	90[a]	80[a]	60[a]	N/A	N/A	NA
Accuracy @ Max Range (m)	0.9	1.6	0.2	0.2	0.26	10	1.5	N/A	N/A	NA
Modulation	Chirp	Hopping OFDMA	DSSS	GFSK	OFDM	AM	Chirp	GFSK	OOK	GFSK
Localization Method	TDoA	TDoA	Signal Strength	Phase-based	Active Reflection	AoA	Multiband Reflection	N/A	N/A	N/A
Carrier Frequency (GHz)	2.1-3.2[c]	2.4	2.4	2.4	0.9/2.4 (T/R)	0.9	0.9/2.4/5.8	2.4	0.92	2.4
Localization Bandwidth (MHz)	8-80[c]	12.5-100	N/A	80	80	N/A	0.5MHz	N/A	N/A	N/A
TX EIRP (dBm)	0.37	N/A	N/A	N/A	N/A	N/A	N/A	-11.2[d]	-15.8	-23.4
Energy per Localization (µJ)	9.7	16.5	N/A	62.5 (TX+RX)	18	552	N/A	N/A	N/A	N/A
Localization Time	250µs	24-192ms	N/A	5.3ms	300µs	N/A	7ms[e]	N/A	N/A	N/A
Plug-in/Battery	No	Yes	No	Yes	Yes	Yes	Yes	No	Yes	Yes
XTAL	No	No	No	Yes	Yes	No	Yes	No	Yes	Yes
Tag & Antenna Size (mm)	1.5×6.1 & 10×5	10×28 & N/A	Small Sys. N/A	Small Sys. N/A	Small Sys. N/A	6.1×6.4 & 2.1 & N/A	11.8×7.5× & N/A	17×27×3 & Included	37×40 & 23×23	3.8×4.5 & 3.5×3.5
Tag Weight (mg)	20	316	N/A	N/A	N/A	102	N/A	N/A	N/A	N/A
Tag Active Power	32.9 mW	10.2mW	0.92µW	5.3mW	62.8 mW	1.6mA*V_DD[f]	93 µW	2.6mW	N/A	606µW

a. 2/3-D Localization; b. LOS Measurements; c. 2.9GHz and 16MHz are used in TDoA Measurement; d. Output power; e. per chirp time; f. V_DD is unknown

Figure 35.2.6: Comparison table with prior low-power localization ICs/systems and mm-scale transmitters.

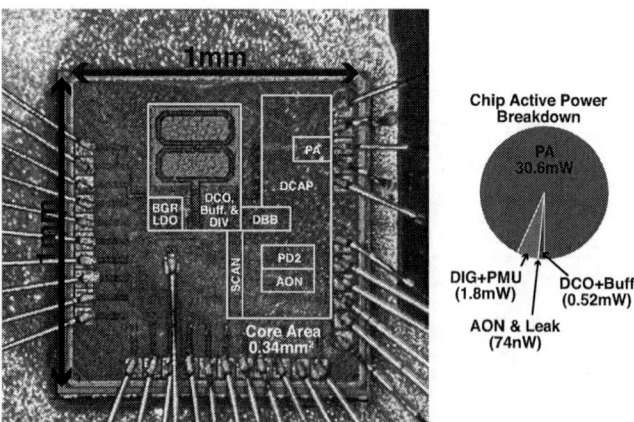

Figure 35.2.7: Die micrograph and power breakdown.

• 2026 IEEE International Solid-State Circuits Conference

979-8-3315-8937-0/26 $31.00 © 2026 IEEE

ISSCC 2026 / SESSION 35 / LOW POWER WIRELESS TRANSCEIVERS FOR LOCALIZATION AND COMMUNICATIONS / 35.3

35.3 A 0.052mm² Blocker-Tolerant Non-Uniform Current Sub-Sampling Receiver with a Discrete-Time FIR/IIR Filter Enabling 56dB Rejection in 28nm CMOS

Mostafa Ayesh[1,2], Mike Shuo-Wei Chen[2]

[1]now with Marvell, Irvine, CA
[2]University of Southern California, Los Angeles, CA

Abstract

This paper presents a 0.052mm² blocker-tolerant 18-to-22GHz non-uniform current sub-sampling (NUCSS) receiver featuring NU DT FIR/IIR filtering and equalized Gm-cells, addressing aliasing, noise folding, and alias/non-alias band blockers. Fabricated in 28nm CMOS, this prototype achieves up to 56dB blocker rejection, >14dB higher blocker rejection than conventional NU DT FIR RX, -29.1dB EVM for a 64-QAM 150MS/s signal centered at 20GHz, and B_{1dB} of 5.8dB while consuming 20mW power.

Wireless sub-sampling receivers (RXs) often offer a reduced implementation cost, i.e. low power and small area, by directly sampling the input voltage of the RF signal at a much lower clock frequency while relying on a desired harmonic aliasing for signal downconversion and reconstruction [1]. This helps in avoiding a significantly higher area and power overhead owing to the required different LO phases' generation and distribution at high frequencies in mixer-based receivers [2]. However, such receivers suffer from two main drawbacks: (1) aliasing and (2) noise folding. They require the desired signal to be strictly bandpass filtered which would otherwise result in destructive aliasing if blockers exist at other harmonics of the sub-sampling frequency. A non-uniform sub-sampling (NUSS) technique was presented in [3] to avoid such destructive in-band aliasing. It spreads the blocker energy that appeared at those unwanted harmonics to out-of-band (OOB) frequencies after sampling, i.e. so-called spectral alias spreading. However, those OOB spectral aliases can still result in a large signal swing and may even saturate the following ADC input. An NU DT FIR receiver [4] was introduced to address this limitation and improve NUSS blocker tolerance, but its notch-based filtering provides only moderate improvement in the blocker tolerance rather than the broader bandpass roll-off needed blocker rejection. Additionally, the NU DT FIR receiver does not address the fundamental noise folding issue. To overcome this and improve non-alias-band blocker tolerance, this work introduces: (1) a non-uniform direct RF current sub-sampling (NUCSS) technique that mitigates noise folding, enhances sampling switch linearity, preserves spectral alias spreading, and provides roll-off bandpass filtering with additional frequency nulls; (2) a non-uniform discrete-time (NU DT) FIR/IIR filter after NUCSS to suppress OOB folded blockers, relax ADC dynamic range and sampling-rate requirements, reconstruct NU samples, and convert them to a uniform grid; and (3) an equalized Gm-cell (Eq-Gm) to de-emphasize low-frequency integrator gain and improving linearity. A 28nm CMOS proof-of-concept prototype with reconfigurable blocker-rejection schemes achieves 41/56 dB alias-band rejection at 2.5/5GHz offsets from a 20GHz carrier, nearly 14dB higher than the conventional NU DT FIR receiver [4] for a 20dB SNDR with 150MHz baseband. Combined with a higher-gain LNA and current-mode sampling, the NUCSS RX attains ~5dB lower noise figure than [4], demonstrating significant performance improvement.

Figure 35.3.1 compares the proposed non-uniform current sub-sampling (NUCSS) technique with existing sub-sampling approaches. As shown in Fig. 35.3.1(A), a conventional uniform-voltage-sub-sampling (USS) RX input is sampled on an evenly spaced time grid at rate f_s, producing spectral replicas (alias bands) at multiples of f_s, with the desired band centered at f_c. The presence of strong blockers in alias bands results in their folding back into baseband with the desired signal, causing destructive aliasing. Thus, a USS receiver relies heavily on analog anti-alias (AA) filtering before sub-sampling, adding significant implementation overhead. On the other hand, non-uniform voltage sub-sampling (NUSS) RX [3], shown in Fig. 35.3.1(B), mitigates this by dithering sampling instants with pre-designed NU shifts, known as NU sequence, spreading alias-band blockers (e.g., those located at $7f_s$ and $9f_s$ in this example) OOB while preserving the desired band (e.g., one located at $8f_s$). However, rejection degrades for blockers in non-alias bands and OOB aliases can still produce large signal swings that can potentially saturate the subsequent ADC. Uniform current sub-sampling (UCSS) RX, another approach shown in Fig. 35.3.1(C), improves rejection via the Sinc-shaped response of current integration, adding spectral nulls at multiples of $1/T_{int}$, the integration period, and providing bandpass roll-off rejection. However, since the current integration periods are still uniformly spaced, alias-band blockers are again allowed to fold into baseband on top of the desired signal. While the integration process reduces the severity of this aliasing compared to USS RX, the problem is not eliminated and can still degrade the receiver performance. To combine the benefits of current-mode integration and non-uniform sampling, the proposed NUCSS RX, shown in Fig. 35.3.1(D), enables clean baseband downconversion of the desired signal while folding alias-band blockers out-of-band, with a stronger suppression than NUSS RX.

After NU voltage/current sub-sampling, a DT filter can reconstruct the signal onto a uniform grid and provide additional AA filtering before the ADC. This offers several advantages: (1) attenuating OOB aliases to reduce output swing and enable further decimation, thereby relaxing ADC dynamic range and sampling rate; (2) avoiding destructive in-band aliasing from both alias-band and non-alias-band blockers; and (3) programmability to tune frequency nulls/stopbands to system needs. Figure 35.3.2 compares NUCSS with NU DT

FIR/IIR against conventional NUSS with NU DT FIR. The NUSS + NU DT FIR response [4] shows limited non-alias band rejection due to extra NU aliases, whereas NUCSS + NU DT FIR/IIR improves non-alias band rejection by +15dB. The frequency response features both NU nulls at alias frequencies (integer multiples of f_s) and Sinc-function nulls at multiples of $1/T_{int}$. It is worth noting that NU sampling instants or integrating clock pulses must be quantized to a time grid T_G, introducing passbands at multiples of $1/T_G$. Since T_G can be designed much finer than the subsampling period (8× finer here), adjacent passbands lie far away (e.g., $8f_s$), unlike uniform DT filters which inevitably place passbands at multiples of f_s. These closer passbands in uniform DT FIR limit blocker attenuation and force reliance on analog AA filtering, whereas NU filters relax AA design by pushing passbands farther apart. Figure 35.3.2 also compares voltage and current sampling. Transfer-function analysis shows current sampling can provide higher gain than voltage sampling across the main and side Sinc lobes. Noise analysis shows current sampling having less total integrated output noise than voltage sampling when only considering the sampling switch thermal noise. However, current sampling turns out to have higher total integrated output noise than the KT/C noise when considering the Gm-cell noise. This noise depends on the switch ON resistance R_{on}, the Gm cell output resistance r_o, the transconductance of the Gm-cell G_m and the integration time T_{int}. Overall, Fig. 35.3.2 shows that current sampling can provide +9dB higher SNR than voltage sampling, mainly due to the signal gain from integration. The integration time T_{int} can be determined by center frequency f_c, required location of Sinc nulls, signal bandwidth, and target SNR.

Figure 35.3.3 shows the proposed NUCSS receiver and its non-uniform clocking scheme. The signal path begins with a two-stage LNA, followed by eight equalized Gm-cells that convert the RF input voltage into RF current. Each Gm-cell includes: (1) RC degeneration to suppress low-frequency gain, reducing total integrated output noise since the input is centered at f_c; and (2) a tunable 6-bit coefficient to set FIR/IIR filter weights. The RF currents are integrated by non-uniformly time-interleaved samplers, which also reduce sampling-induced crosstalk. The resulting voltages on the integrating caps C_h form eight taps, realizing an NU DT filter in the charge domain. A charge-sharing switch network connects these integrated voltages to a summing capacitor C_{sum}, performing the summation for the DT filter output. This summation is triggered by CLK_{sum}, a summation/decimation clock that captures the NU DT filter output once every eight integration cycles. Resetting the summing capacitor yields an FIR filter, while preserving charge across cycles enables an IIR filter. The output is buffered to provide an analog baseband signal, then quantized by an asynchronous SAR ADC. An on-chip delay-locked loop (DLL) generates 64 uniformly spaced phases divided into 8 groups. One phase from each group is selected via multiplexers controlled by the pre-designed NU sequence (S_1 to S_8) to form the eight NU clocks. The pulse width of the selected clock phases ($CLK_{NU1}...CLK_{NU8}$) defines the desired integration period T_{int}.

Figure 35.3.4 highlights a key issue in current sub-sampler implementation. Since NUCSS operates on a Sinc side lobe rather than the main lobe, a low-frequency signal coupled to the Gm-cell input can be integrated with excessive gain, saturating the integrator output and desensitizing it to the desired signal around f_c. To mitigate this and suppress low-frequency noise, an equalized Gm-cell with RC degeneration is proposed. At low frequencies, resistive degeneration attenuates the signal, while at high frequencies, the capacitor bypasses the resistor to restore transconductance. The transistor-level design, its transfer function, and the resulting integrator response are shown in Fig. 35.3.4. The non-uniformly integrated voltages act as DT filter taps; these voltages on integration capacitors are combined with a summing capacitor to form the DT filter output, which is then held for SAR quantization. This charge-domain DT filter reduces area, power, and linearity/noise degradation. With tunable integration and summing capacitors, it also serves as a variable gain amplifier. Figure 35.3.4 (bottom) shows the timing diagram of NU integration, DT filter operation, and asynchronous 6-bit SAR conversion. Thanks to the OOB attenuation from the NU DT filter, 8× decimation is achieved — i.e., the ADC quantizes only once every eight integration cycles — relaxing both dynamic range and speed requirements.

The proposed NUCSS DT FIR/IIR receiver is fabricated in 28nm CMOS, occupying an active area of 0.052mm² and consuming 20mW from a 1V supply. Figure 35.3.5 shows measured frequency responses under different integration periods (T_{int}), and filter coefficients. The NUCSS DT FIR/IIR filter achieves multiple programmable notches in both alias bands

592 • 2026 IEEE International Solid-State Circuits Conference

979-8-3315-8937-0/26 $31.00 © 2026 IEEE

ISSCC 2026 / February 18, 2026 / 4:25 PM

(~41dB) due to NU operation and non-alias bands (~56dB) due to the Sinc nulls, including the LNA bandpass response centered at f_c, providing stronger attenuation than conventional USS and NUSS receivers. In a two-tone blocker test with a blocker and a signal at 17.5GHz and 20GHz respectively, the SNDR is measured over ~150MHz BW, the NUCSS receiver improves blocker tolerance by ~14dB over NUSS + NU DT FIR for a 20dB SNDR target, validating its rejection capability. In a modulated signal test, a 64-QAM test further confirms in-band blocker suppression, achieving −29.1dB EVM in the constellation diagram. Figure 35.3.6 summarizes the NUCSS RX performance against state-of-the-art mm-wave receivers, highlighting its smallest area, lowest power, and reconfigurable notches in both alias and non-alias bands.

Acknowledgement:
The authors like to thank SRC CUbiC for funding support.

References:
[1] J. Cheng, N. Qi, P. Y. Chiang and A. Natarajan, "A 1.3mW 0.6V WBAN-Compatible Sub-Sampling PSK Receiver in 65nm CMOS," *ISSCC*, San Francisco, CA, USA, 2014, pp. 168-169. https://doi.org/10.1109/ISSCC.2014.6757385

[2] M. Pashaeifar, L. C. N. De Vreede and M. S. Alavi, "A Millimeter-Wave Front-End for FD/FDD Transceivers Featuring an Embedded PA and an N-Path Filter Based Circulator Receiver," *IEEE RFIC*, Denver, CO, USA, 2022, pp. 11-14. https://doi.org/10.1109/RFIC54546.2022.9863209

[3] C. Yang, M. Ayesh, A. Zhang, T. -F. Wu and M. S. -W. Chen, "A 29-mW 26.88-GHz Non-Uniform Sub-Sampling Receiver Front-End Enabling Spectral Alias Spreading," *IEEE RFIC*, Los Angeles, CA, USA, 2020, pp. 87-90. https://doi.org/10.1109/RFIC49505.2020.9218365

[4] M. Ayesh, S. Mahapatra, C. Yang and M. S. -W. Chen, "A 0.072mm² 18-to-21GHz Non-Uniform Sub-Sampling Receiver with a Non-Uniform Discrete-Time FIR Filter Achieving 42dB Blocker Rejection in 28nm CMOS," *ISSCC*, San Francisco, CA, USA, 2024, pp. 92-94, doi: 10.1109/ISSCC49657.2024.10454550. https://doi.org/10.1109/ISSCC49657.2024.10454550

[5] C. Yang et al., "A Blocker-Tolerant Receiver with VCO-Based Non-Uniform Multi-Level Time-Approximation Filter with −36dB EVM in 28nm CMOS," *ISSCC*, San Francisco, CA, USA, 2025, pp. 1-3. https://doi.org/10.1109/ISSCC49661.2025.10904562

[6] C. Yang, S. Su and M. S. -W. Chen, "Millimeter-Wave Receiver with Non-Uniform Time-Approximation Filter," in *IEEE JSSC*, vol. 58, no. 5, pp. 1201-1211, May 2023. https://doi.org/10.1109/JSSC.2023.3243044

[7] P. Song and H. Hashemi, "mm-Wave Mixer-First Receiver with Selective Passive Wideband Low-Pass Filtering," in *IEEE JSSC*, vol. 56, no. 5, pp. 1454-1463, May 2021. https://doi.org/10.1109/JSSC.2021.3063726

[8] S. Krishnamurthy and A. M. Niknejad, "10-35GHz Passive Mixer-First Receiver Achieving +14dBm in-band IIP3 for Digital Beam-forming Arrays," *IEEE RFIC*, Los Angeles, CA, USA, 2020, pp. 275-278. https://doi.org/10.1109/RFIC49505.2020.9218301

[9] L. Zhang and M. Babaie, "A 23-to-29GHz Receiver with mm-Wave N-Input-N-Output Spatial Notch Filtering and Autonomous Notch-Steering Achieving 20-to-40dB mm-Wave Spatial Rejection and -14dBm In-Notch IP1 dB," *ISSCC*, San Francisco, CA, USA, 2022, pp. 82-84. https://doi.org/10.1109/ISSCC42614.2022.9731108

[10] R. Garg et al., "A 28GHz 4-Element MIMO Beam-Space Array in 65nm CMOS with Simultaneous Spatial Filtering and Single-Wire Frequency-Domain Multiplexing," *ISSCC*, San Francisco, CA, USA, 2020, pp. 80-82.

https://doi.org/10.1109/ISSCC19947.2020.9063120

[11] S. Mondal, R. Singh and J. Paramesh, "A Reconfigurable 28/37GHz Hybrid-Beamforming MIMO Receiver with Inter-Band Carrier Aggregation and RF-Domain LMS Weight Adaptation," *ISSCC*, San Francisco, CA, USA, 2018, pp. 72-74. https://doi.org/10.1109/ISSCC.2018.8310189

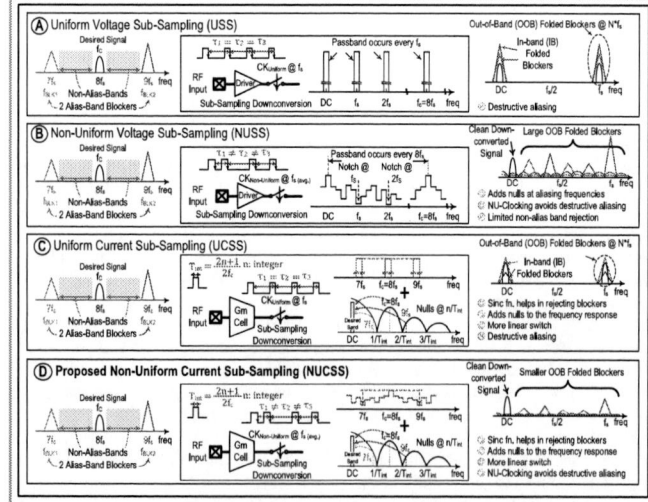

Figure 35.3.1: Comparing uniform and non-uniform voltage sub-sampling techniques vs. uniform and proposed non-uniform current sub-sampling techniques.

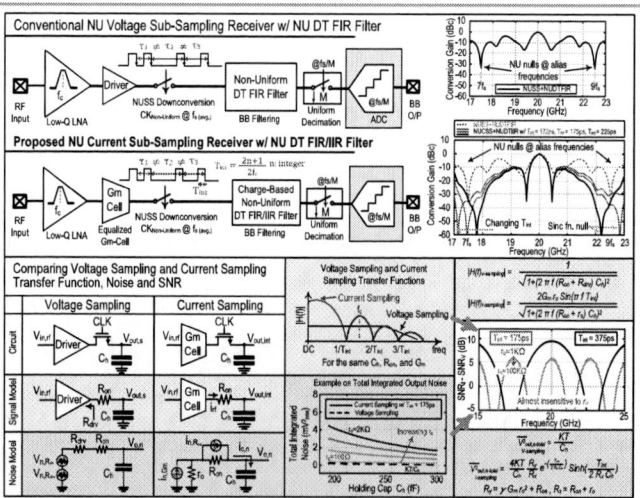

Figure 35.3.2: Proposed non-uniform current sub-sampling (NUCSS) receiver vs. conventional non-uniform voltage sub-sampling receiver (NUSS) and signal and noise analysis.

DIGEST OF TECHNICAL PAPERS • 593

ISSCC 2026 / SESSION 35 / LOW POWER WIRELESS TRANSCEIVERS FOR LOCALIZATION AND COMMUNICATIONS / 35.3

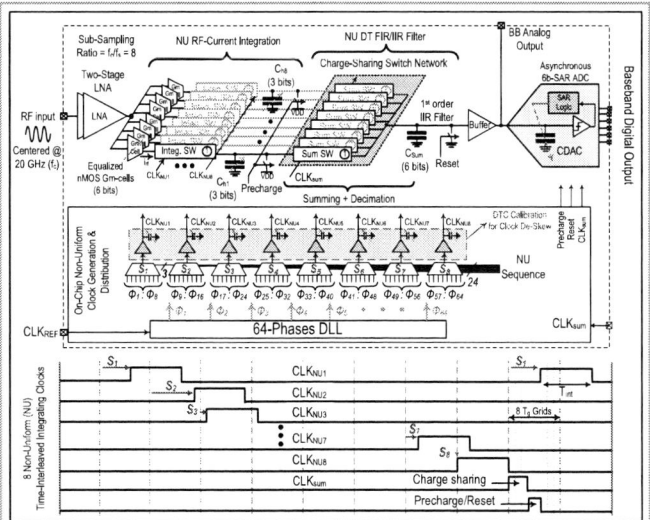

Figure 35.3.3: Block-level architecture of the NUCSS + NU DT IIR/FIR Receiver with clocking diagram.

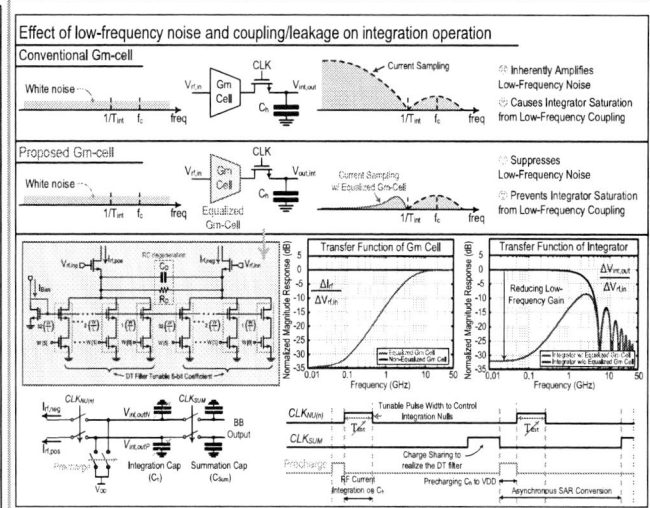

Figure 35.3.4: Implementation of the equalized Gm cell used for current sub-sampling front-end.

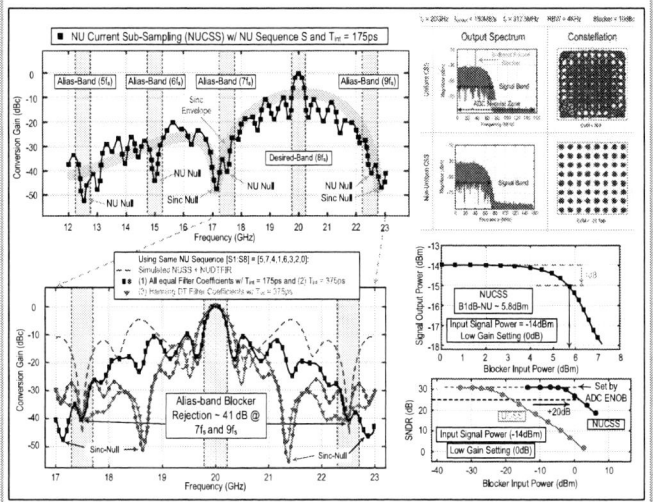

Figure 35.3.5: Measured receiver frequency response with different modes, modulated-signal test, B1dB and SNDR comparing NUCSS w/ NU DT IIR vs. UCSS w/ U DT FIR.

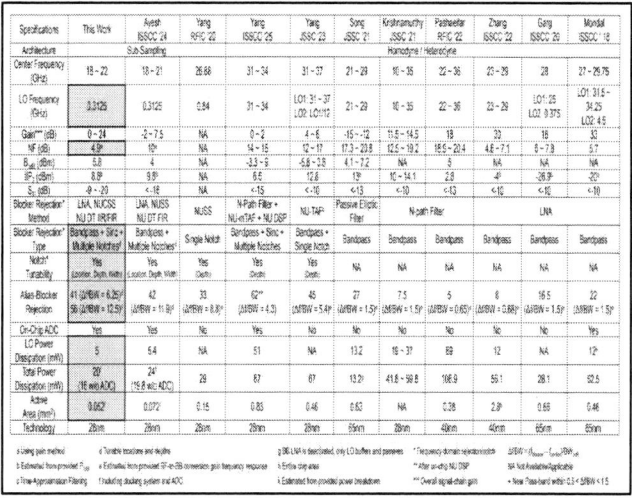

Figure 35.3.6: Performance summary and comparison with state-of-the-art mm-wave receivers.

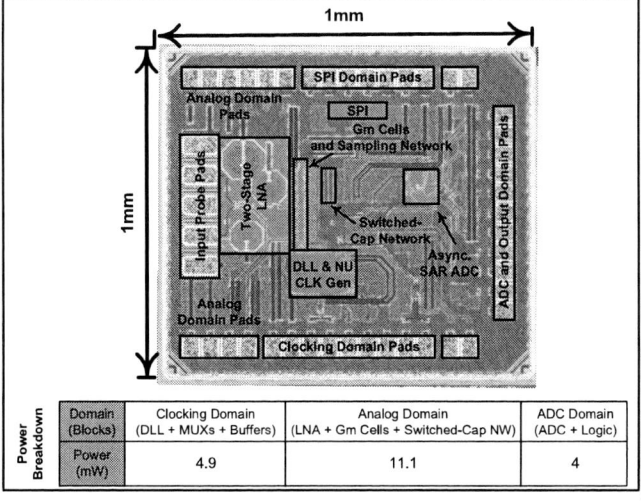

Figure 35.3.7: Die micrograph and power breakdown.

Power Breakdown	Domain (Blocks)	Clocking Domain (DLL + MUXs + Buffers)	Analog Domain (LNA + Gm Cells + Switched-Cap NW)	ADC Domain (ADC + Logic)
	Power (mW)	4.9	11.1	4

• 2026 IEEE International Solid-State Circuits Conference

979-8-3315-8937-0/26 $31.00 © 2026 IEEE

ISSCC 2026 / SESSION 35 / LOW POWER WIRELESS TRANSCEIVERS FOR LOCALIZATION AND COMMUNICATIONS / 35.4

35.4 A NearLink 2.0 Compliant Dual-Band RF Transceiver for Smart Wireless Personal Audio Applications

Rui Yu[1], Supeng Liu[1], Xuesong Chen[1], Tantan Zhang[1], Wei Khuen Chan[1], Deyong Hu[1], Haohong Yu[1], Wang Yao[1], Theng Tee Yeo[1], Peilin Wu[2], Xingpeng Zhao[2], Zhan Guo[2]

[1]Huawei Technologies, Shenzhen, China, [2]HiSilicon Technologies, Shanghai, China

Abstract

This work presents a NearLink 2.0-compliant dual-band RF transceiver. Compared with traditional BT/BLE counterparts, this design achieves receiver figures of merit (FOM) of 188.7dB in the 2.4GHz band and 185.2dB in the 5GHz band, exceeding the state-of-the-art by >4dB. Furthermore, it demonstrates a 30.9% 2.4GHz-transmitter-system power efficiency with exceptional EVM performance: 3.3% in 3Mb/s BT Enhanced Data Rate mode and 3.64% in 4M-16-QAM NearLink 2.0 mode (16Mb/s).

Smart wireless personal audio devices (TWS earbuds, AI/AR glasses etc.) are evolving into indispensable AI gateways, where performance gains (sensing, processing, battery) and next-gen connectivity (speed, latency, stability) form a symbiotic cycle. As AI capabilities expand from health monitoring to immersive AR, these devices will necessitate breakthroughs in wireless protocols to deliver seamless, intuitive, and ubiquitous intelligent experiences. As a next-generation short-distance wireless connectivity system, compared to traditional systems such as Bluetooth/Bluetooth Low Energy (BT/BLE), NearLink 2.0 delivers faster data transmission speeds (up to 16Mb/s with 16QAM@4MHz signal bandwidth), lower latency (sub-ms), support for more simultaneous connected devices (up to 4096 connections), stronger interference resistance (7dB improvement with the help of Polar encoding and adaptive frequency hopping), higher energy efficiency (in term of bit-energy, low standby polling power), and broader frequency band options (2402 to 2480MHz and 5.1 to 5.8GHz), as shown in Fig. 35.4.1. These advancements position NearLink 2.0 as a future-proof solution with commanding competitive strength for smart audio applications. This work presents a NearLink-2.0-compliant dual-band RF transceiver (TRX) to address all aforementioned advantages through comprehensive optimizations in performance, power, and cost. The design achieves receiver (RX) figure-of-merits (FoMs) of 188.7dB (>4dB better than the state-of-the-art) in the 2.4GHz band and 185.2dB in the 5GHz band respectively. Furthermore, it demonstrates 30.9% 2.4GHz-transmitter (TX) system power efficiency with exceptional EVM performance: 3.3% in 3Mb/s-BT Enhanced Data Rate (3Mb/s-BT/EDR3) mode and 3.64% in 4MHz-16-QAM NearLink 2.0 (16Mb/s) mode.

Figure 35.4.1 illustrates the block diagram of the proposed dual-band RF TRX architecture. Two RF I/O pins connect to the 2.4GHz and 5GHz antennas separately. For each frequency bands, TX and RX share the same RF I/O port with a TX-RX switch integrated on-chip. The RX front-end comprises a variable-gain low-noise amplifier (LNA) followed by a mixer for lower-IF downconversion. 1.5MHz IF frequency is chosen for 1MHz and 2MHz signal bandwidth, while 2.5MHz IF frequency is chosen for 4M signal bandwidth. 25% LO signals are generated locally using a frequency divide-by-two circuit. The 2.4GHz and 5.1-to-5.8GHz RX paths share a common RX-IF circuit, which includes a transimpedance amplifier (TIA), a variable-gain amplifier (VGA), and an analog-to-digital converter (ADC), thereby minimizing the chip area. The bandwidth of the RX-IF path is configurable to accommodate different standards. The TX section employs a polar architecture for reduced area and power consumption. Amplitude modulation (AM) is performed at the input of a digital power amplifier (DPA) via the input amplitude control word (ACW). Both DPAs utilize switched-capacitor DPA (SCDPA) to achieve high power efficiency and good linearity. Power detectors are implemented for accurate power level control (with in ±1dB). The architecture integrates an all-digital phase-locked loop (ADPLL) with two digitally controlled oscillators (DCOs) operating in the frequency ranges of 4.8 to 5GHz and 10.36 to 11.7GHz, to support both 2.4GHz and 5GHz band operations. In TX mode, two-point modulation is implemented within the ADPLL to enable phase modulation.

Figure 35.4.2 shows the block diagram of the digital polar TX consisting of digital baseband (DBB), ADPLL and DPA. The DBB operates at 32MHz clock frequency, converting I/Q modulated signals to amplitude and phase data. The amplitude data modulate the DPA directly by controlling the number of active units in the DPA. The digital modulation of the DPA is a zero-order hold function (sinc function) of the amplitude signal, which generates image spectra at the harmonics of the amplitude data sampling frequency. To meet frequency regulation requirements, these image spectra must be suppressed to sufficient low levels. Leveraging the powerful digital processing capabilities of recent CMOS process technologies, an efficient way is to increase the amplitude sampling frequency at a reasonable power consumption cost. Therefore, the amplitude path first converts the sampling clock frequency from 32MHz to $f_{LO}/6$ (i.e., 400.3 to 413.3MHz for the 2402-to-2480MHz frequency band) using an asynchronous FIFO and a symbol rate converter (SRC) module. The SRC output then passes through an interpolation filter to convert the AM sampling clock frequency to $f_{LO}/2$, shifting the image spectra approximately 1.2GHz away from the operating frequency channel. Since the LO is a modulated signal from the ADPLL output, the clock driving the SRC module ($f_{LO}/6$) is not fixed. The phase variation of the SRC clock must be compensated to minimize symbol rate conversion errors, which is critical for achieving lower adjacent-channel leakage power. Phase modulation is implemented in the ADPLL using a two-point modulation scheme. The ADPLL employs a divider-less

architecture for low power consumption. The DCO operates at twice the DPA's operating frequency, which introduces risks of PA pulling. To mitigate this issue, a pulling cancellation algorithm is implemented in the ADPLL. Additionally, digital-to-time converter (DTC) gain calibration, DTC nonlinearity calibration, time-to-digital converter (TDC) gain calibration, and modulation gain calibration are required for optimal TX modulation performance. In BT/NearLink TX, SCDPA is selected to enhance linearity. It consists of 10bit AM code: the most significant (MSB) 6bits are decoded to thermal codes, while the least significant (LSB) 4bits remain in binary format as an optimized configuration for linearity and power efficiency. The on-resistance of the final-stage NMOS and PMOS transistors is precisely matched to improve linearity and efficiency. The OR-Gate driver is employed to provide sufficient drive strength to the final stage. The DPA also incorporates a coupler-based true-power detector, which offers superior detection accuracy compared to voltage-based detectors. The maximum detection error of the power detector is less then 1dB for a Voltage-Standing-Wave-Ratio (VSWR) of 2 and less than 2dB for a VSWR of 3 across all angles.

In the 2GHz RX chain (Fig. 35.4.3a), a single-ended push-pull LNA with a buffer stage is employed to boost the front-end gain and achieve a low noise figure (NF). To mitigate large blockers, a wide-band detector is integrated at the LNA's high-gain mode, dynamically switching the LNA to low-gain mode upon detection of significant out-of-band blockers. The current-mode output of the buffer interfaces with a single balanced passive mixer driven by a 25% duty-cycle RX LO signal. The IF stage comprises a first-order TIA followed by a single opamp dual-pole low-pass filter (LPF, Fig. 35.4.3b) to interface with a 9-bit I/Q ADC. Compared to traditional single-OpAmp-based second-order LPF architectures (e.g., Sallen-Key, multiple feedback), the proposed design demonstrates better noise and reduced die area. For the 5GHz LNA, a two-stage cascode topology with inductive source degeneration is adopted to minimize power consumption while maintaining low NF. A balun, functioning as a transconductance, paired with a passive mixer ensures high linearity. The 5GHz and 2GHz RX share the same IF stage to save die area. In 2GHz RX mode, two divide-by-two frequency dividers, i.e., the PLL divider (generating feedback signal for the ADPLL) and RX-LO divider (generating the RX LO signal), operate simultaneously as shown in Fig. 35.4.3a. The PLL divider output may leak into the RF front-end and be downconverted by the RX LO signal into a DC offset. Without proper synchronization between these two divider outputs, this DC offset will be phase dependent and cannot be eliminated through conventional DC offset calibration. To address this issue, an "LO-sync" scheme shown in Fig. 35.4.3(c-d) is proposed to enforce phase alignment between the divider outputs. For the 5GHz RX, where more stringent timing requirements exist, a synchronous design is unfeasible. Instead, a common divider is utilized for both the ADPLL and RX LO paths, ensuring that any divider leakage into the front-end remains inherently synchronized with the RX LO signal.

This dual-band RF transceiver is fabricated in a 12nm CMOS process and occupies an active area of 2.2mm². The die micrograph is shown in Fig. 35.4.7, and a performance summary is tabulated and benchmarked against prior state-of-the-arts in Fig. 35.4.6. This work achieves the best RX sensitivities across all 2.4GHz legacy BT-BR/EDR and BLE modes, along with competitive RX FoMs. When operating in 2.4GHz NearLink 1Mb/s mode with Polar encoding enabled, both sensitivity and RX FoM improve by over 4dB (Fig. 35.4.4 bottom right). For 5GHz NearLink 1Mb/s mode, superior results (sensitivity of -105.8dBm and FoM of 185.2dB) are still achieved with reasonable power consumption. Sensitivity performance across different channels in the 2.4GHz band is summarized in Fig. 35.4.4 (top left). It should be noted that NearLink 2.0 sensitivity testing requirements are more stringent, assuming a Packet-Error-Rate (PER) threshold of 10% compared to 30% in the BLE standard. PER power scanning plots are provided in Fig. 35.4.4 (bottom left). Image rejection calibration, LNA saturation detection and IP2 calibration give rise to large carrier-to-interference (C/I) margin (>30dB for far bands) relative to the BLE standard, as shown in Fig. 35.4.4 (top right).

Measured DPA efficiencies versus output power are shown in Fig. 35.4.5 (top left). Peak DPA efficiencies of 32%@15.2dBm ad 24%@14.2dBm are achieved for the 2.4GHz and 5GHz bands, respectively, with losses from the TRX-switch, coupler and on-chip harmonic rejection filters considered. Thanks to the inherent high linearity of the SCDPA architecture adopted in this work, superior AM-AM performances are demonstrated in Fig. 35.4.5 (top right), indicating that a digital pre-distortion (DPD) module is actually unnecessary for this

594 • 2026 IEEE International Solid-State Circuits Conference

979-8-3315-8937-0/26 $31.00 © 2026 IEEE

DPA while still achieving good EVM and frequency mask performances (Fig. 35.4.5 bottom left). However, techniques such as the modulation gain calibration in the two-point modulation architecture and pulling cancellation remain essential to maintain good TX performance in the context of high TX system power efficiencies (30.9% and 20.4% for the 2.4GHz and 5GHz bands, respectively). Figure 35.4.5 also present constellation diagrams of BT-EDR3 and NearLink 4M-16-QAM modes, further validating the effectiveness of these techniques.

References:
[1] N. Scolari et al., "A 1mm² Software-Defined Dual-Mode Bluetooth Transceiver with 10dBm Maximum TX Power and -98.2dBm Sensitivity 2.96mW RX Power at 1Mb/s," *ISSCC*, pp. 402-404, Feb. 2024. https://doi.org/10.1109/ISSCC49657.2024.10454304
[2] W. Yang et al., "A +8dBm BLE/BT Transceiver with Automatically Calibrated Integrated RF Bandpass Filter and -58dBc TX HD2," *ISSCC*, pp. 136-137, Feb. 2017. https://doi.org/10.1109/ISSCC.2017.7870298
[3] M. Tamura et al., "A 0.5V BLE Transceiver with a 1.9mW RX Achieving -96.4dBm Sensitivity and 4.1dB Adjacent Channel Rejection at 1MHz Offset in 22nm FDSOI," *ISSCC*, pp. 468-470, Feb. 2020. https://doi.org/10.1109/ISSCC19947.2020.9063021
[4] K. Shibata et al., "A 22nm 0.84mm² BLE Transceiver with Self IQ-Phase Correction Achieving 39dB Image Rejection and On-Chip Antenna Impedance Tuning," *ISSCC*, pp. 398-400, Feb. 2022. https://doi.org/10.1109/ISSCC42614.2022.9731558
[5] M. Ding et al., "A Bluetooth 5 Transceiver with A Phase-Tracking RX and its Corresponding Digital Baseband in 40-Nm CMOS," *IEEE JSSC*, vol. 56, no. 1, pp. 254–266, Jan. 2021. https://doi.org/10.1109/JSSC.2020.3005788
[6] J. Prummel et al., "A 10mW Bluetooth Low-Energy Transceiver with On-Chip Matching," *ISSCC*, pp. 238-240, Feb. 2015. https://doi.org/10.1109/ISSCC.2015.7063014

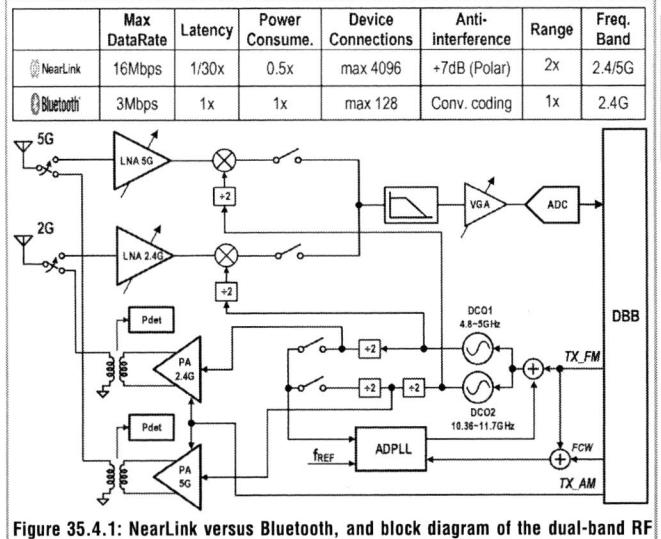

Figure 35.4.1: NearLink versus Bluetooth, and block diagram of the dual-band RF transceiver.

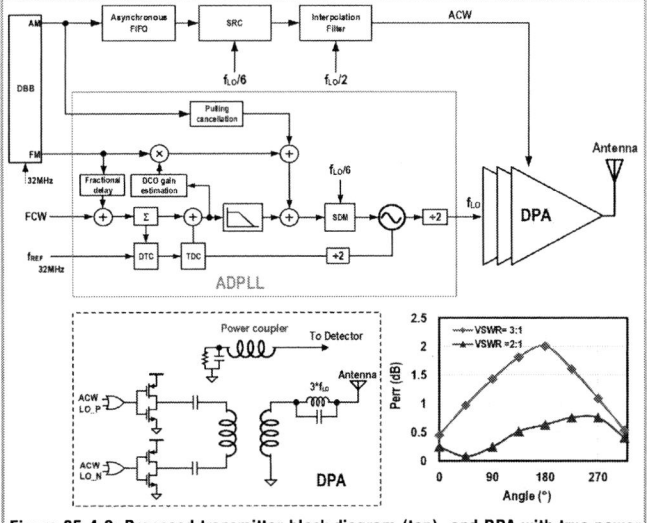

Figure 35.4.2: Proposed transmitter block diagram (top), and DPA with true power detector (bottom).

ISSCC 2026 / SESSION 35 / LOW POWER WIRELESS TRANSCEIVERS FOR LOCALIZATION AND COMMUNICATIONS / 35.4

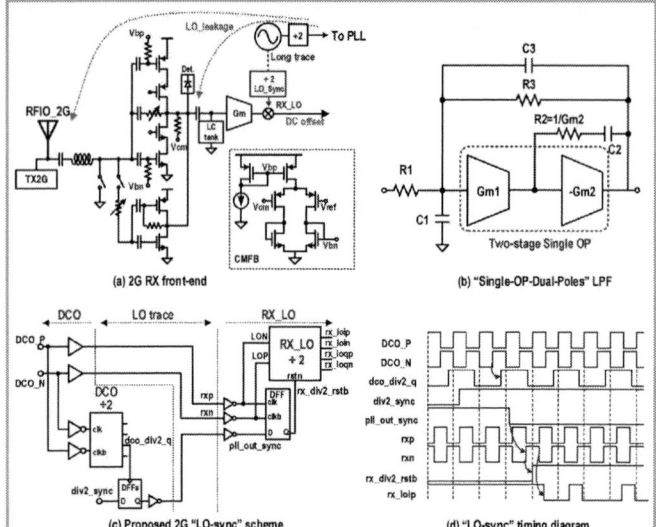

Figure 35.4.3: Proposed 2.4GHz RX core circuit implementation.

(a) 2G RX front-end
(b) "Single-OP-Dual-Poles" LPF
(c) Proposed 2G "LO-sync" scheme
(d) "LO-sync" timing diagram

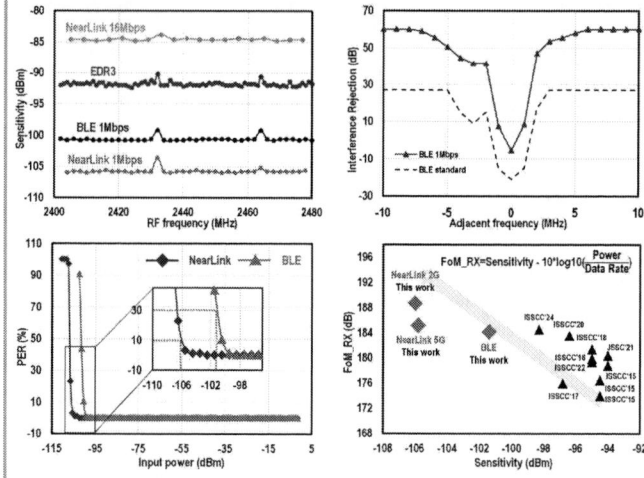

Figure 35.4.4: Measured RX performance (sensitivity, interference rejection, PER & RX Figure of Merit).

Figure 35.4.5: Measured TX performance (DPA efficiency, DPA Power compression, EDR3 frequency spectrum and mask, EDR3 and 4M-16-QAM constellation).

		This Work 2G	This Work 5G	[1] ISSCC'24	[2] ISSCC'17	[3] ISSCC'20	[4] ISSCC'22	[5] JSSC'21	[6] ISSCC'15
Standard		NearLink2.0/BT/BLE		BT/BLE/802.15.4	BT/BLE	BLE	BLE	BT/BLE	BLE
Process (nm)		12		22 FDSOI	55	22 FDSOI	22	40	55
TRX area (mm²)		2.2 (2G+5G)		1.0	2.2	1.9	0.84	1.05	2.9
Max Data Rate (Mbps)		15		3	3	2	2	2	2
Modulation		GFSK/DQPSK D8PSK/BPSK/16QAM		GFSK/DQPSK D8PSK	GFSK/DQPSK D8PSK	GFSK	GFSK	GFSK/DQPSK	GFSK
Bandwidth(Hz)		1M/2M/4M		1M/2M	1M/2M	1M/2M	1M/2M	1M/2M	1M/2M
RX		**2G**	**5G**						
Sensitivity (dBm)	BLE 1Mbps	-101.4	-	-98.3	-96.8	-96.4	-95	-94	-94.5
	BDR	-99.4	-	-96.1	-93.4	-	-	-91	
	NearLink (1Mbps)	-106	-105.8	-	-	-	-	-	-
	EDR3	-92.2	-	-86.7	-87	-	-	-	-
	NearLink (16Mbps)	-84.6	-84.4	-	-	-	-	-	-
C/I & Image (dBc) @ BLE 1Mbps	Image	-41.5	-39#	-33.6	-	-	-39	-	-
	±1MHz	-8.5	-31#	-8.2	-	-4.1	-	-	-
	±2MHz	-47	-49#	-41.5	-51	-36.1	-41	-	-
	±3MHz	-53	-55#	-46.8	-54	-41	-43	-	-
	Power (mW)	5.36	11.6	2.36	12	1.9	3.6	2.3	11.2
FOM_RX(dB)	BLE	184.2	-	184.6	176	183.6	179.4	180.4	174
	BDR	182.1	-	182.3	172.6	-	-	177.4	-
	NearLink	188.7	185.2	-	-	-	-	-	-
Tx	Pout (dBm)	15.2	14.2	10	8	3.3	-2	1.8	0
	FSK Error (%)	1.2	0.36	-	-	-	2.2	2.4	-
DEVM (%)	EDR2/3	2.7/3.3@11dBm	-	6.07@10dBm	5.4@5dBm	-	-	-	-
	16QAM	3.64@9.6dBm	4.34@9dBm	-	-	-	-	-	-
	TX Efficiency (%)	30.9	20.4	15.8	7.9	27.4	15.3	29.5	9.9

Data for 5G NearLink

Figure 35.4.6: Performance summary and benchmark with the state-of-the-arts.

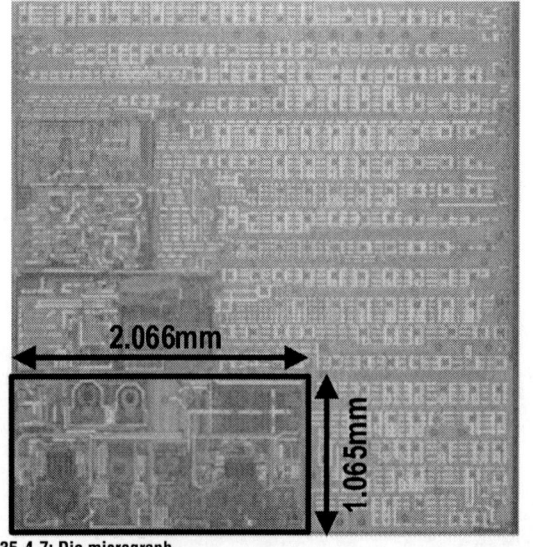

2.066mm
1.065mm

Figure 35.4.7: Die micrograph.

• 2026 IEEE International Solid-State Circuits Conference

ISSCC 2026 / SESSION 35 / LOW POWER WIRELESS TRANSCEIVERS FOR LOCALIZATION AND COMMUNICATIONS / 35.5

35.5 Fully Integrated Backscattered WiFi 802.11b Transmitter with Active Harmonics and Image Rejection for 30dB IRR and 36dB HRR at 0.88µW

Ruiyuan Yang, Karim Ali, Massimo Alioto

National University of Singapore, Singapore, Singapore

Abstract

A sub-µW fully integrated backscattered 802.11b WiFi transmitter is introduced with active image and harmonics rejection for common dense wireless environments. Rejection is achieved via time-domain complex filters, and oversampling with sequential RF switch

Wireless environments are becoming increasingly denser in view of the exponential growth in the number of connected and coexisting devices, making efficient channel utilization a priority especially in unlicensed bands such as 2.4GHz. Pressing demand for lower transmission power, cost and form factor [1-3] has driven the recent evolution of backscattered transmitters [4-18], especially those compliant with existing standards [8-18]. Unfortunately (Fig. 35.5.1), backscattered transmitters traditionally suffer from 1) uncompetitive image rejection ratio (IRR) due to the mixing of the incident wave with the on-chip IF oscillator, and 2) limited harmonics rejection ratio (HRR) due to the abrupt transitions in the reflection coefficient during backscattering. The resulting spurs added to the desired transmitted signal are responsible for major channel utilization deterioration.

Apart from prior art in backscattering without any spectrum control [8-13], IRR is generally improved through SSB/quadrature modulation [14-18]. HRR is partially improved via IF frequency over-clocking and inflexible left/right-most channel allocation to push out the harmonics out of the 2.4GHz band, though at the cost of increased power and uneven/inefficient channel utilization [14-17]. However, these mechanisms cannot mitigate the harmonics caused by discontinuous-phase (PSK) modulations used in WiFi 802.11b backscattered transmitters [14-17], which is the focus of this work considering its suitability for many-device infrastructures with concurrent traffic. Overall, innovation is required in WiFi backscattered transmitters to improve IRR beyond the state of the art of 10 to 18dB [14-17] and tackle the above HRR limitations (unaddressed in prior art), while avoiding the power burden of IF over-clocking and the above channel inflexibility.

In this work, a sub-µW fully integrated WiFi backscattered transmitter with active rejection of both image and harmonics is introduced for densely-connected environments. Rejection without IF frequency over-clocking is achieved by suppressing the image and smoothing reflection coefficient transitions via complex filtering. Power-hungry digital filtering is replaced by time-domain filtering, where progressive (complex) reflection coefficient levels are sequentially enabled by a programmable impedance bank and an accurate delay line. Temperature (and process) variations are compensated via temperature event-driven self-referenced calibration. A 22nm FD-SOI testchip shows 30dB IRR with a 12dB improvement over prior art, and 36dB HRR at 0.88µW. Tested in a real wireless environment, this spur reduction allows an independent TX-RX pair to withstand (same BER) even under up to 35dB backscattered power increase over a baseline lacking active rejection. As further benefit, the allowed power increase extends the backscattered transmitter range by 56× (7×) when compared to a baseline with no spectrum control (quadrature modulation and doubled IF frequency).

In the adopted Passive WiFi backscattered communication scheme (Fig. 35.5.2), the incident wave is sent by an RF source (e.g., shared tone generator, or reverse-whitened BLE signals) at a properly shifted carrier frequency. The latter is then modulated and re-shifted to the targeted channel frequency f_C by the IF frequency f_{IF}, modulating the RF switch impedance connected to the antenna for backscattering. In the proposed architecture, a complex lowpass filter (Fig. 35.5.3) simultaneously 1) rejects the image thanks to its zero response at negative frequencies (where the image lies), and 2) rejects harmonics above its cutoff frequency. A conventional complex filter would draw a power consumption (e.g., 23mW in [19]) much larger than the sub-µW target, and is hence greatly simplified through 1) the Feher's approximation replacing adders/multipliers by a simple 8×8 look-up table [20] (LUT), which reproduces the 8-sample temporal filter response for the 8 possible triplets of 3 subsequent input bits, 2) the elimination of traditional synchronous logic by updating the next-sample RF switch impedance Z of a programmable impedance bank (i.e., desired value of the reflection coefficient $\Gamma=(Z-Z_0)/(Z+Z_0)$, $Z_0=50\Omega$), according to the timing of a delay line. At each symbol period, the latest 3 input bits digitally select one of the 8 complex-response patterns in Fig. 35.5.2, which in turn correspond to different combinations of 8 basic complex loads Z (BCL in Fig. 35.5.2) leading to the desired value. The corresponding impedance Z is generated by the programmable impedance driven by the LUT value in Fig. 35.5.2.

To avoid over-clocking the system for the above 8× symbol oversampling, the 8 subsequent BCLs are enabled in sequence by a delay line, effectively implementing the targeted filter in the time domain. Its sequence of precise enabling pulse widths $1/8 \cdot T_{IF}$ (en_pulse in Fig. 35.5.2) is generated by a pulse generator driven by a digitally controlled oscillator (DCO) in the

enablement (no digital filter). Temperature event-driven calibration adjusts the underlying pulse basis and programmable complex impedance. A 22nm test chip shows 30dB IRR (12dB better than prior art in Fig. 35.5.6), 36dB HRR (not shown priorly).

Fig. 35.5.2). The delay lines for BCL sequencing are also set to $1/8 \cdot T_{DCO}$ to barely disoverlap the various impedance configurations (timing requirements in Fig. 35.5.4). The 5-stage DCO ring oscillator (Fig. 35.5.2) simultaneously provides $f_{IF}=11$MHz (minimum required by 802.11b backscattering) and the delay line stages are a replica of the DCO stages for relatively consistent timing (calibration below).

The above pulse width and cell delay in the delay line in Fig. 35.5.2 are calibrated to $1/8 \cdot T_{IF}$ as in Fig. 35.5.3 to avoid degrading IRR and HRR, acting on the supply voltage of the relevant delay cells in a feedback loop. The pulse width is sensed via two 3-bit counters (counterH and counterL) in Fig. 35.5.2, which are respectively positive and negative edge-triggered and detect whether the pulse is narrower or wider than the target. Their metastability is prevented by delaying the pulse and the counterL clock compared to counter (timing sensing resolution=delay). Counters are clocked 8× faster than f_{IF} via a temporarily-ON calibration DCO, and are enabled by the pulse to be calibrated. Both 11MHz and 88MHz DCOs are calibrated when temperature changes exceed a threshold, adjusting supply and body (similar to [12]). From the analysis in Fig. 35.5.3, when the pulse width is on target, counterH only counts once in each symbol period (details in Fig. 35.5.3), and counterL keeps counting. When the pulse is wider (narrower), this condition is detected by counterL (counterH) as it counts 0 twice and then keeps counting (counterH remains at zero). Similarly, cell delays are calibrated by first converting them to pulse widths (separate pulse generator) and then calibrated as above. Since the programmable impedance is affected by the ON resistance of the transistor switches, their body bias voltage is set according to the sensed temperature (LUT filled at testing time to incorporate process variations).

To assess the effectiveness of the above calibration, Fig. 35.5.4 shows that the EVM degradation across temperatures requires pulse width and cell delay calibration, which keeps their variations under control and keeping the maximum EVM to 11% (well within 802.11b specs). The residual post-calibration timing discrepancy across temperatures added to the worst-case jitter (300ps) is still well below the maximum allowed timing uncertainty maintaining EVM within specs. Similar considerations apply to the inaccuracies in the programmable impedance, as exemplified in Fig. 35.5.4 for the most temperature-sensitive impedance. In detail, pre-calibration variations across temperatures lead to IRR and HRR of 25dB and 27dB, which is already superior to prior art [14-17] thanks to complex filtering. IRR and HRR are further improved to 30dB and 36dB by calibration. Interestingly, IRR and HRR were found to be nearly unaffected by timing inaccuracies, as inaccurate pulses and delays rigidly compress/dilate the temporal filter response, which translate into a mere magnitude scaling factor in the frequency domain. Conversely, EVM was found to be essentially unaffected by impedance changes across temperatures.

Correct packet reception and decoding from commodity WiFi receivers is shown in Fig. 35.5.5, which also shows the backscattered spectrum and the constellation. The benefits of the IRR and HRR improvements were quantified in a real wireless environment, considering that backscattered spurs without spectrum control act as interference on a coexisting transmitter-receiver pair (Fig. 35.5.5). The backscattered radio transmits on the adjacent channel to induce worst-case spur power, which in turn degrades the BER of the TX-RX pair as a proxy for link quality and spectrum utilization (better at lower BER). Experimental results show that the proposed backscattered transmitter allows a WiFi TX-RX pair to maintain the same BER despite an extra 35dB backscattered power, compared to a baseline without active rejection. Such allowable increase in the backscattered power at the same BER clearly quantifies the spectral performance and coexistence capability of the backscattered transmitter. As further comparisons, the proposed transmitter allows the same BER at 20dB (17dB) backscattered power increase compared to a baseline with quadrature modulation for conventional image rejection at minimum $f_{IF}=11$MHz (higher $f_{IF}=22$MHz for conventional harmonics mitigation). This confirms that the proposed active rejection has better spectral performance than spectrum control methods explored in prior art (see below). Interestingly, the higher allowable backscattered power enabled by the IRR and HRR improvement translates into a longer backscattered transmission range by 56× (Fig. 35.5.5).

From Fig. 35.5.6, the 22nm test chip of the proposed backscattered transmitter in Fig. 35.5.7 has the best IRR compared to prior art in WiFi backscattering with an improvement over the prior best by 12dB [16]. Interestingly, the introduction of an active rejection avoids the power burden of IF over-clocking, as confirmed by the adoption of the lowest IF frequency (11MHz) among prior WiFi backscattered transmitters [11-13], [15-17]. The resulting reduction in the IF frequency reduces the overall power consumption down to 0.88μW, which is the lowest reported with a 0.88pJ energy/bit improved by at least 2.8× over [11-13], [15-17]. HRR is also improved by 23dB over a baseline without active rejection (not reported in prior art). Such capabilities are demonstrated in a fully integrated system with no off-chip power combiner or inductor as opposed to [15-17], and with unrestricted channel allocation.

Acknowledgement:
The authors acknowledge the support of the Singapore Ministry of Education (T2EP50125-0009 grant), and GlobalFoundries for chip fabrication.

References:
[1] C. Yang, Z. Wei, H. Gao, C. H. Heng and Y. Zheng, "Towards Ultra-Low Power Transceivers for Pico-IoT," in *IEEE Nanotechnology Magazine*, vol. 18, no. 1, pp. 34-43, Feb. 2024. http://doi.org/10.1109/MNANO.2023.3340391
[2] X. Yu, M. Wei, Y. Yin, Y. Song, Z. Wang, Y. Sun and B. Chi, "A Sub-GHz Low-Power Transceiver with PAPR-Tolerant Power Amplifier for 802.11ah applications," *IEEE RFIC*, Phoenix, AZ, USA, 2015, pp. 231-234. http://doi.org/10.1109/RFIC.2015.7337747
[3] A. Zolfaghari, M. E. Said, M. Youssef, G. Zhang, T. Liu, F. Cattivelli, Y. Syllaios, F. Khan, F. Fang, J. Wang, K. -Y. Li, F. Liao, D. Jin, V. Roussel, D. Lee and F. Hameed, "A Multi-Mode WPAN (Bluetooth, BLE, IEEE 915.4) SoC for Low-Power and IoT Applications," *IEEE Symposium on VLSI Circuits*, Kyoto, Japan, 2017, pp. C74-C75. http://doi.org/10.23919/VLSIC.2017.8008554
[4] Y. Huang, B. Liu, Y. Hou, J. Xu, H. You, A. Hung, S. Ghosh, E. Liu, N. Yang, J. Ma, H. Cai, L. Kondrataviciute, Q. Deng, S. K. Kalia, A. G. Richardson, P. -H. Hsieh, R. Genov and X. Liu, "A Neuroprosthetic SoC with Sensory Feedback Featuring Frequency-Splitting-Based Wireless Power Transfer with 200Mb/s 0.67pJ/b Backscatter Data Uplink and Unsupervised Multi-Class Spike Sorting," *ISSCC*, San Francisco, CA, USA, 2025, pp. 272-274. http://doi.org/10.1109/ISSCC49661.2025.10904677
[5] J. Shen, F. Zhu, Y. Liu, B. Liu, C. Shi, L. Huang, L. Xu, X. Tian and R. Zhang, "A 44μW IoT Tag Enabling 1μs Synchronization Accuracy and OFDMA Concurrent Communication with Software-Defined Modulation," *ISSCC*, San Francisco, CA, USA, 2024, pp. 400-402. http://doi.org/10.1109/ISSCC49657.2024.10454346
[6] C. Yang, Z. Zhang, L. Zhang, Y. Zhang. Z. Li, Y. Luo, G. Pan and B. Zhao, "A 128-Channel 2mmx2mm Battery-Free Neural Dielet Merging Simultaneous Multi-Channel Transmission Through Multi-Carrier Orthogonal Backscatter," *ISSCC*, San Francisco, CA, USA, 2023, pp. 30-32. http://doi.org/10.1109/ISSCC42615.2023.10067688
[7] Y. Zhang, R. Luo, J. Xiong, S. Liang and M. Meng, "A 19μW 200Mb/s IoT Tag Demonstrating High-Definition Video Streaming via a Digital-Switch-Based Reconfigurable 16-QAM Backscatter Communication Technique," *ISSCC*, San Francisco, CA, USA, 2024, pp. 224-226. http://doi.org/10.1109/ISSCC49657.2024.10454414
[8] K. A. Ahmed, R. Yang, P. Salamani, V. Rajanna and M. Alioto, "Single-Antenna Backscattered BLE5 Transmitter with up to 97m Range, 10.6 μW Peak Power for Purely-Harvested Green Systems," *ESSCIRC*, Lisbon, Portugal, 2023, pp. 49-52. http://doi.org/10.1109/ESSCIRC59616.2023.10268708

[9] Z. Chang, Q. Xiao, C. Chen, W. Wang, X. Hu, C. Yang, Z. Li, Y. Luo and B. Zhao, "A Passive Crystal-Less Wi-Fi-to-BLE Tag Demonstrating Battery-Free FDD Communication with Smartphones," *ISSCC*, San Francisco, CA, USA, 2024, pp. 404-406. http://doi.org/10.1109/ISSCC49657.2024.10454302
[10] Z. Chang, Q. Xiao, W. Wang, Y. Luo and B. Zhao, "A Passive Bidirectional BLE Tag Demonstrating Battery-Free Communication in Tablet/Smartphone-to-Tag, Tag-to-Tablet/Smartphone, and Tag-to-Tag Modes," *ISSCC*, San Francisco, CA, USA, 2023, pp. 468-470. http://doi.org/10.1109/ISSCC42615.2023.10067538
[11] Q. Xiao, C. Yang, Y. Zhang, Z. Chang, C. Chen, X. Hu, W. Wang, G. Gu, Y. Luo and B. Zhao, "A Passive Crystal-Less Tag Demonstrating Battery-Free GSM-CW/5G-NR Downlink and BLE-to-BLE/BLE-to-WiFi/BLE-to-WiFi Multi-Channel-Hopping Uplink with Smartphones," *IEEE CICC*, Boston, MA, USA, 2025, pp. 1-3. http://doi.org/10.1109/CICC63670.2025.10983814
[12] L. Lin, K. A. Ahmed, P. S. Salamani and M. Alioto, "Battery-Less IoT Sensor Node with PLL-Less WiFi Backscattering Communications in a 2.5-μW Peak Power Envelope," *IEEE Symposium on VLSI Circuits*, Kyoto, Japan, 2021, pp. 1-2. http://doi.org/10.23919/VLSICircuits52068.2021.9492358
[13] R. Yang, K. A. Ahmed and M. Alioto, "Backscattered Software-Defined Radio for Flexible, Reusable and Upgradeable Transmitters with 34-58 pJ/bit Energy Across Common Standards in 180 nm," *ESSERC*, Bruges, Belgium, 2024, pp. 93-96. http://doi.org/10.1109/ESSERC62670.2024.10719464
[14] S. -K. Kuo, M. Dunna, D. Bharadia and P. P. Mercier, "A WiFi and Bluetooth Backscattering Combo Chip Featuring Beam Steering via a Fully-Reflective Phased-Controlled Multi-Antenna Termination Technique Enabling Operation Over 56 Meters," *ISSCC*, San Francisco, CA, USA, 2022, pp. 1-3. http://doi.org/10.1109/ISSCC42614.2022.9731744
[15] S. -K. Kuo, M. Dunna, H. Lu, A. Agarwal, D. Bharadia and P. P. Mercier, "An LTE-Harvesting BLE-to-WiFi Backscattering Chip for Single-Device RFID-Like Interrogation," *ISSCC*, San Francisco, CA, USA, 2023, pp. 320-322. http://doi.org/10.1109/ISSCC42615.2023.10067815
[16] M. Meng, M. Dunna, H. Yu, S. Kuo, P. -H.P. Wang, D. Bharadia and P. P. Mercier, "Improving the Range of WiFi Backscatter Via a Passive Retro-Reflective Single-Side-Band-Modulating MIMO Array and Non-Absorbing Termination," *ISSCC*, San Francisco, CA, USA, 2021, pp. 202-204. http://doi.org/10.1109/ISSCC42613.2021.9366014
[17] P. -H. P. Wang, C. Zhang, H. Yang, D. Bharadia and P. P. Mercier, "A 28μW IoT Tag That Can Communicate with Commodity WiFi Transceivers via a Single-Side-Band QPSK Backscatter Communication Technique," *ISSCC*, San Francisco, CA, USA, 2020, pp. 312-314. http://doi.org/10.1109/ISSCC19947.2020.9063133
[18] J. Xiong, Y. Zhang, J. Chen, X. Li, J. Zuo, Y. Wang, X. Wang and M. Meng, "A Wearable Backscatter System Featuring Concurrent RF Harvesting and Bidirectional Communication with Commodity BLE Transceivers," *IEEE CICC*, Boston, MA, USA, 2025, pp. 1-3. http://doi.org/10.1109/CICC63670.2025.10983441
[19] S. Lerstaveesin and B. -S. Song, "A Complex Image Rejection Circuit with Sign Detection Only," *ISSCC*, San Francisco, CA, USA, 2006, pp. 1810-1819. http://doi.org/10.1109/ISSCC.2006.1696238
[20] W. Gao, R. Gudipati and K. Feher, "Simple Gaussian Filter Design for FH-SS Applications," *Wireless Access Method and Physical Layer Specifications*, IEEE 802.11, 1995. https://www.ieee802.org/11/Documents/DocumentArchives/1995_docs/1195002_scan.pdf

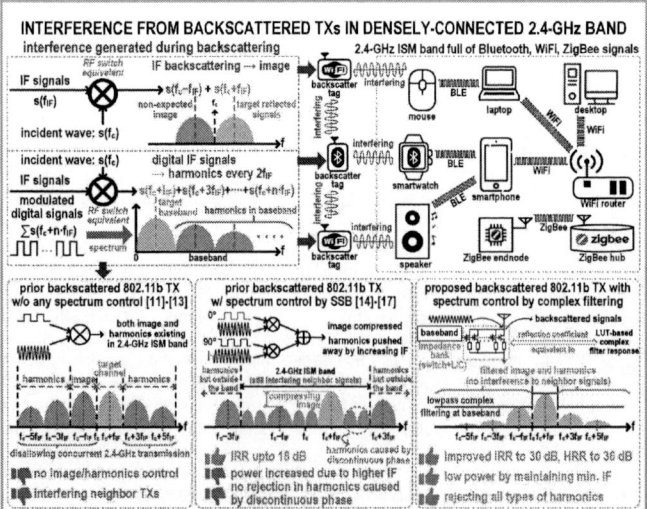

Figure 35.5.1: Motivation for spectrum control in this work and comparison with prior WiFi 802.11b backscattered transmitters.

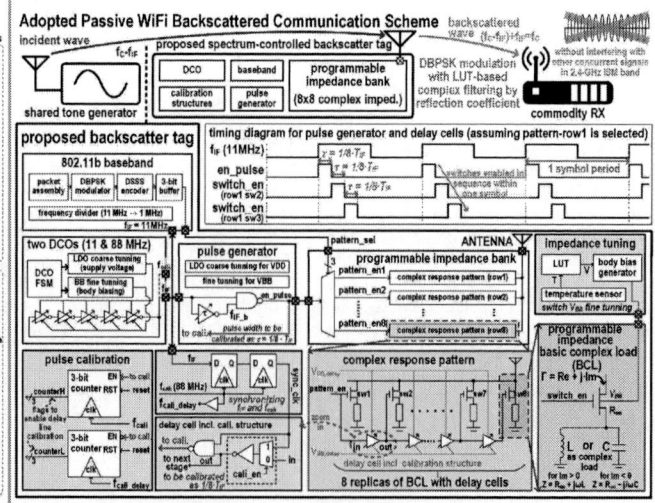

Figure 35.5.2: Passive WiFi backscattering scheme (top), architecture of proposed 802.11b transmitter (bottom).

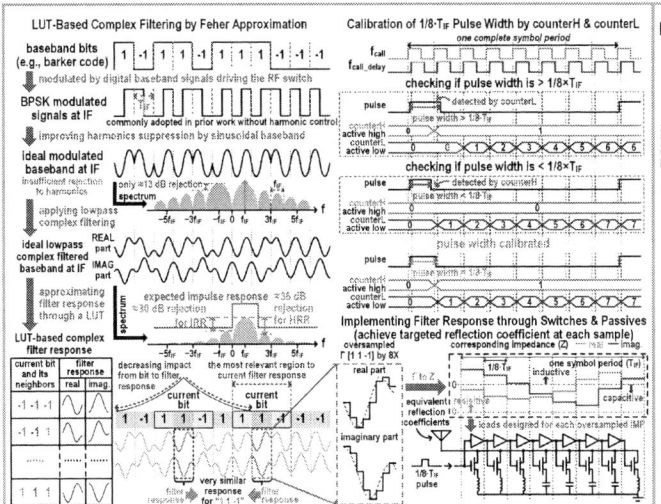

Figure 35.5.3: Proposed architecture of time-domain complex filter (left), delay-line calibration flow (top-right).

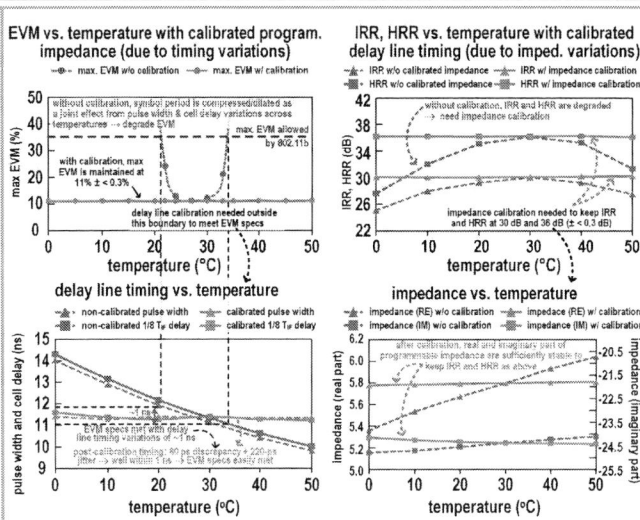

Figure 35.5.4: EVM vs. temperature with delay-line calibration (left), IRR and HRR vs. temperature with impedance calibration (bottom).

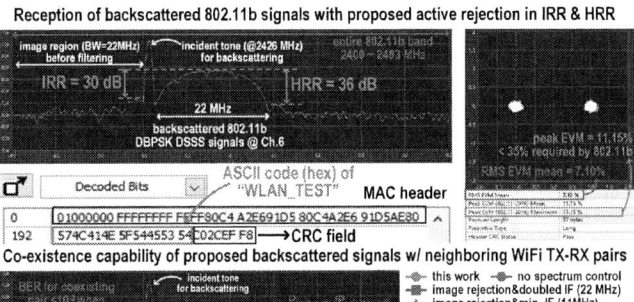

Figure 35.5.5: Measured spectrum and constellation (top), measured spectrum during transfers in a coexisting TX-RX pair (bottom).

	this work [a]	CICC25 [11] [a]	ISSCC23 [15]	ESSERC24 [13] [a]	ISSCC21 [16]	ISSCC20 [17]	VLSI21 [12]
technology (nm)	22 nm	65 nm	65 nm	180 nm	65 nm	65 nm	180 nm
area (mm²)	0.68	0.79	0.43	4	0.41	0.34	1.62
modulations supported	DBPSK	BPSK	QPSK	DQPSK	QPSK	QPSK	DBPSK
on-chip antenna load	programmable impedance	load-switching w/ passives	load-switching w/ passives	load-switching w/ transistor	load-switching w/ passives	load-switching w/ passives	load-switching w/ transistor
max datarate	1 Mbps	1 Mbps	2 Mbps	11 Mbps	2 Mbps	2 Mbps	1 Mbps
no. of antennas for TX	1	1	1	1	1, 4	1	1
range (tag to RX)	20 m	10.5 m	N/R	7 m	23 m (4 antennas) 13 m (1 antenna)	10.5 m	8 m
on-chip oscillator frequency	11 MHz	11 - 70 MHz	8.25 - 63.25 MHz [b]	11 MHz	25 MHz [b]	25 MHz [b]	11 MHz
image control	YES (complex filter)	NO	YES (quadrature mod.)	NO	YES (quadrature mod.)	YES (quadrature mod.)	NO
harmonics control	YES (complex filter)	NO	NO	NO	NO	NO	NO
image rejection ratio (IRR)	30 dB	N/A	N/R (est. 10 dB) [c]	N/A	18 dB	17 dB	N/A
harmonics rejection ratio (HRR)	36 dB	N/A	N/A	N/A	N/A	N/A	N/A
complete TX (packet gen. + backscattering)	YES	YES	NO	YES	NO	NO	YES
fully-integrated	YES	YES	NO (off-chip inductor)	YES	NO (off-chip inductor and combiner)	NO (off-chip combiner)	YES
energy/bit (pJ/bit)	0.88 @ 1 Mbps	110.11 @ 1 Mbps	12.5 @ 2 Mbps	34.1 @ 11 Mbps	19 @ 2 Mbps (4 antennas) 16 @ 2 Mbps (1 antenna)	14 @ 2 Mbps	2.5 @ 1 Mbps
min. power of complete TX	0.88 µW @ 1 Mbps	110.11 µW @ 1 Mbps	25 µW @ 2 Mbps	374.9 µW @ 11 Mbps	38 µW @ 2 Mbps (4 antennas) 32 µW @ 2 Mbps (1 antenna)	28 µW @ 2 Mbps	2.5 µW @ 1 Mbps

[a] configured into 802.11b backscattering mode [b] IF frequency increased to push harmonics away → higher TX power [c] estimated from plots (N/A = not available, N/R = not reported)

Figure 35.5.6: Comparison with state-of-the-art WiFi backscattered transmitters (best performance or unique feature in bold).

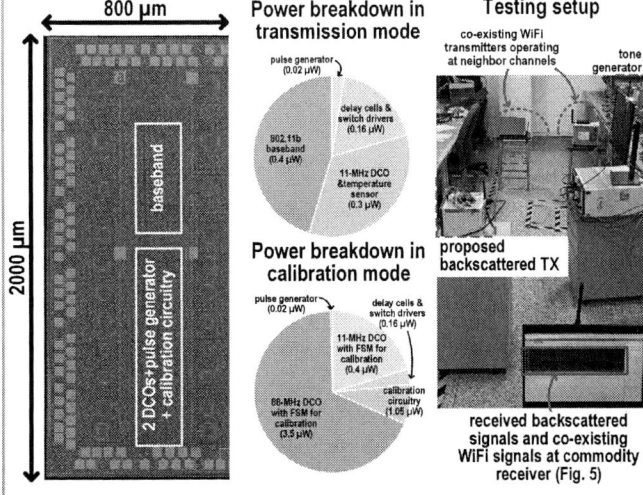

Figure 35.5.7: Die micrograph (left), power breakdown in transmission and calibration modes (middle), testing setup (right).

• 2026 IEEE International Solid-State Circuits Conference

979-8-3315-8937-0/26 $31.00 © 2026 IEEE

ISSCC 2026 / SESSION 36 / NEURAL AND BIOMEDICAL INTERFACES / OVERVIEW

Session 36 Overview: *Neural and Biomedical Interfaces*
MEDICAL SUBCOMMITTEE

Session Chair: Mahsa Shoaran
EPFL, Geneva, Switzerland

Session Co-Chair: Carolina Mora Lopez
imec, Leuven, Belgium

This session highlights innovations in biomedical and neural interface SoCs. It opens with the ReFIND SoC, featuring runtime resolution reconfiguration for >10× AFE power savings, followed by a LoRA-enabled smart-glasses interface with continual on-chip learning and neural commanding. A prototype-based CNN processor achieves patient-independent seizure detection at 16.4nJ/class, while a neuromorphic olfactory chip applies on-chip transfer learning for cross-hospital pulmonary-disease diagnosis. The session continues with a NS-SAR ADC achieving 180.3dB FoM$_{DR}$ for ExG monitoring, a 1024-electrode CIM-based neural interface for hotspot spike tracking, and a hybrid optogenetic stimulator that saves 70.6% optical energy via sub-threshold electrical stimulation. Completing the lineup, advances include a wide-DR retinal prosthesis SoC, a waveguide-based optogenetics neural probe with thermo-optic calibration, a multi-frequency EIT system for 878fps imaging, and a time-to-digital impedance measurement IC for wearable gait mapping.

1:30 PM

36.1 ReFIND: A Resolution-Reconfigurable Bio-Signal Classification SoC Enabling >10× Savings in AFE Power per Channel

Aviral Pandey, University of California, Berkeley, CA

In Paper 36.1, UC Berkeley presents an SoC with 16 reconfigurable AFEs, a feature extractor, and a classifier. Its VCO-ADC trades power for resolution (13 to 10b, 1.77 to 0.12µW/channel), achieving 14× lower power than fixed arrays and enabling runtime resolution scaling for bio-signal classification.

1:55 PM

36.2 A Neural Interface SoC for Smart Glasses with Low-Power Neural Commanding and Efficient LoRA-Enabled On-Chip Learning

Zhiwei Zhong, Northwestern University, Evanston, IL

In Paper 36.2, Northwestern University shows a 65nm ExG SoC for smart glasses performing real-time EOG/EMG/EEG inference at ~1µJ/class. LoRA-based continual learning cuts trainable parameters 54 to 77%, and on-demand compute with delta encoding boosts efficiency and data by 2.1×.

2:20 PM

36.3 A 16.4nJ/Class Patient-Independent Prototype-Based Spatio-Temporal CNN Processor with Forward-Inference-Based Adaptation for Robust and Low-Latency Seizure Detection

Yang Wang, Southern University of Science and Technology, Shenzhen, China

In Paper 36.3, the Southern University of Science and Technology presents a 40 nm prototype-based spatio-temporal CNN processor for patient-independent seizure detection. It achieves 94.3%/94.9% sensitivity/specificity at 16.4nJ/class and <120ms latency, improving to 95.4%/97.9% with zero-shot adaptation at 4.4× lower energy and 3× lower latency.

2:45 PM

36.4 ANP-OT: A 0.17nW/Synapse 0.46pJ/SOP Neuromorphic Olfactory Processor with On-Chip Transfer Learning for Non-Invasive Cross-Hospital Cross-Pulmonary-Disease Diagnosis

Ziyi Cheng,Tsinghua University, Beijing, China

In Paper 36.4, Tsinghua University introduces an olfactory processor with on-chip transfer learning for cross-hospital and cross-disease pulmonary diagnosis. It achieves >94% accuracy across three hospitals and 95.5% lung cancer detection via few-shot adaptation, with 0.46pJ/SOP energy efficiency and 0.17nW/synapse power density.

ISSCC 2026 / February 18, 2026 / 1:30 PM

3:00 PM

36.5 A 185.6dB-FOM$_{DR}$ 180.3dB-FOM$_{SNDR}$ 10.64-NEF NS-SAR-ADC with Calibration-Free 2nd-Order kT/C-Noise Shaping for Wearable ExG Acquisition

Geunha Kim, Daegu Gyeongbuk Institute of Science and Technology, Daegu, Korea

In Paper 36.5, the Daegu Gyeongbuk Institute of Science and Technology describes an NS-SAR ADC with a multi-tasking integrator (MTI) for wearable ExG acquisition. The MTI suppresses quantization and kT/C noise without calibration, maintaining PVT robustness. The ADC achieves 185.6dB-FOM$_{DR}$, 180.3dB-FOM$_{SNDR}$, and 10.64 NEF, enabling low-noise, long-term ExG monitoring.

3:35 PM

36.6 A Sparsity-Aware Neural Interface with CIM-Based Predictive Focused Sampling for Hotspot Spike Tracking

Hui Wu, Westlake University, Hangzhou, China

In Paper 36.6, Westlake University reveals a 1024-electrode neural interface combining low-power panoramic scanning and high-fidelity spike tracking. A real-time CIM engine enables dynamic allocation, achieving 16× resolution boost (10 kHz) and record 0.0011 mm²/ch efficiency in 40 nm CMOS, maximizing spike density per power and area.

4:00 PM

36.7 A 90.7%-Efficiency Hybrid Optogenetic Stimulation System with Sub-Threshold Electrical Stimuli Achieving 70.6% Optical Energy Saving

Joonghoon Kang, Korea University, Seoul, Korea

In Paper 36.7, Korea University shows a hybrid optogenetic stimulator using sub-threshold electrical stimuli to lower optical thresholds and cut energy use. It achieves 90.75% LED efficiency and decaying-exponential irradiance for enhanced neural response, reducing optical energy by 70.6% in ex vivo mouse tests.

4:25 PM

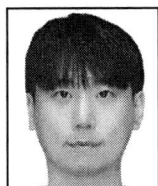

36.8 An 80×80µm²/Pixel 55.48dB-Wide-DR 400-Pixel Subretinal Prosthesis SoC with Power-Aware Light Adaptation and Charge-Recycling Local Dynamic Supply

Kyeongho Eom, Korea University, Seoul, Korea

In Paper 36.8, Korea University presents a light-adaptive subretinal prosthesis featuring power-aware light adaptation with a 55.48dB sensing DR. Pixel clustering with local return electrodes achieves 80×80µm² pixels for higher spatial resolution, while a charge-recycling multi-supply generator cuts stimulation energy by 34%.

4:50 PM

36.9 A CMOS Neural Probe with 1280 Electrodes and 88 Emission Sites Featuring Thermo-Optic Switching and On-Chip Calibration for Dual-Wavelength Optogenetics

Xiaolin Yang, imec, Heverlee, Belgium

In Paper 36.9, imec introduces a waveguide-based optogenetic neural probe with 1,280 electrodes, 384 recording channels, and 88 emission sites using dual-layer silicon photonics. Thermo-optic switches, on-chip calibration, and thermal monitoring enable the highest-density, cell-type-specific optogenetic studies to date.

5:05 PM

36.10 A 43.8-to-662.0µW 27.5-to-878.9fps Frame-Rate-Scalable Duty-Cycled Electrical Impedance Tomography System with MIMO Current-Balancing IA

Haidam Choi, KAIST, Daejeon, Korea

In Paper 36.10, KAIST and NYU Abu Dhabi present a frame-rate-scalable, duty-cycled EIT system that achieves 878.9fps at 662.0µW and 27.5fps at 43.8µW, reducing power by up to 97.5%. A MIMO current-balancing IA halves power, area, and input parasitics, while the FIR filter and digital matched filter suppress unwanted harmonics and reduce settling time.

5:20 PM

36.11 A 0.62µW/sensor 82fps Time-to-Digital Impedance Measurement IC with Unified Excitation/Readout Front-End for Large-Scale Piezo-Resistive Sensor Array

Jiayang Li, University of Bristol, Bristol, United Kingdom, University College London, London, United Kingdom

In Paper 36.11, the University of Bristol presents a fast impedance measurement IC for large-scale piezo-resistive sensor arrays. A differential time-to-digital demodulation and adaptive bias improve efficiency, scanning 253 sensors in 12.2ms (82fps) at 125kHz with 158µW power, 0.5% error, and 71.1dB SNR.

36

DIGEST OF TECHNICAL PAPERS • 599

979-8-3315-8937-0/26 $31.00 © 2026 IEEE

ISSCC 2026 / SESSION 36 / NEURAL AND BIOMEDICAL INTERFACES / 36.1

36.1 ReFIND: A Resolution-Reconfigurable Bio-Signal Classification SoC Enabling >10× Savings in AFE Power per Channel

Aviral Pandey[1], I-Ting Lin[1], Dhruv Vaish[1], Ashwin Rammohan[1], Savit Bhat[1], Jade Pinkenburg[1], Rikky Muller[1,2]

[1]University of California, Berkeley, CA, [2]Weill Neurohub, Berkeley, CA

Abstract

This work presents ReFIND, an SoC integrating together 16 resolution-reconfigurable AFEs, a feature extractor, and a classifier. The VCO ADC architecture trades power for resolution by scaling the integration time, from 13 to 10b at 1.77 to 0.12µW/channel. End-to-end validation shows up to 14× power saving over a fixed array, 3× more than channel selection. ReFIND enables runtime resolution scaling, opening up power-performance optimization in bio-signal classifiers.

Bio-signal recording front ends, which digitize microvolt-level signals from the body, are central to both wearable and implantable neural interfaces. Signals from these front ends are filtered, featurized, and decoded through machine-learning algorithms to the parameter of interest. As decoding tasks become increasingly complex, these front ends must support high channel counts under strict power budgets, constrained by battery capacity and tissue heating limits [1]. Prior studies have shown that, in high-channel-count implants, the analog front-end (AFE) dominates overall power consumption [2]. Because decades of circuit innovation have pushed noise and power efficiency factors (NEF/PEF) close to physical limits, future devices must leverage system-level strategies to further reduce AFE array power.

Prior art suggests several system-level approaches that leverage redundancy in neural recordings to reduce array power. One approach takes advantage of the noise resiliency of machine learning decoders by increasing input-referred noise (IRN) to reduce power consumption at design time [2]. Another approach, referred to as channel selection, dynamically selects which channels to record based on the channel's importance to the classifier [3]. Resolution reconfiguration is an alternative that selects an array or channel-specific resolution based on the impact to classification accuracy at run-time [4]. In this work, we introduce ReFIND (Reconfigurable Frontend Interface for Neural Decoding), an SoC that saves array-level analog power using resolution reconfiguration (Fig. 36.1.1). ReFIND integrates 16 resolution-scalable (13 to 10b) VCO ADCs, a feature extractor, a logistic regression classifier, and setting selection hardware. When validated end-to-end, ReFIND saves up to 14××and 11× power when compared to a traditional array and up to 3× and 2.7× power, when compared to channel selection for seizure decoding and gesture classification, respectively.

Resolution reconfiguration has been suggested before for spike decoding [5] and for human activity recognition [6]. However, prior art has not implemented such AFEs [4,6], or has only scaled ADC resolution without scaling LNA noise, which saves only a fraction of the total power [5]. Other published reconfigurable ADCs achieve power scale factors less than 2× per bit [14,15]. ReFIND's VCO ADCs optimize the resolution to power tradeoff, achieving a scale factor of up to 3.5×, allowing users to select the optimal array resolution or assign a channel dependent resolution at runtime.

The resolution reconfiguration training flow consists of four stages (Fig. 36.1.2): (1) the training dataset is recorded with every channel at maximum resolution, (2) a logistic regression classifier is trained, (3) each channel is assigned an importance score which is mapped to AFE settings, and (4) the training dataset is virtually rerecorded with the new settings and the classifier is retrained. The importance score of a channel is the maximum absolute feature weight for all features corresponding to the channel. User-defined thresholds determine how importance scores map to settings, and an example set of thresholds are shown in Fig. 36.1.2. After setting assignment, noise corresponding to the assigned resolution is injected into the training dataset, and the classifier is retrained to adapt its weights to the added noise. This retraining causes slight changes in model weights (Fig. 36.1.2) but is essential to improving the system accuracy. After the training process, the adjusted weights are loaded onto the chip, and the test dataset is recorded with the new channel settings and classified using the retrained weights for validation.

Figure 36.1.3 shows details of the logistic regression classifier in ReFIND as well as characterization results. The flexible feature-extraction unit outputs spectral band powers, total spectrum power, line length and mean absolute value computed over a variable window. The logistic regression classifier uses the features and the externally programmed weights to first compute the dot product, and then the classified label. Multiple SoCs can be parallelized by summing the dot products off-chip. Figure 36.1.3 shows that the implemented classifier output matches the theoretical fixed-point output, and the classifier achieves an average event sensitivity of 98.8% with an average detection latency of 3.6s on the full CHB-MIT dataset [6], while consuming 38µW at 0.48V.

ReFIND's highest resolution setting was chosen to be 13b to achieve noise floors comparable to state-of-the-art bio-signal recording AFEs. It uses an integrating VCO based ADC that approaches the theoretical maximum scaling of 4× per bit (Fig. 36.1.4). Duty-cycled time integration offers a simple implementation of resolution reconfiguration while holding sample rate constant. Increasing the integration time by 4× increases SNR by 1b while also consuming 4× more power. During integration, the V-to-I stage converts the input voltage to a current, which is converted to a frequency and then accumulated to a phase by the VCO. A VCO quantizer provides integration without large passives. For setting 4 (max resolution), the integration time is the entire cycle; for each lower setting, the integration time is reduced by 4×. Resetting the VCO every cycle requires a brief fixed period where the input is disconnected to mitigate distortion related to the VCO startup. This fixed cost limits the scaling between setting 1 and 2 to 2×.

For the lowest resolution setting, 12b of SQNR are required with an integration time of <125µs. To achieve a high frequency, the VCOs use small devices, allowing the current through the VCO to be set by noise requirements. However, using small VCO devices increases the input referred flicker noise and increases the power consumption of the VCO readout circuitry. To mitigate the increase in flicker noise, the number of stages in the VCO is increased to 31 and the ADC is chopped at the system level. To reduce readout power consumption, a coarse-fine VCO phase decoder samples each VCO only twice per sample period (Fig. 36.1.4) [12].

To further improve the efficiency of the VCO ADCs, the V-to-I converter uses CMOS inputs, as shown in Fig. 36.1.4. The added VCO current flows through the NMOS input pair along with the PMOS current. Current reuse improves overall power efficiency and increases the range of current supplied to each ring oscillator. This allows a higher SQNR for the same DC bias current. Cascode devices are added between the CMOS output and the ring oscillators to account for common mode differences. Split steering is used to drive the CMOS inputs with different common modes allowing for a low 0.8V supply.

While VCO ADCs are a power-efficient architecture, they suffer from distortion due to the nonlinearity of the ring oscillators. To correct this, ReFIND employs a look up table (LUT) based calibration scheme on each channel. A 33-element LUT stores ideal voltage values for 33 calibration codes for each ADC and the final output is linearly interpolated between these 33 calibrated points [13].

The resolution-reconfigurable VCO ADCs efficiently convert input signals up to 25mVptp. However, neural signals have electrode DC offsets up to 100s of mV. To quantize large DC offsets without increasing the dynamic range of each ADC, an offset removal DAC is added to every channel. The 6b DAC removes offsets and uses DEM to remove mismatch in the DAC elements. Gain error between the DAC and ADC is calibrated out.

ReFIND's frontends were validated on the bench to confirm their scaling performance. Figure 36.1.5 shows the ADC output spectrum at peak SNDR and the SNDR and SNR as a function of input amplitude for each setting. The ADCs achieve a noise floor of 13.9, 8.11, 3.6, and 1.83µVrms at a power consumption of 0.12, 0.18, 0.51, and 1.77µW, respectively, corresponding to an NEF of 9.3, 6.6, 5, and 4.7, respectively. While the power scale factor approaches 4× at higher settings, it is limited by the reset time and other fixed power costs, such as leakage, at lower settings.

To validate ReFIND, the system was tested end-to-end on two datasets: patient 16 (chosen for the shortest runtime) of the CHB-MIT seizure dataset [7] and an EMG gesture recognition dataset consisting of 64 channels recorded using 4 chips [8]. The pre-recorded data was played into ReFIND at every setting using DACs and then classified using an off-chip fixed-point-model. Figure 36.1.6 shows accuracy and power scatter plots with randomly sampled thresholds mapping importance scores to settings demonstrating the runtime power-performance tradeoffs achievable with ReFIND. It also compares ReFIND with channel selection, the existing state of the art, and demonstrates how resolution reconfiguration enables a new space of power-performance optimization that offers higher accuracy for less power than channel selection.

For gesture recognition, ReFIND saves on average 11× power for a 3% degradation in accuracy compared to a fixed array, and 2.7× power for equivalent accuracy when compared to channel selection. Lowering AFE resolution across the whole array does degrade

600 • 2026 IEEE International Solid-State Circuits Conference

979-8-3315-8937-0/26 $31.00 © 2026 IEEE

classification accuracy, but channel-specific reconfiguration enables fine-grained optimization of accuracy and power.

For seizure detection, ReFIND saves significant system power by reducing the resolution of all channels. Because seizures are rare events where both the precision and recall of the classifier is important, Fig. 36.1.6 compares the F_1 scores of ReFIND. A fixed array of AFEs at setting 1 achieves the highest power savings of a factor of 14× when compared to the baseline array. For this patient, F_1 score does not degrade with lower resolution, implying that the AFE is over-specified for this application. However, the reconfigurable array in ReFIND allows scaling down the entire array resolution to save system power for such applications.

The 0.95×2mm² ReFIND SoC was fabricated in TSMCs 28nm HPM process and the chip micrograph is shown in Fig. 36.1.7, along with a power breakdown and a comparison to state-of-the-art SoCs incorporating classifiers and AFEs. ReFIND achieves the lowest total power per channel for an integrated solution, the lowest noise AFEs, as well as the lowest AFE power per channel in test mode. The SoC can reduce AFE power by more than an order of magnitude when compared to traditional fixed-resolution arrays. ReFIND enables users to make finer-grained tradeoffs between accuracy and power consumption than channel selection or a fixed array. Random sampling of configurations shows that resolution reconfiguration enables more efficient accuracy to power tradeoff than the existing state of the art.

Acknowledgement:
The authors thank TSMC for chip fabrication, the SRC, JUMP 2.0, Center for Codesign of Cognitive Systems AWD-004311-S2, the NSF GRFP, the Apple Fellowship in Integrated Systems, Weill Neurohub, and the sponsors of BWRC.

References:
[1] P.D. Wolf. "Thermal Considerations for the Design of an Implanted Cortical Brain-Machine Interface (BMI)." *Front. in Neuroengineering*, Ch. 3.
https://pubmed.ncbi.nlm.nih.gov/21204402/
[2] N. Even-Chen et al. "Power-saving design opportunities for wireless intracortical brain-computer interfaces," *Nat. Biomed. Eng.*, Vol. 4, Iss. 10, pp. 984-996.
https://doi.org/10.1038/s41551-020-0595-
[3] U. Shin et al., "NeuralTree: A 256-Channel 0.227-uJ/Class Versatile Neural Activity Classification and Closed-Loop Neuromodulation SoC", *IEEE JSSC*, vol. 57, no. 11, 2022.
http://doi.org/10.1109/JSSC.2022.3204508
[4] A. Pandey et al., "Neural Recording Power Optimization Through Machine Learning Guided Resolution Reconfiguration," 2025, arXiv.
https://doi.org/10.48550/arXiv.2510.22924
[5] S. O'Driscoll et al., "Adaptive Resolution ADC Array for an Implantable Neural Sensor," in *IEEE Transactions on Biomedical Circuits and Systems*, vol. 5, no. 2, pp. 120-130, April 2011, http://doi.org/10.1109/TBCAS.2011.2145418.
[6] L. Galindez et al., "Dynamic Sensor-Frontend Tuning for Resource Efficient Embedded Classification," in *IEEE Journal on Emerging and Selected Topics in Circuits and Systems*, vol. 8, no. 4, pp. 858-872, Dec. 2018, http://doi.org/10.1109/JETCAS.2018.2850451
[7] A. Shoeb, "CHB-MIT Scalp EEG Database." physionet.org, 2010.
http://doi.org/10.13026/C2K01R.

[8] A. Moin et al., "A wearable biosensing system with in-sensor adaptive machine learning for hand gesture recognition," *Nat Electron*, vol. 4, no. 1, pp. 54–63, Dec. 2020, http://doi.org/10.1038/s41928-020-00510-8
[9] M. Shaeri et al., "A 2.46-mm2 Miniaturized Brain-Machine Interface (MiBMI) Enabling 31-Class Brain-to-Text Decoding," *IEEE J. Solid-State Circuits*, vol. 59, no. 11, pp. 3566–3579, Nov. 2024, http://doi.org/10.1109/jssc.2024.3443254
[10] M. Zhang et al., "A Patient-Specific Closed-Loop Epilepsy Management SoC With One-Shot Learning and Online Tuning," *IEEE J. Solid-State Circuits*, vol. 57, no. 4, pp. 1049–1060, Apr. 2022, http://doi.org/10.1109/jssc.2022.3144460
[11] Y. Wang et al., "A Closed-Loop Neuromodulation Chipset With 2-Level Classification Achieving 1.5-Vpp CM Interference Tolerance, 35-dB Stimulation Artifact Rejection in 0.5ms and 97.8%-Sensitivity Seizure Detection," *IEEE Trans. Biomed. Circuits Syst.*, vol. 15, no. 4, pp. 802–819, Aug. 2021, http://doi.org/10.1109/TBCAS.2021.3102261
[12] J. Huang and P. P. Mercier, "A 178.9-dB FoM 128-dB SFDR VCO-Based AFE for ExG Readouts With a Calibration-Free Differential Pulse Code Modulation Technique," in *IEEE Journal of Solid-State Circuits*, vol. 56, no. 11, pp. 3236-3246, Nov. 2021, http://doi.org/10.1109/JSSC.2021.3112635
[13] S. Rao et al., "A Deterministic Digital Background Calibration Technique for VCO-Based ADCs," in *IEEE Journal of Solid-State Circuits*, vol. 49, no. 4, pp. 950-960, April 2014, http://doi.org/10.1109/JSSC.2013.2293753
[14] M. Yip and A. P. Chandrakasan, "A resolution-reconfigurable 5-to-10b 0.4-to-1V power scalable SAR ADC," *ISSCC*, 2011, pp. 190-191, http://doi.org/10.1109/ISSCC.2011.5746277
[15] Y. Liang et al., "An 8-to-12-Bit Resolution-Reconfigurable SAR ADC With Fast-Window-Switching Technique," in *IEEE Transactions on Circuits and Systems II: Express Briefs*, vol. 72, no. 4, pp. 544-548, April 2025. http://doi.org/10.1109/TCSII.2025.3541236

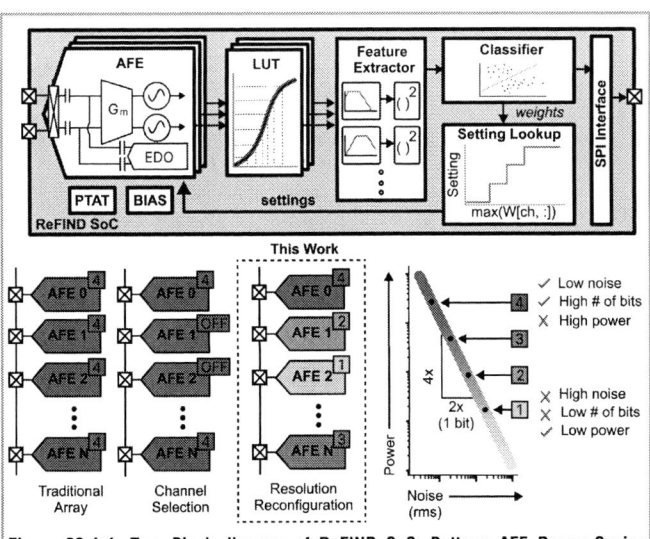

Figure 36.1.1: Top: Block diagram of ReFIND SoC; Bottom: AFE Power Saving approaches.

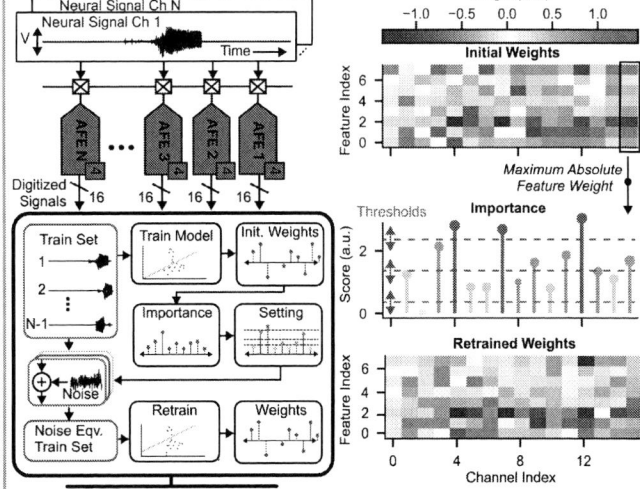

Figure 36.1.2: Left: Training flow block diagram. Right: Example weights, importance scores, and retrained weights, from patient 16 of CHB-MIT.

ISSCC 2026 / SESSION 36 / NEURAL AND BIOMEDICAL INTERFACES / 36.1

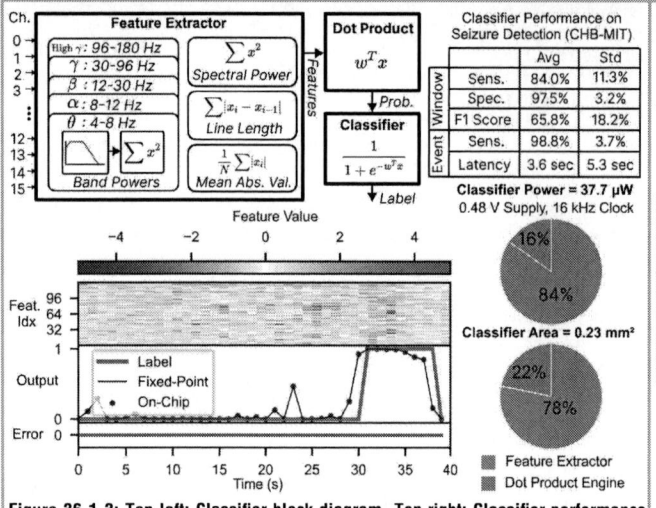

Figure 36.1.3: Top left: Classifier block diagram. Top right: Classifier performance summary. Bottom left: Comparison between fixed point model and silicon implementation, bottom right: Classifier power and area summary.

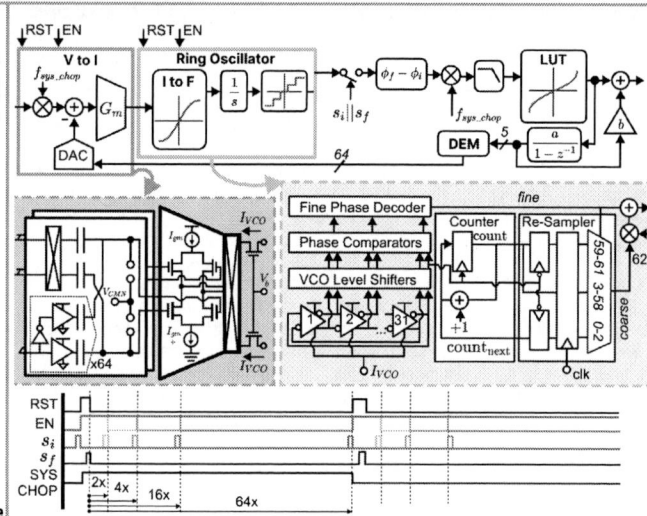

Figure 36.1.4: ADC block diagram showing schematics of key components and a timing diagram demonstrating resolution scaling.

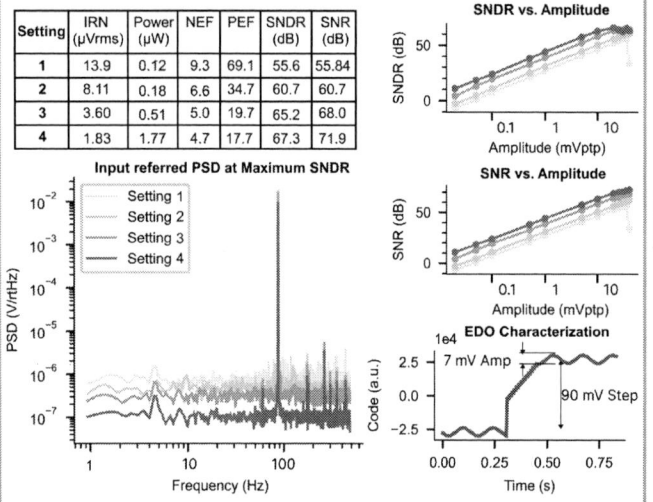

Figure 36.1.5: AFE Performance summary across different settings and example EDO loop settling behavior.

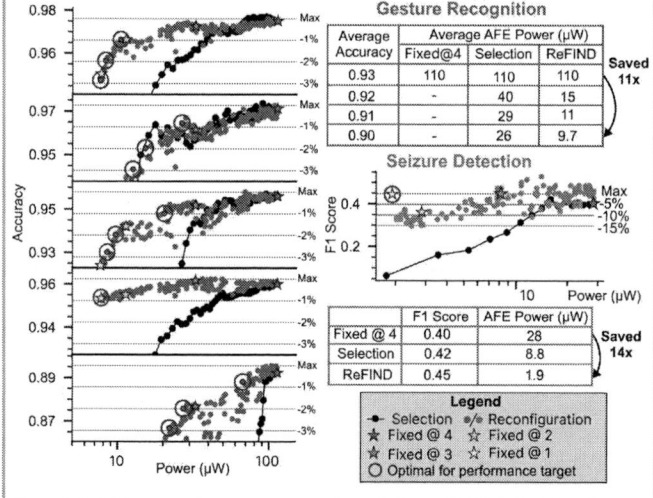

Figure 36.1.6: Power and accuracy comparisons between ReFIND, traditional arrays, and channel selection for the gesture recognition and seizure detection datasets.

SoC Power Breakdown — Total = 73 μW: Digital 94%, AFE 2.7%, Bias 2.8%

	ReFIND	JSSC'24 [9]	JSSC'22 [3]	JSSC'22 [10]	TBioCA21 [11]
Technology (nm)	28	65	65	40	180
Decoding Tasks	Seizure Detect. / Gesture Recog.	Brain-to-text Conversion	Seizure / PD Tremor Detect.	Seizure Detection	Seizure Detection
Integrated Solution	Yes	No	Yes	Yes	No
Supply Voltage (V)	0.8 (A) / 0.48 (D)	1.2	1.2	1.1 (A) / 0.7 (D)	1.5
# of Record. Channels — Train	16	192	256	16	8
# of Record. Channels — Test			64		
Active Area (mm²)	1.9	1.7 (A) / 0.75 (D)	3.48	2.08	2.32 (A) / 3.51 (D)
BW (Hz)	500	10K	500	201	1K
IRN (μVrms)	1.8 (Setting 4)	8.8	3.2	0.85	2.2
SNDR (dB)	67	N/A	49	61	91
AFE Power / Ch. (μW) — Train	1.8	3.4	1.51	—**	1.5
AFE Power / Ch. (μW) — Test	0.13 (Seizure)				
Classifier	Logistic Regression	LDA	Neural Tree	GTCA-SVM	Threshold + LS-SVM
# of Classes	2	31	2	2	2
# of Class. Channels	16	512†	256	16	8
Class. Power* (μW)	37.7	223†	271†‡	—**	1.16
Class. Rate (Hz)	0.5-2 (Seizure)	1 (character/sec)	0.5-4	—**	—**
Tot. Power/Ch. (μW)	4.6 (Seizure)	3.88‡	7.1	—**	1.6†

*Including feature extractor and memory   **Insufficient data   †Calculated from reported power breakdown   ‡Calculated by AFE chip power / ch + Classifier chip power / ch

Figure 36.1.7: Die micrograph, power breakdown and comparison to state-of-the-art.

ISSCC 2026 / SESSION 36 / NEURAL AND BIOMEDICAL INTERFACES / 36.2

36.2 A Neural Interface SoC for Smart Glasses with Low-Power Neural Commanding and Efficient LoRA-Enabled On-Chip Learning

Zhiwei Zhong[*][1], He Yu[*][1], William McGarry[1], Yijie Wei[2], Jie Gu[1]

[1]Northwestern University, Evanston, IL, [2]Texas Instruments, Dallas, TX
*Equally Credited Authors (ECAs)

Abstract

This work presents a 65nm ExG SoC for smart glasses, enabling low-power neural interaction. A 10-ch AFE and on-chip CNN deliver real-time EOG/EMG/EEG inference. On-device continual learning via LoRA fine-tuning cuts trainable parameters by 54 to 77%. An ultra-lower-power nearest-feature neural commanding path was implemented to issue user commands and wake up heavy computation only when needed. Chunked delta encoding and a streaming pipeline reduce memory and expand training data by 2.1×.

Smart glasses powered by augmented reality (AR) and artificial intelligence (AI) are projected as a new market driver for wearable devices, offering a versatile platform for consumer applications [1,2]. However, current smart glasses still fall short of delivering effortless ergonomic human interactions. First, as shown in Fig. 36.2.1, the existing smart glasses interface with users through physical touchpads or voice commands, which require users to perform precise physical movements or to deal with voice-related privacy compromise and sound interference. Second, current smart glasses fail to exploit the available neurological information for creating a human-aware operating environment. Third, existing wearable devices suffer from electrode displacement, body movement, variation of user-specific biological characteristics, causing loss of accuracy or requiring frequent calibration. In this work, we extend the sensing capability of smart glasses with an AI-powered neural interface System-on-Chip (SoC) for context-aware operation. Specifically, we deliver an advanced AI accelerator with efficient on-device learning to classify multimodal neural activities in real time. The contributions of this work are as follows. (1) To overcome the challenges due to the drift of biosignals across sessions and users, this work delivers an on-chip learning solution for real-time training on smart glasses. Also, to overcome the challenges of AI inference and training with highly constrained compute resources, this work incorporates a low-rank adaptation (LoRA) method [3] for hardware efficient fine-tuning. (2) A low-power neural commanding module is introduced to enable low-cost issuing of human command through controlled neural activities without constant running of the power-intensive AI accelerators. (3) To cope with the memory limitation, an on-device data-compression technique with delta encoding is developed, significantly reducing the storage cost for training. (4) A fully integrated system of smart glasses with optimized selection of ExG channels is built and demonstrated through various real-life applications of intelligent neural activity tracking and commanding.

Figure 36.2.2 shows the optimal selection of ExG channels that are around the eye region for different biological sensing requirements. To strike a balance among accuracy, user comfortableness and cosmetics, the final selection of channels are shared EOG/EMG electrodes (HR, HL, Flu, VLd, Vz) and reduced EEG electrodes (T3, T5, O1). The high-level SoC architecture integrates 10 analog front-end (AFE) channels for capturing the bio-signals. In each channel, a two-stage chopper amplifier delivers 40 to 70dB of programmable gain. A 10b SAR ADC digitizes the signals, with a programmable sampling rate ranging from 128Hz to 2kHz. As the reference generators and drivers can dominate AFE's power consumption due to the need for low output impedance to drive the sampling operations of multi-channel ADC simultaneously, a sequential wake-up scheme is incorporated to deliver time-interleaved sampling for different AFE channels so that reference circuitry can be shared and power-gated once the sampling is over, leading to more than 90% reduction of static power for the AFE. The AI core performs end-to-end inference and training fully on-device. The inference modules include FIR filtering, feature extraction, a nearest-feature matcher for neural commanding, and a CNN-based classifier. The training modules comprise data/label collection, LoRA-based fine-tuning, Taylor expansion for cost function approximation, and an additional stochastic rounding scheme for low-precision training at 10b.

As learning is challenging for low-power SoCs, we incorporate an advanced learning solution. Figure 36.2.3 illustrates the hardware computing engine with support for LoRA learning. LoRA leverages projection matrices at lower dimension to enhance fine-tuning efficiency in resource-constrained environments. The LoRA approach integrates low-rank decomposition directly into the forward and backpropagation paths, where the down-projection matrix A and up-projection matrix B are stored in on-chip memory alongside quantized pre-trained weights. As B and A are lower-dimension matrices than the full-weight W, the trainable parameters are significantly reduced, leading to major savings of on-chip memory. By initializing B with pre-trained values rather than all zeros in the conventional method, the system avoids the zero-locking problem in fixed-point training. As shown in Fig. 36.2.3, 10 input buffers are used to receive data from 10 ADCs. A delta encoder is used to compress the data and store it in input memory for training. The computing engine contains 5 processing element (PE) arrays, each incorporating 9 MACs, pooling, activation units, and addition stages for LoRA feedforward. The first PE array is designed for LoRA training. It has a Taylor-expansion unit to approximate nonlinear functions (e.g., exponential functions), a divider for initial error calculation, and a matrix transpose unit for error backpropagation. The feedforward path reads the LoRA matrices A and B from separate memory blocks and performs low-dimensional MAC operations together with the base weight matrix to recover the final weight matrix. For the feedforward process of training, this chip also introduces a LoRA ADD operation that reconstructs and stores the final weight matrix via addition, cutting memory accesses by 25%, and reducing buffer needs and read/write conflicts versus MAC-only designs. Backpropagation computes gradients for A and B via matrix transposes and MAC operations, with per-batch updates handled by an index-controlled scheduler. LoRA reduces the number of trainable parameters by 54 to 77% compared to full fine-tuning.

Figure 36.2.4 illustrates the ExG signal processing techniques that consist of a "neural commanding" feature for issuing low-power hand-voice-free user commands, and data-compression techniques for memory saving for training. To support neural-activity based command operation, we deployed a separate "neural commanding" module that utilizes low-power nearest-feature matching to recognize the human's high-level operational intents. The "neural command" can be pre-defined by the user with easily classified neural activities, e.g., smiling, as compared with the difficult detection tasks such as precise eye tracking. As a result, a low-budget classifier can be used to detect such a command event, which can trigger more powerful CNN computations when necessary. The low-power neural commanding begins with ExG acquisition from a reduced set of m selected channels. A lightweight feature-extraction module processes the incoming signals, and the resulting feature vectors are compared against pre-stored reference vectors using L1 distance. The nearest-feature matcher keeps running until activation criteria are met. Once activated, the corresponding LoRA adapter is selected for further classification. To reduce the memory footprint for training, a chunked delta encoding/decoding pipeline is integrated into the data path. For every 17 data points from ADC outputs, the first data is buffered in 10b format while the rest are compressed via the delta encoder, i.e., storing residuals between successive samples in 5b. The delta encoder is effective due to the continuous nature of ExG signal and leads to 47% memory savings. For end-to-end CNN implementation, input data are typically shifted and duplicated before being sent to the PEs to perform convolution [11]. The stream pipeline performs on-the-fly data shifting, eliminating the need for data duplication. For the chosen data length of 256, this saves 89% memory for data storage. The delta encoding and the stream pipeline increase the storable training data size by 2.1× in the SoC.

A 65nm test chip was fabricated and mounted into smart glasses for real-life demonstration. Figure 36.2.5 shows the demonstrations from the developed smart glasses that support on-device data/label collection, training, and real-time user interaction for three tasks, i.e., visual selection, facial expression and fatigue detections. The LoRA fine-tuning scheduler manages the hyperparameters and data transfer between memory and PE array. A linear-feedback shift register (LFSR)-based shuffling module randomizes data addressing for training, while customized instruction sequences control the progress of training. Stochastic rounding is applied during training to improve convergence speed by approximately 2×. During training, EOG and EMG signals are recorded while the user performs guided actions such as gaze shifts and facial expressions through the display screen on the smart glasses. In our facial-detection demonstration, facial EMGs are classified in real time, and the corresponding emojis are automatically inserted into chat interfaces, enabling expressive, hands-free communication. The 4-class facial detection achieves accuracy of ~86% with 4.7% improvement from using LoRA fine-tuning. For visual selection, the EOG signals are used for object selection. In the demonstration, the camera automatically crops the captured area of the object through EOG-based gaze selection from the user. The image of the object is further classified using an open-source image recognition model. The 9-class visual selection has a gaze tracking accuracy of ~80% with 7% improvement due to LoRA fine-tuning.

Figure 36.2.6 shows more chip measurement results as well as evaluation on public datasets, where the on-chip CNN achieves 84.95% for 8-angle eye-tracking [5], 94.21% for 4-angle eye-tracking [5], 96.41% for 3-angle eye-tracking [4], 85.72% for facial expression recognition [6], and 86.58% for fatigue sensing [7]. The chip is measured with an energy consumption of 1.12µJ/class for CNN inference, 0.0075µJ/class for neural commanding, and 8.6mJ for LoRA based training with 200 epochs. Figure 36.2.6 also shows the comparison table with prior works. Compared to prior fully integrated SoCs with both analog frontend and digital cores which primarily target medical diagnosis or VR, this work not

602 • 2026 IEEE International Solid-State Circuits Conference

979-8-3315-8937-0/26 $31.00 © 2026 IEEE

only extends intelligent neural interface to AI-powered smart glasses but also enables highly efficient online learning for a robust intelligent neural interface. Figure 36.2.7 shows the chip micrograph.

Acknowledgement:
This work was supported by the National Science Foundation under CCF-1846424 and CCF-2208573.

References:
[1] L. -H. Lee and P. Hui, "Interaction Methods for Smart Glasses: A Survey," in *IEEE Access*, vol. 6, pp. 28712-28732, 2018. http://doi.org/10.1109/ACCESS.2018.2831081
[2] L.-H. Lee et al., "Towards augmented reality driven human–city interaction: Current research on mobile headsets and future challenges," *ACM Comput. Surv.*, vol. 54, no. 8, pp. 1–38, 2021. http://doi.org/10.1145/3467963
[3] E. J. Hu et al., "LoRA: Low-Rank Adaptation of Large Language Models," in *Proc. Int. Conf. Learn. Representations* (ICLR), 2022. https://arxiv.org/abs/2106.09685
[4] N. Barbara et al., "Real-Time Continuous EOG-based Gaze Angle Estimation with Baseline Drift Compensation Under Stationary Head Conditions," *Biomedical Signal Processing and Control*, vol. 86, 2023. http://doi.org/10.1016/j.bspc.2023.105282
[5] O. I. Pellico Sánchez et al., "Eye-movement Electrooculography Dataset*", IEEE Dataport*, June 26, 2022. http://doi.org/10.21227/ttw6-ar70
[6] M. Gjoreski et al., "Facial EMG sensing for monitoring affect using a wearable device," *Scientific Reports*, vol. 12, no. 1, p. 16876, 2022. http://doi.org/10.1038/s41598-022-21456-1
[7] J. Min et al., "Driver fatigue detection through multiple entropy fusion analysis in an EEG-based system," *PLoS One*, vol. 12, no. 12, 2017. http://doi.org/10.1371/journal.pone.0188756
[8] H. -J. Lee et al., "A 13.5μW 35-Keyword End-to-End Keyword Spotting System Featuring Personalized On-Chip Training in 28nm CMOS," *ISSCC*, pp. 620-621, 2025. http://doi.org/10.1109/ISSCC49661.2025.10904744
[9] F. Tian et al., "E-NPU: A 34~126nJ/Class Event-Driven Adaptive Neural SoC with Signal-Dynamics-Aware Feature Clustering and Multi-Model In-Memory Inference/Training for Personalized Medical Wearables," *CICC*, 2025. http://doi.org/10.1109/CICC63670.2025.10982760
[10] J. Liu et al., "A High-Accuracy and Energy-Efficient Zero-Shot-Retraining Seizure-Detection Processor with Hybrid-Feature-Driven Adaptive Processing and Learning-Based Adaptive Channel Selection," *ISSCC*, pp. 542-543, 2024. http://doi.org/10.1109/ISSCC49657.2024.10454405
[11] Z. Zhong et al., "A Sub-1μJ/class Headset-Integrated Mind Imagery and Control SoC for VR/MR Applications with Teacher-Student CNN and General-Purpose Instruction Set Architecture," *ISSCC*, pp. 544-545, 2024. http://doi.org/10.1109/ISSCC49657.2024.10454317
[12] U. Shin et al., "A 256-Channel 0.227μJ/class Versatile Brain Activity Classification and Closed-Loop Neuromodulation SoC with 0.004mm2-1.51 μW/channel Fast-Settling Highly Multiplexed Mixed-Signal Front-End," *ISSCC*, pp. 338-339, 2022. http://doi.org/10.1109/ISSCC42614.2022.9731776
[13] C.-W. Tsai et al., "SciCNN: A 0-Shot-Retraining Patient-Independent Epilepsy-Tracking SoC," *ISSCC*, pp. 488-489, 2023. http://doi.org/10.1109/ISSCC42615.2023.10067518

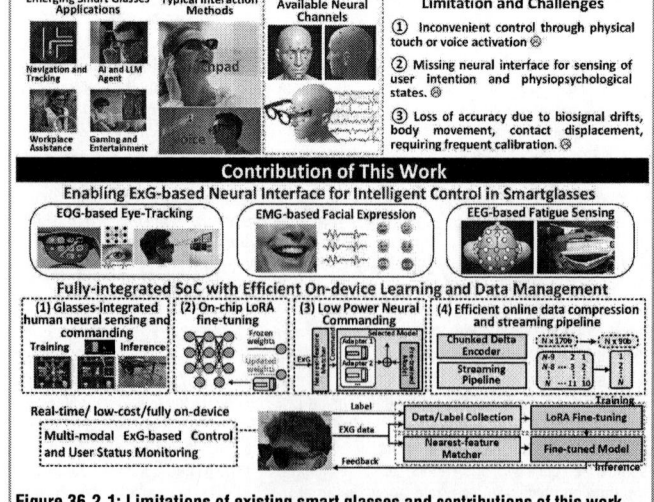

Figure 36.2.1: Limitations of existing smart glasses and contributions of this work.

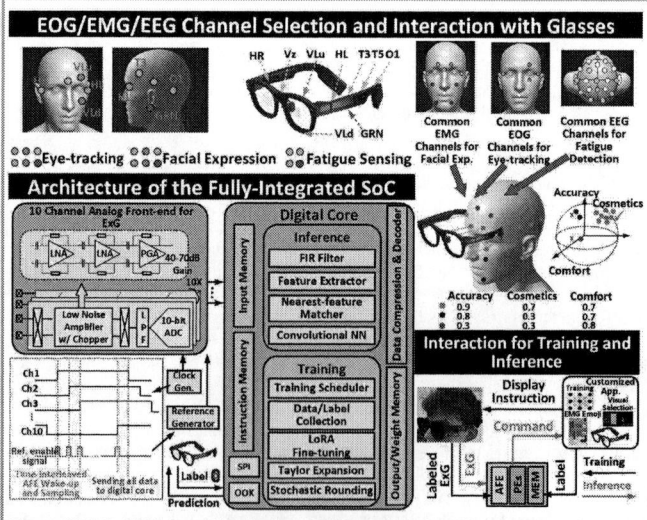

Figure 36.2.2: ExG channel selection with neural training/inference on the smart glasses and block diagram of the fully integrated SoC.

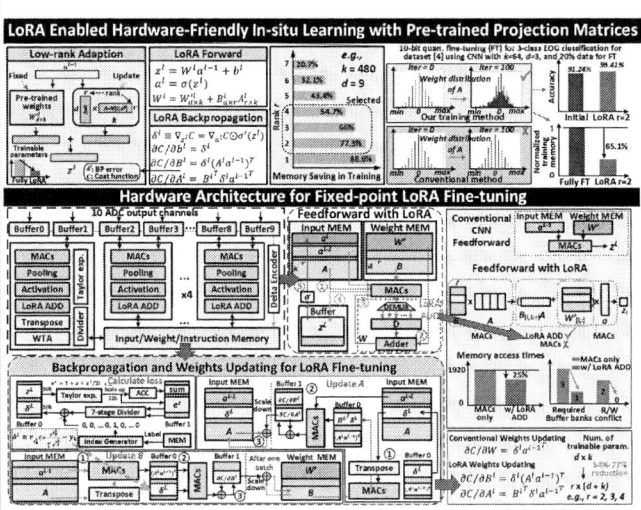

Figure 36.2.3: LoRA-based neural processor architecture and fine-tuning.

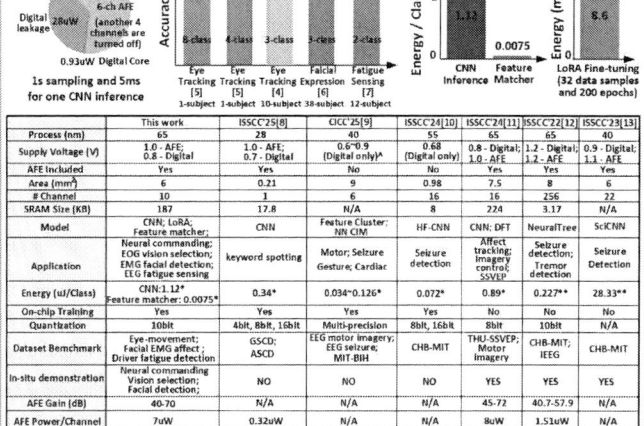

Figure 36.2.4: Hand-voice-free neural commanding using ExG and memory reduction for real time on-chip training.

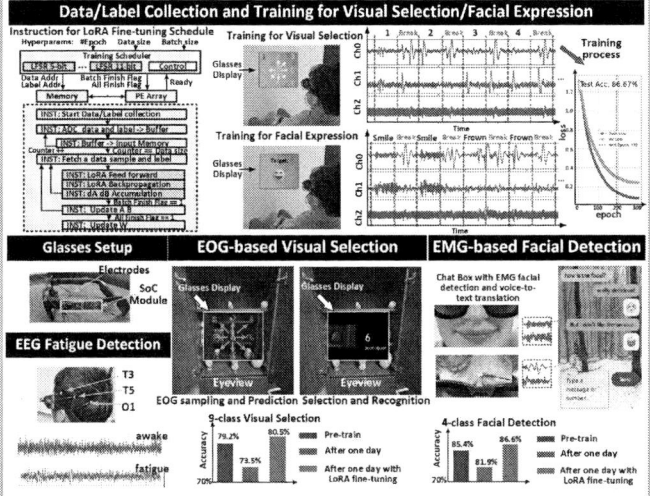

Figure 36.2.5: Demonstration of training and Inference tasks with EOG-based visual selection, EMG-based facial detection and EEG-based fatigue detection.

	This work	ISSCC25[8]	CICC'25[9]	ISSCC'24[10]	ISSCC'24[11]	ISSCC'22[12]	ISSCC'23[13]
Process (nm)	65	28	40	55	65	65	40
Supply Voltage (V)	1.0 - AFE; 0.8 - Digital	1.0 - AFE; 0.7 - Digital	0.6~0.9 (Digital only)^	0.68 (Digital only)	0.8 - Digital; 1.0 - AFE	1.2 - Digital; 1.2 - AFE	0.9 - Digital; 1.1 - AFE
AFE Included	Yes	Yes	No	No	Yes	Yes	Yes
Area (mm²)	6	0.21	9	0.98	7.5	8	6
# Channel	10	1	N/A	16	16	256	22
SRAM Size (KB)	187	17.8	N/A	8	224	3.17	N/A
Model	CNN; LoRA; Feature matcher	CNN	Feature Cluster; NN CIM	HF-CNN	CNN; DFT	NeuralTree	SciCNN
Application	Neural commanding; EOG vision selection; EMG facial detection; EEG fatigue sensing	keyword spotting	Motor; Seizure Gesture; Cardiac	Seizure detection	Affect tracking; Imagery control; SSVEP	Seizure detection; Tremor detection	Seizure Detection
Energy (uJ/Class)	CNN:1.12*; Feature matcher: 0.0075*	0.34*	0.034~0.126*	0.072*	0.89*	0.227**	28.33**
On-chip Training	Yes	Yes	Multi-precision	Yes	No	No	No
Quantization	10bit	4bit, 8bit, 16bit		8bit, 16bit	8bit	10bit	N/A
Dataset Benchmark	Eye-movement; Facial EMG affect; Driver fatigue detection	GSCD; ASCD	EEG motor imagery; EEG seizure; MIT-BIH	CHB-MIT	THU-SSVEP; Motor imagery	CHB-MIT; IEEG	CHB-MIT
In-situ demonstration	Neural commanding Vision selection; Facial detection;	NO	NO	NO	YES	YES	YES
AFE Gain (dB)	40-70	N/A	N/A	N/A	4S-72	40.7-57.9	N/A
AFE Power/Channel	7uW	0.32uW	N/A	N/A	8uW	1.51uW	N/A

*Digital backend only **Analog frontend + digital backend ^A separate chip is served as analog frontend

Figure 36.2.6: Measurement result and performance comparison.

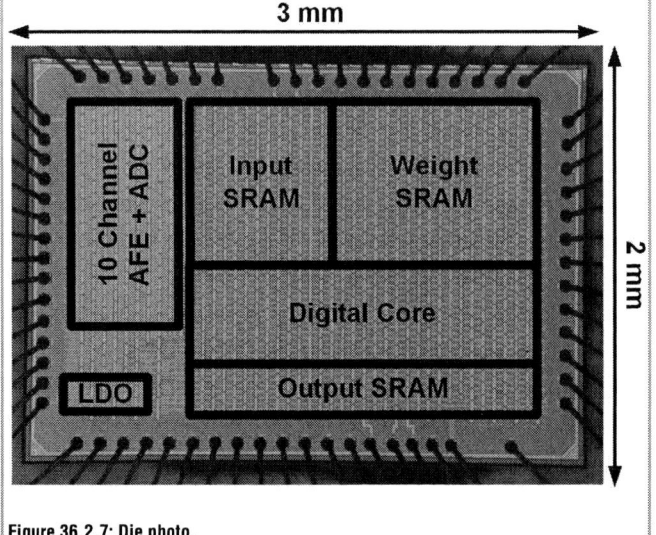

Figure 36.2.7: Die photo.

ISSCC 2026 / SESSION 36 / NEURAL AND BIOMEDICAL INTERFACES / 36.3

36.3 A 16.4nJ/Class Patient-Independent Prototype-Based Spatio-Temporal CNN Processor with Forward-Inference-Based Adaptation for Robust and Low-Latency Seizure Detection

Yang Wang[1], Longyang Lin[1], Jerald Yoo[2], Jiamin Li[1]

[1]Southern University of Science and Technology, Shenzhen, China, [2]Seoul National University, Seoul, Korea

Abstract

We present a patient-independent, prototype-based spatio-temporal CNN processor for seizure detection, achieving high accuracy without patient-specific data, at an energy of 16.4nJ/class and latency of <120ms. An optional forward-inference-based zero-shot adaptation improves robustness to variability and low-quality inputs. The 40nm IC achieves 94.3%/94.9% sensitivity/specificity without patient data, improving to 95.4%/97.9% with adaptation.

For millions of patients worldwide with drug-resistant epilepsy, closed-loop neuromodulation offers a promising solution, providing precise, on-demand intervention upon seizure detection [1-7]. The therapeutic efficacy and practical viability of this approach relies critically on a seizure detection processor capable of simultaneously achieving: 1) high accuracy and minimal latency to exploit the short window (within seconds) between electrical onset and disabling symptoms to intervene seizure evolution; 2) high energy efficiency for chronic use, where the operational duration is proportionally extended by energy efficiency; and 3) low-cost and robust deployment across diverse patient populations considering both intra- and inter-patient variability.

While existing patient-specific seizure detection processors [4-8] deliver high accuracy, they rely on the patient's seizure EEG data for training, which necessitates costly and time-consuming inpatient acquisition due to the low-occurrence of seizure events, and must be repeatedly collected for retraining to address intra-patient pattern drift, further increasing operational burden. To overcome this limitation, zero-shot-retraining systems have been proposed [9], where models are pre-trained on public datasets and retrained at deployment using only the patient's non-seizure EEG data. However, these exhibit limited accuracy, low energy efficiency and high latency due to high computational overhead. While energy efficiency and latency performance were improved in [10] through hybrid feature adaptation and dynamic channel selection, the detection accuracy (retrained with the patient's 1-minute non-seizure data) is still limited. Notably, both methods show strong dependence on the quality of the patient's non-seizure data for retraining, making them vulnerable to artifacts or contaminated recordings, which introduces risks of model degradation and forgetting during retraining, undermining deployment robustness. More recently, a no-patient-data system was proposed [11] that achieves improved accuracy using only dataset-trained models and requires no patient-specific data for deployment. Nonetheless, it incurs high computational costs leading to low energy efficiency and relies on preprocessed frequency-domain features that further increases energy consumption. Furthermore, the lack of model adaptation mechanism makes it susceptible to performance degradation due to intra-patient signal pattern drifting over time.

In this work, we propose a patient-independent prototype-based spatio-temporal CNN (STCNN) processor that enables high-accuracy seizure detection without requiring patient-specific data for low-cost deployment, while achieving low latency and energy consumption. It incorporates an optional forward-inference-based adaptation mechanism that calibrates the model using only the patient's non-seizure EEG, enhancing robustness to inter- and intra-patient variability while maintaining resilience against low-quality adaptation input. Shown in Fig. 36.3.1 (right), the system features: (1) a dual-path (for inter- and intra-patient) learning scheme combining episode training with cross-patient prototypical loss and contrastive learning based on positive and negative pairs to improve feature discrimination; (2) an STCNN processor with receptive-field-driven short window design that minimizes memory access and latency, which is enhanced by energy-efficient temporal and spatial convolution techniques; (3) a forward-inference-driven prototype and distance adaptation mechanism that enhances accuracy using 2-minute non-seizure EEG, avoiding retraining or loss of pre-learned prototype, thereby maintaining robustness against variations and adaptation data quality.

Figure 36.3.2 shows the operational principle of the proposed prototype-based STCNN seizure-detection processor. With 1.9kB parameters and 241.6kMACs, the network backbone employs a lightweight spatio-temporal convolutional structure comprising two depthwise convolutions followed by pointwise convolution for temporal feature extraction, and a depthwise spatial convolution to model inter-channel dependencies. During patient-independent inference, multi-patient training data from public dataset are computed to aggregate embeddings of seizure and non-seizure samples, forming two prototypes that serve as the reference anchors. The incoming extracted embeddings are then classified based on Euclidean distance to these prototypes. For generalization despite inter- and intra-patient variability, a dual-path learning scheme is proposed, where the prototypical episode training improves discriminability through balanced sample selection, while contrastive learning computes a similarity matrix to enforce intra-class consistency and inter-class separation. During the optional adaptation using 2-minute non-seizure EEG from the patient, a forward-inference-driven prototype and distance adaptation approach is proposed,

avoiding conventional backpropagation-based retraining which is susceptible to calibration data quality and incurs high computational overhead. The prototype adaptation uses the patient non-seizure data with the original non-seizure prototype to form a patient-specific negative prototype, enhancing the model's specificity in capturing the patient's non-seizure patterns. The distance adaptation applies a scaling factor, derived based on the distributions of the patient's non-seizure embeddings, to embedding distances while enforcing a safety margin constraint to ensure non-seizure embeddings remain sufficiently closer to the negative prototype than to the positive prototype. Together, this adaptation mechanism improves accuracy while minimizing computational overhead and sensitivity to data quality.

Figure 36.3.3 illustrates the hardware architecture of the STCNN processor. Pre-trained network parameters and configurations are stored in dedicated memory. Incoming EEG data is buffered through a ping-pong FIFO that alternates read and write processes for continuous data flow. A system control unit manages operational modes (for patient-independent classification or adaptation) and coordinates a processing element (PE) scheduler that dynamically allocates computations across eight reconfigurable PEs with a pipelined and layer-wise scheduling strategy for enhanced hardware efficiency. Intermediate feature maps from each layer are post-processed and stored as temporal convolution (TC) boundary partial sums, spatial convolution (SC) context features, or activations for subsequent layers via a configurable feature interconnect block. Upon forming of embeddings, distance calculation to prototypes is performed for classification, which is also used for risk-aware inference interval adaptation. For the optional patient-specific adaptation, non-seizure embeddings are averaged in the prototype adaptation using shift-add operations and fused with the original prototype. Distance adaptation employs shift-based scaling and comparators to adjust decision boundaries under safety margin constraints, all implemented through forward inference without backpropagation overhead.

Figure 36.3.4 shows the receptive-field-driven short window design. By aligning the input segmentation with the receptive field of each TC layer, EEG signals are partitioned into non-overlapping 8-sample window, significantly reducing detection latency and buffer size (4× compared to [10]). Temporal features from each window are cached, with the latest 30 windows aggregated for spatial convolution and projection. As the window size decreases, discarding boundary features in shallow layers [10] incurs greater performance loss due to the rising proportion of discarded features relative to input size. To process the features of adjacent windows with minimized storage or performance penalties, a boundary partial sum (BPS) technique is proposed for temporal convolution. Overlapping Input segments are rearranged and convolved with a fixed kernel, then the previously computed intermediate BPS stored in TCBPS SRAM are added to the current result to construct complete outputs, reducing the MAC count per TC by 51.3%. While short window design lowers latency, it increases computation frequency (triggered by window) of subsequent spatial convolution and projection layers. This is mitigated by the proposed risk-aware spatial convolution with adaptive inference interval, where the ratio of embedding distances to seizure and non-seizure prototypes dynamically adjusts the inference interval. Computation is reduced during stable non-seizure periods to save energy while immediate processing upon short-window aggregation ensures low-latency near seizure onset. This technique reduces the average number of inferences by 27× and overall SRAM access by 30.6%, while maintaining detection sensitivity.

The processor is fabricated in a 40nm CMOS process, consuming 23.08μW and 16.4nJ/classification at 2MHz and 0.58V (Fig. 36.3.7). Evaluated on the CHB-MIT scalp EEG dataset (24 subjects), the system accurately detects both seizure onset and offset when applied to an unseen patient, with one-shot adaptation using the patient's 2-minute non-seizure data further reduces false alarms (Fig. 36.3.5a). Figure 36.3.5(b) summarizes the seizure-detection performance on CHB-MIT, where the direct deployment to unseen patients (no patient data) achieves 94.3% sensitivity and 94.9% specificity, outperforming prior patient-independent designs [11]. With 2-minute prototype adaptation, performance improves to 94.8%/96.7%, and further gains to 95.4%/97.9% are achieved with distance adaptation. Shown in Fig. 36.3.5(c), accuracy (calculated by the average of sensitivity and specificity) is consistently enhanced across all patients by the forward-inference-driven adaptation, with significant improvements for patients exhibiting initial domain drift (e.g., subjects 4, 8, 9, 14, 15). Figure 36.3.6 compares this work with state-of-the-art seizure-detection processors. Among prior patient-independent processors that do not require target-patient training data [9-11], this design achieves the highest sensitivity and specificity in direct deployment, without off-chip feature extraction. When adapted with 0.1% (2-minute) patient-specific non-seizure data, performance further improves to 95.4%/97.9% sensitivity/specificity with robustness to adaptation data quality thanks to the forward-inference-driven approach. Furthermore, the design achieves the lowest reported detection latency and lowest energy per inference, making it well-suited for chronic closed-loop neuromodulation with low-cost and robust deployment across diverse patient population.

Acknowledgment:
This work was funded in part by the National Natural Science Foundation of China (Grant No. 62304099, 62511540072); Shenzhen Science and Technology Innovation Commission (Grant No. JCYJ20230807092359001); Guangdong Grant No. 2023QN10X177. Corresponding author: Jiamin Li.

References:
[1] A. Chua et al., "SPIRIT: A Seizure Prediction SoC with a 17.2nJ/cls Unsupervised Online-Learning Classifier and Zoom Analog Frontends," *IEEE Symposium on VLSI Technology and Circuits*, 2024.
https://doi.org/10.1109/VLSITechnologyandCir46783.2024.10631419
[2] M. Shoaran et al., "Intelligent Neural Interfaces: An Emerging Era in Neurotechnology," *IEEE Custom Integrated Circuits Conference*, 2024.
https://doi.org/10.1109/CICC60959.2024.10529099
[3] D. -Y. Yoon et al., "A 1024-Channel Simultaneous Recording Neural SoC with Stimulation and Real-Time Spike Detection," *Symposium on VLSI Circuits*, 2021.
https://doi.org/10.23919/VLSICircuits52068.2021.9492480
[4] M. Zhang et al., "A One-Shot Learning, Online-Tuning, Closed-Loop Epilepsy Management SoC with 0.97µJ/Classification and 97.8% Vector-Based Sensitivity," *Symposium on VLSI Circuits*, 2021.
https://doi.org/10.23919/VLSICircuits52068.2021.9492429
[5] U. Shin et al., "A 256-Channel 0.227µJ/class Versatile Brain Activity Classification and Closed-Loop Neuromodulation SoC with 0.004mm2-1.51 µW/channel Fast-Settling Highly Multiplexed Mixed-Signal Front-End," *ISSCC*, 2022.
https://doi.org/10.1109/ISSCC42614.2022.9731776
[6] X. Huang et al., "A Closed-Loop Neuromodulation Chipset with 0.0009mm²-0.36µW/Ch Recording Frontend and 0.075mm²-6.76µW Seizure Classification Backend," *Symposium on VLSI Technology and Circuits*, 2025.
https://doi.org/10.23919/VLSITechnologyandCir65189.2025.11074961
[7] M.A. Bin Altaf et al., "A 16-ch patient-specific seizure onset and termination detection SoC with machine-learning and voltage-mode transcranial stimulation," *ISSCC*, 2015.
https://doi.org/10.1109/ISSCC.2015.7063092
[8] S. Qiu et al., "PANDA: A 3.178 TOPS/W Reconfigurable Seizure Prediction ANd Detection Neural Network Accelerator for Epilepsy Monitoring," *Symposium on VLSI Technology and Circuits*, 2025.
https://doi.org/10.23919/VLSITechnologyandCir65189.2025.11075036
[9] C. -W. Tsai et al., "SciCNN: A 0-Shot-Retraining Patient-Independent Epilepsy-Tracking SoC," *ISSCC*, 2023. https://doi.org/10.1109/ISSCC42615.2023.10067518

[10] J. Liu et al., "33.1 A High-Accuracy and Energy-Efficient Zero-Shot-Retraining Seizure-Detection Processor with Hybrid-Feature-Driven Adaptive Processing and Learning-Based Adaptive Channel Selection," *ISSCC*, 2024.
https://doi.org/10.1109/ISSCC49657.2024.10454405
[11] V. Lukito et al., "A No-Patient-Data Seizure Classifier Soc for Real-Time Classification of Seven Seizure Types Using Feature Fusion and Near-Memory Computing," *Symposium on VLSI Technology and Circuits*, 2025.
https://doi.org/10.23919/VLSITechnologyandCir65189.2025.11074982

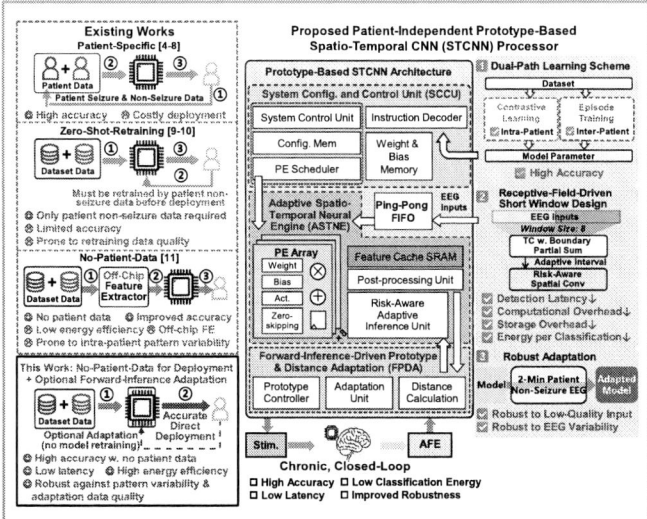

Figure 36.3.1: Limitations of existing works and contributions of this work (left); proposed patient-independent prototype-based STCNN processor.

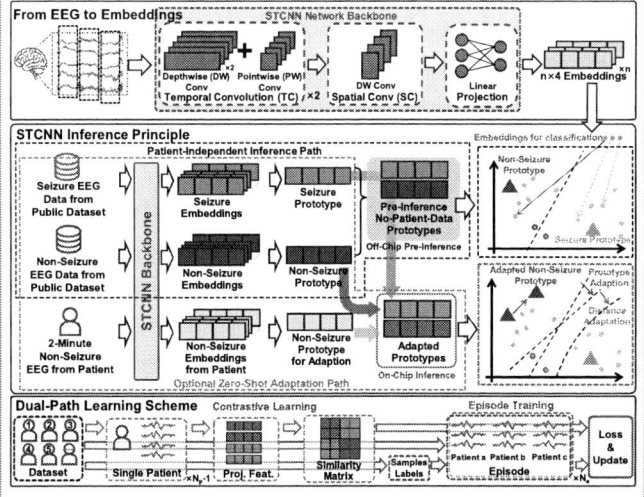

Figure 36.3.2: Principle of the processor with dual-path learning scheme, and optional forward-inference-driven prototype and distance adaptation.

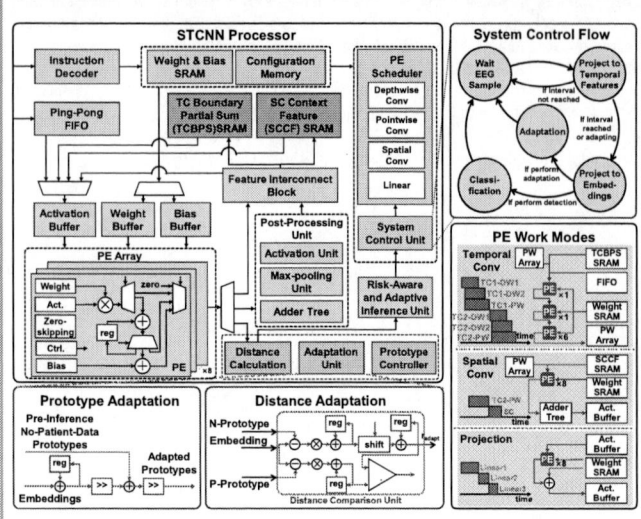

Figure 36.3.3: STCNN processor design.

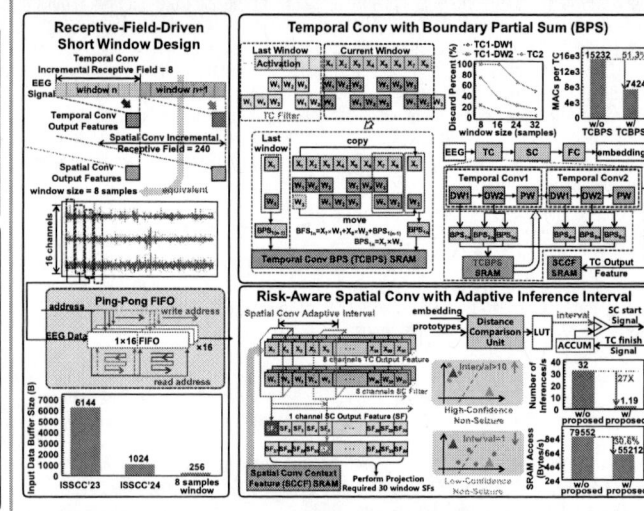

Figure 36.3.4: Receptive-field-driven short window design, with techniques for energy-efficient computations of temporal and spatial convolution.

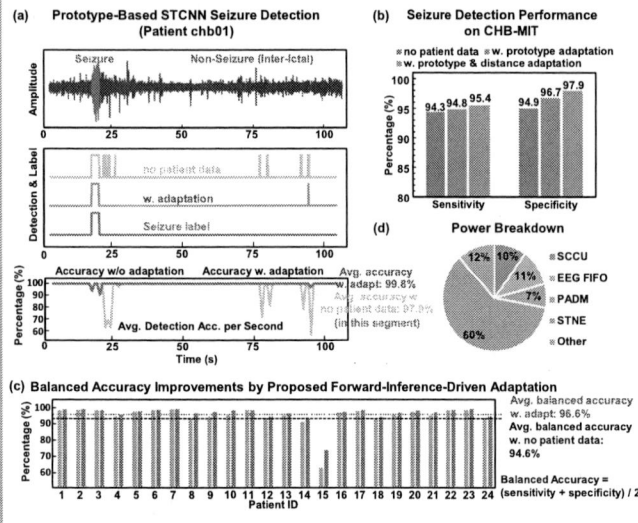

Figure 36.3.5: Measurement results.

	VLSI'21 [4]	ISSCC'22 [5]	VLSI'25 [8]	VLSI'25 [6]	ISSCC'23 [9]	ISSCC'24 [10]	VLSI'25 [11]	This Work
Technology (nm)	40	65	65	28	40	55	65	40
Voltage (V)	1.1 (A) / 0.7 (D)	1.2	0.54-1.2	0.75	0.9	0.68	0.8-1.2	0.58
Frequency (MHz)	1	0.128	0.1 - 10	-	4	2.5	-	2
SRAM (kB)	134	3.17	54	0.091	151.13	8	5.248	2.736
Area (mm²)	1.53#	3.48	2.6	0.075	1.56#	0.98	0.78	0.225
Input Type	Raw EEG	Raw EEG	Raw EEG	Raw EEG	Raw EEG	Raw EEG	Preprocessed FFT	Raw EEG
No. of EEG Channels	16	256	16	64	22	16	19	16
Dataset	CHB-MIT	CHB-MIT	CHB-MIT	CHB-MIT	CHB-MIT / EU	CHB-MIT	CHB-MIT / TUSZ	CHB-MIT
Percent and Type of Test Patient's Data Used for Training	Seizure & Non-Seizure Data					Only Non-Seizure Data		
	0.35	0.8	0.3	0.3	0.1% / 0.1%	0 / 0.05%	0 / 0	0 / 0.1%
Sensitivity (%)	97.8	95.6	99.0	94.93	90.3 / 90.4	78.5 / 90.2	93.07 / 91.63	94.3 / 95.4
Specificity (%)	99.5	96.8	NA	99.55	93.6 / 95.7	74.6 / 94.8	94.23 / 97.04	96.7 / 97.9
Latency (s)	0.74	<1	3.52	-	8.3 / 17.0	0.38	0.0000247*	0.116 / 0.119
Power (µW)	-	453	9.98-843.7	6.76	>1000	25.8	-	23.08
Energy (µJ/class)	0.97##	0.227##	3.57*	0.0085	28.33## / 21.63##	0.072*	4.568	0.016*

\# Calculated from the reported area and ratio.
\## Including AFE and processor, computed as power / sampling rate.
* Computed as power × classification time (0.71ms in this work).
* Computational latency, not detection latency.

Figure 36.3.6: Comparison with the state-of-the-art seizure-detection processors.

Specifications	
Technology	40 nm 1P9M CMOS
Core Area	0.225 mm²
SRAM Size	2.736 kB
Supply Voltage	0.58 V
Frequency	2 MHz
Power Consumption	23.08 µW
Energy / Classification	16.4 nJ

Figure 36.3.7: Chip micrograph.

ISSCC 2026 / SESSION 36 / NEURAL AND BIOMEDICAL INTERFACES / 36.4

36.4 ANP-OT: A 0.17nW/Synapse 0.46pJ/SOP Neuromorphic Olfactory Processor with On-Chip Transfer Learning for Non-Invasive Cross-Hospital Cross-Pulmonary-Disease Diagnosis

Dexuan Huo*[1], Ziyi Cheng*[1], Jilin Zhang[1], Yumeng Jiang[2], Lushuo Zhang[3], Hui Wang[3], Na Ma[2], Zebin Huang[2], Minggui Lin[2], Yunxia Zhao[3], Zhihua Wang[1], Kea-Tiong Tang[4], Hong Chen[1]

[1]Tsinghua University, Beijing, China, [2]Beijing Tsinghua Changgung Hospital, Beijing, China, [3]Hebei Medical University Third Hospital, Shijiazhuang, China
[4]National Tsing Hua University, Hsinchu, Taiwan
*Equally Credited Authors (ECAs)

Abstract

This paper reports an olfactory processor with on-chip transfer learning for cross-hospital and cross-pulmonary-disease diagnosis, which achieves >94% pulmonary disease detection accuracy at 3 different hospitals. This work achieves a 0.46pJ/SOP energy efficiency, and a power density of 0.17nW/synapse.

Many pulmonary diseases have subtle or easily missed symptoms in their early stages [1]. However, the breath composition of early-stage pulmonary disease patients differs from that of healthy individuals [1]. Such differences lead to variations in the response of a multi-channel gas sensor array, generating measurable data deviations. These deviations can be recognized by olfactory processors in electronic nose (E-nose) systems [2-4], enabling early detection of pulmonary diseases. To obtain satisfactory recognition accuracy, some existing olfactory processors incorporate on-chip learning capabilities [2-5]. However, these processors require retraining when deployed in unfamiliar environments and struggle to generalize across different diseases and operational settings. This limits the adaptability of E-nose systems to diverse heterogeneous clinical environments and patient populations. As a result, even for the same pulmonary disease, the recognition accuracy of olfactory processor can vary considerably across different hospitals. Furthermore, these processors lack cross-disease generalization capability, hindering their utility in detecting other types of pulmonary diseases. Additionally, on-chip learning accounts for over 58% of the total power of the chips [3-6], imposing a significant power burden for olfactory processors that need real-time learning to maintain recognition accuracy, as illustrated in Fig. 36.4.1 (middle).

To overcome these challenges, we develop a neuromorphic gas-recognition approach and introduce ANP-OT, an asynchronous neuromorphic olfactory processor, which is the first gas recognition chip with on-chip transfer learning to our knowledge. As shown in Fig. 36.4.1 (bottom), it exploits: 1) bio-inspired SNN with lateral inhibition and dopamine regulation to achieve rapid transfer learning across 3 different hospitals and 2 different types of pulmonary disease, and 2) an asynchronous reconfigurable synaptic weight update mechanism with operator fusion to effectively reduce training power by 62% for synaptic weight update and training power overhead by 76.5 to 86.0%. These key features enable ANP-OT diagnosing pulmonary diseases with 90.7% accuracy at one hospital, achieved by transfer learning gas samples three years ago from a different hospital. ANP-OT also supports few-shot on-chip incremental learning. When combined with transfer learning, it further improves diagnostic accuracy by 4.9 to 15.7%. Also, ANP-OT achieves accuracies of 100% and 87.0% for 10-class and 20-class volatile organic compound (VOC) gases, respectively, using one-shot on-chip incremental learning. The chip has a peak efficiency of 0.46pJ/SOP.

Figure 36.4.2 illustrates the system diagram of the ANP-OT with a three-recurrent-layer SNN topology, and computes the network dynamics in a time-stepped approach. The ANP-OT is composed of four parts: an excitatory layer with 128 excitatory cores, an inhibitory layer with 128 inhibitory cores, a dopamine layer with 32 dopamine cores, and a readout circuit. Within each excitatory core, there is a timestamp generator, a spike generator, and four excitatory neurons (ENs), an SRAM to store excitatory synaptic weights and corresponding addresses, and a weight update circuit. Similarly, each inhibitory core mainly contains an inhibitory timestamp generator, synapse address generator, five LIF inhibitory neurons (INs), and an SRAM to store inhibitory synaptic weights and addresses. Each dopamine core includes synapse address generator, five LIF dopamine neurons (DNs), and an SRAM to store dopamine synaptic weights and addresses. The readout circuit is used to compute the similarity of the spike times of ENs between the test gas and all the trained gases, the category with the greatest similarity is regarded as the inference recognition result.

Figure 36.4.3 presents the SNN enabling lateral inhibition and dopamine regulation that embeds feedforward and recurrent feedback motifs. Every EN together with three INs is regarded as a glomerulus. Once an EN spikes, the weight of excitatory synapses increases, and the spike activity level of the connected inhibitory neurons within a glomerulus arises. When one IN spikes, the weight of inhibitory synapses is reduced, and the spike activity level of connected excitatory neurons in other glomeruli falls. As a result, the difference in spike behavior between ENs is increased, thus a "winner-takes-all" (WTA) effect is achieved. However, during the learning process of similar gases, the SNN may erroneously produce the same winner neuron for both, failing to distinguish between them. To address this issue and avoid the wrong WTA effect, we propose a reward mechanism based on dopamine regulation. This mechanism rewards correct spiking activities of neurons and penalizes incorrect ones based on teaching signals. Therefore, a correctly guided WTA effect is

achieved, thereby realizing >98% accuracy for 20-class recognition with few-shot learning. Furthermore, the SNN model supports transfer learning across different hospitals and diseases. By leveraging gas samples from pulmonary patients at one hospital, ANP-OT achieves an accuracy of 83.6% in diagnosing a different pulmonary disease at another hospital. Subsequently, through incremental learning with a few gas samples from the local hospital, the diagnostic accuracy reaches 91.8% with 8.2% improvement, as shown in Fig. 36.4.3 (bottom right).

Figure 36.4.4 (top left) shows an asynchronous reconfigurable synaptic weight update (ARWU) circuit to achieve efficient update calculation. First, we decompose the weight update function into different calculation operations, each of which is controlled by a control code. The control code (Ctrl. Code) consists of three parts: object selection (O.s.), current calculation (C.c.), and result storage (R.s.). According to the O.s. of the Ctrl. Code, the mux circuit selects the input variables for calculation, and the selected variables are transmitted into the multi-operator fusion circuit to perform calculation according to the C.c. in the Ctrl. Code. If the R.s. in the Ctrl. Code is '1', the calculation result is transmitted into the mux circuit for the next computation until the weight update process is completed. Otherwise, the current calculation result is regarded as the final updated synapse weight. We find the number of control codes determines the training time for a weight update calculation. To reduce the number of control codes, we propose an asynchronous calculation reconstruction representation (ACR) to reorganize the weight update calculation order, enabling independent calculations to be performed simultaneously, thereby reducing the computation cycles by 69.6%. Furthermore, an asynchronous multi-operator fusion (AMOF) circuit is proposed to implement multiple operation paradigms into one operator, as depicted in Fig. 36.4.4 (bottom left). We reconstruct the CORDIC operator and divide it into multiple sub operators, each of which performs simple operations (such as addition, subtraction, multiplication, division, comparison, and so on). These sub operators can operate independently or be dynamically invoked as needed. As a result, multiple operations can be executed in parallel and the entire CORDIC operator circuit is used to perform complex operations such as exponentiation and logarithm. Benefitting from ARWU mechanism with AMOF and ACR, the power consumption for weight update is reduced by 62.0%. Figure 36.4.4 (bottom right) shows that the asynchronous update weight skip mechanism (AUWS) skips 64.8% of redundant computations caused by the uneven convergence speed of synaptic weights during the training process. Additionally, drawing on the observation that synaptic weights tend to polarize during training in winner-take-all SNN models, we have developed an asynchronous weight random compression (AWRC) method. Without losing compression precision for the minimum and maximum synapse weights, this method compresses 8b synapse weights into 2b codes, thus obtaining a 75% reduction in synapse weight storage.

Figure 36.4.5 (top left) shows ANP-OT achieves accuracies of 97.7%, 94.0% and 96.4% in identifying pulmonary diseases by analyzing the gas samples from patients at three different hospitals across diverse areas with on-chip learning respectively. Compared to other algorithms, our chip exhibits an improvement of over 12.4% in accuracy and 7.5% in true positive rate (TPR). Notably, ANP-OT supports on-chip transfer learning across different hospitals. As shown in Fig. 36.4.5 (bottom left), through leveraging transfer learning from gas samples collected in hospital A, ANP-OT achieves an accuracy of 90.7% in identifying pulmonary diseases at hospital C. Based on that, on-chip incremental learning of ANP-OT with a few gas samples from hospital C improve the accuracy rate from 90.7% to 95.6%—only 0.8% lower than the recognition accuracy of 96.4% achieved through local learning in hospital C. This combination of transfer and incremental few-shot learning enables ANP-OT to quickly adapt to new clinical environments with minimal additional training. Moreover, ANP-OT demonstrates robust cross-hospital and cross-disease transfer learning capability. As shown in Fig. 36.4.5 (bottom middle), learning gas samples from pneumonia and bronchial asthma patients at hospital B, ANP-OT attains an accuracy of 82.2% in diagnosing lung cancer at hospital C. After transfer learning, incremental learning with a few local gas samples from hospital C enhances the accuracy to 95.5%. As shown in Fig. 36.4.5 (bottom), the accuracy loss associated with transfer learning across three hospitals and two types of pulmonary diseases is limited to only 0.8 to 5.9% when complemented with on-chip incremental few-shot learning.

ISSCC 2026 / February 18, 2026 / 2:45 PM

Also, using a 20-class volatile organic compounds (VOCs) dataset [7], ANP-OT shows 87% average accuracy with on-chip one-shot learning, outperforming other gas recognition algorithms by over 24.7%. On a 10-class gas sensor sampling dataset [8], it obtains 100% accuracy with one-shot incremental on-chip learning, even under 60% noise, and surpasses the accuracies reported in [2-4] by over 2%, as presented in Fig. 36.4.5 (bottom right). Importantly, the overall accuracy remains stable when sequentially training on all 10 gases, presenting robust one-shot incremental on-chip learning ability [9].

Figure 36.4.6 compares the olfactory processor with other state-of-the-art works. Benefitting from the proposed ARWU with ACR and AMOF methods, AUWS and AWRC mechanisms, ANP-OT consumes only 10nJ/step, achieving 1.39 to 2.64× energy saving compared with that in [3-5], respectively. The training power overhead of the ANP-OT is only 7.5%, which is 90% lower than that in [6]. Figure 36.4.7 shows a die microphotograph and summary. ANP-OT only consumes 7.5 to 9.1nJ/step for inference, which is 45.3% lower than that of [3], and the power density of our processor is 83.7 to 95.8% lower than that of [3-4]. Moreover, the peak energy efficiency is 0.46pJ/SOP, which outperforms 2.3 to 8.1× than that of [3-4].

Acknowledgement:
This work was supported by the National NSFC (Grant No.: 62574117 and 62334014), and partly by National Key R&D Program of China 2018AAA0103100. The corresponding author of this paper is Hong Chen (hongchen@tsinghua.edu.cn).

References:
[1] Chan et al. Engineering synthetic breath biomarkers for respiratory disease. *Nat. Nanotechnol.* 15, 792–800 (2020). https://doi.org/10.1038/s41565-020-0723-4
[2] N. Imam and T. A. Cleland, "Rapid online learning and robust recall in a neuromorphic olfactory circuit," *Nature Machine Intelligence*, vol. 2, no. 3, pp. 181-191, 2020. https://doi.org/10.1038/s42256-020-0159-4
[3] D. Huo et al., "ANP-G: A 28-nm 1.04-pJ/SOP Sub-mm² Asynchronous Hybrid Neural Network Olfactory Processor Enabling Few-Shot Class-Incremental On-Chip Learning," in *IEEE Journal of Solid-State Circuits*, vol. 60, no. 7, pp. 2660-2670, July 2025. https://doi.org/10.1109/jssc.2025.3530513
[4] D. Huo et al., "20.2 A 67µW/Channel, 0.13nW/Synapse/b Nose-on-a-Chip for Noninvasive Diagnosis of Diseases with On-Chip Incremental Learning," *ISSCC*, 2025, pp. 350-351. https://doi.org/doi:%2010.1109/ISSCC49661.2025.10904713
[5] Y. Zhong et al., "PAICORE: A 1.9-Million-Neuron 5.181-TSOPS/W Digital Neuromorphic Processor With Unified SNN-ANN and On-Chip Learning Paradigm," in *IEEE Journal of Solid-State Circuits*, vol. 60, no. 2, pp. 651-671, Feb. 2025. https://doi.org/10.1109/JSSC.2024.3426319
[6] C. Frenkel and G. Indiveri, "ReckOn: A 28nm Sub-mm² Task-Agnostic Spiking Recurrent Neural Network Processor Enabling On-Chip Learning over Second-Long Timescales," *ISSCC*, 2022, https://doi.org/10.1109/ISSCC42614.2022.9731734
[7] CSIRO; A. Berna and S. Trowell, (2015): Electronic nose (FOX) recording of 20 chemicals. v1. CSIRO. Data Collection. https://doi.org/10.4225/08/552C4424EE51E
[8] A. Vergara et al., "On the performance of gas sensor arrays in open sampling systems using Inhibitory Support Vector Machines," *Sensors and Actuators B: Chemical*, vol. 185, pp. 462-477, 2013. https://doi.org/10.1016/j.snb.2013.05.027

[9] D. Huo et al., "A Bio-Inspired Spiking Neural Network with Few-Shot Class-Incremental Learning for Gas Recognition" *Sensors* 2023, 23, 2433. https://doi.org/10.3390/s23052433

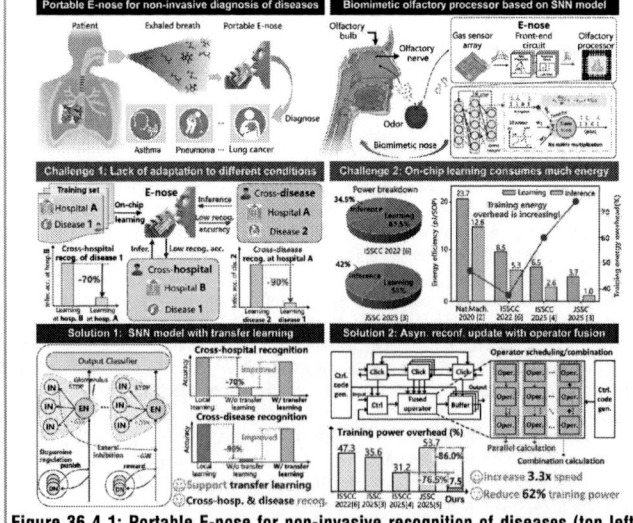

Figure 36.4.1: Portable E-nose for non-invasive recognition of diseases (top left), olfactory processor (top right), challenges (middle), and our solutions (bottom).

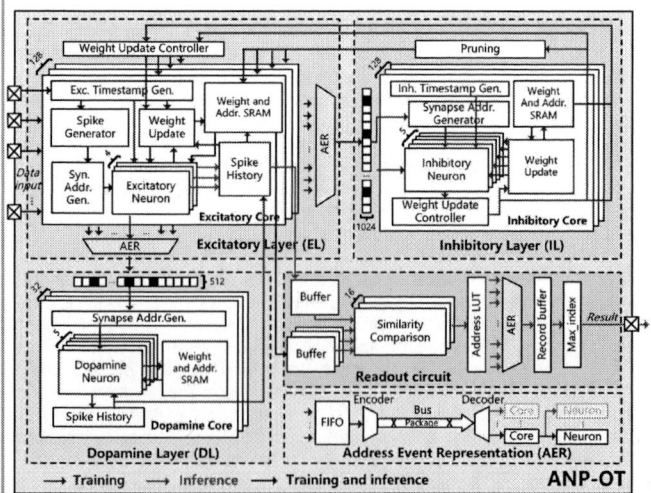

Figure 36.4.2: System diagram of the ANP-OT with on-chip transfer learning, which includes excitatory, inhibitory, and dopamine layers, and readout circuit.

DIGEST OF TECHNICAL PAPERS • 607

979-8-3315-8937-0/26 $31.00 © 2026 IEEE

ISSCC 2026 / SESSION 36 / NEURAL AND BIOMEDICAL INTERFACES / 36.4

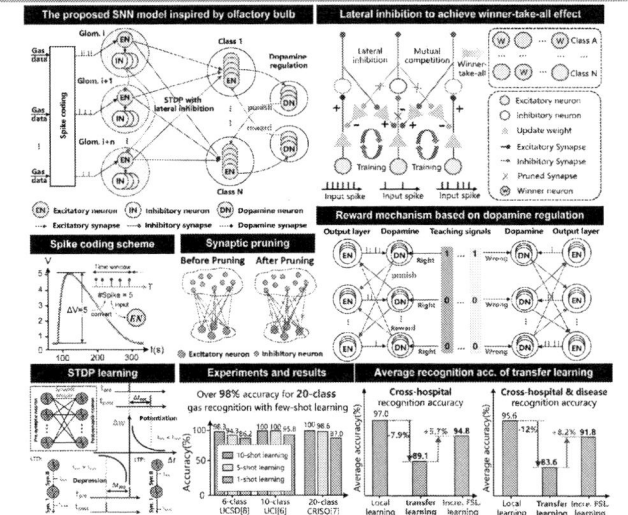

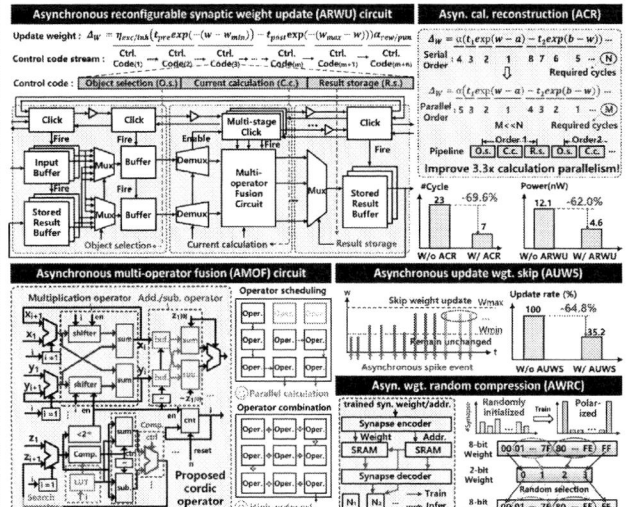

Figure 36.4.3: SNN model (left) with lateral inhibition (top right) and dopamine regulation (middle right), and recognition accuracy (bottom right).

Figure 36.4.4: ARWU approach (top left) with ACR (top right) and AMOF (bottom left), and AUWS (middle right) and AWRC (bottom right) mechanism.

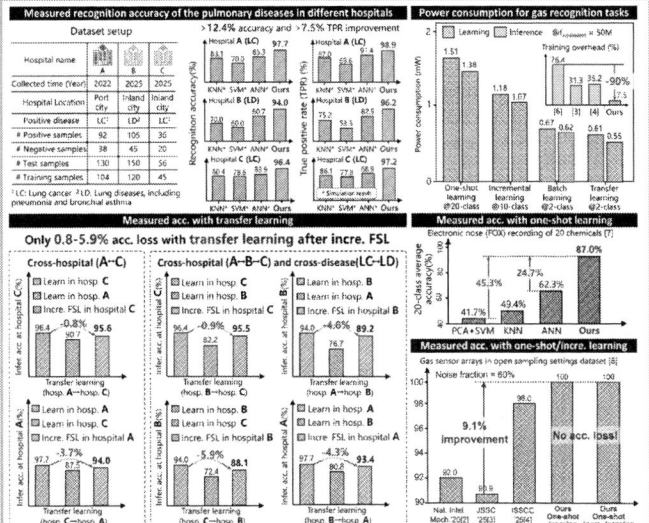

Figure 36.4.5: Benchmarking and measurement results.

	Nat. Mach. Intell.'2020[2]	JSSC'2025[3]	ISSCC'2025[4]	This work
Process	14nm FINFET	28nm CMOS	22nm CMOS	22nm CMOS
Learning engine	SNN	SNN+BPNN	SNN	SNN
On-chip learning	One-shot learning	One/few-shot learning	One/few-shot Incremental Batch learning	One/few-shot learning Incremental learning Batch learning Transfer learning
Voltage(V)	0.75	0.55	0.5	0.7
Size	60mm²	0.63×0.63mm²	1.4×1.8mm²	2.0×1.2mm²
Memory	960kB	2.1kB	58kB	125kB
# Supported channels	72	24	72	128
# Supported identifiable gas	10	10	10	20
Accuracy	92%@10-class VOCs[a]	90.90%@10-class VOCs[a]	98.0%@10-class VOCs[a] 97.7%@2-class LCs[1]	100%@10-class VOCs[a] 87%@20-class CSIRO[b]	97.7%@2-class LCs[1] 94.0%@2-class LDs[2] 96.4%@2-class LCs[3]
Training overhead	N/A	0.35	0.31	0.075
Energy per step (learning/infer.)	N/A	VOCs[a]:13.9nJ/6.3nJ	VOCs[a]:26.4nJ/23.8nJ LCs[1]: 15.6nJ/13.7nJ	VOCs[a]:10nJ/9.5nJ CSIRO[b]:35nJ/32nJ	LCs[1]: 8nJ/7.5nJ LDs[2]: 9.5nJ/8.8nJ LCs[3]: 10.2nJ/9.1nJ
Accuracy loss with transfer learning	N/A	N/A	N/A	0.8~5.9%
Peak energy efficiency	23.6pJ/SOP	1.04pJ/SOP	3.72pJ/SOP	0.46pJ/SOP
FOM[c]: Power density	19.52μW/Synapse	4.08nW/Synapse	1.04nW/Synapse	0.17nW/Synapse

[a] 72-sensor 10-class dataset of VOCs including 10 training samples and 1000 test samples. The test samples are obtained by adding 60% noise to the training samples through data augmentation. [b] 12-sensor 20-class dataset of VOCs including 200 trainings samples and 200 test samples. [1] 14-sensor 2-class dataset of lung cancer patients (92 samples) and normal people (38 samples) in hospital A. [2] 14-sensor 2-class dataset of lung diseases patients (105 samples) and normal people (45 samples) in hospital B. [3] 14-sensor 2-class dataset of lung cancer patients (36 samples) and normal people (20 samples) in hospital C. [c] Power density = (Running power)/(Number of synapses), representing power consumed on each synapse for eliminating the workload dependence.

Figure 36.4.6: Comparison with prior state-of-the-art gas-recognition chips.

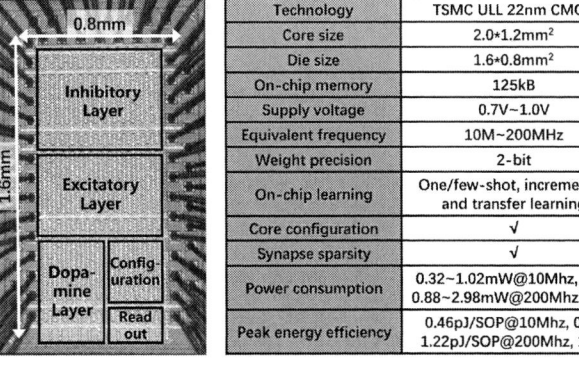

Technology	TSMC ULL 22nm CMOS
Core size	2.0×1.2mm²
Die size	1.6×0.8mm²
On-chip memory	125kB
Supply voltage	0.7V~1.0V
Equivalent frequency	10M~200MHz
Weight precision	2-bit
On-chip learning	One/few-shot, incremental, and transfer learning
Core configuration	√
Synapse sparsity	√
Power consumption	0.32~1.02mW@10Mhz, 0.7V 0.88~2.98mW@200Mhz, 1.0V
Peak energy efficiency	0.46pJ/SOP@10Mhz, 0.7V 1.22pJ/SOP@200Mhz, 1.0V

Figure 36.4.7: Die micrograph and summary table.

• 2026 IEEE International Solid-State Circuits Conference

979-8-3315-8937-0/26 $31.00 © 2026 IEEE

ISSCC 2026 / SESSION 36 / NEURAL AND BIOMEDICAL INTERFACES / 36.5

36.5 A 185.6dB-FOM$_{DR}$ 180.3dB-FOM$_{SNDR}$ 10.64-NEF NS-SAR-ADC with Calibration-Free 2nd-Order kT/C-Noise Shaping for Wearable ExG Acquisition

Geunha Kim, Kyuhyeon Park, Minoo Lee, Jeongyoon Wie, Seokhan Jeong, Yeonjae Shin, Kyoungtae Lee, Jong-Hyeok Yoon, Minyoung Song, Junghyup Lee

Daegu Gyeongbuk Institute of Science and Technology, Daegu, Korea

Abstract

To enable high-fidelity wearable ExG acquisition, this paper presents an NS-SAR ADC that performs 2nd-order shaping of both quantization and kT/C noises through a multi-tasking integrator (MTI) and loop dynamics, without calibration. In addition, correlated-level-shifting (CLS) and Z_{IN}-boosting techniques are used to enhance linearity and Z_{IN}, respectively. Fabricated in 0.18μm CMOS, the ADC achieves 98.3dB SNDR, 103.6dB DR, 3.49μV$_{rms}$ IRN over a 1kHz BW, and Z_{IN} of 168MΩ while consuming 6.3μW.

High-resolution and low-power ExG (ECG, EEG, EMG, and EOG) acquisition systems are essential for emerging wearable healthcare technologies. In wearable environments, motion-induced electrode-skin impedance variation introduces large artifacts. Therefore, these systems must provide a wide input range (IR) exceeding 1V$_{pp}$, achieve input-referred noise (IRN) around 3μV$_{rms}$ within a 1kHz bandwidth, and maintain input impedance (Z_{IN}) above 100MΩ [1-12]. In addition, to enable long-term battery operation, they must deliver exceptional energy efficiency (FOM$_{SNDR}$ >180dB, FOM$_W$ <50fJ/conv.) while consuming less than 10μW [9,10]. However, meeting all these requirements within a single architecture remains highly challenging due to fundamental trade-offs.

Conventional ExG acquisition systems primarily employ continuous-time ΔΣ ADCs (Fig. 36.5.1 top-left) [2-8]. Such architectures readily achieve low IRN and suppress aliasing by using a low-noise amplifier (LNA) or integrator. In addition, because they are inherently immune to kT/C noise at the input, they require only a small input capacitance (C_{IN}) while still maintaining high Z_{IN} through impedance-boosting techniques. However, the LNA and integrator fundamentally limit the IR (<0.5V$_{pp}$), thereby restricting their use in wearable devices and also degrading the achievable dynamic range (DR) and energy efficiency. Alternatively, discrete-time architectures based on noise-shaping SAR (NS-SAR) ADCs (Fig. 36.5.1 top-right) have attracted significant attention for ExG monitoring thanks to their wide IR and high energy efficiency [9,10]. Nevertheless, due to the sampling operation at the input, either a large sampling capacitance (C_{SAR}) or a high oversampling ratio (OSR) is required to meet the target IRN under kT/C noise (V_{sn}) constraints. As a result, this not only increases power consumption but also reduces the effective Z_{IN}.

To overcome such kT/C noise bottlenecks in discrete-time ADCs, several approaches have been proposed, as shown in Fig. 36.5.1 (middle). Pipelined-SAR ADCs employ double-sampling techniques combined with pre-amplifiers to capture kT/C noise (V_{sn1}) and subsequently cancel it by cascading the pre-amplifier with an inter-stage amplifier [13-16]. For NS-SAR ADCs, such cancellation techniques are not easily applicable, since the kT/C noise stored in the comparator directly appears as an all-pass noise component at the digital output [17]. To address this issue, prior work [17] attempted to cancel this noise before the comparison operation. Despite these attempts, the techniques employed in both pipelined-SAR and NS-SAR ADCs remain highly sensitive to gain errors, and thus precise calibration is required, which increases the overall complexity and cost of the ADC.

This paper presents an NS-SAR ADC employing a multi-tasking integrator (MTI), as shown in Fig. 36.5.1 (bottom-right). The MTI provides low IRN by simultaneously shaping quantization (Q_n) and kT/C noise (V_{sn}). Specifically, it pre-stores V_{sn} and combines it with Q_n, thereby shaping both noise sources into 2nd-order through the loop dynamics of the ADC. As a result, in-band kT/C noise is effectively suppressed even with a smaller C_{SAR} while robustness across PVT variations is maintained without any calibration. Moreover, the 2nd-order shaping is achieved using only a single amplifier in the MTI, improving energy efficiency by eliminating the need for an extra stage and its static power. To further enhance performance, a correlated level shifting (CLS) scheme is introduced to mitigate linearity degradation from limited amplifier gain [18,19]. Finally, a Z_{IN} boosting technique that restores the previous input sample onto the C_{SAR} through simple switch control is employed [9-12]. With these techniques, the architecture represents an NS-SAR ADC that meets all the aforementioned requirements for long-term ExG acquisition applications.

Figure 36.5.2 (top) illustrates the overall architecture of the 2nd-order NS-SAR ADC, which consists of a 5b SAR quantizer and the MTI. The system operates in a periodic sequence of input sampling (φ_{s1}, φ_{s2}), kT/C noise storage (φ_{sns}), SAR conversion (φ_{conv}), residue and kT/C noise integration (φ_{int1}, φ_{int2}), and Z_{IN} boosting (φ_{ib}) at a sampling frequency of 128kHz. The single-ended sampling capacitor array (C_{SAR}) has 4pF, implemented with 32 unit capacitors (C_U[31:0]). The input stage uses bottom-plate sampling with a double-sampling scheme, enabling the MTI to pre-store kT/C noise on a dedicated capacitor. CLS is further applied during the 1st integration phase to enhance linearity. To mitigate DAC mismatch errors and non-idealities such as MTI offset and 1/f noise, the system employs dynamic element matching (DEM) and system-level chopping [9]. The chopping frequency is programmable to either one-half or one-quarter of the sampling frequency to maximize Z_{IN}. In addition, Z_{IN} is further enhanced by a boosting technique in which the previous digital

output (D_{out}) is fed back to the C_{SAR}, applying opposite polarity to the top-plate and resetting the bottom-plate in synchronization with the chopping frequency.

To explain the kT/C noise-shaping technique, the z-domain model of the ADC is shown in Fig. 36.5.2 (bottom). The ADC is based on a cascade-of-integrators with feed-forward (CIFF) structure, which is more robust against PVT variations than the EF structure [17,20]. The MTI sequentially performs the kT/C noise storage phase, 1st-integration phase, and 2nd-integration phase through a shared amplifier, thereby saving both power and area. Each residue is passively summed with the input before being processed by the comparator to generate the D_{out}. The sampled kT/C noise (V_{sn}) is integrated together with quantization noise (Q_n) during the 1st integration phase. In the 2nd integration phase, the stored V_{sn} is reset, thereby preventing further accumulation of kT/C noise. As a result, the final D_{out} exhibits 2nd-order noise shaping of both Q_n and V_{sn}.

Figure 36.5.3 shows the detailed operations of the MTI with a simplified single-ended topology for clarity. Figure 36.5.3 (top) illustrates the double-sampling process for storing kT/C noise. During the 1st sampling phase (φ_{s1}), the input voltage (V_{in}) is stored on the bottom plate of the C_{SAR}, while V_{sn} from the 1st sampling switch (φ_{s1}) remains on the top plate of the C_{SAR}. In the subsequent kT/C noise storage phase (φ_{sns}), the storage amplifier (A_{SNS}) amplifies the retained $V_{sn}[n]$, while the residue amplifier (A_R) is disabled for power saving. Finally, at the end of this phase ($t = t_2$), $V_{sn}[n]$ is stored in the storage capacitor (C_{SNS}) before the SAR conversion begins. To prevent A_{SNS} output saturation during the φ_{sns} phase, its gain is set to 5× (~14dB), with a sampling time (t_{sns}) of ~ 250ns.

After the SAR conversion, 2nd-order integration of the residue and kT/C noise is performed, as shown in Fig. 36.5.3 (bottom). The residue voltage on the top plate of C_{SAR} (-Q_n[n]-V_{int1}[n-1]-V_{int2}[n-1]) is integrated by the MTI. Notably, this voltage contains no present kT/C noise. During the 1st integration phase (φ_{int1}), the residue voltage is integrated with the kT/C noise $V_{sn}[n]$ from C_{SNS} (-A_{SNS}·$V_{sn}[n]$) stored during the storage phase. As a result, the 1st residue integration output includes both $V_{sn}[n]$ and $Q_n[n]$. In this phase, both the A_{SNS} and A_R are enabled to improve integration accuracy. To further enhance the MTI open-loop gain, the CLS technique is applied to force the MTI output to return to a common-mode voltage, effectively boosting the open-loop gain. Consequently, the final output at the 1st integration phase, $V_{int1}[n]$, more closely approaches the ideal value compared to the pre-CLS output, $V_{int1}[n]$. During the following 2nd integration phase (φ_{int2}), the 1st integration output across the C_{RS} is integrated into C_{INT2}, which is connected in feedback with the shared amplifier used in φ_{int1}. In this phase, C_{SNS} is reset to prevent the accumulation of additional kT/C noise. Therefore, the system achieves 2nd-order noise shaping, where both Q_n and V_{sn} are shaped.

In practical operation, a portion of the input voltage (ΔV_{in}) is also stored in C_{SNS} during the double-sampling process, as shown in Fig. 36.5.3 (bottom-right). However, the signal transfer function for ΔV_{in} remains all-pass within the bandwidth. As a result, the final digital output corresponds to a linear representation of $V_{in}(t_2)$.

The NS-SAR was implemented in a 0.18μm standard CMOS process. It consumes 6.3μW from a 1V supply and occupies an active area of 0.16mm^2 (Fig. 36.5.7). The measured output power spectral density (Fig. 36.5.4 top-left) demonstrates the effect of kT/C noise shaping in combination with the system-level chopping technique. With a 1.5V$_{pp}$ input at 75Hz, the ADC achieves an SNDR of 98.3dB and an SFDR of 105dB when all techniques are enabled, showing clear improvements compared to the case without these enhancements. The signal bandwidth extends to ~1kHz, beyond which the noise power increases at 40dB/decade, indicating 2nd-order noise-shaping behavior. Under full operation, the peak SNDR of 98.3dB corresponds to a dynamic range of 103.6dB (Fig. 36.5.4 top-right).

The measured input-referred noise over a 1kHz BW is 3.49μV$_{rms}$ with all techniques applied, showing an improvement of 2.25μV$_{rms}$ compared to the case without them (Fig. 36.5.4 bottom-left). Within a 100Hz BW, it is reduced to 1.24μV$_{rms}$, making it suitable for EEG acquisition. Robustness against supply and temperature variations was also verified, with SNDR deviations limited to <1.3dB across a supply range of 0.94 to 1.18V and <0.5dB across a temperature range of -40 to 60°C (Fig. 36.5.4 bottom-right).

608 • 2026 IEEE International Solid-State Circuits Conference

979-8-3315-8937-0/26 $31.00 © 2026 IEEE

Figure 36.5.5 (top-left) shows the measured input impedance, which exceeds 168MΩ at DC and 41.5MΩ at 1kHz. To validate the NS-SAR ADC in biopotential recording, EOG, EMG, EEG, and ECG signals were measured, and the results are shown in Fig. 36.5.5. In the EEG test (Fig. 36.5.5 bottom-left), spectrogram analysis clearly revealed alpha-band activity (8 to 12Hz) under the closed-eye condition. Furthermore, ECG measurements in the presence of large artifacts (Fig. 36.5.5 bottom-right) demonstrate that the ECG waveform can be successfully recovered after high-pass filtering.

Figure 36.5.6 summarizes the measured results and benchmarks them against recently presented works. The design achieves the best FOM$_{SNDR}$, FOM$_{DR}$, and FOM$_W$ in the table, while also demonstrating superior IRN and NEF, with Z_{IN} comparable to other ADCs. Overall, the combination of these features with state-of-the-art efficiency metrics highlights the NS-SAR ADC as a highly suitable solution for low-noise, low-power, high-resolution bio-potential acquisition systems.

Acknowledgement:
This work was supported by the NRF (No. RS-2024-00416319, RS-2025-02303581, RS-2024-00439307) and the IITP (No. RS-2025-02219277), funded by the MSIT. The chip fabrication was supported by the IC Design Education Center (IDEC).

References:
[1] K. Jeong et al., "A 3.47 NEF 175.2dB FOM$_S$ Direct Digitization Front-End Featuring Delta Amplification for Enhanced Dynamic Range and Energy Efficiency in Bio-Signal Acquisition," *ISSCC*, pp. 276-277, Feb. 2025.
https://doi.org/10.1109/ISSCC49661.2025.10904745
[2] H. Chandrakumar and D. Marković, "A 15.2-ENOB 5-kHz BW 4.5-μW Chopped CT ΔΣ-ADC for Artifact-Tolerant Neural Recording Front Ends," *JSSC*, vol. 53, no. 12, pp. 3470-3483, Dec. 2018. https://doi.org/10.1109/JSSC.2018.2876468
[3] C. Lee et al., "A 6.5μW 10kHz-BW 80.4dB-SNDR Continuous-Time ΔΣ Modulator with Gm-Input and 300mV$_{pp}$ Linear Input Range for Closed-Loop Neural Recording," *ISSCC*, pp. 410-411, Feb. 2020. https://doi.org/10.1109/ISSCC19947.2020.9063074
[4] S. Lee et al., "A 0.7V 17fJ/Step-FOM$_W$ 178.1dB-FOM$_{SNDR}$ 10kHz-BW 560mV$_{pp}$ True-ExG Biopotential Acquisition System with Parasitic-Insensitive 421MΩ Input Impedance in 0.18μm CMOS," *ISSCC*, pp. 336-337, Feb. 2022.
https://doi.org/10.1109/ISSCC42614.2022.9731114
[5] T. Seol et al., "A Hybrid Recording System with 10kHz-BW 630mV$_{pp}$ 84.6dB-SNDR 173.3dB-FOM$_{SNDR}$ and 5kHz-BW 114dB-DR for Simultaneous ExG and Biocurrent Acquisition," *ISSCC*, pp. 562-563, Feb. 2024.
https://doi.org/10.1109/ISSCC49657.2024.10454270
[6] S. Lee et al., "A 97dB-PSRR 178.4dB-FOM$_{DR}$ Calibration-Free VCO–ΔΣ ADC Using a PVT-Insensitive Frequency-Locked Differential Regulation Scheme for Multi-Channel ExG Acquisition," *IEEE Symp. VLSI Circuits*, Jun 2024.
https://doi.org/10.1109/VLSITechnologyandCir46783.2024.10631536
[7] J. Huang and P. P. Mercier., "A Distortion-Free VCO-Based Sensor-to-Digital Front-End Achieving 178.9dB FoM and 128dB SFDR with a Calibration-Free Differential Pulse-Code Modulation Technique," *ISSCC*, pp. 386-387, Feb. 2021.
https://doi.org/10.1109/ISSCC42613.2021.9365950
[8] C. Pochet et al., "A 400mV$_{pp}$ 92.3dB-SNDR 1kHz-BW 2nd-Order VCO-Based ExG-to-Digital Front-End Using a Multiphase Gated-Inverted Ring-Oscillator Quantizer," *ISSCC*, pp. 392-393, Feb. 2021. https://doi.org/10.1109/ISSCC42613.2021.9365985

[9] G. Kim et al., "A 1V-Supply 1.85V$_{PP}$-Input-Range 1kHz-BW 181.9dB-FOM$_{DR}$ 179.4dB-FOM$_{SNDR}$ 2nd-Order Noise-Shaping SAR-ADC with Enhanced Input Impedance in 0.18μm CMOS," *ISSCC*, pp. 484-485, Feb. 2023.
https://doi.org/10.1109/ISSCC42615.2023.10067844
[10] Y. Han et al., "An 81.7MΩ-Input-Impedance 179.5dB-FOM$_{SNDR}$ 1.8V$_{PP}$-Input-Range Noise-Shaping-SAR-Based Sensing Frontend with Dynamic Input-Impedance Boosting and Prediction-Assisted Mismatch-Shaping-DEM," *IEEE CICC*, Apr. 2025.
https://doi.org/10.1109/CICC63670.2025.10983204
[11] K. Jeong et al., "A 15.4-ENOB, Fourth-Order Truncation-Error-Shaping NS-SAR-Nested ΔΣ Modulator With Boosted Input Impedance and Range for Biosignal Acquisition," *JSSC*, vol. 59, no. 2, pp. 528-539, Feb. 2024. https://doi.org/10.1109/JSSC.2023.3300928
[12] R. Jiang et al., "A Rail-to-Rail Input NS-Pipelined-SAR ADC With Self-Boosted Input Impedance and Reused Residue Amplifier for Biosignal Acquisition," *JSSC*, 2025, Online early access. https://doi.org/10.1109/JSSC.2025.3590743
[13] R. Kapusta et al., "Sampling Circuits That Break the kT/C Thermal Noise Limit," *JSSC*, vol. 49, no. 8, pp. 1694-1701, Aug. 2014. https://doi.org/10.1109/JSSC.2014.2320465
[14] J. Liu et al., "A 13-bit 0.005-mm 2 40-MS/s SAR ADC With kT/C Noise Cancellation," *JSSC*, vol. 55, no. 12, pp. 3260-3270, Dec. 2020.
https://doi.org/10.1109/JSSC.2020.3016656
[15] M. Zhan et al., "A 0.004mm 2 200MS/s Pipelined SAR ADC with kT/C Noise Cancellation and Robust Ring-Amp," *ISSCC*, pp. 164-165, Feb. 2022.
https://doi.org/10.1109/ISSCC42614.2022.9731599
[16] J. Gao et al., "A 7.9fJ/Conversion-Step and 37.12aF$_{rms}$ Pipelined-SAR Capacitance-to-Digital Converter with kT/C Noise Cancellation and Incomplete-Settling-Based Correlated Level Shifting," *ISSCC*, pp. 346-347, Feb. 2023.
https://doi.org/10.1109/ISSCC42615.2023.10067383
[17] T.-H. Wang et al., "A 13.8-ENOB 0.4pF-C$_{IN}$ 3rd-Order Noise-Shaping SAR in a Single-Amplifier EF-CIFF Structure with Fully Dynamic Hardware-Reusing kT/C Noise Cancelation," *ISSCC*, pp. 374-375, Feb. 2021.
https://doi.org/10.1109/ISSCC42613.2021.9365990
[18] Y. Hu et al., "A 2.87μW 1kHz-BW 94.0dB-SNDR 2-0 MASH ADC Using FIA with Dynamic-Body-Biasing Assisted CLS Technique," *ISSCC*, pp. 410-411, Feb. 2022.
https://doi.org/10.1109/ISSCC42614.2022.9731544
[19] S. Zhang et al., "An 871 nW 96.2 dB SNDR Pipelined Second-Order Noise-Shaping SAR ADC Employing Charge-Efficient CLS-Assisted Residue Amplifier," *IEEE TCAS-I, vol. 72, no. 4, pp. 1510-1521, Apr. 2025. https://doi.org/10.1109/TCSI.2024.3465493
[20] X. Tang et al., "A 13.5-ENOB, 107-μW Noise-Shaping SAR ADC With PVT-Robust Closed-Loop Dynamic Amplifier," *JSSC*, vol. 55, no. 12, pp. 3248-3259, Dec. 2020.
https://doi.org/10.1109/JSSC.2020.3020194

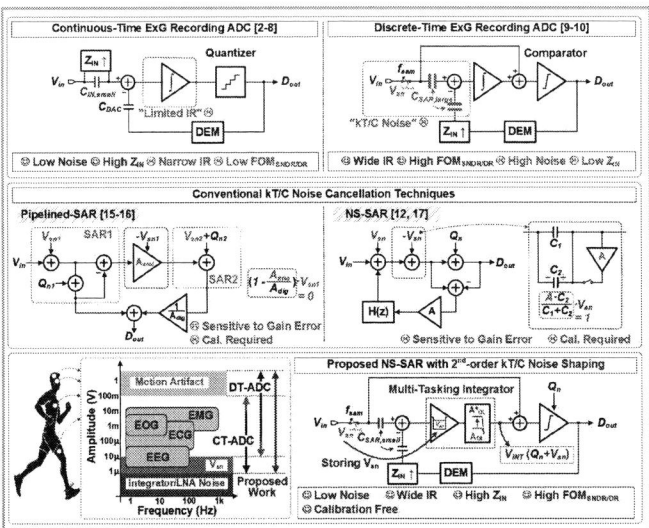

Figure 36.5.1: Conceptual overview of prior ExG recording ADCs (top), kT/C noise cancellation techniques (middle), and the proposed NS-SAR ADC (bottom).

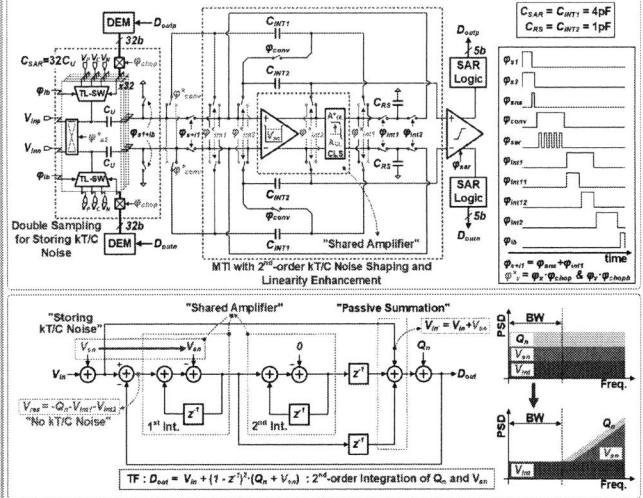

Figure 36.5.2: Proposed NS-SAR ADC architecture (top), timing diagram (top-right), and z-domain model (bottom).

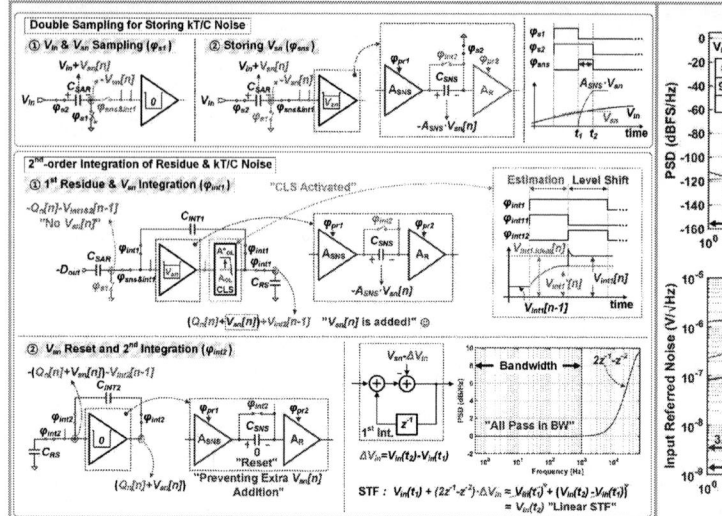

Figure 36.5.3: Operation principle of double sampling for kT/C noise storage (top) and 2nd-order integration of residue and kT/C noise (bottom).

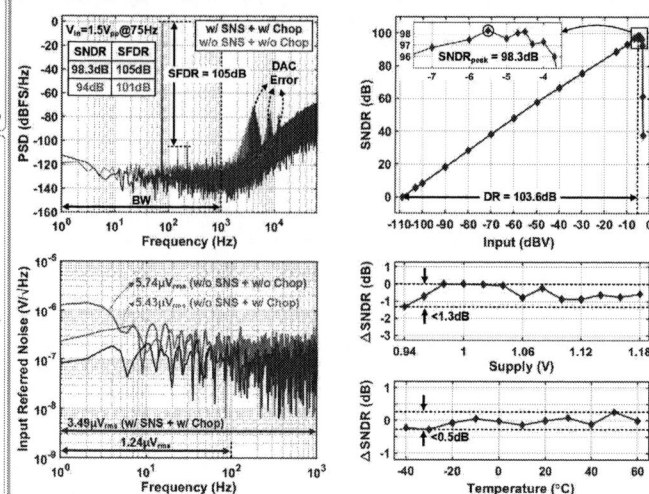

Figure 36.5.4: Measured output PSD (top-left), SNDR (top-right), input-referred noise (bottom-left), and SNDR variations across supply and temperature (bottom-right).

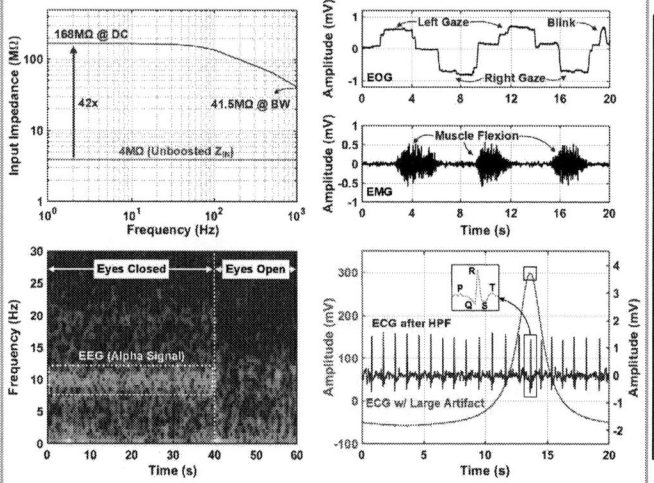

Figure 36.5.5: Measured input impedance (top-left) and ExG measurements: EOG/EMG (top-right), EEG (bottom-left), and ECG with large artifacts (bottom-right).

Parameter	ISSCC'21 [8]	ISSCC'25 [1]	JSSC'24 [11]	JSSC'25 [12]	CICC'25 [10]	ISSCC'23 [9]	This Work
ADC Type	Continuous-Time ADC		Discrete-Time ADC				
Process [nm CMOS]	65	22	65	180	180	180	180
Topology	VCO-based ΔΣM	Δ-Amp + NS-SAR	ΔΣM + NS-SAR	NS-Pipelined SAR	NS-SAR	NS-SAR	NS-SAR
kT/C Noise Suppression	Not Necessary		No	Yes*	No	No	Yes
Total Input Cap. [pF]	2**	20	22	1.536	24	16.2	8
Supply [V]	1.2 / 0.8	1	1.3 / 0.8	1	1 / 0.8	1	1
Area [mm²]	0.075	0.084	0.13	0.12	0.31**	0.12	0.16***
Power [μW]	5.8	4.62	5.2	7	8.2	9.29	6.3
Fs [Hz] / OSR	200k / 100	200k / 100	200k / 200	20k / 10	200k / 100	128k / 64	128k / 64
BW [Hz]	1k	1k	500	1k	1k	1k	1k
IRN [μVrms]	3.56	1.33	3.89	24**	7.14	4.69	3.49
Peak Input [Vpp]	0.4	0.2	0.6	1.96	1.8	1.85	1.5
CMRR [dB]	89	93	83	70	N.A.	93	>81
SNDR/DR [dB]	92.3 / 92.3	91.8 / 92.9	94.5 / 95.8	87.3 / 89.2	98.6 / 101.1	99.1 / 101.6	98.3 / 103.6
Zin@DC/BW [MΩ]	60 / 50	220 / 15	208 / 31.5	1000 / 150	81.7 / 73.5	>30 / >24	168 / 41.5
NEF	9.31	3.47	13.87	77.11	24.83	17.36	10.64
FOMSNDR [dB]	174.7	175.2	174.3	168.8	179.5	179.4	180.3
FOMDR [dB]	174.7	176.2	175.6	170.8	182	181.9	185.6
FOMW [fJ/conv.]	86.1	72.6	119.8	184.8	58.9	63	46.9

FOM$_{SNDR}$ = SNDR + 10log(BW/Power) ; FOM$_{DR}$ = DR + 10log(BW/Power) ; FOM$_W$ = Power/(2·BW·2$^{(SNDR-1.76)/6.02}$)
* DAC and gain cal. required   ** Estimated   *** 65nm CMOS implementation: 0.08mm²

Figure 36.5.6: Performance summary and comparison with state-of-the-art works.

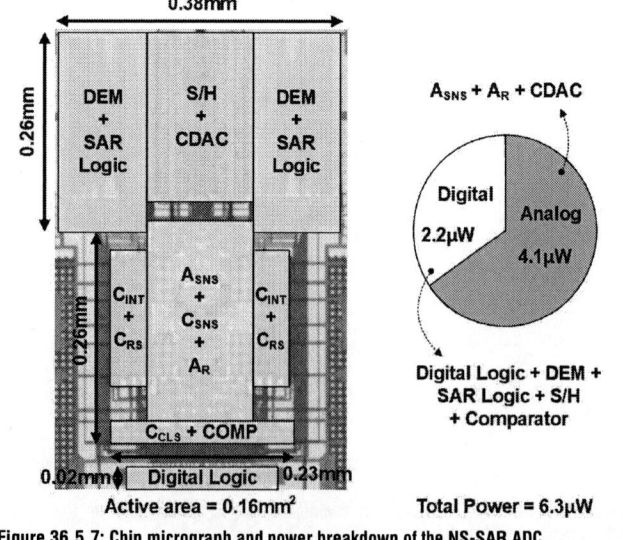

Figure 36.5.7: Chip micrograph and power breakdown of the NS-SAR ADC.

• 2026 IEEE International Solid-State Circuits Conference

ISSCC 2026 / SESSION 36 / NEURAL AND BIOMEDICAL INTERFACES / 36.6

36.6 A Sparsity-Aware Neural Interface with CIM-Based Predictive Focused Sampling for Hotspot Spike Tracking

Hui Wu[*1], Fengshi Tian[*2], Jinbo Chen[1], Zhipeng Liao[1], Hongyong Zhang[1], Xing Liu[1], Wenjun Zou[1], Sirui Cheng[1], Zhao Zhang[3], Jianlong Xu[4], Chi-Ying Tsui[2], Kwang-Ting Tim Cheng[2], Jie Yang[1], Mohamad Sawan[1]

[1]Westlake University, Hangzhou, China, [2]Hong Kong University of Science and Technology, Hong Kong, China
[3]Institute of Semiconductors, Chinese Academy of Sciences, Beijing, China, [4]Soochow University, Suzhou, China
*Equally Credited Authors (ECAs)

Abstract

This 1024-electrode neural interface solves the power, area, and bandwidth bottleneck by combining low-power panoramic scanning for hotspot prediction with high-fidelity spike tracking on active sites. The dynamic allocation is guided by a real-time compute-in-memory (CIM) engine, providing a 16× resolution boost (10kHz) and 0.0011mm²/ch efficiency in 40nm CMOS. The design maximizes spike density to lower power and silicon area.

The pursuit of high-density neural interfaces is fundamentally limited by strict implant power, area, and bandwidth budgets, a constraint exacerbated by the disparity between uniformly distributed hardware and the sparse, non-uniform nature of neural activity [1-6]. Conventional dedicated-per-channel readouts scale poorly in power and area [7-9], while uniform multiplexed scanning wastes critical resources on non-spike signals [10-14]. Some approaches attempt to mitigate this by sub-sampling large electrode arrays with a limited set of readout channels [15-21]. However, these methods often rely on static or slowly adapting techniques such as manual electrode selection [15,16], indirect activity indicators [17,18], or offline training to identify active regions [19,20]. These solutions still face challenges, including frequent recalibration due to chronic electrode drift. While the event-based spatially zooming architecture [22] offers exceptional power efficiency, its scalability is limited by a large silicon footprint. Moreover, the inherently event-driven operation may fail to capture low-amplitude, densely clustered, or rapidly successive spikes, thereby compromising the waveform fidelity essential for reliable spike sorting.

To address these challenges, we present a 1024-electrode sparsity-aware neural interface that employs a predictive focused sampling scheme to optimize resource allocation. First, the architecture introduces a dual-mode scanning strategy, combining low-power 8b panoramic monitoring with high-fidelity 11b spike tracking. This provides a 16× temporal resolution boost (10kHz) for active hotspots and achieves a >5.33× data compression, substantially alleviating the wireless transmission burden. Second, this intelligent allocation is enabled by a real-time hotspot predictor, implemented with an energy-efficient compute-in-memory (CIM) core. This ensures robust tracking of dynamic neural patterns and provides inherent resilience to chronic electrode drift. Finally, by amortizing a minimal, time-shared front-end across the 32×32 array, the design is realized in a silicon area of 0.22mm², achieving a high integration density for kilo-electrode recording implants.

Figure 36.6.1 illustrates the architecture of the proposed focused-sampling neural interface (FSNI). The system consists of two dedicated time-division multiplexed analog front-ends (AFEs), each acquiring signals from a 16×32 sub-array to cover the 1024 electrodes. A Column Scanner sequentially drives the array, allowing a 16-to-1 multiplexer within each half to select a subgroup of electrodes for readout by low-noise transconductance amplifiers (TCs). The amplified signals are then integrated and sampled by a track-and-hold (T/H) stage and digitized by a single, power-efficient successive approximation register (SAR) ADC, which dynamically switches between 8b and 11b resolution based on the scheduled slot type.

The intelligence of this sparsity-aware strategy resides in the digital control core. In its default Baseline Scan Slot (BSS) mode, panoramic 8b data is captured at 625Hz and streamed into a Frame Buffer for real-time processing by a Hotspot Predictor. Upon detection of spike-like activity, the Slot Scheduler commands the front-end to immediately up-sample the identified hotspot by activating available Spike-Focused Slots (SFS). The architecture provides three SFSs per column-scan, enabling the simultaneous tracking of up to three distinct hotspots within each 16-row segment, corresponding to a system-wide maximum of 192 simultaneously active SFS channels across the 32 columns. This capacity was deliberately chosen to match neural firing densities and ensure efficient resource allocation [23,24]. During an SFS, the ADC switches to its high-precision 11b mode, capturing the full spike waveform at an effective sampling rate of 10kHz. This 16× enhancement in temporal resolution is a direct result of the column-based scanning, which allows a column to be revisited 16× more frequently than a full-frame scan. To ensure long-term signal fidelity, a Local Offset Average block provides a crucial feedback loop. It mitigates electrode DC offset by periodically averaging non-spike signals across the array to recalibrate and feed corrections back to the analog chain via IDACs.

Figure 36.6.2 details the circuit implementation of the time-division multiplexed (TDM) analog front-end (AFE). The signal chain begins with a transconductance amplifier (TC) that employs a flipped-voltage-follower (FVF) structure to achieve low output impedance, which is critical for current-mode DC offset feedback. A pair of Gm-boosting OTAs (A1) eliminates the level shift of conventional source followers while stabilizing the common-mode voltage with sufficient headroom. The TC output current is amplified by a 4× current mirror and integrated onto the track-and-hold (T/H) capacitor during the sampling phase. In the subsequent hold phase, the T/H amplifier first executes a charge-transfer operation, actively moving the stored charge to the hold capacitor to provide additional gain and isolating the

electrode from the ADC input; a wide-swing OTA buffer then ensures accurate holding and rapid settling for quantization. Two AFEs share one T/H and a single SAR ADC, improving hardware efficiency under multiplexed operation. The TC–T/H pair inherently performs a windowed integration, simultaneously acting as a charge-domain gain stage and a first-order sinc filter that amplifies the signal, suppresses in-band thermal noise, and attenuates out-of-band noise to prevent aliasing. Programmable gain modes of 1×/2×/4× extend the ADC's dynamic range to resolve both low-amplitude spikes and large baseline shifts, while the ADC itself supports dual-resolution operation: an 8b mode consuming 6.8µW for panoramic baseline scans and an 11b mode consuming 51.2µW for spike-focused slots, enabling power-proportional precision scaling. To further reduce electrode loading in high-speed multiplexing (SFS mode), a ping-pong reservoir (PPR) pre-charges a capacitor one slot ahead of sampling, then transfers the charge to the input capacitor followed by a brief electrode correction, effectively boosting the equivalent input impedance to 84MΩ and relieving instantaneous electrode current stress. Finally, a digital offset-cancellation loop averages non-spike samples and injects compensating currents via an IDAC at the TC output, suppressing quasi-static electrode offsets and keeping the input baseline within the ADC's linear range.

Figure 36.6.3 illustrates the digital core, which leverages a 32×32×8b multi-mode compute-in-memory (MM-CIM) engine to perform power-efficient hotspot prediction. To overcome the memory-wall bottleneck that dominates the energy consumption of conventional digital processors, the design embeds the core computational task—local average calculation for adaptive thresholding—directly within the memory array. This is achieved through a hierarchical, massively parallel architecture detailed in the MM-CIM Engine Organization diagram. Each of the 32 columns integrates a dedicated 4:2 compressor adder tree, enabling simultaneous in-situ partial sum (PSUM) accumulation of analog bitline summations. The column-level PSUMs from a selected 16×16 window are then aggregated by a second-level adder tree. This in-situ computation entirely obviates the need for 256 costly memory read cycles, which is the primary source of the 7.7× reduction in core computation energy. The hotspot prediction operates as a four-phase, single-pass pipeline for real-time analysis. First, an 8b data frame is written into the CIM array (Phase 1). Next, the CIM engine computes the window average in-situ, generating an adaptive threshold (Phase 2). This threshold is then used to binarize the pixel data, creating an activity mask (Phase 3). Finally, a dedicated hardware block calculates the centroid of the activated region, extracting its precise coordinates to close the event-driven feedback loop (Phase 4). This architecture improves computational density by 1.5× and enables the low-latency analysis essential for predictive focused sampling.

Figure 36.6.4 summarizes the measured performance of the FSNI. Compared with a non-CIM baseline, the design reduces energy by 87%, area by 40%, and memory by 20%, highlighting the benefit of in-memory processing. Benchmarking with an open-source dataset shows that the focused-sampling scheme preserves spike sorting accuracy comparable to raw full-rate data, confirming waveform integrity [25]. Measured spectra demonstrate that the dual-mode SAR ADC achieves 41.39dB SNDR (6.58b ENOB) in BSS mode at 625Hz and improves to 48.89dB SNDR (7.83b ENOB) in SFS mode at 10kHz, enabling power-scalable precision. The input-referred noise is 8.7µVrms over the 300Hz-to-10kHz AP band with a flat-band density of 40nV/√Hz. Input impedance measurements show a boost from 300kΩ in BSS mode to 84MΩ in SFS mode using the ping-pong reservoir, corresponding to a 280× improvement that minimizes electrode loading during high-speed sampling. These results validate the system's ability to balance panoramic monitoring with high-resolution spike capture while maintaining low noise and high fidelity.

The system was further validated through in-vitro experiments using stem-cell-derived neuronal cultures interfaced with the 32×32 active electrode array. As shown in Figure 36.6.5, the sparsity-aware neural interface was integrated with a custom readout platform, where an external MCU provided electrode scanning and data streaming. The electrical characteristics of the active electrodes, including transistor I–V behavior and frequency-dependent impedance, confirm stable operation and sufficient bandwidth for neuronal activity recording. Spatial activity maps across successive frames demonstrate that the digital hotspot predictor reliably localizes spike-rich regions, enabling adaptive activation of spike-focused slots. Corresponding neural traces show that the system dynamically switches from low-rate panoramic baseline monitoring (BSS mode, 625Hz) to high-rate, high-SNR capture of localized spikes (SFS mode, 10kHz), thereby validating the predictive focused-sampling strategy in a biologically relevant environment.

Figure 36.6.6 compares the FSNI with prior neural implant systems, and Fig. 36.6.7 presents chip micrographs with area and power breakdowns. Our system demonstrates a 1024-electrode sparsity-aware neural interface in 40nm CMOS that achieves a total chip area of 0.22mm². By combining CIM-based predictive focused sampling with a power-scalable TDM+SAR architecture, this work achieves both high channel density and power-efficient operation for kilo-electrode recording implants.

Acknowledgment:

This work was supported by STI2030-Major Projects (No. 2022ZD0208805), "Pioneer" and "Leading Goose" Research and Development Program of Zhejiang (No. 2024C03002), National Natural Science Foundation of China (No. W2431058), and was partially conducted by ACCESS – AI Chip Center for Emerging Smart Systems, supported by the InnoHK initiative of the Innovation and Technology Commission of the Hong Kong Special Administrative Region Government. Work was conducted through the Integrated-on-Chips Brain-Computer Interfaces Zhejiang Engineering Research Center, Hangzhou, China. The corresponding authors are Jie Yang and Mohamad Sawan (yangjie@westlake.edu.cn, sawan@westlake.edu.cn).

References:

[1] R. Muller et al., "24.1 A miniaturized 64-channel 225μW wireless electrocorticographic neural sensor," *ISSCC*, Feb. 2014, pp. 412–413. https://doi.org/10.1109/ISSCC.2014.6757492

[2] D. Kleinfeld et al., "Can One Concurrently Record Electrical Spikes from Every Neuron in a Mammalian Brain?," *Neuron*, vol. 103, no. 6, pp. 1005-1015, Sept. 2019. https://doi.org/10.1016/j.neuron.2019.08.011

[3] C. Yang et al., "Neural Dielet 2.0: A 128-Channel 2mm×2mm Battery-Free Neural Dielet Merging Simultaneous Multi-Channel Transmission Through Multi-Carrier Orthogonal Backscatter," in *IEEE Transactions on Biomedical Circuits and Systems*, vol. 19, no. 1, pp. 226-237, Feb. 2025. https://doi.org/10.1109/TBCAS.2024.3416728

[4] Tsai, D. et al., A very large-scale microelectrode array for cellular-resolution electrophysiology. *Nat Commun* 8, 1802 (2017). https://doi.org/10.1038/s41467-017-02009-x

[5] M. Shoaran et al., "Intelligent Neural Interfaces: An Emerging Era in Neurotechnology," *IEEE Custom Integrated Circuits Conference*, 2024, pp. 1-7. https://doi.org/10.1109/CICC60959.2024.10529099

[6] Trautmann et al., Large-scale high-density brain-wide neural recording in nonhuman primates. *Nat Neurosci* 28, 1562–1575 (2025). https://doi.org/10.1038/s41593-025-01976-5

[7] X. Yang et al., "A Highly-Integrated 1536-Channel Quad-Shank Monolithic Neural Probe in 55nm CMOS for Full-Band Raw-Signal Recording," 2024 *IEEE Symposium on VLSI Technology and Circuits*, 2024, https://doi.org/10.1109/VLSITechnologyandCir46783.2024.10631430

[8] M. Jang et al., "A 1024-Channel 268-nW/Pixel 36×36 μm²/Channel Data-Compressive Neural Recording IC for High-Bandwidth Brain–Computer Interfaces," in *IEEE Journal of Solid-State Circuits*, vol. 59, no. 4, pp. 1123-1136, April 2024. https://doi.org/10.1109/JSSC.2023.3344798

[9] D. -Y. Yoon et al., "A 1024-Channel Simultaneous Recording Neural SoC with Stimulation and Real-Time Spike Detection," *Symposium on VLSI Circuits*, 2021 https://doi.org/10.23919/VLSICircuits52068.2021.9492480

[10] X. Huang et al., "A 256-Channel Actively-Multiplexed μECoG Implant with Column-Parallel Incremental ΔΣ ADCs Employing Bulk-DACs in 22-nm FDSOI Technology," *ISSCC*, 2021, pp. 200-202. https://doi.org/10.1109/ISSCC42614.2022.9731630

[11] M. A. Shaeri et al., "MiBMI: A 192/512-Channel 2.46mm² Miniaturized Brain-Machine Interface Chipset Enabling 31-Class Brain-to-Text Conversion Through Distinctive Neural Codes," ISSCC, 2024, pp. 546-547. https://doi.org/10.1109/ISSCC49657.2024.10454533

[12] J. Chen et al., "A Neuron-Inspired 0.0032mm²–1.38μW/Ch Wireless Implantable Neural Interface with Direct Multiplexing Front-End and Event-Driven Spike Detection and Transmission," *IEEE Custom Integrated Circuits Conference*, Apr. 2024, https://doi.org/10.1109/CICC60959.2024.10529097

[13] N. S. K. Fathy et al., "A 0.00179 mm²/Ch Chopper-Stabilized TDMA Neural Recording System With Dynamic EOV Cancellation and Predictive Mixed-Signal Impedance Boosting," in *IEEE Transactions on Biomedical Circuits and Systems*, vol. 18, no. 4, pp. 908-922, Aug. 2024. https://doi.org/10.1109/TBCAS.2024.3366649

[14] W. Choi et al., "A 1,024-Channel, 64-Interconnect, Capacitive Neural Interface Using a Cross-Coupled Microelectrode Array and 2-Dimensional Code-Division Multiplexing," *IEEE Symposium on VLSI Technology and Circuits*, 2023, https://doi.org/10.23919/VLSITechnologyandCir57934.2023.10185425

[15] N. Zeng et al., "A Wireless, Mechanically Flexible, 25μm-Thick, 65,536-Channel Subdural Surface Recording and Stimulating Microelectrode Array with Integrated Antennas," *IEEE Symposium on VLSI Technology and Circuits*, 2023, https://doi.org/10.23919/VLSITechnologyandCir57934.2023.10185321

[16] C. M. Lopez et al., "A 16384-electrode 1024-channel multimodal CMOS MEA for high-throughput intracellular action potential measurements and impedance spectroscopy in drug-screening applications," *ISSCC*, 2018, pp. 464-465. https://doi.org/10.1109/ISSCC.2018.8310385

[17] H. -S. Lee et al., "A Multi-Channel Neural Recording System With Neural Spike Scan and Adaptive Electrode Selection for High-Density Neural Interface," in *IEEE Transactions on Circuits and Systems I: Regular Papers*, vol. 70, no. 7, pp. 2844-2857, July 2023. https://doi.org/10.1109/TCSI.2023.3268686

[18] Cartiglia et al., A 4096 channel event-based multielectrode array with asynchronous outputs compatible with neuromorphic processors. *Nat Commun* 15, 7163 (2024). https://doi.org/10.1038/s41467-024-50783-2

[19] Saboo et al., Unsupervised machine-learning classification of electrophysiologically active electrodes during human cognitive task performance. *Sci Rep* 9, 17390 (2019). https://doi.org/10.1038/s41598-019-53925-5

[20] J. Choi et al., "Optimal Adaptive Electrode Selection to Maximize Simultaneously Recorded Neuron Yield," *Advances in Neural Information Processing Systems*, Oct. 2020. https://dl.acm.org/doi/10.5555/3495724.3496241

[21] J. -H. Cha et al., "A Reconfigurable Sub-Array Multiplexing Microelectrode Array System With 24,320 Electrodes and 380 Readout Channels for Investigating Neural Communication," *ISSCC*, 2022, pp. 342-343. https://doi.org/10.1109/ISSCC42614.2022.9731590

[22] J. Xu et al., "Event-Based Spatially Zooming Neural Interface IC with 10nW/Input Reconfigurable-Inverter Fabric and Input-Adaptive Quantization," *ISSCC*, 2025, pp. 274-275. https://doi.org/10.1109/ISSCC49661.2025.10904733

[23] Steinmetz et al., Distributed coding of choice, action and engagement across the mouse brain. *Nature* 576, 266–273 (2019). https://doi.org/10.1038/s41586-019-1787-x

[24] Pachitariu et al., Spike sorting with Kilosort4. *Nat Methods* 21, 914–921 (2024). https://doi.org/10.1038/s41592-024-02232-7

[25] Brochier et al., Massively parallel recordings in macaque motor cortex during an instructed delayed reach-to-grasp task. *Sci Data* 5, 180055 (2018). https://doi.org/10.1038/sdata.2018.55

[26] X. Huang et al., "A 3072-Channel Neural Readout IC with Multiplexed Two-Step Incremental-SAR Conversion and Bulk-DAC-Based EDO Compensation in 22nm FDSOI," *IEEE Symposium on VLSI Technology and Circuits*, 2024, https://doi.org/10.1109/VLSITechnologyandCir46783.2024.10631524

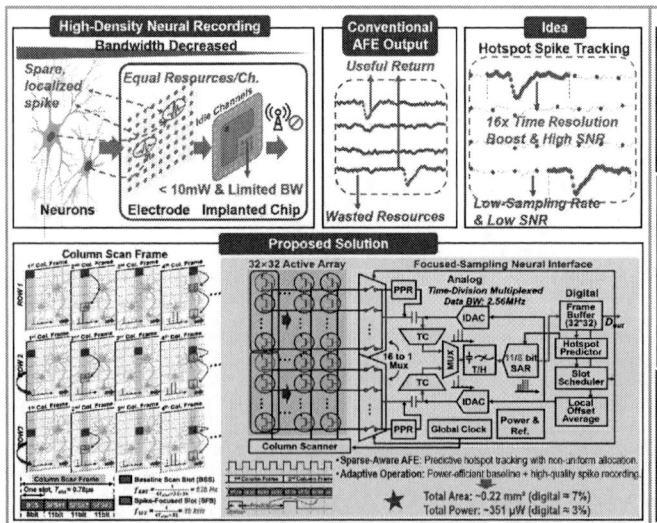

Figure 36.6.1: Conceptual overview and architecture of the proposed sparse-aware neural interface with 16× temporal resolution boost for hotspot spike tracking.

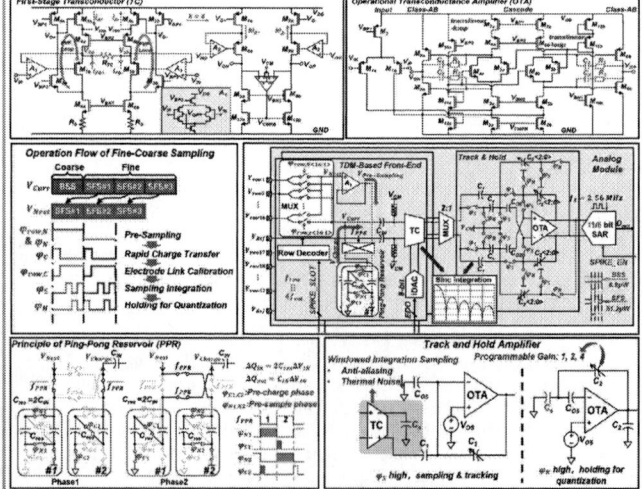

Figure 36.6.2: Key circuits enabling the focused-sampling scheme, including the FVF-based TC, Class-AB OTA, Ping-Pong Reservoir, and programmable-gain T/H.

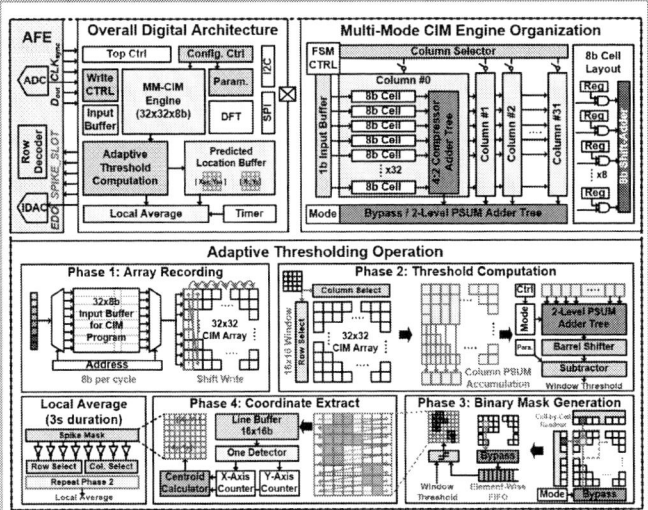

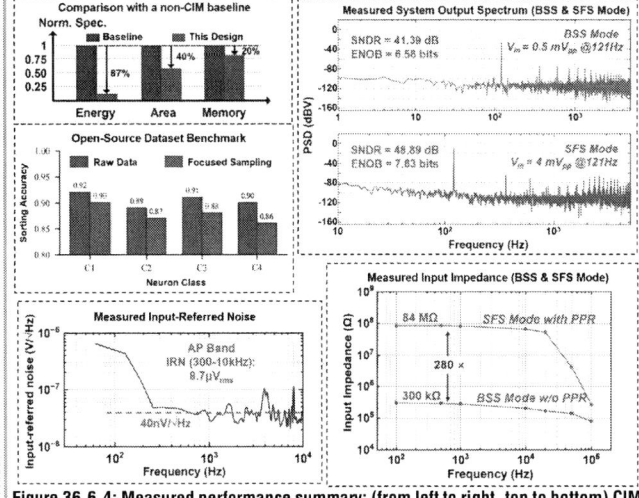

Figure 36.6.3: Digital core featuring a multi-mode Compute-in-Memory (CIM) engine that executes a four-phase adaptive thresholding algorithm to predict spike hotspots.

Figure 36.6.4: Measured performance summary: (from left to right, top to bottom) CIM-core comparison, spike sorting benchmark, input-referred noise, system output spectrum, and input impedance.

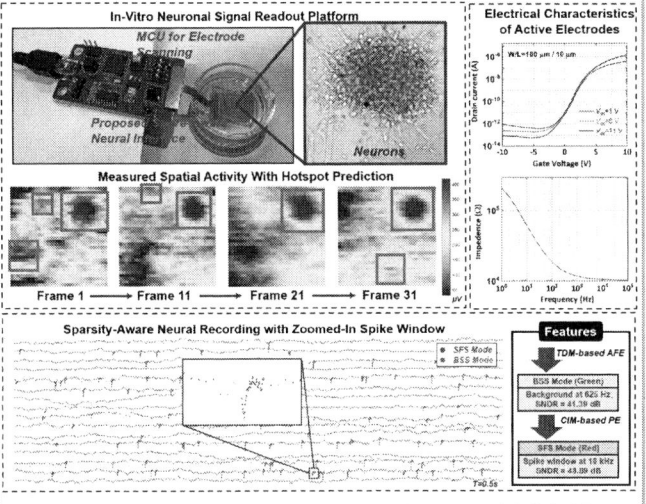

Figure 36.6.5: In-vitro system demonstration showing successful digital hotspot prediction and dual-mode (BSS/SFS) spike capture from a live neuronal culture.

Figure 36.6.6: Performance summary and comparison with state-of-the-art works.

	THIS WORK	ISSCC'25 Xu [22]	VLSI'24 Huang [26]	CICC'24 Chen [12]	NC'24 Cartiglia [18]	TCASI'23 Lee [17]	ISSCC'22 Cha [21]
TECHNOLOGY [nm]	40	65	22	40	180	180	110
SUPPLY VOLTAGE [V]	1.2	0.7 – 3.3	0.5/0.8	1.2/0.6	1.8	1.8	-
ARCHITECTURE	TDM+SAR	INV-based AFE	Two-step I-SAR	TDM+ASC	LC-ADC	AMP+SAR	TDM+SAR
RECORDING MODALITY	SPIKEs	SPIKEs	LFPs & SPIKEs	SPIKEs	SPIKEs	LFPs & SPIKEs	SPIKEs
ELECTRODE TYPE	Active	Passive	Active	Active	Passive	Passive	Passive
ELECTRODE SELECTION	Hotspot Prediction	Spatial Zooming	No	No	No	MUA[†] Scan	Random
ELEC. # / MAX CH. #	1024/192	64/64	3072/3072	32/32	4096/4096	32/12	24320/380
ELEC.-to-REC. RATIO	18.75%	100%	100%	100%	100%	37.5%	1.56%
BANDWIDTH / CH. [kHz]	10	20	7.5	5.2	10	14	20/10/5
ENOB [BITS]	7.83[a]	7.68	-	-	6[*]	12	-
SNDR [dB]	48.89[a]	48	<49.6	<47.2	37.2	-	-
PWR / CHANNEL [μW]	1.83	0.004[c]	0.438	1.38	0.8	10.7	80.79
AREA / CHANNEL [mm²]	0.0011	0.04[d]	0.0004	0.0032	0.0023	0.24	0.02
IRN [μV$_{rms}$]	8.7 @300-10kHz	7.3 @1k-20kHz	6.38 @1-7.5kHz	2.9 @1-1kHz	19.04 @500-3kHz	3.41 @10-10kHz	5.4 @300-10kHz
INPUT IMPEDANCE [GΩ]	0.084[b]	>1 @2kHz	-	-	-	-	-

-: Not reported
[a] Measured from the FFT of the system in SFS mode.
[b] Input impedance measured in spike-focused slot (SFS) mode.
[c] For an average spike rate of 25 spikes/second and 1/32 spatial sparsity.
[d] Estimated from chip micrograph.
[*] Estimated
[†] MUA: Multi-unit activity

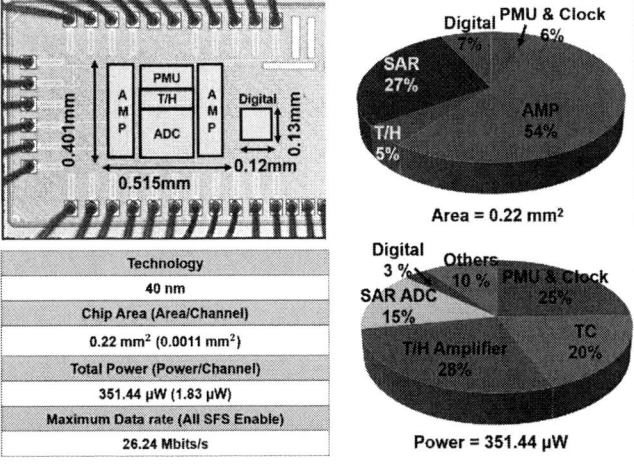

Figure 36.6.7: Die micrograph, performance summary, and breakdown of the 1024-electrode neural interface in 40nm CMOS.

ISSCC 2026 / SESSION 36 / NEURAL AND BIOMEDICAL INTERFACES / 36.7

36.7 A 90.7%-Efficiency Hybrid Optogenetic Stimulation System with Sub-Threshold Electrical Stimuli Achieving 70.6% Optical Energy Saving

Joonghoon Kang[1], Hyeonhee Roh[2], Kyeongho Eom[1], Hyun-Su Lee[1], Hojae Chon[1], Maesoon Im[2], Hyung-Min Lee[1]

[1]Korea University, Seoul, Korea, [2]Korea Institute of Science and Technology, Seoul, Korea

Abstract

This work presents a hybrid optogenetic stimulation system with sub-threshold electrical stimuli that lower the optical threshold and reduce energy consumption. It achieves 90.75% LED driving efficiency through efficient capacitor charge–discharge operation and generates decaying-exponential irradiance to enhance neural responses. *Ex vivo* experiments in mice validate equivalent neural activation with a 70.6% optical energy saving compared to conventional optogenetics.

Neural stimulation has been extensively studied in implantable medical devices for treating neurological disorders. Conventional neural stimulation has utilized electrical stimuli, but stimulation current may lead to unintended neural activation due to stimulus artifact, resulting in poor spatial selectivity [1]. To overcome this limitation, optogenetic stimulation has been studied to enable cell-specific activation and high selectivity [2]. Nevertheless, conventional optogenetics requires high optical energy for providing sufficient irradiance to target cells, which imposes stringent energy constraints in implantable systems [3-8].

Figure 36.7.1 (top) summarizes these limitations of conventional stimulation methods, underscoring the necessity for an efficient stimulation scheme. In this work, we propose a hybrid optogenetic stimulation system that takes advantages of cell-specific optogenetic modulation while employing sub-threshold electrical stimuli to reduce the required LED driving energy, thereby achieving high stimulation precision, efficiency, and efficacy. Figure 36.7.1 (bottom) presents three key advantages of the proposed hybrid stimulation system to enhance both stimulation efficiency and efficacy by combining optical and electrical modalities: (1) A non-responsive sub-threshold electrical stimulus, which is applied prior to optogenetic stimulation, lowers the optical threshold level for neural activation, enabling efficient optogenetics with reduced irradiance and LED driving energy (E_{LED}). (2) An efficient DC-to-capacitor charger enables high-efficiency energy storage and transfer, driving the LED efficiently [9]. Furthermore, the DC-to-capacitor charger does not rely on ac input, which can be disrupted by loosely coupled inductive links, thereby ensuring stable operation [10,11]. (3) Instead of conventional rectangular-shape irradiance [12,13], the hybrid optogenetic system adopts decaying-exponential irradiance waveforms from LEDs, which enhance neural responses even with equal or lower irradiance, further reducing the optical energy demand and improving optogenetic efficacy.

Figure 36.7.2 shows the proposed hybrid optogenetic system, including an optogenetic switched-capacitor stimulator (OSCS), a sub-threshold electrical stimulator (STES), and a hybrid stimulus configurator. The system operates with a DC supply voltage (V_{DD}=4V) which can be available in wirelessly power implantable devices [14]. The OSCS adopts an efficient inductor-assisted four-channel capacitor charger with a digitally tunable target voltage (V_{TG}) to accommodate a wide range of LED driving voltages. The STES employs a 6b wide-swing current-steering digital-to-analog converter (I-DAC) with an H-bridge to realize biphasic stimulation. It operates the electrical stimulation current below the threshold activation level. Hybrid stimulus configurator orchestrates the operation by managing the sequential charging of OSCS and synchronizing the timing of the OSCS with the STES to realize coordinated hybrid stimulation. Upon control commands, the configurator provides a charging signal (S_{CH}) to sequentially charge capacitor voltage (V_{CAP}) up to V_{TG}. Then, the selected capacitor is discharged to drive LED, emitting the required irradiance. The optogenetic stimulation time (T_{LED}) is sufficiently short (<2ms) compared to the optogenetic stimulation period, thereby allowing the capacitors of all channels to be charged during the remaining time.

Figure 36.7.3 (top) shows the schematics of a pulse width modulator (PWM) and a charge status detector (CSD) in OSCS. The PWM generates a PWM signal (S_{PWM}) for capacitor charging. When S_{CH} goes high, the counter starts, a PWM current (I_{PWM}) flows into the capacitor DAC (C-DAC), and C-DAC voltage (V_{CD}) ramps up until the Schmitt trigger toggles, inverting S_{PWM} and incrementing the 7b counter. Two C-DACs with fixed total capacitance, driven by upper 5b of the 7b counter outputs ($D_{i,0.4}$ and $DB_{i,0.4}$), maintain S_{PWM} frequency, while a stop logic halts the counter when $D_{i,0.4}$ reaches the maximum value. The CSD with two comparators enables charging by generating the gate-driver enabling signal (S_{EN}) only when V_{CAP} is lower than both a slope pulse voltage (V_{SP}) and V_{TG}, which is generated by a resistor-string DAC (R-DAC). Figure 36.7.3 (middle) illustrates the efficient charge-discharge operation with PWM and CSD. S_{PWM} is processed through a non-overlapping clock generator to avoid shoot-through, while the gate enabler clamps both gate inputs when S_{EN} = 0. The charger utilizes a compact-volume inductor (4.7µH with 180mΩ DCR and 2×1.25×0.85mm³) with 10µF capacitors. A current limiter below the LED suppresses surge discharging. Figure 36.7.3 (bottom-right) shows the measured charging waveform. V_{CAP1} starts charging from a residual voltage (V_{RSD}) of 2.5V to V_{TG} = 3.8V, and the charging efficiency can be calculated as the ratio of stored capacitor energy to input energy. Figure 36.7.3 (bottom) shows the STES with the 6b wide-swing I-DAC and H-bridge for generating biphasic pulses. I-DAC output transistors operate in deep triode region with 60mV clamp, lowering voltage headroom.

Charging and discharging efficiencies of OSCS were measured as shown in Fig. 36.7.4 (top). Discharging efficiency was defined as the ratio of energy delivered to the LED to energy consumption in capacitors. When a capacitor was charged from V_{RSD} = 2.5V to V_{TG}, high charging efficiency of 94.05% was achieved at V_{TG} = 3.8V thanks to constant inductor current and linearly accumulating losses, while stored energy increases with V_{CAP}^2, yielding higher efficiency at higher V_{TG}. The peak discharging efficiency reaches 99.14% of at V_{TG} = 3.2V, while higher V_{TG} leads to larger LED current and switch I^2R losses, decreasing the efficiency. Then, the overall LED driving efficiency, defined as the product of charging and discharging efficiencies, peaked at 90.75% when V_{TG} = 3.6V, achieving 5.65% higher efficiency than [7,8]. Moreover, the LED driving efficiencies remained high above 89% for $V_{TG} \geq 3.4$V, covering various LED driving conditions for efficient optogenetics [5,7,8]. Figure 36.7.4 (bottom-left) shows the hybrid stimulation waveforms: a 5ms non-responsive sub-threshold cathodic stimulation is first applied to lower the neural activation threshold, followed by a 400µs LED pulse that provides the main optogenetic stimulation. The OSCS sequentially charges four capacitors from V_{RSD} = 2.5V to V_{TG} = 3.8V within 600µs, while V_{TG} of each channel can be programmable through R-DAC. The energy consumption of STES (1.2µJ) was minimally set compared to the total stimulation energy (18.14µJ). Figure 36.7.4 (bottom-right) compares rectangular and decaying exponential irradiance waveforms and their irradiance values. At V_{TG} = 3.2V, the decaying exponential I_{LED} shows an initial peak current, while its total energy equals that of rectangular I_{LED} (14.5mA). Although both waveforms consume the same total energy with identical irradiance, the different shape of irradiance can further improve the optogenetic efficacy, activating target cells with smaller irradiance (and smaller LED energy consumption), which was quantitatively evaluated through *ex vivo* experiments.

Figure 36.7.5 (top) describes the *ex vivo* experimental setup to validate the optogenetic efficacy improvement. Three optogenetic conditions were compared: (1) optogenetics with rectangular irradiance (optogenetic-only), (2) hybrid optogenetics with rectangular irradiance and sub-threshold electrical stimuli (hybrid–rectangular), and (3) hybrid optogenetics with decaying exponential irradiance and sub-threshold electrical stimuli (hybrid–exponential) from the proposed system. *Ex vivo* experiments were conducted using retinas from ChR2(H134R)-expressing retinal degeneration 10 (*rd10*) mice to determine activation (spike) thresholds. Thresholds for optogenetic-only and electrical-only stimulation were first identified through incremental stimulation sweeps. Then, a sub-threshold electrical stimulus, set below the electrical activation threshold, was applied as assistive stimuli to optogenetics. With the assistive stimuli applied, the hybrid stimulation threshold, defined as the optical intensity required to evoke neural activation (≥50% spike probability), was measured for both rectangular and decaying-exponential irradiance waveforms through incremental stimulation sweeps. Finally, the total stimulation energy consumed at each threshold condition was calculated and compared. The *ex vivo* results in Fig. 36.7.5 (bottom) include raster plots from a retinal cell, where spike thresholds were identified under three optogenetic conditions based on spike probability. The stimulation energy graphs with three cells indicate that although activation thresholds vary across cells, hybrid optogenetics consistently requires smaller energy for driving LED than optogenetic-only. As shown in the normalized energy graph, hybrid–rectangular achieved 55.8% saving in LED driving energy, demonstrating higher efficacy of hybrid optogenetics. Furthermore, hybrid–exponential achieved additional 36.7% saving compared to hybrid–rectangular thanks to its unique decaying exponential shape of irradiance, further enhancing the overall hybrid optogenetic efficacy with up to 70.6% energy saving.

Figure. 36.7.6 summarizes the performance of the hybrid optogenetic system and compares it with state-of-the-art optogenetic stimulators. The proposed system achieves up to 90.75% LED driving efficiency, surpassing prior works [7,8], while enabling fast per-channel capacitor charging within 40µs (2.5 to 3.8V). The efficacy improvement of the hybrid optogenetics was validated through *ex vivo* experiments, achieving the same level of neural activation with 70.6% energy saving for LED driving compared to conventional optogenetics, enhancing both system efficiency and optogenetic efficacy. Figure 36.7.7 shows the die micrograph fabricated in 130nm CMOS process with an active area of 3.04mm².

612 • 2026 IEEE International Solid-State Circuits Conference

979-8-3315-8937-0/26 $31.00 © 2026 IEEE

ISSCC 2026 / February 18, 2026 / 4:00 PM

Acknowledgement:

This work was supported by Ministry of Trade, Industry & Energy (RS-2022-00154983) and National Research Foundation (RS-2024-00398460), Korea. Chip fabrication and EDA tool were supported by IC Design Education Center, Korea.

References:

[1] Y. Son *et al.*, "Effects on retinal stimulation of the geometry and the insertion location of penetrating electrodes," *IEEE TNSRE*, vol. 31, pp. 3803–3812, Sep. 2023. https://doi.org/10.1109/TNSRE.2023.3317496

[2] J.-A. Sahel *et al.* "Partial Recovery of Visual Function in a Blind Patient after Optogenetic Therapy," *Nat. Med.*, vol. 27, no. 7, pp. 1223–1229, Jul. 2021. https://doi.org/10.1038/s41591-021-01351-4

[3] Y. Jia *et al.*, "A mm-Sized Free-Floating Wirelessly Powered Implantable Optical Stimulation Device," *IEEE TBioCAS*, vol. 13, no. 4, pp. 608-618, Aug. 2019. https://doi.org/10.1109/TBCAS.2019.2918761

[4] T. Yousefi *et al.*, "An Implantable Optogenetic Neuro-Stimulator SoC with Extended Optical Pulse-Width Enabled by Supply-Variation-Immune Cycled Light-Toggling Stimulation," *IEEE TBioCAS*, vol. 16, no. 4, pp. 557-569, Aug. 2022. https://doi.org/10.1109/TBCAS.2022.3198911

[5] L. Zhao *et al.*, "A Wireless Implantable Opto-Electro Neural Interface ASIC for Simultaneous Neural Recording and Stimulation," *IEEE CICC*, pp. 1–2, Apr. 2023. https://doi.org/10.1109/CICC57935.2023.10121181

[6]] G. Gagnon-Turcotte *et al.*, "An Adaptive and Autonomous System-On-Chip with Data Analysis for μs-Latency Closed-Loop Optogenetics," *IEEE BioCAS*, pp. 1-4, Oct. 2023. https://doi.org/10.1109/BioCAS58349.2023.10388520

[7] L. Zhao *et al.*, "A Miniature Neural Interface Implant with a 95% Charging Efficiency Optical Stimulator and an 81.9dB SNDR ΔΣM-Based Recording Frontend," *IEEE ISSCC*, pp. 558–560, Feb. 2024. https://doi.org/10.1109/ISSCC49657.2024.10454382

[8] L. Zhao *et al.*, "NeuroFlare: An mm³-Scale Wireless Neural Interface Device with Simultaneous Neural Recording and Optical Stimulation," *IEEE JSSC*, vol. 60, no. 9, pp. 3342-3354, Sep. 2025. https://doi.org/10.1109/JSSC.2025.3532646

[9] K. Eom *et al.*, "A 92%-Efficiency Inductor-Charging Switched-Capacitor Stimulation System with Level-Adaptive Duty Modulation and Offset Charge Balancing," *IEEE JSSC*, vol. 59, no. 5, pp. 1521-1531, May 2024. https://doi.org/10.1109/JSSC.2023.3334753

[10] H.-M. Lee *et al.*, "A Power-Efficient Switched-Capacitor Stimulating System for Electrical/Optical Deep Brain Stimulation," *IEEE JSSC*, vol. 50, no. 1, pp. 360–374, Jan. 2015. https://doi.org/10.1109/JSSC.2014.2355814

[11] Y. Jia et al., "A Trimodal Wireless Implantable Neural Interface System-on-Chip," *IEEE TBioCAS*, vol. 14, no. 6, pp. 1207–1217, Dec. 2020. https://doi.org/10.1109/TBCAS.2020.3037452

[12] A. C. Thompson *et al.*, "Hybrid Optogenetic and Electrical Stimulation for Greater Spatial Resolution and Temporal Fidelity of Cochlear Activation," *J. Neural Eng.*, vol. 17, no. 5, Art. no. 056046. Oct. 2020. https://doi.org/10.1088/1741-2552/abbff0

[13] H. Roh *et al.*, "Enhanced Optogenetic Stimulation of Retinal Ganglion Cells With Assistive Electric Stimulation for Low Optical Power Artificial Vision," *IEEE TNSRE*, vol. 33, pp. 1958-1968, May 2025. https://doi.org/10.1109/TNSRE.2025.3568864

[14] T. Lu and S. Du, "A Single-Stage Regulating Voltage-Doubling Rectifier for Wireless Power Transfer," *IEEE SSCL*, vol. 6, pp. 29-32, Jan. 2023. https://doi.org/10.1109/LSSC.2023.3239691

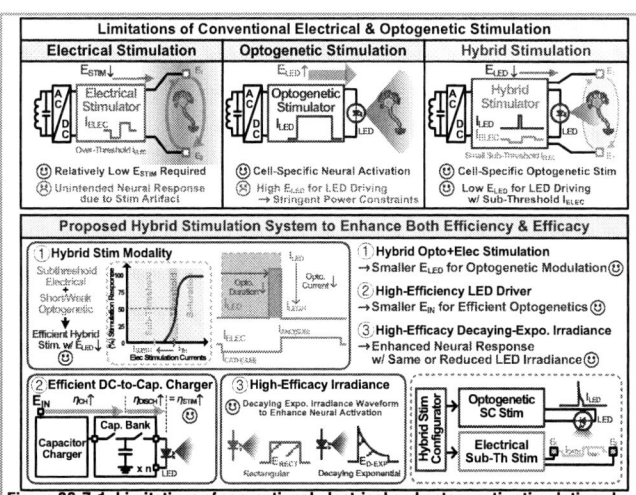

Figure 36.7.1: Limitations of conventional electrical and optogenetic stimulation along with the proposed hybrid stimulation solution to overcome both limitations (top); advantages of the hybrid stimulation system that enhances both efficiency and efficacy (bottom).

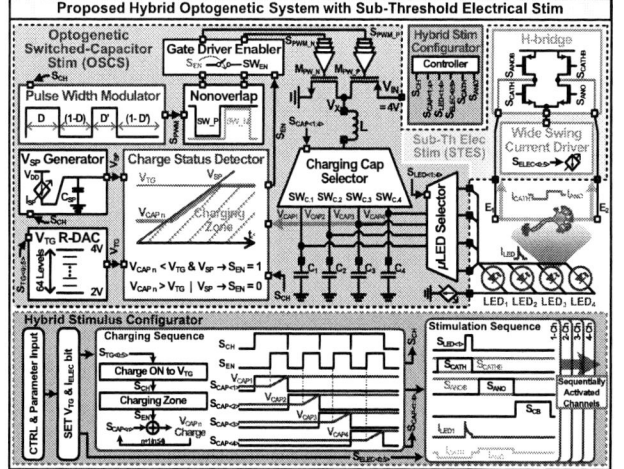

Figure 36.7.2: Hybrid optogenetic system with sub-threshold electrical stimuli including an optogenetic switched-capacitor stimulator (OSCS), a sub-threshold electrical stimulator (STES), and a hybrid stimulus configurator (top); operation timing diagram (bottom).

36

DIGEST OF TECHNICAL PAPERS • 613

979-8-3315-8937-0/26 $31.00 © 2026 IEEE

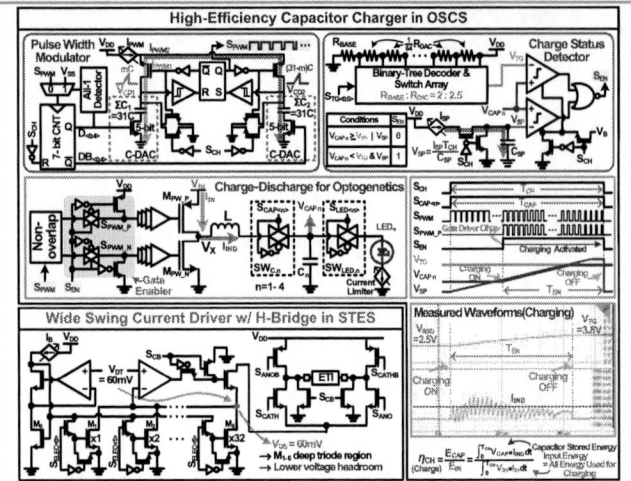

Figure 36.7.3: Schematics of the pulse width modulator and charge status detector in OSCS (top); efficient charge-discharge operation (middle) and measured charging waveforms (bottom-right); schematics of the wide swing current driver in STES (bottom-left).

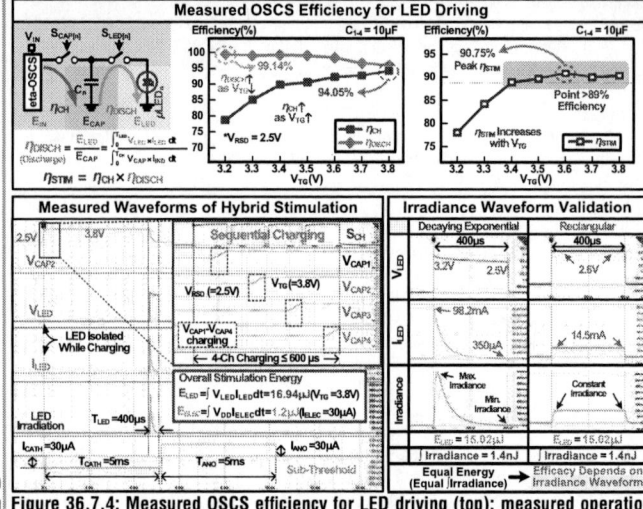

Figure 36.7.4: Measured OSCS efficiency for LED driving (top); measured operation waveforms of hybrid stimulation (bottom-left); and irradiance waveform validation comparing decaying exponential and rectangular LED driving (bottom-right).

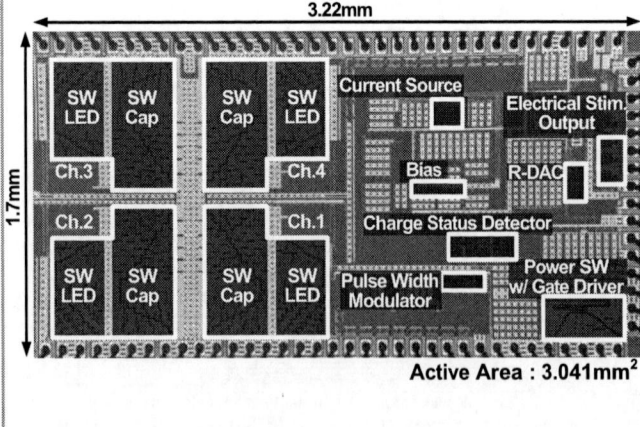

Figure 36.7.5: *Ex vivo* experiments for optogenetic efficacy validation (top); experimental results from mice with ChR2(H134R) opsin expression demonstrating that the hybrid optogenetics needs 70.6% lower energy to induce same neural activation (bottom).

Publication	TBioCAS 22 [4]	CICC 23 [5]	BioCAS 23 [6]	ISSCC 24 [7]	This Work
Technology	130nm CMOS	180nm CMOS	130nm CMOS	180nm CMOS	130nm CMOS
Chip Area (mm²)	3 x 2	2 x 1	2 x 2	1.5 x 1	3.22 x 1.7
Supply Voltage (V)	3.3	1.8	1.3	1.2	4
Stimulation Modality	Optogenetics	Optogenetics	Optogenetics	Optogenetics	Hybrid Optogenetics (w/ Sub-Th Elec Stim)
Optogenetics Topolgy	Toggling CCS	VB-SCS	CCS	LC-SCS	OSCS +Sub-Th CCS
Stimulus Shape	Pulse Train (5kHz, 20%Duty)	Decaying Exponential	Rectangular	Decaying Exponential	Decaying Exponential
Charging Cap (µF)	4.7	100X2	N.A.	22	10
Cap Charging Efficiency (%)	< 50 [A]	< 50 [A]	N.A.	86.4 [D]	94.05
LED Driving Efficiency (%)	< 50 [B]	< 50 [C]	N.A.	85.1 [D]	90.75
Charging Time (µs)	133 [B]	9.36 s [B]	N.A.	1 s	40 (2.5 → 3.8V)
Stim Duration (ms)	5000	1-10	10	1-4	0.4-2
LED Driving Voltage (V)	2.9-3.3	3.6	External	3.6	2.5-3.8
# of Stim Channel	4	1	2	1	4
Area / Ch (mm²)	0.375 [B]	0.1036 [B]	0.05 [B]	0.8 [B]	0.754
Integrated LED Driver	O	O	X (External)	O	O
Optogenetic Efficacy Improvement	X	X	X	X	O (70.64% E_LED Saving)
Verification on Neural Cells	*Ex vivo*	*In vivo*	*In vivo*	*In vivo*	*Ex vivo*
Opsin Compatibility	Specific (Slow-Kinetics)	Universal	Universal	Universal	Universal

[A] Structural limitations of direct charging topology (≤50%), [B] Estimated from the reported data
[C] Assuming discharging efficiency ≈99% (losses from switch R_{ON} & wiring), [D] Estimated from its journal paper [8]

Figure 36.7.6: Performance summary and comparison of the proposed hybrid optogenetic system with state-of-the-art optogenetic stimulators.

Figure 36.7.7: Die micrograph of the hybrid optogenetic system.

ISSCC 2026 / SESSION 36 / NEURAL AND BIOMEDICAL INTERFACES / 36.8

36.8 An 80×80μm²/Pixel 55.48dB-Wide-DR 400-Pixel Subretinal Prosthesis SoC with Power-Aware Light Adaptation and Charge-Recycling Local Dynamic Supply

Kyeongho Eom[1], Hyun-Su Lee[1], Minju Kim[2], Maesoon Im[2], Hyung-Min Lee[1]

[1]Korea University, Seoul, Korea, [2]Korea Institute of Science and Technology, Seoul, Korea

Abstract

We propose a light-adaptive retinal prosthesis (LARP) SoC, which employs power-aware light adaptation achieving a wide light-sensing DR of 55.48dB. Pixel clustering enables a compact pixel size of 80×80μm², enhancing the stimulation spatial resolution. The charge-recycling local dynamic supply scheme enables up to a 34% reduction in stimulation energy. Image projection tests and ex vivo experiments using mouse retinal cells demonstrate the effectiveness of the LARP SoC.

Numerous studies have investigated direct retinal cell stimulation as a strategy to better understand the retinal network and restore vision [1,2]. A retinal prosthesis (RP) system-on-chip (SoC) elicits visual perception by electrically stimulating retinal cells, representing a promising therapeutic option for age-related macular degeneration (AMD) and retinal pigment degeneration. Other approaches, such as optogenetics [3] and visual cortex stimulation [4], are still under investigation to restore visual function.

Figure 36.8.1 (top-left) shows the advantages of subretinal prostheses (Sub-RPs) as well as their existing challenges. Since the Sub-RPs are positioned toward the eye's visual field, they can directly capture images through on-chip photodiodes (PDs) [5-10], eliminating the need for an external camera and excessive wireless data transmission that epiretinal prostheses require [11-13]. Moreover, Sub-RPs take advantage of the residual natural processing capability of the retina by stimulating bipolar cells [1]. Despite these benefits, prior Sub-RP systems have not overcome three key challenges: (1) limited dynamic range (DR), (2) low stimulation efficiency, and (3) low spatial resolution. As depicted in Fig. 36.8.1 (right), we propose the following three solutions to address these challenges of prior Sub-RP systems: light adaptation, on-chip charge recycling, and compact pixel clustering.

While the human visual system has an overall light adaptation (LA) range of 120dB, its instantaneous DR within a single scene is limited to approximately 40dB [14]. Although prior works [15,16] explored LA schemes, they were not fully implemented in SoCs. We propose a light-adaptive Sub-RP SoC to enhance the DR of light sensing. In RP SoCs with continuous stimuli, the stimulator power consumption becomes the dominant factor for heat dissipation and battery life. While adiabatic supply [17-20] or dynamic voltage scaling (DVS) [21,22] has been employed to improve stimulation efficiency, these approaches are difficult to scale for a large number of channels or supply levels, and are thereby not suitable for Sub-RPs. An alternative method in [23] cannot operate standalone since it relies on an additional multi-output rectifier [24] that can provide at most three supply voltages within a limited operating range. In this work, we propose a local DVS method capable of generating six more supply voltages with high efficiency regardless of the load condition. Finally, to further enhance patients' visual acuity, the spatial resolution of stimulation should be improved. To achieve this, we aggressively reduce the pixel area through a pixel clustering and compact circuit configuration.

The 400-pixel light-adaptive retinal prosthesis (LARP) SoC consists of a 400-pixel array and a global controller (GC), as depicted in Fig. 36.8.1 (bottom-left). The GC integrates a bias generator, a serial-to-parallel interface (SPI), and a finite state machine (FSM) while adopting the power-aware light adaptation (PALA) engine and charge-recycling auto-level-sorting multi-supply (CAM) generator. The SPI receives external data to configure the bias voltage (V_{REF}) and current (I_B) for the bias generator as well as timing signals ($S_{G<4:1>}$, S_{SEN}, S_{CATH}, S_{ANO}, S_{CB}) for FSM. The PALA engine and CAM generator are involved in LA and the supply voltage generation for local DVS, respectively. The 400-pixel array is organized into 100 four-pixel clusters. In each cluster, $S_{G<4:1>}$ selects one pixel as the active electrode, while the remaining three pixels serve as local return electrodes. The retinal cells are stimulated through the active electrode, whereas the return electrodes provide the current return path, thereby minimizing current scattering [6,8,10,23].

Figure 36.8.2 (top) shows the detailed schematic of the 4-pixel cluster, including the exposure-controlled light sensor (ECLS), the local dynamic supply (LDS) pixel stimulator, and the LDS controller. The ECLS measures light intensity by counting how many times the PD voltage (V_{PD}) discharges to a reference voltage (V_{TH}) over a specific exposure time (T_{SEN}). To minimize light-sensing mismatch between pixels, the parasitic capacitances of four PDs in each cluster are matched with parasitic equalization. A 6b counter performs this metering, and the exposure time is determined by a signal (S_{SEN}) that is generated by the global power tracker (GPT) in GC. The value of T_{SEN} directly controls the light sensitivity of the ECLS.

The LDS pixel stimulator receives $S_{G<4:1>}$ to determine the active electrode within the cluster and S_{CATH}, S_{ANO}, and S_{CB} to define the cathodic, anodic, and charge-balancing phases, respectively. The LDS pixel stimulator takes the outputs $S_{<5:0>}$ and $S_{B<5:0>}$ from ECLS and applies them to current-steering digital-to-analog converters (I-DACs) to generate the cathodic and anodic stimulation current (I_{STIM}). Then, I_{STIM} flows through the circuit-under-pad (CUP) to the active and return electrodes as determined by group signals, stimulating the retinal cells before returning to the half-rail voltage, V_{HF}. The anodic and cathodic LDS controllers monitor the electrode voltage (V_{EL}) during anodic and cathodic stimulation, respectively. When the supply of I-DAC lacks sufficient voltage headroom, the anodic or cathodic LDS controller increments a 2b counter to select one of four supply voltages, V_{DD1-3} or V_{DD} in the anodic phase, and V_{SS1-3} or V_{SS} in the cathodic phase, and provides it to the pixel stimulator as V_{DD_PX} or V_{SS_PX}.

Figure 36.8.2 (bottom) depicts the operation of the PALA engine and its sub-block, GPT. The PALA engine estimates the stimulation power consumption (P_{TOT}) and compares it to a reference power (P_{REF}) to perform LA. PALA adjusts the light sensitivity to keep P_{TOT} close to P_{REF}: sensitivity is increased when $P_{TOT} < P_{REF}$ and decreased when $P_{TOT} > P_{REF}$. To support monopolar stimulation, a V_{HF} generator provides a stable half-rail voltage (V_{HF}), while triggering a pulse signal (S_{SW}) whenever V_{HF} reaches a specific reference voltage (V_{REF}). On the rising edges of S_{SW}, switching signals (S_1 and S_2) are alternately turned on, connecting the top and bottom plates of the capacitor C_{HF} to V_{DD} and V_{HF}, respectively (or V_{HF} and V_{SS}), to maintain V_{HF} within the range of V_{REF} to $V_{DD} - 2V_{REF}$.

The GPT estimates P_{TOT} from the S_{SW} during cathodic stimulation, as the pulse (T_{SW}) period is inversely proportional to the total current. Then, a control signal (S_{STOP}) can be determined by monitoring the number of pulses ($D_{LA<5:0>}$) over 16 cathodic phases and comparing them to the reference values ($D_{LA\_REF<5:0>}$). If $S_{STOP} = 0$, the rising edge of S_{RST} generated at the end of the 16 cathodic phases increments the 6-bit up-down counter. Conversely, if $S_{STOP} = 1$, the rising edge of S_{RST} decrements the counter value. The counter's output is then fed into the FSM that generates timing signals to modulate the length of T_{SEN}. When T_{SEN} is extended, the ECLS's sensitivity increases, while a shorter T_{SEN} makes it less sensitive.

Figure 36.8.3 (left) depicts the operation of the CAM generator, which uses three compact 10μF capacitors (C_A, C_B, and C_C) to generate six additional supply voltages (V_{DD1-3} and V_{SS1-3}). During the cathodic phase, the bottom plates of C_{A-C} are connected to ground, charging the capacitors to voltages V_A, V_B, and V_C, respectively. The auto-level-sorter (ALS) then arranges V_{A-C} in descending order to provide the highest voltages for stimulation first. For example, as depicted in Fig. 36.8.3, the voltages are sorted as $V_C > V_B > V_A$. In this phase, each pixel cluster sequentially operates with a local DVS voltage (V_{SS_PX}) by connecting it to the highest sorted capacitor voltage first and then downwards. Conversely, during the anodic phase, ALS arranges V_{A-C} in ascending order. The pixel clusters are then sequentially supplied by V_{DD_PX} in ascending order of the sorted capacitor voltage levels, starting with the lowest voltage.

The waveforms in Fig. 36.8.3 (right) highlight the three main features of the CAM generator. 1) Local DVS: CAM applies DVS by providing a different customized supply voltage to each pixel cluster, thereby saving stimulation energy. 2) Charge recycling: by charging C_{A-C} in the cathodic phase and discharging them in the anodic phase, the charge recycling principle allows V_{A-C} to be maintained at specific levels without additional charging or regulation modules. 3) Reliable voltage supply: thanks to ALS, V_{A-C} are reliably sorted into the proper order, ensuring they can be used as stable DVS supplies for the pixels without error. Figure 36.8.3 (middle-right and bottom) depicts the circuit diagram in which ALS connects V_{A-C} to V_{DD1-3} and V_{SS1-3} according to the stimulation phase and the relative order of the capacitor voltage levels.

Stimulation functionality of the LARP SoC was verified with measured results in Fig. 36.8.4 (left). V_{EL} and V_{A-C} waveforms were measured from multiple pixels when the CAM and LDS are applied. The glitches visible in $V_{EL<1:6>}$ occur when the supply voltage changes. The ALS waveforms visually confirm the real-time sorting function of the ALS scheme, which ensures a properly ordered supply. The bottom waveforms show the system operation and performance under various light conditions, resulting in 34% energy saving in dark conditions compared to the fixed-supply system.

The relationships between I_{STIM} and illuminance were measured at different T_{SEN} values, as shown in Fig. 36.8.4 (top-right). A DR was achieved up to 36.12 dB with $T_{SEN} = 5$ms. By applying the PALA, the light-sensing DR can be extended to 55.48dB, which is 19.68dB larger than the prior work that reported the highest DR [23]. Figure 36.8.4 (bottom-right) displays reconstructed images of alphabet patterns (S and C) obtained through the stimuli of all pixels. The left images captured without the PALA scheme are difficult to recognize under both bright and dark conditions. In contrast, the right images, which were obtained using the PALA scheme, are fully recognizable.

ISSCC 2026 / February 18, 2026 / 4:25 PM

Figure 36.8.5 (top) illustrates the *ex vivo* experiment setup with the LARP SoC. To verify the effectiveness of the LARP stimulation, we applied stimuli from both the LARP SoC and the commercial stimulator (STG-4008) to retinal cells of a wild-type mouse and an *rd10* mouse, and activation results were compared. As shown in Fig. 36.8.5 (bottom), the raster plots and the average number of spikes from 7 trials confirm similar stimulation effects across all retinal cells, ensuring effective stimulation with LARP SoC.

Figure 36.8.6 summarizes the performance and compares the LARP SoC with state-of-the-art sub-RP SoCs. With the proposed PALA scheme and ECLS, the highest DR of 55.48dB was achieved. Our LARP SoC uses a 5V supply to ensure sufficient stimulation compliance voltage, resulting in higher static power per pixel, excluding stimulation power, than [8] and [10]; however, the dominant stimulation power was reduced by up to 34% using the CAM and LDS with a monopolar stimulation scheme. Figure 36.8.7 shows the chip photo and pixel layout, demonstrating a 12.4% pixel area reduction compared to the smallest reported work in [7] and a 36% reduction compared to the recent work in [23].

Acknowledgement:
This work was supported by National Research Foundation (RS-2024-00398460, RS-2025-09322969), Korea. Chip fabrication and EDA tool were supported by IC Design Education Center, Korea.

References:

[1] M. Im. and S. I. Fried, "Indirect Activation Elicits Strong Correlations Between Light And Electrical Responses in ON But Not OFF Retinal Ganglion Cells," *The Journal of Physiology*, vol. 593, no. 16, pp. 3577–3596, Aug. 2015. http://doi.org/10.1113/JP270606
[2] Y. J. Yoon *et al.*, "Retinal Degeneration Reduces Consistency of Network-Mediated Responses Arising in Ganglion Cells to Electric Stimulation," *IEEE TNSRE*, vol. 28, no. 9, pp. 1921–1930, Sep. 2020. http://doi.org/10.1109/TNSRE.2020.3003345
[3] T. Yousefi *et al.*, "Closed-Loop 100-Channel Highly-Scalable Retinal Implant with 1.02µW Analog ED-Based Adaptive-Threshold Spike Detection and Poisson-Coded Temporally Distributed Optogenetic Stimulation," *IEEE ISSCC*, pp. 550-552. Feb. 2024. http://doi.org/10.1109/ISSCC49657.2024.10454460
[4] J. Lee *et al.*, "A Wireless Neural Stimulator IC for Cortical Visual Prosthesis," *Symp. VLSI Circuits*, pp. 1-2, Jun. 2023. http://doi.org/10.23919/VLSITechnologyandCir57934.2023.10185375
[5] A. Rothermel *et al.*, "A 1600-pixel Subretinal Chip with DC-free Terminals and ±2V Supply Optimized for Long Lifetime and High Stimulation Efficiency," *IEEE ISSCC*, pp. 144-602, Feb. 2008. http://doi.org/10.1109/ISSCC.2008.4523098
[6] C. -Y. Wu *et al.*, "CMOS 256-Pixel/480-Pixel Photovoltaic-Powered Subretinal Prosthetic Chips with Wide Image Dynamic Range and Bi/Four-Directional Sharing Electrodes and Their *Ex Vivo* Experimental Validations with Mice," *IEEE TCAS-I*, vol. 67, no. 10, pp. 3273-3283, Oct. 2020. http://doi.org/10.1109/TCSI.2020.2976716
[7] J. H. Park *et al.*, "1225-Channel Localized Temperature-Regulated Neuromorphic Retinal-Prosthesis SoC with 56.3nW/Channel Image Processor," *IEEE ISSCC*, pp. 508-509, Feb. 2020. http://doi.org/10.1109/ISSCC19947.2020.9063134
[8] D. -H. Choi *et al.*, "A 4.49nW/Pixel Light-to-Stimulus Duration Converter-Based Retinal Prosthesis Chip," *IEEE TBioCAS*, vol. 15, no. 6, pp. 1140-1148, Dec. 2021. http://doi.org/10.1109/TBCAS.2021.3128418
[9] A. Akinin *et al.*, "An Optically-Addressed Nanowire-Based Retinal Prosthesis with 73% RF-to-Stimulation Power Efficiency and 20nC-to-3µC Wireless Charge Telemetering," *IEEE ISSCC*, pp. 276-278, Feb. 2021. http://doi.org/10.1109/ISSCC42613.2021.9365750

[10] D. -H. Choi and D. -W. Jee, "A 1984-Pixels, 1.26 nW/Pixel Retinal Prosthesis Chip with Time-Domain In-Pixel Image Processing and Bipolar Stimulating Electrode Sharing," *IEEE JSSC*, vol. 58, no. 10, pp. 2757-2766, Oct. 2023. http://doi.org/10.1109/JSSC.2023.3284460
[11] M. Monge *et al.*, "A Fully Intraocular 0.0169mm²/Pixel 512-Channel Self-Calibrating Epiretinal Prosthesis in 65nm CMOS," *IEEE ISSCC*, pp. 296-297, Feb. 2013. http://doi.org/10.1109/ISSCC.2013.6487742
[12] K. Chen, Y. -K. Lo and W. Liu, "A 37.6mm² 1024-Channel High-Compliance-Voltage SoC for Epiretinal Prostheses," *IEEE ISSCC*, pp. 294-295, Feb. 2013. http://doi.org/10.1109/ISSCC.2013.6487741
[13] W. Lemaire *et al.*, "Retinal Stimulator ASIC Architecture Based on a Joint Power and Data Optical Link," *IEEE JSSC*, vol. 56, no. 7, pp. 2158-2170, Jul. 2021. http://doi.org/10.1109/JSSC.2020.3045141
[14] A. Darmont, *High Dynamic Range Imaging: Sensors and Architectures*, Bellingham, WA, USA: SPIE Press, 2012. http://doi.org/10.1117/3.903927
[15] S. Moll *et al.*, "System Design of a Physiological Ambient Illumination Adaptation for Subretinal Stimulator," *IEEE EMBC*, pp. 1962-1965, Jul. 2020. http://doi.org/10.1109/EMBC44109.2020.9175818
[16] K. Eom and H. -M. Lee, "An Average Amplitude Regulation Scheme for Ambient Illuminance Adaptation in Retinal Prosthesis," *IEEE SSCL*, vol. 7, pp. 347-350, 2024. http://doi.org/10.1109/LSSC.2024.3491166
[17] S. Ha *et al.*, "A Fully Integrated RF-Powered Energy-Replenishing Current-Controlled Stimulator," *IEEE TBioCAS*, vol. 13, no. 1, pp. 191-202, Feb. 2019. http://doi.org/10.1109/TBCAS.2018.2881800
[18] Y. Park *et al.*, "A Wireless Adiabatic Stimulator System with Current-Mode Power Reception and Stimulus Current Regulation Achieving Precise Charge Delivery and Electrode Scalability for Miniaturized Electroceuticals," *IEEE ISSCC* Feb. 2025. http://doi.org/10.1109/ISSCC49661.2025.10904531
[19] K. Cui *et al.*, "An Energy-Efficient Wireless Power Receiver with One-Step Adiabatic-Supply Generating for Implantable Electrical Stimulation Applications," *IEEE TBioCAS*, vol. 18, no. 5, pp. 1112-1122, Oct. 2024. http://doi.org/10.1109/TBCAS.2024.3379208
[20] Y. You *et al.*, "A High-Voltage-Compliant 86% Peak Efficiency Current-Mode Stimulator with Dynamic Voltage Supply for Implantable Medical Devices," *IEEE JSSC*, vol. 60, no. 7, pp. 2606-2618, Jul. 2025. http://doi.org/10.1109/JSSC.2024.3505059
[21] K. Eom *et al.*, "A 10-V Tolerant Dual-Mode Neural Stimulation System with Self-Sustaining Dynamic Supply and Error-Resilient Digital Stimulus Odometer," *IEEE JSSC*, vol. 60, no. 9, pp. 3268-3282, Sep. 2025. http://doi.org/10.1109/JSSC.2025.3542022
[22] Z. Luo *et al.*, "A Digitally Dynamic Power Supply Technique for 16-Channel 12 V-Tolerant Stimulator Realized in a 0.18-µm 1.8-V/3.3-V Low-Voltage CMOS Process," *IEEE TBioCAS*, vol. 11, no. 5, pp. 1087-1096, Oct. 2017. http://doi.org/10.1109/TBCAS.2017.2713122
[23] K. Eom *et al.*, "A Stimulus-Scattering-Free Pixel-Sharing Sub-Retinal Prosthesis SoC with 35.8dB Dynamic Range Time-Based Photodiode Sensing and Per-Pixel Dynamic Voltage Scaling," *IEEE ISSCC*, pp. 480-481, Feb. 2023. http://doi.org/10.1109/ISSCC42615.2023.10067387
[24] H. -S. Lee *et al.*, "A 90.8%-Efficiency SIMO Resonant Regulating Rectifier Generating 3 Outputs in a Half Cycle with Distributed Multi-Phase Control for Wirelessly-Powered Implantable Devices," *IEEE ISSCC*, pp. 448-449, Feb. 2024. http://doi.org/10.1109/ISSCC49657.2024.10454403

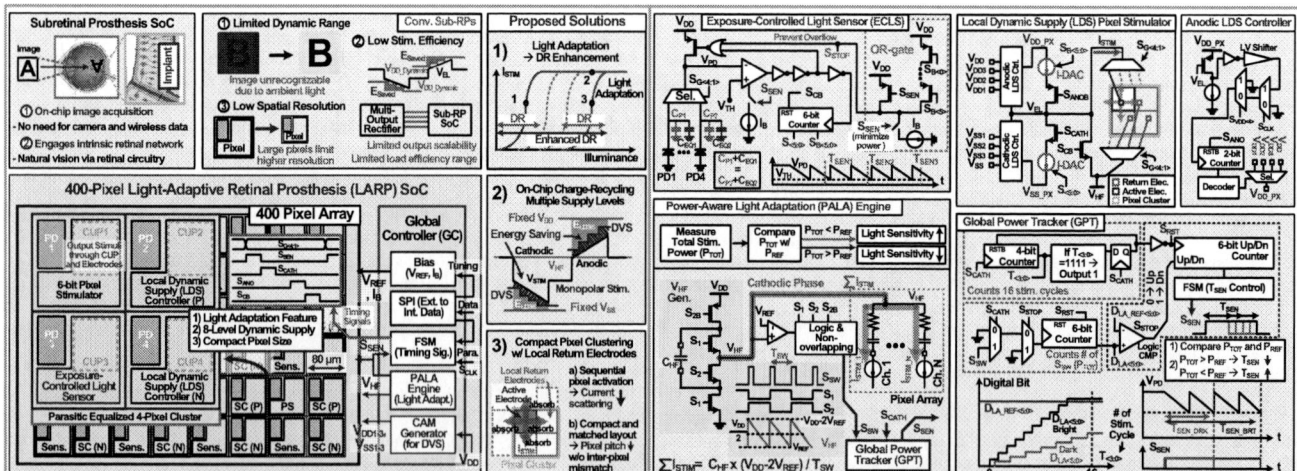

Figure 36.8.1: Advantages of the subretinal prosthesis (sub-RP) system-on-chip (SoC) (top-left), three main challenges of conventional sub-RP SoCs (top-middle), proposed three solutions (right), and a conceptual diagram of the 400-pixel light-adaptive retinal prosthesis (LARP) SoC (bottom-left).

Figure 36.8.2: Schematic diagrams of the exposure-controlled light sensor (ECLS) (top-left), local dynamic supply (LDS) pixel stimulator (top-middle), LDS controller (top-right), and a conceptual diagram of the power-aware light adaptation (PALA) engine (bottom-left) and its sub-block, global power tracker (GPT) (bottom-right).

DIGEST OF TECHNICAL PAPERS • 615

979-8-3315-8937-0/26 $31.00 © 2026 IEEE

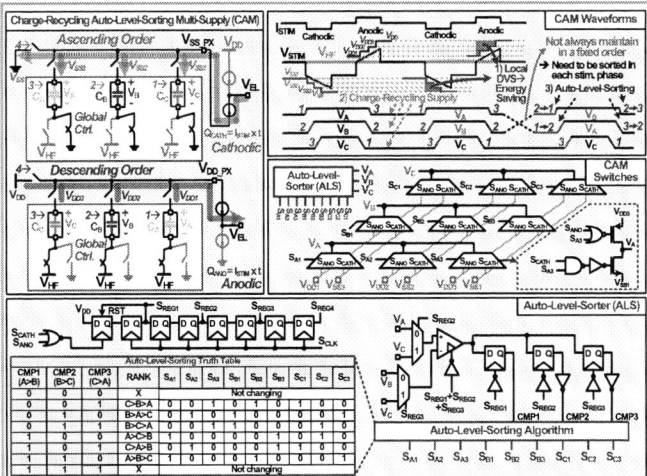

Figure 36.8.3: Charge-recycling auto-level-sorting multi-supply (CAM) scheme and its operation; a conceptual diagram of the CAM (top-left), its operation waveforms (top-right), switch configuration (middle-right), and a conceptual diagram of its sub-block, auto-level-sorter (ALS) (bottom).

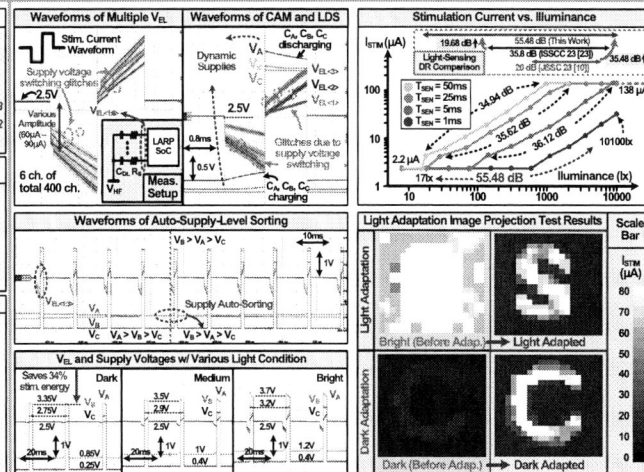

Figure 36.8.4: Measurement results of the LARP SoC, including V_{EL} and supply voltage waveforms with CAM and LDS (top-left), auto-supply-level sorting (middle), and various light conditions (bottom-left); light-to-stimulus relationship and comparison (top-right), and reconstructed images with and without light adaptation (bottom-right).

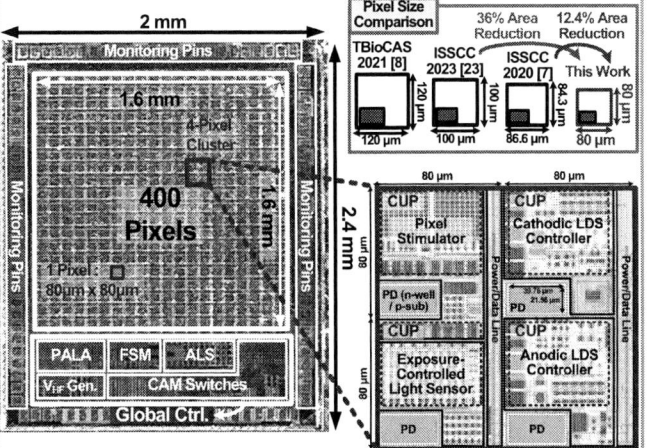

Figure 36.8.5: *Ex vivo* experiments and results: the experimental setup including stimulation parameters and electrode configuration (top) and raster plots with average spike responses from *wild-type* and *rd10* mouse retinal cells comparing the LARP SoC to the commercial stimulator (bottom).

Publication		TCAS-1 20 [6]	ISSCC 20 [7]	TBioCAS 21 [8]	ISSCC 23 [23]	JSSC 23 [10]	This Work
	Technology	180nm CIS	180nm CMOS	180nm CMOS	250nm CMOS	180nm CMOS	250nm CMOS
General	Supply Voltage	1.6V	0.6/±1.6V	1V	5V	1V	5V (Total 8 Supply Levels w/ CAM)
	Chip Area (mm²)	3.2 x 3.2	3.334 x 3.146	2.5 x 2.93	2.4 x 4	3.6 x 3.5	2 x 2.4
	# of Electrode	256	1225	288	505	1984	400
	Pixel Area (μm²)[A]	100 x 100	84.3 x 86.6	120 x 120	100 x 100	110 x 110	80 x 80
	Static Power / Pixel	960 nW	56.3 nW	4.49 nW	335 nW	1.26 nW	60.4 nW
	Power Saving Scheme	No	No	No	DVS w/ multi-output rectifier[B]	No	Charge-Recycling Multi-Supply
	Clock Frequency	0.8 kHz	2 kHz	32 - 64 Hz	1 MHz	1.088 kHz	100 kHz
Electrode	Electrode Location	Sub-retina	Sub-retina	Sub-retina	Sub-retina	Sub-retina	Sub-retina
	Solution for Stimulus Scattering	Local Charge Return	Edge Extraction	Local Charge Return	Local Charge Return	Local Charge Return, Edge Extraction	Local Charge Return
Light Sensor	Light Sensor	Analog	Digital (4b)	Analog (Light-to-duration)	Digital (6b)	Analog (Light-to-duration)	Exposure-Ctrl (6b) + Digital (6b)
	Light input Range	400 - 7400 lx	250 - 2500 lx	600 - 6000 lx	130 - 19000 lx	600 - 6000 lx	17 - 10100 lx
	Light Adaptation Scheme	No	No	No	No	No	Power-Aware Light Adaptation
	Light Sensing Dynamic Range	24.9 dB (Simulated)	24 dB	25.5 dB	35.8 dB	20 dB	55.48 dB
Stim.	I_{STIM} Range	9 μA	0.6/0.8/1mA	Pulse-Width-Ctrl	2.5 - 160 μA	Pulse-Width-Ctrl	2.2 μA - 138 μA
	Stimulation Strategy	Sequential	Simultaneous	Sequential	Sequential	Sequential	Sequential
Implant	Sensing/Stim Tuning	No	No	No	Yes	No	Yes
	Verification on Retinal Cells	*Ex vivo* (rd1 mouse)	No	*Ex vivo*	*Ex vivo* (wt mouse)	*Ex vivo* (wt mouse)	*Ex vivo*[C] (wt & rd10 mouse)

[A] Based on the electrode pitch. [B] Utilized 3-output regulating rectifier in [24] [C] Performed comparison with the commercial stimulator

Figure 36.8.6: Performance summary and comparison with state-of-the-art sub-RP SoCs.

Figure 36.8.7: Photograph of the 400-pixel LARP SoC (left), detailed layout of the 4-pixel cluster (bottom-right), and pixel size comparison (top-right) demonstrating the 80×80μm² pixel with 12.4% area reduction compared to the smallest pixel from prior work.

ISSCC 2026 / SESSION 36 / NEURAL AND BIOMEDICAL INTERFACES / 36.9

36.9 A CMOS Neural Probe with 1280 Electrodes and 88 Emission Sites Featuring Thermo-Optic Switching and On-Chip Calibration for Dual-Wavelength Optogenetics

Xiaolin Yang*, Pieter Neutens*, Alexis Humblet, Chutham Sawigun, John O'Callaghan, Thijs Geurts, Zheyi Li, Yaroslav Gubin, Koen De Munck, Harrie A. C. Tilmans, Anabel De Proft, Barundeb Dutta, Carolina Mora Lopez

imec, Heverlee, Belgium
*Equally Credited Authors (ECAs)

Abstract

We present a waveguide-based optogenetics neural probe with 1,280 electrodes, 384 recording channels, and 88 emission sites enabled by dual-layer silicon photonics. Thermo-optic switches are controlled by a custom photonic driver IC, while integrated tap couplers with photodiodes and a temperature sensor provide on-chip calibration and thermal monitoring. The probe achieves a large number of recording and emission sites, enabling reliable large-scale, cell-type-specific optogenetic studies.

Studying brain function at the single-cell level requires tools that offer both high-resolution recording and cell-type specificity [1]. High-density CMOS neural probes [2-5] provide excellent spatiotemporal resolution for recording neuronal activity but lack the ability to distinguish cell types based on molecular identity. Optogenetics [6], by contrast, complements this limitation by enabling targeted control of genetically defined neuronal populations. Combined, these techniques offer a powerful approach for dissecting neural circuit function. Different approaches have been proposed to enable neural recording and optogenetics in the same implantable device (Fig. 36.9.1, left). Optrodes [7,8] integrate optical fibers with wire electrodes for direct implantation in target brain regions. They offer reliable fabrication and sufficient light delivery for stimulation but are not scalable to large numbers of sites, provide limited spatial resolution, and cannot be directly integrated with CMOS technology. Optogenetic probes incorporating µLEDs or OLEDs [9-12] improve spatial resolution and enable CMOS integration, yet they demand high on-shank power, show limited uniformity across sites, and deliver limited light intensity [11]. Waveguide (WG) based probes [13-22] address some of these challenges by supporting higher light intensities and CMOS compatibility while shifting the power consumption away from the shank. However, they remain constrained by limited scalability, spatial resolution, and severe variability in light delivery arising from propagation losses along the long WGs and fabrication-induced imperfections. In this work, we introduce a WG-based optogenetic neural probe that integrates CMOS electronics with post-processed dual-layer silicon photonics and low-impedance electrodes (Fig. 36.9.1, right). This platform enables high-resolution recording together with high-intensity, dual-wavelength optical stimulation. Compared to previous works, its key innovation is the incorporation of heater-tunable WGs, optical calibration based on on-chip tap couplers, and temperature sensing, which together provide site-to-site uniformity and thermal control for reliable, reproducible optotagging.

The high-level architecture of the neural probe is shown in Fig. 36.9.2. The recording subsystem comprises 1280 switchable electrodes in the shank, which can be selectively connected to 384 low-noise, AC-coupled readout channels. These channels are time-multiplexed at 30kS/s per channel and digitized by 24 12b SAR ADCs. The photonic subsystem for optogenetics couples light from a polarization-maintaining fiber array into single-mode WGs via out-of-plane grating couplers. The optical power is then distributed through 6-level thermo-optic switch (SW) trees to 44 ESs per wavelength. By combining dual-layer photonics with spatial-division multiplexing, the probe delivers light to 2×44 ESs along the shank. For calibration, ~1% of the output power is tapped via directional and grating couplers and measured by 88 integrated photodiodes (PDs) with capacitive transimpedance amplifiers (C-TIAs) and 48:1 time-multiplexed ADCs. Eight additional PDs track the dark current, while a temperature sensor near the shank provides real-time monitoring through a 12b SAR ADC to limit tissue heating during stimulation. The ADCs employ a split capacitive DAC with thermometer-coded MSBs, a redundant bit, and binary-weighted LSBs with reference scaling, achieving improved linearity in a compact area. On-chip power management, including BGRs, LDOs and bias generation, together with a digital control unit for configuration and data serialization, greatly reduces the need for external components. To deliver the high currents (up to 10mA) required by the thermo-optic SW heaters, the system incorporates a custom driver ASIC comprising an array of 12 10b current DACs (IDACs) time-multiplexed to 96 HV output stages.

Figure 36.9.3 (top-left) details the photonic SW, implemented as a symmetric Mach-Zehnder interferometer (MZI) with an integrated TiN heater. Due to process-induced phase variations, nominally identical MZIs exhibit different responses and therefore require calibration. During on-chip heater calibration, the PD outputs are recorded while sweeping the heater power across 9 equidistant points, followed by a finer 7-point scan around the first minimum (Fig. 36.9.3, top-right). Curve fitting then determines the heater powers needed to switch the optical signal between the upper and lower WGs. This procedure is repeated for 90 heaters. SW thermal efficiency is limited by process integration constraints. Figure 36.9.3 (bottom) shows the schematics of the C-TIAs and the temperature sensor. The TIA employs a two-stage OTA with a class-AB rail-to-rail output stage and cascoded Miller compensation (C_{CM1}, C_{CM2}), directly driving the ADC at 9.6kHz. Thick-oxide input transistors minimize gate leakage, while a programmable gain (10-70MΩ) and a 200Hz, 1/48 duty-cycled synchronous reset (Φ_{RST}) enable the TIA to handle up to 100nA input current, ensuring accurate operation across a wide optical-power range. The PD consists of a 10×10µm² metal-defined aperture

over an NWell in a lightly doped P-type SOI substrate. The local interconnect and silicidized area are minimized to enhance light coupling, while a surrounding guard ring reduces spurious light detection. An ultra-low-leakage (ULL) switch between the PD and TIA enables per-channel gain and noise characterization. The temperature sensor is implemented as a vertical PNP-based PTAT voltage generator followed by a 10× amplifier to match the dynamic range of the ADC. It covers a 20-to-50°C range with 0.5°C resolution, achieved using chopper stabilization at 2.4kS/s. Chopping-induced ripples are suppressed by a 1st-order gm-C low-pass filter (300Hz cutoff) built from 2 subthreshold OTAs with 20nA tail current each. The OTAs are configured in unity-gain feedback to minimize low-frequency distortion without additional linearization [17].

The optogenetics probe was fabricated in a 130nm CMOS technology. The post-CMOS fabrication process flow (Fig. 36.9.4, top) adds dual-layer SiN photonics with reflector mirrors, thermo-optic heaters, and low-impedance TiN electrodes on top of the CMOS, using oxide planarization, TSV-like connections, and VIA steps before final passivation, dicing, and packaging. The probe base occupies 66.9mm², while the implantable shank is 10mm long and 70µm wide. The photonic driver IC was fabricated in a 180nm HV technology and occupies 17.8mm². The chip micrographs are shown in Fig. 36.9.4 (bottom).

The electrical performance of the fabricated probe (Fig. 36.9.5) is measured for the 384 recording channels, 96 PDs, 96 C-TIAs, and temperature sensor through the ADCs and digital interface. The recording channels have a gain of 99.2±1.0V/V and input-referred noise of 6.0±0.7µV$_{rms}$ and 4.7±0.8µV$_{rms}$ in the action-potential (300Hz to 10kHz) and local-field-potential (0.5Hz to 1000Hz) bands, respectively. The measured C-TIA gain is 9.9±0.4MΩ for the lowest gain setting, with an input-referred noise of 21.3±5.6pA$_{rms}$. At the highest gain setting (not shown), the gain is 69.1±3.8MΩ with an input-referred noise of 5.0±0.5pA$_{rms}$. The temperature sensor was characterized against a thermistor reference over a 25 to 55°C range for 3 dies. After a 2-point calibration, the measured temperature error is within ±0.25°C.

The optical characterization of the probe was performed using the photonic driver IC to control the thermo-optic SW trees. The PD and heater performances were first evaluated with on-chip test structures. The responsivity of 4 PDs (Fig. 36.9.6, top-left) was measured at a 0.6V reverse bias. The devices exhibited average responsivities of 0.138A/W at 450nm and 0.255A/W at 638nm, corresponding to quantum efficiencies of 37.9% and 49.6%, respectively. The response of the complete photodetection chain (PD+TIA+ADC) to optical power is shown in Fig. 36.9.6 (top-middle). For this measurement, the optical fiber was directly aligned to a single PD and two TIA gain settings were tested. The detection chain demonstrates reliable operation for optical powers up to ~0.4µW (λ=430nm) and ~0.2µW (λ=638nm). The total power consumption of the optogenetics probe is 23.4mW at 1.8V with all circuits enabled. The photonic driver consumes 121.9mW (24.2mW from the IDACs at 1.8V and 97.7mW from the HV output stages at 10V), while the heater array consumes 27.1mW per ES.

Figure 36.9.6 (bottom-left) shows the resistance histogram of 90 heaters, with an average of 1544Ω and a standard deviation of 4.7Ω, demonstrating excellent process uniformity and repeatability. After full calibration of the optical-SW tree, each ES on the shank can be selected by applying the correct current to 6 thermo-optic SWs. Fig. 36.9.6 (right-top) shows a bright-field microscope image of the SW tree and its operation at 450nm when selecting detector 1 (ES1) and detector 44 (ES2). Figure 36.9.6 (bottom-right) displays the shank with all even-numbered ESs activated alternatively, demonstrating programmable optical ESs selection across the full shank length. To characterize the power uniformity of the 450nm ESs, the output from each ES was collected using a 100µm core, 0.39NA multimode fiber. Figure 36.9.6 (bottom-middle) shows the histogram of the measured powers across all 44 ESs, demonstrating excellent uniformity along the shank. Residual variations arise from propagation loss and WG width adjustments used to suppress evanescent coupling between neighboring structures. However, because the PD response maintains a fixed relation to the ES output power, these variations can be compensated by adjusting the laser power to achieve a more uniform distribution along the shank. The average emission power is 6.1µW, corresponding to a power density of 282mW/mm² at the grating coupler surface.

616 • 2026 IEEE International Solid-State Circuits Conference

979-8-3315-8937-0/26 $31.00 © 2026 IEEE

Figure 36.9.7 compares the CMOS WG-based optogenetic probe with state-of-the-art neural probes featuring optical stimulation. In addition to offering the highest number of recording and ESs, it integrates on-chip optical calibration and temperature sensing, ensuring uniform light delivery and reliable thermal control. Optical switching is performed on chip using thermal phase shifters (TPSs), while a co-integrated photonic driver IC provides the required control currents, eliminating the need for bulky external drivers. The recording channels and C-TIAs provide low-noise performance, while the temperature sensor achieves high accuracy. Following calibration, our probe demonstrates the lowest ES output variation reported among optogenetics probes. Moreover, since WGs do not require local drivers, the shank power consumption remains negligible compared to µLED-based approaches. These features make this probe a scalable and reliable platform for large-scale, cell-type-specific optogenetic studies *in vivo*.

References:

[1] G. Buzsáki *et al.* Tools for probing local circuits: high-density silicon probes combined with optogenetics. *Neuron* 86, 92-105 (2015). https://doi.org/10.1016/j.neuron.2015.01.028

[2] G.N. Angotzi *et al.*, 2025. Multi Shank 1024 Channels Active SiNAPS Probe for Large Multi Regional Topographical Electrophysiological Mapping of Neural Dynamics. Advanced Science, 12(16), p. 2416239. https://doi.org/10.1002/advs.202416239

[3] C. Mora Lopez *et al.*, "A Neural Probe With Up to 966 Electrodes and Up to 384 Configurable Channels in 0.13 µm SOI CMOS," in *IEEE Transactions on Biomedical Circuits and Systems*, vol. 11, no. 3, pp. 510-522, June 2017. http://doi.org/10.1109/TBCAS.2016.2646901

[4] S. Wang *et al.*, "A Compact Quad-Shank CMOS Neural Probe With 5,120 Addressable Recording Sites and 384 Fully Differential Parallel Channels," in *IEEE Transactions on Biomedical Circuits and Systems*, vol. 13, no. 6, pp. 1625-1634, Dec. 2019. http://doi.org/10.1109/TBCAS.2019.2942450

[5] X. Yang *et al.*, "A Highly-Integrated 1536-Channel Quad-Shank Monolithic Neural Probe in 55nm CMOS for Full-Band Raw-Signal Recording," *IEEE Symposium on VLSI Technology and Circuits*, 2024, http://doi.org/10.1109/VLSITechnologyandCir46783.2024.10631430

[6] C.K. Kim *et al.*, "Integration of optogenetics with complementary methodologies in systems neuroscience." *Nature Reviews Neuroscience* 18, no. 4 (2017): 222-235. https://doi.org/10.1038/nrn.2017.15

[7] J. Petrovic *et al.*, 2023. Fiber-based optrode with microstructured fiber tips for controlled light delivery in optogenetics. *Journal of Neural Engineering*, 20(3), p.036007. http://doi.org/10.1088/1741-2552/accecf

[8] D. Budai *et al.*, 2018. A novel carbon tipped single micro-optrode for combined optogenetics and electrophysiology. *PLoS One*, 13(3), p.e0193836. https://doi.org/10.1371/journal.pone.0193836

[9] F. Wu *et al.*, "Monolithically integrated µLEDs on silicon neural probes for high-resolution optogenetic studies in behaving animals." *Neuron* 88, no. 6 (2015): 1136-1148. https://doi.org/10.1016/j.neuron.2015.10.032

[10] M. Vöröslakos, *et al.*, HectoSTAR µLED Optoelectrodes for Large-Scale, High-Precision In Vivo Opto-Electrophysiology. *Adv. Sci.* 2022, 9, 2105414. https://doi.org/10.1002/advs.202105414

[11] A. Taal *et al.* Optogenetic stimulation probes with single-neuron resolution based on organic LEDs monolithically integrated on CMOS. *Nat Electron* 6, 669–679 (2023). https://doi.org/10.1038/s41928-023-01013-y

[12] A.M. Clark *et al.* An optrode array for spatiotemporally-precise large-scale optogenetic stimulation of deep cortical layers in non-human primates. *Commun Biol* 7, 329 (2024). https://doi.org/10.1038/s42003-024-05984-2

[13] F. Wu *et al.*. An implantable neural probe with monolithically integrated dielectric waveguide and recording electrodes for optogenetics applications. *J Neural Eng.* 2013 Oct;10(5):056012. http://doi.org/10.1088/1741-2560/10/5/056012

[14] L. Hoffman *et al.*, High-density optrode-electrode neural probe using SixNy photonics for in vivo optogenetics. *IEEE International Electron Devices Meeting*, 2016, p. 29.5.1– 29.5.4. http://doi.org/10.1109/IEDM.2015.7409795

[15] K. Kampasi *et al.* Dual color optogenetic control of neural populations using low-noise, multishank optoelectrodes. *Microsyst Nanoeng* 4, 10 (2018). https://doi.org/10.1038/s41378-018-0009-2

[16] A. Mohanty *et al.* Reconfigurable nanophotonic silicon probes for sub-millisecond deep-brain optical stimulation. *Nat Biomed Eng* 4, 223–231 (2020). https://doi.org/10.1038/s41551-020-0516-y

[17] P. Neutens *et al.*, "Dual-wavelength neural probe for simultaneous opto-stimulation and recording fabricated in a monolithically integrated CMOS/photonics technology platform," *IEEE International Electron Devices Meeting*, 2023, http://doi.org/10.1109/IEDM45741.2023.10413839

[18] A. Lakunina *et al.* "Neuropixels Opto: Combining high-resolution electrophysiology and optogenetics." bioRxiv (2025). https://doi.org/10.1101/2025.02.04.636286

[19] F.D. Chen *et al.* Implantable silicon neural probes with nanophotonic phased arrays for single-lobe beam steering. *Commun Eng* 3, 182 (2024). https://doi.org/10.1038/s44172-024-00328-8

[20] F.D. Chen *et al.* Implantable nanophotonic neural probes for integrated patterned photostimulation and electrophysiological recording. *npj Biosensing* 2, 15 (2025). https://doi.org/10.1038/s44328-025-00024-3

[21] D.A. Roszko *et al.*, "Foundry-fabricated dual-color nanophotonic neural probes for photostimulation and electrophysiological recording." *Neurophotonics* 12, no. 2 (2025): 025002-025002. http://doi.org/10.1117/1.NPh.12.2.025002

[22] W.D. Sacher *et al.*, "Visible-light silicon nitride waveguide devices and implantable neurophotonic probes on thinned 200 mm silicon wafers," *Opt. Express* 27, 37400-37418 (2019). https://doi.org/10.1364/OE.27.037400

[23] C. Sawigun and W. A. Serdijn, "A modular transconductance reduction technique for very low-frequency Gm-C filters," *IEEE International Symposium on Circuits and Systems*, 2012, pp. 1183-1186, http://doi.org/10.1109/ISCAS.2012.6271445

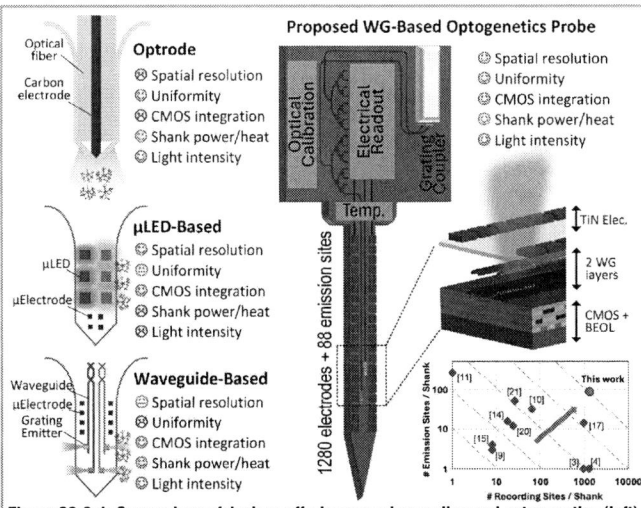

Figure 36.9.1: Comparison of devices offering neural recording and optogenetics (left). Concept of the proposed CMOS WG-based optogenetics probe with 1280 electrodes for neural recording and 88 emission sites for optical stimulation (right).

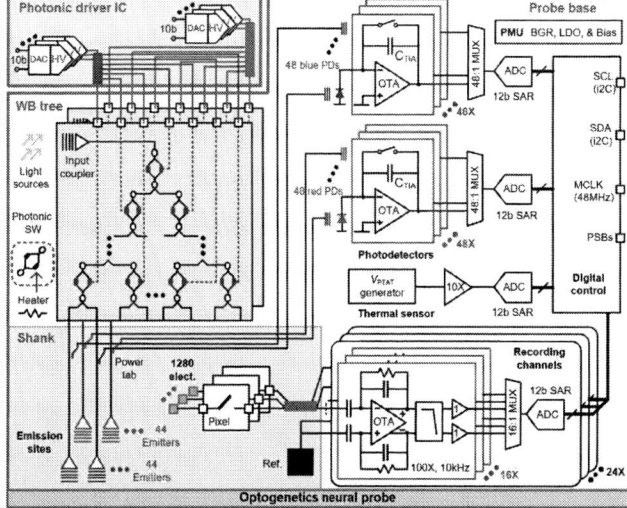

Figure 36.9.2: High-level architecture of the optogenetics neural probe and the photonics driver IC.

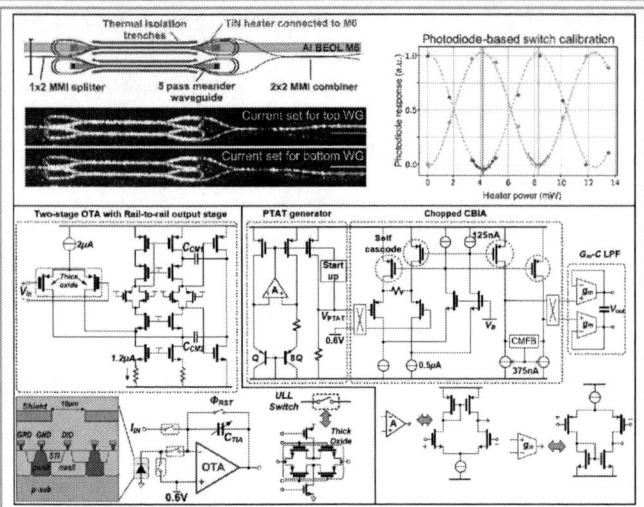

Figure 36.9.3: Detailed photonic SW, and measured PD response vs. heater power during calibration (top); PD implementation and schematics of the C-TIA and temperature sensor (bottom).

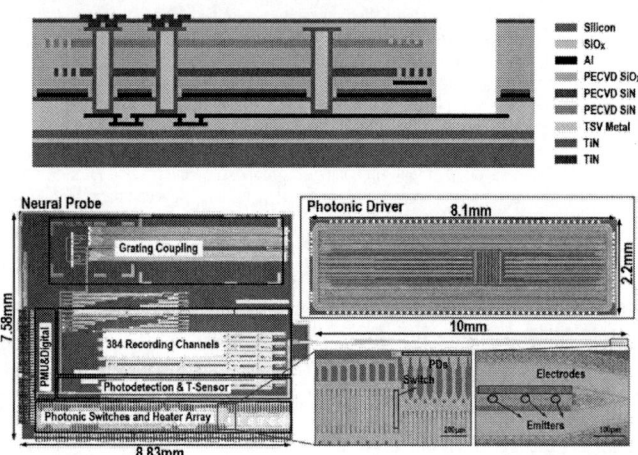

Figure 36.9.4: Simplified cross-section of the 200mm CMOS-photonics integration process to fabricate the WG-based optogenetics probe (top); micrographs of the fabricated probe and photonics driver IC (bottom).

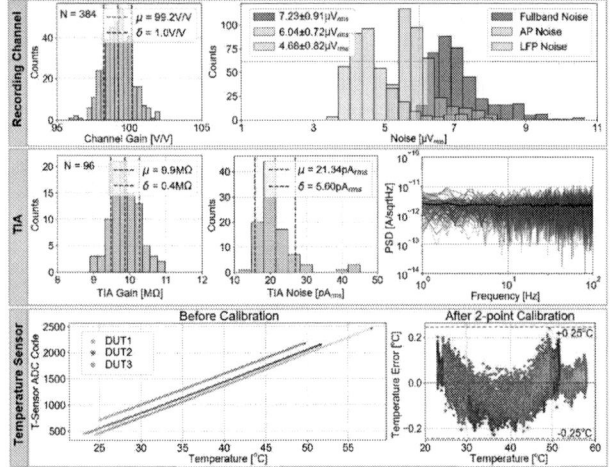

Figure 36.9.5: Measurements of the recording subsystem: end-to-end recording-channel gain and input-referred noise, C-TIA gain and input-referred noise, temperature sensor output vs. temperature before calibration, and temperature error after a 2-point calibration.

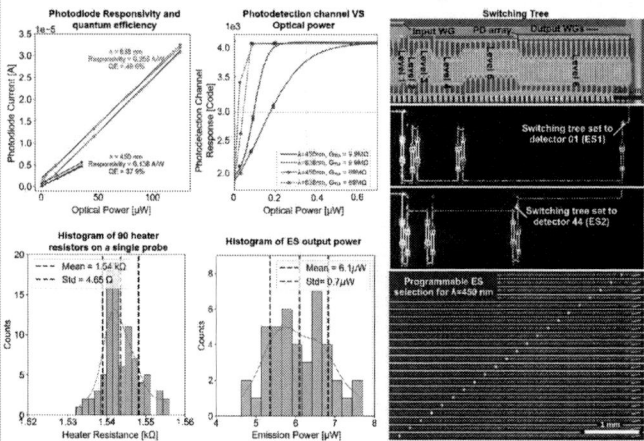

Figure 36.9.6: Measurements of the optical subsystem: PD and heater characterization, photodetection channel response, ES output power histogram, operation of the SW tree at 450nm, and demonstration of the optical ES selection along the shank.

	Vöröslakos et al. (2022)	Neutens et.al (2023)	Taal et al. (2023)	Clark et.al (2024)	Chen et al. (2025)	Rozko et al. (2025)	This work
CMOS Technology	–	130nm	–	–	–	–	130nm, 180nm HV
Probe Specifications							
No. of Shanks	4	1	4	10×10 (Array)	1	1	1
Shank Dimension	6mm x 144µm	10mm x 70µm	6mm x 100µm	2.5mm x 120µm	4mm x 100µm	6.09mm x 70µm	10mm x 70µm
# Electrodes per Shank	64	960	--	--	18	26	1280
Electrode Pitch	N/A	20µm	--	--	80µm	188µm	15µm
No. of Rec. Channels	Off-chip	384	--	--	Off-chip	Off-chip	384
# ESs per Shank	32	14	258	1	16	52	88
ES Pitch	40µm	100µm	24.5µm	400µm	80µm	188µm	100µm
ES Source	µLED	WG	OLED	µLED	WG	WG	WG
Wavelength	470nm	450nm, 638nm	500nm, 615nm	450nm	488nm	473nm, 638nm	450nm, 638nm
Opt. SW Mechanism	External LED driver	On-chip TPS	On-chip CMOS switch	External LED driver	External laser alignment	External laser alignment	On-chip TPS & Photonic Driver IC
Optical Calibration	No	No	No	No	No	No	Yes
No. of Cal. Channels	--	--	--	--	--	--	88 w/ on-chip PD
Temperature Sensor	No	No	No	No	No	No	Yes
Electical & Optical Specifications							
Supply	4V	1.2V/1.8V	7V	5V	--	--	1.2V/1.8V/10V
Rec. Channel Gain	--	50-3000 V/V	--	--	--	--	99.2V/V
Rec. Channel Noise	--	5.24µV$_{rms}$ (AP) 5.05µV$_{rms}$	--	--	--	--	6.0µV$_{rms}$ (AP) 4.7µV$_{rms}$ (LFP)
C-TIA Gain	--	--	--	--	--	--	9.9-69.1MΩ
C-TIA Noise	--	--	--	--	--	--	5.6-21.3pA$_{rms}$
Max. Temp Error	--	--	--	--	--	--	±0.25°C (2-point cal)
Driver Current Range	--	--	--	--	--	--	0-10.2mA
ES Output Density	0.6mW·mm^{-2}	>10mW·mm^{-2}	0.25mW·mm^{-2}	3.79mW·mm^{-2}	84.2mW·mm^{-2}	25.6mW·mm^{-2}	282mW·mm^{-2} ***
ES Output Variation*	14.9%	N/A	55%	54.9%	30.5%**	281%	11.5%
Shank Power	300µW/ES	N/A	49µW/ES (615nm) 245µW/ES (500nm)	--	--	--	<1µW
Total Power	N/A	15mW	--	N/A	--	--	23.4mW (probe) 121.9mW (driver)

* Coeff. of variation = SD/mean; ** Estimated from provided data; *** Calculated at the grating coupler surface;
N/A: Not available; TPS: Thermal phase shifter; AP: action potential; LFP: local field potential;

Figure 36.9.7: Comparison with recent state-of-the-art optogenetic neural probes.

ISSCC 2026 / SESSION 36 / NEURAL AND BIOMEDICAL INTERFACES / 36.10

36.10 A 43.8-to-662.0µW 27.5-to-878.9fps Frame-Rate-Scalable Duty-Cycled Electrical Impedance Tomography System with MIMO Current-Balancing IA

Haidam Choi*[1], Gichan Yun*[1], Ji-Hoon Suh[2], Seonwoo Park[1], Donghyeon Yi[1], Sohmyung Ha[3], Minkyu Je[1]

[1]KAIST, Daejeon, Korea, [2]University of California, San Diego, CA, [3]New York University Abu Dhabi, Abu Dhabi, United Arab Emirates
*Equally Credited Authors (ECAs)

Abstract

This paper presents a frame-rate-scalable duty-cycled electrical impedance tomography system to reduce average power consumption. A proposed multiple-input multiple-output current-balancing IA halves power consumption, area, and input parasitics. Digital matched filters with harmonic-canceling FIR filters significantly reduce settling time, improving system throughput. The system achieves 878.9fps at 662.0µW and 27.5fps at 43.8µW.

Electrical impedance tomography (EIT) systems have been explored beyond their traditional imaging applications, such as lung ventilation monitoring [1–4], extending into human-computer interface, including gesture recognition [5,6]. Despite its high potential, adapting the EIT system into portable and real-time interactive platforms remains challenging, primarily due to two critical bottlenecks: high power consumption and low frame rates (Fig. 36.10.1 (top middle)). First, conventional EIT systems sequentially inject an excitation current through electrode pairs and record the resulting voltages on the remaining electrodes (Fig. 36.10.1 (top left)). Since the recorded signal amplitude is proportional to the current amplitude of the current generator (CG), injecting tens to hundreds of µA of current is necessary to measure small impedance changes, which dominates the system's power consumption. Additionally, multiple wide-bandwidth amplifiers are required in the readout front-end (RFE) for recording the modulated input signals, further increasing power consumption. As a result, conventional EIT systems typically dissipate mW-level power [1,7], restricting their use in battery-powered wearable platforms. Second, conventional EIT systems employ I/Q demodulation, which requires a low-pass filter (LPF) with a large time constant to suppress unwanted harmonics, limiting the achievable frame rate. Previous efforts have implemented pre-demodulation techniques that down-convert the signal to DC [2] or intermediate frequency (IF) [8] before the front-end amplifier, thereby reducing the power consumption of the RFE. However, these approaches still suffer from a long settling time. Alternatively, the IF-sampling with digital matched filter mitigates the settling issue and achieves higher system throughput, but at the expense of increased power consumption due to intensive digital filtering [9]. Nonetheless, none of these approaches addresses the dominant power overhead arising from continuous current injection of the CG.

This work presents a frame-rate-scalable duty-cycled EIT system, which achieves good power efficiency with a high frame rate. Figure 36.10.1 (bottom) shows the overall system architecture with the following features: 1) duty-cycled operation of the CG and RFE, which are fully off in the sleep phase and powered on with fast settling in the active phase, reducing the average power consumption significantly; 2) FIR filters and digital matched filters to suppress unwanted harmonics and reduce settling time; 3) a multiple-input multiple-output (MIMO) current-balancing instrumentation amplifier (CBIA) with embedded switching matrix, reducing both area and power consumption compared to conventional CBIA designs.

Figure 36.10.2 illustrates the main operation of the proposed duty-cycled EIT system. During the wake-up phase, all analog circuits are initially powered on to ensure a stable operation before data acquisition. Once stabilized, the entire system is activated during the data acquisition phase. After acquiring data for one image frame, it is transferred to the SPI protocol, while the entire CG and RFE are powered down. Then, the system enters sleep phase, where only the bandgap reference (BGR) and clock generator remain active, thereby reducing overall average power consumption.

When switching the CG connection between the electrodes, the high-pass filter (HPF) exhibits a settling behavior due to abrupt DC voltage changes. Using a switch connected in parallel with the resistor of the HPF can reduce the settling time by turning it on and off at the zero-crossing points of the input signal [10]. However, ensuring the switch toggles at the zero-crossing points is challenging due to the phase shift of the impedance interface. To address this, the current generator is temporarily connected to the internal resistor (R_S) during channel switching ($0.5 \cdot T_{CG}$), ensuring that the HPF conserves the signal exactly at its zero-crossing points, as shown in Fig. 36.10.2 (top right). When the output switches of CG are disconnected from electrodes, the switch of HPF is shorted to the reference voltage (V_{REF}). This prevents the transient voltages from being sampled by the input capacitors.

After HPF, the signal is amplified by the MIMO CBIA and transferred to the signal demodulation chain (Fig. 36.10.2 (top left)). Here, an IF frequency-shifting chopper is employed to shift the input signal's frequency to an IF (f_{IF}). Due to the square-wave chopper signal, the signal after the IF chopping contains large IF-converted harmonics, which typically induce significant measurement errors. To overcome this, we propose to use an 11-tap FIR filter implemented in the digital domain with an LPF characteristic and nulls that are exactly placed at these harmonics (Fig. 36.10.2 (top left)). The IF chopping also up-converts low-frequency interferences, such as DC offsets, flicker noise, and residual DC settling-induced drifts, to one of the nulls of the FIR filter, allowing them to be effectively

removed. Most of the aliasing components outside the Nyquist band are also placed at the nulls, while high-order harmonics that could alias into f_{IF} are mostly attenuated by the LPF characteristics of the PGA stage. After waiting for a few initial samples, the digital matched filter begins I/Q demodulation and accumulation over an integer multiple of f_{IF} cycles, which corresponds to a multiple of 8 samples. Unlike prior off-chip implementation [5], the digital matched filter is fully integrated on-chip and operates with only a small number of samples, significantly enhancing system throughput in a power-efficient manner. The measured waveforms of the FIR filter output, along with a detailed timing diagram, are shown in Fig. 36.10.2 (bottom), demonstrating the fast-settling and harmonic-canceling characteristics of the filter. After the active phase, the system enters sleep mode with a duty-cycling ratio adjustable between 63.4% and 2.0%.

To further enhance system power efficiency, we propose a MIMO CBIA structure, achieving significant reductions in power consumption and area (Fig. 36.10.3). In conventional CBIA, two flipped voltage followers (FVFs) and a resistor in the transconductance (TC) stage convert the input voltage into a current, which then flows through the output resistor via the transimpedance (TI) stage to generate the output voltage. Instead, the proposed MIMO CBIA structure configures each single-ended TC and TI stage to share AC ground nodes ($V_{TC,S}$ and $V_{TI,S}$) across multiple channels by connecting both ends of the resistors within each single-ended stage. By doing so, the number of circuit components is halved compared to the conventional CBIA (Fig. 36.10.3 (top)). With the AC ground node shared across stages, the signal gain can be calculated as R_{TI}/R_{TC}, where the common-mode signal included in the output signals is removed through differential operation in the PGA following the IF frequency-shifting chopper. Each stage adopts a passing-through IA, eliminating the current mirror, thereby resolving input-dependent noise [11]. Note that the two TC stages corresponding to the electrodes currently connected to the CG are temporarily powered down ($D_{TC}[i]$ operation) to prevent saturation of the TC stage. Furthermore, a switching matrix is embedded in the MIMO CBIA to reconfigure the channel connections according to the CG sequence. The switching matrix is located between the TC and TI stages, which are low-impedance nodes, eliminating the need for large input switches. Along with the switching matrix operation, PGAs with an inverted V-shaped gain configuration are implemented to relax the dynamic range requirement [2]. Moreover, unlike conventional EIT systems (Fig. 36.10.1 (top middle)) that require two FVFs per electrode, the proposed design uses only one FVF per electrode, reducing the parasitic capacitance at the input. Consequently, the MIMO CBIA reduces power consumption, area, and input parasitics by half while preserving gain and noise performance and supporting multi-channel scalability.

The frame-rate-scalable duty-cycled EIT system is fabricated in 180nm CMOS and occupies a total chip die area of 7.5mm² (Fig. 36.10.7). The measured output spectrum plots of the ADC and the FIR filter are shown in Fig. 36.10.4 (top). The measurement was conducted using a 50Ω resistor with 50µA$_{pk}$ CG current, and the data were continuously acquired to validate the transfer function of the FIR filter. It is clearly shown that the IF-converted harmonics and up-converted DC components are observed at the ADC output, but are effectively suppressed by the notches of the FIR filter. The input-referred noise (IRN) of the RFE is measured as 8.31mΩ/√Hz at 878.9fps frame rate with duty-cycling operation. Figure 36.10.4 (bottom left) shows the total average power and noise for different numbers of samples accumulated by the digital matched filter. The accumulation of 16 samples was chosen as the operating mode to provide sufficient noise performance and power efficiency. Figure 36.10.4 (bottom right) illustrates the performance across different frame rates. It demonstrates that power consumption is scalable with the frame rate, while the IRN is maintained under 174.1mΩ$_{rms}$, enabling flexible operation of the EIT system in terms of the frame rate and power consumption, depending on the desired performance specified by the target application. The measured accuracy using a resistive-mesh phantom [12] achieved a mean of 98.8% and a minimum of 96.1% compared to the theoretical values.

Figure 36.10.5 (top) shows the impedance measurement results for a 60mm-diameter dish with an object (or two objects) placed in a saline solution at different positions. The images were reconstructed using an open-source software (EIDORS) [13], and the imaging results clearly demonstrate the functionality of the prototype IC. Figure 36.10.5 (bottom) also shows gesture recognition results based on the data acquired at the wrist from four human subjects. Impedance measurements were performed for six hand gestures and four pinch

618 • 2026 IEEE International Solid-State Circuits Conference

979-8-3315-8937-0/26 $31.00 © 2026 IEEE

ISSCC 2026 / February 18, 2026 / 5:05 PM

gestures with $100\mu A_{pk}$ CG current, and the data were classified using the algorithm presented in [6], achieving 99.7% accuracy for hand gestures and 84.1% accuracy for pinch gestures.

Figure 36.10.6 summarizes the performance of the frame-rate-scalable duty-cycled EIT system and compares it with state-of-the-art EIT systems. By implementing the duty-cycled operation, the proposed system effectively reduces the total average power consumption from $1753.6\mu W$ to $662.0\mu W$ at 878.9fps with a 63.4% duty-cycling ratio (Fig. 36.10.7). It can be further decreased to $43.8\mu W$ at 27.5fps with a 2.0% duty-cycling ratio, where the standby power is $33.1\mu W$ and only $10.7\mu W$ is for the active operation even with a $100\mu A_{pk}$ CG current. As a result, the duty-cycled EIT system enables a highly power-efficient, power-frame-rate scalable design, making it optimal for battery-powered wearable platforms as well as other power-constrained applications.

Acknowledgement:
The chip fabrication was supported by the IC Design Education Center (IDEC), Korea. This work was supported by the Ministry of Science and ICT (MSIT), Korea, under Grants RS-2025-02218624 and IITP-2025-RS-2020-II201461.

References:
[1] M. Kim et al., "A 1.4mΩ-sensitivity 94dB-dynamic-range electrical impedance tomography SoC and 48-channel Hub SoC for 3D lung ventilation monitoring system," *ISSCC*, pp. 354-355, Feb. 2017. http://doi.org/10.1109/ISSCC.2017.7870407
[2] B. Liu et al., "A 13-Channel 1.53-mW 11.28-mm² Electrical Impedance Tomography SoC Based on Frequency Division Multiplexing with 10× Throughput Reduction," *ISSCC*, pp. 370-371, Feb. 2019. http://doi.org/10.1109/ISSCC.2019.8662352
[3] H. Choi et al., "A Fully Dynamic 1st-Order Δ-ΔΣ Modulator with a 468mV$_{pp}$ Input Range for Electrical Impedance Tomography Systems," *IEEE Symp. VLSI Technol. Circuits*, pp. 1–2, Jun. 2024. http://doi.org/10.1109/VLSITechnologyandCir46783.2024.10631480
[4] J. Li et al., "A 1.76 mW, 355-fps, Electrical Impedance Tomography System With a Simple Time-to-Digital Impedance Readout for Fast Neonatal Lung Imaging," *IEEE JSSC*, vol. 60, no. 2, pp. 603–614, Feb. 2025. http://doi.org/10.1109/JSSC.2024.3434638
[5] Y. Wu et al., "A Human–Machine Interface Using Electrical Impedance Tomography for Hand Prosthesis Control," *IEEE TBioCAS*, vol. 12, no. 6, pp. 1322–1333, Dec. 2018. http://doi.org/10.1109/TBCAS.2018.2878395
[6] A. Kyu et al., "EITPose: Wearable and Practical Electrical Impedance Tomography for Continuous Hand Pose Estimation," *CHI Conference on Human Factors in Computing Systems*, pp. 1–10, May 2024. http://doi.org/10.1145/3613904.3642663
[7] J. Lee et al., "A 9.6-mW/Ch 10-MHz Wide-Bandwidth Electrical Impedance Tomography IC With Accurate Phase Compensation for Early Breast Cancer Detection," *IEEE JSSC*, vol. 56, no. 3, pp. 887–898, Mar. 2021. http://doi.org/10.1109/JSSC.2020.3032723
[8] H. Ko et al., "Ultralow-Power Bioimpedance IC With Intermediate Frequency Shifting Chopper," *IEEE TCAS-II*, vol. 63, no. 3, pp. 259–263, Mar. 2016. http://doi.org/10.1109/TCSII.2015.2483258
[9] S.-J. Kweon et al., "An 8MHz 31.25kS/s Impedance-Monitoring IC Based on IF-Sampling Architecture with a Band-Pass Delta-Sigma ADC," *IEEE Symp. VLSI Circuits*, pp. 1–2, Jun. 2021. http://doi.org/10.23919/VLSICircuits52068.2021.9492406

[10] J. Lee et al., "30-fps SNR equalized electrical impedance tomography IC with fast-settle filter and adaptive current control for lung monitoring," *IEEE ISCAS*, pp. 109–112, May 2016. http://doi.org/10.1109/ISCAS.2016.7527182
[11] Q. Pan et al., "A 97.3-dB SNR Bioimpedance AFE With −84-dB THD Segmented-ΔΣM Sinusoidal Current Generator and Passing-Through Instrumentation Amplifier," *IEEE JSSC*, vol. 60, no. 4, pp. 1411–1422, Apr. 2025. http://doi.org/10.1109/JSSC.2024.3516040
[12] G. Hahn et al., "A simple method to check the dynamic performance of electrical impedance tomography systems," *Physiol. Meas.*, vol. 21, no. 1, pp. 53–60, Feb. 2000. http://doi.org/10.1088/0967-3334/21/1/307
[13] A. Adler and W. R. B. Lionheart, "Uses and abuses of EIDORS: an extensible software base for EIT," *Physiol. Meas.*, vol. 27, no. 5, pp. S25–S42, Apr. 2006. https://doi.org/10.1088/0967-3334/27/5/S03

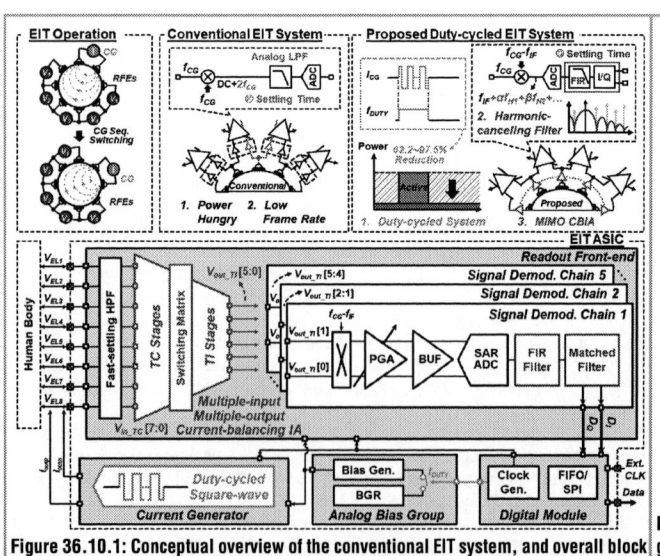

Figure 36.10.1: Conceptual overview of the conventional EIT system, and overall block diagram and main features of the proposed duty-cycled EIT system.

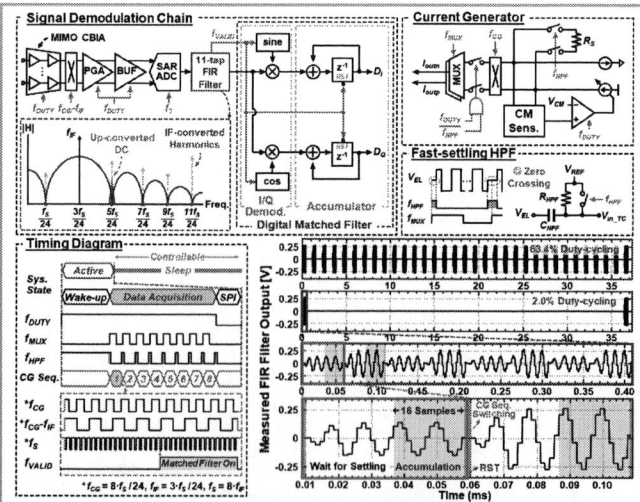

Figure 36.10.2: Structure and operation of the duty-cycled EIT system with current generator and signal demodulation chain (top), detailed timing diagram (bottom left), and measured FIR filter output with different duty-cycling ratios (bottom right).

ISSCC 2026 / SESSION 36 / NEURAL AND BIOMEDICAL INTERFACES / 36.10

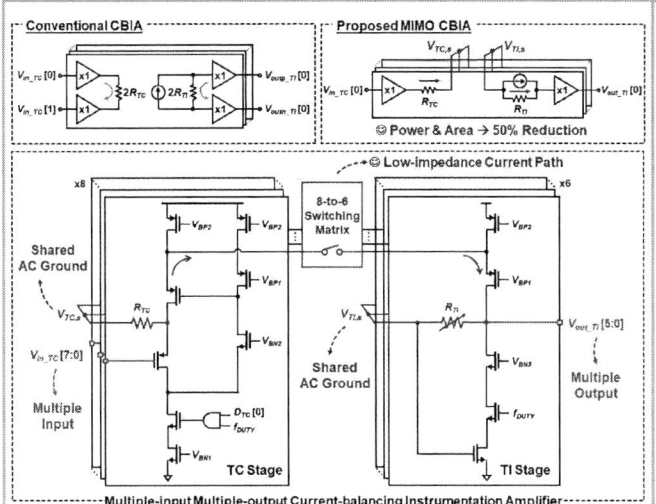

Figure 36.10.3: Detailed schematic of the multiple-input multiple-output (MIMO) current-balancing instrumentation amplifier (CBIA) with an embedded switching matrix.

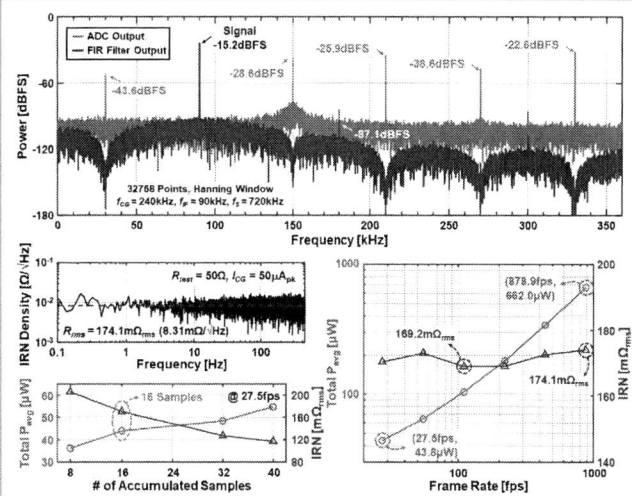

Figure 36.10.4: Measured output spectrum plots of the ADC and the FIR filter (top), IRN density (middle left), total average power and IRN over the number of accumulated samples (bottom left), and over the frame rate (bottom right).

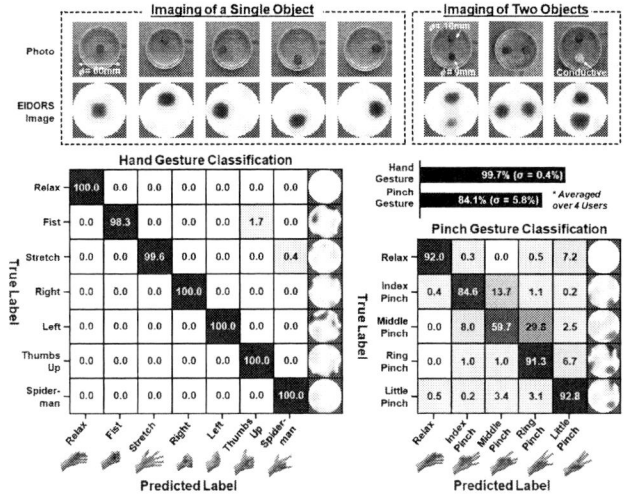

Figure 36.10.5: Imaging results with object(s) in saline solution (top), classification results for hand gestures (bottom left) and pinch gestures (bottom right).

	ISSCC'17 [1]	TBioCAS'18 [5]	ISSCC'19 [2]	VLSI'24 [3]	JSSC'25 [4]	*This Work	
Process (nm)	65	180	130	180	65	180	
Supply (V)	1.2	±1.65	1	1.8	1 / 1.8	1.8	
# of Electrodes (N_{elec})	48	8	16	16	16	8	
# of RFE Ch. (N_{RFE})	a48	1	13	d16	1	5	
Excitation Signal	Pseudo-sine	Pseudo-sine	Pseudo-sine	Square	Pseudo-sine	Square	
CG Frequency (kHz)	10 – 256 (20 steps)	25 – 500	15.625 – 125 (4 steps)	31.25 – 250	100 – 500	30 – 240 (4 steps)	
CG Amplitude (μA_{pk})	50 – 500	Up to 500	10 – 100	5 – 150	Up to 350	5 – 100	
Duty-cycled Operation	X	X	X	X	X	O	
Total Power (μW)	6960	b340000	1530	660.8	1760	662.0	43.8
CG Power (μW)	145	N/A	c538.6	406.4	c1571	83.0	2.2
RFE Power/Ch. (μW)		N/A	c76.3	15.9	172	109.2	1.7
Max. Frame Rate (fps)	70	N/A	5	81.4	355	878.9	27.5
Demodulation Method	Analog I/Q	Digital I/Q (off-chip)	Analog I/Q	SS	T-D	Digital I/Q (on-chip)	
Application	3D-Lung Ventilation	Gesture Recognition	Lung Ventilation	Lung Ventilation	Lung Ventilation	Gesture Recognition	

a Active electrode   b Including off-chip components   c Calculated from reported value in the paper
d Using two ASICs   e Reporting values correspond to a CG frequency of 240 kHz

Figure 36.10.6: Performance summary and comparison with state-of-the-art EIT systems.

Figure 36.10.7: Die micrograph and power breakdown with different duty-cycling ratios.

• 2026 IEEE International Solid-State Circuits Conference

ISSCC 2026 / SESSION 36 / NEURAL AND BIOMEDICAL INTERFACES / 36.11

36.11 A 0.62µW/sensor 82fps Time-to-Digital Impedance Measurement IC with Unified Excitation/Readout Front-End for Large-Scale Piezo-Resistive Sensor Array

Jiayang Li[1,2], Qingyu Zhang[2], Sohmyung Ha[2,3], Andreas Demosthenous[2], Dai Jiang[2], Yu Wu[2]

[1]University of Bristol, Bristol, United Kingdom, [2]University College London, London, United Kingdom, [3]New York University Abu Dhabi, Abu Dhabi, United Arab Emirates

Abstract

This paper presents a fast impedance-measurement IC for large-scale piezo-resistive sensor arrays. It features a unified differential time-to-digital demodulation architecture that reads out impedance directly through the excitation circuit. A pre-saturation adaptive bias technique further improves power efficiency. The chip scans 253 sensors in 12.2ms (82fps) at 125kHz, consuming 158µW (7.5nJ/sensor). With loads from 20Ω to 500kΩ, it achieves 0.5% error and up to 71.1dB SNR.

Frailty increases sharply with age and is the major cause of falls, leading to severe injury and fatality. While fall risks can be effectively predicted using gait analysis, it is often limited in laboratory settings. Piezo-resistive crossbar sensor arrays (Fig. 36.11.1 (top)) can be used towards lab-grade, yet wearable insoles. However, several challenges remain: 1) high sensor density (>250 per insole); 2) high temporal resolution required (50 to 100 frames per second (fps)); 3) wide piezo-resistive range from tens of Ω to hundreds of kΩ, while ohm-level sensitivity is required; 4) ultra-low power operation that lasts days [1]. These pose great challenges for hardware design. Voltage-domain methods using dc current sources [2] or Wheatstone bridges [3] get easily slowed by parasitic capacitances from large arrays and have limited sensitivity to small resistance due to their ADC-like structures, while oscillator-based methods [4], though faster, are sensitive to parasitic capacitances and PVT variations. The time-to-digital (T-D) impedance readout method (Fig. 36.11.1 (bottom left)) based on the cross-over point detection of V_m and V_{ref}, offer high-speed acquisition with almost constant sensitivity across a wide load range and is intrinsically robust against parasitic capacitances [5-6]. However, this prior art is power-demanding compared to other approaches [2-4]. Its major limitation in power (Fig. 36.11.1 (bottom left)) lies in the current driver's large bias current to provide high linear and accurate excitation across wide load ranges. Merely delivering 2 µA$_{pp}$ demands >4 times (8.7µA) the consumption as reported in [7]. Besides, static current is required in the recording amplifier to provide sufficient bandwidth, dynamic range, and noise performance (a static current of 130µA was reported in [5]). Lastly, prior art T-D methods were single-ended, making them susceptible to interference and noise.

To address these limitations, this paper proposes a T-D method (Fig. 36.11.1 (bottom right)) that features: 1) a unified excitation/readout structure that allows directly recording from the voltage excitation circuit, hence replacing the current driver and eliminating the need for static-current-consuming recording amplifiers; 2) a transconductance (TC) voltage driver for the unified excitation/readout structure employs a pre-saturation adaptive biasing, which dynamically adjusts the driver's current driving capability to adapt various load ranges, improving power efficiency without sacrificing performance; 3) a differential T-D demodulation method that further reduces power and circuit complexity. The IC scans 253 piezo-resistive sensors embedded in a single insole, via time-multiplexing at a rate of 12.2ms/frame, resulting in 82 fps. It consumes 158µW in total, achieving a power usage of 0.62µW/sensor (7.5nJ/sensor). It has an average measurement error of 0.5% and a maximum 71.1dB SNR with load resistances from 20Ω to 500kΩ.

The simplified overall block diagram of the IC is shown in Fig. 36.11.2. It comprises four main modules. (1) A unified front-end comprising the TC voltage driver and the T-D readout. The TC voltage driver excites the load with a sinusoidal voltage while mirroring the load current to the dynamic current comparator of the T-D readout. Note that this unified front-end, unlike conventional impedance-readout ICs [4-7] uses fixed-current excitation and voltage-recording architectures; it eliminates the need for additional recording amplifier circuits, reducing both power and hardware complexity. With a programmable dc current offset added, the following T-D readout converts the differential analog current signal into time-coded pulses; (2) a time-to-digital converter (TDC), which includes a pulse counter and a 6-phase 64MHz phase-locked loop (PLL), translates the time-coded pulses into the digital impedance data; (3) a sinusoidal signal generator for generating the required excitation waveform; and (4) a digital module for control and communication. Connected to the prototype insole, the IC addresses 253 piezo-resistive sensors arranged in a crossbar array. Each sensor is measured individually via time-multiplexing. The TC voltage driver routes the sinewave to the target sensor one by one by selecting the row and column through the multiplexer (MUX), thereby capturing the whole foot plantar pressure map.

The sinusoidal signal generator employs the direct digital synthesis, clocked by the PLL. It provides a programmable output amplitude from 15.6mVpp to 1Vpp in 64 steps at a frequency from 125kHz to 1MHz. It also contains a look-up table (LUT), a co-prime current DAC (I-DAC) with dynamic element matching (DEM), and a transimpedance (TI) filter, which removes the offset and harmonics. The co-prime I-DAC, which is based on the co-prime segmentation concept [6] (Fig. 36.11.3 (bottom left)), achieves a total quantization step by 16 times 17 (equivalent to an 8b DAC) with lower hardware cost compared to conventional thermal coding DACs.

The schematic of the TC voltage driver with pre-saturation adaptive bias is shown in Fig. 36.11.3 (top). The differential output, V_{out}, tracks the sinusoidal input, V_{in}, fed from the sinusoidal signal generator, while driving the resistor sensor through its flipped-voltage-follower topology [8]. The load current is supplied by transistors M1 and M2 regulated by the feedback loop, and is mirrored by M3 and M4 to the input of the current comparator (I_{comp}^+ and I_{comp}^-) for T-D demodulation. To extend dynamic range, the mirror ratios of M3/M1 and M4/M2 are programmable from 1 to 25 in five steps. The power bottleneck in this architecture is the static bias current overhead required to maintain linear and accurate excitation across a wide range of loads. To address this, a pre-saturation adaptive-bias technique is proposed (Fig. 36.11.3 (top right)). In this bias block, the currents, I_1 and I_2, provided by M1 and M2 are mirrored to I_3 and I_4 via M5 and M6. When either mirrored current exceeds a threshold set by ($I_{bias} - I_{limit}$), the adaptive-bias loop engages and generates an event-driven additional current, I_{adp}, which is added to the TC voltage driver's constant bias via V_b. This maintains a low quiescent current under light load conditions while expanding the linear drive capability at heavier load conditions with high power efficiency. In addition, unlike conventional adaptive bias approaches [9,10], where the bias current continuously tracks the load and introduces ripple, the bias updates only when the threshold is crossed, thereby minimizing unnecessary bias modulation. Measured THD versus load current (Fig. 36.11.3 (bottom right)) shows 3 increase in the linear output-current range compared to a driver without the pre-saturation adaptive bias. The TC voltage driver consumes 74µA quiescent current, increasing to an average of 172µA with 200µA$_{pp}$ load current.

The schematic of the T-D readout of the unified front-end is shown in Fig. 36.11.4 (top). It includes a dynamic current comparator and two 6b I-DACs, which provide a known threshold offset (I_{os}^+ and I_{os}^-) into the differential input, I_{comp}^+ and I_{comp}^-. After the comparator extracts the critical period N1, subsequently, period N0 and N2, the phase (θ) and magnitude (I_m) of the measured signal can be computed (Fig. 36.11.4 (bottom left)). Despite similarities in the T-D computing in [6], N1 in this work is extracted with a differential current. This differential approach also improves robustness and eliminates the need for differential-to-single-ended conversion, thereby reducing power consumption and circuit complexity. The dynamic comparator is a modified strong-arm latch [11] clocked by the on-chip PLL (CLK_{PLL}). The 6 clock phases are interleaved across successive cycles to increase temporal resolution. Six cycles are used per measurement, yielding the longest conversion time of 48µs (= 6/125kHz) and an array frame rate of 82fps (= 1/12.2ms) for all 253 sensors. Although the comparator is dynamic, dc conduction paths exist from the input mirrors and I-DACs to ground when the clock signal, CLK, is high, degrading the efficiency. Thus, an asynchronous feedback logic is added to address this issue (Fig. 36.11.4 (top-left)). The rising edge of CLK is triggered by the rising edge of CLK_{PLL}. After the comparison, an XOR gate generates a reset signal that immediately pulls CLK to low, which turns off the dc path and significantly increases power efficiency in turn. This T-D readout consumes 28µW at 64MHz. Fig. 36.11.4 (bottom right) shows an example of measured waveforms when the IC drives a 15kΩ load. When the excitation V_{out} is programmed to be smaller, the time-coded pulses N1 measured at the comparator's output become shorter.

The IC was fabricated in a 65nm CMOS process. With a 1.2V supply, it has a power consumption of 158µW (under $I_{load} \leq 55µA_{pp}$ conditions), operating at an excitation frequency of 125kHz. Total five chips were measured with a group of resistive loads from 20Ω to 500kΩ at 125 kHz (Fig. 36.11.5 (top)), spanning an 88dB input dynamic range. An average resistance error (resistance is calculated from the measured phase and amplitude) of 0.5% was obtained across all different resistance loads and chips (Fig. 36.11.5 (top left)). The error slightly increases when the loads are greater than 100kΩ. SNR was evaluated in the time domain by repeating each measurement 5,000 times and SNR = 20×log ($R_{average}/R_{std}$), where $R_{average}$ is the average resistance across the 5,000 measurements and R_{std} is the standard deviation [5,6]. A maximum SNR of 71.1 dB was achieved. The SNR decreases at high load resistances due to the reduced current amplitude. Images of the foot plantar pressure map were obtained using the IC and the prototype insole (Fig. 36.11.5 (bottom)). A custom reconstruction algorithm, similar to [12], is developed to address the cross-talk issue in the crossbar sensor arrays. The reconstructed pressure maps clearly distinguish different gait phases (single-leg neutral, double-leg neutral, forefoot, and rearfoot). A comparison with prior works is shown in Fig. 36.11.6. This IC achieves the lowest power per sensor and a comparable SNR, while providing a superior frame rate of 82fps, scanning 253 sensors. The Schreier FoM was used to evaluate the performance. This work achieves the best FoM among entries in the table. Figure 36.11.7 shows the chip micrograph and a detailed power breakdown pie chart. It has a total die area of $1\times1.7mm^2$, resulting in a die area of $0.0067mm^2$/sensor.

Acknowledgement:
This work was supported by Rosetrees Trust UK Charity and King-Cullimore Charitable Trust.

References:
[1] L. Burnie *et al.*, "Commercially available pressure sensors for sport and health applications: A comparative review," *Foot*, vol. 56, pp. 102046–102046, Sep. 2023. https://doi.org/10.1016/j.foot.2023.102046
[2] H. Xin *et al.*, "A 0.34-571nW All-Dynamic Versatile Sensor Interface for Temperature, Capacitance, and Resistance Sensing," *ESSCIRC*, 2019, pp. 161-164. https://doi.org/10.1109/ESSCIRC.2019.8902918
[3] S. Pan and K. A. A. Makinwa, "A Wheatstone Bridge Temperature Sensor with a Resolution FoM of 20fJ.K2," *ISSCC*, 2019, pp. 186-187. https://doi.org/10.1109/ISSCC.2019.8662337
[4] S. Han *et al.*, "A Time-Domain Multi-Channel Resistive-Sensor Interface IC With High Energy Efficiency and Wide Input Range," in *IEEE Transactions on Biomedical Circuits and Systems*, vol. 19, no. 2, pp. 291-299, April 2025. https://doi.org/10.1109/TBCAS.2025.3526813
[5] J. Li *et al.*, "A 1.76 mW, 355-fps, Electrical Impedance Tomography System With a Simple Time-to-Digital Impedance Readout for Fast Neonatal Lung Imaging," in *IEEE Journal of Solid-State Circuits*, vol. 60, no. 2, pp. 603-614, Feb. 2025. https://doi.org/10.1109/JSSC.2024.3434638
[6] J. Li *et al.*, "A 30MHz Wideband 92.7dB SNR 99.6% Accuracy Bioimpedance Spectroscopy IC Using Time-to-Digital Demodulation with Co-Prime Delay Locked Sampling," *ISSCC*, 2025, pp. 572-573. https://doi.org/10.1109/ISSCC49661.2025.10904781
[7] K. Kim *et al.*, "Design of Sub-10-µW Sub-0.1% THD Sinusoidal Current Generator IC for Bio-Impedance Sensing," in *IEEE Journal of Solid-State Circuits*, vol. 57, no. 2, pp. 586-595, Feb. 2022. https://doi.org/10.1109/JSSC.2021.3100716
[8] J. Li *et al.*, "An 89.3% Current Efficiency, Sub 0.1% THD Current Driver for Electrical Impedance Tomography," in *IEEE Transactions on Circuits and Systems II: Express Briefs*, vol. 70, no. 10, pp. 3742-3746, Oct. 2023. https://doi.org/10.1109/TCSII.2023.3294753
[9] Y. Huang *et al.*, "Nano-Ampere Low-Dropout Regulator Designs for IoT Devices," in *IEEE Transactions on Circuits and Systems I: Regular Papers*, vol. 65, no. 11, pp. 4017-4026, Nov. 2018. https://doi.org/10.1109/TCSI.2018.2851226
[10] S. Ye *et al.*, "A 36V Current-Balancing Instrumentation Amplifier with ±24V Input Range, 5.6MHz BW, and 140dB CMRR at All Gain Settings," *ISSCC*, 2025, https://doi.org/10.1109/ISSCC49661.2025.10904534
[11] B. Razavi, "The StrongARM Latch [A Circuit for All Seasons]," in *IEEE Solid-State Circuits Magazine*, vol. 7, no. 2, pp. 12-17, Spring 2015. https://doi.org/10.1109/MSSC.2015.2418155
[12] S. Müller *et al.*, "Cross-Talk Compensation in Low-Cost Resistive Pressure Matrix Sensors," *IEEE International Conference on Mechatronics*, 2019, pp. 232-237. https://doi.org/10.1109/ICMECH.2019.8722925

[13] Y. Luo *et al.*, "23.2 A 70µW 1.19mm2 Wireless Sensor with 32 Channels of Resistive and Capacitive Sensors and Edge-Encoded PWM UWB Transceiver," *ISSCC*, 2020, pp. 346-347. https://doi.org/10.1109/ISSCC19947.2020.9063079
[14] D. Seo *et al.*, "An RC Delay-Based Pressure-Sensing System With Energy-Efficient Bit-Level Oversampling Techniques for Implantable IOP Monitoring Systems," in *IEEE Journal of Solid-State Circuits*, vol. 58, no. 10, pp. 2745-2756, Oct. 2023. https://doi.org/10.1109/JSSC.2023.3286796
[15] X. Feng *et al.*, "A 72-Channel Resistive-and-Capacitive Sensor-Interface Chip With Noise-Orthogonalizing and Pad-Sharing Techniques," in *IEEE Journal of Solid-State Circuits*, vol. 59, no. 3, pp. 702-715, March 2024. https://doi.org/10.1109/JSSC.2023.3344587

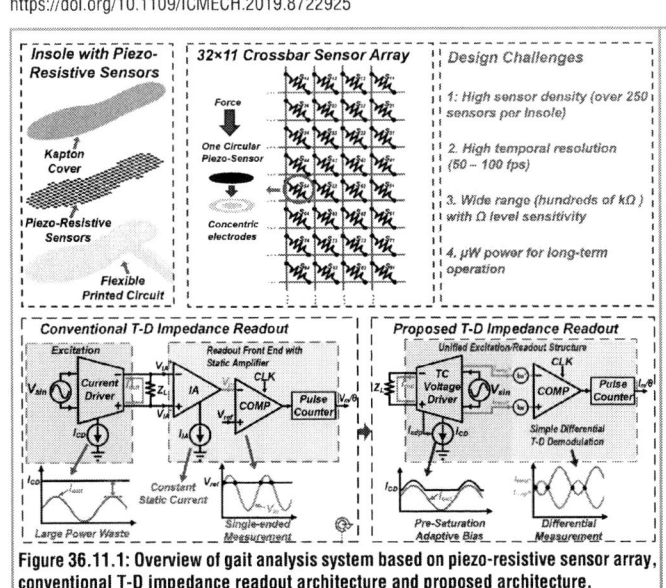

Figure 36.11.1: Overview of gait analysis system based on piezo-resistive sensor array, conventional T-D impedance readout architecture and proposed architecture.

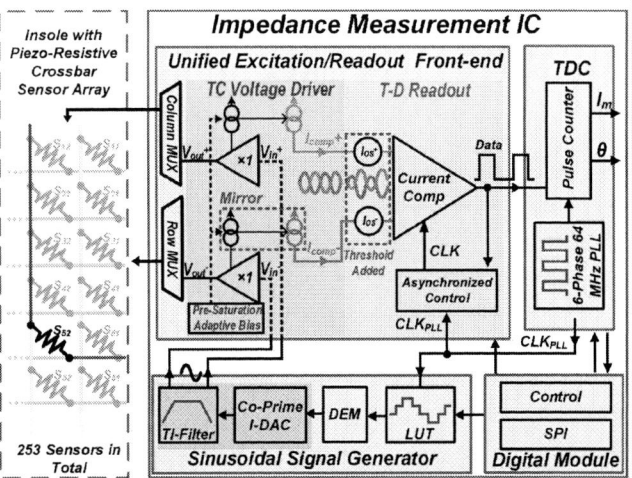

Figure 36.11.2: Simplified block diagram of impedance-measurement IC with unified T-D excitation/readout front end.

ISSCC 2026 / SESSION 36 / NEURAL AND BIOMEDICAL INTERFACES / 36.11

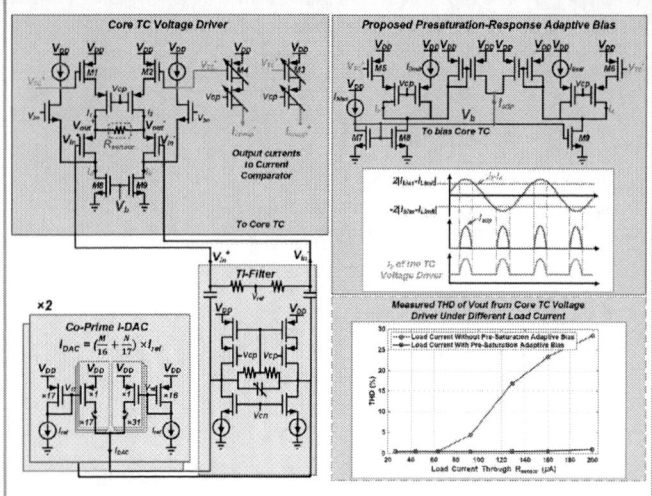

Figure 36.11.3: Schematic and working principle of the TC voltage driver with pre-saturation adaptive bias; measured THD with and without the pre-saturation adaptive bias; schematic of the co-prime I-DAC and TI filter.

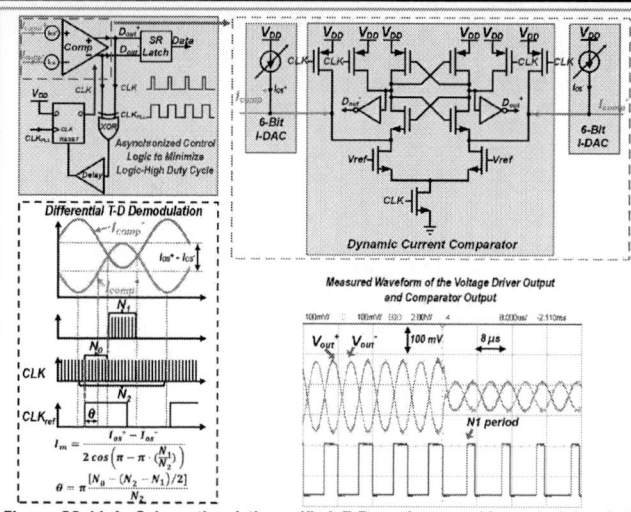

Figure 36.11.4: Schematic of the unified T-D readout, working principle of the differential T-D demodulation, and asynchronized control logic and the measured waveform.

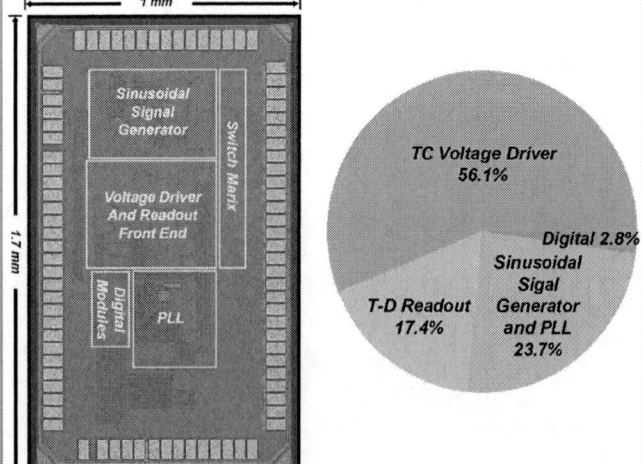

Figure 36.11.5: Measured error and SNR with resistive loads from 20Ω to 500kΩ at 125kHz and image maps of different loading conditions and gait phases.

Parameters	ISSCC'20[13]	JSSC'23[14]	JSSC'24[5]	JSSC'24[15]	This Work
Technology (nm)	130	180	65	65	65
Supply (V)	1.2	1.8	1.8/1.0	1.2	1.2
Readout Method	Voltage	Voltage	T-D	Voltage	T-D
Num. of Sensors	32	1	208[a]	72	253
Input Range (Ω)	249 k	2.2k–4.4k	N/A	<910 k	20–500k
Total Power (µW)	70	12.79	1760	53	158[b]
Power/Sensor (µW)	2.2	12.79	8.46	0.74	0.62
Energy/Sensor (nJ)	N/A	147.3	23.7	83.3	7.5
Chip Area/Sensor (mm²)	0.14	2.52	0.0091	0.06	0.0067
Max Conversion Time (ms)	N/A	11.52	2.81/frame	37.5/measure 112.5/frame[c]	0.048/measure 12.2/frame
Max SNR (dB)	77.7[c]	71.0[c]	52.7	70.0[c]	71.1
ENOB	11.4	10.3	7.3	10.1	10.3[d]
FoM[e]	N/A	79.3	68.9	80.8	92.3

[a]Equivalent from number of measurements per frame
[b]Pre-saturation adaptive bias is not triggered
[c]Calculated from the paper through: SNR = 20×log₁₀ (Input Range/Resolution)
[d]Calculated through: ENOB=log₂(R_average/(2√2×R_std))
[e]FoM=SNR (dB)+10log₁₀(Sampling Frequency (kHz) /Total Power (mW)), where Sampling Frequency= Num. of Sensors/ Max Conversion Time per frame (ms)

Figure 36.11.6: Comparison with prior work.

Figure 36.11.7: Chip micrograph and power breakdown.

• 2026 IEEE International Solid-State Circuits Conference

ISSCC 2026 / SESSION 37 / MEMORY INTERFACE / OVERVIEW

Session 37 Overview: *Memory Interface*
MEMORY SUBCOMMITTEE

Session Chair: Dongkyun Kim
SK hynix, Icheon, Korea

Session Co-Chair: Hye-Ran Kim
Samsung Electronics, Anyang, Korea

This session showcases state-of-the-art high-speed and low-power transmitter and receiver architectures tailored for next-generation memory interfaces: spanning a broad range of techniques, from NRZ, PAM3, to PAM4 signaling to equalization-free, slope-sampling, and self-training schemes. Many techniques focus on enhancing robustness against ISI, supply noise, and reflections in multi-drop or crosstalk on channels. Several techniques employ advanced clocking, calibration, and encoding methods to relax timing and jitter constraints. Power efficiency is also emphasized, with designs achieving sub-pJ/b energy consumption. Overall, this body of work points to more efficient and resilient die-to-die memory links.

1:30 PM

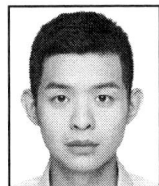

37.1 A 72Gb/s/pin Single-Ended Driver-Cooperative Coded PAM3 Transceiver with Asymmetric Data-Dependent Equalization and Bias-Peaking for Chiplets and Memory Interfaces
Hongzhi Wu, Southern University of Science and Technology, Shenzhen, China
In Paper 37.1, Southern University of Science and Technology (SUSTech) introduces a 72Gb/s/pin PAM3 transceiver with TIA termination, including driver-cooperative encoding, asymmetric data-dependent equalization, and bias-peaking.

1:55 PM

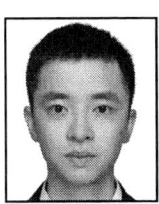

37.2 A 47.0Tb/s/mm 112Gb/s/pin PAM4 Single-Ended Transceiver Featuring 4-Aggressor Crosstalk Cancellation and Supply-Noise Tolerance for Short-Reach Memory Interfaces
Qian Liu, Nanjing University, Nanjing, China
In Paper 37.2, Nanjing University, T-Head (Shanghai) Semiconductor, and Interdisciplinary Research Center for Future Intelligent Chips present a 112Gb/s/pin PAM4 single-ended transceiver in 28nm CMOS for short-reach memory interfaces with crosstalk cancellation & supply-noise-tolerant clock distribution network.

2:20 PM

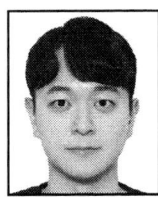

37.3 A 2nm All-Digital 14.4Gb/s/pin LPDDR6 PHY with Quarter-Rate Clocking Architecture and Multi-Level FIFO-Based Speculative DFE
Yoonjae Choi, Samsung Electronics, Yongin, Korea
In Paper 37.3, Samsung Electronics presents a 2nm all-digital LPDDR6 PHY achieving 14.4Gb/s/pin, including a quadrature clocking with QEC to enhance jitter sensitivity, and a multi-level FIFO-based speculative DFE to relax timing constraints.

ISSCC 2026 / February 18, 2026 / 1:30 PM

2:45 PM

37.4 A 100Gb/s, 1.92pJ/b Aggregate 4-Lane Single-Ended NRZ Transceiver with 15dB Far-End Crosstalk Cancellation via On-Chip Feature Extraction and Classification in 16nm FinFET

Xiaohui Lin, Oregon State University, Corvallis, OR

In Paper 37.4, Oregon State University presents a low-power (1.92pJ/b), high-speed (100Gb/s), equalizer-free, 4-lane single-ended transceiver that investigates supervised learning.

3:00 PM

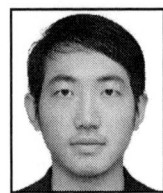

37.5 A 16Gb/s/pin 0.51pJ/b Single-Ended NRZ Transceiver with Distributed Dual-Loop V_{DDQ}-Ripple Compensation and Dynamic Clock Duty-Cycle Calibration for Memory Interfaces

Yuchen He, Fudan University, Shanghai, China

In Paper 37.5, Fudan University introduces a 16 Gb/s/pin 0.51 pJ/b single-ended NRZ transceiver with distributed dual-loop compensation for VDDQ ripple suppression and dynamic duty-cycle calibration for robust clocking.

3:35 PM

37.6 A 0.092pJ/b and 7.7fJ/b/dB Cross-Self-Referenced Slope-Sampling Receiver with Long-Tail ISI Robustness for Next-Generation Low-Power Memory Interfaces

Chanheum Han, Kwangwoon University, Seoul, Korea

In Paper 37.6, Kwangwoon University introduces a cross-self-referenced slope-sampling RX that is highly robust against long-tail ISI without equalization for next-generation low-power memory interfaces.

4:00 PM

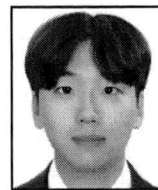

37.7 A 12.8Gb/s Parallel Receiver with a One-Way Self-Training Scheme for Equalizing ISI and Reflections in Multi-Drop Memory Interfaces

Ji-Won Moon, Pohang University of Science and Technology, Pohang, Korea

In Paper 37.7, Pohang University of Science and Technology and Samsung Electronics introduce a 12.8Gb/s/pin receiver with a one-way self-training scheme for highly reflective multi-drop memory interfaces.

4:25 PM

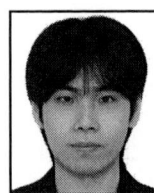

37.8 A 0.87pJ/b 17Gb/s/pin Parallel Receiver with a Local DQS Recovery for a Supply-Noise-Tolerant DQS Distribution in High-Performance NAND Flash Interfaces

Byeong-Chan Kim, Pohang University of Science and Technology, Pohang, Korea

In Paper 37.8, Pohang University of Science and Technology and Samsung present a 0.87pJ/b, 17Gb/s/pin parallel receiver with local DQS recovery and a dual-DQS tree for reduced clock power and mitigated power supply-induced jitter.

4:50 PM

37.9 A 14Gb/s/pin 0.163pJ/b DQ Receiver for HBM with Baud-Rate Phase Tracking Loop Supporting Background Offset Calibration

Jeonghyeon Lee, Yonsei University, Seoul, Korea

In Paper 37.9, Yonsei University and ONE semiconductor present a compact 14Gb/s/pin HBM DQ receiver with a baud-rate phase tracking loop and background offset calibration based on a time window phase detector.

37

DIGEST OF TECHNICAL PAPERS • 623

979-8-3315-8937-0/26 $31.00 © 2026 IEEE

ISSCC 2026 / SESSION 37 / MEMORY INTERFACE / 37.1

37.1 A 72Gb/s/pin Single-Ended Driver-Cooperative Coded PAM3 Transceiver with Asymmetric Data-Dependent Equalization and Bias-Peaking for Chiplets and Memory Interfaces

Hongzhi Wu, Xuxu Cheng, Yangyi Zhang, Xiongshi Luo, Zhenghao Li, Weitao Wu, Quan Pan

Southern University of Science and Technology, Shenzhen, China

Abstract

This paper presents a 72Gb/s/pin PAM-3 single-ended transceiver in 28nm CMOS. Driver-cooperative coding enhances signal integrity and reduces signaling power by ≥46%. Asymmetric data-dependent equalization and bias peaking mitigate PAM-3 different edge ISI and switching jitter, enlarging eye area by 1.9× and 2.8×. A $1/_8$-rate forwarded-clock multiplication scheme enables low-power and wide-range de-skew. The transceiver supports a 72Gb/s/pin data rate, with power efficiencies of 0.7pJ/b at 48Gb/s and 1.17pJ/b at 72Gb/s.

The rapid growth of AI and cloud-computing applications drives demand for high-speed and low-power single-ended (SE) interfaces for chiplets and memories [1-6]. To increase SE interface data rates, PAM3 signaling is emerging as a promising candidate [1-3]: offering a 1.5× bandwidth efficiency over NRZ and better noise and crosstalk immunity than PAM4. However, compared to NRZ, PAM3's higher ISI sensitivity and switching-jitter (SWJ) degraded eye width require the use of asymmetric equalization schemes. Additionally, to improve SE link energy efficiency, reducing the signaling power of the conventional termination scheme shown in Fig. 37.1.1(top left) is a feasible solution [7-10]. Although the AC-coupled unterminated RX shown in Fig. 37.1.1(top middle) lowers signaling power, it does so at the cost of compromised signal integrity (SI) [7-8]. To balance power and SI, transimpedance amplifier (TIA) or V_M termination schemes are proposed [9-10]. In [9], the TIA termination, combined with the proposed Di-code technique, enables low signaling power transmission. In addition, a TIA common-mode (CM) voltage calibration is employed to minimize the extra signaling power caused by TRX CM mismatch. However, this approach lacks TIA impedance calibration and the Di-code technique does not support PAM3, thereby limiting SI and energy efficiency. Although the V_M termination employs a floating middle-level signaling scheme and achieves a low signaling-power transmission for PAM3 [10], it lacks a practical RX architecture for full validation. Moreover, the additional circuitry required to generate the V_M level at the RX side may introduce extra power consumption and degrade the RX's front-end bandwidth.

This paper presents a 72Gb/s/pin SE driver-cooperative coded PAM3 TRX with asymmetric data-dependent equalization and a bias-peaking technique. As shown in Fig. 37.1.1(top right), the proposed low-power driver-cooperative coding scheme with an impedance-calibrated TIA termination achieves higher signal integrity and reduces signaling power by 46% compared to conventional schemes. To equalize channel ISI and to mitigate SWJ, pre-emphasis asymmetric data-dependent equalization and bias-peaking are proposed, achieving a 1.9 and 2.8× eye area improvement. Moreover, the proposed $1/_8$-rate forwarded clock multiplication scheme enables low-power and wide-range de-skew operation.

Figure 37.1.1(bottom) shows the TRX architecture. In the TX, an on-chip PRBS generator outputs parallel data streams, which are input into a driver-cooperative encoder, generating HO, HE, LO, LE for low-power PAM3 modulation. The encoded signals are processed by the equalization logic to generate main cursor signals, and $E_{H/L}/EM_{H/L}$ equalization signals, which are serialized by 8:1 MUX. On-chip feedback equalization is designed to extend the serializer's bandwidth. A tailless push-pull driver with bias-peaking suppresses PAM3 SWJ and enlarges the eye opening. In the RX, a shunt-feedback TIA is used for termination, without consuming static current. A CM calibration loop ensures that the TIA's CM voltage tracks the TX's CM voltage across PVT variations, preventing static short current from TRX CM voltage mismatch. An impedance tuning module compensates for impedance variation and supports impedance relaxation by increasing the RX's termination impedance for low-power operation. An active inductor and a stacked T-coil are integrated in the TIA to boost its output bandwidth by 36.2%. The TIA's output signals are then processed by the slicer and converted into digital signals. These digital signals are then deserialized and decoded for off-chip BER evaluation. In the TX's clock path, the TX generates $1/_8$-rate forwarded clock CK_{OUT} and the RX's clock path receives this clock signal and adjusts its phase via a hybrid de-skew, then drives an injection-locked oscillator and a passive mixer-based multiplier to produce 4-phase clocks for the RX's data path.

Figure 37.1.2(top) shows the driver-cooperative coding scheme and asymmetric data-dependent equalization. As shown in Fig. 37.1.2(top left), a dedicated driver-cooperative encoder is used to implement this scheme: independently generating pull-up (H=0), pull-down (L=1), and disable signals (H=1, L=0), as shown in Fig. 37.1.2(top right). The subsequent push-pull driver outputs the 2/0 levels of PAM3 based on the TX CM voltage (VCM_{TX}), while the 1 level is created by turning off the pull-up and pull-down branches of the driver. This push-pull driver lowers signaling current by 37.5%. Moreover, compared to voltage-mode (VM) drivers, the proposed driver reduces TX output node's parasitics and pre-driver size, ensuring higher bandwidth and lower dynamic power. To further reduce power and mitigate the ISI differences between PAM3 transition edges [11], an asymmetric pre-emphasis equalizer is proposed, and its encoding logic is shown in Fig. 37.1.2(bottom left). A low-speed encoder placed before 8:1 MUX selectively generates control signals $E_{H/L}$

and $EM_{H/L}$ for major and minor edge transitions, reducing equalization path activation time by 56% and lowering signaling power. Equalization coefficients are set by current DACs with 0.5mV$_{pp}$ resolution. As shown in Fig. 37.1.2(bottom middle), applying separate coefficients to major and minor edges enlarges the minimum eye area by 1.9× at 60Gb/s over a 2m cable. Figure 37.1.2(bottom right) compares the power of the proposed coding and equalization with conventional schemes: the proposed scheme reduces signaling power by 46%, before considering the VM driver's higher pre-driver power. Greater energy-efficiency benefits can be achieved with increased equalization.

To further enhance the TRX's bandwidth, a bias-peaking technique is proposed, which provides bandwidth extension, SWJ reduction, improves TX return loss (RL), and energy efficiency over conventional designs. As shown in Fig. 37.1.3(top left), prior schemes place the coupling capacitor at the output node [12]: this increases output parasitics; thus, degrading RL and requiring increased drive strength from preceding stages. T-coil solutions [13] improve RL but incur a large area overhead and do not reduce power. In contrast, the proposed bias-peaking scheme, as shown in Fig. 37.1.3(top right), couples the capacitor C_1 and C_2 to bias transistors M_2 and M_3 instead of the output node, while ensuring M_2 and M_3 operate in saturation to maintain the output impedance; thus, enhancing RL while reducing pre-driver power. A delay-adjust module is inserted in the feed-forward path to improve equalization precision and robustness to PVT variation. The added bias-peaking path slightly limits the 8:1 MUX output nodes' bandwidth, which is compensated by a feedback equalizer. The inverters INV_1/INV_2 placed in the feedback equalizer provide a 3-bit programmable pull-up/pull-down strength, enabling edge-rate mismatch calibration and further improving the TX eye quality. Measured results validate the RL and eye diagram enhancements of this bias-peaking technique. As shown in Fig. 37.1.3(bottom left), conventional designs suffer from degraded RL and faster high-frequency performance degradation. In contrast, the proposed scheme maintains good RL without degradation. As shown in Fig. 37.1.3(bottom right), without the proposed technique, the measured 72Gb/s TX eye exhibits slower edge transitions and poor quality eye. In contrast, with bias-peaking enabled, SWJ is reduced and the minimum eye area increases by 2.8×.

To ensure good SI and static current reduction under PVT variations, on-chip impedance tuning is adopted. As shown in Fig. 37.1.4(top left), a resistor switch array with CM bias is used for impedance tuning. In addition, the TIA front-end input impedance is maintained at 50Ω while maximizing the TIA bandwidth by lowering the feedforward gain A of the Gm stage, reducing the value of R_F, and employing an active inductor together with a T-coil. As shown in Fig. 37.1.4(top right), the input impedance tuning range is 45–60Ω: an impedance setting above 50Ω enables relaxed termination, thus reducing link power by up to 9%.

Forwarded clock schemes are widely used in parallel interfaces. A well-optimized clock architecture ensures low jitter and reduces power overhead. As shown in Fig. 37.1.4(middle left), conventional $1/_2$- or $1/_4$-rate clocking schemes lack jitter-filtering capability, resulting in degraded clock jitter performance, which reduces the RX slicer's decision margin and degrades its bit error rate (BER) performance [14]. In addition, due to their higher operating frequencies, 1/2- and 1/4-rate schemes generally consume more power than 1/8-rate designs. To address power and jitter, this work proposes a $1/_8$-rate forwarded-clock multiplication scheme as shown in Fig. 37.1.4(middle right), which enables low-power operation with improved jitter performance. Figure 37.1.4(bottom) shows the clock path implementation. The clock multiplier uses a passive mixer to reduce power consumption. The de-skew path uses a hybrid-delay scheme, which uses digitally-controlled delay lines (DCDLs), phase interpolators (PIs), and phase selectors to achieve coarse and fine tuning. As shown in Fig. 37.1.4 (bottom right), the fine-tuning range for both DCDLs and PIs is 93ps, with a 427fs step size. The phase selector performs coarse tuning; it expands the tuning range by 4× to 372ps.

The TRX is fabricated in 28nm CMOS. Measurements were performed using an oscilloscope and a BER tester (BERT). Figure 37.1.5(top) shows measured TX eyes at 40 and 48GBaud. At 40GBaud, the NRZ and PAM3 eyes are 234mV × 0.8UI and 95mV × 0.48UI; they are 241mV × 0.72UI and 89mV × 0.44UI at 48GBaud. Figure 37.1.5(bottom) shows the bathtub curves: without on-chip equalization and bias peaking the link fails at 60Gb/s and 72Gb/s. Whereas, with on-chip equalization, a 10^{-12} BER is achieved with a 0.22UI horizontal margins

ISSCC 2026 / February 18, 2026 / 1:30 PM

at 60Gb/s, and 0.14UI at 72Gb/s. Activating bias peaking improves these margins to 0.35UI at 60Gb/s and 0.28UI at 72Gb/s. The TRX consumes 84mW at 72Gb/s (1.17pJ/b) and occupies 0.0087mm².

Figure 37.1.6 summarizes the TRX performance, which achieves 72Gb/s by using asymmetric data-dependent equalization and bias-peaking. Moreover, the TIA-terminated driver-cooperative coding and $1/8$-rate forwarded clocking schemes significantly reduce link power and demonstrate excellent energy efficiency: 0.7pJ/b at 48Gb/s and 1.17pJ/b at 72Gb/s.

Acknowledgement:
The work is supported by the National Natural Science Foundation of China under Grant 92573101 and in part by the Science and Technology Plan of Shenzhen under Grant KJZD20231023100159002 and KJZD20240903100208012. Corresponding author: Quan Pan.

References:
[1] K. Kim et al., "A 0.275pJ/b 42Gb/s/pin Clock-Referenced PAM3 Transceiver Tolerant to Supply Noise, Reference Offset and Crosstalk for Chiplets and Short-Reach Memory Interfaces," *ISSCC*, pp.394-395, 2025.
https://doi.org/10.1109/ISSCC49661.2025.10904729
[2] J. Yang et al., "A 35.4Gb/s/pin 16Gb GDDR7 with a Low-Power Clocking Architecture and PAM3 IO Circuitry," *ISSCC*, pp. 232-234, 2024.
https://doi.org/10.1109/ISSCC49657.2024.10454560
[3] S.-Y. Cho et al., "A 16Gb 37Gb/s GDDR7 DRAM with PAM3-Optimized TRX Equalization and ZQ Calibration," *ISSCC*, pp. 242-243, 2024.
https://doi.org/10.1109/ISSCC49657.2024.10454354
[4] M.-S. Lin et al., "A 32Gb/s 10.5Tb/mm 0.6pJ/b UCIe-Compliant Low-Latency Interface in 3nm Featuring Matched-Delay for Dynamic Clock Gating," *ISSCC*, pp. 586-588, 2025. https://doi.org/10.1109/ISSCC49661.2025.10904767
[5] Z. Wang et al., "A 64Gb/s/wire 10.5Tb/s/mm/Layer Single-Ended Simultaneous Bi-Directional Transceiver with Echo and Crosstalk Cancellation for a Die-to-Die Interface in 28nm CMOS," *ISSCC*, pp. 588-590, 2025.
https://doi.org/10.1109/ISSCC49661.2025.10904631
[6] K. Seong et al., "A 4nm 48Gb/s/wire Single-Ended NRZ Parallel Transceiver with Offset-Calibration and Equalization Schemes for Next-Generation Memory Interfaces and Chiplets," *ISSCC*, pp. 250-252, 2024.
https://doi.org/10.1109/ISSCC49657.2024.10454481
[7] Y. Nishi et al., "A 0.190-pJ/bit 25.2-Gb/s/wire Inverter-Based AC-Coupled Transceiver for Short-Reach Die-to-Die Interfaces in 5-nm CMOS," *IEEE JSSC*, vol. 59, no. 4, pp. 1146-1157, Apr. 2024. https://doi.org/10.1109/JSSC.2023.3338478
[8] S. Kim et al., "A 0.458-pJ/bit 24-Gb/s/pin Capacitively Driven PAM-4 Transceiver With PAM-Based Crosstalk Cancellation for High-Density Die-to-Die Interfaces," *IEEE JSSC*, vol. 59, no. 11, pp. 3730-3740, Nov. 2024. https://doi.org/10.1109/JSSC.2024.3401213
[9] H. Park et al., "A 0.385-pJ/bit 10-Gb/s TIA-Terminated Di-Code Transceiver with Edge-Delayed Equalization, ECC, and Mismatch Calibration for HBM Interfaces," *ISSCC*, pp. 452-453, 2022. https://doi.org/10.1109/ISSCC42614.2022.9731740

[10] C. Han et al., "A VM-Terminated PAM-3 Transmitter Using Floating Middle Level With Enhanced Signal Integrity and Energy Efficiency for Low-Power Memory Interfaces," *IEEE TCAS-I*, vol. 72, no. 6, pp. 2653-2663, June 2025.
https://doi.org/10.1109/TCSI.2025.3552405
[11] B. Kim et al., "A 42Gb/s Single-Ended Hybrid-DFE PAM-3 Receiver for GDDR7 Memory Interfaces," *ISSCC*, pp. 398-399, 2025.
https://doi.org/10.1109/ISSCC49661.2025.10904551
[12] J. Jin et al., "A 4-nm 16-Gb/s/pin Single-Ended PAM-4 Parallel Transceiver With Switching-Jitter Compensation and Transmitter Optimization," *IEEE JSSC*, vol. 59, no. 1, pp. 184-195, Jan. 2024. https://doi.org/10.1109/JSSC.2023.3319637
[13] J.-H. Park et al., "A 32Gb/s/pin 0.51pJ/b Single-Ended Resistor-less Impedance-Matched Transmitter with a T-Coil-Based Edge-Boosting Equalizer in 40nm CMOS," *ISSCC*, pp. 410-412, 2023. https://doi.org/10.1109/ISSCC42615.2023.10067552
[14] H. Xu et al., "A 112-Gb/s PAM-4 Retimer Transceiver With Jitter-Filtering Clocking Scheme and BER Optimization Technique in 28-nm CMOS," *IEEE JSSC*, vol. 60, no. 7, pp. 2305-2318, July 2025. https://doi.org/10.1109/JSSC.2025.3555383
[15] H. Park et al., "A 3-bit/2UI 27Gb/s PAM-3 Single-Ended Transceiver Using One-Tap DFE for Next-Generation Memory Interface," *ISSCC*, pp. 382-384, Feb. 2019. https://doi.org/10.1109/ISSCC.2019.8662462
[16] Y. Nishi et al., "A 0.297-pJ/Bit 50.4-Gb/s/Wire Inverter-Based Short-Reach Simultaneous Bi-Directional Transceiver for Die-to-Die Interface in 5-nm CMOS," *IEEE JSSC*, vol. 58, no. 4, pp. 1062-1073, Apr. 2023.
https://doi.org/10.1109/JSSC.2022.3232024
[17] Y. Kwon et al., "A 33-Gb/s/Pin 1.09-pJ/Bit Single-Ended PAM-3 Transceiver With Ground-Referenced Signaling and Time-Domain Decision Technique for Multi-Chip Module Memory Interfaces," *IEEE JSSC*, vol. 58, no. 8, pp. 2314-2325, Aug. 2023. https://doi.org/10.1109/JSSC.2023.3250706
[18] J. Seo et al., "A 20-Gb/s/pin 0.0024-mm² Single-Ended DECS TRX with CDR-less Self-Slicing/Auto-Deserialization to Improve Tolerance on Duty Cycle Error and RX Supply Noise for DCC/CDR-less Short-Reach Memory Interfaces," *ISSCC*, pp.456-457, 2022. https://doi.org/10.1109/ISSCC42614.2022.9731763

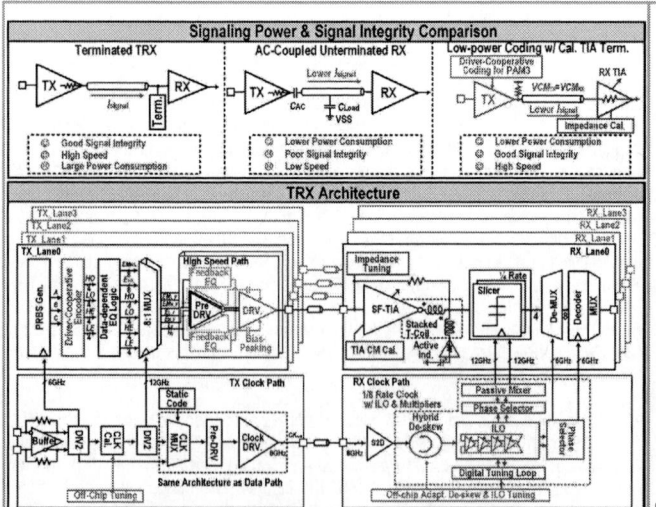

Figure 37.1.1: Comparison of signaling power and signal integrity across different termination schemes and TRX architecture.

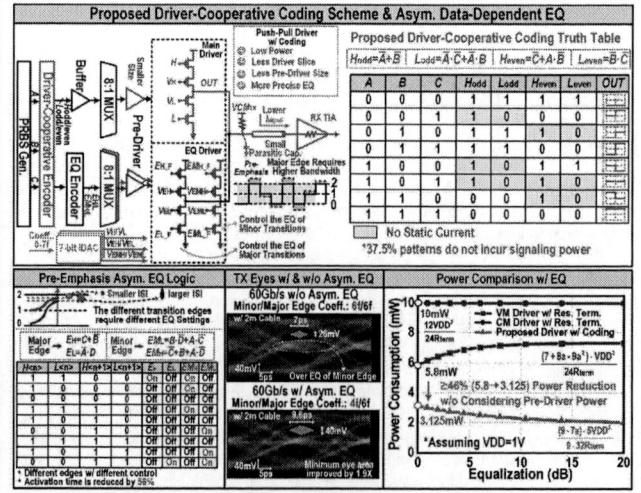

Figure 37.1.2: Proposed driver-cooperative coding scheme, asymmetric data-dependent EQ and its coding logic; TX output eyes w/ & w/o asymmetric EQ, and power comparison.

37

DIGEST OF TECHNICAL PAPERS • 625

979-8-3315-8937-0/26 $31.00 © 2026 IEEE

ISSCC 2026 / SESSION 37 / MEMORY INTERFACE / 37.1

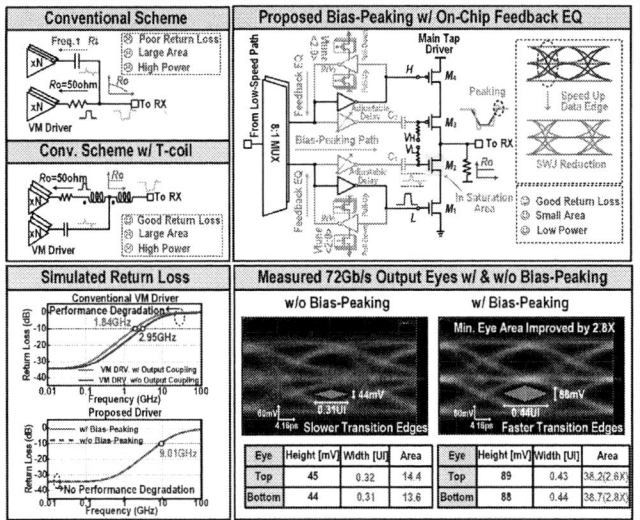

Figure 37.1.3: Schematic of conventional schemes; bias peaking with on-chip feedback equalization; simulated return loss, measured output eyes with & without bias-peaking.

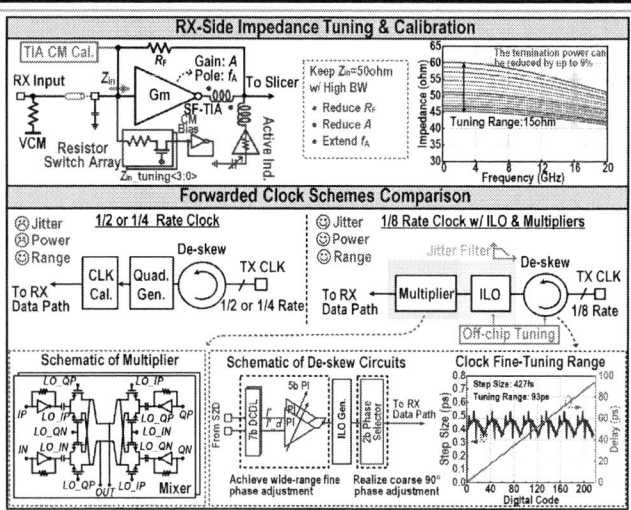

Figure 37.1.4: Schematic of impedance tuning and its tuning range; forwarded-clock scheme comparison; multiplier and de-skew circuit schematics; simulated de-skew range.

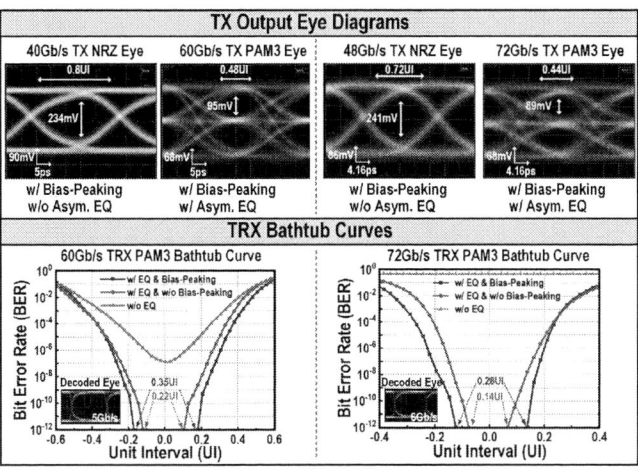

Figure 37.1.5: Measured 40 and 48GBaud TX output eye diagrams; measured TRX bathtub curves and decoded RX output eyes.

	ISSCC'19 [15]	ISSCC'22 [18]	ISSCC'23 [12]	JSSC'23 [17]	ISSCC'24 [6]	ISSCC'25 [4]	ISSCC'25 [1]	ISSCC'25 [5]	This Work		
Technology	28nm	28nm	4nm FinFET	28nm	4nm FinFET	3nm FinFET	28nm	28nm	28nm		
Signaling	PAM3	NRZ	PAM4	PAM3	NRZ	NRZ	PAM3	NRZ	NRZ/PAM3		
Package	Off-Chip	On-chip	Off-Chip	Off-Chip	Off-Chip	Off-Chip	On-chip	On-chip	Off-Chip		
Equalization	DFE	No	FS-FFE	FFE+DFE	Passive EQ+DFE	FFE	Passive EQ+DFE	No	Bias-Peaking+ Data-Dependent EQ		
Driver Type	VM (N-N)	VM (P-N)	VM (P-N)	Charge Bump	VM(P-N)	VM (P-N)	VM (N-N)	VM (N-N)	Push-Pull (P-N)		
SWJ Compensation	No	–	Yes	No	–	–	No	–	Yes		
Data Rate (Gb/s/Pin)	27	20	16	33	48	32	64	48	48	60	72
Insertion Loss @ Nyquist Freq. (dB)	6.5	2.5*	NA	NA	NA	4.6	8.5	2.8	5.4	8.5	9.5
Forwarded Clock Freq.	–	10GHz 1/2 Rate	NA	–	24GHz 1/2 Rate	16GHz 1/2 Rate	7GHz 1/4 Rate	–	4GHz 1/8 Rate	5GHz 1/8 Rate	6GHz 1/8 Rate
Energy Efficiency (pJ/b)	1.03	1.24	1.04	1.09	0.67	0.6	0.275*	1.21	0.7	0.82	1.17
Area(mm²)	0.0135	0.0024	0.0073**	0.006	NA	NA	0.001	0.039	0.0087		

*Clock Path Partly Excluded
**Estimated by Die Photo

Figure 37.1.6: Performance summary and comparison table.

Die Photos & Power Breakdown

Active Area Per Lane: 0.0041um²

TX Die Micrograph

Total: 36mW@72Gb/s
TX Power Breakdown

- Driver 8.3%
- Encoder 19.5%
- Pre-Driver 8.4%
- Clock Path 36%
- MUX 27.8%

Active Area Per Lane: 0.0046um²

RX Die Micrograph

Total: 48mW@72Gb/s
RX Power Breakdown

- AFE 25%
- Slicer & Demux 25%
- Decoder 12.5%
- Clock Path 37.5%

Figure 37.1.7: Die photos and TRX power breakdown.

ISSCC 2026 / SESSION 37 / MEMORY INTERFACE / 37.2

37.2 A 47.0Tb/s/mm 112Gb/s/pin PAM4 Single-Ended Transceiver Featuring 4-Aggressor Crosstalk Cancellation and Supply-Noise Tolerance for Short-Reach Memory Interfaces

Qian Liu[1], Yiheng Hui[1], Yuanlin Nong[1], He Ma[1], Yiwei Xu[1], Hao Hu[1], Jinpeng Zhu[1], Quan Wang[2], Lei Wang[2], Li Du[1,3], Yuan Du[1,3]

[1]Nanjing University, Nanjing, China, [2]T-Head (Shanghai) Semiconductor Co., Ltd., Shanghai, China
[3]Interdisciplinary Research Center for Future Intelligent Chips (Chip-X), Suzhou, China

Abstract

This paper presents a five-lane 112Gb/s/pin PAM4 single-ended transceiver in 28nm CMOS for high-density short-reach memory interfaces. Low-power triple-equalization, 4-aggressor shape-fitting crosstalk cancellation and supply-noise-tolerant clock distribution networks are involved to improve the signal integrity. The design demonstrates robust performance against severe crosstalk and supply noise, achieving an energy efficiency of 0.52pJ/b and an edge density of 47.0Tb/s/mm.

The rapid advancement of high-performance computing and AI promotes the evolutions of short-reach memory interfaces (e.g. HBM), die-to-die and chiplet links (e.g. UCIe); these interfaces utilize single-ended (SE) data transmissions for high-bandwidth I/O density. However, SE signaling is vulnerable to interference and noise: e.g. crosstalk (XT) and supply noise (SN). As the per-pin data rate exceeds 20Gb/s, using higher-level pulse-amplitude modulation (PAM) becomes essential: PAM4 is used for 22Gb/s GDDR6X [1] and PAM3 used for 42Gb/s GDDR7 [2]. However, the reduced signal-to-noise ratio (SNR) introduces greater signal-integrity challenges. The transmitter (TX) in [3] uses a feed-forward equalizer (FFE) and crosstalk cancellation (XTC) to cancel far-end crosstalk (FEXT) for 3D-staggered interposer channels, but it is limited to 4Gb/s NRZ. A 24Gb/s/pin PAM4 transceiver (TRX) [4] uses a PAM-based XTC scheme, but does not address SN. The clock-referenced PAM3 TRX proposed in [5] tolerates both SN and XT, but only two aggressors are considered, which does not address the high-density scenarios in multi-layer interposers where more than two adjacent aggressors can impair the victim signal [6].

To address these problems, this paper presents a 112Gb/s/pin PAM4 single-ended TRX, featuring four-aggressor XTC and SN tolerance. Figure 37.2.1 shows the top-level architecture of the implemented TRX and the on-chip 2mm channels. The TRX includes 5 SE data lanes and 1 differential clock lane. Each data lane sends 112Gb/s data using the proposed energy-efficient triple-equalization (TEQ) driver and termination scheme. The four-aggressor shape-fitting XTC is integrated into the driver. The receiver (RX) uses single-ended-to-differential (S2D) and two-stage buffers to drive the slicers. The clock lane shares the same 4:1 multiplexer, driver and termination as those of the data lanes to transmit the differential quarter-rate forwarded clock. Injection-locked oscillators (ILOs) are used as an I/Q-phase generator (IQG) in the clock lane's RX; per-pin de-skew circuits in the data-lane RXs are also based on ILOs. The 14GHz four-phase clocking architecture comprises of a transmission-line (TL) based global clock-distribution network (CDN) and per-lane local CDNs, both of which are SN insensitive; thus, minimizing power-supply-induced jitter (PSIJ). Moreover, the on-chip 2mm silicon-interposer-mimicked channels are designed as a two-layer structure with a horizontal pitch of 2.35μm and a vertical pitch of 1.75μm: exhibiting high channel density and severe four-adjacent-aggressor XT. The insertion loss (IL) of the channel is 3.9dB at Nyquist (28GHz); the worst case IL-to-FEXT ratio for the same-layer aggressors is 22.4dB, and for the different-layer aggressors is 25.5dB.

Figure 37.2.2 shows the low-power TEQ driver and termination scheme. The relatively low IL of short-reach channels requires less equalization (EQ) and impedance matching for improved power efficiency. First, a small-size high-output-impedance driver is implemented to significantly decrease the power consumption of the driver and the preceding pre-driver, and to minimize the output-node capacitance. Second, the resistor-feedback amplifier (RFA)-based termination in the RX side offers not only a moderate level of impedance matching to reduce reflections, but also a signal amplification to enhance received signal swing. Compared to double- and single-sided resistor termination schemes, the RFA termination has an advantage in both power efficiency and signal integrity. Third, although the on-chip channel loss is only 3.9dB, the aggregate loss, including driver, channel, pad + ESD capacitance (~130fF at each side) and termination, is up to 12.8dB at 28GHz, which requires some EQ. However, rather than a power-hungry TX FFE or an RX continuous-time linear equalizer (CTLE), we use three types of lightweight equalizers targeting different frequency points, resulting in a smooth response within the Nyquist frequency. The passive RC equalizer and active inductor attenuate DC content and also provide gain at 5 and 15GHz. Further, a gain above 20GHz is achieved via the passive inductor in the RFA, which is implemented by middle-layer thin metal to reduce its footprint. Consequently, a 9.1dB signal-loss improvement is realized by TEQ within a 0.5mW power drawn from the active inductor. The overall power consumption of the TEQ driver and termination is <3.8mW.

Figure 37.2.3 displays the proposed four-aggressor shape-fitting XTC combined within the driver. The XT-compensated signal is coupled, via a capacitor, to superimpose it onto the transmitted signal. However, unlike off-chip channel cases [7], the FEXT noise produced by on-chip short-reach channels manifests as changes in polarity, delay and pulse shape. An XTC signal with an unmatched delay or ill-fitting shape cannot fully compensate for the XT noise. Conversely, the maximum level of XTC can be achieved by a delay-matched and shape-fitted XTC signal. In the proposed XTC, delay matching is realized by the digitally controlled capacitor, while shape fitting is achieved by tuning the rise and fall rates of the XTC signal. The measured eye shmoos based on recovered data at RX in the right of Figure 37.2.3 show that the proposed XTC expands the three PAM4 eyes at BER ≤ 10⁻⁹ from completely closed to having openings of 160 × 40, 200 × 60, and 200mUI × 60mV.

Figure 37.2.4 demonstrates the proposed SN-tolerant global and local CDNs. The global CDN is composed of coplanar transmission lines (TLs) and impedance-matched current-mode-logic (CML) drivers. Compared to a conventional inverter-chain CDN the CML driver has a significantly enhanced robustness against SN. The input capacitance of each lane can be absorbed as part of the capacitive components of the TL. With the effect of wire inductance, the distributed clocks maintain sharp edges without increasing power consumption [8]. On the other hand, the local CDN is implemented with inverter chains due to area constraints. However, the clock buffer in the local CDN is SN insensitive due to two mechanisms: (1) the back-to-back (B2B) inverters connected between the differential outputs exhibit an inverse polarity in SN sensitivity to that of the forward (F) inverters. Hence, the overall SN sensitivity decreases as the B2B:F inverter size ratio increases, leading to a reduced PSIJ. (2) The supply noise compensator (SNC) further reduces the residual PSIJ by modulating the inverter delay in the opposite direction to V_{DD} delay modulation [9]. Post-layout simulation shows that the PSIJ reduces from 3.1 to 0.5ps for the case of a 100mV$_{pp}$ 200MHz SN when the B2B:F inverter size ratio is 1:2 and SNC is enabled. Clock-pattern eye diagrams are also measured under a clean V_{DD_CLK} and a noisy V_{DD_CLK}: a 5% degradation of eye width is observed when injecting 100mV$_{pp}$ 200MHz SN onto V_{DD_CLK}.

Measured RX eye shmoos under various configurations of EQ, XTC and SNC are presented in Fig. 37.2.5. To evaluate both XTC and SN tolerance of the proposed TRX, apart from the aggregate 12.8dB signal loss considering both circuit and channel, all the four XT aggressors are activated and a 100mV$_{pp}$ 200MHz SN is also injected onto V_{DD_CLK}. In the first configuration, equalizers (except passive inductor), shape-fitting XTC and SNC are all deactivated, resulting in no observable eye opening at BER ≤ 10⁻⁹. In the second scheme, TEQ is fully enabled while XTC and SNC remain off, and the eye openings improve but remain invisible at BER ≤ 10⁻⁹. In the third scheme, SNC is activated but XTC is not, there are still no effective eyes at BER ≤ 10⁻⁹. Conversely, in the fourth configuration where XTC is activated but SNC is not, the measured eye openings are 120 × 30, 120 × 40, and 120mUI × 50mV. In the final scheme, both XTC and SNC are enabled and the measured eyes are open with 120 × 40, 160 × 60, and 160mUI × 60mV.

The prototype chip, including the proposed TRX and on-chip test channels, is fabricated in 28nm CMOS. The BER and eye-openings are measured via the on-chip PRBS7 pattern generator and checker, and an eye-opening monitor. A differential 14GHz clock is generated by an arbitrary waveform generator and fed into the TX. To emulate a noisy power supply environment, supply noise is injected onto V_{DD_CLK} via capacitive coupling on the PCB. Figure 37.2.6 demonstrates a performance comparison table with prior works for short-reach interfaces, the die micrograph and an area summary of active circuits including clock and data paths. Compared to prior works, the implemented TRX shows effective 4-aggressor XTC and SN tolerance for 112Gb/s/pin PAM4 signals, with a 0.52pJ/b energy efficiency, and an die-shore density of up to 47.0Tb/s/mm (including clock lanes).

Figure 37.2.7 summarizes the measured performance of shape-fitting XTC at different aggressor settings and SNC performance at various frequencies of 100mV$_{pp}$ SN. It shows that the proposed XTC provides up to a 0.16UI extension for the four-aggressor worst case tested. Results show that the SNC is effective below 800MHz, but less effective at higher frequencies due to SNC bandwidth limitations. The test-chip power breakdown is also shown in Fig. 37.2.7.

Acknowledgement:
This work was supported by the National Key Research and Development Program of China under Grant No. 2021YFA0717700 and the Key Research and Development Program of Jiangsu Province under Grant BE2023020-3.

ISSCC 2026 / February 18, 2026 / 1:30 PM

References:

[1] T. M. Hollis et al., "An 8Gb GDDR6X DRAM Achieving 22Gb/s/pin with Single-Ended PAM4 Signaling," *ISSCC*, pp. 348-350, 2021.
http://doi.org/10.1109/ISSCC42613.2021.9365925

[2] B. Kim et al., "A 42Gb/s Single-Ended Hybrid-DFE PAM-3 Receiver for GDDR7 Memory Interfaces," *ISSCC*, pp. 398-400, 2025.
http://doi.org/10.1109/ISSCC49661.2025.10904551

[3] H.-G. Ko, S. Shin, J. Oh, K. Park and D. -K. Jeong, "An 8Gb/s/μm FFE-Combined Crosstalk-Cancellation Scheme for HBM on Silicon Interposer with 3D-Staggered Channels," *ISSCC*, pp. 128-130, 2020. http://doi.org/10.1109/ISSCC19947.2020.9063162

[4] S. Kim et al., "A 0.458-pJ/bit 24-Gb/s/pin Capacitively Driven PAM-4 Transceiver With PAM-Based Crosstalk Cancellation for High-Density Die-to-Die Interfaces," *IEEE JSSC*, vol. 59, no. 11, pp. 3730-3740, Nov. 2024. http://doi.org/10.1109/JSSC.2024.3401213

[5] K. Kim et al., "A 0.275pJ/b 42Gb/s/pin Clock-Referenced PAM3 Transceiver Tolerant to Supply Noise, Reference Offset and Crosstalk for Chiplets and Short-Reach Memory Interfaces," *ISSCC*, pp. 394-396, 2025.
http://doi.org/10.1109/ISSCC49661.2025.10904729

[6] Y. Nishi et al., "A 0.297-pJ/Bit 50.4-Gb/s/Wire Inverter-Based Short-Reach Simultaneous Bi-Directional Transceiver for Die-to-Die Interface in 5-nm CMOS," *IEEE JSSC*, vol. 58, no. 4, pp. 1062-1073, Apr. 2023.
http://doi.org/10.1109/JSSC.2022.3232024

[7] W. Wu et al., "A 64Gb/s/pin PAM4 Single-Ended Transmitter with a Merged Pre-Emphasis Capacitive-Peaking Crosstalk-Cancellation Scheme for Memory Interfaces in 28nm CMOS," *ISSCC*, pp. 240-242, 2024.
http://doi.org/10.1109/ISSCC49657.2024.10454320

[8] F. O'Mahony et al., "A Low-Jitter PLL and Repeaterless Clock Distribution Network for a 20Gb/s Link," *IEEE Symp. VLSI Circuits*, pp. 29-30, 2006.
http://doi.org/10.1109/VLSIC.2006.1705296

[9] T. S. Sandhu and K. El-Sankary, "Supply-Insensitive Digitally Controlled Delay Lines for 3-D IC Clock Synchronization Architectures," *IEEE VLSI Sys.*, vol. 27, no. 6, pp. 1480-1484, June 2019. http://doi.org/10.1109/TVLSI.2019.2894104

[10] H. Park et al., "A 0.385-pJ/bit 10-Gb/s TIA-Terminated Di-Code Transceiver with Edge-Delayed Equalization, ECC, and Mismatch Calibration for HBM Interfaces," *ISSCC*, pp. 452-454, 2022. http://doi.org/10.1109/ISSCC42614.2022.9731740

[11] Q. Liu et al. "A 0.90-Tb/s/in 1.29-pJ/b Wireline Transceiver with Single-Ended Crosstalk Cancellation Coding Scheme for High-Density Interconnects," *IEEE JSSC*, vol. 58, no. 8, pp. 2326-2336, Aug. 2023. http://doi.org/10.1109/JSSC.2023.3261125

[12] J. Lee et al., "A 246-fJ/b 13.3-Tb/s/mm Single-Ended Current-Mode Transceiver with Crosstalk Cancellation for Shield-Less Short-Reach Interconnect," *IEEE Symp. VLSI Tech. and Circuits*, 2024.
http://doi.org/10.1109/VLSITechnologyandCir46783.2024.10631466

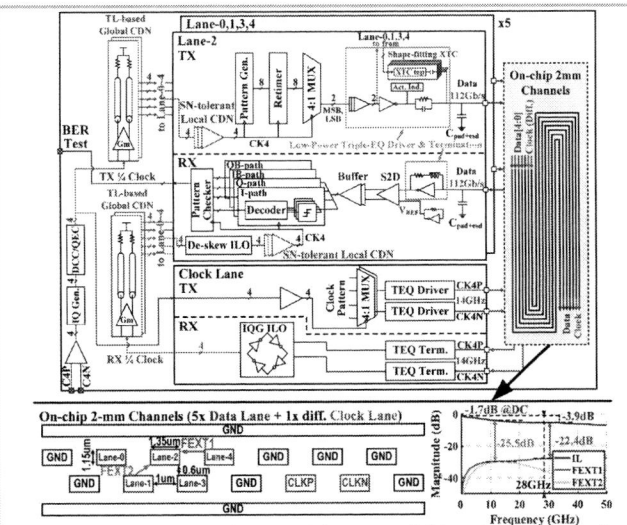

Figure 37.2.1: Top-level architecture of the proposed 5-lane transceiver (top) and characteristics of the on-chip channels (bottom).

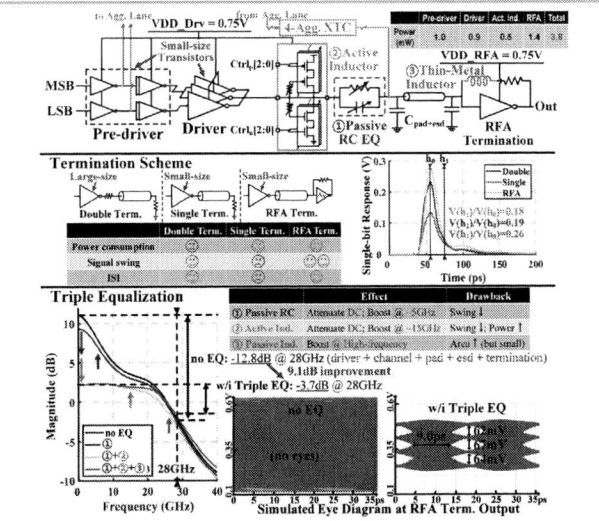

Figure 37.2.2: Schematic of low-power triple-EQ driver and RFA termination (top), comparisons of termination schemes (middle) and details of the triple-EQ (bottom).

DIGEST OF TECHNICAL PAPERS • 627

979-8-3315-8937-0/26 $31.00 © 2026 IEEE

ISSCC 2026 / SESSION 37 / MEMORY INTERFACE / 37.2

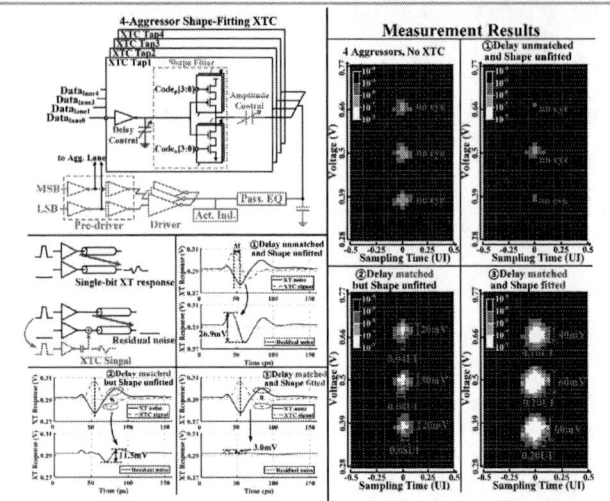

Figure 37.2.3: Schematic of 4-aggressor shape-fitting XTC (top-left), simulated FEXT responses (bottom-left) and measured eyes (right) for various schemes of delay and shape.

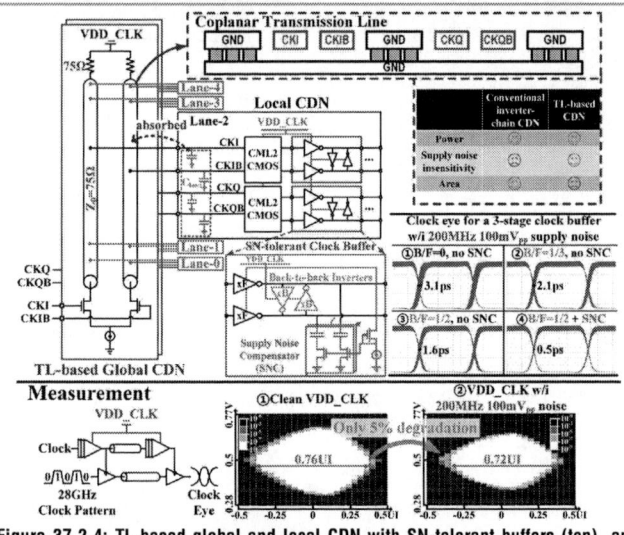

Figure 37.2.4: TL-based global and local CDN with SN-tolerant buffers (top), and measured clock eyes under 200MHz 100mV$_{pp}$ SN (bottom).

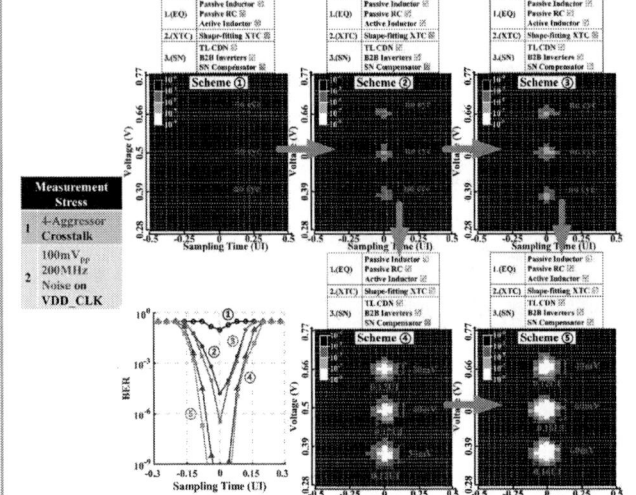

Figure 37.2.5: Measured eye diagrams and BER curves for various configurations of EQ, XTC and SNC, under 4-aggressor XT and 200MHz 100mV$_{pp}$ SN.

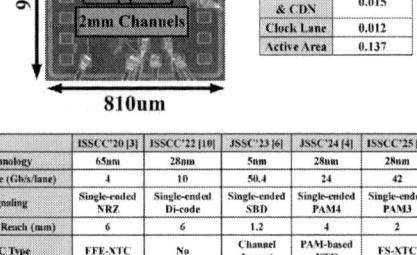

Block	Area (mm²)
5x TX	0.049
5x RX	0.061
Clock Path & CDN	0.015
Clock Lane	0.012
Active Area	0.137

	ISSCC'20 [3]	ISSCC'22 [10]	JSSC'23 [6]	JSSC'24 [4]	ISSCC'25 [5]	This Work
Technology	65nm	28nm	5nm	28nm	28nm	28nm
Data Rate (Gb/s/lane)	4	10	50.4	24	42	112
Signaling	Single-ended NRZ	Single-ended Di-code	Single-ended SBD	Single-ended PAM4	Single-ended PAM3	Single-ended PAM4
Channel Reach (mm)	6	6	1.2	2	2	2
XTC Type	FFE-XTC	No	Channel Layout	PAM-based XTC	FS-XTC	Shape-Fitting XTC
# of XTC Aggressor	4	/	/	2	2	4
Supply Noise Tolerance	No	No	No	No	Yes	Yes
Energy Efficiency (pJ/b)	1.5	0.39	0.30	0.46	0.28	0.52
Area (mm²/lane)	0.0023	0.0006	0.011	0.0010	0.0012	0.027
Edge Density (Tb/s/mm)	8	2	2.14	12	9.16*	47.0*

* Including clock lanes.

Figure 37.2.6: Die micrograph, area allocation, and comparisons with prior works for short-reach interfaces.

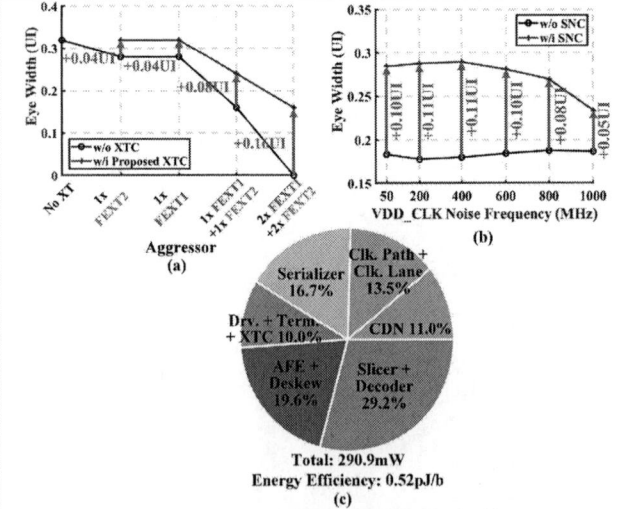

Figure 37.2.7: Measured eye widths at (a) various settings of XT aggressors, (b) various frequencies of 100mV$_{pp}$ SN, and (c) pie-chart showing the TRX power breakdown.

• 2026 IEEE International Solid-State Circuits Conference

979-8-3315-8937-0/26 $31.00 © 2026 IEEE

ISSCC 2026 / SESSION 37 / MEMORY INTERFACE / 37.3

37.3 A 2nm All-Digital 14.4Gb/s/pin LPDDR6 PHY with Quarter-Rate Clocking Architecture and Multi-Level FIFO-Based Speculative DFE

Yoonjae Choi, Daero Kim, Myunggon Kim, Sangmin Lee, Hansol Kang, Gwangwon Lee, Yoonhyung Kim, Sungsik Kang, Hyun Cho, Boyoung Kang, Kyoungwon Lee, Jaegeun Song, Jinho Choi, Shinyoung Yi, Billy Koo, Kwanyeob Chae, Hyo-Gyuem Rhew

Samsung Electronics, Yongin, Korea

Abstract

This paper presents a 2nm all-digital LPDDR6 PHY that achieves 14.4Gb/s/pin, increasing bandwidth while optimizing power. Key innovations include quadrature clocking, with QEC, to reduce jitter sensitivity, and a multi-level FIFO-based speculative DFE to relax timing constraints in standard-cell implementation. An efficiency mode, employing clock gating, dramatically reduces power. A reconfigurable serializer supports both 1:6 and 1:8 DFIs, ensuring high bus efficiency and backward compatibility.

Bandwidth demands of low-power double-data-rate (LPDDR) interfaces continues to grow to support AI, augmented reality (AR), and high-resolution multimedia workloads [1-3]. To meet demand, the PHY's internal clock frequency is aggressively scaled; resulting, in increased jitter sensitivity and power dissipation of the high-speed clock buffers, which are the most power-hungry blocks in the PHY. In LPDDR6 [4], two independent 24b sub-channels are adopted to boost data bandwidth, further multiplying the number of clock buffers and exacerbating power dissipation. At the same time, dynamic voltage-and-frequency scaling (DVFS) expands the operating frequency range of the PHY, making power optimization across day-of-use (DoU) scenarios more difficult. Furthermore, mobile PHYs are optimized for both peak-bandwidth and medium-rate operation under strict power budgets, so a conventional maximum-frequency-centric design often leads to excessive power.

This work proposes a 2nm all-digital LPDDR6 PHY achieving 14.4Gb/s/pin, by introducing a quarter-rate clocking scheme with quadrature-error correction (QEC) to reduce frequency of the clock and jitter sensitivity, while also reducing clocking power. A multi-level FIFO-based speculative decision-feedback equalizer (DFE) is proposed to relax the DFE's stringent timing requirements for the feedback paths in all-digital standard-cell implementations . Furthermore, an efficiency mode, with a clock-gating scheme, is introduced to selectively deactivate the secondary sub-channel and suppress unnecessary clock toggling, substantially reducing power dissipation from power-hungry clock buffers in the PHY. Finally, a reconfigurable serializer is implemented to support both 1:6 and 1:8 DDR PHY interfaces (DFIs), ensuring high bus efficiency and backward compatibility.

Figure 37.3.1 shows the overall architecture of the proposed all-digital LPDDR6 PHY, which is implemented using digital standard cells. In the clock paths, a divider-based quadrature-clock generator (QCG) is used to generate four-phase clocks (I, Q, IB, QB); four digital delay lines are used in the QEC for the four-phase clocks. The calibrated quarter-rate quadrature clocks are distributed via high-speed buffers to each bit slice. In the receiver, a hybrid offset calibration that combines g_m and resistor tuning is employed to guarantee a wide-offset tuning range across the wide DVFS range. Two high-speed current-mode logic (CML) receivers are used for the speculative DFE implementation, and a multi-level FIFO structure ensures timing robustness. In addition, the power domain of the PHY is partitioned into high- and low-speed regions for further PHY power optimization. To enable this, level shifters are employed at the power-domain boundary. The reconfigurable serializer provides 12:1 or 16:1 serialization, to support 1:6 or 1:8 DFI compatibility.

Conventional LPDDR PHYs double the internal clock frequency to double data throughput, but this increases jitter and clock-buffer power. To mitigate these effects, the proposed design adopts quarter-rate quadrature clocking. However, small mismatches among the four clock phases will degrade jitter tolerance. To address this, quadrature-error correction (QEC) training is introduced. Figure 37.3.2 illustrates QEC training in the proposed LPDDR6 PHY. During training, the four phases are compared in pairs. A 4:2 multiplexer sequentially selects phase pairs, and dedicated pre-processing logic [5] measures the phase differences (I–Q, Q–IB, IB–QB, and QB–I). After measuring all phase differences, the average phase difference among the four-phase clocks is calculated to obtain the ideal spacing of 90°. Then, three of the four delay lines are digitally tuned to have the same delay as the average phase difference. The delay of the I-phase delay line is fixed to the median value of delay range to act as a reference for other delay lines and to achieve negative-delay compensation. This approach effectively equalizes phase spacing. Inverter-based phase interpolators embedded in the digital delay lines provide sub-LSB phase adjustment to ensure accuracy across process and voltage variation.

To alleviate the timing constraint of the DFE, a multi-level FIFO-based speculative DFE is proposed and implemented in this work. In a conventional speculative DFE, shown in Fig. 37.3.3(top), the critical path of the DFE is relaxed to one MUX and one flip-flop propagation delay by unrolling the DFE loop. However, it is hard to meet the 1UI feedback timing in a standard-cell-based design as the data rate increases. Figure 37.3.3(bottom) shows the proposed multi-level FIFO-based speculative DFE. It utilizes the zero-cycle read methodology [5], where the FIFO output is available in the same cycle as the read request. The DFE data selection is performed in the first-level (L1) FIFO, where data is deserialized before being sampled in the second-level (L2) FIFO. Meanwhile, both the L1 and L2 FIFOs share the same clock edge through the zero-cycle methodology, effectively flattening the timing path. This eliminates the need for a dedicated feedback MUX in the critical loop, increasing the available slack by nearly a half unit interval. The DFE timing constraint can be considerably alleviated in this way. Moreover, the relaxed timing path, due to the zero-cycle methodology also increases robustness against supply variations, which is crucial for DVFS . Figure 37.3.3(bottom right) compares the maximum data rate of the conventional and the proposed speculative DFEs. The proposed speculative DFE achieves 14.4Gb/s operation at 0.78V, while the conventional approach fails to operate under this condition.

From a system-architecture point of view, a 1:6 DFI can maximize the bus configuration efficiency with 24b DQ signals, as shown in Fig. 37.3.4(top right). The LPDDR6 specification [4] supports both 1:6 and 1:8 DFIs, since system compatibility requires 1:8 as well. 12:1 and 16:1 serialization ratios are required for 1:6 and 1:8 DFIs. To enable this, a phase-counter-based reconfigurable serializer is adopted in this work. Figure 37.3.4(top) shows the reconfigurable serializer and its timing diagrams. It consists of five 4:1 serializers and supports both 16:1 and 12:1 serialization using the phase counter, which counts the phases to distinguish them. In the 16:1 serialization mode for 1:8 DFI, the phase counter cycles from 0 to 3, fully enabling all the 4:1 serializers and producing a full 16:1 output. On the other hand, in the 12:1 serialization mode for 1:6 DFI, the phase counter cycles from 0 to 2, and the preceding four 4:1 serializers operate effectively as 3:1 serializers. Figure 37.3.4(top right) shows timing diagrams that illustrate how the phase counter ensures proper data alignment in both modes. This phase-counter-based serialization method allows efficient reuse of existing serializer blocks with minimal overhead and avoids the need for two separate serializer paths.

Mobile workloads often exhibit bursty traffic, with long idle periods when peak bandwidth is unnecessary: efficiency mode is introduced in LPDDR6 interfaces to reduce DRAM power. In normal mode two sub-channels operate to maximize data bandwidth. If data traffic is low, then the efficiency mode is enabled to reduce DRAM power consumption by deactivating the secondary sub-channel. The proposed PHY enhances this mode by additionally gating the high-speed clock path for the inactive sub-channel, preventing unnecessary toggling in the buffer tree. Figure 37.3.4(bottom) shows the efficiency mode configuration in the proposed LPDDR6 PHY and the normalized power dissipation according to the operation mode, demonstrating the power benefits of efficiency mode with clock gating. If efficiency mode is enabled, then the primary sub-channel remains active while the secondary sub-channel enters standby or deep-sleep mode. The efficiency mode reduces PHY power dissipation by 39% and 29% in read and write modes, respectively. Additional clock gating reduces PHY power dissipation by 46%, 48%, and 59% in write, read, and idle modes, respectively. This demonstrates that PHY-side clock gating is critical to fully exploit the efficiency mode, because the clock-distribution network continues to consume substantial power even when DRAM activity is low.

The proposed all-digital LPDDR6 PHY is fabricated in a 2nm multi-bridge-channel field-effect-transistor (MBCFET) CMOS technology. Figure 37.3.5 shows the measured read and write Shmoo plots at 12.2Gb/s before and after the QEC training. Measurement results show that the read and write valid window margins are improved by 8.3% and 41.6%, respectively, after calibration. This confirms that the fine-grain digital QEC provides robust quadrature clock phase alignment across DVFS conditions, directly reducing clock jitter impact. To demonstrate the maximum performance of the proposed LPDDR6 PHY, internal loopback measurements are performed. Figure 37.3.6(top) shows the measured internal read and write Shmoo plots: 14.4Gb/s operation is achieved at 0.78V. Due to external DRAM limitations, read and write operations with 2-rank LPDDR6 DRAM are validated through system-level tests at 12.8Gb/s. The measured read and write Shmoo plots are shown in Fig. 37.3.6(bottom). The read and write valid window margins are 0.3 and 0.4UI, respectively. These results confirm that the PHY achieves robust data integrity at near-maximum LPDDR6 speeds, while maintaining margin under real DRAM operation. A chip micrograph of the proposed all-digital LPDDR6 PHY and a comparison with prior work is shown in Fig. 37.3.7: illustrating the integration of serializer, quadrature clocking blocks, and multi-level FIFO DFE in a compact digital layout.

ISSCC 2026 / February 18, 2026 / 1:30 PM

References:

[1] J.-H. Baek et al., "A 16Gb 12.7Gb/s/pin LPDDR5-Ultra-Pro DRAM with 4-Phase Self-Calibration and AC-Coupled Transceiver Equalization in a 5th-Generation 10nm DRAM Process," *ISSCC*, pp. 510-512, 2025.
https://doi.org/10.1109/ISSCC49661.2025.10904794

[2] Y. Seo et al., "A 1a-nm 1.05V 10.5Gb/s/pin 16Gb LPDDR5 Turbo DRAM with WCK Correction Strategy, a Voltage-Offset-Calibrated Receiver and Parasitic Capacitance Reduction," *ISSCC*, pp. 246-248, 2024.
https://doi.org/10.1109/ISSCC49657.2024.10454381

[3] K. Chae et al., "An 8nm All-Digital 7.3Gb/s/pin LPDDR5 PHY with an Approximate Delay Compensation Scheme," *IEEE Symp. VLSI Circuits*, pp. 96-97, 2019.
https://doi.org/10.23919/VLSIC.2019.8777959

[4] "LPDDR6 Standard", JESD209-6, *JEDEC*, July 2025. https://www.jedec.org/standards-documents/docs/jesd209-6

[5] K. Chae et al., "A 4nm 1.15TB/s HBM3 Interface with Resistor-Tuned Offset-Calibration and In-Situ Margin-Detection," *ISSCC*, pp. 406-407, 2023.
https://doi.org/10.1109/ISSCC42615.2023.10067736

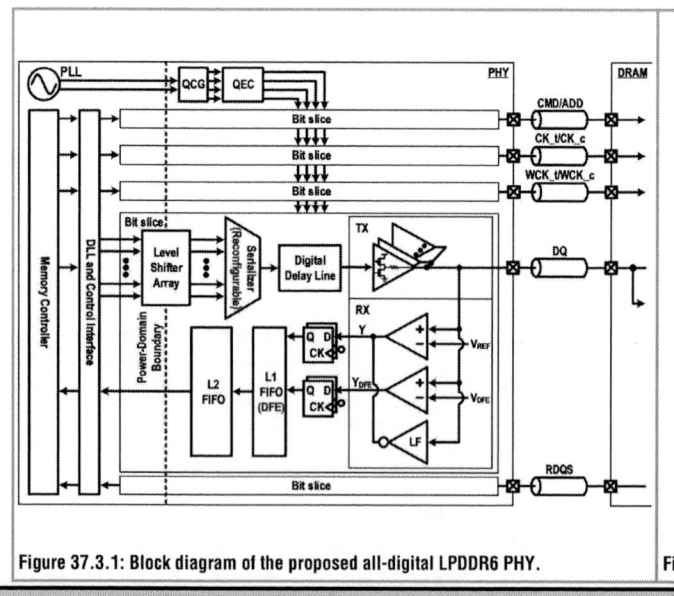

Figure 37.3.1: Block diagram of the proposed all-digital LPDDR6 PHY.

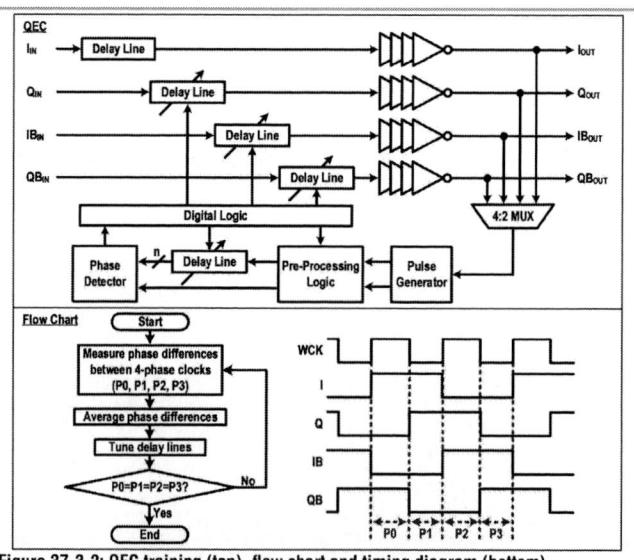

Figure 37.3.2: QEC training (top), flow chart and timing diagram (bottom).

37

DIGEST OF TECHNICAL PAPERS • 629

979-8-3315-8937-0/26 $31.00 © 2026 IEEE

ISSCC 2026 / SESSION 37 / MEMORY INTERFACE / 37.3

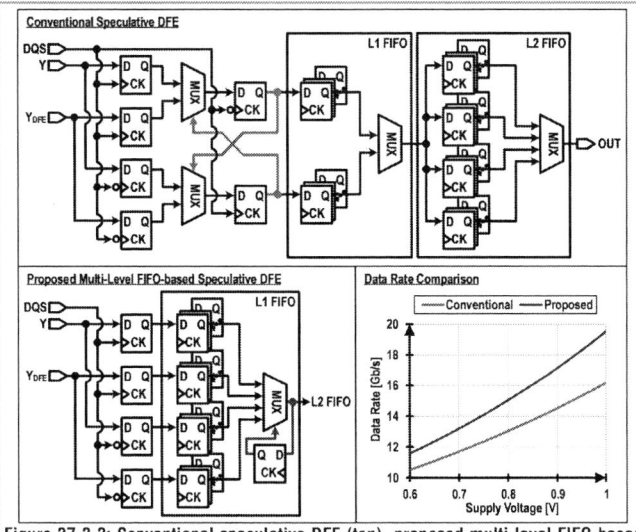

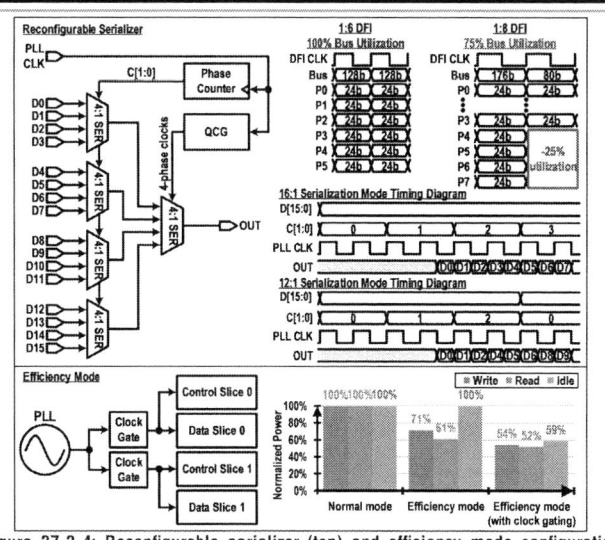

Figure 37.3.3: Conventional speculative DFE (top), proposed multi-level FIFO-based speculative DFE (bottom left), and data rate comparison (bottom right).

Figure 37.3.4: Reconfigurable serializer (top) and efficiency mode configuration (bottom).

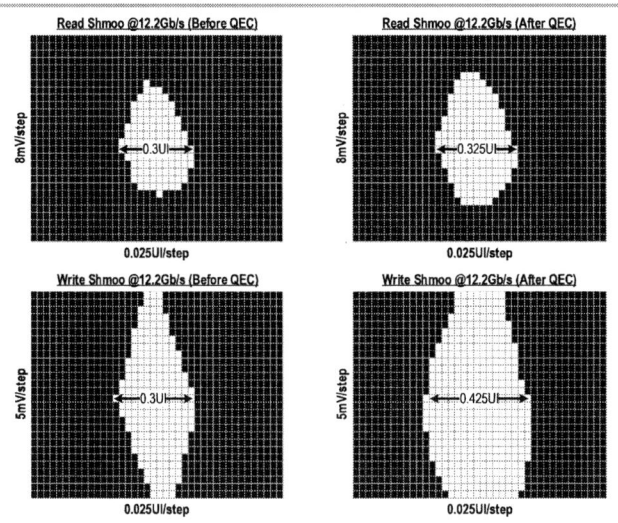

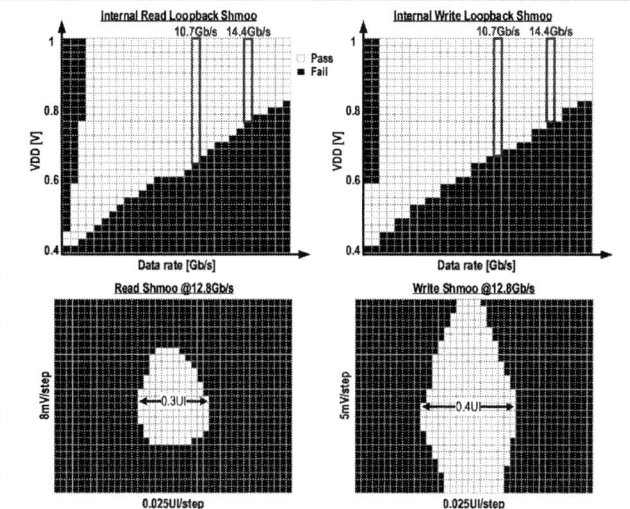

Figure 37.3.5: Measured read/write Shmoo plots at 12.2Gb/s before and after QEC training.

Figure 37.3.6: Measured internal read/write loopback Shmoo plots (top) and measured read/write Shmoos at 12.8Gb/s (bottom).

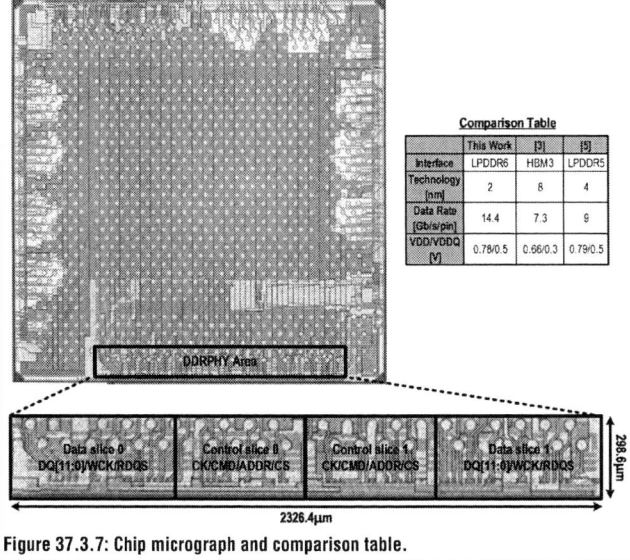

Figure 37.3.7: Chip micrograph and comparison table.

• 2026 IEEE International Solid-State Circuits Conference

979-8-3315-8937-0/26 $31.00 © 2026 IEEE

ISSCC 2026 / SESSION 37 / MEMORY INTERFACE / 37.4

37.4 **A 100Gb/s, 1.92pJ/b Aggregate 4-Lane Single-Ended NRZ Transceiver with 15dB Far-End Crosstalk Cancellation via On-Chip Feature Extraction and Classification in 16nm FinFET**

Xiaohui Lin, Ramin Javadi, Bella Bose, Tejasvi Anand

Oregon State University, Corvallis, OR

Abstract

A low-power (1.92pJ/bit), high-speed (100Gb/s), equalizer-free, 4-lane single-ended transceiver that leverages machine learning principles for crosstalk compensation is proposed. The transceiver uses feature extraction and on-chip classification. It achieves more than 0.2UI horizontal opening for BER < 10^{-12}. The classification compensates the crosstalk from 3 aggressors and -15dB far-end crosstalk power sum. The classification only accounts for 7% of the total transceiver power.

Data-driven AI applications demand increasing bandwidth for processor-to-HBM or processor-to-module interconnects [1]. To meet this required higher throughput, multiple single-ended links are routed in close proximity to each other: resulting in electromagnetic coupling (crosstalk) between these densely packed interconnects. Crosstalk degrades signal integrity, posing a major challenge to achieving <10^{-12} BER. Prior crosstalk cancellation or reduction techniques, see Fig. 37.4.1, can be categorized into three categories: (1) wire-engineering methods, (2) bus-encoding schemes, and (3) active crosstalk cancellation using cross-coupled equalizers. Wire-engineering methods such as ground shielding [2] and wire swizzling [3] incur routing overhead at the PCB or package level; thus, limiting data density. Bus encoding, such as Fibonacci encoding [4], enhances signal integrity at the expense of coding complexity and additional pin overhead arising from redundant encoding. The active crosstalk cancellation (XTC) approach, using cross-coupled equalizers, adds a high-bandwidth cancellation circuit, operating at the data rate, into the transmitter or receiver front-end, resulting in increased loading and reduced energy efficiency. While prior work has shown cross-coupled equalizer-based crosstalk cancellation for just 2-lane high-speed (>20Gb/s) off-chip links [5], the energy inefficiency, due to the front-end loading of the XTC circuit and the long-distance routes of equalizer outputs across lanes, becomes more pronounced when crosstalk cancellation is required for four or more lanes; as is required for memory interfaces. This work proposes an equalizer-free approach that employs machine learning (ML) principles to compensate for crosstalk in a 4-lane off-chip wireline link. Although prior work adopted ML for wireline links, compensation was limited to single-channel imperfections: that is, inter symbol interference (ISI) [6-9]. This work investigates an ML approach to first learn the cross-channel coupling responses, revealing the inherently deterministic nature of the coupling effects that exist between channels, and then compensates for the crosstalk. Figure 37.4.1 presents the proposed approach of employing ML principles to compensate for crosstalk, where the receiver extracts the features of the received data from the victim and aggressor lanes and provides it to the classifier. The decision-tree classifier, first trained in supervised learning mode using PRBS data patterns, is then synthesized on-chip for classification. Classification is performed in the digital domain at the receiver back-end (low frequency: 781.25MHz), which results in energy-efficient crosstalk compensation; the classifier has a latency of only 6UI. The proposed 4-lane 25Gb/s single-ended transceiver, with an aggregate data rate of 100Gb/s, compensates for far-end crosstalk (FEXT) with a power sum up to 15dB; achieving a <10^{-12} BER with an energy efficiency of 1.92pJ/b, including clock distribution power, in a 16nm FinFET process. While the proposed transceiver is a 4-lane link, when compared to prior high-speed 2-lane links operating at a similar or higher data rate and when compensating similar FEXT, the proposed transceiver achieves a 31% lower energy/bit compared to [10] and 34% compared to [11].

Figure 37.4.2 shows the proposed 4-lane transceiver architecture, where the conventional receiver front-end crosstalk cancellation is replaced with feature extraction and classification in the receiver. All 4 lanes of the transmitter are identical, each with its own programmable PRBS generator. The PRBS generator can be initialized with different seeds to generate un-correlated data across 4-lanes, which provides a more representative distribution of coupled data transition events. The PRBS generator is followed by a 32:1 serializer and then a source-series-terminated (SST) output driver. Each transmitter lane has its own duty cycle corrector (DCC) and a programmable-delay line to de-skew phase mismatch among channels. The receiver is a half-rate architecture: its front-end consists of a T-coil that compensates for ESD and termination parasitics, followed by a feature extraction block, which consists of 2×3 slicers (even and odd, with each capturing three features) per lane. Feature extraction is followed by six 2:32 de-serializers per lane to reduce the processing data rate for the combined decision-tree classifier for all 4 lanes. Each lane supports 5 decision trees, which are trained on different channels; a preprogrammed classifier is selected, depending on channel conditions, via the SELECT signal. There are no equalizers such as FFE or CTLE present in the proposed transceiver as the ISI of channel will also be learned and compensated by the decision-tree classifier along with crosstalk.

Figure 37.4.3 illustrates the supervised learning process for all four-lane decision-tree classifiers. The training setup is modeled in MATLAB, where four PRBS-23 sources with various starting seeds drive four parallel lanes with practical channel models that include both insertion loss and inter-lane crosstalk. At the receiver front-end, each lane contains a feature extraction block with three slicers. These slicers compare the sampled incoming waveform against three threshold voltages (V_{TH1}, V_{TH2} and V_{TH3}) to extract first-order features (F1, F2 and F3). Received signal temporal information from both the victim and aggressor lanes is captured by recording several past UIs, the present UI, and a few future UIs forming a feature vector. The feature vectors from all four lanes are merged into a composite vector that represents both the channel impairments on each signal and the crosstalk coupling between them. This combined feature vector is then provided to the classifier for training. The original transmitted 4-lane PRBS data serves as labels for supervised learning. The classifiers build a hierarchical decision structure that maps the received information to the transmitted bits (labels), learning the per-lane channel impairments and crosstalk across all lanes. The trained decision-tree classifiers are then evaluated by analyzing the accuracy using a PRBS checker. System parameters, including the number of past/future UIs considered and the threshold voltages used for feature extraction, are iteratively adjusted through exhaustive search based on the measured accuracy. The classifiers are subsequently retrained with the updated parameters, and this loop continues until the accuracy converges to its highest achieved value among all the training conditions; resulting in an optimized decision-tree configuration for the given channel and crosstalk conditions. Finally, the classifiers generated in MATLAB are converted into Verilog code for on-chip synthesis. The test channel used shows that the training loop converges to a solution using 3 past and 2 future UIs from both the victim and aggressor lanes. To ensure a loose fit, the trained decision-tree model was evaluated across varying data rates, data patterns and channel models before being finalized, and therefore enabling it to compensate for a range of crosstalk.

Figure 37.4.4 depicts the classifier operation when compensating for the crosstalk. For simplicity, the decision-tree classifier is trained on a two-lane transceiver scenario. Both lanes experience mutual crosstalk, but in this example lane 1 is designated as the victim and lane 2 as the aggressor. Figure 37.4.4 shows the trained lane-1 decision tree. For each UI, three features, based on the three threshold voltages, are extracted from the victim at the current (F_{V1}, F_{V2} and F_{V3}) and past UI (F_{V4}, F_{V5} and F_{V6}), as well as from the aggressor at current and past UIs (F_{A1} - F_{A6}); these features are combined and provided to the decision tree. Consider the time instance (highlighted in yellow) when the victim's 0 is corrupted by crosstalk from the aggressor's transition to 1. The decision tree begins classification at the root node using feature F_{V1}, representing the lowest threshold voltage of the victim lane. With $F_{V1}=1$, it checks the middle-threshold feature F_{V2}. Since $F_{V2}=1$, the classifier examines the aggressor using F_{A1} and F_{A6}, corresponding to the current and past UI. With $F_{A1}=1$ and $F_{A6}=0$, a 0→1 transition is detected on the aggressor. Then, the classifier checks F_{V6} and F_{V5} to extract the victim's previous bit. With both being 0 and with $F_{A3}=1$, the classifier determines that the aggressor transitioned and no change occurred on the victim. The decision tree outputs 0, which represents its final decision based on the decisions taken, as previously described. This example shows that the transmitted bit, 0, on the victim lane would have been detected as a 1, due to crosstalk, if only the midpoint (V_{TH2}) sample was used. By incorporating current- and past-UI features from both lanes, the decision tree accurately detects the aggressor transition and outputs the correct decision. In this example, the decision tree uses 12 features and has 9 leaves, but for a more realistic channel like the one used to validate this work, the decision tree leverages 72 features (3×4×6) and has more than a hundred leaves.

A prototype 4-lane single-ended transceiver, using the proposed ML approach, is fabricated in a 16nm FinFET process. Figure 37.4.5 shows the measured Tx near-end eye diagrams for the four lanes at 25Gb/s/pin with PRBS7 data. Note that to measure the Tx outputs on the oscilloscope, a different and longer PCB channel was used due to the spacing required for the SMA connectors; the ISI for this PCB channel is characterized using a single-bit pulse response, shown in Fig. 37.4.5(top left). To validate that the decision-tree has learned the channel ISI and is now able to compensate for it, bathtub plots for all 4 transceiver (Tx & Rx) lanes were measured by turning on one lane at a time with and without the classifier. Measurements show that a >0.2UI improvement in the horizontal eye opening is achieved at <10^{-12} BER when the classifier is enabled. Each lane's measured channel response has a ~8dB insertion loss. The results presented show that the classifier compensates for the channel's ISI non-idealities as well. Figure 37.4.6 demonstrates the resilience of the proposed ML approach from crosstalk by enabling 1, 2 and all aggressor lanes: the crosstalk

630 • 2026 IEEE International Solid-State Circuits Conference

979-8-3315-8937-0/26 $31.00 © 2026 IEEE

interference translates to a power sum FEXT between -23.5 to -15dB for 1 to 3 aggressors. Notably, with classification OFF, the victim lane is unable to achieve a 10^{-12} BER in the presence of one aggressor and the received data was full of errors in the presence of two and more aggressors. With the classifier enabled a ~0.1UI eye opening is achieved at 10^{-12} BER for 1 aggressor, and a ~0.05UI opening for 3 aggressors. Figure 37.4.7 presents the performance comparison with prior work. The proposed equalizer-free transceiver achieves an energy efficiency of 1.92pJ/b operating at a 100Gb/s aggregate data rate. The decision-tree based 4-lane classifiers consume 13.6mW: accounting for 7% of the total transceiver power. The die micrographs for the Tx and Rx, and the power breakdown chart is shown in Fig. 37.4.7.

Acknowledgement:
This work was supported in part by the Center for Ubiquitous Connectivity (CUbiC), by the Semiconductor Research Corporation (SRC), and the Defense Advanced Research Projects Agency (DARPA) under the JUMP 2.0 program. Additional support in part by CDADIC. We would like to thank Intel for chip fabrication, and Tektronix for the measurement equipment.

References:
[1] A. Gholami et al., "AI and Memory Wall," *IEEE Micro*, vol. 44, no. 3, pp. 33-39, May-June 2024. https://doi.org/10.1109/MM.2024.3373763
[2] J. Zhang and E. G. Friedman, "Effect of shield insertion on reducing crosstalk noise between coupled interconnects," *IEEE ISCAS*, pp. 529-532, 2004. https://doi.org/10.1109/ISCAS.2004.1329325
[3] P. Gupta and A. B. Kahng, "Wire swizzling to reduce delay uncertainty due to capacitive coupling," *Intl. Conf. on VLSI Design*, pp. 431-436, 2004. https://doi.org/10.1109/ICVD.2004.1260960
[4] Q. Liu et al., "A 0.90-Tb/s/in 1.29-pJ/b Wireline Transceiver With Single-Ended Crosstalk Cancellation Coding Scheme for High-Density Interconnects," *IEEE JSSC*, vol. 58, no. 8, pp. 2326-2336, Aug. 2023. https://doi.org/10.1109/JSSC.2023.3261125
[5] L. Zhong et al., "A 2×56 Gb/s 0.78-pJ/b PAM-4 Crosstalk Cancellation Receiver with Active Crosstalk Extraction Technique in 28-nm CMOS," *IEEE JSSC*, vol. 59, no. 9, pp. 3008-3020, Sept. 2024. https://doi.org/10.1109/JSSC.2024.3387355
[6] Y. Chun and T. Anand, "An ISI-Resilient Data Encoding for Equalizer-Free Wireline Communication—Dicode Encoding and Error Correction for 24.2-dB Loss With 2.56 pJ/bit," IEEE JSSC, vol. 55, no. 3, pp. 567-579, Mar. 2020. https://doi.org/10.1109/JSSC.2019.2959487
[7] Z. Wang et al., "A Machine Learning Inspired Transceiver with ISI-Resilient Data Encoding: Hybrid-Ternary Coding + 2-Tap FFE + CTLE + Feature Extraction and Classification for 44.7dB Channel Loss in 7.3pJ/bit," *IEEE Symp. VLSI Circuits*, 2021. https://doi.org/10.23919/VLSICircuits52068.2021.9492510
[8] R. Javadi and T. Anand, "A 0.055pJ/bit/dB 42Gb/s PAM-4 Wireline Transceiver with Consecutive Symbol to Center (CSC) Encoding and Classification for 26dB Loss in 16nm FinFET," *IEEE CICC*, 2025. https://doi.org/10.1109/CICC63670.2025.10982737
[9] R. Javadi et al., "A 3.2pJ/b 0.068pJ/b/dB 25Gb/s NRZ Wireline Transceiver with 3-Tap FFE and Random Forest Classification for Compensating 47dB Loss in 16nm FinFET," *IEEE Symp. VLSI Tech. & Circuits*, 2025. https://doi.org/10.23919/VLSITechnologyandCir65189.2025.11075116

[10] L. Zhong et al., "A 112Gb/s/pin Single-Ended Crosstalk-Cancellation Transceiver with 31dB Loss Compensation in 28nm CMOS," *ISSCC*, pp. 134-136, 2024. https://doi.org/10.1109/ISSCC49657.2024.10454508
[11] H. Wu et al., "A 2 × 24 Gb/s Single-Ended Transceiver With Channel-Independent Encoder-Based Crosstalk Cancellation in 28-nm CMOS," *IEEE JSSC*, vol. 60, no. 8, pp. 2959-2970, Aug. 2025. https://doi.org/10.1109/JSSC.2024.3507855
[12] H.-G. Ko et al. "An 8Gb/s/µm FFE-Combined Crosstalk-Cancellation Scheme for HBM on Silicon Interposer with 3D-Staggered Channels," *ISSCC*, pp. 128-130, 2020. https://doi.org/10.1109/ISSCC19947.2020.9063162
[13] Y. Nishi et al., "A 0.297-pJ/Bit 50.4-Gb/s/Wire Inverter-Based Short-Reach Simultaneous Bi-Directional Transceiver for Die-to-Die Interface in 5-nm CMOS," *IEEE JSSC*, vol. 58, no. 4, pp. 1062-1073, April 2023. https://doi.org/10.1109/JSSC.2022.3232024

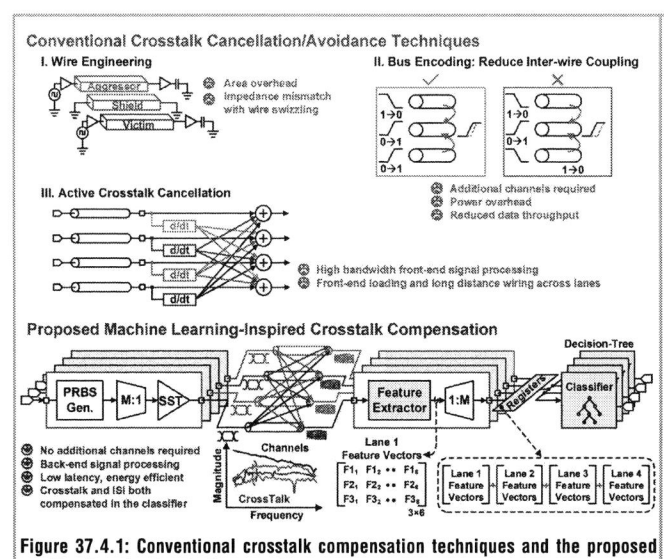

Figure 37.4.1: Conventional crosstalk compensation techniques and the proposed machine learning-inspired approach.

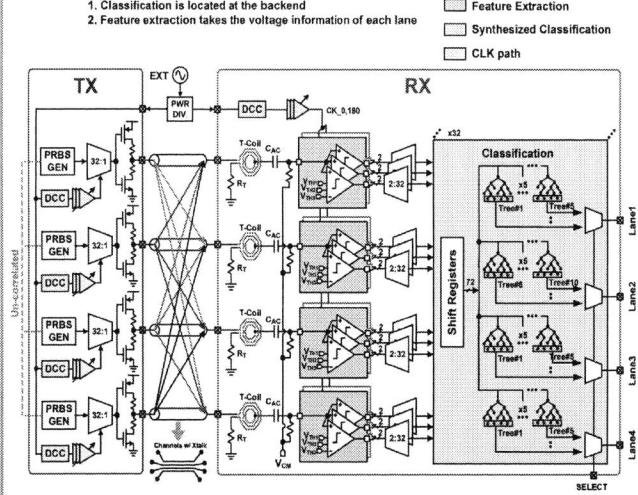

Figure 37.4.2: Proposed 4-lane ML-inspired equalizer-free transceiver architecture for crosstalk compensation.

ISSCC 2026 / SESSION 37 / MEMORY INTERFACE / 37.4

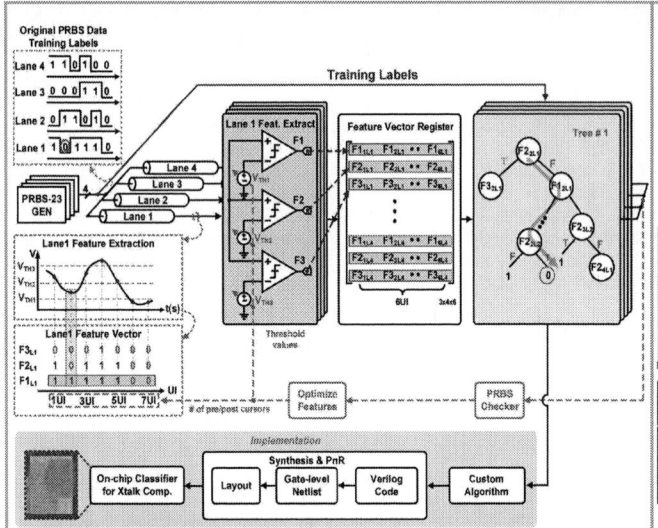

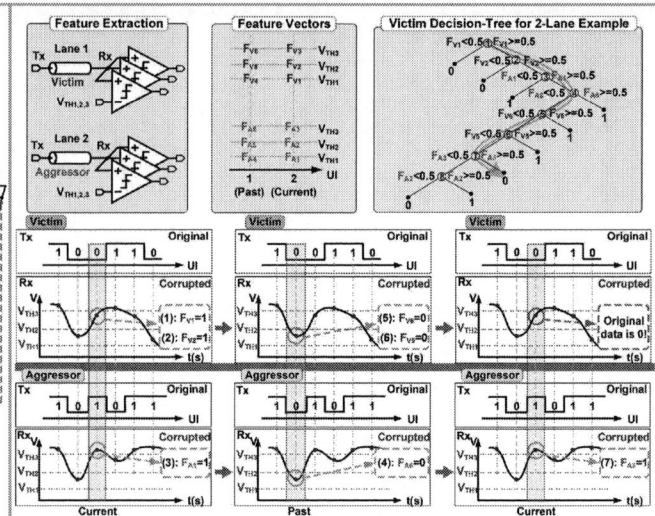

Figure 37.4.3: Decision-tree training process using supervised learning approach.

Figure 37.4.4: An example showing how crosstalk is compensated in a 2-lane coupled channel.

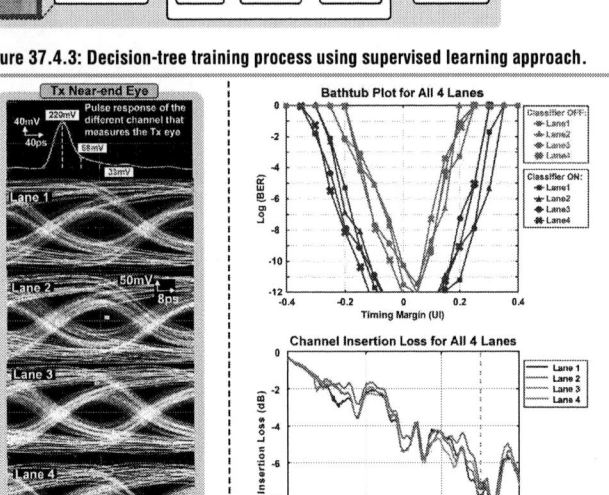

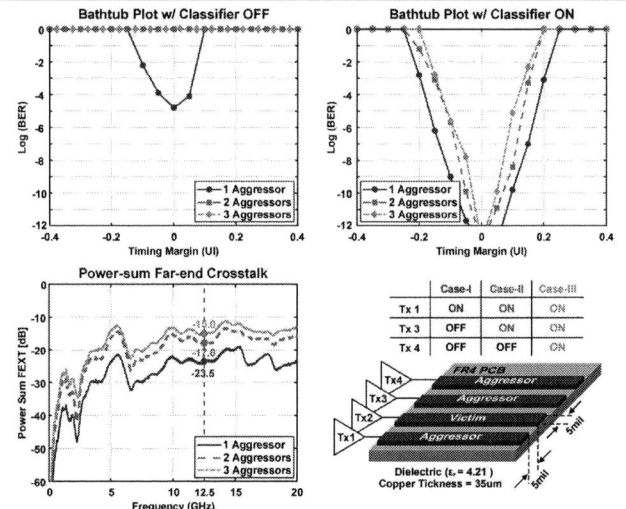

Figure 37.4.5: Measured Tx near-end eye diagram at 25Gb/s/pin, Tx + Rx bathtub plot for each lane without aggressors and with classifier ON/OFF at 25Gb/s/pin with PRBS7 data.

Figure 37.4.6: Measured bathtub plot with multiple aggressors with classifier ON/OFF, at 25Gb/s/pin with PRBS7 data, and 4-lane channel far-end crosstalk profile.

	This work	Off-Chip Links			On-Chip Links	
		Zhong ISSCC' 2024 [10]	Wu JSSC' 2024 [11]	Liu JSSC' 2023 [4]	Ko ISSCC' 2020 [12]	Nishi JSSC' 2023 [13]
Technology	16nm	28nm	28nm	28nm	65nm	5nm
Signaling	Single-ended	Single-ended	Single-ended	Single-ended	Single-ended	Single-ended
Modulation	NRZ	NRZ & PAM4	Orthogonal NRZ	NRZ	NRZ	NRZ SBD
# of Lanes	4	2	2	3+1	6	28
Crosstalk Compensation/ Cancellation	Machine Learning	CTLE 4-tap Tx FFE 4-tap Rx FFE	CTLE 3-tap FS-FFE	Fibonacci Coding	FFE-combined XTC	Ground Shielding
Data Rate [Gb/s/pin]	25	112	24	10	4	50.4
Loss at Nyquist [dB]	7.3	22	20	2	10.2	5.4
FEXT Power Sum at Nyquist [dB]	-15	-14	-10	-4.7	-28.9	-42.4
Power [mW]	70†	192	309.7	139	38.7	36.6	210
Energy Efficiency [pJ/b]	0.7†	1.92	2.77	2.9	1.29	1.5	0.297

†: Exclude clock distribution power from Tx and Rx

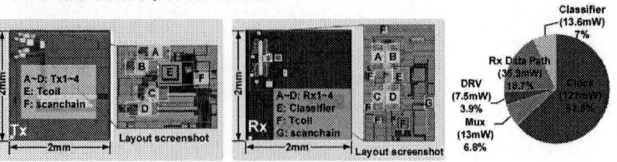

Figure 37.4.7: Comparison table, power breakdown, and die micrograph.

• 2026 IEEE International Solid-State Circuits Conference

ISSCC 2026 / SESSION 37 / MEMORY INTERFACE / 37.5

37.5 A 16Gb/s/pin 0.51pJ/b Single-Ended NRZ Transceiver with Distributed Dual-Loop V_{DDQ}-Ripple Compensation and Dynamic Clock Duty-Cycle Calibration for Memory Interfaces

Yuchen He*[1], Yunbin Luo*[1], Yunzhengmao Wang*[1], Tengyue Yi[2], Chen Jiang[1], Chixiao Chen[1], Qi Liu[1], Ming Liu[1], Wenning Jiang[1]

[1]Fudan University, Shanghai, China, [2]ZhangJiang Laboratory, Shanghai, China
*Equally Credited Authors (ECAs)

Abstract

A 16Gb/s/pin 0.51pJ/b single-ended NRZ transceiver with distributed dual-loop compensation (DDLC) for V_{DDQ}-ripple suppression and dynamic duty-cycle calibration (DDCC) for robust clocking is presented. The DDLC locally regulates V_{DDQ} with a shared low-power LDO, reducing the required decoupling capacitor by 91.4%, while improving horizontal and vertical eye margins by 61% and 36%. Demonstrated on five channels, the proposed DDLC scales naturally to larger DQ counts for high-density memories.

As AI and chiplet technologies proliferate, the demand for high efficiency, high density, and high-speed data transmission between processors and memory continues to surge. Advanced packaging technologies such as silicon bridges, interposers, and through-silicon vias (TSVs) mitigate channel loss and ease equalization requirements [1-8]. While these signal integrity (SI) challenges are mitigated, power integrity (PI), impacted by IR drops, is increasingly critical for high-density and high-speed interfaces [9-11].

Among power delivery networks (PDNs), the data-driver supply (V_{DDQ}) domain in the transmitter (TX) is most critical, as high-speed data-dependent current toggling causes V_{DDQ} ripples. These ripples directly couple into the transmitted data, via the TX driver, degrading the SI of both the active DQ and neighboring DQs. Conventional stabilization techniques rely on power-hungry drivers or large embedded decoupling capacitors, reducing power and area efficiency [9-11]. We introduce a distributed dual-loop compensation (DDLC) technique that provides local V_{DDQ}-ripple suppression for each TX. In addition, dynamic duty-cycle calibration (DDCC) is used to automatically regulate the RX clock's duty cycle to near 50%.

Figure 37.5.1 shows a block diagram for the proposed single-ended 16Gb/s NRZ transceiver. In the DQ TX, five data channels share a common 0.4V V_{DDQ} supplied by a low-power low-dropout regulator (LDO). An NMOS-over-NMOS (N-N) main driver is employed for energy efficiency, while ripples are suppressed using the proposed DDLC. A reflection-canceling driver (RCD) is included for TX equalization [3]; the main driver and RCD are programmable. A differential source-synchronous strobe-clock (DQS) architecture is adopted with an external 8GHz clock. Clock skew is mitigated using global and local digitally controlled delay lines (DCDLs) in the RX. On-chip pseudo-random binary-sequence (PRBS) generation and error checking are integrated for testability.

Figure 37.5.2 illustrates the proposed DDLC for mitigating V_{DDQ} ripple. In this design, the V_{DDQ} LDO uses a conventional architecture [12] with a limited bandwidth and a 50µA quiescent current. IR drops across the long power bus and the narrow bandwidth of LDO makes V_{DDQ} sensitive to simultaneous switching of the five DQs: 0→1 transitions induce a high-speed transient ripple on V_{DDQ}. Furthermore, long-term toggling further leads to an underdamped V_{DDQ} response. As a result of V_{DDQ} ripple, voltage noise and clock jitter are introduced into the transmitted signal; degrading its eye diagram, as shown in Figure. 37.5.2(bottom right). Since the ripple is data-dependent and occurs within tens of picoseconds of each transition, conventional on-chip or off-chip regulators cannot respond quickly enough. To address this, the proposed DDLC is implemented locally within each TX: it consists of a fast feed-forward loop and a slow feedback loop as shown in Figure. 37.5.2 (top). The fast loop predicts V_{DDQ} ripple based on the pre-driver data (D_{PRE}). A data-tracking pulse is generated whenever D_{PRE} toggles 0→1, which is pulse-width (PW) and delay adjusted into a final compensation signal (V_{FC}), which drives an NMOS to inject a compensation current into V_{DDQ} locally. As shown in Figure. 37.5.2 (top), the PW is generated through inverter-based delay cell and a Nor-gate. As the delay cell senses the process, voltage and temperature (PVT) variation, the PW and latency of V_{FC} are adapted for different corners without additional tuning. In a slow corner (e.g., SS and high temperature), wider V_{FC} pulses provide larger compensation currents, while in a fast corner (e.g., FF and low temperature), narrower pulses suffice.

Fast compensation cannot fully suppress V_{DDQ} ripple; thus, the residual uncompensated ripple accumulates and manifests as slow V_{DDQ} wander, which is mitigated by the slow loop compensation. A ripple sense amplifier extracts the low-frequency ripple components, amplifies them 50×, and generates a control voltage (V_{SC}). With its AC-coupled input and intrinsic low-pass response, the ripple sense amplifier filters out both high-speed ripple and the V_{DDQ} DC offset. V_{SC} regulates compensation current via the PMOS stacked above the fast-loop NMOS. The sense amplifier is designed with a 60dB open loop gain and a 100MHz gain bandwidth product (GBW) to cover only slow variations; it consumes 140µA. As shown in Figure. 37.5.2 (bottom left), DDLC stabilizes V_{DDQ} without latency, since compensation is inherently data driven. Therefore, no additional training time or data overhead is required. With DDLC, each DQ TX requires only a 1.5pF local decoupling capacitor (C_D) to suppress the V_{DDQ} ripple to within ±10mV. While demonstrated on five channels, the scheme can scale to larger DQ counts, as each TX integrates a local DDLC

and a shared low-power LDO with appropriate far-end decoupling and a global slow-ripple sense amplifier. The proposed DDLC increases demands on V_{DD}, but the V_{DD} supply is more robust than V_{DDQ}; furthermore, digital logic supplied by V_{DD} is more resilient to power-supply noise. As a result, DDLC only has a slight impact on V_{DD} and overall DQ TX performance.

Conventional duty-cycle calibration (DCC) [13,14] uses a pseudo-differential charge pump (PDCP) to convert duty-cycle error into voltage, followed by pulse-width adjustment. To improve loop gain, [15] introduced a low-pass filter (LPF) to shift the PDCP input common-mode, enhancing the integration slope and reducing detection dead-zone. However, PDCPs require two well-matched current sources (I_P/I_N) and draw static current when V_{FP}/V_{FN} converge. To overcome these drawbacks, this work replaces the PDCP with a fully-differential dynamic amplifier (FDDA) as shown in Figure 37.5.3 (right), which consumes only dynamic power. The gate voltages (V_{FP}/V_{FN}) are generated by a narrow-bandwidth LPF (NB-LPF), which extracts the common-mode of the differential clock. The duty-cycle information is translated into a corresponding voltage by the FDDA. The conversion loop gain is controlled by the integration time, which is set to $16T_{CKIN}$, T_{CKIN} = main-clock (CK_{IN}) period, yielding linear integration. A dynamic comparator [16], which is clocked at the same frequency as the integration clock (CK_S), detects the integrated outputs (V_{OP}/V_{ON}). The decision results are used to control the delay cells in the pulse-width adjustment block. The proposed DDCC achieves an almost zero detection dead-zone by leveraging integration time control and comparator optimization. Unlike conventional DCCs, the presented DDCC requires a single current source and is immune to current mismatch. Once the duty-cycle error calibration converges, duty-cycle correction codes are locked to save power. Foreground calibration cancels residual loop offsets from the NB-LPF, FDDA, and comparator. During offset calibration (os_en=1), the NB-LPF input is shorted to $CK_{IN}/2$. The comparator decisions are captured by an external FPGA, which tunes the offset-correction voltage (V_{CAL}) that is applied to the comparator's auxiliary input pair [16]. Offset calibration can be executed via sideband signals, which periodically update the control codes to compensate for offset drift.

Figure 37.5.4 (left) shows simulated Monte-Carlo results for the proposed DDCC, which corrects a wide duty-cycle error range and remains robust across PVT. Figure 37.5.4(right) shows the measured eye diagrams at 10^{-12} BER with and without DDCC for different data rates. The presented scheme enlarges the horizontal eye opening from 0.68 to 0.84UI at 8Gb/s and from 0.29 to 0.53UI at 14.4Gb/s (LPDDR6 standard) for a PRBS7 pattern.

The prototype chip is implemented in 28nm CMOS and integrates 10 SE DQ channels and one differential DQS channel. Five DQs are powered by an on-chip LDO for V_{DDQ} with a 32pF far-end decoupling capacitor and a 1.5pF local decoupling capacitor for the proposed DDLC. For comparison, the other five DQs are powered directly from an off-chip V_{DDQ} supply, require 465pF of on-chip decoupling capacitance, nearly 12× larger than required for the LDO+DDLC DQs. At 16Gb/s, the measured eye diagrams for 10^{-12} BER and the measured bathtub curves are shown in Fig. 37.5.5. The horizontal and vertical eye margins reach 0.50UI and 237.5mV with DDLC, an improvement of 0.19UI and 62.5mV compared to the off-chip reference. Using a V_{DD} of 1V and V_{DDQ} of 0.4V, the 160Gb/s prototype consumes 81.6mW, achieving a 0.51pJ/b power efficiency: the power breakdown is shown in Fig. 37.5.5. The prototype chip's measured eye diagrams at 14.4Gb/s at 85°C and 105°C for 10^{-10} BER are shown in Fig. 37.5.5 (bottom): both cases have maximum eye margins of 0.47UI and 250mV. Figure 37.5.6 summarizes the transceiver performance and compares it with prior work at similar data rates. The presented DDLC enables robust power integrity using a compact decoupling capacitor, achieving competitive eye margins, energy efficiency, and high bandwidth density.

Figure 37.5.7 shows the die micrograph and block details. Each DQ channel with local DDLC and 1.5pF V_{DDQ} decoupling capacitor occupies 0.003mm². The shared slow V_{DDQ}-ripple sense amplifier and low-power LDO with a 32pF far-end decoupling capacitor occupy 0.0095mm², and the differential DQS occupies 0.0043mm².

Acknowledgement:
This work was supported by the National Key Research and Development Program of China (No.2022YFB4401301). The corresponding author is Wenning Jiang.

ISSCC 2026 / February 18, 2026 / 1:30 PM

References:

[1] Y. Nishi et al., "A 0.190-pJ/bit 25.2-Gb/s/wire Inverter-Based AC-Coupled Transceiver for Short-Reach Die-to-Die Interfaces in 5-nm CMOS," *IEEE JSSC*, vol. 59, no. 4, pp. 1146-1157, April 2024. https://doi.org/10.1109/JSSC.2023.3338478

[2] D. T. Melek et al., "A 0.29pJ/b 5.27Tb/s/mm UCIe Advanced Package Link in 3nm FinFET with 2.5D CoWoS Packaging," *ISSCC*, pp. 590-592, 2025. https://doi.org/10.1109/ISSCC49661.2025.10904754

[3] K. Seong et al., "A 4nm 32Gb/s 8Tb/s/mm Die-to-Die Chiplet Using NRZ Single-Ended Transceiver with Equalization Schemes and Training Techniques," *ISSCC*, pp. 114-116, 2023. https://doi.org/10.1109/ISSCC42615.2023.10067477

[4] J. Jin et al., "A 4nm 16Gb/s/pin Single-Ended PAM4 Parallel Transceiver with Switching-Jitter Compensation and Transmitter Optimization," *ISSCC*, pp. 404-405, 2023. https://doi.org/10.1109/ISSCC42615.2023.10067738

[5] H. Shin et al., "A 15-Gb/s PAM-3 Transceiver with Hybrid Equalization and Time-Domain Decoder for High-Bandwidth-Memory Interfaces," *IEEE JSSC*, 2025. https://doi.org/10.1109/JSSC.2025.3557795

[6] J. Yun et al., "A Single-Ended Impedance-Matched Transmitter with Single Ring-Oscillator-Based Time-Domain ZQ Calibration for Memory Interfaces," *IEEE JSSC*, vol. 59, no. 9, pp. 2971-2982, Sept. 2024. https://doi.org/10.1109/JSSC.2024.3375858

[7] C. Han, et al., "A 25.2Gb/s/pin NRZ/PAM-3 Dual-Mode Transmitter with Embedded Partial DBI Achieving a 133% I/O Bandwidth/Pin Efficiency and 19.3% DBI Efficiency," *ISSCC*, pp. 248-250, 2024. https://doi.org/10.1109/ISSCC49657.2024.10454326

[8] J. Yang et al., "A 35.4Gb/s/pin 16Gb GDDR7 with a Low-Power Clocking Architecture and PAM3 IO Circuitry," *ISSCC*, pp. 232-234, 2024. https://doi.org/10.1109/ISSCC49657.2024.10454560

[9] S. S. Kudva et al., "A switching linear regulator based on a fast-self-clocked comparator with very low probability of meta-stability and a parallel analog ripple control module," *IEEE CICC*, 2018. https://doi.org/10.1109/CICC.2018.8357021

[10] J. Jung et al., "A 4ns Settling Time FVF-Based Fast LDO Using Bandwidth Extension Techniques for HBM3," *IEEE ASSCC*, 2023.
\ https://doi.org/10.1109/A-SSCC58667.2023.10348011

[11] X. Chen et al., "Reference-Noise Compensation Scheme for Single-Ended Package-to-Package Links," *ISSCC*, pp. 126-128, 2020.
https://doi.org/10.1109/ISSCC19947.2020.9063140

[12] C.-J. Park et al., "External Capacitor-Less Low Drop-Out Regulator With 25 dB Superior Power Supply Rejection in the 0.4–4 MHz Range," *IEEE JSSC*, vol. 49, no. 2, pp. 486-501, Feb. 2014. https://doi.org/10.1109/JSSC.2013.2289897

[13] Y.-J. Jung et al., "A low jitter dual loop DLL using multiple VCDLs with a duty cycle corrector," *IEEE Symposium on VLSI Circuits*, pp. 50-51, 2000.
https://doi.org/10.1109/VLSIC.2000.852848

[14] J. S. Humble et al., "A Clock Duty-Cycle Correction and Adjustment Circuit," *ISSCC*, pp. 2132-2141, 2006.
https://doi.org/10.1109/ISSCC.2006.1696273

[15] D. Kwon et al., "A 1.1V 6.4Gb/s/pin 24-Gb DDR5 SDRAM with a Highly-Accurate Duty Corrector and NBTI-Tolerant DLL," *ISSCC*, pp. 27-29, 2023.
https://doi.org/10.1109/ISSCC42615.2023.10067651

[16] M. Miyahara et al., "A low-noise self-calibrating dynamic comparator for high-speed ADCs," *IEEE ASSCC*, pp. 269-272, 2008.
https://doi.org/10.1109/ASSCC.2008.4708780

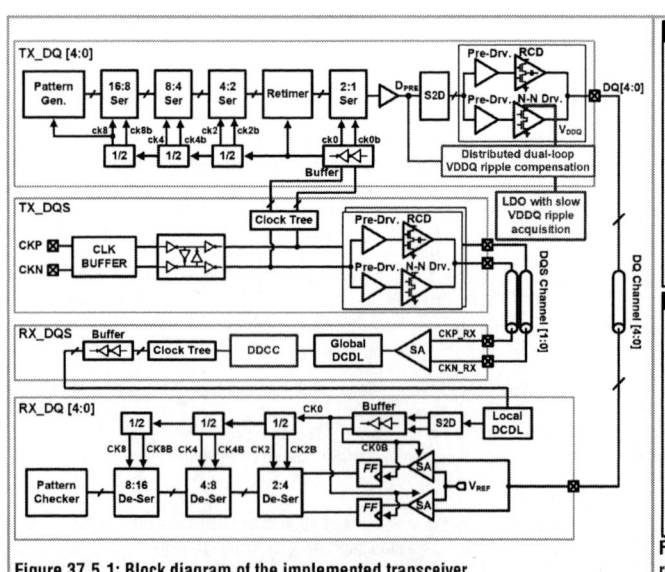

Figure 37.5.1: Block diagram of the implemented transceiver.

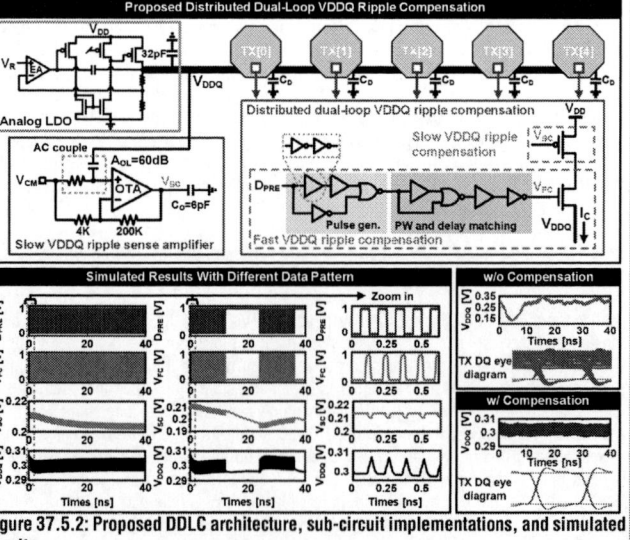

Figure 37.5.2: Proposed DDLC architecture, sub-circuit implementations, and simulated results.

DIGEST OF TECHNICAL PAPERS • 633

979-8-3315-8937-0/26 $31.00 © 2026 IEEE

ISSCC 2026 / SESSION 37 / MEMORY INTERFACE / 37.5

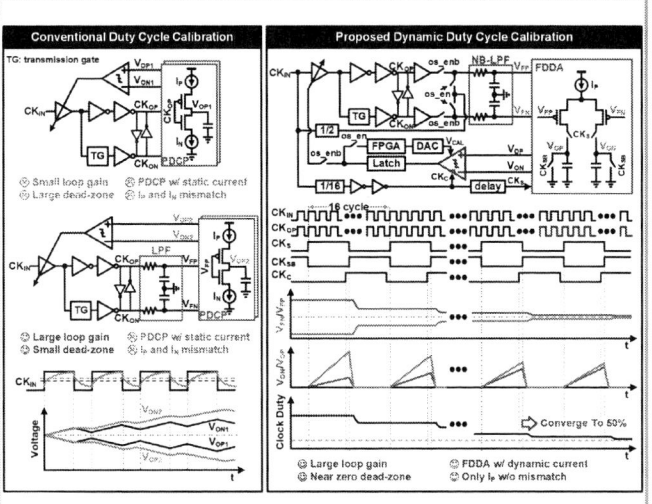

Figure 37.5.3: Comparison between conventional duty-cycle calibration and the presented dynamic duty-cycle calibration.

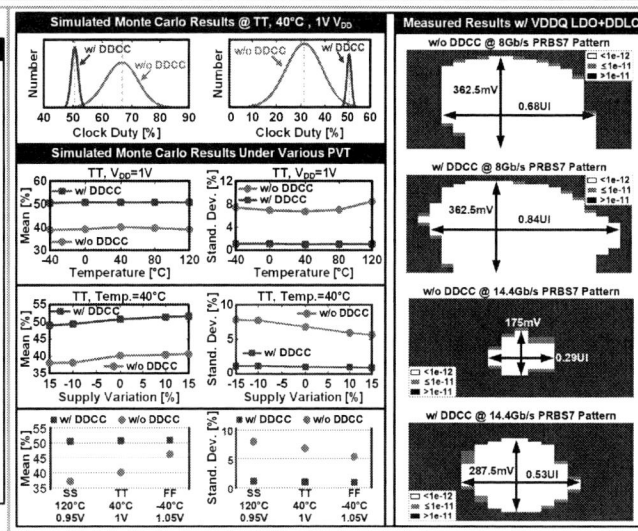

Figure 37.5.4: Monte-Carlo simulation results of clock duty cycle with proposed DDCC across PVT, and measured eye diagrams with and without DDCC for different data rates.

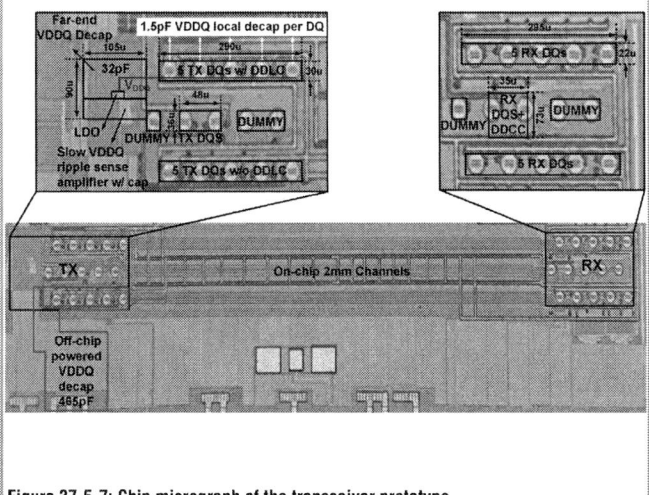

Figure 37.5.5: 16Gb/s measured results: eye diagrams under normal ambient temperature, bathtub curves, and power efficiency breakdown. 14.4Gb/s eye diagrams at 85 and 105°C.

	This work	JSSC'24 Yun	JSSC'25 Shin	ISSCC'25 Melek	ISSCC'23 Jin	ISSCC'24 Han
Architecture	TRX	TX	TRX	TRX	TRX	TX
Technology	28nm	65nm	28nm	3nm	4nm	65nm
VDD/VDDQ [V]	1.0/0.4	1.0/0.4	1.0/1.0	1.2/0.75/0.45	0.75/0.6	-/1.0
Data Rate [Gb/s/pin]	16	12	15	16	16	25.2
Signaling	NRZ	NRZ	PAM3	NRZ	PAM4	PAM3
Efficiency [pJ/bit]	0.35[a]/0.51[b]	1.145	0.274	0.29	0.764[c]/1.046[d]	1.13 @21Gb/s
Channel Loss [dB] @Nyquist Rate	-2.43	-10.3	-4.8	-3.92	-4.65	-2.9
Area [mm²/pin]	0.0049[*]/0.0092[**]	0.115	0.0135	0.015[#]	0.00726[c]	0.0196
Ser./Des. Factor	16:1/1:16	32:1	8:1/1:8	8:1/1:8	NA	32:1
Eye Width @BER	0.5UI @1e-12	0.38UI @1e-12	0.3UI @1e-12	0.325UI @1e-15	0.37UI @1e-12	NA

a: only DQ I/O w/ VDDQ LDO; b: 160Gb/s prototype IC including DQS and DQ I/O w/ VDDQ LDO;
c: only DQ I/O; d: 64Gb/s prototype IC including DQS and DQ I/O; #: Estimated from die photo;
*: 1 DQ I/O + VDDQ LDO with VDDQ decap; **: 1 DQ I/O + VDDQ LDO with VDDQ decap +1 DQS.

Figure 37.5.6: Summary of measured performance results and comparison with prior similar speed designs.

Figure 37.5.7: Chip micrograph of the transceiver prototype.

• 2026 IEEE International Solid-State Circuits Conference

979-8-3315-8937-0/26 $31.00 © 2026 IEEE

ISSCC 2026 / SESSION 37 / MEMORY INTERFACE / 37.6

37.6 A 0.092pJ/b and 7.7fJ/b/dB Cross-Self-Referenced Slope-Sampling Receiver with Long-Tail ISI Robustness for Next-Generation Low-Power Memory Interfaces

Ki-Soo Lee*, Chanheum Han*, Joo-Hyung Chae

Kwangwoon University, Seoul, Korea
*Equally Credited Authors (ECAs)

Abstract

We present a cross-self-referenced slope-sampling RX that is robust against long-tail ISI, without equalization, for next-generation low-power memory interfaces. By determining received data based on the slope of the input signal, the proposed architecture enhances resilience to accumulated post-cursor ISI. Fabricated in a 65-nm CMOS process, the prototype achieves a 0.092pJ/b energy efficiency at 12.5Gb/s and a 7.7fJ/b/dB FoM at 12Gb/s, while occupying only 0.001mm².

As power-hungry data-centric applications are widely employed, the need for energy-efficient high-bandwidth communication between processing units and off-chip memory is increasing. Consequently, low-power double-data-rate (LPDDR) dynamic random-access memory (DRAM) is being adopted in data centers and edge devices due to its energy efficiency. As data rates continue to rise, across LPDDR generations, channel loss becomes more severe; thereby, resulting in increased long-tail post-cursor inter-symbol interference (ISI). Equalization techniques can mitigate this issue but with an impact on power and area. Figure 37.6.1(top) summarizes prior equalization studies for mitigating long-tail post-cursor ISI. A finite impulse response (FIR)-based multi-tap decision-feedback equalizer (DFE) [1] is able to cancel multiple post-cursor ISI without amplifying noise, but at a significant area and power overhead; especially as the number of taps increases. Furthermore, the feedback path has stringent timing constraints, making high-speed operation more challenging. An infinite impulse response (IIR)-based DFE [2] is a power-efficient and compact implementation that mitigates long-tail ISI. However, it suffers from difficulty in coefficient optimization and still incurs feedback latency, which limits its applicability for high-speed systems. To effectively compensate for long-tail ISI, a combination of a continuous-time linear equalizer (CTLE) and a low-frequency equalizer (LFEQ) has been proposed [3]. This approach reduces long-tail ISI with low latency, but it amplifies high-frequency noise and is sensitive to PVT variations. For area- and power-constrained memory interfaces, there is a need for simpler solutions that minimize complexity while providing sufficient ISI compensation. We propose a low-power receiver (RX) that utilizes a cross-self-referenced slope-sampling technique, which is robust to long-tail ISI without using equalization. The proposed slope-sampling architecture achieves a 0.092pJ/b energy efficiency at 12.5Gb/s and a 7.7fJ/b/dB FoM at 12Gb/s, while occupying an active area of only 0.001mm².

Figure 37.6.1(bottom) compares a conventional voltage-based and the proposed slope-based sampling architectures. For the voltage-sampling method, the data decision is determined by comparing the input signal with a reference voltage (V_{REF}). However, in V_{SS}-terminated LPDDR systems, the input common-mode voltage is low, necessitating the use of a PMOS-input front-end buffer, but the PMOS's poor performance degrades sampling performance [4]. In addition, the vertical sampling margin is severely degraded due to the accumulated long-tail post-cursor ISI, which makes accurate sampling difficult. Sampling performance is highly dependent on V_{REF}; hence, training is required to optimize its level. In contrast, the proposed slope-sampling architecture determines data based on the sign and magnitude of the input signal's slope. This improves vertical sampling margin, particularly during transitions, enhancing robustness against accumulated post-cursor ISI. Additionally, it inherently performs level shifting toward an optimized common-mode voltage (V_{CM}), thereby enhancing sampling performance with an NMOS-input front-end. Since the adoption of V_{CM} eliminates the need for a V_{REF} generator and training process, the proposed structure reduces both power and area overhead and enables a faster DRAM startup. In other words, the slope-based approach offers a simpler and more robust solution without equalization, which is particularly well-suited for low-power, high-speed memory interfaces.

Figure 37.6.2(top) presents a comparison of the long-tail ISI robustness between voltage- and slope-sampling architectures. The impulse response of a lossy channel is modeled as an exponentially decaying function [5], where post-cursor ISI extends across multiple unit intervals (UIs) in high-attenuation environments. In a discrete-time expression of the channel output, the voltage-sampling method accumulates post-cursor ISI components directly, leading to degraded signal integrity. In contrast, slope-based sampling benefits from partial cancellation of ISI components from adjacent symbols because the coefficient values across consecutive UIs are similar; thus, this mechanism suppresses long-tail ISI. To validate this, a mathematical analysis of a 1000UI PRBS-7 data pattern was performed: the mean absolute and standard deviation values of accumulated post-cursor ISI under varying attenuation coefficients (α) and number of post-cursors were evaluated, as illustrated in Fig. 37.6.2(bottom). As the post-cursor tail's span increases, the ISI for voltage-based sampling accumulates while ISI for slope-based sampling accumulates less, due to the effective cancellation between the adjacent post-cursor ISIs. As α decreases, the ISI tail decays more slowly, and the accumulated post-cursor ISI is exacerbated for the voltage-based method. However, the slope-based method achieves lower mean absolute and standard deviation values of the accumulated post-cursor ISI. Consequently, the slope-based sampling provides improved immunity for long-tail post-cursor ISI of high-loss channel environments.

Figure 37.6.3(top left) shows a block diagram of the proposed cross-self-referenced slope-sampling RX: consisting of data and clock (CK) paths. The data path includes two slope tracers (for odd and even paths), which extract the slope components of the input signal (RX_{IN}) and shift the common-mode level to V_{CM}. The outputs of the slope tracers, ST_{ODD} and ST_{EVEN}, are then sampled by hysteresis-controlled samplers. The cross-self-reference structure feeds the output of each slope tracer as the reference voltage for the sampler on the other side, enabling stable data comparison and decision without requiring a dedicated V_{REF}. The slope tracer operates in two phases: a reset phase and a sampling phase, as shown in Fig. 37.6.3(top right). When CK_{ST} is high (reset phase), the output is initialized to the V_{CM}. When CK_{ST} is low (sampling phase), the slope tracer extracts the slope component of the RX_{IN} signal by sampling the voltage difference and shifting the common-mode level to V_{CM}. The sampled slope information is then fed into the sampler. A peak-to-peak swing at the sampler input of this slope-sampling method becomes larger than that of the single-ended signaling, improving sampling reliability. The hysteresis-controlled sampler enhances robustness of data decisions, particularly during consecutive identical data where the slope magnitude is small, as shown in Fig 37.6.3(bottom). In such consecutive identical data, the input signal change becomes smaller, causing the vertical voltage margin to approach zero. To improve decision stability for these conditions, the hysteresis current (I_{HYS}) is dynamically adjusted based on previous data decisions, using a feedback path that feeds the sampler's output to the sampler of the other path, as in a DFE. During consecutive identical data, hysteresis contributes to retaining the previous state; thereby, preventing errors caused by a reduced slope magnitude. In contrast, data transition inherently offers a larger slope magnitude, enabling more reliable sampling. The coefficient value (W_{HYS}) allows flexible adjustment of the hysteresis strength, enabling the RX to adapt to various operating conditions.

Figure 37.6.4 illustrates the advantages of the proposed cross self reference architecture. In Fig. 37.6.4(top left), the asymmetric input path leads to voltage disturbances at the virtual ground and output nodes of the sampler. These fluctuations are asymmetrically coupled back to the input through gate-source (C_{GS}) and gate-drain (C_{GD}) capacitances, resulting in mismatched kickback noise and a dynamic offset. While a decoupling capacitor (C_D) can suppress this noise, it degrades settling behavior and increases area overhead. Alternatively, reducing transistor size to minimize kickback exacerbates the intrinsic offset due to device mismatch. In contrast, as shown in Fig. 37.6.4(top right), the proposed cross-self-reference structure symmetrizes the input paths, resulting in a balanced kickback. This symmetry inherently suppresses dynamic offsets, allowing reliable operation without requiring a large decoupling capacitor. The proposed scheme also enhances immunity to simultaneous-switching-output (SSO) noise. During the reset and sampling phases, common-mode noise is coupled similarly to both output nodes (ST_{ODD} and ST_{EVEN}) due to a high-pass filtering formed by the switch's on-resistance and internal capacitance, as shown in Fig. 37.6.4(bottom left). These common-mode components are canceled in the subsequent sampling stage, as validated by simulations with a power-distribution network (PDN) [6]. Similarly, power-supply noise, which perturbs the input common-mode level, is inherently suppressed by the proposed structure. As power-supply noise is symmetrically coupled to the ST_{ODD} and ST_{EVEN} nodes, it is also canceled during sampling. Simulations using a 50mV and 100MHz supply noise [7] demonstrate that the proposed method maintains robust operation under this noise condition, as shown in Fig. 37.6.4(bottom right).

Figure 37.6.5 presents the proposed RX measurement results across various data rates and channel lengths. Figure 37.6.5(top) shows the simulated single-bit response with accumulated long-tail post-cursor ISI, spanning from the 1st to 20th UI, for various data rates and channels. Figure 37.6.5(bottom) shows the measured bathtub curves for varying channel lengths and data rates. The horizontal margins at 10^{-12} BER are 0.244 and 0.177UI for a 9.05" channel at 12 and 12.5Gb/s. At 12Gb/s, the bathtub curves over the 11.25" and 13.45" channels exhibited horizontal margins of 0.205 and 0.022UI. The 13.45" channel's horizontal margins at 10 and 10.5Gb/s are 0.144 and 0.083UI at 10^{-12} BER.

Figure 37.6.6 presents a performance summary of the proposed cross-self-referenced RX and a comparison to recent single-ended NRZ RXs for memory interfaces and long-tail ISI mitigation RXs. The proposed RX achieves 12Gb/s across a 12.2dB channel loss and 12.5Gb/s for a 10.4dB loss, all without the need for equalization circuits. It also demonstrates an energy efficiency and FoM superior to prior work, achieving 0.092pJ/b at 12.5Gb/s and 7.7fJ/b/dB at 12Gb/s.

ISSCC 2026 / February 18, 2026 / 1:30 PM

Figure 37.6.7 presents a die micrograph of the prototype chip, along with the power and area breakdown pie charts. The proposed RX is fabricated in a 65nm CMOS process and occupies an active area of only 0.001mm².

Acknowledgement:
This work was supported by Samsung Research Funding & Incubation Center of Samsung Electronics (No. SRFC-IT2401-06)

References:
[1] T. Toifl et al., "A 2.6 mW/Gbps 12.5 Gbps RX with 8-tap switched-capacitor DFE in 32 nm CMOS," *IEEE JSSC*, vol. 47, no. 4, pp. 897–910, Apr. 2012. https://doi.org/10.1109/JSSC.2012.2185342
[2] S. Son et al., "A 2.3-mW, 5-Gb/s low-power decision-feedback equalizer receiver front-end and its two-step, minimum bit-error-rate adaptation algorithm," *IEEE JSSC*, vol. 48, no. 11, pp. 2693–2704, Nov. 2013. https://doi.org/10.1109/JSSC.2013.2274904
[3] S. Parikh et al., "A 32 Gb/s wireline receiver with a low-frequency equalizer, CTLE and 2-tap DFE in 28 nm CMOS," *ISSCC*, pp. 28–29, 2013. https://doi.org/10.1109/ISSCC.2013.6487622
[4] H. Lee et al., "A 16.8 Gbps/channel single-ended transceiver in 65 nm CMOS for SiP-based DRAM interface on si-carrier channel," *IEEE JSSC*, vol. 50, no. 11, pp. 2613–2624, Nov. 2015. https://doi.org/10.1109/JSSC.2015.2466469
[5] Y.-U. Jeong et al., "A 0.85-pJ/b 16-Gb/s/pin single-ended transmitter with integrated voltage modulation for low-power memory interfaces," *IEEE JSSC*, vol. 58, no. 9, pp. 2659–2667, Sep. 2023. https://doi.org/10.1109/JSSC.2023.3269765
[6] C. Han et al., "A V_M-terminated PAM-3 transmitter using floating middle level with enhanced signal integrity and energy efficiency for low-power memory interfaces," *IEEE Trans. Circuits Syst. I*, vol. 72, no. 6, pp. 2653–2663, Jun. 2025. https://doi.org/10.1109/TCSI.2025.3552405
[7] S. Kim et al., "A 15-Gb/s single-ended NRZ receiver using self-referenced technique with 1-tap latched DFE for DRAM interfaces," *IEEE Trans. Circuits Syst. II*, vol. 70, no. 1, pp. 101–105, Jan. 2023. https://doi.org/10.1109/TCSII.2022.3208280
[8] I.-M. Yi et al., "A time-based receiver with 2-tap decision feedback equalizer for single-ended mobile DRAM interface," *IEEE JSSC*, vol. 53, no. 1, pp. 144–154, Jan. 2018. https://doi.org/10.1109/JSSC.2017.2746698
[9] S.-M. Lee et al., "An 8nm 18Gb/s/pin GDDR6 PHY with TX bandwidth extension and RX training technique," *ISSCC*, pp. 338–340, 2020. https://doi.org/10.1109/ISSCC19947.2020.9062937
[10] T. Kim, et al., "A low-voltage area-efficient TSV I/O for HBM with data rate up to 15Gb/s featuring overlapped multiplexing driver, ISI compensators and QEC," *IEEE Symp. VLSI Circuits*, Jun. 2023. https://doi.org/10.23919/VLSITechnologyandCir57934.2023.10185328
[11] J. Choi et al., "A single-ended NRZ receiver with gain-enhanced active-inductive CTLE and reference-selection DFE for memory interfaces," *IEEE JSSC*, vol. 59, no. 4, pp. 1261–1270, Apr. 2024. https://doi.org/10.1109/JSSC.2024.3358335
[12] S. Shahramian et al., "A 16Gb/s 1 IIR + 1 DT DFE compensating 28dB loss with edge-based adaptation converging in 5µs," *ISSCC*, pp. 410–411, 2016. https://doi.org/10.1109/ISSCC.2016.7418081
[13] J. Lee et al., "A 0.1-pJ/b/dB 1.62-to-10.8-Gb/s video interface receiver with jointly adaptive CTLE and DFE using biased data-level reference," *IEEE JSSC*, vol. 55, no. 8, pp. 2186–2195, Aug. 2020. https://doi.org/10.1109/JSSC.2020.2987690

[14] B. Ye et al., "A 2.29-pJ/b 112-Gb/s wireline transceiver with RX four-tap FFE for medium-reach applications in 28-nm CMOS," *IEEE JSSC*, vol. 58, no. 1, pp. 19–29, Jan. 2023. https://doi.org/10.1109/JSSC.2022.3223052
[15] R. Tang et al., "A 112-Gb/s PAM-4 receiver with ultra-fine gain-adjustment CTLE and novel sample-and-reset slicer in 28-nm CMOS," *IEEE ESSERC*, pp. 297–300, 2024. https://doi.org/10.1109/ESSERC62670.2024.10719519

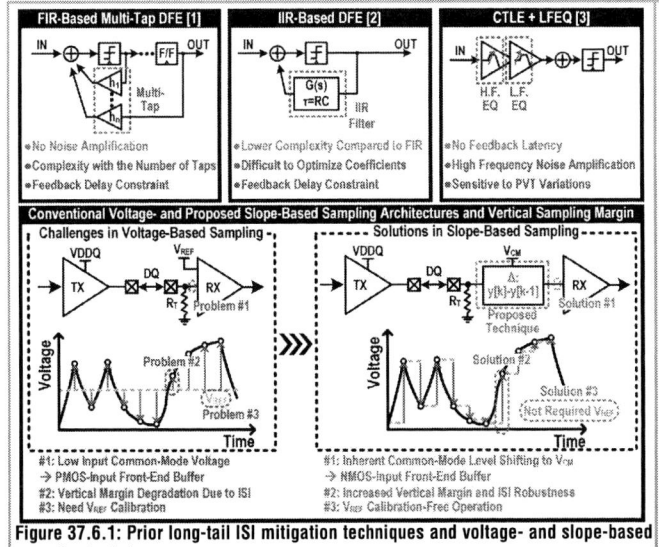

Figure 37.6.1: Prior long-tail ISI mitigation techniques and voltage- and slope-based sampling techniques.

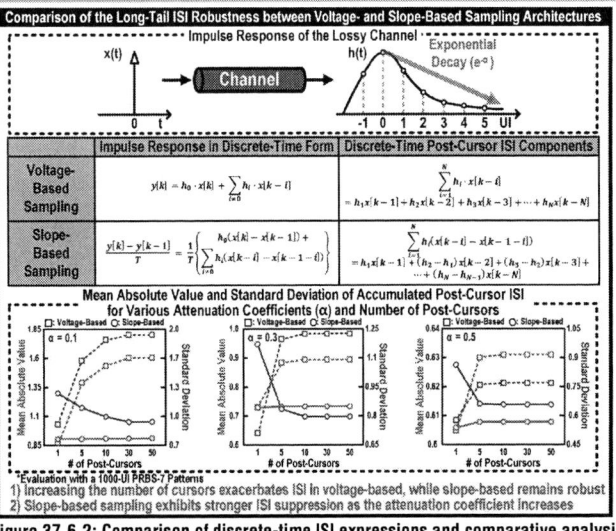

Figure 37.6.2: Comparison of discrete-time ISI expressions and comparative analysis of accumulated post-cursor ISI.

37

DIGEST OF TECHNICAL PAPERS • 635

979-8-3315-8937-0/26 $31.00 © 2026 IEEE

ISSCC 2026 / SESSION 37 / MEMORY INTERFACE / 37.6

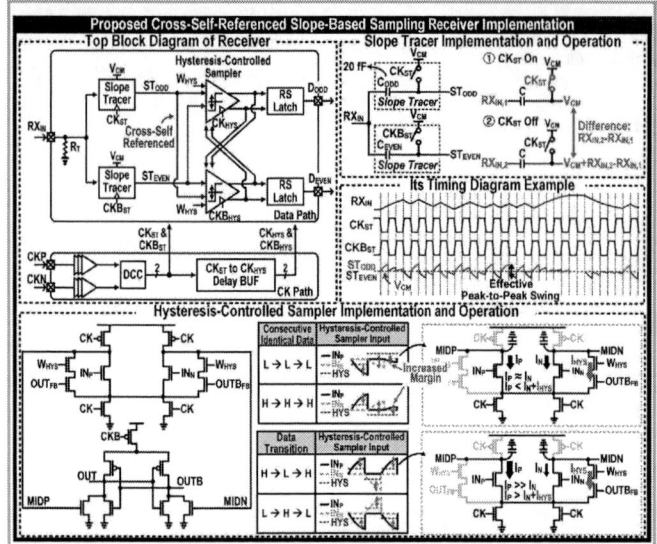

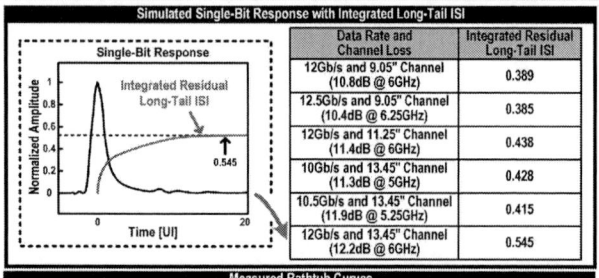

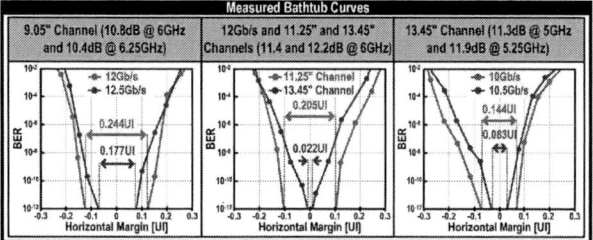

Figure 37.6.3: Cross-self-referenced slope-based sampling RX implementation.

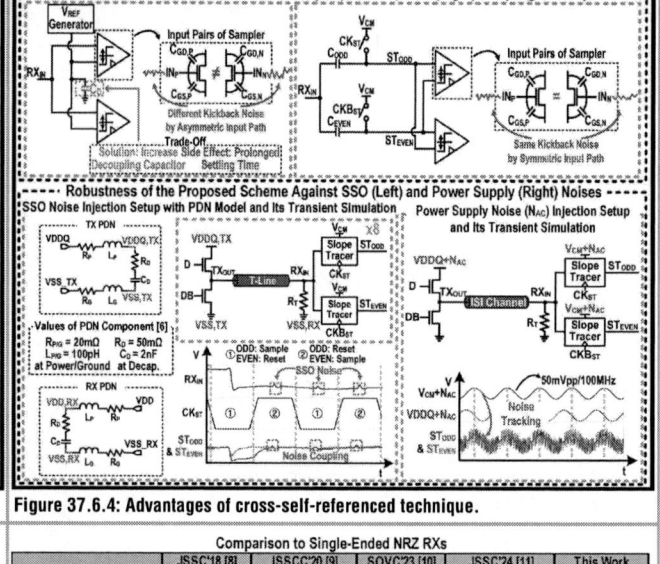

Figure 37.6.4: Advantages of cross-self-referenced technique.

Figure 37.6.5: Simulated single-bit response with integrated long-tail ISI and measured bathtub curves.

Simulated Single-Bit Response with Integrated Long-Tail ISI

Data Rate and Channel Loss	Integrated Residual Long-Tail ISI
12Gb/s and 9.05" Channel (10.8dB @ 6GHz)	0.389
12.5Gb/s and 9.05" Channel (10.4dB @ 6.25GHz)	0.385
12Gb/s and 11.25" Channel (11.4dB @ 6GHz)	0.438
10Gb/s and 13.45" Channel (11.3dB @ 5GHz)	0.428
10.5Gb/s and 13.45" Channel (11.9dB @ 5.25GHz)	0.415
12Gb/s and 13.45" Channel (12.2dB @ 6GHz)	0.545

Comparison to Single-Ended NRZ RXs

	JSSC'18 [8]	ISSCC'20 [9]	SOVC'23 [10]	JSSC'24 [11]	This Work	
Technology [nm]	65 (CMOS)	8 (FinFET)	65 (CMOS)	28 (CMOS)	65 (CMOS)	
Sampling Component	Voltage	Voltage	Voltage	Voltage	Slope	
Reference Voltage	Required	Required	Not Required	Required	Not Required	
Data Rate [Gb/s]	12.5	18	15	18	12	12.5
Channel Loss [dB]	14	10	N/A	15	12.2	10.4
Equalizer	2-Tap DFE	CTLE + 1-Tap DFE	1-Tap DFE	CTLE + 1-Tap DFE	Not Required	
Power [mW]	2.72*	N/A	8.54	2.44*	1.13*	1.15*
Energy Efficiency [pJ/b]	0.22*	N/A	0.57	0.14*	0.094*	0.092*
FoM [fJ/b/dB]	16*	N/A	N/A	9*	7.7*	8.8*
Area [mm²]	0.004*	4.15	0.001*	0.0003*	0.001*	

*: Only RX Core

Comparison to Long-Tail ISI Mitigation RXs

	ISSCC'16 [12]	JSSC'20 [13]	JSSC'23 [14]	ESSERC'24 [15]	This Work	
Technology [nm]	28 (FDSOI)	65 (CMOS)	28 (CMOS)	28 (CMOS)	65 (CMOS)	
Signaling	DS NRZ	DS NRZ	DS PAM-4	DS PAM-4	SE NRZ	
Data Rate [Gb/s]	16	10.8	112	112	12	12.5
Channel Loss [dB]	28	34	20.8	12	12.2	10.4
Equalizer	1-Tap IIR + 1-Tap DFE	CTLE + 2-Tap DFE	CTLE + 4-Tap FFE	CTLE	Not Required	
Power [mW]	15.8*	13.31*	128.24*	108.34	1.13*	1.15*
Energy Efficiency [pJ/b]	0.99*	1.23*	1.15*	0.97	0.094*	0.092*
FoM [fJ/b/dB]	35*	36*	55*	81	7.7*	8.8*
Area [mm²]	0.008*	0.174	0.081	0.322	0.001*	

*: Only RX Core

Figure 37.6.6: Performance summary and comparison to recent single-ended NRZ and long-tail ISI mitigation RXs.

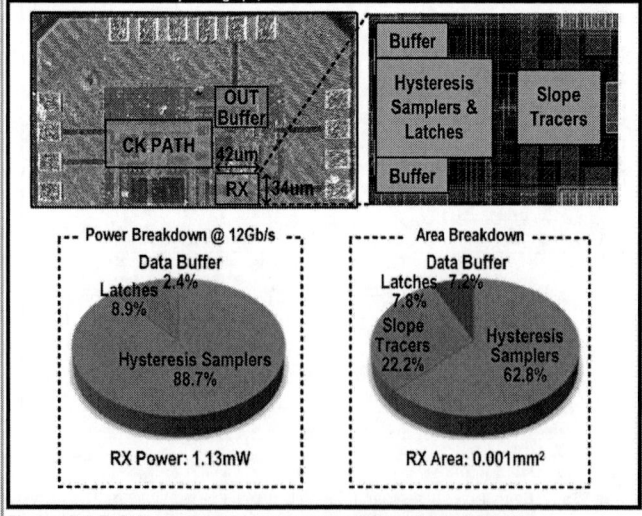

Figure 37.6.7: Chip micrograph, power and area breakdown pie charts.

• 2026 IEEE International Solid-State Circuits Conference

ISSCC 2026 / SESSION 37 / MEMORY INTERFACE / 37.7

37.7 A 12.8Gb/s Parallel Receiver with a One-Way Self-Training Scheme for Equalizing ISI and Reflections in Multi-Drop Memory Interfaces

Ji-Won Moon[1], Taehyeon Kim[1], Minwook Kim[1], Hyeonwoo Seong[1], Jewon Lee[1], Jeongbin Park[1], Dongjun Park[1], Jaehoon Lee[1], Jung-June Park[2], Chiweon Yoon[2], Jae-Yoon Sim[1], Seon-Kyoo Lee[1]

[1]Pohang University of Science and Technology, Pohang, Korea, [2]Samsung Electronics, Hwaseong, Korea

Abstract

This paper presents a 12.8Gb/s/pin receiver with a one-way self-training scheme for highly-reflective multi-drop memory interfaces. DFE tap coefficients are directly derived from a single-pulse response, eliminating controller interaction during training. Implemented in 40nm CMOS, the receiver reopens a closed eye in a 4-drop channel at BER < 10^{-12}. Reusing existing circuitry, the design achieves 102.4Gb/s throughput with only a 6.74% hardware overhead and a 0.64pJ/b energy efficiency.

AI and data-center applications demand memory systems with both high capacity and unprecedented performance, driving their evolution. Interface topologies evolved into distinct regimes: performance-critical systems such as GDDR and LPDDR utilize point-to-point connections for maximum signal integrity. Even mainstream DRAM, which traditionally used multi-drop topologies, has been practically limited to two drops per channel since DDR4 to maintain signal integrity [1]. On the other hand, high-capacity solid-state drives (SSDs) rely on multi-drop topologies with four or more memory chips per lane (Fig. 37.7.1) while also targeting DRAM-like data rates. The multi-drop and high-bandwidth requirements result in a critical bottleneck, as signal reflections from multiple stubs severely degrade signal integrity: causing complete eye closure as data rates approach 12.8Gb/s. A multi-tap decision-feedback equalizer (DFE) can be used to compensate for both inter-symbol interference (ISI) and channel reflections in high-speed memory interfaces. However, conventional DFE training methods require a long training time to converge the tap coefficients, which becomes a significant bottleneck. A conventional controller-driven interactive method has the memory controller sweep the memory chip's DFE coefficients while repeatedly performing write/read operations: the resulting pass/fail statistics determine the set of tap weights that achieve the best eye opening. While the sweep time for a single device is only proportional to the number of DFE taps, the number of tap-weight levels, and the time required to apply each tap-weight update, multi-drop topologies common in high-capacity memory applications exacerbate the problem. Each memory chip observes a different channel response and must be trained independently, so the system-level training time grows with the number of dies and drops. This work proposes a one-way self-training receiver that precisely extracts channel characteristics from a single-bit pulse (SBP) with minimal hardware overhead, allowing for the DFE tap coefficients to be directly derived; thereby, eliminating the need for an iterative algorithm.

Figure 37.7.2(top left) illustrates the goal of the proposed channel characteristic extraction. The SBP captures the ISI and reflection components at each DQS sampling point. Since the objective of this work is to suppress them via a DFE, it is necessary to extract the post-cursor amplitude to determine the optimal DFE coefficients. The proposed method utilizes an SBP as a training signal to extract the post-cursor magnitudes for each DQS sampling point. This is achieved by storing the height of the pulse response and the 1's voltage level as digital codes from the on-chip V_{REF} generator (DAC). The difference between the 1's voltage level and the sampled height directly represents the post-cursor magnitude, and half of this value becomes the target tap coefficient for the DFE. A negative-going pulse is intentionally employed for this training to ensure that all reflection artifacts are accurately captured; since, for some signaling schemes like low-voltage swing terminated logic (LVSTL), reflections can cause the response to undershoot below ground, a voltage range that cannot be resolved by the positive-voltage on-chip DAC. By inverting the training pulse, the entire response, including such negative excursions, are effectively transposed into the DAC dynamic range. This allows for the accurate characterization of all post-cursors, as illustrated by the successful capture of the reflection at the 4th post-cursor (PC4). Figure 37.7.2(top right) shows the block diagram of the proposed extraction scheme. A training pulse generated by the memory controller is applied to the CTLE's positive input via the DQ line, while the DAC is initialized to its maximum code to drive the CTLE's negative input, initiating a sequential downward sweep to digitize the entire pulse response. The SBP is sampled by the DQS via a half-rate architecture, and the comparator output is shifted into a series of flip-flops that feed the polarity detector. This approach avoids additional loading on the DFE's timing-critical feedback path. The shifted data is then latched by an update signal. Subsequently, a V_{REF}-change signal prompts the down counter to decrement the 7b DAC code by one for the next step in the downward sweep. The extraction procedure begins by setting the V_{REF} code to its maximum value and then sampling the SBP. Since the SBP amplitude is initially lower than V_{REF} at all sampling points, all comparator outputs are 0 in cycle #1 and are latched into the polarity detector by the update signal. After lowering V_{REF} by one LSB and allowing it to settle, another SBP is applied, and the process is repeated. In cycle #2, the SBP at the 4th sampling point becomes higher than V_{REF}, and its comparator output flips to 1. When this 1 is shifted and latched, the polarity detector generates a 0→1 transition for that sampling point. This detection event triggers the capture of the current DAC code into the code register, representing the height of the 4th post cursor. By repeating the training sequence until V_{REF} sweeps its full range, a 0→1 transition is eventually detected for the 9 sampling points, which are the 1's voltage level point and the 8 post-cursors. The

corresponding DAC code for each point is then stored in the code register as its height information. The extracted height codes are then used for the subsequent DFE tap calibration.

Figure 37.7.3(top) illustrates the calibration procedure for the first tap. To establish the calibration target, the DAC applies half of the 1's voltage level code and half of the height<1> code to the positive and negative inputs of CTLE. The voltage difference between these two nodes corresponds to half of the post-cursor<1> magnitude, which becomes the calibration target for the DFE tap current source. The tap's current source is then adjusted until this voltage difference between the CTLE's output V+ and output V− is nullified, establishing the precise tap coefficient required to cancel the post-cursor. As shown in Fig. 37.7.3, this calibration is achieved using an up/down counter. Since the V+ input is initially higher than the V−, the comparator output is 1. This output drives the up/down counter to increase the tap coefficient, while the current path is fixed by the tap's sign bit, which encodes the polarity of the 1st post-cursor. As the tap coefficient increases, more current is injected into the V− node, raising its voltage. Once the V− voltage surpasses V+ one, the comparator output flips to 0, and the counter then decreases the tap coefficient. Consequently, the comparator output toggles between 1 and 0, and the tap coefficient converges to the target value corresponding to half of 1st post-cursor. The same procedure is applied to the remaining 7 taps using the respective DAC codes obtained from the channel characteristic extraction. Figure 37.7.3(bottom) details the proposed polarity-controlled DFE tap architecture, which supports two distinct operational modes. In the calibration mode, the DFE tap's sign bit is used to fix the current path to one side. This allows the calibration of the tap's current strength to be independent of the decision data. This independence from decision data is important because it lets the feedback loop adjust the DFE tap's current strength without being perturbed by changing decision data. Following calibration, all DFE taps switch from calibration mode to normal operation mode. In normal operation, the tap current flows to one side of the CTLE output or the other, depending on the previous decision data, and its polarity is selectively reversed by the tap's sign bit according to the calibrated sign of the corresponding post-cursor. A key advantage of this dual-mode architecture is that it provides this compensation without introducing any additional performance-degrading delay into the timing-critical feedback loop.

Figure 37.7.4 shows the functional top-level block diagram of the proposed architecture and the sequence of the one-way calibration scheme. As the diagram indicates, the proposed receiver architecture is composed of blocks already present in a standard memory interface, with the addition of the extraction logic, to digitize the SBP, and the tap calibration logic, to determine the DFE coefficients. The proposed scheme leverages existing RX circuitry, 8-DQ Rx with DFE and a DAC; thus, allowing direct reuse of building blocks and facilitating easy adoption in standard memory interfaces. By using training pulses to sample the SBP, the proposed method reuses the existing RX comparator and DAC to perform SBP digitization, eliminating the need for an extra ADC and reducing hardware overhead. The training pulses needed for extraction can be added to the existing controller training flow in typical DC-coupled memory interfaces. The logic for each step is mainly built from registers and counters, keeping the additional digital complexity low. The one-way calibration proceeds in three sequential steps. The first step is offset tap calibration, where initial offset of the front-end differential pairs is canceled. Next, during channel characteristic extraction, SBP training is executed to digitize the channel response into a set of DAC codes. Finally, in the DFE tap calibration stage, these extracted DAC codes are used to directly calculate and set the coefficients for each DFE tap, taps 1-8, in sequence. The proposed scheme operates as a one-way calibration without back-and-forth interaction between the memory chip and the controller, directly addressing the prohibitive training time of conventional interactive calibration schemes.

The proposed parallel receiver is fabricated in 40nm CMOS. Figure 37.7.5 summarizes the measurement results at 12.8Gb/s, showing that the proposed one-way self-training scheme achieves sufficient timing and voltage margins on the highly reflective 4-drop channel. Consequently, the measured BER eye is completely closed when the DFE is disabled, rendering data recovery impossible, but it is successfully opened when the calibrated 8-tap DFE is enabled. The measured bathtub curve shows timing margins of 0.23UI at 10^{-12} BER and 0.29UI at 10^{-9}. Eye diagram plot in Fig. 37.7.5 indicates a vertical eye opening of approximately 75mV at 10^{-12} BER, confirming that the proposed scheme effectively compensates for the severe channel distortions. Figure 37.7.6 summarizes the performance of this work in comparison to previous work. Using the proposed one-way calibrated equalization, a fully closed eye in a 4-drop environment is successfully opened at 12.8Gb/s, and an aggregate throughput of 102.4Gb/s is achieved with 8 parallel DQs. The design also demonstrates a 0.64pJ/b energy efficiency. The chip die micrograph, along with power and area breakdown, is shown in Fig. 37.7.7. By reusing existing on-chip building blocks, the proposed one-way calibration scheme accomplished with a minimal hardware overhead of only 6.74% for the additional training circuits. This approach also ensures compatibility with conventional memory interface protocols.

Acknowledgement:
This work was supported in part by Samsung Electronics Co., Ltd., and in part by the Korea Institute for Advancement of Technology (KIAT) grant funded by the Korea Government (MOTIE) (RS-2024-00401466, HRD Program for Industrial Innovation). The EDA tools were provided by the IC Design Education Center (IDEC), Korea.

References:
[1] JESD79-4B:JEDEC Standard DDR4 SDRAM, June. 2017.
[2] K. Gharibdoust. et al., "A 7.5mW 7.5Gb/s mixed NRZ/multi-tone serial-data transceiver for multi-drop memory interfaces in 40nm CMOS," *ISSCC*, pp. 180-181, 2015.
http://doi.org/10.1109/ISSCC.2015.7062985
[3] H.-W. Lim et al., "A 5.8-Gb/s Adaptive Integrating Duobinary DFE Receiver for Multi-Drop Memory Interface," *IEEE JSSC*, vol. 52, no. 6, pp. 1563-1575, June 2017.
http://doi.org/10.1109/JSSC.2017.2675923
[4] S. Lee et al., "A 7.8Gb/s/pin 1.96pJ/b compact single-ended TRX and CDR with phase-difference modulation for highly reflective memory interfaces" *ISSCC*, pp. 272-273, 2018.
http://doi.org/10.1109/ISSCC.2018.8310289
[5] K. Chung et al., "A 12-Gb/s Single-Ended Transmitter with Echo-Canceling FFE for Multi-Drop Bus in 28nm CMOS" *IEEE Symp. VLSI Circuits*, pp. C7-5, 2025.
http://doi.org/10.23919/VLSITechnologyandCir65189.2025.11075070

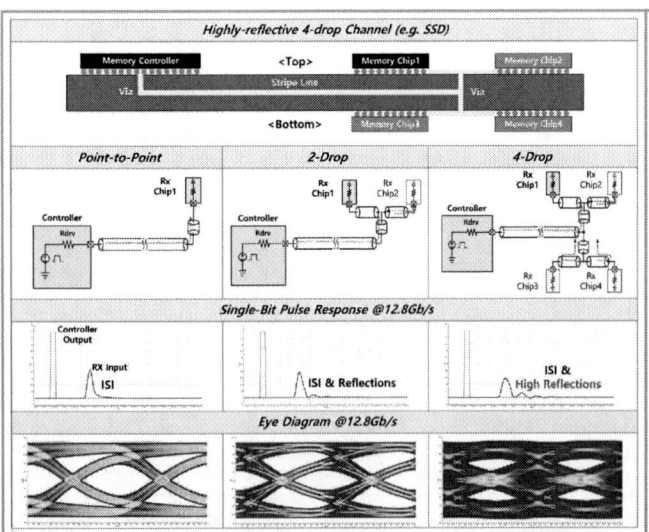

Figure 37.7.1: Point-to-point, 2-drop, and 4-drop channels with single-bit pulse response and eye diagrams at 12.8Gb/s.

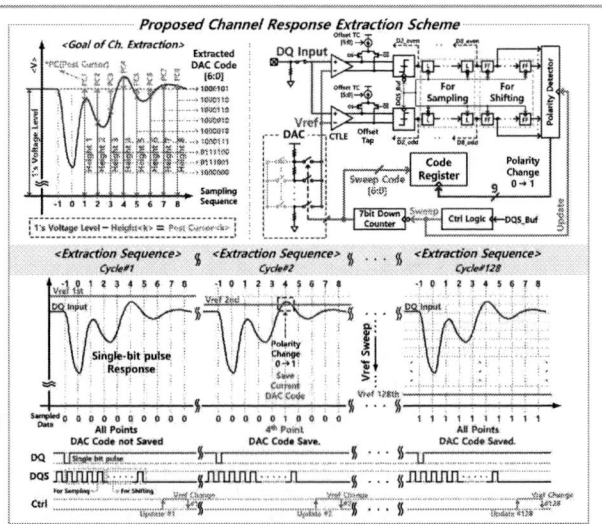

Figure 37.7.2: Proposed channel response extraction scheme using a single-bit pulse response.

ISSCC 2026 / SESSION 37 / MEMORY INTERFACE / 37.7

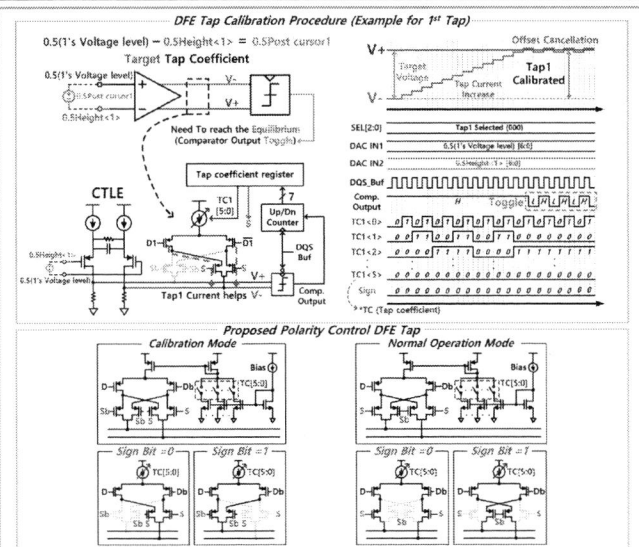

Figure 37.7.3: Proposed DFE tap calibration procedure and DFE tap architecture.

Figure 37.7.4: Functional top-level block diagram of proposed receiver architecture and overall one-way calibration sequence.

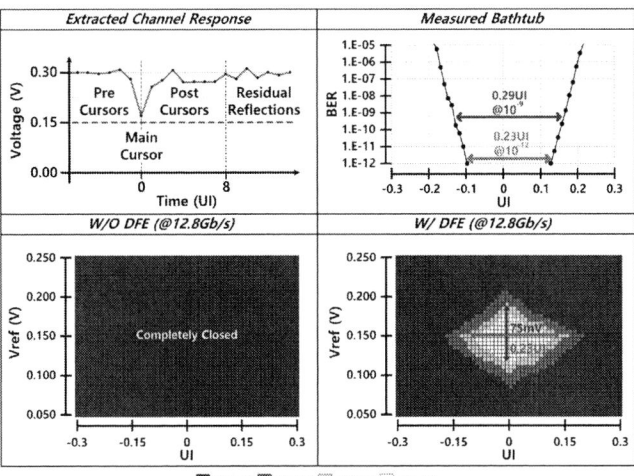

Figure 37.7.5: Measured SBP, bathtub curve, and eye diagrams at 12.8Gb/s.

	This Work	[2]ISSCC 2015	[3]JSSC 2017	[4]ISSCC 2018	[5]VLSI 2025
Process	40nm	40nm	45nm	65nm	28nm
Supply Voltage (V)	1.0	0.9	1.1	1.0/0.9	1.0
Architecture	RX	TRX	RX	TRX	TX
Signaling	Single-ended NRZ	Differential	Differential	Single-ended PDM*	Single-ended NRZ
Compatibility with Existing Systems	Yes	No	-	No	Yes
Equalization	Self-Cal. DFE	-	Integrating Duobinary DFE	-	EC-FFE**
# of Multi Drop	4 drops	4 drops	4 drops	1 stub	1 stub
Adaptation	Ch. Characteristic Extraction	No	SS-LMS	No	No
Data Rate (Gb/s/pin)	12.8	3.75	2.9	7.8	12
Throughput (Data Rate*#DQ)	102.4 (8DQ)	7.5 (2DQ)	5.8 (2DQ)	7.8 (1DQ)	12 (1DQ)
Energy Efficiency (pJ/b)	0.64	1	2.45	1.96	1.52
Active Area/lane	0.018	0.015	0.087	0.0078	0.01

*PDM: Phase-difference modulation   **EC-FFE: Echo-Canceling FFE

Figure 37.7.6: Comparison of key metrics with prior multi-drop memory interface works [2]–[5].

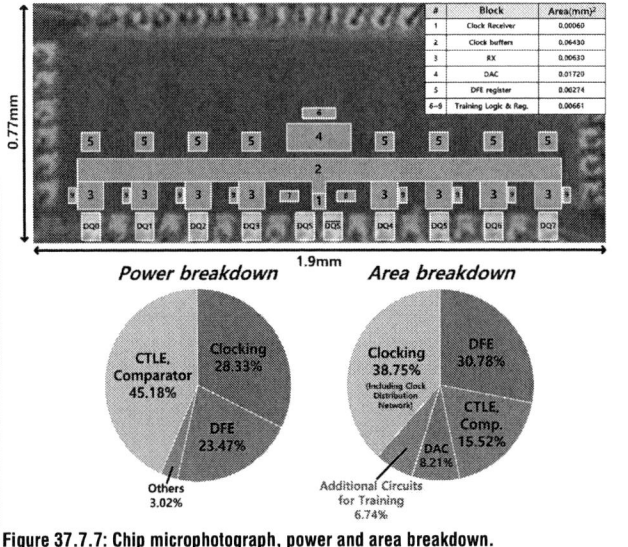

Figure 37.7.7: Chip microphotograph, power and area breakdown.

• 2026 IEEE International Solid-State Circuits Conference

ISSCC 2026 / SESSION 37 / MEMORY INTERFACE / 37.8

37.8 A 0.87pJ/b 17Gb/s/pin Parallel Receiver with a Local DQS Recovery for a Supply-Noise-Tolerant DQS Distribution in High-Performance NAND Flash Interfaces

Byeong-Chan Kim[1], Kyongsu Lee[1], Dongjun Park[1], Jung-June Park[2], Chiweon Yoon[2], Jae-Yoon Sim[1], Seon-Kyoo Lee[1]

[1]Pohang University of Science and Technology, Pohang, Korea, [2]Samsung Electronics, Hwaseong, Korea

Abstract

This work presents a 0.87pJ/b, 17Gb/s/pin parallel receiver with local DQS recovery and a dual-DQS tree for reduced clock power and mitigated power supply-induced jitter (PSIJ). The local DQS recovery time-multiplexes a CML and an inverter tree ($1/8$:$7/8$), enabling a hidden lock to the CML tree and subsequently generating a PSIJ-filtered DQS from the inverter tree with low power consumption. Fabricated in 40nm CMOS, it achieves <10^{-12} BER under 100mV$_{PP}$ supply noise, while reducing clocking power by 70%.

The rapid advancement of AI and large-scale data processing has increased the demand for higher data rates in memory interfaces. While modern DRAM supports multi-Gb/s throughput, conventional NAND interfaces remain limited by their DQ-to-DQS delay-matched architecture. As shown in Fig. 37.8.1, this structure provides strong immunity to power supply-induced jitter (PSIJ), but it precludes the use of advanced equalization techniques such as voltage-domain decision-feedback equalization (DFE), which is essential for high-speed data rates. To achieve data rates comparable to DRAM, the recent NAND interface specification supports DQ-to-DQS delay-unmatched architectures [1,2]. However, these architectures are more susceptible to PSIJ due to timing skew between DQ and DQS. Since NAND interfaces operate without a continuous clock, real-time tracking methods such as clock and data recovery (CDR) cannot be used to establish the optimal DQS-to-DQ timing relationship (t_{DQS2DQ}); instead, this timing must be determined by an initial training sequence. In this case, at the start of a data burst, the DQS signal begins toggling instantly, drawing a large transient current and inducing severe power supply fluctuations, see Fig. 37.8.1(bottom left). Consequently, the resulting PSIJ distorts this pre-trained t_{DQS2DQ}. This ultimately causes data sampling failures and significant degradation in the BER.

One common approach, to mitigate this issue, is to use low-dropout regulators (LDOs) to suppress supply noise [3-5]. However, due to their limited bandwidth, LDOs cannot respond quick enough to compensate for the voltage droop that occurs at the beginning of data bursts. A command-aware LDO is proposed in [6] to address this limitation, but this technique is not suitable for NAND interfaces, which operate based on asynchronous time-triggered protocols. Alternatively, current-mode-logic (CML) drivers offer better immunity to PSIJ and maintain a constant DQS distribution delay (t_{DQS}) under supply fluctuations, but they suffer from large static-power consumption [5,7-9]. Furthermore, since NAND typically transmits data in minimum 4kB bursts, continuous CML operation throughout the entire burst becomes very power inefficient.

To address these limitations, this work proposes a 17Gb/s/pin parallel receiver, featuring local DQS recovery (DQSR) and a dual-DQS tree architecture. Figure 37.8.1(bottom right) shows the proposed scheme's operation. At the beginning of a data transmission, when a voltage droop occurs, the receiver samples data using the DQS delivered from the low-jitter CML tree during the first $1/8$ of the burst length (BL). After the supply voltage stabilizes, the CML tree is disabled, and the inverter-based clock tree delivers DQS to the local PLL's input. The PLL generates a 4-phase DQS signal, which is used for data sampling during the remaining $7/8$ of the BL; during which, the local PLL operates with a low bandwidth to filter the PSIJ on the DQS generated by the inverter tree. By operating the CML tree with a 1:7 on:off duty cycle, the static-power consumption of the CML drivers is significantly reduced. This alternating DQS distribution scheme is enabled by a hidden fast-locking PLL, transparent to the external interface. Triggered during the initial CML tree sampling phase, the PLL acquires lock within the short $1/8$ BL interval, enabling a seamless handover to the recovered DQS without affecting data sampling.

Figure 37.8.2(top) shows the proposed parallel receiver architecture. The dual-DQS tree delivers a 4-phase f÷2 DQS through the CML tree and a 1-phase f÷2 DQS through the inverter tree, where f=8.5GHz. The local DQSR consists of a DLL, a PLL with an embedded frequency-locked loop (FLL), and four multiplexers that select between the two DQS sources. PLLs generally require long lock times so they are not ideal for memory burst-mode operations [10]. To overcome this limitation, the embedded FLL performs frequency acquisition before PLL locking, and an adaptive-bandwidth PLL enables fast locking with a high bandwidth and jitter filtering with a low bandwidth. To compensate for PVT-induced variations in t_{DQS} between the CML and inverter trees, a DLL is employed to align their phases. To ensure safe switching of sampling clocks, flip-flops are inserted in all four clock phases to synchronize the asynchronous multiplexer-control signals (SEL_{SAMP}). The control signals SEL_{REF}, SEL_{SAMP}, and SEL_{BW} for the local DQSR are generated by a state machine operating with an f÷16 DQS derived from the inverter tree.

The timing diagram for the proposed receiver is shown in Fig. 37.8.2(bottom). The receiver switches the sampling DQS within each 4kB burst using a 1:7 duty cycle. When a data transfer begins, the receiver samples data using DQS from the CML tree. Simultaneously, the FLL initiates frequency acquisition. After FLL convergence, a high-bandwidth PLL

eliminates residual frequency and phase errors between DQS from the CML tree and the PLL output. Meanwhile, the DLL compensates for skew between the inverter and CML tree outputs. These loops complete locking within the initial $1/8$ of the BL. Once locked, the PLL transitions into a low-bandwidth mode to suppress jitter, and the reference clock for the DQSR is switched from the CML tree to the inverter tree. During the remaining $7/8$ of the BL, the PLL provides a jitter-filtered DQS to the DQ samplers. Maintaining continuous PLL operation is challenging due to voltage and temperature (VT) drift during long data transfers. To compensate, the PLL increases its loop bandwidth and switches the reference clock back to the CML tree at the beginning of each BL to re-lock. This ensures a consistent sampling phase between bursts despite operating point drift. For long bursts, such as sequential write operations longer than 16kB, where the frequency remains stable, the FLL's frequency acquisition is skipped. In contrast, if a new data burst starts, the receiver re-engages the FLL to acquire the DQS frequency.

Figure 37.8.3(top) illustrates the architecture of the proposed FLL-embedded PLL. To support a wide frequency range, the PLL's digitally controlled oscillator (DCO) includes a 5b coarse inverter-strength control and an 8b fine capacitor bank. To match the injection signal slope with the internal DCO signal, an identical delay cell is placed before the DCO's multiplexer. The SEL logic generates the multiplexer control signal (0 for delay-line mode, 1 for oscillator mode) and provides a clock for SAR updates. The FLL performs frequency locking prior to PLL operation, thereby greatly reducing PLL lock time despite the use of a bang-bang phase detector (BBPD). As a result, the CML-tree operation time is minimized, enabling a $1/8$ duty-cycled CML-tree operation, which considerably reduces power consumption. To ensure proper PLL operation at high-reference frequencies, the BBPD output is first stored in an up/down counter running at f÷4 and is then sampled by the digital loop filter (DLF) at f÷16. The up/down counter is reset after each DLF update to ensure clean loop dynamics. Additionally, multiplexers are inserted by the BBPD to switch its input clocks between FLL and PLL modes, and in the DCO path to select between the SAR code during FLL operation and the DLF code during PLL operation.

The proposed FLL-embedded PLL operates in two phases, as shown in the timing diagram in Fig. 37.8.3. During FLL operation, a CML-based DQS (CML_I & CML_{IB}) passes through injection delay cells, traverses the DCO ($FLL_I \rightarrow DCO_Q \rightarrow DCO_{IB} \rightarrow DCO_{QB} \rightarrow DCO_I$), and is compared to the original FLL_I to update the SAR code. Because of the delay stabilization requirements, after code updates, each step takes three reference cycles, and a full FLL search requires 42 cycles, which is sufficient for lock completion within $1/8$ of BL. After frequency locking, the PLL aligns the DCO phase to the CML_I phase. To minimize the initial phase error for fast locking, the BBPD CMP and REF signals are updated to align DCO_{QB} and CML_I, which are phase aligned due to injection-delay-cell characteristics.

When the PLL's reference clock changes, the phase difference between the two clocks should be minimized, as a large phase difference would act as a step input to the PLL, causing a significant disturbance in its internal state. To mitigate this, the DLL that aligns the phase of the CML and inverter tree is shown in Fig. 37.8.4(top). The DLL incorporates both coarse and fine delay lines to accommodate PVT variation. A down-sweep search followed by SAR locking is employed to avoid stuck lock failure. After locking, the DLL tracks the delay of the CML tree while it is still active and retains the delay code when the CML tree is off, as no reference clock is available during that period. Figure 37.8.4(bottom left) presents the CML-driver schematic. A constant-g_m bias is used to enhance immunity to PSIJ. Periodic on/off switching of the CML driver disturbs the bias voltage due to charge feedthrough through parasitic capacitances: particularly the gate-drain capacitance (C_{gd}) of the current-source MOSFET, resulting in fluctuations in the CML tree's t_{DQS}. To mitigate this issue, a charge-injection switch is added to suppress charge feedthrough during on-transitions. Figure 37.8.4(bottom right) shows the DQ receiver (RX) architecture, which consists of a DC-gain-controllable continuous-time linear equalizer (CTLE) and four samplers for quadrature sampling.

The proposed parallel receiver is fabricated in a 40nm CMOS process and operates up to 17Gb/s/pin using a 1.1V supply voltage. To verify the key achievement of the proposed DQSR, robust supply-noise tolerance, a BER testing is performed under sinusoidal supply

638 • 2026 IEEE International Solid-State Circuits Conference

979-8-3315-8937-0/26 $31.00 © 2026 IEEE

noise conditions. Figure 37.8.5 summarizes the measurement results, comparing the proposed scheme with conventional inverter tree and CML tree modes based on their bathtub curves and power consumption. The power consumption is 11.2mW per DQ receiver, including the DQSR circuit, and 29mW for DQS distribution. The inverter tree fails to achieve <10⁻¹² BER, but the proposed scheme does, which demonstrates robust noise immunity and a timing margin comparable to the CML tree. Most importantly, robust performance is achieved with a 70% reduction in DQS-distribution power compared to the CML tree. Figure 37.8.6 summarizes the performance comparison to prior work, and Fig. 37.8.7 shows the die micrograph and the power breakdown of the proposed receiver.

Acknowledgement:
This work was supported in part by Samsung Electronics Co., Ltd., and in part by the Korea Institute for Advancement of Technology (KIAT) grant funded by the Korea Government (MOTIE) (RS-2024-00401466, HRD Program for Industrial Innovation). EDA tools are supported by the IC Design Education Center (IDEC), Korea.

References:
[1] JESD230G: JEDEC Standard NAND Flash Interface Interoperability, Oct. 2024. http://doi.org/unknown
[2] K. Yanagidaira et al., "A 1Tb 3b/cell 3D-Flash Memory with a 29%-Improved-Energy-Efficiency Read Operation and 4.8Gb/s Power-Isolated Low-Tapped-Termination I/Os," *ISSCC*, pp. 506-507, 2025. http://doi.org/10.1109/ISSCC49661.2025.10904509
[3] E.-H. Chen et al., "A 212.5Gb/s DSP-Based PAM-4 Transceiver with 50dB Loss Compensation for Large AI System Interconnects in 4nm FinFET," *ISSCC*, pp. 136-137, 2025. http://doi.org/10.1109/ISSCC49661.2025.10904601
[4] M. Cusmai et al., "A 224Gb/s sub pJ/b PAM-4 and PAM-6 DAC-Based Transmitter in 3nm FinFET," http://doi.org/10.1109/ISSCC49657.2024.10454558
[5] S.-Y. Cho et al., "A 16Gb 37Gb/s GDDR7 DRAM with PAM3-Optimized TRX Equalization and ZQ Calibration," *ISSCC*, pp. 242-244, 2024. http://doi.org/10.1109/ISSCC49657.2024.10454354
[6] J. Kim et al., "A command-aware hybrid LDO for advanced HBM interfaces with 150 µA quiescent current and 20 pF on-chip capacitor achieving sub-10 mV voltage droop in 400 ps settling time," *ISSCC*, pp. 166–167, 2025. http://doi.org/10.1109/ISSCC49661.2025.10904806
[7] S.-H. Kim et al., "A 24Gb 42.5Gb/s GDDR7 DRAM with Low-Power WCK Distribution, an RC-Optimized Dual-Emphasis TX, and Voltage/Time-Margin-Enhanced Power Reduction," *ISSCC*, pp. 508-510, 2025. http://doi.org/10.1109/ISSCC49661.2025.10904689
[8] J. Yang et al., "A 35.4Gb/s/pin 16Gb GDDR7 with a Low-Power Clocking Architecture and PAM3 IO Circuitry," *ISSCC*, pp. 232-234, 2024. http://doi.org/10.1109/ISSCC49657.2024.10454560
[9] K.-S. Ha et al., "A 7.5Gb/s/pin LPDDR5 SDRAM With WCK Clocking and Non-Target ODT for High Speed and with DVFS, Internal Data Copy, and Deep-Sleep Mode for Low Power," *ISSCC*, pp. 378-380, 2019. http://doi.org/10.1109/ISSCC.2019.8662509
[10] Y. Shin et al., "A Digital-PLL-Based Quadrature Clock Generator for a Low-Power and Jitter-Filtering-Capable Clock Distribution Scheme in High-Speed DRAM Interfaces," *IEEE JSSC*, vol. 60, no. 2, pp. 509-518, 2025. http://doi.org/10.1109/JSSC.2024.3433026

[11] J. Seo et al., "An 850µW 2-to-5GHz Jitter-Filtering and Instant-Toggling Injection-Locked Quadrature-Clock Generator for Low-Power Clock Distribution in HBM Interfaces," *ISSCC*, pp. 396-398, 2025. http://doi.org/10.1109/ISSCC49661.2025.10904589
[12] Y. Jung et al., "A Supply-Noise-Induced Jitter-Cancelling Clock Distribution Network for LPDDR5 Mobile DRAM featuring a 2nd-order Adaptive Filter," *ISSCC*, pp. 458-459, 2022. http://doi.org/10.1109/ISSCC42614.2022.9731682
[13] J. Kim et al., "A 4 × 32 Gb/s 1.8 pJ/bit Collaborative Baud-Rate CDR With Background Eye-Climbing Algorithm and Low-Power Global Clock Distribution," *IEEE JSSC*, vol. 60, no. 8, pp. 2751-2764, 2025. http://doi.org/10.1109/JSSC.2025.3532963
[14] X. Chen et al., "Reference-Noise Compensation Scheme for Single-Ended Package-to-Package Links," *ISSCC*, pp. 126-128, 2020. http://doi.org/10.1109/ISSCC19947.2020.9063140

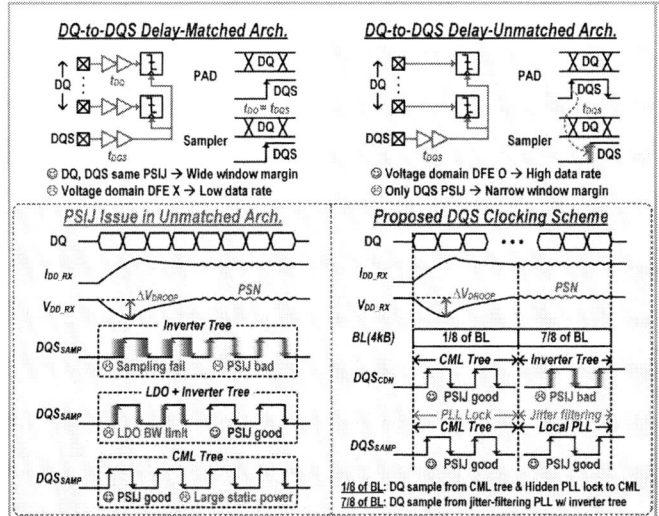

Figure 37.8.1: Proposed DQS clocking scheme for suppressing PSIJ in a delay-unmatched architecture.

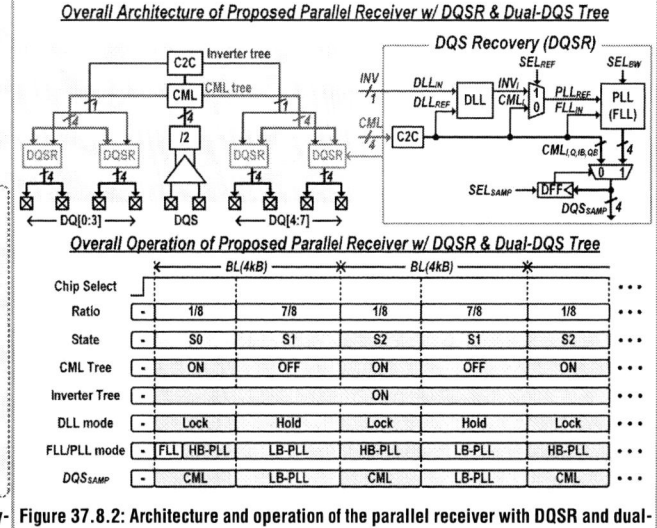

Figure 37.8.2: Architecture and operation of the parallel receiver with DQSR and dual-DQS tree.

ISSCC 2026 / SESSION 37 / MEMORY INTERFACE / 37.8

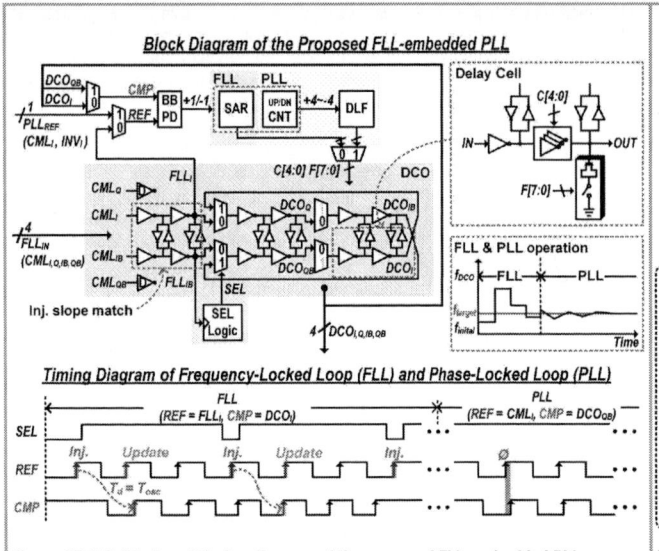

Figure 37.8.3: Block and timing diagram of the proposed FLL-embedded PLL.

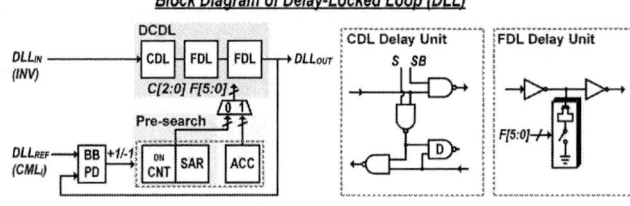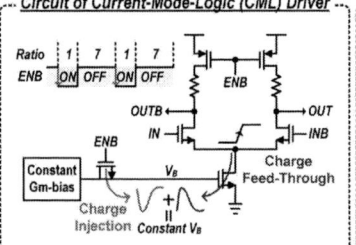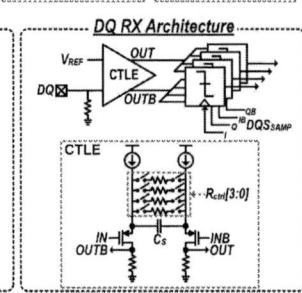

Figure 37.8.4: DLL block diagram, CML driver circuit, and DQ RX architecture.

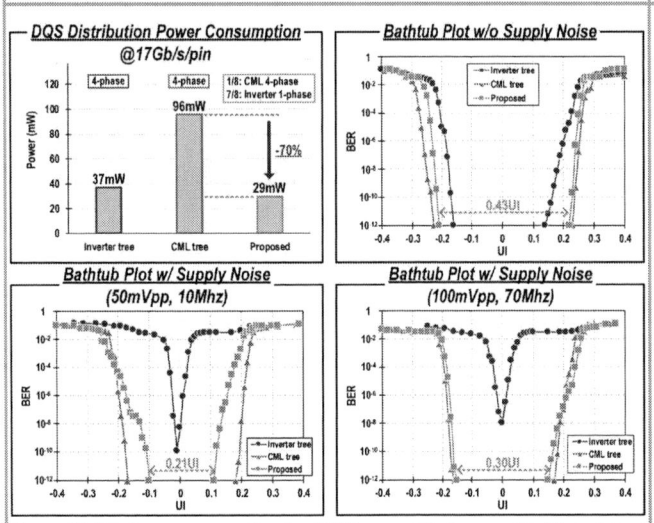

Figure 37.8.5: Power consumption for DQS-tree architectures and bathtub plots at 17Gb/s/pin.

Performance Summary and Comparison with Prior Work

	This work	*ISSCC'25 [11]	*ISSCC'25 [6]	*ISSCC'22 [12]	JSSC'25 [13]	ISSCC'20 [14]
Process [nm]	40	40	40	28	28	16
Supply [V]	1.1	-	1.15	1	-	-
Architecture	DQSR + Dual-DQS tree	Injection Locked QCG	Command-Aware LDO	Adaptive Gain Control of VCDL	FDIV + PI CDR	Ref-Noise Comp. Loop
Data Rate [Gb/s/pin]	17	[a] 10	[a] 10	[a] 12.8	32	25
Aggregate Data Rate [Gb/s]	136	N/A	N/A	N/A	128	100
Equalization	CTLE	N/A	N/A	N/A	CTLE 1-tap DFE	-
PSIJ Immunity	Yes	Yes	Yes	Yes	No	Yes
Initial Voltage Droop Tolerance	Yes	No	Yes	No	No	No
Support Instant Toggling	Yes	Yes	Yes	Yes	No	Yes
Active Area [mm²]	[b] 0.014	0.007	0.014	0.11	[b] 0.105	[b, c] 0.02
RX Clocking Power Efficiency [mW/Gb/s]	0.21	N/A	N/A	N/A	0.65	-
Energy Efficiency [pJ/bits]	0.87	N/A	N/A	N/A	1.81	[c] 1.65

N/A: not applicable, -: not reported
*Same motivation, but not a receiver (BER test excluded) [a] 4-phase clocking assumption [b] per lane [c] transceiver

Figure 37.8.6: Performance summary and comparison with prior PSIJ suppression schemes and parallel receivers.

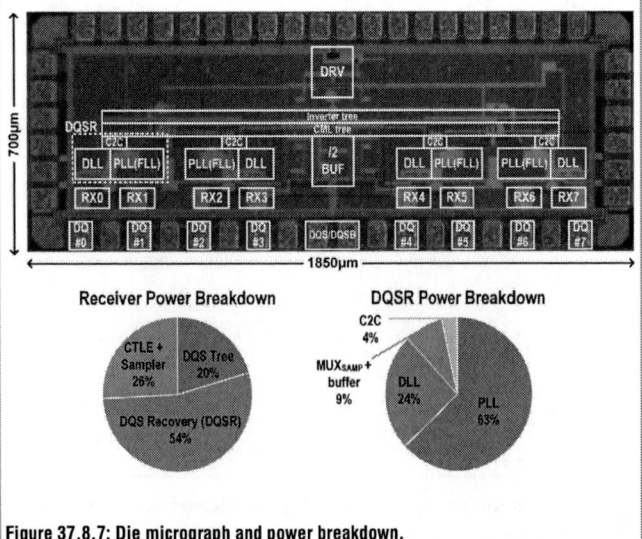

Figure 37.8.7: Die micrograph and power breakdown.

• 2026 IEEE International Solid-State Circuits Conference

ISSCC 2026 / SESSION 37 / MEMORY INTERFACE / 37.9

37.9 A 14Gb/s/pin 0.163pJ/b DQ Receiver for HBM with Baud-Rate Phase Tracking Loop Supporting Background Offset Calibration

Jeonghyeon Lee[1], Ho-Joon Shin[1], Han-Gon Ko[2], Kwanseo Park[1]

[1]Yonsei University, Seoul, Korea, [2]ONEsemiconductor, Gyeonggi, Korea

Abstract

This paper presents a compact 14Gb/s/pin HBM DQ receiver that continuously tracks sampling phase and detector offset in live data, using a baud-rate phase tracking loop with a background-calibrated time-window phase detector. Rise/fall edge-triggered decision logic decouples phase error from detector offset, keeping the sampling-clock lock point of the DCDL fixed under drift and thereby eliminating the need for retraining. A 28nm prototype achieves 0.163pJ/b energy efficiency in 0.00547mm² at <10⁻¹² BER with a 0.295UI sampling margin. High-frequency jitter tolerance reaches 0.188UI at 14Gb/s.

The shift in high-performance computing and memory systems toward multi-chip module packaging, short-reach die-to-die links, and standardized chiplet interfaces is driving requirements for increased bandwidth and I/O density at a low energy per bit [1-6]. Unlike long-reach serial I/Os, high-density memory interfaces typically avoid clock and data recovery and instead rely on source-synchronous clocking and periodic training to save area and power [7-10]. However, as high-bandwidth memories (HBM) scale to thousands of I/O lanes, PVT drift, aging, SSN/PSIJ, and package/board coupling cause lane-to-lane variations and time-varying offsets. These effects cannot be fully suppressed by initial training alone [11-14]. As a result, HBM DQ links face three practical challenges: (1) per-lane V_{REF} generation and trimming, and comparator-offset cancellation consume a large area and do not address time-dependent residual offsets; (2) the DQS-DQ phase relationship drifts while live data is transmitted; thus, reducing eye margins if not continuously tracked; and (3) the dense channel pitch exacerbates crosstalk while simultaneous-switching noise results in supply-induced margin losses for single-ended links. The third challenge is primarily addressed by signaling and packaging choices and is not the focus of this work. To address these challenges, prior work employed V_{REF} tracking or frequent retraining [3,8-10]. V_{REF}-tracking loops often have limited bandwidth due to a complex feedback loop; whereas, retraining interrupts live data transmission and becomes increasingly costly in wide interfaces, where thousands of lanes must periodically suspend traffic for training. Motivated by these limitations, this work focuses on challenges (1) and (2) – offset wander and phase drift during live data transmission – and introduces a baud-rate phase-tracking loop with background offset calibration that continuously aligns the sampling phase and detector threshold on live data.

Figure 37.9.1 shows the overall block diagram of the proposed DQ receiver, with a baud-rate phase-tracking loop and background offset-calibration logic. D_{in} represents the 14Gb/s single-ended data input, and CLK_{in} is the 7GHz differential clock inputs. A clock divider (CLK DIV) converts the half-rate clock into quarter-rate 4-phase clocks, CLK0, CLK90, CLK180, and CLK270: skew between the 4-phase clocks is manually adjusted. A digitally controlled delay line (DCDL) adjusts the quarter-rate clock delays via a 6b DCDL control word (DCW) to achieve an optimal sampling point. Fine phase adjustment is performed by a time-window-based phase detector (TWPD) that integrates D_{in} over the interval between the rising edges of CLK0 and CLK90 and generates a 1-bit digital output T by comparing the resulting integrated voltage with an internal decision threshold. This threshold is defined as the effective input level at which the differential TWPD outputs, V_{OUTP} and V_{OUTN}, become equal; in the absence of internal offset, it coincides with the DC reference level V_{REF} of D_{in}. However, internal TWPD offsets due to process and temperature variation result in threshold voltage variation. To explicitly identify data transitions on D_{in} within the CLK0-CLK90 window, rising-edge-validation (REV) and falling-edge-validation (FEV) circuits generate 1-bit flags R and F, respectively: R = 1 (F = 1) indicates that a 0→1 (1→0) transition occurred between the two clock edges, while R = F = 0 indicates no transition in the window. For an ideal, offset-free TWPD, using only R would be sufficient for phase tracking; however, observing both R and F enables the background offset-calibration logic to distinguish true sampling-phase error from detector offset by checking whether T changes sign with transition polarity. The digital control logic therefore processes the patterns of (R, F, T) and, depending on whether T behaves consistently with phase error or appears stuck high/low due to offset, adjusts either the calibration-control word (CCW) to cancel TWPD offset or the DCW to realign the sampling phase.

Figure 37.9.2 illustrates TWPD operation and possible outputs for different offset conditions. In the TWPD circuit of Fig. 37.9.1, C_{Din} denotes the capacitance connected to D_{in} and is adjusted by CCW, while C_{REF} denotes the capacitance connected to the reference node REF. When no data transition occurs within the CLK0-CLK90 integration window, the TWPD output T simply indicates whether D_{in} is above or below the internal decision threshold over that interval, producing a binary value of 0 or 1 accordingly. When a data transition occurs inside the window and no detector offset is present ($C_{Din} = C_{REF}$), T becomes a function of the sampling-clock phase relative to D_{in}. If the sampling clock lags D_{in} and a rising transition occurs within the window, D_{in} spends more time above the threshold than below it over the integration interval, so T resolves to 1; for a falling transition under the same lagging condition, D_{in} spends more time below the threshold and T resolves to 0. When CLK leads D_{in}, then T = 0 if the rising transition occurs since D_{in} spends more time low than high;

conversely, T = 1 if a falling transition occurs. Thus, under offset-fee conditions the combination of R, F, and T reveal the phase offset polarity; thus, enabling precise DCW adjustment. On the other hand, when an offset is present, the TWPD output is influenced by offset. If C_{Din} is smaller than C_{REF}, T is likely to remain at 0 irrespective of whether a transition occurs earlier or later than the optimal point. Conversely, if C_{Din} is larger than C_{REF}, T is likely to remain at 1 regardless of the transition timing. In this regime, the observed combinations of R, F, and T deviate from those expected for pure phase error and instead reveal the polarity of the detector offset. The calibration logic interprets these biased patterns to iteratively adjust CCW until $C_{Din} \approx C_{REF}$ and T recovers its expected dependence on transition polarity and clock phase, after which the phase-tracking loop can use the same signals to refine DCW.

Figure 37.9.3 details the background offset calibration logic using REV, FEV, and TWPD, interpreted in the voltage variation of D_{in} and the TWPD threshold. The panels plot sampling-clock lock points and rise/fall early/late scenarios for three conditions: $C_{Din} = C_{REF}$, $C_{Din} < C_{REF}$, and $C_{Din} > C_{REF}$. When $C_{Din} = C_{REF}$, the TWPD threshold aligns with the common level of D_{in}; a nominal lock occurs where A = B and C = D, and rise/fall decisions of T are complementary. On the other hand, if the threshold is displaced by offset, collapsing the rise/fall decisions and biasing the early/late classification even when the sampling phase is positioned at the offset-free lock point. Specifically, for $C_{Din} < C_{REF}$ (threshold shifted high), a rise transition is misclassified as late and a fall transition as early at the $C_{Din} = C_{REF}$ lock point; for $C_{Din} > C_{REF}$ (threshold shifted low), a rise is misclassified as early and a fall as late. Near this point, T tends to take the same value for both transition polarities, providing a distinct marker of detector offset. Consequently, the combinations of R, F, and T becomes inconsistent with pure phase error and instead encodes the offset polarity; the synthesized digital logic exploits these signatures to adjust CCW until rise/fall symmetry – and correct early/late labeling – is restored.

Figure 37.9.4 presents the simulated phase tracking behavior for initial optimal clock phase and demonstrates the effectiveness of background offset calibration. In the simulations, three offset conditions are imposed in the TWPD: offset 1 from 0 to 100ns, offset 2 from 100ns to 300ns, and offset 3 from 300ns to 600ns. As a result, with background offset calibration, offset cancellation is achieved by adjusting CCW to 8, 15, and 0 for offsets 1, 2, and 3, respectively, without introducing phase offsets. Therefore, the DCW lock point remains fixed at 30, corresponding to the optimal sampling point. On the other hand, without background offset calibration, the TWPD fails to track offset transitions, causing deviations from the optimal lock point.

The proposed 4-lane DQ receiver is fabricated in 28nm CMOS and occupies 0.00547mm², as shown in Fig. 37.9.7; synthesized digital logic is shared across all lanes. Figure 37.9.5 shows the measured bathtub curve at 14Gb/s, the recovered clock jitter histogram at 14Gb/s, and jitter tolerance plot at 12Gb/s and 14Gb/s. The bathtub curve demonstrates that the proposed receiver achieves <10⁻¹² BER, with a timing margin of 0.295UI. Moreover, with offset calibration, the TWPD locks at the optimal sampling point, yielding the largest margin. The measured jitter histogram exhibits 13.4ps peak-to-peak jitter and 1.982ps RMS jitter: jitter tolerance is measured for <10⁻¹² BER at 12Gb/s for a comparison. High-frequency jitter tolerance (0.294UI) is improved by 345% with background offset calibration, compared to no calibration (0.066UI). At 14Gb/s, the high-frequency jitter tolerance is measured to be 0.188UI. Total power consumption is 2.28mW at 14Gb/s for an energy efficiency of 0.163pJ/b. These results are achieved with compact per-lane hardware with a centralized controller to amortize its area overhead; the baud-rate phase tracking loop uses a low-resolution DCDL and simple decision logic, while the background offset-calibration loop adds only small capacitive trims and a few registers.

Figure 37.9.6 summarizes the results and compares the proposed DQ receiver with other memory interfaces: the proposed receiver achieves better energy efficiency and compact area due to its baud-rate phase-tracking loop and background offset calibration.

ISSCC 2026 / February 18, 2026 / 1:30 PM

Acknowledgement:
This work was supported in part by Samsung Electronics Company Ltd., and in part by the National Research Foundation of Korea (NRF) grant funded by the Korea government (MSIT) (No. RS-2025-00554635). The EDA tool was supported by the IDEC, Korea.

References:
[1] H. Ko et al., "A 370-fJ/b, 0.0056 mm²/DQ, 4.8-Gb/s DQ Receiver for HBM3 with a Baud-Rate Self-Tracking Loop," *IEEE Symp. VLSI Circuits*, pp. C94-C95, 2019. http://doi.org/10.23919/VLSIC.2019.8778082

[2] D. Kang et al., "A 21-Gb/s Duobinary Transceiver for GDDR Interfaces with an Adaptive Equalizer," *IEEE JSSC*, vol. 57, no. 10, pp. 3083-3093, Oct. 2022. http://doi.org/10.1109/JSSC.2022.3170439

[3] Y. Choi et al., "A 25-Gb/s Single-Ended PAM-4 Receiver with Time-Windowed LSB Decoder for High-Speed Memory Interfaces," *IEEE JSSC*, vol. 58, no. 7, pp. 2005-2015, July. 2023. http://doi.org/10.1109/JSSC.2022.3231654

[4] S.-Y. Cho et al., "A 16Gb 37Gb/s GDDR7 DRAM with PAM3-Optimized TRX Equalization and ZQ Calibration," *ISSCC*, pp. 242-244, 2024. http://doi.org/10.1109/ISSCC49657.2024.10454354

[5] B. Kim et al., "A 42Gb/s Single-Ended Hybrid-DFE PAM-3 Receiver for GDDR7 Memory Interfaces," *ISSCC*, pp. 398-400, 2025. http://doi.org/10.1109/ISSCC49661.2025.10904551

[6] J. So et al., "A 0.3pJ/b 32Gb/s/Pin Single-Ended PAM-4 Receiver with a Delay-Less Capacitive-Feedback Equalizer," *ISSCC*, pp. 402-404, 2025. http://doi.org/10.1109/ISSCC49661.2025.10904567

[7] S. Lee et al., "A 7.8 Gb/s/pin, 1.96 pJ/b Transceiver with Phase-Difference-Modulation Signaling for Highly Reflective Interconnects," *IEEE TCAS-I*, vol. 67, no. 6, pp. 2114-2127, Jun. 2020. http://doi.org/10.1109/TCSI.2020.2969472

[8] Y. Kwon et al., "A 33-Gb/s/Pin 1.09-pJ/Bit Single-Ended PAM-3 Transceiver with Ground-Referenced Signaling and Time-Domain Decision Technique for Multi-Chip Module Memory Interfaces," *IEEE JSSC*, vol. 58, no. 8, pp. 2314-2325, Aug. 2023. http://doi.org/10.1109/JSSC.2023.3250706

[9] H. Park et al., "A 3-bit/2UI 27Gb/s PAM-3 Single-Ended Transceiver Using One-Tap DFE for Next-Generation Memory Interface," *ISSCC*, pp. 382-384, 2019. http://doi.org/10.1109/ISSCC.2019.8662462

[10] H. Li et al., "A 100Gb/s-8.3dBm-Sensitivity PAM-4 Optical Receiver with Integrated TIA, FFE and Direct-Feedback DFE in 28nm CMOS," *ISSCC*, pp. 190-192, 2021. http://doi.org/10.1109/ISSCC42613.2021.9365802

[11] T. M. Hollis et al., "An 8Gb GDDR6X DRAM Achieving 22Gb/s/pin with Single-Ended PAM4 Signaling," *ISSCC*, pp. 348-350, 2021. http://doi.org/10.1109/ISSCC42613.2021.9365925

[12] J. Yang et al., "A 35.4Gb/s/pin 16Gb GDDR7 with a Low-Power Clocking Architecture and PAM3 IO Circuitry," *ISSCC*, pp. 232-234, 2024. http://doi.org/10.1109/ISSCC49657.2024.10454560

[13] S. -Y. Cho et al., "A 16-Gb 37-Gb/s GDDR7 DRAM With PAM3-Optimized TRX Equalization and ZQ Calibration," *IEEE JSSC*, vol. 60, no. 1, pp. 184-196, Jan. 2025. http://doi.org/10.1109/JSSC.2024.3472463

[14] K. Sohn et al., "A 1.2 V 20 nm 307 GB/s HBM DRAM With At-Speed Wafer-Level IO Test Scheme and Adaptive Refresh Considering Temperature Distribution," *IEEE JSSC*, vol. 52, no. 1, pp. 250-260, Jan. 2017. http://doi.org/10.1109/JSSC.2016.2602221

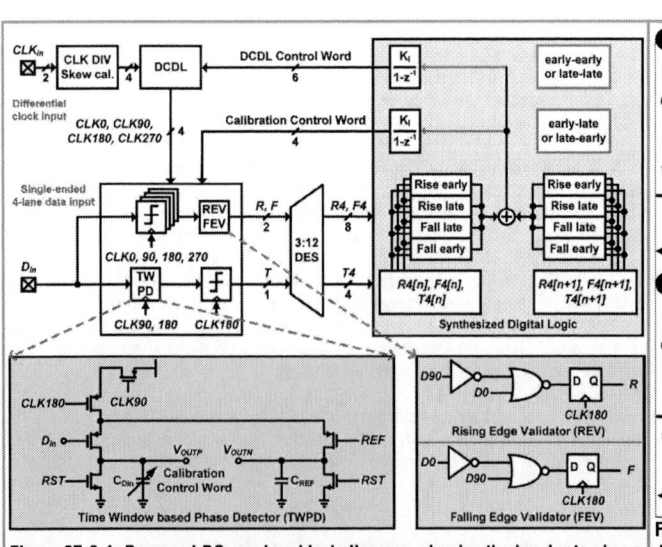

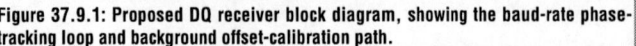

Figure 37.9.1: Proposed DQ receiver block diagram, showing the baud-rate phase-tracking loop and background offset-calibration path.

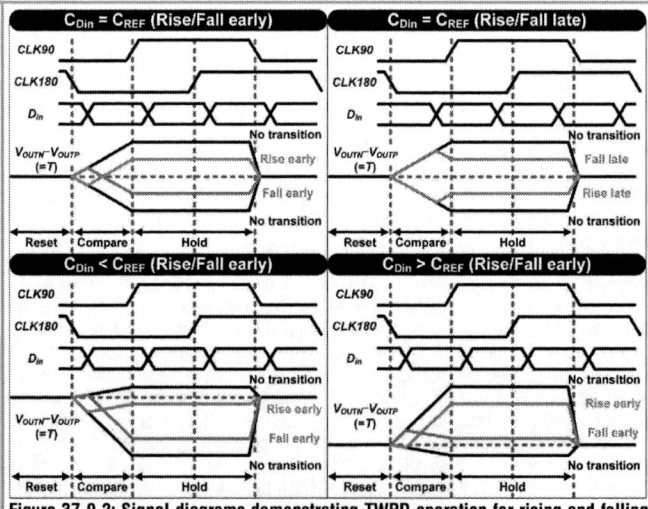

Figure 37.9.2: Signal diagrams demonstrating TWPD operation for rising and falling data transitions under three detector-offset conditions: $C_{Din} = C_{REF}$, $C_{Din} < C_{REF}$, and $C_{Din} > C_{REF}$.

DIGEST OF TECHNICAL PAPERS • 641

979-8-3315-8937-0/26 $31.00 © 2026 IEEE

ISSCC 2026 / SESSION 37 / MEMORY INTERFACE / 37.9

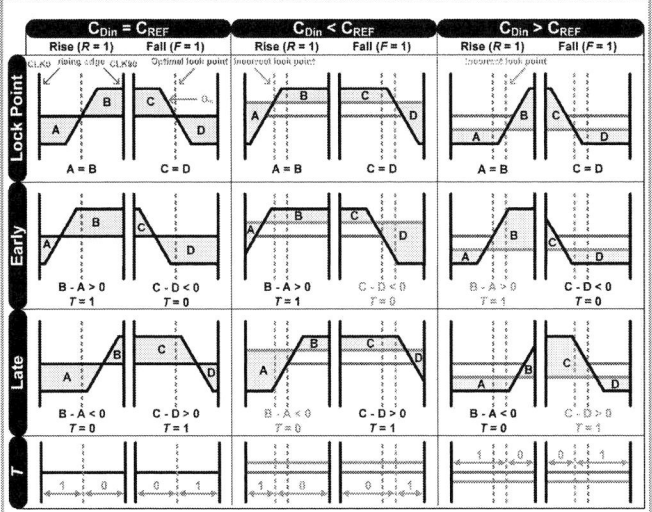

Figure 37.9.3: Background offset-calibration logic operation, using REV, FEV, and TWPD from the perspective of voltage variation.

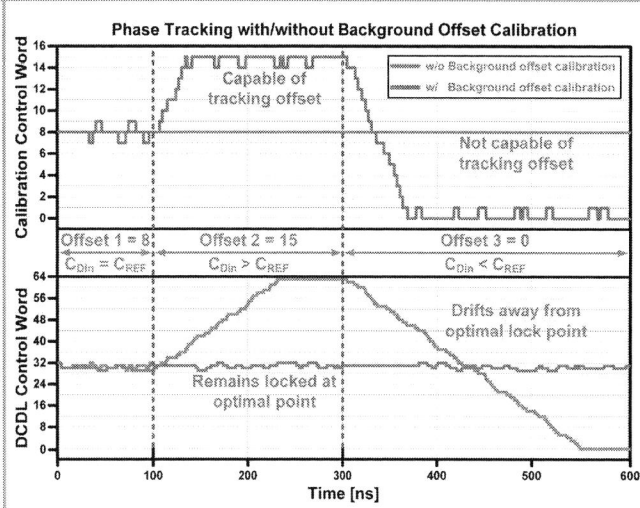

Figure 37.9.4: Simulated phase tracking behavior for an initial optimal clock phase and the effect of background offset calibration.

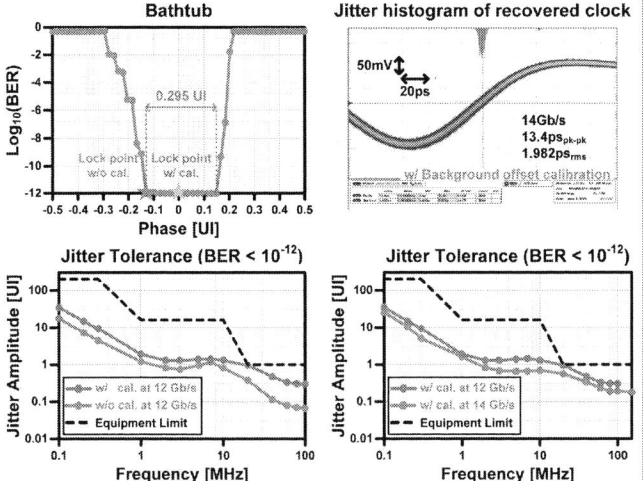

Figure 37.9.5: Measured recovered-data BER bathtub, recovered-clock jitter histogram at 14 Gb/s, and jitter tolerance plots.

	VLSI'19 [1]	TCAS-I'20 [7]	JSSC'22 [2]	JSSC'23 [3]	JSSC'23 [8]	This Work
Technology [nm]	65	65	28	28	28	28
Key Feature	DLL-based Baud-Rate	Phase-Difference Modulation	Adaptive EQ with Clock Recovery	Time-windowed LSB Decoder	Time-Domain Decision Technique	Baud-Rate Self-Tracking w/ Offset Cal.
Clocking Scheme	Quarter rate	Half rate	Quarter rate	Half rate	Quarter rate	Quarter rate
Jitter Improvement	X	O	X	X	O	O
Data Rate [Gb/s]	4.8	7.8	21	25	33	14
Clock Rate [GHz]	1.2	3.9	5.25	6.25	5.5	3.5
Area [mm²]	0.0056	0.0024	0.0437	0.008	0.00408	0.00547
Power [mW]	1.78	4.43	15.96	8.41	16.8	2.28
Power Efficiency [pJ/b]	0.370	0.568	0.760	0.336	0.51	0.163

Figure 37.9.6: Performance summary and comparison to prior memory-interface receiver.

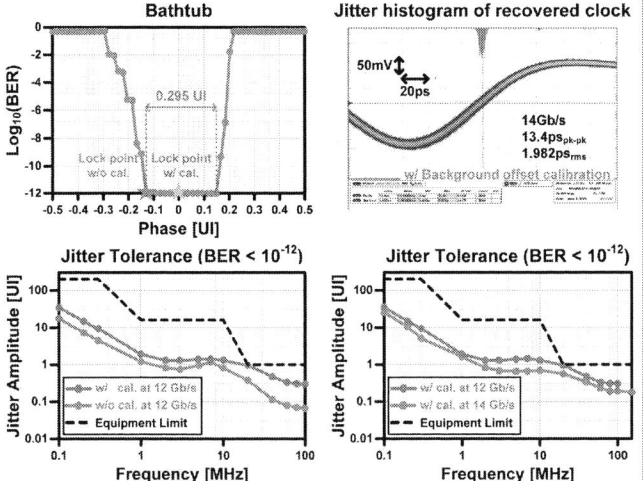

	Block	Area [µm²]	Power [µW]
A	Clocking	179	503
B	DCDL	372	406
C	Sampler	115	426
D	TWPD	17.1	2.10
E	DES	588	829
F	Digital	4200	112
	Total	5471	2278

Figure 37.9.7: Chip micrograph with table summarizing area and power breakdown for each block.

• 2026 IEEE International Solid-State Circuits Conference

ISSCC 2026

Tutorials

Forums

Evening Events

Short Course

Index of Authors

Executive Committee

Program Committee

ISSCC 2027 Call for Papers

Timetable

Conference Space Layout

ISSCC 2026 / TUTORIALS

Sunday, February 15, 8:30 AM

Tutorials

There are a total of 10 tutorials this year on 10 different topics. Each tutorial, selected through a competitive process within each subcommittee of the ISSCC, presents the basic concepts and working principles of a single topic. These tutorials are intended for non-experts, graduate students and practicing engineers who wish to explore and understand a new topic. The coordinators for the 2026 ISSCC Tutorials are (in the same order as the tutorials): Joey Sankman (T1), Yong Ki Lee (T2), Mahsa Shoaran (T3), Jae-sun Seo (T4), Andreas Suess (T5), Hajime Shibata (T6), Gunther Lehmann (T7), Wu-Hsin Chen (T8), Zeynep Deniz (T9), Dmytro Cherniak (T10).

Sudip Shekhar
ISSCC Tutorials Chair

8:30 AM
T1: Fundamentals of Energy-Efficient LDO Regulator Designs
Hyun-Sik Kim
KAIST, Daejeon, Korea

Low-dropout (LDO) regulators are essential for providing stable, ripple-free power with a compact footprint. However, the unavoidable dropout voltage required to maintain adequate regulation performance results in significant power loss, often causing LDO regulators to be perceived as inefficient powe-management elements. This tutorial investigates the fundamental trade-off between dropout voltage and regulator performance by exploring the operational principles and analytical frameworks governing this relationship. We discuss advanced design strategies that effectively overcome this traditional trade-off, enabling extremely low dropout voltages without compromising critical performance metrics. Additionally, we examine design techniques aimed at achieving ultra-low quiescent current consumption. The tutorial concludes by comparatively discussing the strengths and limitations of various state-of-the-art, energy-efficient LDO topologies.

Hyun-Sik Kim is an Associate Professor at KAIST, Daejeon, Korea. He received his Ph.D. degree in electrical engineering from KAIST in 2014. His research interests include analog integrated circuit design, with a particular focus on display drivers and power-management circuits. He has authored or co-authored 90+ peer-reviewed journal and conference papers, including 19 JSSC papers, 20 ISSCC papers, 17 VLSI Symposium papers, and 6 CICC papers. He served as a Guest Editor for IEEE SSC-L and IEEE JESTPE, and is currently serving as a Technical Program Committee (TPC) member for IEEE ISSCC and A-SSCC. Prof. Kim was Chair of the Power Management Technical Subcommittee for IEEE CICC, was Program Chair for PwrSoC 2025, and has been appointed as an IEEE SSCS Distinguished Lecturer (DL) for the term 2024–2026.

8:30 AM
T2: Fundamentals of Cryptography for Chip Designers: Classical to Post-Quantum
Thomas Pöppelmann
Infineon Technologies, Neubiberg, Germany

The advent of quantum computers poses a significant threat to digital communication systems. Now that the NIST post-quantum cryptography (PQC) standardization process has concluded, support for PQC is needed not only in high-end security systems but also in general purpose devices. During this tutorial we discuss how state-of-the-art classical cryptography works today and introduce the basics of PQC. We describe challenges faced by chip architects and IC designers when adding support for the new cryptographic schemes in silicon devices. Additionally, we examine techniques for secured implementation of PQC and how PQC might drive requirements for larger memories and faster communication interfaces in constrained devices like embedded microcontrollers.

Thomas Pöppelmann is a Senior Principal Engineer at Infineon Technologies AG. He is leading the Security Innovation team and working as Platform Security Architect for MCU, IoT, and Automotive. His main area of work is the definition of security architectures and the development of concepts for secured cryptographic modules, security standardization, and innovation projects. His research interests are security architecture, physical protection of cryptographic implementations, post-quantum cryptography, and practical lattice-based cryptography. In 2015 he obtained his PhD (Dr.-Ing.) on practical lattice-based cryptography under the supervision of Prof. Dr.-Ing. Tim Güneysu at Ruhr-University Bochum.

8:30 AM
T3: Design Techniques for Robust and Energy-Efficient Biomedical Readouts
Taekwang Jang
ETH Zurich, Zurich, Switzerland

As wearable and implantable electronics typically operate under stringent energy constraints, their energy efficiency is a primary concern. At the same time, ensuring reliable and robust acquisition of small signals in the presence of various types of noise and interference is essential. In this tutorial, I will introduce the fundamentals and recent design techniques of analog and mixed-signal circuits for energy-constrained biomedical interface systems.

Taekwang Jang received his B.S. and M.S. in EE from KAIST, and Ph.D. from the University of Michigan. In 2008-2013, he worked at Samsung Electronics. Currently, he is an associate professor at ETH Zurich and is leading the Energy-Efficient Circuits and Intelligent Systems group.

He focuses on circuits and systems for energy-constrained applications. Essential components such as an analog interface, energy harvester, PLL, and power/data converters are his primary interests. He received the SSCS New Frontier Award and ISSCC Jan Van Vessem Award for Outstanding European Paper. He is a TPC for the ISSCC, an AE for the JSSC, and an SSCS DL.

DIGEST OF TECHNICAL PAPERS • 643

979-8-3315-8937-0/26 $31.00 © 2026 IEEE

10:30 AM
T4: Fundamentals of Compute-in-Memory: From Circuits to Systems
Jaydeep Kulkarni
The University of Texas at Austin, Austin, TX

Compute-in-memory (CIM) designs are being actively researched to address the "Memory Wall" bottleneck in modern computing systems. This tutorial will begin by discussing system design considerations to tackle data-movement challenges with CIM, then focus on the core principles for performing arithmetic and logic computations using both analog and digital approaches. The design trade-offs, prospects, and challenges of future CIM applications will also be covered.

Jaydeep Kulkarni is an associate professor and a fellow of the Silicon Labs endowed chair at the University of Texas at Austin. His research focuses broadly on IC design. He has been a distinguished lecturer for IEEE SSCS, CAS, and ED societies and a TPC/AE at various conferences and journals.

10:30 AM
T5: Fundamentals of Image Sensors: From Photon to Image
Min-Woong Seo
Samsung Electronics, Hwaseong, Korea

This tutorial introduces the main signal path of an image sensor, from light capture to readout, conversion, and image output. It highlights how design choices impact performance, using examples such as rolling vs. global shutter architectures and column vs. pixel-parallel ADCs. The session balances technical depth with broader context, aiming to engage not only image sensor specialists and enthusiasts but also engineers from other integrated circuit fields. By connecting fundamental concepts to practical design considerations, the tutorial provides a clear perspective on how sensor architecture influences system behavior and imaging outcomes.

Min-Woong Seo received his Ph.D. degree from Shizuoka University, Hamamatsu, Japan, in 2012, with a dissertation focused on low-noise, HDR CMOS image sensors using high-performance ADCs. In 2018, he joined the Semiconductor R&D Center (SRDC) at Samsung Electronics, Hwaseong, Korea. He has been serving on the TPC of EI since 2018 and of the ISSCC since 2024. He has also been a steering committee member of the Korean Image Sensor Society (KISS) since 2024, a board member of the International Image Sensor Society (IISS) since 2025, and an advisory board member of Image Sensors (IS) Asia since 2025.

10:30 AM
T6: Principles and Practices of High-Speed Digital-to-Analog Converter Design
Shiyu Su
University of Waterloo, Waterloo, Canada

High-speed DACs (>GS/s) are crucial not only as performance-limiting building blocks for wideband ADCs but also as key enablers for various emerging communication and computing systems, such as software-defined/DSP-based transceivers and quantum computers. This tutorial offers a comprehensive introduction to the design of high-speed DACs, beginning with system context and fundamental concepts, and progressing to architectural and circuit-level design trade-offs. Additionally, we will delve into advanced techniques for enhancing bandwidth and linearity, including time/frequency interleaving, operation in higher Nyquist zones, mismatch error shaping/mapping/scrambling, and digital predistortion. Finally, we will cover essential layout techniques and strategies.

Shiyu Su is currently an Assistant Professor at the University of Waterloo. He received his Ph.D. degree from the University of Southern California. He is a Senior Member of the IEEE and serves as a Technical Program Committee Member for ISSCC and DAC. He was the recipient of the IEEE SSCS Predoctoral Achievement Award for 2017–2018 and a co-recipient of the Best Student Paper Award (First Place) at the RFIC 2022. He was also a Ming Hsieh Institute Scholar from 2019 to 2020.

1:30 PM
T7: Embedded Memory and Logic Circuit Design in Technologies Beyond FinFET
Zheng Guo
Intel, Portland, OR

With the advent of Gate-All-Around (GAA) technologies, the discretization of transistor width known to FinFET technologies is no longer required. This enables additional flexibility to optimize embedded memory and logic designs through extensive DTCO. Furthermore, power supply from the backside is becoming available and opens additional design-optimization opportunities. This tutorial will focus on the opportunities and challenges in embedded memory and logic circuit design in technologies beyond FinFET.

Zheng Guo received his B.S. degree in Computer Engineering from UIUC in 2003, and M.S. and Ph.D. degrees in electrical engineering from the UC Berkeley in 2005 and 2009, respectively. He joined Intel in 2010, where he is currently a Sr. Principal Engineer and leads embedded-memory-technology/design pathfinding and development.

ISSCC 2026 / TUTORIALS

Sunday, February 15, 1:30 PM

1:30 PM
T8: Interference Mitigation Techniques in Wireless Communication Systems
Negar Reiskarimian
Massachusetts Institute of Technology, Cambridge, MA

The pursuit of ubiquitous connectivity and the evolution of wireless communication technologies such as 5G and beyond and IoT have increased the number of connected devices rapidly and have spurred a growing demand for front-end designs that can operate across a wide frequency spectrum for various communication standards. Furthermore, the growing number of devices creates increasing interference in the environment, from unknown in-band and out-of-band blockers to self-interference in full-duplex and radar systems, and spatial blockers in phased-array and MIMO systems. In this tutorial, a review of circuit and system design techniques for mitigating various types of interference is presented and state-of-the-art performance is summarized.

Negar Reiskarimian received the B.S. and M.S. degrees in electrical engineering from Sharif University of Technology, Iran, in 2011 and 2013, and the M.Phil. and Ph.D. degrees in electrical engineering from Columbia University, New York, NY, USA, in 2017 and 2020. She is currently an Associate Professor with the Department of Electrical Engineering and Computer Science, Massachusetts Institute of Technology, Cambridge, MA, USA. Her research interests include RF and millimeter-wave circuits and systems for a variety of applications. She is currently serving on the TPCs of ISSCC, RFIC and IMS. She was co-recipient of several awards, including the ISSCC 2024 Jack Kilby Outstanding Student Paper Award and 2024 RFIC Best Student Paper Award (1st place).

3:30 PM
T9: Clocking and CDR Techniques for High-Performance Wireline Transceiver
Wei-Zen Chen
National Yang Ming Chiao Tung University, Hsinchu, Taiwan

This tutorial will begin with an overview of jitter budgets for high-performance, DSP-based transceivers, followed by an introduction to the design of high-precision clock generators and CDR circuits. Topics will include low-noise phase-locked loops (PLLs), clock distribution networks, and calibration techniques for clocks and time- interleaved analog-to-digital converters (ADCs). The tutorial will also highlight recent advancements on baud-rate phase detectors and cooperative techniques between CDR and equalizers.

Wei-Zen Chen (M'00–SM'11) is a professor at the Institute of Electronics and the Department of Electronics and Electrical Engineering at National Yang Ming Chiao Tung University (NYCU). He has served as an Associate Editor of IEEE Solid-State Circuits Letters, a Guest Editor of the IEEE JSSC, IEEE OJ-SSCS, and a Technical Program Committee (TPC) member for the IEEE Asian Solid-State Circuits Conference (A-SSCC) and the IEEE Custom Integrated Circuits Conference (CICC). Currently, he is the APAC Chair of the ISSCC regional committee. Professor Chen has published over 100 peer-reviewed journal and conference papers. His research interests include mixed-signal integrated circuits for wireless and wireline communication systems, with a focus on high-speed interconnects, optical communication, and radar sensing systems

3:30 PM
T10: Doherty Power Amplifier: Fundamentals and Recent Advances
Taiyun Chi
Rice University, Houston, TX

Doherty PA architecture has gained growing interest due to its superior efficiency when amplifying modulated signals. This tutorial will begin with Doherty PA fundamentals, along with a few intuitive insights often overlooked in textbooks and research articles. We will then highlight several key ingredients in achieving first-pass Doherty PA design success. Finally, we will delve into recent demonstrations featuring bandwidth extension, deep back-off efficiency enhancement, and size reduction.

Taiyun Chi is an Associate Professor at Rice University. His group received the 2024 ISSCC Lewis Winner Outstanding Paper Award, 2021 CICC Best Student Paper Award, and multiple Best Paper Award Finalists at IMS (2024, 2021) and RFIC (2023, 2022). He was also a recipient of the NSF CAREER Award.

DIGEST OF TECHNICAL PAPERS • 645

979-8-3315-8937-0/26 $31.00 © 2026 IEEE

Forum 1:
Power Efficient Circuits and Systems for Next-Gen Agentic AI and Robotics

Organizers:

Co-Organizers: **Patrik Arno,** *STMicroelectronics, Grenoble, France*
Mahmut Ersin Sinangil, *Nvidia, Santa Clara, CA*
Eric Wang, *TSMC, Hsinchu, Taiwan*
Sugako Otani, *Renesas Electronics, Tokyo, Japan*
Santosh Ghosh, *Nvidia, Hillsboro, OR*

Champions: **Tanay Karnik,** *Intel, Hillsboro, OR*
Makoto Ikeda, *University of Tokyo, Tokyo, Japan*

Jun-Seok Park
S.LSI, Samsung Electronics, Hwaseong, Korea

Frank Prämaßing
Infineon Technologies Austria AG, Villach, Austria

The evolution of AI is now moving toward a new frontier—Agentic AI and Physical AI. Agentic AI represents a shift toward goal-directed, autonomous behavior, while Physical AI refers to embodied systems that interact directly with the physical world. Accordingly, emerging trends in robotics are revolutionizing industrial automation, digitalization, and sustainability. This shift is not merely a transformation in software paradigms but also demands fundamental innovations in hardware design and semiconductor architecture. Real-time computation for agent-based systems, memory-compute integration, and interfaces between sensors, actuators, and edge computing are becoming critical areas that pose new challenges and opportunities for next-generation chip design. This is a forum to hear expert opinions on the crucial aspects of robotics and on the efforts and breakthroughs at the cutting edge of the industry from a circuit and system perspective. We discuss a comprehensive trend of pioneering architectures, systems, and circuits designed to address the power- performance efficiency challenges associated with Agentic and Physical AI.

Robotics and Physical AI: The Next Frontier
Kamesh Medepalli, Infineon, San Jose, CA

Robotics and Physical AI will create the next Industrial revolution, transforming the quality of our lives fundamentally with Humanoids and Autonomous Mobile Robots. In the first half of this talk, we explain why it is the Next Frontier for our Industry, backing it up with mega trends, advancements in AI and simulations and compute power. The second half of the talk focuses on WHAT it means in opening new semiconductor opportunities and innovations required in Sensors, Connectivity (Wired and Wireless), Smart Actuators, Power etc. as well including Security, Safety and Software as key System pillars. We also highlight the importance of Digital Twin based Simulations to close the Sim2Real gap and advance Physical AI algorithms in Humanoid Robots.

Kamesh Medepalli is currently SVP and Global Head of Systems, Research and Innovation for the Connected Secure Systems (CSS) Division of Infineon. He's also on Infineon's Technical Advisory Board, responsible for leading the long-term technology vision for Robotics and Physical AI. Prior to that, he was the CTO and co-founder of the IoT Compute and Wireless (ICW) Business Line at Infineon through Infineon/Cypress/Broadcom acquisitions. He has 25+ years of experience in Systems research and product development spanning Wireless, AI/ML, Networking, Software and System Architecture and held key roles at Bell Labs, Bell Core, Beceem and Broadcom. Kamesh received his PhD in Electrical Engineering from Stanford University where he also served as a Consulting Professor. He has more than 100 patents issued or pending and publications in the general areas of Communications & Compute.

From VLSI through Algorithms: Optimizing the Computing Stack for Agentic AI
Brucek Khailany, NVIDIA, Austin, TX

Agentic AI is rapidly transforming architecture, VLSI implementation, and circuit design tradeoffs for HW, with a need to innovate on domain-specific architectural features and SW/HW co-design for energy efficiency. In this talk, we highlight NVIDIA's technology innovations for AI, from circuits to numerical algorithms to the entire datacenter. We also present research into AI accelerator micro-architecture, SW/HW co-design for quantization, and recent testchips targeting Physical AI.

Brucek Khailany is Sr Director of Research for Accelerators & VLSI at NVIDIA, leading research in AI accelerators, VLSI design methodologies, EDA, and quantum computing. Previously, he was a Co-Founder at Stream Processors Inc. and received a PhD from Stanford Univ and a BSE from the Univ of Michigan.

ISSCC 2026 / February 15, 2026 / 8:00 AM

Agentic AI on Mobile Devices
Chuo-Ling Chang, MediaTek, Hsinchu, Taiwan

Agentic AI, characterized by goal-directed, autonomous behavior, is set to transform mobile computing. However, deploying sophisticated AI models for agentic decision-making on resource-constrained mobile devices presents significant challenges. This presentation will discuss system considerations and innovations for on-device Agentic AI that is personal, prompt, proactive, private, and integrating text, image, speech, and audio modalities—paving the way for truly intelligent mobile experiences.

Chuo-Ling Chang is Assistant General Manager of the Computing and AI Technology Group at MediaTek Inc. Previously, he oversaw engineering for Google's AI Edge Infrastructure. He earned his PhD from Stanford University's Information Systems Laboratory and his BSEE from National Taiwan University.

Revolutionizing Long-Term Memory in AI: New Horizons with High-Capacity and High-Speed Storage
Jun Deguchi, Kioxia, Yokohama, Japan

As AI evolves, memory functionality becomes increasingly essential. The mainstream approach to AI memory management is "extract then store," where potentially useful information is filtered and abstracted prior to being stored. We propose an alternative paradigm: "store then on-demand extract," which comprehensively preserves memory and context, enabling targeted information extraction when needed. This talk introduces the advantages and challenges of the proposed paradigm, and future prospects.

Jun Deguchi received his Ph.D. in 2006. He serves as the group manager of a research team working on AI-related technologies, spanning algorithm/software development to system/hardware design. He has served as a Subcommittee/TPC Vice Chair for A-SSCC and as a TPC Member/Far East Chair for ISSCC.

Architecting Silicon for Intelligence and Autonomy in Embodied AI
Arijit Raychowdhury, Georgia Institute of Technology, Atlanta, GA

This talk presents a cross-layer approach to silicon design for embodied AI, enabling real-time learning, perception, and decision-making under tight power constraints. It highlights compute-in-memory accelerators, neuromorphic designs, and memory-centric SoCs with hybrid RRAM/SRAM arrays. Case studies include reconfigurable dataflow architectures and bio-mimetic chips, illustrating how silicon innovation can deliver the efficiency and adaptability needed for embodied intelligence.

Arijit Raychowdhury is the Steve W. Chaddick School Chair and Professor of ECE at Georgia Tech. His work spans energy-efficient VLSI, hardware accelerators, and mixed-signal ICs. He leads the DARPA/SRC CoCoSys Center and is an IEEE Fellow with 350+ papers and 30+ patents.

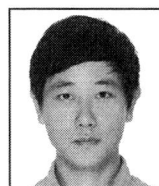

High-Performance DC-DC Conversion for Next-Gen Computing: From Cloud to Edge
Wanyuan Qu, Zhejiang University, Hangzhou, China

With the rapid development of generative AI and AI-powered robotics, the power demand of high-performance processors has grown explosively. Aiming at the demand for high-performance power management technology, this talk will report and analyze the main technical solutions in this field in industry and academia, summarize the advantages and bottlenecks of existing technologies, and explore the technological trends and development of high-density power management chips.

Wanyuan Qu received his Ph.D. degree from Korea Advanced Institute of Science and Technology, South Korea. In 2017, he joined Zhejiang University where he is now a Professor. He has been granted with nine U.S. patents and his research include high-performance dc–dc converters and amplifiers.

On the Move - Semiconductor Technologies for Motion Control in Humanoid Robotics
Mark Nils Muenzer, Infineon, Munich, Germany

Motion will distinguish not only the ability of humanoids to fulfill tasks, but define how it interacts with the environment. In human robotics it is not sufficient to perform a movement with the desired speed and direction, but smart actuators must be synchronized, precise, energy efficient, safe and secure. This unique set of requirements drives semiconductor innovation in multiple areas. The presentation therefore covers the complete motor control loop and the involved semiconductors.

Mark Muenzer has worked for more than 25 years in the semiconductor industry. In his current role as Fellow for Motor Control Solutions at Infineon he is responsible for identifying future motor control trends and pushing the required semiconductor innovations forward.

Hardware Security Challenges for Agentic AI
Ingrid Verbauwhede, KU Leuven, Leuven, Belgium

The deployment of AI and Agentic AI comes with multiple security and privacy challenges. One of the essential requirements is the protection of the data and the models, and being able to operate on the private data with the private models without breaking the security guarantees. To support "confidential computing" there are fundamentally two options. One is the use of a trusted execution environment (TEE) where computations are isolated at the hardware level. We will focus on intrinsic design conflicts to combine performance and scalability with security. A second option is fully homomorphic encryption (FHE). In this setup, the processor remains oblivious to the actual data being processed because the data and/or models are encrypted. We will focus on the implementation challenges of the FHE calculations in hardware.

Ingrid Verbauwhede is a professor in the research group COSIC at the KU Leuven in Belgium and co-founder of the start-up company Belfort. At COSIC, she leads the hardware security team. Her main expertise includes system and architecture design, embedded system, ASIC and FPGA design to make electronic circuits more secure. She is a fellow of IEEE and of IACR, and a member of the Royal Academy of Belgium for Sciences and Arts. She received two advanced ERC grants in 2016 and 2021 respectively. Her most recent awards are the 2023 IEEE Donald O. Pederson Solid-State Circuits Award and the 2024 EDAA Achievement Award. In 2025, Ingrid Verbauwhede received the FWO Excellence Prize Dr. A. De Leeuw-Damry-Bourlart in Applied Sciences. Ingrid Verbauwhede is also a co-founder of the start-up company Belfort, which is based on the results of her advanced ERC grants.

DIGEST OF TECHNICAL PAPERS • 647

979-8-3315-8937-0/26 $31.00 © 2026 IEEE

Forum 2:
Electrical and Optical Links Towards 400G+ Connectivity

Organizers:

Masum Hossain
Carleton University,
Ottawa, Canada

Tamer Ali
Mediatek,
Irvine, CA

Co-Organizers: **Kenny Hsieh,** *TSMC, Hsinchu, Taiwan*
Jay Im, *AMD, San Jose, CA*

Champions: **Wei-Zen Chen,** *National Yang Ming Chiao*
Tung University, Hsinchu, Taiwan
Bill Redman-White, *Top-IC, Southampton,*
United Kingdom

This forum aims to bring the latest developments towards achieving high bandwidth density through electrical and optical connectivity. The forum will include standards, state-of-the-art electrical and optical solutions and applications. The forum will also present the development of the technology platforms by foundries to showcase their latest photonic technology and packaging solutions.

Data Centre Interconnect: 400G, 800G, and Terabit Pluggable Optics
Andy Bechtolsheim, Arista, Santa Clara, CA

The talk will discuss the transition to 448 Gbps electrical signaling and its impact on next-generation AI interconnect architectures. It highlights limitations in PCB loss, crosstalk, channel impairments, equalization requirements, and SNR constraints, and will review Co-Packaged Copper (CPC) and Co-Packaged Optics (CPO) architectures for reducing electrical losses and improving signal integrity.

Andreas "Andy" Bechtolsheim is Chairman, Chief Development Officer and Co-Founder of Arista Networks. Previously, Andy was a Co-Founder and Chief System Architect at Sun Microsystems, responsible for next generation server, storage, and network architectures. Andy received a M.S. in Computer Engineering from Carnegie Mellon University in 1976 and was a doctoral student in Computer Engineering at Stanford University from 1977 to 1982. He has been honored with a Fulbright scholarship, the Stanford Entrepreneur Company of the year award, the Smithsonian Leadership Award for Innovation, the EY 2015 US Entrepreneur of the Year Award (together with Jayshree Ullal) and is a member of the National Academy of Engineering.

Analog and Mixed-Signal Approaches for Low-Complexity Coherent Optical Links
Pavan Hanumolu, University of Illinois (UIUC), Urbana, IL

As data rates surpass 400 Gb/s per wavelength, IMDD systems face limits from optical impairments and bandwidth constraints. Coherent links offer superior spectral efficiency but depend on power-hungry DSP. This talk presents low-complexity coherent architectures using analog and mixed-signal techniques—such as carrier phase recovery and analog-domain phase and polarization tracking—to reduce DSP overhead and enable energy-efficient coherent receivers that extend optical interconnect scalability beyond IMDD.

Pavan Hanumolu is a professor in the Department of Electrical and Computer Engineering at the University of Illinois, Urbana-Champaign. His research interests include integrated circuit implementation of wireline communication systems. He served as the Editor-in-Chief of the IEEE Journal of Solid-State Circuits and is an IEEE Fellow.

State-of-the-Art 200+Gbps Electrical-Optical Interconnects
E-Hung Chen, MediaTek, Hsinchu, Taiwan

In the AI era, high-speed interconnects are critical to sustaining the exponential growth in computing performance. This talk will begin with an overview of the evolution of high-speed serial links, highlighting the trends of electrical links beyond 200 Gb/s. Key design challenges and potential solutions for next-generation interconnects will be discussed. The talk will also cover the latest advancements in silicon photonics, which offer several benefits compared to electrical solutions.

E-Hung Chen received the B.S. degree from National Taiwan University, and the M.S. and Ph.D. degrees from UCLA. He was with Rambus, USA from 2011 to 2014. In 2014, he joined MediaTek, Taiwan, where he is a senior manager and leads the design of high-speed interconnects for data center networking.

ISSCC 2026 / February 15, 2026 / 8:00 AM

Latency Optimized Silicon Photonic I/O
Ganesh Balamurugan, Celestial AI, Santa Clara, CA

High-performance AI systems demand interconnects that overcome the bandwidth, power, and reach limitations of electrical I/O. This talk presents a thermally robust photonic fabric optimized for power and latency. We highlight how in-package optical transceivers enable resource disaggregation, delivering enhanced system performance for AI/ML workloads.

Ganesh Balamurugan (PhD, University of Illinois at Urbana-Champaign, 2004) has over 20 years of experience in high-speed wireline communication circuits and systems. Previously a Principal Engineer at Intel Labs, he is currently a Fellow at Celestial AI working on optical compute interconnects.

Emerging Low-Latency Optical Connectivity
Nikola Nedovic, Nvidia, Santa Clara, CA

The rapid growth of AI workload has accelerated proliferation of optical I/O in data centers and extended its integration into scale-up computing nodes. This presentation reviews the requirements imposed by the emerging applications on these optical interconnects, including bandwidth, energy, and latency, and how they drive the development towards co-packaged optics (CPO) and wide-and-slow architectures.

Nikola Nedovic is a principal research scientist at NVIDIA Corp. where he works on system and circuit design for high-speed links. His research interests include optical and electrical wireline communications, from device modeling and CMOS circuit design, to system design and modeling.

400Gb/s Coherent Optics
Xin Yin, IMEC, Leuven, Belgium

AI's explosive growth is driving optical transceivers beyond the limits of both IMDD and coherent technologies, pushing toward 400G+ per lane and beyond. This talk explores advanced electronics (TIAs, drivers, CDRs, AMUX/ADeMUX, ADC/DACs) and photonic devices enabling next-gen designs. We will examine coherent optics and low-cost coherent-lite schemes for high-capacity links, and highlight how EPIC co-design and co-integration deliver scalable, energy-efficient optical interconnects for the future.

Xin (Scott) Yin, Scientific Director at imec Belgium and professor at Ghent University, is a leading expert in electronic-photonic integrated circuits (EPICs). He has authored 250+ publications in this field and received the GreenTouch 1000x Award for research leadership and innovation.

Scaling AI with Light: 3D Photonic Solutions for Next-Gen Connectivity
Ritesh Jain, Light matter, Mountain View, CA

Silicon photonics has emerged as a key technology to break through the limits of traditional electrical interconnects to meet the growing demands of modern computing and AI.

This presentation will describe the role of 3D photonic solutions, such as PassageTM, to enable higher energy efficiency and bandwidth for advanced AI workloads. It will cover the development of next-generation photonic interconnect solutions, the engineering and packaging challenges and the approaches required for broader adoption. The discussion will include photonic chip architectures, integration strategies, and ecosystem considerations required to realize these solutions.

The session will highlight how 3D photonics and its implementation is enabling the roadmap to a scalable end to end 400G+ connectivity and transforming AI hardware capabilities beyond Moore's law.

Ritesh Jain is Senior Vice President, Engineering & Operations at Lightmatter leading cross-functional engineering organizations responsible for driving innovation and delivering products. Previously as VP at Intel, Ritesh led hardware development for data centers. He holds a master's in semiconductor packaging and during the two plus decades of experience in the semiconductor industry, he drove several generations of data center products into high volume deployments while driving major technology transitions and leading global engineering teams.

Silicon Photonics Platform for Next Generation Data Communication Technologies
Chih-Tsung Shih, TSMC, Hsinchu, Taiwan

As copper and wireline Serdes hits a bandwidth wall beyond 400G, Silicon Photonics becomes a strong candidate for next generation wireline communication. This talk covers a 3D integrated silicon photonics platform tailored to meet the increasing demands of next-generation data communication applications. It does so by integrating a PIC and EIC in compute or switch chips. This talk presents an overview of the platform, key integration challenges, and the performance of key photonic devices, as well as roadmap for technology development in the next few years.

Chih Tsung Shih is a Technical Manager in the Silicon Photonics Devices team at tsmc, where he focuses on advancing integrated photonic devices and PDK to enable the next generation of high-speed data communication technologies.

Prior to joining TSMC, he joined the Industrial Technology research Institute (ITRI), and his work was on the design and package of optical passive devices for long-haul and metro tele-communication from 2001 to 2010.

In 2010, he received the Ph.D degree in Institute of Photonics Technologies (IPT) from NTHU, where his research focused on design and analysis of Silicon-based Microring.

DIGEST OF TECHNICAL PAPERS • 649

979-8-3315-8937-0/26 $31.00 © 2026 IEEE

ISSCC 2026 / FORUM 3

Forum 3:
Powering the Future of AI, HPC, and Chiplet Architectures: From Dies to Package and Rack

Organizers:

Xin Zhang
IBM T. J. Watson Research Center
Yorktown Heights, NY

Kazuki Fukuoka
Renesas Electronics
Tokyo, Japan

Co-Organizers: **Ping-Hsuan Hsieh,** *National Tsing Hua University, Hsinchu, Taiwan*
Monodeep Kar, *IBM T. J. Watson Research Center, Yorktown Heights, NY*
Srividhya Venkataraman, *AMD, Santa Clara, CA*
Rinkle Jain, *Nvidia, Santa Clara, CA*
Nicolas Butzen, *Intel, Hillsboro, OR*

Champions: **Meng-Fan Chang,** *National Tsing Hua University & TSMC, Hsinchu, Taiwan*
Makoto Nagata, *Kobe University, Kobe, Japan*

As AI, HPC, and chiplet-based architectures push the boundaries of performance and integration, intelligent and efficient power management becomes more critical than ever. This forum will explore state-of-the-art power delivery techniques across the full compute stack—from die and package to system and rack levels.

Topics will include power conversion and integrity trends in data-center processors, package-integrated voltage regulators, energy-efficient control strategies, and power converter topologies. Advances in integrated power devices, package-embedded passive technologies, and power integrity optimization for high-bandwidth DRAM will also be discussed. The forum will address emerging challenges posed by AI, HPC and chiplet architectures, as well as the impact of these innovations on system efficiency, scalability, and sustainability.

Integrated Voltage Regulator Solutions to enable 5kW GPUs
Kaladhar Radhakrishnan, Intel, Chandler, AZ

AI scaling laws are driving a rapid increase in datacenter GPU power, which is projected to surpass 5KW before the end of the decade. Evolutionary MBVR based solutions are unlikely to support the current density and bandwidth demands of these high-power systems. In this talk, we will explore the development of IVR solutions that deliver high current density and bandwidth. We will look at different IVR architectures, and the passive technologies required to enable them.

Kaladhar Radhakrishnan joined Intel in 2000 after he received his Ph.D. in Electrical Engineering from UIUC. He is an Intel Fellow and a Power Delivery Architect with the Technology Development group at Intel. His areas of expertise are IVRs, advanced packaging and passives technologies.

Power Delivery Trends and Demands in Data Center AI Processors
Houle Gan, Google, Sunnyvale, CA

In this presentation we will discuss the latest trend in demand and technology directions for Google's ultra-high power data center processors, including topics such as vertical power delivery and its high-volume adoption, power integrity design of highly complex physical power delivery network, and chip-to-datacenter power management.

Houle Gan received the Ph.D. degree in Electrical and Computer Engineering from Purdue University, West Lafayette, IN, USA, in 2010. He is currently a Tech Lead Manager at Google, where he leads the power delivery and power management solutions of Google data center platforms.

Energy Efficiency and Power Management Techniques for AI Accelerators
Luca Benini, Università di Bologna and ETH Zürich, Zurich, Switzerland

The goal is to provide an in-depth review on advanced techniques for boosting energy efficiency and managing power in AI accelerators. The presentation will cover the following topics: (1) boosting efficiency for AI workloads - leveraging specialization, (2) managing idleness and heterogeneity in accelerated systems - multi-stage power management (3) Using AI for managing AI - approaches for data-driven, AI-based power management.

Luca Benini holds the chair of digital Circuits and systems at ETHZ and is Full Professor at the Università di Bologna. His research interests are in energy-efficient parallel computing systems, smart sensing micro-systems and machine learning hardware. He is a Fellow of the IEEE and of the ACM

3D Vertical Power Delivery for AI and Chiplet Systems

Yan Lu, Tsinghua University, Beijing, China

Current density in 3D-integrated high-performance computing/AI systems is extremely high at sub-1V supply. To efficiently and swiftly bridge the gap between 48-V bus voltage and 0.8-V load supply, switched-capacitor converters, resonant converters, and hybrid converters have been heavily investigated in recent years. Meanwhile, advanced 2.5D/3D packaging technologies enable more design flexibility and topology innovations for PoL vertical power delivery. In this talk, we will analyze the challenges and introduce the state-of-the-art DC-DC design examples for the "last-millimeter" vertical power delivery.

Yan Lu received his PhD degree in Electronic and Computer Engineering from HKUST in 2013. From 2014 to 2024, he was with the Institute of Microelectronics, University of Macau. In 2024, He joined the Department of Electronic Engineering, Tsinghua University, Beijing, China, as a Full Professor. His research interests include high-density power converters, integrated voltage regulators, wireless power transfer circuits and systems.

Progress and Scaling in Integrated Power Devices

Sameer Pendharkar, TI, Dallas, TX

Efficient power conversion is essential for the increasing demands of enterprise, automotive, and industrial systems. Scaling of integrated power devices has improved power density and efficiency to support Moore's law of increasing logic density. Silicon remains the primary power technology, while wide band-gap materials enable further enhancement. This talk reviews the scaling of integrated power device architectures, highlighting how silicon and gallium nitride technologies address the challenges of integrating power, analog, and logic in SoC and SiP solutions.

Sameer Pendharkar is Vice President of Process Technology Development and Senior Fellow (emeritus) at Texas Instruments. He joined TI in 1996 and lead the development and execution of technology and foundational IP roadmaps across TI businesses. He has published over 100 technical papers, holds more than 300 U.S. patents, and is a senior member of IEEE.

Integrated Magnetics for Datacenter Applications

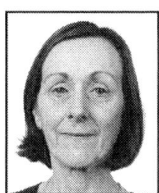

Maeve Duffy, University of Galway, Galway, Ireland

Demands for increasing power and fast response has always presented a challenge to magnetic component design for high performance computing power supplies. The emergence of AI/ML has amplified this challenge, driving new developments in magnetic materials and miniaturized inductor structures. This talk will focus on state of art developments in PCB and silicon integrated magnetics, and their application in new circuit topologies to enable high efficiency power delivery to AI computing loads.

Maeve Duffy completed the PhD in Electronic Engineering in 1992. After four years as a researcher in the Tyndall National Institute, she secured an academic position in the University of Galway where she has continued her research in magnetic component design for power electronics applications.

Packaging-Aware Design and Optimization of HBM Power Delivery in AI Platforms

Daihyun Lim, Samsung, San Jose, CA

High Bandwidth Memory (HBM) systems for AI accelerators demand stringent power integrity as bandwidth and transient currents increase. This work presents a hierarchical PDN and HBM-aware optimization to maintain a well-controlled impedance profile and suppress SSN. A co-design flow using 3D packaging-aware SI/PI simulations ensures stable operation under dynamic rail noise. Measurements confirm sub-milliohm targets, fast transient response, and robust HBM performance for next-generation AI platforms.

Daihyun Lim received the B.S. degree in Electrical Engineering from Seoul National University in 1999 and the S.M. and Ph.D. degrees in Electrical Engineering and Computer Science from MIT in 2004 and 2008. He joined Samsung Electronics in 2023 as a Master (VP of Technology) leading HBM I/O design. He worked at IBM ASIC group (2008–2017) and Nokia Network Infrastructure (2017–2022) as DMTS. He received the 2017 VLSI Symposium Most frequently cited paper award and interested in memory interfaces, signal/power integrity, and optical transceivers.

Powering up Heterogeneously Integrated Multi-Die / Multi-Chiplet Systems

Farhana Sheikh, Intel, Hillsboro, OR

This talk reviews power delivery challenges, emerging solutions, and co-optimization techniques in 2.5D/3D heterogeneously integrated chiplet-based systems. Topics include power distribution across heterogeneous modules, optimization of PDNs across chiplet boundaries, thermal management due to non-uniform power density, on-package voltage regulation, and substrate-centric power routing. Role of TSVs, micro-bumps, and interposers is explored in maintaining power integrity while supporting inter-die communication.

Farhana Sheikh is a Principal Engineer at Intel Corporation. Previously, she co-led Altera's Technology Pathfinding organization. Her research focuses on 2.5D/3D heterogeneous integration, die-to-die interfaces and 3D HI AI/wireless systems. She holds 22 patents, co-authored 60 papers, and won multiple best paper awards. She is an IEEE Fellow.

ISSCC 2026 / FORUM 4

Forum 4:
The Race for 6G FR3 (7-24GHz): From Network Deployment to System Integration and Breakthrough Technology

Organizers:

Co-Organizers: **Chi-Hang (Ivor) Chan,** *University of Macau, Taipa, Macau*
Konstantinos Manetakis, *CSEM, Neuchâtel, Switzerland*
Negar Reiskarimian, *Massachusetts Institute of Technology, Cambridge, MA*
Yun Wang, *Fudan University, Shanghai, China*

Champions: **Danielle Griffith,** *Texas Instruments, Dallas, TX*
Matteo Bassi, *Infineon Technologies AG, Villach, Austria*

Wu-Hsin Chen
Qualcomm,
San Diego, CA

Yuanjin Zheng
Nanyang Technological University,
Singapore, Singapore

As we are finally entering the 6G era, this forum will provide an overview of the challenges and opportunities for the 6G FR3 (7 to 24 GHz). It will delve into system-level considerations, front-end module architectures, spectrum sharing and interference cancellation, offering insights into the next generation of cellular technology. Additionally, the forum will address design techniques for FR3, such as direct ADC sampling, PA linearity and efficiency improvement with Doherty architecture or digital predistortion. It will also discuss process technologies for highly efficient FEMs across different parts of the FR3 bands, evaluating the pros and cons of candidate technologies like GaN-Si. Other critical emerging technologies, such as joined communication and sensing, and ambient IoT, will also be covered. Finally, the use of AI/ML to enhance FR3 network performance will also be explored.

System Consideration and Coexistence with other Incumbent Services
Shawn Tsai, Mediatek, San Diego, CA

6G system is expected to coexist with various services in the same or adjacent frequency across different bands, including FR3. Those services can be scientific use, unlicensed operation, satellite, radar, or combinations thereof. For service deconfliction, 6G system shall consider technology enablers encompassing static/semi-static/near real-time spectrum allocations, radio resource multiplexing, medium access control, integrated functions. Pros and cons are analyzed with outlook into the incoming 6G.

Shawn Tsai received his Ph.D. in Electrical Engineering from Purdue in 2000. From 2000 to 2020, he was with Ericsson for CDMA and IEEE 802.16 standards, Huawei's 4G Research, and Qualcomm Modem System Engineering. Since Feb 2020, he has been a Director at MediatTek working on 5G Advanced and 6G research.

FR3 FEM Architectures: Heterogeneous Integration from Antenna to RF Circuitry
Li Liu, Qualcomm, San Diego, CA

This talk will first discuss the spectrum allocation for FR3 band (7-24 GHz), system requirement, as well as coex requirement. It will then discuss the challenges of the FEM to support such wide bandwidth, especially in the UE devices. Recent progress of the active and filter components, as well as substrate technology, and thermal mitigation in the FEM will be reviewed. The choice of the front end module architecture can be very different for bands below and above 14GHz.

Li Liu is a Senior Director of Technology at Qualcomm, leading transceiver designs. He worked at Triquint (2001-2004) prior to Qualcomm. He got his B.Eng from University of Science and Technology of China in 1996 and M.Eng from National University of Singapore in 1999. He holds 49 U.S. patents.

Spectrum Sensing, Sharing and Interference Cancellation Technologies
Arun Natarajan, Yale University, New Haven, CT

The FR3 spectrum (7–24 GHz) offers new opportunities for 6G by combining wide bandwidths with favorable propagation. These frequencies enable high-rate links and fine-resolution sensing, motivating joint communication and sensing architectures. This talk discusses narrowband and wideband FR3 transceiver designs for MIMO and phased arrays, highlighting frequency-agile, energy-efficient front-ends that unify connectivity and awareness in next-generation wireless systems.

Arun S. Natarajan is a Professor of ECE at Yale University. He received his Ph.D. from Caltech. His research focuses on analog/RF/mm-wave circuits and systems for communication and sensing, and he has contributed extensively to these areas over more than two decades in industry and academia.

FR3 System Analysis, Testbed and Hardware Challenges

Gary Xu, Samsung, Plano, TX

7-24 GHz, referred to as FR3, has emerged as a promising new spectrum for 6G. We present our system analysis for both BS and UE, and eXtreme-MIMO (X-MIMO) system, a 256-transceiver X-MIMO testbed with 1024 antennas at 7 GHz band. We conducted outdoor field tests to validate the capacity, coverage and scalability. Emerging technologies, such as digital envelope tracking (DET) and GaN-on-Si PAs for energy efficiency, and new use cases such as integrated sensing and communication will be presented.

Gary Xu is a vice president at Samsung Research America, where he leads 6G technology development. His recent research topics include eXtreme massive MIMO for FR3, AI for wireless, ISAC and advanced duplexing systems. Before joining Samsung, he worked at Nokia and then Texas Instruments.

6G FR3 RF Front End Modules and Power Amplifiers for Mobile Applications

Florinel Balteanu, Skyworks Solutions, Irvine, CA

Future 6G networks will rely on the effective use of FR3 frequencies, spanning 7-15 GHz. This presentation will explore current designs for transitioning to 6G RF front-end modules for RF cellular technologies, with particular focus on the power amplifier. It will address the challenges posed by 6G implementation while also preparing for the next phase toward FR3. It will address the challenges posed by 6G implementation while also preparing for the next phase toward FR3.

Florinel Balteanu received the Ph.D. degree in electrical engineering from Transylvania University, Brasov, Romania, in 1995. From 1992 to 1993, he was a Fulbright Visiting Scholar at Stanford University, Stanford, CA. He is presently a Technical Director with Skyworks Solutions Inc., Irvine.

Direct RF Sampling ADCs for FR3

Junhua Shen, Analog Devices, Wilmington, MA

The emerging 6G wireless standards in FR3 demands highly flexible radio architectures. Direct RF sampling ADCs offer a compelling solution by enabling wideband digitization with much reduced analog front end complexity. This talk will explore the design challenges and innovations in building high-speed, medium-resolution ADCs capable of sampling directly at FR3 frequencies. Key topics include operation speed, front-end linearity, and interleaving techniques to achieve the required bandwidth and dynamic range. The presentation will also highlight recent advances in ADC architectures that balance performance, power, and integration for next-generation RF systems.

Junhua Shen received the Ph.D degree in EE from Columbia University in 2010. Since Feb. 2010, he has been working at Analog Devices serving technical and managerial roles in multiple product lines, most recently as a technical lead in the high speed converter group. His current research interest is high speed converters. He has published extensively and holds 20 issued US patents. Dr. Shen was a co-recipient of the 2013 JSSC Best Paper Award and a Senior Member of IEEE.

Exploration of 6G FR3 for Coverage, Capacity, and Sensing in Edge AI Era

Kenichi Okada, Institute of Science Tokyo, Tokyo, Japan

In this presentation, the mid-upper 6G band (FR3, 7-24GHz) is investigated in terms of coverage, capacity, and sensing. Comparative link budgets at 2GHz, 13GHz, and 28GHz illustrate array-gain requirements for 500m cells and the role of massive MIMO with spatial duplexing to boost the wireless capacity. The discussion further extends to sensing integration and long-range 6G links including NTN communications.

Kenichi Okada (Fellow, IEEE) received his degrees from Kyoto University. He joined Tokyo Institute of Technology in 2003 and is now a Professor at the Institute of Science Tokyo. His research focuses on ULP/mmW/THz wireless transceiver and mixed-signal circuit design.

Ambient IoT: from Concept to Practicality

David Wentzloff, Everactive, UMICH, Ann Arbor, MI

Ambient-IoT promises pervasive connectivity for batteryless devices, providing the data pipeline for Physical AI. Energy harvesting in the FR3 band presents unique opportunities that will help propel the adoption of A-IoT devices. This talk will introduce the 3GPP plans for A-IoT, and how ICs can meet the 3GPP requirements with RF rectifiers, passive beamforming, and ultra-low power radios.

David Wentzloff is a Professor of Electrical Engineering and Computer Science at the University of Michigan. His research focuses on ultra-low power RFICs. In 2012, he co-founded Everactive, a fabless semiconductor company developing ultra-low power wireless SoCs.

Site-Specific MIMO Channel Optimization in FR3

Sofie Pollin, KU Leuven ESAT and imec, Heverlee, Belgium

The 6G FR3 (7–24 GHz) spectrum offers new flexibility between coverage and capacity, requiring site-specific optimization of antennas, frequency bands, and front-end architectures. This talk explores how AI/ML enhances MIMO design and run-time control in FR3. We present multi-band architectures that balance bandwidth, spatial streams, and RF-chain reuse. Site-specific modeling shows how learned channel predictors and interference-aware beam management enable adaptive, AI-native FR3 transceivers.

Sofie Pollin (SMIEEE) is a full professor at KU Leuven focusing on wireless communication systems since 2012. She is affiliated with imec since 2002. Research interests include cell-free networks, integrated sensing and communication, MIMO, FR3 and data-driven network optimisation.

Forum 5:
Analog for AI and AI for Analog:
What the Analog/RF People Can Do and Leverage in the AI Era

Organizers:

Co-Organizers: **Masoud Babaie,** *Delft University of Technology, Delft, The Netherlands*
Taiyun Chi, *Rice University, Houston, TX*
Jun Yang, *Southeast University, Nanjing, China*
Didem Turker Melek, *Cadence Design Systems, San Jose, CA*

Champions: **Mingoo Seok,** *Columbia University, New York, NY*
Kostas Doris, *NXP, Eindhoven, The Netherlands*

Jiayoon Ru
Peking University, Beijing, China

Vanessa Chen
Carnegie Mellon University, Pittsburgh, PA

As AI advances rapidly, digital circuits such as GPUs work at its foundation, yet analog and mixed-signal circuits remain essential to build AI infrastructure and complement AI computing. At the same time, AI starts to influence the way ICs are modeled, designed, optimized, and packaged. The forum intends to bring the analog community together for a one-stop shop to have a comprehensive understanding and discussion of our impact on AI, and the impact of AI on our jobs and potentially on the way we solve analog problems. By bridging these perspectives, the forum will explore challenges and opportunities for analog/RF/mixed-signal engineers and researchers to shape next-generation AI systems and to leverage AI for improving our design and productivity.

Analog for AI and AI for Analog: Enabling Next Generation AI Datacenters
Tom Gray, NVIDIA, Durham, NC

AI applications on GPU systems have exploded with an almost unlimited appetite for compute creating a critical need for analog circuit research addressing scaling and power reduction. We will explore what is ultimately needed for deployment of these systems including circuits for compute, electrical and photonic interconnect, memory systems, packaging/stacking, power delivery, and thermal design as well as how the AI enabled by these datacenters enables the design and implementation of future circuits and systems.

C. Thomas Gray holds a B.S. from Mississippi College and M.S./Ph.D. from North Carolina State University. He has held various positions at IBM, Cadence, Artisan/ARM, and Nethra working on DSP and high-speed serial link communication systems. In 2011, he joined NVIDIA, where he is Senior Director of Circuit Research, leading activities related to high-speed electrical signaling, photonics, security circuits, low-energy and resilient memories, circuits for machine learning, power delivery, and thermal management.

What Role Can Analog Computation Play in Next-generation AI Systems
Naveen Verma, Princeton University, Princeton, NJ

Fundamental circuit operation shows that analog compute achieves superior efficiency to digital in lower dynamic range regimes. Low precision compute, a critical trend for neural networks, thus makes analog a powerful pathway. But, there are key gaps between fundamental and practical circuit operation, at the scales demanded by AI systems. This talk takes the lens of practical architectures for scaled-up system-level efficiency, which invalidates some analog approaches while illuminating others.

Naveen Verma is the Ralph H. and Freda I. Augustine Professor of ECE at Princeton, and co-founder of EnCharge AI. His research focuses on advanced sensing and computing systems, and he is recipient of numerous teaching and research awards, including several best-paper awards with his students.

Algorithm Architecture Co-Design for Analog In-Memory Computing
Hsinyu Tsai, IBM, San Jose, CA

Large Language Models (LLMs) have gained significant traction in real-world AI applications. However, LLMs demand large weight capacity, efficient computing, and high-throughput data communication. We will discuss how non-volatile memory, analog mixed-signal design, system architecture, and workloads impact efficiency and performance of Analog In-Memory Computing. Through circuit simulations and hardware-aware training, we demonstrate near-software accuracies in both simulation and hardware.

Sidney Tsai leads the Analog AI team at the IBM Research Center in San Jose. Since 2020, her team has successfully demonstrated the training and inference of Deep Neural Networks in analog memory. Dr. Tsai earned her Ph.D. in EECS from MIT, specializing in next-generation logic and memory.

ISSCC 2026 / February 19, 2026 / 8:00 AM

Open-Source AI for Analog Correction: From RF Power Amplifiers to Energy-Efficient Silicon
Chang Gao, TU Delft, Delft, The Netherlands

AI offers a new paradigm for correcting analog/RF non-idealities, but challenges in design, benchmarking, and deployment hinder its adoption. This talk argues for an open-source approach to bridge these gaps. We introduce OpenDPD, a framework for PA linearization that has enabled co-design of optimized mixed-precision & sparse AI models and AI-DPD hardware accelerators. We will explore this path from algorithm to silicon and preview its extension to correct other circuits like ADCs and PLLs.

Chang Gao is an Assistant Professor in the Department of Microelectronics at TU Delft, leading the Lab of Efficient Machine Intelligence (EMI). His research focuses on energy-efficient digital AI hardware design for real-time edge AI inference and learning for embodied intelligence.

Challenges in AI-Based Analog Design
Behzad Razavi, UCLA, Los Angeles, CA

For AI to operate as a good analog designer, it must climb five levels. First, it must be able to analyze circuits correctly, a challenge if AI does not have access to circuit simulators and/or device models. Second, it must gain deep intuition, a formidable undertaking for trainers of AI. Third, it must be able to design a given circuit topology for certain specs while drawing upon intuition and not just formulas. Fourth, it must be able to select from a wide set of topologies for a give function and optimize beyond specs. Fifth, it must innovate.

Behzad Razavi is a professor at UCLA. He has published more than 200 papers and nine books, and received nine IEEE best paper awards and six teaching and education awards. His books have been published in seven languages. He received the IEEE Pederson Award in Solid-State Circuits and was recognized as a top author in the 50-year and 75-year histories of the ISSCC.

State-of-the-art Microwave Device Modeling Enabled by Artificial Intelligence and Machine Learning
Jianjun Xu, Keysight, Santa Rosa, CA

This talk reviews some powerful and practical Artificial Intelligence and Machine Learning (AI/ML) technologies for applications in traditional RF and Microwave design and beyond. After a very brief overview of the AI/ML landscape, we focus on Artificial Neural Networks (ANNs) and provide several key examples of modern ANN applications to electronic, electro-thermal, and electro-chemical device modeling and device characterization to illustrate the substantial benefits and generality of present techniques. The talk concludes with a discussion of the potential of AI/ML technologies to address and solve future challenging and important RF and Microwave design problems, e.g., for 6G.

Jianjun Xu is a Senior Machine Learning Engineer at Keysight Laboratories, Keysight Technology, Inc., in Santa Rosa CA. His ANN research has been integrated into leading commercial simulators and measurement-based design flows, including transistor characterization and nonlinear modeling of GaAs and GaN FETs, cryogenic CMOS devices, lithium-ion battery models, TCAD-to-circuit links, and more. Dr. Xu received the Ph.D. Degree in Electrical Engineering from Carleton University, Ottawa, Canada, in 2004, and is a frequent technical reviewer for the IEEE on topics of AI/ML and ANNs.

Toward Agile and Intelligent Analog/RF IC Design Automation
David Z. Pan, UT Austin, Austin, TX

Analog and RF IC design has long been a labor-intensive, iterative process requiring manual topology generation, device sizing, and layout. This talk presents recent advances in AI-driven agile and intelligent analog/RF IC design automation. We will showcase techniques spanning from automated topology generation to device sizing and to layout synthesis, as well as surrogate modeling and inverse design. Our ultimate vision is to enable a fully automated, end-to-end analog/RF IC design flow, from high-level specifications to layout.

David Pan is the Silicon Labs Endowed Chair Professor at UT Austin. His research includes electronic design automation and AI. He has published over 540 research papers and 10 US patents. He has won SRC Technical Excellence Award and over 20 Best Paper Awards. He is a Fellow of IEEE, ACM, and SPIE.

AI-Empowered Analog/RF IC Design
Ye Lu, Fudan University, Shanghai, China

This talk will demonstrate product-level AI-enabled design tools and discuss their technical foundations. We first introduce Multi-agent Reinforcement Learning and demonstrate its capability for automated optimization of complex analog circuits and knowledge transfer between technologies. Afterwards, a generative design approach is proposed for RFIC design, and a fully automated RFIC-GPT tool is developed. This tool enables the complete Spec-to-Layout automation for RFIC IP design, reducing the design efforts from hours/days to seconds/minutes.

Ye Lu obtained his Ph.D. from the University of Pennsylvania in 2011, and he had been with Intel and Qualcomm from 2011-2019. Dr. Lu joined Fudan University as a professor in early 2020, and his current research interest includes DTCO and AI-EDA. Dr. Lu is also a co-founder of RFIC-GPT.

Handling the Design Complexities of 2D, 2.5D, and 3D Microsystem Assemblies Possessing Heterogeneous Functionality Using AI-Based EDA Technology
John Damoulakis, Cadence Design Systems, San Jose, CA

EDA tools play a critical role in remedying the design, manufacturing, and packaging complexities of advanced analog microelectronic circuitries. EDA tools empowered with artificial intelligence and machine-learning, centered on recent advances in large language models and companion analytics, have proven to be the catalyst alleviating complexities and raising productivity - accelerating analog microsystems turn-around-time. This talk presents AI-based models for complex analog microsystems offering high payoff in high-performance and ultra-low-power computing.

John Damoulakis is a Sr. Director at Cadence Design Systems addressing the intersection of microelectronics and knowledge-based algorithms. He holds a Ph.D. from Rice University and MS in EE and ME from Technical University of Athens, Greece. He has authored numerous technical papers and patents and recipient of the Franklin's Institute Levy medal.

DIGEST OF TECHNICAL PAPERS • 655

979-8-3315-8937-0/26 $31.00 © 2026 IEEE

ISSCC 2026 / FORUM 6

Forum 6:
Calibration and Dynamic Matching Techniques for High-Performance Data Converters

Organizers:

Lucien Breems
NXP Semiconductors
Eindhoven, The Netherlands

Shanthi Pavan
IIT Madras
Chennai, India

Co-Organizers: **Shiyu Su**, *University of Waterloo, Waterloo, Canada*
Nima Maghari, *University of Florida, Gainesville, FL*
Amy Whitcombe, *Intel, Santa Clara, CA*
Seung-Tak Ryu, *KAIST, Daejeon, Korea*

Champions: **Jens Anders**, *University of Stuttgart, Stuttgart, Germany*
Mike Chen, *University of Southern California,*
Los Angeles, CA

Digital assistance through calibration and dynamic matching is a key enabler in modern data converters, enhancing robustness, reducing analog complexity through digital processing, and enabling extreme levels of bandwidth and resolution. These matching techniques correct a wide range of errors — including offset, gain mismatches, dynamic variations, frequency-dependent distortions, and time skew — to ensure high-precision operation. This forum presents a comprehensive overview of calibration strategies and dynamic matching algorithms employed across various data converter architectures: high-speed ADCs, high-speed DACs, continuous-time pipeline ADCs, noise-shaping ADCs, and time-domain ADCs. Looking ahead, the integration of AI and machine learning into calibration workflows holds promise for even greater adaptability and performance.

Data Converter Calibration: Motivation, Evolution, and General Aspects
Boris Murmann, University of Hawai'i, Honolulu, HI

This presentation develops an overarching framework for the subsequent architecture-specific explorations on data converter calibration. It begins by identifying the compelling reasons for calibration, which are typically linked to improving accuracy, robustness and efficiency. These aspects are then projected onto use-case examples to highlight the historical progression of analog, digital and hybrid approaches. Throughout the discussion, we will consider system aspects that often emerge as the ultimate criteria for a calibration scheme's practicality.

Boris Murmann is a Professor of ECE at the University of Hawai'i at Mānoa. His research is in mixed-signal integrated circuit design, including sensor interfaces and data conversion. He is a fellow of the IEEE and currently serves at the Editor-in-Chief of the Journal of Solid-State Circuits.

Background Calibration of High-Speed Pipelined ADCs: Methods, Limitations and Practical Considerations
Huseyin Dinc, Analog Devices, Durham, NC

The design of pipelined ADCs required for demanding wide-band applications is only possible through background calibration of many impairments inherent in analog circuits. However, reliance on calibrations has also resulted in considerable complexity, and increased the possibility of undesirable behavior. In this talk, we will cover the limitations of several calibration techniques and methods to mitigate them. We will touch on strategies for achieving high speed and low power.

Huseyin Dinc received the Ph.D. degree from the Georgia Institute of Technology. Since 2008, he has been with Analog Devices Inc., working on high-speed ADCs. He was the co-recipient of the 2020 ISSCC Outstanding Paper Award. His technical interests include digitally-assisted analog-circuit design.

High-Linearity in Delta-Sigma ADCs using DEM, Calibration, and Architectural Innovation
Maurits Ortmanns, University of Ulm, Ulm, Germany

Sources of nonlinearity in delta-sigma converters are reviewed, and various DEM techniques that address static and dynamic mismatches in order to achieve high linearity are compared. The limitations of DEM in wideband converters are discussed, alongside an overview of calibration techniques for achieving high linearity over wide bandwidths. Converters that achieve superior nonlinearity at an architectural level are also covered. State-of-the-art examples are used for demonstration purposes.

Maurits Ortmanns is a professor at the University of Ulm with a research focus on ADCs. He authored >350 IEEE journal and conference papers. He served as ISSCC TPC member, regional chair, and analog subcom chair. He is Associate Editor for JSSC, Distinguished Lecturer and AdCom member of SSCS.

ISSCC 2026 / February 19, 2026 / 8:00 AM

Ensuring Robustness in MASH / Continuous-Time Pipeline ADCs: Necessity of Calibration
Mitsuya Fukazawa, Renesas Electronics, Tokyo, Japan

MASH and continuous-time pipeline (CTP) ADCs have emerged as promising high-efficient architecture. Despite the theoretical advantages, these architectures inevitably suffer performance degradation due to circuit nonidealities, environmental variations. Consequently, calibration becomes a fundamental requirement for ensuring robustness, yield, and long-term reliability. This talk provides a comprehensive overview and comparison of the calibration techniques essential for MASH and CTP ADCs.

Mitsuya Fukazawa received his Ph.D. from Kobe Univ., Japan, 2008. He joined Renesas Electronics in 2008, and he has been engaged in development of ADC and various mixed-signal circuits. Dr. Fukazawa was a Technical Program Committee Member of Symp. on VLSI Tech. & Circuits from 2018 to 2023.

Calibration for high-speed, high-resolution DACs
Gabriele Manganaro, Mediatek, Woburn, MA

Traditional calibration techniques compensate for static nonidealities. However, as clock rates and signal frequencies continue to rise, dynamic errors are becoming increasingly significant contributors to output spectral purity. In this presentation, an overview of some of the most interesting and innovative recent developments in this area is provided.

Gabriele Manganaro received the Dr.Eng. and Ph.D. degrees in Electronics from the University of Catania, Italy. He worked at Texas Instruments, National Semiconductor and Analog Devices. He joined MediaTek in 2021. He coauthored more than 70 papers and 3 books and has been granted 20 U.S. patents.

Machine Learning-Based Calibration of Analog-to-Digital Converters
Maarten Molendijk, NXP Semiconductors, Eindhoven, The Netherlands

ADC calibration algorithms are paramount in modern IC design to relax or overcome analog limitations. While digital-assisted analog and digital post-correction techniques are widespread in literature, a relatively new approach – machine learning(ML)-based calibration – has the potential to calibrate highly complex errors but is still largely unexplored. In this forum, an overview of ML-based calibration methods and their challenges is presented followed by an outlook on future directions.

Maarten Molendijk received the M.Sc. degree in EE (cum laude) from Eindhoven University of Technology (TU/e) in 2022. He joined NXP, where he is working on ADC calibration, and is pursuing a Ph.D. degree with the IC Group at TU/e. His interest include analog circuit modeling and machine learning.

Calibration Techniques for High-Speed Time-Interleaved Time-Domain ADCs
Sam Palermo, Texas A&M University, College Station, TX

Time-domain ADCs utilize delay-based quantization for speed advantages and low-area in advanced technologies. However, performance limitations arise due to gain, offset, and radix errors present in the voltage-to-time and time-to-digital converters. Moreover, high-speed time-interleaved ADC performance is sensitive to gain, skew, bandwidth, and offset mismatch between the channels. This talk provides an overview of calibration techniques for robust operation of these converters.

Samuel Palermo is a Professor in the Electrical and Computer Engineering Department of Texas A&M University. His research interests include high-speed electrical and optical interconnect architectures, high-speed data converters, radiation-hardened electronics, and AI computing hardware.

Panel Discussion
Kostas Doris, NXP Semiconductors, Eindhoven, The Netherlands

This forum explores the latest advances in calibration techniques across a broad spectrum of both established and emerging data converter architectures. Following seven in-depth technical presentations, this panel will bring together leading experts and the audience in a dynamic and thought-provoking discussion. The goal is to identify common threads and key differences among the presented approaches, examine their implications for system architecture and technology, and assess their relevance to real-world applications. Expect a lively exchange of ideas that challenges assumptions and sparks new insights into the future of data converter design.

Kostas Doris is a Fellow at NXP Semiconductors and part-time Professor at the Technical University of Eindhoven. He pioneered the automotive industry's first mm-wave RFCMOS radar and currently leads NXP's innovation activities in the areas of mm-wave transceivers in silicon processes, and advanced packaging. Kostas has authored numerous publications and patents and actively collaborates with European universities. He has held leadership roles in ISSCC, RFIC, Benelux RF and EURAD conferences.

DIGEST OF TECHNICAL PAPERS • 657

979-8-3315-8937-0/26 $31.00 © 2026 IEEE

ISSCC 2026 / EVENING EVENT / STUDENT RESEARCH PREVIEW (SRP)

Evening Event: Student Research Preview (SRP)

Co-Chair: Jerald Yoo
Seoul National University, Korea

Co-Chair: Mondira Pant
Intel, MA

The Student Research Preview (SRP) will highlight selected student research projects in progress. The SRP consists of 26 sixty-second presentations followed by a Poster Session by graduate students from around the world, which have been selected on the basis of a short submission concerning their ongoing research. Selection is based on the technical quality and innovation of the work. This year, the SRP will be presented in three theme sections: 1) RF, mmWave TRX, 2) ML/AI/Security, and 3) Power, Wireline and Sensing.

The SRP will include an inspirational lecture by Dr. Kinam Kim (Samsung). The SRP begins at 8:00 pm on Sunday, February 15th. It is open to all ISSCC registrants.

Invited Talk: "My Half-Century Journey of Semiconductor Innovation: Key Contributions, Lessons Learned and Insights for Future Engineers"

Dr. Kinam Kim
Senior Advisor and Former CEO,
Samsung Electronics

Among the many scientific and technological achievements of the 20th century, no single invention has arguably influenced our lives as profoundly as transistors or semiconductors. Since the onset of the computer age in the 1960s, we have experienced a series of remarkable transformations such as the transition to personal computers and mobile age, and today, we stand on the cusp of another revolution: digital revolution through AI.

As a semiconductor industry leader as well as an engineer, I would like to share selected insights and lessons learned from almost half-century tenure in the semiconductor industry.

This presentation will begin with a brief introduction, followed by highlights of four key contributions to semiconductor industry among more than 40 years' involvement in semiconductor technologies and products, and some words of guidance and encouragement to the next generation of young semiconductor engineers and researchers.

Main theme of the presentation, several important key technologies, to which I am considered to have contributed in the advancement of semiconductors: RCAT (Recessed Channel Array Transistor) for DRAM, 3-D vertical NAND for NAND flash, MBCFET (Multi Bridged Channel FET) or simply as known as GAA (Gate-All-Around) for logic and HBM (High Bandwidth Memory) for AI, will be discussed with focus on the motivations behind these inventions and how we developed.

Dr. Kinam Kim has greatly contributed to the semiconductor industry on many fronts for more than 45 years since he joined Samsung in 1981. His contributions, leadership, and strategic vision have paved the way for Samsung's role as a semiconductor technology innovator in the industry.

Dr. Kinam Kim currently serves as Senior Advisor of Samsung Electronics. Until 2023, Dr. Kim was the Chairman of SAIT (Samsung advanced institute of Technology), CEO of Samsung Electronics, overseeing global operations of all semiconductor related business.

Under Dr. Kim's leadership as CEO, Samsung's memory business has been fortified on solid foundations with world-leading technological capabilities to include first-to-the market products in DRAM, High Bandwidth Memory (HBM), 3D-NAND and Solid State Drives (SSD). Samsung's System LSI business has also brought innovations for products such as mobile application processors (AP), modem chipsets, CMOS image sensors, and display driver ICs for a wide range of end markets. Samsung Foundry has been an innovator in advanced logic processes and specialty technologies, with industry's first breakthroughs. Notable examples are high-k metal gate (HKMG) transistor (32nm), FinFET process technologies (14nm, 10nm, and 8nm), and EUV (Extreme Ultra Violet)-based process technologies (7nm and below), as well as the world's first 3nm process technology on Gate-All-Around (GAA) transistor architecture.

As a scholar and engineer, Dr. Kim has published more than 480 papers in world class journals, including in Nature and Science, and holds more than 360 patents. His contributions (RCAT DRAM, 3D V-NAND, MBCFET and HBM) to the semiconductor industry have been recognized, and some of the latest recognitions include "IEEE Robert Noyce Medal" in 2025, "IMEC Lifetime Achievement Award" in 2017 and "Flash Memory Summit Lifetime Achievement Award" in 2016. In Korea, Dr. Kim was recognized as the National technology leader in science by receiving the highest honor and recognition by the President of Korea, "The Korea Science & Technology Award" in 2019.

Session 1: RF, mmWave TRX

Session Co-Chair
Hao Gao
Southeast University

Session Co-Chair
Pasqualina Fragneto
ST Microelectronics

Session 1: **RF, mmWave TRX**

Session 1 highlights wideband and high-performance transceiver technologies spanning 24–101.8 GHz for wireless communications, radar, and synchronization. Presentations include 57–71 GHz CMOS phased-array transceivers with aperture-tuning antennas, a 256-element 28 GHz 5G NR wirelessly powered active relay, and chiplet-based CMOS+GaN phased-array transmitters delivering high PSAT, power gain, and peak PAE. The session also features a time-division full-duplex transceiver with strong self-interference rejection, active true-time-delay circuits for beamforming, and a 60 GHz multistatic radar transceiver chipset enabling long-range wireless synchronization. Advances in frequency generation are showcased through Class-F magnetic-coupled multi-phase oscillators, wide-tuning-range quad-mode VCOs, and a dual-slope sampling PLL achieving low RMS jitter and wide lock-in range.

EE1.1
Minghao Fan
Institute of Science Tokyo

EE1.2
Shu Date
Institute of Integrated Research

EE1.3
Dongfan Xu
Institute of Science Tokyo

EE1.4
Eunchae Jo
Chonnam National University

EE1.5
R. Matthew Ma
High5 Semiconductor (Hong Kong) Limited
& The Hong Kong University

EE1.6
Haoran Wang
UESTC

EE1.7
Chawin Khongprasongsiri
University College Dublin

EE1.8
Matoi Ono
The University of Tokyo

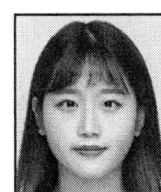

EE1.9
Hyogyoung An
UNIST

ISSCC 2026 / EVENING EVENT / STUDENT RESEARCH PREVIEW (SRP)

Session 2: ML/AI/Security

Session Co-Chair
Kyuho Jason Lee
Yonsei University

Session Co-Chair
Pei-Yun Tsai
National Taiwan University

Session 2: **ML/AI/Security**

Session 2 presents energy-efficient hardware for intelligent, adaptive, and secure computing. The session includes a GALS neuromorphic processor supporting hybrid ANN-SNN operation with on-chip learning, a heterogeneous CGRA SoC for machine learning and sparse tensor algebra, and an MLC RRAM compute-in-memory macro achieving ultra-high TOPS/W efficiency. Additional works address reliability and security, including a post-quantum secure authentication engine with integrated TRNG, an analog-to-cipher converter, and soft-output threshold-guided CRC decoding. The session also highlights system-level intelligence through a spatiotemporal 3D graphics reconstruction processor and a runtime-reconfigurable biomarker and control unit, emphasizing adaptability, robustness, and application-driven AI hardware design.

EE2.1
Zixuan Shen
Huazhong University
of Science and Technology

EE2.6
Haruki Okada
University of Osaka

EE2.2
Yuchen Mei
Stanford University

EE2.7
Zeynep Kizilates
Boston University

EE2.3
Zhen Kong
Southern University
of Science and Technology

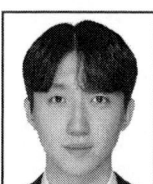

EE2.8
Jae-Young Kim
KAIST

EE2.4
Yi Yang
Southeast University

EE2.9
Jack Burd
UCLA

EE2.5
Seoyoon Jang
MIT

ISSCC 2026 / February 15, 2026 / 8:00 PM

Session 3: Power, Wireline and Sensing

Session Co-Chair
Sijun Du
Delft University of Technology

Session Co-Chair
Philex Fan
National Cheng Kung University

Session 3: **Power, Wireline and Sensing**

Session 3 focuses on integrated solutions for efficient energy delivery, high-speed data links, and precision sensing. Power management presentations include a single-input, multi-isolated-output DC-DC converter, a voltage-mode-boosted current-mode wireless power interface with high PCE, and an optically powered and controlled gate driver for high-voltage SiC MOSFETs. The session also features ISI-robust wireline transceivers and a JTOL-enhanced clock-and-data-recovery circuit with real-time error correction. Advances in sensing are highlighted by a 3D-stacked InGaAs-InP/CMOS SPAD flash LiDAR, a 256-channel wireless neural recording SoC with adaptive compression, and an integrated complex bioimpedance analyzer for organ-on-chip applications.

EE3.1
Yichen Zhang
University of Tokyo

EE3.2
Yating Zou
EPFL

EE3.3
Anish Mondal
University of Pennsylvania

EE3.4
Chanheum Han
Kwangwoon University

EE3.5
Halil Yildirim
EPFL

EE3.6
Ho-Joon Shin
Yonsei University

EE3.7
Wenxian Gu
East China Normal
University

EE3.8
Christian Ziegler
TU Braunschweig

DIGEST OF TECHNICAL PAPERS • 661

979-8-3315-8937-0/26 $31.00 © 2026 IEEE

ISSCC 2026 / EVENING EVENT / STUDENT RESEARCH PREVIEW (SRP)

Poster Session

Poster Session Co-Chair
Nandish Mehta
Nvidia

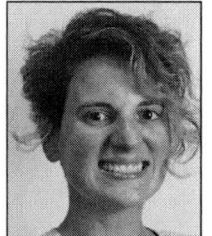

Poster Session Co-Chair
Rabia Tugce Yazicigil
Boston University

SRP ORGANIZING COMMITTEE

Co-Chair:	**Jerald Yoo**, *Seoul National University, Korea*
Co-Chair:	**Mondira (Mandy) Pant**, *Intel, MA*
Advisor:	**Anantha Chandrakasan**, *Massachusetts Institute of Technology, MA*
Advisor:	**Jan Van der Spiegel**, *University of Pennsylvania, Philadelphia, PA*
Media/Publications:	**Laura Fujino**, *University of Toronto, Toronto, Canada*
A/V:	**Trudy Stetzler**, *Halliburton, Houston, TX*

COMMITTEE MEMBERS

Utsav Banerjee
IISC, India

Woo-Seok Choi
Seoul National University, Korea

Zeynep Deniz
IBM Research, NY

Sijun Du
*Delft University of Technology,
The Netherlands*

Philex Fan
*National Cheng Kung University,
Taiwan*

Pasqualina Fragneto
ST Microelectronics, Italy

Antoine Frappé
University of Lille, France

Hao Gao
Southeast University, China

Preet Garcha
Texas Instruments, TX

Bongjin Kim
KAIST, Korea

Jaydeep Kulkarni
University of Texas at Austin, TX

Matthias Kuhl
University of Freiburg, Germany

Hyunjoo Jenny Lee
KAIST, Korea

Kyuho Jason Lee
Yonsei University, Korea

Jiamin Li
*Southern University of Science and
Technology, China*

Xilin Liu
University of Toronto, Canada

Nandish Mehta
Nvidia, CA

Takuji Miki
Kobe University, Japan

Noriyuki Miura
Osaka University, Japan

Phillip Nadeau
Analog Devices, MA

Mondira Pant
Intel, MA

Chutham Sawigun
imec, Belgium

Filip Tavernier
KU Leuven, Belgium

Pei-Yun Tsai
National Taiwan University

Rabia Tugce Yazicigil
Boston University, MA

Jerald Yoo
Seoul National University, Korea

Lian Zhang
Apple, CA

ISSCC 2026 / February 15, 2026 / 1:30 PM

CAREER PANEL: Chip In for Change: Career Insights for Young Designers in an AI-First, Eco-Conscious World

Organizers:

Mondira Pant
Intel, Boxboro, MA

Wanghua Wu
Samsung, San Jose, CA

Elpida Karapepera
*University of Washington
Seattle, WA*

Aishwarya Natarajan
*Hewlett Packard Labs
Palo Alto, CA*

Alana Dee
*University of Washington
Seattle, WA*

Ben Keller
Nvidia, Santa Clara, CA

Dilara Caygara
Boston University, Boston MA

Emily Naviasky
IBM, Yorktown Heights, NY

Jiamin Li
*SUSTech
University of Shenzhen, China*

Kamala Raghavan Sadagopan
Qualcomm, San Diego, CA

Kwantae Kim
Aalto University, Espoo Finland

Marziyeh Rezaei
*University of Washington
Seattle, WA*

Najme Ebrahimi
*Northeastern University
Boston, MA*

Shanshan Xie
Intel, Austin, TX

Sukanya S. Meher
Synopsys, Austin, TX

Yingying Fan
*Washington University
St. Louis, MO*

Yu-Chi Lin
UC Berkeley, CA

Trudy Stetzler
IEEE SSCS, Houston, TX

As AI transforms the global tech landscape and environmental challenges demand urgent innovation, today's young IC designers are stepping into a field with unprecedented potential and responsibility. This panel explores how the next generation of engineers can actively *"chip in"* to drive meaningful change through their work. From AI-accelerated tools to energy-efficient architectures, panelists will share how technological advancement and sustainability are converging—and how early-career professionals can align their skills, values, and ambitions in this new era of purpose-driven design. It aims to offer practical career advice, personal stories, and a candid look at how to thrive at the intersection of cutting-edge innovation and global impact.

Moderator:

Elpida Karapepera, *University of Washington, Seattle, WA*

Elpida Karapepera is an ECE Ph.D. student, working on Mixed-Signal and RF IC design at the University of Washington, Seattle, where she is advised by Professor Chris Rudell. Elpida was awarded the University of Washington's 2022 "Robert E. Rushmer Fellowship" and recognized as a "Next Gen Circuit Designer" at ISSCC 2022. Her industry experience includes roles as an Analog Design Engineer and internships at BOSCH Research in Sunnyvale, CA (2023), and at Qualcomm in Santa Clara, CA (2024 and 2025), where she contributed to Analog and Mixed-Signal IC Design projects. Elpida is a core member of the Women in Circuits Committee, organizing multiple ISSCC events since 2023. She currently serves as the Student Lead for the Next Generation Designer Workshop for ISSCC 2026.

DIGEST OF TECHNICAL PAPERS • 663

979-8-3315-8937-0/26 $31.00 © 2026 IEEE

ISSCC 2026 / SPECIAL EVENT / CAREER PANEL

Panelists

Sashi Obilisetty, Executive Director, *Synopsys, USA*

Sashi Obilisetty is currently Executive Director, R&D in the Technology Products Group at Synopsys. Sashi has been instrumental in enabling several Synopsys technologies in the design, verification and post-silicon space. She was one of the key leaders involved in enabling AI in Synopsys products. Between 2020-2021, she was the Chief Architect for Silicon Solutions at Google Cloud. Prior to joining Synopsys in 2009, Sashi founded 2 EDA startups. She was the Marie Pistilli Women in EDA Award Winner in 2024 and the Synopsys YWCA TWIN nominee in 2018. Sashi has a Bachelors in Electronics and Communication Engineering from Birla Institute of Technology, Mesra and a Master's in Computer Engineering from UMASS, Amherst.

Carolina Mora Lopez, Scientific Director, *imec, Belgium*

Carolina Mora Lopez received her Ph.D. degree in Electrical Engineering in 2012 from the KU Leuven, Belgium, in collaboration with imec, Belgium. From 2012 to 2018, she worked at imec as a researcher and analog designer focused on interfaces for neural-sensing applications. She is currently the Scientific Director and team leader of the biomedical circuits team at imec. Her research interests include analog and mixed-signal circuit design for sensor, bioelectronics and neural interfaces. Carolina is a senior IEEE member and serves on the technical program committee of the ISSCC conference.

Pei-Yun Tsai, Professor, *National Taiwan University, Taiwan*

Pei-Yun Tsai received B.S., M.S., and Ph.D. degrees in electrical engineering from the National Taiwan University, Taipei, Taiwan, in 1994, 1996, and 2005, respectively. From 1996 to 2000, she worked at ASUStek, Taipei. From 2006, she joined the Department of Electrical Engineering, National Central University, Taoyuan, Taiwan, where she has been a Professor since 2016. From 2022 to 2024, she was on a secondment to Tron Future Technology, a leading Defense/Aerospace Tech company in Taiwan, for developing synthetic aperture radar systems. Since 2024, she is with the Graduate School of Advanced Technology, National Taiwan University. Her research interests include digital signal processing algorithms and VLSI design for wireless communication systems, radar systems, and biomedical systems.She has served on numerous Technical committees of top-tier conferences linclude DISP, ASPS, CASS and has served as a guest editor of IEEE SSC-L and IEEE JSSC in 2024. She has received both the Future Tech Award and National Innovation Award in 2022 and 2023 and is a recipient of the 1st Asian Solid-State Circuit Conference Student Design Contest Outstanding Award in 2005.

Srabanti Chowdhury, Professor, *Stanford University, USA*

Prof. Srabanti Chowdhury specializes in the wideband gap (WBG) and ultra-wide bandgap (UWBG) materials and device engineering. Her research focuses on energy-efficient system architecture for power and RF applications, particularly emphasizing thermal management. She earned her M.S. in June 2008 and Ph.D. in December 2010 in Electrical and Computer Engineering from the University of California, Santa Barbara. In recognition of her outstanding work on diamond integration with GaN and SiC, resulting in very low thermal boundary resistances for thermal management, Prof. Chowdhury received the 2023 Technical Excellence Award from the Semiconductor Research Society (SRC). She is the recipient of the 2025 Quantum Device Award for her contribution to Vertical GaN devices and Phonon matching interfaces for thermal boundary resistance lowering. Her contributions to the field encompass 8 book chapters, 150 journal papers, 150 conference presentations, and 27 issued patents. Actively engaged in IEEE conference committees, including IRPS and VLSI Symposium, she serves on the executive committee of IEDM. She became an IEEE fellow in 2024 for her contributions to wide bandgap semiconductor devices and technology.

EE2: Generative AI for Silicon Design:
Mastering Complexity, Democratizing Design, and Building Trust

Organizer:

Alfred Yeung
AMD, Santa Clara, CA

Co-Organizers:

Kaushik Vaidyanathan, *OpenAI, San Francisco, CA*
Julian Tham, *Infineon, San Jose, CA*
Huichu Liu, *Meta Platforms, Sunnyvale, CA*
Cynthia Hsu, *Sandisk, Fremont, CA*
Ping-Hsuan Hsieh, *Nation Tsing Hua University, Tsinchu, Taiwan*
Konstantinos Manetakis, *CSEM, Neuchâtel, Switzerland*
Alberto Valdes-Garcia, *IBM, Yorktown Heights, NY*
Pasqualina Fragneto, *ST Microelectronics, Agrate Brianza, Italy*
Negar Reiskarimian, *MIT, Cambridge, MA*
CM Hung, *Mediatek, Tsinchu, Taiwan*

Generative AI is rapidly transforming chip design, enabling automation in circuit synthesis, layout, and verification. These advancements promise greater efficiency, but also raise critical questions: What are the fundamental limits of AI-driven automation? Can AI truly match human intuition and creativity? And will it democratize or centralize chip design in the long run? To explore these questions, join us for an engaging evening event featuring a panel of experts at the forefront of AI and VLSI design and an interactive game where we see how far AI in VLSI has come. Through a mix of real-world challenges, live audience interaction, and expert commentary, we'll assess where we are with GenAI in chip design and where we're heading.

Moderator:

Siddharth Garg, *NYU, New York, NY*

Siddharth Garg is currently the Institute Associate Professor of ECE at NYU Tandon, where he leads the EnSuRe Research group (https://wp.nyu.edu/ensure_group/). Prior to that he was in Assistant Professor also in ECE from 2014-2020, and an Assistant Professor of ECE at the Unversity of Waterloo from 2010-2014. His research interests are in machine learning, cyber-security and computer hardware design. He received his Ph.D. degree in Electrical and Computer Engineering from Carnegie Mellon University in 2009, and a B.Tech. degree in Electrical Enginerring from the Indian Institute of Technology Madras. In 2016, Siddharth was listed in Popular Science Magazine's annual list of "Brilliant 10" researchers. Siddharth has received the NSF CAREER Award (2015), and paper awards at the IEEE Symposium on Security and Privacy (S&P) 2016, USENIX Security Symposium 2013, at the Semiconductor Research Consortium TECHCON in 2010, and the International Symposium on Quality in Electronic Design (ISQED) in 2009. Siddharth also received the Angel G. Jordan Award from ECE department of Carnegie Mellon University for outstanding thesis contributions and service to the community. He serves on the technical program committee of several top conferences in the area of computer engineering and computer hardware, and has served as a reviewer for several IEEE and ACM journals.

Panelists Statements

Amit Gupta, *Siemens EDA, Saskatoon, Canada*

Generative AI marks a fundamental shift in EDA, but truly transforming chip design requires industrial-grade AI that delivers expertise, accelerated design, and verification through both deep in-tool capabilities and broader cross-tool agentic workflows. At Siemens EDA, we champion an open ecosystem for multi-vendor flows, simultaneously elevating design quality and productivity for all. Ultimately, the most powerful designs will rely on the synergy of human creativity guiding these agents, ensuring that intuition drives innovation while algorithms handle the complexity.

Amit Gupta is a technology executive and serial entrepreneur with over two decades of leadership at the forefront of semiconductor design and AI innovation. As Senior Vice President & GM, AI Strategy and Solido Custom IC Division, Amit directs the company's overarching AI strategy and its Custom IC division. Under his leadership, his teams develop cutting-edge AI technologies for Siemens EDA products, and solutions for custom IC design and verification. Amit's corporate leadership is built on a foundation of proven entrepreneurial success. In 2005, he founded Solido Design Automation and, as President & CEO, grew it into the definitive market leader for AI-based semiconductor design tools, leading to its strategic acquisition by Siemens in 2017. Previously, he founded Analog Design Automation in 1999, which pioneered breakthroughs in circuit design and was acquired by Synopsys in 2004. Amit holds degrees in Electrical Engineering and Computer Science with Great Distinction from the University of Saskatchewan.the University of Saskatchewan.

ISSCC 2026 / EVENING EVENTS / EE2

Panelists Statements

Matheus Trevisan Moreira, *Meta, Sunnyvale, CA*

Generative AI is transforming silicon design by enabling unprecedented levels of automation through agents with reasoning and generative capabilities that can invoke tools for grounding and validation. The first wave of commercial AI agents has focused on automating design-flow tasks traditionally performed by humans, with the longer-term promise of defining and executing complete design flows and discovering novel optimization techniques. However, siloed development and the proprietary nature of silicon data limit progress, and realizing the true boundaries of AI-driven automation requires a collaborative ecosystem that enables shared development and innovation across the industry.

Matheus Trevisan Moreira began his career as a professor in Brazil before transitioning to the technology industry, where he has held key technical leadership roles across startups, Apple, and Meta. He currently serves as a tech lead at Meta, driving the development of advanced silicon architectures and accelerators for AI workloads. Matheus is the author of over 100 peer-reviewed scientific publications, holds 12 patents, and has received multiple awards from IEEE and ACM for his contributions to computing and electronic design. His current interests focus on applying AI to accelerate and optimize silicon design, and on architecting systems for next-generation AI applications.

Husni Habal, *Infineon Munich, Munich, Germany*

Generative AI has the potential to augment, not replace, analog layout synthesis by assisting designers in handling multi-objective complexity when constraints are explicit, workflows are guided, and verification is rigorous.

The methodology must be implementation-driven: designers retain extensive control over placement and routing (P&R) parameters and constraint sets, while AI proposes candidate solutions subject to human review and sign-off.

Verification must extend beyond conventional physical checks to include unit-test-style validations after individual tasks. Constraint satisfaction and feasibility screening (with deterministic routines or theorem-provers) can be used before constraint application. Trust and safety follow from a verification-first stack (DRC, LVS, post-layout parasitic extraction and modeling, etc), audit trails, and conservative defaults. Any AI-generated element that cannot be verified is not production-ready.

In the first generation of generative-AI-assisted design tools, large language model (LLM) copilots can serve as alternative human-machine interfaces, design intent into programmable commands and API calls across procedural module generators, layout tools, and routing engines, while exposing parameters and actions for review before execution.

Husni Habal works on analog design flow automation at Infineon Technologies in Munich, Germany. His work centers on procedural layout generation for analog ICs, with emphasis on constraint-aware placement and routing and on applying numerical methods and AI/ML in production design flows. He developed a reinforcement learning approach augmented by relational graph neural networks to encode circuit structure and positional/layout constraints, improving generalization across diverse circuit topologies and accelerating layout generation. He received his doctorate from the Technical University of Munich on constraint-based, layout-driven sizing of analog circuits. He participates in collaborative research projects funded by Germany's Federal Ministry of Education and Research (BMBF) as well as European calls in his field.

Azalia Mirhoseini, *Google, Palo Alto, CA*

Accelerated Design Cycles: AI will dramatically reduce chip design timelines within the next 3-5 years, reducing what currently takes months or years to weeks or days.From Fabless to Design-less: The industry will shift from "fabless" semiconductor companies to "design-less" models, where AI-powered platforms offer end-to-end chip design as a service. AI companies can focus on the application layer and outsource the design of the chip that fuels it.Cambrian Explosion of Custom Chips: The proliferation of specialized AI models and the massive scale of AI inference across various applications including robotics, AR/VR, space exploration, and autonomous driving will drive an explosion of custom chip designs, each optimized for specific workloads.Hardware as AI's Strategic Moat: For frontier AI labs, the ability to co-evolve models alongside the custom hardware that runs them will become increasingly advantageous, making vertical integration of hardware and software a defining moat.

Azalia Mirhoseini is an Assistant Professor of Computer Science and founder of Scaling Intelligence Lab at Stanford University. Her lab develops scalable and self-improving AI systems and methodologies towards the goal of advancing artificial general intelligence. She also spends time at Google DeepMind as a Senior Staff Scientist. Prior to Stanford, she spent several years in industry AI labs, including Google Brain and Anthropic. Her past work includes Mixture-of-Experts (MoE) neural architectures, now commonly used in leading generative AI models; AlphaChip, a pioneering work on deep reinforcement learning for layout optimization used in the design of advanced chips like Google AI accelerators (TPUs) and data center CPUs; and research on inference-time scaling laws. Her research has been recognized through the MIT Technology Review's 35 Under 35 Award, the Best ECE Thesis Award at Rice University, publications in flagship venues such as Nature, and coverage by various media outlets, including MIT Technology Review, IEEE Spectrum, The Verge, The Times, ZDNet, VentureBeat, and WIRED.

EE3: The Augmented Human –
Will Chips in Our Brain Enhance Our Cognitive Abilities?

Organizer:

Rikky Muller
UC Berkeley, Berkeley, CA

Co-Organizers:

Hidehiro Shiga, *KIOXIA, Yokohama, Japan*
Carolina Mora Lopez, *imec, Leuven, Belgium*
SungWon Chung, *Neuralink, Fremont, CA*
Mutsumi Hamaguchi, *Sharp Corporation, Nara, Japan*
Augusto Ximenes, *CogniSea, Seattle, WA*

Can the brain of the future be enhanced with semiconductor technologies? As we deepen our understanding of neural function and continue to miniaturize and integrate advanced electronics, exciting possibilities emerge, such as expanding cognitive capacity through implanted memory, enhancing our vision to perceive unseen wavelengths through visual prostheses, or accelerating thought processes with integrated computational support akin to a co-processor for the brain. While these ideas may lie far beyond current capabilities, this session invites visionary thinkers and cross-disciplinary experts to examine the scientific, technological, and ethical foundations of such a future.

Moderator:

Azita Emami, *Caltech, Pasadena, CA*

Azita Emami is the Andrew and Peggy Cherng Professor of Electrical Engineering and Medical Engineering, Director of Center for Sensing to Intelligence (S2I), and Vice Chair of the Faculty at the California Institute of Technology (Caltech). She received her M.S. and Ph.D. degrees in Electrical Engineering from Stanford University in 1999 and 2004 respectively, and her B.S. degree from Sharif University of Technology in 1996. From 2004 to 2006 she was with IBM T. J. Watson Research Center before joining Caltech in 2007. She was the Executive Officer (Department Head) for Electrical Engineering at Caltech from 2018 to 2024. Her current research interests include integrated circuits and systems, integrated photonics, wireless wearable and implantable devices, neural recording, neural stimulation, sensing, drug delivery, and high-speed data communication systems. She is currently a returning SSCS Distinguished Lecturer, and serves on the ISSCC Medical Subcommittee.

ISSCC 2026 / EVENING EVENTS / EE3

Panelists Statements

Extending the Mind Without Losing Ourselves
Timothy Denison, *Oxford University, Oxford, United Kingdom*

Neurotechnology sits on a long continuum of tools humans have used to extend their natural capabilities. Coffee helps us stay awake beyond our biological rhythms; eyeglasses allow many of us to see more clearly than "perfect" unaided vision; calculators and search engines offload memory and computation. Seen this way, technologies that interface with the nervous system are not a radical break from the past, but a potentially powerful next step in augmenting human performance.Pragmatism and care, however, are essential. Our current understanding of the nervous system and state of neurotechnology suggests that it is often easier to suppress, block, or remove neural function than to reliably enhance it – although this can provide significant therapeutic benefit. Especially in medicine, "simply" restoring lost capabilities in many diseases remains profoundly difficult. Claims of broad cognitive or perceptual enhancement should therefore be met with healthy skepticism, and new interventions approached with humility. Neurotechnology also carries real risks, including unintended plasticity, long-term side effects, or subtle tradeoffs that only emerge over time.For these reasons, continued investment in treating disease should remain a priority. When individuals suffer from conditions such as Parkinson's disease, epilepsy, or chronic pain, the risk–benefit calculus is clearer, the ethical justification stronger, and the societal value unambiguous. Progress here is also likely to generate the scientific insight and safeguards required for any future enhancement applications.Societal concerns are equally important. Unequal access or high cost could exacerbate existing inequalities, and designers must consider unintended consequences. As recent history shows, technology can amplify inherent human biases. Neurotechnology should aim to help people recognize these biases, and correct for them, rather than reinforce cognitive shortcuts.Handled thoughtfully, neurotechnology can complement human strengths without undermining autonomy or equity. The path forward should be cautious, disease-focused, and ethically grounded—yet still optimistic about responsibly expanding what humans are able to do.

Timothy Denison holds a joint appointment in Engineering Science and Clinical Neurosciences at Oxford, where he explores the fundamentals of physiologic closed-loop systems in collaboration with the MRC Brain Network Dynamics Unit. Tim also serves as an advisor to several governments and industry boards on the field of translational medical devices;. in particular, helping define strategies for mapping scientific discovery to product development roadmaps within the regulatory and economic constraints of medical systems. Prior to Oxford, Tim was a Technical Fellow at Medtronic PLC and Vice President of Research & Core Technology for the Restorative Therapies Group, where he helped oversee the design of next generation neural interface and algorithm technologies for the treatment of chronic neurological disease. In 2015, he was elected to the College of Fellows for the American Institute of Medical and Biological Engineering (AIMBE), and in 2024 a Fellow of the Royal Academy of Engineering. He has an MS and PhD from MIT in electrical engineering, and an AB in Physics and MBA from the University of Chicago.

Sensing Minds: Implantable Neural Interfaces as Catalysts for Neural-Circuit Discovery
Milin Zhang, *Tsinghua University, Beijing, China*

Implantable neural interfaces are developing from experimental prototypes into commercial sensing platforms that can fundamentally change how we interrogate neural circuits and interact with machines. This talk will discuss that advances in implantable sensor technologies offer spatial, temporal, and cell–type specificity for long-term bidirectional access to the living brain. Such sensors accelerated the study on neural circuit mapping of perception, memory, and decision–making, and the emergence of adaptive, personalized human–machine symbioses: high–bandwidth prosthetics, cognitive augmentation, and intuitive multimodal interfaces. I will highlight current technical trend and demonstrations, showing the potential in deepen our scientific understanding of brain function and reshape the landscape of human augmentation by using implantable neural interfaces.

Milin Zhang is an associate professor in the department of Electronic Engineering, Tsinghua University. She received the B.S. and M.S. degrees in electronic engineering from Tsinghua University, Beijing, China, in 2004 and 2006, respectively, and the Ph.D. degree in the Electronic and Computer Engineering Department, Hong Kong University of Science and Technology (HKUST), Hong Kong. After finishing her doctoral studies, she worked as a postdoctoral researcher at the University of Pennsylvania (UPenn). She joined Tsinghua University in 2016. Her research interests include designing of smart sensors, sensor interface circuit and system design for biomedical applications. She serves and has served as the TPC member of ISSCC, CICC, A-SSCC and CASS, the Senior Associate Editor (SAE) of TCAS-II, Associate Editor (AE) of TBioCAS. She is the Chapter chair of the SSCS Beijing chapter. She is the Distinguished Lecturer of CASS and IEEE WiE. She has received the Best Paper Award of the BioCAS Track of the 2014 International Symposium on Circuits and Systems (ISCAS), the Best Paper Award (1st place) of the 2015 Biomedical Circuits and Systems Conference (BioCAS), the best student paper award (2nd place) of ISCAS 2017, the Best Paper Award of ICM 2024.

Panelists Statements

Homo Technologicus - Revisited
Jan Rabaey, *UC Berkeley, Berkeley, CA*

As history has amply demonstrated, there is no doubt that humanity has always used and employed technology to augments its capabilities. There is no reason to believe that this would not be the case for cognitive activities. Already today, we are using technology to augment the way we perceive the world around us or interpret that information. Think about in-ear translation, for instance. As was stated recently in WIRED magazine: "Brain Gear is the Hot New Wearable". Yet, to create a real merger between biology and technology still requires huge progress at many fronts and is clearly not for tomorrow, especially since we do not understand very well how cognitive functions are performed in the brain. Reliability, robustness, longevity, security, and privacy are other major hurdles to overcome, often surpassing the realm of technology.

Jan Rabaey is Professor Emeritus and Professor in the Graduate School in the EECS Department the University of California at Berkeley, after being the holder of the Donald O. Pederson Distinguished Professorship at the same institute for over 30 years. He is a founding director of the Berkeley Wireless Research Center (BWRC) and the Berkeley Ubiquitous SwarmLab, and has served as the Electrical Engineering Division Chair at Berkeley twice. From 2019 until 2025, he also served as the CTO of the System-Technology Co-Optimization (STCO) Division of IMEC, Belgium.Prof. Rabaey has made high-impact contributions to a number of fields, including low power integrated circuits, advanced wireless systems, mobile devices, sensor networks, and ubiquitous computing. Some of the systems he helped envision include the infoPad (a forerunner of the iPad), PicoNets and PicoRadios (IoT avant-la-lettre), the Swarm (IoT on steroids), Brain-Machine interfaces and the Human Intranet.He is the primary author of the influential "Digital Integrated Circuits: A Design Perspective" textbook that has served to educate hundreds of thousands of students all over the world. He is the recipient of numerous awards including the 2025 IEEE James H. Mulligan Jr Education Medal and the 2025 IEEE CASS John Choma Education Award, is a Life Fellow of the IEEE, and has been involved in a broad variety of start-up ventures.

High-Bandwidth Interface to the Brain
Dongjin (DJ) Seo, *Neuralink, Austin, TX*

The brains of the future will be enhanced with semiconductor technologies. The more interesting question is how (whether in, on, around, or away from brains) and when.

Dongjin (DJ) Seo is the President and Co-Founder of Neuralink, where he leads the engineering and operations behind the world's most advanced brain-computer interfaces. Under his leadership, Neuralink has transitioned from early research to successful human clinical trials, aiming to restore autonomy to those with unmet medical needs and unlock new dimensions of human potential.

Prior to Neuralink, DJ received his PhD in EECS and Neuroscience from UC Berkeley. There, he invented Neural Dust, a breakthrough ultrasonic wireless implant technology that demonstrated the feasibility of monitoring neural activity through microscopic, untethered sensors. His pioneering work in this field earned him recognition as one of MIT Technology Review's 35 Innovators Under 35. DJ holds a BS in EE from Caltech.

With over 15 years of experience at the intersection of biology and engineering, he specializes in implantable systems, low-power microelectronics, and distributed wireless sensor networks.

ISSCC 2026 / SHORT COURSE / CIRCUITS FOR OPTICAL SUBSYSTEMS: COMMUNICATIONS AND BEYOND

Short Course:
Circuits for Optical Subsystems: Communications and Beyond

Organizer: *Daniel Friedman*
IBM Thomas J. Watson Research Center
Yorktown Heights, NY

Agenda

Time:	Topic:
8:00 AM	Breakfast
8:25 AM	**Introduction by Chair** *Daniel Friedman, IBM Thomas J. Watson Research Center, Yorktown Heights, NY*
8:30 AM	**Introduction to Optical Communication Systems: From VCSELs to Photonics to Coherent Solutions** *Peter Ossieur, imec, Ghent, Belgium*
10:00 AM	Break
10:30 AM	**VCSEL-Based Solutions: Components, Circuits, and Integration** *Enrico Temporiti, Marvell, Pavia, Italy*
12:15 PM	Lunch
1:20 PM	**Silicon Photonics-Based Solutions: Components, Circuits, and Integration** *Firooz Aflatouni, University of Pennsylvania, Philadelphia, PA*
2:50 PM	Break
3:20 PM	**Emerging Optical Applications and Circuit Approaches** *Ali Hajimiri, California Institute of Technology, Pasadena, CA*
4:50 PM	Conclusion

Introduction

Optical subsystems are taking on growing importance, going beyond traditional critical roles in long- and medium-distance high data rate communication applications toward ever-expanding use cases within compute racks as well as in emerging optical array applications. Circuit and integration techniques supporting optical designs are critical enablers of this growth. Progress in this area is even more critical given the challenges facing increasing per-lane data rates for electrical designs coupled with the seemingly insatiable growth in demand for bandwidth to support data center and AI accelerator workloads. This short course is designed to explore these interface circuits. The first presentation provides an introduction to optical communication subsystems, including a discussion of coherent designs. The second lecture focuses on VCSEL-based solutions, including reviewing optical component characteristics, circuit techniques, and packaging strategies enabling integration in ultra-high bandwidth contexts. The third lecture explores silicon photonics-based design, again reviewing critical system components, here including source lasers, modulators, couplers, detectors, and integration strategies. Finally, the last lecture introduces emerging applications for optical subsystems, including optical phased array-based designs.

OUTLINE

8:30 AM
SC1: Introduction to Optical Communication Systems: From VCSELs to Photonics to Coherent Solutions
Peter Ossieur, imec, Ghent, Belgium

This presentation starts off with a thorough introduction to optical link architectures and their important figures of merit, covering intensity modulated, direct detect, and coherent transmission approaches. Then, as a first means to realize an optical link, direct modulation relying on VCSELs is addressed. This is followed by external modulation with particular focus on Silicon Photonics and InP platforms. The lecture ends with a detailed treatment of dual-polarization, IQ modulation and coherent detection.

Peter Ossieur is a scientific director at imec leading optical interconnect R&D, and holds a part time position as professor of high-speed opto-electronics at Ghent University, Belgium. Prof. Ossieur obtained the PhD degree from Ghent University, Belgium, 2005, where he continued to work as a postdoctoral researcher until 2009. From 2009 till 2017, he worked at Tyndall National Institute, Ireland, eventually as Senior Staff Researcher. His research interests are focused upon high-speed electronic and photonic integrated circuits targeting optical and electrical transceivers. He has authored or co-authored over 200 conference and journal publications. Currently, he is serving on the ISSCC wireline subcommittee.

OUTLINE

10:30 AM
SC2: VCSEL-Based Solutions: Components, Circuits, and Integration
Enrico Temporiti, Marvell, Pavia, Italy

This lecture explores VCSEL-based solutions for short-reach optical interconnects, essential for intra-datacenter connectivity and increasingly critical in AI architectures. After reviewing key optical components, we examine high-speed driver circuits and integration strategies for compact, scalable transmitters. The talk progresses from fundamentals to state-of-the-art advancements, concluding with opportunities and challenges amid relentless bandwidth growth and evolving datacenter demands.

Enrico Temporiti earned his MSc in Electronic Engineering from the University of Pavia in 1999. In 2000 he joined STMicroelectronics, working in R&D and product development. Since 2019 he has been with Marvell, where he serves as Sr. Director of Engineering, leading electro-optical PHY development.

1:20 PM
SC3: Silicon Photonics-Based Solutions: Components, Circuits, and Integration
Firooz Aflatouni, University of Pennsylvania, Philadelphia, PA

In this lecture, monolithic and hybrid electronic-photonic integration approaches are discussed and design approaches for key passive and active photonic devices such as single-mode and multi-mode waveguides, couplers, delay-lines, photodiodes, and modulators are presented. Laser sources and packaging of silicon photonic chips with lasers are discussed and typical noise sources in electronic photonics systems are introduced. Finally, examples of electronic-photonic systems such as optical interconnects using wavelength-division-multiplexing for data-centers and photonic compute modules are presented.

Firooz Aflatouni received the Ph.D. degree in Electrical Engineering from the University of Southern California. He was a post-doctoral scholar in the Department of Electrical Engineering at Caltech before joining the University of Pennsylvania where he is a Professor of Electrical and Systems Engineering. Firooz received the Bell Labs Prize in 2020, the ONR Young Investigator Program Award in 2019, the NASA Early-Stage Innovation Award in 2019, and the 2015 IEEE Benjamin Franklin Key Award. He is a fellow of Optica, and has served as a Distinguished Lecturer of the SSCS and on several IEEE conference program committees.

3:20 PM
SC4: Emerging Optical Applications and Circuit Approaches
Ali Hajimiri, California Institute of Technology, Pasadena, CA

In this presentation, we will discuss emerging optical applications and novel solutions enabling the co-integration of photonics and electronics. We will explore free-space applications such as imaging and projection, as well as the requirements and recent progress in photonic solutions for squeezed light. We will then introduce a subtractive photonic platform that leverages standard CMOS electronic circuits, utilizing the dielectric between metal lines as optical components capable of guiding light at visible and infrared wavelengths.

Ali Hajimiri is Bren Professor of Electrical Engineering and Medical Engineering at California Institute of Technology (Caltech) where he is the Director of the Microelectronics Laboratory, and the Co-Director of the Space Solar Power Project. He is a Fellow of National Academy of Inventors and a Fellow of the IEEE. Hajimiri holds close to 200 issued patents in the field of high-speed and high-frequency integrated circuits for sensors, photonics, biomedical devices, and communication systems. He co-founded Axiom Microdevices Inc. in 2002, which was later acquired by Skyworks Inc. in 2009. He is a recipient of several prestigious awards, including the Feynman Prize for Excellence in Teaching.

979-8-3315-8937-0/26 $31.00 © 2026 IEEE

Gap in pagination due to formatting issues.

Pages 672-679

ISSCC 2026 EXECUTIVE COMMITTEE

CONFERENCE CHAIR
Edith Beigné
Meta
Menlo Park, CA

ITPC EWAA
(Europe, West Asia, Africa)
REGIONAL CHAIR
Jens Anders
University of Stuttgart
Stuttgart, Germany

PRESS COORDINATOR
Shahriar Mirabbasi
University of British
Columbia
Vancouver, Canada

EXHIBITION CHAIR
Eric Karl
Intel

**PAST CONFERENCE
CHAIR AND ADCOM REP**
Eugenio Cantatore
Eindhoven University of
Technology
Eindhoven,
The Netherlands

ITPC EWAA
(Europe, West Asia, Africa)
REGIONAL VICE-CHAIR
Qinwen Fan
Delft University of
Technology
Delft, The Netherlands

**EDUCATIONAL EVENTS
CHAIR**
Ali Sheikholeslami
University of Toronto
Toronto, Canada

**STEERING COMMITTEE
CHAIR**
Anantha Chandrakasan
Massachusetts Institute of
Technology
Cambridge, MA

ITPC AM (Americas)
REGIONAL CHAIR
Danielle Griffith
Texas Instruments
Dallas, TX

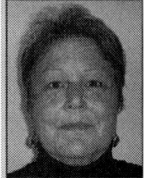

**DIRECTOR OF
OPERATIONS**
Melissa Widerkehr
Widerkehr and Associates
Lewes, DE

PROGRAM CHAIR
Keith Bowman
Qualcomm
Raleigh, NC

ITPC AM (Americas)
REGIONAL VICE CHAIR
Chris Rudell
University of Washington
Seattle, WA

DIRECTOR OF FINANCE
John Weinmann
Rochester, NY

PROGRAM VICE-CHAIR
Marian Verhelst
KU Leuven
Heverlee, Belgium

DEMO SESSION CHAIR
Patrick Mercier
University of California,
San Diego
La Jolla, CA

WEB SITE AND A/V CHAIR
Trudy Stetzler
Halliburton
Houston, TX

PAST PROGRAM CHAIR
Thomas Burd
Advanced Micro Devices
Santa Clara, CA

SRP CHAIR
Jerald Yoo
Seoul National University
Seoul, Korea

SOCIAL MEDIA CHAIR
Vito Giannini
L&T Semiconductor
Technologies
Austin, TX

ITPC APAC (Asia Pacific)
REGIONAL CHAIR
Wei-Zen Chen
National Yang Ming Chiao
Tung University
Hsinchu, Taiwan

ADCOM REPRESENTATIVE
Jan van der Spiegel
University of Pennsylvania
Philadelphia, PA

WIC REPRESENTATIVE
Farhana Sheikh
Intel
Hillsboro, OR

ITPC APAC (Asia Pacific)
REGIONAL VICE-CHAIR
Kousuke Miyaji
Shinshu University
Nagano, Japan

**DIRECTOR OF
PUBLICATIONS**
Laura Fujino
University of Toronto
Toronto, Canada

**INVITED SESSIONS
CHAIR**
Vivek De
Intel
Hillsboro, OR

ISSCC 2026 INTERNATIONAL TECHNICAL PROGRAM COMMITTEE

Technical Editors & Multi-Media Coordinator

Jason H. Anderson
University of Toronto
Toronto, Canada

Dustin Dunwell
Alphawave IP
Toronto, Canada

Glenn Gulak
University of Toronto
Toronto, Canada

Shahriar Mirabbasi
University of British Columbia
Vancouver, Canada

Leonid Belostotski
The University of Calgary
Calgary, Canada

Vincent Gaudet
University of Waterloo
Waterloo, Canada

James W. Haslett
The University of Calgary
Calgary, Canada

Samantha Murray
StarIC
Toronto, Canada

Multi-Media Coordinator and Digest Editor

David Halupka
StarIC
Toronto, Canada

PROGRAM CHAIR: Keith Bowman, Qualcomm, Raleigh, NC

PROGRAM VICE CHAIR: Marian Verhelst, KU Leuven, Heverlee, Belgium

---

Analog Subcommittee

Chair: **Viola Schaffer,** Texas Instruments, Freising, Germany

Ippei Akita
AIST, Tsukuba, Japan

Mei-Chen Chuang
TSMC, Hsinchu, Taiwan

Drew Hall
University of California at San Diego
La Jolla, CA

Tim Piessens
KU-Leuven ESAT-MICAS, Leuven, Belgium

Jens Anders
University of Stuttgart
Stuttgart, Germany

Chinwuba Ezekwe
Robert Bosch, Sunnyvale, CA

Chuan-Hung Hsiao
MediaTek, Hsinchu, Taiwan

Caspar van Vroonhoven
Analog Devices, Ismaning, Germany

Harijot Singh Bindra
University of Twente
Enschede, The Netherlands

Qinwen Fan
Delft University of Technology
Delft, The Netherlands

Minkyu Je
KAIST, Daejeon, Korea

Yihan Zhang
HKUST, Hong Kong, China

Edoardo Bonizzoni
University of Pavia, Pavia, Italy

Danielle Griffith
Texas Instruments, Dallas, TX

Ka-Meng Lei
University of Macau, Taipa, Macau

Data Converters Subcommittee

Chair: **Jan Westra,** Broadcom, Bunnik, The Netherlands

Lucien J. Breems
NXP Semiconductors
Eindhoven, The Netherlands

Wan Kim
Samsung Electronics, Hwaseong, Korea

Shanthi Pavan
IIT Madras, Chennai, India

Amy Whitcombe
Intel, Santa Clara, CA

Chi-Hang (Ivor) Chan
University of Macau, Taipa, Macau

Chin-Yu Lin
Mediatek, Hsinchu, Taiwan

Seung-Tak Ryu
KAIST, Daejeon, Korea

Il-Min Yi
Gwangju Institute of Science and
Technology, Gwangju, Korea

Vanessa Chen
Carnegie Mellon University, Pittsburgh,
PA

Nima Maghari
University of Florida, Gainesville, FL

Hajime Shibata
Analog Devices, Toronto, Canada

Ewout Martens
imec, Leuven, Belgium

Shiyu Su
University of Waterloo, Waterloo, Canada

Pieter Harpe
Eindhoven University of Technology,
Eindhoven, The Netherlands

Shahrzad Naraghi
ANAFLASH, Sunnyvale, CA

Xiyuan Tang
Peking University, Beijing, China

ISSCC 2026 INTERNATIONAL TECHNICAL PROGRAM COMMITTEE

Digital Architectures & Systems (DAS) Subcommittee
Chair: Rahul Rao, IBM India, Bangalore, India

Mark A. Anders
Intel Hillsboro, OR

Francesco Conti
University of Bologna
Bologna, Italy

Giuseppe Desoli
STMicroelectronics S.p.A. Italy,
Cornaredo, Italy

Jie Gu
Northwestern University Evanston, IL

Sumanth Gururajarao
MediaTek, Austin TX

Ji-Hoon Kim
Hanyang University, Seoul, Korea

Sugako Otani
Renesas Electronics, Tokyo, Japan

Jun-Seok Park
S.LSI, Samsung Electronics
Hwaseong, Korea

Nathaniel Pinckney
Nvidia, Austin, TX

Priyanka Raina
Stanford University, Stanford, CA

Soojung Ryu
Seoul National University
Seoul, Korea

Pei-Yun Tsai
National Taiwan University, Taipei, Taiwan

Srividhya Venkataraman
AMD, Santa Clara, CA

Paul Whatmough
Qualcomm, Boston, MA

Zhengya Zhang
University of Michigan, Ann Arbor, MI

Jun Zhou
University of Electronic Science and
Technology of China,
Chengdu, China

Digital Circuits (DCT) Subcommittee
Chair: Huichu Liu, Meta Platforms, Sunnyvale, CA

Sylvain Clerc
CEA LIST, Grenoble, France

Eric Jia-Wei Fang
Mediatek, Hsinchu, Taiwan

Kazuki Fukuoka
Renesas Electronics, Tokyo, Japan

Ping-Hsuan Hsieh
National Tsing Hua University
Hsinchu, Taiwan

Dongsuk Jeon
Seoul National University, Seoul, Korea

Jaydeep Kulkarni
The University of Texas at Austin
Austin, TX

Joachim Rodrigues
Lund University, Lund, Sweden

Visvesh Sathe
Georgia Institute of Technology
Atlanta, GA

Jae-sun Seo
Cornell Tech, New York, NY

Mahmut Ersin Sinangil
Nvidia, Santa Clara, CA

Carlos Tokunaga
Intel, Hillsboro, OR

Kaushik Vaidyanathan
OpenAI, San Francisco, CA

Bo (Angela) Wang
Singapore University
of Technology and Design,
Singapore, Singapore

Masanao Yamaoka
Hitachi, Tokyo, Japan

Alfred Yeung
AMD, Santa Clara, CA

ISD Subcommittee
Chair: Bruce Rae, STMicroelectronics, Edinburgh, United Kingdom

Jan Bogaerts
Gpixel, Antwerpen, Belgium

Leonardo Gasparini
Fondazione Bruno Kessler, Trento, Italy

Mutsumi Hamaguchi
Sharp Corporation, Tenri, Japan

Seunghoon Ko
Kwangwoon University, Seoul, Korea

Masaki Sakakibara
Sony Semiconductor Solutions
Corporation, Atsugi, Japan

Min-Woong Seo
Samsung Electronics Semiconductor R&D
Center, Hwaseong, Korea

Sanshiro Shishido
Panasonic Holdings Corporation,
Kadoma, Japan

Andreas Suess
Google, Mountain View, CA

Augusto Ximenes
CogniSea, Inc., Seattle, WA

Medical Subcommittee
Chair: Rikky Muller, University of California, Berkeley, Berkeley, CA

SungWon Chung
Neuralink, Fremont, CA

Azita Emami
California Institute of Technology,
Pasadena, CA

Venugopal Gopinathan
Analog Devices, Wilmington, MA

Taekwang Jang
ETH Zurich, Zurich, Switzerland

Hyung-Min Lee
Korea University, Seoul, Korea

Yu-Te Liao
National Yang Ming Chiao Tung University
Hsinchu, Taiwan

Carolina Mora-Lopez
imec, Leuven, Belgium

Michiel Pertijs
Delft University of Technology
Delft, The Netherlands

Mahsa Shoaran
EPFL, Geneva, Switzerland

Bo Zhao
Zhejiang University, Hangzhou, China

Memory Subcommittee
Chair: John Wuu, Advanced Micro Devices, Fort Collins, CO

Juang-Ying (Justin) Chueh
Etron Technology, Taipei, Taiwan

Chunmeng Dou
Chinese Academy of Sciences
Beijing, China

Zheng Guo
Intel, Portland, OR

Thomas Hein
Micron, Munich, Germany

Hua-Ling (Cynthia) Hsu
Sandisk, Milpitas, CA

Vita Pi-Ho Hu
National Taiwan University
Taipei, Taiwan

Takashi Ito
Renesas, Tokyo, Japan

Monodeep Kar
IBM Thomas J. Watson
Research Center
Yorktown Heights, NY

Bongjin Kim
KAIST, Daejeon, Korea

Dongkyun Kim
SK hynix, Icheon, Korea

Hye-Ran Kim
Samsung Electronics
Anyang, Korea

Saekyu Lee
EnCharge AI, Denver, CO

Gunther Lehmann
Infineon Technologies AG
Neubiberg, Germany

Xueqing Li
Tsinghua University
Beijing, China

Luigi Pilolli
Openchip & Software
Technologies, Rome, Italy

Shyh-Shyuan Sheu
ITRI, Hsinchu, Taiwan

Hidehiro Shiga, KIOXIA,
Yokohama, Japan

Eric Wang
TSMC, Hsinchu, Taiwan

Jun Yang
Southeast University
Nanjing, China

ISSCC 2026 INTERNATIONAL TECHNICAL PROGRAM COMMITTEE

Power Management Subcommittee

Chair: Bernhard Wicht, University of Hannover, Hannover, Germany
Vice-Chair: Saurav Bandyopadhyay, Texas Instruments, Dallas, TX

Sally Amin
Apple, Cupertino, CA

Patrik Arno
STMicroelectronics
Grenoble, France

Nicolas Butzen
Intel, Hillsboro, OR

Lin Cheng, University of Science
and Technology of China
Hefei, China

Sijun Du
Delft University of Technology
Delft, The Netherlands

Yuan Gao
Southern University of Science
and Technology
Shenzhen, China

Jianping Guo
Sun Yat-sen University
Guangzhou, China

Chen-Yen Ho
MediaTek, Hsinchu, Taiwan

Sung-Wan Hong
Sogang University, Seoul, Korea

Cheng Huang
Iowa State University, Ames, IA

Mo Huang
University of Macau, Taipa,
Macao, China

Rinkle Jain
Nvidia, Santa Clara, CA

Xiaocheng Jing
Nvidia, Santa Clara, CA

Shusuke Kawai
Toshiba, Kawasaki, Japan

Xugang Ke
Zhejiang University
Zhejiang, China

Dongsu Kim
Samsung Electronics
Hwaseong, Korea

Hyun-Sik Kim
KAIST, Daejeon, Korea

Hanh-Phuc Le
UCSD, La Jolla, CA

Xun Liu
The Chinese University
of Hong Kong (Shenzhen)
Shenzhen, China

Kousuke Miyaji
Shinshu University
Nagano, Japan

Frank Prämaßing
Infineon Technologies Austria AG
Villach, Austria

Joseph (Joey) Sankman
Analog Devices, Fort Collins, CO

Xin Zhang
IBM T. J. Watson Research Center
Yorktown Heights, NY

RF Subcommittee

Chair: Brian Ginsburg, Texas Instruments, Dallas, TX
Vice-Chair: Masoud Babaie, Delft University of Technology, Delft, The Netherlands

Sudipto Chakraborty
IBM T.J. Watson Research Center,
Yorktown Heights, NY

Dmytro Cherniak
Marvell Technology
Santa Clara, CA

Taiyun Chi
Rice University, Houston, TX

Wooyeol Choi
Seoul National University
Seoul, Korea

Jeremy Dunworth
Qualcomm Technologies, Inc.,
San Diego, CA

Xiang Gao
Zhejiang University
Zhejiang, China

Hiroshi Hamada
NTT, Kanagawa, Japan

Ruonan Han
Massachusetts Institute
of Technology
Cambridge, MA

Yu-Li Hsueh
MediaTek, Hsinchu, Taiwan

Ping Lu
NVIDIA, Seattle, WA

Danilo Manstretta
University of Pavia, Pavia, Italy

Omeed Momeni
UC Davis, Davis, CA

Jiayoon Ru
Peking University, Beijing, China

Mina Shahmohammadi
NXP Semiconductors
Delft, The Netherlands

Atsushi Shirane
Institute of Science Tokyo
Tokyo, Japan

Teerachot Siriburanon
University College Dublin Dublin,
Ireland

Henrik Sjöland
Lund University & Ericsson Lund,
Sweden

Mikko Varonen
VTT Technical Research
Centre of Finland Ltd.
Espoo, Finland

Yun Wang
Fudan University
Shanghai, China

Dihang Yang
Broadcom, Irvine CA

Jun Yin
University of Macau, Taipa,
Macau, China

Heein Yoon
Ulsan National Institute of
Science and Technology
Ulsan, Korea

Security Subcommittee

Chair: Takeshi Sugawara, The University of Electro-Communications, Tokyo, Japan

Utsav Banerjee
Indian Institute of Science
Bengaluru, India

Pasqualina Fragneto
STMicroelectronics, Agrate, Italy

Santosh Ghosh
Nvidia, Hillsboro, OR

Daniel Holcomb
University of Massachusetts, Amherst
Amherst, MA

Chiraag Juvekar
Apple, San Carlos, CA

Yong Ki Lee
Samsung Electronics, Suwon, Korea

Leibo Liu
Tsinghua University, Beijing, China

Thomas Poeppelmann
Infineon Technologies
Neubiberg, Germany

Shreyas Sen
Purdue University, West Lafayette, IN

ISSCC 2026 INTERNATIONAL TECHNICAL PROGRAM COMMITTEE

Technology Directions Subcommittee

Chair: Alyosha Molnar, Cornell University, Ithaca, NY

Firooz Aflatouni
University of Pennsylvania
Philadelphia, PA

Uygar Avci
Intel, Hillsboro, OR

Joseph Bardin
Google & UMass Amherst
Goleta, CA

Giorgio Ferrari
Politecnico di Milano, Milano, Italy

Koji Inoue
Kyushu University, Fukuoka, Japan

Kyeongha Kwon
KAIST, Daejeon, Korea

Noriyuki Miura
Osaka University, Osaka, Japan

Daniel H. Morris
OpenAI, San Francisco, CA

Chris Rudell
University of Washington, Seattle, WA

Fabio Sebastiano
Delft University of Technology
Delft, The Netherlands

Kaushik Sengupta
Princeton University, Princeton, NJ

Minyoung Song
DGIST, Daegu, Korea

Guy Torfs
Ghent University, Gent, Belgium

Albert Wang
Apple, Cupertino, CA

Kaiyuan Yang
Rice University, Houston, TX

Wireless Subcommittee

Chair: Chih-Ming Hung, MediaTek, Hsinchu, Taiwan

Alyssa Apsel
Cornell University, Ithaca, NY

Wu-Hsin Chen
Qualcomm, San Diego, CA

Hao Gao
Southeast University
Nanjing, China

Jose Luis Gonzalez-Jimenez
CEA-Leti, Grenoble, France

Giuseppe Gramegna
imec, Leuven, Belgium

Kuo-Ken Huang
Everactive, San Jose, CA

Konstantinos Manetakis
CSEM, Neuchâtel, Switzerland

Hyun-Chul Park
Samsung, Hwaseong, Korea

Raja Pullela
Maxlinear, Irvine, CA

Negar Reiskarimian
Massachusetts Institute of Technology,
Cambridge, MA

Shahriar Shahramian, Nokia – Bell
Labs, New Providence, NJ

Julian Tham
Infineon, San Jose, CA

Alberto Valdes-Garcia
IBM Research, Yorktown Heights, NY

Zhiwei Xu
Zhejing University, Zhejiang, China

Yun Yin
Fudan University, Shanghai, China

Yuanjin Zheng
Nanyang Technological University,
Singapore, Singapore

Alireza Zolfaghari
Broadcom, Irvine, CA

Wireline Subcommittee

Chair: Thomas Toifl, Cisco Systems, Thalwil, Switzerland

Tamer Ali
Mediatek, Irvine, CA

Wei-Zen Chen
National Yang Ming
Chiao Tung University
Hsinchu,Taiwan

Zeynep Toprak Deniz
IBM Research, Yorktown Heights, NY

Masum Hossain
Carleton University, Ottawa, Canada

Kenny Hsieh
TSMC, Hsinchu, Taiwan

Jay Im
AMD, San Jose, CA

Didem Turker Melek
Cadence Design Systems, San Jose, CA

Ahmed Mostafa
Marvell, Santa Clara, CA

Peter Ossieur
Imec, Univ. Ghent, Ghent, Belgium

Quan Pan
Southern University of Science
and Technology
Shenzhen, China

Ben Rhew
Samsung, Hwaseong, Korea

ISSCC 2026 ITPC EWAA (EUROPE, WEST ASIA, AFRICA) REGIONAL SUBCOMMITTEE

ITPC EWAA (Europe, West Asia, Africa) Regional Chair
Jens Anders
University of Stuttgart, Stuttgart, Germany

ITPC EWAA (Europe, West Asia, Africa) Regional Vice-Chair
Qinwen Fan
Delft University of Technology, Delft, The Netherlands

ITPC EWAA (Europe, West Asia, Africa) Regional Secretary
Frank Prämaßing
Infineon Technologies Austria AG, Villach, Austria

Patrik Arno
STMicroelectronics, Grenoble, France

Masoud Babaie
Delft University of Technology, Delft, The Netherlands

Harijot Singh Bindra
University of Twente, Enschede, The Netherlands

Jan Bogaerts
Gpixel, Antwerpen, Belgium

Edoardo Bonizzoni
University of Pavia, Pavia, Italy

Lucien J. Breems
NXP Semiconductors, Eindhoven, The Netherlands

Dmytro Cherniak
Marvell Technology, Santa Clara, CA

Sylvain Clerc
CEA LIST, Grenoble, France

Francesco Conti
University of Bologna, Bologna, Italy

Giuseppe Desoli
STMicroelectronics S.p.A. Italy, Cornaredo, Italy

Sijun Du
Delft University of Technology, Delft, The Netherlands

Giorgio Ferrari
Politecnico di Milano, Milan, Italy

Pasqualina Fragneto
STMicroelectronics, Agrate, Italy

Leonardo Gasparini
Fondazione Bruno Kessler, Trento, Italy

Jose Luis Gonzalez-Jimenez
CEA-Leti, Grenoble, France

Giuseppe Gramegna
imec, Leuven, Belgium

Pieter Harpe
Eindhoven University of Technology
Eindhoven, The Netherlands

Thomas Hein
Micron, Munich, Germany

Taekwang Jang
ETH Zurich, Zurich, Switzerland

Gunther Lehmann
Infineon Technologies AG, Neubiberg, Germany

Konstantinos Manetakis
CSEM, Neuchâtel, Switzerland

Danilo Manstretta
University of Pavia, Pavia, Italy

Ewout Martens
imec, Leuven, Belgium

Carolina Mora-Lopez
imec, Leuven, Belgium

Peter Ossieur
Imec, University of Ghent, Ghent, Belgium

Michiel Pertijs
Delft University of Technology, Delft, The Netherlands

Tim Piessens
KU-Leuven ESAT-MICAS, Leuven, Belgium

Luigi Pilolli
Openchip & Software Technologies, Roma, Italy

Thomas Poeppelmann
Infineon Technologies, Neubiberg, Germany

Bruce Rae
STMicroelectronics, Edinburgh, United Kingdom

Joachim Rodrigues
Lund University, Lund, Sweden

Viola Schaffer
Texas Instruments, Freising, Germany

Fabio Sebastiano
Delft University of Technology, Delft, The Netherlands

Mina Shahmohammadi
NXP Semiconductors, Delft, The Netherlands

Mahsa Shoaran
EPFL, Geneva, Switzerland

Teerachot Siriburanon
University College Dublin, Dublin, Ireland

Henrik Sjöland
Lund University & Ericsson,Lund, Sweden

Thomas Toifl
Cisco Systems, Thalwil, Switzerland

Guy Torfs
Ghent University, Ghent, Belgium

Mikko Varonen
VTT Technical Research Centre of Finland Ltd.
Espoo, Finland

Caspar van Vroonhoven
Analog Devices, Ismaning, Germany

Jan Westra
Broadcom, Bunnik, The Netherlands

Bernhard Wicht
University of Hannover, Hannover, Germany

ISSCC 2026 APAC (ASIA PACIFIC) REGIONAL SUBCOMMITTEE

ITPC APAC (Asia Pacific) Regional Chair
Wei-Zen Chen
National Yang Ming Chiao Tung University, Hsinchu, Taiwan

ITPC APAC (Asia Pacific) Regional Secretary
Jun Yin
University of Macau, Taipa, Macau, China

ITPC APAC (Asia Pacific) Regional Vice-Chair
Kousuke Miyaji, Shinshu University, Nagano, Japan

Ippei Akita
AIST, Tsukuba, Japan

Utsav Banerjee
Indian Institute of Science
Bengaluru, India

Chi-Hang (Ivor) Chan
University of Macau, Taipa
Macau, China

Lin Cheng
University of Science and
Technology of China
Hefei, China

Wooyeol Choi
Seoul National University
Seoul, Korea

Mei-Chen Chuang
TSMC, Hsinchu, Taiwan

Juang-Ying (Justin) Chueh
Etron Technology, Taipei, Taiwan

Chunmeng Dou
Chinese Academy of Sciences
Beijing, China

Eric Jia-Wei Fang
Mediatek, Hsinchu, Taiwan

Kazuki Fukuoka
Renesas Electronics
Tokyo, Japan

Hao Gao
Southeast University
Nanjing, China

Xiang Gao
Zhejiang University
Zhejiang, China

Yuan Gao
Southern University of Science
and Technology
Shenzhen, China

Jianping Guo
Sun Yat-sen University
Guangzhou, China

Hiroshi Hamada
NTT, Atugi, Japan

Mutsumi Hamaguchi
Sharp Corporation, Tenri, Japan

Chen-Yen Ho
MediaTek, Hsinchu, Taiwan

Sung-Wan Hong
Sogang University, Seoul, Korea

Chuan-Hung Hsiao
MediaTek, Hsinchu, Taiwan

Kenny Hsieh
TSMC, Hsinchu, Taiwan

Ping-Hsuan Hsieh
National Tsing Hua University
Hsinchu, Taiwan

Yu-Li Hsueh
MediaTek, Hsinchu, Taiwan

Vita Pi-Ho Hu
National Taiwan University
Taipei, Taiwan

Mo Huang
University of Macau, Taipa
Macao, China

Chih-Ming Hung
MediaTek, Hsinchu, Taiwan

Koji Inoue
Kyushu University
Fukuoka, Japan

Takashi Ito
Renesas, Tokyo, Japan

Minkyu Je
KAIST, Daejeon, Korea

Dongsuk Jeon
Seoul National University
Seoul, Korea

Shusuke Kawai
Toshiba, Kawasaki, Japan

Xugang Ke
Zhejiang University
Zhejiang, China

Bongjin Kim
KAIST, Daejeon, Korea

Dongkyun Kim
SK hynix, Icheon, Korea

Dongsu Kim
Samsung Electronics
Hwaseong, Korea

Hye-Ran Kim
Samsung Electronics
Anyang, Korea

Hyun-Sik Kim
KAIST, Daejeon, Korea

Ji-Hoon Kim
Hanyang University
Seoul, Korea

Wan Kim
Samsung Electronics
Hwaseong, Korea

Seunghoon Ko
Kwangwoon University
Seoul, Korea

Kyeongha Kwon
KAIST, Daejeon, Korea

Hyung-Min Lee
Korea University, Seoul, Korea

Yong Ki Lee
Samsung Electronics
Suwon, Korea

Ka-Meng Lei
University of Macau, Taipa
Macau

Xueqing Li
Tsinghua University
Beijing, China

Yu-Te Liao
National Yang Ming
Chiao Tung University
Hsinchu, Taiwan

Chin-Yu Lin
Mediatek, Hsinchu, Taiwan

Leibo Liu
Tsinghua University
Beijing, China

Xun Liu
The Chinese University
of Hong Kong (Shenzhen)
Shenzhen, China

Noriyuki Miura
Osaka University, Osaka, Japan

Sugako Otani
Renesas Electronics
Tokyo, Japan

Quan Pan
Southern University of Science
and Technology
Shenzhen, China

Hyun-Chul Park
Samsung, Hwaseong, Korea

Jun-Seok Park
S.LSI, Samsung Electronics
Hwaseong, Korea

Shanthi Pavan
IIT Madras, Chennai, India

Rahul Rao
IBM India, Bangalore, India

Ben Rhew
Samsung, Hwaseong, Korea

Jiayoon Ru
Peking University, Beijing, China

Seung-Tak Ryu
KAIST, Daejeon, Korea

Soojung Ryu
Seoul National University
Seoul, Korea

Masaki Sakakibara
Sony Semiconductor Solutions
Corporation
Atsugi, Japan

Min-Woong Seo
Samsung Electronics
Hwaseong, Korea

Shyh-Shyuan Sheu
ITRI, Hsinchu, Taiwan

Hidehiro Shiga
KIOXIA, Yokohama, Japan

Atsushi Shirane
Institute of Science Tokyo
Tokyo, Japan

Sanshiro Shishido
Panasonic Holdings Corporation
Kadoma, Japan

Minyoung Song
DGIST, Daegu, Korea

Takeshi Sugawara
The University of
Electro-Communications
Tokyo, Japan

Xiyuan Tang
Peking University, Beijing, China

Pei-Yun Tsai
National Taiwan University
Taipei, Taiwan

Bo (Angela) Wang
Singapore University of
Technology and Design
Singapore, Singapore

Eric Wang
TSMC, Hsinchu, Taiwan

Yun Wang
Fudan University
Shanghai, China

Zhiwei Xu
Zhejing University
Zhejiang, China

Masanao Yamaoka
Hitachi, Tokyo, Japan

Jun Yang
Southeast University
Nanjing, China

Il-Min Yi
Gwangju Institute of Science
and Technology
Gwangju, Korea

Yun Yin
Fudan University
Shanghai, China

Heein Yoon
Ulsan National Institute of
Science and Technology
Ulsan, Korea

Yihan Zhang
HKUST, Hong Kong, China

Bo Zhao
Zhejiang University
Hangzhou, China

Yuanjin Zheng
Nanyang Technological
University
Singapore, Singapore

Jun Zhou
University of Electronic Science
and Technology of China,
Chengdu, China

ISSCC 2026 ITPC AM (AMERICAS) REGIONAL SUBCOMMITTEE

ITPC AM (Americas) Regional Chair
Danielle Griffith
Texas Instruments, Dallas, TX

ITPC AM (Americas) Regional Vice-Chair
Chris Rudell
University of Washington, Seattle, WA

ITPC AM (Americas) Regional Secretary
Carlos Tokunaga
Intel, Hillsboro, OR

Firooz Aflatouni
University of Pennsylvania
Philadelphia, PA

Tamer Ali
Mediatek, Irvine, CA

Sally Amin
Apple, Cupertino, CA

Mark A. Anders
Intel, Hillsboro, OR

Alyssa Apsel
Cornell University, Ithaca, NY

Uygar Avci
Intel, Hillsboro, OR

Saurav Bandyopadhyay
Texas Instruments, Dallas, TX

Joseph Bardin
Google & UMass Amherst
Goleta, CA

Nicolas Butzen
Intel, Hillsboro, OR

Sudipto Chakraborty
IBM T. J. Watson Research
Center, Yorktown Heights, NY

Vanessa Chen
Carnegie Mellon University
Pittsburgh, PA

Wu-Hsin Chen
Qualcomm, San Diego, CA

Taiyun Chi
Rice University, Houston, TX

SungWon Chung
Neuralink, Fremont, CA

Zeynep Toprak Deniz
IBM Research
Yorktown Heights, NY

Jeremy Dunworth
Qualcomm Technologies, Inc.
San Diego, CA

Azita Emami
California Institute
of Technology
Pasadena, CA

Chinwuba Ezekwe
Robert Bosch, Sunnyvale, CA

Santosh Ghosh
Nvidia, Hillsboro, OR

Brian Ginsburg
Texas Instruments, Dallas, TX

Venugopal Gopinathan
Analog Devices
Wilmington, MA

Jie Gu
Northwestern University
Evanston, IL

Zheng Guo
Intel, Portland, OR

Sumanth Gururajarao,
MediaTek, Austin TX

Drew Hall
University of California
at San Diego
La Jolla, CA

Ruonan Han
Massachusetts Institute of
Technology, Cambridge, MA

Daniel Holcomb
University of Massachusetts,
Amherst
Amherst, MA

Masum Hossain
Carleton University
Ottawa, Canada

Hua-Ling (Cynthia) Hsu
Sandisk, Milpitas, CA

Cheng Huang
Iowa State University, Ames, IA

Kuo-Ken Huang
Everactive, San Jose, CA

Jay Im
AMD, San Jose, CA

Rinkle Jain
Nvidia, Santa Clara, CA

Xiaocheng Jing
Nvidia, Santa Clara, CA

Chiraag Juvekar
Apple, San Carlos, CA

Monodeep Kar
IBM Thomas J. Watson
Research Center
Yorktown Heights, NY

Jaydeep Kulkarni
The University of Texas
at Austin
Austin, TX

Hanh-Phuc Le
UCSD, La Jolla, CA

Saekyu Lee
EnCharge AI, Denver, CO

Huichu Liu
Meta Platforms, Sunnyvale, CA

Ping Lu
NVIDIA, Seattle, WA

Nima Maghari
University of Florida
Gainesville, FL

Didem Turker Melek
Cadence Design Systems
San Jose, CA

Alyosha Molnar, Cornell
University, Ithaca, NY

Omeed Momeni
UC Davis, Davis, CA

Daniel H. Morris
OpenAI, San Francisco, CA

Ahmed Mostafa
Marvell, Santa Clara, CA

Rikky Muller
University of California,
Berkeley
Berkeley, CA

Shahrzad Naraghi
ANAFLASH, Sunnyvale, CA

Nathaniel Pinckney
Nvidia, Austin, TX

Raja Pullela
Maxlinear, Irvine, CA

Priyanka Raina
Stanford University
Stanford, CA

Negar Reiskarimian
Massachusetts Institute
of Technology
Cambridge, MA

Joseph (Joey) Sankman
Analog Devices, Fort Collins, CO

Visvesh Sathe
Georgia Institute of Technology
Atlanta, GA

Shreyas Sen
Purdue University
West Lafayette, IN

Kaushik Sengupta
Princeton University
Princeton, NJ

Jae-sun Seo
Cornell Tech, New York, NY

Shahriar Shahramian
Nokia – Bell Labs
New Providence, NJ

Hajime Shibata
Analog Devices
Toronto, Canada

Mahmut Ersin Sinangil
Nvidia, Santa Clara, CA

Shiyu Su
University of Waterloo,
Waterloo, Canada

Andreas Suess
Google, Mountain View, CA

Julian Tham
Infineon, San Jose, CA

Kaushik Vaidyanathan
OpenAI, San Francisco, CA

Alberto Valdes-Garcia
IBM Research
Yorktown Heights, NY

Srividhya Venkataraman
AMD, Santa Clara, CA

Albert Wang
Apple, Cupertino, CA

Paul Whatmough
Qualcomm, Boston, MA

Amy Whitcombe
Intel, Santa Clara, CA

John Wuu
Advanced Micro Devices
Fort Collins, CO

Augusto Ximenes
CogniSea, Inc.
Seattle, WA

Dihang Yang
Broadcom, Irvine CA

Kaiyuan Yang
Rice University, Houston, TX

Alfred Yeung
AMD, Santa Clara, CA

Xin Zhang
IBM T. J. Watson
Research Center
Yorktown Heights, NY

Zhengya Zhang
University of Michigan
Ann Arbor, MI

Alireza Zolfaghari
Broadcom, Irvine, CA

ISSCC 2027 Call for Papers

IEEE INTERNATIONAL SOLID-STATE CIRCUITS CONFERENCE
SUNDAY – THURSDAY, FEBRUARY 14-18, 2027
SAN FRANCISCO MARRIOTT MARQUIS HOTEL, SAN FRANCISCO, CA

ISSCC 2027 CONFERENCE THEME:
TRUSTED SUSTAINABLE SILICON INTELLIGENCE FROM EDGE TO CLOUD

Innovative and original papers are solicited in subject areas including (but not limited to) the following:

ANALOG: Circuits with analog-dominated innovation; amplifiers, comparators, oscillators, filters, references; nonlinear analog circuits; digitally assisted analog circuits; sensor interface circuits; MEMS sensor/actuator interfaces, analog circuits in sub-10nm scaled technologies.

DATA CONVERTERS: Nyquist-rate and oversampling A/D and D/A converters; embedded and application-specific A/D and D/A converters; time-to-digital converters. Focus is on innovative and emerging converter architectures and the integration advantages within larger systems, rather than the figure of merit (FoM) of isolated converters. Submissions should consider converter peripherals, such as input, reference, and clock drivers. Calibration techniques should focus on innovation, robustness, and practical usability.

DIGITAL CIRCUITS, ARCHITECTURES, & SYSTEMS*: Digital circuits, building blocks, and architectures with hardware (HW) optimization of complete systems (monolithic, chiplets, 2.5D, and 3D) for microprocessors, micro-controllers, application processors, graphics processors, automotive processors, processors for machine learning (ML) and artificial intelligence (AI), and system-on-chip (SoC) processors. Digital systems and accelerators for communications, video, and multimedia, cloud and datacenter applications, optimization processors and accelerators, reconfigurable systems, near- and sub-threshold systems, and emerging applications. Digital circuits and architectures for intra-chip communication, clock distribution, soft-error, and variation-tolerant design, power management (e.g., voltage regulators, adaptive digital circuits, digital sensors), and digital clocking circuits (e.g., PLLs, DLLs) for processors. Digital circuits and systems, including near-memory and in-memory computation and HW optimizations for new ML models, such as transformers, graph and spiking neural networks, and hyper-dimensional computing.

IMAGE SENSORS & DISPLAYS: Image sensors; vision sensors, including event-based and computer sensors; LiDAR, time-of-flight, and depth sensing; machine learning and edge computing for imaging applications; display drivers, touch sensing, haptic displays, and interactive display and sensing technologies for AR/VR.

MEDICAL: Medical devices; biomedical sensors and SoCs; brain-computer interfaces, neural interfaces, and closed-loop systems; wearable, implantable, and ingestible devices; ultrasound and medical imaging; medical optical microsystems; body area networks; wireless power transfer and communication to implantable devices; machine learning and edge computing for medical applications; combinatorial innovation, including sensor fusion for disruptive clinical outcomes.

MEMORY: Static, dynamic, and non-volatile memories for stand-alone and embedded applications; memory/SSD controllers; high-bandwidth I/O interfaces for memories; memories based on phase-change, magnetic, spin-transfer-torque, ferroelectric, and resistive materials; array architectures and circuits to improve low-voltage operation, power, reliability, performance, and fault tolerance; application-specific circuit enhancements within the memory subsystem; in-memory-computing and near-memory-computing macros for AI or other applications.

POWER MANAGEMENT: Power management, power delivery, and control circuits; switched-mode power converter ICs using inductive, capacitive, piezoelectric, and hybrid techniques; LDO/linear regulators; power delivery for heterogenous integrated systems, and 2.5/3D-heterogeneous integrated power delivery; gate drivers; wide-bandgap (GaN/SiC) designs; isolated and wireless power converters; envelope supply modulators; energy-harvesting circuits and systems; power-management circuits for automotive and other harsh environments, robotics, LED drivers, and LiDAR.

RF CIRCUITS & WIRELESS SYSTEMS**: Complete solutions and building blocks at RF, mm-Wave, and THz frequencies for receivers, transmitters, frequency synthesizers, RF filters, transceivers, SoCs, and wireless SiPs incorporating multiple chiplets. Innovative circuits, systems, design techniques, heterogeneous packaging solutions, etc. for established and emerging wireless standards as well as future systems or novel applications, such as sensing, radar, imaging, satellite communications, and those improving spectral and energy efficiency.

SECURITY: Chips demonstrating cryptographic accelerators, smart-card security, trusted/confidential computing, security circuits (e.g., PUFs, TRNGs, side-channel and fault-attack countermeasures, circuits and sensors for attack detection and prevention), security for resource-constrained systems, secure micro-processors, secure memories, analog/mixed-signal circuit security (e.g., secure ADC/DAC, RF, sensors), secure supply chains (e.g., hardware trojan countermeasures, trusted microelectronics), security for/with emerging technologies, core circuit-level techniques for logical/physical-level security, secure system integration for specific applications, and secure design methodologies.

TECHNOLOGY DIRECTIONS: Emerging and novel IC, system, and device solutions in various areas, such as integrated photonics, silicon electronics-photonics integration; quantum devices for metrology, sensing, computing, etc.; flexible, stretchable, foldable, printable, and 3D electronic systems; biomedical sensors for cellular and molecular targets; wireless power transfer at-distance (e.g., RF and mm-wave, optical, ultrasonic); ICs for space applications and other harsh environments; novel and unconventional platforms for computing and machine learning, including analog and mixed signal techniques; integrated meta-materials, circuits in alternative device platforms (e.g., carbon, organic, superconductor, spin, etc.). Chip-scale autonomous microsystems and microrobots.

WIRELINE: Receivers/transmitters/transceivers for wireline systems, including backplane transceivers, copper-cable links, chip-to-chip communications, die-to-die interconnects, on-chip/on-package links, high-speed interfaces for memory; optical links and silicon photonics; exploratory I/O circuits for advancing data rates, bandwidth density, power efficiency, equalization, robustness, adaptative capabilities, and design methodologies; building blocks for wireline transceivers, such as AGCs, analog frontends, ADC/DAC/DSPs, TIAs, equalizers, clock generation and distribution circuits, including PLLs/DLLs, clock recovery, line drivers, and hybrids.

*This category will be reviewed by either the Digital Circuits Subcommittee or the Digital Architectures & Systems Subcommittee.

**This category will be reviewed by either the RF Subcommittee or the Wireless Subcommittee.

Deadline for Electronic Submission of Papers: Wednesday, September 2, 2026 • 3:00 PM Eastern Daylight Time (19:00 GMT)

STUDENT INITIATIVES
Graduate students are invited to participate in opportunities to showcase ongoing work and exchange experiences with other students and researchers from academia and industry. These include the Student Research Preview and the Silkroad Award (to a first-time student presenting author of a regular paper from an emerging region in the Far East).

Further information including submission procedures, formats, student initiatives and deadlines can be found at http://www.isscc.org

ISSCC 2026 TIMETABLE

ISSCC 2026 Conference Timetable and Session Room Assignments

FORUMS

F1: Power Efficient Circuits and Systems for Next-Gen Agentic AI and Robotics 8:00 am Salon 7	F2: Electrical and Optical Links Towards 400G+ Connectivity 8:00 am Salon 9

TUTORIALS

T1: Fundamentals of Energy-Efficient LDO Regulator Designs 8:30 am Salon 8	T2: Fundamentals of Cryptography for Chip Designers: Classical to Post-Quantum 8:30 am Salons 1-6	T3: Design Techniques for Robust and Energy-Efficient Biomedical Readouts 8:30 am Salon 8	T4: Fundamentals of Compute-in-Memory: From Circuits to Systems 10:30 am Salons 10-15	T5: Fundamentals of Image Sensors: From Photon to Image 10:30 am Salon 8	T6: Principles and Practices of High-Speed Digital-to-Analog Converter Design 10:30 am Salons 1-6	T7: Embedded Memory and Logic Circuit Design in Technologies Beyond FinFET 1:30 pm Salons 1-6	T8: Interference Mitigation Techniques in Wireless Communication Systems 1:30 pm Salons 10-15	T9: Clocking and CDR Techniques for High-Performance Wireline Transceiver 3:30 pm Salons 1-6	T10: Doherty Power Amplifier: Fundamentals and Recent Advances 3:30 pm Salons 10-15

SUNDAY, FEBRUARY 15th

Events Below in Bold Box are Included with your Conference Registration

AFTERNOON EVENT

CAREER PANEL: Chip In for Change: Career Insights for Young Designers in an AI-First, Eco-Conscious World
1:30 pm
SoMa Room

EVENING EVENTS

Bingo Networking Event (Open to All) 4:00 pm SoMa Room	EE1: Student Research Preview: Short Presentations with Poster Session 8:00 pm Golden Gate Ballroom

---

MONDAY FEBRUARY 16th

SESSION 1: Plenary Session - Invited Papers - 8:30 AM   Salons 6 - 10

SESSION 2: Processors 1:30 pm Salon 7	SESSION 3: Wearable and Wireless Biomedical Systems 1:30 pm Salon 8	SESSION 5: Sub-THz and mm-Wave Phased Arrays and Beamformers 1:30 pm Salons 1-6	SESSION 7: Image Sensors and Ranging 1:30 pm Salons 10-15	SESSION 8: Die-to-Die and High-Speed Electrical Transceivers 1:30 pm Salon 9	SESSION 9: Wireless Power 3:35 pm Foothill C (Second Floor)
	SESSION 4: Analog Techniques & Amplifiers 3:35 pm Salon 8	SESSION 6: Exploratory Receiver Architectures from GHz to THz 3:35 pm Salons 1-6			

3:00 PM to 8:00 PM – Book Displays • Corporations/Institution Exhibition
5:00 PM to 7:00 PM – Demonstration Session • 5:30 PM – Author Interviews • Social Hour

---

TUESDAY, FEBRUARY 17th

SESSION 10: Digital Processing and Circuit Techniques 8:00 am Salon 7	SESSION 11: Pipeline and Ultra-High-Speed Data Converters 8:00 am Salon 8	SESSION 12: Frequency Synthesizers and VCOs 8:00 am Salon 9	SESSION 13: Circuits for AI and AI for Circuits 8:00 am Salons 1-6	SESSION 15: DRAM, SRAM, and Non-Volatile Memories 8:00 am Salons 10-15	SESSION 16: Energy Harvesting, Piezo and Chargers 8:00 am Foothill C (Second Floor)
			SESSION 14: Unusual Interconnects and Other Uses for Light 10:05 am Salons 1-6		
SESSION 17: Highlighted Chip Releases for AI 1:30 pm - Salon 9	SESSION 19: High-Voltage, Isolated and Display Power 1:30 pm Salons 1-6	SESSION 20: RF Transceiver Subsystems from cm-Wave to THz 1:30 pm Salon 7	SESSION 21: Sensor Interfaces 1:30 pm Salon 8	SESSION 22: Circuits in Extreme Environments 1:30 pm - Salons 10-15	SESSION 24: Displays 3:35 pm Foothill C (Second Floor)
SESSION 18: Technology and Circuits for Domain-Specific Accelerators 3:35 pm - Salons 10-15				SESSION 23: Next-Generation Optical Transceivers 3:35 pm - Salon 9	

9:30 AM to 1:30 PM; and from 3:00 PM to 8:00 PM – Book Displays • Corporations/Institution Exhibition
5:00 PM to 7:00 PM – Demonstration Session • 5:30 PM – Author Interviews • Social Hour

EVENING EVENTS

EE2: Generative AI for Silicon Design: Mastering Complexity, Democratizing Design, and Building Trust 8:00 pm Salon 9	EE3: The Augmented Human - Will Chips in Our Brain Enhance Our Cognitive Abilities? 8:00 pm Salon 8

WEDNESDAY, FEBRUARY 18th

SESSION 25: Hardware Security 8:00 am Salons 1-6	SESSION 26: Compute Power and Supply Modulators 8:00 am Salons 10-15	SESSION 27: Frequency Generators, Multipliers and Modulators 8:00 am Salon 7	SESSION 28: Innovations from Outside the (ISSCC) Box 8:00 am Salon 8	SESSION 30: Compute-in-Memory 10:05 am Salon 9
			SESSION 29: Biochemical Sensors for Life Sciences and Agriculture 10:05 am Salon 8	
SESSION 31: AI Accelerators 1:30 pm Salon 9	SESSION 32: Low-Power Noise-Shaping ADCs 1:30 pm - Salon 7	SESSION 34: Integrated Radar and UWB Transceivers from Microwave to Sub-THz 1:30 pm - Salon 8	SESSION 36: Neural and Biomedical Interfaces 1:30 pm Salons 1-6	SESSION 37: Memory Interface 1:30 pm Salons 10-15
	SESSION 33: Time-Varying Circuit Techniques from RF to mm-Wave 3:35 pm - Salon 7	SESSION 35: Low Power Wireless Transceivers for Localization and Communications 3:35 pm - Salon 8		

5:30 PM – Author Interviews

THURSDAY FEBRUARY 19th

SHORT COURSE: Circuits for Optical Subsystems: Communications and Beyond 8:00 am Salons 10-15	F3: Powering the Future of AI, HPC, and Chiplet Architectures: From Dies to Package and Rack 8:00 am Salon 8	F4: The Race for 6G FR3 (7-24GHz): From Network Deployment to System Integration and Breakthrough Technology 8:00 am - Salons 1-6	F5: Analog for AI and AI for Analog: What the Analog/RF People Can Do and Leverage in the AI Era 8:00 am Salon 9	F6: Calibration and Dynamic Matching Techniques for High-Performance Data Converters 8:00 am Salon 7

CONFERENCE SPACE LAYOUT

Lower B2 Level - Yerba Buena Ballroom

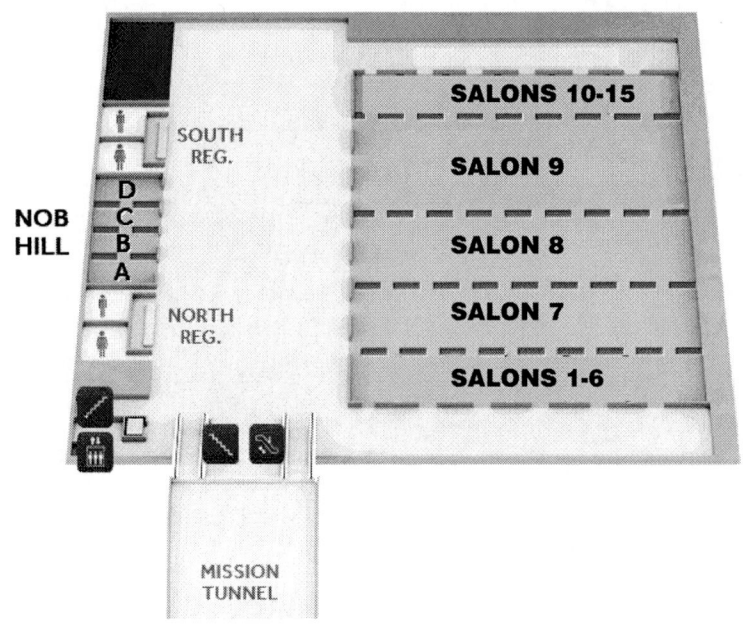

B2 Level - Golden Gate Hall

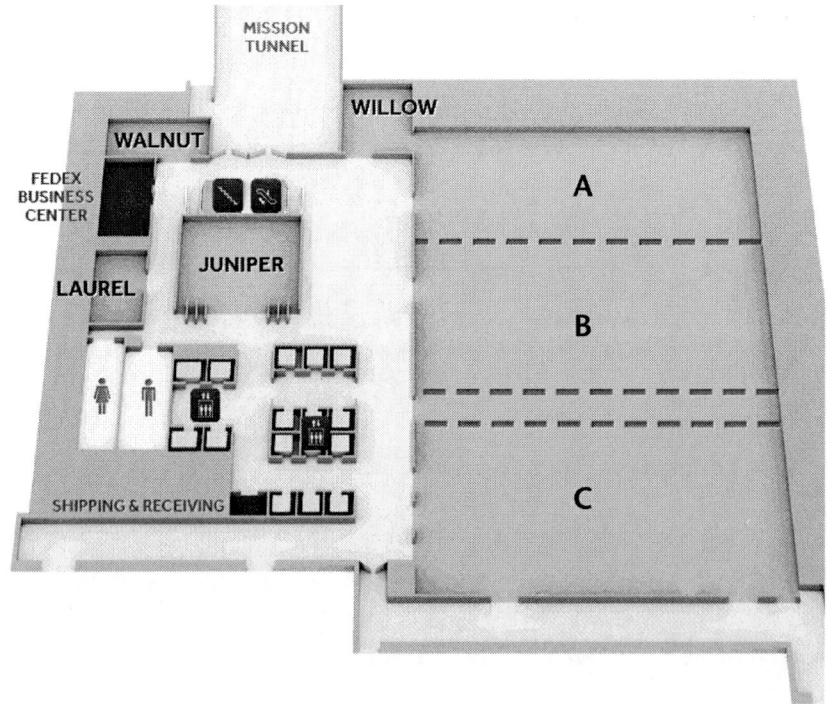

FOOTHILL C - SECOND FLOOR

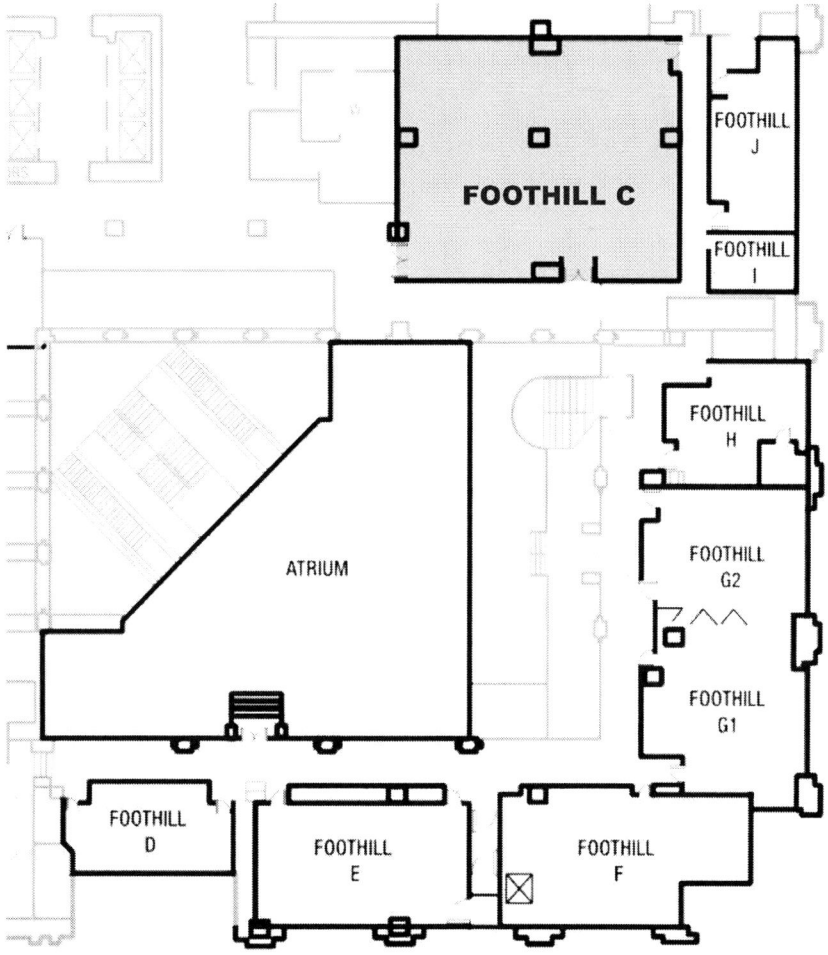

INDEX TO AUTHORS

A

Abe, Yudai 200
Abolmagd, Hany 588
Adaikkalavan, Ramasamy 42
Adebiyi, Jide Yinka 66
Aflatouni, Firooz 242
Agaësse, J.-F. 300
Agarwal, Akshit 508
Agarwal, Amit 528
Aguirre, Pablo 366
Ahn, Gukchae 44
Akinwande, Deji 66
Akso, Emre 488
Al-Hashimi, Bashir M. 482
Al-Rawhani, Mohammed 112
Albano, Domenico 398
Alcalde, Reinaldo E. 508
Ali, Karim 294, 596
Ali, Tamer 404
Alioto, Massimo 248, 596
Alioto, Massimo Bruno 294
Alizad, Sina Haji 366
Allam, Muhamed Fouad 94
Allen, Mark 490
Alvarez-Fontecilla, Enrique 202
Anand, Tejasvi 630
Anders, Mark A. 528
Antal, Sarthak 428
Ara[i]oglu, Arda 192
Ariki, Takuya 254
Ariyarathna, Viduneth 578
Arthanto, Yashael Faith 44
Artz, Patrick 104
Asci, Cihan 202
Asero, Claudio 398
Aseron, Paolo 424
Ashok, Maitreyi 250
Assmann, Andreas 112
Assous, Myriam 308
Ayesh, Mostafa 592
Ayodhyawasi, M. 300

B

Babaie, Masoud 208, 388, 562
Bae, ChangHyun 264
Bae, Daehee 128
Bae, Jaewan 44
Bae, Jeongyeol 578
Bae, Jooyoung 182
Bae, Junhyun 292
Bae, Junsang 116, 118
Bae, Sung-il 44
Baek, Goeun 480, 578
Baek, Sanghoon 258
Baek, Seung Ho 268
Baek, Seungcheol 44
Baek, Seungjae 342
Bahmani, Faramarz 144
Bai, Fujun 524
Bai, Jyun-Cheng 520, 526
Bajoria, Shagun 192
Baks, Christian W. 386
Ballo, Andrea 248, 294
Banerjee, Utsav 432
Bao, Hongyu Bruce 246
Bardsley, Scott 202
Bartholomew, Kevin 138
Bayle, Mathias 254
Bè, Gabriele 194
Bejarano-Carbo, Andrea 590
Beltran, Francisco Cardenas 354
Benites, Jorge Lagos 198
Bernabé, Stéphane 308
Bernard, Christian 308

Bertulessi, Luca 194
Bhamidipati, Sirisha 254
Bharadia, Dinesh 588
Bhat, Mohmad Aasif 572
Bhat, Savit 600
Bhatia, Sneha 254
Bhattacharya, Tinish 186
Bi, Xiaojun 400
Bijinapally, Kamalakanth 424
Bishop, Robert 202
Blaauw, David 590
Black, Ryan 386
Bo, Xiaochen 180
Boecker, Charles 138
Boesch, T. 300
Bolatkale, Muhammed 192
Bonfanti, Andrea Giovanni 194
Bose, Bella 630
Bose, Soumya 134
Bosi, Alessandro 398
Boutafa, Laura 308
Bowman, Keith 172
Branca, Xavier 112
Breems, Lucien 192
Brian, Michele 120
Brunsilius, Janet 202
Bulzacchelli, John F. 386
Bunsen, Keigo 200
Bushimata, Yuto 326
Byeon, Sangyeon 270
Byun, Sanho 420
Byun, Sung-Jae 128
Byun, Young-Yong 264

C

Cai, Fanxun 582
Cai, Hao 512, 536
Cai, Tianyi 376
Cai, Yuancheng 96
Cammarota, Rosario 424
Cantoni, Adalberto 202
Cao, Chenyao 146
Cao, Hung Van 168
Cao, Jun 174
Cao, Lixuan 566
Cao, Nianzheng 52
Cao, Peng 322
Cao, Qiankai 176
Cao, Yue 524
Cao, Yuluan 434
Casas, Jeremy 424
Catalino, Chris 52
Ceroni, Alessia 194
Cha, Jaehoon 270
Cha, Jinyoup 270
Chae, Heeyoung 44
Chae, Joo-Hyung 634
Chae, Kwanyeob 628
Chakraborty, Sudipto 386
Challagundla, AppaRao 424
Chan, Chi Hou 106
Chan, Chi-Hang 86
Chan, Chun-Kun 56
Chan, Lando 438
Chan, Wei Khuen 594
Chandrakasan, Anantha P. 250
Chang, Cheng-Feng 516
Chang, Chia-Ming 56
Chang, Chin-Yu 490
Chang, Huaichung 170
Chang, Jeongtaek 48, 590
Chang, Ken 398
Chang, Leland 52
Chang, Mau-Chung Frank 354
Chang, Meng-Fan 516, 520, 526

Chang, Min-Hua 64
Chang, Muya 548
Chang, ShenKai 56
Chang, Shih-Chieh 526
Chang, Tsung-Yung Jonathan 260
Chang, Yen-An 260
Chang, Young-Uk 264
Chang, Yung-Chang 56
Chapman, Eric 42
Charbonnier, Benoit 308
Charbonnier, Jean 308
Chatarasi, Prasanth 52
Chaves, Jose M. Rojas 424
Chawla, N. 300
Chen, Allen 144
Chen, Barry 170
Chen, Bu 122
Chen, Candy 42
Chen, Chang-Yuan 516
Chen, Changjin 460
Chen, Chia-Ping 56
Chen, Chin-Yi 254
Chen, Chixiao 240, 632
Chen, Christopher 354
Chen, Dihu 162
Chen, Fuzhan 246
Chen, Gong 576
Chen, Gregory K. 178
Chen, Hao 144
Chen, Hong 606
Chen, Hsinchen 170
Chen, Jhih-Wei 354
Chen, Jiawen 482
Chen, Jinbo 610
Chen, Jinghong 498
Chen, Jixin 348, 358
Chen, Junjie 218
Chen, Ko-Chi 526
Chen, Kuan-Chang 402
Chen, Kuan-Chun 260
Chen, Linchien 256
Chen, Mike Shuo-Wei 592
Chen, Minhan 138
Chen, Peng 348
Chen, Po-Wei 54
Chen, Qin 108
Chen, Tao 170
Chen, Wei-Chih 136
Chen, Weinan 434
Chen, Wenhua 350
Chen, Xi 176, 396, 512
Chen, Xibi 580
Chen, Xin 108
Chen, Xinyang 536
Chen, Xuesong 594
Chen, Yan 350
Chen, Yen-Ming 136
Chen, Yi 542
Chen, Yi-Syuan 56
Chen, Yi-Ting 136
Chen, Yifei 342
Chen, Ying 578
Chen, Yong 442
Chen, Yong-Tai 54
Chen, Yu-Chi 136
Chen, Yutang 162
Chen, Zaize 472
Chen, Zhe 358
Chen, Ziteng 448
Cheng, Bojun 314
Cheng, Chia-Yuan 56
Cheng, Chia-Yuan 256
Cheng, Depeng 108
Cheng, Hsin-Ping 56
Cheng, Jinhui 524

Cheng, Kai 442
Cheng, Kwang-Ting 532
Cheng, Kwang-Ting Tim 610
Cheng, Lin 280, 330, 452, 458, 460
Cheng, Nai-Chen 136
Cheng, Quan 390
Cheng, Sirui 610
Cheng, Ting-Yu 508
Cheng, Xuxu 150, 624
Cheng, Ziyi 606
Chenna, Vinay 234, 236
Chhetri, Ghanshyam 172
Chi, Baoyong 152, 220, 222
Chi, Hankyu 270
Chi, Miock 44
Chi, Taiyun 92, 344
Chiang, Ming-Hsuan 56
Chiang, Po-Han 56
Chien, Shihchieh 578
Chighine, Kevin 500
Chih, Hung-Wei 56
Chih, Yu-Der 260
Chin, Woojin 540
Chiou, Pei-Chen 136
Chiu, Taiwei 228
Cho, Gihwan 128
Cho, Hyun 628
Cho, Hyunjun 430
Cho, Joo-Mi 324, 332
Cho, Nara 44
Cho, S. 302
Cho, Wonhee 424
Choe, Yeounghwan 44
Choi, Dooseok 578
Choi, Haejung 372
Choi, Haidam 618
Choi, Hyeon-Ji 324, 456
Choi, Jaehyuk 116, 118, 124
Choi, Jinho 628
Choi, Jinyong 268
Choi, Jiwon 540
Choi, JongMoon 268
Choi, Juncheol 364
Choi, Junkyeong 44
Choi, Michael 372
Choi, Myunghoon 44
Choi, Sungpill 44
Choi, Won Ho 268
Choi, Won-Jong 364
Choi, Woo-Seok 148
Choi, Yong-Suk 128
Choi, Yoonjae 628
Choi, Youngkil 410
Choi, Yuseon 534
Chon, Hojae 612
Chou, Iris Ying 184
Chou, Kuan-Ting 136
Chou, Lin 64
Chou, Tan-Li 260
Chowdhury, Antroy Roy 404
Chu, Chenyuan 556
Chuang, Harry 260
Chuang, Ming-Han 154
Chun, Jung-Hoon 116, 118, 124
Chung, Jaehyun 578
Clemencon, Vincent 112
Clemons, Jason 548
Clymore, Christopher J. 488
Codega, Nicola 398
Cohen, Matthew 52
Collins, Steven 112
Cooman, Adam 198
Cordoba, Cyril 168
Coudyzer, Gertjan 406
Covington, William 176

• 2026 IEEE International Solid-State Circuits Conference

INDEX TO AUTHORS

Crafton, Brian 520
Craninckx, Jan 198
Criss, Russell 178
Crocherie, Axel 112
Crols, Sander 502
Crumley, Paul 52
Cui, Han 576
Cui, Kai 164
Cui, Yilu 552
Cuskelly, Lachlan 354
Cyrusian, Sasan 398

D

Dai, Jun 378
Dai, Steve 548
Dartizio, Simone Mattia 210, 212
Datsko, Benjamin 232
Datta, Kishalay 334
David, Jean-Baptiste 500
Davidson, Alfred 100
Davies, Andy 386
Davoodi, Mehdi 398
Dawson, Geraldine 494
Dayal, Pranav 578
Dayanik, Batu 144
De, Vivek 436
Dehos, Cédric 500
Demirci, Tugba 314
Demosthenous, Andreas 620
Demsky, Kevin 386
Deng, Heqi 562
Deng, Juncheng 576
Deng, Wei 152, 220, 222
Desai, Shaishav 138
Desoli, G. 300
Devgan, Anirudh 26
Diao, Yumei 204
Dikopoulos, Evangelos 232
Dillon, Chris 202
Ding, Lingke 428
Ding, Shixiang 498
Ding, Yifan 310
Ding, Yifang 96
Ding, Yixiao 490
Ding, Yong 448
Ding, Yuhan 466
Do, Hyungrok 266
Dogiamis, Georgios C. 580
Dong, Jun 512
Dong, Liang 158
Dong, Pingcheng 532
Dou, Chunmeng 514, 522, 524
Driel, Willem van 282
Du, Haoran 536
Du, Jieqiong 354
Du, Li 626
Du, Naike 582
Du, Sijun 68, 72, 160, 280, 282, 284, 288, 320, 336
Du, Xincheng 472
Du, Xingyu 490
Du, Yuan 626
Du, Yucheng 512
Du, Yukan 448
Duan, Zhouchi 146
Duda, Kevin 42
Dunna, Manideep 588
Duong, Cuong Manh 430
Dutta, Barundeb 616
Dutton, Neale A. W. 112

E

Ekman, Jeremy 386
Elbadry, Mohamed Abdelrahman 388

Eleraky, Mohamed 340, 346, 564
Eliezer, Oren 578
Elkholy, Ahmed 174
Elmenshawi, Ahmed 360
ElShater, Ahmed 404
Ema, So 90
Emami, Azita 508
Englund, Dirk 384
Enthoven, Luc 388
Eom, Kyeongho 612, 614
Erickson, Emma 386

F

Fabiano, Ivan 398
Fadila, Ashbir Aviat 476
Fagotti, Damiano 210, 212
Fahimnia, Mehrdad 144
Fakkel, Niels 388, 562
Fan, Qinwen 78, 380
Fang, Eric Jia-Wei 170
Fang, Lele 86
Fang, Wenkai 94
Fang, Xiao 582
Fang, Yidong 344
Fang, Yuan 144
Farhoodfar, Arash 398
Fei, Zheyuan 448
Feng, Lichen 518
Feng, Shuo Sarah 246
Feng, Xiaodi 240
Feng, Xiaoyu 60
Feng, Xinhe 126
Ferrari, Victor 52
Ferrer, Florencia 366
Feygin, Gennady 578
Flynn, Michael P. 232
Fojtik, Matthew 548
Francese, Pier Andrea 386
Franiatte, Rémi 308
Frank, David J. 386
Friedman, Daniel J. 386
Fu, Gaoming 512
Fu, Haotian 314
Fu, Jiamu 316
Fu, Renjie 582
Fu, Yang 398
Fu, Yushen 204
Fu, Zhigang 582
Fuguet, César 308
Fujihara, Yasuyuki 254
Fujimura, Susumu 254
Fujisawa, Yui 506
Fujita, Naoya 262
Fukuoka, Kazuki 168

G

Gaddam, Sudheer 402
Gade, Srinivas Pavan Kumar 42
Gai, Weixin 140
Galbraith, Bob 52
Gallucci, Stefano 210, 212
Gambarelli, Serge 500
Gao, Bin 228
Gao, Hanghang 514
Gao, Hao 358
Gao, Wei 492
Gao, Weichen 60
Garampazzi, Marco 398
Garg, Adesh 174
Garimella, Lakshminarasimha Sastry 100
Gaucher, Brian P. 386
Ge, Yu 458
Gebreyohannes, Fikre 172
Geng, Xiang 374

Geng, Xinli 204
Geng, Xinlin 466
Gerfers, Friedel 104
Geurts, Thijs 616
Ghittori, Nicola 398
Ghorbanpoor, Mohsen 564
Ghosh, Archisman 428
Ghozzy, Sherif 230
Giles, Hope 35
Giorgetti, Daniele 120
Gira, Gabriele 398
Giunco, Fabio 398
Giustolisi, Gianluca 248
Go, Jonghyun 128
Golder, Anupam 424
Gomez, Carrel de 476
Gong, Huijing 424
Gong, Xin-Ce 374
Gonzalez, Chris 52
Gooding, Tom 52
Gourdouparis, Marios 70
Grassi, Alberto 144
Grasso, Alfio Dario 248
Gray, C. Thomas 396
Greensky, James 424
Groen, Eric 138
Gu, Jiangyuan 58, 546
Gu, Jie 176, 602
Gu, Mingyang 196
Gu, Tingyi 244
Gu, Wenxian 84
Gu, Ye 406
Gu, Yucong 582
Guan, Ningzi 498
Gubin, Yaroslav 616
Gui, Xiaoyan 146
Guidry, Matthew 488
Guille, Olivier 308
Guillorn, Michael 52
Guo, An 512
Guo, Hao 92, 106
Guo, Jianping 162, 278
Guo, Luyi 482
Guo, Mingqiang 558
Guo, Ruiqi 58, 546
Guo, Yixin 524
Guo, Zhan 594
Guo, Zijun 106
Gurbaxani, Rishabh 208
Gurumurthy, Girishankar 256
Gutierrez, Christopher N. 424

H

Ha, Gyeongmin 420
Ha, Kyung-Soo 264
Ha, Sangwoo 312
Ha, Sohmyung 618, 620
Hajri, Basma 172
Hall, Drew A. 366
Hall, Duncan 112
Halli, Ramesh 256
Ham, Junhee 44
Hammoud, Ali 232
Han, Chanheum 634
Han, Jeongwon 364
Han, Jinho 402
Han, Jun 426
Han, Ruonan 250, 384, 580
Han, S. 302
Han, Seungjun 364
Han, Shinhee 258
Han, Shousheng 454
Han, Su-Hyun 128
Han, Xiangdong 434
Han, Yang 108

Han, Yiming 66
Han, Yuyang 384
Han, Zhongze 514, 522, 524
Handa, Takaya 254
Hanke, Chris 138
Hao, Zhenqi 228
Hara, Yusaku 168
Hardeman, Gilbert 192
Harris, Isaac B. 384
Hasan, Md Nazmul 588
Hasebe, Kazunori 200
Haselhorst, Tom 386
Hashemi, Hossein 234, 236
Hashimoto, Masanori 390
Hattori, Genma 90
Hayashi, Koichiro 254
He, Long 108
He, Longzhen 114
He, Shitu 152
He, Xiyu 190, 196
He, Yongjun 468
He, Yuchen 632
He, Yukun 146
He, Yuming 70
Hedayati, Hiva 402
Hekmatshoartabari, Bahman 52
Hella, Mona M. 360
Helleputte, Nick Van 502
Hensley, Mike 202
Heo, Hyungseok 44
Heo, Sanghyuk 266
Herbas, Daniel L. 398
Higashi, Yumi 254
Hilkens, E. 300
Hong, Binwen 400
Hong, Dongyeon 440
Hong, Gi-Moon 266
Hong, Kieop 124
Hong, Seongyon 540
Hong, Sung-Wan 324, 332, 456
Hong, Tao 122
Hong, Wei 348, 358
Hong, Zhiliang 76, 322, 368
Honma, Naoki 90
Hoover, Kathy 42
Horii, Kohei 326, 328
Hou, Longxiang 466
Hou, Zhang 434
Hsieh, Chih-Cheng 516, 520, 526
Hsieh, Kenny Cheng-Hsiang 136
Hsieh, Le-Jung 520, 526
Hsieh, Shih-Wei 56
Hsu, Chen-Hsing 64
Hsu, Chen-Kai 202
Hsu, Cynthia 254
Hsu, Hung-Hsi 516, 520, 526
Hsu, Lien-Feng 56
Hsu, Steven K. 528
Hsu, Ting-Hao 520, 526
Hsu, Ying-Tuan 232
Hsueh, Sung S.-Y. 170
Hu, Bo 536
Hu, Deyong 594
Hu, Hao 626
Hu, Hao-Tao 106
Hu, Hongyang 514, 522
Hu, Sanming 96
Hu, Yang 58, 316, 546
Hu, Yaolong 92
Hu, Yizhe 576
Hu, Yong 384
Hu, Yu-Jia 526
Huang, Bo-Jr 170
Huang, Chao-Tsung 54
Huang, Cheng 68, 158, 244

INDEX TO AUTHORS

Huang, Jie-Ren 136
Huang, Junwei 454, 462
Huang, Leilei 498
Huang, Mo 276, 286, 290, 446, 450
Huang, Po-Hao 56
Huang, Qiao 330
Huang, Qijing 548
Huang, Siyu 196
Huang, Stan 56
Huang, Tzu-Yuan 340
Huang, Weiwei 458
Huang, Wen-Hung 136
Huang, Xijie 532
Huang, Yao-Wei 64
Huang, Yen-Che 516
Huang, Yi 184
Huang, Yu-Jie 136
Huang, Yulong 314
Huang, Yunbo 442
Huang, Zebin 606
Huang, Zhangcheng 122
Huang, Zhi 174
Huang, Zhikai 504
Huang, Zhiqiang 470
Huang, Zhiwen 140
Hui, Yiheng 626
Humblet, Alexis 616
Hung, Chao-Jung 260
Huo, Dexuan 606
Hur, Joonhoi 578
Hursey, Josh 52
Hutchinson, George Higgins 186
Hwang, JongTae 264
Hwang, Jung-Hye 124
Hwang, Sang-Joon 264
Hwang, SangJoon 268, 272
Hwang, Seonwoo 270
Hwang, Shih-Arn 170
Hwang, Sohee 258

I

III, Roy H. Olsson 490
III, Thomas H. Greer 396
Iizuka, Tetsuya 468
Ilamurugan, Vinoth 192
Im, Maesoon 612, 614
Inagawa, Takahiro 90
Inoue, Ken 386
Irita, Takahiro 168
Ishibashi, Yushi 506
Ishida, Hideaki 554
Ishihara, Hiroaki 326, 328
Ishizaki, Yuki 254
Ismail, Yousr 174
Ito, Masamichi 128
Ito, Takashi 262
Iyer, Arvindh 144

J

Jacob, Philip 52
Jacquot, Jean-François 500
Jahagirdar, S. 304
Jain, Ajaypat 578
Jain, Radhika 52
Jain, Shubham 52
Jamal, Jismal 484
Jang, Jieun 266
Jang, Jinhun 268
Jang, Minsoo 268
Jang, Yonghwan Harold 578
Jantzi, Stephen 398
Jaussi, James 134
Javadi, Ramin 630
Je, Minkyu 618

Je, Sangeun 44
Jeon, H. 302
Jeon, Jaeho 430
Jeon, Jin-Yong 364
Jeon, S. 302
Jeon, Sehyug 342
Jeon, Taehoon 118
Jeon, Taeyoung 44
Jeon,, Gyunam 270
Jeong, H. 302
Jeong, Hyeongsoo 266
Jeong, Hyun-Woo 456
Jeong, Jaehun 128
Jeong, Ji Yong 120
Jeong, Seokhan 608
Jeong, Sera 270
Jeong, Suheon 542
Ji, Junghwan 270
Ji, Yichao 452, 458, 460
Jia, Haikun 152, 220, 222
Jia, Tianyu 310, 538
Jia, Yaoyao 66
Jiang, Chen 240, 632
Jiang, Dai 620
Jiang, Fuze 504
Jiang, Haijun 524
Jiang, Jianqiang 158, 244
Jiang, Jinghao 524
Jiang, Junmin 336
Jiang, Keyao 538
Jiang, Linrui 66
Jiang, Shunmin 288
Jiang, Wenning 632
Jiang, Xiping 524
Jiang, Xuhao 108
Jiang, Yumeng 606
Jiang, Ziwei 350
Jiao, Junkai 582
Jiao, Tianhui 512
Jie, Lu 190, 196, 378, 552
Jimenez, José Luis Gonzalez 500
Jin, Ji 452
Jin, Jinxuan 400
Jin, Ruibin 398
Jin, Young-Jae 44
Jin, Zhenghao 310
Jing, Zixi 478
Jo, Hyeonsu 578
Jo, Hyunje 44
Jo, Wooyoung 312, 540
Jo, Yun-Rae 420
Jo, Yurim 312
John, Deepesh 42
Joo, Sunghwan 264
Joo, Yongsuk 270
Josselin, Vincent 308
Jou, Yucheun Kevin 56
Ju, Chi-Cheng 56
Juan, Bo-Cheng 64
Junaid, Ammaar 100
Jung, C. 302
Jung, In 268, 272
Jung, Jaehong 250
Jung, Jinwook 52
Jung, Jonghoon 258
Jung, Jueun 46
Jung, Jun Won 144
Jung, Junghoon 128
Jung, Sang-Hoon 272
Jung, Seungjae 272
Jung, Woojoong 292, 372
Jung, Y. 302
Jung, Yooseok 266

K

Ka, Dongyoon 266
Kadaveru, Sreevatsank 588
Kalogerakis, Georgios 396
Kalzhan, Zhamaliddin 540
Kam, Dongyun 544
Kamei, Tatsuya 168
Kanagawa, Naoaki 254
Kaneko, Tohru 554
Kang, Bogyeong 440
Kang, Boyoung 628
Kang, Byungjun 270
Kang, Gyuseong 258
Kang, Hansol 628
Kang, Jaeyeol 266
Kang, Joonghoon 612
Kang, Jubin 124
Kang, Kidong 578
Kang, Kyeongpil 266
Kang, KyuChang 272
Kang, Seonghyeon 578
Kang, Shin-haeng 264
Kang, Sungmoon 44
Kang, Sungsik 628
Kang, Xilong 536
Kankuppe, Anirudh 198
Kano, Masahiro 254
Kanybek, A. 302
Kao, Tony 402
Kao, Yu-Sheng 520, 526
Kapusta, Ron 202
Kar, Monodeep 52
Karahan, Emir Ali 230
Karam, Victor 398
Karmakar, Angshuman 428
Kashirin, Alexander 178
Kashmiri, Mahdi 402
Katayama, Yasushi 200
Kato, Sena 90, 392
Kato, Soichi 200
Kato, Yosuke 254
Kaus, Jonathan 386
Kawai, Shusuke 326, 328
Kazemkhani, Shayan 402
Ke, Han-Tzung 136
Keller, Ben 548
Kelly, Dan 202
Kethareswaran, Lalith 424
Khailany, Brucek 548
Khanna, Devrishi 398
Khellah, Muhammad 176
Khiarak, Mehdi N. 398
Khorami, Ata 588
Khwa, Win-San 516, 520
Ki, Myoungoh 44
Kikkawa, Toshiyuki 200
Kim, Bongjin 182
Kim, Boram 270
Kim, Bumjun 128
Kim, Byeong-Chan 638
Kim, Byeongcheol 534
Kim, Chang Won 268
Kim, Daehoon 44
Kim, Daero 628
Kim, Daesun 272
Kim, Dohui 258
Kim, Donggeon 272
Kim, Donghan 44
Kim, Donghyeon 364
Kim, Donghyuk 116
Kim, Dongkyun 266
Kim, Doyoon 480, 578
Kim, Duhyeong 424
Kim, Eunseo 44
Kim, G. 302

Kim, Geunha 608
Kim, H. 302
Kim, Hangyeol 542
Kim, Hongseok 292
Kim, Hongyun 44
Kim, Hun-Seok 590
Kim, Hyeong-Joon 418
Kim, Hyun-Sik 412, 416, 418
Kim, Hyungsoo 266, 270
Kim, Hyunho 44
Kim, Hyunsung 44
Kim, J. 302
Kim, Jaehee 50
Kim, Jeong-Hun 332
Kim, Ji-Young 264
Kim, Jieun 266
Kim, Jinguk 268
Kim, Jinseok 44
Kim, Jinyeon 264
Kim, Jonghwan 270
Kim, Jonghyuk 268, 272
Kim, Jonghyun 578
Kim, Joo-Young 430, 542
Kim, Joohwan 264
Kim, Joonsuk 578
Kim, Jooseong 372
Kim, JuHwan 44
Kim, Junseong 578
Kim, Junsoo 272
Kim, Kahyun 148
Kim, Kangjoo 410
Kim, Kiheung 268
Kim, Kihyun 412, 418
Kim, Kyu-hyoun 52
Kim, Kyunghoon 270
Kim, Kyunghwan 272, 480, 578
Kim, Kyuseong 258
Kim, Kyuyoung 266
Kim, M. 302
Kim, Mijoung 258
Kim, Minchang 266
Kim, Minju 614
Kim, Minkyung 116
Kim, Mino 266
Kim, Minseo 44
Kim, Minsu 292
Kim, Minwoo 292
Kim, Minwook 636
Kim, Myunggon 628
Kim, S. 302
Kim, Sang Young 402
Kim, Sang Yun 268
Kim, Sangjin 312, 534
Kim, Seong-Jin 116, 118, 124
Kim, Seongjin 270
Kim, Seongjung 578
Kim, Seulgi 270
Kim, Seung-Goo 44
Kim, Seung-Sik 128
Kim, Siwoo 420
Kim, Suk Lae 272
Kim, Suksan 128
Kim, Sungjoo 578
Kim, Sunkwon 410
Kim, Taehyeon 636
Kim, Taewan 342
Kim, Taeyeon 578
Kim, W. 302
Kim, Wan 440
Kim, Wan Jong 578
Kim, Yeseul 628
Kim, Yong-Min 264
Kim, Yongjik 44
Kim, Yongjun 128, 272
Kim, Yoonhyung 628

INDEX TO AUTHORS

Kim, Youngtaek 270
Kimura, Hiroshi 144
Kitamura, Kei 254
Kitani, Tomofumi 254
Knag, Phil C. 178
Ko, Han-Gon 640
Ko, Hyeongjun 270
Ko, Seungpil 258
Ko, Youngwoon 364
Kocaman, Namik 144
Kodama, Takuyo 254
Komma, Demba 590
Kong, Hao 184
Kong, Xiangyu 184
Kong, Zhen 126, 390
Koo, Billy 628
Koo, Si-Gyoung 128
Kornfield, Julia A. 508
Koswatta, Siyu 52
Kota, Kishore 398
Kouchi, Toshiyuki 254
Krishnamurthy, Ram 528
Krishnamurthy, Ram K. 178
Krishnamurthy, Sashank 134
Krishnaswamy, Harish 100
Krithivasan, Sarada 52
Kuang, Honglin 426
Kuang, Jian-Jun 374
Kumar, Limitha 68
Kumar, Neelotpala 66
Kumar, Nitish 64
Kumar, Raghavan 424
Kummari, Shekher 366
Kundu, Suparna 428
Kunihiro, Kazuaki 476
Kuo, Hsin-Hung 136
Kuo, Sheng-Po 56
Kuo, Shih-Kai 588
Kwak, Yong-Sik 440
Kwon, Daehan 266
Kwon, Daehyun 268
Kwon, Dongseok 186
Kwon, Hanbyeol 266
Kwon, Hye-Jung 268
Kwon, Hyukbin 128
Kwon, Kyeongha 364
Kwon, Sooncheol 266
Kwon, Soonhyun 48, 544
Kwon, Taehyun 410
Kwon, Yongil 410

L

L'Bahy, Hassan 202
Lacaita, Andrea Leonardo 194, 210, 212
Lai, Sheng-Tsung 136
Lake, Dan 424
Lalwaney, Poornima 424
Lan, Juntao 220, 222
Lancaster, John David 52
Lang, Tian-Chen 374
Lassalle-Balier, Remy 366
Latham, Alex 366
Lau, Pak Tao Alan 246
Lau, Pak-Kim 578
Law, Duncan 42
Law, Man-Kay 114
Lee, Benjamin G. 396
Lee, Chan-Ho 456
Lee, Chanhee 48
Lee, Chia-Fu 260
Lee, Ching-Yen 414
Lee, Dongbeom 266
Lee, Dongkeon 268
Lee, Dongsoo 544
Lee, Eunseok 250, 384

Lee, Gangsik 270
Lee, Gwangwon 628
Lee, Haesuk 264
Lee, Hangil 258
Lee, Hyun-Su 612, 614
Lee, Hyunchang 258
Lee, Hyung-Min 292, 612, 614
Lee, J. 302
Lee, Jae-kyu 128
Lee, Jae-Yeol 410
Lee, Jae-Youl 420
Lee, Jaebong 44
Lee, Jaeho 258
Lee, Jaehoon 264, 636
Lee, Jaehyung 268
Lee, JaeKyung 264
Lee, Jeonghyeon 640
Lee, Jewon 636
Lee, Jihyun 420
Lee, Jingu 312
Lee, Jinseop 364
Lee, Jiseok 270
Lee, Jongmi 440
Lee, Jongmyeong 266
Lee, Jongwoo 410
Lee, Joonggeun 578
Lee, Joonhee 578
Lee, Jooseok 342
Lee, Joungwoo 44
Lee, Jun-Gi 412
Lee, Junghyup 584, 608
Lee, K. 302
Lee, Keonho 270
Lee, Ki-Soo 634
Lee, Kyongsu 638
Lee, Kyoungtae 584, 608
Lee, Kyoungwon 628
Lee, Kyuho Jason 46
Lee, Minoo 608
Lee, Minseob 578
Lee, Nayeong 540
Lee, Po-Hao 260
Lee, Saekyu 52
Lee, Sang-Gug 364
Lee, Sanggwon 128
Lee, Sangho 46
Lee, Sanghoon 270
Lee, Sangmin 628
Lee, Sangsung 578
Lee, Sangyong 268
Lee, Seon-Kyoo 636, 638
Lee, Seongseop 266
Lee, Seungho 266
Lee, Seungjun 268
Lee, Sunghyuck 372
Lee, Sungjun 578
Lee, Sungkwon 270
Lee, Sunkyu 258
Lee, Woncheol 578
Lee, Woo-Nyoung 420
Lee, Wooram 102, 352
Lee, Yong-Chan 332
Lee, Yongsun 264
Lee, Yoona 148
Lee, Youna 128
Lee, Younggeun 44
Lee, Youngjoo 48, 50, 544
Lee, Youngki 578
Lee, Youngsik 264
Lee, Yunho 292
Lei, Hao 582
Lei, Ka-Meng 80
Lei, Mingqian 204
Lekuch, Scott 386
Lele, Ashwin Sanjay 520

Leon, Ana Sonia 178
Leung, Ka Nang 278
Leung, Michael 398
Levantino, Salvatore 194, 210, 212
Levin, A. 304
Li, B. 302
Li, Bingrui 190, 556
Li, Bo 442
Li, Chang-Yi 136
Li, Chushan 448
Li, Dong 518
Li, Duo 476
Li, Haihua 80
Li, Hanyue 198
Li, Haoran 472
Li, Haoyuan 390
Li, Hongou 538
Li, Humiao 126
Li, Jason 254
Li, Jiamin 126, 390, 604
Li, Jiaming 228
Li, Jiawei 474
Li, Jiaxiang 566
Li, Jiayang 620
Li, Jinben 358
Li, Jinge 472, 474
Li, Liangwei 184
Li, Lin 228
Li, Mao 436
Li, Ming-Che 428
Li, Mingxuan 310
Li, Mingyuan 578
Li, Peng 512
Li, Qiufeng 390
Li, Sensen 344
Li, Shenggao 136
Li, Simon 138
Li, Vincent 172
Li, Weizeng 514, 522
Li, Wuhua 448
Li, Xiao 184
Li, Xiaoyu 576
Li, Xiayang 68
Li, Xueqi 228
Li, Xuyang 498
Li, Yang 214, 224
Li, Yicheng 566
Li, Yida 126
Li, Yiqi 176
Li, Yixi 218
Li, Yuanfang 144
Li, Zhao 202
Li, Zhenghao 150, 624
Li, Zhenhao 158, 244
Li, Zhensheng 558
Li, Zheyi 616
Li, Zhi 514, 522
Li, Zhouzheng 228
Li, Ziyang 582
Li1, Lianming 108
Liang, Dingqi 582
Liang, Guirong 254
Liang, Jin 138
Liang, Luhong 532
Liang, Xiqing 570
Liang, Yitao 180
Liang, Yuan 390
Liao, Yu-Te 64
Liao, Zhipeng 610
Lim, Daihyun 264
Lim, Gyu-Wan 416
Lim, Hyun-Wook 410, 420
Lim, Hyungsun 578
Lim, Kyungtae 128
Lim, Yong 440

Lin, Guan-Yu 64
Lin, Hon-Jarn 260
Lin, I-Ting 600
Lin, Jeff 256
Lin, Jiaming 536
Lin, Lishan 180
Lin, Longyang 126, 390, 604
Lin, Ming-Chieh 260
Lin, Ming-Hung 56
Lin, Minggui 606
Lin, Mu-Shan 136
Lin, Qingxuan 470
Lin, Shu-Ping 64
Lin, Wei-Shuo 136
Lin, Xiaohui 630
Lin, Yalong 400
Lin, Yen-Hua 516
Lin, Yi-Jie 64
Lin, Yifan 162
Lin, Ying-Sheng 414
Lin, Yu-En 526
Lin, Yu-Shiang 178
Lin, Zhen 96
Lin, Zhicheng 106
Lin, Ziyi 152
Linnhoff, Sebastian 104
Lipson, Samuel 42
Liss, Andrew 176
Liu, Bo 512, 536
Liu, Chang 152
Liu, Di 384
Liu, Gang 336
Liu, Hangxing 504
Liu, Jett 56
Liu, Jian 142, 218
Liu, Jiang 70
Liu, Jiayao 368
Liu, Kunyang 438
Liu, Leibo 184, 434
Liu, Lianbo 344
Liu, Liyuan 142, 218
Liu, Mengyao 498
Liu, Ming 122, 240, 514, 522, 524, 632
Liu, Qi 122, 240, 524, 632
Liu, Qian 626
Liu, Qilong 192
Liu, Ren-Shuo 516, 520, 526
Liu, Shen-luan 154
Liu, Shih-Yang 532
Liu, Supeng 594
Liu, Wen 138
Liu, Xiao 126
Liu, Xiaosen 216
Liu, Xing 610
Liu, Xinning 512
Liu, Xuanzhi 524
Liu, Xuejiao 532
Liu, Xun 336
Liu, Yanchao 214, 224
Liu, Yao-Hong 70
Liu, Yen-Jen 64
Liu, Yi 478
Liu, Yichen 216
Liu, Yongpan 60
Liu, Yu 532
Liu, Yunlong 518
Liu, Yuqi 434
Liu, Zeguo 458
Liu, Zezheng 476
Liu, Zhaokai 134
Liu, Zhichao 512
Lo, Chung-Chuan 520, 526
Lo, Wei-Chung 516
Lombard, Christian 500
Lombardo, Domenico Maria 78

INDEX TO AUTHORS

Long, Yucheng 576
Long, Zhijun 204
Lopes, Ward 396
Lopez, Carolina Mora 502, 616
Lou, Liheng 576
Lou, Tsung-Han 520, 526
Lovitt, Travis 398
Lu, Cheng-Han 260
Lu, Cheng-Hsun 50
Lu, Chien-Yu 170
Lu, Hao 66
Lu, Haowei 204
Lu, Ping 138
Lu, Po-Yen 54
Lu, Pong-Fei 52
Lu, Siuchuang Ivan 578
Lu, Tianqi 72, 160
Lu, Xiaoyu 348
Lu, Yan 164, 454, 462
Lu, Yue 498
Lu, Yuri 498
Lu, Zhichao 532
Luo, Chenjie 350
Luo, Hao 204
Luo, Haoyang 180, 190, 556
Luo, Jiaming 356
Luo, Peng 532
Luo, Qing 514, 522
Luo, Xiongshi 150, 624
Luo, Xun 568, 570
Luo, Yunbin 632
Luo, Yuxuan 376
Luo, Zhiren 458
Luong, Howard Cam 478
Lutz, Martin 52
Lv, Hangbing 524
Lyu, Liangjian 84
Lyu, Zhichao 126

M

Ma, He 626
Ma, Heng 380
Ma, Na 606
Ma, Rui 220
Ma, Ruitao Matthew 246
Ma, Songchen 532
Ma, Yuan 100
Ma, Yufei 310
Maangat, Simar 402
Machida, Shiro 168
Magod, Raveesh 176
Mai, Hanning 120
Mair, Hugh 170, 256
Maiyuran, Subramaniam 42
Mak, Pui-In 80, 442, 472, 474
Makinwa, Kofi A. A. 370
Malhotra, Seema 254
Malhouitre, Stéphane 308
Mandal, G. 304
Manetakis, Konstantinos 564
Mannari, Alberto 52
Mao, Junfa 568
Marani, Davide 120
Markuli , Nereo 198
Martens, Ewout 198
Martin, Fabrice 112
Martinelli, Fulvio 398
Martino, Matias Di 494
Martins, Rui 80
Martins, Rui P. 86, 114, 276, 286, 290,
446, 450, 472, 474, 558
Masilamanai, Indu 52
Mathaikutty, Deepak A. 528
Mathew, Sanu 424
Mathew, Sanu K. 436

Matsumoto, Tomohiro 200
Maurel, Vincent 500
Maurice, Lisa 52
Mayeda, Jill 90, 392
Mazzanti, Andrea 484
McGarry, William 602
Mehta, Nandish 396
Mello, Shalini De 548
Mellot, Pascal 112
Mendizabal, Laurent 308
Meng, Qianqi 358
Meng, Xiemei 178
Mercier, Patrick 288, 588
Midoh, Yoshihiro 506
Min, Qingqing 452
Min, Y. 302
Minami, Naoyuki 254
Ming, Xin 374
Miral, Nimesh N. 398
Mishra, Umesh K. 488
Miura, Noriyuki 506
Miura, Tomohiro 262
Miyahara, Masaya 392
Miyazaki, Daisuke 200
Miyazaki, Koutaro 326, 328
Mo, Yaowu 130
Moertl, Daniel 386
Moleri, Riccardo 210, 212
Momtaz, Afshin 144, 174
Monaco, Enrico 398
Mondal, Imon 572
Mondal, Susnata 134
Moon, Byeong-Taek 480, 578
Moon, Ji-Won 636
Moon, Kyoung-Jun 440, 578
Moon, Un-Ku 554
Moon, Youngjin 312
Morimoto, Masao 262
Morioka, Sumio 90
Mourik, Patrick van 192
Mueller, Josef 398
Mukkamala, Raj S. 508
Muller, Rikky 600
Munck, Koen De 616
Muralidharan, Sriram 360
Murata, Kentaro 90
Myko, André 308

N

Na, Daehoon 148
Na, H. 302
Na, Ki-Heon 264
Na, Sewhan 410, 420
Naffziger, Samuel 42
Nagarajan, Amrit 52
Nagata, Makoto 506
Nagata, Shun 200
Nagata, Shunya 262
Nair, Indira 52
Naito, Takahiro 200
Naka, Atsushiro 90
Nakamura, Daisuke 262
Nakano, Hiroyuki 168
Nakhkoob, Behrooz 402
Nallaparaju, Kalyan 138
Nam, InCheol 272
Nam, Sang-Yun 324
Naman, Harshit 428
Namkoong, Jin 402
Nani, Claudio 398
Narasimha, Rahul 590
Narukiyo, Yasuto 392
Naveed, Muhammad Zahid 248
Nazemi, Ali 174
Nedovic, Nikola 396

Nejad, Aboozar Ghorbani 588
Nett, Ryan 52
Neutens, Pieter 616
Newman, Dianne K. 508
Nguyen, Bo 402
Ni, Ronghua 214, 224
Nidhi, Nitin 144
Niitsu, Kiichi 438
Nishi, Yoshi 396
Nisi, Fabrizio De 120
Nitto, Daiki 506
Niu, Gaoqiang 390
Niu, Shengpu 406
Nivarthi, Karthik 424
Nogamida, Takeru 200
Nong, Yuanlin 626
Noori, Ari 386
Nussbaum, Pascal 564

O

O'Callaghan, John 616
Ochiai, Soichi 120
Oh, ChiSung 264
Oh, Hyoung-Seok 292
Oh, Jinseok 266
Oh, Jinwook 44
Oh, Jungju 44
Oh, Jungjun 534
Oh, Kwang-Seok 364
Oh, Minwook 266
Oh, Y. 302
Ojima, Naoki 254
Ok, Jiheon 420
Okada, Kenichi 476
Okamoto, Koichi 120
Okamura, Yusuke 120
Omirzakhov, aisarbek 242
Ommen, Hendrik Benjamin van 388
Oncu, Ahmet 506
Onda, Tomoya 168
Oo, Kaung Myat San 202
Ookuma, Naoki 254
Ou, Tincheng 472
Ouyang, Keqing 204

P

Pabba, Kaustubh 100
Pal-Singh, S. 300
Pan, Dongfang 330
Pan, Quan 150, 624
Pan, Sining 82, 552
Pan, Tao 202
Panahandeh, Mohamadamin 508
Pandey, Aviral 600
Pandey, Vaibhav 402
Pang, Di 532
Parisi, Angelo 198
Park, B. 302
Park, Chan-Hong 480, 578
Park, Chansoo 584
Park, Dongjun 636, 638
Park, Geonho 480, 578
Park, Gunho 544
Park, Gwangtae 312, 534
Park, Ha-Jung 148
Park, Hyun-Chul 480, 578
Park, Hyunjun 292
Park, Hyunsu 270
Park, J. 302
Park, Jemin 272
Park, Jeongbin 636
Park, Jeongje 266
Park, Jihye 372
Park, Joonhong 266

Park, Juho 230
Park, Junchul 420
Park, Jung-June 636, 638
Park, Kwanseo 640
Park, Kyuhyeon 608
Park, Kyungdam 364
Park, Minsoo 270
Park, Minsu 584
Park, Sanggyu 44
Park, Seonghyeok 124
Park, Seonwoo 618
Park, Seungwon 342
Park, Sunghyun 44
Park, Tae Jin 272
Park, Youngseok 272
Park, Yousung 416, 418
Parthasarathy, P. 304
Parthasarathy, Vinay 42
Pasdast, Gerald 134
Pasquo, Alessio Di 398
Pastorelli, Cedric 112
Patil, Nishant 100
Patrice, Damien Saint 308
Peng, Ji 576
Peng, Jida 282
Peng, Wenyu 68, 282
Peng, Yatao 474
Peng, Yimai 172
Perenzoni, Daniele 120
Perenzoni, Matteo 120
Perkins, Gina 66
Piao, Canxing 118
Pinckney, N. 298
Pinckney, Nathaniel 548
Pinkenburg, Jade 600
Piotto, Lorenzo 484
Pirbazari, Mahmoud M. 484
Pirmoradi, Ali 242
Privitera, Marco 248
Proft, Anabel De 616
Pu, Dong 224
Putra, Adiwena 430

Q

Qi, Dekui 130
Qian, He 228
Qian, Liwen 498
Qian, Yun 96
Qiao, Ning 314
Qin, Yubin 58
Qiu, Junyi 134
Qu, Tianxiang 76, 368
Qu, Wanyuan 448
Quarta, Gabriele 120

R

Race, Kylan 424
Rae, Bruce R. 112
Raha, Arnab 528
Rahman, Wahid 404
Ramapragada, Krishna Sai Tarun 432
Ramirez, Daniel 386
Rammohan, Ashwin 600
Ramos, Ernesto Zamora 424
Ran, Liang 126
Randhawa, Kavi 52
Rangarajan, Sundar 42
Ranjan, Ashish 52
Ravenhill, P. 300
Ray, Subhajit 386
Rayudu, Sai Krishna 578
Reddy, Sushmitha 138
Reick, Kevin 52
Rekhi, Angad 396

• 2026 IEEE International Solid-State Circuits Conference

INDEX TO AUTHORS

Ren, Guanjing 130
Ren, Hongyu 442
Ren, Huanyu 468
Ren, Shengdao 448
Ren, Wenjie 310
Ren, Zhibin 52
Renukaswamy, Pratap 198
Rey-Losada, Daniel 202
Rhew, Ben 372
Rhew, Hyo-Gyuem 628
Ricci, Luca 194
Rider, Scot 52
Riel, Heike 16
Riggelen, Margriet van 388
Rizzini, Daniele Lodi 210, 212
Robinson, Mike 402
Rocca, Francesco Paolo Mattioli Della 120
Rocco, Michele 194
Roewer, Thomas 52
Rogers, Gregory Eric 578
Roh, Hyeonhee 612
Roh, Wonjong 118
Rosno, Pat 386
Rosseel, G. 298
Rossoni, Michele 210, 212, 476
Ruffino, Andrea 386
Rui, Jiaqing 84
Rutten, Robert 192
Ryu, Je-Min 264
Ryu, Junha 312
Ryu, Kiljun 44
Ryu, Seongyoung 410
Ryu, Wonoh 128

S

Sadhu, Bodhisatwa 386
Saeidi, Hooman 94
Saif, Marco 504
Sakai, Hiroyuki 476
Sako, Mario 254
Salik, Marina 138
Salvi, Pietro 210, 212
Samajdar, Ananda 52
Samori, Carlo 194
Sang, Haoyang 114
Sanjaya, Pranata W. 260
Santana, Lucas Moura 198
Sapiro, Guillermo 494
Satterfield, Dave 52
Sawada, Yohei 262
Sawan, Mohamad 610
Sawigun, Chutham 616
Sayeed, Abdullah 428
Schaal, Marcel 52
Schmerbeck, Timothy J. 386
Scholz, Philipp 104
Scouten, Shawn 398
Sebastiano, Fabio 388, 562
Sen, Sanchari 52
Sen, Shreyas 428
Senger, Rob 52
Sengupta, Kaushik 94, 230
Seo, Daehwan 264
Seo, Jin-O 44
Seo, Min-Woong 128
Seo, Young-Hun 268, 272
Seok, Hyun-Gi 578
Seok, Mingoo 436
Seol, Hoseok 268
Seol, Ji-Hwan 268
Seong, Hyeonwoo 636
Serdijn, Wouter 70
Shallal, Aws 138
Shan, Kexin 84
Shan, Weiwei 512

Shan, Xiaoyu 442
Shang, Dechun 350
Shang, Minghao 280, 460
Shao, Yen-Tung 516
Shao, Zijian 94
Sheikh, Kashif M. 202
Shekhar, Sudip 588
Shen, Dawei 190
Shen, Junzhe 514, 522
Shen, Shen 402
Shen, Xiaohan 240
Shen, Xinyu 218
Shen, Yi 590
Shen, Yili 376
Shen, Yizhu 96
Shen, Zhongqiu 76
Sheng, Chaodi 400
Sheng, Maojia 228
Sheng, Yuguo 504
Shi, C.-J. Richard 84
Shi, Chengyao 70
Shi, Chunqi 498
Shi, Dan 80
Shi, Linxu 582
Shih, Ming-En David 56
Shim, Heesung 128
Shim, Jaechul 258
Shim, Kyu-Ha 264
Shim, Yeongseok 128
Shimazaki, Yasuhisa 168
Shin, D. 302
Shin, Deokha 128
Shin, Ho-Joon 640
Shin, Hoon 268
Shin, Hyunjin 258
Shin, Seunghun 182
Shin, Seunghwa 418
Shin, Taekyun 266
Shin, Wongyu 44
Shin, Yeonjae 608
Shinohara, Hirofumi 438
Shiomi, Jun 506
Shirane, Atsushi 90, 392
Shivnaraine, Ravi 138
Shokrolahzade, Ehsan 562
Shoobi, Amirreza 242
Shui, Hanyue 434
Shum, Kam Man 106
Shyh-Shyuan, Sheu 526
Si, D. 302
Si, Xin 512, 536
Siby, Emil 192
Sideris, Constantine 506
Silberman, Joel 52
Siligaris, Alexandre 500
Silla, Mark 42
Sim, Jae-Yoon 636, 638
Sin, Sai-Weng 454, 558
Singer, Larry 202
Singh, Abhishek 112
Singh, Jaswinder 256
Singh, Nawab 158
Singh, Teja 42
Singh, Ullas 144
Sinha, Sandipan 256
Siriburanon, Teerachot 482
Skende, A. 298
Smith, Alan 42
Snell, Bryce 386
So, Jeonggyu 540
Sohn, Young-Soo 264
Son, Insang 124
Son, Jihun 584
Son, Juhee 578
Son, Muyoung 542

Son, Sieon 44
Son, Young-Suk 364
Song, C. 302
Song, Eunji 270
Song, G. 302
Song, Heewoong 266
Song, Hong-Joo 268
Song, J. 302
Song, Jae-Joon 272
Song, Jaegeun 628
Song, Jaihyuk 128
Song, Jeongeun 148
Song, Minyoung 584, 608
Song, Sanquan 396
Song, Tae-Gyun 420
Song, Taeha 268
Song, Uijong 410
Song, Young Guen 264
Song, Zhibang 454
Soong, Ruei-Chen 354
Sosio, Marco 398
Soulie, M. 300
Spirito, Marco 562
Srinivasan, S. 304
Srinivasan, Viji 52
Srirambhatla, Vasantha 424
Stanzione, Stefano 70
Staszewski, Robert Bogdan 482
Stauth, Jason 334
Steiner, Michael 424
Stepko, Alexander 94
Still, Greg 52
Stojanovic, Vladimir 404
Stoppa, David 120
Strohman, Ryan 590
Strukov, Dmitri 186
Su, Jian-Wei 526
Su, Yu 146
Su, Yumin 92
Su, Yutong 546
Sugihara, Hiroyuki 128
Suh, Ji-Hoon 618
Suh, Junseuk 578
Suh, Kiseok 258
Sullivan, Charles 334
Sun, Guangyu 512
Sun, Huanfa 146
Sun, Lingling 356
Sun, Mingqian 368
Sun, Nan 196, 378, 552
Sun, Quan 170
Sun, Xuan 158, 244
Sun, Yi 426
Sun, Yiyang 538
Sun, Yuxuan 466
Sundaram, Sriram 42
Surana, Saurabh 144
Suresh, Vikram 424
Suriyasak, Chetphilin 506
Suzuki, Atsuya 200
Suzuki, Yoshinao 254

T

Tagawa, Hironori 438
Takahashi, Yuya 90, 392
Takasaki, Mika 200
Takaya, Satoshi 326
Takehara, Satoshi 554
Takiguchi, Kenichiro 262
Talamala, Bala Prasad 424
Tam, Sai-Wang 354
Tambe, Thierry 548
Taminiau, Tim Hugo 388
Tan, Nick Nianxiong 82
Tan, Yonghao 532

Tan, Zhichao 448
Tanaka, Hirofumi 66
Tanaka, Shinji 262
Taneja, Sachin 424
Tang, Adrian 354
Tang, Dawei 358
Tang, Hongxin 570
Tang, Jianshi 228
Tang, Junyao 158
Tang, Kea-Tiong 516, 520, 526, 606
Tang, Miaoyu 512
Tang, Minzhe 476
Tang, Siyuan 358
Tang, Wei 50, 232
Tang, Xiyuan 180, 190, 556
Tang, Yuxiang 76
Tang, Zhong 82
Tang, Zihao 450
Tangirala, Shankar 138
Tao, Guanren 590
Tao, Weichen 576
Tao, Yunsong 196, 378
Tatani, Keiji 120
Taufour, V. 300
Taupin, Sophie 112
Tavernier, Filip 502
Tell, Stephen G. 396, 548
Tellez, Gustavo 52
Temporiti, Enrico 398
Tessitore, Alex 138
Thasari, Kumar 144
Thimmaiah, Jayanth M 254
Thivin, Mathieu 112
Thonnart, Yvain 308
Thuaire, Herve 112
Tian, Fengshi 610
Tian, Li 204
Tian, Meijun 102
Tien, Chao-Jen 354
Tien, Jen-Chun 516, 520, 526
Tien, Kevin 386
Tilmans, Harrie A. C. 616
Timmerwilke, John 386
Tokunaga, Carlos 178
Tolaib, Islam 340
Tombolan, Giacomo 194
Tomita, Kazutoshi 200
Tong, Lin 512
Tong, Zhiguo 462
Torabi, Mohammadamin 144
Toth, Nandor G. 370
Tran, Trong-Hieu 170
Tripathy, S. 304
Trivedi, Manish 256
Trochut, Severin 112
Trotta, Giovanni Rocco 210
Tsai, C.-J. 170
Tsai, Chien-Chun 136
Tsai, Frank W. 254
Tsai, Ping-Yuan 56
Tsai, Rick 10
Tsai, Tsung-Hsien 136
Tsai, Yao-Hung 154
Tsai, Yueh-Feng 414
Tschanz, James W. 178
Tsen, Chia-Jung 260
Tseng, Mai 520, 526
Tseng, Yu-Cheng 56
Tsui, Chi-Ying 532, 610
Tu, Fengbin 532

U

Uddin, Syed Mohammad Ashab 352
Uggu, Viswanath 56
Um, Soyeon 312

INDEX TO AUTHORS

Um, Youngdo 268
Underwood, Devin 386

V

V, Aravinth 254
V, Indra K 254
Vachon, Robert 172
Vaish, Dhruv 600
Valluri, Sravanth 424
Vamvakos, Socrates 138
Vartak, Adish 424
Vaucher, Cicero S. 208
Vecchi, Federico 484
Vélard, Rémi 308
Vemulapalli, Hanish 42
Venkataramani, Swagath 52
Venkatesan, Rangharajan 548
Veraa, Brian 52
Verbauwhede, Ingrid 428
Verhelst, Marian 502
Vu, Roxanne 138

W

Wadatsumi, Takuya 506
Wakahara, Kohei 168
Wan, Jiapeng 96
Wan, Rentao 436
Wan, Yuting 84
Wang, Adam 504
Wang, Alex 144
Wang, Baochuang 280
Wang, Biao 558
Wang, Bin 184
Wang, Bo 114
Wang, Bohan 514, 522
Wang, Cheng 204, 406
Wang, Chia-Yu 260
Wang, Chih-Ming 56
Wang, Chuhui 278
Wang, Ericbill 170
Wang, Hanning 184, 434
Wang, Hedi 60
Wang, Hua 340, 346, 504, 564
Wang, Huanyu 58, 546
Wang, Hui 606
Wang, Jenny 254
Wang, Jian 578
Wang, Jinchen 384
Wang, Jingpeng 190
Wang, Jingyi 122
Wang, Kaihang 214, 224
Wang, Keping 350
Wang, Lei 626
Wang, Li 442
Wang, Linfang 514, 522
Wang, Nan 368
Wang, Pengcheng 582
Wang, Qian 130
Wang, Quan 626
Wang, Ruohan 498
Wang, Shixuan 582
Wang, Song 426, 524
Wang, Tiansu 288
Wang, Ting-Yu 54
Wang, Wei 140
Wang, Weibo 350
Wang, Weixiao 376
Wang, Wen 424
Wang, Wenqian 476
Wang, Xing 512
Wang, Xu 356
Wang, Yan 216
Wang, Yang 58, 316, 546, 604
Wang, Yih 260

Wang, Yiman 514, 522
Wang, Yiming 76
Wang, Yuan 180, 556
Wang, Yuanfei 290
Wang, Yunfan 590
Wang, Yunzhengmao 632
Wang, Zhao 114
Wang, Zheng 466
Wang, Zhifei 140
Wang, Zhihua 152, 220, 222, 358, 606
Wang, Zijie 582
Wang, Zongnan 556
Watanabe, oshihisa 254
Watanabe, Shunya 90
Wei, Kang 176
Wei, Shangjie 146
Wei, Shaojun 58, 316, 546
Wei, Yijie 602
Wei, Yuhong 498
Wen, Chin-Hua 136
Wen, Jincai 356
Wen, Yixuan 466
Weng, Chih-Hui 260
Whang, Sunjoo 312
Wie, Jeongyoon 608
Wilding, Dominik 104
Wilkerson, Chris 424
Wilkins, Paul 202
Willenborg, Scott M. 386
Winckel, Steven Van 198
Wittenhagen, Enne 104
Won, Bokyeon 272
Won, Sanggyeong 258
Woo, Seunghan 272
Woo, Sunsik 364
Woodward, Sandra 52
Wormald, Luke D. 232
Wu, Chih-Ling 516
Wu, Dabin 228
Wu, Defa 512
Wu, Ella 254
Wu, Hongzhi 150, 624
Wu, Huaqiang 228
Wu, Hui 610
Wu, J.J. 260
Wu, Jing 498
Wu, Lianbo 582
Wu, Lin 518
Wu, Mengze 366
Wu, Nanjian 142, 218
Wu, Peilin 594
Wu, Ping-Sheng 516, 520
Wu, Shuxian 490
Wu, Wanghua 578
Wu, Wei 178
Wu, Weitao 150, 624
Wu, Xiaonan 280
Wu, Xiaoyuan 498
Wu, Xing 84
Wu, Xu 108
Wu, Yihao 182
Wu, Yu 620
Wu, Yue 334, 474
Wu, Yuxia 356
Wu, Zihan 180
Wu, Zuoguo 134

X

Xia, Haiyang 108
Xia, Xiaoyue 106
Xiang, Longxi 368
Xiang, Xujiang 316
Xianyu, Haishu 228
Xiao, Junyue 358
Xiao, Qijing 376

Xie, Feng 566
Xie, Kenan 350
Xie, Qian 466
Xie, Shanshan 178
Xie, Wenao 114
Xing, Chaoyang 552
Xing, Yannan 314
Xiong, Yuang 476
Xu, Aolin 400
Xu, Bocheng 180
Xu, Dacheng 400
Xu, Dingxin 476
Xu, Dongfan 476
Xu, Hongtao 566
Xu, Jianlong 610
Xu, Jiawei 76, 322, 368
Xu, Jinglong 340
Xu, Kai 482
Xu, Li 396
Xu, Ningsheng 240
Xu, Richard 130
Xu, Ruohuang 310
Xu, S. 304
Xu, Shizhe 96
Xu, Shufan 438
Xu, Weiwei 330, 452, 458
Xu, Wende 58
Xu, Yilin 582
Xu, Yiqing 142
Xu, Yiwei 626
Xu, Zhuo 472
Xu, Ziang 400

Y

Yada, Nobuhiro 168
Yada, Satish 178
Yadav, Ajay 402
Yagishita, Yuki 200
Yamada, Shuhei 134
Yamaki, Yuzuru 554
Yamanaka, Sho 168
Yamashita, Ryuji 254
Yamauchi, Kazuki 254
Yan, Xinming 536
Yanaka, Kotaro 392
Yanase, Keishi 506
Yang, Bingzheng 568, 570
Yang, Bohan 434
Yang, Chia-Hsiang 414
Yang, Dezhen 314
Yang, H. 302
Yang, Hanlin 478
Yang, Huazhong 60
Yang, Jaehyeok 270
Yang, Jiachang 556
Yang, Jiacheng 450
Yang, Jianguo 524
Yang, Jianxin 286, 446
Yang, Jiawei 96
Yang, Jiaxin 546
Yang, Jie 218, 610
Yang, Jinjiang 434
Yang, Jun 512, 536
Yang, Jun-Hyeok 372
Yang, Kaiyuan 92
Yang, Renxu 314
Yang, Ruiyuan 294, 596
Yang, Shaoqi 576
Yang, Shiheng 466
Yang, Shu-Chun 136
Yang, Sung-gi 342
Yang, Tianze 566
Yang, Wei 228
Yang, Wen 478
Yang, Xiang 254

Yang, Xiaolin 616
Yang, Xiaolong 58
Yang, Xuecheng 76
Yang, Ya-Tang 520, 526
Yang, Yang 498
Yang, Yi 204, 512
Yang, Ying-Yu 64
Yang, Youming 180
Yang, Yunqi 582
Yang, Yunzhe 284, 320
Yang, Zhen 426
Yang, Zhengke 126, 390
Yang, Zhiyong 254
Yang, Zhizhan 472
Yang, Zunsong 442
Yao, Peng 228
Yao, Shun 490
Yao, Wang 594
Ye, Bingyi 140
Ye, Haifeng 498
Ye, Le 310
Ye, Shen 76
Ye, Tianchen 140
Ye, Xiangjun 310
Ye, Xiuzhu 582
Ye, Zhenkai 400
Ye, Zhongxin 278
Ye, Zonglin 466
Yeck, Mark 386
Yeh, Chao-Yang 170
Yeh, Po-Yu 56
Yeh, Shau-Hua 54
Yeh, Yao-Kai 516, 520, 526
Yennampelli, Nataraj 424
Yeo, Theng Tee 594
Yi, Donghyeon 618
Yi, Shinyoung 628
Yi, Tengyue 632
Yim, Sungsoo 272
Yin, Jun 462, 472, 474, 482
Yin, Shouyi 58, 316, 546
Yin, Shuying 434
Yin, Xin 406
Yin, Yun 566
Yingling, Daniel 172
Yoo, Changsik 268, 272
Yoo, Hoi-Jun 312, 534, 540
Yoo, Jerald 390, 604
Yoo, Junghyun 46
Yoo, Kyungwoo 578
Yoo, Sangmin 440, 578
Yoo, Seungjae 542
Yoo, Sun-Young 128
Yoo, Sungmin 372
Yoo, Sunwoo 544
Yook, Byungho 480, 578
Yoon, Chiweon 636, 638
Yoon, Hyunchul 272
Yoon, In-Soo 254
Yoon, Jae-Sung 44
Yoon, Jong-Hyeok 584, 608
Yoon, Jungmin 266
Yoon, Juyeong 44
Yoon, Sangsic 266
Yoon, Sung-Woo 268
Yoshizawa, Satoshi 200
You, De-Qi 516, 520, 526
You, Jungtaek 266
You, Shujuan 228
You, Xiaohu 108
Youn, Yelim 440
Young, Steve 590
Yu, Chang-Hyo 44
Yu, Chao 348
Yu, Dunshan 140

INDEX TO AUTHORS

Yu, Guodong 350
Yu, Haofeng 126
Yu, Haohong 594
Yu, He 602
Yu, Hwayeal 292
Yu, Luqi 348
Yu, Rui 594
Yu, Taewoo 578
Yu, Xiaohua 214, 224
Yu, Xiaopeng 82
Yu, Xinglong 426
Yu, Yichen 588
Yu, Yucheng 348
Yu, Zhehao 242
Yu, Zhewen 454, 462
Yuan, Chia-Hung 56
Yuan, Fengen 474
Yuan, Jinyi 280, 452
Yuan, Luyao 576
Yuasa, Keito 90
Yue, Chih-Yen 526
Yue, Chik Patrick 246
Yue, Zhiheng 58, 316, 546
Yun, Daeho 266
Yun, Ghangmin 46
Yun, Gichan 618
Yun, Jiwon 388
Yun, Sangbu 48
Yune, Sungwoong 430

Z

Zalani, Vidhi 52
Zanoletti, Gabriele 194
Zeng, Jianping 346
Zeng, Xiaoyang 368
Zeng, Yujie 146
Zha, Wenfeng 514, 522
Zhan, Mingtao 190, 378
Zhan, Xiangxun 472
Zhang, Aodong 378
Zhang, Aoyang 184, 434
Zhang, Bo 374, 516, 520
Zhang, Bodong 478
Zhang, Daxu 476
Zhang, Dong 532
Zhang, Fuyao 446
Zhang, Gaojing 348
Zhang, Guangzhong 582
Zhang, Guohe 146
Zhang, Guoqi 282
Zhang, Hao 378
Zhang, Haoming 468
Zhang, Hongyang 466
Zhang, Hongyong 610
Zhang, Huajun 78, 380
Zhang, Hui 582
Zhang, Jian 216
Zhang, Jilin 606
Zhang, Jing 406
Zhang, Lushuo 606
Zhang, Miao 78
Zhang, Nuo 202
Zhang, Qian 524
Zhang, Qingtian 228
Zhang, Qingyu 620
Zhang, Rui 278
Zhang, Runxi 498
Zhang, Shengxin 130
Zhang, Shiwei 222
Zhang, Shun 108
Zhang, Tantan 594
Zhang, Xiang 348
Zhang, Xin 176, 244, 320
Zhang, Yangyi 150, 624
Zhang, Yibin 350

Zhang, Yihan 290
Zhang, Yongle 330
Zhang, Yuncheng 476
Zhang, Zhao 142, 218, 610
Zhang, Zhaoyu 142
Zhang, Zhengya 48, 50, 232
Zhang, Zhi-Jun 54
Zhang, Zhishuai 196
Zhang, Zhiyuan 290
Zhang, Zhongyuan 96
Zhang, Zhuo 314
Zhang, Ziang 108
Zhao, Bo 160, 376
Zhao, Cankun 434
Zhao, Chen 144
Zhao, Guangshu 114
Zhao, Kangjie 498
Zhao, Lei 158
Zhao, Liang 126, 532
Zhao, Linran 66
Zhao, Menglian 448
Zhao, Weisheng 582
Zhao, Xiaoxi 68
Zhao, Xingpeng 594
Zhao, Yifan 426
Zhao, Yunxia 606
Zhao, Ziyan 498
Zheng, Hongzhao 314
Zhong, Jiajun 204
Zhong, Kangping 246
Zhong, Liwen 102
Zhong, Yi 196, 378, 552
Zhong, Zhiwei 602
Zhong, Ziyang 276
Zhou, Ching 52
Zhou, Feichi 126
Zhou, Jia 354
Zhou, Jiaqi 538
Zhou, Jie 568
Zhou, Jonathan 230
Zhou, Kaiwen 76, 368
Zhou, Peigen 358
Zhou, Qi 80
Zhou, Qiang 92, 344
Zhou, Rui 358
Zhou, Wen 578
Zhou, Yue 314
Zhou, Zeyu 576
Zhou, Zhidao 514, 522
Zhu, Haipeng 204
Zhu, Haiyang 202
Zhu, Jianfeng 184
Zhu, Jinpeng 626
Zhu, Junyu 514, 522
Zhu, Min 96, 434
Zhu, Minji 354
Zhu, Wenping 434
Zhu, Wenshuo 158, 244
Zhu, Wentao 204
Zhu, Xiaoge 204
Zhu, Xiaowei 348
Zhu, Yan 86
Zhu, Yuyang 468
Zhu, Zhangming 82, 518
Zhu, Zhaochen 470
Zhu, Zhongyao 454
Zhuang, Yi-Chang 170
Ziegler, Matthew 52
Zimmer, Brian 396
Zou, Wenjun 610
Zou, Yu 120
Zou, Zihan 536
Zuo, Fengguo 524

2026 IEEE International Solid-State Circuits Conference (ISSCC 2026)

San Francisco, California, USA
15-19 February 2026

Pages 1-341

IEEE Catalog Number: CFP26ISS-POD
ISBN: 979-8-3315-8937-0

**Copyright © 2026 by the Institute of Electrical and Electronics Engineers, Inc.
All Rights Reserved**

Copyright and Reprint Permissions: Abstracting is permitted with credit to the source. Libraries are permitted to photocopy beyond the limit of U.S. copyright law for private use of patrons those articles in this volume that carry a code at the bottom of the first page, provided the per-copy fee indicated in the code is paid through Copyright Clearance Center, 222 Rosewood Drive, Danvers, MA 01923.

For other copying, reprint or republication permission, write to IEEE Copyrights Manager, IEEE Service Center, 445 Hoes Lane, Piscataway, NJ 08854. All rights reserved.

*** This is a print representation of what appears in the IEEE Digital Library. Some format issues inherent in the e-media version may also appear in this print version.

IEEE Catalog Number: CFP26ISS-POD
ISBN (Print-On-Demand): 979-8-3315-8937-0
ISBN (Online):   979-8-3315-8936-3
ISSN: 0193-6530

Additional Copies of This Publication Are Available From:

Curran Associates, Inc
57 Morehouse Lane
Red Hook, NY 12571 USA
Phone: (845) 758-0400
Fax: (845) 758-2633
E-mail: curran@proceedings.com
Web: www.proceedings.com

Pagination in this book matches the original digital media

SUNDAY through THURSDAY / FEBRUARY 15, 16, 17, 18, and 19, 2026

2026 IEEE INTERNATIONAL

2026
DIGEST
OF
TECHNICAL
PAPERS

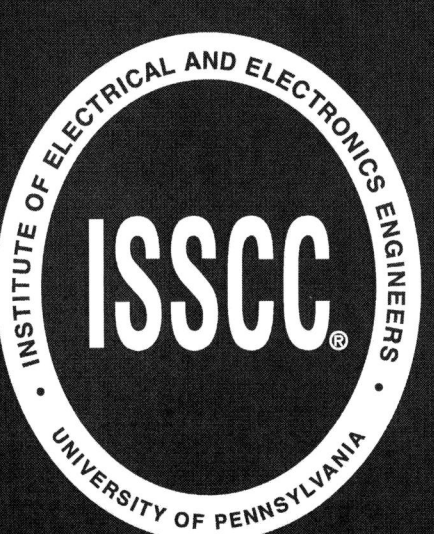

VOLUME SIXTY-NINE
Online ISSN 0193-6530

SOLID-STATE CIRCUITS CONFERENCE

IEEE SOLID-STATE CIRCUITS SOCIETY

2026 IEEE INTERNATIONAL
SOLID-STATE CIRCUITS CONFERENCE

DIGEST OF TECHNICAL PAPERS

First Edition

February 2026

979-8-3315-8937-0/26 $31.00 © 2026 IEEE

2026 IEEE International Solid-State Circuits Conference

DIGEST OF TECHNICAL PAPERS

LAYOUT IN THE UNITED STATES OF AMERICA
by S³ iPublishing, Inc.
Lisbon Falls, Maine

VOLUME 69

Library of Congress Number 81-644810
Online ISSN 2376-8606

Publisher and Managing Editor: Laura C. Fujino
Technical Editors: Jason H. Anderson, Leonid Belostotski, Dustin Dunwell, Vincent Gaudet,
Glenn Gulak, James W. Haslett, David Halupka, Shahriar Mirabbasi, Samantha Murray

TABLE OF CONTENTS

REFLECTIONS..4
FOREWORD..5
AWARDS...23

PAPER SESSIONS

1   Plenary - Invited Papers...8
2   Processors...40
3   Wearable and Wireless Biomedical Systems................................62
4   Analog Techniques & Amplifiers...74
5   Sub-THz and mm-Wave Phased Arrays and Beamformers.............88
6   Exploratory Receiver Architectures from GHz to THz....................98
7   Image Sensors and Ranging...110
8   Die-to-Die and High-Speed Electrical Transceivers.....................132
9   Wireless Power..156
10 Digital Processing and Circuit Techniques..................................166
11 Pipeline and Ultra-High-Speed Data Converters..........................188
12 Frequency Synthesizers and VCOs..206
13 Circuits for AI and AI for Circuits.......................................226
14 Unusual Interconnects and Other Uses for Light.........................238
15 DRAM, SRAM, and Non-Volatile Memories.................................252
16 Energy Harvesting, Piezo and Chargers.....................................274
17 Highlighted Chip Releases for AI...296
18 Technology and Circuits for Domain-Specific Accelerators.........306
19 High-Voltage, Isolated and Display Power..................................318
20 RF Transceiver Subsystems from cm-Wave to THz.....................338
21 Sensor Interfaces...362
22 Circuits in Extreme Environments..382
23 Next-Generation Optical Transceivers......................................394
24 Displays..408
25 Hardware Security..422
26 Compute Power and Supply Modulators...................................444
27 Frequency Generators, Multipliers, and Modulators....................464
28 Innovations from Outside the (ISSCC) Box...............................486
29 Biochemical Sensors for Life Sciences and Agriculture.............496
30 Compute-in-Memory...510
31 AI Accelerators..530
32 Low-Power Noise-Shaping ADCs...550
33 Time-Varying Circuit Techniques from RF to mm-Wave.............560
34 Integrated Radar and UWB Transceivers..................................574
    from Microwave to Sub-THz
35 Low Power Wireless Transceivers..586
    for Localization and Communications
36 Neural and Biomedical Interfaces..598
37 Memory Interface..622

TUTORIALS

TUTORIALS 1-10..643

FORUMS

F1 Power Efficient Circuits and Systems for....................................646
    Next-Gen Agentic and Robotics

F2 Electrical and Optical Links Towards 400G+ Connectivity.........648

F3 Powering the Future of AI, HPC, and Chiplet Architectures:......650
    From Dies to Package and Rack

F4 The Race for 6G FR3 (7-24GHz): From Network Deployment....652
    to System Integration and Breakthrough Technology

F5 Analog for AI and AI for Analog: What the Analog/RF...............654
    People Can Do and Leverage in the AI Era

F6 Calibration and Dynamic Matching Techniques.......................656
    for High-Performance Data Converters

SPECIAL EVENTS

EE1 Student Research Preview: 658
     Short Presentations with Poster Session

     CAREER PANEL: Chip In for Change: Career Insights............663
     for Young Designers in an AI-First, Eco-Conscious World

EE2 Generative AI for Silicon Design: Mastering Complexity,........665
     Democratizing Design, and Building Trust

EE3 The Augmented Human – Will Chips in Our Brain.................667
     Enhance Our Cognitive Abilities?

SHORT COURSE

SC Circuits for Optical Subsystems:...670
    Communications and Beyond

EXECUTIVE COMMITTEE...680

INTERNATIONAL TECHNICAL PROGRAM COMMITTEE...................681

ITPC EWAA SUBCOMMITTEE..685

ITPC APAC SUBCOMMITTEE..686

ITPC AM SUBCOMMITTEE...687

2027 CALL FOR PAPERS...688

CONFERENCE TIMETABLE...689

CONFERENCE SPACE LAYOUT...690

INDEX TO AUTHORS

Reflections

This year, ISSCC 2026 marks my 36th Conference. It has been yet another very challenging year for me. As has been done the past few years, with everyone pulling together, we again had our paper-selection process virtually, as scheduled, and what you see in front of you, is the result of decades of continuous iterative refinement of the submission process and information processing. What has become the norm, we are providing an e-Digest which will include all 3 pages for each paper, and will continue to be included in the Digest download and in IEEE Xplore.

Once more, we have a technical editorial group (listed below) under the direction of a managing editor (Laura Chizuko Fujino). Again, we emphasize full technical and language editing of most papers, as the need dictates.

In recognition of the huge amount of work leading to the Digest open before you, I wish to acknowledge a great many individuals: Keith Bowman, Marian Verhelst, Edith Beigne, Anantha Chandrakasan, the 13 Subcommittee Chairs, members of the ITPC, and all of the authors, for their individual contributions; Brad Phillips, and MiraSMART Conferencing, for Web-based and other preparatory support, including continuing improvement and facilitation of the paper-review and pre-voting process in a continuing double-blind world, as well as preparation and implementation of the Conference platform; Stephen Bonney, and S³ iPublishing, for author and Session-Chair interaction, for figure layout, for paper formatting, and for general assistance; Melissa Widerkehr, for general interfacing, problem solving, and coordination; Jason Anderson (University of Toronto), Leo Belostotski (University of Calgary), Dustin Dunwell (Alphawave), Vincent Gaudet (University of Waterloo), Glenn Gulak (University of Toronto), David Halupka (StarIC), James Haslett (University of Calgary), Shahriar Mirabbasi (University of British Columbia), and Samantha Murray (StarIC) as the technical editors for their heroic effort on the traditional very challenging schedule. My sincere thanks to you all!

Be well, and stay safe!

Laura Chizuko Fujino
ISSCC Director of Publications and Presentations

February 2026

ISSCC 2026 Program Assembly Meeting

Foreword

Advancing AI with IC & SoC Innovations

Keith Bowman
Qualcomm, Raleigh, NC
ISSCC International Technical Program Chair

Welcome to ISSCC 2026, *the foremost global conference for advances in integrated-circuit (IC) and system-on-chip (SoC) designs!* We assembled an exciting program with the latest IC and SoC innovations and developments, perspectives from industry and Solid-State Circuits Society (SSCS) leaders, a wide range of insightful educational events, and opportunities to meet with friends and colleagues from around the world. An enhanced online platform contains presentation slides and recordings for all plenaries, technical sessions, and educational events with the recordings made available the week after the conference.

In the past decade, we have witnessed dramatic improvements in Artificial Intelligence (AI) hardware, software, and models, paving the way for AI applications that are shaping many facets of our everyday life. With IC and SoC designs as the foundation of AI, the theme for this year's conference is *"Advancing AI with IC & SoC Innovations."* ISSCC 2026 opens with four world-renowned plenary speakers. Three plenary speakers highlight this theme with valuable perspectives on the enormous impact of AI on the semiconductor industry, the importance of energy-efficient architectures to enable future AI capabilities, employing agentic AI in design automation and methodologies, and the development of the next-generation IC and SoC design engineers. A fourth plenary speaker provides an insightful overview of the current state of quantum computing and outlines the milestones toward a large-scale, fault-tolerant quantum computer.

This year, marked the third consecutive year with a record number of ISSCC submissions, reaching 1,025 submissions and representing a 12% increase over last year and 66% higher than the 10-year average from 2014 to 2023. The ISSCC International Technical Program Committee (ITPC), composed of 213 experts covering a broad range of expertise from analog circuits to digital systems, selected 265 outstanding technical papers for presentation, including eight invited papers. Academic and industry affiliations represent 66% and 20% of the papers, respectively, with 2% from research institutes and 12% from joint industry, academic, and/or research institute collaborations. These papers are organized across 37 sessions, spanning a wide range of IC and SoC designs. Again, for this year, to accommodate the well-deserved increase in accepted papers, a 6ᵗʰ parallel track on the second level of the hotel in Foothill C is held during the Monday afternoon, Tuesday morning, and Tuesday afternoon sessions. Continuing with previous years, there are two invited half-sessions, one highlighting four groundbreaking chip releases from industry (Session 17) and the other describing recent developments in four "outside-the-box" topics (Session 28), each with future-looking implications and opportunities for the SSCS community. Both in-person and remote attendees have access to an online asynchronous chat to ask questions to authors about their papers. In-person attendees may also ask questions about the papers during the session and at the author interviews at the end of each day. Select papers are displayed during live Demonstration Sessions on Monday and Tuesday evenings, where attendees can observe the functional circuits and systems while engaging with the speakers.

ISSCC 2026 continues the rich tradition of offering a broad range of education events available both in person and on demand. There are ten tutorial sessions on Sunday with live presentations and Q&A for those in person or on-demand recordings and online asynchronous Q&A for remote attendees. The tutorials provide an opportunity to ramp up your knowledge on new topics from leading experts. While tutorials describe the fundamental concepts for specific topics, the full-day short course offers an in-depth learning experience from four experts in one subject. The ISSCC 2026 short course focuses on circuits and systems for optical communications and emerging applications. In addition, six forums provide a deep dive into important contemporary topics, highlighting the latest research, development, and trends in these areas. Each full-day forum consists of eight or nine technical presentations by leading experts in the field. Forum topics this year include circuits and systems for agentic AI and robotics, electrical and optical links, power management for AI and general high-performance computing, wireless communications for 6G FR3, the impact of analog design on AI and vice versa, and data-converter calibration and matching techniques. The forums and the short course are offered as in-person events with a live recording that is made available online for in-person and remote attendees with the corresponding registration. For 3ʳᵈ- and 4ᵗʰ-year undergraduate students and starting graduate students, ISSCC 2026 Circuit Insights cover the fundamentals of six circuit design topics during a live event on Saturday and recorded for a later release via the SSCS/ISSCC YouTube channel.

ISSCC 2026 has an exciting line-up of evening events, including networking and mentorship opportunities, a student research review, exhibitions, special-topic panels, demo sessions, and social hours. Women in Circuits (WiC) continue their successful networking and bingo event on Sunday late afternoon. On Sunday evening, the Student Research Preview (SRP) includes an inspirational lecture and then 26 sixty-second student research presentations, followed by a poster session. On Monday and Tuesday, the Corporate/Institution Exhibition provides a valuable connection for ISSCC attendees to the industry's leading companies and research institutions. On Tuesday, the conference offers two evening panels. The first panel debates the key opportunities and tradeoffs for employing AI in silicon design and the limits of AI-driven automation. The second panel debates the scientific, technological, and ethical considerations for the future possibility of augmenting human brains with semiconductor chips. In addition, a Demonstration Session for many technical papers and an attendee Social Hour follows at the end of the technical paper sessions on Monday and Tuesday evenings. Finally, I encourage you to attend and talk with the speakers at the author interviews on Monday, Tuesday, and Wednesday, directly after the close of the technical paper sessions.

The high quality of the ISSCC 2026 program is the result of the exceptional voluntary work of the ITPC members. I am deeply grateful to the ITPC members for their dedication, creativity, and leadership over the past year to create an exciting and enlightening ISSCC program. Starting in March 2025, the ITPC members proposed and organized the tutorials, forums, and panels, reviewed and selected the technical and invited papers from over one-thousand submissions, organized and attended countless regional, subcommittee, and program committee meetings, provided detailed feedback to all authors on submissions and final drafts, and compiled all associated materials for the Press Kit, Advance Program, and Digest of Technical Papers.

I am especially grateful to the Subcommittee Chairs for their commitment and leadership in these tasks: Viola Schaffer (Analog), Jan Westra (Data Converters), Rahul Rao (Digital Architectures & Systems), Huichu Liu (Digital Circuits), Bruce Rae (Image Sensors & Displays), Rikky Muller (Medical), John Wuu (Memory), Bernhard Wicht (Power Management), Brian Ginsburg (RF), Takeshi Sugawara (Security), Alyosha Molnar (Technology Directions), Chih-Ming Hung (Wireless), and Thomas Toifl (Wireline). I am also thankful for the hard work and leadership of the Subcommittee Vice Chairs: Saurav Bandyopadhyay (Power Management) and Masoud Babaie (RF). My special thanks go to the leadership of the Regional Committees: Wei-Zen Chen, Kousuke Miyaji, and Jun Yin for Asia Pacific, Jens Anders, Qinwen Fan, and Frank Prämaßing for Europe, West Asia, and Africa, and Danielle Griffith, Jacques "Chris" Rudell, and Carlos Tokunaga for the Americas. I am sincerely grateful to Marian Verhelst (ITPC Vice Chair) and her collaboration throughout the year to drive new initiatives, assemble this program, and coordinate the work of the Vision Committee. I deeply appreciate and admire Tom Burd (ISSCC 2025 ITPC Chair and ISSCC 2026 Conference Vice Chair), who taught me immeasurable lessons about the role of ITPC Chair such as managing difficult situations, analyzing complex information for informed decisions, and driving valuable initiatives to improve the conference. It has been an honor to collaborate with Tom.

DIGEST OF TECHNICAL PAPERS • 5

979-8-3315-8937-0/26 $31.00 © 2026 IEEE

ISSCC 2026 / FOREWORD

ISSCC is the product of many individuals who, year after year, work behind the scenes over the course of the year to arrange and support all aspects of the conference program. I sincerely thank Melissa Widerkehr for her valuable support with conference operations and arrangements. I am deeply grateful to Brad Phillips and MiraSMART Conferencing for their assistance in managing nearly all aspects of the conference from their powerful online platform, including the electronic manuscript submissions, reviews and pre-voting, final voting, assembly of the Advance Program, and final manuscripts, presentation slides, and recordings. Brad and his team created multiple enhancements to expand the online platform capabilities for our ITPC and authors, including new features to enable author feedback to all authors and a new blind-review process to reduce the overhead for Subcommittee Chairs and allow more time for paper reviews by ITPC members, while receiving a record-setting number of submissions. I deeply appreciate Steve Bonney and S3 iPublishing in all aspects of layout and formatting the Press Kit, Advance Program, and Digest with the most ISSCC publications in history. I am also sincerely grateful to the Technical Editors, who maintain the highest quality for ISSCC publications: Jason Anderson, Leo Belostotski, Dustin Dunwell, Vincent Gaudet, Glenn Gulak, James Haslett, Shahriar Mirabbasi, and Samantha Murray, and to David Halupka, who serves as both a Technical Editor and a Multi-Media Coordinator, which is a critical role for in-person and on-demand content.

Thank you to Shahriar Mirabbasi (Press Coordinator) for coordinating the process of providing timely and valuable content to press members, to Vito Giannini (Social Media Chair) for growing ISSCC's online outreach, and to Jerald Yoo (SRP Chair) for highlighting the in-progress research of students. ISSCC maintains a strong focus on highly relevant and valuable educational events. In these efforts, I gratefully acknowledge the work of Ali Sheikholeslami (Education Chair and Circuit Insights Organizer), Dan Friedman (Short-Course Chair), Yvain Thonnart (Short-Course Vice Chair), Sudip Shekhar (Tutorials Chair), and Stefano Pellerano (Forums Chair) for their coordination and organization of these insightful events on important contemporary topics. I thank Patrick Mercier for organizing the Demonstration Sessions to allow authors to highlight their functional circuit or system with highly-effective networking opportunities. Thank you to Vivek De (Invited Sessions Committee Chair) for leading the newly formed Invited Sessions Committee, to Hugh Mair (Industry Chair) for organizing the Invited Industry Session and driving industry engagement, and to Kaushik Sengupta (Out-of-the-Box Chair) for organizing the highly acclaimed Invited "Outside-the-Box" Session. Thanks go to Farhana Sheikh for the excellent programming of the Women-in-Circuits (WiC) event, Trudy Stetzler for managing the ISSCC website throughout the year, and to John Weinmann for his financial oversight for ISSCC. A major thank you to Eric Karl (Exhibition Chair) for coordinating an exciting Corporate/Institution Exhibition, providing a valuable connection for ISSCC attendees to the industry's leading companies and research institutions.

My deepest gratitude and sincere admiration go to Laura Fujino (Director of Publications) for the numerous roles that she performs in organizing the ISSCC program, going above and beyond in so many tasks, and resolving a plethora of continuous issues throughout the year. She encouraged and guided me through numerous complex situations. Furthermore, I greatly appreciate her willingness to embrace new initiatives to improve the conference. Laura's immense contributions to ISSCC include coordinating the Call for Papers, the new features for the MiraSMART online platform, the process for paper submission, review, and selection, Advance Program assembly, Digest editing and assembly, pre-conference materials, awards, and plenary-session management. *Simply put, Laura is the foundation of ISSCC!* For me personally, the highlight of serving as the ITPC Chair is my many interactions and collaborations with Laura. In preparations and during the conference, Laura and her fantastic team of University of Toronto graduate student volunteers *(i.e., The Saratoga Group)* consistently resolve unexpected challenges. Over the years, I have witnessed remarkable problem-solving capabilities by Laura and her team of student volunteers. These graduate students are behind every presentation at ISSCC, including the loading and updating of slides at speaker rehearsals and conference sessions, managing the slides during the presentations and Q&A, helping speakers to learn the presentation setup, and generally resolving any technical issues for the speakers. If the session is running smoothly, this is a result of the hard work and dedication of these students. *Laura and her graduate student volunteers are exceptional!*

I sincerely thank Anantha Chandrakasan (Plenary Committee Chair) for providing valuable insights and guidance throughout the process of organizing the Plenary Session. I greatly appreciate Anantha's willingness to help resolve difficult challenges as needed. I am also thankful for the advice and guidance of Eugenio Cantatore and Jan van der Spiegel as AdCom representatives on the Executive Committee. They provide excellent feedback on decisions to improve ISSCC.

I am deeply grateful to Edith Beigné (ISSCC Conference Chair) for her continuous support, advice, and leadership during my role as ITPC Vice Chair and ITPC Chair for the past two years. I consulted with Edith numerous times across a wide range of ISSCC topics, including many complex and difficult decisions. She consistently provided insightful perspectives and sound thinking to reach a resolution.

Finally, I want to express my sincere appreciation to the contributors of the ISSCC content: authors, speakers, panelists, and moderators. *Your effort and dedication make ISSCC the foremost global conference for advances in IC and SoC designs!*

I hope you enjoy the conference!

Keith A. Bowman, ISSCC 2026 Technical Program Chair

This page intentionally left blank.

Session 1 Overview: *Plenary*
INVITED PAPERS

Chair: Edith Beigné
Meta, Menlo Park, CA
ISSCC Conference Chair

Associate Chair: Keith Bowman
Qualcomm, Raleigh, NC
ISSCC International Technical Program Chair

8:30 AM
FORMAL OPENING OF THE CONFERENCE

The Plenary Session starts with welcoming remarks and introduction from the Conference Chair, Edith Beigné, followed by the International Technical Program Chair, Keith Bowman, providing an overview of ISSCC 2026.

The Plenary Session features four distinguished keynote speakers, who are leaders and pioneers in their domain, covering together a broad spectrum of our industry. An Awards Ceremony takes place after the first two Plenary talks to recognize major technical and professional accomplishments presented by the IEEE, Solid-State Circuits Society (SSCS), and ISSCC.

The first plenary **"Advancing Horizons for AI: Perspectives on Semiconductor Innovations"** is by **Rick Tsai**, Vice Chairman and CEO of MediaTek. This presentation describes the enormous impact of artificial intelligence (AI) on the semiconductor industry. While AI's demand for datacenter processing continues to increase at a meteoric rate, this plenary highlights the importance of scalable and energy-efficient architectures for moving a portion of the AI compute from the cloud to edge devices for decreasing energy consumption, lowering costs, protecting user privacy, and improving user experiences. This effort requires innovations in advanced packaging, power delivery, thermal management, high-bandwidth memory, interconnects, and wireless communications. Furthermore, this talk describes the opportunities for cross-layer optimizations across silicon, hardware, software, and algorithms to maximize energy efficiency in future AI systems.

The second plenary **"Quantum Computing – Toward Large-Scale Fault-Tolerant Quantum Computing"** is by **Heike Riel**, IBM Fellow and Head of Science of Quantum and Information Technologies at IBM Research. This presentation introduces quantum computing as a potentially transformative era in computation with the capability of solving certain classes of mathematical problems that are intractable for classical systems. This plenary provides an insightful overview of the current state of quantum computing and covers the requirements for practical quantum systems across the computing stack, from quantum bits (qubits) and quantum processors to control electronics, software, and algorithms. This talk describes a roadmap to realize a commercially viable quantum application, including breakthroughs in error correction, modularity, and integration of quantum accelerators with classical high-performance computing systems.

The third plenary **"Powering the AI Supercycle: Design for AI and AI for Design"** is by **Anirudh Devgan**, President and CEO of Cadence. With AI driving a rapid growth in demand for computing and scalability across all platforms, from datacenters to edge devices, this presentation describes the opportunity of applying agentic AI to electronic design automation (EDA) solutions to deliver increasingly complex systems. This plenary introduces a framework to categorize the application of AI to integrated circuit (IC) design with real-world examples. The framework describes increasing levels of autonomy from optimization AI for improving key metrics such as performance, power, and area across a vast parameter space (e.g., floor planning, place and route, etc.) to fully agentic AI, representing the pinnacle of AI-driven EDA, for managing the entire design lifecycle from initial specification to final system design.

The fourth plenary **"Empowering the Next Wave of Silicon Engineers"** is by **Hope Giles**, Vice President of Hardware Technologies at Apple. This presentation elucidates the significant benefits in product performance, power efficiency, capabilities, and user experiences from custom application-specific system-on-chip (SoC) designs with a full-stack optimization approach. Then, the plenary highlights the limited number of skilled silicon designers as a major challenge facing the future semiconductor industry. To address this critical problem, this talk declares a call to action for deep industry-academic collaborations to motivate, educate, and empower the next generation of silicon design engineers with encouraging examples of improving undergraduate electrical engineering class enrollment from strong industry-academic partnerships.

We hope you find these presentations informative, inspiring, and motivating!

ISSCC 2026 / February 16, 2026 / 8:30 AM

8:50 AM

1.1   Advancing Horizons for AI: Perspectives on Semiconductor Innovations

Rick Tsai, *Vice Chairman and CEO, MediaTek, Hsinchu, Taiwan*
AI is redefining the semiconductor landscape. Exponential growth in compute, bandwidth, and power efficiency is driving the proliferation of agentic and physical AI. The paradigm is shifting from performance alone to breakthroughs in scalable and energy-efficient architectures and cross-layer system innovation with close cooperation across silicon, design, heterogeneous hardware integration, and software. This plenary explores the key vectors of advanced packaging, power delivery, thermal management, high-bandwidth memory, interconnects, and wireless communication that will shape increasingly intelligent, robust AI systems for the decade ahead.

9:20 AM

1.2   Quantum Computing – Toward Large-Scale Fault-Tolerant Quantum Computing

Heike Riel, *IBM Fellow and Head of Science of Quantum & Information Technologies, IBM Research, Rüschlikon, Switzerland*
Quantum computing is advancing rapidly, offering the potential to transform computational paradigms and tackle problems that are intractable for classical systems. Realizing practical quantum systems demands the development of an entirely new computing stack—from qubits and quantum processors to components, wiring and control electronics, transpilers, error handling, software, algorithms, and the compute architecture to integrate quantum and classical resources at scale.

This plenary will provide an overview of the current state of quantum computing and outline the milestones on the path toward building a large-scale, fault-tolerant quantum computer by 2029. The significant improvements in hardware, software, and system integration, pushing the performance of quantum computing to reach quantum utility and advance toward quantum advantage, will be presented. Breakthroughs in error correction and a modular approach indicated in the IBM Quantum Roadmap outline a clear path to quantum systems using 200 logical qubits and 100 million quantum operations by 2029 and 2,000 logical qubits a few years later. Furthermore, advances in quantum algorithms, combined with the integration of quantum systems and high-performance computing (HPC), will unlock powerful synergies—accelerating the timeline for applications previously considered to be far in the future.

9:50 AM
ISSCC, SSCS, IEEE AWARD PRESENTATIONS

10:15 AM
BREAK

10:45 AM

1.3   Powering the AI Supercycle: Design for AI and AI for Design

Anirudh Devgan, *President and CEO, Cadence, San Jose, CA*
The AI supercycle is rapidly increasing demand for compute performance and scalability across all levels, from data centers to edge devices. By 2030, the semiconductor total addressable market (TAM) is projected to reach $1.2T with electronic systems at $5.2T. With silicon designs surpassing 200 billion transistors and chiplet-based architectures becoming common, traditional electronic-design-automation (EDA) workflows are insufficient. This plenary discusses how agentic AI can be applied to complex silicon and system design tasks through multi-agent orchestration and iterative reasoning, outlining a framework for its use in EDA solutions, highlighting some of the challenges, and exploring future directions. From optimizing power, performance, and area of silicon to managing data centers, AI-powered solutions are critical for the engineers designing the next generation of AI infrastructure.

11:15 AM

1.4   Empowering the Next Wave of Silicon Engineers

Hope Giles, *Vice President, Hardware Technologies, Apple, Cupertino, CA*
Developing custom chips for specific applications enables fundamentally better products. As part of a full-stack optimization approach to product design, application-specific systems on chip (SoCs) have been one of the keys to delivering game-changing performance, power efficiency, capabilities, and user experiences. This plenary provides examples of how custom silicon makes delivering innovative breakthroughs in products possible and how techniques like design modularity help manage the SoC scaling challenge. As emerging applications, including artificial intelligence (AI), transform both silicon demand and our design methodologies, we need to reimagine how we motivate and educate the next generation of silicon engineers to address these challenges. We will share our experiences in industry/university/student interactions and how the industry can revitalize these programs. The future of custom silicon depends on all of us investing in the next wave of engineers who will create tomorrow's innovative products.

11:45 AM
PRESENTATION TO PLENARY SPEAKERS

11:50 AM
CONCLUSION

979-8-3315-8937-0/26 $31.00 © 2026 IEEE

ISSCC 2026 / SESSION 1 / PLENARY / 1.1

1.1 Advancing Horizons for AI: Perspectives on Semiconductor Innovations

Rick Tsai

Vice Chairman and CEO, MediaTek, Hsinchu, Taiwan

Abstract

AI is redefining the semiconductor landscape. Exponential growth in compute, bandwidth, and power efficiency is driving the proliferation of agentic and physical AI. The paradigm is shifting from performance alone to breakthroughs in scalable and energy-efficient architectures and cross-layer system innovation with close cooperation across silicon, design, heterogeneous- hardware integration, and software. This plenary paper explores the key vectors of advanced packaging, power delivery, thermal management, high-bandwidth memory, interconnects, and wireless communication that will shape intelligent robust AI systems for the decade ahead.

1.0 The Growth of AI, Unleashing Opportunities and the IC Ecosystem

The pervasive integration of artificial intelligence (AI) across every industry and aspect of daily life is fundamentally enabled by the semiconductor sector, which is projected to reach a trillion-dollar valuation by 2030 [1]-[4], as shown in Figure 1.1.1. This AI evolution is forcing a complete redesign of the entire computational stack, from foundational hardware like advanced packaging [5] and high-bandwidth memory (HBM) [6] to the distributed compute fabric spanning cloud data centers and the robotic edge [7]-[11]. As these hardware and software technologies converge into a transformative, intelligent ecosystem [12], new bottlenecks continually emerge and present significant engineering challenges that must be overcome to support future technological generations.

For example, AI's demand for computing power, particularly in data centers, is growing at a meteoric rate that even outpaces Moore's Law. While the computing demand for AI previously doubled every two years, this demand now doubles every two to four months, with training compute for top-tier AI models growing 4~5 times per year, as shown in Figure 1.1.2. High-performance infrastructure GPUs have leveraged architectural improvements to deliver a 100× and 1000×breakthrough in energy efficiency and throughput over the past decade, respectively [13]. This pace of progress, however, is unsustainable. Future systems will need to be up to a billion times more powerful, and that is unattainable by simply improving silicon design in isolation or with established IC technologies [14]-[15].

Instead, meeting this challenge requires a novel strategy. Rather than a myopic focus on better chips, the industry must embrace a comprehensive co-evolution of hardware, software, and algorithms all optimized at the system level.

1.1 Challenges in the Cloud

Today's cloud providers use advanced-packaging techniques to heterogeneously integrate (HI) different classes of domain-specific processors and accelerators [16]-[19], high-bandwidth memory (HBM), and conventional compute [20]. It is an engineering challenge also confronting an unsustainable energy consumption trend. The cost to train large-scale AI models has increased significantly since 2016, as shown in Figure 1.1.3. If this trend continues, global data-center electricity use is set to more than double to ~945 TWh by 2030 [21]-[22], as shown in Figure 1.1.4. AI's electricity demand could even reach approximately 4.4% of the world's total supply by 2035 [4], as shown in Figure 1.1.5. This reality makes energy efficiency the single most critical semiconductor engineering challenge [9].

Projections show a cumulative $6.7 trillion in data-center capital expenditure (capex) by 2030, with the vast majority, around $5.2 trillion, being AI-oriented [1]-[2]. Independent trends are calling for a compound annual growth rate of approximately 21% through 2029 [23]. This spending directly fuels innovation across the entire hardware stack, including AI accelerators, HBM and advanced packaging, ultra-high-speed networking (transitioning from 800G to 1.6T optics), advanced cooling, and clean-power solutions [24].

These massive financial and technical requirements are not obstacles, rather these constraints define the engineering frontier for the next decade and where innovation will thrive. The core challenge is to improve critical metrics like efficiency (tokens/J), latency (sub ms), and reliability, turning today's limitations into tomorrow's breakthroughs.

1.2 Moving to the Edge

Beyond the cloud, a significant amount of AI compute is also moving directly into edge devices as shown in Figures 1.1.6. and. 1.1.7. On-device AI is scaling fast for four main reasons:

1. It offloads some of the cloud's computational load (especially inference)

2. It reduces service costs

3. It protects user privacy

4. It reduces latency to under 10ms for a vastly improved user experience (UX).

Given these benefits, the market is set to grow dramatically. GenAI-capable smartphones are expected to hit ~730 million annual units by 2028, and AI PCs are forecast to reach ~167 million units by 2027 [23], with dedicated Neural Processing Units (NPUs) becoming standard hardware.

As a result, trillions of daily local AI inferences will power everything from speech and vision to personal copilots that sync with cloud data when needed [25]. However, such unprecedented levels of on-device intelligence will require a new class of hardware. While the cloud continues to handle the heavy computational load of training and large-scale inference, the edge is driving a surge in analog and microcontroller components, advanced sensors, optoelectronics, and sophisticated power management chips [4]. Smart edge devices will not just run AI models; they will fine-tune the models, learning from local data to constantly improve and personalize the user's experience right on the device itself [26].

Often referred to as Physical AI [26]-[27], the integration of AI in physical devices, such as robots, autonomous vehicles, or smart machines, will enable them to act on their surroundings, make decisions, and adapt their behavior in real time [28]. Overall, the path forward is through cross-layer co-design, a holistic strategy in which algorithms, software, and hardware architecture are designed together, from the ground up. It is about optimizing the entire system cohesively [29]-[30], resulting in a balanced fabric: global intelligence in the cloud, low-latency adaptation at the edge, and private instantaneous response on personal devices [31]-[33]. Once again, semiconductors are central to such a significant technological build-out, with AI serving as the primary catalyst for system-level innovation and investment. The engineering challenge is significant, but the payoff is tangible: a trustworthy assistant beside every worker and in every robot, with latency, privacy, and cost steadily improving as the cloud-to-edge fabric matures. This is not a distant vision; it is a healthy and hopeful technology cycle that is already well underway [34]-[35].

2.0 Integrating AI into Applications, Enriching Life through Technology Innovation

Two complementary overview examples in this section provide a glimpse into the application benefits and distinct engineering challenges resulting from AI.

2.1 Agentic AI and Physical AI Augmenting Humans in Autonomous Vehicles and Beyond

Agentic AI and physical AI are complementary paths to augmenting human capability. For instance, applications like autonomous driving benefit from a combination of physical AI and software agents that unify otherwise disparate human-machine interfaces (HMI) with autonomous control, sensing, and actuation [34]-[38]. In these kinds of applications, AI agents will lead autonomous reasoning and interaction, while multi-modal physical AI will support safe perception, control, and actuation.

Over time, experts anticipate that the prevailing human-centric paradigm will shift toward a bot centric one. In this context, the term "bot" includes both physical and non-physical AI (i.e., robots and agents, respectively). In turn, the technical differentiations between entities such as robots and vehicles will likely blur, and the distinctions will become primarily functional, contingent upon specific use cases and operational domains, replacing the rigid and historical definitions. The same AI agents may simultaneously be goal-directed copilots that can plan and execute multi-step tasks, while also conversing with occupants in a range of natural and appropriate emotions. A modern example of AI agents is found in the case of OEMs that embed large-model assistants into In-Vehicle Infotainment (IVI) stacks (e.g., Mercedes-Benz integrating ChatGPT via Azure OpenAI in MB's UX). While agents manage human interaction, physical AI in a vehicle is the embodied counterpart. It handles perception from sensors, planning and performing control loops that actuate steering, braking, powertrain, and executing comfort functions. For safety purposes, functional safety systems are governed by ISO 26262, further strengthened by cybersecurity (ISO/ SAE 21434), OTA/software-update governance (UNECE R155/R156), safety of the intended function (SOTIF), and the overarching framework of IEC 61508.

ISSCC 2026 / February 16, 2026 / 8:50 AM

Projection of navigation commands and traffic-aware advanced driver assistance system (ADAS) cues into the driver's view (e.g., Continental's AR-HUD waveguide systems), crash avoidance, driver drowsiness and attention warning (DDAW), and in-cabin monitoring of wellness will be a safety baseline. Managing conversation length between the agent and the driver, visual load, and AR cue density under driving state will require new regulations.

2.2 Handling the Data

To support adequate data traffic within the vehicle's units, functional safety and security challenges run rampant. In a bot-centric paradigm, a centralized training and distributed-deployment methodology can simultaneously reduce communication data bandwidth, lower compute latency, and improve scalability. The cloud handles compute-intensive pretraining and fine-tuning, while edge devices execute compact and fine-tuned models for low-latency, low-cost decision making (e.g., via federated learning in IoT scenarios [32]). To balance data access across the cloud edge synergy, leveraging hierarchical model deployment, bandwidth-aware model partitioning for inference, adaptive compression and quantization of features, and life-long learning of edge devices can substantially reduce the required data traffic [39]. Because of the need for real-time responsivity, physical AI is pushing compute to the edge under bounded-latency constraints. For this reason, vehicles employ Ethernet networks with the IEEE Time-Sensitive Networking (TSN) suite for time synchronization (802.1AS) and time-aware scheduling (802.1Qbv) to guarantee latency and jitter budgets on Ethernet. At the same time, ISO 26262 for E/E functional safety [36], complemented by ISO 21434, regulates the automotive domain, while different categories of standards regulate cybersecurity. Meanwhile, cars are now fully connected mobile environments for work, business, and entertainment. Today's vehicles are packed with high-performance multimedia systems that offer real-time online interaction, immersive entertainment and gaming, and connectivity through both terrestrial (5G, 6G) and non-terrestrial networks (NTN) [37]. The car's own system, as well as a passenger's many independent mobile devices, are all expected to directly communicate and cooperate as part of a local device cloud [38]. Delivering this smooth, integrated user experience requires a perfect blend of Agentic AI and Physical AI [39]-[41].

From a security and privacy perspective, reinforcement learning with human feedback (RLHF)- trained models may exhibit reward hacking, goal misalignment, and power-seeking behavior [40]. Strengthening governance of training data is essential to cover privacy, consent, and transparency [41]. Several frameworks already guide AI development. These include:

- The NIST AI Risk Management Framework (RMF), which provides methods to identify, assess, and manage AI risks.

- The EU AI Act, which sets legal requirements for deploying AI in the EU.

- ISO/IEC 42001, which offers guidance for organizations to structure policies and controls to manage AI.

To further bridge the security gap, existing frameworks must be extended beyond digital risk management to include cyber-physical safeguards.

2.3 Data-Center Expansion and the Transition from IC Design to Multiphysics Design/ System-Technology Co-Optimization (DTCO and STCO)

As mentioned, multiple inter-related trends emerge in the context of data centers, requiring DTCO and STCO of distinct engineering disciplines. The mix of larger models, tighter synchronization, and denser racks now sets the real constraints.

In this area, MediaTek is enabling the rapid infrastructure compute growth ("overwhelming compute") by solving difficult technical obstacles with multiple leading technologies and through key technology partnerships [42].

2.3.1 Thermal Challenges

The large electrical power consumed by high-density servers leads to difficult thermal management challenges, as heat must leave at the device surface or the rack will exceed safe junction limits and facility Power Usage Effectiveness (PUE) targets. As such, traditional air-cooling technologies are pushed past their physical limits, and more effective liquid-based solutions like direct liquid cooling (DLC) are being embraced. DLC is where a coolant, typically water, circulates through cold plates (heat exchangers) mounted on top of hot components such as CPUs/GPUs, or flows among them in a 3-D heterogeneous integration system via micro-fluid cooling channels, absorbing heat at the source and transporting it away [43]. Additionally, with the more aggressive immersion cooling, entire servers or other components are submerged into a tank filled with a dielectric fluid. Finally, to enhance efficiency and eliminate wasteful overcooling, AIOps (AI for IT Operations) algorithms use predictive analytics on IT workload patterns to forecast cooling needs and dynamically adjust airflow, water temperature, and pump speeds to manage hotspots [44].

2.4 Solutions on Each Level

Heterogeneously integrating chiplets and multi-die system-in-package (SiP) technologies have become mainstream. Systems-on-wafer (SoW) are quickly emerging, consisting of three-dimensional ICs (3DIC), custom HBM, and interposer/substrates aimed to minimize resistive copper electrical communication between the compute and the memory parts. Training clusters now operate as one tightly coupled computer, so memory proximity and die-to-die bandwidth dominate node efficiency. In that context, die-to-die communication is implemented with extra-short reach interfaces having data rates per pin doubling almost every three to four years across different I/O standards, including UCIe. Inside each node, short electrical hops win on cost and density. Across the nodes, copper loss limits the reach at higher baud rates, shifting the bottleneck to the channel rather than the SerDes core [45]-[47].

At a server level, cloud architecture is evolving. The traditional 3-tier architecture of a data center, consisting of core, aggregation, and edge layers, is disaggregating to a highly distributed structure. In these applications, high-bandwidth communication is required between hundreds or more servers simultaneously. Therefore, as interconnect rates increase to meet different distance/reach/channel loss and other physical requirements, distinct data interface solutions are required. DSP-based serial links beyond 200Gb/s have been demonstrated [45]-[46], with unabated throughput demand soon moving to 448Gb/s while improving energy efficiency. Beyond package, connectors, and PCB technologies, SerDes designers implement signal processing and circuit innovations to close the gap [47].

Another approach is offered by silicon photonics and ASICs co-integration, which offers throughput beyond the limitations of electrical interconnects. Co-packaged optics (CPO) have intrinsic physical advantages in terms of data rate, bandwidth density, energy efficiency, communication latency, and reduced crosstalk. General optical connectivity consists of electrical integrated circuit (EIC) dies, including optical driver, transimpedance amplifier (TIA), and potentially retimer and DSP. Photonic integrated circuits (PIC) may also be used for modulators, photodetectors, optical mux/demux, polarization control, and an optional laser source [48].

System-level solutions are also employed. At the board level, a hierarchical power management structure distributes a high voltage (e.g., 48V to 54V) across the servers, which is stepped down to 12V by intermediate bus conversion (IBC). In turn, the IBCs distribute supplies to smart power stages (SPS) on the backside of the packages. At the substrate and die level, complex power delivery networks (PDNs) are developed to manage varying loads and voltage drops in 3DICs. Techniques include through-silicon vias (TSVs), backside power delivery, dedicated integrated voltage regulation (IVR), and hierarchically tied top level supply sources [49], as shown in Figure 1.1.8.

Finally, in a data-based economy, the resilience of data centers is critical. It requires built-in physical security. Zero Trust architectures are being adopted widely integrating physical, cyber, and operational controls. AI infrastructure raises the blast radius of any fault or breach, so designs must assume compromise and contain it at each boundary.

3.0 Enabling Technologies

As we examine the underlying enabling technologies more closely, it becomes possible to identify some key goals and trends that make this vision a practical engineering reality. Performance per watt and bytes moved per joule now decide architecture more than peak TOPS does. Going forward, process technology scaling will remain an important component of computing and communication growth. However, performance scaling will require an optimal blend of traditional device miniaturization and multi-chip heterogeneous integration (HI).

3.1 Device Scaling

Device miniaturization involves shrinking transistors on individual dies, supported by new materials [50]-[51], processes, and transistor architectures like cFETs. HI assembles multiple chiplets using 2.5D, 3D, and 3.5D integration. This connects chips horizontally with interposers (organic, silicon, or glass) and silicon bridges or stacks them vertically. Prominent examples include TSMC's CoWoS™ and SoIC™, Intel's EMIB™ and Foveros™, and substrate-less alternatives like Nvidia's Chip-on-Wafer-on-PCB [52]. Successfully implementing these advanced techniques demands deep expertise in system-level engineering, DTCO/STCO, and close partnerships between designers, fabs, and suppliers. This ecosystem, therefore, relies on emerging EDA tools that support complex multi-physics modeling and design. The new wins come from codesign across logic, memory, package, and cooling, rather than from any single knob.

While process scaling continues to support the speed and power efficiency of data interfaces, the more pressing challenge is overcoming physical channel impairments and maintaining signal integrity. MediaTek is an industry leader in high-speed data interfaces through the co-design of circuits, systems, and signal processing. For example, the company maximized copper channel bandwidth efficiency through high-order modulation schemes like PAM4. For long-reach, ultra-high data rate links, however, optical solutions are a superior alternative due to the vast bandwidth and physical channel isolation of fiber. Established solutions like Active Optical Cables (AOCs) offer high bandwidth and lower energy consumption for offboard connections. Pushing this trend further, emerging

979-8-3315-8937-0/26 $31.00 © 2026 IEEE

ISSCC 2026 / SESSION 1 / PLENARY / 1.1

technologies are integrating optical links closer to the silicon processing units to reduce energy and latency. These include Near Package Optics (NPO) and Co-Packaged Optics (CPO), the latter of which is entering mainstream adoption. MediaTek has demonstrated excellent results in CPO [53]. In practice, designers pick copper for short reach, optics for rack-scale reach, and then codesign the cutover to meet energy per bit limits.

3.2 5G to 6G

As wireless communication transitions from 5G to 6G [38], MediaTek has focused on achieving ultra-low power (ULP) consumption, embedding AI-native intelligence, and enabling ubiquitous coverage. Among them, energy efficiency is crucial for managing thermal constraints and battery life in handhelds and wearables, which in turn enable higher practical data rates and new form factors for immersive applications. Devices like smart glasses, for example, have far more stringent limitations on size, weight, and heat dissipation than smartphones. The challenge is compounded by the fact that high receiver performance drives peak power consumption.

To keep the average power within acceptable limits, energy consumption must be significantly reduced during periods of inactivity or low data rates, such as when the device is performing routine housekeeping tasks (e.g., control channel monitoring, beam management, or idle paging). Wearables set the limiting case, so radios must idle at microwatts and wake fast without dropping link margin.

A defining feature of 6G will be its AI-native architecture that enables agentic AI on edge devices to collaborate across the entire network, as shown in Figure 1.1.9. A good example of edge AI application is in high-resolution video processing, where edge deployment reduces the large data volume streamed without loss of quality. This allows one to intelligently optimize the trade-off between wireless link budget and the energy required for local computation. To handle these demanding workloads, energy-efficient hardware like domain-specific accelerators with adaptable Digital Compute-in-Memory (DCIM) engines has proven highly effective. Built on the latest lithography nodes, their architecture is optimized to reduce external memory access and efficiently handle AI-specific tasks like weight switching. This approach achieves aggressive energy and area efficiencies, making powerful AI feasible on edge devices [54-55]. The goal is to move bits only when needed and compute where the joule is cheapest.

3.3 Neuromorphic Approaches

Looking at the future, cognitive architectures, such as the human brain's dual cognitive system architecture, can help service autonomous edge devices achieve systematic generalization. In the brain, system one handles routine tasks unconsciously, while system two is responsible for complex problem-solving consciously [56]. A working-memory-based planning approach has been proposed for robots [57]. To achieve brain-like energy efficiency, neuromorphic computing is also a promising approach, although its scalability remains a challenge. It is suggested to provide higher-level coding abstraction for programming neuromorphic hardware, along with proper compilation methods for mapping spiking neural networks (SNNs) to hardware architectures. A general model description standard (such as the ONNX format for neural network models) is needed to improve its adoption rate. In addition, identifying key interfaces (e.g., high-speed-sensor integration) and enhancing broad sensor compatibility should be considered when designing a neuromorphic system [58]. These proposals try to raise abstraction, keep determinism, and hold energy within edge budgets.

3.4 Multiprotocol Systems

In the future, users will move seamlessly between different networks (cellular, WiFi, satellite/NTN) without losing connection or experiencing service degradation. In this context, AI can predict when a user is about to move out of WiFi range and preemptively switch to cellular without user intervention. MediaTek's intelligent device edge cloud (device and Radio Access Network (RAN)) concept [59] enables local compute collaboration and allows secure context sync across devices participating in a personal device (edge) cloud. With such technologies, support becomes possible for a new generation of applications that require ultra-high data rates without mandating massive processing, storage or power resources at the devices themselves. Smooth handoff only works if policy, power, and latency agree across links, so control moves closer to the device. MediaTek has taken a leading role in the system design, standardization and ongoing evolution of 5G NTN (satellite), both for NR NTN and IoT NTN, to 6G NTN. Integrating satellite and terrestrial mobile networks to offer pervasive connectivity across the world will enable a new era of innovative digital services and significantly contribute to the United Nations Sustainable Development Goals. Compared with proprietary satellite communication technologies, 6G non-terrestrial network (NTN) technology based on a 3GPP open standard can leverage the economies of scale from the existing global mobile cellular ecosystem to bring satellite communication mainstream [60]. 6G NTN will be native to the 6G physical/protocol layer design, allowing joint optimization of technologies for terrestrial networks and non-terrestrial networks [61]. Direct-to-device reach then becomes a software and RF problem on a common handset bill of materials, not a separate terminal class.

4.0 Ecosystem and Partnerships

Addressing large-scale technological challenges requires productive ecosystems that lower barriers to contribution and foster strategic partnerships. Historically, openness has been a powerful catalyst for innovation. In software, Linux became the backbone of the internet, while Android created a massive mobile ecosystem. In hardware, Tesla's open patents spurred the growth of the EV industry. And in content, Wikipedia's open license made it the world's largest encyclopedia.

This principle of open innovation is helping AI's rapid ascent. Many foundational tools (like TensorFlow and PyTorch) and influential models (such as LLaMA and Mistral) are open source. However, the industry would also benefit from moving beyond open code to embrace open model weights and training data. Similarly, open systems and standardization will be even more important to design and build high-performance systems in areas such as heterogeneous integration in order to accelerate innovation. Looking further ahead, the AI landscape will likely consist of a strategic mix of open and closed approaches as players balance the speed of innovation with commercial value capture.

To power this dynamic and hybrid AI ecosystem, MediaTek delivers a computing foundation that scales from edge to the cloud, as shown in Figure 1.1.10. We meet diverse needs, from small-scale on device AI (MDLA) to large-scale acceleration in data center ASICs. Supported by flexible business models, a resilient supply chain, and wireless technologies like Wi-Fi and 6G, MediaTek is a leader in developing and deploying AI.

Conclusions

The semiconductor industry is entering an era of system-level, efficiency-driven innovation. The immense challenges posed by AI's energy consumption and physical interconnect limits are going to define the next decade of engineering breakthroughs.

The future of compute and communication must be a cohesive, intelligent fabric stretching from massive, disaggregated data centers to private, responsive AI on personal devices and cognitive physical AI with full autonomy. Ultimately, this vision is only possible through a cohesive system where thermal, power and signal integrity, and data flow are co-optimized within a new generation of hardware that is both powerful and efficient.

The ultimate goal is to deliver a future of ubiquitous, trustworthy AI that enhances human productivity, safety, and daily life.

References:
[1] T. Pröttel, "WSTS Semiconductor Market Forecast," WSTS, June 2025. https://www.wsts.org/76/103/WSTS-Semiconductor-Market-Forecast-Spring-2025
[2] SIA, "America Projected to Triple Semiconductor Manufacturing Capacity by 2032," Semiconductor Industry Association, May 2025. https://www.semiconductors.org/america-projected-to-triple-semiconductor-manufacturing-capacity-by-2032-the-largest-rate-of-growth-in-the-world/
[3] O. Burkacky, J. Dragon, N. Lehmann, "The Semiconductor Decade: A trillion-dollar industry," McKinsey & Company, April 2022. https://www.mckinsey.com/industries/semiconductors/our-insights/the-semiconductor-decade-a-trillion-dollar-industry
[4] SRC, "Microelectronics Advanced Packaging Technologies Roadmap (MAPT), Full Report," Semiconductor Research Corporation, 2025. https://srcmapt.org/
[5] C.M. Hung, "Semiconductor Chip Design in a Legoland," IEEE Asian Solid-State Circuits Conference (A-SSCC), Haikou, China: IEEE, pp. 1-4, 2023. doi: 10.1109/A-SSCC58667.2023.10347923.
[6] J. Song, "AI Revolution Driven by Memory Technology Innovation," IEEE International Solid State Circuits Conference (ISSCC), San Francisco, CA: IEEE, pp. 26-36, 2025. doi: 10.1109/ISSCC49661.2025.10904790.
[7] K.-H.L. Loh, "Fertilizing AIoT from Roots to Leaves," IEEE International Solid-State Circuits Conference (ISSCC), San Francisco, CA: IEEE, pp. 15-21, 2020. doi: 10.1109/ISSCC19947.2020.9062950.
[8] L. Su, S. Naffziger, "Innovation for the Next Decade of Compute Efficiency," IEEE International Solid-State Circuits Conference (ISSCC), San Francisco, CA: IEEE, pp. 8-12, 2023. doi: 10.1109/ISSCC42615.2023.10067810.
[9] E. Karl, J.-S. Park, "Forum 2: Energy-efficient AI-computing Systems for Large-language Models," IEEE International Solid-State Circuits Conference (ISSCC), San Francisco, CA: IEEE, pp. 593-596, 2025. doi: 10.1109/ISSCC49657.2024.10454551.
[10] J. Alben, "Computing in the Era of Generative AI," IEEE International Solid-State Circuits Conference (ISSCC), San Francisco, CA: IEEE, pp. 26-28. 2024. doi: 10.1109/ISSCC49657.2024.10454562.
[11] N. Shahriari, "AI Era Innovation Matrix," IEEE International Solid-State Circuits Conference (ISSCC), San Francisco, CA: IEEE, pp. 10-15. 2025. doi: 10.1109/ISSCC49661.2025.10904705.
[12] L.-B. Tan, "Fueling Semiconductor Innovation and Entrepreneurship in the Next Decade," IEEE International Solid-State Circuits Conference (ISSCC), San Francisco, CA: IEEE, pp. 29-33, 2024. doi: 10.1109/ISSCC49657.2024.10454565.

[13] W. Dally, "Hardware for Deep Learning," 35th Hot Chips Symposium, Stanford University, 2023. doi: 10.1109/HCS59251.2023.10254716.

[14] S. Lie, "Wafer-Scale AI: GPU Impossible Performance," IEEE Hot Chips 36 Symposium (HCS), Stanford, CA: IEEE, pp. 1-71, 2024. doi: 10.1109/HCS61935.2024.10664673.

[15] B.-S. Liang, "AI Computing Design Trends for LLMs in the Generative AI Era," International Symposium on Circuits and Systems, London, UK: IEEE, 2025. doi: 10.17023/6csg-2812.

[16] S. Xu, C. Ramakrishnan, "Inside Maia 100," IEEE Hot Chips 36 Symposium (HCS), Stanford, CA: IEEE, pp. 1-17, 2024. doi: 10.1109/HCS61935.2024.10665248.

[17] J. Coburn, et al., "Meta's Second Generation AI Chip: Model-Chip Co-Design and Productionization Experiences," Annual International Symposium on Computer Architecture (ISCA), ACM, pp. 1689-1702, 2025. doi: 10.1145/3695053.3731409.

[18] A. Smith, et al., "AMD Instinct™ MI300 Series Modular Chiplet Package – HPC and AI Accelerator for Exa-Class Systems," IEEE International Solid-State Circuits Conference (ISSCC), San Francisco, CA: IEEE, pp. 490-492, 2024. doi: 10.1109/ISSCC49657.2024.10454441.

[19] R. Prabhakar, "SambaNova SN40L RDU: Breaking the Barrier of Trillion+ Parameter Scale Gen AI Computing," IEEE Hot Chips 36 Symposium (HCS), Stanford, CA: IEEE, pp. 1-24, 2024. doi: 10.1109/HCS61935.2024.10664717.

[20] Y.-T. Yang, C.-M. Hung, "Heterogeneous Integration in Co-Packaged Optics," IEEE Journal on Emerging and Selected Topics in Circuits and Systems, 2025. doi: 10.1109/JETCAS.2025.3590744.

[21] IEA, "AI Is Set to Drive Surging Electricity Demand from Data Centers While Offering the Potential to Transform How the Energy Sector Works," International Energy Agency (IEA), April 2025. https://www.iea.org/news/ai-is-set-to-drive-surging-electricity-demand-from-data-centres-while-offering-the-potential-to-transform-how-the-energy-sector-works

[22] BloombergNEF, "AI Data Centers Fuel Quicker Growth in Power Demand," Bloomberg, September 2025. https://www.bloomberg.com/professional/insights/commodities/ai-data-centers-fuel-quicker-growth-in-power-demand/

[23] Dell'Oro., "Data Center Capex to Grow at 21 Percent CAGR Through 2029," Dell'Oro Group, August 2025. https://www.delloro.com/news/data-center-capex-to-grow-at-21-percent-cagr-through-2029/

[24] MediaTek, "MediaTek Enhances Flagship AI Performance with Dimensity 9400+ Mobile Platform," Press Release, HsinChu, Taiwan: MediaTek, April 2025.

[25] Albogami, N. N., "Enhancing Energy Efficiency in AI Systems Through Edge AI and Federated Learning," Scientific Reports, 2025.

[26] D. Rus, "From Chips to Thoughts: Building Physical Intelligence Into Robotic Systems," IEEE International Solid-State Circuits Conference (ISSCC), San Francisco, CA: IEEE, pp. 16-22, 2025. doi: 10.1109/ISSCC49661.2025.10904576.

[27] J. Chae, S. Lee, J. Jang, S. Hong, K.-J. Park, "A Survey and Perspective on Industrial Cyber-Physical Systems (ICPS): From ICPS to AI-Augmented ICPS," IEEE Transactions on Industrial Cyber-Physical Systems, pp. 257-272, 2023. doi: 10.1109/TICPS.2023.3323600.

[28] G. Manganaro, "Rethinking Mixed-Signal IC Design," IEEE European Solid-State Electronics Research Conference (ESSERC), Bruges, Belgium: IEEE, pp. 552-556, 2024. doi: 10.1109/ESSERC62670.2024.10719447

[29] K.-H. Loh, , "Enabling Generative AI: Innovations and Challenges in Semiconductor Design Technologies," Symposium on VLSI Technology and Circuits (VLSI Technology and Circuits), Kyoto, Japan: IEEE, pp. 1-6, 2025. doi: 10.23919/VLSITechnologyandCir65189.2025.11074926.

[30] A. Nalamalpu, "Accelerating the Future of Overwhelming Compute," Semicon Taiwan, Taipei, Taiwan: Semi, 2025.

[31] R. Tsai, "AI for Everyone, From Edge to Cloud," Computex, Taipei, Taiwan: Computex, 2025.

[32] K. Fukuoka, E.J.-W. Fang, "Forum 6: Evolution of the Software-Defined Vehicle: Navigating Smart Cockpits and In-Car Hardware," IEEE International Solid-State Circuits Conference (ISSCC), San Francisco, CA: IEEE, pp. 642-644, 2025. doi: 10.1109/ISSCC49661.2025.10904587.

[33] P. Schiefer, "The Crucial Role of Semiconductors in the Software-Defined Vehicle," IEEE International Solid-State Circuits Conference (ISSCC), San Francisco, CA: IEEE, pp. 37-41, 2025. doi: 10.1109/ISSCC49661.2025.10904625.

[34] C.-M. Hung, et al., "Toward Automotive Surround-View Radars," IEEE International Solid State Circuits Conference (ISSCC), San Francisco, CA: IEEE, pp. 162-164, 2019. doi: 10.1109/ISSCC.2019.8662489.

[35] M. Kim, P. Pinyoanuntapong, B. Kim, W. Saad, D. Calin, "Edge vs. Cloud: How Do We Balance Cost, Latency, and Quality for Large Language Models Over 5G Networks?" IEEE Wireless Communications and Networking Conference (WCNC), Milan, Italy: IEEE, pp. 1-6, 2025. doi: 10.1109/WCNC61545.2025.10978177.

[36] R. Debouk, "Overview of the Second Edition of ISO 26262: Functional Safety — Road Vehicles," Journal of System Safety, pp. 13-21, 2019. doi: 10.56094/jss.v55i1.55.

[37] I.-K. Fu, , et al., "Satellite and Terrestrial Network Convergence on the Way Toward 6G," IEEE Wireless Communications, pp. 6-8, 2023. doi: 10.1109/MWC.2023.10077212.

[38] N. Tenny, B. Kim, M. Shariat, A. Hsu, D. Calin, "Ambient Computing for 6G Era," IEEE Wireless Communications, pp. 10-12, 2025. doi: 10.1109/MWC.2025.10944645.

[39] M.-E. Shih, , et al., "NVE: A 3nm 23.2TOPS/W 12b-Digital-CIM-Based Neural Engine for High Resolution Visual-Quality Enhancement on Smart Devices," IEEE International Solid-State Circuits Conference (ISSCC), San Francisco, CA: IEEE, pp. 360-362, 2024. doi: 10.1109/ISSCC49657.2024.10454482.

[40] R. Ngo, , "The Alignment Problem From a Deep Learning Perspective," ICLR, 2024. doi: 10.48550/arXiv.2209.00626

[41] J. King, J., "Rethinking Privacy in the AI Era," Stanford Human-centered artificial intelligence (HAI), 2024.

[42] NVLink., "NVIDIA and MediaTek Collaborate to Create a Personal AI Supercomputer," Mediatek.com, January 2025. https://www.mediatek.com/tek-talk-blogs/nvidia-and-mediatek-collaborate-to-create-a-personal-ai-supercomputer

[43] Z. Wu, et al., "Topology Optimization for Embedded Cooling of Multiple and Transient Workloads in 3D Semiconductor Packages," IEEE Intersociety Conference on Thermal and Thermomechanical Phenomena in Electronic Systems (ITherm), IEEE, 2025.

[44] H. Liu, A. Aljbri, J. Song, C. Hua, "Research Advances on AI-powered Thermal Management for Data Centers," Tsinghua Science and Technology, pp. 303-314, 2022. doi: 10.26599/TST.2021.9010019.

[45] E.-H. Chen, et al., "A 212.5Gb/s DSP-Based PAM-4 Transceiver with 50dB Loss Compensation for Large AI System Interconnects in 4nm FinFET," IEEE International Solid-State Circuits Conference (ISSCC), San Francisco, CA: IEEE, pp. 1-3, 2025. doi: 10.1109/ISSCC49661.2025.10904601.

[46] A. Mostafa, et al., "A 2.2pJ/b 212.5Gb/s PAM-4 Transceiver with \>46dB Reach in 5nm FinFET," IEEE International Solid-State Circuits Conference (ISSCC), San Francisco, CA: IEEE, pp. 138-140, 2025. doi: 10.1109/ISSCC49661.2025.10904591.

[47] T. Ali, et al., "56/112Gpsps Wireline Transceivers for Next Generation Data Centers on 7nm FINFET CMOS technology," IEEE Custom Integrated Circuits Conference (CICC), Austin, TX: IEEE, pp. 1-6, 2021. doi: 10.1109/CICC51472.2021.9431430.

[48] M. Mehta, "An AI Compute ASIC with Optical Attach to Enable Next Generation Scale-Up Architectures," 36th Hot Chips Symposium, Stanford University, pp. 1-30, 2024. doi: 10.1109/HCS61935.2024.10664787.

[49] A. Veloso, "Backside Power Delivery: Game Changer and Key Enabler of Advanced Logic Scaling and New STCO Opportunities," International Electron Devices Meeting (IEDM), San Francisco, CA: IEEE, pp. 1-4, 2023. doi: 10.1109/IEDM45741.2023.10413867.

[50] K. Zhang, "Semiconductor Industry: Present & Future," IEEE International Solid-State Circuits Conference (ISSCC), San Francisco, CA: IEEE, pp. 10-15, 2024. doi: 10.1109/ISSCC49657.2024.10454358.

[51] Y.-J. Mii, "Semiconductor Industry Outlook and New Technology Frontiers," IEEE International Electron Devices Meeting (IEDM), San Francisco, CA: IEEE, pp. 1-6, 2024. doi: 10.1109/IEDM50854.2024.10873484.

[52] S. Jangam, S.S. Iyer, "Silicon-Interconnect Fabric for Fine-Pitch (≤10 μm) Heterogeneous Integration," IEEE Transactions on Components, Packaging and Manufacturing Technology, pp. 727-738, 2021. doi: 10.1109/TCPMT.2021.3075219.

[53] Trendforce, "MediaTek Partners with Ranovus to Enter Niche Market, Expands into Heterogeneous Integration Co-Packaged Optics Industry," Trendforce.com, March 2024. https://www.trendforce.com/news/2024/03/21/news-mediatek-partners-with-ranovus-to-enter-niche-market-expands-into-heterogeneous-integration-co-packaged-optics-industry/

[54] A. Varma, et al., "A 4nm 3.4GHz Tri-Gear Fully Out-of-Order ARMv9.2 CPU Subsystem Based 5G Mobile SoC," IEEE International Solid-State Circuits Conference (ISSCC), San Francisco, CA: IEEE, pp. 36-38, 2024. doi: 10.1109/ISSCC49657.2024.10454394.

[55] S.-W. Hsieh, et al., "MAE: A 3nm 0.168mm² 576MAC Mini AutoEncoder with Line-Based Depth-First Scheduling for Generative AI in Vision on Edge Devices," IEEE International Solid State Circuits Conference (ISSCC), San Francisco, CA: IEEE, pp. 414-416, 2025. doi: 10.1109/ISSCC49661.2025.10904763.

[56] D. Kahneman, "Thinking, Fast and Slow," Farrar, Straus and Giroux, 2013.

[57] S. Behnke, (s.d.), "Towards Conscious Service Robots," arXiv, 2501.15198. doi: 10.48550/arXiv.2501.15198.

[58] D. Kudithipudi, et al., "Neuromorphic Computing at Scale," Nature, pp. 801-812, 2025. doi: 10.1038/s41586-024-08253-8.

[59] B. Kim, D. Calin, N. Tenny, M. Shariat, M. Fan, "Device Centric Distributed Compute, Orchestration and Networking," IEEE Wireless Communications, pp. 6-8, 2023. doi: 10.1109/MWC.2023.10251878.

[60] I.-K. Fu, "5G NTN Technology for Direct Satellite Access: From Smart Phone to Automotive," Forum 2: Wireless Communication Technology for Space Applications, IEEE International Solid State Circuits Conference (ISSCC), San Francisco, CA: IEEE, 2025.

[61] C.-S. Yang, et al., "Toward 6G Sustainable Mobile Communications," IEEE Wireless Communications, pp. 44-50, 2025. doi: 10.1109/MWC.012.2400062.

ISSCC 2026 / SESSION 1 / PLENARY / 1.1

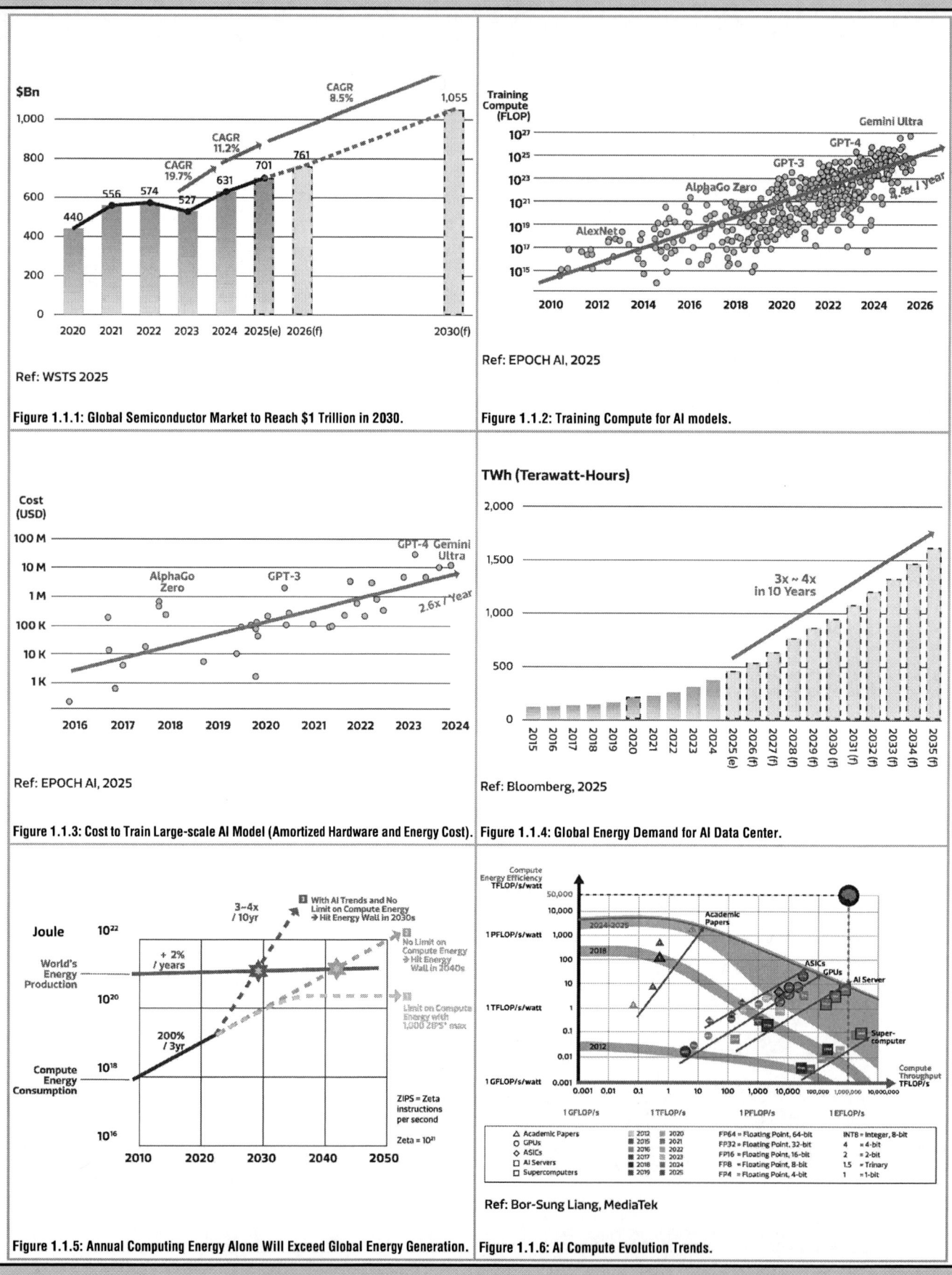

Figure 1.1.1: Global Semiconductor Market to Reach $1 Trillion in 2030.

Figure 1.1.2: Training Compute for AI models.

Figure 1.1.3: Cost to Train Large-scale AI Model (Amortized Hardware and Energy Cost).

Figure 1.1.4: Global Energy Demand for AI Data Center.

Figure 1.1.5: Annual Computing Energy Alone Will Exceed Global Energy Generation.

Figure 1.1.6: AI Compute Evolution Trends.

ISSCC 2026 / February 16, 2026 / 8:50 AM

Figure 1.1.7: AI Compute - From Cloud to Edge.

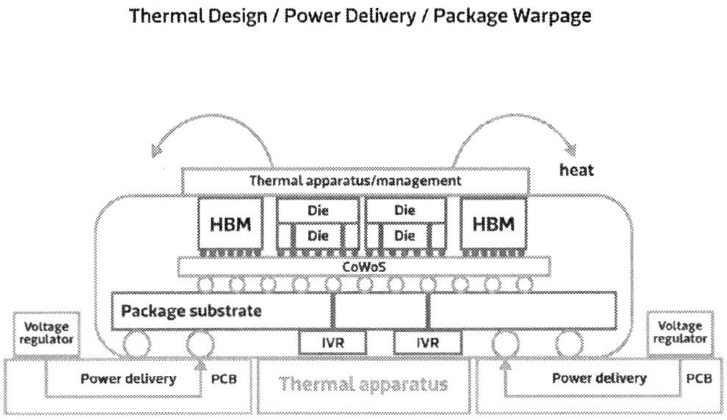

Figure 1.1.8: Technology Challenges in 3.5D Advance Packaging.

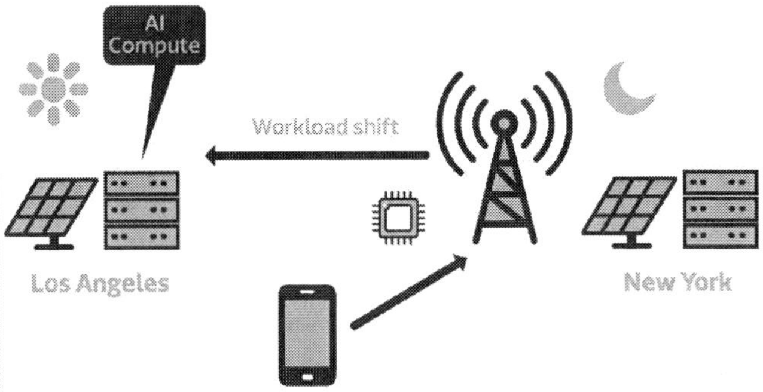

Figure 1.1.9: Mobile Communications for Sustainable AI - Carbon-aware Computation Workload Shift.

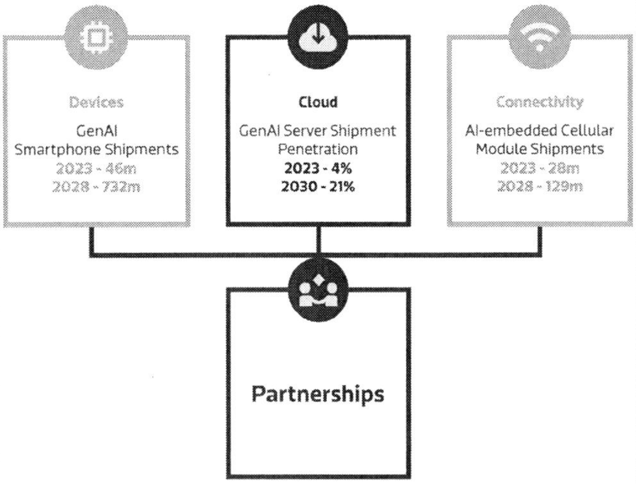

Figure 1.1.10: Cloud Server, Edge Device and Connectivity Modules for AI Compute.

DIGEST OF TECHNICAL PAPERS • 15

979-8-3315-8937-0/26 $31.00 © 2026 IEEE

ISSCC 2026 / SESSION 1 / PLENARY / 1.2

1.2 Quantum Computing – Toward Large-Scale Fault-Tolerant Quantum Computing

Heike Riel

IBM Fellow / Head of Science of Quantum
& Information Technologies, IBM Research, Rüschlikon, Switzerland

Abstract

Quantum computing is advancing rapidly, offering the potential to transform computational paradigms and tackle problems that are intractable for classical systems. Realizing practical quantum systems demands the development of an entirely new computing stack—from qubits and quantum processors to components, wiring and control electronics, transpilers, error handling, software, algorithms, and the compute architecture to integrate quantum and classical resources at scale.

This plenary will provide an overview of the current state of quantum computing and outline the milestones on the path toward building a large-scale, fault-tolerant quantum computer by 2029. The significant improvements in hardware, software, and system integration, pushing the performance of quantum computing to reach quantum utility and advance toward quantum advantage, will be presented. Breakthroughs in error correction and a modular approach indicated in the IBM Quantum Roadmap outline a clear path to quantum systems using 200 logical qubits and 100 million quantum operations by 2029 and 2,000 logical qubits a few years later. Furthermore, advances in quantum algorithms, combined with the integration of quantum systems and high-performance computing (HPC), will unlock powerful synergies—accelerating the timeline for applications previously considered to be far in the future.

1. Introduction

In recent decades, tremendous effort and investment have been dedicated to advancing classical computing, enabling increasingly sophisticated calculations and the processing of vast amounts of data. Despite these achievements, certain complex mathematical problems remain fundamentally intractable for classical computers, regardless of their scale. Quantum computing, based on the principles of quantum mechanics, offers a transformative computational paradigm capable of solving specific problems that are beyond the reach of classical systems.

Current research explores quantum algorithms for a wide range of mathematical problems, including quantum system simulations [1] linear algebra, [2] factoring, [3] optimization, [4] graph problems, and search [5], with the potential to revolutionize science and technology. Many interesting applications in science and business are based on these mathematical problems. So far substantial progress has been made across theory, hardware, and software, driving the field of quantum computing forward. Different technologies to create quantum bits (qubits) are currently explored and developed, each offering certain advantages and challenges. Qubit technologies include implementations based on atomic energy levels like neutral atoms [6] or trapped ions [7], solid-state approaches such as superconducting [8] and spin qubits [9], or photonic qubits [10] for optical-based systems. Among these platforms, superconducting qubits have emerged as a leading technology, demonstrating significant maturity in terms of scalability, fidelity, and operational speed. Quantum processors with more than 100 qubits and error rates of $\sim5\times10^{-4}$ are experimentally realized, solving mathematical problems that are infeasible to simulate classically by brute force [11]. Advances in quantum software and theoretical frameworks continue to lower the barriers to algorithm implementation, enabling new computational capabilities. As these developments accelerate, the first practical benefits of quantum computing are within reach. However, achieving mathematically proven speedups over state-of-the-art classical methods will require quantum circuits comprising hundreds of millions to billions of gates. As quantum information is fragile and susceptible to noise, quantum error correction will be required.

To realize the full potential of quantum computing, several critical challenges must be addressed before practical, large-scale deployment becomes feasible. These hurdles span across physics, engineering, algorithms, and software development. Major technical challenges include scaling quantum systems while simultaneously improving qubit quality—specifically coherence times, noise resilience, and gate fidelity. Demonstrating quantum error correction and building large-scale fault-tolerant quantum systems are essential milestones. The major breakthroughs ahead involve realizing quantum hardware capable of executing fault-tolerant computations at scale and delivering an end-to-end, commercially viable quantum application. The IBM Quantum Development and Innovation Roadmap (Figure 1.2.1) outlines a path toward this goal, aiming to run quantum programs with hundreds of logical qubits and millions of quantum gates by the end of the decade. Further details of the roadmap, with particular emphasis on the quantum hardware, will be described below.

2. Qubit Hardware: Superconducting Qubits

The main building blog of our quantum processors is the superconducting qubit, which is based on a Josephson junction [12] shown in Figure 1.2.2. The Josephson junction consists of two superconducting electrodes separated by a thin insulating barrier (Figure 1.2.2a) magnified view). To reduce susceptibility to charge noise, the junction is capacitively shunted. Functionally, the junction behaves as a nonlinear inductor (Figure 1.2.2b), creating an anharmonic energy spectrum in which quantum energy levels are unequally spaced (Figure 1.2.2c). This anharmonicity enables the isolation of two specific levels that define the computational basis states $|0\rangle$ and $|1\rangle$, allowing selective qubit control and minimizes leakage to higher excited states. The energy separation between these states corresponds to approximately 240mK, equivalent to a microwave frequency near 5GHz. These qubits are operated at cryogenic temperatures of 10-to-20mK, a regime nearly free of dissipation, enabling stable quantum behavior.

A crucial parameter for the stability of a quantum system is the coherence time. It is the time over which a quantum state remains usable for computation before it loses its quantum properties due to noise and interactions with the environment. Typically, it is characterized by two main parameters: the relaxation time T_1 which describes the time for the qubit to decay from the excited state $|1\rangle$ to the ground state $|0\rangle$ and the dephasing time T_2 that describes the time over which the relative phase between $|0\rangle$ and $|1\rangle$ remains stable.

Superconducting qubits are controlled through sequences of precisely shaped microwave pulses that implement quantum circuits. Over the past decade, superconducting qubits have emerged as the leading platform for quantum computing due to several key advantages. They offer extremely fast gate operations, typically in the range of a few tens (for single qubit gate operations) to a few hundred nanoseconds (for two-qubit gate operations), combined with coherence times (T_1: energy relaxation and T_2: dephasing) currently in the range of 300-to-400µs across all qubits, as e.g. measured in IBM's *Heron* processor (Figure 1.2.3a,b), and up to several milliseconds for individual test devices [13]. Faster operations significantly reduce exposure to decoherence, allowing deeper circuits—i.e., more gate operations—to be executed within the coherence window. Continuous improvements in materials, device design, and fabrication processes have enabled steady increases in coherence times. In parallel, high-fidelity gate operations are achieved, with two-qubit gate error rates reaching $\sim5\times10^{-4}$ (Figure 1.2.3c). Achieving such low error rates is essential for implementing quantum error correction, which requires thresholds below approximately 10^{-3}. Furthermore, qubit parameters such as frequency, anharmonicity, and coupling strength can be engineered by design, providing critical flexibility for performance optimization. Significant progress has also been achieved for neutral-atom [14] and trapped-ion qubits [15]. Although trapped-ion qubits can exhibit coherence times of several minutes, their gate operation times are very long up to several hundreds of microseconds with overall speed decreasing with increasing number of qubits thus significantly limiting overall compute speed.

Besides speed, another important advantage of superconducting qubits is their compatibility with classical microwave electronics for control and readout, which simplifies integration and supports scalability. The planar architecture of these qubits further facilitates scaling and aligns well with semiconductor processing technologies. Today, state-of-the-art 300mm semiconductor fabrication processes are leveraged to manufacture the latest superconducting quantum processors, including *Nighthawk* and *Loon* (Figure 1.2.4). The continuous improvements and performance gains achieved highlight superconducting qubits as a preferred qubit technology for scalable quantum hardware delivering high quality circuits and computation speed.

3. Performance Improvements for Quantum Systems

To objectively measure and compare quantum computer performance—and to guide systematic improvements—well-defined metrics are essential. A holistic framework was introduced based on three key dimensions: scale, quality, and speed [16].

Scale refers to the number of qubits available in a system. While qubit count is an important indicator of hardware capability, it is insufficient on its own; meaningful performance evaluation requires consideration of circuit execution accuracy.

Quality captures the ability of a quantum processor to execute complex circuits with high fidelity. This metric is influenced by factors such as coherence times, gate error rates, crosstalk, and qubit connectivity. For processor scale beyond 100 qubits, measures such as layer fidelity and error per layered gate (EPLG) have been introduced [17]. Layer fidelity quantifies the accuracy of executing an entire layer of gates across many qubits, while EPLG tracks error rates per gate layer, offering deeper insight into performance for large circuits and error-mitigated algorithms.

Speed is another important metric and is characterized by Circuit Layer Operations Per Second (CLOPS), which measures how efficiently a quantum system can execute workloads

[16]. CLOPS accounts for gate execution rates, repetition frequency, and classical control overhead, providing a practical benchmark for throughput in real-world applications. Together, these metrics—scale, quality, and speed—form a comprehensive framework for evaluating quantum computing performance and guiding the development of scalable, high-fidelity quantum hardware.

3.1 Scaling Quantum Processors: from Tens to Thousands of Qubits

Scaling the number of qubits and quantum systems is a prerequisite to enable the potential of quantum computing and surpassing the capabilities of classical computation. Over the past several years, significant scientific and engineering challenges have been addressed to realize quantum processors integrating more than 1,000 qubits on a single chip [18]. Initial test devices were demonstrated with increasing qubit counts from 5-to-53, culminating in the release of a first commercial processor with 27 qubits in 2019. Achieving these milestones required overcoming numerous technical hurdles, such as, e.g., the development of the heavy-hex lattice architecture [19] suppressing crosstalk between qubits, improvements in fabrication yield through advanced process control and post-processing techniques [20], and the implementation of multiplexed readout schemes to minimize the number of cryogenic amplifiers.

Breaking the 100-qubit barrier represented a major step forward, as it introduced unprecedented complexity in control and I/O integration. The *Eagle* processor (Figure 1.2.1), which achieved this milestone, relied on a superconducting multi-layer wiring architecture combined with advanced packaging solutions (Figure 1.2.5a). Unlike conventional CMOS back-end-of-line wiring, this approach employs superconducting metals to construct low-loss transmission lines. Multiple metallization layers, separated by planarized oxide, are stacked to form well-isolated microwave interconnects. The integration of an interposer featuring multi-level wiring and superconducting through-substrate vias (TSVs) enabled high-fidelity signal transmission, a critical requirement for processors exceeding 100 qubits (Figure 1.2.5b). This packaging paradigm, compatible with 10mK operation, was subsequently scaled for a quantum processor, which incorporates 1121 qubits, pushing the limits of both fabrication yield and system integration (Figure 1.2.1). These advances in architecture, materials, and packaging represent key enablers for the continued scaling of quantum processors toward fault-tolerant quantum computing.

The significant learnings achieved in this development culminated in the *Heron* processor (Figure 1.2.3a) with several fundamental improvements implemented. It introduces a flux-tunable coupler architecture within a heavy-hex lattice topology, enabling dynamic control of inter-qubit interactions to minimize idle crosstalk to be negligible, while supporting high-fidelity two-qubit gates. This design is optimized for scaling up large-scale systems in a modular approach. The qubit count ranges from 133-to-156 qubits and advanced fabrication techniques, two-level system (TLS) defect mitigation, and multiplexed readout to improve yield and coherence are applied. Figure 1.2.6 shows the continuous improvement of two-qubit gate errors for the different quantum processors demonstrated since 2018. It highlights the significant benefit of the tunable-coupler architecture achieving a best two-qubit gate error (open-colored symbols) of ~5×10⁻⁴ and a significant drop of the median value (filled colored symbols) compared to the former quantum processor types. Also, the median coherence times (T1, T2) were improved to between 300 and 400µs. Performance benchmarking using layer fidelity and EPLG demonstrates substantial improvements over architectures without tunable couplers: EPLG reduced by almost 10× to ~3.2×10⁻³ for full-device layers.

Also the speed of quantum computation measured in CLOPS improved significantly driven by updates to the quantum system software stack, optimization of data movement, and runtime.

Furthermore, parametric compiling was introduced that iterative circuits need to be compiled only once if only parameters change. These updates were key to achieve a CLOPS of over 330,000 CLOPS in 2025 compared to 2,700 in 2023 (Figure 1.2.7). These advances enable execution of utility-scale workloads utilizing error mitigation, with circuit depths approaching 5,000 two-qubit gates [21] while runtime per data point decreased by more than 100× in the last two years.

3.2 System-Level Engineering for Scalable Quantum Platforms

Recent advances in system engineering have focused on three critical domains. First, cryogenic platforms and infrastructure have evolved from conventional dilution refrigerators to modular, large-volume cryostats with enhanced cooling power and port density, supporting multi-processor integration and high I/O throughput. Second, cryogenic electronics and RF hardware have been co-designed to reduce wiring complexity and latency, incorporating cryo-CMOS control circuits [22], low-noise amplification chains, and optimized RF signal generation for high-fidelity qubit operations under stringent thermal budgets. Third, signal delivery and wiring innovations—including high-density flexible printed circuits, multiplexed readout architectures, and superconducting interconnects—have replaced bulky coaxial cables, enabling scalable microwave routing with minimal crosstalk and thermal load [23]. Together, these developments form the backbone of advanced system architectures, addressing the challenges of control, packaging, and integration [24]. These engineering advancements combined with a modular architectural approach on all levels are essential to building systems with millions of qubits for scalable fault-tolerant quantum computers.

4. Modularity

To scale beyond the number of qubits on a single quantum processor, a modular approach utilizing microwave links is used. This modularity is the key for building large-scale fault-tolerant quantum systems that span multiple chips and multiple cryostats. Achieving high-fidelity, large-scale quantum systems requires three levels of modularity as sketched in Figure 1.2.8a: (1) M-type (short-range): partitioning a Quantum Processing Unit (QPU) into multiple chips with minimal impact on gate speed and fidelity; (2) L-couplers (long-range): connections within a cryogenic environment to overcome I/O bottlenecks and enable complex topologies; (3) on-chip non-local C-couplers for LDPC error correction codes. Combining these different coupling schemes is essential to achieve large-scale fault-tolerant quantum computing.

The fabricated m-coupler shown in Figure 1.2.8b, provides high-bandwidth, high-fidelity links to interconnect quantum chips, implementing dense modularity to effectively extend chip size. This approach requires ultra-low-loss, low-cross-talk connections that are sufficiently short to maintain single-mode behavior. The inter-chip spacing must be comparable to intra-chip qubit distances to preserve performance and minimize latency. This type of coupler is demonstrated in the *Crossbill* processor in which three chips with 160 qubits each are connected into a single package for a total of 480 qubits. For long-range, the L-couplers are developed [25]. They connect quantum modules located up to one meter apart within the same cryogenic system demonstrated with the *Flamingo* processor shown in Figure 1.2.8c. Coupling within the same module is achieved by C-couplers and demonstrated recently by *Loon*. These link distant physical qubits within the same quantum chip or module to facilitate complex error correction codes as explained below.

5. Quantum Processors for Quantum Advantage and Fault Tolerance

Another parameter that is important to achieve quantum advantage and fault tolerance is the qubit connectivity. It describes the number of direct interactions between qubits in a quantum processor, and thus, defines which qubits can perform two-qubit gates without additional routing operations. The more neighbors a qubit can interact with, the higher is its connectivity reducing the need for SWAP gates and lowering circuit depth. A SWAP gate exchanges the quantum states of two qubits, acting like a reversible switch to move information between them. High connectivity in a quantum processor is essential for both quantum advantage and fault tolerance because it directly impacts circuit efficiency, error rates, and scalability.

A new chip architecture was designed and built (*Nighthawk*) transitioning from a heavy-hex lattice used before to a square-lattice architecture. This increases qubit connectivity from three to four neighbors, enabling more efficient circuit layouts and reducing the number of gates required for qubit routing. The improved connectivity supports lower circuit depth, which translates into reduced error accumulation and enhanced algorithmic performance.

Using error-mitigation techniques, *Nighthawk* is projected to support the calculation of quantum circuits of approximately 5,000 gate operations in 2025, scaling to 15,000 by 2028 (Figure 1.2.1). These improvements allow users to run more complex quantum algorithms at similar gate scales, accelerating progress toward practical quantum advantage and achieve higher degree connectivity, important for large-scale fault-tolerant systems [13].

6. Handling of Quantum Errors

Errors in quantum systems originate from multiple sources, including decoherence, imperfect gate operations, crosstalk, measurement inaccuracies, leakage into non-computational states, and material or fabrication defects. Consequently, error handling is a fundamental requirement for scalable quantum computing and is addressed through three complementary strategies: error suppression, error mitigation, and error correction.

Error suppression encompasses techniques that proactively reduce error occurrence at the hardware and control levels, thereby improving the intrinsic fidelity of quantum operations. An example is the use of advanced pulse-shaping methods to minimize leakage and phase errors. Error mitigation, in contrast, seeks to counteract the impact of errors after circuit execution. These techniques aim to improve computational accuracy without employing full error-correcting codes, relying instead on statistical and algorithmic approaches to compensate for noise in the measured results [26]. Quantum error correction (QEC) is a fundamental mechanism for achieving fault-tolerant quantum computation, wherein logical information is encoded across multiple physical qubits to ensure computational integrity even in the presence of errors [27]. By introducing structured redundancy, the system can detect and correct faults without compromising the outcome of the computation. This concept parallels classical error correction, where information is encoded with redundancy to enable error detection and recovery through consistency checks. However, QEC is significantly more challenging than classical error correction as quantum information cannot be copied, and additionally there are two types of errors, bit flips and phase flips to be identified and corrected.

ISSCC 2026 / SESSION 1 / PLENARY / 1.2

In QEC the quantum information is encoded across multiple physical qubits—called logical qubits—and specialized gate operations treat this ensemble as a single, effectively error-resilient logical unit. A defined set of operations and syndrome measurements—collectively referred to as an error-correcting code—enables the detection and correction of errors without collapsing the encoded quantum state. According to the quantum threshold theorem, fault-tolerant quantum computation is achievable only if the physical error rates of the hardware fall below a code-dependent threshold, ensuring that logical error rates can be suppressed arbitrarily by increasing the code distance.

7. Scalable Fault-Tolerant Quantum Systems

Achieving the full promise of quantum computing requires systems capable of executing circuits with hundreds of millions of gates across hundreds of qubits while actively correcting errors to prevent propagation [28]. This necessitates a fault-tolerant quantum computer.

There are six essential criteria to be met for scalable, reliable quantum architectures:

- Fault-tolerance – Logical error rates suppressed to enable meaningful algorithms;
- Addressability – Individual logical qubits can be prepared and measured throughout computation;
- Universality – Support for a complete set of quantum instructions;
- Adaptivity – Real-time decoding of measurements to inform subsequent operations;
- Modularity – Hardware organized into replaceable, quantum-connected modules;
- Efficiency – Execution of useful algorithms with practical physical resources.

A new error-correction architecture proposed, satisfies these criteria and provides a roadmap for scaling toward large-scale, error-corrected quantum systems [29]. This architecture integrates fault-tolerant logical qubits, modular hardware design, and adaptive control to enable reliable execution of deep circuits, forming the foundation for quantum-centric supercomputing.

As indicated in section 6, logical states are encoded as entangled superpositions of multi-qubit basis states, and error detection is achieved through syndrome-extraction circuits, which identify error signatures without collapsing the encoded information. So far, the QECs proposed had significant limitations, requiring very low error rates and inefficient scaling behavior. A significant breakthrough was achieved, demonstrating a fault-tolerant quantum memory based on quantum low-density parity-check (qLDPC) codes, specifically the bivariate bicycle (BB) family with low qubit overhead and high error threshold [30]. The initial implementation, termed the *gross code,* encodes 12 logical qubits into 144 data qubits (orange and blue circles), supplemented by 144 syndrome qubits (red and green circles) to check for errors, for a total of 288 physical qubits (Figure 1.2.9a). Each vertex has six incident edges including four short-range edges (pointing north, south, east and west) and two long-range edges. This construction achieves error-correction performance comparable to the surface code while requiring approximately 10× fewer qubits, significantly improving resource efficiency. The former state-of-the-art QEC surface code would require ~3,000 physical qubits. Realizing the gross code requires a 6-degree connectivity and implemented on a 2D chip necessitates therefore long-range connectivity patterned after the symmetry of a 3D torus—imagine the 2-dimensional qubit grid shown in Figure 1.2.9a, rolled up and connected to a torus. With the 4-degree connectivity of the square lattice *Nighthawk* chip, two additional connections are required that go to distant qubits in the lattice as indicated in Figure 1.2.9a.

Beyond memory protection, fault-tolerant quantum computing requires a universal gate set operating on encoded qubits. The universal gate set comprises Clifford gates, which run on the encoded information quickly and with low overhead, complemented by at least one non-Clifford gate, such as the T-gate that is harder to realize. Clifford gates are unitary operations that map Pauli operators to other Pauli operators (e.g. Hadamard gate: creates a superposition and swaps X and Z bases; CNOT gate: entangles two qubits); they preserve the structure of the stabilizer states. The T-gate is a single-qubit rotation around the Z-axis by π/4 and does not belong to the Clifford group. Its fault-tolerant implementation is expensive because it requires magic state injection and entanglement protocols to produce high-fidelity states from multiple noisy copies. Real-time error correction relies on a decoder, a classical system that processes syndrome data, updates error models, and outputs corrected states during computation. Finally, the architecture must be modular and scalable, supporting large qubit arrays and high-throughput error correction to enable practical, fault-tolerant quantum algorithms.

On this foundation, fault-tolerant logical processing units (LPUs) for qLDPC codes are built using generalized lattice surgery (see Figure 1.2.9b) which refers to performing logical operations by merging and splitting patches of qubits in a 2D lattice. LPUs are connected to the quantum memory and enable logical measurements via low-weight checks with minimal qubit overhead and support stabilizer-based operations such as Clifford gates, state preparation, and measurement. Recent work extends these concepts termed the *two-gross code*, which offers increased error resilience. The combined quantum memory and LPU form a modular building block within the scalable architecture (see Figure 1.2.9b).

To enable universal computation across modules, adapter concepts that bridge logical quantum information between distinct modules are introduced. These adapters leverage symmetries of qLDPC codes and can incorporate inter-module microwave L-couplers, first demonstrated in the Flamingo platform (see Figure 1.2.8c). The latest designs characterize baseline inter-module measurement instructions and establish a framework for modular, fault-tolerant quantum computing. Universal quantum computation is achieved by combining stabilizer-based logical operations with magic state factories that generate, distill, and consume magic states for implementing non-Clifford gates. Figure 1.2.10 illustrates the key milestones and critical technical achievements required to develop the fundamental building blocks and to assemble a modular fault-tolerant architecture composed of multiple interconnected modules, each hosting 12 logical qubits and an integrated magic state factory.

The quantum memory architecture with C-couplers enabling long-range connections between distant qubits on the chip is demonstrated with Loon (Figure 1.2.10). The increased qubit connectivity allows to implement the graph of the error correction code. The donut required by the qLDPC code can be mapped onto the qubits of *Loon*. Additional layers offer connecting distant qubits to demonstrate 6-degree connectivity beyond nearest neighbors for experiments with high-rate qLDPC codes. Physically long couplers are integrated, maintaining high-quality gates and long coherence times. Building on this, *Kookaburra* will be demonstrated in 2026, introducing the first processor module integrating qLDPC memory with an LPU comprising around 100 qubits for logical operations (Figure 1.2.10). Finally, *Cockatoo* (2027) will enable inter-module entanglement through universal adapters, advancing modular fault-tolerant quantum computing. Finally in 2028, the use of magic-state injection with multiple modules will be demonstrated in *Starling* and will be scaled to a system capable of running one hundred million gates on 200 logical qubits in 2029 as shown in the Roadmap (Figure 1.2.1). The impact of quantum error correction (QEC) on quantum computing will be transformative, as it is the key enabler for moving from noisy intermediate-scale quantum (NISQ) devices to fault-tolerant quantum computers capable of running arbitrarily long and complex algorithms reliably.

7. Quantum Development Roadmap

Building a completely new stack of computing requires the development of an entire ecosystem. This endeavor can be supported by embracing and sharing a roadmap. Therefore, IBM Quantum released already in 2021 a roadmap that shows how the hardware will evolve and how the software tools and services are developed on top. Figure 1.2.1 shows the 2025 updated IBM Quantum Development and Innovation Roadmap, presenting new processors and capabilities that will pave the way to quantum advantage and fault tolerance as described before. The steps required to realize useful quantum computing are laid out. Recent revisions to the roadmap project a path to 2033 and beyond. The successful delivery of the milestones creates trust in the technology and are the basis for continued progress. The *Nighthawk* quantum processor will be the basis for continuously improving the number of gate operations from 5k in 2025 to 10k in 2027, 15k in 2028, and connecting qubits and quantum processors with quantum links. This culminates in *Starling*—a system capable of running quantum circuits with 100 million gates on 200 logical qubits in 2029. The plan is to extend the system by 2033, demonstrating *Blue Jay*, a quantum system capable of running circuits with a billion gates on 2,000 logical qubits. These large-scale fault-tolerant quantum computers will unlock a new era of algorithmic complexity and application discovery.

8. Quantum-Centric Supercomputing: A Heterogeneous Architecture for Hybrid Workflows

Quantum-centric supercomputing integrates QPUs with CPUs and GPUs in a heterogeneous architecture to accelerate computation by leveraging the strengths of both quantum and classical systems. This approach requires careful consideration of latency, parallelism (quantum and classical), and instruction allocation between processors. Emerging quantum algorithms increasingly exploit multiple quantum circuits executed in parallel, combined with concurrent classical operations, necessitating scalable and parallel circuit execution alongside advanced classical computation [31].

To enable these hybrid workflows, quantum systems are being engineered for tight integration with high-performance computing (HPC) environments. This includes performant software for circuit generation and manipulation, and middleware for dynamic execution of quantum-classical routines. By exploiting synergies between quantum and classical resources, quantum-centric supercomputing provides a foundation for algorithms that partition subroutines across QPUs, CPUs, and GPUs, paving the way for practical quantum acceleration and are the foundation for the future of computing.

9. Quantum Applications

Certain classes of mathematical problems exhibit characteristics that make them amenable to quantum computation, including some that are intractable for classical algorithms. For specific cases there is theoretical evidence that quantum algorithms can achieve exponential or otherwise significant speedups outperforming their classical counterparts. Not all problems that are computationally challenging for classical systems will benefit from quantum acceleration; quantum speed up is problem-dependent and requires careful algorithmic and hardware considerations. Of particular interest are tasks that admit a quantum algorithm with exponential speedup. Known examples of tasks with such a polynomial quantum speed up include simulation of quantum many-body systems [32], number theoretic problems such as integer factoring [3], solving certain types of linear systems [2], applications in topological data analysis [33], and computing topological invariants [34]. Of very high interest are simulations of quantum many-body systems due to their numerous scientific and industrial applications. Quantum computers can provide a significant advantage for highly complex problems, such as high-precision ground-state simulations of strongly interacting electron systems, which are critical in quantum chemistry. Additionally, they offer advantages in calculating dynamic behaviors far from equilibrium including reaction pathways, and simulating the time evolution of complex quantum systems such as spin-chain Hamiltonians. Solving these problems with a guaranteed super polynomial speed up assumes an architecture of fault tolerance.

Also, heuristic methods play a critical role in evaluating the capabilities and potential benefits of quantum computing. Although these approaches lack formal complexity-theoretic guarantees, they are valuable because they enable rigorous error analysis and provide empirical evidence of quantum performance relative to classical baselines. In optimization, heuristics are frequently employed since the performance of an algorithm on a specific instance is generally unknown a priori. Optimization problems are pervasive across science and industry; therefore, any improvement over state-of-the-art classical algorithms using quantum techniques could have substantial impact. Such improvements may manifest across multiple dimensions, including solution quality, diversity of solutions, time-to-solution, and cost-to-solution, thereby influencing both theoretical research and real-world applications [4]. While quantum optimization algorithms will not necessarily improve the performance for all problems, they can improve performance for some problem instances and thus, improve our overall capabilities in optimization. Quantum computing will not necessarily accelerate all problems, therefore it is crucial to understand exactly where advantages might be found. This needs to be probed theoretically, and empirically.

Achieving quantum advantage – defined as solving an information processing task with a quantum computer offering superior efficiency, cost-effectiveness, or accuracy compared to a sole classical computation and the correctness of the output rigorously validated – may not require error correction but could be already possible utilizing error mitigation approaches. These techniques can outperform classical methods for problems at the 100-qubit scale with thousands of two-qubit gates. It was demonstrated that a noisy quantum computer with more than 100 qubits and around 3000 gates can produce accurate expectation values outside of brute force classical computation [11]. Higher qubit connectivity and further improved hardware fidelity continuously increasing the number of gate operations as described before, enable deeper circuits, and thus, more complex observables to be calculated. Based on these developments, it is anticipated that even in the absence of full fault tolerance, hybrid quantum-classical approaches will play a crucial role in the realization of quantum advantage.

The potential applications of quantum computing in science and industry are manifold promising accelerated innovation and computational capabilities beyond classical limits combined with far-reaching economic impact.

10. Conclusions

The progress in Quantum Computing described in this paper demonstrates that we are at the beginning of a transformative era in computation. While quantum mechanics has shaped science for over a century, its application to scalable quantum computing is only starting to unfold. IBM's roadmap toward fault-tolerant quantum computing within the next decade provides a clear trajectory, yet the long-term potential—over the next 20-to-50 years—remains difficult to predict. Quantum-centric supercomputing and modular architectures may redefine computational paradigms, enabling breakthroughs across science and industry. The innovations presented here lay the foundation for systems that will evolve to large-scale, fault-tolerant quantum computers, marking the start of a new chapter in information technology!

Acknowledgement:
The author gratefully acknowledges the entire IBM Quantum team for their dedication and contributions to advancing quantum computing hardware, software, and theory. Their works in achieving the milestones and innovations are summarized and presented here.

References:
[1] D. S. Abrams, S. Lloyd, "Quantum algorithm providing exponential speed increase for finding eigenvalues and eigenvectors", *Phys. Rev. Lett.*, vol. 83, pp.5162-5165, 1999, DOI: doi.org/10.1103/PhysRevLett.83.5162.
[2] A. W. Harrow, A. Hassidim, Seth Lloyd, "Quantum algorithm for linear systems of equations", *Phys. Rev. Lett.*, vol. 103, 150502, 2009. DOI: doi.org/10.1103/PhysRevLett.103.150502.
[3] P. W. Shor, "Algorithms for quantum computation: Discrete logarithms and factoring", *In Proc. 35th Annual Symposium on Foundations of Computer Science IEEE*, pp.124-134, 1994.
[4] A. Abbas et al., "Challenges and opportunities in quantum optimization", *Nature Reviews Physics*, vol. 6, pp.718–735, 2025, DOI: doi.org/10.1038/s42254-024-00770-9.
[5] L.K. Grover, "A fast quantum mechanical algorithm for database search", In

Proceedings of the 28th Annual AM Symposium on theory of Computing STOC '96 (ACM, New York), pp. 212-219, 1996.
[6] T. Macris, "Quantum Computing with Neutral Atoms", In: Jang-Jaccard, J., Caroff, P., Blezinger, E., Mulder, V., Mermoud, A., Lenders, V. (eds), *Quantum Technologies*. Springer, Cham, 2026, DOI: doi.org/10.1007/978-3-031-90727-2_3.
[7] C. Hempel, "Trapped-Ion Quantum Computers," in: Jang-Jaccard, J., Caroff, P., Blezinger, E., Mulder, V., Mermoud, A., Lenders, V. (eds), *Quantum Technologies*. Springer, Cham, 2026, DOI: doi.org/10.1007/978-3-031-90727-2_3.
[8] https://quantum.cloud.ibm.com/computers?type=Heron
[9] G. Burkard et al., "Semiconductor spin qubits", *Rev. Mod. Phys.*, vol.95, p.025003, 2023, DOI: doi.org/10.1103/RevModPhys.95.025003.
[10] H. Aghaee Rad, et al. "Scaling and networking a modular photonic quantum computer," *Nature* vol. 638, pp.912–919, 2025, DOI: doi.org/10.1038/s41586-024-08406-9.
[11] Y. Kim et al., "Evidence for the utility of quantum computing before fault tolerance," *Nature*, vol.618, pp.500–505, 2023, DOI: doi.org/10.1038/s41586-023-06096-3.
[12] J. M. Gambetta, J. M. Chow and M. Steffen, "Building logical qubits in a superconducting quantum computing system", *npj Quantum Information*, vol.3, p.2, 2017, DOI: doi.org/10.1038/s41534-016-0004-0.
[13] https://www.ibm.com/quantum/blog/qdc-2025
[14] NC. Chiu, et al. "Continuous operation of a coherent 3,000-qubit system," *Nature*, vol.646, pp.1075–1080, 2025, DOI: doi.org/10.1038/s41586-025-09596-6.
[15] A. Ransford et al. "Helios: A 98-qubit trapped-ion quantum computer", arXiv:2511.05465 [quant-ph], 2025, DOI: doi.org/10.48550/arXiv.2511.05465.
[16] A. Wack et al., "Scale, Quality, and Speed: Three Key Attributes to Measure the Performance of Near-Term Quantum Computers", arXiv:2110.14108 [quant-ph], 2021, DOI: doi.org/10.48550/arXiv.2110.14108.
[17] D. C. McKay et al., "Benchmarking Quantum Processor Performance at Scale," arXiv:2311.05933 [quant-ph], 2023, DOI: doi.org/10.48550/arXiv.2311.05933.
[18] https://www.ibm.com/quantum/blog/quantum-roadmap-2033
[19] C. Chamberland et al., "Topological and Subsystem Codes on Low-Degree Graphs with Flag Qubits", *Phys. Rev. X*, vol.10, p. 011022, 2020, DOI: doi.org/10.1103/PhysRevX.10.011022.
[20] J. B. Hertzberg, E. J. Zhang, S. Rosenblatt, et al., "Laser-annealing Josephson junctions for yielding scaled-up superconducting quantum processors", *npj Quantum Information*, vol.7, p.129, 2021, DOI: doi.org/10.1038/s41534-021-00464-5.
[21] R. C. Farrell, et al. "Digital quantum simulations of scattering in quantum field theories using W states," arXiv:2505.03111v2, 2025, DOI: doi.org/10.48550/arXiv.2505.03111.
[22] D. Underwood, et al. "Using Cryogenic CMOS Control Electronics to Enable a Two-Qubit Cross-Resonance Gate," *PRX QUANTUM*, vol.5, 010326, 2024, DOI: doi.org/10.1103/PRXQuantum.5.010326.
[23] N. Masluk, presentation at CEC/ICMC 2023, Cern, 2023, https://indico.cern.ch/event/1168385/contributions/5356125/attachments/2680218/46492 07/2023-07-13_Nicholas_Masluk,_CEC_ICMC,_Qubit_Cryogenic_Infrastructure.pdf
[24] https://newsroom.ibm.com/2023-12-04-IBM-Debuts-Next-Generation-Quantum-Processor-IBM-Quantum-System-Two,-Extends-Roadmap-to-Advance-Era-of-Quantum-Utility
[25] K. Heya et al. "Randomized Benchmarking of a Remote cnot Gate Via a Meter-Scale Microwave Link", *Phys. Rev. Lett.*, vol.135, p.200801, 2025, DOI: doi.org/10.1103/xx24-r7q6.
[26] K. Temme et al. "Error mitigation for short-depth quantum circuits", *Phys. Rev. Lett.*, vol. 119, p.180509, 2017, DOI: doi.org/10.1103/PhysRevLett.119.180509.
[27] S. B. Bravyi, A. Y. Kitaev, "Quantum codes on a lattice with boundary", quant-ph/9811052, 1998, DOI: doi.org/10.48550/arXiv.quant-ph/9811052.
[28] S. Bravyi et al. "The future of quantum computing with supercodnncuting qubits," J. Appl. Phys. 132, 160902, 2022.
[29] T. J. Yoder et al. "Tour de gross: A modular quantum computer based on bivariate bicycle codes," arXiv:2506.03094v1, 2025, DOI: doi.org/10.48550/arXiv.2506.03094.
[30] S. Bravyi, et al. "High-threshold and low-overhead fault-tolerant quantum memory," *Nature*, vol.627, pp.778–782, 2024, DOI: doi.org/10.1038/s41586-024-07107-7.
[31] Y. Alexeev et al. "Quantum-centric Supercomputing for Materials Science: A Perspective on Challenges and Future Directions", *Future Generation Computer Systems*, vol.160, pp.666–710, 2024, DOI: doi.org/10.1016/j.future.2024.04.060.
[32] S. Lloyd, "Universal quantum simulators", *Science*, vol.273, pp.1073-1078, 1996, DOI: 10.1126/science.273.5278.1073.
[33] S. Ubaru, et al. "Quantum topological data analysis with linear depth and exponential speedup", arXiv:2108.02811, 2021, DOI: doi.org/10.48550/arXiv.2108.02811.
[34] D. Aharonov, V. Jones, and Z. Landau "A polynomial quanutm algorithm for approximating the Jones polynomial", *Algorithmica*, vol.55, pp.395-421, 2009, DOI: doi.org/10.1007/s00453-008-9168-0.

ISSCC 2026 / SESSION 1 / PLENARY / 1.2

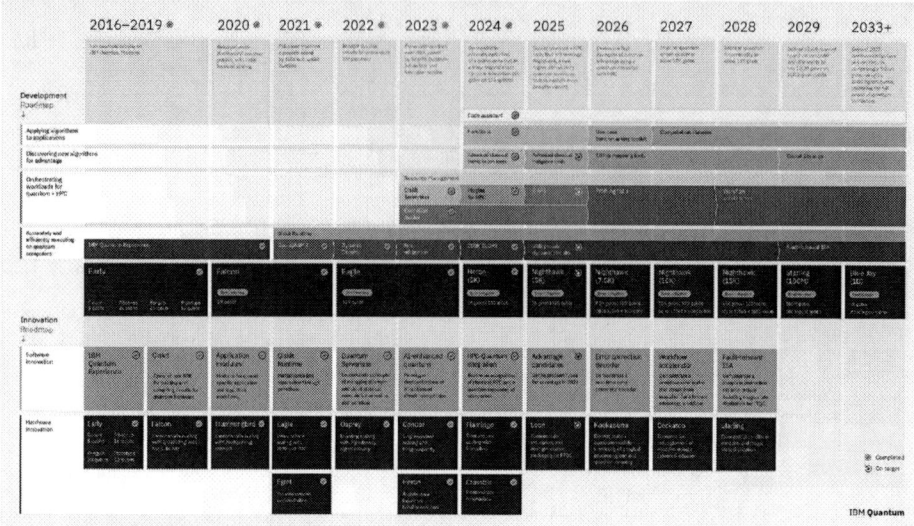

Figure 1.2.1: IBM Quantum Innovation and Development Roadmap (Updated November 2025).

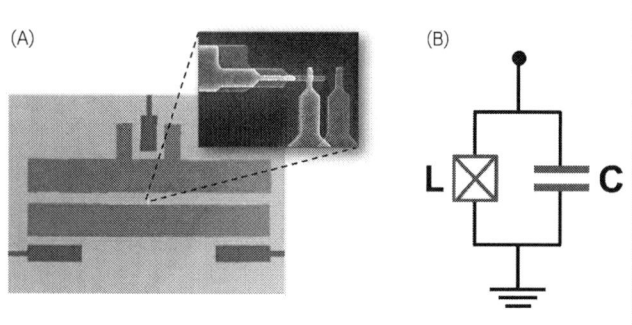

Figure 1.2.2a: Scanning electron micrograph of a superconducting qubit Josephson junction. The blue colored area indicates the shunt capacitor. The junction is created by two overlapping perpendicular metal fingers separated by an oxide and possesses an area of ~100nm x 100nm.

Figure 1.2.2b: Sketch of the equivalent circuit of the Josephson junction comprising of a non-linear inductor and a capacitance acting as non-linear oscillator.

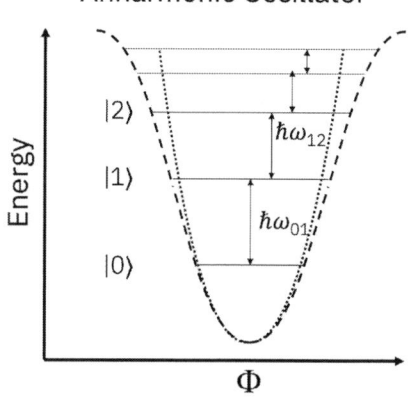

Figure 1.2.2c: An anharmonic energy-level diagram illustrating the ground and first excited states that define the qubit's computational basis, along with higher excited states exhibiting progressively smaller level spacings—an essential feature for selective qubit control.

Figure 1.2.3a: IBM Heron processor with 156 qubits in a heavy-hex qubit layout and tunable couplers achieving negligible cross-talk.

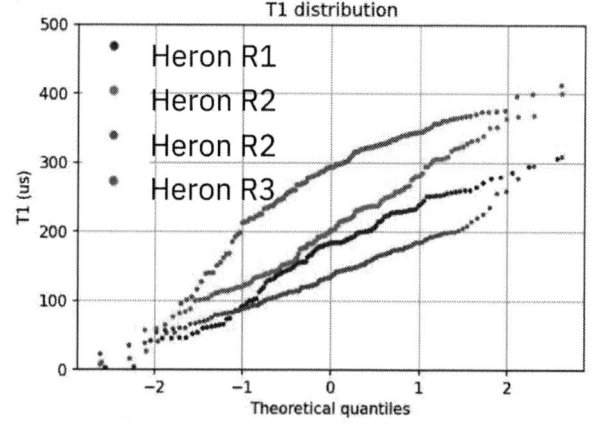

Figure 1.2.3b: Distribution of T1 coherence times of Heron quantum processors. R1, R2, and R3 refer to revisions within the same processor family, representing incremental design and performance improvements. The quantum processor shows significant T1 improvement with median T1 up to 300 µs.

ISSCC 2026 / February 16, 2026 / 9:20 AM

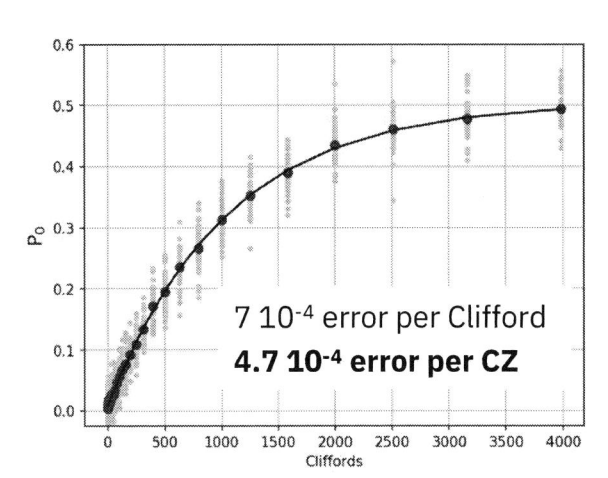

Figure 1.2.3c: Quality measurement error per layered gate (EPLG) for Heron R3 with best gates below 5x10⁻⁴.

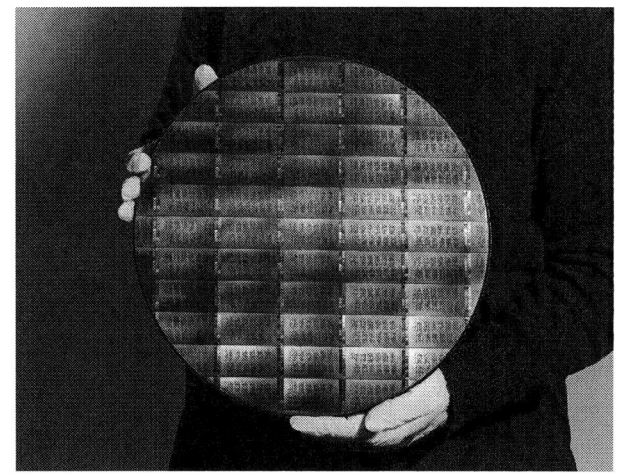

Figure 1.2.4: A 300-mm wafer incorporating Nighthawk quantum processors was fabricated at NY CREATES' Albany NanoTech Complex using advanced semiconductor manufacturing technology.

Qubit plane
Transmon qubits attached to chip via bump bonds

Readout plane
Readout resonators wired through connections.

Wiring plane
Through substrate vias (TSV) providing connections through planes.

Interposer chip
Leverages CMOS packaging techniques.

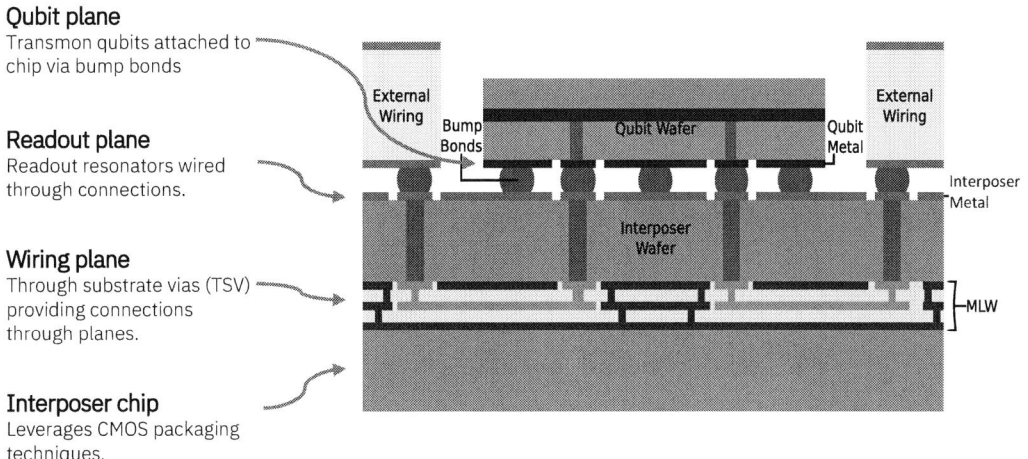

Figure 1.2.5: The architecture for processors with >100 qubits leverages advanced 3D packaging techniques comprising several planes for the qubits, the microwave readout resonators, the wiring plane, attached to an interposer.

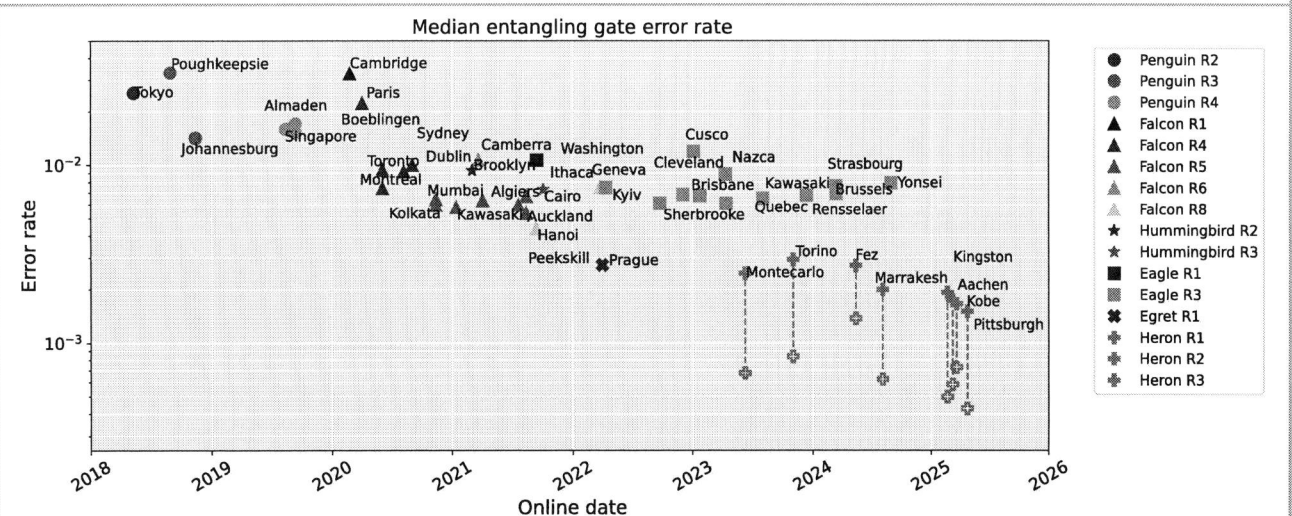

Figure 1.2.6: The road to improved gate errors: Median two-qubit error rate on a logarithmic scale versus the year of quantum processor release. The error rates continuously decrease, and a significant improvement was achieved with the Heron architecture using tunable couplers reaching error rates needed for error correction.

DIGEST OF TECHNICAL PAPERS • 21

979-8-3315-8937-0/26 $31.00 © 2026 IEEE

ISSCC 2026 / SESSION 1 / PLENARY / 1.2

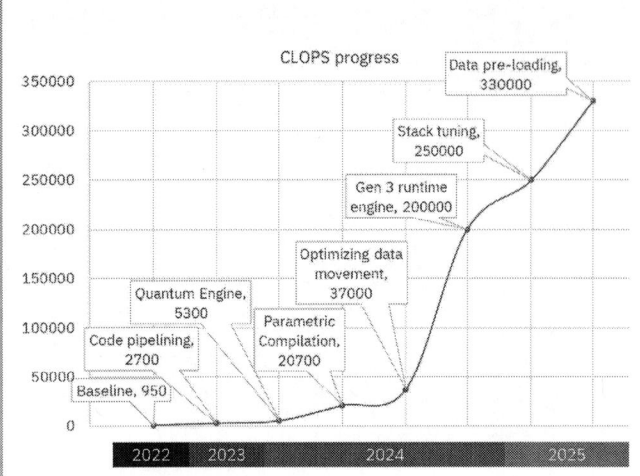

Figure 1.2.7: Speed as measured in circuit layer operations per second (CLOPS) has continuously increased.

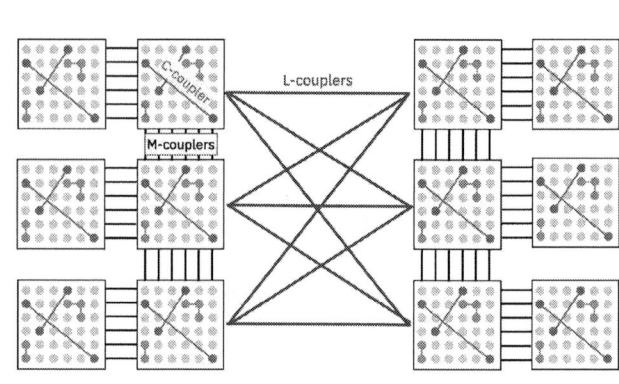

Figure 1.2.8a: Sketch of a modular quantum architecture with M-, L-, and C-couplers.

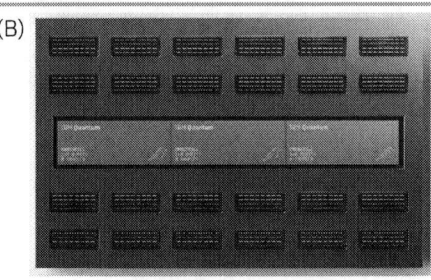

Figure 1.2.8b: M-couplers are interconnects between chips, in this case across 3x160 qubit chips connected into a single package for a total of 480 qubits.
Figure 1.2.8c: L-couplers: Flamingo, which connects two Heron chips with four connectors measuring up to one meter long.

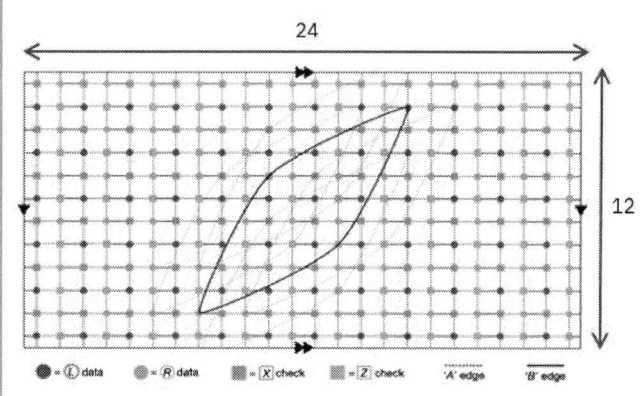

Figure 1.2.9a: The gross code encodes 12 logical qubits into 144 data qubits, along with an additional 144 syndrome check qubits.

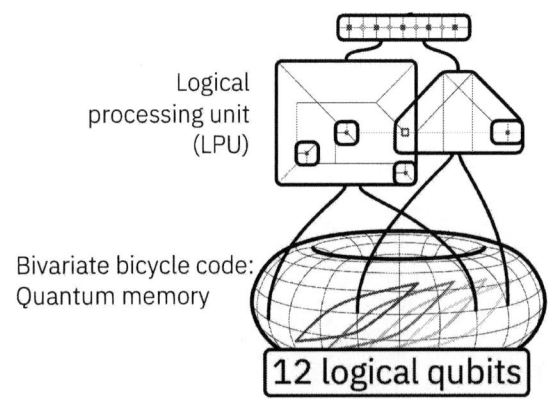

Figure 1.2.9b: A complete module called Kookaburra consisting of a logical processing unit (LPU) connected with a quantum memory to be demonstrated in 2026. Long C-couplers and real-time decoding are required. The LPU comprises around 100 qubits for logical operations.

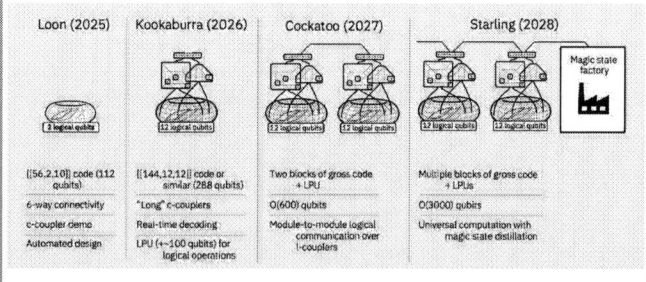

Figure 1.2.10: Roadmap to achieve a fault-tolerant architecture based on modules with 12 logical qubits and a magic state factory using a modular approach. The units with 12 logical qubits are connected with microwave L-couplers. Finally a magic state factory is connected to achieve universality [29].

ISSCC AWARDS

2025 LEWIS WINNER AWARD FOR OUTSTANDING PAPER

"A 9.05-to-37.0GHz LO Generator with Magnetic Mode Switching and Tuning-Free Octave-Bandwidth Common-Mode Resonator Achieving >190.7dBc/Hz FoM "

https://doi.org/10.1109/ISSCC49661.2025.10904644

Hao Guo, Yaolong Hu, Taiyun Chi

Rice University, Houston, TX

2025 ANANTHA P. CHANDRAKASAN DISTINGUISHED-TECHNICAL-PAPER AWARD

"A 212.5Gb/s DSP-Based PAM-4 Transceiver with 50dB Loss Compensation for Large AI System Interconnects in 4nm FinFET "

https://doi.org/10.1109/ISSCC49661.2025.10904601

E-Hung Chen[*1], Henry Park[*2], Mohammed Abdullatif[*2], Miguel Gandara[2], Ahmed ElShater[2], Amr Khashaba[2], Shih-Hao Huang[1], Tsz-Bin Liu[1], Atharav Atharav[2], Joonyeong Lee[2], Qaiser Nehal[2], Mohamed Megahed[2], Yusang Chun[2], Cheng-En Shieh[1], Vidhan Jolly[2], SoonWon Kwon[2], Hsin-Ta Chien[1], Ke-Chung Wu[1], Cheng-En Liu[1], Peng Yan[2], Po-Jui Li[1], Chun-Han Chen[1], Tzu-Shun Lin[1], Pei-Chieh Liu[1], Tamer Ali[2]

[1]MediaTek, Hsinchu, Taiwan, [2]MediaTek, Irvine, CA
*Equally Credited Authors (ECAs)

2025 JAN VAN VESSEM AWARD FOR OUTSTANDING EWAA PAPER

"A Cryo-BiCMOS Controller for $^9Be^+$-Trapped-Ion-Based Quantum Computers "

https://doi.org/10.1109/ISSCC49661.2025.10904696

Peter Toth[1], Paul E. Shine[1], Sebastian Halama[2], Yerzhan Kudabay[1], Kaoru Yamashita[1,3], Hiroki Ishikuro[3], Christian Ospelkaus[2], Vadim Issakov[1]

[1]Technische Universitat Braunschweig, Braunschweig, Germany
[2]Leibniz University Hannover, Hannover, Germany
[3]Keio University, Yokohama, Japan

2025 TAKUO SUGANO AWARD FOR OUTSTANDING APAC PAPER

"An 850µW 2-to-5GHz Jitter-Filtering and Instant-Toggling Injection-Locked Quadrature-Clock Generator for Low-Power Clock Distribution in HBM Interfaces "

https://doi.org/10.1109/ISSCC49661.2025.10904589

Jeongbeom Seo[1], Yoonseo Cho[1,2], Yuhwan Shin[1,2], Jaehyouk Choi[1]

[1]Seoul National University, Seoul, Korea
[2]KAIST, Daejeon, Korea

2025 JACK KILBY OUTSTANDING STUDENT PAPER AWARD

"A Blocker-Tolerant mm-Wave Low-Noise Amplifier Utilizing Doherty Active Load Modulation for Linearity Enhancement "

https://doi.org/10.1109/ISSCC49661.2025.10904630

Hao Yu, Lianbo Liu, Sensen Li

University of Texas, Austin, TX

2025 JACK KILBY OUTSTANDING STUDENT PAPER AWARD

"A 70dB SNDR 80MHz BW Filter-Embedded Pipeline-SAR ADC Achieving 172dB FoMS with Progressive Conversion and Floating-Charge-Transfer Amplifier "

https://doi.org/10.1109/ISSCC49661.2025.10904746

Siyu Huang, Zhishuai Zhang, Xiyu He, Mingyang Gu, Yunsong Tao, Yi Zhong, Nan Sun, Lu Jie

Tsinghua University, Beijing, China

2025 ISSCC AWARD FOR OUTSTANDING FORUM PRESENTER

"Dataflow Optimization and Data Sparsity Management for ML Accelerators "

Marian Verhelst

KU Leuven, Heverlee, Belgium

2025 ISSCC AWARD FOR OUTSTANDING FORUM PRESENTER

"Innovation Trends in Depth Sensing and Imaging: Enabling Technologies and Core Building Blocks "

David Stoppa

Sony Europe, Trento, Italy

2025 DEMONSTRATION SESSION CERTIFICATE OF RECOGNITION

"An 8.62µW 75dB-DRSoC End-to-End Spoken-Language-Understanding SoC with Channel-Level AGC and Temporal-Sparsity-Aware Streaming-Mode RNN"

Sheng Zhou[1], Zixiao Li[1], Tobi Delbruck[1], Kwantae Kim[2], Shih-Chii Liu[1]

[1]University of Zurich and ETH Zurich, Zurich, Switzerland
[2]Aalto University, Espoo, Finland

2025 DEMONSTRATION SESSION CERTIFICATE OF RECOGNITION

"A 2.2pJ/b 212.5Gb/s PAM-4 Transceiver with >46dB Reach in 5nm FinFET"

A. Mostafa[1], A. Hassan[1], A. Hsu[2], A.K. Singh[3], C.-H. Wu[4], C.-R. Yang[1], D. Prabakaran[3], D. Storaska[5], D. Zhou[1], D. Visani[1], E. Hsiao[1], F. Chu[1], F. Khan[1], F. Lu[1], G. Cui[1], G. Wang[1], J. Natonio[5], J. Deng[1], J. Ding[1], J. Guo[1], J. Gu[1], J. Zang[1], L. Jiang[1], K.-M. Lu[4], M. Hasan[1], M. Kelly[6], M. H. Kashani[2], M. Gambhir[1], M. R. Patoju[3], M. Singh[1], M. Shannon[5], M. Yang[1], P. Liu[1], P. Ramakrishna[3], R. Chen[4], R. Ho[7], S. N. Shahi[8], S. Sivakumar[1], S. Xu[7], X. Yang[1], X. Han[1], Y.-P. Su[4], Z. Adal[1], Z. Guo[1], Z. Li[1], Z. Yu[1], Z. Yan[1], H. Wang[1], K. Chang[1], D. Pandya[1], H. Qian[1]

[1]Marvell, Santa Clara, CA; [2]Marvell, Toronto, Canada
[3]Marvell, Bangalore, India; [4]Marvell, Zhubei, Taiwan
[5]Marvell, Fishkill, NY; [6]Marvell, Boston, MA
[7]Marvell, Burlington, VT; [8]Marvell, Kanata, Canada

ISSCC AWARDS

ISSCC 2026 SILKROAD AWARD

"A Sub-Threshold All-nMOS Reconfigurable PUF with Secure Configuration Selection for Stable 6-Bits/Cell"

Shufan Xu, K. Liu, L. Chan, H. Tagawa, H. Shinohara, K. Niitsu

Kyoto University, Kyoto, Japan

ISSCC 2025 STUDENT-RESEARCH PREVIEW (SRP) POSTER AWARD

"Simultaneous Ultrasound Powering and Communication with Continuous Backscatter and Adaptive Modulation"

Conghao Gao

Southern University of Science and Technology (SUSTech), Shenzhen, China

"A High-Power-Density D-Band Phased-Array Transceiver in 65nm CMOS for 6G UE Module"

Yudai Yamazaki

Institute of Science Tokyo

"A 40.68MHz Dual-Output Wireless Power Transfer System Achieving 149.7mW Maximum Power and 51.2% Peak E2E Efficiency with 8mm-Diameter RX Coil for Bio-Implants"

Tianqi Lu

TU Delft, The Netherlands

IEEE SOLID-STATE CIRCUITS SOCIETY AWARDS

2024 JOURNAL OF SOLID-STATE CIRCUITS BEST PAPER AWARD

"8-λ × 50 Gbps/λ Heterogeneously Integrated Si-Ph DWDM Transmitter"

https://doi.org/10.1109/JSSC.2023.3344072

Cooper S. Levy, Zhe Xuan, Jahnavi Sharma, Duanni Huang, Ranjeet Kumar, Chaoxuan Ma, Songtao Liu, Jinyong Kim, Xinru Wu, Tolga Acikalin, Haisheng Rong, Ganesh Balamurugan, James E. Jaussi

Intel, Hillsboro, OR

2024 JOURNAL OF SOLID-STATE CIRCUITS BEST STUDENT PAPER

"A Wireless, Multicolor Fluorescence Image Sensor Implant for Real-Time Monitoring in Cancer Therapy"

https://doi.org/10.1109/JSSC.2024.3435736

Micah Roschelle[1], Rozhan Rabbani[1], Surin Gweon[1], Rohan Kumar[1], Alex Vercruysse[1], Nam Woo Cho[1], Matthew H. Spitzer[1], Ali M. Niknejad[1], Vladimir M. Stojanović[1], Mekhail Anwar[2]

[1]University of California at Berkeley, Berkeley, CA
[2]University of California at San Francisco, San Francisco, CA

2025 IEEE JOURNAL OF SOLID-STATE CIRCUITS TEST-OF-TIME

"Low-Power CMOS Digital Design"

https://doi.org/10.1109/4.126534

Anantha P. Chandrakasan, S. Sheng, Robert W. Brodersen

University of California, Berkeley, CA

2025 IEEE SOLID-STATE CIRCUITS SOCIETY DISTINGUISHED SERVICE AWARD

To: Jeremy Cosson-Martin, *AMD, Toronto, Canada*

2025 IEEE SOLID-STATE CIRCUITS SOCIETY INDUSTRY IMPACT AWARD

To: Taku Umebayashi, *Sony Semiconductor Solutions, Kanagawa, Japan*

2025 IEEE SOLID-STATE CIRCUITS SOCIETY NEW FRONTIER AWARD

To: Lin Cheng, *University of Science and Technology of China, Hefei, China*

2025 IEEE SOLID-STATE CIRCUITS SOCIETY JAMES D. MEINDL INNOVATORS AWARD

To: Anthony Patrick Mauro, *UCSD Futures and Extended Studies, San Diego, CA*

2026 IEEE SSCS PREDOCTORAL ACHIEVEMENT PRIZE IN MEMORY OF THE LATE WILLY SANSEN

To: Ziyuan Wen, *Rice University, Houston, TX*

2026 IEEE SSCS PREDOCTORAL ACHIEVEMENT PRIZE IN MEMORY OF THE LATE KENNETH C (KC) SMITH

To: Zihan Wu, *Peking University, Beijing, China*

2025 IEEE SSCS CHAPTER AWARDS

2025 IEEE SSCS CHAPTER OF THE YEAR AWARD
Lahore Section APS/CAS/MTT/SSCS Joint Chapter (Pakistan)

2025 SSCS STUDENT BRANCH CHAPTER OF THE YEAR AWARD
Alexandria University SSCS Student Branch Chapter (Egypt)

2025 SSCS STUDENT BRANCH CHAPTER OF THE YEAR AWARD
Saintgits College of Engineering-Kottayam SSCS Student Branch Chapter (India)

2025 SSCS CHAPTER WITH BEST EDUCATIONAL PROGRAM AWARD
Kolkata Section SSCS Chapter (India)

2025 CHAPTER WITH BEST PRE-UNIVERSITY OUTREACH AWARD
Central Illinois Section SSCS Chapter (USA)

2025 CHAPTER WITH BEST PRE-UNIVERSITY OUTREACH AWARD
San Diego APS/CAS/EDS/MTTS/SSCS Joint Chapter (USA)

ISSCC 2026 / February 16, 2026 / 9:50 AM

2025 WILEY-IEEE PRESS PROFESSIONAL BOOK AWARD

Bernhard Wicht

University of Hannover, Hannover, Germany

"For his book Design of Power Management Integrated Circuits"

AIME 2026 JOHN FRITZ MEDAL

Mau Chung Frank Chang

University of California, Los Angeles

"For pioneering contributions and innovations in the development of heterojunction and system-on-chip technologies with unprecedented functionality, bandwidth and reconfigurability"

IEEE TECHNICAL FIELD AWARDS

2026 IEEE DONALD O. PEDERSON AWARD IN SOLID-STATE CIRCUITS

Jan M. Rabaey

University of California at Berkeley, Berkeley, CA

"For seminal and visionary contributions to the VLSI implementation of low-energy signal processing and communication systems."

2026 IEEE Fellows

Thomas Burd
AMD, Santa Clara, CA
"For contributions to energy-efficient and high-performance microprocessor design."

Sudipto Chakraborty
IBM T. J. Watson Research Center, Yorktown Heights, NY
"For contributions to ultra-low power circuits and systems."

Debendra Das Sharma
Intel, Santa Clara, CA
"For leadership in establishing PCI-Express, CXL, and UCIe industry-standards, fostering innovation in computing"

Anthony Ghiotto
IMS Research Center, Bordeaux INP
"For contributions to substrate integrated waveguide technologies."

Jaydeep Kulkarni
University of Texas at Austin, TX
"For contributions to low-power SRAM and compute-in-memory circuit technologies."
Princeton University, Princeton, NJ

Antonio Liscidini
University of Toronto, Toronto, Canada
"For contributions to circuit design for wireless communication radios."

Pedram Mohseni
Case Western Reserve University, Cleveland, OH
"For contributions to biosensors and bioelectronics for mitigating debilitating."

Yusuke Oike
Sony Semiconductor Solutions, Kanagawa, Japan
"For contributions to CMOS image sensors with stacked device technologies."

Ada Poon
Stanford University, Stanford, CA
"For contributions to wireless power transfer and implantable devices."

Jafar Savoj
Apple, Cupertino, CA
"For contributions to the design of high-speed wireline transceiver circuits and systems."

Shahriar Shahramian
Bell Laboratories, Nokia, Murray Hill, NJ
"For the development of millimeter-wave phased-arrays and D-bank communication links."

Farhana Sheikh
Intel, Hillsboro, OR
"For contributions to digital signal processing and 2.5D/3D heterogeneous integration"

Lawrence Tse
LTLT Consulting, LLC, Fremont, CA
"For leadership in design and commercialization of wireless and wireline data communication systems."

Leo de Vreede
Delft University of Technology, Delft, The Netherlands
"For contributions to mixed-signal high frequency device characterization and digitally assisted wireless transmitters."

Huaqiang Wu
Tsinghua University, Beijing, China
"For contributions to resistive random-access memory and compute-in-memory technologies."

Geoffrey Choh-Fei Yeap
TSMC, Hsinchu, Taiwan
"For leadership and contribution in development of advanced energy-efficient logic technology ."

Jerald Yoo
Seoul National University, Seoul, Korea
"For contributions to biomedical circuits and systems and body-area-network."

Mohammad Hossein Zarifi
University of British Columbia, Vancouver, Canada
"For contributions to applied electromagnetics and advanced materials for cutting-edge sensing and communication devices."

Zhengya Zhang
University of Michigan, Ann Arbor, MI
"For contributions to the design of high-performance and energy-efficient digital signal processors."

DIGEST OF TECHNICAL PAPERS • 25

979-8-3315-8937-0/26 $31.00 © 2026 IEEE

ISSCC 2026 / SESSION 1 / PLENARY / 1.3

1.3 Powering the AI Supercycle: Design for AI and AI for Design

Anirudh Devgan, President and CEO

Cadence, San Jose, CA

Abstract

The AI supercycle is rapidly increasing demand for compute performance and scalability across all levels, from data centers to edge devices. By 2030, the semiconductor total addressable market (TAM) is projected to reach $1.2T with electronic systems at $5.2T. With silicon designs surpassing 200 billion transistors and chiplet-based architectures becoming common, traditional electronic-design-automation (EDA) workflows are insufficient. This plenary discusses how agentic AI can be applied to complex silicon and system design tasks through multi-agent orchestration and iterative reasoning, outlining a framework for its use in EDA solutions, highlighting some of the challenges, and exploring future directions. From optimizing power, performance, and area of silicon to managing data centers, AI-powered solutions are critical for the engineers designing the next generation of AI infrastructure.

1.1 Background

The evolution of integrated circuit (IC) design has been propelled by decades of innovation, moving from manual layout efforts to today's highly automated, data-driven workflows. Electronic design automation (EDA) tools have enabled engineers to manage increasing complexity through computational software and optimization methods. The current landscape features a significant escalation in system scope; advanced system-on-chip (SoC) designs now incorporate around 400 silicon intellectual property (IP) blocks, a substantial increase from the block counts of the early 2000s [1]. High-performance computing for artificial intelligence (AI) is driving growth in logic and memory beyond typical reticle limits. TSMC has predicted [2] growth from 3.3 reticle with chip-on-wafer-on-substrate (CoWoS) to over 40X reticle with system-on-wafer (SoW-X) (Figure 1.3.1). Concurrently, transistor density has soared, with chips like NVIDIA's Blackwell architecture integrating over 200 billion transistors [3] (Figure 1.3.2).

Sustaining the trajectory described by Moore's Law has become progressively more challenging due to physical, economic, and engineering barriers at advanced nodes. Today's ICs are central to interconnected systems, from edge devices to data centers, further expanding the design landscape. This explosion in scale and complexity has driven the industry to look at AI not only as a design target but also as a foundational enabler for the design process.

As seen in Figure 1.3.3, this trend is set to accelerate with the shift from building AI infrastructure to the emergence of physical AI—pervasive autonomous driving, drones, and robotics. This embodied AI will demand even more infrastructure to train models and deliver intelligence to the edge. The semiconductor total available market (TAM) is projected to reach $1.2T by 2030, with the electronic system TAM reaching $5.2T [4]. The AI era provides a new set of tools to address this complexity, transforming how human expertise interacts with automated flows and forging collaborative systems capable of reasoning, learning, and autonomously executing complex engineering tasks.

1.2 Motivation

The demand for enhanced performance, efficiency, and new capabilities continues to push IC design to its limits. The scale of modern chips, with billions of transistors, and the need for rapid design cycles render traditional workflows insufficient. Historically, progress in EDA and system design has relied on three principles for scaling. The first is hardware-based, while the others are primarily in the software domain (Figure 1.3.4).

- **Accelerated Computing:** The use of multi-core CPUs, GPUs, and cloud clusters for managing massive computational loads.

- **Algorithmic Innovations:** Advanced algorithms, such as dynamic programming and divide-and-conquer, remain essential for addressing NP-complete challenges in SoC assembly.

- **Abstraction:** Higher-level abstractions, from register-transfer level (RTL) to high-level languages, have enabled designers to manage larger system blocks and improve productivity.

Agentic AI promises the next leap, operating at a layer that captures human intent while automating the orchestration of design and verification flows. A significant challenge is "grounding" the AI's knowledge in the physical realities of silicon. A large language model (LLM) might generate syntactically correct RTL, but without an understanding of timing closure, power consumption, or manufacturability, the output is of limited use. Effective grounding requires integrating the AI with EDA tools, allowing it to test its outputs, analyze the results, and learn from the feedback loop between the logical and physical domains.

This is not just a speculative trend. As shown in Figure 1.3.5, industry analysis estimates that we are at an inflection point where over half of advanced-node chips are designed using some form of AI-enabled tools [5]. It is essential for us as an industry, and EDA providers specifically, to understand and address the application of agentic AI to IC design.

1.3 Summary, Scope, and Contributions

This paper presents a technical exploration of agentic AI as applied to the IC domain. We introduce a structured framework that categorizes AI's role in IC design into five distinct levels of increasing autonomy, illustrating capabilities at each level with real-world examples. Furthermore, we discuss key challenges and offer a forward-looking perspective on the directions that will shape the path toward fully autonomous design.

2. The Five Levels of Agentic AI for IC Design

Agentic AI in IC design can be structured as a staged progression akin to the automation levels in autonomous systems (Figure 1.3.6). Each level is defined not only by autonomy but also by the technical sophistication of its models, integration with design environments, and impact on EDA workflows. Grounded in developments across algorithms, language models, and multi-agent orchestration, these levels represent an actionable roadmap for leveraging AI in IC design.

2.1 AI Infrastructure

A key architectural enabler is an infrastructure that enables access to LLMs, compute resources, data, and tools. The heterogeneous nature of IC design environments prevalent across the industry requires a flexible architecture that does not compromise on security. An example is the Cadence® Joint Enterprise Data and AI (JedAI) Platform (Figure 1.3.7 and Figure 1.3.8). Cadence JedAI is a data and AI orchestration platform that abstracts the underlying differences in EDA tools, repositories, and infrastructure, allowing seamless integration of LLM-powered agents within existing design workflows. It achieves this through standardized APIs and context management, enabling secure, controlled access to engineering datasets and knowledge stores, whether deployed on-premises, in the cloud, or in hybrid settings.

The Cadence JedAI platform's architecture manages data ingestion, indexing, and retrieval across multiple silos using vector databases and modular connectors, enabling both language-driven queries and automated agent workflows to access the latest documentation, reference flows, and project state. This approach is essential given the diversity of design assets, proprietary file formats, and vendor-specific environments found in advanced IC development. By separating model hosting, data storage, and inference services, the platform can optimize resource allocation—dynamically using on-premises compute for sensitive workloads while elastically scaling cloud resources for annotation, burst training, or large-scale inference tasks.

Security is a critical consideration: Cadence JedAI platform enforces fine-grained permission models, network isolation, and governed access controls to ensure that proprietary IP and sensitive project data remain protected. Critical operations involving user and tool data are auditable, and all LLM access to data can be restricted to vetted contexts or approved users. This mitigates risks associated with prompt injection, data leaks, or cross-project contamination.

This infrastructure approach enables scalable AI, supporting dynamic onboarding of custom knowledge from customer corpora, robust security policies for data access, and efficient compute utilization across diverse and evolving design environments.

2.2 Level 1: Optimization AI

Optimization AI leverages advanced machine learning (ML) techniques to address well-defined challenges within traditional EDA workflows, significantly enhancing or accelerating tasks that were previously reliant on computationally intensive numerical algorithms or manual adjustments. By integrating reinforcement learning (RL) and supervised learning models, optimization AI focuses on improving key metrics such as performance, power, area (PPA), and resource utilization through the systematic exploration of vast parameter spaces.

Reinforcement learning agents interact dynamically with EDA tools such as place-and-route (P&R), synthesis, and simulation engines. These agents utilize historical data from prior design iterations to train reward or cost functions, enabling them to make informed

26 • 2026 IEEE International Solid-State Circuits Conference

decisions that optimize design outcomes. Techniques like Bayesian optimization, multi-objective Pareto exploration, and policy gradient RL are frequently employed to navigate the complex tradeoffs inherent in chip design. For example, RL-based approaches have been shown to automate tasks like floor planning and resource balancing, reducing power consumption by as much as 15% and achieving up to 10-20× faster verification regressions.

Recent advancements in ML [6] further demonstrate the potential of optimization AI. By employing operator learning frameworks and physics-informed neural networks, ML models can accelerate fundamental algorithms used in thermal simulation and optimization—critical components of three-dimensional integrated circuit (3D-IC) design. These models replace traditional numerical methods, such as finite difference or finite element solvers, with surrogate neural operators that predict temperature fields directly from design configurations. This approach not only eliminates the need for expensive simulation data but also enables real-time analysis of complex thermal patterns, achieving speedups of over 70× in optimization workflows. Additionally, hybrid frameworks that combine ML predictions with iterative refinement methods, such as the Generalized Minimal Residual (GMRES) method, ensure both efficiency and trustworthiness, thereby addressing the accuracy limitations of standalone ML models.

Optimization AI is redefining traditional EDA, enabling faster, more efficient, and more reliable design processes that meet the growing demands of modern semiconductor development.

Industrial Examples:
- RL to optimize digital implementation, automating floor planning, resource balancing, and recipe selection (e.g., Cadence Cerebrus® Studio).

- ML models to prune regression suites, drastically reducing simulation cycles needed for coverage closure (e.g., Cadence Verisium™ AI-Driven Verification Platform).

- AI for analog layout generation, applying knowledge extracted from prior tapeouts to route devices and wires, handle constraint-driven placement, and speed up analog migration tasks (e.g., Cadence Virtuoso® Studio).

- RL-based multi-disciplinary analysis optimization (MDAO) for return loss and crosstalk isolation improvement on package/printed circuit board (PCB) designs (e.g., Cadence Optimality™ Intelligent System Explorer).

2.3 Level 2: Conversational LLM
At this level, AI systems act as language-based interfaces to static knowledge bases, transforming how designers access and utilize complex information. By integrating retrieval-augmented generation (RAG) with LLMs, these systems enable natural language queries on tool documentation, application notes, corporate wikis, and project-specific resources. This approach lowers access barriers and accelerates learning by making specialized knowledge easily accessible through conversational commands.

RAG systems enhance efficiency by chunking large knowledge bases into smaller, semantically meaningful segments, which are stored in a vector database. When a query is made, the system retrieves the most relevant chunks, ensuring the LLM processes only the most pertinent information. For example, a designer would receive precise, contextually relevant guidance drawn from tool documentation and notes.

Advancements in reasoning models further extend these capabilities, enabling tasks like code generation and debugging. For instance, a designer could request, "Generate a Tcl script to buffer all high fan-out nets," and the system would produce a tailored, syntactically correct script. This is achieved through fine-tuned LLMs trained on proprietary languages like Tcl or Cadence SKILL® language. Customer-specific knowledge can be integrated into the vector database, enabling organizations to tailor the system to their specific workflows. This collaborative approach enhances the system's utility and fosters continuous improvement.

Industrial Example:
Designer queries "How to create a flexible H-tree clock structure in Innovus?" LLM returns specific code snippets, explanations, or step-by-step instructions (Figure 1.3.9).

2.4 Level 3: Complex Reasoning
Complex reasoning transitions from static knowledge retrieval to dynamic, tool-driven co-design and debugging. It introduces context-awareness and iterative interaction with live EDA data and application programming interfaces (APIs). At this level, AI systems leverage closed-loop reasoning to autonomously interrogate project states, issue modifications, and refine actions based on structured feedback from tools.

Agents operating at this level interact with EDA tools through APIs, such as Tcl for digital implementation or a SKILL API for custom/analog design. These agents dynamically query the design state, perform modifications, and validate results. An agent could analyze timing reports, identify critical paths, and iteratively adjust placement or buffering strategies using Tcl commands, re-evaluating the design after each modification. In analog design, SKILL-based agents can refine layouts by adjusting device placements or routing to meet performance constraints, guided by feedback from simulation tools.

A defining feature of Level 3 is the iterative refinement process, where agents use context-aware reasoning to adapt their actions based on the tool's outputs. When debugging a high fan-out net, an agent can analyze prior buffering attempts, assess their effectiveness, and propose alternative strategies. This iterative approach ensures convergence toward optimal solutions while addressing evolving design requirements.

Reasoning models play a critical role in enabling these capabilities. Fine-tuned on proprietary Tcl or SKILL languages, these models can autonomously generate tool-specific scripts, validate their syntax and functionality, and iterate based on feedback. For example, a reasoning model might generate a Tcl script to optimize dynamic power in a circuit, run the script, and refine it if the initial results do not meet design objectives. This capability reduces manual intervention, allowing engineers to focus on higher-level design challenges.

Level 3 complex reasoning AI combines advanced reasoning models, robust tool APIs, and iterative refinement to enable dynamic, context-aware co-design and debugging. This level transforms EDA workflows by automating complex tasks and enhancing productivity, while keeping the designer in control of the overall process.

Industrial Examples:
- Formal Verification: Write System Verilog Assertions (SVA), run proof engines, analyze proof failures, and suggest revised assertions or variables to isolate elusive bugs.

- Place-and-Route: Identify and fix paths from P&R tool reports, recommend ECOs, iteratively script placement sets, generate power supply patterns (Figure 1.3.10), and check timing or leakage impacts in an automated loop.

- Analog Design: Update migrated testbenches with new specifications extracted from design documents, recognize key structures and parasitics, identify placement and routing issues, and suggest changes to the circuit/layout.

2.5 Level 4: Agentic Workflows
Level 4 extends beyond single-task automation to orchestrated, multi-agent, and multi-domain workflows. At this level, AI systems leverage interconnected agents to collaboratively achieve outcomes that span multiple design stages and tools, enabling a seamless flow of tasks across heterogeneous, multi-vendor environments. This orchestration is facilitated by standardized protocols, such as the model context protocol (MCP) and agent-to-agent (A2A) communication frameworks, which support live context transfer and state sharing between agents (Figure 1.3.11).

Level 4 introduces a "flow planner" agent, which decomposes high-level design goals into a series of subtasks and delegates them to specialized agents. Each agent operates autonomously within its domain, yet remains interconnected with others to ensure consistency and alignment with the overarching design objectives. This hierarchical approach is particularly critical for managing the complexity of system-on-chip (SoC) designs and mixed-signal integration. By breaking down complex workflows into manageable subtasks and coordinating their execution, Level 4 AI mirrors the hierarchical methodologies traditionally used in programming and IC design. This not only enhances scalability but also builds trust in the system by delivering effective and predictable results.

The orchestration of these agents relies on robust context management and hierarchical design management principles. Agents share live design states and intermediate results, enabling iterative refinement and inter-tool collaboration. An analog migration agent tasked with adapting a circuit from 3nm to 2nm might first interact with an IP catalog agent to identify suitable starting points. Once a candidate design is selected, the migration agent ensures electrical correctness and simulation readiness. Subsequently, an optimization agent fine-tunes the design to meet performance targets, while a layout agent generates the physical implementation. Throughout this process, agents exchange data and feedback, ensuring that each stage builds upon the results of the previous one.

Another example of Level 4 in action is digital subsystem design implementation. In this workflow, a design planning agent partitions the subsystem into blocks, assigns pin placements, and shapes the layout. Block implementation agents then optimize each block in parallel, exploring multiple solutions to maximize performance. An integration agent evaluates the top solutions from each block, combining them into a cohesive floorplan. Finally, a design closure agent ensures that the design meets all constraints, iterating as

ISSCC 2026 / SESSION 1 / PLENARY / 1.3

necessary to achieve the desired results. This multi-agent collaboration significantly accelerates the design process while maintaining high-quality outcomes.

Industrial Examples:

- Analog Migration: Agents migrate a schematic to a different PDK, refine the mapping until the new circuit is functional, then center the design for the target technology, significantly automating analog porting (Figure 1.3.12).

- Front-End Design: RTL creation and validation (Figure 1.3.13), automated debug of regression failures, and proposals for fixes (Figure 1.3.14).

- Subsystem Design: Coordinating block partitioning, implementation, integration, and closure, using a pool of concurrent agents for block implementation and timing convergence.

- PDK Integration: LLM agents parse foundry documentation and automate reference flow setup, reducing integration time from months to weeks. Based on experimental data from our own silicon design group, this can reduce what used to be an 8 – 10-week task to just 1 week using an AI-based flow.

2.6 Level 5: Fully Agentic AI

Level 5 represents the pinnacle of AI-driven design automation. A top-level orchestrator "silicon agent" manages hundreds of subordinate agents across the entire design lifecycle. This lifecycle spans from initial specification to system-level optimization. At this stage, AI systems coordinate complex, multi-domain workflows, integrating architectural, logic, analog, packaging, and system validation processes into cohesive agent teams.

The silicon agent serves as the central orchestrator, dynamically allocating tasks to specialized agents while maintaining a global view of the design state. This requires robust state management, enabling the orchestrator to track the progress of each agent, resolve dependencies, and ensure consistency across the workflow. For example, during the design of a large SoC, the silicon agent might coordinate agents responsible for RTL generation, functional validation, timing closure, power optimization, and physical design, ensuring that each stage aligns with the overall design objectives.

A critical enabler of Level 5 systems is global context awareness, which allows the silicon agent to make informed decisions based on the current state of the design and its requirements. This is achieved through secure data lakes, toolchains, model repositories, and error recovery. The data lake serves as a centralized repository for design data, simulation results, and historical insights, providing agents with the information they need to operate effectively. Toolchains and model repositories ensure that agents have access to the latest algorithms, libraries, and design rules, enabling them to adapt to evolving design challenges. Error recovery ensures the system can detect and address errors at any stage of the workflow, whether they arise from tool limitations, design rule violations, or unexpected interactions between agents. This requires advanced diagnostic capabilities, which identify the root cause of an issue and either resolve it autonomously or escalate it to a human designer with detailed recommendations.

Level 5 systems also emphasize domain-specific tracing, explainability, and alignment with user goals and policies. Each agent is equipped with mechanisms to trace its actions and decisions, providing a clear audit trail that enhances transparency and trust. For example, an agent responsible for analog layout optimization might document the tradeoffs it considered when selecting device placements or routing strategies, enabling designers to understand and validate its decisions. Explainability is further enhanced by embedding domain-specific knowledge into agents, allowing them to provide context-aware justifications for their actions.

The integration of these capabilities enables Level 5 systems to codify and automate workflows across diverse domains. For instance, in a mixed-signal design, the silicon agent might coordinate teams of agents to generate architectural specifications, synthesize RTL, optimize analog circuits, and validate system-level performance. Each team operates autonomously within its domain, yet collaborates seamlessly with others to achieve the overall design objectives. This hierarchical organization mirrors the structure of human design teams, enabling Level 5 systems to tackle the most complex design challenges with unprecedented efficiency and accuracy.

Industrial Examples:

- SoC Prototyping: A silicon agent parses marketing specifications, generates high-level block diagrams, decomposes into implementation/verification subprojects, and automatically generates and validates RTL, constraints, software drivers, and testbenches.

- Derivative Generation: For consumer chip lines, such an agent can create and validate multiple SoC variants (e.g., CPU or memory refresh SKUs) from a single high-level specification and previous design databases, reducing engineering cycle time and errors.

- System Co-Design: A system agent performs real-time optimization across chip, package, and board, responding to power, area, or cost constraints detected at the system level.

3. Challenges in Agentic AI for Integrated Circuit Design

3.1 Data Scarcity and Quality

Progress in agentic AI is significantly constrained by the quantity, diversity, and accessibility of data. Unlike fields such as natural language processing (NLP), where large, open datasets are readily available, the semiconductor industry faces unique challenges due to the proprietary and siloed nature of design data [7]. Intellectual property (IP) protection concerns often prevent data sharing, even within a single organization, while inconsistent data management practices further hinder the creation of cohesive training datasets. Analog design exemplifies these challenges, as much of the expertise resides in undocumented workflows, unmanaged testbenches, and the tacit knowledge of experienced designers.

Data heterogeneity is another barrier. A single SoC encompasses specifications, RTL, netlists, physical layouts, and various reports, all stored in incompatible formats (Figure 1.3.16). Normalizing information into a unified format requires significant engineering investment, particularly when dealing with legacy data or vendor-specific file structures. Additionally, the scarcity of negative data—failed runs or bugs—limits the ability of models to learn from mistakes, which is critical for robust AI systems.

To address these challenges, several technical approaches are being explored:

- **Leveraging Existing Data More Effectively**: Capturing expert debugging sessions, scripts, and design intent through focused events or sandboxes can supplement datasets. For example, prompt indexing of multi-generational IP data, combined with cosine similarity checks and linking English specifications to structured RTL, can bridge gaps in documentation. Testbenches that ensure functional and performance requirements can also serve as valuable data sources for training.

- **Parameter-Efficient Fine-Tuning (PEFT)**: Techniques like low-rank adaptation (LoRA) enable efficient customization of large models on domain-specific data, such as netlists, timing reports, or power analysis results, without requiring full retraining. This approach reduces computational overhead while maintaining model accuracy.

- **Federated Learning**: By training models on decentralized datasets within secure enclaves, federated learning allows organizations to aggregate improvements across silos without exposing proprietary data. This method is particularly promising for industries where data sharing is restricted by IP concerns.

- **Reasoning Models**: These models excel at learning expert workflows from limited examples, making them well-suited for domains where data is scarce [8]. For instance, reasoning models can infer design intent or debug strategies from a small set of annotated examples, reducing the dependency on large datasets.

- **Synthetic Data Generation**: Generating representative RTL snippets, analog layouts, or simulation data can augment small datasets while preserving confidentiality. For example, synthetic data can simulate edge cases or rare design scenarios that are underrepresented in real-world datasets. Synthetic data generation is also a valuable technique for testing the quality of AI assistant or chatbot queries on static documentation, as it enables the creation of large amounts of question-answer pairs.

- **Vendor-Supplied Fine-Tuned Models**: EDA vendors may play a critical role by providing base software with pre-tuned models tailored to specific domains or customer design catalogs. These models can serve as a starting point for further customization, reducing the barrier to entry for adopting AI in design workflows.

3.2 Security Considerations

The integration of agentic AI into IC design workflows introduces several security challenges that must be addressed to ensure the integrity and confidentiality of data, including the design content itself, as well as the underlying process technology and the included IP blocks. One significant risk lies in the models themselves. A stolen model could be reverse-engineered to reveal proprietary information embedded during training, such as design patterns or sensitive IP. Additionally, a compromised model could be manipulated to insert hardware trojans or other malicious modifications, potentially bypassing traditional verification processes and introducing vulnerabilities into the final design.

To mitigate these risks, all LLM operations and data processing must be conducted within secure environments. These environments may include on-premises infrastructure or vendor-owned private clouds, configured with private endpoints and governed by explicit agreements with cloud providers to prevent data leakage. When using foundry-specific data for AI training, explicit permissions must be obtained, and all activities must adhere to strict data handling policies and non-disclosure agreements (NDAs). This ensures compliance with industry standards and protects sensitive design information.

Proactive security measures ("red-teaming") are essential to identify and address potential vulnerabilities. AI agents interacting with external data are particularly susceptible to prompt injection attacks, where malicious actors embed hidden instructions within documents [9]. To defend against such attacks, a fine-grained permission layer must be implemented within the LLM infrastructure. This layer enforces access controls based on user entitlements, ensuring that the AI system can only access data and run actions explicitly authorized by the user. This approach prevents AI from inadvertently accessing and exposing confidential information.

In addition to these measures, robust audit trails and monitoring systems should be established to track AI interactions and detect anomalous behavior. For example, logging all queries and responses can help identify unauthorized access attempts or unusual patterns of activity. These logs can also serve as a valuable resource for post-incident analysis, enabling organizations to refine their security protocols and improve resilience against future threats.

As more comprehensive and capable agentic AI systems are used, it is essential to build a foundation of security and trust from the beginning, to test and verify that the individual agents are secure on a smaller scale. When a complex system of agents begins to tackle a full SoC, it is much more challenging to verify the security of the overall environment. Start small and safe before scaling.

3.3 Infrastructure Requirements

Deploying agentic AI requires a careful investment in computational infrastructure. Training a foundation model from scratch is a resource-intensive process, often costing millions of dollars and requiring access to thousands of GPUs [10] [11] (Figure 1.3.17 and Figure 1.3.18). The deployment of these models across diverse engineering environments—ranging from on-premises systems to public and hybrid clouds—introduces additional complexities, particularly in multi-vendor workflows where interoperability and data security are critical.

To address these challenges, a tiered strategy is proposed, beginning with RAG systems for static data. These allow a project to easily incorporate design- or tool-specific static data (reference documentation, organizational guidelines, best-known-methods, ...) for the AI to access when answering designer questions, without needing to invest in retraining a model. Fine-tuning or full pre-training can then be applied selectively, based on the specific requirements of the application, with ongoing evaluation of accuracy and infrastructure costs to ensure efficiency (Figure 1.3.19). It is important to think carefully about the return on investment when building a custom model. The rate of innovation in the broader AI industry is delivering increasingly better and more capable models every quarter. If the primary medium of work is natural language, a RAG can be an excellent way to get a good return on investment, taking advantage of the latest industry models while using the RAG for static environment-specific information. When tackling more domain-specific areas, especially with low propensity data, a fine-tuned or custom model can deliver better results.

Several additional areas of focus are critical for building a robust and effective infrastructure:

- **Efficient Models**: The development and deployment of smaller, more efficient models are essential to reduce inference costs and hardware requirements. Models such as OpenAI's GPT-OSS and NVIDIA's Nemotron demonstrate that high performance can be achieved with reduced parameter counts, lowering the computational burden. Techniques like model quantization, which reduces the precision of model weights, and model distillation, which transfers knowledge from a large model to a smaller one, can further optimize inference latency. These optimizations are particularly important for interactive design environments, where real-time responsiveness is critical for effective collaboration between human designers and AI systems.

- **Hybrid Cloud Architecture**: A hybrid cloud approach offers a practical solution for balancing security and computational demands. Sensitive data, such as proprietary design files or foundry-specific process design kits (PDKs), can be processed on-premises to ensure confidentiality. Meanwhile, computationally intensive tasks, such as large-scale simulations or model fine-tuning, can be offloaded to the cloud, leveraging its elastic scalability. This approach allows organizations to optimize resource utilization while maintaining control over critical data.

- **EDA-Vendor Owned Cloud**: Deploying agentic AI within an EDA-vendor-owned cloud environment provides an additional layer of security and standardization. Agents are trained and deployed on company-specific data within a secure, managed infrastructure, ensuring that no data is shared between customers. This approach simplifies deployment, as the vendor can pre-configure the environment to meet the specific needs of the design tools and workflows, while also providing centralized updates and maintenance.

3.4 Trust and Adoption

The success of agentic AI depends on the willingness of human designers to trust and adopt it. Overcoming hesitation to cede control over critical decisions requires a strategy focused on transparency, reliability, and collaboration. To build trust, AI systems must not be opaque black boxes. Key principles include:

- **Explainability (XAI)**: AI systems must provide clear and actionable justifications for their decisions. If an agent modifies a design, it should reference specific reports, constraints, or metrics. This could include timing analyses, power estimates, or design rule violations that necessitated the change. Explicit reasoning by the AI enables designers to verify its logic and validate its outputs, building trust and confidence. Explainability is particularly critical in debugging workflows, where understanding the rationale behind an AI-generated solution can help identify and address underlying issues. Techniques such as decision trees, saliency maps, or rule-based explanations can be integrated into the system to enhance explainability.

- **Reliability and Predictability**: AI-driven tools must produce consistent and predictable results. A lack of determinism can erode confidence and make debugging difficult. AI models should incorporate mechanisms to minimize stochastic behavior, such as using fixed random seeds during training and inference, or employing ensemble methods to stabilize their outputs. Additionally, rigorous validation and benchmarking against known datasets can ensure that the system performs reliably across a range of scenarios. Repeatability has been a critical component for the success of modern EDA tools, and the reason designers today are able to leverage automation to tackle larger and larger designs. For agentic AI to enable the next wave of innovation, it must live up to the same.

- **Human-in-the-Loop**: This framework keeps designers in control, allowing them to review, modify, and approve AI suggestions. This positions the AI as an intelligent assistant, fostering a collaborative environment. This iterative interaction allows designers to guide the AI's actions, ensuring alignment with project goals and constraints. It also provides a safety net for early-stage adoption, enabling designers to gradually increase their reliance on AI as they gain confidence in its capabilities. Over time, as trust is established, the scope of automation can be expanded to include more complex and autonomous workflows.

The presented five-stage approach, grounded in current and emerging tool infrastructure, sets a practical path for increasing AI integration toward eventual autonomous design, while maintaining checkpoints for verification, explainability, and human oversight. Each level builds upon the previous, introducing increasingly sophisticated capabilities that transform traditional EDA workflows into intelligent, interconnected systems. As with any automation, agentic AI requires respect and understanding for successful adoption. A design team that is not comfortable with AI might be best served by experimenting with the initial levels—optimization AI and conversational LLM—thereby building confidence for the more advanced stages.

The journey is comparable to the evolution of self-driving cars. Incremental advancements in automation build trust and capability. Just as early levels of autonomous driving, such as lane assist and emergency braking, keep the driver in control, the initial levels of agentic AI focus on augmenting human designers with tools like reinforcement learning and conversational AI. As the levels progress, the system takes on more responsibility, with Level 4 introducing orchestrated multi-agent workflows and Level 5 achieving full autonomy, akin to a self-driving car navigating complex environments without human intervention. The self-driving car analogy underscores the importance of trust and gradual adoption in this journey. Just as drivers must build confidence in autonomous vehicles, designers must trust AI systems to handle critical tasks. This trust is cultivated through transparency, explainability, and incremental improvements in capability, ensuring that human designers remain in control while benefiting from the productivity gains of agentic AI.

4. Future Directions

4.1 Enhancing Agentic Workflows

The future of agentic AI involves expanding the library of specialized agents across many domains and optimizations. This includes developing agents tailored for specific challenges, such as power optimization, thermal management, or advanced analog layout synthesis. A critical enhancement is enabling customers to contribute their domain-specific knowledge by integrating proven scripts, methodologies, and workflows into secure, private vector databases. A customer could upload a library of Tcl scripts for timing closure or SKILL routines for analog layout adjustments, which the AI system could then index and utilize. This approach allows the AI to adapt to the unique requirements of individual teams or projects, creating a feedback loop where the system becomes progressively more aligned with user needs. Incorporating mechanisms for version control and validation ensures that contributed knowledge remains accurate and up to date, further enhancing the reliability of agentic workflows.

4.2 Collaborative Ecosystems

The full potential of agentic AI will be realized through collaborative ecosystems that foster interoperability and standardization. This requires industry-wide collaboration to establish standards for data sharing and agent communication. Adoption of open protocols like MCP and A2A will provide a common language for agents and tools to interact, regardless of their vendor origin. This interoperability fosters a robust ecosystem where startups and established companies alike can contribute specialized agents or tools, accelerating innovation and allowing designers to explore and optimize new physics or types of designs. The adoption of common data formats and APIs ensures that agents can operate across heterogeneous environments, reducing integration overhead and enabling more unique, specialized workflows.

4.3 Training and Upskilling Designers

The integration of agentic AI requires an IC design workforce prepared to collaborate with these new systems. Organizations must invest in training programs that help designers develop skills in AI, like prompt engineering and custom workflow development. An agentic system can be framed as a virtual assistant that takes on more complex tasks as trust is established. For new engineers, the agent can provide verbose explanations of its tasks, serving as a real-time learning tool that readies new engineers even faster. However, this only increases the importance of learning the fundamentals of engineering and physics. AI, abstraction, and automation can help designers tackle larger challenges, but a working understanding of the underlying principles is crucial to the engineering mindset required for success. Designers must retain the ability to critically evaluate AI outputs, ensuring that automation complements, rather than replaces, their expertise.

4.4 Leveraging Emerging AI Trends

The industry must continuously leverage emerging trends. This has always been true for EDA, leveraging the latest in compute and algorithmic developments, and it remains true for AI. A significant trend is the shift toward smaller, more efficient models, including domain-specific small language models (SLMs) tailored for targeted applications [12]. An SLM trained specifically on analog circuit design could outperform general-purpose models in tasks such as parasitic extraction or layout migration. World models and other architectures trained on real-world physics and data show promise for certain engineering problems [13]. These models could be applied to tasks such as thermal simulation or electromagnetic interference analysis, where traditional methods are computationally expensive. In parallel, hybrid AI approaches that combine traditional, deterministic optimization techniques with probabilistic, AI-driven methods will allow this new technology to build on the foundation of decades of good engineering and computer science. Reinforcement learning could be used to explore design spaces, while deterministic solvers ensure that the final solution meets all constraints. These hybrid approaches leverage the strengths of both AI and traditional engineering methods, enabling more robust and efficient workflows. In general, the closer the design task is to natural language, the more it benefits from these modern LLM-based technologies, while deeper backend implementation and analysis tasks could be a better fit for techniques like neural networks [6]

4.5 Integration Across Domains

A key challenge is the integration of disparate design domains. Developing a unified framework connecting digital, analog, and system-level workflows is essential for cross-domain optimization. Initial deployments are most effective where language-based processes are prevalent, such as from product definition to the start of verification. Contemporary RF IC design is inherently multiphysics and multi-domain, where high-frequency electromagnetic effects, substrate coupling, and package-level interactions are not peripheral but instead constitute primary constraints within the design space. 3D-IC design also presents a very broad state-space to explore, implement, and verify. Agentic AI frameworks present a robust methodology for redefining architectural, optimization, and verification processes in these multifaceted environments. The biggest potential emerges when this integration extends across the entire system hierarchy, enabling optimization for end-system metrics, such as tokens generated per watt for an AI data center [14], all the way from the integrated circuit to the data center design and operation.

4.6 Towards Full Autonomy

The long-term vision for agentic AI is the development of fully autonomous design systems capable of orchestrating the entire design lifecycle. Creating high-level silicon agents that manage hundreds of specialized sub-agents and deploying them at scale. These orchestrators must implement dynamic decision-making systems that adapt in real time to changing requirements, such as updated design constraints or new performance targets. A silicon agent might coordinate agents responsible for RTL synthesis, timing closure, and analog layout, dynamically reallocating resources based on the progress of each task. The framework must include mechanisms for human oversight, allowing designers to intervene, adjust parameters, or restart processes as needed. Autonomy must not come at the expense of control or accountability. The development of digestible, explainable decision-making frameworks will be essential to build trust in fully autonomous systems, enabling designers to understand and validate the actions of the silicon agent. By addressing these challenges, the industry can move closer to realizing the vision of fully autonomous design, where human designers focus on high-level strategy while AI systems handle the complexities of implementation.

5. Conclusion

The AI super cycle offers a dual opportunity: designing powerful chips for AI and using AI to transform the design process. The shift to agentic AI is crucial, as it promises to automate complex workflows and boost productivity. This journey, from isolated tools to autonomous, integrated design ecosystems, requires overcoming data sharing, security, and integration challenges. It demands not only technological advancement but also a fundamental change in design approaches, requiring industry collaboration, open standards, and workforce upskilling. By leveraging accelerated compute, simulation, optimization algorithms, and AI trends, we can create intelligent systems that partner with human designers. The future of design is collaborative and interconnected; it's time for the industry to build the agentic frameworks that will define the next generation of technological achievement.

References:

[1] F. Schirrmeister, "Design Complexity in the Golden Age of Semiconductors," 2023. https://semiengineering.com/design-complexity-in-the-golden-age-of-semiconductors/

[2] TSMC, "TSMC Unveils Next-Generation A14 Process at North America Technology Symposium," 24 April 2025. [Online]. https://pr.tsmc.com/english/news/3228

[3] NVIDIA, "NVIDIA Blackwell Architecture," 2024. [Online]. https://www.nvidia.com/enus/data-center/technologies/blackwell-architecture/

[4] IDC, "2030 Semiconductor Market Outlook, 25Q2 Update, Doc #US53381325," 2025. https://my.idc.com/getdoc.jsp?containerId=US5338132

[5] B. O'Donnell, "Chip Design Hits AI Crossover Point," TECHnalysis, 29 April 2025. [Online]. https://www.technalysisresearch.com/blogs/2025%20Blogs/april_29_2025_blog.html

[6] X. Yu et al, "DeepOHeat-v1: Efficient Operator Learning for Fast and Trustworthy Thermal Simulation and Optimization in 3D-IC Design," IEEE Transactions on Components, Packaging and Manufacturing Technology, no. Early Access, 2025. https://arxiv.org/abs/2504.03955

[7] M. Grupen-Shemansky, P. Apte, and M. da Silva, "Smart Manufacturing, Smart Data-AI, and Future of Computing: Technology Communities Make Critical Linkages on Integrating AI," in SEMI, 10 September 2024. [Online]. https://www.semi.org/en/blogs/smart-manufacturing-smart-data-ai-and-the-future-of-computing-technology-communities-make-critical-linkages-on-integrating-ai

[8] C. Deng et al, "ScaleRTL: Scaling LLMs with Reasoning Data and Test-Time Compute for Accurate RTL Code Generation," in ACM/IEEE Symposium on Machine Learning for CAD (MLCAD), 2025. DOI: 10.1109/MLCAD65511.2025.11189212

[9] M. Mudryi et al, "The Hidden Dangers of Browsing AI Agents," arXiv, 19 May 2025. [Online] https://arxiv.org/abs/2505.13076

[10] B. Cottier et al, "The rising costs of training frontier AI models," 03 June 2024. [Online]. https://arxiv.org/abs/2405.21015

[11] J. Sevilla and E. Roldán, "Training Compute of Frontier AI Models Grows by 4-5x per Year," 28 May 2024. [Online]. https://epoch.ai/blog/training-compute-of-frontier-ai-modelsgrows-by-4-5x-per-year

[12] P. Belcak et al, "Small Language Models are the Future of Agentic AI," 2 June 2025. [Online]. https://arxiv.org/abs/2506.02153v2

[13] J. Schmidhuber and D. Ha, "World Models," 27 March 2018. [Online]. https://arxiv.org/abs/1803.10122v4

[14] S. Nadella, Interviewee, World Economic Forum. [Interview]. January 2025. https://www.youtube.com/watch?v=rUO7H7OtW3E

[15] Anthropic, "Anthropic.com," 25 November 2024. [Online]. https://www.anthropic.com/news/model-context-protocol

[16] Google, "Google Developer Blog," 9 April 2025. [Online]. https://developers.googleblog.com/en/a2a-a-new-era-of-agent-interoperability/

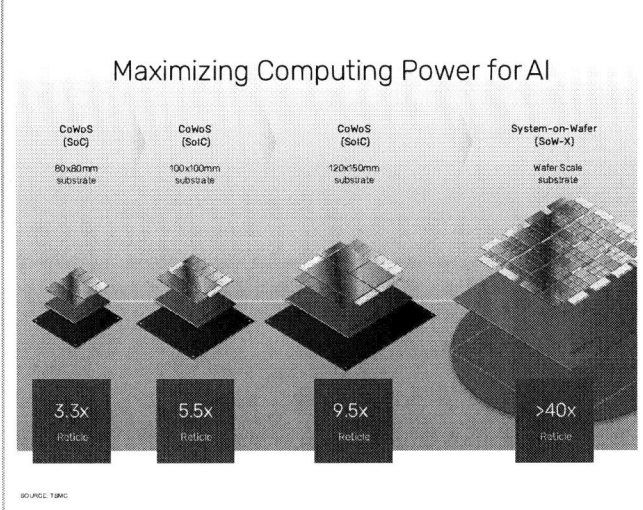

Figure 1.3.1: TSMC roadmap for beyond-reticle growth of compute systems.

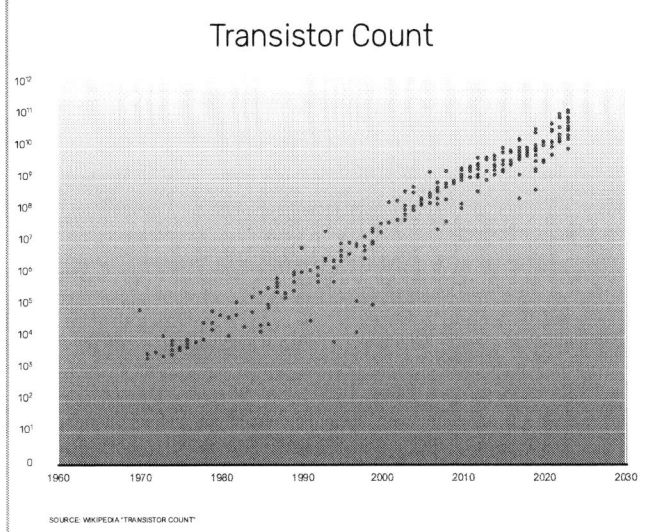

Figure 1.3.2: Growth of transistor count over time.

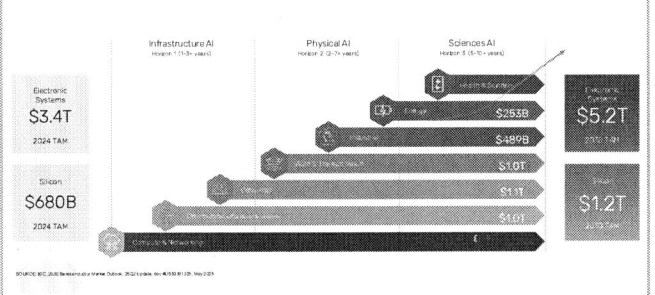

Figure 1.3.3: Silicon and system TAM projection.

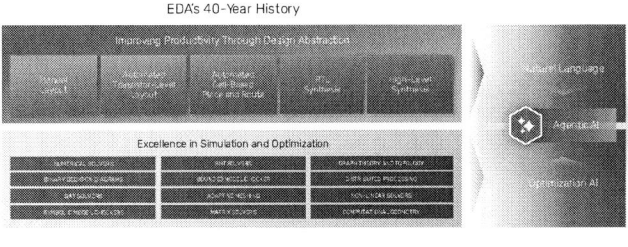

Figure 1.3.4: Continuing evolution of EDA.

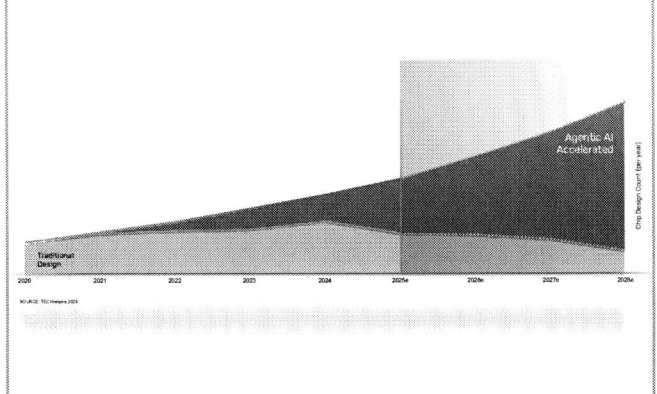

Figure 1.3.5: AI-driven chip design growth.

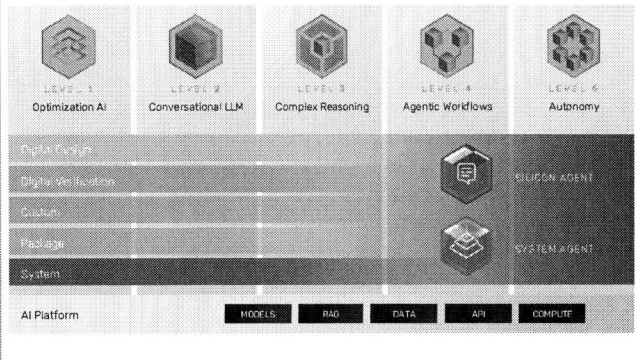

Figure 1.3.6: The journey to autonomous design.

ISSCC 2026 / SESSION 1 / PLENARY / 1.3

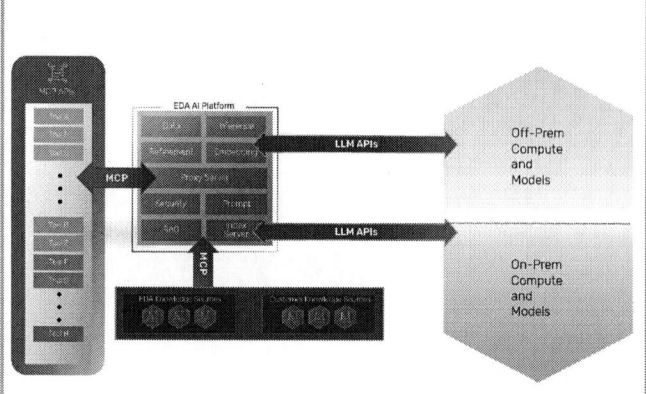

Figure 1.3.7: EDA AI platform standalone.

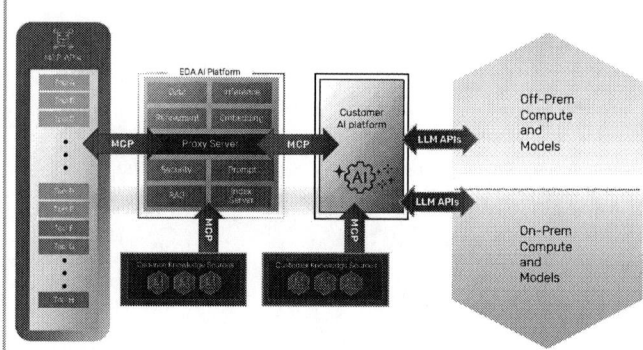

Figure 1.3.8: EDA and customer AI platform co-existence.

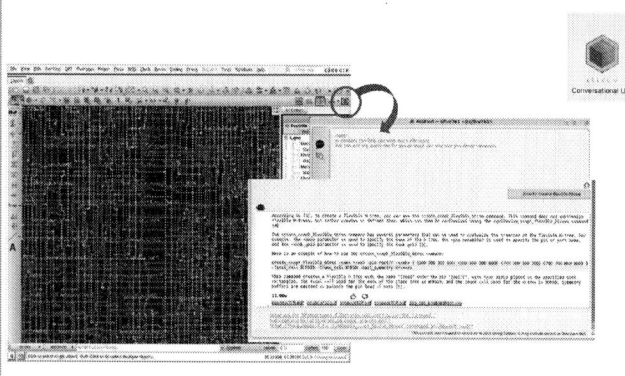

Figure 1.3.9: Example of Level 2: Conversational LLM.

Figure 1.3.10: Example of Level 3: Complex reasoning for power grid generation.

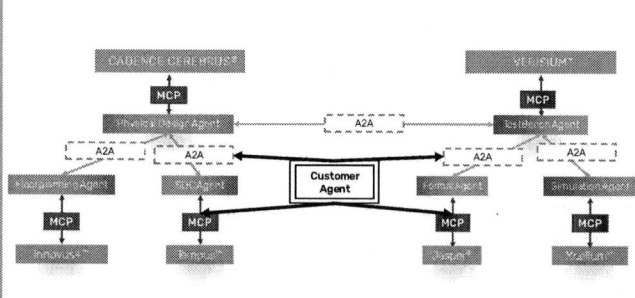

Figure 1.3.11: Using industry-standard protocols to connect agents and tools.

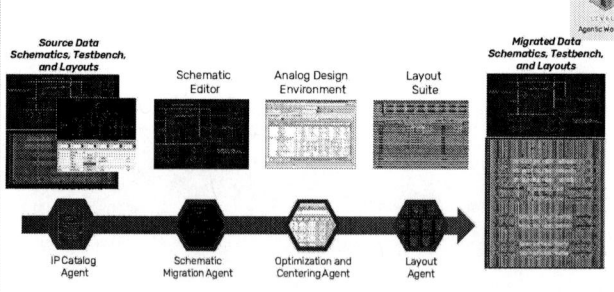

Figure 1.3.12: Example of Level 4: Agentic workflows for custom design migration.

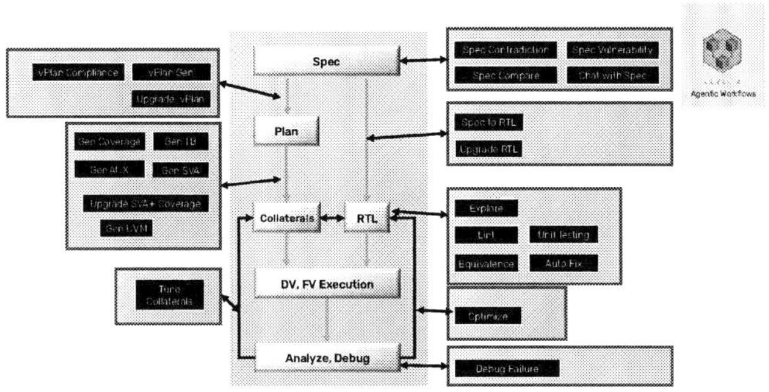

ISSCC 2026 / February 16, 2026 / 10:45 AM

Figure 1.3.13: Example of Level 4: Agentic workflows for front-end silicon design.

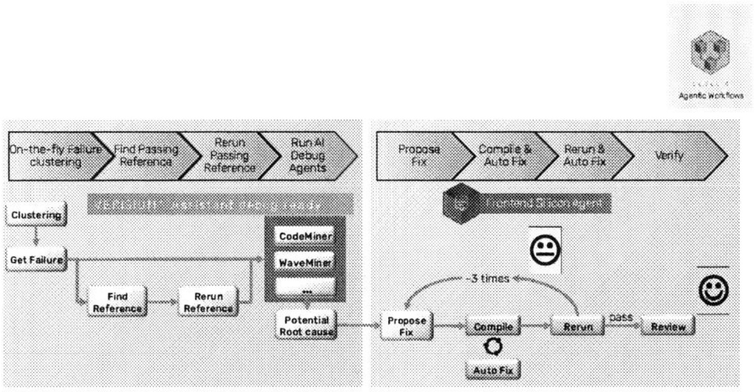

Figure 1.3.14: Using agentic AI for regression failures.

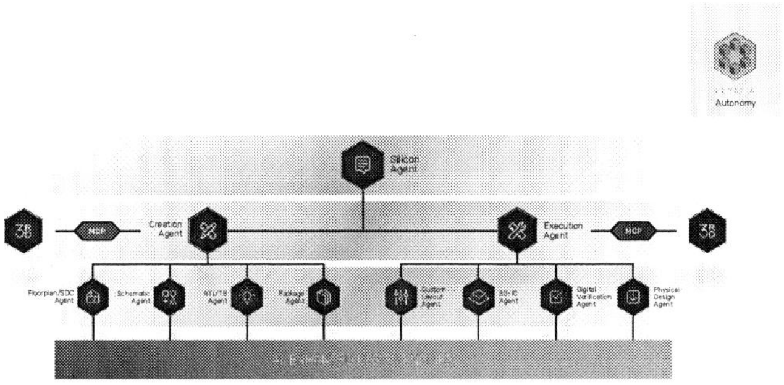

Figure 1.3.15: Level 5: System of agents for autonomous design.

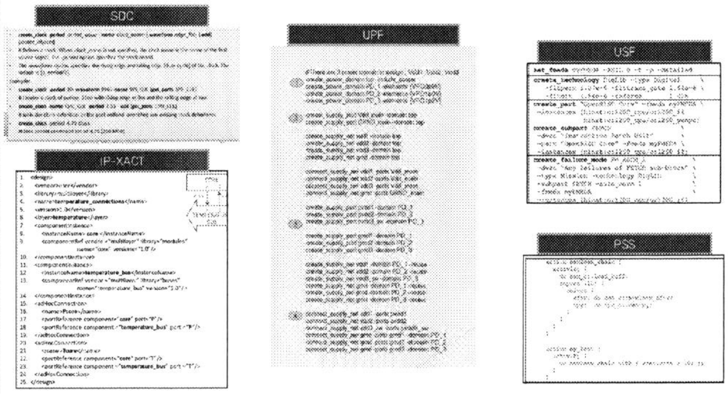

Figure 1.3.16: Tool-specific languages.

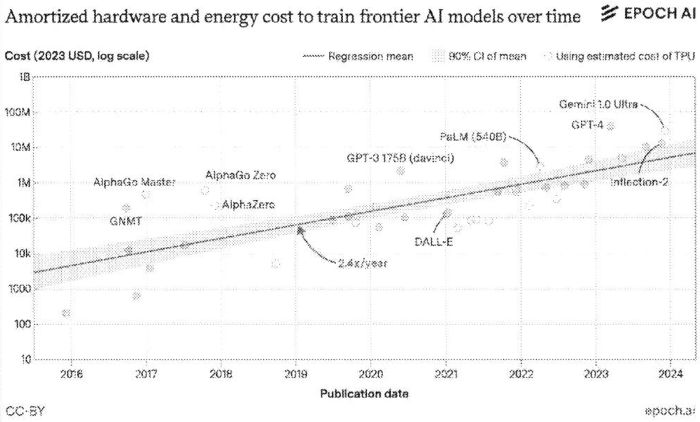

Figure 1.3.17: Hardware and energy cost to train frontier AI models.

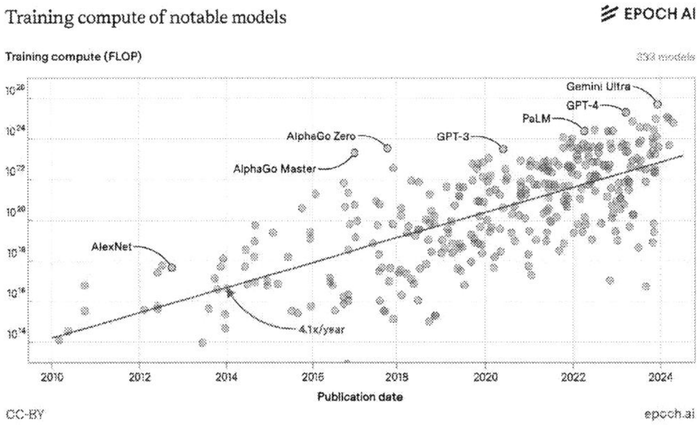

Figure 1.3.18: Training compute of notable models.

ISSCC 2026 / SESSION 1 / PLENARY / 1.3

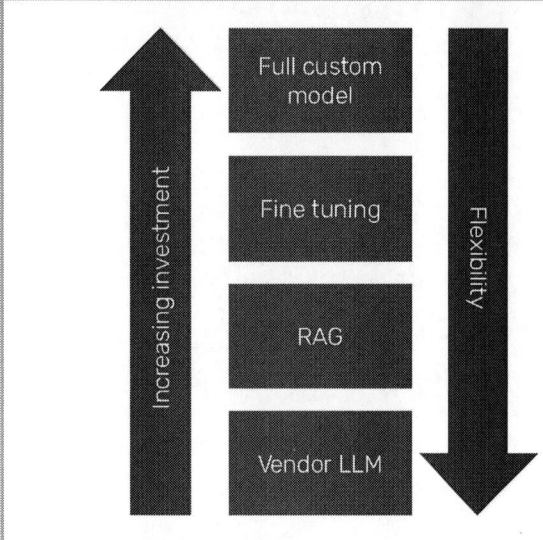

Figure 1.3.19: Infrastructure tradeoffs.

ISSCC 2026 / SESSION 1 / PLENARY / 1.4

1.4 Empowering the Next Wave of Silicon Engineers

Hope Giles

Vice President, Hardware Technologies
Apple, Cupertino, CA

Abstract

Developing custom chips for specific applications enables fundamentally better products. As part of a full-stack optimization approach to product design, application-specific systems-on-chips (SoCs) have been one of the keys to delivering game-changing performance, power efficiency, capabilities, and user experiences. This plenary provides examples of how custom silicon enables innovative product breakthroughs and how techniques like design modularity help manage the SoC scaling challenge. As applications, such as artificial intelligence (AI), transform both silicon demand and design methodologies, we need to reimagine how we motivate and educate the next generation of silicon engineers to address these challenges. We will share our experiences in interactions between industry, universities, and students, showing how industry engagement can re-energize these programs. The future of custom silicon depends on all of us investing in the next wave of engineers who will create tomorrow's innovative products.

Introduction

The semiconductor industry faces a fundamental challenge: the future of custom silicon innovation depends not just on architecture and circuit innovation along with advanced process technologies, but also on having enough skilled engineers working on custom silicon. Industry and universities both play crucial roles in developing these innovations and in building the engineering talent pipeline. In this paper, we will present Apple's journey from being a consumer of commodity silicon to becoming one of the world's largest technology companies that designs its own chips, illustrating the strategic importance of custom silicon. We look at the evolving requirements to execute on the vertical integration required by custom SoCs, and how that is going to influence the skills and educational approaches for the silicon engineering workforce of tomorrow. In particular, we highlight the importance of the industry-university partnership, and how our industry needs to increase engagement with academia to address the skills and needs of future engineers. We introduce some of the key aspects of industry-university interactions based on our experiences with Apple's New Silicon Initiative (NSI) started in 2019, and show some of the positive results that have been achieved in that timeframe. The semiconductor industry is rapidly changing, and it requires dynamic and collaborative approaches between industry and academia to empower the next wave of silicon engineers.

The Evolution to Custom Silicon

Custom-designed SoCs are at the heart of breakthrough technologies across every major product category from low-power consumer electronics to the high-performance supercomputers powering today's artificial intelligence (AI) revolution. Building SoCs that are optimized for specific applications has enabled our industry to achieve unprecedented performance, efficiency, and user experiences that continue to transform the world.

Apple's entry into SoC design began with the A4 chip in 2010, marking a strategic shift towards vertical integration. This decision was driven by the recognition that breakthrough products require optimization across the entire technology stack, from silicon architecture to software implementation, ultimately enabling the delivery of great user experiences. Figure 1.4.1 shows how custom silicon designed at Apple has expanded across all products, from the extremely power-efficient H-series chips used in AirPods to the high-performance M-series chips used in Mac, iPad, and Apple Vision Pro.

The transition from general-purpose to application-specific integrated circuits (ICs) reflects broader industry trends as well. As applications become increasingly specialized, custom silicon and bespoke accelerators offer the primary path to achieving differentiated performance, power efficiency, and area efficiency that is tuned for each product and application [1].

SoCs have evolved to be highly customized; providing not only the general CPU and GPU functions for mobile computing devices, but also the accelerators that enable performance and energy efficiency for key features. For example, the Neural Engine and Neural Accelerators in A19 Pro (Figure 1.4.2) lets the chip run everyday AI tasks efficiently while also handling much heavier, high-throughput AI workloads at significantly higher speeds. Software frameworks running on the A19 Pro can choose the best compute unit depending on the AI-driven task. This bespoke design provides the opportunity to create experiences that would have not been otherwise possible. CPUs, GPUs, and accelerators, as their own discrete devices would be high performance chips, but when they are integrated into one SoC, they provide integrated capabilities that exceed the performance of the sum of the individual blocks.

Apple uses a scalable architecture that allows all Apple silicon to deliver outstanding performance in each device class with industry-leading power efficiency. This is accomplished through a modular design approach where the blocks in the ICs are specifically optimized per functionality and workloads. The M4 Max SoC shown in Figure 1.4.3 is designed for highly compute intensive applications. These SoCs build on the scalable architecture with a fabric and wider unified memory subsystem to address the demanding data requirements. The blocks in the SoCs are mindfully chosen and carefully tuned to customer workloads to make them a key feature of high-performance laptop computers.

A deep understanding of the full-stack system requirements is crucial for future engineers, allowing them to optimize chip designs to find the right balance between performance, power, and area constraints, and effectively add accelerators that incorporate specific algorithms needed for that functionality. In the following sections, we will provide some examples of how Apple has implemented these modular designs.

Case Studies in Full-Stack Optimization

Apple Vision Pro with M5/R1 Co-Architecture

Apple Vision Pro demonstrates the power of purpose-built silicon for emerging applications (Figure 1.4.4). The system employs dual processors: M5 for traditional computing workloads and R1 for real-time sensor processing, achieving sub-12ms motion-to-photon (outside action to eyes) latency. This latency requirement could not be met by any existing off-the-shelf solution, but was necessary to deliver an excellent user experience for spatial computing. Figure 1.4.5 shows how the M5 and the R1 separate the different workloads in their designs.

One significant challenge of any mobile system is power. It takes a lot of energy to move and process the pixel data for many cameras and high-resolution displays. Solving this required full-stack design. Power was reduced by using the custom-designed R1 chip, which processes input from 12 cameras, five sensors, and six microphones, and streams new images to the displays within 12 milliseconds to create a real-time view of the physical world. Power was also reduced by using foveated rendering to use more pixels in the zone of focus and fewer outside the area of focus.

Unified Memory Architecture

The M-series and A-series chips implement a unified memory architecture (UMA) that eliminates traditional CPU-GPU memory bottlenecks by sharing tightly integrated on-package DRAM (Figure 1.4.6), greatly improving efficiency by eliminating the need for data to be copied between two separate memory pools.

A unified memory architecture enables each of the chip's subsystems to access a single pool of memory, which allows iPhone, Mac, iPad, and Apple Vision Pro to run large AI models completely on device. It fuels the CPU, GPU, and Neural Engine with data as well, offering higher multithreaded performance in apps, faster graphics performance in creative apps and games, and faster AI performance running models on the Neural Accelerators in the GPU and CPU or the Neural Engine.

Bringing UMA to a performance mobile SoC required co-design of silicon architecture, chip packaging, operating system, and application frameworks, demonstrating the need for engineers who understand the entire technology stack and the needs of the end user experience.

Ultra-Low Power Design for Wearables

The low-power and extremely compact form-factor of Apple Watch and AirPods exemplify the challenges of designing for extremely low power within volume constraints. Successfully designing these products required from-the-ground-up optimized silicon architecture and circuit design for their specific use cases, achieving all-day battery life in severely space-constrained form factors. Achieving performance at the lowest power levels requires a clean sheet and a dedicated design mindset.

Some of these improvements can be achieved by focusing on performance and power optimization. This involves predictive power management using workload analysis and aggressive dynamic voltage and frequency scaling. These techniques help provide efficient performance by minimizing the power consumed by unused or heavily utilized blocks on the chip. Clock gating provides substantial power savings, along with power gating and entire power rail gating, which eliminates both switching and DC power from leakage. Additionally, whole memory subsystems are custom designed to employ retention modes and power-optimized SRAMs minimize standby power.

ISSCC 2026 / SESSION 1 / PLENARY / 1.4

Although many of these techniques have been used in SoC designs, they highlight the power of system-level understanding for engineers of today and tomorrow. By understanding exactly how the SoCs will be used in the system, they can leverage these and other techniques in a custom design to deliver outstanding performance while still meeting strict limits on battery life and form factor.

The Full-Stack Engineering Imperative

The examples described here highlight a fundamental shift in how ICs are designed, with optimization of the design across the stack (Figure 1.4.7) to achieve the performance, power, and form-factors required. IC design has now evolved from being just about building a great stand-alone processor or SoC to optimizing across the stack.

To achieve maximum impact modern silicon engineers should possess:
- Systems-level thinking: Understanding how silicon decisions impact overall product performance and user experience
- Cross-domain expertise: Navigating trade-offs between analog, digital, algorithm, and system-level considerations
- Application awareness: Deep knowledge of target applications to make optimal architectural decisions and design tradeoffs
- Integration skills: Ability to work across traditional engineering boundaries.
- Deep expertise: Concentrated focus in design areas that bring innovation, experience, and expertise

The Future of Custom Silicon Design

Our industry is experiencing a proliferation of custom SoCs [1]. Many software-focused technology companies are developing highly specialized custom chips, while AI is driving a silicon renaissance across our industry. To address the demands for high performance along with the application-specific requirements, the potential growth in new chip designs is expected to rise rapidly over the next decade [1]. Complexity of these SoCs has increased incredibly over the last two decades, reaching over 100 billion transistors per leading edge SoC today, with scaling continuing to increase transistor count [2].

Some of this increase is reflected in the growth in the number of transistors of Apple SoCs (Figure 1.4.8). The requirements for higher performance and more capabilities, have led to three orders of magnitude growth in transistor count over a decade and a half from 190 million in the original A4 in 2010 to 184 billion in the M3 Ultra in 2025.

The demand for more custom ICs and the increase in the size of these SoCs is going to drive a need for increased design productivity and more IC engineers across our industry. The next two sections discuss how a combination of these can develop to meet the demands for custom silicon.

Design Productivity

Automated design tools are a critical part of the IC design infrastructure. The use of electronic design automation (EDA) has enabled the industry to increase the size and number of the ICs being built, without proportional increases in development time. Our industry needs a significant scaling of these capabilities to meet the needs of custom IC designs in the next decade. AI and machine learning (ML) are beginning to contribute to productivity efficiency across our industry. AI and ML in IC design methodologies could provide a significant improvement in productivity and capabilities to build even larger, more complex ICs [1][3]. Advancing these techniques into EDA tools will enable IC designers across the spectrum to significantly enhance their productivity, efficiency, and also the breadth of their skills and creativity, potentially boosting innovation in our industry.

Today, AI-based tools are already being used in the industry for physical design, place-and-route, and verification. On-going research and development in industry and academia will make these tools more powerful, as well as create the capabilities to apply them further in the design of digital and analog circuits. The use of AI-based tools in the verification and sign-off flow will also be a significant efficiency improvement. However, there is still a long way to go. More sophisticated tools with the capability to simulate and emulate across the stack will have a significant role in the future of chip design, taking into consideration the application requirements along with the microarchitecture development. This is an area that will benefit from investments in academic research to advance the state-of-the-art, as well as in the development and training of graduates with the skills to drive these capabilities into future design flows and methodologies. Engineers entering the workforce who are more familiar with AI will bring fresh perspectives to our industry.

Industry-University Collaboration

Across the industry, the raw demand for silicon design is growing, as more companies explore the performance advantages of developing their own ICs. Demands for chip designers are rapidly beginning to outpace the growth in the workforce [3], [4]. This difference is on track to limit the industry's capability to expand product lines and could potentially result in stifled innovation. In short, this could impact the future of our industry.

A 2023 analysis by the Semiconductor Industry Association (SIA) projects a shortfall of about 30,000 engineers by 2030 [5]. The skills required by these engineers will keep evolving with more demands for full-stack engineering and more knowledge across the design optimization stack, along with the utilization of AI. All of these will need to be addressed for the workforce of the future.

At Apple we have seen encouraging results from deepening industry/academic partnership in silicon engineering, particularly with the "New Silicon Initiative" (NSI), an in-depth engagement that Apple began in 2019. Through our NSI work and academia partnerships, we have learned that the most effective workforce development requires active industry participation in partnership with universities. The more that students interact with technical leaders and recent graduates from our industry directly in their classrooms and labs, the more they are motivated and appreciate the impact of silicon engineering and design.

The following are some approaches that we have found effective in motivating students and elevating their experiences in IC design:

A. Curriculum Enhancement
- Industry partnerships with faculty and departments to enhance the syllabuses of IC design and the pre-requisite courses, and to build curricular pathways for students to study IC design.
- Guest lectures by industry engineers that provide visibility into the interesting technical challenges in industry and potential solutions.
- Real-world design projects with industry mentorship, from circuit design to silicon tapeout, PCB design, algorithm development, test, debug, and validation. These are highly motivating for engineering students and provide them invaluable experience [6], [7].
- Industry-sponsored senior design projects where students get to engage with industry engineers who participate in project reviews and provide guidance.
- Knowledge exchange on the best practices in education and research is an on-going process, especially with workshops and conferences targeted towards curriculum development where examples of what works or does not work can be reviewed.
- Internship programs within the industry that provide meaningful hands-on experiences.

B. Investment in Faculty Development
A strong and well-educated undergraduate program is not possible without faculty that are experts in their fields and are exploring the cutting-edge IC design. Attracting top-tier faculty to IC academic research requires sustained industry investment. To achieve this, it is important that industry builds strong relationships with professors. Some of the areas that are important include:
- Industry partnerships with faculty that highlight the value of the collaboration within the university and raise awareness of the significance of the research areas. To this end, joint research initiatives and collaboration between the industry and academia are important, especially with new faculty as they build their research labs and recruit graduate students.
- To have top graduate students with the IC design expertise, especially in areas of digital and analog design, it is increasingly critical for the industry to identify and fund research in those areas. Continued engagement with graduate students with regular check-ins during their graduate years helps to guide them during their research, making both the students and their research more valuable to the industry.
- IC technology is becoming complex and more difficult to access by universities on their own. The full-stack environment requires direct engagement with multiple experts. Industry sabbatical programs can provide academics with the exposure to state-of-the-art technologies, and visibility into the challenges and areas of interest for industry, and help academics understand how they can translate these into elevated learning experiences for their students.
- Further motivation can be provided through faculty career development opportunities from industry collaboration, as well as recognition programs that highlight academic contributions to industry innovation.

C. IC-Specific Student Engagement
- To ensure students are well-prepared for the future of IC design, it's important for academia to continually update the curriculum to incorporate emerging and relevant technologies in this rapidly evolving field. The industry can help to guide the curriculum development in those cases and suggest curriculum pathways for IC design.
- Exposure to real-world design challenges provides students visibility into what is needed to be successful as IC designers. Industry sponsorship of IC design and tapeout classes that take projects from concept to actual packaged silicon and eventually working demos of the application [6]-[9]. Both undergraduates and graduate students can make meaningful contributions to these complex projects. Design reviews and mentoring IC design course projects by industry engineers, enables students to learn what the key priorities are during a design project, and what they need to do to prepare for a career in industry.

- State-of-the-art design tools and access to silicon fabrication are important for students to gain the skills and experience necessary to become strong contributors to the workforce. Towards this objective, there is a need for industry support in identifying and providing the design tools as well as access to reliable and rapid turnaround silicon processes.
- Programs like these are resource intensive, and universities need support to build the infrastructure to make them happen. For example, industry support for hiring and training teaching assistants and lab assistants is important to the success of programs that include IC tapeout, silicon bringup, and IC validation.
- Industry and academia should work together to show a clear program pipeline development from undergraduate to employment in the industry.

Six Year Engagement Results

Figure 1.4.9 shows the change in enrollment in junior-level undergraduate electrical engineering classes at a leading U.S. university. The NSI partnership at this school began in academic year (AY) 2019/2020, and the enrollments in these classes have gone up nearly 3x over the last six years. Figure 1.4.10 shows how the enrollment in IC design pathway courses over this time at three leading U.S. universities has increased after starting NSI engagements. While correlation does not equal causation, these are excellent results and show what is possible with industry and academia collaboration.

Call to Action

The semiconductor industry is at a pivotal moment of rapid transformation, presenting an unprecedented opportunity to redefine workforce development for custom silicon engineers. We can no longer rely on traditional, passive approaches. To meet tomorrow's full-stack design challenges for highly complex custom ICs, we must proactively cultivate a new generation of engineers, equipped with the cutting-edge skills they need to thrive. Rising to the challenge together to build a future-ready workforce requires:
- Reimagined engineering education that emphasizes full-stack thinking and integration of real-world applications and algorithms.
- Increased industry engagement in university programs and curriculum revitalization.
- Sustained investment in faculty development and academic research in silicon engineering and design methodologies.
- Cross-industry collaboration to address workforce challenges.

Conclusion

Apple's experience in silicon development demonstrates that breakthrough products can be achieved through custom silicon and full-stack engineering. Across the industry, the raw demand for custom silicon is growing, with more and more companies designing their own ICs. Custom ICs drive product differentiation, with SoCs growing more complex, heterogeneous, and feature-rich every year. But workforce and design tool limitations are rapidly emerging as constraints that will limit the potential these chips will be able to achieve. The demand for chip designers is outpacing the growth in the workforce and, eventually, this will begin to limit the capability of our industry.

The future of semiconductor innovation depends on the ability to educate and train engineers who can think across traditional boundaries. Our industry must act collectively to ensure that the next generation of engineers are equipped to meet these challenges. Apple's experience partnering with universities over the past six years with the New Silicon Initiative is an example of what can be done. In our engagements with universities, we have seen 2x to 3x increases in the number of undergraduates in the IC curriculum when partnering with leading U.S. universities. These numbers can be greater with more industry involvement and more direct engagement with universities and students.

The seeds the industry plants today in education and university research will determine whether the semiconductor industry can continue its trajectory of transformative growth and innovation, or whether workforce limitations will constrain the collective future. Our industry must act together to turn around these trends. The time to return to the classroom is now.

References:
[1] Fortune Business Insights, "Application Specific Integrated Circuit (ASIC) Market Size, Share & Industry Analysis, and Regional Forecast, 2025 – 2032", November 2025, https://www.fortunebusinessinsights.com/application-specific-integrated-circuit-market-104779
[2] Mark Liu, Philip Wong, "The Path to a 1 Trillion Transistor GPU", IEEE Spectrum, July 2024, https://doi.org/10.1109/MSPEC.2024.10589682
[3] Danny Crichton, "The Looming Labor Crisis in Chip Design", Lux Capital, August 2024, https://www.luxcapital.com/content/the-looming-labor-crisis-in-chip-design

[4] Ramiro Palma, Raj Varadarajan, Jimmy Goodrich, Thomas Lopez and Aniket Patil, "The Growing Challenge of Semiconductor Design Leadership", , BCG, November 2024, https://www.semiconductors.org/wp-content/uploads/2022/11/2022_The-Growing-Challenge-of-Semiconductor-Design-Leadership_FINAL.pdf
[5] "Chipping Away", Semiconductor Industry Association, July 2023, https://www.semiconductors.org/wp-content/uploads/2023/07/SIA_July2023_ChippingAway_website.pdf
[6] Peter Kinget, "Teaching IC Circuit Design: From Concepts to Testing a Fabricated Chip", IEEE Sol. St. Circuits Mag., 2023, https://doi.org/10.1109/mssc.2023.3283976
[7] Behzad Razavi, "Education of Chip Designers at a Large Scale: A Proposal", IEEE Solid. State Circuits Mag., 16, 2024, https://doi.org/10.1109/mssc.2024.3395073
[8] Lucy Revina, E. Gao, K. Ho, D. Lovell, K. Pister, B. Nikolic, "Taping Out Three Class Chips per Semester in Intel 16 Technology", Hot Chips 2025. https://doi.org/10.1109/hcs66204.2025.11154390
[9] Viansa Schmulbach, Jason Kim, Ethan Gao, et al, "NeCTar and RASoc: Tale of Two Class SoCs for Language Model Interference and Robotics in Intel 16, HotChips 2024, https://doi.org/10.1109/hcs61935.2024.10665203

ISSCC 2026 / SESSION 1 / PLENARY / 1.4

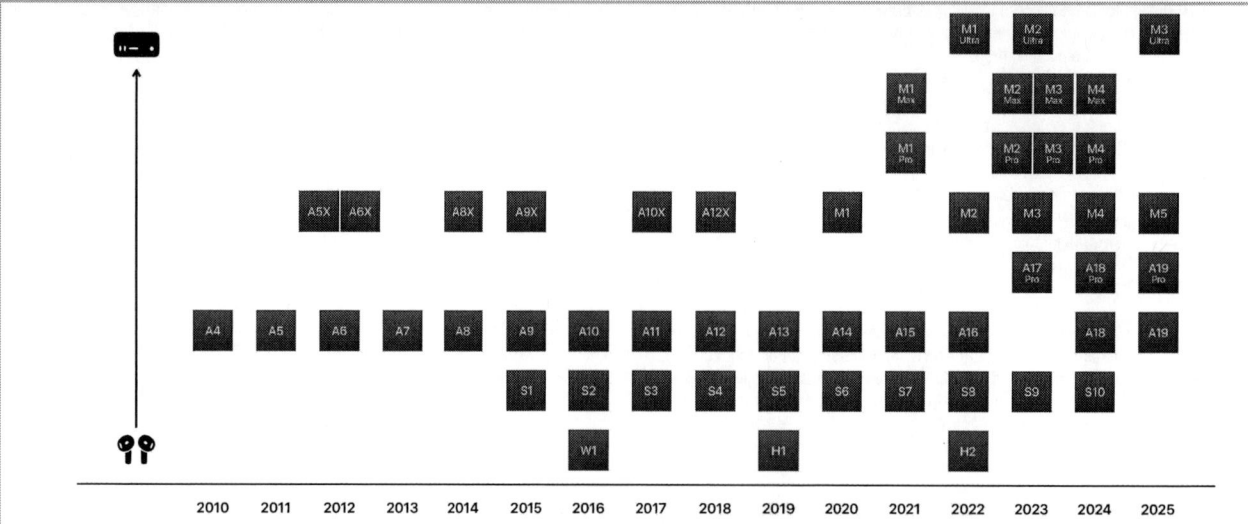

Figure 1.4.1: Evolution of Apple's custom silicon over the last decade and a half across a broad range of power/performance capabilities. W/H-series: AirPods; S-series: Apple Watch and HomePod family; A-series: iPhone, iPad, Apple TV, and Studio Display; Ax-series: iPad; M-series: iPad, Apple Vision Pro, and Mac laptop and desktop computers.

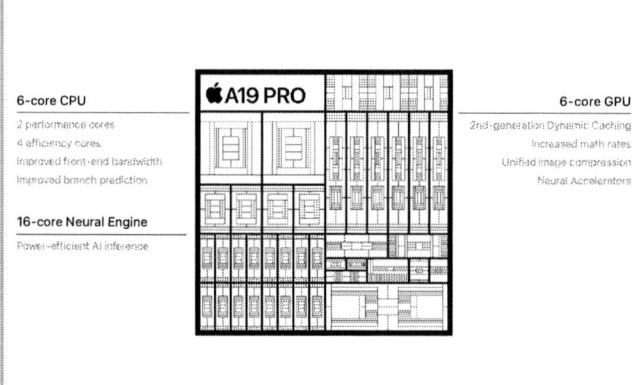

Figure 1.4.2: Block diagram of recently released Apple A19 Pro SoC.

Apple M4 Max

Compute functions
Including CPU, graphics

Specialized accelerators
Including media codecs, neural network acceleration, computer vision

Communication functions

Interfaces
Including memory, IO

Sensors

Figure 1.4.3: Die photo of the M4 Max, which is designed for compute intensive performance-centric applications. The specialized functions in SoC are listed.

Figure 1.4.4: Apple Vision Pro with the M5 and R1 chips. M5 manages traditional computing workloads while R1 manages real-time sensor processing.

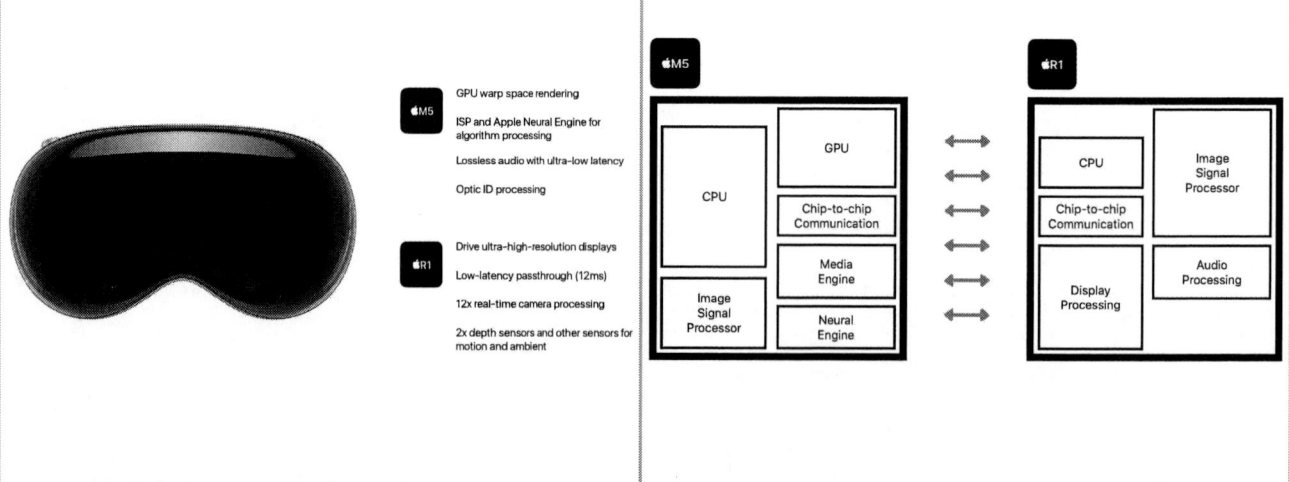

Figure 1.4.5: The two custom silicon chips on Apple Vision Pro, showing how the M5 chip and the R1 sensor chip manage different functions.

ISSCC 2026 / February 16, 2026 / 11:15 AM

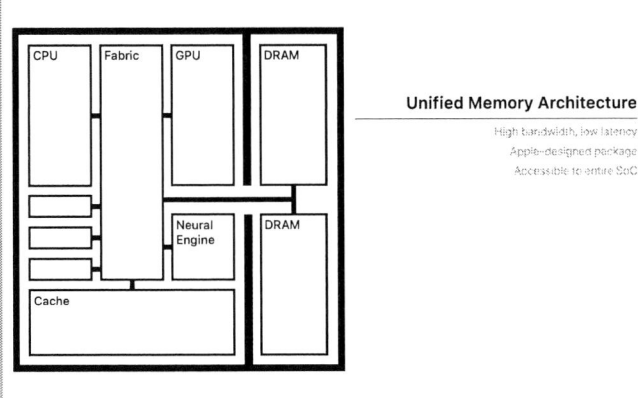

Unified Memory Architecture

High bandwidth, low latency
Apple-designed package
Accessible to entire SoC

Figure 1.4.6: Block diagram of Unified Memory Architecture in a mobile SoC

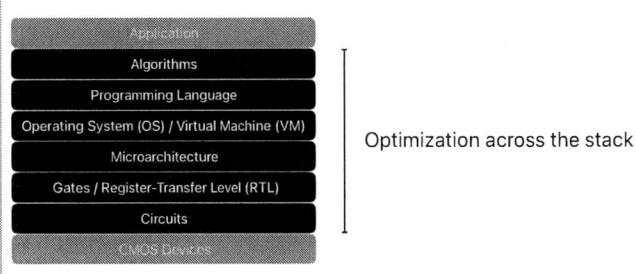

Optimization across the stack

Figure 1.4.7: Full-Stack engineering showing the skills needed for building optimized SoCs, from physics to the systems application.

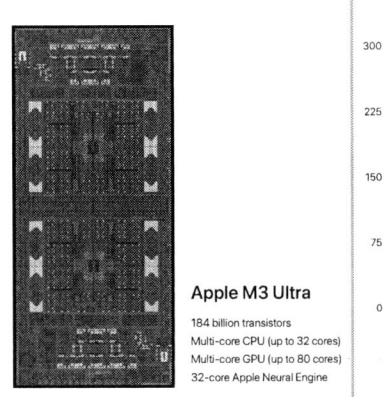

Apple A4

190 million transistors
Single-core CPU
Single-core GPU

Apple M3 Ultra

184 billion transistors
Multi-core CPU (up to 32 cores)
Multi-core GPU (up to 80 cores)
32-core Apple Neural Engine

Figure 1.4.8: The growth in size of Apple's custom SoCs from 190 million transistors in the A4 in 2010 to 184 billion transistors in the M3 Ultra in 2025.

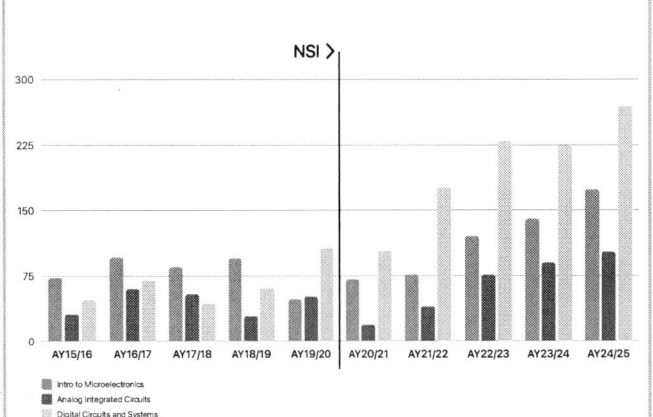

Figure 1.4.9: Enrollment in key digital, analog, and microelectronics classes over the last 10 years at a leading U.S. university. The increase in enrollment following the start of the NSI engagement is significant.

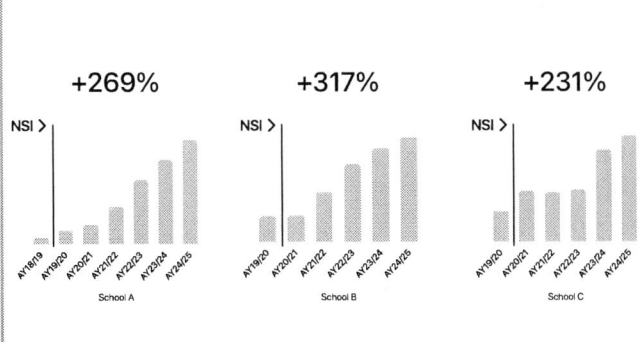

Figure 1.4.10: The increase in undergraduate IC design course enrollment at three leading U.S. universities over the last six years, following the start of the NSI engagement.

DIGEST OF TECHNICAL PAPERS • 39

979-8-3315-8937-0/26 $31.00 © 2026 IEEE

Session 2 Overview: *Processors*
DIGITAL ARCHITECTURES AND SYSTEMS SUBCOMMITTEE

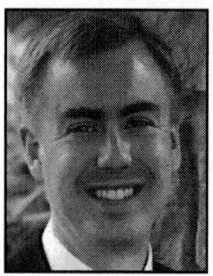

Session Chair: Mark Anders
Intel, Hillsboro, OR

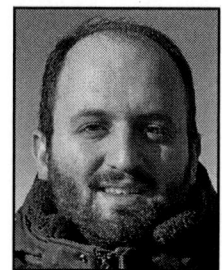

Session Co-Chair: Francesco Conti
University of Bologna, Bologna, Italy

This session showcases processor architectures that range from general-purpose GPUs and NPUs to accelerators that are customized for specific end-applications. The papers highlight advances in 3D stacking and 2.5D chiplet integration enabling heterogeneous, high-performance, energy-efficient processors and customized architectures for channel error correction, end-to-end driving, 3D/4D Gaussian splatting and diffusion model processors.

1:30 PM

2.1 AMD Instinct MI350 Series GPUs: CDNA 4-Based 3D-Stacked 3nm XCDs and 6nm IODs for AI applications

Ramasamy Adaikkalavan, AMD, Austin, TX

In Paper 2.1, AMD introduces a 3nm 3D-stacked GPU for AI applications. Compared to their prior generation, it offers 1.9× peak theoretical performance and 1.5× HBM bandwidth and capacity at higher power efficiency within the same area footprint.

1:55 PM

2.2 A Quad-Chiplet AI SoC with Full-Chip Scalable Mesh Over 16Gb/s UCIe-Advanced Die-to-Die Interface for Large-Scale AI Inferencing

Chang-Hyo Yu, Rebellions, Seongnam, Korea

In Paper 2.2, Rebellions showcases a quad-chiplet AI SoC for large-scale AI inferencing. The chiplet integrates an NPU implemented in 4nm with HBM3E modules and integrated silicon capacitors for stable power delivery. The LLM accelerator achieves 56.8TPS on LLaMA v3.3 70B with single-batch 2k/2k input/output sequences.

2:20 PM

2.3 A 71.3mJ/Frame End-to-End Driving Processor with Flexible Heterogeneous Core Orchestration via Sparsity Reasoning

Jueun Jung, Ulsan National Institute of Science and Technology, Ulsan, Korea, Yonsei University, Seoul, Korea

In Paper 2.3, the Ulsan National Institute of Science and Technology presents an end-to-end driving processor with flexible heterogeneous core orchestration via sparsity reasoning. Fabricated in 28nm, the 10.5mm² core operates at 50 to 250MHz with a power range of 97.6 to 735.6mW and 71.3mJ/frame.

2:45 PM

2.4   UniC-Vision: A 14.4Gb/s 7.3pJ/b Universal Vision Transformer OFDM Channel Estimation Accelerator for B5G/6G AI-RAN

Sangbu Yun, Korea Advanced Institute of Science and Technology, Daejeon, Korea,
                Pohang University of Science and Technology, Pohang, Korea

In Paper 2.4, Korea Advanced Institute of Science and Technology presents a 28nm 6.35mm² AI-RAN (radio access network) channel estimation accelerator for next-generation communication, achieving a high throughput of 14.4Gb/s with a high energy efficiency of 7.3pJ/b.

3:00 PM

2.5   A 1.1mm², 14.4ns, 13.1pJ/b Forward Error Correction with Ordered-Statistics Post Processing for Ultra-Reliable and Low-Latency Communications

Cheng-Hsun Lu, University of Michigan, Ann Arbor, MI

In Paper 2.5, the University of Michigan presents a 16nm, 1.1mm² two-stage forward-error-correction decoder combining soft-decision Chase and order-1 ordered-statistics decoding, achieving 14.4ns average latency, 13.1pJ/b energy efficiency and 15.2Gb/s throughput for ultra-reliable low-latency communication, improving worst-case latency by five orders of magnitude over prior designs.

3:35 PM

2.6   Spyre: An Inference-Optimized Scalable AI Accelerator for Enterprise Workloads

Matthew Cohen, IBM Research, Yorktown Heights, NY

In Paper 2.6, IBM introduces Spyre, a scalable, power-efficient AI accelerator optimized for inference, featuring advanced power management, flexible low-precision support, and a high-throughput memory subsystem, all within a single-slot PCIe card. Manufactured in 5nm CMOS, its 330mm² die integrates 26B transistors and delivers 32% higher throughput and 2-to-3× better energy efficiency compared to leading GPUs.

4:00 PM

2.7   Tiamat: A 98-to-134ms/Step Transformer-Based Diffusion Model Processor Supporting Classifier-Free Guidance for Image Generation

Po-Yen Lu, National Tsing Hua University, Hsinchu, Taiwan

In Paper 2.7, National Tsing Hua University proposes Tiamat, a 16nm 98-134ms/step transformer-based diffusion model processor demonstrating text-to-image capabilities on DiT-XL/2 and PixArt-α. Tiamat achieves peak compute of 7.37TOPS with an energy efficiency of 4.99TOPS/W in a die area of 8mm² and reduces external memory access by ~67%.

4:25 PM

2.8   MADiC: A 3nm 7.4TOPS/mm², 17.4TOPS/W Generative Diffusion Accelerator Enabled by Hardware-Compiler Co-Optimization of Memory Hierarchy and Operator Parallelism

Shih-Wei Hsieh, MediaTek, Hsinchu, Taiwan

In Paper 2.8, MediaTek presents MADiC, their 3nm 0.338mm² diffusion accelerator for generative image editing on edge devices. MADiC achieves 7.4TOPS/mm² and 17.4TOPS/W when running a diffusion ConvNet model through design-time memory optimization, cross-loop feature map allocation, and concurrent preloading when resizing.

4:50 PM

2.9   A 0.24mJ/Frame Quadratic Interpolation 4DGS Processor with Recursive Computation Reuse and Tree-Based Parallel-Rendering

Xiaolong Yang, Tsinghua University, Beijing, China

In Paper 2.9, Tsinghua University introduces a 28nm 3.65mm² 4D Gaussian Splatting (4DGS) processor that uses quadratic frame interpolation and tree-based parallel-rendering to achieve energy of 0.24mJ/frame for rendering.

5:15 PM

2.10   A 1286fps 0.39mJ/Frame Modeling/Rendering Unified 3D GS Processor with Locality-Optimized Computation and Reconfigurable Architecture

Hedi Wang, Tsinghua University, Beijing, China

In Paper 2.10, Tsinghua University presents a 28nm 4.76mm² reconfigurable 3D Gaussian Splatting (3DGS) processor architecture with a locality-optimized fine-grained rendering array and workflow achieving a throughput of 1286fps at 0.39mJ/frame.

ISSCC 2026 / SESSION 2 / PROCESSORS / 2.1

2.1 AMD Instinct MI350 Series GPUs: CDNA 4-Based 3D-Stacked 3nm XCDs and 6nm IODs for AI applications

Ramasamy Adaikkalavan[1], Alan Smith[1], Teja Singh[1], Sundar Rangarajan[1], Eric Chapman[1], Samuel Naffziger[2], Subramaniam Maiyuran[3], Candy Chen[4], Sriram Sundaram[1], Mark Silla[1], Duncan Law[5], Kathy Hoover[1], Samuel Lipson[1], Kevin Duda[2], Vinay Parthasarathy[6], Deepesh John[1], Hanish Vemulapalli[1], Srinivas Pavan Kumar Gade[7]

[1]AMD, Austin, TX, [2]AMD, Fort Collins, CO, [3]AMD, Folsom, CA, [4]AMD, Shanghai, China, [5]AMD, Markham, Canada, [6]AMD, San Diego, CA, [7]AMD, Hyderabad, India

Abstract

The AMD Instinct MI350 Series GPU, based on a 4th-generation AMD CDNA architecture, features eight stacked accelerator complex dies (XCD), two I/O dies (IOD) interconnected with AMD Infinity Fabric, and interfaces for eight stacks of 12-hi HBM3E memory. The MI350 Series GPU delivers over 3× generational inference performance increase by improving energy efficiency, increasing compute throughput, supporting new lower-precision FP6/FP4 datatypes, and increasing both memory bandwidth and capacity.

The AMD Instinct™ MI350 Series GPUs establish a high level of performance for generative AI and training applications within datacenters. The AMD Instinct MI350 Series GPU, based on a 4th-generation AMD CDNA™ 4 architecture, features eight stacked accelerator complex dies (XCD), two I/O dies (IOD) interconnected with proprietary AMD Infinity Fabric™, and interfaces for eight stacks of 12-hi HBM3E memory. The heterogeneous 3D-stacked architecture facilitates integrating more compute, memory capacity, and bandwidth into a CoWoS-S module, surpassing the limitations of traditional 2D implementations. The XCD die is fabricated in TSMC's performance-enhanced 3nm FinFET process (N3P) for improved density and performance per Watt. The IOD is fabricated in TSMC's mature 6nm FinFET (N6) process for improved yield [1]. The product line features the AMD Instinct MI350X GPU (maximum total board power (TBP) of 1kW) and the AMD MI355X GPU (maximum TBP of 1.4kW). The AMD Instinct MI355X GPU delivers over 3× generational inference performance increase by improving energy efficiency, increasing the compute throughput, supporting new lower-precision FP6/FP4 datatypes, and increasing both memory bandwidth and capacity [5] [7].

The overarching goal for the AMD CDNA 4 architecture is to deliver a generational performance throughput increase with enhanced energy efficiency for AI workloads compared to the AMD CDNA 3 architecture used in the prior generation [6]. The AMD Instinct MI350 achieves this by integrating more computational hardware within a similar footprint to the previous generation, while also boosting power efficiency as shown in Fig. 2.1.1. The 3nm process technology scaling, combined with a power-efficient standard-cell library choice, provided an optimized performance per Watt solution to deliver the increased compute. Power efficiency gains are accomplished through design, architecture and physical design improvements to reduce effective capacitance per operation (C_{AC}/op) and improved frequency. To support the increased compute needs, the IOD is architected to deliver a 1.5× peak HBM bandwidth and capacity at more power-efficient frequencies compared to the MI300X as shown in Fig. 2.1.1.

Figure 2.1.2 shows the AMD MI355X XCD die which accommodates around 30% more transistors within roughly the same area footprint as the MI350X via several advancements across the GFX implementation. This improved transistor density is enabled by the shift from 5nm to 3nm process technology, the standard-cell library choice and two additional routing layers (15-layer metal to 17-layer metal). Unlocking the full technology process scaling improvement required close collaboration between design teams, TSMC and EDA vendors. PD methodology innovations like custom standard-cell optimizations, synthesis, place and route (SAPR) tool optimizations to improve utilization and SRAM efficiency improvements helped drive the density further. The increased transistor density, combined with area reduction through hardware reuse across select datatypes, enables the doubling of execution hardware resources for 16b and 8b formats. The improved hardware capacity delivers a 1.9× improvement in peak theoretical performance for these existing machine learning-focused data types. Additionally, this higher transistor density allows for the introduction of FP6 and FP4 precision formats, delivering a 3.85× improvement in peak theoretical performance relative to the FP8 peak of the previous generation as shown in Fig. 2.1.4 [3].

Dynamic power dominates the total power profile of compute heavy AI workloads and the largest levers for driving dynamic power reduction are to lower the voltage of operation at ISO-frequency and to reduce the effective switching capacitance per operation (C_{AC}/op). In addition to the frequency boost from process technology scaling, XCD targets a 5% technology process-neutral GPU device-level frequency increase focusing on the optimal operating frequencies for the high computational throughput workloads. A robust standard-cell (stdcell) library with a combination of single row and custom multi-row stdcells, careful cell palette selection, innovations in timing flows and methodology contributed to frequency improvement at these critical operating points [2]. The increased compute throughput in the AMD MI355X combined with higher gate density in 3nm drives an increase in current density, increasing the voltage supply droop. Voltage droop on the XCD is mitigated through a combination of advanced di/dt-based adaptive clocking, decoupling capacitance, architectural instruction throttlers and improved power delivery networks.

The XCD targeted up to 30% reduction in process-neutral C_{AC}/op over the prior design for critical workloads through strong cross-disciplinary collaboration across microarchitecture, RTL, physical design and SoC teams. Detailed power breakdowns across RTL features and pipelines were used to identify power inefficiencies in design across a range of targeted workloads that are critical for AI performance. A breakdown of the XCD C_{AC} is shown in Fig. 2.1.5. Efforts focused on minimizing sequential and clocking power overhead through strategic pipeline balancing, the use of low-power flops, and multi-bit flop banking optimization. Enhanced CTS methodologies, improved clock gating efficiency and use of improved activity-based clock gating cells further drove clock power reduction. The design also emphasized careful floorplaning and the use of machine learning-based flows to reduce wire C_{AC}. The combined improvements in frequency, C_{AC} and process technology scaling outlined above have enhanced power efficiency, enabling 2× increase in compute throughput with less than 2× the power.

A key design goal for the IOD in the AMD CDNA 4 architecture was to extend the leadership in HBM bandwidth (BW) and capacity of the MI300X, while minimizing uncore power consumption, thereby freeing up more power for XCD compute. The architecture now features two larger IODs manufactured on TSMC's N6 process rather than four smaller IODs as in the prior generation, as shown in Fig. 2.1.3. This change enables greater power efficiency by reducing the chip-to-chip (C2C) power. The AMD MI355X features a redesigned scalable interconnect layout topology aimed at reducing power consumption associated with data movement. The CDNA 4 architecture also widens the data pipeline to deliver peak HBM bandwidth at $V_{min,}$ which lowers BW per Watt.

Considerable effort was dedicated to increasing interconnect routing density and minimizing power usage through advanced custom wire engineering. The selection of interconnect repeater topology, including segment length, NDR options, routing pattern and drive strengths was carefully optimized to ensure optimal performance per Watt at V_{min}. A combination of floorplan and layout topology improvements, architecture improvements to deliver peak BW at power efficient frequencies and PD power optimization delivers up to 1.3× better HBM read BW/Watt compared to the MI300X [4].

The AMD Instinct MI350 Series delivers on a product strategy which provides up to 1.9× peak theoretical performance improvement and 1.5× HBM bandwidth and capacity at improved power efficiency for next-generation AI workloads. The voltage reduction at constant GPU frequency combined with C_{AC}/op and throughput improvement on the XCD and BW per Watt improvement on the IOD delivers more than 3× generational inference performance improvement for broad AI use cases as shown in Fig. 2.1.6 [5]. The AMD MI355X also delivers high inference throughput for larger models compared to the competition, as shown in Fig. 2.1.7 [7]. This was possible through holistic architectural innovation, collaborative optimization between RTL, physical design and technology teams, and a focus on energy efficiency at every stage of the design process.

Acknowledgement:
The authors acknowledge the efforts of the AMD design teams worldwide whose contributions were instrumental to the MI350 design.

References:
[1] L. T. Su et al., "Multi-Chip Technologies to Unleash Computing Performance Gains Over the Next Decade," *IEDM*, pp. 1.1.1-1.1.8, 2017. https://doi.org/10.1109/IEDM.2017.8268306
[2] T. Singh et al., "Zen 5": The AMD High-Performance 4nm x86-64 Microprocessor Core," *ISSCC*, pp. 44-45, 2025. https://doi.org/10.1109/ISSCC49661.2025.10904529
[3] MI350-005: Based on calculations by AMD Performance Labs in May 2025 for the AMD Instinct™ MI355X and MI350X GPUs to determine the peak theoretical precision performance when comparing FP16, FP8, FP6 and FP4 datatypes with Matrix vs. AMD Instinct MI325X, MI300X, MI250X and MI100 GPUs. Server manufacturers may vary configurations, yielding different results.

[4] MI350-036 (n.d.). Based on testing by AMD Performance Labs in June 2025, on the AMD Instinct MI350X vs. AMD Instinct MI300X GPUs, using an AMD proprietary synthetic micro-benchmark to determine the memory read bandwidth for each GPU, and an AMD internal tool to collect per watt performance in a time-series format at the 1ms timescale, for each GPU. Server manufacturers may vary configurations, yielding different results. Results may vary based on use of the latest drivers and optimizations.

[5] MI350-042 (n.d.). Based on AMD measurements as of 6/62025 of the text-generated offline inference throughput for Llama 3.1-405B chat model on an 8x GPU AMD Instinct MI355X platform running 8x TP1 (8 copies of model on 1 GPU) with (FP4) compared to an 8x GPU AMD Instinct MI300X platform running 2x TP4 (2 copies of model on 4 GPUs) with (FP8). Tests were conducted using a synthetic dataset with different combinations of 128 and 2048 input/output tokens. Server manufacturers may vary configurations, yielding different results. Performance may vary based on the use of the latest drivers and optimizations.

[6] A. Smith and V. Alla, "AMD Instinct MI300X: A Generative AI Accelerator and Platform Architecture," *IEEE Hot Chips Symp.*, 2024. https://doi.org/10.1109/HCS61935.2024.10664659

[7] MI350-038: Based on testing by AMD internal labs as of 6/6/2025 measuring text generated throughput for LLaMA 3.1-405B model using FP4 datatype. Test was performed using input length of 128 tokens and an output length of 2048 tokens for AMD Instinct™ MI355X 8xGPU platform compared to NVIDIA B200 HGX 8xGPU platform published results. Server manufacturers may vary configurations, yielding different results. Performance may vary based on use of latest drivers and optimizations.

MI350-039: Based on Lucid automation framework testing by AMD labs as of 6/6/2025, measuring text generated throughput for LLaMA 3.1-405B model using FP4 datatype. Test was performed using 4 different combinations of input/output lengths (128/2048) to achieve a mean score of tokens per second for AMD Instinct™ MI355X 4xGPU platform compared to NVIDIA DGX GB200 4xGPU platform. Server manufacturers may vary configurations, yielding different results. Performance may vary based on use of latest drivers and optimizations.

MI350-040: Based on testing (tokens per second) by AMD internal labs as of 6/6/2025 measuring text generated online serving throughput for DeepSeek-R1 chat model using FP4 datatype. Test was performed using input length of 3200 tokens and an output length of 800 tokens with concurrency up to 64 looks, serviceable with 30ms ITL threshold for AMD Instinct™ MI355X 8xGPU platform median total tokens compared to NVIDIA B200 HGX 8xGPU platform results. Server manufacturers may vary configurations, yielding different results. Performance may vary based on use of latest drivers and optimizations.

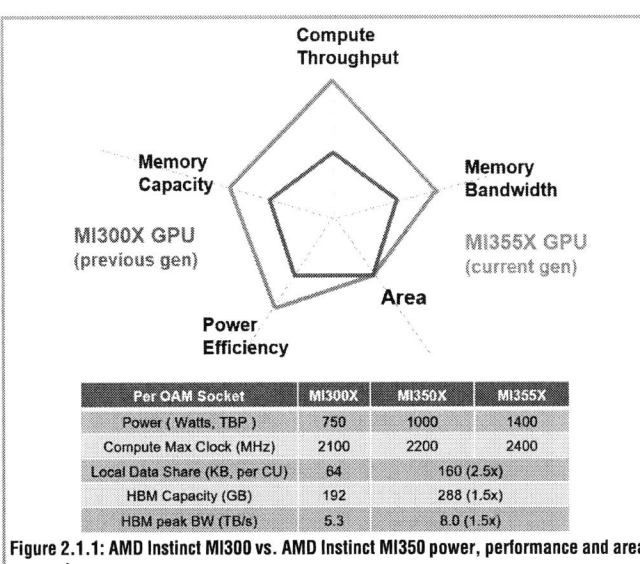

Per OAM Socket	MI300X	MI350X	MI355X
Power (Watts, TBP)	750	1000	1400
Compute Max Clock (MHz)	2100	2200	2400
Local Data Share (KB, per CU)	64	160 (2.5x)	
HBM Capacity (GB)	192	288 (1.5x)	
HBM peak BW (TB/s)	5.3	8.0 (1.5x)	

Figure 2.1.1: AMD Instinct MI300 vs. AMD Instinct MI350 power, performance and area comparison.

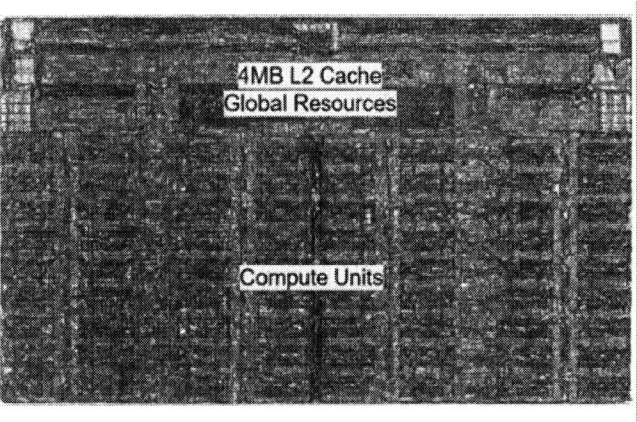

Figure 2.1.2: AMD Instinct MI350 XCD.

ISSCC 2026 / SESSION 2 / PROCESSORS / 2.1

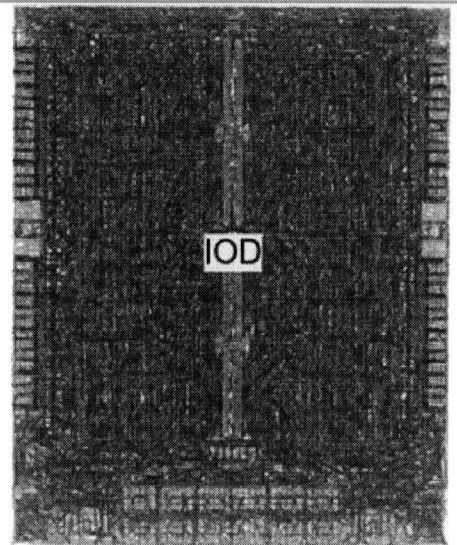

Figure 2.1.3: AMD Instinct MI350 IOD.

Figure 2.1.4: Generational peak theoretical performance improvement.

Computation	FLOPS/clock/CU		Peak Theoretical		MI355X Peak Speedup Over MI300X[1]
	MI300X	MI355X	MI300X	MI355X	
Matrix FP16/BF16	2048	4096	1.3 PF	2.5 PF	1.9x
Matrix FP8	2048	4096	2.6 PF	5 PF	1.9x
Matrix INT8	4096	8192	2.6 POPs	5 POPs	1.9x
Matrix MXFP6	NA	16384	NA	10 PF	New to MI350
Matrix MXFP4	NA	16384	NA	10 PF	New to MI350

Peak theoretical performance without sparsity

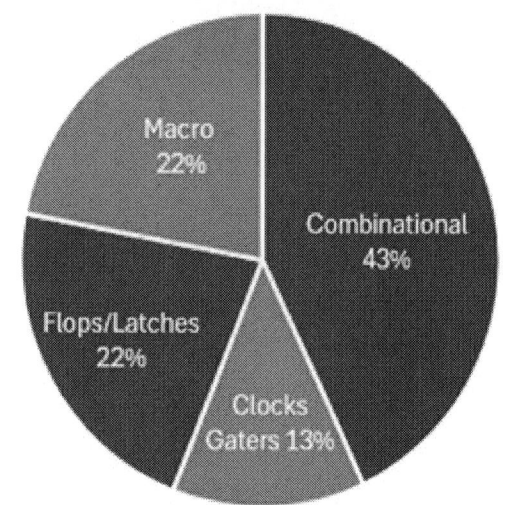

Figure 2.1.5: AMD Instinct MI350 XCD C_{ac} breakdown.

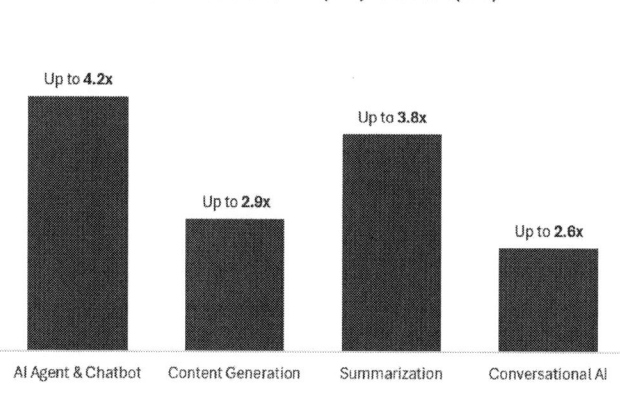

Figure 2.1.6: Inference throughput comparison between AMD Instinct MI355X (FP4) and AMD Instinct MI300X(FP8).

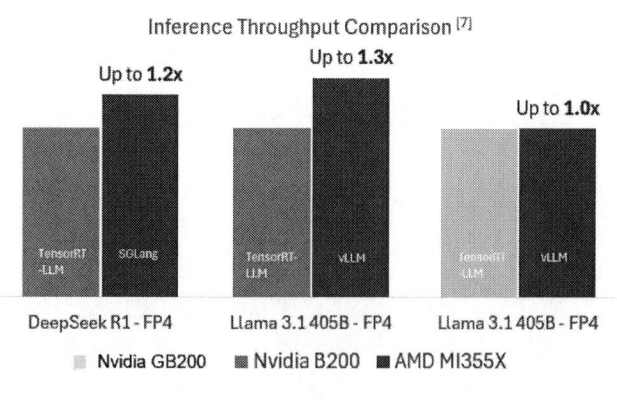

Figure 2.1.7: Inference throughput comparison between AMD Instinct MI355X vs. Nvidia B200 and Nvidia GB200.

- 2026 IEEE International Solid-State Circuits Conference

ISSCC 2026 / SESSION 2 / PROCESSORS / 2.2

2.2 A Quad-Chiplet AI SoC with Full-Chip Scalable Mesh Over 16Gb/s UCIe-Advanced Die-to-Die Interface for Large-Scale AI Inferencing

Chang-Hyo Yu, Jaewan Bae, Jinseok Kim, Hongyun Kim, Wongyu Shin, Jae-Sung Yoon, Young-Jae Jin, Jungju Oh, Jaebong Lee, Eunseo Kim, Miock Chi, Sungpill Choi, Donghan Kim, Hyunje Jo, Hyunho Kim, Hyungseok Heo, Hyunsung Kim, Seung-Goo Kim, Myunghoon Choi, Sangeun Je, Junhee Ham, Juyeong Yoon, Yashael Faith Arthanto, Sung-il Bae, Sanggyu Park, Joungwoo Lee, Heeyoung Chae, Kiljun Ryu, Yongjik Kim, Jin-O Seo, Nara Cho, Taeyoung Jeon, Gukchae Ahn, Myoungoh Ki, Junkyeong Choi, Seungcheol Baek, Daehoon Kim, JuHwan Kim, Sungmoon Kang, Sieon Son, Minseo Kim, Yeounghwan Choe, Younggeun Lee, Sunghyun Park, Jinwook Oh

Rebellions, Seongnam, Korea

Abstract

A 4nm-based quad-chiplet with an advanced packaged LLM accelerator achieving 56.8TPS on LLaMA v3.3 70B with single-batch 2k/2k input/output sequences. The architecture combines chiplet-based design, low-latency die-to-die interfaces, unified mixed-precision compute, holistic synchronization, and HBM3E with advanced power schemes to sustain bandwidth, capacity and thermal stability. The SoC integrates four NPU chiplets, four HBM3E modules, and four ISCs for stable power delivery.

This paper presents an NPU architecture optimized for rapidly evolving, large-scale LLMs such as GPT-4.5, Claude 4, Gemini 2.5, and Llama 4, addressing two primary computational challenges. The compute-bound prefill phase requires a massive amount of floating-point operations, exacerbated by increasing model complexity and prompt engineering, while the memory-bound decoding phase is limited by KV cache bandwidth, becoming more severe with context windows extending to millions of tokens. To overcome these challenges, we propose a scalable chiplet-based architecture that defines the minimum system granularity, enabling the construction of a large, virtually monolithic system. Modular chiplets support both scale-up and scale-out expansion, allowing growth from single-die setups to multi-chiplet clusters. The design also accommodates I/O chiplets for seamless integration of additional compute or memory resources, mitigating compute and memory bottlenecks while establishing a robust foundation for next-generation LLMs.

We present a quad-chiplet configuration as the foundational building block, delivering 56.8TPS on Llama 3.3 70B [1] with a single-batch 2k/2k input/output sequence in a single-chip setup. The architecture combines chiplet-based design, a low-latency die-to-die interface with memory load–store semantics, and a unified compute and interconnect fabric that scales efficiently across chiplet and chip boundaries. Holistic synchronization ensures high utilization across model phases, while HBM3E integration with advanced power schemes sustains bandwidth, capacity and thermal stability. Together, these innovations establish a robust platform that not only addresses the continuous rise in compute density and design scale but also provides the scalability required for future system expansion.

The AI SoC integrates four identical NPU chiplets, four HBM3E modules, and four integrated silicon capacitors (ISCs), as illustrated in Figs. 2.2.1 and 2.2.7. Each NPU chiplet employs three UCIe channels, with topology-aware die sliding and rotation applied to ensure seamless vertical and horizontal mesh continuity. The ISCs feature multiple power domains, actively supplying sufficient capacitance to both the NPU and HBM3E.

The SoC interfaces with the host system via two PCIe Gen5 ×16 links, supporting SRIOV and peer-to-peer connectivity. A command processor compiles descriptors and manages task scheduling both cooperatively and independently across all NPU chiplets, in conjunction with a full-chip synchronization manager and DMA engines. The SoC delivers 2PFLOPS (FP8) of compute capability through a unified multiple- and mixed-precision arithmetic unit [2]. By eliminating the need for separate arithmetic units across different precision levels, the design improves compute density while supporting per-operand configurability for diverse precisions, enabling resource-efficient mixed-precision execution. Complementing this, the SoC integrates 512MB of on-chip SRAM organized in a multi-cycle, multi-bank architecture: 256MB high-density scratchpad memory providing 128TB/s aggregated bandwidth, and 256MB of shared memory delivering 64TB/s aggregated bandwidth. This balanced memory hierarchy supports both low-latency buffering and high-bandwidth shared access, sustaining throughput under demanding workloads. The combined compute and memory resources are defined as a Neural Core, grouped into Neural Core Clusters. These clusters communicate across the full-chip interconnect, which extends seamlessly over chiplet boundaries via UCIe-based links. Moreover, the architecture is explicitly designed to enable scalability beyond package boundaries, providing a pathway to future system expansion without sacrificing performance efficiency.

The data transfer across all chiplets is orchestrated through four key architectural components shown in Figs. 2.2.1 and 2.2.2: a full-chip spanning mesh network, a configurable multi-engine DMA, a synchronization mechanism, and a proprietary die-to-die interface. Each chiplet integrates 16 Neural Cores and 64MB of shared memory, organized into an 8×4 granular mesh interconnected via 64 routers and supported by a DMA subsystem with 8 execution engines. Leveraging the die-to-die interface, the system seamlessly extends into a unified virtual mesh topology, scaling to 256 routers across the entire multi-chiplet configuration.

To dynamically adapt to diverse AI workloads, traffic-aware prioritization and QoS controls govern real-time resource allocation. Fully interleaved shared-memory access paths and multi-path routing deliver up to 4× peak per-node bandwidth, while a virtual synchronization network embedded in the mesh minimizes dependency resolution latency, as illustrated in the green path of Fig. 2.2.2. In clustered configurations, groups of eight Neural Cores and 32MB of shared memory are connected through a customized mesh NoC comprising three logically independent channels (D, R, and C). The D-channel handles bulk data transfers, while the R- and C-channels serve requests, responses, and control messages. Routers employ XY-routing with turn-model constraints and weighted round-robin arbitration, where adjustable weights enable fine-grained QoS tuning from a system-wide perspective. Distributed shared memory is partitioned into 16 slices across NoC nodes, with logical mapping managed through address-decode tables (ADT) in the network interface unit (NIU) for efficient interleaving. To improve yield and manage design complexity associated with increasing die and chip size, a harvesting scheme is supported at both the cores and shared memory levels, with the mesh isolating defective nodes from the system, as illustrated in the red-dashed region of Fig. 2.2.2.

The multi-engine DMA exploits both local/remote HBM3E and shared memory, sustaining up to 2.6TB/s per-DMA bandwidth. Per-task QoS features further reduce long-tail latency and mitigate dependency bottlenecks, streamlining execution across heterogeneous workloads. Meanwhile, the synchronization mechanism operates through a dedicated virtual channel over the mesh, eliminating direct peer-to-peer dependencies between Neural Cores, DMAs and synchronization managers. Each chiplet incorporates a synchronization manager with hardwired sync control for scalability over the previous scheme [3], coordinating cross-chiplet operations either centrally or autonomously to minimize inter-entity interactions and enhance efficiency across chiplet boundaries.

The chiplet interface employs a UCIe-based latency-optimized protocol, achieving 11ns FDI-to-FDI latency at 16Gbps, while fully utilizing the available port bandwidth. By extending load–store memory semantics across dies, the interconnect enables the multi-chip system to operate as a virtually monolithic unit with inherent support for both scale-up and scale-out expansion. To ensure reliability in data center environments, the architecture incorporates a configurable switching that trades marginal performance for significant improvements in MTBF and MTTF, thereby reducing downtime and lowering MTTR to enhance overall system availability. Testability, often constrained in multi-die systems, is improved through multi-layer loopback paths that extend beyond the UCIe specification. These include transaction-level and channel-level modes, as well as inbound-to-outbound loopback with optional address remapping, offering comprehensive diagnostic coverage. As shown in Fig. 2.2.5, bathtub and eye-diagram measurements confirm high signal fidelity and wide margins, with error-free operation verified at 16Gbps during the bring-up phase. Remarkably, these results were achieved within two months of silicon arrival, demonstrating the robustness and maturity of the chiplet interface design.

With a thermal design power (TDP) of 600W, the system encounters transient surges exceeding twice the nominal level, posing significant challenges for power integrity (PI). The lock-step activation and deactivation of multiple Neural Cores, while essential for functional correctness, induces large transient currents (high di/dt) that trigger voltage droops and threaten stability. To address this, we employ a hardware-based staggering technique, skewing the start-up sequences of individual Neural Cores. As shown in Fig. 2.2.4, this approach markedly reduces peak transient current: the synchronous method (top-left) exhibits steep current slopes and severe voltage noise, whereas the staggered method (top-middle) achieves gentler slopes and improved supply stability. In addition, power throttling and smoothing mechanisms, applied at both the Neural Core and inter-chiplet granularity, dynamically regulate instruction execution rate and issue count over a moving time window to further suppress transient fluctuations. In parallel, the high-transient HBM3E traffic, comparable in intensity to NPU compute bursts, places additional stress on the power delivery network. To address this, we integrate a ISC die, embedding distributed on-chip capacitance across the V_{DD} rails of both the HBM3E and PHY. This reinforcement effectively suppresses traffic-induced voltage fluctuations, as demonstrated in Fig. 2.2.5, where impedance profiles with ISC integration show a clear reduction in resonance peaks versus the baseline. Through these combined architectural and physical design measures, we achieved stable HBM3E operation, validating both EOM (UI) and performance metrics within just three weeks of silicon bring-up. Collectively, these innovations establish a robust PI solution, enabling reliable operation under extreme transient workloads.

Our architectural innovations establish a petaflop-class LLM solution that not only matches but surpasses the capabilities of leading commercial accelerators. Figure 2.2.6 presents the measurements during bring-up up to 1.7× higher power efficiency compared to an H200 [5], and emulator-correlated performance estimations at the design-target configuration indicate highly promising efficiency gains, further underscoring the competitiveness of our approach. These results demonstrate the practicality of our approach in delivering scalable, high-efficiency AI acceleration that can serve as a compelling alternative to existing GPU-based solutions. This design serves as a foundational building block for larger AI systems, supporting both scale-up and scale-out through chiplet expansion. To further enhance scalability, we are developing I/O and memory expander chiplets as shown in Fig. 2.2.7 connected through the scalable mesh over the die-to-die interfaces. This scalable architecture supports cross-node/rack LLM systems for trillion-parameter models with million-token contexts, reshaping large-scale AI economics.

Acknowledgement:
The authors gratefully acknowledge the enthusiastic collaboration and invaluable support from the entire Rebellions, and the Samsung Foundry members for their exceptional assistance and contribution to the successful completion of the advanced packaging.

References:
[1] Llama 3.3 70B, <https://www.llama.com/docs/model-cards-and-prompt-formats/llama3_3/>, <https://huggingface.co/meta-llama/Llama-3.3-70B-Instruct>, accessed: November 2025. https://www.llama.com/docs/model-cards-and-prompt-formats/llama3_3/
[2] Rebellions Inc., "REBEL-Quad, <https://rebellions.ai/peta-scale-soc-for-massive-ai-serving-rebel-quad/>, accessed: November 2025.
https://rebellions.ai/peta-scale-soc-for-massive-ai-serving-rebel-quad
[3] C. Yu et al., "ATOMUS: A 5nm 32TFLOPS/128TOPS ML System-on-Chip for Latency Critical Applications," *ISSCC*, pp. 42-43, 2024.
https://doi.org/10.1109/ISSCC49657.2024.10454509
[4] NVIDIA Corp., "NVIDIA H100". https://www.nvidia.com/en-us/data-center/h100/
[5] NVIDIA Corp., "NVIDIA H200". https://www.nvidia.com/en-us/data-center/h200/

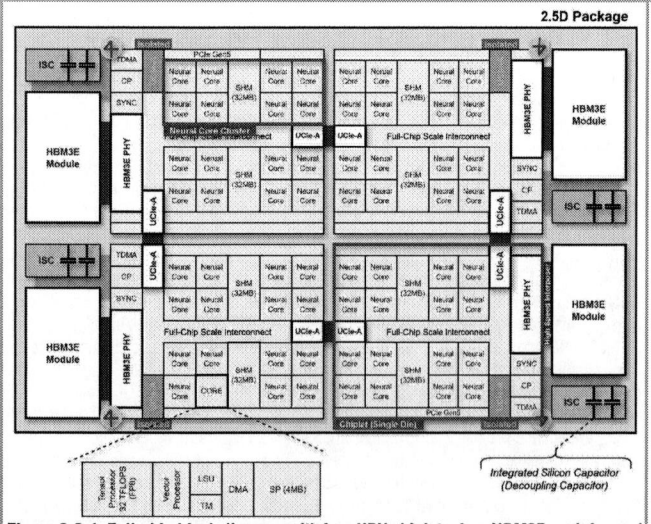

Figure 2.2.1: Full-chip block diagram with four NPU chiplets, four HBM3E modules and four ISCs.

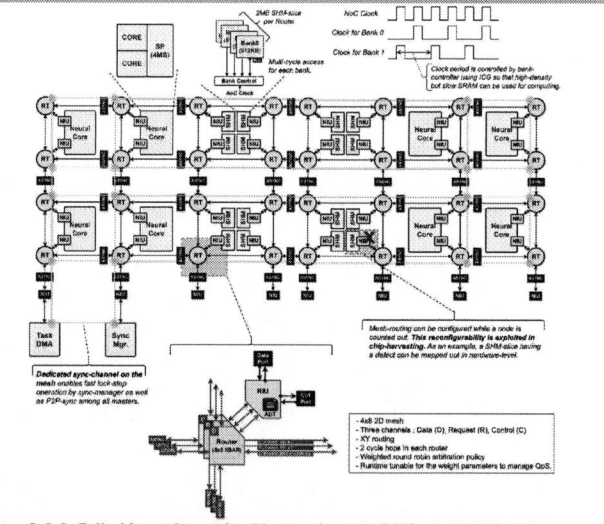

Figure 2.2.2: Full-chip scale mesh with neural cores, DMA, synchronization manager connected with three logically independent channels.

ISSCC 2026 / SESSION 2 / PROCESSORS / 2.2

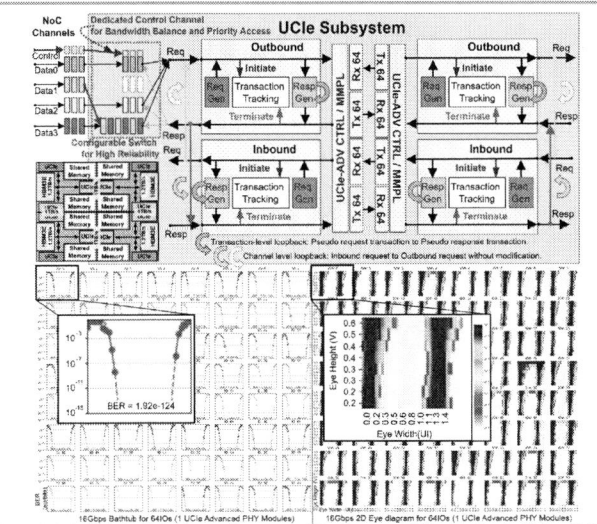

Figure 2.2.3: Chiplet interface subsystem with advanced UCIe die-to-die links. Bathtub and eye diagram snapshots at target speed during silicon bring-up.

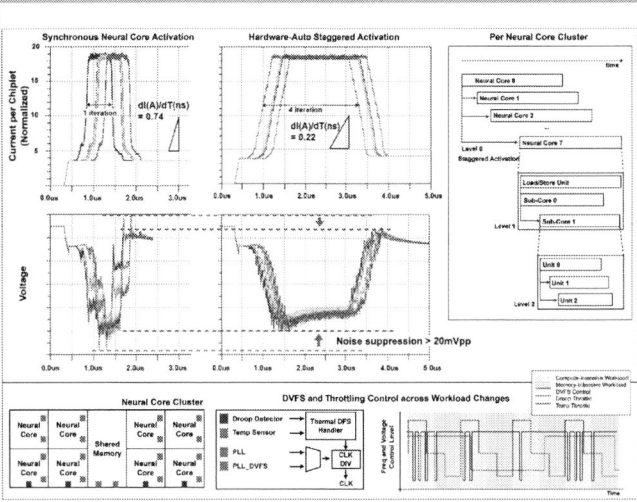

Figure 2.2.4: Hardware-based staggering technique and core-group closed-loop control for voltage droop and temperature, with per-cluster DVFS control loop.

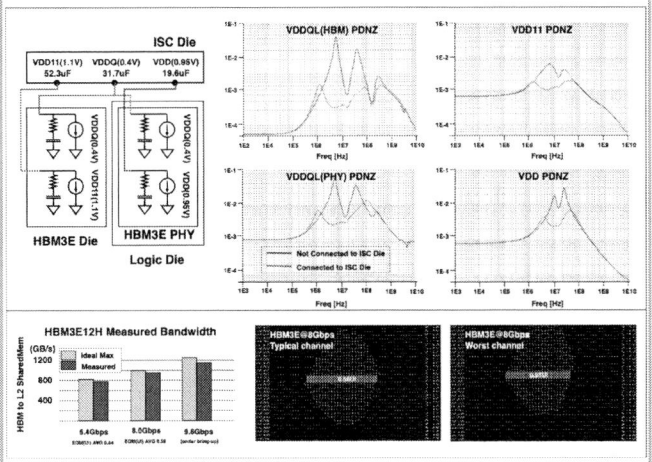

Figure 2.2.5: ISC for power integrity enhancement of HBM3E operation. Measured bandwidth performance and eye diagram; the eye diagram at the target speed (9.6Gbps) will be updated after silicon bring-up.

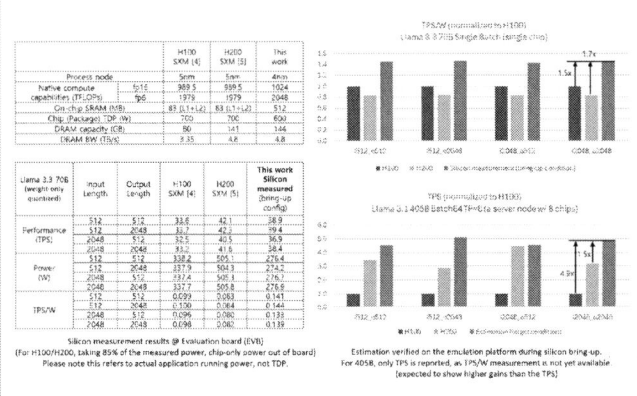

Figure 2.2.6: Performance (TPS) and power efficiency (TPS/W) comparison on silicon vs. flagship AI processors. Weight-only quantization applied due to DRAM limits on the 70B model.

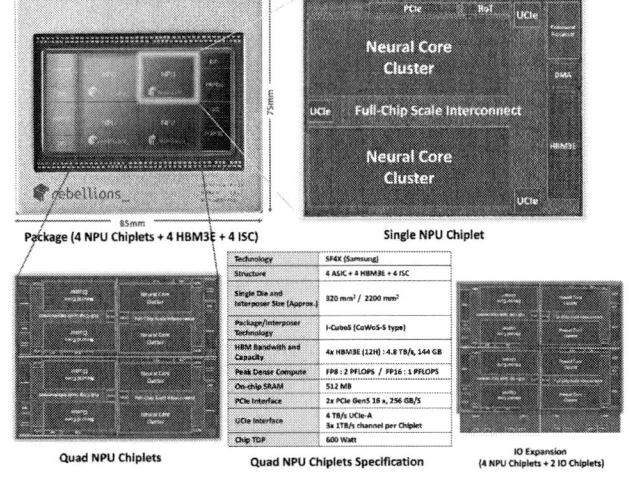

Figure 2.2.7: Package photo with four NPU chiplets, four HBM3E modules and four ISC dies. Various package configurations are being developed in parallel with the NPU chiplet.

• 2026 IEEE International Solid-State Circuits Conference

979-8-3315-8937-0/26 $31.00 © 2026 IEEE

ISSCC 2026 / SESSION 2 / PROCESSORS / 2.3

2.3 A 71.3mJ/Frame End-to-End Driving Processor with Flexible Heterogeneous Core Orchestration via Sparsity Reasoning

Jueun Jung*[1,2], Sangho Lee*[1], Junghyun Yoo[2], Ghangmin Yun[2], Kyuho Jason Lee[2]

[1]Ulsan National Institute of Science and Technology, Ulsan, Korea, [2]Yonsei University, Seoul, Korea
*Equally Credited Authors (ECAs)

Abstract

A multi-modal end-to-end driving processor is proposed with 4 features: 1) a sparsity reasoning unit to maximize sparsity exploitation, 2) a flexible sparse-dense heterogeneous architecture with a sparsity-aware adaptive core orchestrator to maximize core utilization, 3) an energy-efficient segmented aggregation network, and 4) a long-/short-term memory unit to minimize external memory access (EMA) of temporal attention. It achieves 10.3fps at 71.3mJ/frame, consuming 218× less energy than a state-of-the-art driving SoC.

Recently, multimodal sensor-fusion end-to-end driving (EED) models that integrate a CNN and transformer have demonstrated remarkable achievements. By unifying CNN-based feature extraction with transformer-based spatio-temporal correlation learning across tokens in the Bird's-Eye-View (BEV) space, these models enable breakthrough improvements in driving performance [1-4]. However, these EED models introduce three key challenges in hardware acceleration caused by fusing multi-sensor/modal/temporal information (Fig. 2.3.1). First, EED models suffer from drastic sparsity fluctuations across layers, where transformer layers fused with multi-modal data exhibit no structural patterns for strongly related tokens, causing sparsity vanishing and hindering the benefit of sparsity exploitation. Second, prior sparse-dense accelerators [5-7] showed that heterogeneous architectures can efficiently handle both sparse and dense tensors. However, their core configuration is fixed to a predetermined dense/sparse ratio (R_{MAC}), which limits peak utilization to a narrow sparsity range. Such enforced sparsity leads to significant accuracy loss in sensor-fusion models [8], where sparsity patterns vary across modalities. Third, temporal attention inspired by the human-memory structure – attending past-frame keys (K) and values (V) with current queries (Q) – is widely employed in EED models and sequential frame processing models (e.g., video) [4, 9-16], delivering significant accuracy improvements. However, the massive memory demand of long-/short-term data, along with the resulting attention score (S), probability (P), and weights (A), makes on-chip storage infeasible, thereby requiring off-chip access (e.g., >12 MB per stored temporal frame). This single layer incurs substantial external memory access (EMA), accounting for 34.1% of the total EMA.

The proposed processor addresses these challenges through the following 4 key features: 1) a sparsity reasoning unit (SRU) that predicts and generates sparsity by referencing past-future correlations to maximize sparsity exploitation in the current frame, 2) a flexible sparse-dense heterogeneous architecture with a sparsity-aware adaptive core orchestrator that maximizes core utilization, 3) a N-to-N segmented aggregation network (SAN) that enables dynamic multi-core aggregation and improves scalability, and 4) a long-/short-term memory unit (LSTMU) that preserves region-of-interest (ROI)-based critical tokens, while progressively discarding irrelevant temporal memories for efficient temporal attention.

Figure 2.3.2 illustrates the overall chip architecture of the proposed processor, which consists of two Reasoning Engine (RE) clusters, a global memory core, and a top controller. Each RE cluster comprises four dense-matrix cores (DMCs) with an 8×32 Processing Element (PE) array and eight sparse-matrix cores (SMCs) with a 1×32 PE array. These heterogeneous cores are organized into flexible processing groups by the core orchestrator for efficient acceleration under varying sparsity conditions. First, the Sparsity Reasoning Unit (SRU) reuses sparsity information from previous frames to estimate the workload of the current frame. From the predicted future frame's occupancy map of the EED model, the SRU derives an important mask that filters out irrelevant tokens, maximizing sparsity exploitation across the entire network. Next, the core orchestrator distributes computations to each core according to the predicted tile-level sparsity and flexibly aggregates the results through the SAN. Finally, the LSTMU in the global memory core progressively discards redundant memory entries in the large long-/short-term buffers for temporal attention, enabling efficient temporal reasoning.

Figure 2.3.3 illustrates the detailed operation of the past-future guided sparsity reasoning. To maximize the utilization of heterogeneous cores, the SRU performs both sparsity speculation and sparsity generation by exploiting inter-frame similarity between past frames and the estimated future frames. For workload allocation in the next layer (L+1) of the current frame (F_t), the SRU reuses sparsity information from the previous frame (F_{t-1}), while simultaneously performing sparsity speculation for the next frame (F_{t+1}) based on the statistics of the current-layer output. For each tile, the SRU only stores the channel-wise sparsity mode (1b index per channel) and the required R_{MAC}, thereby maintaining high inter-frame consistency, while reducing the memory demand for statistics by 73.9%. In addition, to address the sparsity vanishing issue in transformer layers, the SRU introduces object probability-based sparsity generation. A large portion of tokens in the BEV map carry low importance, while certain objects exhibit strong global correlations. Leveraging the object probability from the previous frame together with the predicted future occupancy maps, the SRU identifies probable object paths and generates an importance mask. As a result, the SRU eliminates computations for unmasked tokens without degrading driving score, reducing the computational load of transformer layers by 34.3%.

Figure 2.3.4 illustrates the operation and detailed architecture of the core orchestrator within the RE cluster. The sparse-dense heterogeneous architecture separates and processes input data into dense and sparse tensors. However, while DMCs exhibit fixed throughput due to unit-based processing grouped as a column for data reuse, the SMCs show highly variable throughput depending on sparsity owing to the extreme zero-skipping mechanism. Consequently, the optimal workload allocation to the sparse-dense core configuration varies depending on the polarity of sparsity, but conventional heterogeneous architectures with predetermined core organizations [5-7] cannot adapt to the variation, degrading hardware utilization. To overcome the constraint of such a predetermined core organization, the proposed core orchestrator flexibly forms processing groups of DMCs and SMCs. For effective data allocation with the physically limited core count, the core orchestrator coordinates the processing groups through dynamic tile reordering. First, using the tile-wise R_{MAC} predicted by the SRU, each tile is classified into either heterogeneous mode ($1 \le R_{MAC} < 32$) or single mode ($R_{MAC} < 1$ or $R_{MAC} > 32$) based on boundary criteria. Next, the tiles deemed heterogeneous are further sorted through min-max pairing, where "min tiles" require more SMCs while "max tiles" require more DMCs. The remaining cores are then filled with single-mode tiles, providing a compact packing of workloads. The rearranged tiles are processed through flexible processing groups, and their outputs are aggregated into a unified result. However, previous reconfigurable aggregation networks relied on large multiplexers to support aggregation among cores [17]. To further enhance the efficiency of core orchestration, the proposed SAN replaces these multiplexers with three 2-to-2 switches that support N-stage aggregations by a reconfigurable datapath, reducing power consumption and area by 89.9% and 83.2%, respectively, compared to the baseline [17]. Moreover, the linear complexity per stage improves scalability across multiple cores. Consequently, the core orchestrator reduces total processing time by 39.6% through flexible core allocation and reconfigurable aggregation that exploits tile-level sparsity characteristics.

Figure 2.3.5 illustrates the Temporal Attention Pipeline (TAP) and the details of the LSTMU. Temporal attention requires sequential accumulation of row-wise summations for softmax normalization and column-wise accumulation for P to track the importance (usage frequency) of past K-V pairs. However, usage frequency is computed only after the completion of the attention probability calculation, which necessitates storing massive amounts of intermediate data (S and P) on-chip, thereby causing significant EMA. The LSTMU eliminates EMA for intermediate data through a query and attention weight (QA) cache and a two-stage softmax. First, the LSTMU stores only the row-wise normalization factors of S. Then, by streaming K and V in block units, it sequentially computes P, the column-wise accumulated usage frequency, and A by removing EMA for both scores and probabilities. Furthermore, to reduce redundant attention operations on less relevant tokens in short-term memory, the LSTMU performs winner-take-all memory pruning. Since an object is repeatedly referenced over frames, its tokens are frequently present in recently accessed memory; temporal similarity of frequent tokens during feature matching (i.e., temporal attention) is associated with the same object with high probability. Therefore, after completing processing of F_t, the LSTMU generates an ROI mask using the object probability map, groups tokens into ROI/non-ROI categories, reorders the K-V pairs accordingly, and updates them in external short-term memory. When computing temporal attention for F_{t+1}, the usage frequency for block streamed K-V is input to the Usage-Adaptive Pruning Engine (UAPE). If the input belongs to an ROI Group or ID within the pruning boundary, it bypasses to the LTM/STM Update BUF. UAPE sorts the inputs with their ID outside the pruning boundary into a table and stores only top-k tokens to the Update BUF to preserve non-ROI tokens with high similarity. As a result, the LSTMU with TAP and progressive memory pruning reduces EMA for temporal attention by 92.8%.

Figure 2.3.6 presents the benchmark results of the proposed processor, evaluated on the Town05 Long benchmark [18] in the CARLA simulator [19], a state-of-the-art closed-loop driving platform. The benchmark reports the driving score, which combines a route completion and infraction score, along with a comparison table. Unlike previous accelerators [7, 17, 20-21] that utilize only partial models or sensors for specific stages of autonomous driving, this work processes the entire end-to-end driving pipeline, encompassing multi-sensor/modal/temporal information. The proposed processor achieves a driving score of 69.6 on the Town05 Long benchmark [18]. Compared to a GPU implementation [22], it delivers comparable throughput, while consuming 544× less power. Relative to the state-of-the-art autonomous driving SoC [23], it demonstrates an average of 2.67× higher

46 • 2026 IEEE International Solid-State Circuits Conference

979-8-3315-8937-0/26 $31.00 © 2026 IEEE

processing speed, achieving real-time performance for multi-sensor fusion (>10fps), while reducing energy per frame by 218×. As shown in Fig. 2.3.7, the proposed chip is fabricated in 28nm CMOS technology and occupies 10.5mm². In conclusion, the proposed processor successfully realizes real-time end-to-end driving with an energy consumption of 71.3mJ/frame.

References:
[1] H. Shao et al., "Safety-Enhanced Autonomous Driving Using Interpretable Sensor Fusion Transformer," *Proc. of Machine Learning Research*, vol. 205, pp. 726–737, 2023. https://doi.org/10.48550/arXiv.2207.14024
[2] K. Chitta et al., "TransFuser: Imitation With Transformer-Based Sensor Fusion for Autonomous Driving," *IEEE Trans. on Pattern Analysis and Machine Intelligence*, vol. 45, no. 11, pp. 12878–12895, 2023. https://doi.org/10.1109/TPAMI.2022.3200245
[3] H. Shao et al., "LMDrive: Closed-Loop End-to-End Driving with Large Language Models," *CVPR*, pp. 15120–15130, 2024. https://doi.org/10.1109/CVPR52733.2024.01432
[4] H. Shao et al., "ReasonNet: End-to-End Driving with Temporal and Global Reasoning," *CVPR*, pp. 13723–13733, 2023. https://doi.org/10.1109/CVPR52729.2023.01319
[5] D. Han and A. P. Chandrakasan, "MEGA.mini: A Universal Generative AI Processor with a New Big/Little Core Architecture for NPU," *ISSCC*, pp. 416-418, 2025. https://doi.org/10.1109/ISSCC49661.2025.10904514
[6] D. Han et al., "MetaVRain: A 133mW Real-Time Hyper-Realistic 3D-NeRF Processor with 1D-2D Hybrid-Neural Engines for Metaverse on Mobile Devices," *ISSCC*, pp. 50–52, 2023. https://doi.org/10.1109/ISSCC42615.2023.10067447
[7] S. Moon et al., "T-REX: A 68-to-567μs/Token 0.41-to-3.95μJ/Token Transformer Accelerator with Reduced External Memory Access and Enhanced Hardware Utilization in 16nm FinFET," *ISSCC*, pp. 406–408, 2025. https://doi.org/10.1109/ISSCC49661.2025.10904793
[8] S. Sun et al., "AlterMOMA: Fusion Redundancy Pruning for Camera-LiDAR Fusion Models with Alternative Modality Masking," *NeurIPS*, vol. 37, pp. 46653–46679, 2024. https://doi.org/10.48550/arXiv.2409.17728
[9] Y. Hu et al., "Planning-oriented Autonomous Driving," *CVPR*, pp. 17853–17862, 2023. https://doi.org/10.1109/CVPR52729.2023.01712
[10] B. Jiang et al., "VAD: Vectorized Scene Representation for Efficient Autonomous Driving," *ICCV*, pp. 8306–8316, 2023. https://doi.org/10.1109/ICCV51070.2023.00766
[11] Z. Li et al., "BEVFormer: Learning Bird's-Eye-View Representation From LiDAR-Camera via Spatiotemporal Transformers," *IEEE Trans. on Pattern Analysis and Machine Intelligence*, vol. 47, no. 3, pp. 2020–2036, 2025. https://doi.org/10.1109/TPAMI.2024.3515454
[12] H. Cheng et al., "XMem: Long-Term Video Object Segmentation with an Atkinson-Shiffrin Memory Model," *ECCV*, pp. 640–658, 2022. https://doi.org/10.1007/978-3-031-19815-1_37
[13] M. Bekuzarov et al., "XMem++: Production-level Video Segmentation From Few Annotated Frames," *ICCV*, pp. 635–644, 2023. https://doi.org/10.1109/ICCV51070.2023.00065
[14] S. Wang et al., "Exploring Object-Centric Temporal Modeling for Efficient Multi-View 3D Object Detection," *ICCV*, pp. 3598–3608, 2023. https://doi.org/10.1109/ICCV51070.2023.00335
[15] H. Cheng et al., "Putting the Object Back into Video Object Segmentation," *CVPR*, pp. 3151–3161, 2024. https://doi.org/10.1109/CVPR52733.2024.00304

[16] E. Song et al., "MovieChat: From Dense Token to Sparse Memory for Long Video Understanding," *CVPR*, pp. 18221–18232, 2024. https://doi.org/10.1109/CVPR52733.2024.01725
[17] J. Jung et al., "LSPU: A Fully Integrated Real-Time LiDAR-SLAM SoC with Point-Neural-Network Segmentation and Multi-Level kNN Acceleration," *ISSCC*, pp. 370–372, 2024. https://doi.org/10.1109/ISSCC49657.2024.10454374
[18] A. Prakash et al., "Multi-Modal Fusion Transformer for End-to-End Autonomous Driving," *CVPR*, pp. 7073–7083, 2021. https://doi.org/10.1109/CVPR46437.2021.00700
[19] A. Dosovitskiy et al., "CARLA: An Open Urban Driving Simulator," *CoRL*, vol. 78, 2017. https://doi.org/10.48550/arXiv.1711.03938
[20] P. Dong et al., "A 28nm 0.22μJ/Token Memory-Compute-Intensity-Aware CNN-Transformer Accelerator with Hybrid-Attention-Based Layer-Fusion and Cascaded Pruning for Semantic-Segmentation," *ISSCC*, pp. 408–410, 2025. https://doi.org/10.1109/ISSCC49661.2025.10904499
[21] X. Feng et al., "A Scalable BEV Perception Processor for Image/Point Cloud Fusion Applications Using CAM-Based Universal Mapping Unit," *IEEE JSSC*, vol. 60, no. 3, pp. 1002–1013, 2025. https://doi.org/10.1109/JSSC.2024.3514733
[22] NVIDIA A100 Tensor Core GPU, https://www.nvidia.com/en-us/data-center/a100, Accessed: Sep. 2025. https://www.nvidia.com/en-us/data-center/a100
[23] NVIDIA Jetson Orin, https://www.nvidia.com/en-us/autonomous-machines/embedded-systems/jetson-orin, Accessed: Sep. 2025. https://www.nvidia.com/en-us/autonomous-machines/embedded-systems/jetson-orin

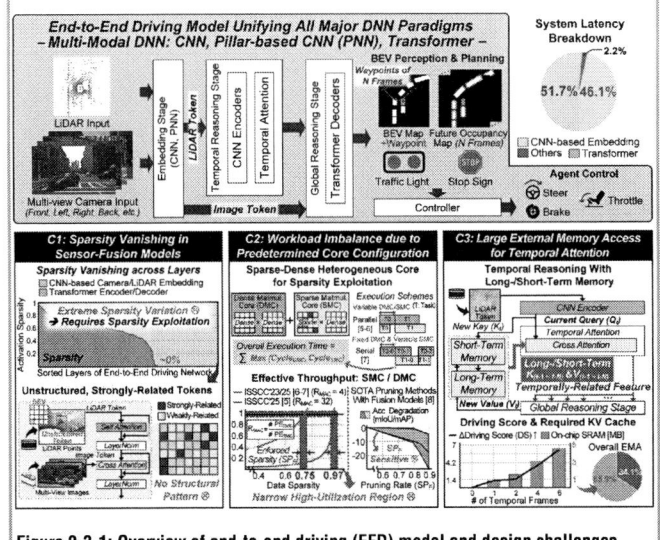

Figure 2.3.1: Overview of end-to-end driving (EED) model and design challenges.

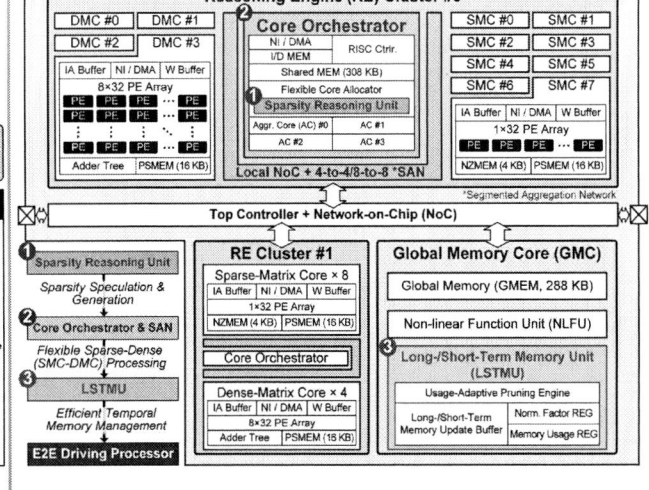

Figure 2.3.2: Overall chip architecture.

ISSCC 2026 / SESSION 2 / PROCESSORS / 2.3

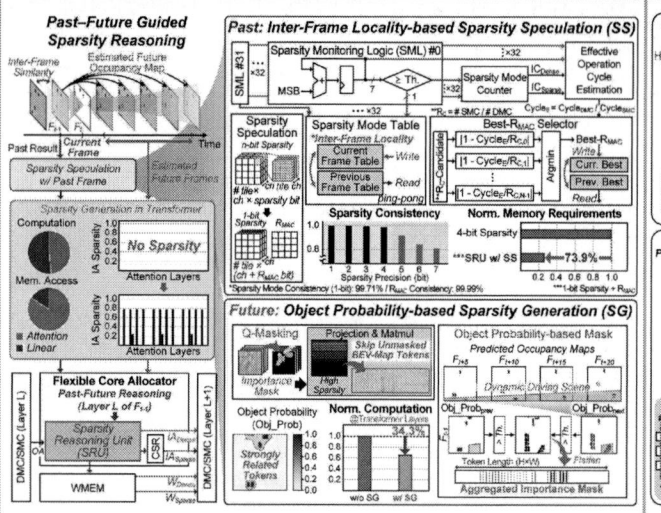

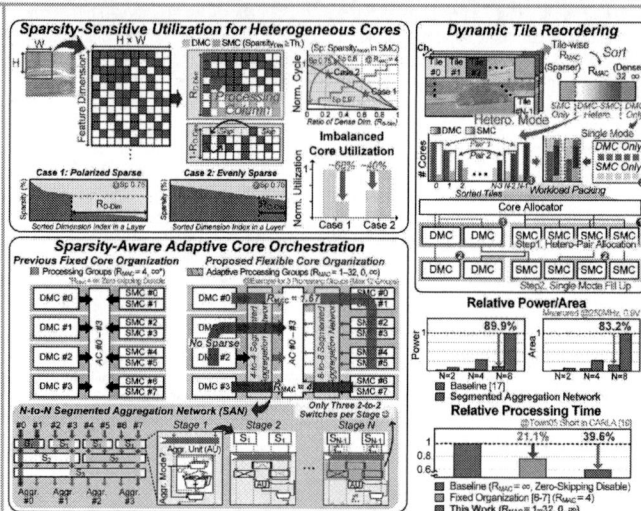

Figure 2.3.3: Details of operations and architecture of the sparsity reasoning unit (SRU) with past-future guided sparsity speculation and generation.

Figure 2.3.4: Core orchestrator and segmented aggregation network with sparsity-aware adaptive core orchestration and dynamic tile reordering.

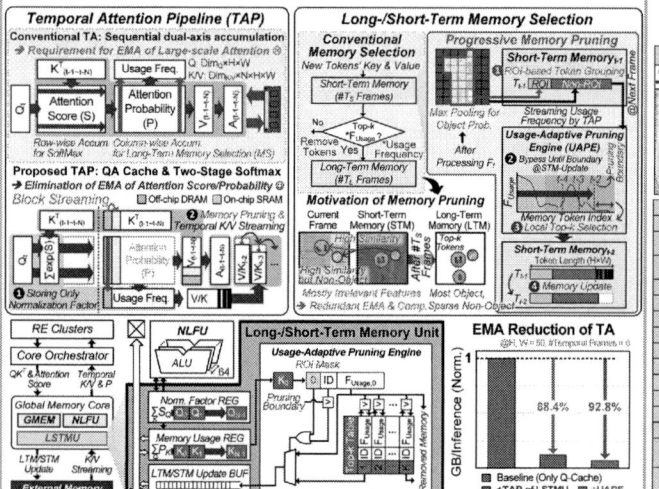

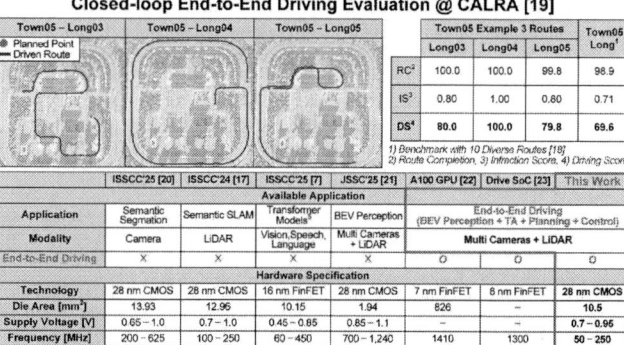

Figure 2.3.5: Long-/short-term memory unit (LSTMU) with temporal attention pipeline and progressive memory pruning.

Figure 2.3.6: Measurement results and performance comparison table.

Closed-loop End-to-End Driving Evaluation @ CALRA [19]

	Town05 Example 3 Routes			Town05 Long[1]
	Long03	Long04	Long05	
RC[2]	100.0	100.0	99.8	96.9
IS[3]	0.80	1.00	0.80	0.71
DS[4]	80.0	100.0	79.8	69.6

1) Benchmark with 10 Diverse Routes [18]
2) Route Completion, 3) Infraction Score, 4) Driving Score

	ISSCC'25 [20]	ISSCC'24 [17]	ISSCC'25 [7]	JSSC'25 [21]	A100 GPU [22]	Drive SoC [23]	This Work
Available Application							
Application	Semantic Segmentation	Semantic SLAM	Transformer Models[5]	BEV Perception	End-to-End Driving (BEV Perception + TA + Planning + Control)		
Modality	Camera	LiDAR	Vision,Speech, Language	Multi Cameras + LiDAR	Multi Cameras + LiDAR		
End-to-End Driving	X	X	X	X	O	O	O
Hardware Specification							
Technology	28 nm CMOS	28 nm CMOS	16 nm FinFET	28 nm CMOS	7 nm FinFET	8 nm FinFET	28 nm CMOS
Die Area [mm²]	13.93	12.96	10.15	1.94	826	–	10.5
Supply Voltage [V]	0.65 – 1.0	0.7 – 1.0	0.45 – 0.85	0.85 – 1.1	–	–	0.7 – 0.95
Frequency [MHz]	200 – 625	100 – 250	60 – 450	700 – 1,240	1410	1300	50 – 250
On-chip SRAM [KB]	3,280	914	1,320	160	–	–	1,352
Data Type	INT8	INT16/INT11	INT4/8/16	INT8	INT4 – FP64	INT8 – FP32	INT16
SoC Power [mW]	160 – 970	349.6	7.12 – 152.5	68 – 264	400,000	60,000	88.3 – 689.1
End-to-End Driving Performance							
Frame Rate [frame/second]	–	–	–	2.78[6]	14.01	3.86	10.32[7,8]
Frame Energy [mJ/frame]	–	–	–	95.0[6]	28,546	15,539	71.3[7,8]

5) Image/Text Classification, Translation, Speech-to-Text, 6) Only BEV Perception Stage, 7) EMA is included, 8) @250MHz, 0.95V

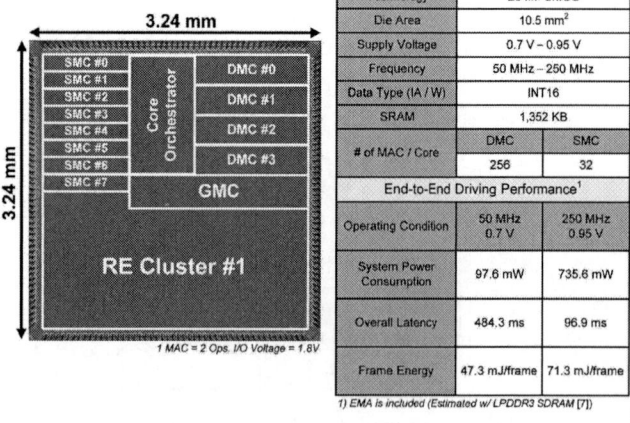

Technology	28 nm CMOS	
Die Area	10.5 mm²	
Supply Voltage	0.7 V – 0.95 V	
Frequency	50 MHz – 250 MHz	
Data Type (IA / W)	INT16	
SRAM	1,352 KB	
	DMC	SMC
# of MAC / Core	256	32
End-to-End Driving Performance[1]		
Operating Condition	50 MHz 0.7 V	250 MHz 0.95 V
System Power Consumption	97.6 mW	735.6 mW
Overall Latency	484.3 ms	96.9 ms
Frame Energy	47.3 mJ/frame	71.3 mJ/frame

1 MAC = 2 Ops, I/O Voltage = 1.8V
1) EMA is included (Estimated w/ LPDDR3 SDRAM [7])

Figure 2.3.7: Chip micrograph and performance summary.

ISSCC 2026 / SESSION 2 / PROCESSORS / 2.4

2.4 UniC-Vision: A 14.4Gb/s 7.3pJ/b Universal Vision Transformer OFDM Channel Estimation Accelerator for B5G/6G AI-RAN

Sangbu Yun[1,2], Chanhee Lee[1,2], Soonhyun Kwon[1], Jeongtaek Chang[3], Zhengya Zhang[3], Youngjoo Lee[1]

[1]Korea Advanced Institute of Science and Technology, Daejeon, Korea, [2]Pohang University of Science and Technology, Pohang, Korea, [3]University of Michigan, Ann Arbor, MI

Abstract

This work presents an AI-RAN channel estimation accelerator for next-generation communications, offering hyper reliability, low latency, and universal frequency range coverage, thereby meeting B5G/6G requirements. Utilizing algorithm-hardware co-optimization and architecture optimizations, the prototype achieves a throughput of 14.4Gb/s at 7.3pJ/b in 28nm CMOS, surpassing the state-of-the-art channel estimation systems by 7.7 to 18.9dB in reliability and 16-to-39× in throughput.

Future wireless systems beyond 5G and 6G (B5G/6G) aim to meet the stringent requirements of hyper-reliable low-latency communication (HRLLC), imposing unprecedented constraints on wireless physical layer (PHY) design. 6G involves a broad frequency range (FR), including sub-6GHz (FR1), mm-Wave (FR2), and cm-Wave (FR3), along with ultra-massive MIMO and wide bandwidth, which introduces new physical phenomena and severe workload [1]. Under such reliability and latency-hungry conditions, however, conventional channel estimation techniques offer insufficient data rate and limited quality of essential channel information for subsequent computations, which prevents the system from meeting HRLLC requirements. An artificial intelligence radio access network (AI-RAN) has been recognized as a promising solution for the B5G/6G paradigm, capable of overcoming the inherent limitations of traditional algebra-based solutions [2]. For channel estimation, a deep-learning approach using a Vision Transformer (ViT) has been proposed to bridge this gap [3]. As shown in Fig. 2.4.1, the mean squared error (MSE) performance indicates that the AI-RAN solution provides more accurate channel state information (CSI) with improved robustness to channel variation. However, its substantial computational and memory footprint imposes critical challenges for real-world applications. To this end, we present UniC-Vision: a ViT channel estimation accelerator for future AI-RAN systems, offering hyper reliable and real-time performance with universal FR coverage.

As shown in Fig. 2.4.1, our work not only achieves superior MSE performance compared to traditional estimators and the baseline model [3], but also meets the tight latency demands. The first key requirement of HRLLC is ≤0.1ms end-to-end (E2E) PHY processing and ≤15μs for channel estimation [1], a goal hindered by external memory access (EMA), which constitutes a significant portion of latency. To mitigate this, we first introduce channel-aware compression (CAC), a wireless channel-oriented optimization that reduces weight parameters by 95%. This compression enables the entire pre-trained model to be loaded onto on-chip SRAM, thus eliminating EMA. Furthermore, we present relaxed self-attention (RSA), which simplifies computation by eliminating external processor access (EPA). EPA is required for nonlinear operations, which are costly to implement on-chip and cause additional latency from off-chip communication. By mitigating this overhead, RSA achieves an 83% latency reduction. For an efficient universal estimator for B5G/6G's broad spectrum, we employ mode switching, utilizing different patch embedding dimensions for FR1 and FR2/3 to optimize performance for each spectrum. Furthermore, a quad-patch architecture is designed to maximize the resource efficiency of multi-mode design, boosting throughput by 5× over a naïve realization.

Figure 2.4.2 shows the overall architecture of the proposed UniC-Vision. The system comprises a processing element (PE) cluster, an aggregation unit, a buffer cluster, a global parameter memory (GPM), and a top controller. The PE cluster consists of 40 reconfigurable SIMD lanes. Each PE is equipped with five fixed-point multipliers and a 5-to-1 adder tree (AT). A flexible 8-way adder (Flex-8A) reduces SIMD results from each lane. The variable fractional word lengths (FWL) of Flex-8A are managed by a mode switching agent (MSA). The aggregation unit supports several operator types to accomplish ViT inference. To enable E2E layer fusion, the buffer cluster utilizes purpose-built buffers for the demodulation reference signal (DMRS), CSI tokens, and key-value (K-V) pairs. This design adds only 720B of overhead. The GPM module contains 31KB of weight and 3KB of bias memory. It also integrates a parameter casting unit (PCU) to reshape trained parameters for each instruction. The PCU includes a tokenizer to generate SIMD CSI tokens on the fly, a broadcast (Bcast) unit to feed weight-stationary parameters to a partitioned PE cluster for quad-patch processing, and a stationary controller to access SRAM banks in various stationary patterns. A top controller with a RISC-based domain-specific processor (DSP), instruction memory (IMEM), a decoder, and AXI-stream/SPI interfaces controls the system.

Figure 2.4.3 depicts compression methods in CAC and the approximation techniques used in RSA, resulting in E2E layer fusion and zero external access throughout the entire process. The proposed CAC first leverages the independent and identically distributed (i.i.d.) nature of in-phase (I) and quadrature (Q) channels, reusing parameters across I and Q components by assuming activation symmetry, resulting in a non-lossy compression. Second, replacing least-square (LS) and bilinear interpolation with [4] improves estimation quality but incurs an extra 70% memory cost. The OFDM channel grid has two major axes: time and frequency, which are affected by different sources of noise, the Doppler effect, and selective fading, respectively. Hence, the interpolation fully connected (FC) layer is decomposed into two orthogonal axes, saving 93% of the extra memory. Third, by removing nonlinear dependencies and redundant biases using RSA, particular weight tensors can be merged offline, further reducing memory usage by 11%. Lastly, by analyzing the trade-off between the hidden dimensions and estimation accuracy, measured in MSE loss, another 41% of the parameters are reduced. Moreover, the dimension calibration further eliminates computational and memory overhead of input and output FC layers of the ViT encoder block, and also eliminates inverse patch embedding layer which forms an output activation into a result channel grid.

Another solution for removing external access is RSA, which consists of three approximation schemes. RSA first exploits structural patterns in the attention map of wireless channel matrices to eliminate resource-intensive max-tree operations: instead of calculating maxima to relax overflow [5], RSA subtracts predetermined pseudo-max values identified through offline head-wise Monte Carlo analysis. The pseudo-max uses only 5% of the area and 3% of the power of the max-tree and reduces instruction counts in the multi-head-attention (MHA) part by 17%. Second, the softmax operator is simplified via a first-order Taylor expansion and an integral upper bound, eliminating internal nonlinear look-up tables and dividers, resulting in a further 73% latency reduction. Lastly, compute-intensive layer normalization and GeLU activation are replaced by batch normalization and ReLU activation based on [6] and [7], leveraging the fixed token length and a simple model structure thanks to CAC and RSA. Consequently, CAC results in a 95% storage reduction, and RSA leads to 83% of the MHA latency as depicted in Fig. 2.4.3, even when external processor latency is omitted. Note that the optimizations result in at most 5% MSE loss from the baseline.

Figure 2.4.4 shows the mode switching and quad-patch processing mechanism of the proposed system, which depends on the target FR. The full-patch mode features the maximum patch embedding size (40), capturing finer-grained activations and global patterns with full-sized tokens. Hence, the full-patch mode enjoys the full expressive power of the neural network, offering superior MSE performance in complex profiles. In the FR2/3 bands, the predominance of line-of-sight components results in increased sparsity and reduced complex patterns [8]. Therefore, a simpler quad-patch mode is employed thanks to reduced nonlinearity and increased local patterns. The quad-patch mode reduces the number of parameters from 25.4K to 9.4K with a half patch size (20) and corresponding hidden dimensions. As in Fig. 2.4.4, the patch size for each mode has been examined through trade-off analysis. The datapath of the Flex-8A has been designed to process four patches per clock cycle, maximizing utilization. In detail, the halved patch first produces doubled patch embeddings, then four embeddings are fed into the PE cluster, resulting in halved processing latency to estimate equally sized physical resource blocks (PRBs). Corresponding weight and bias parameters for each patch are managed by the PCU within the GPM module. Furthermore, the critical path delay in the datapath can be shortened, as a 40-to-1 adder tree can be reconfigured into two 20-to-1 adder trees. The path delay can be reduced by 31% since the adders in the latter part of the effective tree utilize a large bitwidth to avoid overflow and underflow. As a result, quad-patch mode throughput can be increased 2.5× from full-patch mode, while saving 29.7% dynamic power consumption from inactivated SRAM regions as well.

Figure 2.4.5 shows the measurement results of the prototype UniC-Vision channel estimation accelerator. In terms of MSE with true channel coefficients, even with various cost optimizations, this work successfully minimizes the loss to 5% of the baseline single-precision ViT model [3]. The prototype operates with a 12b fixed-point 2's complement number system, showing negligible degradation compared to the FP32 equivalent, by applying iterative quantization-aware training and fine-tuning to achieve a loss of 1E-4 vs. the FP32 model. The universality of the proposed system is verified via MSE evaluation across a wide frequency range and successfully adapts to varying channel profiles without retraining for each FR. An efficiency comparison with state-of-the-art (SOTA) wireless systems has been made [9-13], and this work offers up to 5.4× better energy efficiency than the latest channel estimation fabric [9]. Another comparison with the SOTA channel estimator [4,14], including FPGA designs, shows that the chip surpasses others in both reliability and throughput. Note that pilot overhead becomes more essential for bandwidth

48 • 2026 IEEE International Solid-State Circuits Conference

ISSCC 2026 / February 16, 2026 / 2:45 PM

utilization in B5G/6G, where the prototype incurs only 4.76% pilot overhead, whereas [9] uses a full OFDM symbol as a pilot vector (100%). Considering the pilot efficiency, it outperforms [9] by 16-40× and 9-30× in throughput and energy efficiency, respectively. Figure 2.4.6 shows system settings for the evaluations, link-level evaluation by integrating other baseband signal processing, and a comparison table with the SOTA fabrics. Link-level simulation reveals ≤1.7 dB loss from a system with perfect CSI and ≥3.6 dB gain from a conventional LS estimator, confirming the potential of this work for future HRLLC systems. Fabricated in 28nm CMOS, the chip operates at up to 250MHz with a 1.0 V supply, dissipating 104.6mW, delivering a steady throughput of 14.4Gb/s, and achieving an energy efficiency of 7.3pJ/b, respectively. Figure 2.4.7 shows the chip micrograph and summary. This work offers 0.28-to-0.7μs E2E latency per PRB with up to 24dB estimation gain, and universal spectrum coverage, paving a roadmap for next-generation AI-RAN communications.

Acknowledgement:
This work was supported by the NRF (No. 2022R1A2C2092521, RS-2025-02264052) and IITP (RS-2024-00398449) grants funded by the Korea government (MSIT). The chip fabrication and EDA tools were supported by IDEC, Korea.

References:
[1] B. Ji et al., "Several Key Technologies for 6G: Challenges and Opportunities," *IEEE Comm. Magazine*, vol. 5, no. 2, pp. 44-51, 2021. https://doi.org/10.1109/MCOMSTD.001.2000038
[2] N. A. Khan and S. Schmid, "AI-RAN in 6G Networks: State-of-the-Art and Challenges," *IEEE Open Journal of the Communications Society*, vol. 5, pp. 294-311, 2024. https://doi.org/10.1109/OJCOMS.2023.3343069
[3] F. Liu et al., "CE-ViT: A Robust Channel Estimator Based on Vision Transformer for OFDM Systems," *IEEE GLOBECOM*, pp. 4798-4803, 2023. https://doi.org/10.1109/GLOBECOM54140.2023.10436847
[4] A. Sharma et al., "Low Complexity Deep Learning Augmented Wireless Channel Estimation for Pilot-Based OFDM on Zynq System on Chip," in *IEEE TCAS-I*, vol. 71, no. 5, pp. 2334-2347, 2024. https://doi.org/10.1109/TCSI.2024.3371780
[5] S. Kim et al., "I-BERT: Integer-only BERT Quantization," arXiv preprint arXiv:2101.01321, 2021. https://arxiv.org/abs/2101.01321
[6] Z. Yao, Y. Cao, Y. Lin, Z. Liu, Z. Zhang and H. Hu, "Leveraging Batch Normalization for Vision Transformers," *ICCV Workshops*, pp. 413-422, 2021. https://doi.org/10.1109/ICCVW54120.2021.00050
[7] I. Mirzadeh et al., "ReLU Strikes Back: Exploiting Activation Sparsityl in Large Language Models," arXiv preprint arXiv:2310.04564, 2023. https://doi.org/10.48550/arXiv.2310.04564
[8] Q. Xue et al., "A Survey of Beam Management for mmWave and THz Communications Towards 6G," *IEEE Communications Surveys & Tutorials*, vol. 26, no. 3, pp. 1520-1559, 2024. https://doi.org/10.1109/COMST.2024.3361991
[9] O. Castañeda et al., "A Resolution-Adaptive 8 mm² 9.98 Gb/s 39.7 pJ/b 32-Antenna All-Digital Spatial Equalizer for mmWave Massive MU-MIMO in 65nm CMOS," *ESSCIRC*, pp. 247-250, 2021. https://doi.org/10.1109/ESSCIRC53450.2021.9567843
[10] P. -J. Chen et al., "A 142Mw 6.4Gbps Massive MU-MIMO RSMA Detector for Next-Generation Communication Systems," *IEEE Symp. VLSI Circuits*, 2025. https://doi.org/10.23919/VLSITechnologyandCir65189.2025.11075000

[11] T. Lee et al., "A 131mW 6.4Gbps 256×32 Multi-User MIMO OTFS Detector for Next-Gen Communication Systems," *ISSCC*, pp. 46-48, 2024. https://doi.org/10.1109/ISSCC49657.2024.10454410
[12] Y. Zhang et al., "BayesBB: A 9.6Gbps 1.61ms Configurable All-Message-Passing Baseband-Accelerator for B5G/6G Cell-Free Massive-MIMO in 40nm CMOS," *ISSCC*, pp. 48-50, 2024. https://doi.org/10.1109/ISSCC49657.2024.10454287
[13] W. Tang et al., "A 1.8Gb/s 70.6pJ/b 128×16 link-adaptive near-optimal massive MIMO detector in 28nm UTBB-FDSOI," *ISSCC*, pp. 224-226, 2018. https://doi.org/10.1109/ISSCC.2018.8310265
[14] S. A. Ul Haq, S. J. Darak and A. K. Gizzini, "Low Complexity Deep Learning Aided Channel Estimation Architecture for Vehicular Networks," *IEEE ISCAS*, 2024. https://doi.org/10.1109/ISCAS58744.2024.10557931
[15] 3GPP, "Study on Channel Model for Frequencies From 0.5 to 100 GHz," TR 38.901, ver. 18.1.0, Rel-18, 2023, <https://www.etsi.org/deliver/etsi_tr/138900_138999/138901/18.00.00_60/tr_138901v180000p.pdf>.
[16] J. Hoydis et al., "Sionna: An Open-Source Library for Next-Generation Physical Layer Research," arXiv preprint arXiv:2203.11854, 2023. https://arxiv.org/abs/2203.11854

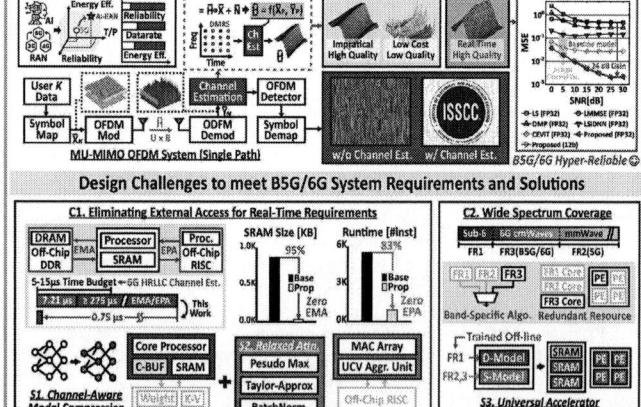

Figure 2.4.1: AI-RAN OFDM channel estimation accelerator solutions to meet B5G/6G system requirements.

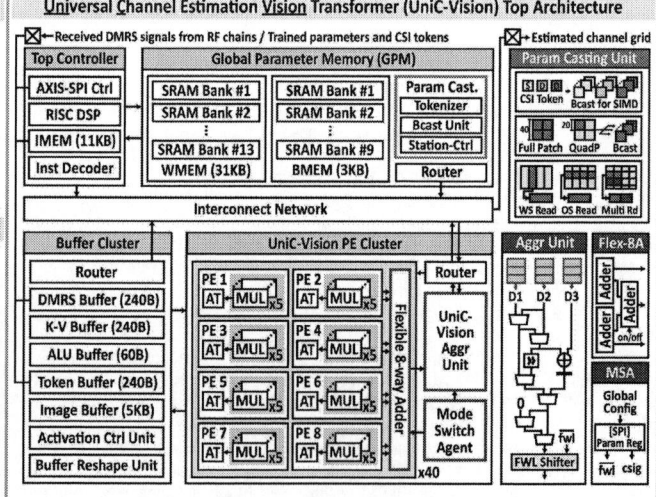

Figure 2.4.2: System architecture of the proposed UniC-Vision OFDM channel estimation accelerator.

ISSCC 2026 / SESSION 2 / PROCESSORS / 2.4

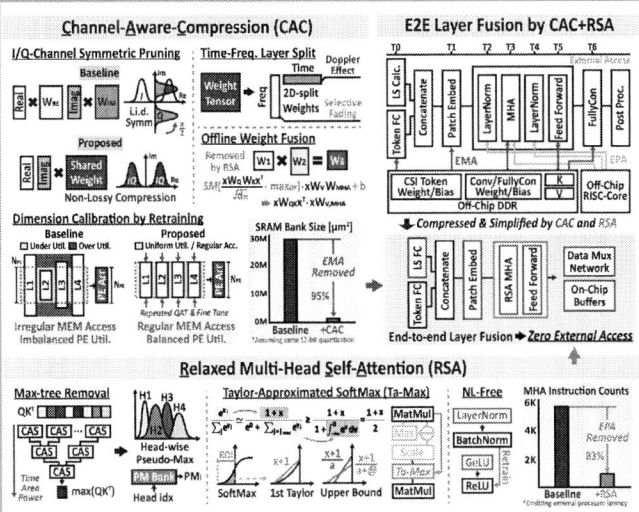

Figure 2.4.3: CAC and RSA algorithm-hardware co-optimization eliminating external access for real-time processing.

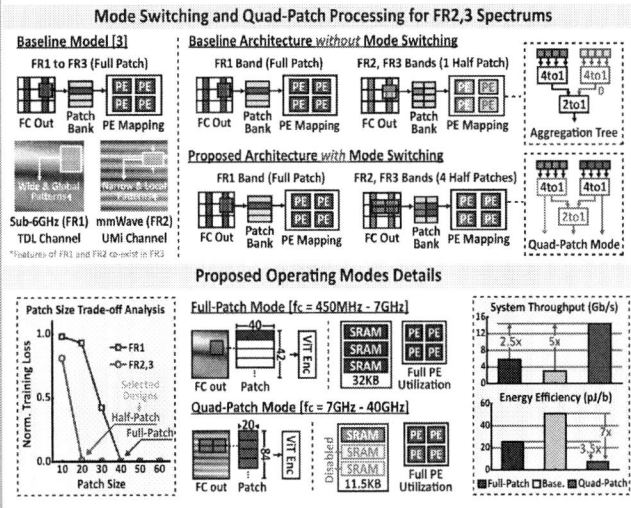

Figure 2.4.4: Mode switching to support FR2 and FR3 bands with the maximum utilization of the on-chip PE.

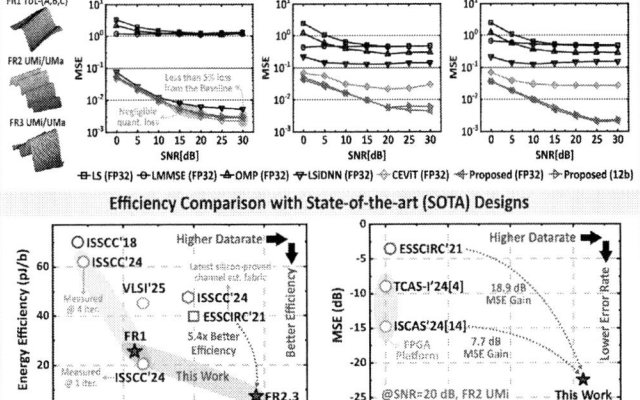

Figure 2.4.5: Channel estimation performance evaluation and efficiency analysis with SOTA wireless system fabrics, and channel estimation circuits.

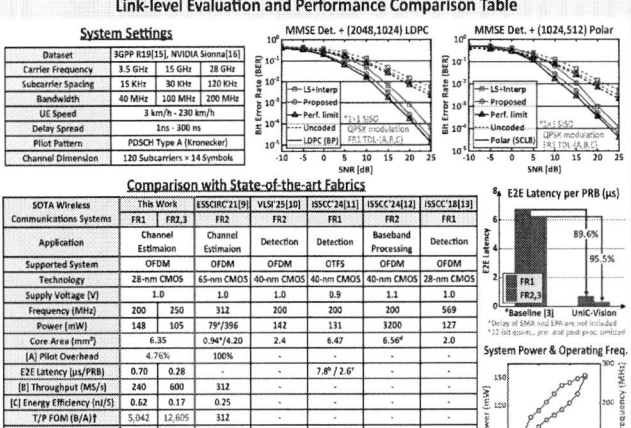

Figure 2.4.6: Link-level performance evaluation and comparison with silicon-proven SOTA wireless communications system.

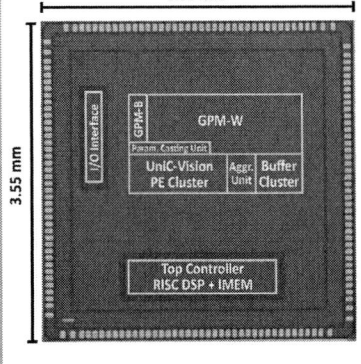

Technology	28-nm CMOS	
Application	Channel Estimation	
Supply Voltage	0.65 - 1.1 V	
Max. Frequency	50 - 275 MHz	
Die Area	3.55 mm × 3.55 mm	
On-chip Memory	51 KB	
Data Precision	Fixed point 12-bit	
Pilot Overhead	4.76%	
System Performance[a]		
Frequency Range	FR1	FR2,3
Channel MSE Gain[b]	24.6 dB	18.9 dB
Latency per RB	0.70 μs	0.28 μs
Power	148 mW	105 mW
Throughput	5.8 Gb/s	14.4 Gb/s
Energy Efficiency	25.5 pJ/b	7.3 pJ/b

[a]Measured at 1.0 V supply voltage
[b]Channel estimation MSE gain over least square with 2D-bilinear interpolation.

Figure 2.4.7: Chip micrograph and summary.

• 2026 IEEE International Solid-State Circuits Conference

ISSCC 2026 / SESSION 2 / PROCESSORS / 2.5

2.5 A 1.1mm², 14.4ns, 13.1pJ/b Forward Error Correction with Ordered-Statistics Post Processing for Ultra-Reliable and Low-Latency Communications

Cheng-Hsun Lu[1], Wei Tang[1], Jaehee Kim[2], Youngjoo Lee[3], Zhengya Zhang[1]

[1]University of Michigan, Ann Arbor, MI, [2]Pohang University of Science and Technology, Pohang, Korea, [3]Korea Advanced Institute of Science and Technology, Daejeon, Korea

Abstract

We present a two-stage FEC decoder that combines soft-decision Chase decoding with order-1 ordered statistics decoding, offering robust error correction for URLLC. A 1.1mm², 16nm decoder prototype with optimized Gaussian elimination and early exits achieves an average latency of 14.4ns and a worst-case latency of 132.6ns, improving upon prior work by 53× and five orders of magnitude, respectively. The prototype delivers 15.2Gbps at 13.1pJ/b, outperforming state-of-the-art URLLC FEC decoders.

Ultra-reliable and low-latency communication (URLLC) [1] is a key enabler for 6G wireless networks, targeting sub-millisecond latencies and high reliability, with block error rate (BLER) below 10⁻⁵ for applications from autonomous systems to extended reality. Forward error correction (FEC) is essential in URLLC, as it ensures reliable and rapid data transmission over imperfect wireless channels without relying on lengthy retransmissions. In designing FEC codes, long codes can provide stronger error protection, but their long decoding latencies are incompatible with URLLC. Consequently, short codes (≤128b) are preferred, demanding powerful decoders to maintain high reliability [2]. Near-maximum-likelihood (near-ML) decoders such as Guessing Random Additive Noise Decoding (GRAND) [3] and Ordered Statistics Decoding (OSD) [4] have emerged as promising solutions for short codes, as illustrated in Fig. 2.5.1. GRAND is primarily practical for high code rates (code rate refers to the ratio of information bits to total transmitted bits) suitable for good channel conditions and less protection. However, its complexity becomes prohibitive for the moderate to low code rates required for a wide range of channel conditions and enhanced protection in URLLC. In contrast, OSD is better suited for these rates, but faces implementation challenges due to the complexity of high-order OSD and latency in sequential processing for sorting, Gaussian elimination (GE), and candidate re-encoding.

Recent research combining hard-decision decoding (HDD) and order-3 OSD achieves excellent BLER and an average latency of 773ns [5], but the complexity of order-3 OSD results in a worst-case latency of 28.9ms, with an information throughput of 83Mbps at 350pJ/b. This work presents a new two-stage decoder, employing a soft-decision Chase decoder and an order-1 OSD for more efficient, complementary error correction. Implemented in a 1.1mm² 16nm silicon prototype, the architecture lowers the OSD order from 3 to 1 and uses an efficient iterative GE with pre-sorting and early exits. Applied to the Bose-Chaudhuri-Hocquenghem (BCH) (127,64) code, our decoder achieves 14.4ns average and 132.6ns worst-case latency, improving upon the state-of-the-art [5] by 53× and five orders of magnitude, respectively. The prototype delivers 15.2Gbps information throughput at 13.1pJ/b, outperforming the latest URLLC FEC decoders [5], [14].

For a (n, k) binary linear block code, k message bits are encoded into an n-bit codeword using a systematic generator matrix $G = [I_k | P]$, where I_k is the $k \times k$ identity matrix and P is a $k \times (n-k)$ matrix. The $1 \times k$ binary message u is encoded as $c = uG$. After transmission over a noisy channel, the received vector r consists of n soft values known as log-likelihood ratios (LLRs). OSD operates in two stages: preprocessing and reprocessing as shown in Fig. 2.5.1. In preprocessing, LLRs in r are sorted by reliability (magnitude), from most reliable bits (MRBs) to least reliable bits (LRBs). The columns of G are permuted accordingly. Since the permutation changes the systematic form of G, GE is used to restore a new systematic matrix, G'. Then in reprocessing, candidate codewords are generated by flipping MRBs and re-encoding. Specifically, order-t candidates are created by flipping all possible combinations of t MRBs ($C(k,t)$ patterns) and encoding them using G'. Each candidate codeword is then compared to r using weighted Hamming distance (WHD), and the candidate with the smallest WHD is inverse permuted as the decoding result. Order-t decoding includes all candidates from order 0 up to t; as t approaches k, OSD approaches ML decoding. In practice, setting $t \le 3$ is sufficient to achieve near-ML performance for short codes.

While OSD provides highly reliable decoding, its latency is high for two reasons, as shown in Fig. 2.5.2: 1) as decoding order increases, the number of candidates grows rapidly, e.g., in decoding the BCH (127,64) code, each order increment can require an order-of-magnitude or more increase in complexity; 2) OSD relies on GE in preprocessing, a sequential operation that incurs significant sorting and control overhead, e.g., GE can account for over 60% of the latency in order-1 OSD. To minimize the impact of OSD's long latency, we employ OSD as a post-processor in a two-stage decoder, illustrated for decoding BCH codes in Fig. 2.5.3. Our approach uses a first-stage Chase decoder [6], a classic soft-decision decoder implemented in parallel for low latency. The OSD post-processor is activated only if the Chase decoder fails, ensuring most codewords are decoded quickly, while high reliability is still maintained. For instance, when decoding the BCH (127,64) code at 5dB SNR, the Chase decoder fails 0.07% of the time, so the OSD is infrequently activated and the average two-stage decoding latency remains close to that of the first-stage Chase

decoder. Nevertheless, the OSD is essential for correcting residual errors and achieving the target BLER of 10⁻⁵. A key advantage of the two-stage architecture is the ability to use a lower OSD order. In this work, we adopt Chase + order-1 OSD, which outperforms single-stage order-2 OSD in BLER above 5dB, while achieving a 156× reduction in average latency, as shown in Fig. 2.5.3. The efficiency arises because Chase and OSD offer complementary strengths: Chase flips LRBs to fix likely errors, while order-1 OSD acts as a safety net by flipping MRBs and re-encoding to cover patterns missed by Chase.

While the two-stage decoder effectively reduces average latency, worst-case latency remains critical for URLLC applications. Addressing this requires further optimization of the order-1 OSD. In order-1 OSD, reprocessing stage latency can be effectively reduced through parallel candidate generation and processing. However, the preprocessing stage remains constrained by GE, as pivot selection and row swapping are inherently sequential [7]. A recent approach called reduced GE [8] lowers GE complexity by applying it only to a submatrix, excluding columns associated with the MRBs, but this adds overhead for selection, indexing, and bookkeeping. Another approach called iterative basis update [9] divides bits into two sets: Set A, with k bits (initially message bits), and Set B, with the remaining (n-k) bits (initially parity bits). In each iteration, if the least reliable bit in Set A is less reliable than the most reliable bit in Set B, the corresponding columns in the generator matrix G are swapped and row operations are performed. This continues until no further swaps are possible, resulting in Set A comprising the k MRBs. The approach avoids complete column reordering and reduces the number of swaps and row operations, but still requires per-step search of minimum and maximum in Sets A and B, respectively, increasing control complexity and latency. To address this, we designed a pre-sorted basis update (PSBU), illustrated in Fig. 2.5.4. PSBU is a modified iterative basis update that begins with a single pre-sorting of bits in both sets, tagging each bit with its reliability, column index, and usage status (whether it has already been checked). In each iteration, the unused bit in Set A with the lowest reliability and the unused bit in Set B with the highest reliability are selected by referencing the index and usage tags, eliminating the need for additional sorting. Before swapping the two columns, a pivot check is performed to ensure the next pivot element in the generator matrix is non-zero. If not, the unused bit in Set B with the next highest reliability is considered for swapping, and the process continues. Furthermore, PSBU updates only the P part of the G matrix as the I_k part remains an identity matrix after row operations and is not needed for re-encoding, further reducing complexity and latency. Figure 2.5.4 compares full GE, reduced GE, and PSBU. Reduced GE incurs substantial area and frequency penalties due to higher control complexity, resulting in a higher latency than full GE. In contrast, PSBU reduces latency by over 75% and area by 39% compared to full GE.

The two-stage decoder prototype implementation is illustrated in Fig. 2.5.5. In the first stage, a Chase decoder is designed for 127b BCH codes. It processes 127 LLRs at a time, ranks the bits by reliability, and identifies the four LRBs. Flipping 0 to 4 of these LRBs generates 16 test patterns (TPs), which are decoded in parallel across 16 lanes. Among the successful candidates, the result with the minimum WHD is selected. If none of the TPs decode successfully, the OSD post-processor is engaged. In the second stage, an order-1 OSD is employed to post-process 127b words that could not be decoded in the first stage. The design incorporates four early exits to minimize latency as shown in Fig. 2.5.5: exit 1 for already correct codewords, exit 2 if a TP is a correct codeword, exit 3 if Chase decoding succeeds, and exit 4 if order-1 OSD succeeds. By reduced-order OSD, efficient PSBU, and multiple early exits, the average latency is reduced to 13.7 clock cycles, and the worst-case latency to 126 cycles for BCH (127,64) at 5dB SNR and BLER of 10⁻⁵.

The prototype is fabricated in Intel 16, occupying a total silicon area of 1.1mm², with the Chase decoder and order-1 OSD utilizing 0.8mm² and 0.3mm², respectively, as illustrated in Fig. 2.5.7. Operating at room temperature, 0.9V supply voltage and 950MHz, the decoder achieves an information throughput of 15.2Gbps and an energy efficiency of 13.1pJ/b when decoding the BCH (127,64) code. BLER performance and latency for two sample codes, BCH (127,64) and BCH (127,85), are presented in Fig. 2.5.7. The design is compared with state-of-the-art FEC decoders in Fig. 2.5.6. Conventional LDPC and polar codes typically feature longer code lengths [10]-[12], resulting in decoding latencies that are unsuitable for URLLC. The GRAND decoder [13] offers superior energy efficiency but is only practical

---

50 • 2026 IEEE International Solid-State Circuits Conference

979-8-3315-8937-0/26 $31.00 © 2026 IEEE

for high-rate codes. It requires 7dB SNR to reach a BLER of 10^{-5}, and its latency increases by three orders of magnitude for lower-rate codes, making it unsuitable for URLLC. The BOSS decoder [14] achieves BLER of 10^{-5} at 5.2dB SNR, but it is limited to a fixed low code rate of 0.11, resulting in information throughput below 1Gbps. The two-stage HDD BCH decoder + order-3 OSD [5] achieves BLER of 10^{-5} at 4dB SNR, but it has a long worst-case latency of 28.9ms, which is impractical for URLLC. Our Chase + order-1 OSD outperforms the BOSS decoder in error correction, while delivering 22× higher information throughput at 3.7× lower energy. It supports a broad range of code rates suitable for URLLC, maintaining a short average decoding latency of 14.4ns and a worst-case latency of 132.6ns – 53× lower average latency and over five orders of magnitude lower worst-case latency than the state-of-the-art HDD + order-3 OSD decoder [5], with 183× higher information throughput and 26× lower energy.

Acknowledgment:
This work was supported in part by the Center for Ubiquitous Connectivity (CUbiC), one of the seven centers sponsored by the Semiconductor Research Corporation (SRC) and DARPA under the Joint University Microelectronics Program 2.0 (JUMP 2.0). Chip fabrication was provided by the Intel University Shuttle Program.

References:
[1] G. Pocovi et al., "Achieving Ultra-Reliable Low-Latency Communications: Challenges and Envisioned System Enhancements," *IEEE Network*, vol. 32, no. 2, pp. 8-15, 2018. https://doi.org/10.1109/MNET.2018.1700257
[2] C. Yue et al., "Efficient Decoders for Short Block Length Codes in 6G URLLC," *IEEE Communications Magazine*, vol. 61, no. 4, pp. 84-90, 2023. http://doi.org/10.1109/MCOM.001.2200275
[3] K. R. Duffy et al., "Capacity-Achieving Guessing Random Additive Noise Decoding," *IEEE Trans. on Information Theory*, vol. 65, no. 7, pp. 4023-4040, 2019. http://doi.org/10.1109/TIT.2019.2896110
[4] M.P.C. Fossorier et al., "Soft-Decision Decoding of Linear Block Codes Based on Ordered Statistics," *IEEE Trans. on Information Theory*, vol. 41, no. 5, pp. 1379-1396, 1995. http://doi.org/10.1109/18.412683
[5] J. Kim et al., "Hybrid Ordered Statistics Decoding of Short-Length BCH Codes for URLLC Systems: Theoretical Analysis and Decoder Implementation," *IEEE TCAS-I*, vol. 72, no. 11, pp. 6528-6540, 2025. http://doi.org/10.1109/TCSI.2025.3570034
[6] D. Chase, "Class of Algorithms for Decoding Block Codes with Channel Measurement Information," *IEEE Trans. on Information Theory*, vol. 18, no. 1, pp. 170-182, 1972. http://doi.org/10.1109/TIT.1972.1054746
[7] A. Rupp et al., "A Parallel Hardware Architecture for fast Gaussian Elimination over GF(2)," *IEEE Symp. on Field-Programmable Custom Computing Machines*, pp. 237-248, 2006. http://doi.org/10.1109/FCCM.2006.12
[8] M. Fossorier et al., "Modified OSD Algorithm with Reduced Gaussian Elimination," *IEEE Communication Letters*, vol. 28, no. 8, pp. 1755-1759, 2024. http://doi.org/10.1109/LCOMM.2024.3416515
[9] X. Li et al., "Iterative Basis Update for Ordered Statistics Decoding of Linear Block Codes," *IEEE Communication Letters*, vol. 28, no. 9, pp. 1981-1985, 2024. http://doi.org/10.1109/LCOMM.2024.3425388
[10] M. Milicevic et al., "A Multi-Gb/s Frame-Interleaved LDPC Decoder with Path-Unrolled Message Passing in 28-nm CMOS," *IEEE TVLSI*, vol. 26, no. 10, pp. 1908-1921, 2018. http://doi.org/10.1109/TVLSI.2018.2838591

[11] C.-F. Teng et al., "An Ultra-Low Latency 7.8–13.6 pJ/b Reconfigurable Neural Network-Assisted Polar Decoder with Multi-Code Length Support," *IEEE Symp. VLSI Circuits*, 2020. http://doi.org/10.1109/VLSICircuits18222.2020.9163022
[12] D. Kam et al., "A 1.1μs 1.56Gb/s/mm² Cost-Efficient Large-List SCL Polar Decoder Using Fully-Reusable LLR Buffers in 28nm CMOS Technology," *IEEE Symp. VLSI Circuits*, pp. 204-205, 2022. http://doi.org/10.1109/VLSITechnologyandCir46769.2022.9830317
[13] A. Riaz et al., "A Sub-0.8pJ/b 16.3Gbps/mm² Universal Soft-Detection Decoder Using ORBGRAND in 40nm CMOS," *ISSCC*, pp. 432-434, 2023. http://doi.org/10.1109/ISSCC42615.2023.10067519
[14] D. Kam et al., "A 21.9ns 15.7 Gbps/mm² (128,15) BOSS FEC Decoder for 5G/6G URLLC Applications," *ISSCC*, pp. 50-52, 2024. http://doi.org/10.1109/ISSCC49657.2024.10454363

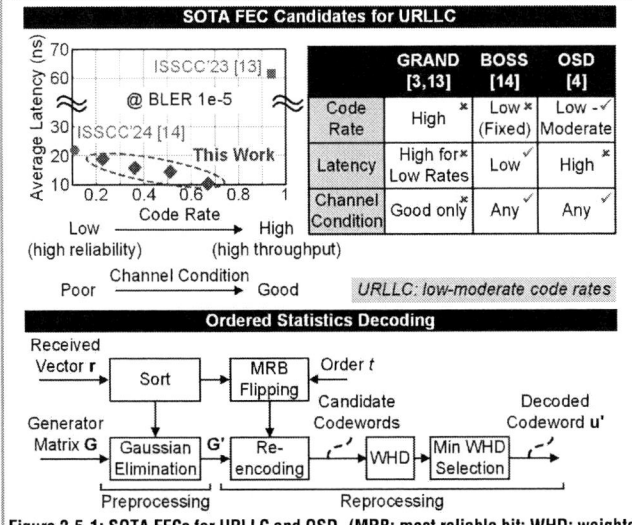

Figure 2.5.1: SOTA FECs for URLLC and OSD. (MRB: most reliable bit; WHD: weighted Hamming distance).

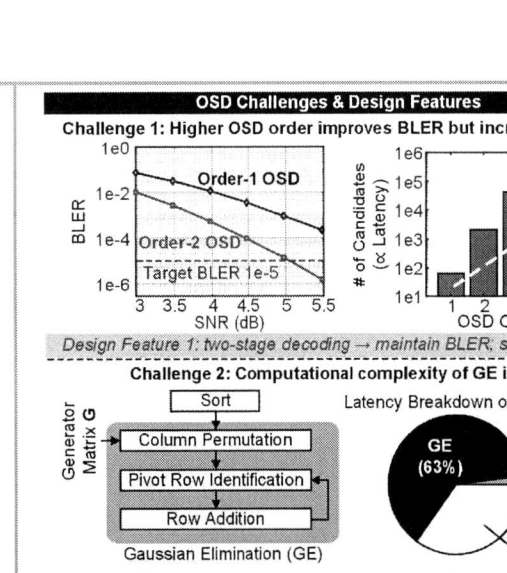

Figure 2.5.2: Design challenges for OSD hardware implementations and design features.

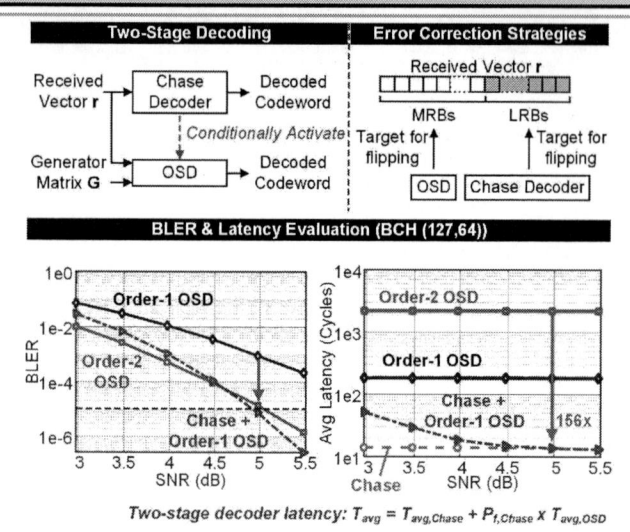

Figure 2.5.3: Two-stage decoding and BLER and latency evaluations. ($P_{f,Chase}$: failure probability of Chase decoder).

Two-stage decoder latency: $T_{avg} = T_{avg,Chase} + P_{f,Chase} \times T_{avg,OSD}$

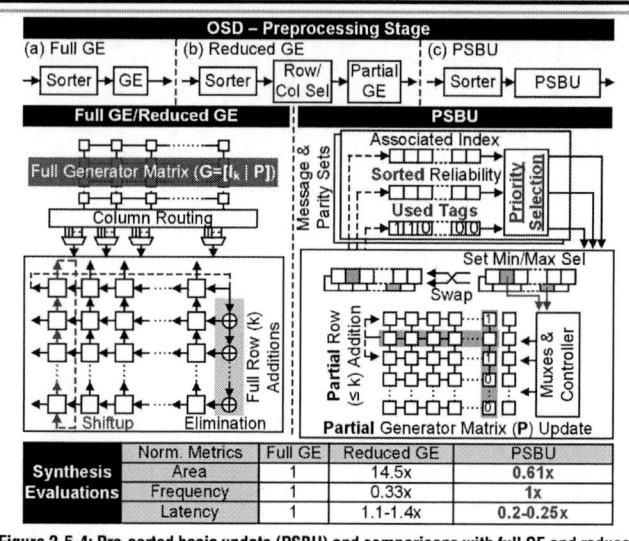

Figure 2.5.4: Pre-sorted basis update (PSBU) and comparisons with full GE and reduced GE.

Figure 2.5.5: Architecture block diagram and latency evaluation.

Latency Evaluation (BCH (127,64) @ SNR 5dB)

(1) Two-stage decoding; (2) PSBU; (3) Multi-exits

Exit Ratios — Exit 3 (95.16%), Exit 2 (3.99%), Exit 1 (0.78%), Exit 4 (0.07%)

Worst Case: 2219, 214, 126, 126 — 94.3%
Average: 2200, 14.14, 14.07, 13.69 — 99.4%

126 cycles = 132.6ns @ 950MHz
13.69 cycles = 14.4ns @ 950MHz

Figure 2.5.6: Comparison of state-of-the-art FECs.

	This Work	ISSCC'23 [13]	ISSCC'24 [14]	TCAS1'25 [5]	TVLSI'18 [10]	VLSI'20 [11]	VLSI'22 [12]
Technology	16nm	40nm	28nm	28nm	28nm	40nm	28nm
Code	BCH	CA-Polar/CRC	BOSS	BCH	LDPC	Polar	Polar
Decoding Algorithm	Chase + Order-1 OSD	ORBGRAND	BOSS Decoding	HDD + Order-3 OSD	MS	RNN-BP	SCL
Code Length	127	Up to 256	128	127	672	Up to 256	1024
Code Rate	Any	High Rate Only (0.94)	Low Rate Only (0.11)	Fixed Rate (0.5)	0.5	Fixed Rate (0.5)	Fixed Rate (0.5)
Voltage (V)	0.9	1	0.95	0.9	0.9	0.9	1.05
Frequency (MHz)	950	90	590	200	150	225	413
Decoder Area (mm²)	0.8 (Chase) 0.3 (OSD)	0.4	0.37	0.49	1.99	0.18	0.59
SNR (dB) @ BLER 1e-5	5*	7	5.2	4*	5.2	5.8	2.75
Info Throughput (Gbps)	15.2*	4.3	0.68+	0.083*	4.85+	0.48+,**	0.47+
Average Latency (ns)	14.4*	61.3	21.9	773*	555	270**	1100
Worst Case Latency (ns)	132.6*	1657000	21.9	28902000*	793	270**	1100
Energy Efficiency (pJ/b)	13.1*	1.14	48.6+	350*	82.4+	23.4+,**	219+

*: BCH (127,64); **: Code Length = 128; +: Info Throughput = Raw Throughput x Code Rate

Figure 2.5.7: Die photo and measurement results.

• 2026 IEEE International Solid-State Circuits Conference

979-8-3315-8937-0/26 $31.00 © 2026 IEEE

ISSCC 2026 / SESSION 2 / PROCESSORS / 2.6

2.6 Spyre: An Inference-Optimized Scalable AI Accelerator for Enterprise Workloads

Matthew Cohen[1], Monodeep Kar[1], Swagath Venkataramani[1], Viji Srinivasan[1], Brian Veraa[2], Matthew Ziegler[1], Nianzheng Cao[1], Ashish Ranjan[1], Joel Silberman[1], Michael Guillorn[1], Sandra Woodward[3], JohnDavid Lancaster[1], Josh Hursey[4], Kyu-hyoun Kim[1], Alberto Mannari[5], Amrit Nagarajan[1], Ananda Samajdar[1], Bahman Hekmatshoartabari[1], Bob Galbraith[3], Ching Zhou[1], Dave Satterfield[1], Greg Still[6], Gustavo Tellez[1], Indira Nair[1], Indu Masilamanai[2], Jinwook Jung[1], Kavi Randhawa[2], Marcel Schaal[1], Martin Lutz[1], Paul Crumley[1], Philip Jacob[1], Prasanth Chatarasi[1], Radhika Jain[1], Saekyu Lee[1], Sanchari Sen[1], Sarada Krithivasan[1], Scot Rider[1], Shubham Jain[1], Siyu Koswatta[1], Thomas Roewer[1], Tom Gooding[3], Victor Ferrari[7], Vidhi Zalani[1], Zhibin Ren[1], Kevin Reick[2], Lisa Maurice[8], Chris Gonzalez[9], Chris Catalino[2], Ryan Nett[2], Pong-Fei Lu[1], Rob Senger[1], Leland Chang[1]

[1]IBM Research, Yorktown Heights, NY, [2]IBM Infrastructure, Austin, TX, [3]IBM Infrastructure, Rochester, MN, [4]IBM Research, Rochester, MN, [5]IBM Research, Zurich, Switzerland, [6]IBM Infrastructure, Research Triangle Park, NC, [7]IBM Research, Sao Paulo, Brazil, [8]IBM Research, Austin, TX, [9]IBM Infrastructure, Yorktown Heights, NY

Abstract

Spyre is a scalable, power-efficient AI accelerator product for enterprise workloads. Featuring 32 AI cores, mixed-precision support, and LPDDR5 memory, it fits in a single-slot PCIe form factor and scales over a standard PCIE fabric. Optimized for inference workloads, Spyre achieves 2-to-3× better power/performance than GPUs on encoder-class models and scales up to 4 or more devices for large generative models.

Enterprise AI workloads require power/performance efficiency to enable broad deployment in a variety of servers with standard PCIe slots. Power-optimized circuit design and advanced power management are thus key to maximizing performance in a single-slot power budget. Scalable, multi-accelerator operation is also critical to aggregate performance across PCIe slots to support larger models while meeting aggressive throughput and latency SLAs. In this work, we present Spyre [1], an AI accelerator product for enterprise workloads that meets these requirements across a broad application space – satisfying the stringent reliability and security requirements of mainframe, on-prem infrastructure and cloud servers. Spyre scales over a PCIe fabric, which removes the need for high-power GPUs, while delivering leading-edge performance on enterprise use cases. Spyre features 32 AI cores, 9 customized micro-controllers, PCIe Gen 5 ×16, 16 independent LPDDR5 interfaces (to address 128GB of DRAM), DMA/RDMA engines and advanced power management features.

Spyre AI cores are divided into two corelets, each containing a 2D SIMD-systolic array for operations like matrix multiplication and convolution as well as two 1D vector arrays for activation/normalization functions and quantize/dequantize operations. The mixed-precision engines (MPE), similar to [2, 3], support fp16, fp8 (e4m3/e5m2), int8, and int4 arithmetic to enable customers to run Granite [6] and other foundation models at various levels of quantization. The 2D MPE array supports n:4 (n=1,2,3) structured sparsity by provisioning 4× bandwidth for activations and reusing the local MPE register file for weight indices [4]. Each core contains a 2MB scratchpad memory divided into 4 banks, where consecutive 128B aligned addresses are mapped to adjacent banks to avoid conflicts.

The AI cores are interconnected by a bi-directional ring (Fig. 2.6.1). When splitting work across sequence/batch, a software-managed multicast protocol streams weights and activations. These strategies reduce ring latency, especially for distant cores. When AI operations are distributed across cores by splitting across input channels, groups of cores perform reductions by using independent ring segments to minimize traffic overlap, aided by a dedicated lightweight streaming ring for partial sums. The ring includes stops for the high-speed memory interface (HMI) and core management unit (CMU). The HMI feeds into a memory crossbar (MCX) that connects to 16 LPDDR5 channels @ 6.4Gbps to deliver 204GB/sec peak to the cores. LPDDR5 optimizes the capacity-to-bandwidth ratio at low cost per bit, while still providing sufficient memory bandwidth for targeted applications. MCX also has ports for the DMA and RDMA engines, which move data over PCIe between Spyre and the host and between multiple Spyres, respectively, and the micro-controller subsystem (MCS), which is a collection of 9 customized micro-controllers that handle chip management and other tasks. The MCS controls job execution in the AI cores via the CMU and directly dispatches work to the DMA/RDMA engines, while simultaneously ensuring user isolation. The SoC supports fully pipelined DMA/RDMA input transfers, execution, and DMA/RDMA output transfers, allowing these stages to overlap and enable different phases of independent inferences to run in parallel. Similarly, CMU pre-fetches program data for the next operation, while the current operation is still executing, further enhancing parallelism and reducing idle time.

The Spyre SoC is integrated into a single-slot PCIe card alongside 8 dual-channel LPDDR5 modules and power management control loop circuits. Communicating through a PCIe switch, multiple Spyres form a local group with lowest latency inter-communication, which allows for multi-card operation to aggregate memory bandwidth for performance on Generative AI (GenAI) models. In a z17 mainframe deployment, up to eight of these cards can be deployed in a server I/O drawer and up to 96 Spyres could be supported in a system (Fig. 2.6.1).

An overarching design constraint for Spyre is the power budget of a single-slot PCIe card with no auxiliary power connector. An important design decision to reduce power was to adopt a multiple power domain strategy with a reduced supply voltage for the high activity regions on the chip. A higher voltage domain is employed for timing-critical and less active regions. Figure 2.6.2 illustrates the chip floorplan, highlighting the 0.55V domain for the high activity and densely placed compute regions that occupy ~42% of the die area. Closing the physical design at 0.55V for the compute regions required higher effort to meet the additional timing margins and standard-cell library restrictions associated with low voltage operation. However, the overall power savings provided a strong return-on-investment. The ring connecting the cores was assigned to a 0.75V domain so that the ring frequency could be maximized. While smaller-track standard-cell libraries can be more power-efficient at higher voltages, we found that at 0.55V, a larger 7-track library led to lower overall power than the smaller 6-track library, likely because the lower inherent drive strength of the 6-track library required the use of additional buffering and higher drive strength cells in comparison to the 7-track library.

A key power reduction technique in this work was the use of specific stimulus vectors files from a set of key client workloads within the logic synthesis and physical design flow for high accuracy dynamic power analysis and optimization. As an example, when software stack optimizations advanced more than originally expected and achieved higher compute utilization, the core cycle time was opportunistically tuned to a lower, more optimal frequency to minimize power-limited core throttling. By applying this power analysis methodology during the physical design phase, we achieved ~8% power savings and ~6% cell area reduction compared to relying solely on post-silicon frequency scaling (Fig. 2.6.2).

Aggressive power management is important to extract the best performance across AI models when operating under platform-specific power budgets. Previous solutions [5] focus on a single control loop that regulates the consumed power at a fixed target operating under a loop bandwidth. Figure 2.6.3 shows the peak power constraints for a PCIe card across time constants for two system deployment scenarios S_A and S_B, which have strict peak power requirements at longer and shorter time constants, respectively. Single-loop control would require keeping the set-point of the loop at the worst-case peak power at T_{LONG}. To mitigate underutilization, a dual-loop control is introduced, which enables the peak power to be regulated at two different set-points and two different time constants. A dedicated IIR filter-based control loop with programmable coefficients manages power at smaller time constants. The loop set-point, which is platform-specific, is configured through the firmware. An MCS micro-controller controls the long time-constant loop, as it requires storing more than 100K current samples since a new current sample is received every 1μs. A hierarchical 2-step averaging scheme is used to: 1) fit the power history in the 16KB MCS micro-controller cache, 2) absorb the jitter introduced by the asynchronous events serviced by the MCS core, and 3) avoid aliasing issues introduced by traditional undersampling techniques.

The 5b stall rate generated by individual control loops is merged by computing the maximum value, which is then forwarded to the cores. Core throttling is implemented using a sigma-delta modulator that stalls the matrix engines, similar to the approach described in [5]. Figure 2.6.4a illustrates the improvement in inference throughput for use cases relevant to transaction processing and fraud detection, two key enterprise workloads. Compared to single-loop control, dual-loop control enables a higher peak current than the long-term average, resulting in up to a 28% improvement in throughput. Figure 2.6.4b presents normalized throughput as a function of frequency for both single and dual-loop operations on BERT-Large with a sequence length of 256. While dual-loop operation consistently outperforms single-loop control, a performance roll-off is observed beyond the intended operating frequency. This degradation is due to power-limited throttling in the digital filter and/or MCS loops. These results underscore the importance of accurate pre-silicon power modeling to prevent runtime scaling issues and to optimize the design for the target operating frequency. Figure 2.6.4c shows the power trace captured during inference with and without the second loop active. The bottom waveform with the second loop enabled shows the current reaching higher short-term peaks, which allows inference to complete faster.

A single Spyre device can efficiently support encoder models, which are often used inside financial transactions with enterprise-scale throughput and millisecond latency requirements such as for fraud detection and anti-money laundering applications. GenAI models such as Granite capitalize on multiple Spyre devices working in concert to meet customer performance targets. Instead of using a proprietary interconnect, the Spyre device uses a standard PCIe fabric to scale inference workloads within a server. Figure 2.6.5 shows two views of the performance of a Granite v3.3 8B model running on up to 8 Spyre devices. This instance of the model was configured to run a prompt of 1024 tokens, a batch size of 1, with precision set to fp16, and generate 8 tokens. Figure 2.6.5a shows the normalized scaling of inter-token latency (ITL) for this model, which demonstrates good scalability using the PCIe fabric. Figure 2.6.5b shows example power waveforms when using 1, 2, and 4 Spyre devices. Each of the 8 tokens generated can be observed from the waveform, with the time-to-first-token (TTFT) exhibiting a distinct performance profile due to the additional computational demands. As more Spyre devices are added, the time required to produce each token decreases. This improvement is the result of both smaller computational kernels and increased aggregate memory bandwidth at the cost of some added communication to synchronize data in tensor-parallel execution.

The Spyre SoC (Fig. 2.6.7) is fabricated in a 5nm, 14-metal level technology. It measures 330mm², contains 26B transistors, and has over 64MB of embedded SRAM. Logic operates at 0.55V and 0.75V, with the cores and ring running at up to 1.2GHz. Spyre offers 98/157/315/629 TOPS (dense) of compute performance per card for fp16/fp8/int8/int4 precisions and achieves ~2-to-3× power/performance improvement relative to GPU solutions on internal workloads exercising encoder-class models. Scaling up to 4 or more cards meets TTFT/ITL latency targets for generative AI models.

References:
[1] C Berry, "IBM Telum II processor and IBM Spyre Accelerator chip for AI," *IEEE Hot Chips Symp.*, 2024.
<https://hc2024.hotchips.org/assets/program/conference/day1/04_HC2024.IBM.CBerry.final.pdf>.
[2] A. Agrawal et al., "A 7nm 4-Core AI Chip with 25.6TFLOPS Hybrid FP8 Training, 102.4TOPS INT4 Inference and Workload-Aware Throttling," *ISSCC*, pp. 144-146, 2021. https://doi.org/10.1109/ISSCC42613.2021.9365791
[3] RaPiD: AI Accelerator for Ultra-low Precision Training and Inference. https://doi.org/10.1109/ISCA52012.2021.00021
[4] N. Wang et al., "Deep Compression of Pre-trained Transformer Models," *NeurIPS, pp. 14140 - 14154, 2022.* https://dl.acm.org/doi/10.5555/3600270.3601298
[5] M. Kar et al., "A Software-Assisted Peak Current Regulation Scheme to Improve Power-Limited Inference Performance in a 5nm AI SoC," *ISSCC*, pp. 254-256, 2024. https://doi.org/10.1109/ISSCC49657.2024.10454301
[6] IBM Granite Documentation - <https://www.ibm.com/granite/docs/>.

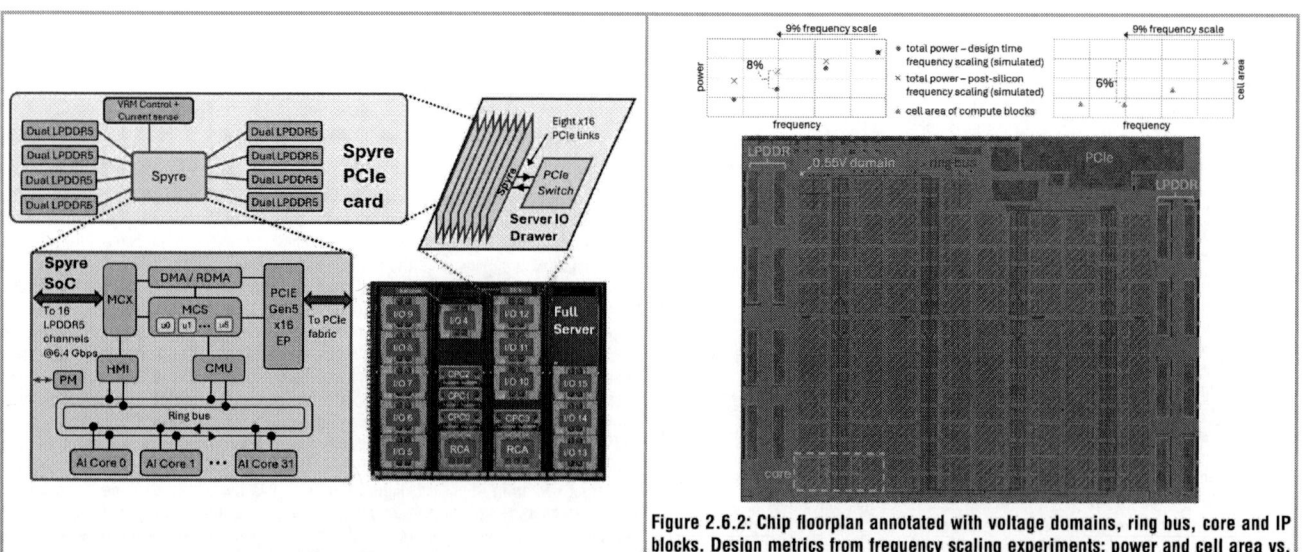

Figure 2.6.1: Spyre chip, card and system diagrams.

Figure 2.6.2: Chip floorplan annotated with voltage domains, ring bus, core and IP blocks. Design metrics from frequency scaling experiments: power and cell area vs. frequency.

ISSCC 2026 / SESSION 2 / PROCESSORS / 2.6

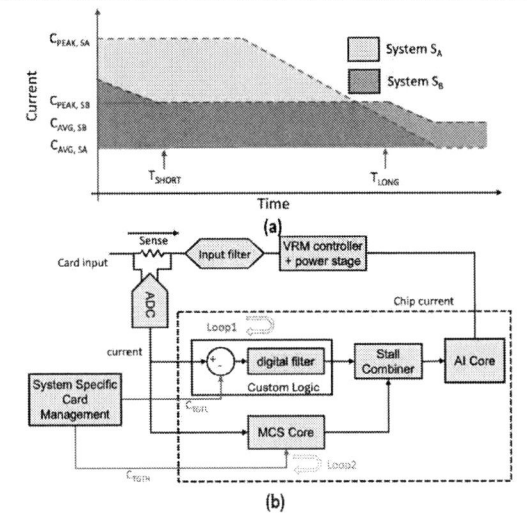

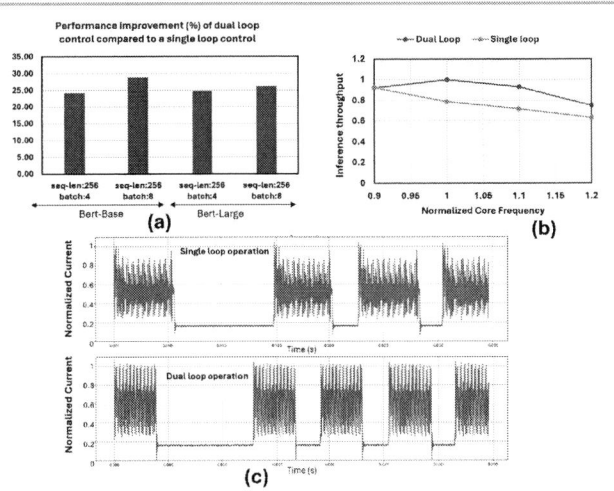

Figure 2.6.3: (a) Peak power specification across different systems; (b) Dual loop control for peak current management.

Figure 2.6.4: (a) Performance enhancement with dual-loop control; (b) performance against core frequency; (c) measured waves with and without a micro-controller-based loop.

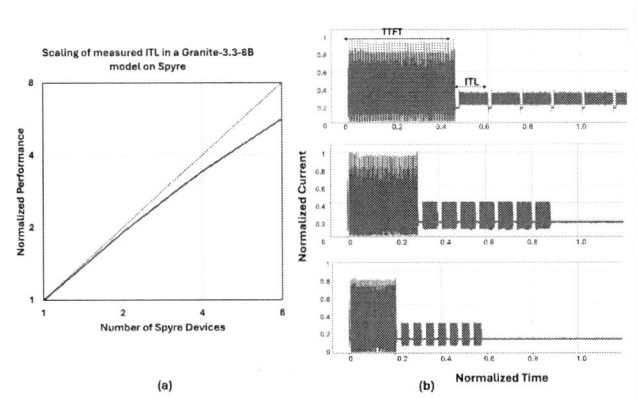

Figure 2.6.5: (a) Measured normalized performance (1/ITL) as a function of Spyre instance count; (b) Inference with 1, 2, and 4 Spyres, identical time scales, showing TTFT and ITL improvements.

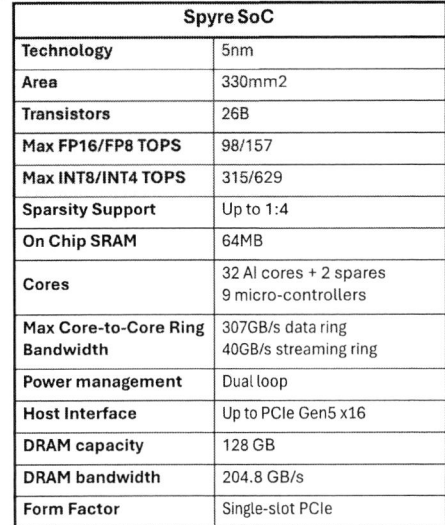

Spyre SoC	
Technology	5nm
Area	330mm2
Transistors	26B
Max FP16/FP8 TOPS	98/157
Max INT8/INT4 TOPS	315/629
Sparsity Support	Up to 1:4
On Chip SRAM	64MB
Cores	32 AI cores + 2 spares 9 micro-controllers
Max Core-to-Core Ring Bandwidth	307GB/s data ring 40GB/s streaming ring
Power management	Dual loop
Host Interface	Up to PCIe Gen5 x16
DRAM capacity	128 GB
DRAM bandwidth	204.8 GB/s
Form Factor	Single-slot PCIe

Figure 2.6.6: Spyre technical specifications.

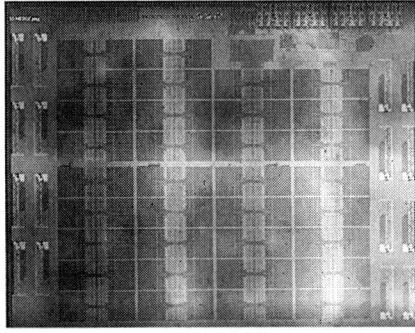

Figure 2.6.7: Photos of the die and the PCIe card.

• 2026 IEEE International Solid-State Circuits Conference

979-8-3315-8937-0/26 $31.00 © 2026 IEEE

ISSCC 2026 / SESSION 2 / PROCESSORS / 2.7

2.7 Tiamat: A 98-to-134ms/Step Transformer-Based Diffusion Model Processor Supporting Classifier-Free Guidance for Image Generation

Po-Yen Lu, Po-Wei Chen, Shau-Hua Yeh, Zhi-Jun Zhang, Yong-Tai Chen, Ting-Yu Wang, Chao-Tsung Huang

National Tsing Hua University, Hsinchu, Taiwan

Abstract

A 16nm FinFET transformer-based diffusion model processor chip is fabricated for supporting class-conditional DiT-XL/2 and text-to-image PixArt-α with 98ms and 134ms generation time per step with 7.37TOPS and 1477mW at 400MHz. Classifier-free guidance

The rapid growth of generative AI drives the demand for real-time, high-quality image generation on edge devices. Recently, transformer-based diffusion models (DMs) [1-2] have emerged as strong candidates, delivering superior image quality with 1.5-to-4.3× smaller model sizes compared to traditional U-Net-based models [3-4], as depicted in Fig. 2.7.1. These models replace the U-Net backbone with stacked transformer blocks and usually enhance generation quality through classifier-free guidance (CFG) [5], a technique that combines positive samples (PSs) and negative samples (NSs). In addition, the emerging microscaling (MX) data format [6] applies a shared scale across the elements in one block, improving efficiency for multiply-accumulate (MAC) energy and data traffics while maintaining quality without model fine-tuning or re-training. However, even with the MX format, a single step of transformer-based DMs still requires up to 911.4MB of external memory access (EMA) due to the large number of parameters and the two-sample inferences for CFG, creating a major bottleneck for real-time deployment on bandwidth-constrained devices. Furthermore, directly converting floating-point (FP) residual connections into MX format would cause severe quality degradation, leading to the demand of large buffers to store FP feature maps (FMs). To address these issues, we present Tiamat, a memory-efficient transformer processor optimized for real-time, high-quality DM inference with CFG. Tiamat introduces three key features: 1) a CFG batch difference processing (CBDP) flow reducing weight EMA by 50% and MAC energy by 14%; 2) a weight-reordered quantization (WRQ) combined with two-stage block-aware FM alignment (TSBFA) further cutting weight EMA by 37%; 3) a hybrid MX blocking datapath (HMBD) and two-stage quantization (TSQ) to preserve FP-level generation quality, while lowering on-chip buffer requirements by 37%.

Figure 2.7.2 illustrates the proposed CBDP flow. For a single-sample design, supporting CFG requires loading weights for both PS and NS, resulting in 910MB of weight EMA. Instead, batching the two samples reduces this EMA by 50%, but increases on-chip buffer demand up to 1594KB for storing both FMs. To improve this trade-off, CBDP leverages the similarity between FMs of PS and NS. Since the input latent (X_t) is identical for both samples, the resulting PSFM ($Z_{t,i}^+$) and NSFM ($Z_{t,i}^-$) remain highly correlated across all transformer blocks, with relative differences below 20%. This enables storing one FM, $Z_{t,i}^+$, as a reference, while representing the other, $Z_{t,i}^-$, as the difference, $\delta_{t,i}$, with reduced precision. Evaluations show that $\delta_{t,i}$ can be stored in MXINT5 format with only a 0.05 increase in Fréchet inception distance (FID). To keep inference correctness for non-linear functions (NLFs), we temporarily recover the PSFM, $Z_{t,i}^+$, by first adding the reference back, then applying the NLF, and finally restoring the difference $\delta_{t,i}$ by subtracting the reference again. Accordingly, we devised two specialized cores: a reference processing core (RPC) for computing of $Z_{t,i}^+$ and a difference processing core (DPC) for computing of $\delta_{t,i}$. This parallel RPC/DRC architecture enables on-the-fly $Z_{t,i}^+$ recovery and $\delta_{t,i}$ restoration without storage overheads. Compared with the batching baseline, the proposed CBDP reduces on-chip FM buffer usage and MAC energy by 14% each. Figure 2.7.2 also shows the overall chip architecture. It consists of a reference sub-system with eight RPCs and a difference sub-system with eight DPCs for CBDP. Each sub-system also incorporates a block-aware FM alignment (BFA) unit and a two-stage quantization (TSQ) unit for supporting TSBFA and HMBD, respectively.

Figure 2.7.3 shows the proposed WRQ and TSBFA for weight quantization. MX quantization applies weight blocking along the input channels (iChs) for efficient matrix multiplication. However, the large variation of weights across iChs could lead to truncation errors, hindering narrow bit-width quantization. We observed that the weight magnitudes stay consistent along output channels (oChs) for transformer-based DMs. Therefore, we devised WRQ, which reorders iChs for weights so that elements within each block exhibit similar magnitudes. Specifically, we use the average along oChs to serve as the representative for each iCh and generate a new iCh order by sorting these representatives. The weight matrices are then quantized offline based on the new orders. With this approach, WRQ preserves MXINT8 generation quality even with MXINT5 precision, thereby reducing EMA by 37%. However, to preserve the correctness of MACs and residual connections, FMs must be aligned with the new orders on-the-fly, for which we propose TSBFA as an efficient solution. A naïve approach would deploy multiplexers (MUXs) to handle all possible cases in one cycle, demanding 1152 9-to-1 MUXs to align one FM tile, eight tokens for eight channels, to the 1152 alignment buffers (ABufs) in our chip. In contrast, we devised a block-aware FM alignment (BFA) unit that considers 32 ABufs in one MX block as a group and aligns at

batching reduces external memory access (EMA) for weights by 50%, and weight-reordered quantization brings another 37% reduction. Hybrid microscaling blocking reduces feature-map buffer size by 37%, while approaching floating-point quality.

most one channel, instead of eight, to each group/block. It reduces the complexity overhead to only 36 8-to-1 and 1152 2-to-1 MUXs, with latency from block conflicts mitigated by a two-stage (TS) alignment scheme: the first stage pre-aligns the FM tile, according to an offline-determined intermediate order, to reduce block conflicts, and the second stage finishes alignment with nearly no latency. The TS alignment reduces the latency by 12.4%, with only 0.3% overhead compared to the naïve approach. Overall, TSBFA complements WRQ while reducing the total gate count of processing cores by 8% versus a direct MXINT8 design.

Figure 2.7.4 depicts the proposed HMBD and TSQ for FM quantization. We observed that the default channel-wise MX blocking, denoted as C32, results in poor image quality and significant quantization errors. This stems from the large variance across channels in the residual connections of transformer-based DMs. In contrast, the variance across tokens is relatively small, motivating the use of a hybrid MX blocking strategy to preserve image fidelity without requiring an on-chip buffer of up to 2,528KB for storing FP residual connections. Specifically, HMBD applies token-wise MX blocking, denoted as T32, to store residual connections in FM buffers, and converts them on-the-fly to the C32 format before matrix multiplications to facilitate fixed-point channel reduction. However, this approach alone would need to collect multiple tokens until a full T32 block is available, demanding a large collection buffer of up to 295KB. We further devise TSQ to address this issue. The output FM tile – consisting of eight tokens for eight channels – is first directly quantized using an intermediate blocking format of eight tokens and four channels, denoted as T8C4, without the need of collection. Then a second-stage quantization converts the format to T32 by reading in token-first order and transposing T8C4 blocks, reducing the collection buffer to just 2KB with only 0.48% of latency overhead. Although T8C4 blocks require higher MXINT11 precision to preserve small quantization errors – raising the FM buffer size from 1,372KB to 1,594KB – the overall buffer size is still reduced by 37% compared to using FP residual buffers, thanks to the small collection buffer enabled by TSQ. HMBD and TSQ also preserve the generation quality with only 0.2 increase in FID. Together with CBDP, the reduction ratio further goes up to 46% for using MXINT5 FMs in DPC.

Figure 2.7.5 presents the measurement and inference results of Tiamat. At 1.05V and 400MHz, Tiamat delivers a peak computing throughput of 7.37TOPS with an energy efficiency of 4.99TOPS/W, achieving generation time of 98ms/step for class-conditional image synthesis with the DiT-XL/2 model and 134ms/step for text-to-image synthesis with the PixArt-α model. At 0.66V and 100MHz, Tiamat reaches its peak energy efficiency of 11.64TOPS/W, corresponding to generation energy of 75mJ/step and 99mJ/step for DiT-XL/2 and PixArt-α, respectively. With the proposed EMA optimization techniques, the system energy – including DDR4 [7] access – is well contained at 121.9mJ/step and 159.2mJ/step. Inference results across different diffusion steps demonstrate the trade-off between generation energy and image quality, highlighting the flexibility of this chip for real-world applications.

Figure 2.7.6 shows the comparison table with DM processors [8-11]. This chip is optimized for advanced transformer-based DMs and MX datapath to achieve fast and realistic image generation with CFG. It achieves 3.90-to-7.48× faster generation time per step per prompt compared to [8-10] and comparable generation time per inference compared to [11], which runs on a step-distilled model. In particular, it generates one high-fidelity image with 15 to 25 diffusion steps in 1.47 to 3.35 seconds for real-time interaction. Also, the image quality is close to original FP models without fine-tuning even when EMA is greatly reduced by 68% and 66% for DiT-XL/2 and PixArt-α, respectively. The chip photograph and specifications are given in Fig. 2.7.7. The Tiamat processor delivers memory-efficient and datapath-optimized acceleration for transformer-based DM inference with CFG support, facilitating real-time and high-quality image generation on edge devices.

Acknowledgment:

This work was supported by NSTC. The authors would like to thank the guidance from Meng-Fan Marvin Chang, Win-San Khwa, Ping-Sheng Wu, and TSMC colleagues, the support from TSMC University Shuttle Program, and EDA tool support from TSRI.

References:

[1] W. Peebles et al., "Scalable Diffusion Models with Transformers," *ICCV*, pp. 4172-4182, 2023. https://doi.org/10.1109/ICCV51070.2023.00387

[2] J. Chen et al., "PixArt-α: Fast Training of Diffusion Transformer for Photorealistic Text-to-Image Synthesis," *ICLR*, 2024. https://arxiv.org/abs/2310.00426

[3] R. Rombach et al., "High-Resolution Image Synthesis with Latent Diffusion Models," *IEEE CVPR*, pp. 10674-10685, 2022. https://doi.org/10.1109/CVPR52688.2022.01042

[4] D. Podell et al., "SDXL: Improving Latent Diffusion Models for High-Resolution Image Synthesis," *ICLR*, 2024. https://arxiv.org/abs/2307.01952

[5] J. Ho et al., "Classifier-Free Diffusion Guidance," *arXiv preprint arXiv:2207.12598*, 2022. https://arxiv.org/abs/2207.12598

[6] B. D. Rouhani et al., "Microscaling Data Formats for Deep Learning," *arXiv preprint arXiv:2310.10537*, 2023. https://arxiv.org/abs/2310.10537

[7] N. P. Jouppi et al., "Ten Lessons from Three Generations Shaped Google's TPUv4i: Industrial Product," *ACM/IEEE ISCA*, 2021. https://doi.org/10.1109/ISCA52012.2021.00010

[8] R. Guo et al., "A 28nm 74.34TFLOPS/W BF16 Heterogenous CIM-Based Accelerator Exploiting Denoising-Similarity for Diffusion Models," *ISSCC*, pp. 362-364, 2024. https://doi.org/10.1109/ISSCC49657.2024.10454308

[9] Y. Jing et al., "A 22nm 60.81TFLOPS/W Diffusion Accelerator with Bandwidth-Aware Memory Partition and BL-Segmented Compute-in-Memory for Efficient Multi-Task Content Generation," *ISSCC*, pp. 616-617, 2025. https://doi.org/10.1109/ISSCC49661.2025.10904573

[10] D. Han et al., "MEGA.mini: A Universal Generative AI Processor with a New Big/Little Core Architecture for NPU," *ISSCC*, pp. 416-417, 2025. https://doi.org/10.1109/ISSCC49661.2025.10904514

[11] S. Kim et al., "EdgeDiff: 418.4mJ/Inference Multi-Modal Few-Step Diffusion Model Accelerator with Mixed-Precision and Reordered Group Quantization," *ISSCC*, pp. 410-411, 2025. https://doi.org/10.1109/ISSCC49661.2025.10904594

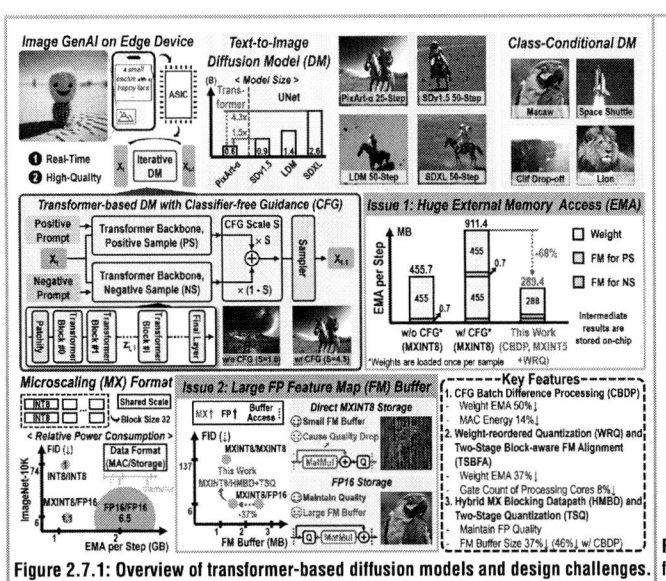

Figure 2.7.1: Overview of transformer-based diffusion models and design challenges.

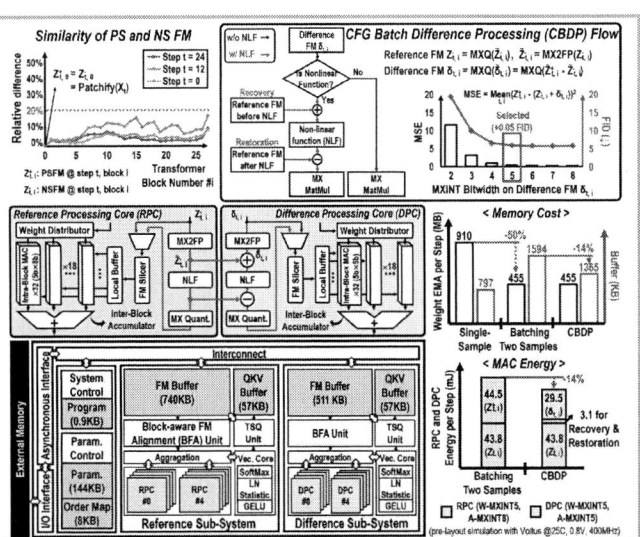

Figure 2.7.2: Proposed classifier-free guidance batch difference processing (CBDP) flow and overall system architecture.

ISSCC 2026 / SESSION 2 / PROCESSORS / 2.7

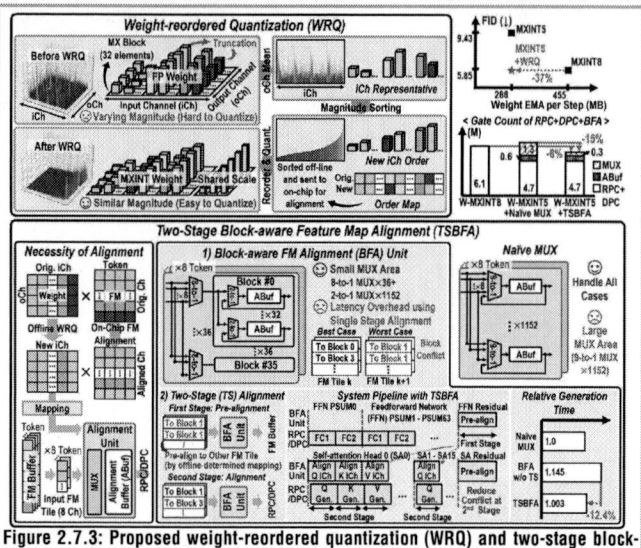

Figure 2.7.3: Proposed weight-reordered quantization (WRQ) and two-stage block-aware feature map alignment (TSBFA).

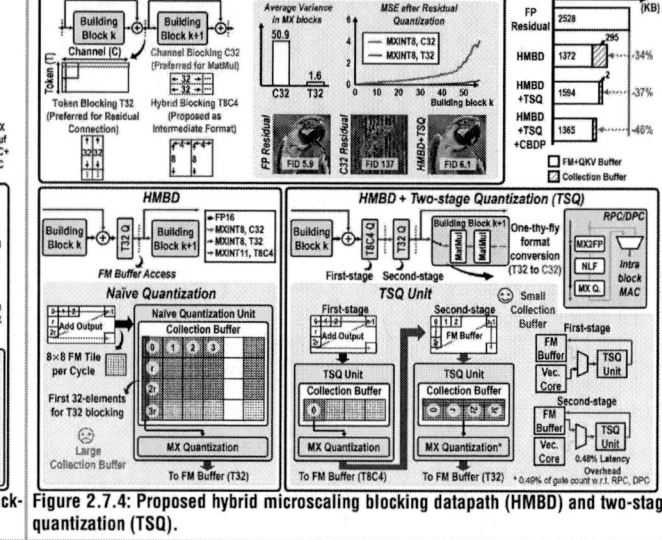

Figure 2.7.4: Proposed hybrid microscaling blocking datapath (HMBD) and two-stage quantization (TSQ).

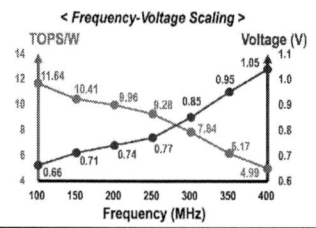
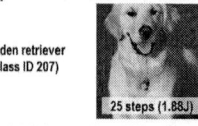

Model	DiT-XL/2	PixArt-α
Task	Class-Conditional Image Generation	Text-to-Image Generation
Building Blocks	Self-attention Feedforward Network	Self-attention Cross-attention Feedforward Network
Dataset	ImageNet (256×256)	MS-COCO (256×256)
FID (25 steps)	6.12 (FP: 5.85)	36.47 (FP: 37.17)
Generation Time w/ CFG (two prompts)[1]	98 ms/step	134 ms/step
Generation Energy w/ CFG (two prompts)[2]	75 mJ/step	99 mJ/step
System Energy w/ CFG (two prompts)[2,3]	121.9 mJ/step	159.2 mJ/step

[1] Best Performance, 400 MHz, 1.05 V. [2] Best Efficiency, 100 MHz, 0.66 V. [3] 1.3nJ per 64-bit access via DDR4 [7]

Figure 2.7.5: Measurement and inference results of Tiamat.

	ISSCC'24 [8]	ISSCC'25 [9]	ISSCC'25 [10]	ISSCC'25 [11]	This Work
Process (nm)	28	22	28	28	16
Supply Voltage (V)	0.60 - 1.00	0.60 - 0.90	0.57 - 0.90	0.68 - 1.00	0.66 - 1.05
Frequency (MHz)	50 - 540	180 - 478	10 - 250	50 - 250	100 - 400
Die Area (mm²)	3.67	12.96	8.20	20.25	8
Precision	INT10/16, FP16/BF16	INT4/8/12, FP6, BF16	W: INT4/8/12/16 IA_MAC: DFXP6 IA_exp: DFP12	INT4/8/12/16 + Group Quant.	W: MXINT5 IA_RPC: MXINT8 IA_DPC: MXINT5
Peak Performance (TOPS)	6.64 (BF16) 4.42 (FP16)	9.71 (Hybrid-precision)	2.05 (Hybrid IA)	31.30 (INT4) / 7.60 (INT8) 3.05 (INT12) / 1.95 (INT16)	7.37
Energy Efficiency (TOPS/W)	74.34 (BF16) 67.89 (FP16)	49.74 - 60.81 (Hybrid-precision)	12.90 - 38.10 (Hybrid IA)	34.40 (INT4-GQ) 8.60 (INT8-GQ)	4.99 - 11.64
Image Diffusion Model	SDv1.5	SDv1.5	SDv1.5	SDXL-Turbo	DiT-XL/2 PixArt-α
Backbone	UNet	UNet	UNet	UNet	Transformer Transformer
Guidance Support	Naïve[1]	Naïve[1]	Naïve[1]	Naïve[1]	Intrinsic Intrinsic
Generation Time per Step-Prompt (ms)	367[2]	261[2]	267	-	49 67
Generation Time per Inference (sec)	18.38 (50 steps, 1 prompt)	13.09 (50 steps, 1 prompt)	13.35 (50 steps, 1 prompt)[3]	2.12 (1 step, 1 prompt)[4] 7.33 (4 steps, 1 prompt)[4]	1.47 (15 steps) 2.45 (25 steps) 2.01 (15 steps) 3.35 (25 steps)

[1] Two-pass generation for two prompts (positive/negative).
[2] Projected from the reported 50-step generation time in [8] and [9].
[3] Projected from the reported per-step generation time in [10].
[4] Including encoder and decoder.

Figure 2.7.6: Performance comparison table and EMA optimization summary.

Chip Specifications

Technology	TSMC 16nm FinFET	
Supply Voltage	0.66 - 1.05 V	
Die Area	4 mm × 2 mm (8 mm²)	
Core Frequency	100 - 400 MHz	
SRAM Size	1518 KB	
	RPC #0-7 (Weight, IA)	DPC #0-7 (Weight, IA)
Precision	MXINT5, MXINT8	MXINT5, MXINT5
Peak Throughput[1,2]	3.68 TOPS	3.68 TOPS

System Performance

Condition	100 MHz, 0.66 V	400 MHz, 1.05 V
Core Power	158 mW	1477 mW
Efficiency	11.64 TOPS/W	4.99 TOPS/W

Diffusion Model Performance

Model	DiT-XL/2	PixArt-α
Task	Class-Conditional Image Generation	Text-to-Image Generation
Generation Time w/ CFG (two prompts)[2]	98 ms/step	134 ms/step
Generation Energy w/ CFG (two prompts)[3]	75 mJ/step	99 mJ/step

[1] 1MAC = 2 OPs
[2] Best Performance, 400 MHz, 1.05 V
[3] Best Efficiency, 100 MHz, 0.66 V

Figure 2.7.7: Chip micrograph and performance summary.

• 2026 IEEE International Solid-State Circuits Conference

ISSCC 2026 / SESSION 2 / PROCESSORS / 2.8

2.8 MADiC: A 3nm 7.4TOPS/mm², 17.4TOPS/W Generative Diffusion Accelerator Enabled by Hardware-Compiler Co-Optimization of Memory Hierarchy and Operator Parallelism

Shih-Wei Hsieh, Yi-Syuan Chen, Ping-Yuan Tsai, Ming-Hung Lin, Chia-Yuan Cheng, Lien-Feng Hsu, Po-Hao Huang, Hung-Wei Chih, Po-Han Chiang, Chia-Ming Chang, Ming-Hsuan Chiang, Chia-Hung Yuan, Sheng-Po Kuo, Viswanath Uggu, Chun-Kun Chan, Ming-En David Shih, Yu-Cheng Tseng, Hsin-Ping Cheng, Stan Huang, Chia-Ping Chen, ShenKai Chang, Chih-Ming Wang, Po-Yu Yeh, Jett Liu, Yung-Chang Chang, Chi-Cheng Ju, Yucheun Kevin Jou

MediaTek, Hsinchu, Taiwan

Abstract

This work presents MADiC, a 3nm 0.338mm² diffusion accelerator for generative image editing on edge devices. MADiC features design-time L1/L2 memory optimization, cross-loop feature map allocation with write-over-read relaxation, and activation-aware input preloading with concurrent resizing. Operating at 0.575V and 546MHz, it delivers 7.4TOPS/mm² and 17.4TOPS/W when running the Diffusion ConvNet model.

Generative diffusion models have become the mainstream approach for photo-realistic image synthesis, delivering exceptional quality across a range of applications including super-resolution and style transfer [1-3]. However, a significant gap remains in translating these capabilities to edge devices, where compact area and limited power budgets are challenged by the substantial computational and bandwidth demands inherent to these models. Recent algorithmic advances have begun to address these challenges. Diffusion Transformers (DiT) [4] leverage transformer architectures for better scalability, while step distillation [5, 6] reduces denoising steps from hundreds to one, reducing complexity by 50-100×. Diffusion ConvNets (DiCo) [7, 8] and architectural distillation [9] eliminate costly attention operations and further reduce model complexity, achieving an additional 3-to-10× reduction. These advances narrow the gap for diffusion model deployment; however, significant hardware challenges persist as DiCo-based models require multi-scale feature connections and squeeze-and-excitation (SE) mechanisms [10] in each block. Both induce large memory bandwidth burdens and increased latency on typical AI accelerators, resulting in low operational intensity and utilization for partial computation loops.

Increasing on-chip memory enables more data reuse, which improves operational intensity and alleviates memory bottlenecks in these loops. However, as technology moves to more advanced nodes, SRAM area scaling has increasingly lagged behind logic, as shown in Fig. 2.8.1. Our empirical study shows that from 28nm to 3nm, the SRAM area shrink at only 0.44× the rate of logic. This discrepancy poses a significant challenge for efficiently allocating on-chip memory to reduce external memory access (EMA) in area-constrained edge devices. In addition to memory bottlenecks, hardware utilization drops when input data fetching and resizing are performed sequentially with convolution, causing idle hardware and limited performance. To address these challenges, we present MADiC, a 3nm mini accelerator for Diffusion ConvNets, featuring: 1) design-time L1/L2 memory size optimization for minimizing SRAM area cost at a fixed EMA target, 2) cross-loop feature map (FM) storage allocation with write-over-read relaxation to reduce EMA, and 3) activation-aware input preloading and concurrent resizing enable operator parallelism to improve utilization. As a result, MADiC delivers 7.4TOPS/mm² and 17.4TOPS/W when running the DiCo model.

Figure 2.8.2 shows the overall architecture of MADiC, which includes a tensor core, a DMA module, 128KB L2 and 384KB L1 memory for activations, instructions, and weights, a preload controller, and two data streaming modules – the cross-loop feature map (FM) streamer and the activation L1 interconnection – that bridge the memory system and the tensor core datapath. The tensor core comprises a convolution (Conv) core with 2304 MAC units, a vector unit for element-wise operations, and a resizer for scaling and mean operations. Each unit's input and output connect to either the dual-mode activation L1 or the cross-loop FM streamer via the activation interconnection. The dual-mode activation L1 provides either dedicated bandwidth for parallel access or increased capacity for single-client access. The cross-loop FM streamer manages activation data across computation loops and selectively localizes frequently reused activations in L2 memory. Control registers, instructions, and weights are configured by a binary file generated by the accompanying compiler. At design time, the compiler serves as a design space optimizer, evaluating architectural choices and optimizing on-chip memory size for the target model pool. Once the memory size is fixed, the compiler performs memory allocation optimization, binary file generation, and power and latency estimation at runtime.

Figure 2.8.3 illustrates the design space exploration workflow that optimizes on-chip memory size under a predefined EMA target. MADiC executes a network by dividing it into inner loops, each representing a fused stack of layers for depth-first execution. During each inner loop, temporary FMs are stored in activation L1, while cross-loop FMs are exchanged via L2 memory or DRAM. Increasing L1 enables deeper inner-loop fusion and reduces external memory exchanges, while increasing L2 allows more cross-loop activations to remain on-chip. Both strategies lower EMA, leading to an area allocation tradeoff between L1 and L2. Additionally, since the benefits of scaling L1 or L2 are non-linear and depend strongly on the model topology, operation types, and tensor dimensions, we perform design space exploration over a pool of models, including both the main target and other representative models, to avoid overfitting to specific workloads. The exploration evaluates

the weighted sum of EMA for a set of (L1, L2) combinations, where area is calculated as the sum of L1 and L2 sizes normalized by their respective memory densities. Each evaluated (L1, L2) pair is plotted on the EMA-area graph, with the Pareto front showing the most efficient options. The final design is chosen as the Pareto-optimal point with the smallest area that satisfies the EMA constraint, minimizing the costly SRAM area in advanced nodes with minimal power overhead. Unlike traditional cache hierarchies that favor smaller lower-level caches without model- or hardware-specific optimization, our compiler-driven, design-time flow analyzes hardware-specific reuse and compiler-identified fusion to jointly optimize L1 and L2 for the target model pool. This approach outperforms typical baseline choices in both area and power, with 11% less area and 27% lower EMA than a naïve L1:L2=1:4 allocation.

Figure 2.8.4 shows our cross-loop FM allocation strategy, which maximizes L2 utilization by allocating FMs at a fine-grained, line-based level. Each FM is defined by its lifetime, line count, line size, and total accesses. The L2 bandwidth contribution of each FM is calculated as its allocated size times its access count. Our objective is to maximize total L2 bandwidth and minimize EMA, which can be formulated as an Integer Linear Programming (ILP) problem. This formulation includes two key constraints: the non-overlap constraint ensures that FMs with overlapping lifetimes are assigned to distinct addresses in L2, while the non-overflow constraint keeps the total allocation within the available L2 size. We solve the problem optimally using the CP-SAT solver [11]. While previous works [12–14] have primarily focused on finding feasible allocations, our formulation explicitly optimizes for bandwidth. To further reduce memory waste, we apply Write-over-Read (WoR) relaxation by analyzing lifetimes at the line level within each FM. Instead of assigning a single lifetime to each FM, we calculate a minimum line offset K_{ij} between FM_i and FM_j, which represents the smallest safe distance that allows their memory regions to partially overlap. K_{ij} is derived from the allocation and deallocation times of each line, and is integrated into the non-overlap constraint to enable tighter L2 memory usage. We design a light-weight cross-loop FM streamer to support the line-based allocation strategy. The streamer maintains a table for each FM with the base address, line size, and total allocated lines in L2. For each data access, it uses a counter table to track the current line count and computes the L2 address as (base + current line count × line size). If the current line count exceeds the allocated lines, the access is redirected to DRAM. We compare a greedy method, which heuristically places the FM with the highest access count first, to our optimal solution, both with and without WoR relaxation. The optimal solution without WoR improves L2 bandwidth by 21% over the greedy method, while enabling WoR relaxation provides an additional 15% improvement. From an area perspective, to achieve the same EMA target as the optimal method with WoR, the greedy method and the optimal method without WoR require 48KB and 16KB more L2 usage, respectively.

Figure 2.8.5 illustrates activation-aware input preloading and the concurrent resizing mechanism to improve hardware and L1 bandwidth utilization. In standard processing, input loading, convolution, and resizing are executed sequentially, resulting in underutilized compute units and memory bandwidth. To address this, input preloading initiates data transfer from DRAM to L1 at the earliest time possible, loading input data into L1 whenever write bandwidth is available. A preload sync signal is added to ensure that data dependencies are maintained. Since preloading extends the lifetime of input activations in L1, additional space is required for pre-allocation. Concurrent resizing further enables the resizer to operate in parallel with preloading and convolution. To support simultaneous access, L1 is partitioned into two banks, allowing the resizer and Conv core to operate without contention. The architecture supporting these features includes a preload controller and a dual-mode activation L1. The preload controller monitors write requests from the Conv core to L1 and arbitrates access between DMA and Conv core on a per-cycle basis. A global sync unit coordinates the preload operation and tensor core for correct sequencing. When concurrent resizing is disabled, the dual-mode activation L1 operates in stack mode, unifying two groups of activation SRAMs as a larger bank to increase overall capacity. During concurrent resizing, the groups operate independently to provide separate bandwidth for the resizer and Conv core. This design minimizes area overhead by reusing existing SRAM resources, eliminating the need for additional dedicated memory. The compiler enables input preloading and concurrent resizing by default, and selectively disables them only if L1 memory demand exceeds available space. For the DiCo model, enabling input preloading and concurrent resizing reduces resizer time by 41% and input loading time by 45%.

Figure 2.8.6 presents the measurement results and comparison table. Fabricated in 3nm technology, MADiC occupies 0.338mm² with a total of 512KB on-chip memory. When evaluated on the DiCo model, MADiC achieves a generation energy of 0.011J/inference and a latency of 0.116s, enabling low energy consumption and fast inference for end-to-end diffusion-based image editing. Notably, running DiCo model on MADiC with 8b activations and weights maintains high output quality, with negligible difference compared to FP32 outputs. Operating at 0.575V, MADiC achieves an area efficiency of 7.4TOPS/mm² and an energy efficiency of 17.4TOPS/W, outperforming previous diffusion processors [15-17] that do not rely on inter-timestep redundancy in both metrics. Figure 2.8.7 shows the chip micrograph and measurement results for additional non-diffusion benchmark models, demonstrating MADiC's ability to sustain high efficiency across a wide range of AI workloads.

References:

[1] L. Sun et al., "Pixel-Level and Semantic-Level Adjustable Super-Resolution: A Dual-Lora Approach," *IEEE CVPR*, pp. 2333-2343, 2025. http://doi.org/10.1109/cvpr52734.2025.00223
[2] R. Xu et al., "StyleSSP: Sampling StartPoint Enhancement for Training-free Diffusion-based Method for Style Transfer," *IEEE CVPR*, pp. 18260-18269, 2025. http://doi.org/10.1109/cvpr52734.2025.01702
[3] R. Rombach et al., "High-Resolution Image Synthesis with Latent Diffusion Models," *IEEE CVPR*, pp. 10684-10695, 2022. http://doi.org/10.1109/cvpr52688.2022.01042
[4] W, Peebles et al., "Scalable Diffusion Models with Transformers," *ICCV*, pp. 4195-4205, 2023. http://doi.org/10.1109/iccv51070.2023.00387
[5] Y. Kim et al., "Autoregressive Distillation of Diffusion Transformers," *IEEE CVPR*, pp. 15745-15756, 2025. http://doi.org/10.1109/cvpr52734.2025.01468
[6] T. Yin et al., "One-Step Diffusion with Distribution Matching Distillation," *IEEE CVPR*, pp. 6613-6623, 2024. http://doi.org/10.1109/cvpr52733.2024.00632
[7] Y, Tian et al., "DiC: Rethinking Conv3x3 Designs in Diffusion Model," *IEEE CVPR*, pp. 2469-2478, 2025. http://doi.org/10.1109/cvpr52734.2025.00236
[8] Y, Ai et al., "DiCo: Revitalizing ConvNets for Scalable and Efficient Diffusion Modeling," arXiv:2505.11196, 2025. http://doi.org/10.48550/arXiv.2505.11196
[9] B, Chen et al., "Adversarial Diffusion Compression for Real-World Image Super-Resolution," *IEEE CVPR*, pp. 28208-28220, 2025. http://doi.org/10.1109/cvpr52734.2025.02627
[10] J. Hu et al., "Squeeze-and-Excitation Networks," *IEEE CVPR*, pp. 7132-7141, 2018. http://doi.org/10.1109/cvpr.2018.00745
[11] Google OR Tools: CP-SAT Solver, Accessed Sept. 2025. http://developers.google.com/optimization/cp/cp_solver/
[12] M. Mass et al., "TelaMalloc: Efficient On-Chip Memory Allocation for Production Machine Learning Accelerators," *ACM ASPLOS*, pp. 127-137, 2023. http://doi.org/10.1145/3567955.3567961
[13] M. D. Moffitt, "MiniMalloc: A Lightweight Memory Allocator for Hardware-Accelerated Machine Learning," *ACM ASPLOS*, pp. 238-252, 2023. http://doi.org/10.1145/3623278.3624752

[14] M. Scherer et al., "Deeploy: Enabling Energy-Efficient Deployment of Small Language Models on Heterogeneous Microcontrollers," *IEEE TCAD*, pp.4009-4020, 2024. http://doi.org/10.1109/tcad.2024.3443718
[15] S. Kim et al., "EdgeDiff: 418.4 mJ/Inference Multi-Modal Few-Step Diffusion Model Accelerator with Mixed-Precision and Reordered Group Quantization," *ISSCC*, pp.410-412, 2025. http://doi.org/10.1109/isscc49661.2025.10904594
[16] D. Han et al., "MEGA.mini: A Universal Generative AI Processor with a New Big/Little Core Architecture for NPU," *ISSCC*, pp.416-418, 2025. http://doi.org/10.1109/isscc49661.2025.10904514
[17] S-W. Hsieh et al., "MAE: A 3nm 0.168mm² 576MAC Mini AutoEncoder with Line-Based Depth-First Scheduling for Generative AI in Vision on Edge Devices," *ISSCC*, pp. 414-416, 2025. http://doi.org/10.1109/isscc49661.2025.10904763
[18] M-E. Shih et al., "NVE: A 3nm 23.2 TOPS/W 12b-Digital-CIM-Based Neural Engine for High-Resolution Visual-Quality Enhancement on Smart Devices," *ISSCC*, pp. 360-362, 2024. http://doi.org/10.1109/isscc49657.2024.10454482
[19] R. Guo et al., "A 28nm 74.34 TFLOPS/W BF16 Heterogenous CIM-Based Accelerator Exploiting Denoising-Similarity for Diffusion Models," *ISSCC*, pp. 362-364, 2024. http://doi.org/10.1109/isscc49657.2024.10454308
[20] Y. Qin et al., "A 52.01 TFLOPS/W Diffusion Model Processor with Inter-Time-Step Convolution-Attention-Redundancy Elimination and Bipolar Floating-Point Multiplication," *IEEE Symp. VLSI Circuits*, pp. C304-C305, 2024. http://doi.org/10.1109/vlsitechnologyandcir46783.2024.10631322
[21] J-S. Park et al., "An On-Device Generative AI Focused Neural Processing Unit in 4nm Flagship Mobile Soc with Fan-Out Wafer-Level Package," *ISSCC*, pp. 286-288, 2025. http://doi.org/10.1109/isscc49661.2025.10904722

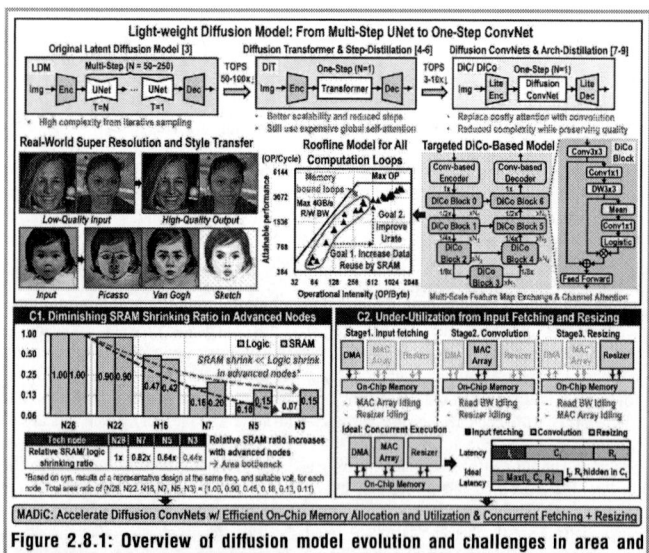

Figure 2.8.1: Overview of diffusion model evolution and challenges in area and utilization for efficient deployment.

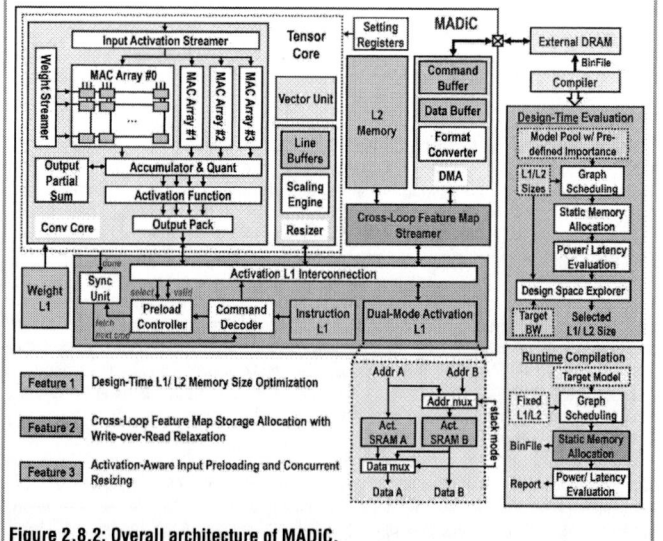

Figure 2.8.2: Overall architecture of MADiC.

ISSCC 2026 / SESSION 2 / PROCESSORS / 2.8

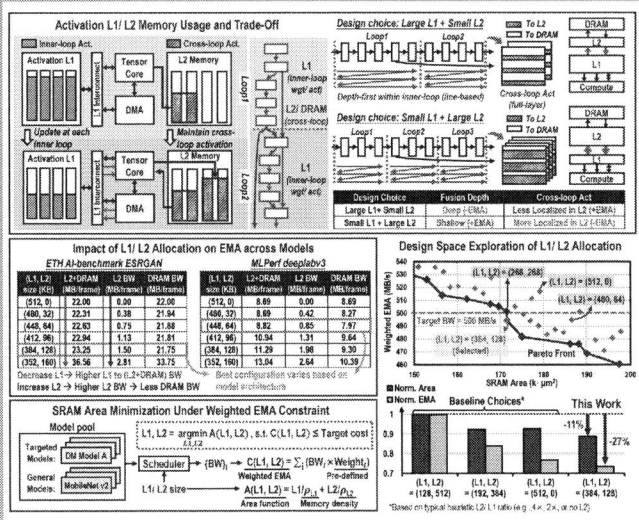

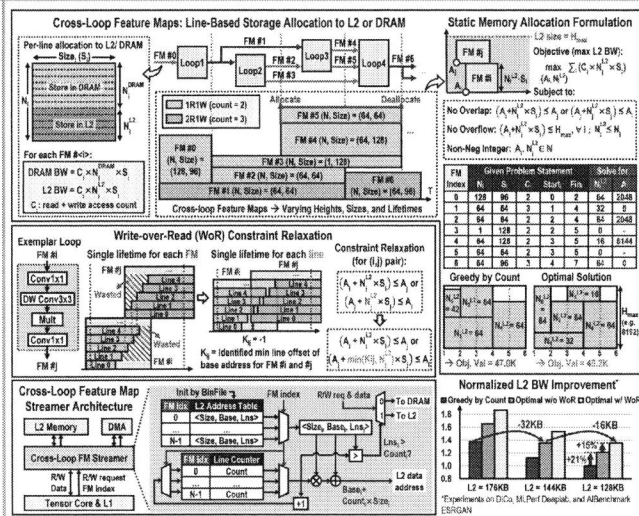

Figure 2.8.3: Design-time L1/L2 memory size optimization for minimizing SRAM cost.

Figure 2.8.4: Cross-loop data storage allocation and Write-over-Read (WoR) relaxation.

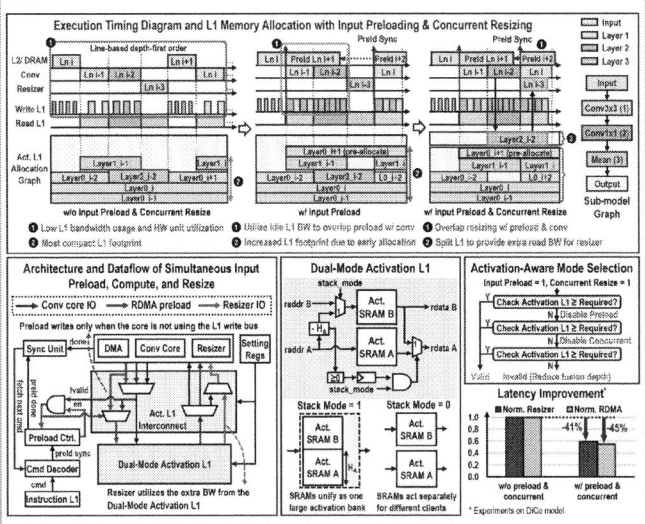

Figure 2.8.5: Activation-aware input preloading and concurrent resizing.

Figure 2.8.6: Measurement results and comparison table.

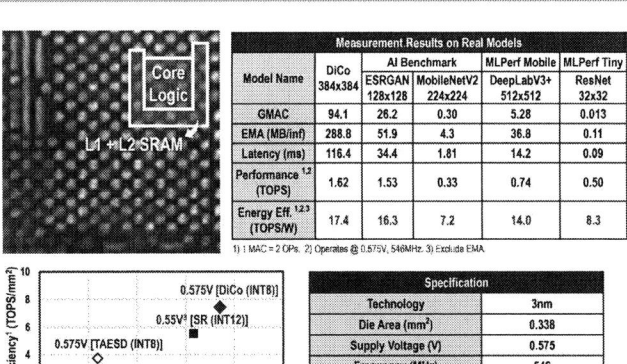

Figure 2.8.7: Chip micrograph and performance summary.

• 2026 IEEE International Solid-State Circuits Conference

ISSCC 2026 / SESSION 2 / PROCESSORS / 2.9

2.9 A 0.24mJ/Frame Quadratic Interpolation 4DGS Processor with Recursive Computation Reuse and Tree-Based Parallel-Rendering

Xiaolong Yang, Yang Wang, Wende Xu, Yubin Qin, Huanyu Wang, Ruiqi Guo, Zhiheng Yue, Jiangyuan Gu, Shaojun Wei, Yang Hu, Shouyi Yin

Tsinghua University, Beijing, China

Abstract

4D Gaussian Splatting (4DGS) has widespread applications in fields such as VR, AR and industrial simulation. However, 4DGS suffers from significant memory requirements, redundant computations and low PE utilization. This paper introduces adaptive quadratic interpolation, recursive computation reuse and tree-based parallel rendering to tackle these challenges. The processor in the paper achieves a rendering energy of 0.24mJ/frame and improves energy efficiency by 6.11× on the D-Nerf dataset.

4D Gaussian Splatting (4DGS) is a pivotal technology for rendering dynamic scenes, with widespread applications in fields such as VR, AR and industrial simulation [1-5]. Unlike its static counterpart, 3D Gaussian Splatting (3DGS) [6-8], the attributes of 4DGS, like position, shape, and color, are functions of time, enabling photorealistic dynamic renderings [9-11]. The 4DGS workflow consists of two stages: pre-processing and rendering. In the pre-processing stage, a large number of 4DGS parameters are loaded to compute 2DGS parameters based on the frame time and the camera pose of the frame. These 2DGS parameters are then passed to the rendering stage. The rendering stage is a two-step process: opacity computation and pixel rendering. The opacity of the kth 2DGS on pixel (i, j) is computed as $\alpha_{i,j,k}=\alpha_k \exp(a_k x_i^2+b_k x_j y_j+c_k y_j^2)$, where $(\alpha_k, a_k, b_k, c_k)$ are the 2DGS parameters, and $(x_i=i-x_g, y_j=j-y_g)$ represents the pixel coordinates relative to the 2DGS center. These opacities are passed to subsequent pixel rendering. The pixel color is obtained by iteratively accumulating 2DGS. At iteration k for pixel (i,j), color c^k and opacity $\alpha_{i,j,k}$ of the kth 2DGS are used to update pixel color and transparency as $C_{i,j,k+1}=C_{i,j,k}+T_{i,j,k}\alpha_{i,j,k}c^k$ and $T_{i,j,k+1}=T_{i,j,k}(1-\alpha_{i,j,k})$. The accumulation stops at iteration k if $T_{i,j,k}$<threshold, and different pixels stop at different iterations. Typically, the memory access and computation overhead of 4DGS is substantially higher than that of 3DGS. For instance, the size of a 4DGS can be up to 6GB, which is 100× larger than the typical 60MB of a 3DGS [12]. This overhead necessitates the design of a time and energy-efficient specialized processor for 4DGS.

Designing such a processor faces three key challenges. (1) The pre-processing stage suffers from a severe memory bottleneck, as loading 4DGS parameters for each frame to compute 2DGS parameters accounts for 89.11% of the total memory access overhead. Furthermore, due to the motion of 4DGS and the camera pose across frames, 96.66% of the 2DGS parameters change across frames, preventing the direct reuse of 2DGS parameters from the previous frame. (2) The opacity computation of the rendering stage suffers from O(N²) redundant computations for N×N pixels, which constitute 59.15% of the rendering computation. The repeated computation of ax_i^2 for pixels in the ith row and cy_j^2 for pixels in the jth column creates explicit redundancies, and the recursive relationship $bx_iy_{j+1}= bx_i(y_j+1)=bx_iy_j+bx_i$ between adjacent pixels creates implicit redundancies. (3) The pixel rendering of the rendering stage exhibits a highly imbalanced workload because different pixels stop rendering at different iterations. This results in a low PE utilization of only 28.1%. Moreover, the pixel rendering process has a recursive dependency, preventing the utilization of idle PEs.

This paper introduces a 4DGS processor with three key features to address these challenges. (1) An adaptive quadratic frame interpolation (AQFI) unit replaces memory-intensive pre-processing by generating 2DGS parameters for intermediate frames via quadratic interpolation of neighbor frames, reducing memory access overhead by 62.1%. (2) A fused recursive opacity computation (FROC) unit reuses the explicitly redundant computations of ax_i^2 and cy_j^2 across the ith row and the jth column, recursively computes the implicitly redundant computations of bx_iy_j between neighbor pixels, reducing the rendering computation by 59.15%. (3) A tree-based adaptive-precision parallel pixel rendering (TAPR) core breaks dependencies by transforming the traditional serial pixel rendering into a tree-based parallel pixel rendering, thus enabling the use of idle PEs, increasing PE utilization by 2.61×.

Figure 2.9.2 illustrates the overall architecture of the processor, comprising a pre-processing unit, an AQFI unit, 16 FROC units, a TAPR core, a top-level controller and 204KB SRAM. The AQFI chooses to use the pre-processing unit to compute 2DGS parameters or selects a suitable interpolation rate to obtain 2DGS parameters through memory-efficient quadratic interpolation of neighbor frames, reducing memory access. The FROC unit caches reusable intermediate results through a register file and computes (in parallel) the opacities of 8 pixels using one vertical recursion unit and eight fused horizontal recursion units, eliminating redundant computations. The TAPR core contains 8 tree-based pixel rendering PE lines. Each PE line includes 8 merge PEs and 8 bottom PEs, which can, in parallel, complete FP16 accumulations of 16 2DGS or FP8 accumulations of 32 2DGS, improving the PE utilization.

Figure 2.9.3 details the AQFI, reducing memory access overhead by replacing memory-intensive pre-processing with quadratic interpolation of neighbor frames. Quadratic interpolation, which uses three frames for interpolation, can fit variable curve motion of 2DGS across frames well. The quadratic interpolation function is $f(y_1,y_2,y_3)=$ $-1/8\times y_1+3/4\times y_2+3/8\times y_3$. Loading 2DGS params y_t, y_{t+2}, and y_{t+4} from frames t, t+2 and t+4, the quadratic interpolation result for frame t+1 and t+3 is $f(y_{t+4},y_{t+2},y_t)$ and $f(y_t,y_{t+2},y_{t+4})$. Compared to reusing the previous frame and linear interpolation, quadratic interpolation reduces the interpolation error by 99.9% and 94.2%, respectively. The AQFI adaptively determines frame interpolation rates of 2DGS according to their motion speeds. If the 2DGS's inter-frame change $S_4=|y_{t+4}-y_t|+|y_t-y_{t-4}|$<threshold, 4× interpolation (interpolate 3/4 frames) is performed. Otherwise, if $S_2=|y_{t+4}-y_{t+2}|+|y_{t+2}-y_t|$<threshold, 2× interpolation (interpolate 1/2 frames) is performed. 78.2% of 2DGS use 4× interpolation, 12.7% use 2× interpolation, and only 9.1% need to be pre-processed. 2DGS need to be sorted by depth to determine occlusion relationships. The AQFI collects partial depth-sorted results D_1 and D_2 for the 2DGS of frame t+1 from depth-sorted results of frame t and frame t+2. If there remain 2DGS of frame t+1 not in D_1 and D_2 (<0.1% probability), they are sorted by depth to get D_3. Then, D_1, D_2 and D_3 are merged to get the complete depth-sorted result for frame t+1. The AQFI reduces the sorting overhead from O(NLog(N)) to O(N). The AQFI reduces the energy of memory access by 62.1% and pre-processing time by 59.8%.

Figure 2.9.4 details the FROC, reducing redundant computations by reusing ax_i^2 and cy_j^2 across the ith row and the jth column, recursively computing bx_iy_j between neighbor pixels. The opacity for the pixel (i,j) is computed as $\alpha_{i,j}=\alpha\times \exp(ax_i^2)\times \exp(bx_iy_j)\times \exp(cy_j^2)$. To eliminate explicit redundancy, the FROC reuses computations of $x_i=i-x_g$ and $\exp(ax_i^2)$ for pixels in the ith row, and reuses computations of $y_j=j-y_g$ and $\exp(cy_j^2)$ for pixels in the jth column. To reduce the implicit redundancy, the FROC recursively computes $\exp(bx_iy_j)$. First, starting from $\exp(bx_1y_1)$, the first column is computed through the vertical recursion $\exp(bx_{i+1}y_1)=\exp(b(x_i+1)y_1)=\exp(bx_1y_1)\times \exp(by_1)$. Next, starting from $\exp(bx_iy_1)$, each row is computed through the horizontal recursion $\exp(bx_iy_{j+1})=\exp(bx_i(y_j+1))=\exp(bx_iy_j)\times \exp(bx_i)$. After obtaining $\exp(ax_i^2)$, $\exp(cy_j^2)$ and $\exp(bx_iy_j)$, the opacity is computed as $\alpha_{i,j}=\alpha\times \exp(ax_i^2)\times \exp(bx_iy_j)\times \exp(cy_j^2)$, where the first two multiplications can be fused to the horizontal recursion of $\exp(bx_iy_j)$ as $h_{i,j}=\alpha\times \exp(ax_i^2)\times \exp(bx_iy_j)$. $h_{i,j}$ also satisfies the horizontal recursion $h_{i,j+1}=h_{i,j}\times \exp(bx_i)$. After fusion, the opacity is computed via $\alpha_{i,j}=h_{i,j}\times \exp(cy_j^2)$. For N×N pixels, the FROC reduces total multiplication and addition operations from 11N² to 2N²+7N, and exponent operations from N² to 3N. The FROC reduces the total rendering computation overhead by 59.15% and improves the throughput by 2.14×.

Figure 2.9.5 details the TAPR, improving the PE utilization by converting the traditional serial pixel rendering process into a tree-based parallel pixel rendering. In serial pixel rendering, (C_{k+1},T_{k+1}) depends on (C_k,T_k). The TAPR breaks the dependency by transforming the serial pixel rendering into a tree-based parallel pixel rendering. First, N 2DGSs are divided into segments with two 2DGS in each segment. Bottom PE (BPE) accumulates each segment $[k,k+1]$ in parallel using $C[k,k+1]=\alpha_k c_k+(1-\alpha_k)\alpha_{k+1}c_{k+1}$ and $T[k,k+1]=(1-\alpha_k)(1-\alpha_{k+1})$. Then, a merge PE (MPE) parallelly and hierarchically merges the results of BPEs. MPE merges segments $[m,r]$ and $[r,n]$ using $C[m,n]=C[m,r]+T[m,r]C[r+1,n]$ and $T[m,n]=T[m,r]T[r+1,n]$. When workload imbalance occurs, the TAPR uses idle PEs for parallel computation, increasing PE utilization by 2.61×. N 2DGS can be divided into the former segment $[0,r]$ and the latter segment $[r+1,N]$. The final pixel color is $C[0,N]=C[0,r]+T[0,r]C[r+1,N]$. As r increases, $C[0,r]$ approaches the final color and $T[0,r]C[r+1,N]$ quickly approaches 0. The value of r is determined such that $T[0,r]$<threshold. Due to characteristics of FP addition, low-precision can be used for the computation of $C[r+1,N]$ without significantly affecting the addition result. The PE line contains 8 BPEs and 8 MPEs. By sharing the resources of the adder and multiplier, one FP16 PE can be efficiently reconfigured into two FP8 PEs. The PE operates at FP16 precision for the segment $[0,r]$ and FP8 precision for the segment $[r+1,N]$, improving throughput by 1.43×. The TAPR improves throughput by 3.59×.

Figure 2.9.6 shows the measurement results. The processor consumes 63.1 to 699.2mW of power with a voltage of 0.60 to 1.0V at 110 to 500MHz. The peak energy efficiency is 16.27TFLOPS/W, achieved at 0.65V with a 175MHz frequency. Evaluated on the D-NeRF Dataset [13], the processor achieves 1043.6fps with 699.2mW power consumption at 500MHz, 2.80× faster than the state-of-the-art (SOTA) rendering processor [20]. At 175MHz, it maintains 365.3fps for real-time rendering, while consuming 87.7mW and achieving a rendering energy of 0.24mJ/frame. The rendering energy is 26.4× less than the

58 • 2026 IEEE International Solid-State Circuits Conference

979-8-3315-8937-0/26 $31.00 © 2026 IEEE

SOTA NeRF processor [18], and 2.42× less than the SOTA 3DGS processor [20]. On D-NeRF, Immersive, and Neu3D datasets [13-15], the processor reaches 6.23×, 6.18×, 6.31× speedup, and 6.11×, 6.04×, 6.22× energy efficiency improvement. Figure 2.9.7 shows a summary and die photo of the processor fabricated in 28nm CMOS technology with an area of 3.65mm².

Acknowledgement:
This work was supported in part by the NSFC under Grant 62125403, Grant 92464302, Grant 92164301 and Grant 62304121; in part by the National Key R&D Project under Grant 2023YFB4403100; in part by the National Key Research and Development Program under Grant 2021ZD0114400; in part by the National Science and Technology Major Project under Grant 2022ZD0115201; Beijing S&T Project Z221100007722023; in part by the Beijing National Research Center for Information Science and Technology; and in part by the Beijing Advanced Innovation Center for Integrated Circuits. (Corresponding authors: Yang Wang and Shouyi Yin).

References:
[1] M. Kim et al., "4D Gaussian Splatting in the Wild with Uncertainty-Aware Regularization," *NeuriPS*, 2024. http://doi.org/10.48550/arXiv.2411.08879
[2] Y.-H. Huang et al., "SC-GS: Sparse-controlled Gaussian Splatting for Editable Dynamic Scenes," *IEEE CVPR*, 2024. http://doi.org/10.1109/CVPR52733.2024.00404
[3] H. Li et al., "GIFStream: 4D Gaussian-Based Immersive Video with Feature Stream," *IEEE CVPR*, 2025. http://doi.org/10.1109/CVPR52734.2025.02027
[4] Q. Hu et al., "4DGC: Rate-Aware 4D Gaussian Compression for Efficient Streamable Free-Viewpoint Video," *IEEE CVPR*, 2025. http://doi.org/10.1109/CVPR52734.2025.00090
[5] G. Wu et al., "4D Gaussian Splatting for Real-Time Dynamic Scene Rendering," *IEEE CVPR*, 2024. http://doi.org/10.48550/arXiv.2412.20720
[6] B. Kerbl et al., "3D Gaussian Splatting for Real-Time Radiance Field Rendering," *ACM Trans. Graph.*, vol. 42 no. 4, 2023. http://doi.org/10.1145/3592433
[7] J. Lee et al., "GSCore: Efficient Radiance Field Rendering via Architectural Support for 3D Gaussian Splatting," *ACM ASPLOS*, pp. 497–511, 2024. http://doi.org/10.1145/3620666.3651385
[8] L. Wu et al., "GauSPU: 3D Gaussian Splatting Processor for Real-Time SLAM Systems," *ACM/IEEE MICRO*, pp. 1562-1573, 2024. http://doi.org/10.1109/MICRO61859.2024.00114
[9] Y. Duan et al., "4D-Rotor Gaussian splatting: Towards Efficient Novel View Synthesis for Dynamic Scenes," *ACM SIGGRAPH Conf.*, 2024. http://doi.org/10.1145/3641519.3657463
[10] Z. Li et al., Spacetime Gaussian Feature Splatting for Real-Time Dynamic View Synthesis," *IEEE CVPR*, 2024. http://doi.org/10.1109/CVPR52733.2024.00813
[11] Z. Yang et al., "Real-Time Photorealistic Dynamic Scene Representation and Rendering with 4D Gaussian Splatting," *ICLR*, 2024. http://doi.org/10.48550/arXiv.2310.10642
[12] X. Zhang et al., "MEGA: Memory-Efficient 4D Gaussian Splatting for Dynamic Scenes," *ICCV*, 2025. http://doi.org/10.48550/arXiv.2410.13613
[13] A. Pumarola et al., "D-NeRF: Neural Radiance Fields for Dynamic Scenes," *CVPR*, 2021. http://doi.org/10.48550/arXiv.2011.13961

[14] M. Broxton et al., "Immersive Light Field Video with a Layered Mesh Representation," *ACM Trans. Graph.*, vol. 39, no. 4, pp. 86:1–86:15, 2020. http://doi.org/10.1145/3386569.3392485
[15] T. Li et al., "Neural 3D Video Synthesis From Multi-View Video," *CVPR*, 2022. http://doi.org/10.48550/arXiv.2103.02597
[16] Z. Ye et al., "Gaussian blending Unit: An Edge GPU Plug-In for Real-Time Gaussian-Based Rendering in AR/VR," *IEEE HPCA*, pp. 353-365, 2025. http://doi.org/10.1109/HPCA61900.2025.00036
[17] J. Ryu and al., "NeuGPU: A 18.5mJ/iter Neural-Graphics Processing Unit for Instant-Modeling and Real-Time Rendering with Segmented-Hashing Architecture," *ISSCC*, pp. 372-374, 2024. http://doi.org/10.1109/ISSCC49657.2024.10454276
[18] G. Park et al., "Space-Mate: A 303.5mW Real-Time Sparse Mixture-of-Experts-Based Nerf-SLAM Processor for Mobile Spatial Computing," *ISSCC*, pp. 374-376, 2024. http://doi.org/10.1109/ISSCC49657.2024.10454487
[19] S. Song et al., "IRIS: A 8.55mJ/frame Spatial Computing SoC for Interactable Rendering and Surface-Aware Modeling with 3D Gaussian Splatting," *ISSCC*, pp. 56-57, 2025. http://doi.org/10.1109/ISSCC49661.2025.10904521
[20] Feng et al., "1.78mJ/frame 373fps 3D GS Processor Based on Shape-Aware Hybrid Architecture Using Earlier Computation Skipping and Gaussian Cache Scheduler," *ISSCC*, pp. 54-55, 2025. http://doi.org/10.1109/ISSCC49661.2025.10904813

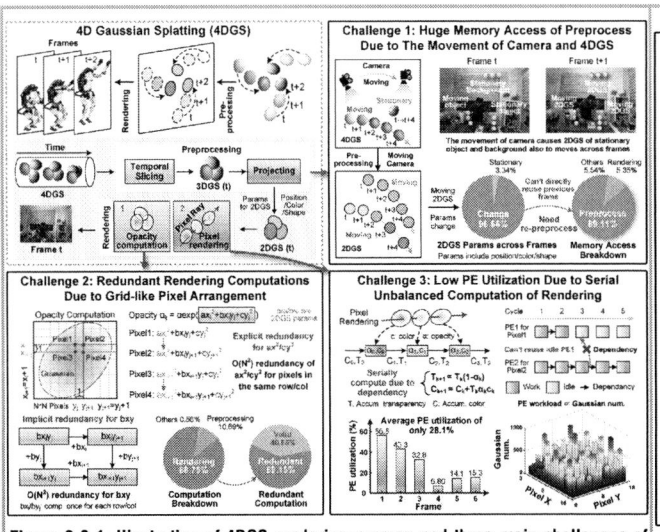

Figure 2.9.1: Illustration of 4DGS rendering process and three main challenges of efficient 4DGS rendering.

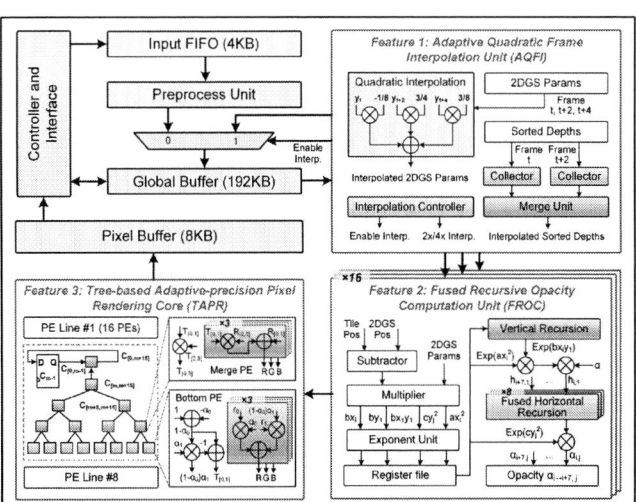

Figure 2.9.2: The overall architecture of the 4DGS processor.

ISSCC 2026 / SESSION 2 / PROCESSORS / 2.9

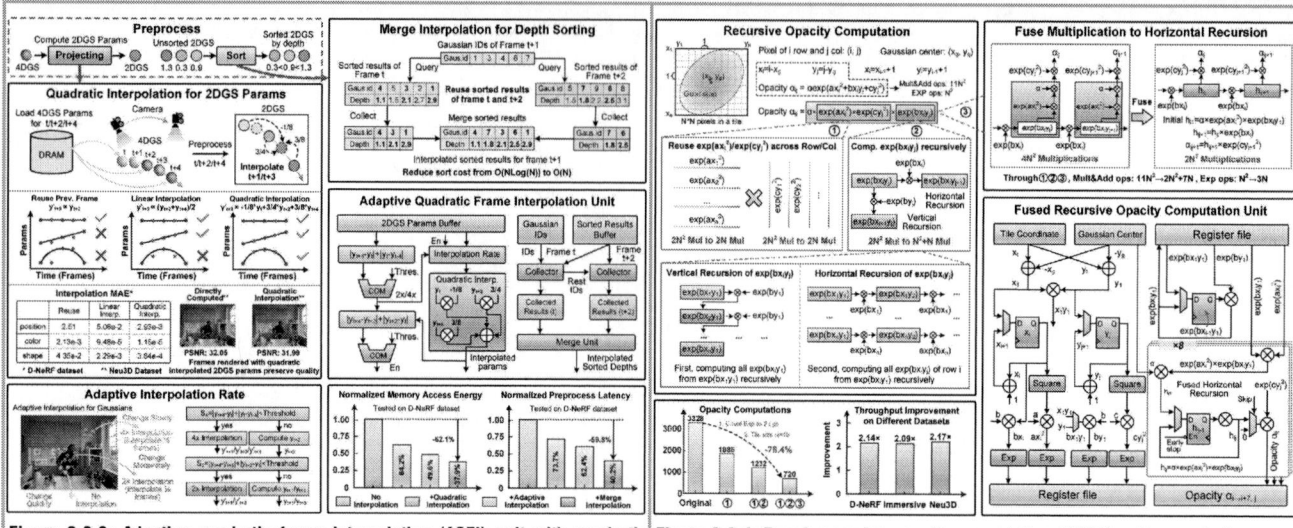

Figure 2.9.3: Adaptive quadratic frame interpolation (AQFI) unit with quadratic interpolation of neighbor frames.

Figure 2.9.4: Fused recursive opacity computation (FROC) unit recursively reuses redundant computations.

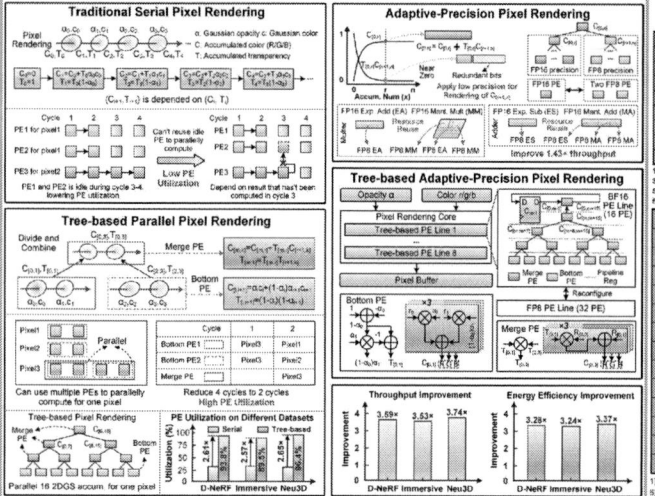

Figure 2.9.5: Tree-based adaptive-precision pixel rendering (TAPR) core with tree-based parallel pixel rendering.

Figure 2.9.6: Measurement results and comparison with state-of-the-art processors.

Model	HPCA'25[16]	ISSCC'24[17]	ISSCC'24[18]	ISSCC'25[19]	ISSCC'25[20]	This Work
	3DGS[2]	NeRF[7]	NeRF[3]	3DGS[8]	3DGS[8]	4DGS[5]
PSNR(dB)	33.26	30.32	/	25.8	33.98	34.05
Technique (nm)	28	28	28	28	28	28
Die Area (mm²)	1.78	20.25	20.25	20.25	2.43	3.65
Supply Voltage (V)	/	0.68-0.9	0.7-0.9	0.7-0.9	0.65-1.23	0.60-1.0
Frequency (MHz)	1000	100-200	50-200	200	150-700	110-500
Data Type	FP16	Mixed FP16/FP8	Mixed FP16/FP8	FP16	FP16	Mixed FP16/FP8[4]
On Chip SRAM (KB)	63	2112	1348	1743	187.4	204
Power (mw)	780	297.8-728.4	118.6-655.7	584	46.5-564	63.1-699.2
Energy Efficiency (TFLOPS/W)	/	4.46	2.78	/	6.65	16.27[5][6]
Throughput (FPS)	172.0	36.7-73.5	13.1-52.6	142.1	80-373	229.6[3]-1043.6[5][6]
Area Efficiency (FPS/mm²)	96.6	1.81-3.63	0.62-2.60	7.02	32.9-153	62.9[3][6]-285.9[4][6]
Energy per Frame (mJ/frame)	4.53	8.11-9.91	6.34-8.76	3.9	0.58-1.78	0.24[2][3][6]-0.67[4][6]

Figure 2.9.7: Chip micrograph and performance summary.

	Specifications
Technology	28nm CMOS
Die Area	3.65 mm²
On Chip SRAM	204 KB
Voltage	0.60-1.0 V
Frequency	110-500 MHz
Power	63.1-699.2 mW
Data Type	Mixed FP16/FP8[1]
Energy Efficiency	16.27[2][3][6] TFLOPS/W
Throughput	229.6[3]-1043.6[2][4] FPS
Area Efficiency	62.9[3]-285.9[2][4] FPS/mm²
Rendering Energy	0.24[2][3][6]-0.67[2][4][6] mJ/frame
PSNR — D-NeRF	34.05 (-0.04[5] loss)
PSNR — Immersive	29.05 (-0.15[5] loss)
PSNR — Neu3D	31.97 (-0.08[5] loss)

1) Mixed FP16/FP8 for TAPR, FP16 for other units.
2) Evaluated on the D-NeRF dataset.
3) Highest efficiency point, 0.65V, 175MHz.
4) Highest performance point, 1.0V, 500MHz.
5) Compared with videos rendering with FP32 precision.
6) Include all on-chip components. DRAM is not included.

• 2026 IEEE International Solid-State Circuits Conference

979-8-3315-8937-0/26 $31.00 © 2026 IEEE

ISSCC 2026 / SESSION 2 / PROCESSORS / 2.10

2.10 A 1286fps 0.39mJ/Frame Modeling/Rendering Unified 3D GS Processor with Locality-Optimized Computation and Reconfigurable Architecture

Hedi Wang*, Xiaoyu Feng*, Weichen Gao, Huazhong Yang, Yongpan Liu

Tsinghua University, Beijing, China
*Equally Credited Authors (ECAs)

Abstract

A modeling/rendering unified 3D GS processor is proposed with: 1) A locality-aware dynamic fine-grained rendering engine for reduced redundant computation. 2) A locality-optimized unified rendering workflow to reduce EMA. 3) A unified reconfigurable architecture for neural modeling and Gaussian rendering with minimal area overhead. It achieves 3.4× higher rendering throughput and 74.1% lower energy per frame than SOTA 3D GS accelerators, and an orders-of-magnitude reduction in modeling latency.

3D Gaussian Splatting (3D GS) [1] is widely adopted in applications such as virtual reality due to its realistic modeling and efficient rendering. However, traditional 3D GS modeling relies on backpropagation-based optimization with up to 30,000 iterations, leading to slow modeling and limited generalization. To enable faster modeling, feedforward GS modeling employs neural networks to generate a Gaussian model in a single inference [2-3]. This approach improves generalization and reduces modeling time from tens of minutes to seconds, enabling rapid creation, sharing, and viewing of 3D models on mobile devices. Nevertheless, feedforward GS modeling introduces different Gaussian distributions and dataflows, posing new challenges for traditional 3D GS processors, as shown in Fig. 2.10.1. First, feedforward pixel-aligned modeling produces smaller, more localized Gaussians, resulting in redundant coarse-grained rasterization. Traditional accelerators [4-5] rely on static coarse-grained rendering arrays. However, when processing small Gaussians, many computing units remain idle as they fall outside the Gaussian coverage, leading to low utilization and high redundancy. Second, irregular Gaussian shapes and positions necessitate a gather step to assign each tile, introducing significant external memory access (EMA) overhead. Although 3D Gaussian storage can be optimized with existing techniques [6], the intermediate 2D Gaussians are repeatedly written to and read from external memory, accounting for 50.6% of the total memory access. Finally, feedforward 3D GS requires hardware support for heterogeneous multiply-accumulate operations and rasterization dataflows for modeling and rendering, respectively. Adopting a heterogeneous architecture to accommodate both would incur additional area overhead.

To address these challenges, this paper presents a modeling/rendering unified 3D GS processor with three key features, as illustrated in Fig. 2.10.2: 1) A locality-aware Dynamic Fine-Grained Rendering Engine (DFGRE) that employs multiple parallel Micro-Clusters (MCs), each dynamically assigned to render one Gaussian. This design reduces redundancy and improves utilization for localized Gaussians. 2) A Locality-Optimized Unified Rendering (LOUR) workflow, which utilizes a Sliding Window-Based (SWB) canvas to establish a one-to-one correspondence between each Gaussian and a window, ensuring rendering occurs exclusively within a single window. This eliminates the gather step and the associated 2D Gaussian memory access. 3) A modeling/rendering Unified Reconfigurable Architecture (URA) that shares memory, controllers, and logic arrays to support both operational modes, improving area efficiency and hardware reusability.

The processor integrates a DFGRE, an SWB canvas, and a top controller. The DFGRE consists of a Scanline-Based Traversal Unit (SBTU) for Gaussian boundary detection, eight MCs – each configurable as four Gaussian PEs (GPEs) or two lines of Neural PEs (NPEs) – and a Dynamic Mapping Crossbar (DMC). Results are written to the SWB canvas, which contains 256 FP accumulators and a 32-bank output buffer. Additionally, 122KB of SRAM is used to store Gaussian parameters and images during rendering, as well as features, weights, and output during modeling.

Figure 2.10.3 illustrates the detailed architecture of the DFGRE. In static coarse-grained rendering, the entire array processes only one Gaussian per cycle using pixel-level parallelism, which leads to high redundancy and low utilization when dealing with small, localized Gaussians. In contrast, the DFGRE employs fine-grained computing units and Gaussian-level parallelism to reduce redundant computations. Gaussian parameters are fetched from a 4-bank Gaussian memory and stored in 8 Gaussian buffers, each assigned to one MC. Each MC contains 4 GPEs that compute the influence of each Gaussian on pixels based on position, covariance, opacity and color. Results are written to RGBA FIFOs and forwarded to the DMC. The DMC employs periodic priority scheduling to map Gaussian distributions to the 16-bank SWB canvas for accumulation, minimizing collision and blocking. Owing to the sort-free Gaussian algorithm [7], Gaussians at different depths can be processed out of order. Finally, pixel color and opacity are accumulated on the SWB canvas. Throughout this process, each MC fetches the next Gaussian during idle cycles, extracts its boundary, and rasterizes a 2×2 Pixel Group (PG) within that boundary. The MC proceeds to the next PG within the same boundary and only fetches a new Gaussian after all pixels in the current boundary have been traversed. Furthermore, for accurate boundary detection, the DFGRE incorporates an SBTU that extracts scanline-based boundaries instead of using Axis-Aligned Bounding Boxes (AABBs). The SBTU locates the ellipse where the transparency α equals a predefined threshold and determines the valid pixel ranges per row.

The dynamic scheduler and SBTU consume only 13.7% of total power and 12.0% of area. Compared to a 16×16 static array, the DFGRE using AABB reduces computational load by 89.2% and improves efficiency by 8.4×. Adding the SBTU further reduces computation by 58.1% and enhances efficiency by 63.6%, without compromising image quality.

Figure 2.10.4 depicts the LOUR workflow with the SWB canvas. Traditional Tile-Based Rendering (TBR) employs a discrete workflow. Since a Gaussian may span multiple tiles, all Gaussians are first projected globally, followed by a gather step to assign them to tiles. The intermediate 2D Gaussians are written to external DDR memory and repeatedly read back during the subsequent per-tile rasterization, accounting for 50.6% of the total EMA. In contrast, LOUR uses a unified workflow that exploits the spatial locality of Gaussians. It replaces tiles with sliding windows and establishes a one-to-one mapping between Gaussians and windows based solely on the position projection. Since each Gaussian is rendered exclusively within a single window, LOUR performs continuous per-window feature projection and rasterization without needing to store intermediate 2D Gaussian results in DDR. Consequently, the gather step and its associated 2D Gaussian memory access overhead are eliminated. However, using naïve LOUR alone can cause cropping of edge and large Gaussians, resulting in degraded rendering accuracy. To address this, the SWB canvas was introduced in hardware. The processor rasterizes a 16×16 sliding window at a time, which contains an 8×8 core region. Only Gaussians whose centers lie in the core region are mapped to the sliding window, while their contributions to the padding region are also rendered, thereby mitigating edge cropping. The SWB canvas uses 4 ping-pong banks for alternating read, write, and compute tasks. The window moves in 8-pixel steps row and column-wise, ultimately traversing the full image. On the software side, a Scale-Penalized Gaussian (SPG) method was employed. During training, over-scaled Gaussians are identified, and their average scale is penalized through the loss function, mitigating large Gaussian cropping. Together, the SWB canvas and SPG method recover 5.36dB in rendering quality. By eliminating the 2D Gaussian memory accesses, LOUR reduces EMA by 42.6% and 34.1% on the Synthetic NeRF and RE10K datasets, respectively.

Figure 2.10.5 illustrates the unified reconfigurable architecture for modeling and rendering. Both workflows share certain similarities: sparse inputs, broadcast of Gaussian parameters or weights, floating-point units and an output-stationary paradigm. Leveraging these commonalities, we designed a unified hardware architecture that accommodates both dataflows. In rendering mode, Gaussians are fetched from SRAM, buffered, and broadcast to an MC. PGs within coverage are traversed, and results are accumulated into the output buffer via the DMC. In modeling mode, the same hardware is reconfigured for outer-product matrix multiplication. Input features and sparse weights are read line-wise from SRAM and buffered. Each line of weights iterates through its index-value pairs one by one, broadcasting to the NPE lines for multiplication with input features until a preconfigured effective weight limit is reached. The process reuses the same traversal and data-fetching path as Gaussian rendering. At the computation level, the modeling configuration uses 16 lines of NPEs, each with 16 NPEs reconfigured from 2 GPEs. The DMC routes results to the 16-bank output buffer, where partial sums are accumulated using 256 FP adders – reusing the rendering-mode accumulation structure. Moreover, since modeling mode permits the use of FP8 precision, each neural multiply-accumulator (comprising one NPE and one adder) supports FP16/FP8 reconfiguration. By reusing most storage and compute resources, the URA reduces area overhead by 40.2% compared to a heterogeneous architecture. Additionally, by leveraging the sparse data path and reconfigurable precision, URA achieves up to 4.04× higher throughput and 8.26× better energy efficiency than dense FP16 computation in modeling.

Fabricated in 28nm CMOS technology, the unified 3D GS processor occupies 4.76mm² and operates at 80-680MHz with 0.58-1.01V supply. Figure 2.10.6 presents the measurement results and a comparison table. In rendering mode, the processor supports models from both optimization-based (on the Synthetic NeRF dataset [8]) and feedforward (on the RE10K dataset [9]) methods. At 680MHz, it achieves a throughput of 1286fps with a power consumption of 507mW, representing an 11.7× improvement over existing NeRF accelerators [10-11] and a 3.4× gain compared to 3D GS accelerators [5, 12]. At 80MHz, it consumes only 0.15mJ per frame, which is 74.1% lower than the state-of-the-art rendering accelerator [5]. The sort-free Gaussian algorithm provides the foundation for out-of-order

ISSCC 2026 / February 16, 2026 / 5:15 PM

rendering, while the performance improvements in rendering mode are realized through the DFGRE and LOUR hardware architectures. The unified processor also supports feedforward 3D GS modeling. The feedforward algorithm reduces modeling time from several minutes to 1.2s, and the processor's support for sparse and FP8-quantized computations further cuts inference time to 0.30s, which achieves an order-of-magnitude reduction in modeling latency compared to optimization-based modeling accelerators [11-12]. Moreover, the unified reconfigurable architecture minimizes the additional area overhead required to support feedforward modeling. In modeling mode, the processor consumes only 1067mW at 680MHz, which is 0.30% of the power consumption of a GPU. The die photo and chip summary are provided in Fig. 2.10.7. In conclusion, we have introduced a modeling/rendering unified 3D GS processor capable of performing both 3D GS rendering and feedforward 3D GS modeling with high speed and energy efficiency, making it suitable for mobile devices.

Acknowledgement:
This work was supported in part by the National Science and Technology Major Project (Grant No. 2021ZD0114402); the National Natural Science Foundation of China under grant number 92267203; and the Beijing Major Science and Technology Project under contract No. Z241100004224016. The corresponding author is Xiaoyu Feng.

References:
[1] B. Kerbl et al., "3D Gaussian Splatting for Real-Time Radiance Field Rendering," *ACM Trans. Graph.*, vol. 42, no. 139, 2023. https://doi.org/10.1145/3592433
[2] D. Charatan et al., "PixelSplat: 3D Gaussian Splats from Image Pairs for Scalable Generalizable 3D Reconstruction," *IEEE CVPR*, pp. 19457-19467, 2024. https://doi.org/10.1109/CVPR52733.2024.01840
[3] Y. Chen et al., "MVSplat: Efficient 3D Gaussian Splatting from Sparse Multi-view Images," *ECCV*, pp. 370-386, 2024. https://doi.org/10.1007/978-3-031-72664-4_21
[4] J. Lee et al., "GSCore: Efficient Radiance Field Rendering via Architectural Support for 3D Gaussian Splatting," *ACM ASPLOS*, pp. 497-511, 2024. https://doi.org/10.1145/3620666.3651385
[5] X. Feng et al., "1.78mJ/Frame 373fps 3D GS Processor Based on Shape-Aware Hybrid Architecture Using Earlier Computation Skipping and Gaussian Cache Scheduler," *ISSCC*, pp. 54-55, 2025. https://doi.org/10.1109/ISSCC49661.2025.10904813
[6] P. Papantonakis et al., "Reducing the Memory Footprint of 3D Gaussian Splatting," *ACM on Computer Graphics and Interactive Techniques*, vol. 7, no. 16, pp. 1-17, 2024. https://doi.org/10.1145/3651282
[7] Q. Hou et al., "Sort-free Gaussian Splatting via Weighted Sum Rendering," *ICLR*, 2025. https://doi.org/10.48550/arXiv.2410.18931
[8] B. Mildenhall et al., "NeRF: Representing Scenes as Neural Radiance Fields for View Synthesis," *Commun. of the ACM*, vol. 65, no. 1, pp. 99-106, 2021. https://doi.org/10.1145/3503250
[9] T. Zhou et al., "Stereo magnification: learning view synthesis using multiplane images," *ACM Trans. Graph.*, vol. 37, no. 65, 2018. https://doi.org/10.1145/3197517.3201323
[10] D. Han et al., "MetaVRain: A 133mW Real-Time Hyper-Realistic 3D-NeRF Processor with 1D-2D Hybrid-Neural Engines for Metaverse on Mobile Devices," *ISSCC*, pp. 50-52, 2023. https://doi.org/10.1109/ISSCC42615.2023.10067447
[11] J. Ryu et al., "NeuGPU: A 18.5 mJ/Iter Neural-Graphics Processing Unit for Instant-Modeling and Real-Time Rendering with Segmented-Hashing Architecture," *ISSCC*, pp. 372-374, 2024. https://doi.org/10.1109/ISSCC49657.2024.10454276

[12] S. Song et al., "IRIS: A 8.55mJ/frame Spatial Computing SoC for Interactable Rendering and Surface-Aware Modeling with 3D Gaussian Splatting," *ISSCC*, pp. 56-57, 2025. https://doi.org/10.1109/ISSCC49661.2025.10904521

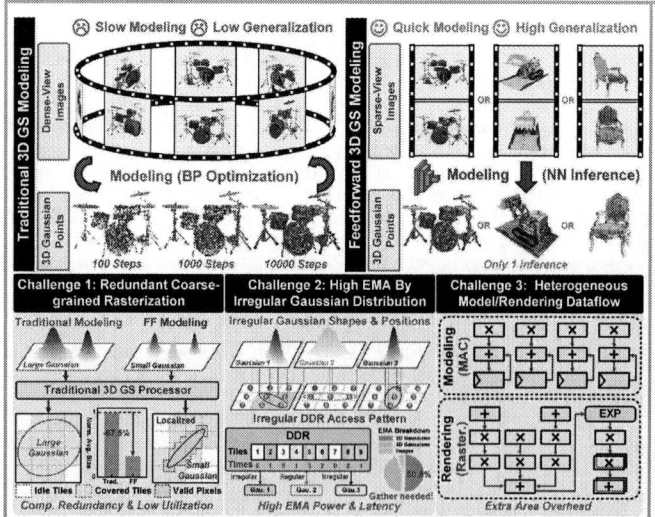

Figure 2.10.1: Overview of feedforward 3D Gaussian splatting modeling and challenges of traditional 3D GS processor.

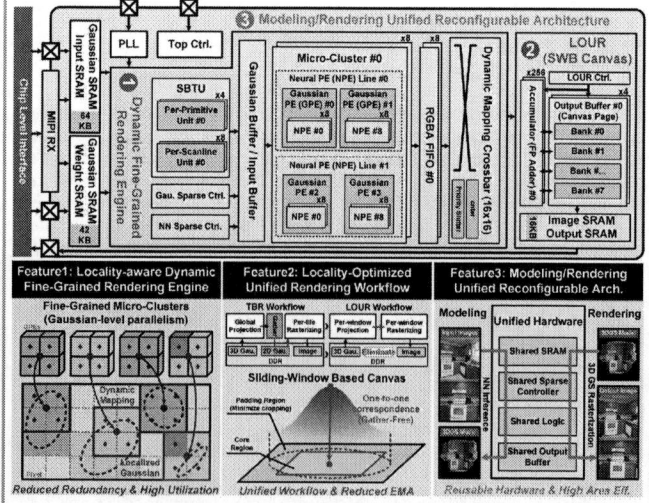

Figure 2.10.2: Overall chip architecture and key features of the proposed modeling/rendering unified 3D GS processor.

ISSCC 2026 / SESSION 2 / PROCESSORS / 2.10

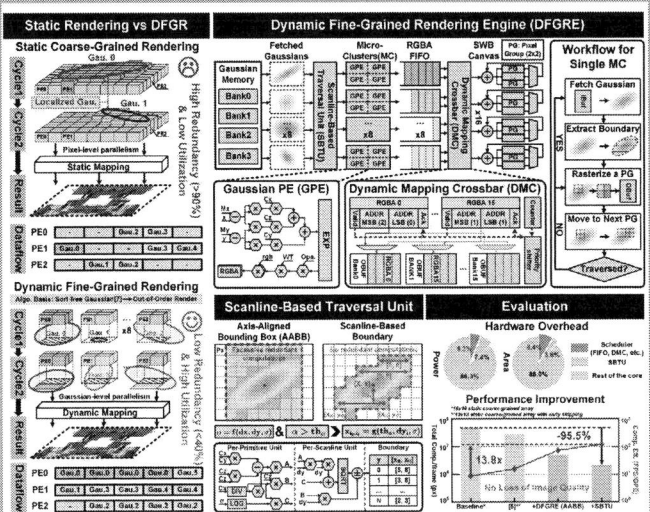

Figure 2.10.3: Locality-aware dynamic fine-grained rendering engine featuring a scanline-based traversal unit for reduced redundant computation.

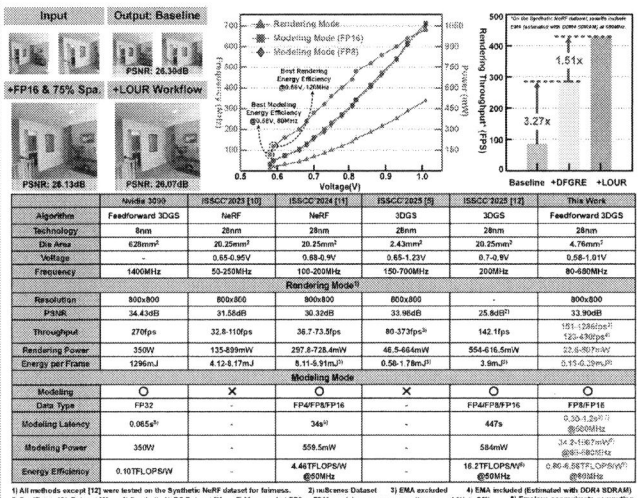

Figure 2.10.4: Locality-optimized unified rendering workflow adopting sliding window-based canvas and scale-penalized gaussians to reduce EMA.

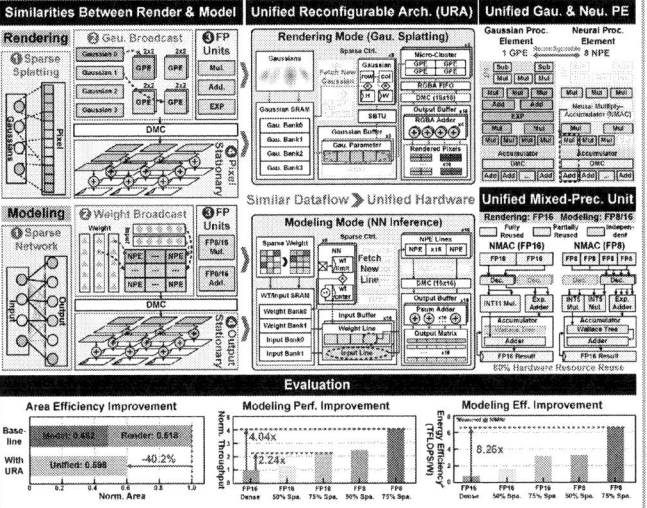

Figure 2.10.5: Unified reconfigurable architecture for neural modeling and Gaussian splatting rendering via exploitation of dataflow similarity.

Figure 2.10.6: Measurement results and performance comparison table.

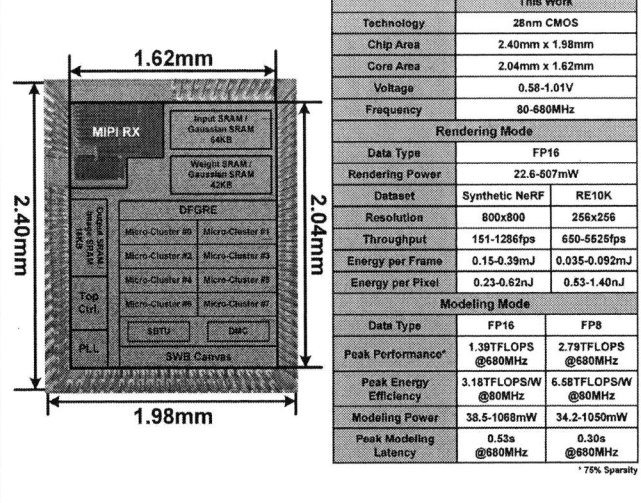

Figure 2.10.7: Die photo and chip summary.

- 2026 IEEE International Solid-State Circuits Conference

ISSCC 2026 / SESSION 3 / WEARABLE AND WIRELESS BIOMEDICAL SYSTEMS / OVERVIEW

Session 3 Overview: *Wearable and Wireless Biomedical Systems*

MEDICAL SUBCOMMITTEE

Session Chair: Hyung-Min Lee
Korea University, Seoul, Korea

Session Co-Chair: Venugopal Gopinathan
Analog Devices, Wilmington, MA

Biomedical systems continue to integrate and improve various functionalities for a wide range of wearable and wireless applications. The first paper describes a wireless multiple bio-signal sensing SoC, followed by a blood pressure monitoring transceiver using near-field RF reflection. A hybrid resonant pulse-train transmitter for ultrasound wireless power transfer comes next, followed by another ultrasound transmitter with autonomous power regulation. The final paper showcases a wireless power and bidirectional data transfer system for miniaturized implants.

62 • 2026 IEEE International Solid-State Circuits Conference

979-8-3315-8937-0/26 $31.00 © 2026 IEEE

ISSCC 2026 / February 16, 2026 / 1:30 PM

3

1:30 PM

3.1 A Multimodal Biosensing System-on-Chip with Integrated Wireless Transceiver and Power Management for Stress Monitoring

Lin Chou, National Yang Ming Chiao Tung University, Hsinchu, Taiwan

In Paper 3.1, National Yang Ming Chiao Tung University describes a highly integrated SoC in 180nm CMOS with multiple front-ends allowing ECG, PPG, GSR, and ECH sensor data to be acquired and transmitted through a wireless link, together with an on-chip PMU. The chip consumes only 125.8µW of AFE power.

1:55 PM

3.2 A Near-Field RF Reflection Transceiver ASIC for Continuous Unobtrusive Blood Pressure Monitoring

Yiming Han, University of Texas, Austin, TX

In Paper 3.2, University of Texas at Austin proposes a near-field RF reflection transceiver to measure systolic and diastolic blood pressure using the changes in electrical characteristics of the arterial wall and blood within 4.9/3.2mmHg, implemented in 65nm CMOS.

2:20 PM

3.3 A Battery-Powered Hybrid Resonant Pulse-Train Generator with Adaptive Frequency Tracking and Residual Energy Recycling for Ultrasonic Implants

Xiayang Li, Delft University of Technology, Delft, The Netherlands

In Paper 3.3, Delft University of Technology presents a battery-powered hybrid resonant pulse-train generator for ultrasonic wireless power transfer to implantable medical devices. The 180nm BCD ASIC provides an output up to 25V and achieves 68.6% C_p-loss reduction and 21.7% end-to-end efficiency.

2:45 PM

3.4 An Ultrasound-Powering TX with Standing-Wave Peak Tracking Employing Adiabatic Power Sensing Achieving 82% Power-Tracking Accuracy and <90ms Settling Time for Brain Implants

Marios Gourdouparis, imec, Eindhoven, The Netherlands, TU Delft, Delft, The Netherlands

In Paper 3.4, imec presents an ultrasound transmitter ASIC in 65nm CMOS that regulates the power delivered to an implant using autonomous on-chip standing wave peak tracking. It consumes only 43µW for power tracking and improves power delivery up to 2.7× with a tracking accuracy of 82%.

3:00 PM

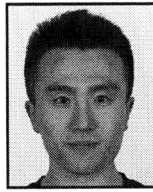

3.5 A Simultaneous Wireless Power and Full-Duplex Data Transfer System Over a Single Inductive Link Achieving 17/3.4Mb/s and 61.1% Efficiency for Miniature Biomedical Implants

Tianqi Lu, Delft University of Technology, Delft, The Netherlands

In Paper 3.5, Delft University of Technology introduces a wireless power and data transfer system in 180nm BCD for miniaturized biomedical implants, achieving 17Mb/s uplink and 3.4Mb/s downlink over a single 40.68MHz inductive link with 61.1% end-to-end power transfer efficiency.

DIGEST OF TECHNICAL PAPERS • 63

979-8-3315-8937-0/26 $31.00 © 2026 IEEE

ISSCC 2026 / SESSION 3 / WEARABLE AND WIRELESS BIOMEDICAL SYSTEMS / 3.1

3.1 A Multimodal Biosensing System-on-Chip with Integrated Wireless Transceiver and Power Management for Stress Monitoring

Min-Hua Chang*[1], Lin Chou*[1], Yen-Jen Liu[1], Chen-Hsing Hsu[1], Ying-Yu Yang[1], Bo-Cheng Juan[1], Guan-Yu Lin[1], Yi-Jie Lin[1], Yao-Wei Huang[1], Nitish Kumar[2], Shu-Ping Lin[2], Yu-Te Liao[1]

[1]National Yang Ming Chiao Tung University, Hsinchu, Taiwan, [2]National Chung Hsing University, Taichung, Taiwan
*Equally Credited Authors (ECAs)

Abstract

A highly integrated SoC incorporates ECG, PPG, GSR, and ECH biosignal acquisition interface circuits with a 920MHz, 2kb/s, 1.1µW WuRx, a 2.4GHz, 1Mb/s Tx, and a PMU with <5µs recovery time under 600µA-to-50mA load steps in 0.18µm CMOS. A shared ECG/ECH AFE reaches $0.32\mu V_{rms}$ IRN and 130dB CMRR. The PPG circuits attain a 113.9dB DR using a precharge scheme, while the GSR AFE achieves a 99dB DR with $3.1pA_{rms}$ IRN. The biosensing interface consumes only 125.8µW.

Psychological stress monitoring (PSM) has emerged as a critical challenge in modern health assessment. Fingers are an ideal sensing site due to their high gland density [1], strong sympathetic responsiveness [2], and convenient accessibility for bio-signal acquisition. Current commercial smart rings [3] primarily rely on photoplethysmography (PPG), offering limited stress estimation. Recent multimodal SoCs have extended sensing coverage, such as a behind-the-ear design [4] that integrates multiple modalities but lacks chemical sensing and requires off-chip Bluetooth Low Energy (BLE), and an adhesive patch [5] that combines chemical and physiological sensing but incurs high power costs (>800 µW). These works highlight the potential of multimodal integration, but still leave room for a compact, low-power solution tailored to comprehensive PSM. This work presents a PSM SoC that integrates galvanic skin response (GSR), electrocardiogram (ECG), PPG, and electrochemical (ECH) sensing with on-chip wireless connectivity and power-management circuits. This design supports the extraction of key stress biomarkers, including skin conductance level/response (SCL/SCR), heart rate variability (HRV), peripheral oxygen saturation (SpO$_2$), pulse arrival time (PAT), and cortisol levels in sweat [6].

Figure 3.1.1 illustrates the circuit diagram of the wireless multimodal sensing SoC, including optical, electrical, and electrochemical measurements. The implemented ECG and ECH circuits share a single instrumentation amplifier (IA), enabling the simultaneous acquisition of both signals while reducing power and area compared to separate readouts. Low-frequency motion artifacts and DC drifts are extracted from the integrator of the DC servo loop (DCSL). The GSR circuit includes both SCR and SCL readout paths, corresponding to short-term and long-term stress, respectively, and uses a current compensation technique [7] to accommodate a wide range of skin impedance changes (10kΩ to 10MΩ). The PPG circuits consist of ambient light (AL) compensation through the AL-DAC, a coarse DC current removal loop using the DC-DAC, and fine current digitization to enhance the readout range (200µA) in the presence of large interference (e.g., AL current of 50µA). This design proposes a swapping chopper with a pre-charge scheme that mitigates signal-dependent distortion caused by photodiode (PD) parasitics. Additionally, a 920MHz wake-up receiver (WuRx) applies an automatic offset calibration loop (AOCL) [8] for rapid offset correction over environment variations and receives control information at 2kb/s from the interrogator, while a 2.4GHz all-digital phase-locked loop (ADPLL)-based transmitter (Tx) synchronizes and serializes multimodal sensing data and transmits it at 1Mb/s to an external reader. The design also incorporates a power-management unit (PMU) with adaptive gain control, which achieves high efficiency and minimal overshoot/undershoot during both high and low load transients.

Figure 3.1.2 presents the key circuits of the ECG and ECH readouts. The ECG readout comprises a low-noise current-balancing IA (CBIA), a programmable gain amplifier (PGA), a DCSL, a positive feedback loop (PFL) for input impedance boosting, and a voltage-controlled oscillator-based delta-sigma ADC [9]. The CBIA enhances its common-mode rejection ratio (CMRR) [10,11], while its gain is precisely defined by the resistor ratio $2R_2/R_1$. To improve the buffering capability under small R_1, the CBIA transconductance stage combines a flipped voltage follower (FVF) to maintain sufficient gain [12]. DCSL is employed to suppress DC drift caused by electrodes. Since the integrator of the DCSL works as a low-pass filter (LPF), low-frequency motion artifacts are captured before being converted to high-frequency components [4]. The low-frequency motion signal is sent to the PGA and digitized by the ADC, then subtracted from the ECG output to enable precise extraction of the ECG signal. As a result, an 11dB attenuation for DC drifts is achieved compared to the one without subtraction. For ECH mode, the current input, sensed through the midpoint of R_1, appears as the common-mode signal of the CBIA. The resulting output voltage is subsequently fed into the ECH PGA and LPF. Finally, a feedback loop incorporating a V-to-I converter is implemented to improve linearity. Figure 3.1.2 also illustrates the gain between the ECG and ECH channels via Monte Carlo simulations. The average isolations of 79dB from ECG to ECH and 104dB from ECH to ECG indicate that the ECG and ECH signals do not interfere with each other due to the high inherent CMRR of the CBIA.

Figure 3.1.3 (top) illustrates the automatic feedback adjustment mechanism that extends the operating range of the GSR sensing system. When the input SCL signal (I_{SEN}) exceeds the range and reaches the upper or lower bound, the compensation current (I_{comp}) detector asserts $Trig_{up}$ or $Trig_{low}$ to 1. It then determines the magnitude (5b D_{Icomp}) and direction (1b I_{sign}) of I_{comp}, thereby dynamically generating an I_{comp} to shift the input back into range. A delay margin mechanism is incorporated to mitigate instability caused by T_{delay} in the LPF of the SCL readout path. A delay counter ensures sufficient time between updates, allowing each adjustment to be applied only after the previous one has taken effect, thereby preventing redundant triggers and oscillation. With the feedback adjustment mechanism, the SCL operating range is extended from ±0.2µA to ±3.3µA under a 1MΩ skin resistance. Figure 3.1.3 (bottom) illustrates the schematic, operating principles, and control waveforms of the PPG analog front-end (AFE). The chopper-swapping scheme has been adopted to remove the AL effect [4,13]. A pre-charging method (PCM) with a buffer (B_1) compensates for PD and trace parasitic capacitance (C_{PD}) effects [4]. However, when switch S_1 is turned on by Φ_{in}, this approach suffers from charge sharing between C_{PD} and the integrating capacitance (C_{INT}) due to different PD bias voltages (V_{PD}) in the pre-charging phase (V_X) and the signal integration phase ($V_{REF1} - I_{INT2}/G_m$), causing signal-dependent distortion. In addition, the offset voltages between the integrator and buffer amplifier increase the deviation of V_{PD}, which makes the charge-sharing effect more severe. Furthermore, the extra buffer must deliver sufficient current to charge C_{PD}, leading to increased power consumption. The enhanced pre-charging scheme (phase 2b) utilizes only the integrating operational amplifier (OPA) as a unity-gain buffer to pre-charge the C_{PD}, thereby establishing a constant V_{PD} in both phase 2b and 3, which prevents the charge-sharing issue. The reconfigurable pulse repetition frequency (PRF) adjusts for different settling times of the PD based on LED light intensity. Unlike previous designs [13] that used a fixed C_{INT} and required a power-hungry OPA to drive C_{INT}, this design reconfigures C_{INT} according to PRF to lower power consumption dynamically. Furthermore, an AL-DAC is added to cancel slow-changing AL currents, reducing the strict requirements of the OPA's output voltage range and power consumption. Consequently, the integrator output (V_{OUT}) is connected to two comparators with thresholds set at 0.9 and 0.3V. The compare signal (Φ_{CAL}) is fed back to ensure proper OPA operations and manages the AL-DAC to prevent saturation.

A WuRx [8] is adopted to minimize standby power. Upon detecting a wireless trigger, the WuRx activates the PMU for the readout circuits accordingly. Additionally, the AOCL [14] accelerates comparator calibration by initializing both capacitor arrays to their maximum capacitance and gradually reducing the capacitance on one side, thereby providing high gain at large offsets and minimizing the number of steps required for fast offset correction to overcome temperature and environmental interferences. The WuRx achieves a sensitivity of −57dBm and consumes 1.1µW at a data rate of 2kb/s. The 2.4GHz Tx employs an ADPLL [15] with a Class-C digitally controlled oscillator (DCO), achieving low phase noise and a 20kHz resolution. A two-point modulation scheme further preserves signal integrity despite bandwidth limits. The Tx adopts 1Mb/s GFSK modulation, which meets the required transmission mask of BLE. An adaptive-gain pulse-frequency modulation (PFM) buck converter obtains both high efficiency and fast transient response. Using DC resistor (DCR) sensing, the loop adjusts the gain according to load conditions, which reduces switching loss at light loads and accelerates the response at heavy loads. Zero-current detection (ZCD) and current-limit signals enable dynamic shutdown, minimizing quiescent current (3.7µA). The converter achieves 82.5% efficiency at 600µA load, and the efficiency exceeds 90% once the load current is greater than 5mA, reaching a peak of 94.2% at 30mA. Measurements show less than 10mV/40mV overshoot/undershoot and less than 5µs recovery for load steps from 600µA to 50mA.

The design was fabricated in a 0.18µm CMOS process. Figure 3.1.4 shows the measurement results of the multimodal sensing signal acquisition circuits. The ECG readout achieves a maximum CMRR of 130dB. The input-referred noise (IRN) is $0.32\mu V_{rms}$ (0.5 to 150Hz) and $24.9pA_{rms}$ (0.01 to 20Hz) for the ECG and ECH channels, respectively. The PPG channel obtains a dynamic range of 113.9dB with an 11b DAC and 50µA AL current tolerance. The impact of C_{PD} is presented. Even when applying only a 1mV offset between V_{PRE} and V_{REF1}, the PCM exhibits a C_{PD}-dependent output signal, whereas the enhanced method remains immune. The SCR readout path achieves an IRN of $3.1pA_{rms}$ at I_{comp} = 0µA and $23.8pA_{rms}$ at I_{comp} = ±3.1µA, resulting in a dynamic range of 99dB. Additionally, the SCR readout path achieves an SFDR of 52.5dB. Figure 3.1.5 shows the bio-signal acquisition results of PPG, ECG, motion detection, GSR, and cortisol concentration. PAT is derived as the interval between the R-peak of the ECG and the rising edge of the PPG signal. At stress stimulation, the SCL exhibits a gradual increase, while the SCR demonstrates transient fluctuations. In addition, a customized nanobrush electrode [16] enhances sweat cortisol detection (50 to 250ng/ml), achieving an R^2 linearity of 0.99. In this design, the biosensing circuits only consume 125.8µW. Figure 3.1.6 presents a performance summary and comparison with state-of-the-art multimodal [4,5,17-19] and bio-sensing designs [20-23]. Compared to prior works in the summary table, this work achieves good noise-power trade-offs, a wide dynamic range, and full integration of GSR, ECG, PPG, and sweat cortisol level sensing with embedded wireless connectivity and power management. The low power consumption, compact form factor, and broad biomarker coverage make it well-suited for continuous and comprehensive wearable PSM. Figure 3.1.7 shows the micrograph of the multimodal biosensing SoC.

Acknowledgement:
This work was supported in part by the National Science and Technology Council, Taiwan, under Grants 114-2628-E-A49-005, 112-2221-E-A49-143-MY3, and 114-2640-B-005-001. The authors would also like to acknowledge the Taiwan Semiconductor Research Institute for their assistance with chip fabrication.

References:
[1] N.A. Taylor and C.A. Machado-Moreira, (2013). Regional variations in transepidermal water loss, eccrine sweat gland density, sweat secretion rates and electrolyte composition in resting and exercising humans. *Extreme physiology & medicine*, 2(1), 4. https://doi.org/10.1186/2046-7648-2-4
[2] A.A. Alian and K.H. Shelley, (2013). "Photoplethysmography: Analysis of the pulse oximeter waveform". In *Monitoring technologies in acute care environments: a comprehensive guide to patient monitoring technology* (pp. 165-178). New York, NY: Springer New York. https://doi.org/10.1007/978-1-4614-8557-5_19
[3] Oura Health Oy, "Stress & Resilience," Accessed: Oct. 28, 2025. [Online]. Available: https://ouraring.com/stress. https://ouraring.com/stress
[4] H. Kim *et al.*, " A Behind-The-Ear Patch-Type Mental Healthcare Integrated Interface with 275-Fold Input Impedance Boosting and Adaptive Multimodal Compensation Capabilities," *ISSCC*, 2023, http://doi.org/10.1109/ISSCC42615.2023.10067723
[5] J. Cho et al., "An Adhesive Interposer-Based Reconfigurable Multi-Sensor Patch Interface with On-Chip Application Tunable Time-Domain Feature Extraction," *ISSCC*, 2024, pp. 554-555. http://doi.org/10.1109/ISSCC49657.2024.10454293
[6] G. Taskasaplidis et al., "Review of Stress Detection Methods Using Wearable Sensors," in *IEEE Access*, vol. 12, pp. 38219-38246, 2024. http://doi.org/10.1109/ACCESS.2024.3373010
[7] Y. -J. Lin et al., "A 96 dB Input Dynamic Range Galvanic Skin Response Readout IC with 3.5 pArms Input-Referred Noise for Mental Stress Monitoring," in *IEEE Transactions on Biomedical Circuits and Systems*. http://doi.org/10.1109/TBCAS.2025.3573614
[8] J. Moody et al., "Interference robust detector-first near-zero power wake-up receiver," *IEEE Journal of Solid-State Circuits*, vol. 54, no. 8, pp. 2149–2161, Aug. 2019. http://doi.org/10.1109/JSSC.2019.2912710
[9] W. Zhao et al., "A 0.025-mm2 0.8-V 78.5-dB SNDR VCO-Based Sensor Readout Circuit in a Hybrid PLL-ΔΣM Structure," *IEEE JSSC*, vol. 55, no. 3, pp. 666-679, Mar. 2020. http://doi.org/10.1109/JSSC.2019.2959479
[10] Z. Hoseini et al., "Current Feedback Instrumentation Amplifier with Built-In Differential Electrode Offset Cancellation Loop for ECG/EEG Sensing Frontend," in *IEEE Transactions on Instrumentation and Measurement*, vol. 70, pp. 1-11, 2021. http://doi.org/10.1109/TIM.2020.3031205

[11] A. Worapishet et al., "Generalized Analysis of Random Common-Mode Rejection Performance of CMOS Current Feedback Instrumentation Amplifiers," in *IEEE Transactions on Circuits and Systems I: Regular Papers*, vol. 62, no. 9, pp. 2137-2146, Sept. 2015. http://doi.org/10.1109/TCSI.2015.2411794
[12] R. Carvajal et al., "The flipped voltage follower: a useful cell for low-voltage low-power circuit design," *IEEE Transactions on Circuits and Systems I: Regular Papers*, vol. 52, no. 7, pp. 1276–1291, 2005. http://doi.org/10.1109/TCSI.2005.851387
[13] Q. Lin et al., "A 119dB Dynamic Range Charge Counting Light-to-Digital Converter for Wearable PPG/NIRS Monitoring Applications," in *IEEE TBioCAS*, vol. 14, no. 4, pp. 800-810, Aug. 2020. http://doi.org/10.1109/TBCAS.2020.3001449
[14] Y.-W. Huang and Y.-T. Liao, "A 62nW, 920MHz Wake-up Receiver with Automatic Offset Calibration Technique," *IEEE Asia Pacific Conference on Circuits and Systems*, 2024, pp. 432–436. http://doi.org/10.1109/APCCAS62602.2024.10809001
[15] P. Chen et al., "A 529-µW Fractional-N All-Digital PLL Using TDC Gain Auto-Calibration and an Inverse-Class-F DCO in 65-nm CMOS," in *IEEE Transactions on Circuits and Systems I: Regular Papers*, vol. 69, no. 1, pp. 51-63, Jan. 2022. http://doi.org/10.1109/TCSI.2021.3094094
[16] Shu-Ping Lin et al., "Nano-Brush Structure for Rapid Label-Free Differentiation of Alzheimer's Disease Stages and Direct Capture of Neuron-Derived Exosomes from Human Blood Plasma" in *ACS Applied Materials & Interfaces*, 2023. http://doi.org/10.1021/acsami.3c12766
[17] U. Ha et al., "A 25.2mW EEG-NIRS Multimodal SoC for Accurate Anesthesia Depth Monitoring," *ISSCC*, 2017, pp. 450–451. http://doi.org/10.1109/ISSCC.2017.7870455
[18] Y.-S. Shu et al., "A 4.5mm² Multimodal Biosensing SoC for PPG, ECG, BIOZ and GSR Acquisition in Consumer Wearable Devices," *ISSCC*, pp. 400-401, 2020. http://doi.org/10.1109/ISSCC19947.2020.9063112.
[19] T. Seol et al., "A Hybrid Recording System with 10kHz-BW 630mVpp 84.6dB-SNDR 173.3dB-FOMSNDR and 5kHz-BW 114dB-DR for Simultaneous ExG and Biocurrent Acquisition," *ISSCC*, 2024, pp. 562–563. http://doi.org/10.1109/ISSCC49657.2024.10454270
[20] M. Roham et al., "A Wireless IC for Time-Share Chemical and Electrical Neural Recording," *ISSCC*, 2009, pp. 430-431,431a. http://doi.org/10.1109/ISSCC.2009.4977492
[21] S. -Y. Lu et al., "A Wireless Multimodality System-on-a-Chip with Time-Based Resolution Scaling Technique for Chronic Wound Monitoring," *ISSCC*, 2021, pp. 282–283. http://doi.org/10.1109/ISSCC42613.2021.9365992
[22] mpatica, "User Manuals," Accessed: Oct. 28, 2025. [Online]. Available: https://www.empatica.com/manuals. https://www.empatica.com/manuals
[23] M. Konijnenburg et al., "A Multi(bio) sensor acquisition system with integrated processor, power management, 8x8 LED drivers, and simultaneously synchronized ECG, BIO-Z, GSR, and two PPG readouts," *IEEE J. Solid-State Circuits*, vol. 51, no. 11, pp. 2584-2595, Nov. 2016. http://doi.org/10.1109/JSSC.2016.2605660

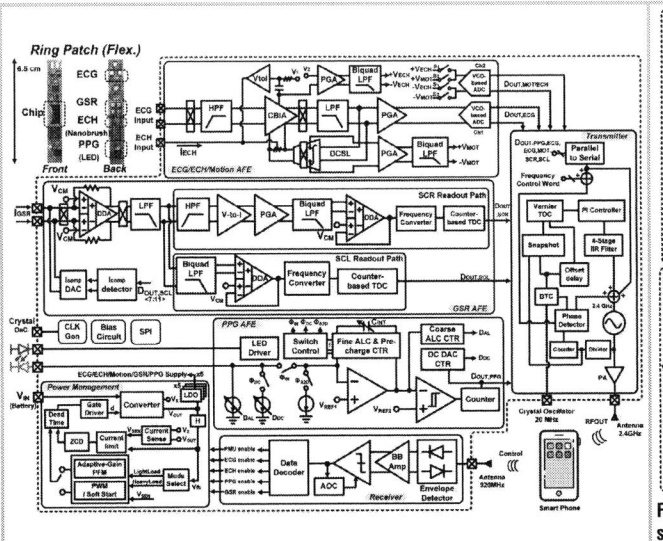

Figure 3.1.1: Block diagram of overall multimodality biosensing system.

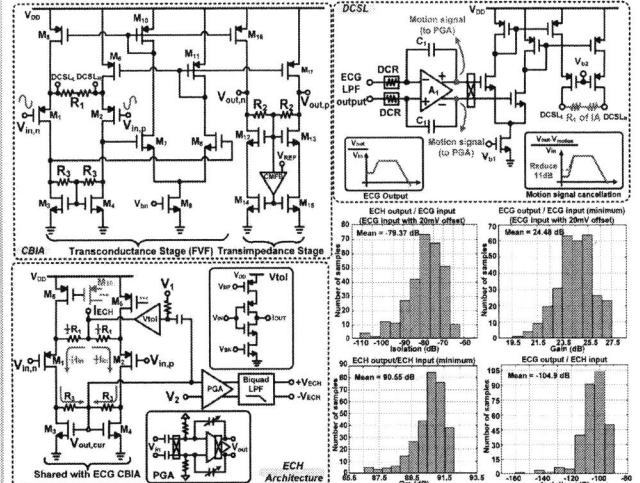

Figure 3.1.2: Schematic and transfer function of the CBIA and DCSL in the ECG, schematic of ECH, and Monte Carlo simulation results of ECG and ECH gain and isolation.

ISSCC 2026 / SESSION 3 / WEARABLE AND WIRELESS BIOMEDICAL SYSTEMS / 3.1

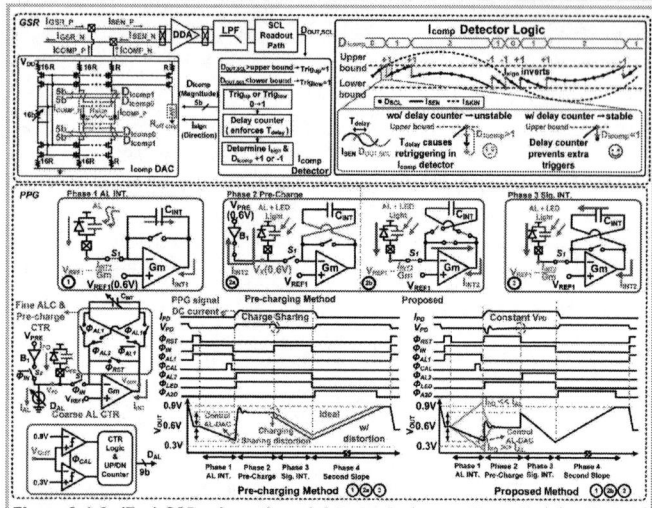

Figure 3.1.3: (Top) GSR schematic and detector logic operating principle. (Bottom) Schematic and timing diagram of the chopper swapping scheme in PPG readout interface.

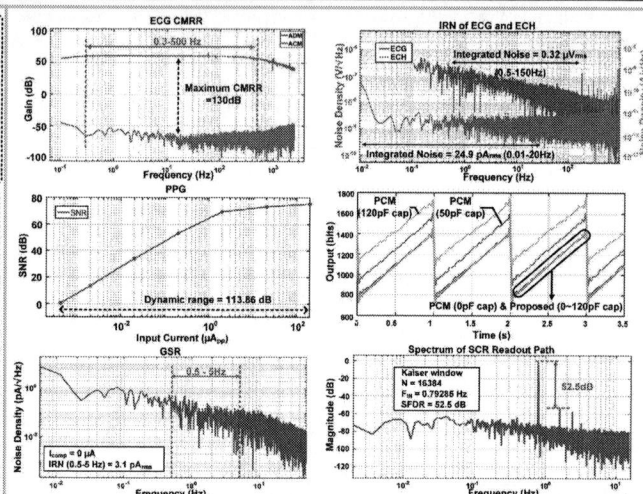

Figure 3.1.4: (Top) Measured CMRR, IRN of ECG AFE, IRN of ECH AFE. (Middle) Measured dynamic range, pre-charging methods of PPG AFE. (Bottom) Output noise of GSR AFE and spectrum of SCR path.

Figure 3.1.5: (Top) Measured waveforms of PPG, ECG, PAT, motion detection. (Bottom) GSR human stress experiment result and cortisol measurement result with SEM image of Nanobrush sensor.

	ISSCC'23 [4]	ISSCC'24 [5]	ISSCC'17 [17]	ISSCC'20 [18]	ISSCC'24 [19]	This work
Technology (nm)	180	180	65	55	180	180
Supply (V)	1.8	1.8/3	1.2/3.3	0.9/1.8/2.8	0.8	1.2/3.3
Modality	ExG/PPG/GSR/BIOZ/tVNS	ExG/PPG/BIOZ	Chemo-R/ECH/ExG/PPG/BIOZ	EEG/NIRS	PPG/ECG/BIOZ/GSR	ExG/PPG	ECG/ECH/Motion Detect/PPG/GSR
Wireless Rx/Tx	X	X	X	X	X	O
PMU	O	O	X	X	X	O
ECG (ExG)						
Bandwidth (Hz)	1000	-	0.5-250	-	10k	0.3-500
Gain (dB)	40	-	20-54	-	-	21-61
IRN (μVrms)	0.22 (0.5-300Hz)	0.27 (0.5-125Hz)	0.44 (0.5-100Hz)	0.73 (0.5-150Hz)	12 (~10kHz)	0.32 (0.5-150Hz)
CMRR (dB)	91	-	110	118	79	130
Power (μW)	54	72	7.8*	192.6	13.5 (include PPG)	24
PPG						
Input Current (μA)	100	265	-	200	40	200
AL Current (μA)	-	-	-	40	-	50
DR (dB)	102	108	-	130	114	113.86
LED Duty Cycle (%)	0.25	-	-	0.625	-	0.744
Power (μW)	59 (w/o DC can.)	155	-	72	13.5 (include ExG)	39.4
ECH			ISSCC'09 [20]	ISSCC'21 [21]	This work	
Input Range (μA)	-	105.9	0.00001-0.01	±0.75	±6.14	± 8
Noise (pArms/√Hz)	-	-	-	0.8*	2**	5.6
DR (dB)†	-	158.5	100*	88.4	129.7‡	116.12
Power (μW)	-	586	-	76	49	16.8
GSR			E4 wristband [22]	JSSC'16 [23]	This work	
Bandwidth (Hz)	DC-4	-	-	DC-4	DC-4 / 0.5-4	DC-5 / 0.5-5
Sensitivity	-	-	-	900 pS/code	100 pA	30 pA/code
DR (dB)	-	-	-	80	87	99.28
IRN	0.37Ωrms (0.1-4Hz)	-	-	-	36 pArms (0.5-4Hz)	3.1 pArms (0.5-5Hz)
Power (μW)	2 (Hybrid sensing)	-	-	-	56.4	43.2

* Estimated ** Resolution †20log(Input range/pArms) ‡ Operational range

Figure 3.1.6: Performance summary and comparison with state-of-the-art biosensing ICs.

Figure 3.1.7: Die microphotograph and power breakdown.

4.67 mm — 6.848 mm

Power Breakdown of Biosensers
Total = 125.8 μW

GSR 43.2μW · ECG 24μW (with ADC) · ECH 16.8μW (with ADC) · Motion 2.4μW · PPG 39.4μW

WuRx: 1.1μW
Tx: 6.2mW (intermittent)
PMU: 1.5mW (include buffer & bias circuit)

• 2026 IEEE International Solid-State Circuits Conference

ISSCC 2026 / SESSION 3 / WEARABLE AND WIRELESS BIOMEDICAL SYSTEMS / 3.2

3.2 A Near-Field RF Reflection Transceiver ASIC for Continuous Unobtrusive Blood Pressure Monitoring

Yiming Han, Neelotpala Kumar, Linran Zhao, Jide Yinka Adebiyi, Linrui Jiang, Hao Lu, Gina Perkins, Hirofumi Tanaka, Deji Akinwande, Yaoyao Jia

University of Texas, Austin, TX

Abstract

This work presents continuous, unobtrusive, and clinically accurate BP monitoring in a fully wearable form factor using an near-field RF reflection TRx ASIC in 65nm CMOS. The NRR approach enables noncontact monitoring with comfort and long-term wear. The optimized TRx achieves 98.6nV$_{RMS}$ IRN, while SAR-based DC-offset cancellation and closed-loop phase calibration mitigate Rx leakage and phase error for clinical accuracy. The ASIC is thoroughly validated in human studies achieving accurate BP extraction within 6mmHg mean absolute error.

Continuous, unobtrusive, and clinically accurate blood pressure (BP) monitoring provides crucial diagnostic value for cardiovascular disease. Conventional singular BP measurement using sphygmomanometers, though clinically accurate, is uncomfortable and unable to capture continuous BP trends, which are far stronger predictors of cardiovascular mortality. These limitations have motivated intensive research into wearable approaches such as ultrasound, photoplethysmography (PPG), electrocardiogram (ECG), and bioimpedance (BioZ). However, each approach still faces challenges, such as discomfort, skin-tone bias, or limited low integration, highlighting the need for a new monitoring paradigm that solves these challenges while preserving clinical accuracy.

In Fig. 3.2.1, an ultrasound-based approach achieves accurate BP estimation by tracking arterial wall displacement with high resolution [1,2] but requires gel-mediated coupling to reduce acoustic impedance, causing discomfort and limiting it to short-term use. High transducer driving voltages further increase power and hinder integration. The measurement of pulse transit time (PTT) between fiducial points using two modalities of PPG, ECG, and BioZ, provides a noninvasive marker for BP [3-6]. While they use µW-level sensing front ends that are easy to integrate, ECG and BioZ electrodes require intimate contact and are sensitive to the skin-electrode interface, such as oil and sweat. PPG operates without intimate contact, enabling comfort and long-term use [7], but suffers from skin-tone bias and shallow penetration, limiting accuracy. To overcome these limitations, we introduce near-field RF reflection (NRR), which leverages electromagnetic (EM) boundary modulation effects on the antenna. When an antenna is placed near the wrist, the arterial wall behaves as a pressure-dependent dielectric and the blood column as a conductive medium, together forming the effective near-field EM boundary [8]. As BP changes, usually within a 4Hz bandwidth (BW), (1) vessel geometry and wall stiffness vary, changing the dielectric properties; (2) the volume and velocity of the blood flow change, altering the conductivity. These BP-driven dielectric and conductive variations modulate the near-field EM boundary of the antenna, altering the antenna reflection coefficient, which manifests as changes in the amplitude and phase of reflected signals [9,10]. Therefore, when the antenna transmits an RF carrier, the amplitude and phase variations of the reflected signal encode the BP. Specifically, during systole, arterial expansion increases wall stiffness and blood flow, producing stronger reflections with larger amplitude and phase shifts. During diastole, relaxation decreases wall stiffness and blood flow, yielding weaker reflections.

In this paper, we present a realization of continuous, unobtrusive, and clinically accurate BP monitoring in a wearable design (i.e., smartwatch) through the proposed NRR approach, with successful validation in human studies. To implement NRR, we designed a transceiver (TRx) ASIC that operates with a circulator and a patch antenna positioned above the wrist, monitoring 4Hz-BP-encoded amplitude and phase variations of the reflected signal. In the ASIC, a quadrature clock generator provides a 2.4GHz clock for the transmitter (Tx), which excites the patch antenna through the circulator. The amplitude and phase-modulated reflections from the antenna are routed back to a direct-conversion receiver (Rx) for quadrature demodulation. The resulting demodulated quadrature signals are then processed for BP extraction. However, the accurate measurement of weak reflected signals is hindered by strong DC offsets from carrier reflection and leakage, and phase errors from Tx–Rx clock misalignment. We employ a SAR-based DC-offset cancellation (DCOC) to suppress DC offsets and a closed-loop phase calibration (CLPC) to ensure Tx-Rx clock alignment. Our work provides three distinct advantages over existing BP monitoring approaches: (1) noncontact sensing for comfort and long-term wear; (2) the low-power, low-voltage TRx ASIC facilitating easy integration to wearable devices; and (3) the DCOC and CLPC for clinically accurate BP estimation.

In the NRR TRx AISC (Fig. 3.2.2), a voltage-controlled oscillator (VCO) generates a 2.4GHz signal buffered by a CML-to-CMOS buffer, producing differential $LO_{TX\_P/N}$. A Gilbert mixer and LC filter convert $LO_{TX\_P/N}$ into a differential 2.4GHz sine wave, $Tx_{OUTP/N}$, with adjustable power (-40 to -18dBm). An off-chip balun transfers $Tx_{OUTP/N}$ to a single-ended carrier, V_{RF_OUT}, delivered to the antenna via a circulator that isolates Tx–Rx leakage. Reflections from the antenna return to the Rx, where a low-noise amplifier (LNA) amplifies, band-pass filters, and converts them to differential outputs, $V_{LNA\_P/N}$, driving a current-domain quadrature mixer. The mixer consists of a G_M stage that converts $V_{LNA\_P/N}$ into currents, $I_{RF\_P/N}$, and mixer switches that are controlled by quadrature local oscillation (LO) clocks from a delay-locked loop (DLL), and transimpedance amplifiers (TIAs) that filter and convert the demodulated

currents into voltages. Since G_M outputs and TIA inputs are low-impedance nodes to virtual ground, mixer switches operate with stable V_{GS} and near-zero V_{DS}, ensuring constant on-resistance for high linearity and negligible noise. The TIA outputs are amplified by a baseband (BB) amplifier and digitized by a 10b SARADC.

To enable SAR-based DCOC, the SAR logic within the ADC is reused, driving two 9b differential IDACs that inject cancellation currents at the BB amplifier feedback nodes. To enable CLPC, the Q-path mixer outputs, $V_{MIX\_QP/N}$, are amplified, generating a control voltage, V_{DEL}, for a voltage-controlled delay buffer (VCDB) that adjusts the mixer clock phase to align Tx and Rx clocks. In our NRR approach, flicker noise requires special attention, unlike in general-purpose RF applications. The LNA and G_M contribute negligible thermal noise folded from 2.4GHz to the Rx input-referred noise (IRN) in the <4Hz equivalent noise BW (ENBW). However, the TIA and BB amplifier introduce high flicker noise that dominates the Rx IRN in the <4Hz ENBW and overlaps with NRR signals. Thanks to the Rx structure, equipped with a 32dB mixer gain and an 8dB LNA gain, the flicker noise is suppressed by 40dB, achieving an ultra-low noise floor without chopping.

In Fig. 3.2.3, the DC offsets originate from impedance mismatches among the antenna, circulator, and TRx ASIC. The impedance mismatches offer carrier reflection and Tx–Rx leakage that, after amplification and demodulation, appear as large DC offsets, $I_{DC\_IP/N}$, at the BB amplifier. With a total Rx gain of 60dB, only 2mV_{PP} leakage can saturate the Rx, making it impractical to monitor the NRR signal. DCOC is proposed to solve this issue, starting from the I-path offset cancellation. The SAR comparator compares the BB amplifier differential outputs, $V_{BB\_IP/N}$. The resulting SAR logic signals drive the 9b IDAC, whose output currents, $I_{DCOC\_IP/N}$, cancel $I_{DC\_IP/N}$. This reduces the DC offset at the BB amplifier output to <4mV, equivalent to <0.4mV at the input with 20dB BB amplifier's gain. After cancellation, the IDAC code is latched to maintain $I_{DCOC\_IP/N}$, preserving the low-frequency NRR signal. After I-path cancellation, a 2:1 multiplexer shifts to the Q-path for the same process, before normal digitization resumes. This approach achieves effective DCOC with only two IDACs and no extra logic.

The phase error, φ_{ERR}, arises from Tx–Rx clock misalignment between the phase of V_{RF_OUT}, φ_{RF_OUT}, and the phase of the I-path clock ($LO_{0°}$), φ_{LO}. Without phase calibration, the phase of mixer outputs includes the desired phase shift between φ_{RF_OUT} and the phase of $V_{LNA\_P/N}$, φ_{LNA}, as well as the undesired φ_{ERR}. Our CLPC removes φ_{ERR} by first connecting the circular to a 50Ω load, before normal BP monitoring, so that the mixer outputs represent only φ_{ERR}. $V_{MIX\_QP/N}$ is amplified to generate V_{DEL}, which tunes the VCDB to adjust φ_{LO} until $V_{MIX\_QP/N}$ reaches 0V. This achieves φ_{ERR} removal because ideal Tx-Rx clock alignment ($\varphi_{ERR} = 0$) maximizes I-path mixer output, $V_{MIX\_IP/N}$, and drives $V_{MIX\_QP/N}$ to zero. V_{DEL} is stored on capacitor C_1 to maintain the phase alignment during normal BP monitoring.

In Fig. 3.2.4, the measured Tx S_{22} is –12dB at 2.4GHz with a minimum of –13.8dB at 60MHz away, while the Rx S_{11} is –18.4dB at 2.4GHz with a minimum of –19dB at 20MHz offset. Despite the frequency offsets, they remain sufficiently small to demonstrate effective impedance matching at 2.4GHz. The Tx spectrum shows a 2.425GHz carrier tunable from –40 to –18dBm with only 2.5MHz offset from the 2.4GHz target. The PSD of the Tx shows a noise floor of –83dBm with a resolution bandwidth (RBW) of 100kHz, indicating a low noise floor of –133dBm/Hz. The IRN spectral density shows a low noise of 98.6nV_{RMS} integrated over 0.1-to-4Hz BW. Figure 3.2.5 shows the human experimental setup of the NRR-based smartwatch prototype. A patch antenna is positioned above the wrist without skin contact to acquire NRR signals, while a commercial BP monitor (Finapres NOVA with finger cuff and PPG sensor) provides baseline transient BP. Before normal BP monitoring, DCOC and CLPC are enabled. The measured transient waveform of DCOC shows the cancellation of >600mV DC offset into <4mV at the BB amplifier output. The transient response of CLPC shows V_{DEL} stabilizing at ~550mV upon completion, which leads to $V_{MIX\_QP/N}$ reduced to <8mV and $V_{MIX\_IP/N}$ reaching its maximum differential swing, indicating effective compensation for φ_{ERR}.

Figure 3.2.6 presents the results of the human experiment after DCOC and CLPC. In the Valsalva protocol, the subject performed three 10s Valsalva maneuvers separated by 10s rest. The transient I/Q NRR waveforms closely tracked the temporal patterns of the baseline BP waveform. The systolic BP (SBP)/ diastolic BP (DBP) extracted from the NRR signal

66 • 2026 IEEE International Solid-State Circuits Conference

979-8-3315-8937-0/26 $31.00 © 2026 IEEE

ISSCC 2026 / February 16, 2026 / 1:55 PM

matched the baseline with a mean absolute error (MAE) of 6/4mmHg. SBP/DBP variations captured the onset, strain, and release phases of each Valsalva. In the ice-bath protocol, the subject immersed the hand in 4°C water for 120s followed by 60s rest. The transient I/Q NRR signals showed clear differences between rest and immersion, indicating vasoconstriction-induced BP variation. The extracted SBP/DBP followed the baseline within 4.9/3.2mmHg MAE, and the immersion-induced BP rise was clearly observed. These results confirm clinical-level accuracy of the NRR TRx for continuous BP monitoring under both rapid and slow variations. Figure 3.2.7 shows the ASIC micrograph in 65nm standard CMOS, with a summary of key parameters and benchmarking. We demonstrate a fully wearable system for continuous, unobtrusive, and clinically accurate BP monitoring, with low power enabled by DCOC, CLPC, and optimized TRx designs. Moreover, the system is validated in human studies.

Acknowledgement:
We acknowledge support from NSF ECCS (2239915), NIH NHLBI (1R01HL177816), the AFRL RISING Center, and the NSF Graduate Research Fellowship.

References:
[1] Sai Zhou *et al.*, "Clinical validation of a wearable ultrasound sensor of blood pressure," *Nat. Biomed. Eng*, vol. 9, pp. 865–881, 2025.
https://doi.org/10.1038/s41551-024-01279-3
[2] Muyang Lin *et al.*, "A fully integrated wearable ultrasound system to monitor deep tissues in moving subjects," *Nat Biotechnol*, vol. 42, pp. 448–457, 2024.
https://doi.org/10.1038/s41587-023-01800-0
[3] T. Seol *et al.*, "A Hybrid Recording System with 10kHz-BW 630mVPP 84.6dB-SNDR 173.3dB-FOMSNDR and 5kHz-BW 114dB-DR for Simultaneous ExG and Biocurrent Acquisition," *ISSCC*, 2024, pp. 562-563.
https://doi.org/10.1109/ISSCC49657.2024.10454270
[4] H. Kim *et al.*, "A Behind-The-Ear Patch-Type Mental Healthcare Integrated Interface with 275-Fold Input Impedance Boosting and Adaptive Multimodal Compensation Capabilities," *ISSCC*, 2023, https://doi.org/10.1109/ISSCC42615.2023.10067723
[5] C. S. Park *et al.*, "A 145.2dB-DR Baseline-Tracking Impedance Plethysmogram IC for Neckband-Based Blood Pressure and Cardiovascular Monitoring," *ISSCC*, 2022, https://doi.org/10.1109/ISSCC42614.2022.9731712
[6] D. Kireev *et al.*, "Continuous cuffless monitoring of arterial blood pressure via graphene bioimpedance tattoos," *Nat. Nanotechnol*, vol. 17, pp. 864–870, 2022.
https://doi.org/10.1038/s41565-022-01145-w
[7] H. Mohammadi *et al.*, "Cuff-less blood pressure monitoring via PPG signals using a hybrid CNN-BiLSTM deep learning model with attention mechanism," *Sci Rep*, vol. 15, pp. 22229, 2025. https://doi.org/10.1038/s41598-025-07087-2
[8] X. Hui and E.C. Kan, "Monitoring vital signs over multiplexed radio by near-field coherent sensing," *Nat Electron*, vol. 1, pp. 74–78, 2018. https://doi.org/10.1038/s41928-017-0001-0
[9] N. Kumar *et al.*, "Toward Clinically Accurate Continuous Blood Pressure Smart Watch Biosensors Using Equitable Bioimpedance Modality and AI," *IEEE MTT-S International Microwave Biomedical Conference*, 2025,
https://doi.org/10.1109/IMBioC63524.2025.10989672
[10] X. Hui et al., "Multi-Point Near-Field RF Sensing of Blood Pressures and Heartbeat Dynamics," in *IEEE Access*, vol. 8, pp. 89935-89945, 2020.
https://doi.org/10.1109/ACCESS.2020.2993994

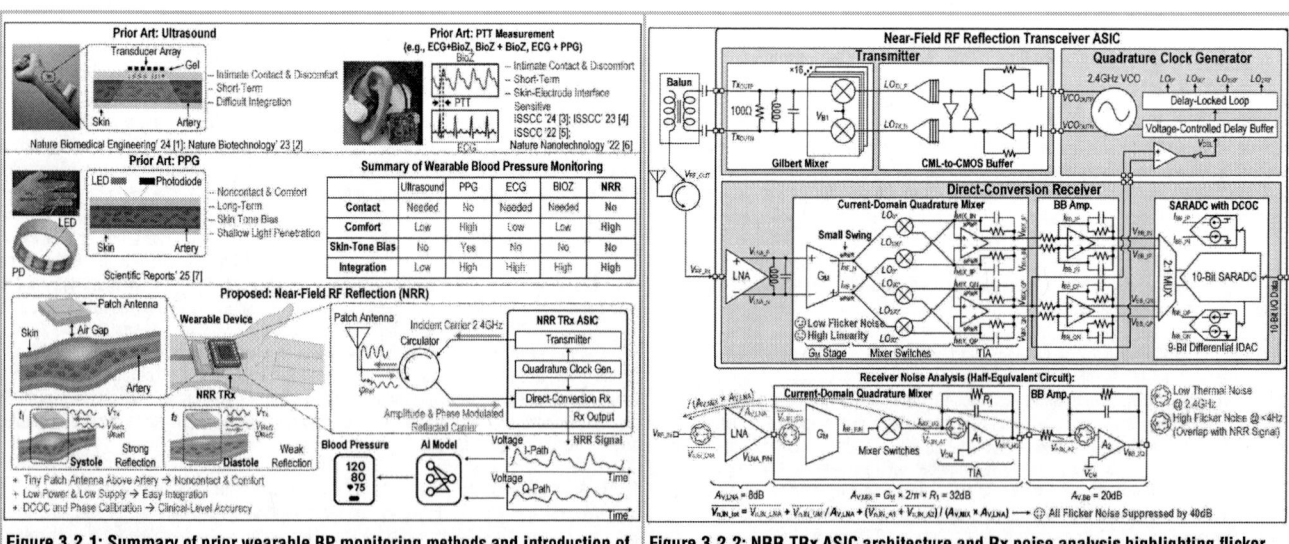

Figure 3.2.1: Summary of prior wearable BP monitoring methods and introduction of our NRR architecture.

Figure 3.2.2: NRR TRx ASIC architecture and Rx noise analysis highlighting flicker noise suppression consideration.

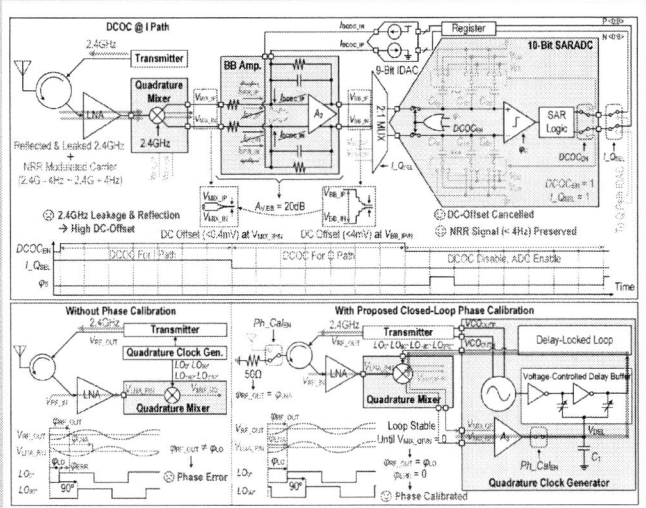

Figure 3.2.3: NRR challenges and solutions: SAR-based DCOC for DC offset removal and CLPC for phase correction.

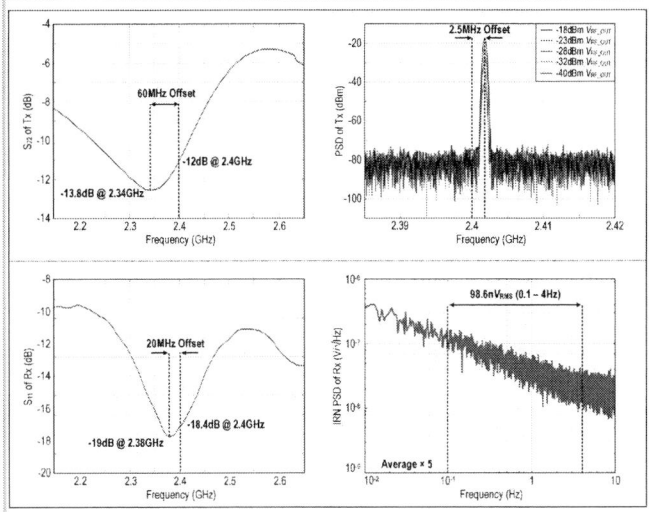

Figure 3.2.4: Measured S_{22} parameter and output PSD of Tx, along with the measured S_{11} parameter and IRN PSD of Rx.

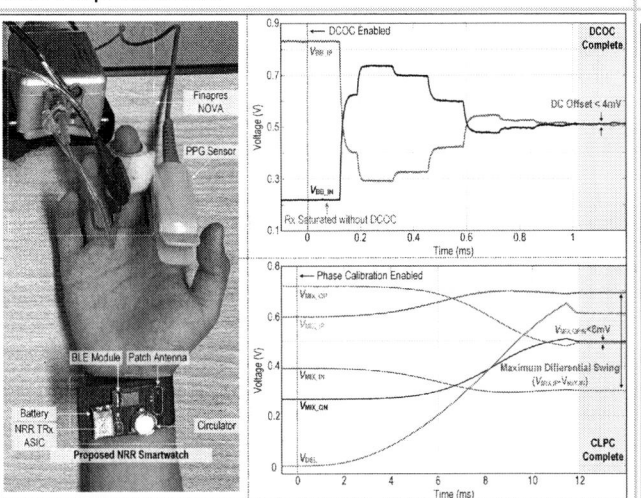

Figure 3.2.5: Human experiment setup of the NRR smartwatch prototype synchronized with a commercial BP monitor, with the measured transient waveforms of DCOC and CLPC.

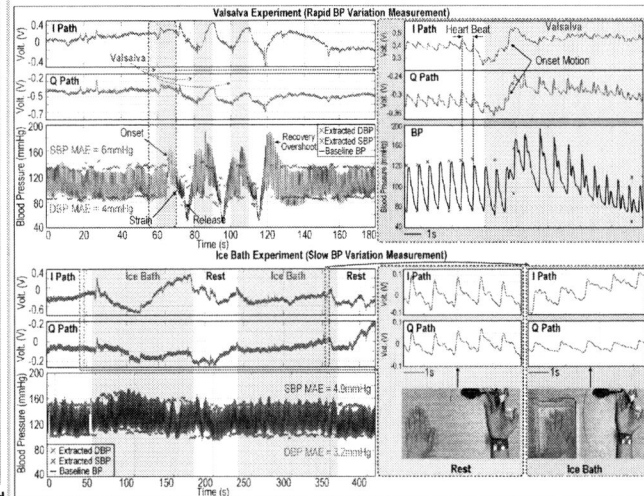

Figure 3.2.6: Human experiment results in Valsalva and ice-bath protocols, both demonstrating accurate NRR-extracted SBP/DBP in comparison with baseline BP.

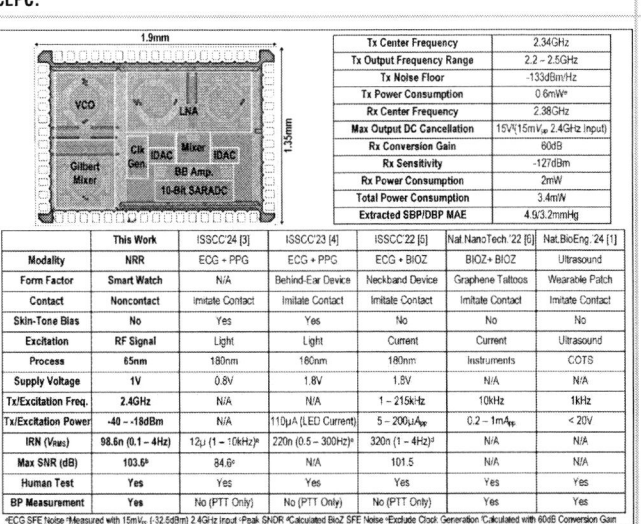

	Tx Center Frequency	2.34GHz
	Tx Output Frequency Range	2.2 - 2.5GHz
	Tx Noise Floor	-133dBm/Hz
	Tx Power Consumption	0.6mW*
	Rx Center Frequency	2.38GHz
	Max Output DC Cancellation	15V*(15mV$_{pp}$ 2.4GHz Input)
	Rx Conversion Gain	60dB
	Rx Sensitivity	-127dBm
	Rx Power Consumption	2mW
	Total Power Consumption	3.4mW
	Extracted SBP/DBP MAE	4.9/3.2mmHg

	This Work	ISSCC'24 [3]	ISSCC'23 [4]	ISSCC'22 [5]	Nat.NanoTech.'22 [6]	Nat.BioEng.'24 [1]
Modality	NRR	ECG + PPG	ECG + PPG	ECG + BIOZ	BIOZ+ BIOZ	Ultrasound
Form Factor	Smart Watch	N/A	Behind-Ear Device	Neckband Device	Graphene Tattoos	Wearable Patch
Contact	Noncontact	Imitate Contact	Imitate Contact	Imitate Contact	Imitate Contact	Imitate Contact
Skin-Tone Bias	No	Yes	Yes	No	No	No
Excitation	RF Signal	Light	Light	Current	Current	Ultrasound
Process	65nm	180nm	180nm	180nm	Instruments	COTS
Supply Voltage	1V	0.8V	1.8V	1.8V	N/A	N/A
Tx/Excitation Freq.	2.4GHz	N/A	N/A	1 - 215kHz	10kHz	1kHz
Tx/Excitation Power	-40 - -18dBm	N/A	110μA (LED Current)	5 - 200μA$_{pp}$	0.2 - 1mA$_{pp}$	< 20V
IRN (V$_{RMS}$)	98.6n (0.1 - 4Hz)	12μ (1 - 10kHz)^a	220n (0.5 - 300Hz)^b	320n (1 - 4Hz)^d	N/A	N/A
Max SNR (dB)	103.6^b	84.6^a	N/A	101.5	N/A	N/A
Human Test	Yes	Yes	Yes	Yes	Yes	Yes
BP Measurement	Yes	No (PTT Only)	No (PTT Only)	No (PTT Only)	Yes	Yes

*ECG SFE Noise *Measured with 15mV$_{pp}$ (-32.5dBm) 2.4GHz input. *Peak SNDR *Calculated BioZ SFE Noise *Exclude Clock Generation *Calculated with 60dB Conversion Gain

Figure 3.2.7: Chip micrograph with a summary of key ASIC parameters and benchmarking table.

ISSCC 2026 / SESSION 3 / WEARABLE AND WIRELESS BIOMEDICAL SYSTEMS / 3.3

3.3 A Battery-Powered Hybrid Resonant Pulse-Train Generator with Adaptive Frequency Tracking and Residual Energy Recycling for Ultrasonic Implants

Xiayang Li*[1], Wenyu Peng*[1], Limitha Kumar[1], Xiaoxi Zhao[1], Cheng Huang[2], Sijun Du[1]

[1]Delft University of Technology, Delft, The Netherlands, [2]Iowa State University, Ames, IA
*Equally Credited Authors (ECAs)

Abstract

This work presents a battery-powered hybrid resonant pulse-train generator for ultrasonic wireless power transfer (US-WPT) in implantable medical devices. The ASIC integrates adaptive resonance tracking, burst-mode power regulation, a switched-capacitor–resonant architecture, and residual energy harvesting. Fabricated in 0.18μm BCD, it achieves programmable US power, 68.6% C_p-loss reduction, and 21.7% end-to-end efficiency.

Ultrasonic wireless power transfer (US-WPT) is an attractive solution for non-invasive energy delivery to miniaturized implantable medical devices (IMDs), thanks to its deep tissue penetration and excellent biocompatibility [1]. Delivering sufficient power at depths beyond 10mm requires large piezoelectric transducers (PZTs) driven by high-voltage (HV) pulse-trains to generate intense ultrasonic (US) bursts [2-3]. However, existing pulser designs face two critical challenges: (1) low efficiency under continuous HV pulsing due to repeated charging of the transducer capacitance, and (2) poor adaptability to different transducers or resonance drift caused by process and environmental variations. Most prior works target single-pulse excitation for imaging, leaving the need for an efficient, battery-powered HV transmitter (TX) for US-WPT largely unmet. Meanwhile, although many US-WPT receiver (RX) designs have recently been proposed [4-8], the lack of an adaptable and power-regulated transmitter highlights the urgent need for a new architecture that can sustain high-intensity pulse-train excitation while dynamically tracking transducer resonance and regulating the output power.

The core challenge of continuous HV pulsing arises from the intrinsic behavior of the PZT transducer, which can be modelled as a parallel capacitance (C_p) and a mechanical RLC branch [9]. While the mechanical current (I_m) dictates the acoustic output, significant dynamic loss occurs from repeatedly charging C_p, scaling with $f \cdot C_p \cdot V^2$. High efficiency is only achieved when the driving pulses are strictly aligned with the resonant frequency (f_r) and the voltage (V_{TX}) remains strictly in-phase with I_m. In practice, process variations, environmental factors, and packaging cause the actual f_r to drift, so fixed-frequency driving leads to power cancellation and efficiency degradation. Conventional class-D pulsers (Fig. 3.3.1 top-middle) provide simple continuous pulsing rely on a preset frequency and a dedicated HV rail (V_{HV}), suffering from low conversion efficiency and dynamic losses due to C_p hard-charging [10]. Resonant pulsers (Fig. 3.3.1, top-right) improve efficiency by leveraging resonant energy transfer, similar to a buck-boost converter, to generate half-sinusoidal HV pulses from a battery supply, reducing $f \cdot C_p \cdot V^2$ loss [11]. However, they only support discrete single-pulse generation and still depend on a preset pulse width, lacking adaptability to frequency drift and limiting their use as power TXs. These limitations call for a new architecture that unifies the advantages of both approaches: sustaining continuous pulse-train excitation for high US output, minimizing dynamic losses through resonant energy transfer, and adaptively tracking f_r to maintain efficiency under varying conditions.

To address these challenges, this work presents a hybrid resonant high-voltage pulse-train generator (Fig. 3.3.1, bottom). The proposed design introduces four key innovations: (1) adaptive resonance tracking, which detects and dynamically locks onto the actual resonant frequency of the PZT during operation, ensuring in-phase excitation and sustaining efficiency under process or environmental variations; (2) continuous burst-mode excitation with programmable power regulation, where both the pulse amplitude and pulse count can be tuned to maintain a desired output power level, unlike prior resonant pulsers limited to discrete single pulses; (3) a hybrid switched-capacitor (SC) resonant architecture, where the SC booster works with the resonant pulser to reduce conduction loss and deliver higher output power from a battery supply, enabling efficient drive of large PZTs; and (4) residual energy harvesting, where leftover mechanical energy is reclaimed after each burst and recycled to further enhance system efficiency. Together, these innovations enable a compact, battery-compatible pulse-train TX ASIC that provides efficient, adaptive, and power-scalable ultrasonic excitation for next-generation IMDs.

The operating principle of the ASIC is illustrated in Fig. 3.3.2. At the system level, the chip is controlled by two external digital signals: EN and BU. EN enables the ASIC, while BU defines the timing and duration of each US burst. By adjusting the frequency and pulse width of BU, different burst patterns can be flexibly programmed. Each burst consists of four modes: M_0 (Idle), M_1 (f_r Detecting), M_2 (Pulsing), and M_3 (Energy Harvesting). These modes correspond to different sequences of power stage configurations, or phases, shown in top-right corner of Fig. 3.3.2, with the associated configurations ($\Phi_0-\Phi_3$) detailed at the bottom.

The power stage integrates a PZT, a resonant pulser, and an SC booster, comprising three 5V switches ($S_{C1}-S_{C3}$), five HV Laterally Diffused MOS (LDMOS) switches (S_1-S_5), an inductor (L) and a flying capacitor (C). The hybrid SC-inductive topology boosts the input

voltage of the resonant pulser to $2 \cdot V_{BAT}$, shortening the inductor charging time and thus improves efficiency at the same output power. This faster charging also extends the applicable range of transducer frequencies, as higher-frequency PZTs allow less charging time per pulse. In Φ_0 (Open), PZT is left open to sustain its intrinsic resonance while C is charged by V_{BAT}. In Φ_1 (Inductive), V_{BAT} and C are stacked to energize L with $2 \cdot V_{BAT}$ (solid path). The reverse current path (dotted) returns energy from L back to the battery when $I_L<0$. Meanwhile, S_5 grounds the PZT to enable zero-crossing detection ($I_m=0$), providing a reliable reference for cycle-accurate frequency tracking. In Φ_2 (Resonant), C is recharged by V_{BAT} while energy in L is transferred to the PZT (solid path). At the end of each pulse, the same configuration allows reverse transfer when $I_L<0$, with the inductor energized by the PZT. An "intense" variant of Φ_2 phase is implemented in M_2 to boost the delivered energy by turning on S_1 and S_4. In Φ_3 (Ground), the PZT is shorted to ground, preventing energy cancellation during the negative half-cycle of vibration.

The detailed burst operation is illustrated in Fig. 3.3.2 (top-right). When either EN or BU is low, the system stays in the low-power mode M_0. Once both signals go high, the system enters M_1, where a single half-sinusoidal pulse ($\Phi_1 \rightarrow \Phi_2 \rightarrow \Phi_1 \rightarrow \Phi_3$) excites a weak vibration on the PZT, V_{TX}. By sensing the zero-crossing moments of V_{TX}, the resonant frequency f_r is estimated. The system then enters M_2, where continuous HV pulses are applied at the measured f_r. The sequence $\Phi_1 \rightarrow \Phi_2 \rightarrow \Phi_0 \rightarrow \Phi_2 \rightarrow \Phi_1 \rightarrow \Phi_3$ extends each pulse width by holding V_{TX} at its peak during Φ_0, increasing the input energy per pulse. When BU goes low, the system transitions to M_3, in which the residual mechanical energy is harvested through the reverse sequence $\Phi_0 \rightarrow \Phi_2 \rightarrow \Phi_1 \rightarrow \Phi_3$ and partially returned to the battery. The system then reverts to M_0.

The system architecture is shown in Fig. 3.3.3. It consists of a power stage, a sensor block, a digital controller, and on-chip clocking blocks. The sensor block includes two zero-crossing detectors (ZCDs) and a peak detector (PD) to monitor V_{TX} and V_R. The ZCDs provide timing information, while the PD identifies amplitude peaks; both signals are digitized for resonance estimation and real-time frequency tracking. A 100MHz on-chip voltage-controlled oscillator (VCO) and a time-to-digital converter (TDC) further quantify the resonance period. These sensing outputs are fed to a finite state machine (FSM), which coordinates the four operating modes (M_0-M_3), while a configuration controller sequences the power-stage configurations ($\Phi_0-\Phi_3$) based on the FSM commands and sensor feedback. As shown in Fig. 3.3.3 (top-right), the system applies HV pulses at resonance while maintaining phase alignment between V_{TX} and I_m to maximize energy transfer. The pulse amplitude is tuned by adjusting the inductor charging time T_c, thereby regulating acoustic output power. High-voltage switching is enabled by bootstrap drivers (Fig. 3.3.3, bottom-left), where a stacked capacitor C_{BS} provides gate overdrive for HV LDMOS devices, enabling efficient switching of S_1, S_4, and S_{C1}.

The prototype ASIC was fabricated in a 0.18μm BCD process with an active chip area of 4.92mm² (Fig. 3.3.7). Electrical measurements were first performed using a 1nF off-chip capacitor load, with a 0.68μH off-chip inductor for the resonant pulser and a 1μF off-chip capacitor for the SC booster (Fig. 3.3.4). Because a pure capacitor does not exhibit mechanical resonance, the operating frequency was externally provided by an FPGA. As shown in Fig. 3.3.4 (top-left), the chip achieves continuous pulse-train generation at 2MHz, where resonant power transfer enables soft-charging quasi-square pulses. The output amplitude is programmable from 5 to 25V; at 24.6V, the chip consumes 0.38W power and achieves a maximum C_p loss reduction of 68.6% (Fig. 3.3.5, bottom-left).

For ultrasonic evaluations, a 12×12×1.02mm³ PZT-5A transducer was used as the load, with its resonant frequency characterized at around 2.2MHz by an impedance analyzer. Driven by an external BU signal, the ASIC generates bursts with a 2kHz pulse repetition frequency (PRF) and 10% duty cycle (Fig. 3.3.4, top-right). A zoomed-in waveform of a single burst (Fig. 3.3.4, middle) demonstrates the full operation sequence, including f_r-detection (M_1), pulsing (M_2), and energy harvesting (M_3). The detailed waveforms for each mode are further shown in Fig. 3.3.4 (bottom). In M_1, after initialization by resetting the power stage, a single half-sinusoidal pulse is applied for zero-crossing detection to estimate f_r. In M_2, the measured pulse-train period is 456ns, confirming accurate resonance tracking at around 2.2MHz. In M_3, the residual mechanical energy is recycled, with the PZT's kinetic energy decaying rapidly.

68 • 2026 IEEE International Solid-State Circuits Conference

979-8-3315-8937-0/26 $31.00 © 2026 IEEE

Figure 3.3.5 presents the acoustic measurement setup and results. Tests were conducted in a water tank using the same 12×12×1.02mm³ PZT transducer, while a fiber-optic hydrophone on a 3-axis stage scanned the acoustic field. At the focal depth of 13mm, the measured distribution (Fig. 3.3.5, top-right) shows a peak acoustic pressure of 0.392MPa at V_{TX} = 8.8V. Leveraging on-chip power regulation, the ASIC can program different output intensities, and the end-to-end (E2E) efficiency is obtained as the ratio of the received acoustic power at the focal spot to the electrical input power. The received acoustic power is evaluated by integrating the acoustic intensity over a 1mm² focal area, which corresponds to both the effective beam spot and the aperture size of practical implant receivers, providing a representative metric for E2E efficiency. As shown in Fig. 3.3.5 (bottom-right), the system achieves a peak E2E efficiency of 21.7%, which corresponds to a peak acoustic pressure of 0.392MPa at the focal spot. Compared to prior works (Fig. 3.3.6), the proposed ASIC shows higher efficiency under adaptive resonant pulse-train operation, while uniquely supporting output power regulation and residual energy harvesting, making it a promising US-WPT solution for IMDs.

Acknowledgement:
The authors would like to express their gratitude to Tiago Costa and Michiel Pertijs from TU Delft for their invaluable advice and technical support throughout this work.

References:
[1] J. Charthad *et al.*, "A mm-Sized Implantable Medical Device (IMD) With Ultrasonic Power Transfer and a Hybrid Bi-Directional Data Link," in *IEEE Journal of Solid-State Circuits*, vol. 50, no. 8, pp. 1741-1753, Aug. 2015.
https://doi.org/10.1109/JSSC.2015.2427336
[2] J. Choi *et al.*, "An Energy-Replenishing Ultrasound Pulser with 0.25CV2f Dynamic Power Consumption," *ISSCC*, 2021, pp. 486-487.
https://doi.org/10.1109/ISSCC42613.2021.9365826
[3] P. Guo *et al.*, "A Single-Inductor-Based High-Voltage Transmit Beamformer for Wearable Ultrasound Devices Achieving 88% fCV² Power Reduction," *ISSCC*, 2025,
https://doi.org/10.1109/ISSCC49661.2025.10904513
[4] Y. -S. Luo *et al.*, "Ultrasonic Power/Data Telemetry and Neural Stimulator With OOK-PM Signaling," in *IEEE Transactions on Circuits and Systems II: Express Briefs*, vol. 60, no. 12, pp. 827-831, Dec. 2013. https://doi.org/10.1109/TCSII.2013.2286000
[5] T. C. Chang *et al.*, "A 30.5mm³ fully packaged implantable device with duplex ultrasonic data and power links achieving 95kb/s with <10⁻⁴ BER at 8.5cm depth," *ISSCC*, 2017, pp. 460-461. https://doi.org/10.1109/ISSCC.2017.7870460
[6] Piech *et al.* A wireless millimetre-scale implantable neural stimulator with ultrasonically powered bidirectional communication. *Nat Biomed Eng* 4, 207–222 (2020).
https://doi.org/10.1038/s41551-020-0518-9
[7] J. Charthad *et al.*, "A mm-Sized Wireless Implantable Device for Electrical Stimulation of Peripheral Nerves," in *IEEE Transactions on Biomedical Circuits and Systems*, vol. 12, no. 2, pp. 257-270, April 2018. https://doi.org/10.1109/TBCAS.2018.2799623
[8] C. Lee *et al.*, "DustNet: A Network of Time-Division Multiplexed Ultrasonic Implants with 16-Level ASK Backscatter Modulation," *ISSCC*, 2025, pp. 582-583.
https://doi.org/10.1109/ISSCC49661.2025.10904679
[9] J. Choi *et al.*, "Energy-Efficient High-Voltage Pulsers for Ultrasound Transducers," in *IEEE Transactions on Circuits and Systems II: Express Briefs*, vol. 68, no. 1, pp. 19-23, Jan. 2021. https://doi.org/10.1109/TCSII.2020.3040548

[10] H. Rivandi and T. L. Costa, "A 2D Ultrasound Phased-Array Transmitter ASIC for High-Frequency US Stimulation and Powering," in *IEEE Transactions on Biomedical Circuits and Systems*, vol. 17, no. 4, pp. 701-712, Aug. 2023.
https://doi.org/10.1109/TBCAS.2023.3288891
[11] I. Bellouki *et al.*, "A Resonant High-Voltage Pulser for Battery-Powered Ultrasound Devices," *ISSCC*, 2024, pp. 106-107.
https://doi.org/10.1109/ISSCC49657.2024.10454286
[12] M. Gourdouparis *et al.*, "An Ultrasound-Powering TX with a Global Charge-Redistribution Adiabatic Drive Achieving 69% Power Reduction and 53° Maximum Beam Steering Angle for Implantable Applications," *ISSCC*, 2024, pp. 102-103.
https://doi.org/10.1109/ISSCC49657.2024.10454411
[13] E. So *et al.*, "An RF-Ultrasound Relay for Adaptive Wireless Powering Across Tissue Interfaces," in *IEEE Journal of Solid-State Circuits*, vol. 57, no. 11, pp. 3429-3441, Nov. 2022. https://doi.org/10.1109/JSSC.2022.3171233
[14] T. Costa *et al.*, "An Integrated 2D Ultrasound Phased Array Transmitter in CMOS With Pixel Pitch-Matched Beamforming," in *IEEE Transactions on Biomedical Circuits and Systems*, vol. 15, no. 4, pp. 731-742, Aug. 2021.
https://doi.org/10.1109/TBCAS.2021.3096722
[15] K. Chen *et al.*, "Ultrasonic Imaging Transceiver Design for CMUT: A Three-Level 30-Vpp Pulse-Shaping Pulser With Improved Efficiency and a Noise-Optimized Receiver," in *IEEE Journal of Solid-State Circuits*, vol. 48, no. 11, pp. 2734-2745, Nov. 2013.
https://doi.org/10.1109/JSSC.2013.2274895

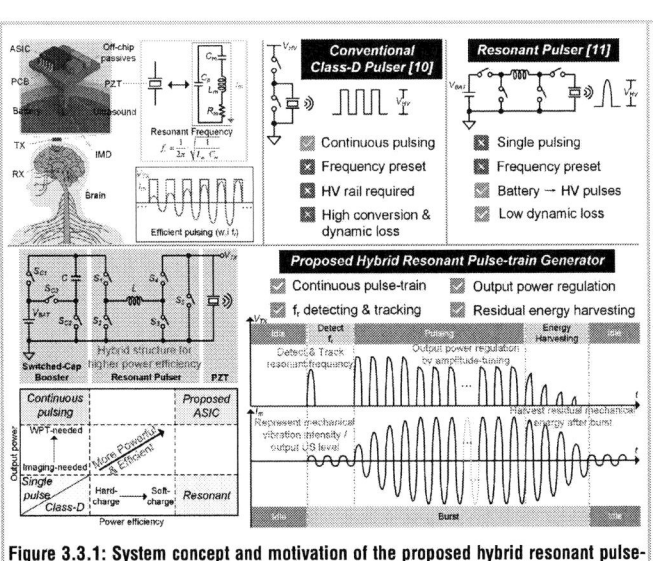

Figure 3.3.1: System concept and motivation of the proposed hybrid resonant pulse-train generator.

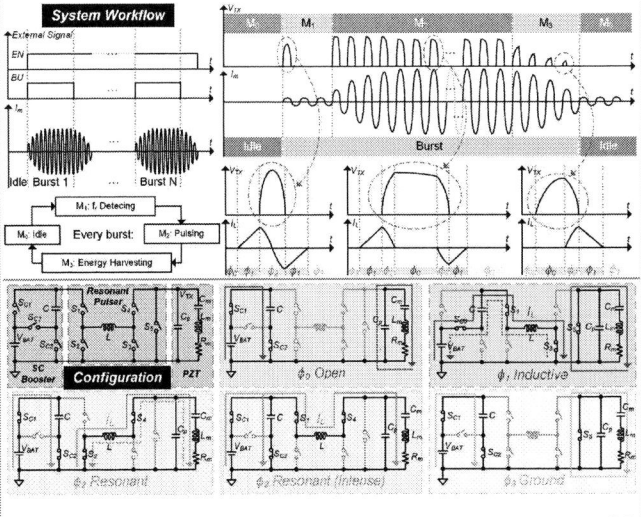

Figure 3.3.2: Operating principle and power-stage configurations of the proposed ASIC.

ISSCC 2026 / SESSION 3 / WEARABLE AND WIRELESS BIOMEDICAL SYSTEMS / 3.3

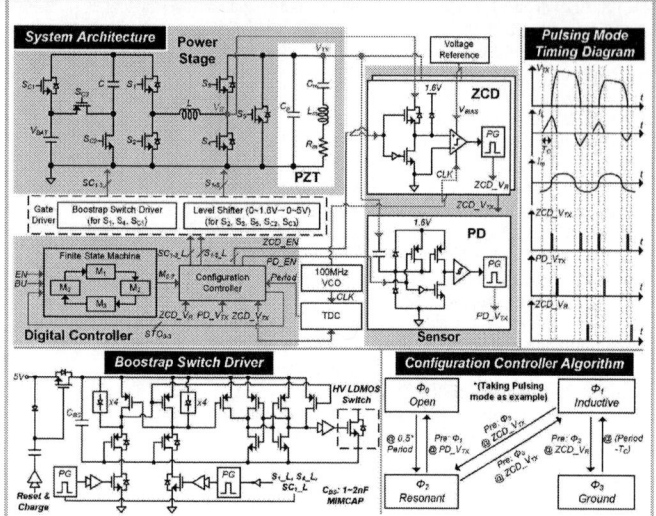

Figure 3.3.3: System architecture and circuit implementation of the proposed ASIC.

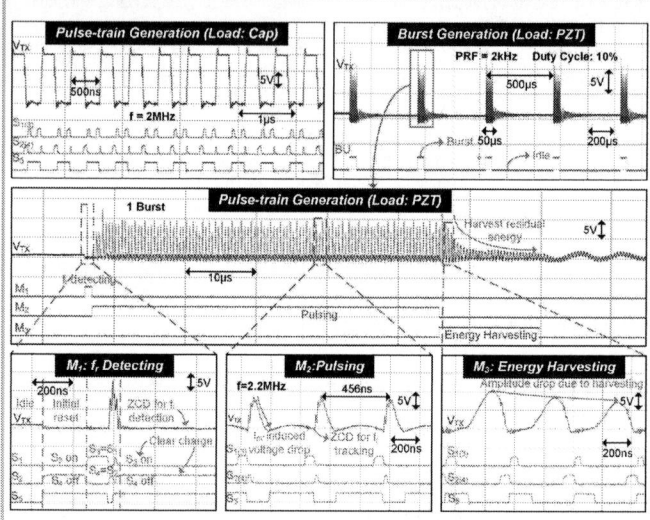

Figure 3.3.4: Electrical measurement results of both capacitor and PZT load.

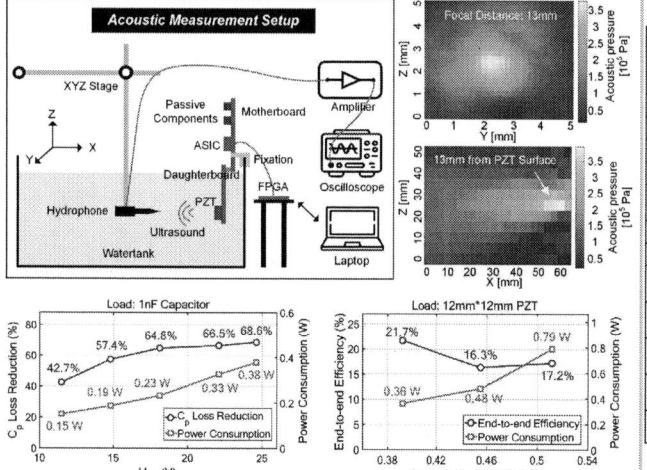

Figure 3.3.5: Acoustic measurement setup and results; C_p loss reduction and end-to-end efficiency analysis.

Comparison with State-of-the-art TX ASIC for Ultrasonic Applications

Reference	This work	Gourdouparis, ISSCC' 2024 [12]	So, JSSC' 2022 [13]	Costa, TbioCAS' 2021 [14]	Bellouki, ISSCC' 2024 [11]	Choi, ISSCC' 2021 [2]	Chen, JSSC' 2013 [15]
Application	Powering/Stimulation				Imaging		
Process	180nm BCD	65nm	180nm BCD	180nm	180nm BCD	180nm BCD	180nm BCD
Transducer	PZT	PMUT	PZT	PZT	CMUT	PZT	CMUT
Frequency	0.5-2.5MHz	7.8MHz	850kHz	8.4MHz	2.5MHz	1MHz	3.3MHz
f_r tracking	Yes	No	No	No	No	No	No
Driving voltage	5-25V (Adjustable)	4.8V	4.5V	5V	10-30V (Adjustable)	30V	30V
HV supply	Not Needed	No HV TX	External	No HV TX	Not Needed	External	External
Transducer load C_p	200pF-1.5nF	2.5pF	125pF	N/A	120pF	820pF	40pF
Measured C_p loss reduction	68.6%	46%	N/A	N/A	63%	73.1%	38%
Powering method	Pulse train generation	Beamforming	Beamforming	Beamforming	N/A		
Output power regulation	Yes	No	No	No	N/A		
Residual energy harvesting	Yes	No	No	No	N/A		
End-to-end efficiency	21.7%*	N/A	0.86%*	N/A	N/A	N/A	1%**

*Efficiency = Acoustic power @ focal spot / Electrical power from supply
**Efficiency = Acoustic power @ TX surface / Electrical power from supply

Figure 3.3.6: Comparison table of the proposed ASIC with prior state-of-the-art TX designs for ultrasonic applications.

Figure 3.3.7: Chip micrograph.

- 2026 IEEE International Solid-State Circuits Conference

ISSCC 2026 / SESSION 3 / WEARABLE AND WIRELESS BIOMEDICAL SYSTEMS / 3.4

3.4 An Ultrasound-Powering TX with Standing-Wave Peak Tracking Employing Adiabatic Power Sensing Achieving 82% Power-Tracking Accuracy and <90ms Settling Time for Brain Implants

Marios Gourdouparis[1,2], Chengyao Shi[1], Jiang Liu[1], Yuming He[1], Stefano Stanzione[1], Wouter Serdijn[2], Yao-Hong Liu[1]

[1]imec, Eindhoven, The Netherlands, [2]TU Delft, Delft, The Netherlands

Abstract

An ultrasound powering TX ASIC for brain implants with autonomous on-chip standing-wave peak tracking for RX power (PDL) regulation is presented. With a proposed adiabatic power-sensing scheme, the TX consumes 43µW for power tracking, and the system achieves a settling time of <90ms while using the standing-wave peak-tracking FSM. The TX can achieve PDL improvement of up to 2.7× with a power-tracking accuracy of 82%.

State-of-the-art intracortical brain-computer interfaces rely on implants on the cortex surface to achieve selective neural recording and stimulation [9]. Wireless powering replaces wires penetrating the dura (a membrane above the cortex), thus enabling safe and long-term sub-dural implantation (Fig. 3.4.1). To increase the powering-link efficiency, two-stage approaches involving a relay cranial implant have recently been presented [1,6], with ultrasound (US) serving as the "trans-dural" wireless powering modality thanks to its low attenuation in tissue, relaxed safety limits on power density and beam-steering capabilities. For trans-dural powering, the distance (d) between the cranial and subdural implants is up to 1cm (Fig. 3.4.1), while the implant diameter (D) should not exceed several mm to minimize the invasiveness of skull-opening surgery. Transducers with resonant frequencies in the MHz range are preferred [6], since their higher TX pressure sensitivity (kPa/V) maximizes power delivered to the load (PDL) over the small powering distance. Thus, d can be smaller than the Rayleigh distance $D^2/(4\lambda)$, meaning that the RX can be in the US near field (Fresnel zone). Acoustic reflections in this region create "standing waves", maximizing PDL at distances approximately equal to integer multiples of $\lambda/2$ and making it dependent on both distance and frequency. However, distance variations can occur during the implantation surgery but also due to the vibrations caused by breathing, heartbeat or head movements. Displacements by around 100mm [7] are close to the distance between peak and valley of the US standing wave of 4 to 4.5MHz, varying PDL by more than 2×. With a fixed ultrasound driver frequency (f_D), RX power is unregulated against such displacements [1], since the frequencies corresponding to the peak of the standing wave are shifted when the distance changes. Thus, there is a need to tune f_D and ensure operation at the peak of the standing wave for maximum PDL. This process is standing wave peak (SWP) tracking and can be characterized by the power-tracking accuracy, meaning how close to the optimal value PDL is regulated with varying distance. Within the narrowband f_D tuning range for SWP tracking, the transducer's pressure sensitivity remains unchanged and PDL varies only due to the standing waves. SWP tracking could be achieved with a global feedback loop as in [2,3], with PDL monitoring on the subdural implant and a wireless uplink of the feedback data to the power TX where it would be used for f_D tuning. However, such an approach increases complexity and power consumption on the subdural implant since it requires PDL monitoring as well as a dedicated driver circuit and transducer for the uplink of the power feedback [2,3]. Also, the wireless feedback does not allow for fast f_D settling (<100ms) needed to compensate for displacements caused by fast movements such as the heartbeat. To overcome these limitations, this work proposes a local TX f_D tuning loop for fast and autonomous SWP tracking that maximizes and regulates PDL with a high tracking accuracy and low power.

For near-field US powering, the medium acts as a lossy acoustic transmission line, with a change in frequency affecting the equivalent impedance seen by the TX (ZTX). PZT transducers with a resonant frequency at 4.3MHz are chosen as a trade-off between pressure sensitivity and attenuation in tissue. The standing-wave effect adds a magnitude and phase modulation on top of the ZTX profile. Thus, complex ZTX sensing has been used as the TX local feedback signal for f_D tuning [4]. Since voltage mode drivers are preferred for driving US transducers thanks to their power efficiency [13], the latter conventionally requires current sensing, with high resolution and low phase distortion necessary to accurately measure ZTX and maximize PDL. However, direct current sensing with a resistor in the power path [4] large enough to output a voltage signal $V(I)$ with high SNR leads to significant TX power loss, while bulky off-chip current sensing components exceed the cranial implant form factor constraints [5]. Alternatively, integrated current sensors with TIA [11] have constraints on the amplifier BW and power consumption for low phase distortion at high US frequencies. However, around resonance, ZTX is dominated by a resistive part [10] and PDL sensing can be simplified to active power monitoring. To enable low-power and high-accuracy power-tracking for 4.3MHz US without the constraints of ZTX sensing, an adiabatic power-sensing scheme is proposed.

The TX system architecture is shown in Fig. 3.4.2 (bottom left), with the full system comprised of an adiabatic driver with a V_{DD} of 4.8V and an f_D tuning loop. In this work, a 5-level adiabatic driver is adopted (Fig. 3.4.2 top left), with 3 capacitors connected to the intermediate nodes V2, V3, V4 and switches S1-S5 turning on and off sequentially, driving the US link with a 5-level 4.3MHz periodic waveform. Inside the f_D tuning loop, an ADC samples V3 from the adiabatic driver and provides the digitized voltage value V_p to a "SWP tracking FSM". Based on V_p, the FSM determines the frequency control word (FCW) that

serves as the input of a frequency locked loop (FLL). Based on the FCW value, the FLL controls the frequency of a 12-phase DCO that provides the driving signals for S1-S5 in the 4-to-5MHz region.

In the adiabatic power-sensing scheme, the capacitors of the adiabatic TX are used as power-tracking elements. In adiabatic drivers [1,6], these capacitors serve to recycle part of the charge that would be dissipated by the load and thus improve driving efficiency of US transducers. At steady state in each US period, in the first half cycle the capacitors sequentially charge the US link load, while in the second half cycle there is sequential charge transfer from the load to the capacitors. With on-chip capacitors in the same range as the load capacitance, the voltage across these capacitors will vary due to the charge transfer. The voltages are decreased during the charging phase of the TX output (VTX) and increased during the discharging phase of VTX, with the example of V3 varying between $V3_L$ and $V3_H$ (Fig. 3.4.2 top right). For two US frequencies f_A and f_B, if V3 at f_B is lower than V3_L at f_A, it means that at f_B there is a larger charge transfer from the cap of node V3 and the US link. This corresponds to a higher absolute value of output current ITX while the US link is connected to V3. Since the US link load is approximately resistive, this translates to higher PDL. Thus, V3 is monitored in the adiabatic power sensing scheme, and the system settles to the f_D of minimum $V3_L$ which maximizes PDL. In this way, direct PDL tracking with a high tracking accuracy can be achieved at TX with simple voltage sampling. For sampling, a S/H and 8b SAR-ADC running at 125S/s are used. For proper closed-loop operation, $V3_L$ should be sampled when it has settled so the S&H switch opens in sync with S5 opening, once every SWP tracking FSM cycle of 8ms. The adiabatic power sensing has negligible power consumption (43µW) and requires no current sensing or external components.

Prior art [4] performs f_D tuning for SWP tracking with a blind frequency sweep. This results in long f_D settling time (seconds) and the system cannot compensate for fast movements or heartrate vibrations and requires power consuming and long-latency memory access to store all sensing values. The proposed f_D tuning loop, instead, is fully autonomous, achieves frequency settling in <90ms and without accessing memory, thanks to the proposed SWP tracking FSM, as shown in Fig. 3.4.3. The FSM uses a differential approach and a pair of frequencies f_1 and f_2 that differ by a fixed amount df. df is selected to be in the same range as the frequency difference between consecutive peak and valley of the standing wave, that for trans-dural US powering at 4.3MHz is in the range of 40 to 60kHz, and thus df is set to 64kHz. Depending on the sampled voltage difference ΔV_p between f_1 and f_2, the FCW is updated accordingly, and the pair of f_1 and f_2 moves across the spectrum with a step of df_2 until ΔV_p is smaller than a pre-defined value lim (i.e., $\Delta V_p < lim$). Then an extra operation is performed to check if V_p is smaller than the value at f_2 the middle frequency of f_1 and f_2. This is an indication of a local minimum of V_p, and the FSM locks the FCW at this f_D, maximizing PDL. This continues until there is a new distance change and a voltage difference is detected to update the loop. Compared to conventional MPPT algorithms like hill climbing [12], the SWP tracking FSM takes advantage of the well-known periodicity of the power over frequency due to the standing wave and achieves power tracking with a higher gain (Fig. 3.4.3 right). Parameters df, df_2 and lim are programmable to allow for a wide range of US f_D and powering distances. df_2 is set to 4kHz as a tradeoff between tracking accuracy and settling speed.

With the use of the FLL, the FSM accurately sets the DCO frequency by setting the FCW. FLL adopts a 4kHz reference clock to allow a df_2 step of 4kHz. A 12-phase differential DCO is implemented to create driving signals for the adiabatic driver switches S2-S4 with duration $T_D/12$, where T_D is the US period. A leakage-based delay cell is used [8] as well as supply independent current bias with a self-start-up capability. DCO f_D tuning is achieved with resistor tuning of the current bias. The FLL tunes the DCO's 6b resistor fine bank, targeting a resolution of 2kHz to ensure a low frequency error. The fine bank range is designed to be 120kHz to ensure loop settling by having at least 2 standing-wave peaks within this range. The 4b coarse bank value is programmable, with a coarse resolution of 60kHz so that it is smaller than the fine bank range and a coarse range of 1MHz to compensate for DCO PVT variations and allow for the use of 4-to-5MHz transducers.

The chip was fabricated in 65nm CMOS, and the micrograph is shown in Fig. 3.4.7. Acoustic measurements are performed in water with all 64 TX units driving a single 6×6mm² PZT

70 • 2026 IEEE International Solid-State Circuits Conference

979-8-3315-8937-0/26 $31.00 © 2026 IEEE

transducer, with a 6×6mm² RX PZT connected to a 50Ω load. By sweeping f_D and monitoring the digitized $V3_L$ and the RX AC voltage, the correlation of $V3_L$ with PDL is demonstrated (Fig. 3.4.4), validating the adiabatic power sensing scheme. The FLL can be settled within 8ms (Fig. 3.4.4). The closed-loop operation shows that the SWP tracking FSM tunes f_D in <90ms, and PDL is improved up to 2.7×(Fig. 3.4.5). PDL can be regulated to more than 82% of the maximum open loop value, when varying the powering distance but also when varying the initial f_D of the loop. Figure 3.4.6 summarizes the performance and compares it with state-of-the-art US drivers. Autonomous on-chip SWP tracking is performed on the TX with high PDL improvement and power-tracking accuracy while settling >50× faster than [4] and consuming only 43µW for SWP tracking, most of which is due to FLL operation.

Acknowledgement:
Funded from European Research Council (grant No. 101001448).

References:
[1] E. So et al. "12Mb/s 4×4 Ultrasound MIMO Relay with Wireless Power and Communication for Neural Interfaces." *ISSCC*, 2024.
https://doi.org/10.1109/ISSCC49657.2024.10454377
[2] M.L. Wang et al. "A Wireless Implantable Closed-Loop Electrochemical Drug Delivery System." *IEEE Transactions on Biomedical Circuits and Systems* (2024).
https://doi.org/10.1109/TBCAS.2024.3507022
[3] D. Shmilovitz et al. "Noninvasive control of the power transferred to an implanted device by an ultrasonic transcutaneous energy transfer link." *IEEE Transactions on Biomedical Engineering* 61.4 (2013): 995-1004.
https://doi.org/10.1109/TBME.2013.2280460
[4] H. Vihvelin et al. "Compensating for tissue changes in an ultrasonic power link for implanted medical devices." *IEEE Transactions on Biomedical Circuits and Systems* 10.2 (2015): 404-411. https://doi.org/10.1109/TBCAS.2015.2421823
[5] N. Guo et al. "Design of an automatic resonance frequency tracking chip for power ultrasonic transducer with a center frequency below 100 kHz." *IEEE Sensors Journal* 23.9 (2023): 9848-9858. https://doi.org/10.1109/JSEN.2022.3232155
[6] M. Gourdouparis et al. "An Ultrasound-Powering TX with a Global Charge-Redistribution Adiabatic Drive Achieving 69% Power Reduction and 53° Maximum Beam Steering Angle for Implantable Applications." *ISSCC*, 2024.
https://doi.org/10.1109/ISSCC49657.2024.10454411
[7] A. Lecchini-Visintini et al. "The pulsing brain: state of the art and an interdisciplinary perspective." *Interface Focus* 15.1 (2025): 20240058.
https://doi.org/10.1098/rsfs.2024.0058
[8] M. Ding et al. "A 0.7-V 0.43-pJ/cycle wakeup timer based on a bang-bang digital-intensive frequency-locked-loop for IoT applications." *IEEE Solid-State Circuits Letters* 1.2 (2018): 30-33. https://doi.org/10.1109/LSSC.2018.2810602
[9] D.-Y. Yoon et al., "A 1024-Channel Simultaneous Recording Neural SoC with Stimulation and Real-Time Spike Detection," *IEEE Symp. VLSI Circuits*, June 2021.
https://doi.org/10.23919/VLSICircuits52068.2021.9492480
[10] J. Charthad et al. "A mm-sized implantable medical device (IMD) with ultrasonic power transfer and a hybrid bi-directional data link." *IEEE Journal of Solid-State Circuits* 50.8 (2015): 1741-1753. https://doi.org/10.1109/JSSC.2015.2427336

[11] S. Agarwal et al. "A Current-Source-Free Constant-Current Wireless Adiabatic Neural Stimulator Achieving a 5.5-27.7 x Improved RF-to-Electrode Stimulation Efficiency Factor." *IEEE Symposium on VLSI Technology and Circuits*, 2024.
https://doi.org/10.1109/VLSITechnologyandCir46783.2024.10631437
[12] Y. Yang et al. "Dynamic improvement of series–series compensated wireless power transfer systems using discrete sliding mode control." *IEEE Transactions on Power Electronics* 33.7 (2017): 6351-6360. https://doi.org/10.1109/TPEL.2017.2747139
[13] Y. Zhang et al. "Integrated circuits for medical ultrasound applications: Imaging and beyond." *IEEE Transactions on Biomedical Circuits and Systems* 15.5 (2021): 838-858.
https://doi.org/10.1109/TBCAS.2021.3120886

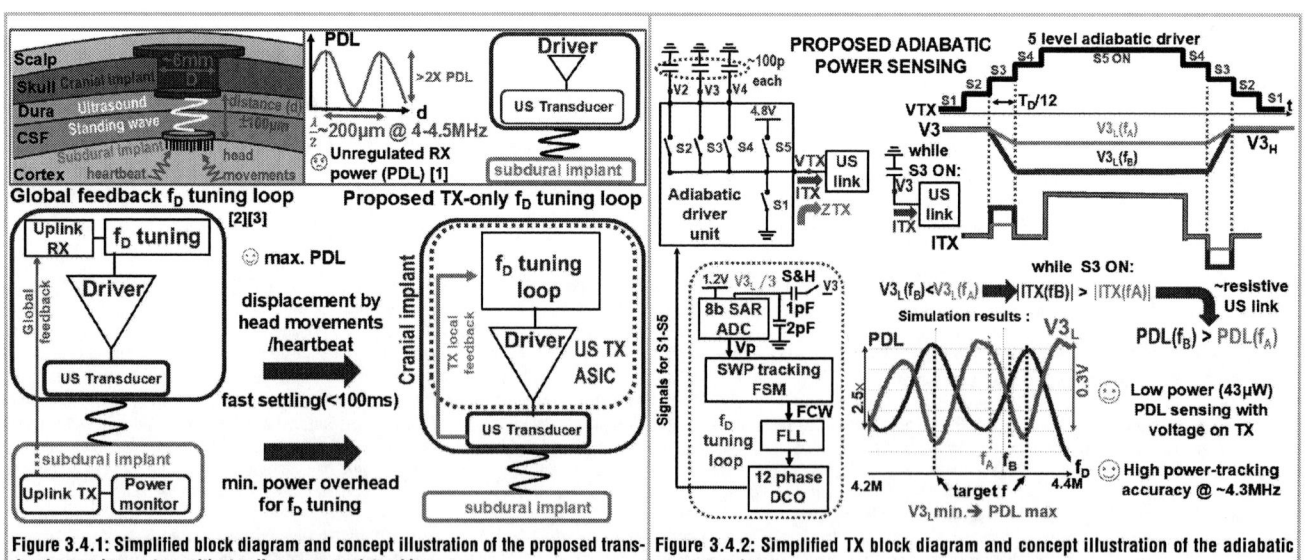

Figure 3.4.1: Simplified block diagram and concept illustration of the proposed trans-dural powering system with standing wave peak tracking.

Figure 3.4.2: Simplified TX block diagram and concept illustration of the adiabatic power sensing.

ISSCC 2026 / SESSION 3 / WEARABLE AND WIRELESS BIOMEDICAL SYSTEMS / 3.4

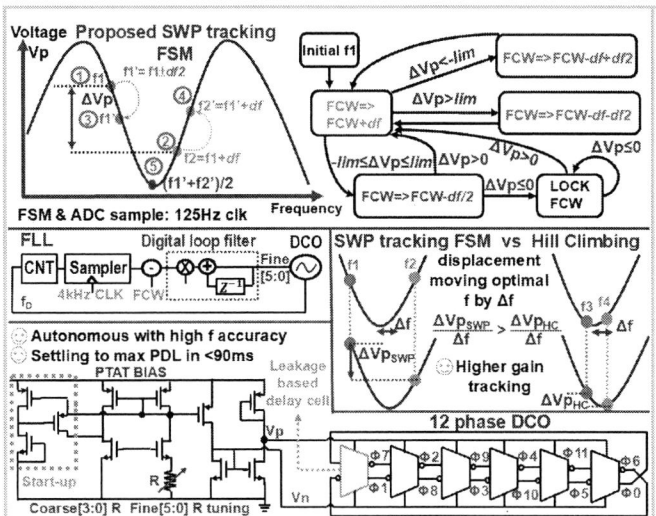

Figure 3.4.3: Simplified schematic of the frequency-tuning loop and concept illustration of the SWP tracking FSM.

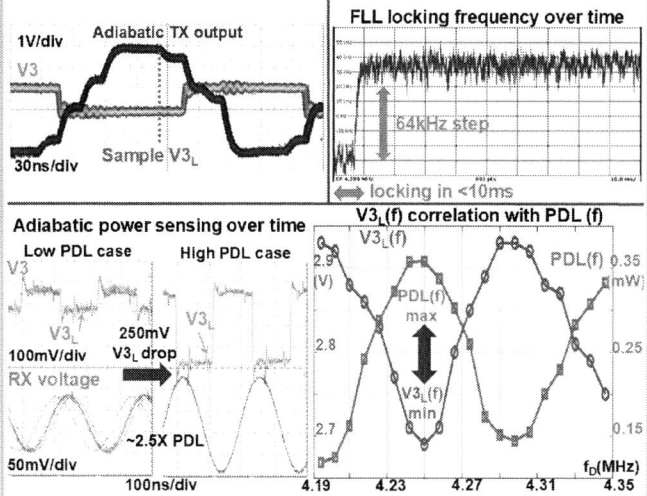

Figure 3.4.4: Measured TX output, voltage V3 to be sampled, FLL operation and acoustic measurements showing the correlation between sampled V3($V3_L$) and PDL.

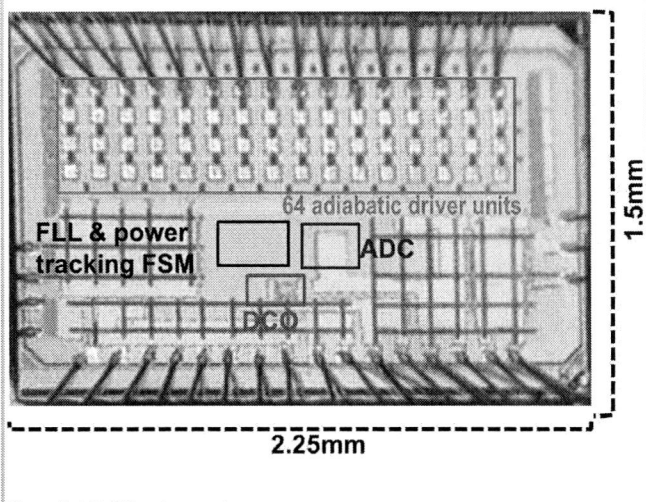

Figure 3.4.5: Acoustic measurements for PDL with and without the proposed SWP tracking, with varying f_D or powering distance d, as well as transient frequency measurement.

Figure 3.4.6: Summary and comparison table of state-of-the-art ultrasound powering systems with power regulation.

	This work	[4]Vihvelin, TBIOCAS'15	[2] Wang, TBioCAS'24	[3] Shmilovitz TBME' 13	[1] So, ISSCC'24	[6] Gourdouparis iSSCC' 24
CMOS process	65nm	Discrete	180nm	180nm	180nm BCD	65nm
Frequency	4.3MHz	1.3MHz	1MHz	720kHz	850kHz	7.8MHz
Autonomous closed-loop power regulation	YES	YES	YES	YES	NO	NO
Data link required for power regulation	NO	NO	YES	YES	-	-
Standing wave peak (SWP) tracking	YES	YES	NO	NO	NO	NO
On-chip SWP tracking	YES	NO	NO	NO	NO	NO
PDL improvement by SWP tracking	Up to 2.7X	Up to 2.5X	-	-	-	-
Power-tracking accuracy	82%	80%	-	-	-	-
Power loss for SWP tracking	43µW	Large (series R losses)[A]	-	-	-	-
settling time	<90ms	>5sec	-	-	-	-

[A]Estimated to be in the mW range, sensing R=209 Ω in series with transducer of 100-1K Ω under several Volts of supply.

Figure 3.4.7: Chip micrograph.

• 2026 IEEE International Solid-State Circuits Conference

979-8-3315-8937-0/26 $31.00 © 2026 IEEE

ISSCC 2026 / SESSION 3 / WEARABLE AND WIRELESS BIOMEDICAL SYSTEMS / 3.5

3.5 A Simultaneous Wireless Power and Full-Duplex Data Transfer System Over a Single Inductive Link Achieving 17/3.4Mb/s and 61.1% Efficiency for Miniature Biomedical Implants

Tianqi Lu, Sijun Du

Delft University of Technology, Delft, The Netherlands

Abstract

A single-link wireless power and full-duplex data transfer system is presented. Operating at 40.68MHz, it delivers >50mW maximum output power with 61.1% peak end-to-end efficiency while supporting 17Mb/s uplink using resonance-length keying (RLK) and 3.4Mb/s downlink using instant-carrier-suspension keying (Instant-CSK). The design removes ASK/LSK trade-offs and achieves an aggregate data rate of half the carrier frequency using 180nm TX/RX chips and 15-/8-mm-diameter coils.

Wireless power and data transfer (WPDT) holds great promise for next-generation implantable medical devices (IMDs), such as brain-machine interfaces and artificial retina [1], which require WPDT systems delivering high-speed data (>10Mb/s [2,3]) and continuous power (>50mW [3,4]) in small form factors. Traditional WPDT adopts separate data links in addition to the power link, supporting up-to-Gb/s data rates with techniques such as ultrawideband (UWB) [5]. However, that requires additional hardware and a larger form factor.

A single-link solution, where both power and data share one inductive link, eliminates additional antennas and cross-coupling. However, existing designs still fall short of IMD requirements for power and data (Fig. 3.5.1 [7-21]). Many support only simplex data transfer [9-20], while IMDs require both a precise downlink to modulate bioactivities and a high-speed uplink to transmit neural data. Reported bidirectional schemes usually achieve <1Mb/s [6-8,21], constrained by their modulation methods. For downlink, amplitude-shift keying (ASK) is commonly adopted, but limited by speed inversely proportional to the TX tank quality factor (Q) [9] and coupling-sensitive modulation depth (MD). Though frequency-shift keying (FSK) removes the Q-limitation for downlink, it sacrifices link efficiency and occupies extra frequency bands [9,15]. For uplink, load-shift keying (LSK) is typically used, which interrupts RX resonance and degrades power transfer [18]. In addition, the ASK/LSK interference prevents full-duplex operation with high data rate [7], necessitating time-division half-duplex instead.

To address these limitations, this paper presents a single-link, full-duplex WPDT system (Fig. 3.5.1 bottom), operating at 40.68MHz carrier frequency (f_c) that delivers continuous output power >50mW and supports full-duplex data transfer with an aggregate data rate >20Mb/s ($f_c/2$) with no extra off-chip hardware. For the uplink, resonance-length keying (RLK) is proposed, which enables $0.5 \cdot f_c$-rate data transfer without interrupting power delivery. Complementing this, instant-carrier-suspension keying (Instant-CSK) is proposed for the downlink, which provides instant carrier suspension/recovery and coexists seamlessly with RLK, establishing full-duplex operation with an aggregate rate around $0.5 \cdot f_c$. Together, these cycle-accurate modems overcome the ASK/LSK trade-offs tied to tank-Q and support compact IMDs requiring stable power and high-rate bidirectional links.

Figure 3.5.2 (top) shows the principle of the RLK for uplink data transfer. The RX uses a non-residual resonant-current-mode (RCM) rectifier, that alternates between two phases: resonance (Φ_1) and charging (Φ_2) [22]. In Φ_1, the RX coil (L_{RX}) resonates with C_{RX} to build up current (I_{LRX}) for a given duration; when I_{LRX} reaches its peak, the rectifier switches to Φ_2, bypassing C_{RX} and transferring the stored energy in L_{RX} to C_{OUT}. This $\Phi_1+\Phi_2$ cycle then repeats. Because Φ_1 and Φ_2 present distinct reflected impedances to the TX, setting the duration of one $\Phi_1+\Phi_2$ cycle to different multiples of the carrier period (T_c) directly modulates the TX coil current (I_{LTX}). As shown in Fig. 3.5.2 (top-right), two successive $1 \cdot T_c$ cycles (total $\approx 2 \cdot T_c$) lead to increased I_{LTX} amplitude, while a single $2 \cdot T_c$ cycle decreases it. These two cases can be mapped to binary symbols "0" and "1", respectively, forming the baseline RLK modulation. Each bit spans $2 \cdot T_c$ on average, yielding an uplink rate of $0.5 \cdot f_c$, while crucially preserving uninterrupted power delivery.

Figure 3.5.2 (bottom) shows the principle of the proposed instant-carrier-suspension keying (Instant-CSK) for downlink data transfer. The TX employs a class-D power amplifier (PA) augmented with a carrier-suspension switch, M_{CS}. With M_{CS} on, the PA operates in its normal ON state; turning off M_{CS} places the PA into a carrier-suspension (CS) state. During CS, M_{NT} remains on and M_{PT} is off; M_{CS} is turned off precisely when most resonant energy is in C_{TX}, leaving only minimal TX coil current (I_{LTX}) and negligible energy received by RX. After $2 \cdot T_c$, turning M_{CS} back on instantly recovers full-scale resonance by releasing the stored energy in C_{TX}, avoiding the gradual build-up required in conventional designs [7,10, 23]. By modulating the duty ratio between ON and CS states as 20/2 for "1" and 16/2 for "0", the baseline Instant-CSK downlink is established (Fig. 3.5.2 bottom-right). It reduces the power delivered to loads (PDL) by 10% (i.e., 90% Power Index (PI), where PI = PDL/PDL$_{w/oCSK}$) due to the average 18/2 duty ratio, and reallocates part of the uplink data rate to the downlink. In return, it removes the Q-limitation of conventional ASK and achieves an ON-OFF-keying-like MD without interfering with RLK uplink. To further enhance performance, a multi-level Instant-CSK is introduced: instead of two duty ratios (20/2, 16/2), four levels (24/2, 20/2, 16/2, 12/2) encode 2-bit symbols ("11", "10", "01", "00"). This doubles the average downlink rate while keeping the average duty ratio at 18/2, maintaining

90% PI and 18.3Mb/s (=$0.5 \cdot f_c \cdot 90\%$) uplink rate. Two optimization approaches are enabled by adding more Instant-CSK levels: (1) increasing the average duty to raise PI and uplink rate at the same downlink rate; or (2) fixing the average duty at 18/2 to accelerate downlink, achieving aggregate downlink/uplink throughput beyond $0.5 \cdot f_c$.

Figure 3.5.3 (top) shows the architectures of the TX and RX chips. In the TX, M_{PT} and M_{NT} are 5V MOSFETs, while M_{CS} is a 36V lateral-diffusion MOSFET to withstand the high V_{CTX} during the CS state. A 40.68MHz clock (clk$_{TX}$) drives the class-D PA, with the ON/CS duty controlled by the Instant-CSK modulator based on downlink input data (D_{DL}). An RLK demodulator extracts the uplink data. In the RX, 5V devices (M_{N1}, M_{N2}, M_P) form the RCM rectifier. A switching controller selects Φ_1 duration based on uplink input data (D_{UL}) via the RLK modulator. It also reports the LC energy level to the Instant-CSK demodulator for counter-based downlink decoding and the RLK modulator for D_{UL} holding during CS intervals. Both chips output recovered signals (D_{ULDEM}, D_{DLDEM}) for bit error rate (BER) analysis. During each CS interval, the RX holds D_{UL} while the TX skips updating D_{ULDEM}, ensuring proper uplink in full-duplex. Figure 3.5.3 (bottom-left) highlights the TX's RLK demodulator, which derives an amplitude-scaled and DC-shifted signal V_{MIDL} from V_{MID} (V_{MID} also reflects RLK modulation as I_{LTX}). An amplifier further amplifies the derivative in V_{MIDL} amplitude envelope with a shifted linear range around V_{MIDL} valleys (controlled by reference V_{ENV}). This enables cycle-accurate detection with logic timing referenced to clk$_{TX}$.

Given that the RLK demodulator samples V_{MID} periodically, long runs of "0"s (i.e., six consecutive $1 \cdot T_c$ cycles) can saturate the V_{MID} amplitude, leading to negligible envelope derivative and possible bit errors. To address this, two advanced RLK modes, low-BER and phase-position (PP) modes, are proposed (Fig. 3.5.3 bottom-right). The low-BER mode encodes data "1" and "0" using $2 \cdot T_c$-cycle×2 and ($2 \cdot T_c$-cycle×1+$1 \cdot T_c$-cycle×2), respectively. Since both cases include the $2 \cdot T_c$-cycle, V_{MID} amplitudes stabilize below the always-$1 \cdot T_c$-cycle extreme, avoiding saturation. This yields larger MD and lower BER, at the cost of halving the data rate. To recover rate, the PP mode further uses one $2 \cdot T_c$-cycle as a distinct symbol to be placed in every four T_c, and two bits of data are sent simultaneously. By doing so, V_{MID} remains sensitive (unsaturated) to RLK modulation, while it achieves the same data rate as the baseline RLK mode. The implemented RLK demodulator remains validated for these advanced RLK modes.

Both TX and RX chips are fabricated in a 180nm BCD process, occupying active areas of 0.15 and 0.18mm², respectively (Fig. 3.5.7). The TX and RX coils have inductance of 289nH and 97nH, and diameters of 15mm and 8mm, respectively. Figure 3.5.4 (top-left) shows the measured RLK waveform. It is observed that the $1 \cdot T_c$-cycle×2 and $2 \cdot T_c$-cycle×1 RX operations respectively increase and decrease V_{MID} amplitudes on TX. Figure 3.5.4 (top-right) shows the measured Instant-CSK waveform. The ON/CS duty is adjusted between 12/2 and 8/2, which represent downlink data "1" and "0", respectively. During CS intervals, RX receives negligible energy; the TX recovers steady power transfer immediately after it returns to ON states. Figure 3.5.4 (bottom-left) demonstrates the baseline Instant-CSK downlink modem in full-duplex. The TX ON states of $12 \cdot T_c$ and $8 \cdot T_c$ modulate downlink data of "1" and "0", respectively, which are correctly demodulated by RX (indicated by D_{DLDEM}), achieving an average data rate of 3.4Mb/s and a latency up to $12 \cdot T_c$. Figure 3.5.4 (bottom-right) shows the RLK uplink modem waveform in full-duplex. The RX operates the baseline RLK at 17Mb/s with $2 \cdot T_c$ latency. The uplink is activated during TX ON states, which are part of the downlink, demonstrating full-duplex.

Figure 3.5.5 shows the measured power and data transfer performance of the full-duplex WPDT system. At a coil separation distance, D_{COIL}, of 3mm, the full-duplex system achieves the peak end-to-end (E2E, from TX input to RX output) efficiency at 61.1%. It obtains a maximum output power (delivered to load R_{OUT}), P_{OUT}, of 73mW as the input voltage of TX, V_{IN}, reaches 5V. When disabling the data transfer, the system realizes a maximum P_{OUT} of 104mW. The relations between P_{OUT} and both downlink data rate and RX operation mode are illustrated in Fig. 3.5.5 (top right). Due to CS intervals, P_{OUT} drops if the baseline downlink data rate increases (RX in $1 \cdot T_c$ cycles). On RX, always-$1 \cdot T_c$-cycle case and always-$2 \cdot T_c$-cycle case yield two extreme P_{OUT} levels, confirming the RLK load-variation principle (Fig. 3.5.2). Three RLK modes report P_{OUT} levels between two extremes. The BER is measured using PRBS 7, 15, and 31 data patterns. The downlink reports a BER below 10^{-7} under three coupling conditions, and the uplink reports a BER below 10^{-5} except the weakest coupling condition. The low-BER and PP modes of RLK significantly improve BER compared to the baseline RLK. Figure 3.5.5 (bottom right) shows two figure-of-merit (FoM) comparisons. The system reports the highest FoM_1 with no extra hardware, indicating high bandwidth efficiency [20]. It also exhibits the highest FoM_2, indicating efficient power delivery with high output power.

Figure 3.5.6 shows a comparison table. The proposed WPDT system achieves full-duplex data transfer with much higher data rates compared to the state-of-the-art full-duplex design. It features small chip areas and small coil form factors without extra hardware. Its strong compatibility between the power and data links brings both high output power and high E2E efficiency.

References:
[1] C. Lee et al., "DustNet: A Network of Time-Division Multiplexed Ultrasonic Implants with 16-Level ASK Backscatter Modulation," *ISSCC*, 2025, pp. 582-583. https://doi.org/10.1109/ISSCC49661.2025.10904679
[2] W. Choi et al., "A 1,024-Channel, 64-Interconnect, Capacitive Neural Interface Using a Cross-Coupled Microelectrode Array and 2-Dimensional Code-Division Multiplexing," *IEEE Symposium on VLSI Technology and Circuits*, 2023, https://doi.org/10.23919/VLSITechnologyandCir57934.2023.10185425
[3] Q. Lin et al., "Advances and Challenges in Integrated Circuits for Electrochemical Sensing: Enabling Next-Generation Biomedical and Molecular Applications," in *IEEE Transactions on Biomedical Circuits and Systems*. https://doi.org/10.1109/TBCAS.2025.3589027
[4] K. Eom et al., "A 10-V-Tolerant Dual-Mode Neural Stimulation System with Self-Sustaining Dynamic Supply and Error-Resilient Digital Stimulus Odometer," in *IEEE Journal of Solid-State Circuits*, vol. 60, no. 9, pp. 3268-3282, Sept. 2025. https://doi.org/10.1109/JSSC.2025.3542022
[5] C. Ding et al., "A 49.8-mm² IR-UWB Transmitter With Co-Designed Power Amplifier and Antenna for Neural Implants With Extended Transmission Range," in *IEEE Journal of Solid-State Circuits*, vol. 60, no. 9, pp. 3174-3188, Sept. 2025. https://doi.org/10.1109/JSSC.2025.3531234
[6] Y. Jia et al., "A Dual-Band Wireless Power Transmission System for Evaluating mm-Sized Implants," in *IEEE Transactions on Biomedical Circuits and Systems*, vol. 13, no. 4, pp. 595-607, Aug. 2019. https://doi.org/10.1109/TBCAS.2019.2915549
[7] J. Lee et al., "A Wireless Power and Synchronized Full-Duplex Data Transceiver IC with 400 Kbps Bidirectional Data Rate Using a Single Inductive Link for Low-Power Systems," *Symposium on VLSI Technology and Circuits*, 2025, https://doi.org/10.23919/VLSITechnologyandCir65189.2025.11074871
[8] Y. -T. Hsiao et al., "An RFID-Inspired One-Step Packaged Multimode Bio-Analyzer with Vacuum Microfluidics for Point-of-Care Diagnostics," *ISSCC*, 2025, pp. 352-353. https://doi.org/10.1109/ISSCC49661.2025.10904714
[9] Y. Park et al., "An Enhanced-Frequency-Splitting-Based Wireless Power and Data Transfer System Achieving 60.2% End-to-End Efficiency and 1 Mb/s Data Rate with a Sub-cm RX Coil for Miniaturized Implants," *ISSCC*, 2025, https://doi.org/10.1109/ISSCC49661.2025.10904522
[10] H. -S. Lee and H. -M. Lee, "A Power-Efficient Envelope-Detector-Less Amplitude-Shift-Keying Forward Telemetry for Wirelessly Powered Biomedical Devices," in *IEEE Transactions on Biomedical Circuits and Systems*, vol. 19, no. 2, pp. 374-384, April 2025. https://doi.org/10.1109/TBCAS.2024.3427396
[11] H. Roh et al., "A Smart Contact Lens System with 433MHz Wireless Power and Data Transfer at a Modulation Index Down to 0.02%," *IEEE Symposium on VLSI Technology and Circuits*, 2024, https://doi.org/10.1109/VLSITechnologyandCir46783.2024.10631553

[12] Q. Zhuang et al., "A 6.78-MHz Wireless Power and Data Transfer System Achieving Simultaneous 48.6% End-to-End Efficiency and 4.0-Mb/s Forward Data Delivery With Interference-Free Rectifier," in *IEEE Journal of Solid-State Circuits*, vol. 60, no. 9, pp. 3283-3293, Sept. 2025. https://doi.org/10.1109/JSSC.2025.3541290
[13] Y. Liu et al., "A 13.56-MHz Single-Input Dual-Output Wireless Power and Data Transfer System for Bio-Implants," in *IEEE Journal of Solid-State Circuits*, vol. 59, no. 8, pp. 2557-2567, Aug. 2024. https://doi.org/10.1109/JSSC.2024.3372430
[14] H. Kim et al., "DiTTO: A Distance Adaptive Over 100-mW Wireless Power Transfer System With 1.695-Mb/s Uplink Telemetry and a Shared Inductor Two-Output Regulating Rectification," in *IEEE Journal of Solid-State Circuits*, vol. 59, no. 8, pp. 2568-2580, Aug. 2024. https://doi.org/10.1109/JSSC.2024.3372418
[15] Y. Park et al., "A Frequency-Splitting-Based Wireless Power and Data Transfer IC for Neural Prostheses with Simultaneous 115mW Power and 2.5Mb/s Forward Data Delivery," *ISSCC*, 2021, pp. 472-473. https://doi.org/10.1109/ISSCC42613.2021.9365781
[16] S. -W. Hong, "A 13.56MHz Current-Mode Wireless Power and Data Receiver with Efficient Power Extracting Controller and Energy-Shift Keying Technique for Loosely Coupled Implantable Devices," *ISSCC*, 2020, pp. 486-487. https://doi.org/10.1109/ISSCC19947.2020.9062923
[17] J. Pan et al., "Simultaneous Transmission of Up To 94-mW Self-Regulated Wireless Power and Up To 5-Mb/s Reverse Data Over a Single Pair of Coils," in *IEEE Journal of Solid-State Circuits*, vol. 54, no. 4, pp. 1003-1016, April 2019. https://doi.org/10.1109/JSSC.2018.2888884
[18] S. Ha et al., "Energy Recycling Telemetry IC With Simultaneous 11.5 mW Power and 6.78 Mb/s Backward Data Delivery Over a Single 13.56 MHz Inductive Link," in *IEEE Journal of Solid-State Circuits*, vol. 51, no. 11, pp. 2664-2678, Nov. 2016. https://doi.org/10.1109/JSSC.2016.2600864
[19] M. Kim et al., "A 13.56-MHz Wireless Power and Data Transfer System With Current-Modulated Energy-Reuse Back Telemetry and Energy-Adaptive Voltage Regulation," in *IEEE Journal of Solid-State Circuits*, vol. 58, no. 2, pp. 400-410, Feb. 2023. https://doi.org/10.1109/JSSC.2022.3207549
[20] Y. -P. Lin and K. -T. Tang, "An Inductive Power and Data Telemetry Subsystem with Fast Transient Low Dropout Regulator for Biomedical Implants," in *IEEE Transactions on Biomedical Circuits and Systems*, vol. 10, no. 2, pp. 435-444, April 2016. https://doi.org/10.1109/TBCAS.2015.2447526
[21] C. -H. Cheng et al., "A Fully Integrated 16-Channel Closed-Loop Neural-Prosthetic CMOS SoC With Wireless Power and Bidirectional Data Telemetry for Real-Time Efficient Human Epileptic Seizure Control," in *IEEE Journal of Solid-State Circuits*, vol. 53, no. 11, pp. 3314-3326, Nov. 2018. https://doi.org/10.1109/JSSC.2018.2867293
[22] T. Lu and S. Du, "A Three-Phase Regulating Resonant-Current-Mode Rectifier with Bypass-Capacitor Residual-Free Charging for Wireless Power Transfer," in *IEEE Journal of Solid-State Circuits*. https://doi.org/10.1109/JSSC.2025.3597069
[23] A. Trigui et al., "Generic Wireless Power Transfer and Data Communication System Based on a Novel Modulation Technique," in *IEEE Transactions on Circuits and Systems I: Regular Papers*, vol. 67, no. 11, pp. 3978-3990, Nov. 2020. https://doi.org/10.1109/TCSI.2020.3010308
[24] M. Kiani and M. Ghovanloo, "A Figure-of-Merit for Designing High-Performance Inductive Power Transmission Links," in *IEEE Transactions on Industrial Electronics*, vol. 60, no. 11, pp. 5292-5305, Nov. 2013. https://doi.org/10.1109/TIE.2012.2227914

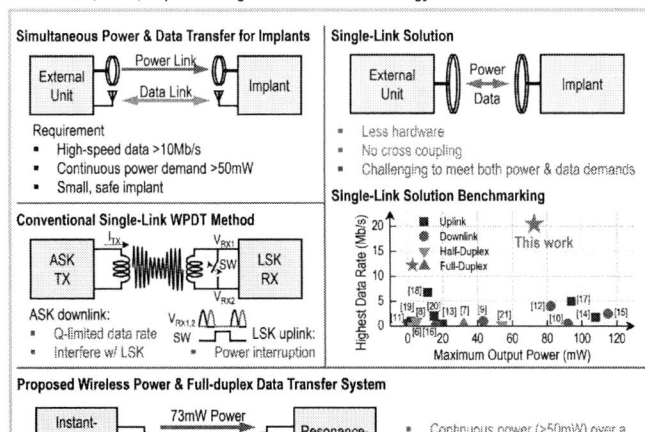

Figure 3.5.1: Requirements for simultaneous power and data transfer in bio-implants, limitations of conventional single-link WPDT, and the proposed full-duplex solution.

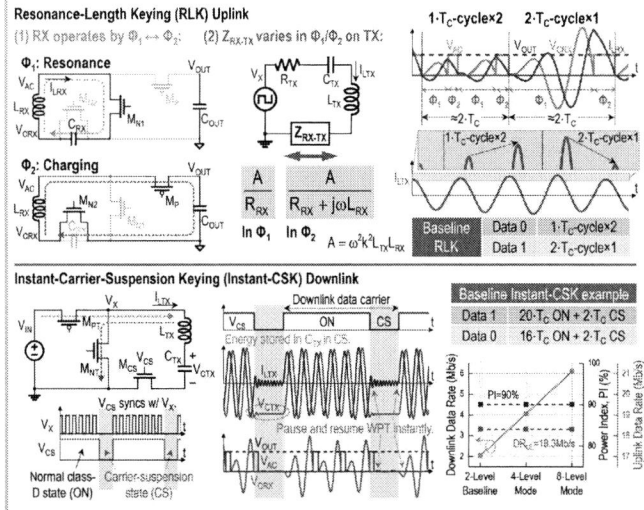

Figure 3.5.2: Operating principle of the proposed full-duplex WPDT system with resonance-length keying uplink and instant-carrier-suspension keying downlink.

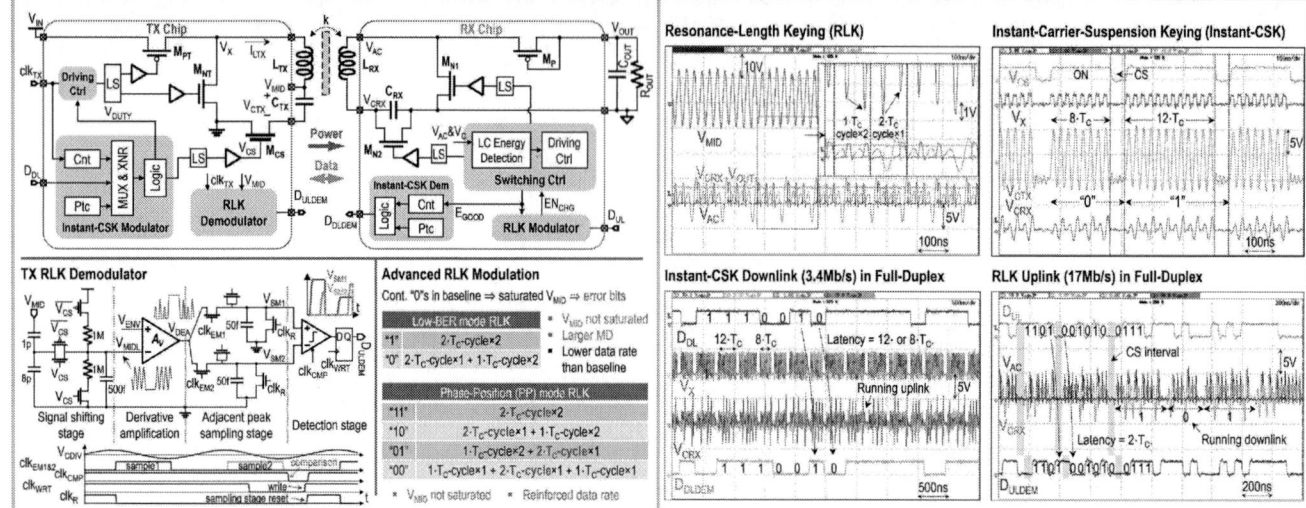

Figure 3.5.3: System diagram of the full-duplex WPDT, circuit implementation of the RLK demodulator, and advanced RLK modulation modes.

Figure 3.5.4: Measured waveforms of RLK and Instant-CSK operation, Instant-CSK downlink and RLK uplink modem in the full-duplex system.

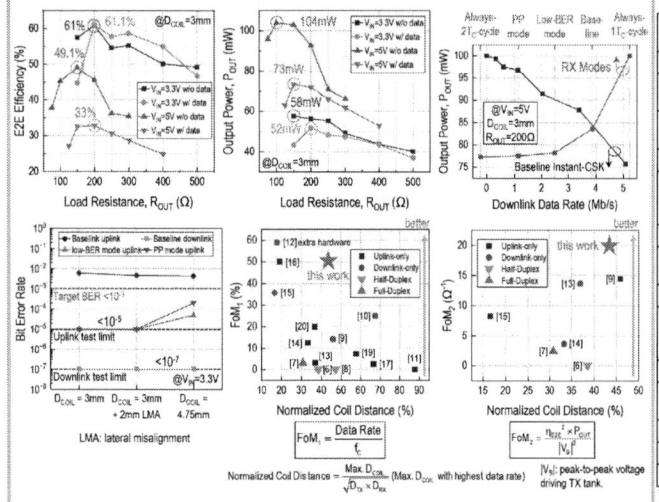

Figure 3.5.5: Measured power and data transfer performance of the WPDT system, with figure-of-merit (FoM) benchmarking.

	JSSC'24 [13]	JSSC'24 [14]	TBioCAS'24 [10]	JSSC'25 [12]	ISSCC'25 [9]	ISSCC'25 [8]	VLSI'25 [7]	This work
Data Directionality	Uplink	Uplink	Downlink	Downlink	Downlink	Half-duplex	Full-duplex	Full-duplex
Data Rate (Mb/s)	0.424	1.695	0.5	4	1	downlink 0.22 / uplink 0.9	downlink 0.4 / uplink 0.4	downlink 3.4 / uplink 17
Carrier Frequency, f_C (MHz)	13.56	13.56	2	6.78	6.8~7.2	700	13.56	40.68
Technology (nm)	65	65	250	180	180	180	180	180
Chip Area (mm²) Total / Active	TX - / RX -, 0.15 / 0.15	TX - / RX -, Ext. / 0.2	TX Ext. / RX 1, Ext. / 0.52	TX Ext. / 0.52	TX 1.71 / RX 2.25, 10.2	TX Ext., RX 2.25	TX 4.6, RX 4.6	TX 0.72 / RX 0.8, 0.15 / 0.18
Extra Hardware*	1 coil	No	No	4 caps	No	No	No	No
Coil Diameter (mm) D_{TX}, D_{RX}	35, 30	30, 30	61, 35	45, 35	12.5, 9.5	TX 3, RX 1.5	TX 35, RX 30	TX 15, RX 8
Norm. D_{COIL} (%)	36.93	33.33	67.52	17.78	45.88	47.14	30.86	43.36
Modulation Methods	LSK	DCK	ASK	Harmonic	FSK	downlink ASK / uplink LSK	downlink ODSK / uplink LSK	downlink Instant-CSK / uplink RLK
Bit Error Rate	2×10⁻⁸	2.2×10⁻⁷	<10⁻⁴	10⁻⁶	<10⁻⁷	-	10⁻⁶	downlink <10⁻⁷ / uplink <10⁻⁵
Output Power, P_{OUT} (mW)	20	108	92	82	43.4	5	32	73 w/ data / 104 w/o data
E2E Efficiency (%)	62.7	58.4	36	48.6	60.2	>10%	34.8	61.1
FoM₁ [20] (%)	3.13	12.5	25	59	14.29	0.16	2.95	50
FoM₂ [24] (Ω⁻¹)	13.65	3.68#	-	-	14.44	-	2.42#	19.72

*for WPDT use, excluding resonant link components and output capacitors. #estimated from paper. Reported results was obtained through air.

Figure 3.5.6: Comparison of recently reported WPDT systems.

Figure 3.5.7: Chip micrographs.

Session 4 Overview: *Analog Techniques & Amplifiers*
ANALOG SUBCOMMITTEE

Session Chair: Chinwuba Ezekwe
Robert Bosch, Sunnyvale, CA

Session Co-Chair: Ka-Meng Lei
University of Macau, Macau

The Analog Techniques & Amplifiers session showcases advancements in precision analog design, highlighting circuits that push the performance boundaries of accuracy, power efficiency, and compactness. The first two papers describe Class-D amplifiers that improve efficiency over a wide range of signal levels while achieving state-of-the-art linearity performance. The third paper demonstrates a compact and low-voltage-compatible current-pulse-injection crystal oscillator. The three final papers present innovative high-order TC compensation techniques to improve the accuracy of voltage and current references.

ISSCC 2026 / February 16, 2026 / 3:35 PM

3:35 PM

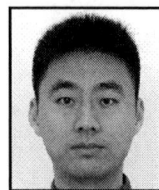

4.1 A 0.64mA, -108.2dB THD+N Class-D Amplifier with Neural-Assisted Pre-Reconfiguration for Smart Power Optimization

Kaiwen Zhou, Fudan University, Shanghai, China

In Paper 4.1, Fudan University presents a Class-D amplifier that leverages the assistance of a neural network to improve efficiency by up to 18% at small-signal levels while achieving a THD+N of –108.2dB and the highest FoM$_{THD+N}$ of 3,747.

4

4:00 PM

4.2 A -102.2dB THD+N, 92% Efficiency, 1.08mW Idle Power Digital-Input Class-D Amplifier with Power-Adaptive Techniques and Dual-Edge Pulse-Width Adjustment Modulator

Miao Zhang, TU Delft, Delft, The Netherlands

In Paper 4.2, Delft University of Technology introduces a digital-input Class-D amplifier for portable audio that improves efficiency by up to 20% at small signal levels by combining digital-pulse-width modulation with power-adaptive techniques. It achieves –102.2dB THD+N, 117.8dB DR, and 92% peak efficiency.

4:25 PM

4.3 A 0.6V 9.4µW 1,892µm² Current-Pulse-Injection Crystal Oscillator Featuring Capacitively Biased Amplifier with 242.2dBc/Hz PN FoM @1kHz Offset

Haihua Li, University of Macau, Macau, China, UM Hetao IC Research Institute, Shenzhen, China

In Paper 4.3, the University of Macau showcases a compact current-pulse-injection 12/16MHz crystal oscillator featuring a capacitively biased amplifier that can support operation down to 0.5V. The XO core occupies 1,892µm² in the 65nm process, and consumes 9.4µW (16MHz) at 0.6V V$_{DD}$, with a FoM of 242.2dBc/Hz.

4:50 PM

4.4 A 2.1-to-3.7ppm/°C Bandgap Voltage Reference with a Current-Domain TC Compensation and ±0.06% Inaccuracy from -40°C to 125°C in 130nm CMOS

Zhong Tang, Xidian University, Xi'an, China

In Paper 4.4, Xidian University describes a bandgap voltage reference that uses a capacitively-biased-diode-based super PTAT current generator to compensate high-order TC in 130nm CMOS. It achieves a TC of 2.1 to 7.1ppm/°C from -40 to 125°C after a simple batch trim.

5:05 PM

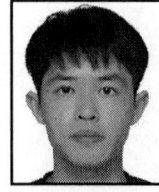

4.5 A 1ppm/°C and ±0.066% 3σ Accuracy Bandgap Reference with Temperature-Adaptive PTAT Scaling

Jiaqing Rui, East China Normal University, Shanghai, China

In Paper 4.5, East China Normal University presents a bandgap reference with temperature-adaptive PTAT Scaling to compensate for high-order non-linearity. The bandgap reference in 180nm CMOS has a TC of 0.508ppm/°C across –40 to 125°C and a 3σ inaccuracy of ±0.042% after a single trim.

5:20 PM

4.6 An Integrated Voltage and Current Reference Together Achieving 5.7 and 9.1ppm/°C from -40 to 125°C

Lele Fang, University of Macau, Macau, China

In Paper 4.6, the University of Macau reveals an integrated voltage and current reference featuring sub-ranging compensation with seamless transitions. The voltage and current references, fabricated in 65nm CMOS, achieve an average TC of 5.7ppm/°C and 9.1ppm/°C, respectively, across –40 to 125°C after batch trimming.

DIGEST OF TECHNICAL PAPERS • 75

979-8-3315-8937-0/26 $31.00 © 2026 IEEE

ISSCC 2026 / SESSION 4 / ANALOG TECHNIQUES & AMPLIFIERS / 4.1

4.1 A 0.64mA, -108.2dB THD+N Class-D Amplifier with Neural-Assisted Pre-Reconfiguration for Smart Power Optimization

Kaiwen Zhou, Yuxiang Tang, Zhongqiu Shen, Xuecheng Yang, Tianxiang Qu, Yiming Wang, Shen Ye, Zhiliang Hong, Jiawei Xu

Fudan University, Shanghai, China

Abstract

This work presents a high-precision, power-efficient Class-D audio amplifier (CDA) that addresses switching losses and reconfiguration-induced nonlinearity through neural-network–assisted pre-reconfiguration. Consequently, this CDA achieves the lowest THD+N

of -108.2dB and a very high FoM$_{THD+N}$ of 3747, corresponding to a 48% improvement over the state-of-the-art in the accompanying comparison table. This CDA also attains an ultra-low quiescent current of 0.64mA and highest efficiency of 66% at 10mW.

In Class-D audio amplifiers (CDAs), power efficiency is typically quoted near full-scale output. In practice, audio signals are highly dynamic, with speech containing syllabic pauses and inter-word gaps, and music alternating between soft and loud passages. As a result, the amplifier operates for a substantial portion of time at low output levels, where power-stage switching loss dominates and efficiency collapses (Fig. 4.1.1, top left) [1]. These switching losses also limit the minimum achievable quiescent current (I_Q), making the improvement of low-power efficiency and the reduction of I_Q essential for portable, battery-powered audio systems. To address this challenge, various reconfiguration techniques have been explored, including dual-supply operation [2], spread-spectrum modulation [3], and dynamic power-stage activation [4]. However, these approaches often entail trade-offs between power efficiency and linearity. Beyond circuit-level techniques, control strategies also critically influence reconfiguration effectiveness (Fig. 4.1.1, top right). For example, envelope tracking can substantially improve efficiency, but frequent mode transitions degrade linearity and can induce click and pop noise [2,3]. Hold-time insertion alleviates this by deferring switching decisions, yet often results in unnecessary power under dynamic signals [5]. In contrast, the "delay for decision" strategy [6] provides a more favorable balance by postponing transitions until the signal trend is established, enabling zero-crossing switching to minimize interference and power loss. However, its buffer-based delay limits use in time-sensitive applications such as active noise cancellation (ANC). These constraints highlight the need for an adaptable and intelligent reconfiguration scheme that can preserve high linearity and efficiency without incurring latency penalties.

This work presents a high-precision, power-efficient CDA (Fig. 4.1.1, bottom) that addresses switching losses and reconfiguration-induced nonlinearity through: (1) a neural-network-assisted predictive control for zero-delay, linearity-preserving pre-reconfiguration; (2) a layer-reuse topology with clock gating to implement a compact, energy-efficient predictor; (3) a dynamic loop enabling wide switching-frequency spreading while maintaining stability; and (4) zero-crossing power-stage slicing to further reduce switching loss and improve linearity. Consequently, this CDA delivers a THD+N of -108.2dB at a quiescent current of only 0.64mA, achieving an FoM$_{THD+N}$ of 3747, which surpasses the state-of-the-art in the comparison table in Fig. 4.1.6 by 48%.

Figure 4.1.2 illustrates the neural prediction architecture and operation. The key idea is to use previously digitized input peaks to forecast the upcoming peak one half-period ahead, enabling zero-crossing reconfiguration in advance without incurring signal delay while preserving linearity by minimizing mode transitions. Due to multi-frequency components and large amplitude variations of audio signals, accurately predicting future waveforms using linear circuitry is challenging. Neural networks, known for their superior nonlinear function approximation and temporal sequence prediction capabilities, offer a promising solution for predictive reconfiguration. However, embedding a neural network in a CDA imposes stringent constraints on silicon area and power consumption. To overcome these challenges, an ultra-low-power, compact predictive network with high-accuracy inference is developed. As shown in Fig. 4.1.2, the input signal is digitized by a 7-bit SAR ADC operating at 1MS/s. Each half-period of the input waveform is defined as a cycle, and a zero-crossing detector identifies the cycle boundaries and triggers the neural prediction process. During each cycle, the signal peak is captured by a peak detector and stored in memory for the previous four cycles (T_n to T_{n-3}). Upon detection of a new zero crossing, these stored peaks are fed into the neural prediction network to estimate the upcoming peak at T_{n+1}. The predicted peak amplitude is then quantized into four discrete levels according to predefined thresholds, generating a classification code that guides the pre-reconfiguration of analog CDA.

To minimize the area and power consumption of the prediction network, a layer-reuse topology with integrated clock gating is adopted, as illustrated in Fig. 4.1.2. A single pipelined multiply-accumulate layer comprising 8 neurons is time-multiplexed to sequentially process all network layers, with weights and biases retrieved from on-chip memory. This serialized architecture significantly reduces hardware overhead. The network, trained on 50,000 randomly generated dual-tone samples and evaluated on 1,000 test samples, achieves a classification accuracy of 83.5%. Since prediction is triggered only once per zero-crossing event (41 cycles at a 16MHz clock), clock gating disables the network when during idle periods, thereby substantially reducing dynamic power. Compared

with a fully parallel implementation in digital back-end simulations, the proposed architecture reduces silicon area by 53% and lowers the quiescent current from 2mA to 103µA, while maintaining real-time, low-power, and high-accuracy prediction with minimal hardware cost.

The predicted amplitude is quantized into four levels and used for pre-reconfiguration of the analog CDA, as shown in Fig. 4.1.3. Two techniques, dynamic loop and zero-crossing slicing, are employed to minimize switching loss while preserving loop linearity and stability. In dynamic loop control (Fig. 4.1.3, bottom left), the PWM switching frequency f_{SW} is constrained by the loop stability criterion $f_{SW} > \pi \cdot f_{UG}$, where f_{UG} is the unity-gain bandwidth of the CDA loop. Simple frequency spreading may violate this limit, resulting in degraded linearity or even loop instability. To address this, the proposed CDA dynamically reconfigures the loop bandwidth from 75kHz to 150kHz by adjusting a resistive attenuator reused from the anti-PWM aliasing filter [7]. This enables ultra-wide f_{SW} adaption from 250kHz to 650kHz, enabling pronounced reduction in switching losses. In conventional envelope tracking, f_{SW} shifting occurs when the differential feedback current $I_{FB} = I_{FB+} - I_{FB-}$ is nonzero, which perturbs the PWM duty cycle and consequently introduces nonlinearity and audible transition pops. Owing to zero-crossing pre-reconfiguration, where the I_{FB} approaches zero, abrupt changes in f_{SW} introduce no duty-cycle interference, thereby minimizing the instantaneous loop error and suppressing pops. If the predicted level is inaccurate, the selected loop gain becomes suboptimal for power-stage linearization, raising THD. Nevertheless, the loop remains stable, and because decisions are refreshed at each zero crossing, the system rapidly returns to the optimal mode in the subsequent cycle.

The power stage (Fig. 4.1.3) consists of 4 parallel units. For signals below -18dBFS, only ×1 unit is activated to minimize switching loss. As the amplitude increases to -18dBFS, -12dBFS, and -6dBFS, ×2, ×3, and ×4 units are enabled, respectively, to reduce conduction loss. As illustrated in Fig. 4.1.3 (bottom right), slicing transients introduce variations in Δr_{on}, which can distort V_{out} when the load current I_{load} fluctuates. However, by aligning slicing events with the input zero crossing, where the load current is momentarily zero, such distortion is also inherently suppressed. When the prediction is wrong, the efficiency is not optimal but the CDA remains stable. Since predictions are correct in the majority of cases, the long-term average efficiency is improved.

The CDA was fabricated in a 0.18µm CMOS process, occupying an active area of 2.2mm² (Fig. 4.1.7). Figure 4.1.4 compares the measured performance of conventional envelope tracking and neural-network-assisted prediction under a dual-tone input signal (1V$_{rms}$, 500Hz + 2kHz). The oscilloscope waveforms of input and control code (Fig. 4.1.4, top) indicate that envelope tracking responds only after instantaneous amplitude crosses predefined thresholds, resulting in frequent mode transitions. In contrast, neural prediction enables zero-crossing pre-reconfiguration, thereby notably reducing switching events. The output spectrum (Fig. 4.1.4, bottom) demonstrates significant performance gains: the prediction method achieves an intermodulation distortion (IMD) of -113.3dBc, improving by 18.8dB over the envelope-tracking baseline (-94.5dBc). Furthermore, the spurious-free dynamic range (SFDR) is enhanced from 72.1dBc to 98.4dBc (+26.4dB). Consequently, all spurs are suppressed below click-and-pop hearing threshold [8].

Figure 4.1.5 (top) compares the spectral performance under correct and worst-case prediction scenarios. For a -3dBFS input with correct prediction (-6 to 0dBFS), the CDA achieves a THD+N of -108.2dB. In the worst case, corresponding to the bottom-right cell of the confusion matrix in Fig. 4.1.2, the -3dBFS input is severely underestimated (<-18dBFS), and THD+N degrades to -72.6dB due to insufficient loop gain and conduction capacity. Despite this, the output remains stable without oscillation or latch-up, confirming system robustness. Figure 4.1.5 (bottom) compares output efficiency with and without pre-reconfiguration. In the switching-loss-dominated region, reconfiguration yields up to 18.6% efficiency improvement. At higher output power, both curves converge, with ~2% loss from -12dBFS to -6dBFS due to conduction-loss dominance, indicating that the threshold may be set lower. For low-power operation, Fig. 4.1.6 (top-left) shows the quiescent current breakdown: by suppressing switching loss, the total quiescent current is reduced from 1.1mA to 0.64mA, achieving a 42% reduction.

76 • 2026 IEEE International Solid-State Circuits Conference

979-8-3315-8937-0/26 $31.00 © 2026 IEEE

ISSCC 2026 / February 16, 2026 / 3:35 PM

Figure 4.1.6 compares the proposed CDA with state-of-the-art designs. The scatter plot (top-right) shows the measured minimum THD+N versus FoM$_{THD+N}$, where this work achieves a THD+N of −108.2dB and an FoM$_{THD+N}$ of 3747, corresponding to a 48% improvement over prior results. The summary table (bottom) further highlights that, enabled by neural-assisted pre-reconfiguration, the CDA attains the highest linearity, an ultra-low quiescent current of 0.64mA, and 66% efficiency at 10mW—leading among the compared designs. The wide reconfigurable switching frequency (250 to 650kHz) also provides headroom for EMI mitigation. As a neural-network-assisted CDA, it achieves 83.5% dual-tone prediction accuracy and enables zero-crossing pre-reconfiguration, which is not feasible with analog-only control. When evaluated on an audio dataset with richer spectral content, prediction accuracy drops to 75.8%. However, system functionality is maintained and can be restored using higher-capacity or sequence-aware models (e.g., deep networks or RNNs) at modest area and power overhead. This design establishes a new benchmark for battery-powered high-fidelity audio.

Acknowledgement:
This work was supported by the State Key Laboratory of Integrated Chips and Systems, Fudan University.

References:
[1] Y. Qiu et al., "A 0.4-mA-Quiescent-Current, 0.00091%-THD+N Class-D Audio Amplifier with Low-Complexity Frequency Equalization for PWM-Residual- Aliasing Reduction," *IEEE JSSC*, vol. 57, no. 2, pp. 423-433, Feb. 2022. https://doi.org/10.1109/JSSC.2021.3093309
[2] M. Berkhout et al., "A 4Ω 2.65W Class-D Audio Amplifier with Embedded DC-DC Boost Converter, Current Sensing ADC and DSP for Adaptive Speaker Protection," *IEEE JSSC*, vol. 48, no. 12, pp. 2952-2964, Dec. 2013. https://doi.org/10.1109/JSSC.2013.2284692
[3] L. Guo et al., "A 101 dB PSRR, 0.0027% THD + N and 94% Power-Efficiency Filterless Class D Amplifier," *IEEE JSSC*, vol. 49, no. 11, pp. 2608-2617, Nov. 2014. https://doi.org/10.1109/JSSC.2014.2359913
[4] L. Dooper, M. Berkhout, "A 3.4 W Digital-In Class-D Audio Amplifier in 0.14 µm CMOS," *IEEE JSSC*, vol. 47, no. 7, pp. 1524-1534, July. 2012. https://doi.org/10.1109/JSSC.2012.2191683
[5] Analog Devices Inc., 28V Digital Input, Class-DG Amplifier with IV$_{SENSE}$, Ultra-Low I$_Q$, and Brownout Prevention, [Online]. Available: https://www.analog.com/media/en/technical-documentation/data-sheets/max98397.pdf.
[6] Texas Instruments, TAS6684-Q1 - 45V, 13A Digital Input 4-Channel Automotive Class-D Audio Amplifier with Current Sense and Real-time Load Diagnostics, [Online]. Available: https://www.ti.com/lit/ds/symlink/tas6684-q1.pdf.
[7] K. Zhou et al., "A 0.81mA, -105.2dB THD+N Class-D Audio Amplifier with Capacitive Feedforward and PWM-Aliasing Reduction for Wide-Band-Effective Linearity Improvement," *ISSCC*, pp. 380-382, Feb. 2024. https://doi.org/10.1109/ISSCC49657.2024.10454485
[8] Texas Instruments, Click and Pop Measurement Technique, [Online]. Available: https://www.ti.com/lit/an/slea044/slea044.pdf.
[9] S. Chien et al., "A Low Quiescent Current, Low THD+N Class-D Audio Amplifier with Area-Efficient PWM-Residual-Aliasing Reduction," *IEEE JSSC*, vol. 53, no. 12, pp. 3377-3385, Dec. 2018. https://doi.org/10.1109/JSSC.2018.2873613

[10] W. Wang, and Y. Lin, "A 0.0004% (−108dB) THD+N, 112dB-SNR, 3.15W Fully Differential Class-D Audio Amplifier with Gm Noise Cancellation and Negative Output-Common-Mode Injection Techniques," *ISSCC*, pp. 58-60, Feb. 2018. https://doi.org/10.1109/ISSCC.2018.8310182
[11] W. Wang, and Y. Lin, "A 118 dB PSRR, 0.00067% (-103.5 dB) THD+N and 3.1 W Fully Differential Class-D Audio Amplifier with PWM Common Mode Control," *IEEE JSSC*, vol. 51, no. 12, pp. 2808-2818, Dec. 2016. https://doi.org/10.1109/ISSCC.2016.7417921
[12] W. Sun et al., "A 121dB DR, 0.0017% THD+N, 8× Jitter-Effect Reduction Digital-Input Class-D Audio Amplifier with Supply-Voltage-Scaling Volume Control and Series-Connected DSM," *ISSCC*, pp. 486-488, Feb. 2022. https://doi.org/10.1109/ISSCC42614.2022.9731791

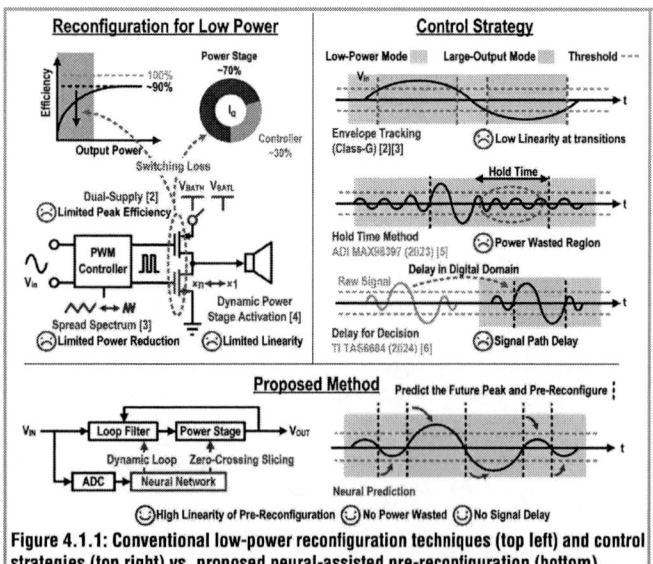

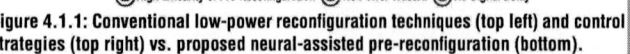

Figure 4.1.1: Conventional low-power reconfiguration techniques (top left) and control strategies (top right) vs. proposed neural-assisted pre-reconfiguration (bottom).

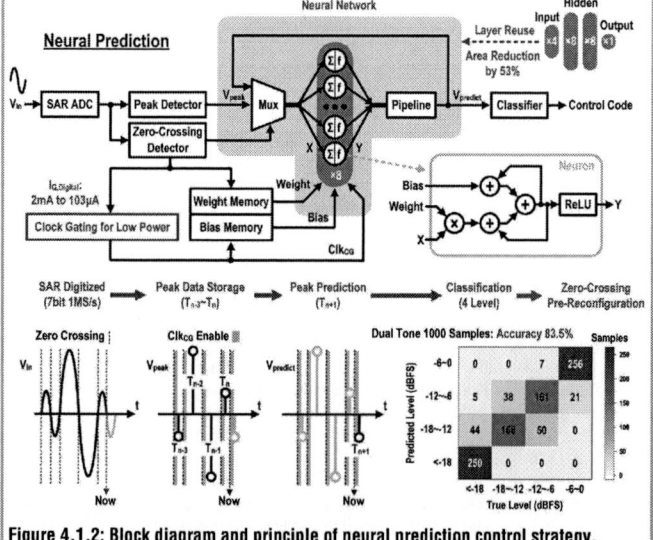

Figure 4.1.2: Block diagram and principle of neural prediction control strategy.

DIGEST OF TECHNICAL PAPERS • 77

ISSCC 2026 / SESSION 4 / ANALOG TECHNIQUES & AMPLIFIERS / 4.1

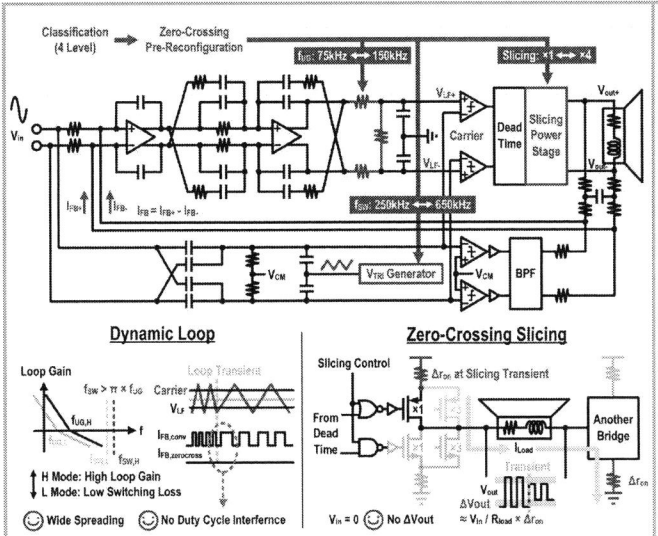

Figure 4.1.3: Schematic and principle of dynamic loop and zero-crossing slicing.

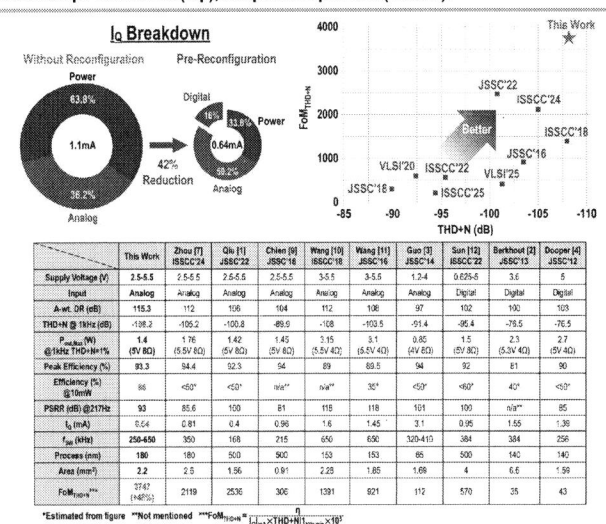

Figure 4.1.4: Dual-tone measurement of envelope tracking and neural prediction: Oscilloscope waveforms (top); Respective spectrum (bottom).

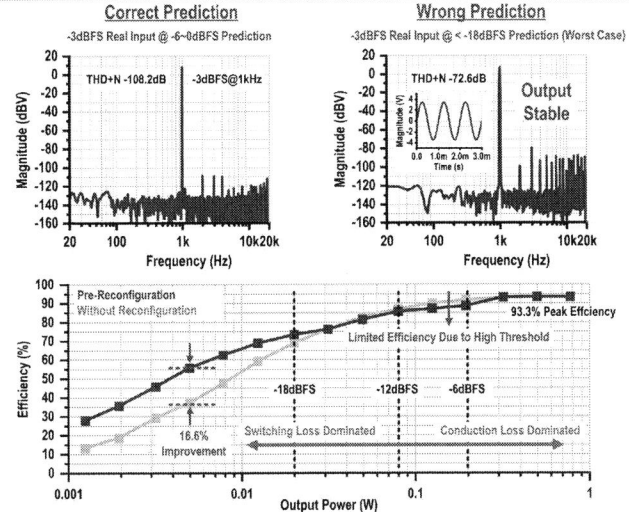

Figure 4.1.5: Measured spectrum under correct and worst-case prediction scenarios (top). Efficiency with and without pre-reconfiguration (bottom).

Figure 4.1.6: I_Q breakdown with and without pre-reconfiguration (top left). Scatter plot of state of the art (top right). Measured performance summary (bottom).

	This Work ISSCC'24	Zhou [7] JSSC'22	Qiu [1] JSSC'22	Chien [9] JSSC'16	Wang [10] ISSCC'18	Wang [11] JSSC'16	Guo [3] JSSC'14	Sun [12] ISSCC'22	Berkhout [2] JSSC'13	Dooper [4] JSSC'12
Supply Voltage (V)	2.5-5.5	2.5-5.5	2.5-5.5	2.5-5.5	3.5-5	3-5.5	1.2-4	0.625-5	3.6	5
Input	Analog	Analog	Analog	Analog	Analog	Analog	Analog	Digital	Digital	Digital
A-wt. DR (dB)	115.3	112	106	104	112	108	102	102	100	103
THD+N @ 1kHz (dB)	-108.2	-105.2	-100.8	-89.9	-108	-103.5	-91.4	-95.4	-76.5	-76.5
$P_{out,Max}$ (W) @1kHz THD+N<1%	1.4 (5V 8Ω)	1.76 (5.5V 8Ω)	1.42 (5V 8Ω)	1.46 (5V 8Ω)	3.15 (3.5V 4Ω)	3.1 (5.5V 4Ω)	0.85 (4V 8Ω)	1.5 (5V 8Ω)	2.3 (3.3V 4Ω)	2.7 (5V 4Ω)
Peak Efficiency (%)	93.3	94.4	92.3	94	89	89.5	94	92	81	90
Efficiency (%) @10mW	86	<50*	<50*	n/a**	n/a**	35*	<50*	<50*	40*	<50*
PSRR (dB) @217Hz	93	85.6	100	81	118	118	101	100	n/a**	85
I_Q (mA)	0.64	0.81	0.4	0.96	1.6	1.45	3.1	0.95	1.55	1.38
f_{sw} (kHz)	250-650	350	168	215	650	650	320-410	384	384	256
Process (nm)	180	180	500	500	153	153	65	500	140	140
Area (mm²)	2.2	2.5	1.56	0.91	2.26	1.85	1.69	4	6.5	1.59
FoM_{THD+N}***	3747 (+48%)	2119	2536	306	1391	921	112	570	35	43

*Estimated from figure **Not mentioned ***$FoM_{THD+N} = \frac{P_{out,Max} \times THD + N \vert_{1\%,max}}{I_Q} \times 10^3$

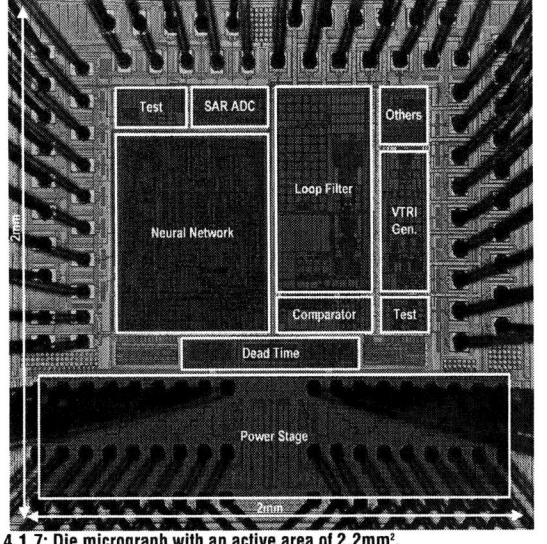

Figure 4.1.7: Die micrograph with an active area of 2.2mm².

ISSCC 2026 / SESSION 4 / ANALOG TECHNIQUES & AMPLIFIERS / 4.2

4.2 A -102.2dB THD+N, 92% Efficiency, 1.08mW Idle Power Digital-Input Class-D Amplifier with Power-Adaptive Techniques and Dual-Edge Pulse-Width Adjustment Modulator

Miao Zhang, Huajun Zhang, Domenico Maria Lombardo, Qinwen Fan

TU Delft, Delft, The Netherlands

Abstract

This paper presents a digital-input Class-D headphone amplifier that achieves low noise and low idle power simultaneously. Three key techniques enable this: a power-adaptive 3-level IDAC that scales power with signal level, an assistant IDAC that relaxes the first integrator's power consumption, and a dual-edge PWA modulator that minimizes loop filter swing. Fabricated in 0.18μm CMOS, it achieves 117.8dB DR, –102.2dB THD+N, 0.6mA idle current, and 92% peak efficiency when driving a 16Ω load.

Portable audio devices, such as true wireless stereo (TWS) headphones and hearing aids, increasingly rely on digital-input amplifiers to deliver high-quality sound while consuming minimal power to extend battery life. Class-D amplifiers (CDAs) are particularly advantageous thanks to their inherent efficiency compared to their Class AB counterparts [1,2]. Moreover, it saves system complexity and cost compared to Class-G/H topologies, which require either multiple supply rails or envelope tracking [3,4]. However, a prior Class-D headphone amplifier [5], which achieves high peak efficiency (93%) and low idle power (1.23mW), suffers from limited linearity (-93dB). To achieve high linearity, low noise, and low idle power simultaneously, [6] and [7] introduce multiple operating modes, which could potentially introduce audible artifacts. Additionally, since audio signals typically have a relatively high crest factor (10dB to 20dB) [8], the power consumption of the amplifier at small signal levels significantly impacts battery lifetime. This paper presents a digital-input, filter-less CDA that achieves low noise, low THD+N, and low idle power simultaneously, as well as optimized small-signal power efficiency with three key techniques: 1) A power-adaptive 3-level current-steering DAC (IDAC) that dynamically scales the power consumption according to the signal input. 2) An assistant IDAC (A-IDAC) that relaxes the loop filter's power consumption while ensuring good linearity. 3) A power-efficient dual-edge pulse width adjustment (DE-PWA) modulator that eliminates the need for power-consuming triangle wave generators and comparators and minimizes signal swing for high linearity. Fabricated in a standard 0.18μm CMOS process, the CDA achieves a 117.8dB dynamic range (DR) and a peak THD+N of –102.2dB at 1kHz signal. It consumes 0.6mA from a 1.8V supply in the idle state, while achieving a peak efficiency of 92%, and up to 20% small-signal efficiency improvement when driving a 16Ω load.

The system block diagram is shown in Fig. 4.2.1, which essentially adopts a digital-input direct PWM architecture similar to [6]. The 32-bit up-sampled digital audio input is truncated to 6 bits by a 6th-order digital ΔΣ modulator, which is then converted into a 1.5-bit digital PWM signal (D_{IN}) with a carrier frequency of 960kHz and a 30.72MHz master clock (T_{MCLK}=32.6ns) defining its timing resolution. While D_{IN} directly drives the 3-level main IDAC (M-IDAC) after a digital delay block (DDB), it is also decoded into INP/N in a BD modulation fashion [9] to drive the DE-PWA modulator, forming a feedforward path such that the loop filter only needs to process the residual error signals. Both operations will be explained in detail later. The rest of the CDA analog loop consists of a 3rd-order loop filter, a common-mode (CM) stable 3-level output stage [5], and feedback resistors (R_{FB}). A pre-distortion module [9] is employed before the ΔΣ modulator to compensate for the 3rd-order harmonic distortion inherent to the digital PWM modulation. The A-IDAC, controlled by D_{IN} and the output stage driving signals, supports the first loop filter OP1 when high output current is required. Thanks to the 3-level M-IDAC, the noise of the CDA in idle state is mostly determined by the noise from R_{FB} and OP1, where R_{FB} is chosen to be ~2.4kΩ as a compromise between noise and power. Compared to a multi-bit digital-input CDA [5], direct PWM architecture offers simplicity but is sensitive to clock jitter [6]. Fortunately, in the targeted applications, sub-ps clocks [10] are available, which allows a >116dB SNDR.

In idle or small-signal conditions, D'_{IN} stays mostly at level 0 with ±1 as occasional narrow pulses. Therefore, the M-IDAC does not need to output current to the CDA most of the time, but still consumes power continuously. To mitigate this, the degeneration resistor R_D, employed to suppress the M-IDAC's flicker noise, is increased by 64× when D'_{IN}=0 and restored when D'_{IN}=±1, significantly lowering M-IDAC's power consumption during idle and small-signal operations as shown in Fig. 4.2.2 (left). However, the biasing of current sources (V_{G1} and V_{G4}), filtered by on-chip RC networks to reduce bias noise, is inevitably affected by this switching operation through parasitic capacitances (e.g. $C_{GS1,4}$), disturbing the output current. To ensure sufficient settling (5τ over PVT), the transition of R_D back to its nominal value is advanced by T_{ADV}=T_{MCLK} relative to the transition of the M-IDAC controlled by switches (S_P, S_N, S_C). R_D switches only when D'_{IN}=0 lasts at least one T_{ADV}, it remains unchanged under clipping conditions. By employing this power-adaptive scheme, the M-IDAC power consumption becomes automatically adjustable with the input signal, ranging from 69μW to 440μW from idle to -1dBFS. Idle switching activities due to digital DSM quantization noise cause a non-zero M-IDAC idle power. To minimize glitches and ISI, S_P, S_N, S_C are implemented by matched NMOS transistors and are driven by overlapping gate signals.

The A-IDAC, together with the M-IDAC output currents (I_{INP}, I_{INN}) and feedback currents (I_{FBP}, I_{FBN}), are illustrated in Fig. 4.2.2. Since both the M-IDAC and output stage operate in a 3-level manner, $I_{INP/N}$ can largely cancel $I_{FBP/N}$. However, $I_{FBP/N}$ would lag $I_{INP/N}$ due to the DE-PWA delay (~100 ns in this design), forcing a large net current into the first integrator around OP1 and causing degraded linearity due to large signal swing. To mitigate this, a T_{DDB}=3T_{MCLK} has been added to the M-IDAC path (Fig. 4.2.1) to roughly compensate for the DE-PWA delay such that $I_{FBP/N}$ and $I_{INP/N}$ are center-aligned, reducing the integrator swing by ~8×, and thus significantly improving linearity with the same C_1 (2×32pF). Still, IR drop across the power stage transistors reduces the amplitude of the CDA output PWM, and consequently the amplitude of $I_{FBP/N}$. The closed-loop CDA compensates by widening the feedback pulses, which introduces a short timing misalignment with $I_{INP/N}$, as shown in Fig. 4.2.2 (top-right). This short residual misalignment forces OP1's output stage to source or sink large instantaneous currents, demanding high bias. To alleviate this, the A-IDAC [11] is activated during these intervals to supply currents (I_{ADP}, I_{ADN}), thereby reducing the burden on OP1. Since the A-IDAC is only activated around the PWM edges, it adopts the same power-adaptive scheme as the M-IDAC to minimize its own power consumption. With its assistance, the output stage bias current of OP1 is reduced to 128μA, which would otherwise need to be 1.5mA to ensure the same linearity. Meanwhile, the A-IDAC consumes only 29μA during the idle state.

The DE-PWA is inspired by a single-edge PWA (SE-PWA) in [6], which employs two voltage-controlled delay cells (VCDLs) to adjust the falling edges of BD-modulated PWM signals that drive an output stage. As a result, $I_{FBP/N}$ are naturally aligned with $I_{INP/N}$ at the rising edge but are misaligned at the falling edge (Fig. 4.2.3 top-right), leading to large swing at OP1 output, degrading linearity. To mitigate this issue, a DE-PWA consisting of four VCDLs is designed, which modulates both the rising and falling edges to minimize signal swing. Figure 4.2.3 (top-right) showcases the processing of INN: its rising and falling edges trigger VCDL4 and VCDL2, generating pulses R_N and F_N whose widths are controlled by loop filter outputs V_{INT3P} and V_{INT3N}, respectively. Digital logic then reconstructs INN_D, with rising and falling edges defined by the falling edges of R_N and F_N. The same applies to the INP path with swapped control signals. Finally, INN_D and INP_D drive the output stage to generate the complementary 3-level PWM signals SWP/N. With no differential signal, each VCDL would introduce a CM delay T_{DCM} (Fig. 4.2.3 bottom-right), which is compensated by T_{DDB} as explained earlier. To process the same amount of baseband differential signal as SE-PWA, where one VCDL adds a delay (T_d) and the other subtracts T_d both at the falling edge, each VCDL in a DE-PWA only adds or subtracts $T_d/2$ at both the rising and falling edge (Fig. 4.2.3 top-right). This provides a further 2× reduction in OP1 swing thanks to 0.5× misaligned pulse width. To accommodate the CDA analog loop errors over PVT, the VCDL is designed to handle a maximum delay range of ~120ns, and an ~1.05V CM voltage V_{CM3} is chosen at the loop filter output. To further reduce power, the VCDLs adopt a dynamic scheme, as shown in Fig. 4.2.3 (bottom-left): it is activated by an enable signal en_i just before the rising edges of INN, similar to the degeneration control in the IDACs, and enters a low-power mode immediately after the inverter I_0 is triggered. With this dynamic VCDL, the entire DE-PWA modulator consumes only 45μW in the idle state, compared to 142μW without the power-adaptive technique. To mitigate process variation, the loop filter capacitors and the DE-PWA current source are 2-bit trimmable.

The prototype CDA was fabricated in a standard 0.18μm CMOS process and occupies an active area of 0.72mm². Both the 24-bit digital input and output measurements are handled by an APx555 audio analyzer, with an FPGA implementing the interpolation filter, pre-distortion, DSM, and digital PWM. The CDA drives a 16Ω + 33μH load emulating a speaker. Figure 4.2.4 (right) shows the measured output FFT spectra when delivering a 1kHz sinewave. A THD+N of –102.2dB is achieved at –12dBFS. At –80dBFS input, an A-weighted SNR of 37.8dB is measured, indicating a dynamic range of 117.8dB. Figure 4.2.4 (left) shows the measured THD+N versus output power, with peak THD+N of –102.2dB and –110.0dB for 1kHz and 6kHz inputs, respectively. Disabling the A-IDAC results in a significant degradation of linearity, as expected. The measured maximum output power (P_{MAX}) with 1% THD is 71mW. Figure 4.2.5 (top) shows a peak efficiency of 92%. For typical music playback, where average output power is 0.01 to 0.1P_{MAX} [8], the power-adaptive technique improves small-signal efficiency by 10 to 20%. During idle, the CDA's power consumption is 1.08mW with the power-adaptive technique enabled, which would otherwise

78 • 2026 IEEE International Solid-State Circuits Conference

979-8-3315-8937-0/26 $31.00 © 2026 IEEE

increase to 2.27mW. Figure 4.2.5 (bottom-left) plots total analog power with and without the power adaptive techniques used in M-IDAC, A-IDAC, and DE-PWA modulators. The idle power breakdown in Fig. 4.2.5 (bottom-right) shows the output stage as the largest contributor, followed by OP1.

Figure 4.2.6 compares this work with state-of-the-art digital-input headphone amplifiers under the 1.8V supply. It has achieved the lowest power consumption during idle as well as at 0.1mW output power. Compared to its Class-D counterpart [5], it improves THD+N by 9dB and DR by 4.8dB. Lastly, it achieves the highest FoM_{THD+N} and FoM_{DR} in the comparison chart.

References:
[1] S. -H. Wen et al., "A -105dBc THD+N (-114dBc HD2) at 2.8VPP Swing and 120dB DR Audio Decoder with Sample-and-Hold Noise Filtering and Poly Resistor Linearization Schemes," *ISSCC*, pp. 294-295, 2019. https://doi.org/10.1109/ISSCC.2019.8662456
[2] S. -H. Wen et al., "A −117dBc THD (−132dBc HD3) and 126dB DR Audio Decoder with Code-Change-Insensitive RT-DEM Algorithm and Circuit Technique for Relaxing Velocity Saturation Effect of Poly Resistors," *ISSCC*, pp. 482-484, 2022. https://doi.org/10.1109/ISSCC42614.2022.9731634
[3] A. Lollio et al., "Class-G Headphone Driver In 65nm CMOS Technology," *ISSCC*, pp. 84-85, 2010. https://doi.org/10.1109/JSSC.2010.2076450
[4] S. -H. Wen et al., "A -108dBc THD+N, 2.3mW Class-H Headphone Amplifier with Power-Aware SIMO Supply Modulator," *ISSCC*, pp. 384-386, 2024. https://doi.org/10.1109/ISSCC49657.2024.10454450
[5] A. Matamura et al., "An 82mW ΔΣ-Based Filter-Less Class-D Headphone Amplifier with -93dB THD+N, 113dB SNR and 93% Efficiency," *ISSCC*, pp. 432-434, 2021. https://doi.org/10.1109/ISSCC42613.2021.9365773
[6] W. -H. Sun et al., "A 121dB DR, 0.0017% THD+N, 8× Jitter-Effect Reduction Digital-Input Class-D Audio Amplifier with Supply-Voltage-Scaling Volume Control and Series-Connected DSM," *ISSCC*, pp. 486-488, 2022. https://doi.org/10.1109/ISSCC42614.2022.9731791
[7] J. -H. Cho et al., "A Boosted 3.5W, -81.6dB THD+N, 92.6% Total Efficiency, Battery-Powered Class-D Audio Amplifier with True-Zero-Switching Achieving Quiescent Current of 0.9mA," *IEEE Symp. VLSI Circuits*, pp. 1-3, 2025. https://doi.org/10.23919/VLSITechnologyandCir65189.2025.11074806
[8] S. -H. Chien et al., "A 0.41mA Quiescent Current, 0.00091% THD+N Class-D Audio Amplifier with Frequency Equalization for PWM-Residual-Aliasing Reduction," *ISSCC*, pp. 352-354, 2020. https://doi.org/10.1109/ISSCC19947.2020.9062932
[9] M. Berkhout et al., "Audio at Low and High Power," *ESSCIRC*, pp. 40-49, 2008. https://doi.org/10.1109/ESSCIRC.2008.4681788
[10] SiTime, SiT1605 Datasheet, Accessed: Sep. 2025. https://www.sitime.com/support/resource-library/datasheets/sit1605-datasheet
[11] S. Pavan et al., "Power Reduction in Continuous-Time Delta-Sigma Modulators Using the Assisted Opamp Technique," *JSSC*, vol. 45, no. 7, pp. 1365-1379, July 2010. https://doi.org/10.1109/JSSC.2010.2048082
[12] Texas Instruments, TAC5242 Datasheet, Accessed: Sep. 2025. https://www.ti.com/document-viewer/tac5242/datasheet

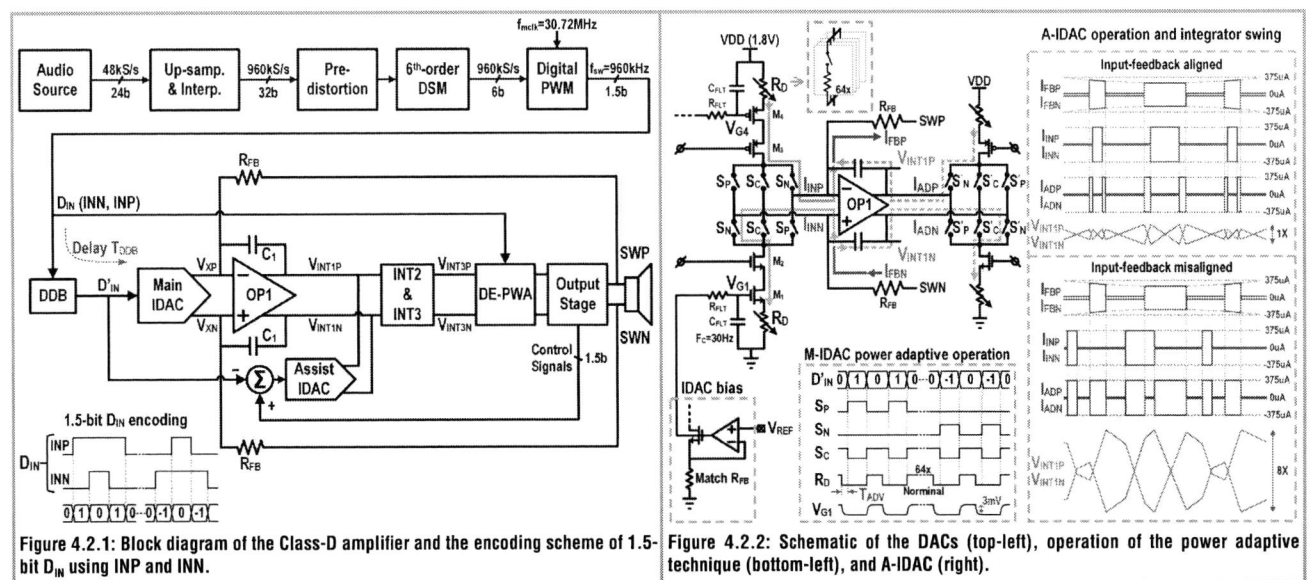

Figure 4.2.1: Block diagram of the Class-D amplifier and the encoding scheme of 1.5-bit D_{IN} using INP and INN.

Figure 4.2.2: Schematic of the DACs (top-left), operation of the power adaptive technique (bottom-left), and A-IDAC (right).

ISSCC 2026 / SESSION 4 / ANALOG TECHNIQUES & AMPLIFIERS / 4.2

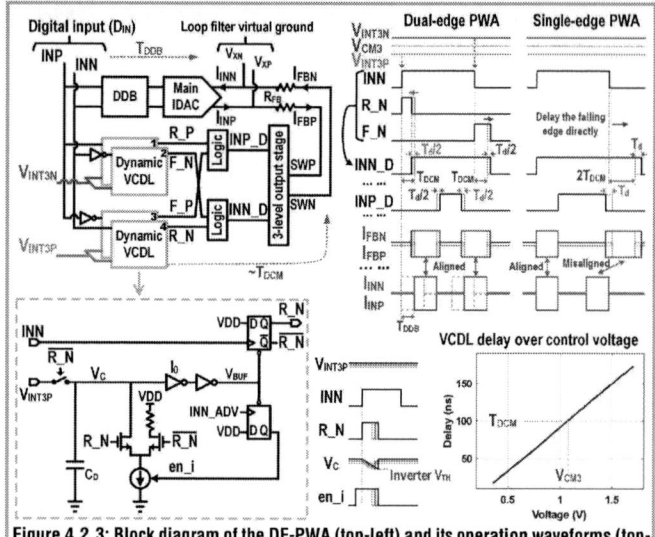

Figure 4.2.3: Block diagram of the DE-PWA (top-left) and its operation waveforms (top-right). The dynamic VCDL (bottom-left) and its transfer curve (bottom-right).

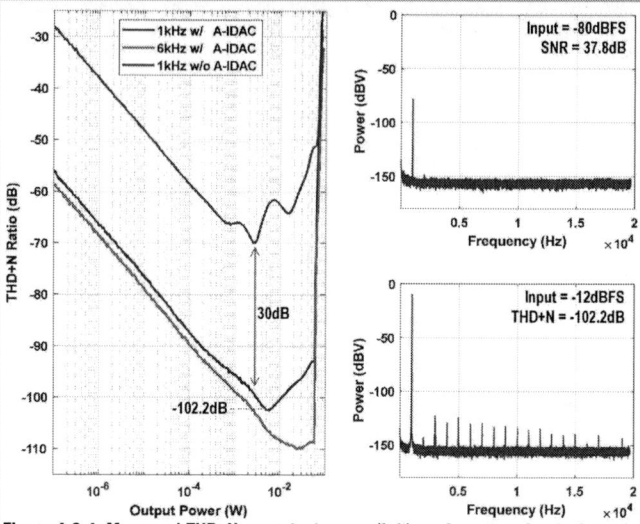

Figure 4.2.4: Measured THD+N vs. output power (left), and measured output spectra (right) (128 points, 8× averaged).

Figure 4.2.5: Measured efficiency (top), analog power consumption (bottom-left), and idle power breakdown (bottom-right).

	This work	Matamura ISSCC'21 [5]	Wen ISSCC'24 [4]	TI TAC5242 [12]	Wen ISSCC'22 [2]	Wen ISSCC'19 [1]
Process (nm)	180	40	28	—	28	153
Topology	IDAC + class-D	IDAC + class-D	IDAC + class-H	—	IDAC + class-AB	IDAC + class-AB
Supply (V)	1.8	1.8	1.8	1.8	1.8	1.8
Load (Ω)	16	16	16	16	16	16
P_{IDLE} (mW) [1]	1.08	1.23	6.3	—	8.28	5.58
$P_{0.1mW}$ (mW) [2]	1.2	1.41	7.7	—	—	—
P_{MAX} (mW) [3]	71	86	62	62.5	62	62
Efficiency (%)	92	93	—	—	—	—
PSR (dB) [4]	93.1	94	—	118 [5]	—	—
THD+N (dB) [6]	-102.2	-93.2	-108	—	-117	-105
A-wt. DR (dB)	117.8	113	126	115	126	120
FoM_{THD+N} [7]	1975	625	—	—	—	—
FoM_{DR} [8]	1190	611	—	—	—	—

[1] Including DACs and drivers
[2] Measured with 0.1mW output power
[3] Maximum output power with 1% THD+N
[4] Measured with 217Hz supply noise
[5] Lowest THD+N measured at 1kHz
[6] Extracted from plot
[7] FoM_{THD+N} = Efficiency/($I_{Q,mA}$×THD+N×100)
[8] FoM_{DR} = Efficiency×DR/($I_{Q,mA}$×1000)

Figure 4.2.6: Comparison with state-of-the-art digital-input headphone amplifiers.

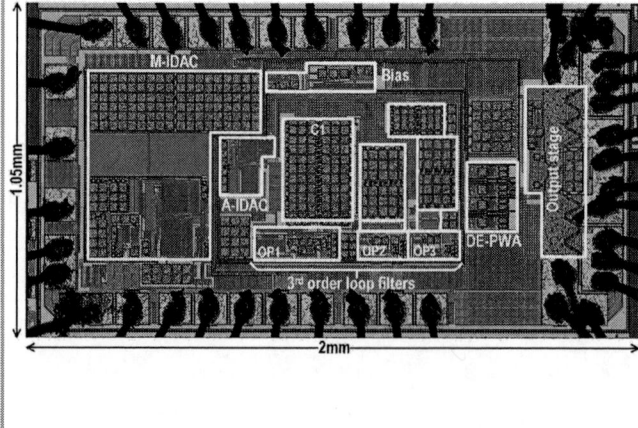

Figure 4.2.7: Die micrograph.

• 2026 IEEE International Solid-State Circuits Conference

ISSCC 2026 / SESSION 4 / ANALOG TECHNIQUES & AMPLIFIERS / 4.3

4.3 A 0.6V 9.4µW 1,892µm² Current-Pulse-Injection Crystal Oscillator Featuring Capacitively Biased Amplifier with 242.2dBc/Hz PN FoM @1kHz Offset

Haihua Li[1,2], Dan Shi[1], Qi Zhou[1], Rui Martins[1], Pui-In Mak[1], Ka-Meng Lei[1]

[1]University of Macau, Macau, China, [2]UM Hetao IC Research Institute, Shenzhen, China

Abstract

We report a current-pulse injection-based Crystal Oscillator (XO). A capacitively biased amplifier, powered by the stacked capacitor power supply (SCPS), delivers a short boosted current pulse into the crystal to limit the amplifier's shoot-through current while upholding the XO's phase noise. Fabricated in 65nm CMOS, the XO core occupies 1,892µm². The 16MHz XO consumes 9.4µW at 0.6V V_{DD}, associated with a phase noise of –137.9dBc/Hz @1kHz offset, corresponding to a superior FoM of 242.2dBc/Hz.

The explosive growth of Internet-of-Things (IoT) applications in the artificial intelligence era requires resource-constrained IoT nodes to handle surging data volumes for edge computing. Low-power crystal oscillators (XOs) serve as a critical enabler by delivering stable clock signals that support reliable, real-time data communication and processing at the edge [1]. The mainstream architecture for realizing MHz-range XOs is the Pierce configuration [2-4], which involves a transconductance amplifier (G_M) to compensate for the energy loss within the crystal (Fig. 4.3.1, top-left). While it offers superior frequency stability (Δf) and compactness, the G_M suffers from low energy efficiency due to its Class-A/AB operation and shoot-through current in the deep-submicron process. A duty-cycling technique to turn on/off the G_M periodically could reduce the average power consumption, at the expense of jeopardizing the phase noise (PN) (–120dBc/Hz @1kHz offset) [5]. An amplifier with a stacking technique can reuse the current and boost the effective g_m to improve the PN. However, it penalizes the oscillation swing (<0.4V_{pp}) and requires a high V_{DD} (3.3V), associated with a large area consumption (0.09mm²) for the AC-coupling capacitors [6]. Recently, a 96MHz XO co-integrated with a PLL using a self-aligned pulse injection (PI) driver has been proposed [7]. Yet, the XO alone consumes 2.4mW, primarily attributable to the undesired dynamic power dissipation on the load capacitor (C_L) during the injection, and relies on the VCO to generate the requisite injection pulses.

To this end, this work proposes a current-pulse injection XO based on a capacitively biased amplifier (CBA) with capacitor-boosting technique. The stack capacitor injects the boosted current pulse into the crystal to maintain the oscillation. As the total injection energy is pre-defined by the capacitor and only half of the cycle is enabled, this design shrinks the amplifier's shoot-through current and improves the energy efficiency by 2.9× compared with the traditional Class-AB Pierce XO. Fabricated in 65nm CMOS, the XO core occupies 1,892µm² (5,470µm² with C_L). With a 16MHz crystal, it attains a power consumption of 9.4µW at 0.6V V_{DD}, while being capable of delivering a 0.38V_{pp} oscillation swing and a PN of –137.9dBc/Hz (at 1kHz offset). The Δf over the ranges of voltage (0.5 to 0.7V) and temperature (–20 to 100°C) variations are ±1.8 and ±10ppm, respectively (See Figure 4.3.5). We also obtain consistent performances from the 12MHz crystal, showing the versatility of the proposed XO.

The proposed XO includes an inverter as an amplifier to deliver the current into the crystal, a stacked-capacitor power supply (SCPS) to boost the voltage swing and power the amplifier, and a buffer to generate the switching signals for the SCPS (Fig. 4.3.2). The XO operates in two modes: constant mode ($MD=V_{DD}$) with a fixed V_{DD} for XO startup, analogous to the traditional XO, and capacitively biased (CB) mode ($MD=GND$) with the XO powered by the SCPS. We can divide the operation in CB mode into 2 phases in one cycle. In the charging phase (Φ_1), where V_{XO-} decreases from the maximum towards the minimum, $C_{S1/2}$ in the SCPS stores the charges for the boosting operation in the next phase, and the crystal is disconnected from the G_M. In the delivery phase (Φ_2), where V_{XO-} increases from the minimum towards the maximum, the G_M is connected to the crystal, and SCPS switches the configuration of $C_{S1/2}$ to power the G_M. During this phase, a drain current (I_D) delivers the required energy to the crystal to compensate for its resistive loss. A bypass capacitor (C_F=60fF) is implemented on the supply of the G_M to limit the transient current and protect the circuit.

The central idea of this design is to utilize a CBA to deliver current into the crystal to sustain its oscillation while upholding the energy efficiency [8,9]. At the onset of enabling the XO, MD is enabled and the supply of the G_M is connected to V_{DD} to initiate the oscillation. During this mode, a passive resistor R_F (20kΩ) serves as a feedback element to define the operating point. After the crystal garners sufficient energy to clock the buffer, MD turns off, switching the supply of the G_M to the SCPS. With the CBA architecture, the passive R_F with reasonable sizing would introduce a current path between V_{XO+} and V_{XO-}, degrading the resultant crystal's motional current (I_M). Hence, to suppress this current while maintaining a path for self-biasing in the CB mode, we replace the feedback element with a pseudo-resistor (formed by $M_{P3/4}$) with an equivalent resistance of 3.1MΩ.

The SCPS delivers the requisite current pulse to the G_M and sustains the oscillation while reducing the unnecessary power consumption. During Φ_1, both $C_{S1/2}$ are charged to pose a voltage difference of V_{DD}. Note that the amplifier's supply is disconnected from the capacitor

in this phase. When the minimum of V_{XO-} arrives, the control signals invert (i.e., Φ_2 enables). Correspondingly, the top plate of C_{S1} is connected to the bottom plate of C_{S2}. Due to the charge conservation, it instantaneously elevates the voltage on C_{S2}'s top plate to ~2×V_{DD}, boosting the available current delivered into the G_M. At the same time, the C_{S2}'s top plate is connected to the G_M. Hence, it can deliver a boosted transient current pulse into the G_M to provide energy to the crystal.

Inspired by the PI technique, this architecture shortens the energy injection period into a brief pulse (<500ps) and delivers it into the crystal. Compared with prior voltage PI works with wider pulse width (~3ns), this brief pulse induces less influence on the phase distortion on the V_{XO} waveform, thereby preserving the PN. In addition, the M_{N1}, driven by V_{XO-}, provides a down-to-GND current path and replaces the pull-down driver in the prior PI XO, leading to a less abrupt phase shift on the V_{XO-} and upholding the PN. In the steady state, Φ_2 is enabled for half of the cycle to provide a current path (I_D) for the charge from the SCPS to flow into the G_M. At the onset of Φ_2, V_{GS} of M_{P1} ($V_{XO-} – V_{DD}$) is at its minimum, ensuring a maximum I_D to pass through the crystal rapidly (Fig. 4.3.3 top-left). As the switching on the SCPS introduces non-linear behaviour on G_M, we can derive the G_M's effective transconductance based on averaging its transient $g_m(t)=[\partial I_D(t)/\partial t]/[\partial V_{GS}(t)/\partial t]$ over the entire oscillation period ($g_{m,avg}$) [9], as shown in the right of Fig. 4.3.3. The derived $g_{m,avg}$ is 1.9mS, 10× larger than that of the traditional Pierce XO (195µS) with a 4× smaller current budget (3.4µA). The interpretation of the current flow is depicted in the bottom-left of Fig. 4.3.3.

Subsequently, as V_{XO-} rises, M_{P1} and M_{N1} enter the sub-threshold region. However, as most of the charges from the SCPS have been delivered during the peak of I_D, the shoot-through current through M_{P1} and M_{N1} is substantially reduced, a hallmark feature of the CBA architecture. Eventually, when V_{XO-} reaches the maximum, the SCPS enters the charging phase (Φ_1), whereas the G_M is disconnected from the SCPS and the crystal, resulting in V_G (amplifier's input) being maintained at its maximum. Compared with the constant mode, the CBA could reduce the shoot-through current as there is a fixed amount of charge available from the SCPS, which decays after the injection pulse of I_D. As such, the peak current through M_{N1} is decreased by 4.2×, ensuring the energy transmission efficiency from the SCPS to the crystal (Fig. 4.3.3, Right).

During the operation, an AC-coupled buffer converts the sinusoidal V_{XO-} into square waves ($\Phi_{1/2}$) to control the switches in the SCPS. According to the impulse-sensitivity-function theory, the injection during the low-ISF regions with a narrow pulse width will introduce minimum PN to the system, requiring a delicate design on the buffer to control the timing of $\Phi_{1/2}$ [7,10]. Hence, as shown in Fig. 4.3.2, the buffer stage includes a self-bias AC-coupled stacked inverter in the forefront to convert the sinusoidal input to a square wave with a low power consumption (5.2µW). An RC delay of 1/4 cycle provides the necessary phase shift to delay the square wave and enable injection in the low-ISF regions, preserving the PN in the CB mode. The forefront inverter has a gain of >16dB and a 3dB bandwidth of >27MHz across –20 to 100°C, with a simulated power consumption of 3.2µW (in nominal condition). The simulated XO power consumption is 9.1µW in the CB mode, whereas the power consumption of the G_M (P_{GM}) is 1.8µW (Fig. 4.3.3, right) to maintain an I_M of 138µA. In contrast, the total power consumption in the constant mode is 14.6µW, whereas P_{GM} is 8.5µW to deliver an I_M of 175µA; the P_{GM}-to-I_M^2 conversion efficiency in CB Mode improves by 2.9× compared to the Pierce architecture.

We fabricated the proposed CB XO in the 65nm CMOS process. The total area, including the on-chip C_L of 4pF, is 5,470µm², where the XO core only occupies 1,892µm² (Fig. 4.3.7); this area efficiency is attributed to the current-pulse injection scheme that eliminates the need for large-size transistors and bypass capacitors to suppress the noise, unlike previous implementations. We tested the XO with a standard commercial 16MHz crystal (Abracon ABM8W), with a nominal V_{DD} of 0.6V. MD is first enabled for 3ms to operate the XO in the constant mode and generate the oscillation signal for the buffer to detect the input clock (Fig. 4.3.4, Left). After this, MD is disabled and the XO transits to CB mode. The average power consumption from 5 samples is 9.4µW (Fig. 4.3.5, top). Between –20 and 100°C, the XOs exhibit a Δf of ±10ppm, associated with a maximum power consumption of 12.8µW (@100°C), showing the excellent frequency stability of the proposed design despite having

80 • 2026 IEEE International Solid-State Circuits Conference

979-8-3315-8937-0/26 $31.00 © 2026 IEEE

a low power consumption and a reduced V_{DD} (Fig. 4.3.5, bottom). The Δf under a V_{DD} variation between 0.5 and 0.7V is 3.9ppm, whereas the power varies between 5.5 and 17.3μW. Benefitting from the stack-capacitor powering scheme, the XO can achieve a steady-state swing of 0.3 to 0.42V_{PP} across V_{DD} variation, with a nominal value of 0.38V_{PP} (Fig. 4.3.4, Right). Consequently, the XO achieves a superior PN of –137.9dBc/Hz @1kHz offset, corresponding to an FoM of 242.2dBc/Hz (Fig. 4.3.6, Top). We also tested the XO with the 12MHz quartz crystal, and the results are consistent with the 16MHz one (nominal power: 8.4μW, Δf across temperature: ±15ppm, PN: –138.3dBc/Hz @1kHz offset, FoM: 240.7dBc/Hz).

Figure 4.3.6 (bottom) benchmarks the CB XO against the state-of-the-art MHz-level XOs [11-14]. Leveraging the CBA with the SCPS, this XO achieves a power consumption of 9.4/8.1μW (16/12MHz), one of the lowest in the reported XO, and maintains a premier PN of –137.9/–138.3dBc/Hz at 1kHz offset, despite having a low V_{DD} of 0.6V. It can deliver a 0.38V_{PP} swing to drive the subsequent buffer/logic gate without penalizing their performance. In addition, the XO core only occupies 1,892μm² (excluding the C_L), highlighting its cost/area effectiveness for the IoT nodes for the edge computing scenario.

Acknowledgement:
The work is funded by The Macau Science and Technology Development Fund (004/2023/SKL, 0149/2022/A3, and 0005/2024/RIC), the University of Macau (MYRG-GRG2024-00125-IME and MYRG-GRG2025-00025-IME), and Hetao SZ-HK S&T Innovation Cooperation Zone Project (HTHZQSWS-KCCYB-2023030). Corresponding author: Ka-Meng Lei.

References:
[1] M. Alrowaily et al., "Secure Edge Computing in IoT Systems: Review and Case Studies," *IEEE/ACMSEC*, pp. 440-444, Oct. 2018.
https://doi.org/10.1109/SEC.2018.00060
[2] K. -M. Lei, et al., "A 0.4V 4.8μW 16MHz CMOS Crystal Oscillator Achieving 74-Fold Startup-Time Reduction Using Momentary Detuning," *IEEE ISCAS*, pp. 1-4, May. 2017.
https://doi.org/10.1109/ISCAS.2017.8051002
[3] H. Li et al., "A 12-/13.56-MHz Crystal Oscillator with Binary-Search-Assisted Two-Step Injection Achieving 5.0-nJ Startup Energy and 45.8-μs Startup Time," *IEEE JSSC*, vol. 59, no. 2, pp. 464-475, Feb. 2024. https://doi.org/10.1109/JSSC.2023.3300589
[4] W. Kruiskamp, "A Fully Differential 40 MHz Switched-Capacitor Crystal Oscillator with Fast Start-Up," *ESSCIRC*, pp. 397-400, Sep. 2022.
https://doi.org/10.1109/ESSCIRC55480.2022.9911475
[5] S. Iguchi et al., "93% Power Reduction by Automatic Self Power Gating (ASPG) and Multistage Inverter for Negative Resistance (MINR) in 0.7V, 9.2μW, 39MHz Crystal Oscillator," *VLSI*, pp. C142-C143, Jun. 2013.
https://ieeexplore.ieee.org/document/6578640
[6] S. Iguchi et al., "A 39.25MHz 278dB-FoM 19μW LDO-Free Stacked-Amplifier Crystal Oscillator (SAXO) Operating at I/O Voltage," *ISSCC*, pp. 100-101, Feb. 2016.
https://doi.org/10.1109/ISSCC.2016.7417926
[7] C. Livanelioglu et al., "A 4.6 GHz 63.3 fsrms PLL-XO Co-Design Using a Self-Aligned Pulse-Injection Driver Achieving– 255.2 dB FoMJ Including the XO Power and Noise," *ISSCC*, pp. 342-344. Feb. 2025. https://doi.org/10.1109/isscc49661.2025.10904568

[8] X. Tang et al., "A 13.5-ENOB, 107-μW Noise-Shaping SAR ADC with PVT-Robust Closed-Loop Dynamic Amplifier," *IEEE JSSC*, vol. 55, no. 12, pp. 3248-3259, Sep. 2020. https://doi.org/10.1109/JSSC.2020.3020194
[9] R. S. A. Kumar et al., "Analysis and Design of a Discrete-Time Delta-Sigma Modulator Using a Cascoded Floating-Inverter-Based Dynamic Amplifier," *IEEE JSSC*, vol. 57, no. 11, pp. 3384-3395, Nov. 2022. https://doi.org/10.1109/JSSC.2022.3171790
[10] A. Hajimiri et al., "A General Theory Of Phase Noise In Electrical Oscillators," *IEEE JSSC*, vol. 33, no. 2, pp. 179-194, Feb. 1998. https://doi.org/10.1109/4.658619
[11] K.-M. Lei et al., "A Regulation-Free Sub-0.5V 16/24MHz Crystal Oscillator for Energy-Harvesting BLE Radios with 14.2nj Startup Energy and 31.8pw Steady-State Power," *ISSCC*, pp. 52-54, Feb. 2018. https://doi.org/10.1109/ISSCC.2018.8310179
[12] Z. Xu et al., "Ultralow-Power Class-C Complementary Colpitts Crystal Oscillator," *IEEE SSC-L*, vol. 3, pp. 274-277, Aug. 2020.
https://doi.org/10.1109/LSSC.2020.3014048
[13] Z. Cai et al., "A 16MHz XO with 17.5μs Startup Time Under 10⁴ppm-ΔF Injection Using Automatic Phase-Error Correction Technique," *ISSCC*, pp. 66-67, Feb. 2023. https://doi.org/10.1109/ISSCC42615.2023.10067675
[14] R. Luo et al., "A 0.5-V 6.14- μW Trimming-Free Single-XO Dual-Output Frequency Reference with [5.1-nJ, 120-μs] XO Start-Up and [8.1-nJ, 200-μs] Successive-Approximation-Based RTC Calibration," *ISSCC*, pp. 58-60, Feb. 2024. https://doi.org/10.1109/ISSCC49657.2024.10454426

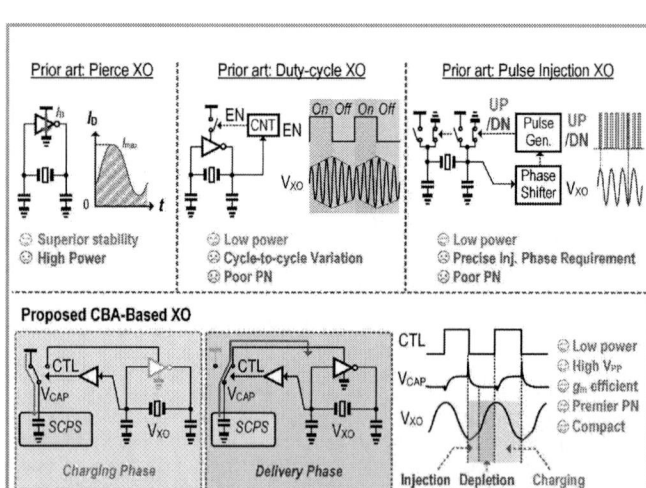

Figure 4.3.1: Comparison of the existing MHz-range XO architecture and the proposed XO based on a capacitively biased amplifier, featuring low power consumption, high energy-efficiency, and competitive phase-noise performance.

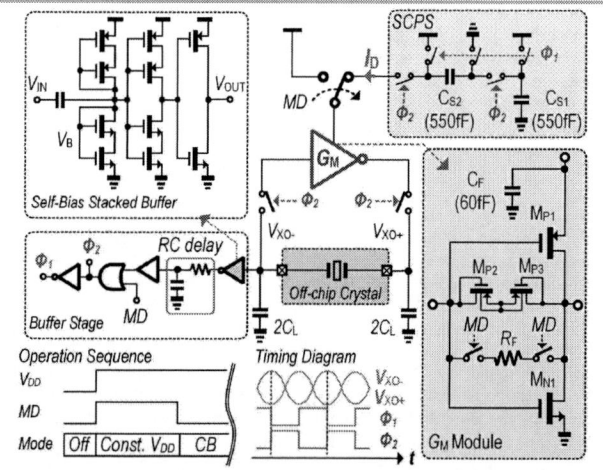

Figure 4.3.2: The schematic of the proposed XO with CBA. It primarily includes a G_M, SCPS, buffer, load capacitors, and off-chip crystal. The bottom of the figure illustrates the operation sequence during the startup and the timing diagram of critical waveforms during the operation.

ISSCC 2026 / SESSION 4 / ANALOG TECHNIQUES & AMPLIFIERS / 4.3

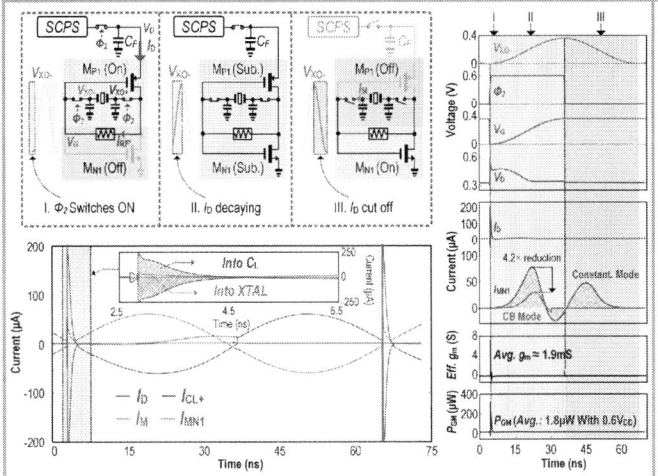

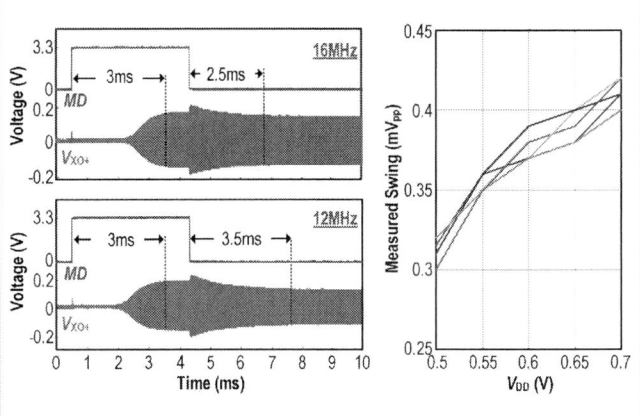

Figure 4.3.3: (Left) The detailed operation states according to V_{XO}. and the relevant current flows. (Right) Simulation result of the CB XO, showing the critical signals during the operation, as well as the effective transconductance of G_M with its transient P_{GM}. The CBA helps reduce the peak shoot-through current by 4.2×.

Figure 4.3.4: (Left) Measured time-domain waveforms of the 12 and 16MHz XO, illustrating the startup sequence and the steady-state swing (measured with an active probe). (Right) Measured oscillation swing versus the supply voltage from 5 samples.

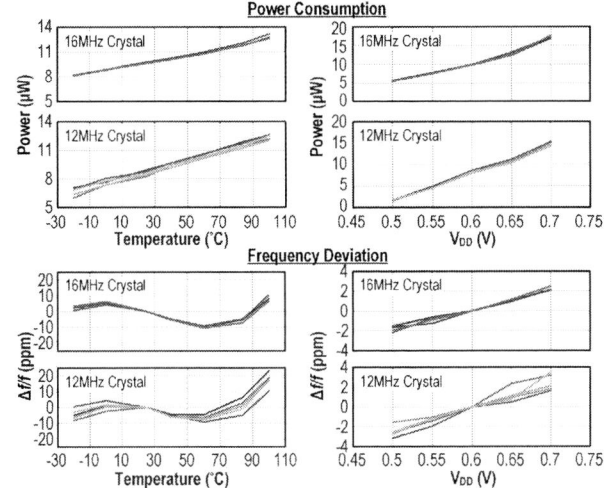

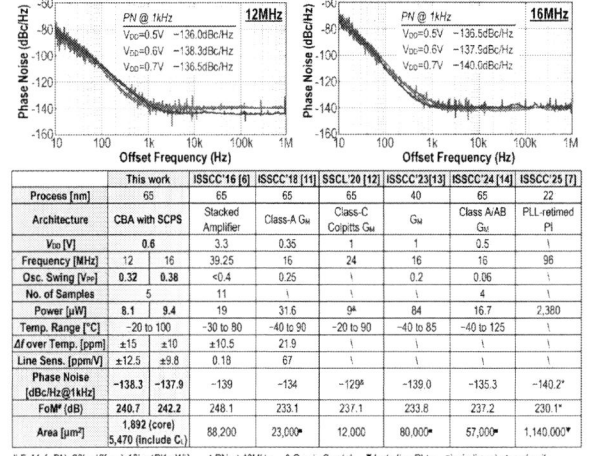

Figure 4.3.5: Measured power consumption (top) and frequency inaccuracy (bottom) of the XO with 12/16MHz crystals across temperature (–20 to 100°C) and supply voltage (0.5 to 0.7V) from 5 samples.

Figure 4.3.6: (Top) Measured PN of the 12/16MHz XO versus V_{DD} (0.5 to 0.7V) in CB mode. (Bottom) Performance summary and comparison with the state-of-the-art MHz-range XOs.

	This work		ISSCC'16 [6]	ISSCC'18 [11]	SSCL'20 [12]	ISSCC'23[13]	ISSCC'24 [14]	ISSCC'25 [7]
Process [nm]	65		65	65	65	40	65	22
Architecture	CBA with SCPS		Stacked Amplifier	Class-A G_M	Class-C Colpitts G_M	G_M	Class A/AB G_M	PLL-retimed PI
V_{DD} [V]	0.6		3.3	0.35	1	1	0.5	1
Frequency [MHz]	12	16	39.25	16	24	16	16	96
Osc. Swing [V_{PP}]	0.32	0.38	<0.4	0.25	\	0.2	0.06	\
No. of Samples	5		11	\	\	\	4	\
Power [µW]	8.1	9.4	19	31.6	9&	84	16.7	2,380
Temp. Range [°C]	–20 to 100		–30 to 80	–40 to 90	–20 to 90	–40 to 85	–40 to 125	\
Δf over Temp. [ppm]	±15	±10	±10.5	21.9	\	\	\	\
Line Sens. [ppm/V]	±12.5	±9.8	0.18	67	\	\	\	\
Phase Noise [dBc/Hz@1kHz]	–138.3	–137.9	–139	–134	–129&	–139.0	–135.3	–140.2*
FoM# [dB]	240.7	242.2	248.1	233.1	237.1	233.8	237.2	230.1*
Area [µm²]	1,892 (core) 5,470 (include C_L)		88,200	23,000■	12,000	80,000■	57,000■	1,140,000▼

FoM: [-PN+20log(f/f_{offset})-10log(P/1mW)]; * PN at 48MHz; & Quartz Crystal; ▼ Including PLL; ■including startup circuit

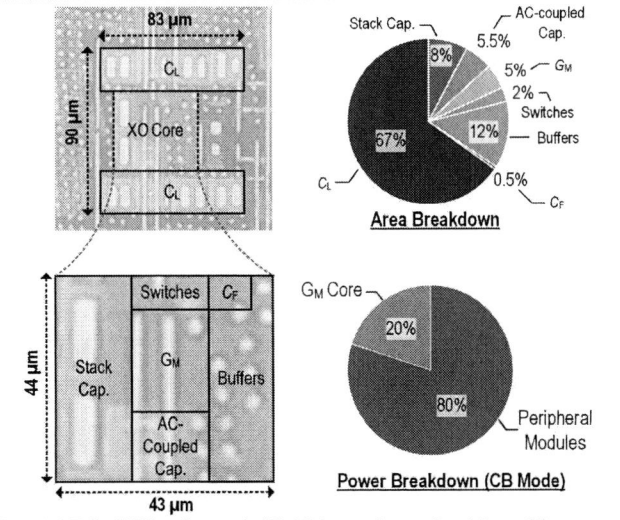

Figure 4.3.7: (Left) Chip micrograph. (Right) Area and power breakdown of the proposed XO.

• 2026 IEEE International Solid-State Circuits Conference

979-8-3315-8937-0/26 $31.00 © 2026 IEEE

ISSCC 2026 / SESSION 4 / ANALOG TECHNIQUES & AMPLIFIERS / 4.4

4.4 A 2.1-to-3.7ppm/°C Bandgap Voltage Reference with a Current-Domain TC Compensation and ±0.06% Inaccuracy from -40°C to 125°C in 130nm CMOS

Zhong Tang[1], Sining Pan[2], Xiaopeng Yu[3], Nick Nianxiong Tan[4], Zhangming Zhu[1]

[1]Xidian University, Xi'an, China, [2]Tsinghua University, Beijing, China, [3]Zhejiang University, Hangzhou, China, [4]Vango Technologies, Hangzhou, China

Abstract

This paper presents a CMOS bandgap reference (BGR) with a current-domain high-order TC compensation by using a capacitively-biased-diode-based super-PTAT current bias. It achieves a TC of 2.1 to 7.1ppm/°C from -40 to 125°C after a batch trim, which is on par with state-of-the-art BGRs with a 1-point trim. After a 2-point trim, it achieves a TC that ranges from 2.1 to 3.7ppm/°C and an inaccuracy of ±0.06% (3σ) from -40 to 125°C. Its simple architecture also results in 2× higher energy efficiency.

BJT-based bandgap references (BGRs) are essential building blocks for measurement and metering chips. Their 1st-order temperature coefficient (TC) can be minimized by adding a PTAT voltage (ΔV_{BE}) to a CTAT voltage (V_{BE}) in the correct proportions. However, due to the nonlinear term ($\propto T \cdot \ln(T)$) in V_{BE}, high-order TCs remain, which limit the accuracy of BGRs in standard CMOS processes [1-9]. To avoid this, a nonlinear compensation voltage V_{NL} can be added to the BGR output (Fig. 4.4.1 top). This can be generated by a nonlinear PTAT voltage ΔV_{BE} [1-3], piecewise-linear compensation [4,5], and other nonlinear voltage-generation circuits [6,7]. However, such methods usually result in larger noise and spread, which limit the energy efficiency and accuracy of the BGR. Although better accuracy can be achieved with a digital look-up-table (LUT) and a temperature sensor [8], the compensated result is only available in the digital domain. As mentioned in [9], the curvature in V_{BE} can also be suppressed by using a strongly temperature-dependent bias current ($\propto T^N$). However, generating such a super-PTAT current requires either translinear circuits based on high-quality BJTs [10], or V_{PTAT}/R current- generation circuits based on a resistor with a large negative TC (~-0.33%/°C) [11], which are not available in most standard CMOS processes.

In this work, a CMOS BGR with a current-domain high-order TC compensation is presented. The curvature of V_{BE} is compensated by a tunable super- PTAT bias current through a capacitively-biased-diode (CBD)-based current generator. Implemented in a standard 130nm CMOS process, the proposed BGR achieves a TC of 2.1 to 7.1ppm/°C from -40 to 125°C after a simple batch trim, which is on par with state-of-the-art precision BGRs with a 1-point trim [3]. After a 2-point trim, it achieves a TC that ranges from 2.1 to 3.7ppm/°C and an inaccuracy of ±0.06% (3σ) from -40 to 125°C. Compared to state-of-the-art precision BGRs [2-6], its simple architecture results in 2× higher energy efficiency.

Figure 4.4.1 (bottom) shows the diagram of the proposed current-mode BGR with a current-domain TC compensation. The reference output V_{REF} is generated by adding an amplified PTAT voltage ΔV_{BE} to a CTAT voltage V_{BE}. Apart from the classic PTAT bias current I_{PTAT}, a super-PTAT bias current I_{BIAS} is also injected into the emitter of the BJT Q_3, compensating for the curvature of V_{BE} directly. Compared with other voltage-domain TC compensation schemes [2-7], in which the voltage error is linearly added to the BGR, this current-domain TC compensation is less sensitive to errors, as any variation in I_{BIAS} will only lead to a logarithmic change in V_{BE}. The main challenge is then how to generate such a super-PTAT bias current in a simple way.

To avoid the use of BJT-based translinear circuits, a CBD-based super-PTAT bias generation approach is proposed that only requires a few transistors and capacitors (Fig. 4.4.1 bottom-right). As in [12], the capacitor C (=6.2pF) is first pre-charged to V_{DD} during the reset (Φ_{RST}) phase. Then it is discharged via a diode-connected PMOS M_D. After a short settling time (tens of ns), the current through M_D is proportional to $kT/q \cdot (C/t)$. Since the on-chip MIM capacitor C has a negligible TC, the current is PTAT when a fixed discharging time t is used [13-15]. However, if t decreases with temperature, the temperature dependency of the current will be significantly increased, thus achieving super-PTAT behavior. Moreover, due to the diode characteristic, this current is also insensitive to the initial voltage on the capacitor C, resulting in a good power-supply rejection (PSR) [12]. After discharging, the current of M_D is sampled and replicated via the current-mirror transistor M_1 to bias the BJT. By tuning the 1st-order TC of the discharging time t, the TC of the super-PTAT I_{BIAS} can be tuned to compensate for the curvature of V_{BE}. In this approach, the TC compensation is finally accomplished in the time-domain, which simplifies the analog design.

As in [16], the discharging time t can be defined by an on-chip RC delay. To compensate the 2nd-order TC in V_{BE}, t should have a large negative TC (~-0.35%/°C) [11]. Unfortunately, no standard CMOS resistor can provide this temperature characteristic. Instead, this is achieved by using logic gates to subtract a positive TC delay t_2 from a smaller negative TC delay t_1 ($t_1>t_2$, Fig. 4.4.2 top). Figure 4.4.2 (bottom) shows the diagram of the tunable negative-TC delay-generation circuit. It consists of two modified RC polyphase filters (PF$_1$ and PF$_2$) and two inverter-based comparators (COMP$_1$ and COMP$_2$) [16]. The polyphase filters are implemented with unsilicided N-poly resistors (R_N=46.8kΩ, TC=~-0.1%/°C), silicided N-poly resistors (R_P=14.8kΩ, TC=~0.28%/°C) and MIM capacitors C_1 and C_3 (=3.1pF). As in [16], the extra capacitors C_2 and C_4 (=C_1=C_3) prevent the occurrence of beyond-the-rail output spikes, thus allowing the comparators to be realized with core

devices without over-voltage stress issues. In this work, the auxiliary comparator COMP$_1$ with a relaxed speed is used to detect the zero-crossing (ZC) moment of PF$_2$ (V_{AP}=V_{AN}). Its output (SEL) is then used to select the input of the main comparator COMP$_2$. Before the ZC moment of PF$_2$, the input of the COMP$_2$ is connected to PF$_2$. After that, it is switched to PF$_1$. The time interval t between the ZC moments of the two PFs can then be extracted from the output of COMP$_2$. Since COMP$_2$ is reused for both PFs, its switching delay and offset are cancelled, making the time interval t purely defined by the RC time constants. To save power in the inverter-based comparator, reset switches are used to pull V_{AP}/V_{AN} and V_{BP}/V_{BN} up/down to V_{DD}/ground, respectively, shortly after the comparator switches [16]. In case the high-order TCs of the BJT are not well-modeled, the TC of the discharging time t and thus the BGR's curvature is adjustable by a 4-bit tunable R_N.

The generated negative TC delay is used to control the CBD-based super-PTAT bias generator. However, the CBD bias cannot provide a stable output during the reset and discharging phases (Φ_{RST} and Φ_{DC}). To solve this issue, a ping-pong technique [12] is adopted to provide a continuous-time bias current for the PNP Q_3 in the BGR core (Fig. 4.4.3). To mitigate the impact of the leakage current, the bias current is refreshed at f=200kHz. With a discharging time of 165ns, the simulated super-PTAT I_{BIAS} is about 9µA at room temperature and increases to 16µA at 125°C. For simplicity, I_{BIAS} is also mirrored to bias the error amplifier A in the BGR core. Thanks to the simple structure of the CBD-based bias generator, it can easily start up as long as a clock is available, thus avoiding additional start-up circuits, which are usually necessary for conventional BGRs. The PNP Q_1 and Q_2 have an emitter ratio of 15 (unit size 5µm×5µm), which defines the PTAT bias current (I_U=$\Delta V_{BE}/R_1$). It is further mirrored to R_2 and Q_3, thus generating the curvature-compensated output V_{REF}=2R_2/R_1·ΔV_{BE}+V_{BE}. To ensure that the emitter current of Q_3 is dominated by the super PTAT I_{BIAS}, the normal PTAT current (2I_U=3.6µA) is much smaller than I_{BIAS}. R_2 is 6-bit tunable, so that the PTAT voltage can be tuned to compensate for the 1st-order TC variation. To reduce the offset and 1/f noise of the error amplifier, it is chopped at 200kHz. In addition, the current mirrors are cascoded to improve the PSR. Their mismatch is mitigated by using dynamic element matching (DEM).

Implemented in a standard 130nm CMOS process, the BGR occupies an active area of 0.1mm² (Fig. 4.4.7). It generates a 1.187V output V_{REF} with a 3.3V analog supply and a 1.2V digital supply for the delay-generation circuit. To mitigate measurement errors and suppress the chopping and DEM ripples, the output is followed by an on-chip chopped buffer and then filtered by an external 1µF capacitor. Figure 4.4.4 (top-left) shows the measured power consumption of the BGR over temperatures. At room temperature, it consumes 91µA and 10µA in the analog and digital domains, respectively. Due to the super-PTAT bias current, the analog current increases to 177µA at 125°C. Figure 4.4.4 (top-right) shows the measured output noise of the BGR. In the frequency band of interest (0.1Hz to 10Hz), the noise can be significantly suppressed by chopping and DEM, resulting in an integrated noise of 1µVrms. Changing the analog supply from 2.4V to 3.6V only changes the BGR output by 200µV, corresponding to a power-supply sensitivity (PSS) of less than 0.016%/V (8 samples, Figure 4.4.4 bottom-left). Varying the digital supply from 1.15V to 1.25V causes ±0.06% error in V_{REF} due to the residual supply dependency of the RC delay, which can be mitigated by the LDO. By injecting a 100mVp AC signal to the analog supply, it shows a PSR of -90dB at 10Hz.

To verify the high-order TC compensation of the proposed BGR, a typical sample is first characterized from -40 to 125°C. As shown in Fig. 4.4.4 (bottom-right), the residual high-order TC of the BGR can be tuned by adjusting R_N in the RC delay-generation circuit. With the optimal trimming code, it achieves a residual TC of 2.6ppm/°C. The same high-order TC trimming code is applied to all the samples. 16 samples from one batch with DIL ceramic package were characterized over temperature. Without any individual trimming, the BGR has an untrimmed inaccuracy of ±0.26%(3σ) from -40 to 125°C (Fig. 4.4.5, top-left). The TCs of different samples vary from 2.1ppm/°C to 7.1ppm/°C with an averaged value of 4.6ppm/°C (Fig. 4.4.5, top-right). The inaccuracy is reduced to ±0.13%(3σ) after a 1-point offset trim at room temperature. To further improve the accuracy, 2-point calibration at 20°C and 95°C can be used to trim the 1st-order TC by tuning R_2, the residual TC is then reduced to 2.1 to 3.7ppm/°C with an averaged value of 2.7ppm/°C (Fig. 4.4.5, bottom). The BGR then achieves an inaccuracy of ±0.06%(3σ) from -40 to 125°C. To investigate the

82 • 2026 IEEE International Solid-State Circuits Conference

979-8-3315-8937-0/26 $31.00 © 2026 IEEE

effect of packaging stress, 16 samples from the same batch were characterized in SOP plastic packaging, resulting in a small shift in the optimal trimming code. Thanks to the added polymide layer in the die, it achieves a similar inaccuracy of ±0.25%(3σ) and a TC of 2.9 to 8.1ppm/°C with a batch trim. After 2-point calibration for the 1ˢᵗ-order TC and nominal value trim, the inaccuracy can be further improved to ±0.05%(3σ) from -40 to 125°C.

Figure 4.4.6 summarizes the performance of this work and compares it to state-of-the-art precision BGRs in standard CMOS processes. Thanks to the direct current-domain TC compensation scheme, this work achieves the lowest untrimmed TC, which is on par with the state-of-the-art precision BGR with a 1-point trim [3]. It also achieves low noise of 1μVrms from 0.1Hz to 10Hz while consuming 101μA, improving the energy efficiency by 2× due to its simple structure.

Acknowledgement:
This work was supported by the National Natural Science Foundation of China under Grant 62522409. The corresponding author is Zhangming Zhu (zmyh@263.net).

References:
[1] C. Palmer et al., "A Curvature Corrected Micropower Voltage Reference," *ISSCC*, New York, NY, USA, 1981, pp. 58-59. https://doi.org/10.1109/ISSCC.1981.1156266
[2] G. Ge et al., "A Single-Trim CMOS Bandgap Reference with a 3σ Inaccuracy of ± 0.15% From –40°C to 125°C," *IEEE JSSC*, vol. 46, no. 11, pp. 2693-2701, Nov. 2011. https://doi.org/10.1109/JSSC.2011.2165235
[3] J.-H. Boo et al., "A Single-Trim Switched Capacitor CMOS Bandgap Reference with a 3σ Inaccuracy of +0.02%, –0.12% for Battery-Monitoring Applications," *IEEE JSSC*, vol. 56, no. 4, pp. 1197-1206, April 2021. https://doi.org/10.1109/JSSC.2020.3044165
[4] K. Chen et al., "A 1.16-V 5.8-to-13.5-ppm/°C Curvature-Compensated CMOS Bandgap Reference Circuit with a Shared Offset-Cancellation Method for Internal Amplifiers," *IEEE JSSC*, vol. 56, no. 1, pp. 267-276, Jan. 2021. https://doi.org/10.1109/JSSC.2020.3033467
[5] H. Tian et al., "A 1.9μVrms 7.7ppm/°C ADC Reference with 20mA Output Current and Single-Trim Inaccuracy of ±0.03%(3σ) from -40°C to 125°C," *ESSCIRC*, Lisbon, Portugal, 2023, pp. 93-96. https://doi.org/10.1109/ESSCIRC59616.2023.10268754
[6] S. Huang et al., "A Sub-1 ppm/°C Bandgap Voltage Reference with High-Order Temperature Compensation in 0.18-μm CMOS Process," *IEEE TCAS-I*, vol. 69, no. 4, pp. 1408-1416, April 2022. https://doi.org/10.1109/TCSI.2021.3139908
[7] X. Liao et al., "A 3.0 μVrms, 2.4 ppm/°C BGR With Feedback Coefficient Enhancement and Bowl-Shaped Curvature Compensation," *IEEE TCAS-I*, vol. 71, no. 5, pp. 2424-2433, May 2024. https://doi.org/10.1109/TCSI.2024.3373788
[8] G. Maderbacher et al., " A Digitally Assisted Single-Point-Calibration CMOS Bandgap Voltage Reference with a 3σ Inaccuracy of ±0.08% for Fuel-Gauge Applications," *ISSCC*, San Francisco, CA, USA, 2015, pp. 1-3. https://doi.org/10.1109/ISSCC.2015.7062946
[9] G. C. M. Meijer et al., "A New Curvature-Corrected Bandgap Reference," *IEEE JSSC*, vol. 17, no. 6, pp. 1139-1143, Dec. 1982. https://doi.org/10.1109/JSSC.1982.1051872
[10] I. M. Filanovsky et al., "BiCMOS Cascaded Bandgap Voltage Reference," Proceedings of the *39th Midwest Symposium on Circuits and Systems*, Ames, IA, USA, 1996, pp. 943-946 vol.2. https://doi.org/10.1109/MWSCAS.1996.588110
[11] C. Falconi et al., "Low Cost Curvature Correction of Bandgap References for Integrated Sensors," *Sensors and Actuators A*: Physical, Volume 117, Issue 1, 2005, pp 127-136. https://doi.org/10.1016/j.sna.2004.05.030

[12] M. Eberlein et al., "A 40nW, Sub-1V Truly 'Digital' Reverse Bandgap Reference Using Bulk-Diodes in 16nm FinFET," *A-SSCC*, pp. 99-102, Nov. 2018. https://doi.org/10.1109/ASSCC.2018.8579306
[13] X. Wu et al., "A Sub-1V 14b 5.8nW/Hz BW/Power-Scalable CT Sensor Interface with a Frequency-Controlled Current Source Achieving a 225× Scalable Range," *ISSCC*, San Francisco, CA, USA, 2025, pp. 1-3. https://doi.org/10.1109/ISSCC49661.2025.10904802
[14] Z. Tang et al., "A Sub-1 V Capacitively Biased BJT-Based Temperature Sensor with an Inaccuracy of ±0.15 °C (3σ) from -55 °C to 125 °C," *IEEE JSSC*, vol. 58, no. 12, pp. 3433-3441, Dec. 2023. https://doi.org/10.1109/JSSC.2023.3308554
[15] H. Eum et al., "A Sub-1-V Capacitively-Biased Voltage Reference with an Auto-Zeroed Buffer and a TC of 18-ppm/°C," *IEEE TCAS-II*, vol. 72, no. 1, pp. 8-12, Jan. 2025. https://doi.org/10.1109/TCSII.2024.3454348
[16] Z. Tang et al., "A 14-b BW /Power Scalable Sensor Interface with a Dynamic Bandgap Reference," *IEEE JSSC*, vol. 59, no. 12, pp. 4077-4087, Dec. 2024. https://doi.org/10.1109/JSSC.2024.3471820

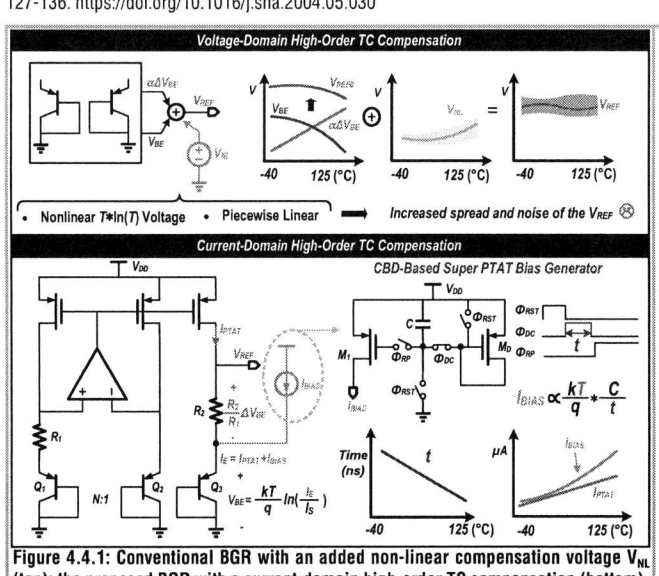

Figure 4.4.1: Conventional BGR with an added non-linear compensation voltage V_NL (top); the proposed BGR with a current-domain high-order TC compensation (bottom).

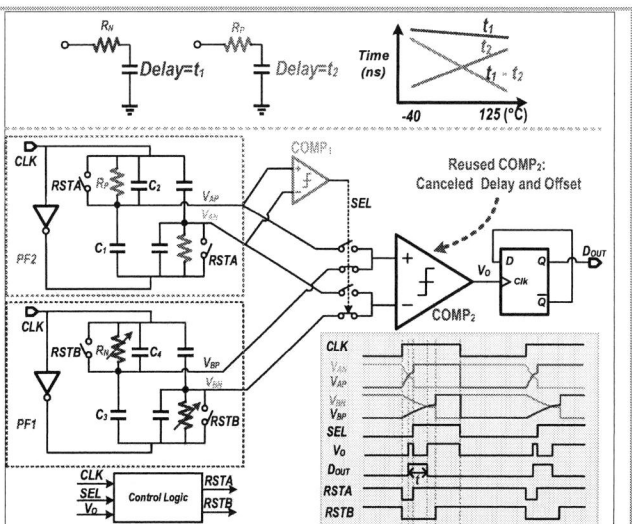

Figure 4.4.2: Concept of the on-chip negative TC delay generation (top) and its circuit implementation (bottom).

ISSCC 2026 / SESSION 4 / ANALOG TECHNIQUES & AMPLIFIERS / 4.4

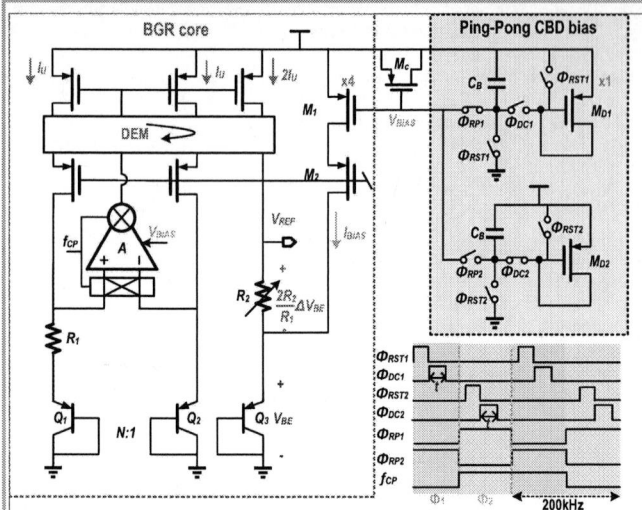

Figure 4.4.3: Circuit and timing diagrams of the current-mode BGR with a ping-pong super-PTAT bias.

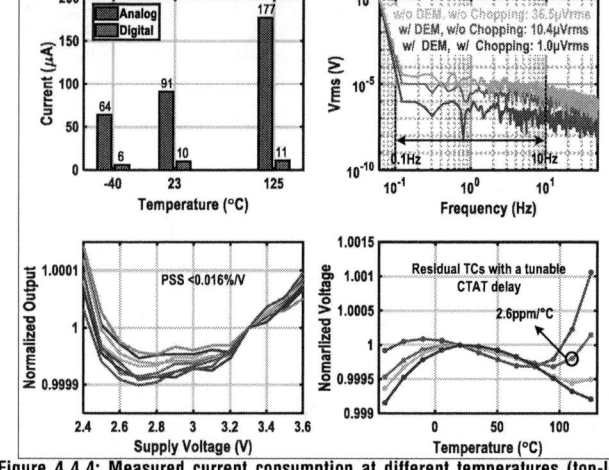

Figure 4.4.4: Measured current consumption at different temperatures (top-left); measured output noise with different DEM and chopping settings (top-right); measured power supply sensitivity (PSS) of 8 samples (bottom-left); measured residual TCs with different CTAT delay times of one typical sample (bottom-right).

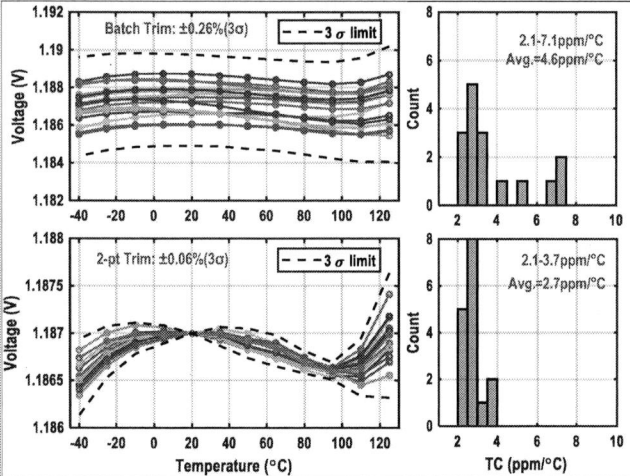

Figure 4.4.5: Measured inaccuracy of 16 samples in ceramic package after a batch trim (top-left) and their residual TCs (top-right); inaccuracy after a 2-point trim (bottom-left) and their residual TCs (bottom-right).

	This work	JSSC'11 [2]	JSSC'21 [3]	TCAS-I'22 [6]	JSSC'21 [4]	ESSCIRC'23 [5]	ISSCC'15 [8]
Technology (nm)	130	160	180	180	130	40	130
Area (mm²)	0.10	0.12	0.38	0.256	0.08	0.18	0.034
Curvature Correction	Super PTAT Current	Non-Linear Voltage	Non-Linear Voltage	Non-Linear Voltage	Piecewise Linear	Piecewise Linear	Digital LUT
Supply (V)	2.4-3.6/1.2	1.8±10%	1.8±10%	3.2-3.7	3.3	1.4-1.8	1.5
Bandgap Voltage (V)	1.187	1.0875	1.14	2.14	1.16	1.2	1.215
Temp. range (°C)	-40 to 125	-40 to 125	-40 to 125	-25 to 125	-40 to 150	-40 to 125	-40 to 120
3σ Inaccuracy (Trim Point)	±0.26% (0) ±0.13% (1) ±0.06% (2)	±0.75% (0) ±0.15% (1)	±0.21% (0) -0.02%/ +0.12% (1)	N/A	±1.62% (0)	±0.35% (0)* ±0.1% (1)*	±0.08% (1)
TC (ppm/°C) (Trim Point)	2.1-7.1 (0) 2.1-3.7 (2)	5-12 (1)	3.2-5.5 (1)	7.4-20.3 (0) 0.7-1.53 (3)	5.78-13.5 (1)	6.5-9.6 (1)	7 (1)
Current (µA)	91/10	55	17	409	120	56	N/A
Noise (µV) @BW	1@10Hz	6.1@10Hz	56@2.5Hz	15@1kHz	175.5@10Hz	1.9@10Hz	N/A
Line Sensitivity (%/V)	0.016	N/A	N/A	0.0146	0.03	0.19	N/A
PSR (dB)	-90@10Hz	-74@DC	-76dB@1Hz	-63.4@10Hz	-82@10Hz	-77@10Hz	N/A
Efficiency (µA×µV²/Hz)**	10.1	204.7	21324.8	92.0	369603	20.2	N/A
Samples	16 Ceramic + 16 Plastic	61	18	6	7	10	13 Ceramic

*Estimated from figures;   **Efficiency=Current × Noise²/Hz

Figure 4.4.6: Performance summary and comparison with the state-of-the-art.

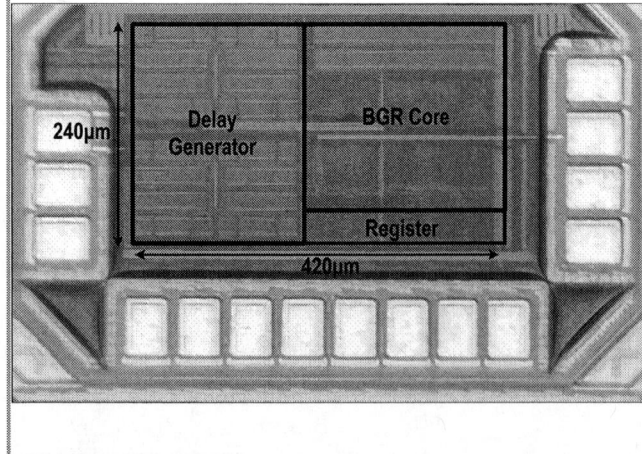

Figure 4.4.7: Die micrograph.

• 2026 IEEE International Solid-State Circuits Conference

ISSCC 2026 / SESSION 4 / ANALOG TECHNIQUES & AMPLIFIERS / 4.5

4.5 A 1ppm/°C and ±0.066% 3σ Accuracy Bandgap Reference with Temperature-Adaptive PTAT Scaling

Jiaqing Rui[1], Liangjian Lyu[1], Yuting Wan[1], Kexin Shan[1], Wenxian Gu[1], C.-J. Richard Shi[2], Xing Wu[1]

[1]East China Normal University, Shanghai, China, [2]University of Washington, Seattle, WA

Abstract

This paper presents a process-independent, curvature-compensated bandgap reference. By introducing a temperature-adaptive duty-cycled resistor to scale the PTAT voltage, the design effectively compensates for CTAT nonlinearity without introducing offset errors. Fabricated in 0.18µm CMOS, the prototype chip occupies 0.068mm² and consumes 15µA. After two-point trimming, it achieves an average temperature coefficient of 1ppm/°C and a 3σ output accuracy within ±0.066%.

A bandgap voltage reference (BGR) with low temperature drift and high accuracy is crucial for high-precision circuit systems. Therefore, current research focuses on reducing the temperature coefficient (TC) and improving voltage accuracy [1-7]. To address these design challenges, this work proposes a process-independent curvature compensation method, by utilizing a duty-cycled resistor (DCR) to generate a nonlinear proportional-to-absolute-temperature (PTAT) voltage. Ultimately, an optimal low temperature drift of 0.51ppm/°C is achieved, along with high voltage accuracy of ±0.041% (3σ) for single-point trimming and ±0.066% (3σ) for two-point trimming.

Figure 4.5.1 (a) reveals that the output characteristics of a BGR typically exhibit one or more of the following errors: PTAT error (i.e., slope error) and non-PTAT errors (including offset and curvature errors). In BGR circuits without high-order compensation, non-PTAT errors mainly originate from the offset voltage of the error amplifier and current mirror mismatch. By adopting chopping or auto-zeroing techniques, and replacing the MOSFET current mirror with a resistor-based current mirror, non-PTAT errors can be almost completely suppressed. When only PTAT errors remain in the circuit, the correlation between temperature drift errors and voltage errors among chips follows a certain linear relationship: if the voltage error at a specific temperature point is calibrated to zero, the temperature drift error can be simultaneously reduced to zero. Based on this characteristic, an inaccuracy of less than ±0.1% can be achieved [6]; furthermore, relying on the above principle, the automatic calibration function of the BGR can be further realized [2]. But, due to the lack of temperature compensation, their temperature drift typically exceeds 10ppm/°C [1], which is unsatisfactory for high-precision applications.

However, the temperature compensation strategies of traditional BGRs introduce new error terms in practical applications. Current mainstream temperature compensation schemes are divided into piecewise compensation and curvature compensation. As shown in Fig. 4.5.1 (b), in the piecewise compensation architecture, each compensation interval may generate error components due to process variation and device mismatch. Although such errors can be eliminated through segment-by-segment trimming, this significantly increases the chip trimming cost, making it difficult to implement [5]. For the curvature compensation scheme, as shown in Fig. 4.5.1 (c), the introduced compensation term usually leads to additional offset; especially when using the nonlinearity of MOSFETs to construct the compensation circuit, such as [3], a non-zero threshold voltage at 0°K is inevitably introduced. In addition, the temperature-dependent curvature characteristic of carrier mobility is limited by specific process nodes, which makes it difficult for the curvature compensation effect to meet expectations when the circuit is implemented in different processes, restricting its applicability in advanced processes.

Figure 4.5.1 (d) presents the principle of the proposed curvature compensation scheme in this work. To address the defects of the aforementioned traditional temperature compensation schemes, the inherently existing V_{PTAT} ($V_T \cdot \ln(N)$) in the BGR core is utilized to construct a nonlinear PTAT voltage for achieving curvature compensation. $V_T \cdot \ln(N)$ is zero at 0°K, therefore the offset error is eliminated. As shown in Fig. 4.5.1 (d), unlike the output voltage V_{BG} of a traditional BGR (which is generated by superposing voltage complementary to absolute temperature (V_{CTAT}) and α-times V_{PTAT}, i.e., $V_{BG} = V_{CTAT} + \alpha \cdot V_{PTAT}$), this work proposes to achieve curvature compensation by designing V_{PTAT} with a temperature-dependent amplification coefficient α(T).

To intuitively analyze the influence of the nonlinear component of V_{CTAT} on V_{BG}, we perform a Taylor expansion of V_{CTAT} with respect to temperature and approximate it as a quadratic function: $V_{CTAT} = V_{Eg0} - AT^2 - BT$, where V_{Eg0} is the semiconductor bandgap voltage at 0°K, and A and B are coefficients related to specific process parameters. Based on this model, if the linear V_{PTAT} is multiplied by the equally linear α(T) through circuit design to generate a nonlinear PTAT voltage (i.e., $\alpha(T) \cdot V_{PTAT} = AT^2 + BT$) with flexibly adjustable coefficients for each temperature term, perfect compensation for the curvature term of V_{CTAT} can be achieved, and finally the BGR output voltage is stabilized at $V_{BG} = V_{Eg0}$. To achieve the above curvature compensation goal, reference [7] proposed a scheme of constructing a voltage amplification network using resistors with different temperature coefficients. However, this scheme relies on diffused resistors with low precision and large mismatch, and it is difficult to achieve good matching with the most widely used polysilicon resistors. Therefore, this work

proposes to use a duty-cycled resistor (DCR) to generate the temperature coefficient required for compensation. As shown in Fig. 4.5.2 (a), the DCR R_x is controlled by a clock signal CLK_{DCR} with a duty cycle of D, and its equivalent resistance can be configured as $R_1/(2-D)$ by adjusting the switching duty cycle D, thereby making the amplification coefficient α linearly related to temperature. The CLK_{DCR} generation circuit is show in Fig. 4.5.2 (b). First, a ramp signal is generated using V_{BG}, and this ramp signal is compared with the k-times V_{PTAT} ($V_{PTAT,CMP}$); the comparator then outputs the duty cycle signal CLK_{DCR}. Figure 4.5.2 (c) shows the schematics of V_{RAMP} and the comparator used in the DCR clock generator. Figure 4.5.2 (c) deduces the expression of the nonlinear PTAT voltage α(T)·V_{PTAT}. It can be seen from the formula that R_S can be set as a knob to adjust the first-order term of V_{PTAT}, and R_R can be set as a knob to adjust the second-order term of V_{PTAT}.

The schematic diagram of the BGR core circuit in this work is shown in Fig. 4.5.3 (a). To minimize the non-PTAT errors introduced by current mirror and amplifier, a resistor-based current mirror and chopping amplifier are adopted in this BGR. All bias currents and bias voltages are generated by a β multiplier circuit with a negative feedback loop to improve the power supply rejection (PSR) of the bias circuit. To reduce the inaccuracy of BGR caused by process variation, this work integrates two resistor digital-to-analog converters (RDACs) to calibrate the linear term and nonlinear term of V_{BG} respectively. As shown in Fig. 4.5.3 (b), since the variation of code1 can be equivalent to the variation of code2, we analyze the relationship between the quadratic term trimming code2 and the linear term trimming code1. The simulation results in Fig. 4.5.3 (b) also verify this characteristic: during the trimming process, one code can be used to derive the other, which provides support for the subsequent implementation of the BGR single-point trimming scheme.

The curvature-compensated BGR proposed in this work is implemented based on a 0.18µm CMOS process. Figure 4.5.7 (a) shows the chip micrograph, with an area of 0.068mm². Under a 1.8V power supply, the current consumption is 15µA. Figure 4.5.4 (a) presents the measurement results of the V_{BG} temperature characteristics, comparing the differences between the modes with and without curvature compensation, within the temperature range from -40 to 125°C. The measured results show that after adopting the curvature compensation scheme proposed in this work, the optimal temperature coefficient of V_{BG} is optimized to 0.51ppm/°C; compared with the optimal TC without compensation, this scheme achieves a 25-fold improvement in temperature stability. Figure 4.5.4 (b) shows the measured noise spectrum, with an integrated noise of 44µV$_{rms}$ (0.1Hz to 1kHz). Figure 4.5.4 (c) shows the measured spectrum of PSR, with the test frequency ranging from 1Hz to 10kHz. The results show that the PSR is approximately −85dB at 10Hz and approximately −63dB at 1kHz. Figure 4.5.4 (d) shows the measured line regulation (LR), which is 41.95ppm/V within the power supply voltage range of 1.5 to 3.3V.

The curvature-compensated BGR chip designed in this work supports two trimming modes: one is single-temperature-point trimming at room temperature, and the other is two-temperature-point trimming at 0°C and 80°C (the latter further optimizes the temperature drift accuracy). Figure 4.5.5 (a) presents the output voltage-temperature characteristic curves of 10 test chips before trimming, and the curves calibrated to minimize temperature drift under a 1.8V power supply. Figure 4.5.5 (b) shows the optimal trimming codes for each sample, and linear fitting is performed. Based on the relationship between these codes, single-parameter trimming is carried out. Figure 4.5.5 (c) shows the results of two-point trimming using the code values on the fitting curve: the average temperature drift of the chips is 1.02ppm/°C, and the average output voltage error (3σ) is ±0.066%. Figure 4.5.5 (d) shows the performance after single-temperature-point trimming (at 20°C). Among them, the average output voltage error (3σ/µ) is ±0.042%, and the average temperature drift of the chips is 1.86ppm/°C.

Figure 4.5.6 systematically summarizes the core performance parameters of the proposed curvature-compensated BGR and the comparison of the most advanced low-temperature-drift and high-accuracy BGR schemes. The proposed compensation scheme achieves independent adjustment of the linear component and nonlinear component of V_{BG}, thus achieving good process compatibility. An average temperature drift of 1.86ppm/°C and an output voltage accuracy of ±0.042% can be achieved through single-temperature-point trimming. With two-temperature-point trimming, the temperature drift can be further

ISSCC 2026 / February 16, 2026 / 5:05 PM

optimized to an average of 1.02ppm/°C. The proposed BGR also achieves high PSR and low noise. Figure 4.5.7 (a) shows the micrograph, and Fig. 4.5.7 (b) presents the power consumption breakdown of the proposed chip. Figure 4.5.7 (c) shows the leading advantages of this work.

Acknowledgement:
This work was supported by the STI2030 Major Projects (Grant Nos. 2021ZD0202200 and 2021ZD0202202) and the National Natural Science Foundation of China (Grant No. 62204085). Corresponding author: Liangjian Lyu.

References:
[1] Y. Li et al., "Compact PNP BJT-Based Temperature Sensor and Sub-1-V Bandgap Reference for SoC Applications in 4-nm FinFET," *IEEE JSSC*, vol. 60, no. 8, pp. 2842-2853, Aug. 2025. http://doi.org/10.1109/JSSC.2024.3524245
[2] C.-W. U et al., "Sub-µW Auto-Calibration Bandgap Voltage Reference with 1σ Inaccuracy of ± 0.12% Within – 40°C to 120°C," *IEEE JSSC*, vol. 59, no. 2, pp. 540-550, Feb. 2024. http://doi.org/10.1109/JSSC.2023.3294996
[3] S. Huang et al., "A Sub-1 ppm/°C Bandgap Voltage Reference with High-Order Temperature Compensation in 0.18-µm CMOS Process," *IEEE TCAS-I*, vol. 69, no. 4, pp. 1408-1416, April 2022. http://doi.org/10.1109/TCSI.2021.3139908
[4] J. -H. Boo et al., "A Single-Trim Switched Capacitor CMOS Bandgap Reference with a 3σ Inaccuracy of +0.02%, –0.12% for Battery-Monitoring Applications," *IEEE JSSC*, vol. 56, no. 4, pp. 1197-1206, April 2021. http://doi.org/10.1109/JSSC.2020.3044165
[5] K. Chen et al., "A 1.16-V 5.8-to-13.5-ppm/°C Curvature-Compensated CMOS Bandgap Reference Circuit with a Shared Offset-Cancellation Method for Internal Amplifiers," *IEEE JSSC*, vol. 56, no. 1, pp. 267-276, Jan. 2021. http://doi.org/10.1109/JSSC.2020.3033467
[6] G. Maderbacher et al., "A Digitally Assisted Single-Point-Calibration CMOS Bandgap Voltage Reference with a 3σ Inaccuracy of ±0.08% for Fuel-Gauge Applications," *ISSCC*, pp. 1-3, 2015. http://doi.org/10.1109/ISSCC.2015.7062946
[7] K. Leung et al., "A 2-V 23-µA 5.3-ppm/°C curvature-compensated CMOS bandgap voltage reference," *IEEE JSSC*, vol. 38, no. 3, pp. 561-564, March 2003. http://doi.org/10.1109/JSSC.2002.808328

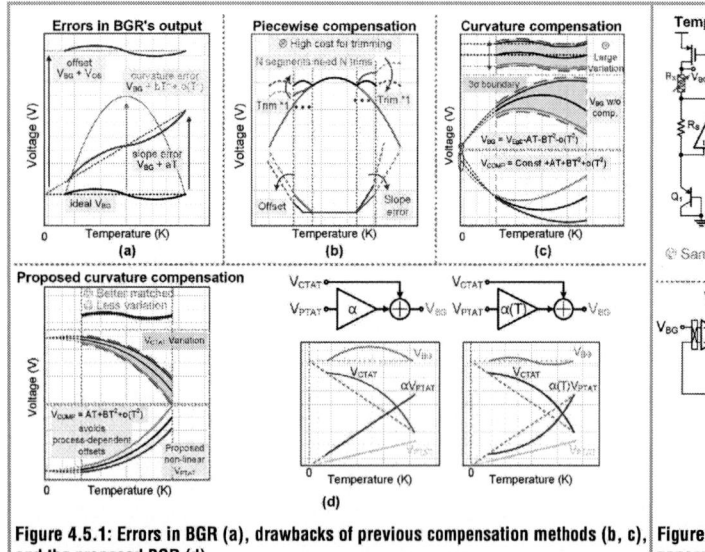

Figure 4.5.1: Errors in BGR (a), drawbacks of previous compensation methods (b, c), and the proposed BGR (d).

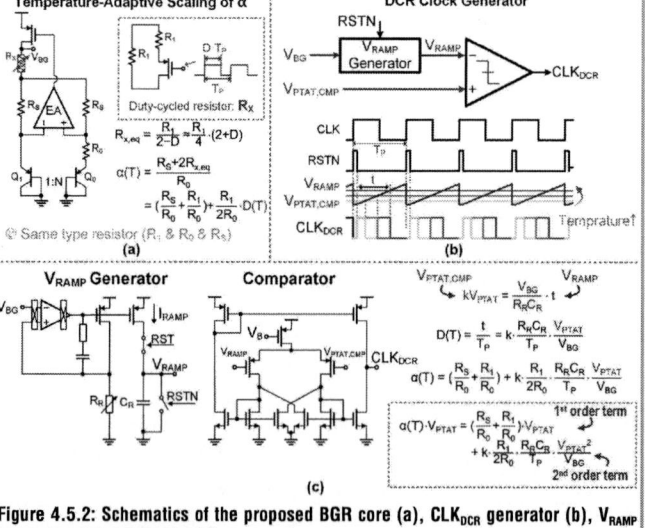

Figure 4.5.2: Schematics of the proposed BGR core (a), CLK_DCR generator (b), V_RAMP generator and comparator (c).

DIGEST OF TECHNICAL PAPERS • 85

979-8-3315-8937-0/26 $31.00 © 2026 IEEE

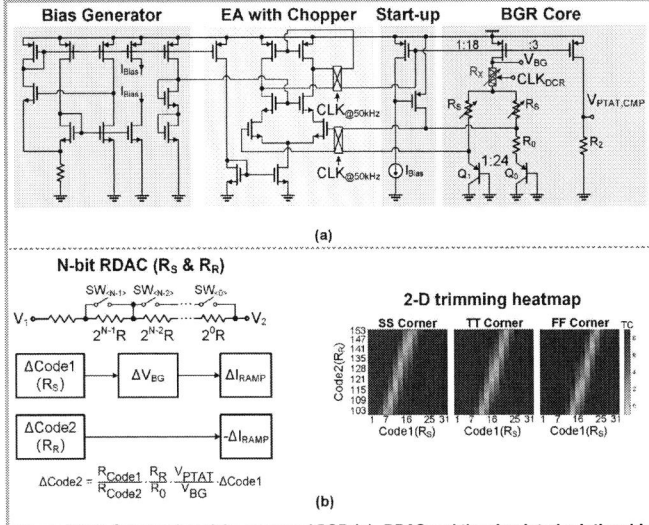

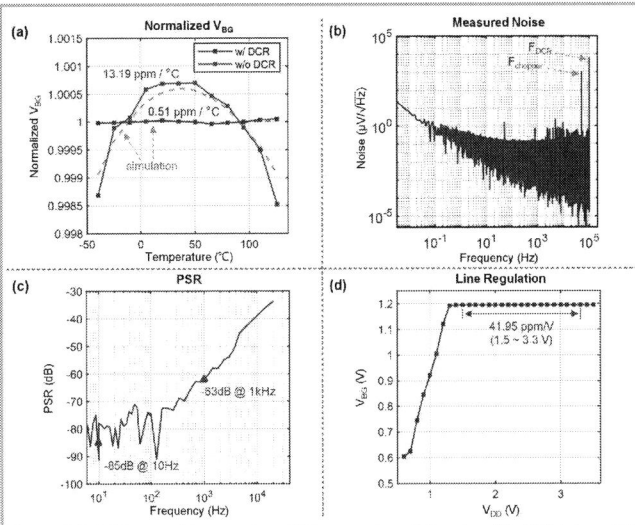

Figure 4.5.3: Schematics of the proposed BGR (a), RDAC and the simulated relationship between trimming codes (b).

Figure 4.5.4: Measured temperature drift comparison (a), noise (b), PSR (c), and LR (d) of the proposed BGR.

Figure 4.5.5: Measured V_{BG} of 10 chips (a), relationship of trimming codes (b), and histograms of V_{BG} (c, d).

Figure 4.5.6: Performance summary and comparison with state-of-the-art bandgap references.

Parameter	This work	JSSC'25[1]	JSSC'24[2]	TCASI'22[3]	JSSC'21[4]	JSSC'21[5]	ISSCC'15[6]
Technology (nm)	180	4	65	180	180	130	130
Supply Voltage (V)	1.5 ~ 3.3	> 0.9	1.2 ~ 2.5	3.2 ~ 3.7	1.8±10%	1.3 ~ 3.3	1.5±10%
Total Current (µA)	15	49.5	0.81	409	17	120	10.5
Compensation	curvature	/	/	curvature	curvature	piecewise	/
Temp. Range (°C)	-40 ~ 125	-10 ~ 110	-40 ~ 125	-25 ~ 125	-40 ~ 125	-40 ~ 150	-40 ~ 120
TC (ppm/°C)	0.51 (best) / 0.61 ~ 1.28 (2pt) / 0.61 ~ 2.8 (1pt)	9 ~ 45	18.6 ~ 28.8	0.611A / 0.7 ~ 1.557B	3.2 ~ 5.5	2.3 ~ 6.3A / 5.78 ~ 13.5B	7
Output Voltage (V)	1.221	0.45	1.001	2.14	1.1419	1.16	1.215
3σ Inaccuracy (%)	±0.066 (2pt) / ±0.042 (1pt)	±1.1	±0.04A / ±0.36B		+0.02, -0.12	1.62	+0.08
Line Regulation (%/V)	0.0042	/	0.13	0.0146	/	03	/
PSR (dB)	-85@10Hz	-40@DC	-58@10Hz		-63.4@10Hz / -56.5@1kHz	-76@DC	-82@10Hz
Noise @ BW	44 µV @0.1-1k Hz	/	457 µV @0.1-10 Hz	15 µV @1~k Hz	56 µV @1~1k Hz	175.5 µV @0.1~10 Hz	/
Trim parameters & points	1-parameter 2-pt trim	2-parameter 2-pt trim	1-parameter 2-pt trim	3-parameter 2-pt trim	1-parameter 1-pt trim	/	1-parameter 1-pt trim
Area (mm²)	0.068	0.006	0.04	0.256	0.38	0.08	0.034

A simulation result, B measurement result.

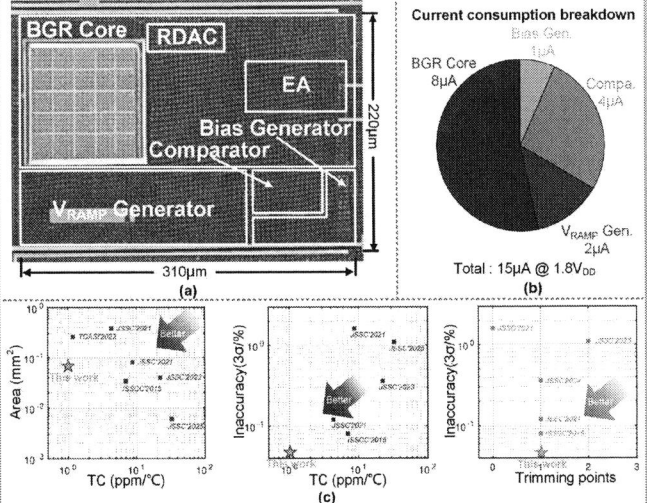

Figure 4.5.7: Die micrograph (a), current consumption breakdown (b), and performance comparison (c).

ISSCC 2026 / SESSION 4 / ANALOG TECHNIQUES & AMPLIFIERS / 4.6

4.6 An Integrated Voltage and Current Reference Together Achieving 5.7 and 9.1ppm/°C from -40 to 125°C

Lele Fang, Yan Zhu, Rui P. Martins, Chi-Hang Chan

University of Macau, Macau, China

Abstract

This work implements an integrated voltage and current reference circuit, which has low temperature drift. By ensuring the reuse of the main circuit, we propose a current-mirror-based seamless transition technique, which can be used to achieve sub-ranging compensation. Eventually, this work realizes 3rd-order compensation in two domains simultaneously, resulting in temperature coefficients of 5.7ppm/°C and 9.1 ppm/°C for voltage and current references, respectively.

The accuracy of high-precision meters, sensor readout, and control systems relies heavily on the stability of their voltage and current references (VR and CR). To accommodate broad scenarios while preserving a small form factor, integrating VR and CR with high temperature-insensitivity in a compact area is desirable. The VR or CR is typically derived from its accurate VR or CR counterparts via a resistor [1-3]. Despite being area-efficient, this approach inevitably limits the temperature stability of the sequentially generated voltage/current reference due to the added temperature-dependent component. For instance, the measured temperature coefficient (TC) of CR in [1] is 283ppm/°C, which is an order of magnitude higher than that of its VR.

A sub-ranging TC compensation for CR is presented in [4], as illustrated in Fig. 4.6.1(top). It partitions the temperature range and eliminates the 1st-order TC within each sub-range, achieving an exponential reduction in temperature drift. It also employs a bridge-resistor-based seamless transition (BRST) technique to ensure the continuity of the compensation curve between temperature sub-ranges. The bridge resistor between R_1 and R_2 determines the TC control parameter (α), and I_{REF} becomes independent of α at sub-ranging temperature (T_X) due to the same algebraic form involving α in both numerator and denominator. A seamless and process-insensitive transition is also achieved, owing to the consistent R_1/R_2 relationship in RBST and process-invariant V_{PTAT}/V_{CTAT}. However, this concurrently locks the PTAT/CTAT ratio within the core (area/power-dominated) circuitry, limiting its sub-ranging TC correction to a single domain. A straightforward solution to decouple the CR and VR TC compensation is to separate the core unit circuitry. However, such an approach introduces significant overhead due to the multiple bulky BJTs. This work presents a 3rd-order TC-compensated voltage and current integrated references circuit in 65nm, achieving average TCs of 5.7ppm/°C and 9.1ppm/°C for VR and CR, respectively, over a temperature range of -40 to 125°C within 0.16mm^2 area. This work also has a seamless transition characteristic, and without the requirement of chip-by-chip trimming.

To facilitate a compact solution, a current-mirror-based seamless transition (CMST) scheme is proposed, as shown in Fig. 4.6.1(bottom). This approach unlocks the correlation between the core unit and the seamless sub-ranging transition control in BRST, enabling an independent TC compensation for CR and VR with shared core circuity. Unlike RBST, which relies on the bridge resistor to correlate α among V_{PTAT}/V_{CTAT} and R_1/R_2 for T_X facilitation, the proposed CMST associates α with its mirrored copy. Except for expressing in the fraction form, I_{REF} can be decomposed as the weighted sum of I_{PTAT} and I_{CTAT}. Based on the additive form, the α can be canceled out when $I_{PTAT}=I_{CTAT}$ at T_X by assigning α for I_{PTAT} and (1-α) for I_{CTAT}. Therefore, a 2nd-order compensation for I_{REF} is realized by configuring an appropriate α within the ranges of $T<T_X$ and $T>T_X$, respectively. Since the α is tuned via scaling the current mirror source from the core circuitry, a similar scheme can be applied independently for VR with a distinct TC control parameter (β). Eventually, both the core circuitry and the temperature sensor for the TC sub-ranging operation can be shared for VR and CR implementation.

Building on the concept given in Fig. 4.6.1, Fig. 4.6.2 illustrates the overall architecture of the integrated VR and CR (VCR) and its sub-ranging operation. A typical structure with three current branches, employing two error amplifiers, resistor R_1 and R_2, and BJTs, generates the I_{PTAT} and I_{CTAT}. They can be expressed as $I_{CTAT}=V_{BE1}/R_1$ and $I_{PTAT}=ΔV_{BE}/R_2$, where $ΔV_{BE}=V_{BE2}-V_{BE1}$. Such a configuration sets I_{PTAT} equal to I_{CTAT} at T_X. Since T_X is primarily determined by the ratios of $ΔV_{BE}/V_{BE1}$ and R_1/R_2, it is inherently process-insensitive [5]. This design entails three temperature ranges to facilitate outstanding TC while simultaneously achieving a compact VCR thanks to the CMST approach, as detailed next. By simply mirroring the I_{PTAT} with dedicated width ratios (m, n) and comparing with the mirrored CTAT current, multiple transition temperatures (T_{X1} and T_{X2}) can be obtained, enabling the two-bit temperature sensor for 3rd-order TC compensation of both VR and CR. The process-insensitive width ratios ensure that the generated T_{X1} and T_{X2} inherit the process-stable characteristic, which is verified through 10k Monte-Carlo simulations. The standard deviation (σ) of T_{X1} and T_{X2} is < 0.86°C, which is a value comparable to the $σ_{TX} = 0.74°C$ reported in [4]. Note that the process stability of T_{X1} and T_{X2} helps the designer avoid additional trimming. The decisions of cross-detector (T_{X1_SEL}, T_{X2_SEL}) identify three sub-ranges and control the seamless pair #A & #B (for V_{REF}) and #C & #D (for I_{REF}) to sweep their corresponding control factors (a_{1-2}, b_{1-2}, and c_{1-2}, d_{1-2}), facilitating the 1st-order TC cancellation within each of the three regions. The seamless transition is achieved by simultaneously adjusting the mirrored PTAT and CTAT current units by the same amount. For instance, in seamless pair #A, a unit increase in the PTAT array (m*I_{PTAT_u}) corresponds to a unit decrease in the CTAT array (I_{CTAT_u}), and vice versa. The control factor for each array corresponds to the ratio of the number of enabled units to the total number of units, and the total number of active units across both the PTAT and CTAT arrays remains constant. Following the above operation, m*I_{PTAT_u} and I_{CTAT_u} in seamless pair #A are given coefficients 'a' and '1-a', respectively. Since I_{CTAT_u} and m*I_{PTAT_u} are equal at T_{X1}, T_{X1_SEL}-controlled switching of a_1 and a_2 does not induce an abrupt voltage change in V_{REF}.

The shared I_{PTAT}, I_{CTAT} generator employs a BJT emitter area ratio and R_1/R_2 values of 8 and 11.7, respectively. The error amplifiers (A_1, A_2) adopt the folded-cascode topology, as illustrated in Fig. 4.6.3 (top-left). The error amplifiers are chopped at 10KHz to remove the TC nonlinearity arising from the amplifier offset [6]. Note that the chopper technique also suppresses 1/f noise from error amplifiers. To originate T_{X1} and T_{X2}, the main current is mirrored to three copies with width ratios m=1.175 and n=0.825. These currents are applied to identical resistive loads (R_{T1-3}), and their corresponding IR drop voltages (V_1, V_2, and V_3) are compared with dedicated comparators realizing the two-bit temperature sensor for T_{X1_SEL} and T_{X2_SEL} generation, as shown in Fig. 4.6.3 (top-right). The hysteresis comparator architecture is employed to mitigate noise-induced flickering of the comparator output as the temperature approaches T_{X1} or T_{X2} [7]. The hysteresis window meets the noise immunity requirements, minimizing impact on TCs of both VR and CR. Figure 4.6.3(middle-left) presents the control logic of seamless sub-ranging operation; the PTAT, CTAT weight factors (a, b, c, d) are decoded to determine the number of activated PTAT and CTAT current units in the seamless pair A~D. T_{X1_SEL} and T_{X2_SEL} control MUX selection to achieve independent controls of three regions for VR and CR. The implementation of the seamless pair is presented in Fig. 4.6.3(bottom). The width ratios of the current mirror units correspond to the previously denoted m and n values. The current mirror array is realized using a hybrid encoding scheme, where the lower 4 bits and upper 8 bits employ binary and thermometer coding, respectively. This hybrid coding scheme, combined with the elongated channel length of mirror transistors, improves the mismatch across all arrays.

The proposed 3rd-order TC-compensated VCR is fabricated in a 65nm CMOS process. It occupies 0.16mm^2 as shown in Fig. 4.6.7, and consumes 58µA from a 2V supply at 25°C. The power consumption from VR and CR is 40µA and 42µA, respectively, when their counterparts are off. The shared circuitry accounts for 60% and 41% of the total area and power consumption, respectively. Figure 4.6.4 (top) demonstrates that the measured best TCs of V_{REF} and I_{REF} reach 1.8ppm/°C and 4.8ppm/°C, reduced by 82% and 83% compared to their raw values without sub-ranging compensation. T_{X1_SEL} and T_{X2_SEL} transit from 0 to 1 at 12°C and 65°C (Fig. 4.6.4(bottom-left)). Within a supply voltage range of 1.5V to 2.5V, the line regulations of I_{REF} and V_{REF} are 0.16%/V and 0.15%/V, respectively.

Figure 4.6.5 presents the measurement result of V_{REF} and I_{REF} from 20 chips in three process corners (TT:12, FF:4, SS:4). These chips are measured over a temperature range of -40 to 125°C, with a supply voltage of 2V and a load voltage of 0.8V for current reference. After the weight factor (a_{1-2}, b_{1-2}, and c_{1-2}, d_{1-2}) for all chips is determined through batch trimming, the average TCs of V_{REF} and I_{REF} reach 5.7ppm/°C and 9.1ppm/°C with the proposed CMST 3rd-order sub-ranging compensation, respectively. Across the 20 samples, the µ and σ/µ of V_{REF} are 1000mV and 0.35%, respectively, while the µ and σ/µ of I_{REF} are 18.25µA and 0.44% respectively.

Figure 4.6.6 compares the proposed low-TC hybrid voltage/current reference with prior works. The proposed V&CR scheme facilitates a non-sequential voltage and current reference generation scheme, thus allowing individual compensation for VR and CR and simultaneously a low TC characteristic for both references. Furthermore, it also enables individual power-down control of VR and CR, which is unique compared with prior V&CR designs. Figure 4.6.7 shows the chip micrograph.

Acknowledgement:

This work was supported in part by the Guangdong-Hong Kong-Macau Joint Laboratories (GDSTC) by FDCT under Grant File No.: 001/2024/COP and Grant File No.: 004/2023/SKL, and in part by the university of Macau Research under Grant MYRG-GRG2023-00100-IME.

References:

[1] Y. Ji, C. Jeon, H. Son, B. Kim, H. -J. Park and J. -Y. Sim, " A 9.3nW All-In-One Bandgap Voltage and Current Reference Circuit," *ISSCC*, San Francisco, CA, USA, 2017. https://doi.org/10.1109/ISSCC.2017.7870280

[2] I. -F. Lin, Y. -C. Tsai, H. -L. Lin and Y. -T. Liao, "An 18-nW CMOS Current and Voltage Reference Circuit with Low Line Sensitivity and Wide Temperature Range," in *IEEE SSCL*, vol. 7, pp. 179-182, 2024. https://doi.org/10.1109/LSSC.2024.3407583

[3] L. Wang and C. Zhan, "A 0.7-V 28-nW CMOS Subthreshold Voltage and Current Reference in One Simple Circuit," in *IEEE TCAS-1*, vol. 66, no. 9, pp. 3457-3466, Sept. 2019. https://doi.org/10.1109/TCSI.2019.2927240

[4] P. Park, J. Lee and S. Cho, "A PVT-Insensitive Sub-Ranging Current Reference Achieving 11.4-ppm/ ° C From -20 °C to 125 °C," in *IEEE JSSC*, vol. 59, no. 12, pp. 4057-4067, Dec. 2024. https://doi.org/10.1109/JSSC.2024.3450950

[5] B. Yousefzadeh, S. Heidary Shalmany and K. A. A. Makinwa, "A BJT-Based Temperature-to-Digital Converter with ±60 mK (3 σ) Inaccuracy From –55 °C to +125 °C in 0.16-µm CMOS," in *IEEE JSSC*, vol. 52, no. 4, pp. 1044-1052, April 2017. https://doi.org/10.1109/JSSC.2016.2638464

[6] G. Ge, C. Zhang, G. Hoogzaad and K. A. A. Makinwa, "A Single-Trim CMOS Bandgap Reference with a 3σ Inaccuracy of ±0.15% from –40°C to 125°C," in *IEEE JSSC*, vol. 46, no. 11, pp. 2693-2701, Nov. 2011. https://doi.org/10.1109/JSSC.2011.2165235

[7] C. -W. U, C. Liu, R. P. Martins and C. -S. Lam, "A 1 V Supply, 740 nW, 8.7 ppm/°C Bandgap Voltage Reference with Segmented Curvature Compensation," in *IEEE TCAS-I*, vol. 70, no. 12, pp. 4755-4766, Dec. 2023. https://doi.org/10.1109/TCSI.2023.3301736

[8] J. Lee and S. Cho, "A 1.4-µW 24.9-ppm/°C Current Reference with Process-Insensitive Temperature Compensation in 0.18-µm CMOS," in *IEEE JSSC*, vol. 47, no. 10, pp. 2527-2533, Oct. 2012. https://doi.org/10.1109/JSSC.2012.2204475

[9] K. Chen, L. Petruzzi, R. Hulfachor and M. Onabajo, "A 1.16-V 5.8-to-13.5-ppm/°C Curvature-Compensated CMOS Bandgap Reference Circuit with a Shared Offset-Cancellation Method for Internal Amplifiers," in *IEEE JSSC*, vol. 56, no. 1, pp. 267-276, Jan. 2021. https://doi.org/10.1109/JSSC.2020.3033467

[10] C. -W. U, M. -K. Law, R. P. Martins and C. -S. Lam, "Sub-µW Auto-Calibration Bandgap Voltage Reference with 1σ Inaccuracy of ± 0.12% Within - 40°C to 120°C," in *IEEE JSSC*, vol. 59, no. 2, pp. 540-550, Feb. 2024. https://doi.org/10.1109/JSSC.2023.3294996

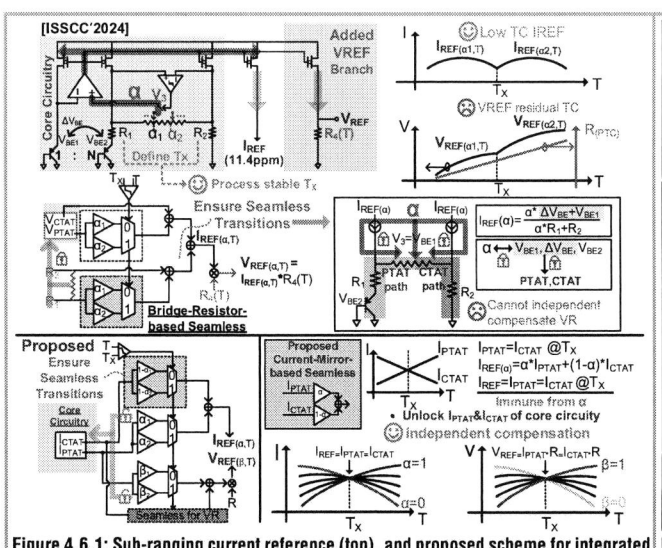

Figure 4.6.1: Sub-ranging current reference (top), and proposed scheme for integrated sub-ranging voltage/current reference (bottom).

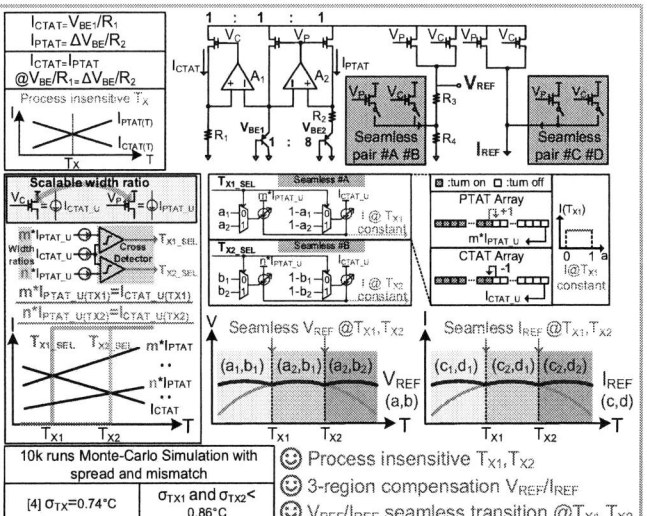

Figure 4.6.2: The overall architecture of the 3rd-order sub-ranging voltage/current reference.

ISSCC 2026 / SESSION 4 / ANALOG TECHNIQUES & AMPLIFIERS / 4.6

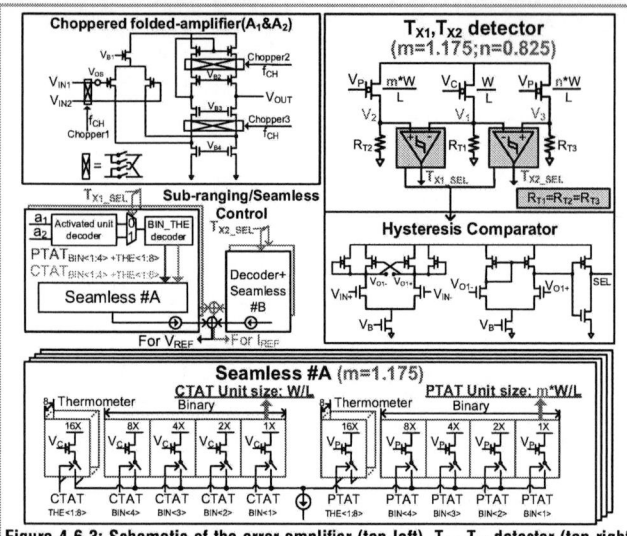

Figure 4.6.3: Schematic of the error amplifier (top-left), T_{X1}, T_{X2} detector (top-right), seamless/sub-ranging control blocks (middle-left), and seamless pair (bottom).

Figure 4.6.4: Measured I_{REF} with and without sub-ranging (top-left), V_{REF} with and without sub-ranging (top-right), T_{X1_SEL}, T_{X2_SEL} (bottom-left), and line regulation of I_{REF} and V_{REF} (bottom-right).

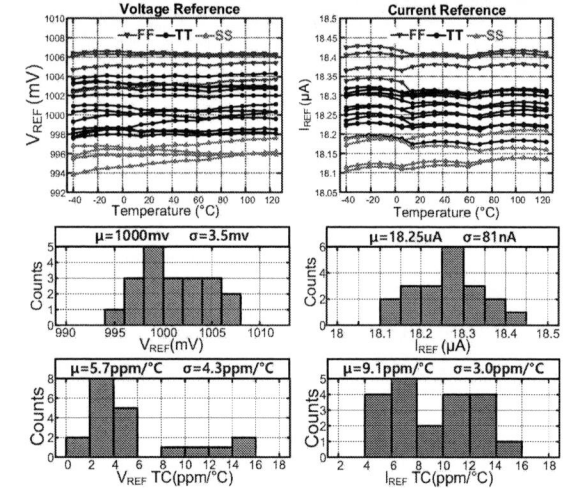

Figure 4.6.5: Measured I_{REF} and V_{REF} variation across 20 chips from 3 process corners: measured V_{REF} (top-left), measured I_{REF} (top-right), histograms of the average V_{REF} and its TC (bottom-left), and histograms of the average I_{REF} and its TC (bottom-right).

	ISSCC' 17 [1]	SSCL 24 [2]	JSSC 24 [4]	JSSC 12 [8]	JSSC 21 [9]	JSSC 24 [10]	This Work
Process(nm)	180	180	180	180	130	65	65
Type	V&CR	V&CR	CR	CR	VR	VR	V&CR
V&CR Generation Scheme	Sequential	Sequential	/	/	/	/	Parallel
V_{DD}(V)	1.3-1.8 [a]	0.8-1.8	1.3-2.4	1.2-3	2-3.3	1.2-2.5	1.5-2.5
Supply Current(µA)	N.A (9.3nW)	N.A (18.51nW)	43.5	27 [c]	120	2.18 [c]	58
Temperature Range(°C)	0-110	-40-130	-20-125	0-100	-40-150	-40-120	-40-125
Best/Average I_{REF} TC(ppm/°C)	283/ N.A	N.A/ 264	7.8/ 11.4	18.4/ 24.9	/	/	4.8/ 9.1
Best/Average V_{REF} TC(ppm/°C)	26/ N.A	N.A/ 124	/	/	5.7/ 8.75	18.6/ 22.3	1.8/ 5.7
Output Current (µA)	0.00664	0.02043	10.3	7.81	/	/	18.25
Output Voltage (mV)	1238	270	/	/	1160	1001.1	1000
I_{REF} Line regulation(%/V)	1.16	0.094	0.036	0.13	/	/	0.16
V_{REF} Line regulation(%/V)	0.08	0.011	/	/	0.03	0.126	0.15
TC Trimming	YES [b]	Batch Trimming	Batch Trimming	No	No	Auto Trimming	Batch Trimming
PSRR of VR (dB@Hz)	-46@100	-66@100	/	/	-82@10	-58@10	-65@10
Samples	10	13	45	10	7	10	20
Area(mm²)	0.055	0.025	0.08	0.123 [c]	0.08	0.04	0.16

a: Estimated from figure
b: Trimmed but the method is not available
c: Including all required blocks

Figure 4.6.6: Performance summary and comparison.

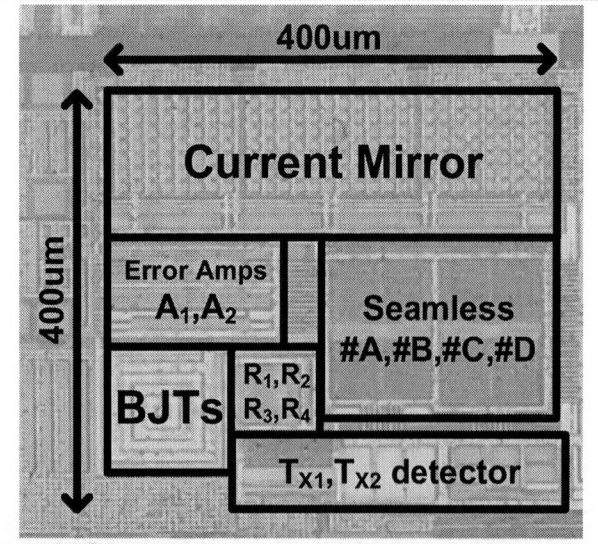

Figure 4.6.7: Chip micrograph.

• 2026 IEEE International Solid-State Circuits Conference

Session 5 Overview: *Sub-THz and mm-Wave Phased Arrays and Beamformers*

WIRELESS SUBCOMMITTEE

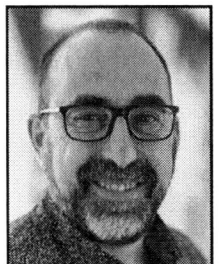

Session Chair: José Luis González-Jiménez
CEA-Leti
Grenoble, France

Session Co-Chair: Hao Gao
Southeast University
Nanjing, China

This session presents four papers addressing innovations in phased arrays and beamformers for mm-wave communications. The first paper demonstrates a TRX IC for picosatellite collective phased-arrays operating in the 1.7-to-1.8GHz band. The second paper is a phased-array TX operating at 28GHz that extends signal interception resilience with eavesdroppers in the same beam as the wanted user. The third paper presents a 60GHz dual-surface active TRX relay with +/-45° beam-steering capabilities that can operate as a reflective or transmissive intelligent surface, and support bidirectional links as well. Finally, the last paper of the session demonstrates a single channel TRX operating at D-band featuring an on-chip dual-polarization antenna and a double-polarization meta-surface supporting full-duplex bi-directional links.

ISSCC 2026 / February 16, 2026 / 1:30 PM

1:30 PM

5.1 A Formation Flight Phased-Array Transceiver for Spatial Power Combining and Distributing Architectures in Direct-to-Device-Communication Satellite Constellations

Keito Yuasa, Institute of Science Tokyo, Tokyo, Japan

In Paper 5.1, The Institute of Science of Tokyo in collaboration with Microwave Factory, Interstellar Technologies and Iwate University introduces a novel FDD transceiver architecture for large-scale, formation-flying satellite arrays that wirelessly receive a common LO signal from a gateway satellite. The 65nm IC integrates a receiver and a transmitter operating in the 1.7GHz band for the up-link and in the 1.8GHz band for the downlink, consuming 206mW and 135mW, respectively. The proposed transceiver enables OTA 256-QAM signals as LTE.

1:55 PM

5.2 A 28GHz Frequency-Diverse Sub-Array TX with Secret Phase Keys and Antenna Subset Modulation for Eavesdropping-Resilient Wireless Communication

Qiang Zhou, Rice University, Houston, TX

In Paper 5.2, the Rice University team demonstrates for the first time a phased-array transmitter circuit that extends existing techniques with a key-based frequency-diverse antenna-switching sequence that effectively suppresses the possibility of signal interception for eavesdroppers placed in the desired user beam. The 22nm CMOS SOI IC operates in the 28GHz band and incorporates 12 channels sharing the same input LO and low IF inputs. They are integrated in a PCB implementing a linear phased-array antenna. Each channel can radiate up to 15.5dBm at saturation. A 0.3Gb/s link using 64-QAM signal is demonstrated with close to 100% of EVM for the eavesdropper compared to the unaffected desired user.

2:20 PM

5.3 SPARTA: A Scalable, Programmable and Active mm-Wave Dual-Surface Reflect/Transmit Array with Integrated Gain and Phase Control Allowing Bidirectional Signal Routing Capability for Robust and Reconfigurable mm-Wave Networks at 60GHz

Muhamed Fouad Allam, Princeton University, Princeton, NJ,

In Paper 5.3, Princeton University presents an active relay array able to spatially route an incident signal from any angle to any other angle, acting as a reflective or transmissive surface. The dual-surface array can also operate in a bidirectional mode. The dual surface is composed of a couple of linear arrays of 23 ICs, each one integrating a dual-channel amplify-and-forward chain with 2-stage LNA, 360 degree 8-bit digitally controlled phase shifters, and a 3-stage PA, consuming 336mW each and offering up to 7.7dBm of saturated output power. In board splitters and through-pcb-substrate vias allow to redirect the mm-wave signals from the antennas of one side of the PCB to the circuits on the other side, enabling reconfiguring the surface as a reflecting array or as a transmission array. It supports 0.5Gb/s in either transmission or reflection mode over a double 2m distance link.

2:45 PM

5.4 A 140GHz Full-Duplex CMOS Transceiver with Metasurface-Integrated Self-Interference-Cancelling Antenna Supporting 16Gb/s 16-QAM Dual-Mode Bidirectional Communication

Jiawei Yang, Southeast University, Nanjing, China, Purple Mountain Laboratories, Nanjing, China

In Paper 5.4, Southeast University and Purple Mountain Laboratories demonstrate a full-duplex link using the same antenna for both directions. The link is based on a single-channel transceiver IC with integrated antenna incorporating self-interference techniques, combined with a meta-surface to increase the antenna gain. The IC, fabricated in a 40nm CMOS process, features a dual polarization on-chip antenna with more than 49dB of self-interference cancellation over a band from 132 to 150GHz. A link in FD mode with 16Gb/s data-rate based on 16-QAM modulation is demonstrated over 1m.

DIGEST OF TECHNICAL PAPERS • 89

979-8-3315-8937-0/26 $31.00 © 2026 IEEE

ISSCC 2026 / SESSION 5 / SUB-THz AND mm-WAVE PHASED ARRAYS AND BEAMFORMERS / 5.1

5.1 A Formation Flight Phased-Array Transceiver for Spatial Power Combining and Distributing Architectures in Direct-to-Device-Communication Satellite Constellations

Keito Yuasa[1], Yuya Takahashi[1], Shunya Watanabe[1], Sena Kato[1], So Ema[2], Genma Hattori[2], Atsushiro Naka[3], Sumio Morioka[3], Takahiro Inagawa[3], Kentaro Murata[4], Naoki Honma[4], Jill Mayeda[1], Atsushi Shirane[1]

[1]Institute of Science Tokyo, Tokyo, Japan, [2]Microwave Factory, Kanagawa, Japan, [3]Interstellar Technologies, Hokkaido, Japan, [4]Iwate University, Iwate, Japan

Abstract

Conventional satellites are costly and unreliable. We propose a pico-satellite formation as a virtual phased array to reduce costs and improve reliability. A low-power FDD transceiver wirelessly receives an LO signal from a gateway, eliminating power-hungry PLLs. Precise beamforming is achieved by calibrating path differences and using SRFA technology. This work enables future low-cost, highly reliable satellite communications.

Conventional satellite communication has relied on large, monolithic satellites orbiting in Low Earth Orbit (LEO) at altitudes of 500 km (Fig. 5.1.1). Notably, systems like BlueWalker, developed by AST, require large deployable 8m×8m antennas to achieve sufficient gain. This leads to extremely high launch costs and creates a vulnerable single point of failure. If a failure occurs, the entire satellite can become non-operational, as replacing damaged parts is impossible.

To overcome these limitations, satellite communication systems based on large-scale phased arrays composed of tens of thousands of pico-satellites in a formation flight in Fig. 5.1.1 have been proposed. This formation-flying concept creates a massive antenna aperture by coherently combining signals in space from numerous small, low-cost satellite elements, without any electrical connections. Specifically, the proposed phased array consists of more than 15,000 pico-satellites in a formation flight at an altitude of 500km with an equivalent aperture area of 13m×13m or larger.

As summarized in the table in Fig. 5.1.1, this approach offers significant advantages in launch cost and robustness. First, the distributed nature of the system allows for partitioned launches using multiple small, inexpensive rockets. Furthermore, the system is highly resilient to failures. If a satellite malfunctions, the array can be reconfigured to compensate for the loss, ensuring system continuity.

Phased arrays in formation flights operation relies on a gateway satellite (GS) to act as a relay for the feeder link. In the uplink (UL) operation, an RF signal at 1.7GHz Band 3 as LTE from a user equipment (UE), such as a smartphone, is received by the pico-satellites. Concurrently, the GS transmits a 24GHz LO signal to the pico-satellites, which upconverts the received signal to a 25.7GHz IF for the uplink. This IF signal is then transmitted from each pico-satellite toward the GS. When the IF signals from all pico-satellites are coherent, they are power-combined in space at the GS's receiver. The IF signal received by the GS is then converted in the Q/V band and sent to the base station (BS) via the feeder link. Conversely, in the downlink (DL) operation, data transmitted in the Q/V band is converted to 25.8GHz by the GS and then sent to the pico-satellites. Similar to the UL operation, a 24GHz LO signal is also transmitted from the GS. In this case, power is distributed from the GS, enabling each pico-satellite to receive the signal. Consequently, the received IF signal is downconverted to a 1.8GHz RF signal as LTE. This downconverted RF signal is then beamformed towards the UE on the ground.

To realize the proposed system, it is necessary to perform power combining and distribution in space among tens of thousands of satellites without any physical connections. This requires achieving precise phase control between the pico-satellites and with the GS. A key challenge in formation flight is that for each wirelessly linked satellite, significant phase errors arise from propagation path differences between the IF and LO signals. However, as shown in Fig. 5.1.2, a conventional transceiver integrates a dedicated VCO and PLL on each satellite to generate its local signal. This approach requires tens of thousands of PLLs, resulting in enormous power consumption for the entire system. Furthermore, the frequency offsets and phase noise from each PLL make strict frequency synchronization among satellites impossible, thus hindering coherent beamforming.

The proposed formation-flying system achieves high-precision frequency and phase synchronization among tens of thousands of satellite transceivers with low power consumption by adopting an in-space wireless combining method where the local oscillator (LO) signal is transmitted from the GS. This architecture obviates the need for complex and heavy inter-satellite links for phase synchronization, contributing to lower power consumption and reduced weight for each satellite element. The 24GHz LO and 25.8GHz IF signals transmitted from the GS are received by a common antenna and separated by an on-chip divider. By supplying this external LO (24GHz) to the mixers in both the UL and DL paths, we eliminate the need for a PLL on each satellite, achieving significant power reduction.

Figure 5.1.3 describes the technique for generating high-quality circular polarization (CP) and the high-precision beamforming method using Over-the-Air (OTA) calibration.

The conventional method of generating it with a single element requires precise control of the element's amplitude and phase. This approach suffers from the degradation of the Axial Ratio (AR), especially at wide angles.

To address this challenge, we propose a Sequential Rotation Feed Array (SRFA) technique with integrated IC to control the phase shift and combining the power in space that uses a 2×2 subarray as a fundamental unit in Fig. 5.1.3. For instance, to transmit Right-Hand Circular Polarization (RHCP), the four elements are fed with signals progressively phase-shifted by -90°. For Left-Hand Circular Polarization (LHCP), the phase progression is reversed. This technique achieves a good axial ratio of under 3dB for RHCP over a wide scan angle of ±40° without requiring precise control of individual elements. This ensures stable communication quality even during beam steering.

In the proposed system, the propagation path lengths for the LO/IF signals from the GS to each satellite vary, resulting in phase offsets (φLOn, φIFn) in the signals received at each element, so first the distance between the GS and satellite array is fixed and then OTA measurements are performed by transmitting the reference LO and IF signals from the GS. The combined phase difference of the LO and IF paths up to the mixer, (φLO + φIFn), is measured for each element. This measurement creates a phase error map that accounts for the different propagation path lengths and φpsn, the n-th element phase shift is set based on the following equation:

$$\varphi psn = \varphi SRFA + \varphi Beam + \varphi LO + \varphi IFn$$

where φSRFA is the phase term for SRFA-based circular polarization, φBeam is the phase progression required for the desired beam angle, φLOn for the LO path phase, φIFn for the IF path phase. Through this process illustrated in Fig. 5.1.3, path-induced phase differences from the in-space distribution of the LO and IF signals can be compensated, while also applying the required phase shifts for beamforming. As a result, high-precision beamforming is achieved across the entire system.

The design of the mixer and its peripheral circuitry is key to achieving low power consumption and miniaturization in the proposed transceiver. In particular, since the power of the LO signal supplied wirelessly from an external source is limited, a highly efficient mixer that operates with extremely low LO input power is essential. Figure 5.1.4 shows the architecture of the UL/DL mixers proposed in this work. For the UL mixer, we adopted a passive mixer, which offers excellent low-distortion characteristics. For the DL mixer, we employed a passively operating self-heterodyne mixer.

To use the transmitted LO signals, the UL and DL systems utilize different LO paths. For the UL, an on-chip LC-based directional coupler separates the LO and IF signal paths. In the UL path, this directional coupler separates the 25.7GHz IF signal generated by the mixer and amplified by the IF amplifier from the 24GHz LO signal supplied by a high-gain three-stage LO amplifier, achieving high isolation. This allows the LO and IF amplifier configurations to be independently optimized, enabling significant reductions in circuit area and power consumption.

The directional coupler achieves high isolation of 30dB at the target frequencies. This ensures high isolation between the mixer's LO leakage (P1) and the coupler's port (P3) that feeds into the three-stage amplifier. For isolation against the 25.7GHz signal leaking into the DL path, the power divider achieves a wideband isolation of -34.4dB (Fig. 5.1.4). This successfully suppresses potential oscillations within the circuit.

For the DL path, the 25.8GHz IF and 24GHz LO signals share the same path. The two-stage amplifier is designed to amplify both of these frequency bands, eliminating the need for separate IF and LO amplifiers. This reduces the number of components and reduces the power consumption. The DL mixer achieves a high conversion gain of over 30dB with a low LO input power range of -29dBm to -20dBm, demonstrating the low-power LO requirement for this system.

Figure 5.1.5 shows the measured UL and DL beam patterns and phase shift characteristics. The results confirm that the beam direction can be electronically steered by digitally controlling the on-chip 8-bit phase shifters. Good beam steering is observed over a wide range of ±25° for both uplink and downlink. This demonstrates that by using phase calibration to compensate for phase errors caused by IF/LO path differences, the beam direction can be effectively controlled. This successfully demonstrates in-space power combining for the UL and in-space power distribution for the DL.

Next, the communication performance of the system is presented. A 15MHz bandwidth 64-QAM and 256-QAM modulation signals of LTE were used. The Error Vector Magnitude (EVM) achieved values of -28.3dB (256QAM) for the UL, and -29.7dB (256QAM) for the downlink. These results prove that the proposed in-space wireless combining scheme and transceiver architecture are fully applicable for high-speed data communication using modulated signals.

Figure 5.1.6 presents a table comparing the performance of this work with other state-of-the-art LTE transceivers and Sub-6GHz phased array TRX. This is the fully integrated phased-array transceiver for pico-satellites intended for formation flight. Despite its small IC area of 1.73×2.29mm² and low power consumption of 206mW (UL) and 135mW (DL), the proposed transceiver enables OTA 256-QAM signals as LTE. Furthermore, the IC supports precise beamforming, integrating a 1TX/1RX chain equipped with a high-resolution 8-bit phase shifter. Additionally, by supplying the LO from an external source, we frequency-converted the RF signal to a K-band IF signal and achieved a successful demonstration of its in-space combining and distribution.

Acknowledgement:
This work is partially supported by the MIC (JPJ000254), MIC/SCOPE (JP192203002, JP192103003, JP235003011), and VDEC in collaboration with Cadence Design Systems, Inc., Mentor Graphics, Inc., and Keysight Technologies Japan, Ltd.

References:
[1] Z. Xu, G. Chen, R. Fernandez, Y. Gao and R. Tafazolli, "Enhancement of Direct LEO Satellite-to-Smartphone Communications by Distributed Beamforming," in *IEEE Transactions on Vehicular Technology*, vol. 73, no. 8, pp. 11543-11555, Aug. 2024. https://doi.org/10.1109/TVT.2024.3379017
[2] R. Deng, B. Di and L. Song, "Ultra-Dense LEO Satellite Based Formation Flying," in *IEEE Transactions on Communications*, vol. 69, no. 5, pp. 3091-3105, May 2021. https://doi.org/10.1109/TCOMM.2021.3058370
[3] N. Klemmer *et al.*, "A 45nm CMOS RF-to-Bits LTE/WCDMA FDD/TDD 2×2 MIMO base-station transceiver SoC with 200MHz RF bandwidth," *ISSCC*, San Francisco, CA, USA, 2016, pp. 164-165. https://doi.org/10.1109/ISSCC.2016.7417958
[4] K. Lim *et al.*, "A 65-nm CMOS 2×2 MIMO Multi-Band LTE RF Transceiver for Small Cell Base Stations," in *IEEE JSSC*, vol. 53, no. 7, pp. 1960-1976, July 2018. https://doi.org/10.1109/JSSC.2018.2824300
[5] X. Lei *et al.*, "A 4 × 4 5–6-GHz Analog Beamforming Wi-Fi Transceiver Front End for Fiber to the Room in 55-nm CMOS," in *IEEE TMTT*, Vol 73, Issue 9, Sept. 2025. https://doi.org/10.1109/TMTT.2025.3565564
[6] Jeong, J.-C., Shin, D., Ju, I. and Yom, I.-B. (2013), "An S-Band Multifunction Chip with a Simple Interface for Active Phased Array Base Station Antennas," *ETRI Journal*, 35: 378-385. https://doi.org/10.4218/etrij.13.0112.0400

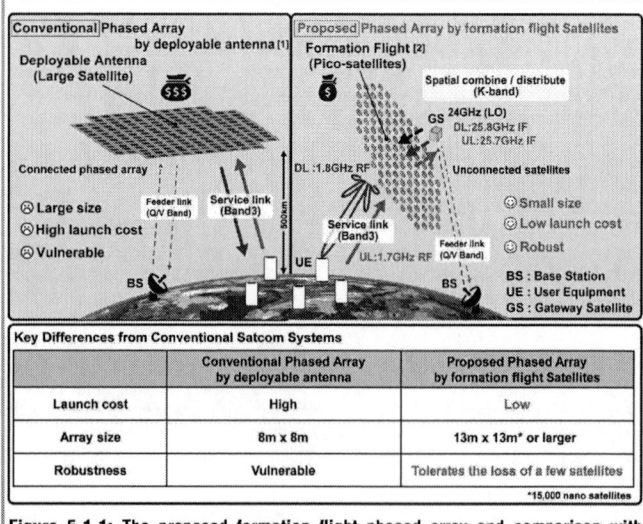

Figure 5.1.1: The proposed formation flight phased array and comparison with conventional deployable phased array.

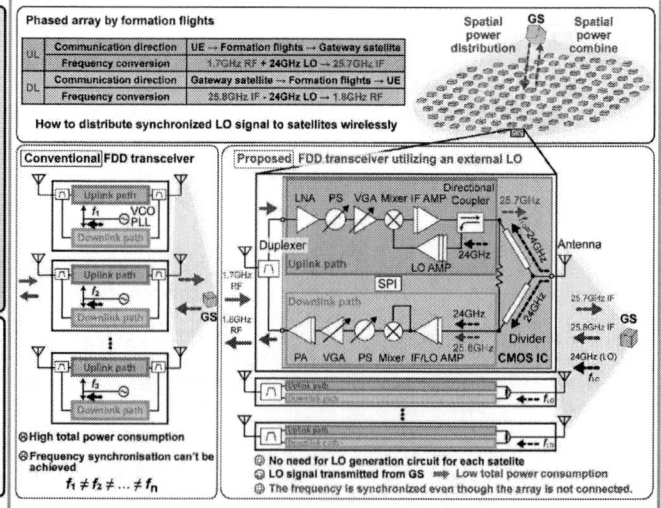

Figure 5.1.2: A transceiver architecture for satellite formation flight utilizing an external LO signal.

ISSCC 2026 / SESSION 5 / SUB-THz AND mm-WAVE PHASED ARRAYS AND BEAMFORMERS / 5.1

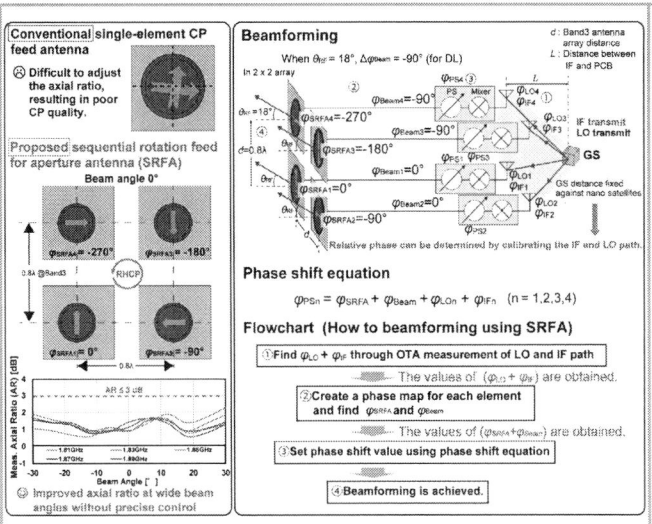

Figure 5.1.3: Beamforming using an antenna array with sequential rotation feed in formation flight satellites.

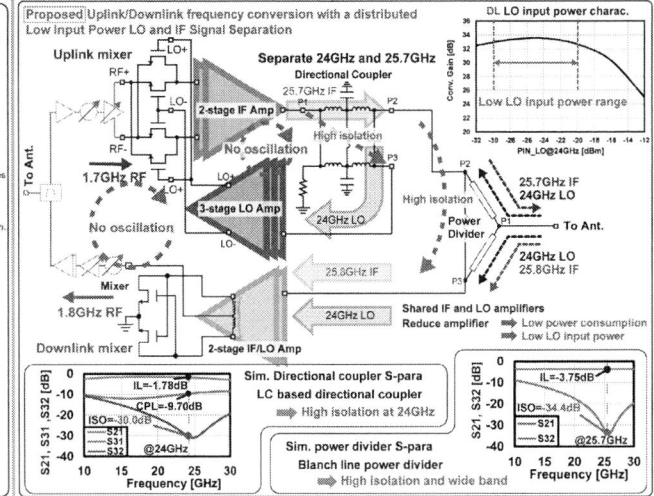

Figure 5.1.4: Proposed up and downlink transceiver architectures for a distributed low LO input power, IF separation, compact design and low-power communication systems.

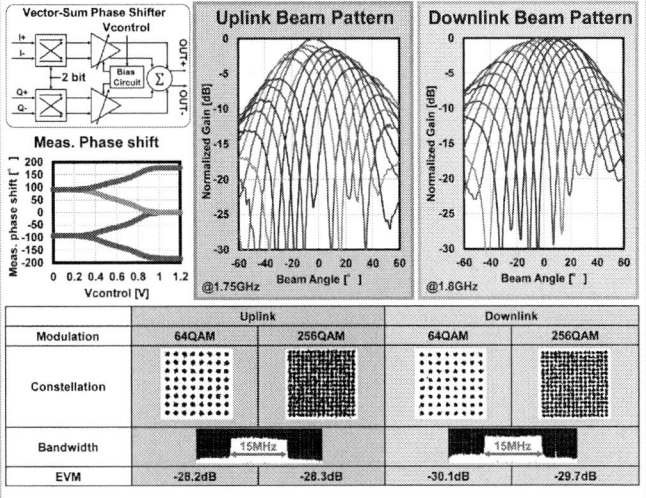

Figure 5.1.5: Measured OTA beam pattern, EVM, and constellations for 2×2 array module.

		This work	ISSCC 2016[3]	JSSC 2018[4]	TMTT 2025[5]	ETRI 2013[6]
Process		65nm CMOS	45nm CMOS	65nm CMOS	55nm CMOS	0.5um pHEMT
Application		Satcom	LTE/WCDMA	LTE	Wi-Fi	LTE
Operation Band		UL:1.7GHz DL:1.8GHz	2.6GHz (0.4-4GHz)	1.4-2.7GHz(MHB) (0.68-6GHz))	5-6GHz	1.8-3.2GHz
Bandwidth		15/20MHz	200MHz	10MHz	40/80/160MHz	N/A
Number of TRX		1TX/1RX	2TX/2RX	16TX/16RX	4TX/4RX	1TRM
Die Area		1.73x2.29mm²	13x13mm²	5x5.6mm²	3.15x5.85mm²	4x4mm²
Supply voltage		1.2V	N/A	N/A	1.2/2.5V	5V
Uplink	Conv. Gain	18.6dB	N/A	Max >93dB Min <-2dB	33.1dB	26±0.5dB
	NF	4.8dB	2dB	3.2dB	4.3dB	4dB
	IIP3	-25.4dBm	-4dBm	> -2dBm	12.1dBm	N/A
	EVM	-28.4dB*	N/A	< -31.1dB** @-50dBm	-48.8dB**	N/A
	Power cons.	206mW	200mW	< 480mW	108mW	500mW(TX or RX)
Downlink	Conv. Gain	29.8dB	N/A	N/A	27.5dB	26±0.5dB
	Psat	0dBm	N/A	N/A	21.8dBm	12dBm***
	Pout	-5dBm	0dBm	-0.7dBm	27dBm	N/A
	EVM	-29.6dB*	-46.0dB**	< -31.1dB**	-45dB**	N/A
	Power cons.	135mW	500mW	< 750mW	560mW 140mW/1path	500mW(TX or RX)
Phase Shift		RX:8bit, 1.42° TX:1.41°	N/A	N/A	4bit	6bit
Frequency Conversion		UL IF : 25.7GHz DL IF : 25.8GHz	Baseband	Baseband	N/A	N/A
LO Supply Method		Free space distribute 24GHz	IC Internal PLL 0.4-4GHz	IC Internal PLL 5.4-12GHz	N/A	N/A

TRM : transmit / receive module *OTA measurement **Conducted measurement ***P1dB

Figure 5.1.6: Performance comparison with state-of-the-art works.

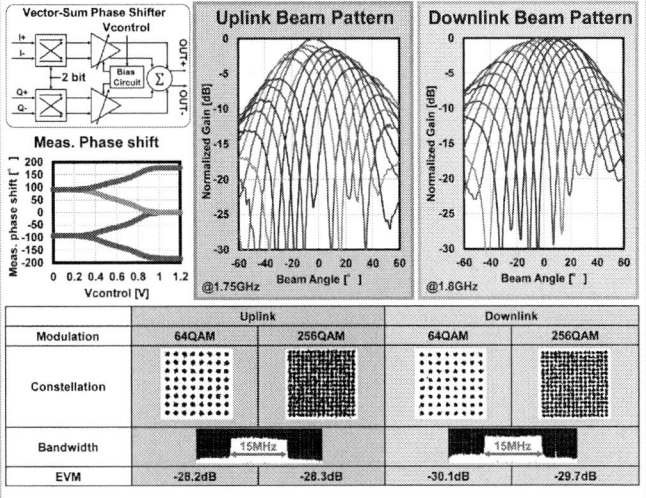

Figure 5.1.7: Die micrograph.

• 2026 IEEE International Solid-State Circuits Conference

ISSCC 2026 / SESSION 5 / SUB-THz AND mm-WAVE PHASED ARRAYS AND BEAMFORMERS / 5.2

5.2 A 28GHz Frequency-Diverse Sub-Array TX with Secret Phase Keys and Antenna Subset Modulation for Eavesdropping-Resilient Wireless Communication

Qiang Zhou, Yumin Su, Hao Guo, Yaolong Hu, Kaiyuan Yang, Taiyun Chi

Rice University, Houston, TX

Abstract

Prior works on physical-layer eavesdropping protection only addressed off-axis Eve. This paper presents the first TX to defend against both main-beam and off-axis Eves. Incorporating three key components: frequency-diverse sub-arrays, secret phase keys, and

antenna subset modulation, the 28GHz TX demonstrates highly scrambled constellations for all Eves, while the legitimate RX's reception remains intact (~6% EVM$_{rms}$), with only 2% DC power overhead compared to a standard phased-array TX.

Physical-layer security has attracted growing interest within the wireless communication community [1-14]. It can be either combined with conventional cryptography for enhanced, "two-factor" defenses [15] or used to reduce encryption workload in energy-constrained systems [16,17]. The latter is particularly attractive for mm-wave communication, where real-time, multi-Gb/s encryption/decryption incurs significant power consumption and latency overhead [18].

Eavesdropping is one of the most common security threats in wireless communication. To mitigate such attacks, several physical-layer security techniques have been proposed [6-10]. One example is antenna subset modulation (ASM) [6], which randomizes the active antenna selection for each transmitted symbol (e.g., antennas #1, #4 for symbol S_1 and #2, #3 for symbol S_N) (Fig. 5.2.1, top left). This randomization scrambles the received I/Q constellation for eavesdroppers (Eve) placed off-axis (i.e., outside the intended direction) while preserving signal fidelity for the legitimate receiver (Bob). Related techniques [7,8] have demonstrated similar intentional off-axis distortion (Fig. 5.2.1, top middle).

Despite these advances, existing methods suffer from a common limitation: they only address "off-axis eavesdropping" but remain vulnerable to adversaries located in the main-beam direction. In fact, such "main-beam eavesdropping" attack has been successfully demonstrated in [19] (Fig. 5.2.1, top right), where a small reflector placed in the main-beam direction deflects a small portion of the signal toward a high-gain Eve, allowing Eve to fully recover Alice's information without disrupting Bob's reception.

This work aims to develop a physical-layer security approach capable of defending against both main-beam and off-axis Eves, enabling secure communication only to Bob, located in the intended direction and at the intended distance. To achieve this, we introduce three key components: (1) frequency-diverse sub-array (FDSA), (2) secret phase keys, and (3) ASM. In our TX architecture (Fig. 5.2.1, bottom), the baseband or IF input is first split into K sub-arrays. Within each sub-array, the signal is modulated onto N equally spaced carrier frequencies (f_1, $f_2 = f_1 + \Delta f$, ..., $f_N = f_{N-1} + \Delta f$), creating a multi-tone signal, similar to frequency diverse array radars [20,21]. Next, secret phase keys (φ), which are implemented as distinct phase shifts at $f_1 - f_N$ and shared only between Alice and Bob, are applied within each sub-array. The resulting signals are then transmitted through antennas, which are randomly activated on a per-symbol basis through ASM.

On the receiver side, the received signal is the superposition of multi-tone signals transmitted from all sub-arrays (Fig. 5.2.2, top left). The t-th received symbol at frequency f_n (denoted as $S_{RX,t,n}$) can be expressed as $S_{RX,t,n} = S_t \cdot \sum_{k=1}^{K} a_{t,k,n} e^{j\varphi_{k,n}}$ where S_t is the t-th transmitted symbol, $a_{t,k,n}$ indicates n-th antenna state in k-th sub-array (1 for on, 0 for off via ASM), and $\varphi_{k,n}$ is the applied secret phase key. For each symbol, M out of N antennas are active per sub-array, i.e., $\sum_{n=1}^{N} a_{t,k,n} = M$, with independent antenna selections across sub-arrays.

For Bob, signal recovery is straightforward (Fig. 5.2.2, top right): similar to how an OFDM receiver processes a multi-tone signal [22], he first performs FFT-based downconversion. Next, using the pre-shared keys, he applies a phase compensation factor $\zeta_n = K/\sum_{k=1}^{K} e^{j\varphi_{k,n}}$ to each bin. Finally, he coherently combines the phase-corrected signals, resulting in an accurate reconstruction of the original transmitted symbol.

In contrast, Eve's summation is incoherent, leading to scrambled constellations (Fig. 5.2.2, middle right). The distortion is quantified by an error factor $\alpha(t,\varphi)$, which arises from the secret phase keys (shared only with Bob) and the random antenna selections across symbols. To overcome this distortion, with P-bit phase keys, Eve would need to search a brute-force space of $2^{P \cdot K \cdot N}$ possibilities.

It is important to note that all three components: FDSA, secret phase keys, and ASM, are critical to achieving eavesdropping resilience (Fig. 5.2.2, bottom left). First, the FDSA architecture distributes secret phase keys across frequencies. Without this frequency diversity, all phase keys would collapse onto the same frequency, causing Bob to experience the same constellation scrambling as Eve. Second, the secret phase keys act as a physical-layer encryption scheme. While the distance between main-beam Eve and Bob (ΔR) creates a phase shift at each frequency, an intelligent Eve could exploit this deterministic distance-phase relationship to reconstruct the signal. The proposed secret keys, known only to Bob and that can be updated dynamically, make this compensation infeasible. Third, ASM introduces symbol-level one-to-many randomness, preventing Eve from learning a fixed channel response. The simulated constellations for Bob and Eve (Fig. 5.2.2, bottom right) clearly show that Eve's received symbols are highly scrambled with a one-to-many randomization, while Bob's remain unaffected.

As proof of concept, we implement a TX prototype at 28GHz in a 22nm CMOS process that incorporates two sub-arrays ($K = 2$), each consisting of six channels ($N = 6$), with three channels randomly activated per symbol ($M = 3$) and a 50MHz offset between adjacent channels ($\Delta f = 50$MHz). The top-level schematic is shown in Fig. 5.2.3. Each channel includes an IF multiplier for frequency shifting by Δf, followed by a 10-bit vector-modulator-based phase shifter, a 5-bit VGA, an up-conversion mixer, an LO buffer, and a two-stage PA. A high-speed true random number generator (TRNG) adopting the frequency-collapse concept [23] is also integrated to enable per-symbol antenna subset randomization.

The proposed system architecture requires high-speed antenna switching for ASM and low-overhead frequency shifting for diverse frequency generation. For this, we incorporate two circuit techniques (Fig. 5.2.3, bottom right).

First, a MUX-based biasing scheme is implemented for the PA, mixer, and LO buffer. The MUX selects the nominal bias when the channel is on or sets the bias to 0 when off. This avoids loss from RF switches at the PA output in conventional ASM implementation [6]. Furthermore, the inactive channels are completely turned off with bias set to 0, eliminating their power consumption. The MUX is placed at the center tap of the transformer-based matching network (i.e., the common-mode terminal) [24,25], allowing the use of a large transistor (120μm/20nm) to reduce its on-resistance and the resulting on/off time. The simulated on/off time is only 0.4ns, sufficiently short for symbol-level switching.

Second, a cascaded frequency multiplier chain generates the multi-tone signals at IF (i.e., f_{IF}, $f_{IF} + \Delta f$, ..., $f_{IF} + 5\Delta f$). The IF input is first converted to differential I/Q signals through a three-stage RC-CR polyphase filter [26], followed by a pair of IF analog multipliers that perform single sideband (SSB) up-conversion (Fig. 5.2.3, bottom right). The cascode transistors in the multipliers operate in the small-signal region (not as switches in mixers) to suppress undesired harmonics. The differential I/Q outputs at $f_{IF} + \Delta f$ are cascaded to subsequent stages to generate $f_{IF} + 2\Delta f$ and so on. The gain mismatches between channels are calibrated by tunable buffers after the multipliers. This architecture requires only a single Δf input and enables scalable diverse frequency generation with minimal hardware overhead.

In the measurements, we first characterize the TX performance by sequentially probing all 12 channels. In continuous-wave (CW) measurements with $f_{IF} = 7$GHz, $f_{LO} = 21$GHz, and $\Delta f = 50$MHz (Fig. 5.2.4, left), each channel achieves consistent gain, and all undesired harmonics, IF feedthrough, and images are below -21.3dBc across 12 channels (Fig. 5.2.4, top left). A gradual degradation in harmonic suppression is observed along the chain (from -39.1dBc in the 2nd channel to -21.3dBc in the 6th channel). This is mainly due to the differential imbalance of I and Q, which can be readily addressed by introducing single-ended calibration in the tunable buffer. The CW power sweep demonstrates 13.0-to-14.2dBm OP_{1dB} and 14.4-to-15.5dBm P_{SAT} over 12 channels (Fig. 5.2.4, bottom left). Next, we evaluate the TX using a 400MSym/s 64-QAM signal. At -25.1dB EVM$_{rms}$, it achieves 8.7dBm P_{avg} (Fig. 5.2.4, top right). All 12 channels demonstrate consistent 8.1-to-8.9dBm P_{avg} under modulation (Fig. 5.2.4, bottom right).

Next, we conduct over-the-air (OTA) measurements using a packaged TX module with on-board patch antennas (Fig. 5.2.7). To evaluate the I/Q symbol scrambling against both main-beam and off-axis Eves, we repeatedly transmit a fixed symbol (I, Q) = (1, 0) and record the received symbols at the far-field across all possible antenna subset selections. This measurement is performed for Bob (0°, 0.7m), main-beam Eves (0°, 0.5m and 0.6m), and off-axis Eves (-30° to 30° with 10° step, 0.7m), with and without the secret phase keys (Fig. 5.2.5). To quantify the scrambling performance, we define EVM$_{rms,PLS}$ as the minimum achievable EVM$_{rms}$ for collected symbols over all possible ASM selections. Without applying the phase keys, main-beam Eves show very low EVM$_{rms,PLS}$ after a straightforward distance-phase compensation. In contrast, when the phase keys are applied, main-beam Eves' received symbols become highly scrambled. For example, as shown for Eve$_1$ (0°, 0.5m) in Fig. 5.2.5 (bottom left), EVM$_{rms,PLS}$ increases from 5.5% to 95.9%, which prevents her from recovering any useful information. Importantly, Bob's received symbols remain unaffected (EVM$_{rms,PLS}$ = 2.8%) with correct phase compensation.

Finally, we perform OTA modulation tests using 50MSym/s 16- and 64-QAM signals (Fig. 5.2.5, bottom right). The measured EVM$_{rms}$ for Bob is 5.8% and 6.3%, respectively, confirming that Bob's signal fidelity is unaffected under the proposed scheme. In contrast, the constellations cannot be demodulated at all for Eves at all other distances or angles (i.e., Eve$_1$ and Eve$_2$). While only 50MHz modulation bandwidth is demonstrated in this work, our idea can be readily extended to larger Δf values, which would support wider modulation bandwidths.

Compared with existing TXs designed for physical-layer eavesdropping prevention (Fig. 5.2.6), our proposed TX, featuring FDSA, secret phase keys, and ASM, (1) demonstrates defense against main-beam Eve for the first time, (2) incurs minimal overhead (with power consumption <0.1% for TRNG and 2% for IF multiplier), and (3) achieves competitive TX performance while integrating an on-chip true-random entropy (i.e., TRNG). Given its low-overhead physical-layer protection, scalable architecture, and compatibility with standard phased arrays, the proposed TX provides a practical and efficient solution for securing mm-wave wireless communication.

Acknowledgement:
The authors would like to thank GlobalFoundries for chip fabrication, Keysight and UT-Austin Circuits and Electromagnetics (UT-ACE) Lab for measurement equipment support, and members of the Rice Integrated Systems and Electromagnetics (RISE) Lab for insightful technical discussions.

References:
[1] M. Bloch and J. Barros, *Physical-Layer Security: From Information Theory to Security Engineering.* Cambridge: Cambridge University Press, 2011. https://doi.org/10.1017/CBO9780511977985
[2] A. Mukherjee, S. A. A. Fakoorian, J. Huang and A. L. Swindlehurst, "Principles of Physical Layer Security in Multiuser Wireless Networks: A Survey," in *IEEE Communications Surveys & Tutorials*, vol. 16, no. 3, pp. 1550-1573, 2014. https://doi.org/10.1109/SURV.2014.012314.00178
[3] N. Yang, L. Wang, G. Geraci, M. Elkashlan, J. Yuan and M. Di Renzo, "Safeguarding 5G Wireless Communication Networks Using Physical Layer Security," in *IEEE Communications Magazine*, vol. 53, no. 4, pp. 20-27, April 2015. https://doi.org/10.1109/MCOM.2015.7081071
[4] Y. Zou, J. Zhu, X. Wang and L. Hanzo, "A Survey on Wireless Security: Technical Challenges, Recent Advances, and Future Trends," in *Proceedings of the IEEE*, vol. 104, no. 9, pp. 1727-1765, Sep. 2016. https://doi.org/10.1109/JPROC.2016.2558521
[5] Y. Liu, H. -H. Chen and L. Wang, "Physical Layer Security for Next Generation Wireless Networks: Theories, Technologies, and Challenges," in *IEEE Communications Surveys & Tutorials*, vol. 19, no. 1, pp. 347-376, 2017. https://doi.org/10.1109/COMST.2016.2598968
[6] Q. Zhou, Y. He, Y. Hu, X. Zhang, K. Yang and T. Chi, "A Millimeter-Wave Antenna Subset Modulation Transmitter Array for Low-Overhead Physical-Layer Security Against Wireless Eavesdropping Attacks," in *IEEE TMTT*, vol. 73, no. 6, pp. 3630-3643, June 2025. https://doi.org/10.1109/TMTT.2024.3487894
[7] N. S. Mannem et al., "A 25–34-GHz Eight-Element MIMO Transmitter for Keyless High Throughput Directionally Secure Communication," in *IEEE JSSC*, vol. 57, no. 5, pp. 1244-1256, May 2022. https://doi.org/10.1109/JSSC.2021.3135481
[8] X. Lu, S. Venkatesh, B. Tang and K. Sengupta, "Space-Time Modulated 71-to-76GHz mm-Wave Transmitter Array for Physically Secure Directional Wireless Links," *ISSCC*, San Francisco, CA, USA, 2020, pp. 86-88. https://doi.org/10.1109/ISSCC19947.2020.9062929
[9] A. Babakhani, D. B. Rutledge and A. Hajimiri, "A Near-Field Modulation Technique Using Antenna Reflector Switching," *ISSCC*, San Francisco, CA, USA, 2008, pp. 188-605. https://doi.org/10.1109/ISSCC.2008.4523120
[10] Y. Ju, Y. Zhu, H. -M. Wang, Q. Pei and H. Zheng, "Artificial Noise Hopping: A Practical Secure Transmission Technique with Experimental Analysis for Millimeter Wave Systems," in *IEEE Systems Journal*, vol. 14, no. 4, pp. 5121-5132, Dec. 2020. https://doi.org/10.1109/JSYST.2020.2976852
[11] N. S. Mannem, E. Erfani, T. -Y. Huang and H. Wang, "A mm-Wave Frequency Modulated Transmitter Array for Superior Resolution in Angular Localization Supporting Low-Latency Joint Communication and Sensing," in *IEEE JSSC*, vol. 58, no. 6, pp. 1572-1585, June 2023. https://doi.org/10.1109/JSSC.2022.3212207

[12] Q. Zhou, Y. He, K. Yang and T. Chi, "Exploring PUF-Controlled PA Spectral Regrowth for Physical-Layer Identification of IoT Nodes," *ISSCC*, San Francisco, CA, USA, 2021, pp. 204-206. https://doi.org/10.1109/ISSCC42613.2021.9365941
[13] Z. Shaikhanov, F. Hassan, H. Guerboukha, D. Mittleman, and E. Knightly, "Metasurface-in-the-Middle Attack: From Theory to Experiment," *2022 ACM Proceedings of the 15th ACM Conference on Security and Privacy in Wireless and Mobile Networks (WiSec)*, New York, NY, USA, 2022, pp. 257–267. https://doi.org/10.1145/3507657.3528549
[14] Q. Zhou, Y. He, K. Yang and T. Chi, "Physical-Layer Identification of Wireless IoT Nodes Through PUF-Controlled Transmitter Spectral Regrowth," in *IEEE TMTT*, vol. 72, no. 2, pp. 1045-1055, Feb. 2024. https://doi.org/10.1109/TMTT.2023.3305055
[15] X. Wang, P. Hao, and L. Hanzo, "Physical-Layer Authentication for Wireless Security Enhancement: Current Challenges and Future Developments," in *IEEE Communications Magazine*, vol. 54, no. 6, pp. 152-158, June 2016. https://doi.org/10.1109/MCOM.2016.7498103
[16] A. Yasar and R. T. Yazicigil, "Physical-Layer Security for Energy-Constrained Integrated Systems: Challenges and Design Perspectives," in *IEEE Open Journal of the Solid-State Circuits Society*, vol. 3, pp. 262-273, 2023. https://doi.org/10.1109/OJSSCS.2023.3327326
[17] Q. Rui et al., "Lightweight Machine Learning and Embedded Security Engine for Physical-Layer Identification of Wireless IoT Nodes," *2024 IEEE International Conference on Communications (ICC)*, Denver, CO, USA, 2024, pp. 2865-2870. https://doi.org/10.1109/ICC51166.2024.10622365
[18] M. Alioto, "Trends in Hardware Security: From Basics to ASICs," in *IEEE Solid-State Circuits Magazine*, vol. 11, no. 3, pp. 56-74, 2019. https://doi.org/10.1109/MSSC.2019.2923503
[19] J. Ma et al., "Security and Eavesdropping in Terahertz Wireless Links," *Nature*, vol. 563, no. 7729, pp. 89–93, Nov. 2018. https://doi.org/10.1038/s41586-018-0609-x
[20] P. Antonik, M. C. Wicks, H. D. Griffiths and C. J. Baker, "Frequency Diverse Array Radars," *2006 IEEE Conference on Radar*, Verona, NY, USA, 2006, pp. 3. https://doi.org/10.1109/RADAR.2006.1631800
[21] P. F. Sammartino, C. J. Baker and H. D. Griffiths, "Frequency Diverse MIMO Techniques for Radar," in *IEEE Transactions on Aerospace and Electronic Systems*, vol. 49, no. 1, pp. 201-222, Jan. 2013. https://doi.org/10.1109/TAES.2013.6404099
[22] X. Lin et al., "5G New Radio: Unveiling the Essentials of the Next Generation Wireless Access Technology," in *IEEE Communications Standards Magazine*, vol. 3, no. 3, pp. 30-37, Sep. 2019. https://doi.org/10.1109/MCOMSTD.001.1800036
[23] K. Yang, D. Fick, M. B. Henry, Y. Lee, D. Blaauw and D. Sylvester, "A 23Mb/s 23pJ/b Fully Synthesized True-Random-Number Generator in 28nm and 65nm CMOS," *ISSCC*, San Francisco, CA, USA, 2014, pp. 280-281. https://doi.org/10.1109/ISSCC.2014.6757434
[24] X. Zhang, H. Guo and T. Chi, "A Millimeter-Wave Four-Way Doherty Power Amplifier with Over-GHz Modulation Bandwidth," in *IEEE JSSC*, vol. 59, no. 12, pp. 3898-3914, Dec. 2024. https://doi.org/10.1109/JSSC.2024.3453321
[25] Y. Hu, X. Zhang and T. Chi, "A 28-GHz Hybrid Beamforming Transmitter with Spatial Notch Steering Enabling Concurrent Dual Data Streams for 5G MIMO Applications," in *IEEE JSSC*, vol. 59, no. 10, pp. 3378-3391, Oct. 2024. https://doi.org/10.1109/JSSC.2024.3399220
[26] H. Wang, H. Guo, X. Zhang and T. Chi, "A Packaged D-Band Transmitter with a Multifeed Lens Antenna Achieving 25.3dBm Single-Element EIRP for 2-D Scalable Arrays," *IEEE CICC*, Boston, MA, USA, 2025, pp. 1-3. https://doi.org/10.1109/CICC63670.2025.10983759

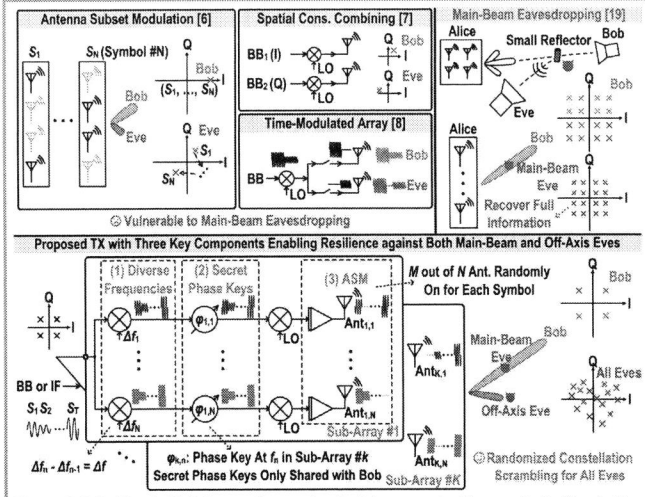

Figure 5.2.1: (Top left) Prior works on physical-layer protection against off-axis Eve. (Top right) Main-beam Eve attack, which cannot be addressed by existing schemes. (Bottom) Proposed TX architecture enabling protection against all Eves.

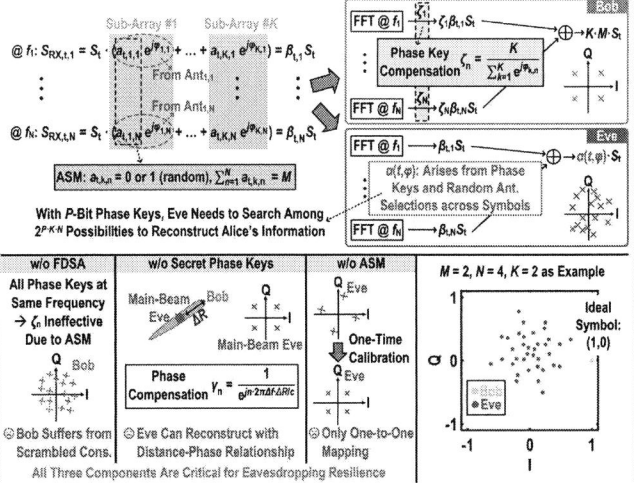

Figure 5.2.2: (Top) Bob successfully reconstructs the transmitted symbols, while Eve's received symbols are scrambled. (Bottom left) Missing any one of the three components breaks security protection. (Bottom right) Simulated symbol scrambling.

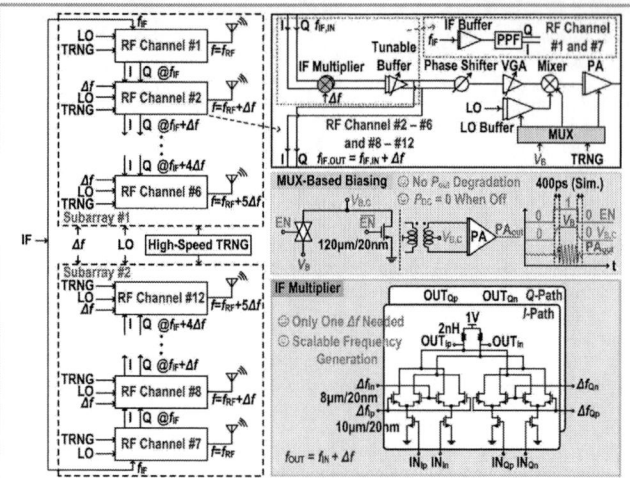

Figure 5.2.3: (Left) TX prototype with two sub-arrays (*K*=2), three out of six channels randomly activated per sub-array (*M*=3, *N*=6), and Δ*f* = 50MHz. (Top right) Block diagram of each channel. (Bottom right) Proposed MUX-based biasing scheme and IF analog multiplier.

Figure 5.2.4: Probing-based measurements. (Left) CW testing results: gain, harmonic rejection, and power sweep across all 12 channels. (Right) Modulation testing results of the first channel and a summary across all 12 channels at -25dB EVM$_{rms}$.

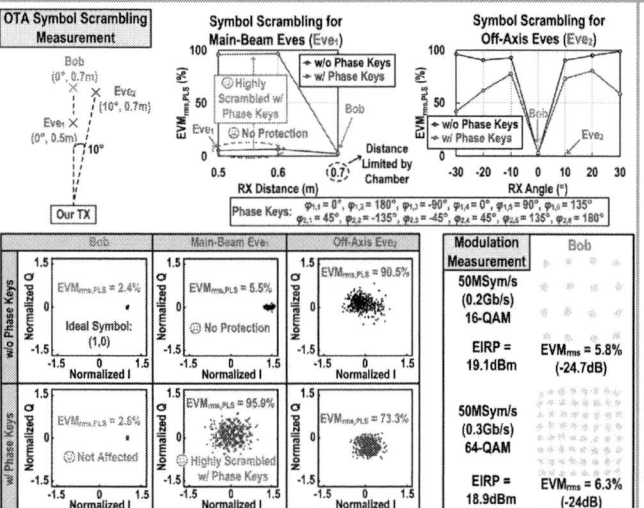

Figure 5.2.5: OTA measurement results: symbols are scrambled for both main-beam and off-axis Eves, while Bob's received symbols remain unaffected.

Figure 5.2.6: Comparison with recently reported physical-layer security schemes for preventing wireless eavesdropping attacks.

		This Work		TMTT 2025 [6]	JSSC 2022 [7]	ISSCC 2020 [8]
Frequency		28 GHz		28 GHz	30 GHz	73 GHz
Technology		22-nm CMOS SOI		45-nm CMOS SOI	45-nm CMOS SOI	65-nm CMOS
Architecture		FDSA + Secret Phase Keys + ASM		Antenna Subset Modulation (ASM)	Spatial Constellation Combining (SCC)	Time-Modulated Array
Antenna		On-PCB Patch		On-PCB Patch [a]	On-PCB Patch	On-PCB Patch
Number of Antennas		12 (6 Radiating)		8 (4 Radiating)	8 (8 Radiating)	4 (1 Radiating)
Randomness Entropy		On-Chip TRNG (True Randomness)		On-Chip TRNG (True Randomness)	Off-Chip Temporal Swapping (Periodic)	Off-Chip Chirping (Periodic)
Off-Axis Eve Prevention?		Yes		Yes	Yes	Yes
Main-Beam Eve Prevention?		Yes		No	No	No
P_{SAT} per Channel		15.5 dBm		14.6 dBm	18 dBm	9 dBm
DC Power		149 mW per Channel at P_{SAT}		282 mW per Channel at P_{SAT}	300 mW per Channel at P_{SAT}	49 mW per Channel
Chip Size		5.2 mm × 2.7 mm (12 Channels)		3 mm × 1.9 mm (8 Channels)	5.5 mm × 2.3 mm (8 Channels)	1.4 mm × 0.6 mm (2 Channels)
OTA Modulation Performance	Modulation Scheme	16-QAM	64-QAM	64-QAM	64-QAM	QPSK
	Data Rate	0.2 Gbps	0.3 Gbps	1.2 Gbps	3 Gbps	0.4 Gbps
	EVM$_{rms}$	5.8%	6.3%	7%	7.1%	4%
Overhead		TRNG (<0.1% P_{DC}) and IF Multiplier (2% P_{DC})		TRNG and RF Switch Matrix (4.5 dB P_{SAT} Degradation) [b]	Additional Baseband/ Digital Backend for Modulated Signal and Temporal Swapping	Additional Baseband/ Digital Backend for Chirping

[a] λ/4 Spacing with Electromagnetic Bandgap
[b] Graphically Estimated

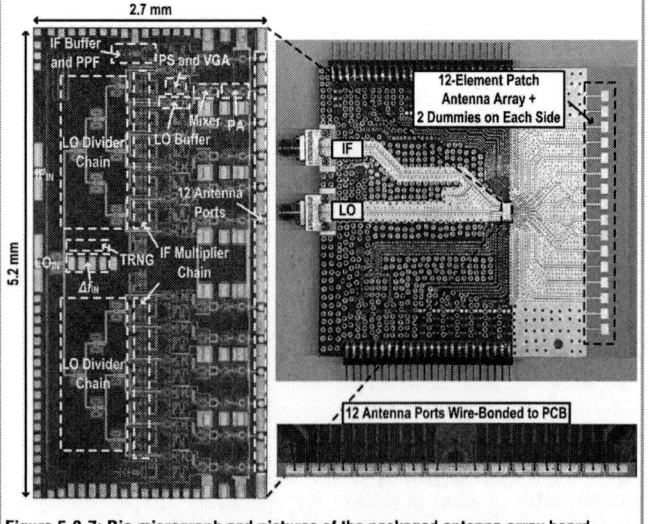

Figure 5.2.7: Die micrograph and pictures of the packaged antenna array board.

ISSCC 2026 / SESSION 5 / SUB-THz AND mm-WAVE PHASED ARRAYS AND BEAMFORMERS / 5.3

5.3 SPARTA: A Scalable, Programmable and Active mm-Wave Dual-Surface Reflect/Transmit Array with Integrated Gain and Phase Control Allowing Bidirectional Signal Routing Capability for Robust and Reconfigurable mm-Wave Networks at 60GHz

Muhamed Fouad Allam[1]*, Wenkai Fang[1]*, Zijian Shao[1]*, Hooman Saeidi[2], Alexander Stepko[1], Kaushik Sengupta[1]

[1]Princeton University, Princeton, NJ, [2]Qualcomm, San Diego, CA
*Equally Credited Authors (ECAs)

Abstract

We present SPARTA, the first dual-surface mm-wave active surface enabling RF signal processing (gain and beamforming) on both sides with multiple transmit and reflect configurations, including simultaneous reception and multi-beam re-transmission in either direction across seven modes. The surface integrates 46 flip-chip bonded custom designed CMOS RFICs (23 per side) operating at 60 GHz on a 14-layer laminate and demonstrates NLOS wireless link closure in both reflect and transmit modes.

Millimeter-wave (mm-wave) networks offer the potential for high-speed gigabit-per-second wireless links with spatial multiplexing and low latency. However, their performance is highly sensitive to physical channel disruptions such as blockages, propagation variability, and fading, making reliable connectivity a significant challenge [1-4]. While relays and Reconfigurable Intelligent Surfaces (RIS) have been proposed as scalable and energy-efficient solutions to dynamically reconfigure the wireless environment by steering incident beams toward desired directions, the dual path losses from access point (AP) to RIS and RIS to user equipment (UE) create fundamental limitations in received signal strength at the receiver [1,4]. While previous works have demonstrated custom-designed CMOS chips wire-bonded to substrates as small-scale demonstrations for active gain-enhanced RIS with and without frequency conversion [6-9] or time-modulated and reconfigurable arrays [10-14], scalable arrays that allow simultaneous programmability of transmit and reflect modes on demand for a reconfigurable wireless network have not been demonstrated. Furthermore, to create a dynamic network (Figure 5.3.1), the surface needs to allow directing the received signal to be transmitted to any node distributed across 360 degrees of azimuth. Here, we present SPARTA that demonstrates a dual-surface mm-wave active surface that allows RF signal processing on both surfaces allowing multiple transmit and reflect configurations including simultaneous reception and multi-beam re-transmission in either directions on demand, with a total of 7 modes of operation. The surface is built with 46 flip-chip-bonded CMOS ICs implemented in 65nm CMOS technology (23 on either side) on a 14-layer laminate (EM89BK 1078 and EM890K 3313 core). Each IC consists of a dual-channel amplify-and-forward chain with 2-stage LNA, 360 degree 8-bit digitally controlled phase shifters, and a 3 stage PA. Each chip processes incoming radiation from both surfaces through series patch antennas, amplifiers, and phase shifts, and re-radiates in a full-duplex manner with another set of transmit antenna arrays realized on the laminate. Each channel can operate in reflect (input and output signals on the same side), transmit (input signals go through the vertical transition and output at the opposite side of the surface), or simultaneous reflect-transmit modes. Such a flexible platform—through programmable manipulation of the amplitude and phase of the incident wavefront—enables dynamic beamforming for NLOS communication, obstacle avoidance, interference mitigation, and efficient aperture reuse.

The architecture and functionality of this complex array assembly is illustrated in Fig. 5.3.1. The incident radiation on the front surface is captured by the series patch antenna array and is channelized to the front and back surfaces simultaneously. The signal is then processed by the CMOS ICs on either surface, which includes amplification, phase control, and power amplification before being re-transmitted through another set of series-fed patch antenna arrays. This happens simultaneously across 23 CMOS ICs on either side of the package, leading to a surface that allows a programmable reflect/transmit configuration across both sides with arbitrary beam configurations. The total board measures 11.14cm×23.85cm with 14 layers of the laminate measuring 1.6mm thickness. The PCB and the CMOS IC are co-designed to allow a seamless transition from the RF path on the board to the RFIC and back to the board for re-transmission. Each CMOS IC constitutes a dual-channel forward-phase-relay path. The front incident RF signal from the antenna is split into two paths—the first path is processed by the first channel on the front CMOS IC, and the second path goes through a vertical via acting as a vertical t-line designed to be processed by the first channel of the back chip. Similarly, to process an incident signal from the back, the back antenna reception path is split feeding to the second channels of the CMOS ICs on the back and the front. On the transmit side, the power amplifiers (PAs) are combined from the two channels and re-transmitted on one side. This allows several modes of operation: 1) active front-surface RIS (incident signal from front and radiation on front), 2) active back-surface RIS (incident signal from back and radiation from back), 3) active front-surface TIS (front incident signal, back re-radiation), 4) active back-surface TIS (back incident signal, front re-radiation), 5) active front-surface RIS/TIS (front incident signal, front and back re-radiation), 6) back-surface RIS/TIS (back incident, simultaneous front and back re-radiation), 7), simultaneous front-back RIS/TIS (back and front incident, simultaneous front and back re-radiation).

Figure 5.3.2 shows the connection of the CMOS IC on the front and back surface to the two antennas on two surfaces. Each CMOS RFIC includes two chains of amplify and phase

control including LNA, phase shifter and PA in each channel. To allow the RIS mode, each front/back antenna is connected to the corresponding front-back surface CMOS RFIC. To also allow the TIS mode, each of the antennas are also connected to the other side CMOS RFIC (to the second RF channel), as shown in the figure. The interface, packaging and transition are carefully designed in the packaging to maintain impedance matching and minimize reflections and coupling. When scaled up, the entire architecture consists of 46 CMOS RFICs with 23 flip-chipped on either side allowing high directional beamforming in azimuth. In a 5G mm-wave communication network, the predominant mode of beamforming is in azimuth, and hence, 1D-beamforming is not just functional but mostly desired for more seamless deployment. Each channel dissipates 168mW.

The chip architecture and the circuit schematics are shown in Fig. 5.3.2. Each LNA comprises of two stages of cascode amplifiers with inductive loads. To allow the surface to act as a backscatter, the second stage of the LNA includes a switch that can be modulated at Gb/s rates. The LNA chain has a simulated peak gain of 9.8dB and input $IP_{in,1dB}$ of -14dBm at 60GHz. The outputs are then fed through a transformer-based quadrature hybrid to generate I and Q signals. The signals are processed through an 8-bit phase vector-based phase shifter to allow full 360° coverage with a resolution of 5.625 degrees. The differential outputs are processed through a chain of three PAs, and then finally converted into single-ended through a balun and connected to the output. The PA chain has a simulated peak gain of 14dB and a P_{sat} of 7.7dBm. The entire chain has a simulated gain of 17dB, allowing the surface to boost the incoming radiation significantly.

The ICs are co-designed with the packaging as shown in Fig. 5.3.3. The ICs are flip-chipped onto the surface, and the RF signal path from the antenna through the vertical transition, off-chip matching networks, buried transmission lines and the bumps are carefully designed. Figure 5.3.3 shows the patch antenna array realized on the front and back layer of the 14-layer laminate. The simulated gain of a series patch is 11.7dBi (~25dBi for the 2D array) demonstrating radiation efficiency of 83% and an input matching of below -10dB nearly across 52 to 62GHz. The simulated radiation patterns are shown in the figure. Both the RX/TX antennas are connected to a low-loss buried transmission line of length 9.3mm with a simulated loss of 0.9dB/cm. The figure also shows the vertical power splitter and associated matching structure. Incident power from the RX antenna is evenly divided between two GCPW (grounded coplanar waveguide) lines—one routed on the same layer and the other coupled through a vertical via to the opposite side of the board as shown in the figure. The multiple metal pads around the via and ground-plane openings are optimized for least insertion loss and balanced power split. The performance of the power splitter demonstrates insertion losses of ~1dB at 60GHz. A matching structure is inserted between the splitter and signal bumps to minimize insertion loss between the RX antenna and the chip LNA. After amplification and phase shifts within the TRX chip, the two RF output paths are combined and routed to the TX antenna via a buried transmission line. This buried transition provides routing flexibility and reserves surface space for surface-mount capacitors and resistors which enhances stabilization. A critical point of consideration in this design is the coupling between the RX and TX paths, that can cause instability due to the gain of the CMOS RFIC. The antennas and the electromagnetic passives were carefully optimized to reduce surface-wave excitation, and spurious coupling (due to transitions). Collectively, the loop gain was reduced below -16dB throughout the frequency range to ensure stable operation, and remove the phase distortions (Fig. 5.3.3).

Figure 5.3.4 shows the measurement results of SPARTA. First, we characterized the system with an incident RIS signal to demonstrate the capability of amplification and beamforming. A TX/RX system is set up with horn antennas at 60GHz as shown in the figure. We fix the TX at an angle and optimize the reception power at the RX by dynamically reconfiguring the phase control on SPARTA. To allow for 8-bit phase control optimization across 23 channels, we employ a genetic algorithm (population size=69, generation count=30) to search for the optimal codes through receiver feedback. In the figure, we demonstrate a TIS and RIS mode experiments (TX angle=135°, RX=90°) and (TX angle=-60°, RX=45°). As can be seen from the measured radiation patterns, SPARTA successfully beamforms towards the RX, allowing a 17dB power enhancement compared to when SPARTA is switched off. To allow for precise measurement of the radiation patterns from the board (and not the background), we employ on/off modulation on SPARTA allowing a frequency shift of the backscattered signals. This mode is only for demonstrating the patterns, and not required when the beam is optimized towards the RX.

Figure 5.3.5 shows a wireless link closure at 57GHz where the link dies without SPARTA and comes alive with 9dB increase in SNR when SPARTA is on and optimized for the TIS mode. Comparison with the state-of-the-art is shown in Fig. 5.3.6 [5-9]. This is a reconfigurable RIS/TIS surface allowing multiple modes of operation and realized with custom CMOS RFICs packaged and co-designed with mm-wave passives and antennas on the laminate. The chip micrograph and the entire SPARTA system is shown in Fig. 5.3.7.

Acknowledgement:
The authors would like to acknowledge the funding agencies: Army Research Office (W911NF2110314, W911NF2410111), Air Force Office of Scientific Research (FA9550-23-1-0176), Office of Naval Research (N00014-23-1-2592), DURIP funding (N00014-23-1-2332), and National Science Foundation (CNS-2211617, CNS-2148271, CNS-2402782). Authors also would like to thank Bert Harrop from the Princeton University Physics Department for professional assistance with packaging, and Princeton ECE IMRL for design support and discussions. The corresponding author is Zijian Shao (zs9193@princeton.edu).

References:
[1] E. Basar et al., "Wireless Communications Through Reconfigurable Intelligent Surfaces," *IEEE Access*, vol. 7, pp. 116753-116773, Aug. 2019.
https://doi.org/10.1109/ACCESS.2019.2935192
[2] J. -B. Dore et al., "Technology Roadmap for Beyond 5G Wireless Connectivity in D-band," *2nd 6G Wireless Summit (6G SUMMIT)*, pp. 1-5, March 2020.
https://doi.org/10.1109/6GSUMMIT49458.2020.9083890
[3] M. Elkhouly et al., "Fully Integrated 2D Scalable TX/RX Chipset for D-Band Phased-Array-on-Glass Modules," *ISSCC*, pp. 76-78, Feb. 2022.
https://doi.org/10.1109/ISSCC42614.2022.9731626
[4] S. Basharat et al., "Reconfigurable Intelligent Surfaces: Potentials, Applications, and Challenges for 6G Wireless Networks," *IEEE Wireless Communications*, vol. 28, no. 6, pp. 184-191, Dec. 2021. https://doi.org/10.1109/MWC.011.2100016
[5] B. A. Abdelmagid, B. Lin and H. Wang, "A D-Band 2D-Scalable 4×4 Active Reflective Relay with Orthogonally Polarized on-Chip TX/RX Antennas and in-Front-End Common-Centroid Fast Azimuth/Elevation Angle-of-Arrival Detection," *ISSCC*, 2025, pp. 206-208.
https://doi.org/10.1109/ISSCC49661.2025.10904706
[6] S. Nooshabadi et al., "A 28-GHz, Multi-Beam, Decentralized Relay Array," *IEEE JSSC*, vol. 58, no. 5, pp. 1212-1227, May 2023. https://doi.org/10.1109/JSSC.2023.3251898
[7] A. Fikes et al., "Programmable Active Mirror: A Scalable Decentralized Router," *IEEE TMTT*, vol. 69, no. 3, pp. 1860-1874, March. https://doi.org/10.1109/TMTT.2020.3042516
[8] N. M. Monroe et al., "Electronic THz Pencil Beam Forming and 2D Steering for High Angular-Resolution Operation: A 98 × 98-Unit 265GHz CMOS Reflectarray with In-Unit Digital Beam Shaping and Squint Correction," *ISSCC*, pp. 1-3, Feb. 2022.
https://doi.org/10.1109/ISSCC42614.2022.9731671
[9] S. Venkatesh, H. Saeidi, X. Lu and K. Sengupta, "Active Tunable Millimeter-Wave Reflective Surface across 57-64 GHz for Blockage Mitigation and Physical Layer Security," *IEEE RFIC*, Denver, CO, USA, 2022, pp. 63-66.
https://doi.org/10.1109/RFIC54546.2022.9863076

[10] X. Lu, S. Venkatesh, B. Tang and K. Sengupta, "Space-Time Modulated 71-to-76GHz mm-Wave Transmitter Array for Physically Secure Directional Wireless Links," *ISSCC*, 2020, pp. 86-88.
https://doi.org/10.1109/ISSCC19947.2020.9062929
[11] S. Venkatesh, X. Lu, B. Tang, K. Sengupta, "Secure Space–Time-Modulated Millimeter-Wave Wireless Links that are Resilient to Distributed Eavesdropper Attacks," *Nature Electronics*, vol. 4, no 11, Nov. 2021, pp.827-36. https://doi.org/10.1038/s41928-021-00664-z
[12] C. R. Chappidi, X. Lu, X. Wu and K. Sengupta, "Antenna Preprocessing and Element-Pattern Shaping for Multi-Band mmWave Arrays: Multi-Port Transmitters and Antennas," *IEEE JSSC*, vol. 55, no. 6, pp. 1441-1454, June 2020.
https://doi.org/10.1109/JSSC.2020.2967545
[13] X. Lu, C. R. Chappidi, X. Wu and K. Sengupta, "Antenna Preprocessing and Element-Pattern Shaping for Multi-Band mmWave Arrays: Multi-Port Receivers and Antennas," *IEEE JSSC*, vol. 55, no. 6, pp. 1455-1470, June 2020.
https://doi.org/10.1109/JSSC.2020.2967544
[14] B. Sadhu et al., "A 24-to-30GHz 256-Element Dual-Polarized 5G Phased Array with Fast Beam-Switching Support for >30,000 Beams," *ISSCC*, 2022, pp. 436-438.
https://doi.org/10.1109/ISSCC42614.2022.9731778

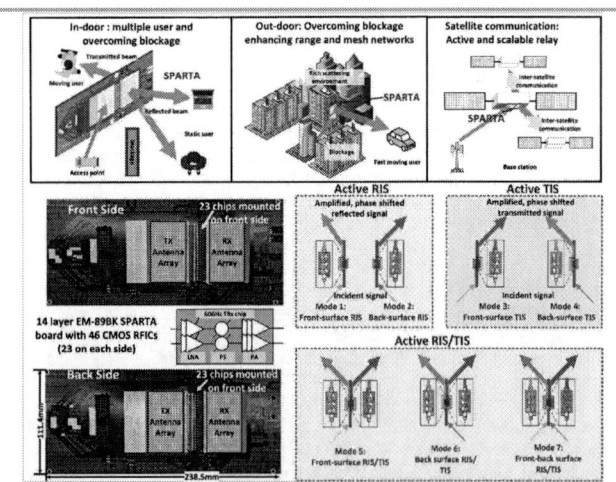

Figure 5.3.1: SPARTA: A scalable, programmable and active mm-wave dual-surface reflect/transmit array with 46 custom-designed CMOS RFICs operating at 60GHz with amplify, beamform and relay, packaged on both sides of a 14-layer EM89BK laminate allowing 7 modes of operation.

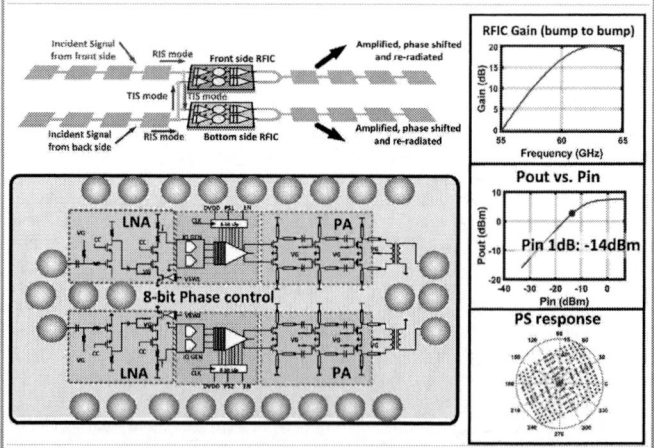

Figure 5.3.2: Schematics of the dual-channel CMOS RFIC integrating a two stage LNA, 8-bit phase shifter, and three-stage PA operating at 60GHz, with each chip capable of processing RF incident signals from both sides of the board, allowing the multi-mode configuration.

ISSCC 2026 / SESSION 5 / SUB-THz AND mm-WAVE PHASED ARRAYS AND BEAMFORMERS / 5.3

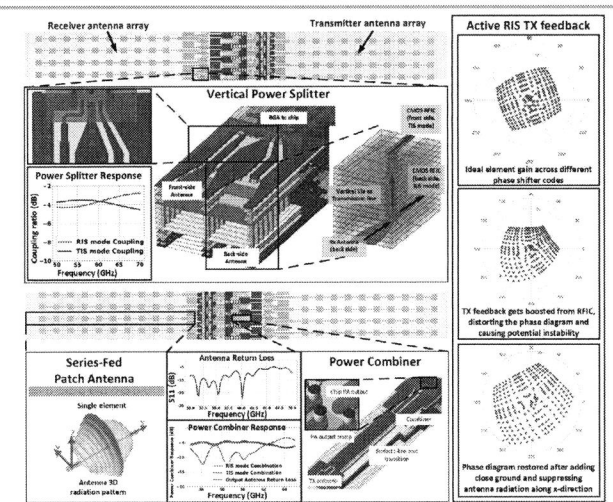

Figure 5.3.3: Packaging and chip co-design showing input and output antenna interfaces, vertical transitions, power combiner on PCB, and improved phase control with optimized packaging and antenna designs.

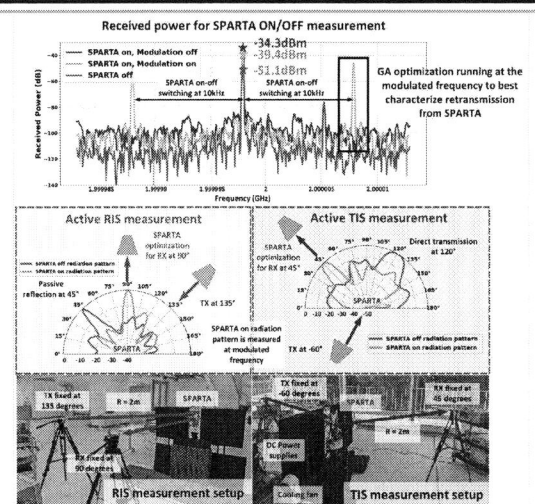

Figure 5.3.4: Measured wireless links at 57GHz with SPARTA optimization for both RIS and TIS modes with the measurements demonstrating 17dB enhancement due to on-board application and beamforming by SPARTA.

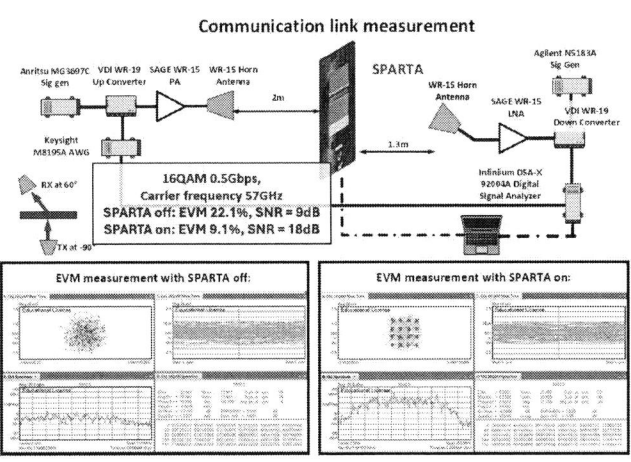

Figure 5.3.5: Wireless link closing demonstrated in the TIS mode at 57GHz with 16-QAM and 0.5Gb/s and the link collapses when SPARTA is off.

	This work	ISSCC 2025 [5]	JSSC 2023 [6]	TMTT 2021 [7]	ISSCC 2022 [8]	RFIC 2022 [9]
Frequency	60 GHz	120 GHz	28 GHz	25 GHz	260 GHz	60 GHz
System Architecture	Dual-surface reconfigurable RIS/TIS with active RFICs	Active reflect relay	Active reflect relay	Active reflect relay	Passive reflect relay	Active reflect relay
Modes of operation	7 modes (combination of RIS and TIS modes)	1 (RIS)	1 (RIS)	1 (RIS)	1 (RIS)	1 (RIS)
Array size	Each side: 23 (packaged CMOS ICs) x 10 (series patch)	4x4	1x4	1x4	98x98	2x2
CMOS Technology Node	65nm CMOS	22nm CMOS	65nm CMOS	65nm CMOS	22nm CMOS	65nm CMOS
Packaging	23 CMOS ICs flip chipped on either side of a dual-surface EM889BK laminate with full digital control	One chip with integrated antennas	One chip with patch antennas on Rogers	One chip mounted with patch antennas on PCB	196 CMOS ICs with integrated antennas	2 CMOS ICs wire-bonded to antennas on PCB
Beamforming architecture	360° with 8-bit resolution	360° with 5-bit resolution	True-time delay with 1.7 ns range	True-time delay + phase shifter	1-bit phase scanning	360° with 5-bit resolution
Scanning range	±60° (1D)	±35° (2D)	±10° (1D)	±5° (1D)	±60° (2D)	±45° (2D)
Element CMOS RFIC/antenna gain	21dB (CMOS), 11.7dB	23.6dB	5.6dB	26dB	-3dB	15dB
Bandwidth	7 GHz	NA	NA	NA	NA	7 GHz
Element P$_{SAT}$	7.7dBm	6.8dBm	3dBm	11dBm	NA	4.2dBm
Element power consumption	336mW	100mW	472mW	80mW (only LO + baseband)	0.1mW (digital)	55mW
Wireless measurements	RIS and TIS modes: 0.5Gbps, 16-QAM, 2m	RIS mode: 6 Gbps, 64-QAM, 1.3m	RIS mode: 0.625 Gbps, 32-QAM, 0.75m	RIS mode: 0.27 Gbps, 64-QAM, 5m	No modulation	RIS mode: 0.4 Gbps, QPSK

Figure 5.3.6: Comparison with state-of-the-art RIS systems demonstrating the first reconfigurable RIS/TIS modes of operation.

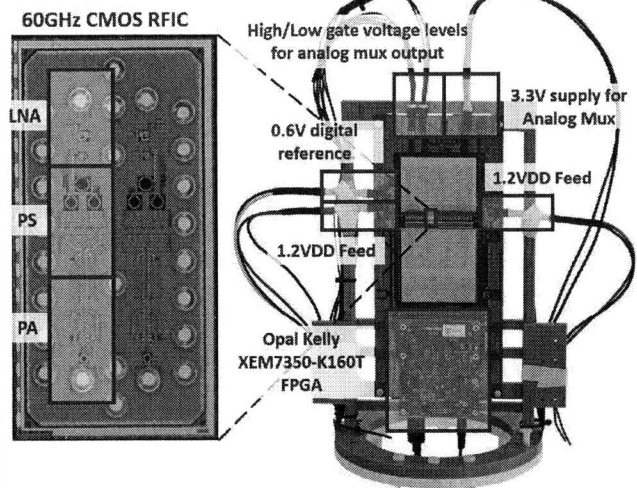

Figure 5.3.7: CMOS RFIC micrograph and SPARTA system (with 46 packaged custom-designed CMOS RFICs on both sides) showing its digital control and supplies.

ISSCC 2026 / SESSION 5 / SUB-THz AND MM-WAVE PHASED ARRAYS AND BEAMFORMERS / 5.4

5.4 A 140GHz Full-Duplex CMOS Transceiver with Metasurface-Integrated Self-Interference-Cancelling Antenna Supporting 16Gb/s 16-QAM Dual-Mode Bidirectional Communication

Jiawei Yang[1,2], Yizhu Shen[1,2], Shizhe Xu[1], Zhongyuan Zhang[1], Zhen Lin[1,2], Yifan Ding[1], Yun Qian[1], Jiapeng Wan[1], Yuancheng Cai[2], Min Zhu[1,2], Sanming Hu[1,2]

[1]Southeast University, Nanjing, China, [2]Purple Mountain Laboratories, Nanjing, China

Abstract

This paper presents a 140GHz CMOS transceiver for both full-duplex and frequency-division duplex communication. By proposing self-interference (SI)-cancelling antenna and dual-polarized metasurface, the design achieves measured EIRP of 26.8dBm and SI suppression of >49dB, therefore providing enough SINR for transceivers. A 16Gb/s 16QAM simultaneous bidirectional communication over 1m is demonstrated, achieving the highest data rate reported for full-duplex dual-mode transceivers in the Figure 5.4.6 comparison table.

A full-duplex (FD) terahertz (0.1-to-10THz) transceiver (TRX) is very promising to efficiently support emerging 6G application scenarios including immersive holographic communications, collaborative industrial robotics, tactile internet, etc. However, an FD THz transceiver seriously suffers from the strong self-interference (SI) from the local transmitter (TX) to the co-located receiver (RX), which severely degrades RX performances and even directly blocks the RX. Effective mitigation of SI requires multiple cancellations spanning from antenna, RF, even to digital domains [1]. Among these, the antenna-domain cancellation directly determines the linearity and dynamic range of the subsequent RF and analog circuits, thereby influencing overall system capabilities.

Conventional antenna-domain self-interference-cancelling (SIC) techniques, however, involve significant performance compromises, as shown in Fig. 5.4.1. Spatially separated dual-antenna configurations achieve substantial SI suppression through physical decoupling, but necessitate considerable layout area, increasing system size and cost [2,3]. Alternatively, reciprocal structures such as electrical-balance duplexers [4] and non-reciprocal devices [5,6] including circulators inherently introduce an additional insertion loss of at least 3dB, directly degrading both TX output power and RX noise figure (NF). Multi-feed antenna structures avoid the intrinsic 3dB loss, but require duplicated TX and RX chains equipped with variable-gain amplifiers (VGAs) and phase shifters (PSs) to calibrate amplitude and phase for effective cancellation [7,8]. This structure redundancy results in increased power consumption, large chip area, and additional digital control complexity. Furthermore, most existing antenna- and RF-domain SIC techniques operate effectively only within narrow bandwidths (typically within 2GHz), which constrains achievable modulation bandwidth and ultimate data rates. In addition, many integrated SIC antennas suffer from low antenna gain (often below 5dBi) and severely limit the TX effective isotropic radiated power (EIRP) and the available RX gain, therefore resulting in diminished signal-to-interference-plus-noise ratio (SINR), reduced communication range, and data throughput.

To address the abovementioned challenges, this paper presents a dual-mode dual-polarization D-band transceiver for bidirectional high-speed communication. The key innovations include: 1) A three-port dual-polarized on-chip antenna that leverages structural symmetry to create inherent voltage nulls at the RX ports during TX operation, achieving high differential-mode (DM) isolation without tuning or active components; 2) An integrated slow-wave coupler that suppresses common-mode (CM) leakage by an additional 20dB, effectively mitigating inherent CM coupling; 3) A dual-polarization metasurface which boosts antenna gain by ~25dB, significantly improving TX EIRP and RX gain. 4) One SIC-and-gain enhancement method that combines high inherent isolation with metasurface gain. These innovations substantially improve the key performance index including TX-RX SI suppression, EIRP and SINR, and also enables robust support for dual-mode operation, i.e., both frequency-division duplex (FDD) and FD communication modes, without recourse to power-hungry RF-domain or complex digital-domain cancellation. Fabricated in 40nm CMOS, the transceiver demonstrates an antenna-domain isolation exceeding 49dB covering 132 to 150GHz and a peak EIRP of 26.8dBm. A wireless link demonstration validates simultaneous bidirectional 16-QAM communication at 16Gb/s over a 1-meter distance in both FDD and FD modes, without employing any RF or digital SIC.

Figure 5.4.1 shows the proposed dual-mode dual-polarization D-band transceiver for simultaneous bidirectional high-speed communication system. The system consists of two identical TRX chips in a face-to-face configuration. Each chip includes independent ×8 multiplier chains to generate local oscillator (LO) signals for the TX and RX paths, enabling flexible and reconfigurable FDD or FD operation. The transmitted signal from one chip is radiated through a dual-polarized SIC antenna, propagating through a dual-polarization metasurface to significantly enhance the EIRP. An identical configuration rotated by 90° is employed at the opposite TRX. The signal is captured by the dual-polarized metasurface, and then amplified by the RX front-end. Thanks to the wideband high SI suppression of the dual-polarized SIC antenna and high gain of the metasurface, the desired signal experiences substantial gain while the SI leakage remains highly suppressed, and therefore results in a significantly improved SINR, enabling robust high-speed communication without additional cancellation stages.

A dual-polarized SIC antenna is one of the most important part in the system, as shown in Fig. 5.4.2. The on-chip antenna is implemented as a three-port ring slot. The TX signal is fed at a single port, exciting a linearly polarized electromagnetic wave. Because of the structural symmetry, the two differential RX feed points are situated at voltage nulls during TX operation, providing inherent high DM isolation for the desired DM RX signal. Despite the structural symmetry, practical implementation introduces symmetric parasitic couplings that can result in significant CM leakage, deteriorating linearity of the RX. This issue is mitigated by an integrated slow-wave coupler, which provides an additional rejection of 20dB specifically for CM signals. The differential DM signals are with low-loss transmission. Full-wave simulations show that the antenna achieves a DM isolation greater than 44dB and a CM isolation exceeding 35dB, and the CM isolation is being further improved by 20dB through the integration of the slow-wave coupler. The on-chip antenna is fabricated on a CMOS substrate of 100μm, then covered with a 550μm quartz superstrate, and finally mounted on a printed circuit board (PCB) as metal reflector. A three-layer dual-polarized metasurface is positioned 25mm above the chip. This planar metasurface consists of 8 types of discrete phase-shifting unit cells, and each unit cell is specially designed with a specific phase shift to collimate the radiated electromagnetic wave into a highly directive pencil beam. This topology significantly enhances antenna gain and improves spatial power combining for both TX and RX modes. Full-wave simulation demonstrates that the substantial improvement afforded by the dual-polarized 140GHz metasurface is validated by the TX mode gain increasing from -2dBi to 23.9dBi, and the RX mode gain improving from -0.55dBi to 25.8dBi.

The detailed schematics of the RF front-end circuits are depicted in Fig. 5.4.3. Two wideband LO generation chains are implemented to ensure robust power across the band. Each chain employs an odd-harmonics-recycling quadrupler followed by a push-push doubler, combining an amplitude-phase coordinating technique [9] to enhance output power, conversion efficiency, and even-harmonic rejection. This structure enables high efficiency, broad bandwidth, and high output power simultaneously, delivering an output power of 0dBm across the 126-to-154GHz frequency range. The TX path features a gain-boosted up-conversion mixer [10] that drives a four-stage cross-coupled neutralized power amplifier (PA). The design employs lossy neutralization techniques to achieve gain enhancement and utilizes asymmetrically coupled transmission lines to form wideband baluns, thereby improving output power and bandwidth. Simulation results indicate a saturated output power (P_{sat}) of 5dBm, a power gain of 24.4dB, and an output 1dB compression point (OP1dB) of 0.4dBm. The RX path comprises a four-stage transformer-coupled low-noise amplifier (LNA), optimized for noise performance and linearity, followed by a double-balanced Gilbert-cell down-conversion mixer. A lossy embedded gain-boosting network is adopted to simultaneously achieve optimal noise and gain performance. The simulated RX chain exhibits a peak gain of 18.5dB and a minimum NF of 8.5dB.

The TRX is fabricated in a 40nm CMOS process, with a compact area of 1.82mm², as shown in Fig. 5.4.7. The chip is covered by a 700×700×550μm³ quartz superstrate, and then mounted on a PCB. A dual-polarized metasurface, fabricated using a standard PCB process with two-layer RO5880, is aligned above the chip with a height of 25mm. Over-the-air (OTA) measurement results are summarized in Fig. 5.4.4. By integrating the metasurface, the TX achieves a measured peak EIRP of 26.8dBm at 140GHz, marking a 27.1dB improvement over the EIRP of -0.3dBm without metasurface. The corresponding TX gain reaches 44dB, also representing a 25dB enhancement compared to the configuration without metasurface. On the RX side, the RX gain is measured as 42.8dB, significantly higher than that of 16.8dB without metasurface. The measured minimum NF, extracted using the gain method, is 9dB at 141GHz.

To experimentally validate the high-speed communication capabilities of the proposed system, a bidirectional wireless link is established between two TRXs spaced 1m apart, as shown in Fig. 5.4.5. In FDD mode, the system achieves a data rate of 12Gb/s using QPSK modulation, with a measured SINR of 13.29dB and a BER of 2.75×10⁻⁵. With 16-QAM modulation, the link supports 16Gb/s at an SINR of 16.61dB. In FD mode, the system demonstrated simultaneous bidirectional operation, achieving 8Gb/s and 12Gb/s using QPSK modulation with SINRs of 15.13dB and 11.53dB, respectively. With 16QAM, the system attains data rates of 8Gb/s and 16Gb/s, corresponding to SINRs of 16.07dB and 14.92dB.

96 • 2026 IEEE International Solid-State Circuits Conference

979-8-3315-8937-0/26 $31.00 © 2026 IEEE

Figure 5.4.6 compares this work with state-of-the-art mm-wave and THz FD systems. Without employing any additional RF or digital SIC, this work achieves a high antenna-domain SI suppression of >49dB over the 132-to-150GHz band, and a data rate of 16Gb/s for simultaneous bidirectional 16-QAM communication at 1m wireless distance with dual modes (both FDD and FD modes), recording the highest data rate for a dual-mode bidirectional communication system among the state-of-the-art publications in the comparison table (Figure 5.4.6).

Acknowledgement:
This work was supported in part by the China Postdoctoral Science Foundation under Grant 2025M770564, and in part by the National Natural Science Foundation of China under Grant 62022023 (corresponding author: Sanming Hu).

References:
[1] A. Nagulu et al, "Doubling Down on Wireless Capacity: A Review of Integrated Circuits, Systems, and Networks for Full Duplex," *Proceeding of IEEE*, vol. 112, no. 5, pp. 405-432, August 2024. http://doi.org/10.1109/JPROC.2024.3438755.
[2] T. Dinc et al., "A 60GHz CMOS Full-Duplex Transceiver and Link with Polarization-Based Antenna and RF Cancellation," *IEEE JSSC*, vol. 51, no. 5, pp. 1125-1140, May 2016. http://doi.org/10.1109/JSSC.2015.2507367.
[3] B. A. Abdelmagid, "A D-Band 2D-Scalable 4×4 Active Reflective Relay with Orthogonally Polarized on-Chip TX/RX Antennas and in-Front-End Common-Centroid Fast Azimuth/Elevation Angle-of-Arrival Detection," *ISSCC*, pp. 206-208, Feb. 2025. http://doi.org/10.1109/ISSCC49661.2025.10904706.
[4] N. V. Thienen et al, "Bidirectional Communication Circuits for a 120-GHz PMF Data Link in 40-nm CMOS," *IEEE JSSC*, vol. 53, no. 7, pp. 2023-2031, July 2018. http://doi.org/10.1109/JSSC.2018.2822714.
[5] T. Dinc et al, "A 28GHz Magnetic-Free Non-Reciprocal Passive CMOS Circulator Based on Spatio-Temporal Conductance Modulation," *ISSCC*, pp. 294-295, Feb. 2017. http://doi.org/10.1109/ISSCC.2017.7870377.
[6] M. Kucharski et al, "Monostatic and Bistatic G-Band BiCMOS Radar Transceivers with On-Chip Antennas and Tunable TX-to-RX Leakage Cancellation," *IEEE JSSC*, vol. 56, no. 3, pp. 899-913, March 2021. http://doi.org/10.1109/JSSC.2020.3041045.
[7] T. Chi et al, "A 64GHz Full-Duplex Transceiver Front-End with an On-Chip Multifeed Self-Interference-Canceling Antenna and an All-Passive Canceler Supporting 4Gb/s Modulation in One Antenna Footprint," *ISSCC*, pp. 76-78, Feb. 2018. http://doi.org/10.1109/ISSCC.2018.8310191.
[8] C. Wang et al, "A Sub-THz Full-Duplex Phased-Array Transceiver with Self-Interference Cancellation and LO Feedthrough Suppression," *IEEE JSSC*, vol. 59, no. 4, pp. 978-992, April 2024. http://doi.org/10.1109/JSSC.2024.3353067.
[9] Z. Lin et al, "A 12.4% Efficiency, 11dBm Psat, Odd-Harmonics-Recycling, 62-to-92GHz CMOS Frequency Quadrupler Using an Amplitude-Phase Coordinating Technique," *ISSCC*, pp. 350-352, Feb. 2024. http://doi.org/10.1109/ISSCC49657.2024.10454527.
[10] J. Wan et al, "A 136-GHz Compact Gain-Boosted Mixer with Wideband Balance-Compensated Baluns for 42-Gbps Communications," *IEEE TCAS-I*, vol. 72, no. 4, pp. 1587-1597, April 2024. http://doi.org/10.1109/TCSI.2024.3476128.
[11] X. Chen et al, "A 140GHz Transceiver with Integrated Antenna, Inherent-Low-Loss Duplexing and Adaptive Self-Interference Cancellation for FMCW Monostatic Radar," *ISSCC*, pp. 80-82, 2022. http://doi.org/10.1109/ISSCC42614.2022.9731637.

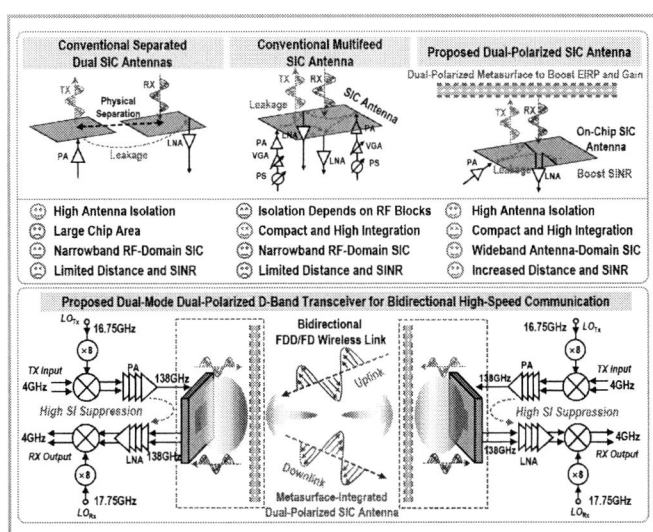

Figure 5.4.1: The traditional SIC architectures, and the proposed dual-mode dual-polarized D-band transceiver for bidirectional high-speed communication.

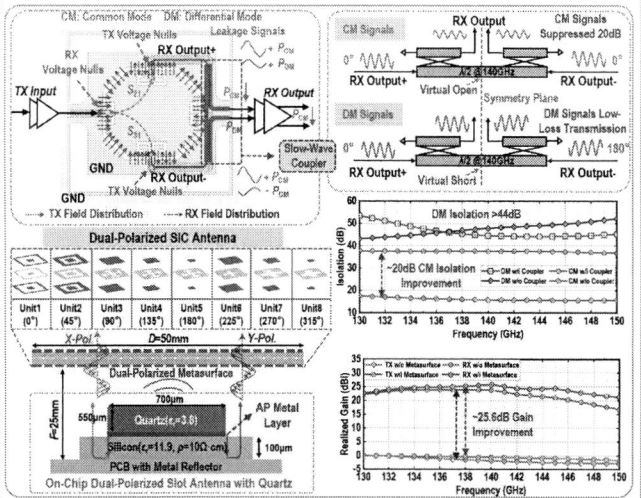

Figure 5.4.2: Detailed analysis and simulation results of the proposed dual-polarized SIC antenna with metasurface.

979-8-3315-8937-0/26 $31.00 © 2026 IEEE

ISSCC 2026 / SESSION 5 / SUB-THz AND mm-WAVE PHASED ARRAYS AND BEAMFORMERS / 5.4

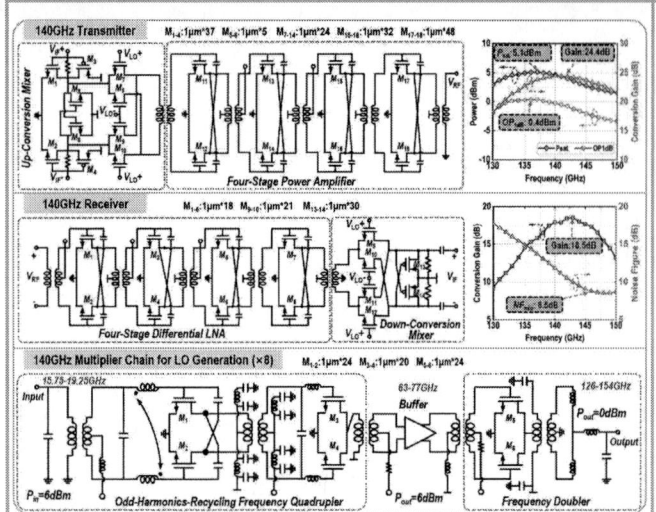

Figure 5.4.3: Schematic and key performances of the D-band transmitter, receiver, and ×8 frequency multiplier chain.

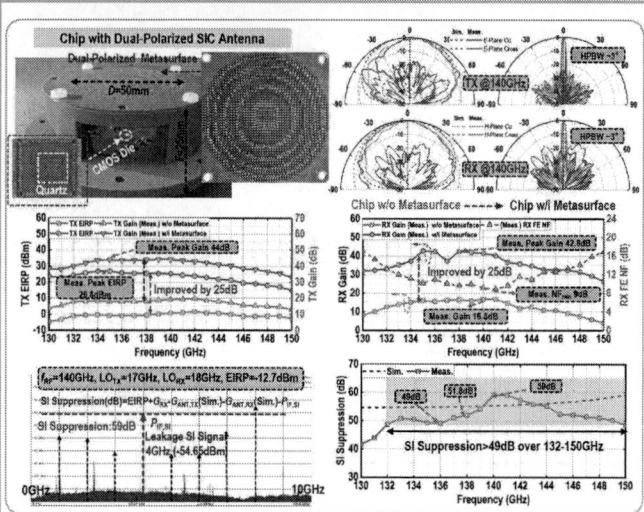

Figure 5.4.4: TRX Implementation and OTA measurement results including radiation patterns, TX EIRP and gain, RX gain and NF, and TX-RX SI suppression.

Figure 5.4.5: High-speed single-channel wireless communication demonstrations for FDD and FD modes over 1m between two TRXs.

Reference	This Work	JSSC 2024[8] K. Okada	JSSC 2018[7] H. Wang	JSSC 2016[2] T. Dinc	JSSC 2018[4] P. Reynaert	ISSCC 2022[11] R. Han
Technology	40nm CMOS	65nm CMOS	45nm CMOS SOI	65nm CMOS	40nm CMOS	65nm CMOS
Frequency (GHz)	135-141	88-136	60-75	120	134-148	134-148
Antenna Domain SIC Architecture	One SIC Antenna	One SIC Antenna	One SIC Antenna	Two Separated SIC Antennas	Electrical-Balance Duplexer	One SIC Antenna with Duplexer
Polarization	Dual	Dual	Dual	Dual	Single	Dual
Integration	TRX + On-Chip Ant. + Metasurface	TRX +Off-Chip Ant.	RF Blocks +On-Chip Ant.	TRX +Off-Chip Ant	TRX + On-Chip Ant.	TRX + On-Chip Ant. + 3D-Printed Lens
Without Additional RF/Digital SIC?	Yes	No	No	No	Yes	No
SI Suppression (dB)	>49 (132-150GHz)	>60* (2GHz BW)	>35 >60*(1GHz BW)	>65* (1GHz BW)	>30	>33.3*
Peak EIRP (dBm)	26.8	19**	18.5**	18.3	/	25.2
RX NF$_{min}$ (dB)	9	15	4.8	4.52	/	12.9
Wireless Distance (m)	1	0.015	0.5	0.7§	2*(Wireline)	2.15 (FMCW)
Duplexing Mode	Dual	Single	Single	Single	Single	Single
Duplexing	FDD	FD	FD	FD	FD§	FD	FD
Modulation	16QAM	16QAM	16QAM	16QAM	BPSK	FSK	N/A (FMCW Radar)
Data Rate (Gb/s)	16	16	6	4	1	6.2	
SINR (dB)	16.5	14.9	16.7	15.9	7.2	/	
P$_{DC}$ (mW)	330	600×8	410	361	73	405
Chip Area (mm²)	1.82	5.7	7.29	4.42	2.75	3.1

*The isolation including RF SIC circuit with narrow band. #Polymer microwave fiber link. ** The saturated one-way TX output power.
** The total saturated TX output power is 9dBm calculated from one-way TX output power of 0dBm, and the antenna array gain is estimated of 10dBi from figures. §FD chip to horn antenna with instruments communication.

Figure 5.4.6: Performance summary and comparison with state-of-the-art FDD/FD systems.

Figure 5.4.7: Die micrograph of the FD/FDD TRX chip.

• 2026 IEEE International Solid-State Circuits Conference

ISSCC 2026 / SESSION 6 / EXPLORATORY RECEIVER ARCHITECTURES FROM GHz TO THz / OVERVIEW

Session 6 Overview: *Exploratory Receiver Architectures from GHz to THz*

WIRELESS SUBCOMMITTEE

Session Chair: Alberto Valdes-Garcia
IBM Research, Yorktown Heights, NY

Session Co-Chair: Giuseppe Gramegna
imec, Leuven, Belgium

RF receiver architectures continue to evolve toward higher frequencies, data rates, and greater versatility. This session features five innovative papers that push the boundaries of receiver IC design. The first paper presents a CMOS full-duplex RF canceller based on an N-path filter. It is followed by an RF spectrum sensor that exploits the bandpass response arising from constructive and destructive interference in a programmable array. Next is a 10GS/s sampled 2×2 receiver beamforming front-end, followed by a 436-to-472GHz four-element IF beamforming phased array. The session concludes with a fully connected MIMO receiver front-end for 5G FR2, covering the 24.25-to-29.5GHz and 37-to-40GHz bands.

98 • 2026 IEEE International Solid-State Circuits Conference

979-8-3315-8937-0/26 $31.00 © 2026 IEEE

ISSCC 2026 / February 16, 2026 / 3:35 PM

3:35 PM

6.1 Full-Duplex RF Canceler Achieving Wideband High-SI-Power Low-Noise Cancellation Through A Novel N-Path-Filter-Based Architecture and ML-Based Canceler Configuration

Lakshminarasimha Sastry Garimella, Columbia University, New York, NY

In Paper 6.1, Columbia University presents an 11.9mm² 65nm CMOS Full Duplex (FD) RF canceler that utilizes an N-path-filter architecture with filter stacking, a sub-cycle rotation technique in a rotary clock-path frequency scheme and an ML canceller algorithm. The 0.2-to-1GHz FD RF canceller achieves self-interferer (SI) suppression of 17/22dB across 200/140MHz BW when operating at 675MHz and -3dBm SI power handling with 0.7/1dB NF degradation for 150/200MHz SI BW

4:00 PM

6.2 A Phased-Array-Inspired Broadband RF Signal Processor for 2-to-32GHz Spectrum Sensing

Liwen Zhong, Pennsylvania State University, State College, PA

In Paper 6.2, Pennsylvania State University demonstrates a 5mm² 22nm FDSOI RF spectrum sensor that exploits the bandpass response resulting from the constructive and destructive interference of a programmable dispersion-engineered array. The sensor operates in the 2-to-32GHz range with 3GHz resolution, 1.2GHz frequency step and draws 101mW.

4:25 PM

6.3 A 2×2 10GS/s TTD BF Receiver Utilizing Charge-Based Summation with 10.5GHz Bandwidth and SNDR/SFDR with 49.5dB/57.5dBc in 22nm CMOS.

Enne Wittenhagen, TU Berlin, Berlin, Germany

In Paper 6.3, TU Berlin describes a 0.3mm² 22nm FDSOI 10GS/s sampled 2×2 receiver beamforming front-end (FE). The time-interleaved FE enables steering in azimuth and elevation using a 200ps range/14.3ps LSB true-time-delay employing passive charge-based summation with 10.5GHz signal bandwidth. The FE RX achieves an SFDR/SNDR >57.5dBc/49.5dB with 52mW/channel.

4:50 PM

6.4 A 436-to-472GHz 4-Element IF Beamforming Phased-Array Receiver in 65nm CMOS

Hao Guo, City University of Hong Kong, Hong Kong, China

In Paper 6.4, City University of Hong Kong shows a 436-to-472GHz 4-element IF beamforming phased-array with a two-step mixing architecture and using an on-chip antenna array with 10dBi gain over 91GHz 3dB bandwidth. The 7.68mm² 65nm CMOS beamformer demonstrates 36GHz bandwidth, ±35° scanning range and 27.5dB noise figure at 211mW per element.

5:15 PM

6.5 A 26/28/37/39GHz Reconfigurable Fully Connected MIMO Receiver Front-End with On-Chip Diplexer Achieving 52-to-70dB Blocker Rejection

Xuhao Jiang, Southeast University, Nanjing, China, Purple Mountain Laboratories, Nanjing, China

In Paper 6.5, Southeast University introduces a 24.25-to-29.5 / 37-to-40GHz fully connected MIMO receiver (RX) front-end for 5G FR2 with 52-to-70dB inter-band blocker rejection and up to -5dBm blocker tolerance. The RX is integrated in 65nm CMOS (0.7mm² and 99.67mW per channel), supports concurrent dual-band or multi-stream single-band operation, and achieves 4.8Gb/s 64-QAM non-contiguous inter-band carrier aggregation.

DIGEST OF TECHNICAL PAPERS • 99

979-8-3315-8937-0/26 $31.00 © 2026 IEEE

ISSCC 2026 / SESSION 6 / EXPLORATORY RECEIVER ARCHITECTURES FROM GHz TO THz / 6.1

6.1 Full-Duplex RF Canceler Achieving Wideband High-SI-Power Low-Noise Cancellation Through A Novel N-Path-Filter-Based Architecture and ML-Based Canceler Configuration

Lakshminarasimha Sastry Garimella, Alfred Davidson, Nishant Patil, Yuan Ma, Ammaar Junaid, Kaustubh Pabba, Harish Krishnaswamy

Columbia University, New York, NY

Abstract

We present a novel RF canceller that utilizes (i) a novel N-path-filter-based architecture that uses filter stacking to achieve passive tap combining leading to a lower NF degradation, (ii) a sub-cycle rotation technique that doubles the largest frequency shift possible in a rotary clock-path passive-frequency-shifting technique enabling wideband SIC, and (iii) an ML-based canceler configuration algorithm that requires only a limited number of canceller measurements.

Self-interference cancellation (SIC) for full-duplex (FD), frequency-division duplex (FDD), and co-site interference mitigation is drawing significant interest from commercial and defense applications alike. In defense applications, enabling FD operation in arrays (Fig. 6.1.1) enables multifunctional operation across communication, radar and electronic warfare (EW), concurrent use of threat detection and reactive jamming for EW, and conducting blue-force communications with the same hardware [1], resulting in Microelectronics Commons (CHIPS ACT)-funded research into this topic [2]. While arrays potentially require N² RF/analog cancellers across every TX-RX pair, a well-architected array can limit the need to cancellers only at the TX-RX boundary where the antenna coupling is most severe (Fig. 6.1.1). The need for multiple RF/analog cancellers necessitates architectures that provide high RF SIC over large bandwidths (BWs), with small footprint, high SI power handling, low power consumption and low noise figure (NF) degradation. In this work, we present a novel frequency-domain equalization (FDE) RF canceller that utilizes (i) a novel N-path-filter-based architecture that uses filter stacking to achieve passive filter tap combining leading to a lower NF degradation, (ii) a sub-cycle rotation technique that doubles the largest frequency shift possible in a near-zero-power rotary clock-path passive frequency-shifting technique for N-path filters enabling wideband SIC, and (iii) a machine-learning-based canceler characterization and configuration algorithm where the model accurately predicts the exact RF canceler filter response for any possible canceler setting in the presence of tap non-idealities from a limited number of canceller measurements. The 0.2-to-1GHz 65nm CMOS FD RF canceller achieves (i) wideband SI suppression of up to 17/22dB across 200/140MHz BW when operating at 675MHz (1.7× more fractional bandwidth (FBW) compared to [3]), (ii) -3dBm SI power handling (2.5× more than [3]), while (iii) producing only 0.7/1dB NF degradation for 150/200MHz SIC BW (1.3 to 1.6dB less than [3]), thus advancing the state of the art along all axes.

Time-domain equalization (TDE) and FDE are the two common architectures for RF cancelers, utilizing broadband delay taps and frequency-separated bandpass filters respectively. Figure 6.1.1 compares their trade-offs. TDE cancelers have two degrees of freedom (DoFs) per tap (delay and gain). FDE cancelers based on N-path filters [3] feature four DoFs per tap, namely filter center frequency, Q, gain and phase. For wideband SIC, TDE typically results in higher filtering loss (Fig. 6.1.1), defined as difference between the sum of total power in each of the taps and the total power of the channel replicated, as broadband delay taps destructively interfere to recreate the channel. FDE however breaks the SIC BW into smaller slices which are then replicated using non-overlapping filter taps, resulting in lower filtering loss. Hence, less gain is required to overcome the overall losses in the canceller, lowering power consumption and added noise, and improving power handling.

A challenge with FDE based on N-path filters is the need for a unique synthesizer for each filter to set its center frequency. Rotary clocking has been proposed in [3] as a passive approach to achieve frequency shifts in N-path filters from a common synthesizer, where the clocks driving each path of an N-path filter are rotated periodically to shift the center frequency by the rate at which clock phase changes. If we have P clock rotations in Q cycles, the frequency shift achieved is $f_c P/NQ$. To achieve shifts larger than f_c/N, $P/Q>1$ is required, which increases the quantization loss of the filter [3]. In this work, we introduce a sub-cycle phase rotation technique (Fig. 6.1.2), where the phase can be rotated twice a clock cycle (i.e., at both rising and falling clock edges) allowing up to $2f_c/N$ largest frequency shift without an increase in quantization loss. This essentially doubles the SIC BW achievable. Figure 6.1.2 compares the measurement results for the two cases described above, increasing the largest frequency shift from approximately 100MHz to 200MHz.

Figure 6.1.2 also contrasts two potential architectures for an FDE canceller – the traditional [3] where the variable Gm cells for programmable gain follow the filter taps and the taps are combined in the current domain, and the proposed where the gain cells precede the filter. The advantage of the proposed architecture is that the broadband noise of the Gm cells is filtered by the taps, significantly reducing the NF degradation, but this requires a means of passive summation of the filter taps. In this work, we propose a filter stacking technique for passive summation. Multiple N-path filters are independently charged during their charging phase but have their capacitors stacked on top of each other in the discharging phase to provide a voltage summation at the output. Figure 6.1.2 shows the

behavior for a two-filter-stacked case. Capacitors of the bottom filter are always connected to ground while the capacitors of the top filter alternate between ground and the output of the bottom filter. Proper clocking needs to be performed to ensure the correct capacitor is stacked as both the filters operate at different center frequencies. Figure 6.1.2 also shows a comparison between individual filter measurements and their combined measurement from stacking. The combined response has a slightly higher loss compared to ideal summation due to the presence of parasitics at the output of the bottom filter. This loss reduces as the path capacitance (Q) increases.

Figure 6.1.3 shows the block diagram and chip photo of the 65nm CMOS 3.5mm×3.4mm FDE-based RF canceler having 12 filter taps, with each tap being a two-port 8-path rotary-clocked filter. Each RF canceler filter has 3-bit phase shifting realized by staggering the clocks of the switches at the two ports. These filters are driven using two-stage thick-oxide inverter-based Gm cells with tunable shunt resistors to individually control the gain and output impedance. This provides a 6-bit gain control for each tap. The programmable output impedance sets the quality factor (Q) of the filter, providing up to an equivalent 6-bit Q tunability. However, this output impedance tuning cannot cover the required Q range from 15 to 50, and thus separate taps are designed for different ranges of Q. The 12 taps are divided into 4 each of high-Q, mid-Q and low-Q taps with varying path capacitance. Due to an increase in parasitic loss as the number of filters stacked increases, a set of 4 high-Q, 4 mid-Q and 2 low-Q filters are stacked together. Lower stacking is used for low-Q filters as their parasitic loss is higher due to lower path capacitance. These stacked filters drive tunable 6-bit thick-oxide inverter-based Gm cells that are co-designed into the partial noise-canceling LNA (similar to [3]) and induce the cancellation current into the RX while providing additional gain control. A 20-bit shift register stores the sequence for phase rotation in the tap N-path filters, and is used to drive a MUX that selects the clock phase that drives each switch of the N-path filters providing a minimum frequency shift of $f_c/20N$ and a maximum frequency shift of $2f_c/N$.

Figure 6.1.4 shows the measured reconfigurability of a single tap while the tap is driven using 750MHz clocks, demonstrating the 4 DoFs. Figure 6.1.5 shows the number of tunable parameters, leading to a very large number of canceller configurations to be measured. This work focuses on creating a model to predict the filter response along with tap non-idealities from a limited set of canceller measurements. It is known that an N-path filter can be modeled as a second-order bandpass filter around the center frequency. However, for wideband SIC, each tap should be modeled for a wider BW (about 80 to 100MHz in this work). We have found that the measured N-path filter taps can be accurately modeled as a sum of two second order bandpass filters with high accuracy (30dB). Interestingly, for different configurations of the same tap, the ratio of gains (G_1/G_2), Qs (Q_1/Q_2) and the difference of phases ($f_1 - f_2$) were almost constant. Learning these ratios for each of the 12 taps reduced complexity of the model from 7 variables (f_c, G_1, Q_1, f_1, G_2, Q_2, f_2) to 4 variables (f, G, Q, f). Using the tap model developed above, 1000 different configurations are measured for each tap which are then divided into 600 train-plus-cross-validation and 400 test data points. A polynomial-kernel-based non-linear ridge-regression model is used with the hyper-parameter being the ridge parameter. As every tap behaves differently, each tap has its own model developed using its train data. Figure 6.1.5 shows the training and test percentage errors for a high-Q filter for each of the 4 output parameters defining the tap response. The test prediction error is about 0.65% for G and Q, 0.8% for f_c and about 5% for f. Figure 6.1.5 also shows an example test measurement and its corresponding predicted filter response. The prediction residue is about 26dB lower in power on average for all the test configurations. The closed-loop iterative canceller optimization algorithm extensively uses the ML model developed here. A one-shot initial configuration and gain selection is achieved assuming the FDE response as the weighted sum of the tap responses and minimizing the measured SI. This is performed by employing a greedy mechanism that iteratively selects, for each tap, the configuration which yields the largest SI reduction. This is done with a constraint on the sum of all tap gains, which in turn limits the NF degradation. The ML model developed above can only predict the filter response non-idealities but not the filter summation losses that result from stacking. Then, an iterative gain adaptation is then undertaken to reduce the residual SI by iterating over gain using gradient descent while keeping other parameters the same and measuring the residual SI.

ISSCC 2026 / February 16, 2026 / 3:35 PM

Figure 6.1.6 shows the measured SIC for a circulator-based antenna interface operating at 675MHz. An SIC of 17/22dB is achieved for BWs of 200/150MHz (29.6/22.2% fractional BW). Figure 6.1.6 also shows the achieved SIC across various BWs. The NF degradation ranges from 0.1dB to 1dB for 50-to-200MHz SIC BWs. SIC improves SI power handling (the SI power level that produces 1dB gain compression in the RX) from -21dBm to -3dBm, an 18dBm improvement. The RF canceler power consumption per DoF is 2.4mW. Figure 6.1.7 compares the state-of-the-art cancelers to this work. Notably, this work has advanced the state of the art on all relevant metrics for FD, achieving the highest RF canceler complexity, the highest RF canceler delay, the widest SIC BW (absolute and fractional), the lowest NF degradation and the highest SI power handling among the devices shown in the comparison chart, resulting in a significant step towards practical FD array-based systems.

References:
[1] K. E. Kolodziej, B. T. Perry and J. S. Herd, "In-Band Full-Duplex Technology: Techniques and Systems Survey," in *IEEE TMTT,*
https://ieeexplore.ieee.org/abstract/document/8642523, vol. 67, no. 7, pp. 3025-3041, July 2019. doi: 10.1109/TMTT.2019.2896561.
[2] https://nstxl.org/opportunity/microelectronics-commons-call-for-projects/
[3] S. Garimella et al., "Frequency-Domain-Equalization-Based Full-Duplex Receiver with Passive-Frequency-Shifting N-Path Filters Achieving >53 dB SI Suppression Across 160 MHz BW," in *IEEE RFIC*, San Diego, CA, 2023, pp. 225-228.
doi: 10.1109/RFIC54547.2023.10186132.
https://ieeexplore.ieee.org/abstract/document/10186132.
[4] D. Dosluoglu, K. -D. Chu, D. Pena-Colaiocco, I. Zhao, V. Sathe and J. C. Rudell, "A Reconfigurable Digital Beamforming V-Band Phased-Array Receiver," in *IEEE ESSCIRC*, Milan, Italy, 2022, pp. 493-496, doi: 10.1109/ESSCIRC55480.2022.9911486.
https://ieeexplore.ieee.org/abstract/document/9911486.
[5] A. Nagulu et al., "Full-Duplex Receiver with Wideband Multi-Domain FIR Cancellation Based on Stacked-Capacitor, N-Path Switched-Capacitor Delay Lines Achieving >54dB SIC Across 80MHz BW and >15dBm TX Power-Handling," in *ISSCC*, San Francisco, CA, 2021, pp. 100-102. doi: 10.1109/ISSCC42613.2021.9365947.
https://ieeexplore.ieee.org/abstract/document/9365947.
[6] X. Li et al., "A Highly Tunable RF Self-Interference Canceler with Near Real-Time Machine-Learning Augmented Adaptation for Full-Duplex Radio Applications," in *IEEE ESSERC*, Bruges, Belgium, 2024, pp. 436-439, doi:
10.1109/ESSERC62670.2024.10719413.
https://ieeexplore.ieee.org/abstract/document/10719413.
[7] X. Ma et al., "A Compact Full-Duplex Receiver with Wideband Multi-Domain Hilbert-Transform-Equalization Cancellation Based on Multi-Stage APFs Achieving 65dB SIC Across 120MHz BW," in *ISSCC*, San Francisco, CA, 2025, pp. 1-3. doi: 10.1109/ISSCC49661.2025.10904506.
https://ieeexplore.ieee.org/abstract/document/10904506.

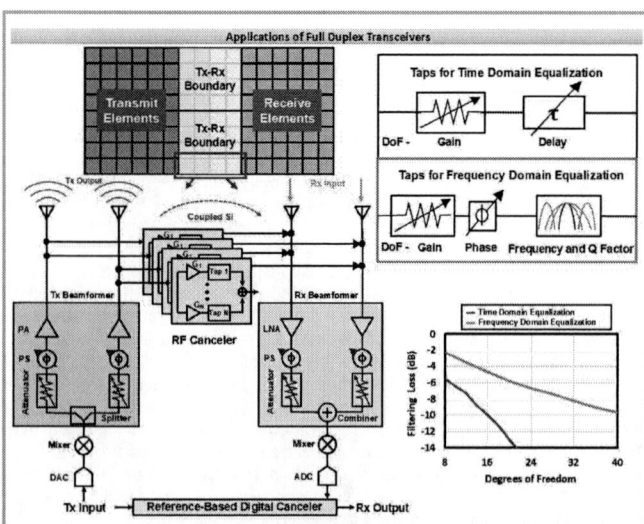

Figure 6.1.1: Motivation for RF cancelers in full duplex transceiver arrays along with system level comparison for time domain vs frequency domain equalization.

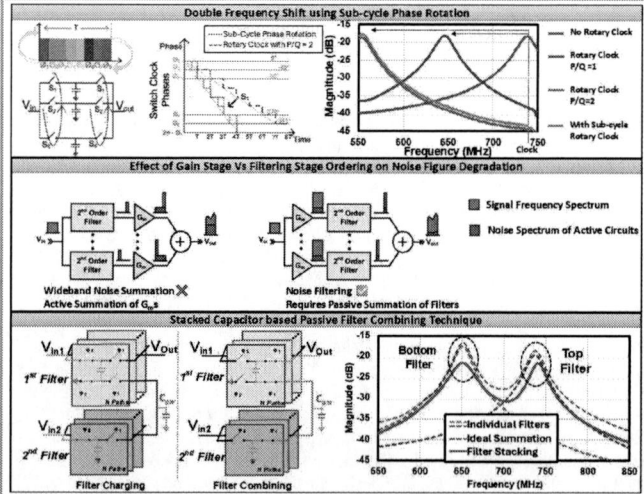

Figure 6.1.2: Sub-cycle phase-rotation-based rotary clocking achieving double frequency shift. Filtering and gain-stage ordering for minimizing noise degradation. Proposed stacked-capacitor-based passive filter combining technique.

979-8-3315-8937-0/26 $31.00 © 2026 IEEE

ISSCC 2026 / SESSION 6 / EXPLORATORY RECEIVER ARCHITECTURES FROM GHz TO THz / 6.1

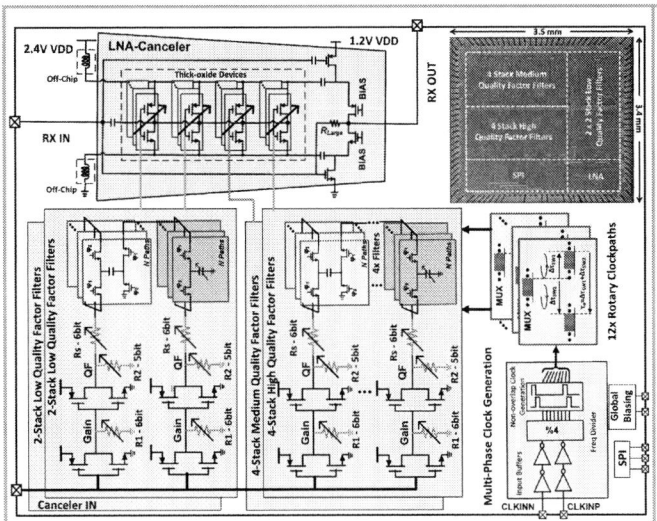

Figure 6.1.3: Block diagram of implemented FD receiver with FDE-based RF canceler with 12 taps of rotary clocked 8-path filters. Chip micrograph.

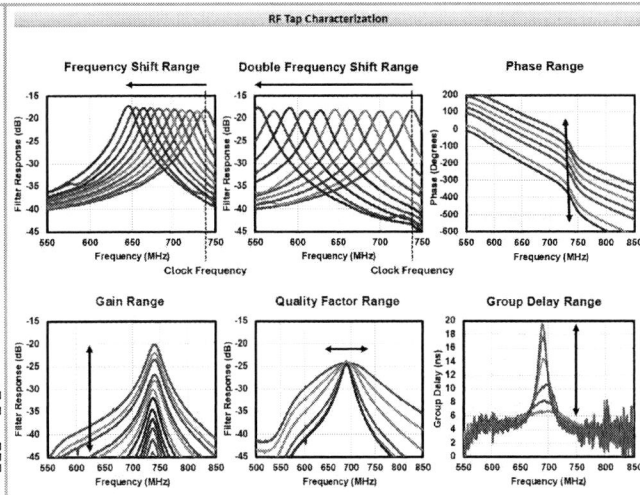

Figure 6.1.4: Measurement results for a single RF tap in the FD receiver having tunability in 4 degrees of freedom, namely center frequency, quality factor, gain and phase.

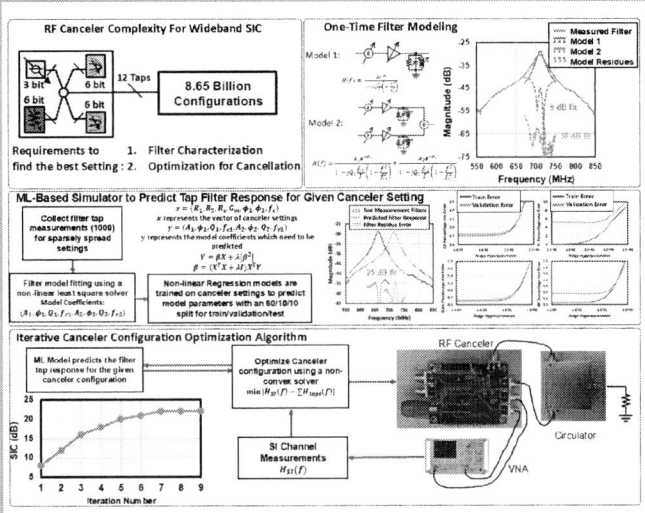

Figure 6.1.5: ML-based canceler characterization and closed-loop optimization algorithm.

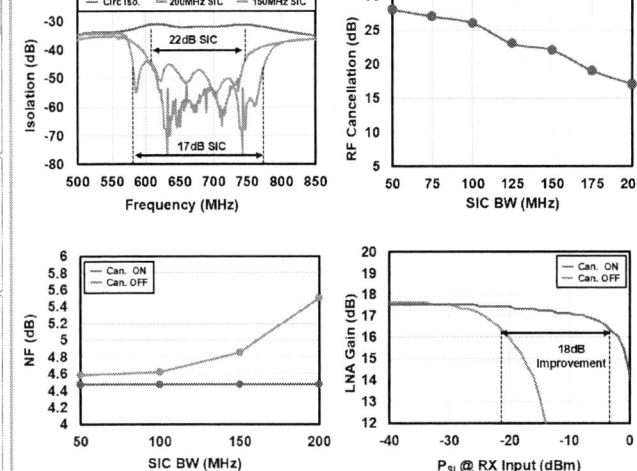

Figure 6.1.6: Cancelation measurements along with noise figure degradation and TX SI induced receiver 1dB compression point with and without cancellation.

	ISSCC 2021 [5]	RFIC 2023 [3]	ESSERC 2024 [6]	ISSCC 2025 [7]	This Work
Chip Architecture	RX with SIC in RF and BB domains using stacked-capacitor switched-capacitor FIR filters and an embedded LNTA-canceler	RX with frequency domain equalization based RF SIC using passive frequency shifting in N-path filters and time domain equalization based BB SIC	A Highly Tunable RF Self-Interference Canceler with Near Real-Time Machine-Learning Augmented Adaptation for Full-Duplex Radio Applications	A Compact Full-Duplex Receiver with Wideband Multi-Domain Hilbert-Transform Equalization Cancellation Based on Multi-Stage APFs	Full-Duplex RF Canceler with Passive Summation of N-Path Filter Taps and Machine-Learning Based Canceler Characterization
Canceler Architecture	TDE	FDE	TDE	TDE	FDE
Integrated SIC Domains	RF + BB	RF + BB	RF	RF + BB	RF
RF Canceler Delay	0 to 7.75ns	0 to 1ns*	2ns	1 to 12ns*	0 to 20ns*
RF Complex Canceler Gain	No	Yes	Yes	Yes	Yes
Number of taps in RF Canceler	16	8	6	2	12
Number of DoFs in RF Canceler	32 (2 per tap)	32 (4 per tap)	24 (6 per tap)	12 (3 per tap)	48 (4 per tap)
SIC from RF Canceler* (Not including Ant. ISO.)	30dB for 40MHz (3%) @ 800MHz; 25dB for 40MHz (8.7%) @ 460MHz	27dB for 120MHz (16.2%) @ 740MHz; 24dB for 160MHz (22.2%) @ 720MHz	32 dB for 40MHz (1.6%) @ 2.5GHz	23.9dB for 80MHz (3%) @ 2.6GHz; 34dB for 120MHz (4.6%) @ 2.6GHz	28dB for 50MHz (7.4%) @ 675MHz; 22dB for 150MHz (22.2%) @ 675MHz; 17dB for 200MHz (29.6%) @ 675MHz
RF Canceler NF Degradation*	2.3dB for 40MHz Canceler BW (frac. BW = 5%)	3.2dB for 120MHz Canceler BW (frac. BW = 16.2%)	1.8dB* (frac. BW = 1.6%)	1.2dB* (frac. BW = 3%)	0.8dB for 150MHz Canceler BW (frac. BW = 22.2%); 1dB for 200MHz Canceler BW (frac. BW = 29.6%)
SI Power Handling Referenced to RX Input*	-7dBm	-12 dBm	N/R	-14 dBm	-3 dBm
RF Canceler Power	3.7mW per DoF	1.8mW per DoF	1.625mW per DoF	2.88mW per DoF	2.5mW per DoF
Technology	65nm CMOS	65nm CMOS	40nm CMOS	40nm CMOS	65nm CMOS
Active Area	7.2mm²	10.9mm²	1.8mm²	0.5mm²	11.9mm²

* - Narrowband delay due to bandpass filter(s).
* - Fractional BW is calculated from the frequency of measured isolation.
* - High power mode metrics where applicable
* - The NF degradation calculated by assuming a RX NF of 6dB.
* - The NF degradation calculated by assuming an RX NF of 4.4dB; work mentions noise contribution is dominated by RF taps.
N/R – Not reported

Figure 6.1.7: Detailed comparison to state-of-the-art full-duplex receivers.

ISSCC 2026 / SESSION 6 / EXPLORATORY RECEIVER ARCHITECTURES FROM GHz TO THz / 6.2

6.2 A Phased-Array-Inspired Broadband RF Signal Processor for 2-to-32GHz Spectrum Sensing

Liwen Zhong, Meijun Tian, Wooram Lee

Pennsylvania State University, State College, PA

Abstract

This paper presents a phased-array-inspired RF signal processor for ultra-broadband spectrum sensing. The proposed architecture exploits the bandpass response resulting from the constructive and destructive interference of a programmable dispersion-engineered array, implemented with broadband true-time delay lines and programmable phase shifters. The proposed design is fabricated in a 22nm FDX process, achieving a measured scanning range of 2 to 32GHz with a power consumption of 101mW.

Frequency Range 3 (FR3), which covers the frequency range from 7.125GHz to 24.25GHz, is a strong candidate for 6G wireless communications due to a favorable coverage-capacity trade-off. To fully utilize the FR3 spectrum with existing mobile services at sub-6GHz and the FR2 mm-wave spectrum, innovative temporal and spatial spectrum sharing is essential across the entire frequency band, requiring energy-efficient, ultra-broadband, real-time spectrum sensing [1]. Spectrum occupancy can be measured by directly digitizing RF signals and then performing a Fast Fourier Transform (FFT) using an analog-to-digital converter (ADC) followed by a digital signal processor (DSP) [2]. However, ultra-broadband (>30GHz) spectrum sensing requires excessively high-speed ADCs and DSPs, leading to significant design complexity and power dissipation. To reduce the required speed of ADCs and DSPs for higher energy efficiency, several analog preprocessing techniques have been proposed for spectrum sensing with a limited frequency scan range below 10GHz [3-6].

This paper presents an ultra-broadband RF signal processor which can sense the spectrum of 2 to 32GHz. Inspired by phased arrays, the proposed RF signal processor leverages the constructive and destructive interference of a programmable dispersion-engineered array, implemented with broadband true-time delay lines and digitally controlled wideband phase shifters. The operation principle of the proposed signal processor can be understood by considering its duality with a phased-array receiver (Fig. 6.2.1). In a phased-array receiver (Fig. 6.2.1 left), phase shifters compensate for the phase differences among received signals across the antenna elements at a fixed frequency ω. The phase difference between the two adjacent antenna elements is given by $\omega\Delta\tau$, where $\Delta\tau$ is the arrival time difference, which varies with the steering angle θ. By appropriately adjusting phase-shifter settings, the received signals from all antenna elements for a desired steering angle θ are constructively combined, enabling spatial filtering. In the proposed signal processor (Fig. 6.2.1 right), the phase shifters similarly compensate for the phase differences among the amplifier elements. However, the phase differences vary with the input frequency ω (instead of a steering angle θ) for a fixed time delay $\Delta\tau$ given by a delay line. Consequently, the input frequency ω corresponding to the peak output amplitude can be selected by adjusting phase-shifter settings, thereby achieving spectral filtering. The peak frequency is given by $\omega_{peak} = (\Delta\varphi + 2n\pi)/\Delta\tau$, showing the peak frequency linearly changing with $\Delta\varphi$, where $\Delta\varphi$ is the phase difference between two adjacent phase shifters. By controlling $\Delta\varphi$, the proposed analog processor sweeps the center frequency of a "pencil-like" narrow passband linearly over a wide frequency range for spectrum sensing, while maintaining constant gain and resolution bandwidth. A preliminary implementation of a similar concept using process design kit (PDK) transistors and lossless inductors and capacitors with simulation results was reported in [7]. However, the realization of low-loss true-time delay lines and precise phase shifters with bandwidths exceeding 30GHz remains a significant challenge for the development of phased-array-inspired RF signal processors in silicon.

To address the challenges mentioned above, Fig. 6.2.2 shows the proposed RF signal processor architecture designed using a 22nm FD-SOI process. To reduce the chip area of the true-time delay, a path-sharing delayed-signal combiner architecture is adopted [7] (Fig. 6.2.2 top). A single-ended RF input is first converted to a balanced differential signal by an active balun, which comprises a common-source (CS) amplifier, a common-gate (CG) amplifier, and an output buffer (Fig. 6.2.3 top left). An input-matching network comprising L_1 and C_1, placed before the shared input terminal of the CS and CG amplifiers, achieves an input return loss better than -10dB over the 2.7GHz to 40GHz frequency range. Inductive peaking is used at the first stage of the active balun to compensate for the gain roll-off in subsequent stages, maintaining a flat overall gain. The differential output of the active balun is applied to a broadband transformer-based 3-stage poly-phase network (Fig. 6.2.3 bottom) to generate in-phase (I) and quadrature (Q) signals [8]. Two cascaded bandwidth extension stages reduce phase and amplitude imbalance over a wide bandwidth by equalizing the highpass and lowpass responses of the THRU and CPL ports in the first stage. The simulated I/Q amplitude and phase mismatch are maintained below 1dB and 5°, respectively, over the frequency range of 7 to 40GHz. The simulated insertion loss of the poly-phase network is approximately 6dB. The I/Q signals are distributed to 10 vector-modulator-based phase shifters through input transmission lines (Fig. 6.2.2 bottom). Each vector modulator consists of two 5-bit variable-gain amplifiers (VGAs). The vector modulators combine the I and Q signals with programmable gain and polarity, achieving precise 360° phase control (Fig. 6.2.3 top right). The implemented 5-bit vector modulator reports a measured phase

step of 11.25° with <2° RMS phase error and <1dB RMS gain error over 7 to 35GHz. The digital control settings of the 10 vector modulators are selected from a 32-state look-up table. The phase difference between adjacent elements $\Delta\varphi$ is swept from 0° to 348.75° with an 11.25° step, corresponding to 32 different pass bands. The output currents of the vector modulator array are combined on the shared delay combining network (Fig. 6.2.4 top left). The wideband delay line is designed using a differential coplanar waveguide implemented with the two topmost metal layers (LB and QB) stacked for lower ohmic loss. The capacitance loading from VGAs is absorbed as a part of an artificial transmission line. Three additional shunt capacitors, implemented using vertical natural capacitors (VNCAP) with capacitance equal to VGA loading, are periodically placed in the delay line between every two vector modulators to increase delay per area and improve broadband delay flatness. Each delay segment exhibits approximately 280ps of delay, folded to form an "S" shape for a compact chip area.

To address higher propagation loss with a longer delay path and higher frequencies in the shared delay combining network, a broadband loss compensation technique using distributed negative conductance is proposed (Fig. 6.2.4 top right). One hundred and nine small-sized active negative conductance units are uniformly distributed along the delay line to equalize loss, while maintaining a constant group delay up to 35GHz. Each negative conductance unit is implemented using a cross-coupled NMOS differential pair with R-C degeneration. The R-C degeneration provides stronger negative conductance at higher frequencies to compensate for frequency-dependent propagation loss over the delay line. Each negative conductance unit consumes a DC power of 230μW. The proposed distributed negative conductance units minimize the difference in propagation loss between the input and output signal paths, ensuring that every vector modulator experiences the same propagation loss from the input to the output (Fig. 6.2.4, middle). Loss equalization across the 10 vector modulator paths is verified in the gain measurement by turning on one vector modulator at a time. (Fig. 6.2.4 bottom) The distributed negative conductance reduces the loss variation across the vector modulators to less than 5dB over the 5-to-30GHz frequency range.

The proposed broadband signal processor is fabricated in a 22nm FD-SOI process with a chip area of 5mm² and a DC power consumption of 101mW (Fig. 6.2.7). The measured frequency response of the proposed RF signal processor is shown in Fig. 6.2.5 using only one single-ended output port with the other output port terminated with 50Ω for a simple VNA calibration. For the uniform amplitude setting across the vector modulators, the measured single-ended peak gain is -5dB with a 3dB bandwidth of 3.5GHz for frequency state 9 corresponding to the peak frequency of 11.5GHz (Fig. 6.2.5 top left). The measured side lobe suppression is 13dB, which is close to that of an ideal uniform linear array. Amplitude tapering is applied by adjusting the gain of each vector modulator according to the Taylor tapering function to achieve a high side lobe suppression of 23dB (Fig. 6.2.5 top right). With amplitude tapering for the same frequency code, the measured peak gain and 3dB bandwidth are -6dB and 4.2GHz, respectively. Twenty-six of the total 32 frequency states are selected for frequency sweeping, while the remaining six states are excluded to avoid grating lobes. The single-ended gain for 26 frequency states is measured without and with amplitude tapering (Fig. 6.2.5 middle left and right). The peak frequency is linearly tuned with a constant frequency step of 1.2GHz. The peak magnitude is approximately -5dB with < ± 1.5dB variation at frequencies from 2 to 32GHz without tapering. The 3dB bandwidth is approximately 3 ± 1GHz with a frequency-shifted uniform pattern. With amplitude tapering, the average measured side-lobe suppression is improved from 11.7 to 22.8dB for 26 frequency states, as shown in Fig. 6.2.5, at the cost of a slightly degraded average peak gain of -7dB and an increased resolution bandwidth of 4.4GHz.

The noise figure (NF) is measured using the Y-factor method with a Keysight 346CK01 1-to-50GHz noise source and a Keysight N9042B signal analyzer equipped with a built-in preamplifier. The differential output signal is converted into a single-ended signal using a broadband Marki EBAL-0050 50GHz Balun before being sent to the signal analyzer. The NF for the selected five frequency states corresponding to the peak frequencies of 5GHz, 10GHz, 15GHz, 20GHz, and 25GHz is shown in the top left of Fig. 6.2.6. The measured NF at these frequency states ranges from 13.4dB to 14.9dB, which is close to the simulation results. The linearity is measured for three different frequency states, as shown in the top right of

ISSCC 2026 / February 16, 2026 / 4:00 PM

Fig. 6.2.6. The measured IP1dB values are -15dBm, -15.2dBm, and -17.7dBm at the frequency states corresponding to the peak frequencies of 5, 15, and 25GHz, respectively. The performance of the proposed RF signal processor is summarized in Fig. 6.2.6. This work presents an analog signal processor for precise spectrum sensing with a frequency-scan range exceeding 30GHz.

Acknowledgement:
This work was supported by the National Science Foundation under Award 2318759. The authors would like to thank the GlobalFoundries University Partnership Program for fabrication support.

References:
[1] I. F. Akyildiz, W.-y. Lee, M. C. Vuran, and S. Mohanty, "A Survey on Spectrum Management in Cognitive Radio Networks," *IEEE Communications Magazine*, vol. 46, no. 4, pp. 40-48, 2008. http://doi.org//10.1109/MCOM.2008.4481339
[2] R. Sharma, R. Shrestha and S. K. Sharma, "Hardware-Efficient and Short Sensing-Time Multicoset-Sampling Based Wideband Spectrum Sensor for Cognitive Radio Network," in *IEEE TCAS I: Regular Papers*, vol. 70, no. 3, pp. 1298-1310, March 2023. http://doi.org//10.1109/TCSI.2022.3223356
[3] N. -S. Kim and J. M. Rabaey, "A Dual-Resolution Wavelet-Based Energy Detection Spectrum Sensing for UWB-Based Cognitive Radios," in *IEEE TCAS-I*, vol. 65, no. 7, pp. 2279-2292, July 2018. http://doi.org//10.1109/TCSI.2017.2781542
[4] Y. Wang, G. J. Mendis, J. Wei-Kocsis, A. Madanayake and S. Mandal, "A 1.0-8.3 GHz Cochlea-Based Real-Time Spectrum Analyzer with Δ-Σ-Modulated Digital Outputs," in *IEEE TCAS I: Regular Papers*, vol. 67, no. 9, pp. 2934-2947, Sept. 2020. http://doi.org//10.1109/TCSI.2020.2990364
[5] P. Sepidband and K. Entesari, "A CMOS Real-Time Spectrum Sensor Based on Phasers for Cognitive Radios," *IEEE TMTT*, vol. 66, no. 3, pp. 1440-1451, Mar. 2018. http://doi.org//10.1109/RFIC.2016.7508304
[6] M. Kitsunezuka, H. Kodama, N. Oshima, K. Kunihiro, T. Maeda, and M. Fukaishi, "A 30-MHz-2.4-GHz CMOS Receiver with Integrated RF Filter and Dynamic-Range-Scalable Energy Detector for Cognitive Radio Systems," *IEEE JSSC*, vol. 47, no. 5, pp. 1084-1093, 2012. http://doi.org//10.1109/JSSC.2012.2185531
[7] L. Zhong, M. Abbasi, S. M. A. Uddin, and W. Lee, "Broadband Frequency-Domain Analog Processor for Spectrum Sensing with 20 GHz Scan Range," *IEEE TCAS II: Express Briefs*, vol. 70, no. 5, pp. 1759-1763, 2023. http://doi.org//10.1109/TCSII.2023.3260088
[8] J. S. Park and H. Wang, "A Transformer-Based Poly-Phase Network for Ultra-Broadband Quadrature Signal Generation," *IEEE IMS*, pp. 1-4, 2015 http://doi.org//10.1109/MWSYM.2015.7167078

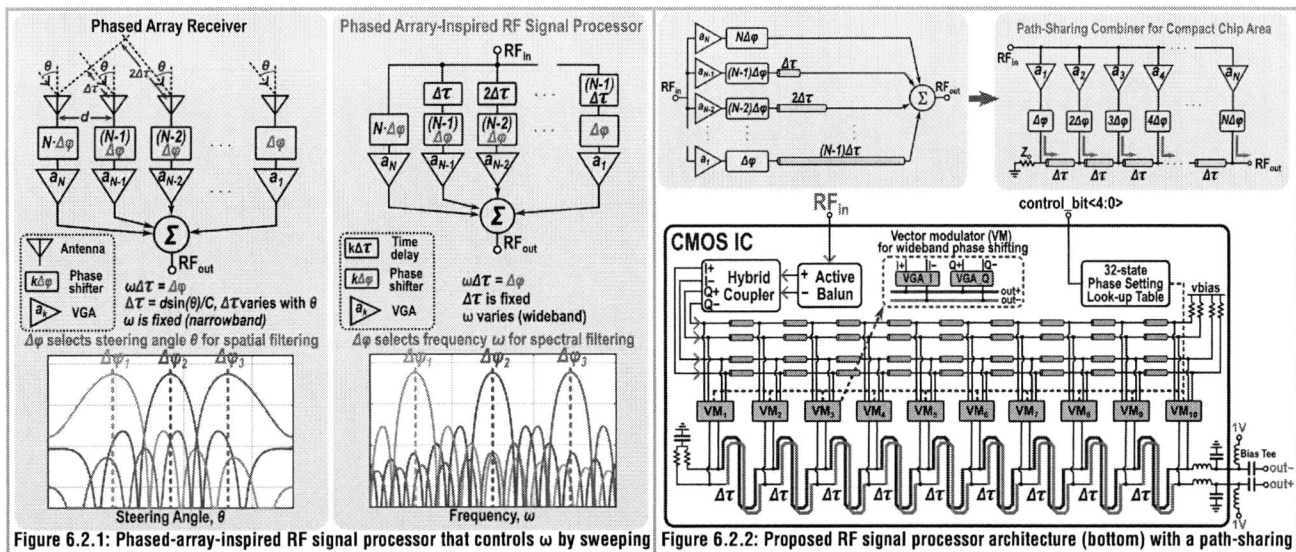

Figure 6.2.1: Phased-array-inspired RF signal processor that controls ω by sweeping Δφ (right), compared to a phased-array receiver that controls θ by sweeping Δφ (left).

Figure 6.2.2: Proposed RF signal processor architecture (bottom) with a path-sharing delayed combiner for a compact chip area (top).

DIGEST OF TECHNICAL PAPERS • 103

979-8-3315-8937-0/26 $31.00 © 2026 IEEE

ISSCC 2026 / SESSION 6 / EXPLORATORY RECEIVER ARCHITECTURES FROM GHz TO THz / 6.2

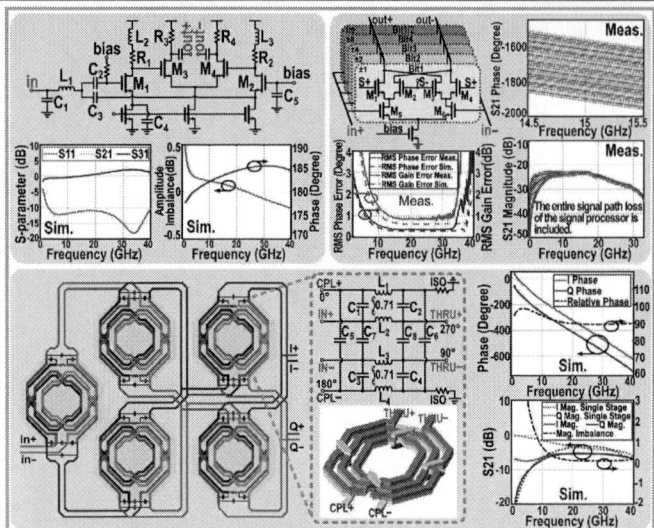

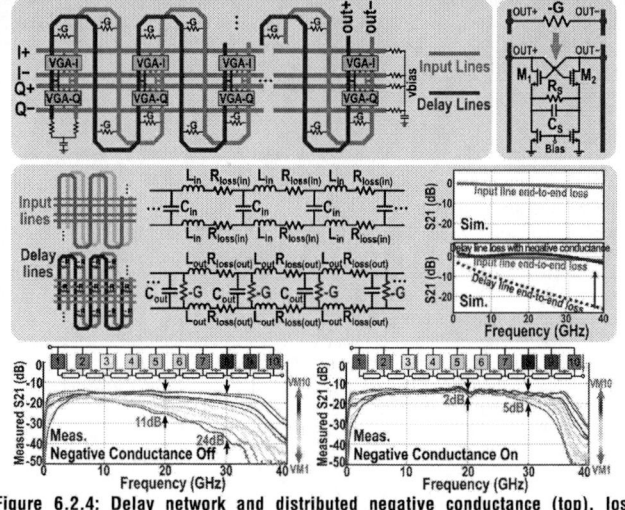

Figure 6.2.3: Circuit details and performance of the active balun (top left), vector modulator (top right), and I/Q coupler (bottom).

Figure 6.2.4: Delay network and distributed negative conductance (top), loss equalization (middle), and measured single-element S21 with negative conductance on/off (bottom).

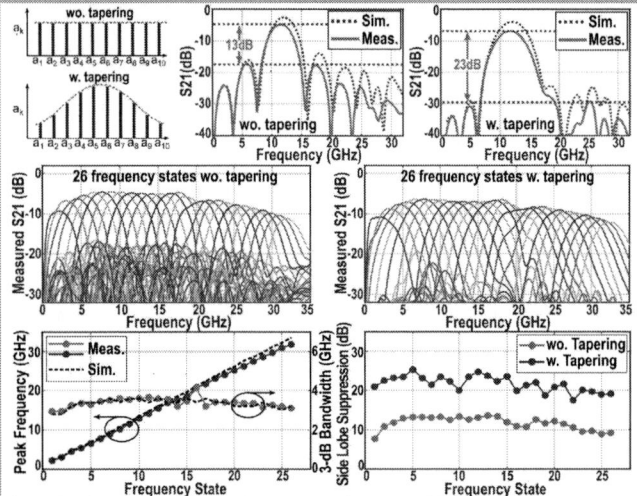

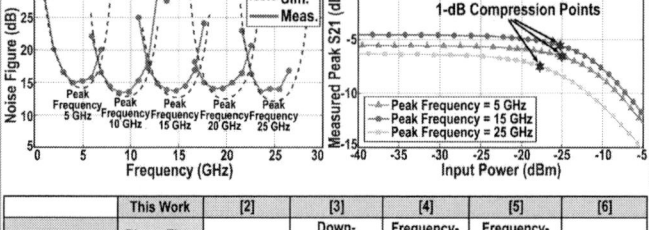

Figure 6.2.5: Measured S21 for frequency state 9 (top). S21 for 26 states with/without tapering (middle), peak frequency, resolution bandwidth, and side-lobe suppression vs. frequency state (bottom).

Figure 6.2.6: Measured noise figure (top left) and linearity (top right). Comparison table with prior art based on different spectrum sensing methods (bottom).

Method	This Work	[2]	[3]	[4]	[5]	[6]
	Phase-Time Delay Array	ADC + DSP	Down-Conversion & LO Sweeping	Frequency-Space Mapping	Frequency-Time Mapping	Filter Bank
Frequency Range	2-32 GHz	0-2.2 GHz	3.1-10.6 GHz	1.0-8.3 GHz	57-354 MHz	0.03-2.4 GHz
Sensing BW	30 GHz	2.2 GHz	7.5 GHz	7.3 GHz	297 MHz	2.4 GHz
Resolution BW	3±0.5 GHz	-NA-	132 MHz	100-200 MHz	27 MHz	0.2-30 MHz
Frequency Step	1.2 GHz	-NA-	132 MHz	~140 MHz	27 MHz	Continuous
Uniform BW	Yes	-NA-	Yes	No	No	No
Uniform Gain	Yes	-NA-	Yes	No	No	Yes
Core Area	5.0 mm²	6.7 mm²	2.5 mm²	8.0 mm²	1.1 mm²	2.3 mm²
Technology	22nm FDSOI	90nm CMOS	65nm CMOS	65nm CMOS	180nm CMOS	90nm CMOS
Power Consumption	101 mW	-NA-	26.4-47.9 mW	418 mW	20 mW	30-44 mW

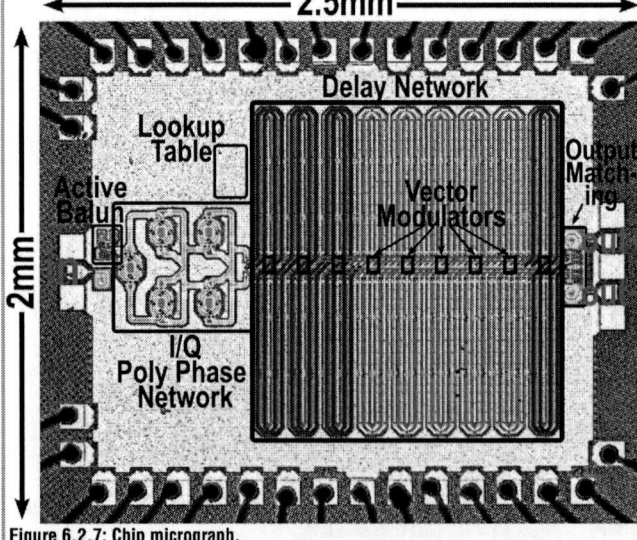

Figure 6.2.7: Chip micrograph.

- 2026 IEEE International Solid-State Circuits Conference

979-8-3315-8937-0/26 $31.00 © 2026 IEEE

ISSCC 2026 / SESSION 6 / EXPLORATORY RECEIVER ARCHITECTURES FROM GHz TO THz / 6.3

6.3 A 2×2 10GS/s TTD BF Receiver Utilizing Charge-Based Summation with 10.5GHz Bandwidth and SNDR/SFDR with 49.5dB/57.5dBc in 22nm CMOS.

Enne Wittenhagen, Dominik Wilding, Patrick Artz, Sebastian Linnhoff, Philipp Scholz, Friedel Gerfers

TU Berlin, Berlin, Germany

Abstract

This work presents an ultra-compact 10GS/s sampled beamforming (BF) receiver front-end (FE) for a 2×2 antenna array, enabling both azimuth and elevation steering in 22nm FDSOI CMOS. The 4× time-interleaved FE features a 200ps range/14.3ps LSB true-time-delay employing a TaH-based BF with passive charge-based summation with 10.5GHz signal bandwidth, eliminating power-hungry summing amplifiers. The FE receiver achieves an SFDR/SNDR >57.5dBc/49.5dB with only 52mW/channel and a Schreier FOM of 153dB.

Next-generation wireless communication links demand both high bandwidth and high throughput. Consequently, techniques leveraging high aggregated bandwidth, large-scale MIMO antenna arrays, and direct RF transceiver architectures are gaining importance to address the performance and energy-efficiency trade-off [1-5]. However, the limited signal bandwidth necessitates an increase in spectral efficiency and consequently in the signal-to-noise ratio (SNR) to support the transmission of large data-rates per antenna.

A beamforming (BF) receiver with an N×N antenna array utilizes spatial signal processing to coherently combine signals from multiple antennas, enhancing reception in the desired direction while suppressing interference. This enhances SNR, spectral efficiency, and link reliability in wireless systems [6].

True-Time-Delay (TTD) operation in phased arrays ensures a uniform frequency response without beam squinting, enabling wide signal bandwidths [7]. However, this approach relies on digital BF, which is power- and area-intensive as a dedicated ADC per antenna element is required [8].

Active analog-based switched-capacitor TTD BF solutions have been presented in [2,3]. But these works suffer from limited signal bandwidth and SNR/SFDR while demanding power-hungry BF summing amplifiers at sample rates below 3GS/s. To address these limitations, a 2×2 10GS/s sampled-BF receiver with a TTD-based phase-tunable track-and-hold (TaH) and passive charge summation is demonstrated, achieving 7× bandwidth and >12dB SNDR improvement over the state-of-the-art (SOTA). This enables robust reception of high data rates over a 5GHz signal bandwidth, obtaining an excellent Schreier figure-of-merit (FOM) of 153dB.

Figure 6.3.1 shows the proposed 10GS/s 4× time-interleaved (TI) BF receiver for a 2×2 antenna array. Due to the λ/2 antenna spacing and the angle of incidence, each antenna element receives the incoming wave with a relative delay $\Delta T_{i,j}$ determined by the array geometry [6]. Each antenna input is isolated by a highly linear pseudo-differential front-end buffer (FE) to protect the RF front-end (not part of this work) from input impedance variations and kickback introduced by the sampling architecture, improving both the receiver linearity and FE signal bandwidth.

Key innovations of the proposed receiver are a highly accurate phase-aligned sampling of each RF input, combined with an area-efficient and highly energy-efficient signal summing architecture. A beamforming clock and phase generator, driven by an external f_{EXT}=10GHz clock, provides the required sampling phases with time step sizes (LSB step size) of 14.3ps and 200ps total range, while the FE configuration and control is managed via an SPI. By individually altering the sampling instant of the respective $RX_{i,j}$ input using the corresponding clock $\Phi_{BF,i,j}$, the beam is steered to different angles to account for the signal time difference $\Delta T_{i,j}$. Considering a center frequency f_c of 3.5GHz and a θ_0 steering to -90°, a time difference between $\Phi_{BF,1,1}$ and $\Phi_{BF,2,2}$ of 200ps is required for Φ_0=45°, as depicted in Fig. 6.3.1. Clock-phase inversion steers the beam toward +90°, thereby enabling the full degree of freedom. To improve FE settling (linearity) and simplify the sample-based BF operation, 4×TI is utilized in this work.

The operating principle of the sampled-BF receiver is illustrated in Fig. 6.3.2, where each antenna receives an input signal with an inter-element delay $\Delta T_{i,j}$, such that: $V_{IN,1,1}(\Phi_{BF,1,1})=V_{IN,i,j}(\Phi_{BF,i,j}+\Delta T_{i,j})$. Each FE buffered input signal is top-plate sampled onto the capacitor $C_{BF,i,j}$ with respect to V_{CM} of 325mV using the appropriate clock phase within each of the four TI TaH slices.

The FE buffer is implemented as a push-pull source follower (SF), operating from a 0.85V core power supply, which doubles g_M/I_D enhancing the FE linearity [9]. A low output impedance of 5Ω allows the BF sampling capacitance $C_{BF,i,j}$=500fF to be sized to simultaneously satisfy kT/C noise, settling (linearity), and maximum BF gain requirements, while supporting signal bandwidth beyond 10GHz.

A 4-bit R-2R DAC is employed to control the operating point of the source follower across PVT variations. A single FE buffer drives the 4×TI TaH slices, with only one slice active at a time, thereby enhancing overall power efficiency. The FE buffer supports an input signal swing of 500mV$_{pp,DIFF}$ with an output common-mode of 425mV, compensating for the offset introduced by charge injection from the switches M_3 and M_4.

Clock bootstrapping for M_3 and M_4 ensures high linearity (>60dBc at full scale) and wide signal bandwidth, despite a tracking duration of only T_S=50ps [10]. Furthermore, switch M_4 remains always active outside the summing-phase, enabling fast and accurate signal tracking.

After the sampling phase, the input signal is stored on $C_{BF,i,j}$ according to the steering angle. Clock Φ_{SUMP} initiates the summation phase by connecting all the $C_{BF,i,j}$ capacitors in series via switch M_5 to ideally produce a voltage gain of 12dB (if $C_{PAR,i,j}$=0 and no capacitor mismatch). A design trade-off exists between voltage gain ($A_{OUT,ALL}/A_{IN,1,1}$) and BF gain ($A_{OUT,ALL}/A_{OUT,1,1}$). Adding a switch before the BE buffer avoids directly charging C_{BE} (option 1, Fig. 6.3.2), leading to a 0.6dB BF gain improvement while suffering from signal gain / SNR loss. Thus, option 2 in Fig. 6.3.2 is realized, enhancing the overall FE voltage gain and thereby RX SNR considering a C_{BE} =35fF (BE buffer + routing). Finally, the summed voltage corresponds to $V_{SUM}\approx[C+(C_{PAR,1,1}+C_{BE})]/[(C/4)+C_{PAR,EFF}]$. The passive summing approach eliminates the need for an active summing amplifier, enabling high linearity and low noise, wide bandwidth, low power consumption, and minimal area overhead at the cost of a slightly reduced BF gain.

The proposed BF sampler serves as the input FE for a complete receiver with an integrated ADC. A stacked source-follower back-end buffer (BE) isolates the envisioned final quantizer, while serving as an open summing node with low input capacitance.

Source followers M_7 and M_8 (in Fig. 6.3.2) employ active transistor bulk boosting by connecting the bulk nodes to the SF output to achieve a high signal swing (up to 1.4V$_{pp,DIFF}$) with only -0.3dB SF gain loss while maintaining high linearity. Furthermore, by driving the gates of M_6 and M_9 with the SF output, the input capacitive load of the SF is reduced by 50%, maximizing the FE voltage gain. The 4×TI differential outputs are multiplexed via transmission-gate (TG), to minimize the number of pads and measurement IO buffers.

A low-jitter multi-phase clock generator is depicted in Fig. 6.3.3. The buffered fully differential f_{EXT}=10GHz (1/(2T_S)) input clock generates 4×TI clocks, each producing 4 BF and 2 summing clock pulses. An on-chip AC-coupled, self-biased inverter regenerates fast clock edges driving a delay chain, which comprises powers-of-two inverters to provide a 14.3ps delay LSB step size [11]. The measured time delay step size is strictly monolithic, with a DNL and INL well below 0.5 LSB (bottom left of Fig. 6.3.3).

With only 7 inverters, a TTD network is realized covering exactly one period of delay (100ps), resulting in a low jitter clock generation to maximize the SNR at high frequencies. The 8 delayed clocks are selected using a TG MUX network. The TI sampling clocks are generated using dedicated ring counters, each producing 4×TI 2.5GHz clock signals with 100ps pulse width, switching TGs to propagate the appropriately delayed signals. An additional period delay $T_{EXT} = 1/f_{EXT}$ (coarse delay) is obtained by selecting the next clock period T_{EXT} with the next trigger signal. In summary, this enables a total delay range of 200ps for full angular resolution.

Due to time-interleaving in the BF front-end, precise time-skew (TS) calibration is needed, achieved via capacitive DAC load tuning. The TS step size is ~27fs with a 7-bit resolution (>3ps range). Similarly to the sampling clocks, also the summing clocks Φ_{SUMP} and Φ_{SUMN} are generated by a ring counter, resulting in a 2T_S (100ps) duration for the summing phase at 2.5GS/s.

To characterize the BF receiver IC, realized in a 22nm FDSOI CMOS process, each of the TI outputs is analyzed using a DCA-X sampling oscilloscope. Spectral tests (incl. SNDR) use an SMA100B with dividers and baluns, while FE bandwidth, gain, and phase are measured with an AWG M8194A. The realized BF receiver achieves an input bandwidth beyond 10.5GHz, enabling sub-sampling up to the H-band, as shown in Fig. 6.3.4. The spectral results prove an uncalibrated SFDR of 37.4dBc, while after TI offset- and gain mismatch as well as TS calibration, the SFDR improves to 57.5dBc. For all measurement results shown, a single set of TS coefficients is used. Uncritical TI offset and gain mismatch errors are calibrated offline. High-frequency two-tone tests exhibit excellent linearity after calibration with interleaving spurs below -69dBFS (-61dBc). The SFDR and SNDR remains above 57.5dBc and 49.5dB (single lane: 58dBc and 51.2dB), respectively, up to 6GHz input signals. A sweep of the input signal amplitude at low (10MHz) and at Nyquist frequency (i.e. 5GHz) shows >75dBFS SFDR (limited by the measurement setup) at an amplitude of -12dB.

Figure 6.3.5 illustrates the spectral performance when summing two different waves $f_{RF,1,1}=f_{RF,2,2}=2.28$GHz and $f_{RF,1,2}=f_{RF,2,1}=3.96$GHz achieving 55.8dBFS SFDR. Furthermore, the normalized array factor (AF) for elevation and azimuth sweep at an f_c of 3.5GHz is shown. Here, the BF receiver is controlled from -90° to 90° in 30° steps, with θ_0 or Φ_0 being 45° (maximum delay setting). An accurate steering to θ_0 or Φ_0 is achieved by sweeping the angle at the input. In addition, the input angle is kept constant at ±90°, 0°, and 45°, and the angle of the BF receiver is swept, revealing the 14.3ps quantized phase steps of the TTD clock, validating the precise sampling phase control against all target angles. An input frequency sweep up to Nyquist frequency exhibits no beam squint, demonstrating the wideband FE operation. The BF receiver provides a flat BF gain of 7.2 to 7.8dB across all angles and input frequencies. Furthermore, as the input signal is already sampled, the performance requirements of the succeeding ADC are significantly relaxed.

Figure 6.3.6 summarizes the performance parameters compared to a SOTA BF receiver. The proposed receiver surpasses the SOTA in terms of signal bandwidth by 7×, the SNDR by more than 12dB (i.e. >2bit), while improving the FOM by more than 15dB. The total power consumption of the 4-channel receiver is 208mW (i.e. 52mW/channel), corresponding to a Schreier and Walden FOM of 153.9dB and 79.6fJ/conv.

Figure 6.3.7 shows the micrograph of the chip with a core area of 630×470µm² for 2×2 BF RX front-end. The measurement setup uses on-wafer probing of all differential RF input and output signals while the differential 10GHz clock, SPI, and power pads are connected using bondwires.

References:
[1] E. Ghaderi, C. Puglisi, S. Bansal and S. Gupta, "10.8 A 4-Element 500MHz-Modulated-BW 40mW 6b 1GS/s Analog-Time-to-Digital-Converter-Enabled Spatial Signal Processor in 65nm CMOS," *ISSCC*, pp. 186-188, Feb. 2020. https://doi.org/10.1109/ISSCC19947.2020.9063106
[2] K. Spoof, M. Zahra, V. Unnikrishnan, K. Stadius, M. Kosunen and J. Ryynänen, "A 0.6–4.0 GHz RF-Resampling Beamforming Receiver With Frequency-Scaling True-Time-Delays up to Three Carrier Cycles," *IEEE SSCL*, vol. 3, pp. 234-237, Jul. 2020. https://doi.org/10.1109/LSSC.2020.3012654
[3] Q. Xu et al., "A TTD-Based Fast Precise Localization Enabled by Passive-Active Signal Combiner with Negative-Capacitance Stabilized RAMP," *IEEE JSSC*, vol. 60, no. 9, pp. 3202-3217, Sept. 2025. https://doi.org/10.1109/JSSC.2025.3546958
[4] A. Kharalkar et al., "An 8-Element 800MHz BW 0.083mm²/element Scalable Current-Mode True-Time-Delay Analog Combiner for Low SWaP-C Antenna Arrays," *RFIC*, pp. 399-402, Jun. 2025. https://doi.org/10.1109/RFIC61188.2025.11082954

[5] E. Wittenhagen et al., "An 11-Bit 12 GS/s Beam-Forming Receiver ADC for a 2x2 Antenna Array utilizing True Time-Delay with 68 dBc SFDR and 55 dB SNDR," *ISCAS*, pp. 1-5, May 2024. https://doi.org/10.1109/ISCAS58744.2024.10558495
[6] C. A. Balanis, "Antenna Theory: Analysis and Design," John Wiley and Sons, Hoboken, New Jersey, 2016.
https://www.wiley.com/en-us/Antenna+Theory%3A+Analysis+and+Design%2C+4th+Edition-p-9781118642061
[7] K. Spoof, V. Unnikrishnan, M. Zahra, K. Stadius, M. Kosunen and J. Ryynänen, "True-Time-Delay Beamforming Receiver with RF Re-Sampling," *IEEE TCAS I*, vol. 67, no. 12, pp. 4457-4469, Dec. 2020. https://doi.org/10.1109/TCSI.2020.3005475
[8] S. Jang, J. Jeong, R. Lu and M. P. Flynn, "A 16-Element 4-Beam 1 GHz IF 100 MHz Bandwidth Interleaved Bit Stream Digital Beamformer in 40 nm CMOS," *IEEE JSSC*, vol. 53, no. 5, pp. 1302-1312, May 2018. https://doi.org/10.1109/JSSC.2018.2791483
[9] A. M. A. Ali et al., "A 12b 18GS/s RF Sampling ADC with an Integrated Wideband Track-and-Hold Amplifier and Background Calibration," *ISSCC*, pp. 250-252, Feb. 2020. https://doi.org/10.1109/ISSCC19947.2020.9063011
[10] A. M. Abo and P. R. Gray, "A 1.5 V, 10-bit, 14 MS/s CMOS pipeline analog-to-digital converter," *IEEE VLSI*, pp. 166-169, Jun. 1998. https://doi.org/10.1109/VLSIC.1998.688071
[11] L. Kull et al., "A 10-Bit 20-40 GS/S ADC with 37 dB SNDR at 40 GHz Input Using First Order Sampling Bandwidth Calibration," *IEEE VLSI*, pp. 275-276, Jun. 2018. https://doi.org/10.1109/VLSIC.2018.8502268

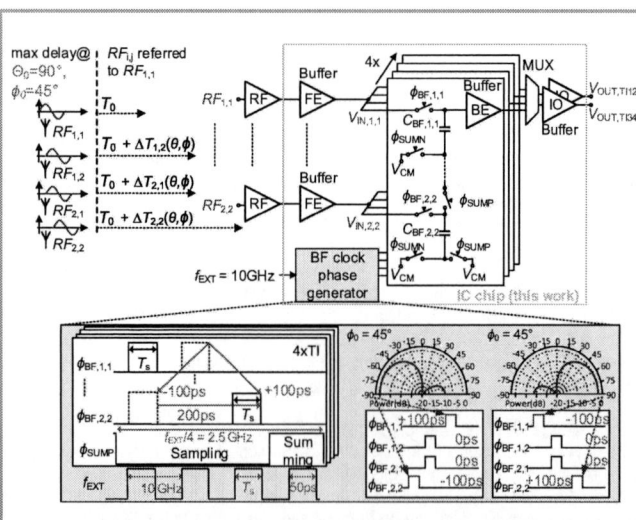

Figure 6.3.1: Proposed sampled beamformer solution for a 2×2 antenna array operating at 10GS/s utilizing 14.3ps/step and 200ps total sample phase clock tuning.

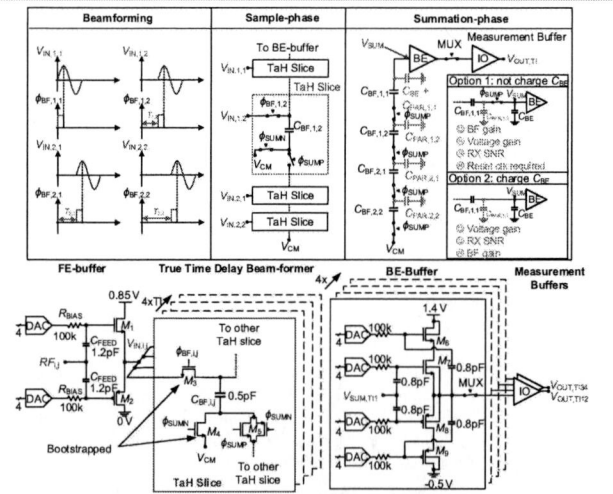

Figure 6.3.2: Simplified concept of the charge-based sample and summation phases. Schematic details of the FE-buffer, TTD beamformer and active body-driven stacked SF BE-buffer.

ISSCC 2026 / SESSION 6 / EXPLORATORY RECEIVER ARCHITECTURES FROM GHz TO THz / 6.3

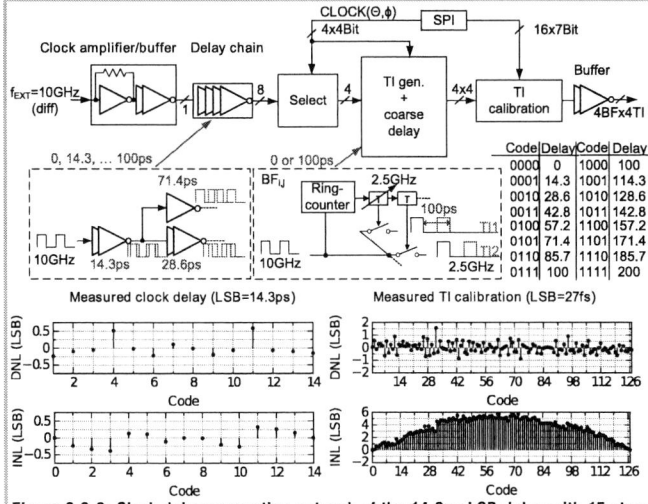

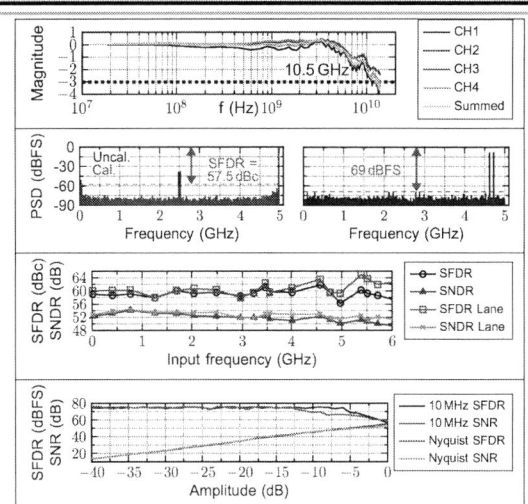

Figure 6.3.3: Clock delay generation network of the 14.3ps LSB delay with 15 steps yielding in 200ps total delay range and TI calibration with ~27fs LSB. Measured DNL and INL of the clock generation and TI calibration.

Figure 6.3.4: Measured ($\Phi_0=45°$, $\theta_0=0°$): input bandwidth, Nyquist- and two-tone spectrum (2048 points, uncal/cal), SFDR and SNR across f_{IN}, and SFDR/SNR across the amplitude.

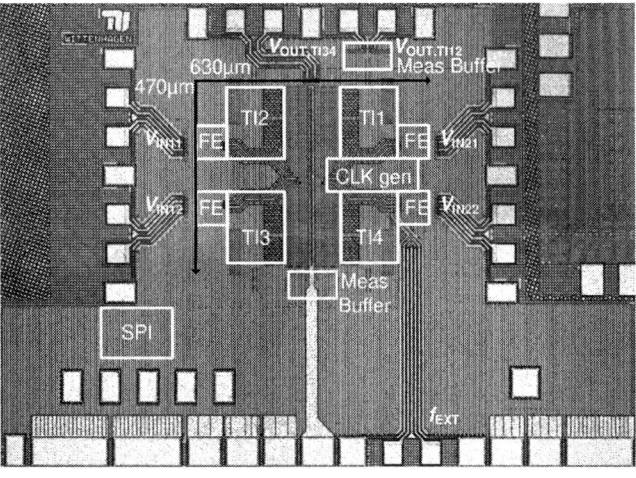

Figure 6.3.5: Measured channel linearity, normalized array factor of an external and internal phase shift (PS), across f_{IN} (no beam squint), and the gain across the angle (other angle 45°) and f_{IN}.

	ISSCC 2020 [1]	SSCL 2020 [2]	JSSCC 2025 [3]	RFIC 2025 [4]	**This Work**
Technology (nm)	65	28	65	65	**22 FDSOI**
# Elements	1x4	1x2	1x2	1x8	**2x2**
BF-Combination	VTC	Charge active	Charge active	Current mode	**Charge passive**
Bandwidth (GHz)	0.5	0.07	1.5[3]	0.85[3]	**10.5**
f_S (GS/s)	1	0.6-4	3	1.7	**10**
f_{IN} (GHz)	0.5	-	1.2	0.8	**5**
SFDR (dBc)	38	-	33	52	**57.5**
SNDR (dB)	29.9	-	32.8	37.3	**50.1**
BF Gain (dB)	11.75	-	5.75	18	**7.5**
Power/Channel (mW)	10[1]	35	37.3	7.5	**52[4] 18.7(FE) 11.6(TaH) 7(CLK) 14.7(BE)**
Supply					
Area (mm²)	0.31	1.2	0.45	0.66	**0.3**
FOM_Schreier[5] (dB)	130.9	-	135.8	138.8	**153.9**
FOM_Walden[6] (fJ/conv)	1566	-	681	589.5	**79.58**

[1]Including ADC   [2]no FE-buffer   [3] $f_S/2$
[4]excluding measurement buffer   [5]FOM_Schreier=SNDR+10log₁₀(f_S/(2Power))
[6]FOM_Walden=Power/($f_S\ 2^{ENOB}$)

Figure 6.3.6: Comparison with state-of-the-art beamforming receiver.

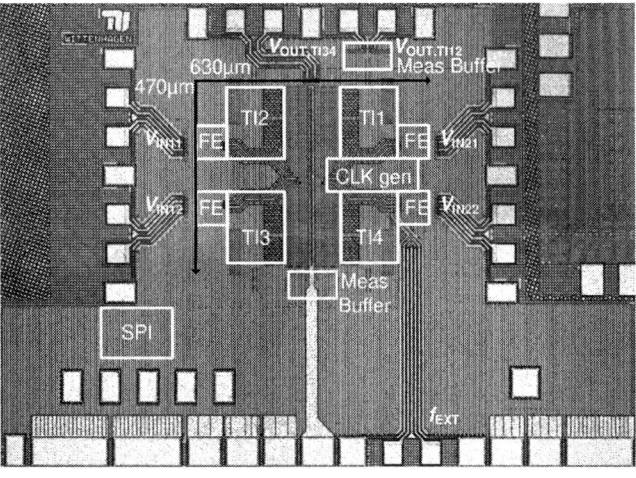

Figure 6.3.7: Die micrograph of the 2×2 10GS/s beamforming receiver.

- 2026 IEEE International Solid-State Circuits Conference

979-8-3315-8937-0/26 $31.00 © 2026 IEEE

ISSCC 2026 / SESSION 6 / EXPLORATORY RECEIVER ARCHITECTURES FROM GHz TO THz / 6.4

6.4 A 436-to-472GHz 4-Element IF Beamforming Phased-Array Receiver in 65nm CMOS

Hao Guo, Hao-Tao Hu, Zhicheng Lin, Zijun Guo, Xiaoyue Xia, Kam Man Shum, Chi Hou Chan

City University of Hong Kong, Hong Kong, China

Abstract

This paper presents a 436-to-472GHz 4-element IF beamforming phased-array receiver in 65nm CMOS. A two-step mixing architecture reduces layout overhead. The LO frequency multiplier chain provides >-1dBm over 180 to 214GHz to drive the THz mixer. The 6-bit

phase shifter covers 51 to 66GHz with 2.5° phase error. The on-chip stacked patch antenna array enables 10dBi gain and 91GHz 3dB gain-bandwidth. The receiver achieves ±35° scanning range, 36GHz system operating bandwidth and 27.5dB noise figure.

The terahertz (THz) spectrum (300GHz to 3THz) offers tremendous potential for ultra-high-speed wireless communication and high-resolution sensing. CMOS phased arrays are key enablers for THz systems, providing beam steering and enhanced sensitivity [1-3]. However, implementing THz phased-array systems remains challenging. The RF phase-shifting architecture is demanding to realize at THz frequencies, since the design of active phase shifters is difficult, while passive phase shifters will inevitably introduce significant insertion loss. In LO phase-shifting architectures, the presence of frequency multipliers imposes stringent phase-accuracy requirements on the phase shifters. The digital phase-shifting architecture requires multiple high-speed ADCs and DACs, increasing power consumption and design complexity. In contrast, the IF phase-shifting architecture avoids these limitations and has emerged as a promising approach for THz phased-array systems. However, it still presents several critical challenges above 300GHz: 1) layout complexity due to the mismatch in size between compact THz antenna arrays (<500µm pitch for 0.5λ spacing) and large-area low-frequency IF circuits; 2) limited system bandwidth imposed by both LO and IF circuits; and 3) the difficulty of designing wideband on-chip phased-array antennas, where broadband slot antennas require silicon lenses that constrain beam steering, while conventional lens-less patch antennas have narrow bandwidths.

This work demonstrates a 436-to-472GHz, four-element, IF-beamforming phased-array receiver in 65nm CMOS. To enhance scalability, a two-step mixing scheme is presented. Each channel integrates a broadband high-power sub-THz frequency multiplier chain to drive the THz mixer and extend the receiver operating bandwidth. A 6-bit phase shifter achieves 51-to-66GHz bandwidth by suppressing leakage introduced by transistor off-state capacitance. A wideband on-chip patch antenna array with a quartz superstrate achieves a 3dB gain-bandwidth of 91GHz through the introduction of three resonance modes. Measurements show a 36GHz system operating bandwidth with a ±35° beam-scanning range and a noise figure of 27.5dB.

As frequencies increase into the THz band, the wavelength reduction results in decreased inter-element spacing within antenna arrays, placing stringent constraints on the layout area available for active circuits. In contrast, IF circuits operating at lower frequencies occupy significantly larger chip area, posing critical challenges for compact array integration. To address these challenges, a scalable IF beamforming architecture is presented, employing a two-step mixing scheme. Figure 6.4.1 illustrates the block diagram of the 436-to-472GHz 4-element phased-array receiver. The receiver integrates an on-chip patch antenna array with 10dBi gain and 91GHz 3dB gain-bandwidth to capture the incoming THz signal. A THz subharmonic mixer (SHM) down-converts the RF signal to the first IF at 60GHz, using a broadband high-power LO signal generated by the multiplier chain. Subsequently, a millimeter-wave low-noise amplifier (LNA) and a 6-bit phase shifter (PS) with full 360° tuning range are employed to improve the noise figure and enable phase control. By increasing the operating frequency of the LNA and PS to 60GHz, the layout space occupied by the IF circuits is significantly reduced. The footprint of the IF circuit along the array direction only occupies 0.2mm, alleviating the problem of insufficient layout space. Furthermore, by placing the LO and IF circuits on each side of the antenna, the issue of inadequate layout space is further addressed, making the architecture highly suitable for THz 1-D phased-array integration. Finally, a fundamental mixer downconverts the signal from the first IF at millimeter-wave frequency to the second IF at a lower frequency for baseband processing.

Figure 6.4.2 illustrates the LO generation and mixer circuit in the presented phased-array receiver. An external $LO_{1/4}$ signal is frequency-multiplied to generate a high-power 200GHz LO signal that drives a second-order SHM for THz down-conversion. The SHM topology inherently reduces LO frequency requirements [4-10], enabling practical LO generation. To support wideband operation of the mixer, a broadband, high-power LO chain is required. The sub-THz frequency doubler is optimized by creating an AC short at LO frequency at the gate terminal, which enhances both conversion gain and output power, as shown in Fig. 6.4.2. This AC short is implemented using a λ/4 stub at the LO frequency, which outperforms the λ/2 stub with resonant capacitors employed in [11-13]. A driver chain comprising a three-stage differential PA and a frequency doubler efficiently drives the sub-THz doubler. The gate of the input-stage frequency doubler connects a λ/4 stub at $LO_{1/2}$ to enhance both CG and P_{out}. Subsequently, the output signal is distributed through a 4-way Wilkinson power

divider, followed by a balun. The balun grounds the center tap of the single-ended coil, eliminating lossy DC-blocking capacitors before the PA. Simulations show the presented LO multiplier chain delivers maximum P_{out} of 6.7dBm, with P_{out} exceeding -1dBm over 180 to 214GHz, meeting the LO power requirements of the mixer.

Figure 6.4.3 shows the IF chain of the presented phased-array receiver. The millimeter-wave LNA and buffer amplify the first IF signal from the THz SHM to improve the noise figure. A 6-bit 360° switched-type phase shifter achieves beam steering, followed by a fundamental double-balanced mixer that downconverts the signal to the second IF. The 45° phase-shifter cell operates in two states: phase-delay and bypass. In conventional designs, the bypass state suffers from abrupt phase change due to the equivalent capacitance C_{eq}, limiting the bandwidth. By shorting C_{eq} with transistors M1 and M2, the proposed design extends the bandwidth from 30 to 80GHz. The 90° cell operates in phase-delay and phase-advance states. In conventional phase-delay states, the off-state capacitance of the switch introduces strong leakage from the highpass network, causing large phase deviations. To address the issue, an additional CMOS switch M3 is introduced to break the leakage path in the highpass network, reducing leakage by 11.5dB compared with a conventional design, enabling 40-to-80GHz bandwidth. For the 180° cell, two additional switches (M4, M5) are introduced to suppress the leakage from the lowpass network in the phase-advance state. Measurement results show the 6-bit phase-shifter circuit (Fig. 6.4.3) achieves 360° phase coverage from 51 to 66GHz with a minimum phase error of 2.5°.

Conventional on-chip patch antennas suffer from narrow bandwidth due to low-profile geometry. To address this, a broadband stacked-patch antenna is presented. The antenna consists of an upper patch in the M10 layer and a lower patch in the M9 layer. A dual-slot configuration on the upper patch enables resonant tuning and inter-patch coupling control, while capacitive vias coupling reduces antenna size. A 300µm-thick quartz superstrate enhances radiation efficiency, and a surrounding GND wall suppresses coupling to adjacent components. Figure 6.4.4 shows the simulated surface current distributions at the three dominant resonant frequencies. The stacked configuration provides multi-resonant behavior that supports wideband operation. The simulated S11 of a single antenna remains below -10dB from 410GHz to 510GHz, corresponding to a relative bandwidth of 21.7%. The radiation efficiency is 38%. A 4-element array achieves 10dBi simulated peak radiation gain in the boresight direction and 3dB gain-bandwidth of 91GHz, while maintaining active S11 below -8dB over ±35° scan angles.

The presented 4-element phased-array receiver is implemented in a 65nm CMOS process with a chip area of 7.68mm² (Fig. 6.4.7) and a total power consumption of 842mW. First, the received signal spectrum shows no spurious interference (Fig. 6.4.5). Second, the transmitter is calibrated using a PM-5 power meter to enable accurate measurement of the equivalent isotropic conversion gain (EICG) and equivalent isotropic noise figure (EINF) of the receiver. With the IF fixed at 7GHz and RF swept, the measured 3dB bandwidth of EICG spans 436 to 472GHz. With the LO fixed at 50GHz and RF swept, a minimum EINF of 12dB is measured, corresponding to a receiver NF of 27.5dB using simulated antenna directivity. Finally, the radiation patterns are measured at 440, 450, 460, and 470GHz with IF fixed at 10GHz, demonstrating a ±35° beam-steering angle.

Figure 6.4.6 summarizes the state-of-the-art sub-THz and THz receivers. Prior work on phased-array receivers focuses on the sub-THz band (<300GHz), while THz receivers are limited to fixed-beam designs. This work represents a demonstration of a CMOS phased-array receiver operating above 300GHz. The performance is enabled by the efficient two-step mixing architecture, compact circuit integration, and broadband on-chip antennas, paving the way toward scalable, low-cost CMOS THz phased-array systems.

Acknowledgement:

This work was supported by the Hong Kong Research Grants Council Collaborative Research Fund under Grant C1009-22G. The authors thank Dr. Ka Fai Chan and Dr. Bao-Jie Chen for their assistance with chip measurements. The corresponding author is Chi Hou Chan.

106 • 2026 IEEE International Solid-State Circuits Conference

979-8-3315-8937-0/26 $31.00 © 2026 IEEE

References:

[1] S. Kato et al., "A 256-Element Ka-Band CMOS Phased-Array Receiver Using Switch-Type Quadrature-Hybrid-First Architecture for Small Satellite Constellations," *ISSCC*, pp. 1-3, Feb. 2025. http://doi.org/10.1109/ISSCC49661.2025.10904607

[2] Z. Zhang et al., "A 77GHz Hybrid TDMA-MIMO Phased-Array Radar with 186m Detection Range and 3cm Range Resolution," *ISSCC*, pp. 1-3, Feb. 2025. http://doi.org/10.1109/ISSCC49661.2025.10904816

[3] B. Liu et al., "A 132-to-148GHz CMOS 4TX-4RX FMCW Radar Transceiver Array with Cavity-Backed Antenna-in-Package Achieving 28dBm EIRP," *ISSCC*, pp. 1-3, Feb. 2025. http://doi.org/10.1109/ISSCC49661.2025.10904524

[4] I. Momson, S. Lee, S. Dong, and K. O. Kenneth, "425-to-25-GHz CMOS-Integrated Downconverter," *IEEE SSCL*, vol. 4, pp. 80-83, Mar. 2021. http://doi.org/10.1109/LSSC.2021.3067192

[5] K.-S. Choi, K.-M. Kim, D. R. Utomo, I.-Y. Lee and S.-G. Lee, "A Fully Integrated 490-GHz CMOS Receiver Adopting Dual-Locking Receiver-Based FLL," *IEEE JSSC*, vol. 57, no. 9, pp. 2626-2639, Sept. 2022. http://doi.org/10.1109/JSSC.2022.3159656

[6] Y. Zhu, P. R. Byreddy, S. Dong, K. O. Kenneth, and W. Choi, "A 430GHz CMOS Concurrent Transceiver Pixel Array for High Angular Resolution Reflection-Mode Active Imaging," *ISSCC*, pp. 1-3, Feb. 2022. http://doi.org/10.1109/ISSCC42614.2022.9731698

[7] H. Guo, K. Guo, Z. Lin, K. Man Shum, K. Fai Chan and C. Hou Chan, "A 460-GHz Receiver Using Second-Order Subharmonic Mixer in 65-nm CMOS," *IEEE TMTT*, vol. 73, no. 4, pp. 2440-2452, April 2025. http://doi.org/10.1109/TMTT.2024.3471676

[8] J. Bott et al., "A 335-407-GHz SiGe-Based Subharmonic Mixer Using a Fully Integrated LO Generation," *IEEE JSSC*, vol. 34, no. 6, pp. 675-678, June 2024. http://doi.org/10.1109/LMWT.2024.3389061

[9] C. Mangiavillano, A. Kaineder, K. Aufinger, and A. Stelzer, "A 1.42-mm² 0.45–0.49 THz Monostatic FMCW Radar Transceiver in 90-nm SiGe BiCMOS," *IEEE TTST*, vol. 12, no. 6, pp. 592-602, Nov. 2022. http://doi.org/10.1109/TTHZ.2022.3208069

[10] W. Choi, Z. Ahmad, A. Jha, J.-Y. Lee, I. Kim and K. O. Kenneth, "410-GHz CMOS Imager Using a 4th Sub-Harmonic Mixer with Effective NEP of 0.3 fW/Hz 0.5 at 1-kHz Noise Bandwidth," *IEEE VLSI*, pp. C302-C303, Jun. 2015. http://doi.org/10.1109/VLSIC.2015.7231299

[11] H. -C. Lin and G. M. Rebeiz, "A 200-245 GHz Balanced Frequency Doubler with Peak Output Power of +2 dBm," *IEEE CSICS*, pp. 1-4, Oct. 2013. http://doi.org/10.1109/CSICS.2013.6659189

[12] H. Guo, K. Guo, Z. Lin, K. M. Shum and C. H. Chan, "A 190-217-GHz Frequency Multiplier Chain with 13.2 dB Conversion Gain in 65-nm CMOS," *IEEE TCAS-II*, vol. 71, no. 12, pp. 4859-4863, Dec. 2024. http://doi.org/10.1109/TCSII.2024.3449631

[13] B. Cetinoneri, Y. A. Atesal, A. Fung and G. M. Rebeiz, "W-Band Amplifiers with 6-dB Noise Figure and Milliwatt-Level 170–200-GHz Doublers in 45-nm CMOS," *IEEE TMTT*, vol. 60, no. 3, pp. 692-701, March 2012. http://doi.org/10.1109/TMTT.2011.2165964

[14] S. Li, Z. Zhang, B. Rupakula and G. M. Rebeiz, "An Eight-Element 140-GHz Wafer-Scale IF Beamforming Phased-Array Receiver with 64-QAM Operation in CMOS RFSOI," *IEEE JSSC*, vol. 57, no. 2, pp. 385-399, Feb. 2022. http://doi.org/10.1109/JSSC.2021.3102876

[15] M. Elkhouly et al., "Fully Integrated 2D Scalable TX/RX Chipset for D-Band Phased-Array-on-Glass Modules," *ISSCC*, pp. 76-78, Feb. 2022. http://doi.org/10.1109/ISSCC42614.2022.9731626

[16] I. Abdo et al., "A Bi-Directional 300-GHz-Band Phased-Array Transceiver in 65-nm CMOS with Outphasing Transmitting Mode and LO Emission Cancellation," *IEEE JSSC*, vol. 57, no. 8, pp. 2292-2308, Aug. 2022. http://doi.org/10.1109/JSSC.2022.3179166

[17] D. Simic, K. Guo, and P. Reynaert, "A 420-GHz Sub-5-µm Range Resolution TX–RX Phase Imaging System in 40-nm CMOS Technology," *IEEE JSSC*, vol. 56, no. 12, pp. 3827-3839, Dec. 2021. http://doi.org/10.1109/JSSC.2021.3111152

[18] Q. Zhong, W. Choi, and K. O. Kenneth, "Terahertz Even-Order Subharmonic Mixer Using Symmetric MOS Varactors," *IEEE JSSC*, vol. 56, no. 2, pp. 355-366, Feb. 2021. http://doi.org/10.1109/JSSC.2020.3024177

[19] K. Guo and C. H. Chan, "A 0.68-THz Receiver with Third-Order Subharmonic Mixing in 65-nm CMOS," *IEEE JSSC*, vol. 59, no. 8, pp. 2469-2480, Aug. 2024. http://doi.org/10.1109/JSSC.2024.3371162

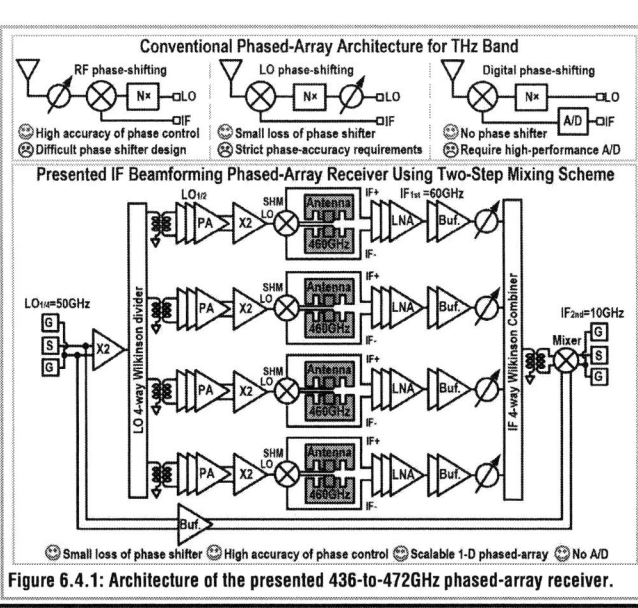

Figure 6.4.1: Architecture of the presented 436-to-472GHz phased-array receiver.

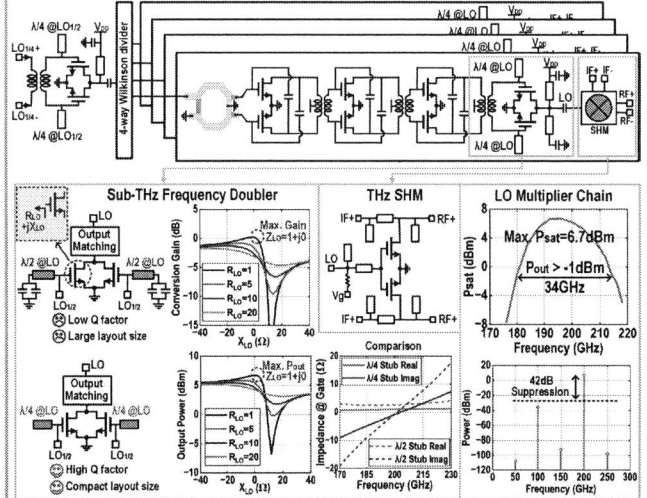

Figure 6.4.2: Schematic of LO multiplier chain with the optimized frequency doubler and subharmonic mixer.

ISSCC 2026 / SESSION 6 / EXPLORATORY RECEIVER ARCHITECTURES FROM GHz TO THz / 6.4

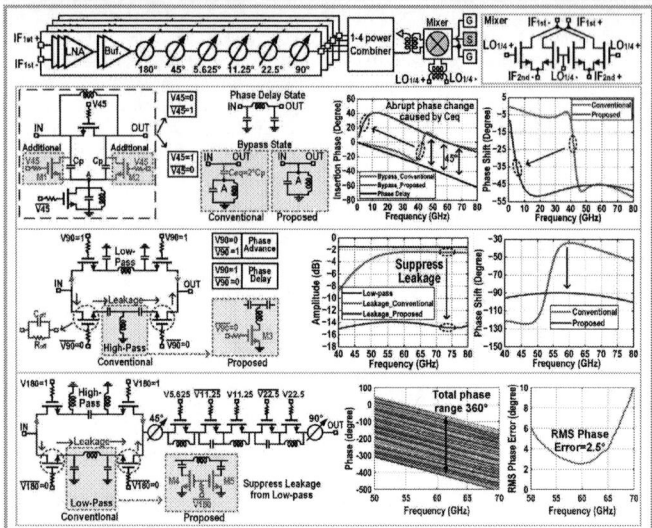

Figure 6.4.3: IF chain including proposed phase-shifter design, and measurement results.

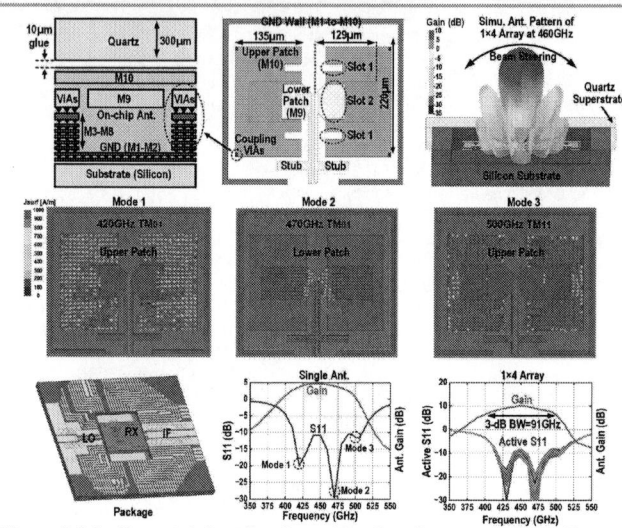

Figure 6.4.4: Presented broadband on-chip phased-array antenna with quartz superstrate.

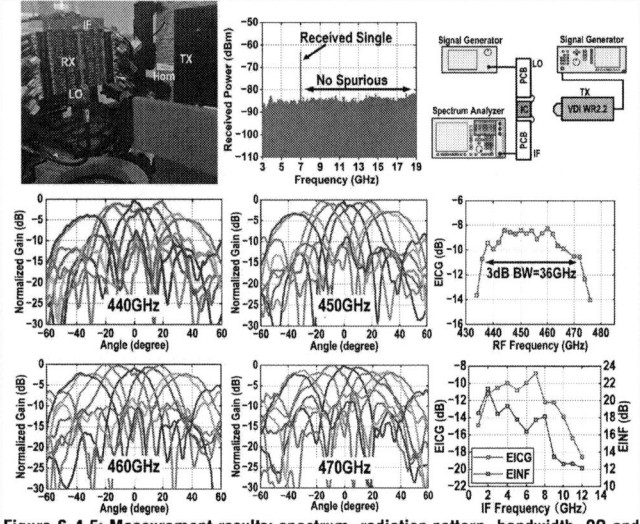

Figure 6.4.5: Measurement results: spectrum, radiation pattern, bandwidth, CG and NF.

Figure 6.4.6: Performance comparison with the state-of-the-art receivers operating at sub-THz and THz frequency band.

Ref.	This work	JSSC 2022 [14]	ISSCC 2022 [15]	JSSC 2022 [16]	JSSC 2021 [17]	JSSC 2021 [18]	JSSC 2024 [19]
Frequency (GHz)	436-472	<300			>300		
		139-155	130-164	256	420	552	682
Scan Range (°)	±35	±35	±20	±16	Fixed Beam	Fixed Beam	Fixed Beam
Element Number	4	4×2	16×16	4	1	1	1
Beamforming Architecture	IF Beamforming	IF Beamforming	RF Beamforming	LO Beamforming	N/A	N/A	N/A
Operating BW (GHz)	36	16	34	N/A	20E	N/A	2E
Mixing Method	2nd SHM + Fund. Mixer	LNA+ Fund. Mixer	No Mixer	2nd SHM	2-step Fund. Mixer	4th SHM	3rd SHM
Ant. Type	Patch	Patch	Glass RDL Hybrid	On-PCB Vivaldi	Dipole With Lens	Dipole	Slot With Lens
NF (dB)	27.5-36.7	6.4-7.5	7.5-10	20C	27	35	28.4
P_{DC}/ Element (mW)	211	145	200	750	601	N/A	183
Chip Area (mm²)	7.68	25.38	4.46S	4.17S	2.88	0.8	1.31
Process	65nm CMOS	45nm RFSOI CMOS	130nm SiGe BiCMOS	65nm CMOS	40nm CMOS	65nm CMOS	65nm CMOS

E Estimated from the curve   C Simulation results without antenna loss   S Size of single chip containing single element

Figure 6.4.7: Die micrograph of the 436-to-472GHz 4-element IF-beamforming phased-array receiver.

• 2026 IEEE International Solid-State Circuits Conference

ISSCC 2026 / SESSION 6 / EXPLORATORY RECEIVER ARCHITECTURES FROM GHz TO THz / 6.5

6.5 A 26/28/37/39GHz Reconfigurable Fully Connected MIMO Receiver Front-End with On-Chip Diplexer Achieving 52-to-70dB Blocker Rejection

Xuhao Jiang*[1,2], Qin Chen*[1,2], Ziang Zhang[1,2], Haiyang Xia[2], Shun Zhang[1], Yang Han[1], Xin Chen[2], Depeng Cheng[2], Long He[2], Xu Wu[1,2], Lianming Li[1,2], Xiaohu You[1,2]

[1]Southeast University, Nanjing, China, [2]Purple Mountain Laboratories, Nanjing, China
*Equally Credited Authors (ECAs)

Abstract

A 24.25-to-29.5 / 37-to-40GHz fully connected MIMO receiver (RX) front-end for 5G FR2 that enables 52-to-70dB calibration-free inter-band blocker rejection and up to -5dBm blocker tolerance is introduced. A compact on-chip diplexer and band-selective network are integrated in the RF chain to reject undesired signals before the mixer. The RX can support concurrent dual-band or multi-stream single-band operation, and achieve 4.8Gb/s 64-QAM non-contiguous inter-band carrier aggregation operation.

Fully-connected (FC) MIMO architectures have recently gained attention due to their benefits of reduced power consumption and chip area overhead [1-5]. To meet the evolving demand for full coverage of multiple non-contiguous 5G millimeter-wave bands, most MIMO receivers employ the Hartley architecture, which enables image selection for multi-band operation and achieves an image-rejection-ratio (IRR) exceeding 30dB [6,7]. However, it is very challenging for the Hartley-based MIMO receivers to maintain accurate phase resolution across wide RF bandwidths. To address this, the Cartesian phase-shifting technique has been introduced [8], in which the quadrature mixers replace the traditional IQ generator and eliminate the explicit phase shifters. Nevertheless, the overall IRR performance remains bandwidth-limited due to the high-precision I/Q calibration requirements. Moreover, as the noise and blockers from the image band propagate to the mixers through the wideband RF chain, the system NF and blocker tolerance are constrained. To overcome these limitations, this work presents a multi-band FC MIMO RX front-end that incorporates an on-chip diplexer and the band-selective network before the mixers, and the concurrent dual-band operation, i.e. non-contiguous inter-band carrier aggregation (CA) and frequency-domain multiplexed MIMO, can be supported. In particular, the diplexer splits the RF signal into two narrowband paths, thereby simplifying the phase-shifter design and improving the RX linearity. With measurements, the RX front-end can cover 5G frequency range 2 (FR2) bands of 24.25 to 29.5GHz (low-band, LB) and 37 to 40GHz (high-band, HB), while achieving inter-band blocker rejection up to 70dB.

Figure 6.5.1 (bottom) shows the block diagram of the FC MIMO phased-array RX architecture with the illustrated dual-stream operation. The chip adopts an RF phase-shifting architecture with two configurable modes: (1) single-band mode for the LB or HB spatially-multiplexed MIMO operation, and (2) dual-band mode for the concurrent inter-band CA and frequency-multiplexed MIMO operation, as illustrated in Fig. 6.5.1 (top right). Each antenna element employs a shared multi-band LNA followed by an on-chip diplexer, which separates the input signal into LB and HB paths. To enable multi-beam spatial filtering for MIMO, each LB and HB path comprises two independent amplitude-phase control branches, in which three stages of variable-gain amplifiers (VGAs) and a passive phase shifter are integrated. Thanks to the diplexer, the VGA/phase-shifter chain processes the signals from a single band, thereby relaxing its bandwidth and linearity performance requirements while inherently isolating the out-of-band blockers. The LB and HB paths for the same stream are then combined and downconverted to the intermediate frequency (IF). Instead of using a wideband Wilkinson power combiner (WPC) to merge the LB and HB paths, a transformer-based band-selective network (TFBN) is employed while providing an out-of-band notch [9]. Consequently, potential image interference is eliminated before the downconverter input. To simplify the LO generation, the LB and HB signals are configured as the mixer lower and upper sidebands, and by setting the IF frequency to 5 to 7GHz, the LO frequency range is reduced to 30 to 36GHz.

Figure 6.5.2 shows the key circuit schematics of the RX, highlighting the multi-band LNA, on-chip diplexer, and TFBN. By introducing a triple-winding transformer (TWTF)-based amplifier and a current-reused gm-boosted common-gate (CG) amplifier, the LNA can simultaneously cover LB and HB to suppress the overall RX NF. The TWTF input network provides multiple resonances for the wideband impedance matching and establishes a noise-canceling path for NF reduction [10,11], while the CG amplifier peaks gain in HB to stagger the overall frequency response. As a result, the LNA consumes 16mW and achieves <4dB NF over 20 to 40GHz with <1dB gain ripple. Figure 6.5.2 (top middle) illustrates the on-chip diplexer, which consists of a TWTF and two magnetically coupled resonators (MCRs). With the TWTF, the input signal is split into LB and HB, and subsequently the LB path is capacitively coupled to its MCR, whereas the HB path employs the direct coupling. Note that each MCR absorbs the parasitic input capacitances of the following VGA while providing the impedance transformation and single-ended-to-differential conversion. To maintain low in-band insertion loss while rejecting the out-of-band undesired signals, both MCRs are designed for the LB and HB operation, respectively, resulting in in-band impedance matching and out-band impedance mismatching at the VGA input (Fig. 6.5.2 middle center). Compared with single-band MCRs, the diplexer achieves the high out-of-band rejection performance with two mechanisms: (1) both LB and HB paths show high-order bandpass responses that can sharpen out-of-band roll-off, and (2) with the mutual loading effect between LB and HB paths, multiple transmission zeros are introduced, which can be designed carefully to eliminate the undesired signals. As demonstrated in Fig. 6.5.2 (bottom middle and right), the diplexer achieves <6.5dB insertion loss and >19dB inter-band rejection across both LB and HB, with an ultra-compact area of 0.017mm². Note that the diplexer is sensitive to its load variations, and thus its following VGAs adopt a current-steering structure to stabilize input impedance across various gain states and bias conditions [12,13]. As illustrated in Fig. 6.5.2 (top right), the TFBN consists of two current-combining transformer-based signal paths, and the band-selection operation is achieved by enabling or disabling the preceding VGA. With simulations, the TFBN achieves 8-to-13dB and 6-to-11dB inter-band rejection at LB and HB, respectively.

As a proof-of-concept, an FC RX front-end prototype is fabricated in a 65nm bulk CMOS process. Figure 6.5.3 presents its on-wafer measurement results. With a 6GHz IF, the RX achieves 25.5/28dB peak conversion gain with 4.7-to-5.4 / 5.1-to-6.1dB NF in the LB/HB. The measured IP1dB is -23/-25dBm at the maximum VGA gain and improves to -9/-10dBm at lower gain states, indicating the RX robust blocker tolerance benefits. The inter-band rejection is characterized by the IRR measurements, and it achieves 52 to 70dB across both operating bands without additional calibration. With 5-to-7GHz IF, the amplitude and phase control for the MIMO beamforming are evaluated with a 32GHz LO frequency and corresponding 25-to-27GHz LB and 37-to-39GHz HB. The measured gain tuning range exceeds 37dB in 0.5dB step, and corresponding RMS gain/phase errors are <0.2dB/1.5° and <0.35dB/3.9° in LB and HB, respectively. With the same IF range, the phase shifter covers a full 360° range with 5.625° step, the LB (HB) RMS phase/gain errors are <1.0°/0.63dB (1.2°/0.5dB).

Figure 6.5.4 (top) shows the on-wafer single-element RX measurements using 600MSym/s 256-QAM modulated signals. Without blockers, the RX achieves a minimum EVM_{RMS} of about -35dB, and by adjusting RX VGA gain states the measured EVM maintains <-30dB (3%) with the input power levels ranging from -48 to -22dBm. To demonstrate the RX blocker rejection capability, the RX is injected by the desired 27/39GHz signals of -35/-37dBm and a single-tone blocker at the image frequency with a wideband Wilkinson power combiner (WPC). The measurement results show that the RX can endure inter-band blockers up to -5dBm while maintaining EVM below 3%, confirming its robust inter-band blocker-tolerance benefits.

To demonstrate the MIMO and CA functionality, over-the-air (OTA) measurements are performed by using a four-element phased-array PCB module, which integrates four FC RX front-end chips and a 2dBi patch antenna array covering 24 to 40GHz with half-wavelength spacing at 40GHz. Moreover, the IF output streams from each chip are combined through a PCB 4:1 WPC network. With measurements, the RX module can support a beam steering range of -50° to 50°. By enabling single-element RX in the dual-band mode, Fig. 6.5.4 (bottom) shows its EVM measurements when receiving an inter-band CA signal at a 0.6m OTA distance. The CA signal consists of two independently modulated component carriers (CC1, CC2) at 27 and 39GHz. With the 400MSym/s 64QAM modulation, the EVM_{RMS} of CC1 and CC2 are -27.2 and -32.6dB, respectively, demonstrating successful inter-band CA reception. Note that the EVM difference of CC1 and CC2 is attributed to the signal quality of the available signal sources.

As shown in Fig. 6.5.5, the OTA MIMO measurements are conducted by enabling four elements in two scenarios: dual-band frequency-multiplexed MIMO at LB&HB (Fig. 6.5.5 top) and single-band spatially-multiplexed MIMO at HB (Fig. 6.5.5 bottom). For dual-band MIMO, the transmitting LB and HB beams are realized with two horn antennas, which are placed at incident angles of 0° and -20° relative to the RX array. With the 600MSym/s 64-QAM signals, when the HB and LB beams are received concurrently/individually, the measured EVM_{RMS} are -27.06/-28.18dB and -28.9/-29.47dB for the LB and HB beams, respectively. The slight EVM degradation observed in concurrent multi-band beams could be due to the crosstalk between the PCB IF routings for two received streams. To demonstrate the single-band MIMO operation, the two beams are aligned with the spatial nulls of each other, e.g. incident angles of -30° and 10°. As indicated in the Fig. 6.5.5 (bottom), the measured spatial null exceeds 20dB at 39GHz without the beam tapering. With 200MSym/s 16-QAM signals, the EVM_{RMS} of the two beams achieves -19.51/-20.03dB, consistent with theoretical expectations.

Figure 6.5.6 summarizes and compares this work with state-of-the-art multi-band receivers for 5G NR FR2 applications. Benefiting from the on-chip diplexer and TFBN, this work features an excellent calibration-free inter-band blocker rejection with competitive NF and IP1dB. Moreover, the RX front-end can support multiple operating modes, enabling both MIMO and inter-band CA with an FC-tile architecture, while dissipating 0.1W per element per stream. The die micrograph is shown in Fig. 6.5.7, with a total chip area of 1.79mm×1.54mm.

Acknowledgement:
This work was supported by the National Key Research and Development Program of China under Grant No. 2023YFB4403803, and in part by the Major Key Project of PCL.

References:
[1] S. Mondal et al., "A 28/37GHz Scalable, Reconfigurable Multi-Layer Hybrid/Digital MIMO Transceiver for TDD/FDD and Full-Duplex Communication," *ISSCC*, pp. 82-84, Feb., 2020. http://doi.org/10.1109/ISSCC19947.2020.9063167
[2] M. Yang et al., "K/Ka-Band Hybrid-Packaged Four-Element Four-Beam Phased-Array Transmitter and Receiver Front-Ends with Optimized Beamforming Passive Networks," *IEEE JSSC*, vol. 59, no. 10, pp. 3142–3155, Oct. 2024. http://doi.org/10.1109/JSSC.2024.3391893
[3] Y. Hu et al., "A 28-GHz Hybrid Beamforming Transmitter with Spatial Notch Steering Enabling Concurrent Dual Data Streams for 5G MIMO Applications," *IEEE JSSC*, vol. 59, no. 10, pp. 3378-3391, Oct. 2024. http://doi.org/10.1109/JSSC.2024.3399220
[4] O. Hassan et al., "An Eight-Channel 15–55 GHz Dual-Beam Receive Phased-Array Beamformer IC with 2.9–4.2 dB NF for Multiband 5G Operation," *IEEE TMTT*, vol. 73, no. 1, pp. 661-673, Jan. 2025. http://doi.org/10.1109/TMTT.2024.3421642
[5] E. Naviasky et al., "A 71-to-86GHz Packaged 16-Element by 16-Beam Multi-User Beamforming Integrated Receiver in 28nm CMOS," *ISSCC*, pp. 218-220, Feb. 2021. http://doi.org/10.1109/JSSC.2021.3118641
[6] M. -Y. Huang et al., "A 24.5–43.5-GHz Ultra-Compact CMOS Receiver Front End with Calibration-Free Instantaneous Full-Band Image Rejection for Multiband 5G Massive MIMO," *IEEE JSSC*, vol. 55, no. 5, pp. 1177-1186, May 2020. http://doi.org/10.1109/JSSC.2019.2959495
[7] J. Pang et al., "A Power-Efficient 24-to-71 GHz CMOS Phased-Array Receiver Utilizing Harmonic-Selection Technique Supporting 36dB Inter-Band Blocker Rejection for 5G NR," *ISSCC*, pp. 434-436, Feb. 2022. http://doi.org/10.1109/ISSCC42614.2022.9731619
[8] S. Mondal et al., "A Reconfigurable 28/37GHz Hybrid-Beamforming MIMO Receiver with Inter-Band Carrier Aggregation and RF-Domain LMS Weight Adaptation," *ISSCC*, pp. 72-74, Feb. 2018. http://doi.org/10.1109/ISSCC.2018.8310189
[9] Q. Chen et al., "A Compact Reconfigurable 24-29.5/38-43.5GHz Phased Array Transceiver Front-End with Self-Interference Rejection and Wideband IF Supporting TDD/FDD Operation," *IEEE CICC*, pp. 1-3, Apr. 2025. http://doi.org/10.1109/CICC63670.2025.10983676
[10] H. Xu et al., "A 5-to-16GHz Reconfigurable Quadrature Receiver with 50% Duty-Cycle LO and IQ-Leakage Suppression," *ISSCC*, pp. 88-90, Feb. 2024. http://doi.org/10.1109/ISSCC49657.2024.10454493
[11] J. Ke et al, "A 52–73-GHz LNA with Tri-Coupled Transformer for Gm Boosting and Enhanced Noise Canceling," *IEEE JSSC*, vol. 59, no. 3, pp. 668–676, Mar. 2024. http://doi.org/10.1109/JSSC.2023.3340300

[12] Y. Yu et al., "A 22.4-to-30.7GHz Phased-Array Receiver with Beam-Pattern Null-Steering and Beam-Tracking Techniques Achieving >30.2dB OTA-Tested Spatial Rejection," *ISSCC*, pp. 94-96, Feb. 2024. http://doi.org/10.1109/ISSCC49657.2024.10454530
[13] S. Shakib et al., "A Wideband 28GHz Power Amplifier Supporting 8×100MHz Carrier Aggregation for 5G in 40nm CMOS," *ISSCC*, pp. 44-45, Feb. 2017. http://doi.org/10.1109/ISSCC.2017.7870252
[14] Y. Yu et al., "A 26/28/39-GHz Reconfigurable Phased-Array Receiver Front-End with Built-In Calibration Technique for 5G New Radio," *IEEE JSSC*, vol. 60, no. 2, pp. 382-393, Feb. 2025. http://doi.org/10.1109/JSSC.2024.3425889

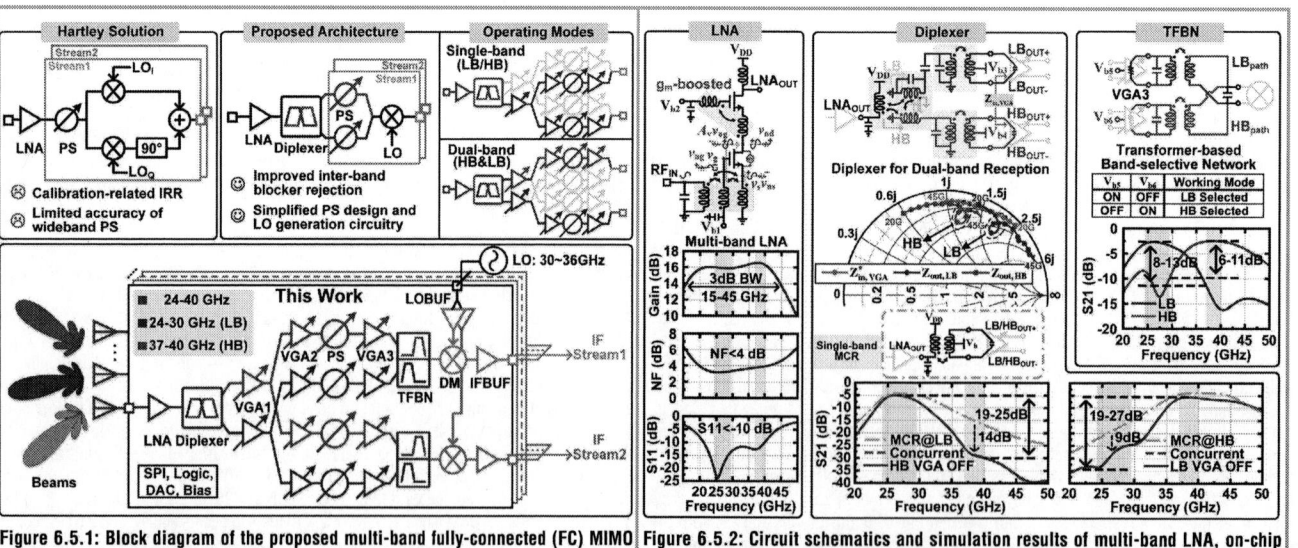

Figure 6.5.1: Block diagram of the proposed multi-band fully-connected (FC) MIMO phased-array RX with built-in diplexer.

Figure 6.5.2: Circuit schematics and simulation results of multi-band LNA, on-chip diplexer and transformer-based band-selective network (TFBN).

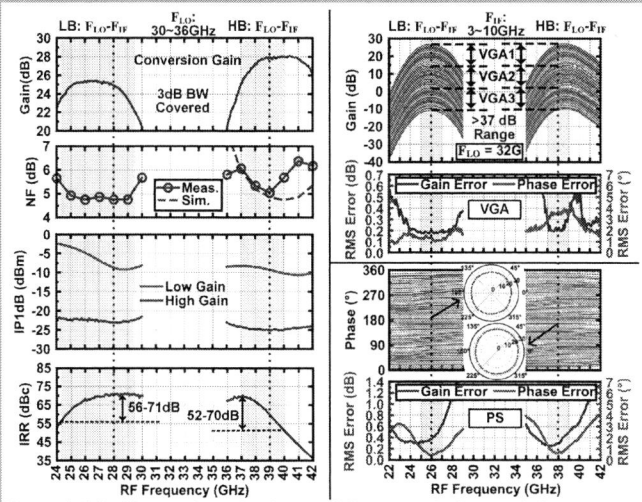

Figure 6.5.3: Measured single-element RX front-end characteristics including conversion gain, NF, IP1dB, IRR performance (left), and amplitude/phase control blocks measurement results(right).

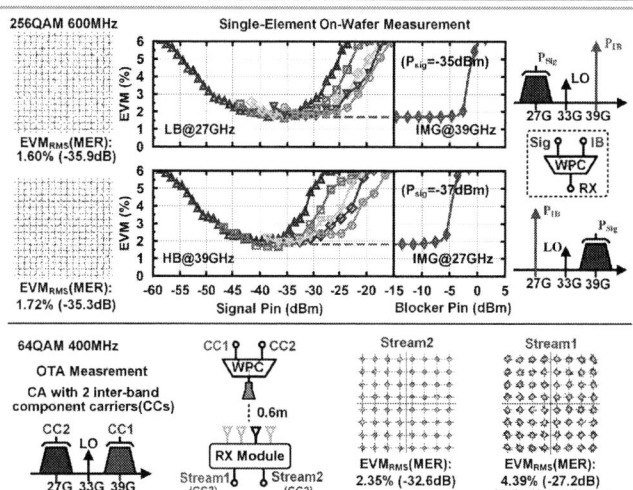

Figure 6.5.4: Measured single-element RX modulation performance with different gain states and inter-band blocker input power (top), and constellations and EVM results of concurrent inter-band CA mode (bottom).

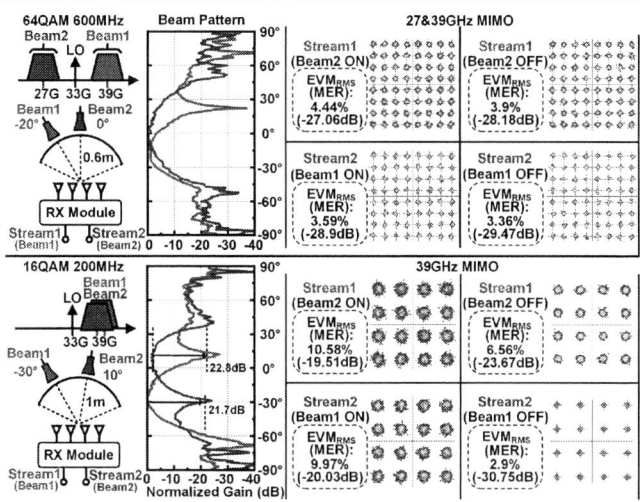

Figure 6.5.5: Measured four-element RX phased-array module beam patterns and constellations of MIMO in the dual-band mode (top) and the single-band mode (bottom).

	This Work	[14] JSSC'25	[9] CICC'25	[4] TMTT'25	[7] ISSCC'22	[1] ISSCC'20	[6] JSSC'20	[8] ISSCC'18
Functionality	FC MIMO/CA RX FE	Phased Array RX FE	Phased Array T/R FE	FC MIMO RX FE	Phased Array RX FE	FC MIMO TRX	Phased Array RX	FC-HBF MIMO/CA RX
Technology	65nm CMOS	65nm CMOS	65nm CMOS	90nm SiGe BiCMOS	65nm CMOS	65nm CMOS	45nm CMOS SOI	65nm CMOS
IF Freq. (GHz)	5-7	4-6	4.5-9.5	N/A	8	0.1	3-5	3.75-4.5
LO Freq. (GHz)	30-36	10-12	30-36	N/A	16-26	28/37	25-50	30-36
RF Freq. (GHz)	24.25-29.5 / 37-40	24-30 / 36-40	24.25-29.5 / 38-43.5	15-55	24 25-71	26.5-29 / 35.25-38	24.5-43.5	27-29.75 / 35-38.75
RX Gain (dB)	25.5 / 28	20.1 / 19.7	21 / 20	25	15	44 / 37	35.2	33 / 26.5
NF (dB)	4.7-5.4 / 5.1-6.1	5.3-7.5 / 4.7-8.4	5.2-6.8 / 6.3-8.5	2.9-5.2	3.6-8.0	7.9-9.6 / 8.8-9.9	5.3-7.4	5.7-8.4 / 8.5-9.7
IP1dB (dBm)	-23/-9§	-25/-10§	-31 / -32.8	-21 / -17	-34	-31.8	-29 / -22	-27/-8§	-30 / -23
Inter-Band Rejection (dB)	56-71¶ / 52-70§	>43	13-40# / 14-41#	N/A	>36	>28	30-56	10/35* / 17/33* @28G @37G
RX Power/ Chan. (mW)	99.67*	130	109.9	185*	32-75	98.75*	60	52.5*
Area/Chan. (mm²)	0.7	1.92	1.1	1.6	1.2	1.05	0.77	0.46

*Per element per stream §With gain control #Fixed IF at 6-GHz ^After digital calibration

Figure 6.5.6: Performance comparison of 5G NR FR2 mm-Wave phased-array receivers.

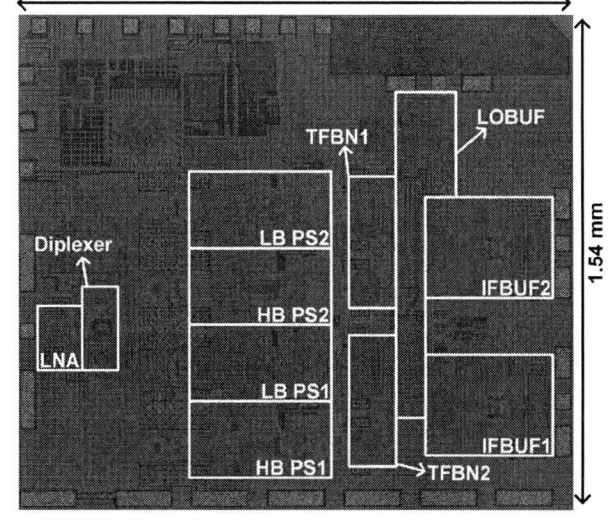

Figure 6.5.7: Die micrograph.

ISSCC 2026 / SESSION 7 / IMAGE SENSORS AND RANGING / OVERVIEW

Session 7 Overview: *Image Sensors and Ranging*
IMAGE SENSORS & DISPLAYS SUBCOMMITTEE

Session Chair: Augusto Ximenes
CogniSea, Seattle, WA

Session Co-Chair: Andreas Suess
Google, Mountain View, CA

The Image Sensors & Ranging session is characterized by a mix of classic and emerging challenges. Classic challenges focus on hardware integration, such as realizing highly integrated Systems-on-Chip (SoCs) and developing intensity imagers that combine high resolution with low noise. New challenges center on integration and efficiency, including the on-chip integration of intelligence for signal processing, the efficient use of on-chip memory for histogramming in LiDARs, and the development of beyond-silicon sensing layers.

The first paper presents a complete LiDAR SoC for consumer applications, this is followed by a compute-in-memory SoC for on-chip inference. Two high-resolution dToF sensors implementing on-chip non-destructive compressed histogramming come next, followed by a low-power SWIR depth sensor based on Ge-on-Si SPADs. A novel approach to LiDAR image processing is introduced afterwards, with a sensor performing on-chip object segmentation using an Ising machine. On-sensor intelligent signal processing is also the topic of the following paper, implementing a fully analog vision SoC.for AI-driven applications. The session concludes with two high-resolution, ultra-low-noise image sensors for consumer applications.

1:30 PM

7.1 54×42 LiDAR 3D-Stacked System-On-Chip with On-Chip Point Cloud Processing and Hybrid On-Chip/Package-Embedded 25V Boost Generation

Neale A. W. Dutton, STMicroelectronics, Edinburgh, United Kingdom

In Paper 7.1, STMicroelectronics presents a 54×42 resolution 3D-stacked SPAD-based LiDAR receiver SoC in 65/40nm CMOS with all-digital Time of Flight pixels, point-cloud processing, and a hybrid on-chip/package-embedded 25V boost converter. Operating at 60fps, the system achieves <1cm ranging error up to 9.6m while consuming 153mW for the receiver SoC.

1:55 PM

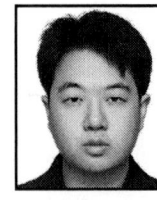

7.2 VoxCAD: A 0.82-to-81.0mW Intelligent 3D-Perception dToF SoC with Sector-Wise Voxelization and High-Density Tri-Mode eDRAM CIM Macro

Haoyang Sang, University of Macau, Macau, China

In Paper 7.2. University of Macau presents a 3D Perception dToF SoC for efficient on-chip inference utilizing Sector Wise Voxelization and a high-density Tri-Mode eDRAM Macro enabling memory, CIM, and XNOR capability.

2:20 PM

7.3 A Multi-Range, Multi-Resolution LiDAR Sensor with 2,880-Channel Modular Survival Histogramming TDC and Delay Compensation Using Double Histogram Sampling

Minkyung Kim, Sungkyunkwan University, Suwon, Korea

In Paper 7.3 Sungkyunkwan University presents a 720×192 dToF sensor with a 2,880-channel modular survival histogram TDC (shTDC) array. The shTDC supports non-destructive accumulation while concentrating depth on active bins, reducing memory by 75% with precision comparable to one-step hTDC.

ISSCC 2026 / February 16, 2026 / 1:30 PM

2:45 PM

7.4   A 480×320 CMOS LiDAR Sensor with Tapering 1-Step Histogramming TDCs and Sub-Pixel Echo Resolvers

Wonjong Roh, Sungkyunkwan University, Suwon, Korea, SolidVue, Seongnam, Korea

In Paper 7.4, Sungkyunkwan University presents a 480×320 SPAD-based solid-state LiDAR sensor fabricated in a 90nm BSI CIS process, with tapering 1-step histogramming TDCs and sub-pixel echo resolvers (SPER) that enable 2× finer time resolution of 500ps and 4× higher image resolution with 85% memory reduction, achieving 3.8cm precision at 100m and 4.1cm accuracy under 110klux sunlight at 25fps.

3:00 PM

7.5   A 26.0mW 30fps 400x300-pixel SWIR Ge-SPAD dToF Range Sensor with Programmable Macro-Pixels and Integrated Histogram Processing for Low-Power AR/VR Applications

Matteo Perenzoni, Sony Semiconductor Solutions Europe, Trento, Italy

In Paper 7.5, Sony Semiconductor Solutions presents a 10µm-pitch Ge-on-Si SPAD sensor featuring a 400×300 array operating at room temperature for low-power SWIR dToF ranging. The sensor integrates a programmable macro-pixel architecture and on-chip histogram processing to enable selective activation. It consumes 26mW when operating 500 macropixels of 3×3 SPAD at 30fps.

7

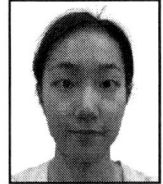

3:35 PM

7.6   A 128×96 Multimodal Flash LiDAR SPAD Imager with Object Segmentation Latency of 18µs Based on Compute-Near-Sensor Ising Annealing Machine

Jingyi Wang, Fudan University, Shanghai, China

In Paper 7.6, Fudan University outlines a 128×96 multimodal Flash LiDAR SPAD Imager utilizing an on-chip Ising compute engine realizing object segmentation.

4:00 PM

7.7   A Fully Reconfigurable Hybrid SPAD Vision Sensor with 134dB Dynamic Range Using Time-Coded Dual Exposures

Kieop Hong, Ulsan National Institute of Science and Technology, Ulsan, Korea, Sogang University, Seoul, Korea

In Paper 7.7, Ulsan National Institute of Science and Technology introduces an intensity/event-based hybrid SPAD sensor characterized by 134dB dynamic range in both working modes and low event noise.

4:25 PM

7.8   A 55nm Intelligent Vision SoC Achieving 346TOPS/W System Efficiency via Fully Analog Sensing-to-Inference Pipeline

Zhengke Yang, Southern University of Science and Technology, Shenzhen, China

In Paper 7.8, Southern University of Science and Technology develops a fully analog intelligent vision SoC in 55nm process that achieves end-to-end analog datapath from sensing to multi-layer inference, delivering 346TOPS/W system efficiency and 8791TOPS/W peak MAC efficiency. The system energy efficiency of 2.489TOPS/W for classification and object detection of five classes from PASCAL VOC 2007 achieves mAP@0.5 of 20.5%.

4:50 PM

7.9   A 1.09e⁻-Random-Noise 1.5µm-Pixel-Pitch 12MP Global-Shutter-Equivalent CMOS Image Sensor with 3µm Digital Pixels Using Quad-Phase-Staggered Zigzag Readout and Motion Compensation

Sanggwon Lee, Samsung Electronics, Hwaseong, Korea

In Paper 7.9, Samsung Electronics describes a 12MP global-shutter equivalent CMOS image sensor having a 1.5µm pixel pitch read by a 2×2-shared cluster-parallel ADC achieving 1.09e⁻ rms.

5:15 PM

7.10   A 200MP 0.61µm-Pixel-Pitch CMOS Imager with Sub-1e⁻ Readout Noise Using Interlaced-Shared Transistor Architecture and On-Chip Motion Artifact-Free HDR Synthesis for 8K Video Applications

Guanjing Ren, SmartSens Technology, Shanghai, China

In Paper 7.10, SmartSens Technology presents a 1/1.28-inch 200MP stacked BSI CMOS image sensor with a 0.61µm pixel pitch and sub-electron read noise, featuring a novel interlaced-shared transistor 2×2 pixel architecture with dual conversion gain, on-chip motion-artifact-free HDR synthesis, and self-CG ratio calibration, enabling 8K HDR video with high dynamic range and low power consumption.

DIGEST OF TECHNICAL PAPERS • 111

979-8-3315-8937-0/26 $31.00 © 2026 IEEE

ISSCC 2026 / SESSION 7 / IMAGE SENSORS AND RANGING / 7.1

7.1 54×42 LiDAR 3D-Stacked System-On-Chip with On-Chip Point Cloud Processing and Hybrid On-Chip/Package-Embedded 25V Boost Generation

Neale A. W. Dutton[1], Xavier Branca[2], Mathieu Thivin[2], Steven Collins[1], Herve Thuaire[2], Fabrice Martin[2], Vincent Clemencon[2], Mohammed Al-Rawhani[1], Duncan Hall[1], Axel Crocherie[1], Cedric Pastorelli[2], Sophie Taupin[2], Severin Trochut[2], Abhishek Singh[1], Andreas Assmann[1], Bruce R. Rae[1], Pascal Mellot[2]

[1]STMicroelectronics, Edinburgh, United Kingdom, [2]STMicroelectronics, Grenoble, France

Abstract

We report a 54×42-resolution SPAD-based LIDAR receiver SoC leveraging 3D-stacked 65/40nm technology with high-frequency TDC and 2 counters for ToF and intensity exposure, with peripheral memories used for image signal processing (ISP) frame stores, point cloud generation & HDR management. A hybrid on-chip & package-embedded 25V boost converter is described integrating SPAD supply generation synchronized with sensor frame timing. The sensor is integrated in a LIDAR module for 9.6m ranging with <1cm error.

Time of Flight (ToF) or Light Detection and Ranging (LIDAR) devices are in widespread use for scene mapping and machine vision tasks. ToF receivers sensor arrays utilising CMOS single-photon avalanche diodes (SPAD) and time-to-digital converters (TDC) offer high frequency all-digital operation. Longer distance range devices are found in automotive and specialist applications utilizing large histogram generation memories at the periphery of the sensor outside the imaging plane [1]. Short to medium range devices are prevalent in mobile, xR devices, security, robotics. For those applications, monolithic front-side illuminated (FSI) devices likewise have histogram generation at the periphery [2] so suffer from a scalability limitation to higher resolutions, whereas this is overcome by those leveraging wafer-to-wafer 3D-stack technologies that place circuits under back-side illuminated (BSI) SPADs with in-pixel TDCs, counters or compact histograms [3-5] but these tradeoff resolution and distance range. This paper overcomes that tradeoff with a combined approach of a 54×42 resolution SPAD-based LIDAR receiver System-on-Chip (SoC) leveraging 3D-stacked 65/40nm technology with high-frequency TDC and 2 counters under the pixel for ToF and intensity exposure, and peripheral memories used for image signal processing (ISP) frame stores, point cloud generation & HDR management. Furthermore, the SoC and LIDAR module package integrates power generation synchronized with sensor frame timing, for the 25V SPAD supply facilitating battery-powered operation.

Figure 7.1.1 provides a block diagram of the LIDAR receiver SoC. The sensor array comprises 220×198 BSI SPADs. 4 SPADs are connected in a 2×2 macropixel so the native resolution of the sensor output is 110×96 with an additional test row for a 110×98 total physical array. The 10.17µm SPADs are wafer-to-wafer stacked on logic wafer with 4 front-end circuits, a flash time to digital converter (TDC) and two integrating counters. The macropixel digital column-parallel output is sequentially read into embedded frame store memories and processed by the point-cloud ISP chain. There is a secondary small 3×2 array for an optical zero-distance reference channel. Distance computation at 60fps has default binning 2×2 and cropping for 54×42 resolution. Flexible digital binning (3×3, 4×4, etc.) offers a range of lower depth map resolutions while decreasing processing overhead in the ISP chain and increasing signal to obtain longer distance range. A frame timing engine controls the sequencing of the array exposure and readout, ToF operation and the ISP operations. ToF clock generation is performed by a PLL and programmable multi-frequency clock divider outputting from 8.0MHz to 1GHz to the pixel arrays. The pixel receives a burst of clock pulses and can skip clock pulses to produce a delay versus the VCSEL pulse. The clock generation block embeds programmable delays for VCSEL pulse generation versus the high frequency clock to the TDC's to calibrate on-die and package signal chain delays removing global ToF distance offsets from the point cloud data. The flash TDC resolution is simply the ToF clock period: from 1ns and in binary multiples thereof (2ns, 4ns, etc.). On a SPAD event pulse the flash TDC triggers the respective counter and is immediately ready for the next photon. This is an advantage in high ambient versus other TDC architectures which limit photon throughput [5-7]. The SoC has SPI communications controller and low-voltage differential signaling (LVDS) trigger interfaces to either a 1 or a 2 channel companion laser-driver IC (LDIC). A micro-controller unit (MCU) controls and synchronizes system operation of both the SoC and LDIC over SPI interface. There are three on-chip frame memories for ToF range, ToF signal and non-ToF ambient which form the on-chip point cloud. Two output data paths access the computed point cloud either up to 60fps via a 1Gb/s D-PHY single lane interface, or solely over I3C for lower frame rate applications with combined control and data over the same 2-wire interface. A second PLL is embedded for MIPI D-PHY clock generation.

Figure 7.1.2 details the imaging operation and timing. There are three primary pixel operating modes: sequential histogram ToF [4,8-10], sequential multi-frequency ToF [3,11-13] and non-ToF intensity. For all three modes, the sensor field of view is split into two halves to perform a dual scan operation. The timing engine performs the ping-pong array operation of upper half-global shutter exposure (54×21 points) and simultaneous lower half readout and processing, then vice versa in the next exposure. Two sequential upper/lower exposures form one sub-frame. In sequential histogram ToF mode, the ToF clock is iteratively delayed versus the laser trigger. The frame rate is a function of the required total histogram bins and exposure time per sub-frame, up to maximum 32 sub-frames. In multi-frequency ToF mode, two counters provide partial range information, so to fully compute a distance measurement the frame timing engine supports up to 7 ToF sub-frames per output frame at different frequencies, and a final 8th sub-frame dedicated to a non-ToF ambient exposure.

The pixel reads out directly to the point cloud image signal processing (ISP) engine and results are stored in the ISP frame memories at up to 480 sub-frames per second for point cloud output up to 60fps. The exposure time in sequential histogram ToF mode is the same for all sub-frames whereas in the multi-frequency mode the sub-frame exposure times and frequency selection may be fixed or varied depending on system tuning and settings. A high dynamic range management (HDRM) process is performed on a frame basis by the MCU in parallel to the ongoing exposure, accessing the frame memories for saturated pixels and subsequently dynamically switching on/off SPADs on a per-macropixel basis. HDRM has two key benefits: mitigating time-domain pile-up distortion effects of close targets (a signal rate dependent depth shift) to preserve depth linearity and managing SPAD load current for close targets which is critical for thermal management and control of power generation in package.

The SoC embeds power generation for the high-voltage supply (VHV 19-to-25V range) for the SPAD detectors in a hybrid on-die/in-package design in Fig. 7.1.3 for integration in battery powered applications. There are two configurations of the power management. The first boost converter configuration comprises low-voltage (2.8 to 3.3V) controller on the 40nm SoC and package (or board) embedded high-voltage (≤25V) MOS, tank capacitor, Schottky diode, sense resistor and inductor. The controller loop comprises linear current DAC, and a controller with resistive feedback, load current sense feedback and synchronization from the frame timing engine. The SPAD load during an exposure is lightly loaded in dark conditions, but always > 0A, so the loop is in continuous conduction mode (CCM). However, as shown in the Fig. 7.1.3 frame timing diagram there is a short ~10µs period between sub-frames where all SPADs are off so there is always zero load in this condition and so the loop would enter undesired discontinuous conduction mode (DCM). The advantage of integrating the boost controller on the receiver SoC is it allows timing synchronization between controller and the LIDAR frame timing engine to both avoid DCM and to control load transients for under/overshoots at starts and ends of exposures shown as the timing signal "DCDC Shutter" in Figs. 7.1.2 & 7.1.3. This signal disables the CCM controller output for this system inter-sub frame transition time, blanking the external MOS driver output while relying solely on the tank capacitor to provide charge. Consequently, this approach transforms the DCM-induced >3V overshoot (in simulation) at end of sub-frame exposure to a 5µs duration passive discharge undershoot which is signal dependent but systematic on every sub-frame. The second configuration is for external VHV supply and the controller is powered down.

The receiver SoC is embedded in a compact LIDAR module measuring 12.8×6.1×4.6mm³ (0.359cm³) with a F#1.3 receiver lens, dual-channel LDIC and two VCSELs to generate images, point clouds and ranging measurement. The hybrid power converter discrete components are embedded in the optical module package near the SoC die and the overall DC/DC implementation sustains maximum 60mA average current. The DC/DC boost converter is measured with 2.8V input and 20V output achieving 55% efficiency at <10mA load rising to 65% efficiency at 60mA maximum load. Load regulation is 67mV at 30mA SPAD array consumption. The point cloud images are captured with a 54° by 42° field of view (FOV) where each pixel spans 1°×1°. The average ToF modulation frequency employed in the multi-frequency ToF mode measurements is 15.63MHz. Figure 7.1.4 illustrates the sensor output of two scenes indoor and outdoor with 54×42 resolution in multi-frequency ToF mode. The ToF depth map & non-ToF ambient are both direct from the sensor. Both outdoor and indoor images of ToF amplitude are normalized by exposure time & scaled for the HDRM SPAD selection for pixels with N enabled of 16 SPADs, scaling by $16/(N \cdot T_{exposure})$. The signal confidence and reflectance images are generated in software post processing. The range and accuracy for the center of the image are shown in Fig. 7.1.5 from 5cm to 9.6m measured in the dark at 60fps, for 3 square Lambertian target with reflectances: 3% black, 13% dark grey and 88% white. The accuracy is <1cm across the full distance range for all targets. Both range measurement and image capture were performed with optical peak power of 4.2W at 940nm wavelength. The measured power consumption is 153mW of the LIDAR receiver SoC, and 432mW power consumption of the full LIDAR system at 60fps including all power generation, the VCSEL driver IC and VCSELs.

Figure 7.1.6 shows a performance summary and comparison to the state of the art. The sensor presented has competitive accuracy and distance range to the compared works while the power also remains competitive with the addition of point cloud data processing. Figure 7.1.7 shows a microscope image of the LIDAR SoC die.

112 • 2026 IEEE International Solid-State Circuits Conference

979-8-3315-8937-0/26 $31.00 © 2026 IEEE

ISSCC 2026 / February 16, 2026 / 1:30 PM

References:

[1] O. Kumagai et al., "A 189×600 Back-Illuminated Stacked SPAD Direct Time-of-Flight Depth Sensor for Automotive LiDAR Systems," *ISSCC*, 2021, pp. 110-111. https://doi.org/10.1109/ISSCC42613.2021.9365961

[2] N.A.W. Dutton et al., "A time-correlated single-photon-counting sensor with 14GS/s histogramming time-to-digital converter," *ISSCC*, 2015. https://doi.org/10.1109/ISSCC.2015.7062997

[3] B. Kim et al., "A 48×40 13.5mm Depth Resolution Flash LiDAR Sensor with In-Pixel Zoom Histogramming Time-to-Digital Converter," *ISSCC*, 2021, pp. 108-109. https://doi.org/10.1109/ISSCC42613.2021.9366022

[4] F. Mattioli Della Rocca et al., "A 128 × 128 SPAD Motion-Triggered Time-of-Flight Image Sensor With In-Pixel Histogram and Column-Parallel Vision Processor," in *IEEE Journal of Solid-State Circuits*, vol. 55, no. 7, pp. 1762-1775, July 2020. https://doi.org/10.1109/JSSC.2020.2993722

[5] P. Padmanabhan et al., "A 256×128 3D-Stacked (45nm) SPAD FLASH LiDAR with 7-Level Coincidence Detection and Progressive Gating for 100m Range and 10klux Background Light," *ISSCC*, 2021, pp. 111-112. https://doi.org/10.1109/ISSCC42613.2021.9366010

[6] V. Sesta, et al. "Range-Finding SPAD Array with Smart Laser-Spot Tracking and TDC Sharing for Background Suppression," in *IEEE Open Journal of the Solid-State Circuits Society*, vol. 2, pp. 26-37, 2022. https://doi.org/10.1109/OJSSCS.2021.3116920

[7] E. Manuzzato et al., "A 64x64 Pixel Flash LiDAR SPAD Imager with Distributed Pixel-to-Pixel Correlation for Background Rejection, Tunable Automatic Pixel Sensitivity and First-Last Event Detection Strategies for Space Applications," *ISSCC*, 2022, pp. 96-97. https://doi.org/10.1109/ISSCC42614.2022.9731622

[8] K. Hatakeyama et al., "A Hybrid Indirect ToF Image Sensor for Long-Range 3D Depth Measurement under High Ambient Light Conditions," *IEEE Symposium on VLSI Technology and Circuits*, 2022, pp. 46-47. https://doi.org/10.1109/VLSITechnologyandCir46769.2022.9830139

[9] R. K. Henderson et al., "A 256×256 40nm/90nm CMOS 3D-Stacked 120dB Dynamic-Range Reconfigurable Time-Resolved SPAD Imager," *ISSCC*, 2019, pp. 106-107. https://doi.org/10.1109/ISSCC.2019.8662355

[10] S. Zhuo et al., "Solid-State dToF LiDAR System Using an Eight-Channel Addressable, 20W/Ch Transmitter, and a 128x128 SPAD Receiver with SNR-Based Pixel Binning and Resolution Upscaling," *IEEE Custom Integrated Circuits Conference*, 2022. https://doi.org/10.1109/CICC53496.2022.9772823

[11] A. Payne et al., "A 512×424 CMOS 3D Time-of-Flight image sensor with multi-frequency photo-demodulation up to 130MHz and 2GS/s ADC," *ISSCC*, 2014, pp. 134-135. https://doi.org/10.1109/ISSCC.2014.6757370

[12] C. S. Bamji et al., "1Mpixel 65nm BSI 320MHz demodulated TOF Image sensor with 3µm global shutter pixels and analog binning," *ISSCC*, 2018, pp. 94-95. https://doi.org/10.1109/ISSCC.2018.8310200

[13] C. Tubert et al., "4.6µm Low Power Indirect Time-of-Flight Pixel Achieving 88.5% Demodulation Contrast at 200MHz for 0.54MPix Depth Camera," *IEEE European Solid State Circuits Conference*, 2021, pp. 135-138. https://doi.org/10.1109/ESSCIRC53450.2021.9567878

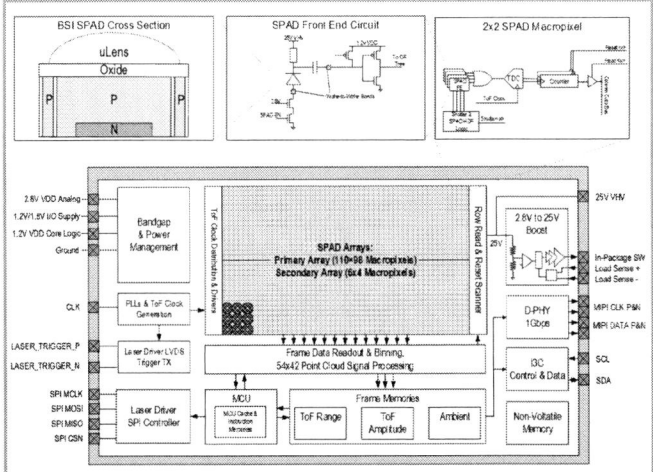

Figure 7.1.1: LiDAR sensor System-on-Chip (SoC). The sensor has native 110×98 macropixel resolution and with default 2×2 binning and cropping provides 54×42 point cloud resolution. The array is split into two for simultaneous upper half exposure and lower half readout, and vice versa.

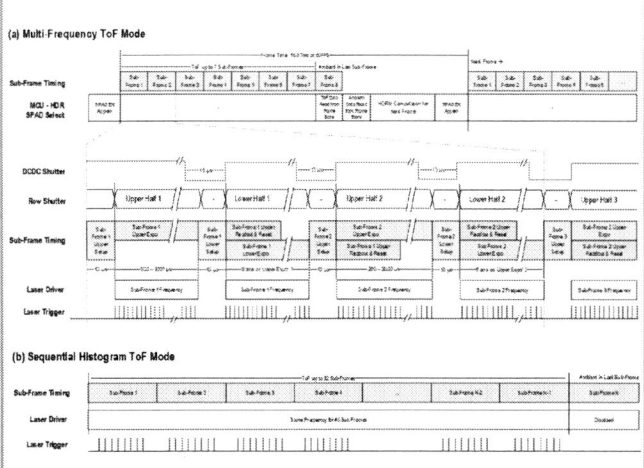

Figure 7.1.2: SOC Timing of (a) Multi-Frequency ToF Mode (b) Sequential Histogram ToF Mode. Non-ToF intensity mode is the same as the last sub-frame ambient exposure in both other modes.

979-8-3315-8937-0/26 $31.00 © 2026 IEEE

ISSCC 2026 / SESSION 7 / IMAGE SENSORS AND RANGING / 7.1

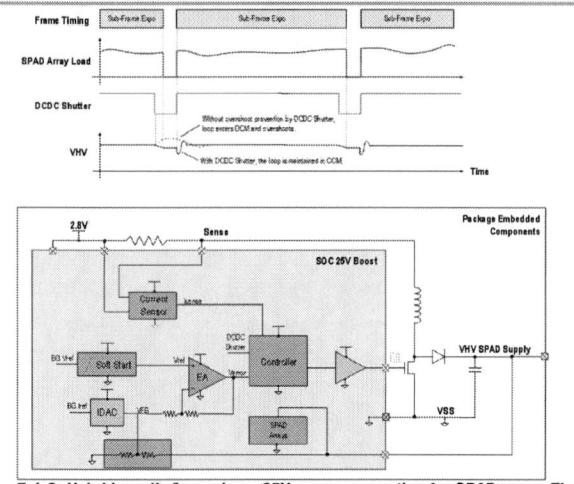

Figure 7.1.3: Hybrid on-die/in-package 25V power generation for SPAD array. Timing diagram shows the synchronization with the frame timing controller. The block diagram illustrates the primary configuration, the second configuration is not shown and uses external VHV supply and the controller is powered down.

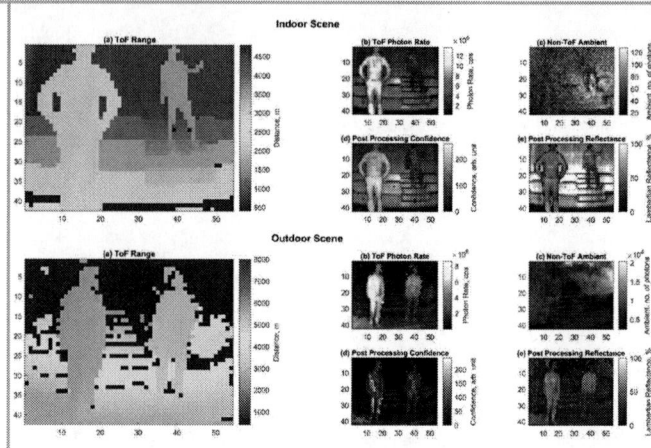

Figure 7.1.4: Multi-frequency ToF mode 54×42 output of indoor and outdoor scenes with 4.2W peak optical power. Each set displays (a) ToF Depth, pixels with low confidence show as black (b) ToF Rate = scaled ToF Amplitude data for Exposure & HDRM SPAD patterning (c) Non-ToF Ambient (d) Software Post Processed Confidence (e) Software Post Processed ToF Reflectance.

Figure 7.1.5: Ranging distance sweep measurement in the dark: Linear ranging 5cm to 9.6m and accuracy <1cm across that range. Beyond 9.6m, the ISP quantizes to the max range of 9.6m.

Reference	This Work	ISSCC'21 [1]	ISSCC'22 [3]	JSSC'20 [4]	ISSCC'19 [9]	CICC'22 [10]	VLSI'22 [8]	ISSCC'18 [12]	ESSERC'23 [13]
Detector	SPAD	SPAD	SPAD	SPAD	SPAD	SPAD	ITOF PD	ITOF PD	ITOF PD
CMOS Node	65/40nm	90/40nm	110nm FSI	40nm FSI	65/40nm	180nm BCD	110nm BSI	65nm BSI	65/40nm
Detector Pixel Pitch	10.17µm	10µm	27µm	20µm	9.6µm	25µm	5.8µm	3.2 µm	4.8 µm
Physical Resolution	220x196	189x600	96x120	256x256	256x256	128x128	640x480	1024x1024	804x672
Output Resolution	54x42	168x63	48x40	128x128	64x64	32x32	640x480	1024x1024	804x672
Depth Points	2,268	10,584	1,920	16,384	4,096	1,024	307k	1M	540k
ToF Exposures	≤7	1	5	3	2	32	3	>2	>2
ToF Range	5cm – 9.6m	200m	5cm-0.8m	45m	0.5m - 3.5m	50m	15m	0.5m – 30m	0.4m – 7m	0.25m - 4m
ToF Accuracy	≤ 1cm (dark)	≤ 30cm	1.35cm	40cm	2.1cm	17cm	15cm	≤ 6cm at 10m	≤ 0.25cm at 4m	≤1cm at 4m
ToF Frequency Range	8-1000MHz	1000MHz	50-400MHz	31-500MHz	50M / 2GHz	Not reported	~20MHz (*)	320MHz	200MHz
Depth Processing	On-chip	On-chip	External	External	External	External	External	Partially On-chip	Partially On-chip
Power	153mW RX SOC at 60 FPS	1192mW RX SOC at 20 FPS	840mW RX at unknown FPS	185mW RX at 30 FPS	77.6mW RX at 30 FPS	227mW System at 10 FPS	Not reported	650mW RX SOC at 30 FPS	300mW System at 30 FPS
SPAD Power Generation	Hybrid on-die/in-package	External	External	External	External	External	N/A	N/A	N/A

Figure 7.1.6: Table of comparison of selected ToF image sensors. (* = estimated)

Figure 7.1.7: Die photomicrograph. Both primary and secondary BSI SPAD arrays are visible as black rectangles. A backside metal shield covers the die.

• 2026 IEEE International Solid-State Circuits Conference

ISSCC 2026 / SESSION 7 / IMAGE SENSORS AND RANGING / 7.2

7.2 VoxCAD: A 0.82-to-81.0mW Intelligent 3D-Perception dToF SoC with Sector-Wise Voxelization and High-Density Tri-Mode eDRAM CIM Macro

Haoyang Sang*[1], Zhao Wang*[1], Longzhen He[1], Guangshu Zhao[1], Wenao Xie[2], Bo Wang[3], Rui P. Martins[1], Man-Kay Law[1]

[1]University of Macau, Macau, China, [2]KAIST, Daejeon, Korea, [3]Hamad Bin Khalifa University, Doha, Qatar
*Equally Credited Authors (ECAs)

Abstract

VoxCAD, an intelligent low-power dToF SoC for end-to-end 3D perception applications, is presented with three features: 1) LiDAR dToF sensing-integrated 2D-ROI guided point cloud construction for reducing sensor transfer energy and latency; 2) coarse-fine sector-wise voxelization with a central-computation unit for reducing on-chip memory size and EMAs; and 3) reconfigurable tri-mode eDRAM CIM macros with hybrid 1T1C/2T1C cells for supporting 3D transformers at high area efficiency and storage density.

The rapid growth of the electric-vehicle (EV) and smart robotics market drives demand for full self-driving (FSD), advanced driver-assistance systems (ADAS) and autonomous navigation. The foundation of such systems lies in end-to-end environment perception from 2D/3D sensors and efficient execution of AI tasks. Figure 7.2.1 presents a conventional 3D perception flow, beginning with a range sensor for retrieving object distances, followed by the transferring of the sensor data with user datagram protocol (UDP)/serial interfaces (I/Fs) to a preprocessor for point-cloud construction. The voxelized point clouds are then stored in external memory, and finally processed by CPUs/GPUs for AI tasks such as 3D object detection, segmentation, and classification. Yet, CPUs/GPUs lack application-specific optimizations resulting in low energy efficiency and high latency.

Recently, several application-specific digital processors have emerged to improve energy efficiency and latency on 3D perception tasks [1-3], but still suffer from three critical challenges. First, the sensor data transfer via UDP/serial I/Fs dominates performance, using 60 to 72% of system energy and accounting for 51 to 65% of latency [1]. Even though on-chip sensor integration in [4–6] reduces transfer costs, this approach can only demonstrate 2D imaging and simple AI tasks (e.g., face recognition). Also, the implemented single-photon avalanche diode (SPAD)-based vision SoC with 2D/3D imaging capability in [7] still suffers from a limited energy efficiency and latency due to the fully digital implementation. Furthermore, despite the improved energy efficiency and latency offered by 3D stacking [6,8], such techniques can still incur a high fabrication and packaging cost. Second, voxelization workloads require massive external memory access (EMA), which can degrade system performance by up to 7× in [1]. To reduce EMA, [2] stores voxels fully on-chip, but inevitably demands an impractically large on-chip memory (5.9Mbit). Third, although mainstream 3D transformers allow higher accuracy, they can introduce an excessive hardware cost due to the 2× increase in parameters [9,10], not to mention the underutilization and low energy efficiency of CNN-specific processors due to different data reusability on mixed CNN/attention operations [1-3].

To resolve the above challenges, this work presents an ultra-low power intelligent SoC, called VoxCAD, with key features including: 1) sensing-integrated 2D region-of-interest (ROI)-guided point cloud construction for reducing sensor data transfer latency, energy, and EMAs; 2) sector-wise voxelization with a central-computation unit (CCU), enabling full on-chip voxel management with lower memory requirements and EMA reduction; and 3) reconfigurable tri-mode eDRAM compute-in-memory (CIM) macros based on hybrid 2T1C and 1T1C cells to maximize the storage density and area efficiency, while supporting both 2D CNN and attention workloads.

Figure 7.2.2 depicts the SoC with a 128×1 line SPAD direct time-of-flight (dToF) sensor. It supports 256×128 2D intensity and 512×128 3D distance output via scanning operation with galvo mirror. The dToF operates at 600MHz with an on-chip PLL, or 200MHz with an external clock to support always-on tasks. The tightly coupled preprocessor receives distance data with ultra-low latency and energy, and then performs point cloud construction and radius outlier removal (ROR) denoising. This work employs four CCUs for point cloud voxelization and feature extraction. Each CCU includes a point-to-voxel unit, a programmable SIMD, a run-length encoder (RLE), and a look-up table (LUT). The point-to-voxel units support point cloud voxelization through quantization, rounding, and translation lookaside buffer (TLB)-based address translation. The eight reconfigurable eDRAM CIM macros (516Kbit total) support high-density storage and energy efficient computation. A RISC host core with two global 8KB SRAMs handle system control and data buffering, while a 64b 2D mesh network-on-chip (NoC) interconnects all processing modules.

Figure 7.2.3 presents the 2D/3D dToF sensing-integrated preprocessor and 2D-ROI guided point cloud construction flow. Each dToF pixel integrates a SPAD, an analog front-end, and an up/down counter. In 2D mode, the sensor delivers 128×256 8b intensity maps based on photon counting at 20 frame per second (fps) with on-chip PLL, or at 1fps in ultra-low-power mode. In 3D mode, the up/down counting scheme with coarse-fine coding resolves the depth information [11], yielding a 9b distance resolution. The preprocessor streams up to 1152b/cycle of intensity/distance data into eDRAM macros, where the lightweight 2D CNN identifies regions of interest (ROI) such as cars, pedestrians, obstacles, and other objects. We divide the intensity data into multiple azimuth sectors based on rotation angles.

Only the ROI-flagged azimuth sectors require distance measurement, while the unflagged ones are skipped for reducing point cloud scale. We construct the measured distance into point clouds with ROR denoising. Measurement results demonstrate that this sensing-integrated ROI-guided flow reduces transfer energy and latency by more than 99% compared to UDP solutions, and suppresses 82% of raw points with only a 2.8% mean average precision (mAP) loss. It further achieves a 70 and 77% reduction in the system's end-to-end inference latency and energy consumption, respectively.

Figure 7.2.4 illustrates the CCU with sector-wise voxelization flow. To reduce the on-chip storage requirement, VoxCAD employs a coarse-to-fine sector division strategy. Initially, we partition the point clouds into 'M' coarse azimuth sectors, with each further subdivided into 'N' fine sectors along distance bins. However, the non-uniform point cloud density across near and far distances can still compromise the on-chip memory requirement. This work employs logarithmic distance bins to balance memory allocation among the near-field dense points and far-field sparse points. Each fine sector is processed sequentially in a streaming manner to reduce the on-chip memory size. Specifically, first we quantize and round the points to voxel indices, and forward the results to the TLB. TLB previews the sector, and constructs a voxel table to record point counts per voxel. Guided by the voxel table, the RISC core allocates contiguous SRAM blocks for non-empty voxels, which are then processed by the SIMD to obtain both the mean position and cluster features. Finally, we transfer the voxel-level features to the eDRAM CIM macros for higher-dimensional feature extraction. Measurement results show that this sector-wise voxelization flow eliminates the need for a large on-chip buffer, reduces EMA of activation by nearly 99%, and achieves an overall EMA reduction of up to 87% at a division level of 'M=8' and 'N=6' with only 1% mAP loss. This offers a trade-off between sector granularity and model parameter EMA overhead.

Figure 7.2.5 shows the reconfigurable tri-mode eDRAM macros, integrating buffers, 256×258 1T1C storage cells, 16×128 2T1C compute cells, a word-line (WL) decoder, WL drivers, bit-line (BL) drivers, local sense amplifiers (LSAs), global sense amplifiers (GSAs), and 4 SAR-ADCs. Each macro comprises 16×128 local cell arrays (LCAs), with each containing 16×2 1T1C cells, one 2T1C cell, and one LSA. It supports three operation modes: 1) CIM mode for matrix multiplication; 2) memory mode for tensor storing; and 3) XNOR mode for similarity checking. In CIM mode, the index buffer stores the non-zero value index, with the all-zero value operations skipped for achieving high area and energy efficiency. Non-zero weights are stored in 1T1C cells, then loaded to the local bit-line (LBL) and further amplified by LSA. Non-zero activation controls the gate of 2T1C cells to charge or discharge the coupled capacitor. Afterwards, the LCAs perform intra-inter charge sharing (for single-bit activation, multi-bit weight, and multi-channel addition), and the SAR-ADCs convert the final voltage into digital values. CCU supports shift-add operations for multi-bit accumulation. The simultaneous operation of multiple LCAs contributes to high utilization for high area efficiency. For memory mode, we write the data to 1T1C cells via global bit-line (GBL), and readout by LSA and GSA. We first configure half macros to memory for storing Q/K/V tensors, and later reconfigure them to CIM to reduce data movement for attention layers. In XNOR mode, weight and activation are represented in a complementary format. The gate of 2T1C is always on, and we load the activation on WL & WLB to readout the stored complementary weight in capacitors. The bitwise XNOR operation is accomplished by capacitor charging. LCAs then perform charge sharing to obtain multi-tensor similarity and readout by ADCs. The system initializes with low power always-on mode, and it performs similarity checking between 2 adjacent intensity frames at 1fps. If the similarity does not exceed the threshold, the system remains in low-power mode to conserve energy, and performs 2D-ROI detection and 3D detection otherwise. By using high-density 1T1C cells for storing and 2T1C cells for computing, the macro achieves a 9.2× storage density increase compared to 2T1C-only macro [12]. Further, the intra/inter charge sharing with sparsity exploration contributes to a 34.2× area efficiency improvement compared to previous works [12,13].

Figure 7.2.7 illustrates the 130nm chip photo, summary table, and power measurement setup. The processing modules operate at a supply voltage from 1.1 to 1.7V, with an operating frequency from 10 to 250MHz. The dToF operates at 600MHz with an on-chip PLL, and switches to ultra-low power always-on mode with a 200MHz external clock, based

114 • 2026 IEEE International Solid-State Circuits Conference

979-8-3315-8937-0/26 $31.00 © 2026 IEEE

on temporal similarity checking between two adjacent intensity frames at 1fps with CIM macros. Figure 7.2.6 presents the measured SoC detection results and benchmarks this work with the state of the art. Using the Kitti dataset and our measured point clouds, this work achieves 80.3 and 75.6 mAP on the dynamic sparse voxel transformer (DSVT) [10] backbone, respectively. With a 97.5% reduction in system energy and a 91.2% reduction in latency, this work also demonstrates a 101-to-406× density FoM improvement compared to digital point cloud processors [1-3], and a 12.7-to-153× improvement over prior DRAM macros [12,13] through the integration of high-density eDRAM macros. It also exhibits an end-to-end energy efficiency improvement of 3.51 to 6.85×, and a latency reduction of 2.4 to 22.9× compared to digital point cloud processors [1-3] through the co-optimization of sensor and processor. Compared to previous sensor-integrated smart vision SoCs [4-7], it offers an energy efficiency improvement of 9.5 to 60.5× and a latency reduction of 2.1 to 26.1× using the eDRAM CIM macro with sparsity exploration.

Acknowledgement:
The authors thank Macau FDCT (0004/2023/AKP and 004/2023/SKL), University of Macau (MYRG-GRG2024-00173-IME and MYRG-CRG2024-00028-IME), and Qatar Research Development and Innovation Council (ARG02-0421-240247) for financial support.

References:
[1] X. Feng et al. "A 28-nm energy-efficient sparse neural network processor for point cloud applications using block-wise online neighbor searching," *IEEE Journal of Solid-State Circuits* 59.9 (2024): 3070-3081. https://doi.org/10.1109/JSSC.2024.3386878
[2] S. Lim et al., "Hawkeye: A Point Cloud Neural Network Processor With Virtual Pillar and Quadtree-Based Workload Management for Real-Time Outdoor BEV Detection," *IEEE J. Solid-State Circuits*, vol. 60, no. 3, pp. 990-1001, March 2025, https://doi.org/10.1109/JSSC.2024.3508873
[3] X. Feng et al., "A Scalable BEV Perception Processor for Image/Point Cloud Fusion Applications Using CAM-Based Universal Mapping Unit," *IEEE J. Solid-State Circuits*, vol. 60, no. 3, pp. 1002-1013, March 2025, https://doi.org/10.1109/JSSC.2024.3514733
[4] M. Lefebvre and D. Bol, "MANTIS: A Mixed-Signal Near-Sensor Convolutional Imager SoC Using Charge-Domain 4b-Weighted 5-to-84-TOPS/W MAC Operations for Feature Extraction and Region-of-Interest Detection," *IEEE J. Solid-State Circuits*, vol. 60, no. 3, pp. 934-948, March 2025, https://doi.org/10.1109/JSSC.2024.3484766
[5] X. Yang et al., "A Bio-Inspired Spiking Vision Chip Based on SPAD Imaging and Direct Spike Computing for Versatile Edge Vision," *IEEE J. Solid-State Circuits,* vol. 59, no. 6, pp. 1883-1898, June 2024, https://doi.org/10.1109/JSSC.2023.3340018
[6] H. Song et al., "A 120 Frames/s CMOS Image Sensor With 8.19 TOPS/W Computing-In-Pixel for Energy-Efficient Low-Latency Face Detection," *IEEE J. Solid-State Circuits*, https://doi.org/10.1109/JSSC.2025.3592980
[7] L. Millet et al., "A 5500-frames/s 85-GOPS/W 3-D Stacked BSI Vision Chip Based on Parallel In-Focal-Plane Acquisition and Processing," *IEEE J. Solid-State Circuits*, vol. 54, no. 4, pp. 1096-1105, April 2019, https://doi.org/10.1109/JSSC.2018.2886325
[8] M.F. Amir et al., "3-D Stacked Image Sensor With Deep Neural Network Computation," in *IEEE Sensors Journal*, vol. 18, no. 10, pp. 4187-4199, 15 May, 2018, https://doi.org/10.1109/JSEN.2018.2817632
[9] I. Misra et al., "An end-to-end transformer model for 3D object detection," *IEEE/CVF Int. Conf. Comput. Vis.*, Oct. 2021, pp. 2906–2917, https://doi.org/10.1109/ICCV48922.2021.00290

[10] Z. Han et al., "DSVT: Dynamic sparse voxel transformer with rotated sets," *IEEE/CVF Conf. Comput. Vis. Pattern Recognit.*, June 2023, pp. 13445–13454, https://doi.org/10.1109/CVPR52729.2023.01299
[11] S. Park et al., "An 80×60 Flash LiDAR Sensor with In-Pixel Histogramming TDC Based on Quaternary Search and Time-Gated Δ-Intensity Phase Detection for 45m Detectable Range and Background Light Cancellation," *ISSCC*, 2022, pp. 98-199, https://doi.org/10.1109/ISSCC42614.2022.9731112
[12] D. Kim et al., "DPIM: A 19.36 TOPS/W 2T1C eDRAM Transformer-in-Memory Chip with Sparsity-Aware Quantization and Heterogeneous Dense-Sparse Core," *IEEE European Solid-State Electronics Research Conference*, 2024, pp. 141-144, https://doi.org/10.1109/ESSERC62670.2024.10719539
[13] S. Hong et al., "Dyamond: A 1T1C DRAM In-memory Computing Accelerator with Compact MAC-SIMD and Adaptive Column Addition Dataflow," *IEEE Symposium on VLSI Technology and Circuits*, 2024, https://doi.org/10.1109/VLSITechnologyandCir46783.2024.10631334

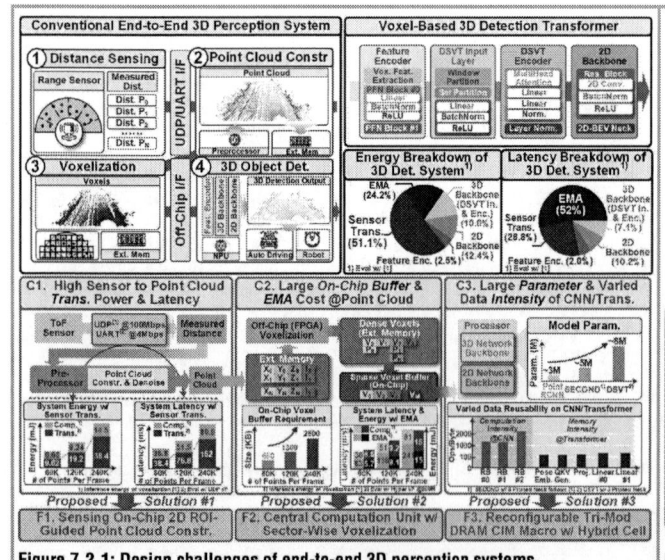

Figure 7.2.1: Design challenges of end-to-end 3D perception systems.

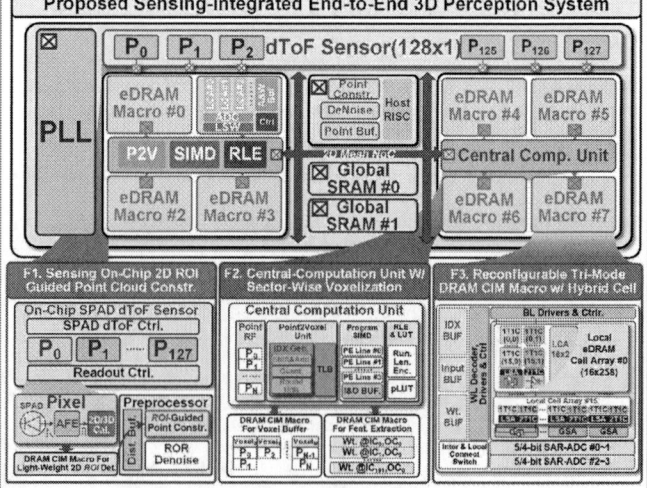

Figure 7.2.2: Intelligent 3D perception dToF SoC overview with three key features.

ISSCC 2026 / SESSION 7 / IMAGE SENSORS AND RANGING / 7.2

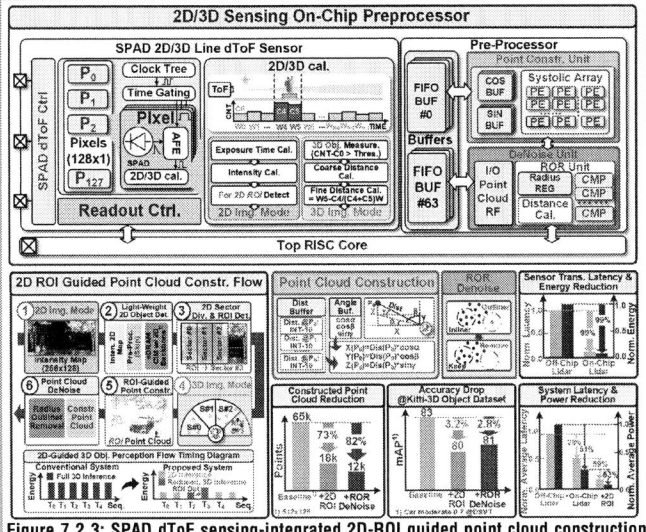

Figure 7.2.3: SPAD dToF sensing-integrated 2D-ROI guided point cloud construction flow.

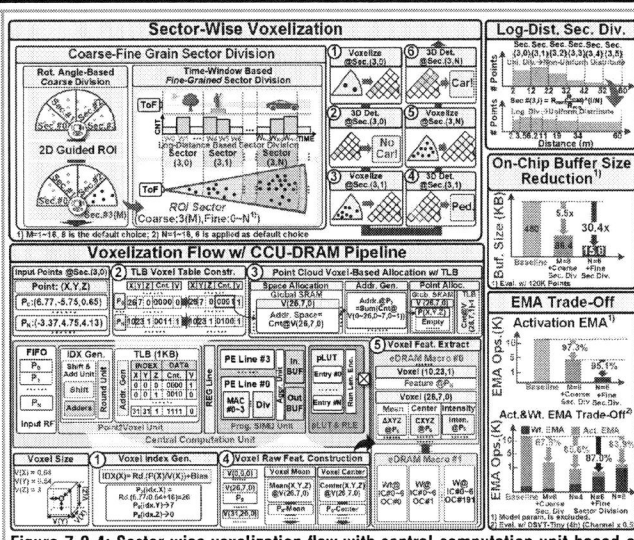

Figure 7.2.4: Sector-wise voxelization flow with central-computation unit based on coarse-fine sector division.

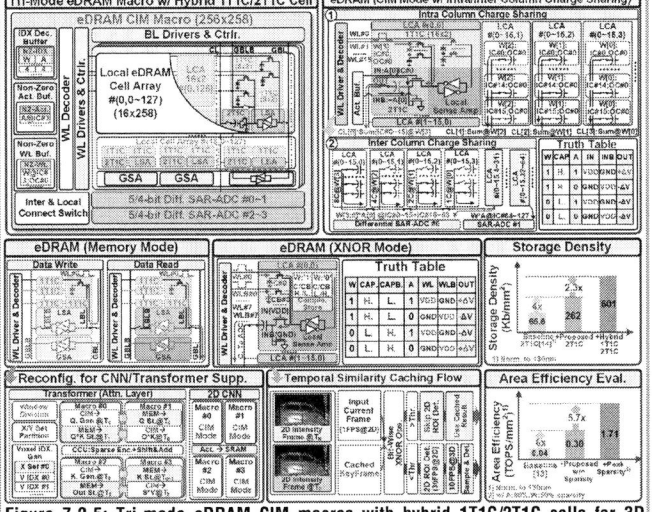

Figure 7.2.5: Tri-mode eDRAM CIM macros with hybrid 1T1C/2T1C cells for 3D transformer and similarity checking.

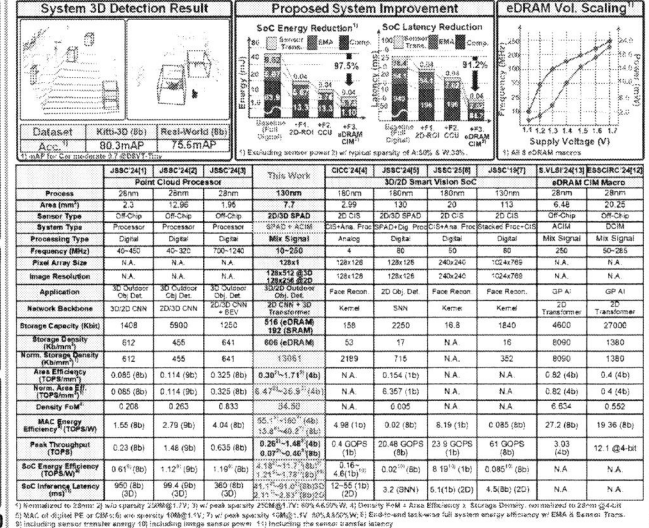

Figure 7.2.6: Measurement results and performance comparison table.

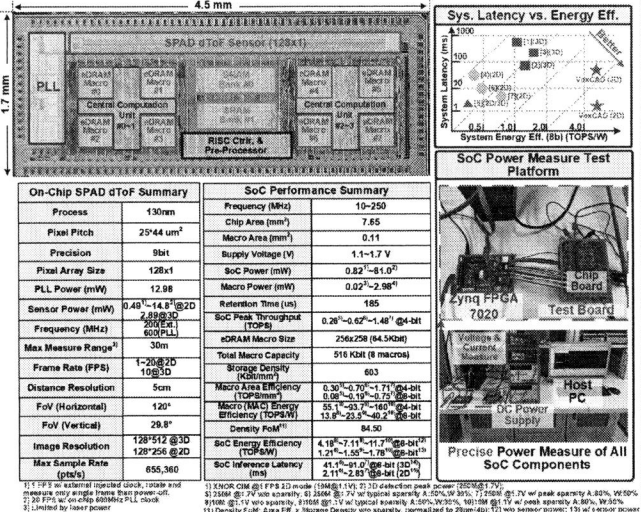

Figure 7.2.7: Chip photo, performance summary and system power measurement setup.

• 2026 IEEE International Solid-State Circuits Conference

979-8-3315-8937-0/26 $31.00 © 2026 IEEE

ISSCC 2026 / SESSION 7 / IMAGE SENSORS AND RANGING / 7.3

7.3 A Multi-Range, Multi-Resolution LiDAR Sensor with 2,880-Channel Modular Survival Histogramming TDC and Delay Compensation Using Double Histogram Sampling

Minkyung Kim[1], Junsang Bae[1], Donghyuk Kim[1], Jung-Hoon Chun[1,2], Seong-Jin Kim[3], Jaehyuk Choi[1,2]

[1]Sungkyunkwan University, Suwon, Korea, [2]SolidVue, Seongnam, Korea, [3]Sogang University, Seoul, Korea

Abstract

We present a 720×192 dToF sensor with a 2,880-channel modular survival histogram TDC (shTDC) array. The shTDC supports non-destructive accumulation while concentrating depth on active bins, reducing memory by 75% with precision comparable to one-step hTDC. Modular hTDCs enable multi-range and multi-resolution imaging to 69.6m, while double histogram sampling cancels pixel-to-shTDC offsets. The prototype achieves 0.13% precision and 0.12% accuracy.

Light detection and ranging (LiDAR) is a key 3D-perception modality for automotive, robotics, and AR/VR applications. In particular, solid-state LiDAR, free of mechanical laser scanners, enables compact and monolithic integration. The floor for reliable recognition is set by high spatial resolution and edge/detail preservation. However, resolution scaling is limited by three key factors. First, on-chip histogram memory and column-parallel histogramming time-to-digital converters (hTDCs) become area-dominant at small pixel pitch. Consequently, existing sensors remain capped at 400×112, while a 70°×20° field-of-view (FOV) with 0.1° angular resolution demands about 700×200 pixels [1]. Furthermore, the 400-channel TDC architecture of existing solutions forces single-row readout, resulting in reduced frame rate. Recently, priority-based histogram allocation and successive time-of-flight (ToF) searching (zoom/time-gating) have been reported to reduce memory footprint. However, at step transitions the reused memory is reinitialized, discarding prior accumulation and thereby degrading precision compared to full-range one-step accumulation at the same laser emission counts [2, 3]. Second, resolution scaling is further constrained by uniform macro pixels. At long-range, the signal-to-background ratio (SBR) degrades, necessitating macro-pixels that combine multiple single-photon avalanche diodes (SPADs), yet SBR is not uniform across the range; applying macro-pixels uniformly penalizes near-range high-resolution depiction. Third, resolution scaling is further limited by position-dependent delay offsets. A rolling-scan architecture with column-parallel hTDCs inevitably suffers from pixel-to-TDC delays. When multiple hTDCs are placed within a column to support multi-row readout for higher frame rates, their positions cause depth inconsistencies due to position-dependent delay variations. Therefore, realizing high-resolution LiDAR demands: (i) preservation of accumulated histograms with memory steered to signal-dense regions for area efficiency, (ii) flexible adjustment of pixel granularity and TDC range to distance/scene conditions, and (iii) cancellation position-dependent delays.

In this paper, we present a 720×192 high-resolution direct ToF sensor. The sensor supports multi-row scanning with 2,880-channel hTDCs, making it compatible with rolling-scan solid-state LiDAR. The sensor realizes three key ideas: (i) survival histogram TDC (shTDC), (ii) multi-range and multi-resolution readout with modular shTDCs, and (iii) delay compensation using double histogram sampling (DHS). A shTDC scheme maintains accumulation non-destructively while progressively promoting memory only in regions of high activity, delivering one-step-equivalent precision with high area efficiency. For multi-range and multi-resolution operation, near-range (high-SBR) pixels run in a high-resolution mode where each modular hTDC covers a short-range window, while long-range pixels are configured as macro-pixels with a selectable number of SPADs per pixel. Using the shTDC modules, the effective TDC dynamic range is partitioned and extended up to 69.6m, enabling multi-range operation. Finally, dedicated delay-sampling pixels and shTDCs accumulate histograms of pixel-to-shTDC propagation delays. The DHS then subtracts the delay, eliminating position-dependent offsets and enhancing depth consistency across the FOV. The sensor provides multi-range up to 69.6m and multi-resolution up to 720×192, achieving 0.13% depth precision and 0.12% depth accuracy while reducing memory usage by 75% compared to conventional one-step hTDC-based sensors.

Figure 7.3.1 compares histogramming strategies for hTDCs. In one-step hTDC, a histogram is accumulated across the entire detection range; counts are accumulated throughout acquisition, inherently supporting multi-echo detection, but making the required histogram memory prohibitive for high-resolution arrays. In two-/multi-step hTDC, the histogram memory is reused over successively narrowed windows. At each step, the histogram is destructively reset, clearing counts instead of carrying them forward. Thus, more laser shots are needed to achieve the same precision, especially at low SBR. To address these limitations, we propose the shTDC, which preserves full-range accumulation while steering memory capacity to active echo regions. A total of 128 bins are initially divided into 16 groups of 8 bins each with 2b/bin capacity. When saturation occurs, group activity is ranked, and only the upper half is retained while memory from the others is reassigned. The preserved groups are progressively expanded in bit depth—from 2b to 4b at the first saturation, and ultimately 8b at the next—without resetting counts. This preserves accumulation continuity where signals reside, maintains multi-echo capability, and achieves one-step-level precision with 1/4 the histogram memory at the same binning.

Figure 7.3.2 (left) illustrates the sensor architecture, comprising a 720×192 SPAD array, modular configuration logic for resolution and range selection, and 2,880-channel modular shTDCs operating in rolling-shutter (multi-row) scan. The array is partitioned into 4×4 SPAD clusters; each cluster is connected to 16 modular shTDCs, and each SPAD interfaces with a 3-transistor, nMOS-only SPAD analog front end (AFE). Within each cluster, SPADs can be grouped into macro-pixels or separated to support multi-resolution and multi-range operation with three configurations. In the high-resolution (HR) mode, single-SPAD pixels within the cluster operate independently, with one modular shTDC per pixel providing a measurement range up to 19.2m. In the double-extension (DE) mode, 1×2-SPAD macro-pixels are employed; two series-connected modular shTDCs per pixel extend the range to 36m, including a 2.4m overlap between two hTDC modules. In the quadruple-extension (QE) mode, 2×2-SPAD macro-pixels are employed, where four series-connected modular shTDCs per pixel extend the range to 69.6m. This per-cluster resolution selection preserves near-range detail while ensuring long-range detection, as illustrated by the multi-resolution depth images in Fig. 7.3.2 (top-right). Multi-range mode selection is determined from the previous frame's range estimates; the system periodically executes measurements in QE mode to refresh long-range coverage. The rightmost column consists of delay-sampling pixels driven by a digital pulse $TRIG_D$ distributed through a double-sampling (DS) buffer. The same shTDC chain used for depth sampling accumulates a histogram of the pixel-to-shTDC propagation delay. The DHS subtracts this delay, eliminating position-dependent offsets.

Figure 7.3.3 illustrates the shTDC with reconfigurable histogram memory, which enables non-destructive accumulation while concentrating histogram depth on active time-bin groups. Upon a pixel-output TRIG, the TDC front end samples the 8 phases of 125MHz clock signals from a global PLL, providing 1ns time bins. A 62.5MHz synchronizer then delivers the sampled 16-bin timestamps to the hTDC logic. Both the hTDC and the SRAM bank run at 62.5MHz. A 3b counter in the SRAM controller sequentially scans addresses 0–7; at each clock, 16 time-stamp bins are accumulated at the selected address, completing a 128-bin sweep in eight cycles. The SRAM bank is organized into 8 addresses, each containing 2 bin groups (16 groups total). Each group entry stores HIST[15:0] (eight adjacent time bins, 2b/bin), a bit-slice order identifier (MO) specifying the histogram depth, and a surviving peak-group identifier (PA) specifying the time-bin group. Accumulation begins uniformly at 2b/bin (MO = $00_{(2)}$). When any bin overflows, the reconfiguration controller in hTDC logic ranks group activity, records four peak addresses, retains those peaks together with four adjacent redundancy groups (eight survivors), and clears non-survivor capacity to reassign it as upper-bit storage. During this reassignment, both MO and PA are updated according to the stage (e.g., MO = $01_{(2)}$ at the first reconfiguration, MO = $10_{(2)}/11_{(2)}$ at the second, with PA updated to the final peak address). Previously accumulated counts are preserved, and the overflowed-bin index is tracked so that upper-bit histogramming continues at the same bin. At the first reconfiguration, the per-bin depth is promoted from 2b to 4b while retaining eight survivors. If overflow recurs at the 4b stage, the survivor set is pruned to two peak groups (first and second echoes), and the released capacity is assigned to their most-significant slices, yielding up to 8b depth. Consequently, each single-echo histogram—16-bins with up to 8b depth—is physically distributed across 8 bin groups, with the bit slices arranged from the least significant to the most significant. It preserves non-destructive accumulation, concentrates depth on signal-bearing bins, and achieves the precision of one-step hTDC with far less memory at the same shot budget. The shTDC supports two-echo detection per module providing up to eight-echo detection in the case of four series-connected module.

A prototype sensor was fabricated using a 90nm BSI CIS process and integrated into a VCSEL-based solid-state LiDAR platform for evaluation. Figure 7.3.4 presents both outdoor and indoor point-cloud depth images. Depth values are computed with sub-centimeter resolution using an on-chip FIR filter. Outdoors, a building beyond 35m is captured in the DE mode, whereas a statue and a staircase are captured in the HR mode. Indoors, high-resolution imaging preserves contours; compared with cluster-merged macro-pixels, higher point density yields more faithful circular geometry. Figure 7.3.5 (top) presents one-point depth measurements on a 90%-reflectivity whiteboard under 80klx ambient light. The sensor achieves a range of 54m with a depth precision of 9.2cm, and demonstrates a depth accuracy of 8.9cm at 27.2m. Within each SPAD cluster, position-dependent delay variation is dominated by the physical placement of the connected modular shTDCs and amounts up to 2ns. The DHS effectively removes the offset, as shown in Fig. 7.3.5 (bottom). Figure 7.3.6 presents chip characteristics and a comparison with state-of-the-art LiDAR sensors. In the table, this work achieves the highest resolution of 720×192 among solid-state LiDARs, enabled by the largest hTDC array—2,880 column-parallel channels—realized with an area-efficient modular survival-histogram architecture. The multi-range and multi-resolution depth imaging enables accurate depth acquisition across scenes with widely varying SBR, while DHS eliminates pixel-to-shTDC delay offsets. Chip micrograph is shown in Fig. 7.3.7.

Acknowledgement:
This work was supported in part by the Ministry of Trade, Industry & Energy (MOTIE), Korea (No. 1415187379) and in part by the MOTIE and Korea Institute for Advancement of Technology (KIAT) through the "International Cooperative R&D program" (No. P0028389). The EDA tool was supported by IC Design Education Center (IDEC), South Korea.

References:
[1] W. Roh *et al.*, "A High-Resolution Solid-State LiDAR Sensor With Reconfigurable Histogramming Time-to-Digital Converter and Filter for Depth Refinement," *IEEE JSSC*, pp. 1-17, June 2025. http://doi.org/10.1109/JSSC.2025.3571446
[2] M. Kim *et al.*, "A 320 × 240 CMOS LiDAR sensor with 6-transistor nMOS-Only SPAD analog front-end and area-efficient priority histogram memory," *ISSCC*, pp. 120-121, Feb. 2024. http://doi.org/10.1109/ISSCC49657.2024.10454449.
[3] S. Park *et al.*, "An Asynchronous 160×90 Flash LiDAR Sensor with Dynamic Frame Rates of 5 to 250fps Based on Pixelwise ToF Validation via a Background-Light-Adaptive Threshold," *ISSCC*, pp. 116-117, Feb. 2025. http://doi.org/10.1109/ISSCC49661.2025.10904624.
[4] O. Kumagai et al., "A 189 × 600 back-illuminated stacked SPAD direct time-of-flight depth sensor for automotive LiDAR systems," *ISSCC*, pp. 110-111, Feb. 2021. http://doi.org/10.1109/ISSCC42613.2021.9365961.
[5] T. Yui *et al.*, "A 25M Points/s Back-Illuminated Stacked SPAD Direct Time-of-Flight Depth Sensor with Equivalent Time Sampling for Automotive LiDAR," *IEEE Symp. VLSI Circuits*, June 2025. http://doi.org/10.23919/VLSITechnologyandCir65189.2025.11075078.
[6] C. Zou *et al.*, "A 256×192-pixel 30fps automotive direct time-of-flight LiDAR using 8× current-integrating-based TIA, hybrid pulse position/width converter, and intensity/CNN-guided 3D inpainting," *ISSCC*, pp. 114-115, Feb. 2024. http://doi.org/10.1109/ISSCC49657.2024.10454461.
[7] S. Zhuo *et al.*, "Solid-State dToF LiDAR System Using an Eight-Channel Addressable, 20-W/Ch Transmitter, and a 128 × 128 SPAD Receiver With SNR-Based Pixel Binning and Resolution Upscaling," *IEEE JSSC*, pp. 757-770, Mar. 2023. http://doi.org/10.1109/JSSC.2022.3227078.

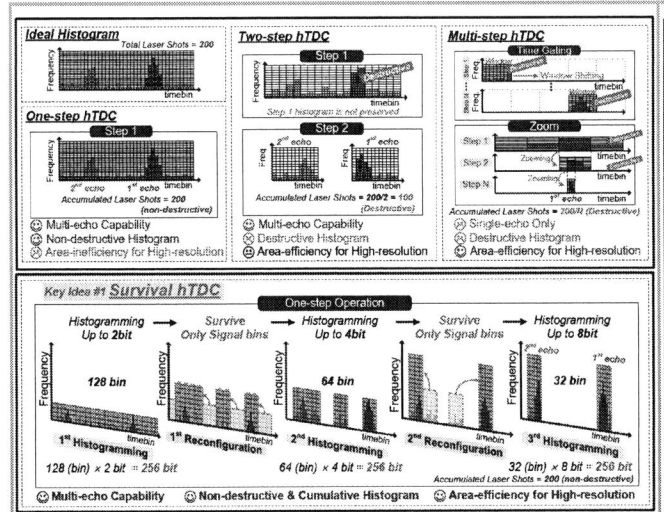

Figure 7.3.1: Comparison of various histogramming TDCs: one-step hTDC, two-step hTDC, multi-step hTDC (time gating and zoom), and survival hTDC.

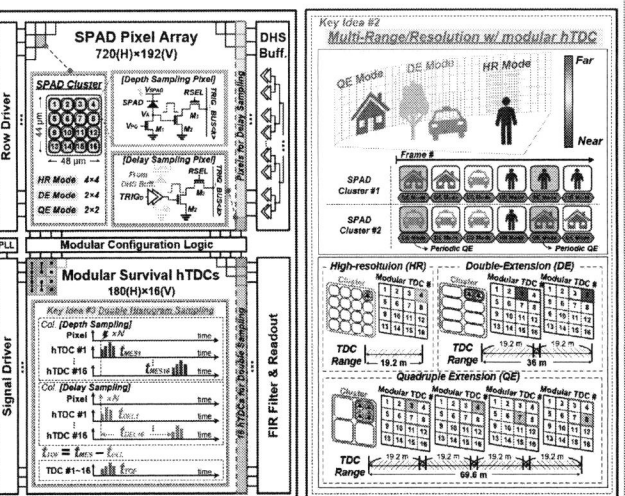

Figure 7.3.2: Overall architecture and operating principles of double histogram sampling (left) and multi-range/resolution configuration with modular shTDC (right).

ISSCC 2026 / SESSION 7 / IMAGE SENSORS AND RANGING / 7.3

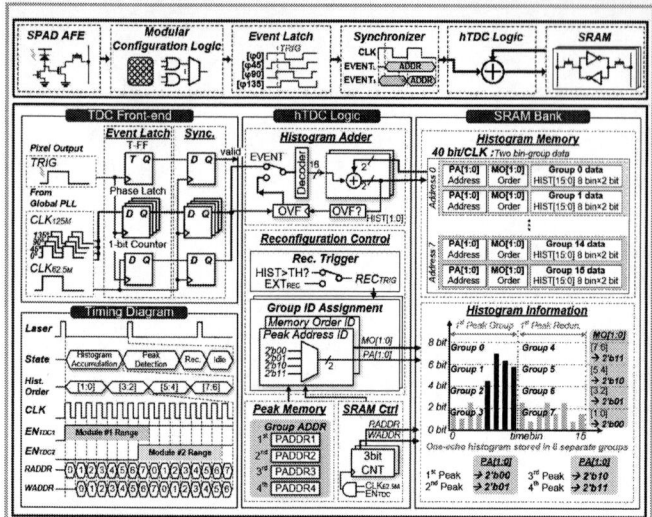

Figure 7.3.3: Signal chain of the sensor (top), and block diagrams of the TDC front-end, hTDC logic, SRAM bank, and its timing diagram (bottom).

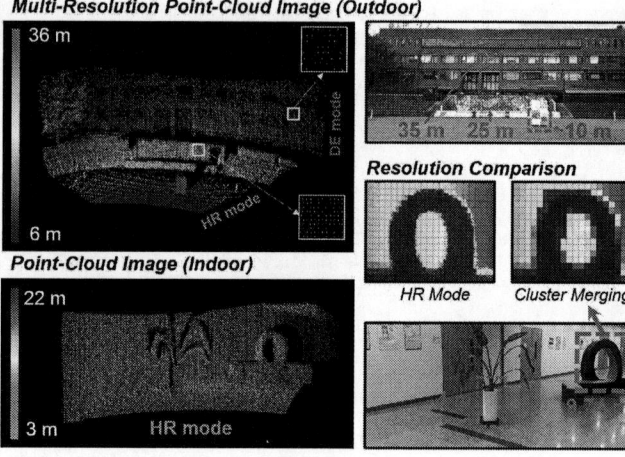

Figure 7.3.4: Point-cloud depth images for outdoor (top) and indoor (bottom) scenes; resolution comparison between high-resolution and cluster-merging.

One-point Depth Measurement Results

Double Histogram Sampling Results

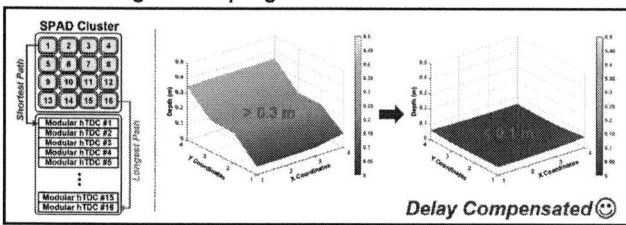

Figure 7.3.5: One-point depth measurement results under 80klux background light (top), double histogram sampling results (bottom).

Figure 7.3.6: Performance summary and comparison with state of the art.

Parameter	ISSCC'24 [6]	VLSI'25 [5]	ISSCC'21 [4]	JSSC'23 [7]	ISSCC'24 [2]	JSSC'25 [1]	This Work
Process	65nm	90 / 40 nm	90 / 40 nm	180 nm	110 nm	90 nm	**90 nm**
Optical System	Scanning	Scanning	Scanning	Scanning	Scanning	Scanning	**Multi-row Scanning**
	Mechanical	MEMS	Solid-State	Solid-State	Solid-State	Solid-State	**Solid-State**
Field-of-View (H × V)	N/A	120° × 26°	25.2° × 9.45°	36° × 36°	15° × 17°	80° × 30°	**80° × 30°**
Pixel Array (H × V)	Off-chip SiPM	1572 × 105	600 × 189	128 × 128	120 × 160 / 240 × 320	200 × 56 / 400 × 112	**720 × 192**
Image resolution (H × V)	256 × 192	2400 × 520	168 × 63	128 × 128	240 × 320	400 × 112	**Multi-res. up to 720 × 192**
Number of hTDC Channels	4	N/A	192	32	120	400	**2880**
Transistors/AFE (Type)	84 (CMOS)	> 30 (CMOS)	31 (CMOS)	21 (CMOS)	6 (nMOS-Only)	5 (nMOS-Only)	**3 (nMOS-Only)**
Frame rate	30 fps	20 fps	20 fps	1 fps	40 / 10 fps	20 / 10 fps	**15 fps**
Maximum Distance	160 m - 240 m	300 m	150 - 200 m	15 m	48 m	200 m	**Multi-range up to 69.6 m**
TDC bit depth	14 bit	N/A	12 bit	10 bit	9 bit	10 / 9 bit	**7 bit to 9 bit**
Depth Precision	15 cm	8 cm	N/A	N/A	8.8 cm	10 cm	**9.2 cm**
Depth Accuracy	1 cm	17 cm	15 - 30 cm	20 cm	5.7 cm	12 cm	**8.9 cm**
TDC resolution	100 ps	333 ps	1 ns	N/A	625 ps	1 / 2 ns	**1 ns**
Double Histogram Sampling	X	X	X	X	X	X	**O**

Figure 7.3.7: Chip micrograph of the CMOS LiDAR sensor.

ISSCC 2026 / SESSION 7 / IMAGE SENSORS AND RANGING / 7.4

7.4 A 480×320 CMOS LiDAR Sensor with Tapering 1-Step Histogramming TDCs and Sub-Pixel Echo Resolvers

Wonjong Roh*[1,2], Canxing Piao*[2], Taehoon Jeon[1], Junsang Bae[1], Jung-Hoon Chun[1,2], Seong-Jin Kim[3], Jaehyuk Choi[1,2]

[1]Sungkyunkwan University, Suwon, Korea, [2]SolidVue, Seongnam, Korea, [3]Sogang University, Seoul, Korea
*Equally Credited Authors (ECAs)

Abstract

A 480×320 CMOS LiDAR sensor using tapered 1-step hTDCs and SPER for memory-efficient operation is presented. The hTDC accumulates histograms with 1ns resolution over the full range and applies 500ps resolution only at the echoes, while SPER maps echoes to individual SPADs, enabling 4× image resolution. Implemented with 2×2 SPAD macro-pixels and shared AFEs (2.5T/SPAD) in a 90nm BSI CIS process, the sensor achieves 3.8cm precision at 100m and 4.1cm accuracy under 110klx sunlight at 25fps.

Light detection and ranging (LiDAR) is indispensable for 3D imaging in autonomous vehicles, robotics, and surveillance systems. For reliable object recognition, LiDAR faces two key challenges: fine time resolution for depth accuracy and high image resolution for detailed spatial mapping. Fine time resolution requires a large number of histogram bins, while high image resolution demands histograms for more pixels. Conventional 1-step hTDCs therefore face scalability issues, as improving either time or image resolution proportionally increases memory demand [1]. Previously, a QVGA LiDAR sensor was reported [2], but insufficient histogram memory forced time-division multiplexing, operating only one-quarter of the channels at a time, limiting frame rate and scalability. Although the multi-step hTDC architecture enabled memory reduction, the histogramming process was inevitably divided across laser repetitions for echo detection and for fine ToF acquisition. This division caused failures in detecting weak echoes and made the design vulnerable to strong ambient light. Another work reported a 400×112 LiDAR sensor with reconfigurable hTDCs [3]. To achieve high image resolution, a single-channel hTDC was reconfigured into two channels, enabling simultaneous histogramming of two SPADs within a macro pixel. However, this came at the cost of halving the time resolution and the frame rate. Recently, equivalent-time sampling (ETS) with multi-phase clock oversampling was introduced, extracting fine timing information from coarsely accumulated histograms [4]. However, when observing distinct objects within a macro-pixel at the SPAD level, achieving fully fine resolution remains challenging, leaving some room for further improvement. Achieving both fine time and image resolution under constrained memory remains an open challenge.

In this paper, we propose a 480×320 CMOS LiDAR sensor that employs tapering 1-step histogramming TDCs (hTDCs) and sub-pixel echo resolvers (SPER). The sensor employs 2×2 SPAD macro-pixels with shared AFEs (2.5T/SPAD) and row-scanning operation. The tapering hTDC forms a 1ns basis histogram over the full detection range and applies 2× time resolution (500ps) only within the time windows where echoes are detected. Within the same window, per-SPAD intensity counting is performed. Based on per-SPAD intensity at each echo, SPER maps echoes to individual SPADs to achieve a 4× increase in image resolution. An intensity-aware depth-refining filter (IA-DRF) further enhances time-of-flight (ToF) estimation by selecting filter coefficients according to echo intensity. Fabricated in a 90nm BSI CIS process, the sensor achieves a worst-case precision of 3.8cm (0.04%) at 100m and a worst-case accuracy of 4.1cm under 110klx sunlight at 25fps.

Figure 7.4.1 (top) shows the conceptual operation of the proposed tapering 1-step hTDC with SPER. The sensor integrates four SPADs (2×2) grouped into a single macro-pixel. The tapering 1-step hTDC first accumulates a basis histogram with 1ns resolution over the entire detection range using the macro-pixel while performing real-time echo detection. Once an echo is detected, the corresponding time window is subdivided by accumulating a tapered histogram with a 2× finer time resolution (500ps) over the basis histogram. This provides finer time resolution only at the detected echoes. In the same window, per-SPAD activations are counted to obtain the intensity of each SPAD. SPER compares these per-SPAD intensities with a SPER threshold, and only SPADs exceeding the threshold are identified as strong contributors to that echo. Through this process, echoes are mapped to individual SPADs, enabling sub-pixel resolution. Figure 7.4.1 (bottom) shows a comparison of histogram memory capacity with a conventional 1-step hTDC. Conventionally, achieving 2× finer time resolution and 4× higher image resolution requires proportional increases in histogram memory capacity by 2× and 4×, respectively. In contrast, the proposed tapering 1-step hTDC combined with SPER achieves 2× finer time resolution and 4× higher image resolution by requiring only additional memory to store the tapered histograms and per-SPAD intensities of four echoes. This reduces the overall histogram memory by 85% (5.6×) compared to the conventional design.

Figure 7.4.2 depicts the sensor architecture and key components. The sensor operates in a row-scanning manner and employs a 240×160 macro-pixel array, where each macro-pixel consists of 2×2 SPADs. Each SPAD is connected to a dedicated transistor for higher excess voltage and shares three transistors for quenching, pull-down, and selection with an adjacent-row SPAD that is not activated simultaneously. Thus, the AFE uses 2.5 transistors per SPAD, enhancing the fill factor and photon detection efficiency (PDE). The pull-up circuit is arranged in a column-parallel structure. Each of the four TDC front ends (FEs) is assigned to a SPAD in a macro-pixel. Each TDC FE uses the pixel-output TRIG to sample a 125MHz

clock and four 250MHz clocks with 45° phase differences from a global PLL. This enables photonic-event detection with 500ps resolution [6]. The output transitions of the TRIG-toggled flip-flop are used to mark the validity of the sampled phase value. The 5b phase information and VALID signals of the four SPADs are synchronized at 125MHz and passed to the phase decoder in the tapering hTDC. For the tapering, an echo-address memory and an address counter are used. The echo-address memory stores up to four echo addresses across laser repetitions. Only when the current address counter matches one of the stored echo addresses, the phase information is classified as an echo-induced event and mapped to 500ps time bins; otherwise, it is assigned to 1ns time bins. The event counter, which accelerates echo histogram accumulation, sums events from the four SPADs that fall into the same time bin and accumulates them in the histogram memory. The tapering hTDC also includes a real-time echo detector, SPER logic, and a threshold generator. The SRAM bank for histogram memory is divided into three parts: (i) a basis histogram memory with 720 time bins at 1ns resolution (6b/bin); (ii) a tapered histogram memory with 32 time bins at 500ps resolution (6b/bin) for four echoes; and (iii) a per-SPAD intensity memory for four echoes (4×12b/SPAD). The sensor periphery includes a charge pump to internally supply the SPAD supply voltage V_{SPAD}, a system controller, an IA-DRF for ToF acquisition, and a MIPI interface.

Figure 7.4.3 presents the sensor timing diagram and the operations for final ToF acquisition. The acquisition proceeds row-wise, while filtering and readout are pipelined to process the preceding row. Before the laser shot, background light (BGL) is measured to estimate the BGL event rate of each pixel. Afterward, with the laser active, a basis histogram with 1-ns resolution is accumulated concurrently with real-time echo detection. The echo threshold is generated by adding redundancy to the BGL event rate within the echo window. The echo threshold is then adaptively updated to reflect the increasing BGL count over successive laser repetitions. Once an echo is detected in real time, a tapered histogram with 2× finer time resolution of 500ps is accumulated over the 1ns basis histogram within the corresponding 16ns window, with each histogram storing up to 6b and combining to 7b. In the same window, per-SPAD intensity counting is also performed. Real-time echo detection continues until up to four echoes are identified. Conventional multi-step hTDCs detect echoes only once after a fixed number of laser repetitions and discard all prior accumulated data [2]. In contrast, the proposed tapering 1-step hTDC detects echoes in real time while retaining previously accumulated data. This enables early capture of strong echoes and late detection of weak echoes, which multi-step hTDCs would miss, all without data loss. Once histogramming is complete, SPER compares per-SPAD intensities with a threshold and assigns SPADs exceeding it to the corresponding echo. This process maps echoes from the macro-pixel level down to individual SPADs, enabling sub-pixel resolution. For example, in Fig. 7.4.3, the per-SPAD intensities of SPADs 1–3 exceed the threshold at Echo 0, whereas SPAD 0 exceeds the threshold at Echo 1. Consequently, Echo 0 is assigned to SPADs 1–3, while Echo 1 is assigned to SPAD 0. The SPER threshold derives from the same BGL event rate used for the echo threshold, divided across individual SPADs. SPER threshold is then scaled in proportion to the number of laser repetitions after each echo was detected. To extract the ToF of each mapped echo, an IA-DRF is applied. Histogram shapes corresponding to varying intensities are predetermined and stored in the LUT. For each echo, the most appropriate shape is selected based on its intensity, and the final ToF is determined as the point showing the highest correlation with the predicted shape [3]. Finally, the ToF values of each echo, combined with the SPER results, yield the per-SPAD ToF within the macro-pixel.

The prototype sensor with a 480×320 SPAD array was fabricated in a 90nm BSI CIS process and integrated into a VCSEL-based solid-state LiDAR system for evaluation. Figure 7.4.4 shows the depth-measurement results up to 100m and the corresponding test environment. The measurement was performed under 110klx sunlight using 1000 samples with a 90% reflectivity target at 25fps. When using the tapering 1-step scheme, a worst-case precision of 3.8cm (0.04%) was achieved at 100m, representing a 45% improvement over the conventional 1-step, which showed 6.9cm (0.07%). The worst-case accuracy was 4.1cm with tapering and 6.3cm with the conventional scheme. Figure 7.4.5 (top) presents the 3D point-cloud image and the 480×320 infrared (IR) intensity image under 90klx sunlight. Figure 7.4.5 (bottom) shows the color-mapped depth image with a camera-captured image, demonstrating the sub-pixel resolving capability of SPER. Without SPER, the vertical barrier

118 • 2026 IEEE International Solid-State Circuits Conference

979-8-3315-8937-0/26 $31.00 © 2026 IEEE

ISSCC 2026 / February 16, 2026 / 2:45 PM

is represented only at the macro-pixel level, and several inter-column spaces appear as a solid wall. With SPER applied, the barrier contour is more clearly represented at the SPAD level, the previously occluded inter-barrier space is resolved, and the vehicle outline appears more precisely. The surveillance camera, difficult to discern without SPER, is also represented at the SPAD level with well-defined contours. A performance summary and comparison with state-of-the-art LiDAR sensors is provided in Fig. 7.4.6. Compared to prior solid-state LiDAR sensors, this work achieves the highest image resolution of 480×320 with compact AFEs requiring only 2.5T/SPAD, leveraging SPER. In addition, thanks to tapering, the design attains the finest time resolution of 500ps and achieves the best depth precision of 3.8cm (0.04%) at 100m under 110klx sunlight. A chip photograph is shown in Fig. 7.4.7.

Acknowledgement:
This work was supported in part by the Technology Innovation Program (00419971) funded by the Ministry of Trade Industry & Energy (MOTIE), in part by the Technology Innovation Program (20025946) funded by the MOTIE, Korea, in part by Institute of Information & Communications Technology Planning & Evaluation (IITP) grant funded by the Korea government (MSIT) (RS-2025-02217613).

References:
[1] O. Kumagai *et al.*, "7.3 A 189×600 Back-Illuminated Stacked SPAD Direct Time-of-Flight Depth Sensor for Automotive LiDAR Systems," *ISSCC*, pp. 110-111, Feb. 2021. http://doi.org/10.1109/ISSCC42613.2021.9365961
[2] M. Kim *et al.*, "6.11 A 320x240 CMOS LiDAR Sensor with 6-Transistor nMOS-Only SPAD Analog Front-End and Area-Efficient Priority Histogram Memory," *ISSCC*, pp. 120-121, Feb. 2024. http://doi.org/10.1109/ISSCC49657.2024.10454449
[3] W. Roh *et al.*, "A High-Resolution Solid-State LiDAR Sensor With Reconfigurable Histogramming Time-to-Digital Converter and Filter for Depth Refinement," *IEEE JSSC*, vol. 60, no. 10, pp. 3665-3681, Oct. 2025. http://doi.org/10.1109/JSSC.2025.3571446
[4] T. Yui *et al.*, "A 25M Points/s Back-Illuminated Stacked SPAD Direct Time-of-Flight Depth Sensor with Equivalent Time Sampling for Automotive LiDAR," *IEEE Symp. VLSI Technology and Circuits*, June 2025. http://doi.org/10.23919/VLSITechnologyandCir65189.2025.11075078
[5] S. Zhuo *et al.*, "Solid-State dToF LiDAR System Using an Eight-Channel Addressable, 20-W/Ch Transmitter, and a 128 × 128 SPAD Receiver With SNR-Based Pixel Binning and Resolution Upscaling," *IEEE JSSC*, vol. 58, no. 3, pp. 757-770, Mar. 2023. http://doi.org/10.1109/JSSC.2022.3227078
[6] C. Niclass *et al.*, "A 100-m Range 10-Frame/s 340 ×96-Pixel Time-of-Flight Depth Sensor in 0.18-μm CMOS," *IEEE JSSC*, vol. 48, no. 2, pp. 559-572, Feb. 2013. http://doi.org/10.1109/JSSC.2012.2227607

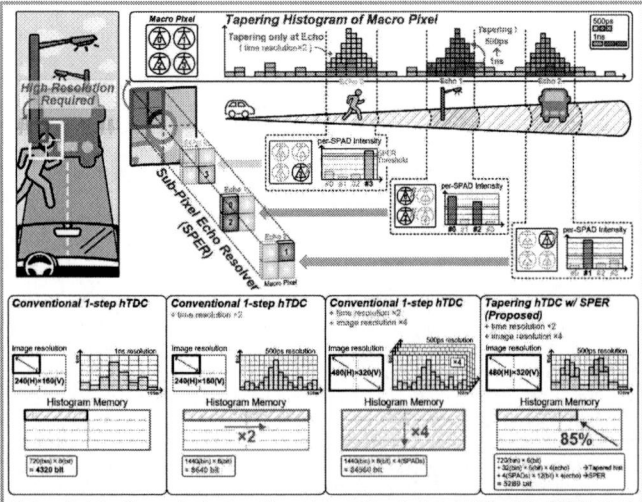

Figure 7.4.1: Conceptual operation of tapering 1-step hTDC with SPER (top). Comparison of histogram memory usage between conventional 1-step hTDC and proposed tapering hTDC (bottom).

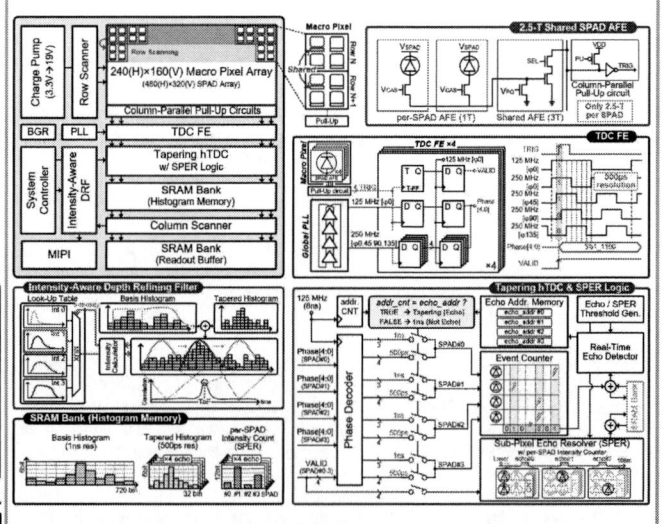

Figure 7.4.2: Overall sensor architecture and circuit diagram of key components.

DIGEST OF TECHNICAL PAPERS • 119

979-8-3315-8937-0/26 $31.00 © 2026 IEEE

ISSCC 2026 / SESSION 7 / IMAGE SENSORS AND RANGING / 7.4

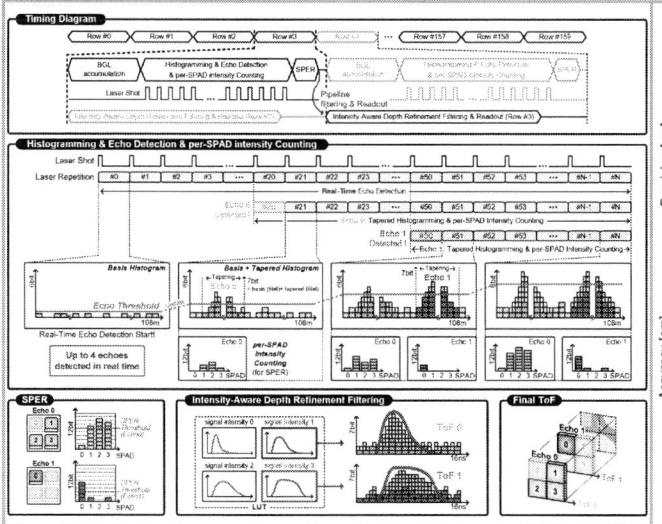

Figure 7.4.3: Timing diagram and operation principle of the tapering hTDC with SPER and IA-DRF.

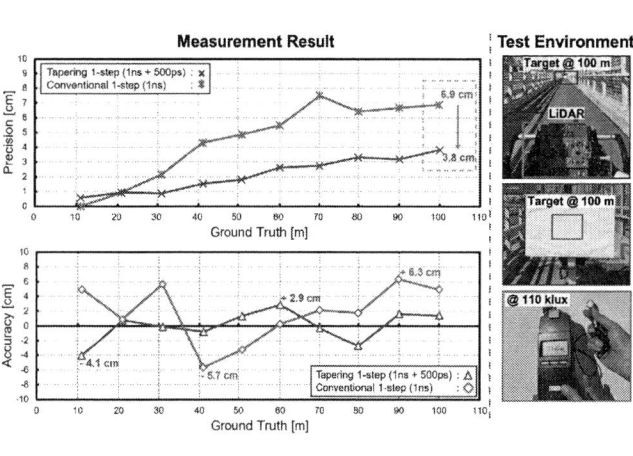

Figure 7.4.4: Depth measurement results under 110klx sunlight; depth was measured using 1000 samples per point with 90% reflectivity target (operating @ 25fps).

Figure 7.4.5: Captured images from the LiDAR sensor (operating @ 25fps): Point-cloud image and IR intensity image under 90-klx sunlight (top). Color mapped depth image, and the camera-captured image (bottom).

	This Work	ISSCC'21 [1]	JSSC'23 [5]	ISSCC'24 [2]	JSSC'25 [3]	VLSI'25 [4]
Process	90 nm	90 nm / 40 nm	180 nm	110 nm	90 nm	90 nm / 40nm
LiDAR System	Solid-State	Solid-State	Solid-State	Solid-State	Solid-State	Mechanical
Laser Wavelength	940 nm	905 nm	940 nm	940 nm	940 nm	940 nm
Pixel Array (H × V)	240 × 160	168 × 63	128 × 128	160 × 120	200 × 56	35 × 520
Image Resolution (H × V)	480 × 320	168 × 63	-	320 × 240	400 × 112	2400 × 520
Transistor/AFE	2.5-T	31-T	21-T	6-T	5-T	-
Frame Rate	25 fps	20 fps	1 fps	40 / 10 fps	20 / 10 fps	20 fps
Distance Range	108 m	150 – 200 m	15 m	48 m	150 / 200 m	300 m
hTDC Architecture	Tapering 1-Step	Conventional 1-Step	Off-Chip Histogram	2-Step	Conventional 1-Step	Conventional 1-Step
hTDC Time Resolution	500 ps	1 ns	-	625 ps	1 ns	333 ps (Effective)
Depth Precision (Relative)	3.8 cm (0.04%)	-	-	8.8 cm (0.2%)	10 cm (0.18%)	8cm (0.03%)
Depth Accuracy	4.1 cm	15 – 30 cm	20 cm	5.7 cm	17 cm	17 cm
Background Light	110 klx	117 klx	0.5 klx	100 klx	135 klx	120 klx
Target Reflectivity	90%	95%	10%	White	90%	10%

Figure 7.4.6: Performance summary and comparison with state-of-the-art LiDAR sensors.

Figure 7.4.7: Chip photograph.

- 2026 IEEE International Solid-State Circuits Conference

ISSCC 2026 / SESSION 7 / IMAGE SENSORS AND RANGING / 7.5

7.5 A 26.0mW 30fps 400x300-pixel SWIR Ge-SPAD dToF Range Sensor with Programmable Macro-Pixels and Integrated Histogram Processing for Low-Power AR/VR Applications

Matteo Perenzoni[1], Koichi Okamoto[1,2], Fabrizio De Nisi[1], Daniele Perenzoni[1], Ji Yong Jeong[1], Yu Zou[1], Hanning Mai[1], Gabriele Quarta[1], Francesco Paolo Mattioli Della Rocca[1], Yusuke Okamura[1,2], Davide Marani[1], Michele Brian[1], Daniele Giorgetti[1], Soichi Ochiai[2], Keiji Tatani[2], David Stoppa[1]

[1]Sony Semiconductor Solutions Europe, Trento, Italy, [2]Sony Semiconductor Solutions, Atsugi, Japan

Abstract

This paper presents a 400×300-pixel Ge-SPAD SWIR/NIR sensor enabling very-low-power operation: the flexible macropixel allocation scheme and high-rate multievent time-to-digital converters, combined with up to 100 simultaneous histograms having autoexposure control allows to turn on the sensor only when and where needed, with resilience to DCR. The sensor power consumption is 26mW when operating 500 macropixels of 3×3 SPAD at 30fps.

Direct time-of-flight (dTOF) range sensors have reached maturity in the last years, thanks to high-performing and low-noise SPADs while 3D-stacking technology has enabled complex on-chip processing [1-17]. Still, rejection of ambient light remains a challenge, and for wearables such as augmented reality/virtual reality (AR/VR) devices, power consumption of dTOF sensors in the hundreds-mW range is one order of magnitude more than required.

One of the most efficient solutions to reduce overall TOF system power is to selectively illuminate only the necessary regions within the field of view, shining laser dots only 'where' it is needed and only for the minimum required exposure time to get sufficient SNR. In [1], a software/hardware co-designed 128×80 SPAD array featuring 16-channel rolling-shutter operation coupled to addressable TX controlled by external powerful computational board proved such a concept, reducing overall TOF system power. However, this dToF sensor architecture has very limited and coarse addressability, and fully relies on external processing to implement dynamic RX/TX pattern reconfigurability. Another way to improve TOF system performance under strong sunlight conditions is to shift TX wavelength from near-infrared (NIR) to short-wave infrared (SWIR) region, where the solar spectrum is weaker and higher laser power is allowed within the eye-safety regime. Recently several Ge-on-Si SPADs (Ge-SPAD) have been reported [2-4] as an economically viable technology to address SWIR detection; however, their high dark count rate (DCR) noise makes it challenging to exploit such devices for large pixel array operating without cooling. In [6], pixel-to-pixel correlation is adopted for further mitigation of background light at the expense of increased pixel complexity. SPAD iToF architectures have also been proposed [7] to both suppress BGL and reduce sensor power.

This work uses a room-temperature SWIR Ge-SPAD device to mitigate the ambient light contribution, with a highly flexible macro-pixel (MP) allocation architecture and integrated histogram processing to reduce the total TOF system power consumption. Indeed, flexible MP allocation is fundamental when working with scanned dot pattern illuminators to spatially and temporally optimize the power consumption. Additional benefits include relaxed eye-safety constraints and high adaptability to different laser illuminator types. A known drawback of Ge-SPADs is their higher DCR, which is managed in this work by a signal chain designed to efficiently sustain a continuous rate: a short deadtime frontend, coincidence detection, high bandwidth array-to-periphery connections, and zero-latency multi-event TDC.

The architecture of Fig. 7.5.1 features an array of 400×300 pixels with a programmable MP allocation scheme that delivers the detected photons' triggers to 206 column event filtering/gating circuits and time-to-digital converters (TDC), which in turn are multiplexed towards 100 digital histograms. The histograms are processed by a dedicated control logic and there are two microprocessors: CPU1 with 32kB memory for housekeeping and a more powerful CPU2 with 256kB memory for high-complexity tasks. A MIPI interface outputs either raw histograms or processed data, while I2C interfaces are used to read/write configurations and control external peripherals; a LVDS interface delivers fast pulses for an external laser driver. A PLL and a DLL provide stable clocks and PVT-tracking biases for delay lines.

The MP allocation architecture, shown in Fig. 7.5.2, connects the pixels hierarchically to the TDCs through vertical buses (VBUS) using three levels L1, L2, and L3. L1 simply merges the signals of enabled pixels in a 2×2 arrangement: if a pixel is disabled, its contribution is irrelevant, therefore enabled pixels define the MP shape. L2 can be enabled via a local memory, and if so, they merge the signals of 2×2 L1 effectively identifying the whole MP. To consider various MP boundaries, L2 are overlapped but neighbors are not expected to be simultaneously active because of separation between laser dots. L3 are instead needed to route the signals to the VBUS: to manage congestion, each L3 can be programmed to deliver to 1 out of 4 VBUS via a local horizontal bus. To reduce the number of vertical buses, L3 anyway combines the outputs of 2x2 L2 with the assumption that only one of them will be active. This scheme allows defining MPs with any shape and position up to 4×4 pixels (only the 4×4-pixel MP is constrained by being aligned to an L2) and it can accommodate as many MP as the available histograms. The L3 bus selection enables up to 4 vertically aligned MPs, while borrowing available VBUS from neighbors. An example is shown in

Fig. 7.5.2. To reach the total number of MPs, the on-chip logic can store up to 1600 pre-programmed MP sizes and positions, while running acquisitions in 16 subsequent groups up to 100 MPs each; however, with custom firmware, the embedded CPU can override and freely reconfigure the array and sequence.

The detailed signal chain from photon to histogram is depicted in Fig. 7.5.3. The pixel frontend biases the SPAD with a time-gated active quenching circuit to sustain high rates without paralysis: hold-off time is configurable at chip-level down to a minimum of ~5ns. The received pulse can either be locally counted by an 8b ripple counter, or delivered to the 4 neighboring L1 both as a calibrated-width pulse up to 2ns and as an edge using a toggle flip-flop. If the pixel is disabled, the SPAD is off, and no pulses nor edges are produced. At the L1, pixel signals' edges are combined with a balanced XOR preserving timing, while pulses are used to detect coincidence if desired. Similarly, at L2 circuit, if enabled, edges are again combined with a balanced XOR and pulses checked for coincidence. The result is demultiplexed towards the selected VBUS by the L3, where edges are delivered with high-bandwidth differential tristate buffers. At column level, the differential signal is converted to standard logic and filtered by an optional column-level deadtime, gating, and coincidence circuitry. The resulting edge stream is fed into the TDC delay line which is composed by a 20-tap PVT-locked and calibrated delay line covering 1 TDC clock cycle, selectable from 105MHz, 210MHz, and 350MHz to achieve 476ps, 238ps and 143ps resolution respectively. To ensure uniformity of all TDCs, at startup and after DLL locking, a calibration is performed by injecting the clock as input: the delay-line is digitally fine-tuned until the edge coincides with the last tap, and the calibration value is stored in a local register.

The delay line sits at the boundary between analog and digital circuitry: after its synchronization, freezing the stream status at clock edge, a synthesized digital circuitry grabs the 20 delay-line taps and updates the histogram, allowing the TDC to operate with zero latency. The position of the 20 bins in the histogram is identified by a coarse clock counter that spans the 280 bins in 14 cycles, defining the TDC and histogram's full scale: this scheme ensures performance at minimum energy, avoiding power-hungry ring oscillators [6] or high-speed clocks. A shared processing unit continuously scans the histograms performing an approximate calculation of the SNR of a potential detected peak: if a programmable threshold is exceeded, the digital logic reacts by deactivating the corresponding MP, TDC, and histogram, thus saving the associated power. At the end of the exposure, the histograms can either be delivered to MIPI as raw or processed on-chip to extract the distance using background estimation, subtraction and matched filter peak detection.

The sensor is fabricated in a 22nm CMOS technology, 3D-stacked with a specialized Ge-SPAD process with 10μm-pitch pixels (micrograph of Fig. 7.5.7), and thoroughly characterized electrically and optically. The Ge-SPAD device features a DCR=5.7MHz at room temperature and a PDE = 5.1% at 1300nm with optical microlenses, making it suitable for low power dToF operation.

Figure 7.5.4 shows the overall timing and distance performance of the sensor. The left side depicts the signal chain timing performance at nominal clock of 210MHz both in dark and when stimulating the sensor with a diffused 1300nm pulsed laser with ~400ps pulse width and peak average of ~40mW, while activating 10 groups of 50 MPs of 3×3-pixel size simultaneously, and the respective TDCs. The code-density test in dark conditions gives a linearity of the TDC after calibration bounded by a DNL = [-0.35,0.32] and an INL = [-0.66,0.78] across all 206 TDCs of a typical chip, seamlessly connecting the coarse histogram portions. The 1300nm pulsed laser measurement shows the high robustness to >10× temperature-induced DCR increase, preserving the detected peak height without suffering compression. The right side of Fig. 7.5.4 concerns the distance performance using a 25×15 field-of-view (FOV) optics at F# = 1.6, with the collimated 1300 nm laser: at mid-distance of 4.4m, a single macropixel receives ~10% of the overall laser energy, i.e. ~4mW/dot. The 3×3-pixel MP receiving the spot peak is selected and distance is extracted relying on the on-chip peak extractor, measuring up to 10m with a precision <3cm.

The acquisition of a scene is pictured in Fig. 7.5.5: the 905nm laser with 2.5ns pulse and 75W peak power used in this measurement is not patterned, and therefore the scene is

120 • 2026 IEEE International Solid-State Circuits Conference

979-8-3315-8937-0/26 $31.00 © 2026 IEEE

ISSCC 2026 / February 16, 2026 / 3:00 PM

sampled at a predefined uniform MP arrangement with an estimated peak power of less than 5mW/dot. With the same FOV as above, a total of 500 MPs having 3×3-pixel size are acquired in 10 groups of 50 MPs, each exposing for 1.5ms performing 22.7k repetitions, resulting in 30fps including histogram processing and stream-out. Due to the specific laser limitations, pulse emissions have been decimated to ~750 laser pulses (one every 30 repetitions) with the received energy scaling accordingly, while dark counts are continuously accumulated. In these conditions the total sensor consumption amounts to 26.0mW: the breakdown of Fig. 7.5.5 emphasizes the optimization performed on the analog and digital signal chain, realized in particular with extensive on-chip supply power-gating, where individual blocks can be completely turned off achieving a very high power efficiency. The power consumption can be further reduced either at system level or with the integrated CPU by exploiting the unique sensor architecture, e.g., MP grouping by exposure time, region of interest, incremental histograms, etc., therefore reducing the operating duty-cycle and taking advantage of the power gating.

The sensor main KPI are listed in the table of Fig. 7.5.6 together with the relevant literature, demonstrating the effectiveness of the low-power design with the lowest sensor power consumption while achieving competitive ranging performance.

Acknowledgement:
The authors would like to thank F. Campi and N. Broseghini for the measurements support, and S.Yoshida, S. Shimada, A. Matsuzaki, S. M. L. Loo, Y. Otake, and T. Wakano for their support and advice regarding the SPAD pixel development. The authors would also like to acknowledge the continuous support from the members of Taiwan Semiconductor Manufacturing Company, Sony Semiconductor Solutions, and Sony Semiconductor Manufacturing.

References:
[1] Y. Wu et al., "dToF LIDAR System Using Addressable Multi-Channel VCSEL Transmitter, 128x80 SPAD Sensor, and ML-Based Object Detection for Adaptive Beam-Steering", *IEEE CICC*, April 2023. https://doi.org/10.1109/CICC57935.2023.10121184
[2] S. Yoshida et al., "10 μm Pitch Ge-on-Si SPAD Pixel Array with PDE of 33.8% at 1300 nm and 23.3% at 1550 nm under Room Temperature Environment", *IEDM, 2025.*
[3] Y. Benhammou et al., "Germanium on silicon SPAD 32×32 pixel array in 3D-stacked technology for SWIR applications", *IISW*, 2023. https://doi.org/10.60928/a28x-pby2
[4] C.-E. Chen et al., "First demonstration of room temperature SWIR flash LiDAR using a 160 × 116 Ge-on-Si SPAD array", *CLEO*, 2025. https://doi.org/10.1364/CLEO_AT.2025.PD103_7
[5] P.-Y. Taloud et al., "A 1.2K dots dToF 3D Imaging System in 45/22nm 3D-stacked BSI SPAD CMOS", *ISSW*, 2022. https://www.imagesensors.org/Past%20Workshops/2022%20ISSW/1145%20-%201215_Pierre-Yves_Taloud.pdf
[6] E. Manuzzato et al., "A 64x64-pixel flash LiDAR SPAD imager with distributed pixel-to-pixel correlation for background rejection, tunable automatic pixel sensitivity and first-last event detection strategies for space applications," *ISSCC*, pp. 96-97, Feb. 2022. https://doi.org/10.1109/ISSCC42614.2022.9731622
[7] H.-S. Choi et al., "SPAD flash LiDAR with chopped analog counter for 76m range and 120klx background light", *ISSCC*, pp. 118-119, Feb. 2025. https://doi.org/10.1109/ISSCC49661.2025.10904623

[8] A.R. Ximenes, et al., "A 256×256 45/65nm 3D-stacked SPAD-based direct TOF image sensor for LiDAR applications with optical polar modulation for up to 18.6dB interference suppression", *ISSCC*, pp. 96-97, 2018. https://doi.org/10.1109/ISSCC.2018.8310201
[9] S. W. Hutchings et al., "A Reconfigurable 3-D-Stacked SPAD Imager with In-Pixel Histogramming for Flash LIDAR or High-Speed Time-of-Flight Imaging", *IEEE JSSC*, vol. 54, no. 11, pp. 2947-2956, 2019. https://doi.org/10.1109/JSSC.2019.2939083
[10] R. K. Henderson et al., "A 256×256 40nm/90nm CMOS 3D-Stacked 120dB Dynamic-Range Reconfigurable Time-Resolved SPAD Imager", *ISSCC*, pp. 106-107, 2019. https://doi.org/10.1109/ISSCC.2019.8662355
[11] P. Padmanabhan et al., "A 256×128 3D-Stacked (45nm) SPAD FLASH LiDAR with 7-Level Coincidence Detection and Progressive Gating for 100m Range and 10klux Background Light", *ISSCC*, pp. 111-112, 2021. https://doi.org/10.1109/ISSCC42613.2021.9366010
[12] O. Kumagai et al., "A 189x600 Back-illuminated stacked SPAD direct Time-of-Flight depth sensor for automotive LiDAR systems", *ISSCC*, pp. 110-111, 2021. https://doi.org/10.1109/ISSCC42613.2021.9365961
[13] T. Ohkubo et al., "A Color Image Sensor Using 1.0-μm Organic Photoconductive Film Pixels Stacked on 4.0-μm Si Pixels for Near-Infrared Time-of-Flight Depth Sensing", *IEEE IEDM*, 2024. https://doi.org/10.1109/IEDM50854.2024.10873486
[14] I. Gyongy et al., "A Direct Time-of-Flight Image Sensor with In-Pixel Surface Detection and Dynamic Vision", *IEEE J. of Selected Topics in Quantum Electronics*, Vol. 30, No. 1, pp. 1-11, Jan./Feb. 2024. https://doi.org/10.1109/JSTQE.2023.3238520
[15] J. Jang *et al.*, "A 336 x 240 Backside-Illuminated 3D-Stacked 7μm SPAD for LiDAR Sensor with PDE 28% at 940nm and under 0.4% Depth Accuracy Up to 10m", *IEEE Symp. VLSI Technology and Circuits*, 2024. https://doi.org/10.1109/VLSITechnologyandCir46783.2024.10631458
[16] T. Yui *et al.*, "A 25M Points/s Back-Illuminated Stacked SPAD Direct Time-of-Flight Depth Sensor with Equivalent Time Sampling for Automotive LiDAR," *IEEE Symp. VLSI Technology and Circuits*, 2025. https://doi.org/10.23919/VLSITechnologyandCir65189.2025.11075078
[17] S. Park *et al.*, "An Asynchronous 160x90 Flash LiDAR Sensor with Dynamic Frame Rates of 5 to 250fps Based on Pixelwise ToF Validation via a Background-Light-Adaptive Threshold", *ISSCC*, pp. 116-117, 2025. https://doi.org/10.1109/ISSCC49661.2025.10904624

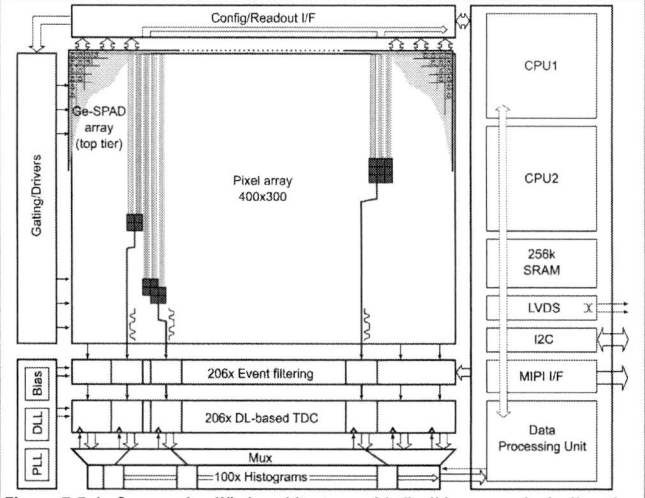

Figure 7.5.1: Sensor simplified architecture with flexible macropixel allocation: highlighted in red is the photon counting datapath, in blue is the photon timing datapath.

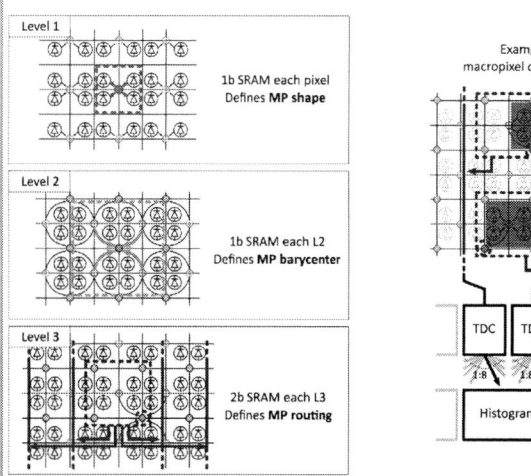

Figure 7.5.2: Architecture of the macropixel allocation with three hierarchical levels, showing a configuration with a possible conflict, resolved by use of different vertical buses.

ISSCC 2026 / SESSION 7 / IMAGE SENSORS AND RANGING / 7.5

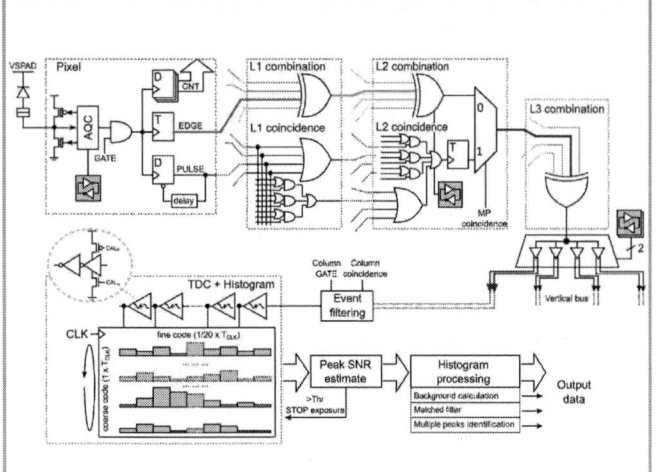

Figure 7.5.3: Signal path from SPAD to output: pixel, levels, and vertical bus in the pixel array; event filtering, TDC, and histogram at column level; processing logic at chip level.

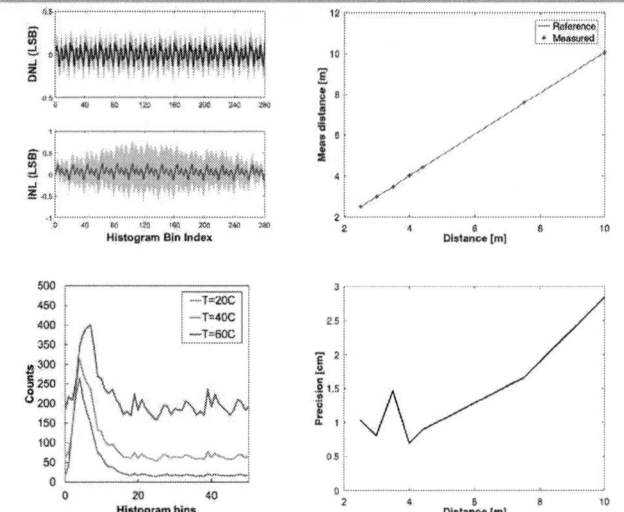

Figure 7.5.4: Timing and distance performance of the sensor: DNL/INL of the multievent TDC, resilience to DCR levels up to 60°C, and distance sweep measurements at 25°C.

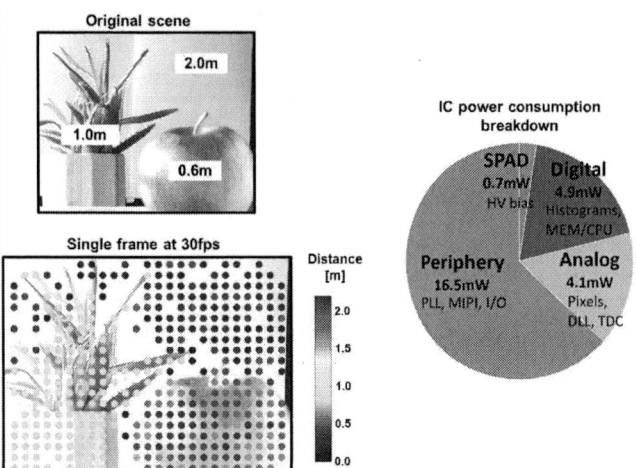

Figure 7.5.5: Sample scene at 25°C with 940nm laser, using a configuration with 500 MPs electronically scanned in 10 groups at 30fps, and power consumption breakdown.

	Unit	This work	[1]	[2]	[4]	[5]	[6]	[7]
Technology	nm	22	N.A.	22	22	22	110 CIS	110
SPAD Technology	-	Ge-on-Si BSI 65nm	Si	Ge-on-Si BSI 65nm	Ge-on-Si BSI 45nm	Si BSI 45nm	Si FSI	Si BSI
SPAD pixel array	-	400x300	128x80	165x140	160x116	320x240	64x64	64x64
VEX / VBD	V	3 / 18.2	3 / 19.5	2 / 18.08	3 / ~21V	2 / 17.5	6.6 / N.A.	2 / N.A.
SPAD PDE/PDP	%	5.1 PDE@1300nm	3.6 PDP@940nm	33.8 PDE@1300nm	~31 PDE@1300nm	>12 PDE@940nm	54.5 PDP@450nm	14.3 PDP@850nm
SPAD DCR@RT	cps	5.7M	10	22.7M	11.16M	0.8	388	N.A.
Chip size	mm²	6.16x4.65	4.6x4.0	4.2x3.5	5x4	NA	NA	2.9x2.9
Pixel pitch	µm	10	20	10	10	10	48	32
Laser wavelength	nm	905/1300	940	1380	1310/1550	940	905	850
TDC resolution	ps	143 - 475	100	833	N.A	250 - 350	250-10k	NA, IToF
TDC #bins	-	280	N.A.	N.A.	N.A.	59	16b	N.A.
Number of TDCs	-	100 – 206*	80	N.A.	N.A.	1200	4096	N.A.
On-chip distance extraction	-	Y	Y	N	N	Y	N	N
Frame rate	-	30	10	30	4	30	25	50
Distance range	m	10	15	18	2.5	5	8.2	75
Depth precision	-	<0.3%	<5cm	<1%	NA	<1cm	<27cm	6.4cm
Sensor power	mW	26.0	N.A.	N.A.	N.A	284 (TX+RX)	206	39.4

* 206 total available TDCs, 100 maximum parallel operation

Figure 7.5.6: Comparison table.

Figure 7.5.7: Chip micrograph.

• 2026 IEEE International Solid-State Circuits Conference

979-8-3315-8937-0/26 $31.00 © 2026 IEEE

ISSCC 2026 / SESSION 7 / IMAGE SENSORS AND RANGING / 7.6

7.6 A 128×96 Multimodal Flash LiDAR SPAD Imager with Object Segmentation Latency of 18µs Based on Compute-Near-Sensor Ising Annealing Machine

Jingyi Wang*, Tao Hong*, Bu Chen*, Zhangcheng Huang, Qi Liu, Ming Liu

Fudan University, Shanghai, China
*Equally-Credited Authors (ECAs)

Abstract

This paper presents a 128×96 multimodal flash LiDAR SPAD imager integrated with a compute-near-sensor Ising-model annealing processor for dynamic object segmentation, framed as a combinatorial optimization problem (COP). The system enables real-time updates of Hamiltonian coefficients using a high-bandwidth output-while-write scheme, and accelerates Ising spin iterations by exploiting multimodal SPAD information, achieving a segmentation latency of 18µs.

Single-photon avalanche diode (SPAD) imagers offer single-photon sensitivity, picosecond-level timing resolution, and intrinsic time-of-flight depth sensing capabilities, enabling robust 2D/3D perception in low-light and high-dynamic-range scenarios. These attributes render SPADs highly suitable for real-time sensing applications including autonomous driving, 3D mapping, and augmented reality (AR) [1-3]. In many scenarios, object segmentation is critical for accurate perception, tracking, and decision-making processes [4,5]. However, conventional architectures suffer from prolonged data transmission between sensors and processors [6-10], compounded by the limited speed of edge processors when executing neural network algorithms under power-constrained conditions. This results in excessively long image segmentation latency [11,12], posing challenges for tracking high-speed moving targets, particularly 3D dynamic targets with drastically increased data volumes. To address these limitations, prior research has integrated processors and sensors onto a single chip, significantly enhancing sensor-processor bandwidth [13-16]. Separately, studies have explored the use of Ising machines for solving combinatorial optimization problems (COP) such as image segmentation to achieve accelerated processing [17-19]. To enable rapid segmentation of dynamic objects, this work presents a compute-near-sensor architecture that co-integrates a 128×96 Ising annealing processor with a 128×96 flash-mode LiDAR SPAD imager. The bandwidth bottleneck between the sensor and processor is mitigated through on-chip high-bandwidth interconnections, and the Ising machine, which is specialized in solving COPs, is employed to execute the image segmentation task.

Figure 7.6.1 illustrates the operating principle of the proposed architecture and presents a comparison with traditional schemes. In conventional architectures, latency incurred by massive data transmission between SPAD sensors and processors undermines the ability of segmentation tasks to track high-speed moving targets. By contrast, in the proposed scheme, data from the SPAD array is rapidly transmitted to the Ising machine for processing through ultra-high on-chip bandwidth, thus overcoming the data bandwidth bottleneck inherent in traditional architectures. Furthermore, by concurrently feeding intensity information and dynamic vision sensing (DVS) information from the multimodal SPAD into the Ising machine in an "output-while-write" mode, acceleration of image segmentation tasks is achieved via the Ising machine.

Figure 7.6.2 presents the architecture of the chip and the schematic diagram of the SPAD pixel. The chip comprises a 128×96 pixel array, a Hamiltonian coefficient computation (HCC) circuit, a 128×96 Ising spin array, a region-of-interest (ROI) generation circuit, a readout-control circuit, and a microcontroller unit (MCU). The pixel front-end is designed to support three operational modes: 2D intensity, 2D DVS, and 3D detection. In the 2D DVS mode, during the first half-cycle, the counter increments by 1 each time the SPAD device is triggered (P_D). At the midpoint of the cycle, the chip outputs 2D intensity information, denoted as D_{INT}. Concurrently, the counter result is inverted via the INV signal. During the second half-cycle, the counter continues to increment by 1 with each SPAD trigger. At the end of the full cycle, the counter output D_{DVS} corresponds exactly to the differential value between the two half-cycles. In the 3D direct time-of-flight (dToF) mode, the SPAD pulse P_T triggers the time-to-digital converter (TDC) ring oscillator (RO). The RO drives the counter until the TDCSTOP signal disables the counter, thereby generating the ToF result D_{TIME} [15]. In this work, under the 2D DVS mode, both the intensity information D_{INT} and differential information D_{DVS} are read out row-by-row and written into the HCC circuit. First, 128 comparators are used to parallelize the binarization of D_{INT} against an externally configured intensity threshold, with the output labeled as h_{INT}. Subsequently, the same set of comparators is utilized to binarize the absolute values of D_{DVS}, with the output labeled as h_{DVS}; this is followed by denoising the D_{DVS} data using a programmable 3×3 convolution kernel. The resulting h_{INT} and h_{DVS} signals are written into the Ising machine immediately upon generation.

Figure 7.6.3 (top) illustrates the operating principle of the Ising machine. The Ising machine solves COP by finding the minimum Hamiltonian value through an annealing mechanism. The process begins with the initialization of all spins. Each spin cell (SC) then calculates its local variable F_i and updates its spin state σ_i based on F_i, aiming to minimize its local Hamiltonian H_i. To avoid trapping in local minima, spin flips are enforced by introducing a nonlinear probabilistic flip (NPF) with a random probability P. Through continuous iteration of the UPDATE and FLIP phases, the Ising machine gradually converges to the global minimum of the Hamiltonian. In this work, the DVS information h_{DVS} from the SPAD sensor is adopted as the initial state of the spin array, which significantly accelerates the annealing process of the Ising machine. To prevent global oscillations, spins are not flipped simultaneously. Instead, 2×2 spin units are grouped into a macro, and the four spin cells within each of the 64×48 macros are flipped probabilistically in four sequential phases (from SC1 to SC4). The circuit diagram of the spin cell is shown in Fig. 7.6.3 (lower left). It reads the spin states σ_U, σ_D, σ_L, σ_R and coefficients J_U, J_L from adjacent spin cells. Together with locally stored parameters J_D, J_R, h_{INT}, and h_{DVS}, it performs Hamiltonian calculation, state update, and spin flipping. Figure 7.6.3 (lower right) illustrates the chip's workflow. The Ising machine operates cyclically on four spin cells within each of the 64×48 macros. After 75 cycles, the Hamiltonian converges to its minimum value, completing the image segmentation of dynamic targets in the SPAD array. At a clock frequency of 50MHz, the entire segmentation process takes 18µs. The segmentation results can be directly output, or alternatively, the region of interest can be computed on-chip from the segmentation data. The SPAD array can then be configured to activate only the ROI for 3D dToF detection, which significantly reduces both power consumption and data output latency.

Figure 7.6.4 illustrates the workflow of the entire Ising machine for performing image segmentation tasks. In the LOAD stage, the initial spin array can be configured to a fixed value or set to h_{DVS} which is derived from the DVS image. At the onset of Ising processing, the system energy is high, corresponding to a "high-temperature" state. As iterations progress, the flipping frequency of the spin array gradually decreases. As the spin values of the Ising machine become increasingly stable, they eventually exhibit a high degree of consistency with the contours of dynamic targets. The randomness required for spin flipping is generated by two components: an on-chip linear feedback shift register (LFSR) array with configurable seeds, and an external global temperature coefficient $T[t]$. Over time, the temperature sequence $T[t]$ evolves, with the proportion of "0" values gradually increasing. This corresponds to the annealing process moving toward a lower-energy state. Subsequently, the segmentation result is transmitted to the ROI computation circuit. This circuit first receives spin-state data row-by-row, applies a 3×3 convolution for denoising, and accumulates the denoised results using row and column adder trees. The accumulated data are then binarized and deburred to generate horizontal and vertical binary representations. By right-shifting and XORing these binary sequences, a two-hot encoding is produced; the positions of "1"s in this encoding mark the start and end indices of the target region, enabling precise ROI extraction.

Figure 7.6.5 presents the experimental results, including real-scene picture, 2D intensity image, 2D DVS images (both pre- and post-denoising), the Hamiltonian convergence curve, spin-array state evolution, and 3D ROI detection results. The system executes 300 total iterations, with each spin cell undergoing 75 iterations, ultimately converging to state [D] at the minimum energy level. The final segmentation result retains only the region corresponding to the moving car while suppressing static background objects, thus enabling efficient dynamic-object segmentation and ROI extraction.

Figure 7.6.6 summarizes the chip's performance and provides a comparison with state-of-the-art SPAD image sensors and Ising processors. This work represents a monolithic integration of a large-format digital Ising machine with a flash-mode LiDAR SPAD sensor, enabling near-sensor object segmentation. Operating at 50MHz, the chip achieves a segmentation latency of 18µs, while simultaneously delivering 2D imaging at 10kfps and 3D dToF sensing at 1kfps. Compared with conventional architectures in Fig. 7.6.6, the proposed design realizes ultra-fast image segmentation by leveraging a compute-near-sensor architecture and Ising processing accelerated via multimodal information—thus exhibiting significant advantages in both on-edge image processing and intelligent detection.

Acknowledgement:

This work was supported in part by the National Natural Science Foundation of China under Grant 62374039 and 62235009, in part by the National Key Research and Development Program of China under Grant 2021YFA1200700, and in part by State Key Laboratory of Integrated Chips and Systems (Grant No. SKLICS-Z202405). The corresponding author is Zhangcheng Huang (huangzc@fudan.edu.cn).

References:

[1] S. Park *et al.*, "An Asynchronous 160×90 Flash LiDAR Sensor with Dynamic Frame Rates of 5 to 250fps Based on Pixelwise ToF Validation via a Background-Light-Adaptive Threshold," *ISSCC*, Feb. 2025, pp. 116–117. http://doi.org/10.1109/ISSCC49661.2025.10904624

[2] C. Zou *et al.*, "A 256×192-Pixel 30fps Automotive Direct Time-of-Flight LiDAR Using 8× Current-Integrating-Based TIA, Hybrid Pulse Position/Width Converter, and Intensity/CNN-Guided 3D Inpainting," *ISSCC*, Feb. 2024, pp. 114–115. http://doi.org/10.1109/ISSCC49657.2024.10454461

[3] S.-H. Han *et al.*, "A 160×120 Flash LiDAR Sensor with Fully Analog-Assisted In- Pixel Histogramming TDC Based on Self-Referenced SAR ADC," *ISSCC*, Feb. 2024, pp. 112–113. http://doi.org/10.1109/ISSCC49657.2024.10454470

[4] S. Minaee *et al.*, "Image Segmentation Using Deep Learning: A Survey," *IEEE Trans. Pattern Anal. Mach. Intell.*, pp. 1–1, 2021. http://doi.org/10.1109/TPAMI.2021.3059968

[5] A. S. Mohan and R. Resmi, "Video image processing for moving object detection and segmentation using background subtraction," *International Conference on Computational Systems and Communications*, Dec. 2014, pp. 288–292. http://doi.org/10.1109/COMPSC.2014.7032664

[6] Y. Zhou *et al.*, "Computational event-driven vision sensors for in-sensor spiking neural networks," *Nat. Electron.*, vol. 6, no. 11, pp. 870–878, Nov. 2023, doi: 10.1038/s41928-023-01055-2. http://doi.org/10.1038/s41928-023-01055-2

[7] S. K. Bose *et al.*, "A 51.3-TOPS/W, 134.4-GOPS In-Memory Binary Image Filtering in 65-nm CMOS," *IEEE J. Solid-State Circuits*, vol. 57, no. 1, pp. 323–335, Jan. 2022. http://doi.org/10.1109/JSSC.2021.3098539

[8] T.-H. Hsu *et al.*, "A 0.8 V Multimode Vision Sensor for Motion and Saliency Detection With Ping-Pong PWM Pixel," *IEEE J. Solid-State Circuits*, vol. 56, no. 8, pp. 2516–2524, Aug. 2021. http://doi.org/10.1109/JSSC.2021.3075746

[9] M. Lefebvre *et al.*, "A 0.2-to-3.6TOPS/W Programmable Convolutional Imager SoC with In-Sensor Current-Domain Ternary-Weighted MAC Operations for Feature Extraction and Region-of-Interest Detection," *ISSCC*, Feb. 2021, pp. 118–119. http://doi.org/10.1109/ISSCC42613.2021.9365839

[10] I. Gyongy *et al.*, "A Direct Time-of-Flight Image Sensor with In-Pixel Surface Detection and Dynamic Vision," *IEEE J. Sel. Top. Quantum Electron.*, vol. 30, no. 1: Single-Photon Technologies, pp. 1–11, Jan. 2024. http://doi.org/10.1109/JSTQE.2023.3238520

[11] V. Badrinarayanan *et al.*, "SegNet: A Deep Convolutional Encoder-Decoder Architecture for Image Segmentation," *IEEE Trans. Pattern Anal. Mach. Intell.*, vol. 39, no. 12, pp. 2481–2495, Dec. 2017. http://doi.org/10.1109/TPAMI.2016.2644615

[12] R. Girshick *et al.*, "Rich Feature Hierarchies for Accurate Object Detection and Semantic Segmentation," *IEEE Conference on Computer Vision and Pattern Recognition*, June 2014, pp. 580–587. http://doi.org/10.1109/CVPR.2014.81

[13] X. Yang *et al.*, "A 10 000-Inference/s Bio-Inspired Spiking Vision Chip Based on an End-to-End SNN Embedding Image Signal Enhancement," *IEEE J. Solid-State Circuits*, pp. 1–17, 2025. http://doi.org/10.1109/JSSC.2025.3583310

[14] R. Gomez-Merchan, *et al.*, "Dynamic Vision With Single Photon Detectors: A Discrete DVS Architecture Using Asynchronous Sensor Front-Ends," *IEEE Trans. Circuits Syst. I: Regul. Pap.*, vol. 72, no. 1, pp. 36–49, Jan. 2025. http://doi.org/10.1109/TCSI.2024.3441371

[15] J. Wang *et al.*, "A 32×32 Flash LiDAR SPAD Sensor with Up-to-1kfps Motional Target Detection by Threshold-adaptive 2D Dynamic Vision," *IEEE Custom Integrated Circuits Conference*, Apr. 2024, http://doi.org/10.1109/CICC60959.2024.10529051

[16] F. Mattioli Della Rocca *et al.*, "A 128 × 128 SPAD Motion-Triggered Time-of-Flight Image Sensor With In-Pixel Histogram and Column-Parallel Vision Processor," *IEEE J. Solid-State Circuits*, vol. 55, no. 7, pp. 1762–1775, July 2020. http://doi.org/10.1109/JSSC.2020.2993722

[17] S. Geman and D. Geman, "Stochastic Relaxation, Gibbs Distributions, and the Bayesian Restoration of Images," *IEEE Trans. Pattern Anal. Mach. Intell.*, vol. PAMI-6, no. 6, pp. 721–741, Nov. 1984. https://doi.org/10.1109/TPAMI.1984.4767596

[18] M. Yamaoka *et al.*, "A 20k-Spin Ising Chip to Solve Combinatorial Optimization Problems With CMOS Annealing," *IEEE J. Solid-State Circuits*, vol. 51, no. 1, pp. 303–309, Jan. 2016. http://doi.org/10.1109/JSSC.2015.2498601

[19] N. Mohseni *et al.*, "Ising machines as hardware solvers of combinatorial optimization problems," *Nat. Rev. Phys.*, vol. 4, no. 6, pp. 363–379, May 2022. http://doi.org/10.1038/s42254-022-00440-8

[20] E. Manuzzato *et al.*, "A 64×64-Pixel Flash LiDAR SPAD Imager with Distributed Pixel-to-Pixel Correlation for Background Rejection, Tunable Automatic Pixel Sensitivity and First-Last Event Detection Strategies for Space Applications," *ISSCC*, Feb. 2022, pp. 96–97. http://doi.org/10.1109/ISSCC42614.2022.9731622

[21] J. Vohra *et al.*, "Imager with In-Sensor Event Detection and Morphological Transformations with 2.9pJ/pixel×frame Object Segmentation FOM for Always-On Surveillance in 40nm," *ISSCC*, Feb. 2024, pp. 104–105. http://doi.org/10.1109/ISSCC49657.2024.10454391

[22] T. Takemoto *et al.*, "A 144Kb Annealing System Composed of 9×16Kb Annealing Processor Chips with Scalable Chip-to-Chip Connections for Large-Scale Combinatorial Optimization Problems," *ISSCC*, Feb. 2021, pp. 64–65. http://doi.org/10.1109/ISSCC42613.2021.9365748

[23] S. Xie *et al.*, "Ising-CIM: A Reconfigurable and Scalable Compute Within Memory Analog Ising Accelerator for Solving Combinatorial Optimization Problems," *IEEE J. Solid-State Circuits*, vol. 57, no. 11, pp. 3453–3465, Nov. 2022. http://doi.org/10.1109/JSSC.2022.3176610

[24] J. Bae *et al.*, "CTLE-Ising:A 1440-Spin Continuous-Time Latch-Based isling Machine with One-Shot Fully-Parallel Spin Updates Featuring Equalization of Spin States," *ISSCC*, Feb. 2023, pp. 142–143. http://doi.org/10.1109/ISSCC42615.2023.10067622

[25] Y. Zhou *et al.*, "A Compute-in-Memory Annealing Processor With Interaction Coefficient Reuse and Sparse Energy Computation for Solving Combinatorial Optimization Problems," *IEEE J. Solid-State Circuits*, vol. 59, no. 9, pp. 3094–3105, Sept. 2024. http://doi.org/10.1109/JSSC.2024.3376410

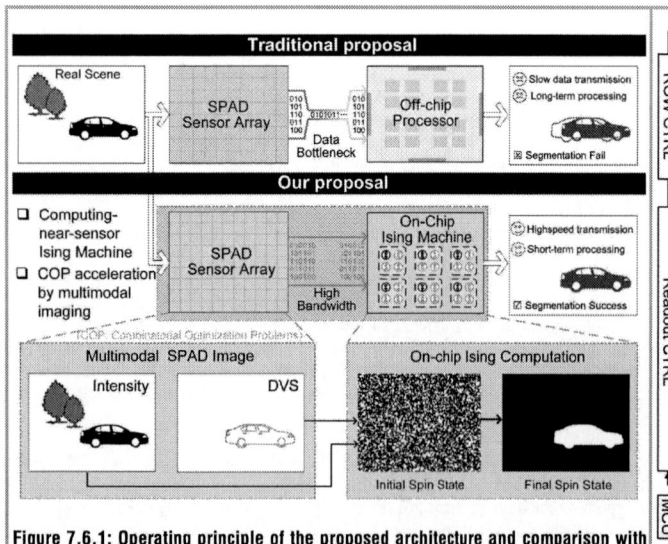

Figure 7.6.1: Operating principle of the proposed architecture and comparison with traditional scheme.

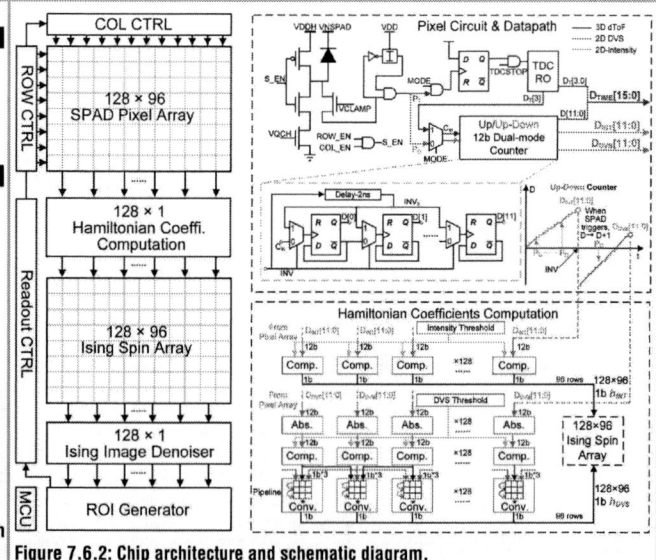

Figure 7.6.2: Chip architecture and schematic diagram.

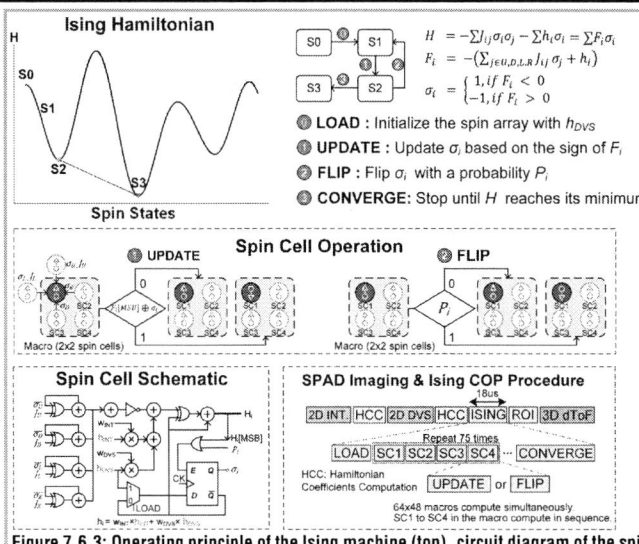

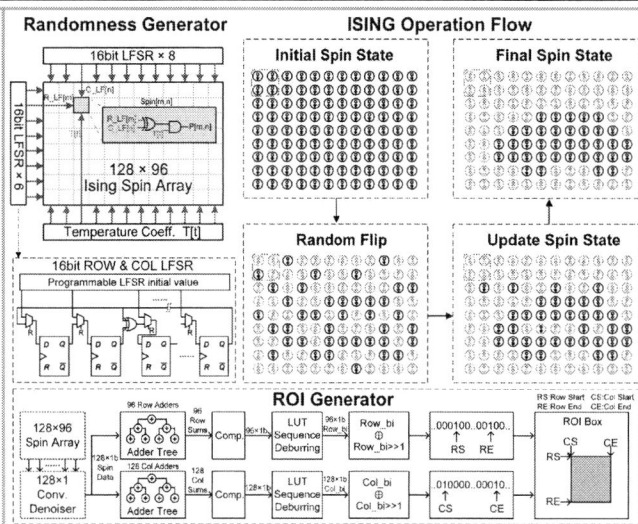

Figure 7.6.3: Operating principle of the Ising machine (top), circuit diagram of the spin cell (lower left), and the chip's workflow (lower right).

Figure 7.6.4: Workflow of the entire Ising machine and diagram of the Randomness Generator and ROI Generator.

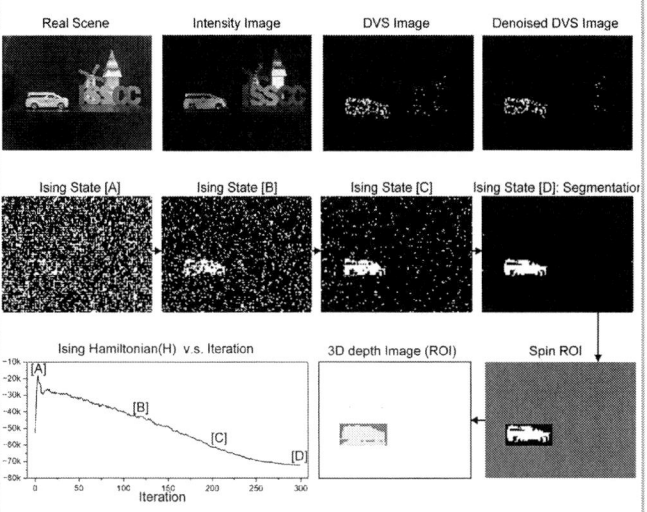

Figure 7.6.5: Experimental results of SPAD imaging, Hamiltonian convergence curve, and segmentation task.

Figure 7.6.6: Comparison table of the chip with state-of-the-art SPAD image sensors and Ising processors.

		ISSCC22[20]	CICC24[15]	JSSC25[13]	ISSCC24[21]	This work
Type		SPAD	SPAD	SPAD	CIS	SPAD
2D/3D Archit.		dToF	Vision/DVS/dToF	Vision	Vision	Vision/DVS/dToF
Technology		110 nm	130 nm	55 nm	40 nm	180 nm
Pixel pitch		48 μm	60 μm	/	4.95 μm	60 μm
Supply Voltage		/	1.2 V/3.3 V	0.95 V-1.2 V	0.8 V/1.6 V	1.8 V/3.3 V
Resolution		64 × 64	32 × 32	128×128	240×320	128 × 96
Power	2D	/	39.63 mW	384.2 mW	258.6 μV@30 fps	65.67 mW[a]
	3D	205.7 mW	42.66 mW	/	/	311.42 mW[a]
Frame rate	2D	/	10 Kfps	100 K spikes	30 fps	10 Kfps[b]
	3D	25	1 Kfps	/	/	1 Kfps[c]
Pixel counter		/	1 × 16 bit	/	/	1 × 12 bit
PDP@V_x		54.5 %@6.6 V	/	/	/	11.7 %@2 V
DCR		388 Hz	/	/	/	4.08 KHz
TDC depth		16-13 bit	16 bit	/	/	15 bit
TDC resolution		250-10000 ps	63-150 ps	/	/	108 ps
Near-Sensor task		/	ROI	SNN Classification	Object Segmentation	Object Segmentation
Time-to-Solution		/	/	/	21.7-25.8 μs	18 μs

a: Total chip power consumption; b: Intensity counting time is 75 μs; c: 1000 frames histogram with 4×4 ROI.

	ISSCC21[22]	JSSC22[23]	ISSCC23[24]	JSSC24[25]	This work
Technology	40 nm	65 nm	65 nm	55 nm	180 nm
# of Spins	16000	6400	1440	900	12288
Randomness Source	On-chip LFSR	Off-chip	Latch Equalization	XNOR-based SRAM	On-chip LFSR Array
Initial Spin states	Fixed	Fixed	Superposed (or random)	Random	Image/Fixed
Spin Circuit	Flip-flop	eDRAM	CMOS Latch	Register	Register
Spin State Representation	Digital	Digital	Analog	Digital	Digital
Spin Interaction	Digital MAC	Analog MAC	Latch Coupling	Digital MAC	Digital MAC
Power/Spin	/	0.11 μW	/	0.08 μW	2.9 μW[d]
Task	Max-cut	Max-cut	Max-cut	Max-cut & Image Segmentation	Object Segmentation

d: The circuit operates at 50 MHz.

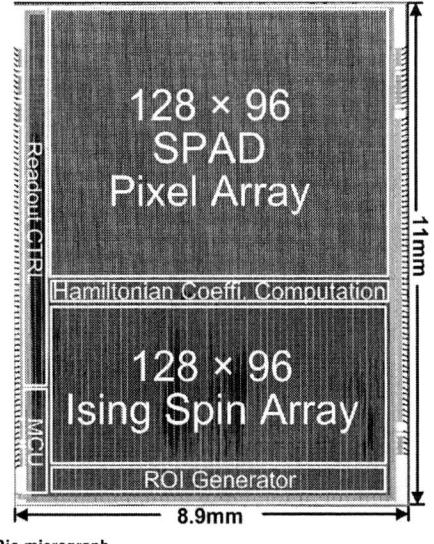

Figure 7.6.7: Die micrograph.

ISSCC 2026 / SESSION 7 / IMAGE SENSORS AND RANGING / 7.7

7.7 A Fully Reconfigurable Hybrid SPAD Vision Sensor with 134dB Dynamic Range Using Time-Coded Dual Exposures

Kieop Hong[1,2], Jubin Kang[1,2], Jung-Hye Hwang[1,2], Insang Son[3], Seonghyeok Park[3], Jung-Hoon Chun[3,4], Jaehyuk Choi[3,4], Seong-Jin Kim[2]

[1]Ulsan National Institute of Science and Technology, Ulsan, Korea, [2]Sogang University, Seoul, Korea, [3]SolidVue, Seongnam, Korea, [4]Sungkyunkwan University, Suwon, Korea

Abstract

This paper presents a fully reconfigurable HDR intensity/event hybrid SPAD sensor. The fully reconfigurable scheme enables both pixel types to be freely blended and operated simultaneously. A time-coded dual exposure encodes saturation time to provide an adaptive secondary exposure, achieving >134dB DR in both modes while reducing quantization error of the time code. Event detection with pixel-wise shot-noise thresholding keeps uniform event noise <0.46% across all illumination conditions.

Recently, hybrid image sensors that combine event-based vision sensors (EVSs) with conventional CMOS image sensors (CISs) have garnered significant attention for various applications, including consumer electronics, machine vision, and robotics, as they can detect motion events and intensity information of fast-moving objects [1-3]. Such imagers typically employ either two distinct pixel types or a time-division multiplexing architecture for event detection and intensity acquisition. By integrating the complementary advantages of both modalities into a single chip without additional optical modules, these hybrid sensors achieve compact form factors and low power consumption while capturing both dynamic and static scenes. Frame-based hybrid sensing architectures, in particular, are highly compatible with post-processing and neural network engines, enabling a broad range of functions such as image deblurring, object tracking, and other advanced vision tasks. However, PPD-based hybrid sensors suffer from inherent mismatches between their pixel types. First, while logarithmic event-vision pixels support more than 120dB of event dynamic range (DR), conventional intensity pixels have limited DR due to reduced full-well capacity and aggressive pixel scaling. Second, despite the excellent noise performance of the PPD devices, their analog front-end for event detection remains vulnerable to noise and PVT variation. SPAD-based hybrid sensors have been introduced to enhance noise robustness, but their DR is intrinsically constrained by the bit depth of in-pixel counters, restricting scalability [4-6]. Although dual-PPD pixels [7,8] and time-code-based SPAD pixels [9-11] for achieving high DR (HDR) have been reported, none of them integrates EVS functionality. Furthermore, a conventional EVS employs a fixed value or a fixed ratio for comparison, resulting in suboptimal thresholds that do not compensate for the trade-off between sensitivity and noise performance in diverse illumination conditions [1-5].

This paper presents a photon-counting/event-detecting hybrid SPAD image sensor in which each pixel encodes its saturation time and autonomously decodes it to control the secondary exposure time, thereby achieving DR enhancement through extrapolation and reducing quantization error. The time-coded dual-exposure scheme is intensively repurposed to subtract the data from two sub-frames for event detection across HDR illumination levels, realizing fully reconfigurable hybrid operation. A column-parallel event detection unit performs shot-noise-based thresholding, which ensures stable sensitivity and noise performance regardless of pixel brightness, while minimizing pixel-level event detection logic to preserve high spatial resolution. The prototype sensor fabricated in a 90nm CMOS process achieves a DR exceeding 134dB in both intensity and event modes, with event noise maintained below 0.46% across the entire DR.

Figure 7.7.1 illustrates the conceptual operating diagram and overall architecture of the hybrid SPAD sensor, which simultaneously provides intra-scene HDR imaging and precise event detection compared to conventional designs with limited DR and noisy event generation. The device consists of a 250×164 SPAD pixel array, where each pixel incorporates an active-recharging analog front-end operating with a minimum dead time of 4ns. Each pixel also contains a single-bit memory to configure operating modes and multiplexers to select corresponding input signals, enabling software-defined operation mode and synchronous frame-based readout feature. Therefore, two different pixel types can be freely blended in the array and simultaneously operated. Additionally, the pixel comprises a 9b up–down counter (UDC) for accumulating or subtracting SPAD pulses, a 4b memory for storing a time code or event data, and a 1b overflow memory for recording illuminance conditions, all controlled by compact combinational logic. Pixel outputs are either read out directly as 14b intensity data via the data-sampling circuit or processed through the event detection logic to produce 1.5b event data. A threshold look-up table (LUT) is preloaded during initialization to provide adaptive thresholds derived from intensity information.

Figure 7.7.2 illustrates the operating principle of the hybrid pixel with the time-coded dual exposures under different illuminance conditions in both modes. The exposure time is divided into two phases, referred to as the up and down periods. During the up phase, the UDC integrates photon pulses until its MSB flips over at T1, defining the primary exposure time. Subsequently, the time-coding clock (CLK_{TC}), applied with logarithmic intervals, encodes the pixel illuminance into the time code at the 4b LSB of the UDC from T1 to T2. In the succeeding down phase, the CLK_{TC} interval sequence is reversed, resetting the time code to zero, which is the inherent decoding procedure to define the secondary exposure

time. At the next rising edge of the CLK_{TC} (T3) after decoding, the UDC clock source switches from CLK_{TC} to the SPAD pulses, allowing accumulation of additional intensity data. The secondary exposure time is always automatically adjusted to be shorter than the primary exposure time, preventing saturation while maximizing count values. During the time coding period from T1 to T3, the SPAD is turned off to reduce activation power. At the transition point (T2), the MSB state of the UDC is latched into the overflow memory to store the pixel's illuminance condition. In intensity mode, if the pixel is under dark conditions (OF = 0), the UDC continues to increment without entering the down phase, collecting sufficient photon counts even beyond the reference level. The UDC is not saturated because it functions as a standard up counter in this case. Under bright conditions (OF = 1), the time code in the lower 4b of the UDC is stored in the memory (MEM_{TC}), and the pixel intensity is obtained by extrapolating this code with the adaptively adjusted secondary exposure counts (CNT_{2nd}), refining the intensity image. A single frame period is divided into several sub-frames to ensure no significant change in illuminance and prevent LED flickering.

In event mode, the upper 4b of the UCD, excluding the MSB, are sampled into the memory (MEM_{EVENT}) at every CLK_{TC} cycle and at the up-down transition point, unless the SPAD is deactivated. This up-motion information is used for event thresholding and detection. The illuminance difference caused by motion between the two phases is directly estimated by the counter in dark conditions, while the secondary exposure counts during the same time as the primary exposure are subtracted from the stored up-motion data in bright conditions. The resulting delta value is compared with the threshold in the column-parallel event detection logic in Fig. 7.7.3 to determine event occurrence. One of the 6b event thresholds in the 16-level LUT is chosen by the up-motion data, which roughly represents the pixel intensity, to provide an optimal threshold reflecting the shot-noise characteristics of the pixel signal across all illuminance conditions. The MSB of the UDC indicates either a positive or a negative event in dark conditions, whereas the carry of the subtractor signifies it in bright conditions. To align the bit width in the 9b subtractor, the MSB is filled with zero, and the lower bits of the up-motion data are padded with $0111_{(2)}$ to form a medium value, which contributes less than 1% error to the event decision. The combinational logic, comprising the subtractor and the comparator, produces the event decision result within a single clock cycle after the data is sampled.

The prototype hybrid vision sensor with 250×164 SPAD pixels was fabricated in a 90nm CMOS process. All digital pixel circuits, including a SPAD device, occupy 30×30μm², which can be further reduced by adopting advanced process nodes. Figure 7.7.4 plots experimental results in both intensity and event modes. In intensity mode, a DR of 134dB was measured, determined by the maximum and dark count rates of the SPAD. The time-coded dual-exposure scheme ensures sufficient photon counts, resulting in an SNR above 32dB in the extrapolated DR with four sub-integration times. The SNR is calculated by the ratio of the average counts and the standard deviation under various illumination conditions given by a light source with 8b intensity control, and its fluctuation is attributed to the adaptively adjusted secondary exposure counts. In event mode, considering an event detection probability above 50%, a conventional SPAD-based EVS is typically constrained to a DR of about 20dB due to counter bit depth limitations. In contrast, the proposed sensor achieves an event DR of more than 134dB with the 9b UDC. Event noise was further evaluated under three different threshold schemes: fixed value, fixed ratio, and 3σ shot-noise-based thresholds. Fixed-value and ratio-based thresholds exhibit steep increases in noise rates above 5% under bright or dark conditions. By contrast, shot-noise-based thresholds maintain a peak noise rate of less than 0.46% uniformly across all illumination levels.

Figure 7.7.5 shows the reconstructed HDR intensity image alongside the corresponding event image, which are configured separately. To form an HDR scene, objects on the left-hand side are under bright illumination from a lamp, while a large panel located at the center blocks light to create a dark condition. The scene DR is set to approximately 100dB, limited by the lens glare artifact, which results in the time code interval being increased by 1.5×. The HDR image is obtained by combining the time-code map and the secondary count map. The secondary count image mitigates quantization errors in the time-code map, enhancing fine details, such as the text on white paper, where neighboring pixels have an identical code. No secondary count is given when the time code is zero. The event map also demonstrates accurate motion extraction in two different illuminations, while maintaining

124 • 2026 IEEE International Solid-State Circuits Conference

979-8-3315-8937-0/26 $31.00 © 2026 IEEE

uniformly low event noise across the entire scene without requiring additional post-processing. The green and red colors represent positive and negative events, respectively. The performance of the hybrid sensor and comparison with state-of-the-art designs are tabulated in Fig. 7.7.6. This work achieves over 130dB DR in both intensity and event modes, surpassing previous approaches with limited intensity or event DR. A die micrograph is shown in Fig. 7.7.7.

Acknowledgement:
This work was supported in part by RS-2021-NR059470 funded by the Ministry of Science and ICT & Future Planning (MSIT, Korea), in part by Institute of Information & Communications Technology Planning & Evaluation (IITP) grant funded by the Korea government (MSIT) (RS-2025-02217613), and in part by Samsung Research Funding & Incubation Center of Samsung Electronics (SRFC-TA 1803-51).

References:
[1] M. Guo et al., "A 3-Wafer-Stacked Hybrid 15MPixel CIS + 1MPixel EVS with 4.6GEvent/s Readout, In-Pixel TDC and On-Chip ISP and ESP Function," *ISSCC*, pp. 90-91, Feb. 2023. http://doi.org/10.1109/ISSCC42615.2023.10067476
[2] A. Niwa et al., "A 2.97μm-Pitch Event-Based Vision Sensor with Shared Pixel Front-End Circuitry and Low-Noise Intensity Readout Mode," *ISSCC*, pp. 94-95, Feb. 2023. http://doi.org/10.1109/ISSCC42615.2023.10067566
[3] K. Kodama et al., "1.22μm 35.6Mpixel RGB Hybrid Event-Based Vision Sensor with 4.88μm-Pitch Event Pixels and up to 10K Event Frame Rate by Adaptive Control on Event Sparsity," *ISSCC*, pp. 92-93, Feb. 2023. http://doi.org/10.1109/ISSCC42615.2023.10067520
[4] H. Lee et al., "All-Digital Event-Based Vision Sensor with Multi-Event Generation for Motion/Vibration-Adaptive Detection," *IEEE JSSC*, vol. 60, no. 4, pp. 1162-1173, Apr. 2025. http://doi.org/10.1109/JSSC.2025.3536184
[5] F.M.D. Rocca et al., "A 128 × 128 SPAD Motion-Triggered Time-of-Flight Image Sensor With In-Pixel Histogram and Column-Parallel Vision Processor," *IEEE JSSC*, vol. 55, no. 7, pp. 1762-1775, Jul. 2020. http://doi.org/10.1109/JSSC.2020.2993722
[6] J. Wang et al., "A 32×32 Flash LiDAR SPAD Sensor with Up-to-1kfps Motional Target Detection by Threshold-adaptive 2D Dynamic Vision," *IEEE CICC,* Apr. 2024. http://doi.org/10.1109/CICC60959.2024.10529051
[7] Y. Sakano et al., "A 132dB Single-Exposure-Dynamic-Range CMOS Image Sensor with High Temperature Tolerance," *ISSCC*, pp. 106-107, Feb. 2020. http://doi.org/10.1109/ISSCC19947.2020.9063095
[8] M.-S. Keel et al., "A 12-Mpixel Automotive Image Sensor with 137-dB Single-Exposure Dynamic Range and 0.55-electron Read Noise by Oversampling-based Noise Reduction," *IEEE Symp. VLSI Technologies and Circuits*, July 2025. http://doi.org/10.23919/VLSITechnologyandCir65189.2025.11074844
[9] J. Ogi et al., "A 250fps 124dB Dynamic-Range SPAD Image Sensor Stacked with Pixel-Parallel Photon Counter Employing Sub-Frame Extrapolating Architecture for Motion Artifact Suppression," *ISSCC*, pp. 146-147, Feb. 2021. http://doi.org/10.1109/ISSCC42613.2021.9365977
[10] B. Park et al., "A 400×200 600fps 117.7dB-DR SPAD X-Ray Detector with Seamless Global Shutter and Time-Encoded Extrapolation Counter," *ISSCC*, pp. 100-101, Feb. 2023. http://doi.org/10.1109/ISSCC42615.2023.10067344

[11] Y. Ota et al., "A 0.37W 143dB-Dynamic-Range 1Mpixel Backside-Illuminated Charge-Focusing SPAD Image Sensor with Pixel-Wise Exposure Control and Adaptive Clocked Recharging," *ISSCC*, pp. 94-95, Feb. 2023. http://doi.org/10.1109/ISSCC42614.2022.9731644

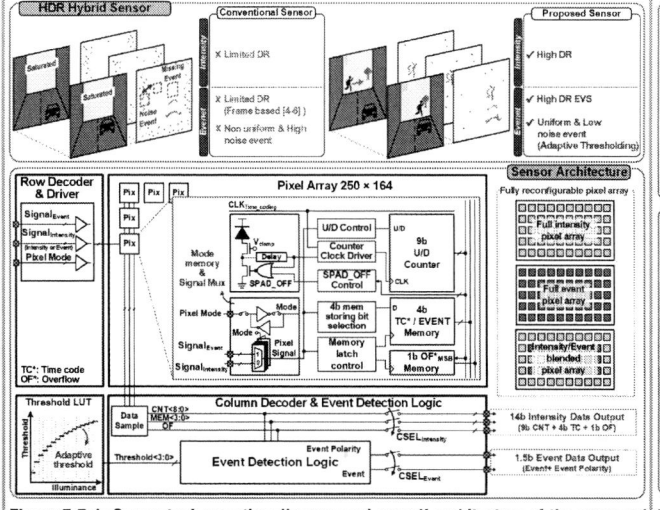

Figure 7.7.1: Conceptual operating diagram and overall architecture of the proposed hybrid SPAD sensor.

Figure 7.7.2: Operating principle of the hybrid pixel with the time-coded dual exposures under different illuminance conditions in both modes.

ISSCC 2026 / SESSION 7 / IMAGE SENSORS AND RANGING / 7.7

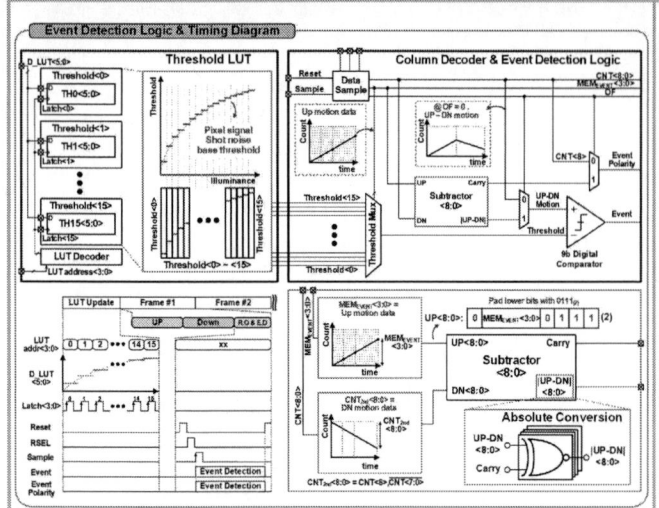

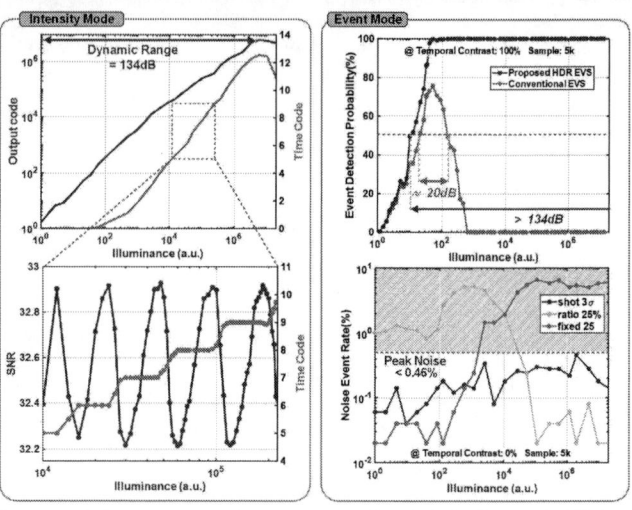

Figure 7.7.3: Schematic and operating principles of column-parallel event detection logic and its timing diagram.

Figure 7.7.4: Experimental results in both intensity (left) and event (right) modes: DR and SNR for intensity mode, and event detection and noise performance for event mode.

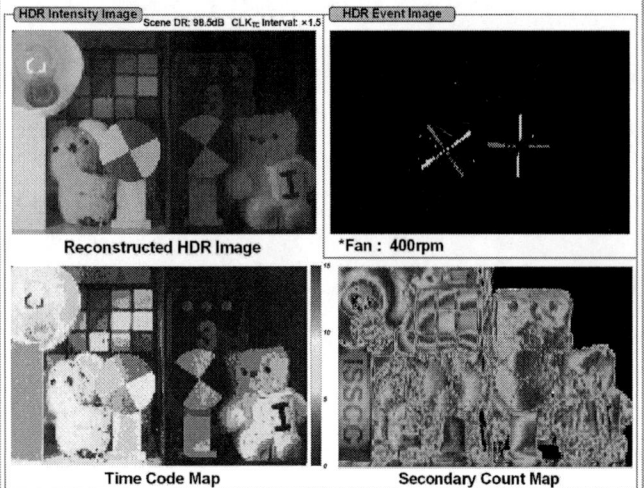

Figure 7.7.5: Captured HDR intensity and event image from the prototype chip.

Figure 7.7.6: Performance comparison with the state-of-the-art hybrid sensors.

		Unit	This Work	JSSC25 [4]	CICC24 [6]	ISSCC23 [1]	ISSCC23 [2]
	Technology	nm	90	110	130	40/65/40	90/22
	Resolution	-	250 x 164	160 x 120	32 x 32	4096 x 3680 / 1032 x 928 (CIS / EVS)	640 x 640
	Maximum Power	mW	90	15.6 (Core power)	39.63	64 (EVS power)	79.8
	Pixel Pitch	μm^2	30 x 30	40.5 x 32	60 x 60	$(8.8)^2$ / $(2.2)^2$ (EVS / CIS)	2.97 x 2.97
	Photo Detector	-	SPAD	SPAD	SPAD	PPD	PPD
Event	Dynamic Range	dB	> 134	< 54 [a]	< 96 [a]	n.a	> 120
Event	Thresholding	-	Adaptive (Pixel wise signal's shot noise based)	Fixed (All pixel has same TH @ single frame)	Adaptive (Pixel wise signal's shot noise based)	Ratio (Log pixel)	Ratio (Log pixel)
Event	Peak noise rate	-	0.46% [b]	n.a	n.a	< 1 Hz	< 0.71 Hz
Event	Max event rate	eps	200M [c]	800M	n.a	4.6G	1412M
Intensity	Dynamic Range	dB	134	54 [a]	96 [a]	73 [a]	72.2
Intensity	Shutter	-	GS	GS	GS	RS	RS
Intensity	Frame Rate	fps	60 ~ 240	10416	10000	18	n.a
Intensity	Pixel Counter or FWC (memory)	-	9b (+ 4b TC/EVENT + 1b overflow)	9b (+ 2b saturation)	16b	10ke⁻	10.6ke⁻
Intensity	DCR or Read noise @ RT	-	50 cps (V_{ex} = 1V)	n.a	n.a	2.2 e⁻	2.6σ

a: [20log(counter bit depth) or 20log(FWC/Read noise)], Calculated from the values reported in the paper
b: (noise motion event frame number / total frame number) x 100 @ no motion condition
c: Limited by IO speed

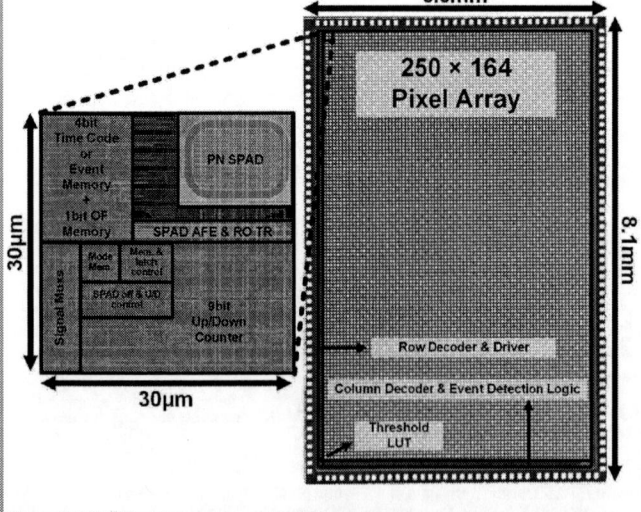

Figure 7.7.7: Chip micrograph and pixel layout.

• 2026 IEEE International Solid-State Circuits Conference

ISSCC 2026 / SESSION 7 / IMAGE SENSORS AND RANGING / 7.8

7.8 A 55nm Intelligent Vision SoC Achieving 346TOPS/W System Efficiency via Fully Analog Sensing-to-Inference Pipeline

Zhengke Yang[1], Haofeng Yu[1], Xiao Liu[1], Zhen Kong[1], Humiao Li[1], Liang Ran[2], Zhichao Lyu[3], Xinhe Feng[2], Liang Zhao[3], Yida Li[1], Jiamin Li[1], Feichi Zhou[1], Longyang Lin[1]

[1]Southern University of Science and Technology, Shenzhen, China, [2]Beijing Pixelcore Technology, Beijing, China, [3]Hefei Reliance Memory, Hefei, China

Abstract

This work presents a fully analog intelligent vision SoC that eliminates both sensor-processor and inter-layer A/D conversions for end-to-end vision. A continuous analog datapath from sensing to multi-layer inference is enabled by a PWM imager, an RRAM-based CIM, and a linearity-recovery analog memory. Fabricated in 55nm CMOS, the chip achieves 11pJ/(pixel·frame) sensing efficiency, 8,791 TOPS/W MAC efficiency, and 346 TOPS/W system-level efficiency across diverse vision tasks.

Artificial intelligence (AI)-enabled vision systems are rapidly expanding at the edge, requiring versatile support for tasks ranging from feature extraction and region-of-interest (RoI) detection to object recognition. Imagers with stacked digital processors and memory [1,2] improve locality but still rely on upfront analog-to-digital (A/D) conversion and incur significant energy and latency overhead from image data transfer. To address this, imagers with near- or in-sensor processing have been presented [3–7] to reduce data traffic and improve efficiency. Prior arts such as charge-domain Log-Haar feature extraction with weak classifiers [3] and in-sensor object segmentation [4] achieve low-energy operation, but remain task-specific and cannot support convolutional neural networks (CNNs). Designs embedding tiny CNN models through mixed-mode [5], optical-digital [6], or static random-access memory (SRAM)-based compute-in-memory (CIM) processing [7] improve energy efficiency, but remain constrained by limited programmability and memory, restricting scalability beyond shallow networks. An imager system-on-chip (SoC) combining near-sensor charge-domain multiply-and-accumulate (MAC) operations with a microcontroller [8] enhances programmability for feature extraction and RoI detection, yet repeated analog-digital conversions degrade system-level efficiency, while limited on-chip memory again prevents support for complex CNNs. Consequently, enhancing imager intelligence is fundamentally constrained by limited on-chip compute and memory capacity, as complex neural networks demand both high computational throughput and large storage, while maintaining energy efficiency and programmability for edge deployment (Fig. 7.8.1).

This work presents a fully analog versatile intelligent vision system that eliminates both sensor-processor and inter-layer A/D conversions while supporting end-to-end vision tasks. The prototype integrates: (1) a pulse-width modulation (PWM) image sensor tightly coupled with embedded resistive random-access memory (RRAM)-based CIM, (2) a continuous fully analog datapath spanning sensing to multi-layer inference, and (3) a linearity-recovery analog memory (LR-AMEM) with built-in voltage-to-time converter (VTC) and self-referenced reset for process, voltage, and temperature (PVT) resilience. Fabricated in 55nm CMOS with foundry RRAM bitcells, this design achieves 11pJ/(pixel·frame) sensing figure-of-merit (FoM), 8,791 TOPS/W peak MAC efficiency, and 346 TOPS/W system-level efficiency, enabling diverse workloads including feature extraction, classification, and object detection on a single chip.

Figure 7.8.1 (bottom) illustrates the conceptual diagram of the fully analog vision system with a continuous time-voltage datapath. The PWM image sensor converts photocurrent into pulse widths, which serve as time-domain inputs to a near-sensor MAC array. The MAC outputs are accumulated in voltage domain (i.e., V_{ACC}). For lightweight tasks such as feature extraction, V_{ACC} can be stored in LR-AMEM and directly digitized by a time-to-digital converter (TDC) to produce the results. For multi-layer inference, V_{ACC} is written into a ping-pong LR-AMEM, which stores the voltage V_{MEM} and directly provides pulse width readout (t_{RBL}) via built-in VTC. The RRAM-based CIM macro then receives t_{RBL} as word-line (WL) input, generating bit-line (BL) voltages which are stored back into the alternate LR-AMEM array for layer-by-layer processing. This seamless AMEM–RRAM cycle maintains fully analog operation across sensing, computation, and memory throughout all layers, eliminating inter-layer A/D conversions and their associated quantization noise while enabling scalable end-to-end vision inference.

Figure 7.8.2 presents the fully analog intelligent vision system architecture. The system integrates a 128×128 PWM pixel array [9], 128×4 near-sensor MAC units [10], an LR-AMEM array (65,536 analog memory cells, 2×32 banks, 32×32 cells/bank), dual 256×1024 foundry 1T1R multi-level-cell (MLC) RRAM CIM macros (1.5Mbit, 3b/cell), a ping-pong dataflow controller, and peripheral circuits including column-parallel 8b TDCs, write-bit-line (WBL) broadcast drivers, and a 512×512 shifter matrix. The system operates in three modes (direct imaging, feature extraction and multi-layer inference). In direct imaging mode, TDCs with 3:1 multiplexing digitize pixel outputs. In feature extraction mode, each pixel column generates 3-row parallel readouts feeding four near-sensor MAC units for real-time 3×3 convolution, producing accumulated voltages V_{ACC} stored directly in a linear AMEM bank before optional TDC digitization. In multi-layer inference mode, feature outputs from near-sensor MAC units are stored in LR-AMEM arrays, which perform internal voltage-to-time conversion to generate time-domain inputs for RRAM CIM macros. The shifter matrix enables flexible routing between LR-AMEM outputs and RRAM inputs. Dual RRAM CIM

macros with shared BLs perform charge-domain MAC operations, with results in voltage domain transferred to alternate LR-AMEM banks via column multiplexers and voltage-summing circuits, preserving analog signals throughout the pipeline. Digitization occurs only at final classification outputs, thus removing intermediate conversion overhead across the entire sensing-to-inference pipeline.

Figure 7.8.3 illustrates the co-design of LR-AMEM and RRAM-based CIM for energy-efficient MAC operations with linearity recovery and PVT resilience. Conventional RRAM-CIM architectures face a fundamental linearity–efficiency trade-off: current-domain designs [11,12] enforce Ohmic linearity through voltage regulation/clamping but suffer area/power overhead and limited accuracy from finite amplifier gain, while charge-domain designs [13,14] avoid voltage clamping yet exhibit intrinsic RC non-linearity, requiring swing restriction or extensive calibration. In this work, our CIM macro accumulates BL charge directly from V_{SL}, performing fully analog MAC operations (i.e. analog input in time, analog weight in RRAM conductance, analog output in voltage) in a single shot, achieving a peak energy efficiency of 8,791 TOPS/W. The RC-induced nonlinearity from this MAC operation is naturally compensated by the exponential readout generated by the built-in VTC of LR-AMEM.

Figure 7.8.3 (top-right) shows the LR-AMEM cell design, which employs a subthreshold-biased comparator (M_P/M_N) that simultaneously enables threshold variation cancellations (TVC) [9] and linearity recovery. For TVC, during reset, M_N is diode-connected by M_{RM}, setting $V_{MEM}=V_{RST}$, where the subthreshold leakage current I_N inherently tracks threshold, temperature, and supply variations, thereby canceling their mismatches and establishing a self-referenced operating point. For linearity recovery, following RRAM accumulation, V_{MAC} is transferred through a switched-capacitor voltage-summing stage (flying capacitor samples V_{RST}, then V_{MAC} pumps the top plate to $V_{RST}+V_{MAC}$) and buffered onto C_{MEM} (~5fF) without signal degradation. During readout, an exponential ramp generated by discharging through a resistor is applied to M_P, producing a rising I_P; when I_P exceeds I_N, V_{READ} rises, triggering RBL discharge to generate a linear MAC-encoded pulse. M_{RV} accelerates reset for high throughput. This exponential–logarithmic transformation intrinsically cancels RC nonlinearity in charge-domain CIM macros, yielding linear time-domain output from LR-AMEM without calibration circuits. Measurements (Fig. 7.8.3, bottom-left) confirm excellent MAC linearity with $R^2=0.9985$ with integral nonlinearity/differential nonlinearity (INL/DNL) of 3.39%/0.57%. Monte Carlo simulations across process corners and local mismatch show that the TVC scheme reduces output variation by 91.5% (11.76× reduction in σ) at mid-scale compared with open-loop storage without reset phase. Temperature measurements from -40 to 80°C demonstrates 83.1% variation reduction, while supply voltage sweeps from 0.5 to 0.7V yields 70.3% variation suppression, confirming robustness across temperature and supply voltage variations. Compared with [15], which uses feedback amplifiers for reducing AMEM variation and mismatch, this scheme achieves the results without amplifier area/power overhead, enabling scalable dense analog memory arrays for multi-layer inference.

Figure 7.8.4 illustrates the system-level analog dataflow from input LR-AMEM through RRAM computation to output LR-AMEM. The 3D feature tensor (H×W×CH) adopts a channel-centric mapping: channels are assigned to LR-AMEM rows for parallel readout, while spatial dimensions (H×W) map across columns and bank indices. This flattened addressing enables single-cycle multi-channel access aligned with RRAM array geometry. The RRAM array stores 4D weight tensors (K×K×CH$_{IN}$×CH$_{OUT}$) in a corresponding mapping, with dimensions K×K×CH$_{IN}$ flattened along array height and CH$_{OUT}$ distributed across columns. This mapping ensures that parallel channel readouts from LR-AMEM are directly fed to RRAM input dimensions, eliminating transpose operations. The timing diagram illustrates a full layer execution: parallel feature pulse readout (green) triggers charge-domain MAC in RRAM (purple), and accumulated bit-line (BL) voltages are written to destination LR-AMEM banks (red). This architecture supports diverse CNN topologies including conventional, depthwise separable convolutions and fully connected layers through flexible readout patterns and programmable weight mapping. Figure 7.8.4 (top-right) presents deployment examples for both classification and detection tasks, with entire weight tensors stored in the RRAM array. This direct tensor-to-hardware mapping achieves high MAC utilization and maintains programmability across arbitrary kernel sizes through configurable readout sequences, enabling end-to-end system energy efficiency of 345.54 TOPS/W from sensing to inference within a single chip.

Figure 7.8.5 demonstrates three edge AI vision applications evaluated on the 55nm test chip. In feature extraction, near-sensor MAC units achieve 15.32% root-mean-square error (RMSE) relative to ideal computation, with a system efficiency of 2.489 TOPS/W limited primarily by sensor exposure time rather than computation. For classification tasks, the system achieves 91.12% accuracy on CIFAR-10 and 77.22% on CIFAR-100. Operating at a 100MHz datapath frequency, this mode delivers 227.9 TOPS/W including both sensing and CNN computation, completing inference in 0.12ms. Object detection using five classes from PASCAL VOC 2007 achieves a mean average precision (mAP) of 20.5% at an intersection-over-union threshold of 0.5 (mAP@0.5), with a peak efficiency of 345.54 TOPS/W and 0.195ms latency, demonstrating versatility from simple filtering to complex detection workloads while maintaining the analog computation advantage throughout the processing pipeline.

Compared with prior intelligent imagers [4-6,8] in Fig. 7.8.6, the 55nm design (Fig. 7.8.7) demonstrates versatility across feature extraction and multi-layer CNNs for end-to-end sensing and inference, with system efficiency outperforming prior art by 75.6×–966×. Compared with prior analog processor [15], this work achieves 15.9× improvement in both MAC and system-level energy efficiency, enabling ultra-high-efficiency, fully integrated intelligent vision processing at the edge.

Acknowledgement:
This work was supported by the National Natural Science Foundation of China (Grant 62274081) and Guangdong Projects (Grant 2023QN10X177, Grant 2025B0101180002). We acknowledge the technical support from SUSTech SME-Pixelcore Neuromorphic In-sensor Computing Joint Laboratory and the SUSTech SME-CIMCube Joint Laboratory. The corresponding author is Longyang Lin (linly@sustech.edu.cn).

References:
[1] M. -W. Seo et al., "A 1.22 E-Rms Temporal Random Noise, 110 Db High Dynamic Range, 2.988µm Pixelpitch 3-Stacked Digital Pixel Sensor With on-Chip Hdr Merger", *Symposium on VLSI Technology and Circuits*, 2025, http://doi.org/10.23919/VLSITechnologyandCir65189.2025.11074898
[2] R. Eki et al., "A 1/2.3inch 12.3Mpixel with On-Chip 4.97TOPS/W CNN Processor Back-Illuminated Stacked CMOS Image Sensor", *ISSCC*, 2021, pp. 154-155. http://doi.org/10.1109/ISSCC42613.2021.9365965
[3] H. Song et al., "A 5.1ms Low-Latency Face Detection Imager with In-Memory Charge-Domain Computing of Machine-Learning Classifiers", *Symposium on VLSI Circuits*, 2021, http://doi.org/10.23919/VLSICircuits52068.2021.9492432
[4] J. Vohra et al., "Imager with In-Sensor Event Detection and Morphological Transformations with 2.9pJ/pixel×frame Object Segmentation FOM for Always-On Surveillance in 40nm", *ISSCC*, 2024, pp. 104-105. http://doi.org/10.1109/ISSCC49657.2024.10454391
[5] T. -H. Hsu et al., "A 0.8V Intelligent Vision Sensor with Tiny Convolutional Neural Network and Programmable Weights Using Mixed-Mode Processing-in-Sensor Technique for Image Classification", *ISSCC*, 2022, http://doi.org/10.1109/ISSCC42614.2022.9731675

[6] X. Wang et al., "A 0.35V 0.367TOPS/W Image Sensor with 3-Layer Optical-Electronic Hybrid Convolutional Neural Network", *ISSCC*, 2024, pp. 116-117. http://doi.org/10.1109/ISSCC49657.2024.10454479
[7] M. Nazhamaiti et al., "Selfputing: A 0.57 µW @ 15 fps Vision Chip with Self-powered In-Pixel Computing and In-Memory Computing for Visual Perception on the Edge", *IEEE European Solid-State Electronics Research Conference*, 2024, pp. 585-588. http://doi.org/10.1109/ESSERC62670.2024.10719556
[8] M. Lefebvre and D. Bol, "MANTIS: A Mixed-Signal Near-Sensor Convolutional Imager SoC Using Charge-Domain 4b-Weighted 5-to-84-TOPS/W MAC Operations for Feature Extraction and Region-of-Interest Detection", *IEEE Journal of Solid-State Circuits*, vol. 60, no. 3, pp. 934-948, March 2025. http://doi.org/10.1109/JSSC.2024.3484766
[9] M. -T. Chung et al., "A 0.5 V PWM CMOS Imager With 82 dB Dynamic Range and 0.055% Fixed-Pattern-Noise", *IEEE Journal of Solid-State Circuits*, vol. 48, no. 10, pp. 2522-2530, Oct. 2013. http://doi.org/10.1109/JSSC.2013.2269857
[10] T. -H. Hsu et al., "A 0.5-V Real-Time Computational CMOS Image Sensor With Programmable Kernel for Feature Extraction", *IEEE Journal of Solid-State Circuits*, vol. 56, no. 5, pp. 1588-1596, May 2021. http://doi.org/10.1109/JSSC.2020.3034192
[11] J. M. Correll et al., "An 8-bit 20.7 TOPS/W Multilevel Cell ReRAM Macro With ADC-Assisted Bit-Serial Processing", *IEEE Journal of Solid-State Circuits*, vol. 60, no. 8, pp. 2995-3008, Aug. 2025. http://doi.org/10.1109/JSSC.2025.3540114
[12] P. Yao et al., "A 28 nm RRAM-Based 81.1 TOPS/mm²/bit Compute-In-Memory Macro with Uniform and Linear 64 Read Channels under 512 4-bit Inputs", *IEEE European Solid-State Electronics Research Conference*, 2024, pp. 577-580. http://doi.org/10.1109/ESSERC62670.2024.10719511
[13] S. Wei et al., "A 28-nm Static-Power-Free Fully Parallel RRAM-Based TD CIM Macro With 1982 TOPS/W/Bit for Edge Applications", *IEEE Solid-State Circuits Letters*, vol. 8, pp. 21-24, 2025. http://doi.org/10.1109/LSSC.2024.3520593
[14] J. -M. Hung et al., "8-b Precision 8-Mb ReRAM Compute-in-Memory Macro Using Direct-Current-Free Time-Domain Readout Scheme for AI Edge Devices", *IEEE Journal of Solid-State Circuits*, vol. 58, no. 1, pp. 303-315, Jan. 2023. http://doi.org/10.1109/JSSC.2022.3200515
[15] J. -O. Seo et al. "ARCHON: A 332.7TOPS/W 5b Variation-Tolerant Analog CNN Processor Featuring Analog Neuronal Computation Unit and Analog Memory", *ISSCC*, 2022, pp. 258-259. http://doi.org/10.1109/ISSCC42614.2022.9731654

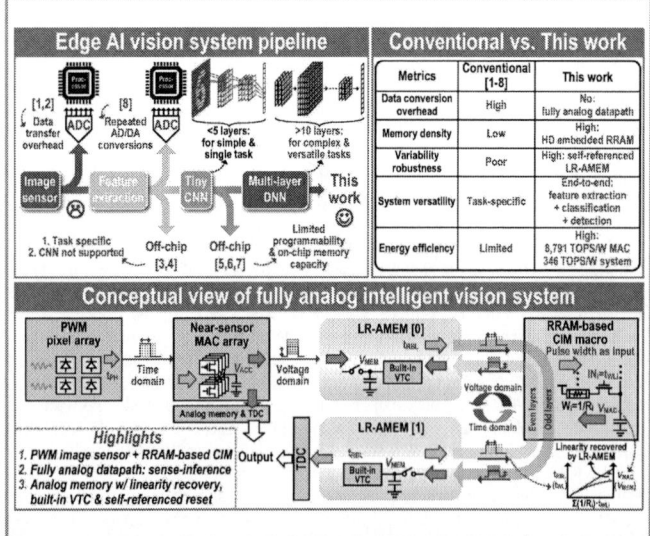

Figure 7.8.1: Motivations and contributions of the fully analog intelligent vision system.

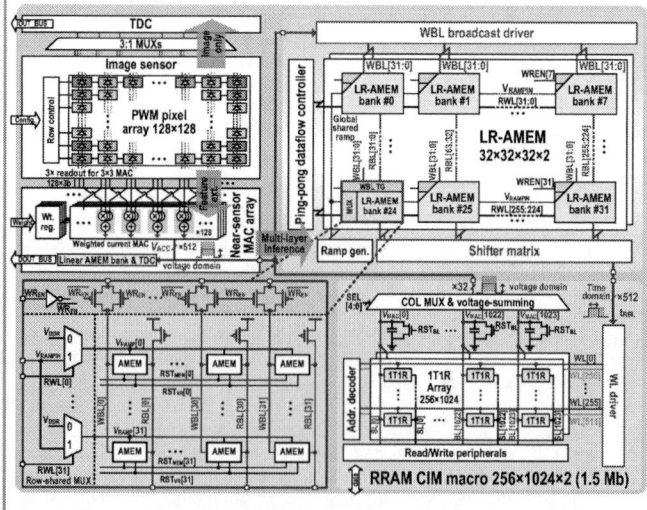

Figure 7.8.2: Overall architecture of the fully analog intelligent vision system.

ISSCC 2026 / SESSION 7 / IMAGE SENSORS AND RANGING / 7.8

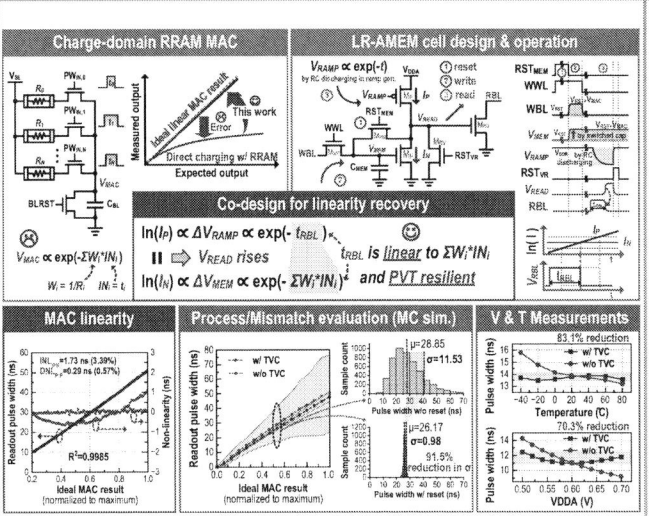

Figure 7.8.3: Co-design of LR-AMEM and RRAM-based CIM.

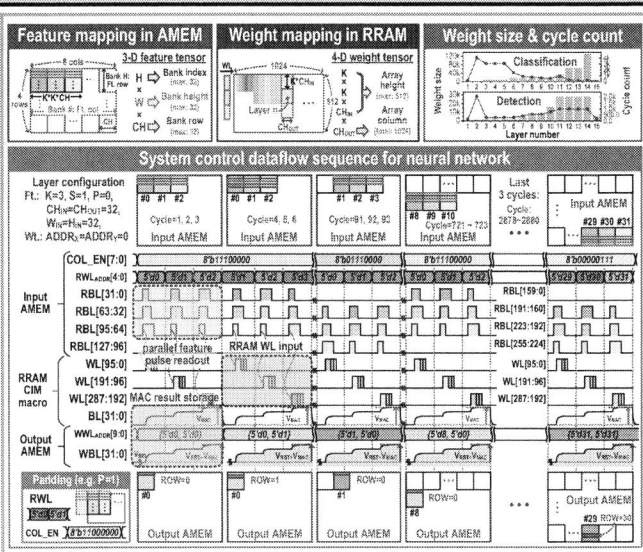

Figure 7.8.4: System-level fully analog dataflow pipeline.

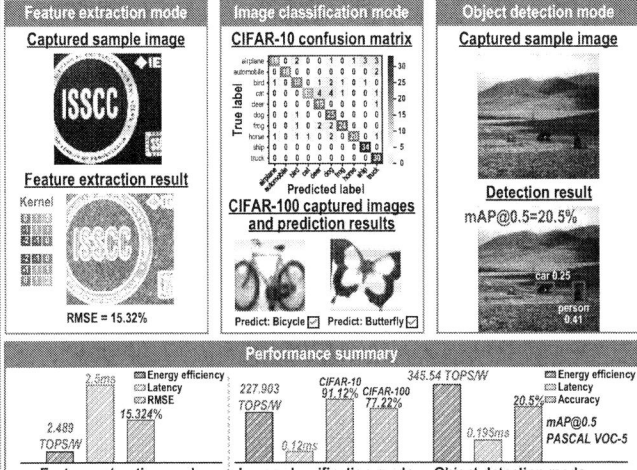

Figure 7.8.5: Measurements of edge intelligent vision applications.

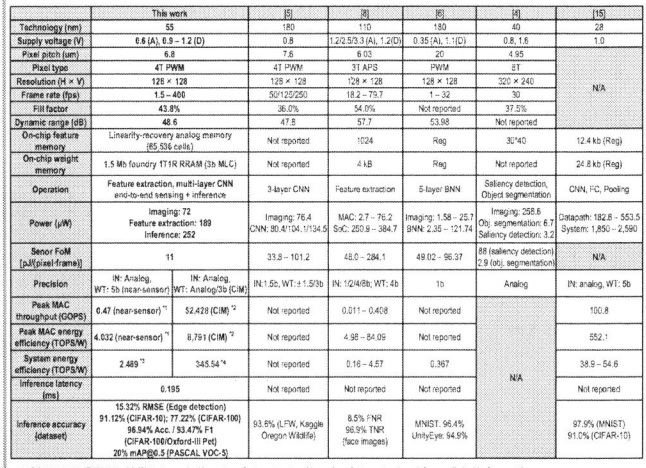

Figure 7.8.6: Comparison with prior intelligent imagers and analog processors.

Figure 7.8.7: Chip micrograph and summary.

• 2026 IEEE International Solid-State Circuits Conference

979-8-3315-8937-0/26 $31.00 © 2026 IEEE

ISSCC 2026 / SESSION 7 / IMAGE SENSORS AND RANGING / 7.9

7.9 **A 1.09e⁻-Random-Noise 1.5μm-Pixel-Pitch 12MP Global-Shutter-Equivalent CMOS Image Sensor with 3μm Digital Pixels Using Quad-Phase-Staggered Zigzag Readout and Motion Compensation**

Sanggwon Lee*, Sun-Young Yoo*, Yeongseok Shim, Yong-Suk Choi, Daehee Bae, Masamichi Ito, Deokha Shin, Si-Gyoung Koo, Sung-Jae Byun, Gihwan Cho, Hyukbin Kwon, Jaehun Jeong, Bumjun Kim, Su-Hyun Han, Youna Lee, Hiroyuki Sugihara, Junghoon Jung, Suksan Kim, Kyungtae Lim, Wonoh Ryu, Yongjun Kim, Seung-Sik Kim, Heesung Shim, Min-Woong Seo, Jae-kyu Lee, Jonghyun Go, Jaihyuk Song

Samsung Electronics, Hwaseong, Korea
*Equally Credited Authors (ECAs)

Abstract

A quad phase-staggered digital pixel sensor with a 1.494μm pixel pitch featuring a 1.09 e-rms RN, delivering a total resolution of 12MP composed of four 3MP sub-frames is presented. A single ADC is shared by a 2×2 pixels sequentially digitized, enabling optical flow extraction using two-stage deep burst super-resolution framework that incorporates a U-Net-based Quad-Bayer Re-mosaic module, extending the 12MP resolution up to 48MP while providing de-noising, motion compensation.

Mobile imaging systems have long relied on rolling-shutter (RS) CMOS image sensors (CISs), which offer compact pixel sizes and high resolution well suited for smartphones [1]. RS operation, however, inevitably causes motion artifacts such as blur, wobble, and partial exposure mismatch, degrading image quality in dynamic scenes. To address these issues, global-shutter (GS) architectures have been explored in charge-domain [2-4], voltage-domain [5-8], and digital-domain [9-11] implementations. Each approach requires additional in-pixel storage, such as capacitors or latches, which increase memory area and consequently enlarge pixel pitch. As a result, the pixel sizes generally exceed near 2μm, limiting the adoption of mobile devices that demand aggressive scaling and power efficiency. Array-parallel analog-to-digital converter (ADC) architectures have also been proposed to reduce power consumption in GS sensors [12,13]. While they improve throughput and shorten analog transfer paths, GS functionality is often restricted to selected regions of interest (ROI) rather than the entire array. Thus, uniform full-array GS capture has yet to be achieved within a mobile-compatible pixel pitch. The realization of a GS sensor full-array operation at a sub-1.5μm pixel pitch would establish a new class of image sensors for mobile applications. Such a device enables compact form factor, reduced power consumption, and high-resolution imaging with robust motion performance, extending GS technology beyond traditional industrial and automotive applications into the mobile domain.

This paper introduces a quad phase-staggered (QPS) digital pixel sensor (DPS) with a 1.494μm pixel pitch, delivering a total resolution of 12 megapixels (MP) composed of four 3MP sub-frames. In the proposed architecture, a single ADC is shared by a 2×2 pixel block, and the four pixels are sequentially digitized with a uniform time interval. Three outputs are stored in frame memories of the bottom plate, while the fourth output is stored in the in-pixel memory of the middle plate, generating four phase-staggered sub-frames. The sequential phase shifts enable optical flow [14] extraction using Deep Burst Super-Resolution (DBSR) algorithm [15]. To further improve image quality, we propose a novel two-stage DBSR framework that incorporates a U-Net-based Quad-Bayer Re-mosaic (QBR) module, referred to as QPS Super-Resolution & Re-mosaic (QPS-SRR) framework, extending the native 12MP resolution up to 48MP while providing de-noising, motion compensation and robustness against edge artifacts. Furthermore, the sub-frame differences inherently provide edge detection capability, enhancing recovery of high-frequency patterns. The QPS design demonstrates that GS-equivalent imaging can be achieved at a 1.494μm pixel pitch, establishing a promising solution for mobile applications that demand compact, power efficient, and motion-robust ultra-high-resolution imaging.

The sensor is fabricated in a 3-stack configuration consisting of a photodiode layer, an ADC/in-pixel memory layer, and a digital/frame memory layer, enabling compact pixel scaling. Figure 7.9.1 shows the chip-level diagram of the 3-stacked QPS DPS along with the schematic of a 2×2 pixel block and its shared ADC connection. On the top plate, the sensor integrates a 12MP array (3872ᴴ×3104ᵛ) with a 1.494μm pixel pitch. Each quad pixel comprises a pinned photodiode (PPD), transfer gate (TG), dual conversion gain (DCG), reset (RST), source follower (SF), and select (SEL) transistors. The four SF outputs converge into a single ADC, realizing a quad-shared ADC topology with zigzag-direction readout. Additional bias transistor (Vp), and a clamp transistor (CLP) with enable control (CLP_EN) prevent the output-node floating, which is critical for reducing signal settling time and ensuring consistent output behavior. The input transistor of the operational transconductance amplifier (OTA), which is the main source of low-frequency noise such as a random telegraph signal (RTS) and flicker noises, is placed on the top layer to allow surface treatment at the channel interface. Two auto-zero capacitors (C_AZ) and bandwidth-limiting capacitor (C_BWL) are also integrated to minimize pixel offset, mismatch, and to suppress sensor noise by reducing the OTA's noise bandwidth. The middle layer contains pixel drivers, logic drivers for controlling data-bus (DBS) signals, ramp buffers, and the remaining array-parallel ADC circuits. Each pixel output is compared against the ramp and stored in SRAM cells as reset and signal data, controlled by RST_EN and SIG_EN. To optimize spatial efficiency, one ADC is shared by a 2×2 pixel block, and groups of 4×16 ADCs are organized into a bank unit. Each bank integrates 64 ADCs and 128 SRAM cells (64 SRAMs for reset and 64 SRAMs for signal). Reset and signal data are further processed in the ADBUS block, where digital correlated double sampling (D-CDS) suppresses circuit variation-induced noise. The bottom layer implements a voltage doubler (DBR), ramp generator, and reference circuit, along with three 3MP frame memories for storing phase-staggered sub-frames, as well as an ISP block for digital processing.

The QPS scheme defines the sequential digitization of four pixels in a 2×2 quad, generating well-controlled temporal offsets that form four sub-frames. Figure 7.9.2 illustrates the timing diagram of the QPS readout operation. Each operation cycle consists of ADC and readout phases. During the ADC phase, each quad phase (QP1 to QP4) is processed in sequence. For a given phase, 3MP are simultaneously digitized by their corresponding shared ADCs, enabling GS capture. Both high- and low-conversion-gain (HCG/LCG) modes are supported to accommodate low- and high-illumination conditions. Each conversion generates reset and signal data, which are stored in SRAM. During the readout phase, the stored data are transferred through the DBS. The readout process currently takes 2.67ms/sub-frame, which can be further reduced by structural optimization of DBS block. Before transmission, bit lines (BL/BLn) are pre-charged (PC) to reduce error rates. Data are then read in a top-down sequence and passed to the DBS block. The quad phases are read sequentially in a zigzag direction, with a constant interval ΔT between sub-frame operations. This produces four sub-frames containing identical spatial but shifted temporal information, enabling optical flow estimation and reconstruction. To support sequential storage and retrieval, QP1–QP3 data are written to the three bottom-layer frame memories, while QP4 data are temporarily stored in the in-pixel memory before sequential release.

Exploiting QPS zigzag readout method at the circuit level, optical-flow-based motion compensation and artifact suppression are achieved by aligning and fusing the QPS sub-frame images. Leveraging all four 3MP images enables reconstruction of a 12MP image and a 48MP image can also be realized according to a pre-defined scale factor using high-resolution techniques. To this end, the DBSR framework is adopted, treating the QPS sub-frame images as burst inputs. First, four temporally shifted 3MP inputs are aligned to a reference image using pixel-wise optical-flow estimation, with the second sub-frame serving as the reference to stabilize motion estimation and integrate motion information across all four sub-frames. When fusing the sub-frames, weights are adaptively assigned local reliability with the reference image emphasized in edge regions and auxiliary sub-frames contributing complementary information in degraded areas. While this conventional DBSR architecture is sufficient for 12MP or 48MP reconstruction, our QPS-SRR framework, which incorporates an additional quality enhancement module, enables improved reconstruction at 48MP. The DBSR framework can be vulnerable to aliasing artifacts due to effective input sampling at two-pixel intervals. To mitigate this limitation, a U-Net-based QBR module is integrated into the second stage [16,17] since the QBR reconstructs from unsampled representations of all QPS sub-pixel images, enhancing robustness against edge artifacts. Accordingly, in this stage, four 12MP images sub-sampled at two-pixel intervals from the 48MP reconstructed image, along with the 12MP QBR output, serve as inputs. A Reference Image Selector (RIS) block determines the optimal reference by evaluating motion intensity from the first-stage DBSR and high-frequency density from the QBR module. When the QBR image is chosen as reference, the reconstruction exhibits enhanced suppression of aliasing artifacts. Figure 7.9.3 illustrates the results of the QPS-SRR. Using the second sub-frame as reference allows full exploitation of motion information across all sub-frames, resulting in a motion compensated image, while the simply combined 12MP image exhibits motion blur. Integration of the U-Net-based QBR further reduces artifacts at edges, a yielding high-resolution image with improved structural fidelity. Moreover, edge information extracted from inter-sub-frame differences highlights motion boundaries and salient features in dynamic scenes. Figure 7.9.4 demonstrates spatiotemporal information extraction, where edges indicate motion regions. Event polarity can be inferred from intensity variations, while brightness fluctuations caused by illumination changes and global motion can be analyzed across inter-sub-frames, enabling more accurate identification of motion boundaries. From a broader perspective, QPS sub-frame images may contribute to higher-level vision tasks such as object recognition, where both motion cues and structural features are jointly leveraged.

Figure 7.9.5 shows the sensor characteristics along with a summary table. The four QP outputs exhibit nearly identical linearity, attributed to the symmetric layout. The average fixed-pattern noise (FPN) is 0.65 e-rms, and random noise is 1.09 e-rms at an analog gain

128 • 2026 IEEE International Solid-State Circuits Conference

979-8-3315-8937-0/26 $31.00 © 2026 IEEE

of 16. With a 1.494µm pixel pitch, the sensor forms a 12MP effective array from four 3MP sub-pixels, consuming 355.5mW of total power. Figure 7.9.6 presents a performance comparison with array-parallel, pixel-parallel and column-parallel ADC-based GS image sensors. Among GS-equivalent sensors, the developed sensor achieves the smallest pixel pitch while demonstrating state-of-the-art noise performance. In particular, the QPS DPS architecture enables a well-balanced trade-off among noise, power, and scalability. Figure 7.9.7 shows chip micrographs of the fabricated device. The proposed sensor architecture represents a promising candidate for extending GS technology specifically into mobile imaging applications.

References:
[1] D. Kim et al., "A 1/1.56-inch 50Mpixel CMOS image sensor with 0.5µm pitch quad photodiode separated by front deep trench isolation," *ISSCC*, pp. 118-119, 2024. https://doi.org/10.1109/ISSCC49657.2024.10454448
[2] Y. Kumagai et al., "Back-illuminated 2.74µm-pixel-pitch global shutter CMOS image sensor with charge-domain memory achieving 10ke- saturation signal," *IEEE IEDM*, pp. 237-240, 2018. https://doi.org/10.1109/IEDM.2018.8614676
[3] M. Kobayashi et al., "A 1.8e-rms temporal noise over 110-dB-dynamic range 3.4µm pixel pitch global-shutter CMOS image sensor with dual-gain amplifiers SS-ADC, light guide structure, and multiple-accumulation shutter," *IEEE JSSC*, vol. 53, no. 1, pp. 219-228, 2018. https://doi.org/10.1109/JSSC.2017.2737143
[4] I. Mizuno et al., "A High Performance 2.5um Charge Domain Global Shutter Pixel," *Proc. Int. IISW*, June 2019. https://doi.org/10.60928/3le4-os80
[5] S.–S. Kim et al., "3-Layer Stacked Voltage-Domain Global Shutter CMOS Image Sensor with 1.8µm-Pixel-Pitch," *IEEE IEDM*, pp. 902-905, 2022. https://doi.org/10.1109/IEDM45625.2022.10019515
[6] H. Shim et al., "A 3-Stacked Hybrid-Shutter CMOS Image Sensor with Switchable 1.2µm-Pitch 50MPixel Rolling Shutter and 2.4µm-Pitch 12.5MPixel Global Shutter Modes for Mobile Applications," *ISSCC*, pp. 114-115, 2025. https://doi.org/10.1109/ISSCC49661.2025.10904571
[7] P. Malinge et al., "2.16µm Back Side Illuminated Voltage Domain Global Shutter CMOS Image Sensor with single silicon layer pixel," *IEEE IEDM*, 2023. https://doi.org/10.1109/IEDM45741.2023.10413802
[8] T. M. Kim et al., "A High-Performance 2.2µm 1-Layer Pixel Global Shutter CMOS Image Sensor for Near-Infrared Applications", *IEEE IEDM*, 2024. https://doi.org/10.1109/IEDM50854.2024.10873463
[9] T. Kainuma et al., "A 25.2Mpixel 120frames/s Full-Frame Global-Shutter CMOS Image Sensor with Pixel-Parallel ADC," *ISSCC*, pp. 122-123, 2025. https://doi.org/10.1109/ISSCC49661.2025.10904642
[10] M.-W. Seo et al., "A 1.22 e-rms Temporal Random Noise, 110 dB High Dynamic Range, 2.988 µm Pixel-Pitch 3-Stacked Digital Pixel Sensor with On-Chip HDR Merger," *in Proc. Symp. VLSI*, 2025. https://doi.org/10.23919/VLSITechnologyandCir65189.2025.11074898
[11] T.–H. Tsai et al., "A 400x400 3.24µm 117dB-Dynamic-Range 3-Layer Stacked Digital Pixel Sensor," *ISSCC*, pp. 120-121, 2025. https://doi.org/10.1109/ISSCC49661.2025.10904717
[12] A. Berkovich et al., "A 3D-Integrated 2-Megapixel Imager with Sparse Capture and Fine-Grain Power Gating," *IEEE IEDM*, 2023.

https://doi.org/10.1109/IEDM45741.2023.10413713
[13] T. Takahashi et al., "A Stacked CMOS Image Sensor with Array-Parallel ADC Architecture," *IEEE JSSC*, vol. 53, no. 4, pp. 1061-1070, 2018. https://doi.org/10.1109/JSSC.2017.2784759
[14] D. Sun et al., "PWC-Net: CNNs for Optical Flow Using Pyramid, Warping, and Cost Volume," *IEEE Conference on Computer Vision and Pattern Recognition workshops*, 2018. https://doi.org/10.48550/arXiv.1709.02371
[15] G. Bhat et al., "Deep Burst Super-Resolution," *IEEE Conference on Computer Vision and Pattern Recognition workshops*, 2021. https://doi.org/10.48550/arXiv.2101.10997
[16] D. Sun et al., "U-Net: Convolutional Networks for Biomedical Image Segmentation," *IEEE Conference on Computer Vision and Pattern Recognition workshops*, 2015. https://doi.org/10.48550/arXiv.1505.04597
[17] Q. Yang et al., "MIPI 2022 Challenge on Quad-Bayer Re-mosaic: Dataset and Report," *European Conference on Computer Vision*, 2022. https://doi.org/10.48550/arXiv.2209.07060

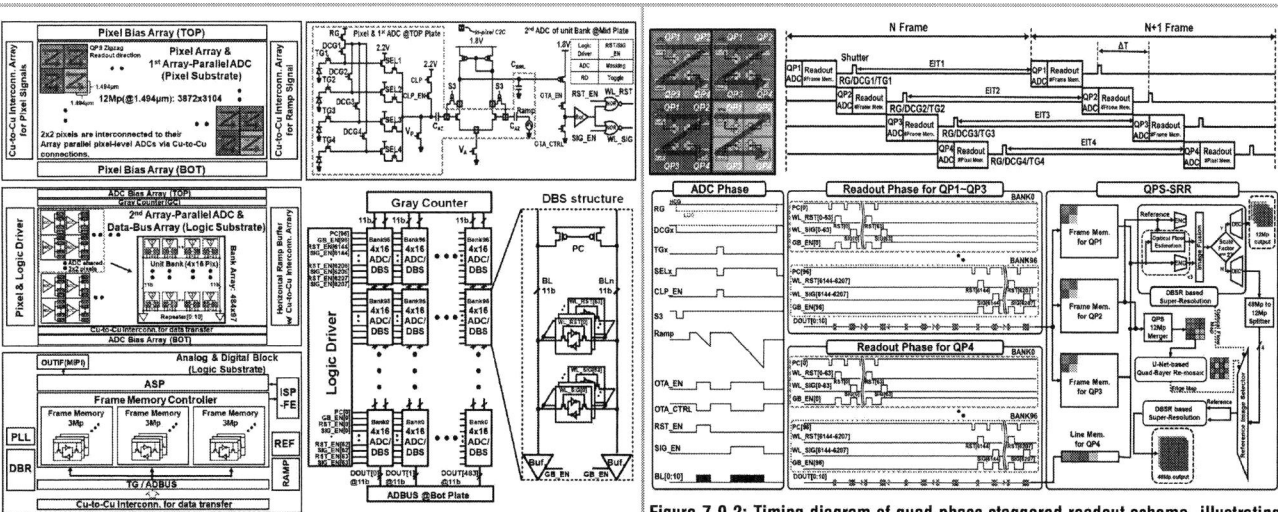

Figure 7.9.1: Chip-level diagram of the proposed 3-stacked quad-phase staggered digital pixel sensor (left), and schematic of a 2×2 pixel block with a shared ADC (right).

Figure 7.9.2: Timing diagram of quad-phase staggered readout scheme, illustrating zigzag selection of 2×2 sub-pixels (top), ADC and readout phases, and data allocation to frame memories and in-pixel memory (bottom).

ISSCC 2026 / SESSION 7 / IMAGE SENSORS AND RANGING / 7.9

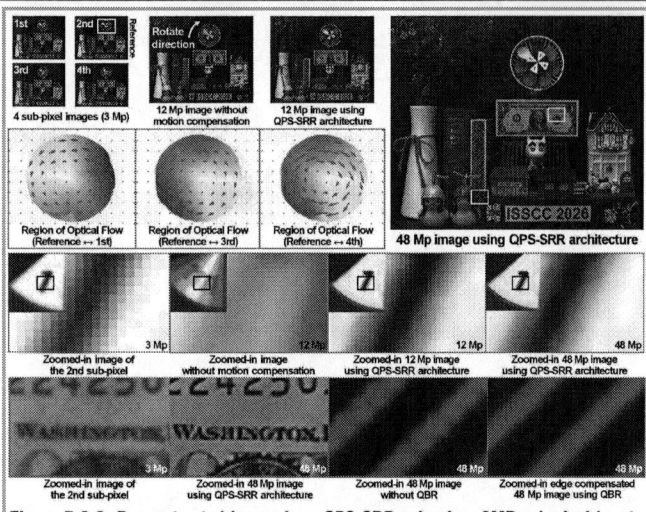

Figure 7.9.3: Reconstructed image from QPS-SRR using four 3MP sub-pixel inputs, achieving super-resolution and motion compensation via DBSR with optical flow, and edge-artifact mitigation using a U-Net-based Quad-Bayer Re-mosaic module.

First sub-pixel image

Second sub-pixel image

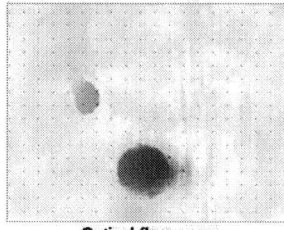

Optical-flow map

Optical-flow driven edge map

Figure 7.9.4: Edge image extracted from the optical flow between consecutive sub-pixel images, containing spatiotemporal information.

Figure 7.9.5: Sensor performance characteristics and summary table.

Parameter		Value
Supply Voltage		2.2V /1.8V /1.0V
Process	TOP	65 nm
	MID	28 nm
	BOT	28 nm
Pixel Pitch		1.494 um
Resolution		12 Mp (3872x3104)
Frame Rate		45 fps
C.G [uV/e]	HCG	254.1
	LCG	63.3
FWC [e⁻] @LCG, AGx1		11.9K
RN [e⁻] @HCG, AGx16		1.09
pFPN [e⁻] @HCG, AGx16		0.65
Power Consumtion [mW]		355.5

Figure 7.9.6: Performance comparison with array-parallel, pixel-parallel and column-parallel ADC image sensors.

Parameter		This work	IEDM'2023 [12]	JSSC'2018 [13]	ISSCC'2025 [9]	VLSI'2025 [10]	ISSCC'2025 [6]
ADC Type		Array-parallel (2x2)	Array-parallel (2x2)	Array-parallel (10x16)	Pixel-parallel	Pixel-parallel	Column-parallel
Process technology [nm]		65/28/28	65/40	90/55	90/40	65/28/28	65/65/28
Supply Voltage [V]		2.2/1.8/1.0	3.3/2.5/1.1	2.9/1.2/1.1	2.9/1.1	2.2/1.8/1.0	2.2/2.2/1.0
Array size		12Mp (3872x3104)	1.9Mp (1400x1400)	4.1Mp (2360x1728)	25.2Mp (6144x4104)	3Mp (1936x1552)	12.5Mp(4096x3072)
Pixel size [μm]		1.494	4.23	4.8	5.94	2.988	2.4
Full well capacity [e⁻]		11.9k	7k	6.7k*	40k	N/A	52k
Frame rate [fps]		45	30	280	120	192	95
Random noise	AGx1 [e]	2.92	11	4.3	2.66	2.3**	N/A
	AGx16 [e]	1.09	N/A	2.4	N/A	1.2**	2.4
ADC resolution [bit]		10	8	12	14	10	10
Power consumption [mW]		355.5	4.2	1340	1545	270.1	650

*Calculated value
**Reported improvement at VLSI presentation

TOP layer
2x2 sub pixel APS Array

MID layer
Driver — ADC & DBS Array — Ramp Buffer

BOT layer
Frame Memory & ISP
DBR — ADBUS & TG — RAMP & BST

Figure 7.9.7: Chip micrographs.

• 2026 IEEE International Solid-State Circuits Conference

ISSCC 2026 / SESSION 7 / IMAGE SENSORS AND RANGING / 7.10

7.10 A 200MP 0.61μm-Pixel-Pitch CMOS Imager with Sub-1e⁻ Readout Noise Using Interlaced-Shared Transistor Architecture and On-Chip Motion Artifact-Free HDR Synthesis for 8K Video Applications

Richard Xu[1], Guanjing Ren[2], Yaowu Mo[2], Dekui Qi[2], Qian Wang[2], Shengxin Zhang[2]

[1]SmartSens Technology, Santa Clara, CA, [2]SmartSens Technology, Shanghai, China

Abstract

We propose a 200MP 0.61μm pixel size CMOS image sensor featuring: 1) an interlaced-shared pixel architecture to achieve a 0.7e- readout noise, 2) an on-chip motion artifact-free HDR synthesis with single-frame dynamic range of 77dB, 3) a DCG ratio self-calibration algorithm to address the CG ratio non-uniformity, and 4) a 22nm high-speed readout circuit suitable for 8K HDR video with an optimal balance of low-power and high-speed performance.

The resolution race is once again gaining momentum. Driven by advancements in pixel technology and ongoing developments in CMOS processes, the demand for higher resolutions, reaching up to 200 megapixels (MP), is increasingly prominent in the latest high-end smartphones [1]. To accommodate the constraints of lens modules and device compartments, the pixel pitch for 200MP sensors typically falls within the sub-0.7 μm range. One of the leading designs for achieving high-performance 200MP sensors with a compact pixel pitch is detailed in reference [2]. This design employs Full Deep Trench Isolation (FDTI) and a 2×4 shared-pixel architecture, allowing for a full well capacity of 40ke-, which surpasses that of a 50MP sensor with the same optical format. However, due to the inherent limitations of the pixel-sharing architecture, the pixel conversion gain is relatively low, resulting in a readout noise level > 1e-, thereby compromising low-light performance. In contrast, a 50MP sensor can achieve sub-1e- noise levels [3], which restricts the usage of the 200MP sensor primarily to tele-photo applications, where low-light performance is less critical. Alternative approaches to developing a 200MP sensor with sub-0.7μm pixel pitch without utilizing FDTI are explored in [4,5]. Unfortunately, these designs exhibit even poorer full-well capacity than the aforementioned reference [2], and the readout noise issues remain unresolved due to constraints on transistor placement within the pixel, which are dictated by the small pixel pitch. Furthermore, the demand for 8K video with high dynamic range (HDR) is surging, fueled by the popularity of short video applications. However, traditional staggered HDR methods often result in motion artifacts or ghosting effects when capturing fast-moving subjects, which detracts from user experience.

To address the above challenges, this paper presents a 1/1.28-inch, 200MP CMOS image sensor featuring a 0.61μm pixel pitch and a readout noise of less than 1 electron, fabricated using the TSMC 40nm/22nm stacked backside illumination (BSI) CMOS process. Figure 7.10.1 illustrates a pixel schematic diagram of a 2×2 shared-pixel architecture that employs an interlaced-shared transistor design, facilitating an ultra-high conversion gain and superior low-light performance. In the depicted 2×2 pixel unit, each row comprises three transistors and four transfer (TX) transistors. The row select (RS) transistor for row N0 is shared with row N1, while the reset (RST) transistor for row N1 is shared with row N2. This interlaced-shared architecture minimizes the number of transistors required, allowing for a compact pixel pitch. The design accommodates an additional Dual Conversion Gain (DCG) transistor positioned between the floating diffusion (FD) node and the RST transistor. When the DCG is in the off state during readout, it reduces the number of transistors and minimizes the P/N junction connections to this node. Furthermore, the interconnect metal routing on the FD is optimized, resulting in a capacitance of 0.58fF. This optimization leads to an impressive conversion gain of 275μV/e, corresponding to a readout noise of 0.7 electrons. This conversion gain is designated as High Conversion Gain (HCG). When the DCG is in the on state during readout, the FD node connects to a metal-oxide-metal (MOM) capacitor shared between rows N1 and N2, serving as a substantial charge reservoir to accommodate all charges from the photodiode (PD) and maintain a high full-well capacity. This configuration, characterized by low conversion gain and high charge capacity, is designated as Low Conversion Gain (LCG). Consequently, this design achieves both HCG for low-light sensing with sub-1 electron readout noise and LCG for enhanced charge capacity, all within a compact 0.61μm pixel pitch without compromising performance. Moreover, since rows N0 and N1 share the same RS transistor, as illustrated in Fig. 7.10.1, the drain of the RST transistor (RSTD) is connected to the row driver, enabling control over whether row N0 or row N1 can be read out without interfering with the other shared rows.

The detailed working principle of the pixel is illustrated in the timing diagram presented in Fig. 7.10.2. This diagram outlines the readout process for two consecutive rows, capturing both HCG and LCG signals. Prior to the readout phase, a reset operation is conducted to simultaneously reset the photodiodes (PDN0 and PDN1). During the readout phase, row N0 is read first. Concurrently, the RST and DCG transistors of row N1 are activated, applying a low voltage from RSTDN1 to the FD node of row N1, thereby disabling the source follower (SF) and preserving the integrity of the signal from row N0 at the output line. Once the readout of row N0 is complete, row N1 is then read. At this time, the RST and DCG transistors of row N0 are activated, with a low voltage from RSTDN0 applied to the FD node of row N0, ensuring that the signal from row N1 remains unaffected at the output line. Furthermore, during the readout of each row, the reset level for LCG is first measured with the DCG in the on state and stored in memory for subsequent processing. This is followed by the measurement of the reset level for HCG, which is then succeeded by the HCG signal level. The DCG transistor is subsequently reactivated to measure the LCG signal level, which is then subtracted from the previously stored LCG reset level for correlated double sampling (CDS) operation. To facilitate this interlaced operation, a specialized design for the row decoder is required, as illustrated in Fig. 7.10.3. Given that a single output line is shared among four rows of the 2×2 pixel unit, the row decoder must employ address connections for the RS, RST, RSTD, and DCG signals in an interlaced and cross-connected pattern. This design ensures that all RS, RST, RSTD, and DCG signals are controlled in the correct sequence, in accordance with the timing requirements and the shared architecture.

The block diagram of the readout circuitry is presented in Fig. 7.10.4. The architecture employs a column parallel single-slope analog-to-digital converter (ADC) design, which effectively reduces power consumption while enhancing noise performance. The comparator features a differential input and single-ended output configuration, consisting of three stages with programmable gain control at the input. To further enhance noise reduction and speed, a programmable noise reduction block is integrated between the first and second stages of the comparator. In the counter/memory readout block, a 14b counter is utilized, enabling rapid counting speeds of up to 2Gcounts/s. Additionally, a multi-channel D flip-flop with reset (DFFR) sequential memory readout configuration is implemented to minimize noise interference with other analog blocks, thereby improving overall noise performance. A top-bottom four half-row multi-channel concurrent readout architecture is employed for both HCG and LCG signals, ensuring high-speed readout capabilities suitable for 8K HDR video modes. Furthermore, by leveraging an advanced 22nm CMOS logic process and operating at a voltage of 0.9V, the readout circuit achieves a desirable balance of low power consumption and high-speed performance.

Furthermore, the HCG and LCG signals can be synthesized within the chip's signal pipeline to create a HDR image. As depicted in the readout timing diagram in Fig. 7.10.2, the HCG and LCG signals are read out with minimal time delay, effectively eliminating motion artifacts or ghosting issues commonly associated with traditional staggered HDR methods. Following the on-chip HDR synthesis, the dynamic range of the final image is enhanced to 77dB. Due to variations in the manufacturing process, the CG ratio for each sensor and color channel can differ from die to die and wafer to wafer, leading to color ratio nonuniformity or color noise, particularly at the knee point during HDR combination. Our CG ratio also fluctuates slightly with changes in gain and exposure time, and in certain conditions, these variations can become noticeable. To address the issue of CG ratio variation, we employ a specialized algorithm known as self-CG ratio calibration. This method operates as follows: an initial CG ratio is calculated and stored in OTP memory during the chip production test. Each time the sensor undergoes changes in gain and exposure, the CG ratio is recalibrated, provided that the HCG and LCG signals are not saturated. The newly calculated ratio replaces the previous one and is utilized to derive the linearization curve for each color channel after combination, resulting in a significantly smoother image with reduced color noise.

The sensor is fabricated using a TSMC 40nm/22nm stacked backside illumination (BSI) CMOS process. A sample 8K video image captured with on-chip motion artifact-free HDR synthesis is presented in Fig. 7.10.5, demonstrating a clear advantage over traditional staggered HDR video, as evidenced by the absence of ghosting artifacts around fast-moving objects. A comprehensive performance summary including pixel size from 0.56 to 0.61μm is provided in Fig. 7.10.6, while Fig. 7.10.7 displays the micrograph of both the sensor and the ASIC die.

In conclusion, we propose a novel interlaced-shared transistor architecture designed for compact pixel sizes, achieving a low noise level of sub-1e- while simultaneously maintaining a high full-well capacity. The sensor is equipped with on-chip motion artifact-free HDR synthesis and a self-CG ratio calibration algorithm, along with a low-power, high-speed readout circuitry that supports 8K HDR video output. This sensor effectively addresses the increasing performance demands for 200MP sensors in high-end smartphone main camera applications.

Acknowledgement:
Thanks to Taiwan Semiconductor Manufacturing Co. for fabricating the chip.

References:

[1] J. S. Thomas et al. "Trends and Developments in State-of-the-Art CMOS Image Sensors", *Int. Image Sensor Workshop*, pp. 10-13, May 2023.
https://imagesensors.org/papers/10.60928/za59-hge8/

[2] J. Yun et al., "A 0.6 μm Small Pixel for High Resolution CMOS Image Sensor with Full Well Capacity of 10,000e- by Dual Vertical Transfer Gate Technology", *Symp. VLSI Technology and Circuits*, pp. 351-352, June 2022.
https://doi.org/10.1109/VLSITechnologyandCir46769.2022.9830254

[3] H. Kim et al., "A 0.64μm 4-Photodiode 1.28μm 50Mpixel CMOS Image Sensor with 0.98e- Temporal Noise and 20Ke- Full-Well Capacity Employing Quarter-Ring Source-Follower", *ISSCC*, Feb. 2023. https://doi.org/10.1109/ISSCC42615.2023.10067732

[4] C. Y. Ai et al., "0.56 μm-pitch CMOS image sensor for high resolution application", *Int. Image Sensor Workshop*, pp. 22-25, May 2023.
https://imagesensors.org/papers/10.60928/4t8q-rmzl/

[5] M. Uchiyamaet al., "A 40/22nm 200MP Stacked CMOS Image Sensor with 0.61μm Pixel", *Int. Image Sensor Workshop*, vol. 1, p. 3, Sep. 2021.
https://imagesensors.org/papers/10.60928/vt74-nsnt/

[6] M. Sugimoto, et al. "High full well capacity and low noise characteristics in 0.6μm pixels via buried sublocal connections in a 2-layer transistor pixel stacked CMOS image sensor", *Int. Image Sensor Workshop*, pp.-R1.2, May 2023.
https://imagesensors.org/papers/10.60928/mpmm-kwbx/

[7] S. Choi, et al. "World smallest 200Mp CMOS image sensor with 0.56μm pixel equipped with novel deep trench isolation structure for better sensitivity and higher CG", *Int. Image Sensor Workshop*, pp. R1.3, May 2023.
https://imagesensors.org/papers/10.60928/sel5-hn0e/

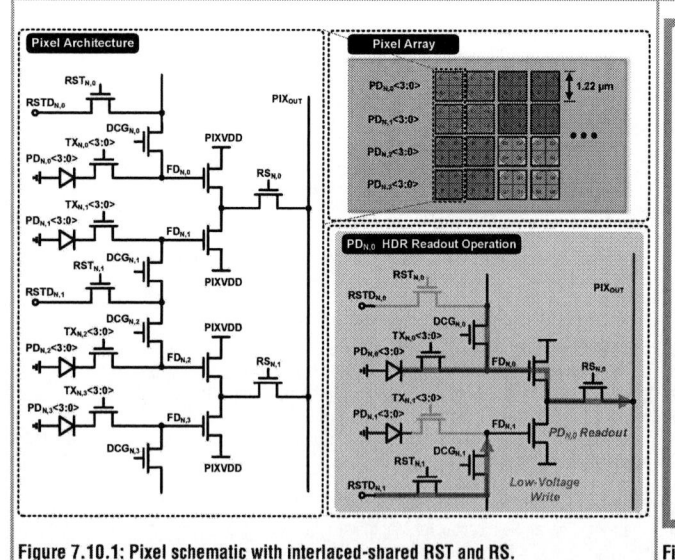

Figure 7.10.1: Pixel schematic with interlaced-shared RST and RS.

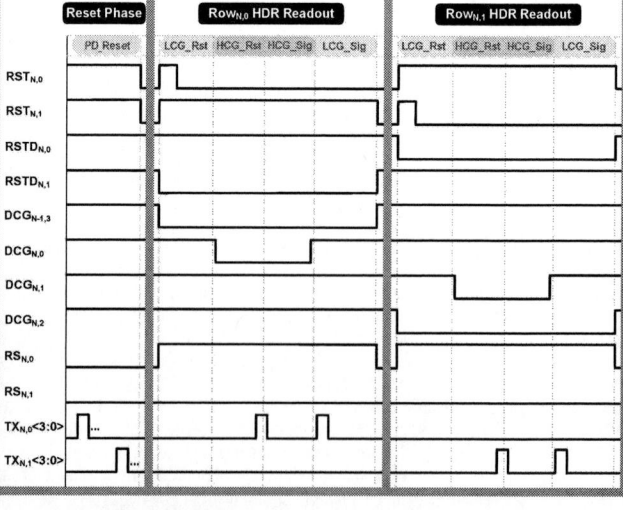

Figure 7.10.2: Pixel readout timing.

ISSCC 2026 / SESSION 7 / IMAGE SENSORS AND RANGING / 7.10

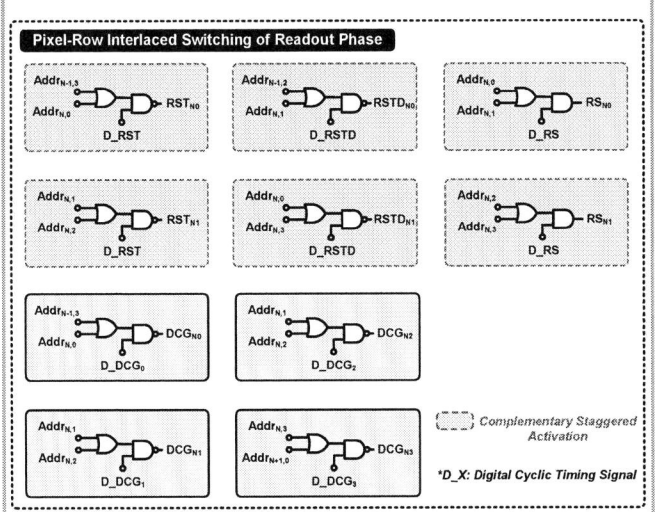

Figure 7.10.3: Special design at the row driver for interlaced-shared architecture.

Figure 7.10.4: Readout circuitry block diagram.

Figure 7.10.5: (a) Regular stagger HDR video, (b) Motion artifact free HDR video.

Figure 7.10.6: Performance summary table.

Parameters	Unit	This work	[2]	[4]	[5]	[6]	[7]
Pixel size	μm	0.61	0.60	0.56	0.61	0.60	0.56
Resolution	MP	200	200	200	200	-	200
FD share	-	2x2	2x4	2x4	2x4	2x2	2x4
FDTI	-	No	Yes	No	No	Yes	Yes
Read Noise @200MP	e-	0.7	1.5	1.6	1.6	0.99	1.6
Linear FWC @200MP	e-	5000	10000	5500	5000	8000	4700
Dynamic range @200MP	dB	77	76	71	70	78.1	69.4
Read Noise @50MP	e-	0.7	-	-	-	-	-
Linear FWC @50MP	e-	10000	-	-	-	-	-
Dynamic range @50MP	dB	83	-	-	-	-	-
Read Noise @12.5MP	e-	1.7 @normal 1.0 @CMS4	-	4.1 @normal 1.6 @LN	-	-	-
Linear FWC @12.5MP	e-	40000	-	72000	-	-	75000
Dynamic range @12.5MP	dB	87 @normal 92 @CMS4	-	85 @normal 93 @LN	-	-	-
HCG/LCG ratio	1	4:1	-	-	-	-	-
RTS@1.0mV	ppm	10	-	-	10	6	1
Dark current	e-/s	1.5	0.8	1.1	1.3	-	-
White pixel	ppm	30	-	-	26	-	-

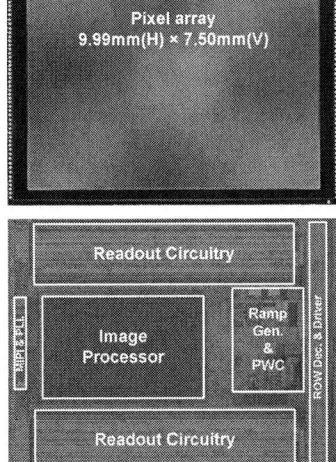

Figure 7.10.7: Chip micrograph.

ISSCC 2026 / SESSION 8 / DIE-TO-DIE AND HIGH-SPEED ELECTRICAL TRANSCEIVERS / OVERVIEW

Session 8 Overview: *Die-to-Die and High-Speed Electrical Transceivers*

WIRELINE SUBCOMMITTEE

Session Chair: Didem Turker Melek
Cadence Design Systems, San Jose, CA

Session Co-Chair: Kenny Hsieh
TSMC, Hsinchu, Taiwan

This session explores advancements in transceivers for die-to-die and high-speed data communication to enable AI (artificial intelligence) and other applications. The use-cases range from low-power, high-bandwidth parallel interconnects with forwarded clock for chip-to-chip, module-to-module, all the way to intra-data-center communication based on CDR. The papers describe cutting-edge transmitter and receiver architectures optimizing energy efficiency and performance. The first three papers (8.1, 8.2 and 8.3) present advanced die-to-die interconnects achieving high bandwidth density, ultra-low power consumption, and modular low-latency designs for next-generation chiplet and data center applications. Paper 8.4 and 8.7 present 112Gb/s NRZ and PAM-4, respectively, single-ended simultaneous bi-directional transceiver featuring crosstalk and reflection cancellations. The fifth paper proposes a reference-less PAM-4 CDR that operates at 112Gb/s and achieves competitive recovered clock jitter of 343fs$_{rms}$. Paper 8.6 demonstrates an ultra-low power, compact 112Gb/s PAM-4 analog RX based transceiver. Multiple techniques to enable lower power such as AFE tuning to realize a DFE tap-2, single PI clocking scheme is introduced with best pJ/b/dB FOM. Paper 8.8 and 8.9 present advancement in 56 and 72Gb/s single-ended simultaneous bi-directional transceiver, respectively, proposing quadrature rate per-lane CDR with QEC/DCC secure clock margin and capacitive peaking leakage cancellation. Paper 8.10 describes an analog-intensive PAM-4 transmitter architecture to improve the data rate, and the first reported above 240Gb/s per channel SerDes TX in 65nm process. Paper 8.11 proposes a low-power 112Gb/s PAM-4 and 168Gb/s PMA-8 TX design with resistor-less source-series-termination segment (SSTS) to achieve at least 1.84× input parasitic capacitance reduction and scaling friendliness.

1:30 PM

8.1   A 48Gb/s/lane 1.24Tb/s/mm UCIe-Compliant Die-to-Die Link Over 30mm Standard Package

Susnata Mondal, Intel, Hillsboro, OR

In Paper 8.1, Intel presents a 22nm UCIe standard package compliant die-to-die PHY at 48Gb/s/lane across 16 lanes, achieving 1.24Tb/s/mm shoreline BW density over a 30mm organic package at 1.2pJ/b, extendable to 56GT/s (1.13pJ/b). Circuit innovations include an impedance-invariant TX CTLE, high-swing N/N driver, compact resonant clocking, double-tail latch w/ improved setup/hold times, and common-mode supply-noise rejection. Compared to prior UCIe-S, it achieves 3× higher data rate and 2.8× higher BW density.

1:55 PM

8.2   A 32Gb/s 12.35Tb/s/mm² 0.36pJ/b UCIe-Like Die-to-Die Interface Featuring Edge-Triggered Transceivers in 3nm with Active LSI Packaging

Wei-Chih Chen, TSMC, Hsinchu, Taiwan

In Paper 8.2, TSMC presents an edge-triggered transceiver (ETT), enabling a 32Gb/s UCIe-compatible chiplet interconnect achieving 10Tb/s/mm shoreline BW when integrated within an active local silicon interconnect (aLSI). The ETT achieves power consumption of 0.07pJ/b and supports compact top-die TX/RX designs while optimizing 0.36pJ/b energy efficiency and 12.35Tbps/mm² area bandwidth density. It demonstrates a 32Gb/s die-to-die link at 0.75V with 20.46ps eye width (65% UI) and 530mV eye height across 64 lanes in the 3nm CMOS.

2:20 PM

8.3   A 0.23pJ/b 24Gb/s Modular D2D Interface With Zero Wake Penalty Clock Gating in 3nm

Ravi Shivnaraine, Microsoft, Sunnyvale, CA

In Paper 8.3, Microsoft presents a 3nm 24Gb/s D2D Interface achieving 0.23pJ/b (analog PHY power) and zero wake penalty clock gating and 0.17pJ/b at 20Gb/s. The PHY achieves a total end-to-end average latency of 4.2ns (~25% analog macro). Measured performance demonstrates a worst-case bit error rate, extrapolated to 1e-18 measured across 1680 lanes over PVT of 12ps$_{pp}$. The modular nature of this work allows flexibility in the construction of D2D links and natively supports asymmetric TX and RX bandwidths.

2:45 PM

8.4   A 112Gb/s/wire Single-Ended Simultaneous Bi-Directional Transceiver with Dynamic Equalizer for Die-to-Die Interface in 28nm CMOS

Zhiwen Huang, Peking University, Beijing, China

In Paper 8.4, Peking University describes 8-lane 112Gb/s/wire NRZ single-ended simultaneous bi-directional transceiver with a 3mm shield-less on-chip channel. A dynamic equalizer is proposed to decouple the bi-directional signals and compensate for insertion loss and crosstalk. Fabricated in 28nm CMOS, the transceiver achieves a BER of less than 1E-14, and an energy efficiency of 1.01pJ/b.

132 • 2026 IEEE International Solid-State Circuits Conference

979-8-3315-8937-0/26 $31.00 © 2026 IEEE

ISSCC 2026 / February 16, 2026 / 1:30 PM

3:00 PM

8.5   A 112Gb/s 0.76pJ/b Reference-less Mixed-Signal PAM-4 CDR in 28nm CMOS

Yiqing Xu, Institute of Semiconductors, Chinese Academy of Sciences, Beijing, China, University of Chinese Academy of Sciences, Beijing, China
In Paper 8.5, Institute of Semiconductors of Chinese Academy of Sciences demonstrates a 112Gb/s mixed-signal reference-less PAM-4 CDR in 28nm CMOS. By proposing the hybrid architecture based on a symmetrical linear PD and a bang-bang PFD, a frequency acquisition controller and a DCC-PEC-merged MPCG, the design issues of reference-less PAM-4 CDR operating at over 100Gb/s data rate are mitigated, achieving 0.76pJ/b efficiency and 0.11UIPP jitter tolerance at BER 1E-12.

3:35 PM

8.6   A 280mW 112Gb/s PAM-4/NRZ Transceiver for Low-Power IOs in 5nm FinFET Technology

Ullas Singh, Broadcom, Irvine, CA
In Paper 8.6, Broadcom presents the design of an ultra-low power, compact 112Gb/s PAM-4 transceiver. Each lane incorporates a quarter rate RX with 3-tap FFE and 11-tap DFE equalization and a half rate TX using 7b SST DAC driver. The transceiver can compensate up to 35dB channel loss at 112Gb/s and achieves BER < 1E-6, dissipating only 280mW including both analog and digital power domains. This work has the best pJ/b/dB FOM as well as the lowest area compared to previously published work at this data rate.

8

4:00 PM

8.7   A 112Gb/s PAM-4 SBD Transceiver with Mismatch-Compensated 2×VDD Hybrid and Two-Step Echo Canceller in 28nm CMOS

Huanfa Sun, Xi'an JiaoTong University, Xi'an, China
In Paper 8.7, Xi'an JiaoTong University describes a 112Gb/s PAM-4 SBD transceiver in 28nm CMOS. It features a novel hybrid with a 2×VDD stacked driver to restore signal swing, a joint delay and slew-rate matching scheme to eliminate dynamic glitches, and a two-step echo canceller to mitigate reflections. Over a 12.7dB channel (equivalent to 24.4dB at 28GHz Nyquist), the transceiver achieves the highest TX swing of 1.27V, a competitive energy efficiency of 1.73pJ/b and a record FoM of 0.14pJ/b/dB among PAM-4 SBD transceivers.

4:25 PM

8.8   A 0.292pJ/b 56Gb/s/wire Capacitively Driven Simultaneous Bidirectional Transceiver with PVT/Mismatch Tracking for XSR and D2D Interfaces in 28nm CMOS

Kahyun Kim, Seoul National University, Seoul, Korea
In Paper 8.8, Seoul National University proposes a low-power (0.292pJ/b), high-bandwidth (20.7Tb/s/mm, 56Gb/s/wire) single-ended capacitively driven simultaneous bidirectional (CDSBD) TRX with PVT tolerance is proposed. Capacitive driving reduces power and self-interference in SBD. AC/DC replica and PVT/mismatch tracking provides PVT robustness without extra hardware. Per-lane CDR and quadrature generator with QEC/DCC secure clock margin. The TRX sustains <1e-12 BER across -6.5 to -11.5dB loss, 0.93-1.2V supply, and 0-80°C temperature variation.

4:50 PM

8.9   A 72Gb/s/pin Single-Ended Simultaneous Bi-Directional Transceiver with C-Peaking Leakage Cancellation and Dual-Loop Hybrid Impedance Calibration for Chiplet Interfaces

Xuxu Cheng, Southern University of Science and Technology, Shenzhen, China
In Paper 8.9, Southern University of Science and Technology presents a 72Gb/s/pin single-ended simultaneous bi-directional (SBD) TRX in 28nm CMOS. Capacitive peaking leakage cancellation (CPLC) suppresses the high-frequency leakage due to the main and hybrid driver mismatch in SBD links by 63%. Dual-loop hybrid impedance calibration reduces coefficients error of hybrid circuits under PVT variations by 92%. The TRX achieves 72Gb/s with eye opening of 0.45UI and 243mV and energy efficiency of 1.5pJ/b.

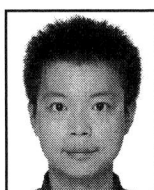

5:05 PM

8.10   A 180-to-240Gb/s Analog-Intensive PAM-4 Transmitter with 0.70pJ/b Analog Power Efficiency in 65nm CMOS

Ziyi Lin, Tsinghua University, Beijing, China
In Paper 8.10, Tsinghua University demonstrates a 180-240Gb/s analog-intensive PAM-4 transmitter in 65nm CMOS process. A three-stage cascaded 2-to-1 analog MUX (AMUX) is employed to reduce the complexity and therefore the parasitic capacitance of both output and internal high-speed nodes. Reconfigurable 4-tap FFE, current injection, and dual T-coil techniques are proposed to enhance the output swing and the bandwidth. The transmitter achieved 0.70 pJ/b analog energy efficiency at 240Gb/s data rate.

5:20 PM

8.11   A 1.59pJ/b 112Gb/s PAM-4 and 1.06pJ/b 168Gb/s PAM-8 Resistor-Less 7-Bit SST DAC-Based Transmitter with 8-Tap FFE in 28nm CMOS

Yao-Hung Tsai, National Taiwan University, Taipei, Taiwan
In Paper 8.11, National Taiwan University proposes a resistor-less 7b SST DAC-based TX with 8-tap FFE and 3 types of segments is presented to achieve low parasitic capacitance, compact area, and scaling friendliness. To reduce inter-symbol-interference jitter of the output of a 4:1 multiplexer, a dual-feedback equalizer is proposed. The TX is fabricated in 28nm CMOS with a maximum output swing of $1.1V_{ppd}$. The power dissipation is 178mW at 112Gb/s PAM-4 and 168Gb/s PAM-8, and energy efficiency is 1.59pJ/b at PAM-4 and 1.06pJ/b at PAM-8.

DIGEST OF TECHNICAL PAPERS • 133

979-8-3315-8937-0/26 $31.00 © 2026 IEEE

ISSCC 2026 / SESSION 8 / DIE-TO-DIE AND HIGH-SPEED ELECTRICAL TRANSCEIVERS / 8.1

8.1 A 48Gb/s/lane 1.24Tb/s/mm UCIe-Compliant Die-to-Die Link Over 30mm Standard Package

Susnata Mondal*[1], Sashank Krishnamurthy*[1], Shuhei Yamada[1], Zhaokai Liu[2], Junyi Qiu[1], Soumya Bose[3], Zuoguo Wu[2], Gerald Pasdast[2], James Jaussi[1], Mozhgan Mansuri[1]

[1]Intel, Hillsboro, OR, [2]Intel, Santa Clara, CA, [3]University of California, Santa Cruz, CA
*Equally Credited Authors (ECAs)

Abstract

This work demonstrates a UCIe-S compliant die-to-die PHY at 48Gb/s/lane across 16 lanes, achieving 1.24Tb/s/mm shoreline BW density over a 30mm organic package at 1.2pJ/b, extendable to 56GT/s (1.13pJ/b). Circuit innovations include an impedance-invariant TX CTLE, high-swing N/N driver, compact resonant clocking, double-tail latch w/ improved setup/hold times, and common-mode supply-noise rejection. Compared to prior UCIe-S, it achieves 3× higher data rate and 2.8× higher BW density.

As system complexity grows, monolithic SoC scaling faces challenges in integration, die size, and technology heterogeneity, motivating chiplet disaggregation that improves yield and cost while enabling IP reuse and heterogeneous integration. Die-to-die (D2D) interconnects with low latency, high bandwidth (BW) density and energy efficiency (EE) are therefore critical. The Universal Chiplet Interconnect Express (UCIe) standard [1] enables interoperability between dies from various vendors over multiple protocols (4-64Gb/s). This work demonstrates a UCIe-compliant physical link (PHY) over standard organic package (pkg), or UCIe-S, operating at the 48Gb/s/lane across all 16 lanes, and further extending to 56Gb/s. This was enabled by several circuit-level innovations: an impedance-invariant passive TX continuous-time linear equalizer (CTLE), a high-swing NMOS-over-NMOS (N/N) driver, coupling-canceled compact resonant clock (ck) distribution, a high-sensitivity double-tail latch (DTL) with improved setup/hold times (t_{SU}/t_H), and common mode (CM) supply-noise (SN) rejection in the RX using the track (trk) signal.

Figure 8.1.1 shows a block diagram of the UCIe-S PHY, consisting of 16 data lanes, quadrature (quad) forwarded clocks (ck_{fw}) at ¼ rate, and a trk/valid signal between the two dies. The dies are mounted on a 5-2-5 organic 2D standard package substrate with 30mm channels routed in two layers, as per UCIe specification. A differential (diff) ¼-rate external reference ck is supplied to each die. Quad phases generated by a coupled-resonator quad hybrid (QH) [2] drive two 8b phase interpolators (PIs) which provide the rotated ck to the 16 TX data lanes and the two quad ck_{fw}, respectively. The RX employs a matched architecture, with both TX and RX incorporating independent per-channel de-skew to compensate for lane-to-lane skew.

A passive CTLE, at the TX output or RX input, can equalize channel loss w/o power overhead when the TX driver directly drives the CTLE impedance (Z). Conventional passive CTLEs [3], however, exhibit strong frequency-dependent Z variation (~150% for 5dB peaking) at the pad, leading to reflections and ISI. A pole in the impedance response aligned with the CTLE zero ($-1/R_1C_1$ in Fig. 8.1.2) causes larger variation at higher equalization gain. This work mitigates the Z-variation by augmenting the series R_1-C_1 network with a shunt R_2-L_2 branch, which introduces an additional zero at $-R_2/L_2$ in both gain and impedance responses. The proposed CTLE therefore achieves higher peaking from two low-frequency zeros while maintaining near-constant output impedance through a compensating low-frequency pole–zero pair (Fig. 8.1.2). While the Z-invariance and peaking benefit holds both in the TX and RX configuration, the CTLE is implemented at the TX side in this work. The additional CTLE inductor is collocated with the TX BW-extension inductor, as the overall response is largely insensitive to their coupling, eliminating area overhead. Compared to a feedforward equalizer (FFE), the CTLE avoids extra data-path power and reduces TX driver power by a factor of $1+\alpha(1-\alpha)\times(R_R/R_T)$, where R_R, R_T and α are RX termination, TX term. and FFE coefficient, respectively. The CTLE also improves eye width (EW) through an enhanced high-frequency response (Fig. 8.1.2). Moreover, a TX FFE does not shape unmodulated low-frequency SN. However, the CTLE attenuates it by its equalization gain, making it beneficial for single-ended (SE) voltage-mode (VM) drivers. While the CTLE, tunable through R_1/R_2, is sufficient, one configurable pre/post-cursor FFE tap (not both as [1]) is implemented but unused.

To reduce ck distribution power and achieve better lane-to-lane matching, the active circuits of all 16 lanes are clustered into two groups (Fig. 8.1.7), with the ck driven symmetrically from a central location. TX outputs are routed to the pads through matched 200µm on-die T-lines, which act as extensions of package channels with negligible loss. Resonant clocking in the TX ck network halves power, with an option to switch to non-resonant mode below 12GHz using center-tap switches to support lower UCIe data rates (Fig. 8.1.3). Given the stringent bump-map area, multiple I/Q inductor pairs are impractical; instead, the global ck with a single I/Q inductor drives all 16 lanes w/o repeaters, enabled by lane clustering. The global ck for ck_{fw} and track lanes, with lighter loading, is distributed in non-resonant mode. A single I/Q pair with matched inductor is challenging to accommodate; so, the I/Q inductors are overlapped, with an '8-shaped' [4] turn introduced in one winding to cancel mutual coupling that would otherwise cause large I/Q gain and phase errors (Fig. 8.1.3). Each TX lane employs per-lane de-skew to compensate up to 0.5UI lane-to-lane variation as required by UCIe-S spec. To meet this range efficiently, a "sub-UI PI" is used instead of inverter-based delay lines. The sub-UI PI is implemented with interpolating strength set to half of

the main path, incurring ~40% higher power only in the first-stage local ck buffer. Furthermore, since the inner lanes away from the die edge incur an additional ~6ps delay relative to the outer lanes, the sub-UI PI interpolates a leading phase for inner lanes and a lagging phase for outer lanes (Fig. 8.1.3). Using separate lead/lag sub-UI PIs reduces its power by ~40% compared to supporting both options in every lane.

The TX 4:1 serializer (Fig. 8.1.3) uses a combinational 1UI pulse generator (PG) and directly combines 4UI data at the high-BW TX output node. Unlike [2], three inverter stages are inserted between the PG and driver to reduce PG power and ck loading. As the pre-drivers buffer a 1UI pulse padded with zeros over the remaining three UIs, they tolerate moderate fanout while remaining ISI-free, unlike full-rate data paths where combining is done before the output. A VM driver is adopted for efficiency, but conventional VM drivers face a tradeoff between low swing in N/N drivers [2] and high power in linearized PMOS-over-NMOS (P/N) drivers, which require upsizing of the driver and preceding stages [5]. In this work, a high-swing N/N driver is introduced that operates at nominal supply and employs a PMOS resistor (~⅓ of the termination resistance) at the pull-up NMOS drain. When the pull-up branch is activated, the voltage across this PMOS resistor pushes the NMOS into the linear region, improving linearity. Simulations show 2.5× higher EE than linearized P/N drivers and 1.5× greater swing than low-swing N/N drivers. The final pre-driver stage uses a higher supply V_{DDH} than IO supply V_{DD}, within the process-nominal supply range, to further aid linearization and mitigate pulse-width distortion inherent to N/N drivers [2].

The ck distribution in the matched RX architecture (Fig. 8.1.4) receives two quad phases of the ck_{fw} from the TX, converts them to diff and further amplifies the four quad phases. Similar to the TX, a resonant buffer with coupling-cancelling I/Q inductors distributes the ck to all 16 RX lanes, followed by per-lane buffering and de-skew. At the RX lane, the SE input data from the TX is converted to diff and amplified through high-BW inverter-based Cherry–Hooper (CH) stages, where g_m stages drive shunt-feedback transimpedance amplifiers; this RX amplifier chain delays the data-path to match the ck-path timing while amplifying the data with minimal jitter amplification. As the TX signal has a low CM voltage (~0.1V), the input g_m stage uses a PMOS device M_p biased by an NMOS current source M_n. A closed-loop DC offset correction (DCOC) minimizes the diff output CM mismatch of the RX data amplifier chain by injecting current into a resistor R at the input. The current injected into R raises the gate voltage of M_p and keeps it in saturation, while a parallel capacitor C extends the BW.

To improve link performance in SE UCIe-S, the RX employs CM noise rejection using the trk signal (Fig. 8.1.4). The low-frequency TX SN, common to both the trk and data paths, is extracted from the trk path using an RC filter. This filtered signal (V_{B-DATA}) is distributed to all 16 lanes and applied to the M_n devices that bias the respective input PMOS M_p of the CH amplifier, thereby canceling the low-frequency TX-SN in the data path. The amplified and delay-matched diff output then drives four slices of a ¼-rate cascaded DTL. The 2-ck-phase DTL [6] improves ck-to-Q delay and sensitivity, which are further enhanced by cascading with a ¼-sized 1-ck-phase DTL driven by a complementary ck phase. This cascaded DTL achieves <20mV$_{pk-pk,diff}$ simulated sensitivity at 56Gb/s or, equivalently, <2ps simulated t_{SU}/t_H with a 400mV$_{pk-pk}$-swing diff signal, improving link timing margin.

Fabricated in 22nm FinFET CMOS with 112.65µm bump pitch, the UCIe-S PHY core occupies 1.24×0.72mm² (Fig. 8.1.7). Two dies were routed across a 30mm on-pkg channel, mounted on a custom PCB supplying external ck and DC power supplies. An on-die PRBS generator produced PRBS15 data at the TX, and the 1/8-rate deserialized RX output was routed off-chip to an external BERT.

For standalone TX characterization, a die on the same package (Fig. 8.1.7) drove a 10mm channel to a high-speed probe, with the probe, cable and channel producing ~5.5dB Nyquist loss, comparable to the 30mm UCIe-S link, for representative TX eye monitoring. With the passive TX CTLE at its optimal setting, the measured 56Gb/s eye was 112mV/10ps at BER=1e-12, while the least-peaking setting (~3dB lower peaking) degraded the eye to 80mV/7.5ps, showing CTLE effectiveness (Fig. 8.1.6).

134 • 2026 IEEE International Solid-State Circuits Conference

979-8-3315-8937-0/26 $31.00 © 2026 IEEE

The link operates at 48Gb/s per lane with all 16 lanes active. At V_{DD}=0.7V (V_{DDH}=0.85V for TX pre-driver), the worst-case bathtub curve across 16 channels shows a 1e-12 EW of 0.26UI (Fig. 8.1.5). BER was measured directly to 1e-12; extrapolation shows 0.15UI at 1e-15. Bathtub curves were obtained by measuring the PI output delay across code and mapping it to UI delay. The link achieves UCIe-S-compliant 48Gb/s/lane data rate at an energy efficiency of 1.2pJ/b (0.45, 0.69, and 0.06pJ/b in TX, RX, and PI+QH, respectively; a PLL is not integrated in this work). The link also extended operation to 56Gb/s with all 16 lanes active, at an EE of 1.13pJ/b (0.44-TX, 0.63-RX, 0.06-PI+QH) with the same V_{DD}/V_{DDH}. Representative bathtub curves for four lanes are shown in Fig. 8.1.5, with a worst-case 1e-12 EW of 0.23UI. Compared to prior D2D NRZ links (Fig. 8.1.6) over organic pkg [7-11], this work achieves the highest demonstrated data rate of 56Gb/s/lane, similar to [9], but with better EE despite using a less advanced process node. The link delivers a high shoreline density of 1.45Tb/mm and is UCIe compliant with 2-layer package routing at 3x the channel length compared to the 4nm design in [10], which uses 3 package layers. Compared to the 16Gb/s UCIe-S compliant design in 3nm CMOS [11], this 48Gb/s UCIe-S compliant design in 22nm FinFET achieves 3× higher data rate and 2.8× higher shoreline density.

Acknowledgment:
The authors thank N. Sanchez, J. Tagumasi, Y. Zhu, E. Calderon, E. J. Hernandez, A. Kumar and S. Tiagaraj for design/layout help; T. Acikalin, T. Prabhakaran, D. Lake, J. Escobar, N. Viswanathan, D. Ye and Z. Qian for package/PCB design/modeling; J. Kennedy for die photo; and G. Campbell, B. Whipker and T. Nguyen for assembly/lab support

References:
[1] *UCIe 3.0 Specification*, 2025. Available: https://www.uciexpress.org/.
https://www.uciexpress.org
[2] S. Mondal et al., "A 4-Ch × 64 Gb/s/Ch NRZ VCSEL-Based Co-Packaged Fiber-Terminated Optical TX and 80-Gb/s Optical Driver," in IEEE Journal of Solid-State Circuits. https://doi.org/10.1109/JSSC.2025.3563073
[3] B. Dehlaghi and A. Chan Carusone, "A 0.3 pJ/bit 20 Gb/s/Wire Parallel Interface for Die-to-Die Communication," in IEEE Journal of Solid-State Circuits, vol. 51, no. 11, pp. 2690-2701, Nov. 2016. https://doi.org/10.1109/JSSC.2016.2596773
[4] L. Fanori, T. Mattsson and P. Andreani, "21.6 A 2.4-to-5.3GHz dual-core CMOS VCO with concentric 8-shaped coils," 2014 IEEE International Solid-State Circuits Conference Digest of Technical Papers (ISSCC), San Francisco, CA, USA, 2014, pp. 370-371. https://doi.org/10.1109/ISSCC.2014.6757474
[5] C. Menolfi et al., "A 112Gb/S 2.6pJ/b 8-Tap FFE PAM-4 SST TX in 14nm CMOS," 2018 IEEE International Solid-State Circuits Conference - (ISSCC), San Francisco, CA, USA, 2018, pp. 104-106. https://doi.org/10.1109/ISSCC.2018.8310205
[6] S. Krishnamurthy et al., "A 4×50Gb/s NRZ 1.5pJ/b Co-Packaged and Fiber-Terminated 4-Channel Optical RX," 2024 IEEE Symposium on VLSI Technology and Circuits (VLSI Technology and Circuits), Honolulu, HI, USA, 2024, pp. 1-2.
https://doi.org/10.1109/VLSITechnologyandCir46783.2024.10631327
[7] J. W. Poulton et al., "A 1.17-pJ/b, 25-Gb/s/pin Ground-Referenced Single-Ended Serial Link for Off- and On-Package Communication Using a Process- and Temperature-Adaptive Voltage Regulator," in IEEE Journal of Solid-State Circuits, vol. 54, no. 1, pp. 43-54, Jan. 2019. https://doi.org/10.1109/JSSC.2018.2875092

[8] K. McCollough, S. D. Huss, J. Vandersand, R. Smith, C. Moscone and Q. O. Farooq, "11.3 A 480Gb/s/mm 1.7pJ/b Short-Reach Wireline Transceiver Using Single-Ended NRZ for Die-to-Die Applications," 2021 IEEE International Solid-State Circuits Conference (ISSCC), San Francisco, CA, USA, 2021, pp. 1-3.
https://doi.org/10.1109/ISSCC42613.2021.9366048
[9] C. F. Poon et al., "A 1.24-pJ/b 112-Gb/s (870 Gb/s/Mm) Transceiver for In-Package Links in 7-nm FinFET," in IEEE Journal of Solid-State Circuits, vol. 57, no. 4, pp. 1199-1210, April 2022. https://doi.org/10.1109/JSSC.2022.3141802
[10] K. Seong et al., "A 4nm 48Gb/s/wire Single-Ended NRZ Parallel Transceiver with Offset-Calibration and Equalization Schemes for Next-Generation Memory Interfaces and Chiplets," 2024 IEEE International Solid-State Circuits Conference (ISSCC), San Francisco, CA, USA, 2024, pp. 250-252. https://doi.org/10.1109/ISSCC49657.2024.10454481
[11] J. Vandersand et al., "A 0.52pJ/bit 0.448Tbps/mm UCIe Standard Package Die-to-Die Transceiver with Low-Latency TX Clock Alignment in 3nm FinFET," 2025 Symposium on VLSI Technology and Circuits (VLSI Technology and Circuits), Kyoto, Japan, 2025, pp. 1-3. https://doi.org/10.23919/VLSITechnologyandCir65189.2025.11075053
[12] G. Gangasani et al., "A 1.6Tb/s Chiplet over XSR-MCM Channels using 113Gb/s PAM-4 Transceiver with Dynamic Receiver-Driven Adaptation of TX-FFE and Programmable Roaming Taps in 5nm CMOS," 2022 IEEE International Solid-State Circuits Conference (ISSCC), San Francisco, CA, USA, 2022, pp. 122-124.
https://doi.org/10.1109/ISSCC42614.2022.9731636

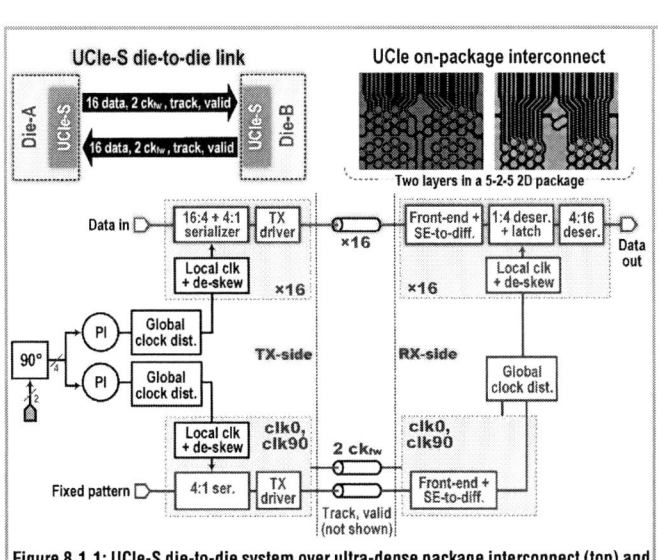

Figure 8.1.1: UCIe-S die-to-die system over ultra-dense package interconnect (top) and transceiver architecture highlighting data/clock paths (bottom).

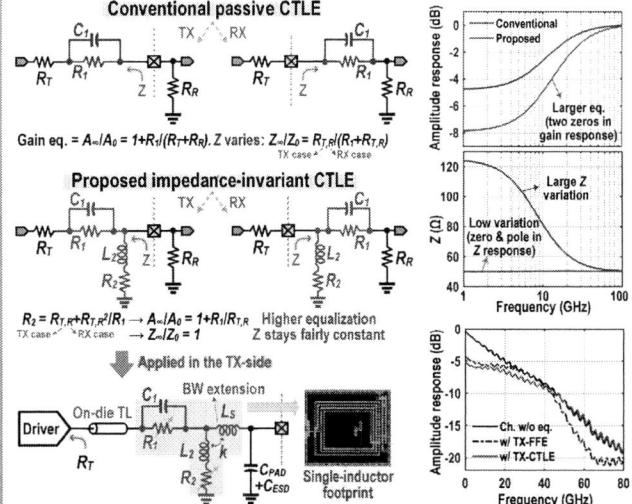

Figure 8.1.2: Proposed impedance-invariant passive CTLE and representative simulations illustrating impedance invariance and channel equalization.

ISSCC 2026 / SESSION 8 / DIE-TO-DIE AND HIGH-SPEED ELECTRICAL TRANSCEIVERS / 8.1

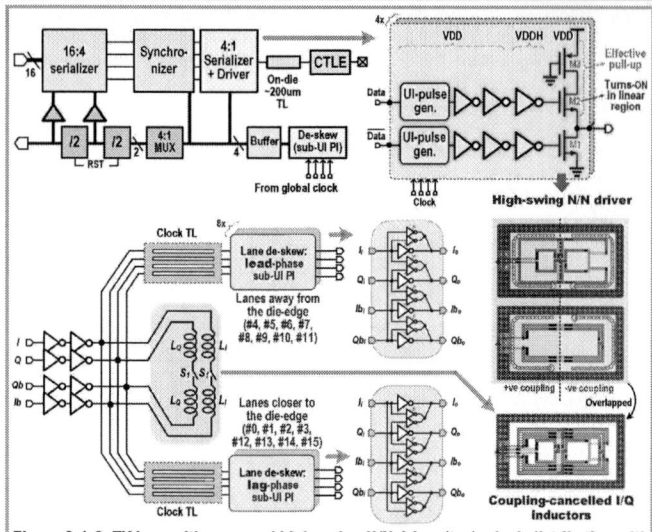

Figure 8.1.3: TX lane with proposed high-swing N/N driver (top); clock distribution with per-lane sub-UI PI and global I/Q overlapped inductors (bottom).

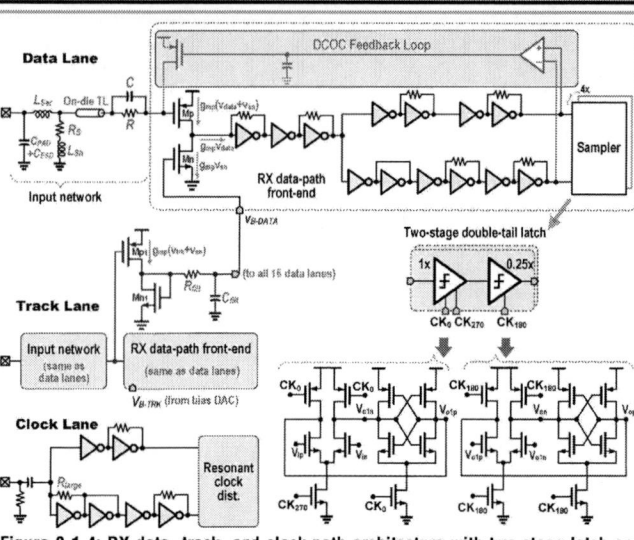

Figure 8.1.4: RX data, track, and clock path architecture with two-stage latch and common-mode supply-noise rejection using the track lane.

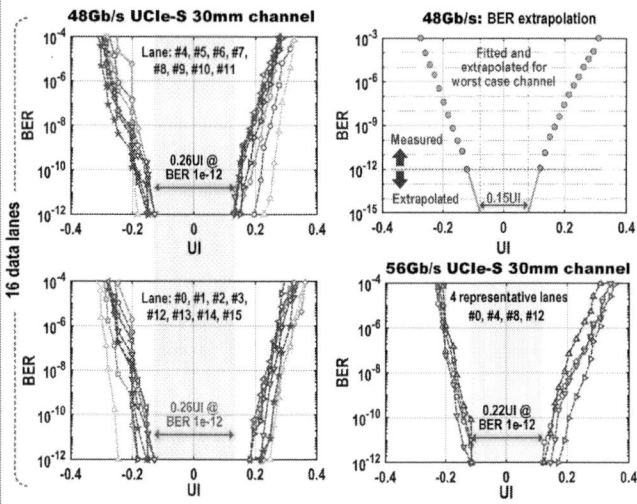

Figure 8.1.5: Measured bathtub plots for UCIe-S D2D interconnect over 30mm standard organic package: 48Gb/s (left, top-right) and 56Gb/s (bottom-right).

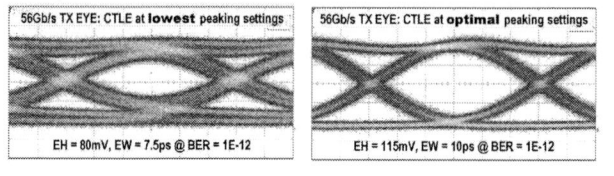

	Poulton, JSSC'19 [7]	McCollough, ISSCC'21 [8]	Poon, JSSC'22 [9]	Gangasani, ISSCC'22 [12]	Seong, ISSCC'24 [10]	Vandersand, VLSI'25 [11]	This Work	
Process	16nm	7nm	7nm	5nm	4nm	3nm	22nm	
Standard	Custom	Custom	Custom	Custom	Custom	UCIe-S	UCIe-S	Custom*
Package type	Organic	Organic	Organic	Organic	Organic	Organic	Organic	
Modulation	NRZ	NRZ	NRZ/PAM4	PAM4	NRZ	NRZ	NRZ	
Bump pitch [μm]	-	130	-	-	130	130	112.65	
Data rate [Gb/s]	25	40	56 (NRZ)/ 112 (PAM4)	113	48	16	48Gb/s	56Gb/s
Channel length [mm]	10 (60 off-pkg)	20	5-30	5-80	10	5-25	30mm	
Shore-line BW [Tb/s/mm]	0.200	0.450	0.870	0.460	1.850	0.448	1.243	1.450
Energy efficiency [pJ/b]	1.17	1.7	1.24	1.55	0.67*	0.52	1.2pJ/b†	1.13pJ/b†

*UCIe-S compliant at 48 Gb/s; 56 Gb/s is an extended data-rate (not part of the current UCIe-S standard). †A PLL is not integrated in this work.

Figure 8.1.6: Measured TX eye at 56Gb/s with CTLE at lowest and optimal peaking (top); performance comparison to custom and UCIe-S standard on-package links (bottom).

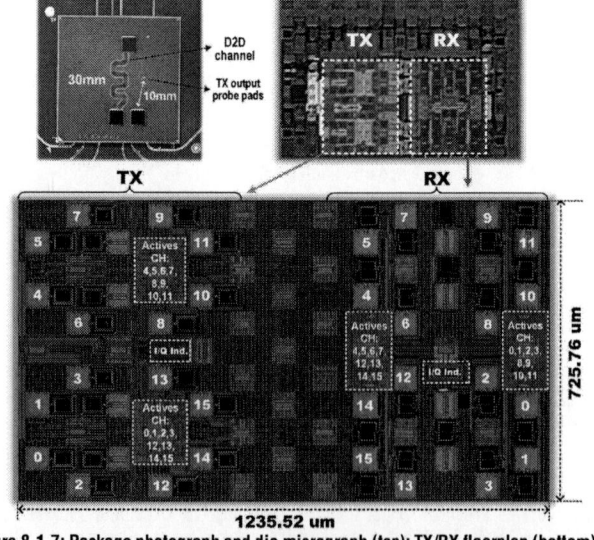

Figure 8.1.7: Package photograph and die micrograph (top); TX/RX floorplan (bottom).

• 2026 IEEE International Solid-State Circuits Conference

ISSCC 2026 / SESSION 8 / DIE-TO-DIE AND HIGH-SPEED ELECTRICAL TRANSCEIVERS / 8.2

8.2 A 32Gb/s 12.35Tb/s/mm² 0.36pJ/b UCIe-Like Die-to-Die Interface Featuring Edge-Triggered Transceivers in 3nm with Active LSI Packaging

Wei-Chih Chen[1], Mu-Shan Lin[1], Chien-Chun Tsai[1], Shenggao Li[2], Wei-Shuo Lin[1], Yu-Jie Huang[1], Nai-Chen Cheng[1], Yu-Chi Chen[1], Wen-Hung Huang[1], Chin-Hua Wen[1], Hsin-Hung Kuo[1], Han-Tzung Ke[1], Jie-Ren Huang[1], Chang-Yi Li[1], Sheng-Tsung Lai[1], Shu-Chun Yang[1], Kuan-Ting Chou[1], Pei-Chen Chiou[1], Tsung-Hsien Tsai[1], Yi-Ting Chen[1], Yen-Ming Chen[1], Kenny Cheng-Hsiang Hsieh[1]

[1]TSMC, Hsinchu, Taiwan, [2]TSMC, San Jose, CA

Abstract

This paper presents edge-triggered transceivers (ETT), enabling a 32Gb/s UCIe-like chiplet interconnect when integrated within an active local silicon interconnect (aLSI). The ETT achieves power consumption of 0.07pJ/b and supports compact top-die TX/RX designs while optimizing 0.36pJ/b energy efficiency and 12.35Tb/s/mm² area bandwidth density. It demonstrates a 32Gb/s die-to-die link at 0.75V with 20.46ps eye width (65% UI) and 530mV eye height across 64 lanes in a 3nm CMOS process.

The rapid growth of data-centric computing, fueled by internet traffic, AI, and GPUs/TPUs, is driving demand for high-bandwidth interconnect solutions. Universal Chiplet Interconnect Express (UCIe) is widely adopted with passive local silicon interconnect (pLSI), prioritizing bandwidth density, energy efficiency, and low latency for die-to-die (D2D) links [1-2]. To enable flexible and efficient extended-range communication, this paper presents a 32Gb/s UCIe-like edge-triggered transceiver (ETT) integrated with an active local silicon interconnect (aLSI). The ETT circuits within the aLSI have an ultra-low power consumption of 0.07pJ/b, addressing thermal concerns in stacked dies. Fabricated using the 3nm CMOS process, this design achieves total link power efficiency of 0.36pJ/b and area bandwidth density of 12.35Tb/s/mm².

This system features 64 lanes, organized into 4 macros, with each macro comprising 16 lanes. Figure 8.2.1 shows the transceiver block diagram spanning two dies: the top-die and the aLSI die. Integrating the ETT within the aLSI significantly reduces the loading capacitance of the top-die transmitter (TX) driver, making a weak TX feasible, which in turn allows a smaller pre-driver and smaller clock buffers. Meanwhile, the ETT output can generate a large swing to eliminate the need for signal amplification on the top-die receiver (RX) path. The ETT circuit in the aLSI thus enables a compact top-die TX/RX design, allowing the bump pitch to be reduced to below 45μm and shortening the PHY depth from 1043μm to 850μm. This further improves power efficiency at 32Gb/s and increases area bandwidth density. The proposed transceiver is compatible with the UCIe protocol, including training sequences, bump map structure, forwarding clock architecture, and matched-delay design.

The ETT integrates a driver, an AC-coupling capacitor (Cac), an amplifier with both negative and positive feedback, and an output stage to enable high-speed and energy-efficient data transmission, as shown in Fig. 8.2.2. The data transmitted through Cac introduces pre-emphasis at the transition edges on the VI node, effectively enhancing signal integrity along the trace. The AC amplitude of the transmitted signal is defined by the capacitive division ratio, Cac / (Cac + CL), which is realized through the neutralization capacitor, enabling adjustable swing levels. The amplifier leverages both positive and negative feedback loops to stabilize the DC voltage level at the VM node across two distinct threshold voltages, effectively mitigating baseline wander and ensuring signal integrity. With optimized sizing, as outlined in the expression, the voltage level is determined by the resistor ratio. In this design, Cac is set to be 180fF for a 1.7mm channel length, while Ra and Rb are set to 2kΩ and 3kΩ, respectively. To ensure a reliable initial state, the start-up circuitry and sequence are designed to establish the correct signal polarity. Prior to the ETT start-up, the negative feedback loop is disabled to conserve power, and VM is tied to ground to maintain a zero output. During start-up, the circuit powers on with VM held in a steady state while awaiting data transmission. A clock lane is utilized to track and absorb the amplifier PVT variations. After N/P transistor sizing adjustment, the generated calibration codes are applied to the data lane to optimize the eye opening.

Figure 8.2.3 depicts the top-die TX architecture. Data and Forwarded Clock lanes share identical structures for matched delay and synchronization using 0, 45, 90, and 135-degree 8GHz quadrature phases. The dynamic logic gate, positioned between the 4:1 MUX and post-driver, improves bandwidth and enables independent adjustment of the crossing point for each branch, improving signal quality. The proposed 16:4 serializer captures signals via clock pulses rather than edges to reduce latency and power. A NAND/NOR-based 1UI pulse generator achieves 25% capture duty clocks, generating 1UI-wide pulses extracted from 4UI-wide data (D0-D3) using quarter-rate clock gating. The duty cycle corrector (DCC) and quadrature error corrector (QEC) are strategically placed midway to drive 8 data lanes, optimizing power efficiency and ensuring clock quality. To enhance DCC linearity and prevent clock disappearance under extreme conditions, adaptive current compensation adjusts bleeding current strength based on Vctrl. Resistors connected to MOS transistors are several kΩ to minimize power consumption, with bleeding path resistance in the range of a few kΩ to avoid interference with normal DCC operation. The single-in, differential-out clocks reduce power consumption and incorporate a low-distortion single-to-differential converter (LDS2D). A source-follower buffer in the feed-forward path compensates for propagation delay and bandwidth differences between the inverter and transmission gate. The cross-latch output stage is replaced with a feed-forward cross-source follower buffer to maintain a 50% duty cycle.

Figure 8.2.4 shows the top-die receiver architecture with matched-delay design to mitigate power supply-induced jitter. For pLSI applications, the RXAMP compensates for longer reach in the interposer [1]. In aLSI, the long-reach metal channel is replaced by ETT within the aLSI, reducing the driving requirements of the top-die TX and the equalization needs of the top-die RX. Consequently, the RXAMP is simplified to a single active-peaking stage with 2b programmable gain peaking. VCM adjustment is accomplished using an MOS array before the RXAMP, which adjusts N/P transistor sizes to adapt to varying VCM levels across process skew corners. The power consumption of the RXAMP without peaking and the sampler is reduced by 90% and 54%, respectively, compared to [1]. The received clock and data are edge-aligned, removing the necessity of 45° phase-interpolated clocks at the TX. Center alignment (0.5UI) across data rates from 8Gb/s to 32Gb/s is achieved through coarse/fine tuning mechanisms. A wide tuning range is provided by a 4b, 6ps resolution coarse-tune DCDL (digitally controlled delay line) in each data lane. Fine resolution is achieved through a 6b, 1/64 UI resolution phase shifter in the clock lane. A 2b speed adjustment mechanism ensures 1UI eye-width scanning on the phase shifter across various data rates. In aLSI interconnects, received data passes through multiple buffer stages, causing lane-to-lane mismatch and common-mode distortion. To address this, de-skew circuits are included in each data lane, along with two common-mode calibration procedures. In the aLSI, the VCM is monitored after ETT amplifier and calibrated via N/P transistor sizing adjustment. In the top-die, the common-mode characteristics of the data path are detected prior to de-serialization. The clock pattern from CKP is duplicated to a replica data lane and incorporates a duty detector. Any common-mode distortion can be calibrated through the VCM tuners on both data and clock lanes.

The delay-locked loop (DLL) block diagram is shown in Fig. 8.2.5. The multi-phase delay line (MPD) consists of 12 inverting stages with shared coarse and fine-tuning controls. Coarse tuning adjusts inverter strength, while fine-tuning employs a monotonic capacitor switching scheme, ensuring 30° equal delay per stage as the phase detector (PD) aligns edges between 0° and 360°. Before the MPD, single-ended clock phases are distributed over long distances for power saving, while LDS2D and DCC generate differential phases and ensure a 50% duty cycle. To minimize the dead zone of the PD, a buffer is meticulously designed and integrated into the data path. It balances the setup and hold times of the sense-amplifier flip-flop (SAFF) and reduces the impact of metastability. The phase accuracy of the phase interpolator (PI) outputs are dependent on the co-design of the MPD and the PI with 5b resolution. To ensure that the multi-phase delay line achieves a wide locking range across various data rates, the design incorporates a mechanism that combines 32 coarse tuning (CT) bands with 32 fine tuning (FT) thermometer steps. The finite state machine (FSM) initiates the FT search from the minimum delay CT band to prevent harmonic locking. Additionally, 8 FT codes on either side of the CT band are allocated as a safe margin. Once the delay aligns with the desired frequency, the CT band is locked, while the FT codes continue to perform on-the-fly tracking. The safe margin is designed to accommodate dynamic voltage (±15mV) and temperature drift (-40 to 125°C). The maximum phase error among 12 phases remains below 1.6° across all corners.

Figure 8.2.6 illustrates the channel model in the aLSI, testing steps and measurement data. Ultra-thick metal wires are used for the routing, optimized for the 32Gb/s UCIe-like link at a 10.5Tb/s/mm shoreline bandwidth density. Key concerns such as routability, IR drop, insertion loss, reflection, and crosstalk are addressed. The aLSI package testing includes three phases: KGD for die validation, KGS for stack functionality, and KGP after full assembly, to comprehensively verify functionality, performance, and reliability. A fully functional 32Gb/s link with sufficient design margin at 0.75V has been demonstrated. Specifically, an aggregate eye width of 20.46ps (65% UI) and an eye height of 530mV across 64 lanes has been achieved in the KGP tests.

Figure 8.2.7 shows a bird's-eye view photograph of the stacked KGS die. The design incorporates four aLSIs to facilitate SoC-to-SoC and SoC-to-IOD chiplet interconnects. Each aLSI integrates four macros, delivering a total of 64 lanes to enable high-bandwidth data transfer. In summary, this paper presents a 32Gb/s per lane chiplet interconnect. The aLSI allows a simplified transceiver design, and a reduced PHY depth at 850μm, allowing the area bandwidth density to be 12.35Tb/s/mm², and an energy efficiency of 0.36pJ/b. Other than the aLSI, the design is fully compatible with the UCIe protocol. A comparative analysis against state-of-the-art technologies is provided in the table.

Acknowledgement:
The authors would like to thank TSMC DTP design, layout, chip implementation and validation teams for their dedication and support.

References:
[1] M. -S. Lin *et al.*, "A 32Gb/s 10.5Tb/s/mm 0.6pJ/b UCIe-Compliant Low-Latency Interface in 3nm Featuring Matched-Delay for Dynamic Clock Gating," *2025 IEEE International Solid-State Circuits Conference (ISSCC)*, San Francisco, CA, USA, 2025, pp. 586-588. https://doi.org/10.1109/ISSCC49661.2025.10904767
[2] D. T. Melek *et al.*, "A 0.29pJ/b 5.27Tb/s/mm UCIe Advanced Package Link in 3nm FinFET with 2.5D CoWoS Packaging," *2025 IEEE International Solid-State Circuits Conference (ISSCC)*, San Francisco, CA, USA, 2025, pp. 590-592. https://doi.org/10.1109/ISSCC49661.2025.10904754
[3] J. Gu *et al.*, "A 32Gb/s 0.36pJ/bit 3nm Chiplet IO Using 2.5D CoWoS Package with Real-Time and Per-Lane CDR and Bathtub Monitoring," *2024 IEEE Symposium on VLSI Technology and Circuits (VLSI Technology and Circuits)*, Honolulu, HI, USA, 2024, pp. 1-2. https://doi.org/10.1109/VLSITechnologyandCir46783.2024.10631527
[4] K. Seong *et al.*, "A 4nm 32Gb/s 8Tb/s/mm Die-to-Die Chiplet Using NRZ Single-Ended Transceiver With Equalization Schemes And Training Techniques," *2023 IEEE International Solid-State Circuits Conference (ISSCC)*, San Francisco, CA, USA, 2023, pp. 114-116. http://doi.org/10.1109/ISSCC42615.2023.10067477
[5] Y. Nishi *et al.*, "A 0.190-pJ/bit 25.2-Gb/s/wire Inverter-Based AC-Coupled Transceiver for Short-Reach Die-to-Die Interfaces in 5-nm CMOS," *2023 IEEE Symposium on VLSI Technology and Circuits (VLSI Technology and Circuits)*, Kyoto, Japan, 2023, pp. 1-2. http://doi.org/10.23919/VLSITechnologyandCir57934.2023.10185334

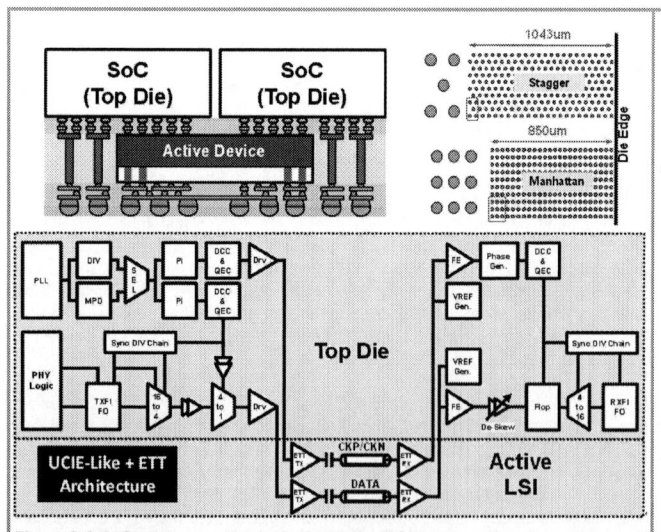

Figure 8.2.1: System overview including active LSI cross-section, bump pitch and arrangement and die-to-die transceiver block diagram.

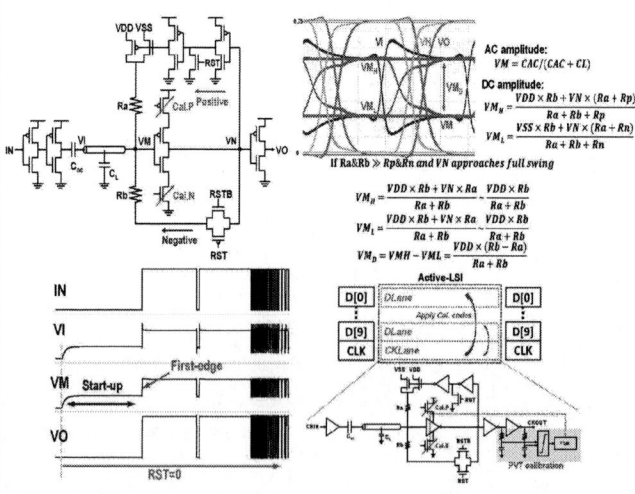

Figure 8.2.2: Edge-triggered transceiver circuit architecture, operations, start-up and PVT skew calibration.

ISSCC 2026 / SESSION 8 / DIE-TO-DIE AND HIGH-SPEED ELECTRICAL TRANSCEIVERS / 8.2

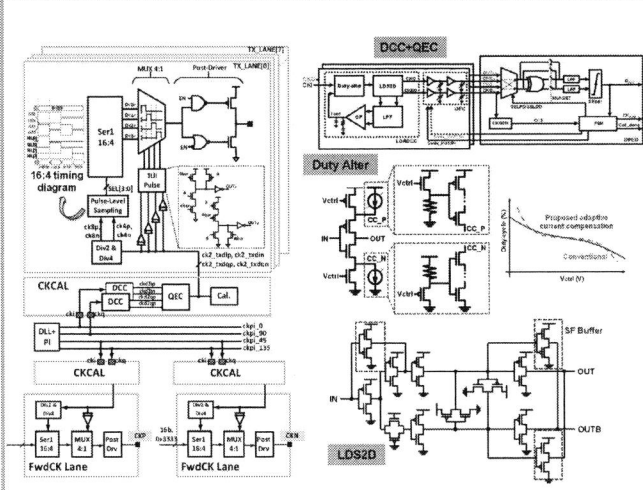

Figure 8.2.3: Top-die transmitter circuit architecture: 1UI-pulse generator, push-pull driver, duty cycle alter, low distortion single-to-differential driver.

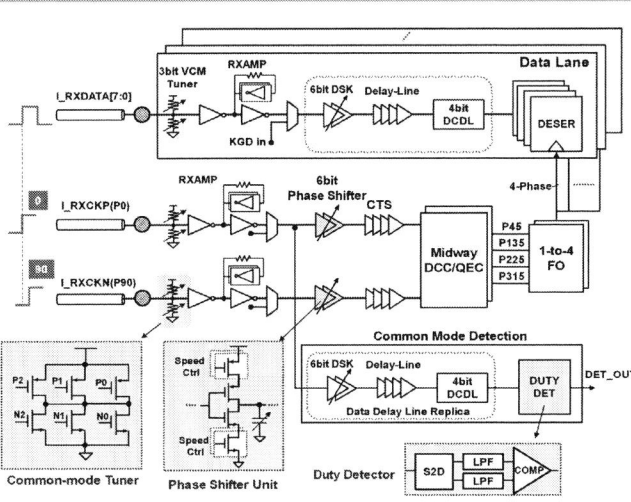

Figure 8.2.4: Top-die receiver circuit architecture: matched-delay clock scheme, de-skew adjustable delay buffers, RX amplifier with common-mode voltage tuner.

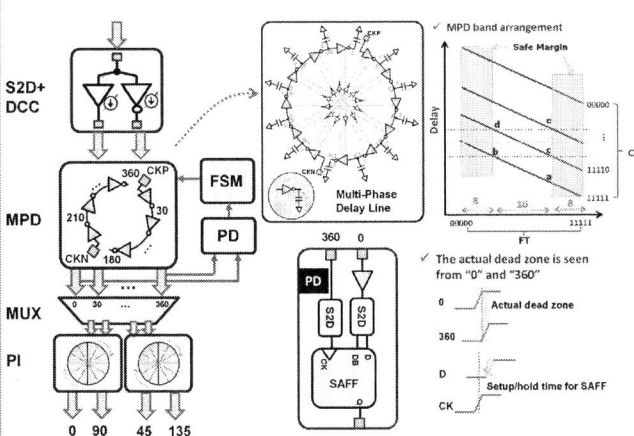

Figure 8.2.5: Delay locked loop circuit architecture: multi-phase delay line, adaptive banks selection by coarse and fine tuning.

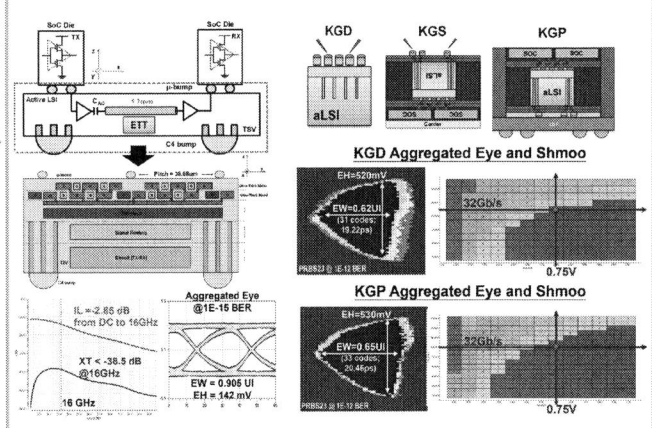

Figure 8.2.6: Channel model of the ETT design, route-ability, IL, crosstalk, eye opening. Testing steps, KGD and KGP aggregated lanes eye diagram and shmoo.

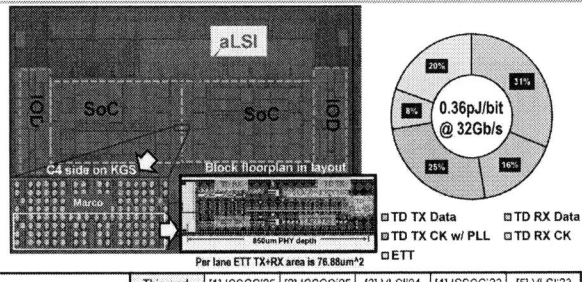

	This work	[1] ISSCC'25	[2] ISSCC'25	[3] VLSI'24	[4] ISSCC'23	[5] VLSI'23
Technology	3nm	3nm	3nm	5nm	4nm	5nm
Module width (um)	388.8	388.8	388.8	300	250	330
Advanced Packages	Active interposer	Passive Interposer	Passive Interposer	Passive Interposer	Passive Interposer	On-chip Channel
Channel Length	1.7mm	1.7mm	1.4mm	2.0mm	3.0mm	1.2mm
Data Rate (Gbps/pin)	32	32	16	32	32	25.2
D2D bump pitch (um)	35	45	45	-	50	55
Energy Efficiency (pJ/bit)	0.36*	0.6*	0.29*	0.36	0.44	0.19**
Shoreline BW (Tbps/mm)	10.5	10.5	5.27	3.84	8.00	5.8**
Area BW density (Tbps/mm²)	12.35	10.07	5.05	4.47	11.43	6.04**
FoM (Tbps/mm)/(pJ/bit)	29.41	17.50	18.17	10.67	18.18	30.53**

*PLL is included   **Extrapolated data

Figure 8.2.7: Die (KGS) and layout photograph, power breakdown and comparison.

• 2026 IEEE International Solid-State Circuits Conference

979-8-3315-8937-0/26 $31.00 © 2026 IEEE

ISSCC 2026 / SESSION 8 / DIE-TO-DIE AND HIGH-SPEED ELECTRICAL TRANSCEIVERS / 8.3

8.3 A 0.23pJ/b 24Gb/s Modular D2D Interface With Zero Wake Penalty Clock Gating in 3nm

Ravi Shivnaraine[1], Charles Boecker[1], Eric Groen[1], Simon Li[1], Ping Lu[1], Socrates Vamvakos[1], Roxanne Vu[1], Chris Hanke[1], Shankar Tangirala[1], Wen Liu[1], Marina Salik[2], Kevin Bartholomew[2], Sushmitha Reddy[2], Alex Tessitore[2], Aws Shallal[2], Kalyan Nallaparaju[2], Jin Liang[2], Minhan Chen[2], Shaishav Desai[1]

[1]Microsoft, Sunnyvale, CA, [2]Microsoft, Raleigh, NC

Abstract

This work presents a 3nm 24Gb/s D2D Interface achieving 0.23pJ/b (analog PHY power) and zero wake penalty clock gating and 0.17pJ/b at 20Gb/s. The PHY achieves a total end-to-end average latency of 4.2ns (~25% analog macro). Measured performance demonstrates a worst-case bit-error rate, extrapolated to 1e-18 measured across 1680 lanes over PVT of 12ps$_{pp}$. The modular nature of this work allows flexibility in the construction of D2D links and natively supports asymmetric TX and RX bandwidths.

Modern AI and compute systems increasingly rely on chiplet architectures and disaggregated designs, driving the need for die-to-die interconnect solutions that deliver high bandwidth, low active power, and minimal latency. Emerging standards such as UCIe [1] provide a framework for standardized die-to-die connectivity. In this work, we propose a modular die-to-die interface that achieves low active power, low latency, and robust clock gating support with zero wake penalty.

The system concept is shown in Fig. 8.3.1. Each D2D interface, a 'Node', consists of a digital interface and an analog hard macro. The digital interface manages traffic flow control (credit based), performs calibration tasks during initialization, houses error correction, and redundancy/repair logic. The analog macro consists of a ring PLL inside a 'Common Macro' plus transmitter and receiver macros, grouped in rows. Each TX/RX macro contains 16 micro-bumps with 14 data lanes, and 2 clock bumps which are shorted for mechanical redundancy – this work does not use repair on the clock channels. Since each TX and RX macro is independent, therefore this architecture also natively supports asymmetric TX and RX memory bandwidths. The clock from the PLL is distributed to each row using a global clock distribution spine, and at each row a junction box, CL_N, generates a clock for each TX macro which is sent to the digital interface. At each row, the system half-rate clock (CLK$_2$) and digital word clock (CLK$_{10}$) are sent to TXs via a per-row clock distribution channel.

Prior art supporting UCIe needs to follow a standardized bump map [2,3], however this presents a few challenges. Die screening, specifically at speed testing during wafer testing of the TX and RX paths is critical to screening parts. Wafer probe testing is a key aspect of manufacturing tests, so probe pads need to be included inside the D2D macro, not just on the periphery. In this work probe pads are included every 2 rows to achieve good power delivery for chip probing. To reduce complexity for analog loopback testing, a single TX data lane and clock is forwarded to the receiver macro and fanned out to all receiver lanes. This enables end-to-end data path testing, at maximum speed for single dies. Due to the rectangular placement of the power bumps for a N/S breakout the power and ground columns can be rotated, and the TX and RX macros can be re-used to allow for re-use in both N/S and E/W configurations. To demonstrate clock gating, a traffic generator is included in the test fixture to model traffic starting and stopping, this can be done at varying duty cycles to measure power vs. activity factor.

Figure 8.3.2 shows the transmit macro, each TX macro takes in a half rate clock and 10th-rate data word clock. Data from the digital interface is grouped into 140b (10b per lane × 14 lanes per macro) and includes a 'Valid' signal, which indicates when the analog macro should stop sending data and halt traffic on the channel. In this design, 'Valid' is not transmitted between states, instead the transmit macro's forwarded clock stops toggling going into idle, and resumes toggling upon idle exit. Data valid is encoded in the first data word using three bits for error protection. Inside the TX macro data is serialized in 2 stages, 1st up to half rate and then a 2:1 mux generates the final high-speed data, and forwarded clock, which is programmable for calibration and initialization purposes. To achieve low idle state power during clock gating both clock and data are stopped, the clock is parked high, corresponding to the last transmitted bit of a 1010 pattern.

The receiver macro (shown in Fig. 8.3.3) has no notion that data has stopped since the data valid is not sent across the interface. Valid framing at the receiver is handled in the data traffic, 3 data bits are encoded with valid in the last data bit to protect against bit errors. The main novelty of the receiver is its lack of complexity. The clock path is intentionally kept electrically short to limit delay excursions during a clock gating event, whereas designs like [3] use a phase interpolator for design simplicity, a CDR is avoided due to clock gating support and to maintain a low power envelope. The receiver path data can be muxed between interposer traffic, and on-die loopback, each receiver lane has a small per-lane skew adjustment, calibrated once at start-up, and is sampled by an even and odd sampler. Common to all receiver lanes is a clock adjustment block, ΔT, which acts as the macro's global sampling adjustment – also calibrated once at start-up. For diagnostic purposes an eye monitor is included, however, to keep hardware at a minimum all lanes are muxed into one eye monitor sampler. Lastly, all data lanes, and eye monitor are de-serialized using a global 2:10 de-serializer. Each receiver lane (and clock) includes an 'async buffer', which is a low-speed buffer with weak pull up and pull down to facilitate a simple open/short package fault detection to mimic functionality of boundary scan on non-interposer-based systems.

Figure 8.3.4 shows the test vehicle which includes two pairs of die-to-die nodes. Two nodes were included to simulate more activity on the power supply, additionally the power delivery network of the test fixture was tuned to mimic the real SoC PDN so clock gating mimics the final application. As shown an integrated packaged device (IPD) is key to achieving low latency clock gating, with an IPD the system with all lanes in both nodes transitioning at the exact same instance in time achieves less than 40mV peak of droop. In a product the IPD is placed on the bottom of the substrate, to avoid being optimistic the test-chip was scaled to achieve a representative impedance to a final large scale SoC. Without IPD the droop performance of the system is significantly degraded to approximately 112mV. Also shown in Fig. 8.3.4, the test fixture also includes a full routing mock-up to include a realistic crosstalk profile of the end application.

Figure 8.3.5 shows the measurement results of the die-to-die link. BER projections are done using a destructive measurement (PRBS15), the ΔT timing offset is swept until the edge of the eye is found, the timing offset is adjusted using coarse codes ~8ps/step and fine codes ~0.8ps/step. Due to a test chip limitation, the clock always has a positive delay relative to the data, so the left edge of the bath-tub curve could not be captured. This limitation stemmed from the usage of an RS latch to compare clock to data; an RS latch was used to make the clock-to-data alignment insensitive to DCD which necessitated the clock always starting at a positive delay relative to the data for correct operation, which introduced this measurement limitation. An XOR gate and RS latch are functionally equivalent for clock alignment, so to achieve a better phase detector gain and average multiple lanes together, an XOR gate is used for production versions. The right edge was captured and compared to the calibrated center value to estimate the peak-to-peak opening, the center value was confirmed by measuring a phase comparator output on an analog test point. The device was able to achieve an eye opening of 12ps at a BER of 1e-18. To demonstrate the clock gating functionality, the average DC current of the link is plotted vs. clock gating percentage of the test fixture traffic generator. Clock gating factor is plotted for 0.1% to 100%. At 0% clock gating the D2D PHY achieves 0.33pJ/b at 24Gb/s (0.895A × 0.75V ÷ 6 macro pairs × 14 lanes per macro × 24Gb/s) and 0.25pJ/b at 20Gb/s. For all activity factor measurements no bit errors were observed. When fully clock gated, the system achieves a power efficiency of 0.05pJ/b.

A comparison to prior art is shown in Fig. 8.3.6, the link operates at a nominal supply of 0.65V for 20Gb/s and 0.75V for 24Gb/s. Bit-error rate is measured down to 1e-13 and is extrapolated down to 1e-18. The measured performance confirms that utilizing tight data and clock delay alignment is critical to achieving both low power and zero wake latency clock gating performance. The PHY achieves an average latency of 4.2ns (end-to-end, including FIFOs, 25% analog latency). The end-to-end latency quoted is approximately equivalent to the latency between FDI interfaces defined in UCIe, however, there is no additional protocol overhead required in this scheme. The bandwidth density achieved for this prototype is 7.4Tb/s/mm (24Gb/s × 12 macro pairs × 14 lanes per macro in 0.542mm).

Figure 8.3.7 shows a die photo of the work in 3nm CMOS, the test fixture consisted of four die-to-die nodes, two pairs connected to an IPD, and two without. This work demonstrates a modular D2D architecture capable of supporting asymmetric TX and RX bandwidths and achieves 0.25pJ/b (0.17pJ/b PHY only) at 20Gb/s and 0.33pJ/b (0.23pJ/b PHY only) at 24Gb/s - power is quoted at 0.75V for 24Gb/s and 0.65V for 20Gb/s. At 24Gb/s, using representative high-density routing between dies, an extrapolated eye opening of 12ps$_{pp}$ is measured.

References:
[1] UCIe Standard Specification, Home | UCIe Consortium.
https://www.uciexpress.org/specifications
[2] M. -S. Lin et al., "36.1 A 32Gb/s 10.5Tb/s/mm 0.6pJ/b UCIe-Compliant Low-Latency Interface in 3nm Featuring Matched-Delay for Dynamic Clock Gating," *2025 IEEE International Solid-State Circuits Conference (ISSCC)*, San Francisco, CA, USA, 2025, pp. 586-588, doi: 10.1109/ISSCC49661.2025.10904767.
https://ieeexplore.ieee.org/document/10904767

[3] D. T. Melek *et al.*, "A 0.29pJ/b 5.27Tb/s/mm UCIe Advanced Package Link in 3nm FinFET with 2.5D CoWoS Packaging," *2025 IEEE International Solid-State Circuits Conference (ISSCC)*, San Francisco, CA, USA, 2025, pp. 590-592, doi: 10.1109/ISSCC49661.2025.10904754. https://ieeexplore.ieee.org/document/10904754
[4] J. Gu *et al.*, "A 32Gb/s 0.36pJ/bit 3nm Chiplet IO Using 2.5D CoWoS Package with Real-Time and Per-Lane CDR and Bathtub Monitoring," *2024 IEEE Symposium on VLSI Technology and Circuits (VLSI Technology and Circuits)*, Honolulu, HI, USA, 2024, pp. 1-2, doi: 10.1109/VLSITechnologyandCir46783.2024.10631527.
https://ieeexplore.ieee.org/document/10631527
[5] K. Seong *et al.*, "A 4nm 32Gb/s 8Tb/s/mm Die-to-Die Chiplet Using NRZ Single-Ended Transceiver With Equalization Schemes And Training Techniques," *2023 IEEE International Solid-State Circuits Conference (ISSCC)*, San Francisco, CA, USA, 2023, pp. 114-116, doi: 10.1109/ISSCC42615.2023.10067477.
https://ieeexplore.ieee.org/document/10067477
[6] Y. Nishi *et al.*, "A 0.190-pJ/bit 25.2-Gb/s/wire Inverter-Based AC-Coupled Transceiver for Short-Reach Die-to-Die Interfaces in 5-nm CMOS," in *IEEE Journal of Solid-State Circuits*, vol. 59, no. 4, pp. 1146-1157, April 2024, doi: 10.1109/JSSC.2023.3338478.
https://ieeexplore.ieee.org/document/10356114

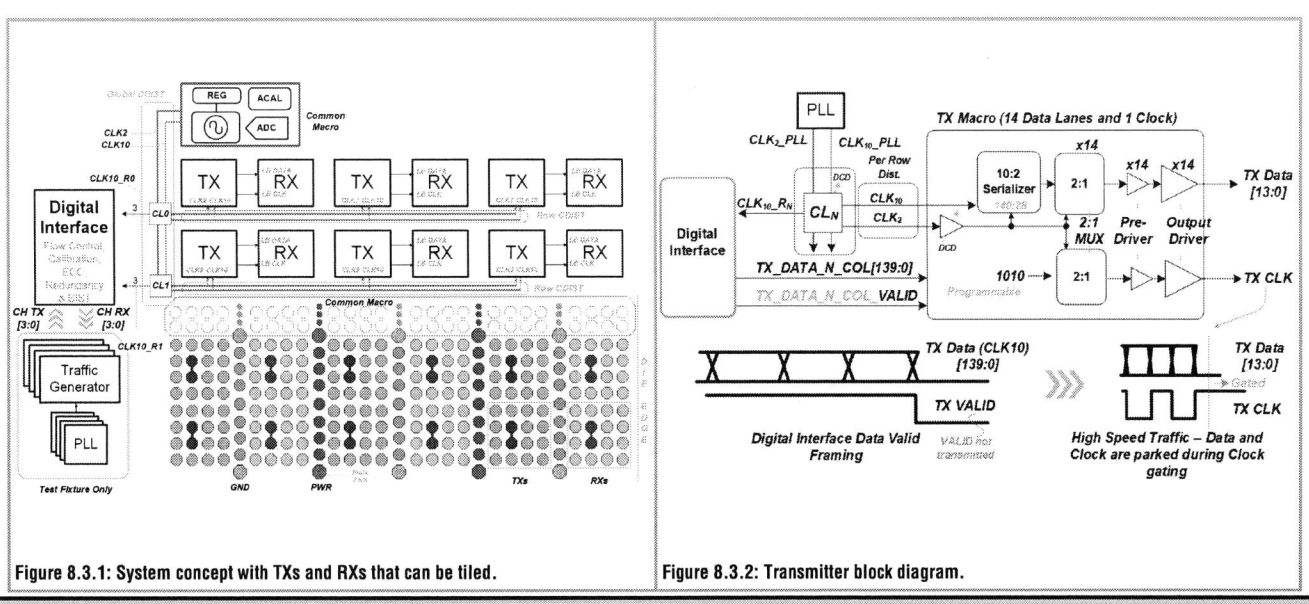

Figure 8.3.1: System concept with TXs and RXs that can be tiled.

Figure 8.3.2: Transmitter block diagram.

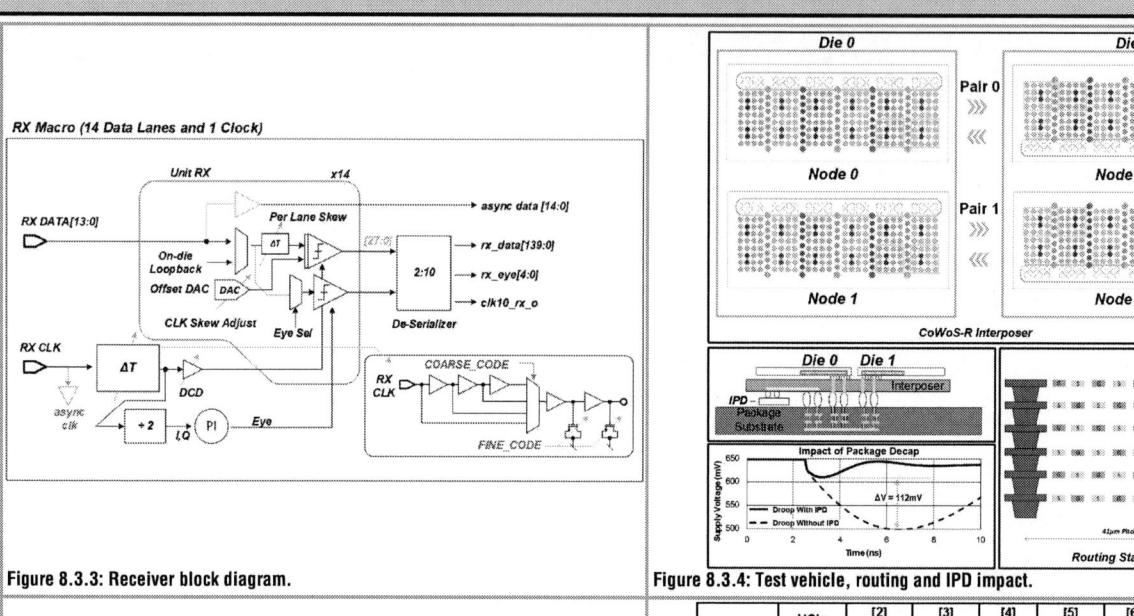

Figure 8.3.3: Receiver block diagram.

Figure 8.3.4: Test vehicle, routing and IPD impact.

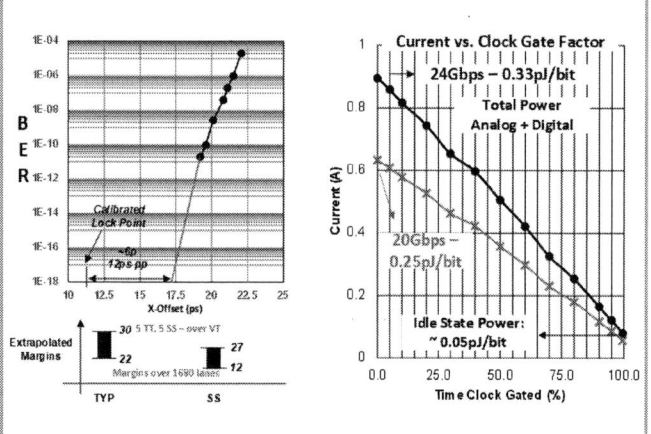

Figure 8.3.5: Measurement results.

Parameter	UCIe Spec	[2] ISSCC '25	[3] ISSCC '25	[4] VLSI '24	[5] ISSCC '23	[6] VLSI '23	This Work
Technology		3nm	3nm	3nm	4nm	5nm	3nm
Module Width	64	64 / Module	64 / Module	36 / Slice	39 / Slice	19 / Module	42 / Slice (Row) Max 11 Rows
Channel	Advanced Package	Si Interposer	Si Interposer	On-chip Channel	Si Interposer	On-Chip Channel	Org. Interposer
Channel Length	< 2mm	1.7mm	1.4mm	2mm	3mm	1.2mm	1.8mm
Bump Pitch	45	45	45	-	50	-	41
Data Rate /pin (Gbps)	32	32	16	32	32	25.2	24
Analog Efficiency (pJ/bit) 24Gbps – 0.75V 20Gbps – 0.65V	0.6	0.6 0.46	0.29	0.36	0.44	0.19	0.226 (24Gbps) 0.17 (20Gbps)
Full System Efficiency Analog PHY + Digital Interface (pJ/bit) 24Gbps – 0.75V 20Gbps – 0.65V	-	-	-	-	-	-	0.33 (24Gbps) 0.25 (20Gbps)
Shoreline BW Density (Tbps/mm)	10.5	10.5	5.27	3.84	8	5.8	7.4
Latency (ns) PHY / PHY + Digital or (FDI-FDI)	-	-	-/3.5	-	-	-	1/4.2

Figure 8.3.6: Comparison to prior art.

Figure 8.3.7: Die photo.

• 2026 IEEE International Solid-State Circuits Conference

ISSCC 2026 / SESSION 8 / DIE-TO-DIE AND HIGH-SPEED ELECTRICAL TRANSCEIVERS / 8.4

8.4 A 112Gb/s/wire Single-Ended Simultaneous Bi-Directional Transceiver with Dynamic Equalizer for Die-to-Die Interface in 28nm CMOS

Zhiwen Huang*[1], Zhifei Wang*[1], Bingyi Ye[2], Tianchen Ye[1], Dunshan Yu[1], Wei Wang[1,3], Weixin Gai[1,3]

[1]Peking University, Beijing, China, [2]East China Normal University, Shanghai, China, [3]Beijing Advanced Innovation Center for Integrated Circuits, Beijing, China
*Equally Credited Authors (ECAs)

Abstract

This work presents an 8-lane 112Gb/s/wire single-ended simultaneous bi-directional transceiver with a 3mm shield-less on-chip channel. A dynamic equalizer is proposed to decouple the bi-directional signals, and compensate for insertion loss and crosstalk.

Fabricated in 28nm CMOS, the transceiver achieves a BER of less than 10^{-14}, and an energy efficiency of 1.01pJ/b.

The growth in artificial intelligence (AI) applications is pushing die-to-die (D2D) interfaces toward higher edge density and lower bit-error rate (BER). Increasing the per-wire data rate and reducing the data channel pitch can improve edge density, albeit at the cost of degrading signal-to-noise ratio (SNR). The degraded SNR requires better equalization to mitigate insertion loss (IL) and crosstalk to achieve a low BER. Simultaneous bi-directional (SBD) signaling and shield-less data channels are adopted in [1] for improved edge density. However, the digital signal processing (DSP)-based equalizer consumes a large amount of power and area. Feed-forward equalizers (FFEs) and decision-feedback equalizers (DFEs) are widely used in D2D interfaces to eliminate inter-symbol interference (ISI) introduced by a lossy channel [2,3], but the static current in an FFE and the feedback timing constraints in a DFE incur power and area efficiency penalties. In view of these drawbacks, this work presents a 112Gb/s/wire single-ended SBD transceiver with a dynamic equalizer for crosstalk and IL, achieving less than 10^{-14} BER, and 1.01pJ/b energy efficiency on a 3mm shield-less channel in 28nm CMOS.

The source-synchronous SBD interconnect is illustrated in Fig. 8.4.1, comprising eight single-ended data lanes and two differential clock lanes. The 32 bits of parallel data are input to each data lane and sent to adjacent victim lanes for near-end crosstalk (NEXT) equalization. The transmitter (TX) serializes input data through a 32:4 multiplexer (MUX), a 4:1 pulse generator (PG), and an N-over-N driver to the output V_O. The TX output impedance is designed to match the channel's characteristic impedance, thus reducing reflection. In addition, each side of the channel is terminated with a TX, so the receiver (RX) input termination is eliminated for power saving. As the TXs work simultaneously, bi-directional signals are coupled, along with crosstalk and ISI, degrading the SNR of the received data. To address this problem, we propose a dynamic equalizer for bidirectional decoupling, crosstalk, and IL equalization. The decoupling signal V_{REP} is generated by a replica PG and driver, which is designed for a lighter load to match the rising and falling edges of the coupled signal. The NEXT equalization (XTEQ) signal V_{XTEQ} is generated by serializing and high-pass filtering data from aggressors (AGGRs). V_{REP} and V_{XTEQ} are sent to the dynamic equalizer and subtracted from V_O to recover the received data. Moreover, the IL equalization (ILEQ) path operates concurrently to eliminate the ISI. The far-end crosstalk (FEXT) is equalized by designing the channel's geometry to have a balanced capacitive and inductive coupling [1,4]. After the previously mentioned equalization, the received signal is resolved into digital data by the slicers and deserialized through an aligner and a demultiplexer (DEMUX). The 28GHz differential clock CK_{IN} is delivered to TXs and feed-forwarded to RXs after passing through the delay-line-based phase rotators (PRs). In each clock domain, both the TX and RX clocks (CK_{TX} and CK_{RX}) are delivered through standing waves for synchronous operation and NEXT equalization alignment [1,5].

The top of Fig. 8.4.2 shows circuit details of the V_{XTEQ} generator, where the input parallel data is from the NEXT aggressor, and the output V_{XTEQ} is sent to the dynamic equalizer. The crosstalk exhibits a high-pass characteristic attributed to channel coupling and is fitted by an RC high-pass filter (HPF). The driver utilizes the same N-over-N structure and power supply as that in TX for better matching. Its pull-up and pull-down current I_D and the coupling capacitor C_{XTEQ} can be tuned to match the NEXT signal (V_{NEXT}) profile. The single-bit response (SBR) is shown in the middle of Fig. 8.4.2. Increasing I_D enhances the V_{XTEQ} amplitude, while increasing C_{XTEQ} accelerates the falling edge of V_{XTEQ}. Through the joint tuning of I_D and C_{XTEQ}, V_{XTEQ} aligns precisely with V_{NEXT}. The emulated eye diagram of the received data is shown at the bottom of Fig. 8.4.2, and the XTEQ improves eye height by approximately 70m.

Figure 8.4.3 shows the schematic of the dynamic equalizer, which leverages the subtraction property of data slicing. The equalizer comprises two stages: the first serves as a sense amplifier, and the second is a clock-controlled cross-coupled latch. The sense amplifier features three differential paths for bi-directional decoupling, ILEQ, and XTEQ. The decoupling path amplifies the difference between the RX input V_{IN} and the decoupling signal V_{REP} through transistors M_1 and M_2, subtracting the coupled TX output. The ILEQ path utilizes the RC high-pass filtered signal through M_3 and M_4, boosting V_{IN}'s high-frequency components like a continuous-time linear equalizer (CTLE). The boost gain can be tuned via switching transistor M_7 by code C_{ILEQ}. The switch is turned on to connect the input transistors M_3 and M_4 to a common-mode voltage, and turned off to enable the input of

high-pass filtered signals. Moreover, since the circuit operates under clock control, no static current exists, rendering it more suitable for low-power D2D applications than conventional CTLEs. The XTEQ path subtracts V_{XTEQ} signals from two adjacent aggressor lanes, denoted as $V_{XTEQ,1}$ and $V_{XTEQ,2}$. $VB_{XTEQ,1}$ is the inverse of $V_{XTEQ,1}$, and is placed on the opposite side of $V_{XTEQ,2}$ for symmetrical loading. The ILEQ performance is modeled by the impulse sensitivity function (ISF) and sampling frequency response [6], as shown at the bottom of Fig. 8.4.3. Stronger equalization yields a narrower ISF window, a higher peaking gain, and a larger sampling bandwidth. The equalizer provides a maximum equalization of 4.2dB at the Nyquist frequency, which is 7.8dB better than the un-equalized slicer.

The top left of Fig. 8.4.4 shows the quadrature clock generator (QCG) in each data lane, which receives half-rate (28GHz) differential clocks to generate quarter-rate quadrature clocks. The first stage of the QCG is delay-matched true single-phase clock D flip-flop (DM-TSPC DFF)-based dividers. The divided quarter-rate clocks (CK4I & CK4Q) are then converted to four quadrature phases. The always-on transmission gates match the delay of the corresponding inverters, and the subsequent quadrature error correction (QEC) circuit [7] calibrates the phase error. The TSPC DFF is adopted for its advantages in speed and power consumption. However, the mismatch between the 0-to-1 and 1-to-0 clock-to-Q delays (t_{cq}) in conventional architectures leads to a duty-cycle error in the divided clocks. To solve this problem, a DM-TSPC DFF is proposed, as shown in the top right of Fig. 8.4.4. A delayed clock CKD is used for the third stage of the DFF, to increase the 1-to-0 t_{cq} and match with 0-to-1 t_{cq}. As shown at the bottom of Fig. 8.4.4, the duty cycle of the divided clocks improves from 57.5% to 49.8% at the typical case, and varies by 1% across process, voltage, and temperature (PVT) corners. The simulated eye width of the TX output shows an improvement of 0.12-unit interval (UI).

The 8-lane 112Gb/s/wire single-ended SBD transceiver with a 3mm shield-less on-chip channel is fabricated in 28nm CMOS technology. The channel is routed using the top metal layer with a pitch of 6.1μm. An electromagnetic solver is used to simulate the channel's frequency response, as shown in the bottom left of Fig. 8.4.5. The IL is 3.1dB at the Nyquist frequency, and the return loss (RL) is 16.1dB. The FEXT is -22.1dB, and the NEXT is -10.1dB, which is only 7dB smaller than the IL. The input 28GHz clocks are generated by an arbitrary waveform generator (AWG); the transmitted data is generated by an on-chip pseudo-random binary sequence (PRBS) generator, and BER is measured by an on-chip checker. The measured bathtub curves are shown at the top of Fig. 8.4.5. The BER is higher than 10^{-5} when both ILEQ and XTEQ are disabled, about 10^{-7} when either ILEQ or XTEQ is enabled, and lower than 10^{-14} (error-free for 10^{14} bits) with 0.16UI eye width when both ILEQ and XTEQ are enabled. The measured power efficiency is 1.01pJ/b, and the simulated power breakdown is shown at the bottom right of Fig. 8.4.5.

Figure 8.4.6 shows the performance comparison. This work utilizes SBD signaling and shield-less channels to achieve a per-wire data rate of 112Gb/s. The RC HPF-based XTEQ and ILEQ ensure a BER of lower than 10^{-14}. Figure 8.4.7 shows the die micrograph and area summary.

Acknowledgement:
This work was supported in part by the Beijing Major Science and Technology Project (Grant No. Z221100007722019).

References:
[1] Z. Wang et al., "A 64Gb/s/wire 10.5Tb/s/mm/Layer Single-Ended Simultaneous Bi-Directional Transceiver with Echo and Crosstalk Cancellation for a Die-to-Die Interface in 28nm CMOS," *ISSCC*, pp. 588-590, Feb. 2025.
http://doi.org/10.1109/ISSCC49661.2025.10904631
[2] M. S. Lin et al., "A 32Gb/s 10.5Tb/s/mm 0.6pJ/b UCIe-Compliant Low-Latency Interface in 3nm Featuring Matched-Delay for Dynamic Clock Gating," *ISSCC*, pp. 586-588, Feb. 2025.
http://doi.org/10.1109/ISSCC49661.2025.10904767
[3] K. Seong et al., "A 4nm 32Gb/s 8Tb/s/mm Die-to-Die Chiplet Using NRZ Single-Ended Transceiver with Equalization Schemes and Training Techniques," *ISSCC*, pp. 114-116, Feb. 2023. http://doi.org/10.1109/ISSCC42615.2023.10067477

979-8-3315-8937-0/26 $31.00 © 2026 IEEE

[4] J. Lee et al., "A 246fJ/b 13.3Tb/s/mm Single-Ended Current-Mode Transceiver with Crosstalk Cancellation for Shield-Less Short-Reach Interconnect," *IEEE Symp. VLSI Technology*, pp. C1-1, Jun. 2024.
http://doi.org/10.1109/VLSITechnologyandCir46783.2024.10631466

[5] G. Li et al., "Standing Wave Based Clock Distribution Technique with Application to a 10×11 Gbps Transceiver in 28 nm CMOS," *IEEE A-SSCC*, pp. 13-5, Nov. 2015.
http://doi.org/10.1109/ASSCC.2015.7387451

[6] M. Jeeradit et al., "Characterizing sampling aperture of clocked comparators," *IEEE Symp. VLSI Technology*, pp. 68-69, Jun. 2008.
http://doi.org/10.1109/VLSIC.2008.4585955

[7] P. J. Peng et al., "A 112-Gb/s PAM-4 Voltage-Mode Transmitter with Four-Tap Two-Step FFE and Automatic Phase Alignment Techniques in 40nm CMOS," *IEEE JSSC*, vol. 56, no. 7, pp. 2123-2131, Jul. 2021. http://doi.org/10.1109/JSSC.2020.3038818

[8] H. G. Ko et al., "An 8Gb/s/μm FFE-Combined Crosstalk-Cancellation Scheme for HBM on Silicon Interposer with 3D-Staggered Channels," *ISSCC*, pp. 128-130, Feb. 2020.
http://doi.org/10.1109/ISSCC19947.2020.9063162

[9] Y. Nishi et al., "A 0.297-pJ/Bit 50.4-Gb/s/Wire Inverter-Based Short-Reach Simultaneous Bi-Directional Transceiver for Die-to-Die Interface in 5-nm CMOS," *IEEE JSSC*, vol. 58, no. 4, pp. 1062-1073, Apr. 2023.
http://doi.org/10.1109/JSSC.2022.3232024

[10] J. Gu et al. "A 32Gb/s 0.36pJ/bit 3nm Chiplet IO using 2.5D CoWoS Package with Real-Time and Per-Lane CDR and Bathtub Monitoring," *IEEE Symp. VLSI Technology*, pp. C19-3, Jun. 2024.
http://doi.org/10.1109/VLSITechnologyandCir46783.2024.10631527

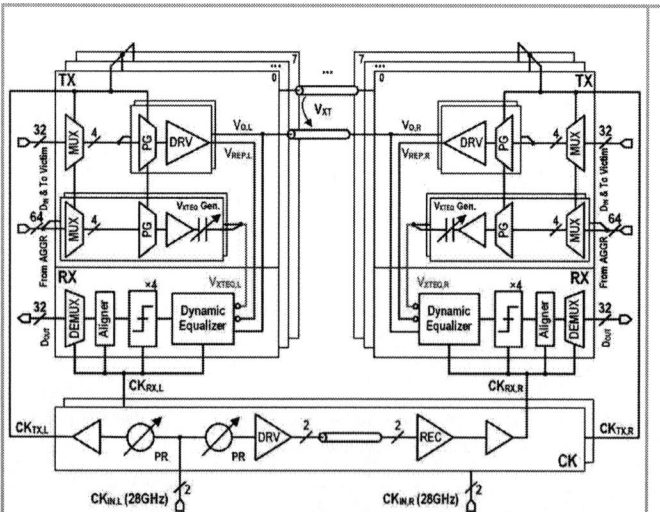

Figure 8.4.1: Source-synchronous simultaneous bi-directional interconnect with the dynamic equalizer for crosstalk and insertion loss.

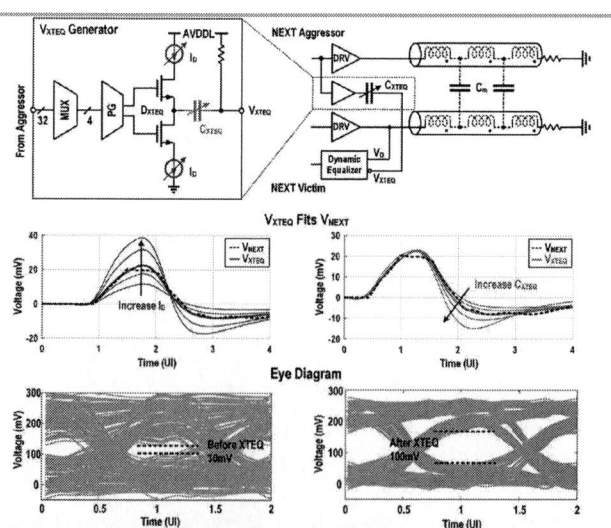

Figure 8.4.2: Crosstalk equalization voltage generator; simulation results of the generator output and the received data eye diagrams.

ISSCC 2026 / SESSION 8 / DIE-TO-DIE AND HIGH-SPEED ELECTRICAL TRANSCEIVERS / 8.4

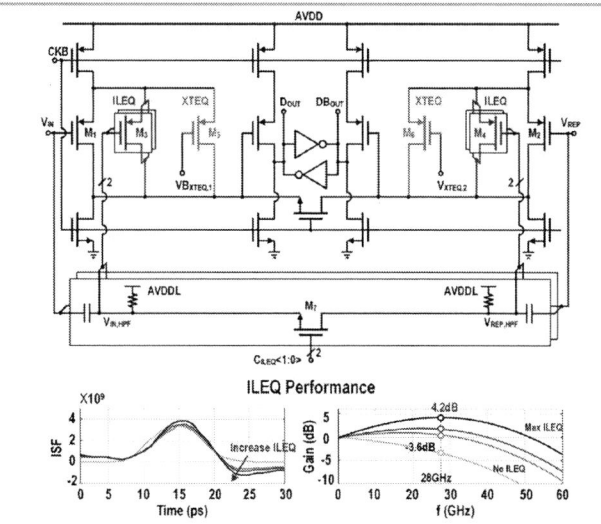

Figure 8.4.3: Dynamic equalizer and the simulated insertion loss equalization performance.

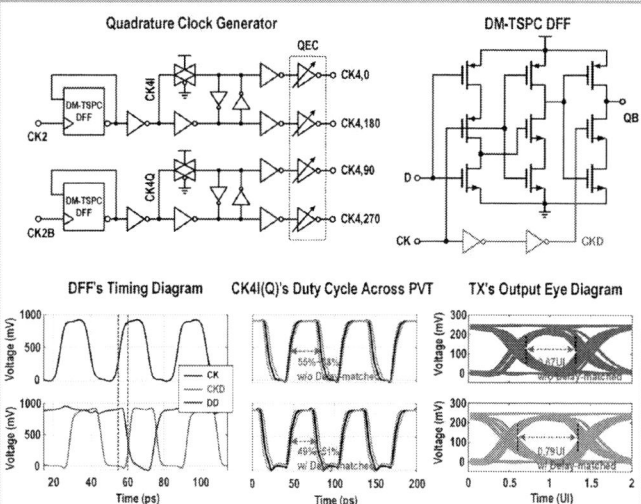

Figure 8.4.4: Quadrature clock generator and duty cycle improvement with the delay-matched TSPC DFF.

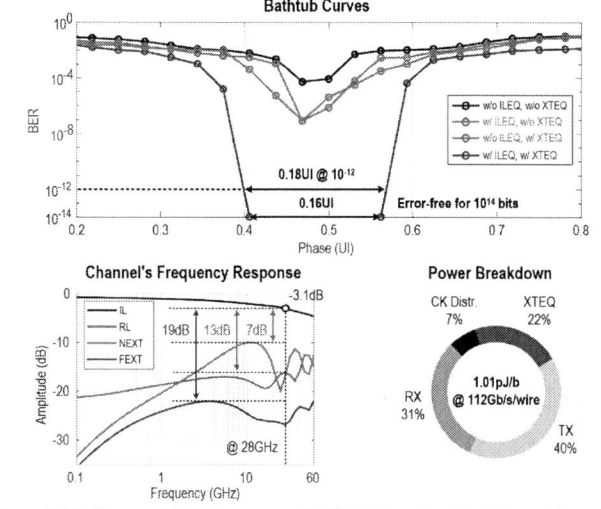

Figure 8.4.5: Measured bathtub curves at 112Gb/s/wire; simulated channel frequency response and power breakdown.

		ISSCC'20 [8]	ISSCC'23 [3]	JSSC'23 [9]	VLSI'24 [10]	ISSCC'25 [1]	This Work
Technology		65nm	4nm	5nm	3nm	28nm	28nm
Data Rate (Gb/s/wire)		4	32	50.4	32	64	112
Modulation		NRZ	NRZ	NRZ Bi-dir	NRZ	NRZ Bi-dir	NRZ Bi-dir
Channel Length (mm)		6	3	1.2	2	3	3
Channel Pitch (um)		0.5	-	-	-	6.1	6.1
Energy Efficiency (pJ/b)		1.5	0.44	0.297	0.36	1.21	1.01
Measured BER		<1e-12	-	<1e-12	<1e-12	<1e-16	<1e-14
Eq.	XT	FIR	No	No	No	FIR	RC HPF
	IL	No	DFE	No	No	No	RC HPF

Figure 8.4.6: Performance comparison with previous works.

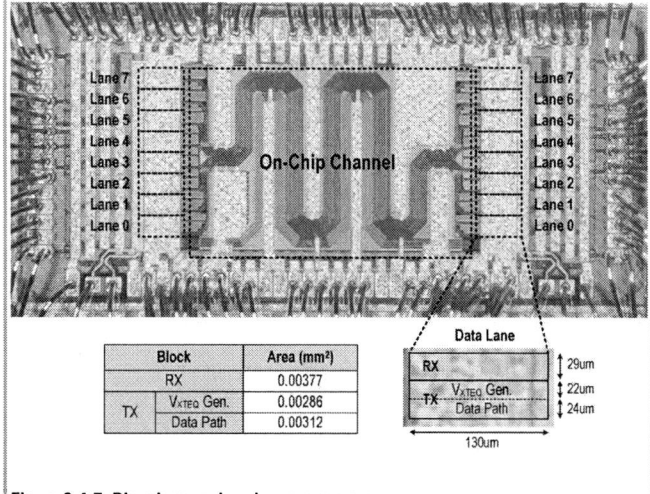

Block		Area (mm²)
RX		0.00377
TX	V_XTEQ Gen.	0.00286
	Data Path	0.00312

Figure 8.4.7: Die micrograph and area summary.

• 2026 IEEE International Solid-State Circuits Conference

979-8-3315-8937-0/26 $31.00 © 2026 IEEE

ISSCC 2026 / SESSION 8 / DIE-TO-DIE AND HIGH-SPEED ELECTRICAL TRANSCEIVERS / 8.5

8.5 A 112Gb/s 0.76pJ/b Reference-less Mixed-Signal PAM-4 CDR in 28nm CMOS

Zhaoyu Zhang*[1,2], Yiqing Xu*[1,2], Jian Liu[1,2], Nanjian Wu[1,2], Zhao Zhang[1,2], Liyuan Liu[1,2]

[1]Institute of Semiconductors, Chinese Academy of Sciences, Beijing, China, [2]University of Chinese Academy of Sciences, Beijing, China
*Equally Credited Authors (ECAs)

Abstract

This paper presents a 112Gb/s mixed-signal reference-less PAM-4 CDR in 28nm CMOS. By proposing the hybrid architecture based on a symmetrical linear PD and a bang-bang PFD, a frequency acquisition controller and a DCC-PEC-merged MPCG, the design issues of reference-less PAM-4 CDR operating at a data rate over 100Gb/s are mitigated. The 28nm CMOS prototype achieves 0.76pJ/b efficiency and 0.11UI_{PP} jitter tolerance at BER $\leq10^{-12}$.

With growing demand for I/O bandwidth in applications like artificial intelligence and high-performance computing, extra-short-reach over-100Gb/s energy-efficient PAM-4 serial transceivers [1-6] are increasingly popular. For the clock-and-data recovery circuits (CDR) in these transceivers, reference-less operation [7-10] is highly desirable to eliminate external crystal references for lower cost and to simplify the clock path for low-power design. Previous state-of-the-art over-100Gb/s PAM-4 CDRs use phase-interpolator (PI)-based digital architectures [11-16], as depicted in Fig. 8.5.1 (top-left). These cannot support reference-less operation, and their digitally intensive nature with multiple high-speed PIs consume high power, especially in low-cost processes like 28nm CMOS. Alternatively, mixed-signal architectures are more energy-efficient and compatible with low-cost CMOS process, as the bang-bang phase detector (BBPD)-based reference-less PAM-4 CDR (bottom left of Fig. 8.5.1) reported in [17-19]. However, BBPD quantization noise degrades recovered clock jitter, reducing the horizontal eye opening of the bit-error rate (BER) bathtub curve at over-100Gb/s rates. Furthermore, divider-based multi-phase clock generators (MPCG) [17,18] are unsuitable for over-100Gb/s operation, as they require the main LC voltage-controlled oscillator (VCO) frequency (F_{VCO}) to be at least twice the MPCG output frequency (F_{MPCG}), and high-speed BBPD logic limits energy efficiency. While using a ring VCO [19] avoids the MPCG issue, its poor phase noise degrades BER at high data rates. Reference [20] presents a reference-less PAM-4 CDR using an asymmetrical linear phase detector (A-LPD), as shown in Fig. 8.5.1 (bottom-left), to eliminate quantization noise at lower power than a BBPD, and operates with F_{VCO} equal to F_{MPCG}. However, its asymmetrical transfer curve results in a narrow linear detection range at speeds over 100Gb/s, causing robustness issues. Additionally, the A-LPD unbalances the loading on each MPCG output phase, introducing phase errors that severely degrade CDR performance. Thus, dummy loads at the MPCG output for balance or separate duty-cycle correctors (DCC) and phase-error correctors (PEC) are needed, which significantly reduce energy efficiency at over-100Gb/s data rates.

To overcome these issues, this paper proposes a 112Gb/s mixed-signal reference-less PAM-4 CDR in 28nm CMOS that scores 0.76pJ/b efficiency and 0.11UI_{PP} jitter tolerance at BER $\leq10^{-12}$, as shown in Fig. 8.5.1 (bottom-right). Our proposed hybrid architecture based on a symmetrical LPD (S-LPD) and a bang-bang phase/frequency detector (BBPFD) enhances robustness at over-100Gb/s compared to prior A-LPD and reduces quantization noise versus BBPD, while maintaining the low-power feature of prior A-LPD. The detection-of-UP-and-DN-occurrence-probability-based frequency acquisition controller [P(UP&DN)-FAC] is devised to enable reference-less operation by using a BBPFD with a CP to control the loop and scan VCO bands, then automatically switches to an LPD-based CDR loop without quantization noise for phase tracking after lock. This also allows the BBPFD [22], which is previously limited to unilateral NRZ detection, to support bilateral PAM-4 detection, ensuring robust reference-less operation. Furthermore, by developing a duty-cycle-corrector and phase-error-corrector merged MPCG (DPM-MPCG), phase error can be eliminated without dummy loads or separate DCC/PEC, and $F_{VCO}=F_{MPCG}$, significantly cutting the power of the clock path.

Figure 8.5.2 (top) illustrates the complete block diagram of our proposed S-LPD and BBPFD hybrid reference-less mixed-signal PAM-4 CDR. Quarter-rate architecture is used for the consideration of lower clock frequency and low power. It mainly consists of four data paths for data receiving, a BBPFD, a proposed P(UP&DN)-FAC, a CP, an S-LPD, an LC-VCO as the main oscillator, a proposed DPM-MPCG, and an analog front-end (AFE). The whole CDR loop can be controlled to operate in frequency-acquisition mode (FAM) or phase-tracking mode (PTM) by P(UP&DN)-FAC based on the lock state, which is detected by estimating the probability of UPDNI signal of the BBPFD output. At FAM, the CP controlled by the BBPFD is enabled for frequency acquisition, and P(UP&DN)-FAC searches the frequency control word band based on the control voltage V_C. At PAM, CP is disabled to make the CDR loop controlled by S-LPD, while the BBPFD, along with the P(UP&DN)-FAC, is kept on to monitor the lock state so that the CDR loop can be relocked if it suffers from some interference or data rate change. The probability of UPDNI is estimated by the P(UP&DN) detector in the P(UP&DN)-FAC, as shown in Fig. 8.5.2(bottom-left). A counter counts the UPDNI signal from BBPFD, while CK_0 is also counted simultaneously. Once the number of UPDNI events reaches a threshold UPDNC, the elapsed clock cycles N_{CLK} are recorded as N_{CLKR}. The probability P(UP&DN) is thus proportional to $1/N_{CLKR}$. To prevent P(UP&DN) from becoming too low to reach UPDNC when the CDR is near locked, CK_0 is continuously

monitored. If N_{CLK} exceeds a predefined value, the CDR is considered locked and automatically switches to LPD mode. UPDNC is programmable and is selected based on simulation and measurement results.

To address the robustness limitations of the A-LPD in [20] at over-100Gb/s, this design employs an S-LPD (bottom left of Fig. 8.5.2) modified from the A-LPD to ensure unbiased phase sampling at over-100Gb/s data rate. Thanks to the utilization of our DPM-MPCG, the phase error issue induced by the unbalanced clock load is eliminated. Thus, we can further reduce the power using the subsampling technique [21], namely, only one edge sampler instead of 4 in the prior BBPD-based CDR, and only one BBPFD logic is used. This does not influence the overall CDR performance because the BBPFD is used only for frequency acquisition and lock state monitoring.

The BBPFD has been used in the NRZ case owing to its simplicity for low-power design [22] (top left of Fig. 8.5.3). By analyzing its logic truth table, we can get that the occurrence probability of UP&DN, P(UP&DN), is non-zero only when the clock frequency f_C is lower than the baud rate of the input data f_D, namely a unilateral frequency detection, which is not robust because such a BBPFD cannot make the CDR relock if the CDR loses lock induced by a positive difference of (f_C-f_D). Interestingly, this is quite different in our PAM-4 case if we directly use the same BBPFD for frequency acquisition of reference-less PAM-4 CDR, as analyzed in Fig. 8.5.3 (top-right). Our BBPFD is performed using the input T_{M_0}, T_{M_1}, and edge sampler output E_0. Since T_{M_0} and T_{M_1} are from the middle slicers in the data path 0 and 1, respectively, all the transition edges of the PAM-4 signal that cross zero are used for the BBPFD without edge selection. The equivalent eye width of the PAM-4 signal detected by the BBPFD is reduced to τ_{EW} ($<1/f_D$), compared to the NRZ eye width of $1/f_D$. This indicates that P(UP&DN) is zero when $f_C>1/\tau_{EW}$. Thus, we can still get a non-zero P(UP&DN) if $f_D<f_C\leq1/\tau_{EW}$. According to the simulation results, our BBPFD can detect a frequency offset of up to 30% with a central f_D of 14GHz. This is quite enough for our design because the tuning range of our LC-VCO is 12.6-15.2GHz. Hence, we can propose to use the simple and power-efficient BBPFD to achieve bilateral frequency detection for robust reference operation with the help of our P(UP&DN)-FAC.

The operation principle of the proposed P(UP&DN)-FAC is detailed in Fig. 8.5.3 (bottom). Upon receiving a reset signal RST, the CDR enters FAM mode with the initial frequency band word of band of 00000, and both the main CP and auxiliary CP are enabled to increase V_C. Band[4:0] increases by 1 once V_C exceeds V_{TH} to switch to the next VCO band; meanwhile, the signal SN and SP generate a pulse to reset V_C to V_{TL} to repeat the same process until the CDR nears the lock state. At that time, N_{CLK} or N_{CLKR} reaches threshold $LOCK_{LMTR}$, indicating a small P(UP&DN), the P(UP&DN)-FAC switches the CDR to PTM while disabling CP. The BBPFD and P(UP&DN)-FAC keep monitoring the lock state. If N_{CLKR} reduces below a threshold $UNLOCK_{LMTR}$ in PTM, indicating a higher P(UP&DN) and loss of lock, the CDR enters FAM again with the initial band of 00000 for relocking. The current of the auxiliary CP is controlled by N_{CLKR} for the smooth switch from FAM to PTM.

Figure 8.5.4 (top) shows our proposed DPM-MPCG, which is based on a ring-current-controlled-oscillator-based (RCCO) PLL (RPLL) improved from [20] with $F_{VCO}=F_{MPCG}$. Prior PEC [24] requires additional delay-tuning circuits, leading to high power overhead. To save power, we propose to introduce an auxiliary current-controlled path within each delay cell of the RCCO, namely our devised phase-error-adjustor-aided delay cell (P-DC), enabling independent delay adjustment and phase error correction inside the RCCO without additional delay-tuning circuits. Inspired by the PEC and DCC technique in [23], phase errors are detected by comparing the DC levels derived from clock pairs with 45° phase spacing. Phase errors are corrected when these levels match each other, indicating aligned phases. Furthermore, we integrate eight duty-cycle adjustors (DCAs) directly into the self-biased buffers (SB-BUFs), avoiding extra power consumption for DCC in the clock path. The loop bandwidth of our RPLL-based DPM-MPCG is several hundreds of MHz, which is more than one order higher than the bandwidth of DCC and PEC. This not only significantly suppresses the phase noise (PN) of RCCO for low recovered clock jitter with low MPCG power, but also keeps the loop stability.

Figure 8.5.4 (bottom-left) shows the PN profile of the 14GHz recovered clock with an rms jitter of 343.3fs with a 112Gb/s PAM-4 PRBS-9 input. The measured JTOL is 0.11 UI_{PP} at a 100MHz jitter frequency, as depicted in Fig. 8.5.4 (bottom-right). Figure 8.5.5 (top) presents

142 • 2026 IEEE International Solid-State Circuits Conference

979-8-3315-8937-0/26 $31.00 © 2026 IEEE

the measured settling process in terms of V_{C_OUT}, the buffered V_C in the CDR (top of Fig. 8.5.2), demonstrating the robustness for reference-less operation. Under a channel loss of 9.1dB at 28GHz, the horizontal eye opening of the bathtub curve is 0.11UI with BER<10^{-12}, as presented in Fig. 8.5.5 (bottom).

This is the first reported reference-less PAM-4 CDR that can operate at over-100Gb/s data rate, and maintains a competitive recovered clock jitter compared to other reference-less PAM-4 CDRs in Fig. 8.5.6. Compared to other over-100Gb/s PAM-4 CDRs in 28nm CMOS, our work achieves the best energy efficiency, and it is also among the best reported to date when comparing with the over-100Gb/s CDR in the advanced FinFET process. Thanks to our linear-PD-based architecture, our work achieves improved JTOL compared to other over-100Gb/s works in Fig. 8.5.6. The energy efficiency of our 28nm CMOS prototype is 0.76pJ/b (Fig. 8.5.7).

Acknowledgement:
The authors would like to acknowledge Sinolink Technologies for the support of test equipments. This work was supported by the National Natural Science Foundation of China (No. 62174153 and 62222409). Corresponding author: Zhao Zhang (zhangzhao11@semi.ac.cn).

References:
[1] B. Ye *et al.*, "A 1.11pJ/b 224Gb/s XSR Receiver with Slice-Based CTLE and PI-Based Clock Generator in 12nm CMOS", *IEEE ISSCC*, pp. 140-142, 2025. https://doi.org/10.1109/ISSCC49661.2025.10904800
[2] B. Ye *et al.*, "A 2.29-pJ/b 112-Gb/s Wireline Transceiver With RX Four-Tap FFE for Medium-Reach Applications in 28-nm CMOS", *IEEE JSSC*, vol. 58, no. 1, pp. 19-29, Jan. 2023. https://doi.org/10.1109/JSSC.2022.3223052
[3] G. Gangasani *et al.*, "A 1.6Tb/s Chiplet over XSR-MCM Channels using 113Gb/s PAM-4 Transceiver with Dynamic Receiver-Driven Adaptation of TX-FFE and Programmable Roaming Taps in 5nm CMOS", *IEEE ISSCC*, pp. 122-124, Jan. 2022. https://doi.org/10.1109/ISSCC42614.2022.9731636
[4] R. Yousry *et al.*, "A 1.7pJ/b 112Gb/s XSR Transceiver for Intra-Package Communication in 7nm FinFET Technology", *IEEE ISSCC*, pp. 180-182, 2021. https://doi.org/10.1109/ISSCC42613.2021.9365752
[5] H. Park et al., "A 4.63pJ/b 112Gb/s DSP-Based PAM-4 Transceiver for a Large-Scale Switch in 5nm FinFET", *IEEE ISSCC*, pp. 5-7, 2023. https://doi.org/10.1109/ISSCC42615.2023.10067613
[6] A. Chowdhury *et al.*, "A 0.9pJ/b 9.8-113Gb/s XSR SerDes with 6-tap TX FFE and AC coupling RX in 3nm FinFet Technology", *IEEE VLSI*, pp. 1-2, 2024. https://doi.org/10.1109/VLSITechnologyandCir46783.2024.10631354
[7] W. Rahman *et al.*, "A 22.5-to-32-Gb/s 3.2-pJ/b Referenceless Baud-Rate Digital CDR With DFE and CTLE in 28-nm CMOS", *IEEE JSSC*, vol. 52, no. 12, pp. 3517-3531, Dec. 2017. https://doi.org/10.1109/JSSC.2017.2744661
[8] X. Zhao *et al.*, "A Reference-Less CDR Using SAR-Based Frequency-Acquisition Technique Achieving 55ns Constant Band-Searching Time and up to 63.64Gb/s/µs Acquisition Speed", *IEEE ISSCC*, pp. 150-152, 2025. https://doi.org/10.1109/ISSCC49661.2025.10904756
[9] Y. Jung, Y. -W. Kim, S. Lee, S. Kang and K. Park, "A 16-30Gb/s 1.03pJ/b Referenceless Baud-Rate CDR with Integrated Pattern Decoding Technique for Fast Frequency Acquisition", *IEEE ASSCC*, pp. 1-3, 2024. https://doi.org/10.1109/A-SSCC60305.2024.10849287

[10] H. -S. Choi et al., "A 14-to-32-Gb/s Deadzone-Free Referenceless CDR with Autocovariance-based Frequency Detector in 40-nm CMOS Technology", *IEEE ASSCC*, pp. 1-3, 2024. https://doi.org/10.1109/A-SSCC60305.2024.10848602
[11] R. Tang *et al.*, "A 112-Gb/s PAM-4 Receiver with Ultra-Fine GainAdjustment CTLE and Novel Sample-and-Reset Slicer in 28-nm CMOS", *IEEE ESSERC*, pp. 297-300, 2024. https://doi.org/10.1109/ESSERC62670.2024.10719519
[12] Y. -P. Lin, P. -J. Peng, C. -C. Lu, P. -T. Shen, Y. -C. Jao and P. -H. Hsieh, "A 2.16pJ/b 112Gb/s PAM-4 Transceiver with Time-Interleaved 2b/3b ADCs and Unbalanced Baud-Rate CDR for XSR Applications in 28nm CMOS", *IEEE ISSCC*, pp. 136-138, 2024. https://doi.org/10.1109/ISSCC49657.2024.10454418
[13] Y. -P. Lin, Y. -C. Jao, W. -H. Hsieh and P. -J. Peng, "A 2.06pJ/b 106.25Gb/s PAM-4 Receiver with 3-Tap FFE and 1-Tap Speculative DFE in 28nm CMOS", *IEEE ISSCC*, pp. 146-148, 2025. https://doi.org/10.1109/ISSCC49661.2025.10904743
[14] R. Shivnraine *et al.*, "A 26.5625-to-106.25Gb/s XSR SerDes with 1.55pJ/b Efficiency in 7nm CMOS", *IEEE ISSCC*, pp. 181-183, 2021. https://doi.org/10.1109/ISSCC42613.2021.9365975
[15] G. Gangasani *et al.*, "A 1.1-pJ/b/Lane, 1.8-Tb/s Chiplet Using 113-Gb/s PAM-4 Transceiver With Equalization Strategy to Reduce Fractionally Spaced 0.5-UI ISI in 5-nm CMOS", *IEEE SSCL*, vol. 8, pp. 33-36, 2025. https://doi.org/10.1109/LSSC.2025.3526877
[16] C. F. Poon *et al.*, "A 1.24-pJ/b 112-Gb/s (870 Gb/s/Mm) Transceiver for In-Package Links in 7-nm FinFET", *IEEE JSSC*, vol. 57, no. 4, pp. 1199-1210, April. 2022. https://doi.org/10.1109/JSSC.2022.3141802
[17] X. Zhao, Y. Chen, L. Wang, P. -I. Mak, F. Maloberti and R. P. Martins, "A Sub-0.25-pJ/bit 47.6-to-58.8-Gb/s Reference-Less FD-Less Single-Loop PAM-4 Bang-Bang CDR With a Deliberate-Current-Mismatch Frequency Acquisition Technique in 28-nm CMOS", *IEEE JSSC*, vol. 57, no. 5, pp. 1358-1371, May. 2022. https://doi.org/10.1109/JSSC.2022.3140778
[18] X. Zhao, Y. Chen, P. -I. Mak and R. P. Martins, "A 0.0285-mm² 0.68-pJ/bit Single-Loop Full-Rate Bang-Bang CDR Without Reference and Separate FD Pulling Off an 8.2-Gb/s/µs Acquisition Speed of the PAM-4 Input in 28-nm CMOS", *IEEE JSSC*, vol. 57, no. 2, pp. 546-561, Feb. 2022. https://doi.org/10.1109/JSSC.2021.3113773
[19] L. Feng, T. Li, X. Zou, X. Xiong and Z. Zhang, "A 6–64-Gb/s 0.41-pJ/Bit Reference-Less PAM4 CDR Using a Frequency-Detection-Gain-Enhanced PFD Achieving 19.8-Gb/s/µs Acquisition Speed", *IEEE TCAS-II*, vol. 72, no. 1, pp. 68-72, Jan. 2025. https://doi.org/10.1109/TCSII.2024.3481436
[20] Z. Zhang *et al.*, "A 64-Gb/s Reference-Less PAM4 CDR with Asymmetrical Linear Phase Detector Soring 231.5-fs$_{rms}$ Clock Jitter and 0.21-pJ/bit Energy Efficiency in 40-nm CMOS", *IEEE VLSI*, pp. 1-2, 2023. https://doi.org/10.23919/VLSITechnologyandCir57934.2023.10185285
[21] Z. Zhang, G. Zhu, C. Wang, L. Wang and C. P. Yue, "A 32-Gb/s 0.46-pJ/bit PAM4 CDR Using a Quarter-Rate Linear Phase Detector and a Self-Biased PLL-Based Multiphase Clock Generator", *IEEE JSSC*, vol. 55, no. 10, pp. 2734-2746, Oct. 2020. https://doi.org/10.1109/JSSC.2020.3005780
[22] K. Park, D. -K. Jeong, "Analysis of frequency detection capability of Alexander phase detector", *Electronics Letters* vol. 56, no. 4, pp. 180-182, Feb. 2020. https://doi.org/10.1049/el.2019.3488
[23] Y. -T. Lin and W. -Z. Chen, "A 50 Gb/s PAM-4 Transmitter with Feedforward Equalizer and Background Phase Error Calibration", *IEEE ASSCC*, pp. 1-2, 2020. https://doi.org/10.1109/A-SSCC48613.2020.9336134
[24] J. Oh and S. Cho, "A 0.001-mm², 1.15–11-GHz Background Quadrature Phase and Duty-Cycle Error Corrector Using a NAND- Based Phase Detector in 28-nm CMOS", *IEEE SSCL*, vol. 7, pp. 247-250, 2024. https://doi.org/10.1109/LSSC.2024.3452280

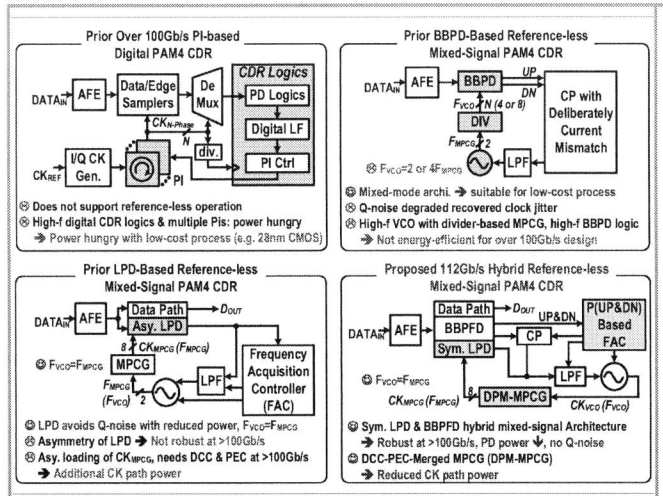

Figure 8.5.1: Review of prior over 100Gb/s digital PAM-4 CDR, prior reference-less PAM-4 CDR, and proposed hybrid reference-less mixed-signal PAM-4 CDR (bottom-right).

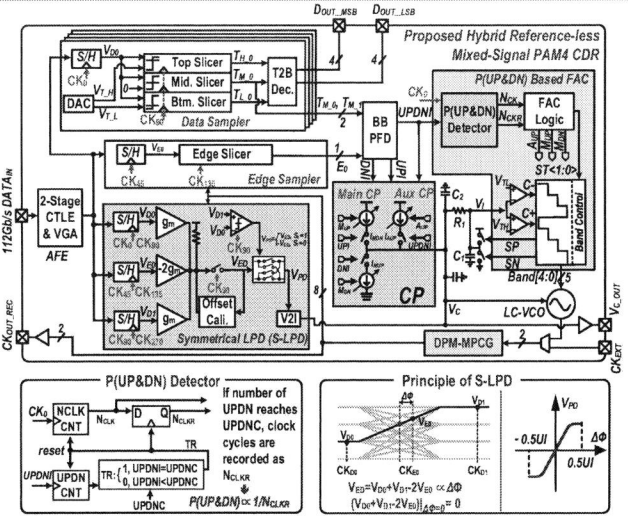

Figure 8.5.2: Block diagram of the proposed hybrid reference-less mixed-signal PAM-4 CDR (top), P(UP&DN) detector (bottom-left) and principle of S-LPD (bottom-right).

ISSCC 2026 / SESSION 8 / DIE-TO-DIE AND HIGH-SPEED ELECTRICAL TRANSCEIVERS / 8.5

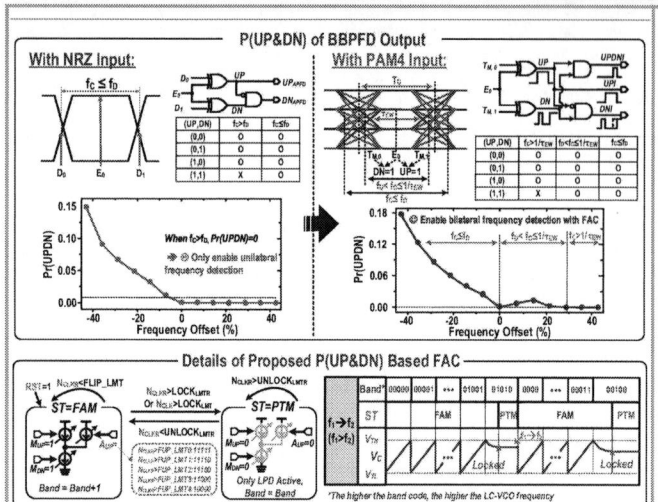

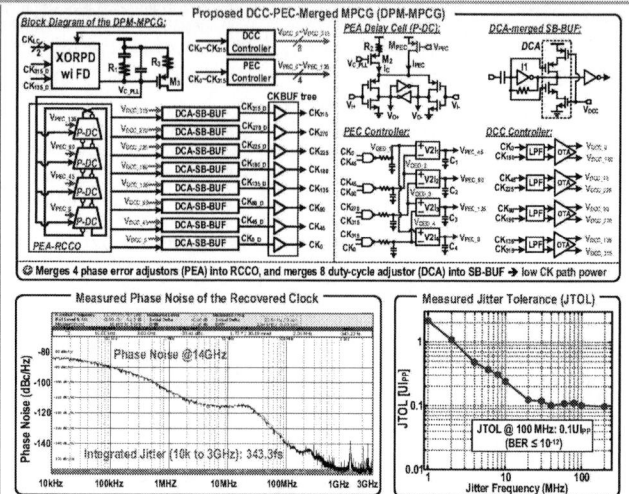

Figure 8.5.3: Principle of the P(UP&DN) generation with NRZ input and PAM-4 input (top) and details of the proposed P(UP&DN)-based FAC (bottom).

Figure 8.5.4: Block diagram of the proposed DPM-MPCG (top), measured phase noise of the recovered clock at 14GHz (bottom-left), and JTOL with 112Gb/s input (bottom-right).

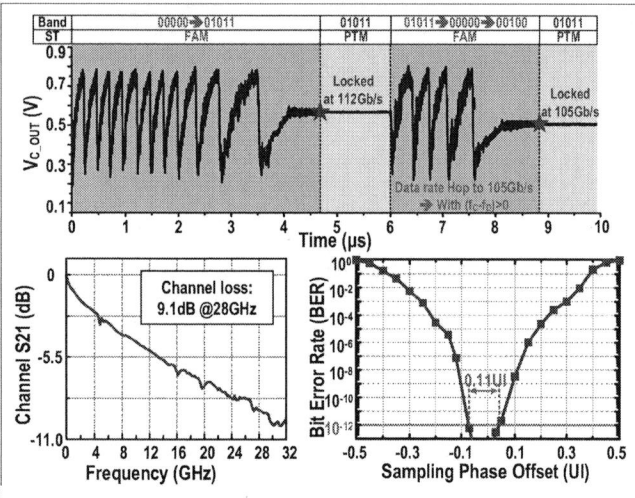

Figure 8.5.5: Measured settling process, channel S21, and bathtub curve for 112Gb/s input data.

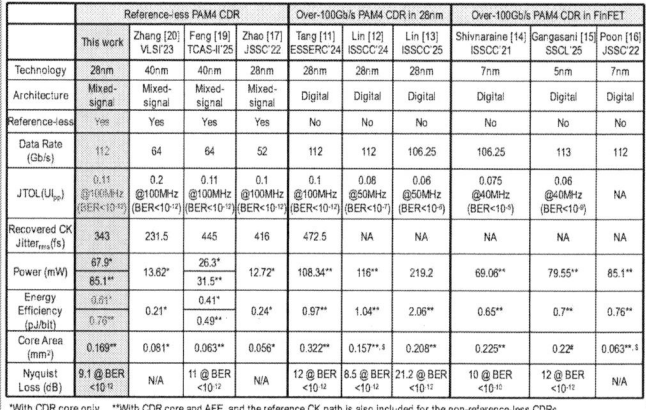

	Reference-less PAM4 CDR				Over-100Gb/s PAM4 CDR in 28nm			Over-100Gb/s PAM4 CDR in FinFET		
	This work	Zhang [20] VLSI'23	Feng [19] TCAS-II'25	Zhao [17] JSSC'22	Tang [11] ESSERC'24	Lin [12] ISSCC'24	Lin [13] ISSCC'25	Shivnaraine [14] ISSCC'21	Gangasani [15] SSCL'25	Poon [16] JSSC'22
Technology	28nm	40nm	40nm	28nm	28nm	28nm	28nm	7nm	5nm	7nm
Architecture	Mixed-signal	Mixed-signal	Mixed-signal	Mixed-signal	Digital	Digital	Digital	Digital	Digital	Digital
Reference-less	Yes	Yes	Yes	Yes	No	No	No	No	No	No
Data Rate (Gb/s)	112	64	64	52	112	112	106.25	106.25	113	112
JTOL(UI$_{pp}$)	0.11 @100MHz (BER<10^{-12})	0.2 @100MHz (BER<10^{-12})	0.11 @100MHz (BER<10^{-12})	0.1 @100MHz (BER<10^{-12})	0.1 @100MHz (BER<10^{-12})	0.06 @50MHz (BER<10^{-7})	0.06 @50MHz (BER<10^{-9})	0.075 @40MHz (BER<10^{-9})	0.06 @40MHz (BER<10^{-9})	NA
Recovered CK Jitter$_{rms}$(fs)	343	231.5	445	416	472.5	NA	NA	NA	NA	NA
Power (mW)	67.9* 85.1**	13.62*	26.3* 31.5**	12.72*	108.34**	116**	219.2	69.06**	79.55**	85.1**
Energy Efficiency (pJ/bit)	0.61* 0.76**	0.21*	0.41* 0.49**	0.24*	0.97**	1.04**	2.06**	0.65**	0.7**	0.76**
Core Area (mm²)	0.169**	0.081*	0.063**	0.056*	0.322**	0.157**,§	0.208**	0.225**	0.22**	0.063**,§
Nyquist Loss (dB)	9.1 @ BER <10^{-12}	N/A	11 @ BER <10^{-12}	N/A	12 @ BER <10^{-12}	8.5 @ BER <10^{-12}	21.2 @ BER <10^{-12}	10 @ BER <10^{-10}	12 @ BER <10^{-12}	N/A

*With CDR core only　**With CDR core and AFE, and the reference CK path is also included for the non-reference-less CDRs
§Estimated from chip photo　♯Whole TRX, area of CDR and AFE are not reported

Figure 8.5.6: Performance summary and comparison.

1. LC-VCO
2. DPM-MPCG with CKBUF
3. LPD, BBPFD, Data & Edge Sampler
4. CP
5. LPF
6. FAC
7. AFE

	LC-VCO	DPM-MPCG, CKBUF	LPD, CP	BBPFD,FAC, Data & Edge Sampler	AFE	Total
Power (mW)	5.4	34.8	4	23.7	17.2	85.1

Energy efficiency @ 112Gb/s: 0.76pJ/bit

Figure 8.5.7: Microscope photograph (top) and power breakdown of the proposed reference-less PAM-4 CDR (bottom).

• 2026 IEEE International Solid-State Circuits Conference

979-8-3315-8937-0/26 $31.00 © 2026 IEEE

ISSCC 2026 / SESSION 8 / DIE-TO-DIE AND HIGH-SPEED ELECTRICAL TRANSCEIVERS / 8.6

8.6   A 280mW 112Gb/s PAM-4/NRZ Transceiver for Low-Power IOs in 5nm FinFET Technology

Ullas Singh[1], Kumar Thasari[1], Nitin Nidhi[1], Arvindh Iyer[1], Saurabh Surana[1], Batu Dayanik[1], Yuanfang Li[1], Mohammadamin Torabi[1], Jun Won Jung[1], Alberto Grassi[1], Mehrdad Fahimnia[1], Hiroshi Kimura[2], Faramarz Bahmani[2], Hao Chen[2], Alex Wang[2], Chen Zhao[2], Yuan Fang[2], Allen Chen[3], Afshin Momtaz[1], Namik Kocaman[1]

[1]Broadcom, Irvine, CA, [2]Broadcom, San Jose, CA, [3]Broadcom, Fort Collins, CO

Abstract

This paper presents the design of an ultra-low power, compact 112Gb/s PAM-4 transceiver. Each lane incorporates a quarter rate RX with 3-tap FFE, 11-tap DFE equalization, and a half-rate TX using a 7b SST DAC driver. The transceiver can compensate up to 35dB channel loss at 112Gb/s and achieves BER < 1E-6, dissipating only 280mW including both analog and digital power domains. This work has the best pJ/b/dB FOM as well as the lowest area versus comparable work at this data rate.

The increased demand in next-generation data centers and high-performance computing has pushed serial link data rates into 100Gb/s territory and beyond. To cope with channel impairments with more than 20dB Nyquist loss at 112Gb/s PAM-4, recently published transceivers [1-4] adopted ADC/DSP-based receivers in advanced FinFET processes. An important set of IO applications are chip-to-chip and chip-to-module links. Such short-to-medium reach links, which are the focus of standards such as CEI-112G-MR and IEEE 100G VSR/SR, have moderate Nyquist channel loss of up to 28dB or less. Typically, in these applications, the power efficiencies of the transmitter (TX) and the receiver (RX) are of primary concern. While the ADC-based receiver solution provides the flexibility of digital equalization, the power/area consumption cannot easily match that of an analog-based receiver solution. This work presents an ultra-low-power (< 2.5pJ/b) and area-efficient non-ADC/DSP based 112Gb/s transceiver architecture targeting short and medium reach channels. It employs fully adaptive 3-tap FFE/11-tap DFE in the receiver (RX) and a 7b DAC-based transmitter (TX) in 5nm FinFET technology.

Although the data path implementation is fully differential to achieve good supply immunity, a simplified single-ended data path implementation of the transceiver block diagram is illustrated in Fig. 8.6.1. The RX input signal goes into the 50Ω termination impedance, which utilizes a bridged T-coil and a shunt inductor to extend the bandwidth. Then, the input signal is passed through AC capacitors to enable direct attachment to an external transmitter. The first stage is a continuous-time linear equalizer (CTLE) that provides peaking controls for Nyquist equalization. This is followed by a variable gain amplifier (VGA) with T-coil bandwidth extension. The VGA drives 4 interleaved sample/hold amplifiers (S/H) feeding into a 3-tap FFE. The receiver has an 11-tap DFE to further equalize the channel loss and reflections. The output currents of all the FFE/DFE taps are aggregated in the summer. The receiver is clocked by a single half-rate phase interpolator (PI) followed by a divide by 2 to generate quarter-rate I/IB/Q/QB clock phases for sampled baud-rate CDR operation to capture the data at the optimal sampling point. The DEMUX block sends de-serialized 40b received data output to the digital block for CTLE/VGA/FFE/DFE adaptation and CDR. The TX has a half-rate 40:1 MUX architecture and a 7b DAC-based source-series terminated (SST) driver. The common PLL utilizes a low jitter, single LC VCO to achieve the full operation range (49-57GHz). It distributes half-rate clocks to RX/TX clock generators.

The CTLE and VGA stages are gm-TIA-based structures, as shown in Fig. 8.6.2. The CTLE stage implements a tunable linear peaking equalizer. The VGA stage has a similar architecture to the CTLE stage with adjustable degeneration resistor (R_m) for gain control. The passive RC network in the Gm stage degeneration gives the advantage of no extra power dissipation, and uniform step sizes throughout the range. The RX peaking and gain ranges should be wide enough to compensate for the targeted link loss across PVT corners. PVT-compensated bias currents are used to achieve the desired loss compensation capability at optimal power efficiency. This approach results in wide common-mode voltage (VCM) variation. However, the final stage summer, which drives the data slicers, requires well-controlled input and output VCM for the best slicer performance in terms of low noise and sensitivity. Thus, the reference voltages for VGA/CTLE CMFB loops are generated by replica bias blocks which mimic the entire RX chain comprising the FFE, S/H and summer circuits. Both CTLE and VGA VCM voltages are adapted over PVT corners to ensure tightly controlled slicer input VCM voltages while maintaining good CTLE/VGA linearity.

After the VGA, a 3-tap FFE is implemented, one pre-cursor, one post-cursor, and the main tap using interleaved S/H amplifiers [5]. The addition of an FFE provides the precursor equalization that a CTLE/DFE cannot offer. Both FFE and DFE outputs are summed into the output of the summer stage (Fig. 8.6.1). The first two DFE taps are power hungry and subject to stringent timing closure for higher data rates. In this design, the first DFE tap (H1) is not necessary because the FFE post-cursor tap can provide a similar ISI reduction at 1UI. For cancelling ISI at 2UI, the second DFE tap (H2) feedback loop must settle within 2UI, i.e. 35.6ps for 112G PAM-4. Meeting this stringent timing over PVT would result in significant power overhead. Therefore, various CTLE peaking controls are used concurrently to cancel the H2 ISI and eliminate the need for the second DFE tap. CTLE frequency control components impact the ISI at multiple UI points, not just at a single UI as in a FFE/DFE. The table in Fig. 8.6.2 categorizes various CTLE frequency controls based on their impact at different ISI locations. For example, inductive peaking control ($L_{deQ(ctle)}$) has a strong

impact on ISI at 1UI, but it has an opposite effect on ISI at 2UI. Conversely, source degeneration peaking cap ($C_{m(ctle)}$) typically has its strongest impact on H2 and no effect on H1. By using the combinations of these controls, the ISI at the second UI is cancelled with minimal impact to other ISI locations. These controls are part of the equalization algorithm that are adapted together with FFE/DFE using standard LMS optimization.

All other DFE taps (3-11) are realized by direct feedback using CMOS latch circuits. Taps 3-7 are fixed taps while Taps 8-11 serve as floating taps, which are programmable to compensate for reflections. The slicer is a two-stage self-timed comparator with a single clock phase to minimize clock load. Each summer consists of an NMOS CML main amplifier and DFE/FFE tap differential pairs sharing a calibrated load resistor. It is inductively peaked to improve the bandwidth. The summer output VCM drops as DFE/FFE tap weights are increased. A CMFB loop as well as a bleeding current scheme into the load resistor is used to continuously regulate the summer output VCM across PVT variations such as during a temperature ramping test.

The detailed TX block diagram is depicted in Fig. 8.6.3. To reduce the power of the TX, aggressive sizing for the driver is implemented by increasing the resistance ratio between the active switch and an on-chip passive resistor. This reduces capacitive loading down the chain for both clock and data paths. The entire TX runs from a single 0.8V supply. The TX can track any frequency difference from the incoming data rate and the PLL frequency using a phase interpolator (PI). Moreover, it also has duty-cycle correction circuits. A 2b thermometer, 5b-binary driver segmentation scheme has been implemented to mitigate excessive DNL transitions due to MSB segment switching and improve linearity. At the digital RTL interface, a 40b-wide data stream is sent into a 40:1 MUX consisting of a 40:2 MUX driving a 2:1 MUX. The pre-driver receives full-rate data from the final 2:1 MUX, which utilizes both edges of the half-rate clock. The pre-drivers are actively peaked to realize power-efficient high fan-out stages, and at the same time reduce ISI. Active peaking employs an inverter with pass-gate-based resistor feedback (Fig. 8.6.3) At the driver output; the bandwidth is extended using a T-coil. The driver is also equipped with a continuous-time high-pass filter implemented with a capacitive path parallel to the main path. This high-pass filter sharpens the TX output edges and improves the bandwidth further.

In the common PLL, a fractional-N multi-modulus divider with a 3rd-order delta-sigma modulator is implemented to provide clocking for various data rates with a single reference frequency. Tuned clocking has proved to be a very power efficient way to distribute clocks [1-5]. The half-rate clock from the PLL is distributed to the lanes via differential transmission lines implemented as standing wave resonators. Clocks are driven by open-drain CML drivers supplied through the center-tap of the shunt inductors that terminate the structure.

Conventional RX clocking implementation (Fig. 8.6.4) first reduces the I/Q clock speed using a CML divider and uses two separate phase interpolators to control the phase of I and Q clocks. This scheme suffers from dynamic phase mismatches and non-linearity during frequency tracking due to dual PI usage. In the proposed RX clocking design, PLL global clock channel drives a single inductively tuned PI, followed by CMOS dividers eliminating the need for dual PI usage. The detailed schematic of the PI is shown in Fig. 8.6.4. The architecture is similar to a twin INL cancelling PI [6]. The phase control codes of twin PI are offset by 45° and summed after a buffer. Thus, the INL errors are out of phase with each other and are cancelled out.

The transceiver is fabricated in 5nm FinFET technology and tested against multiple 100Gb/s standards. The measurements are performed with long package routings mimicking the worst-case scenario. For a link with insertion loss of 35dB at 28GHz for 112Gb/s PAM-4 (Fig. 8.6.5), the transceiver achieves raw BER better than 1E-6. The receiver easily meets the CEI-112G-MR jitter tolerance mask with *0.2UI* margin due to low latency and high CDR bandwidth, which is a benefit of this architecture where CDR loop latency is less than ADC/DSP based receivers. The measured TX 112Gb/s eye diagram in Fig. 8.6.5 shows an *RLM* of 0.991, *J4U* of 73mUI, *Jrms* of 8.85mUI, *EOJ* of 13.5mUI, and exhibits an SNDR better than 37.5dB providing significant margin compared to MR standards. The PLL has a lock range from 49 to 57GHz, with measured jitter of 100fs$_{rms}$. The transceiver consumes 250mW for analog and 280mW overall from a single 0.8V supply per RX/TX. The FOM, as

144 • 2026 IEEE International Solid-State Circuits Conference

979-8-3315-8937-0/26 $31.00 © 2026 IEEE

measured by pJ/b/dB, is 0.07 which is among the lowest compared to published solutions with similar data rates and processing [1-5]. The die area is 0.33mm² for analog and 0.43mm² overall per RX/TX, which is also one of the lowest compared to prior art. A chip micrograph of four transceivers with the common PLL is shown in Fig. 8.6.7.

References:
[1] J. Im et al., "A 112Gb/s PAM-4 Long-Reach Wireline Transceiver Using a 36-Way Time-Interleaved SAR-ADC and Inverter-Based RX Analog Front-End in 7nm FinFET", ISSCC, pp. 116-117, Feb. 2020. https://doi.org/10.1109/ISSCC19947.2020.9063081
[2] T. Ali et al, "A 460mW 112Gb/s DSP-Based Transceiver with 38dB Loss Compensation for Next-Generation Data Centers in 7nm FinFET Technology", ISSCC, pp.118-119, Feb. 2020. https://doi.org/10.1109/ISSCC19947.2020.9062925
[3] M. LaCroix et al., "A 116Gb/s DSP-Based Wireline Transceiver in 7nm CMOS Achieving 6pJ/b at 45dB Loss in PAM-4/Duo-PAM-4 and 52dB in PAM-2", ISSCC, pp.132-133, Feb. 2021. https://doi.org/10.1109/ISSCC42613.2021.9366030
[4] Z. Guo et al., "A 112.5Gb/s ADC-DSP-Based PAM-4 Long-Reach Transceiver with>50dB Channel Loss in 5nm FinFET", ISSCC, pp. 116 117, Feb. 2022. https://doi.org/10.1109/ISSCC42614.2022.9731650
[5] B. Zhang et al., "A 112Gb/s Serial Link Transceiver With 3-tap FFE and 18-tap DFE Receiver for up to 43dB Insertion Loss Channel in 7nm FinFET Technology", ISSCC, pp.108-110, Feb.2023. https://doi.org/10.1109/ISSCC42615.2023.10067657
[6] Z. Wang et al., "A 65nm CMOS, 3.5-to-11GHz, Less-Than-1.45LSB-INLpp, 7b Twin Phase Interpolator with a Wideband, Low-Noise Delta Quadrature Delay-Locked Loop for High-Speed Data Links", ISSCC, pp.292-294, Feb.2022. https://doi.org/10.1109/ISSCC42614.2022.9731649

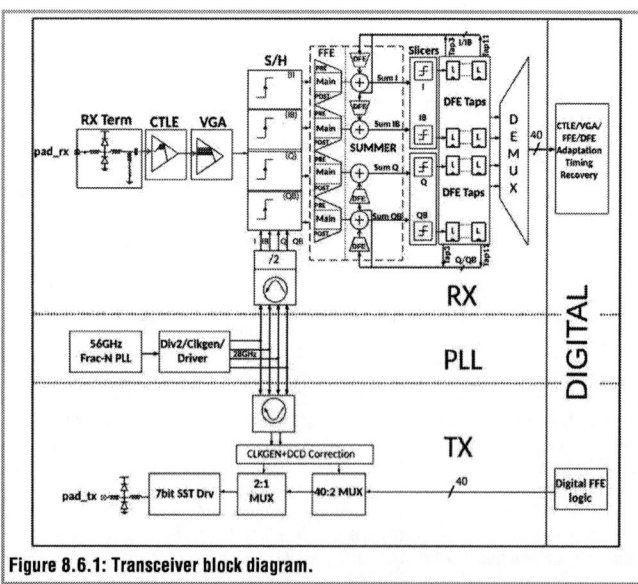

Figure 8.6.1: Transceiver block diagram.

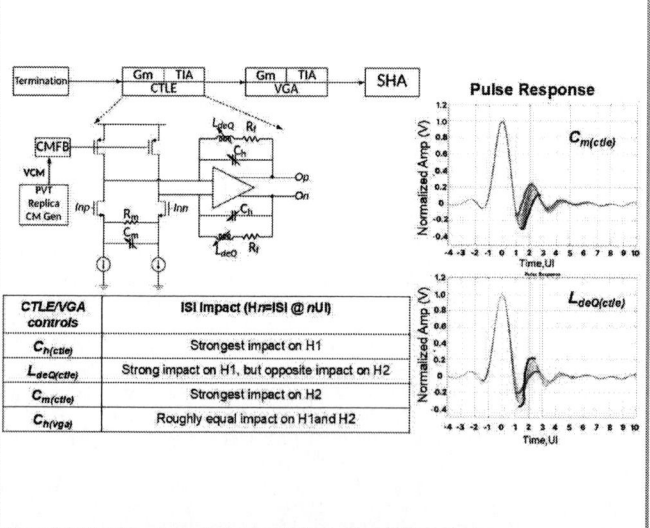

Figure 8.6.2: RX implementation and ISI impact of AFE parameters.

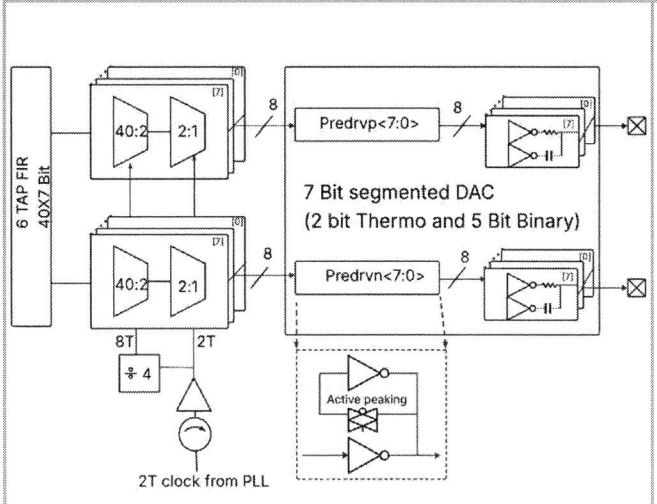

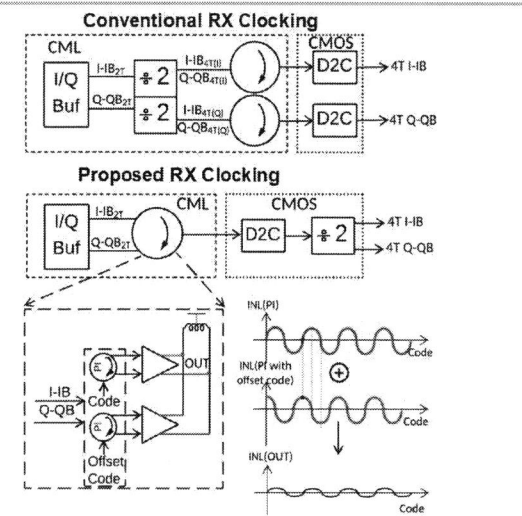

Figure 8.6.3: Transmitter block diagram.

Figure 8.6.4: Conventional and proposed RX clocking with INL cancelled phase interpolator circuit.

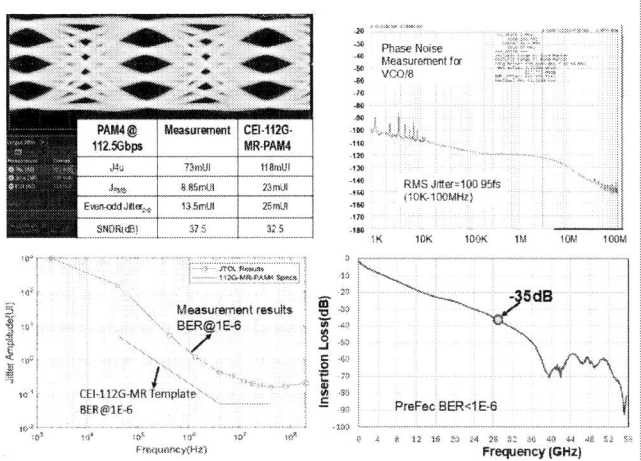

Figure 8.6.5: Measured transmitter PAM-4 eye diagram at 112.5Gb/s, PLL phase noise, jitter tolerance for 112Gb/s PAM-4 and equalized channel insertion loss.

Design	[1]	[2]	[3]	[4]	[5]	This work
Technology	7nm	7nm	7nm	5nm	7nm	5nm
Data Rate	112Gbps	112Gbps	112Gbps	112Gbps	112Gb/s	112Gbps
RX Topology	7-bit ADC	7-bit ADC	7-bit ADC	7-bit ADC	Analog CTLE/FEE/DFE	Analog CTLE/FEE/DFE
TX Topology	7-bit DAC	7-bit DAC	7-bit DAC	7.5-bit DAC	7-bit DAC	7-bit DAC
RX DFE tap	1	1	2	1	18	11
RX FFE tap	31	24	25	30	3	3
TX FFE tap	4	6	7	6	6	6
TX RLM	NA	0.99	NA	0.96	0.99	0.991
TX SNDR	NA	36dB	NA	36dB	37dB	37.5dB
Channel Loss	37.5dB (BER<1e-8)	38.9dB (BER<5e-7)	45dB (BER<1e-5)	50dB (BER<1e-6)	43dB (BER<1e-5)	35dB (BER<1e-6)
AFE+DSP Power/lane	602mW*	460mW*	662mW	504mW	690mW	256*280mW
FOM(pJ/Bit/dB)	0.14+DSP	0.106+DSP	0.131	0.09	0.143	0.07
Area mm²/lane	0.41*	0.385*	0.53*	0.49	0.47*/0.63	0.33*/0.43

*: Analog only

Figure 8.6.6: Performance comparison 112G PAM-4 transceivers.

Figure 8.6.7: Chip photograph of the transceiver.

- 2026 IEEE International Solid-State Circuits Conference

ISSCC 2026 / SESSION 8 / DIE-TO-DIE AND HIGH-SPEED ELECTRICAL TRANSCEIVERS / 8.7

8.7 A 112Gb/s PAM-4 SBD Transceiver with Mismatch-Compensated 2×VDD Hybrid and Two-Step Echo Canceller in 28nm CMOS

Huanfa Sun, Shangjie Wei, Yu Su, Chenyao Cao, Yujie Zeng, Yukun He, Zhouchi Duan, Guohe Zhang, Xiaoyan Gui

Xi'an JiaoTong University, Xi'an, China

Abstract

A 112Gb/s PAM-4 simultaneous bidirectional (SBD) transceiver in 28nm CMOS is presented. It features a hybrid with a 2×VDD stacked driver to restore signal swing, a joint delay and slew-rate matching scheme to eliminate dynamic glitches, and a two-step echo canceller to

mitigate reflections. The transceiver achieves BER < 1E-10 over a 12.7dB loss channel (equivalent to 24.4dB at Nyquist), with 1.73pJ/b energy efficiency and an FoM of 0.14pJ/b/dB.

The relentless growth in data traffic demands higher I/O bandwidth density, making simultaneous bidirectional (SBD) signaling attractive for doubling channel throughput and halving equivalent channel loss, positioning it as a promising solution for energy-efficient 112Gb/s communication over medium-reach channels. However, prior SBD transceivers face significant trade-offs. ADC-based solutions [1] can tolerate high channel loss at the cost of prohibitive power consumption due to the digital intensive echo cancellers (ECs). SBD transceivers using wide linear range (WLR) hybrids [2] suffer from large echo-cancellation residuals and output swing loss, handling only low channel loss of 5.6dB, and thus fails to meet the >20dB medium-reach (MR) link budget required by 100G OIF-CEI standards [3]. R-gm-based hybrids [4] are typically limited to lower-speed NRZ signaling up to 32Gb/s SBD. Further, the inherent output swing loss of ~34% and dynamic glitches of resistive hybrids [5], plus the eye-height compression of more than 1/3 in PAM-4 signaling, severely degrade the signal integrity of the transmission link. This paper presents a 112Gb/s PAM-4 SBD transceiver (TRX) in 28nm CMOS with a hybrid circuitry employing a 2×VDD stacked driver to restore the signal swing and an adjustable output impedance to counteract process variations. A joint delay and slew-rate matching scheme is introduced to eliminate dynamic glitches incurred by the misalignment between the main- and replica driver, while a power-efficient, two-step EC mitigates the long-latency reflections. The SBD TRX achieves robust operation over a 12.7dB loss channel (equivalent to an insertion loss of 24.4dB at 28 GHz Nyquist), achieving an energy efficiency of 1.73pJ/b and the best FoM (0.14pJ/b/dB) among PAM-4 SBD TRXs.

Figure 8.7.1 shows the architecture block diagram of the SBD TRX, comprising four primary blocks: a 1/4-rate transmitter (TX), a half-rate receiver (RX), a hybrid circuit, and an EC. An on-chip PRBS generator provides 4MSB and 4LSB data streams, which are retimed and then serialized into the full-rate differential data by a 4:1 multiplexer (MUX). Meanwhile, a precise replica of the MSB and LSB streams is routed to the EC for echo reconstruction and cancellation. The hybrid, composed of adjustable delay lines, pre-drivers, 2×VDD stacked high-swing output drivers, and an on-chip termination network, couples the TX signal onto the channel while actively cancelling out the outbound signal at the receiver port (RXA), delivering the far-end inbound signal to the RX front-end with minimized local TX interference. In the RX path, the inbound signal passes through a continuous-time linear equalizer (CTLE) to compensate for the channel loss, and a variable-gain amplifier (VGA) further boosting the signal amplitude. An EC generates a delayed and scaled replica of the outbound signal and subtracts it from the received signal, to suppress the residual echoes. Finally, the echo-cancelled signal is sampled and demuxed, with the sampling clock generated by the clock and data recovery (CDR) circuitry. An external 14GHz clock is split into two paths: the divide-by-two 7GHz quadrature (I/Q) clock in the TX, and the 14GHz I/Q clock for the phase interpolator (PI) in the RX, respectively.

High-speed SBD TRXs with PAM-4 modulation impose stringent linearity requirements on the hybrid circuitry. A resistor-based hybrid has drawn considerable interests in PAM-4 signaling transmissions due to its inherent high linearity and low power overhead [5]. Nevertheless, a key drawback of the resistive hybrid is the inherent signal attenuation, which reduces the outbound TX signal amplitude and attenuates the inbound received signal, both leading to a significant degradation of the signal-to-noise ratio. To overcome this limitation, we propose a stacked driver with a 2×VDD supply, which effectively doubles the driver's output amplitude. As shown in Fig. 8.7.2(a), the hybrid circuitry consists of a main driver, a replica driver and a resistive network, which includes a main path resistor (R_{dr}), a replica path resistor (R_{rep}), and a hybrid resistor (R_h). To provide 50Ω output impedance matching and minimum outbound signal residual at the local RX node (RXA/B in Fig. 8.7.2(a)), the required values of R_{dr} and R_{rep} are plotted as a function of R_h in Fig. 8.7.2(b). As R_h increases, R_{rep} is required to increase linearly, while R_{dr} decreases slightly to maintain a 50Ω impedance. As plotted in Fig. 8.7.2(c), this approach is theoretically limited by a maximum inbound-signal amplitude of $0.64V_{pp}$ at node RXA/B with a 0.9V supply in 28nm CMOS for instance, even with large R_h values (>550Ω). The proposed stacked driver employs a level-shifting pre-driver to drive the stacked PMOS and NMOS output stages. In addition, variable MOS resistors are integrated into the PMOS and NMOS paths, which calibrate the resistive network, in combination with R_{dr} and R_{rep} in the main and replica driver, respectively. The drain-to-source voltage (V_{ds}) of each transistor in the stacked output stage is maintained below 1V, preventing device breakdown. For the design case with $R_h = 100Ω$ in Fig. 8.7.2(c),

the stacked driver boosts the equivalent output swing from 0.54 to $1.08V_{pp}$. Another critical challenge for the resistive hybrid is its susceptibility to process-induced resistor variation. Simulations in Fig. 8.7.2(d) show that a 20% deviation of all resistors can degrade the self-interference cancellation (SIC) by up to 8dB. To address this, the variable MOS resistor arrays are calibrated to compensate for process variations. With a 12.7dB channel loss and a 10% resistor mismatch, the residual outbound signal at node RXA/B is suppressed by more than 3.6dB, and a 5.3dB improvement in SIC is achieved even with a 20% mismatch.

Moreover, the transition-edge-induced glitches at the RX node (Fig. 8.7.3(a)) in the resistive hybrid design arise from mismatched charge/discharge dynamics between the main- and replica-driver paths, and thus severely corrupt the inbound signal integrity since they appear at random positions in the inbound data. Mitigating these glitches requires precise matching of both delay and slew-rate between the two paths. As illustrated in Fig. 8.7.3(b), this work employs a joint delay and slew-rate matching scheme. A voltage-controlled delay line (VCDL) is embedded in both the main and replica paths for delay matching, providing an adjustable delay range of ±5ps with a 0.4ps resolution. In addition, the slew-rate mismatch between the main and replica paths, caused by unequal driving current and disparate loadings, is compensated through a coarse-fine tuning scheme. A dummy capacitor in the replica path provides coarse matching to mimic the main path load, while varactors at the pre-driver stage of both the main and replica path enable fine-tuning of the slew rate. Simulation results in Fig. 8.7.3(c) show that with this joint matching scheme, the residual glitches are suppressed from 135mV to 48mV, improving the SIC from 18.9dB to 26.02dB. The table in Fig. 8.7.3 summarizes the cancellation improvement from the proposed matching scheme. Figure 8.7.3(d) and (e) compare the eye diagrams of the 56Gb/s inbound signal at the RX node after a 12.7dB channel loss in SBD operation. With the joint matching scheme, the eye height (EH) improves from 23 to 154mV, and the eye width (EW) is extended from 0.17 to 0.54UI.

In SBD links, reflections from impedance discontinuities at interfaces generate echoes mingled with the inbound signal, which also degrade signal integrity. Conventional long-latency echo cancellers suffer from substantial power consumption, with delay modules in the mixed-signal EC consuming up to 0.38pJ/b [4] and ADC-based EC solutions requiring up to 2pJ/b [1]. This work presents a high-speed, low-power echo canceller (EC) employing a two-step architecture, as illustrated in Fig. 8.7.4(a), which partitions the total echo delay path into coarse and fine stages. The coarse stage processes data at a lower rate, enabling the use of small-sized D flip flops (DFFs) for reduced power consumption. It consists of 31 DFFs and provides a programmable delay of 128UI in total, by increment of 4UI. The fine-tuning stage utilizes a 4:1 MUX for phase selection, allowing it to reuse the TX clock generation. Signals tapped from the selected coarse delay node are fed into the 4:1 MUX, which performs fine-delay selection with 1UI resolution, with the timing diagram depicted in Fig. 8.7.4(b). A 4-tap echo cancellation (EC) scheme is implemented, consisting of a 1-tap near-end (NE) canceller and a 3-tap far-end (FE) canceller. Reduced DFF size and improved delay-chain efficiency yield substantial power savings. The replicated echoes from the delay chain, with the weighting determined by coefficients N1–N4, are subtracted from the incoming signal by the EC summer at the RX. As shown in Fig. 8.7.4(c), the time-domain reflectometry (TDR) measurement shows that the channel exhibits a 22.5mV near-end echo, with far-end echoes of up to 13mV at 56UI delay. Simulations with the 4tap EC, shown in Fig. 8.7.4(d), verify that the residual voltage is reduced from 49.2 to 29.4mV and the SIC is improved from 26.2 to 31.3dB.

The transceiver was fabricated in a 28nm CMOS process. Figure 8.7.5(a) and (b) show the measured eye diagrams at the TX output, driving 1.27V and 1.29V differential swings for PAM-4 and NRZ signaling, respectively. For the 112Gb/s PAM-4 SBD measurement, two transceivers communicate over a channel with 12.7dB insertion loss, as shown in Fig. 8.7.5(c), equivalent to 24.4dB at the 28GHz Nyquist frequency for 112Gb/s PAM-4 MR uni-direction transmission. Figure 8.7.5(e) presents the measured bathtub curves. With the proposed mismatch compensation and EC enabled, a bit-error rate (BER) of 1E-10 is achieved. For the 56Gb/s NRZ SBD measurement, the transceivers were tested over a 23.7dB loss channel as presented in Fig. 8.7.5(d). An eye margin up to 0.11UI is observed at a 1E-10 BER, as shown in Fig. 8.7.5(f).

Figure 8.7.6 compares the performance with prior SBD transceivers, highlighting that this work operates at 112Gb/s SBD with 12.7dB channel loss at 14GHz Nyquist frequency, and achieves 1.73pJ/b energy efficiency. This work features the best equalization power efficiency (FoM) of 0.14pJ/b/dB compared to prior SBD arts with ≥5dB channel loss, and the highest TX output swing of 1.27V_{ppd} and inbound swing at RX of 1.08V_{ppd} at similar data rates. Further, compared to 112Gb/s uni-direction transceivers for MR applications, this work exhibits the best energy efficiency of 1.73pJ/b and an FoM of 0.072pJ/b/dB with an equivalent channel loss of 24.4dB at the 28GHz Nyquist frequency. Figure 8.7.7 shows the die micrograph.

Acknowledgement:
This work was supported by the National Natural Science Foundation of China (NSFC) under Grant 62174132 and 92573110. The authors appreciate valuable discussions with Dr. Bingyi Ye. Corresponding author: Xiaoyan Gui (xy.gui@xjtu.edu.cn).

References:
[1] R. Farjadrad et al., "An Echo-Cancelling Front-End for 112Gb/s PAM-4 Simultaneous Bidirectional Signaling in 14nm CMOS," ISSCC, pp. 194-196, Feb. 2021. https://doi.org/10.1109/ISSCC42613.2021.9365852
[2] Y. Lee et al., "An 80-Gb/s PAM-4 Simultaneous Bidirectional Transceiver With Hybrid Adaptation Scheme," IEEE TCAS-II, vol. 70, no. 8, pp. 2884-2888, Aug. 2023. https://doi.org/10.1109/TCSII.2023.3253679
[3] Implementation Agreement OIF-CEI-05.2. Accessed: Aug. 2024. [Online]. https://www.oiforum.com/wp-content/uploads/OIF-CEI-05.2.pdf
[4] Y.-H. Fan et al., "A 32-Gb/s Simultaneous Bidirectional Source-Synchronous Transceiver With Adaptive Echo Cancellation Techniques," IEEE JSSC, vol. 55, no. 2, pp. 320-332, Feb. 2020. https://doi.org/10.1109/JSSC.2019.2956369
[5] Y. Nishi et al., "A 0.297-pJ/Bit 50.4-Gb/s/Wire Inverter-Based Short-Reach Simultaneous Bi-Directional Transceiver for Die-to-Die Interface in 5-nm CMOS," IEEE JSSC, vol. 58, no. 4, pp. 1062-1073, Apr. 2023. https://doi.org/10.1109/JSSC.2022.3232024
[6] Y. Tomita et al., "A 20Gb/s Bidirectional Transceiver Using a Resistor-Transconductor Hybrid," ISSCC, pp. 2102-2111, Feb. 2006. https://doi.org/10.1109/ISSCC.2006.169627
[7] Z. Wang et al., "A 64Gb/s/wire 10.5Tb/s/mm/Layer Single-Ended Simultaneous Bi-Directional Transceiver with Echo and Crosstalk Cancellation for a Die-to-Die Interface in 28nm CMOS," ISSCC, pp. 588-590, Feb. 2025. https://doi.org/10.1109/ISSCC49661.2025.10904631
[8] T. Ali et al., "A 460mW 112Gb/s DSP-Based Transceiver with 38dB Loss Compensation for Next-Generation Data Centers in 7nm FinFET Technology," ISSCC, pp. 118-120, Feb. 2020. https://doi.org/10.1109/ISSCC19947.2020.9062925
[9] B. Ye et al., "A 2.29pJ/b 112Gb/s Wireline Transceiver with RX 4-Tap FFE for Medium-Reach Applications in 28nm CMOS," ISSCC, pp. 118-120, Feb. 2022. https://doi.org/10.1109/ISSCC42614.2022.9731591
[10] L. Zhong et al., "A 112Gb/s/pin Single-Ended Crosstalk-Cancellation Transceiver with 31dB Loss Compensation in 28nm CMOS," ISSCC, pp. 134-136, Feb. 2024. https://doi.org/10.1109/ISSCC49657.2024.10454508
[11] Y.-P. Lin et al., "A 2.06pJ/b 106.25Gb/s PAM-4 Receiver with 3-Tap FFE and 1-Tap Speculative DFE in 28nm CMOS," ISSCC, pp. 146-148, Feb. 2025. https://doi.org/10.1109/ISSCC49661.2025.10904743

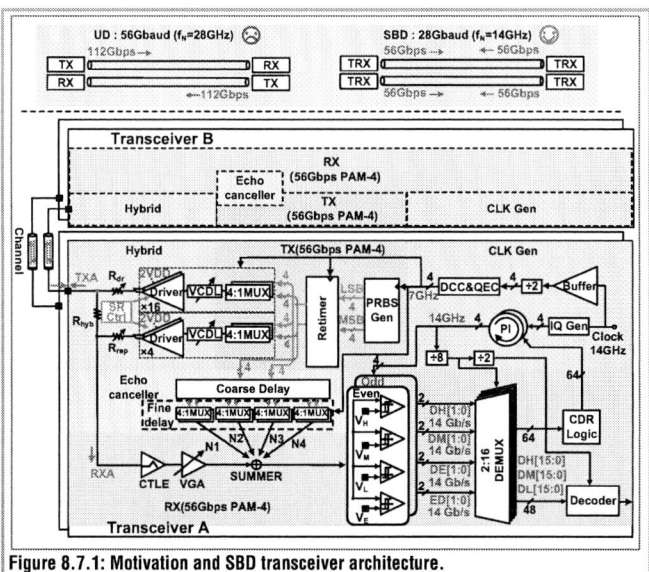

Figure 8.7.1: Motivation and SBD transceiver architecture.

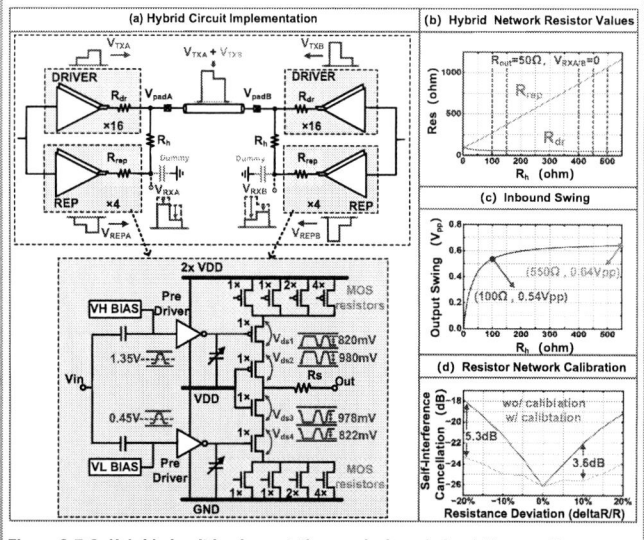

Figure 8.7.2: Hybrid circuit implementation, analysis and simulation results.

ISSCC 2026 / SESSION 8 / DIE-TO-DIE AND HIGH-SPEED ELECTRICAL TRANSCEIVERS / 8.7

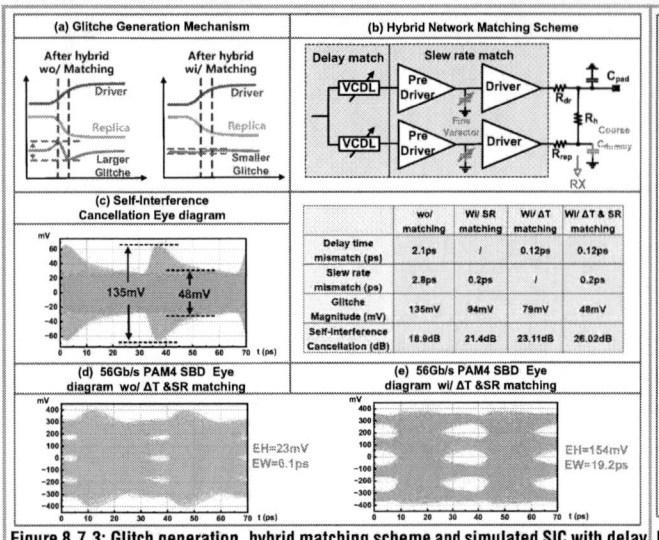

Figure 8.7.3: Glitch generation, hybrid matching scheme and simulated SIC with delay and SR matching.

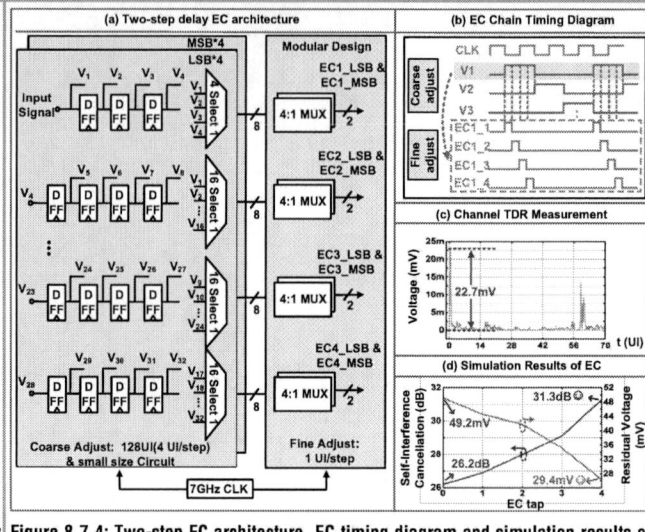

Figure 8.7.4: Two-step EC architecture, EC timing diagram and simulation results of the EC.

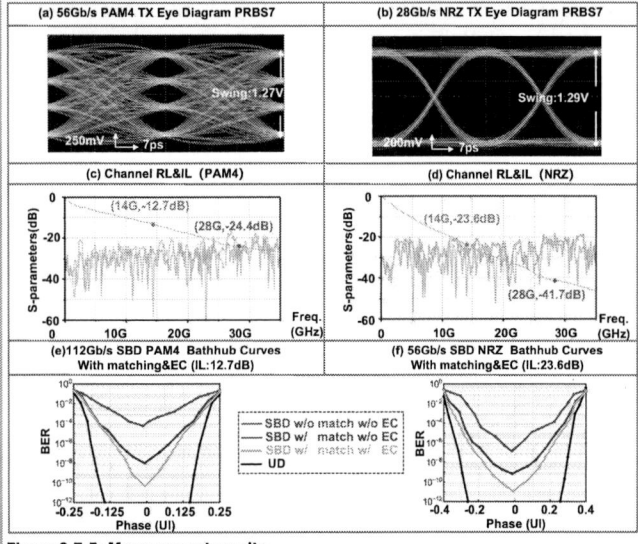

Figure 8.7.5: Measurement results.

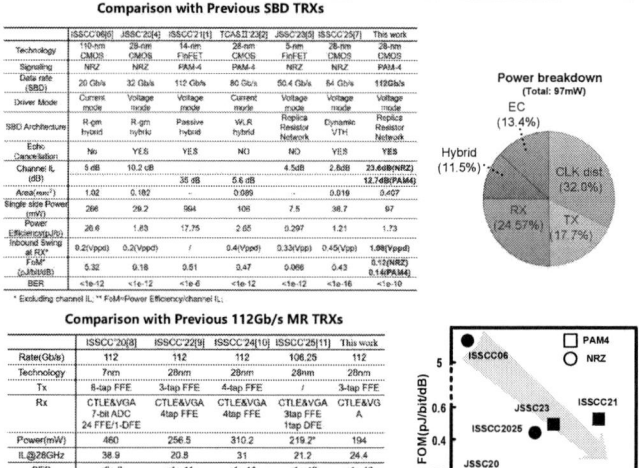

Figure 8.7.6: Performance comparisons with prior works and power breakdown.

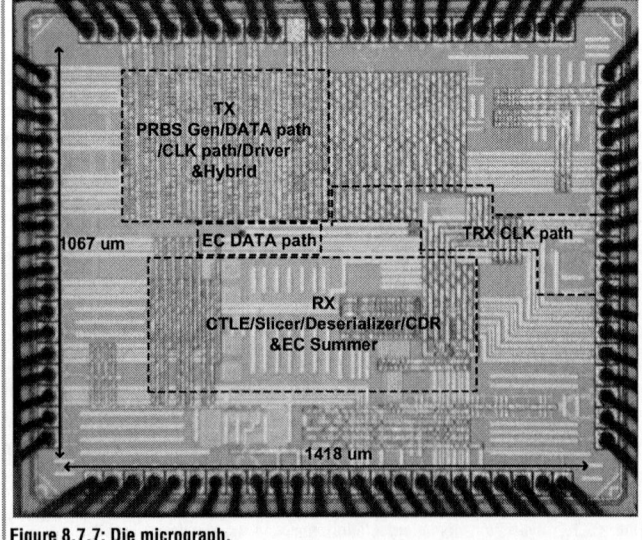

Figure 8.7.7: Die micrograph.

• 2026 IEEE International Solid-State Circuits Conference

979-8-3315-8937-0/26 $31.00 © 2026 IEEE

ISSCC 2026 / SESSION 8 / DIE-TO-DIE AND HIGH-SPEED ELECTRICAL TRANSCEIVERS / 8.8

8.8 A 0.292pJ/b 56Gb/s/wire Capacitively Driven Simultaneous Bidirectional Transceiver with PVT/Mismatch Tracking for XSR and D2D Interfaces in 28nm CMOS

Kahyun Kim, Yoona Lee, Daehoon Na, Ha-Jung Park, Jeongeun Song, Woo-Seok Choi

Seoul National University, Seoul, Korea

Abstract

A low-power (0.292pJ/b), high-bandwidth (56Gb/s/wire) single-ended capacitively driven simultaneous bi-directional (CD-SBD) TRX with PVT tolerance is proposed. Capacitive driving reduces power and self-interference in SBD. AC/DC replica and PVT/mismatch tracking provides PVT robustness without extra hardware. Per-lane CDR and a quadrature generator with QEC/DCC, secure clock margin. The TRX sustains <1e-12 BER across -6.5 to -11.5dB loss, 0.93-1.2V supply, and 0-80°C temperature variation.

Low power consumption and high bandwidth are two major priorities in XSR and die-to-die (D2D) interfaces. Figure 8.8.1(top) shows 3 signaling candidates to achieve 2× bandwidth at the same channel density as in XSR. Capacitively driven PAM-4 (CD-PAM-4) is power-efficient and provides inherent equalization (EQ) as it uses capacitive driving [1-3]. However, the eye height (EH) is only 1/3 of CD-NRZ [4], leading to a degraded SNR. PAM-4 further degrades the eye margin due to switching jitter (SWJ) [5] and complicates CDR [6]. On the other hand, simultaneous bidirectional signaling (SBD) enables simpler CDR and is free from SWJ [7], when using NRZ. However, in RC-dominant channels (e.g., silicon interposers for D2D), the R component causes resistive voltage division, making the inbound (IB) signal markedly smaller than the outbound (OB) signal [8], thereby resulting in harsher self-interference. Furthermore, SBD requires additional EQ and is vulnerable to mismatch between the replica and main driver [9].

In this work, capacitively driven simultaneous bidirectional signaling (CD-SBD) is proposed. CD-SBD delivers the same EH as CD-NRZ, while preserving the low driving power and EQ advantages of capacitive driving. In addition, the OB and IB amplitudes remain equal, since the swing size is determined by the capacitive ratio between driver and channel, rather than by resistive voltage division. The ease of CDR in SBD is also preserved. Nevertheless, the inherent replica mismatch issue of SBD still remains, especially under PVT variations. Figure 8.8.1(bottom left) shows the frequency-dependent mismatch [10] that arises between the main and replica driver. Prior works attempted to address this, but were limited by PVT variations. Reference [8] suffers from both self-interference and PVT. Reference [12] enhances SBD signal integrity, but at the cost of huge hardware overhead. An SBD mismatch adaptation method has been suggested in [11], but it only adjusts gain magnitude and requires external sweeping, making it vulnerable to VT drift. In this paper, a CD-SBD transceiver with an AC/DC replica incorporating PVT/mismatch tracking is proposed. This work suggests to decompose the mismatch into AC (i.e., high-frequency) and DC (i.e., low-to-mid frequency) components. That is, rather than using a single replica and dummy RC, the OB signal is reproduced by combining AC and DC replicas. Proposed edge-sample-based PVT/mismatch tracking enables real-time adaptation of those two weights (Wac, Wdc in Fig. 8.8.1) under VT drift. This enables: (i) good power efficiency by employing CD-SBD and removing dummy RC, (ii) PVT/mismatch tolerance by adapting AC/DC swings in background, and (iii) no additional hardware cost, as it reuses edge data for per-lane CDR.

Figure 8.8.2(top) shows the overall architecture of the proposed 56Gb/s/wire CD-SBD transceiver. The AFE integrates a capacitive driver, AC/DC replica paths, and a hybrid circuit that subtracts the AC/DC replica signal and a crosstalk cancellation (XTC) signal. Each of the 8 data channels incorporates a per-lane CDR, whose information is also used to track AC/DC mismatch. Two clock-forwarding channels are implemented, one per direction, with the clock driven capacitively as data and received via a TIA. In addition to the global CDR, a DLL-based quadrature generator is employed, incorporating a quadrature-error corrector (QEC) and duty-cycle corrector (DCC) with a quadrature-phase detector (QPD). Figure 8.8.2(left) shows the circuit implementation of the AC/DC replica, and corresponding hybrid circuit with XTC. In the replica path, a small inverter delivers a DC signal, while a small MOSCAP extracts the AC components. With peak-to-peak swing resolutions of about 5mV and 1.5mV, the AC/DC replicas are designed to cover gains of about 19dB and 16dB, respectively. Besides subtracting the emulated AC/DC components of the OB signal, the hybrid circuit also cancels out the crosstalk from 2 adjacent channels. Unlike the band-pass behavior in resistively driven interconnects, crosstalk (XT) in CD exhibits a low-pass nature [13]. To mimic the low-pass filtered XT signal, the XTC signal is constructed from the SBD - AC of the two adjacent channels. Since IB is already attenuated through the channel, it inherently resembles the low-pass filtered XT signal. Figure 8.8.2 (bottom) shows how the CD-SBD signal is composed of IB, OB, and XT components, and is restored by removing the XTC, AC, and DC components.

Figure 8.8.3 illustrates the operation of PVT/mismatch tracking. While conventional edge-sample-based CDR uses only edge and IB data, proposed tracking also utilizes OB data that is already known. Figure 8.8.3(top left) shows the case with a DC mismatch. If the replica DC weight is set too large so that '11' DC components are over-subtracted in the hybrid compared to the actual OB signal, the IB transition edge is pulled toward '0'. Thus, the edge is sampled as '0'. Conversely, if the replica DC weight is set too small and the '11' DC

component is under-subtracted, the IB edge is pulled toward '1', and is sampled as '1'. That is, DC mismatch can be tracked using the edge data when the OB data is either '00' or '11'. Similarly, Fig. 8.8.3(bottom left) shows the case with an AC mismatch. When the OB data is '01', an over-estimated AC weight pulls the IB edge towards '0'. Conversely, when the OB data is '10', an over-estimated AC weight pulls the IB edge towards '1', and is sampled as '1'. Figure 8.8.3 (top right) shows the pattern filter table. While the per-lane CDR operates as convention using IB and edge data, the proposed CD-SBD additionally exploits the OB data to track AC/DC mismatch in the hybrid under PVT variation. Figure 8.8.3 (bottom right) compares the RC variation tolerance between the conventional and the proposed schemes. With conventional SBD using a replica and dummy RC, a -20% RC variation in the replica path reduces the eye height (EH) to about 40%. In contrast, the proposed CD-SBD with PVT/mismatch tracking restores EH to nearly 100% under the same condition.

Figure 8.8.4 illustrates the proposed DLL-based quadrature generator with QEC and DCC function. Since the clock is forwarded using a single lane in this work, it is crucial to correct duty cycle and quadrature error to ensure clock quality [14]. Figure 8.8.4(top) shows the block diagram of the quadrature generator and the circuit implementation of the QPD. The durations of Q1-Q4 in the timing diagram correlate with duty and quadrature errors as follows. If the quadrature delay is less than 90°, Q2 and Q4 increase while Q1 and Q3 decrease, and vice versa. Similarly, if the duty cycle modulated by DCC exceeds 50%, Q1 and Q2 increase while Q3 and Q4 decrease, and vice versa. Thus, the sign (+/-) of (Q2-Q3) + (Q4-Q1) indicates the quadrature error polarity, while the sign of (Q2-Q3) + (Q1-Q4) indicates the duty cycle error polarity. The proposed QPD converts (Q2-Q3) and (Q1-Q4) into complementary voltage pairs. Each quadrature interval (Q1-Q4) is translated into pull-up (PU) and pull-down (PD) currents, and the resulting charge is integrated onto the corresponding node (Fig. 8.8.4, top right). Thanks to the complementary PU/PD networks, PN mismatch is canceled between the two nodes, enabling accurate comparison of (Q2-Q3) and (Q1-Q4). Next, the comparators detect the sign of (Q2-Q3) + (Q4-Q1) and (Q2-Q3) + (Q1-Q4), thereby the quadrature phase error and the duty cycle error are precisely detected. The proposed DLL-based quadrature generator with QEC/DCC achieves <0.3% duty cycle error and 230fs quadrature phase error.

The prototype chip is fabricated in a 28nm CMOS process. Since real process variations and mismatch are difficult to quantify and reproduce across chips, intentional variations were introduced into 8 channels to explicitly demonstrate the effects of process variation and mismatch, as illustrated in Fig. 8.8.5 (top). Channels are configured as on-chip metal lines, and additional R and C are inserted in the middle of lanes 3-8 with various amounts to emulate process variation in RC-dominant channels for D2D interfaces [15]. These introduce up to 3dB of additional insertion loss. Furthermore, to realize mismatch between the main and replica drivers, additional R is inserted to the main drivers of the lane 2, 4, 6, and 8, resulting in 2dB of additional loss. All 8 channels are implemented with 2mm long, 0.4μm wide, and 1.8μm pitch. The baseline channel, without additional RC components or driver mismatch, exhibits an insertion loss of -6.5dB at the Nyquist frequency of 14GHz, while the worst-case channel shows an insertion loss of -11.5dB. The worst-case FEXT from adjacent channels is -18.2dB at Nyquist, for all 8 channels.

Figure 8.8.5 (bottom) shows the measured bathtub curves and eye diagram of the proposed transceiver. Bottom left illustrates the bathtub curves of CH0 when (i) CDR, PVT/mismatch tracking, and XTC are all disabled, (ii) only CDR and PVT/mismatch tracking are enabled, and (iii) all are enabled. When all are disabled, the eye margin is measured as 0UI at a BER of 10[-12], whereas enabling all improves it to 0.32UI. With PVT/mismatch tracking enabled, all channels exhibit BER below 10[-12]. The bottom right shows the eye diagram of CH0, with the vertical and horizontal eye margin of 84mV and 0.56UI, respectively, at a BER of 10[-12].

Figure 8.8.6 presents the measured performance of PVT/mismatch tracking under process variation, voltage drift, and temperature drift, respectively. To evaluate process variation tracking, CDR and PVT/mismatch tracking were applied in CH0 and, using the same parameters, the eye diagram was measured in CH3, where process variation was emulated by clock skew, driver mismatch, and RC variation with different channel loss. Without CDR and tracking, the optimum sampling point deviates by ~0.25UI, with the vertical and horizontal eye margin of 28mV and 0.16UI, respectively, although BER at the sampling point is unmeasurable. With CDR and tracking enabled, the sampling point is restored to the optimum position, and the eye margin improves to 80mV and 0.28UI. Similarly, when supply voltage (VDDA, VDDQ) drifts from 1.0V to 0.93V, the eye shifts by ~0.5UI, and the voltage margin reduces to 24mV. After applying CDR and tracking, the eye margin recovers to 64mV and 0.28UI. Figure 8.8.6(right) shows the measured eye height and eye width under supply voltage variation from 0.93V to 1.2V, and ambient temperature variation from 0 to 80°C. Figure 8.8.7 summarizes the TRX performance compared with other state-of-the-art D2D TRXs. Per-lane CDR and PVT/mismatch tracking incur 15.6% area and ($\alpha \times 16.8$)% power overhead, where α denotes the adjustable CDR loop active factor used to save power for low-frequency VT tracking. The area and power overhead of the proposed tracking itself is negligible. This work demonstrates PVT-variation tolerance and apower efficiency of 0.292pJ/b by proposing: (i) CD-SBD with (ii) AC/DC hybrid, (iii) PVT/mismatch tracking, and (iv) DLL-based quad-gen. The die photograph shows the active area of 0.01697mm².

Acknowledgement:
This work was supported in part by Samsung Electronics Co., Ltd., and by Institute of Information & communications Technology Planning & Evaluation (IITP) under the artificial intelligence semiconductor support program to nurture the best talents (IITP-2025-RS-2023-00256081) grant funded by the Korea government(MSIT). The EDA tool was supported by the IC Design Education Center(IDEC), Korea. Woo-Seok Choi (wooseokchoi@snu.ac.kr) is the corresponding author.

References:
[1] S. Lee, J. Yun and S. Kim, "A 78.8fJ/b/mm 12.0Gb/s/Wire Capacitively Driven On-Chip Link Over 5.6mm with an FFE-Combined Ground-Forcing Biasing Technique for DRAM Global Bus Line in 65nm CMOS," 2022 IEEE International Solid-State Circuits Conference (ISSCC), San Francisco, CA, USA, 2022, pp. 454-456. https://doi.org/10.1109/ISSCC42614.2022.9731653
[2] Y. Nishi et al., "A 0.190-pJ/bit 25.2-Gb/s/wire Inverter-Based AC-Coupled Transceiver for Short-Reach Die-to-Die Interfaces in 5-nm CMOS," in IEEE Journal of Solid-State Circuits, vol. 59, no. 4, pp. 1146-1157, April 2024. https://doi.org/10.1109/JSSC.2023.3338478
[3] D. Walter et al., "A source-synchronous 90Gb/s capacitively driven serial on-chip link over 6mm in 65nm CMOS," 2012 IEEE International Solid-State Circuits Conference, San Francisco, CA, USA, 2012, pp. 180-182. https://doi.org/10.1109/ISSCC.2012.6176902
[4] S. Kim et al., "A 0.458-pJ/bit 24-Gb/s/pin Capacitively Driven PAM-4 Transceiver With PAM-Based Crosstalk Cancellation for High-Density Die-to-Die Interfaces," in IEEE Journal of Solid-State Circuits, vol. 59, no. 11, pp. 3730-3740, Nov. 2024. https://doi.org/10.1109/JSSC.2024.3401213
[5] J. Jin et al., "A 4-nm 16-Gb/s/pin Single-Ended PAM-4 Parallel Transceiver With Switching-Jitter Compensation and Transmitter Optimization," in IEEE Journal of Solid-State Circuits, vol. 59, no. 1, pp. 184-195, Jan. 2024. https://doi.org/10.1109/JSSC.2023.3319637
[6] Y. -P. Lin, P. -J. Peng, C. -C. Lu, P. -T. Shen, Y. -C. Jao and P. -H. Hsieh, "A 2.16-pJ/b 112-Gb/s PAM-4 Transceiver With Time-Interleaved 2-b/3-b ADCs and Unbalanced Baud-Rate CDR for XSR Applications in 28-nm CMOS," in IEEE Journal of Solid-State Circuits. https://doi.org/10.1109/JSSC.2025.3562885

[7] R. J. Drost and B. A. Wooley, "An 8-Gb/s/pin simultaneously bidirectional transceiver in 0.35- μm CMOS," in IEEE Journal of Solid-State Circuits, vol. 39, no. 11, pp. 1894-1908, Nov. 2004. https://doi.org/10.1109/JSSC.2004.835837
[8] Y. Nishi et al., "A 0.297-pJ/Bit 50.4-Gb/s/Wire Inverter-Based Short-Reach Simultaneous Bi-Directional Transceiver for Die-to-Die Interface in 5-nm CMOS," in IEEE Journal of Solid-State Circuits, vol. 58, no. 4, pp. 1062-1073, April 2023. https://doi.org/10.1109/JSSC.2022.3232024
[9] Y. Tomita, H. Tamura, M. Kibune, J. Ogawa, K. Gotoh and T. Kuroda, "A 20-Gb/s Simultaneous Bidirectional Transceiver Using a Resistor-Transconductor Hybrid in 0.11-μm CMOS," in IEEE Journal of Solid-State Circuits, vol. 42, no. 3, pp. 627-636, March 2007. https://doi.org/10.1109/JSSC.2006.891719
[10] C. Yuan, A. Naguib and S. Shekhar, "On the Design of Low-Power Hybrids for Full Duplex Simultaneous Bidirectional Signaling Links," in IEEE Transactions on Circuits and Systems I: Regular Papers, vol. 67, no. 4, pp. 1413-1422, April 2020. https://doi.org/10.1109/TCSI.2019.2962359
[11] Y. -H. Fan et al., "A 32-Gb/s Simultaneous Bidirectional Source-Synchronous Transceiver With Adaptive Echo Cancellation Techniques," in IEEE Journal of Solid-State Circuits, vol. 55, no. 2, pp. 439-451, Feb. 2020. https://doi.org/10.1109/JSSC.2019.2956369
[12] R. Farjadrad et al., "An Echo-Cancelling Front-End for 112Gb/s PAM-4 Simultaneous Bidirectional Signaling in 14nm CMOS," 2021 IEEE International Solid-State Circuits Conference (ISSCC), San Francisco, CA, USA, 2021, pp. 194-196. https://doi.org/10.1109/ISSCC42613.2021.9365852
[13] J. Lee, W. Lee and S. Cho, "A 2.5-Gb/s On-Chip Interconnect Transceiver With Crosstalk and ISI Equalizer in 130 nm CMOS," in IEEE Transactions on Circuits and Systems I: Regular Papers, vol. 59, no. 1, pp. 124-136, Jan. 2012. https://doi.org/10.1109/TCSI.2011.2161394
[14] M. -S. Lin et al., "A 32Gb/s 10.5Tb/s/mm 0.6pJ/b UCIe-Compliant Low-Latency Interface in 3nm Featuring Matched-Delay for Dynamic Clock Gating," 2025 IEEE International Solid-State Circuits Conference (ISSCC), San Francisco, CA, USA, 2025, pp. 586-588. https://doi.org/10.1109/ISSCC49661.2025.10904767
[15] M. Choi, J. -Y. Sim, H. -J. Park and B. Kim, "An Approximate Closed-Form Channel Model for Diverse Interconnect Applications," in IEEE Transactions on Circuits and Systems I: Regular Papers, vol. 61, no. 10, pp. 3034-3043, Oct. 2014. https://doi.org/10.1109/TCSI.2014.2327275
[16] Z. Wang et al., "A 64Gb/s/wire 10.5Tb/s/mm/Layer Single-Ended Simultaneous Bi-Directional Transceiver with Echo and Crosstalk Cancellation for a Die-to-Die Interface in 28nm CMOS," 2025 IEEE International Solid-State Circuits Conference (ISSCC), San Francisco, CA, USA, 2025, pp. 588-590. https://doi.org/10.1109/ISSCC49661.2025.10904631
[17] J. Gu et al., "A 32 Gb/s 0.36 pJ/bit 3 nm Chiplet IO Using 2.5-D CoWoS Package With Real-Time and Per-Lane CDR and Bathtub Monitoring," in IEEE Journal of Solid-State Circuits, vol. 60, no. 4, pp. 1289-1298, April 2025. https://doi.org/10.1109/JSSC.2025.3545483
[18] K. Seong et al., "A 4nm 48Gb/s/wire Single-Ended NRZ Parallel Transceiver with Offset-Calibration and Equalization Schemes for Next-Generation Memory Interfaces and Chiplets," 2024 IEEE International Solid-State Circuits Conference (ISSCC), San Francisco, CA, USA, 2024, pp. 250-252. https://doi.org/10.1109/ISSCC49657.2024.10454481

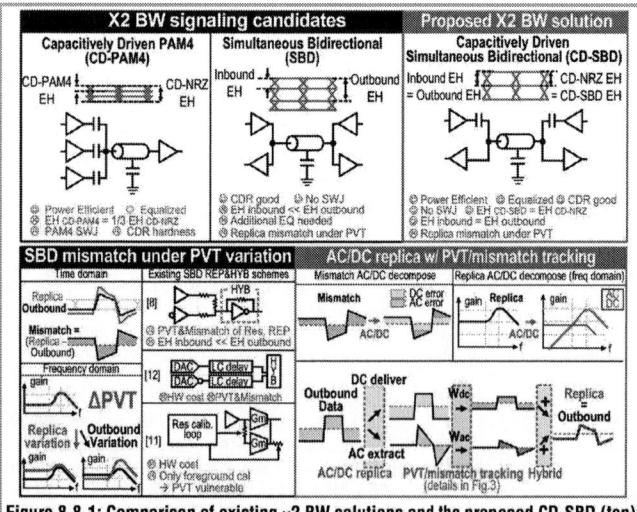

Figure 8.8.1: Comparison of existing ×2 BW solutions and the proposed CD-SBD (top); replica mismatch problem of SBD under PVT variation and the proposed AC/DC replica with PVT/mismatch tracking (bottom).

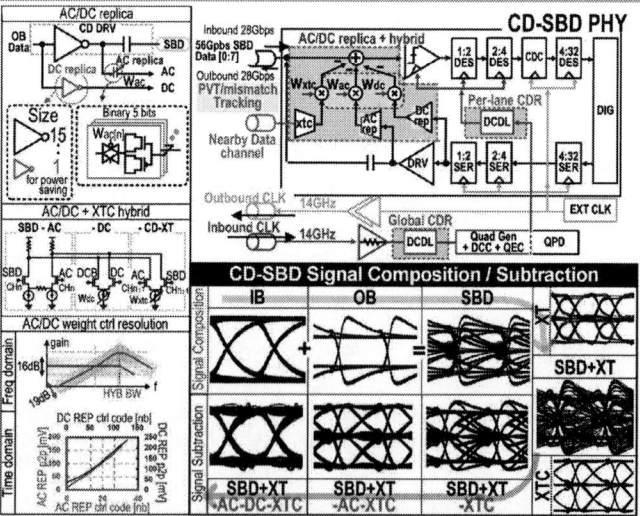

Figure 8.8.2: Proposed CD-SBD TRX (top right); circuit implementation of the AC/DC replica and hybrid (left); signal composition/subtraction principle in CD-SBD (bottom).

ISSCC 2026 / SESSION 8 / DIE-TO-DIE AND HIGH-SPEED ELECTRICAL TRANSCEIVERS / 8.8

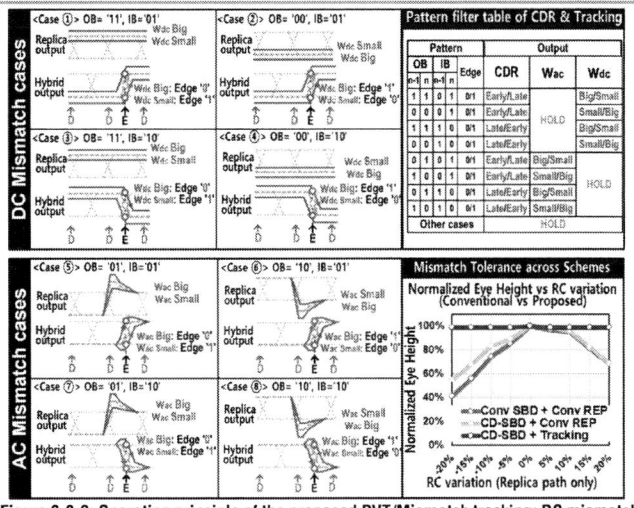

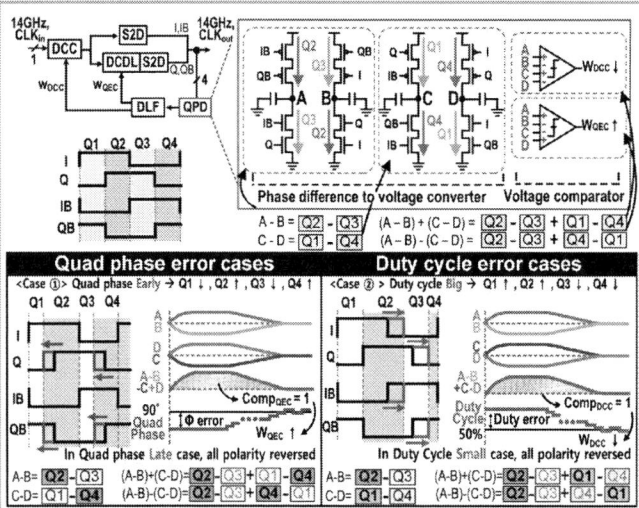

Figure 8.8.3: Operating principle of the proposed PVT/Mismatch tracking: DC mismatch cases (top left), AC mismatch cases (bottom left), pattern filter table (top right) and comparison of mismatch tolerance.

Figure 8.8.4: Proposed quadrature generator with QEC/DCC: block diagram (top left), circuit implementation of QPD (top right), error detection in QEC/DCC (bottom).

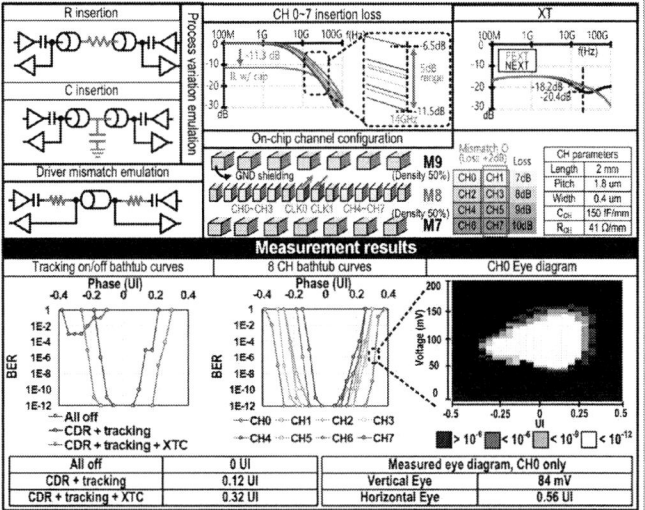

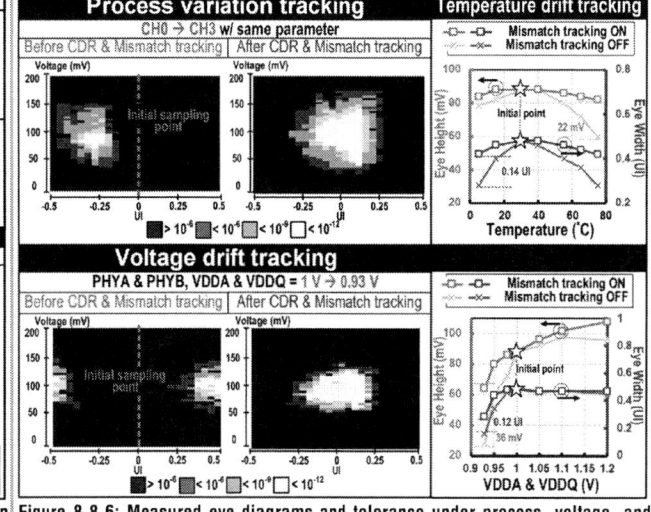

Figure 8.8.5: Channel setup with injected mismatch, process, channel loss variation and channel parameters (top); measured BER bathtub curves and eye diagram.

Figure 8.8.6: Measured eye diagrams and tolerance under process, voltage, and temperature variations.

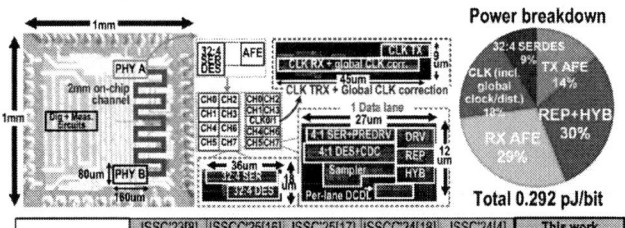

Power breakdown

Total 0.292 pJ/bit

Technology	JSSC'23[8]	ISSCC'25[16]	JSSC'25[17]	ISSCC'24[18]	JSSC'24[4]	This work
	5nm FinFET	28nm CMOS	3nm FinFET	4nm FinFET	28nm CMOS	28nm CMOS
Data rate (Gb/s/pin)	50.4	64	32	48	24	56
Signaling	SBD	SBD	NRZ	NRZ	CD-PAM4	CD-SBD
Channel Loss @ Nyq. Freq. [dB]	-5.4	-2.8	-2.4	-3	-4.3	-6.5 ~ -11.5
PVT variation tracking — CLK skew	X	X	O	X	X	Per-lane CDR, QEC+DCC+4φgen AC/DC replica w/ PVT/Mismatch tracking
PVT variation tracking — Data mismatch	X	X	X	X	X	
XTC	X	O	X	X	O	O
Area (µm²)	11326*	11820*	7000	67833	1010	1697.5**
Energy efficiency (pJ/bit)	0.297	1.21	0.36	0.67	0.458	0.292***

* per-lane area including 2 PHY for SBD.
** per-lane active area of 8 data & 2 clock lanes, including 2 PHY for SBD.
  Calculated by dividing the total active area (including global clock & 2 clock TRX lanes, excluding dig) by 8 data lanes.
*** per-lane power of 8 data & 2 clock lanes, including 2 PHYs for SBD.
  Calculated by dividing the total PHY power (including global clock & 2 clock TRX lanes, excluding dig) by 8 data lanes.

Figure 8.8.7: Chip microphotograph, power breakdown and comparison table.

• 2026 IEEE International Solid-State Circuits Conference

ISSCC 2026 / SESSION 8 / DIE-TO-DIE AND HIGH-SPEED ELECTRICAL TRANSCEIVERS / 8.9

8.9 A 72Gb/s/pin Single-Ended Simultaneous Bi-Directional Transceiver with C-Peaking Leakage Cancellation and Dual-Loop Hybrid Impedance Calibration for Chiplet Interfaces

Xuxu Cheng, Hongzhi Wu, Zhenghao Li, Weitao Wu, Xiongshi Luo, Yangyi Zhang, Quan Pan

Southern University of Science and Technology, Shenzhen, China

Abstract

This paper presents a 72Gb/s/pin single-ended simultaneously bi-directional (SBD) TRX in 28nm CMOS. Capacitive peaking leakage cancellation (CPLC) suppresses the high-frequency leakage due to the main and hybrid driver mismatch in SBD links by 63%. Dual-loop hybrid impedance calibration reduces the coefficient errors of the hybrid circuits under PVT variations by 92%. The TRX achieves a 72Gb/s data rate with an eye-opening of 0.45UI and 243mV and an energy efficiency of 1.5pJ/b.

The explosive growth of artificial intelligence has accelerated the development of high-performance computing and high-bandwidth interconnects. Single-ended parallel links, like chiplet interfaces, are favored for their high bandwidth density and low bit-error rate (BER) [1-5]. Simultaneous bi-directional (SBD) links have shown potential to double the system throughput [6-7]. However, the design of high-speed bi-directional links is challenging. The top-left of Fig. 8.9.1 shows the single-ended SBD interconnect. Signals are simultaneously coupled to both sides of the channel and then decoupled by the hybrid circuits. As shown in the top-right of Fig. 8.9.1, the signal integrity of the SBD link is compromised by two main challenges: (1) high-frequency leakage due to the mismatch between main and hybrid drivers [8]; and (2) the residual outbound signal caused by imprecise coefficients of the hybrid circuits [6]. Consequently, these effects shrink the eye opening and limit the overall link performance. This work presents a 72Gb/s/pin single-ended SBD transceiver (TRX) with capacitive-peaking leakage cancellation (CPLC) to suppress the high-frequency leakage by 63%. A dual-loop hybrid impedance calibration scheme adaptively adjusts the hybrid coefficient, achieving accurate cancellation of the outbound signal with residual error less than 5mV.

The overall TRX architecture is shown in the bottom of Fig. 8.9.1. In the transmitter (TX), the pattern generator outputs 1/8-rate low-speed data, which is fed to an 8:4 multiplexer (MUX) to generate the outbound signal D4 and inversed outbound signal D4B. These two signals are serialized by 4:1 MUXs and distributed to the output stage by the slew rate (SR)-controlled pre-drivers. The output stage employs voltage mode (VM) drivers to maximize signal swing to improve the signal-to-noise ratio (SNR) of the SBD link. 24 main-driver slices and 6 hybrid-driver slices are connected through hybrid resistor R_H, with a CPLC driver to suppress the high-frequency leakage. The dual-loop hybrid impedance calibration technique calibrates the output impedance and hybrid impedance of drivers. The receiver (RX) receives the output signal of the hybrid circuit and amplifies it by the inverter-based analog front-end (AFE). To mitigate the channel inter-symbol interference (ISI), the RX adopts a single-ended track and hold feed-forward equalizer (TAH FFE) with feedthrough cancellation that eliminates the feedthrough error during the holding phase. The output of the TAH FFE is sampled by the quarter-rate slicer, deserialized by the 1:2 DEMUX, and finally fed into the pattern checker for BER measurement.

Figure 8.9.2 shows the conventional leakage cancellation scheme and the proposed CPLC scheme. The SBD signal fed to the hybrid is superimposed by the inbound and outbound signals. The relative timing of the inbound and outbound signal transition edges depend on the channel delay. In the worst case, the high-frequency leakage adds directly at the center of the inbound signal eye. The conventional leakage cancellation technique (top-left of Fig. 8.9.2) using the tunable LC delay line [8] aligns edges of the inbound and outbound signals, thereby limiting the residual leakage to the edge regions of the inbound signal to improved eye-opening. However, it introduces significant jitter to the inbound signal, and the implementation of the passive LCs occupy large area. FIR-based leakage cancellation in the middle-left of Fig. 8.9.2 utilizes digital signal processing (DSP) to detect the leakage and adjust cancellation coefficients to minimize residual leakage. However, since the leakage is primarily high-frequency noise, the FIR-based solution only eliminates the leakage partially. To suppress high-frequency leakage, a VM hybrid circuit with CPLC is proposed as shown in the top-right of Fig. 8.9.2. The CPLC driver extracts high-frequency components from the input signal to generate reversed leakage signal $V_{Leakage,B}$, which is then fed forward to the hybrid's output for the cancellation of $V_{Leakage}$. To enable precise elimination of leakage, the strength of pull-up and pull-down of CPLC driver is controlled by the 6b switchable header and footer. Additionally, a series inductor is adopted to broaden the bandwidth and reduce the leakage. The simulated waveforms of leakage are shown in the middle-right of Fig. 8.9.2. By using the series inductor and CLPC, the amplitude of the leakage signal is suppressed by 63% (from 148mV to 55mV). The bottom of Fig. 8.9.2 shows the simulated eye diagrams of the output signal of the hybrid circuit. The eye is completely closed without the series inductor and CPLC. The eye of the output signal is slightly opened with an eye height of 38mV and eye width of 0.23UI by introducing the series inductor. When the CPLC is further enabled, the eye-opening is improved to an eye height of 134mV and an eye width of 0.65UI. The eye-opening is improved by 6 times, ensuring sufficient SNR for SBD links.

The accuracy of the hybrid coefficients across PVT variations is critical for effective cancellation of the outbound signal. Thus, hybrid coefficient adaptation is required to improve the robustness of SBD links. In the VM hybrid circuit, the hybrid coefficients are determined by the output impedance ratio between the main driver and the hybrid driver, along with an appropriate hybrid resistor R_H value [6]. As shown in the top-left of Fig. 8.9.3, conventional impedance calibration uses an impedance control loop shared by the main and hybrid drivers to match their impedance to a precise off-chip resistor, thereby ensuring an accurate ratio between them. However, the fixed output impedance of drivers fails to compensate for the R_H variation, resulting in incomplete cancellation of the outbound signal. To address this issue, this work introduces a dual-loop hybrid impedance calibration scheme that calibrates the impedance of hybrid driver according to the impedance of R_H and main driver. As depicted in the top-right of Fig. 8.9.3, Loop#1 calibrates the pull-up (PU) driver and pull-down (PD) driver branches to a high-precision off-chip resistor, generating the bias voltage V_{BP} and V_{BN} for the replica main driver. In Loop#2, two operational amplifiers adjust the bias voltage of the hybrid driver based on the impedance of the replica main driver and R_H. This process forces the node V_{RX} to remain at VDD/2 for both 'high' and 'low' states of the outbound signal, thereby achieving cancellation of the outbound signal. The bottom-left of Fig. 8.9.3 shows the simulation results of the conventional impedance calibration scheme and the proposed dual-loop hybrid impedance calibration scheme when varying R_H. With the conventional scheme, the residual hybrid error is up to 60mV as R_H varies by 20%. In contrast, the proposed scheme ensures that the impedance of the hybrid driver tracks R_H variation, minimizing the error to below 5mV, reducing by 92% compared to the conventional scheme. Transient simulation waveforms shown in the bottom-middle of Fig. 8.9.3 further confirm that the proposed calibration scheme significantly reduces residual error compared to the conventional approach. The simulated output impedance is shown in the bottom-right of Fig. 8.9.3. The output impedance can be adjusted within a range from 38 to 63Ω by configuring the number of active driver slices, thereby enabling impedance matching to the channel.

To mitigate ISI in the channel, a TAH FFE is employed in the RX side, as illustrated in the top of Fig. 8.9.4. The TAH FFE is implemented using an inverter-based design to reduce power consumption while achieving high data rate. A single-ended transmission gate is adopted for the track and hold switch due to its wide input and output common-mode range and strong immunity to clock feedthrough. However, during the holding phase, data feedthrough occurs if the input signal D_{in} changes, thereby affecting the holding signal. The feedthrough can be alleviated by using the differential signals, but with high power consumption. To address this issue, a feedthrough cancellation (FTC) tap is introduced in the RX TAH FFE. As shown in the timing diagram in the top-right of Fig. 8.9.4, taking the THIB path as an example, during the holding phase, the transition edge between data symbols D2 and D3 feeds through to THIB, which introduces an error on the holding value. When the post tap (THQ) is enabled, the timing margin for the FFE operation is 1.5UI; during this period, the error is further amplified. The FTC tap is introduced to eliminate the feedthrough by using the THQB signal. Simulation results are shown in the bottom of Fig. 8.9.4. The optimal clock timing is set for TAH operation. The eye-opening of the TAH signal is degraded by the ISI and feedthrough. With the post tap of RX FFE enabled only, the ISI is eliminated but the feedthrough remains. With the FTC tap further enabled, the feedthrough is eliminated, and the eye-height is improved from 233 to 285mV. The overshooting is outside the TAH FFE timing margin, and thus it has little impact on the link performance.

The proposed SBD TRX is fabricated in 28nm CMOS and evaluated using a PCB. Figure 8.9.5 shows the measurement results. In the uni-directional (UD) mode, the TX output eye exhibits a height of 189mV and a width of 20.7ps at 36Gb/s, which is 214mV and 25.7ps at 32Gb/s. In the bi-directional mode, the TRX is measured over an off-chip channel with an insertion loss of 6dB at 18GHz. The measured single-channel RX bathtub curve at data rate of 72Gb/s (36Gb/s in each direction) is shown in the top-middle of Fig. 8.9.5. The TRX reaches error-free margin of 0.2UI at 1e-12 BER when the CPLC and RX TAH FFE are turned off. The error free margin is extended to 0.39UI with the CPLC enabled; it is further extended to 0.45UI with the TAH FFE enabled. As shown in the top-right of Fig. 8.9.5, with both the CPLC and TAH FFE enabled, the RX internal eye height is improved from 131mV to 244mV. Measurement results of the SBD TRX at data rate of 64Gb/s are shown in the bottom-right of Fig. 8.9.5. With the CPLC and TAH FFE enabled, the error free margin at 1e-12 BER is extended from 0.22UI to 0.52UI, and the eye height is increased from 158mV to 263mV. Figure 8.9.7 shows the die photo and power breakdown.

The TRX performance summary and comparison with the prior works are shown in Fig. 8.9.6. With the proposed CPLC, dual-loop hybrid impedance calibration and TAH FFE, the SBD TRX achieves the data rate of 72Gb/s/pin and the vertical eye-opening of 244mV, which compares favorably among prior works.

Acknowledgement:
The work is supported by the Science and Technology Plan of Shenzhen under Grant KJZD20231023100159002 and KJZD20240903100208012, and in part by the SUSTech High-Level Special Funds under Grant G03034K007. Corresponding author: Quan Pan.

References:
[1] K. Seong et al., "A 4nm 32Gb/s 8Tb/s/mm Die-to-Die Chiplet Using NRZ Single-Ended Transceiver With Equalization Schemes And Training Techniques," *ISSCC*, pp. 114–116, Feb. 2023. http://doi.org/10.1109/ISSCC49657.2024.10454481
[2] K. Seong et al., "A 4nm 48Gb/s/wire Single-Ended NRZ Parallel Transceiver with Offset-Calibration and Equalization Schemes for Next-Generation Memory Interfaces and Chiplets," *ISSCC*, pp. 250–252, Feb. 2024. http://doi.org/10.1109/ISSCC49657.2024.10454481
[3] Y. Nishi *et al.*, "A 0.190-pJ/bit 25.2-Gb/s/wire Inverter-Based AC-Coupled Transceiver for Short-Reach Die-to-Die Interfaces in 5-nm CMOS," *IEEE JSSC*, vol. 59, no. 4, pp. 1146-1157, April 2024. http://doi.org/10.1109/JSSC.2023.3338478
[4] D. T. Melek et al., "A 0.29pJ/b 5.27Tb/s/mm UCIe Advanced Package Link in 3nm FinFET with 2.5D CoWoS Packaging," *ISSCC*, pp. 590-592, Feb. 2025. http://doi.org/10.1109/ISSCC49661.2025.10904754
[5] M. -S. Lin et al., "A 32Gb/s 10.5Tb/s/mm 0.6pJ/b UCIe-Compliant Low-Latency Interface in 3nm Featuring Matched-Delay for Dynamic Clock Gating," *ISSCC*, pp. 586-588, Feb. 2025. http://doi.org/10.1109/ISSCC49661.2025.10904767
[6] Y. Nishi et al., "A 0.297-pJ/Bit 50.4-Gb/s/Wire Inverter-Based Short-Reach Simultaneous Bi-Directional Transceiver for Die-to-Die Interface in 5-nm CMOS," *IEEE JSSC*, vol. 58, no. 4, pp. 1062–1073, Apr. 2023. http://doi.org/10.1109/JSSC.2022.3232024
[7] Z. Wang et al., "A 64Gb/s/wire 10.5Tb/s/mm/Layer Single-Ended Simultaneous Bi-Directional Transceiver with Echo and Crosstalk Cancellation for a Die-to-Die Interface in 28nm CMOS," *ISSCC*, pp. 588–590, Feb. 2025. http://doi.org/10.1109/ISSCC49661.2025.10904631
[8] R. Farjadrad et al., "An Echo-Cancelling Front-End for 112Gb/s PAM-4 Simultaneous Bidirectional Signaling in 14nm CMOS," *ISSCC*, pp. 194–196, Feb. 2021. http://doi.org/10.1109/ISSCC42613.2021.9365852
[9] Y.-H. Fan et al., "A 32-Gb/s Simultaneous Bidirectional Source-Synchronous Transceiver With Adaptive Echo Cancellation Techniques," *IEEE JSSC*, vol. 55, no. 2, pp. 439–451, Feb. 2020. http://doi.org/10.1109/JSSC.2019.2956369

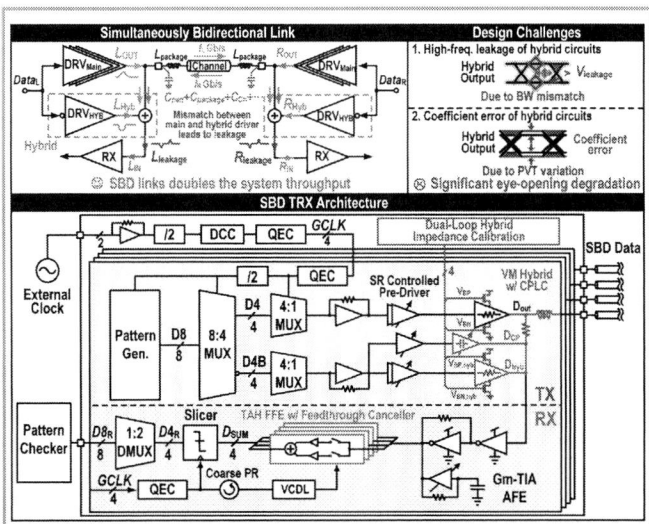

Figure 8.9.1: SBD link and its design challenges (top); SBD transceiver architecture (bottom).

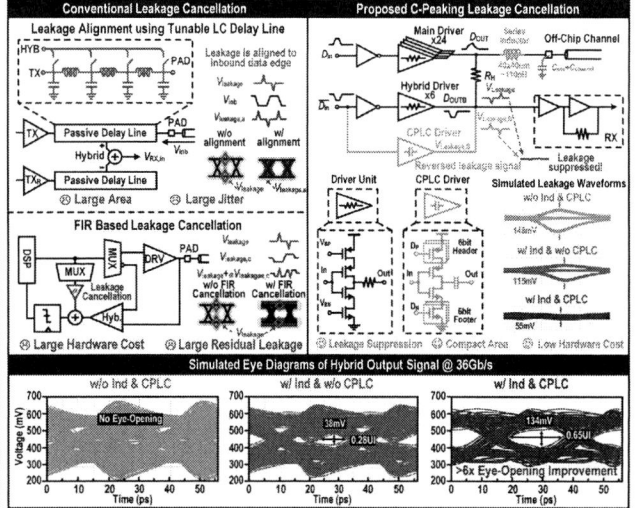

Figure 8.9.2: Conventional leakage cancellation (top-left) and proposed CPLC (top-right); simulated eye diagrams of the SBD links (bottom).

ISSCC 2026 / SESSION 8 / DIE-TO-DIE AND HIGH-SPEED ELECTRICAL TRANSCEIVERS / 8.9

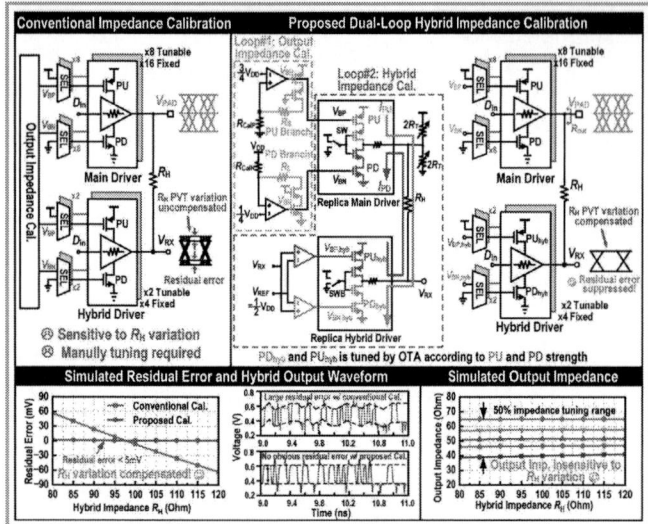

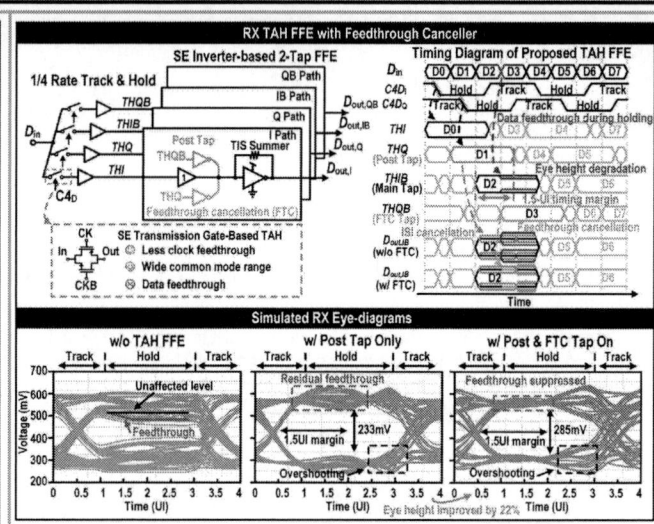

Figure 8.9.3: Conventional impedance calibration (top-left); proposed dual-loop hybrid impedance calibration scheme (top-right); simulation results (bottom).

Figure 8.9.4: Proposed RX TAH FFE with feedthrough cancellation and corresponding timing diagram (top); simulation results of TAH FFE (bottom).

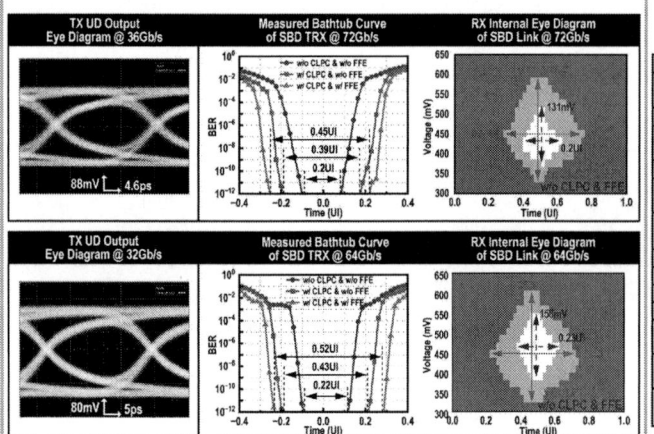

Figure 8.9.5: Measured TX output eye diagrams (left), RX bathtub curves (middle) and RX internal eye diagrams (right).

	JSSC'20 [9]	JSSC'23 [6]	ISSCC'23 [1]	ISSCC'24 [2]	JSSC'25 [3]	ISSCC'25 [7]	This Work
Process	28nm CMOS	5nm FinFET	4nm FinFET	4nm FinFET	3nm FinFET	28nm CMOS	28nm CMOS
Supply	0.9	0.75	0.9	0.9	0.75	0.9/0.45	0.9
Data Rate Per Pin (Gb/s/pin)	32	50.4	32	48	32	64	72
Signaling	SBD NRZ	SBD NRZ	NRZ	NRZ	NRZ	SBD NRZ	SBD NRZ
Energy Efficiency (pJ/bit)	1.83	0.30	0.44	0.67	0.36	1.21	1.50
Channel Loss @ f_{Nyq} (dB)	4.4	2	3.9	3	2.4	2.8	6
Channel Type	Off-Chip	On-Chip	Off-Chip	Off-Chip	Off-Chip	On-Chip	Off-Chip
Equalization	CTLE	No	DFE	C-Peaking	No	No	2-Tap RX FFE
Hybrid Impedance Cal.	No	No	-	-	-	-	Yes
Leakage Cancellation	No	No	No	No	No	No	C-Peaking
Eye Height (mV)	20	NA	170	88	NA	80	244 @ 72Gb/s / 263 @ 64Gb/s
Eye Width (UI)	0.4	0.56	0.53	0.58	0.34	0.64	0.44 @ 72Gb/s / 0.52 @ 64Gb/s
Core Area / Lane (mm²)	0.182	0.004	0.004	0.067	0.34	0.019	0.012

Figure 8.9.6: Performance summary and comparison with prior works.

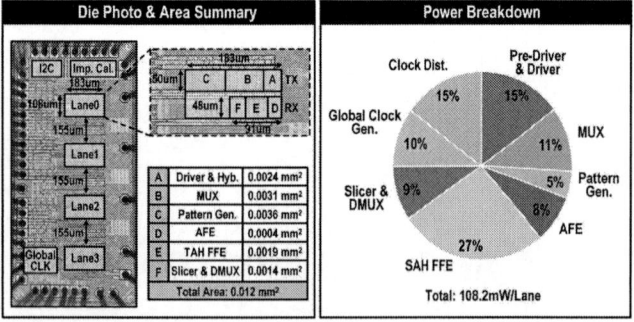

Figure 8.9.7: Die micrograph, area summary, and power breakdown.

• 2026 IEEE International Solid-State Circuits Conference

ISSCC 2026 / SESSION 8 / DIE-TO-DIE AND HIGH-SPEED ELECTRICAL TRANSCEIVERS / 8.10

8.10   A 180-to-240Gb/s Analog-Intensive PAM-4 Transmitter with 0.70pJ/b Analog Power Efficiency in 65nm CMOS

Ziyi Lin, Haikun Jia, Wei Deng, Shitu He, Chang Liu, Zhihua Wang, Baoyong Chi

Tsinghua University, Beijing, China

Abstract

This paper presents a 180-240Gb/s analog-intensive PAM-4 transmitter in 65nm CMOS process. A three-stage cascaded 2-to-1 analog MUX (AMUX) is employed to reduce the complexity and therefore the parasitic capacitance of both output and internal high-speed nodes. Reconfigurable 4-tap FFE, current injection, and dual T-coil techniques are proposed to enhance the output swing and the bandwidth. The transmitter achieves 0.70pJ/b analog energy efficiency at a 240Gb/s data rate.

The rapid growth of applications such as artificial intelligence (AI) and high-performance computing (HPC) has driven strong demand for high-speed transceivers (TRXs) operating beyond 200Gb/s per channel. Several transmitter (TX) implementations in advanced CMOS FinFET processes exceeding 200Gb/s have been reported in [1]-[4]. However, design challenges still exist to pursue further higher data rates and higher energy efficiency. First, either DAC-based or analog-FFE-based TXs require a large driver array, which introduces excessive capacitance and thus limits the bandwidth. Second, unlike the output nodes where inductive peaking is commonly used to enhance the bandwidth, the bandwidth of internal nodes is less considered in literature, which would become a significant bandwidth bottleneck at data rates above 200Gb/s. Third, the conventional low-pass inverter-based clock delivery path faces a stringent jitter performance requirement and consumes a large amount of power.

To address those challenges, we propose an analog-intensive PAM-4 TX architecture with several key circuits to improve the speed and energy efficiency in this work. First, a three-cascaded 2-to-1 analog MUX (AMUX) is employed to reduce the complexity and therefore the parasitic capacitance of both the output and internal high-speed nodes. Second, thanks to the AMUX's ability to preserve the amplitude information, unlike traditional analog-FFE-based TXs, we can partially push the reconfigurable 4-tap FFE operation to low-speed nodes, thus reducing the parasitic capacitance at the output nodes. Third, current injection and dual T-coil techniques are proposed to enhance the output swing and the bandwidth. At last, a transformer-based jitter filtering clock distribution scheme together with replica timing loops is proposed to generate the required low-jitter clocks for MUXs. Leveraging the proposed analog-intensive techniques, a 180-240Gb/s PAM-4 TX with 0.70pJ/b analog energy efficiency is achieved in a 65nm CMOS process.

The architecture of the proposed PAM-4 TX is shown in Fig. 8.10.1. An on-chip PRBS generator produces 256 parallel MSB and LSB signals, which are serialized through 256:8 digital serializers and retimed by FFE retimers. The resulting 14Gb/s single-ended data streams are then converted into differential signals. The MSB and LSB signals are combined in CML adders to generate PAM-4 signals while simultaneously realizing a 2UI-spaced FFE, where 2UI-spaced data are summed by weight in the analog domain. Eight differential PAM-4 signals are multiplexed by two cascaded 2:1 AMUX stages, producing two differential PAM-4 signals at 112Gb/s. In the 3rd AMUX stage, two differential PAM-4 signals are serialized into full-rate differential data with an embedded 2-tap FFE, which is delivered to the output through a dual T-coil network. MUXs in the analog domain are favorable for high-speed design due to the faster switching speed of analog switches in current paths. The bandwidth can be further expanded by inserting inductors. A transformer-based jitter-filtering CML clock distribution network is employed to provide clocks for the 3 AMUX stages. The 56GHz differential input clocks are externally supplied and buffered for the output stage, while 28GHz and 14GHz are generated by on-chip CML dividers. Two 360° PIs are implemented to meet the timing requirements between adjacent AMUX stages. In addition, replica-based timing loops are proposed to automatically adjust the clock phases across different stages.

In the conventional 2-stage AMUX with an inherent 2-tap FFE reported in [5]-[7], 2 differential PAM-4 signals are first converted into 2 interlaced return-to-zero (RZ) signals. An RZ summer then combines these signals to produce full-rate differential data, introducing two additional full-rate nodes, creating a bandwidth bottleneck. In this work, the 2-stage structure is improved by merging them into a single-stage output driver, as shown in Fig. 8.10.2. With only one full-rate node in the merged AMUX, the bandwidth requirement is relaxed, and inductors can be more easily incorporated for bandwidth extension. The inherent 2-tap FFE is preserved without duplicating the array and introducing excessive capacitance. The coefficients and polarity of the FFE can be continuously tuned through the analog bias voltages V_{BP} and V_{BN}. Moreover, depending on the relative phase between clock and data, two different FFE configurations can be realized [8], as illustrated in the bottom-left of Fig. 8.10.2. Two staggered data streams, D1 and D2, are sampled by CKP and CKN, respectively. In CONFIG-A, the first half-UI of D1 and D2 is sampled as the main-tap. Meanwhile, the information of the second half-UI is preserved and summed with the next data as the post-tap, thereby forming a 2-tap FFE of main and post-tap. In CONFIG-B, the first half-UI of D1 and D2 is preserved as the pre-tap, while the second half-UI is sampled

as the main-tap. The pre-tap information is summed with the last data, yielding a 2-tap FFE of main and pre-tap. In addition, a 2-UI spaced FFE is implemented in the digital domain using retimers. By combining two segments of 2-tap FFE, an overall 4-tap FFE is achieved. Depending on the configuration in analog FFE, the 4-tap FFE can be realized either with 3 post-taps or with 1 pre-tap plus 2 post-taps. Compared with conventional 4-tap FFE, the proposed approach lowers retimer and output-array complexity, thereby saving power and improving speed. Part of the FFE array is shifted to low-stage stages thanks to the analog-domain operation, thereby reducing the capacitive load in high-speed stages. Simulation results show that, compared with a 2-tap FFE, the proposed architecture achieves a 14% eye height and a 9% eye width improvement at 224Gb/s PAM-4 operation.

In the proposed 2-to-1 output stage, part of the DC current is consumed for RZ generation without contributing to the output swing. This lowers the common-mode voltage and drives the output transistor into the linear region, thereby degrading both output swing and termination impedance. To address this, a pair of injection current sources is introduced at the output node to raise the common-mode level and boost the output swing without increasing the supply voltage, as shown in Fig. 8.10.3. The size of the current source is carefully optimized. Sufficient currents should be provided with a large output resistance while not introducing excessive capacitance. With an appropriate injection current, a 50Ω output resistance can be achieved. The low output resistance of the driver transistors and current sources would decrease the overall output resistance under a lower or higher injection current, respectively. An optimized DC gain and bandwidth can also be achieved under the appropriate current. However, the injection current source introduces significant parasitic capacitance. To mitigate this, a dual T-coil network is proposed. Compared with the conventional T-coil network, the peaking inductor is split into two magnetically coupled inductors that resonate with the additional capacitance introduced by the current source. Using a high inductance density stacked inductor topology, the dual T-coil occupies only 160×70μm². Compared with the conventional single T-coil network, the proposed network achieves a higher bandwidth of 63.5GHz with in-band gain flatness within 1.5dB.

A transformer-based clock distribution network is proposed to provide clocks for the AMUX stages, as illustrated in Fig. 8.10.4. A pair of half-rate differential clocks is externally supplied. The transformer-based clock path covers the 45-60GHz input frequency range, corresponding to a 180-240Gb/s data rate. Two CML dividers operating at 45-60GHz and 22.5-30GHz, respectively, generate lower-speed clocks. Two PIs are incorporated to satisfy the timing requirements between adjacent AMUX stages. The clock path adopts a band-pass characteristic to filter the clock jitter and deliver sufficient gain while minimizing power consumption. To adaptively tune the two PIs, a replica-based timing loop is proposed, whose block diagram is shown in the middle-left of Fig. 8.10.4. The purpose of the loop is to align the relative phase of CK_{S2} with CK_{S1} to ensure optimal sampling of D2 by CK_{S2}. MUX replicas are used, with inputs tied to V_{DD} and V_{SS}, respectively, and sampled by CK_{S1}. The replica output, D_{REF}, is therefore a clock signal with the same phase as D2. D_{REF} is then fed into a quadrature error detector (QED) and compared with the clock CK_{DIV}, which operates at the same frequency as D_{REF} and has a fixed delay from CK_{S2}. The delay is designed to be approximately 0.25UI of D2. When D_{REF} and CK_{DIV} are in quadrature, CK_{S2} samples D2 at its midpoint. The QED outputs, DETP and DETN, are compared, and the results are counted and passed through a low-pass filter. The loop subsequently adjusts the phase of the 8b PI until DETP equals DETN, indicating that D_{REF} and CK_{DIV} are properly quadrature-aligned. Both simulation and measurement results confirm this behavior, showing a sinusoidal relationship between the PI phase and the QED output. The measured output eye diagram presents optimal quality near the zero-crossing point.

The proposed analog-intensive PAM-4 TX has been fabricated in a 65nm CMOS technology, occupying an area of 1.7mm×1.1mm including pads, with a core area of 0.52mm². The die micrograph is shown in Fig. 8.10.7. The analog power excluding digital serializers is 169mW (0.70pJ/b) from a 1.2V supply when operating at 240Gb/s PAM-4. The entire chip consumes 307mW, largely due to the power-hungry digital serializers and PRBS generator in 65nm CMOS, which can be easily scaled down using smaller process nodes. The performance of the TX is measured through a probe, cables, DC blocks, and a sampling scope, with approximately 4dB loss at 56GHz. The eye diagram is equalized using a 3-tap scope FFE. The measured PRBS-15 eye diagrams in Fig. 8.10.5 show the operating range of 180-

240Gb/s. The injection current technique effectively boosts the output swing, achieving 440mV with a 99.8% relative level mismatch (RLM). The PAM-4 eye diagrams with and without the on-chip 4-tap FFE validate its effectiveness. A performance summary and comparison with prior works are provided in Fig. 8.10.6. Despite being implemented in a planar 65nm CMOS process, this work achieves the highest reported data rate and best analog energy efficiency. The area consumption of the proposed TX is also competitive.

Acknowledgement:
The authors would like to thank the Wuhan National Optoelectronics Innovation Center and the Beijing Innovation Center for Future Chip for their measurement support.

References:
[1] J. Kim *et al.*, "A 224Gb/s DAC-Based PAM-4 Transmitter with 8-Tap FFE in 10nm CMOS," *ISSCC*, pp. 126-128, Feb. 2021.
http://doi.org/10.1109/ISSCC42613.2021.9365840
[2] J. Q. Wang *et al.*, "A 2.69pJ/b 212Gb/s DSP-Based PAM-4 Transceiver for Optical Direct-Detect Application in 5nm FinFET," *ISSCC*, pp. 123-125, Feb. 2024.
http://doi.org/10.1109/ISSCC49657.2024.10454275
[3] M. Cusmai *et al.*, "A 224Gb/s sub pJ/b PAM-4 and PAM-6 DAC-Based Transmitter in 3nm FinFET," *ISSCC*, pp. 126-128, Feb. 2024.
http://doi.org/10.1109/ISSCC49657.2024.10454558
[4] A. Mostafa *et al.*, "A 2.2pJ/b 212.5Gb/s PAM-4 Transceiver with >46dB Reach in 5nm FinFET," *ISSCC*, pp. 138-140, Feb. 2025.
http://doi.org/10.1109/ISSCC49661.2025.10904591
[5] H. Ramon *et al.*, "A 100-GS/s Four-to-One Analog Time Interleaver in 55-nm SiGe BiCMOS," *IEEE JSSC*, vol. 56, no. 8, pp. 2539-2549, Aug. 2021.
http://doi.org/10.1109/JSSC.2021.3057575
[6] Z. Lin *et al.*, "A 112Gb/s Analog-MUX-based PAM-4 Transmitter with Inherent 2-tap FFE in 65nm CMOS," *ISCAS*, pp. 1-4, May 2025.
http://doi.org/10.1109/ISCAS56072.2025.11043913
[7] R. Tang *et al.*, "A Low-Latency 200Gb/s PAM-4 Heterogeneous Transceiver in 0.13μm SiGe BiCMOS and 28nm CMOS for Retimed Pluggable Optics," *ISSCC*, pp. 594-596, Feb. 2025. http://doi.org/10.1109/ISSCC49661.2025.10904643
[8] F. Chen *et al.*, "A 56-Gbaud 7.3-Vppd Linear Modulator Transmitter with AMUX-Based Reconfigurable FFE and Dynamic Triple-Stacked Driver in 130-nm SiGe BiCMOS," *CICC*, pp. 1-2, Apr. 2024,. http://doi.org/10.1109/CICC60959.2024.10529032
[9] X. Zheng *et al.*, "A 50–112-Gb/s PAM-4 Transmitter With a Fractional-Spaced FFE in 65-nm CMOS," *IEEE JSSC*, vol. 55, no. 7, pp. 1864-1876, July 2020.
http://doi.org/10.1109/JSSC.2020.2987712
[10] M. Choi *et al.*, "An Output-Bandwidth-Optimized 200Gb/s PAM-4 100Gb/s NRZ Transmitter with 5-Tap FFE in 28nm CMOS," *ISSCC*, pp. 128-130, Feb. 2021. http://doi.org/10.1109/ISSCC42613.2021.9366012
[11] K. Sheng *et al.*, "A 128Gb/s PAM-4 Transmitter with Programmable-Width Pulse Generator and Pattern-Dependent Pre-Emphasis in 28nm CMOS," *ISSCC*, pp. 120-122, Feb. 2023. http://doi.org/10.1109/ISSCC42615.2023.10067407

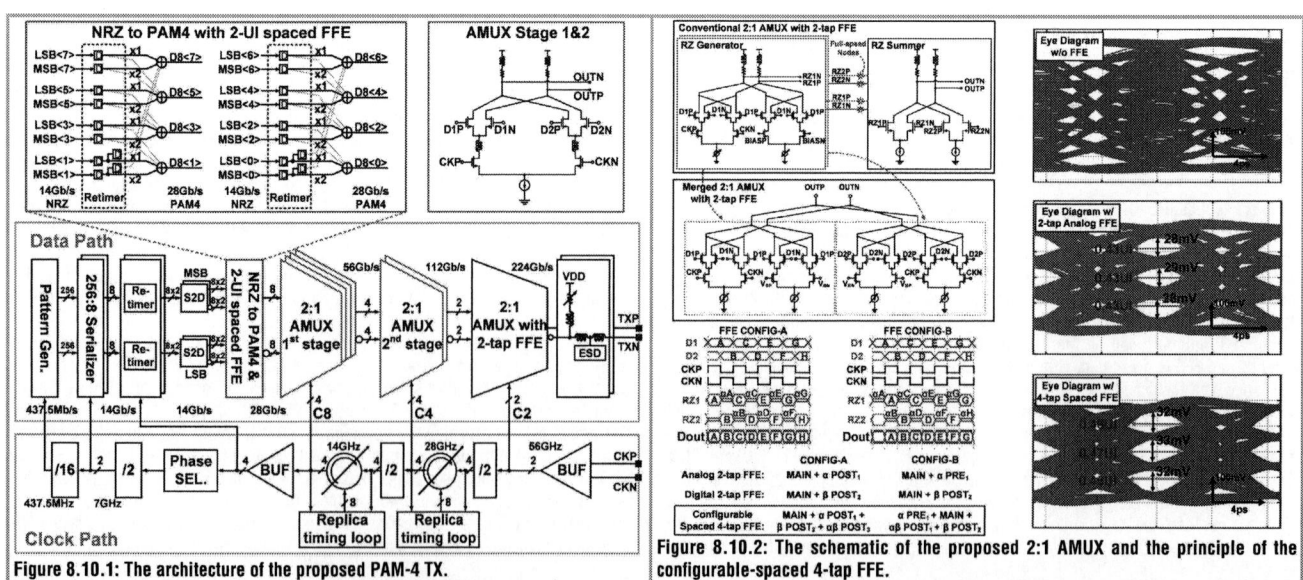

Figure 8.10.1: The architecture of the proposed PAM-4 TX.

Figure 8.10.2: The schematic of the proposed 2:1 AMUX and the principle of the configurable-spaced 4-tap FFE.

ISSCC 2026 / SESSION 8 / DIE-TO-DIE AND HIGH-SPEED ELECTRICAL TRANSCEIVERS / 8.10

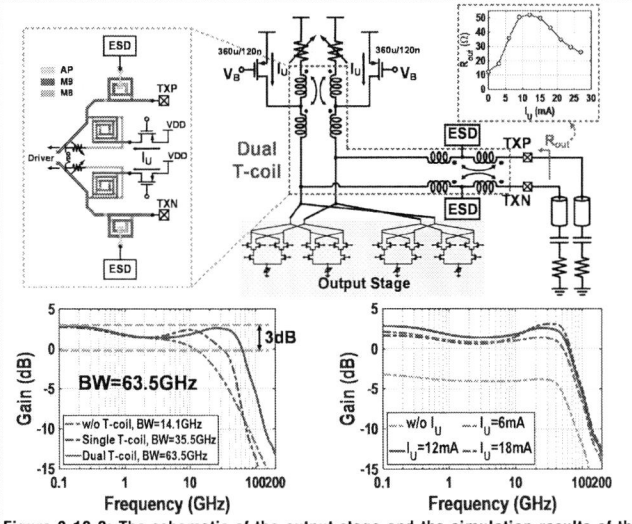

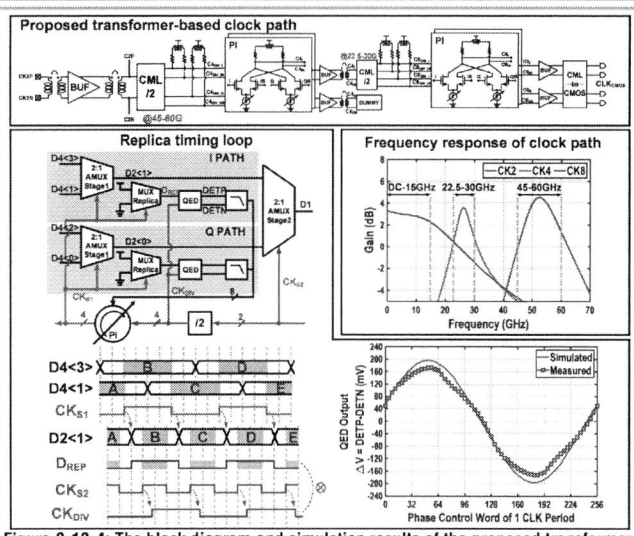

Figure 8.10.3: The schematic of the output stage and the simulation results of the frequency response.

Figure 8.10.4: The block diagram and simulation results of the proposed transformer-based clock distribution network and replica-based timing loop.

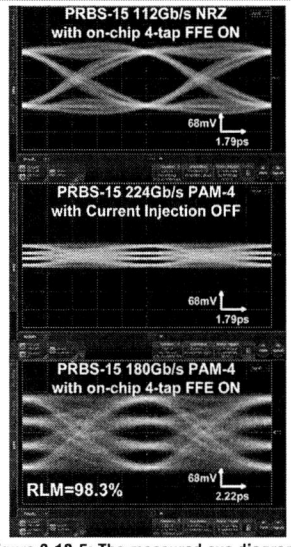

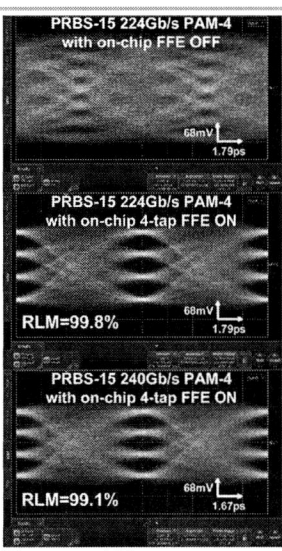

Figure 8.10.5: The measured eye diagrams of the proposed TX.

	This work	ISSCC'21[1]	ISSCC'24[2]	ISSCC'24[3]	ISSCC'25[4]	JSSC'20[9]	ISSCC'21[10]	ISSCC'23[11]
Technology	65nm CMOS	10nm CMOS	5nm CMOS	3nm CMOS	5nm CMOS	65nm CMOS	28nm CMOS	28nm CMOS
Architecture	Half-rate	Quarter-rate.	Quarter-rate	Quarter-rate	Quarter-rate	Quarter-rate	Quarter-rate	Quarter-rate
No. of FFE Tap	4	8	N/A	9	10	2	5	4
Serialization Ratio	512:1	64:1	128:1	64:1	64:1	N/A	256:1	64:1
Modulation	PAM-4	PAM-4	PAM-4	PAM-4	PAM-4	PAM-4	PAM-4	PAM-4
ESD	YES	YES	YES	YES	YES	YES	YES	NO	N/A
Data Rate (Gb/s)	180-240	224	212	224	212	112	180	200	128
RLM (%)	99.8	99	98	97	N/A	99.7	99	99	96
Output Swing (mV)	440	1000	780	1000	560**	600**	260	340	840
Supply Voltage (V)	1.2	0.85/1/1.5	0.65/0.9/4/1.2	0.75/0.9/1.2	N/A	1.2	N/A	N/A	N/A
Analog Power (mW)	169*	340	282	206	182	N/A	N/A	N/A	N/A
Total Power (mW)	307	473	N/A	233	N/A	243	826	926	157
Analog Efficiency (pJ/b)	0.70*	1.7	1.3	0.92	0.86	N/A	N/A	N/A	N/A
Total Efficiency (pJ/b)	1.3	2.1	N/A	1.04	N/A	2.2	4.6	4.6	1.4
Active area (mm²)	0.52	0.088	0.28	0.10	0.29	0.69	0.43	0.137

*Excluding digital serializers
**Estimated from eye diagram

Figure 8.10.6: Comparison table with prior works.

Figure 8.10.7: The die micrograph of the proposed TX.

• 2026 IEEE International Solid-State Circuits Conference

ISSCC 2026 / SESSION 8 / DIE-TO-DIE AND HIGH-SPEED ELECTRICAL TRANSCEIVERS / 8.11

8.11 A 1.59pJ/b 112Gb/s PAM-4 and 1.06pJ/b 168Gb/s PAM-8 Resistor-Less 7-Bit SST DAC-Based Transmitter with 8-Tap FFE in 28nm CMOS

Yao-Hung Tsai, Ming-Han Chuang, Shen-Iuan Liu

National Taiwan University, Taipei, Taiwan

Abstract

A resistor-less 7b SST DAC-based TX with 8-tap FFE and 3 types of segments is presented to achieve low parasitic capacitance, compact area, and scaling friendliness. To reduce inter-symbol-interference jitter at the output of a 4:1 multiplexer, a dual-feedback equalizer is proposed. The TX is fabricated in 28nm CMOS with a maximum output swing of $1.1V_{ppd}$. The power dissipation is 178mW at 112Gb/s PAM-4 and 168Gb/s PAM-8, and energy efficiency is 1.59pJ/b at PAM-4 and 1.06pJ/b at PAM-8.

To support higher serial-link data rates, >100Gb/s transmitters (TXs) [1-8] often employ digital-to-analog converters (DACs) [3-8] for multi-level pulse-amplitude-modulated (PAM) signaling. Current-mode logic (CML) DAC-based TXs [3], [4] are adopted due to their wide bandwidth and good supply noise rejection. Source-series termination (SST) DAC-based TXs [5-8] have been developed due to their low power, large swing, and high linearity characteristics.

The top of Fig. 8.11.1 shows several SSTSs with digitally controlled redundancies to calibrate the output resistance over process, voltage, and temperature (PVT) variations. The top left of Fig. 8.11.1 shows a double-stacked SSTS, which consists of a double-stacked inverter and a passive resistor [5-7]. Both the header and the footer of the inverter are used for calibration. An unstacked SSTS consists of N plus M slices, as shown in the top middle of Fig. 8.11.1, where N and M represent the numbers of the enabled and redundant ones, respectively. Each slice contains NAND/NOR gates, an unstacked inverter, and a passive resistor [8]. For calibration, the redundant slices are used. However, the double-stacked inverter, the redundant slices, and the NAND/NOR gates cause additional parasitic capacitance and power. The passive resistor occupies large area and is not scaling-friendly. To solve these issues, the proposed resistor-less SSTS is shown in the top right of Fig. 8.11.1. It consists of an unstacked inverter and digitally controlled transmission gates (TGs), which are controlled by two K-bit binary calibration codes, $Cal_{P,N}$ [K−1:0].

To demonstrate that the resistor-less SSTS can further reduce input parasitic capacitance, the comparison of the 7b DACs realized by the unstacked SSTSs and the resistor-less ones is given. For unstacked SSTSs, to neglect the redundant slices, N = 1 and M = 0, and NAND/NOR gates are not considered. For the resistor-less SSTSs, K = 5. The widths of $M_{1,2}$ and $M_{P0,N0}$ are multiplied by the scaling factor (SF). The aspect ratios of M_1 and M_{P0} are the same, and those of M_2 and M_{N0} are the same. For a given SF, the passive resistor R_{res} and the 5b TGs are adjusted to keep the output resistance at about 50Ω. Assume the differential DACs can cancel even-order harmonics. The output voltage of the 7b differential DAC is expressed as $V_{out} = 2(m_1x + m_3x^3 + m_5x^5)$, where $x = D_{in}/127 - 0.5$, $-0.5 \le x \le 0.5$, D_{in} is a decimal value of the DAC input code $[B_6, B_5, ..., B_0]$, $0 \le D_{in} \le 127$, and $m_{1,3,5}$ are coefficients. The bottom left of Fig. 8.11.1 shows the simulation results of m_3. When the DAC input code varies, m_3 causes the third-order harmonic distortion (HD3), where m_3 is contributed by the code-dependent resistance variation of the inverters and TGs. As the SF increases, $|m_3|$ from both the unstacked SSTSs and the inverters of resistor-less SSTSs decreases, and those m_3 are negative. For the TGs of the resistor-less SSTSs, those m_3 are positive. Thus, for SF = 1, the combined m_3 from the inverters and the TGs of the resistor-less SSTSs is close to zero. For SF = 1.84, m_3 from the unstacked SSTSs is also close to zero. The bottom right of Fig. 8.11.1 shows the calculated and simulated INL_{max} of two DACs realized by the unstacked SSTSs and the resistor-less SSTSs, respectively. Both the unstacked SSTSs with SF = 1.84 and the resistor-less SSTSs with SF = 1 can reach $INL_{max} \approx 0.1$ LSB, but the required widths of the resistor-less SSTSs are reduced by 1.84 times which improves both area and input parasitic capacitances. If the redundant slices of the unstacked SSTSs [8] are further considered, additional area and input parasitic capacitance are contributed. Assume the double-stacked SSTSs have a 2× larger inverter size compared with unstacked one. That is, the size of resistor-less SSTSs is 3.68 times smaller than the double-stacked ones. The TGs contribute the positive HD3 to cancel the negative HD3 from the inverters, thereby further reducing input parasitic capacitance. This lowers the power of the previous stage of the inverter of SSTSs. By using the TG instead of the passive resistor, the ESD-induced voltage stress on TG should be considered.

The proposed resistor-less 7b SST DAC-based TX with 8-tap FFE is shown in the top of Fig. 8.11.2. It consists of pseudo-random binary sequence (PRBS) generators, an 8-tap lookup-table (LUT)-based finite-impulse-response (FIR) filter [9], the 7b DAC, the output impedance calibration (OIC) circuit, a CML-to-CMOS level converter, a single-ended-to-differential (S2D) converter, a quadrature-locked-loop (QLL), a resistor-less phase interpolator (PI), and two dividers. This 7b DAC consists of 64:4 multiplexers (MUXes), S2Ds, 4:1 MUXes, the proposed dual-feedback equalizers (DFEQs), resistor-less SSTSs, T-coils, and electrostatic discharge (ESD) diodes. In [7], to reduce the number of segments, two types of segments with the aspect ratios of 32W/L and W/L are adopted. Although the self-loading effect is mitigated, the calibration between the two types of segments is needed to improve the matching issue. In this work, three types of segments A, B, and C are shown in a single-ended representation in the bottom of Fig. 8.11.2. Segment A consists of an inverter controlled by the bit $\bar{B_i}$, $2 \le i \le 6$, and TG_0. TG_0 is realized by 5 binary-weighted TGs which are controlled by two 5b binary codes, $Cal_{P,N}$[4:0]. The segment B is controlled by the bit $\bar{B_1}$. The aspect ratios of $TG_{0,1,2}$ are identical. The segment C is controlled by the bit $\bar{B_0}$. $TG_{3,4}$ are similar to TG_0 except that they are controlled by two 4b binary codes. The width ratios of $M_{P0,P6-P9}$ and $M_{N0,N6-N9}$ are also given in Fig. 8.11.2. Since the SF is no more than 2 among the segments A, B, and C, the device mismatch is kept low, and the inter-segment calibration is not required. In summary, the proposed 7b DAC is realized by 31-segment A, 1-segment B, and 1-segment C. Consequently, the total number of the segments is reduced from 127 to 33, mitigating the self-loading effect. The impact of loading mismatch between B_{0-2} can be mitigated by adding dummies on $B_{0,2}$.

A 4:1 MUX, a parasitic capacitance C_X, and the proposed DFEQ are illustrated in the top of Fig. 8.11.3. To suppress inter-symbol interference (ISI) caused by internal capacitive loads of V_X and V_Y, the DFEQ is used to boost the signal transition and reduce their DC signal amplitudes. Since there is a delay time between V_{out} and V_X, it is equivalent to performing the on-chip de-emphasis for V_X [10]. To lower C_X, the size of INV₁ should be small. However, it may introduce ISI jitter for V_Y and V_{out}. By adding INV$_{4,5}$, the de-emphasis is realized for V_Y. The simulated eye diagrams of V_{out} without and with INV$_{3-5}$, respectively, for the resistor-less SSTS are shown in the bottom of Fig. 8.11.3. Without the DFEQ, the jitter of V_{out} is quite large. With the DFEQ, the simulated peak-to-peak jitter of V_{out} is reduced to 0.06UI. The transition time from 10% to 90% is equal to 0.52UI.

The OIC circuit is shown in the top of Fig. 8.11.4. It consists of a replica of segment A, two external resistors, R_{ext}, two comparators, and calibration logic. To calibrate the output resistance at the two boundaries of the differential DAC input code, two reference voltages of $0.75V_{DD}$ and $0.25V_{DD}$ are set. The up-branch resistance R_U is only contributed by M_{P0-P5} instead of M_{P0-P5} and M_{N1-N5}. If M_{N1-N5} exist, they are almost turned off. Similar to the down-branch resistance R_D. The R_U and R_D will be calibrated as $R_U = R_U = R_{ext}/3$. The output resistances within the boundaries are slightly larger than those at the boundaries, which is verified by simulation. While considering the process corners, the full range of DAC output voltage, and the temperature from −40°C to 120°C, the simulated output resistance of the resistor-less SST DAC ranges from 46 to 57Ω after calibration. The proposed 6b resistor-less PI is shown in the top of Fig. 8.11.4, which is modified from [11]. It consists of 2 and 16 MUXes in the first stage STG₁ and the second stage STG₂, respectively, 16 resistor-less SSTSs, an inverter with a feedback TG, and a binary-to-thermometer (B2T) decoder. With STG₁ and, STG₂ the interpolated phases are coarsely and finely tuned, respectively. Thanks to a TG feedback inverter, the swing of V_X is reduced to lower drain-to-source resistance variation of the SSTSs to improve linearity. Compared with [11], two modifications are made. One is that the number of the MUXes is reduced from 32 to 2 in STG₁ to lower the capacitive loads for the quadrature clocks, $CK_{I,IB,Q,QB}$. The other is that TGs are used instead of passive resistors to save area.

To generate quadrature-phase clocks, the QLL is shown in the bottom left of Fig. 8.11.4. It consists of two positive delay lines (PDLs), four negative delay lines (NDLs), two buffers (BUFs), a quadrature phase detector (QPD) [12], and an operational transconductance amplifier (OTA). It adopts both the PDLs and NDLs to extend the phase error correction range of the analog loop only. The PDL consists of four inverters and two PMOS varactors. Similarly, the NDL consists of four inverters and two NMOS varactors. The quadrature errors among the $CK_{I,IB,Q,QB}^{14G}$ are detected by the QPD and the OTA to control PDLs and NDLs. This work adopts 2 PDLs in $CK_{I,IB}^{14G}$ path, and 4 NDLs in $CK_{Q,QB}^{14G}$ path. In [12], no delay line is used in the $CK_{I,IB}^{14G}$ path, and 2 NDLs are used in the $CK_{Q,QB}^{14G}$ path. By comparing this work and [12], their common-mode delay between $CK_{I,IB}^{14G}$ and $CK_{Q,QB}^{14G}$ are similar. Tunable differential-mode delay between 2 PDLs and 4 NDLs is increased compared with that of 2 NDLs only. The simulated differential-mode delay range is improved from 29ps to 85ps, corresponding to an output frequency range from 17.5–11.5GHz to 18–7GHz with no more than 0.03UI timing error. The PDL can not only cancel the intrinsic delay of the NDL but also increase the correction range.

154 • 2026 IEEE International Solid-State Circuits Conference

979-8-3315-8937-0/26 $31.00 © 2026 IEEE

Figure 8.11.5 shows the measured eye diagrams at 56Gbaud with 8-tap FFE coefficients. For 56Gb/s PAM-2 with PRBS-15 and insertion loss (IL) of 10dB, the eye width (EW) and eye height (EH) are equal to 0.6UI and 150mV, respectively. For 112Gb/s PAM-4 with QPRBS-15 and an IL of 2.7dB, EW and EH are equal to 0.24UI and 110mV, respectively. The relative level mismatch (RLM) is equal to 0.991. For 144.48Gb/s PAM-6 with PRBS-15 and IL of 2.7dB, EW and EH are equal to 0.12UI and 33mV, respectively. The RLM is equal to 0.975. For 168Gb/s PAM-8 with PRBS-15 and IL of 2.7dB, EW and EH are equal to 0.056UI and 22mV, respectively. The RLM is equal to 0.967.

Figure 8.11.6 shows the measured INL and DNL are within +0.58/−0.49 LSB and within +0.24/−0.68 LSB, respectively. The measured S_{11} is below −10dB from DC to 33GHz. The comparison table in the bottom of Fig. 8.11.6. Figure 8.11.7 shows normalized single-bit responses (SBRs), simulated signal-to-noise-and-distortion ratio (SNDR) and spurious-free dynamic range (SFDR), a die photo, and power breakdown.

Acknowledgement:
The authors would like to thank the Taiwan Semiconductor Research Institute, Hsinchu, for fabricating this chip. The authors also thank the Intelligent & Sustainable Medical Electronics Research Fund of National Taiwan University, the National Science and Technology Council, Taipei, Taiwan, and the TSMC University Shuttle Program for their support.

References:
[1] K. Sheng, W. Gai, Z. Feng, H. Niu, B. Ye, and H. Zhou, "A 128Gb/s PAM-4 transmitter with programmable-width pulse generator and pattern-dependent pre-emphasis in 28nm CMOS," in *IEEE Int. Solid-State Circuits Conf. (ISSCC) Dig. Tech. Papers*, Feb. 2023, pp. 120–122. https://doi.org/10.1109/ISSCC42615.2023.10067407
[2] J. Yang, E. Song, S. Hong, D. Lee, S. Lee, H. Im, T. Shin, and J. Han, "A 100Gb/s 1.6Vppd PAM-8 transmitter with high-swing 3+1 hybrid FFE taps in 40nm," in *IEEE Int. Solid-State Circuits Conf. (ISSCC) Dig. Tech. Papers*, Feb. 2023, pp. 122–124. https://doi.org/10.1109/ISSCC42615.2023.10067452
[3] J. Kim, S. Kundu, A. Balankutty, M. Beach, B. C. Kim, and S. T. Kim, "A 224-Gb/s DAC-based PAM-4 quarter-rate transmitter with 8-tap FFE in 10-nm FinFET," *IEEE J. Solid-State Circuits*, vol. 57, no. 1, pp. 6–20, Jan. 2022. https://doi.org/10.1109/JSSC.2021.3108969
[4] M. Cusmai, N. Familia, E. Kuperberg, M. Nashash, D. Gottesman, and Z. Marcus, "A 0.92-pJ/b PAM-4 and 0.61-pJ/b PAM-6 224-Gb/s DAC-based transmitter in 3-nm FinFET," *IEEE J. Solid-State Circuits*, vol. 60, no. 1, pp. 23–34, Jan. 2025. https://doi.org/10.1109/JSSC.2024.3456672
[5] T. Ali, E. Chen, H. Park, R. Yousry, Y. M. Ying, and M. Abdullatif, "A 460 mW 112 Gb/s DSP-based transceiver with 38dB loss compensation for next-generation data centers in 7 nm FinFET technology," in *IEEE Int. Solid-State Circuits Conf. (ISSCC) Dig. Tech. Papers*, Feb. 2020, pp. 118–120. https://doi.org/10.1109/ISSCC19947.2020.9062925
[6] M. A. Kossel, V. Khatri, M. Braendli, P. A. Francese, T. Morf, and S. A. Yonar, "An 8b DAC-based SST TX using metal gate resistors with 1.4 pJ/b efficiency at 112 Gb/s PAM-4 and 8-tap FFE in 7 nm CMOS," in *IEEE Int. Solid-State Circuits Conf. (ISSCC) Dig. Tech. Papers*, Feb. 2021, pp. 130–131. https://doi.org/10.1109/ISSCC42613.2021.9365784

[7] T. Dickson, Z. Deniz, M. Cochet, M. Kossel, T. Morf, and Y. H. Choi, "A 72GS/s, 8-bit DAC-based wireline transmitter in 4nm FinFET CMOS for 200+Gb/s serial links," *IEEE J. Solid-State Circuits*, vol. 58, no. 4, pp. 1074–1086, April 2023. https://doi.org/10.1109/JSSC.2022.3228632
[8] Y. Perelman, Z. Toroker, D. Kumar, E. Maday, N. Familia, and T. Carbone, "A 116-Gb/s PAM4 0.9-pJ/b transmitter with eight-tap FFE in 5-nm FinFET," *IEEE J. Solid-State Circuits*, vol. 59, no. 7, pp. 2260–2271, Jul. 2024. https://doi.org/10.1109/JSSC.2024.3351372
[9] E. Groen, C. Boecker, M. Hossain, R. Vu, S. D. Vamvakos, and H. Lin, "10-to-112-Gb/s DSP-DAC-based transmitter with flex clocking architecture," *IEEE J. Solid-State Circuits*, vol. 56, no. 1, pp. 30–42, Jan. 2021. https://doi.org/10.1109/JSSC.2020.3036981
[10] S.-J. Bae, Y.-S. Sohn, T.-Y. Oh, S.-H. Kwak, D.-M. Kim, D.-H. Kim, Y.-S. Kim, Y.-S. Yang, S.-Y. Doo, J.-I. Lee, S.-Y. Bang, S.-Y. Park, K.-W. Yeom, J.-Y. Lee, H. Park, W.-S. Kim, H.-J. Yang, K.-I. Park, J. S. Choi, and Y.-H. Jun, "A 40nm 7Gb/s/pin single-ended transceiver with jitter and ISI reduction techniques for high-speed DRAM interface," in *Symp. VLSI Circuits Dig. Tech. Papers*, June 2010, pp. 193–194. https://doi.org/10.1109/VLSIC.2010.5560300
[11] B. Razavi, "The design of a phase interpolator," *IEEE Solid-State Circuits Mag.*, vol.15, no. 4, pp. 6–10, Fall 2023. https://doi.org/10.1109/MSSC.2023.3315653
[12] Z. Wang and P. R. Kinget, "A very high linearity twin phase interpolator with a low-noise and wideband delta quadrature DLL for high-speed data link clocking," *IEEE J. Solid-State Circuits*, vol. 58, no. 4, pp. 1172–1184, Apr. 2023. https://doi.org/10.1109/JSSC.2022.3197061

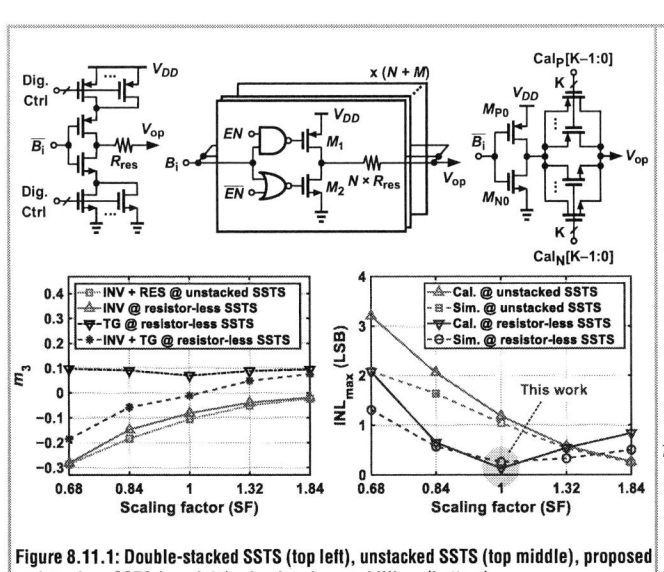

Figure 8.11.1: Double-stacked SSTS (top left), unstacked SSTS (top middle), proposed resistor-less SSTS (top right), simulated m_3 and INL_{max} (bottom).

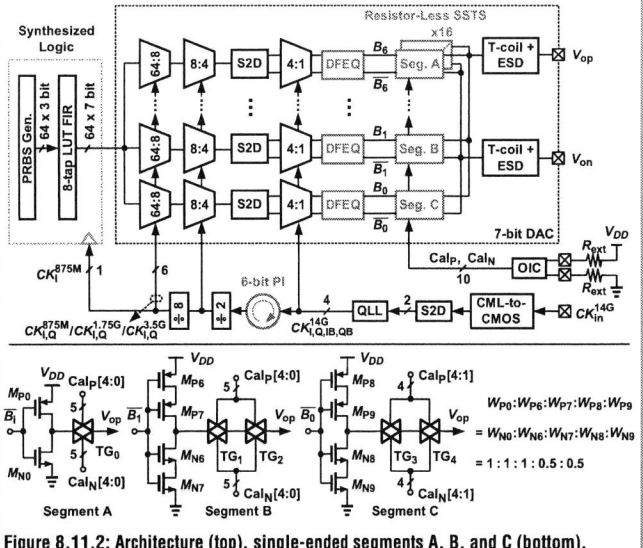

Figure 8.11.2: Architecture (top), single-ended segments A, B, and C (bottom).

ISSCC 2026 / SESSION 8 / DIE-TO-DIE AND HIGH-SPEED ELECTRICAL TRANSCEIVERS / 8.11

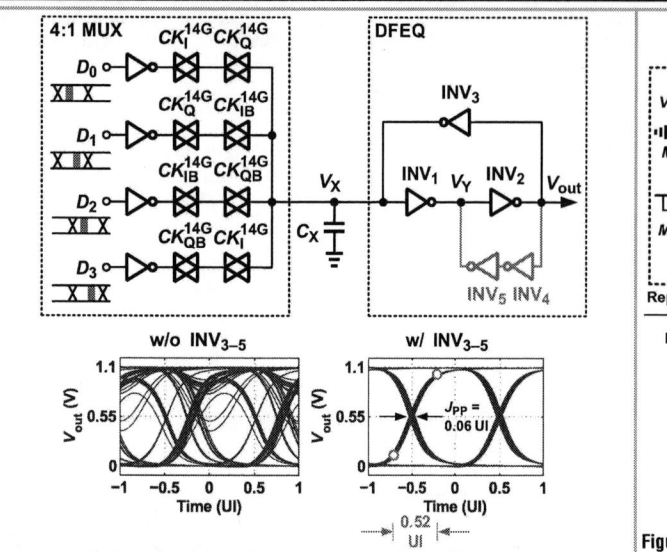

Figure 8.11.3: The 4:1 MUX with DFEQ and simulated V_{out} of DFEQ.

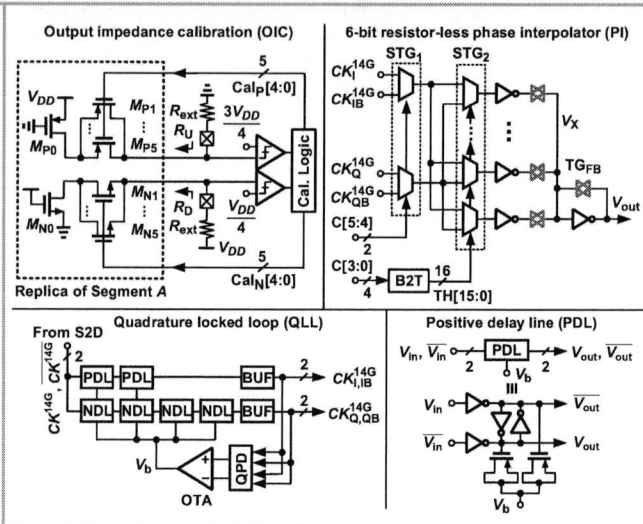

Figure 8.11.4: OIC (top left), 6-bit resistor-less PI (top right), QLL (bottom left), and PDL (bottom right).

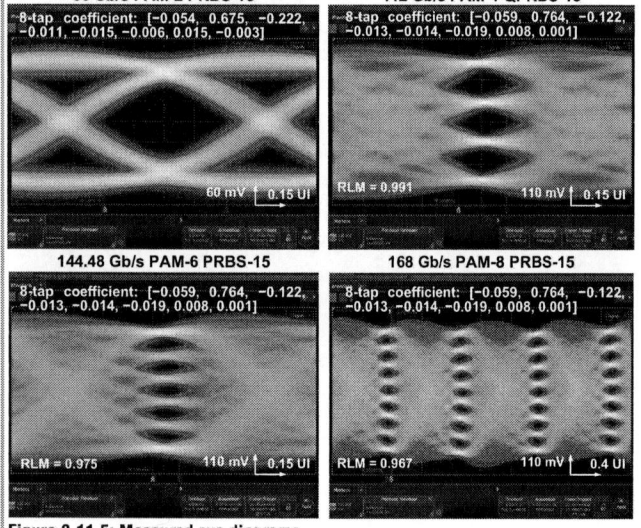

Figure 8.11.5: Measured eye diagrams.

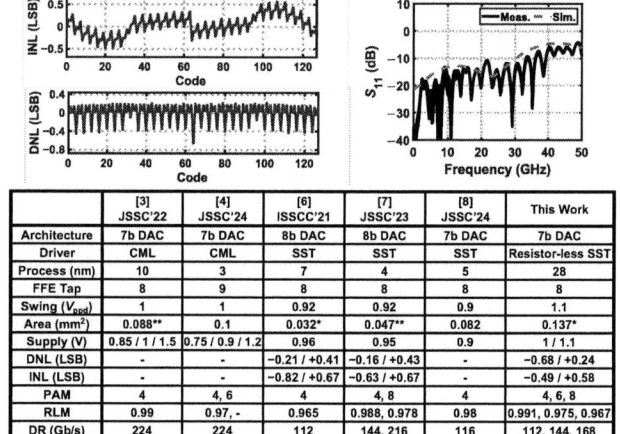

Figure 8.11.6: Measured INL / DNL (top left), measured and simulated S_{11} (top right), and comparison table (bottom).

	[3] JSSC'22	[4] JSSC'24	[6] ISSCC'21	[7] JSSC'23	[8] JSSC'24	This Work
Architecture	7b DAC	7b DAC	8b DAC	8b DAC	7b DAC	7b DAC
Driver	CML	CML	SST	SST	SST	Resistor-less SST
Process (nm)	10	3	7	4	5	28
FFE Tap	8	9	8	8	8	8
Swing (V_{ppd})	1	1	0.92	0.92	0.9	1.1
Area (mm²)	0.088**	0.1	0.032*	0.047**	0.082	0.137*
Supply (V)	0.85 / 1 / 1.5	0.75 / 0.9 / 1.2	0.96	0.95	0.9	1 / 1.1
DNL (LSB)	-	-	-0.21 / +0.41	-0.16 / +0.43	-	-0.68 / +0.24
INL (LSB)	-	-	-0.82 / +0.67	-0.63 / +0.67	-	-0.49 / +0.58
PAM	4	4, 6	4	4, 8	4	4, 6, 8
RLM	0.99	0.97, -	0.965	0.988, 0.978	0.98	0.991, 0.975, 0.967
DR (Gb/s)	224	224	112	144, 216	116	112, 144, 168
Power (mW)	473*	233, 168	157*	288**, 287**	104.4	178*
Pwr Eff. (pJ/b)	2.11*	1.04, 0.75	1.4*	2**, 1.33**	0.9	1.59*, 1.23*, 1.06*

* Exclude PLL, **Exclude DSP + PLL

Figure 8.11.7: Normalized single-bit responses, simulated SNDR and SFDR, die photo, and power breakdown.

Building Block	Power (mW)	
Resistor-less SSTS*	10.5	(5.8%)
S2D, 4:1 MUX, DFEQ*	75	(42%)
8:4 MUX*	17.3	(9.7%)
64:8 MUX*	6.3	(3.5%)
CML-to-CMOS, S2D, QLL, PI, ÷8, ÷2*	21	(11.8%)
OIC*	2.1	(1.2%)
Synthesized logic**	46	(25.8%)
Total	178	(100%)

*Analog: 1.1 V, **Digital: 1 V

• 2026 IEEE International Solid-State Circuits Conference

Session 9 Overview: *Wireless Power*
POWER MANAGEMENT SUBCOMMITTEE

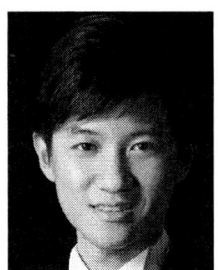

Session Chair: Chen-Yen Ho
MediaTek
Hsinchu, Taiwan

Session Co-Chair: Jianping Guo
Sun Yat-Sen University
Guangzhou, China

Wireless power transfer systems are evolving towards higher integration and multifunctionality to address the demands of complex application scenarios. The papers in this session focus on architectural innovations such as phase-shifted time-multiplexing, multi-output regulating rectification, single-transmitter-dual-receiver power management, and networked receiver coordination. These advancements enable simultaneous power and dual-uplink data transmission over a single link, multi-channel or multi-receiver independent outputs, adaptive power allocation, and cross-unit energy sharing.

3:35 PM

9.1   A Single-Power-Link 13.56MHz Wireless Power and Data Transfer System with Synchronized Phase-Shifted Time-Multiplexing Dual Uplinks for Implantable Voltammetry

Zhenhao Li, Iowa State University, Ames, IA

In Paper 9.1, Iowa State University and University of Georgia present a wireless power and data transfer (WPDT) system for implantable voltammetry sensors. Phase-shifted time-multiplexing control is proposed to enable simultaneous transmission of power and two types of data (power control signal and sensor data) via a single power link. The dual-uplink feature was verified in measurements showing no obvious efficiency degradation, with a fast dynamic response and a >99.9% sensor data linearity.

9

4:00 PM

9.2   A 91%-Efficiency Single-Stage Bipolar Quad-Output Regulating Rectifier with Event-Driven Output Power Enhancement via Coil-Reused DC-DC for Wireless Power Transfer

Tianqi Lu, Delft University of Technology, Delft, The Netherlands

In Paper 9.2, Delft University of Technology and Zhejiang University present a 91%-efficiency bipolar quad-output regulating rectifier with event-driven power enhancement for wireless power transfer. Using only five switches, it delivers four regulated outputs. A coil-reused DC-DC mode sustains heavy loads beyond RX input limits, achieving 173mW maximum output power and fast transient recovery.

4:25 PM

9.3   An Output-Domain-Independent Single-Transmitter-Dual-Receiver Wireless Power Transfer System with Detuned-Tank and Time-Multiplexing Control for Adaptive Power Distribution

Yutang Chen, Sun Yat-Sen University, Guangzhou, China

In Paper 9.3, Sun Yat-Sen University presents a highly integrated single-transmitter-dual-receiver (STDR) wireless power transfer system. With the proposed detuned-tank and time-multiplexing control, this STDR system achieves adaptive power distribution and two independent output domains, feathering load regulations of 0.07 and 0.26 mV/mA, 82.2% peak efficiency, and 1099mW maximum output power.

4:50 PM

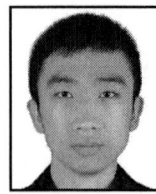

9.4   A Multi-Coil Scalable Energy-Shared Wireless Power Receiver Network for Distributed Time-Division-Multiplexing Somatosensory Cortex Stimulation

Kai Cui, Tsinghua University, Beijing, China

In Paper 9.4, Tsinghua University presents a system-level scalable resonant-current-mode wireless power receiver (RX) network for miniaturized distributed somatosensory cortex stimulation. By interconnecting the power outputs of all RX cells within the network, the stimulator in each cell can access the total received power of the entire network, in a time-division-multiplexing manner.

ISSCC 2026 / SESSION 9 / WIRELESS POWER / 9.1

9.1 A Single-Power-Link 13.56MHz Wireless Power and Data Transfer System with Synchronized Phase-Shifted Time-Multiplexing Dual Uplinks for Implantable Voltammetry

Zhenhao Li[1], Junyao Tang[1], Nawab Singh[2], Jianqiang Jiang[1], Lei Zhao[1], Wenshuo Zhu[1], Xuan Sun[1], Liang Dong[2], Cheng Huang[1]

[1]Iowa State University, Ames, IA, [2]University of Georgia, Athens, GA

Abstract

A 13.56MHz wireless power and data transfer (WPDT) system for implantable voltammetry sensor is presented. With the synchronized phase-shifted time-multiplexing technique, the system uses a single power link to support simultaneous power and two types of data transmission via dual-uplinks with neither additional coils nor efficiency degradation. The single-power-link operation is verified in silicon, measuring a 60% peak end-to-end efficiency, a fast dynamic response, and a >99.9% data linearity.

Implantable voltammetry systems play an important role in tracking biochemical activities in living organisms [1]. In implantable applications, batteries are generally not preferred due to the size, weight, risk of leakage, and the need for periodical surgical replacements. Wireless power transfer systems thus can provide a safer and long-lasting solution. To achieve higher power efficiency and reduce heat dissipation, a wireless uplink is necessary for the external transmitter (TX) to obtain feedback from the receiver (RX) to close the global loop and optimize its power transmission based on the coupling condition and loading at the implant. Wirelessly powered implantable sensors also need to retrieve digital data or analog signals with a data/signal uplink from RX to TX. Supporting two uplinks for different purposes is thus required, which poses design challenges especially with low power budget for implantable devices. Existing wireless power and data transfer (WPDT) systems typically provide only one uplink, mostly for raw data transmission only [2-4], leaving the TX operated in open loop considering maximum power transfer at worst-case scenarios, which results in energy waste and excessive heat in the coils and the implant. Some designs only provide the power uplink, with no data transmission capability [5-7]. Moreover, the uplink in previous works is often utilized with an additional coil [3] or a separated high-frequency telemetry path [4], which increases implant volume and packaging complexity or introduces cross-coupling that disturbs the power link, reducing efficiency and dynamic performance. Some WPDT systems do provide two data links, but with only one uplink while the other one is a downlink from TX to RX specifically for stimulation applications [3,4,8] or toggling between different types of the data uplink [9].

To address the needs, the paper introduces a single-power-link WPDT system with dual-uplink capability: a power uplink for global TX power regulation, and a data uplink designed for cyclic voltammetry (CV) applications, which is one of the most widely used methods for chemical sensing, with the goal of simultaneous power and dual-link data transfer without degrading power efficiency and dynamic performance.

This system is developed in conjunction with an implantable CV sensor, enabling long-term monitoring of chemical properties, for example, antibody and antigen detection, as shown in Fig. 9.1.1. The proposed system addresses the following key challenges: **1)** Efficient power transfer and fast dynamic performance. With the power uplink for closed-loop operation, the TX can adjust its power immediately based on the RX needs to avoid overpowering and heating up the coils or the implant. **2)** Dual uplinks for sensor and voltage regulation data transmission. To reduce power overhead, instead of quantizing the sensor analog information in RX and transmitting digital data, a time-based mechanism is proposed. To transmit both types of data, a synchronized phase-shifted time-multiplexing technique is proposed to process and combine the data at RX, transmit through load-shift-keying (LSK) modulation. Then, the data will be sensed, differentiated and recovered at TX. **3)** Addressing potential data conflicts and errors. This includes start-up alignment for initial phase shifting and clock correction and realignment. To address these challenges, we designed both RX and TX chips with dedicated circuits for rectification, sensing, synchronization, and data modulation, as detailed in the following sections.

RX chip Design (Fig. 9.1.2): For power conversion, an active rectifier is designed with delay compensation [8] for efficiency enhancement. A comparator is used to generate the voltage regulation request (D_{PWR_PRE}) indicating an "overpower" status when V_{FB} (representing V_0 through a fixed resistor ratio) exceeds V_{REF} for TX to reduce its transmitted power when triggered, and a current sink is used as a shunt regulator for local voltage regulation in RX. For CV sensing, the RX chip is equipped with three on-chip electrodes: the working, reference, and counter electrode (WE, RE, and CE, respectively). By applying a scanning voltage to WE with a fixed reference to RE, the current through CE contains chemical information. This current is then sensed and converted into time-based pulses (current to frequency, I2f), whose frequency is further divided by 4 before being output as the sensor data (D_{SEN_PRE}). The clock in RX (CLK_{RX}) is important as it needs to be synchronized with the TX for data recognition, so it is extracted from the AC power inputs $V_{AC}s$ and down scaled by a 1/64 divider as CLK_{RX}. Both D_{PWR_PRE} and D_{SEN_PRE} (with the "$\_PRE$" denoting pre-synchronization) are then synchronized to the rising and falling edge of CLK_{RX}, converted into LSK signals as D_{PWRRX} and D_{SENRX}, respectively, and transmitted by pulsing on the LSK switch briefly. The carrier $V_{AC}s$ will be interrupted during LSK signal transmission, but because its duration and timing are initiated by and known to RX, the counters will be corrected by adding predicted counts so that the CLK_{RX} remains unaffected.

TX Chip Design (Fig. 9.1.3): For DC-AC power conversion, a Class-D power stage is designed to operate in either High-Power (HP) full-bridge mode or Low-Power (LP) half-bridge mode to drive the LC tank, controlled by D_{PMODE}. This D_{PMODE} originates from the RX, indicating RX being "overpowered". The RX generates D_{PWR_PRE}, synchronizes it as D_{PWRRX}, transmits it via LSK, and the TX recovers it as D_{PWRTX}, which directly triggers D_{PMODE}. Whenever a D_{PMODE} request is detected, the TX will switch into the LP mode for a fixed duration, which reduces a predictable amount of transmitted energy for each triggering and achieves reliable power regulation with minimized complexity. Please note that the power stage switches at 13.56 MHz, which is from an always active CLK, so not to be confused with CLK_{TX} introduced later for data identification. For data extraction, as shown in Fig. 9.1.3, the currents through the power stage will be sensed by an integrated current sensor, and the impedance change due to RX LSK signals will be recognized. To improve the detectability, a modulation depth (MD) enhancement module is designed, extracting and amplifying the voltage difference due to impedance shifts. The MD enhancement circuit captures the raw current signal V_{SENSE} and compares it with a slowly varying baseline voltage V_{CAP} that is obtained through a capacitor acting as a low-pass filter. Because the intensity of HP and LP modes is different, two current sensors and MD enhancement modules are designed. The data identification will be discussed in the following section.

System Design (Fig. 9.1.4): During start up, the RX controller will bypass any LSK requests until V_{FB} crosses V_{REF} for the first time. Then the first LSK signal will be generated and transmitted, ensuring the first data is always D_{PWR}, which is used as a reference for both synchronization and data identification at TX. Enforcing D_{PWR} as the first transmitted LSK symbol also guarantees that subsequent D_{SEN} are generated only after the supply voltage for CV sensor (V_0) has reached its desired level to ensure normal sensor operation. The CLK_{RX} and CLK_{TX} have a matching frequency (1/64 of 13.56 MHz divided from CLK) but a phase shift of 90°, which not only provides the maximum detection margin but also prevents data from overlapping at the edges, thereby minimizing data ambiguity. At start up, both CLK_{RX} and CLK_{TX} stay high without switching. When the 1st D_{PWRRX} is generated at RX and received at TX as D_{PWRTX}, both CLK_{RX} and CLK_{TX} switch to low after ½ and ¼ of a cycle (creating a 180° and 90° delay), respectively, and start switching afterwards. This initiates a 90° phase shift, and D_{PWRTX} and D_{SENTX} align at the center of the high and low levels of CLK_{TX}, respectively. The phase shift between RX generating and TX receiving an LSK signal is negligible due to the 64x faster carrier frequency.

To maximize detectability and avoid error accumulation, both CLK_{RX} and CLK_{TX} will realign to 90° phase shift whenever an LSK signal is generated by RX and received by TX. This ensures that even interruptions occur, the system can automatically restore synchronization with a tolerance of up to half the detection window. To address potential data conflicts, at RX, D_{PWRRX} and D_{SENRX} are aligned to the rising and falling of CLK_{RX}, respectively, within a window of 4 carrier clocks. If a data misses the window, it is held and aligned to the next available edge. This ensures zero conflicts, and the size of the detection window is determined based on the settling time of TX LSK signal reception. However, time errors of at most a CLK_{RX} cycle may be introduced during this process. To minimize the impact of this error and ensure >99.9% linearity and <0.2% current sensing accuracy, a 2-ms averaging window is applied at readout. The averaging window could be set longer if better accuracy is needed, and will not affect the operation in our application as a CV scanning can take seconds to complete.

Both chips are fabricated in 0.18μm standard CMOS. Fig. 9.1.5 measures load transient responses and system behaviors during simultaneous power and dual-uplink data transmission, verifying the single-power-link dual-uplink WPDT operation. Measured output linearity is shown in Fig. 9.1.6, which is greater than 99.9%. Figure 9.1.6 also includes measured RX time to TX output codes. As for the current sensing accuracy, fractional Allan deviation calculated with measured TX data under different RX input frequencies (as sensor signals after I2f conversion) shows a consistent trend (white frequency noise), with the target specification of 0.2% successfully achieved at a 2-ms averaging time. As also shown in Fig. 9.1.6, up to 20.76% end-to-end (E2E) efficiency enhancement is observed by enabling the closed-loop global power control, and the impact of sensor data on power efficiency is negligible. Figure 9.1.7 shows a micrograph of both TX and RX chips along with a comparison against state-of-the-art WPDT systems. To the best of our knowledge, the

158 • 2026 IEEE International Solid-State Circuits Conference

979-8-3315-8937-0/26 $31.00 © 2026 IEEE

proposed system is the only one that can transmit power with both user data/signal and global closed-loop control (dual uplink) simultaneously via a single power link, while maintaining a decent E2E efficiency.

Acknowledgement:
This work is supported in part by USDA Agriculture Research Service under Project 5030-32000-228-017-S, and in part by NSF under Project 2338697

References:
[1] S.-N. Hosseini, P.S. Das, V.K. Lazarjan, G. Gagnon-Turcotte, K. Bouzid, and B. Gosselin, "Recent Advances in CMOS Electrochemical Biosensor Design for Microbial Monitoring: Review and Design Methodology," *IEEE Transactions on Biomedical Circuits and Systems*, vol. 17, no. 2, pp. 202-228, April 2023. http://doi.org/10.1109/TBCAS.2023.3252402
[2] M. Kim, H.-S. Lee, J. Ahn, and H.-M. Lee, "A 13.56-MHz Wireless Power and Data Transfer System With Current-Modulated Energy-Reuse Back Telemetry and Energy-Adaptive Voltage Regulation," *IEEE Journal of Solid-State Circuits*, vol. 58, no. 2, pp. 400-410, Feb. 2023. http://doi.org/10.1109/JSSC.2022.3207549
[3] Y.-K. Lo et al., "A Fully Integrated Wireless SoC for Motor Function Recovery After Spinal Cord Injury," *IEEE Transactions on Biomedical Circuits and Systems*, vol. 11, no. 3, pp. 497-509, June 2017. http://doi.org/10.1109/TBCAS.2017.2679441
[4] R. Ranjandish, K. Ture, F. Maloberti, C. Dehollain, and A. Schmid, "All Wireless, 16-Channel Epilepsy Control System with Sub-µW/Channel and Closed-Loop Stimulation Using a Switched-Capacitor-Based Active Charge Balancing Method," *ESSCIRC*, pp. 226-229, Sept. 2018. http://doi.org/10.1109/ESSCIRC.2018.8494252
[5] C. Huang, T. Kawajiri, and H. Ishikuro, "A 13.56-MHz Wireless Power Transfer System With Enhanced Load-Transient Response and Efficiency by Fully Integrated Wireless Constant-Idle-Time Control for Biomedical Implants," *IEEE Journal of Solid-State Circuits*, vol. 53, no. 2, pp. 538-551, Feb. 2018. http://doi.org/10.1109/JSSC.2017.2767181
[6] T. Lu and S. Du, "A 13.56MHz Wireless Power Transfer System with Hybrid Voltage-/Current-Mode Receiver and Global Digital-PWM Regulation Achieving 150% Transfer Range Extension and 72.3% End-to-End Efficiency," *ISSCC*, pp. 450-451, Feb. 2024. http://doi.org/10.1109/ISSCC49657.2024.10454547
[7] J. Tang, L. Zhao and C. Huang, "A Wireless Power Transfer System with Up-to-20% Light- Load Efficiency Enhancement and Instant Dynamic Response by Fully Integrated Wireless Hysteretic Control for Bioimplants," *ISSCC*, pp. 470-471, Feb. 2021. http://doi.org/10.1109/ISSCC42613.2021.9365859
[8] A. Akinin et al., "An Optically Addressed Nanowire-Based Retinal Prosthesis With Wireless Stimulation Waveform Control and Charge Telemetering," *IEEE Journal of Solid-State Circuits*, vol. 56, no. 11, pp. 3263-3273, Nov. 2021. http://doi.org/10.1109/JSSC.2021.3113648
[9] Y. Liu, Y. Yao, and W.-H. Ki, "A 13.56-MHz Single-Input Dual-Output Wireless Power and Data Transfer System for Bio-Implants," *IEEE Journal of Solid-State Circuits*, vol. 59, no. 8, pp. 2557-2567, Aug. 2024. http://doi.org/10.1109/JSSC.2024.3372430
[10] Y. Park et al., "A Frequency-Splitting-Based Wireless Power and Data Transfer IC for Neural Prostheses with Simultaneous 115mW Power and 2.5Mb/s Forward Data Delivery," *ISSCC*, pp. 472-473, Feb. 2021. http://doi.org/10.1109/ISSCC42613.2021.9365781

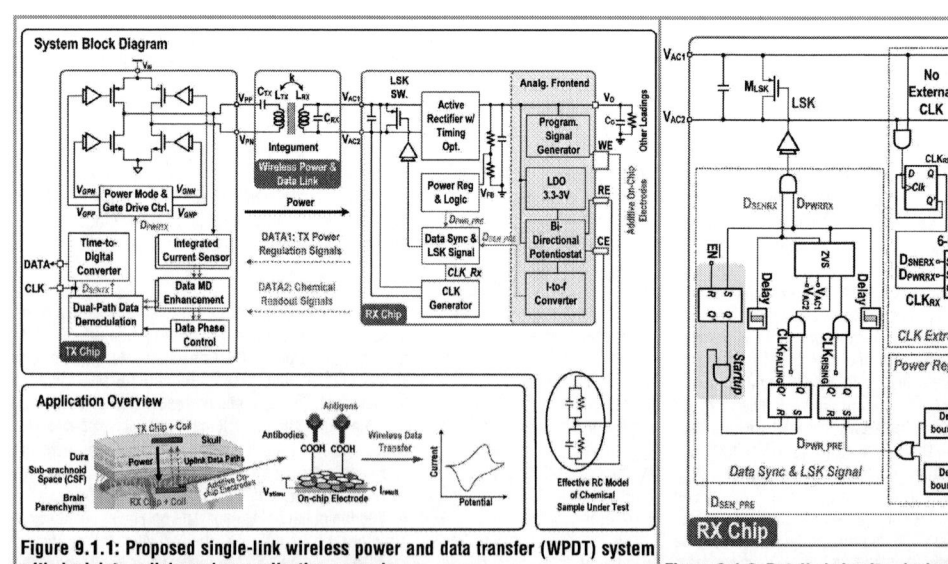

Figure 9.1.1: Proposed single-link wireless power and data transfer (WPDT) system with dual data uplinks and an application example.

Figure 9.1.2: Detailed circuitry design of the RX chip.

ISSCC 2026 / SESSION 9 / WIRELESS POWER / 9.1

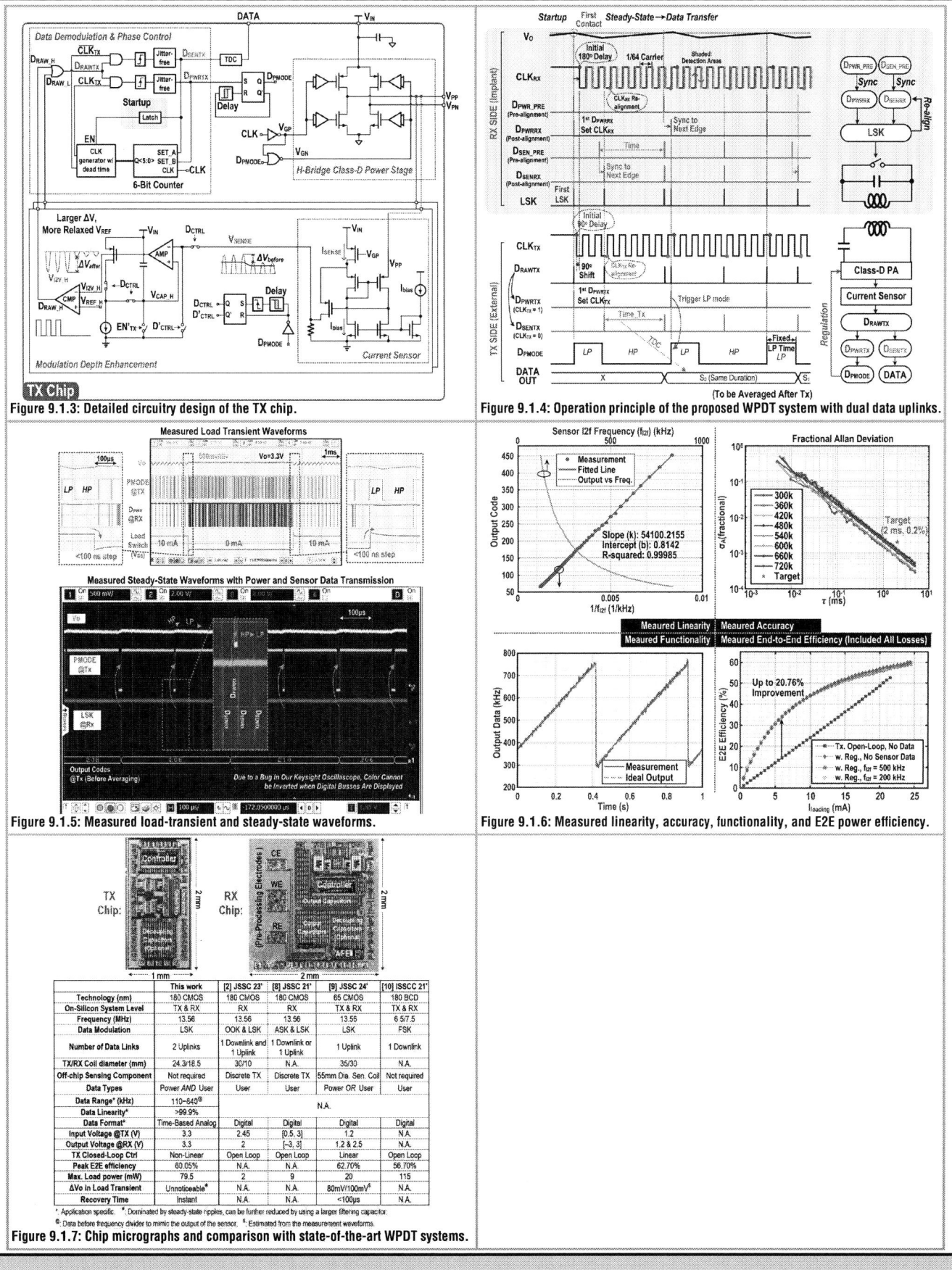

Figure 9.1.3: Detailed circuitry design of the TX chip.

Figure 9.1.4: Operation principle of the proposed WPDT system with dual data uplinks.

Figure 9.1.5: Measured load-transient and steady-state waveforms.

Figure 9.1.6: Measured linearity, accuracy, functionality, and E2E power efficiency.

Figure 9.1.7: Chip micrographs and comparison with state-of-the-art WPDT systems.

• 2026 IEEE International Solid-State Circuits Conference

ISSCC 2026 / SESSION 9 / WIRELESS POWER / 9.2

9.2 A 91%-Efficiency Single-Stage Bipolar Quad-Output Regulating Rectifier with Event-Driven Output Power Enhancement via Coil-Reused DC-DC for Wireless Power Transfer

Tianqi Lu[1], Bo Zhao[2], Sijun Du[1]

[1]Delft University of Technology, Delft, The Netherlands, [2]Zhejiang University, Hangzhou, China

Abstract

A bipolar quad-output regulating rectifier for wireless power transfer (WPT) is presented. Using only five power switches, it provides four independently regulated bipolar outputs. A coil-reused DC-DC mode sustains operation under heavy-load conditions beyond the RX input power limit. The rectifier achieves up to 173mW output power, 91% peak efficiency, and fast transient recovery.

Wireless power transfer (WPT) is a promising approach for non-invasive powering of biomedical implants such as neural interfaces and retinal prostheses. These advanced applications, involving large-array neural recording and multi-channel stimulation, impose stringent requirements on the WPT system. First, it must deliver multiple high-power (tens of mW) DC outputs across distinct voltage domains to supply diverse functional blocks [1-4]. Second, bipolar supplies are highly desired for neural stimulators, eliminating the need for charge pumps and full-bridge inverters [5,6]. In addition, fast load-transient recovery and event-driven heavy-load capability are critical, as implants in active mode can suddenly draw hundreds of mW for bio-modulation or wireless data transmission [4,7].

To achieve multiple regulated outputs, many designs have been proposed based on a typical full-bridge rectifier by adding additional high-side branches [1,4,8] (Fig. 9.2.1, top right). However, these lack bipolar outputs and rely on many power switches. A dual-output voltage doubler was reported in [3], providing two regulated outputs with only two switches (Fig. 9.2.1, mid left), but it remains unipolar. A dedicated bipolar-output full-bridge rectifier [9] achieves bipolar outputs, yet requires six switches, leading to low efficiency and poor hardware utilization. A bipolar current-mode rectifier using only two switches was proposed in [5] (Fig. 9.2.1, bottom left), but its bipolar outputs share a single current loop and cannot be regulated independently under variable loads. Moreover, none of the above designs can deliver higher-than-input output power to handle potential heavy loads, exposing limited operational range dependent on the RX input power levels. A DC-DC energy recycling approach [4] enhances this capability by reusing the RX coil as an inductor to transfer energy from a reservoir to a DC output, but it only supports positive regulated outputs.

To overcome these limitations, this paper proposes a single-stage bipolar quad-output regulating rectifier with coil-reusing DC-DC-enhanced heavy-load ability. The power stage uses only five switches (Fig. 9.2.1, bottom right), with the output mid-node connected to ground. M_{P1} and M_{P2} form the two positive-side branches; M_{N1} and M_{N2} constitute the two negative-side branches. Switch S_1 is activated during DC-DC operation. The rectifier delivers three stably regulated outputs, including two positive outputs, V_{OUTP1} and V_{OUTP2}, and one negative output, V_{OUTN1}. In addition, an energy reservoir (e.g., a high-density super capacitor) is employed in the negative voltage domain (V_{STO}), which can either be a regulated output or a temporary power supply. Through inverting buck–boost operation with the RX coil reused as the DC-DC inductor, V_{STO} can provisionally supply the other three outputs under event-driven heavy loads. As a result, the system maintains four bipolar regulated outputs in steady state, with three outputs sustaining heavy-load operation even when RX input power is insufficient.

The proposed rectifier operates in two modes: RX and DC-DC. In the RX mode (Fig. 9.2.2, top), S_1 is continuously ON, while M_{P1}, M_{P2}, M_{N1}, and M_{N2} function as active diodes depending on output voltage levels. When all four output voltages (V_{OUTP1}, V_{OUTP2}, V_{OUTN1} and V_{STO}) are below the regulated levels, the rectifier enters the "all-ON" case with all active diodes enabled. In every two resonant periods, it cyclically operates in the sequence of $\Phi_{R1} \rightarrow \Phi_{R4} \rightarrow \Phi_{R2} \rightarrow \Phi_{R3}$, with each phase lasting for half a period, to charge V_{OUTP2}, V_{STO}, V_{OUTP1} and V_{OUTN1}, respectively, as shown in Fig. 9.2.3 (top left). If an output voltage is sufficient, its corresponding phase will be skipped. When both outputs on the same polarity are high, the rectifier allows the L_{RX}–C_{RX} tank to freely resonate in these spare phases. In addition, a 0X phase (Φ_{0X}) will be introduced to freewheel L_{RX} if all four outputs are sufficiently charged. Figure 9.2.2 (bottom left) illustrates the DC-DC mode, which relocates energy from V_{STO} to the other three outputs by reusing the existing power stage and the RX coil (L_{RX}) as a DC-DC inductor, requiring no extra components. Φ_{D1} is the L_{RX}-energizing phase, while Φ_{D2}, Φ_{D3}, and Φ_{D5} are the L_{RX}-deenergizing phases delivering power to V_{OUTP2}, V_{OUTP1}, and V_{OUTN1}, respectively. Notably, since charging V_{OUTN1} (Φ_{D5}) requires a L_{RX} current (I_{LRX}) in the other direction, an I_{LRX}-flipping phase, Φ_{D4}, is inserted, which relies on the resonant capacitor C_{RX} to fast flip I_{LRX} with low loss. Hence, the operation phase sequences to sustain the three outputs are: $\Phi_{D1} \rightarrow \Phi_{D2}$ for charging V_{OUTP2}, $\Phi_{D1} \rightarrow \Phi_{D3}$ for charging V_{OUTP1}, and $\Phi_{D1} \rightarrow \Phi_{D4} \rightarrow \Phi_{D5}$ for charging V_{OUTN1}, as shown in Fig. 9.2.3 (top right). The system workflow is presented in Fig. 9.2.2 (bottom right). Based on the hysteresis regulation of output voltages, the rectifier operates in the RX mode under normal loads; if any of V_{OUTP1}, V_{OUTP2}, and V_{OUTN1} experience undershoots due to suddenly heavier loads, the DC-DC mode will be triggered till all outputs return within their hysteresis windows.

Figure 9.2.3 (bottom) shows the system architecture in a 6.78MHz resonant WPT configuration. The five power switches, M_{P1}, M_{P2}, M_{N1}, M_{N2}, and S_1, and the 0X switch, S_{0X}, are implemented with 5V MOSFETs. M_{P1}, M_{P2}, M_{N1}, and M_{N2} employ actively biased gate and body voltages. Each of the four output voltages is monitored by a dedicated output controller. For example, V_{OUTP1} controller scales V_{OUTP1} through a multi-level resistive divider, whose three mid-nodes are compared with a reference V_{REF}, forming a hysteresis window with three levels: V_{OUTP1} is normally regulated between the upper two levels in the RX mode (indicated by EN_{RXP1}), while hitting the lowest threshold triggers the DC-DC mode (indicated by EN_{DCP1}). Notably, the V_{STO} controller uses only high and middle thresholds since V_{STO} serves as an energy reservoir that sustains the other three outputs during heavy loads while remaining regulated under normal operation.

When the rectifier operates in RX mode, 16 phase combinations arise from the four output states, each requiring a different delay-compensation strength for M_{P1}, M_{P2}, M_{N1}, and M_{N2} to achieve soft active-diode switching. Conventional fixed [10] and adaptive [11,12] delay-compensation schemes suffer from either a fixed compensation strength or slow convergence. To address this limitation, we propose a state-based delay-compensation scheme that dynamically selects the appropriate compensation strength for each active diode based on the rectifier state during the preceding half-resonant period. For example, prior to entering Φ_{R1}, the rectifier may have been operating in Φ_{R3}, Φ_{R4}, or the L_{RX}-C_{RX} free-resonance state. Accordingly, in Φ_{R1}, only the matching compensation branch in M_{P2}'s active-diode comparator is enabled with the appropriate strength. This ensures that soft switching is maintained across RX phase transitions. In DC-DC mode, constant-on-time control is applied to energize L_{RX} (Φ_{D1}), while adaptive delay compensation is used to achieve soft switching during the L_{RX}-deenergizing phases and I_{LRX}-flipping phase. Notably, all NMOS devices in the proposed system are triple-well devices with their bodies tied to V_{OUTN1}, while the chip substrate remains connected to ground, ensuring compatibility with system-on-chip (SoC) integration.

The proposed rectifier was fabricated in a 0.18μm BCD process, occupying a silicon area of 2.84mm² (Fig. 9.2.7). In measurement, the transmitter (TX) generates a driving signal (V_S) with a peak-to-peak voltage up to 5V. The TX and RX coils have 1.137μH and 166nH inductance, and 29mm and 16mm diameters, respectively. The coils are placed coaxially with a distance of 10.2mm. Figure 9.2.4 (top) shows two measured steady-state RX waveforms. V_{OUTP1}, V_{OUTP2}, V_{OUTN1} and V_{STO} are regulated at 2.2V, 1.1V, 2.2V, and 1.8V, respectively, with ≤100mV ripples. Figure 9.2.4 (bottom) shows zoomed-in steady-state RX waveforms under different phase combinations, confirming near-optimal active-diode switching. Figure 9.2.5 (top-left) shows the load-transient waveform when R_{P1} varies between 1kΩ and 300Ω, with $R_{P2}=R_{N1}=1$kΩ. No undershoots or overshoots is observed at the transients, while the cross-regulation is also unnoticeable. Figure 9.2.5 (top right) shows the load-transient waveform when R_{P2} varies between 1kΩ and 300Ω, with $R_{P1}=R_{N1}=1$kΩ. The rectifier still shows negligible recovery time and unnoticeable cross-regulation at the transients. Figure 9.2.5 (bottom left) shows the load-transient waveform under weaker coupling conditions, and R_{P2} varies between 5kΩ and 200Ω. It is observed that V_{OUTP2} presents undershoots at the transients and under the heavier-R_{P2} condition, which trigger the DC-DC mode; the DC-DC mode further recovers V_{OUTP2} back to the preset RX hysteresis window. The zoomed-in DC-DC waveforms are shown in Fig. 9.2.5 (mid bottom) under different DC-DC phase combinations. Figure 9.2.5 (bottom right) shows the power conversion efficiency (PCE) of the proposed rectifier in the RX mode, with $R_{N1}=100$Ω and $R_{STO}=1$kΩ, under two conditions: (1) varying R_{P1} and $R_{P2}=50$Ω, and (2) varying R_{P2} and $R_{P1}=100$Ω. The peak PCE of 91% is obtained at the maximum output power of 173mW. Figure 9.2.6 shows the comparison table. The proposed rectifier realizes four regulated outputs using only five power switches, while a DC-DC mode enhances its heavy-load capability, resulting in both high PCE and high maximum output power.

References:

[1] Q. Zhuang et al., "A 6.78MHz Single-Stage Regulating Rectifier with Dual Outputs Simultaneously Charged in a Half Cycle Achieving 92.2% Efficiency and 131mW Output Power," *ISSCC*, pp. 188-189, Feb. 2025.
https://doi.org/10.1109/ISSCC49661.2025.10904779

160 • 2026 IEEE International Solid-State Circuits Conference

979-8-3315-8937-0/26 $31.00 © 2026 IEEE

[2] H.-S. Lee, K. Eom, and H.-M. Lee, "A 90.8%-Efficiency SIMO Resonant Regulating Rectifier Generating 3 Outputs in a Half Cycle with Distributed Multi-Phase Control for Wirelessly-Powered Implantable Devices," *ISSCC*, pp. 448-449, Feb. 2024. https://doi.org/10.1109/ISSCC49657.2024.10454403

[3] T. Lu, K.A.A. Makinwa, and S. Du, "A Single-Stage Dual-Output Regulating Voltage Doubler for Wireless Power Transfer," *IEEE Journal of Solid-State Circuits*, vol. 59, no. 9, pp. 2922-2933, Sept. 2024. https://doi.org/10.1109/JSSC.2024.3378675

[4] F.-B. Yang, D.-H. Yao, and P.-H. Chen, "A Quad-Mode Structure-Reconfigurable Regulating Rectifier With Shared-Inductor DC–DC Energy Recycling in a Wireless Power Receiver," *IEEE Journal of Solid-State Circuits*, vol. 59, no. 2, pp. 574-582, Feb. 2024. https://doi.org/10.1109/JSSC.2023.3298720

[5] Y. Park, C. Kim, and M. Je, "A Wireless Adiabatic Stimulator System with Current-Mode Power Reception and Stimulus Current Regulation Achieving Precise Charge Delivery and Electrode Scalability for Miniaturized Electroceuticals," *ISSCC*, pp. 576-577, Feb. 2025. https://doi.org/10.1109/ISSCC49661.2025.10904531

[6] S. Agarwal, G. Pillonnet, H. Lu, N.S. Kassem Fathy, and P.P. Mercier, "A Current-Source-Free Constant-Current Wireless Adiabatic Neural Stimulator Achieving a 5.5-27.7x Improved RF-to-Electrode Stimulation Efficiency Factor," *IEEE Symp. VLSI Circuits*, 2024, pp. 1-2, June 2024. https://doi.org/10.1109/VLSITechnologyandCir46783.2024.10631437

[7] H. S. Gougheri, P. Graybill, and M. Kiani, "A Dual-Output Reconfigurable Shared-Inductor Boost-Converter/Current-Mode Inductive Power Management ASIC With 750% Extended Output-Power Range, Adaptive Switching Control, and Voltage-Power Regulation," *IEEE Transactions on Biomedical Circuits and Systems*, vol. 13, no. 5, pp. 1075-1086, Oct. 2019. https://doi.org/10.1109/TBCAS.2019.2937253

[8] Z. Luo, J. Liu, and H. Lee, "A 90%-Efficiency 40.68MHz Single-Stage Dual-Output Regulating Rectifier with ZVS and Synchronous PFM Control for Wireless Powering," *ISSCC*, pp. 454-455, Feb. 2023. https://doi.org/10.1109/ISSCC42615.2023.10067331

[9] C. Huang, C. Zhan, X. Bai, and Y. Lu, "A Single-Stage Bipolar-Output Regulating Rectifier With Negligible Cross-Regulation for Wireless Display," *IEEE Journal of Solid-State Circuits*, vol. 59, no. 9, pp. 2983-2994, Sept. 2024. https://doi.org/10.1109/JSSC.2024.3385337

[10] Y. Chen, Y. Luo, Y. Lin, L. Shao, D. Chen, and J. Guo, " A Wireless Power Transfer System with Up-to-27.9% Efficiency Improvement under Coupling Coefficient Ranging from 0.1 to 0.39 Based on Phase-Shift/Time-Constant Detection and Hybrid Transmission Power Control," *ISSCC*, pp. 452-453, Feb. 2024. https://doi.org/10.1109/ISSCC49657.2024.10454528

[11] L. Cheng, W.-H. Ki, Y. Lu, and T.-S. Yim, "Adaptive On/Off Delay-Compensated Active Rectifiers for Wireless Power Transfer Systems," *IEEE Journal of Solid-State Circuits*, vol. 51, no. 3, pp. 712-723, March 2016. https://doi.org/10.1109/JSSC.2016.2517119

[12] C. Huang, T. Kawajiri, and H. Ishikuro, "A Near-Optimum 13.56 MHz CMOS Active Rectifier With Circuit-Delay Real-Time Calibrations for High-Current Biomedical Implants," *IEEE Journal of Solid-State Circuits*, vol. 51, no. 8, pp. 1797-1809, Aug. 2016. https://doi.org/10.1109/JSSC.2016.2582871

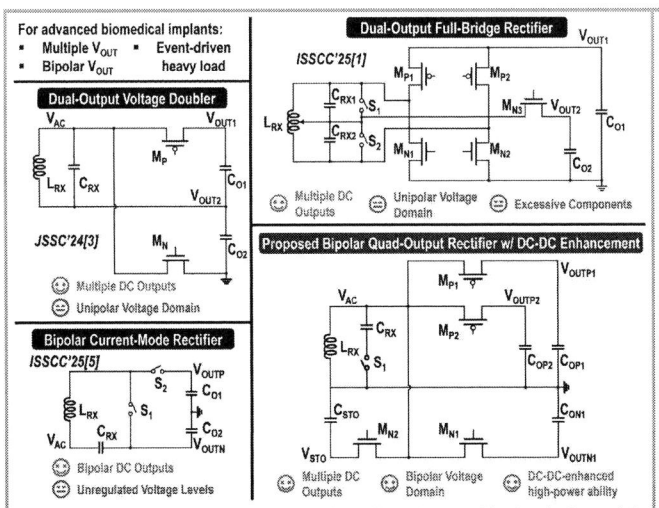

Figure 9.2.1: Conventional multi-output and bipolar-output rectifier topologies and the proposed single-stage bipolar quad-output rectifier with DC-DC-enhanced high-power ability for biomedical wireless power transfer applications.

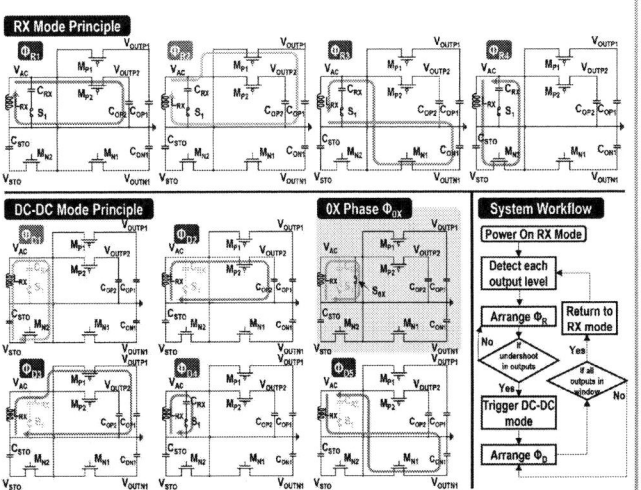

Figure 9.2.2: Operating principle of the proposed rectifier in wireless power receiver (RX) mode and DC-DC mode, and system workflow.

ISSCC 2026 / SESSION 9 / WIRELESS POWER / 9.2

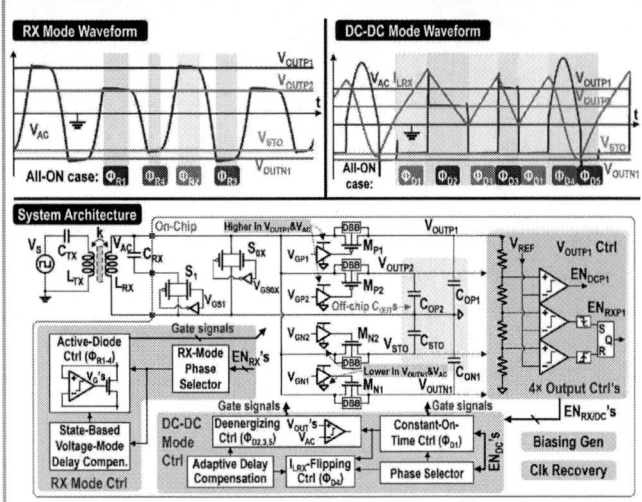

Figure 9.2.3: Operating waveform of the proposed rectifier in RX and DC-DC modes (top) and the system architecture (bottom). (DBB: dynamic body bias)

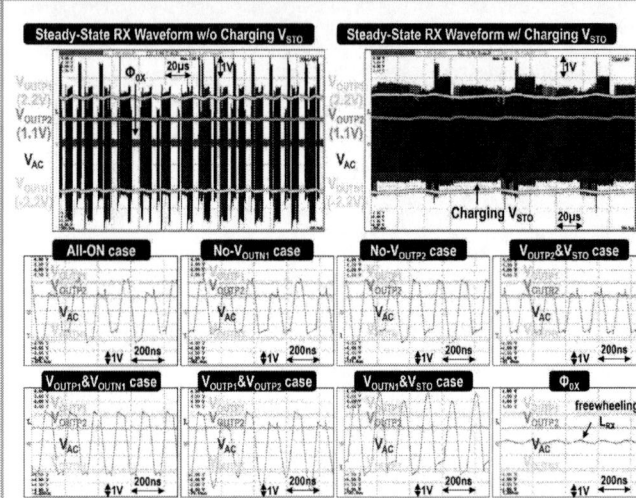

Figure 9.2.4: Measured steady-state waveform of the proposed rectifier in the RX mode (top) and zoomed-in waveforms with different phase combinations (bottom).

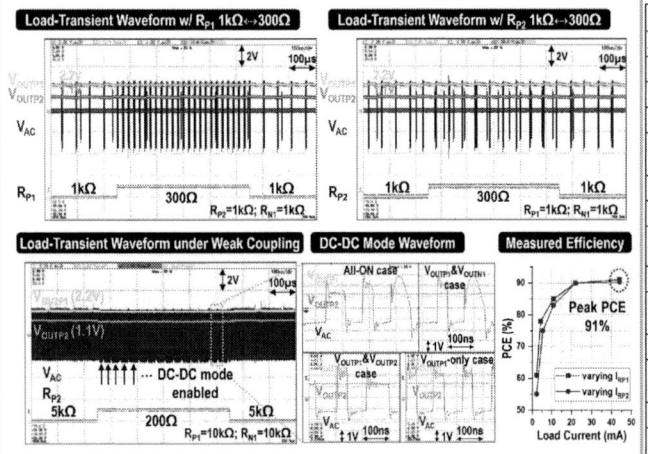

Figure 9.2.5: Measured load-transient waveforms, DC-DC mode waveform, and power conversion efficiency. (R_{P1}, R_{P2}, R_{N1} represent the load resistance at V_{OUTP1}, V_{OUTP2}, and V_{OUTN1}; I_{RP1} and I_{RP2} represent the corresponding load currents.)

	JSSC'23 [4]	JSSC'24 [9]	JSSC'24 [3]	ISSCC'25 [5]	ISSCC'25 [1]	This work
Technology	180nm CMOS	180nm BCD	180nm BCD	180nm BCD	180nm CMOS	180nm BCD
Frequency	6.78MHz	6.78MHz	6.78MHz	6.78MHz	6.78MHz	6.78MHz
LC Tank Topology	Parallel	Series	Parallel	Series	Parallel	Parallel
Chip Area	2.32mm²	8.47mm²	0.77mm²	3.24mm²	1.63mm²	2.84mm²
Number of Power Switches	6	6	2	2	5	5
Number of Outputs	2	2	2	2	2	4
Bipolar Output	No	Yes	No	Yes	No	Yes
Output Voltages	5V, 3.7V	5V, -5V	3.3V, 1.8V	1V, -1V	3.3V, 1.6V	2.2V, 1.1V, -1.8V, -2.2V
Output Ripples	35mV	45mV	100mV*	100mV*	≤75mV	≤100mV
Load-Transient Recovery Time	negligible	12µs	negligible	n/r	negligible	negligible
Cross Regulation	unnoticeable	unnoticeable	unnoticeable	n/r	unnoticeable	unnoticeable
DC-DC Enhancing Operation	Yes	No	No	No	No	Yes
Peak PCE	91.8%	86.4%	92.9%	72.5%	92.2%	91%
Maximum Output Power	300mW	5W	90.5mW	n/r	131mW	173mW

*estimated from papers. n/r: not reported.

Figure 9.2.6: Comparison table.

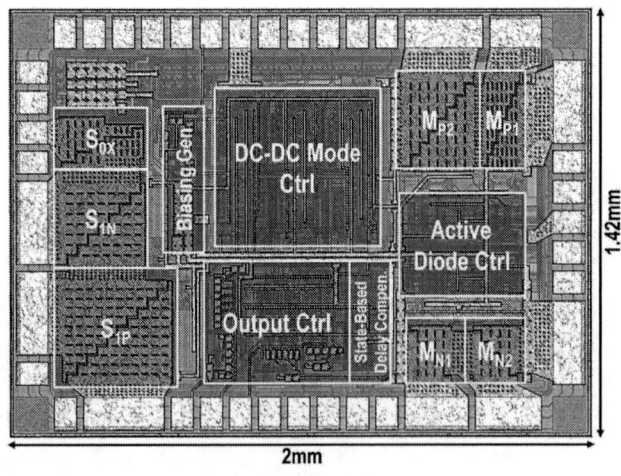

Figure 9.2.7: Chip micrograph.

• 2026 IEEE International Solid-State Circuits Conference

ISSCC 2026 / SESSION 9 / WIRELESS POWER / 9.3

9.3 An Output-Domain-Independent Single-Transmitter-Dual-Receiver Wireless Power Transfer System with Detuned-Tank and Time-Multiplexing Control for Adaptive Power Distribution

Yutang Chen, Yifan Lin, Dihu Chen, Jianping Guo

Sun Yat-Sen University, Guangzhou, China

Abstract

This paper presents an output-domain-independent single-transmitter-dual-receiver wireless power transfer system. The proposed detuned-tank and time-multiplexing control enable adaptive power distribution, preventing power conflicts caused by different receiver impedance (Z_{RX}) and coupling coefficient (k). The system achieves two output domains over k ranging from 0.14 to 0.28, with excellent load regulation of 0.07/0.26mV/mA, a peak efficiency of 82.2% and a maximum output power of 1099mW.

With the growing number of portable, wearable, and implantable devices, multi-output wireless power transfer (WPT) systems have become increasingly important [1-6]. Previous solutions typically realize multiple outputs within a single receiver (RX) and cannot simultaneously deliver multiple outputs across different RX devices (i.e., cannot achieve independent output domains), which limits their application scenarios. To power multiple RXs, multi-transmitter (TX)-multi-RX systems have been proposed [7,8]. However, the use of multiple TX tanks can lead to magnetic field cancellation, resulting in higher power consumption. Clearly, a single-TX-multiple-RX WPT system provides a simple and effective solution for achieving independent output domains, as illustrated in Fig. 9.3.1 (top left).

To realize a single-TX-dual-RX (STDR) system, multi-frequency resonating compensation (MFRC) networks have been proposed in [9,10], as illustrated in Fig. 9.3.1 (top right). The MFRC creates low impedance at two frequencies: the TX resonant tank and the first RX (RX1) are tuned to the fundamental frequency ω_0, while the MFRC and the second RX (RX2) are tuned to a higher frequency ω_1 ($\geq 2\omega_0$). This allows both RXs to receive power simultaneously. However, MFRC is unsuitable for MHz-frequency WPT systems with narrow ISM bands, and switching losses increase at high frequencies. An alternative STDR-WPT system was proposed in [11], in which a rectifier followed by a DC-DC buck converter is adopted at RX to achieve voltage regulation. However, the cascaded power stage reduces overall power efficiency. Moreover, the aforementioned works [9-11] were implemented with discrete components, limiting integration.

The single-stage reconfigurable resonant regulating (R^3) rectifier [12] can significantly improve efficiency, but it introduces power conflicts in STDR-WPT systems. As shown in Fig. 9.3.1 (middle left), the 0X mode presents a low RX impedance Z_{RX}, corresponding to a high reflected impedance Z_{RL}, whereas the 1X mode exhibits the opposite behavior. Consequently, an RX operating in 0X mode absorbs most of the TX power, but this power is largely wasted in freewheeling and does not contribute to the output V_O. Conversely, the RX in 1X mode receives insufficient input power, resulting in poor load regulation, particularly under heavy-load conditions. To address this power conflict, an open-mode RX for isolated DC-DC conversion was proposed [13], which operates similarly to the R^3 rectifier in WPT systems. However, the open-mode RX cannot function in a single-RX WPT system due to the absence of a freewheeling phase. In addition, the coupling coefficient k also affects Z_{RL} and can exacerbate power conflicts. When both RXs operate in 1X mode, power is preferentially delivered to the higher-k RX, resulting in poor regulation of the RX with lower k.

In this work, a STDR-WPT system with detuned-tank control (DTC) and time-multiplexing control (TMC) is proposed to achieve adaptive power distribution, as illustrated in Fig. 9.3.1 (middle right). The proposed DTC mitigates the impact of Z_{RX} by controlling the GaN FET (EPC2012C) while retaining the freewheeling path necessary for output regulation. Meanwhile, the proposed TMC, enabled via RX-RX communication, alleviates the influence of the coupling coefficient k. As shown in Fig. 9.3.1 (bottom), the DTC introduces two detuned 0X/1X (D0X/D1X) modes, where a capacitor C_{DT} is series-connected into the LC tank to increase Z_{RX} through the added impedance Z_{DT}. Consequently, the reflected impedance Z_{RL} in the detuned mode decreases, reducing the power received by the RX. Additionally, in D0X mode, the input current continues to freewheel, ensuring correct operation in both single-and dual-RX systems. Since Z_{RL} is also affected by k, C_{DT} is chosen as half of C_{RX} to ensure a reasonable power distribution under extreme conditions. Assuming RX1 operates under the lowest coupling coefficient (k_1) and RX2 under the highest ($k_2=2k_1$), the Z_{RL} of RX1 in 1X mode remains higher than that of RX2 in D0X mode. In this way, a sufficiently large Z_{DT} compensates for the influence of k.

Figure 9.3.2 (left) describes the circuit implementations of the proposed RX chip, which integrates an R^3 rectifier, a current sensor, a feedback resistor chain, a GaN driver for M_N, two detectors for input current amplitude and V_O, and a controller. In the DTC, the signal V_{RM} determines whether the rectifier operates in full-bridge rectification or freewheeling, while the signal V_{DTU} turns M_N on or off to detune the LC tank. When the TMC is inactive, the RX will only switch between 1X and D0X modes. Since Z_{RL} in 1X mode is higher than in D0X mode, TX power preferentially flows to the 1X-mode RX, ensuring a reasonable power distribution.

Additionally, the TMC will be enabled via RX-RX communication when both RXs operate in 1X mode simultaneously. The workflows of the RXs are shown in Fig. 9.3.2 (right), and the operation principle of the proposed system is illustrated in Fig. 9.3.3. In period I, V_{O1} in the low-k RX1 rises in 1X mode, while V_{O2} in the high-k RX2 decreases in D0X mode. With the proposed DTC, no power conflict occurs. In period II, when RX2 transitions from D0X mode to 1X mode, TX power preferentially flows to RX2 due to its large k and high Z_{RL}, causing V_{O1} to decrease unexpectedly in 1X mode and degrading load regulation. To address this issue, the proposed RX chip employs a voltage detector to sense power conflict induced by k, with the reference voltage V_{TH1} set below the hysteresis window. In period III, once V_{O1} falls below V_{TH1}, RX-RX communication is enabled. A 2μs blanking signal V_{BLK1} is inserted into V_{RM1}, briefly forcing RX1 into 0X mode. As Z_{RL} is large in 0X mode, the input current i_{RX1} increases, while i_{RX2} in RX2 decreases. In period IV, the amplitude detector in high-k RX2 senses the reduction in i_{RX2}, thereby enabling the TMC. An 8μs clock signal V_{CLK} with 25% duty cycle is inserted into V_{DTU}. Consequently, RX2 switches between D1X and 1X modes: when RX2 is in D1X mode, most power is delivered to RX1, raising V_{O1}; when RX2 is in 1X mode, power flows to V_{O2}. The TMC thus eliminates the power conflict caused by k and ensures proper regulation for both RXs. In period V, if V_{O1} and V_{O2} reach the upper hysteresis window, both RXs enter D0X mode.

In the adopted TMC, D1X mode is chosen over D0X mode for two reasons. First, D1X mode has the lowest Z_{RL} among the four operation modes in the proposed WPT system, allowing low-k RX1 to receive high power. Second, although the input power transferred to high-k RX2 is insufficient, it can still be delivered to V_{O2}. In the clock generator, a relatively long period is preferred to suppress ringing and instability of i_{RX1} and i_{RX2} during mode switching, providing steady and sufficient power delivery. Meanwhile, a 25% duty cycle is selected to charge V_{O1} for a longer time, thereby improving the load regulation. Figure 9.3.2 (bottom) shows the proposed amplitude detector enabled by signal V_{EN}. This detector is activated under the following conditions: 1) the RX is in 1X mode (i.e., $V_{HY}=1$); 2) V_O is within the hysteresis window (i.e., $V_{DVO}=0$); and 3) the input current is sufficient (i.e., $V_{AVG}>V_{TH3}$, resetting V_{DIRX}). Additionally, an RC low-pass filter is incorporated to suppress narrow mis-triggered pulses generated by the input current ringing.

Figure 9.3.4 shows the measured steady-state waveforms under different load currents. In Fig. 9.3.4 (top), with 50mA I_{L1} and 165mA I_{L2}, both outputs V_{O1} and V_{O2} are properly regulated at 4V using the proposed DTC. Since their 1X modes rarely overlap, the TMC remains inactive. In Fig. 9.3.4 (middle), when I_{L1} increases to 70mA, the original duration of 1X mode becomes insufficient to support the load, causing V_{O1} to drop. When V_{O1} falls below V_{TH1}, RX-RX communication is triggered, activating the TMC in the high-k RX2. This enables adaptive power distribution and simultaneous charging of both outputs, thereby eliminating the impact of k. Figure 9.3.4 (bottom) compares the output voltages with and without the TMC, showing that the TMC significantly improves load regulation, especially under heavy-load conditions.

Figure 9.3.5 (top) shows the measured load transient response waveforms, where I_{L2} is fixed at 165mA, while I_{L1} switches between 10mA and 102mA. Both outputs maintain correct regulation, with unnoticeable voltage droop or recovery time. As depicted in Fig. 9.3.5 (bottom left), with proposed DTC and TMC, the load regulation in the low-k RX1 is significantly improved, especially under heavy load, decreasing from 1.27mV/mA to 0.07mV/mA. The power efficiency is shown in Fig. 9.3.5 (bottom right), with a peak efficiency of 82.2%, and a maximum output power of 1099mW.

Figure 9.3.6 summarizes the performance and compares it with state-of-the-art WPT systems. To the best of our knowledge, this is the first reported output-domain-independent STDR-WPT system with highly integrated RX chips, in which power conflicts caused by Z_{RX} and k in conventional topologies are eliminated, and adaptive power distribution is achieved through the proposed DTC and TMC strategies. The measurement results show that two independent output domains can be well generated, with good load regulation of 0.07/0.26 mV/mA. The measured peak efficiency is 82.2%, and the maximum output power is 1099mW. The output voltages recover instantly, with negligible voltage droop during load transients. Figure 9.3.7 shows the RX chip micrograph, occupying an area of 0.9×1.75mm². The TX and RX coils are designed with inductances of 2.2μH and 551nH, and parasitic resistances of approximately 620mΩ and 310mΩ, respectively. The measured k ranges from 0.28 to 0.14 as the coil distance varies from 4 to 10mm.

Acknowledgement:
This work was partly supported by the Key Area R&D Program of Guangdong Province under Grant 2022B0701180001, and the National Natural Science Foundation of China under Grant 62274189. Corresponding author: Jianping Guo.

References:
[1] J. Lin et al., "A Single-Stage Dual-Output Regulating Rectifier With Hysteretic Current-Wave Modulation," *IEEE JSSC*, vol. 56, no. 9, pp. 2770-2780, Sep. 2021. https://doi.org/10.1109/JSSC.2021.3071221

[2] Y. Chen et al., "A 2-W, 90%-Efficiency Single-Stage Dual-Output Wireless Power Receiver with 0.1 to 700-mA Output Current Range Through Dynamic Delay Compensation and Bootstrap Adaptive Body Biasing Circuit," *A-SSCC*, pp. 1-3, Nov. 2023. https://doi.org/10.1109/A-SSCC58667.2023.10348004

[3] F. Wang et al., "A Hybrid-Resonance Single-Stage Dual-Output Rectifier With High Voltage Difference for Wireless Power Transfer System," *A-SSCC*, pp. 1-3, Nov. 2024. https://doi.org/10.1109/A-SSCC60305.2024.10848581

[4] H.-S. Lee et al., "A 90.8%-Efficiency SIMO Resonant Regulating Rectifier Generating 3 Outputs in a Half Cycle with Distributed Multi-Phase Control for Wirelessly-Powered Implantable Devices," *ISSCC*, pp. 448-449, Feb. 2024. https://doi.org/10.1109/ISSCC49657.2024.10454403

[5] T. Lu and S. Du, "A 92.3%-Efficiency Switching-Mode Dual-Output Regulating Rectifier With Improved Link Adaptability for Wireless Power Transfer," *IEEE JSSC*, vol. 60, no. 7, pp. 2354-2366, Jul. 2025. https://doi.org/10.1109/JSSC.2025.3540596

[6] Q. Zhuang et al., "A 6.78MHz Single-Stage Regulating Rectifier with Dual Outputs Simultaneously Charged in a Half Cycle Achieving 92.2% Efficiency and 131mW Output Power," *ISSCC*, pp. 188-189, Feb. 2025. https://doi.org/10.1109/ISSCC49661.2025.10904779

[7] H. Qiu et al., "A 6.78-MHz Coupling Coefficient Sensorless Wireless Power Transfer System Charging Multiple Receivers With Efficiency Maximization by Adaptive Magnetic Field Distributor IC," *IEEE TCAS-I*, vol. 71, no. 2, pp. 974-983, Feb. 2024. https://doi.org/10.1109/TCSI.2023.3340681

[8] X. Wang et al., "Multioutput Wireless Charger for Drone Swarms With Reduced Switch Requirements and Independent Regulation Capability," *IEEE TIE*, vol. 71, no. 5, pp. 4883-4895, May 2024. https://doi.org/10.1109/TIE.2023.3277116

[9] Z. Zhang et al., "Multiple-Frequency Resonating Compensation for Multichannel Transmission of Wireless Power Transfer," *IEEE TPEL*, vol. 36, no. 5, pp. 5169-5180, May 2021. https://doi.org/10.1109/TPEL.2020.3027916

[10] Y. Gong et al., "Selected-Interharmonic-Injected Pulse Density Modulation for One-to-Many WPT Systems," *IEEE TPEL*, vol. 39, no. 9, pp. 11784-11793, Sep. 2024. https://doi.org/10.1109/TPEL.2024.3402209

[11] M. Fu et al., "Megahertz Multiple-Receiver Wireless Power Transfer Systems With Power Flow Management and Maximum Efficiency Point Tracking," *IEEE TMTT*, vol. 65, no. 11, pp. 4285-4293, Nov. 2017. https://doi.org/10.1109/TMTT.2017.2689747

[12] J. Ge et al., "A 6.78-MHz 79.5%-Peak-Efficiency Wireless Power Transfer System using a Wireless Mode-Recognition Technique and a Fully-On/off Class-D Power Amplifier," *ISSCC*, pp. 446-447, Feb. 2024. https://doi.org/10.1109/ISSCC49657.2024.10454467

[13] J. Jiang et al., "A Single-Link Multi-Domain-Output (SLiMDO) Isolated DC-DC Converter with Passive Magnetic Flux Sharing for Local Energy Distribution and Rx Behavior Sensing-Based Global Power Modulation," *ISSCC*, pp. 530-531, Feb. 2025. https://doi.org/10.1109/ISSCC49661.2025.10904792

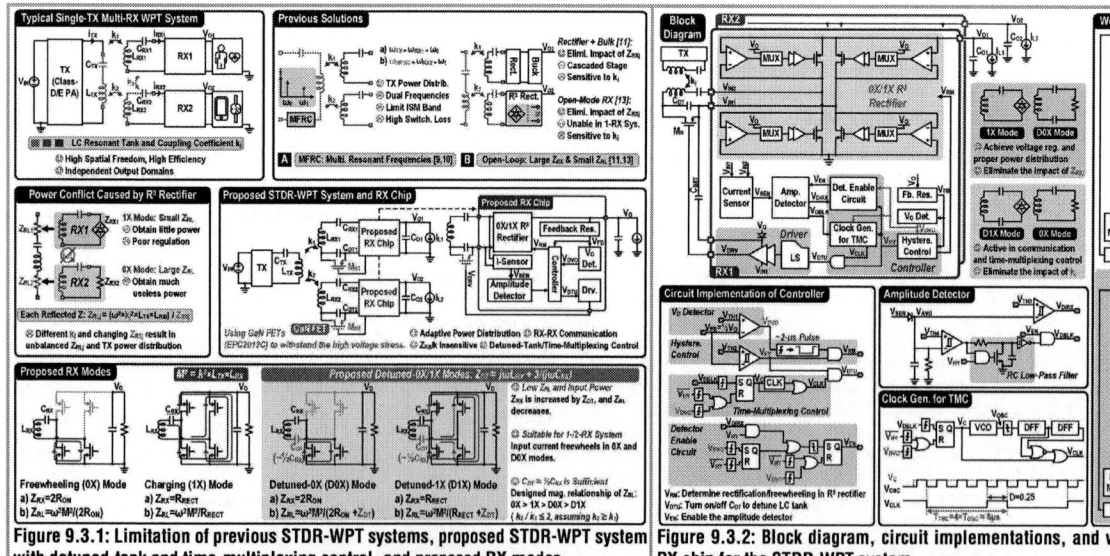

Figure 9.3.1: Limitation of previous STDR-WPT systems, proposed STDR-WPT system with detuned-tank and time-multiplexing control, and proposed RX modes.

Figure 9.3.2: Block diagram, circuit implementations, and workflow of the proposed RX chip for the STDR-WPT system.

ISSCC 2026 / SESSION 9 / WIRELESS POWER / 9.3

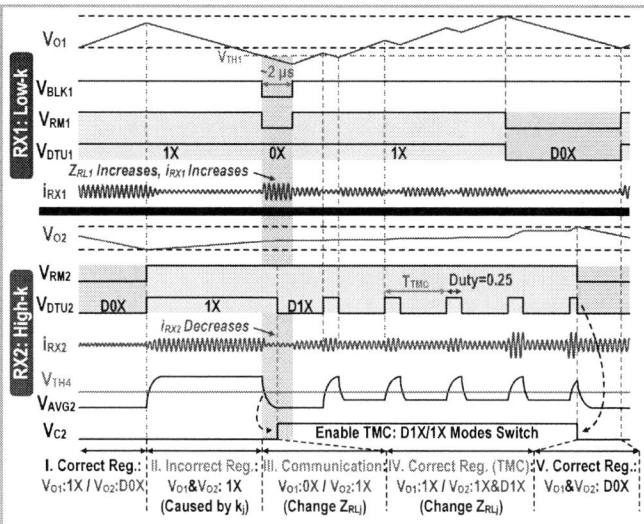

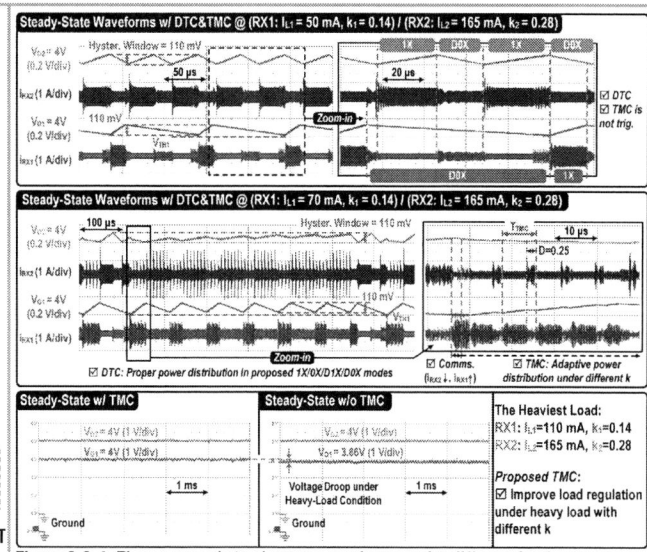

Figure 9.3.3: Timing diagram and operation principle of the proposed STDR-WPT system, illustrated under a low-k RX1 and high-k RX2 condition.

Figure 9.3.4: The measured steady-state waveforms under different load currents.

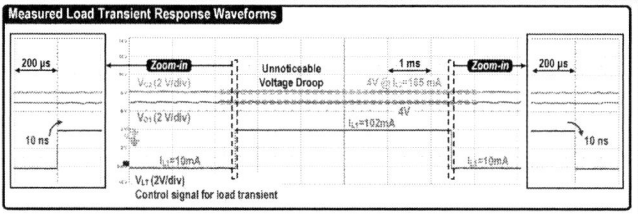

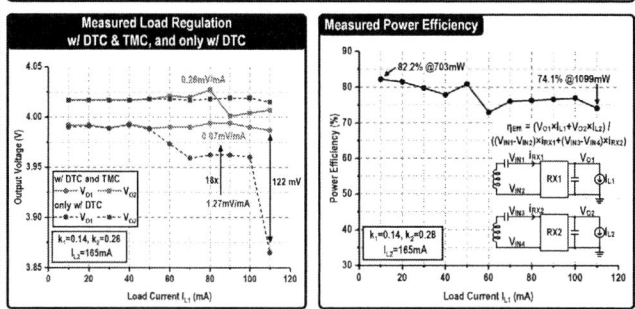

Figure 9.3.5: The measured load transient response, load regulation, and power efficiency.

	[1] JSSC 2021	[4] ISSCC 2024	[5] JSSC 2025	[6] ISSCC 2025	This work
Technology	180nm CMOS	250nm CMOS	180nm CMOS	180nm CMOS	180nm CMOS
RX Tank	Series	Parallel	Parallel	Parallel	Series
RX Structure	SSDO R^3 Rectifier	SIMO R^3 Rectifier	Dual-Output Voltage Doubler	SSDO R^3 Rectifier	Detuned R^3 Rectifier
Frequency (MHz)	6.78	2	6.78	6.78	6.78
# of Outputs	2	3	2	2	2
Independent Output Domains	1	1	1	1	2
RX-RX Comms.?	No	No	No	No	Yes
Adaptive Power Distribution	None	None	None	None	Detuned-Tank/ Time-Multiplexing Ctrl.
V_O (V)	1.8/3.3	1-4.5	1.8/3.3	1-3.3	4
Max. P_O (mW)	1020	135.53	171	131	1099
Trans. V_O Droop	Unnoticeable	N/A	Unnoticeable	Unnoticeable	Unnoticeable
Recovery Time	Instant	N/A	Instant	Instant	Instant
Load Reg. (mV/mA)	0.23	N/A	N/A	N/A	0.07/0.26
Peak Efficiency (%)	91.9	90.82	92.3	92.2	82.2
Range of k	N/A	N/A	N/A	N/A	0.14-0.28

Figure 9.3.6: Performance summary and comparison with the state-of-the-art WPT systems.

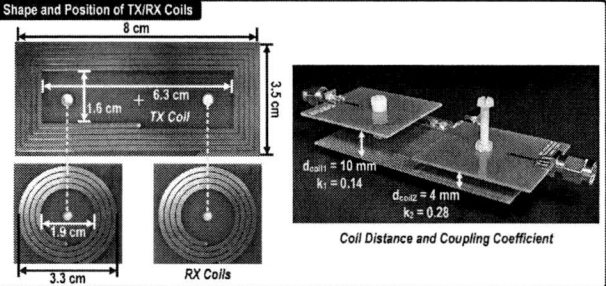

6.78-MHz Resonant Link		
	TX Tank	RX Tank
L_{TX}	2.2 µH	L_{RX}	551 nH
C_{TX}	250 pF	C_{RX}	970 nF
\	\	C_{DT}	470 pF
R_{TX}	620 mΩ	R_{RX}	310 mΩ

Figure 9.3.7: Die micrograph of RX chip, and information of TX/RX tanks.

• 2026 IEEE International Solid-State Circuits Conference

ISSCC 2026 / SESSION 9 / WIRELESS POWER / 9.4

9.4 A Multi-Coil Scalable Energy-Shared Wireless Power Receiver Network for Distributed Time-Division-Multiplexing Somatosensory Cortex Stimulation

Kai Cui, Yan Lu

Tsinghua University, Beijing, China

Abstract

This work presents a system-level scalable wireless power receiver (RX) network for miniaturized distributed somatosensory cortex stimulation. By interconnecting the power outputs of all RX cells in the network, the stimulator in each cell can access the total received power of the whole network, with a time-division-multiplexing stimulation manner. Thus, each cell owns a multiple-input single-output wireless power RX network without additional RX coils, with improved efficiency and robustness.

Brain-computer interfaces (BCIs) hold great promise for restoring motor and sensory function in individuals with severe limb disabilities. Recent advances have enabled intuitive somatosensory feedback, such as touch, vibration, and force sensations, from contact events in brain-controlled bionic limbs, significantly improving their usability and user acceptance [1, 2]. Delivering precise tactile feedback requires distributed micro-stimulation across large cortical neuron networks [3], often demanding hundreds of independently driven stimulus channels placed across multiple neural sites. Minimizing device dimensions to reduce foreign body reactions and inflammation further complicates power delivery, making wireless power transfer (WPT) indispensable. Near-field resonant inductive powering is widely used; however, the small size of the receiver (RX) coil in typical two-coil links results in a very low power transfer efficiency [4-7], along with limited transmission distance and poor robustness to frequent micromotions of the brain. Recent efforts have improved robustness using an omnidirectional wireless power reception, but the required 3D coil geometries increase the RX volume [8]. Three-coil configurations, which add a relay coil in the implant plane, can enhance efficiency and extend range by partially compensating for misalignments [3, 9-11], and have thus become widely adopted. Yet, due to the large size mismatch between the external transmitter (TX) and miniature RX coils, gains from the relay alone remain limited and typically must be combined with additional system- and circuit-level optimizations. In this work, we present a scalable system-level wireless power RX network tailored for miniaturized distributed stimulation. By interconnecting the power outputs of all RX cells in the network, the stimulation site in each cell can access the total received power of the entire network through a time-division multiplexing (TDM) stimulation scheme. As a result, every cell effectively benefits from a multiple-input single-output wireless power RX network without requiring additional RX coils, achieving improved efficiency and enhanced robustness.

Figure 9.4.1 (top) illustrates the distributed BCI concept powered by the proposed RX network. The network is implemented on a flexible PCB substrate that interconnects scalable RX cells, enabling energy sharing and distribution across all cells while preserving their free-floating characteristics. The RX network offers three key features: (1) Interconnecting the power-stage outputs of all RX cells allows each cell to access the total received power of the entire network; (2) Even under severe coil misalignment, an individual cell can still receive energy through the network from other better-aligned cells, thereby imporving robustness; (3) Each cell can initiate leader-follower control of other cells' power stages through the control interconnect, ensuring proper regulation of output power. This work demonstrates power transfer and sharing from a single TX to a three-cell chip-based RX network, and the architecture readily scales to ab arbitrary number of cells (N) without loss of generality.

Typically, the electrical stimulus pulses occupy only a small duration of the stimulation priod, and thus this work employs TDM stimulation to maximize each RX cell's power. Figure 9.4.1 (bottom) shows a wireless micro-stimulation system block diagram incorporating the proposed network. In multi-site TDM stimulation for distributed BCIs, different cells activate by detecting unique IDs via a forward data link. During a cell's enable phase, its stimulation function activates, and the RX network delivers power to its stimulus load via the interconnect. Thus, each cell consistently achieves multi-coil powering. As illustrated in Fig. 9.4.2 (top), in most conventional wireless power systems for distributed BCIs, each RX is powered solely by a single coil, limiting the available received power. Moreover, variations in coil orientation and position readily reduce received power, resulting in inconsistent output levels. In contrast, the wireless power RX network allows each cell to harvest energy from all networked coils, ensuring that all cells maintain consistent output power ranges. These features significantly enhance the stability and reliability of the stimulation power supply.

Figure 9.4.2 (bottom) shows the system architecture and control timing of the cell chip in the proposed RX network. In this work, a resonant current-mode (RCM) WPT scheme is adopted. In RCM operation, the LC-tank circuit undergoes multiple resonance cycles to progressively accumulate energy, which is then directly transferred to the output once the inductor is fully energized. Note that RCM WTP is ideally suited for multi-coil powering, as the resonant coil in charging phase can be regarded as a current source, and connecting multiple coils in parallel will not diminishes their effectiveness [8]. Key components of the RX cell include an RCM power stage, an RCM controller, a hysteresis comparator, and a mode-selection circuit. The output voltage is regulated through a hysteresis feedback control loop, which configures the RCM power stage into 1X and 0X modes via the hysteresis comparator. In 1X mode, the RX charges the output using RCM operation, while in 0X mode, both PMOS and NMOS transistors are turned. Since the equivalent input impedance of the power stage in 0X mode is very large, the AC current I_{AC} and the AC voltage V_{AC} are nearly negligible. The operation principle is as follows: the feedback voltage V_{FB} is compared with a reference voltage to generate the mode-control signal MD. When the output voltage V_{OUT} exceeds the upper bound of the hysteresis window, the RX enters 0X mode, temporarily halting power delivery to the output. When V_{OUT} falls below the lower bound of the hysteresis window, the RX return to 1X mode.

In the proposed wireless power RX network, any stimulus-activated cell shares the power of all other cells in the network to. In general, as shown in Fig. 9.4.3 (top), the power delivered to the load can be only adjusted by the local hysteresis feedback loop in each cell. In this way, the complexity of the control circuits design can be simplified in the system level. However, due to the difficulty of achieving good matching among key modules such as the feedback voltage divider, bandgap reference, and hysteresis comparator across different chips, offsets occur in the hysteresis window. This reduces system stability and increases output ripple. Overall, the lack of direct output power control poses safety risks, which is unacceptable for implantable medical devices. To address this issue, a control interconnection is introduced to transfer the control signal MD form the stimulus-activated cell to others, and thus all cells can share the same hysteresis feedback loop, which can be regarded as a global leader-follower hysteresis control. Note that only one hysteresis comparator is enabled, and the power loss can be reduced. In the charging phase during each RCM cycle, an exact zero-current switching (ZCS) is typically required to prevent the revise current form output to the ground once the energy of the coil is completely released. However, as shown in Fig. 9.4.3 (bottom), the frequent micromotions of the brain induce fluctuations in the power delivered to the RX LC tank, significantly altering the duration of the charging phase, which in turn varies the timing of the zero current. Therefore, the ZCS controller should promptly track the zero-current instant to further optimize efficiency. For this purpose, a simple fast zero-current tracking technique is proposed based on the common delay-line controlled ZCS. During normal operation, a dynamic comparator DCMP controls a 32b bidirectional shift register to adjust the 32b delay line with a slow digital feedback loop, and the output of DCMP will switch between '0' and '1' periodically in the steady state. If the angel/position varies, the large changing of the zero-current timing will be sensed by the 3b monitor after three consecutive RCM cycles. The signal Q[0:2] of '000' indicates that the high-side PMOS transistor was turned off too late, while '111' indicates an early turn-off. In this case, the high-side switch is forced into diode conduction mode over one resonant cycle, and this results in an optimal duration (reflected by t_{DIO}) for the charging phase. Meanwhile, t_{DIO} is quantitated by the delay line and stored in the shift register, which can be directly used in the following RCM cycles. Thus, the fast calibration is achieved.

In this work, a 3-cell RX network powered by a 3-coil link mentioned in [9] is mainly chosen to verify the functionality and the performance. The TX is implemented with a half-bridge Class-D power amplifier supplied by a 1.8V DC source. The 34mm TX coil has an inductance of 47.4nH with a quality factor (Q) of 202, while the 24mm relay coil exhibits an inductance 42.4nH and a Q of 131. For the 2mm RX coil in each cell, the simulated inductance is 22.59nH with a Q of 18.8. To verify the effectiveness of the power and control interconnection, the stimulus load of Cell 1 (emulated by the resistor R_{L1}) is activated, and the output voltage V_{OUT} is regulated using the proposed leader-follower hysteresis control, as illustrated in Fig. 9.4.4 (top). It can be clearly observed that all RCM power stages are synchronously controlled, periodically switching between 0X and 1X modes to regulate the output voltage V_{OUT}. Additionally, the resonant and charging phases in RCM wireless power receiving cycle are also distinctly visible in Fig. 9.4.4 (bottom). The measured waveforms of the TDM procedure is shown in Fig. 9.4.5 (top), demonstrating that the network power is delivered to different loads in different phases without cross regulation. Figure 9.4.5 (bottom) presents the power performance analysis of the proposed wireless power RX network. As the number of cells in the RX network increases, both the output power and the overall power transfer efficiency (PTE) improve substantially. In particular, when

ISSCC 2026 / February 16, 2026 / 4:50 PM

expanding to a 3-cell network yields approximately a 3× increase in output power compared to a single cell, while the efficiency improves by nearly 2.7×. Fig. 9.4.6 summarizes and compares the proposed RX network with prior work. The chip in each cell is fabricated in a 0.18μm BCD process. Photos of the TX and RX modules, along with RX and TX chip micrographs and the implant mock-up are shown in Fig. 9.4.7.

Acknowledgement:
This work was supported in part by the National Natural Science Foundation of China (92573201).

References:
[1] G. Valle et al., "Tactile Edges and Motion via Patterned Microstimulation of the Human Somatosensory Cortex," *Science*, vol. 387, no. 6731, pp. 315-322, Jan. 2025. https://doi.org/10.1126/science.adq5978
[2] C. M. Greenspon et al., "Evoking Stable and Precise Tactile Sensations via Multi-Electrode Intracortical Microstimulation of the Somatosensory Cortex," *Nat. Biomed. Eng*, vol. 9, no. 7, pp. 935–951, June 2025. https://doi.org/10.1038/s41551-024-01299-z
[3] A.-H. Lee et al., "Patterned Electrical Brain Stimulation by a Wireless Network of Implantable Microdevices," *Nat Commun*, vol. 15, no. 10093, pp. 1-14, Nov. 2024. https://www.nature.com/articles/s41467-024-54542-1
[4] W. Biederman et al., "A Fully-Integrated, Miniaturized (0.125mm²) 10.5mW Wireless Neural Sensor," *IEEE JSSC*, vol. 48, no. 4, pp. 960–970, Apr. 2013. https://doi.org/10.1109/JSSC.2013.2238994
[5] A. Khalifa et al., "The Microbead: A 0.009 mm³ Implantable Wireless Neural Stimulator," *IEEE TBCAS*, vol. 13, no. 5, pp. 971-985, Oct. 2019. https://doi.org/10.1109/TBCAS.2019.2939014
[6] C. Kim et al., "A 3 mm × 3 mm Fully Integrated Wireless Power Receiver and Neural Interface System-on-Chip," *IEEE TBCAS*, vol. 13, no. 6, pp. 1736-1746, Dec. 2019. https://doi.org/10.1109/TBCAS.2019.2943506
[7] H. Rahmani et al., "A Wirelessly Powered Reconfigurable FDD Radio with On-Chip Antennas for Multi-Site Neural Interfaces," *IEEE JSSC*, vol. 56, no. 10, pp. 3177-3190, Oct. 2021. https://doi.org/10.1109/JSSC.2021.3076014
[8] J.-H. Kim et al., "A Programming-Free Three-Dimensional Resonant Current-Mode Wireless Receiver with Real-Time Link-Adaptivity and a 0.904cm³ Receiver Coil for Implantable Systems," *ISSCC*, pp. 580-581, Feb. 2025. https://doi.org/10.1109/ISSCC49661.2025.10904511
[9] S.A. Mirbozorgi et al., "Robust Wireless Power Transmission to mm-Sized Free-Floating Distributed Implants," *IEEE TBCAS*, vol. 11, no. 3, pp. 692–702, Jun. 2017. https://doi.org/10.1109/TBCAS.2017.2663358
[10] V.W. Leung et al., "A CMOS Distributed Sensor System for High-Density Wireless Neural Implants for Brain–Machine Interfaces," *ESSCIRC*, pp. 230–233, Sep. 2018. https://doi.org/10.1109/ESSCIRC.2018.8494335
[11] G.L. Barbruni et al., "A Frequency-Switching Inductive Power Transfer System for Wireless, Miniaturised and Large-Scale Neural Interfaces," *IEEE TBCAS*, vol. 18, no. 3, pp. 679-690, June 2024. https://doi.org/10.1109/TBCAS.2024.3359481

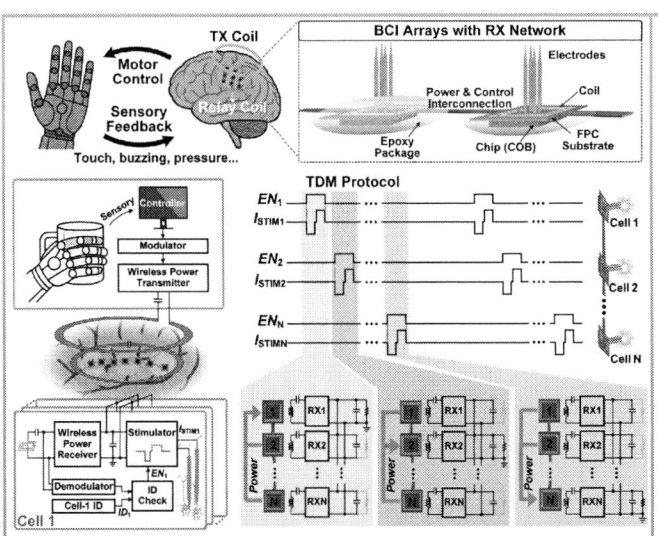

Figure 9.4.1: Concept of the proposed multi-coil scalable wireless power RX network (top). Working mechanism of the RX network based on the TDM stimulation (bottom).

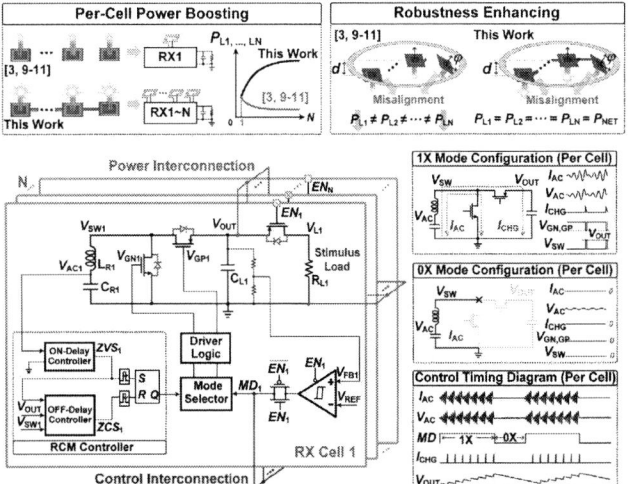

Figure 9.4.2: Main contributions of this work compared with the prior arts (top). System architecture of the RX network with the configurations and the control timing diagram of per cell (bottom).

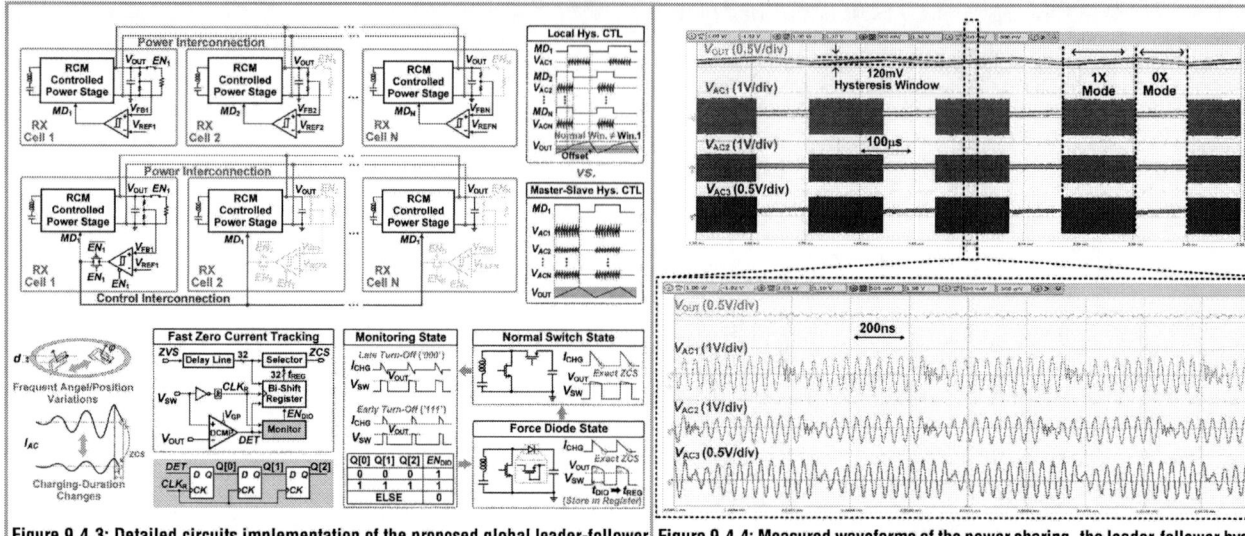

Figure 9.4.3: Detailed circuits implementation of the proposed global leader-follower hysteresis feedback control (top) and fast zero current tracking (bottom).

Figure 9.4.4: Measured waveforms of the power sharing, the leader-follower hysteresis control, and the detailed RCM wireless power receiving for a 3-cell RX network.

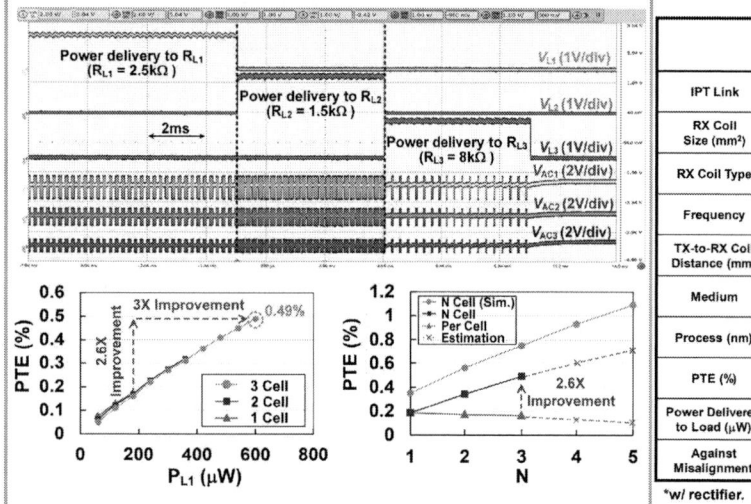

Figure 9.4.5: Measured waveforms of the TDM procedure for a 3-cell RX network (top). Measured PTE performance (bottom).

	JSSC 2013 [4]	TBCAS 2019 [5]	TBCAS 2017 [9]	ESSCIRC 2018 [10]	TBCAS 2024 [11]	This Work
IPT Link	2-Coil	2-Coil	3-Coil	3-Coil	3-Coil	3-Coil
RX Coil Size (mm²)	0.25x0.5	0.3x0.3	1x1	0.5x0.5	0.2x0.2	2x2
RX Coil Type	On-chip spiral	On-chip spiral	WWC	On-chip spiral	On-chip spiral	FPCB spiral
Frequency	1.5 GHz	1.18 GHz	60 MHz	915 MHz	433.92 MHz	27.12 MHz
TX-to-RX Coil Distance (mm)	1	6.6	16	8	14	16
Medium	Air	Beef	Air	Liquid phantom	Tissue	Air
Process (nm)	65	130	N/A	65	180	180 BCD
PTE (%)	0.021	0.0019*	2.4	0.019~0.047	0.013	0.49* (@3 Cell) 0.17* (@1 Cell)
Power Delivered to Load (µW)	10.5	55.5	1300	95~235	1970	600 (@3 Cell) 200 (@1 Cell)
Against Misalignment	No	No	Yes	Yes	Yes	Yes

*w/ rectifier.

Figure 9.4.6: Performance summary and comparison with prior arts.

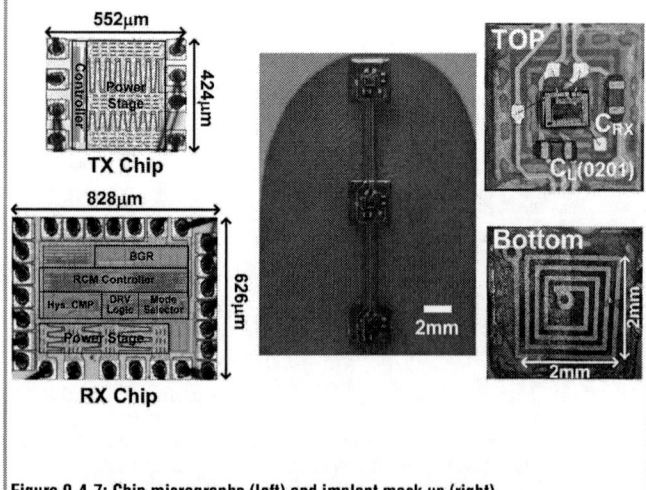

Figure 9.4.7: Chip micrographs (left) and implant mock-up (right).

ISSCC 2026 / SESSION 10 / DIGITAL PROCESSING AND CIRCUIT TECHNIQUES / OVERVIEW

Session 10 Overview: *Digital Processing and Circuit Techniques*
DIGITAL CIRCUITS SUBCOMMITTEE

Session Chair: Visvesh S. Sathe
Georgia Institute of Technology
Atlanta, GA

Session Co-Chair: Ping-Hsuan Hsieh
National Tsing Hua University
Hsinchu, Taiwan

Digital circuits continue to enable an increasingly broader range of applications. The first paper describes a Software-Defined Vehicle (SDV)-driven automotive 3nm SoC to achieve ASIL-D. The second paper reports energy-reliability optimizations for a mobile CPU. The next three papers explore circuit techniques for energy-efficient digital datapaths, clocking and voltage guardband reduction. In the second half of the session, advances in 3D integration are demonstrated through a 2nm-3nm hybrid-bonded die-stacked DNN processor, following by four papers that report recent advances in optimization solvers.

8:00 AM

10.1 A 3nm, 400TOPS, 1080k DMIPS SoC with Chiplet Support for ASIL D Automotive Cross-Domain Applications
Shiro Machida, Renesas Electronics, Tokyo, Japan
In Paper 10.1, Renesas presents a 312mm^2 SDV-driven automotive 3nm SoC with Freedom From Interference (FFI) functional safety techniques to achieve ASIL D. The system operates at a core voltage of 0.765V and an APU overdrive voltage of 0.865V, and comprises a 32-core APU cluster achieving 2.7GHz and 1,080k DMIPS, a 24-core NPU cluster at 1.066GHz and 400TOPS, and 51.2GB/s inter-chiplet bandwidth with UCIe.

8:25 AM

10.2 A Dynamic Performance Augmentation in a 3nm-Plus Mobile CPU
Chien-Yu Lu, MediaTek, Hsinchu, Taiwan
In Paper 10.2, MediaTek describes a dynamic performance augmentation technique for a mobile CPU, featuring boosting-duty control and an adaptive thermal cooler to boost CPU maximum performance. The 3nm-plus multicore CPU occupies 24.69mm^2 and boosts F_{max} to 4.4GHz, achieving a score of 3917 in the GeekBenchv6 single-core benchmark on a flagship smartphone, with the adaptive performance boosting circuitry consuming 0.078% of total power and 0.0122% CPU area.

8:50 AM

10.3 A 2nm Clock-Edge Architecture for Processor Clock-Power Reduction
Yimai Peng, Qualcomm, Raleigh, NC
In Paper 10.3, Qualcomm presents a 2nm NPU matrix-multiplication unit occupying 0.170mm^2 featuring dual-edge-triggered flip-flops and clock-gating circuits with an adaptive clock duty-cycle controller of 675μm^2, realizing 39-to-40% lower clock power and a total dynamic power reduction of up to 15%.

9:15 AM

10.4 A 0.008mm^2 16-to-1600MHz All-Digital Fractional Divider Using AUX-DLL for Background LMS-Based DTC Calibration
Ahmed Elkholy, Broadcom, Irvine, CA
In Paper 10.4, Broadcom presents a 7nm fractional divider with a replica-free least-mean-square-based digital-to-time converter background calibration using an auxiliary delay-locked loop. The proposed FDIV consumes 10mW at 1.6GHz, 0.9V and provides an output frequency range of 16 to 1600MHz with worst-case integrated jitter of 350fs$_{rms}$, occupying 0.008mm^2.

ISSCC 2026 / February 17, 2026 / 8:00 AM

9:30 AM

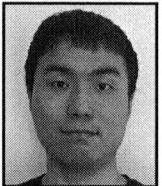

10.5 Proactive Power Management-Based Supply Regulation with Online Learning for Variation-Tolerant Workload-Aware Droop Mitigation in 28nm CMOS

Xi Chen, Northwestern University, Evanston, IL

In Paper 10.5, Northwestern University presents integrated proactive power management for droop mitigation by combining a neural droop management unit, integrated high-speed power converter, and an online learning engine to combat the PDN and workload variations. The 28nm SoC, comprising a CPU operating at 1.2GHz and an accelerator operating at 600MHz under a nominal 1V supply, occupies 2.58mm² and achieves a peak converter efficiency of 91.7% across an output voltage range of 0.5 to 1.2V and an average of 79mV droop reduction and 48.5× throttling reduction.

10:05 AM

10.6 A Hybrid-Bonded 12.1TOPS/mm² 56-Core DNN Processor with 2.5Tb/s/mm² 3D Network on Chip

Phil C. Knag, Intel, Hillsboro, OR

In Paper 10.6, Intel presents a 3D stacked DNN processor leveraging hybrid bonding of Intel 18A and Intel 3 dies in a 3D network on chip design. The processor occupies 2.74mm² and achieves peak bandwidth of 7.0Tb/s, peak AI performance density of 12.1TOPS/mm² at 1.1V, 1.205GHz, and peak dynamic energy efficiency of 16.1TOPS/W at 0.5V, 280MHz.

10:30 AM

10.7 A 28nm Mode-Reconfigurable CAM-CIM Hybrid Complete 3-SAT Solver Supporting Conflict-Driven Clause Learning with 100% Solvability

Zihan Wu, Peking University, Beijing, China

In Paper 10.7, Peking University presents a 28nm complete 3-SAT solver supporting Conflict-Driven Clause Learning (CDCL) with a mode-reconfigurable CAM-CIM hybrid architecture achieving 4.0-to-8.3μs solution time for SATLIB uf/uuf50-218. The design occupies 0.65mm² and consumes 10.4 to 11.2mW, on average, at 0.95V, 185MHz.

10:55 AM

10.8 COBI: A Degree-of-56 Column-Bipartite Densely Connected Digital Ising Chip with 8b Spin Coefficients

Yihao Wu, University of California, Santa Barbara, CA

In Paper 10.8, the University of California, Santa Barbara, and KAIST present a 65nm digital Ising chip using a column-bipartite topology for solving combinatorial optimization problems. The chip occupies 0.483mm² with a spin area of 7547μm² and achieves sub-72ns solution time for various problems using 64 densely connected spin processing elements with 8b spin coefficients, consuming 518μW at 1.2V, 111.11MHz.

11:20 AM

10.9 SharpSAT: A Heuristic-Learning-Based SAT Accelerator Achieving 0.8μs/16.1μs Solution Time in SAT/UNSAT Cases

Aoyang Zhang, Tsinghua University, Beijing, China

In Paper 10.9, Tsinghua University presents a 28nm heuristic-learning-based SAT solver chip that uses a clause learner to efficiently explore the search space, achieving solution times of 0.8μs for SAT cases and 16.1μs for UNSAT cases having 50 variables and 218 clauses. The chip occupies 0.78mm² and operates at 375MHz and 0.9V, consuming 99.1mW.

11:45 AM

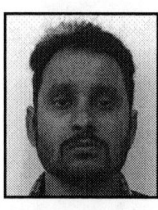

10.10 PCIM-SAT: A 55nm Probabilistic K-SAT Solver with p-Bit-Based Parallel-Variable Update on a Mixed-Signal Compute-in-Memory Architecture

Tinish Bhattacharya, University of California, Santa Barbara, CA

In Paper 10.10, the University of California, Santa Barbara presents a K-SAT solver with p-bit-based parallel-variable update on a mixed-signal CIM architecture. The 55nm chip achieves 5.5μs mean solution time and 265nJ mean solution energy for 50-variable and 218-clause problems operating at 1.3V, 100MHz and occupies 0.42mm².

DIGEST OF TECHNICAL PAPERS • 167

979-8-3315-8937-0/26 $31.00 © 2026 IEEE

ISSCC 2026 / SESSION 10 / DIGITAL PROCESSING AND CIRCUIT TECHNIQUES / 10.1

10.1 A 3nm, 400TOPS, 1080k DMIPS SoC with Chiplet Support for ASIL D Automotive Cross-Domain Applications

Shiro Machida[1], Kazuki Fukuoka[1], Tomoya Onda[1], Nobuhiro Yada[1], Hiroyuki Nakano[1], Sho Yamanaka[1], Hung Van Cao[2], Yusaku Hara[1], Takahiro Irita[1], Kohei Wakahara[1], Cyril Cordoba[3], Tatsuya Kamei[1], Yasuhisa Shimazaki[1]

[1]Renesas Electronics, Tokyo, Japan, [2]Renesas Design Vietnam, Ho Chi Minh, Vietnam, [3]Renesas Electronics Europe, Paris, France

Abstract

This paper presents a 3nm SoC, designed for software defined vehicles, integrating various functions for zone-based computing. The chip includes a 1,080kDMIPS APU, 400TOPS NPU and 51.2GB/s inter-chip bandwidth. Functional safety is ensured by RegionID-based FFI

over UCIe. Power efficiency is achieved with fine-gating across 90+ power domains with enhanced IR-drop control and ASIL D-compliant DCLS. Hierarchical mCPGs reduce clock latency. The chip supports low-power modes such as Sentry and CPD.

In recent years, the Software-Defined Vehicle (SDV) has emerged as a paradigm in the automotive industry. The SDV features a shift from domain-based to zone-based computing, supported by an Electrical/Electronic (E/E) architecture that enables seamless expansion across various applications. This includes most in-vehicle applications and computing domains. For instance, autonomous functions cover advanced driver assistance systems (ADAS) and vehicle control, while cockpit functions encompass in-vehicle infotainment (IVI), instrument clusters, and gateways that connect to external networks. These diverse applications depend on the underlying SDV architecture, which offers a flexible and scalable E/E framework. This framework simplifies the integration and addition of new functions, enabling rapid deployment of advanced vehicle services. Conventional vehicles and their E/E architectures are typically domain specific and exhibit limited scalability, making the addition of new applications complex and slow. Conversely, the SDV architecture overcomes these issues by offering a modular, software-centric framework that supports quick deployment and flexible reconfiguration of in-vehicle functions. However, developing a SDV architecture presents new challenges for system-on-chip (SoC) design, as shown in Fig. 10.1.1. The rising computational requirements from multiple concurrent applications, along with real-time safety demands, call for high-performance, energy-efficient SoCs that support heterogeneous processing units and scalable chiplet extensions. Additionally, integrating autonomous and cockpit functionalities requires careful partitioning and scheduling to ensure deterministic behavior and meet the requirements of automotive safety standards. This paper introduces a SoC design that aims to address these challenges. The proposed architecture focuses on optimizing computing resources, supporting heterogeneous workloads, and achieving real-time performance, enabling efficient integration of autonomous and cockpit systems within SDVs. Results demonstrate improved scalability and compliance with automotive safety standards. Consequently, the single chip achieved Application Processing Unit (APU) operation at 2.7GHz and Neural Processing Unit (NPU) operation at 1.066GHz. Furthermore, in-chiplet operation reached an inter-chip data transfer rate of 51.2GB/s.

Figure 10.1.2 provides an overview of this work using a function block diagram. The implemented NPU delivers over 400TOPS, while the APU achieves 1080k DMIPS, along with conventional SoC features such as DRAM bandwidth and high-speed I/O, including 10Gbps Ethernet and camera and display interfaces. The proposed design meets the overall requirements of the SoC. This integration concept allows the chip to support the SDV E/E architecture, integrating multiple applications, including autonomous driving, central computing, in-vehicle infotainment, and gateways, all on a single chip, while enabling software scalability. Additionally, UCIe (Universal Chiplet Interconnect Express) and PCIe inter-die connections support chiplet configurations by accommodating extra dies, thereby promoting future scalability. From a performance architecture standpoint, the larger chip scale and increased number of cores present a key challenge: maintaining low latency between clusters and the latency between cores and memory, which is essential for optimal processing performance. To address this, the bus architecture is designed to reduce both the physical distance between clusters and the latency between cores and DRAM.

In automotive SoCs, meeting functional safety requirements is crucial when adding functionality with chiplets, as simply connecting multiple dies via UCIe does not guarantee ASIL D compliance. To address this, we propose a method that combines the standard UCIe connection with RegionID-based mechanisms to achieve Freedom from Interference (FFI) by preventing access conflicts to hardware resources among multiple concurrent applications. Fig. 10.1.3 depicts our approach, showing the proposed UCIe connection alongside the FFI architecture, which ensures secure access control through the RegionID concept. However, conventional UCIe lacks a mechanism to transmit the RegionID between dies. To overcome this, the RegionID is converted and mapped into an existing physical address space, then encoded into parameters such as physical function bits and other unused fields within the UCIe interface. This enables transmission to the System Memory Management Unit (sMMU) and Real-Time (RT) cores, facilitating the practical implementation of FFI across dies. Additionally, to maintain adequate inter-die communication performance on this chip, the architecture is designed to sustain adequate bandwidth from the processor to the memory bus. Performance testing of the UCIe showed a sufficiently wide eye opening on the transmission lines, reaching 51.2GB/s and demonstrating performance close to the memory access speed limit.

Meeting the requirements of SDVs requires continuous growth in both chip performance and functionality. Fig. 10.1.4 shows the trend of NPU performance and area. In this work, compared to Generation A, the chip area has increased, and the NPU area has expanded by 1.5×. Adopting a 3nm process reduces delay per gate, but the larger logic scale increases clock latency from the shared clock source to each circuit's flipflops. To maintain a 400ps timing window for flipflops under various conditions, the maximum permissible clock latency is limited to 2578ps, accounting for on-chip variation, aging, and guard bands. Initial estimates indicated that the NPU's clock latency reached 3600ps, surpassing the limit in all worst conditions. To address this, the Clock Pulse Generator (CPG), conventionally located at each module's entry point, has been divided, with mini-CPGs (mCPGs) now positioned at the entry points of submodule hierarchies. This structural refinement successfully reduces the clock latency to 2440ps, satisfying timing requirements. Deploying mCPGs across many subdivided modules, however, makes synchronizing test clocks difficult, hindering zero-defect operation, which is essential for automotive applications. To overcome this, the chip integrates a tuning circuit within a hierarchical CPG architecture, comprising a main CPG, an always-on mCPG, and mCPGs located in power-off regions. These power-off mCPGs are further subdivided and assigned to each NPU. Conventionally, the test clock was generated via a separate path from the user clock, increasing circuit complexity and making timing closure more difficult as chip size grew. The proposed design uses a shared clock path for both user and test clocks, solving these issues. Additionally, to eliminate phase delays caused by upper-level mCPGs during test clock selection, the same clock source is used to synchronize both upper and lower mCPGs even in test mode. This approach enables unified tuning of all mCPGs as a single-phase system, which is crucial for zero-defect operation.

Recent automotive SoCs need to support different power modes to manage energy efficiently based on specific conditions. These modes include Full Run, Deep Stop, and Cyclic Run. Additionally, updates have added Sentry Mode and Child Presence Detection (CPD) mode. Fig. 10.15 outlines the proposed approach for managing power. Sentry Mode acts as a security system that monitors parked vehicles, and CPD is a safety feature that observes activity inside the vehicle during parking, issues alerts, and records video if suspicious activity occurs. During chip operation, these modes require activating the camera and sensors in Deep Stop mode with minimal power consumption. The chip manages this by controlling the NPU, APU, and other components, such as the DRAM interface, graphics processing unit, and camera serial interface, with precise granularity. The following section explains how this has been enabled.

Figure 10.1.6 shows the proposed power control method To manage the increasing number of modules caused by the adoption of multi-purpose automotive SoCs and the higher power demands driven by performance improvements, the design uses power gating across over 90 distinct power domains. This enables precise power regulation, from a few milliwatts to several tens of watts, depending on the operational conditions. Additionally, due to higher current densities caused by process miniaturization and performance scaling, it is essential to shut off power within each domain while maintaining equal or lower IR drop levels to ensure reliable operation and prevent voltage instability. To control IR-drop behavior, the chip architecture separates the power switch (PSW) into two parts: a ring-type PSW (Ring PSW) and a row-type PSW (Row PSW). When power is restored, the Ring PSW is analog-controlled to suppress rush current [8], gradually charging the virtual power rail (VVDD) from the outer edge, thus reducing noise interference with adjacent domains. Next, the Row PSW, embedded within the standard cell rows, is activated to evenly reduce impedance across the power domain. Compared to the earlier ADAS product that only used the Ring PSW, the proposed CPU Dual Core Lock Step (DCLS) achieves a 0.87× reduction in IR drop under 2× power density conditions. Moreover, the DCLS power gating system must meet ASIL D safety standards. To prevent dependent failures within the DCLS power switch circuitry, the master and checker components are independently controlled by separate power switch controllers (PSWC) and power switches (PSW). This configuration allows fault detection in lockstep mode, even if a voltage drop occurs due to a failure on one side's power switch. Additionally, each power switch's gate signal is monitored through a loopback mechanism that issues an OFF-detection signal in case of failure. Since a malfunction of a power switch can cause the loss of DCLS functionality - particularly in the comparator region - a digital voltage monitor (DVMON) continuously tracks voltage drops. DVMON has been enhanced from traditional designs to ensure stable power supply detection with minimal temperature drift, using multiple ring oscillators with different temperature characteristics. Furthermore, a current-starved oscillator is adopted to reduce detection variation caused by aging, resulting in a 1.4mV improvement in aging tolerance. The total variation in detection ranges from –1.9mV to +3.2mV from day 1 to the end-of-life, while [9] reports the variation at the fresh condition.

Figure 10.1.7 shows the micrograph of the chip. The essential architecture and circuits are fabricated on one chip using 3nm CMOS technology. The SHMOO plots demonstrate that both NPU and APU achieve their performance targets.

References:
[1] Vivek Bhan, "TECHNOLOGY UPDATES AT ELECTRONICA", ELECTRONICA, 2024, <https://presse.hbi.de/pub/Renesas/Presskits/electronica2024/R-Car_X5H_Press_Conf_Presentation_final2.pdf>.
[2] Pierrick Boulay et al.,"The Present and Future of Automotive ADAS through Sensing, Advanced Computing and Memory technology," Yole, 2023, <https://medias.yolegroup.com/uploads/2023/03/analyst-thursday-march-30-automotive-adas.pdf>.
[3] Rama Venkatasubramanian et al., "A 16nm 3.5B+ Transistor >14TOPS 2-to-10W Multicore SoC Platform for Automotive and Embedded Applications with Integrated Safety MCU, 512b Vector VLIW DSP, Embedded Vision and Imaging Acceleration," *ISSCC*, pp. 52-53, 2020. http://doi.org/10.1109/ISSCC19947.2020.9062915
[4] Katsushige Matsubara et al., "A 12nm Autonomous Driving Processor with 60.4 TOPS, 13.8 TOPS/W CNN Executed by Task-Separated ASIL D Control," *ISSCC*, pp. 56-57, 2021. http://doi.org/10.1109/JSSC.2021.3120191
[5] M. Ditty, "NVIDIA Orin System on Chip" *IEEE Hot Chips Symp.*, 2022. http://doi.org/10.1109/HCS55958.2022.9895609
[6] Kenichi Shimada et al.,"A 33kDMIPS 6.4W Vehicle Communication Gateway Processor Achieving 10Gbps/W Network Routing, 40ms CAN Bus Start-Up and 1.4mW Standby Power," *ISSCC*, pp. 240-241, 2023. http://doi.org/10.1109/ISSCC42615.2023.10067585
[7] NXP Semiconductors, "S32N55 Vehicle Super–Integration Processor Fact Sheet," Rev. 0, 2024, Accessed on Aug 28, 2025, <https://www.nxp.com/docs/en/fact-sheet/S32N55FS.pdf>.
[8] K.Fukuoka et al., "Power Management Features of 1.5GHz Dual-Core Application Processor and LTE Capable Baseband Processor Named R-Mobile U2," *IEEE Micro*, vol. 6, pp. 26-36, 2013. http://doi.org/10.1109/MM.2013.109
[9] T.Uemura et al., "A 28nm Fully Digital Voltage Monitor with 16.5uV/°C Accuracy and 0.8mV Quantized Error from -40 to 160°C for ISO26262 ASIL-D capable MCU," *ASSCC*, pp. 129-132, 2019. http://doi.org/10.1109/A-SSCC47793.2019.9056911

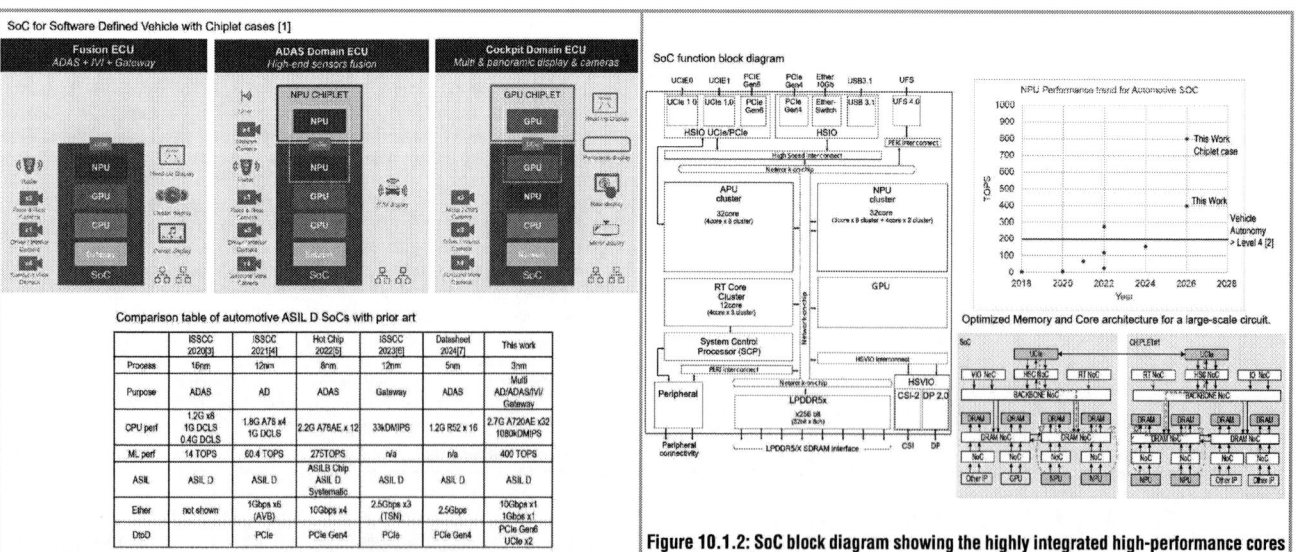

Figure 10.1.1: What the automotive SoC aims to achieve for SDVs.

Figure 10.1.2: SoC block diagram showing the highly integrated high-performance cores on a single chip.

ISSCC 2026 / SESSION 10 / DIGITAL PROCESSING AND CIRCUIT TECHNIQUES / 10.1

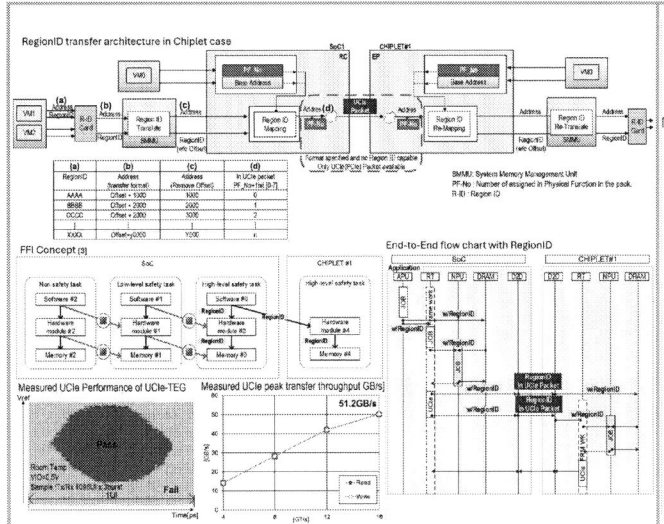

Figure 10.1.3: ASIL D capable FFI architecture with RegionID for chiplet.

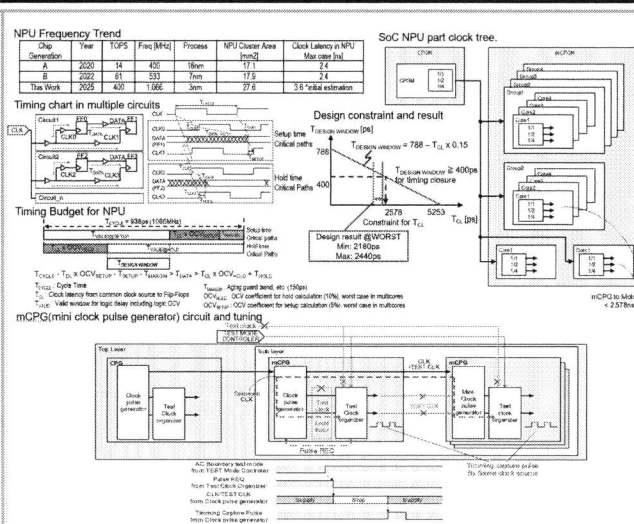

Figure 10.1.4: Clock distribution for the advanced large-scale and high-performance chip enabled by the layer tuning method.

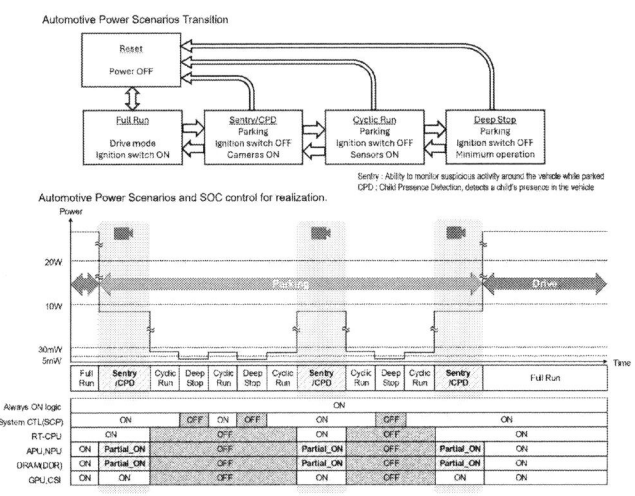

Figure 10.1.5: Automotive power scenarios enabled by multiple power-saving operation modes.

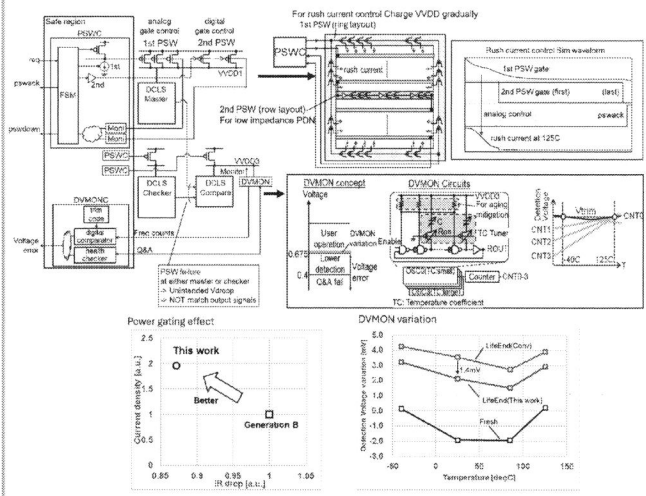

Figure 10.1.6: Hybrid power gating for compensating voltage-drop in a high-power consumption chip with ASIL D capability.

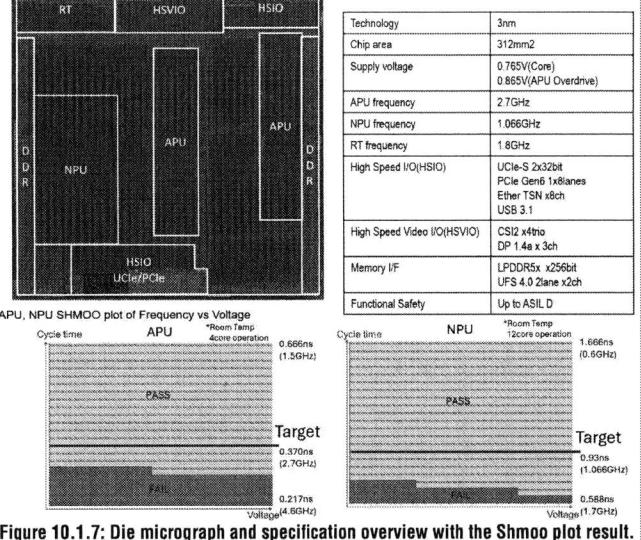

Figure 10.1.7: Die micrograph and specification overview with the Shmoo plot result.

• 2026 IEEE International Solid-State Circuits Conference

979-8-3315-8937-0/26 $31.00 © 2026 IEEE

ISSCC 2026 / SESSION 10 / DIGITAL PROCESSING AND CIRCUIT TECHNIQUES / 10.2

10.2 A Dynamic Performance Augmentation in a 3nm-Plus Mobile CPU

Chien-Yu Lu[1], Bo-Jr Huang[1], Sung S.-Y. Hsueh[1], Trong-Hieu Tran[1], Eric Jia-Wei Fang[1], Chao-Yang Yeh[1], Quan Sun[2], Tao Chen[2], Hsinchen Chen[2], Huaichung Chang[1], C.-J. Tsai[1], Yi-Chang Zhuang[1], Barry Chen[1], Ericbill Wang[1], Hugh Mair[2], Shih-Arn Hwang[1]

[1]MediaTek, Hsinchu, Taiwan, [2]MediaTek, Austin, TX

Abstract

This work presents dynamic mobile-performance augmentation (DMPA), featuring boosting-duty control (BDC) and an adaptive thermal cooler (ATC) to augment CPU maximum performance. The BDC uses an aging sensor (A-sensor) to leverage an aging budget for performance boosting. For thermal, the ATC integrates temperature sensors (T-sensors) across the CPU and works with multiple sensors and a throttler to optimize clock speed effectively cooling hotspots. The DMPA offers a score of 3917 in the GeekBenchv6 single-core benchmark on a flagship smartphone, consuming 0.078% or less total power, while occupying only 0.0122% of the CPU area.

Significant performance improvements in flagship smartphones and tablets are fulfilling the demands of high-end gaming, achieving a rich display with a stable high frame rate per second (fps), and effective thermal control. These demands drive CPUs to offer better performance and power efficiency through multi-core architectures that enable parallel processing for diverse workloads. However, sequential tasks within high fps gaming, such as logic processing, AI computation, and physics simulation, still rely on CPU single-core performance (SCP). SCP becomes the critical bottleneck for the most demanding frames and determines overall gameplay smoothness. In gameplay, the 1%-low fps is considered a measure of the most demanding frames that determines gaming smoothness more so than average fps. In Fig. 10.2.1, when the game runs above 80fps, it requires more SCP to improve the 1%-low fps, even when the average fps target is met. A 5% increase in SCP improves the 1%-low fps by 4.1%. Consequently, flagship SoCs are aggressively enhancing CPU maximum SCP by generations [1][2]. Increasing the CPU maximum clock frequency and voltage to boost SCP results leads to challenges in thermal control [3][4][5] and aging degradation [6][7]. As shown in Fig. 10.2.1, the CPU anchor cell models the speed degradation after high-temperature operation life (HTOL) at various maximum CPU operation voltages (V_{max}). When V_{max} increases by 10%, the speed degradation due to HTOL rises by over 80%, which limits the CPU SCP.

In this work, a dynamic mobile-performance augmentation (DMPA) approach is implemented in an 3nm-plus CPU, consisting of eight single-thread cores as part of the 5G mobile flagship SoC. The DMPA scheme maximizes achievable SCP through boosting-duty control (BDC) and a hardware adaptive thermal cooler (ATC) along with the CPU performance cores. The BDC utilizes the available aging budget with a digital aging sensor (A-sensor) to boost the CPU highest operation point (OPP0) to an ultimate performance point (OPP0+), occupying 0.0024% of the CPU area and consuming negligible power in daily use. The ATC deploys digital temperature sensors (T-sensors) spread within the CPU to efficiently detect hotspots. The T-sensor provides accurate temperature readings with less than 2.5°C error and works with multiple sensors in the throttler to adjust clock speed in a 25μs response time for effectively cooling the CPU. With the DMPA, the CPU enhances maximum SCP up to 4.5% with less than 0.078% additional power and area contribution of 0.0122%. It achieves a leading score of 3917 in the GeekBenchv6 single-core benchmark on a flagship smartphone, up from a base score of 3762 without the DMPA.

Figure 10.2.2 illustrates the CPU within the SoC, featuring octa cores that comply with the ARMv9.4 instruction set in three performance gears. The first gear includes a single ARM C1-Ultra High-Performance (HP) core, operating at 4.21GHz and with a speed boost of +3.3% to +4.5% with DMPA. The second gear comprises three ARM C1-Premium Balanced-Performance (BP) cores, running at 3.5GHz with a boost of +2% to +3%. The third gear consists of four C1-Pro High-Efficiency (HE) cores, optimized for energy-efficient computing, operating at 2.7GHz. This tri-gear configuration offers better peak performance over an equal split between BP and HE cores, while achieving a more competitive balance of area and power efficiency than an equal HP and HE (or BP) split for better user experience in daily low-power operation.

In FinFET technology, the device's contact poly pitch (CPP) significantly impacts power, performance and area (PPA) at the physical layers and thus needs optimization for PPA tradeoffs. A narrower CPP leads to a better area efficiency, but suffers higher leakage, increased coupling effects, and potential process reliability issues. Fig. 10.2.3 compares the power simulation of the smallest CPP (48nm) and the optimized CPP (54nm) in the HP core under high-performance conditions. Since the optimized CPP offers better power efficiency, it allows the CPU to achieve maximum performance with less thermal penalty. The aging margin of the CPU is assessed for sustainable performance until the end-of-life (T_{EOL}). Fig. 10.2.3 shows the measured performance-to-supply voltage (V_{DD}) curve of the HP core. The performance target is set at OPP0 (F_{max}, V_{max}) for T_{EOL}. This condition offers sufficient aging margin at the start-of-life (T_0) with V_{DD} set to V_0. F_{max} also considers other margins like temperature margin for the worst thermal fluctuations and process margin for variations. To achieve an ultimate performance (F_{max+}) from the OPP0 to the OPP0+, the aging margin headroom at T_0 is used to boost the voltage V_0 to V_{max}. The temperature margin can be reduced dynamically by effective thermal control. And, the process margin gain is determined by post-silicon binning with adequate yield. In Fig. 10.2.4, the HTOL quantifies

the total aging budget under (F_{max}, V_{max}) to ensure the CPU reliability at T_{EOL}. During the period from T_0 to T_{EOL}, the available aging budget can be converted into the equivalent boosting-duty time (T_{BD}) that allows the CPU to increase V_0 to V_{max} for higher performance.

The BDC is designed to manage the aging budget for each CPU's boosting request. It includes a digital ring oscillator (ROSC)-based A-sensor, which acts as a timer to indicate whether the aging budget is due at T_{BD}. The A-sensor uses series-stacking and parallel-connected devices as a unified inverter to achieve better aging sensitivity, as shown in Fig. 10.2.4. In the BDC, the A-sensor is enabled to capture ROSC counts (RO) during a 10μs sampling window (T_{SW}) at a 100ms software-polling interval. It checks if the RO meets the boosting-duty threshold counts (RO_{BD}), which represents the aged speed target at the boosting-duty of A-sensor counts. The 10μs/100ms sampling/polling ratio ensures minimal aging impact on the A-sensor during regular checks. If the RO is equal to or greater than the RO_{BD}, the boost flag is set to 1, indicating the CPU's OPP0 is eligible for boosting upon request. If the RO is less than RO_{BD}, the flag is set to 0, denying any boosting request.

The RO is calibrated at initial T_0 across various voltages within the CPU sign-off range. The RO_{BD} is calculated based on the aging model characterized for various voltage/temperature using 84 corner samples during HTOL validation. Since RO varies with temperature/voltage, the voltage sensor (V-sensor) and T-sensor are needed for the final RO_{BD} calculation using a lookup table and a two-point interpolation. Initially, the boost flag is disabled when the CPU powers on. Then, regular RO polling enables the boost ready flag if the RO count is higher than RO_{BD}. Once the CPU requests the performance, the boost flag allows the speed boosting. In the boosting procedure, the A-sensor acts as an aging timer during the boosting period, with the software polling interval accelerating down to 1ms to prevent over-duty.

As the performance is boosted, an effective thermal control is required to sustain performance against thermal spikes, Fig. 10.2.5 shows the digital T-sensors, which consist of leakage-dominated delay cells in a ROSC implemented in the CPU core. The natural logarithm of the T-sensor speed is linearly related to the inverse of absolute temperature [8], assuming a constant threshold voltage (Vth) for the leakage cell. This allows measurement of temperature from T-sensor speed counts. However, as the CPU scales V_{DD} for power efficiency, Vth changes due to Drain Induced Barrier Lowering (DIBL), may cause a voltage-dependent error. Fig. 10.2.5 illustrates the Differential-Voltage-based Temperature Compensation (DVTC) scheme, which places two abutting T-sensors on the CPU core boundary: one (T_{SV}) at the core's scaling V_{DD}, another (T_{FV}) at a fixed V_{DD} source from the system. Since they share the same hotspot, the error from different voltages can be quantified at run-time to calibrate other T-sensors across the CPU cores where a fixed V_{DD} is not available. After process calibration is done at single temperature/voltage condition, the T-sensor is measured to achieve a temperature error of +2.5°C to -1.6°C with DVTC from -40°C to 100°C, verified in a 25μs response time across 40% to 120% V_{DD}, covering all CPU sign-off conditions. The DVTC improves the worst T-sensor error by 5.7°C compared to no voltage-dependent compensation.

Figure 10.2.6 illustrates the DMPA, which consists of the BDC and ATC to augment CPU maximum performance. The BDC leverages the A-sensor to utilize the available aging budget, allowing an optimized performance point OPP0+. For sustainable performance against thermal challenges, the ATC uses T-sensors to monitor CPU temperature with a 25μs response time and employs a hardware temperature throttler to reduce CPU temperature by slowing down the clock when it exceeds the throttling temperature threshold (T_{thr}). T_{thr} is adaptively determined by on-die process sensors (P-sensors) and leakage sensors (LKG-sensors) [3], which predict total CPU power and convert it to an expected thermal jump for T_{thr}. Additionally, current sensors (I-sensors) [9] and V-sensors provide power for the throttler to adjust the clock using a tunable Proportional-Integral-Derivative (PID) control.

Figure 10.2.6 presents the measured run-time clock speed during peak performance when the HP core runs Geekbenchv6 benchmarks. The DMPA further boosts HP core performance from 4.21GHz to 4.4GHz. Two test cases at the same initial temperature (T_{thr} -10°C) with and without the ATC are demonstrated. Without the ATC, the software-based cooler with 1ms response time controls the thermal state for system stability and results in a

ISSCC 2026 / February 17, 2026 / 8:00 AM

temperature damping over 8°C when thermal throttling. That causes a corresponding performance degradation. With the ATC, the temperature damping is well controlled within 2°C when throttling and thus the performance degradation is mitigated.

The die photo in Fig. 10.2.7 shows the 3nm-plus multi-core CPU with an area of 24.69mm². This work presents the DMPA, featuring the BDC and ATC to augment CPU maximum performance. The BDC utilizes an A-sensor to leverage an aging budget as duty for performance boosting, occupying only 0.0024% of the CPU area and consuming negligible power in daily use. For thermal management, the ATC integrates the T-sensors across the CPU and works with multiple sensors in the throttler to adjust clock speed in a 25μs response time, effectively cooling CPU hotspots. The ATC occupies only 0.0098% of the CPU area and consumes 0.048% of the total CPU power at most. Overall, the DMPA offers the HP core a maximum performance enhancement up to 4.5% and the BP core by up to 3%, consuming less than 0.078% total power while occupying only 0.0122% of the CPU area. With the DMPA, the CPU achieved a score of 3917 in the GeekBenchv6 single-core benchmark on a flagship smartphone, a 4.1% improvement over its score of 3762 without the augmentation.

Acknowledgement:
The authors thank Alfred Tsai, Albert He, Pi-Cheng Chen, TC Tsai, Rex Che-Yuan Liu, Eden Tsai, Lauren Yang, Chih-Hsuan Chang, Po-Yang Hsu, Cheng-Yuh Wu, Yungching Chen, Yuwen Tsai, and Alex Chiou, Mediatek, Hsinchu, Taiwan, for their support on this work.

References:
[1] A. Varma et al., "A 4nm 3.4GHz Tri-Gear Fully Out-of-Order ARMv9.2 CPU Subsystem-Based 5G Mobile SoC," *ISSCC*, pp. 36-38, 2024.
https://doi.org/10.1109/ISSCC49657.2024.10454494
[2] C. -Y. Lu et al., "Run-Time Power Management System by on-Die Power Sensor with Silicon Machine Learning-Based Calibration in a 3nm Octa-Core CPU," *ISSCC*, pp. 160-162, 2025. https://doi.org/10.1109/ISSCC49661.2025.10904564
[3] B. -J. Huang et al., "A 5G Mobile Gaming-Centric SoC with High-Performance Thermal Management in 4nm FinFET," *ISSCC*, pp. 40-42, 2023.
https://doi.org/10.1109/ISSCC42615.2023.10067271
[4] H. S. Ozturk et al., "A Dual VDD-Temperature Sensor Employing Sensor Fusion with 2.4°C, 9mV (±3) Inaccuracy in 65nm CMOS," *ISSCC*, pp. 170-172, 2025.
https://doi.org/10.1109/ISSCC49661.2025.10904772
[5] B. -S. Lien et al., "A 0.65V 900μm² BEoL RC-Based Temperature Sensor with ±1°C Inaccuracy from −25°C to 125°C," *ISSCC*, pp. 68-70, 2024.
https://doi.org/10.1109/ISSCC49657.2024.10454423
[6] T. Webel et al., "Dynamic Guard-Band Features of the IBM zNext System," *ISSCC*, pp. 158-159, 2025. https://doi.org/10.1109/ISSCC49661.2025.10904615
[7] D. Yingling et al., "A 3nm Adaptive Clock Duty-Cycle Controller for Mitigating Aging-Induced Clock Duty-Cycle Distortion," *ISSCC*, pp. 258-260, 2024.
https://doi.org/10.1109/ISSCC49657.2024.10454421
[8] Z. Tang et al., "A 1770-μm² Leakage-Based Digital Temperature Sensor With Supply Sensitivity Suppression in 55-nm CMOS," *IEEE JSSC*, vol. 55, no. 3, pp. 781-793, 2020.
https://doi.org/10.1109/JSSC.2019.2952855
[9] C. -Y. Lu et al., "A Fully Digital Current Sensor Offering Per-Core Runtime Power for System Budgeting in a 4nm-Plus Octa-Core CPU," *ISSCC*, pp. 260-262, 2024.
https://doi.org/10.1109/ISSCC49657.2024.10454415

[10] Y. Duan et al., "A PVT-Robust 5.5GHz Fractional-N Cascaded RO-Based Digital PLL with Voltage-Domain Feedforward Noise Cancellation," *ISSCC*, pp. 324-326, 2025.
https://doi.org/10.1109/ISSCC49661.2025.10904636

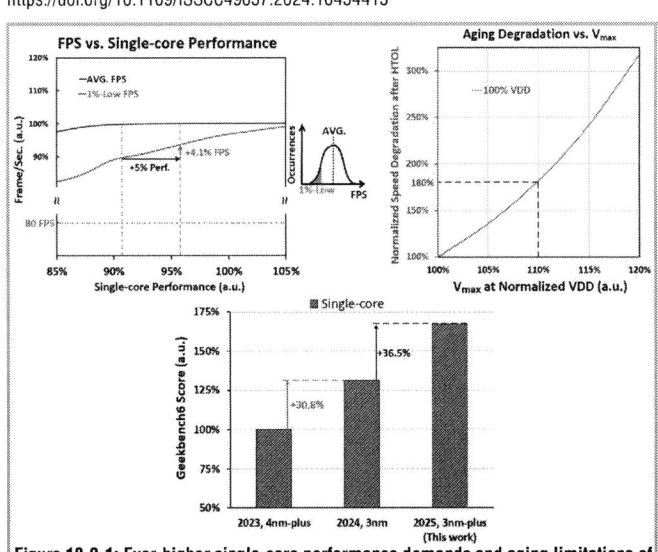

Figure 10.2.1: Ever-higher single-core performance demands and aging limitations of maximum voltage in 3nm-plus mobile flagship smartphone CPU.

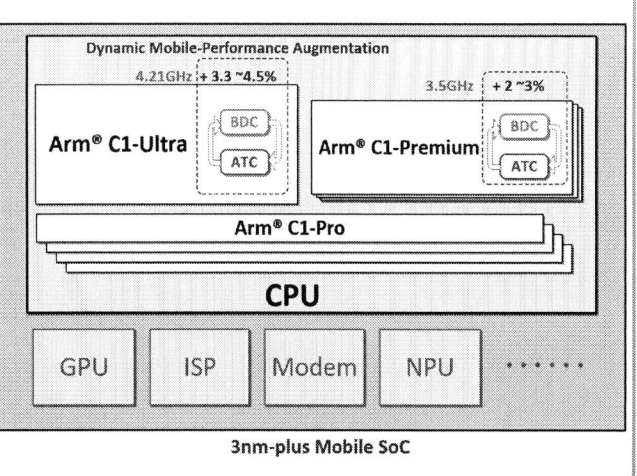

Figure 10.2.2: The DMPA of the CPU with BDC/ATC block diagram and architecture in 3nm-plus SoC.

ISSCC 2026 / SESSION 10 / DIGITAL PROCESSING AND CIRCUIT TECHNIQUES / 10.2

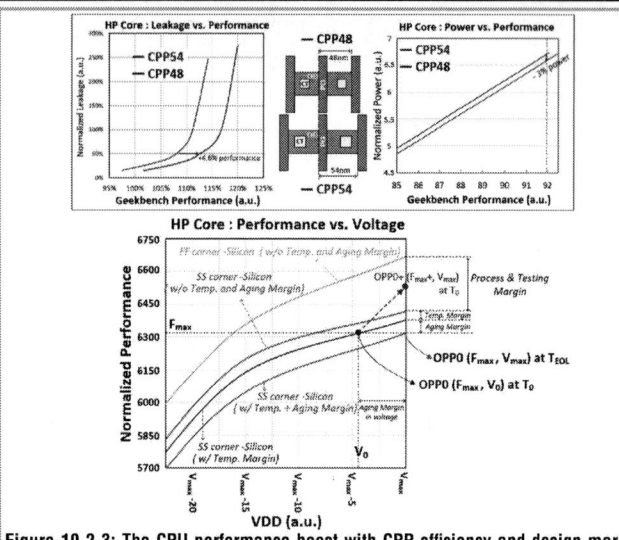

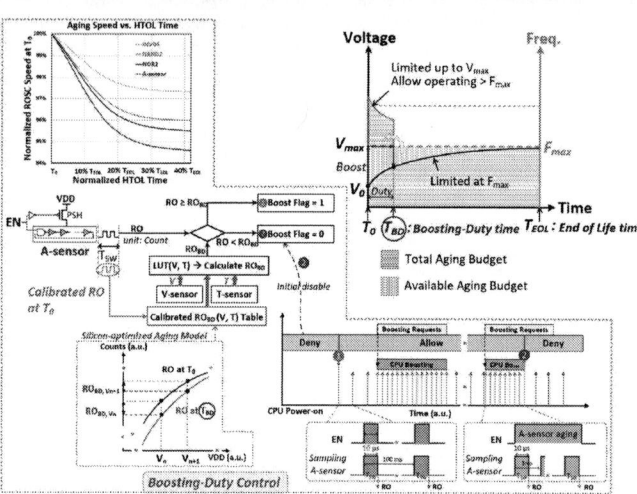

Figure 10.2.3: The CPU performance boost with CPP efficiency and design margin optimization to achieve higher OPPO+.

Figure 10.2.4: The proposed Boosting-Duty Control (BDC) with a digital A-sensor to leverage available aging-budget in CPU.

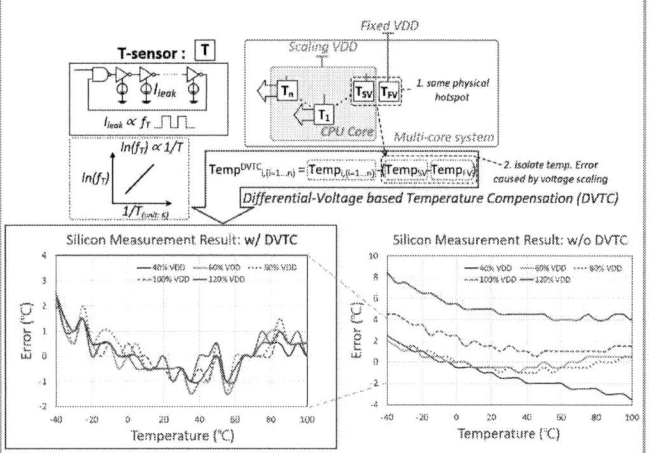

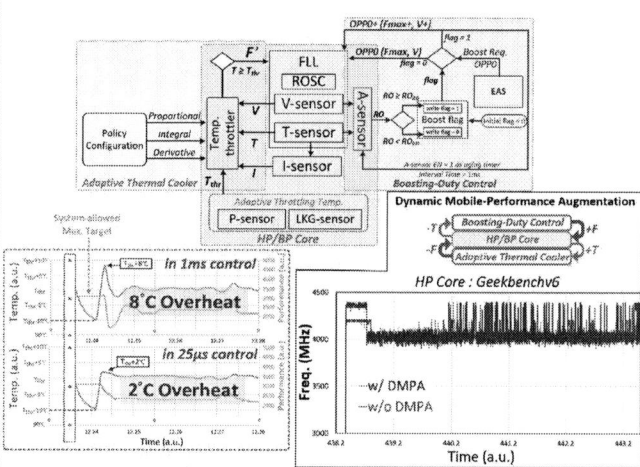

Figure 10.2.5: The digital T-sensor with proposed Differential-Voltage-based Temperature Compensation (DVTC) achieving fast and accurate temperature sensing in silicon for the CPU.

Figure 10.2.6: The Dynamic Mobile-Performance Augmentation (DMPA) contains the ATC and BDC for the CPU with silicon measurement results.

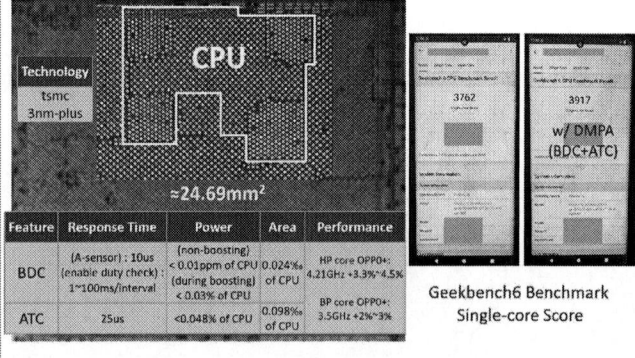

Figure 10.2.7: The 3nm-plus CPU die-photo with DMPA feature summary and on-phone benchmark demonstration.

• 2026 IEEE International Solid-State Circuits Conference

979-8-3315-8937-0/26 $31.00 © 2026 IEEE

ISSCC 2026 / SESSION 10 / DIGITAL PROCESSING AND CIRCUIT TECHNIQUES / 10.3

10.3 A 2nm Clock-Edge Architecture for Processor Clock-Power Reduction

Yimai Peng[1], Daniel Yingling[1], Basma Hajri[2], Robert Vachon[1], Fikre Gebreyohannes[2], Vincent Li[3], Ghanshyam Chhetri[4], Keith Bowman[1]

[1]Qualcomm, Raleigh, NC, [2]Qualcomm, Cork, Ireland, [3]Qualcomm, San Diego, CA, [4]Qualcomm, Bangalore, India

Abstract

A 2nm clock-edge architecture (CEA) for an NPU matrix-multiplication unit (MXU) features dual-edge-triggered (DET) flip-flops, DET clock-gating circuits, and an adaptive clock duty-cycle controller to achieve iso-performance as a conventional (CNV) design while operating at half the clock frequency. Silicon measurements of the CEA MXU demonstrate ~39-40% lower clock power and a total dynamic power reduction of ~7-15%, depending on the workload, as compared to the CNV MXU at iso-throughput.

In conventional processors, pipeline registers with single-edge-triggered (SET) flip-flops (FFs) only transition data at one clock edge (e.g., rising edge) with no data movement at the opposite clock edge (e.g., falling edge). Since pipeline data movement only occurs at one clock edge, conventional designs waste half of the dynamic clock power from a microarchitecture perspective. In contrast, dual-edge-triggered (DET) FFs [1]-[17] transition data at both clock edges, enabling the processor to operate at half the clock frequency (F_{CLK}) for the same throughput (TP) as a conventional design (Fig. 10.3.1, top), and as a result, significantly reducing the processor clock power. Although proposed more than four decades ago [1], DET FFs have not been adopted as pipeline registers in commercial processors due to four key challenges. First and foremost, clock duty-cycle distortion (DCD) negatively impacts processor performance, where the shortest clock-phase delay (i.e., high-phase or low-phase delay) limits the performance. A second key challenge is the lack of reset and design-for-test (DFT) functionality in prior DET FF designs [1]-[17], while maintaining similar timing and low clock power as compared to production-level SET FFs. Third, the prior-art DET clock-gating circuit (CGC) [18] induces cycle-to-cycle clock DCD, resulting in performance degradation. Fourth, production processors require compatible design flows for DET FFs and CGCs. This paper presents a clock-edge architecture (CEA) for a neural processing unit (NPU) matrix-multiplication unit (MXU) in a 2nm Gate-All-Around (GAA) CMOS technology [19], featuring an adaptive clock duty-cycle controller (DCC) [20]-[21] to mitigate clock DCD for maximum performance, a low-clock-power DET FF with reset and DFT capabilities and similar timing as an SET FF, and a DET CGC to minimize cycle-to-cycle clock DCD.

The test chip (Figs. 10.3.1 and 10.3.7) integrates two separate industry-level NPU MXUs for conventional (CNV) and CEA designs to provide a direct comparison. Each MXU contains 1,024 multiply-accumulate units (MACs) with each MAC performing four 8b×8b multiplications and an accumulation per cycle. An on-die built-in self-test (BIST) engine executes MXU test patterns and verifies the expected outputs. The CNV MXU with SET FFs and CGCs employs standard design tools and flows. Since production design flows are not available for DET FFs and CGCs, the CEA MXU requires a unique physical-design (PD) methodology. The CEA MXU replicates the CNV MXU PD while swapping the SET FF and CGC cells with customized DET FF and CGC cells. After the PD-cell swapping, logical gate-level simulations verify proper connections and functionality without performing any timing analysis for the CEA MXU.

As required for the unique CEA MXU implementation, the proposed DET FF (Fig. 10.3.2) provides reset and DFT capabilities and similar timing as an SET FF with a focus on low clock power, while avoiding the overheads of designs based on pulsed latches [12]-[13], true-single-phase clocks [14]-[16], or C-elements [17]. From a logical perspective, the SET FF contains an input MUX for DFT followed by two series-connected latches. The proposed DET FF consists of an input MUX, two parallel latches, and an output MUX. Based on the DFT shift-enable signal (SHIFT), the input MUX selects between the FF input data (D) and the DFT scan-in signal (SIN). When the FF input clock (CLK) is logically low, the negative latch is transparent to the input-MUX output while the positive latch is opaque to the input-MUX output, holds the current value at node pn1, and drives the FF output (Q) as selected by the output MUX. When CLK is logically high, the functionality reverses for the positive and negative latches with the output MUX selecting the negative latch to drive Q. Asserting the reset signal (RST) forces both latches logically low (i.e., nodes pn1 and nn1 equal 0V) and Q to 0V while disabling the pull-up path at the input-MUX output to prevent a potential short-circuit path from the input MUX to the transparent latch. In comparison to the SET FF with 12 clock transistors toggling at F_{CLK}, the DET FF contains 16 clock transistors switching at $\frac{1}{2}F_{CLK}$ (i.e., equivalent of 8 clock transistors at F_{CLK}), resulting in a projected ~33% clock-power reduction at the FF level. From circuit simulations with extracted layout (Fig. 10.3.7), the DET FF enables ~30% power reduction with a 2% data activity and an equivalent worst-case FF power with 100% data activity as compared to the SET FF. The proposed DET FF provides the same worst-case CLK-to-Q delay and a slightly higher worst-case setup time (i.e., 0.3% of the ideal clock-phase delay), while maintaining a negative hold-time margin as compared to the SET FF.

For an SET CGC, a logically high enable (EN) results in the CGC clock input (CLK$_{IN}$) driving the CGC clock output (CLK$_{OUT}$), and a logically low EN forces CLK$_{OUT}$ low to prevent clock toggling. In contrast, the DET CGC requires a different function. When gating the DET CGC with a logically low EN, CLK$_{OUT}$ maintains the same logic or voltage level to avoid clock toggling (Fig. 10.3.1, top). With a logically high EN, CLK$_{OUT}$ transitions voltage levels either from 0V to the supply voltage (V_{DD}) or from V_{DD} to 0V when CLK$_{IN}$ transitions voltage levels, where the CLK$_{IN}$ and CLK$_{OUT}$ transitions may change in the same or opposite directions. The prior-art DET CGC design [18] consists of a DET FF and a 2-input XOR-logic gate (Fig. 10.3.7). The CGC CLK$_{IN}$ drives the DET FF CLK, and the DET FF Q represents CLK$_{OUT}$. The CGC EN and CLK$_{OUT}$ drive the two XOR inputs with the XOR output connecting to the DET FF D. The XOR gate either inverts or non-inverts CLK$_{OUT}$ to the DET FF D with EN logically high or low, respectively. Although this prior design achieves the proper DET CGC functionality, the DET FF imbalanced clock path induces CLK-to-Q delay variations across the combinations of rising and falling CLK and D transitions, resulting in a cycle-to-cycle clock DCD, and thus, degrading performance based on the number of clock-path CGCs.

The proposed DET CGC (Fig. 10.3.2) minimizes the cycle-to-cycle clock DCD by integrating two latches, which are controlled by the CGC EN, rather than the CGC CLK$_{IN}$, to avoid the imbalanced clock path. When EN is logically low, latch #1 is opaque to hold CLK$_{OUT}$ at the same voltage level while latch #2 is transparent to set a comparison signal (CP) of CLK$_{IN}$ and CLK$_{OUT}$ as the inverted XOR output of CLK$_{OUT}$ and the inverted CLK$_{IN}$ (CKB). When CLK$_{OUT}$=CLK$_{IN}$, CP=0V, and when CLK$_{OUT}$ ≠ CLK$_{IN}$, CP=V_{DD}. When EN is logically high, latch #2 is opaque to hold CP and the XNOR output drives the transparent latch #1 and eventually CLK$_{OUT}$, thus, CLK$_{OUT}$ transitions whenever CLK$_{IN}$ transitions. The XNOR-logic gate performs a MUX function of non-inverted and inverted CLK$_{IN}$ based on CP to drive CLK$_{OUT}$. With a traditional MUX-logic gate, the inverted CLK$_{IN}$ requires an additional inverter as compared to CLK$_{IN}$. The XNOR gate minimizes the input-to-output delay difference between non-inverting and inverting CLK$_{IN}$ transitions. As a result, the CLK$_{IN}$-to-CLK$_{OUT}$ delays are similar across the different combinations of rising or falling CLK$_{IN}$ or CLK$_{OUT}$ transitions to minimize the cycle-to-cycle CGC-induced clock DCD. From circuit simulations with extracted layout (Fig. 10.3.7), the proposed DET CGC reduces the cycle-to-cycle clock DCD by ~2.6× as compared to the prior-art DET CGC design, resulting in only a 0.12% performance degradation per clock-path CGC.

The DCC [20]-[21] integration into the clock path (Figs. 10.3.1 and 10.3.3) is a fundamental requirement to maximize the CEA MXU performance by mitigating the clock DCD from the phase-locked loop (PLL), the overall clock path from the PLL to the clock-leaf nodes driving the MXU FFs, and transistor aging. Since clock DCD appears similar across the MXU clock-leaf nodes, the DCC duty-cycle monitor (DCM) measures only eight clock-leaf nodes to guide a duty-cycle adjuster (DCA[1]) to maintain ~50%-50% duty cycle for the clock-leaf nodes. To validate the DCC functionality, a second duty-cycle adjuster (DCA[0]) intentionally induces clock DCD to represent the impact of aging. Silicon test-chip measurements (Fig. 10.3.3) demonstrate the CEA MXU TP degradation vs. the clock-leaf duty cycle with the DCC duty-cycle correction disabled. As an example with a 60%-40% duty cycle, the CEA MXU requires a 20% longer clock period to satisfy the timing margin, resulting in ~17% reduction in F_{CLK} and TP. After enabling the DCC duty-cycle correction, DCA[1] adjusts the CEA MXU duty cycle to ~50%-50% at the clock-leaf nodes to maximize the CEA MXU TP. Since the DCA resolution is ~0.4% of the ideal clock-phase delay, the duty cycle at the clock-leaf nodes may slightly deviate from the 50%-50% target, resulting in a small performance loss.

While enabling the DCC duty-cycle correction, silicon measurements (Fig. 10.3.4) compare the minimum V_{DD} (V_{MIN}) for CNV and CEA MXUs versus TP. CEA achieves the same TP as CNV while operating at half F_{CLK}. The CEA V_{MIN} is higher than the CNV V_{MIN} by 4mV, 4mV, and 9mV for the supply-voltage-scaled (SVS), nominal, and turbo operating conditions, respectively. This slightly larger V_{MIN} may result from (i) Longer DET FF setup time as compared to an SET FF, (ii) Small cycle-to-cycle clock DCD from the proposed DET CGC, and (iii) DCC DCA resolution error. For 40 parts across SVS, nominal, and turbo conditions (Fig. 10.3.4), the CEA MXU V_{MIN} distribution closely tracks the CNV MXU V_{MIN} distribution, where the average CEA V_{MIN} is ~5-10mV higher than the average CNV V_{MIN}.

Silicon measurements (Fig. 10.3.5) demonstrate the CEA MXU power reduction by operating at half F_{CLK}. At iso-V_{DD} and iso-TP, CEA enables an MXU clock-power reduction of ~39-40%

172 • 2026 IEEE International Solid-State Circuits Conference

979-8-3315-8937-0/26 $31.00 © 2026 IEEE

at SVS, nominal, and turbo. Since the total dynamic power reduction depends on the magnitude of the data-switching power, the measurements evaluate two workloads: (i) Peak performance with a worst-case MXU data-switching activity and (ii) Typical with realistic MXU data-switching activity. At the nominal operating condition with CNV and CEA at iso-V_{DD}, CEA enables a total dynamic power reduction of 9% and 17% for the peak-performance and typical workloads, respectively. From SVS to turbo conditions, the CEA total dynamic power reduction ranges from ~9-19%, depending on the workload. To capture the effect of the higher CEA V_{MIN}, the CEA MXU power measurements (Fig. 10.3.5, bottom-right) apply a higher V_{DD} based on the ΔV_{MIN} between CEA and CNV MXUs. Accounting for this ΔV_{MIN} effect, the CEA total dynamic power reduction ranges from ~7-8% for a peak-performance workload to ~16-17% for a typical workload across SVS, nominal, and turbo conditions. Across 40 parts, while capturing the ΔV_{MIN} effect for each part, the average CEA total dynamic power reduction ranges from ~7% (peak performance) to ~15% (typical) at iso-TP from SVS to turbo (Fig. 10.3.6).

Acknowledgement:
The authors sincerely thank Bhaskar Mittal, Nirmal Nanaware, Abhijith Ubhayakar, and Ganesh Murugesan for chip implementation, Wenjing Song for test-chip measurements, and Paul Penzes and Jason (Shih-Hsin) Hsu for encouragement and support.

References:
[1] S. H. Unger, "Double-Edge-Triggered Flip-Flops," *IEEE Trans. on Computers*, vol. C-30, no. 6, pp. 447-451, 1981. https://doi.org/10.1109/TC.1981.1675811
[2] S.-L. Lu and M. Ercegovac, "A Novel CMOS Implementation of Double-Edge-Triggered Flip-Flops," *IEEE JSSC*, vol. 25, no. 4, pp. 1008-1010, 1990. https://doi.org/10.1109/4.58294
[3] M. Afghahi and J. Yuan, "Double-Edge-Triggered D-Flip-Flops for High-Speed CMOS Circuits," *IEEE JSSC*, vol. 26, no. 8, pp. 1168-1170, 1991. https://doi.org/10.1109/4.90071
[4] A. Gago et al., "Reduced Implementation of D-Type DET Flip-Flops," *IEEE JSSC*, vol. 28, no. 3, pp. 400-402, 1993. https://doi.org/10.1109/4.210012
[5] R. Hossain et al., "Low Power Design Using Double Edge Triggered Flip-Flops," *IEEE TVLSI*, vol. 2, no. 2, pp. 261-265, 1994. https://doi.org/10.1109/92.285754
[6] R. P. Llopis and M. Sachdev, "Low Power, Testable Dual Edge Triggered Flip-Flops," *ACM/IEEE ISLPED*, pp. 341-345, 1996. https://doi.org/10.1109/LPE.1996.547536
[7] A. G. M. Strollo et al., "Analysis of Power Dissipation in Double Edge-Triggered Flip-Flops," *IEEE TVLSI*, vol. 8, no. 5, pp. 624-629, 2000. https://doi.org/10.1109/92.894168
[8] N. Nedovic and V. G. Oklobdzija, "Dual-Edge Triggered Storage Elements and Clocking Strategy for Low-Power Systems," *IEEE TVLSI*, vol. 13, no. 5, pp. 577-590, 2005. https://doi.org/10.1109/TVLSI.2005.844302
[9] C. Oh et al., "Timing Analysis of Dual-Edge-Triggered Flip-Flop Based Circuits with Clock Gating," *IEEE ICICDT*, pp. 59-62, 2009. https://doi.org/10.1109/ICICDT.2009.5166265
[10] M. Alioto et al., "DET FF Topologies: A Detailed Investigation in the Energy-Delay-Area Domain," *IEEE ISCAS*, pp. 563-566, 2011. https://doi.org/10.1109/ISCAS.2011.5937627
[11] A. Bonetti et al., "Automated Integration of Dual-Edge Clocking for Low-Power Operation in Nanometer Nodes," *ACM Trans. Design Automation*, vol. 22, no. 62, 2017. https://doi.org/10.1145/3054744

[12] J. Tschanz et al., "Comparative Delay and Energy of Single Edge-Triggered & Dual Edge-Triggered Pulsed Flip-Flops for High-Performance Microprocessors," *ACM/IEEE ISLPED*, pp. 147-152, 2001. https://doi.org/10.1109/LPE.2001.945391
[13] M. W. Phyu et al., "Power-Efficient Explicit-Pulsed Dual-Edge Triggered Sense-Amplifier Flip-Flops," *IEEE TVLSI*, vol. 19, no. 1, pp. 1-9, 2011. https://doi.org/10.1109/TVLSI.2009.2029116
[14] Y. Lee et al., "A Fully Static True-Single-Phase-Clocked Dual-Edge-Triggered Flip-Flop for Near-Threshold Voltage Operation in IoT Applications," *IEEE Access*, vol. 8, pp. 40232-40245, 2020. https://doi.org/10.1109/ACCESS.2020.2976773
[15] Z. Wang et al., "Low-Power Redundant-Transition-Free TSPC Dual-Edge-Triggering Flip-Flop Using Single-Transistor-Clocked Buffer," *IEEE TVLSI*, vol. 31, no. 5, pp. 706-710, 2023. https://doi.org/10.1109/TVLSI.2023.3251286
[16] Y. K. Maheshwari and M. Sachdev, "Low-Power, Low-Energy, Static, Contention-Free, TSPC Dual-Edge Triggered Flip-Flops," *IEEE Access*, vol. 13, pp. 66040-66050, 2025. https://doi.org/10.1109/ACCESS.2025.3560850
[17] S. Lapshev and S. M. R. Hasan, "New Low Glitch and Low Power DET Flip-Flops Using Multiple C-Elements," *IEEE TCAS-I*, vol. 63, no. 10, pp. 1673-1681, 2016. https://doi.org/10.1109/TCSI.2016.2587282
[18] R. P. Llopis, "Electronic Circuit with Dual Edge Triggered Flip-Flop," *U.S. Patent 6,137,331*, 2000. https://patents.google.com/patent/US6137331A/en
[19] D. Jeong et al., "Product Performance Aware 3rd Generation GAA Platform Transistor Design with Extreme Small Local Layout Effect and Transistor Variation," *IEEE Symp. VLSI Tech.*, 2024. https://doi.org/10.1109/VLSITechnologyandCir46783.2024.10631329
[20] D. Yingling et al., "A 3nm Adaptive Clock Duty-Cycle Controller for Mitigating Aging-Induced Clock Duty-Cycle Distortion," *ISSCC*, pp. 258-260, 2024. https://doi.org/10.1109/ISSCC49657.2024.10454421
[21] D. Yingling et al., "An Adaptive Clock Duty-Cycle Controller to Mitigate Aging-Induced Clock Duty-Cycle Distortion in Automotive and IoT Processors," *IEEE JSSC*, 2025. https://doi.org/10.1109/JSSC.2025.3578413

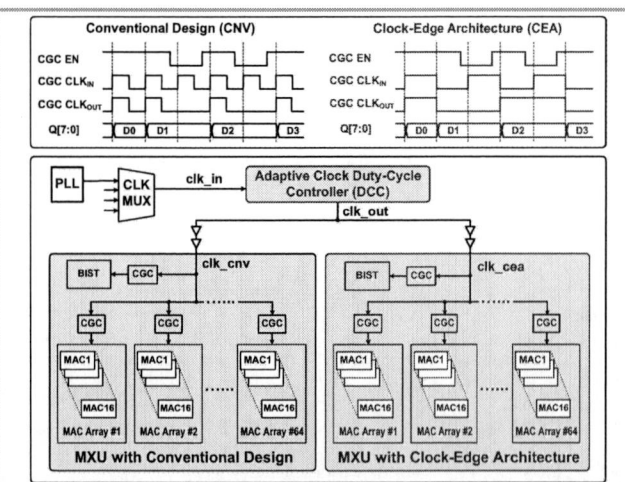

Figure 10.3.1: Conventional (CNV) design and clock-edge architecture (CEA) timing diagrams with a clock-gating circuit (CGC) and test-chip block diagram, containing the adaptive clock duty-cycle controller (DCC) and two NPU MXUs with CNV and CEA designs.

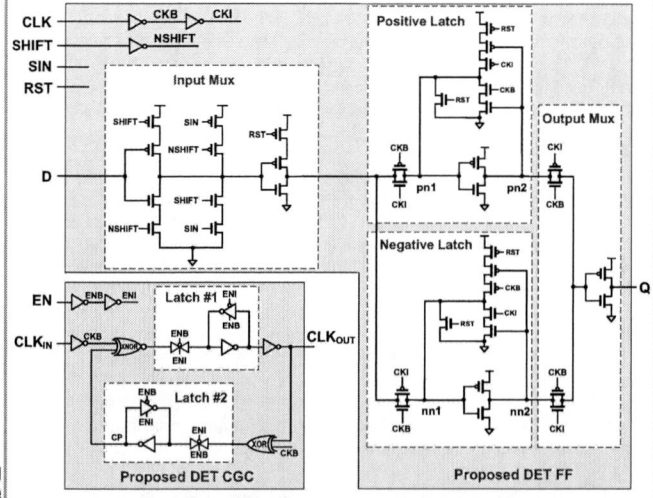

Figure 10.3.2: CEA dual-edge-triggered (DET) flip-flop (FF) and DET CGC circuit schematics.

ISSCC 2026 / SESSION 10 / DIGITAL PROCESSING AND CIRCUIT TECHNIQUES / 10.3

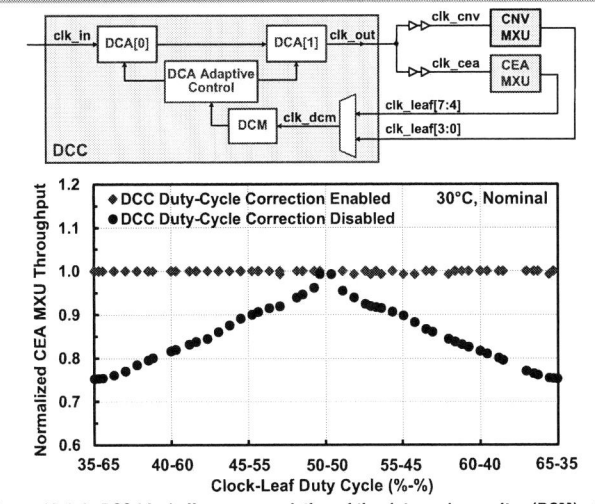

Figure 10.3.3: DCC block diagram, consisting of the duty-cycle monitor (DCM), duty-cycle adjuster (DCA), and DCA adaptive control, and measured normalized CEA MXU throughput (TP) vs. the clock-leaf duty cycle with DCC duty-cycle correction disabled.

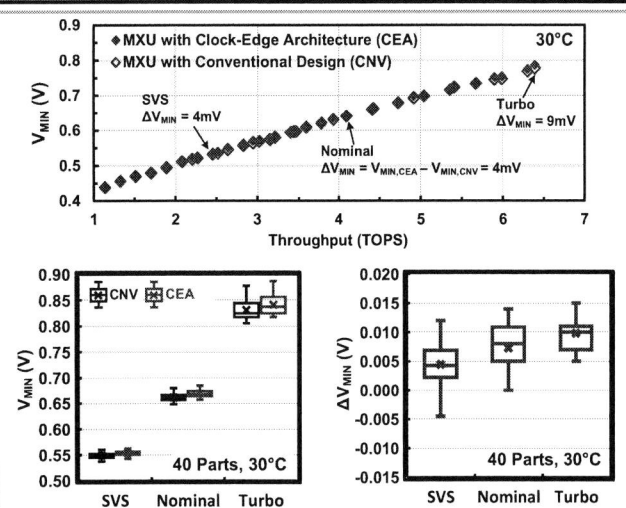

Figure 10.3.4: Measured MXU minimum V_{DD} (V_{MIN}) for CNV and CEA vs. TP, MXU V_{MIN} distribution for CNV and CEA vs. operating condition, and ΔV_{MIN} (=$V_{MIN,CEA}$ − $V_{MIN,CNV}$) distribution vs. operating condition.

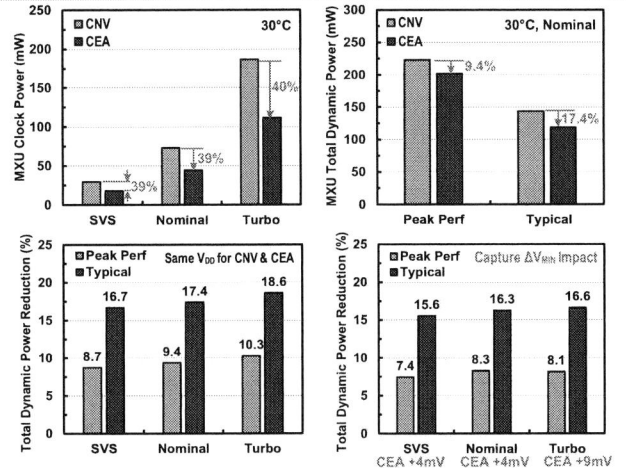

Figure 10.3.5: Measured MXU clock power vs. operating condition, MXU total dynamic power vs. workload, and CEA MXU total dynamic power reduction vs. operating condition with CNV and CEA at iso-V_{DD} and with higher CEA V_{DD} to capture the ΔV_{MIN} impact in Fig. 10.3.4.

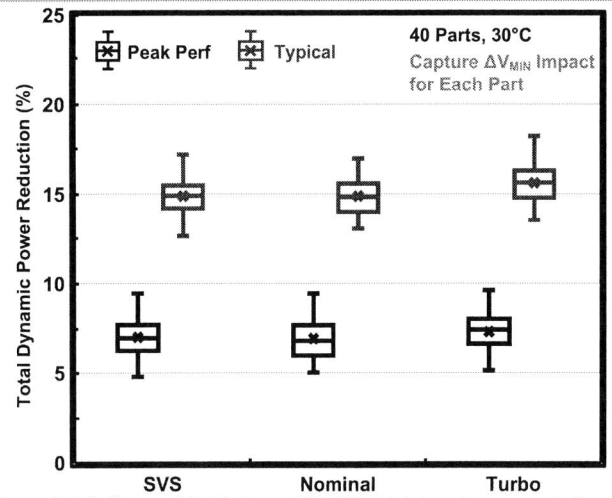

Figure 10.3.6: Measured distributions of CEA MXU total dynamic power reduction vs. operating condition for peak-performance and typical workloads across 40 parts while capturing the ΔV_{MIN} impact for each part.

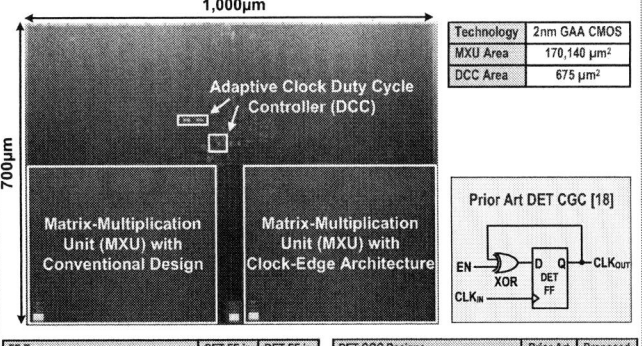

FF Types		SET FF in CNV MXU	DET FF in CEA MXU
Transistor Count		38	43
Normalized Power *	Data Activity = 2%	1.00	0.70
	Data Activity = 100%	2.39	2.39
Worst-Case Setup Time (ps)		13	20
Worst-Case CLK-to-Q Delay (ps)		23	23
Worst-Case Hold Time (ps)		-9	-7

DET CGC Designs		Prior Art [18]	Proposed
Transistor Count		48	44
Normalized Power *	Clock Gated	0.69	2.17
	Clock Ungated	3.15	2.47
TP Degradation per Clock-Path CGC		0.31%	0.12%

Simulation @ V_{DD} = 0.7V, 25°C, TT,
$F_{CLK,SET}$ = 500MHz, $F_{CLK,DET}$ = 250MHz
*Normalized to SET FF Power with Data Activity = 2%

Technology	2nm GAA CMOS
MXU Area	170,140 µm²
DCC Area	675 µm²

Figure 10.3.7: Test-chip microphotograph, characteristics, and comparisons of the SET FF vs. the proposed DET FF and the prior-art DET CGC [18] vs. the proposed DET CGC.

• 2026 IEEE International Solid-State Circuits Conference

ISSCC 2026 / SESSION 10 / DIGITAL PROCESSING AND CIRCUIT TECHNIQUES / 10.4

10.4 A 0.008mm² 16-to-1600MHz All-Digital Fractional Divider Using AUX-DLL for Background LMS-Based DTC Calibration

Ahmed Elkholy, Yousr Ismail, Zhi Huang, Adesh Garg, Ali Nazemi, Jun Cao, Afshin Momtaz

Broadcom, Irvine, CA

Abstract

An all-digital high-performance standalone fractional divider (FDIV) is presented. It leverages a robust replica-free least-mean square (LMS)-based digital-to-time converter (DTC) background calibration using a compact auxiliary delay locked loop (AUX-DLL). The proposed FDIV provides a wide output frequency range (16-1600MHz) with very fine frequency resolution (50b) and achieves worst-case integrated jitter performance of 350fs$_{rms}$, while occupying only 0.008mm².

Modern system-on-chips (SoCs) demand compact, multi-output, low-power and low-jitter clock generators. Traditional approaches using multiple phase-locked loops (PLLs), increase silicon area, power consumption and system complexity. As an effective alternative, standalone open-loop fractional dividers (FDIVs) have been introduced to generate multiple clock frequencies from a single PLL [1]. An FDIV typically integrates a multi-modulus divider (MMD), a digital-to-time converter (DTC), and a Delta-Sigma ($\Delta\Sigma$)-based digital controller. The $\Delta\Sigma$ modulator dithers the MMD division ratio to produce fractional division, while the DTC cancels phase quantization noise (P_Q) introduced by $\Delta\Sigma$ dithering. This approach considerably reduces power and area overhead compared to multiple PLL solutions. For accurate FDIV quantization noise cancellation (QNC), DTC gain (K_{DTC}) needs to match the input clock period (T_{IN}). This requires precise background calibration of DTC gain for robust jitter performance across process, voltage, and temperature (PVT) variations.

Two techniques to realize DTC gain calibration have been presented in the literature, namely delay matching [1-3], and least-mean square (LMS) using auxiliary PLL [4]. Prior delay matching calibration techniques [1-3] relied on an extra re-timer stage at MMD output to create a T_{IN} time reference. A feedback loop then attempts to match the delay of two paths to set $K_{DTC} \approx T_{IN}$ as shown in Fig. 10.4.1. Specifically, [1] added a complimentary replica DTC after the main, comparing it against the T_{IN} path to ensure $P_Q K_{DTC} + (1-P_Q)K_{DTC} \approx K_{DTC} \approx T_{IN}$. In [2], the re-timer stage was added to the main DTC path and compared against an offset replica, aiming for $T_{IN} + P_Q K_{DTC} \approx (1+P_Q)K_{DTC}$, thus $K_{DTC} \approx T_{IN}$. In [3], DTC delay ($P_Q K_{DTC}$) is directly compared to a reconfigurable path using a multiplexer, where the re-timer stage is bypassed (when $P_Q=0$) or provide T_{IN} delay (when $P_Q=1$). Delay matching calibration techniques [1-3] practically suffer from limited accuracy due to the random and systematic mismatches between the two paths, which require very careful layout and tedious offset calibration steps [5]. In contrast, [4] used an auxiliary PLL for accurate and robust LMS calibration, similar to traditional DTC-based digital PLLs [6]. This, however, incurred significant power and area penalties, making FDIV less appealing compared to simply using a fractional-N digital PLL. In this work, we propose a high performance standalone FDIV leveraging a robust replica-free LMS-based DTC background calibration using a compact auxiliary delay locked loop (AUX-DLL). The proposed FDIV provides a wide output frequency range (16-1600MHz) with very fine frequency resolution (50b) and achieves worst-case integrated jitter performance of 350fs$_{rms}$, while occupying a small footprint of 0.008mm².

To demonstrate the basic operation of the proposed FDIV and how the background calibration works, Fig. 10.4.1 depicts a simple example of a divide by 4.25 operation using a 1st-order $\Delta\Sigma$ modulator. A division ratio of 4.25 ($N_I=4$ and $\alpha=1/4$) is realized by dividing the input by 4 for three cycles and by 5 for one cycle in a repetitive manner. Any deviation in DTC gain from T_{IN} causes P_Q to leak to the output, appearing as deterministic jitter. By comparing CLK$_{OUT}$ edges to an ideal clock, the absolute deterministic jitter (ADJ) can be expressed as $ADJ \approx P_Q(T_{IN}-K_{DTC})$. Period deterministic jitter (PDJ) represents CLK$_{OUT}$ period fluctuations, which depend on the time difference between two consecutive jittery edges (i.e. 1st time difference of ADJ) and can be expressed as $PDJ=(1-z^{-1})ADJ \approx P_Q(1-z^{-1})(T_{IN}-K_{DTC}) \approx F_Q(T_{IN}-K_{DTC})$, where F_Q is $\Delta\Sigma$ frequency quantization noise. We propose to leverage a digital DLL to detect PDJ accurately. This DLL comprises a phase detector (PD), digitally controlled delay line (DCDL), and an accumulator (ACC). The DLL ensures CLK$_{OUT}$ and CLK$_{DLY}$ are separated precisely by 1T$_{CLK}$ on average sense, and PD output (Err) detects residual PDJ due to DTC gain error (Err - $F_Q(T_{IN}-K_{DTC})$). By correlating Err with F_Q, an LMS algorithm can precisely converge to the optimum DTC gain calibration factor (K_G).

Figure 10.4.2 presents a detailed block diagram of the proposed FDIV. A reconfigurable 1st/2nd-order $\Delta\Sigma$ modulator with 44b input generates 4-level $\Delta\Sigma_{OUT}$ which is added to a 6b integer division control (N_I). The resulting dithered division signal (N_{DIV}) controls an extended range MMD, seamlessly dividing the input clock frequency (F_{IN}) by $N_I + \alpha$. The MMD output (CLK$_{MMD}$) feeds the DTC, which compensates for P_Q after truncation from 44b to 9b. The DTC is followed by a post divider ($\div L$), implemented simply using a cascaded divide-by-2s, a MUX, and re-timer. Hence, the FDIV can generate a wide range of output frequency, $F_{OUT}=F_{IN}/[(N_I+\alpha)\times L]$. DTC output (CLK$_{DTC}$) also drives DCDL and PD of the AUX-DLL. The PD measures the sign of the phase error between CLK$_{DTC}$ and CLK$_{DLY}$, and the generated Err signal is fed to a ACC$_D$ after scaling by a programmable digital gain factor (K_{DLL}) to control DLL bandwidth. The accumulator output (K_D) is truncated to 7b using a 2nd-order $\Delta\Sigma$

modulator to control 7b thermometer-coded current digital-to-analog converter (IDAC$_D$). A 1st-order RC post filter then suppresses the shaped quantization noise from IDAC$_D$, generating a control voltage (V_D) to tune DCDL towards delay lock condition $T_{DCDL}= T_{CLK}$. Once the DLL is locked, Err is correlated with F_Q using a mux-based correlator. The LMS correlator output (LMS$_{ERR}$) is fed to ACC$_G$ after scaling by a programmable digital gain factor (K_{LMS}). K_{LMS} must be carefully selected, considering the tradeoff between LMS convergence time and accuracy. LMS accumulator output (K_G) fine tunes DTC gain K_{DTC} to match T_{IN} using 2nd-order $\Delta\Sigma$ modulator, IDAC$_G$, and 1st-order RC post filter, a similar implementation to the DCDL tuning path. When LMS converges, Err is no longer correlated with F_Q and DTC cancels P_Q correctly.

The DTC is implemented using a cascade of 8 identical digitally controlled delay cells (DCDCs) as shown in Fig. 10.4.2. Delay tuning is achieved by adjusting the load capacitance of the first inverter via a 6b capacitor bank, while gain is tuned by filter output voltage V_G. This multi-stage implementation ensures fast rise/fall times, enabling high-frequency operation beyond 1.6GHz, and reduces DTC sensitivity to supply/thermal noise. To enhance DTC linearity, a segmented control scheme [1] distributes the desired delay equally among all DCDCs. $P_Q[8:3]$ controls the 63-unit binary capacitor bank in all 8 DCDCs, while $P_Q[2:0]$ drives a binary-to-thermometer (B2T) decoder. Each B2T output controls a unit capacitor in the corresponding DCDC, reducing DTC integral non-linearity (INL) to <3ps. Fig. 10.4.2 also details the 7b unary IDACs, designed for high linearity and minimal quantization noise. High linearity is critical to reduce folding of shaped $\Delta\Sigma$ quantization noise, and fine resolution minimizes filter requirements for compact implementation. A two-dimensional (2D) (8×16) B2T decoder, using even/odd local decoder cells and a zigzag switching scheme, ensures only one-line changes at a time (C-Reg change at the start of the column, A-Reg change at the end of the column, otherwise R-Reg change). This minimizes glitches, differential non-linearity (DNL) and digital activity.

The proposed FDIV is designed to cover a versatile range of applications with a wide input frequency range (8-32GHz), and wide output frequency range (16-1600MHz). The current-starved DTC tracks the broad range of T_{IN}. By utilizing a post-divider ($\div L$), the operating frequency range of the DTC and DLL is constrained to one octave (0.8-1.6GHz), which limits the required DCDL delay range. Despite this, designing a wide-frequency digital DLL remains challenging due to potential false locking. The correct locking condition requires CLK$_{DLY}$ Edge-0 to lock to CLK$_{DTC}$ Edge-1, preventing stuck locking (CLK$_{DLY}$ Edge-0 attempting to lock to CLK$_{DTC}$ Edge-0), or harmonic locking (CLK$_{DLY}$ Edge-0 locking to CLK$_{DTC}$ Edge-2, Edge-3, etc.). As shown in Fig. 10.4.3, conventional bang-bang (BB)-PD has a limited input delay range, making them susceptible to these false locking scenarios, which would compromise DTC gain calibration. To address this, a tri-state phase and frequency detector (PFD) with a dedicated reset path is integrated before the BBPD. The reset pulse is generated from the DCDL mid-point tap, incurring no additional overhead. This ensures CLK$_{DLY}$ Edge-0 is consistently compared against CLK$_{DTC}$ Edge-1, extending PD input delay range to +/-T$_{CLK}$. This robust phase detection scheme eliminates potential DLL false locking, ensuring reliable operation across the entire 0.8-to-1.6GHz output frequency range. This also enables a simplified FDIV usage scenario with automatic locking on-startup, requiring no internal signal probing or user intervention.

Fabricated in a 7nm CMOS process, the proposed FDIV with LMS-based DTC background calibration using AUX-DLL occupies an active area of 0.008mm². It generates a wide range of output frequencies from 16MHz to 1.6GHz with very fine resolution (50b). Power consumption at 1.6GHz is 10mW from a single 0.9V supply. Fig. 10.4.4 shows measured phase noise for integer- and fractional-N mode with integrated jitter [100kHz-to-100MHz] of 352fs$_{rms}$ for fractional, and 282fs$_{rms}$ for integer (limited by input clock jitter). Fig. 10.4.5 (top-left) shows the FDIV output spectrum at 0.9GHz, with excellent spurious performance of -69dBc. The worst-case spur performance of -50dBc ($\alpha=2^{-17}$), is limited by DTC INL. Fig. 10.4.5 (top-right) illustrates measured spurs and integrated jitter versus fractional division factor (α) and supply voltage. At 0.9V supply, the FDIV achieved 0.3ps$_{rms}$ jitter with calibrated DTC QNC, compared to 1.65ps$_{rms}$ without DTC QNC. If the supply voltage varies from 0.9V, the FDIV maintains the excellent jitter performance demonstrating the effectiveness and robustness of the proposed background DTC calibration technique, while if DTC calibration is frozen, the jitter degrades as DTC gain deviates from T_{IN}. Fig. 10.4.5 (bottom) presents

DLL and LMS measured results of 550 parts from corner-wafers. Histograms of 10b $IDAC_G$ and $IDAC_D$ codes are centered around mid-code with a relatively small standard deviations (63.5LSB for $IDAC_G$ and 31.5LSBs for $IDAC_D$), indicating ample margin for voltage and temperature (VT) and input frequency variations. Fig. 10.4.6 summarizes the performance of the proposed FDIV and compares it to prior works. The proposed FDIV generates the highest output frequency (5× higher than [2-4]), and excellent jitter performance (4× better than [1]). The proposed LMS calibration using AUX-DLL is very robust and practical compared to delay matching techniques [1-3], while occupying 10× smaller area than [4] which used AUX-PLL. Furthermore, AUX-DLL can be easily leveraged to compensate for DTC INL with minimum overhead.

References:
[1] A. Elkholy et al., "A 20-to-1000MHz ±14ps Peak-to-Peak Jitter Reconfigurable Multi-Output All-Digital Clock Generator Using Open-Loop Fractional Dividers in 65nm CMOS," *ISSCC*, pp. 272-273, 2014. http://doi.org/10.1109/ISSCC.2014.6757431
[2] Y. Yu et al., "A 0.024mm² All-Digital Fractional Output Divider with 257fs Worst-Case Jitter Using Split-DTC-Based Background Calibration," *ISSCC*, pp. 168-169, 2025. http://doi.org/10.1109/ISSCC49661.2025.10904505
[3] C. -Y. Lin et al., "A 0.008mm² 1.5mW 0.625-to-200MHz Fractional Output Divider with 120fs_rms Jitter Based on Replica-DTC-Free Background Calibration," *ISSCC*, pp. 412-414, 2021. http://doi.org/10.1109/ISSCC42613.2021.9365821
[4] Y. Yang et al., "A 10-to-300MHz Fractional Output Divider with -80dBc Worst-Case Fractional Spurs Using Auxiliary-PLL-Based Background 0th/1st/2nd-Order DTC INL Calibration," *ISSCC*, pp. 228-229, 2023. http://doi.org/10.1109/ISSCC42615.2023.10067785
[5] A. Elkholy et al., "Low-Jitter Multi-Output All-Digital Clock Generator Using DTC-Based Open Loop Fractional Dividers," *IEEE JSSC*, vol. 53, no. 6, pp. 1806-1817, 2018. http://doi.org/10.1109/JSSC.2018.2817602
[6] D. Tasca et al., "A 2.9–4.0-GHz Fractional-N Digital PLL With Bang-Bang Phase Detector and 560-fs_rms Integrated Jitter at 4.5-mW Power," *IEEE JSSC*, vol. 46, no. 12, pp. 2745-2758, 2011. http://doi.org/10.1109/JSSC.2011.2162917

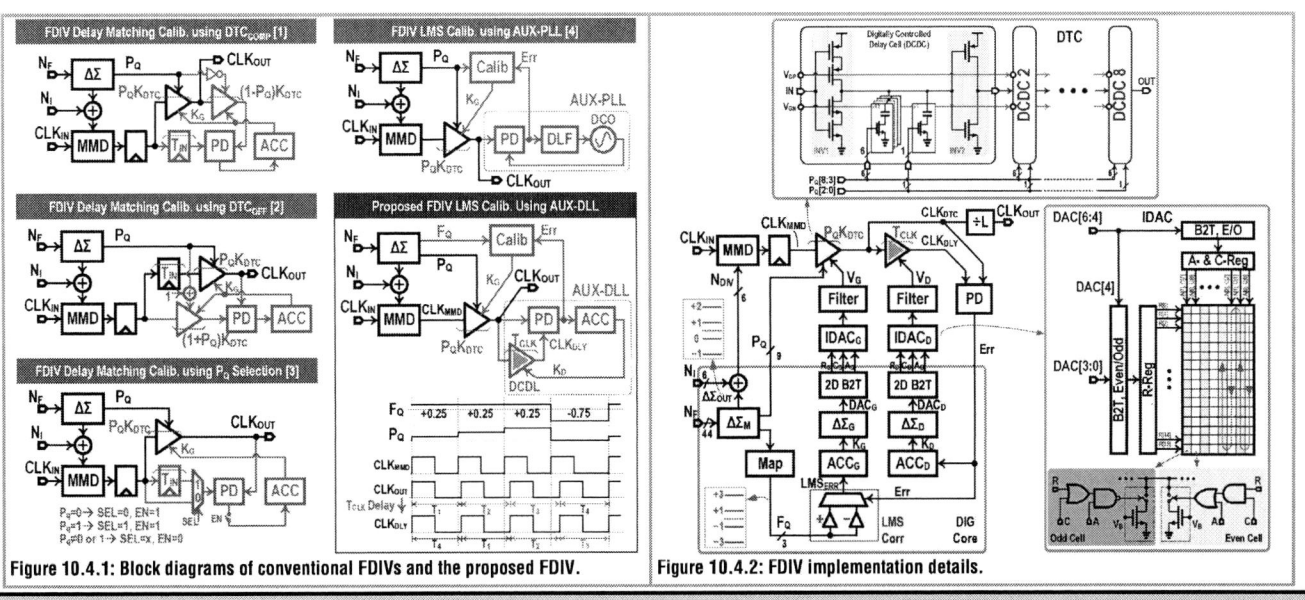

Figure 10.4.1: Block diagrams of conventional FDIVs and the proposed FDIV.

Figure 10.4.2: FDIV implementation details.

ISSCC 2026 / SESSION 10 / DIGITAL PROCESSING AND CIRCUIT TECHNIQUES / 10.4

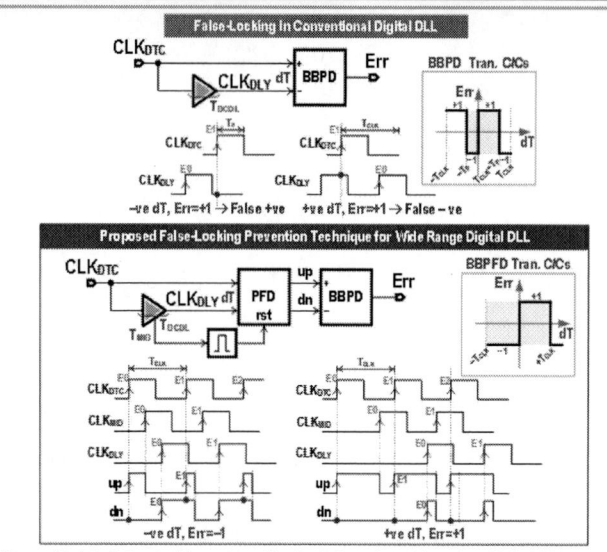

Figure 10.4.3: False-locking prevention technique for wide-range digital DLL.

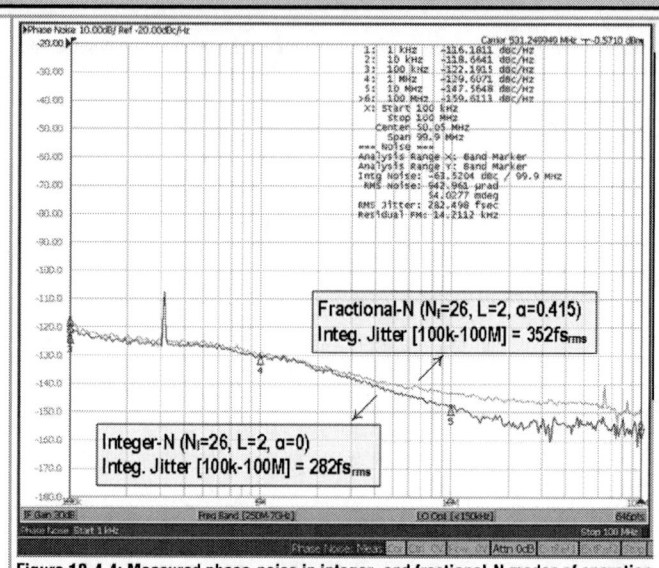

Figure 10.4.4: Measured phase-noise in integer- and fractional-N modes of operation.

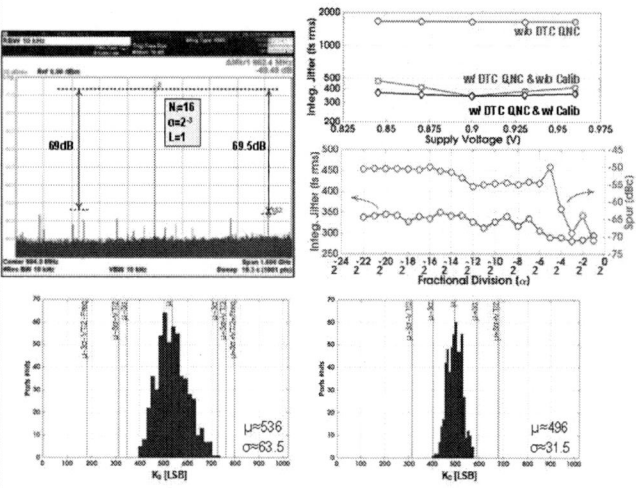

Figure 10.4.5: Measured spurs and integrated jitter across fractional division factor (α) and supply voltage (top), measured IDAC_G and IDAC_D codes of 550 parts (bottom).

	ISSCC'14 [1]	ISSCC'25 [2]	ISSCC'21 [3]	ISSCC'23 [4]	This Work
Technology	65nm	28nm	90nm	28nm	7nm
Supply [V]	0.9	0.9	1.0	0.9	0.9
F_IN Range [GHz]	5	5	5	8	8-32
F_OUT Range [MHz]	20-1000	2.5-250	0.6-200	100-300	16-1600
Freq. Resolution	21-bit	23-bit	21-bit	25-bit	50-bit
Instant. Switching	Yes	Yes	Yes	Yes	Yes
RMS Jitter [fs]	1440 [10k-40M]	257 [10k-20M]	340 [100-30M]	316 [10k-20M]	350 [10k-100M]
Worst Spur [dBc] Normalized to 1GHz	-49	-59	-50	-60	-49
Power Consumption	3.2mW @1GHz	3.4mW @123MHz	1.5mW @192MHz	1.25mW @50MHz	10mW @1.6GHz
Power Efficiency [mW/GHz]	3.2	27.6	7.8	25	6.25
Architecture	FDIV Delay Matching Calib. using DTC_COMP	FDIV Delay Matching Calib. using DTC_OFF	FDIV Delay Matching Calib. using P_Q Sel.	FDIV LMS Calib. using AUX-PLL	FDIV LMS Calib. using AUX-DLL
Robustness & Tolerance to Mismatches	No (DTC offset delay, replica mismatch, etc)	No (DTC Offset Delay, Replica mismatch, etc)	No (DTC Offset Delay, MUX Mismatch, etc)	Yes (Measurements of 1 Part)	Yes (Validated by Measurements of 550 Parts)
DTC INL Calib.	No	Yes	No	Yes	No
Area [mm²]	0.017	0.024	0.008	0.084	0.008

Figure 10.4.6: Performance summary and comparison with state-of-the art.

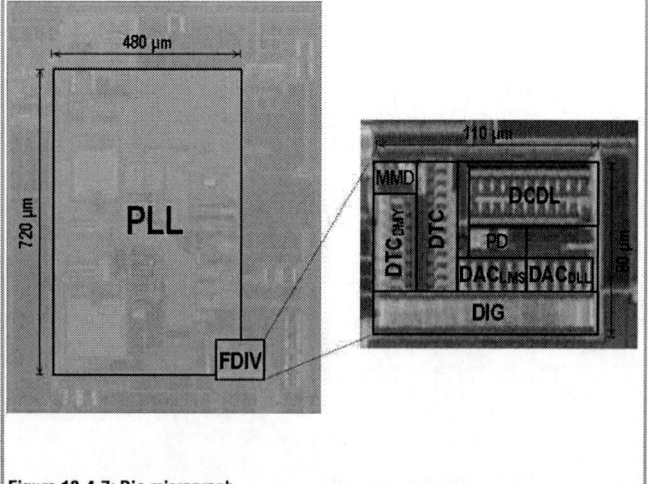

Figure 10.4.7: Die micrograph.

- 2026 IEEE International Solid-State Circuits Conference

ISSCC 2026 / SESSION 10 / DIGITAL PROCESSING AND CIRCUIT TECHNIQUES / 10.5

10.5 Proactive Power Management-Based Supply Regulation with Online Learning for Variation-Tolerant Workload-Aware Droop Mitigation in 28nm CMOS

Xi Chen[1], Andrew Liss[1], William Covington[1], Qiankai Cao[1], Yiqi Li[1], Kang Wei[2], Raveesh Magod[3], Muhammad Khellah[4], Xin Zhang[5], Jie Gu[1]

[1]Northwestern University, Evanston, IL, [2]Texas Instruments, Dallas, TX, [3]Indian Institute of Technology Madras, Chennai, India, [4]Intel, Hillsboro, OR
[5]IBM T. J. Watson Research Center, Yorktown Heights, NY

Abstract

A 28nm SoC solution with integrated proactive power management for droop mitigation is demonstrated combining a neural droop management unit, integrated high speed power converter, and an online learning engine to combat the PDN and workload variations. The 28nm test chip integrated with CPU and accelerators achieves 59% worst-case droop reduction, 48× throttling reduction, and >91% regulator peak efficiency, reducing performance degradation from prior fixed-model or throttling-only schemes.

Accelerator-enriched microprocessors with highly dynamic workload cause significant supply droops. Recent works address droop challenges through a variety of schemes including reactive [1], or proactive [2], [3] clock throttling, unified clock and power regulation [4], machine learning (ML)-based power management (PM) [5], software-assisted workload management [6]. However, there are still several unaddressed challenges, as shown in Fig. 10.5.1. First, the inevitable package-level impedance variations and die-level Process-Voltage-Temperature (PVT) variations cause a significant uncertainty of the droop magnitude and frequency due to the deviation of power delivery network (PDN) impedance from the pre-silicon design. Prior works using pre-determined PDN models suffer from inaccurate droop mitigation or costly post-silicon characterization efforts [3]. Second, the state-of-the-art countermeasures using clock throttling introduces significant performance penalties, e.g. up to 30% in the recent report [1]. In addition, the unresolved supply undershoot or overshoot from clock throttling cause reliability and signal integrity degradation to the chips. Finally, integrated power converters, e.g. DC-DC convertors, face an efficiency-speed trade-off, i.e. large off-chip inductors offer high efficiency but suffer from slow transient response, while small in-package/on-die inductors respond faster at the cost of efficiency loss. To address the above challenges, this work proposes a holistic solution utilizing both advanced ML computation and new circuit designs. The contributions of this work are summarized as follows: (1) To minimize performance loss from clock throttling, this work develops a workload-aware proactive power management which predicts upcoming droops and proactively adjusts the regulator to achieve near-optimal droop regulation with minimal performance loss. (2) To address the efficiency and speed tradeoff of power converters, this work employs a dual-inductor buck converter achieving both high efficiency and high speed. (3) To overcome the impact of the variation, an on-device online learning optimizer is developed to track the PDN and work-load dependent droop characteristics. The proposed technique is applied to a SoC integrating CPU, CNN and transformer accelerators, demonstrating the versatility of the proposed droop mitigation methods.

Figure 10.5.2 shows the system-on-chip architecture of the learning-enabled droop mitigation scheme. The SoC consists of main processors including a RISC-V CPU, a CNN and a transformer accelerator serving as workload for droop management, a runtime Neural Droop Management Unit (NDMU), an online learning engine (OLE), and an integrated DC-DC converter with both off-chip and on-chip inductors. The DC-DC converter takes 1.8V IO supply and generates digital core supply from 0.5-1.2V by conventional feedback control along with an enhanced NDMU-guided feedforward control. Fig. 10.5.2 also depicts the proposed NDMU's internal architecture, which comprises four sub-modules. (1) an Activity Embedding Module (AEM) monitors processors' signals related to critical switching activities (e.g., instruction, memory accesses, operation types, etc.) from the CPU and accelerators, converting them into a representative activity vector via an embedding matrix. (2) a Power Predictor Module (PPM) feeds the embedded activity vector into a bank of pipelined regression operators to compute a running sum over the last programmable N cycles (e.g., N=5) of predicted power in parallel, reducing critical-path delay by ~5× vs. a serial predictor [5]. (3) a Droop Predictor Module (DPM) uses a neural network (NN) model, with inputs from the predicted power and recently sensed voltages over the last i cycles (i is programmable) to forecast upcoming voltage droops. The NN is initialized with pre-trained model and is capable of performing on-device online learning as a fine-tuning operation to adapt to silicon and workload variations. (4) a Mitigation Modulator (MM) receives upcoming droop events from DPM and issues activity commands to the regulator. The MM determines the droop event levels based on droop prediction and generates the feedforward control signal for the regulator to provide the charge compensation that is adaptive to the workload change, leading to proactive mitigation of the predicted droop. An error detection module is used to halt the on-going feedforward boost or activate the throttling-based guardband by monitoring the real-time supply in the rare event of misprediction. A 5b flash ADC-based voltage meter is included to provide real-time supply voltage measurement data to support online learning and testing with a total of 32KB on-chip memory for recording. The droop prediction results are also stored in memory for testing purposes.

As shown in Fig. 10.5.3, use of large off-chip inductors for higher efficiency in power converters leads to slow current slopes with steeper first-droop and longer recovery times. This work proposes an additional feedforward droop mitigation scheme to boost the speed

of operation. The main control loop is through a conventional feedback loop consisting of an error amplifier, compensator, PWM generator, and timing control circuits. When a transient load event is predicted (indicated by the LT_ACT signal from the NDMU), the regulation is taken over by the feedforward path. Boost signals including both Main_Ctrl and AUX_Ctrl from the NDMU trigger a feedforward boost controller that includes an adaptive switch modulator and a handover control to provide 2-stage regulation with off-chip inductor and on-chip inductor. Simulation results under a 500mA load step show that relative droop and settling time are reduced by 7.5× and 4.3×, respectively, compared to the baseline. The 2-stage adaptive boost strategy is detailed as follows. At the first stage, LT_ACT triggers a swap from PWM to a single-on switch until the main inductor current reaches the predicted workload. PWM status is monitored during the swap to determine whether to trigger an on-switch or utilize the existing PWM on-phase. At the end, a pre-bias-based handover circuit gives the control back to the PWM loop. At stage 2, an on-die ~3nH auxiliary inductor is rapidly ramped up to the workload current, while the main inductor lags due to its slower slope. Then, a programmable PWM generator gradually reduces auxiliary current to maintain a constant total current of both inductors, and a shunt switch removes residual oscillation once current returns to zero.

Figure 10.5.4 shows the simulated runtime droop prediction for CPU and AI accelerators which exhibit ~5mV mean error across different RISC-V programs or Transformer/CNN layers. Since the worst-case droop (e.g., >5% of nominal supply) in SoC only occurs in a small fraction of runtime (<20% across benchmarks), continuous operation of the NDMU at full activity would incur non-negligible overhead. To reduce this overhead, a dual-mode NDMU is implemented: The high-performance mode is activated under droop risk, enabling cycle-wise monitoring, while the low-power mode leverages less frequent prediction during near-steady-state operation, e.g., intra-layer execution of CNN. To deal with variation impact in either PDN or workload, this work proposed an online transfer learning scheme which captures such variations by fine-tuning the prediction model with an online learning engine (OLE). Real-time supply and operation signals during the droop events are stored into a data memory while the mitigation is paused. Instead of re-training the entire network, most layers of the droop predictor NN are frozen, and only the last one or two layers are fine-tuned through the backpropagation with the mini batch from learning pool. The online learning scheme can be activated either by a customized pilot program that exercises the PDN of the chip, or at an early stage of real execution, e.g., the 1st layer of AI models to capture real workload features.

Figure 10.5.5 shows the measurement results from a test chip fabricated in a 28nm process. Real-time droop measurements as well as the activation of clock gating signals are recorded from voltage meters and plotted for both CPU and accelerator operations, with three test conditions (1) original droop without mitigation, (2) prediction with pre-trained models, (3) prediction with online learning. As shown in Fig. 10.5.5, the model with online learning can effectively cut down the droop misprediction cases from pre-trained model, reducing the droop magnitude by 79mV on average over 100K run cycles of Mibench and VGG-16 test cases. The recorded clock gating signals show that the proposed NDMU with OLE can remove most of the clock throttling operations, in the operation of transformer accelerator. To evaluate the targeted variation tolerance, four different package configurations including QFN60, QFN64, and QFN80 packages with different bond wire lengths and different numbers of bond wires were tested for PDN variation, as shown in Fig. 10.5.5. The OLE improves droop prediction accuracy by 32% to 42% over pre-trained models, with only 8 epochs of model update. Over 100k cycles of VGG-16 operation, the worst-case droop magnitude has been reduced by 47% to 59% as measured from the four package configurations. Fig. 10.5.5 also shows the trend of clock throttling-only operation (as in prior works) vs. the proposed scheme at various voltage margins. As voltage is reduced from 1V without lower margins, a significant increase of clock throttling with high performance loss is observed, while the proposed NDMU/OLE with proactive power management sustains low performance loss by reducing clock gating cycles from 24% to ~1%. The power overhead of the NDMU/OLE is measured to be up to 2.4% of the SoC power in the high performance model, but is reduced to 0.64% in the low power model.

Figure 10.5.6 presents additional measured power efficiency and performance gain of the proposed droop mitigation scheme. The integrated DC-DC achieves a peak efficiency of

176 • 2026 IEEE International Solid-State Circuits Conference

979-8-3315-8937-0/26 $31.00 © 2026 IEEE

91.7% across an output voltage range of 0.5-1.2 V with 0.5% efficiency loss from the auxiliary feedforward circuit. 6 CPU benchmarks, 4 AI models are measured with over 1 million cycles showing average 79.3mV droop magnitude reduction and an average of 48.5× reduction of clock throttling event. A comparison table is given in Fig. 10.5.6 highlighting the contributions of this work over prior droop mitigation methods. Unlike prior clock-throttling schemes that incur a significant performance penalty as well as the presence of supply droops, the proposed scheme has significantly reduced the performance loss by up to 23×. While all prior works rely on pre-trained models or chip calibration, the proposed scheme enables online learning which delivers high tolerance to PDN and workload variation, reducing the worst-case droops by up to 59% across a range of chip configurations. Fig. 10.5.7 shows the die micrograph and the packaging variations used for the measurements.

Acknowledgement:
This work is supported in part by SRC through the Texas Analog Center of Excellence, UT, Dallas (Task ID: 3160.039) and NSF ECCS-2435138.

References:
[1] T. Webel et al., "Dynamic Guard-Band Features of the IBM zNext System," *ISSCC*, pp. 158-159, 2025. http://doi.org/10.1109/ISSCC49661.2025.10904615
[2] V. K. Kalyanam et al., "A Proactive System for Voltage-Droop Mitigation in a 7-nm Hexagon™ Processor," *IEEE JSSC*, vol. 56, no. 4, 2021. http://doi.org/10.1109/JSSC.2020.3043786
[3] W. Shan et al., "Proactive Voltage Droop Mitigation Using Dual-Proportional-Derivative Control Based on Current and Voltage Prediction Applied to a Multicore Processor in 28nm CMOS," *ISSCC*, pp. 256-258, 2024. http://doi.org/10.1109/ISSCC49657.2024.10454398
[4] C.-H. Huang et al., "A Single-Inductor 4-Output SoC with Dynamic Droop Allocation and Adaptive Clocking for Enhanced Performance and Energy Efficiency in 65 nm CMOS," *ISSCC*, 2021. http://doi.org/10.1109/ISSCC42613.2021.9365760
[5] X. Chen et al., "Proactive Power Regulation with Real-time Prediction and Fast Response Guardband for Fine-grained Dynamic Voltage Droop Mitigation on Digital SoCs," *IEEE Symp. VLSI Circuits*, 2023. http://doi.org/10.23919/VLSITechnologyandCir57934.2023.10185397
[6] M. Kar et al., "A Software-Assisted Peak Current Regulation Scheme to Improve Power-Limited Inference Performance in a 5nm AI SoC," *ISSCC*, pp. 254-256, 2024. http://doi.org/10.1109/ISSCC49657.2024.10454301
[7] S. J. Kim et al., "A 0.5-1V Input Event-Driven Multiple Digital Low-Dropout-Regulator System for Supporting a Large Digital Load," *IEEE Symp. VLSI Circuits*, pp. C128-C129, 2019. http://doi.org/10.23919/VLSIC.2019.8778117
[8] C.-H. Huang et al., "Improving SIMO-Regulated Digital SoC Energy Efficiencies Through Adaptive Clocking and Concurrent Domain Control," *IEEE JSSC*, vol. 57, no. 1, pp. 90-102, 2022. http://doi.org/10.1109/JSSC.2021.3102603
[9] T. Hu, M. Huang et al., "A 12V-to-1V Quad-Output Switched-Capacitor Buck Converter with Shared DC Capacitors Achieving High Peak Efficiency and Power Density," *ISSCC*, pp. 184–186, 2023. http://doi.org/10.1109/ISSCC42615.2023.10067463
[10] G. Cai et al., "A compact 12V-to-1V Hybrid Resonant Switched-Capacitor Parallel Inductor (ReSC-PL) Buck Converter with High Peak Efficiency," *ISSCC*, pp. 198-199, 2023. http://doi.org/10.1109/ISSCC42615.2023.10067303

[11] S. Kim et al., "A Monolithic Top-Metal and C4 Planar Spiral Inductor-Based Integrated Buck Voltage Regulator on 16nm-Class CMOS," *ISSCC*, pp. 270–272, 2024. http://doi.org/10.1109/ISSCC49657.2024.10454473
[12] K. Joshi et al., "A Fully Integrated Multi-Phase Voltage Regulator with Enhanced Light-Load Efficiency and Autonomous Mode Transition in 3nm FinFET CMOS," *ISSCC*, pp. 382–384, 2025. http://doi.org/10.1109/ISSCC49661.2025.10904723
[13] X. Chen et al., "A 65-nm Proactive Power Management Technique with Real-Time Machine Learning Engine for Droop Prediction and Mitigation on Microprocessors," *IEEE JSSC*, vol. 60, no. 6, pp. 2170-2181. http://doi.org/10.1109/JSSC.2024.3479273
[14] H. Han et al., "A Parallel-SC Hybrid Buck Converter Designed Using VCR-Aware Topology Optimizer for High Efficiency And Current-Density FoM," *ISSCC*, pp. 464–466, 2024. http://doi.org/10.1109/ISSCC49657.2024.10454535
[15] Y.-H. Kao et al., "A Ripple-Less Buck Converter with Sub −21.94 dB EVM for 5G Low Earth Orbit Application," *ISSCC*, pp. 498–500, 2024. http://doi.org/10.1109/ISSCC49657.2024.10454514
[16] Z. Ahmed et al., "A Fully Integrated Buck Voltage Regulator in 16nm with In-Package Air-Core Inductor Featuring Digital Computational Control for Fast Transient Responses," *IEEE Symp. VLSI Circuits*, pp. C18–3, 2025. http://doi.org/10.23919/VLSITechnologyandCir65189.2025.11075218
[17] X. Chen et al., "A 65-nm Proactive Power Management Technique With Real-Time Machine Learning Engine for Droop Prediction and Mitigation on Microprocessors," *IEEE JSSC*, vol. 60, no. 6, pp. 2170-2181, 2024. http://doi.org/10.1109/JSSC.2024.3479273

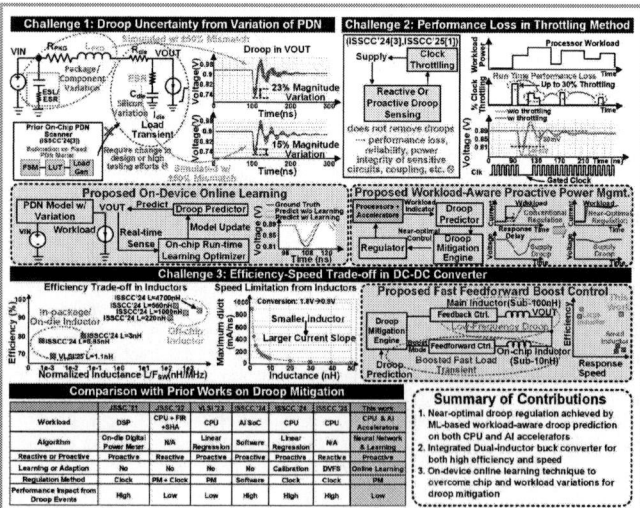

Figure 10.5.1: Existing challenges in droop mitigation and key contributions of the proposed regulation framework.

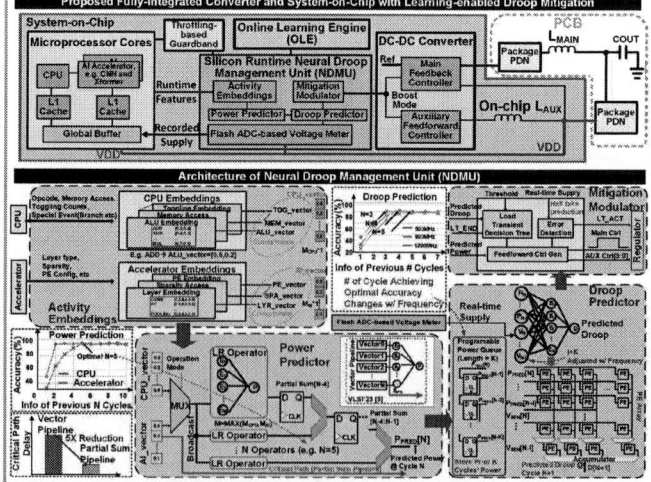

Figure 10.5.2: System-level diagram of the proposed proactive supply regulation scheme with online learning, including the architecture of the Neural Droop Management Unit (NDMU).

ISSCC 2026 / SESSION 10 / DIGITAL PROCESSING AND CIRCUIT TECHNIQUES / 10.5

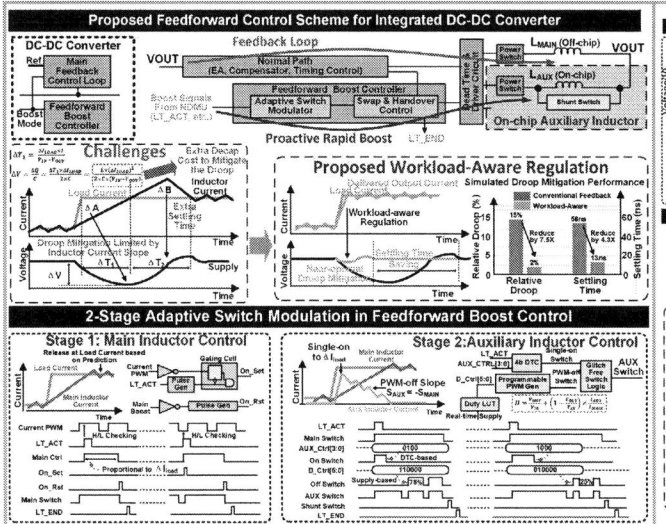

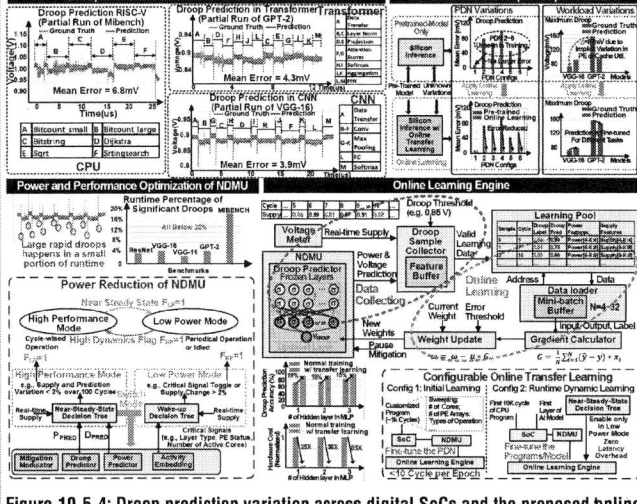

Figure 10.5.3: Proposed feedforward control of the DC–DC converter employing dual-switch modulation for rapid droop mitigation.

Figure 10.5.4: Droop prediction variation across digital SoCs and the proposed online transfer learning scheme for adaptive model update.

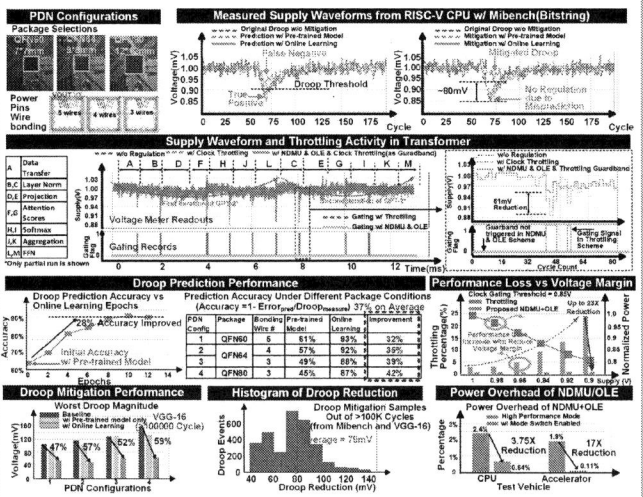

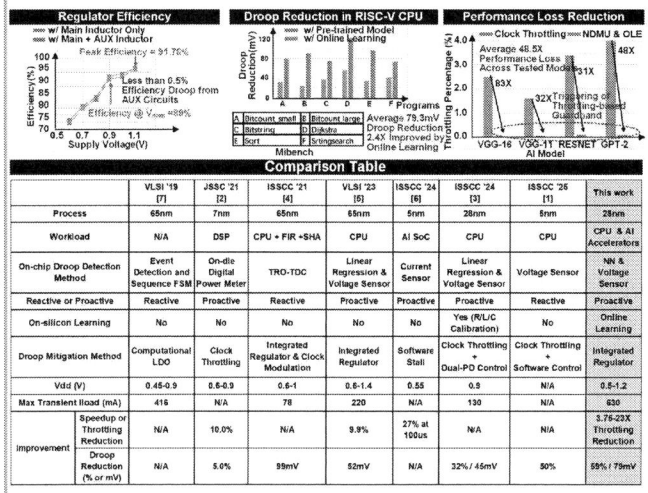

Figure 10.5.5: Measured performance of the Neural Droop Management Unit (NDMU) and Online Learning Engine (OLE).

Figure 10.5.6: Measured regulator efficiency and mitigation performance across benchmarks, with comparison to prior works.

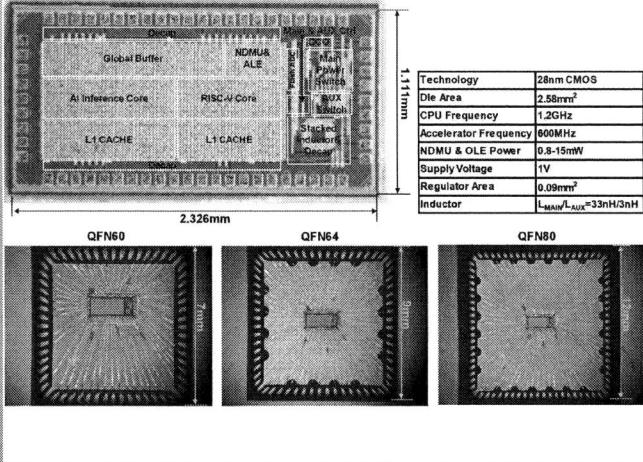

Technology	28nm CMOS
Die Area	2.58mm²
CPU Frequency	1.2GHz
Accelerator Frequency	600MHz
NDMU & OLE Power	0.8-15mW
Supply Voltage	1V
Regulator Area	0.09mm²
Inductor	L_{MAIN}/L_{AUX}=33nH/3nH

Figure 10.5.7: Die micrograph and packaging/bonding configurations of the fabricated chip.

• 2026 IEEE International Solid-State Circuits Conference

ISSCC 2026 / SESSION 10 / DIGITAL PROCESSING AND CIRCUIT TECHNIQUES / 10.6

10.6 A Hybrid-Bonded 12.1TOPS/mm² 56-Core DNN Processor with 2.5Tb/s/mm² 3D Network on Chip

Phil C. Knag[1], Gregory K. Chen[1], Shanshan Xie[2], Satish Yada[1], Wei Wu[1], Yu-Shiang Lin[1], Alexander Kashirin[1], Xiemei Meng[2], Russell Criss[1], Ana Sonia Leon[3], Carlos Tokunaga[1], Ram K. Krishnamurthy[1], James W. Tschanz[1]

[1]Intel, Hillsboro, OR, [2]Intel, Austin, TX, [3]Intel, Santa Clara, CA

Abstract

A manycore DNN processor leverages hybrid bonding in a 14×4×2 mesh network on chip to increase 3D bandwidth between cores, accelerators, and SRAM to 2.5Tb/s/mm². Memory throughput is improved by 39% compared to an equivalent 2D design with the same memory capacity. The 56 16×16 INT8 systolic array accelerators are stacked on top of 56 RISC-V cores to enable a peak AI performance density of 12.1TOPS/mm². The DNN processor scores a latency of 53μs on the image classification benchmark from MLPerf.

AI and machine-learning workloads feature dense data movement and high-bandwidth communication between cores, accelerators, and memory. Reducing the power and performance overhead of this data movement is key for energy-efficient operation. As die area increases, cross-die communication can become a critical bottleneck, limiting workload performance. 3D stacking addresses this communication challenge, but achieving the full throughput benefits of 3D requires high-bandwidth connectivity between dies, such as that enabled by hybrid-bonded interconnect (HBI). In manycore processors, HBI enables both wide memory access from parallel processing cores and lower latency to shared distributed memory. 3D networks on chip (NoCs) provide low-latency access to memory on both top and bottom dies, with cores agnostic to 3D memory design. Further, creating a general-purpose manycore bottom die allows design extensibility, with the base die used either stand-alone or augmented with application-specific top dies matching a common 3D NoC and HBI footprint. For DNN applications, multiply accumulate (MAC) operations can be offloaded to accelerators in the top die, while the bottom die orchestrates 3D data movement.

A manycore 3D DNN processor (M3DProc) is implemented with 9μm HBI connecting 56 RISC-V core tiles in an Intel 3 bottom die to 56 DNN accelerator tiles in a 12.1TOPS/mm² Intel 18A top die (Fig. 10.6.1). Cores and DNN accelerators access data from a shared 5.25MB SRAM that is distributed across the stacked dies. A 14×4×2 mesh 3D NoC (M3DNoC) delivers memory accesses with 2.5Tb/s/mm² throughput across the HBI. M3DNoC improves memory throughput by up to 39% compared to 2D NoCs spanning an identical memory capacity. M3DProc has a peak dynamic energy efficiency of 16.1TOPS/W excluding leakage at 0.5V, 280MHz. The chip scores a latency of 53μs on the image classification benchmark from MLPerf Tiny.

Manycore processors are designed to maximize compute density for increased core count and greater on-chip physical address space, often eliminating features such as out-of-order operation and coherent shared memory [1][2]. In M3DProc, the 56 bottom tiles each contain a 32b RISC-V (RV32) core, modified from open-source RTL [3]. The cores use the 5.25MB on-chip memory as a partitioned global address space, giving all cores access to memory in all top and bottom tiles. The 4-entry fully associative data (D$) and instruction (I$) caches occupy only 0.5% of the bottom tile area to maximize total shared memory. Write-through store operations to 64b cache lines eliminate modified data tracking and simplify memory control.

M3DNoC provides low-latency data transfer between compute elements and memory, while also allowing for seamless design extensibility (Fig. 10.6.2). M3DProc size can be increased via simple abutment of tiles on both the top and bottom dies, which share the same area footprint. M3DNoC has a 14×4×2 3D mesh topology which spans both top and bottom dies as well as connecting off-chip through the chip bridge. It performs dimension-order routing with two separate physical channels for requests and responses. M3DNoC uses wormhole flow control with header packets containing addresses followed by one or more 64b data flits. Routers use valid/ready signaling with single-cycle latency to reduce in-network register requirements. Input and output port latches work in tandem to support full-throughput operation [4], while requiring 50% fewer sequential elements compared to a 2-entry FIFO. The data latches are clock gated (CG) by the valid and ready signals to reduce active idle power.

3D NoCs must handle communication between layers in the 3D design, which is challenging since uncorrelated process variation on the two dies induces clock skew that impacts timing margin [5]. Balancing clock paths with post-silicon de-skewing and reducing the length of divergent clock paths with numerous 3D clock connections [6] have been proposed to address these issues. To simplify design construction, increase modularity, and facilitate timing closure, the M3DNoC HBI interface uses a forwarded-clock design, which alleviates timing constraints by allowing clock delay to scale together with data between dies. While this incurs a latency penalty when crossing between dies, DNN workloads are more throughput oriented, making this a good tradeoff for power, performance, and ease of design. A credit-based clock-crossing circuit synchronizes data to the local clock as it crosses the 3D interface, using request and acknowledge HBI pads accompanying the data in each direction, with an 8-entry asynchronous FIFO enabling full bandwidth across the 3D interface.

The top die pairs with the RISC-V base die and M3DNoC to offer application-specific acceleration. In this design, each of the 56 top tiles contains a DNN accelerator that implements convolution (Conv) with bit accuracy to a Tensorflow Lite software baseline (Fig. 10.6.3). Computation consists of matrix-matrix multiplication performed with a 16×16 INT8 systolic array, followed by bias addition, batch normalization, and activation function units. Conv is divided into operations for 4×4 height (h) by width (w) output feature map patches with 16 output channels (k) and is either distributed spatially or processed serially by accelerators. Cores initiate DNN computation through memory-mapped communication, and accelerators signal completion via an interrupt (IRQ), enabling concurrent RISC-V and accelerator processing. The memory-mapped registers also hold Conv scalar input parameters such as tensor pointers and layer dimensions. Input (X), output (Y) data, and weights (W) are stored channel-first, with INT8 values packed into 64b words. Per-channel parameters for element-wise operations are also read from memory. The accelerator transposes and lowers input activations before systolic array computations, and the output register file has capacity for two output feature map patches for double buffering. The accelerator latency is s×r×c+34 cycles, where s and r are the filter dimensions, and c is the number of input channels.

For high-throughput data movement, each of the 112 tiles contains a direct memory access (DMA) engine that supports memory copy operations for up to 2kB of contiguous memory at a time. DMA is used for Conv operations across multiple accelerators with the same input feature map patch. Burst DMA transfers are transported over the NoC as a single packet, with one header flit followed by up to 256 data flits. Similar to the DNN accelerator, the DMA is configured via memory-mapped registers and returns an interrupt to the core when complete. Core-to-core interrupts are also used to create hardware threads and perform barrier synchronization. Any core can create a thread by sending a function pointer, input parameter pointer, and interrupt to an idle follower core. The follower core's interrupt handler executes the function and returns an interrupt to the leader core. Leader cores perform a 'join' operation to wait for thread completion and complete a barrier synchronization by performing 'join' for all active threads.

M3DProc is implemented with an Intel 3 [7] bottom die and an Intel 18A [8] top die, connected by 9μm HBI (Fig. 10.6.7). It is part of a larger test vehicle with a shared power rail, so dynamic energy is reported by subtracting out the measured full-chip leakage energy at each data point. M3DProc is measured across a wide range of voltages and frequencies (Fig. 10.6.4) for a peak TOPS workload, achieving a peak performance of 33.0TOPS at 1.1V, 1.205 GHz, where one MAC is 2 OPs. The chip was measured down to 0.5V where the peak dynamic energy efficiency is 16.1TOPS/W at 280MHz. M3DNoC throughput is measured for 2D and 3D random uniform access patterns across a 2.6MB address space, where traffic is generated by RISC-V cores with varying transfer sizes. The 2D traffic pattern accesses all 56 bottom tiles, while the 3D test accesses 28 tiles in each of the top and bottom dies. Larger transfer sizes amortize the overheads of core operation; however, 2D traffic saturates at a lower transfer size due to NoC congestion. 3D communication benefits from higher bisection bandwidth and improves throughput by up to 39%. Energy for 2D and 3D DMA memory copy operations is measured across supply voltages for nearest-neighbor communication between all 112 tiles. For 2D operation, all tiles communicate with their neighboring tile on the same die, while for 3D operation, each tile communicates with the tile stacked directly above/below across the HBI interface. The DMA energy measurement has multiple components including core DMA initialization, NoC transfer, and core interrupt handling of the DMA completion interrupt. Measurements show no energy penalty for 3D DMA transfers.

Performance is measured for the image classification benchmark from MLPerf Tiny, which performs inference for CIFAR-10 using ResNet-8 [9]. Since each processor core runs an independent thread, the design enables flexible workload mappings that are optimized for different metrics such as latency and energy. Two alternative mappings are written in C++, compiled with GCC, and run with weights in on-chip memory (Fig. 10.6.5). A parallel mapping optimizes for the best end-to-end latency by parallelizing each DNN layer across as many cores as possible. It processes layers sequentially with barrier synchronization between layers to resolve core-to-core dependencies. Independent layers are computed in parallel, such as Conv layers 3 and 5, and 'Add' operations are fused with Conv layers 2 and 4. 'Add' and 'Pool' operations contribute significantly to the inference latency since they contain many calculations performed in the RISC-V cores. The parallel mapping scores a latency of 53µs on the MLPerf image classification benchmark [10]. An alternative dataflow workload mapping optimizes for lowest inference energy by reducing cross-chip communication, trading off end-to-end latency. This dataflow mapping allocates a subset of cores for each DNN layer, passing results between layers in a pipelined fashion [11]. Conv layer 7 and the fully connected layer are processed in the same core since neither of these operations limit overall throughput. By confining layer computation to a smaller portion of the chip communication overheads are reduced, improving throughput by 1.5× to 27.4k inferences/sec at 1.1V. Energy is also reduced by 1.5× to as low as 15.6µJ/inference at 0.5V.

Figure 10.6.6 compares M3DProc to previous works on HBI interfaces and manycore DNN processors. M3DNoC employs higher HBI signal density than previous works, enabling a 3D throughput density of up to 2.5Tb/s/mm². The combination of 3D stacking and process scaling enables M3DProc to achieve higher compute density than previous manycore DNN processors.

References:
[1] A. Rovinski et al., "A 1.4 GHz 695 Giga Risc-V Inst/s 496-Core Manycore Processor With Mesh On-Chip Network and an All-Digital Synthesized PLL in 16nm CMOS," *IEEE Symp. VLSI Circuits*, pp. C30-C31, 2019. http://doi.org/10.23919/VLSIC.2019.8778031
[2] S. Riedel et al., "MemPool: A Scalable Manycore Architecture With a Low-Latency Shared L1 Memory," *IEEE Trans. on Computers*, vol. 72, no. 12, pp. 3561-3575, Dec. 2023. http://doi.org/10.1109/TC.2023.3307796
[3] M. Gautschi et al., "Near-Threshold RISC-V Core With DSP Extensions for Scalable IoT Endpoint Devices," *IEEE TVLSI*, vol. 25, no. 10, pp. 2700-2713, Oct. 2017. http://doi.org/10.1109/TVLSI.2017.2654506
[4] G. Michelogiannakis and W. J. Dally, "Elastic Buffer Flow Control for On-Chip Networks," *IEEE Trans. on Computers*, vol. 62, no. 2, pp. 295-309, 2013. http://doi.org/10.1109/TC.2011.237
[5] P. Vivet et al., "A 4×4×2 Homogeneous Scalable 3D Network-on-Chip Circuit With 326 MFlit/s 0.66 pJ/b Robust and Fault Tolerant Asynchronous 3D Links," *IEEE JSSC*, vol. 52, no. 1, pp. 33-49, 2017. http://doi.org/10.1109/JSSC.2016.2611497
[6] T. Wu et al., "A 3D Integrated Prototype System-on-Chip for Augmented Reality Applications Using Face-to-Face Wafer Bonded 7nm Logic at <2µm Pitch with up to 40% Energy Reduction at Iso-Area Footprint," *ISSCC*, pp 210-212, 2024. http://doi.org/10.1109/ISSCC49657.2024.10454529
[7] W. Hafez et al., "An Intel 3 Advanced FinFET Platform Technology for High Performance Computing and SOC Product Applications," *IEEE Symp. VLSI Tech.*, pp. 1-2, 2024. http://doi.org/10.1109/VLSITechnologyandCir46783.2024.10631513
[8] K. Fischer et al., "Intel 18A Platform Technology Featuring RibbonFET (GAA) and PowerVia for Advanced High-Performance Computing," *IEEE Symp. VLSI Tech.*, pp. 1-3, 2025. http://doi.org/10.23919/VLSITechnologyandCir65189.2025.11075006
[9] C. Banbury et al., "MLPerf Tiny Benchmark," *Neural Information Processing Systems*, 2021. http://arxiv.org/abs/2106.07597
[10] MLCommons, "Benchmark MLPerf Inference: Tiny | MLCommons V1.3 Results," [Online]. Available: <https://mlcommons.org/benchmarks/inference-tiny/>.

[11] R. Prabhakar et al., "SambaNova SN40L: A 5nm 2.5D Dataflow Accelerator with Three Memory Tiers for Trillion Parameter AI," *ISSCC*, pp. 288-290, 2025. http://doi.org/10.1109/ISSCC49661.2025.10904578
[12] J. Wuu et al., "3D V-Cache: the Implementation of a Hybrid-Bonded 64MB Stacked Cache for a 7nm x86-64 CPU," *ISSCC*, pp. 428-429, 2022. http://doi.org/10.1109/ISSCC42614.2022.9731565
[13] D. Niu et al., "184QPS/W 64Mb/mm² 3D Logic-to-DRAM Hybrid Bonding with Process-Near-Memory Engine for Recommendation System," *ISSCC*, pp. 462-464, 2022. http://doi.org/10.1109/ISSCC42614.2022.9731694
[14] A. Smith et al., "AMD Instinct™ MI300 Series Modular Chiplet Package – HPC and AI Accelerator for Exa-Class Systems," *ISSCC*, pp. 490-492, 2024. http://doi.org/10.1109/ISSCC49657.2024.10454441
[15] C. -H. Yu et al., "ATOMUS: A 5nm 32TFLOPS/128TOPS ML System-on-Chip for Latency Critical Applications," *ISSCC*, pp. 42-44, 2024. http://doi.org/10.1109/ISSCC49657.2024.10454509
[16] P. A. Hager et al., "Metis AIPU: A 12nm 15TOPS/W 209.6TOPS SoC for Cost- and Energy-Efficient Inference at the Edge," *ISSCC*, pp. 212-214, 2024. http://doi.org/10.1109/ISSCC49657.2024.10454395
[17] A. S. Cassidy et al., "IBM NorthPole: An Architecture for Neural Network Inference with a 12nm Chip," *ISSCC*, pp. 214-215, 2024. http://doi.org/10.1109/ISSCC49657.2024.10454451
[18] S. M. Lee et al., "RNGD: A 5nm Tensor-Contraction Processor for Power-Efficient Inference on Large Language Models," *ISSCC*, pp. 284-286, 2025. http://doi.org/10.1109/ISSCC49661.2025.10904727

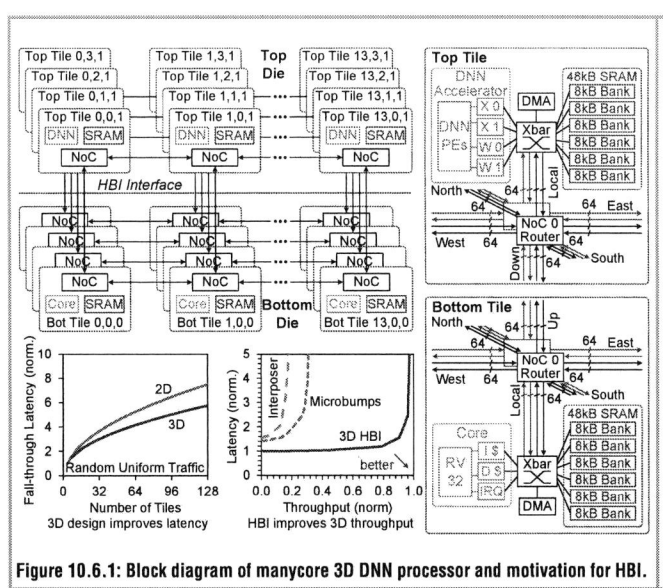

Figure 10.6.1: Block diagram of manycore 3D DNN processor and motivation for HBI.

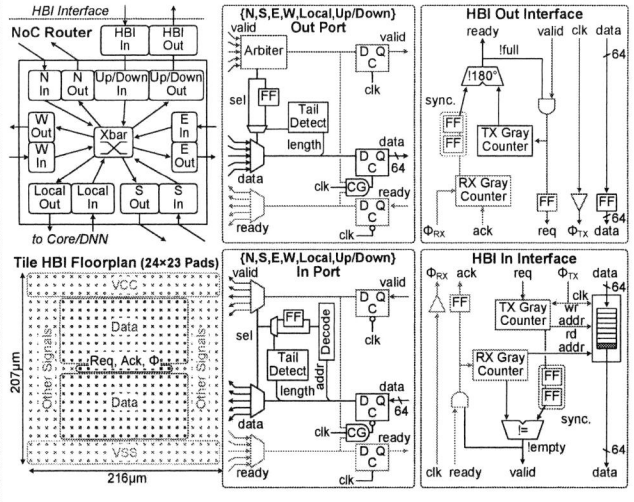

Figure 10.6.2: Mesh 3D network on chip and hybrid-bonded interface.

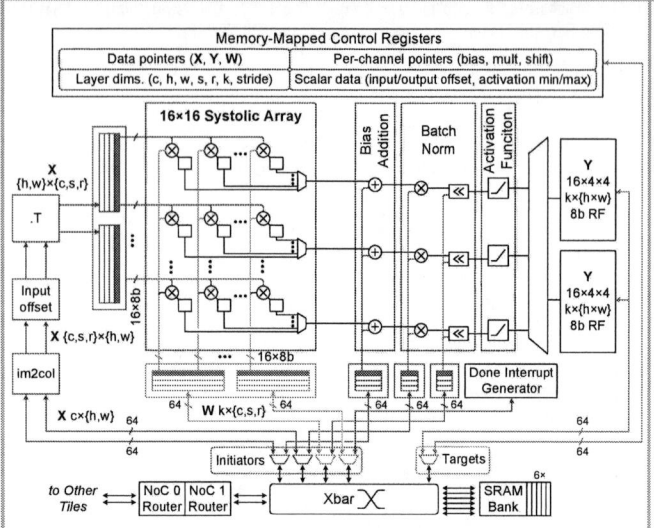

Figure 10.6.3: Block diagram of DNN accelerator in the top tile.

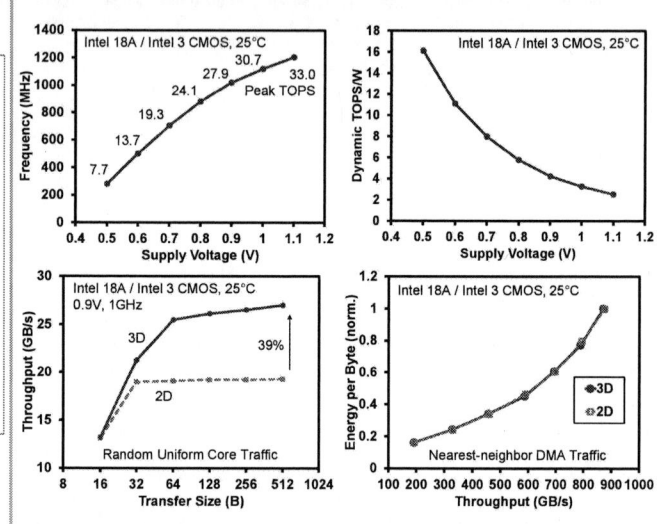

Figure 10.6.4: Chip measurements and comparison of 2D vs. 3D traffic.

Image Classification Results, 25°C

Mapping	Parallel		Dataflow	
Voltage, V	0.5	1.1	0.5	1.1
Latency, µs	260	53	1,687	329
Throughput, Inf / s	3.8k	18.8k	**5.3k**	**27.4k**
Dynamic Energy, µJ / Inf	23.2	161	**15.6**	**111**

Figure 10.6.5: Image classification measurements for parallel and dataflow mappings.

Hybrid Bonded Chips	ISSCC22 [12]	ISSCC22 [13]	ISSCC24 [6]	ISSCC24 [14]	This Work
Top Process	7nm	25nm	7nm	5nm	Intel 18A
Bottom Process	7nm	55nm	7nm	6nm	Intel 3
Bonding Technology	HBI 9µm	HBI 3µm	HBI <2µm	HBI 9µm	HBI 9µm
Area, mm²	81	602.22	15.17	4× ½Reticle	2.74
Voltage, V	-	1.1, 1.2	0.58-0.69	-	0.5-1.1
Frequency, GHz	-	0.15, 0.3	0.4-0.6	2.1	0.280-1.205
3D Signal Count	-	-	33.7 k	1.4 M*	21.5 k
Peak 3D Bandwidth, Tb/s	16	11.04	6.6†	-	7.0
3D Signal Density, / mm²	-	-	2.2 k	-	**7.9 k**
Peak BW Density, Tb/s/mm²	0.020	0.018	0.44	-	**2.5**

*Includes both signal and power hybrid bonded connections, †6.1Tb/s local memory + 512Gb/s SMEM

DNN Processors	ISSCC24 [14]	ISSCC24 [15]	ISSCC24 [16]	ISSCC24 [17]	ISSCC25 [11]	ISSCC25 [18]	This Work
Process	5nm, 6nm	5nm	12nm	12nm	5nm	5nm	Intel 18A, Intel 3
Area, mm²	4× ½Reticle	149	144	795	2× 600	653	2.74
Voltage, V	-	-	0.68	0.8	0.75	-	0.5-1.1
Frequency, GHz	2.1	-	0.800	0.4-0.5	-	1.0	0.280-1.205
SRAM Capacity, MB	256*	64†	52‡	224§	520	256	5.25
Data Precision	BF16	INT8	INT8	INT8	BF16	INT8	INT8
Peak Perf., TOPS/TFLOPS	1307	128	209.6	200	640	512	33.0
Comp. Density, TOPS/mm²	-	0.86	1.46	0.25	2.13‡	0.78	**12.1**

*Memory side cache capacity, †32MB L2 + 2×16MB L0, ‡32MB L2 + 4×(4MB L1 + 8Mb IMC), §192MB + 32MB Frame Buffer, ‡BF16 scaled to 4 INT8 ops

Figure 10.6.6: Comparison tables to hybrid bonded chips and manycore DNN processors.

Figure 10.6.7: Top and bottom die photos of 3D stacked DNN processor.

• 2026 IEEE International Solid-State Circuits Conference

ISSCC 2026 / SESSION 10 / DIGITAL PROCESSING AND CIRCUIT TECHNIQUES / 10.7

10.7 A 28nm Mode-Reconfigurable CAM-CIM Hybrid Complete 3-SAT Solver Supporting Conflict-Driven Clause Learning with 100% Solvability

Zihan Wu, Xiyuan Tang, Lishan Lin, Youming Yang, Haoyang Luo, Bocheng Xu, Yitao Liang, Xiaochen Bo, Yuan Wang

Peking University, Beijing, China

Abstract

The K-SAT problem is NP-complete and costly on von Neumann machines. Several ASIC solvers have been proposed to mitigate this, but they rely on inefficient crossbar mapping, overlook community structures and lack adaptability to formula size. This work presents an ASIC complete solver with conflict-driven clause learning (CDCL), employing a mode-reconfigurable CAM–CIM hybrid architecture. Fabricated in 28nm CMOS, it achieves 4.3-to-5.1× speedup and 1.4× energy reduction over a state-of-the-art ASIC complete solver.

The Boolean satisfiability (K-SAT) problem is a fundamental combinatorial problem. Its objective is to find a satisfying (SAT) assignment for a Boolean formula F(x) in conjunctive normal form (CNF), or prove it is unsatisfiable (UNSAT). K-SAT finds applications in many areas, such as formal verification [1], synthesis [2], and AI planning [3]. According to complexity theory [4], K-SAT is NP-complete for all K$\geq$3 and can be reduced to 3-SAT with only polynomial overhead. In contrast, 2-SAT belongs to P and can be solved in polynomial time. Thus, 3-SAT has become the canonical benchmark for K-SAT hardness analysis. However, due to its NP-complete nature, solving 3-SAT problems on von Neumann machines incurs substantial time and energy costs. To address this, several ASIC SAT solvers have been developed using continuous-time dynamics [5,6], computing-in-memory [7-10], oscillator [11], and Ising machine [12]. While these designs show promising performance on uniform 3-SAT benchmarks, they overlook several key structural properties of K-SAT formulas, limiting their efficiency in broader application scenarios. 1) Prior SAT solvers typically adopt a crossbar topology to map the entire input formula F(x) [5-11]. It requires $O(n^2)$ processing elements (PE), while each clause contains only a few literals. As a result, merely a small fraction of PEs are set '1' for literal-clause associations, while the majority are set '0' and do not participate in inference. As the number of variables grows, the density of '1' decreases, making the entire F(x) mapping increasingly inefficient. 2) Real-life F(x) exhibit abundant community structures [13] that can be exploited by conflict-driven clause learning (CDCL) [14], which significantly accelerates solving. As a widely adopted framework in modern software SAT solvers [15,16], CDCL relies on structured Boolean constraint propagation (BCP) over implication graph (IG) to enable non-chronological backtracking. However, prior works flatten BCP results without preserving community structures, leading to chronological schemes such as Davis-Putnam-Logemann-Loveland (DPLL). 3) Fixed-size macros operate efficiently only in a narrow formula-size range, whereas SAT formulas vary widely in size across scenarios. When the macro size exceeds the formula size, only a portion of the PEs are active and the rest remain idle, leading to macro underutilization. Conversely, if the formula is oversized, only part of the F(x) can be mapped, leading to massive crosschecking and external memory access that degrades system efficiency.

In light of this, this work proposes a complete CDCL 3-SAT solver. As shown in Fig. 10.7.2, the system consists of 4 cores, a top controller, an initial sequence scheduler (ISS), a global result combiner (GRC), and readout/write-in units. Each core integrates a content-addressable-memory-based propagation unit (CAM-PU) and a compute-in-memory-based learning unit (CIM-LU), forming a producer-consumer paradigm to support CDCL. The CAM-PU stores the large and static input F(x) in binary-encoded format, avoiding crossbar expansion and reducing PE overhead from $O(n^2)$ to $O(n\log n)$. It performs propagation and decision making to generate a structured IG, which is consumed by CIM-LU for conflict analysis and learnt clause generation. Unlike the input F(x), learnt clauses do not affect formula satisfiability but provide critical guidance for inference. To achieve real-time tracking of all learnt clauses, a crossbar CIM macro supporting parallel clause update is employed. Since the learnt clause set is small and dynamically maintained, the macro is lightweight with low area overhead. To accommodate varying F(x) sizes, the solver adopts a mode-reconfigurable 4-core architecture. For small F(x), each core replicates the entire F(x) and explores different branches in split mode. For large F(x), the system switches to merge mode, partitioning the F(x) across 4 cores for concurrent inference. Prototyped in 28nm CMOS, this work achieves a 4.3-to-5.1× speedup and 1.4-to-1.7× energy reduction over a state-of-the-art ASIC complete solver on uf/uuf50-218 benchmarks. The design supports CDCL, showing excellent performance on both uniform and structured formulas.

Figure 10.7.3 illustrates the architecture of the CAM-PU. Each clause in F(x) is mapped onto three CAM tiles, each storing one 9b literal composed of 1b polarity (0 for positive, 1 for negative) and an 8b variable index (e.g., $\neg x_8$ is encoded as 1_00000111). Each tile contains two 9T NAND CAM arrays with dedicated match lines (ML) for polarity and variable index. In the precharge phase (EVAL=0), all MLs are low. When EVAL is asserted, both arrays compare input search lines (SL) with stored data, raising ML only on an exact match. To reduce unnecessary searches, a SAT-gating scheme disables entries of satisfied clauses, preventing redundant lookups and cutting search power by 2.1× compared with the baseline. The CAM-PU proceeds with propagation and decision making in a literal-watching loop. Each iteration first pushes a new literal into the CAM for matching. The match results are sent to the clause analyzer (CA) for clause state update: if both polarity and variable match, the clause is satisfied; if only the variable matches, the literal is false. The CA counts false literals to determine its state, and forwards it to the token mixer for clause state merging. If conflicts occur, it awakens the CIM-LU to backtrack; otherwise, new unit clauses are added to the watching list for BCP. When the list is exhausted, a heuristic/minor clause is injected to make a new decision, resuming the BCP loop. For the clause fetched from the watching list, its three literals are read from 3 CAMs and sent to the literal engine. A literal sorter extracts the propagation constraint (i.e., the forced literal in unit clauses, the entire conflict clause, or the chosen branch literal in heuristic/minor clauses) and stores it in the IG buffer for conflict analysis.

Figure 10.7.4 shows the architecture of the CIM-LU, which performs CDCL using the implication graph (IG) constructed from both input formula F(x) and learnt clauses G(x). The solver alternates between IG construction and IG reduction based on the combined clause states. When no conflict is detected, it enters the IG construction phase. In this phase, each new IG entry is stamped with a decision level (DL), which increments only when the entry is derived from a heuristic or minor clause (i.e., decision making) and remains unchanged for literals inferred through BCP. This bottom-up process encodes assignment dependencies for CDCL. When a conflict is detected, the CIM-LU switches to IG reduction. Starting from the conflict clause, it recursively traverses the IG to trace the assignment reason of each literal: literals assigned at the current DL are further expanded, while those from earlier DLs are collected into the learnt clause. This traversal stops when only one literal remains at the current DL, marking it as the first unique implication point (FUIP). The learnt clause consists of the negations of literals from earlier DLs encountered during the traversal and FUIP, and the solver backtracks non-chronologically to the second highest DL among those negations. The learnt clause is stored in a 50×32 SRAM-based CIM macro, where each column represents a clause and each row corresponds to a variable. Each PE contains 2 SRAMs for positive and negative literals, along with logic for clause state evaluation. Clause satisfaction is determined via wired-OR logic: if an assignment matches any stored literal, the corresponding SAT line is pulled high. Other clause states are derived from the number of unassigned literals using a position-encoded counter (PEC) [9]. The PEC is driven by SW, the combinational result of the stored value and variable assignment (AGN). Each PEC right-shifts when SW=1 and holds when SW=0. The first PEC is initialized with <100>, and after a 50-stage cascade, the final PEC outputs a 3b position code (PC) indicating the number of unassigned literals. Notably, the CIM macro is reset only upon conflicts, as satisfied clauses remain SAT during propagation and decision making. This conflict-aware mechanism avoids unnecessary refresh operations and reduces power by 1.54× compared to conventional dynamic logic.

Figure 10.7.5 illustrates the mode-reconfigurable architecture. The solver supports two modes that adapt to formula size. Split mode targets small F(x), where the entire F(x) is replicated on 4 cores. The initial sequence scheduler (ISS) partitions the solution space into 32 sub-branches, where each sub-branch corresponds to an initial sequence with 5 literals. Each core loads one initial sequence and verifies its satisfiability. If UNSAT, the core returns to ISS for a new sequence; if SAT, the core raises a global SAT flag. When all sequences are exhausted without finding a solution, ISS declares global UNSAT. Merge mode targets large F(x), where each core stores part of the formula clauses and performs local inference. In each iteration, all cores send their inference results to the global result combiner, which determines the next action. If no conflict occurs, the combiner selects a clause from the watching list, assigns a satisfying literal, and broadcasts it to all cores for synchronized clause state updating. Upon conflict, each core performs parallel local IG reduction to identify its local FUIP and DL under the coordination of the combiner. The combiner then aligns four local FUIPs to identify the deepest one as the global FUIP, which guides the generation of learnt clause and non-chronological backtracking. This iterative process continues until a solution is found or UNSAT is proven.

Figure 10.7.6 shows the chip measurements. Fabricated in 28nm CMOS, the solver occupies 0.65mm^2 and operates at 45-185MHz with 0.65-0.95V supply. In split mode, 1000 uniform SATLIB uf/uuf50-218 benchmarks are measured. At 185 MHz and 0.95V, the solver achieves average solution times of 4.0µs (SAT cases) and 8.3µs (UNSAT cases) with 100% solvability, consuming 10.4mW on average. Compared to the state-of-the-art complete

ISSCC 2026 / February 17, 2026 / 8:00 AM

solver [9], it delivers a 4.3-to-5.1× speedup and achieves 1.4-to-1.7× energy reduction on the same test set. In merge mode, 150 SAT and 46 UNSAT structured formulas (872 clauses, 200 variables) are measured. At 185MHz and 0.95V, the solver achieves average solution times of 91.8µs (SAT cases) and 846.0µs (UNSAT cases) with 100% solvability, consuming 11.2mW on average. On the same formulas, an ablation study shows that CDCL yields ~45× speedup over the baseline without CDCL, demonstrating its efficiency in exploiting formula implications to accelerate solving. These results confirm the proposed solver's versatility from compact to large-scale K-SAT problems.

Acknowledgement:
This work was supported by the National Key Research and Development Program of China (Grant No. 2024YFB4405501). The corresponding authors are Xiyuan Tang and Yuan Wang (xitang@pku.edu.cn, wangyuan@pku.edu.cn).

References:
[1] Prasad, Mukul et al., "A Survey of Recent Advances in SAT-Based Formal Verification." *International Jour. on Software Tools for Technology Transfer*, pp. 156-173, 2005. https://doi.org/10.1007/s10009-004-0183-4
[2] Bloem, Roderick et al., "SAT-Based Synthesis Methods for Safety Specs," *International Conf. on Verification, Model Checking, and Abstract Interpretation*, 2014. https://doi.org/10.1007/978-3-642-54013-4_1
[3] Schreiber, Dominik, "Lilotane: A Lifted SAT-Based Approach to Hierarchical Planning," *Journal of Artificial Intelligence Research*, pp. 1117-1181, 2021. https://doi.org/10.1613/jair.1.12520
[4] Cook, Stephen A. "The Complexity of Theorem-Proving Procedures," *Logic, Automata, and Computational Complexity: The Works of Stephen A. Cook*, pp. 143-152, 2023. https://doi.org/10.1145/3588287.3588297
[5] Chang, Muya et al., "An Analog Clock-Free Compute Fabric Base on Continuous-Time Dynamical System for Solving Combinatorial Optimization Problems," *IEEE CICC*, 2022. https://doi.org/10.1109/CICC53496.2022.9772850
[6] Zhang, Qiaochu et al., "A Stochastic Analog SAT Solver in 65nm CMOS Achieving 6.6us Average Solution Time with 100% Solvability for Hard 3-SAT Problems." *IEEE Symp. VLSI Circuits*, 2024. http://doi.org/10.1109/VLSITechnologyandCir46783.2024.10631503
[7] Kim, Daehyun et al., "A 32.5 mW Mixed-Signal Processing-In-Memory-Based k-SAT Solver in 65nm CMOS with 74.0% Solvability for 30-Variable 126-Clause 3-SAT Problems," *ISSCC*, pp. 28-30, 2023. http://doi.org/10.1109/ISSCC42615.2023.10067570
[8] Xie, Shanshan et al., "Snap-SAT: A One-Shot Energy-Performance-Aware All-Digital Compute-in-Memory Solver for Large-Scale Hard Boolean Satisfiability Problems," *ISSCC*, pp. 420-422, 2023. http://doi.org/10.1109/ISSCC42615.2023.10067380
[9] Wu, Zihan et al., "SKADI: A 28nm Complete K-SAT Solver Featuring Dual-Path SRAM-Based Macro and Incremental Update with 100% Solvability," *ISSCC*, pp. 614-615, 2025. http://doi.org/10.1109/ISSCC49661.2025.10904686
[10] Bhattacharya, Tinish et al., "A Fully Integrated Mixed-Signal Compute-In-Memory Accelerator for Solving Arbitrary Order Boolean Satisfiability Problems," *IEEE Symp. VLSI Circuits*, 2025. http://doi.org/10.23919/VLSITechnologyandCir65189.2025.11074791
[11] Dikopoulos, Evangelos et al., "A Physics-Inspired Oscillator-Based Mixed-Signal Optimization Engine for Solving 50-Variable 218-Clause 3-SAT Problems with 100% Solvability and 31.7µs Solution Time," *ISSCC*, pp. 446-447, 2025. http://doi.org/10.1109/ISSCC49661.2025.10904814

[12] Su, Yuqi et al., "A Reconfigurable Ising Machine for Boolean Satisfiability Problems Featuring Many-Body Spin Interactions," *IEEE CICC*, 2023. http://doi.org/10.1109/CICC57935.2023.10121303
[13] Ansótegui, Carlos et al., "The Community Structure of SAT Formulas," *International Conference on Theory and Applications of Satisfiability Testing*, pp. 410-423, 2012. https://doi.org/10.1007/978-3-642-31612-8_31
[14] Joao Marques-Silva, Karem A. Sakallah, "GRASP: A Search Algorithm for Propositional Satisfiability," *IEEE Trans. on Computers*, vol. 48, no. 5, pp. 506-521, 2002. https://doi.org/10.1109/12.769433
[15] Eén, Niklas, Niklas Sörensson, "An Extensible SAT-Solver," *International Conference on Theory and Applications of Satisfiability Testing*, pp 502-518, 2003. https://doi.org/10.1007/978-3-540-24605-3_37
[16] Gilles Audemard, Laurent Simon, "Predicting Learnt Clauses Quality in Modern SAT Solvers," *International Joint Conf. on Artificial Intelligence*, pp. 399-404,, 2009. https://www.ijcai.org/Proceedings/09/Papers/074.pdf
[17] Shim, Chaeyun et al., "VIP-Sat: A Boolean Satisfiability Solver Featuring 5× 12 Variable In-Memory Processing Elements with 98% Solvability for 50-Variables 218-Clauses 3-SAT Problems," *ISSCC*, pp. 486-487, 2024. http://doi.org/10.1109/ISSCC49657.2024.10454397

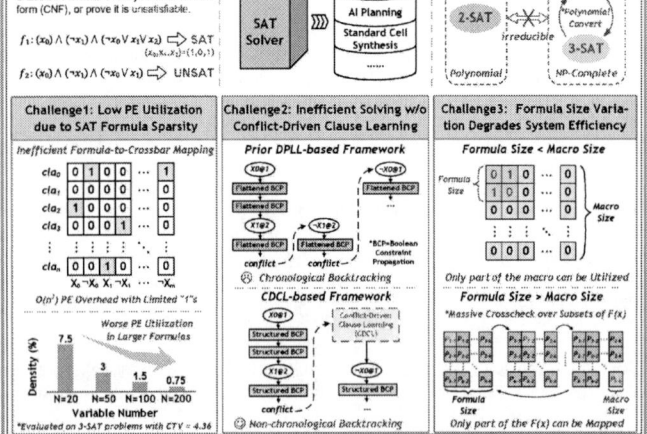

Figure 10.7.1: Overview of the K-SAT concept and the challenges of efficient SAT inference in ASIC.

Figure 10.7.2: The overall architecture of the proposed CDCL K-SAT solver and three pivotal features.

DIGEST OF TECHNICAL PAPERS • 181

979-8-3315-8937-0/26 $31.00 © 2026 IEEE

ISSCC 2026 / SESSION 10 / DIGITAL PROCESSING AND CIRCUIT TECHNIQUES / 10.7

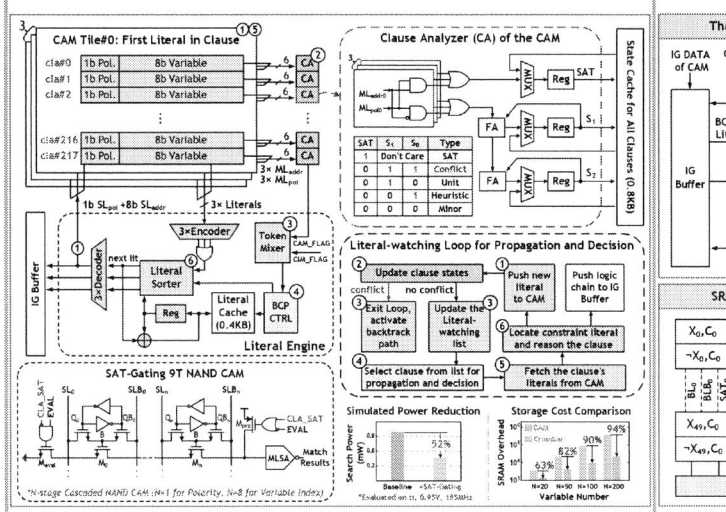

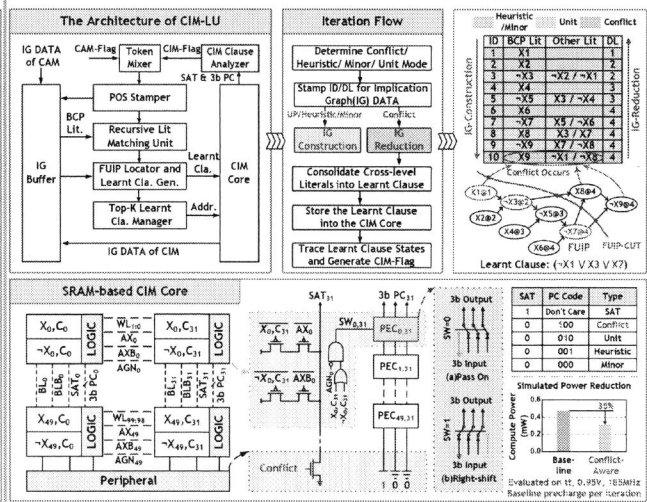

Figure 10.7.3: The schematic of CAM-based propagation unit (CAM-PU) and its operation flow.

Figure 10.7.4: CIM-based learning unit (CIM-LU) for conflict analysis and learned clause management.

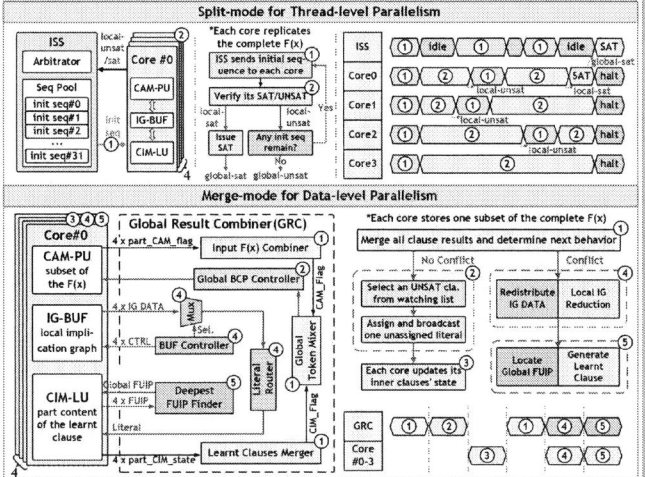

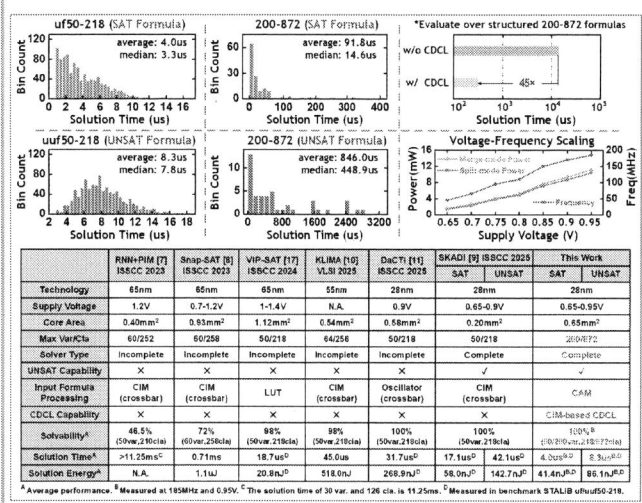

Figure 10.7.5: Mode-reconfigurable architecture: the split-mode for thread-level parallelism on small formulas, and the merge-mode for data-level parallelism on large formulas.

Figure 10.7.6: Measurement results and the comparison with state-of-the-art ASIC K-SAT solvers.

	RNN+PIM [7] ISSCC 2023	Snap-SAT [8] ISSCC 2023	VIP-SAT [17] ISSCC 2024	KLIMA [16] VLSI 2025	DaCTi [11] ISSCC 2025	SKADI [9] ISSCC 2025		This Work	
						SAT	UNSAT	SAT	UNSAT
Technology	65nm	65nm	65nm	55nm	28nm	28nm		28nm	
Supply Voltage	1.2V	0.7-1.2V	1-1.4V	N.A.	0.9V	0.65-0.9V		0.65-0.95V	
Core Area	0.40mm²	0.93mm²	1.12mm²	0.54mm²	0.58mm²	0.20mm²		0.65mm²	
Max Var/Cla	60/252	60/258	50/218	64/256	50/218	50/218		200/872	
Solver Type	Incomplete	Incomplete	Incomplete	Incomplete	Incomplete	Complete		Complete	
UNSAT Capability	✗	✗	✗	✗	✗	✓		✓	
Input Formula Processing	CIM (crossbar)	CIM (crossbar)	LUT	CIM (crossbar)	Oscillator (crossbar)	CIM (crossbar)		CAM	
CDCL Capability	✗	✗	✗	✗	✗	✗		CIM-based CDCL	
Solvability*	46.5% (50var,210cla)	72% (60var,258cla)	98% (50var,218cla)	98% (50var,218cla)	100% (50var,218cla)	100% (50var,218cla)		100% (50/200var,218/872cla)	
Solution Time*	>11.25ms⁶	0.71ms	18.7us⁵	45.0us	31.7us⁰	17.1us⁰	42.1us⁰	4.0us⁰·⁵	8.3us⁰·⁵
Solution Energy*	N.A.	1.1uJ	20.8nJ⁰	518.0nJ	268.9nJ⁰	58.0nJ⁰	142.7nJ⁰	41.4nJ⁸·⁰	86.1nJ⁸·⁰

*ᴬ Average performance. ᴮ Measured at 185MHz and 0.95V. ᶜ The solution time of 30 var. and 126 cla. is 11.25ms. ᴰ Measured in benchmark STALIB ufuuf50-218.

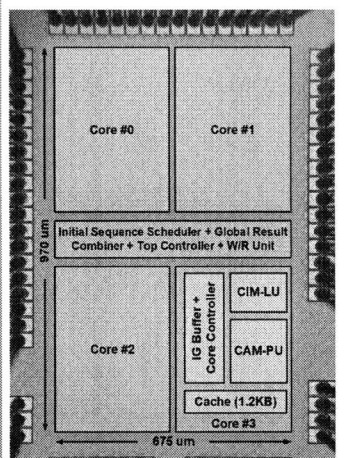

Chip Summary

Technology	28nm HPC+
Core Size	0.97×0.68mm²
Supply Voltage	0.65-0.95V
Frequency	45-185MHz
Power	1.4-10.4mW@split 1.6-11.2mW@merge
Completeness	Complete
Implication Learning Method	Conflict-driven Clause Learning (CDCL)
Solvability	100%

Performance Summary

Benchmark		Solution Time^A	Energy^A
50-218^B (split mode)	SAT	4.0us	41.4nJ
	UNSAT	8.3us	86.1nJ
200-872^C (merge mode)	SAT	91.8us	1.0uJ
	UNSAT	846.0us	9.5uJ

^A Measured at 185MHz@0.95V.
^B 50-218 are uniform formulas.
^C 200-872 are structured formulas.

Figure 10.7.7: Die micrograph and performance summary of the proposed solver across different formula sizes.

• 2026 IEEE International Solid-State Circuits Conference

ISSCC 2026 / SESSION 10 / DIGITAL PROCESSING AND CIRCUIT TECHNIQUES / 10.8

10.8 COBI: A Degree-of-56 Column-Bipartite Densely Connected Digital Ising Chip with 8b Spin Coefficients

Yihao Wu[1], Jooyoung Bae[1], Seunghun Shin[2], Bongjin Kim[2]

[1]University of California, Santa Barbara, CA, [2]Korea Advanced Institute of Science and Technology, Daejeon, Korea

Abstract

We present a 65nm digital Ising chip with an advanced column-bipartite topology, featuring densely connected spins with a degree of 56 and 8b coefficients for mapping and solving computationally intensive combinatorial optimization problems. A fabricated Ising test chip with 64 spin processing elements with spin coefficients stored in flexible embedded SRAM demonstrates solving practical problems, such as MIMO detection for a wireless network, with a fast solution time (<72ns).

Combinatorial optimization problems (COPs) are prevalent in various industries, including logistics, networks, finance, drug discovery and wireless communications. Many of these problems are NP-hard and scale poorly on conventional CPUs and GPUs. Ising machines provide a promising alternative by mapping COP variables onto Ising spins, where spin interactions encode the energy function. The system then searches for low-energy states of the Ising Hamiltonian, typically using energy gradient descent combined with stochastic methods, such as annealing to escape local minima. Quantum annealers [1] have demonstrated the feasibility of large-scale Ising machines, but their applications are limited due to the extremely low operating temperature (<20mK) and high-power consumption (15-25kW). In contrast, CMOS Ising machines have emerged as viable approaches, achieving low power consumption and good scalability while operating at room temperature [2-6]. However, prior CMOS Ising machines suffer from poor problem-mapping capabilities, limiting their practical applications, due to fixed and low-degree hardware topologies, such as lattice and King's graphs [1-3]. A continuous-time Ising machine with embedded DRAM [4] implemented a King's graph topology with the degree of 8 (allowing up to 8 interconnects per spin) and 4b coefficients. An SRAM-based macro implemented a network of spins in an SRAM macro using an enhanced Chimera topology [5], inspired by a quantum annealer [1]. However, its precision is limited to ternary due to the limited spatial connectivity within the macro. A recent m-Zephyr topology [6] further advances connectivity and achieves a degree of 24 through cross-dimensional connections, yet it requires computationally intensive pre-processing for problem mapping. Due to the limited hardware topologies, prior works have primarily focused on showcasing simple benchmarks, such as Max-Cut problems, which can be mapped into any hardware topology regardless of connectivity and precision. In this work, we introduce a digital Ising chip with an advanced column-bipartite topology, featuring densely connected spins with a degree of 56 and 8b coefficients for mapping and solving practical problems, such as community and MIMO detection, with densely connected spins and high coefficient precisions. The digital Ising chip enhances reconfigurability by introducing flexibility in mapping spin coefficients to embedded memory. The fabricated 65nm test chip implements 64 densely connected spin processing elements (PEs) with 8b spin coefficients and demonstrates solving community and MIMO detection problems with a fast solution time (<72ns). Fig. 10.8.1 provides an overview of this work, highlighting its main design features with the operating principles of the Ising machine.

Figure 10.8.2 illustrates the overall architecture of a 65nm test chip, comprising 64 spin PEs arranged in a two-dimensional array, along with detailed circuit building block schematics and a spin layout. A PE integrates four modules: (i) a 512b 8T-SRAM for storing sixty-four 8b spin coefficient values; (ii) a near-memory MAC comprising eight multipliers, an adder tree, and an accumulator for accumulating gradients; (iii) a register for storing a binary spin state with a zero detector; (iv) a bus controller for orchestrating spin connectivity. Note that a MAC computes a gradient accumulation by multiplying eight 8b spin coefficients and the incoming eight binary spin states, adding them using an adder tree, and accumulating them. Then, the inverted sign bit of the accumulated gradient value is used to update the spin state. When the accumulated gradient value is zero, the spin state is inverted to introduce extra randomness, improving the chance of escaping local minima and reaching better sub-optimal solutions. A bus controller embedded in each spin PE plays a key role in implementing the column-bipartite topology by routing an updated spin state output to all other spins by connecting it to horizontal and vertical 8b spin buses distributed throughout the Ising core. In each cycle, all eight spin state outputs from a single column are connected to all other spins via buses. It takes eight clock cycles to complete a full cycle of spin interactions, thereby maximizing parallelism and achieving a significant speedup.

Figure 10.8.3 (top) shows the sequential activation of columns, where spins in an active column transmit their spin state outputs to vertical buses and spins in inactive columns receive incoming spin states from horizontal buses. Fig. 10.8.3 (bottom) illustrates detailed operating sequences of a spin in an activated and an inactive column. In each cycle, spins in an active column (in transmit or Tx mode) enable their bus controller switches to broadcast their spin state outputs to other spins via vertical and horizontal spin buses. Meanwhile, spins in inactive columns (in receive or Rx mode) compute interactions with eight spins in active columns by receiving their transmitted spin states and multiplying them with interaction coefficients retrieved from local 8T-SRAM via local bitlines (LBL). Eight

element-wise multiplied results are then added (via an adder tree), accumulated, and then used to update a spin state at a spin register.

Figure 10.8.4 (top) shows measurement results of easy Max-Cut problems with a predefined ground state encoded as a clean monster image. Measured transient spin state maps (where white and black pixels indicate spin states +1 and -1) showcase fast convergence. It takes two cycles to reach convergence with the stabilized Ising Hamiltonian of -1792 at a ground state. Note that the top and bottom row transient spin state maps result from the same Max-Cut problem, sharing the same coefficients but with different initial spin states. Their converged spin maps appear inverted, but both represent the ground state with the minimum Ising Hamiltonian value. Fig. 10.8.4 (bottom) illustrates hard Max-Cut problems with random spin interaction coefficients, whose values range between -127 and +127. Measured results show that the chip achieves a solution time more than ~$10^7\times$ faster than the baseline CPU using the Metropolis algorithm. Statistics of the measured results (with 1000 samples) indicate that the Ising chip achieves lower mean Hamiltonians with less variation. Note that this work achieves better suboptimal results (with lower Hamiltonians), while requiring only eight clock cycles for convergence.

Figure 10.8.5 illustrates problem mappings and evaluation results for two computationally intensive problems with versatile and practical applications, requiring dense connections and high coefficient precision for solving them using the Ising chip. Each problem is mapped to the test chip with two different spin coefficient configurations, one with J coefficients only and the other with both J and h coefficients, where J and h indicate interaction and bias coefficients, respectively. Fig. 10.8.5 (top-right) shows the measured results for a 64-node community detection problem. A complex graph is embedded into the chip and initialized with random spin states. After the first run, spins form a clear bipartition. For the second run, we reprogram the J coefficient matrices by detaching all inter-community links by setting their corresponding J coefficients to zero, while keeping intra-community couplings. The chip then operates on each group of spins, bisecting them, resulting in four communities. Spin state map snapshots and their corresponding visualized graph views show clusters emerging and stabilizing through bisections. If more partitions are desired, we repeat the same process (reprogramming the J coefficients) as needed. Fig. 10.8.5 (bottom right) shows the measured results for a 24×24 BPSK MIMO detection problem for wireless communications. In the MIMO communication system, multiple antennas transmit and receive simultaneously, while a BPSK modulation constrains each transmitted symbol to one of the two possible data symbols. The MIMO detector's objective is to recover the original transmitted symbols from the received signals that are linearly mixed by the channel and corrupted by additive noise. Then, we directly map the MIMO detection problem to the Ising chip with the column-bipartite densely connected topology, where couplings between spins capture channel interactions and bias terms capture the influence of received data. A 24×24 BPSK MIMO detection problem is segmented and mapped to fit into the Ising chip with eight spins allocated for six active columns, while a column is allocated for h bias coefficients for active spins. Measured transient Hamiltonian traces show a faster time to solution than the baseline, which uses the CPU to run the Metropolis algorithm. Note that the column-bipartite topology is readily scalable for solving higher-order modulation (e.g., QPSK, 16-QAM), with the expectation of a significant reduction in latency, which is a critical performance bottleneck for next-generation wireless networks.

Figure 10.8.6 presents a comparison table that compares this work with the state of the art in digital and mixed-signal implementations. A die micrograph and a summary table of this work, fabricated with a 65nm process, are shown in Fig. 10.8.7. A spin PE occupies 120.75×62.5μm², and the measured power consumption is 518μW at 1.2V when solving hard Max-Cut problems with random coefficients.

Acknowledgement:

This work was supported by the National Research Foundation of Korea (NRF) grant (RS-2025-00516009/RS-2025-02264052), and the National Research Council of Science and Technology (NST) grant (GTL24042-201) funded by the Korea Government (MIST). This work was also supported by the Korea Advanced Institute of Science and Technology (KAIST) C2 Research Project.

References:

[1] Johnson, M., Amin, M., Gildert, S. et al., "Quantum Annealing with Manufactured Spins," *Nature*, pp. 194–198, 2011. http://doi.org/10.1038/nature10012

[2] S. Xie et al., "Ising-CIM: A Reconfigurable and Scalable Compute Within Memory Analog Ising Accelerator for Solving Combinatorial Optimization Problems," *IEEE JSSC*, vol. 57, no. 11, pp. 3453-3465, Nov. 2022. http://doi.org/10.1109/JSSC.2022.3176610

[3] J. Bae et al., "CTLE-Ising:A 1440-Spin Continuous-Time Latch-Based Ising Machine with One-Shot Fully-Parallel Spin Updates Featuring Equalization of Spin States," *ISSCC*, pp. 142-144, 2023. http://doi.org/10.1109/ISSCC42615.2023.10067622

[4] J. Song et al., " A Variation-Tolerant In-eDRAM Continuous-Time Ising Machine Featuring 15-Level Coefficients and Leaked Negative-Feedback Annealing," *ISSCC*, pp. 490-492, 2024. http://doi.org/10.1109/ISSCC49657.2024.10454272

[5] J. Bae et al., "e-Chimera: A Scalable SRAM-Based Ising Macro with Enhanced-Chimera Topology for Solving Combinatorial Optimization Problems Within Memory," *ISSCC*, pp. 286-288, 2024. http://doi.org/10.1109/ISSCC49657.2024.10454340

[6] Y. Wu et al., "m-Zephyr: A Digital In-Memory Ising Chip with 240 Spins Featuring Enhanced Connectivity Based on a Modified 3D Zephyr Topology," *IEEE Symp. VLSI Tech.*, 2025. http://doi.org/10.23919/VLSITechnologyandCir65189.2025.11075071

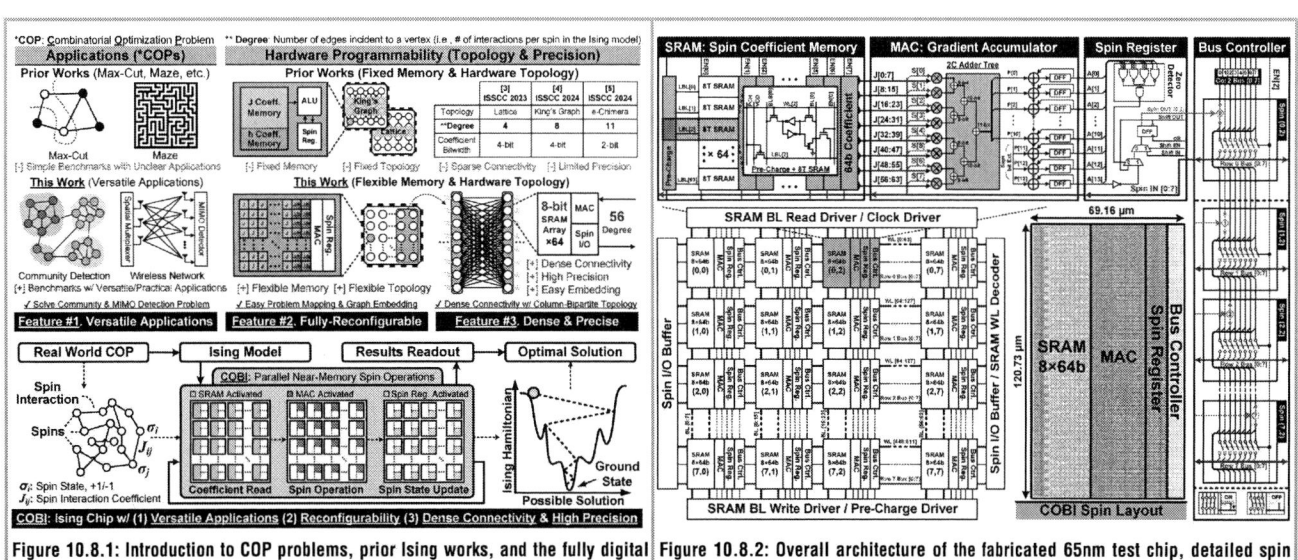

Figure 10.8.1: Introduction to COP problems, prior Ising works, and the fully digital Ising chip (COBI) with reconfigurability, dense connectivity, and high precision.

Figure 10.8.2: Overall architecture of the fabricated 65nm test chip, detailed spin structure, and a single spin layout.

ISSCC 2026 / SESSION 10 / DIGITAL PROCESSING AND CIRCUIT TECHNIQUES / 10.8

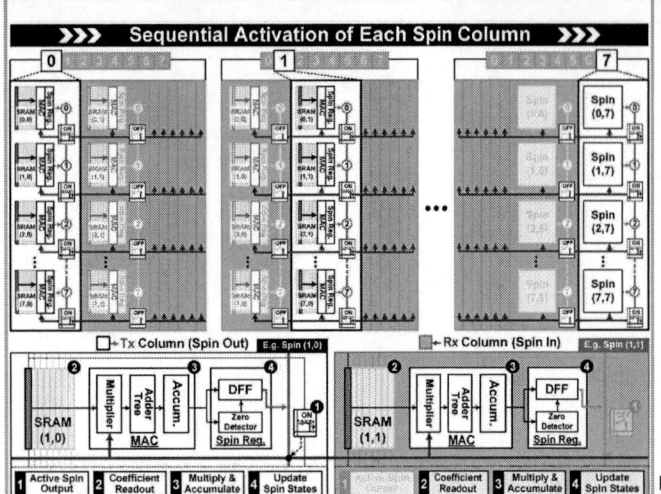

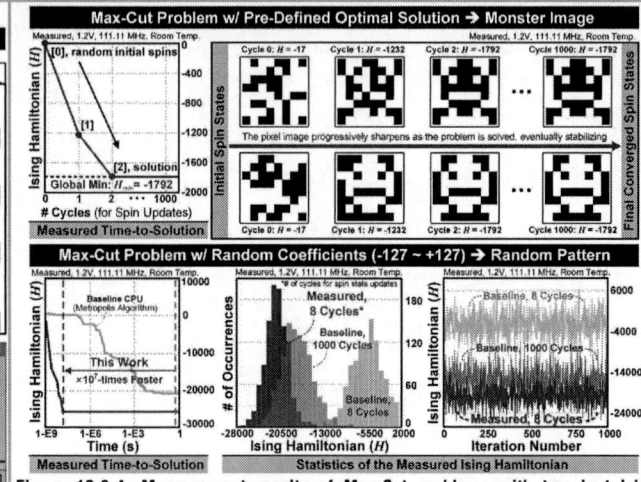

Figure 10.8.3: Sequential activation of spin column: Transmit (Tx) and receive (Rx) operations, coefficient readout, and MAC operation.

Figure 10.8.4: Measurement results of Max-Cut problems with transient Ising Hamiltonians (top-left) and transient spin maps (top-right), and hard Max-Cut problems with transient Ising Hamiltonians (bottom-left) and statistical distributions (bottom-right).

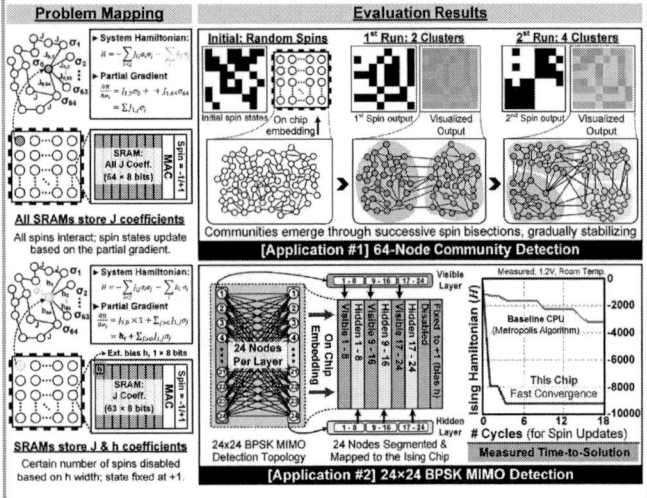

Figure 10.8.5: Mapping (left) and evaluated results (right) of 64-node community detection (top) and 24×24 BPSK MIMO detection (bottom) problems.

	JSSC 2022 [2]	ISSCC 2023 [3]	ISSCC 2024 [4]	ISSCC 2024 [5]	VLSI 2025 [6]	This Work
Technology (Circuit Type)	65nm CMOS (Mixed-Signal)	65nm CMOS (Analog)	65nm CMOS (Mixed-Signal)	65nm CMOS (Mixed-Signal)	65nm CMOS (Digital)	65nm CMOS (Digital)
Operating Temperature	Room Temp. (300K)	Room Temp. (300K)	Room Temp. (300K)	Room Temp. (300K)	Room Temp. (300K)	Room Temp. (300K)
Computing Type	Discrete-Time Mixed-Signal	Continuous-Time Analog	Continuous-Time Mixed-Signal	Continuous-Time Mixed-Signal	Discrete-Time Digital MAC	Discrete-Time Digital MAC
Memory-Centric Computing	Near-Memory Computing	Near-Memory Computing	Near-Memory Computing	Within-Memory Computing	Near-Memory Computing	Near-Memory Computing
Spin Memory	eDRAM Cell	CMOS Latch	eDRAM Cell	SRAM Cell	Register	Register
Spin State Representation	Digital (Binary State)	Analog (Latch Voltage)	Digital (Binary State)	Analog (Latch Voltage)	Digital (Binary State)	Digital (Binary State)
Spin Interaction	Mixed-Signal MAC	Latch Coupling Via Switches	Mixed-Signal MAC	Latch Coupling Via Switches	Digital MAC	Digital MAC
Graph Topology	King's Graph	Lattice	King's Graph	e-Chimera	m-Zephyr	Column Bipartite
# of Neighbors (Degree of Graphs)	8	4	8	11	24	56
Coeff. SRAM R/W During Operation	Required (via SRAM I/O)	Directly Read (No Write)	Required (via SRAM I/O)	Required (via SRAM I/O)	Directly Read (No Write)	Directly Read (No Write)
Coefficient Bitwidth	1-4bit	1bit	4bit	2bit	7bit	8bit
Supply Voltage	0.9-1.2V	0.75-1.05V	0.9-1.2V	0.8-1.4V	0.6-1.4V	0.8-1.4V
Tested Problems	Max-Cut Only					Max-Cut, MIMO, Community Detection
Time-to-Solution (TTS)	51200 ns (12800 Cycles)	<20ns	<20.7ns	<100 ns	<150ns (10 Cycles)	<72 ns* (8 Cycles)

*TTS of the proposed chip is measured at 1.2 V and 111.11 MHz while solving hard Max-Cut problems with all random coefficients

Figure 10.8.6: A comparison table with the state-of-the-art CMOS Ising machines.

Technology	65nm LP CMOS
Test Chip	Fully-Reconfigurable Digital Ising Chip
Core Circuit	64× Spin Processing Element Array
Topology (Degree)	Column-Bipartite Dense Connections 56 (# of interactions per spin)
Core/Spin Area	966×500 μm² / 120.75×62.5 μm²
SRAM	64 × 512b (Store 64×64 8-bit Spin Coefficients)
Supply Voltage	Core: 0.8 V - 1.2 V, I/O: 3.3 V
Clock Freq.	111.11 MHz
Time-to-Solution (TTS)	<72 ns*
Measured Power Consumption	518 μW @ 1.2 V

*TTS of the proposed chip is measured at 1.2 V and 111.11 MHz while solving hard Max-Cut problems with all random coefficients

Figure 10.8.7: Die micrograph and a summary table.

ISSCC 2026 / SESSION 10 / DIGITAL PROCESSING AND CIRCUIT TECHNIQUES / 10.9

10.9 SharpSAT: A Heuristic-Learning-Based SAT Accelerator Achieving 0.8µs/16.1µs Solution Time in SAT/UNSAT Cases

Yi Huang, Hao Kong, Iris Ying Chou, Bin Wang, Xiangyu Kong, Jianfeng Zhu, Liangwei Li, Xiao Li, Hanning Wang, Aoyang Zhang, Leibo Liu

Tsinghua University, Beijing, China

Abstract

We present SharpSAT, a heuristic-learning SAT accelerator that achieves fast solution times of 0.8µs for SAT and 16.1µs for UNSAT cases. Our design integrates: a fast clause learning unit that prunes the search space by deriving constraints from conflicts; a dual-BCP unit that accelerates implication propagation; and a heuristic variable decider introducing non-determinism to efficiently traverse the solution space. A 28nm prototype demonstrates 21.11×/2.61× (SAT/UNSAT) speedup over prior accelerators.

The Boolean satisfiability (SAT) problem, a proven NP-complete problem, sits at the core of modern computational tasks, from circuit verification and cryptographic analysis to AI planning [1]. A SAT instance asks for a set of Boolean assignments that make all clauses in a conjunctive normal form (CNF) evaluate to TRUE. Conventionally, solving the SAT problem using software can be time-consuming, and several ASIC accelerators have been implemented to improve performance [2-7]. However, prior accelerators [3-7] based on local search algorithms are incomplete solvers [8]: they may find satisfying assignments but cannot prove unsatisfiability (UNSAT) cases in applications like formal verification, gene sequencing, and cryptographic protocol verification [9-11]. SKADI [2] moves toward completeness by implementing a DPLL accelerator. However, this complete solver requires exploring a large search space, often necessitating the traversal of numerous candidate assignments. It also suffers from slow Boolean constraint propagation (BCP) and employs a rigid, deterministic exploration strategy that misses the diversification benefits of randomized assignment, as used in [8]. These factors collectively lead to significant slowdowns, resulting in long runtimes for large problems [9-11].

To alleviate these drawbacks, we develop SharpSAT, a complete SAT accelerator that uses heuristic learning to refine the search space by deriving clauses from prior searches [12-14], which largely reduces the solution time in both SAT and UNSAT scenarios (Fig. 10.9.1). The key attributes of the SharpSAT approach are: (1) a fast clause learning unit that tracks the causes of each assignment and promptly generates a learned clause upon conflict, guiding the search and avoiding redundant variable flips; (2) a dual-BCP unit that accelerates the BCP steps in the complete algorithm; (3) a smart variable assignment strategy which introduces randomness in the solution space search without sacrificing completeness; (4) a 28nm CMOS prototype that achieves an average solution time of 0.8µs in SAT cases and 16.1µs in UNSAT cases with 50 variables and 218 clauses. It offers 70.67×/6.17× speedups over software, MINISAT [13], on a CPU and 21.11×/2.61× speedups over a prior complete accelerator [2].

Figure 10.9.2 shows the overall architecture and the design operation flow. SharpSAT comprises an assignment management unit, a clause calculation unit, and a clause learner. The assignment management unit chooses decision variables using a PRNG-based decider and maintains a history for backtracking. The clause learner implements clause learning and backtracking. The clause calculation unit, which is responsible for BCP, conflict, and SAT detection, contains 384 clause processing elements (PEs), a subset of which store learned clauses depending on the CNF size. Each PE contains a 128b register encoding 64 variables (x_0-x_{63}) with 2b codes (e.g., the code 10 indicates that the clause contains $\neg x_0$, whereas 00 indicates ABSENT). Similarly, the current-assignment register also uses 128b to encode the assignment of x_0-x_{63} (e.g., the 2b 10 denotes x_0 = FALSE, while 00 denotes UNDEF). Accordingly, the BCS logic in each clause PE uses bitwise XOR between the current assignment and the clause encoding to evaluate clause satisfiability, and it uses the OR of each 2b encoding to identify variables for BCP and to determine if a conflict should be reported. For each variable, the AND result of the 2b XOR between assignment and clause shows whether this variable is satisfied, and the AND of the OR within the two bits of assignment and the OR within the two bits of clause indicates whether the variable present in current clause is assigned. With these two per-variable results, subsequent OR-reduction and an encoder yield the clause-level satisfiability, conflict, and BCP outputs. Consequently, the deployed dual-BCP arbiter issues either two BCP implications or one conflict to the clause learner. The overall operation flow of SharpSAT is as follows: once the CNF file is programmed into the clause PEs, the accelerator enters the Assign state, in which the heuristic decider selects a variable and assigns a value. It then enters the BCP state, in which the clause calculation unit repeatedly identifies implied variables and propagates them until the problem is satisfied (and the calculation terminates), a conflict is encountered, or no further propagation is available. If a conflict is encountered, the clause learner is activated to derive a learned clause and trigger a backtrack, or to report UNSAT and terminate. After learning a clause, the state returns to BCP. If no additional propagation is available in the BCP state, control transitions back to the Assign state.

Figure 10.9.3 (right) illustrates the workflow of our hardware clause learning and backtrack control, and the bottom panel depicts the hardware implementation of the key components. The clause learning and backtrack control are driven by the BCP/Conflict/SAT (BCS) signals

from the clause calculation unit. In each cycle, the clause learner inspects the BCS signals and selects the corresponding computation. While last-UIP clause learning [12] is traditionally performed in a backward manner (top of Fig. 10.9.3), we reformulate it as a forward VarCause construction, as shown in the workflow. In our forward algorithm, we maintain two register banks, one is the VarLevelMem in which each row encodes the variables at certain Decision Level (DL) in a bit-vector form (e.g., if the first row's third and fifth bits are set, then x_2 and x_4 have DL = 0), and each column encodes the DL of a particular variable in a bit-vector form. The other bank is the VarCauseMem, in which each row encodes the CauseSet of a certain variable in a bit-vector form (e.g., if the first row's third and fifth bits are set, then the value of x_0 depends on x_2 and x_4). Thus, the CauseSet records the variable decisions that lead to the current variable's assignment. Whenever the clause learner receives a new decision, it treats the variable as self-caused: it writes the variable into the VarLevelMem at the current DL (thereby setting its DL to the current level) and initializes its CauseSet to itself, then increments the DL. When a BCP signal arrives, the learner treats the new variable as implied by other assignments in the triggering clause. It sets the variable's DL to the current level and forms its CauseSet as the union of the Σ set (variables whose DL is less than the current DL) in the BCP clause and the CauseSets of the ∧ set (variables whose DL is equal to the current DL) in the BCP clause. Since we have already computed all the CauseSets of the variables in the BCP clause previously, simply fetching the CauseSets of the Σ set and vector-ORing them with the ∧ set can yield the CauseSet of the current BCP variable. Conflict construction follows the same procedure as BCP, and the CauseSet of the added conflict node becomes the learned clause. After obtaining the learned clause, we choose the largest DL in the Conflict clause using an arbiter as the backtrack level and we use this level to flush the two memories and reset the assignment history to the current variable assignment.

Figure 10.9.4 depicts the deployed dual-BCP technique and the PRNG-based heuristic decider. The dual-BCP arbiter collects BCP results from 384 clauses and selects two BCP results to deliver to the clause learner for VarCause construction. It also arbitrates a single conflict result among the 384 clauses for clause learning. Issuing two BCP variables from distinct clauses does not affect the correctness of VarCause construction and therefore does not distort exploration of the solution space. We implement the dual-BCP logic using use two arbiters with reversed priority (c_0-c_{63} and c_{63}-c_0). For the variable decider, we combine the benefits of both local search and complete search by introducing a heuristic decider. First, we generate a set of initial assignments through a polynomial-time analyzer for the last assignment registers, similar to an incomplete algorithm. For assigning a non-assigned variable, if the initial assigned value met the requirement of a certain clause, the variable value is preserved. Otherwise (e.g., when we backtrack to lower levels and we need to re-assign its value), we generate its value randomly using an on-chip Geffe generator. For BCP variables (containing backtrack), we simply adopt the value calculated by the clause calculation unit. Randomizing non-assigned decisions enables non-deterministic traversal of the solution space rather than a fixed order, improving exploration efficiency in a manner reminiscent of local search algorithms. Fig. 10.9.4 (bottom) shows the benefit of both techniques.

Figure 10.9.5 shows the measured results of SharpSAT. We first conducted tests on 1000 SAT and 1000 UNSAT 3-SAT cases (50-218 clauses) from SATLIB. SharpSAT achieves speedups of 21.11× (SAT) and 2.61× (UNSAT) over the prior complete SAT solver [2], with 100% solvability while running at 375MHz and 0.9V, consuming 99.1mW. Although a CDCL-based solver is more complicated than a DPLL solver due to learning clause storage and a complex clause learner, SharpSAT delivers comparable energy in SAT cases (only 1.38× higher) compared to the DPLL solver owing to faster convergence enabled by clause learning, dual-BCP and the randomized decider. We also analyze the evolution of the total clause count during computation: in SAT cases, fewer than 20 learned clauses are added before a solution is found, whereas in UNSAT cases, the full budget of 384 clauses is exercised before UNSAT is reported. Fig. 10.9.6 provides a comparison against prior works, and Fig. 10.9.7 shows the 28nm test-chip die micrograph and a summary table. SharpSAT implements 64 variables and 384 clauses in the 0.78mm² core area.

ISSCC 2026 / February 17, 2026 / 8:00 AM

Acknowledgement:
This work is supported by the National Key R&D Program of China
(No. 2024YFB3108102). Corresponding author: Aoyang Zhang
(email: aoyang@tsinghua.edu.cn).

References:
[1] J. Marques-Silva, "Practical Applications of Boolean Satisfiability," *Workshop on Discrete Event Systems*, pp. 74-80, 2008. http://doi.org/10.1109/WODES.2008.4605925
[2] Z. Wu et al., "SKADI: A 28nm Complete K-SAT Solver Featuring Dual-Path SRAM-Based Macro and Incremental Update with 100% Solvability," *ISSCC*, pp. 614-616, 2025. http://doi.org/10.1109/ISSCC49661.2025.10904686
[3] E. Dikopoulos et al., "A Physics-Inspired Oscillator-Based Mixed-Signal Optimization Engine for Solving 50-Variable 218-Clause 3-SAT Problems with 100% Solvability and 31.7µs Solution Time," *ISSCC*, pp. 446-447, 2025. http://doi.org/10.1109/ISSCC49661.2025.10904814
[4] Q. Zhang et al., "A Stochastic Analog SAT Solver in 65nm CMOS Achieving 6.6µs Average Solution Time with 100% Solvability for Hard 3-SAT Problems," *IEEE Symp. on VLSI Circuits*, 2024. http://doi.org/10.1109/VLSITechnologyandCir46783.2024.10631503
[5] C. Shim et al., "VIP-SAT: A Boolean Satisfiability Solver Featuring 5×12 Variable In-Memory Processing Elements with 98% Solvability for 50-Variables 218-Clauses 3-Sat Problems," *ISSCC*, pp. 486-488, 2024. http://doi.org/10.1109/ISSCC49657.2024.10454397
[6] D. Kim et al., "A 32.5mw Mixed-Signal Processing-In-Memory-Based K-Sat Solver in 65nm CMOS With 74.0% Solvability for 30-Variable 126-Clause 3-SAT Problems," *ISSCC*, pp. 28-30, 2023. http://doi.org/10.1109/ISSCC42615.2023.10067570
[7] S. Xie et al., "Snap-SAT: A One-Shot Energy-Performance-Aware All-Digital Compute-In-Memory Solver for Large-Scale Hard Boolean Satisfiability Problems," *ISSCC*, pp. 420-421, 2023. http://doi.org/10.1109/ISSCC42615.2023.10067380
[8] H. Holger et al, "Local Search Algorithms for SAT: An Empirical Evaluation," *J. Autom. Reason.*, vol 24, pp. 421–481, 2000. http://doi.org/10.1023/A:1006350622830
[9] A. Kuehlmann, V. Paruthi, F. Krohm and M. K. Ganai, "Robust Boolean Reasoning for Equivalence Checking and Functional Property Verification," *IEEE TCAD*, vol. 21, no. 12, pp. 1377-1394, Dec. 2002. http://doi.org/10.1109/TCAD.2002.804386
[10] A. Armando and L. Compagna, "SAT-Based Model-Checking for Security Protocols Analysis," *Int. J. Inf. Secur.*, vol 7, pp.3–32, January 2008. http://doi.org/10.1007/s10207-007-0041-y
[11] D. He et al., "Optimal Algorithms for Haplotype Assembly from Whole-Genome Sequence Data," *Bioinformatics*, vol 26, no. 12, pp. i183–i190, June 2010. http://doi.org/10.1093/bioinformatics/btq215
[12] J. P. Marques-Silva and K. A. Sakallah, "GRASP: A Search Algorithm for Propositional Satisfiability," *IEEE Trans. on Computers*, vol. 48, no. 5, pp. 506-521, 1999. http://doi.org/10.1109/12.769433
[13] N. Eén et al., "An Extensible SAT-Solver," *Conf. on Theory and Applications of Satisfiability Testing*, pp. 502-518, 2003. http://doi.org/10.1007/978-3-540-24605-3_37
[14] A. Biere et al., "CaDiCaL, Kissat, Paracooba, Plingeling and Treengeling Entering the SAT Competition 2020," *Proc. of SAT Competition 2020 - Solver and Benchmark Descriptions*, vol. B-2020-1, pp. 50-53, 2020. https://hdl.handle.net/10138/318450

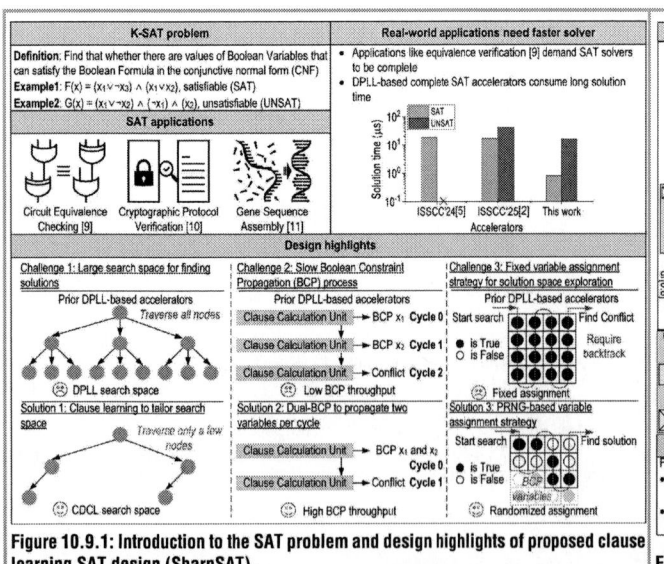

Figure 10.9.1: Introduction to the SAT problem and design highlights of proposed clause learning SAT design (SharpSAT).

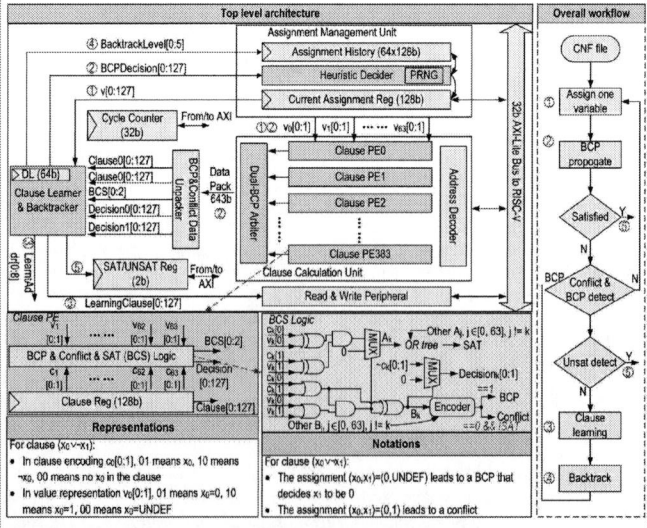

Figure 10.9.2: The top-level architecture and workflow of SharpSAT.

DIGEST OF TECHNICAL PAPERS • 185

979-8-3315-8937-0/26 $31.00 © 2026 IEEE

ISSCC 2026 / SESSION 10 / DIGITAL PROCESSING AND CIRCUIT TECHNIQUES / 10.9

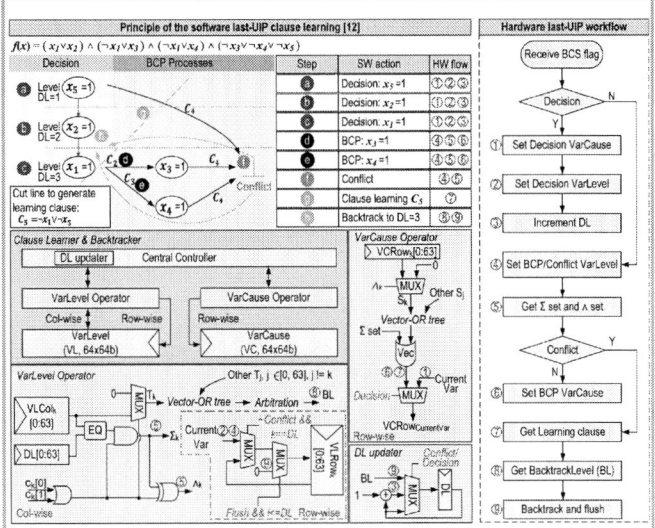

Figure 10.9.3: Proposed clause learner and its workflow.

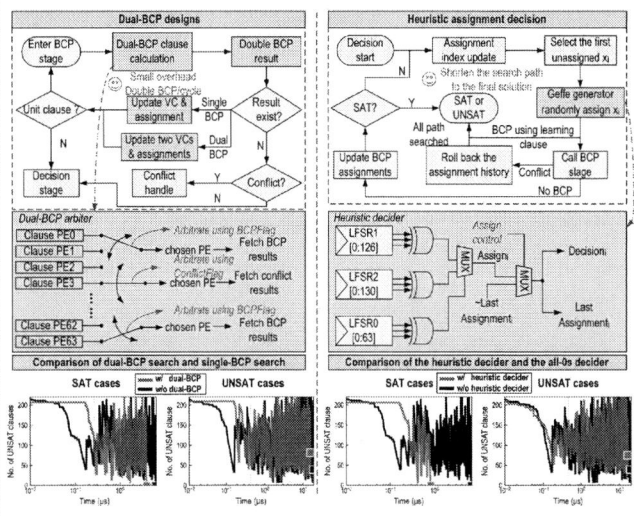

Figure 10.9.4: Design of the proposed dual-BCP unit and PRNG-based heuristic decider.

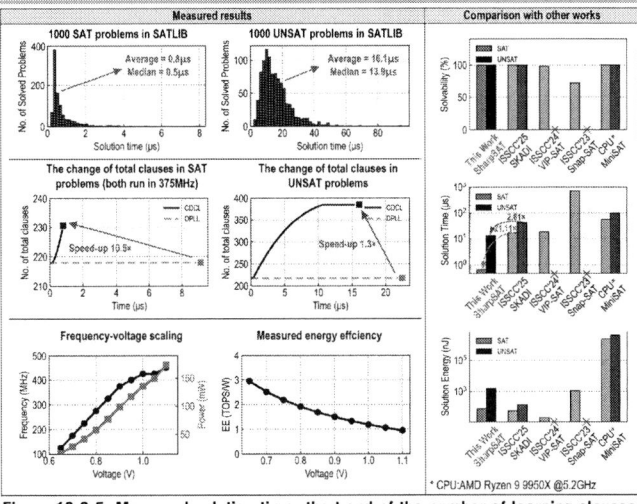

Figure 10.9.5: Measured solution time, the trend of the number of learning clauses, energy efficiency, frequency voltage plot, and comparison with prior ASICs and an HPC CPU.

	This Work	ISSCC 2025 [2]	ISSCC 2025 [3]	VLSI 2024 [4]	ISSCC 2024 [5]	ISSCC 2023 [6]	ISSCC 2023 [7]
Technology	28nm	28nm	28nm	65nm	65nm	65nm	65nm
Computing Domain	Digital	Digital	Mixed-Signal	Analog	Digital	Mixed-Signal	Digital
Computation Mechanism	Near-Memory Digital PE Array	In-Memory Digital PE Array	Mixed-Signal Oscillator Array	Stochastic CT+DT	In-Memory Digital PE Array	In-Memory Mixed-Signal	In-Memory Digital Macro
Supply Voltage	0.65-1.1V	0.65-0.9V	0.9V	1.0V	1.1-1.4V	1.1-1.3V	0.7-1.2V
Core Area	0.78mm²	0.20mm²	0.58mm²	0.37mm²	1.12mm²	0.40mm²	0.93mm²
Variables/Clauses	50/218+166 learning clauses	50/218	50/218	20/91	50/218	60/252	60/258
CTV	4.36	4.36	4.36	4.55	4.36	4.2	4.3
Accuracy	100%	100%	100%	100%	100%	N/A	100%
Solver Type	Complete	Complete	Incomplete	Incomplete	Incomplete	Incomplete	Incomplete
UNSAT Capability	√	√	×	×	×	×	×
Heuristic Learning Ability	√	×	×	×	×	×	×
Solvability (# of Var./Clauses)	100% (50var, 218cla)	100% (50var, 218cla)	100% (50var, 218cla)	100% (20var, 91cla)	98% (50var, 218cla)	31.5% (60var, 252cla)	72% (60var, 258cla)
SAT Solution Time	0.8µs*	17.1µs	31.7µs	6.6µs	18.7µs	125000µs	710µs
UNSAT Solution Time	16.1µs*	42.1µs	×	×	×	×	×
SAT Solution Energy	80.1nJ*	58.0nJ	268.9nJ	11.0nJ	20.8nJ	N/A	1100nJ
UNSAT Solution Energy	1599.3nJ*	142.7nJ	×	×	×	×	×

* Evaluated using 1000 SAT/UNSAT cases @ 0.9V, 375MHz

Figure 10.9.6: Comparison with state-of-the-art ASIC SAT solvers.

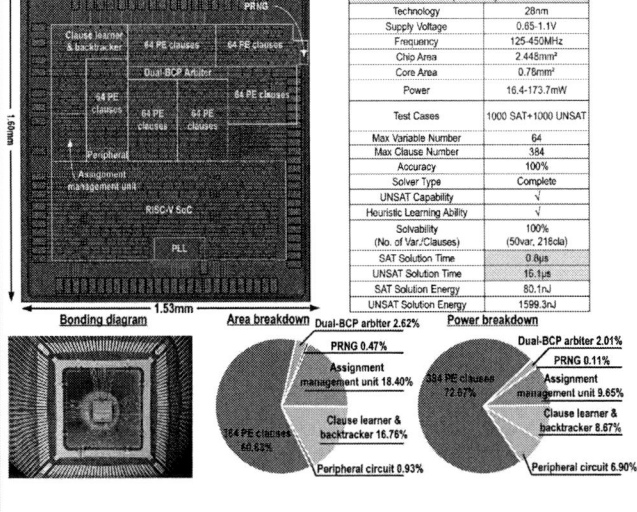

Chip summary	
Technology	28nm
Supply Voltage	0.65-1.1V
Frequency	125-450MHz
Chip Area	2.448mm²
Core Area	0.78mm²
Power	16.4-173.7mW
Test Cases	1000 SAT+1000 UNSAT
Max Variable Number	64
Max Clause Number	384
Accuracy	100%
Solver Type	Complete
UNSAT Capability	√
Heuristic Learning Ability	√
Solvability (No. of Var./Clauses)	100% (50var, 218cla)
SAT Solution Time	0.8µs
UNSAT Solution Time	16.1µs
SAT Solution Energy	80.1nJ
UNSAT Solution Energy	1599.3nJ

Figure 10.9.7: Chip micrograph, summary table, and area/power breakdown.

• 2026 IEEE International Solid-State Circuits Conference

ISSCC 2026 / SESSION 10 / DIGITAL PROCESSING AND CIRCUIT TECHNIQUES / 10.10

10.10 PCIM-SAT: A 55nm Probabilistic K-SAT Solver with p-Bit-Based Parallel-Variable Update on a Mixed-Signal Compute-in-Memory Architecture

Tinish Bhattacharya, George Higgins Hutchinson, Dongseok Kwon, Dmitri Strukov

University of California, Santa Barbara, CA

Abstract

We present a 55nm mixed-signal K-SAT solver with parallel-variable update algorithm for improved convergence and a compute-in-memory fabric that maps arbitrary-order K-SAT instances without preprocessing. It solves 50-variable 3-SAT (satisfiable) problems in 5.5µs with 100% solvability at 265nJ. A custom 11T-SRAM crossbar for one-shot analog gradient computation and ADC-less gradient-based sampling with probabilistic-bit circuits enable 4-to-13× speedup over prior ASICs.

Boolean Satisfiability (K-SAT, $K{\geq}3$) is an NP-complete problem that arises across nearly all domains of science and engineering like cryptography, computational biology, AI planning and EDA [1]. A K-SAT instance with N variables and M clauses is expressed as a conjunction (AND) of clauses, each being a disjunction (OR) of up to K literals (Boolean variables or their complements, Fig. 10.10.1, top). Solving SAT, i.e., finding satisfying assignment for all clauses, is computationally expensive on any hardware, which has motivated efforts towards designing SAT-solving hardware accelerators [2-7]. Complete K-SAT solvers based on the DPLL algorithm [2] can prove unsatisfiability of instances, but their decision-tree search scales poorly with problem size compared to other algorithms. A fully digital WalkSAT-based solver [3] exploited SRAM-based In-Memory Computing (IMC) for parallel clause evaluation but computed break-values (B_i, Fig. 10.10.1, top) sequentially, thereby limiting its performance. A mixed-signal IMC solver [4,5] improved parallelism by computing clause and break-values in one shot using SRAM-based IMC, but updated only one variable per iteration, limiting its navigational efficiency. Solvers updating multiple variables in parallel using chromatic block updates [6] or continuous-time (CT) dynamics [7,8] have demonstrated fast convergence but are typically restricted to 3-SAT problems. Solving higher-order (K>3)-SAT problems on such hardware [6-8], as well as other accelerators that are limited to second-order interactions [9,10], requires introducing additional variables and clauses, which inflates the search space exponentially and degrades performance [11].

This paper introduces a probabilistic SAT-solving algorithm, that operates directly in the high-order K-SAT space and leverages the bipartite clause-variable relationship to update multiple variables per Monte Carlo step with high navigational efficiency. It modifies the sigmoidal probability function $\sigma(\cdot)$ of Gibbs sampling-based solvers by scaling it with a Heaviside step function $\delta(\cdot)$ of the variable's make-value M_i. This restricts updates to *promising* variables in unsatisfied clauses. Variables accepted by this function are updated independently in parallel, while a rejection-free mechanism ensures progress by updating all *promising* variables with non-zero M_i, if no candidate is accepted. This heuristic outperforms state-of-the-art (SOTA) algorithms such as WalkSAT and the Digital Annealer (when simulated on a third-order Ising formulation of 3-SAT), with performance gains increasing with problem size (Fig. 10.10.1, left).

This work advances the field of SAT-solving hardware by making four key contributions (Fig. 10.10.1): (1) a mixed-signal in-memory computing architecture that is order-invariant and implements a novel SAT solver capable of updating multiple variables in each time step without prior preprocessing, (2) A custom 11T SRAM crossbar array that allows computation of both make and break-values of all variables (effectively the gradients) in a single cycle, (3) dynamic comparator-based Probabilistic-bit (p-bit) circuits on the variable side, enabling massively parallel sampling informed by the analog gradients computed in the crossbar, eliminating the need for ADCs or analog-domain reductions, (4) multi-threaded pipeline that exploits the decoupled orthogonal accesses to the crossbar array to launch three independent and parallel solvers, improving convergence by 1.7× at similar energy. Our prototype fabricated in 55nm, outperforms SOTA solvers, including continuous-time implementations in more advanced 28nm [8] and discrete-time solvers [4] by 4-13× on uniform random 3-SAT instances having up to 50 variables and on practically relevant K-SAT instances by 2-14×.

The algorithm implementation (Fig. 10.10.2, right) involves the following circuit blocks (Fig. 10.10.2, left): a 256×128 bidirectional 11T-SRAM array; clause evaluators; variable/clause registers and drivers; satisfiability (SAT) evaluators; make-value filter; p-bits, a 128b Pseudo Random Number Generator (PRNG); and a 128b wide 4-stage RNG buffer chain. In the first step, the K-SAT problem is loaded into the memory array such that each row corresponds to a clause and each column to a literal. Depending on whether x_i or $\sim x_i$ is present in clause C_j, $SRAM_{j,2i}$ or $SRAM_{j,2i+1}$ is set to 1 respectively, otherwise both are 0. In the second step, all variables are initialized to random binary values, using the PRNG. Subsequently the solver attempts to find the solution by iteratively executing the following three functions in sequence: (i) Forward Pass (FP), where literal-values are applied to the crossbar and all clauses are evaluated in parallel; (ii) Backward Pass (BP), where clause signals are applied to the crossbar array to compute the make and break values of all variables in parallel. The make-value filters detect the variables with $M_i > 0$, whereas the p-bits for each variable output a 1 with a probability proportional to the gain of the variable ($= M_i - B_i$). Simultaneously, the SAT evaluator checks if problem is satisfied, (iii) Variable Update (VU), which checks if there are *promising* variables with $M_i > 0$ and accepted by p-bit, in which case all such variables are flipped, otherwise all with $M_i > 0$ are flipped. Existence of at least one variable with $M_i > 0$ is guaranteed at this stage, since no variable with $M_i > 0$ implies the solution has been reached.

On the schematic level (Fig. 10.10.3), each coupling cell in a row consists of a 6T SRAM, 2T forward switch and 3T backward switch. During the forward pass, drivers apply variable-proportional voltages to the RWLFs, while at the same time pull-up pFETs connected to the RBLFs on the clause side are turned on. The forward switches that are ON, represent literals in respective clauses that have been set to 1 (TRUE). As a result, the voltage divider formed by the pFET and the nFETs in the forward switches generates a voltage on RBLF$_j$ inversely proportional to the number of true literals in clause C_j. StrongARM latch-based voltage sense amplifiers (VSAs), referenced to off-chip analog reference voltages ref0 and ref1, are used to detect whether a clause has zero (S_j=1) or at most one (O_j=1) true literal. Additional logic derives signal Z, indicating clauses with exactly one true literal. The SAT evaluator uses tree of OR gates to check if all clauses are satisfied. In the Backward Pass, clause signals S_j and Z_j are applied to RWLBS$_j$ and RWLBZ$_j$, respectively. The make-value of variable x_i is evaluated by accumulating contribution from clauses with S_j=1 and that contain its zero-valued literal, on PXM$_i$. The break-value of variable x_i is computed by accumulating contribution from clauses with Z_j=1 and that contain its one-valued literal on PXB$_i$ [12]. The peripheral pull-up pFETs and MUXes ensure for each variable that a voltage-divider on only the RBLBS (RBLBZ) line of its zero- (one-) valued literal is activated and its output is connected to PXM$_i$ (PXB$_i$). VSAs referenced to ref2 are used to detect $M_i > 0$. Probabilistic bits are implemented using a StrongARM latch with three differential pairs: (i) T_1 and T_3, (ii) T_2 and T_4, (iii) $T_{5:12}$ and $T_{13:20}$. While both (i) and (ii) discharge nodes P and Q during evaluation phase (PEN=1), with differential current proportional to PXB$_i$ – PXM$_i$, the latter along with its p-type input common-source amplifiers extends support for rail-to-rail input range. Each branch in (iii) is a 4b binary-weighted current switching DAC and the pair contributes a differential current proportional to refp*(R_p –R_n), where R_p, R_n are the uniform random numbers corresponding to binary bits $rp_{3:0}$ and $rn_{3:0}$. Ignoring offset voltage, transconductance variations in the different pairs, the output of the p-bit is one if (PXB$_i$ – PXM$_i$) > refp×R_0, where $R_0 = R_n - R_p$ is also a uniform random number. The differential voltage (PXB$_i$ – PXM$_i$) is a saturating nonlinear function of the variable's gain (ΔG_i=M_i-B_i), leading to an approximately sigmoidal dependence of the p-bit's output probability and the gain. Each p-bit receives 8 random bits from the RNG buffer chain with each of the four buffer stages supplying random bits for a total of 16 p-bits.

The experimental results for probabilistic-bits (Fig. 10.10.4, left) reveal a relatively tight spread in the characteristics across all p-bits, arising primarily from the offset voltage variations in the circuit's differential pairs. (The results also indicate that the slope of the p-bit function, i.e., the solver's temperature parameter T, can be dynamically adjusted by tuning the reference bias (refp), a feature that can be exploited in more advanced algorithms with annealing). While prior SAT accelerators [4] have been restricted by structural hazards in the IMC fabric to single-thread operation, our circuit can interlace three independent solver threads for the same SAT problem in its three-stage pipeline (Fig. 10.10.4, right). While the problem graph mapped to the SRAM array is reused, three sets of variable and clause registers are required for each thread to independently maintain its variable and clause statuses, respectively.

The solver's measured results on common SAT benchmarks show that multi-thread parallelism improves convergence by 1.7×, while maintaining similar energy consumption (Fig. 10.10.5). Independent threads parallelly explore different regions of the energy landscape, thereby increasing the ensemble's probability of reaching a solution early. Our chip's performance scaling is compared with several prior solvers not requiring preprocessing. Compared to prior SOTA discrete-time incomplete solvers [4], our solver is 8-to-13× faster and 2-to-5× more energy efficient on uniform random 3-SAT problems, whereas it is 2-to-14× faster on practically relevant higher-order K-SAT problems. Compared to a continuous-time solver in a more advanced 28nm node [8] our solver is 4-to-5.6× faster with comparable energy efficiency. Compared to a purely digital DPLL-based complete solver [2] also in 28nm, our chip achieves a 3× speedup. Modeling in [2] further suggests a 84× speedup on larger problems.

186 • 2026 IEEE International Solid-State Circuits Conference

979-8-3315-8937-0/26 $31.00 © 2026 IEEE

In summary, PCIM-SAT can natively map arbitrary-order K-SAT problems and accelerate efficient multi/parallel-variable update heuristics. Leveraging SRAM-based mixed-signal in-memory computing and a new probabilistic SAT-solving algorithm, it achieves 100% solvability on 50-variable 3-SAT (satisfiable) instances in 5.5µs, while consuming 265nJ (Fig. 10.10.6). The locally dense yet globally sparse structure of K-SAT problems, combined with the use of p-bits, enables seamless scalability to multi-tile architectures [15], where all graph computations can be performed locally within each tile, eliminating the need for analog reductions across tiles and limiting inter-tile communication to simple digital exchanges of variable and clause values.

Acknowledgement:
This work is supported by the Defense Advanced Research Projects Agency (DARPA) under Air Force Research Laboratory (AFRL) contract no FA8650-23-3-7313. The authors thank Giacomo Pedretti, Thomas Van Vaerenbergh, Raymond Beausoleil, John Paul Strachan, Ignacio Rozada and other members of the DARPA QuICC team for valuable discussion and feedback.

References:
[1] Marques-Silva, Joao. "Practical Applications of Boolean Satisfiability," *IEEE International Workshop on Discrete Event Systems*, 2008.
https://doi.org/10.1109/WODES.2008.4605925
[2] Wu, Zihan et al., "SKADI: A 28-nm Complete K-SAT Solver Featuring Bidirectional In-Memory Deduction and Incremental Updating," *IEEE JSSC*, 2025.
https://doi.org/10.1109/JSSC.2025.3598289
[3] S. Xie et al., "Snap-SAT: A One-Shot Energy-Performance-Aware All-Digital Compute-in-Memory Solver for Large-Scale Hard Boolean Satisfiability Problems," *ISSCC*, pp. 420-421, 2023. https://doi.org/10.1109/ISSCC42615.2023.10067380
[4] Bhattacharya, Tinish et al., "A Fully Integrated Mixed-Signal Compute-In-Memory Accelerator for Solving Arbitrary Order Boolean Satisfiability Problems," *IEEE Symp. VLSI Circuits*, 2025. https://doi.org/10.23919/VLSITechnologyandCir65189.2025.11074791
[5] Bhattacharya, Tinish et al., " KLIMA: Low-latency mixed-signal In-Memory Computing accelerator for solving arbitrary-order Boolean Satisfiability," *IEEE Hot Chips Symp*, 2025.
https://doi.org/10.1109/HCS66204.2025.11154396
[6] C. Shim et al., "VIP-Sat: A Boolean Satisfiability Solver Featuring 5× 12 Variable In-Memory Processing Elements with 98% Solvability for 50-Variables 218-Clauses 3-SAT Problems," *ISSCC*, pp. 486-488, 2024.
https://doi.org/10.1109/ISSCC49657.2024.10454397
[7] Q. Zhang et al., "A Stochastic Analog SAT Solver in 65nm CMOS Achieving 6.6µs Average Solution Time with 100% Solvability for Hard 3-SAT problems," *IEEE Symp. VLSI Circuits*, 2024. https://doi.org/10.1109/VLSITechnologyandCir46783.2024.10631503
[8] Dikopoulos, Evangelos et al., "A Physics-Inspired Oscillator-Based Mixed-Signal Optimization Engine for Solving 50-Variable 218-Clause 3-SAT Problems with 100% Solvability and 31.7µs Solution Time," *ISSCC*, pp. 446-447, 2025.
https://doi.org/10.1109/ISSCC49661.2025.10904814
[9] Tsukamoto, Sanroku et al., "An Accelerator Architecture for Combinatorial Optimization Problems," *Fujitsu Sci. Tech. Jour.*, vol. 53, no. 5,pp. 8-13, 2017.
https://www.fujitsu.com/global/documents/about/resources/publications/fstj/archives/vol 53-5/paper02.pdf

[10] Grimaldi, Andrea et al., "Spintronics-Compatible Approach to Solving Maximum-Satisfiability Problems with Probabilistic Computing, Invertible Logic, and Parallel Tempering," *Physical Review Applied*, vol. 17, no. 2, Feb. 2022.
https://doi.org/10.1103/physrevapplied.17.024052
[11] Bybee, Connor et al., "Efficient Optimization with Higher-Order Ising Machines," *Nature Communications*, vol 14, 2023. https://doi.org/10.1038/s41467-023-41214-9
[12] Bhattacharya, Tinish et al., "Computing High-Degree Polynomial Gradients in Memory," *Nature Communications*, vol. 15, 2024. https://doi.org/10.1038/s41467-024-52488-y
[13] SAT Benchmarks to Assess Quantum-Inspired Solvers, William Regli, Gianni Mossi, Mohammad Taghi Hajiaghayi, Humberto Munoz Bauza, Ian Joseph Whitehouse, Kiarash Banihashem, Peyman Jabbarzade, Taylor Henry Paul, 2025. Licensed under CC BY 4.0.
https://github.com/UMD-ARLIS/QuICC-SAT-Datasets.git
[14] H. H. Hoos, T. Stützle, "SATLIB: An Online Resource for Research on SAT," *SAT*, pp. 283-292, 2000. https://www.cs.ubc.ca/~hoos/SATLIB/benchm.html
[15] Bhattacharya, Tinish et al., "HO-FPIA: High-Order Field-Programmable Ising Arrays with In-Memory Computing," *IEEE Computer Society Annual Symposium on VLSI*, pp. 252-259, 2024. https://doi.org/10.1109/ISVLSI61997.2024.00054
[16] Patel, Saavan et al., "Logically Synthesized and Hardware-Accelerated Restricted Boltzmann Machines for Combinatorial Optimization and *Integer Factorization*," *Nature Electronics*, vol 5, pp. 92-101, 2022. https://doi.org/10.1038/s41928-022-00714-0

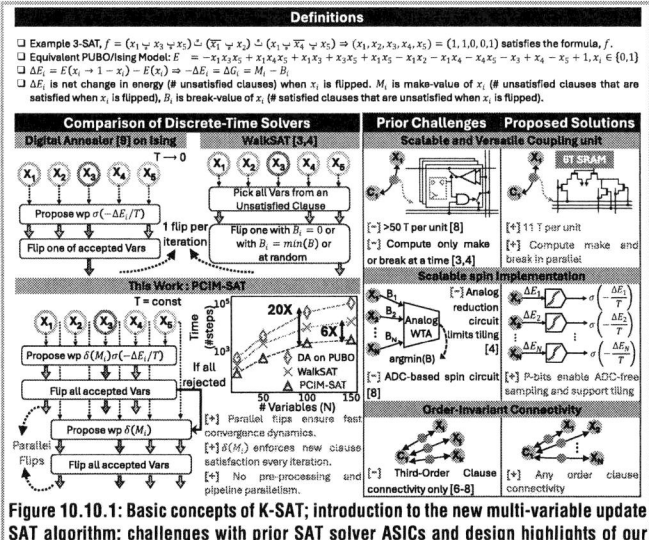

Figure 10.10.1: Basic concepts of K-SAT; introduction to the new multi-variable update SAT algorithm; challenges with prior SAT solver ASICs and design highlights of our solver architecture.

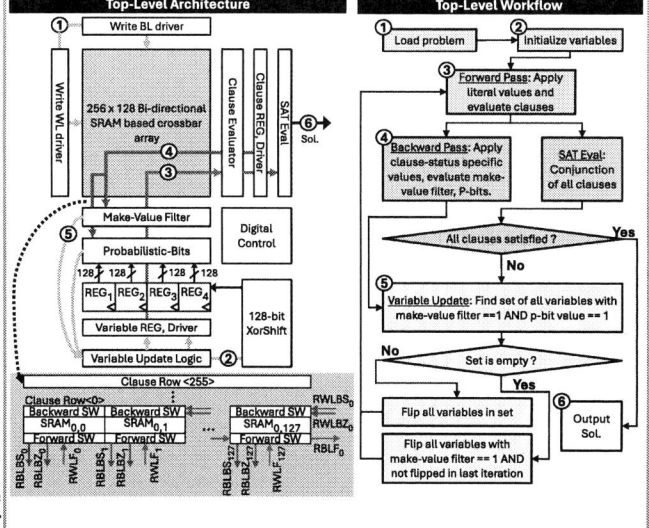

Figure 10.10.2: Top-level architecture and workflow of proposed solver.

ISSCC 2026 / SESSION 10 / DIGITAL PROCESSING AND CIRCUIT TECHNIQUES / 10.10

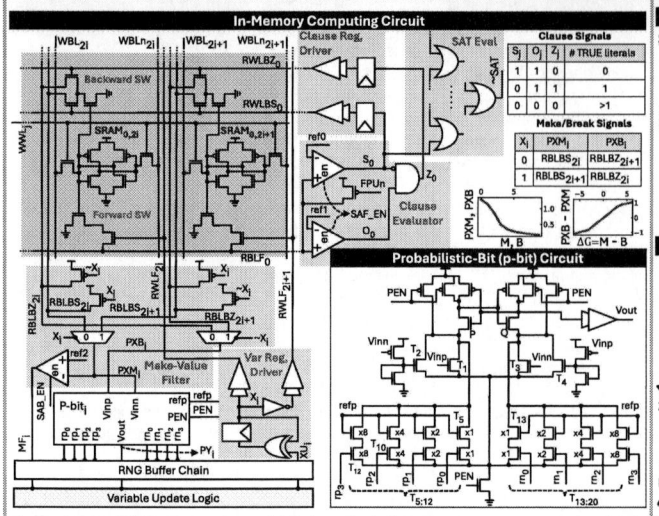

Figure 10.10.3: Detailed schematic of the mixed-signal in-memory computing fabric and probabilistic bit.

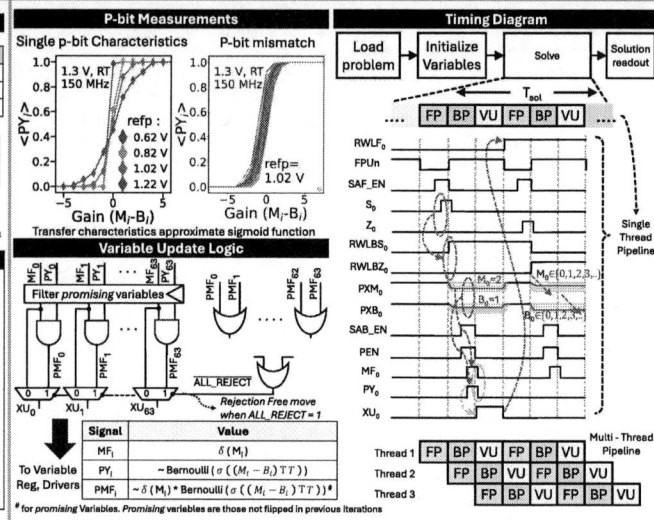

Figure 10.10.4: Measurement results of the probabilistic-bits; schematic of variable update logic and timing diagram of solver.

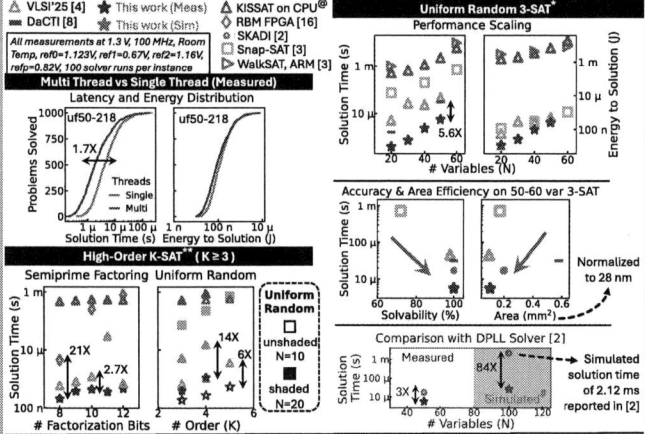

@ KISSAT on CPU: C language implementation of KISSAT on AMD EPYC7373X@3.0/3.73 GHz, *Uniform Random 3-SAT: N=20, 50 has 1000 instances each from uf20-91/uf50-218 sets in SATLIB [14], N=30, 40 has 1000 custom generated instances each. **High-Order K-SAT: (A) Semiprime Factoring: batch-07 in [13], (B) Uniform Random: 100 custom generated instances for each (N,K) pair. # Our solver's simulation done with macro extended to map N=100, M=430 size SAT problems.

Figure 10.10.5: Measurement results of PCIM-SAT on different SAT benchmarks and comparison with other solvers. Multi-thread mode implemented for comparison.

	ISSCC 2023 Snap-SAT [3]	ISSCC 2024 VIP-SAT [6]	VLSI 2025 [4]	ISSCC 2025 SKADI [2]	ISSCC 2025 DaCTI [8]	This Work
Technology	65nm	65nm	55nm	28nm	28nm	55 nm
Technique	Column wise In-memory OR	PE Array	WalkSAT, in-memory	DPLL Solver	Dynamic CT Injection	PCIM-SAT, In-memory
Architecture	Digital	Digital	Mixed	Digital	Mixed	Mixed
Memory Type (#transistors)	SRAM (6T)	Near-Memory	SRAM (10T)	SRAM (~20T)	Flip-Flop (>50T)	SRAM (11T)
Chip Area (mm²)	0.93	1.115	0.544	0.2	0.58	0.42
Pre-processing	NO	YES	NO	NO	NO	NO
Flips per iteration	Single	Multiple	Single	NA	Multiple	Multiple
Arbitrary Order	Yes (up to 128)	No (3)	Yes (up to 64)	Yes (up to 50)	No (3)	Yes (up to 64)
# Variables / Clauses	20/86	60/258	20/91 50/218	50/218	20/91 50/218	20/91 50/218
# Problems Tested	1000[D]	100[C]	100[C]	1000[C]	1000[C]	1000[C]
Solvability[A] (%)	NA	72[H]	100[H] 98[H]	100[G]	100[H] 100[H]	100[H] 100[H]
Solution Time[B] (us)	70	713	7[E] 18.7[E]	5.12 45 17.1	1.6 31.7	0.39 5.5
Solution Energy[C] (nJ)	100	1098	2.1[E] 20.8[E]	59 518 58.0	7.8 268.9	11.8[F] 265[F]

A Number of SAT instances that can be solved in a given time, B Mean time/energy to find a solution across all instances in the set, C SATLIB Benchmark [14], D Benchmark not publicly available, E energy and latency incurred during off-chip preprocessing not taken into account, F (Mean Energy) = P x (Mean solution time), where P is the average power consumption of chip while solving SAT instances not including off-chip DACs used to supply reference biases ref0, ref1, ref2, refp, G Reported on satisfiable instances, H Reported on satisfiable instances. Incomplete/heuristic algorithms cannot prove unsatisfiability.

Figure 10.10.6: Feature and performance comparison with prior SAT solver ASICs.

Chip Summary

Prototype	PCIM-SAT
Technology	55 nm CMOS
Core Area	0.42 mm²
Supply	1.3 V
Frequency	100 MHz
Max Variables	64
Max Clauses	256
Max Order	64
Accuracy	100%@
Solvability	100%#
Mean Solution Time	0.39 us (20 var 3-SAT*)
	5.5 us (50 var 3-SAT**)
Mean Solution Energy	11.8 nJ (20 var 3-SAT*)
	265 nJ (50 var 3-SAT**)

@ Correctness of the solved problem, # Our solver achieves 100% solvability on all tested Uniform random 3-SAT (satisfiable) instances, * all instances from uf20-91 set in SATLIB [14], ** all instances from uf50-218 set in SATLIB [14].

Figure 10.10.7: Die micrograph and performance summary.

ISSCC 2026 / SESSION 11 / PIPELINE AND ULTRA-HIGH-SPEED DATA CONVERTERS / OVERVIEW

Session 11 Overview: *Pipeline and Ultra-High-Speed Data Converters*

DATA CONVERTERS SUBCOMMITTEE

Session Chair: Shahrzad Naraghi
ANAFLASH, Sunnyvale, CA

Session Co-Chair: Chin-Yu Lin
Mediatek, HsinChu, Taiwan

This session features eight data converters that leverage pipelined architectures and ultra-high-speed A/D and D/A techniques. The first part of the session highlights both continuous-time and discrete-time pipelined ADCs, achieving over 60dB SNDR across bandwidths of several hundred MHz. Novel residue amplification schemes, time-domain signal processing techniques, and calibrations enable high-performance A/D conversion. The second part of this session demonstrates advanced ultra-high-speed data converters. The first two works introduce massive time-interleaved ADCs with moderate resolution, designed to support DSP-based wireline receivers. The final two papers present wide-bandwidth ADC and DAC architectures that enable direct-RF-sampling wireless infrastructure.

8:00 AM

11.1 A 14b 400MS/s TDC-Assisted Pipelined-SAR ADC with Rail-to-Rail Input VTC and Background Time-Domain Error Calibration

Jingpeng Wang, *Peking University, Beijing, China*

In Paper 11.1, Peking Univ. presents a 14b 400MS/s TDC-assisted pipelined-SAR ADC, achieving 67.0dB SNDR and 80.0dB SFDR at Nyquist rate, consuming 4.6mW. The performance is enabled by a novel linearized rail-to-rail input VTC and background time-domain error calibration.

8:25 AM

11.2 A 28nm CMOS SAR-Based Continuous-Time Pipeline ADC with 103dB SFDR and 270MHz Bandwidth Using NCF and DAC Error Calibration

Qilong Liu, *NXP Semiconductors, Eindhoven, The Netherlands*

In Paper 11.2, NXP semiconductors introduces a SAR-based continuous-time pipeline ADC achieving 69.5dB SNDR and 103.0dB SFDR over 270MHz bandwidth, exceeding state-of-the-art by 20dB. The noise-cancellation filter and DAC error calibrations result in the wideband high spectral purity.

8:50 AM

11.3 A 500MS/s 12b Pipe-SAR ADC Using a Triple-Cascode FIA with Virtual Supply Extension

Michele Rocco, *Politecnico di Milano, Milan, Italy*

In Paper 11.3, Politecnico di Milano proposes a triple-cascode single-stage FIA for a 12b 500MS/s pipeline-SAR ADC. This ADC realizes an SNDR of 63.4dB and 81dB SFDR at Nyquist rate, consuming 6.64mW with on-chip digital calibration.

188 • 2026 IEEE International Solid-State Circuits Conference

979-8-3315-8937-0/26 $31.00 © 2026 IEEE

ISSCC 2026 / February 17, 2026 / 8:00 AM

9:15 AM

11.4 A 13b 500MS/s 94dB-SFDR Resistive-Input Pipelined-SAR ADC with Linear and Efficient Current-Buffer-Based Integrating Sampler

Xiyu He, Tsinghua University, Beijing, China

In Paper 11.4, Tsinghua Univ. presents a 13b 500MS/s resistive-input pipelined-SAR achieving 67.5dB SNDR and 94.5dB SFDR at Nyquist rate. The highest reported SFDR among discrete-time ADCs operating above 200MS/s is enabled by a current-buffer-based integrating sampler and a floating charge transferrer (FCT).

10:05 AM

11.5 A Compact 7b 175GS/s Linearized Time-Interleaved Slope ADC with Switched Input Buffers

Ewout Martens, imec, Leuven, Belgium

In Paper 11.5, Imec introduces a 7b 175GS/s 2048× time-interleaved slope ADC achieving a low energy per sample of less than 2.16pJ while occupying only 0.063mm², which is the smallest ultra-high-speed ADC reported to date.

11

10:30 AM

11.6 An 8b 20GS/s Time-Interleaved ADC with 2.6mW 1GS/s Hybrid Voltage/Time-Domain Sub-ADC in 12nm FinFET

Daisuke Miyazaki, Sony Semiconductor Solutions, Atsugi, Japan

In Paper 11.6, Sony Semiconductor Solutions presents an 8b 20GS/s time-interleaved ADC achieving an SNDR of 37dB, consuming 135mW with input and reference buffers included. The performance is enabled by hybrid voltage/time-domain conversion and a new common-mode feedback scheme in the sample-and-hold circuit.

10:55 AM

11.7 A 12b 12GS/s Two-Way Interleaved Pipeline ADC with Integrated Broadband RF VGA in 5nm

Haiyang Zhu, Analog Devices, Wilmington, MA

In Paper 11.7, Analog Devices introduces a 12b 12GS/s 2× interleaved pipeline ADC with integrated broadband RF VGA, achieving 49.0dB SNDR and 70.0dB SFDR with a wide programmable input range. The proposed flash early-trigger and embedded SHA enable an ultra-high-speed 6GS/s sub-ADC.

11:20 AM

11.8 A 14b 20GS/s RF-Sampling DAC Achieving 70.4dBc IMD3 up to 8.9GHz

Haowei Lu, Sanechips Technology, Shenzhen, China

In Paper 11.8, Sanechips demonstrates a 14b 20GS/s RF-sampling DAC, achieving 70.4dBc IMD3 up to an 8.9GHz input and a maximum output power of 7dBm. The proposed ISI-minimizing dynamic element matching (ISIM-DEM) mitigates the output distortion induced by inter-symbol-interference (ISI).

DIGEST OF TECHNICAL PAPERS • 189

979-8-3315-8937-0/26 $31.00 © 2026 IEEE

ISSCC 2026 / SESSION 11 / PIPELINE AND ULTRA-HIGH-SPEED DATA CONVERTERS / 11.1

11.1 A 14b 400MS/s TDC-Assisted Pipelined-SAR ADC with Rail-to-Rail Input VTC and Background Time-Domain Error Calibration

Jingpeng Wang[1], Bingrui Li[1], Haoyang Luo[1], Mingtao Zhan[2], Xiyu He[2], Dawei Shen[1], Lu Jie[2], Xiyuan Tang[1]

[1]Peking University, Beijing, China, [2]Tsinghua University, Beijing, China

Abstract

This paper presents a 14b 400MS/s Pipelined-SAR ADC utilizing a front-end TDC with a linearized rail-to-rail input VTC. A background calibration engine corrects VTC gain and time-domain offset errors by monitoring digital output statistics with minimal analog overhead. A replica-biasing-based ring amplifier improves energy efficiency and robustness. The 28nm chip achieves 67dB Nyquist SNDR at 4.6mW with a 6.3fJ/conv-step FoMw and 173.5dB FoMs.

The proliferation of 5G/6G communications and high-speed wireline systems creates a relentless demand for ADCs achieving both high resolution (≥14b) and sampling rates well into the hundreds of MS/s. The pipelined-SAR architecture has become a popular solution, effectively balancing power efficiency with the required conversion speed. However, the sequential bit-cycling of the frontend SAR often becomes the primary speed bottleneck. To break this latency barrier, the time-domain (TD) quantizer, which employs a voltage-to-time converter (VTC) and a time-to-digital converter (TDC), offers a compelling alternative for realizing multi-bit/cycle conversion. Yet, the advantage of TD quantization is weakened when applied at the frontend, as the conventional dynamic integrator or discharging-based VTC exhibits a severely limited linear input range, usually smaller than $1.2V_{pp}$ [1]–[4]. Compromises have been made by relegating the TD quantizer to a backend role to reduce the VTC input swing [5], [6], but this method fails to alleviate the timing pressure on the first stage. A 5b/cycle 2-step architecture introduced in [7] shrinks the VTC input by a coarse TD quantization. The reduced VTC input swing at the 2nd-stage TD converter enables 12b linearity. However, the input swing of the ADC is still constrained to $0.8V_{pp}$, limiting its dynamic range. Another challenge is that the VTC gain and time-domain offsets are sensitive to PVT variations. Prior works have addressed this with either foreground calibration that fails to track environmental change [2]–[5], [7], or with background calibrations that introduce extra analog hardware complexity [8], [9]. Beyond the TDC, the settling speed and power efficiency of the residue amplifier present another critical bottleneck. To confront these issues, this work introduces a 14b 400MS/s TDC-assisted pipelined-SAR ADC featuring a linearized rail-to-rail input VTC. A background statistics-based calibration engine is constructed to correct major time-domain errors by monitoring only digital outputs, avoiding extra analog complexity. Additionally, a ring amplifier with replica biasing is introduced to realize high speed, energy efficiency, and PVT robustness simultaneously. The prototype ADC achieves 69.2dB SNDR at a low input frequency and 67dB SNDR at the Nyquist input. Fabricated in 28nm, it consumes 4.6mW of power, yielding a 6.3fJ/conv.-step FoMw and 173.5dB FoMs at the Nyquist input.

As depicted in Fig. 11.1.1, the pipelined-SAR employs a 3-then-4 bit per cycle TDC-assisted frontend ADC, an 18× residue amplifier, and a 10b backend SAR ADC. To break the speed bottleneck of conventional SAR logic, the frontend replaces SAR bit-cycling with high-speed TD quantizers that produce multiple bits in one conversion cycle. In the first cycle (Cyc1), a 2b coarse TDC combined with an MSB arbiter resolves the 3 coarse bits. Following the residue subtraction on the CDAC, a 3b fine TDC with the reused MSB arbiter resolves the 4 fine bits in Cyc2. The fine TDC LSB is generated from the 4× phase-interpolated unit delay ($T_{LSB2}=\tau/4≈3ps$), while the coarse TDC LSB is derived from a 2τ-delay block ($T_{LSB1}=2\tau≈24ps$), guaranteeing an accurate 8× time-reference scaling factor (T_{LSB1}/T_{LSB2}). It allows the same VTC to be adopted during the two TD conversions, obviating VTC gain calibration as required in [7]. To increase the fine TDC redundancy, the intermediate taps $T_{PI}<7:5>$ of the phase interpolator (PI) are delayed by a full delay unit τ, extending quantization range. With a single-side sampling capacitor of 1.7pF, the entire 2-step TD quantization completes in approximately 500ps, halving the conversion time of a comparable 7b SAR, which is critical to achieving the 400MS/s sample rate of the ADC.

A background statistics-based calibration engine ensures the PVT robustness of the TD frontend by correcting 3 key error sources: VTC gain, VTC offset ($T_{os,vtc}$), and P/N-side PI offsets ($T_{os,pi,p}/T_{os,pi,n}$), as illustrated in Fig. 11.1.2. The engine leverages inter-stage redundancy to extract error signatures. First, for the VTC gain error, if it is too large, the TDC overestimates the input, leading to an over-approximated CDAC voltage. Thus, the polarity of differential CDAC voltage will be changed twice in two TD conversions ("two crossings"), i.e., coarse cycle MSB (MSB_1) ≠ fine cycle MSB (MSB_2), MSB_2 ≠ backend MSB ($MSB_{backend}$), together with a backend overrange flag. Conversely, if the VTC gain is too small, under-approximation can happen ("no crossing"), manifesting as $MSB_1=MSB_2$ and $MSB_2=MSB_{backend}$. As shown in Fig. 11.1.2 (top-right), it only requires simple combinational logic and a counter, followed by a charge pump to tune the VTC's current bias and correct its gain. Second, the multiple offset sources are hierarchically calibrated to improve calibration stability. Due to the folding operation of the TDC, the total equivalent offset ($T_{os,eq}$) presented to the backend ADC depends on MSB_2, which selects one of two signal paths. If $MSB_2=1$: $T_{os,eq}=T_{os,vtc}+T_{os,pi,p}$, and if $MSB_2=0$: $T_{os,eq}=-T_{os,vtc}+T_{os,pi,n}$. Therefore, two parallel calibration loops use the sign of $MSB_{backend}$ to drive the long-term average of $T_{os,eq}$ to 0 for both MSB_2 states. This forces $T_{os,pi,p}$ and $T_{os,pi,n}$ to cancel $T_{os,vtc}$. During normal operation, $T_{os,vtc}$ is intentionally left uncalibrated and canceled by the other two offsets, thereby largely improving the multi-loop calibration stability. The third auxiliary loop for $T_{os,vtc}$ is activated only when $T_{os,eq}$ becomes excessively large that saturates the fine stage. The calibration scheme offers significant advantages over conventional methods. Prior approach for VTC gain tuning relies on dedicated analog replicas and feedback loops, which introduce their own mismatch errors and hardware overhead [9]. Other solutions employ redundant TDC stages to detect and correct offset but do not address VTC gain variations [8]. In contrast, the introduced calibration engine provides comprehensive correction for both TD gain and offset errors by leveraging the digital output statistics of the ADC with minimal auxiliary analog hardware.

To accommodate the $1.8V_{pp}$ rail-to-rail input swing during the coarse cycle, the VTC is enhanced with a linearization technique (Fig. 11.1.3 top). A conventional integrator-based VTC suffers from severe expanding nonlinearity for large inputs, as one input transistor enters the cutoff region, severely increasing the discharge time. To overcome this limitation, the presented VTC introduces a level-shifted parallel input branch driven by a source follower. It provides a DC-shifted version of the input, ensuring the auxiliary branch remains active and provides a discharging current even when the main input transistor is driven into cutoff. Moreover, the source follower introduces a compressive nonlinearity that provides first-order cancellation of the expanding nonlinearity from the main input pair. Meanwhile, the background calibration loop actively controls the source follower's bias current, which tunes the level-shift voltage to adjust the overall VTC gain. With these features, the designed VTC preserves a linear V-T transfer characteristic even at a rail-to-rail input. Post-extraction simulation shows that the VTC realizes over 32dB SNDR across all PVT corners at a $1.8V_{pp}$ input, robustly meeting the coarse conversion requirement. During the fine TD conversion, the VTC linearity is further improved due to the 8×-shrunk input swing. As a result, the realized SFDR of the entire TD frontend exceeds 50dB.

To further improve the PVT robustness and energy efficiency, a replica-biased ring amplifier is introduced (Fig. 11.1.3 bottom). Conventional self-biased ring amplifiers benefit from a simple inverter-based structure but suffer from an ill-defined biasing point [10]. The current-biased ring amplifiers show good PVT robustness [11]–[13], but still need a dedicated global common-mode feedback (CMFB) loop which degrades speed. A floating inverter amplifier (FIA) output stage achieves intrinsic CMFB without extra circuitry [14]–[17], but its quenching nature makes it undesirable for high-speed applications. In this work, A1's biasing is established through auto-zeroing, while a replica-based loop generates PVT-stable bias points for A2. The A2 bias current is defined by ($V_{CM1}-V_{CM2}$)/R_{dz}, and its two output common-mode voltages are set to V_{CM2} and V_{CM1}, respectively. V_{CM2} and V_{CM1} are generated through an on-chip resistive voltage divider. For the output stage A3, an FIA with bias enhancement technique is applied, eliminating the need for global CMFB. Compared with conventional self-biased ring amplifiers that use the IR drop of R_{dz} to generate a dead zone [18], this work leverages the FIA's inherent self-quenching to create the dead zone at its supply. Thus, the IR drop of R_{dz} can be utilized to boost the output stage overdrive instead. This bias enhancement technique largely accelerates the initial slewing, enabling FIA usage in high-speed RAs. The correlated level shifting (CLS) technique is adopted to enhance gain and output swing [19]. A split-FIA output stage optimizes the speed-noise trade-off in the estimate phase and level shift phase, respectively [20]. At the level-shift phase, the enabled FIA branch is optimized for larger output impedance to further lower the dominant pole frequency and filter out high-frequency noise. The post-extraction simulation shows that the ring amplifier SDR is over 59dB across all corners, confirming its robustness.

Fabricated in 28nm CMOS, the ADC occupies an active area of 0.016mm² (Fig. 11.1.7). One-time foreground calibration is adopted to address the capacitor mismatches and the inter-stage gain (defined by capacitor ratio). During measurement, the on-chip background calibration engine corrects TD error. With a sample rate of 400MS/s and a differential input swing of $1.8V_{pp}$, the ADC achieves an SNDR/SFDR of 69.2/81.7dB at a low input frequency and 67/80dB at the Nyquist frequency. The total power consumption is 4.59mW from a 1V supply, including the on-chip calibration engine (Fig. 11.1.4). The robustness of the TD error calibration is verified by measuring the real-time backend SAR code when a supply fluctuation is manually induced (Fig. 11.1.5). With calibration enabled, the backend SAR code does not overrange, demonstrating the fast convergence of the implemented calibration scheme without interrupting ADC operation. The performance variations under ±6% supply and -20~80℃ of 3 samples are also measured. With the background calibration on, the SNDR variations are within 2dB and 2.2dB, respectively. A comparison with state-of-the-art single-channel ADCs highlights the presented design in achieving a highly competitive combination of speed, resolution, and power efficiency.

Acknowledgement:
This work is supported in part by National Key R&D Program of China (2022YFB4401900) and the NSFC (62274005). The corresponding author is Xiyuan Tang.

References:

[1] A. S. Yonar *et al.*, "An 8b 1.0-to-1.25GS/s 0.7-to-0.8V Single-Stage Time-Based Gated-Ring-Oscillator ADC with 2× Interpolating Sense-Amplifier-Latches," *2023 IEEE International Solid-State Circuits Conference (ISSCC)*, San Francisco, CA, USA, 2023, pp. 1-3. https://doi.org/10.1109/ISSCC42615.2023.10067745

[2] M. Zhang, Y. Zhu, C. -H. Chan and R. P. Martins, "16.2 A 4× Interleaved 10GS/s 8b Time-Domain ADC with 16× Interpolation-Based Inter-Stage Gain Achieving >37.5dB SNDR at 18GHz Input," *2020 IEEE International Solid-State Circuits Conference - (ISSCC)*, San Francisco, CA, USA, 2020, pp. 252-254. https://doi.org/10.1109/ISSCC19947.2020.9062986

[3] K. Ohhata, "A 2.3-mW, 1-GHz, 8-Bit Fully Time-Based Two-Step ADC Using a High-Linearity Dynamic VTC," in *IEEE Journal of Solid-State Circuits*, vol. 54, no. 7, pp. 2038-2048, July 2019. https://doi.org/10.1109/JSSC.2019.2907401

[4] A. Whitcombe *et al.*, "22.3 A 76mW 40GS/s 7b Time-Interleaved Hybrid Voltage/Time-Domain ADC with Common-Mode Input Tracking," *2024 IEEE International Solid-State Circuits Conference (ISSCC)*, San Francisco, CA, USA, 2024, pp. 392-394. https://doi.org/10.1109/ISSCC49657.2024.10454355

[5] M. Zhang, C. -H. Chan, Y. Zhu and R. P. Martins, "3.5 A 0.6V 13b 20MS/s Two-Step TDC-Assisted SAR ADC with PVT Tracking and Speed-Enhanced Techniques," *2019 IEEE International Solid-State Circuits Conference - (ISSCC)*, San Francisco, CA, USA, 2019, pp. 66-68. https://doi.org/10.1109/ISSCC.2019.8662350

[6] J. Muhlestein, S. Leuenberger, H. Sun, Y. Xu and U. -K. Moon, "A 73dB SNDR 20MS/s 1.28mW SAR-TDC using hybrid two-step quantization," *2017 IEEE Custom Integrated Circuits Conference (CICC)*, Austin, TX, USA, 2017, pp. 1-4. https://doi.org/10.1109/CICC.2017.7993701

[7] H. Zhao, M. Zhang, Y. Zhu, C. -H. Chan and R. P. Martins, "17.1 A 2x-Interleaved 9b 2.8G8S/s 5b/cycle SAR ADC with Linearized Configurable V2T Buffer Achieving >50dB SNDR at 3GHz Input," *2023 IEEE International Solid-State Circuits Conference (ISSCC)*, San Francisco, CA, USA, 2023, pp. 1-3. https://doi.org/10.1109/ISSCC42615.2023.10067627

[8] J. Liu, A. Shabra, S. Ho, G. Manganaro and M. Shuo- Wei Chen, "A 16GS/s 10b Time-domain ADC using Pipelined-SAR TDC with Delay Variability Compensation and Background Calibration Achieving 153.8dB FoM in 4nm CMOS," *2024 IEEE Symposium on VLSI Technology and Circuits (VLSI Technology and Circuits)*, Honolulu, HI, USA, 2024, pp. 1-2. https://doi.org/10.1109/VLSITechnologyandCir46783.2024.10631373

[9] H. Liang *et al.*, "A 32GS/s 8b Time-Interleaved Hybrid ADC with Self-Detection Offset Calibration, DLL-Based TLSB PVT Variation Calibration and VTC Gain Self-Tracking," *2025 IEEE Custom Integrated Circuits Conference (CICC)*, Boston, MA, USA, 2025, pp. 1-3. https://doi.org/10.1109/CICC63670.2025.10983486

[10] B. Hershberg *et al.*, "A 4-GS/s 10-ENOB 75-mW Ringamp ADC in 16-nm CMOS With Background Monitoring of Distortion," in *IEEE Journal of Solid-State Circuits*, vol. 56, no. 8, pp. 2360-2374, Aug. 2021. https://doi.org/10.1109/JSSC.2021.3053893

[11] Y. Cao *et al.*, "24.2 A 14b 1GS/s Single-Channel Pipelined ADC with A Parallel-Operation SAR Sub-Quantizer and A Dynamic-Deadzone Ring Amplifier," *2025 IEEE International Solid-State Circuits Conference (ISSCC)*, San Francisco, CA, USA, 2025, pp. 430-432. https://doi.org/10.1109/ISSCC49661.2025.10904766

[12] Y. Lim *et al.*, "9.2 A 2.08mW 64.4dB SNDR 400MS/s 12b Pipelined-SAR ADC using Mismatch and PVT Variation Tolerant Dynamically Biased Ring Amplifier in 8nm," *2024 IEEE International Solid-State Circuits Conference (ISSCC)*, San Francisco, CA, USA, 2024, pp. 170-172. https://doi.org/10.1109/ISSCC49657.2024.10454422

[13] M. Zhan, L. Jie and N. Sun, "17.5 A 10mW 10-ENOB 1GS/s Ring-Amp-Based Pipelined TI-SAR ADC with Split MDAC and Switched Reference Decoupling Capacitor," *2023 IEEE International Solid-State Circuits Conference (ISSCC)*, San Francisco, CA, USA, 2023, pp. 272-274. https://doi.org/10.1109/ISSCC42615.2023.10067475

[14] X. Tang *et al.*, "An Energy-Efficient Comparator With Dynamic Floating Inverter Amplifier," in *IEEE Journal of Solid-State Circuits*, vol. 55, no. 4, pp. 1011-1022, April 2020. https://doi.org/10.1109/JSSC.2019.2960485

[15] X. Tang *et al.*, "A Bandwidth-Adaptive Pipelined SAR ADC With Three-Stage Cascoded Floating Inverter Amplifier," in *IEEE Journal of Solid-State Circuits*, vol. 58, no. 9, pp. 2564-2574, Sept. 2023. https://doi.org/10.1109/JSSC.2023.3268719

[16] S. Ye *et al.*, "9.1 A 2mW 70.7dB SNDR 200MS/s Pipelined-SAR ADC with Continuous-Time SAR-Assisted Detect-and-Skip and Open-then-Close Correlated Level Shifting," *2024 IEEE International Solid-State Circuits Conference (ISSCC)*, San Francisco, CA, USA, 2024, pp. 168-170. https://doi.org/10.1109/ISSCC49657.2024.10454412

[17] S. Song, T. Kang, S. Lee and M. P. Flynn, "A 150-MS/s Fully Dynamic SAR-Assisted Pipeline ADC Using a Floating Ring Amplifier and Gain-Enhancing Miller Negative-C," *2023 IEEE Symposium on VLSI Technology and Circuits (VLSI Technology and Circuits)*, Kyoto, Japan, 2023, pp. 1-2. https://doi.org/10.23919/VLSITechnologyandCir57934.2023.10185377

[18] Y. Lim and M. P. Flynn, "A 100 MS/s, 10.5 Bit, 2.46 mW Comparator-Less Pipeline ADC Using Self-Biased Ring Amplifiers," in *IEEE Journal of Solid-State Circuits*, vol. 50, no. 10, pp. 2331-2341, Oct. 2015. https://doi.org/10.1109/JSSC.2015.2453332

[19] B. R. Gregoire and U. -K. Moon, "An Over-60 dB True Rail-to-Rail Performance Using Correlated Level Shifting and an Opamp With Only 30 dB Loop Gain," in *IEEE Journal of Solid-State Circuits*, vol. 43, no. 12, pp. 2620-2630, Dec. 2008. https://doi.org/10.1109/JSSC.2008.2006312

[20] A. ElShater *et al.*, "3.7 A 10mW 16b 15MS/s Two-Step SAR ADC with 95dB DR Using Dual-Deadzone Ring-Amplifier," *2019 IEEE International Solid-State Circuits Conference - (ISSCC)*, San Francisco, CA, USA, 2019, pp. 70-72. https://doi.org/10.1109/ISSCC.2019.8662400

[21] H. Zhao and F. F. Dai, "A 0.97mW 260MS/s 12b Pipelined-SAR ADC with Ring-TDC-Based Fine Quantizer for PVT Robust Automatic Cross-Domain Scale Alignment," *2022 IEEE International Solid-State Circuits Conference (ISSCC)*, San Francisco, CA, USA, 2022, pp. 1-3. https://doi.org/10.1109/ISSCC42614.2022.9731702

[22] W. Jiang *et al.*, "A 14b 500 MS/s Single-Channel Pipelined-SAR ADC With Reference Ripple Mitigation Techniques and Adaptively Biased Floating Inverter Amplifier," in *IEEE Journal of Solid-State Circuits*, vol. 58, no. 10, pp. 2709-2721, Oct. 2023. https://doi.org/10.1109/JSSC.2023.3290119

[23] C. Chen, Z. Yuan, P. Cao, J. Xu and Z. Hong, "A 71.5-dB SNDR 475-MS/s Ringamp-Based Pipelined SAR ADC with On-Chip Bit-Weight Calibration," *2024 IEEE Symposium on VLSI Technology and Circuits (VLSI Technology and Circuits)*, Honolulu, HI, USA, 2024, pp. 1-2. https://doi.org/10.1109/VLSITechnologyandCir46783.2024.10631448

[24] X. He, M. Gu, H. Jiang, Y. Zhong, N. Sun and L. Jie, "9.3 A 71dB SNDR 200MHz BW Interleaved Pipe-SAR ADC with a Shared Residue Integrating Amplifier Achieving 173dB FoMs," *2024 IEEE International Solid-State Circuits Conference (ISSCC)*, San Francisco, CA, USA, 2024, pp. 172-174. https://doi.org/10.1109/ISSCC49657.2024.10454431

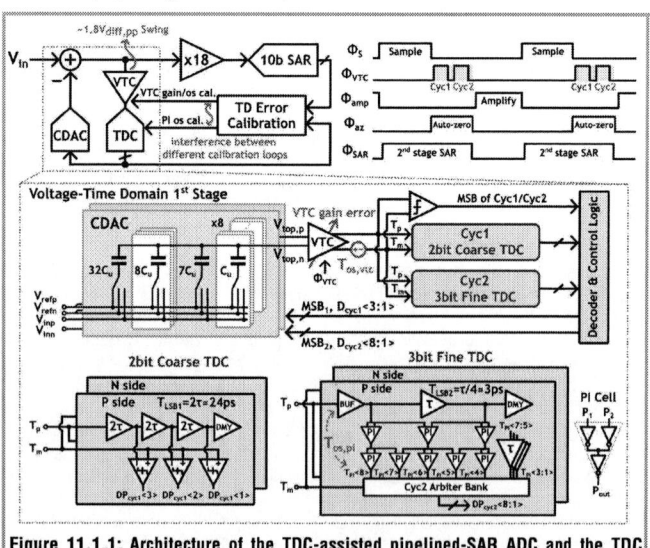

Figure 11.1.1: Architecture of the TDC-assisted pipelined-SAR ADC and the TDC implementation.

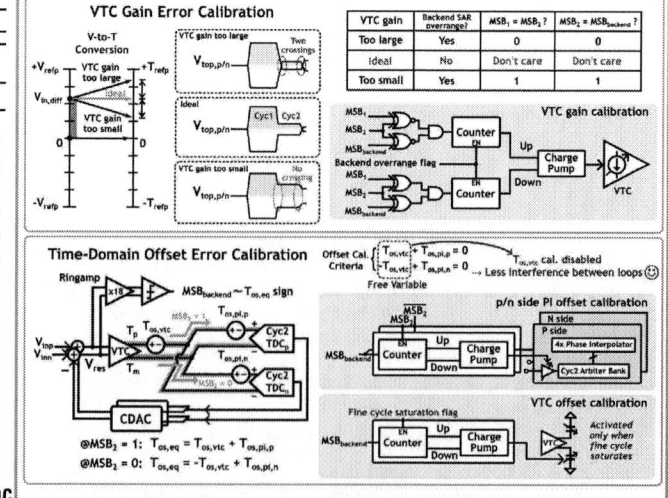

Figure 11.1.2: VTC gain error and time-domain offset error calibration principle.

ISSCC 2026 / SESSION 11 / PIPELINE AND ULTRA-HIGH-SPEED DATA CONVERTERS / 11.1

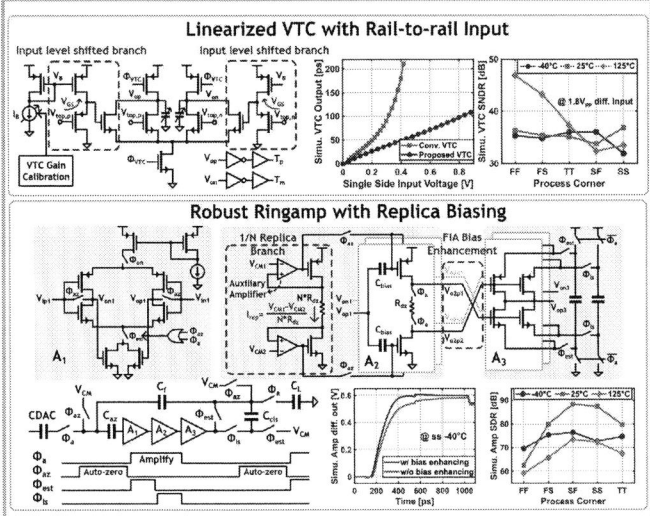

Figure 11.1.3: Linearized VTC with rail-to-rail input and replica-biasing-based ring amplifier details.

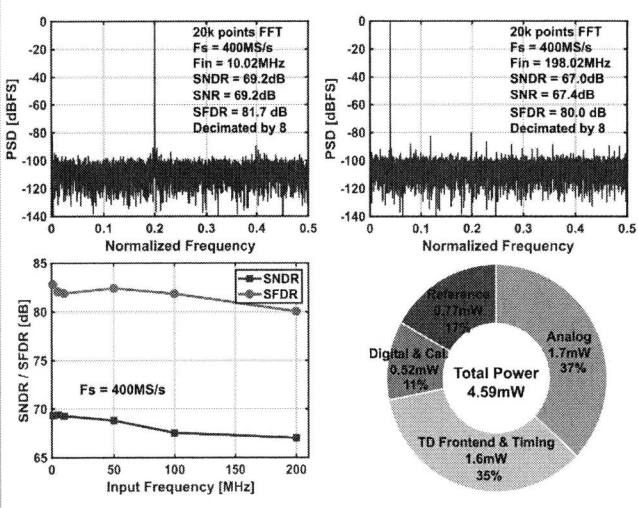

Figure 11.1.4: Measured ADC performance at low/high frequencies and power breakdown.

Figure 11.1.5: Measured real-time calibration result with supply fluctuation and SNDR vs. supply/temperature variation.

Figure 11.1.6: Performance summary and comparison with the state-of-the-art single-channel ADCs with similar specifications.

		ISSCC22 H. Zhao	JSSC23 W. Jiang	VLSI24 C. Chen	ISSCC24 X. He	ISSCC24 Y. Lim	This Work
Architecture		TDC-Assisted Pipeline	Pipelined-SAR	Pipelined-SAR	Pipelined-SAR	Pipelined-SAR	TDC-Assisted Pipeline
Technology		22nm	28nm	22nm	28nm	8nm FinFET	28nm
Supply Voltage [V]		0.8	0.9	1	1	0.9	1
Input Swing [V]		1.25	1.6	-	-	1.4	1.8
Resolution [bit]		12	14	13	14	12	14
Sample Rate [MS/s]		260	500	475	1600	400	400
Bandwidth [MHz]		130	250	237.5	200	200	200
Core Area [mm²]		0.048	0.018	0.038	0.0219	0.0168	0.016
Power [mW]		1.0	6.3	9.9	12.5	2.1	4.6
SNDR@Low Freq. [dB]		67	67	71.5	71.2	64.4	69.2
SNDR@High Freq.[dB]		60.5	64.2	65.9	-	62.8	**67.0**
@Low Freq.	FoM$_W$ [fJ/c-s]	3.1	6.9	6.8	10.5	3.8	4.9
	FoM$_S$ [dB]	174.6	173	175.3	173.2	174.2	**175.6**
@High Freq.	FoM$_W$ [fJ/c-s]	4.3	9.6	13	-	4.6	6.3
	FoM$_S$ [dB]	171.8	170.2	169.7	-	172.6	**173.5**

Figure 11.1.7: Die micrograph.

• 2026 IEEE International Solid-State Circuits Conference

ISSCC 2026 / SESSION 11 / PIPELINE AND ULTRA-HIGH-SPEED DATA CONVERTERS / 11.2

11.2 A 28nm CMOS SAR-Based Continuous-Time Pipeline ADC with 103dB SFDR and 270MHz Bandwidth Using NCF and DAC Error Calibration

Qilong Liu[1], Robert Rutten[1], Muhammed Bolatkale[1], Arda Aralioglu[1], Gilbert Hardeman[1], Emil Siby[1], Patrick van Mourik[1], Vinoth Ilamurugan[2], Shagun Bajoria[1], Lucien Breems[1]

[1]NXP Semiconductors, Eindhoven, The Netherlands, [2]NXP Semiconductors, Delft, The Netherlands

Abstract

This 28nm CMOS SAR-based continuous-time pipeline ADC for automotive receivers achieves -95dBc THD and 103dB SFDR in 270MHz bandwidth, exceeding state-of-the-art by >20dB. Wideband high spectral purity is realized with a highly linear all-pass filter, high-

SFDR coarse ADCs, a fast tone-based frequency-domain noise-cancellation filter calibration and a correlation-based element-wise DAC gain and timing error calibration. Power and area, including LDOs and calibrated filters is 155mW and 1.4mm².

Automotive grade receivers for FMCW radar sensors and high-end car infotainment systems require ADCs with very high linearity (-100dB THD) and spectral purity (>100dB SFDR). A continuous-time (CT) delta-sigma (DS) modulator is the preferred architecture as it combines inherent anti-alias filtering, easy drivability, high dynamic range and superb linearity and spurious-free dynamic range (SFDR) capabilities with high system power efficiency. As a DS modulator is an oversampled feedback system, the largest bandwidth demonstrated with this linearity performance is 120MHz [1]. Multi-stage noise-shaping (MASH) and CT pipeline (CTP) architectures can have substantially higher bandwidths while preserving inherent anti-alias suppression and ease of drivability. However, their spectral purity is an order of magnitude lower compared to state-of-the-art single-loop DS ADCs [2],[3],[4]. A reason for this gap is that a multi-stage ADC requires very precise frequency-dependent matching between analog and digital filters for error cancellation, while a DS loop is based on robust error suppression by means of a high-gain loop filter. This paper presents a 270MHz bandwidth CTP ADC that achieves -95dBc THD and 103dBFS SFDR, same as state-of-the-art CT DS ADCs but with >2× larger bandwidth and advancing SFDR of CTP ADCs by 20dB. Key enablers of this performance are a highly linear all-pass filter (APF), high-SFDR coarse ADC, ISI mitigated DAC, a fast-converging frequency-domain tone-based noise-cancellation filter (NCF) calibration and a time-domain correlation-based DAC gain and timing errors calibration.

The CTP ADC architecture (Fig. 11.2.1) consists of three identical 4b front-end stages and a 6b back-end SAR ADC, all operating at 1.6GS/s. The digital outputs Y_1-Y_3 of the front-end stages are filtered by digital NCFs h_1-h_3 to compensate for the analog quantization noise transfers and combined with back-end ADC output Y_4 to cancel the quantization errors of the front-end stages. Each stage comprises a 4b SAR coarse ADC, for power efficiency, and a 3-1 segmented 4b resistive DAC with 7 unary and 1 binary element for compact area and low noise. A 1st-order RC cross-coupled all-pass filter (APF) compensates for the propagation delay from SAR input buffer to DAC output. Kickback from the SAR ADC S/H circuitry can propagate through the APF, leading to spurs at the ADC output. To prevent this, a unity-gain input buffer is used between the substage input and the SAR ADC. To minimize the offset-induced swing at the residue amplifier output, and the resulting 2nd-order distortion, the comparator of the 4b SAR ADC is calibrated to a residual offset below 10mV. The inter-stage residue amplifier is a 2nd-order single-opamp biquad filter [2] with an inverter-based amplifier which has a moderate gain of 12dB, to relax prefiltering requirements and prevent overload of the ADC in case of near-Fs blockers. As the APF of the first stage is directly at the input of the ADC and the capacitor calibration bank switches of C_{APF} are not located at a virtual ground node, linearity of the switches is critical. The switch design is shown in Fig. 11.2.1 and consists of a bootstrapped thin-oxide NMOS M1 achieving high-linearity (<-100dBc THD). The APF input and output common-mode voltages are at 0.45V (half supply). In ON state, the M1 gate voltage Vg is resistively connected to a 1.35V supply (via thick-oxide PMOS switch M2) and AC coupled to node $V_{2P/N}$, which results in 0.9V Vgs, preventing voltage stress of M1. When the branch is conducting, the M1 gate voltage tracks $V_{2P/N}$ via capacitor C_B to maintain a constant on-resistance. This is achieved by setting the pole of the RC filter (~16MHz) much below the APF pole frequency. In OFF state the gate of M1 is grounded via a thick-oxide NMOS switch M3.

A key challenge to achieve >100dB SFDR with a pipelined ADC architecture is to precisely cancel the large in-band quantization tones that are created by the coarse 4b SARs in the front-end stages. To reduce the intrinsic level of the quantization tones, a 4b dither signal (±1/2 LSB amplitude) is injected at each SAR ADC input (Fig. 11.2.1). With 4b dithering, the spurious performance of the 4b SAR is improved to an 8b ADC equivalent level [5]. The dither is injected by driving the bottom plate of the dither C_{dither} during the charge transfer before the bit cycling. To match the dither level with the quantization level, the input is sampled to the bottom plates of the C_{SAR} array. The reference voltage V_{ref} of the SAR ADC is derived from an LDO supply with a resistive divider. A resistive dual return-to-open [6] DAC has been used for low noise and to mitigate inter-symbol interference. Separate LDOs are deployed in the DAC to isolate the output stage from the input latch driver. Process variability related mismatch and timing errors between the DAC elements limit the ADC non-linearity to ~80dB. In this design, each output bit of stage-1 (D_1-D_8) has a dedicated NCF ($h_{1,1}$-$h_{1,8}$) that is individually calibrated for DAC element gain and timing mismatch. As the digital FIR filters process 1b input signals only, they can be hardware efficient.

The calibration is done at foreground in two phases. In phase 1, the frequency responses of the quantization noise transfers of the front-end stages to the ADC output are extracted and mapped to digital NCFs h_1-h_3, starting with calibration of h_3, then h_2 (with calibrated h_3) and lastly h_1 (with calibrated h_2 and h_3). The 6b back-end SAR ADC does not need calibration. As an example, consider the case where h_3 will be calibrated (x=3). In calibration mode, the ADC input is disconnected, and the dither/tone generator (DTG) generates a full-scale (±1/2 LSB amplitude) square wave test tone with programmable frequency that is applied to the input of stage-3 SAR ADC. Using test tones in a foreground calibration yields fast and accurate convergence, allowing fast recalibration (~1ms) for temperature tracking, in contrast to a (background) PRBS based calibration which can take seconds [7]. The SAR ADC output Y_3 and the back-end ADC output Y_4 are fed to the calibration circuit (Fig. 11.2.2). The tone is mixed down to DC with complex mixers and lowpass filtered to remove noise and undesired harmonics of the square wave. The gain and phase differences between vectors $V_B=I_B+jQ_B$ and $V_F=I_F+jQ_F$ represent the frequency transfer from stage-3 DAC to stage-4 at tone frequency. The control loop of the detector block extracts the correction vector V_k, that equalizes complex vectors V_B and $V_A=V_k*V_F$, where k is the frequency index of the test tone. After convergence, V_k contains the amplitude and phase transfer at the test tone frequency. This is done for 32 frequencies sequentially, evenly distributed in the band from 0 to 0.8GHz (Fs/2). The 32 complex vectors are stored in a register set H_3 which forms the fully extracted frequency transfer. Then, the extracted response H_3 is mapped to FIR NCF coefficient set h_3 by calculating the modified inverse DFT (IDFT) of the frequency responses according to $h_3=H2h*H_3$, where H2h is the (hard coded) modified inverse DFT matrix [8],[9]. For 32 frequency points and a filter length of 20 the IDFT requires 20×32 complex multiplications. As the complex vectors are stored in registers, the matrix multiplications are fully serialized, reducing filter mapping hardware to 1 complex multiplier and 1 adder.

In phase 2, the gain and timing errors of the first stage DAC are extracted and calibrated by adjusting the individual NCFs $h_{1,1}$-$h_{1,8}$. Filters $h_{1,1}$-$h_{1,8}$ are initialized with coefficients h_1 (from phase 1). Figure 11.2.3 shows a conceptional block diagram of the LMS-based gain and timing error calibration loop for an example 2-stage CTP ADC with 2b coarse ADC. The 2b binary output is converted to a 3b thermometer code, for which the order is fully randomized by the barrel shifter with random rotation-based binary-weighted selection (RRBS) [10]. The randomization ensures that on average data streams D_1-D_3 only correlate with the errors of their corresponding DAC element. The DAC elements generate current pulses with timing errors (t_1-t_3) and gain errors (D_1-D_3). Figure 11.2.3 (top) shows example DAC pulses with late and early timing and a gain error. The residue lowpass filter $h_1(s)$ shapes the DAC pulses to smoothed waveforms. Neglecting quantization noise, only the DAC-related gain/timing error contributions appear at the output Y_{out}. Figure 11.2.3 shows the sampled errors in case of late and early timing and in case of a gain error. For late timing, a negative error appears at $i+1$, while for early timing a negative error appears at $i-1$ (and a smaller positive error at $i+1$). Thus, timing errors of DAC element n can be found by correlating Y_{out} with delayed (z^{-1}) and inverted advanced (-z) replicas of D_n. Note that the imbalance between the magnitude of late/early timing errors does not impact the calibration accuracy. Also, convergence of the timing error calibration loop is highly insensitive to the presence of a gain error. After calibrating the timing error, the gain error remains, present at sampling moment i, and can be detected by correlating Y_{out} with D_n. In the actual implementation, timing and gain errors are calibrated simultaneously, even though a timing error also results in an error at sampling moment i that interferes with the gain error detection. As the LMS calibration loop minimizes the timing error and hence this cross-correlation error at i, it does not impact the final gain error calibration convergence and accuracy. The outputs of the correlators in Fig. 11.2.3 are fed to integrators in the calibration loop that converge to a correction vector $Vc_{n,k}$ which holds the accurate gain and timing information of DAC element n. Gain and timing error correction is done by alternating the complex vector V_k (for h_1 estimation) to $V_{n,k}$=$Vc_{n,k}$·V_k and computing $h_{1,1}$-$h_{1,8}$ according to Fig. 11.2.2. The correlation loops are fully serialized, minimizing hardware cost to 1 slice of the correlator block. Total time of calibration phases 1 and 2 at startup is 3ms.

The ADC is fabricated in TSMC 28nm CMOS (Fig. 11.2.7). The linearity measurements are done with 0.51V (-5dBFS) peak differential signal amplitude, which is the signal level that can be delivered by the front-end circuit with -100dBc THD linearity. Figure 11.2.4 shows the output spectrum with a single tone input at 66.1MHz. Measured THD is -95dBc. Figure

192 • 2026 IEEE International Solid-State Circuits Conference

979-8-3315-8937-0/26 $31.00 © 2026 IEEE

11.2.5 shows the measured two-tone spectrum with input signals at 217.1MHz and 233.05MHz. Intermodulation distortion tones are ≤-108dBFS. The measured DR and peak SNDR are 69.5dB in 270MHz BW. The SFDR (excluding noise floor in 1/f noise region) is -103dBFS (limited by HD2). The ADC, including on-chip supply LDOs and on-chip NCFs is 1.2mm². The off-chip calibration hardware (mixers, filters, IDFT, multipliers, registers, etc) is 0.2mm² (estimated). Total power consumption is 155mW. Measured performance is summarized in Fig. 11.2.6.

Acknowledgement:
The authors would like to thank Frank Siebler, Vinay Shenoy, Amol Dhok, Manikandan Panchapakesan, Surabhi B Krishnamurthy, Nico Morskieft, Jagadeesh Anisetti, Hans Brekelmans, Stephane Le Tual, Yann Carminati, Wei Kong, Mohammed Abo Alainein and Max Schuurmans. This result is part of the IPCEI ME/CT and is funded by the Dutch Ministry of Economic Affairs and Climate Policy.

References:
[1] M. Bolatkale *et al.*, "A 28nm 6GHz 2b Continuous-Time ΔΣ ADC with –101 dBc THD and 120MHz Bandwidth Using Digital DAC Error Correction," *2022 IEEE International Solid-State Circuits Conference (ISSCC)*, San Francisco, CA, USA, 2022, pp. 416-418. http://doi.org/10.1109/ISSCC42614.2022.9731602
[2] K. Xing, Y. Zhu, R. P. Martins and C. -H. Chan, "A 320MHz BW NS TD-ADC Assisted C/DT Hybrid Pipelined ADC with SAB-based Residue Amplifying Filter," *2024 IEEE Asian Solid-State Circuits Conference (A-SSCC)*, Hiroshima, Japan, 2024, pp. 1-3. http://doi.org/10.1109/A-SSCC60305.2024.10849229
[3] S. Patil et al., "A 700MHZ-BW –164dBFS/Hz-Small-Signal-NSD 703mW Continuous-Time Pipelined ADC with On-Chip Digital Reconstruction Achieving <-85dBFS HD3 using Digital Cancellation of DAC Errors," *2024 IEEE International Solid-State Circuits Conference (ISSCC)*, San Francisco, CA, USA, 2024, pp. 390-392. http://doi.org/10.1109/ISSCC49657.2024.10454477
[4] H. Zhao et al., "A 0.16mm² 450MHz-BW 72dB-SNDR Continuous-time Pipeline ADC with APF+HPF and APF+FIR Hybrid Delay Alignment Techniques," *2025 IEEE Custom Integrated Circuits Conference (CICC)*, Boston, MA, USA, 2025, pp. 1-3. http://doi.org/10.1109/CICC63670.2025.10983217
[5] R. A. Wannamaker, S. P. Lipshitz, J. Vanderkooy and J. N. Wright, "A theory of nonsubtractive dither," *in IEEE Transactions on Signal Processing*, vol. 48, no. 2, pp. 499-516, Feb. 2000. http://doi.org/10.1109/78.823976
[6] S. Loeda, J. Harrison, F. Pourchet and A. Adams, "A 10/20/30/40 MHz Feedforward FIR DAC Continuous-Time ΔΣ ADC With Robust Blocker Performance for Radio Receivers," *in IEEE Journal of Solid-State Circuits*, vol. 51, no. 4, pp. 860-870, April 2016. http://doi.org/10.1109/JSSC.2016.2519395
[7] Y. Dong et al., "Adaptive digital noise-cancellation filtering using cross-correlators for continuous-time MASH ADC in 28nm CMOS," *2017 IEEE Custom Integrated Circuits Conference (CICC)*, Austin, TX, USA, 2017, pp. 1-4. http://doi.org/10.1109/CICC.2017.7993658
[8] L. Rabiner and R. Schafer, "Recursive and nonrecursive realizations of digital filters designed by frequency sampling techniques," *in IEEE Transactions on Audio and Electroacoustics*, vol. 19, no. 3, pp. 200-207, September 1971. http://doi.org/10.1109/TAU.1971.1162185

[9] Yuheng He, K. Hueske, J. Götze and E. Coersmeier, "A matrix-vector based approach to FFT implementations," *2009 IEEE International Symposium on Signal Processing and Information Technology (ISSPIT)*, Ajman, United Arab Emirates, 2009, pp. 490-494. http://doi.org/10.1109/ISSPIT.2009.5407501
[10] W. -T. Lin and T. -H. Kuo, "A Compact Dynamic-Performance-Improved Current-Steering DAC With Random Rotation-Based Binary-Weighted Selection," *in IEEE Journal of Solid-State Circuits*, vol. 47, no. 2, pp. 444-453, Feb. 2012. http://doi.org/10.1109/JSSC.2011.2168651
[11] J. Edler, M. Runge, S. Linnhoff and F. Gerfers, "A 4.4 GS/s 220 MHz ΣΔ ADC with a Linearized Back-Gate Controlled GmC Filter," *2023 IEEE Symposium on VLSI Technology and Circuits (VLSI Technology and Circuits)*, Kyoto, Japan, 2023, pp. 1-2. http://doi.org/10.23919/VLSITechnologyandCir57934.2023.10185281

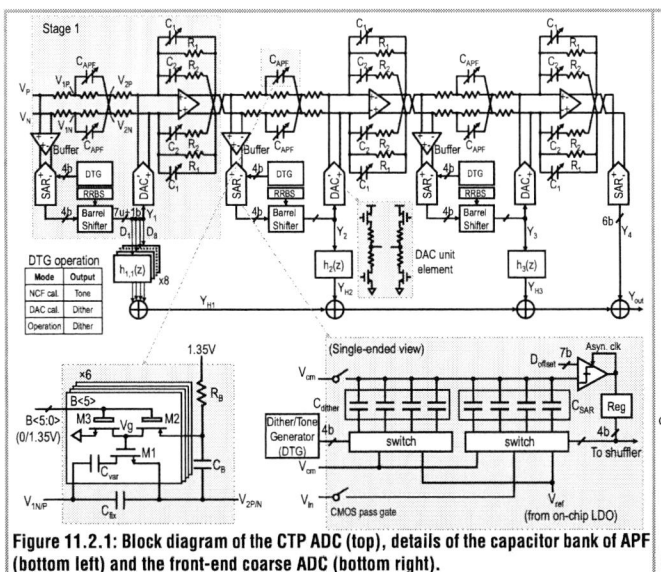

Figure 11.2.1: Block diagram of the CTP ADC (top), details of the capacitor bank of APF (bottom left) and the front-end coarse ADC (bottom right).

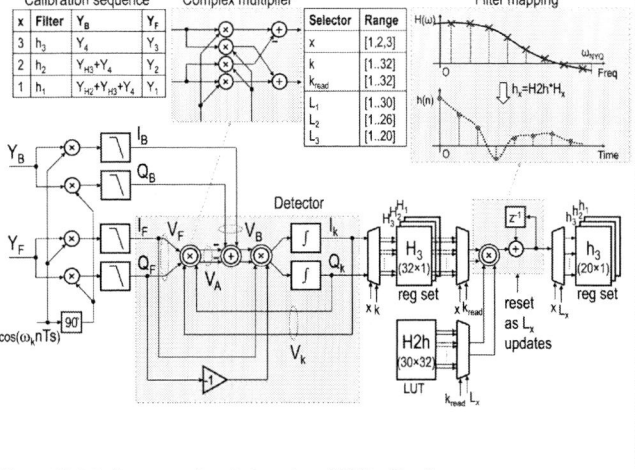

Figure 11.2.2: Frequency-domain tone-based NCF calibration.

ISSCC 2026 / SESSION 11 / PIPELINE AND ULTRA-HIGH-SPEED DATA CONVERTERS / 11.2

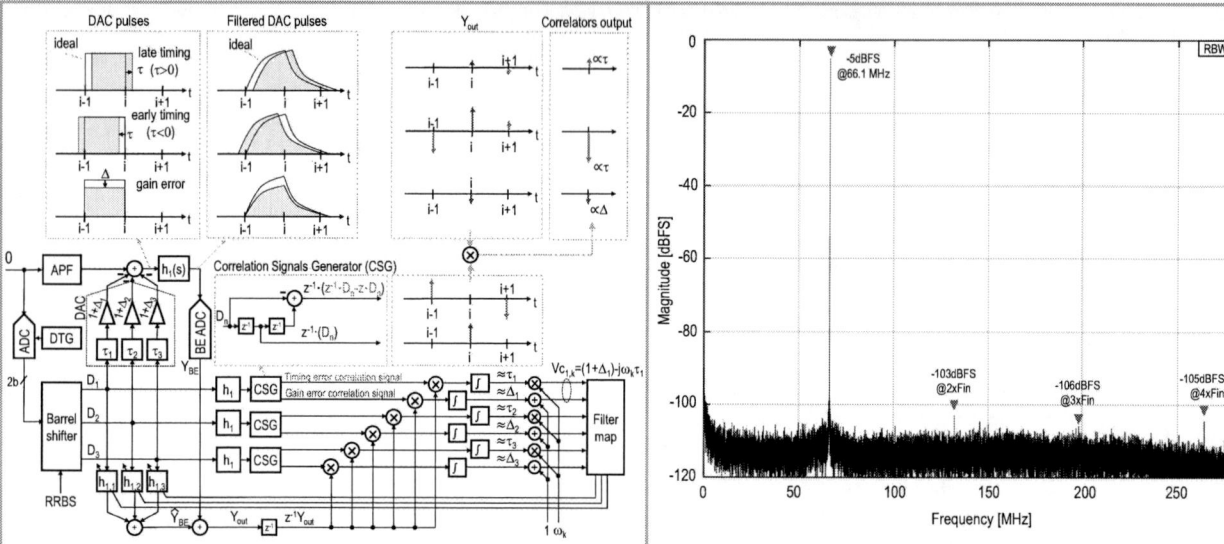

Figure 11.2.3: Time-domain correlation-based DAC gain and timing error calibration.

Figure 11.2.4: Single-tone measurement.

Figure 11.2.5: Two-tone measurement.

Figure 11.2.6: Benchmark of CTP and CTDS ADCs.

	This work	[3]	[2]	[4]	[1]	[11]
Technology (nm)	28	16	65	28	28	22
Architecture	CTP	CTP	C/DTP	CTP	CT DSM	CT DSM
FE / BE	SAR / SAR	Flash / VCO	NS TDC / TI SAR	Flash / VCO	DSM / -	DSM / -
Sub-stage ADC / IG*(V/V)	4b / 4	4.1b / 6	4b / 7	4.1b / 6.7	2b / -	4b / -
NCF estimation	Foreground	Background	Foreground	n.a.	n.a.	n.a.
DAC estimation	Foreground	Foreground	Foreground	n.a.	Background	n.a.
Vdd (V)	0.9/1.5	1/1.8	n.a.	1	0.9/1.5	-0.5/0.9
Full Scale Input (Vdiff-pk)	0.9	1	n.a.	1	0.64	0.73
Fs (GHz)	1.6	6.4	2.4	4	6	4.4
SNDR (dB)	69.5	71	65.4	71.8	72.3	62
Bandwidth (MHz)	270	700	320	450	120	220
Fin@ THD (MHz)	66.1	228	53	21	38.8	2
THD (dBc)	-95	-76.6	-75.4	-76.6	-101.1	-73.6
Fin@IMx (MHz)	217&233	560&580	245&255	386&415	88&94	209&211
IMx, worst case (dBFS)	-108	-79	-77	-74	-111.7	-81.5
SFDR (dBFS)	103	78	76.1	77	114***	75
Power (mW) analog	100	355	56	329	108.8	22
Power (mW) digital	55	348	n.a.	n.a.	n.a.	n.a.
Area (mm²)	1.4	3	0.081**	0.16**	0.13**	0.06**
FoMs_SNDR (dB)	162	161	163	163	162	162

* Inter-stage gain of the first sub-stage: from ADC input to the second sub-stage quantizer input.
** Analog core area only
*** Non-harmonic SFDR

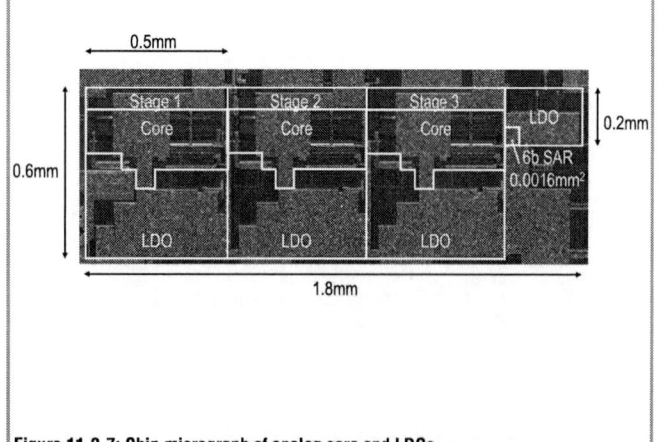

Figure 11.2.7: Chip micrograph of analog core and LDOs.

• 2026 IEEE International Solid-State Circuits Conference

979-8-3315-8937-0/26 $31.00 © 2026 IEEE

ISSCC 2026 / SESSION 11 / PIPELINE AND ULTRA-HIGH-SPEED DATA CONVERTERS / 11.3

11.3 A 500MS/s 12b Pipe-SAR ADC Using a Triple-Cascode FIA with Virtual Supply Extension

Michele Rocco*[1], Gabriele Zanoletti*[1], Alessia Ceroni*[1], Giacomo Tombolan[1], Gabriele Bè[1,2], Luca Ricci[1], Salvatore Levantino[1], Andrea Leonardo Lacaita[1], Luca Bertulessi[1], Carlo Samori[1], Andrea Giovanni Bonfanti[1]

[1]Politecnico di Milano, Milan, Italy, [2]now with Infineon Technologies, Villach, Austria
*Equally Credited Authors (ECAs)

Abstract

This paper presents a 28nm CMOS 500MS/s 12b three-stage pipeline SAR ADC based on a triple-cascode single-stage floating residue amplifier with virtual supply extension to improve its efficiency and linearity. The ADC achieves 63.4dB SNDR and 81.0dB SFDR at 240MHz input across PVT variations, consuming 6.64mW with on-chip digital calibrations and occupying 0.011mm².

Over the past decade, the pipeline-SAR converter has emerged as a hybrid architecture in which each stage is implemented as a SAR ADC. In this context, interstage residue amplifiers (RAs) play a crucial role, as they must meet stringent speed, linearity, and noise requirements. Their design is challenging, and they can account for up to 40% of the overall ADC power consumption [1–4].

Thanks to their energy efficiency and scalability, ring amplifiers have become a widely adopted solution despite some inherent limitations (Fig. 11.3.1, top) [5]. This circuit topology exhibits three high-impedance nodes, leading to a frequency response with three poles [6–7]. The last stage, biased in the subthreshold regime, determines the dominant pole and thereby the gain–bandwidth product (GBWP), while the phase margin is jointly set by the poles of the first and second stages. As a result, in high-speed ADCs the ring amplifier suffers from a bandwidth–stability trade-off: pushing the GBWP close to the internal poles lowers the phase margin, causing overshoot and ringing that may lead to distortion [7–8]. Moreover, a differential ring amplifier requires additional CMFB and biasing circuitry [5], reducing its power efficiency. Finally, the last-stage bias is highly sensitive to mismatch and PVT variations, which calls for digitally intensive nonlinearity calibration [8–10].

To address this last issue, PVT-robust architectures have been proposed, exploiting dynamic dead-zone with current biasing [11] or dynamic current-mirror techniques [12], at the cost of further reducing amplifier efficiency. To improve efficiency, floating inverter amplifier (FIA)-based RAs have been recently introduced using either a hybrid solution with a static inverter followed by two FIAs [3] or employing three FIAs in cascade [4]. However, these architectures show the same bandwidth–stability issue as ring amplifiers, featuring three high-resistive nodes. Sufficient phase margin is reached by carefully sizing the reservoir capacitors and the transistors in the three stages so that the GBWP is pushed below the non-dominant poles at the end of the transient. It follows that, despite recent advances, the design of an RA that combines high efficiency, robust PVT behavior, and stability remains critical in pipeline converters. In this work, we propose a closed-loop triple-cascode FIA (Fig. 11.3.1, bottom). The amplifier, implemented in a 500MS/s 12b 3-stage pipeline-SAR ADC, leverages a virtual-supply extension technique to deliver wide swing at the output node, high linearity (THD=−75dB), strong PVT robustness, and high efficiency, with an FoM (energy noise-power product [13]) of 12.9 nJ·(μV)².

Conceptually, a single-stage cascode FIA can solve the stability-bandwidth issue. It features only a single high-resistive node, which sets the dominant pole, whereas the secondary poles lie at much higher frequencies. Thus, either the GBWP can be increased, or, at a comparable bandwidth, loop stability is relaxed. However, it suffers from a limited output swing, which degrades linearity and prevents stacking more than four transistors between the supply rails, thereby limiting the gain achievable with a single stage. The proposed amplifier avoids these issues by implementing a triple-cascode amplifier with switched battery capacitors C_{bat}. These capacitors act as voltage generators that properly shift the voltage of internal nodes to recover the output-node swing and high linearity. The voltage across every two stacked transistors is V_{dd}, thus virtually extending the supply of the whole stage to 3 V_{dd}, while keeping each node voltage within its maximum rating.

The equivalent common-mode half-circuit, Fig. 11.3.2 (top), illustrates the bias condition of the circuit nodes in the two phases. In the reset phase (Φ_{amp}=0), the two battery capacitors (C_{bat}) and the reservoir one (C_{res}/2) are pre-charged to V_{dd} and the output nodes to V_{cm}=V_{dd}/2. To avoid current consumption, the gates of the innermost cascode transistors $M_{C2,p}$ and $M_{C2,n}$ are tied to V_{dd} and GND, respectively. In the amplification phase (Φ_{amp}=1), the gates of $M_{C2,p}$ and $M_{C2,n}$ are connected to V_{cm}, and C_{res}/2 is tied to the amplifier. In the available technology, both PMOS and NMOS transistors are electrically equivalent for the same (W/L) ratio. Therefore, when the common-mode current discharges C_{res}/2 to V_{dd}−2·ΔV(t), the variations of the $M_{1,n}$ and $M_{1,p}$ sources are the same, i.e. ΔV(t), in opposite directions. Besides, since all devices have equal W/L and carry the same current, they must show the same source voltage variation. Finally, being C_{bat}=C_{res}, for charge conservation the voltage on each C_{bat} is V_{dd}−ΔV(t). Node voltages, shown in red, are thus defined and all devices are well in saturation. The advantage of this solution is further detailed in Fig. 11.3.2 (bottom). In a conventional cascode FIA (left), M_{C1} gate is tied to V_{cm}, pushing M_1 out of the saturation region. Either using a low-V_T transistor M_{C1} or raising its gate voltage helps keep M_1 in saturation, but reduces V_{out} swing. In the proposed scheme (right), C_{bat} acts as a battery. The

M_{C1} gate is set to V_{dd} to keep M_1 in saturation while M_{C1} is also in saturation since its drain is shifted to ≈V_{dd} by C_{bat}. Finally, thanks to the proposed virtual supply extension, it is possible to stack another cascode transistor, M_{C2}, and increase the gain from $(g_m r_o)^2$ to $(g_m r_o)^3$.

Figure 11.3.3 shows the architecture of the pipeline-SAR ADC embedding the proposed RA in a closed-loop configuration with a time diagram of the main signals. The first stage resolves 4 bits, the RA gain of 8 ensures 1b (V_{lsb}/2~60 mV) of interstage redundancy for the absorption of conversion errors. The V_{lsb}/4-amplitude dithering signal for the interstage gain estimation is injected into the capacitive DAC (CDAC) with a dedicated capacitor C_{dith} equal to the unitary capacitance, C_u=10.3fF. To improve linearity, the signal is sampled using bootstrapped switches on the bottom plates of an input single-ended 320fF capacitance (equivalent to 11.5-ENOB), and a loop-unrolled SAR logic [14] has been chosen for a faster quantization using the V_{cm}-based switching algorithm [15]. The second stage of the pipeline-SAR ADC is a downscaled version of the first one. It does not employ bootstrapped switches, and the single-ended CDAC capacitance is reduced to 30fF. The third stage resolves 6 bits using a split-monotonic switching scheme on the 16fF CDAC and a conventional SAR logic. The three stages are asynchronously controlled by implementing a handshake protocol [16–17].

The 14b raw data are initially used to compensate the comparators offset in the first two stages using a redundancy-based algorithm [18–19]. While the pipeline architecture inherently compensates for this error, calibrating the offsets restores the nominal interstage redundancy range and centers the residue around ±V_{lsb}/2 for optimal RA linearity. The data are then corrected for radix errors due to CDAC mismatches and recombined with the interstage gains G_1 and G_2 to compose an 18b word with 4 fractional bits. Thanks to the high linearity of the RA, even a simple on-chip LMS loop can be used to update the interstage gain estimates. Only the radix errors are estimated once off-chip in the foreground using a sine fit optimization algorithm and then stored on chip for bit-weight correction.

The first stage RA has been designed for a THD lower than −60dB at half the output range, which corresponds from simulations to an ENOB=11.7, considering just the effect of the RA non linearity on the quantization noise, and for an input-referred noise comparable to the sampling one. Figure 11.3.4 (top left) shows the simulated THD vs. the supply voltage (±10% variation) for different temperatures (−40 °C to 80 °C) and process corners. Without any trimming, the THD remains below the −60dB constraint. The figure also shows the amplifier's 95% settling time across PVT variations. Under nominal conditions (V_{dd}=1V, TT corner and T=27°C), the RA settles in 250ps (for a target amplification time of 500ps) with a THD=−78dB and an input-referred noise of 116μVrms. Considering a 0.96pJ energy consumption per amplification, the amplifier features a FoM of 12.9nJ·(μV)². The table in Fig. 11.3.4 (top right) compares the proposed amplifier performance with other RAs in literature, showing that the implemented RA achieves the best FoM while operating at 500MS/s. As the 3-stage FIA in [4] our RA does not require auxiliary circuitry and with respect to [3] achieves a 2× better efficiency even working at a 2.5× larger sampling rate.

The prototype has been fabricated in a 28nm CMOS supplied at 1V except for the reference buffer, which is biased at 1.2V. It occupies an active area of 0.011mm² (Fig. 11.3.7). Four chips have been characterized; for each of them, the radix coefficients have been estimated once at nominal supply voltage (V_{dd}=1V) and room temperature, and kept constant over all the following measurements. Figure 11.3.4 (bottom) and Fig. 11.3.5 (top/middle) show the measured ADC performance. Without any RA linearity calibration, the prototype achieves an SNDR of 63.5dB and an SFDR of 79.2dB at low input frequency, and an SNDR of 63.4dB and an SFDR of 81.0dB near Nyquist (Fig. 11.3.4 bottom). Measurements on four chips confirm that performance remains stable across the entire input frequency range (Fig. 11.3.4 bottom and Fig. 11.3.5 top), with SNDR variation limited to 0.9dB over a 250MHz sampling frequency range (Fig. 11.3.5 middle). Measurements in Fig. 11.3.5 (bottom) demonstrate the linearity of the proposed RA across voltage and temperature variations. In the prototype, the RA supply can be varied while keeping the supply of the other blocks constant. The overall ADC performance in terms of SNDR and SFDR is unaffected, highlighting the robustness of the RA.

Figure 11.3.6 provides a detailed power breakdown of the analog blocks and the synthesized digital section. Excluding debug and communication features, the pipelined-SAR ADC consumes a total of 6.64mW, with 2.69mW dissipated by the core and 3.95mW by the digital section. The digital power accounts for the on-chip RA gain estimation (1mW), comparator offset calibration (0.78mW), and pipeline recombination (1.65mW). Since most literature references report these functions as off-chip, we quote them separately for a fair comparison. Based on this, Fig. 11.3.6 also compares the prototype performance with state-of-the-art pipeline-SAR ADCs. The Schreier FoM is 173.1dB when excluding on-chip digital calibrations and corrections but including the reference buffer; it becomes 169.2dB when considering the total power consumption. This FoM, in line with the state of the art, is achieved with stable performance across PVT variations.

Acknowledgment:
The authors thank Infineon Technologies Villach for partially funding this research.

References:
[1] M. Zhan, L. Jie, X. Tang, Y. Zhong and N. Sun, "A 0.004-mm2 200-MS/s Pipelined SAR ADC With kT/C Noise Cancellation and Robust Ring-Amp," *in IEEE Journal of Solid-State Circuits*, vol. 59, no. 7, pp. 2209-2218, July 2024, doi: 10.1109/JSSC.2023.3344461. http://doi.org/10.1109/JSSC.2023.3344461.
[2] Y. Shen et al., "A 12-bit 1.5-GS/s Single-Channel Pipelined SAR ADC With a Pipelined Residue Amplification Stage," *in IEEE Journal of Solid-State Circuits*, vol. 60, no. 1, pp. 260-271, Jan. 2025, doi: 10.1109/JSSC.2024.3412090. http://doi.org/10.1109/JSSC.2024.3412090
[3] S. Ye et al., "A 2-mW 70.7-dB SNDR 200-MS/s Pipelined-SAR ADC Using Continuous-Time SAR-Assisted Detect-and-Skip and Open-Then-Close Correlated Level Shifting," *in IEEE Journal of Solid-State Circuits*, vol. 60, no. 7, pp. 2581-2594, July 2025, doi: 10.1109/JSSC.2024.3497175. http://doi.org/10.1109/JSSC.2024.3497175
[4] X. Tang et al., "A Bandwidth-Adaptive Pipelined SAR ADC With Three-Stage Cascoded Floating Inverter Amplifier," *in IEEE Journal of Solid-State Circuits*, vol. 58, no. 9, pp. 2564-2574, Sept. 2023, doi: 10.1109/JSSC.2023.3268719. http://doi.org/10.1109/JSSC.2023.3268719
[5] B. Hershberg, S. Weaver, K. Sobue, S. Takeuchi, K. Hamashita, and U.-K. Moon, "Ring Amplifiers for Switched Capacitor Circuits," *IEEE J. Solid-State Circuits*, vol. 47, no. 12, pp. 2928–2942, Dec. 2012, doi: 10.1109/JSSC.2012.2217865. http://doi.org/10.1109/JSSC.2012.2217865
[6] K. M. Megawer, F. A. Hussien, M. M. Aboudina and A. N. Mohieldin, "A Systematic Design Methodology for Class-AB-Style Ring Amplifiers," *in IEEE Transactions on Circuits and Systems II: Express Briefs,* vol. 65, no. 9, pp. 1169-1173, Sept. 2018, doi: 10.1109/TCSII.2018.2815705. http://doi.org/10.1109/TCSII.2018.2815705
[7] J. Conrad, P. Vogelmann, M. A. Mokhtar and M. Ortmanns, "Design Approach for Ring Amplifiers," *in IEEE Transactions on Circuits and Systems I: Regular Papers,* vol. 67, no. 10, pp. 3444-3457, Oct. 2020, doi: 10.1109/TCSI.2020.2986553. http://doi.org/10.1109/TCSI.2020.2986553
[8] J. Lagos et al., "A 10.1-ENOB, 6.2-fJ/conv.-step, 500-MS/s, Ringamp-Based Pipelined-SAR ADC With Background Calibration and Dynamic Reference Regulation in 16-nm CMOS," *in IEEE Journal of Solid-State Circuits*, vol. 57, no. 4, pp. 1112-1124, April 2022, doi: 10.1109/JSSC.2021.3133829. http://doi.org/10.1109/JSSC.2021.3133829
[9] B. Hershberg et al., "3.1 A 3.2GS/s 10 ENOB 61mW Ringamp ADC in 16nm with Background Monitoring of Distortion," *2019 IEEE International Solid-State Circuits Conference - (ISSCC),* San Francisco, CA, USA, 2019, pp. 58-60, doi: 10.1109/ISSCC.2019.8662290. http://doi.org/10.1109/ISSCC.2019.8662290
[10] B. Hershberg, N. Markulić, J. Lagos, E. Martens, D. Dermit and J. Craninckx, "A 1-MS/s to 1-GS/s Ringamp-Based Pipelined ADC With Fully Dynamic Reference Regulation and Stochastic Scope-on-Chip Background Monitoring in 16 nm," *in IEEE Journal of Solid-State Circuits,* vol. 56, no. 4, pp. 1227-1240, April 2021, doi: 10.1109/JSSC.2020.3044831. http://doi.org/10.1109/JSSC.2020.3044831

[11] Y. Cao et al., "24.2 A 14b 1GS/s Single-Channel Pipelined ADC with A Parallel-Operation SAR Sub-Quantizer and A Dynamic-Deadzone Ring Amplifier," *2025 IEEE International Solid-State Circuits Conference (ISSCC),* San Francisco, CA, USA, 2025, pp. 430-432, doi: 10.1109/ISSCC49661.2025.10904766. http://doi.org/10.1109/ISSCC49661.2025.10904766
[12] Y. Lim et al., "9.2 A 2.08mW 64.4dB SNDR 400MS/s 12b Pipelined-SAR ADC using Mismatch and PVT Variation Tolerant Dynamically Biased Ring Amplifier in 8nm," *2024 IEEE International Solid-State Circuits Conference (ISSCC),* San Francisco, CA, USA, 2024, pp. 170-172, doi: 10.1109/ISSCC49657.2024.10454422. http://doi.org/10.1109/ISSCC49657.2024.10454422
[13] X. Tang et al., "An Energy-Efficient Comparator With Dynamic Floating Inverter Amplifier," *in IEEE Journal of Solid-State Circuits,* vol. 55, no. 4, pp. 1011-1022, April 2020, doi: 10.1109/JSSC.2019.2960485. http://doi.org/10.1109/JSSC.2019.2960485
[14] G. V. d. Plas and B. Verbruggen, "A 150 MS/s 133 µW 7 bit ADC in 90 nm Digital CMOS," *in IEEE Journal of Solid-State Circuits,* vol. 43, no. 12, pp. 2631-2640, Dec. 2008, doi: 10.1109/JSSC.2008.2006315. http://doi.org/10.1109/JSSC.2008.2006315
[15] Y. Zhu et al., "A 10-bit 100-MS/s Reference-Free SAR ADC in 90 nm CMOS," *in IEEE Journal of Solid-State Circuits,* vol. 45, no. 6, pp. 1111-1121, June 2010, d oi: 10.1109/JSSC.2010.2048498. http://doi.org/10.1109/JSSC.2010.2048498
[16] B. Vaz et al., "16.1 A 13b 4GS/s digitally assisted dynamic 3-stage asynchronous pipelined-SAR ADC," *2017 IEEE International Solid-State Circuits Conference (ISSCC),* San Francisco, CA, USA, 2017, pp. 276-277, doi: 10.1109/ISSCC.2017.7870368. http://doi.org/10.1109/ISSCC.2017.7870368
[17] B. Hershberg et al., "Asynchronous Event-Driven Clocking and Control in Pipelined ADCs," *in IEEE Transactions on Circuits and Systems I: Regular Papers,* vol. 68, no. 7, pp. 2813-2826, July 2021, doi: 10.1109/TCSI.2021.3077881. http://doi.org/10.1109/TCSI.2021.3077881
[18] K. Bunsen, E. Martens, D. Dermit and J. Craninckx, "A Redundancy-Based Background Calibration for Comparator Offset/Threshold and DAC Gain in a Ping-Pong SAR ADC," *in IEEE Transactions on Circuits and Systems II: Express Briefs,* vol. 68, no. 2, pp. 592-596, Feb. 2021, doi: 10.1109/TCSII.2020.3046091. http://doi.org/10.1109/TCSII.2020.3046091
[19] C. Nani et al., "A 5-nm 60-GS/s 7b 64-Way Time Interleaved Partial Loop Unrolled SAR ADC Achieving 35.2dB SNDR up to 32 GHz," *in IEEE Journal of Solid-State Circuits,* vol. 60, no. 4, pp. 1210-1222, April 2025, doi: 10.1109/JSSC.2024.3517333. http://doi.org/10.1109/JSSC.2024.3517333
[20] J. Lagos et al., "A 10.0 ENOB, 6.2 fJ/conv.-step, 500 MS/s Ringamp-Based Pipelined-SAR ADC with Background Calibration and Dynamic Reference Regulation in 16nm CMOS," *2021 Symposium on VLSI Circuits, Kyoto, Japan,* 2021, pp. 1-2, doi: 10.23919/VLSICircuits52068.2021.9492354. http://doi.org/10.23919/VLSICircuits52068.2021.9492354
[21] C. Chen, Z. Yuan, P. Cao, J. Xu and Z. Hong, "A 71.5-dB SNDR 475-MS/s Ringamp-Based Pipelined SAR ADC with On-Chip Bit-Weight Calibration," *2024 IEEE Symposium on VLSI Technology and Circuits (VLSI Technology and Circuits),* Honolulu, HI, USA, 2024, pp. 1-2, doi: 10.1109/VLSITechnologyandCir46783.2024.10631448. http://doi.org/10.1109/VLSITechnologyandCir46783.2024.10631448
[22] Y. Lim et al., "9.2 A 2.08mW 64.4dB SNDR 400MS/s 12b Pipelined-SAR ADC using Mismatch and PVT Variation Tolerant Dynamically Biased Ring Amplifier in 8nm," *2024 IEEE International Solid-State Circuits Conference (ISSCC),* San Francisco, CA, USA, 2024, pp. 170-172, doi: 10.1109/ISSCC49657.2024.10454422. http://doi.org/10.1109/ISSCC49657.2024.10454422
[23] X. He, M. Gu, H. Jiang, Y. Zhong, N. Sun and L. Jie, "9.3 A 71dB SNDR 200MHz BW Interleaved Pipe-SAR ADC with a Shared Residue Integrating Amplifier Achieving 173dB FoMs," *2024 IEEE International Solid-State Circuits Conference (ISSCC),* San Francisco, CA, USA, 2024, pp. 172-174, doi: 10.1109/ISSCC49657.2024.10454431. http://doi.org/10.1109/ISSCC49657.2024.10454431

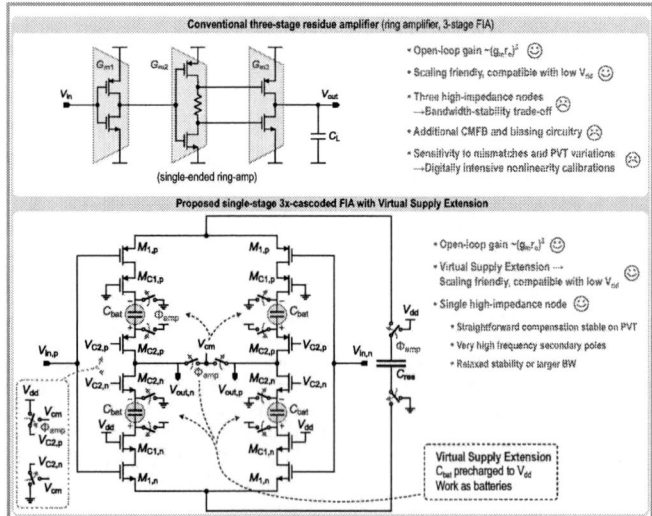

Figure 11.3.1: Comparison of the conventional ring amplifier (top) with the proposed 3×-cascoded FIA with virtual supply extension (bottom).

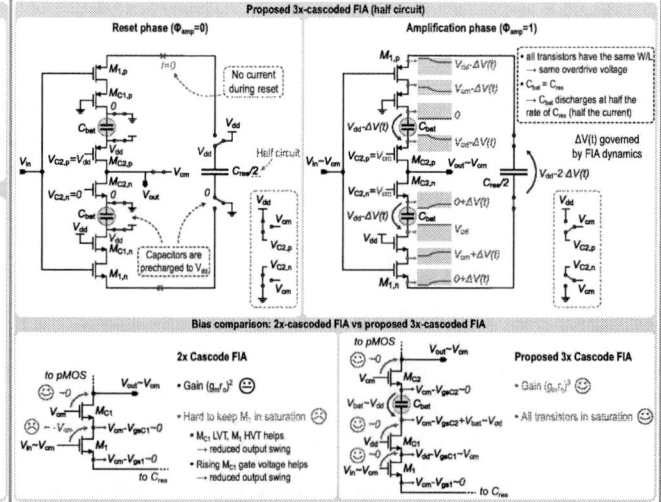

Figure 11.3.2: Operating phases of the proposed 3×-cascoded FIA (top) and bias comparison with the 2×-cascoded FIA (bottom).

ISSCC 2026 / SESSION 11 / PIPELINE AND ULTRA-HIGH-SPEED DATA CONVERTERS / 11.3

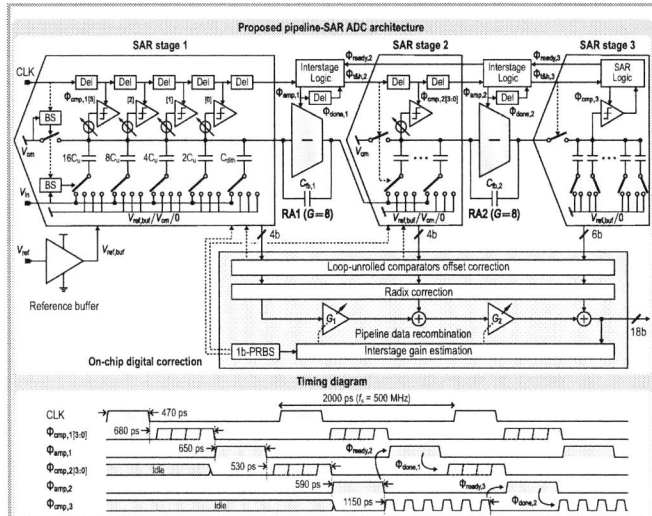

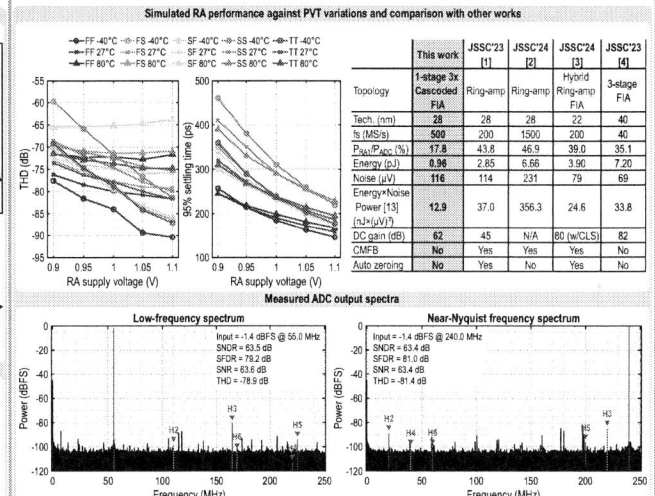

Figure 11.3.3: Proposed pipeline-SAR ADC architecture and timing diagram.

Figure 11.3.4: Simulation results and comparison of the proposed RA (top) and measured spectra of the prototype ADC (bottom).

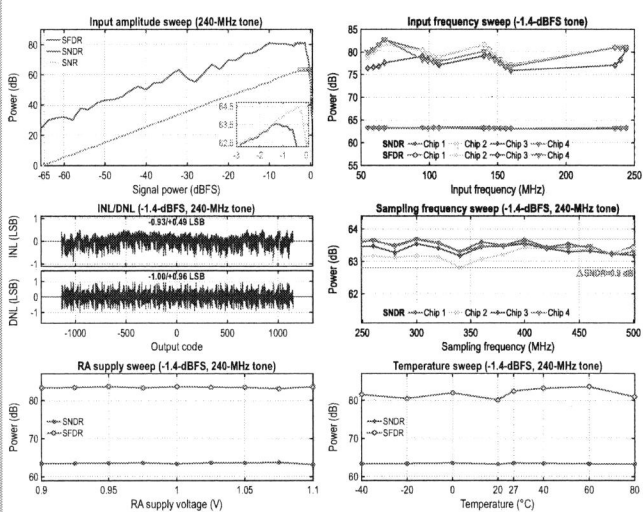

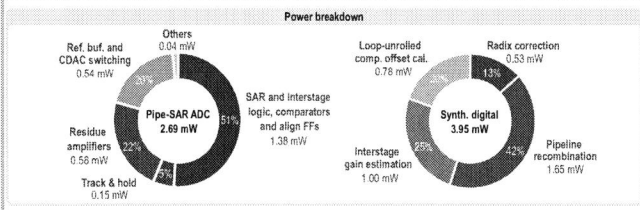

Figure 11.3.5: Measurement results of the prototype pipeline-SAR ADC.

Figure 11.3.6: Power breakdown of the ADC prototype and comparison with state-of-the-art pipeline-SAR ADCs with similar input bandwidth.

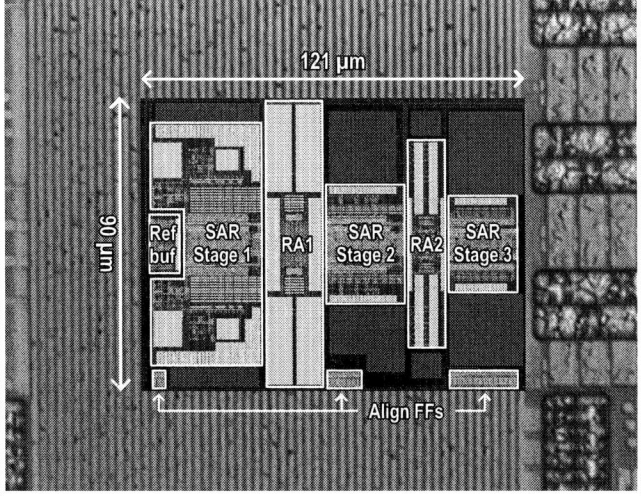

Figure 11.3.7: Die microphotograph with layout details of the pipeline-SAR ADC.

• 2026 IEEE International Solid-State Circuits Conference

979-8-3315-8937-0/26 $31.00 © 2026 IEEE

ISSCC 2026 / SESSION 11 / PIPELINE AND ULTRA-HIGH-SPEED DATA CONVERTERS / 11.4

11.4 A 13b 500MS/s 94dB-SFDR Resistive-Input Pipelined-SAR ADC with Linear and Efficient Current-Buffer-Based Integrating Sampler

Xiyu He, Mingyang Gu, Yunsong Tao, Siyu Huang, Zhishuai Zhang, Yi Zhong, Nan Sun, Lu Jie

Tsinghua University, Beijing, China

Abstract

This work proposes a pipelined-SAR ADC featuring a current-buffer-based integrating sampler that presents a resistive input with good linearity. A floating charge transferrer (FCT) with multi-bit pre-conversion (MB-PC) enables linear residue amplification and accelerates the second-stage conversion. The prototype ADC achieves an SNDR of 67.5dB and an SFDR of 94.5dB under a Nyquist input at 250MHz. This SFDR is the highest reported for discrete-time ADCs operating above 200MS/s.

Wideband ADCs have advanced significantly at both the architecture and circuit levels. As the performance and energy efficiency of ADC cores continue to improve, the frontend often emerges as the system bottleneck. Figure 11.4.1 summarizes the representative ADC frontend structures. Conventional Nyquist ADCs with switched-capacitor (SC) sampling frontends impose a large capacitive load that necessitates a power-hungry input buffer [1-3]. Additional filters are also indispensable to address aliasing, incurring extra power costs. Continuous-time (CT) delta-sigma modulators (DSMs) offer a resistive input that is easy to drive and provides inherent anti-aliasing capability. However, their energy efficiency and bandwidth are limited due to their closed-loop nature [4-5]. The current integrating sampler (CIS) [6-9] offers an alternative to drive the sampling capacitor of Nyquist ADCs, where a transconductance (g_m) cell converts the input voltage into current and integrates it onto the sampling capacitor. The high-impedance transistor gate input eases driving requirements, and the integration provides anti-aliasing filtering. However, the g_m cell suffers from poor linearity and high sensitivity to PVT variations, which has limited previous implementations to a maximum SFDR of 65dB [7].

This work proposes a current-buffer-based integrating sampler, as illustrated in Fig. 11.4.1. The input voltage is converted to a current by a linear resistor, buffered, and integrated onto the sampling capacitor of a pipelined-SAR ADC. Owing to the low input impedance of the current buffer, the voltage-to-current conversion remains highly linear, and the ADC presents a constant resistive input. This approach combines the high efficiency of a Nyquist quantizer with an easy-to-drive resistive input and anti-aliasing capability, while improving input linearity by more than 25dB over prior CIS implementations. The prototype pipelined-SAR ADC achieves 94.5dB SFDR under a Nyquist input at 500MS/s. This represents the highest reported SFDR among discrete-time (DT) ADCs operating above 200MS/s.

The proposed current-buffer-based integrating sampler is detailed in Fig. 11.4.2, enabled by three key techniques:
(1) A gain-boosted continuous-time floating charge transferrer (CT-FCT) implements the current buffer, providing low input impedance and high transconductance linearity;
(2) An MSB pre-conversion technique doubles the effective sampler swing, enabling a rail-to-rail equivalent input signal swing on the CDAC at core supply voltage;
(3) A buffer-connected conversion technique improves linearity by draining the parasitic charge from the buffer output node during the first-stage conversion.

In the proposed scheme, the input voltage (V_{in}) is converted to a current (I_{in}) by R_{in}, then buffered and integrated onto the sampling capacitor (C_{DAC}). A conventional common-gate amplifier could implement the current buffer, but its high noise makes it unsuitable for this frontend application. Recently, the floating charge transferrer (FCT) was proposed in [10] to implement low-noise and power-efficient common-gate amplifiers, where a floating capacitor (C_{res}) suppresses bias noise, and a folded-cascode stage enhances output impedance without degrading swing. In this work, we extend the FCT from a DT amplifier to a CT current buffer. As a DT residue amplifier, the FCT input sees a capacitor, thus its input resistance (R_{FCT}) only needs to satisfy settling requirements. In contrast, the CT-FCT for current buffering encounters a resistive input source, where the input current equals $V_{in}/(R_{in} + R_{FCT})$. Therefore, a small R_{FCT} is essential for maintaining voltage-to-current linearity. Gain-boosting is applied to the input transistors, reducing R_{FCT} by a factor of (A+1) and creating a stable virtual ground at the CT-FCT input. Simulation results confirm that input-current linearity improves by more than 10dB after introducing gain-boosting. Since the sampling switch (φ_1) connects to a virtual ground, its linearity requirement is also significantly relaxed. An auxiliary switch ($\overline{\varphi_1}$) is introduced to maintain a constant input impedance outside the sampling phase.

Two additional factors affect the linearity of the proposed integration sampler. First, the output swing of CT-FCT is limited by the reservoir cap voltage and the $V_{ds,sat}$ of the cascoding transistors, which is conventionally solved by increasing supply voltage and power. In this work, we apply MSB pre-conversion [11] to mitigate this swing limitation: the MSB is resolved and fed back into the C_{DAC} at the midpoint of the sampling phase, which halves the integrator output swing and effectively doubles the available sampler gain. As a result, an equivalent full-swing input signal can be sampled on the C_{DAC} with a core-voltage supply. Second, when the output sampling switch (φ_2) opens, the nonlinear parasitic capacitor (C_{par}) at the buffer output draws error charge from the C_{DAC}, severely degrading sampling linearity.

To address this issue, we propose a buffer-connected conversion technique: the buffer output sampling switch (φ_2) remains closed throughout the sampling and conversion phase, and opens only after the conversion completes. The buffer input sampling switch (φ_1) closes only during the sampling phase, which defines the integration window. In this way, C_{par} remains connected to the C_{DAC} during the first-stage SAR operation, and the voltage on C_{par} is gradually reduced to V_{cm}. Consequently, the nonlinear charge on C_{par} is drained into the C_{DAC}, leaving only negligible error charge at the end of conversion. This effectively eliminates the nonlinearity induced by C_{par}, bringing a 20dB improvement in the measured SFDR (Fig. 11.4.2).

The residue amplifier is a critical block in pipelined-SAR ADCs, largely determining performance and energy efficiency. It requires a wide output swing, high linearity, low noise, and low power. To meet these demands, this work proposes a residue amplifier based on a discrete-time (DT) FCT with multi-bit pre-conversion (MB-PC). This scheme achieves low noise, high efficiency, and enhanced linearity, while also accelerating the second-stage conversion. The FCT in [10] provides low-noise and power-efficient residue amplification, but its output swing is constrained by the floating supply, and its linearity is limited by output-node parasitic capacitance. To relax the swing requirement, a conventional 1b MSB pre-conversion can be applied, as we did in the input buffer, but linearity remains limited since a single-bit operation only halves the output range. To address this limitation, we propose a MB-PC scheme, where a 4b SAR operation occurs in parallel with the amplification. With MB-PC, the amplifier output is reduced to a small residue at the end of amplification, effectively suppressing nonlinearity from output parasitics, similar to the buffer-connected conversion described earlier. The key enabler is that the DT-FCT performs charge-domain amplification with first-order settling. Its output converges rapidly to the final value, and C_{DAC} switching does not interfere with the charge-transfer process. As a result, progressive multi-bit SAR operations introduce only small quantization errors, which can then be absorbed by the redundancy afterwards. Simulation results confirm that MB-PC not only extends the usable equivalent output swing, but also improves linearity beyond what is achievable with conventional MSB pre-conversion. Furthermore, MB-PC overlaps part of the SAR decision process with residue amplification, thereby accelerating the second-stage conversion.

Figure 11.4.3 illustrates the overall architecture of the proposed 2-stage pipelined-SAR ADC featuring the current-integrating sampler. The sampler employs a 330Ω input resistor, a CT-FCT buffer, and a 1.5pF first-stage CDAC. With a 500ps sampling window, this configuration sets an equivalent voltage gain of 1×. The first-stage SAR resolves 5 bits, after which the DT-FCT residue amplifier delivers 16× gain. The second-stage SAR resolves 11 bits in total, including 4 bits performed concurrently during the residue amplification phase and 2 bits of redundancy. The overall resolution of the prototype is therefore 13 bits.

The prototype ADC is fabricated in a 28nm CMOS process and occupies an active area of 0.033mm² (Fig. 11.4.7). The CDAC mismatch is one-time off-chip calibrated in the foreground, and the inter-stage gain is calibrated by background dither injection. Figure 11.4.4 shows the measured output spectra with 50MHz and 250MHz inputs at a sampling rate of 500MS/s. At the Nyquist input frequency, the ADC achieves 67.5dB SNDR and 94.5dB SFDR. Figure 11.4.4 also shows the SNDR and SFDR versus input frequency of three measured devices, demonstrating consistent SFDR above 92dB across the first and second Nyquist zones. The measured DNL and INL are +0.81/-0.50 LSB and +0.80/-0.72 LSB for 13b resolution, respectively. Operating at 500MS/s, the ADC core consumes 3.54mW and the input buffer consumes 7.24mW. Figure 11.4.5 shows the detailed power breakdown. The Schreier Figure-of-Merit (FoM$_S$) is 176.0dB for the ADC core and 171.2dB including the input buffer. Figure 11.4.5 also shows the measured signal transfer function (STF), and the ADC achieves a dynamic range of 69.2dB. Multi-device measurements are conducted across supply and temperature variations (Fig. 11.4.5). The SNDR variation is below 0.2dB under ±5% supply variation and below 1.5dB from -40°C to 80°C, and the SFDR remains above 91.5dB across all PVT conditions. Figure 11.4.6 summarizes the overall performance and compares it with state-of-the-art ADCs. This work achieves the highest SFDR among all published DT ADCs operating above 200MS/s.

196 • 2026 IEEE International Solid-State Circuits Conference

979-8-3315-8937-0/26 $31.00 © 2026 IEEE

Acknowledgement:
This work is supported in part by the NSFC under Grant 62090042, 62374098, 62434004 and 62434005, National Key R&D Program of China (No. 2019YFB2205003), the Beijing National Research Center for Information Science and Technology, the Beijing Innovation Center for Future Chip.

References:
[1] J. Lagos et al., "A 10.0 ENOB, 6.2 fJ/conv.-step, 500 MS/s Ringamp-Based Pipelined-SAR ADC with Background Calibration and Dynamic Reference Regulation in 16nm CMOS," *in 2021 Symposium on VLSI Circuits, Jun. 2021*, pp. 1–2. https://doi.org/10.23919/VLSICircuits52068.2021.9492354
[2] Y. Lim et al., "A 2.08mW 64.4dB SNDR 400MS/s 12b Pipelined-SAR ADC using Mismatch and PVT Variation Tolerant Dynamically Biased Ring Amplifier in 8nm," *in 2024 IEEE International Solid-State Circuits Conference (ISSCC), Feb. 2024*, pp. 170–172. https://doi.org/10.1109/ISSCC49657.2024.10454422
[3] Y. Cao et al., "A 14b 1GS/s Single-Channel Pipelined ADC with A Parallel-Operation SAR Sub-Quantizer and A Dynamic-Deadzone Ring Amplifier," *in 2025 IEEE International Solid-State Circuits Conference (ISSCC), Feb. 2025*, pp. 430–432. https://doi.org/10.1109/ISSCC49661.2025.10904766
[4] M. Bolatkale et al., "A 28nm 6GHz 2b Continuous-Time ΔΣ ADC with –101 dBc THD and 120MHz Bandwidth Using Digital DAC Error Correction," *in 2022 IEEE International Solid-State Circuits Conference (ISSCC), Feb. 2022*, pp. 416–418. https://doi.org/10.1109/ISSCC42614.2022.9731602
[5] Q. Liu et al., "A 5GS/s 360MHz-BW 68dB-DR Continuous-Time 1-1-1 Filtering MASH ΔΣ ADC in 40nm CMOS," *in 2022 IEEE International Solid-State Circuits Conference (ISSCC), Feb. 2022*, pp. 414–416. https://doi.org/10.1109/ISSCC42614.2022.9731789
[6] A. Mirzaei, S. Chehrazi, R. Bagheri, and A. A. Abidi, "Analysis of first-order anti-aliasing integration sampler," *IEEE Transactions on Circuits and Systems I: Regular Papers*, vol. 55, no. 10, pp. 2994–3005, Nov. 2008. https://doi.org/10.1109/TCSI.2008.924127
[7] B. Malki, T. Yamamoto, B. Verbruggen, P. Wambacq, and J. Craninckx, "A 70dB DR 10b 0-to-80MS/s current-integrating SAR ADC with adaptive dynamic range," *in 2012 IEEE International Solid-State Circuits Conference, Feb. 2012*, pp. 470–472. https://doi.org/10.1109/ISSCC.2012.6177095
[8] W. Jiang, Y. Zhu, C.-H. Chan, B. Murmann, S.-P. U, and R. P. Martins, "A 7b 2 GS/s Time-Interleaved SAR ADC with Time Skew Calibration Based on Current Integrating Sampler," *in 2018 IEEE Asian Solid-State Circuits Conference (A-SSCC), Nov. 2018*, pp. 235–238. https://doi.org/10.1109/ASSCC.2018.8579344
[9] D. Li, M. Qian, D. Li, H. Liang, and Z. Zhu, "A 2.6-GS/s 8-bit Time-Interleaved ADC With Fully Dynamic Current Integrating Sampler," *IEEE Solid-State Circuits Letters*, vol. 8, pp. 29–32, 2025. https://doi.org/10.1109/LSSC.2024.3523509
[10] S. Huang et al., "A 70dB SNDR 80MHz BW Filter-Embedded Pipeline-SAR ADC Achieving 172dB FoMs with Progressive Conversion and Floating-Charge-Transfer Amplifier," *in 2025 IEEE International Solid-State Circuits Conference (ISSCC), Feb. 2025*, pp. 318–320. https://doi.org/10.1109/ISSCC49661.2025.10904746
[11] Z. Chen et al., "A 182.3dB FoMs 50MS/s Pipelined-SAR ADC using Cascode Capacitively Degenerated Dynamic Amplifier and MSB Pre-Conversion Technique," *in 2024 IEEE International Solid-State Circuits Conference (ISSCC), Feb. 2024*, pp. 174–176. https://doi.org/10.1109/ISSCC49657.2024.10454297

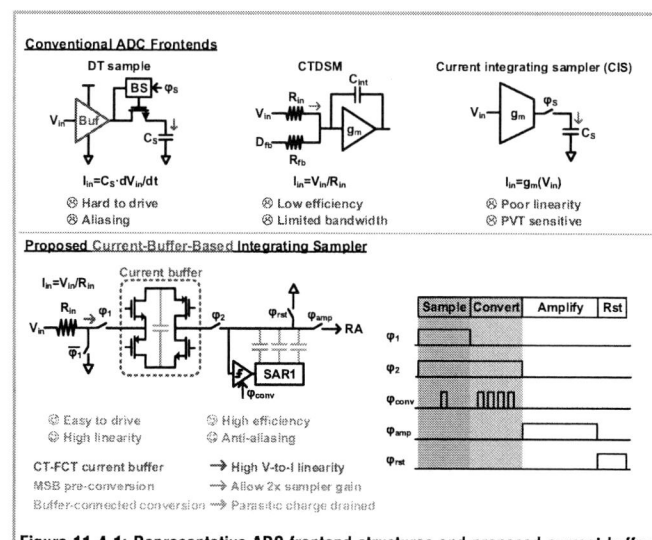

Figure 11.4.1: Representative ADC frontend structures and proposed current-buffer-based integrating sampler.

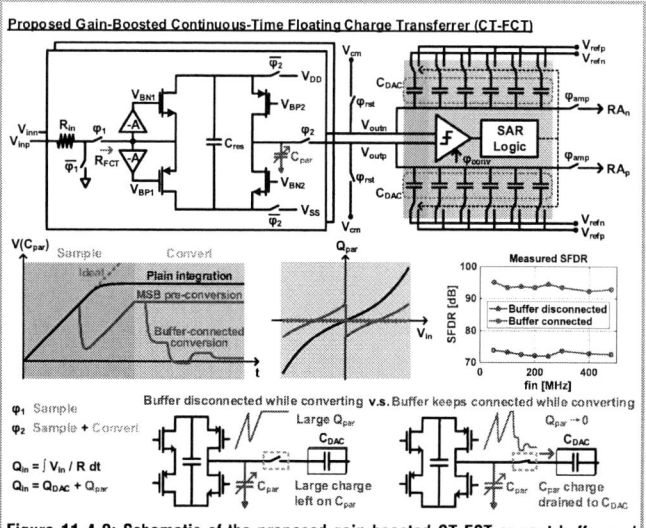

Figure 11.4.2: Schematic of the proposed gain-boosted CT-FCT current buffer and buffer-connected conversion technique.

ISSCC 2026 / SESSION 11 / PIPELINE AND ULTRA-HIGH-SPEED DATA CONVERTERS / 11.4

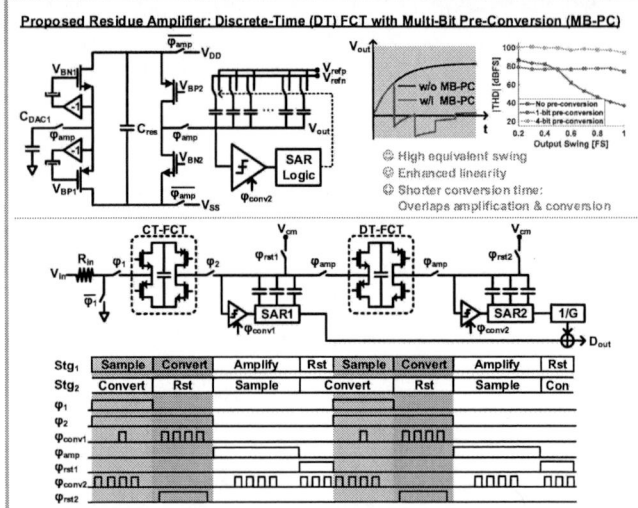

Figure 11.4.3: Schematic of the proposed DT-FCT residue amplifier with multi-bit pre-conversion and overall schematic and timing diagram of the proposed ADC.

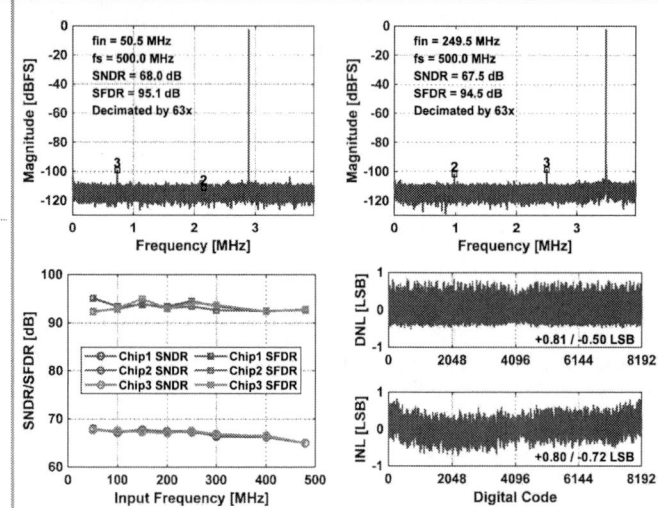

Figure 11.4.4: Measured spectra with 50MHz and 250MHz inputs, performance versus input frequency, and DNL/INL curves.

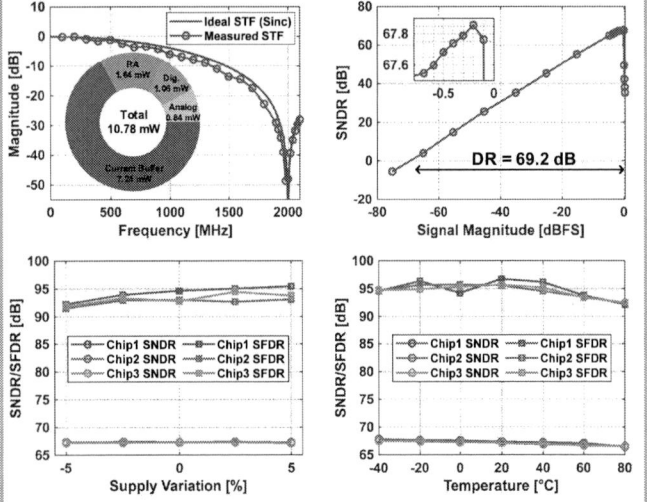

Figure 11.4.5: Measured STF, dynamic range, power breakdown, and multi-device performance under supply and temperature variations.

Figure 11.4.6: Performance summary and comparison with state-of-the-art ADCs.

	This Work		ISSCC12 Malki [7]	ISSCC22 Bolatkale [4]	ISSCC22 Liu [5]	VLSI21 Lagos [1]	ISSCC24 Lim [2]	ISSCC25 Cao [3]
Architecture	CIS Pipe-SAR		CIS SAR	CTDSM	CTDSM	Pipe-SAR	Pipe-SAR	Pipeline
Input Type	Resistor		Gate	Resistor	Resistor	Swi. Cap	Swi. Cap	Swi. Cap
Easy to Drive	Yes		Yes	Yes	Yes	No	No	No
Technology	28nm		40nm	28nm	40nm	16nm	8nm	28nm
Area [mm²]	0.033		0.080	0.13	0.21	0.0084	0.017	0.022
Fs [MHz]	500		80	6000	5000	500	400	1000
BW [MHz]	250		40	120	360	250	200	500
SNDR [dB]	67.5		53.6	71.5	65	62.3	62.8	68.2
SFDR [dB]	94.5		65.1	102.5	79.4	75.5	68.6	85.8
Include Input Buffer	Yes	No	Yes	-	-	No	No	No
Power [mW]	10.8	3.5	6.0	108.8	158	3.3	2.08	15.3
FoM$_S$ [dB]	171.2	176.0	151.8	161.9	158.6	171.1	172.6	173.3

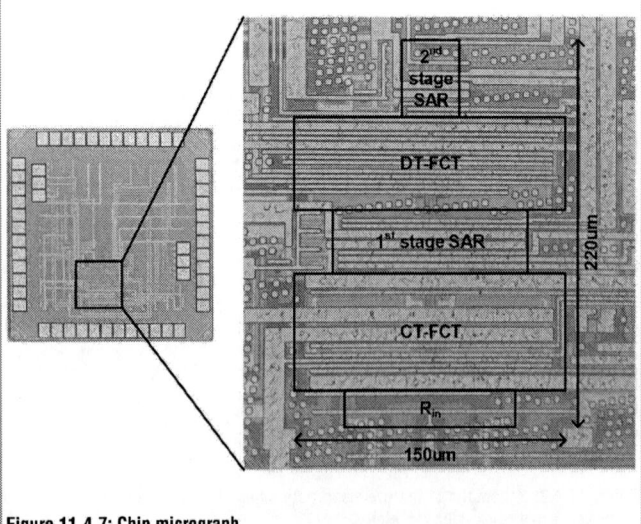

Figure 11.4.7: Chip micrograph.

• 2026 IEEE International Solid-State Circuits Conference

ISSCC 2026 / SESSION 11 / PIPELINE AND ULTRA-HIGH-SPEED DATA CONVERTERS / 11.5

11.5 A Compact 7b 175GS/s Linearized Time-Interleaved Slope ADC with Switched Input Buffers

Ewout Martens, Angelo Parisi, Anirudh Kankuppe, Adam Cooman, Hanyue Li, Steven Van Winckel, Pratap Renukaswamy, Lucas Moura Santana, Jorge Lagos Benites, Nereo Markulić, Jan Craninckx

imec, Leuven, Belgium

Abstract

We present a 2048× time-interleaved slope-ADC implementing 7b, 175GS/s conversion. Samplers with switched buffers are proposed to realize wideband sampling in rank 1 driven by a multi-phase clock generated by a delay line with feedforward coupling. Making the slope nonlinear compensates static nonlinearities of the hierarchical sampling network. With only 0.063mm² in 5nm CMOS, it is the smallest ultra-high-speed ADC reported to date with an excellent energy per sample of less than 2.16pJ.

Recent wireline communication standards like IEEE 802.3db/df/dj increase bit rates to 100Gb/s/lane and beyond. An analog-to-digital converter (ADC) based receiver (RX) offers powerful digital processing to correct imperfections, which benefits from technology scaling in terms of power and area [1]. Realizing ADC-based receivers at these high speeds require ADCs with sampling rates well above 100GS/s, a bandwidth of several 10s of GHz, and a moderate 6-8b resolution. The de facto standard way of implementing these ADCs is to time-interleave (TI) as many successive approximation register (SAR) ADCs as needed, which often results in more than 100 channels [2-6]. Due to the reliance on mismatch-sensitive capacitive digital-to-analog converters (DACs), SAR ADCs benefit only mildly from technology scaling resulting in a rapid increase in ADC area when upscaling the speed, leading to more interconnect parasitics that ultimately limit the bandwidth. Limiting the interleaving factor by increasing the speed of a channel, e.g., with loop-unrolled SARs [7-8], increases complexity, calibration effort and area within a channel. An alternative approach proposed in [9] interleaves a huge number of tiny channels with a slope ADC. The simple channel offers a low area per conversion leading to an overall small area with minimal calibration requirements within the channel. This work demonstrates the scalability of this architecture towards both 5nm CMOS and to ultra-high speeds resulting in a power- and area-efficient solution for a high-speed ADC based on a 3-rank hierarchical sampler and a compact TI slope-ADC array as the back-end.

We introduce a switched-buffer technique for the wideband high-speed samplers. The different phases of the high-speed clock driving these samplers are generated by a delay line with feedforward (FF) couplers. In the 2nd rank of the hierarchy, a common-source/common-drain (CSCD) amplifier provides 2× gain to relax comparator noise. For the slope ADC in the channels, an inverter-based continuous-time (CT) comparator is proposed, providing high gain and bandwidth. Static nonlinearities in the hierarchical sampler are readily compensated by making the slope signal nonlinear. These techniques have been implemented in a 7b 175GS/s ADC.

As shown in Fig. 11.5.1, the first rank of the hierarchical sampler consists of 8 units each operating at 21.875GHz. A source-follower buffer brings the sampled signal to 1 of 8 rank-2 units connected to each rank-1 unit. Each of the 64 rank-2 amplifiers drives a row of 32 rank-3 cells resulting in a TI slope-ADC array of 2048 channels each converting at an 85MHz-rate. A 7b counter is synchronized with the slope signal and both are distributed together to all rank-3 cells. The slope frequency is about 100MHz to meet the condition $T_{slope} < T_{conversion}$ required to enable TI with slope ADCs [9]. Outputs are multiplexed per row resulting in 64 parallel output data streams of 2.73GHz.

A dual symmetrical T-coil configuration [4] provides wideband 50Ω-matching of the 8 parallel input buffers. Electrostatic discharge (ESD) protection diodes are connected to the first T-coil while the second T-coil sees the 1-to-8 interconnection network to the rank-1 input buffers and the on-chip termination resistor. The T-coils have been optimized for a 65GHz bandwidth to drive the 8 rank-1 units exhibiting a 15fF single-ended input capacitance each, and a total equivalent parasitic load of 37fF due to the interconnection network.

The conventional bottom-plate sampler shown in Fig. 11.5.2 reduces nonlinearities by opening the top-plate switch a small time after the bottom-plate sampling switch that sees signal-independent gate and source voltages. However, it requires two switches in series limiting the bandwidth of the sampling circuit. In this work, the top-plate switch is removed from the sampling network reducing its bandwidth and instead the input buffer is switched off when not tracking. By placing the enabling switch in the drain of the transistors of the push-pull input buffer as depicted in Fig. 11.5.2, the resistance of this switch has only a minor impact on the output impedance of the buffer, hence enabling wide bandwidth with no major increase of the load on the clock signal CK_1. Besides, this solution also provides isolation between the rank-1 samplers and eliminates the need for the inverse of the clock signal that is often used to drive the typical complementary top-plate switch to handle a large signal range. Instead of the conventional differential sampling switch, a tripartite switch [10] is adopted. To realize the same differential switch resistance R_{eq}, this configuration halves the load on the critical sampling clock line. Custom-made compact 3-terminal (1 top, 2 bottoms) AC-coupling capacitors (120fF each part) bring the signal to the gates of the push-pull buffer where the bias voltages V_{b1p}, V_{b1n} are set via programmable DACs with required values readily obtained from simulations in our prototype. The buffers see an input signal of 800mV$_{ptp,diff}$.

All clock signals in the ADC are derived from a low-jitter sinusoidal main clock CK_{in} at 21.875GHz as shown in Fig. 11.5.1 (inverter colors indicate strengths). First, coarse tuning of inverter sizes and their loads generates 8 clock phases CK_0 with FF coupling to enable high-speed operation [11]. Then, a programmable delay with fine resolution tunes out skew on the falling edges of CK_0. An off-chip LMS-based calibration loop selects the settings for coarse and fine delay lines using a high-frequency sine wave input signal. Finally, per rank-1 unit, 2 quadrature clocks are combined to generate the sampling clock CK_1 based on the falling edges of the CK_0 clocks as shown in Fig. 11.5.2. AC capacitors make the level-shifted early sampling clock CK_{1eH}. The bias voltages V_{bc} are generated with on-chip DACs, set all to the same value in our reported results.

In a parallel path (see Fig. 11.5.1), two differential clocks CK_0 are selected to generate the sampling clock for rank 2, also set by falling edges eliminating any need for duty cycle control. To minimize differences between the sampling clocks CK_2 in rank 2, a combination of all sampling clocks of 8 rank-2 units CKS_2 is sent to all units, and a one-hot shift register creates a mask M_1 to pick the actual sampling pulse. Selecting different CK_0 clocks and a small programmable delay Δ allows aligning the clocks of rank 1 and 2. Proper alignment (with relaxed margin) results in maximum gain and is done using a sine-wave input.

Figure 11.5.3 shows that the rank-1 output is sampled twice with complementary switches, followed by switching the bottom plates of $C_{s2\#}$ (6/8/2/9fF) so that 2 input signals at different common-mode (CM) levels are presented to the CSCD amplifier [12] to turn it on. The amplification factor is about 2V/V within a time set by a second one-hot shift register generating the pulse M_2 to end amplification and start reset of rank 2. Cross-coupled capacitors over the sampling switches cancel signal feedthrough. Parts of the switching capacitors are made tunable; the settings are found by monitoring the common-mode value of the outputs.

In the rank-3 cells, the signal is sampled differentially onto C_{s3} (20fF). First, the MSB is found as depicted in the timing diagram of Fig. 11.5.3. Then, C_{s3} is placed between the slope signal and one input of the CT comparator with a reference voltage V_{ref} on the other side ensuring the comparator toggles and clocks in the counter value when both inputs are equal to V_{ref}, eliminating any CM-dependent offset. The whole configuration is flipped depending on the MSB. A redundancy of a few LSBs between the 2 phases makes the conversion resilient to MSB errors. A T-switch configuration [13] avoids any leakage from the slope signal C_{s3} when sampling and cross-coupled capacitors eliminate signal feedthrough when converting. As drawn in Fig. 11.5.3, the CT comparator is an inverter with its local supply set by the loop around another inverter with V_{ref} as input which tunes the threshold of the first inverter based on V_{ref} [14]. Around the threshold, the inverter gain is maximized while the g_m of both NMOS and PMOS contribute to a high bandwidth. To speed up MSB detection, the CT comparator is set in a dynamic mode.

After sampling in rank 1 on C_{s1} (20fF), all nonlinearities from the hierarchical sampler are frequency-independent. The TI slope ADC provides an easy way to compensate these static nonlinearities by making the common slope voltage vary nonlinearly with time as depicted in Fig. 11.5.4. The counter keeps running linearly. The sampled voltage in rank 3 of an input voltage V_{IN} consists of a linear part V_{linear} and a nonlinear part $f_{NL}(V_{IN})$. When the slope is linear, that extra term leads to errors in the final digital output D_{OUT}. However, adding a nonlinear part g_{NL} to the slope signal allows to cancel the odd-order nonlinearities completely. As illustrated in Fig. 11.5.4, due to the conversion scheme in rank 3, the nonlinear part of the slope should be set to $-f_{NL}(|V_{IN}|)$ to compensate the odd-order harmonics over the full range. That shape can be extracted from the INL curve of the ADC. For the prototype chip, the nonlinear slope is generated externally and synchronized on-chip with the Gray counter signal CNTR. The counter clock runs at 10.9375GHz, derived from a main clock CK_0.

Figure 11.5.7 shows the chip micrograph, fabricated in a 1P17M 5nm FinFET process with the hierarchical sampler and back-end array occupying only 250×250μm². The slope ADCs are laid out following the patterns of the digital standard cells for maximum compactness. Reported results are obtained after a one-time off-chip calibration of skew, and with digital offset and gain correction per cell. Cables, balun and probe losses have been extracted and de-embedded. The ADC consumes 378mW from different 0.85V supplies. All clocking circuitry dominate the power while the 2048 cells contribute only to 1/4ᵗʰ of the total power.

198 • 2026 IEEE International Solid-State Circuits Conference

979-8-3315-8937-0/26 $31.00 © 2026 IEEE

Figure 11.5.5 shows that the ADC achieves 5.4 to 3.8 ENOB over a 52GHz band. The -3dB-bandwidth is 22GHz and limited by improper package modeling during design. When activating the nonlinear slope compensation, the HD_3 at low frequencies improves from -35 to -61dBc. Compared to the state-of-the-art of (all TI SAR) ultra-high-speed ADCs (see Fig. 11.5.6), this design occupies the smallest area despite realizing one of the highest reported sampling speeds. Furthermore, the conversion energy (= power / f_{sample}) is among the lowest, and includes the complete generation of all phases of the jitter-sensitive high-speed clocks. These results demonstrate that the presented approach avoids a rapid increase in area and power when upscaling ADCs for next-generation digital-intensive wireline links.

Acknowledgment:

The authors would like to thank Pratik Purbey, Stephane Zagrocki and Bert Blockx for layout engineering; Hans Suys for package and PCB design; Luc Pauwels and Nele Van Hoovels for lab support; Ambra Neri for project management; and Bart Moeneclaey, Joris Lambrecht, Jiji Aravind and Peter Ossieur for valuable discussions.

References:

[1] S. Palermo, S. Hoyos, S. Cai, S. Kiran and Y. Zhu, "Analog-to-Digital Converter-Based Serial Links: An Overview," in *IEEE Solid-State Circuits Magazine*, vol. 10, no. 3, pp. 35–47, Summer 2018. http://doi.org/10.1109/MSSC.2018.2844603
[2] G. Li *et al.*, "18.1 A 600Gb/s DP-QAM64 Coherent Optical Transceiver Frontend with 4x105GS/s 8b ADC/DAC in 16nm CMOS," *2024 IEEE International Solid-State Circuits Conference (ISSCC)*, San Francisco, CA, USA, 2024, pp. 338–340. http://doi.org/10.1109/ISSCC49657.2024.10454499
[3] R. L. Nguyen *et al.*, "18.4 A 200GS/s 8b 20fJ/c-s Receiver with >60GHz AFE Bandwidth for 800Gb/s Optical Coherent Communications in 5nm FinFET," *2024 IEEE International Solid-State Circuits Conference (ISSCC)*, San Francisco, CA, USA, 2024, pp. 344–346. http://doi.org/10.1109/ISSCC49657.2024.10454385
[4] D. Pfaff *et al.*, "A 224 Gb/s 3 pJ/bit 40 dB Insertion Loss Transceiver in 3-nm FinFET CMOS," in *IEEE Journal of Solid-State Circuits*, vol. 60, no. 1, pp. 9–22, Jan. 2025. http://doi.org/10.1109/JSSC.2024.3466092
[5] J. Q. Wang *et al.*, "7.1 A 2.69pJ/b 212Gb/s DSP-Based PAM-4 Transceiver for Optical Direct-Detect Application in 5nm FinFET," *2024 IEEE International Solid-State Circuits Conference (ISSCC)*, San Francisco, CA, USA, 2024, pp. 123–125. http://doi.org/10.1109/ISSCC49657.2024.10454275
[6] E. -H. Chen, H. Park, M. Abdullatif *et al.*, "7.1 A 212.5Gb/s DSP-Based PAM-4 Transceiver with 50dB Loss Compensation for Large AI System Interconnects in 4nm FinFET," *2025 IEEE International Solid-State Circuits Conference (ISSCC)*, San Francisco, CA, USA, 2025, pp. 1–3. http://doi.org/10.1109/ISSCC49661.2025.10904601
[7] A. Khairi, *et al.*, "Beyond 200-Gb/s PAM4 ADC and DAC-Based Transceiver for Wireline and Linear Optics Applications," in *IEEE Open Journal of the Solid-State Circuits Society*, vol. 4, pp. 265–276, 2024. http://doi.org/10.1109/OJSSCS.2024.3501975
[8] C. Nani, *et al.*, "A 5-nm 60-GS/s 7b 64-Way Time Interleaved Partial Loop Unrolled SAR ADC Achieving 35.2dB SNDR up to 32 GHz" in *IEEE Journal of Solid-State Circuits*, vol. 60, no. 4, pp. 1210–1222, April 2025. http://doi.org/10.1109/JSSC.2024.3517333
[9] E. Martens *et al.*, "22.5 A 42GS/s 7b 16nm Massively Time-Interleaved Slope-ADC," *2024 IEEE International Solid-State Circuits Conference (ISSCC)*, San Francisco, CA, USA, 2024, pp. 396–398. http://doi.org/10.1109/ISSCC49657.2024.10454361

[10] Benwei Xu, Yuan Zhou and Yun Chiu, "A 23mW 24GS/s 6b Time-interleaved hybrid two-step ADC in 28nm CMOS," *2016 IEEE Symposium on VLSI Circuits (VLSI-Circuits)*, Honolulu, HI, 2016, pp. 1–2. http://doi.org/10.1109/VLSIC.2016.7573535
[11] M. Baert and W. Dehaene, "A 5-GS/s 7.2-ENOB Time-Interleaved VCO-Based ADC Achieving 30.5 fJ/cs," in *IEEE Journal of Solid-State Circuits*, vol. 55, no. 6, pp. 1577–1587, June 2020. http://doi.org/10.1109/JSSC.2019.2959484
[12] J. Fang *et al.*, "A 5-GS/s 10-b 76-mW Time-Interleaved SAR ADC in 28 nm CMOS," in *IEEE Transactions on Circuits and Systems I: Regular Papers*, vol. 64, no. 7, pp. 1673–1683, July 2017. http://doi.org/10.1109/TCSI.2017.2661481
[13] K. Ishida, K. Kanda, A. Tamtrakarn, H. Kawaguchi and T. Sakurai, "Managing leakage in charge-based analog circuits with low-V/sub TH/ transistors by analog T-switch (AT-Switch) and super cut-off CMOS," *Digest of Technical Papers. 2005 Symposium on VLSI Circuits, 2005.*, Kyoto, Japan, 2005, pp. 122–125. http://doi.org/10.1109/VLSIC.2005.1469348

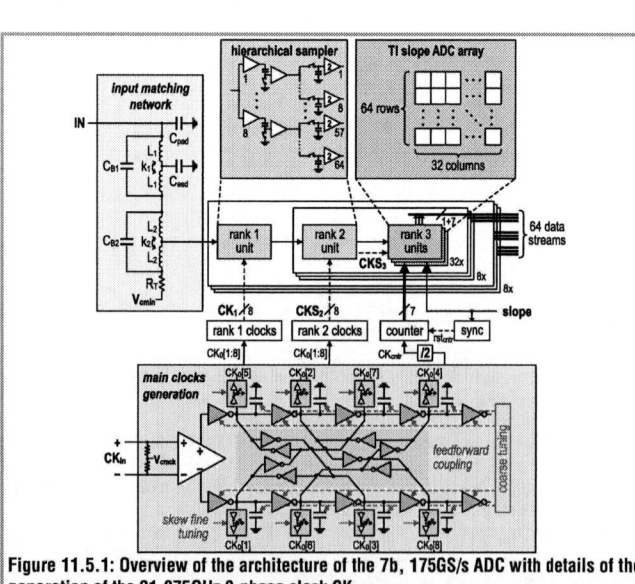
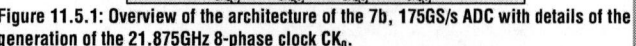

Figure 11.5.1: Overview of the architecture of the 7b, 175GS/s ADC with details of the generation of the 21.875GHz 8-phase clock CK_8.

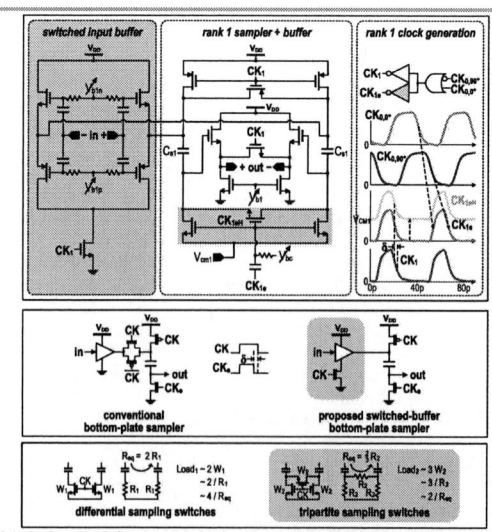

Figure 11.5.2: A unit of the 1st rank of the hierarchical sampler with the switched input buffer.

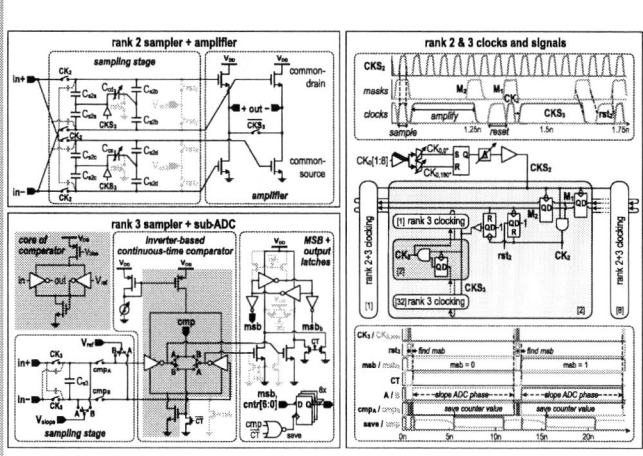

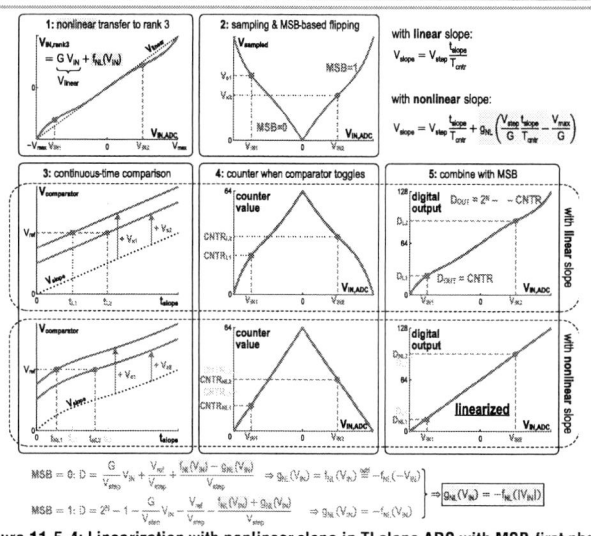

Figure 11.5.3: Details of a unit of the 2nd rank of the hierarchical sampler with a common-source/common-drain amplifier, and of the sub-ADCs in the TI slope ADC array.

Figure 11.5.4: Linearization with nonlinear slope in TI slope ADC with MSB-first phase and slope shifting.

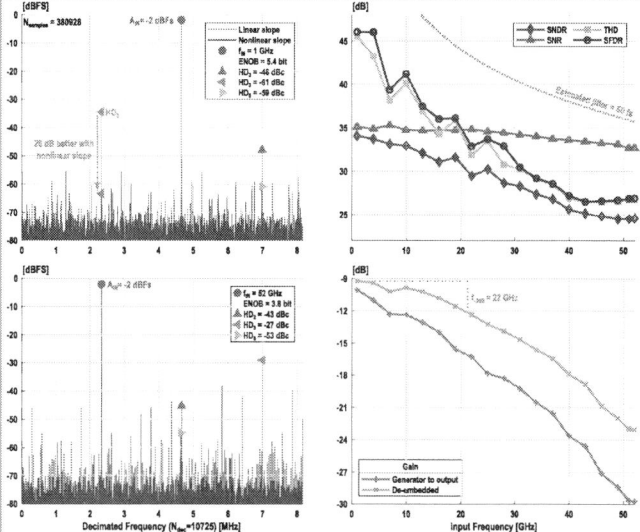

Figure 11.5.5: Main measurement results of the ADC at 175GS/s.

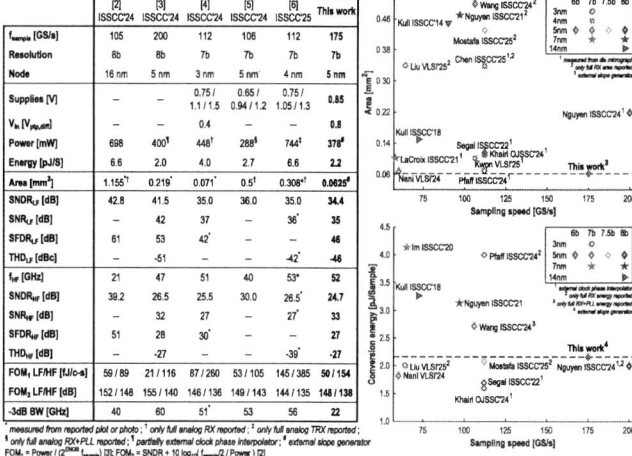

	[2] ISSCC'24	[3] ISSCC'24	[4] ISSCC'24	[5] ISSCC'24	[6] ISSCC'25	This work
f_sample [GS/s]	105	200	112	106	112	175
Resolution	8b	8b	7b	7b	7b	7b
Node	16 nm	5 nm	3 nm	5 nm	4 nm	5 nm
Supplies [V]	–	–	0.75/ 1.1/1.5	0.65/ 0.94/1.2	0.75/ 1.05/1.3	0.85
V_IN [V_pp,diff]	–	–	0.4	–	–	0.8
Power [mW]	698	400†	448†	288§	744§	378§
Energy [pJ/S]	6.6	2.0	4.0	2.7	6.6	2.2
Area [mm²]	1.155*	0.219*	0.071*	0.5†	0.308*†	0.0625§
SNDR_LF [dB]	42.8	41.5	35.0	36.0	35.0	34.4
SNR_LF [dB]	–	42	37	–	36*	35
SFDR_LF [dB]	61	53	42*	–	–	46
THD_LF [dBc]	–	-51	–	–	-42*	-46
f_in [GHz]	21	47	51	40	53*	52
SNDR_HF [dB]	39.2	26.5	25.5	30.0	26.5*	24.7
SNR_HF [dB]	–	32	27	–	27*	33
SFDR_HF [dB]	51	28	30*	–	–	27
THD_HF [dB]	–	-27	–	–	-39*	-27
FOM_1 LF/HF [fJ/c-s]	59/89	21/116	87/260	53/105	145/385	50/154
FOM_2 LF/HF [dB]	152/148	155/140	146/136	149/143	144/135	148/138
-3dB BW [GHz]	40	60	51†	53	56	22

* measured from reported plot or photo ; † only full analog RX reported ; ‡ only full analog TRX reported ;
§ only full analog RX+PLL reported ; ¶ partially external clock phase interpolator ; † external slope generator
FOM$_1$ = Power / (2^{ENOB} f$_{sample}$) [3]; FOM$_2$ = SNDR + 10 log$_{10}$(f$_{sample}$/2 / Power) [2]

Figure 11.5.6: Performance summary and comparison with state-of-the-art ultra-high-speed ADCs implementing the TI SAR architecture.

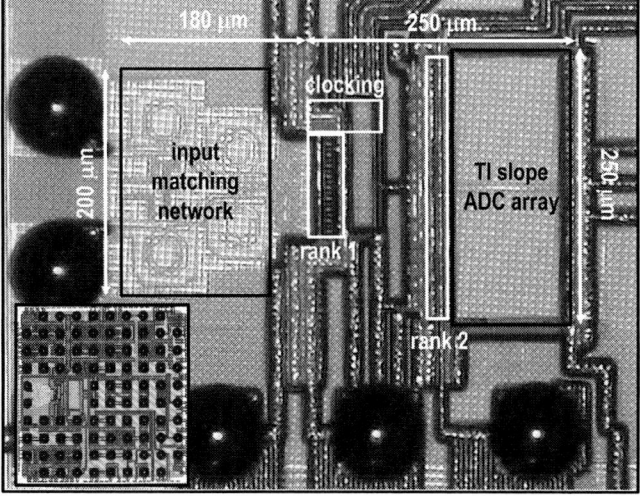

Figure 11.5.7: Chip micrograph.

ISSCC 2026 / SESSION 11 / PIPELINE AND ULTRA-HIGH-SPEED DATA CONVERTERS / 11.6

11.6 An 8b 20GS/s Time-Interleaved ADC with 2.6mW 1GS/s Hybrid Voltage/Time-Domain Sub-ADC in 12nm FinFET

Daisuke Miyazaki, Yuki Yagishita, Mika Takasaki, Shun Nagata, Takeru Nogamida, Yudai Abe, Kazutoshi Tomita, Kazunori Hasebe, Toshiyuki Kikkawa, Satoshi Yoshizawa, Atsuya Suzuki, Soichi Kato, Takahiro Naito, Keigo Bunsen, Tomohiro Matsumoto, Yasushi Katayama

Sony Semiconductor Solutions, Atsugi, Japan

Abstract

A 2.6mW, 1GHz, 42dB-SNDR sub-ADC is presented. A hybrid voltage- and time-domain architecture is employed to avoid the theorical SAR comparator trade-offs and enable power efficiency benefit from the time-domain ADC. Additionally, a hierarchical sample-and-hold circuit with a new common-mode feedback scheme is introduced to enhance signal distribution. A 20GHz time-interleaved ADC, including an input buffer and reference, demonstrates an SNDR of 37dB with a power consumption of 135mW.

High-speed DSP-based wireline transceivers require medium-resolution (6-8b) ADCs that operate at multi-GHz sampling rates while maintaining low power consumption. SAR ADCs are a preferred choice due to their high energy efficiency and robustness for sub-channels in time-interleaving architectures. However, achieving high sampling rates with SAR ADCs necessitates massive interleaving because of the fundamental power-speed trade-off in comparators, which limits channel speed. Such massive interleaving increases design complexity and calibration difficulty. To mitigate these issues, techniques have been reported to accelerate SAR ADC operation by reducing comparator conversion cycles [1]. As an alternative to conventional voltage-domain designs, time-domain architectures offer excellent energy and area efficiency. To achieve medium resolution, several multi-step architectures have been proposed [2-5]. However, significant variations in amplitude and inter-stage gain errors necessitate complex calibration schemes. This paper introduces a low-power sub-ADC that employs a hybrid voltage- and time-domain ADC architecture. Through a large redundancy between the voltage-and time-domain stages, the offset and noise requirements of the SAR comparator can be minimized. This approach avoids the fundamental trade-offs of SAR architectures, which enables both increased speed and reduced power consumption. Furthermore, applying a short time-to-digital conversion (TDC) for LSB quantization enables power efficiency benefits from the time-domain ADC. As a result, the 1GHz sub-ADC achieves a remarkably low power consumption of only 2.6mW while demonstrating an SNDR of 42dB. Additionally, this paper presents a sample-and-hold (SH) circuit featuring a new common-mode feedback (CMFB) scheme, which facilitates high-precision signal distribution. The 20GHz time-interleaved ADC, including the input buffer and reference, demonstrates an SNDR of 37dB at Nyquist with power consumption of 135mW.

Figure 11.6.1 shows the overall block diagram, which consists of an input buffer, four-channel 5GHz sample-and-hold circuits, sub-ADCs, clocking circuits, and a DC reference block. Each of the 4-channel SH circuits drives a 5-channel sub-ADC and is clocked through a programmable delay for skew adjustment. The DC block composed of BGR and reference driver supplies the reference current and voltage to all sub-ADCs.

Figure 11.6.2 shows the sub-ADC of the hybrid voltage- and time-domain architecture. It consists of a 5b asynchronous SAR ADC, a voltage-to-time converter (VTC), an edge detector that sorts pulses based on their arrival timing [1] and an 11-level flash TDC. The SAR ADC employs a standard architecture that performs top-plate sampling using binary-weighted capacitors of the C-DAC and controls the bottom plate terminals of the C-DAC based on the decision results of the comparator. After the 5b SAR operation, the residue voltage is buffered using source follower (SF) and is sampled to the VTC sampling capacitors through a pipeline operation. Simultaneously, an additional trimming C-DAC is activated to adjust the offset of the subsequent VTC and TDC. The trimming code can be easily determined by identifying the code that produces the midpoint output of the TDC in foreground calibration. To improve the fidelity of the VTC, the common-mode ramp technique is applied by charging a capacitor at the input node of the sampling capacitors. The VTC circuit converts voltage information into time information. However, the initial period before the internal slope voltage reaches the first threshold is wasted conversion time, which limits the VTC's maximum operating speed. To address this issue, a NOR gate circuit is added to control the additional current source of the slope generator. The ramp speed is increased until one of the pulses occurs, thereby reducing this wasted time and maintaining high operating speed. Furthermore, a new edge detector is employed to improve the linearity of the TDC. The conventional edge detector based on NAND/NOR circuits suffers from a large delay offset due to the imbalance between the stacking and non-stacking paths of the internal circuit. The technique provides a fully balanced timing-determining (early or late) path, eliminating the asymmetry between early and late delay. The edge detector is implemented using only standard cells by strategically adding dummy transistors and achieves compact layout in good compatibility with the following TDC. For the TDC architecture, a standard flash TDC is used to quantize the LSBs. The residue range and TDC quantization range are 8 LSB and 22 LSB, respectively, providing a large redundancy (22 - 8 = 14). This redundancy minimizes SAR comparator offset and noise requirements, enabling speed-optimized design for substantial power savings. In addition to the offset trimming mentioned earlier, VTC gain calibration can be achieved through a simple sequence. Using a C-DAC, two reference voltages are generated, and the capacitance for

the ramp generator is trimmed to ensure that the difference between the TDC output codes is aligned to 8 LSB.

Figure 11.6.3 illustrates the detailed input network and SH circuit diagram. Inductors are placed before the push-pull linear input buffer, providing a 50Ω input impedance matching for input frequencies up to 10GHz. When a standard SF is used as a unity-gain buffer in an SH circuit such as [6], the common-mode voltage varies due to fluctuations in the V_{th} of the follower transistors. This variation limits the signal amplitude to ensure adequate voltage headroom, making it difficult to maintain a high SNR. The SH circuit enables bottom-plate sampling while allowing control of the common-mode voltage. The differential input signal is sampled at the sampling capacitor (Cs) with a voltage of Vcm1 using the clock signal of S. Vcm1 is the output voltage of the common-mode feedback circuit. During the hold phase, the top plate of Cs is connected to VDD, and the bottom plate voltage is level-shifted and buffered with SFs. The output voltage of the SF is sampled by Cfb every clock cycle, and feedback is applied continuously to ensure that the common-mode voltage reaches the specified voltage of ½ VDD. As a result, variations in V_{th} can be absorbed. Since the voltage of Vcm1 is low, below 0.4 V, the current source of the SF enters the linear region in the sampling phase, causing slow settling in the subsequent hold phase. To avoid this issue, additional current is injected into the current source of SF during the sampling phase, and the output voltage is controlled to be ½ VDD. This ensures that the SF including load current MOS operates in the saturation region at all times, allowing high-speed operation. Total power consumption remains unchanged because the main and additional path are isolated by a power gating switch. The input terminal voltages of the 2nd push-pull SF, coupled with level-shift capacitors are biased to the optimal voltage for low distortion buffering. This 2nd SF drives individual sub-ADC.

The prototype ADC is fabricated using 12nm FinFET CMOS technology, and the active area occupies about 0.23mm² as shown in Fig. 11.6.7. The VTC gain/offset errors in the sub-ADCs and the timing mismatch of the SH clocks are adjusted with a one-time foreground calibration by trimming the capacitor array of the sub-ADC and using a programmable delay of the clock generator. Channel gain mismatch calibration is not performed to avoid power-hungry digital-domain multipliers, which is feasible due to the robust characteristics of the SH and SAR. Figures 11.6.4 and 11.6.5 show the performance of the AC and DC characteristics. The measured output spectra of the ADC are also presented where the FFT point is 1024. There are small spurs remaining, caused by channel mismatch. These spurs do not significantly affect the SNDR. At 20GS/s, the ADC consumes a total of a 135mW and demonstrates a 37dB SNDR and a 46dB SFDR with a Nyquist input. The ADC can operate with supply voltages of 0.8V and 1.8V. Since no special voltages are needed, a simple power supply configuration becomes possible in actual applications. Figure 11.6.6 presents the performance summary and comparison with time-domain or hybrid voltage/time-domain ADCs of similar resolution and sampling speed published in ISSCC or VLSI over the past 5 years. Single-channel performance is also included in this table. The sub-ADC consumes only 2.6mW and achieves a 42dB SNDR with a single supply voltage of 0.8V. This power efficiency is realized through the voltage- and time-domain hybrid architecture. The Schreier Figure of Merit (FoMs) of TI-ADC reaches 145.7dB. The FoMs represents the total system performance, including the input buffer and reference generation circuits, which demonstrates a state-of-the-art performance when compared to other ADCs under similar conditions.

Acknowledgement:
The authors would like to thank H.Yamasaki, K.Hagisaki, H.Sato, T.Nakamura and H.Fujita for circuit implementation and verification. The authors would also like to thank Y. Jingu,Y. Hasegawa and N. Shoji for technical discussion.

References:
[1] A. Whitcombe et al, "A 76mW 40GS/s 7b Time-interleaved Hybrid Voltage/Time-Domain ADC with Common-Mode Input Tracking," *ISSCC*, pp.392-393, Feb 2024. https://doi.org/10.1109/ISSCC49657.2024.10454355
[2] M.Zhang et al., "A 4x Interleaved 10GS/s 8b Time-Domain ADC with 16xInterpolation-Based Inter-Stage Gain Achieving >37.5dB SNDR at 18GHz Input," *ISSCC*, pp.252-253, Feb 2020. https://doi.org/10.1109/ISSCC19947.2020.9062986

ISSCC 2026 / February 17, 2026 / 10:30 AM

[3] H Li et al, "A PVT-Robust 16GS/s 4×TI Time-Domain ADC with Vernier-based Multipath Flash TDC achieving 25.7fJ/c-s FoM in 28nm CMOS," *IEEE Symp. on VLSI Circuits*, C8-4, June 2025. https://doi.org/10.23919/VLSITechnologyandCir65189.2025.11074876

[4] M.M.Ghahramani et al., "A 46GS/s 7-bit Time-Interleaved Time-Domain ADC with Synthesizable Unit ADCs in 16nm FinFET," *IEEE Symp. on VLSI Circuits*, C8-2, June 2025. https://doi.org/10.23919/VLSITechnologyandCir65189.2025.11075030

[5] M.Zhang et al, "A 20GS/s 8b Time-Interleaved Time-Domain ADC with Input-Independent Background Timing Skew Calibration," *IEEE Symp. on VLSI Circuits*, C15-2, June 2021. https://doi.org/10.23919/VLSICircuits52068.2021.9492436

[6] E.Martens et al., "A 42GS/s 7b 16nm Massively Time-Interleaved Slope-ADC," *ISSCC*, pp.396-397, Feb 2024. https://doi.org/10.1109/ISSCC49657.2024.10454361

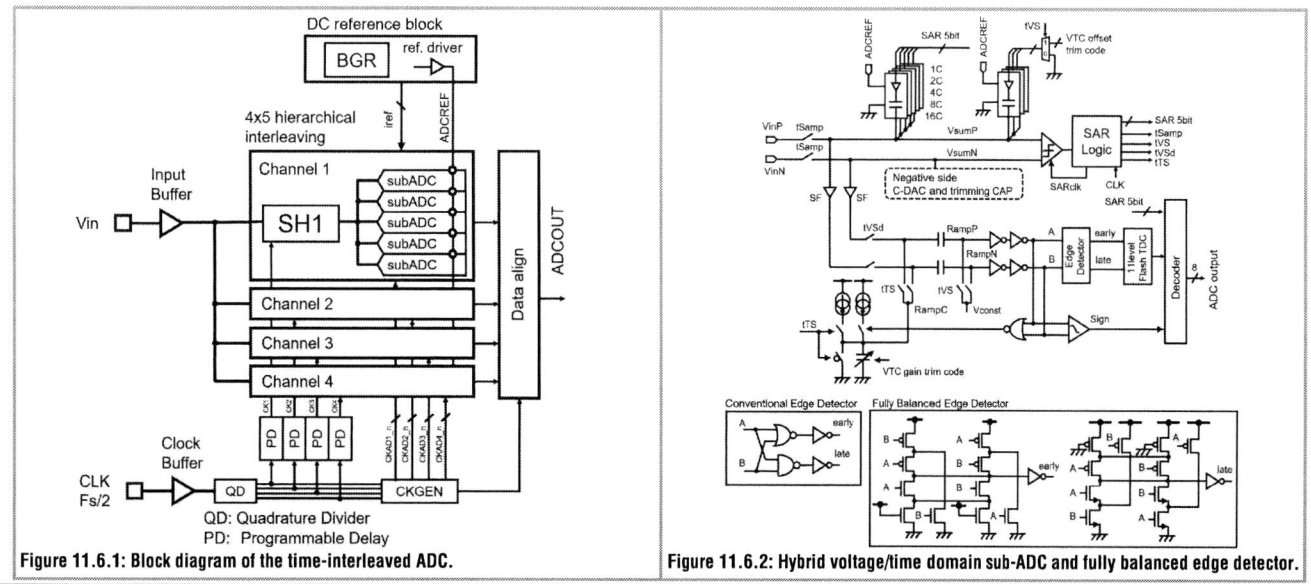

Figure 11.6.1: Block diagram of the time-interleaved ADC.

Figure 11.6.2: Hybrid voltage/time domain sub-ADC and fully balanced edge detector.

DIGEST OF TECHNICAL PAPERS • 201

ISSCC 2026 / SESSION 11 / PIPELINE AND ULTRA-HIGH-SPEED DATA CONVERTERS / 11.6

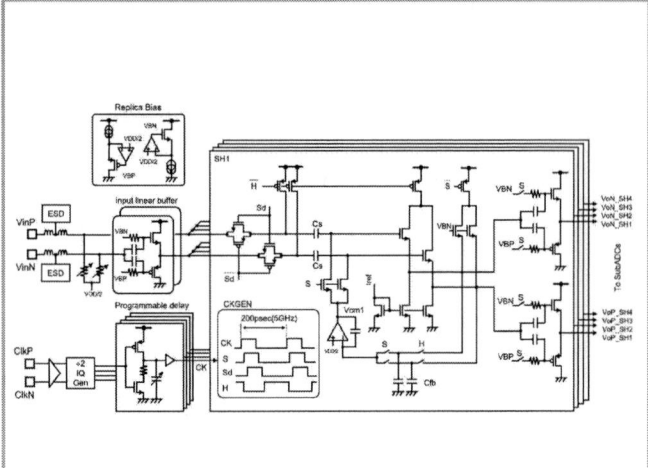

Figure 11.6.3: Detailed circuit implementation of Input matching, input buffer, hierarchical sampler and sampling clock.

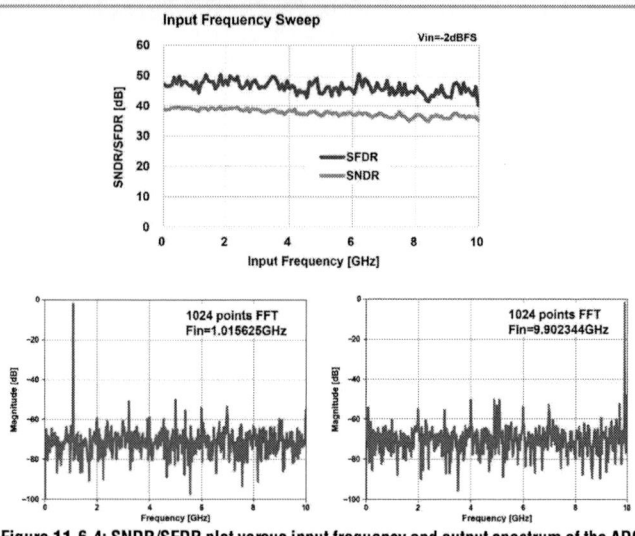

Figure 11.6.4: SNDR/SFDR plot versus input frequency and output spectrum of the ADC with low (1GHz) input and Nyquist (9.9GHz) input frequency.

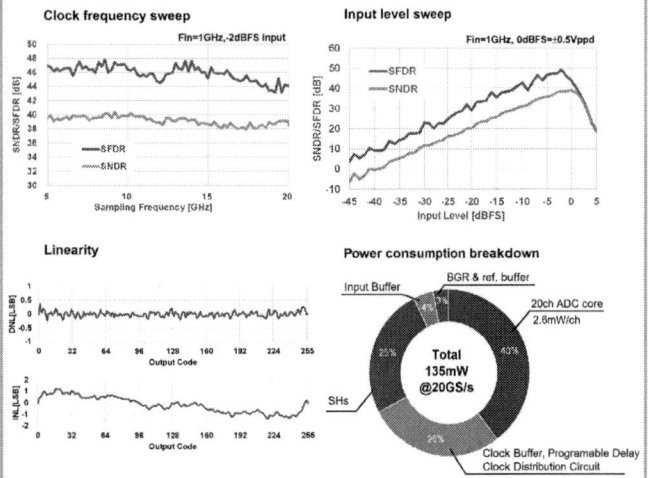

Figure 11.6.5: SNDR/SFDR plot versus clock frequency and input amplitude, linearity, and measured power consumption with power breakdown.

	This work	[1] Whitcombe ISSCC2024	[2] Zhang ISSCC2020	[3] Li VLSI2025	[4] Ghahramani VLSI2025	[5] Zhang VLSI2021
Channel Arch.	SAR/TDC Hybrid	TDC-assisted SAR	2-stage TDC	2-stage TDC	2-stage TDC	2-stage TDC
Tech.	12nmFinFET	22nmFinFET	65nm	28nm	16nmFinFET	65nm
Resolution	8	8	7	8	7	8
Fs [GHz]	20	1	40	10	46	20
SubADC Fs [GHz]	1	-	0.83	2.5	2.875	2.5
# Channel	20	-	48	4	16	8
Supply	0.8V/1.8V	0.8V	0.85V	1.0V	N/A	1.0/1.2V
Power[mW]	135	2.6	76.4	50.8	133	129.9
SFDR@nyq. [dB]	46	52	45	52.8	40.07	52.5
SNDR@nyq. [dB]	37	42	32.3	40.1	33.18	38.8
FoMs@nyq. [dB]	145.7	154.9	146.5	150	145.5	147.7
Area [mm^2]	0.23*	0.006	0.103	0.095	0.085	0.22
Input Buffer**	v	N/A	v	N/A	v	N/A

* T-Coil not included
** Input Buffer
v: include buffer

Figure 11.6.6: Performance summary and comparison with state-of-the-art ADCs.

subADC 1ch (0.006mm²)

SAR ADC — VTC — TDC

Figure 11.6.7: Chip micrograph.

- 2026 IEEE International Solid-State Circuits Conference

ISSCC 2026 / SESSION 11 / PIPELINE AND ULTRA-HIGH-SPEED DATA CONVERTERS / 11.7

11.7 A 12b 12GS/s Two-Way Interleaved Pipeline ADC with Integrated Broadband RF VGA in 5nm

Haiyang Zhu[1], Larry Singer[1], Ron Kapusta[1], Dan Kelly[1], Tao Pan[1], Mike Hensley[2], Scott Bardsley[3], Chris Dillon[3], Daniel Rey-Losada[4], Kashif M. Sheikh[5], Chen-Kai Hsu[1], Zhao Li[6], Janet Brunsilius[4], Enrique Alvarez-Fontecilla[4], Robert Bishop[1], Paul Wilkins[1], Hassan L'Bahy[1], Adalberto Cantoni[1], Nuo Zhang[1], Kaung Myat San Oo[1], Cihan Asci[1]

[1]Analog Devices, Wilmington, MA, [2]Analog Devices, Austin, TX, [3]Analog Devices, Durham, NC, [4]Analog Devices, San Diego, CA, [5]Analog Devices, Ottawa, Canada
[6]Analog Devices, Toronto, Canada

Abstract

A 12b 12GS/s ADC with an integrated broadband RF VGA is implemented in a 5nm FinFET process. This combination achieves a programmable input range of 36dB from 250mV$_{ppd}$ to 16V$_{ppd}$, bandwidth of 8.5GHz, SNDR of 49dB, SFDR of 70dBc (with a 5.1GHz, -1dBFS input tone), and a single interleaving spur (<-90dBc) between DC and F$_S$/2. Two pipeline-stage architectures, "Flash Early-Trigger" and "Embedded SHA", enable 6GS/s in a 12b sub-ADC, which is key to achieve two-way interleaving at 12GS/s.

Wireless communication infrastructure demands highly integrated radio platforms. These systems benefit from direct-RF-sampling receivers but must meet strict requirements: bandwidth, noise spectral density (NSD), spectral purity, performance consistency across a widely adjustable input range (to withstand large blockers), and power consumption. Integrating the RF variable-gain amplifier (VGA) that drives the ADC simplifies system design significantly. This paper presents a 12b 12GS/s ADC with an integrated broadband RF VGA in a 5nm FinFET process. This combination achieves a programmable input range of 36dB from 250mV$_{ppd}$ to 16V$_{ppd}$ (peak-to-peak differential), bandwidth (BW) of 8.5GHz, SNDR of 49dB, SFDR of 70dBc (with a 5.1GHz, -1dBFS input tone at 12GS/s), and an output spectrum with very few spurs. The power consumption is 1.7W including all analog, calibration, and digital functions.

Interleaving (IL) is essential to achieve sampling rates (F$_S$) above 10GS/s [1-6]. However, reducing the IL factor by pushing the sub-ADC sampling rate mitigates the IL overhead (ADC-driver load, clock distribution, etc.) [7]. Furthermore, two-way IL has advantages over N-way IL (N>2) that the literature has not adequately recognized. First, in N-way IL, mismatches in offset, gain, and timing skew across sub-ADCs lead to multiple IL spurs. By contrast, two-way IL produces a single IL spur between DC and F$_S$/2, providing much wider spur-free regions (Fig. 11.7.1 top). SFDR does not reflect this advantage because it considers only the largest spur, ignoring the number and location of spurs that critically impact frequency planning. Second, N-way IL leads to noise humps around multiples of F$_S$/N, caused by flicker noise uncorrelated across sub-ADCs (Fig. 11.7.1 middle). This blocks out some frequency regions due to degraded NSD, which complicates frequency planning. By contrast, in two-way IL, flicker-noise humps appear only near DC and F$_S$/2, regions commonly unused especially in direct RF sampling. The SNR reported in the literature is often averaged over the full spectrum. Particularly, if the output is decimated for measurement convenience. This aliases the noise and results in misleadingly white spectra that hide the humps.

Various IL mismatch calibrations mitigate spurs, but they have shortcomings and do not suppress flicker noise humps outside the (typically narrow) calibration-loop bandwidth. Sub-ADC randomization [6] spreads all IL spurs and flicker noise across the spectrum at the cost of complexity and power. Random chopping of input switches [8] whitens offset spurs and flicker noise but leaves gain and timing skew spurs unmitigated, degrades input BW, and complicates calibration. In this work, several techniques enable 6GS/s in a 12b sub-ADC. This is ≥2× the rate of any published sub-ADC of comparable resolution and enables two-way IL at 12GS/s.

The architecture is shown in Fig. 11.7.1 (bottom). The RF input is amplified by the VGA and alternately sampled by two sub-ADCs. Each is a pipeline ADC comprising a sample-and-hold amplifier (SHA0), nine stages (STAGE1-9), and a 3b flash. A digital block with a microprocessor (μP) and firmware (FW) runs sub-ADC and IL calibrations. Sub-ADC calibrations mitigate stage imperfections, as discussed later. IL calibrations estimate mismatches between the sub-ADCs (offset, gain, and timing skew) via signal autocorrelation. They correct offset and gain digitally; and timing skew by adjusting SHA0's sampling times.

Figure 11.7.2 shows the VGA simplified schematic. It is designed for high and consistent performance from 250mV$_{ppd}$ to 8V$_{ppd}$ (blockers) and to withstand up to 16V$_{ppd}$ inputs (overdrive protection). To tolerate large input voltage, no transistors connect to the inputs except ESD devices. Each segment of a passive binary-segmented attenuator [9] converts the input voltage to a current through a resistor and steers it to a low-impedance VCML node or a virtual ground, depending on the attenuation setting. This keeps a 50Ω single-ended input impedance, independent of setting. A transimpedance amplifier (TIA) converts the current signal back to voltage. The gain is chosen to balance power, noise, and linearity. It is set by the feedback resistor and maintained from 100MHz to 8.5GHz. MN0 and MP0 provide push-pull transconductance (g$_m$), and active cascodes (MN1, A1, MP1, A2) enhance linearity. A buffer drives the sub-ADCs and isolates the TIA from their kickback.

In a conventional pipeline stage [10], during the first half clock cycle, the flash and MDAC track the input simultaneously. In the next half cycle, the flash quantizes the input quickly

to satisfy the metastability-error target before the MDAC amplifies the residue (accurately to meet the ADC resolution). This is impractical in a 12b 6GS/s sub-ADC: both would need to complete within 80ps. Two pipeline-stage architectures are implemented: "Flash Early-Trigger" and "Embedded SHA". They are applied to SHA0+STAGE1 and STAGE2-9, respectively, to optimize the performance and power of individual stages. For MDAC residue amplification, both provide a full half cycle (80ps); for flash regeneration, 40 and 80ps, respectively. For communication applications, 40ps is sufficient to prevent metastability errors from degrading the ADC noise. This criterion is easier to meet than the code-error rate in oscilloscope-type applications.

Figure 11.7.3 depicts the schematic and timing of SHA0 and STAGE1. Flash1 and MDAC1 start tracking from a q1 rising edge. After ~40ps, Flash1 is "early-triggered" (qlatch falling edge) and regenerates within the remaining ~40ps, while MDAC1 continues tracking. Thus, valid Flash1 data is ready when MDAC1 finishes tracking, giving MDAC1 the full 80ps of the subsequent q2 for residue amplification. Since Flash1 and MDAC1 track at different times, a critical challenge is preserving MDAC1's redundancy range. SHA0 is introduced to sample the VGA, so that Flash1 and MDAC1 track the settling SHA0 output instead of the fast-moving VGA output. Flash1's sample still differs from MDAC1's due to incomplete SHA0 settling. However, with the chosen MDAC1 architecture, the ratio of the two samples is nearly signal-independent (due to linear settling), making it equivalent to a Flash1 gain error (Fig. 11.7.3 bottom left). A Flash1 gain calibration handles this error instead of lumping it into comparator offsets, thus saving offset-correction actuator range. First, comparator offsets are calibrated at power-up by shorting their inputs and adjusting actuators for 0.5 average outputs. This holds over temperature. Then, Flash1 gain is background-calibrated by examining the slope through the midpoints of the DAC transitions [11] and adjusting the Flash1 ladder voltages (VREFT_FL, VREFB_FL) for zero slope.

MDAC1 is closed-loop and uses a separate-DAC-capacitor architecture, both with well-known design trade-offs [6]. In addition, they offer two advantages for the SHA0-MDAC1 interface because during charge transfer (q2) the input capacitor C$_S$ is connected to a constant voltage V$_{CM}$. First, the memory of C$_S$ is eliminated. This allows SHA0 to settle linearly during q1, critical for Flash1 gain calibration. Second, while the SHA0 amplifier is disabled to save power during q2, its outputs can be left connected to C$_S$ and overridden by the switch S$_{SH}$. This removes an expensive linear switch between SHA0 and MDAC1, expediting SHA0 settling, improving linearity, and saving power.

MDAC1 employs an inverter-style amplifier (Fig. 11.7.3 bottom right) for speed and low power. Its push-pull topology doubles g$_m$ for a given bias current. All capacitances between transistor gate and drain, including parasitic ones, are absorbed into C$_{FB}$, preserving the feedback factor. The single-stage design ensures closed-loop stability at much lower power than a two-stage one. However, it suffers from low gain and poor non-linearity (NL), which would degrade SNDR. These are addressed via correlation-based inter-stage gain-error (IGE) calibration [5-6] and discrete-time non-linearity calibration (DNLC) [5], both supported by a calibration bit (dither_cal1). The absence of reset time causes inter-stage memory error (IME), which is calibrated via correlation with a delayed dither_cal1. The DAC mismatch and code-dependent reference settling cause INL breaks, also calibrated via DNLC [5]. The sub-ADC calibrations of all stages (that is, comparator offset, Flash1 gain, and MDAC calibrations above) improve the SNDR of each sub-ADC from 25 to 49dB.

Figure 11.7.4 depicts the schematic and timing of STAGE2. In each backend stage (STAGE2-9), an "Embedded SHA" (flip-around, gain=1) is inserted before the MDAC. This adds pipeline depth to double the time for Flash2 regeneration and MDAC2 residue amplification. During q2, Flash2 and SHA2 track the stage input. In the next q1, Flash2 regenerates for 80ps while MDAC2 tracks SHA2's output. As a result, the Flash2 data is ready before MDAC2 begins residue amplification, which leaves 80ps for settling.

STAGE1 used "Flash Early-Trigger" to avoid the noise, linearity, and power penalty of an "Embedded SHA". By contrast, in the backend stages, the costs of the "Embedded SHAs" are manageable. Their noise is attenuated by the gain of all preceding stages. Imperfections such as IGE, NL, and IME are lumped and calibrated together with those of the preceding MDAC, enabling a low-power implementation. Additionally, the "Embedded SHA" avoids

202 • 2026 IEEE International Solid-State Circuits Conference

979-8-3315-8937-0/26 $31.00 © 2026 IEEE

the precisely delayed clock (qlatch) required in "Flash Early-Trigger", reducing complexity and power; and longer regeneration time lowers comparators' power in the backend stages, where they account for a greater portion of the power budget.

Figure 11.7.5 shows measured performance of the VGA-ADC combination with all background calibrations running on-chip. Unless noted, data is taken at 250mV$_{ppd}$ input (0dB attenuation setting). Spectra with a -1dBFS tone at 1.15 and 5.15GHz show a single <-90dBc IL spur. NSD degrades only 0.4dB when sweeping the input level from -15 to -1dBFS, indicating minimal signal-dependent noise. Harmonic distortions are flat across a 30dB VGA input range up to 8V$_{ppd}$ (VGA input adjusted for a constant -1dBFS ADC input).

Figure 11.7.6 compares performance with state-of-the-art ADCs (≥10GS/s, ≥12b). This work is the only one to report an integrated VGA with 36dB input range and 8.5GHz BW. Also, the 6GS/s sub-ADC is ≥2× the rate of any published ≥12b sub-ADC, while achieving a competitive Schreier's FoM (148.5dB). Total power consumption is 1.7W: 200mW from VGA, 1.3W from ADC analog (including bias, reference, and clock), and 200mW from digital + μP. This work invests power for system-level benefits: first, it integrates a VGA; second, it reduces IL spurs dramatically and avoids noise humps by limiting IL to two-way (running the sub-ADCs very fast). Backend stages reuse the same STAGE2 design. Had they been properly scaled, ADC analog power would have improved by 45% and the ADC FoM by 2dB. Figure 11.7.7 shows the die microphotograph.

Acknowledgement:
The authors would like to acknowledge the support and guidance of S. Devarajan; the contributions of J. Gealow, C. Petersen, B. Wilcox, S. Parthasarathy, R. Schubert, R. Miller, A. DeFiore, R. Britton, E. Krommenhoek, D. Casey, K. Burke, A. Mulholland, C. Lay, J. Rioux, B. Luu, D O'Connor, S. Debnath, D. Dereli, D. Debolt, S. Prabhu, T. Parsons, P. Brown, C. Angell, D. McLaurin, H. Yektaii, S. Veluri, T. Gaiser, B. Swahn, A. Deignan, S. Whiston; and all others who supported or contributed to this work.

References:
[1] C.-E. Hsieh, et al., "A Power and Area Efficient 4nm Self Calibrated 12b/16GS/s Hierarchical Time Interleaving ADC," *2025 IEEE International Solid-State Circuits Conference (ISSCC)*, San Francisco, CA. USA, pp. 438-440, 2025. https://doi.org/10.1109/ISSCC49661.2025.10904740
[2] Y. Cao, et al., "A 12GS/s 12b 4× Time-Interleaved Pipelined ADC with Comprehensive Calibration of TI Errors and Linearized Input Buffer," 2024 *IEEE International Solid-State Circuits Conference (ISSCC)*, San Francisco, CA, USA, pp. 388-390, 2024. https://doi.org/10.1109/ISSCC49657.2024.10454350
[3] S. S. Kumar, et al., "A 750mW 24GS/s 12b Time-Interleaved ADC for Direct RF Sampling in Modern Wireless Systems," *2023 IEEE International Solid-State Circuits Conference (ISSCC)*, San Francisco, CA, USA, pp. 270-272, 2023. https://doi.org/10.1109/ISSCC42615.2023.10067793
[4] K.-J. Moon, et al., "A 12-bit 10GS/s 16-Channel Time-Interleaved ADC with a Digital Processing Timing-Skew Background Calibration in 5nm FinFET," *2022 IEEE Symposium on VLSI Technology and Circuits (VLSI Technology and Circuits)*, Honolulu, HI, USA, pp. 172-173, 2022. https://doi.org/10.1109/VLSITechnologyandCir46769.2022.9830208
[5] A. M. A. Ali, et al., "A 12-b 18-GS/s RF Sampling ADC With an Integrated Wideband Track-and-Hold Amplifier and Background Calibration", *IEEE Journal of Solid-State Circuits*, vol.55, no.12, pp. 3210-3224, 2020. https://doi.org/10.1109/JSSC.2020.3023882

[6] S. Devarajan, et al., "A 12-b 10-GS/s Interleaved Pipeline ADC in 28-nm CMOS Technology," in *IEEE Journal of Solid-State Circuits*, vol. 52, no. 12, pp. 3204-3218, Dec. 2017. https://doi.org/10.1109/JSSC.2017.2747758
[7] C.-H. Chan, et al., "The Race for the Extra Pico Second without Losing the Decibel: A Partial-Review of Single-Channel Energy-Efficient High-Speed Nyquist ADCs," *2024 IEEE Custom Integrated Circuits Conference (CICC)*, Denver, CO, USA, pp. 1-8, 2024. https://doi.org/10.1109/CICC60959.2024.10528981
[8] B. Vaz, et al., "A 13Bit 5GS/S ADC with Time-Interleaved Chopping Calibration in 16NM FinFET," *2018 IEEE Symposium on VLSI Circuits*, Honolulu, HI, USA, pp. 99-100, 2018. https://doi.org/10.1109/VLSIC.2018.8502306
[9] D. J. McLaurin, et al., "A Highly Reconfigurable 65nm CMOS RF-to-Bits Transceiver for Full-Band Multicarrier TDD/FDD 2G/3G/4G/5G Macro Basestations," *2018 IEEE International Solid-State Circuits Conference (ISSCC)*, San Francisco, CA. USA, pp. 162–163, 2018. https://doi.org/10.1109/ISSCC.2018.8310234
[10] I. Mehr, et al., "A 55-mW, 10-bit, 40-Msample/s Nyquist-rate CMOS ADC", in *IEEE Journal of Solid-State Circuits*, vol. 35, no. 3, pp. 318-325, Mar. 2000. https://doi.org/10.1109/4.826813
[11] A. M. A. Ali, et al., "A 14-bit 2.5GS/s and 5GS/s RF sampling ADC with background calibration and dither," *2016 IEEE Symposium on VLSI Circuits*, Honolulu, HI, USA, pp. 206-207, 2016. https://doi.org/10.1109/VLSIC.2016.7573537

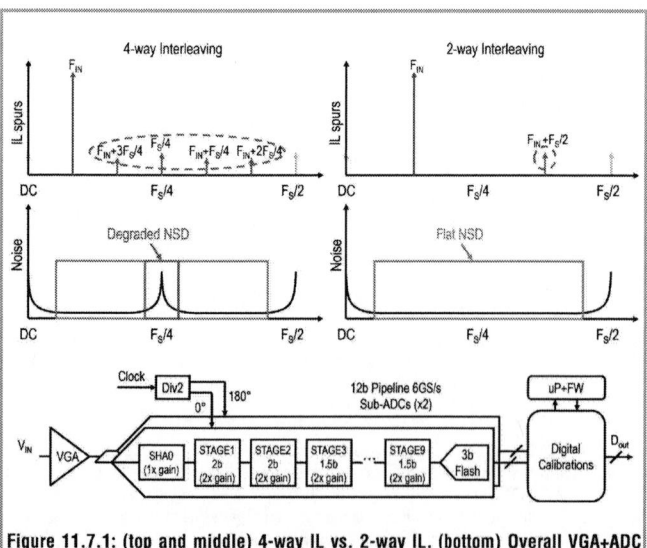

Figure 11.7.1: (top and middle) 4-way IL vs. 2-way IL. (bottom) Overall VGA+ADC architecture.

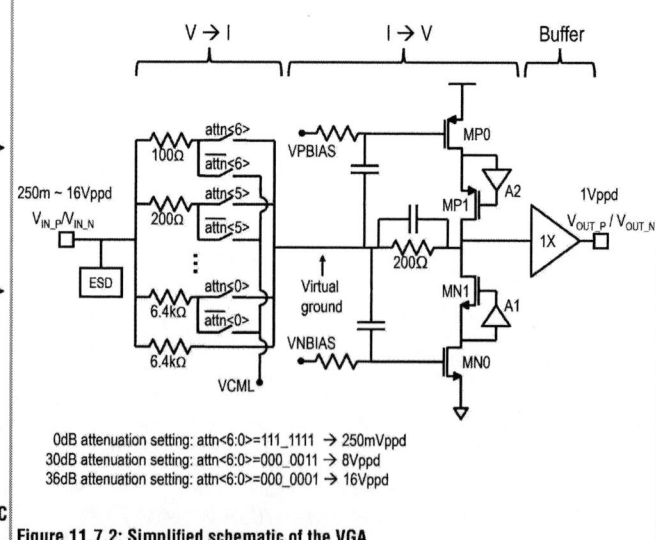

Figure 11.7.2: Simplified schematic of the VGA.

ISSCC 2026 / SESSION 11 / PIPELINE AND ULTRA-HIGH-SPEED DATA CONVERTERS / 11.7

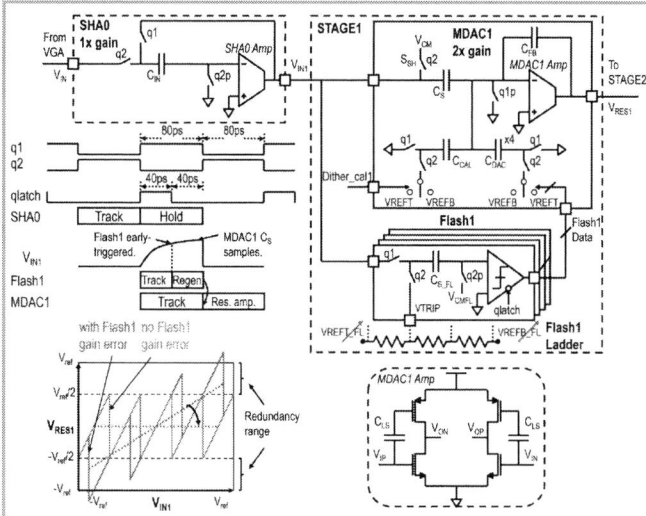

Figure 11.7.3: SHA0 and STAGE1 schematic, "Flash Early-Trigger" timing diagram, Flash1 gain calibration, and MDAC1 amplifier detail.

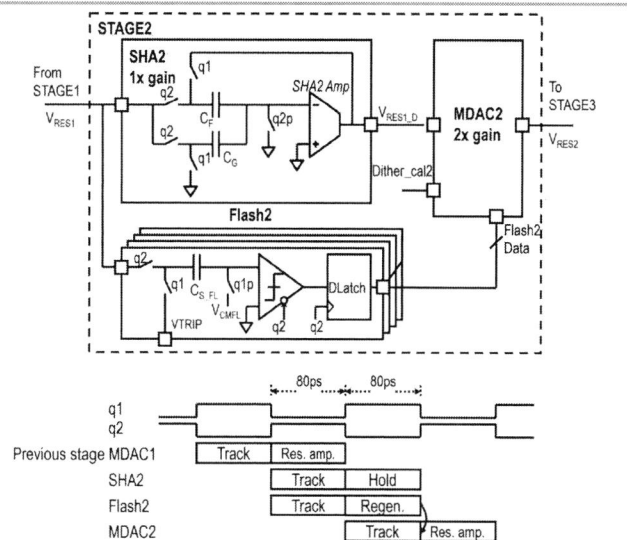

Figure 11.7.4: STAGE2 schematic and "Embedded SHA" timing diagram.

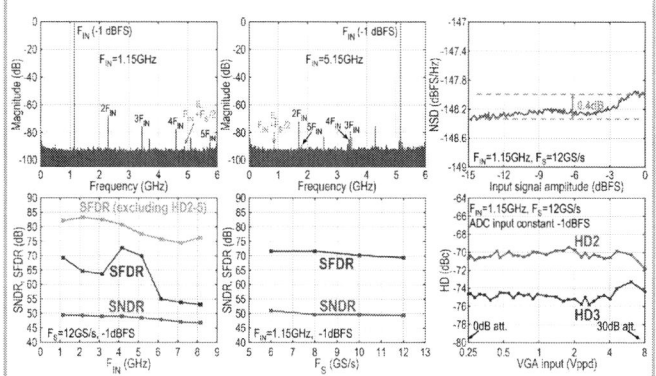

Figure 11.7.5: Measured performance.

Publication	This Work		Hsieh ISSCC 25 [1]	Cao ISSCC 24 [2]	Kumar ISSCC 23 [3]	Moon VLSI 22 [4]	Ali ISSCC 20 [5]		Devarajan ISSCC 17 [6]
	VGA+ADC	ADC							
Sampling Rate (GS/s)	12		16	12	24	10	18	10	10
Sub-ADC Sampling rate (GS/s)	6		1	3	0.3	0.625	2.25	2.5	1.4
Resolution (bits)	12		12	12	12	12	12		12
Architecture	IL Pipe		IL Pipe-SAR	IL Pipe	IL SAR	IL Pipe-SAR	IL Pipe		IL Pipe
Input Frequency (GHz)	5.15		7.4	5.71	7.24	4.8	8	4	4
Full Scale (Vppd)	0.25 (Min) 16 (Max)ª	1	1.1	1	0.7	-	1.4	1.4	1.4
SNDR at Fin (dB)	49.0	52.5ᵇ	47	54.1	46.5	48	48	53.5	55
SFDR at Fin (dBc)	70	-	58	66	76	61	54	65	66
SFDR (excluding HD2-5) at Fin (dBc)	78	-	66	66	76	61	-	-	80
NSD_small-signal (dBFS/Hz)	-148.3	-151.8ᵇ	-155	-	-147.5	-	-157	-157	-157
NSD_large-signal (dBFS/Hz)	-147.9	-151.2ᵇ	-147	-151.9	-147.3	-145	-147.5	-150.5	-152
Power (mW)	1700		1500	570	179.8 (see †)	750	625	1300	2900
FoM_S @ Fin (dB)	144.5	148.5	148	159.3 (see †)	148.5	147	146.4	151.5	147.4
Process Node (nm)	5		4	28	7	5	16		28
Calibration Type	On-chip BG/FGᶜ		On-chip BG/FG	Off-chip	On-chip BG/FG	On-chip BG	On-chip BG		On-chip BG

a: Performance consistent from 0.25 to 8Vppd.
b: Indirectly measured by subtracting VGA noise; ADC and VGA noise are determined by disabling VGA and shorting ADC inputs.
c: Comparator offset calibration is performed once during power-up and holds over temperature. All other calibrations are background.

† : Excluding digital calibration and reference buffers.

Figure 11.7.6: Performance summary and comparison.

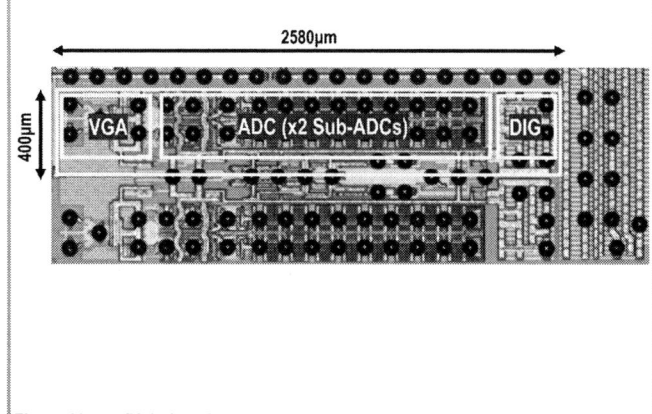

Figure 11.7.7: Die microphotograph.

ISSCC 2026 / SESSION 11 / PIPELINE AND ULTRA-HIGH-SPEED DATA CONVERTERS / 11.8

11.8 A 14b 20GS/s RF-Sampling DAC Achieving 70.4dBc IMD3 up to 8.9GHz

Hao Luo*, Zhijun Long*, Wentao Zhu, Haowei Lu, Yushen Fu, Mingqian Lei, Cheng Wang, Xiaoge Zhu, Yumei Diao, Haipeng Zhu, Jiajun Zhong, Li Tian, Yi Yang, Xinli Geng, Keqing Ouyang

Sanechips Technology, Shenzhen, China
*Equally Credited Authors (ECAs)

Abstract

This work presents a 14b 20GS/s RF-sampling DAC. By introducing a switch driver with process-and-temperature-adaptive (PT-adaptive) level shifting and enhanced low-crosspoint control, the DAC demonstrates an IMD3 of 70.8dBc at 7.2GHz and 16Gs/s, and a maximum

output power of 7dBm. An algorithm, named ISI-minimizing dynamic element matching (ISIM-DEM), is proposed to mitigate output distortion induced by inter-symbol interference (ISI) effect at 20GS/s sampling, achieving 73dBc IMD3 at 8.5GHz.

5G-A and 6G applications depend heavily on data converters with multi-GHz signal bandwidth and ever-increasing carrier frequency, while at the same time the trend of IF/RF integration requires an advanced technology node and good power efficiency. In [1] a mixing-DAC is used to put the carrier in 2nd and 3rd Nyquist zones, which leads to additional duty cycle error and more images in the spectrum, as compared with direct synthesis. Reference [2] uses a capacitive DAC as a direct RF DAC over a bandwidth from 0.5GHz to 8GHz, which cannot support DC signals, and the impact of code-dependent source impedance and of the relatively large data driving load need be considered.

This work presents a 14b 20GS/s RF-sampling DAC achieving a bandwidth from DC to 8.9GHz. By introducing a switch driver with process-and-temperature-adaptive (PT-adaptive) level shifting and enhanced low-crosspoint control, the DAC demonstrates an IMD3 of 70.8dBc at 7.2GHz and 16GS/s, and a maximum output power of 7dBm without negative supply. And an algorithm, named ISI-minimizing dynamic element matching (ISIM-DEM), is proposed to mitigate output distortion induced by inter-symbol interference (ISI) effects at 20GS/s sampling, achieving 70.4dBc IMD3 at 8.9GHz. In addition, a technique of toggle charge compensation is proposed to realize an output phase deviation of <5 deg and an output power control range of >29dB when DAC full-scale attenuates from 0dB to 30dB.

Figure 11.8.1 depicts the overall architecture of the RF DAC. The 14b DAC design is split into 5-3-6 unary-unary-binary segmentation. Sixteen parallel data streams at 1.25GS/s are sent to the decoder that performs data format conversion and serialization. The output 4-phase 5GS/s data is fed to serializer firstly and then serialized up to 20GS/s in the switch driver, which switches current cells in the DAC core. In the DAC core, a source degeneration resistor is added to the current source array and placed above a MOS device to improve matching so that no extra calibration is needed. A 6-level horizontal output tree is placed above the switch array by using M15/M16/AP, which reduces die area without compromising performance.

To compensate for the subsequent PA power fluctuation, DAC output is required with constant phase and linearly changed amplitude over the whole full-scale range. However, data toggles at the switch input inject a relatively constant charge to the DAC output through switch gate-drain cap, leading to amplitude non-linearity and phase shift versus DAC full-scale. In this design, two source-floating MOS capacitors (M3/M4) are introduced to compensate the effect of toggle charge at the DAC output.

Figure 11.8.2 shows the implementation of the data path in our DAC. For the proposed PT-adaptive level-shifter in the switch driver, a V_{sw_on} is firstly generated by presetting the source of a DAC switch replica device, and then buffered to shift the low level of the output data in switch driver. A set of super source followers (used as buffers) are distributed locally across switch driver units, with each buffer incorporating a dynamic current bias circuitry to absorb the driver toggle current. The switch cascode bias (V_{sw_cas}) is generated similarly to the V_{sw_on} (as drawn in Fig. 11.8.1). As a result, constant source potentials are maintained for both the switch and switch cascode devices in the DAC core over process corners and temperatures, ensuring well-controlled headroom for the tradeoff between DAC linearity and output swing.

For the proposed low-crosspoint output stage in switch driver, PM1/PM5 are stacked to improve random skew in pull-up branch, and NM1/PM7 are parallel for robust pulldown across MOS process corners. During fast low-to-high transitions at D_{mid}/Db_{mid}, the PMOS stack node goes to a high-Z state and overshoots above VDD following with a very slow recovery. If D_{mid}/Db_{mid} is switched back to low at different time points during recovery, memory timing skew occurs. PM3/PM4 are introduced to remove the skew, which rapidly pulls the stack node to VDD during recovery and gets turned off at low inputs.

When a DAC bit toggles continuously every clock cycle, its output behavior differs from non-continuous toggling, creating ISI error. To address this, we developed the ISIM-DEM algorithm in the decoder that dynamically shifts the MSB "1" sequence position to minimize continuous bit toggling. As illustrated in Fig. 11.8.3(a), if the "1" sequence of the current cycle is longer than that of the previous cycle, its start position remains unchanged. If shorter, its end position is aligned with the endpoint of the previous cycle. As a result, the

"1-priority zone" of the current cycle is always positioned around the endpoint of the previous cycle, and maximally occupied by "1" sequence, while the "0-priority zone" is protected from "1" sequence intrusion. New spurs may emerge due to the rotational "1" sequence movement among MSB locations with different analog errors. To suppress these spurs, MSB bits are sorted and remapped according to timing/amplitude error, and a randomization option is proposed to control the movement direction of the "1" sequence. When $D_{k-2} \geq D_{k-1} \geq D_k$ or $D_{k-2} \leq D_{k-1} \leq D_k$, the "1" sequence can randomly move left or right without introducing extra ISI compensation error. While [3] describes a similar random direction algorithm to mitigate mismatch and glitch errors, the proposed algorithm further suppresses ISI error by minimizing continuous MSB toggling. Figure 11.8.3(b) shows the implementation block diagram of the ISIM-DEM. Figure 11.8.3(c) compares the SFDR test result at 7.577GHz and 18GS/s, and the ISIM-DEM provides a 12dB improvement in HD3.

Figure 11.8.4 shows the output power and phase versus DAC full-scale attenuation. With toggle charge compensation, <5deg phase deviation and >29dB power control range are measured with 0~30dB attenuation, as compared to >80deg and <20.7dB without this technique. Figure 11.8.5 shows the IMD3, SFDR, fullscale output power and NSD test results as a function of output frequency. At 8.9GHz and 20GS/s, the ISIM-DEM improves SFDR over Nyquist BW and IMD3 by 13.8dB and 4dB, respectively. Figure 11.8.6 shows a performance summary table. At 20GS/s, this DAC consumes 530mW from 0.95V/1.8V analog supply as well as 0.95V digital LDO supply. Figure 11.8.7 shows the die micrograph.

Acknowledgement:
The authors would like to thank the entire team of this project including system, analog, digital, layout, test and CAD for their dedication and support. The corresponding author is Keqing Ouyang (ouyangkeqing@sanechips.com.cn).

References:
[1] C. Erdmann et al., "A 330mW 14b 6.8GS/s dual-mode RF DAC in 16nm FinFET achieving –70.8dBc ACPR in a 20MHz channel at 5.2GHz," *2017 IEEE International Solid-State Circuits Conference (ISSCC)*, San Francisco, CA, USA, 2017, pp. 280-281, doi: 10.1109/ISSCC.2017.7870370. https://ieeexplore.ieee.org/document/7870370

[2] D. Gruber et al., "A 12b 16GS/s RF-Sampling Capacitive DAC for Multi-Band Soft-Radio Base-Station Applications with On-Chip Transmission-Line Matching Network in 16nm FinFET," *2021 IEEE International Solid-State Circuits Conference (ISSCC)*, San Francisco, CA, USA, 2021, pp. 174-176, doi: 10.1109/ISSCC42613.2021.9365744. https://ieeexplore.ieee.org/document/9365744

[3] M. -H. Shen, J. -H. Tsai and P. -C. Huang, "Random Swapping Dynamic Element Matching Technique for Glitch Energy Minimization in Current-Steering DAC," in IEEE Transactions on Circuits and Systems II: Express Briefs, vol. 57, no. 5, pp. 369-373, May 2010, doi: 10.1109/TCSII.2010.2043400. https://ieeexplore.ieee.org/document/5466568

[4] C. Huang et al., "A 16-bit 10-GS/s DAC Achieving >65-dBc SFDR and <−75-dBc IM3 up to the Nyquist in 28-nm CMOS," in *IEEE Journal of Solid-State Circuits*, doi: 10.1109/JSSC.2025.3583372. https://ieeexplore.ieee.org/document/11084876

[5] B. Koo et al., "A 12-bit 16GS/s Single-Channel RF-DAC with Hybrid Segmentation for Digital Back-Off and Code-Dependent Free Switch Driver Achieving –85dBc IMD3 in 5nm FinFET," *2024 IEEE Symposium on VLSI Technology and Circuits (VLSI Technology and Circuits)*, Honolulu, HI, USA, 2024, pp. 1-2, doi: 10.1109/VLSITechnologyandCir46783.2024.10631542. https://ieeexplore.ieee.org/document/10631542

[6] W. -H. Tseng et al., "A 14b 16GS/s Time-Interleaving Oirect-RF Synthesis Oae with T-OEM Achieving -70dBc IM3 up to 7.8GHz in 7nm," *2023 IEEE International Solid-State Circuits Conference (ISSCC)*, San Francisco, CA, USA, 2023, pp. 268-270, doi: 10.1109/ISSCC42615.2023.10067459. https://ieeexplore.ieee.org/document/10067459

[7] H. -Y. Huang and T. -H. Kuo, "A 0.07-mm2 162-mW DAC Achieving >65 dBc SFDR and < −70 dBc IM3 at 10 GS/s With Output Impedance Compensation and Concentric Parallelogram Routing," in *IEEE Journal of Solid-State Circuits*, vol. 55, no. 9, pp. 2478-2488, Sept. 2020, doi: 10.1109/JSSC.2020.2993672. https://ieeexplore.ieee.org/document/9098055

ISSCC 2026 / February 17, 2026 / 11:20 AM

[8] C. -H. Lin et al., "A 16b 6GS/S nyquist DAC with IMD <-90dBc up to 1.9GHz in 16nm CMOS," *2018 IEEE International Solid-State Circuits Conference - (ISSCC), San Francisco, CA, USA*, 2018, pp. 360-362, doi: 10.1109/ISSCC.2018.8310333. https://ieeexplore.ieee.org/document/8310333

[9] S. Su and M. S. -W. Chen, "A 16-bit 12-GS/s Single-/Dual-Rate DAC With a Successive Bandpass Delta-Sigma Modulator Achieving <-67-dBc IM3 Within DC to 6-GHz Tunable Passbands," *in IEEE Journal of Solid-State Circuits*, vol. 53, no. 12, pp. 3517-3527, Dec. 2018, doi: 10.1109/JSSC.2018.2871143. https://ieeexplore.ieee.org/document/8485418

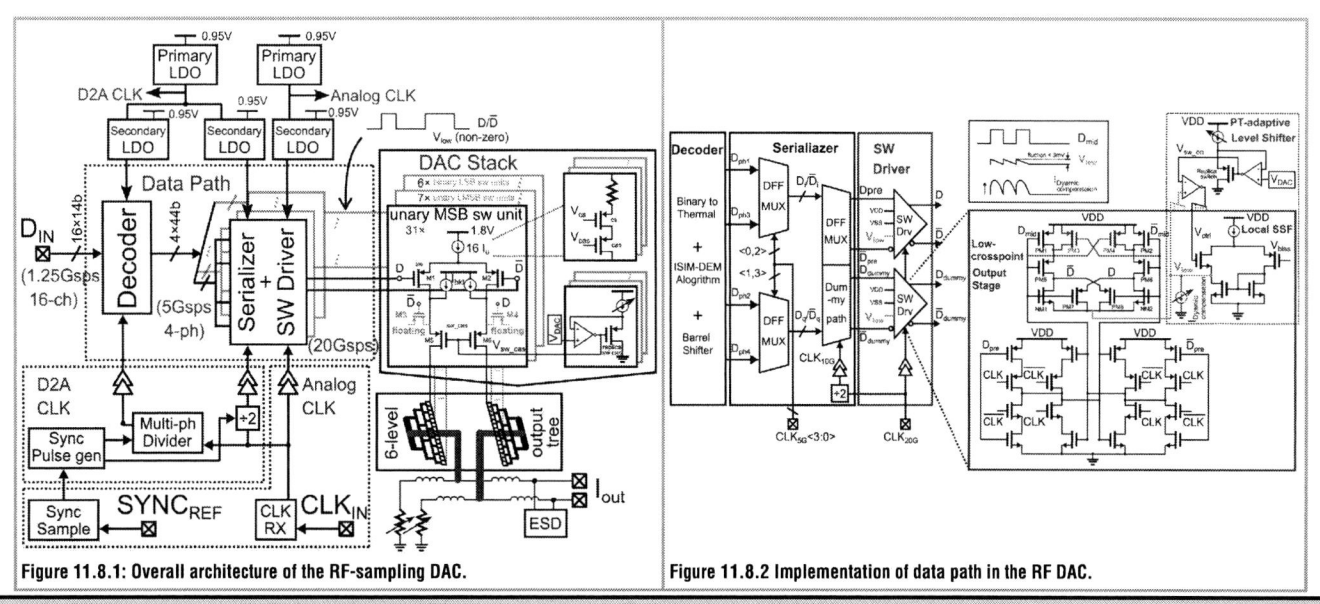

Figure 11.8.1: Overall architecture of the RF-sampling DAC.

Figure 11.8.2 Implementation of data path in the RF DAC.

ISSCC 2026 / SESSION 11 / PIPELINE AND ULTRA-HIGH-SPEED DATA CONVERTERS / 11.8

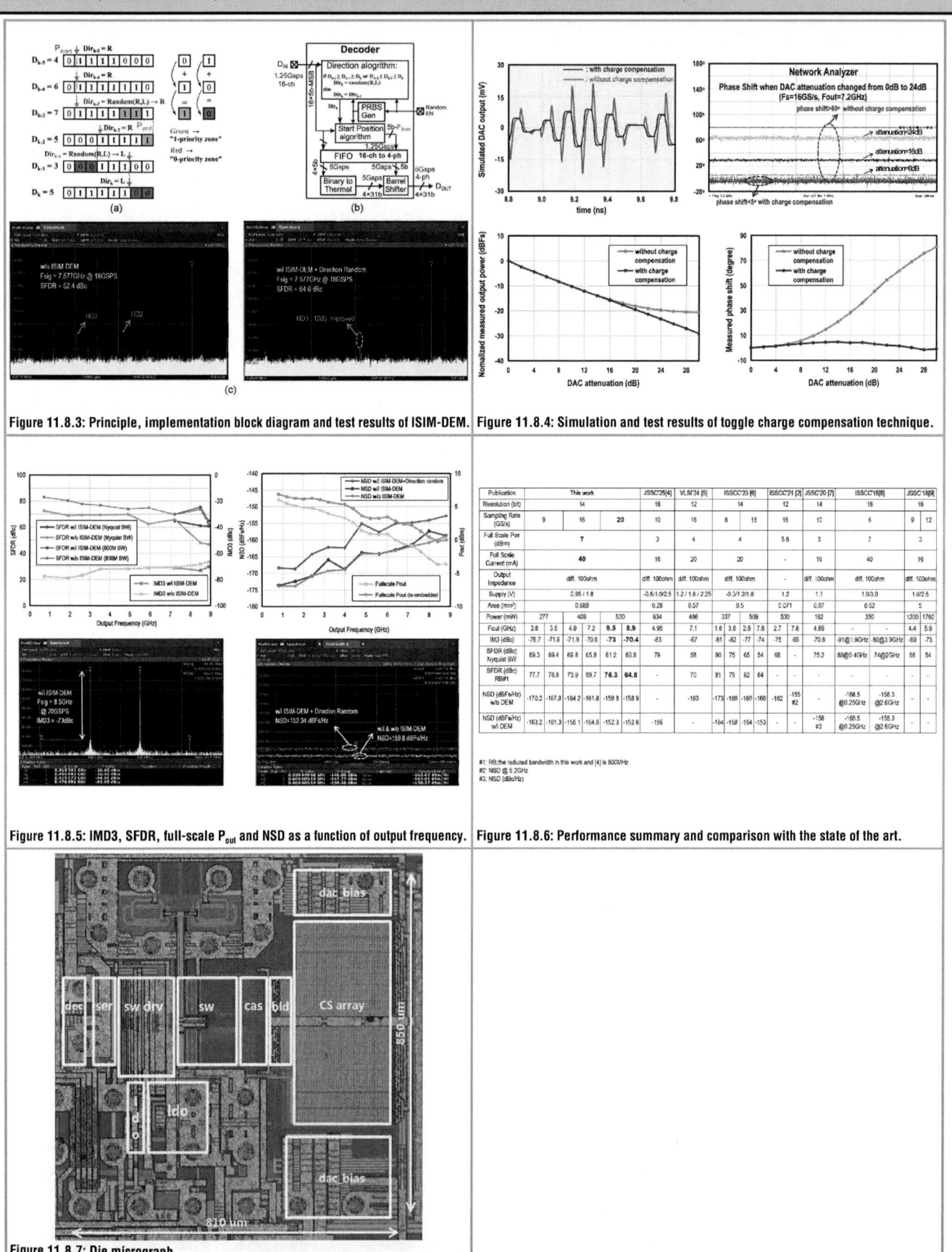

Figure 11.8.3: Principle, implementation block diagram and test results of ISIM-DEM.

Figure 11.8.4: Simulation and test results of toggle charge compensation technique.

Figure 11.8.5: IMD3, SFDR, full-scale P_out and NSD as a function of output frequency.

Figure 11.8.6: Performance summary and comparison with the state of the art.

Figure 11.8.7: Die micrograph.

• 2026 IEEE International Solid-State Circuits Conference

979-8-3315-8937-0/26 $31.00 © 2026 IEEE

ISSCC 2026 / SESSION 12 / FREQUENCY SYNTHESIZERS AND VCOs / OVERVIEW

Session 12 Overview: *Frequency Synthesizers and VCOs*
RF SUBCOMMITTEE

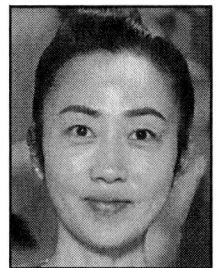

Session Chair: Dmytro Cherniak
Marvell Technology, Santa Clara, CA

Session Co-Chair: Ping Lu
NVIDIA, Seattle, WA

High-performance frequency synthesizers are essential building blocks of any wireless and wireline communication systems. As data rates increase and high-order modulation schemes are adopted, the stringent requirements for low phase noise become increasingly critical to maintain system performance and minimize bit-error rates. This session features nine papers on ultra-low-noise analog and digital PLLs and series-resonance VCOs.

8:00 AM

12.1 A 74fs-Jitter, −59dBc-Spur Fractional-N DPLL Using a Supply-Resilient Time-Amplifying Dual-Ramp DTC
Rishabh Gurbaxani, TU Delft, Delft, The Netherlands
In Paper 12.1, the Delft University of Technology presents a fractional-N DPLL with a supply-resilient time-amplifying dual-ramp DTC, achieving a 74fs jitter and -59dBc fractional spur.

8:25 AM

12.2 A Fractional-N Digital PLL with a Supply-Insensitive DTC Achieving −62dBc Spur and 69fs Jitter Under 10mV$_{pp}$ Sinusoidal DTC Supply Ripple and 6.2mV$_{rms}$ DTC Supply Noise
Damiano Fagotti, Politecnico di Milano, Milan, Italy
In Paper 12.2, Politecnico di Milano presents a fractional-N DPLL with supply-insensitive DTC, achieving a 69fs jitter under 10mVpp sinusoidal supply ripple and 6.2mV$_{rms}$ supply noise.

8:50 AM

12.3 A −66dBc-Worst-Fractional-Spur and 58fs-Jitter Fractional-N Digital PLL Using a Supply-Resilient Pseudo-Differential Inverse-Constant-Slope DTC
Pietro Salvi, Politecnico di Milano, Milan, Italy
In Paper 12.3, Politecnico di Milano presents a fractional-N DPLL with supply-resilient pseudo-differential inverse-constant-slope DTC, achieving a 58fs jitter and -66dBc fractional spur.

206 • 2026 IEEE International Solid-State Circuits Conference

979-8-3315-8937-0/26 $31.00 © 2026 IEEE

ISSCC 2026 / February 17, 2026 / 8:00 AM

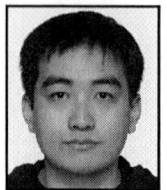

9:15 AM

12.4 A 21.6fs$_{rms}$-Jitter, -260.7dB-FoM Fractional-N PLL Enabled by an Intrinsically Linear Variable-Slope SPD for Quantization-Error Cancellation

Yanchao Liu, Fudan University, Shanghai, China

In Paper 12.4, Fudan University presents a fractional-N PLL with intrinsically linear variable-slope sampling phase detector, achieving a 21.6fs jitter and -260.7dB FoM.

10:05 AM

12.5 A 14GHz Chopper-Refolding Sampling PLL Achieving 33.8fs$_{rms}$ and 80.8dBc Reference Spur with a kT/C-Noise-Cancellation SPD

Yichen Liu, Tsinghua University, Beijing, China

In Paper 12.5, Tsinghua University presents a 14GHz PLL with kT/C noise-cancellation sampling phase detector, achieving a 33.8fs jitter and -257.1dB FoM.

10:30 AM

12.6 A 0.65-to-1V-V$_{DD}$ 10.5-to-11.85GHz Fractional-N Sampling PLL Achieving 71.47fs Integrated Jitter and <-60dBc Near-Integer Fractional Spur in 40nm CMOS

Yixi Li, Institute of Semiconductors, Chinese Academy of Sciences, Beijing, China, University of Chinese Academy of Sciences, Beijing, China

In Paper 12.6, the Chinese Academy of Sciences presents a fractional-N sampling PLL enabled for low-voltage operation, achieving a 71.4fs jitter and below -60dBc fractional spur.

10:55 AM

12.7 A 7.15-to-7.95GHz Magnetically Enhanced Feedforward Waveform-Shaping CMOS Oscillator with Implicit Common-Mode Noise Cancellation Achieving -146.72dBc/Hz PN@1MHz and 190.6dBc/Hz FoM

Rui Ma, Tsinghua University, Beijing, China

In Paper 12.7, Tsinghua University presents a 7.15-to-7.95GHz oscillator with magnetically enhanced feedforward waveform shaping and implicit common-mode noise cancellation, achieving -146.7dBc/Hz phase noise at a 1MHz offset.

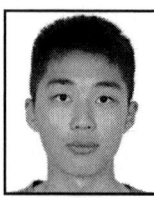

11:20 AM

12.8 A 5.7mW@0.55V-to-50mW@0.9V Deeply Power-Scalable Reconfigurable Series-Resonance/Class-F VCO with Mutual-Inductance Self-Cancellation and Hybrid 8-Shaped Coupling Techniques

Juntao Lan, Tsinghua University, Beijing, China

In Paper 12.8, Tsinghua University presents a power-scalable reconfigurable series-resonance/Class-F VCO utilizing mutual-inductance self-cancellation and hybrid-coupling techniques, achieving -149dBc/Hz phase noise at a 10MHz offset and offering power scalability from 5.7 to 93.1mW.

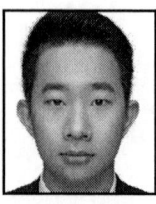

11:35 AM

12.9 A 10.2-to-16.2GHz Dual-Mode-Transformer-Based Wideband Series-Resonance VCO Achieving >201.1dBc/Hz FoM$_T$ at a 10MHz Offset

Yang Li, Fudan University, Shanghai, China

In Paper 12.9, Fudan University presents a 10.2-to-16.2GHz transformer-based series-resonance VCO with dual-mode operation, achieving -145.2dBc/Hz phase noise at a 10MHz offset.

ISSCC 2026 / SESSION 12 / FREQUENCY SYNTHESIZERS AND VCOs / 12.1

12.1 A 74fs-Jitter, −59dBc-Spur Fractional-N DPLL Using a Supply-Resilient Time-Amplifying Dual-Ramp DTC

Rishabh Gurbaxani[1], Cicero S. Vaucher[1,2], Masoud Babaie[1]

[1]TU Delft, Delft, The Netherlands, [2]NXP Semiconductors, Eindhoven, The Netherlands

Abstract

This work presents a fractional-N digital PLL that employs a supply-resilient, time-amplifying dual-ramp DTC, which offers 8× time amplification and maintains its linearity under supply variations. Additionally, a DTC code-randomization technique is included to enhance the DTC gain-calibration accuracy and convergence speed at near-integer channels. The PLL achieves <−59dBc fractional spurs and 74fs rms jitter for a stable supply and 88.5fs rms jitter for a 29mV$_{rms}$, 10MHz-bandwidth supply noise.

Exploiting narrow-input-range phase detectors (PDs), such as bang-bang (BBPD) [1-3] or subsampling [4,5] architectures, improves the jitter–power FoM of PLLs by enhancing the PD linearity and gain, thereby suppressing noise from subsequent loop components. To implement this in fractional-N PLLs, a digital-to-time converter (DTC) [6-12] is typically used to cancel the $\Delta\Sigma$-modulator (DSM) quantization error before the PD, with its gain K_{DTC} estimated by correlating the PD output to the DSM error sequence using a least-mean-square (LMS) calibration technique [6,7]. However, for near-integer channels, PLL dynamics strongly attenuate errors originated from inaccurate K_{DTC} estimation at the PD output, prolonging calibration, reducing K_{DTC} calibration accuracy, and elevating fractional spurs [13]. Frequency-control-word (FCW) subtractive dithering [14] and cascaded-fractional-divider [15] techniques have been proposed to mitigate this issue; however, the former requires an additional calibration to compensate for digitally controlled oscillator (DCO) duty-cycle error, while the latter requires an extra full-range DTC, thereby degrading the PLL jitter.

The second issue is the sensitivity of prior-art variable-slope [6,16], constant-slope (CS) [17,18], and inverse-constant-slope (ICS) [1,14] DTC architectures to dynamic in-band supply voltage variations, which can couple to the PD from other circuits in real-life applications. Given the low-jitter performance of modern DTCs and PDs, PLL bandwidths (BWs) are typically in the MHz range to optimize the PLL jitter-power FoM. Consequently, DTC and PD distortions caused by supply variations in the 10kHz-to-1MHz range cannot be compensated either by the PLL, as they lie within the loop BW, or by K_{DTC} gain calibration, which is too slow to track fast supply changes, leading to significant spur degradation.

This work presents a fractional-N digital PLL (DPLL) that achieves <−59dBc fractional spurs and 74fs rms jitter at near-integer channels by introducing: 1) a time-amplifying dual-ramp (TADR) DTC that offers 8× embedded time amplification to suppress the PD noise and maintains its linearity under in-band supply variations, and 2) a DTC code-randomization technique that enhances the K_{DTC} calibration accuracy and convergence speed at near-integer channels with minimal hardware overhead and jitter penalty.

Figures 12.1.1 and 12.1.2 illustrate the evolution and operating principles of the proposed TADR DTC. As a baseline, the resistor-based ICS DTC from [1] is adopted due to its excellent jitter and linearity performance under stable-supply conditions. The circuit consists of a capacitor (C) precharged to the supply voltage (V_{DD}) before each phase-detection cycle, along with two resistors, R and R/M, that discharge C in two steps.

In the first step, C discharges with a time constant $\tau_{in} = RC$, triggered by an input pulse (IN_P) of width mT_{DCO}, where T_{DCO} is the DCO period and m is a programmable integer determined from the accumulated DSM quantization time error (T_Q). In the second step, the divider rising edge (t_{div}) acts as the DTC input, further discharging C with a time constant $\tau_o = RC/M$. When the capacitor voltage (V_C) falls below the comparator threshold (V_{th}), the comparator generates the DTC output edge (t_o). The phase comparison is then performed between the reference edge (t_{ref}) and t_o using a time-to-digital converter (TDC). In this scheme, the DTC delay (T_{DTC}) equals $\tau_o \ln(V_{DD}/V_{th}) + t_d - (\tau_o/\tau_{in})mT_{DCO}$, where t_d is the comparator delay. The term mT_{DCO}/M cancels the deterministic T_Q and remains largely insensitive to PVT, since T_{DCO} is precisely stabilized by the PLL. Meanwhile, a Type-II PLL adjusts t_{div} to compensate for T_{DTC} time offset: $\tau_o \ln(V_{DD}/V_{th}) + t_d$. However, because this offset is strongly V_{DD} dependent, in-band supply fluctuations directly modulate T_{DTC}, producing spurs that the PLL cannot suppress.

To mitigate this, a dual-ramp DTC is proposed, comprising two ICS DTCs—main and auxiliary—as shown in Fig. 12.1.1b. The main branch retains the original ICS structure with $\tau_{in,M} = RC$ and launches an output rising edge: $t_{o,M} = t_{div} + \tau_o \ln(V_{DD}/V_{th}) + t_d - mT_{DCO}/M$. In the auxiliary branch, the first discharge having time constant $\tau_{in,A} = RC/M$, is triggered by an input pulse (IN_{PD}) spanning the divider and reference rising edges, yielding a pulse width of $t_{div} - t_{ref} = T_Q + t_n$, where t_n denotes the random jitter. Next, t_{div} initiates the discharge of the auxiliary capacitor with the same RC/M time constant, after which the auxiliary comparator generates an output edge: $t_{o,A} = t_{div} + \tau_o \ln(V_{DD}/V_{th}) + t_d - (T_Q + t_n)$. The phase error is now encoded in the time difference between the main and auxiliary outputs. Since the output discharging step is common to both branches, the terms $\tau_o \ln(V_{DD}/V_{th}) + t_d$ are naturally canceled, suppressing supply-induced variations and resulting in-band spurs. Furthermore, since $\tau_o/\tau_{in} = 1/M$ and $\tau_o/\tau_{in,A} = 1$, the condition $m=MT_Q/T_{DCO}$ is preserved, leaving only t_n at the PD output.

Although the dual-ramp DTC reduces supply sensitivity, adding the auxiliary branch doubles the power consumption (P_{DC}) and KT/C-induced jitter power ($\sigma^2_{j,c}$), degrading the jitter-power FoM ($\sigma^2_{j,c}P_{DC}$) by 4×. However, since $\sigma_{j,c}$ scales with the DTC maximum required delay, the PLL locking point is adjusted so that T_Q can assume both positive and negative polarity [20]. This effectively halves the maximum delay, reducing $\sigma^2_{j,c}$ by 4× and restoring the FoM. As shown in Fig. 12.1.2a, the additional negative time shifts are achieved by introducing extra discharge branches and swapping IN_P and IN_{PD} based on T_Q polarity. With the maximum delay halved, all time constants are scaled by reducing the resistors by 2×, while capacitors remain unchanged to preserve DTC jitter.

After achieving supply immunity and preserving the DTC FoM, an implicit time-amplification gain of M is incorporated to suppress TDC quantization noise. As shown in Fig. 12.1.2b, this is achieved by increasing $\tau_o/\tau_{in,M}$ and $\tau_o/\tau_{in,A}$ by a factor of M. This modification increases the maximum delay of both the main and auxiliary branches by M times, raising the KT/C-induced jitter power at the TADR-DTC output by M^2. However, due to the time-amplification factor M, the input-referred KT/C-induced jitter power remains unchanged. Meanwhile, the slopes of the main and auxiliary ramps during the output discharging phase decrease by M, which increases the contribution of comparator noise to the output jitter power by M^2 [13]. Nevertheless, when referred back to the input, this contribution remains unchanged, realizing time amplification with no jitter penalty.

The proposed TADR DTC is integrated into a DPLL, as shown in Fig. 12.1.3a. The phase-frequency detector (PFD) [19] extracts the IN_{PD} pulse, and the T_{DCO}-Scaler [14] provides programmability of the IN_P pulse width. The phase error is extracted using a differential vernier TDC [13], whose multi-bit output signal improves the LMS convergence speed compared to a BBPD [10], and its resolution, determined by difference of delay elements, exhibits resilience to supply variations. The TADR DTC offers a dynamic range of $\pm T_{DCO}$ to cover the quantization error introduced by the MASH 1–1 DSM controlling the PLL division ratio. However, its coarse resolution of T_{DCO}/M ($\approx$16ps at $f_{DCO}\approx$8GHz with M=8) is insufficient to suppress the DSM quantization noise below the PLL in-band noise floor. Therefore, a fine variable-slope DTC [6] is inserted in the reference path, covering the T_{DCO}/M range with an $\approx$125fs resolution. Because the fine DTC has a narrow dynamic range, its nonlinearity has a negligible impact on the PLL spur performance. Its gain, $K_{DTC,f}$, however, is PVT sensitive and must be continuously tracked by a background LMS calibration loop. Monte Carlo simulations in Fig. 12.1.3b show that device mismatch in the TADR DTC causes $\tau_o/\tau_{in,M}$ and $\tau_o/\tau_{in,A}$ to deviate from their nominal values. This results in distinct gain errors for positive and negative T_Q cancellations, since different circuit branches are active in each case. Two additional LMS calibration loops estimate these gain errors.

For the near-integer channels, the PLL dynamics strongly attenuate errors originated from an inaccurate K_{DTC} estimation at the PD output, degrading the LMS loop accuracy and convergence speed. To address this, a pseudo-random 0–1 sequence (R_n) is added to the TADR-DTC codeword in each reference cycle ($1/f_{ref}$), while a scaled version, R_n/M, is subtracted from the fine DTC quantization error residue ($Q_{n,f}$). This shifts the phase error components to $0.5f_{ref}$ and enables accurate and fast calibration even for the near-integer channels, as shown in Fig. 12.1.3c. Figure 12.1.3d shows the DTC code sequences before and after dithering. When $R_n=1$, the TADR DTC generates an additional delay of T_{DCO}/M, and the fine DTC produces the corresponding negative delay, leaving overall T_Q cancellation unaffected. The TADR DTC range thus increases by 6.25% with a negligible linearity impact, while the fine DTC range doubles from T_{DCO}/M to $2T_{DCO}/M$ yet remains sufficiently narrow to maintain its linearity within acceptable limits. Consequently, compared to prior art [14,15], the proposed technique achieves a high-accuracy calibration but with a minimal hardware overhead, while increasing the total DTC range by only 12.5%.

The DPLL, fabricated in 28nm CMOS, occupies 0.2mm^2 of active area (Fig.12.1.7) and consumes 23mW. It operates from 8 to 9.7GHz with f_{ref}=125MHz. Figure 12.1.4 shows the measured output spectra and phase-noise (PN) performance under different conditions.

208 • 2026 IEEE International Solid-State Circuits Conference

979-8-3315-8937-0/26 $31.00 © 2026 IEEE

With dithering disabled and $FCW_{FRAC} = 2^{-11}$, the worst spur of -42.4dBc appears at the 8th harmonic of the fractional frequency, indicating inaccuracies in $K_{DTC,f}$ calibration. Enabling dithering improves the LMS convergence, reducing the worst-case fractional spur to -59dBc. Similar suppression is observed across other in-band fractional-N channels, with spur levels consistently <-59dBc. The rms jitter is 72fs with spurs and 68.2fs without.

Supply resilience was evaluated (Fig. 12.1.5) by injecting both wideband noise and single-tone ripple into the TADR DTC, PFDs, T_{DCO} scaler, and TDC supply. With a 29mV$_{rms}$ wideband noise (10MHz BW), the rms jitter increases only to 88.5fs, confirming the robustness of the TADR DTC. The injected wideband noise also elevates the PLL phase noise even outside the loop bandwidth, as the DTC noise—due to absence of additional IIR filtering—is comparable to the DCO contribution in this region. A 20mV$_{PP}$ ripple at 100kHz produces a -58.9dBc spur, corresponding to a DTC supply pushing factor, K_{sup} = 4ps/V. Compared with prior-art PLLs (Fig. 12.1.6), the jitter, spur, and FoM of this work remain highly competitive, while also demonstrating notable resilience to supply variations.

Acknowledgement:
This work is supported by NXP-TU Delft Industry partnership program TKI2212P02. The authors thank Harro Koning, Chuang Lu, Nenad Pavlovic and Athon Zanikopoulos from NXP Semiconductors.

References:
[1] P. Salvi et al., "A Low-Noise Fractional-N Digital PLL Using a Resistor-Based Inverse-Constant-Slope DTC," *IEEE JSSC*, vol. 60, no. 7, pp. 2619-2631, July 2025. https://doi.org/10.1109/JSSC.2024.3501196
[2] S. M. Dartizio et al., "A 68.6fs$_{rms}$-Total-Integrated-Jitter and 1.56μs-Locking-Time Fractional-N Bang-Bang PLL Based on Type-II Gear Shifting and Adaptive Frequency Switching," *ISSCC*, pp. 386-387, Feb. 2022. https://doi.org/10.1109/ISSCC42614.2022.9731683
[3] A. Santiccioli et al., "A 66-fs-rms Jitter 12.8-to-15.2-GHz Fractional-N Bang–Bang PLL With Digital Frequency-Error Recovery for Fast Locking," *IEEE JSSC*, vol. 55, no. 12, pp. 3349-3361, Dec. 2020. https://doi.org/10.1109/JSSC.2020.3019344
[4] X. Gao et al., "A Low Noise Sub-Sampling PLL in Which Divider Noise is Eliminated and PD/CP Noise is Not Multiplied by N^2," *IEEE JSSC*, vol. 44, no. 12, pp. 3253-3263, Dec. 2009. https://doi.org/10.1109/JSSC.2009.2032723
[5] H. Liu et al., "A 265-μW Fractional-N Digital PLL With Seamless Automatic Switching Sub-Sampling/Sampling Feedback Path and Duty-Cycled Frequency-Locked Loop in 65-nm CMOS," *IEEE JSSC*, vol. 54, no. 12, pp. 3478-3492, Dec. 2019. https://doi.org/10.1109/JSSC.2019.2936967
[6] W. Wu et al., "A 28-nm 75-fsrms Analog Fractional-N Sampling PLL With a Highly Linear DTC Incorporating Background DTC Gain Calibration and Reference Clock Duty Cycle Correction," *IEEE JSSC*, vol. 54, no. 5, pp. 1254-1265, May 2019. https://doi.org/10.1109/JSSC.2019.2899726
[7] D. Tasca et al., "A 2.9–4.0-GHz Fractional-N Digital PLL With Bang-Bang Phase Detector and 560-fsrms Integrated Jitter at 4.5-mW Power," *IEEE JSSC*, vol. 46, no. 12, pp. 2745-2758, Dec. 2011. https://doi.org/10.1109/JSSC.2011.2162917
[8] H. Liu et al., "A 0.98mW fractional-N ADPLL using 10b isolated constant-slope DTC with FOM of −246dB for IoT applications in 65nm CMOS," *ISSCC*, pp. 246-248, Feb. 2018. https://doi.org/10.1109/ISSCC.2018.8310276

[9] W. Wu et al., "A 14nm Analog Sampling Fractional-N PLL with a Digital-to-Time Converter Range-Reduction Technique Achieving 80fs Integrated Jitter and 93fs at Near-Integer Channels," *ISSCC*, pp. 444-446, Feb. 2021. https://doi.org/10.1109/ISSCC42613.2021.9365850
[10] A. Elkholy et al., "A 3.7 mW Low-Noise Wide-Bandwidth 4.5 GHz Digital Fractional-N PLL Using Time Amplifier-Based TDC," *IEEE JSSC*, vol. 50, no. 4, pp. 867-881, Apr. 2015. https://doi.org/10.1109/JSSC.2014.2385753
[11] H. Park et al., "A 365fs$_{rms}$-Jitter and -63dBc-Fractional Spur 5.3GHz-Ring-DCO-Based Fractional-N DPLL Using a DTC Second/Third- Order Nonlinearity Cancelation and a Probability-Density-Shaping ΔΣM," *ISSCC*, pp. 442-444, Feb. 2021. https://doi.org/10.1109/ISSCC42613.2021.9365798
[12] G. Castoro et al., "A 9.25GHz Digital PLL with Fractional-Spur Cancellation Based on a Multi-DTC Topology," *ISSCC*, pp. 82-84, Feb. 2023. https://doi.org/10.1109/ISSCC42615.2023.10067351
[13] Z. Gao et al., "A Low-Spur Fractional-N PLL Based on a Time-Mode Arithmetic Unit," *IEEE JSSC*, vol. 58, no. 6, pp. 1552-1571, June 2023. https://doi.org/10.1109/JSSC.2022.3209338
[14] S. M. Dartizio et al., "A 76.7fs-Integrated-Jitter and −71.9dBc In-Band Fractional-Spur Bang-Bang Digital PLL Based on an Inverse-Constant-Slope DTC and FCW Subtractive Dithering," *ISSCC*, pp. 78-79, Feb. 2023. https://doi.org/10.1109/ISSCC42615.2023.10067719
[15] D. Xu et al., "A 7GHz Digital PLL with Cascaded Fractional Divider and Pseudo-Differential DTC Achieving -62.1dBc Fractional Spur and 143.7fs Integrated Jitter," *ISSCC*, pp. 192-194, Feb. 2024. https://doi.org/10.1109/ISSCC49657.2024.10454284
[16] M. Rossoni et al., "An 8.75GHz Fractional-N Digital PLL with a Reverse-Concavity Variable-Slope DTC Achieving 57.3fsrms Integrated Jitter and −252.4dB FoM," *ISSCC*, pp. 188-190, Feb. 2024. https://doi.org/10.1109/ISSCC49657.2024.10454388
[17] J. Z. Ru et al., "A High-Linearity Digital-to-Time Converter Technique: Constant-Slope Charging," *IEEE JSSC*, vol. 50, no. 6, pp. 1412-1423, June 2015. https://doi.org/10.1109/JSSC.2015.2414421
[18] P. Chen et al., "A 31-μW, 148-fs Step, 9-bit Capacitor-DAC-Based Constant-Slope Digital-to-Time Converter in 28-nm CMOS," *IEEE JSSC*, vol. 54, no. 11, pp. 3075-3085, Nov. 2019. https://doi.org/10.1109/JSSC.2019.2939663
[19] C.-L. Ti et al., "A 2.4-GHz fractional-N PLL with a PFD/CP linearization and an improved CP circuit," *IEEE ISCAS*, pp. 1728-1731, May 2008. https://doi.org/10.1109/ISCAS.2008.4541771
[20] M. Chae et al., "A 65fs$_{rms}$-Jitter and −272dB-FoM$_{jitter,N}$ 10.1GHz Fractional-N Digital PLL with a Quantization-Error-Compensating BBPD and an Orthogonal-Polynomial LMS Calibration," *ISSCC*, pp. 554-556, Feb. 2025. https://doi.org/10.1109/ISSCC49661.2025.10904741
[21] A. Elkholy et al., "A 3.7 mW Low-Noise Wide-Bandwidth 4.5 GHz Digital Fractional-N PLL Using Time Amplifier-Based TDC," *IEEE JSSC*, vol. 50, no. 4, pp. 867-881, Apr. 2015. https://doi.org/10.1109/JSSC.2014.2385753
[22] Z. Gao et al., "A 2.6-to-4.1GHz Fractional-N Digital PLL Based on a Time-Mode Arithmetic Unit Achieving -249.4dB FoM and -59dBc Fractional Spurs," *ISSCC*, pp. 380-382, Feb. 2022. https://doi.org/10.1109/ISSCC42614.2022.9731561
[23] Y. Chen et al., "A Fractional-N Digitally Intensive PLL Achieving 428-fs Jitter and <−54-dBc Spurs Under 50-mVpp Supply Ripple," *IEEE JSSC*, vol. 57, no. 6, pp. 1749-1764, June 2022. https://doi.org/10.1109/JSSC.2021.3123386

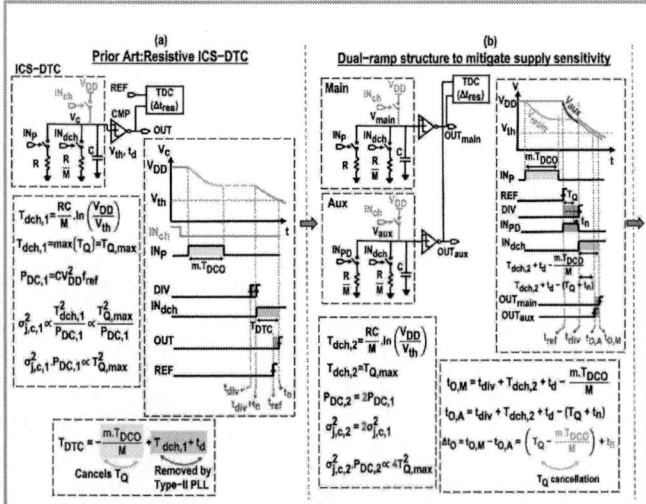

Figure 12.1.1: (a) Operating principle and timing diagram of the prior resistor-based inverse-constant-slope DTC [1]; (b) the dual-ramp DTC.

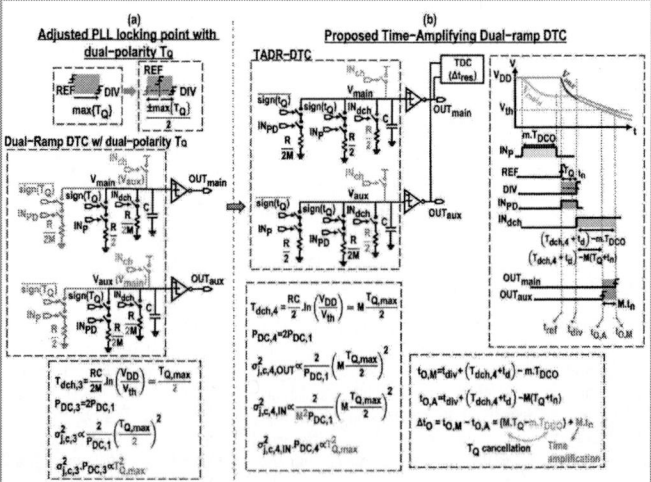

Figure 12.1.2: Evolution of the time-amplifying dual-ramp DTC: (a) Using positive and negative time shifts to reduce DTC noise; (b) employing time amplification to suppress the TDC quantization noise.

ISSCC 2026 / SESSION 12 / FREQUENCY SYNTHESIZERS AND VCOs / 12.1

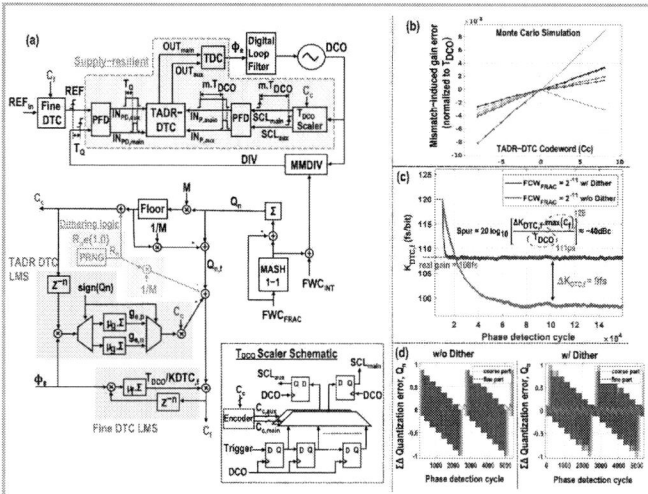

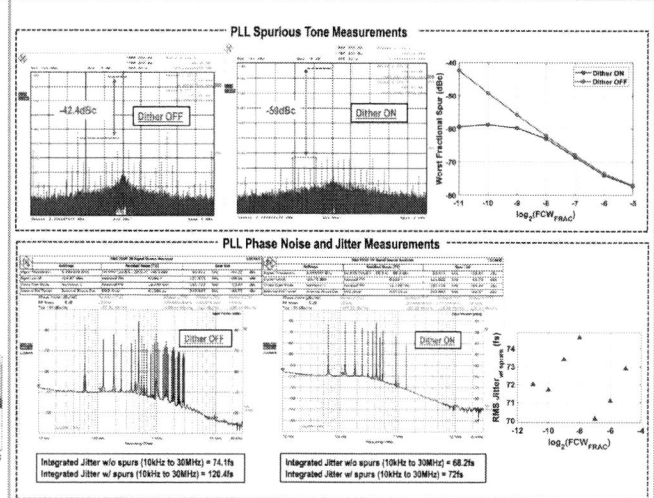

Figure 12.1.3: (a) Simplified block diagram of the proposed digital PLL; (b) simulated TADR-DTC gain error; (c) LMS loop accuracy and convergence; (d) TADR and fine-DTC ranges with and without the randomization technique.

Figure 12.1.4: Measured fractional spurs and phase noise for $FCW_{FRAC} = 2^{-11}$ and worst fractional-spur level and rms jitter across FCW_{FRAC}.

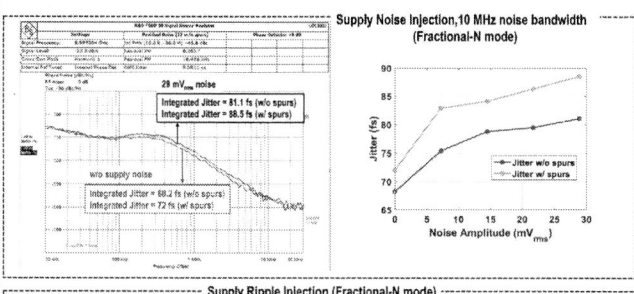

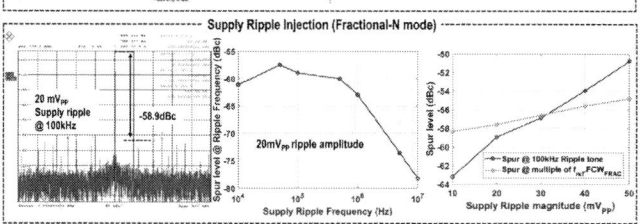

Figure 12.1.5: Measurement results showing robustness of the TADR DTC against wideband supply noise (top) and single-tone ripple injection (bottom).

	Supply-Resilient PLLs			Regular PLLs				
	This Work	Chen [23] JSSC'22	Rossoni [21] VLSI'25	Salvi [1] JSSC'25	Chae [20] ISSCC'25	Xu [15] ISSCC'24	Dartizio [14] ISSCC'23	Gao [22] ISSCC'22
PLL Architecture	DPLL	DPLL	DPLL	BBPLL	QEC-BBPD	DPLL	BBPLL	DPLL
Supply Disturbance Cancellation Method	TADR DTC	Ripple Pattern Estim. & Canc.	PGRO w/ FSL	-	-	-	-	-
Output Freq (GHz)	8 to 9.7	4.47 to 5.14	9 to 10.25	8.75 to 10.25	10 to 11.5	6.5 to 7.5	9.25 to 10.5	2.56 to 4.1
Reference Freq (MHz)	125	50	250	125	100	100	250	40
Supply Ripple Amplitude (mV$_{PP}$)	20	50	3	-	-	-	-	-
Ripple Output Spur# (dBc)	-58.9	-48.9[A3]	-50.2	-	-	-	-	-
Supply Noise Amplitude (mV$_{rms}$)	29	-	2.5	-	-	-	-	-
Jitter w/ Supply Noise (fs)	88.5	-	81	-	-	-	-	-
Jitter w/o supply noise (fs)	72	422	72.2	66.7	65	143.7	76.7	182
K_{sup} (ps/V)	4	9.1	70.8	-	-	-	-	-
Power Dissipation (mW)	22.65	3.25	19.5	16.7	14.4	8.89	17.2	3.48
FoM (dB)	-249.3	-242.4	-249.9	-251.3	-252.2	-247.4	-249.9	-249.4
FoM$_N$ (dB)	-267.9	-262.2	-265.6	-270	-272.2	-265.9	-265.6	-267.7
Fractional Spur (dBc)	-59	-55	-57.9	-63.8	-62.8	-62.1	-71.9	-59
DTC Calibration Type	Gain	Gain (DAC) + DCO Duty cycle	-	Gain + DCO Duty cycle	Polynomial DPD	Gain	Gain + DCO Duty cycle	LUT DPD
DTC Calibration Time (us)	40	64	-	-	262	-	-	-
Active area (mm²)	0.2	0.39	0.27	0.21	0.12	0.23	0.33	0.31
CMOS Process (nm)	28	40	28	28	40	65	28	40

FoM = $10.\log_{10}((Jitter_{w/\,spurs}/1s)^2.(Power/1mW))$ FoM$_N$ = FoM + $10.\log_{10}(f_{TP}/f_{ref})$ [A]Normalized to 9GHz   [B]Known Ripple Frequency

Figure 12.1.6: Performance summary and comparison with prior art.

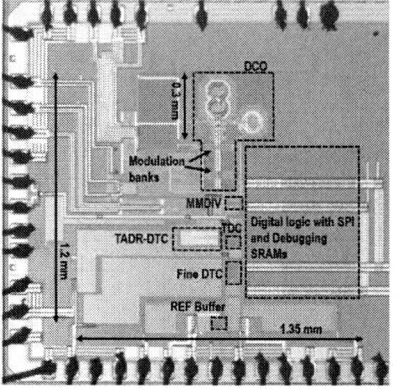

Power and Area Breakdown

Block	Power (mW)	Area (mm²)
REF Buffer	0.25	0.002
Fine DTC	0.3	0.004
TADR-DTC	2.7	0.02
TDC	0.2	0.002
Digital	1.2	0.03#
MMDIV	6	0.002
DCO	12	0.14
Total	22.65	0.2

#Excluding debugging SRAMs

Figure 12.1.7: Die micrograph; breakdown of power consumption and area.

ISSCC 2026 / SESSION 12 / FREQUENCY SYNTHESIZERS AND VCOs / 12.2

12.2 A Fractional-N Digital PLL with a Supply-Insensitive DTC Achieving –62dBc Spur and 69fs Jitter Under 10mV$_{pp}$ Sinusoidal DTC Supply Ripple and 6.2mV$_{rms}$ DTC Supply Noise

Damiano Fagotti, Riccardo Moleri, Michele Rossoni, Daniele Lodi Rizzini, Pietro Salvi, Stefano Gallucci, Giovanni Rocco Trotta, Andrea Leonardo Lacaita, Simone Mattia Dartizio, Salvatore Levantino

Politecnico di Milano, Milan, Italy

Abstract

A 28nm fractional-N digital PLL with a supply-insensitive variable-slope digital-to-time converter (DTC) is presented. Supply insensitivity is achieved by a background calibration loop, without affecting DTC linearity or noise. Measured results demonstrate 65.6fs jitter upon 6.2mV$_{rms}$ supply noise and <–62dBc spurs under 10mV$_{pp}$ supply-ripple over 10k-to-50MHz disturbance frequency range. The PLL covers 8.75-to-10.25GHz tuning range, while achieving a –251dB jitter-power figure of merit at 22.5mW power.

Frequency synthesizers with spurious tones below -60dBc and jitter lower than 100fs are essential for high data rate wireless and wireline transceivers [1,2]. Digital-to-time converter (DTC)-based fractional-N PLLs have emerged as a compelling solution, effectively eliminating the quantization error associated with fractional frequency multiplication [3-7]. Among various DTC topologies, the variable-slope DTC (VS-DTC) is particularly attractive due to its simplicity and superior power–jitter trade-off [8]. Its main drawback—nonlinearity causing large spurs—can be mitigated using the reverse-concavity effect, allowing for simultaneous low jitter and low spurs [3]. However, embedding DTC-based PLLs in modern SoCs remains challenging, because of supply network noise and ripple from voltage regulators. The high sensitivity of VS-DTCs to supply voltage (K$_{VDD}$) makes this a critical issue for reliable operation (Fig. 12.2.1, top). While fully differential techniques improve PSRR [9,10], they add complexity and lack experimental validation for achieving ultra-low jitter under supply disturbances. Calibration loops can suppress periodic ripples [11] but are ineffective against unknown tones or supply noise. Although a recent digital PLL attempted to address this by using a coarse ring-oscillator (RO)-based DTC stage with a frequency-stabilizing loop [12], its narrow bandwidth limits the suppression to low-frequency disturbances. Furthermore, this approach cannot be applied to a full VS-DTC architecture, which offer better noise performance with simpler design.

This work introduces a technique that makes the VS-DTC architecture inherently insensitive to supply disturbances. Validated in 28nm CMOS, fractional-N digital PLL demonstrates robust operation with -62dBc spurs under a 10mV$_{pp}$ sinusoidal DTC-supply ripple and 65.5fs integrated jitter under 6.2mV$_{rms}$ DTC-supply noise.

Figure 12.2.1 illustrates our approach to suppressing the VS-DTC supply sensitivity by counterbalancing two opposing K$_{VDD}$ contributions. A simplified VS-DTC model (bottom-left) shows the core of this principle. The first stage is modelled as an RC-discharge starting from V$_{DD}$. The second stage is an ideal zero-delay CMOS inverter with threshold voltage V$_{TH}$, assumed to increase with V$_{DD}$. Based on this model, the VS-DTC delay can be expressed as $T_{DTC} = \tau_{DTC} \ln(V_{DD}/V_{TH})$, where τ_{DTC} is the RC time constant and is assumed to be independent of V$_{DD}$. As the equation shows, an increase in V$_{DD}$ has two opposite effects on the delay: it increases the numerator, which tends to increase T$_{DTC}$, and it also increases the threshold voltage V$_{TH}$, which tends to reduce T$_{DTC}$. We can achieve a zero K$_{VDD}$ (i.e., $\partial T_{DTC}/\partial V_{DD} = 0$) by carefully balancing these two effects. The model shows this occurs when $(V_{TH}/V_{DD}) = (\partial V_{TH}/\partial V_{DD})$. To meet this condition, we introduce a CMOS inverter stage with a digitally scalable NMOS transistor with size s$_n$ and a PMOS transistor with size s$_p$, as shown in Fig. 12.2.1 (bottom-right). By digitally adjusting s$_n$, we can find a specific size that ensures the condition is met, thereby achieving an optimal operating point with a near-zero K$_{VDD}$. Figure 12.2.1 analyses the two extreme cases of the digitally controllable CMOS inverter: when the PMOS is much larger than the NMOS (s$_p$ >> s$_n$), the PMOS dominates and the inverter threshold voltage V$_{TH}$ approaches V$_{DD}$ - |V$_{Tp}$|, where |V$_{Tp}$| is the PMOS threshold voltage and is independent of V$_{DD}$. Here, $\partial V_{TH}/\partial V_{DD} = 1$, and since $0 < V_{TH}/V_{DD} < 1$, we have $\partial V_{TH}/\partial V_{DD} > V_{TH}/V_{DD}$, meaning the second stage K$_{VDD}$ contribution is larger than of the first stage, resulting in a negative overall K$_{VDD}$. Conversely, when the NMOS is much larger than the PMOS (s$_n$ >> s$_p$), the NMOS dominates the inverter behavior, and V$_{TH}$ approaches the NMOS threshold voltage V$_{Tn}$. Since V$_{Tn}$ is independent of V$_{DD}$, it follows that $\partial V_{TH}/\partial V_{DD} = 0$ and $\partial V_{TH}/\partial V_{DD} < V_{TH}/V_{DD}$. In this scenario, the second stage K$_{VDD}$ contribution is smaller than that of the first stage, resulting in a positive overall K$_{VDD}$. Since the overall K$_{VDD}$ can be either negative or positive depending on the NMOS-to-PMOS size ratio, there must be an intermediate value of s$_n$ that achieves the condition $(\partial V_{TH}/\partial V_{DD}) = (V_{TH}/V_{DD})$. This specific setting effectively nullifies the total K$_{VDD}$. This is confirmed by the simulated curves of $\partial V_{TH}/\partial V_{DD}$ and V$_{TH}/V_{DD}$ in Fig. 12.2.1 (bottom-right) for a standard CMOS inverter, which intersect at the point of zero K$_{VDD}$. Despite the simplifying assumptions of the model, the concept holds: the optimal s$_n$ value in a real-world circuit simply differ from the predicted value. To ensure robustness, a digital calibration algorithm is used to automatically tune s$_n$ to the optimal setting.

The above technique is embedded in the presented supply-insensitive (SI)-DTC, shown in Fig. 12.2.2 (top). It consists of two cascaded coarse and fine VS-DTC stages, driven by S$_{DTC,c}$ and S$_{DTC,f}$, achieving 150fs resolution over a 140ps range. Each coarse cell includes a series MOS, whose gate is driven by a DAC to enable reverse-concavity linearization [3],

while an auxiliary switch and resampling flip-flops suppress memory effects [13]. In the inverter stage, a resistor R$_{PU}$ in series with the PMOS reduces flicker noise [14]. K$_{VDD}$ tuning is achieved by splitting the NMOS into a coarse 11-device thermometric bank controlled by C$_{TH,c}$ and a fine-tuning element comprising a NMOS, whose gate is driven by a DAC, controlled by C$_{TH,f}$. To save hardware resources, the two DAC voltages are generated by a dual-output R-string DAC (bottom-right), sharing the 64 resistor elements between two independently driven switch sets. Post-layout simulations (bottom-right) show the overall K$_{VDD}$ of the DTC chain, including the contribution of both the coarse and fine DTC stages, as a function of C$_{TH,c}$ and C$_{TH,f}$, with the DTC input codes fixed at mid-range (S$_{DTC,c}$ = 64, S$_{DTC,f}$ = 32). In nominal conditions, the zero-K$_{VDD}$ setting occurs for C$_{TH,c}$ between 4 and 5, with precise cancellation obtained by fine tuning C$_{TH,f}$. Figure 12.2.2 (bottom-right) also reports the simulated K$_{VDD}$ versus S$_{DTC,c}$, with C$_{TH,c}$ and C$_{TH,f}$ fixed at the optimal values. Note that K$_{VDD}$ is exactly zero only at the mid-range code, varying almost linearly across S$_{DTC,c}$. Across DTC operation, with S$_{DTC,c}$ swept over its full range, the relevant metric for supply disturbance conversion is the average K$_{VDD}$, i.e., its mid-code value. Simulations in Fig. 12.2.2 (bottom-right) confirm that, despite residual K$_{VDD}$ vs. S$_{DTC,c}$, the PLL spectral degradation due to the DTC-supply disturbances is suppressed by the zero-average K$_{VDD}$. The presented technique does not compromise the DTC performance; nonlinearity remains minimized by the reverse-concavity effect [3], while the DTC jitter (Fig. 12.2.2, bottom-right) mildly increases with C$_{TH,c}$ (<10% increase for the optimum C$_{TH,c}$).

Figure 12.2.3 shows the implemented bang-bang digital-PLL block diagram and the operating principle of the V$_{TH}$-LMS algorithm, which automatically tunes the NMOS size to achieve zero average K$_{VDD}$. A supply-dither generator—consisting of a 1b pseudo-random number generator (PRNG), a CMOS buffer, and a resistive divider—superimposes a perturbation onto the SI-DTC supply voltage, V$_{DD,DTC}$. As illustrated in Fig. 12.2.3 (top-left), the PRNG output d[k], toggling between 0 and 1, is converted into a perturbation of the amplitude A$_{dith}$≈1.3mV through a small resistor R$_1$=20Ω in series to V$_{DD,DTC}$ (with negligible impact on the DTC operation) and a large resistor R$_2$=14kΩ at the buffer output. The induced supply perturbation modulates the DTC delay through K$_{VDD}$, producing an error signal at the BBPD output proportional to K$_{VDD}$·A$_{dith}$·d[k], correlated with d[k]. The sign of the correlation between e[k] and d[k] can be used to extract the sign of K$_{VDD}$, indicating whether the SI-DTC NMOS size must increase or decrease to reach zero K$_{VDD}$. To this aim, the V$_{TH}$-LMS algorithm (bottom) computes this correlation as the product of e[k] and d[k]. After scaling by a gain γ, the result is accumulated to form the control code C$_{TH}$. Its MSBs and LSBs drive the C$_{TH,c}$ and C$_{TH,f}$ controls, respectively. This loop converges to nullify the correlation between e[k] and d[k], driving C$_{TH}$ towards the zero-average K$_{VDD}$ condition. The V$_{TH}$-LMS operates in the background together with the adaptive concavity zeroing (ACZ) algorithm [3], the DTC gain calibration [15], and the DTC-range-reduction technique [15,16], the latter further improving the DTC noise and linearity.

The PLL prototype, implemented in 28nm CMOS (Fig. 12.2.7), occupies 0.28mm² and consumes 22.5mW, excluding the output buffer. Figure 12.2.4 (top) reports the measured PLL performance at an 8.75GHz near-integer channel under an undisturbed DTC supply. The PLL achieves 59.6fs integrated jitter with worst-case fractional spurs below <–64dBc, performance consistently maintained across different fractional-N channels (Fig. 12.2.4, top-right). Figure 12.2.4 (bottom-left) shows the measured K$_{VDD}$ of the SI-DTC versus NMOS control codes with DTC inputs fixed at mid-range. K$_{VDD}$ increases monotonically with NMOS control and changes sign, confirming the effectiveness of the presented compensation scheme. Around the optimum point, obtained by fine-tuning C$_{TH,c}$ and C$_{TH,F}$, the zoomed-in plots show that K$_{VDD}$ decreases from –350fs/mV in the unoptimized case to less than 5fs/mV, with the optimum consistently identified by the V$_{TH}$-LMS algorithm after convergence. The residual sensitivity across DTC input codes (bottom-right) remains below 80fs/mV at its peak, while the measured DTC INL at the optimum is ~142fs$_{pp}$ after ACZ convergence. Figure 12.2.5 shows the PLL performance under a perturbed DTC supply, with and without enabling the V$_{TH}$-LMS. When a 1MHz-BW, 6.2mV$_{rms}$ supply noise is applied (top-left), the PLL jitter is reduced from 714 to 65.5fs, thanks to the convergence of the average K$_{VDD}$ to zero. With a 10mV$_{pp}$, 94kHz sinusoidal disturbance, the V$_{TH}$-LMS suppresses the corresponding spur in the PLL spectrum from –27 to –65.7dBc, while the PLL jitter reduces from 1.2ps to 66fs. A second-harmonic spur at 188kHz is observed due to the nonlinear delay-vs-supply dependence of the DTC, but it remains sufficiently small (below

−64dBc) to negligibly affect the PLL jitter (raising it only by 1.8fs). Sweeping the disturbance frequency across the 10kHz-to-50MHz range demonstrates residual spurs (first and second harmonics) always below −62dBc with integrated jitter below 69fs, demonstrating robust supply immunity across a wide range. Figure 12.2.6 compares this work with recent low-jitter fractional-N PLLs. Among reported designs employing DTC-supply-resilience techniques and targeting sub-100fs jitter, it achieves the lowest integrated jitter under the widest supply disturbances. At the same time, it delivers high FoM, low jitter, and spectral purity, remaining competitive even against PLLs listed in Fig. 12.2.6 that were tested under undistorted supply conditions.

References:

[1] W. Wu, "Low-Jitter Frequency Generation Techniques for 5G Communication: A tutorial," *IEEE Solid-State Circuits Magazine*, vol. 13, no. 4, pp. 44-63, Fall 2021, https://doi.org/10.1109/MSSC.2021.3111430

[2] Y. Hu et al., "Nonlinearity-Induced Spur Analysis in Fractional-N Synthesizers With ΔΣ Quantization Cancellation," *IEEE OJSSC*, vol. 4, pp. 226-237, Oct. 2024. https://doi.org/10.1109/OJSSCS.2024.3476035

[3] M. Rossoni et al., "A Low-Jitter Fractional-N Digital PLL Adopting a Reverse-Concavity Variable-Slope DTC," *IEEE JSSC*, vol. 60, no. 6, pp. 2122-2133, June 2025. https://doi.org/10.1109/JSSC.2024.3469556

[4] P. Salvi et al., "A Low-Noise Fractional-N Digital PLL Using a Resistor-Based Inverse-Constant-Slope DTC," *IEEE JSSC*, vol. 60, no. 7, pp. 2619-2631, July 2025. https://doi.org/10.1109/JSSC.2024.3501196

[5] M. Chae et al., "A 65fs_rms-Jitter and -272dB-FoM_jitter,N 10.1GHz Fractional-N Digital PLL with a Quantization-Error-Compensating BBPD and an Orthogonal-Polynomial LMS Calibration," *ISSCC*, pp. 554-556, Feb. 2025. https://doi.org/10.1109/ISSCC49661.2025.10904741

[6] H. Li et al., "A 27GHz Fractional-N Sub-Sampling PLL Achieving 57.9f_rms Jitter, −249.7dB FoM, and 1.98μs Locking Time Using Polarity-Reversible SSPD," *ISSCC*, pp. 336-337, Feb. 2025. https://doi.org/10.1109/ISSCC49661.2025.10904556

[7] Z. Gao et al., "A Low-Spur Fractional-N PLL Based on a Time-Mode Arithmetic Unit," *IEEE JSSC*, vol. 58, no. 6, pp. 1552-1571, June 2023. https://doi.org/10.1109/JSSC.2022.3209338

[8] A. Santiccioli et al., "Power-Jitter Trade-Off Analysis in Digital-to-Time Converters," *Electron. Lett.*, vol. 53, no. 5, pp. 306–308, Mar. 2017. https://doi.org/10.1049/el.2016.4577

[9] L. Wu et al., "A Power-Efficient Fractional-N DPLL with Phase Error Quantized in Fully Differential-Voltage Domain," *IEEE JSSC*, vol. 56, no. 4, pp. 1254-1264, Apr. 2021. https://doi.org/10.1109/JSSC.2020.3047431

[10] J. Jung et al., "A 55.8-to-64.2GHz, 58.3fs_rms-Jitter, -250.2dB-FoM_J Fractional-N Cascaded PLL in 28nm CMOS," *IEEE Symp. VLSI Circuits*, pp. 1-3, June 2025. https://doi.org/10.23919/VLSITechnologyandCir65189.2025.11075118

[11] Y. Chen et al., "A Fractional-N Digitally Intensive PLL Achieving 428-fs Jitter and <−54-dBc Spurs Under 50-mVpp Supply Ripple," *IEEE JSCC*, vol. 57, no. 6, pp. 1749-1764, June 2022. https://doi.org/10.1109/JSSC.2021.3123386

[12] M. Rossoni et al., "A Fractional-N Digital-PLL Based on a Power-Gated Ring-Oscillator and a Frequency-Stabilizing Loop Achieving 74fs Jitter Under 3mV_pp Supply Ripple," *IEEE Symp. VLSI Circuits*, pp. 1-3, June 2025. https://doi.org/10.23919/VLSITechnologyandCir65189.2025.11074831

[13] A. Santiccioli et al., "A 66-fs-rms Jitter 12.8-to-15.2-GHz Fractional-N Bang–Bang PLL with Digital Frequency-Error Recovery for Fast Locking," *IEEE JSSC*, vol. 55, no. 12, pp. 3349-3361, Dec. 2020. https://doi.org/10.1109/JSSC.2020.3019344

[14] N. Markulic et al., "A 10-bit, 550-fs step Digital-to-Time Converter in 28nm CMOS," *ESSCIR*, pp. 79-82, Sept. 2014. https://doi.org/10.1109/ESSCIRC.2014.6942026

[15] D. Tasca, et al., "A 2.9–4.0-GHz Fractional-N Digital PLL With Bang-Bang Phase Detector and 560-fs_rms Integrated Jitter at 4.5-mW Power," *IEEE JSSC*, vol. 46, no. 12, pp. 2745-2758, Dec. 2011. https://doi.org/10.1109/JSSC.2011.2162917

[16] W. Wu et al., "A 14-nm Ultra-Low Jitter Fractional-N PLL Using a DTC Range Reduction Technique and a Reconfigurable Dual-Core VCO," *IEEE JSSC*, vol. 56, no. 12, pp. 3756-3767, Dec. 2021. https://doi.org/10.1109/JSSC.2021.3111134

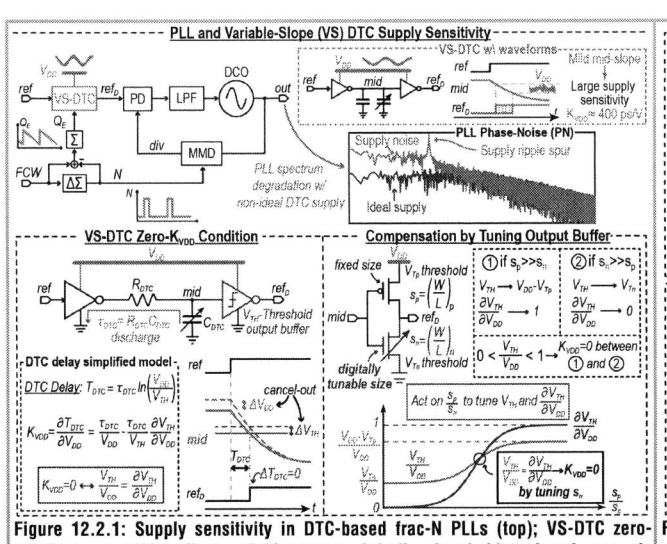

Figure 12.2.1: Supply sensitivity in DTC-based frac-N PLLs (top); VS-DTC zero-sensitivity condition (bottom-left); proposed buffer-threshold tuning for supply resilience (bottom-right).

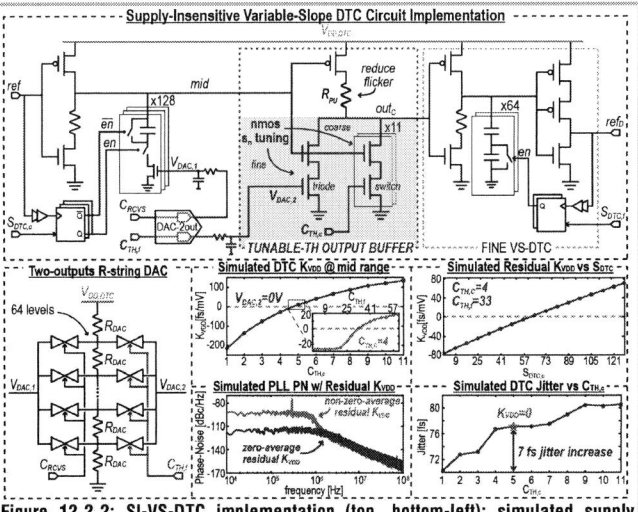

Figure 12.2.2: SI-VS-DTC implementation (top, bottom-left); simulated supply sensitivity and jitter vs. buffer NMOS size, residual sensitivity vs. SDTC, and the PLL PN with zero/non-zero average K_VDD (bottom-right).

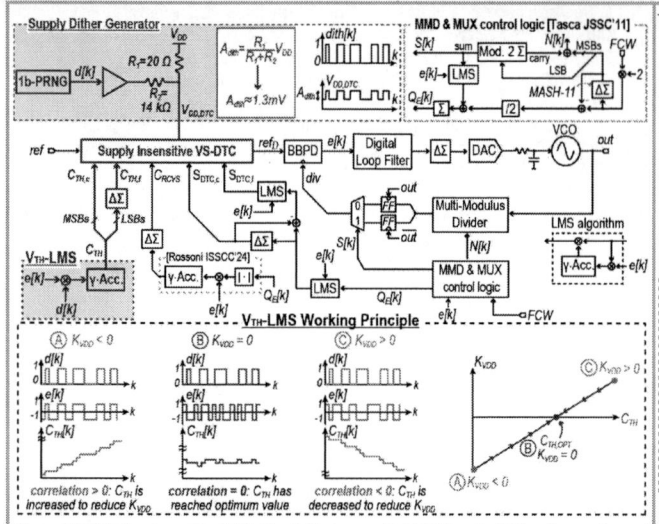

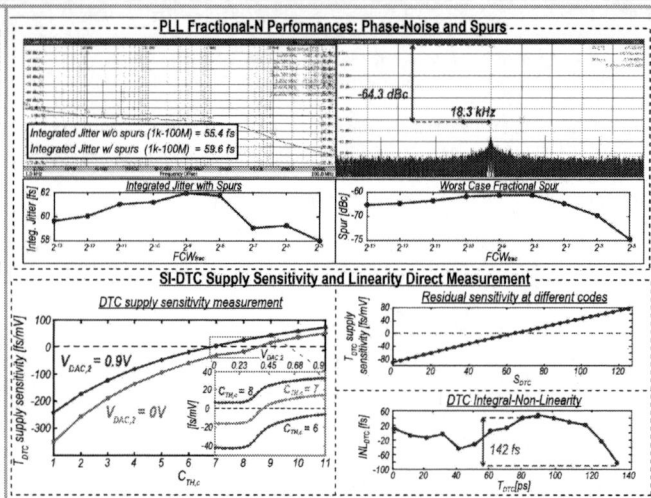

Figure 12.2.3: Implemented PLL block diagram with DTC-Supply Dither Generator and V$_{TH}$-LMS (top); V$_{TH}$-LMS working principle (bottom).

Figure 12.2.4: Measured PLL jitter and frac. spurs vs. frac-N channels (top); measured DTC supply sensitivity vs. C$_{TH}$ (bottom-left), measured residual sensitivity vs. S$_{DTC}$, and measured DTC INL (bottom-right).

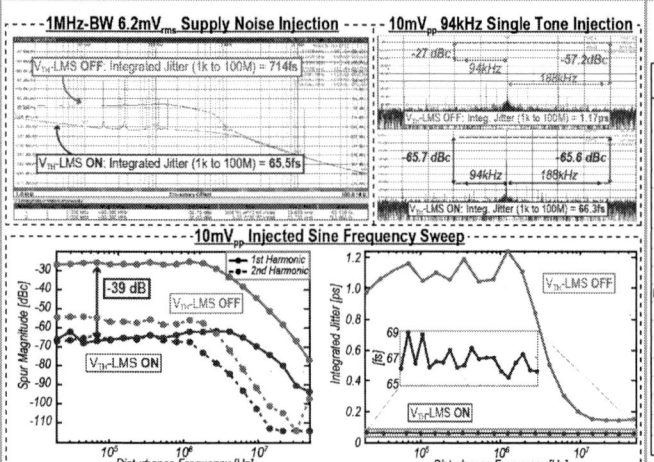

Figure 12.2.5: Measured PLL phase noise with 6.2mV$_{rms}$, 1MHz-BW supply noise (top-left), and spectrum with 10mV$_{pp}$ 94kHz supply ripple (top-right), with/without V$_{TH}$-LMS; output spur and jitter vs. disturbance frequency with/without V$_{TH}$-LMS (bottom).

	PLLs with Supply Sensitivity Reduction Techniques				Conventional PLLs		
	This Work	M. Rossoni VLSI'25 [12]	Y.Chen JSSC'22 [11]	L.Wu JSSC'21 [9]	H.Li ISSCC'25 [6]	M. Rossoni ISSCC'24 [3]	M. Chae ISSCC'25 [5]
PLL Architecture	DPLL + DTC	DPLL + DTC	DPLL + DAC	DPLL + FDVPD	SSPLL +DTC	DPLL + DTC	DPLL + QEC-BB
Supply Sensitivity Cancellation Technique	SI-DTC w/ V$_{TH}$-LMS	PGRO w/ FSL	Ripple Patt. Canc.	Fully-Differential Voltage Domain	N/A	N/A	N/A
Output Frequency [GHz]	8.75 to 10.25	9 to 10.25	4.47 to 5.14	2.99 to 3.5	23.2 to 27	8.75 to 10.25	10 to 11.5
Reference Frequency [MHz]	150	250	50	80	120	250	100
Amplitude Ripple [mV$_{pp}$]	10	3	50	20	N/A	N/A	N/A
Amplitude Noise [mV$_{rms}$]	6.2	2.5	N/A	N/A	N/A	N/A	N/A
Ripple Output Spur*** [dBc]	<-62	<-50.25	<-48.5	-57.4	N/A	N/A	N/A
Fractional Spur [dBc]	-64.3	-57.9	-55	-56.4	-55.2	-63.4	-63
Frac-N Jitter w/o Supply Ripple [fs]	59.6	72.2	422	101	57.9	57.3	65
Frac-N Jitter w/ Supply Noise [fs]	65.5	81	N/A	N/A	N/A	N/A	N/A
Frac-N Jitter w/ Supply Ripple [fs]	<69	<100	428	125	N/A	N/A	N/A
Integration BW [Hz]	1k to 100M	1k to 100M	10k to 30M	10k to 40M	1k to 100M	1k to 100M	1k to 100M
Power Dissipation [mW]	22.5	19.5	3.25	9.2	32.16	17.5	14.4
FoM* [dB]	-251	-249.9	-242.4	-250.3	-249.7	-252.4	-252.2
FoM$_{RSR}$ **[dB]	-239.2	-235.9	-235.4	-241.2	-238.9	-238.4	-242.2
Active Area [mm²]	0.28	0.29	0.39	0.27	0.09	0.21	0.12
CMOS Process [nm]	28	28	40	130	28	28	40

*FoM = 10 log$_{10}$ [(Power/1mW)·(Frac-N Jitter/1s)²]   **FoM$_{RSR}$ = FoM +10 log$_{10}$(F$_{ref}$/10MHz)   ***Normalized to 8.75 GHz

Figure 12.2.6: Comparison table with prior-art PLLs with reduced supply sensitivity (left columns) and conventional low-jitter fractional-N PLLs (right columns).

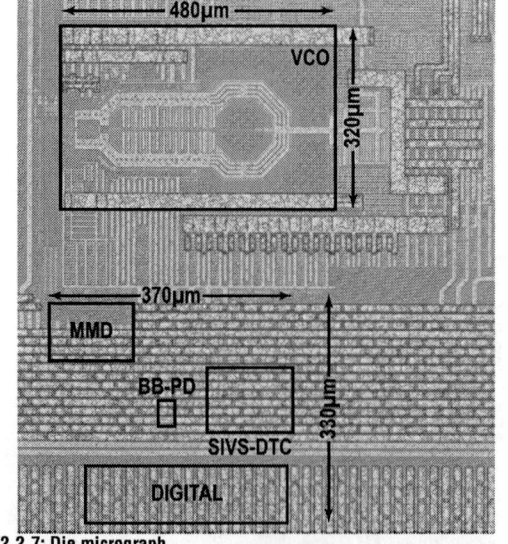

Figure 12.2.7: Die micrograph.

ISSCC 2026 / SESSION 12 / FREQUENCY SYNTHESIZERS AND VCOs / 12.3

12.3 A −66dBc-Worst-Fractional-Spur and 58fs-Jitter Fractional-N Digital PLL Using a Supply-Resilient Pseudo-Differential Inverse-Constant-Slope DTC

Pietro Salvi, Michele Rossoni, Riccardo Moleri, Daniele Lodi Rizzini, Damiano Fagotti, Stefano Gallucci, Andrea Leonardo Lacaita, Simone Mattia Dartizio, Salvatore Levantino

Politecnico di Milano, Milan, Italy

Abstract

A fractional-N digital PLL adopting a pseudo-differential inverse-constant-slope DTC for rejection of supply disturbances is presented. Compared to traditional fractional-N digital PLLs, it requires no additional calibration for supply rejection or additional supply-insensitive blocks. The PLL achieves integrated jitter below 59fs under 10mV$_{pp}$ sinusoidal disturbance applied to the DTC supply in the fractional-N mode, with fractional spurs below -66dBc and the jitter-vs-power figure of merit of -251.4dB.

Modern systems-on-chip for wireless/wireline communications typically integrate multiple functional blocks, including several frequency synthesizers, within a compact silicon area. In this context, sensitivity to supply disturbances—arising either from crosstalk between adjacent sub-blocks or by DC/DC converters and low-dropout regulators in the power supply network—becomes a critical concern. Among frequency synthesizers, digital-to-time-converter (DTC)-based fractional-N phase-locked loops (PLLs) have gained widespread adoption owing to their excellent noise performance and compact footprint. Their noise and spectral purity are largely dictated by the DTC performance, which has motivated several recent solutions aimed at improving DTC linearity and noise [1-6]. A persistent challenge, however, is the high sensitivity of DTCs to supply fluctuations, an inherent consequence of their shallow output slope when generating large delays. This leads to additional spurious tones and increased jitter at the PLL output in the presence of supply disturbances, such as periodic ripples or supply noise (Fig. 12.3.1, top) [7]. Although the supply sensitivity of ring and LC-based oscillators in PLLs has been extensively studied in the literature [8-10], only a few works have addressed the supply sensitivity of DTCs [7,11,12]. Specifically, [11] employs a digital calibration scheme to estimate the supply-ripple pattern, but it is limited to periodic disturbances at specific frequencies. The approach in [12] implements a fully differential voltage-domain phase detector (PD), but it requires a supply-independent DAC to reduce supply sensitivity. The solution proposed in [7] uses digital calibration to compensate for supply-induced delay variations, but it exhibits a limited operating bandwidth.

This work presents a DTC-based fractional-N digital PLL featuring a supply-resilient pseudo-differential inverse-constant-slope DTC, achieving <59fs integrated jitter under 10mV$_{pp}$ sinusoidal supply disturbances across a wide frequency range (up to 40MHz) and under broadband supply noise with 4.7mV$_{rms}$ and a 10MHz bandwidth. The proposed DTC architecture is inherently resilient to both deterministic and random supply-ripple fluctuations, without requiring any dedicated digital calibration.

Figure 12.3.1 (bottom) illustrates that a simple approach to reduce DTC supply sensitivity may be to adopt a pseudo-differential DTC configuration [13-15], in which two DTCs—one on the reference signal path (DTC$_R$) and the other on the divider signal path (DTC$_D$)—are driven with complementary input codes. In this configuration, a disturbance affecting the common supply voltage V_{DD} of the DTCs manifests as a common-mode perturbation at the positive and negative input terminals of the PD, thus not being transferred to the PLL output. However, the validity of this property strongly depends on the DTCs architecture. To analyze this, the DTC delay can be expressed as the sum of a fixed delay term, T_{fix}, and a code-dependent delay term, $S_{dtc}T_{LSB}$, where S_{dtc} is the DTC input code and T_{LSB} is the DTC resolution. Even in the pseudo-differential configuration, if T_{LSB} depends on V_{DD}, a residual effect of the supply disturbance remains at the PD input (bottom-left), limiting the effectiveness of the technique. This occurs because the code-dependent delay terms of the two DTCs, unlike the fixed delay terms, are opposite and thus not cancelled by the differential operation. In variable-slope (VS) DTCs, the dependence of the DTC resolution on the supply rail arises from the fact that the voltage slope of internal signals is code dependent, leading to a code-dependent supply sensitivity and, consequently, a V_{DD}-dependent T_{LSB}. Instead, for constant-slope (CS) DTCs, this behavior stems from the supply sensitivity of the DAC that controls the propagation delay. Additionally, CS-DTCs exhibit worse noise performance than VS-DTCs [16], as they rely on current generators with inherently higher flicker noise, making them unsuitable for low-jitter PLLs. To overcome these limitations, this work adopts the inverse-constant-slope (ICS) DTC topology, originally introduced in [3], to implement the individual DTCs (Fig. 12.3.1 bottom-right). The ICS-DTC provides two main advantages: (i) its resolution is defined as a fraction of the digitally controlled oscillator (DCO) period, T_{DCO}, which is a stable, supply-tolerant time reference within the PLL, thereby eliminating the need for additional supply-independent blocks or biasing circuits; (ii) it can employ resistors instead of current generators, achieving a reduced noise level comparable to that of VS-DTCs [4].

Figure 12.3.2 (left) shows the implemented ICS driving circuitry. Compared to [3,4], modifications in the precharge generator (PG) and front-end circuit were made to enable the pseudo-differential configuration of the ICS-DTC. In the PG, to achieve the independent control of the precharge duration T_{pch} for DTC$_D$ and DTC$_R$, the MUX, which collects T_{DCO} shifted versions of the multi-modulus-divider (MMD) output *mmd*, and the subsequent flip-flops (FFs)—enabling $T_{DCO}/2$ resolution of T_{pch} [3]—were duplicated. An additional PG output signal *hold* was added, which comes after every in_{pch} signal to stop the precharge phase of both DTC$_D$ and DTC$_R$. Successively, in the front-end circuit, three set-reset FFs and a couple of logic gates combine the PG outputs to create the $in_{pch,n}$, $in_{dtc,r}$ and *reset* signals. As can be seen from the waveforms in Fig. 12.3.2 (center), the *hold* signal allows for T_{pch} to be independent of $in_{dtc,r}$, thus guaranteeing correct operation of the ICS-DTC even on the *ref* path where $in_{dtc,r}$ has time-variant distance compared to $in_{pch,r}$ due to fractional-N operation. Figure 12.3.2 (top-right) shows the implemented ICS with resistor-based current branches, which reduce noise (especially flicker) without impairing the DTC linearity [4]. Unlike previous methods [3,4], this work refines the coarse resolution by using a bank of switchable capacitive cells in parallel to a fixed capacitor C_{fix}, which collectively tune the total load capacitance C_{TOT}. This technique, essentially embedding a VS-DTC within the ICS-DTC, eliminates the need for a separate fine DTC, avoiding extra noise, INL, supply sensitivity, and power consumption. Finally, a pull-down logic discharges C_{TOT} between each conversion cycle, avoiding code-dependent current absorption from the supply [4]. Figure 12.3.2 (bottom-center) shows the analytical expression of the ICS-DTC delay T_{dtc}, where S_C and S_F are the coarse and fine control, respectively, $S_{C,max} = 17$, V_{th} is the threshold of the output comparator, R is the resistance discharging C_{TOT} during the precharge phase, K is the resistance ratio between the precharge and the delay phase. As shown by the formula, the coarse resolution is independent of V_{DD}, while the delay T_{dtc} is proportional to S_F, thus validating the effectiveness of integrating a VS-DTC within an ICS-DTC. Additionally, the VS-DTC embedded in the ICS-DTC benefits from the large C_{fix} used in the ICS, which reduces the slope variation, thus improving the VS-DTC linearity [17]. Figure 12.3.2 (bottom-right) shows simulation results of the pseudo-differential resistor-based ICS (DRICS) DTC. The INL was calculated through a best linear fit of the DTC delay over coarse and fine tuning, resulting in a peak-to-peak value of 17fs. The DTC range is 129ps (with K=15 and T_{DCO}=108ps), with a fine resolution of 137fs. The edge phase noise shows a white level of -168dBc/Hz with flicker corner at 94kHz. The residual supply sensitivity, with a peak-to-peak value of 14ps/V, is mainly ascribed to the embedded VS-DTC, as it is almost constant over S_C.

Figure 12.3.3 (top) shows the implemented BBPLL scheme with the DRICS DTC. The MMD is driven by a MASH 1-1 modulator combined with a DTC-range-reduction technique [18]. At the system level, the DRICS is treated as a single DTC, where both DTCs (DTC$_R$ and DTC$_D$) share the same coarse and fine controls, except for the opposite sign. For this reason, the number of LMS algorithms is the same as in [4], with two gain calibrations—for the coarse and fine tuning of the DRICS delay—and two duty-cycle corrections—for the DTC-range-reduction technique and the improved PG-resolution ($T_{DCO}/2$) mechanism. Figure 12.3.3 (bottom-right) shows the measurement setup used for the prototype validation. A separate supply was fed to the DRICS to demonstrate its effectiveness in rejecting supply noise/ripples, and to decouple its sensitivity from that of the other PLL blocks (e.g., PG, MMD, etc.), which are less critical owing to the steepness of their internal signals. The DRICS supply was modulated by a signal generator through an AC-coupling network, while the supply ripple was monitored close to the chip with an oscilloscope, ensuring constant amplitude over a wide frequency range.

The PLL prototype, fabricated in 28nm CMOS (Fig. 12.3.7), occupies an active area of 0.2mm² and consumes 22mW (excluding output driver). It operates at a 125MHz reference clock with an output tuning range from 8.625 to 10.25GHz. At a fractional-N channel 15kHz off the 8.625GHz integer one (Fig. 12.3.4 left), the PLL demonstrates the worst-case fractional spur of -66.6dBc and jitter—integrated from 1kHz to 100MHz—of 57.1fs. Sweeping the fractional-N channel (FCW$_{frac}$ from 2^{-13} to 2^{-5}), fractional spurs always stay below -65.7dBc and integrated jitter never exceeds 57.6fs (Fig. 12.3.4 right). Figure 12.3.5 (top-left) shows the PLL phase-noise spectra when a broadband noise with a 4.7mV$_{rms}$ amplitude and a 10MHz bandwidth is added to the DTC supply. To validate the DRICS effectiveness in rejecting supply modulations, DTC$_R$ can be bypassed (DRICS OFF), while doubling the DTC$_D$ range by reducing K to 7.5. In this condition, the supply noise transfers to the PD input through the large sensitivity of the single DTC$_D$, leading to an integrated jitter of 558.7fs. When the pseudo-differential configuration is active (DRICS ON)—with K=15 in the DTCs—supply noise equally affects the PD inputs and is thus not transferred to the PLL output, leading to an integrated jitter of 57.9fs. Similarly, Fig. 12.3.5 (top-right) demonstrates the effectiveness of the technique under a 10mV$_{pp}$-amplitude sinusoidal disturbance with a 100kHz frequency. When activating the DRICS, the spur induced by the supply modulation decreases from -21.1 to -69.6dBc. Sweeping the sinusoidal disturbance frequency (Fig. 12.3.5 bottom), the presented technique shows 48.5dB spur attenuation within the PLL bandwidth, with supply-ripple spur always below -68.3dbc and integrated jitter never exceeding 59fs. Compared to prior-art supply-resilient fractional-N PLLs (Fig. 12.3.6), this work achieves the lowest integrated jitter and fractional spur, with a figure of merit of -251.4dB. Compared with conventional fractional-N PLLs (Fig. 12.3.6), the presented prototype achieves the lowest in-band phase noise, with -113.2dBc/Hz phase noise at a 10kHz offset.

References:
[1] M. Chae et al., "A 65fs$_{rms}$-Jitter and -272dB-FoM$_{jitter,N}$ 10.1GHz Fractional-N Digital PLL with a Quantization-Error-Compensating BBPD and an Orthogonal-Polynomial LMS Calibration," ISSCC, pp. 554-556, Feb. 2025.
http://doi.org/10.1109/ISSCC49661.2025.10904741
[2] M. Rossoni et al., "An 8.75GHz Fractional-N Digital PLL with a Reverse-Concavity Variable-Slope DTC Achieving 57.3fs$_{rms}$ Integrated Jitter and –252.4dB FoM," ISSCC, pp. 188-190, Feb. 2024. http://doi.org/10.1109/ISSCC49657.2024.10454388
[3] S. M. Dartizio et al., "A 76.7fs-Integrated-Jitter and –71.9dBc In-Band Fractional-Spur Bang-Bang Digital PLL Based on an Inverse-Constant-Slope DTC and FCW Subtractive Dithering," ISSCC, pp. 188–189, Feb. 2023.
http://doi.org/10.1109/ISSCC49657.2024.10454388
[4] P. Salvi et al., "A 66.7fs-Integrated-Jitter Fractional-N Digital PLL Based on a Resistive-Inverse-Constant-Slope DTC," IEEE CICC, pp. 1–2, Apr. 2024.
http://doi.org/10.1109/CICC60959.2024.10529003
[5] H. Liu et al., "An Ultra-Low-Jitter Fast-Hopping Fractional-N PLL With LC DTC and Hybrid-Proportional Paths," IEEE JSSC, vol. 60, no. 3, pp. 785-798, Mar. 2025.
http://doi.org/10.1109/JSSC.2024.3514870
[6] D. Zhang et al., "A 6.65-to-7.75GHz Fractional-N Digital PLL with Analog Pre-Distortion DTC Implementing 2nd/3rd-Order Calibration and Achieving –65.7dBc Fractional Spur and 154fs Integrated Jitter," IEEE CICC, pp. 1-3, Apr. 2025.
http://doi.org/10.1109/CICC63670.2025.10983671
[7] M. Rossoni et al., "A Fractional-N Digital-PLL Based on a Power-Gated Ring-Oscillator and a Frequency-Stabilizing Loop Achieving 74fs Jitter Under 3mVpp Supply Ripple," IEEE Symp. VLSI Circuits, pp. 1-3, June 2025.
http://doi.org/10.23919/VLSITechnologyandCir65189.2025.11074831
[8] Y. Duan et al., "Supply-Noise-Desensitized Techniques for Low Jitter RO-Based PLL Achieving ≤1.6 ps RMS Jitter Within Full-Spectrum Supply Interference," IEEE TCAS-I, vol. 69, no. 12, pp. 4799-4809, Dec. 2022. http://doi.org/10.1109/TCSI.2022.3204655
[9] Y. Chen et al., "A Supply Pushing Reduction Technique for LC Oscillators Based on Ripple Replication and Cancellation," IEEE JSSC, vol. 54, no. 1, pp. 240-252, Jan. 2019.
http://doi.org/10.1109/JSSC.2018.2871195
[10] X. Gui et al., "Low-Supply Sensitivity LC VCOs with Complementary Varactors," IEEE Trans. VLSI Systems, vol. 28, no. 7, pp. 1589-1599, July 2020.
http://doi.org/10.1109/TVLSI.2020.2991765
[11] Y. Chen et al., "A Fractional-N Digitally Intensive PLL Achieving 428-fs Jitter and <–54-dBc Spurs Under 50-mVpp Supply Ripple," IEEE JSSC, vol. 57, no. 6, pp. 1749-1764, June 2022. http://doi.org/10.1109/JSSC.2021.3123386

[12] L. Wu et al., "A Power-Efficient Fractional-N DPLL With Phase Error Quantized in Fully Differential-Voltage Domain," IEEE JSSC, vol. 56, no. 4, pp. 1254-1264, Apr. 2021. http://doi.org/10.1109/JSSC.2020.3047431
[13] A. Ba et al., "A 0.62nJ/b multi-standard WiFi/BLE wideband digital polar TX with dynamic FM correction and AM alias suppression for IoT applications," IEEE RFIC, pp. 308-311, June 2018. http://doi.org/10.1109/RFIC.2018.8428999
[14] D. Xu et al., "A 7GHz Digital PLL with Cascaded Fractional Divider and Pseudo-Differential DTC Achieving -62.1dBc Fractional Spur and 143.7fs Integrated Jitter," ISSCC, pp. 192-194, Feb. 2024. http://doi.org/10.1109/ISSCC49657.2024.10454284
[15] Z. Gao et al., "A 2.6-to-4.1GHz Fractional-N Digital PLL Based on a Time-Mode Arithmetic Unit Achieving -249.4dB FoM and -59dBc Fractional Spurs," ISSCC, pp. 380-382, Feb. 2022. https://doi.org/10.1109/ISSCC42614.2022.9731561
[16] D. Xu et al., "A 6.5-to-8-GHz Cascaded Dual-Fractional-N Digital PLL Achieving –52.79-dBc Fractional Spur with 50-MHz Reference," IEEE JSSC, vol. 60, no. 3, pp. 1043-1055, Mar. 2025. http://doi.org/10.1109/JSSC.2024.3447021
[17] W. Wu et al., "A 28-nm 75-fs$_{rms}$ Analog Fractional-N Sampling PLL With a Highly Linear DTC Incorporating Background DTC Gain Calibration and Reference Clock Duty Cycle Correction," IEEE JSSC, vol. 54, no. 5, pp. 1254-1265, May 2019.
http://doi.org/10.1109/JSSC.2019.2899726
[18] D. Tasca et al., "A 2.9–4.0-GHz Fractional-N Digital PLL with Bang-Bang Phase Detector and 560-fs$_{rms}$ Integrated Jitter at 4.5-mW Power," IEEE JSSC, vol. 46, no. 12, pp. 2745-2758, Dec. 2011. http://doi.org/10.1109/JSSC.2011.212917
[19] H. Liu et al., "A 0.18-µs-Locking-Time Fractional-N PLL with Stochastic Gradient Descent Tuning Curve Fitting, Initial Phase Error Zeroing, and Random DSM Achieving 44.4-fs Jitter at Near-Integer Channel," IEEE CICC, pp. 1-3, Apr. 2025.
http://doi.org/10.1109/CICC63670.2025.10982962

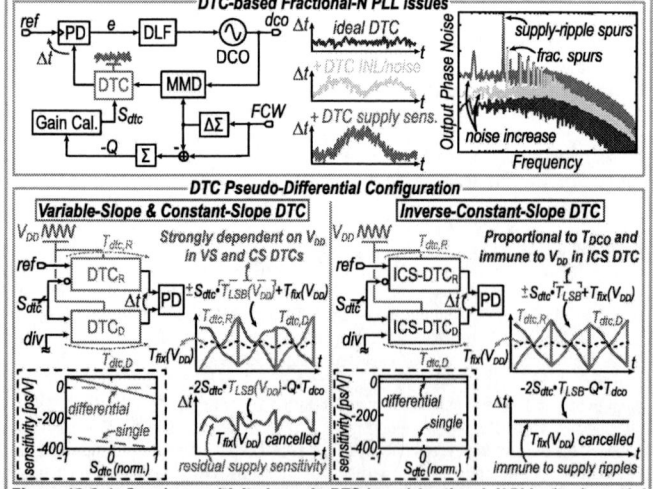

Figure 12.3.1: Supply-sensitivity issue in DTC-based fractional-N PLLs (top); partial mitigation using DTC pseudo-differential configuration (bottom-left), and presented inverse-constant-slope DTC-based solution (bottom-right).

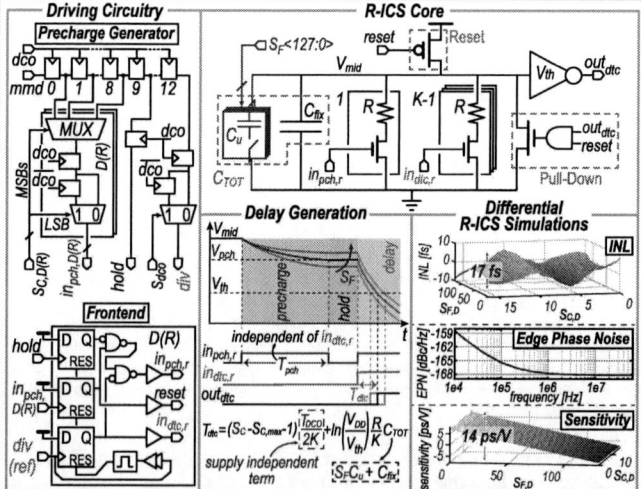

Figure 12.3.2: Implementation of DTC driving circuitry for ICS pseudo-differential operation (left), schematic of resistor-based ICS core (top-right), coarse and fine delay-tuning mechanisms (bottom-center), and simulation results (bottom-right).

ISSCC 2026 / SESSION 12 / FREQUENCY SYNTHESIZERS AND VCOs / 12.3

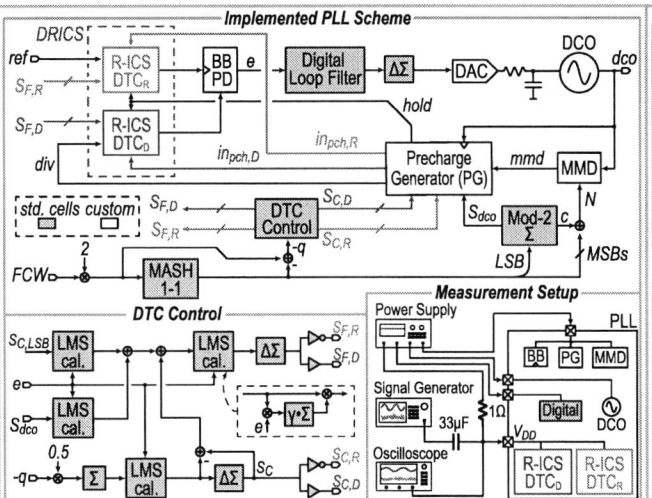

Figure 12.3.3: Diagram of the implemented PLL (top), LMS calibrations and DRICS controls generation (bottom-left), and supply-resilience measurement setup (bottom-right).

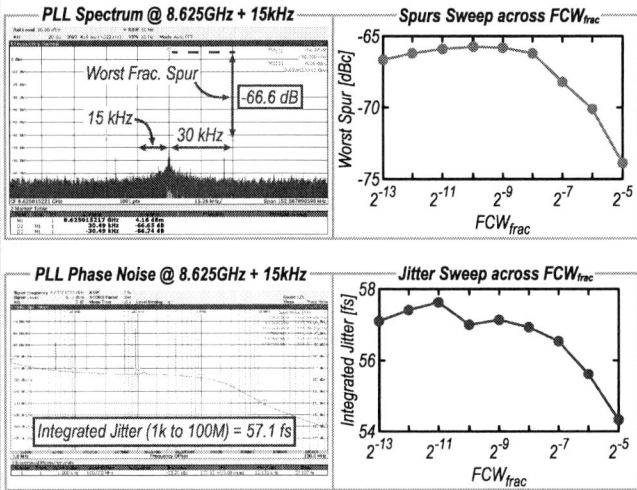

Figure 12.3.4: Measured PLL output spectrum and phase noise for FCW = $69 + 2^{-13}$ (left), and worst-case spur and integrated jitter sweeps across fractional-N channels (right).

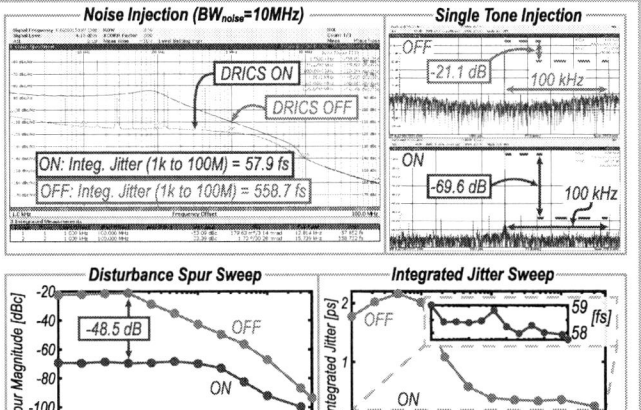

Figure 12.3.5: Measured DRICS effectiveness in rejecting broadband supply noise (top-left) and sinusoidal supply disturbance (top-right), and supply-ripple spur and integrated jitter sweeps across the disturbance frequency when DRICS is ON and OFF (bottom).

	Supply-Resilient frac-N PLLs				Conventional frac-N PLLs		
	This Work	Rossoni [7] VLSI 25	Chen [11] JSSC 22	Wu [12] JSSC 21	Chae [1] ISSCC 25	Salvi [4] CICC 24	Liu [19] CICC 25
Output Freq. [GHz]	8.625 to 10.25	9 to 10.25	4.47 to 5.14	2.99 to 3.5	10 to 11.5	8.75 to 10.25	10.8 to 14.5
Reference Freq. [MHz]	125	250	50	80	100	125	250
Supply Sensitivity Cancellation Technique	Pseudo-Diff. R-ICS	PGRO w/ FSL	Ripple Patt. Cal.	Fully-Diff. Voltage PD	N/A	N/A	N/A
Noise Amplitude [mV$_{rms}$]	4.7	2.5	N/A	N/A	N/A	N/A	N/A
Jitter with Supply Noise [fs]	57.9	80.6	N/A	N/A	N/A	N/A	N/A
Ripple Amplitude [mV$_{pp}$]	10	3	50	32	N/A	N/A	N/A
Jitter with Supply Ripple [fs]	58.4	74	423	115*	N/A	N/A	N/A
Ripple Output Spur** [dBc]	-69.6	-66.6	-56.5	-53.8	N/A	N/A	N/A
Fractional Spur [dBc]	-66.6	-57.9	-55	-55.8	-63	-63.8	-60.9
Jitter w/o Supply Ripple [fs]	57.1	72.2	N/A	101.7	65	66.7	44.4
Integration BW [Hz]	1k to 100M	1k to 100M	10k to 30M	10k to 40M	1k to 100M	10k to 100M	1k to 100M
PN @ 10kHz** [dBc/Hz]	-113.2	-108.8	-86.5	-100.1	-112.6	-109.1	-112.9
Power Dissipation [mW]	22	19.5	3.25	9.2	14.4	16.7	35.1
FoM$_J$ [dB]	-251.4	-249.9	-242.4	-250.2	-252.2	-251.3	-251.6
FoM$_N$ [dB]	-269.8	-265.6	-262.2	-266.4	-272.2	-270	-269.6
Active Area [mm²]	0.2	0.27	0.39	0.27	0.12	0.21	0.22
CMOS Process [nm]	28	28	40	130	40	28	28

*Estimated from Measurements **Normalized to 8.625GHz FoM$_J$ = $10log_{10}[P_{DD}[mW]\cdot(Jitter[s])^2]$ FoM$_N$ = FoM$_J$ + $10log_{10}(f_{REF}/f_{OUT})$

Figure 12.3.6: Performance comparison with prior-art supply-resilient and conventional fractional-N PLLs.

Figure 12.3.7: Die micrograph of the implemented PLL prototype.

214 • 2026 IEEE International Solid-State Circuits Conference

979-8-3315-8937-0/26 $31.00 © 2026 IEEE

ISSCC 2026 / SESSION 12 / FREQUENCY SYNTHESIZERS AND VCOs / 12.4

12.4 A 21.6fs$_{rms}$-Jitter, -260.7dB-FoM Fractional-N PLL Enabled by an Intrinsically Linear Variable-Slope SPD for Quantization-Error Cancellation

Yanchao Liu, Yang Li, Kaihang Wang, Xiaohua Yu, Ronghua Ni

Fudan University, Shanghai, China

Abstract

This work presents a fractional-N PLL with an intrinsically linear variable-slope sampling phase detector for quantization-error cancellation. Linearity, noise, and power efficiency are improved by eliminating the edge-restoration buffer. Combined with the proposed charge-injection-based nonlinearity compensation, the 14GHz prototype PLL achieves 21.6fs rms jitter, -260.7dB FoM, and -67.4dBc worst-case spur in the fractional-N mode with 18.1mW power consumption in a 28nm CMOS process.

The evolution of communication protocols beyond 5G imposes increasingly stringent phase-noise (PN) requirements on local oscillators (LOs) [1]. While rms jitter below 30fs has been demonstrated in several integer-N phase-locked loops (PLLs) [2-4], achieving comparable performance in fractional-N PLLs remains significantly more challenging due to the nonlinearity and additional noise introduced by the quantization-error (QE) cancellation circuitry. Digital-to-time converters (DTCs) are widely employed in fractional-N PLLs for QE cancellation, owing to their finer resolution, simpler implementation, and reduced sensitivity to process, voltage, and temperature (PVT) variations [5-13]. However, the RC-based variable-slope DTCs (VS-DTCs), as shown in Fig. 12.4.1 (top), are susceptible to nonlinearity introduced by the edge-restoration buffer, which arises from the nonlinear relationship between its input slope and inversion delay [5]. When followed by a sampling phase detector (SPD), DTC integral-nonlinearity (INL) translates to voltage ripples, which raise the noise floor and fractional spur. Several prior works have attempted to compensate the DTC INL through pre-distortion using either digital or analog techniques. However, their effectiveness is limited and often comes at the cost of additional calibration circuitry or increased thermal noise [7-9]. Constant-slope DTCs (CS-DTCs), based on current source and digital-to-analog converters (DACs), have been introduced [14] to mitigate the INL problem by enforcing a constant slope at the input of the edge-restoration buffer (Fig. 12.4.1, middle). Nevertheless, the elevated noise and inevitable nonlinearity of the current sources and DACs constrain their suitability for ultra-low-jitter LO applications. Moreover, since an edge-restoration buffer is unavoidable in both VS-DTCs and CS-DTCs, its noise significantly degrades the overall DTC noise performance due to the slow ramp of the input waveform [5]. A merged CS-DTC and SPD has been presented in [15] to eliminate the buffer noise, but the dominant noise and INL sources of the CS-DTC remain.

To mitigate the linearity and noise challenges in fractional-N PLLs, this work presents an intrinsically linear variable-slope sampling phase detector (VS-SPD), which simultaneously achieves QE cancellation and phase-to-voltage conversion with improved linearity and noise performance, as shown in Fig. 12.4.1 (bottom). Exploiting the linear relationship between the RC delay and capacitance value, linear digital-to-time conversion is achieved by controlling the capacitance of CS1 with the digital control word (DCW). Meanwhile, the linearity of the RC delay is preserved, and the noise is significantly reduced through the elimination of the edge-restoration buffer. The linear relationship between the DCW and the RC delay of VS-SPD (t_d) can be derived from the step response of the VS1 falling edge, which is given by $VS1 = VDD \times exp(-t / \tau)$, where $\tau = R \times CS1$. When VS1 drops to a certain level, e.g. V_{REF}, the delay in the VS-SPD is given by $t_d = k \times DCW$, where $k = R \times C_{LSB} \times \ln(VDD / V_{REF})$ and C_{LSB} is the unit capacitance of CS1. Therefore, t_d is a linear function of the DCW. Generated by the delta-sigma modulator (DSM), the DCW represents the expected QE and scales linearly with the edge displacement in CLKDIV (t_{QE}), which is determined by the DSM-generated division ratio. Consequently, t_d also scales linearly with t_{QE}. In the locking mode, an appropriate scaling factor from the gain calibration ensures that t_d cancels t_{QE} in the time domain, so that CLKREF samples exactly when VS1 crosses V_{REF}. From a voltage-domain perspective, the DCW regulates the slope of VS1 such that it reaches V_{REF} exactly at the CLKREF sampling instant. As a result, the sampled voltage consistently equals V_{REF} (Fig. 12.4.1, bottom), leading to a stable VS-SPD output voltage with a complete QE cancellation and minimized ripple. By eliminating the nonlinearity and noise from the edge-restoration buffer, the presented VS-SPD achieves simultaneous improvements in linearity, noise, and power efficiency compared to the conventional DTCs.

The detailed circuit implementation and locking-mode waveforms of the presented VS-SPD are shown in Fig. 12.4.2 (top). The rising edge of the feedback clock from the divider (CLKDIV) triggers the falling edge of VS1. With CLKREF = 1 and SW1 closed, the slope of VS1 is determined by the DCW-controlled bank capacitance (CS1). When the subsequent falling edge of CLKREF opens SW1, the instantaneous voltage of VS1 is sampled and held on CS1. A second sample-and-hold stage, consisting of SW2 and CS2, preserves the sampled voltage over the entire reference period. Enforced by the loop dynamics and gain calibration in the locking mode, the average VS-SPD output voltage (V_{SAMP}) equals V_{REF}, which is defined by the loop filter and is typically set to approximately half of VDD to maximize the SPD linear range. The dynamic INL introduced by memory effects in the switched-capacitor bank and supply is mitigated through two design optimizations: (1) both plates of C_{LSB} are reset to ground after SW2 opens, ensuring an identical and well-defined initial condition for C_{LSB} at the start of each reference period; (2) a replica VS-SPD driven by a complementary DCW is added in parallel with the main VS-SPD (Fig. 12.4.2, middle-left) [10]. This guarantees that the same total capacitance is charged during each reference cycle, producing a periodic, data-independent supply ripple and thereby reducing dynamic INL.

Since the VS-SPD is intrinsically linear, the INL introduced by the on-resistance of the capacitor switch becomes noticeable, even though it is still much smaller than the INL contributed by the edge-restoration buffer in conventional DTCs. As shown in Fig. 12.4.2 (middle-right), the effect of the capacitor switch on-resistance (R_M) in CS1 can be modeled as an effective resistor with a resistance of $R_E = R_M / DCW$. During the sampling phase when SW1 is closed, R_E slows down the discharging of CS1, resulting in a sampled voltage larger than the ideal value (Fig. 12.4.2, bottom-left). The offset voltage introduced by R_M, denoted as V_{RM}, is proportional to R_E, and hence inversely proportional to the DCW (Fig. 12.4.2, bottom-middle). Reducing R_M lowers the INL, but this requires a larger switch size, which inevitably increases parasitic capacitance, leading to higher noise and power consumption. To address this trade-off, a charge-injection-based INL-compensation (CIIC) technique is presented. As shown in Fig. 12.4.2 (bottom-left), a small charge-injection capacitor (C_{INJ}) is added in parallel with CS1. During the sampling phase when SW1 is closed, the bottom plate of C_{INJ} is connected to VDD. Once the sampling phase finishes and SW1 opens, the bottom plate of C_{INJ} is switched to ground. This switching introduces a negative offset voltage (V_{INJ}) at VS1 due to charge redistribution between C_{INJ} and C_E. Since C_E is proportional to the DCW, V_{INJ} is approximately inversely proportional to the DCW (Fig. 12.4.2, bottom-middle). Therefore, V_{RM} and V_{INJ} can largely cancel each other if C_{INJ} is properly chosen. As shown in Fig. 12.4.2 (bottom-right), post-layout simulations confirm that the INL is reduced from 60fs to 9fs after CIIC is applied.

The system architecture of the VS-SPD-based sampling PLL is illustrated in Fig. 12.4.3 (top). The QE present in the rising edge of CLKDIV is cancelled during the sampling procedure in the VS-SPD, leaving a QE-free voltage to the loop filter. Two range reduction techniques are employed in this system, which reduces the delay range requirement of the VS-SPD by a factor of 4: the modified multi-modulus divider (MMD) and DSM [6] in the feedback path, and the reference phase selection (RPS) [11] in the reference path. As shown in Fig. 12.4.3 (top), these techniques enable a four-times higher slew rate in the VS-SPD, which reduces the bottom noise and power consumption while suppressing noise from the subsequent loop filter. Gm in the loop filter is designed to be small to save area with negligible noise contribution, thanks to the high gain provided by the VS-SPD. The sign of the phase error after the VS-SPD is detected by a low-noise comparator and used for loop calibrations. Several least-mean-square (LMS)-based background calibrations are implemented in the custom digital circuit to achieve the required INL level, including the gain calibration for the VS-SPD to cover PVT variations, the duty-cycle correction for the voltage-controlled-oscillator (VCO) output in the modified MMD, and the calibration for the RPS to cancel delay mismatch between two reference phases.

A dual-core VCO based on a circular-topology transformer [16] is implemented to enhance out-of-band noise performance, as shown in Fig. 12.4.3 (left). With three ultra-thick metals (M9) connected in parallel, the simulated quality factor (Q) of L_G at 14GHz is as high as 40 (Fig. 12.4.3, top-middle), achieving significant PN improvement. A low magnetic coupling factor ($\approx$0.25) between L_D and L_G is adopted to reduce the tank impedance at drains and boost voltage swing at gates. The two cores are synchronized by shorting their drain coils with a thick top metal (M8), which suppresses the undesirable low-Q mode. The nested shape-8 inductors (Fig. 12.4.3, bottom-middle) are designed for L_{tail-p} and L_{tail-n} with cancelled magnetic coupling and reduced area. The high source impedance introduced by L_{tail-p} and L_{tail-n} reshapes the drain-source voltage waveforms, creating a plateau region during which the impulse sensitivity function (ISF) equals zero and noise injection from circuits is suppressed, as shown in Fig. 12.4.3 (bottom-right). At 13.9GHz, the free-running VCO measures a PN of -143dBc/Hz and an FoM of 196dBc/Hz at a 10MHz offset.

The prototype sampling PLL is fabricated in a 28nm CMOS process and occupies a core area of 0.34mm². It consumes 18.1mW from a 0.9V supply in the fractional-N mode while generating a 13.1 to 14.5GHz VCO output from a 250MHz crystal oscillator reference. In

214 • 2026 IEEE International Solid-State Circuits Conference

979-8-3315-8937-0/26 $31.00 © 2026 IEEE

the integer-N mode, the measured rms jitter (integrated from 10kHz to 100MHz) is 19.1fs and the reference spur is -73.2dBc (Fig. 12.4.4, top). In the fractional-N mode, the measured rms jitter is 21.6fs and 29.7fs with a corresponding FoM of -260.7dB and -258dB for far-integer and near-integer channels, respectively (Fig. 12.4.4, bottom). As shown in Fig. 12.4.5 (top), the worst-case fractional spur is improved from -64.5 to -67.4dBc when the CIIC is enabled, indicating nearly 3dB of improvement attributed to the CIIC. The near-integer fractional spur across the fractional part of the frequency control word (FCW) is plotted in Fig. 12.4.5 (bottom-left). The measured performance is compared with recent low-jitter fractional-N and integer-N PLLs in Fig. 12.4.5 (bottom-right) and Fig. 12.4.6. The die micrograph and power breakdown table are shown in Fig. 12.4.7. Enabled by the low-noise, intrinsically linear VS-SPD, the presented fractional-N PLL achieves jitter and FoM comparable to prior-art integer-N PLLs, thereby narrowing the long-standing performance gap between the two architectures.

References:
[1] W. Wu, "Low-Jitter Frequency Generation Techniques for 5G Communication: A tutorial," *IEEE Solid-State Circuits Magazine*, vol. 13, no. 4, pp. 44-63, Fall 2021. http://doi.org/10.1109/MSSC.2021.3111430
[2] Y. Zhao et al., "A 20-GHz PLL with 20.9-fs Random Jitter," *IEEE JSSC*, vol. 58, no. 6, pp. 1597-1609, June 2023. http://doi.org/10.1109/JSSC.2022.3225105
[3] H. Liu et al., "A Multireference PLL: Theory and Implementation," *IEEE JSSC*, vol. 59, no. 7, pp. 1981-1994, July 2024. http://doi.org/10.1109/JSSC.2024.3383605
[4] D. Sun et al., "A 13GHz Charge-Pump PLL Achieving 15.8fsrms Integrated Jitter and -98.5dBc Reference Spur," *ISSCC*, pp. 344-346, Feb. 2025. http://doi.org/10.1109/ISSCC49661.2025.10904662
[5] W. Wu et al., "A 28-nm 75-fs$_{rms}$ Analog Fractional- N Sampling PLL with a Highly Linear DTC Incorporating Background DTC Gain Calibration and Reference Clock Duty Cycle Correction," *IEEE JSSC*, vol. 54, no. 5, pp. 1254-1265, May 2019. http://doi.org/10.1109/JSSC.2019.2899726
[6] W. Wu et al., "A 14nm Analog Sampling Fractional-N PLL with a Digital-to-Time Converter Range-Reduction Technique Achieving 80fs Integrated Jitter and 93fs at Near-Integer Channels," *ISSCC*, pp. 444-446, Feb. 2021. http://doi.org/10.1109/ISSCC42613.2021.9365850
[7] S. Levantino et al., "An Adaptive Pre-Distortion Technique to Mitigate the DTC Nonlinearity in Digital PLLs," *IEEE JSSC*, vol. 49, no. 8, pp. 1762-1772, Aug. 2014. http://doi.org/10.1109/JSSC.2014.2314436
[8] M. Rossoni et al., "An 8.75GHz Fractional-N Digital PLL with a Reverse-Concavity Variable-Slope DTC Achieving 57.3fs$_{rms}$ Integrated Jitter and –252.4dB FoM," *ISSCC*, pp. 188-190, Feb. 2024. http://doi.org/10.1109/ISSCC49657.2024.10454388
[9] D. Zhang et al., "A 6.65-to-7.75GHz Fractional-N Digital PLL with Analog Pre-Distortion DTC Implementing 2nd/3rd-Order Calibration and Achieving –65.7dBc Fractional Spur and 154fs Integrated Jitter," *IEEE CICC*, pp. 1-3, Apr. 2025. http://doi.org/10.1109/CICC63670.2025.10983671
[10] A. Santiccioli et al., "A 66-fs-rms Jitter 12.8-to-15.2-GHz Fractional-N Bang–Bang PLL with Digital Frequency-Error Recovery for Fast Locking," *IEEE JSSC*, vol. 55, no. 12, pp. 3349-3361, Dec. 2020. http://doi.org/10.1109/JSSC.2020.3019344
[11] Y. Liu et al., "A 37.5fs-rms Jitter and –254.1dB FoM Fractional-N Sampling PLL with Reference-Phase-Selection and Complementary-DTC Achieving 8× DTC Range Reduction and Zero DTC Delay Offset," *IEEE CICC*, pp. 1-3, Apr. 2025. http://doi.org/10.1109/CICC63670.2025.10983724

[12] H. Liu et al., "An Ultra-Low-Jitter Fast-Hopping Fractional-N PLL With LC DTC and Hybrid-Proportional Paths," *IEEE JSSC*, vol. 60, no. 3, pp. 785-798, Mar. 2025. http://doi.org/10.1109/JSSC.2024.3514870
[13] Z. Ye et al., "A Sub-50-fsrms Jitter Fractional-N CPPLL Based on a Dual-DTC-Assisted Time-Amplifying Phase-Frequency Detector with Cascadable DTC Nonlinearity Compensation Algorithm," *IEEE JSSC*, vol. 59, no. 3, pp. 677-689, Mar. 2024. http://doi.org/10.1109/JSSC.2023.3339679
[14] J. Z. Ru et al., "A High-Linearity Digital-to-Time Converter Technique: Constant-Slope Charging," *IEEE JSSC*, vol. 50, no. 6, pp. 1412-1423, June 2015. http://doi.org/10.1109/JSSC.2015.2414421
[15] G. Jin et al., "A Fractional-N Sampling PLL with a Merged Constant-Slope DTC and Sampling PD," *IEEE JSSC*, vol. 59, no. 8, pp. 2407-2417, Aug. 2024. http://doi.org/10.1109/JSSC.2024.3358564
[16] D. Murphy and H. Darabi, "A 27-GHz Quad-Core CMOS Oscillator with No Mode Ambiguity," *IEEE JSSC*, vol. 53, no. 11, pp. 3208-3216, Nov. 2018. http://doi.org/10.1109/JSSC.2018.2865460

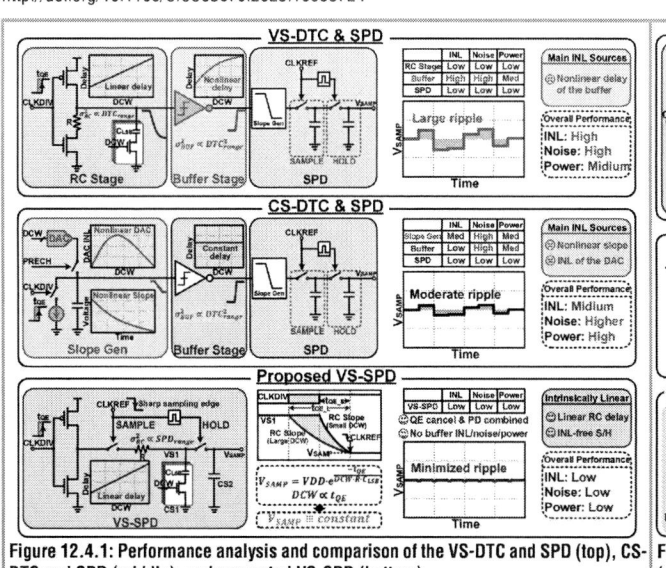

Figure 12.4.1: Performance analysis and comparison of the VS-DTC and SPD (top), CS-DTC and SPD (middle), and presented VS-SPD (bottom).

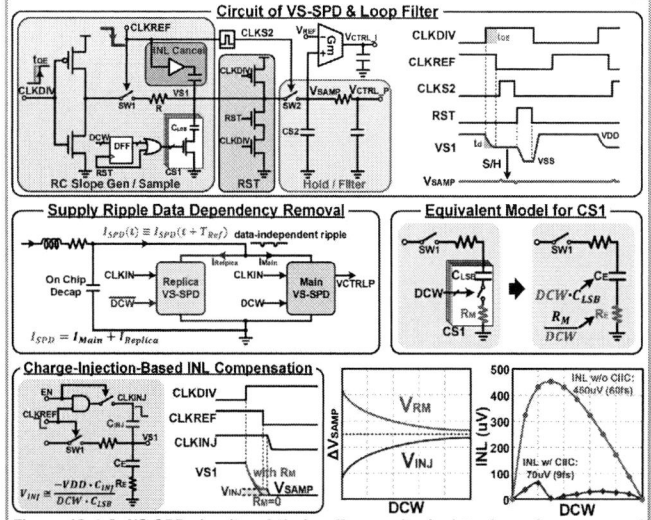

Figure 12.4.2: VS-SPD circuit and timing diagram (top), data-dependency removal (middle-left), and the CIIC circuit and principle (middle-right and bottom).

ISSCC 2026 / SESSION 12 / FREQUENCY SYNTHESIZERS AND VCOs / 12.4

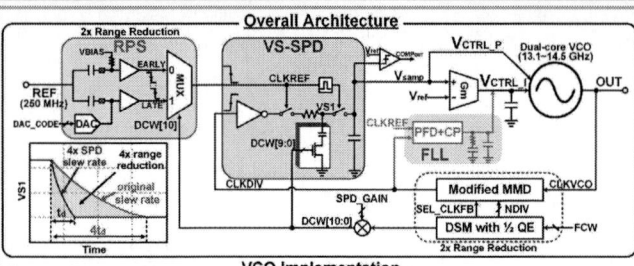

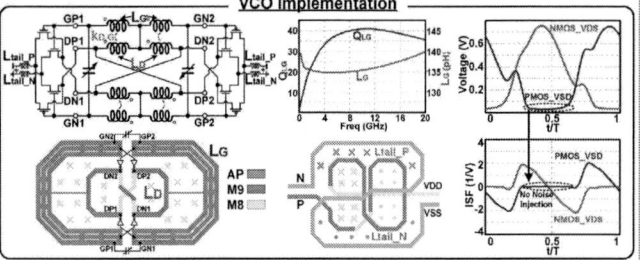

Figure 12.4.3: Overall architecture of the presented PLL (top), the VCO circuit, layout, and simulated results (bottom).

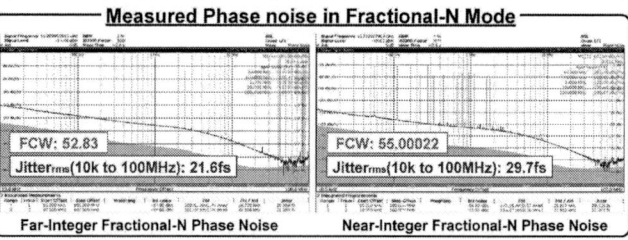

Figure 12.4.4: Measured PN and reference spur in the integer-N channel (top), measured PN in far-integer and near-integer fractional-N channels (bottom).

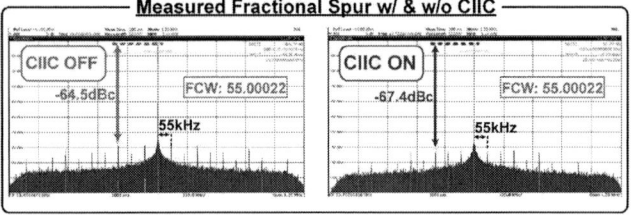

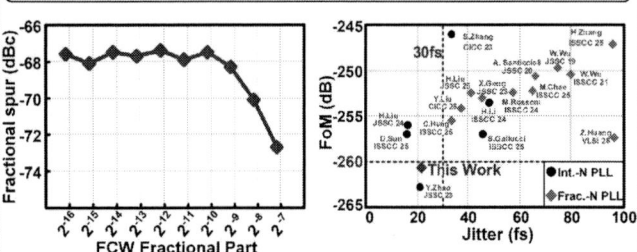

Figure 12.4.5: Measured spectra without and with CIIC (top), fractional spur (bottom-left), and benchmark (bottom-right).

Performance Comparison Table

	This work	Low jitter Fractional-N PLLs						Low jitter Integer-N PLLs		
		ISSCC'25 C.Hung	CICC'25 Y.Liu	JSSC'25 H.Liu	JSSC'23 X.Geng	ISSCC'24 M.Rossoni	ISSCC'21 W.Wu	ISSCC'25 D.Sun	JSSC'24 H.Liu	JSSC'23 Y.Zhao
Technology	28nm	22nm	28nm	28nm	65nm	28nm	14nm	65nm	65nm	28nm
Architecture	Analog SPLL	Analog CPPLL	Analog SPLL	Hybrid PLL	Analog CPPLL	Digital BBPLL	Analog SPLL	Analog CPPLL	Analog SPLL	Analog SPLL
Output Freq. [GHz]	13.1 to 14.5	7.3 to 9.5	13.7 to 15.4	11.1 to 14.9	24 to 28	8.8 to 10.3	5 to 7	13	4.8 to 5.6	20
Ref. Freq. [MHz]	250	80 x 2	250	250	250	250	76.8 x 2	250	100	250
Integer-N Jitter [fs]	19.1	-	28.5	36.8	37.1	-	-	15.8	16.1	20.9
Far-Integer Fractional-N Jitter [fs]	21.6	33.8	37.5	41.3	45.6	-	80	-	-	-
Near-Integer Fractional-N Jitter [fs]	29.7	37.7	-	61.7	-	57.3	93.2	-	-	-
PN@1MHz (dBc/Hz) (Norm. to 13.2GHz)	-125.7	-119.6	-121.2	-119.2	-118.4	-116.4	-109.9	-127.5	-128.1	-129.2
Jitter Integration Bandwidth [Hz]	10k to 100M	10k to 100M	1k to 100M	1k to 100M	1k to 100M	1k to 30M	1k to 100M	10k to 40M	10k to 100M	10k to 40M
Frac. Spur [dBc]	-67.4	-64.8	-64.6	-62.4	-59	-63.4	-66.4*	-	-	-
Ref. Spur [dBc]	-73.2	-101	-69.1	-65.3	-72	-69.4	-66	-98.5	-71	-66
Core Area [mm²]	0.34	0.39	0.25	0.21	0.47	0.21	0.31	0.52	1.1	0.06
Power [mW]	18.1	24.4	27.9	34	23.9	17.5	14.2	79.5	83	12
Far-Integer FoM**[dB]	-260.7	-255.5	-254.1	-252.4	-253	-	-250.4	-257***	-256.7***	-262.8***
Near-Integer FoM**[dB]	-258	-254.6	-	-248.9	-	-252.4	-249.1			

*Normalized to VCO frequency **FoM=10log₁₀[(Jitter_rms/1s)²·(Power/1mW)] ***Measured at integer-N channel

(superscript corrected: $**\text{FoM}=10\log_{10}[(\text{Jitter}_{rms}/1s)^2\cdot(\text{Power}/1mW)]$)

Figure 12.4.6: Performance comparison with low-jitter fractional-N and integer-N PLLs.

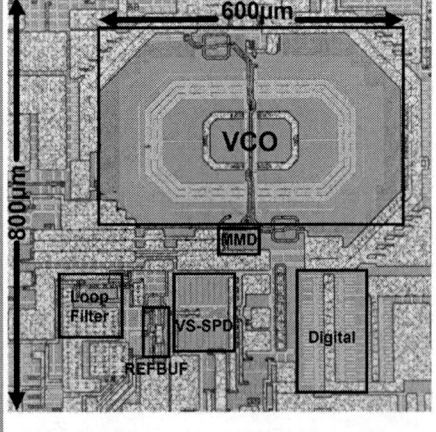

Power @ 13.2GHz (mW)		
VCO	9.2	51%
VS-SPD	2.5	14%
MMD	2.4	13%
REFBUF w/ RPS	1.4	8%
Loop Filter	0.7	4%
Digital	1.9	10%
Total	18.1	100%

Figure 12.4.7: Die micrograph (left) and power breakdown table (right).

ISSCC 2026 / SESSION 12 / FREQUENCY SYNTHESIZERS AND VCOs / 12.5

12.5 A 14GHz Chopper-Refolding Sampling PLL Achieving 33.8fs$_{rms}$ and 80.8dBc Reference Spur with a kT/C-Noise-Cancellation SPD

Yichen Liu, Jian Zhang, Xiaosen Liu, Yan Wang

Tsinghua University, Beijing, China

Abstract

A 14GHz chopper-refolding sampling PLL is implemented in 28nm CMOS, integrating a kT/C-noise-cancellation sampling phase detector (SPD) and a self-injection VCO with harmonic-impedance expansion. The SPD decouples jitter and phase-detection gain, while the chopper-refolding scheme filters flicker noise. The VCO enables a high FoM without manual tuning, achieving 33.8fs$_{rms}$ jitter, -80.8dBc spur, 16.9mW power, and competitive jitter performance among integer-N PLLs.

In high-speed RF direct-sampling systems, a low-jitter phase-locked loop (PLL) is critical for providing precise clock signals to ADCs, thereby enabling accurate signal capture and processing. Such systems impose stringent requirements on PLL performance, particularly in terms of low jitter and low phase noise—both of which are essential for maintaining high dynamic range and signal fidelity [1]. First, the PLL rms jitter must be tightly controlled to ensure it does not degrade the ADC sampling accuracy, with typical target values in the order of tens of femtoseconds. Second, the design must be low power to minimize power consumption, making it suitable for integration in RF systems. This paper presents a chopper-refolding sampling PLL (SPLL) featuring a kT/C-noise-cancellation sampling phase detector (SPD), which achieves 33.8fs$_{rms}$ while consuming 16.9mW of power in a 28nm CMOS process.

Figure 12.5.1 shows the features of the presented chopper-refolding sampling PLL. The conventional slope controller [2,3] uses a reference signal (REF) to enable charging of the sampling capacitor. At the falling edge of the DIV, the integrated voltage on the sampling capacitor is held. In order to reduce the in-band phase noise, the slope controller requires a relatively large phase-detection gain (G_{SPD}) to achieve a sufficiently low input-referred jitter (IRJ), thus necessitating a smaller sampling capacitor for a higher slew rate (SR) [4]. Integrating an excessively high G_{SPD} is infeasible for ensuring an adequate phase-detection range. This leads to long locking times [5] and greater implementation challenges in fractional-N PLLs, where stricter requirements are placed on digital-to-time converters (DTCs) for precise fractional division. This in turn limits the generalizability of the IRJ reduction [6,7], as the approach cannot be readily applied to other PLL types.

To decouple IRJ and G_{SPD}, a kT/C-noise-cancellation SPD is presented in this work. As shown in the lower-left corner of Fig. 12.5.1, by introducing the kT/C-noise-cancellation technique [8,9], the voltage of the reference frequency, after undergoing direct sampling, amplification, and noise cancellation, enables the achievement of low IRJ within an acceptable phase-detection range. Furthermore, to filter out flicker noise, which is more severe especially in advanced processes, a chopper-refolding sampling scheme is presented, as shown on the right of Fig. 12.5.1. The frequency of the DIV is set to twice the reference frequency to sample both edges of the REF, similar to the double sampling in [5]. The sampling operation effectively performs frequency mixing, such that the frequency of the sampled voltage ($V_{sam,p/n}$) is exactly equal to the reference frequency. $V_{sam,p/n}$ retains the flicker noise from the amplifier, and after being converted into current via the transconductance (g_m), it is further superimposed with the flicker noise of the g_m. At this point, since $V_{sam,p/n}$ is at the reference frequency, the sampled current ($I_{sam,p/n}$) is down-converted to baseband by the post-stage chopper; meanwhile, the additional flicker noise is up-converted to the reference frequency. After passing through the loop filter (LF), the sampled current ($I_{sam,p/n}$) is converted to voltage (V_c), while the flicker noise and chopper-induced reference spur are filtered out.

Figure 12.5.2 depicts the circuit implementation and the working principle of the chopper-refolding sampling with a kT/C-noise-cancellation SPD. The SPD comprises a sampling amplifier, an output hold circuit, a chopper, and a g_m. The sampling amplifier employs the kT/C-noise-cancellation technique ensuring low-noise sampling while delivering sufficient gain. During the sampling phase of Φ_1, C_1 and C_2 are reset. Upon reset completion, the residual noise voltage from the switches is stored on the capacitors. During the sampling phase of Φ_2, C_2 continues sampling. At this time, the input amplifier outputs a fixed-slope voltage, which is amplified by amplifier A_1 and then stored on C_2. Here, the amplifier A_1 also functions as a low-pass filter, effectively suppressing noise from the input amplifier. During the amplification phase of Φ_A, the loop around the entire amplifier is closed. C_1 is a 6-bit CDAC configured with a fixed code to output a voltage V_d, compensating for the offset caused by the non-idealities of the system, and it can also be configured as a frac-spur cancellation CDAC presented in [10] for fractional PLLs. When reaching the final steady state, the input of A_2 approaches 0. The output voltage can be derived using the principle of charge conservation. All noise generated during the reset phase is canceled out or rejected in the closed-loop mode. The closed-loop gain (A_{cl}) is set to 4×. The amplified signal is then stored in the subsequent hold circuit using a ping-pong operation. During the chopping phase of Φ_C, the front-stage amplifier is reset, while the rear-stage hold circuit switches simultaneously. The chopper switches to the newly stored voltage output, ensuring continuous current output and suppressing the reference spurs.

The left side of Fig. 12.5.3 details the amplifier in the SPD. The amplifier A_1 in the SPD operates in dual modes. Since A_1 amplifies the output of the dual-sampling slope controller during the Φ_{12} sampling phase, it has a requirement for a high slew rate. Thus, during the sampling phase Φ_{12}, A_1 is configured as an inverter. During the Φ_A closed-loop amplification, A_1 requires lower equivalent input noise and faster settling time, so it is configured in a floating inverter amplifier (FIA) mode [11] to achieve low-noise output in closed-loop amplification. The amplifiers A_2 and A_3 perform closed-loop amplification solely during Φ_A amplification. To obtain a faster closed-loop response, larger reservoir capacitors are used at the initial stage of Φ_A to broaden the amplifier bandwidth. After rapid establishment, part of the reservoir capacitors is disconnected to stabilize the loop. Consequently, A_2 and A_3 are designed as FIAs with a dynamic bandwidth enhancement.

The right side of Fig. 12.5.3 presents the overall architecture of the sampling PLL and the clock generation circuit. The reference clock is fed to the on-chip dual-sampling slope controller via an off-chip balun, undergoes phase detection by the SPD, is converted to current by the g_m after passing through the chopper, and is used to generate a control voltage (V_c) via an on-chip analog loop filter to tune the voltage-controlled oscillator (VCO). The timing generation circuit addresses two key requirements: First, the frequency divider output should not undergo further frequency division; otherwise, the output frequency step of the PLL becomes larger. Second, the sampling instant (i.e., the falling edge of Φ_{12}) should be directly synchronized with the VCO output. This not only requires the frequency divider to perform retiming but also demands that the timing generation circuit minimize the jitter added to the retimed falling edge. Therefore, as depicted, a combinational-logic timing-generation circuit is presented, which cooperates with the retimer to output two edges: $\Phi_{1, div}$ generates the falling edge of the clock Φ_1, and the falling edge of $\Phi_{12, div}$ generates the falling edge of the sampling clock Φ_{12}. By introducing a delay, it is ensured that there is no overlap during each timing switch. The input amplifier is shown at the bottom of Fig. 12.5.3, which is designed as a configurable transimpedance amplifier (TIA) to amplify the input signal while featuring a high output bandwidth. Benefiting from the kT/C-noise-cancellation technique, the sampling capacitor C_1 can be sized small, enabling acceptable power consumption.

Figure 12.5.4 presents the circuits and layout implementation of the presented self-injection VCO with harmonic-impedance expansion. High impedance at twice the oscillation frequency has been proven to effectively reduce phase noise [12,13]. In conventional VCOs, independent tuning at the second-harmonic frequencies is achieved via additional adjustments of the capacitors connected in the common-mode circuit. The presented VCO employs a transformer for voltage gain, achieving a lower equivalent noise figure [14]. By connecting the drain capacitor to the source terminal, the common-mode circuit shown in the lower-left corner of Fig. 12.5.4 is realized, where a self-injection is formed to achieve lower phase noise [15]. The common-mode current after capacitive phase shifting is injected back into the tank at the leakage terminal through the mixer formed by the negative resistance pair, forming self-injection and further suppressing the phase noise of the VCO. Adjusting Cs tunes the single-ended resonant frequency (lower-right corner of Fig. 12.5.4), which indirectly adjusts the bandwidth of the common-mode resonant impedance. Concurrently, adjusting the length (inductance) of the supply feeder L_1 in the layout (see the upper-right corner of Fig. 12.5.4) enables the adjustment of the common-mode resonant frequency. The VCO incorporates a 7-bit switched-capacitor array (SCA) and a varactor (upper-middle part of Fig. 12.5.4). Without the SCA, the simulation results of the VCO show that a figure-of-merit (FoM) value higher than 198dB is achieved within a 29.7% bandwidth (middle part of Fig. 12.5.4), with the corresponding simulation waveforms displayed in the lower-middle part of Fig. 12.5.4.

The presented SPLL was fabricated in a 28nm CMOS process. Figure 12.5.7 presents the die micrograph and the power consumption of each module. The presented PLL dissipates 16.9mW. The 250MHz reference clock is provided by R&S SMA100B. Figure 12.5.5 shows the measured phase noise and reference spur. The measured integrated jitter from 10kHz to 100MHz is 33.8fs. The reference spur is -80.8dBc. Since the minimum divide ratio of the divider in the presented SPLL is 28, the jitter performance across the full frequency range is measured using a 230MHz reference clock. Given the better performance of the VCO at lower frequencies, the measured jitter ranges from 32.66 to 39.26fs across the output

216 • 2026 IEEE International Solid-State Circuits Conference

979-8-3315-8937-0/26 $31.00 © 2026 IEEE

frequency range from 12.88 to 14.72GHz. The comparison table with prior-art integer-N PLLs [5,16-20], shown in Fig. 12.5.6, demonstrates low jitter, low reference spur, high FoM, and small chip area. The jitter performance achieved with the presented techniques demonstrates that the presented SPLL holds considerable potential as a candidate for high-performance ADCs.

Acknowledgment:
Project is supported by the National Key Research and Development Program of China: Design Technology Co-optimization Methodology (grant number: 2023YFB4402700).

References:
[1] X. Tang et al., "Low-Power SAR ADC Design: Overview and Survey of State-of-the-Art Techniques," *IEEE TCAS-I.*, vol. 69, no. 6, pp. 2249–2262, June 2022. https://doi.org/10.1109/TCSI.2022.3166792
[2] M. Mercandelli et al., "A 12.5-GHz Fractional-N Type-I Sampling PLL Achieving 58-fs Integrated Jitter," *IEEE JSSC*, vol. 57, no. 2, pp. 505–517, Feb. 2022. https://doi.org/10.1109/JSSC.2021.3123827
[3] G. Jin et al., "A Fractional- N Sampling PLL with a Merged Constant-Slope DTC and Sampling PD," *IEEE JSSC*, pp. 1–11, Aug. 2024. https://doi.org/10.1109/JSSC.2024.3358564
[4] J. Gong et al., "A Cryo-CMOS PLL for Quantum Computing Applications," *IEEE JSSC*, vol. 58, no. 5, pp. 1362–1375, May 2023. https://doi.org/10.1109/JSSC.2022.3223629
[5] Y. Zhao et al., "A 20-GHz PLL with 20.9-fs Random Jitter," *IEEE JSSC*, vol. 58, no. 6, pp. 1597–1609, June 2023. http://doi.org/10.1109/JSSC.2022.3225105
[6] X. Shen et al., "A 0.65V-V$_{DD}$ 10.4-to-11.8GHz Fractional-N Sampling PLL Achieving 73.8fs$_{rms}$ Jitter, -271.5dB FoM$_N$, and -61dBc in-Band Fractional Spur in 40nm CMOS," *ISSCC*, pp. 338–339, Feb. 2025. https://doi.org/10.1109/ISSCC49661.2025.10904612
[7] Y. Shin et al., "A 76 fs$_{rms}$-Jitter and -65dBc- Fractional-Spur Fractional-N Sampling PLL Using a Nonlinearity-Replication Technique," *ISSCC*, pp. 196–198, Fe.b 2025. https://doi.org/10.1109/ISSCC49657.2024.10454557
[8] M. Zhan et al., "A 0.004mm² 200MS/S Pipelined SAR ADC with kT/C Noise Cancellation and Robust Ring-Amp," *ISSCC*, pp. 164–166, Feb. 2022. http://doi.org/10.1109/ISSCC42614.2022.9731599
[9] J. Liu et al., "A 13-bit 0.005-mm² 40-MS/s SAR ADC with kT/C Noise Cancellation," *IEEE JSSC*, vol. 55, no. 12, pp. 3260–3270, Dec. 2020. http://doi.org/10.1109/JSSC.2020.3016656
[10] D. Liao and F. Dai, "A Fractional-N Reference Sampling PLL with Linear Sampler and CDAC Based Fractional Spur Cancellation," *IEEE JSSC*, vol. 56, no. 3, pp. 694–704, Mar. 2021. https://doi.org/10.1109/JSSC.2020.3033271
[11] X. Tang et al., "A Bandwidth-Adaptive Pipelined SAR ADC with Three-Stage Cascoded Floating Inverter Amplifier," *IEEE JSSC*, vol. 58, no. 9, pp. 2564–2574, Sept. 2023. http://doi.org/10.1109/JSSC.2023.3268719
[12] D. Murphy et al., "A VCO with Implicit Common-Mode Resonance," *ISSCC*, pp. 442–443, Feb. 2015. http://doi.org/10.1109/ISSCC.2015.7063116
[13] H. Guo et al., "A 5.0-to-6.36GHz Wideband-Harmonic-Shaping VCO Achieving 196.9dBc/Hz Peak FoM and 90-to-180kHz 1/f3 PN Corner Without Harmonic Tuning," *ISSCC*, pp. 294–296, Feb. 2021. http://doi.org/10.1109/ISSCC42613.2021.9365761
[14] B. Jafari et al., "A 4.6-6GHz Self-Injection LC Oscillator Exploiting 2nd Harmonic Extraction and Self-Mixing to Achieve 5-35kHz 1/f3 Phase Noise Corner and 201dB FoMT," *IEEE CICC*, pp. 1–3, Apr. 2025. https://doi.org/10.1109/CICC63670.2025.10983692

[15] M. Babaie and R. Staszewski, "A Class-F CMOS Oscillator," *IEEE JSSC*, vol. 48, no. 12, pp. 3120–3133, Dec. 2013. http://doi.org/10.1109/JSSC.2013.2273823
[16] H. Liu et al., "A Multireference PLL: Theory and Implementation," *IEEE JSSC*, vol. 59, no. 7, pp. 1981–1994, July 2024. http://doi.org/10.1109/JSSC.2024.3383605
[17] L. Wang et al., "A Compact 20–24-GHz Sub-Sampling PLL with Charge-Domain Bandwidth Control Scheme," *IEEE JSSC*, vol. 60, no. 3, pp. 768–784, Mar. 2025. http://doi.org/10.1109/JSSC.2024.3488277
[18] X. Kong et al., "A 9-GHz Low-In-Band Noise Sub-Sampling-Chopper PLL with Charge-Share Canceling Technique," *IEEE JSSC*, pp. 1384-1396, Apr. 2025. http://doi.org/10.1109/JSSC.2025.3532504
[19] H. Cao et al., "A 24.6-to-30.6GHz Magnetic-Isolated Sub-Sampling PLL with a Fast-Locking FLL Achieving 64.9fs jitter, –253.3dB FoM$_J$, and –69.1dBc Reference Spur in 65nm CMOS," *IEEE CICC*, pp. 1–3, Apr. 2025.
[20] Y. Huang et al., "A 22.4-25.6GHz Ping-Pong Sub-Sampling PLL Featuring Unified Supply Voltage and Balanced 2nd Harmonic Extraction Achieving 45.8fs$_{rms}$ Jitter and -254.3dB FoM," *IEEE CICC*, pp. 1–3, Apr. 2025. http://doi.org/10.1109/CICC63670.2025.10983523

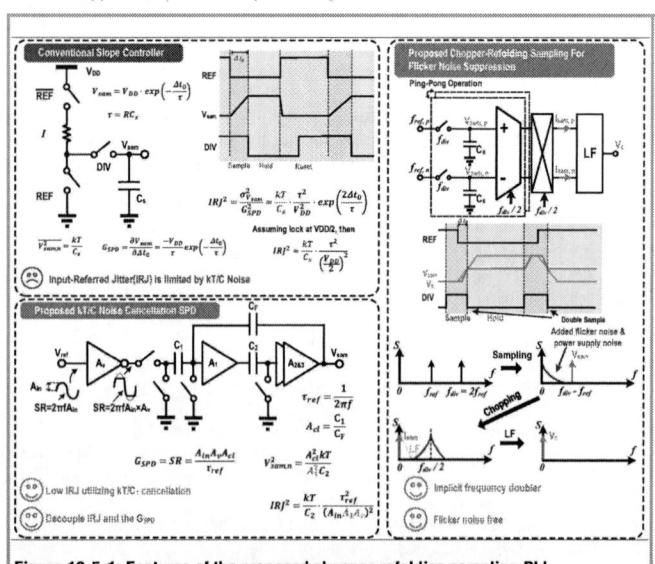

Figure 12.5.1: Features of the proposed chopper-refolding sampling PLL.

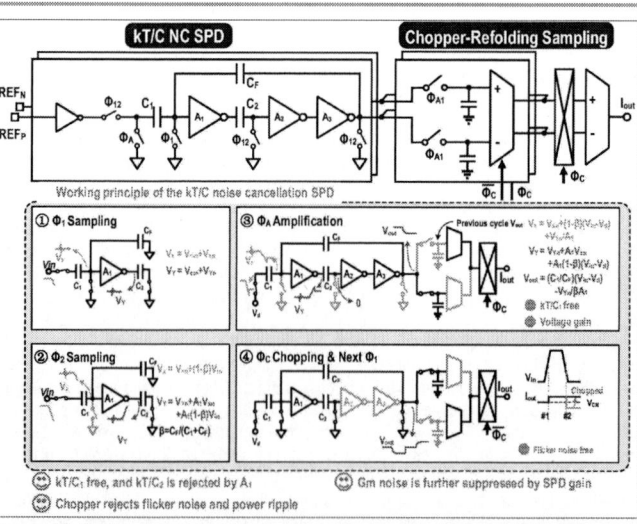

Figure 12.5.2: Circuits and working principle of the presented kT/C NC SPD and chopper-refolding sampling.

ISSCC 2026 / SESSION 12 / FREQUENCY SYNTHESIZERS AND VCOs / 12.5

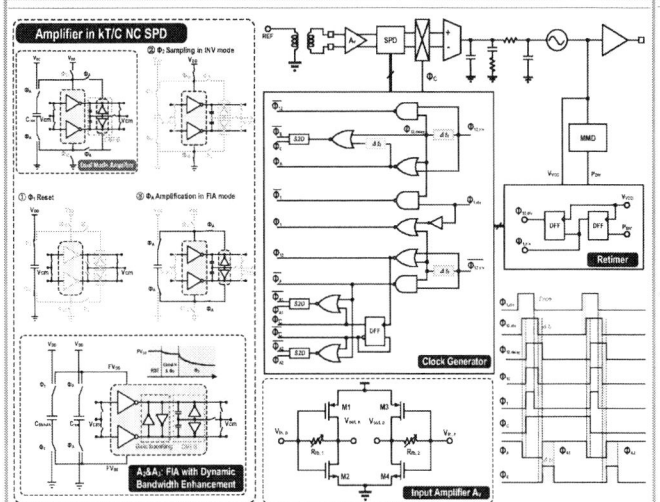

Figure 12.5.3: Details of the amplifier and the overall architecture of the proposed sampling PLL.

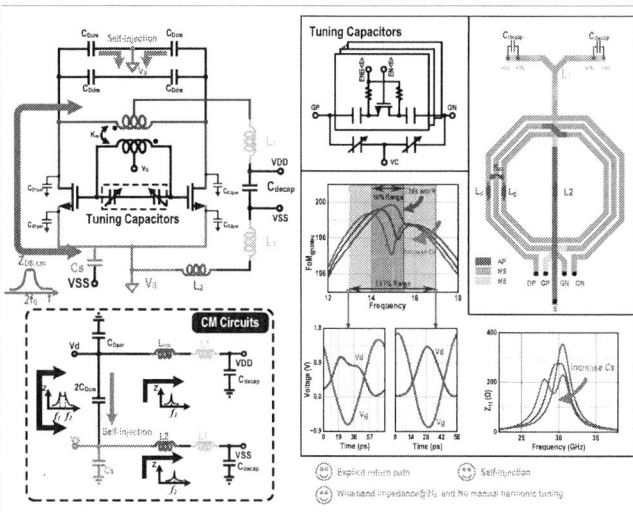

Figure 12.5.4: Circuits and layout of the presented self-injection VCO with harmonic-impedance expansion.

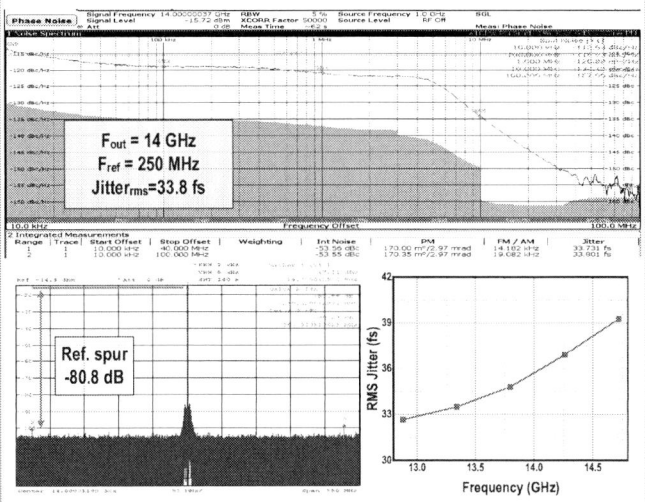

Figure 12.5.5: Measured jitter (top), reference spur (bottom), and jitter performance over the full frequency range.

	This Work	Liu JSSC 2024	Wang JSSC 2025	Kong JSSC 2025	Cao CICC 2025	Huang CICC 2025	Zhao JSSC 2023
Architecture	SPLL	SPLL	SSPLL	SSPLL	SSPLL	SSPLL	SPLL
Process (nm)	28	65	40	65	65	65	28
Ref. Freq (MHz)	250	100	250	100	100	100	250
Ref. Number	1	8	1	1	1	1	1
Freq. Range (GHz)	14	1.8 to 5.6	22	9	27	24	20
RMS Jitter (fs)	33.8	16.1	61.23	49.9	64.9	45.8	20.9
Integ. Range (MHz)	0.01 to 100	0.01 to 40	0.001 to 100	0.001 to 40	0.01 to 100	0.001 to 100	0.01 to 40
Ref. Spur (dBc)	-80.8	-71	-44	-58	-69.1	-65.6	-66
Power (mW)	16.9	83	13.35	7.8	11.2	17.6	12
Area (mm²)	0.3	1.1	0.057	0.64	0.19	0.23	0.06
FoM (dB)	-257.1	-256.7	-253.0	-257.1	-253.3	-254.3	-262.8

Figure 12.5.6: The comparison table with prior-art integer-N PLL.

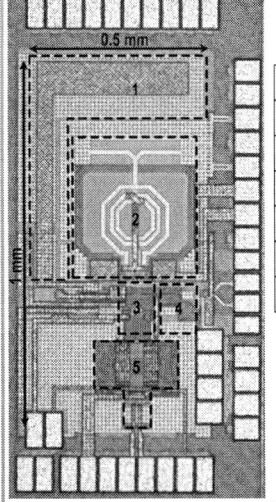

No	Block	Power(mW)	Area(mm²)
1	Loop Filter	0	0.12
2	VCO	11	0.14
3	Chopper&gm	0.5	0.006
4	Div+ClkGen	2.7	0.005
5	kT/C NC SPD	2.7	0.027
	Total	16.9	0.30

Figure 12.5.7: Die micrograph with the active area of 0.30mm², and the power consumption and area of each module.

ISSCC 2026 / SESSION 12 / FREQUENCY SYNTHESIZERS AND VCOs / 12.6

12.6 A 0.65-to-1V-V_{DD} 10.5-to-11.85GHz Fractional-N Sampling PLL Achieving 71.47fs Integrated Jitter and <-60dBc Near-Integer Fractional Spur in 40nm CMOS

Yixi Li[1,2], Junjie Chen[1,2], Xinyu Shen[1], Jie Yang[3], Jian Liu[1,2], Nanjian Wu[1,2], Zhao Zhang[1,2], Liyuan Liu[1,2]

[1]Institute of Semiconductors, Chinese Academy of Sciences, Beijing, China, [2]University of Chinese Academy of Sciences, Beijing, China, [3]Westlake University, Hangzhou, China

Abstract

This paper presents a 0.65-to-1V, 10.5-to-11.85GHz wide-V_{DD}-range (WV) fractional-N sampling PLL. The QE-reduction WV SPD, high-linearity 2-stage DTC, and DCC-aided QE dithering method are proposed to address the issue of poor phase noise at low-V_{DD}, reliability at regular V_{DD}, nonlinearity, and convergence issues at small fractional FCWs. The 40nm prototype achieves 71.47fs jitter, −251.8dB FoM, and <−60dBc near-integer spur.

Dynamically varying supply voltage (V_{DD}) is attractive for digital-centric system on chips (SoC) to meet throughput and power-consumption requirements in different scenarios [1]. To meet the performance requirements of wireless transceivers in the SoCs, the phase-locked loop (PLL) with high spectral purity and high spectrum-utilization efficiency is crucial. Therefore, a fractional-N PLL (FN-PLL) that achieves low jitter and low spur across a wide dynamic voltage operating range (WV) is highly desirable. Several high-performance FN-PLLs operating at regular V_{DD} (≥0.9V) have been reported [2-12], including the digital-to-time-converter (DTC)-based and digital-to-analog-converter (DAC)-based architectures, which are widely adopted to eliminate quantization errors (QEs) generated by ΔΣ modulators (ΔΣMs) from time domain and voltage domain [Fig. 12.6.1(top-left)], respectively. However, at low V_{DD} (LV), their performances degrade significantly. The DTC suffers from degraded phase noise (PN) and stronger trade-off between PN and its linearity at LV [8-11], whereas the integral nonlinearity (INL) of the sampling phase detector (SPD) in the DAC-based FN-PLL degrades significantly due to the reduced voltage headroom at LV [12]. Moreover, to ensure a sufficient dynamic range for QE compensation, the gain of the SPD in the DAC-based approach gets decreased at LV, making it insufficient to suppress in-band phase noise (PN) of the FN-PLL. To overcome these issues, a LV low-jitter FN-PLL has been reported in [13] [Fig. 12.6.1(top-right)]. Yet, the deployment of voltage-booster-like circuits in its LV-C-DTC and LV-SPD introduces significant reliability concerns for regular V_{DD} operation. Furthermore, both LV-C-DTC and LV-SPD exhibit severe INL, necessitating a 16-coefficient hybrid-cascaded digital pre-distortion for INL compensation [13]. This induces convergence issue and significant hardware overhead.

To surmount these issues, we propose a 0.65-to-1V-V_{DD}, 10.5-to-11.85GHz wide-V_{DD}-range FN sampling PLL (WV-FN-SPLL) [Fig. 12.6.1(bottom)] achieving 71.47fs jitter, -251.8dB FoM, and <−60dBc near-integer fractional spur. The 1/2 QE-range reduction technique [10] is utilized to decrease the overall QE range, thereby enhancing the SPD gain and reducing the required delay range of the DTC for in-band PN reduction, particularly at LV. The QE is coarsely compensated by the QE-reduction wide-voltage SPD (QR-WV-SPD) and is further finely compensated by the high-linearity 2-stage DTC. It features: 1) a QR-WV-SPD suitable for the WV operation, proposed to reduce the DTC range with high linearity and high SPD gain, thereby improving the overall in-band PN; 2) high-linearity 2-stage DTC, devised to further release INL and noise folding of the main DTC (M-DTC) over a wide V_{DD} range; 3) our developed duty-cycle-corrector-aided (DCC-aided) QE dithering method to guarantee the reliable convergence of M-DTC INL calibration without requiring an additional dither-elimination DTC. As a result, the architecture achieves high linearity and low PN across a wide V_{DD} range.

The concept of the proposed high-linearity QE-range-reduction SPD is shown in Fig. 12.6.2(top). To balance the hardware complexity and the in-band PN contribution of the M-DTC, the proposed SPD performs coarse compensation for the 2-bit MSB QE. Therefore, the SPD operates with four turn-on times: $1T_{VCO}$, $0.75T_{VCO}$, $0.5T_{VCO}$, and $0.25T_{VCO}$. To achieve the same lock voltage (V_{Lock}) across all cases, the number of charge-current-source (I_U) cells is scaled accordingly as ×3, ×4, ×6, and ×12 while the capacitor (C) remains constant. Indeed, both I_U and C are voltage dependent (denoted as $I_U(V)$ and $C(V)$), especially at LV, which affects the actual V_{Lock} due to charging nonlinearity. Nevertheless, the concept remains valid since all four cases experience identical nonlinear effects along the same voltage trajectory, allowing V_{SPD} to lock at the same V_{Lock} without INL issues in each charging cycle. The proof is provided in Fig. 12.6.2(top-right). Ultimately, high-linearity QE coarse compensation is achieved across a wide V_{DD}, enabling low fractional spurs and suppressed noise folding while the M-DTC range is reduced to $0.25T_{VCO}$, alleviating the trade-off between DTC PN and INL.

The gain of prior SPD in a DAC-based FN-PLL [12] is constrained by the maximum QE range [Fig. 12.6.2(middle-left)], and worsens at LV, leading to inherently poor PN across wide V_{DD}. Unlike prior designs, the proposed SPD features four configurable gains that adapt to the QE magnitude, as illustrated in Fig. 12.6.2(middle-middle). The minimum gain (activated at QE=$1T_{VCO}$) matches that of prior SPDs, while the maximum gain (activated at QE=$0.25T_{VCO}$) offers a 4× improvement. The high-frequency variation of the SPD gain allows it to be approximated at a fixed value within the PLL bandwidth, ensuring stable dynamics. Moreover, the probability density function (PDF) of the MASH1-1 modulator 2-bit MSB

enables the SPD to operate predominantly in a higher-gain mode, allowing for a higher average gain. Consequently, reduced in-band PN is achieved across a wide V_{DD} range. Under identical circuit topologies and at a regular V_{DD}, the equivalent input PN of both SPDs was simulated [Fig. 12.6.2(middle-right)]. The proposed SPD demonstrates a 5.4dB improvement in the in-band PN compared to the prior SPD (with a DAC).

An LV-SPD was proposed in [13], as shown in Fig. 12.6.2(bottom-left). However, its triple-stack PMOS charging path severely restricts voltage headroom, resulting in poor PN and INL. Although the negative-level-boosted inverter (NL-INV) [13] is used to alleviate the voltage headroom issue, the voltage-booster nature of the NL-INV suffers from reliability issues at regular V_{DD}, making such an LV-SPD unsuitable for WV operation. To address this, the WV-SPD is proposed, as described in Fig. 12.6.2(bottom-middle). In the design, complementary switches (SW_{GP}, SW_{GN}) and the switch (SW_D) cooperate to perform sampling. The sampling process consists of four phases: 1) Start of sampling: SW_H and SW_D are off, SW_{GP} is on, SW_{GN} is off. The PMOS gate is biased at V_{CGO}, initiating the PMOS current charging of the capacitor C_S. 2) End of sampling: SW_H and SW_D remain off; SW_{GP} turns off, and SW_{GN} turns on. The PMOS gate voltage switches to V_{DD}, immediately turning-off the charging current, thus holding the sampled voltage V_{SPD}. 3) Hold phase: SW_H turns on; all other switches remain unchanged. Voltage V_{SPD} is transferred to V_H. 4) Reset phase: SW_H turns off; SW_D turns on; SW_{GP} is on, SW_{GN} is off. The PMOS gate returns to V_{CGO}, stabilizing for the next sampling cycle. The proposed WV-SPD uses only one single PMOS current source in the charging path, significantly improving voltage headroom. This allows V_{SPD} to settle rapidly across a wide V_{DD} range, achieving high gain and low PN for WV operation. Furthermore, there is no voltage-booster-like circuit in our WV-SPD, ensuring reliable and robust WV operation. As simulated in Fig. 12.6.2(bottom-right), the INL of the WV-SPD at 0.65V V_{DD} is reduced by more that one order compared to the prior LV-SPD in [13].

A variable-slope DTC (VS-DTC) is a promising candidate for WV operation because it is simple and free of voltage-booster-like circuits. However, a severe trade-off between the figure of merit (FoM) and INL emerges at low V_{DD} [Fig. 12.6.3(top-left)]. A high-linearity 2-stage DTC architecture is proposed to break this trade-off [Fig. 12.6.3(top-middle)]. The convex INL characteristic of the M-DTC and its mechanisms were analyzed in [8]. A nonlinear-compensation DTC (NLC-DTC) is cascaded after the M-DTC to compensate for the M-DTC INL. The NLC-DTC control code |DTC$_{CODE}$| is simply generated by applying an absolute-value operation to the M-DTC code. Furthermore, by adjusting the gain of the NLC-DTC, its dynamic delay T_3 can be configured to precisely cancel the M-DTC INL. Besides, a large-size NMOS in the M-DTC discharge path ensures the discharge resistance is dominated by R_D across a wide V_{DD} range, minimizing delay variation. Consequently, without employing any voltage booster, the proposed high-linearity 2-stage DTC simultaneously achieves low power, low PN, and low INL across a wide V_{DD} range. Simulations show the proposed high-linearity 2-stage DTC improves INL by ~60% with negligible PN degradation over a single M-DTC [Fig. 12.6.3(top-right)].

As shown in Fig. 12.6.3(bottom-left), since the 2-bit MSB of QE is compensated by the SPD, the dither in the M-DTC code is eliminated. Consequently, the DTC nonlinearity only introduces low-frequency disturbances in the VCO control voltage V_P at a small fractional frequency control word (FCW). This results in the failure to extract INL at small FCW. By introducing dither to the M-DTC code, the INL of the M-DTC induces sufficiently high-frequency disturbances in V_P, enabling the comparator loop to extract the INL information even at a small fractional FCW. Thus, the DCC-aided QE dithering method is proposed. Since the phase selection signal SEL-CK$_{FB}$, used in the 1/2 QE range reduction, exhibits high-frequency characteristics, we select the SEL-CK$_{FB}$ as the dither signal for the M-DTC code as illustrated in Fig. 12.6.3(bottom-middle). This enables the extraction of the M-DTC INL at a small fractional FCW, ensuring reliable convergence for the NLC-DTC calibration. Here, the functionality of the DCC-DTC is multiplexed: on one hand, it performs duty-cycle correction for 1/2 QE-range reduction; on the other hand, it cancels the M-DTC dither. The delays of each DTC-stage output are shown in Fig. 12.6.3(bottom-right). It is noteworthy that while dither techniques have been proposed in [7,14], they universally require an extra DTC for dither removal. Figure 12.6.4(left) shows the convergence transient simulation (FCW=73+2^{-11}), which demonstrate reliable convergence at a small fractional FCW.

218 • 2026 IEEE International Solid-State Circuits Conference

979-8-3315-8937-0/26 $31.00 © 2026 IEEE

Figure 12.6.4(right) plots the measured spectrum and PN at the fractional-N (FCW=73+2^{-11}) mode with a 150MHz F_{REF} at regular V_{DD}. The largest fractional spur is −64.85dBc and the integrated jitter from 1kHz to 100MHz is 54.96fs, respectively. Figure 12.6.5(left) plots the measured spectrum and PN at 0.65V V_{DD}. The largest fractional spur is −60.02dBc and the integrated jitter from 1kHz to 100MHz is 71.47fs, respectively. (In both modes, the power supply voltage for the VCO is 0.55V.) Thanks to the SPD gain boosting, wide-voltage techniques, and a low-noise DTC design, the WV-FN-PLL achieves <72fs jitter across a 0.65-to-1.0V V_{DD}. Benefiting from the high linearity of both the SPD and DTC, the worst fractional spur is below −60dBc as shown in Fig. 12.6.4(left). This WV-FN PLL simultaneously supports a wide V_{DD} range from low V_{DD} (0.65V) to regular V_{DD} (1.0V) at F_{OUT} above 10GHz while maintaining jitter below 72fs across the entire voltage range [Fig. 12.6.6]. The 40nm CMOS prototype achieves a competitive FoM$_N$ across both low V_{DD} [15-19] and regular V_{DD}, particularly for over-10GHz F_{OUT} as shown in Fig. 12.6.7.

Acknowledgement:
This work was supported by the National Natural Science Foundation of China (No. 62174153 and 62222409). Corresponding author: Zhao Zhang (zhangzhao11@semi.ac.cn).

References:
[1] T. D. Burd et al., "A Dynamic Voltage Scaled Microprocessor System," *IEEE JSSC*, vol. 35, no. 11, pp. 1571-1580, Nov. 2000. https://doi.org/10.1109/4.881202
[2] H. Liu et al., "An Ultra-Low-Jitter Fast-Hopping Fractional-N PLL with LC DTC and Hybrid-Proportional Paths," *IEEE JSSC*, vol. 60, no. 3, pp. 785-798, Mar. 2025. https://doi.org/10.1109/JSSC.2024.3514870
[3] M. Chae et al., "34.1 A 65fs$_{rms}$-Jitter and −272dB-FoM$_{jitter,N}$ 10.1GHz Fractional-N Digital PLL with a Quantization-Error-Compensating BBPD and an Orthogonal-Polynomial LMS Calibration," *ISSCC*, pp. 554-556, Feb. 2025. https://doi.org/10.1109/ISSCC49661.2025.10904741
[4] Y. Shin et al., "10.5 A 76 fs$_{rms}$-Jitter and -65dBc-Fractional-Spur Fractional-N Sampling PLL Using a Nonlinearity-Replication Technique," *ISSCC*, pp. 196-198, Feb. 2024. https://doi.org/10.1109/ISSCC49657.2024.10454557
[5] J. Kim et al., "A 12.8-15.0-GHz Low-Jitter Fractional-N Subsampling PLL Using a Voltage-Domain Quantization-Error-Cancellation," *IEEE JSSC*, vol. 59, no. 2, pp. 424-434, Feb. 2024. https://doi.org/10.1109/JSSC.2023.3297618
[6] Z. Gao et al., "A 2.6-to-4.1GHz Fractional-N Digital PLL Based on a Time-Mode Arithmetic Unit Achieving -249.4dB FoM and -59dBc Fractional Spurs," *ISSCC*, pp. 380-381, Feb. 2022. https://doi.org/10.1109/ISSCC42614.2022.9731561
[7] S. M. Dartizio et al., "A 76.7fs-Integrated-Jitter and -71.9dBc In-Band Fractional-Spur Bang-Bang Digital PLL Based on an Inverse-Constant Slope DTC and FCW Subtractive Dithering," *ISSCC*, pp. 78-79, Feb. 2023. https://doi.org/10.1109/ISSCC42615.2023.10067719
[8] M. Rossoni et al., "An 8.75GHz Fractional-N Digital PLL with a Reverse-Concavity Variable-Slope DTC Achieving 57.3fs$_{rms}$ Integrated Jitter and -252.4dB FoM," *ISSCC*, pp. 188-189, Feb. 2024. https://doi.org/10.1109/ISSCC49657.2024.10454388
[9] D. Xu et al., "A 7GHz Digital PLL with Cascaded Fractional Divider and Pseudo-Differential DTC Achieving -62.1dBc Fractional Spur and 143.7fs Integrated Jitter," *ISSCC*, pp. 192-193, Feb. 2024. https://doi.org/10.1109/ISSCC49657.2024.10454284
[10] W. Wu et al., "A 16-22 GHz Fractional-N PLL in 8nm FinFET with 68 fs$_{rms}$ Jitter," *IEEE RFIC*, pp. 111-114, June 2025. https://doi.org/10.1109/RFIC61188.2025.11082795

[11] Y. Liu et al., "A 37.5fs-rms Jitter and –254.1dB FoM Fractional-N Sampling PLL with Reference-Phase-Selection and Complementary-DTC Achieving 8× DTC Range Reduction and Zero DTC Delay Offset," *IEEE CICC*, pp. 1-3, Apr. 2025. https://doi.org/10.1109/CICC63670.2025.10983724
[12] G. Jin et al., "A Fractional-N Sampling PLL with a Merged Constant-Slope DTC and Sampling PD," *IEEE JSSC*, vol. 59, no. 8, pp. 2407-2417, Aug. 2024. https://doi.org/10.1109/JSSC.2024.3358564
[13] X. Shen et al., "A 0.65V-V$_{DD}$ 10.4-to-11.8GHz Fractional-N Sampling PLL Achieving 73.8fs$_{rms}$ Jitter, -271.5dB FoM$_N$, and -61dBc in-Band Fractional Spur in 40nm CMOS," *ISSCC*, pp. 338-339, Feb. 2025. https://doi.org/10.1109/ISSCC49661.2025.10904612
[14] Z. Ye et al., "A 6.8-to-14.4GHz Octave-Tuning Fractional-N Charge-Pump PLL with Slide-Dithering-Based Background DTC Nonlinearity Calibration for Near-Integer Fractional Spur Mitigation Achieving 78fs RMS Jitter and -258.6dB FoM$_T$," *IEEE CICC*, pp. 1-2, Apr. 2024. https://doi.org/10.1109/CICC60959.2024.10529020
[15] L. Feng et al., "A 0.45V 0.72mW 2.4GHz Bias-Current-Free Fractional-N Hybrid PLL Using a Voltage-Mode Phase Interpolator in 28nm CMOS," *ISSCC*, pp. 266-268, Feb. 2024. https://doi.org/10.1109/ISSCC49657.2024.10454404
[16] H. Liu et al., "A Sub-mW Fractional-N ADPLL with FOM of -246 dB for IoT Applications," *IEEE JSSC*, vol. 53, no. 12, pp. 3540-3552, Dec. 2018. https://doi.org/10.1109/JSSC.2018.2878836
[17] H. R. Kooshkaki and P. P. Mercier, "A 0.55mW Fractional-N PLL with a DC-DC Powered Class-D VCO Achieving Better than −66dBc Fractional and Reference Spurs for NB-IoT," *IEEE CICC*, pp. 1-4, Apr. 2020. https://doi.org/10.1109/CICC48029.2020.9075944
[18] C.-C. Li et al., "A Compact Transformer-Based Fractional-N ADPLL in 10-nm FinFET CMOS," *IEEE TCAS-I*, vol. 68, no. 5, pp. 3540-3552, May 2018. https://doi.org/10.1109/TCSI.2021.3059484
[19] H. Liu et al., "A 265-mW Fractional-N Digital PLL with Seamless Automatic Switching Sub-Sampling/Sampling Feedback Path and Duty-Cycled Frequency-Locked Loop in 65-nm CMOS," *IEEE JSSC*, vol. 54, no. 12, pp. 3478-3492, Dec. 2019. https://doi.org/10.1109/JSSC.2019.2936967

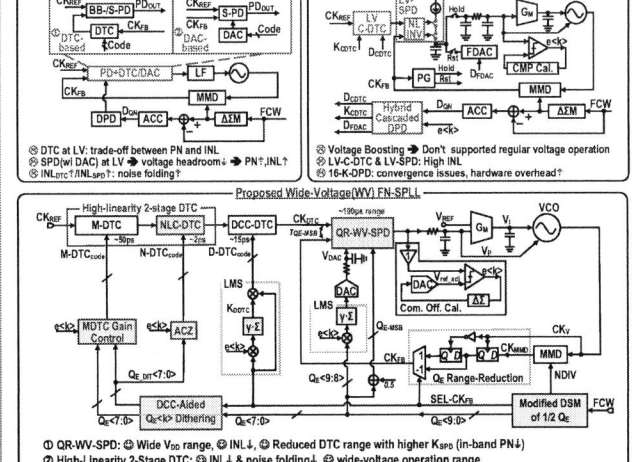

Figure 12.6.1: (top) Review of prior regular-V$_{DD}$ and low-V$_{DD}$ PLLs; (bottom) the overall architecture of the wide-voltage fractional-N sampling PLL.

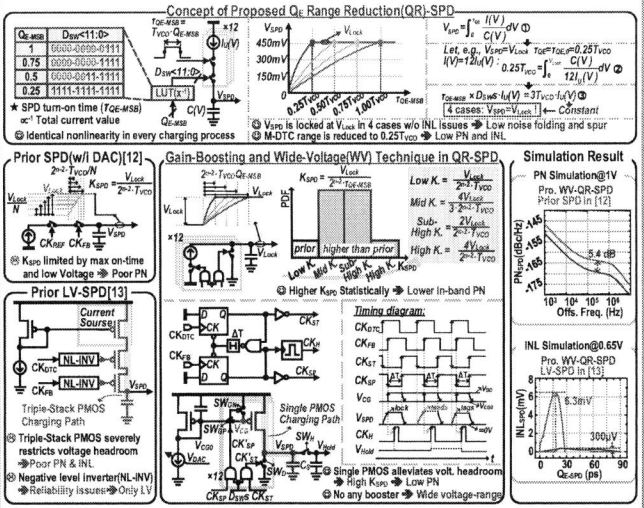

Figure 12.6.2: (top) Concept of the QE-range-reduction SPD; (middle and bottom) comparison between prior SPDs and the proposed WV-SPD.

ISSCC 2026 / SESSION 12 / FREQUENCY SYNTHESIZERS AND VCOs / 12.6

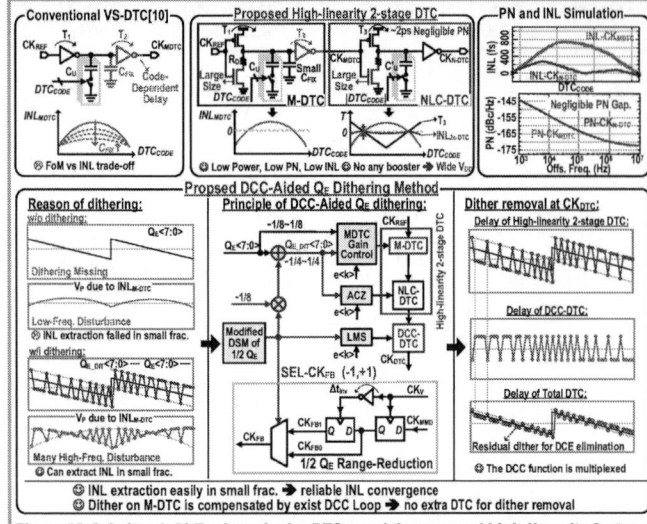

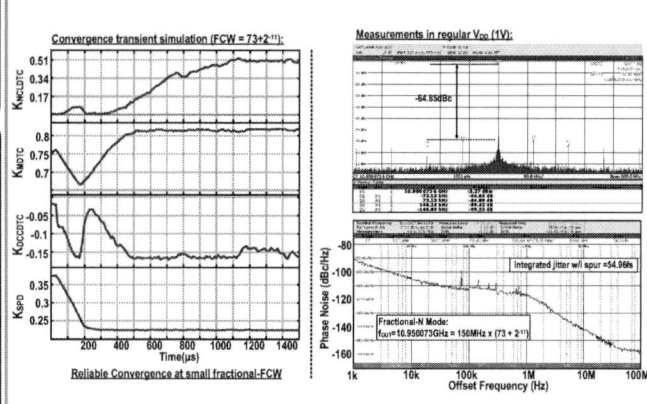

Figure 12.6.3: (top-left) Review of prior DTCs, and the proposed high-linearity 2-stage-DTC (top-right); the proposed DCC-aided QE dithering method (bottom).

Figure 12.6.4: Convergence transient simulation (left); measured spectrum and rms jitter at a regular V_{DD}.

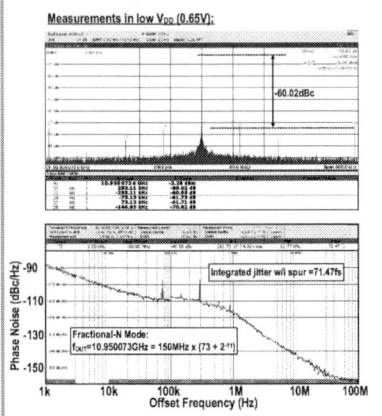

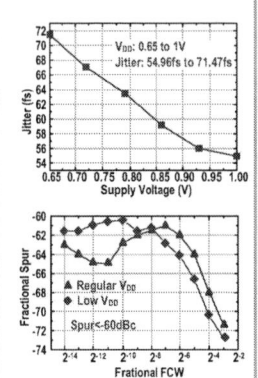

Figure 12.6.5: (left) Measured spectrum and rms jitter at a low V_{DD}; (top-right) measured rms jitter versus V_{DD}; measured fractional spur versus fractional FCW (bottom-right).

Figure 12.6.6: Performance summary and comparison.

	WV-FN-PLL		Low voltage fractional-N PLLs			Regular voltage fractional-N PLLs			
	This Work		ISSCC'25[13] X. Shen	ISSCC'24[15] L. Feng	JSSC'18[16] H. Liu	JSSC'25 [2] H. Liu	ISSCC'25 [3] M. Chae	ISSCC'24 [4] Y. Shin	JSSC'24 [5] J. Kim
Architecture	SPLL		SPLL	Hybrid PLL	DPLL	Hybrid PLL	BBPLL	Digital SPLL	Digital-SSPLL
Key Technology	QR-WV-SPD & NLC-DTC		C-DTC&F-DAC with HC-DPD	VPI	CS-DTC + 4-bit TDC	LC-DTC	QEC-BBPD	DAC with NL Replication	DAC with SCF DPD
CMOS (nm)	40		40	28	65	28	40	40	65
Supply Voltage	0.65	1	0.65	0.55	0.8	N/A	0.9	1	1.2
F_{OUT} (GHz)	10.950073 (10.5 to 11.85)		10.506758 (10.4 to 11.8)	2.4 (2.21 to 2.57)	2.44 (2.0 to 2.8)	11.2500045 (11.1 to 14.8)	10.100012 (10.0 to 11.5)	10.500293 (10.4 to 11.8)	14.00039 (12.8 to 15.0)
F_{REF} (MHz)	150		100.0625	50	26	250	100	150	100
Ref. spur (dBc)	-67.1	-64	-72.9	-66	-72	-65.3	-67.4	-67	N/A
Integrated Jitter (fs)	71.47	54.96	73.8§	470.2§	530§	61.7	65	76	104
In-Band Frac. Spur (dBc)	-60.02	-64.85	-61	-63	-56	-62.4	-63	-65	-58
Power(mW)	11	15.5	13.7	0.85	0.98	34	14.4	15.3	7.3
*FoM(dB)	-252.5	-253.3	-251.3	-247.3	-246	-250.7	-252.2	-250.5	-251
**FoM$_N$(dB)	-271.1	-271.9	-271.5	-264.1	-265.7	-267.2	-272.2	-269.2	-271
Core Area (mm²)	0.378		0.27	0.24	0.23	0.21	0.12	0.17	0.21

*FoM = $20\log(\sigma_{rms})+10\log(P_{DC}/1mW)$, FoM$_N$ = FOM-10log(N), the lower, the better
§integrated jitter without spur

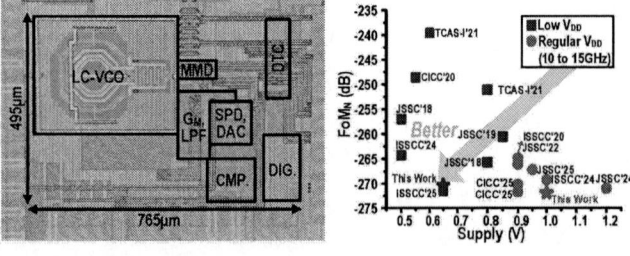

Figure 12.6.7: Die micrograph (left) and performance comparison with prior low-V_{DD} and regular-V_{DD} (over 10GHz) fractional-N PLLs (right).

ISSCC 2026 / SESSION 12 / FREQUENCY SYNTHESIZERS AND VCOs / 12.7

12.7 A 7.15-to-7.95GHz Magnetically Enhanced Feedforward Waveform-Shaping CMOS Oscillator with Implicit Common-Mode Noise Cancellation Achieving -146.72dBc/Hz PN@1MHz and 190.6dBc/Hz FoM

Rui Ma, Wei Deng, Haikun Jia, Juntao Lan, Zhihua Wang, Baoyong Chi

Tsinghua University, Beijing, China

Abstract

Pure microwave sources are vital for precision instrumentation, but conventional schemes are bulky and costly. CMOS oscillators are inexpensive and compact yet limited by phase-noise performance. To address this, we present a magnetically enhanced, feedforward waveform-shaping oscillator with an implicit common-mode noise-cancellation technique. Measurements show –146.72dBc/Hz PN at a 1MHz offset and a 190.6dBc/Hz FoM at a 7.5GHz carrier.

Ultra-low-phase-noise microwave sources are critical for applications such as quantum computing and precision instrumentation. Conventional solutions, including frequency synthesizers and dielectric-resonator oscillators, deliver excellent performance but are bulky, costly, and power hungry, making them unsuitable for large-scale or portable systems [1]. The CMOS technology, with its low cost, process maturity, and high integration capability, offers a promising alternative. However, achieving ultra-low phase noise in CMOS remains challenging due to the inherently low-Q passives.

Multi-core oscillators reduce phase noise theoretically by 10 log(N) through mutual coupling. However, oscillators with more than ten cores often fail to realize the expected improvement due to core-to-core mismatch [2-5]. Recently, series-resonant (SR) oscillators have been introduced, offering a potential 10 log(Q²) phase noise reduction [6,7]. Nevertheless, CMOS implementations of SR oscillators have yet to fully exploit this advantage and still lag behind BiCMOS counterparts in phase-noise performance. As CMOS technology scales, the reduced current-driving capability of transistors poses an additional challenge. Enlarging device sizes to increase current inevitably adds gate–drain parasitics, which lower the output waveform slope and increase distortion, thereby degrading the phase noise of CMOS SR oscillators [8]. Moreover, issues such as common-mode supply noise, over-voltage stress, IQ mismatch [9], and weak coupling—highly susceptible to environmental interference—remain major bottlenecks [10]. Consequently, CMOS oscillators are constrained by a performance ceiling that prevents them from meeting the stringent requirements of ultra-pure microwave sources, while the FoMs of CMOS designs remain limited.

To address these challenges, including inter-core mismatch, supply common-mode noise, and waveform distortion induced by enlarged active-switch dimensions, this work introduces two key circuit innovations, enabling low phase noise, improved robustness, and scalable array operation (Fig. 12.7.1):

(1) The proposed transformer network with a magnetically enhanced feedforward waveform-shaping technique, which compensates the waveform distortion induced by enlarged active-switch dimensions, compensates the voltage dip caused by the capacitive feedthrough, and sharpens the output waveform toward an ideal square shape.

(2) An implicit common-mode noise-cancellation technique, which employs an implicit transformer network to common-mode and differential-mode paths, cancels the injected common-mode supply noise in the LC tank and enhances the oscillator robustness against supply disturbances.

By introducing these two techniques, the proposed CMOS oscillator breaks the phase-noise-performance boundary of existing silicon-based oscillators. The prototype, fabricated in a standard 65nm CMOS process, achieves –146.72dBc/Hz phase noise at a 1MHz offset and an FoM of 190.6dBc/Hz, from a carrier frequency of 7.5GHz.

Figure 12.7.1 illustrates the prototype of the proposed oscillator architecture. The coupling network induces inter-core mismatch in multi-core extensions due to the indirect or inconsistent interconnections. To reduce the mismatch between cores outside the ring cell, a normalized resistor coupling path is introduced (Fig. 12.7.1). With this approach, any two cores—even when not belonging to the same ring cell—can achieve equivalent coupling through the unified synchronization path, thereby ensuring consistent phase alignment and scalable performance.

As shown in the top part of Fig. 12.7.2, prior distributed boosting schemes suffer from asymmetric differential ports and unbalanced magnetic coupling modes, causing systematic mismatches [11]. A collaborative-enhanced, ring-coupled transformer topology has been introduced, which lowers the equivalent inductance (Leq) and increases the resonator Q. Compared with weakly coupled designs, it offers stronger mismatch tolerance, reduced pulling sensitivity, and the effective suppression of mode ambiguity under multi-core scaling.

The bottom part of Fig. 12.7.2 presents the complete oscillator topology together with its operating modes. By employing a high-Q series-resonant (SR) tank in combination with a properly chosen coupling coefficient, the oscillator achieves ultra-low phase noise and an improved FoM. In this topology, the gate inductors (L_G) and drain inductors (L_D) of multiple cores are strongly coupled (K_G, K_D), which simultaneously enhances the effective inductance density, improves the resonator Q, and alleviates mismatch during multi-core synchronization. The transformer coupling coefficient is accurately adjusted by controlling the spacing between the coupled coils. During the differential-mode operation, the in-phase currents (indicated by black arrows) reinforce one another, producing a spatially uniform magnetic field around the ring coil and, thereby, achieving substantial DM boosting.

As illustrated in Fig. 12.7.2 (right), common-mode (CM) supply noise remains a critical limitation for high-performance oscillators. Under standard supply voltages, CMOS oscillators are typically biased in the Class-B mode to avoid device breakdown, which keeps the active devices in partial saturation for a portion of each cycle [4]. In typical cross-coupled-pair-based oscillators, power-supply noise is directly injected into the LC tank through the supply network. Unlike cross-coupled-pair-based oscillators, inverter-based oscillators inherently amplify CM disturbances and inject them into the resonant tank, leading to a slower output waveform slope and increased distortion, which in turn deteriorates phase-noise performance. To overcome this issue, we propose an implicit CM noise-cancellation scheme. In this approach, opposing current flows (red arrows) in the transformer network decouple the gate inductor (L_G) from the drain inductor (L_D), effectively driving the drain-to-gate coupling coefficient toward zero. As a result, the injected common-mode supply noise is suppressed, and the current inversion through the L_G path compensates the distortion, thereby enhancing oscillator robustness and mitigating phase-noise degradation.

As shown in Fig. 12.7.3, a relatively weak coupling between the drain and gate inductors is required to ensure sufficient differential-phase gain and to maintain a properly low impedance seen from the drain. Conventional SR oscillators suffer from severe waveform degradation when the transistor aspect ratio (W/L) is increased to provide sufficient current drive. A larger device W/L introduces significant parasitic gate-to-drain capacitance (CGD). The input signal is then partially injected through this parasitic capacitance into the drain node, which not only reduces the transition slope of the output square waveform but also introduces additional distortion. The resulting deterioration of the waveform degrades the impulse sensitivity function, ultimately limiting both the achievable phase-noise performance and the overall FoM.

To overcome these limitations, we propose a transformer network with magnetically enhanced feedforward waveform shaping, as illustrated in Fig. 12.7.3. To clearly illustrate this technique, the figure presents the proposed dual-core structure and the schematic of two adjacent cores, where a tunable strong mutual inductance is introduced before LD. In this approach, the negative output of the differential SR oscillator is filtered by an LC network and magnetically coupled to the positive output. By tuning the coupling strength in proportion to the level of input-signal injection through C_{GD}, the feedforward distortion at the positive input is effectively suppressed. Furthermore, a dual-core cell implementation is introduced, whose transformer and equivalent circuit are shown in Fig. 12.7.3. The feedback factor of this structure is expressed as $L_G/(2M_1)$. This technique increases the effective output admittance, sharpens the drain-voltage transition, and compensates for parasitic-induced distortion, thereby enabling SR oscillators to trade the additional power consumption for significantly improved phase-noise performance and higher overall system efficiency.

Figure 12.7.4 shows the measured phase noise and FoM at a 1MHz offset under four different control codes. Due to instrumentation limitations, the phase noise measured at a 10MHz offset is constrained by the noise floor and thus excluded from the FoM calculation. Figure 12.7.5 shows the FoM of 190.6dBc/Hz and a phase noise of –146.72dBc/Hz at a 1MHz offset. In comparison with these earlier publications, this CMOS-only design delivers superior phase noise and FoM, markedly extending the performance boundary of the oscillators in Fig. 12.7.5. Furthermore, without calibration or auxiliary techniques, the proposed oscillator array achieves a corner frequency ranging from 170 to 530kHz, which is enabled by its collaborative-enhanced topology.

The proposed CMOS oscillator was fabricated in a 65nm CMOS process. As shown in Fig. 12.7.7, it consumes 1.5 to 2.3W of DC power from a 1.2V supply and occupies a core area of 2mm², which represents modest power and area overheads for instrumentation applications. Figure 12.7.6 compares the proposed design with other ultra-low-phase-noise oscillators. Compared with previous works, the proposed oscillator, implemented in standard CMOS technology, achieves the lowest phase noise and a high FoM at a nominal supply voltage. After normalization to 10GHz, the proposed oscillator outperforms existing silicon-based oscillators in Fig. 12.7.6 by 6.22dB in phase-noise performance. This demonstrates that high-performance microwave sources can be realized at low cost using a scalable CMOS process featuring high area efficiency, without requiring additional LDO circuits. Such an approach is favorable for precision measurement instruments, enabling compactness and seamless system integration.

The proposed CMOS oscillators exhibit the lowest experimentally reported phase noise among fully integrated silicon oscillators in Fig. 12.7.6, with their performance surpassing that of oscillators implemented in BiCMOS technology and even the oscillator in a mainstream III-V commercial implementation.

Acknowledgement:
This work was supported in part by the National Key R&D Program of China under Grant No.2020YFB1807300, in part by the National Natural Science Foundation of China under Grant 62131013, and in part by the Beijing Advanced Innovation Center for Integrated Circuits. Corresponding author: Wei Deng (wdeng@tsinghua.edu.cn).

References:
[1] A. Chu et al., "A 263GHz 32-Channel EPR-on-a-Chip Injection-Locked VCO-array," *ISSCC*, pp. 20–22, Feb. 2023. http://doi.org/10.1109/ISSCC42615.2023.10067623
[2] H. Jia et al., "A 53.6-to-60.2GHz Many-Core Fundamental Oscillator with Scalable Mesh Topology Achieving -136.0dBc/Hz Phase Noise at 10MHz Offset and 190.3dBc/Hz peak FoM in 65nm CMOS," *ISSCC*, pp. 154–156, Feb. 2022. http://doi.org/10.1109/ISSCC42614.2022.9731581
[3] L. Tomasin et al., "A 12-GHz Reconfigurable Multicore CMOS DCO, with a Time-Variant Analysis of the Impact of Reconfiguration Switches on Phase Noise," *IEEE JSSC*, vol. 57, no. 9, pp. 2802–2811, Sep. 2022. http://doi.org/10.1109/JSSC.2022.3167109
[4] Q. Wu et al., "An 11.5-to-14.3GHz 192.8dBc/Hz FoM at 1MHz Offset Dual-Core Enhanced Class-F VCO with Common-Mode-Noise Self-Cancellation and Isolation Technique," *ISSCC*, pp. 146–147, Feb. 2023. http://doi.org/10.1109/ISSCC42615.2023.10067672
[5] X. Zhan et al., "A 22.4-to-26.8GHz Dual-Path-Synchronized Quad-Core Oscillator Achieving -138dBc/Hz PN and 193.3dBc/Hz FoM at 10MHz Offset from 25.8GHz," *ISSCC*, pp. 148–150, Feb. 2023. http://doi.org/10.1109/ISSCC42615.2023.10067277
[6] F. Pepe et al., "On the Remarkable Performance of the Series-Resonance CMOS Oscillator," *IEEE TCAS-I*, vol. 65, no. 2, pp. 531–542, Feb. 2018. http://doi.org/10.1109/TCSI.2017.2727283
[7] A. Franceschin et al., "Series-Resonance BiCMOS VCO with Phase Noise of -138dBc/Hz at 1MHz Offset from 10GHz and -190dBc/Hz FoM," *ISSCC*, pp. 150–151, Feb. 2022. http://doi.org/10.1109/ISSCC42614.2022.9731738
[8] S. Zhang et al., "A Transformer-Based Series-Resonance CMOS VCO," *IEEE JSSC*, vol. 60, no. 2, pp. 529–542, Feb. 2025. http://doi.org/10.1109/JSSC.2024.3433521.

[9] S. Zhang et al., "A Multi-Core Series-Resonance CMOS Oscillator," *IEEE JSSC*, vol. 60, no. 5, pp. 1644–1655, May 2025. http://doi.org/10.1109/JSSC.2025.3529600
[10] J. Guo et al., "A Differential Series-Resonance CMOS VCO with Pole-Convergence Technique Achieving 202.1dBc/Hz FoM$_{TA}$ at 10MHz Offset," *ISSCC*, pp. 332–334, Feb. 2025. http://doi.org/10.1109/ISSCC49661.2025.10904629
[11] Y. Shu et al., "A 3.09-to-4.04GHz Distributed-Boosting and Harmonic-Impedance-Expanding Multi-Core Oscillator with-138.9dBc/Hz at 1MHz Offset and 195.1dBc/Hz FoM," *ISSCC*, pp. 296–298, Feb. 2021. http://doi.org/10.1109/ISSCC42613.2021.9365737

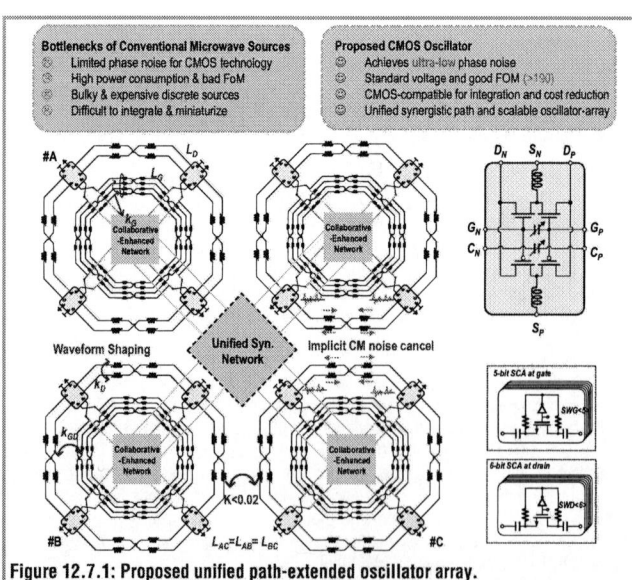

Figure 12.7.1: Proposed unified path-extended oscillator array.

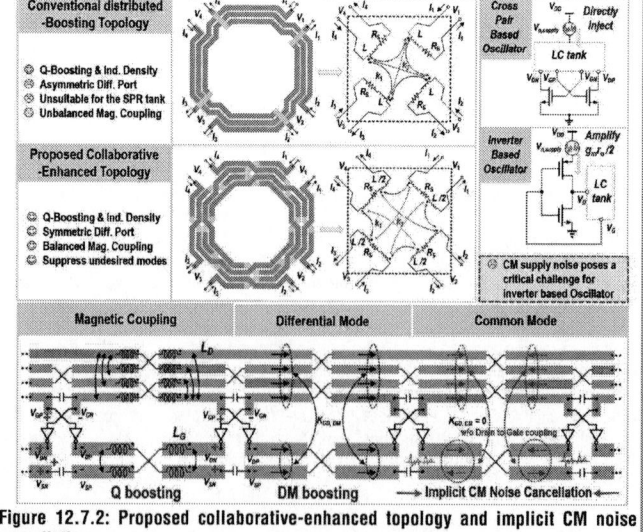

Figure 12.7.2: Proposed collaborative-enhanced topology and implicit CM noise cancellation.

ISSCC 2026 / SESSION 12 / FREQUENCY SYNTHESIZERS AND VCOs / 12.7

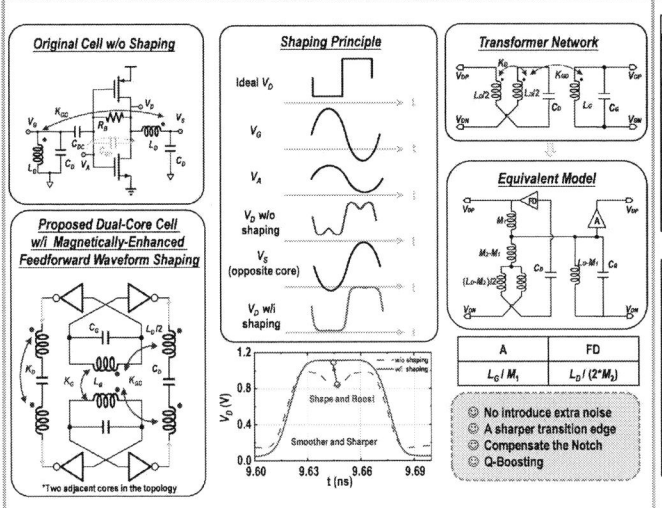

Figure 12.7.3: Proposed magnetically enhanced feedforward waveform shaping (MFWS) technology.

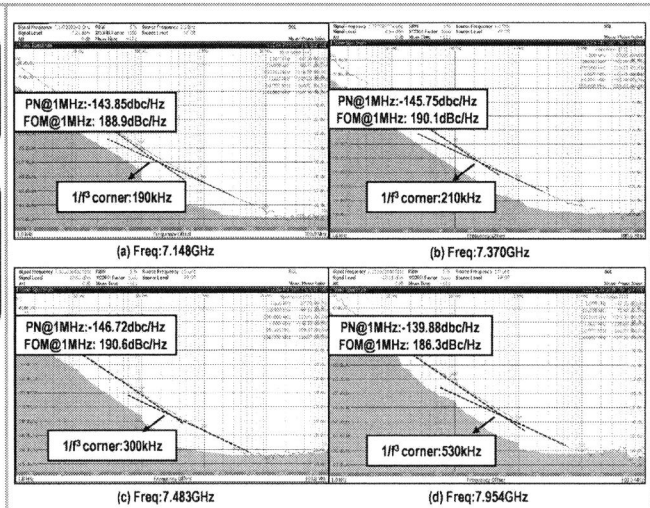

Figure 12.7.4: Measured phase noise results. Phase noise at a 10MHz offset is limited by the noise floor.

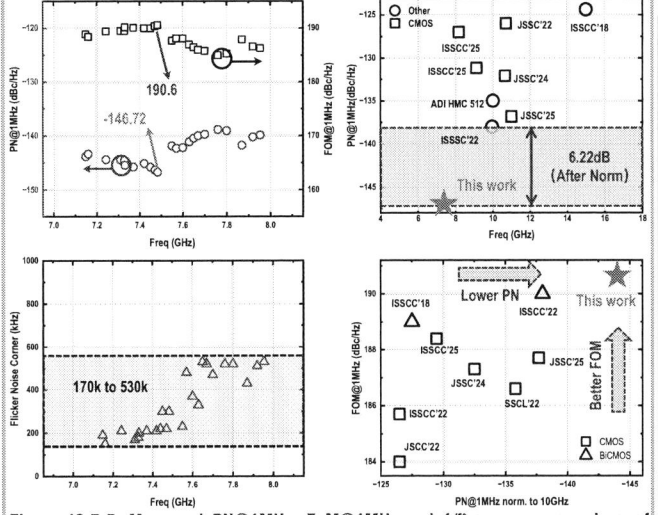

Figure 12.7.5: Measured PN@1MHz, FoM@1MHz and 1/f³ corner; comparison of PN@1MHz and FoM@1MHz with prior oscillators.

	CMOS					Other		
	This Work	JSSC'22	JSSC'24	JSSC'25	ISSCC'25	ISSCC'23	ISSCC'22	Analog Devices HMC512
Technology	65nm CMOS	28nm CMOS	65nm CMOS	65nm CMOS	65nm CMOS	130nm BiCMOS	55nm BiCMOS	GaAs InGaP HBT
Key Topology	Unified Path + Implicit CM Noise Cancellation + MFWS	MultiCore	Series Resonance XFMR			VCO-Array	Series Resonance	N/A
Supply(V)	1.2	1.1	1.2	1	1.2	N/A	1.2	5
Standard Supply?	Yes	No	Yes	No	Yes	N/A	No	N/A
Power(W)	1.5 to 2.3	0.173	0.33	1.2	0.164	4.3	0.6	1.65
Tunning Range (GHz)	7.15 to 7.95 (11%)	10.7 to 14.1 (27%)	10.3 to 11.1 (9%)	11 to 12.2 (10%)	7.65 to 9.14 (18%)	261.3 to 263.6 (1%)	10 to 10.9 (9%)	9.6 to 10.8 (12%)
Frequency (GHz)	7.482	10.7	10.6	11	9.122	263	10	N/A
PN@1MHz (dBc/Hz)	-146.72	-126	-132	-136.82	-131.27	-84*	-138	N/A
PN@1MHz Norm.10GHz (dBc/Hz)	-144.22	-126.5	-132.5	-137.7	-130.48	112.4*	-138	-135
FOM@1MHz (dBc/Hz)	190.6	184	187.4	187.1	188.4	156.2*	190	N/A
Core Area (mm²)	2	3.3	0.43	3.4	0.192	4.2	0.54	N/A

*Estimated from figures FOM=PN+20log₁₀(f₀/Δf)-10log₁₀(P_DC/1mW)

Figure 12.7.6: Comparison with prior-art ultra-low PN oscillators.

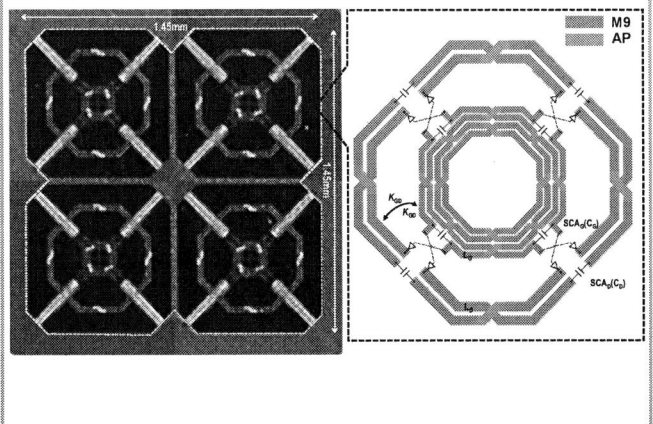

Figure 12.7.7: Die micrograph and the top view of the CMOS oscillator.

ISSCC 2026 / SESSION 12 / FREQUENCY SYNTHESIZERS AND VCOs / 12.8

12.8 A 5.7mW@0.55V-to-50mW@0.9V Deeply Power-Scalable Reconfigurable Series-Resonance/Class-F VCO with Mutual-Inductance Self-Cancellation and Hybrid 8-Shaped Coupling Techniques

Juntao Lan, Wei Deng, Haikun Jia, Shiwei Zhang, Zhihua Wang, Baoyong Chi

Tsinghua University, Beijing, China

Abstract

This work presents a 28nm deeply power-scalable reconfigurable series-resonance/Class-F VCO with mutual-inductance self-cancellation and hybrid 8-shaped coupling techniques, achieving power scalability from 5.7 to 93.1mW. By switching between parallel-resonance (Class-F) and series-resonance modes, the VCO achieves −149dBc/Hz PN and 192.8dBc/Hz FoM at a 10MHz offset. This design enables flexible PN–power scalability for Wi-Fi 7, IoT, and wearable devices.

Voltage-controlled oscillators (VCOs) in next-generation wireless systems are frequently required to exhibit more flexible power-performance scalability to address more complex standards and increasingly diverse application scenarios. Modern communication protocols are not just about high data rates, but rather represent a process where demands evolve continuously over time [1,2]. For instance, Wi-Fi 7 (802.11be) introduces 4096-QAM modulation, enabling higher data rates and better spectral efficiency, with phase-noise (PN) requirements typically around −145dBc/Hz at a 10MHz offset [3]. In applications like base stations or wireless access points, VCOs must support dynamic switching between high- and low-order modulation to meet varying network conditions and protocol requirements. In power-sensitive applications, like IoT devices or mobile electronics, Wi-Fi 7 employs techniques like target wake time (TWT) and orthogonal-frequency-division multiple access (OFDMA) to reduce power consumption and extend battery life. To meet these diverse requirements, VCOs must dynamically switch between low-power (typically 5 to 50mW) and low-phase-noise modes.

In [2], a multi-core VCO with reconfigurable cores was reported, enabling a certain degree of power scalability. However, its power scalability range was limited, and the achievable PN performance remained suboptimal due to core mismatch. To improve PN performance, an SR-VCO was introduced in [4-6], achieving excellent PN but offering little power scalability and a relatively high power floor, making it unsuitable for low-power applications with extended battery life demands.

To address these limitations, this work proposes a deeply power-scalable reconfigurable series-resonance/Class-F VCO, which can dynamically switch between the low-PN and low-power modes. Figure 12.8.1 (right side) shows the simplified block diagram of the proposed VCO, which introduces two key circuit ideas: 1) reconfigurable resonance modes between series resonance (SR) and parallel resonance (PR), and 2) mutual inductance self-cancellation and hybrid 8-shaped coupling techniques aimed at optimizing the EM parameters adaptively for each mode. Two sets of resonance-mode switches (MSWs), located between the gates and drains of two active cores, are used. Since the source nodes in the SR mode exhibit large voltage swings, I/O devices are adopted for the gate-switching transistors to ensure reliability. As a result, this work achieves deep PN-power scalability, providing ultra-low PN in the SR mode and low power in the PR mode, with minor FoM variation. Measurement results demonstrate a minimum power of 5.7mW in the PR mode, and a peak PN of −128.1dBc/Hz at a 1MHz offset, -149 dBc/Hz at a 10MHz offset from a 10GHz carrier in the SR mode.

The top left of Fig. 12.8.2 presents the full schematic of the proposed reconfigurable series-resonance /Class-F VCO operating in two modes. By controlling the resonance MSW, the two ends of the SR path are out of phase when SR mode is enabled. In the PR mode, which works in Class-F, the two ends of the SR path are in phase, making it invisible in the EM environment. The current of the drain inductor flows through a cross path, connecting to the opposite core. The RF current from the gate nodes flows through a shared-edge inductor, forming an 8-shape PR tank with SCA_G, which ensures stable oscillation at ultra-low power.

In both resonance modes, a self-biased inverter is used as the active core. While it does not degrade the PN in the SR mode, it would introduce significant phase noise if directly used in the PR mode. To address this, a common-mode (CM) resonator consisting of a head inductor, SCA_CM, and tail inductor is designed, which effectively improves the PN performance and the FoM in the PR mode. Furthermore, to ensure reliability, I/O devices are used in the differential capacitor array SCA_G. A 500Ω resistor is connected to the substrate of SCA_G for noise isolation, blocking high-frequency noise from entering the resonator through parasitic capacitors C_{db} and C_{sb}. This resistor has an optimal value, requiring a trade-off between thermal noise and isolation performance.

To better illustrate the reconfigurable series-resonance/Class-F scheme, the bottom of Fig. 12.8.2 shows the equivalent circuit for both modes. In the PR mode, the SR path is invisible, and C_{D-PR} resonates with the drain inductance L_{D-PR} near $3\omega_0$ as required by the Class-F operation [7]. However, in the SR mode, C_{D-PR} is not shielded by phase changes, which causes the equivalent circuit of the SR mode to differ slightly from the traditional SPR tank.

By analyzing Z_{in} at f_0, it can be found that C_{D-PR} introduces a pole in Z_{in} together with C_{D-SR}. Since the role of C_{D-PR} is to generate $3\omega_0$ in the Class-F operation, its capacitance value is typically small. As a result, the pole is usually located at a higher frequency than the zero of Z_{in}. By appropriately reducing the capacitance value of C_{D-PR}, the pole can be pushed further away, making its impact on the resonance point negligible in the SR mode. The top right of Fig. 12.8.2 shows the simulated transient waveforms of the gate and drain differential signals for both the SR and PR modes, along with their respective ISF analysis results. The ISF analysis clearly demonstrates that, compared to the PR mode, the SR mode reduces noise sensitivity by approximately 65%.

Figure 12.8.3 illustrates two adaptive tuning techniques—mutual inductance self-cancellation and hybrid 8-shaped coupling—that optimize EM parameters to simultaneously accommodate both resonance modes. One key distinction between the SR and PR modes is the k_m between L_D and L_G. The PR mode using Class-F operation requires a high k_m for effective harmonic shaping, while SR mode necessitates a low k_m to merge poles and eliminate mode ambiguity. As shown on the left side of Fig. 12.8.3, in the SR mode, the transformer is primarily formed by slab inductors L_{D-SR} and L_{G-SR}. After partial co-directional coupling, L_{D-SR} and L_{G-SR} cross over, with the drain inductor extending outward to connect to the SCA_S. After crossing, a mutual-inductance self-cancellation region is formed between L_{D-SR} and L_{G-SR}, effectively reducing k_m to meet the requirements of the SR mode.

As shown in the middle of Fig. 12.8.3, when switching to the PR mode, the hybrid 8-shaped coupling topology is used for the Class-F operation. Since the quality factor Q of L_{G-PR}, Q_{G-PR}, is more critical for the PR mode performance, an edge-shared 8-shaped inductor with Q-boosting functionality is used for L_{G-PR} [8]. However, Q-boosting typically means increasing the inductance of L_{G-PR}, and both modes share the same gate capacitor array SCA_G. To ensure consistent frequency coverage across both modes, there is a trade-off between Q-boosting and L-alignment. To address this, the cross-coupled 8-shaped inductor of L_{D-PR} provides more flexibility in balancing this trade-off. By elongating the middle section of L_{D-PR}, the coupling coefficient between the dual cores, k_m', is reduced, which offers chances for better L-alignment without greatly breaking the Q-boosting effect for L_{G-PR}. While this may sacrifice the Q_{D-PR}, its impact on the PR-mode performance is much less significant. On the right side of Fig. 12.83, the simulated results of inductance, k_m, and Q factors for both modes are shown. As observed, during the mode switching process, the inductance of L_G, which primarily determines the oscillation frequency, changes by only 14pH. Meanwhile, the k_m adaptively switches between the two modes, with k_m being 0.34 in the PR mode and 0.12 in the SR mode. The Q_G factors for both modes remain around 20, achieving a good balance between Q-boosting and L-alignment.

The proposed reconfigurable series-resonance/Class-F VCO was fabricated in a 28nm CMOS process. Measured by R&S FSWP50 phase-noise analyzer, Fig. 12.8.4 shows the measured PN results in the SR mode and the PR mode under a standard supply voltage of 0.9V. In the SR mode, the proposed VCO achieves a PN of −128.1dBc/Hz and −125.2dBc/Hz at a 1MHz offset from a 10 and 11.3GHz carrier, respectively. In the PR mode, the PN is −119.9dBc/Hz and −116.5dBc/Hz at a 1MHz offset from a 10 and 10.9GHz carrier, respectively. Thanks to the CM-resonance design, the 1/f³ corner are lower than 1MHz across the frequency tuning range (FTR) in both modes. More measurement results under a standard supply voltage (0.9V) are exhibited in the top of Fig. 12.8.5. The FTR of the SR mode is from 9.96 to 11.26GHz, with a minor FoM fluctuation of 1.9dB at a 10MHz offset. In the PR mode, the FTR is from 9.95 to 11.12GHz, achieving a peak FoM of 192.1dBc/Hz at a 10MHz offset with a fluctuation of 2.8dB.

The bottom of Fig. 12.8.5 shows the measurement results under varying supply voltages (V_{DD}), representing real scenarios such as power systems with dynamic voltage scaling. In the PR mode, the lowest power consumption of 5.7mW is obtained at a V_{DD} of 0.55V, with a PN and FoM of -136.3dBc/Hz and 192.8dBc/Hz at a 10MHz offset, respectively, from a 10GHz carrier. In the SR mode, a peak power consumption of 93.1mW occurs at a V_{DD} of 1.15V. The lowest PN at a 10MHz offset in the SR mode is -149dBc/Hz under a 0.95V supply from a 10GHz carrier. Across the FTR and power variation, the FoM at 10MHz varies from 188.3 to 192.1 and 187.1 to 192.8dBc/Hz for the SR and PR modes, respectively. It is worth noting that the reported measurement results are based on voltage variations within the

222 • 2026 IEEE International Solid-State Circuits Conference

979-8-3315-8937-0/26 $31.00 © 2026 IEEE

safe operating range of the VCO, and no significant performance degradation or even oscillation failure due to aging or thermal decay was observed during the VCO operation.

Figure 12.8.6 compares this work with other prior-art SR- and PR-VCOs. The proposed reconfigurable series-resonance/Class-F VCO demonstrates the ability to simultaneously achieve competitive ultra-low PN and ultra-low power consumption within a single VCO. Its deeply power scalability range (5.7 to 93.1mW) expected to empower future RF systems to adapt to more diverse range of application scenarios.

Acknowledgement:
This work was supported in part by the National Key R&D Program of China under Grant No.2020YFB1807300, in part by the National Natural Science Foundation of China under Grant 62131013, and in part by the Beijing Advanced Innovation Center for Integrated Circuits. The corresponding author is Wei Deng (wdeng@tsinghua.edu.cn).

References:
[1] A. Liscidini et al. , "A Power-Scalable DCO for Multi-Standard GSM/WCDMA Frequency Synthesizers," *IEEE JSSC*, vol. 49, no. 3, pp. 646-656, Mar. 2014. http://doi.org/10.1109/JSSC.2014.2302292.
[2] L. Tomasin et al., "A 12-GHz Reconfigurable Multicore CMOS DCO with a Time-Variant Analysis of the Impact of Reconfiguration Switches on Phase Noise," *IEEE JSSC*, vol. 57, no. 9, pp. 2802-2811, Sept. 2022. http://doi.org/10.1109/JSSC.2022.3167109.
[3] IEEE Standard for Information technology—Telecommunications and Information Exchange between Systems Local and Metropolitan Area Networks—Specific requirements - Part 11: Wireless LAN Medium Access Control (MAC) and Physical Layer (PHY) Specifications Amendment 2: Enhancements for Extremely High Throughput (EHT)," in *IEEE Std 802.11be-2024 (Amendment to IEEE Std 802.11-2024, as amended by 802.11bh-2024)*, pp.1-1020, 22 July 2025. http://doi.org/10.1109/IEEESTD.2024.11090080.
[4] J. Guo et al., "A Differential Series-Resonance CMOS VCO with Pole-Convergence Technique Achieving 202.1dBc/Hz FoM$_{TA}$ at 10MHz Offset," *ISSCC*, pp. 332-334, Feb. 2025. http://doi.org/10.1109/ISSCC49661.2025.10904629.
[5] S. Zhang et al., "A Multi-Core Series-Resonance CMOS Oscillator," *IEEE JSSC*, vol. 60, no. 5, pp. 1644-1655, May 2025. http://doi.org/10.1109/JSSC.2025.3529600.
[6] S. Zhang et al., "A Transformer-Based Series-Resonance CMOS VCO," *IEEE JSSC*, vol. 60, no. 2, pp. 529-542, Feb. 2025. http://doi.org/10.1109/JSSC.2024.3433521.
[7] M. Babaie and R. B. Staszewski, "A Class-F CMOS Oscillator," *IEEE JSSC*, vol. 48, no. 12, pp. 3120-3133, Dec. 2013. http://doi.org/10.1109/JSSC.2013.2273823.
[8] Q. Wu et al., "An 11.5-to-14.3GHz 192.8dBc/Hz FoM at 1MHz Offset Dual-Core Enhanced Class-F VCO with Common-Mode-Noise Self-Cancellation and Isolation Technique," *ISSCC*, pp. 146-147, Feb. 2023. http://doi.org/10.1109/ISSCC42615.2023.10067672.
[9] J. Chen, et al., "An 8.1-to-9.9GHz Single-Core Pseudo-Series-Resonance Oscillator Achieving -128.7dBc/Hz PN at 1MHz," *ISSCC*, pp. 330-331, Feb. 2025. http://doi.org/10.1109/ISSCC49661.2025.10904613.

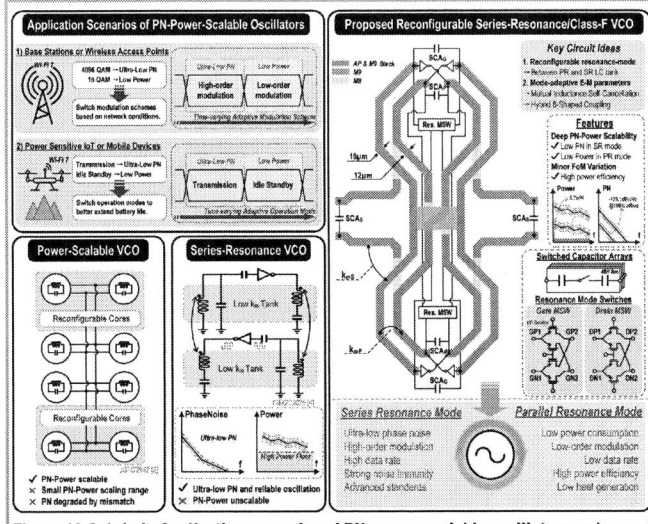

Figure 12.8.1: Left: Application scenarios of PN-power-scalable oscillators and recent VCO techniques. Right: Key ideas and features of the proposed VCO.

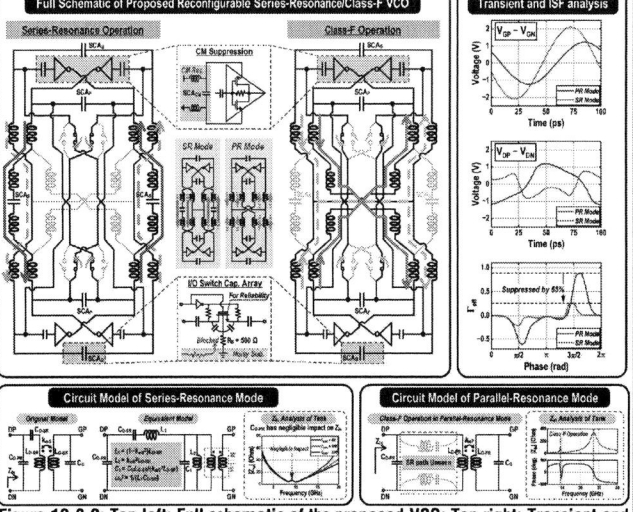

Figure 12.8.2: Top left: Full schematic of the proposed VCO; Top right: Transient and ISF analysis; Bottom: Equivalent circuit models of the SR and PR modes.

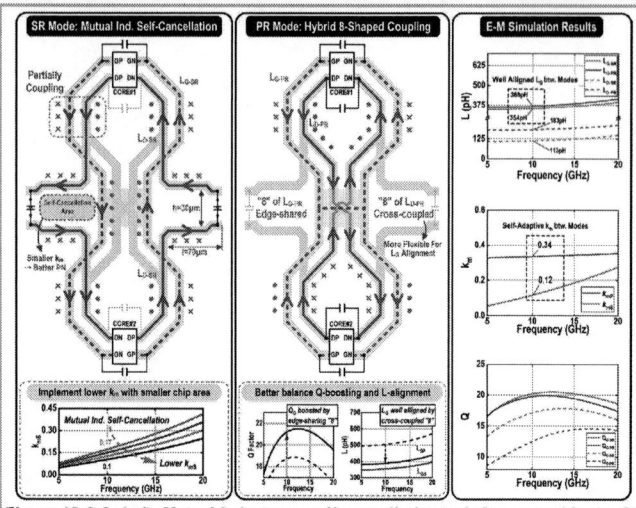

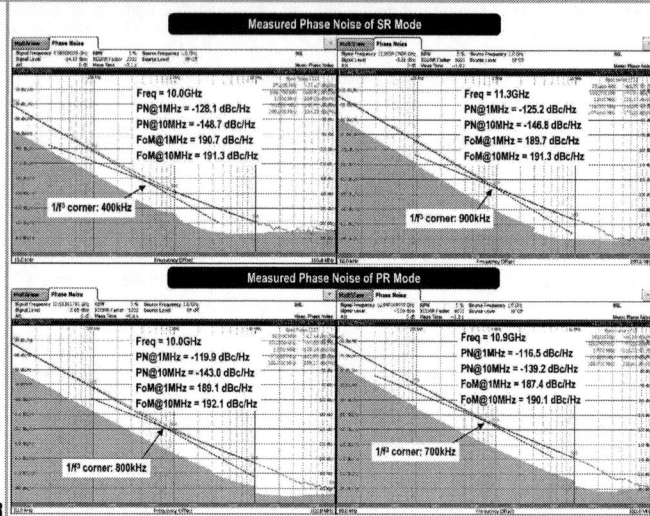

Figure 12.8.3: Left: Mutual inductance self-cancellation techniques used in the SR mode; Middle: Hybrid 8-shaped coupling techniques used in the PR mode; Right: Simulation results of L, k_m, and Q.

Figure 12.8.4: Measured phase noise of the SR and PR modes of the proposed reconfigurable series-resonance/Class-F VCO.

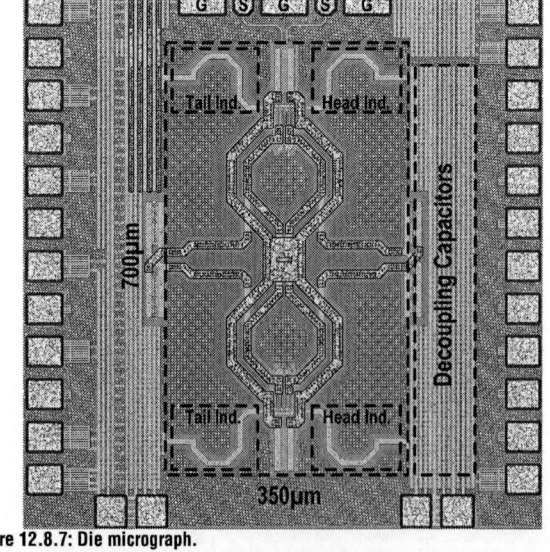

Figure 12.8.5: Measured phase noise and FoM across FTR at a standard supply voltage (0.9V); measured phase noise and FoM across FTR and various supply voltages.

	This Work		JSSCC22 [2] L. Tomasin	ISSCC23 [8] Q. Wu	ISSCC25 [4] J. Guo	ISSCC25 [9] J. Chen
Technology	28nm CMOS		28nm CMOS	65nm CMOS	65nm CMOS	28nm CMOS
Key Technique	Mutual Inductance Self-Cancellation and Hybrid 8-Shaped Coupling		Reconfigurable Multicore	CM-Noise Self Cancellation	Pole- Convergence	Pseudo SR
Resonance Mode	Reconfigurable Resonance Mode		Parallel Resonance	Parallel Resonance	Series Resonance	Series Resonance
	Series Resonance	Parallel Resonance (Class-F)				
No. of Cores (N)	2		8	2	1	1
Supply Voltage [V]	0.9		1.1	-	1.2	0.9
Frequency Tuning Range (FTR) [GHz]	9.96 to 11.26	9.95 to 11.12	10.7 to 14.1	11.5 to 14.3	7.7 to 9.1	8.1 to 9.9
Power [mW]	40.1 to 61.8	9.2 to 15.1	173 (10.7GHz)	5.6 to 10	159 to 164	33.5 to 36.3
PN @ 1MHz [dBc/Hz]	-128.1 to -125.2	-120.1 to -115.4	-126 to -121*	-119.2 to -115.6	-131.9 to -130.6*	-128.7 to -124.0
PN @ 10MHz [dBc/Hz]	-148.7 to -146.7	-143.1 to -138.1	-147 to -145*	-138.9 to -135.7*	-153 to -150.7*	-146.3 to -142.9*
PN @10MHz [dBc/Hz] (normalized to 10GHz)	-148.7 to -147.7	-143.1 to -139	-147.6 to -147.5	-140.1 to -137.8	-152 to -148.4	-144.5 to -141.8
FoM @10MHz [dBc/Hz]	191.4 to 190.2	192.2 to 189.4	184 to 185	191.8 (peak)	190 to 186.4	189.2 to 186.1
Power Scalable?	Yes, under 0.55-to-1.15V VDD 5.7 to 93.1 mW (16.3x)		Yes 36 to 173 mW (4.8x)**	No	No	No
PN scale @ 10MHz across power and FTR	-149 to -136.3		-147 to -139*	-	-	-
FoM scale @ 10MHz across power and FTR	192.1 to 188.3		192.8 to 187.1	184 to 185	-	-
Core Area [mm²]	0.24		3.1	0.065	0.192	0.16

FoM = |PN| + 2Clog10(f₀/Δf) -10log10(P_DC/1mW)
* Read from plot   ** Estimated from reported data

Figure 12.8.6: Performance summary and comparison with prior-art oscillators.

Figure 12.8.7: Die micrograph.

ISSCC 2026 / SESSION 12 / FREQUENCY SYNTHESIZERS AND VCOs / 12.9

12.9 A 10.2-to-16.2GHz Dual-Mode-Transformer-Based Wideband Series-Resonance VCO Achieving >201.1dBc/Hz FoM$_T$ at a 10MHz Offset

Yang Li, Yanchao Liu, Kaihang Wang, Dong Pu, Xiaohua Yu, Ronghua Ni

Fudan University, Shanghai, China

Abstract

This work presents a 28nm CMOS SR-VCO enabled by a dual-mode transformer with magnetic control, achieving a wide tuning range of 10.2 to 16.2GHz (45.2%) and a high FoM$_T$/FoM$_{AT}$@10MHz of 201.1/207.7dBc/Hz. A transformer optimization method is introduced to ensure consistent performance across modes, achieving less than 2dB variation in measured PN and FoM at 10MHz. A sampling PLL incorporating the presented SR-VCO demonstrates rms jitter <18.1fs and FoM$_{JT}$ of -265.8dB.

While series-resonance voltage-controlled oscillators (SR-VCOs) are gaining increased popularity in the communication systems beyond 5G by virtue of their superior phase-noise (PN) performance, their tuning range (TR) is typically limited to below 20% due to the small on/off capacitance ratio of the switched capacitor array, which restricts multi-band operation. This limitation stems from the need for relatively large MOS switches to ensure device reliability under large tank voltage swings. Recently, TRs in the range of 6.2% to 20% have been reported for SR oscillators with single-mode inductors [1-7]. A wider TR of 31.6% has been measured in a SR-VCO employing switched transformer and C$_M$ co-tuning (Fig. 12.9.1, left-bottom) [8]. However, the quality factor (Q) of the inductor is degraded by magnetic mode switching, resulting in a figure-of-merit (FoM) drop from 189 to 185dBc/Hz. In this work, we introduce a dual-mode-transformer-based dual-core SR-VCO that achieves an extended TR with a stable PN and FoM. The presented dual-mode transformer with magnetic control enables switchable inductance without Q degradation, ensuring consistent PN and FoM across the entire TR. Furthermore, a dual-mode-transformer optimization method is presented to achieve consistent inductance ratios and coupling factors across two modes. A consistent inductance ratio improves the oscillator flicker-noise performance, while a consistent coupling factor ensures PN stability across modes. Fabricated in a 28nm CMOS process, a TR of 45.2% with an FoM of 188 to 190dBc/Hz at 10MHz is measured in the prototype wideband SR-VCO, corresponding to a FoM$_T$ exceeding 201.1dBc/Hz over the entire TR. Meanwhile, an analog sampling phase-locked loop (PLL) enabled by this SR-VCO is also implemented and demonstrates an rms jitter of less than 18.1fs across the TR.

The schematic of the presented wideband SR-VCO is shown in Fig. 12.9.1 (right). The dual-mode-transformer-based resonator consists of two coupled tanks: the primary series-resonance (SR) tank (red) and the secondary parallel-resonance (PR) tank (blue). The primary SR tank, connected to the drain nodes (DP1, DN1, DP2, DN2), consists of L$_P$, L$_{P,CM}$, and C$_P$. The secondary PR tank, connected to the gate nodes (GP1, GN1, GP2, GN2), consists of L$_S$, L$_{S,CM}$, and C$_S$. Pole-convergence series resonance is established through coupling between the two tanks (with k$_1$ and k$_2$) [2]. Magnetic mode switching is achieved through the mode switching coils L$_{MS1}$ and L$_{MS2}$ controlled by SW1 and SW2, respectively, (orange) [9], which determine the effective inductance of the two tanks. The tank currents and magnetic flux for both modes are depicted in Fig. 12.9.2 (left). In MODE1, the primary tank current (i$_P$) flows through DP2, L$_P$, C$_P$, DT1, C$_P$, L$_P$, and DN1 on the left and through DP1, L$_P$, C$_P$, DT2, C$_P$, L$_P$, and DN2 on the right. As a result, virtual grounds are formed at DT1 and DT2, rendering L$_{P,CM}$ invisible. Meanwhile, the secondary tank current (i$_S$) flows through the loop GP2, L$_S$, GT1, L$_S$, GN1, C$_S$, GP1, L$_S$, GT2, L$_S$, GN2, C$_S$, and GP2. Virtual grounds are also formed at GT1 and GT2, rendering L$_{S,CM}$ invisible. Therefore, the effective inductance in MODE1 is L$_{P,tank}$=2L$_P$ and L$_{S,tank}$=2L$_S$ for the primary and secondary tanks, respectively. In MODE2, i$_P$ flows through DP2, L$_P$, C$_P$, DT1, L$_{P,CM}$, DT2, C$_P$, L$_P$, and DN2 on the top and through DN1, L$_P$, C$_P$, DT1, L$_{P,CM}$, DT2, C$_P$, L$_P$, and DP1 on the bottom. Differential signals appear across DT1 and DT2, making L$_{P,CM}$ visible. Meanwhile, i$_S$ flows in the top loop GP2, L$_S$, GT1, L$_{S,CM}$, GT2, L$_S$, GN2, and C$_S$, and the bottom loop GN1, L$_S$, GT1, L$_{S,CM}$, GT2, L$_S$, GP1, C$_S$, and GN1. Differential signals appear across GT1 and GT2, making L$_{S,CM}$ visible as well. Therefore, the effective inductance in MODE2 is L$_{P,tank}$=2(L$_P$+L$_{P,CM}$) and L$_{S,tank}$=2(L$_S$+L$_{S,CM}$) for the primary and secondary tanks, respectively. The equivalent circuits for the primary SR and secondary PR tanks in both modes are shown in Fig. 12.9.2 (left-bottom). With proper design of L$_{P,CM}$ and L$_{S,CM}$, the TR of the SR-VCO can be extended, with controlled overlap between the two modes to cover process, voltage, and temperature (PVT) variations.

Mode switching is achieved through magnetic control using L$_{MS1}$ and L$_{MS2}$, as shown in orange in Fig. 12.9.2 (left). In MODE1, SW1 is closed and the figure-8 L$_{MS1}$ is enabled. The drain-to-gate transimpedance (TZ) provided by the circular MODE1 transformer remains unaffected, while the TZ of the figure-8 MODE2 transformer is suppressed, ensuring a higher start-up gain for MODE1. Similarly, in MODE2, SW2 is closed and the circular L$_{MS2}$ is enabled. In this case, the TZ provided by the figure-8 MODE2 transformer remains unaffected, while the TZ of the circular MODE1 transformer is suppressed, yielding a higher start-up gain for MODE2. The simulated TZ for the desired modes is at least 8dB larger than that for the undesired modes across the whole TR, as shown in Fig. 12.9.2 (right-bottom), confirming robust operation of the SR-VCO in the intended mode. It is worth noting that,

due to magnetic flux cancellation, no current is induced in the mode selection coils in either mode, therefore the Q of the primary and secondary tanks in the desired mode is not degraded. The simulated inductance and Q of the primary and secondary tanks in both modes are plotted in Fig. 12.9.2 (right-top). At 13GHz, L$_{P,tank}$ and L$_{S,tank}$ in MODE1/MODE2 are 91/140pH and 181/283pH, respectively, while Q$_{P,tank}$ and Q$_{S,tank}$ are 17/13 and 29/18, respectively. The effective coupling factor between the two tank inductors (k$_{eff}$) in MODE1/MODE2 is 0.158/0.152 (Fig. 12.9.2, right-middle).

Two essential design considerations must be addressed to maintain high performance in the presented dual-mode SR-VCO. First, the resonant frequencies of the primary tank (f$_P$) and the secondary tank (f$_S$) should be as close as possible to suppress the flicker noise up-conversion [1], which requires L$_{P,tank}$·C$_P$=L$_{S,tank}$·C$_S$. Since C$_P$ and C$_S$ remain unchanged, this condition translates to maintaining a consistent L$_{S,tank}$/L$_{P,tank}$ ratio across the two modes. Second, in the pole-convergence technique, the tank impedance, power consumption, and PN varies strongly with k$_{eff}$. As shown in Fig. 12.9.3 (bottom-left), when k$_{eff}$ increases from 0.15 to 0.3, the tank oscillation impedance seen by MOSFET drains (Z$_{drain\_osc}$) increases by a factor of 3.5, theoretically leading to a 5.5dB PN degradation along with a 3.5× reduction in power consumption. Therefore, maintaining a consistent k$_{eff}$ across the two modes is critical for PN stability across the entire TR. Because inductance and coupling factor are inherently interdependent in a transformer design, multiple design iterations are typically required to satisfy both requirements, significantly increasing design complexity. To address this challenge, a four-step dual-mode-transformer optimization method (Fig. 12.9.3, top) is introduced to reliably achieve consistent L$_{P,tank}$/L$_{S,tank}$ and k$_{eff}$ across modes. Before optimization, a fundamental transformer is first designed with rectangular L$_P$ and L$_S$ and straight L$_{P,CM}$ and L$_{S,CM}$ (Fig. 12.9.3, top-left). Co-simulated with C$_P$ and C$_S$ in MODE1, L$_P$ and L$_S$ are sized to target f$_P$ = f$_S$ = f$_{MODE1}$ and k$_{eff}$ ≈0.16, determined by the PN requirement. Without optimization, the simulated L$_{S,tank}$/L$_{P,tank}$ and k$_{eff}$ values in the two modes are misaligned. The optimization process is illustrated in Fig. 12.9.3 (top-right). In step 1, by examining the frequency TR of MODE1 and MDOE2, L$_S$ is bent inward to decrease L$_{S,CM}$ if band overlap is insufficient to cover PVT variations; otherwise, L$_S$ is bent outward to increase L$_{S,CM}$. In step 2, with L$_P$, L$_S$, and L$_{S,CM}$ remaining essentially unchanged, the shape of L$_P$ is adjusted to tune L$_{P,CM}$ until L$_S$/L$_P$ = (L$_S$+L$_{S,CM}$)/(L$_P$+L$_{P,CM}$) is satisfied. In step 3, the straight lines of L$_{P,CM}$ and L$_{S,CM}$ are bent into S-shapes in opposite directions so that the currents on both sides become perpendicular at the middle crossing point, thereby reducing k$_{eff}$ in MODE2 until it aligns with MODE1. In step 4, since the shape adjustment changes L$_{P,CM}$ and L$_{S,CM}$ slightly, the line widths of the S-shaped L$_{P,CM}$ and L$_{S,CM}$ are fine-tuned to restore the L$_S$/L$_P$ = (L$_S$+L$_{S,CM}$)/(L$_P$+L$_{P,CM}$) condition. The values of L$_{S,eff}$/L$_{P,eff}$ and k$_{eff}$ in MODE2 successfully converge to MODE1 after the presented optimization, as shown by the graph in Fig. 12.9.3 (top-right).

While C$_S$ experiences differential signals in both modes and can, therefore, employ a conventional differential switched-capacitor design, C$_P$ encounters different signal conditions in the two modes (Fig. 12.9.3, bottom-right). To ensure high-Q switching and reliable operation for both modes, a new switched-capacitor design is introduced for C$_P$. The biasing voltages and signal waveforms of this design, under both on and off states for the two modes, are illustrated in Fig. 12.9.3(bottom-right). With DT biased at 0.7V and a signal swing below 0.6V, the conduction of parasitic diodes is avoided. A thick-oxide NMOS with a 1.8V gate control is employed to minimize breakdown risk while maintaining reasonable switch on-resistance.

A sampling PLL integrating the presented wideband SR-VCO, a high-gain sampling phase detector, a high-frequency multi-modulus divider, and a proportional-integral loop filter was fabricated in a 28nm CMOS process. The PLL core occupies 0.29mm², with 0.22mm² attributed to the SR-VCO. The PN plots of the free-running SR-VCO at its highest and lowest frequencies are shown in Fig. 12.9.4 (top). The best and worst measured PLL rms jitter (integrated from 10kHz to 100MHz) are 15.4fs at 13GHz and 18.1fs at 10.25GHz, respectively (Fig. 12.9.4, bottom). The measured TRs for MODE1 and MODE2 are 12.8 to 16.2GHz and 10.2 to 13.1GHz, respectively, with a 300MHz overlap, corresponding to a total TR of 45.2% (Fig. 12.9.5, top-left). The measured PN at 1MHz and 10MHz offsets is -126.4 to -122.1dBc/Hz and -148.6 to -146.8dBc/Hz, respectively, over the entire TR (Fig. 12.9.5, top-right). Although affected by flicker noise, the variation of PN at a 1MHz offset remains within 4.3dB. At a 10MHz offset, the PN variation is well controlled below 2dB. The measured FoM and FoM$_T$ at a 10MHz offset are 188 to 190dBc/Hz and 201.1 to 203.1dBc/Hz, respectively (Fig. 12.9.5, bottom-left). The measured rms jitter of the sampling PLL across the entire TR is shown in Fig. 12.9.5 (bottom-right).

The measured performance is compared with recent SR oscillators in Fig. 12.9.6 (top) and Fig. 12.9.7 (right), and the die micrograph is shown in Fig. 12.9.7 (left). Thanks to the presented dual-mode transformer and the optimization method, this design achieves the largest TR as well as leading FoM$_T$ and FoM$_{A,T}$ at a 10MHz offset, with the improvement of 13.6%, 5.4dB, and 3.4dB, respectively, relative to designs in Figs. 12.9.6 and 12.9.7. Consequently, the sampling PLL enabled by this SR-VCO demonstrates rms jitter better than 18.1fs and achieves the most competitive FoM$_{J,T}$ among prior-art 20fs-class PLLs shown in Fig. 12.9.6 (bottom).

References:
[1] J. Chen et al., " An 8.1-to-9.9GHz Single-Core Pseudo-Series-Resonance Oscillator Achieving -128.7dBc/Hz PN at 1MHz," *ISSCC*, pp. 1-3, Feb. 2025. https://doi.org/10.1109/ISSCC49661.2025.10904613
[2] J. Guo et al., "A Differential Series-Resonance CMOS VCO with Pole-Convergence Technique Achieving 202.1 dBc/Hz FoM$_{TA}$ at 10MHz Offset," *ISSCC*, pp. 332-334, Feb. 2025. https://doi.org/10.1109/ISSCC49661.2025.10904629
[3] L. Chen et al., "A 0.06mm² 27.5-to-30GHz Series Resonance VCO with Magnetic Mutual Resistance Achieving 207.2dBc/Hz FoM$_A$ at 10MHz Offset," *IEEE VLSI*, pp. 1-3, Jun. 2025. https://doi.org/10.23919/VLSITechnologyandCir65189.2025.11075196
[4] S. Zhang et al., "A Transformer-Based Series-Resonance CMOS VCO," *IEEE JSSC*, vol. 60, no. 2, pp. 529-542, Feb. 2025. https://doi.org/10.1109/JSSC.2024.3433521
[5] D. Sun et al., "A 13GHz Charge-Pump PLL Achieving 15.8fs Integrated Jitter and -98.5dBc Reference Spur," *ISSCC*, pp. 344-346, Feb. 2025. https://doi.org/10.1109/ISSCC49661.2025.10904662
[6] S. Zhang et al., "A Multi-Core Series-Resonance CMOS Oscillator," *IEEE JSSC*, vol. 60, no. 5, pp. 1644-1655, May 2025. https://doi.org/10.1109/JSSC.2025.3529600
[7] S. Zhang et al., "A Series-Parallel-Resonance Oscillator With a 191.5 dBc/Hz FoM," *IEEE SSCL*, vol. 8, pp. 141-144, 2025. https://doi.org/10.1109/LSSC.2025.3564312
[8] Q. Leng et al., "7.8-to-10.7GHz Reliable-Mode-Switching Series Resonance Oscillator with Bidirectional Inductive-Mode-Pulling Achieving -156.5dBc/Hz Phase Noise and 199.2dBc/Hz FoM$_T$ at 10 MHz Offset in 40-nm CMOS," *IEEE RFIC*, pp. 247-250, Jun. 2025. https://doi.org/10.1109/RFIC61188.2025.11082967
[9] H. Guo et al., "A 9.05-to-37.0GHz LO Generator with Magnetic Mode Switching and Tuning-Free Octave-Bandwidth Common-Mode Resonator Achieving >190.7dBc/Hz FoM," *ISSCC*, pp. 560-562, Feb. 2025. https://doi.org/10.1109/ISSCC49661.2025.10904644
[10] H. Liu et al., "A Multireference PLL: Theory and Implementation," *IEEE JSSC*, vol. 59, no. 7, pp. 1981-1994, Jul. 2024. https://doi.org/10.1109/JSSC.2024.3383605
[11] Y. Zhao et al., "A 20-GHz PLL with 20.9-fs Random Jitter," *IEEE JSSC*, vol. 58, no. 6, pp. 1597-1609, June 2023. https://doi.org/10.1109/JSSC.2022.3225105

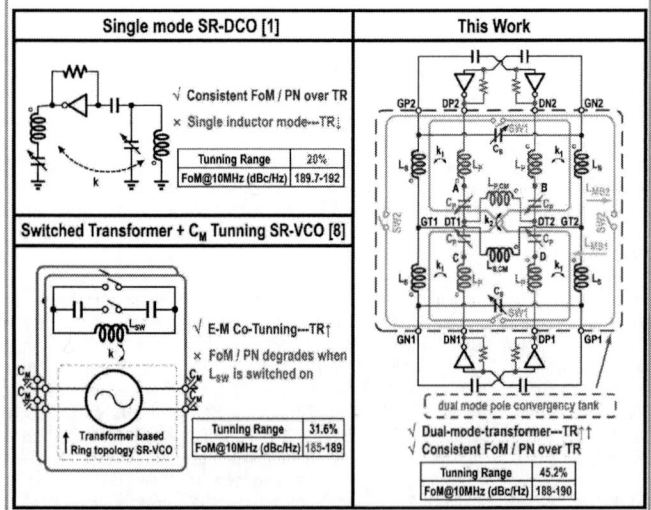

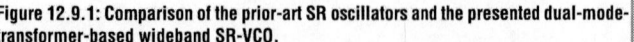

Figure 12.9.1: Comparison of the prior-art SR oscillators and the presented dual-mode-transformer-based wideband SR-VCO.

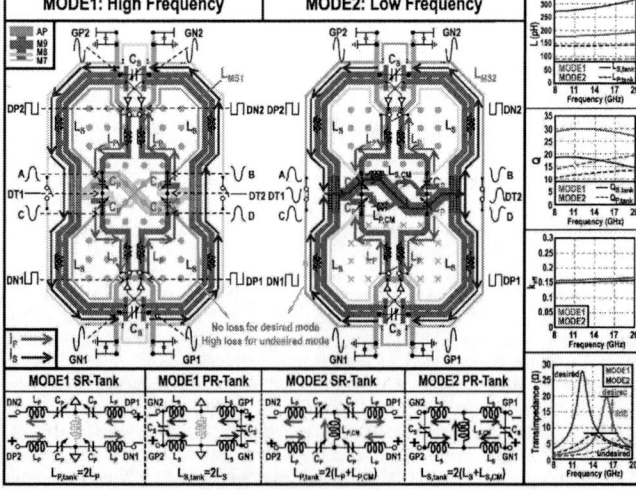

Figure 12.9.2: The layout, waveforms, equivalent circuits, and simulated results for two tanks in two modes of the presented SR-VCO.

ISSCC 2026 / SESSION 12 / FREQUENCY SYNTHESIZERS AND VCOs / 12.9

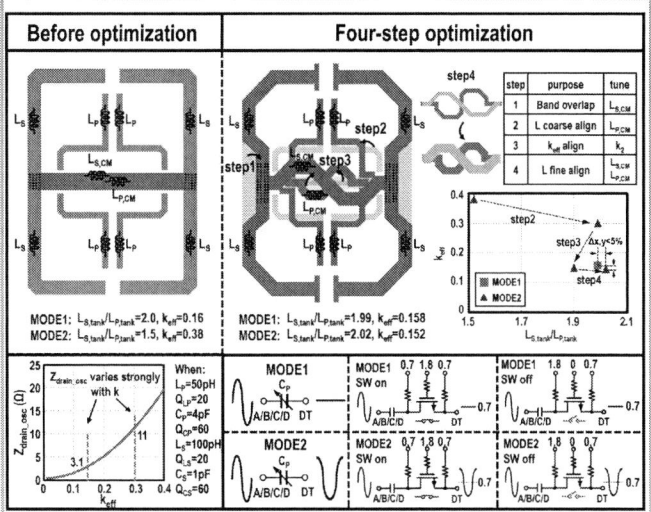

Figure 12.9.3: The four-step optimization (top), simulated Z_{drain_osc} vs. k_{eff} (bottom-left), and dual-mode switched-capacitor design (bottom-right).

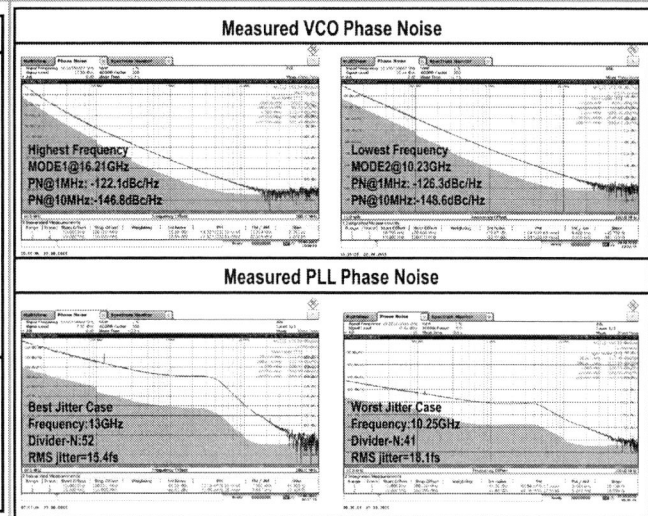

Figure 12.9.4: Measured phase-noise plots for the presented SR-VCO (top) and sampling PLL (bottom).

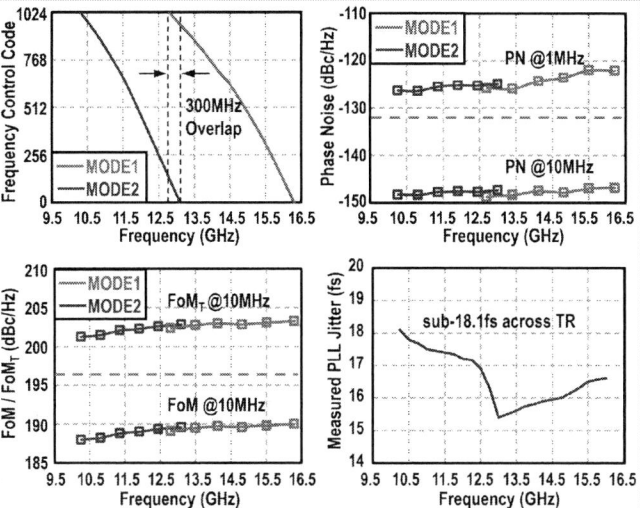

Figure 12.9.5: The measured TR, PN, and FoM/FoM$_T$ for the presented SR-VCO (top and bottom-left). Measured rms jitter for the sampling PLL (bottom-right).

Comparison with prior-art SR oscillators

	This work	ISSCC'25 [1]	ISSCC'25 [2]	VLSI'25 [3]	JSSC'25 [4]	ISSCC'25 [5]	RFIC'25 [8]
Technology	28nm CMOS	28nm CMOS	65nm CMOS	65nm CMOS	65nm CMOS	65nm CMOS	40nm CMOS
Freq. range(GHz) (FTR)	10.2 to 16.2 (45.2%)	8.1 to 9.9 (20%)	7.6 to 9.1 (17.7%)	27.5 to 30 (8.6%)	10.3 to 11.1 (7.5%)	12.6 to 13.4 (6.2%)	7.8 to 12.0 (31.6%)
Supply (V)	0.9	0.9	1.2	-	1.2	1	0.8
Power (mW)	110 to 150	33.5 to 36.3	159 to 164	46	228 to 330	62	350 to 380
Worst PN @1MHz Eq. to 15G (dBc/Hz)	-122.7	-120.4	-124.9	-122.3	-122.2	-121.7	-124.5
Worst PN @10MHz Eq. to 15G (dBc/Hz)	-145.2	-140.4	-145.0	-147.1	-143.1	-143.0	-147.3
Worst FoM@10MHz over TR (dBc/Hz)	188	189.7*	186.4	193.4	183	189*	185
Worst FoM$_T$@10MHz over TR (dBc/Hz)	201.1	195.7	191.4	192.1	180.5	184.8	194.9
Worst FoM$_A$@10MHz over TR (dBc/Hz)	207.7	203.7	198.6	204.3	184.0	189.9	198.4
Area (mm²)	0.22	0.16	0.19	0.06	0.45	0.31	0.46

Comparison with prior-art 20fs-class PLLs

	Technology	Ref (MHz)	Freq. (GHz)	TR (%)	Jitter (fs)	Power (mW)	FoM$_J$ (dB)	FoM$_{A,T}$ (dB)	Ref. Spur (dBc)	Area (mm²)
This Work	28nm CMOS	250	10.25 to 16	43.8	15.4 to 18.1	115 to 155	-253	-265.8	-57**	0.29
ISSCC'25 [5]	65nm CMOS	250	12.75 to 13.25	3.8	15.8	79.53	-257	-248.6	-98.5	0.52
JSSC'24 [10]	65nm CMOS	100	4.8 to 5.6	15.4	16	83	-256.7	-260.5	-71	1.1
JSSC'23 [11]	28nm CMOS	250	20	-	20.9	12	-262.8	-	-66	0.06

*Estimated from the measured PN plot **Mainly due to reference clock coupling through PCB
VCO: FoM=|PN-20log(f/Δf)+10log(Power*1000)| FoM$_T$=FoM+20log(TR/10%) FoM$_{A,T}$=FoM$_T$+10log(1/Area$_{(mm2)}$).
PLL: FoM$_J$=10log(jitter²*Power*1000) FoM$_{J,T}$=FoM$_J$-20log(TR/10%).

Figure 12.9.6: Performance summary and comparison with prior-art SR oscillators and 20fs-class PLLs.

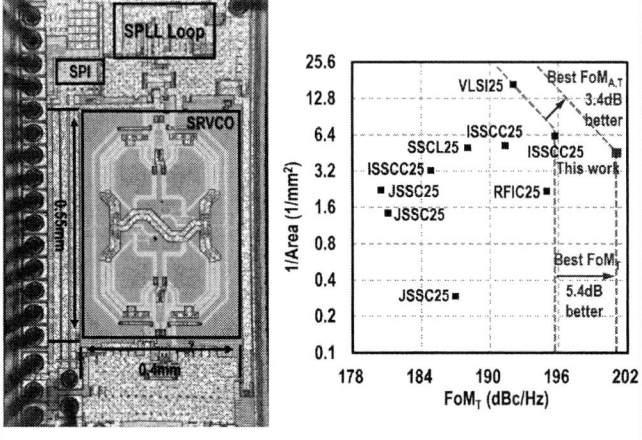

Figure 12.9.7: Die micrograph (left). Benchmarking against prior-art SR oscillators (right).

ISSCC 2026 / SESSION 13 / CIRCUITS FOR AI AND AI FOR CIRCUITS / OVERVIEW

Session 13 Overview: *Circuits for AI and AI for Circuits*
TECHNOLOGY DIRECTIONS SUBCOMMITTEE

Session Chair: Minyoung Song
DGIST, Daegu, Korea

Session Co-Chair: Chris Rudell
University of Washington, Seattle, WA

This session highlights a two-way revolution with AI-driven design that discovers non-intuitive RF/analog architectures beyond traditional human templates and hardware that pushes both compute and optimization into silicon. The first and third papers introduce hybrid compute-in-memory accelerator for large-scale recommendation and a quantum-inspired analog solver for combinatorial optimization, highlighting hardware tailored to AI workloads. The second paper presents an AI-enabled end-to-end RFIC design flow that synthesizes LNAs from specifications to layout. The last two papers present RF front-ends designed with algorithmic inverse and topology-optimization techniques, including an inverse-designed N-path filter and a topology-optimized multilayer wideband mm-wave LNA.

ISSCC 2026 / February 17, 2026 / 8:00 AM

8:00 AM

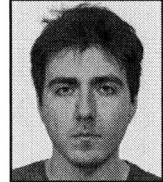

13.1 HYDAR: A 390K QPS, 1574K QPS/W Hybrid Analog/Digital Compute-in-RRAM Accelerator for Efficient Recommendation System

Jiaming Li, Tsinghua University, Beijing, China

In Paper 13.1, a joint development led by Tsinghua University presents a hybrid compute-in-RRAM accelerator for large-scale recommendations. The 36M-RRAM chip reaches 390K QPS at 1,574K QPS/W and scales to a 576M system with 66× higher QPS and 181× higher QPS/W.

8:25 AM

13.2 AI-Enabled End-to-End Design in RFICs with Controllable Architectural Style from 'Classical' to 'Non-Intuitive' for mm-Wave/sub-THz LNAs

Emir Ali Karahan, Princeton University, Princeton, NJ

In Paper 13.2, Princeton University introduces an AI-enabled end-to-end RFIC design flow that synthesizes LNAs from specs to layout with controllable architecture. Measured LNAs span 24 to 90GHz with 16dB peak gain and 3.8-to-6.8dB NF, and another AI-synthesized design shows 30dB and 18dB peaks at 85 and 170GHz.

8:50 AM

13.3 Medusa: A Quantum-Inspired 200-Variable 1016-Clause Analog k-SAT Solver

Luke D. Wormald, University of Michigan, Ann Arbor, MI

In Paper 13.3, the University of Michigan describes a quantum-inspired analog variable k-SAT solver supporting up to 200 variables and 1,016 clauses. On 50-variable 3-SAT it delivers 4.92µs mean solution time and 19.1nJ with 100% solvability and accuracy, improving energy and time by 3.5× and 3× over state-of-the-art solvers.

9:15 AM

13.4 An Inverse-Designed Passively Coupled N-Path Filter with g_m-Boosted Active HBT Switches

Vinay Chenna, University of Southern California, Los Angeles, CA

In Paper 13.4, the University of Southern California presents an inverse-designed, passively coupled N-path filter in BiCMOS using gm-boosted active HBT switches and compact multilayer pixelated passives operating at low-GHz frequencies. It tunes 0.8 to 2.6GHz with >+15dBm P1dB, <5dB NF, and <4dB insertion loss.

9:30 AM

13.5 A Nonintuitively Frequency-Staggered Wideband mm-Wave Low-Noise Amplifier

Vinay Chenna, University of Southern California, Los Angeles, CA

In Paper 13.5, the University of Southern California shows an algorithmic topology-optimization framework that synthesizes non-intuitive, multilayer wideband mm-wave LNAs by co-optimizing active and passive devices directly at a layout level. A 3-stage SiGe HBT LNA validates the efficacy of this approach with 26.3dB gain, 22-to-46GHz 3dB BW, 3.1dB NF, and 18mW.

979-8-3315-8937-0/26 $31.00 © 2026 IEEE

ISSCC 2026 / SESSION 13 / CIRCUITS FOR AI AND AI FOR CIRCUITS / 13.1

13.1 HYDAR: A 390K QPS, 1574K QPS/W Hybrid Analog/Digital Compute-in-RRAM Accelerator for Efficient Recommendation System

Jiaming Li[1], Peng Yao[1], Xueqi Li[1], Zhenqi Hao[1], Dabin Wu[1], Zhouzheng Li[1], Haishu Xianyu[2], Lin Li[3], Shujuan You[4], Taiwei Chiu[5], Maojia Sheng[6], Wei Yang[7], Qingtian Zhang[1], Jianshi Tang[1], He Qian[1], Bin Gao[1], Huaqiang Wu[1]

[1]Tsinghua University, Beijing, China, [2]Beijing Elemem Technology, Beijing, China, [3]Migu Culture Technology, Beijing, China, [4]China Mobile Research Institute, Beijing, China, [5]Xiamen Industrial Technology Research Institute, Fujian, China, [6]Bytedance China, Beijing, China, [7]Huawei Technologies, Shenzhen, China

Abstract

We present the first 28nm hybrid compute-in-RRAM (CiR) accelerator for recommendation system (RecSys) based on the HYDAR framework: (1) DL ADCs enable early termination of non-Top-K calculations. (2) PPSP dataflow boosts throughput for irregular workloads. (3) A coarse-to-fine architecture preserves system recall accuracy. The 36M-RRAM CiR chip achieves 390K QPS with a SOTA 1574K QPS/W. The chips scale out to a 576M system for practical RecSys task, improving by 66x in QPS and 181x in QPS/W.

Recommendation systems (RecSys) play a crucial role in connecting users with massive volumes of content and services, and are widely deployed in e-commerce and streaming platforms. Similarity Vector Search (SVS) serves as the core operator in RecSys, retrieving the Top-K nearest vectors by computing distances between query embeddings and a large base vector set [1]. SVS dominates RecSys' computation time and power consumption, primarily due to external memory access (EMA) overhead [2]. DRAM accelerators with hybrid bonding [3] enhance bandwidth to mitigate EMA but incur high cost and remain limited by DRAM–logic data transfer. NAND TCAM-based accelerators [4] integrate computation within memory arrays to reduce EMA, but suffer from high read latency and limited precision in data and distance representation. To overcome these limitations, this paper proposes HYDAR, a HYbrid Digital-Analog compute-in-RRAM accelerator, enabling high-throughput, energy-efficient, and accurate SVS. Compute-in-RRAM (CiR) has been recognized as a promising approach for deep learning acceleration [5,6], due to its minimized data movement, high storage density, and massive parallelism. However, extending CiR to SVS introduces additional challenges as shown in Fig. 13.1.1: (1) SVS requires only top-matching vectors, but conventional designs compute all distances at equal cost, leading to excessive energy and latency as databases scale; (2) SVS workloads are dynamic and irregular, in contrast to the deterministic dataflow of deep learning, severely degrading processing element (PE) utilization and throughput; (3) the efficiency benefits of analog CiR typically come with precision trade-offs. Thus, an efficient SVS accelerator is required to ensure high-throughput retrieval without compromising recall accuracy.

HYDAR addresses these challenges through co-optimization across CiR PE, hybrid chip design, and multi-chip system architecture as shown in Fig. 13.1.2: (1) CiR PE with Dynamic-Latency ADC (DL-ADC): A multi-bit analog CiR PE integrates DL-ADC for histogram-based SVS, comparing distances against a retrieval threshold in advance and skipping non-Top-K vectors, thereby reducing latency and energy; (2) Prediction-based Preemptive Scheduling Pipeline (PPSP): A hybrid chip mechanism that adapts to dynamic SVS workloads by predicting per-PE runtime, interrupting unbalanced tasks, and inserting short ones to balance workloads, improving utilization and throughput; (3) Two-step coarse-to-fine retrieval architecture: A software–hardware co-designed framework that performs coarse retrieval on CiR PEs for high throughput, followed by fine retrieval on a digital SVS engine, maximizing throughput while preserving recall accuracy. Based on HYDAR framework, a 28nm CiR prototype chip is implemented with 36M RRAM cells across 16 parallel PEs, each comprising a 288×4096 array (shown in Fig. 13.1.2). Queries are broadcast to PEs under the control of a dynamic task scheduler for parallel computation and filtering, and subsequently generate histograms and execute out-of-order committing. The chip supports 65,536 high-dimensional base vectors, delivering 390K queries per second (QPS) at a state-of-the-art 1574K QPS/W. A multi-chip system achieves real-time end-to-end million-scale RecSys with CPU-comparable accuracy (95% Recall@10) and >66.8× throughput improvement.

Figure 13.1.3 details the realization of histogram-based SVS with analog CiR PE, and the DL-ADC design to enable early termination of calculation. A histogram of distance distribution between queries and base vectors is employed to determine the cutoff key (CK) value for Top-K retrieval. Within the Euclidean distance framework, base vectors with distances exceeding the CK are filtered by the dual-mode DL-ADCs that dynamically monitor comparison results, enabling early termination of non-Top-K vectors. Euclidean distance computation can be achieved on 288×4096 CiR arrays, where each 2T2R cell represents a 4-bit dimension, and each column represents a 256-dimension base vector along with a 32-dimensions bias. In this design, a histogram is recorded in each histogram memory following CiR PE during computation, then synchronized to a Cross-PE Histogram Unit (CHU) to merge distributed results for CK generation. Three customized instructions are designed to execute this process as shown in Fig. 13.1.3. In terms of the DL ADC, the SAR-based structure supports Early-Termination (ET) mode that takes a pre-generated CK as input, which is fed into a bit-wise comparator alongside the SAR code generated per cycle. During the process of iteratively adjusting the IDAC to approximate the ADC input current, any bit mismatch indicates a discrepancy between the computation result and the CK. This triggers early termination, which halts the computation and outputs 2-bit vMasks (2'b01 means result < CK, discard the vector; 2'b10 means result > CK, retain the vector). By setting DL-ADC in ET mode, distance calculation and filtering are conducted simultaneously. This ET mechanism is highly effective as database scales, reducing 60% computation time and 71% power consumption on average with only 7% macro area overhead.

Figure 13.1.4 illustrates the proposed Prediction-based Preemptive Scheduling Pipeline for dynamic SVS workloads. Queries undergo parallel computation across different PEs, where the number of base vectors involved in computation typically varies per PE. This introduces difference in computation periods and cross-PE synchronization, which further lead to scheduling stalls and pipeline bubbles. PPSP addresses this issue by adopting a novel continuous and preemptive scheduling with Dynamic Task Scheduler (DTS). DTS monitors and predicts the time stamp of completion of queries execution across PEs. Our preemptive scheduling mechanism enables new tasks to preempt the ongoing tasks that could finish early. This not only eliminates pipeline bubbles, but also facilitates early task completion and PE release for subsequent queries. Upon instruction reception, the task occupies a DTS slot and store its PE/section mask in a task table, and is routed as a sub-task to one of the two outstanding buffers in target PEs. Each PE Status Unit (PSU) tracks the Predicted End Time (PET) of allocated sub-tasks: For I_H and I_T, PET depends on the 1s count in the section mask, a PET factor of 10 (DL-ADC cycles in SAR Mode), and PET of on-going sub-task; For I_C, the factor is 4 (average DL-ADC cycles in ET Mode). DTS also monitors each task's Predicted Critical End Time (PCET), defined as the maximum PET of all sub-tasks. An arbiter in DTS checks PE overlap with ongoing tasks, and switches outstanding buffers to preemptively schedule new tasks, if this reduces new task PET without affecting ongoing task's PCET, improving throughput and reducing latency. Outputs of PEs are processed by Out-of-order Committing Unit: I_T uses CHU-generated CK in parallel comparators with the intermediate results in the per-PE result buffers to conduct CK-based filtering and generate vMasks, avoiding recalculation of distance; while I_C directly retrieves DL-ADC vMasks from PEs. ID generation unit converts vMasks into vector IDs with counters for each PE, while I_H bypasses this stage to write the merged histogram directly into the output buffer. Also, during queries scheduling, DTS pre-allocates address space for each query in a back-end memory allocator. This enables PEs to immediately write results to the output buffer without inter-PE synchronization, enabling fast PE release for new queries. Leveraging the optimizations above, PPSP improves PE utilization to 91%, reduces average query latency by 30%, and increases QPS throughput by 1.82×.

Figure 13.1.5 details the CiR-featured two-step retrieval architecture for large-scale SVS. To improve the systematic accuracy, the architecture integrates a digital fine-retrieval engine to precisely select vectors among the results after high-throughput coarse retrieval. This maintains high recall accuracy in spite of the noisy and low-precision processing within analog CiR. This architecture also expands the base capacity and enables broader parallel coarse retrieval via multi-CiR chip parallelism, while adopting a Thresh-IVF [7] workflow with system pipelining for further throughput improvement. CiR PEs are categorize into three types: 1) Centroid PEs (CPEs) storing cluster centroid coordinates; 2) Sample PEs (SPEs), storing small sets of vectors sampled from each cluster to characterize distribution and generate CK; 3) Full-base PEs (FPEs) storing all base vectors and operating entirely in highly efficient DL-ADC ET Mode, taking up 92.7% of vector storge in the workflow. A CiR-dedicated Thresh-IVF workflow proceeds as follows: 1) Queries are first sent to CPEs, where I_T computes distance between queries and centroids to identify closest clusters. 2) The system routes queries to SPEs of selected clusters, where I_H generates histogram across chips, then generate CK. 3) CK is routed to all FPEs of clusters identified in Step 1, with I_C generating coarse retrieval IDs. This system-level threshold-based coarse-retrieval minimizes the number of filtering results generated by each chip, avoiding redundant ID filtering caused the same Top-K computation on each chip. Finally, a small set of candidate IDs is sent to the digital engine for fine retrieval in FP16 format, reducing system-level memory bandwidth requirement by 97.44%. This enables a system-level four-stage pipeline parallelism, while different tasks are also processed in parallel across distinct PEs within a chip as shown in the bottom middle of Fig. 13.1.5. This multi-chip hierarchical pipeline can achieve a 10.17× latency reduction comparing to traditional CPU-based IVF solution.

Figure 13.1.6 shows the measurement results and comparisons with prior works [3,4,8,9]. The highly efficient CiR chip is fabricated under 28nm technology, and a multi-chip system is built for real-time processing in practical million-scale RecSys. The CiR coarse-retrieval chip delivers 390K QPS and a state-of-the-art 1574K QPS/W in DL-ADC ET mode. The

228 • 2026 IEEE International Solid-State Circuits Conference

979-8-3315-8937-0/26 $31.00 © 2026 IEEE

ISSCC 2026 / February 17, 2026 / 8:00 AM

coarse-to-fine retrieval system integrates 576M RRAM with 16 CiR chips and a FPGA. A million-scale GloVe dataset for RecSys is experimentally deployed, and the system achieves 66.8× higher QPS (with 95% Recall@10) compared to ScaNN [10] solution based on XEON Silver 4310 CPU, with a 181.6× improvement in QPS/W. Figure 13.1.7 shows the die micrograph with chip and system summary.

Acknowledgement:
This work is supported in part by the STI 2030-Major Projects (2021ZD0201200), the NSFC (62422405 and 62495100), the IoT Intelligent Microsystem Center of Tsinghua University - China Mobile Joint Research Institute, the Shanghai Municipal Science and Technology Major Project, and the Beijing Advanced Innovation Center for Integrated Circuits. (Corresponding author is Peng Yao, pyao@mail.tsinghua.edu.cn)

References:
[1] J.-T. Huang *et al.*, "Embedding-based Retrieval in Facebook Search," *Proceedings of the 26th ACM SIGKDD International Conference on Knowledge Discovery & Data Mining*, Aug. 2020, pp. 2553–2561. http://doi.org/10.1145/3394486.3403305
[2] Y. Lee *et al.*, "ANNA: Specialized Architecture for Approximate Nearest Neighbor Search," *IEEE International Symposium on High-Performance Computer Architecture (HPCA)*, Apr. 2022, pp. 169–183. http://doi.org/10.1109/HPCA53966.2022.00021
[3] D. Niu *et al.*, "184QPS/W 64Mb/mm23D Logic-to-DRAM Hybrid Bonding with Process-Near-Memory Engine for Recommendation System," *IEEE International Solid- State Circuits Conference (ISSCC)*, Feb. 2022, pp. 1–3. http://doi.org/10.1109/ISSCC42614.2022.9731694
[4] C.-C. Hsieh *et al.*, "Chip Demonstration of a High-Density (43Gb) and High-Search-Bandwidth (300Gb/s) 3D NAND Based In-Memory Search Accelerator for Ternary Content Addressable Memory (TCAM) and Proximity Search of Hamming Distance," *IEEE Symposium on VLSI Technology and Circuits (VLSI Technology and Circuits)*, June 2023, pp. 1–2. http://doi.org/10.23919/VLSITechnologyandCir57934.2023.10185361
[5] A. S. Lele *et al.*, "A Heterogeneous RRAM In-Memory and SRAM Near-Memory SoC for Fused Frame and Event-Based Target Identification and Tracking," *IEEE Journal of Solid-State Circuits*, vol. 59, no. 1, pp. 52–64, Jan. 2024. http://doi.org/10.1109/JSSC.2023.3297411
[6] K. Ueyoshi *et al.*, "DIANA: An End-to-End Energy-Efficient Digital and ANAlog Hybrid Neural Network SoC," *IEEE International Solid-State Circuits Conference (ISSCC)*, Feb. 2022, pp. 1–3. http://doi.org/10.1109/ISSCC42614.2022.9731716
[7] A. Babenko and V. Lempitsky, "The inverted multi-index," *IEEE Conference on Computer Vision and Pattern Recognition*, June 2012, pp. 3069–3076. http://doi.org/10.1109/CVPR.2012.6248038
[8] F. Tu *et al.*, "TensorCIM: A 28nm 3.7nJ/Gather and 8.3TFLOPS/W FP32 Digital-CIM Tensor Processor for MCM-CIM-Based Beyond-NN Acceleration," *IEEE International Solid-State Circuits Conference (ISSCC)*, Feb. 2023, pp. 254–256. http://doi.org/10.1109/ISSCC42615.2023.10067285
[9] E. Garzón *et al.*, "A 128-kbit Approximate Search-Capable Content-Addressable Memory (CAM) With Tunable Hamming Distance," *IEEE Journal of Solid-State Circuits*, vol. 60, no. 8, pp. 3009–3019, Aug. 2025. http://doi.org/10.1109/JSSC.2025.3529715
[10] R. Guo *et al.*, "Accelerating Large-Scale Inference with Anisotropic Vector Quantization," Dec. 04, 2020, *arXiv*: arXiv:1908.10396. http://doi.org/10.48550/arXiv.1908.10396

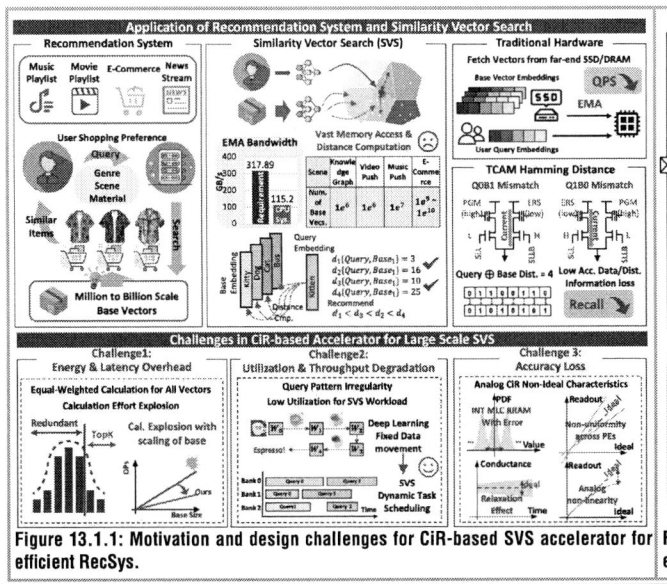

Figure 13.1.1: Motivation and design challenges for CiR-based SVS accelerator for efficient RecSys.

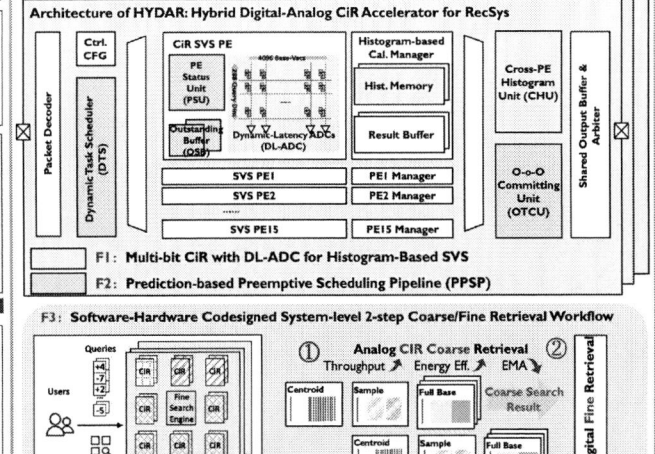

Figure 13.1.2: Overall architecture and key features of the HYDAR, and the CiR-featured end-to-end retrieval system.

ISSCC 2026 / SESSION 13 / CIRCUITS FOR AI AND AI FOR CIRCUITS / 13.1

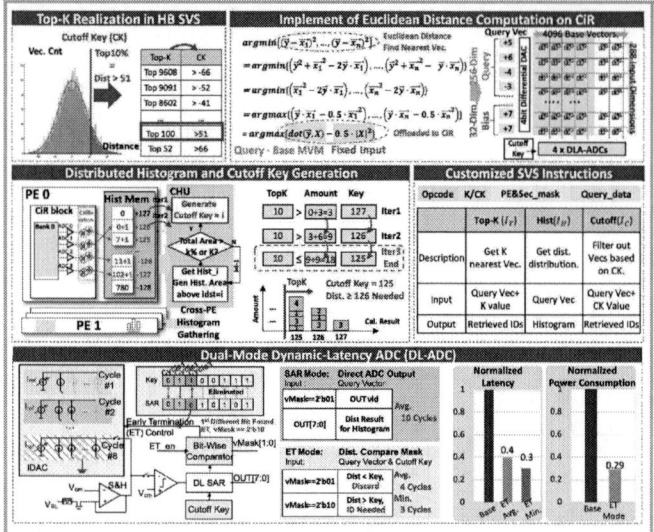

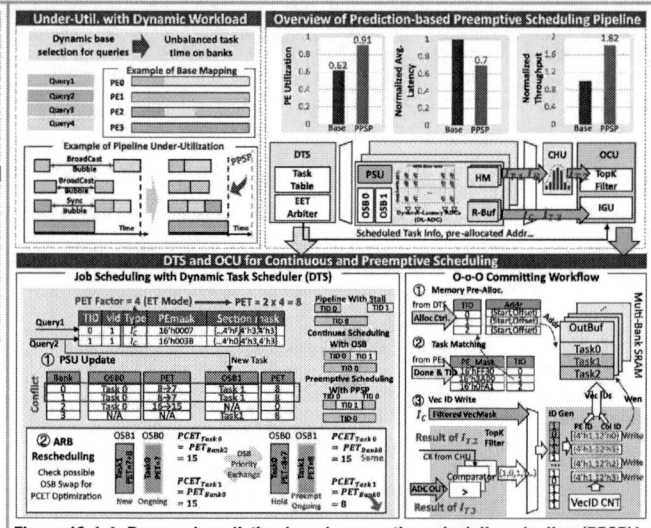

Figure 13.1.3: CiR implementation of histogram-based SVS with Dynamic-Latency ADC.

Figure 13.1.4: Proposed prediction-based preemptive scheduling pipeline (PPSP) for dynamic SVS workload.

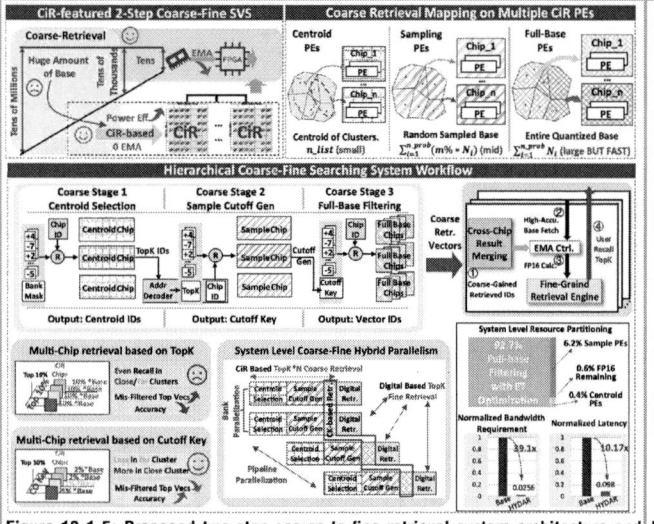

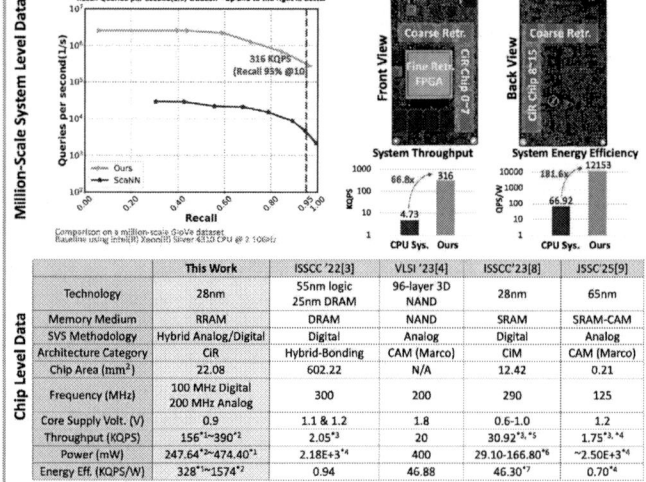

Figure 13.1.5: Proposed two-step coarse-to-fine retrieval system architecture and workflow.

Figure 13.1.6: System and chip performance evaluation and comparison with the state-of-the-art designs.

	This Work	ISSCC '22[3]	VLSI '23[4]	ISSCC'23[8]	JSSC'25[9]
Technology	28nm	55nm logic 25nm DRAM	96-layer 3D NAND	28nm	65nm
Memory Medium	RRAM	DRAM	NAND	SRAM	SRAM-CAM
SVS Methodology	Hybrid Analog/Digital	Digital	Analog	Digital	Analog
Architecture Category	CiR	Hybrid-Bonding	CAM (Marco)	CIM	CAM (Marco)
Chip Area (mm²)	22.08	602.22	N/A	12.42	0.21
Frequency (MHz)	100 MHz Digital 200 MHz Analog	300	200	290	125
Core Supply Volt. (V)	0.9	1.1 & 1.2	1.8	0.6-1.0	1.2
Throughput (KQPS)	156[*1]~390[*2]	2.05[*3]	20	30.92[*3, *5]	1.75[*1, *4]
Power (mW)	247.64[*1]~474.40[*1]	2.18E+3[*4]	400	29.10-166.80[*6]	~2.50E+3[*4]
Energy Eff. (KQPS/W)	328[*1]~1574[*2]	0.94	46.88	46.30[*7]	0.70[*4]

[*1]DL-ADC in SAR mode, [*2]DL-ADC in ET mode, [*3]Normalized to 4bit calculation, [*4] DDR data movement included; [*5]with 4 chiplets MCM, [*6]Per-chiplet result, [*7]At highest power;

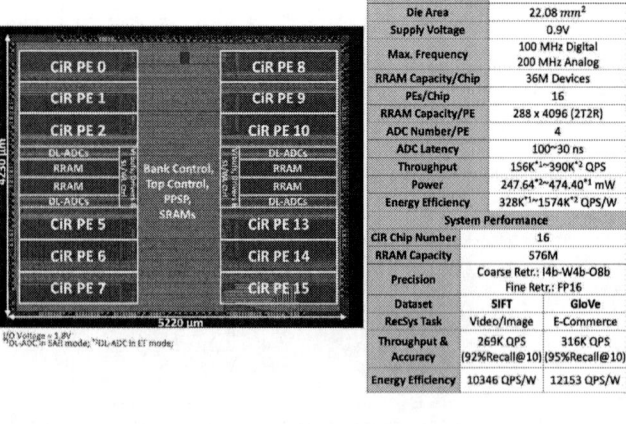

Chip Specifications	
Technology	28nm
Die Area	22.08 mm^2
Supply Voltage	0.9V
Max. Frequency	100 MHz Digital 200 MHz Analog
RRAM Capacity/Chip	36M Devices
PEs/Chip	16
RRAM Capacity/PE	288 x 4096 (2T2R)
ADC Number/PE	4
ADC Latency	100~30 ns
Throughput	156K[*1]~390K[*2] QPS
Power	247.64K[*2]~474.40K[*1] mW
Energy Efficiency	328K[*1]~1574K[*2] QPS/W
System Performance	
CiR Chip Number	16
RRAM Capacity	576M
Precision	Coarse Retr.: I4b-W4b-O8b Fine Retr.: FP16
Dataset	SIFT / GloVe
RecSys Task	Video/Image / E-Commerce
Throughput & Accuracy	269K QPS (92%Recall@10) / 316K QPS (95%Recall@10)
Energy Efficiency	10346 QPS/W / 12153 QPS/W

I/O Voltage = 1.8V
[*1]DL-ADC in SAR mode; [*2]DL-ADC in ET mode;

Figure 13.1.7: Die micrograph and summary.

• 2026 IEEE International Solid-State Circuits Conference

ISSCC 2026 / SESSION 13 / CIRCUITS FOR AI AND AI FOR CIRCUITS / 13.2

13.2 AI-Enabled End-to-End Design in RFICs with Controllable Architectural Style from 'Classical' to 'Non-Intuitive' for mm-Wave/sub-THz LNAs

Jonathan Zhou*[1], Emir Ali Karahan*[2], Juho Park[1], Sherif Ghozzy[1], Kaushik Sengupta[1]

[1]Princeton University, Princeton, NJ, [2]now with Marvell, Irvine, CA
*Equally Credited Authors (ECAs)

Abstract

This paper introduces a unified algorithmic design flow for low-noise amplifiers (LNAs), spanning specifications to layout and integrating topology, architecture, circuit, and electromagnetic (EM) design in one framework. The algorithm supports designer input from classical to unconventional architectures, enhancing interpretability, debugging, and usability. We demonstrate two LNAs covering 24 to 150GHz with state-of-the-art performance, including novel architectures uncovered by the algorithm.

This paper introduces an unified algorithmic design flow for low-noise amplifiers (LNAs) that spans from specifications to layout, integrating topology, architecture, circuit, and electromagnetic (EM) design in an end-to-end framework. The approach incorporates controllable EM interfaces ranging from classical transmission-line shapes to arbitrary pixelated structures, each offering distinct design trade-offs. Unlike prior AI-enabled inverse design methods for EM and passive synthesis [1-3], this framework supports controlled architectural evolution, enabling designer preference, interpretability, and seamless integration with surrounding circuits—facilitating debugging, fault isolation, and broader usability. In addition, this paper also demonstrates unique topology choices for the LNA that emerge as a result of the algorithmic approach that would not have been possible with a classical designer-driven synthesis.

Designing LNAs to meet targets for gain, matching, noise figure, stability, linearity, and power efficiency requires multi-dimensional optimization across architecture, topology, stage count, and device type (e.g., common-emitter/source or common-base/gate), along with EM co-design using lumped or distributed elements. Conventional RFIC design depends on fixed templates and iterative tuning; if unmet, new architectures are introduced, followed by circuit and EM optimization at layout to handle parasitics and coupling. This process is lengthy—often months—and constrained by designer intuition regarding feasible topologies and EM structures.

Here, we propose an end-to-end parasitic and EM-aware algorithmic design flow for RF/mmWave ICs, particularly focusing on LNA design that enables specifications to layout synthesis, allowing: 1) discovery of novel LNA architectures that are extremely challenging to achieve with human designs due to the lack of suitable templates, 2) topology and architecture selection, 3) LNA circuit and parameter optimization, 4) full circuit-EM co-design and synthesis covering both classical and non-intuitive structures, and 5) end-to-end optimization allowing specifications to layout-ready synthesis. We demonstrate the methodology with proof-of-concept LNAs spanning 30 to 160GHz. The first example is a frequency-diplexing LNA achieving near-optimal performance covering all mmWave frequency bands from 24-to-90 GHz, while overcoming the trade-offs associated with bandwidth, efficiency and noise, by employing highly efficient non-intuitive arbitrary-shaped three-port broadband frequency-diplexing matching networks. The LNA achieves an effective bandwidth of 24-to-90GHz with peak gain of 16dB and noise factor (NF) ranging from 3.8 to 5.7dB. It can be switched between a 24-to-55GHz low-frequency mode, a 55-to-90GHz high-frequency mode, or a combined path mode. We also present a fully AI-synthesized LNA using classical transmission-line-based matching networks exhibiting dual resonances with competitive bandwidth, noise figure, and gain, demonstrating peak gains of 30dB and 18dB at 85GHz and 150GHz respectively. To our knowledge, this is the first demonstration of a universal algorithmic flow capable of synthesizing RFICs from specifications to layout, validated through measurement of both classical and non-intuitive topologies.

The overarching vision of this work is summarized in Fig. 13.2.1. Mapping from an RFIC to its specifications is unique, whereas the reverse mapping—from specifications to a realizable RFIC (architecture, topology, circuit, EM, and parameters)—is inherently one-to-many. Multiple designs can satisfy the same specifications, with final choices often shaped by subjective considerations such as explainability, simplicity, area, sensitivity, debugging, and interfacing with ancillary circuits. Thus, controlling the evolution of architecture is essential for the broader adoption of AI-enabled RFIC design. Figure 13.2.2 outlines the proposed flow for LNA design. Specifications and architectural style (classical transmission-line vs. pixelated) serve as inputs. For pixelated architectures, synthesis proceeds in two stages: (1) reinforcement learning (RL) explores optimal architectures, including topology, stage configuration, device sizing, biasing, and interface impedances to meet specifications and (2) inverse EM design generates structures that satisfy target scattering parameters in the chosen "style". We demonstrate two strategies for pixelated synthesis: one employs a forward AI model with metaheuristics to generate EM structures, while the other incorporates transmission-line parameters directly into the RL process, enabling joint circuit–EM optimization. Through this, the designer can enforce their own constraints on topology, architecture, and interpretability, where prior inverse works did not allow for EM structure "type" selection between pixelated or classical [4-6].

Figure 13.2.2 depicts the integration of reinforcement learning (RL) to optimize circuit architecture, topology, parameters, and impedance matching, with the latter subsequently dispatched to the inverse algorithm for optimization and actual structure synthesis. The encoded design space accommodates architectural options (e.g., number of stages), topological variations (e.g., common-emitter, common-base, cascode), device-level parameters (e.g., transistor sizing, biasing), and interface impedances, while also incorporating operating classes and layout-aware constraints for joint circuit and electromagnetic (EM) synthesis. As shown in Figure 13.2.2, the RL engine is formulated as a Markov decision process where the policy network maps design states and performance metrics to a probability distribution over actions, guiding optimization toward multi-objective specifications. The reward function combines hard and soft targets, penalizing deviations from constraints. Circuit parameters are linked with an AI-assisted inverse network that generates compact EM structures consistent with desired S-parameters [7], thereby extending the design space beyond conventional heuristics. Training with ~350,000 circuit–EM examples requires ~24 hours on a 192-core cluster using proximal policy optimization (PPO) [8] with a three-layer multilayer perceptron (MLP) actor of 128 neurons per layer and a three-layer MLP critic of 256 neurons per layer. The training data is generated directly from the model's interactions with a commercial circuit simulator and a custom EM surrogate model, resulting in high accuracy. Data to train the EM surrogate model comes from simulating random examples within the design space in a commercial EM simulator. Once trained, output-stage layouts are synthesized in 5–10 minutes, while a complete low-noise amplifier (LNA) design, including auxiliary circuitry, is achieved within hours. Partitioning circuit and EM synthesis enables efficient end-to-end optimization. It is important to note that the architectural choice between classical and non-intuitive does not come at a cost performance. As can be seen in Fig. 13.2.3, different EM structures (both classical and non-classical) can achieve similar EM performance that can be guided through an algorithm. Compared to [9] for PCB based filter designs, this work avoids the training of any diffusion models, and demonstrates on-chip realization of mmWave/sub-THz passives that are integrated with the circuit synthesis in a tight loop.

AI not only allows synthesis of particular circuits, but also allows canvassing of possible designs in a rapid fashion, allowing us to evaluate the Pareto front achievable in a given process design kit. This is often an overlooked benefit. Figure 13.2.4 demonstrates the RL-based synthesis of LNAs with the proposed methods demonstrating achieved target gains and NF with the minimum power dissipation. It is to be noted that due to the parasitic and EM-aware training process, the reported numbers are the final end-to-end simulated RF performance for the GDS-ready circuit. As can be seen, the simulated performances are competitive with state of the art, demonstrating that the RL algorithm has been able to learn the design space effectively. The example designs in the figure demonstrate that the algorithm can effectively choose circuit topologies, bias them properly for optimal noise performance, add gain stages when needed, and synthesize the optimal impedance as well. More importantly, as shown in the figure, the design allows us to quantify the Pareto front of the design space. The figure shows the trade-off of NF and gain with DC power, and other considerations such as linearity, area, robustness, sensitivity, stability measures can easily be added, allowing a designer to evaluate the PDK and get to the optimal design orders of magnitudes faster.

The paper presents two design examples of LNAs demonstrating the pixelated and classical approaches, particularly focusing on how algorithms can allow synthesis of architectures that would be very challenging for a human designer. The first LNA is designed to cover the mmWave spectrum from 24 to 90GHz for joint sensing and communication, covering 5G mmWave, satellite communication, and radar as a broadband front-end. Broadband design presents key challenges, particularly achieving simultaneous power and noise matching across wide frequencies. Matching networks must provide flat gain while incurring minimal loss and noise penalties. Leveraging arbitrary multi-port synthesis, we introduce a frequency-diplexing LNA that separates matching into two bands: 24 to 55GHz and 55 to 90GHz. The two-stage design employs cascode and common-emitter stages, with input/output networks enabling diplexing. The input directs the 24-to-55GHz band through the upper branch for optimal power/noise match without leakage, while the 55-to-90GHz band is directed oppositely; the output combiner mirrors this behavior. Designing such low-loss broadband networks is difficult due to absent templates, but the algorithmic flow enables their realization. As shown in the measurement results in Fig. 13.2.6, the LNA effectively realizes 16dB gain across a record bandwidth of 24 to 90GHz, with state of the art NF (3.8 to 6.8dB) across the entire spectrum matching very closely with simulations,

230 • 2026 IEEE International Solid-State Circuits Conference

979-8-3315-8937-0/26 $31.00 © 2026 IEEE

dissipating approximately 24mW of power in the low-band mode and 32mW of power in the high-band mode.

To demonstrate the synthesis with classical t-lines, we present a dual peaking LNA between 85GHz and 150GHz allowing for dual purpose usage at radar and 6G frequencies. Designing multi-peaking requires careful tuning to align the matching while providing simultaneous noise-optimal matching, all of which are taken into account during the RL training. The measurement results are seen in Fig. 13.2.6, where the dual peaking LNA realizes a peak gain of 30dB and minimum NF of 5.8dB at 85GHz, and a peak gain of 18dB at 160GHz, with a power dissipation of 63mA.

A table of comparison with state of the art is shown Fig. 13.2.7. The presented LNAs demonstrate state of the art performance and are synthesized for the first time through an end-to-end AI algorithm that allows controllable architectural choice as designer inputs for broad applicability.

Acknowledgement:
The authors would like to acknowledge GlobalFoundries for chip fabrication support, and the funding agencies: Army Research Office (W911NF2110314, W911NF2410111), Air Force Office of Scientific Research (FA9550-23-1-0176), Office of Naval Research (N00014-23-1-2592), DURIP funding (N00014-23-1-2332), and National Science Foundation (CNS-2211617, CNS-2148271, CNS-2402782). Authors also would like to thank Bert Harrop from the Princeton University Physics Department for professional assistance with packaging, and Princeton ECE IMRL for design support and discussions.

References:
[1] V. Chenna and H. Hashemi, "Topology-Optimized Nonintuitive Multilayered mm-Wave Power Amplifiers," *IEEE Radio Frequency Integrated Circuits Symposium (RFIC)*, 2025, pp. 279-282, doi: 10.1109/RFIC61188.2025.11082875. https://doi.org/10.1109/RFIC61188.2025.11082875

[2] E. A. Karahan, Z. Liu and K. Sengupta, "Deep-Learning-Based Inverse-Designed Millimeter-Wave Passives and Power Amplifiers," *IEEE Journal of Solid-State Circuits*, vol. 58, no. 11, pp. 3074-3088, Nov. 2023, doi: 10.1109/JSSC.2023.3276315. http://doi.org/10.1109/JSSC.2023.3276315.

[3] J. Zhou, E. A. Karahan, S. Ghozzy, Z. Liu, H. Jalili and K. Sengupta, "AI-Enabled Design Space Discovery and End-to-End Synthesis for RFICs with Reinforcement Learning and Inverse Methods Demonstrating mm-Wave/sub-THz PAs Between 30 and 120GHz," *IEEE International Solid-State Circuits Conference (ISSCC)*, Feb. 2025, pp. 1-3, doi: 10.1109/ISSCC49661.2025.10904600. http://doi.org/10.1109/ISSCC49661.2025.10904600.

[4] K. Settaluri, A. Haj-Ali, Q. Huang, K. Hakhamaneshi and B. Nikolic, "AutoCkt: Deep Reinforcement Learning of Analog Circuit Designs," *Design, Automation & Test in Europe Conference & Exhibition (DATE)*, 2020, pp. 490-495, doi: 10.23919/DATE48585.2020.9116200. http://doi.org/10.23919/DATE48585.2020.9116200

[5] E.A. Karahan et al., Deep-learning enabled generalized inverse design of multi-port radio-frequency and sub-terahertz passives and integrated circuits. *Nat Commun* 15, 10734 (2024). https://doi.org/10.1038/s41467-024-54178-1.

[6] H. Wang et al., "GCN-RL Circuit Designer: Transferable Transistor Sizing with Graph Neural Networks and Reinforcement Learning," *57th ACM/IEEE Design Automation Conference (DAC)*, 2020, pp. 1-6, doi: 10.1109/DAC18072.2020.9218757. http://doi.org/10.1109/DAC18072.2020.9218757

[7] C. Chu et al., "AI-Assisted Template-Seeded Pixelated Design for Multi-Metal-Layer High-Coupling EM Structures: A Ku-Band 6G FR3 PA in 22nm FDX+," *IEEE/MTT-S International Microwave Symposium - IMS 2025*, 2025, pp. 922-925, doi: 10.1109/IMS40360.2025.11103802. http://doi.org/10.1109/IMS40360.2025.11103802

[8] J. Schulman, F. Wolski, P. Dhariwal, A. Radford, and O. Klimov, "Proximal Policy Optimization Algorithms," arXiv.org, Aug. 28, 2017. https://arxiv.org/abs/1707.06347.

[9] Y. Guo et al., "Dall-EM: Generative AI with Diffusion Models for New Design Space Discovery and Target-To-Electromagnetic Structure Synthesis," *IEEE/MTT-S International Microwave Symposium - IMS 2025*, 2025, pp. 926-929, doi: 10.1109/IMS40360.2025.11103838. http://doi.org/10.1109/IMS40360.2025.11103838.

[10] J. Park and H. Wang, "A 26-to-39GHz Broadband Ultra-Compact High-Linearity Switchless Hybrid N/PMOS Bi-Directional PA/LNA Front-End for Multi-Band 5G Large-Scaled MIMO System," *IEEE International Solid-State Circuits Conference (ISSCC)*, Feb. 2022, pp. 322-324, doi: 10.1109/ISSCC42614.2022.9731651. http://doi.org/10.1109/ISSCC42614.2022.9731651

[11] Z. Zhao et al., "Design and Analysis of a 22.6-to-73.9 GHz Low-Noise Amplifier for 5G NR FR2 and NR-U Multiband/Multistandard Communications," *IEEE Journal of Solid-State Circuits*, vol. 60, no. 9, pp. 3189-3201, Sept. 2025, doi: 10.1109/JSSC.2025.3545463. http://doi.org/10.1109/JSSC.2025.3545463

[12] L. Mendes, J. Silva, N. Lourenço, J. C. Vaz, R. Martins and F. Passos, "Fully Automatically Synthesized mm-Wave Low-Noise Amplifiers for 5G/6G Applications," *IEEE Transactions on Microwave Theory and Techniques*, vol. 73, no. 8, pp. 4828-4841, Aug. 2025, doi: 10.1109/TMTT.2025.3537588. http://doi.org/10.1109/TMTT.2025.3537588

[13] Y. Yu, H. Liu, Y. Wu and K. Kang, "A 54.4–90 GHz Low-Noise Amplifier in 65-nm CMOS," *IEEE Journal of Solid-State Circuits*, vol. 52, no. 11, pp. 2892-2904, Nov. 2017, doi: 10.1109/JSSC.2017.2727040. http://doi.org/10.1109/JSSC.2017.2727040

[14] B. Bae, E. Kim, S. Kim and J. Han, "Dual-Band CMOS Low-Noise Amplifier Employing Transformer-Based Band-Switchable Load for 5G NR FR2 Applications," *IEEE Microwave and Wireless Technology Letters*, vol. 33, no. 3, pp. 319-322, March 2023, doi: 10.1109/LMWC.2022.3218001. http://doi.org/10.1109/LMWC.2022.3218001

[15] C. Han, J. Zhou, Z. Deng, Y. Shu and X. Luo, "A 4.8dB NF, 70-to-86GHz Deep-Noise-Canceling LNA Using Asymmetric Compensation Transformer and 4-to-1 Hybrid-Phase Combiner in 40nm CMOS," *IEEE International Solid-State Circuits Conference (ISSCC)*, Feb. 2023, pp. 24-26, doi: 10.1109/ISSCC42615.2023.10067549. http://doi.org/10.1109/ISSCC42615.2023.10067549

[16] Y. Zhang, X. Tang, Z. Wei, X. Feng and F. Huang, "A 58–110 GHz 4.2 dB Minimum NF CMOS LNA With Broadband Simultaneous Noise and Impedance Matching," *IEEE Microwave and Wireless Technology Letters*, vol. 34, no. 5, pp. 504-507, May 2024, doi: 10.1109/LMWT.2024.3373618. http://doi.org/10.1109/LMWT.2024.3373618

[17] G. D. Filippi, L. Piotto and A. Mazzanti, "A D-Band LNA Exploiting Ultra-Wideband Sixth-Order Matching Networks in SiGe BiCMOS," *IEEE European Solid-State Electronics Research Conference (ESSERC)*, 2024, pp. 705-708, doi: 10.1109/ESSERC62670.2024.10719560. http://doi.org/10.1109/ESSERC62670.2024.10719560

[18] S. Li, T. Chi, D. Jung, T. -Y. Huang, M. -Y. Huang and H. Wang, "An E-Band High-Linearity Antenna-LNA Front-End with 4.8dB NF and 2.2dBm IIP3 Exploiting Multi-Feed On-Antenna Noise-Canceling and Gm-Boosting," *IEEE International Solid-State Circuits Conference - (ISSCC)*, Feb. 2020, pp. 1-3, doi: 10.1109/ISSCC19947.2020.9063008. http://doi.org/10.1109/ISSCC19947.2020.9063008

[19] A. Moradinia, S. G. Rao and J. D. Cressler, "A SiGe HBT D-Band LNA Utilizing Asymmetric Broadside Coupled Lines," *IEEE Microwave and Wireless Technology Letters*, vol. 33, no. 6, pp. 707-710, June 2023, doi: 10.1109/LMWT.2023.3247792. http://doi.org/10.1109/LMWT.2023.3247792

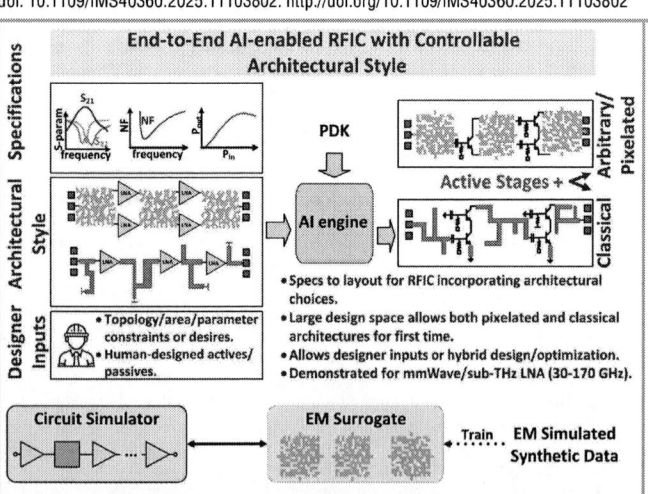

Figure 13.2.1: RFIC synthesis from specifications to layout allowing designer choice of either classical or non-intuitive pixelated architecture for interpretability and debugging. A commercial circuit simulator and EM surrogate are used for accurate results.

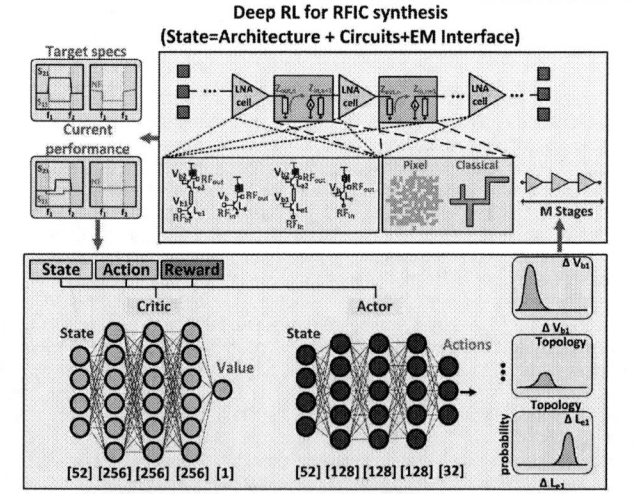

Figure 13.2.2: Deep reinforcement learning algorithm to allow optimization over architecture, circuit topology, parameters and full EM-circuit co-design incorporating classical t-line optimization and pixelated architecture.

ISSCC 2026 / SESSION 13 / CIRCUITS FOR AI AND AI FOR CIRCUITS / 13.2

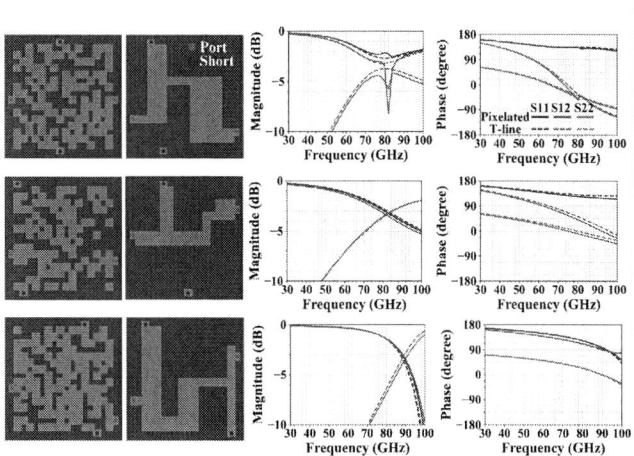

Figure 13.2.3: The same EM performance can be realized by either inverse designed pixelated structures or more 'classical' transmission-line based structures as demonstrated by the example in 90nm SiGe BiCMOS showing near-identical performance across all scattering parameters.

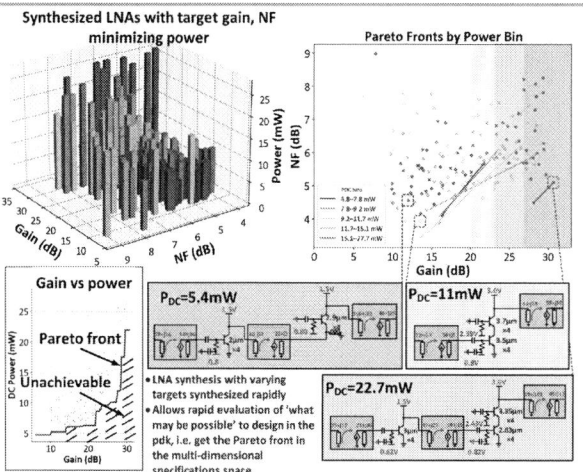

Figure 13.2.4: Algorithmic synthesis approaches can demonstrate the 'Pareto fronts' for RFICs indicating what might be possible to synthesize in the process design kit, as shown in the synthesized examples of LNA achieving varying gain and NF, while minimizing power consumption.

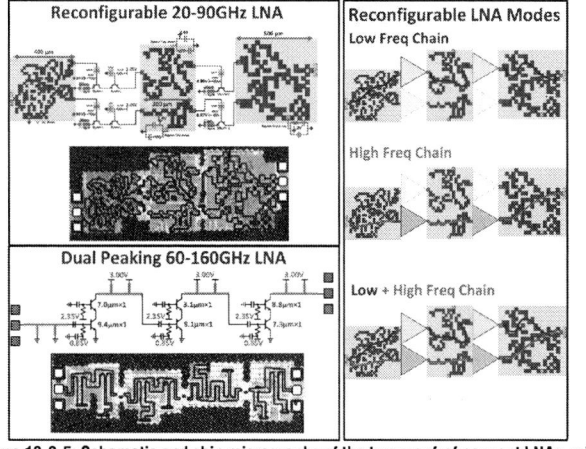

Figure 13.2.5: Schematic and chip micrographs of the two proof-of-concept LNAs, with the first example demonstrating a frequency diplexing architecture (enabling NF minimization) covering all mmWave bands between 20-to-90GHz aided by an inverse designed 3-port structure, and the second demonstrating a dual band LNA with a more classical architecture with peaks at 85 and 150 GHz.

Figure 13.2.6: Measurement results demonstrating the switched-mode frequency diplexing operating from 20-to-90GHz with state-of-the-art NF, and measurements of the dual band LNA.

Reference	This work: 24-90GHz Reconfigurable LNA	[10] Park, ISSCC'22	[11] Zhao, JSSC'25	[12] Mendes, MTT'25	[13] Yu, JSSC '17	[14] Bae, MWTL'23
Technology	90nm SiGe	45nm CMOS SOI	130nm SiGe	65nm CMOS	65nm CMOS	65n CMOS
Design Method	End-to-end AI flow (topology, circuits, designer-input on classical/pixelated)	Manual	Manual	Automated layout & parameter optimization	Manual	Manual
Architecture	Frequency-duplexing LNA with 2-stage (Cascode, Common emitter) as decided by AI	Bi-directional PA/LNA	2-stage with feedback/compensation	2-stage CS cells	4-stage CS cells with transformers	Transformer-based band-switching load
Frequency/BW (GHz)	Combined: 24-90 Mode 1: 24-55 GHz Mode 2: 55-90 GHz	27-38	22.8-73.9	22.3-29.6/30.7	54.4-90	26-30/37-40
Gain (dB)	Mode 1: 14-17, Mode 2: 14-16	17.8	15.2	10/13.9	17.7	13.66/13.04
NF (dB)	Mode 1: 3.8-5, Mode 2: 4-6.8	5.2-7.8	4.06-4.94	3.8/3.47 @28GHz	5.4-7.4	3.84/4.47
DC Power (mW)	Mode 1: 24, Mode 2: 32.4	66	17.5	2.26/4.8	19	11.6
Core Area (mm²)	0.70	0.19	0.06	0.29/0.30	0.37 (with pads)	0.17

Reference	This work: 80-160GHz Dual Peaking LNA	[15] Han, ISSCC'23	[16] Zhang, MWL'24	[17] De Filippi, JSSC'25	[18] Li, ISSCC'20	[19] Moradinia, MWTL'23
Technology	90nm SiGe	40nm CMOS	40nm CMOS	55nm SiGe	45nm CMOS SOI	90nm SiGe
Design Method	End-to-end AI flow (topology, circuits, designer-input on classical/pixelated)	Manual	Manual	Manual	Manual	Manual
Architecture	3-stage Cascode cells with RL-optimized T-lines	4-way noise-canceling combiner with CG cells	3-stage differential CS cells	3-stage Cascode cells	Noise-canceling CG/CS pair	5-stage staggered CE and Cascode cells
Frequency/BW (GHz)	75-95, 150	79-96	70-91 ($S_{21,3dB}$)	105-175	73-88	120-157.6
Gain (dB)	30/18	16.5	16.4	23	16.8	26.5
NF (dB)	5.8 @85GHz	4.8 to 6.5	4.2-6.9	5-6.5	4.8-6.1	7.2
DC Power (mW)	63	25	20.7	60	46	20.6
Core Area (mm²)	0.46	0.06	0.163	0.109	0.63 (with on-chip antenna)	0.69 (with pads)

Figure 13.2.7: Comparison of the LNAs with the state of the art, demonstrating new architectures, record bandwidth, and competitive performance.

ISSCC 2026 / SESSION 13 / CIRCUITS FOR AI AND AI FOR CIRCUITS / 13.3

13.3 Medusa: A Quantum-Inspired 200-Variable 1016-Clause Analog *k*-SAT Solver

Luke D. Wormald, Ying-Tuan Hsu, Evangelos Dikopoulos, Wei Tang, Benjamin Datsko, Ali Hammoud, Zhengya Zhang, Michael P. Flynn

University of Michigan, Ann Arbor, MI

Abstract

A quantum-inspired analog variable *k*-SAT solver supporting up to 200 variables and 1016 clauses. Enabling techniques include make/break feedback, distributed *k*-SAT logic, digital macro coupling, and feedback optimization. The 28nm CMOS prototype achieves a 4.92µs mean solution time and a 19.1nJ mean energy consumption for 50-variable 3-SAT problems with 100% solvability and accuracy, representing 3.5× and 3× improvements in energy and solution time respectively, compared to the state-of-the-art.

Boolean satisfiability (SAT or *k*-SAT, *k* 3) problems are essential combinatorial optimization problems (COPs) for artificial intelligence [1], drug discovery [2], scheduling, and automated design [3]. SAT problems solve for binary variables such that all clauses are true, with each clause comprising variables or their negations (e.g., $x_1 \vee \neg x_2 \vee x_9$). However, SAT problems are NP-Complete, making them difficult for conventional computing. Quantum systems can tackle SAT problems, but are limited by immature technology and restricted qubit connectivity, hindering the ability to address k-SAT problems [4]. Most silicon digital and analog SAT solver architectures are limited to 3-SAT operation (i.e., 3 variables in each clause) [5], [6], require extensive preprocessing [5], or are restricted in problem size (often only 50 variables). Some recent solvers [7], [8] address *k*-SAT (i.e., *k* variables per clause) but are limited in size, hindering their ability to address *k* > 3 problems. [9] implements a larger *k*-SAT solver but does not demonstrate functionality beyond 60 variables. We overcome these obstacles and introduce a quantum-inspired, mixed *k*-SAT solver that is 4× larger than the state-of-the-art [8] while achieving 3.5× better solution times and 3× better energy efficiency. Furthermore, we demonstrate 250ns and 95.7µs solution times for the mixed *k*-SAT encoded Blocks World Planning [10], [11] "anomaly" and "medium" problems respectively, demonstrating the potential of this work to address real-world problems.

Supporting *k*-SAT with large and mixed values of *k* is essential for solving real-world optimization problems without decomposition or costly preprocessing. This work introduces circuit architecture and design innovations (Fig. 13.3.1) to support large and varying *k*, overcome the size and scaling limitations of the state-of-the-art, and improve performance and efficiency: (i) Dual *Make* and *Break* feedback improves system performance and stability, particularly for larger problem sizes. Adding *Break* feedback improves the solution time and accuracy for all problems when compared with only using *Make* feedback. (ii) Distributed clause logic enables a scalable system with all-to-all *k*-SAT connectivity and supports mixed *k* problems. (iii) Digital coupling between spins in separate compute macros enables the solver to scale beyond the practical limits of a single spin array. (iv) Several innovative architectural and layout techniques alleviate the impact of parasitic delays in the feedback system. This is critical as feedback delay limits the speed and scale of the system.

The solver (Fig. 13.3.2) comprises two mixed-signal compute macros, each containing 200 Gated Relaxation Oscillator (GRXO) based spins and supporting up to 508 *k*-SAT clauses. Each macro can function independently as a 200-variable 508-clause machine, or the macros can couple together to tackle *k*-SAT problems of up to 200 variables and 1016 clauses. The solver algorithm is directly implemented in the hardware and requires no pre- or post-processing, unlike [5]. An on-chip RISC-V core configures the mixed-signal macros but takes no part in the solver computation. The mixed-signal hardware operates entirely asynchronously, allowing it to take full advantage of the high-resolution continuous-time feedback and massive parallelism of analog computing. We implement a one-to-one variable-to-spin mapping where each variable is represented by a single GRXO-based spin, eliminating the need for any complex problem embedding. The *k*-SAT clauses form the mixed-signal feedback system, which receives the digital spin states and returns analog feedback to perturb the spins. During a solver run (Fig. 13.3.2), the RISC-V writes the problem information to the clauses and then enables the oscillators. The oscillators and clauses evolve until no errors are detected, after which the sampling system captures the oscillator states to be later read out by the RISC-V.

Each macro in the system comprises a central array of GRXO-based spins and two split banks of clauses with mixed-signal feedback. The GRXO (Fig. 13.3.3) is based on the twin-capacitor relaxation oscillator (RXO) [12]. RXOs are more PVT robust and flexible than oscillators, making them an excellent candidate for dynamical system solvers [13]. A recent SAT solver [6] proposes a GRXO system without continuous oscillation to reduce energy consumption, increase performance, and simplify sampling. This is done by gating the GRXO bias in the absence of any feedback, so that the output remains static. However, [6] relies on additional circuitry, including ADCs and multiple DACs in the feedback path, which increases energy consumption and feedback delay. To address this limitation, we directly inject the analog feedback currents into the GRXO capacitors, thereby increasing the resolution and dynamic range of the feedback while eliminating the need for data converters in the feedback path.

One of the key innovations of this work (Fig. 13.3.1) is the use of opposing make and break feedback mechanisms to improve stability and performance. This approach is inspired by software SAT solvers [14], which count the number of clauses that become satisfied (the make-count) or unsatisfied (the break-count) if a given variable changes state. We adopt this terminology to describe the two feedback mechanisms within our solver. If a clause is unsatisfied, we apply a *Make* current (I_{MAKE}) to each spin comprising that clause, thereby stimulating those spins to satisfy the clause. When a clause is satisfied by more than one variable, no feedback is applied to the associated spins. However, a small *Break* current (I_{BREAK}) is applied to the satisfying spin when a clause is satisfied by one and only one variable. The *Break* feedback subtracts from the *Make* feedback for a given spin, thereby discouraging it from changing state and breaking the satisfaction of the clause. The *Break* feedback guides the system to a final satisfied state and contributes inertia to the system, reducing the risk of unstable oscillations. The current subtractor (Fig. 13.3.3) also acts as a buffer, preventing the *Break* current from depleting charge already integrated onto the active capacitor in the GRXO. The use of *Break* feedback in addition to *Make* feedback improves the mean solution time of the solver by 30× for 50-variable 3-SAT problems.

Distributed clause logic (Fig. 13.3.4) implements *k*-SAT functionality (i.e., with arbitrary clause sizes) and greatly improves system scalability while maintaining all-to-all connectivity. Each clause can evaluate all 200 spins for both clause satisfaction and possible *Break* condition, and comprises 200 programmable clause cells. The clause cells contain logic for variable selection, negation, and feedback, and the analog circuitry to provide feedback to the connected spin. Instead of constructing a 200-input OR gate in a conventional binary tree manner, we cascade 200 two-input OR gates together. Cascading gates drastically reduces the interconnect overhead to just one net connecting between each clause cell for the *Make* logic, at the cost of increased maximum delay and delay variation. Figure 13.3.4 illustrates the distributed logic, where the topmost OR gate determines *Make* and the lower AND/OR logic determines *Break*. The logic required to evaluate the *Break* condition is distributed similarly, resulting in fully decentralized clause logic. Likewise, the analog feedback circuitry is also fully decentralized, with current sources placed in each clause cell. This decentralized design scales linearly with the total number of variables in one dimension and with the total number of clauses in the other.

Inter-macro spin coupling is a key innovation for scaling the size of the SAT solver. Strongly coupling the corresponding spins in different macro locks them in phase. As the system traverses the solution space, updates in one macro propagate to the other. We implement digital spin coupling to minimize delay. In each spin, a state machine evaluates the condition of the spin and its counterpart. When one spin changes state, it immediately forces its counterpart to change state with a hard, digital flip. The state machine in Fig. 13.3.3 prevents undesirable oscillations from rapid state changes. Once a spin state change is issued, the state machine prevents any additional flips until the coupled spin is updated. When spins are coupled, the system effectively shares the clauses between both macros. This work demonstrates the potential for scaling beyond a single compute macro to larger, multi-macro systems and chiplet implementations.

We introduce several circuit and layout techniques to reduce the delay and improve the efficiency of the current feedback. These techniques are essential because the delay limits scalability. First, a low-power active-cascode stage (Fig. 13.3.3) terminates each current summation line (I_{MAKE}, I_{BREAK}), fixing the common-mode voltage of the line and reducing the effect of the large 130fF capacitance by greatly reducing the voltage swing. Second, the mixed-signal feedback system is split into two separate groups of clauses, with half above and half below the GRXOs. Splitting into two halves the parasitic delay of the current summation lines at the cost of doubling the number of active-cascodes for each GRXO. We mitigate the extra cascode power consumption by disabling them when their associated clauses are not needed. Third, a switch placed in the middle of each summation line allows half the parasitic load to be disconnected when the clauses are not in use.

The prototype solver (Fig. 13.3.7) is implemented in 28nm CMOS and has an active core area of 2.59mm² for the two-macro system. No pre- or post-processing is required or utilized. The prototype was evaluated across a wide range of challenging benchmarks (Fig. 13.3.5), including the popular SATLIB library of 3-SAT problems [10] and a library of

232 • 2026 IEEE International Solid-State Circuits Conference

979-8-3315-8937-0/26 $31.00 © 2026 IEEE

ISSCC 2026 / February 17, 2026 / 8:50 AM

k-SAT problems generated in the same uniform random, unforced manner near the SAT phase transition. With the commonly used 50-variable uf50-218 SATLIB benchmark, the prototype achieves a 4.92µs mean time to solution (TTS) and a 19.1nJ mean energy to solution (ETS) (Fig. 13.3.6). This surpasses both the performance and energy efficiency of the previous state-of-the-art [8] (also a *k*-SAT solver) by 3.5× and 3×, respectively. For the 200-variable uf200-860 benchmark, the prototype achieves a 30.9ms TTS, which is less than half the time required for the D-Wave quantum solver [4] for problems of the same size. Among *k*-SAT problems (for k > 3), the prototype demonstrates excellent scaling and achieves a 33.0ms TTS for 100 variable 4-SAT problems. This solver introduces a high-performance, energy-efficient, and scalable architecture to address mixed *k*-SAT problems with up to 200 variables, achieves better than state-of-the-art solution time and energy, and surpasses the performance of quantum systems.

Acknowledgement:
This work was supported by DARPA QuICC.

References:
[1] J. Rintanen, K. Heljanko, and I. Niemelä, "Planning as satisfiability: parallel plans and algorithms for plan search," *Artif. Intell.*, vol. 170, no. 12–13, pp. 1031–1080, 2006. https://doi.org/10.1016/j.artint.2006.08.002
[2] N. Ollikainen, E. Sentovich, C. Coelho, A. Kuehlmann, and T. Kortemme, "SAT-based protein design," *IEEE/ACM International Conference on Computer-Aided Design*, 2009, pp. 128–135. https://ieeexplore.ieee.org/abstract/document/5361301/?casa_token=u-4CGznX9VEAAAAA:tsiHlnZ4PQsN080I6RtKAnvXjcqGroxlGldPG4Rz-DWlPQib3qvGXSqCJdW5LeoXJgGTfWj7
[3] J. P. Marques-Silva and K. A. Sakallah, "Boolean satisfiability in electronic design automation," *Proceedings of the 37th conference on Design automation - DAC '00*, ACM Press, 2000, pp. 675–680. https://doi.org/10.1145/337292.337611
[4] S. Tan *et al.*, "HyQSAT: A Hybrid Approach for 3-SAT Problems by Integrating Quantum Annealer with CDCL," *IEEE International Symposium on High-Performance Computer Architecture (HPCA)*, Feb. 2023, pp. 731–744. https://doi.org/10.1109/HPCA56546.2023
[5] C. Shim, J. Bae, and B. Kim, "VIP-Sat: A Boolean Satisfiability Solver Featuring 5×12 Variable In-Memory Processing Elements with 98% Solvability for 50-Variables 218-Clauses 3-SAT Problems," *IEEE International Solid-State Circuits Conference (ISSCC)*, Feb. 2024, pp. 486–488. https://doi.org/10.1109/ISSCC49657.2024.10454397
[6] E. Dikopoulos, Y. Hsu, L. Wormald, W. Tang, Z. Zhang, and M. P. Flynn, "A Physics-Inspired Oscillator-Based Mixed-Signal Optimization Engine for Solving 50-Variable 218-Clause 3-SAT Problems with 100% Solvability and 31.7µs Solution Time," *IEEE International Solid-State Circuits Conference (ISSCC)*, Feb. 2025, pp. 01–03. https://doi.org/10.1109/ISSCC49661.2025.10904814
[7] T. Bhattacharya, D. Kwon, G. H. Hutchinson, X. Zhang, I. Rozada, and D. Strukov, "A Fully Integrated Mixed-Signal Compute-In-Memory Accelerator for Solving Arbitrary Order Boolean Satisfiability Problems," *Symposium on VLSI Technology and Circuits (VLSI Technology and Circuits)*, IEEE, 2025, pp. 1–3. https://doi.org/10.23919/VLSITechnologyandCir65189.2025.11074791
[8] Z. Wu *et al.*, "SKADI: A 28nm Complete K-SAT Solver Featuring Dual-Path SRAM-Based Macro and Incremental Update with 100% Solvability," *IEEE International Solid-State Circuits Conference (ISSCC)*, Feb. 2025, pp. 614–616. https://doi.org/10.1109/ISSCC49661.2025.10904686

[9] S. Xie *et al.*, "snap-SAT: A one-shot energy-performance-aware all-digital compute-in-memory solver for large-scale hard Boolean satisfiability problems," *IEEE International Solid-State Circuits Conference (ISSCC)*, 2023, pp. 420–422. https://doi.org/10.1109/ISSCC42615.2023.10067380
[10] H. H. Hoos and T. Stützle, "SATLIB: An online resource for research on SAT," *Sat*, vol. 2000, pp. 283–292, 2000. https://www.cs.ubc.ca/~hoos/Publ/sat2000-satlib.pdf
[11] American Association for Artificial Intelligence, Ed., *Proceedings of the Thirteenth National Conference on Artificial Intelligence and the Eighth Innovative Applications of Artificial Intelligence Conference:* August 4 - 8, 1996, AAAI Press [u.a.]. https://aaai.org/proceeding/aaai-13-1996/
[12] M. P. Flynn and S. U. Lidholm, "A 1.2-µm CMOS current-controlled oscillator," *IEEE J. Solid-State Circuits*, vol. 27, no. 7, pp. 982–987, 2003. https://doi.org/10.1109/4.142592
[13] E. Dikopoulos *et al.*, "RXO-LDPC: A Physics-Inspired Relaxation Oscillator-Based Solver Leveraging Six-Body Spin Interactions for Soft Decoding of LDPC Codes," *IEEE J. Solid-State Circuits*, 2025. https://doi.org/10.1109/JSSC.2025.3561780
[14] C. P. Gomes, H. Kautz, A. Sabharwal, and B. Selman, "Satisfiability solvers," *Found. Artif. Intell.*, vol. 3, pp. 89–134, 2008. https://doi.org/10.1016/S1574-6526(07)03002-7

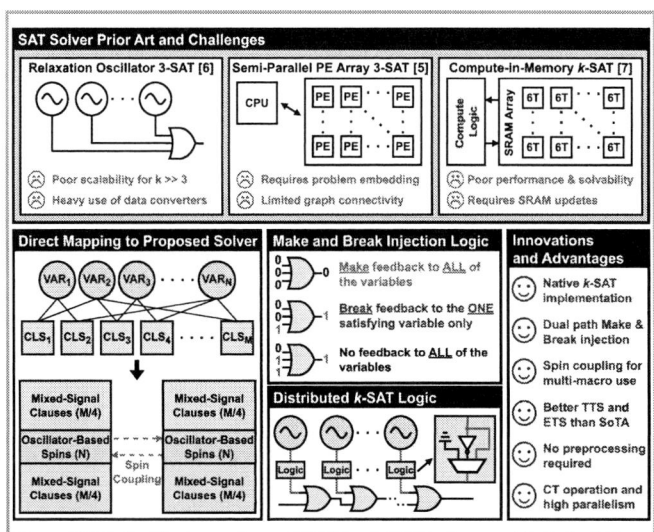

Figure 13.3.1: Architectures of recent 3-SAT and *k*-SAT solvers and their limitations. Overview of key design highlights and innovations of the proposed solver.

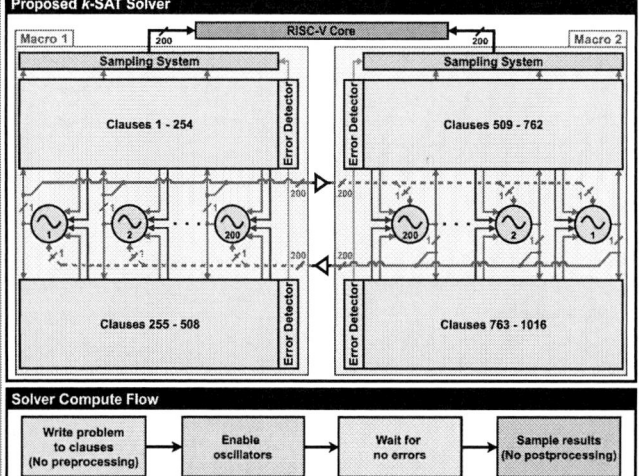

Figure 13.3.2: System architecture of the proposed *k*-SAT solver including spins, clauses, and sampling and explanation of solver compute flow and methodology.

DIGEST OF TECHNICAL PAPERS • 233

979-8-3315-8937-0/26 $31.00 © 2026 IEEE

ISSCC 2026 / SESSION 13 / CIRCUITS FOR AI AND AI FOR CIRCUITS / 13.3

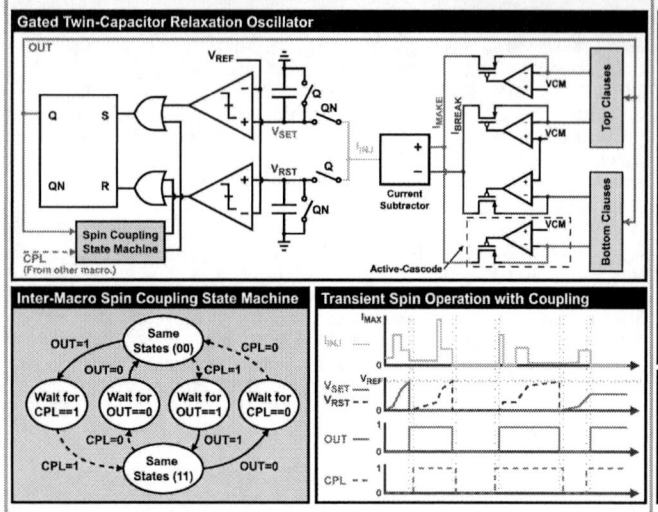

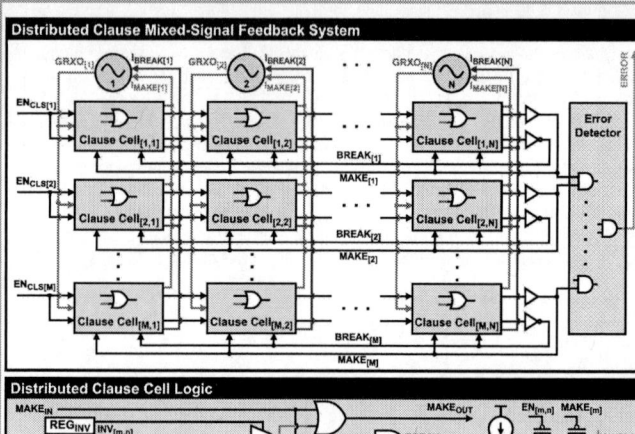

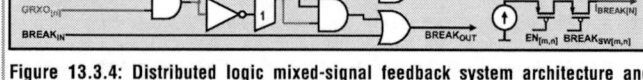

Figure 13.3.3: Gated Relaxation Oscillator (GRXO) spin design with inter-macro coupling logic and transient operation plot.

Figure 13.3.4: Distributed logic mixed-signal feedback system architecture and individual clause cell logic design.

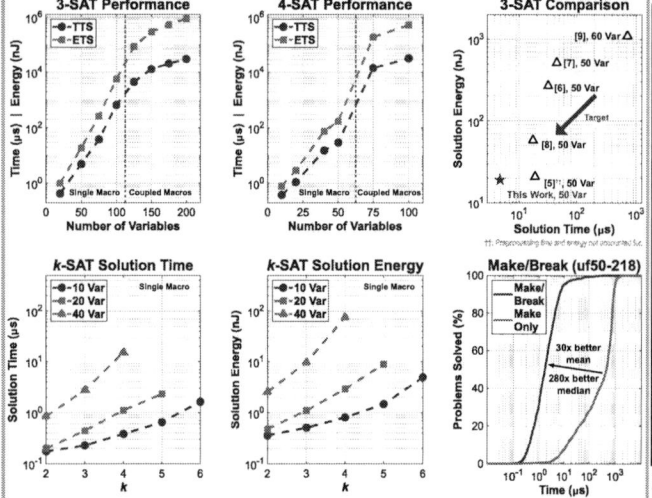

Figure 13.3.5: Plots labeled 3-SAT used the SATLIB tests. 4-SAT and k-SAT plots are based on generated unforced problems at the phase transition.

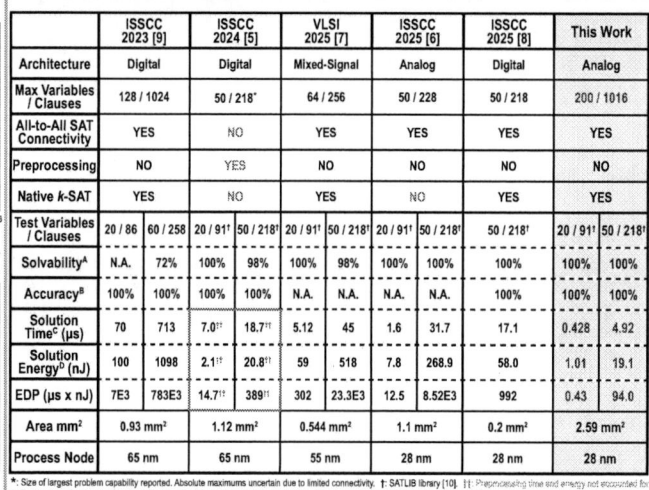

	ISSCC 2023 [9]	ISSCC 2024 [5]	VLSI 2025 [7]	ISSCC 2025 [6]	ISSCC 2025 [8]	This Work
Architecture	Digital	Digital	Mixed-Signal	Analog	Digital	Analog
Max Variables / Clauses	128 / 1024	50 / 218*	64 / 256	50 / 228	50 / 218	200 / 1016
All-to-All SAT Connectivity	YES	NO	YES	YES	YES	YES
Preprocessing	NO	YES	NO	NO	NO	NO
Native k-SAT	YES	NO	YES	NO	YES	YES
Test Variables / Clauses	20 / 86	60 / 258	20 / 91†	50 / 218†	20 / 91†	50 / 218†	20 / 91†	50 / 218†	50 / 218†	20 / 91†	50 / 218†
Solvability^A	N.A.	72%	100%	98%	100%	98%	100%	100%	100%	100%	100%
Accuracy^B	100%	100%	100%	100%	N.A.	N.A.	N.A.	N.A.	100%	100%	100%
Solution Time^C (µs)	70	713	7.0††	18.7††	5.12	45	1.6	31.7	17.1	0.428	4.92
Solution Energy^D (nJ)	100	1098	2.1††	20.8††	59	518	7.8	268.9	58.0	1.01	19.1
EDP (µs x nJ)	7E3	783E3	14.7††	389††	302	23.3E3	12.5	8.52E3	992	0.43	94.0
Area mm²	0.93 mm²	1.12 mm²	0.544 mm²	1.1 mm²	0.2 mm²	2.59 mm²
Process Node	65 nm	65 nm	55 nm	28 nm	28 nm	28 nm

*: Size of largest problem capability reported. Absolute maximums uncertain due to limited connectivity. †: SATLIB library [10]. ††: Preprocessing time and energy not accounted for.
A: The percent of problems solved. B: Percent of solved problems which are correct. C: The mean time required to find a solution. D: The mean energy required to find a solution.

Figure 13.3.6: Performance summary and comparison with other work. Comparison with 3-SAT operation as this is most widely available.

Chip Summary

Technology	28 nm
Active Area	2.59 mm²
Supply	900 mV
Max Variables	200
Max Clauses	1016
Mean Solution Time (s)	uf50-218: 4.92E-6
	uf100-430: 6.64E-4
	uf200-860: 3.09E-2
Mean Solution Energy (J)	uf50-218: 1.91E-8
	uf100-430: 5.69E-6
	uf200-860: 8.83E-4
Pass Rate (Solvability x Accuracy)	uf50-218: 100 %
	uf100-430: 99.5 %
	uf200-860: 80.0 %

Figure 13.3.7: Die micrograph, chip summary, and area breakdown.

- 2026 IEEE International Solid-State Circuits Conference

ISSCC 2026 / SESSION 13 / CIRCUITS FOR AI AND AI FOR CIRCUITS / 13.4

13.4 An Inverse-Designed Passively Coupled N-Path Filter with g_m-Boosted Active HBT Switches

Vinay Chenna, Hossein Hashemi

University of Southern California, Los Angeles, CA

Abstract

A 0.8-to-2.6GHz N-path filter with g_m-boosted HBT switches and inverse-designed non-resonant passive networks is presented to enhance dynamic range and tunability of passively coupled higher-order N-path filters. A hybrid real-binary inverse design algorithm capable of inserting variable-size MIM capacitors in layouts enables synthesis of compact multilayered low-GHz passives. A BiCMOS filter prototype achieves >15dBm P1dB, <5dB noise figure and <4dB insertion loss across the full tuning range.

N-path filters with multiphase, periodically switched filter kernels have gained attention over the past 2 decades for enabling high-Q, frequency-tunable responses not achievable with traditional on-chip passives. Their performance tradeoffs are governed by the CMOS switch $R_{ON}C_{OFF}$ product [1-2]. Larger switches improve linearity and out-of-band (OOB) rejection but increase parasitics, raising CV^2f power and unwanted charge sharing that elevates noise figure (NF) and in-band insertion loss (IL). These CMOS-switch tradeoffs can only improve through technology scaling. Further, a single RC N-path resonator can only yield a 2-pole filter response, necessitating coupling multiple resonators or using higher-order baseband loads [3] to achieve higher-order filtering. State-of-the-art coupling schemes include active [4], passive [5-7] and frequency-translational [8]. Active coupling offers low NF and DC power consumption (P_{DC}) but suffers from poor in-band (IB) linearity and is limited to low frequencies by the g_m cells' bandwidth. In contrast, passive coupling achieves high IB linearity, but the lossy and resonant LC coupling networks cause large IL, NF and P_{DC} while severely restricting tuning range. Frequency translational filters avoid these limitations but require large chip areas to realize the baseband filters and suffer from approximately −25dBc in-/near-band intermixing spurs [9] that fold onto the neighboring channels and can potentially exceed intermodulation or LO feedthrough levels. To address these tradeoffs, this work presents: (1) a g_m-boosted active SiGe HBT switch with negative feedback that boosts linearity without large switches, that consequently improves NF, IL and P_{DC} when compared with using standard CMOS switches; (2) a *non-resonant* inverse-designed passive coupling network enabling a higher-order response with lower loss and a wider tuning range than prior passive coupling approaches; and (3) a hybrid real-binary inverse-design algorithm that introduces variable-size MIM capacitors in 3D pixelated electromagnetic (EM) networks to autonomously synthesize compact multilayered passives operating at low-GHz frequencies.

Figure 13.4.1 presents the concept of active switches. Traditional passive switches suffer from a high $R_{ON}C_{OFF}$ product as the switch size simultaneously sets both R_{ON} and C_{OFF}. To decouple R_{ON} and C_{OFF}, the switch impedance can be the output impedance of a biased emitter/source follower: a physically small device yields low C_{OFF}, while a large ON-state bias current provides high g_m and thus low R_{ON}. Both FETs and HBTs can operate in this mode, but forward-active HBTs offer much higher g_m/I_C (>30) than FET g_m/I_D in saturation (<5), motivating using HBTs. The drawback is sensitivity to the LO high level, which sets the ON-state g_m. In addition, emitter-follower LO-amplitude noise appears at the output and degrades NF, as also observed in top-plate, base-voltage-switched, common-gate HBT N-path filters [10]. To address these drawbacks, a super emitter follower (SEF) switch is presented that: (1) uses negative feedback to boost ON-state output admittance by the loop-gain ($\approx g_{m,2}R_C$) without enlarging device size, thereby also reducing the $R_{ON}C_{OFF}$ product by the loop-gain; (2) improves linearity via the same feedback (3) toggles the *base currents* of the forward-active $Q_{1,2}$ (rather than their base voltages) by inserting passive switches ($Q_{3,4}$) in series with their bases. This eliminates sensitivity to the LO high level and amplitude noise, since the toggled devices no longer operate as transconductances in the ON state but behave like digital switches. SEF switches, therefore, improve linearity, NF, IL, and P_{DC} for a given R_{ON} by using significantly smaller devices than passive FET switches or common-gate HBT switches while leveraging negative feedback to maintain linearity.

Figure 13.4.2 shows the architecture of the passively coupled N-path filter that uses SEF switches. It consists of 2 passively coupled 8-phase bottom-plate-switched N-path resonators. Differential operation doubles the LO duty cycle from 12.5% to 25% by interleaving the {0°, 90°, 180°, 270°} and {45°, 135°, 225°, 315°} sub-resonators within a stage [11]. The stages are coupled by *wideband, non-resonant* passive networks, enabling a higher-order response with wide tuning range while preventing interstage charge sharing. The 8-phase, 25% duty cycle LO phases are generated by a wideband 4-stage RC polyphase filter followed by current-mode logic (CML) logic gates, producing ~400mV pk-pk voltage waveforms, sufficient for the SEF switches and lower than typically required for FET switches. Finally, the SEF transconductances ($Q_{1,2}$) are only ~4.5µm long, confirming the ability to operate with small device sizes. The circuit is implemented in Tower SBC18S5 180nm BiCMOS process with peak simulated f_T/f_{max} of 240/300GHz, comparable to or inferior to 65nm bulk CMOS processes [12], where most prior CMOS N-path filters were implemented.

No known LC networks can realize the wideband coupling network in Fig. 13.4.2 due to a non-straightforward target functionality. Inverse design synthesizes nonintuitive on-chip passives by algorithmically distributing metal and dielectric within a given volume to meet a target response [13-16]. Prior work, limited to mm-Wave frequencies, inverse-designed EM structures that occupy a significant fraction of the wavelength (λ). At low-GHz, the much larger λ makes such structures prohibitively large on chip, rendering direct adoption of mm-Wave approaches infeasible. To keep area manageable at low-GHz, the inverse-design stack in this work uses metal-insulator-metal capacitor (MIMcap) layers alongside the top 3 metal + 2 via layers (Fig. 13.4.3). Earlier on-chip inverse-design flows used purely binary pixels to represent layouts. For MIM layers, a binary representation results in adding/removing fully occupied MIMcap pixels in every iteration. Given the high MIMcap density (≈2-to-4fF/µm²), such abrupt additions/removals cause sudden performance variations that hinder convergence. To address this issue, continuously varying real-valued variables can be used to represent the area of the MIMcap top metal within a given pixel. In this scheme, a value of 0 indicates the absence of a MIMcap, while a value of 1 indicates a MIMcap that fully occupies the pixel. By using real values between 0 and 1, the MIMcap size can scale continuously between these extremes. This method allows optimization algorithms to modify the size of MIMcaps gradually in the layout. New MIMcaps are always introduced with the minimum size. Existing MIMcaps cannot be removed until they reach this minimum size. MIMcap sizes can only increase or decrease incrementally in each iteration, enabling smooth convergence. Therefore, EM structures can now be represented using hybrid real-binary 3D arrays: metal/via layers using binary entries while MIM layer using real-valued entries.

The algorithm in Fig. 13.4.3 synthesizes the coupling network. The design space is a pixelated region with a 50×27 grid resolution with a pixel size of 12.5×12.5µm². A quadrant symmetry constraint reduces the effective optimization domain to 25×14, spanning 6 layers: 3 metal, 2 via, and 1 MIM. Optimization begins with an initial seed. In each iteration, the best-known solution is perturbed by randomly flipping up to 10 binary pixels and incrementing or decrementing up to 2 real-valued variables. The resulting structure is EM simulated, followed by a circuit simulation of the filter in Fig. 13.4.2 but with ideal LOs. Performance is evaluated against a 4-pole Chebyshev response using mean square error (MSE). If the MSE improves, the modification is accepted and the best solution is updated; otherwise, the changes are undone, and a new perturbation is applied in the next iteration. Once the passive network is finalized, the remaining circuit, including the LO path and the N-path resonators (component sizes and biases), is optimized using Virtuoso ADE Assembler to maximize overall filter performance. Figure 13.4.4 shows the final layout obtained from a multilayered spiral seed after ~4 days of hybrid real-binary optimization. MIMcaps are laid out such that, at each location on the optimization grid, the bottom plates connect to metal 5 and the top plates to metal 6.

An N-path filter prototype (Fig. 13.4.7) was fabricated and measured, with key results summarized in Fig. 13.4.5 and performance compared against state-of-the-art in Fig. 13.4.6. Compared to other passive coupling approaches, the design achieves superior tuning range, NF and IL, despite operating at higher frequencies. It is also the most power-efficient among all passive techniques (passively coupled, frequency-translational, and rotary clocking methods), even though it uses HBT switches with CML-based LO generation that consumes static DC power. These results highlight the efficiency advantages of using small HBT switches, which significantly reduce LO-path power consumption.

Compared to active-coupling approaches, the filter in this work achieves >2× the tuning range while operating at >2× the peak LO frequency. This advantage stems from the absence of bandwidth-limiting active g_m-coupling cells in passive coupling schemes. Additionally, the filter exhibits a higher-order response across the entire tuning range when compared to a single standard N-path resonator, demonstrating the effectiveness of inverse design in synthesizing *wideband, non-resonant* coupling networks.

The higher breakdown voltages of HBTs than FETs, combined with negative feedback in SEF switches, contribute to the filter's high linearity. An IB P1dB exceeding +15dBm is maintained across the entire tuning range, consistent with simulations that show tiny gain compression followed by expansion. A notable observation is that the IM3 tones initially rise with a 2dB/dB slope before compressing sharply, well before the fundamental tones begin to compress. Although counterintuitive, this behavior has been previously documented in HBT circuits and is attributed to scenarios where IM3 tone power is dominated by the intermixing of 2nd-order distortion products with the fundamental tones, resulting in an IM3 slope matching that of HD2 [19]. This behavior results in significantly lower IM3 spurs compared to state-of-the-art at high P_{in} levels. Since IIP3 cannot be defined for IM3 slopes of 2dB/dB, an EVM test was performed using a 64-QAM signal with +10dBm peak power, occupying ~ 50% of the filter bandwidth. This test stresses the filter linearity while probing memory effects and verifying that group delay variations across the passband are acceptable. The low EVM confirms that the filter handles high input powers without distorting the signal beyond recognition. The minimal degradation in EVM in the presence of a strong +5dBm OOB blocker @ Δf/BW = 4 demonstrates high OOB linearity. This metric offers more practical insight than traditional single-tone OOB metrics such as OOB B1dB.

Acknowledgement:
The authors thank Tower Semiconductor for chip fabrication. This research was partially supported by the Northrop Grumman Corporation and the National Science Foundation (ECCS-2229535). The authors are grateful to Dr. Tim LaRocca and Dr. Kevin Leong from Northrop Grumman Corporation and Dr. Pingyue Song for valuable technical discussions.

References:
[1] C. Andrews and A. C. Molnar, "Implications of Passive Mixer Transparency for Impedance Matching and Noise Figure in Passive Mixer-First Receivers," *IEEE Transactions on Circuits and Systems I: Regular Papers*, vol. 57, no. 12, pp. 3092-3103, Dec. 2010. https://doi.org/10.1109/TCSI.2010.2052513
[2] D. Yang, C. Andrews and A. Molnar, "Optimized Design of N-Phase Passive Mixer-First Receivers in Wideband Operation," *IEEE Transactions on Circuits and Systems I: Regular Papers*, vol. 62, no. 11, pp. 2759-2770, Nov. 2015. https://doi.org/10.1109/TCSI.2015.2479035
[3] S. Krishnamurthy and A. M. Niknejad, "Synthesis and Design of Enhanced N-Path Filters With 60-dB/decade RF Selectivity," *IEEE Solid-State Circuits Letters*, vol. 3, pp. 522-525, 2020. https://doi.org/10.1109/LSSC.2020.3035962
[4] M. Darvishi, R. van der Zee and B. Nauta, "A 0.1-to-1.2GHz tunable 6th-order N-path channel-select filter with 0.6dB passband ripple and +7dBm blocker tolerance," *IEEE International Solid-State Circuits Conference (ISSCC)*, Feb. 2013. https://doi.org/10.1109/ISSCC.2013.6487686
[5] P. Song and H. Hashemi, "A 13th-order CMOS reconfigurable RF BPF with adjustable transmission zeros for SAW-less SDR receivers," *IEEE International Solid-State Circuits Conference (ISSCC)*, Feb. 2018, pp. 416-418. https://doi.org/10.1109/ISSCC.2018.8310361
[6] C. M. Thomas and L. E. Larson, "Broadband Synthetic Transmission-Line N-Path Filter Design," *IEEE Transactions on Microwave Theory and Techniques*, vol. 63, no. 10, pp. 3525-3536, Oct. 2015. https://doi.org/10.1109/TMTT.2015.2473161
[7] N. Reiskarimian and H. Krishnaswamy, "Design of all-passive higher-order CMOS N-path filters," *IEEE Radio Frequency Integrated Circuits Symposium (RFIC)*, 2015, pp. 83-86. https://doi.org/10.1109/RFIC.2015.7337710

[8] A. Nagulu, M. Yi, Y. Zhuang, S. Garikapati and H. Krishnaswamy, "A 1-to-5GHz All-Passive Frequency-Translational 4th-Order N-path Filter with Low-Power Clock Boosting for High Linearity and Relaxed Pdc-Frequency Trade-Off," *IEEE International Solid-State Circuits Conference (ISSCC)*, Feb. 2023, pp. 378-380. https://doi.org/10.1109/ISSCC42615.2023.10067862
[9] A. Nagulu, Y. Zhuang, M. Yuan, S. Garikapati and H. Krishnaswamy, "A Third-Order Quasi-Elliptic N-Path Filter With Enhanced Linearity Through Clock Boosting," *IEEE Journal of Solid-State Circuits*, vol. 58, no. 12, pp. 3351-3363, Dec. 2023. https://doi.org/10.1109/JSSC.2023.3318357
[10] R. Ying and A. Molnar, "Impedance Transparency and Performance Metrics of HBT-Based N-Path Mixers for mmWave Applications," *IEEE Transactions on Circuits and Systems I: Regular Papers*, vol. 68, no. 5, pp. 2210-2223, May 2021. https://doi.org/10.1109/TCSI.2021.3060644
[11] Z. G. Boynton and A. Molnar, "A 9-31GHz 65nm CMOS Down-Converter with >4dBm OOB B1dB," *IEEE Radio Frequency Integrated Circuits Symposium (RFIC)*, 2020, pp. 279-282. https://doi.org/10.1109/RFIC49505.2020.9218432
[12] B. Razavi, "A 300-GHz Fundamental Oscillator in 65-nm CMOS Technology," *IEEE Journal of Solid-State Circuits*, vol. 46, no. 4, pp. 894-903. https://doi.org/10.1109/JSSC.2011.2108122
[13] E. A. Karahan, Z. Liu and K. Sengupta, "Deep-Learning-Based Inverse-Designed Millimeter-Wave Passives and Power Amplifiers," *IEEE Journal of Solid-State Circuits*, vol. 58, no. 11, pp. 3074-3088, Nov. 2023. https://doi.org/10.1109/JSSC.2023.3276315
[14] J. Zhou, E. A. Karahan, S. Ghozzy, Z. Liu, H. Jalili and K. Sengupta, "AI-Enabled Design Space Discovery and End-to-End Synthesis for RFICs with Reinforcement Learning and Inverse Methods Demonstrating mm-Wave/sub-THz PAs Between 30 and 120GHz," *IEEE International Solid-State Circuits Conference (ISSCC)*, Feb. 2025, pp. 1-3. https://doi.org/10.1109/ISSCC49661.2025.10904600
[15] V. Chenna and H. Hashemi, "Topology-Optimized Nonintuitive Multilayered mm-Wave Power Amplifiers," *IEEE Radio Frequency Integrated Circuits Symposium (RFIC)*, 2025, pp. 279-282. https://doi.org/10.1109/RFIC61188.2025.11082875
[16] V. Chenna and H. Hashemi, "Algorithmic Design of Nonintuitive on-Chip Multilayered Passive Networks," *IEEE/MTT-S International Microwave Symposium - IMS 2025*, pp. 918-921. https://doi.org/10.1109/IMS40360.2025.11103811
[17] M. Darvishi, R. Van der Zee, E. Klumperink and B. Nauta, "A 0.3-to-1.2GHz tunable 4th-order switched gm-C bandpass filter with >55dB ultimate rejection and out-of-band IIP3 of +29dBm," *IEEE International Solid-State Circuits Conference (ISSCC)*, Feb. 2012, pp. 358-360. https://doi.org/10.1109/ISSCC.2012.6177050
[18] M. Khorshidian, S. L. N. Garimella, A. Nagulu and H. Krishnaswamy, "An Inductor-Less All-Passive Higher-Order N-Path Filter Based on Rotary Clocking in N-Path Filters," *IEEE/MTT-S International Microwave Symposium - IMS 2022*, pp. 241-244. https://doi.org/10.1109/IMS37962.2022.9865397
[19] T.-Y. Lee, S. Lee, P. Zampardi and J. Kang, "Large-signal modeling of SiGe HBT for PA applications," *IEEE Bipolar/BiCMOS Circuits and Technology Meeting (BCTM)*, 2010, pp. 94-97. https://doi.org/10.1109/BIPOL.2010.5667936

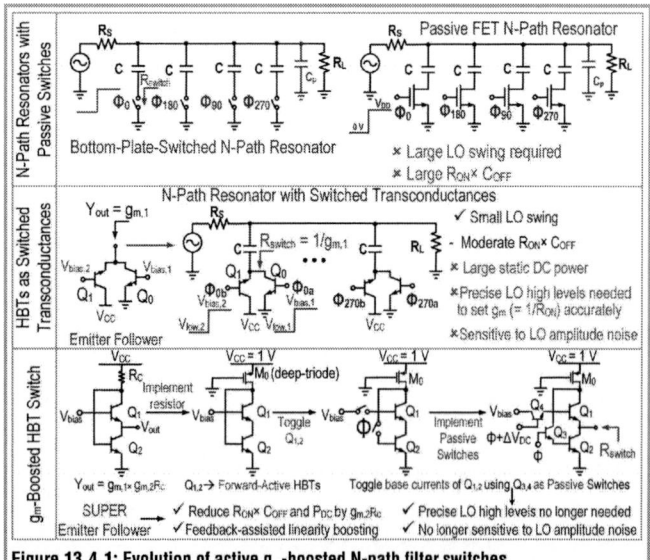

Figure 13.4.1: Evolution of active g_m-boosted N-path filter switches.

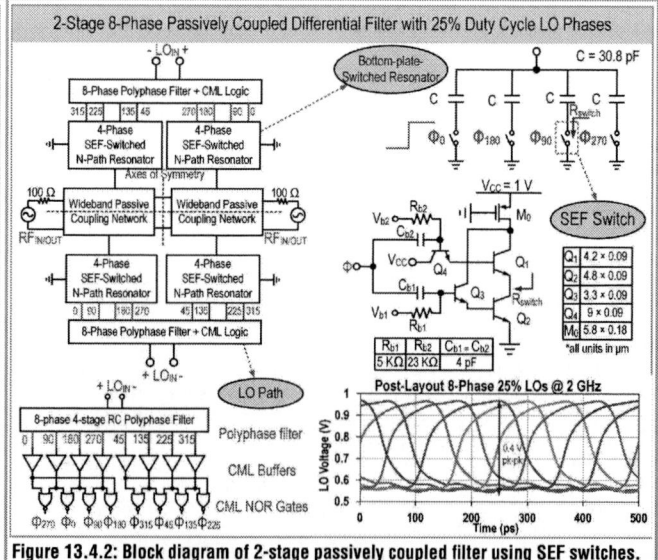

Figure 13.4.2: Block diagram of 2-stage passively coupled filter using SEF switches.

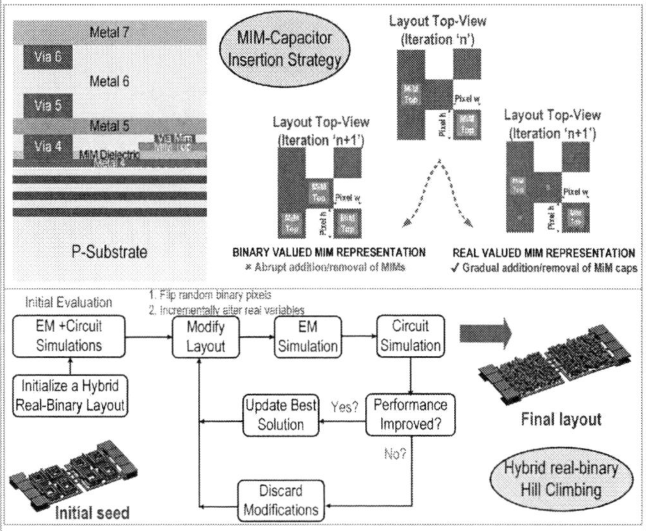

ISSCC 2026 / SESSION 13 / CIRCUITS FOR AI AND AI FOR CIRCUITS / 13.4

Figure 13.4.3: Inverse design of wideband, non-resonant passive coupling network.

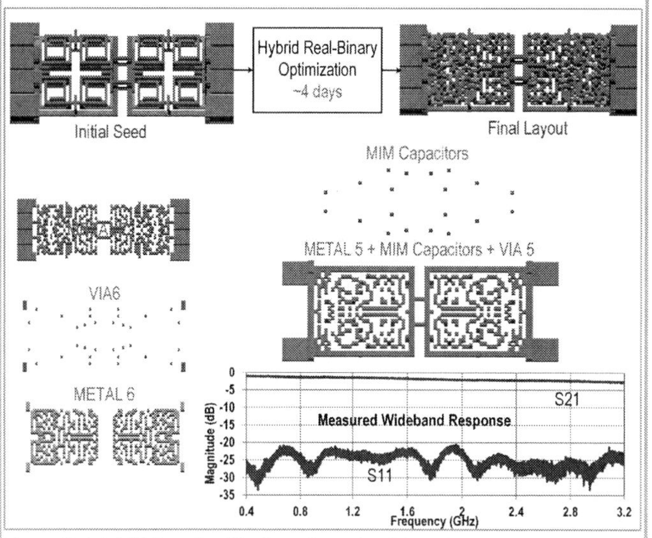

Figure 13.4.4: Initial seed and final coupling network layouts.

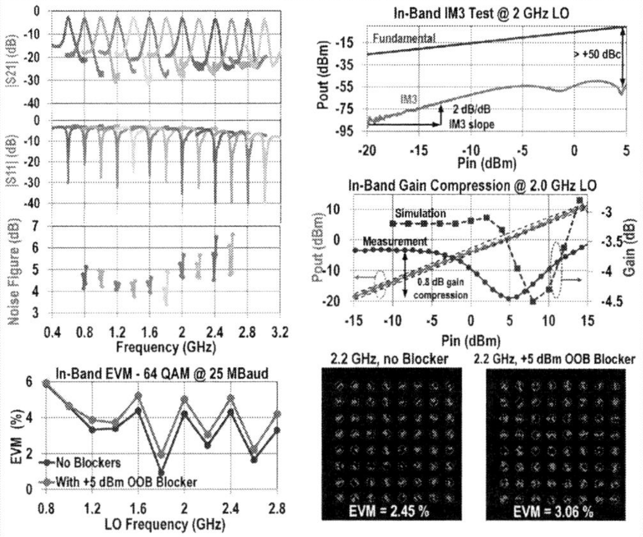

Figure 13.4.5: Key measurement results.

Performance Metric	This work	ISSCC 2018 [5]	RFIC 2015 [7]	TMTT 2015 [6]	ISSCC 2013 [4]	ISSCC 2023 [8]	ISSCC 2012 [17]	IMS 2022 [18]
Architecture	Inverse Designed Passive Coupling +Active Switches	LC-Coupled Filter	CLC-Coupled Filter	Artificial T-Line Coupling	Gyrator-Coupled Filter	Frequency Translational	Switched gm-C Filter	Rotary Clocked Filter
Coupling	Passive	Passive	Passive	Passive	Active	N/A	N/A	N/A
Small Signal Performance Metrics								
Tuning (GHz)	0.8 - 2.6	0.8 - 1.1	0.6 - 0.85	0.1 - 1.6	0.1 - 1.2	1 - 5	0.3 - 1.2	0.2 - 0.8
Gain (dB)	-2.8 to -4.1	-4.6 to -3.8	-4.7 to -6.2	-0.5 to -5	+25	-8.4 to -5	+3.5	+10[b]
BW (MHz)	60	30 – 50	9 – 15	23	8	5 - 80	21	40
IB NF (dB)	3.4 to 4.7	5 to 8.6	8.6	1.5 to 5.4	2.8[a]	5 to 8.5	9.5[a]	3.7 to 6.4[a]
OOB Rejection	> 16 dB (left) & > 20 dB (right)	17 dB	30-50 dB	30-40 dB	59 dB	> 24 dB	59 dB	30-50 dB
IB S11 (dB)	< -20	< -10	< -10	< -25	-5	< -10	N/R	< -10
Linearity and Modulation Metrics								
IB IP1dB	> +15 dBm	+7 dBm	0 dBm	+11 dBm	-23 dBm	+8.8	-4.4	-8.2
IM3 Slope	≤ 2 dB/dB	≥ 3dB/dB	N/R	3dB/dB	N/R	≥ 3dB/dB	N/R	N/R
IB IIP3 (dBm)	N/A	+24	+7	+29	-12	+23	+9	+4.2
IM3 @ +5 dBm IB Inputs	< -45 dBc	-26 dBc*	N/R	N/R	N/R	-36 dBc*	N/R	N/R
Modulation	64 QAM, +10 dBm Pin,peak	N/R	N/R	N/R	N/R	N/R	N/R	N/R
EVM	0.9 to 5.8% (150 Mbps)	N/R	N/R	N/R	N/R	N/R	N/R	N/R
EVM with +5 dBm OOB Blocker	2 to 6 % (150 Mbps)	N/R	N/R	N/R	N/R	N/R	N/R	N/R
Chip Metrics								
Technology	180 nm SiGe BiCMOS	65 nm CMOS	65 nm CMOS	65 nm CMOS	65 nm CMOS	65 nm CMOS	65 nm CMOS	65 nm CMOS
P_{DC} (mW)	72 (constant versus f_{LO})	80-97	75	30 - 200	21-69	40 - 167	17.6	5.5 – 38
$P_{DC}/f_{LO, high}$	27.7 mW/GHz	88 mW/GHz	88 mW/GHz	125 mW/GHz	57.5 mW/GHz	33.4 mW/GHz	21.4 mW/GHz	47.5 mW/GHz
Core Area (mm²)	1.15	1.9	1.21 + off-chip inductors	4.2	0.27	4.6	0.127	0.56

a – buffer noise contribution deembeded, b – passive voltage gain not power gain, * - estimated from measurement plots . N/A – not applicable. N/R – not reported

Figure 13.4.6: Performance comparison with state-of-the-art N-path filters.

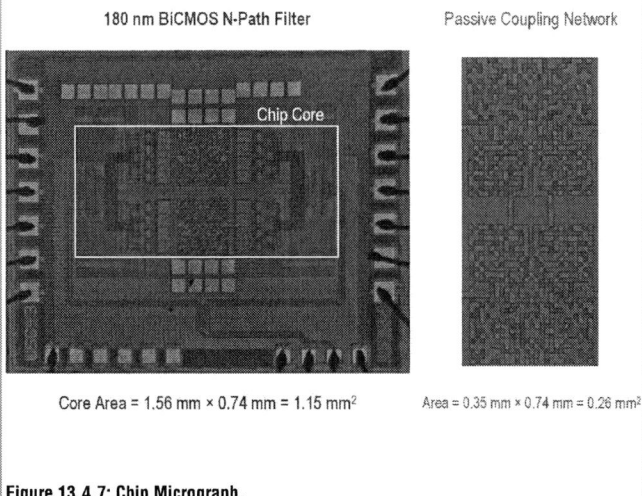

180 nm BiCMOS N-Path Filter

Passive Coupling Network

Chip Core

Core Area = 1.56 mm × 0.74 mm = 1.15 mm²

Area = 0.35 mm × 0.74 mm = 0.26 mm²

Figure 13.4.7: Chip Micrograph.

ISSCC 2026 / SESSION 13 / CIRCUITS FOR AI AND AI FOR CIRCUITS / 13.5

13.5 A Nonintuitively Frequency-Staggered Wideband mm-Wave Low-Noise Amplifier

Vinay Chenna, Hossein Hashemi

University of Southern California, Los Angeles, CA

Abstract

An algorithmic topology optimization framework is presented to autonomously synthesize nonintuitive, multilayered wideband mm-Wave LNAs with arbitrary stage count. Co-optimization of actives and passives directly at a layout level enables improved tradeoffs among metrics like gain, bandwidth, NF, and DC power. A 3-stage SiGe HBT LNA prototype fabricated in a 180nm BiCMOS process achieves 26.3dB gain, 22-to-46GHz 3dB bandwidth, 3.1dB NF and 18mW DC power, experimentally validating the method.

Inverse design and topology optimization have recently emerged as powerful techniques for autonomously synthesizing mm-Wave amplifiers with nonintuitive pixelated layouts that can significantly outperform classical designs based on lumped or distributed elements such as inductors, transmission lines, capacitors, and transformers. Recent machine learning (ML)-based approaches [1–2] have demonstrated amplifier synthesis using single-layer pixelated passives, but these result in large chip areas. In contrast, the exponentially larger design spaces associated with multilayered passives make ML methods impractical, requiring explicit electromagnetic (EM) simulators within the algorithmic loops typically constrained to optimize a single performance metric [3]. All prior efforts have focused exclusively on power amplifiers (PAs), used only 2 stages, and have not demonstrated experimental generalization to multi-stage designs that could enable higher gain. Furthermore, previous multilayered designs have been limited to single-objective optimization. This work advances the state of amplifier synthesis by: (1) Targeting low-noise amplifiers (LNAs) instead of PAs, (2) Presenting a generalized algorithmic strategy capable of handling an arbitrary number of stages, and (3) Simultaneously optimizing multiple objectives including gain (S21), input match (S11), noise figure (NF), DC power consumption (P_{DC}), and stability across multiple frequency points in multilayered designs. A 3-stage, wideband mm-Wave SiGe HBT LNA prototype with nonintuitive frequency staggering is fabricated to experimentally validate the synthesis methodology while achieving superior performance over state-of-the-art traditional LNA designs.

Frequency staggering is a well-established technique for designing wideband amplifiers (Fig. 13.5.1), wherein the frequency responses of individual stages are deliberately engineered so that their cascade yields a flat, broadband gain. Traditional staggered tuning approaches include linearly shifted bandpass stages [5], cascaded combinations of low-pass and high-pass stages [6–7], or hybrids of these techniques. While such strategies are intuitive and human-conceivable, there is no fundamental reason to believe they represent the only or optimal solution. Beyond conventional human intuition, nonstandard passband shapes of individual stages may yield superior overall gain and bandwidth, particularly under practical constraints such as minimized P_{DC}, NF, and S11. The algorithmic methods presented in this work automatically uncover such nonintuitive responses for each stage (Fig. 13.5.1, bottom), achieving wideband performance while satisfying all design constraints across the operating frequency range.

The algorithm used in this work consists of 2 main components: passive network updates and active device size and bias updates. Figure 13.5.2 illustrates the methodology for iteratively updating the passive networks in an N-stage amplifier, which contains N+1 passive networks. Each iteration involves two key steps: selecting which networks to update and evaluating which updates to propagate to the future iterations. To systematically explore the design space, all possible M = (N+1)×(N)/2 unique pairs of the passive networks are enumerated and indexed. In iteration 'n', if n modulo M equals k – 1, the kth pair, comprising networks 'p' and 'q', is selected for update. Both p and q are modified by applying up to 10 random pixel flips to each layout, and the resulting structures are EM simulated using Cadence EMX. Then, circuit simulations are performed of the entire LNA circuit for 3 different cases: one with only p network modified, one with only q network modified, and one with both p and q modified. In each case, the remaining networks are taken from the previous iteration. The simulation yielding the best performance determines whether the pixel modifications to p, q, or both are retained in future iterations. The performance of each circuit is evaluated by a combination of the results obtained from SP, noise and DC analyses using Cadence Spectre.

The procedure for updating transistor sizes and bias voltages is illustrated in Fig. 13.5.3. Similar to the passive update mechanism, the active update also operates cyclically across amplifier stages. First, all transistors are enumerated from left to right and assigned ascending indices. Then, once every 10 passive-network iterations, a specific transistor is selected for update. This selection is determined by verifying that the current passive iteration number n is divisible by 10. The value of $n/10$ is then taken modulo N, where N is the total number of amplifier stages. If this result equals k – 1, then the transistor in the kth stage is chosen for update. Once selected, the transistor undergoes a local sweep of its emitter length and base bias voltage. Specifically, the emitter length is swept over a ±0.5µm range in 0.1µm steps (11 points), and the base voltage is swept over a ±10mV range in

5mV steps (5 points), resulting in a total of 55 circuit simulations. The combination of length and bias that yields the best overall circuit performance is then adopted as the new size and bias setting for the selected transistor. By systematically cycling through both active and passive stages, the algorithms (Figs. 13.5.2-13.5.3) jointly co-optimize transistor and passive networks and ultimately converge to a tapeout-ready LNA layout fully automatically while being scalable to any number of stages, without human intervention.

The algorithms were used to synthesize a mm-Wave LNA targeting the following specifications: (1) in-band gain (S_{21}) of 26 dB, (2) input return loss (S_{11}) below −10dB, (3) NF below 3dB, (4) stability factor (μ) greater than 1, and (5) P_{DC} below 20mW from a 1.2V supply. These constraints were enforced over a full octave bandwidth from 23-to-46GHz. Multiple objectives were combined nonlinearly using sigmoidal weighting functions, which serve as smooth switches to prioritize the worst-performing parameters during optimization, following the approach in [14]. The only manual steps involved were the selection of transistor locations and EM port locations in the layout, along with a set of high-level design hyperparameters: (1) core layout area (1mm × 0.25mm), (2) number of amplifier stages (3 stages with 4 passive networks), (3) pixelation resolution (9 × 9 grid with 5 layers (3 metal + 2 via layers) for each passive network), and (4) pixel dimensions (23µm × 23µm). For these hyperparameters, the size of the design space becomes astronomically large with $13^{4 \times 9 \times 9} = 10^{324}$ different layout configurations [15], reinforcing the importance of requiring a judicious algorithmic strategy to systematically explore the design space. Once these constraints are defined, the core layout of the LNA is fully determined by the algorithms, without any further human intervention.

Figure 13.5.4 presents an example of a low-noise amplifier whose core layout was fully synthesized using the design algorithms and previously described hyperparameters. Starting from a random initial condition, the algorithm completed the full synthesis in under 5 days using 128 threads on an AMD Ryzen Threadripper PRO 7985WX CPU. The fabricated prototype occupies a core area of 0.25mm² and was implemented in Tower Semiconductor's SBC18S5 180nm SiGe BiCMOS process, with a simulated f_t/f_{max} of 240/300GHz, lower than that of a typical 65nm bulk CMOS process [4, 16]. The chip was assembled using chip-on-board wire bonding onto a PCB. Measured small-signal results show a peak gain of 26.3dB, with a 3dB bandwidth spanning from 22-to-46GHz. Within this range, the NF varies from 3.1 to 4.7dB, and the S_{11} remains below −10dB. The measured DC power consumption during small-signal operation is 18mW from a 1.2V supply. The absence of oscillations or unexpected resonances in the measured S-parameters confirms the small-signal stability of the fabricated design. These measurements verify that the LNA either meets or closely approaches the constraints enforced by the optimization objectives.

The large-signal performance of the LNA was evaluated through compression, intermodulation, and error vector magnitude (EVM) measurements, as summarized in Fig. 13.5.5. The LNA exhibits an in-band IIP3 of −9.3dBm at 30GHz and a corresponding input-referred P1dB of −17.9dBm. In EVM tests, the LNA achieves 5.2% EVM for 64-QAM modulation at a data rate of 2.4Gbps, and 8.6% EVM for 16-QAM at 6Gbps. These results confirm the LNA's capability to handle large signal operation while maintaining acceptable linearity and modulation fidelity.

Figure 13.5.6 compares the synthesized LNA against state-of-the-art designs, including both narrowband and wideband LNAs. The synthesized design demonstrates a significantly higher gain–bandwidth product than all the compared works. Among LNAs with wider fractional bandwidths, the synthesized LNA exhibits notably superior output-referred linearity. While its input-referred linearity might appear lower, this is due to its higher gain, making output-referred metrics a more meaningful basis for comparison for LNAs with different gains. Compared to LNAs with narrower bandwidths, the synthesized design achieves comparable output-referred linearity. Additionally, it maintains a competitive NF relative to all state-of-the-art designs.

To quantify how the topology optimization methodology overcomes traditional design tradeoffs, two figures of merit (FoMs) are evaluated. FoM_1, which captures tradeoffs between key small-signal performance metrics, is significantly higher for the synthesized topology-optimized LNA compared to prior works, demonstrating the benefit of exploring

236 • 2026 IEEE International Solid-State Circuits Conference

979-8-3315-8937-0/26 $31.00 © 2026 IEEE

noninuitive, beyond-human design spaces. FoM₂ extends this comparison by also incorporating input-referred linearity, and again, the topology-optimized design outperforms recent state-of-the-art. These results confirm that the presented synthesis approach not only accelerates the design cycle but also unlocks fundamentally superior solutions relative to traditional design methodologies.

In conclusion, this work demonstrated that algorithmic topology optimization can transcend traditional design methods to synthesize wideband mm-Wave LNAs with nonintuitive layouts that achieve superior performance across multiple metrics. By cyclically co-optimizing multilayered passive networks and active device parameters in a fully autonomous manner, the synthesis methodology presented in this work enables faster design cycles and unlocks new regions of the design space that are otherwise inaccessible through manual techniques. A fabricated 3-stage LNA prototype, synthesized entirely using the algorithmic topology-optimization approach, experimentally validates its effectiveness by meeting all design constraints across a wide frequency range of 23-to-46 GHz. The chip micrograph is shown in Fig. 13.5.7.

Acknowledgement:
The authors thank Tower Semiconductor for chip fabrication. This research was partially supported by the National Science Foundation under ECCS-2229535.

References:
[1] E. A. Karahan, Z. Liu and K. Sengupta, "Deep-Learning-Based Inverse-Designed Millimeter-Wave Passives and Power Amplifiers," *IEEE Journal of Solid-State Circuits*, vol. 58, no. 11, pp. 3074-3088, Nov. 2023, doi: 10.1109/JSSC.2023.3276315. https://doi.org/10.1109/JSSC.2023.3276315
[2] J. Zhou, E. A. Karahan, S. Ghozzy, Z. Liu, H. Jalili and K. Sengupta, "AI-Enabled Design Space Discovery and End-to-End Synthesis for RFICs with Reinforcement Learning and Inverse Methods Demonstrating mm-Wave/sub-THz PAs Between 30 and 120GHz," *IEEE International Solid-State Circuits Conference (ISSCC)*, Feb. 2025, pp. 1-3, doi: 10.1109/ISSCC49661.2025.10904600. https://doi.org/10.1109/ISSCC49661.2025.10904600
[3] V. Chenna and H. Hashemi, "Topology-Optimized Nonintuitive Multilayered mm-Wave Power Amplifiers," *IEEE Radio Frequency Integrated Circuits Symposium (RFIC)*, 2025, pp. 279-282, doi: 10.1109/RFIC61188.2025.11082875. https://doi.org/10.1109/RFIC61188.2025.11082875
[4] B. Razavi, "A 300-GHz Fundamental Oscillator in 65-nm CMOS Technology," *IEEE Journal of Solid-State Circuits*, vol. 46, no. 4, pp. 894-903, April 2011, doi: 10.1109/JSSC.2011.2108122. https://doi.org/10.1109/JSSC.2011.2108122
[5] Y. O. Hassan, M. Oveisi, H. Wang and P. Heydari, "A Power-Efficient, F-Band, 6.5-dB NF, Staggered-Tuned, Inverter-Based CMOS LNA for 6G Receivers," *IEEE/MTT-S International Microwave Symposium - IMS 2024*, pp. 333-336, doi: 10.1109/IMS40175.2024.10600407. https://doi.org/10.1109/IMS40175.2024.10600407
[6] O. Kazan and G. M. Rebeiz, "A 10–110 GHz LNA with 19-25.5 dB Gain and 4.8-5.3 dB NF for Ultra-Wideband Applications in 90nm SiGe HBT Technology," *IEEE Radio Frequency Integrated Circuits Symposium (RFIC)*, 2021, pp. 39-42, doi: 10.1109/RFIC51843.2021.9490504. https://doi.org/10.1109/RFIC51843.2021.9490504

[7] W. Wu, X. Bao, S. Chen, Y. Wang and L. Zhang, "A 67.8-to-108.2GHz Power Amplifier with a Three-Coupled-Line-Based Complementary-Gain-Boosting Technique Achieving 442GHz GBW and 23.1% peak PAE," *IEEE International Solid-State Circuits Conference (ISSCC)*, Feb. 2024, pp. 526-528, doi: 10.1109/ISSCC49657.2024.10454503. https://doi.org/10.1109/ISSCC49657.2024.10454503
[8] H. Yu, L. Liu and S. Li, "A Blocker-Tolerant mm-Wave Low-Noise Amplifier Utilizing Doherty Active Load Modulation for Linearity Enhancement," *IEEE International Solid-State Circuits Conference (ISSCC)*, Feb. 2025, pp. 110-112, doi: 10.1109/ISSCC49661.2025.10904630. https://doi.org/10.1109/ISSCC49661.2025.10904630
[9] A. G. Gadelkarim and P. Mercier, "A 3.5mW mm-Wave Low-Noise Active Bandpass Filter Employing an All-Passive Interferer-Cancellation Feedforward Path," *IEEE International Solid-State Circuits Conference (ISSCC)*, Feb. 2025, pp. 1-3, doi: 10.1109/ISSCC49661.2025.10904619. https://doi.org/10.1109/ISSCC49661.2025.10904619
[10] J. Park and H. Wang, "A 26-to-39GHz Broadband Ultra-Compact High-Linearity Switchless Hybrid N/PMOS Bi-Directional PA/LNA Front-End for Multi-Band 5G Large-Scaled MIMO System," *IEEE International Solid-State Circuits Conference (ISSCC)*, Feb. 2022, pp. 322-324, doi: 10.1109/ISSCC42614.2022.9731651. https://doi.org/10.1109/ISSCC42614.2022.9731651
[11] Z. Zhao *et al.*, "Design and Analysis of a 22.6-to-73.9 GHz Low-Noise Amplifier for 5G NR FR2 and NR-U Multiband/Multistandard Communications," *IEEE Journal of Solid-State Circuits*, vol. 60, no. 9, pp. 3189-3201, Sept. 2025, doi: 10.1109/JSSC.2025.3545463. https://doi.org/10.1109/JSSC.2025.3545463
[12] J. -H. Kim, J. -T. Son, J. -T. Lim, H. -W. Choi and C. -Y. Kim, "Ultralow Noise Figure and Broadband CMOS LNA With Three-Winding Transformer and Large Transistor," *IEEE Transactions on Microwave Theory and Techniques*, vol. 72, no. 5, pp. 2734-2744, May 2024, doi: 10.1109/TMTT.2024.3354908. https://doi.org/10.1109/TMTT.2024.3354908
[13] J. Fu, C. Song, Y. Wang and L. Wu, "A 4.4-mW 19–46-GHz Low-Noise Amplifier with Pole-Converging Gain Flattening and Triple-Resonance Input Matching," *IEEE Radio Frequency Integrated Circuits Symposium (RFIC)*, 2024, pp. 311-314, doi: 10.1109/RFIC61187.2024.10600028. https://doi.org/10.1109/RFIC61187.2024.10600028
[14] M. R. Khan, C. L. Zekios, S. Bhardwaj and S. V. Georgakopoulos, "Multiobjective Fitness Functions With Nonlinear Switching for Antenna Optimizations," *IEEE Open Journal of Antennas and Propagation*, vol. 3, pp. 613-626, 2022, doi: 10.1109/OJAP.2022.3178840. https://doi.org/10.1109/OJAP.2022.3178840
[15] V. Chenna and H. Hashemi, "Algorithmic Design of Nonintuitive on-Chip Multilayered Passive Networks," *IEEE/MTT-S International Microwave Symposium - IMS 2025*, pp. 918-921, doi: 10.1109/IMS40360.2025.11103811. https://doi.org/10.1109/IMS40360.2025.11103811
[16] X. Huang, H. Jia, W. Deng, Z. Wang and B. Chi, "A Compact E-Band Load-Modulation Balanced Power Amplifier in 65-nm CMOS," *IEEE Journal of Solid-State Circuits*, vol. 59, no. 10, pp. 3172-3182, Oct. 2024, doi: 10.1109/JSSC.2024.3404610. https://doi.org/10.1109/JSSC.2024.3404610

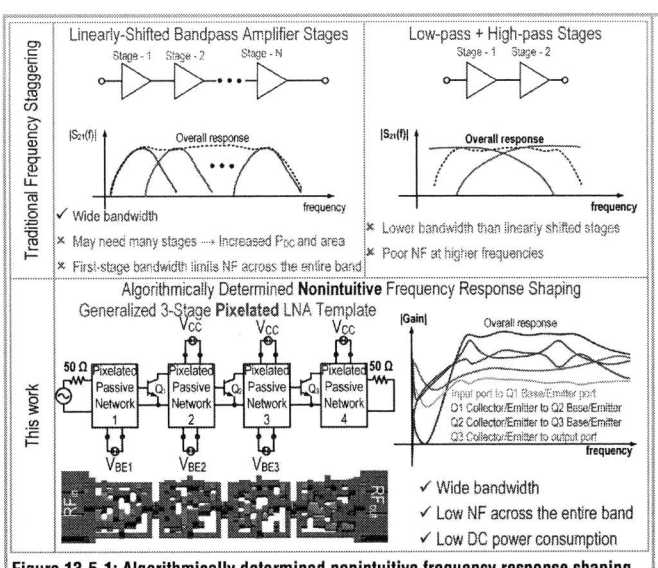

Figure 13.5.1: Algorithmically determined nonintuitive frequency response shaping.

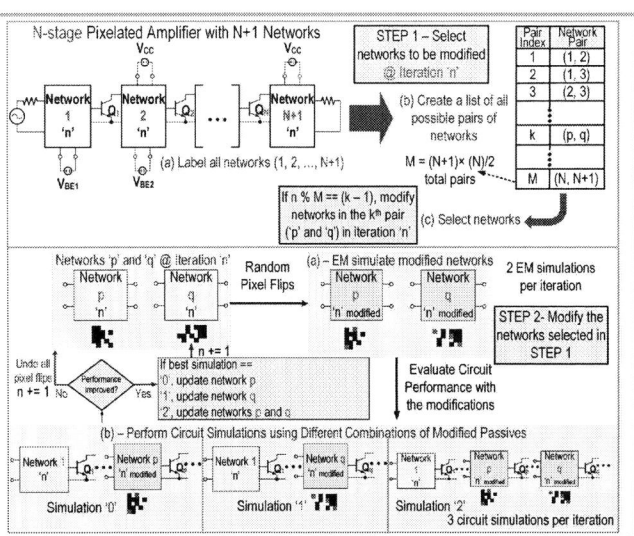

Figure 13.5.2: Strategy to update passive networks for amplifiers with any number of stages.

ISSCC 2026 / SESSION 13 / CIRCUITS FOR AI AND AI FOR CIRCUITS / 13.5

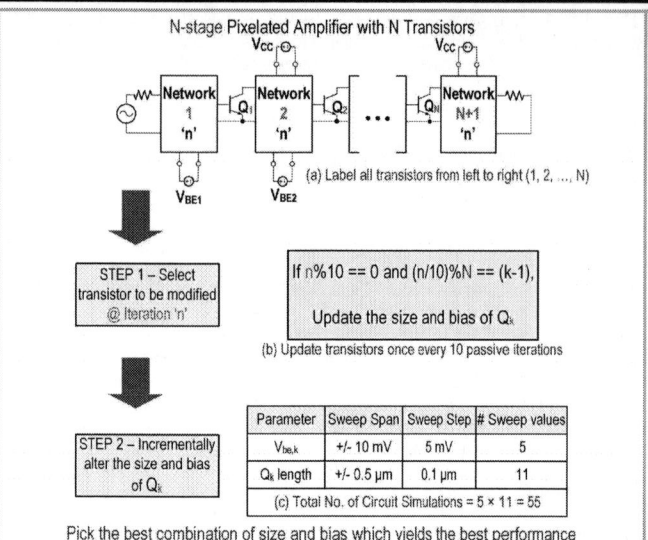

Figure 13.5.3: Cyclic strategy to periodically update the transistor size and bias.

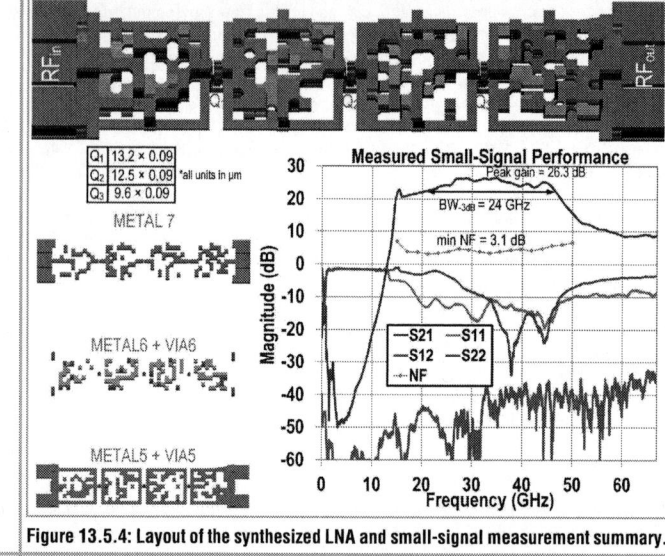

Figure 13.5.4: Layout of the synthesized LNA and small-signal measurement summary.

Gain Compression @ 30 GHz

IP1dB = -17.9 dB
OP1dB = IP1dB + Gain|IP1dB = +7 dBm

IIP3 @ f_c = 30 GHz, Δf = 1 GHz

Fundamental
IIP3 = -9.3 dBm
OIP3 = IIP3 + Gain = +16.6 dBm
IM3

IB Linearity VS Frequency

IIP3 > -10.8 dBm in-band
IP1dB > -17.9 dBm in-band

64 QAM – 400 Mbaud – 2.4 Gbps
EVM = 5.2% = -25.57 dB
$P_{in,avg}$ = -29 dBm, f_{IN} = 30 GHz

16 QAM – 1.5 Gbaud – 6 Gbps
EVM = 8.6% = -21.31 dB
$P_{in,avg}$ = -27.5 dBm, f_{IN} = 30 GHz

Figure 13.5.5: Linearity and modulation measurements.

Performance Metric	This work		ISSCC 2025 [8]	ISSCC 2025 [9]	ISSCC 2022 [10]	JSSC 2025 [11]	TMTT 2024 [12]	RFIC 2024 [13]
Architecture/ Technique	Topology-Optimized 3-Stage Frequency-Staggered LNA		Doherty LNA	LNA BPF	Bidirectional PA/LNA	Common Emitter +Shunt Feedback + Dual inductive peaking	3-winding Transformer	Gain flattening + Triple Resonance Matching
Small Signal Performance Metrics								
f_c (GHz)	34		27.2	24.5	32.5	48.25	38.1	32.5
Peak Gain (dB)	26.3		18.4	11.6	17.6	15.2	21.1	12.4
BW (GHz)	24 (70.6 %)		8.4 (31 %)	1.32 (5.4 %)	11 (33.8 %)	51.3 (106.3 %)	20 (53%)	27 (83.1 %)
G×BW Product	496 GHz		70 GHz	5 GHz	83.5 GHz	295.2 GHz	227 GHz	112.6 GHz
Min NF (dB)	3.1		2.5	3.6	5.2	4.06	2.2	3.4
Linearity and Modulation Metrics								
IB IIP1dB (dBm)	> -17.9		-9.3	> -20.4	> -8.6	> -20*	-17.2ᵃ	> -19.4
IB OP1dB (dBm)	> +7		+9.1	> -8.8	> 9	> -14*	+3.9ᵃ	> -7
IB IIP3 (dBm)	> -10.8		-2.5	> -11.9	> +0.9	> -6.8*	-7.6	N/R
IB OIP3 (dBm)	> +15.2		15.9	> -0.3	> +18.5	> +7*	+13.5	N/R
Modulation	64 QAM	16 QAM	64 QAM	1024 QAM	N/R	N/R	N/R	N/R
Data Rate	2.4 Gbps	6 Gbps	2.4 Gbps	3 Gbps	N/R	N/R	N/R	N/R
EVM (dB)	-25.6	-21.3	-27.13	N/R	N/R	N/R	N/R	N/R
Chip Metrics								
Technology	180 nm BiCMOS		65 nm CMOS	22 nm FDX	45 nm SOI	130 nm BiCMOS	65 nm CMOS	40 nm CMOS
P_{DC} (mW)	18		14.6	3.5	66	17.5	22	4.4
Core Area (mm²)	0.25		> 0.55*	0.07	0.19	0.06	0.16	0.096
Figures of Merit								
FoM_1	546		51.14	4.23	4.14	62.75	177.6	89.8
FoM_2	10.9		6	0.04	0.57	0.63	3.4	1.03

FoM_1 =(Power Gain [lin.] × BW [GHz])/(P_{DC} [mW] × (F-1)) FoM_2 =FoM_1 × IP1dB [mW]
* – estimated from figures, a – IP1dB [dB] estimated as IIP3 – 9.6 dB, N/R – not reported

Figure 13.5.6: Performance comparison with state-of-the-art mm-Wave LNAs.

Topology-Optimized mm-Wave Low-Noise Amplifier

Top-cell bypass capacitors

LNA Core

Top-cell bypass capacitors

LNA Core

Core Area =
1 mm × 0..25 mm =
= 0.25 mm²

Figure 13.5.7: Chip Micrograph.

Session 14 Overview: *Unusual Interconnects and Other Uses for Light*

TECHNOLOGY DIRECTIONS SUBCOMMITTEE

Session Chair: Uygar Avci
Intel, Hillsboro, OR

Session Co-Chair: Guy Torfs
Ghent University, Ghent, Belgium

Advances in photonics and novel interconnects are enabling breakthroughs in bandwidth, energy efficiency, and novel system applications. The session begins with a terahertz-coupled through-silicon-like interconnect module for 3D integration, achieving high energy efficiency at very fast data rates. High-performance optical transmitters follow, leveraging WDM in CMOS SOI and high-efficiency VCSEL drivers in GaN HEMT for parallel wireless links. Next is a sensing technology using a single-chip FMCW LiDAR laser diode driver using a flexible IGZO-based light sensing system with RF harvesting. The session concludes with a self-programmable twin PUF architecture powered by photovoltaic energy harvesting before wafer dicing, providing a unique approach to hardware security.

ISSCC 2026 / February 17, 2026 / 10:05 AM

10:05 AM

14.1 THz-TSI: A 0.33pJ/b 264Gb/s Through-Silicon Interconnect Module for 3D Integration Utilizing Terahertz Coupling

Chen Jiang, Fudan University, Shanghai, China

In Paper 14.1, Fudan University presents a through-silicon interconnect module for 3D integration utilizing terahertz coupling. It achieves a data-rate of 264Gb/s with an energy efficiency of 0.33pJ/b, supports both bi-directional point-to-point and broadcast modes of communication, in addition to reduced fabrication complexity and cost.

10:30 AM

14.2 An 8λ×38Gb/s/λ 106fJ/b Optical WDM Transmitter in 45nm CMOS SOI

Amirreza Shoobi, University of Pennsylvania, Philadelphia, PA

In Paper 14.2, University of Pennsylvania demonstrates an ultra-energy-efficient microring modulator-based 8-channel optical WDM NRZ transmitter with monolithic electronic–photonic integration. The design uses 0.018mm²-per-channel area and has autonomous wavelength-locking with near-zero static power, achieving 304Gb/s aggregate throughput at 106fJ/b efficiency.

10:55 AM

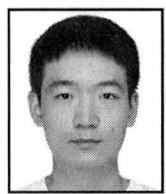

14.3 A Single-Chip Laser Diode Driver with Built-In Frequency-Sweep Linearization for FMCW LiDAR

Wenshuo Zhu, Iowa State University, Ames, IA

In Paper 14.3, Iowa State University develops a laser diode driver with Built-In Frequency-Sweep Linearization for FMCW LiDAR applications. The design achieves this without external interferometers or detectors, in 180nm CMOS, outperforming prior art in integration and shows mobile LiDAR suitability.

14

11:20 AM

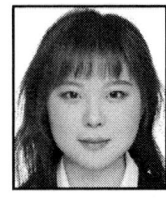

14.4 A 40Gb/s 8mW-OMA 1-to-N VCSEL Driver for Parallel and Wireless Optical Links Using 150nm GaN HEMT

Shuo Sarah Feng, Hong Kong University of Science and Technology, Hong Kong, China

In Paper 14.4, Hong Kong University of Science and Technology introduces a 40Gb/s 1-to-N VCSEL driver built on 150nm GaN HEMT technology with f_T/f_{MAX} of 50GHz/154GHz. It drives VCSEL arrays in series or parallel, delivering 8mW OMA at 40Gb/s NRZ for four VCSELs with 3-tap FFE under 15V, consuming 480mW, and supporting up to 38Gb/s PAM-4.

11:35 AM

14.5 Highly-Integrated Light-Sensing System with RF Harvesting and Transmission in Commercial N-Type IGZO Flexible Technology

Marco Privitera, University of Catania, Catania, Italy, National University of Singapore, Singapore, Singapore

In Paper 14.5, University of Catania and National University of Singapore present a flexible, sticker-like light sensing system based on N-type-only IGZO TFT technology, integrating RF harvesting and backscatter communications. The 2.02mm² system operates at 15.5µW (<1% duty cycle), reuses resistors as light sensors, and demonstrates indoor light sensing with 0.9GHz transmission.

11:50 AM

14.6 Self-Programmable Twin PUFs via Photovoltaic Energy Harvesting During the Pre-Wafer-Dicing Stage

Eunseok Lee, Massachusetts Institute of Technology, Cambridge, MA

In Paper 14.6, MIT proposes a CMOS self-programmable twin PUF formed at the pre-dicing stage via photovoltaic harvesting. Two adjacent PUF macros share entropy during oxide breakdown using a 4-T circuit, achieving 98.4% pairwise consistency with a 0.4% native BER, and can be generated through wafer-level batch processing using a low-cost LED.

DIGEST OF TECHNICAL PAPERS • 239

979-8-3315-8937-0/26 $31.00 © 2026 IEEE

ISSCC 2026 / SESSION 14 / UNUSUAL INTERCONNECTS AND OTHER USES FOR LIGHT / 14.1

14.1 THz-TSI: A 0.33pJ/b 264Gb/s Through-Silicon Interconnect Module for 3D Integration Utilizing Terahertz Coupling

Chen Jiang, Xiaodi Feng, Xiaohan Shen, Chixiao Chen, Qi Liu, Ming Liu, Ningsheng Xu

Fudan University, Shanghai, China

Abstract

A through-silicon interconnect module for 3D integration utilizing THz coupling is presented, which achieves a record high data rate of 264Gb/s and efficiency of 0.33pJ/b. Both bi-directional point-to-point link and broadcast mode are successfully demonstrated. This scheme offers highly competitive bandwidth and density with significantly simpler fabrication and lower cost. Furthermore, its high flexibility enables real-time adjustable interconnect topology to effectively reduce latency.

The 3D integration technology has been widely recognized as a critical catalyst for high-performance computing (HPC) and artificial intelligence (AI) integrated systems, which enables a 10-to-100x memory bandwidth enhancement through innovations like high bandwidth memory (HBM) [1,2]. In the large language model era, an exponential surge in model parameters is placing an unprecedented demand on interconnect performances across the dies. However, current 3D integration, which relies on through-silicon vias (TSVs), faces significant challenges on further scaling. As the TSV density increases, the already costly and complicated fabrication process faces escalating yield and reliability issues, which mandates sub-micron alignment precision to mitigate the risk of electrical failures like opens and shorts, as well as mechanical stress and warpage [3]. To overcome these challenges, wireless through-silicon interconnects (TSI) have been proposed as a reliable, flexible, and scalable alternative solution [4,5]. However, the direct transfer of baseband data using inductive coupling introduces a stringent tradeoff between signal quality and inductor size. Therefore, this method requires extensive die thinning (to below 10μm [5]) to facilitate vertical communication across multiple dies. In this work, a TSV-free terahertz TSI (THz-TSI) utilizing near-field wireless communication technique is proposed. With the vast bandwidth and short wavelengths of the THz waves, exceptionally high data rates are achieved within a compact physical layout, enabling substantial bandwidth and density improvement. In the 3D integration applications, the sub-millimeter-level distances only require low transmission power, thereby circumventing the challenge of high-power signal generation at THz frequencies. Another key advantage of this scheme is its architectural flexibility. High-speed, point-to-point links can be established between any two dies in the stack. Moreover, direction of these links can be dynamically configured to fit the read and write tasks. This bi-directional capability helps to significantly improve the link utilization and reduce latency compared to conventional uni-directional designs [6]. This scheme also incorporates a broadcast mode to efficiently support one-to-many data distribution scenarios. To validate the proposed scheme, a 4-layer stacked THz-TSI prototype was implemented, which supports both NRZ and PAM4 signaling. Operating at a 154GHz carrier frequency, the prototype achieves a peak data rate of 264Gb/s at an energy efficiency of 0.33pJ/b. Furthermore, both bi-directional point-to-point and broadcast communication modes were successfully demonstrated.

The proposed THz-TSI topology is illustrated in Fig. 14.1.1. Two transceiver (TRX) channels are implemented on each die, which share a common VCO for LO generation. The close proximity among dies is leveraged to enable direct VCO coupling to achieve LO synchronization. This approach eliminates the need for power and area consuming frequency synthesizers [7], therefore the overall energy efficiency and interconnect density are significantly enhanced. The proposed THz coupler is based on a circular slot structure, which provides robust and wideband coupling under both point-to-point and broadcast modes. The performance is confirmed by the 5-layer stack simulations (Fig. 14.1.1), which show that, even the channel loss and bandwidth degrade as the distance increases, a 45GHz bandwidth is still maintained under worst-case conditions. To ensure robust performance across diverse operating scenarios, the TRX can be configured for either NRZ or PAM4 signaling to adapt to different SNR conditions. Furthermore, the RX path incorporates adjustable VGA and CTLE stages to compensate for different channel responses.

The reconfigurable THz TRX architecture is shown in Fig. 14.1.2. Switching among different operating modes is realized by a transmission line (TL) to slot mode conversion mechanism and a set of parallel switches. In TX mode ($V_{ct\_rx}=0$), TL_1 resonates with the parasitic cap of M_1 and presents a high impedance, which disconnects the RX from the slot structure. Then, the magnitude and direction of the E-fields in the slot is modulated by the differential inputs, V_{inp} and V_{inn}, forming NRZ or PAM4 signals that are sent to the THz coupler. Simulation results show that the modulator achieves a minimal LO-to-RF modulation loss of 4.2dB and an excellent OP1dB of 3.5dBm while consuming zero dc power. In RX mode ($V_{ct\_rx}=0.9V$, $V_{inp}=V_{inn}=0$), the symmetrical excitation decouples the modulator from the slot, allowing the RF signal to pass to the RX. The RX mixer employs gate-injection of combined LO and RF signals. Leveraging the high gate impedance to boost voltage swing, conversion gain and noise figure (NF) are enhanced. A slot-balun structure with even/odd mode excitation is used for this combination, avoiding the 3dB loss inherent to conventional power combiners [7]. Under a -5dBm LO input, the mixer achieves a low NF of 14.5dB. In OFF mode ($V_{ct\_rx}=V_{inp}=V_{inn}=0$), both the TX and RX are disconnected from the THz channel to minimize the loading effect. With the flexibility provided by the TRX design, the THz-TSI module can be configured for both bi-directional point-to-point and broadcast modes.

The VCO structure and direct coupling mechanism are shown in Fig. 14.1.3. To optimize LO power generation efficiency, the design employs a fundamental oscillator with a self-feeding structure [8]. Mutual coupling among the VCOs arises from standing waves within the slot lines. Shape of these slot lines is designed to balance the coupling strength and radiation loss. The VCO's free-running frequency is adjustable by tuning the gate bias [9]. To support both TRX channels on each die, every VCO generates four LO outputs. The TX path incorporates LO buffers to provide additional power gain and to isolate the VCO from the modulator, which prevents load-pulling effects that could disrupt the oscillation. Simulations indicate that, consuming 32mW power, the VCO delivers 0.2dBm of power to each TX and -1.3dBm to each RX.

A 4-layer stacked prototype, depicted in Fig. 14.1.3, was assembled to validate the THz-TSI design. Wire bonds were selected as a cost-effective solution for providing dc power and SPI control signals. However, production implementations may consider vertical fan out packaging [10] to create a more sophisticated power distribution network. To ensure pad accessibility after die stacking, two variants were fabricated in a 28nm CMOS technology, differing only in pad locations (Fig. 14.1.7). Each die also has duplicated DC and SPI pads on both sides. The 4-layer sample was diced and arranged to allow pad access to all layers. A dummy die was placed on top of the stack to ensure a consistent EM environment for VCO locking. The wireless coupling scheme significantly relaxes alignment tolerances, making low-cost die-attach techniques sufficient for assembly. Limited by our packaging facility capability, all dies are only thinned to 100μm. Further substrate thinning would effectively improve both coupling loss and bandwidth.

The VCO frequencies and spectrum were measured by capturing leaked radiation from the sample with a horn antenna. This method eliminates the need for on-chip THz test structures that would otherwise degrade circuit performance. A spectrum analyzer with a frequency extender was used to analyze the spectrum. With only 1 of the 4 stacked dies powered on, the VCO frequency ranges of each die were measured. Fig. 14.1.4 illustrates the VCO spectrum before and after locking, showing a locked LO frequency of 153.8GHz. Under mutual injection, the relative LO phase of each layer is adjustable by tuning their free-running frequencies [7,11]. With this mechanism, VCOs were tuned to adjust the TRX LO phases to maximize the output signal amplitude for an optimized SNR (Fig. 14.1.4). The measured isolation between the two channels is better than 25dB up to 40GHz (Fig. 14.1.4), therefore the impact of crosstalk is negligible for NRZ/PAM4 signals up to 80Gbaud.

To evaluate the link quality, DC-to-50GHz RF probes were used to send and acquire signals from the sample. An arbitrary waveform generator provided the test patterns, while a sampling oscilloscope was used to analyze the received signal. The measured eye diagrams under different modes and test conditions are shown in Fig. 14.1.5 and Fig. 14.1.6. Since CH1 and CH2 show very symmetrical performances, only CH1 results are reported. First, a point-to-point link was established by configuring L4 as the TX and L3 as the RX, while L1/L2 were deactivated. After de-embedding the frequency response of the probes and cables, the raw signals at chip out were obtained. Leveraging the large bandwidth, peak data rates of 58Gb/s for NRZ and 86Gb/s for PAM4 signaling were achieved in each channel without any equalization. To further increase the data rate, an offline adaptive linear FIR filter was used for equalization. This post-processing boosted the per-channel data rates to 90Gb/s and 132Gb/s for NRZ and PAM4 signaling, respectively. At the peak total data rate of 264Gb/s, the L3 and L4 THz-TSI modules consumed 43.2mW and 44.7mW, respectively. This corresponds to an excellent energy efficiency of 0.33pJ/b. It's worth mentioning that this link is bi-directional. While limited pad access here prevents the demonstration, it was successfully validated on a 2-layer sample with all TX/RX pads accessible. To verify the broadcast capability, L2 was first configured as TX while L1 and L3 as RX. This link successfully demonstrated clear PAM4 eye diagrams at 124Gb/s/ch. In a second test, where L4 broadcasted to L1 and L3, the maximum data rate demonstrated was 80Gb/s/ch. This limitation is due to the longer L4-L1 distance and signal blockage by L3, which degrade the link's SNR and bandwidth. These findings are consistent with the simulation results in Fig. 14.1.1.

A performance summary and comparison table is given in Fig. 14.1.6. Thanks to the large bandwidth achieved by THz coupling, as well as circuit design innovations, this work achieves a record high data rate, energy efficiency and interconnect density among all previously published designs on wireless 3D interconnect. The performance is also highly competitive with state-of-the-art TSV-based solutions in advanced technology nodes. The wireless nature of this technique can significantly simplify 3D integration and fabrication, substantially lowering cost while improving yield and reliability. Furthermore, this highly flexible architecture allows task-based, real-time reconfigurable interconnect topologies with dynamic node and transmission direction control, as well as broadcast capabilities, which will substantially reduce interconnect latency. With these advantages, the proposed THz-TSI technique has great potential to unlock new possibilities for chiplet architectures, and revolutionize future HPC and AI integrated systems.

Acknowledgement:
This work was supported by the National Natural Science Foundation of China under Grant 92473101 and 62488101, by the Science and Technology Commission of Shanghai Municipality under Grant 25JD1400500, by the State Key Laboratory of Integrated Chips and Systems under Grant SKLICS-Z202505, by the Fudan University Nano Information Science Innovation Support Center Project, Ministry of Education of China, and by the Nano Institute of Fudan University. Corresponding author: Chen Jiang (cjiang@fudan.edu.cn).

References:
[1] M. Park *et al.*, "A 192-Gb 12-High 896-GB/s HBM3 DRAM With a TSV Auto-Calibration Scheme and Machine-Learning-Based Layout Optimization," *IEEE J. Solid-State Circuits*, vol. 58, no. 1, pp. 256-269, Jan. 2023. https://doi.org/10.1109/JSSC.2022.3193354
[2] K. Chae *et al.*, "A 4-nm 1.15 TB/s HBM3 Interface With Resistor-Tuned Offset Calibration and In Situ Margin Detection," *IEEE J. Solid-State Circuits*, vol. 59, no. 1, pp. 231-242, Jan. 2024. https://doi.org/10.1109/JSSC.2023.3330485
[3] J. Wang *et al.*, "A short review of through-silicon via (TSV) interconnects: metrology and analysis," *Applied Sciences*, vol. 13, no. 14, Jul. 2023. https://doi.org/10.3390/app13148301
[4] B. J. Fletcher, T. Mak, and S. Das, "A 3D-stacked Cortex-M0 SoC with 20.3Gbps/mm² 7.1mW/mm² simultaneous wireless inter-tier data and power transfer," *IEEE Symp. on VLSI Circuits*, Jun. 2020. https://doi.org/10.1109/VLSICircuits18222.2020.9162824
[5] K. Shiba *et al.*, "A 7-nm FinFET 1.2-Tb/s/mm² 3D-stacked SRAM module with 0.7-pJ/b inductive coupling interface using over-SRAM coil and Manchester-encoded synchronous transceiver," *IEEE J. Solid-State Circuits*, vol. 58, no. 7, pp. 2075–2086, Jul. 2023. https://doi.org/10.1109/JSSC.2022.3224421
[6] B. Jiao *et al.*, "SHINSAI: a 586mm² reusable active TSV interposer with programmable interconnect fabric and 512Mb 3D underdeck memory," *IEEE Int. Solid-State Circuits Conf. (ISSCC)*, Feb. 2025. https://doi.org/10.1109/ISSCC49661.2025.10904819
[7] C. Jiang *et al.*, "A 1.25pJ/b 73Gb/s/ch 210GHz transceiver front-end for wireless 3D IC thru-silicon interface," *IEEE Euro. Solid-State Electron. Res. Conf. (ESSERC)*, Sep. 2025. https://doi.org/10.1109/ESSERC66193.2025.11214044
[8] R. Han *et al.*, "A SiGe terahertz heterodyne imaging transmitter with 3.3 mW radiated power and fully-integrated phase-locked loop," *IEEE J. Solid-State Circuits*, vol. 50, no. 12, pp. 2935 - 2947, Dec. 2015. https://doi.org/10.1109/JSSC.2015.2471847

[9] C. Jiang, A. Cathelin, and E. Afshari, "An efficient 210GHz compact harmonic oscillator with 1.4dBm peak output power and 10.6% tuning range in 130nm BiCMOS," *IEEE Radio Freq. Integr. Circuits Symp.*, May 2016. https://doi.org/10.1109/RFIC.2016.7508284
[10] K. Sung *et al.*, "Vertical Fan Out (VFO) package with enhanced form factor and performances for mobile applications," *IEEE Electronic Components and Technology Conference (ECTC)*, pp. 34-39, May 2024. https://doi.org/10.1109/ECTC51529.2024.00014
[11] B. Razavi, "A study of injection locking and pulling in oscillators," *IEEE J. Solid-State Circuits*, vol. 39, no. 9, pp. 1415–1424, Sep. 2004. https://doi.org/10.1109/JSSC.2004.831608
[12] P. Guan *et al.*, "A fully integrated QPSK/16-QAM D-band CMOS transceiver with mixed-signal baseband circuitry realizing digital interfaces," *IEEE J. Solid-State Circuits*, vol. 59, no. 10, pp. 3123 - 3141, Oct. 2024. https://doi.org/10.1109/JSSC.2024.3432759
[13] S. Callender *et al.*, "A fully integrated 160-Gb/s D-band transmitter achieving 1.1-pJ/b efficiency in 22-nm FinFET," *IEEE J. Solid-State Circuits*, vol. 57, no. 12, pp. 3582–3598, Dec. 2022. https://doi.org/10.1109/JSSC.2022.3208510

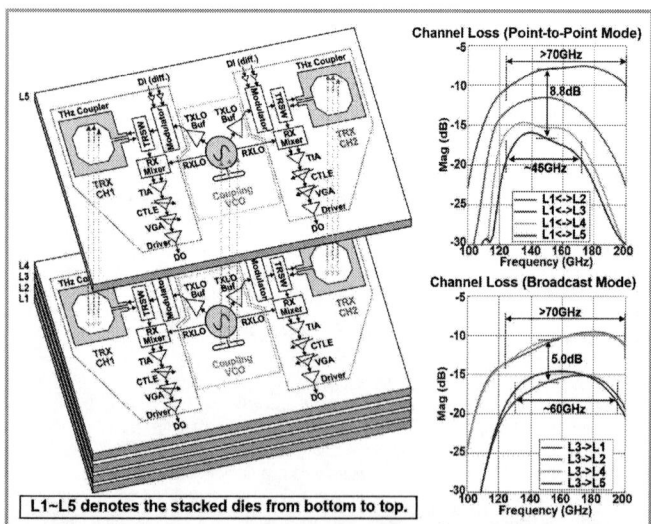

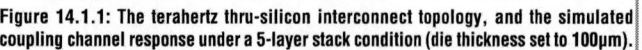

Figure 14.1.1: The terahertz thru-silicon interconnect topology, and the simulated coupling channel response under a 5-layer stack condition (die thickness set to 100µm).

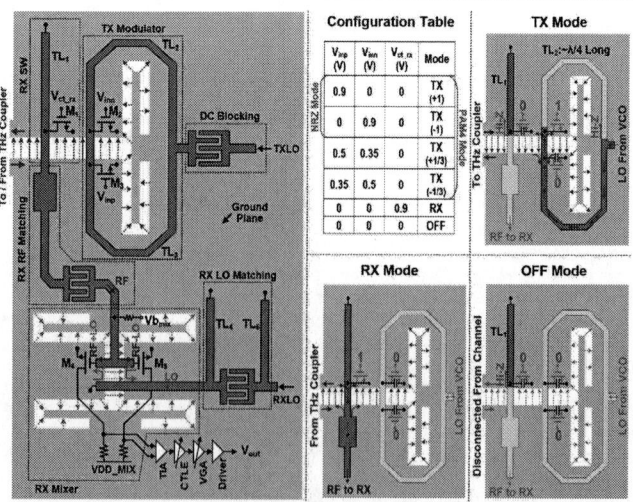

Figure 14.1.2: The reconfigurable terahertz transceiver architecture and its working principle.

ISSCC 2026 / SESSION 14 / UNUSUAL INTERCONNECTS AND OTHER USES FOR LIGHT / 14.1

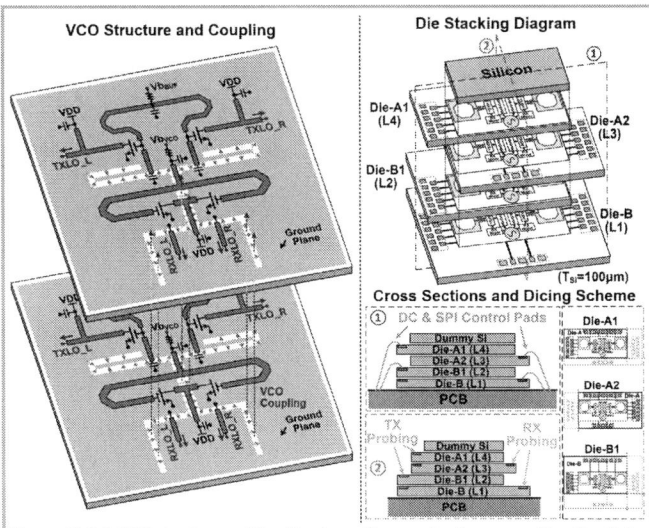

Figure 14.1.3: VCO structure and the direct coupling mechanism, as well as the dicing and stacking scheme of the 4-layer prototype.

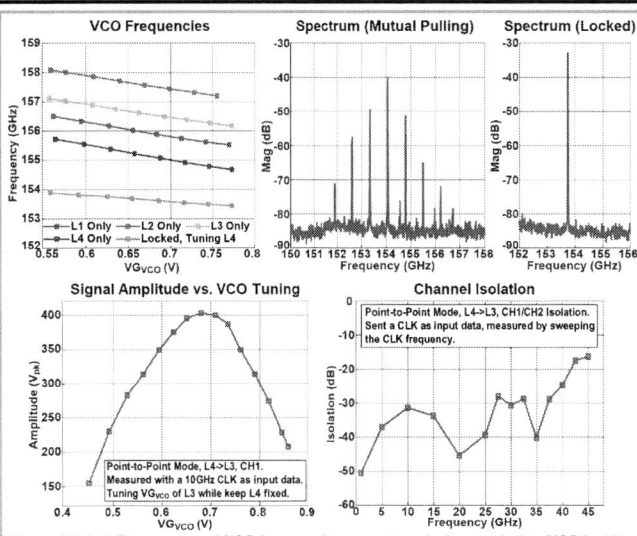

Figure 14.1.4: The measured VCO frequencies, spectrum before and after VCO locking, output signal amplitude optimization, as well as channel isolation.

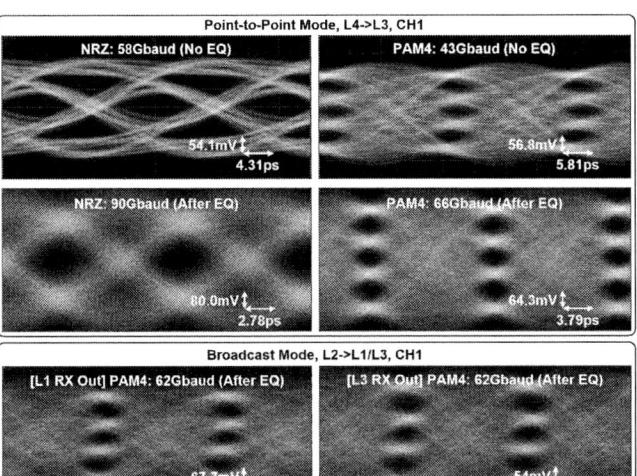

Figure 14.1.5: The measured eye diagrams under different modes and test conditions.

Performance Summary and Comparison

	Link Type	Modulation	Data Rate [Gb/s]	Energy Efficiency [pJ/bit]	Interconnect Density [1] [Tb/s/mm²]	Die Thinning [μm]	Link Direction	Broadcast Mode	Process
JSSC 2023 [1]	TSV+μBump (for HBM3)	NRZ	7.0	Not Reported	0.66 [2]	30	Uni-Direction	No	1z-nm DRAM
JSSC 2024 [2]	TSV+μBump (for HBM3)	NRZ	9.0	0.29	Not Reported	Not Reported	Uni-Direction	No	4nm CMOS
VLSI 2020 [4]	Wireless (Ind. Coupling)	NRZ	1.27	17.0	0.02	70	Uni-Direction	No	65nm CMOS
JSSC 2023 [5]	Wireless (Ind. Coupling)	NRZ	8.5	0.7	0.53	8	Uni-Direction	No	7nm CMOS
ESSERC 2025 [7]	Wireless (Modulated)	QPSK	36.5×2	1.25	0.47	100	Uni-Direction	No	28nm CMOS
		PAM4	71.4	1.22	0.53				
This Work	Wireless (Modulated)	NRZ	90×2	0.49	0.51	100	Bi-Direction	Yes	28nm CMOS
		PAM4	132×2	0.33	0.75				

(1) Interconnect Density = Data Rates / Area Overhead (includes TRX, coupler, TSV, keep out regions, etc).
(2) Estimated based on bump pitch and data rate per pin.

Figure 14.1.6: More measured eye diagrams, as well as the performance summary and comparison table.

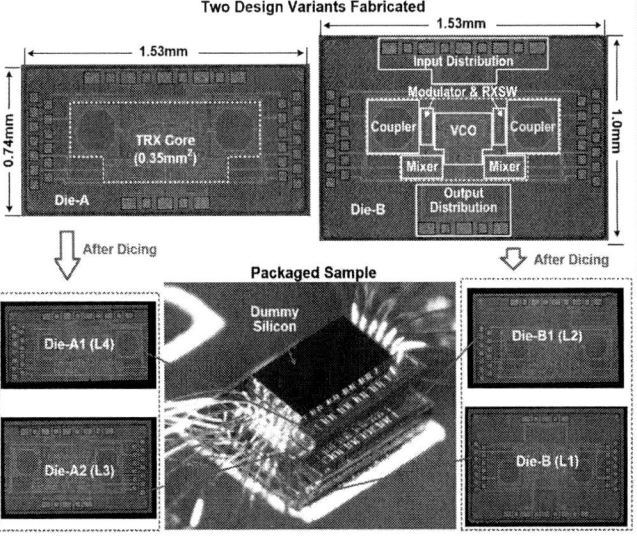

Figure 14.1.7: Die micrographs and the packaged 4-layer sample.

• 2026 IEEE International Solid-State Circuits Conference

979-8-3315-8937-0/26 $31.00 © 2026 IEEE

ISSCC 2026 / SESSION 14 / UNUSUAL INTERCONNECTS AND OTHER USES FOR LIGHT / 14.2

14.2 An 8λ×38Gb/s/λ 106fJ/b Optical WDM Transmitter in 45nm CMOS SOI

Amirreza Shoobi, Kaisarbek Omirzakhov, Zhehao Yu, Ali Pirmoradi, Firooz Aflatouni

University of Pennsylvania, Philadelphia, PA

Abstract

This paper presents an ultra-energy-efficient MRM based 8-channel optical WDM NRZ transmitter, employing a scalable driver topology combined with monolithic electronic–photonic integration, which minimizes packaging parasitics and enables a compact design within a footprint of 0.018mm² per-channel. The chip incorporates an autonomous concurrent wavelength-locking mechanism with near-zero static power consumption, achieving an aggregate data rate of 304Gb/s at 106fJ/b energy efficiency.

High-performance computing systems and modern data centers are experiencing unprecedented compute demand driven by data-intensive applications such as generative artificial intelligence (AI). Optical links have emerged as a transformative interconnect solution within data centers, offering high bandwidth, low latency, and low loss. In particular, wavelength-division multiplexing (WDM) enables scaling in aggregate capacity by utilizing multiple optical carriers with different wavelengths. Despite significant advances in silicon photonic interconnects, energy efficiency and bandwidth density remain critical constraints.

This work presents a compact energy-efficient 8-channel optical WDM NRZ transmitter (OTX) integrated using GlobalFoundries 45CLO CMOS SOI process. The monolithic integration of compact multi-section capacitively tuned micro-ring modulator (MRM) devices, scalable ultra-high energy-efficient wideband MRM drivers, autonomous wavelength locking and tracking systems with a near-zero power consumption, as well as careful mm-wave and optical routing result in a record energy efficiency of 106fJ/b when 8 optical carriers are simultaneously modulated at 38Gb/s/carrier, corresponding to an aggregate data-rate of 304Gb/s. The chip was fabricated with a per-channel footprint of 0.018mm², resulting in an areal bandwidth density of 2.1Tb/s/mm². Figure 14.2.1 (top-left) shows a survey of existing CMOS transmitters and provides a comparison between this work and the state-of-the-art, where the implemented transmitter chip achieves more than 10 times better (energy-efficiency of driver and modulator) x (transmitter active core area) while all 8 channels are operating simultaneously at an aggregate data-rate of 304Gb/s.

Figure 14.2.1 (top right) shows the conventional 4-transistor stacked push-pull topology, that is widely used with some variations in prior works [1-8], which can typically achieve a differential swing of up to 4xVDD. Typically, mid-point transistors (M_9 to M_{12}) are employed to suppress voltage overstress and accelerate the charge–discharge process. Despite excellent performance, this topology often requires utilization of level-shifters to set correct DC values for the two inputs (i.e., $D_{in,L}$ and $D_{in,H}$), which could increase the overall power consumption and latency, decrease the bandwidth, and require careful delay matching between the two input signals.

Figure 14.2.1 (bottom left) shows the schematic of the proposed modulator driver, where the requirement for the two DC levels for the input data, and thus level shifters, is eliminated. In this topology, the differential input is connected to the sources of M_5 and M_6 through a pair of ac-coupling capacitors (C_{in}) [9]. As a result, when D_{in} changes from gnd to VDD, in steady state, the voltage at node S_1 increases from VDD to near 2xVDD, and since M_3 is ON, D_{outp} changes from gnd to near 2xVDD. Similarly, when D_{in} falls from VDD to gnd, D_{outn} changes from gnd to near 2xVDD. Consequently, the proposed driver can provide an output voltage swing of nearly 4xVDD, which is comparable with the output voltage swing generated by the conventional architecture in Fig. 14.2.1 (top right), but at a lower power consumption. Note that the supply voltage for the proposed and the conventional driver (top right of Fig. 14.2.1) is VDD and 2xVDD, respectively. Also note that, based on simulations, the output swing is affected by the total parasitic capacitance, C_p, at nodes S_1 and S_2, and is approximately proportional to $(C_{in}+ C_p)/C_{in}$. Hence, the value of C_{in} must be carefully chosen, given the trade-off between the input capacitance and output swing.

Figure 14.2.1 (bottom right) shows the structure of the MRM used in this work. The design of the MRM is inspired by the device in [6], where a p-n section is used for high-frequency and wide-band modulation, a capacitive section is utilized for fine-tuning and wavelength locking with near-zero static power consumption, and a heater (i.e., a doped silicon resistor) is placed inside the MRM for thermal coarse-tuning if necessary. To ensure 8 MRMs can effectively be used to modulate 8 optical carriers with 1.6nm spacing concurrently, the straight section of the MRM, L_1, is increased with an increment of 30nm across the eight rings. Note that since the free spectral range (FSR) of the MRMs, which is about 8.65nm on average, is smaller than the system bandwidth (i.e., 8 optical carriers with 1.6nm carrier-to-carrier spacing), two consecutive resonances of some MRMs will appear within the range of interest, which has been considered in the frequency planning of the system to minimize the channel-to-channel crosstalk. To study the effect of process variations on the MRM response, the resonance wavelengths of 8 cascaded MRMs (on a chip) were measured for several chips, which is shown in Fig. 14.2.2 (top left). In this case, while the majority of the resonance wavelengths could be aligned with the laser grid using capacitive tuning with a tuning efficiency of 50pm/V and a zero static power consumption, thermal tuning (with an efficiency of 0.2nm/mW) is needed to align the rest of the MRMs. The resonance wavelength of the 8 MRMs after wavelength alignment is also shown in this figure. To lock the wavelengths of MRMs to the corresponding target optical carriers, following the design in [6], a dual sensing, actuation, and memory (DSAM) unit is implemented for each MRM (top right of Fig. 14.2.2) to digitally control the heater and capacitive section of the MRM, while synchronized with a global decision feedback and control (DFC) unit. The DSAM units of the MRMs are selected sequentially using a serial clock (SCK) and data (SDI) for either capacitive tuning/locking or thermal tuning. During the capacitive tuning/locking process, when an MRM is selected, in a feedback loop, a small fraction of the MRM output is photodetected using a SiGe photodiode. The photocurrent is routed to the DFC, where it is converted to a voltage, V_{FB}, and compared with a threshold voltage, V_{TH}, which is set using a 7-bit memory (counter) followed by an R-2R DAC. The comparator output is routed to the DSAM unit serving as the Up/Down control of the 5-bit counter. The feedback loop typically reaches the steady-state after several cycles of the counter clock (CCK), at which point, the final value is stored in the 5-bit counter, keeping the resonance wavelength of the MRM at the target point. This process is sequentially repeated for all 8 MRMs within the chip. Figure 14.2.2 (bottom left) shows the capacitive locking process for 8 MRMs of multiple chips. Figure 14.2.2 (bottom right) shows the measured optical response of an MRM, with a quality factor of around 5700 and an extinction ratio of more than 10dB, resulting in an OMA of nearly 4.5dB when a 4V reverse bias voltage is applied.

The schematic of the modulator driver and pre-driver is shown in Fig. 14.2.3 (top left). Transistors in the pre-driver inverter stages are carefully sized to perform single-ended to differential conversion with a negligible delay mismatch, while minimizing the power consumption. Using two sets of a DC-blocking capacitor and a shunt resistor, the reverse bias voltage across the p-n junction of the MRM can be set independently of the DC voltage at the output of the driver. Assuming an RC equivalent model for the reverse-biased p-n junction of the MRM [6] (with estimated R and C of 40Ω and 22fF, respectively), the output swing of the driver across the MRM is simulated, which is shown in Fig. 14.2.3 (top right), indicating that the driver is capable of providing more than 3.7V swing at high data-rates. Figure 14.2.3 (bottom) shows the block diagram of the monolithically integrated transmitter chip. The input light is coupled into the chip using a grating coupler and routed to the main waveguide bus. Eight MRMs with slightly different circumference (C_{ring}), resulting in different FSRs, are placed on the waveguide bus, each autonomously locked to the wavelength of the corresponding target carrier using its DSAM unit. The input bit-stream is coupled into the chip using an RF probe, asymmetrically distributed (i.e., with different routing lengths) among eight drivers, each placed next to an MRM, and used to modulate all 8 optical carriers simultaneously.

Figure 14.2.4 (top left) depicts the measurement setup of the OTX. Eight outputs of a Thorlabs PRO8000 distributed feedback (DFB) laser bank emitting from 1546nm to 1557.2nm on a 1.6nm grid are individually polarization adjusted, combined using an 8-to-1 combiner, amplified using a polarization maintaining (PM) erbium doped fiber amplifier (EDFA) to about 13dBm/carrier (with a 0.5dB variations), and coupled into the transmitter chip using its input grating coupler (with a ~5dB coupling loss). The output of an arbitrary waveform generator (AWG), generating a PRBS7 waveform at 38Gb/s, is amplified and bias-adjusted (to reach 0 to 1.2V levels) and applied to the input of the transmitter chip, where it is routed to all eight drivers for concurrent operation. All carriers are simultaneously modulated using the 8 MRMs, routed to the output grating coupler on the bus waveguide, and coupled to the output optical fiber. A tunable optical band-pass filter, with an insertion loss of 4dB and a bandwidth of around 0.3nm, is used to select one channel at a time for characterization. The filtered output is amplified using an EDFA, photo-detected using a Gtran GT40 photo-receiver, and routed to a sampling oscilloscope and an error analyzer for BER measurements.

While on-chip electronics are on, the resonance wavelengths of the MRMs are autonomously aligned with and locked to the corresponding optical carriers. Figure 14.2.4 (top right) shows the measured spectrum of the combined optical carriers after the EDFA (in blue) and the measured normalized optical response of the cascaded MRMs after wavelength alignment (in red). Figure 14.2.4 (bottom) shows the measured eye diagrams

242 • 2026 IEEE International Solid-State Circuits Conference

979-8-3315-8937-0/26 $31.00 © 2026 IEEE

of the 8 output channels when simultaneously modulated at 38Gb/s with a BER of less than 1E-12. The measured and simulated per-channel energy efficiency of the driver (including the MRM) for different data rates is shown in Fig. 14.2.5 (top left). The measured BER for different input coupled optical power levels and data rates is shown in the top right and bottom left of Fig. 14.2.5, respectively, showing an error-free operation for data rates up to 38Gb/s. The transmitter output bathtub curves are shown in Fig. 14.2.5 (bottom right). The performance of the implemented transmitter chip is compared with other works in Fig. 14.2.6.

Figure 14.2.7 shows the chip micrograph integrated within a footprint of 0.7mm² with a per-channel core area of 0.018mm², achieving a bandwidth density of 2.1Tb/s/mm². The presented OTX achieves a higher energy efficiency compared to the conventional approaches by utilizing the proposed modulator driver topology, eliminating the need for level shifters and a 2xVDD supply rail, as well as capacitive MRM tuning with near-zero static power consumption.

Acknowledgement:
This work was funded by DARPA PIPES program under contract number HR0011-19-2-0016.

References:
[1] H. Li *et al.*, "A 25 Gb/s, 4.4 V-Swing, AC-Coupled Ring Modulator-Based WDM Transmitter with Wavelength Stabilization in 65 nm CMOS," *IEEE Journal of Solid-State Circuits*, vol. 50, no. 12, pp. 3145-3159, Dec. 2015, doi: 10.1109/JSSC.2015.2470524. https://doi.org/10.1109/JSSC.2015.2470524
[2] C. S. Levy *et al.*, "8-λ × 50 Gbps/λ Heterogeneously Integrated Si-Ph DWDM Transmitter," *IEEE Journal of Solid-State Circuits*, vol. 59, no. 3, pp. 690-701, March 2024, doi: 10.1109/JSSC.2023.3344072. https://doi.org/10.1109/JSSC.2023.3344072
[3] J. Sharma *et al.*, "Silicon Photonic Microring-Based 4 × 112 Gb/s WDM Transmitter With Photocurrent-Based Thermal Control in 28-nm CMOS," *IEEE Journal of Solid-State Circuits*, vol. 57, no. 4, pp. 1187-1198, April 2022, doi: 10.1109/JSSC.2021.3134221. https://doi.org/10.1109/JSSC.2021.3134221
[4] H. Li *et al.*, "A 3-D-Integrated Silicon Photonic Microring-Based 112-Gb/s PAM-4 Transmitter With Nonlinear Equalization and Thermal Control," *IEEE Journal of Solid-State Circuits*, vol. 56, no. 1, pp. 19-29, Jan. 2021, doi: 10.1109/JSSC.2020.3022851. https://doi.org/10.1109/JSSC.2020.3022851
[5] L. Szilagyi *et al.*, "An 8-Lane 58 Gb/s/lane 0.66 pJ/bit Modulator Driver Electrical-IC for a 3-D Integrated Silicon Photonic Transmitter in 22 nm FD-SOI Process," *IEEE Radio Frequency Integrated Circuits Symposium (RFIC)*, 2025, pp. 267-270, doi: 10.1109/RFIC61188.2025.11082778. https://doi.org/10.1109/RFIC61188.2025.11082778
[6] K. Omirzakhov, H. Hao, A. Pirmoradi and F. Aflatouni, "Energy Efficient Monolithically Integrated 256 Gb/s Optical Transmitter With Autonomous Wavelength Stabilization in 45 nm CMOS SOI," *IEEE Journal of Solid-State Circuits*, vol. 60, no. 7, pp. 2522-2531, July 2025, doi: 10.1109/JSSC.2024.3511673. https://doi.org/10.1109/JSSC.2024.3511673
[7] N. Qi *et al.*, "A Monolithically Integrated DWDM Si-Photonics Transceiver for Chiplet Optical I/O," *IEEE Journal of Solid-State Circuits*, doi: 10.1109/JSSC.2025.3585584. https://doi.org/10.1109/JSSC.2025.3585584
[8] M. Raj *et al.*, "Design of a 50-Gb/s Hybrid Integrated Si-Photonic Optical Link in 16-nm FinFET," *IEEE Journal of Solid-State Circuits*, vol. 55, no. 4, pp. 1086-1095, April 2020, doi: 10.1109/JSSC.2019.2960487. https://doi.org/10.1109/JSSC.2019.2960487

[9] T. B. Cho and P. R. Gray, "A 10 b, 20 Msample/s, 35 mW pipeline A/D converter," *IEEE Journal of Solid-State Circuits*, vol. 30, no. 3, pp. 166-172, March 1995, doi: 10.1109/4.364429. https://doi.org/10.1109/4.364429
[10] P. Bhargava *et al.*, "A 256Gbps Microring-Based WDM Transceiver with Error-Free Wide Temperature Operation for Co-Packaged Optical I/O Chiplets," *IEEE Symposium on VLSI Technology and Circuits (VLSI Technology and Circuits)*, 2024, pp. 1-2, doi: 10.1109/VLSITechnologyandCir46783.2024.10631547. https://doi.org/10.1109/VLSITechnologyandCir46783.2024.10631547

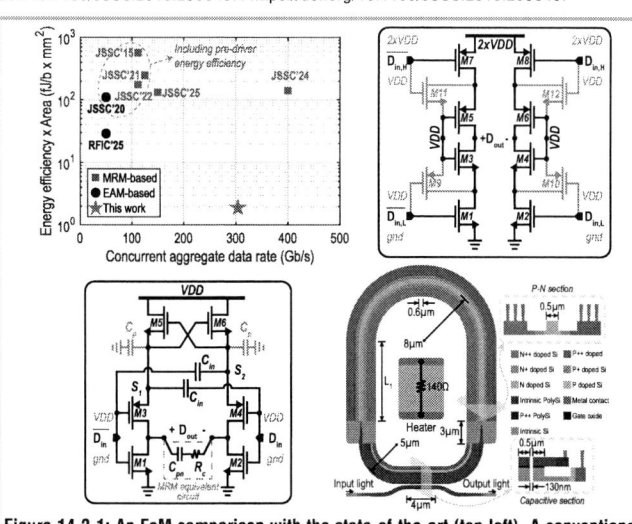

Figure 14.2.1: An FoM comparison with the state-of-the-art (top left). A conventional optical modulator driver circuit diagram (top right) and proposed (bottom left), MRM structure besides PN and capacitive cross-sections (bottom right).

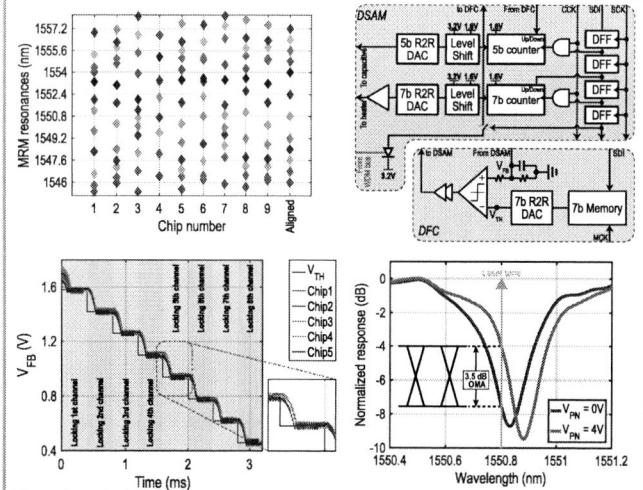

Figure 14.2.2: Optical resonance distribution of 8 cascaded MRMs over multiple chips (top left). Block diagrams of DSAM and DFC units (top right). MRM wavelength locking (bottom left). MRM optical response for different reverse bias voltages (bottom right).

ISSCC 2026 / SESSION 14 / UNUSUAL INTERCONNECTS AND OTHER USES FOR LIGHT / 14.2

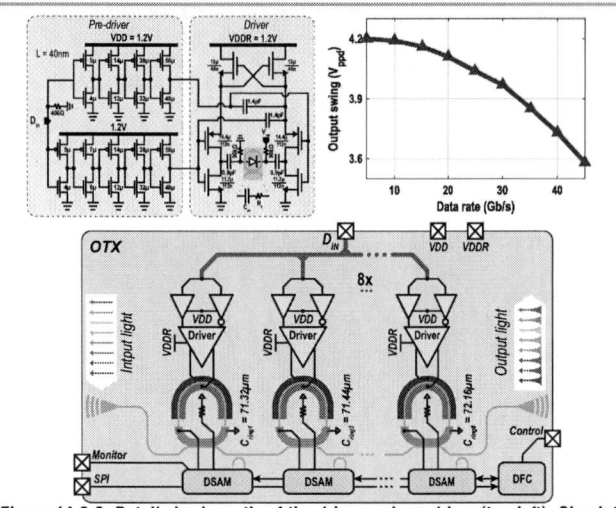

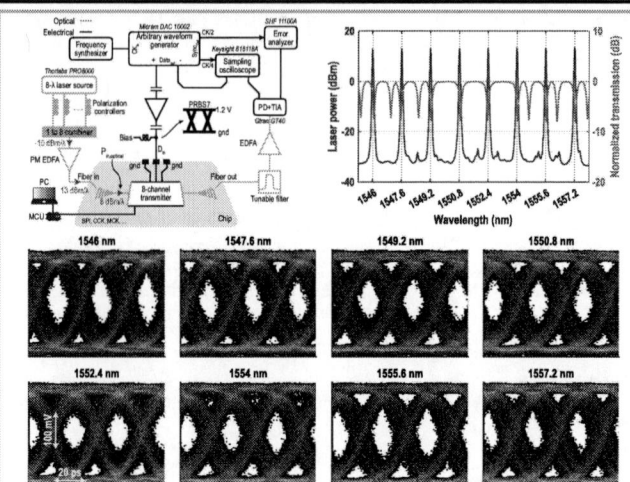

Figure 14.2.3: Detailed schematic of the driver and pre-driver (top left). Simulated driver output swing vs. data rate (top right). The architecture of the transmitter chip (bottom).

Figure 14.2.4: Measurement setup (top left). Optical response of the locked MRM resonances and the spectrum of the combined and amplified laser tones (top right). Measured eye diagrams of 8 channels operating concurrently at 38Gb/s/λ with BER<1E-12 (bottom).

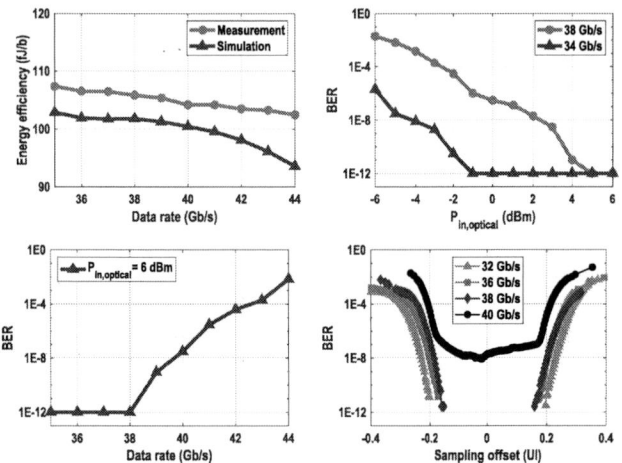

Figure 14.2.5: Energy consumption of the driver+MRM (top left), BER as a function of coupled input optical power (top right), data rate (bottom left), and sampling offset at different data rates (bottom right).

	JSSC'15 [1]	JSSC'24 [2]	JSSC'22 [3]	JSSC'21 [4]	RFIC'25 [5]	JSSC'25 [6]	JSSC'25 [7]	JSSC'20 [8]	VLSI'24 [10]	This work
Technology	65nm CMOS	28nm CMOS	28nm CMOS	28nm CMOS	22nm FDSOI	45nm CMOS SOI	45nm CMOS SOI	16nm FinFET	45nm CMOS SOI	45nm CMOS SOI
Integration	Hybrid	Hybrid	Hybrid	Hybrid	Hybrid	Monolithic	Monolithic	Hybrid	Monolithic	Monolithic
Modulator type	MRM	MRM	MRM	MRM	EAM	MRM	MRM	EAM	MRM	MRM
Operation band	O	O	C	O	C	C	O	L	-	C
Resonance tuning/locking	Thermal	Thermal	Thermal	Thermal	-	Capacitive+thermal	Thermal	-	Thermal	Capacitive+thermal
Signaling	NRZ	NRZ	PAM-4	PAM-4	NRZ	NRZ	NRZ	NRZ	NRZ	NRZ
Output swing/VDD (V_{pp}/V)[1]	2.2/2.4	1.8/2	1.5/2	1.5/2.2	1.87/2	2.4/2.4	2/2	1.8/1.8	2/-	1.8/1.2
Concurrent Laser Sources	5-λ	8-λ	1-λ	1-λ	1-λ	8-λ	3-λ	1-λ	8-λ	8-λ
Aggregate[2] data rate (Gb/s)	125	400	112	112	50	240	150	50	256	304
Energy efficiency (fJ/b)	2472[3](4)	560	1102[3]	1429[3]	760[5]	300[5]	400	1220[5]	740	106
BER	-	1E-12	-	-	1E-12	1E-12	1E-12	1E-12	1E-15	1E-12
Core area (mm²)	0.1	0.25	0.16	0.4	0.038	-	0.33	0.09[6]	-	0.018
BW density (Tb/s/mm²)	-	-	-	-	-	3.3[7]	0.176	-	-	2.1

(1) Single-ended (2) Concurrent lasers x per-channel data rate (3) Including pre-driver
(4) Calculated from the pi chart (5) Excluding level shifters (6) Calculated from chip area
(7) Excluding pre-drivers and level shifters

Figure 14.2.6: Performance summary and comparison with state-of-the-art modulator-based OTXs.

Figure 14.2.7: Chip micrograph and transmitter per-channel core area.

• 2026 IEEE International Solid-State Circuits Conference

ISSCC 2026 / SESSION 14 / UNUSUAL INTERCONNECTS AND OTHER USES FOR LIGHT / 14.3

14.3 A Single-Chip Laser Diode Driver with Built-In Frequency-Sweep Linearization for FMCW LiDAR

Wenshuo Zhu[1], Jianqiang Jiang[1], Xuan Sun[1], Zhenhao Li[1], Tingyi Gu[2], Xin Zhang[3], Cheng Huang[1]

[1]Iowa State University, Ames, IA, [2]University of Delaware, Newark, DE, [3]IBM T. J. Watson Research Center, Yorktown Heights, NY

Abstract

A laser diode driver with supply-intrinsic current shaping for FMCW LiDAR is presented. By combining a fast optical loop and an efficient switching loop, and with the LD fitted model realized in analog circuits, the design achieves closed-loop optical frequency linearization without MZI, external PD, or bulky equipment. Measurements in 180nm CMOS confirm effective frequency-sweep linearization and tunable modulation capability in a compact, highly integrated design for mobile FMCW LiDAR.

The Frequency-Modulated Continuous-Wave (FMCW) Light Detection and Ranging (LiDAR) technology achieves the level of precision desired for applications e.g. Advanced Driver-Assistance Systems, 3D points cloud generation, and autonomous driving [1-2]. FMCW LiDAR offers several advantages: 1) Unlike pulsed LiDAR that measures distance by blasting short light pulses and detecting the round-trip time, FMCW LiDAR transmits a linearly-swept-frequency laser towards the target, and receives the reflected laser mixed with a local oscillator (LO, split from the source). The round-trip delay introduces a frequency offset, known as the beat frequency (f_{beat}), which is directly proportional to the target distance (Fig. 14.3.1, top-left). This coherent detection improves receiver sensitivity, eliminates the need for wide-bandwidth processing circuits, and provides fine depth accuracy even under strong ambient light [3-5]. 2) FMCW LiDAR also enables simultaneous velocity extraction through Doppler shift analysis. 3) FMCW LiDAR employs more gentler optical power rather than high peak pulses, making it inherently eye-safe and highly compatible with chip-level integration [1-3].

Despite these advantages, the actual performance of FMCW LiDAR is limited by the non-linear electro-optical (EO) conversion of the laser diode (LD). Ideally, a linear sweeping of the transmitted optical frequency f_{TX} ($1/\lambda$, where λ is the wavelength) should be ensured with sweeping of the LD current I_{LD} to guarantee that the received optical-frequency f_{RX} is mixed into a constant f_{beat} [4-5]. However, most LDs exhibit a non-linear EO conversion at wide-range tuning (> 1GHz f_{TX} tuning range), leading to sweeping nonlinearity, which distorts the mapping between time and f_{TX}, broadening and shifting the beat spectrum (Fig. 14.3.1, top-right). As a result, the detected f_{beat} no longer represents the true round-trip delay, causing depth and velocity errors. Moreover, the up- and down-chirp responses become inconsistent among detections, further degrading measurement reliability [4]. Therefore, mitigating frequency-sweep nonlinearity is essential for achieving the high depth precision promised by FMCW LiDAR.

Prior linearization works can be classified into open-loop pre-distortion [6-7] and closed-loop EO phase-locked loop (EOPLL) [8-11] approaches (Fig. 14.3.1, bottom). With Iterative-Learning pre-distortion, the laser drive voltage is iteratively updated until the measured frequency sweep converges to the desired linear chirp using a computer. This technique provides a wide tuning range for LD with excellent linearity but relies on bench-top setups with Mach–Zehnder interferometers (MZIs) and external photodetectors. It also requires many cycles of computation-intensive and time-, power-consuming updates. In contrast, EOPLLs place the LD in a closed loop, locking and calibrating f_{beat} to a fixed reference with an MZI. This real-time correction suppresses nonlinearity and eliminates bench-top equipment, however, still requires a bulky MZI and external photodiode (PD), which adds to the total volume, cost and complexity. Although a Butterfly-Packaged LD (e.g., DFB, DBR, and ECL types) typically comes with a built-in PD, it internally couples with the LD and cannot be used without an MZI. Moreover, due to limited PLL loop bandwidth, the modulation rate and locking range are restricted; due to the typical use of a fixed-frequency clock [8-11], the achievable chirp rate is constrained by this reference and hence lacks flexibility for dynamic optimization. A portion of the laser is also used for calibration, reducing the optical intensity for detection.

Here, we present a single-chip LD driver for FMCW LiDAR with dual-loop supply-intrinsic current-shaping (Fig. 14.3.2):

1) A linear fast optical loop regulates the measured f_{TX} to a linearly sweeping reference V_{REF}, shaping I_{LD} to achieve a linear sweeping f_{TX}. The measured f_{TX} is extracted by sensing I_{PD} from the built-in PD, and mapping I_{PD} to the wavelength shift $\Delta\lambda$ (which represents f_{TX}) by an integrated Analog Function Generator (AFG).

2) A switching slow current loop regulates the current from the lower-efficiency linear power stage (I_{LIN}) to a small reference of 10mA, so that the higher-efficiency switching power stage takes over the majority of I_{LD} to enhance system efficiency.

With this new architecture, closed-loop linearization is achieved without any MZI, external PD, PLL, or bulky equipment. The power supply and linearization circuits are further combined, hence "supply-intrinsic". As a result, the level of integration is significantly improved over state-of-the-art. The supported modulation rate, range, and chirp rate are also flexible, enabling possible dynamic optimization to balance the ranging precision and acquisition rate.

Linear optical loop: One key challenge is to measure f_{TX} without an MZI. The internal PD in a butterfly LD directly senses the optical output power (P_{OPT}) and linearly converts it to PD current (I_{PD}). As shown in Fig. 14.3.2 (lower right), I_{LD}, f_{TX}, and P_{OPT} have a one-to-one correspondence: f_{TX} is a non-linear function of I_{LD} determined by the LD characteristic, and cannot be directly obtained without a real-time optical spectrum analyzer; I_{LD}, P_{OPT} and I_{PD} have a linear correlation; I_{PD} can be sensed easily; f_{TX} thus can be recovered from I_{PD} if an inverse function matching the LD characteristic exists, which is achieved by the AFG in this work.

Figure 14.3.3 shows the equation-based fitting model and the corresponding circuits for mapping I_{PD} to $\Delta\lambda$. The measured LD I_{PD}-$\Delta\lambda$ relation is fitted to an equation-based model: $\Delta\lambda = \alpha_1 I_{PD} \times \ln(\alpha_2 I_{PD} + \beta_1)$, which is analytically derived from the semiconductor laser's physical equations [12-13] and should be universal for different LDs with adjustable parameters in the model. It is then realized by the introduced AFG: first, I_{PD} is amplified through the TIA resistor R_T and combined with an offset V_{BIAS}, then converted into a current with a gain of $1/R_S$. This current is then processed by a log converter using a matched BJT pair to generate the logarithmic operation. Finally, the log-converted signal and the amplified I_{PD} are fed into an analog multiplier, implemented with a folded Gilbert cell [14] and differential-to-single-ended output conversion. The parameters α_1, α_2, and β_1 inside the AFG are programmable, allowing adaptation to different LDs and process variation with a one-time calibration. As shown in Fig. 14.3.3 (lower-left), the measured $\Delta\lambda$-I_{PD} curve, the fitted model, and the actual measured AFG output voltage exhibit close agreement, confirming the physical validity of the introduced model and the functionality of the designed AFG. In this prototype, EX-Q938 from ExOptronics is used for demonstration. We have also verified with other LDs to confirm the general applicability.

Once $\Delta\lambda$ is extracted from I_{PD}, an error amplifier (EA) is used to compare it to a linearly sweeping reference V_{REF}, then drive a power PMOS to adjust the current to the LD (I_{LIN}) (Fig. 14.3.2). In this way, a linear optical loop is formed to regulate the AFG output $\Delta\lambda$ to a linearly sweeping V_{REF}, ensuring linear $\Delta\lambda$ sweeping while driving the LD. Please note that, linearizing $\Delta\lambda$ is equivalent to linearizing f_{TX} (Fig. 14.3.3, top-left). As a result, f_{TX} linearization is achieved without using an MZI or external PD. Furthermore, the modulation range and rate are easily adjustable by changing the amplitude and frequency of V_{REF} to optimize the ranging precision and acquisition rate in operation.

Switching current loop: Linear power stage has low efficiency. A switching power stage is thus placed in parallel to supply more efficient I_{SW} to off-load the majority of I_{LD} from I_{LIN}. As shown in Fig. 14.3.2, the I_{LIN} is sensed by a scaled current sensor, then compared with a reference V_{REFLIN} representing a 10mA current by an EA, followed by a PWM controller to adjust the duty-cycle of the switching power stage to deliver enough I_{SW} to result in an averaged 10mA I_{LIN}. In this way, a switching current loop is formed. The "10mA" V_{REFLIN} is selected based on the inductor current ripple of less than 20mA, with which the 18MHz-BW fast linear loop can track and cancel the 4MHz inductor ripple effectively, resulting in a clean I_{LD} and maintaining a good efficiency.

This LD driver was fabricated in standard 180nm CMOS. Figure 14.3.4 shows the I_{LD} waveforms, measured with a 100MHz current probe, under both sweeping and steady-state conditions. With a 100kHz modulation rate, the shaped I_{LD} sweeps between 30mA and 420mA, achieving a wide 390mA range. Figure 14.3.4 also shows a different modulation rate and range of 1kHz and I_{LD} between 350mA and 400mA. In steady-state, the current ripple is well suppressed to ~1mA, a 16× reduction of the inductor current ripple. Please note that the I_{LD} waveforms appear non-linear because they are intentionally shaped to compensate for the LD's EO non-linearity for a linear optical frequency sweep. This driver was further validated in an optical setup (Fig. 14.3.5). The LD output was split by a 2×2 50:50 coupler, with one branch passing through a 1 or 2m fiber to emulate the round-trip propagation between TX and RX, and the other branch serving as the LO. A second coupler mixes the delayed RX and the LO signals to a 3GHz photodetector for measurement and

ISSCC 2026 / February 17, 2026 / 10:55 AM

evaluation. The measured f_{beat} spectrograms at 5-kHz modulation rate (Fig. 14.3.5, left) confirm the effectiveness of the introduced I_{LD} linearization: with linearization, both upward and downward f_{beat} show flat frequency trajectories over the region of interest, whereas without linearization, the f_{beat} changes significantly with time due to the LD EO nonlinearity. The f_{beat} spectra (Fig. 14.3.5, right) shows that this linearization not only sharpens spectral peaks by suppressing nonlinearity-induced broadening, but also enables the system to support a wide range of modulation rates and ranges. With a 2m fiber length, the ranging precision was measured to be 1.16mm. Despite the 18MHz fast-loop bandwidth, the modulation rate is only measured up to 1MHz. This is limited by the LDO-type linear stage, which can only source current, while current sinking relies on the switching power stage with a much lower bandwidth. On the other hand, a push-pull linear stage could fully reveal the fast-loop bandwidth for a much higher modulation rate if needed.

Figure 14.3.6 summarizes a comparison with prior EOPLL-based linearization works. This work eliminates the bulky components, broadens the modulation flexibility, and maintains an mm-level ranging precision. This work paves the way towards highly integrated, battery-powered FMCW LiDAR for mobile applications. Figure 14.3.7 shows the chip micrograph and optical testing setup.

Acknowledgement:
This work is supported by CogniSense under SRC Project 3133.016.

References:
[1] C. Rogers *et al.*, "A universal 3D imaging sensor on a silicon photonics platform," *Nature*, vol. 590, pp. 256–261, 2021. https://doi.org/10.1038/s41586-021-03259-y
[2] P. Bhargava *et al.*, "Fully Integrated Coherent LiDAR in 3D-Integrated Silicon Photonics/65nm CMOS," *IEEE Symposium on VLSI Circuits*, 2019, pp. C262-C263. https://doi.org/10.23919/VLSIC.2019.8778154
[3] N. Li *et al.*, "A Progress Review on Solid-State LiDAR and Nanophotonics-Based LiDAR Sensors," *Laser & Photonics Reviews*, vol. 16, pp. 2100511, 2022. https://doi.org/10.1002/lpor.202100511
[4] B. Behroozpour, P. A. M. Sandborn, M. C. Wu and B. E. Boser, "Lidar System Architectures and Circuits," *IEEE Communications Magazine*, vol. 55, no. 10, pp. 135-142, Oct. 2017. https://doi.org/10.1109/MCOM.2017.1700030
[5] H. Hashemi, "A Review of Silicon Photonics LiDAR," *IEEE Custom Integrated Circuits Conference (CICC)*, 2022, pp. 1-8. https://doi.org/10.1109/CICC53496.2022.9772845
[6] J. Riemensberger *et al.*, "Massively parallel coherent laser ranging using a soliton microcomb," *Nature*, vol. 581, pp. 164–170, 2020. https://doi.org/10.1038/s41586-020-2239-3
[7] Xiaosheng Zhang, Jazz Pouls, and Ming C. Wu, "Laser frequency sweep linearization by iterative learning pre-distortion for FMCW LiDAR," *Opt. Express*, vol. 27, pp. 9965-9974, 2019. https://doi.org/10.1364/OE.27.009965
[8] B. Behroozpour *et al.*, "Chip-scale electro-optical 3D FMCW lidar with 8μm ranging precision," *IEEE International Solid-State Circuits Conference (ISSCC)*, Feb. 2016, pp. 214-216. https://doi.org/10.1109/ISSCC.2016.7417983
[9] A. Binaie, S. Ahasan, and H. Krishnaswamy, "A 65nm CMOS continuous-time electro-optic PLL (CT-EOPLL) with image and harmonic spur suppression for LIDAR," *IEEE Radio Freq. Integr. Circuits Symp. (RFIC)*, 2019, pp. 103–106. https://doi.org/10.1109/RFIC.2019.8701771

[10] K. Kondo and H. Hashemi, "Electro-Optical Phase-Locked Loop Generating Linear Frequency Chirp for FMCW LiDAR," *Conference on Lasers and Electro-Optics (CLEO)*, 2020, pp. 1-2. https://doi.org/10.1364/CLEO_SI.2020.SM1O.5
[11] M. Rezaei, L. Hussein, A. Dee, and S. Moazeni, "An electro-optical synthesizer to generate random chirp rates for secure FMCW LiDAR applications," *Proc. IEEE Radio Freq. Integr. Circuits Symp. (RFIC)*, 2024, pp. 231–234. https://doi.org/10.1109/RFIC61187.2024.10599997
[12] M. Funabashi *et al.*, "Recent advances in DFB lasers for ultradense WDM applications," *IEEE Journal of Selected Topics in Quantum Electronics*, vol. 10, no. 2, pp. 312-320, March-April 2004. https://doi.org/10.1109/JSTQE.2004.826576
[13] A. Asmari, J. Hodgkinson, E. Chehura, S. E. Staines, and R. P. Tatam, "All-electronic frequency stabilization of a DFB laser diode," *Opt. Express*, vol. 25, pp. 11679-11691, 2017. https://doi.org/10.1364/OE.25.011679
[14] J. N. Babanezhad and G. C. Temes, "A 20-V four-quadrant CMOS analog multiplier," *IEEE Journal of Solid-State Circuits*, vol. 20, no. 6, pp. 1158-1168, Dec. 1985. https://doi.org/10.1109/JSSC.1985.1052454

14

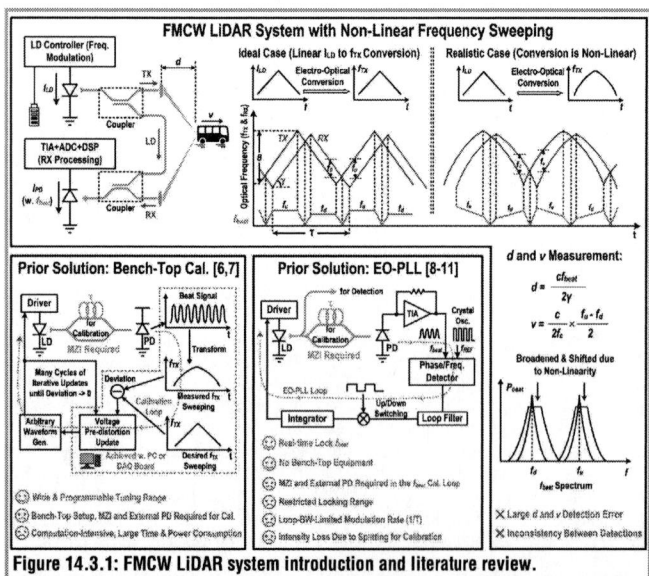

Figure 14.3.1: FMCW LiDAR system introduction and literature review.

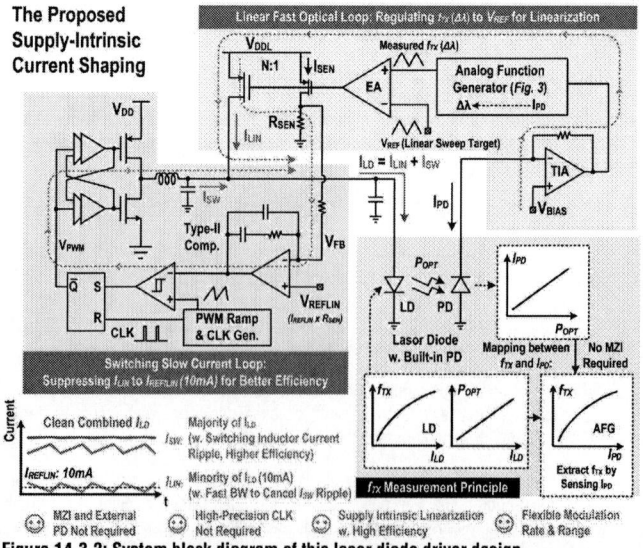

Figure 14.3.2: System block diagram of this laser diode driver design.

DIGEST OF TECHNICAL PAPERS • 245

979-8-3315-8937-0/26 $31.00 © 2026 IEEE

ISSCC 2026 / SESSION 14 / UNUSUAL INTERCONNECTS AND OTHER USES FOR LIGHT / 14.3

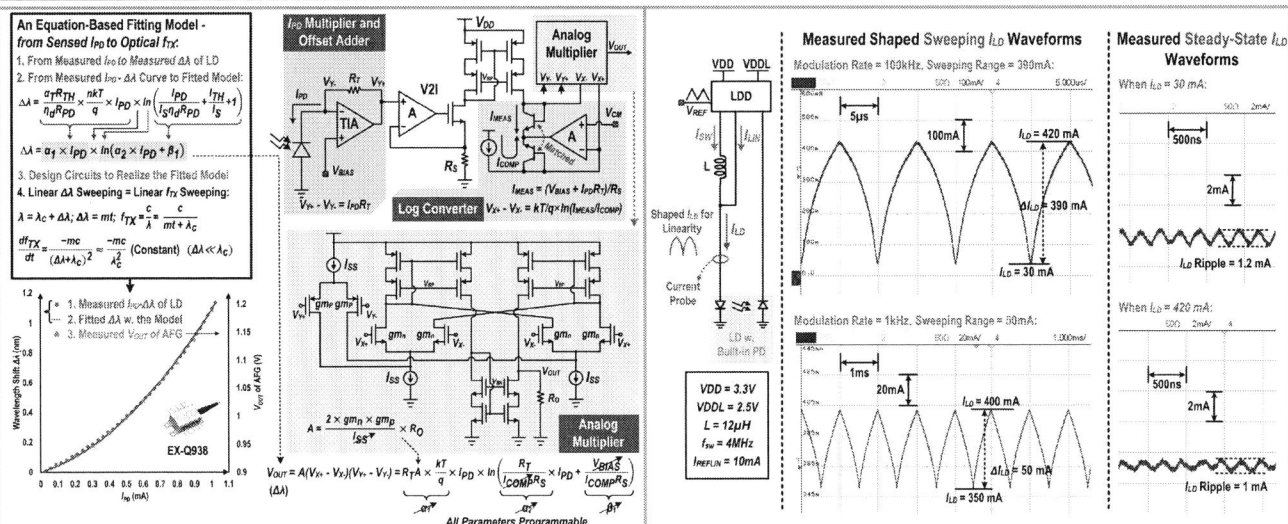

Figure 14.3.3: Design considerations of the analog function generator for I_{PD} to f_{TX} mapping.

Figure 14.3.4: Measured sweeping and steady-state I_{LD} waveforms at different conditions.

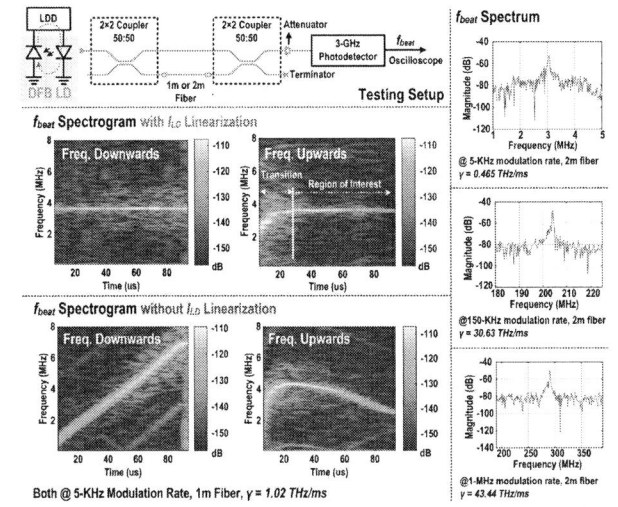

Figure 14.3.5: Measured f_{beat} spectrogram and f_{beat} spectrum at different modulation rates.

	This Work	ISSCC 2016 [8]	RFIC 2019 [9]	CLEO 2020 [10]	RFIC 2024 [11]
Linearization Architecture	Closed-Loop Supply Intrinsic Current Shaping	EOPLL	EOPLL	EOPLL	EOPLL
Technology	180nm CMOS	180nm CMOS	65nm CMOS	Discrete	180nm CMOS
Laser Type	DFB	DBR	DFB	DFB	ECL
Laser Diode Intrinsic Non-linearity $(\delta\gamma/\gamma)$[1]	94%	~8%	~30%	NA	NA
Laser Center Wavelength (λ_0)	1555 nm	1530 nm	1546 nm	1549 nm	1550 nm
MZI Requirement for Linearization	No	Yes	Yes	Yes	Yes
MZI Type	Not Needed	Si-Photonics	Si-Photonics	Discrete Fibers	Discrete Fibers
External Photodiode Requirement for Linearization	No	Yes	Yes	Yes	Yes
High-Precision CLK Requirement	No	Yes	Yes	Yes	Yes
Max. Current Tuning Range (ΔI_{LD})	390 mA	5 mA	NA	NA	50 mA
Max. Delivered Power $(P_{OUT\,Max})$	741 mW	~100 mW	150 mW	NA	NA
Linearization Loop BW	1 MHz	0.7 MHz	0.02 MHz	0.1 MHz	0.004 MHz
Freq. Modulation Range (B)	22 GHz – 102 GHz	122 GHz	16 GHz	10 GHz	1.8/3.6 GHz
Modulation Rate (1/T)	1 KHz – 1 MHz	89 KHz	1 KHz	0.45 KHz	1.7/5 KHz
Supported Chirp Rate (γ)	0.043 – 43.44 THz/ms	21.8 THz/ms	0.032 THz/ms	0.009 THz/ms	0.012/0.018 THz/ms
Chirp Rate Adjustability[2]	Linearly Adjustable	Fixed	Fixed	Fixed	4 Discrete Settings
Ranging Precision @ Range	1.16 mm @ 1 m	4.2 mm @ 1.4 m	0.55 mm @ 2 m	NA	72 mm @ 4 m

[1] Describes how non-linear the Laser Diode $I_{LD} - f_{TX}$ intrinsically is before linearization (related to the quality and cost of the laser diode). Higher intrinsic non-linearity requires more aggressive linearization to return to a linear $I_{LD} - f_{TX}$ conversion.

[2] PLL-based designs typically fix the chirp rate due to the use of a crystal oscillator as a high-presition clock with a fixed f_{beat} for linearization. The proposed design does not have this limitation, thus chirp rate can be adaptively optimized in applications with different ranges.

Figure 14.3.6: Comparison with prior work.

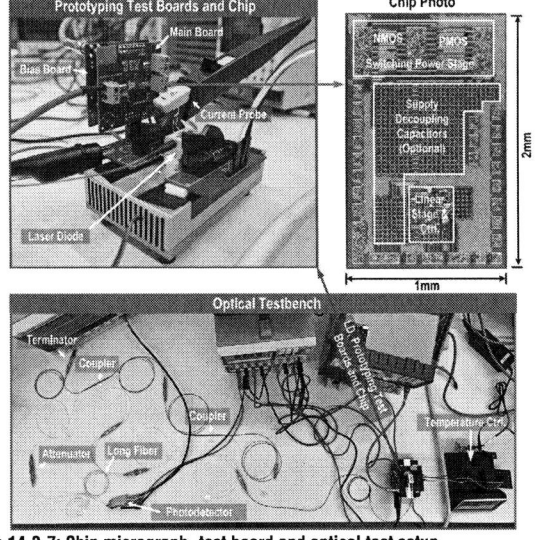

Figure 14.3.7: Chip micrograph, test board and optical test setup.

• 2026 IEEE International Solid-State Circuits Conference

979-8-3315-8937-0/26 $31.00 © 2026 IEEE

ISSCC 2026 / SESSION 14 / UNUSUAL INTERCONNECTS AND OTHER USES FOR LIGHT / 14.4

14.4 A 40Gb/s 8mW-OMA 1-to-N VCSEL Driver for Parallel and Wireless Optical Links Using 150nm GaN HEMT

Shuo Sarah Feng[1], Fuzhan Chen[1], Ruitao Matthew Ma[1,2], Hongyu Bruce Bao[1], Kangping Zhong[3], Pak Tao Alan Lau[3], Chik Patrick Yue[1,2]

[1]Hong Kong University of Science and Technology, Hong Kong, China, [2]High5 Semiconductor, Hong Kong, China, [3]Hong Kong Polytechnic University, Hong Kong, China

Abstract

This paper presents a 40Gb/s 1-to-N VCSEL driver using 150nm GaN HEMT with f_T/f_{MAX} of 50GHz/154GHz. The proposed topology can drive VCSEL arrays in series or parallel without the need of pre-driver, using a common-source main data stage and auxiliary adjustable bias stage for different array configurations. Measured optical eye diagrams show 8mW OMA at 40Gb/s NRZ with 3-tap FFE for 4 VCSELs in series under a 15V supply. The driver consumes 480mW. For PAM-4 input, the driver can support 38Gb/s.

Emerging non-terrestrial network (NTN) links demand high optical modulation amplitude (OMA) to sustain long-reach free-space channels. Conventional NTN links utilize large-footprint optical modulators with large-swing drivers to satisfy the OMA requirement, suffering from high power and area overhead [1-2]. On the other hand, VCSEL array-based optical link emerges as a promising candidate (top of Fig. 14.4.1). VCSEL has advantages of low power consumption, low cost, and small footprint. Moreover, the output swing requirement for VCSEL driver is typically much lower than that for optical modulator driver, leading to lower power overhead. For such application, 1-to-N driver with strong driving capability is highly demanded. And the arrangement strategy of VCSEL array is to be explored to achieve the optimal tradeoff between bandwidth, OMA and power consumption. Besides, 1-to-N driver is also necessitated for clock-and-data recovery (CDR)-less wireline communications as shown in the bottom of Fig. 14.4.1, where the transmitter generates multiple identical optical pulses to facilitate the synchronization of the communication channels.

Recently, taking advantages of its high breakdown voltage and higher power conversion efficiency, GaN emerges as a favored technology for radio-frequency (RF) components such as power amplifiers (PAs) [3-5] and low-noise amplifiers (LNAs) [6-7]. Despite these advances, the potential of GaN devices for optical and wireline transmitters has not yet been explored. Therefore, in this work, we research the feasibility of GaN high-electron-mobility transistor (HEMT) for broadband front-end circuits and propose the first 1-to-N driver in 150nm GaN process for VCSEL array-based optical interconnects. Leveraging the high drain-to-source breakdown voltage of 28V of utilized GaN HEMT, the proposed 1-to-N driver can directly drive up to 10 in-series VCSEL units. Moreover, compared to widely used CMOS and SiGe BiCMOS technologies, GaN HEMT can sustain multi-Gb/s operations even when operating at the full 28V stress. With this merit, the proposed driver can support amplifications of NRZ and PAM-4 signals with data rates beyond 40Gb/s.

Figure 14.4.2 shows the schematic of the proposed 1-to-N driver, consisting of a main data stage (M_1) and an auxiliary bias stage (M_2-M_3). The main stage adopted a common-source topology with input signals fed through the capacitor C_1. Resistor R_1 provides 50Ω matching to the preceding building block. As the adopted GaN HEMT operates in depletion mode, negative gate bias is required for its gate. In this work, this bias voltage is supplied by the external equipment. R_2 and C_{b1} form a low-pass filter to suppress the noise from the external DC input. To properly bias the cathode of VCSEL array, an auxiliary path (M_2-M_3) is implemented. The gate bias voltage V_{G2} for M_2 and M_3 is filtered by bypass capacitor C_{b2}, and two dedicated pads are included to monitor the voltage drop V_{sen} across R_4 at the drain of M_3. This enables temperature-dependent bias control for V_{G2}, ensuring stable VCSEL operation across operating conditions. In this process, source and body terminals of all HEMTs are connected to ground through back vias. To ensure accurate performance evaluation, during design stage, packaging components including bond wires are modeled using compact RLC networks, while the evaluation board (EVB) traces are extracted from HFSS. The bottom left of Fig. 14.4.2 shows the VCSEL equivalent electrical model, including pad capacitance C_p, pad resistance R_p, DBR resistance R_s, junction resistance R_j, and junction capacitance C_j. The bottom right illustrates seven different VCSEL arrays from 1×1 up to 4×1 in both series and parallel configurations for the demonstrations of the proposed 1-to-N driver prototype.

The driver is fabricated in 150nm GaN HEMT with f_T/f_{MAX} of 50GHz/154GHz. Electrical testing is first performed to validate its bandwidth and signal integrity prior to optical characterization. The measured electrical eye diagrams and frequency responses are shown in Fig. 14.4.3. As shown in the top left of Fig. 14.4.3, with 28Gb/s NRZ and 44Gb/s PAM4 inputs from arbitrary waveform generator (AWG), the measured eyes show clear openings with swings of 170mV and 180mV. And then, with 3-tap AWG FFE applied, as shown in the top right of Fig. 14.4.3, the driver can support the amplifications of 48Gb/s NRZ and 54Gb/s PAM-4 signals, with output swings of 130mW and 70mV, respectively. The 3-tap FFE can be easily achieved by the preceding blocks of the driver such as serializers in the future work. The bottom plots of Fig. 14.4.3 demonstrate the measured frequency responses of the driver and compare them with simulation results. The driver achieves −3dB and −6dB bandwidths of 8.5GHz and 11.7GHz, respectively. The measured output return loss (S_{22}) is around −6dB within the operating band. The reason for the poor output reflection

performance is because the output impedance of the driver is designed to be around 100Ω to match the VCSEL array instead of 50Ω. The above results validate the high-speed performance of the proposed GaN driver and confirm agreement between measurement and simulation.

Optical measurements are performed for the driver when connected to VCSEL arrays with seven configurations illustrated in the bottom of Fig. 14.4.2. The optical measurement results are shown in Fig. 14.4.4. For series connection, the average eye width increases with VCSEL count, indicating improved bandwidth. In series chains, the first VCSEL exhibits wider eyes since it is loaded only by its own anode capacitance, whereas downstream devices see both anode and preceding cathode capacitances. The bottom-right plot of Fig. 14.4.4 summarizes the measured average eye width versus VCSEL count for both series and parallel connections. While series operation of the VCSEL array yields progressively wider eye openings, with a measurable improvement of +20.37% in eye width at 4×1 configuration compared to a single device, parallel operation exhibits the opposite behavior, resulting in a 44.35% reduction in eye width under the same 4×1 condition. This behavior can be directly explained by the VCSEL equivalent model in Fig. 14.4.2: the effective load at the driver is dominated by the pad capacitance C_p and the junction capacitance C_j. For an N×1 series connection, the equivalent capacitance scales approximately as $C_{eff,ser,N} \approx (C_p+C_j)/N$, leading to reduced capacitance and improved bandwidth as N increases. Conversely, for 1×N parallel connection, $C_{eff,par,N} \approx (C_p+C_j)×N$, causing the effective capacitance to grow with N and thereby slowing the driver output time constant.

The measured optical performance of various VCSEL array configurations is summarized in Fig. 14.4.5, which includes the corresponding eye diagram for the best-performing case. The top plot shows the maximum average data rate for each configuration, where series connections achieve progressively higher rates, up to a 37.5% improvement with 4×1 VCSELs, while parallel connections exhibit degraded scaling. The bottom eye diagrams of Fig. 14.4.5 illustrate the highest-speed case of a 4×1 series array. At 28Gb/s NRZ and 32Gb/s PAM4 with direct AWG input (no FFE), clear optical eyes are observed. With FFE pre-emphasis applied by the AWG, simulating the equalization from a preceding TX stage, the data rate is extended to 40Gb/s NRZ and 38Gb/s PAM4. These results confirm that the proposed GaN driver can support multi-VCSEL series arrays at high data rates, with system-level equalization enabling further scaling.

Figure 14.4.6 summarizes the measured performance of the proposed driver. To the best of our knowledge, no prior standalone GaN laser drivers have been reported, necessitating performance benchmarks against state-of-the-art GaN PAs as the closest functional analog. As shown in the top comparison table in Fig. 14.4.6, the design achieves an electrical −3dB bandwidth of 8.5GHz and −6dB bandwidth of 11.7GHz, surpassing all previously reported GaN PAs in bandwidth. When benchmarked against complete optical transmitters in the lower table, the driver achieves 40Gb/s NRZ optical data rate while delivering superior OMA of 8.56mW and extinction ratio of 5.19dB, significantly exceeding the output capabilities of CMOS and SiGe-based counterparts. Although optical measurements are demonstrated up to 4-VCSEL series operation, the 28V capability of this process allows extension to 10-VCSEL series arrays for higher optical power, highlighting the scalability of the approach. The full die micrograph and a detailed 3D layout view are presented in Fig. 14.4.7. The chip area is 1.63×1.30mm² with a total area of 2.12mm². The core circuit occupies 1.61mm², which includes the main and auxiliary amplifier stages, while the remaining area is allocated to pads and sealing structures.

Acknowledgement:
This work was supported in part by Hong Kong Research Grants Council through the Areas of Excellence (AoE) Scheme under Grant AoE/E-601/22-R and in part by the General Research Fund (GRF) under Grant 16205522 and Grant 16205023. The authors would like to thank Dynax Semiconductor, Inc. for providing GaN foundry service and fabrication support. Corresponding author: Chik Patrick Yue (eepatrick@ust.hk).

246 • 2026 IEEE International Solid-State Circuits Conference

979-8-3315-8937-0/26 $31.00 © 2026 IEEE

ISSCC 2026 / February 17, 2026 / 11:20 AM

References:

[1] Y. Horst *et al.*, "Tbit/s line-rate satellite feeder links enabled by coherent modulation and full-adaptive optics," *Light Sci. Appl.*, vol. 12, p. 153, Jun. 2023.
https://doi.org/10.1038/s41377-023-01201-7

[2] J. Sim *et al.*, "A 10-Gb/s Wireline Receiver Using Linear Baud-Rate CDR and Analog Equalizer for Free Space Optical Communication Over 10- and 100-m Distances," *IEEE JSSC*, vol. 59, no. 6, pp. 1835–1846, Jun. 2024.
https://doi.org/10.1109/JSSC.2023.3347758

[3] G. Lv, W. Chen, X. Chen, F. Huang, and Z. Feng, "A GaN SLCG-Doherty-Continuum Power Amplifier Achieving >38% 6dB Back-Off Efficiency Over 1.35 to 7.6GHz," *ISSCC*, pp. 90–92, Feb. 2025. https://doi.org/10.1109/ISSCC49661.2025.10904605

[4] G. Lv, W. Chen, X. Chen, F. M. Ghannouchi, and Z. Feng, "A 1.8–5.4-GHz GaN MMIC Distributed Efficient Power Amplifier With Reactance Compensation and Adaptive Biasing," *IEEE TCAS-I*, vol. 71, no. 4, pp. 1531–1543, Apr. 2024.
https://doi.org/10.1109/TCSI.2024.3354923

[5] X. Chen, M. Zhao, W. Chen, and Z. Feng, "A 700-2800MHz Switchless Class-G Power Amplifier with Two-Quadrant Modulation for Back-off Efficiency Improvement," *IEEE IMS*, pp. 410–413, Jun. 2022. https://doi.org/10.1109/IMS37962.2022.9865384

[6] C.-J. Hu, H.-Y. Li, J.-X. Xu, L. Gao and X. Y. Zhang, "A 7.2–29.8 GHz LNA With 1.35-2.67-dB NF Using Coupled-Line-Based Transformers in 0.15- μm GaN-on-SiC Technology," IEEE TCAS-I, vol. 72, no. 10, pp. 5302-5313, Oct. 2025.
https://doi.org/10.1109/TCSI.2025.3554211

[7] H. B. Ahn, H.-G. Ji, Y. Choi, S. Lee, D. M. Kang, and J. Han, "25–31 GHz GaN-Based LNA MMIC Employing Hybrid-Matching Topology for 5G Base Station Applications," *IEEE MWTL*, vol. 33, no. 1, pp. 47–50, Jan. 2023.
https://doi.org/10.1109/LMWC.2022.3201075

[8] D. Gustafsson, J. C. Cahuana, D. Kuylenstierna, I. Angelov, and C. Fager, "A GaN MMIC Modified Doherty PA With Large Bandwidth and Reconfigurable Efficiency," *IEEE T-MTT*, vol. 62, no. 12, pp. 3006–3016, Dec. 2014. https://doi.org/10.1109/TMTT.2014.2362136

[9] S. Krishnamurthy *et al.*, "A 0.9pJ/b 108Gb/s PAM-4 VCSEL-Based Direct-Drive Optical Engine," *ISSCC*, pp. 592–594, Feb. 2025.
https://doi.org/10.1109/ISSCC49661.2025.10904635

[10] K. Nagashima *et al.*, "An Ultra-Compact CPO Transceiver based on a 1060-nm Single-Mode VCSEL Array and Multicore Fibers," *J. Lightwave Technol.*, pp. 1–11, 2025.
https://doi.org/10.1109/JLT.2025.3555591

14

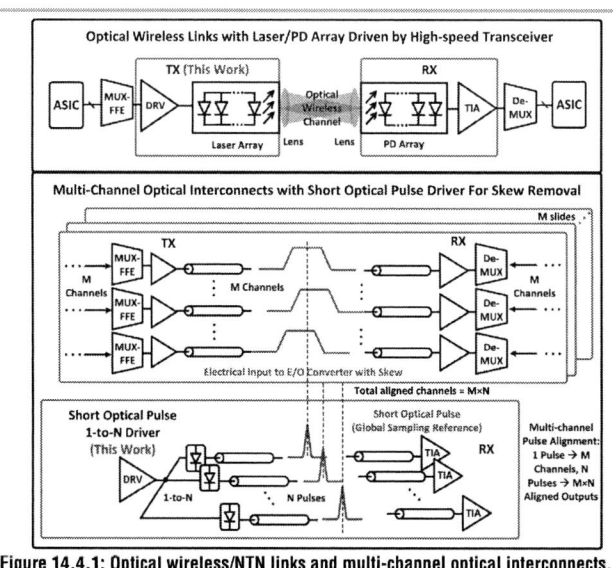

Figure 14.4.1: Optical wireless/NTN links and multi-channel optical interconnects.

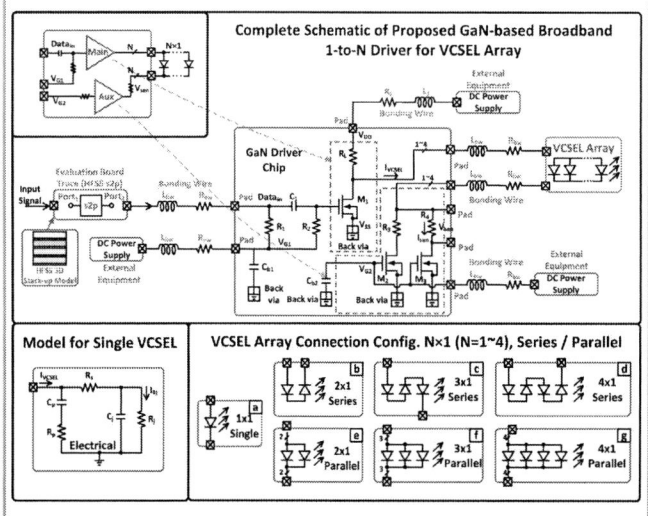

Figure 14.4.2: Schematic of the proposed GaN-based broadband 1-to-N driver.

DIGEST OF TECHNICAL PAPERS • 247

979-8-3315-8937-0/26 $31.00 © 2026 IEEE

ISSCC 2026 / SESSION 14 / UNUSUAL INTERCONNECTS AND OTHER USES FOR LIGHT / 14.4

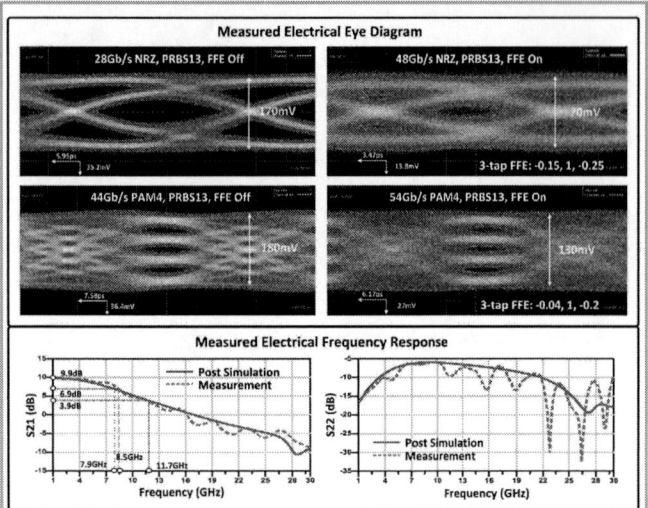

Figure 14.4.3: Measured electrical results: time-domain eye diagrams and frequency-domain S-parameters.

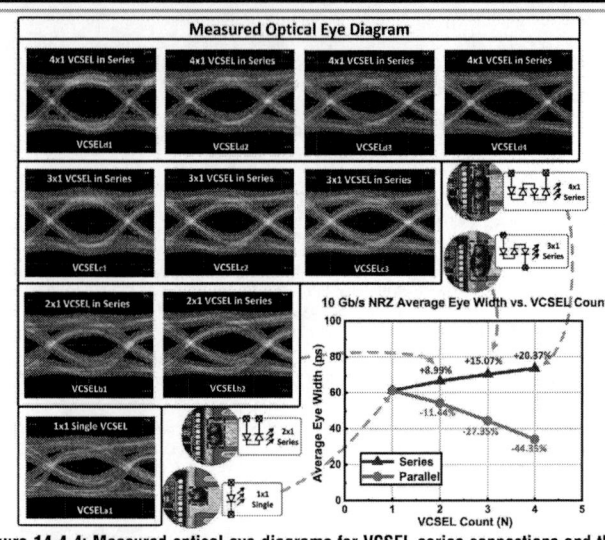

Figure 14.4.4: Measured optical eye diagrams for VCSEL series connections and the eye width versus VCSEL count.

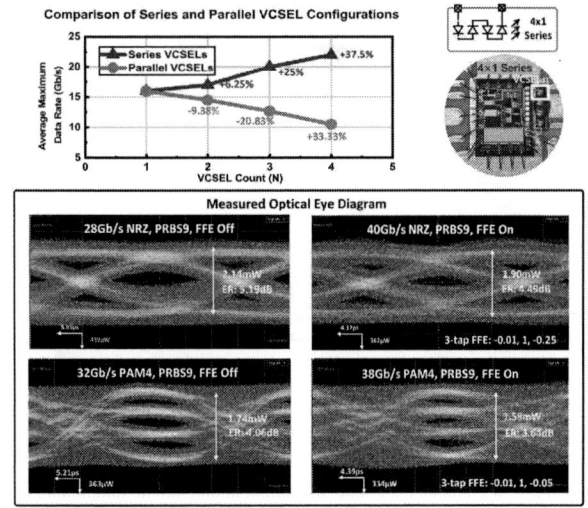

Figure 14.4.5: Maximum data rates comparison of series and parallel VCSEL arrays, and eye diagrams for 4×1 series.

Comparison of Standalone GaN Drivers (Electrical Performance Only)

	[3] ISSCC'25	[4] TCASI'24	[8] TMTT'14	This Work
Technology	250nm GaN	SiGe BiCMOS	250nm GaN	150nm GaN
Supply Voltage (V)	15/30	28	12.5/28	15
Bandwidth (GHz)	6.35 (1.35 to 7.6)	3.6 (1.8 to 5.4)	3 (5.8 to 8.8)	8.5 (DC to 8.5) 11.7 for -6dB BW
Modulation	100MHz LTE	100MHz LTE	20MHz 256QAM	NRZ / PAM4
Electrical Data Rate (Gb/s)	N/A	N/A	N/A	48 for NRZ 54 for PAM4
Gain (dB)	7.8 to 13	10 to 12.5	8.5 to 9	9.7 to 14.2
Area (mm²)	3.36	10.2	8.4	2.12

Comparison of VCSEL-based Optical Transmitters (Driver + VCSEL)

	[9] ISSCC'25	[10] JLT'25	This Work
Technology	22nm FinFET CMOS	SiGe BiCMOS	150nm GaN
VCSEL Bandwidth (GHz)	26.5	16	16
Optical Data Rate (Gb/s)	64 for NRZ 128 for PAM4	25 for NRZ (16 Channel)	40 for NRZ 38 for PAM4
Optical Modulation Amplitude (mW)	3.53*	1.55*	8.56*,**
Extinction Ratio (dB)	N/A	2.3	5.19
Power (mW)	40	2000	480
Area (mm²)	0.19	9	2.12

*3-dB butt coupling loss de-embedded. ** Total power of the 4-VCSEL array.

Figure 14.4.6: Comparison tables: driver-only metrics and full optical transmitters with both driver and VCSEL.

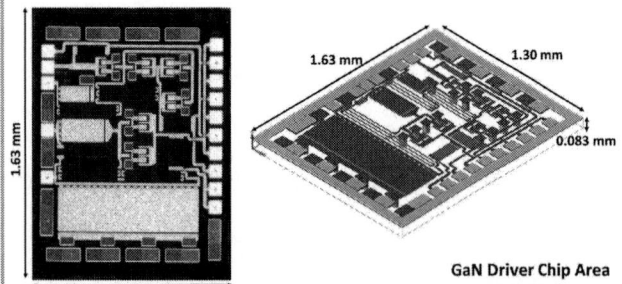

Block		Value (mm²)
Core Area	Active Area — Main Stage	0.84
	Active Area — Aux. Stage	0.48
	Inactive Area	0.29
Pad and Sealing		0.51
Total Chip Area		2.12

GaN Driver Chip Area
- Main Stage 39%
- Aux. Stage 24%
- Inactive Area 14%
- Pad and Sealing 23%

Figure 14.4.7: Die micrograph and 3D layout view of the proposed GaN driver.

• 2026 IEEE International Solid-State Circuits Conference

ISSCC 2026 / SESSION 14 / UNUSUAL INTERCONNECTS AND OTHER USES FOR LIGHT / 14.5

14.5 Highly-Integrated Light-Sensing System with RF Harvesting and Transmission in Commercial N-Type IGZO Flexible Technology

Marco Privitera[1,2], Muhammad Zahid Naveed[1], Andrea Ballo[1,2], Gianluca Giustolisi[1], Alfio Dario Grasso[1], Massimo Alioto[2]

[1]University of Catania, Catania, Italy, [2]National University of Singapore, Singapore, Singapore

Abstract

A sticker-like light sensing system is demonstrated in flexible N type-only IGZO TFT. It reuses RF signals for harvested battery charging and backscattered communications, and repurposes resistors as light sensors. Low-power operation is enabled by techniques for state-of-the-art efficiency in voltage regulation, signal amplification, data conversion and pseudo-dynamic logic. Indoor light sensing and 0.9GHz data transmission are demonstrated with 2.02mm² area at 15.5µW (<1% duty cycle).

The quest for thin and conformable electronics fuels the demand for flexible systems based on IGZO thin-film transistors (TFTs) [1]-[3] in lightweight, bendable and low-cost wearable/patchable wireless sensing systems [4]-[6]. However, the large process minimum feature size and the absence of complementary transistors (N type-only) and other basic devices (e.g., accessible pn junctions) make flexible system integration challenging in commercial IGZO technologies. Indeed, very limited level of integration was shown in prior demonstrations, which 1) focus on sub-systems, 2) suffer from energy provision limitations and lack the necessary on-chip harvesting and flexible battery energy replenishment (without harvesting, few weeks of flexible battery life are achieved at tens of µWs), 3) require off-chip (inductive) passives, and do not demonstrate integrated power management and/or voltage regulation [3], [4], [6].

Another obstacle in system integration is the poor area efficiency in key sub-systems such as 1) amplifiers (sub-mm² OTA [7], [8] to mm² area per OTA [9]-[12]), 2) low-dropout regulators (LDO, <<1 mA/mm² area efficiency [13]). Other fundamental advances are required to improve the currently poor energy efficiency of sub-systems such as data converters (Walden FOM generally >1 nJ/convstep [14]-[19], recently sub-1 nJ/convstep [20]). Such area and energy efficiency limitations prevented radio-sensing integration, as in touch/pressure and light sensors without wireless communications [16], [21]-[27], or RFID and NFC flexible radios with no sensing [4]-[6] and only sub-meter range. As exception, [3] integrates temperature sensing and inductive datalink for 12m range, though at the extremely low datarate of 2kbs at 1.7X/oct range degradation (7m at 4kbps). Its range again falls in the sub-meter range at typical >100-kbps datarates required to keep channel occupancy within customary ms time at hundreds of bits/packed (e.g., as in Zigbee, NR-U/5G, Bluetooth classic and other networks with coexistence capabilities).

In this work, a flexible highly-integrated sensing system is presented from harvesting to battery charging, light sensing and wireless data transfers, enabling sticker-like and bendable form factors for attachment to existing surfaces (e.g., window, curtain, wall). The system architecture uniquely reuses RF incident signals for energy harvesting, flexible battery charging and 0.9GHz backscattered communications. Light sensing is enabled by unconventionally repurposing on-chip resistors and exploiting their photoconductivity, remedying the lack of accessible pn junctions in commercial IGZO technologies (photovoltaic effect). As selected highlights, 2 orders of magnitude improvement over prior flex ICs is demonstrated in 1) OTA power efficiency, 2) LDO area efficiency, 3) wireless transmission energy/bit, while eliminating conventional off-flex chip components (e.g., discrete inductors/capacitors [3]-[6]).

The sticker-like system is demonstrated in an indoor smart lighting system detecting light from outdoor (sticker on window) and indoor artificial lighting (sticker on curtain) for automatic lighting and electrochromic window glass tint adjustment for minimum building power usage at targeted comfort level (Fig. 14.5.1). The light sensors transfer data to a shared receiver (e.g., access point, plugged-in device, PID) by backscattering a 0.9GHz tone intermittently generated by an indoor plug-in device. The adopted time-driven scheme (e.g., sense every 10 minutes) is supported by an (off-chip) 32.7kHz real time clock in flex sensors, as commercially available with power in the tens of nWs.

The system architecture (Fig. 14.5.1) includes 1) a signal path comprising a resistor-based light sensor, OTA amplification, data conversion and backscattered wireless transmission timed by an on-chip FLL, and 2) a power path with a natively 50Ω-matched RF-to-DC converter to harvest ambient RF power from the antenna skipping off-chip matching network, a 3.7V LiPo flex battery charger, and LDO voltage regulation. In the signal path, light sensing is embedded into an OTA via the light-sensitive resistor R_A in Fig. 14.5.1, which would conventionally be shielded from light via metal layer(s) in the adopted technology. Instead, R_A is deliberately left uncovered for full light exposure to leverage its photoconductivity, whereas resistor R_B is covered by metal 2 (and PCB) to generate a light-sensitive voltage divider output. The signal is amplified via non-inverting OTA with light-to-output voltage set only by resistor ratios, suppressing the effect of process corners and unintended partial light leakage via biaxial layout symmetry (both x and y axis). The signal dynamic range (indoor, outdoor) can be adjusted via resistor selection.

Figure 14.5.2 shows the N type-only on-chip inverting current-buffer nested-Miller compensated 3-stage OTA for input amplification, LDO error amplifier and FLL loop in Fig. 14.5.1. Its bias current and the output common-mode voltage are set by transistors M_{1B}-M_{5B}. The M_{1B}-M_{4B} loop (analysis in Fig. 14.5.2) compensates transistor/resistor mistracking across process/temperature corners, which in turn sets the bias voltage drop $RI_{B,OTA}$ across

any resistor R proportional to R_{BIAS}/R (ratiometric), and hence corner/temperature-independent bias point and small-signal gains. The fully-differential input stage is converted to single-ended by converting voltage to current via M_3-M_4 and R_3-R_4, followed by two common-source stages with resistive load. Interestingly, an inverting current buffer Nested Miller compensation network (e.g., C_{C1}, C_{C2}, R_C) is introduced to improve both large and small signal efficiency when driving capacitive loads in the range of 10-to-100 pF (i.e., SAR capDAC). In addition, the use of the inverting current buffer avoids the insertion of non-inverting stages (no-need for sign inversion in the second/third stage of the OTA) and hence it reduces the overall power consumption of the OTA. The proposed ratiometric pre-amplified clocked comparator in Fig. 14.5.2 is used in the 8b SAR ADC and the battery-charge controller in Fig. 14.5.1, and has a low kickback noise of <0.2 LSB. To overcome the limited swing of prior unipolar TFT comparators [11], [24], [28], [29], rail-to-rail output voltage is reinstated via resistor-transistor (RT) logic in digital buffers and output SR-latch. Data conversion is carried out by a binary-weighted 8b SAR ADC (Fig. 14.5.3) with unconventional use of differential capacitive DAC (capDAC) for single ended input, helping to reduce unavoidable charge-injection and clock-feedthrough effects given by the absence of complementary devices in IGZO technology, and hence of transmission gates. The N type-only LDO in Fig. 14.5.1 uses the 3-stage OTA as error amplifier and a pW-power 2-transistor voltage reference (V_{RLDO}~120 mV, 6.9%/V line sensitivity from 0.5-to-4.5 V, and 1,600 ppm/°C temperature coefficient).

In Fig. 14.5.3, the proposed switched-capacitor frequency-locked loop (FLL) locally derives a faster clock (*CLKH*) at frequency f_H for the SAR ADC to support its multi-cycle conversion (and further frequency multiplication to support oversampling/time averaging for sensing noise reduction), while keeping the system clock at lower frequency f_L for lower power. The circuit is based on transistors $M_{1,2}$, which effectively implement two voltage-controlled current sources $I_{1,2}$. The 3-stage OTA in Fig. 14.5.3 sets $V_A=V_B$ for virtual short-circuit, leading to an output frequency $f_H=f_L(C_H/C_L)$ simply set by a capacitive ratio. The real time clock $f_L=32.7$kHz is provided off chip for extra testability. One-time offset compensation is carried out at chip boot time by simply tuning $V_{G1,2}$ until the divided output frequency matches the input frequency (simple counters). The ample 250kHz-to-1.2MHz range of output frequency (sufficient for backscattering, see below) is obtained by introducing pseudo-dynamic voltage-controlled delay cells for the first time in flex ICs, whose N type-only output stage not dominated by RC time constants as opposed to RT logic.

Measurements in Fig. 14.5.4 show the OTA performance. The above reduction in the compensation capacitor C_{C1} allows the proposed OTA to outperform prior figure-of-merit for small-signal (large-signal) power efficiency FOMS (FOML) by >109× (>318×), while reducing area by >5× [7]-[12]. The 78% LDO peak power efficiency is >28% better than the prior best in flex ICs [13], despite the latter's advantageous availability of custom-process complementary TFT transistors. The maximum output current is improved by 6.5× at 4.6× lower area.

The SAR ADC in Fig. 14.5.5 shows a Walden's energy FOM at least 1.21× better than prior best-in-class flex ADCs [14]-[20]. Fig. 14.5.5 shows the output frequency vs. time for the proposed FLL, and adequate locking time of 1ms for low channel occupancy time and a ±2%-accuracy in the obtained output frequency. The 213µW FLL power in steady-state locked operations outperforms prior flex IC oscillators with comparable frequency by >23× [21], [23], [27], [30], while shrinking area by >2×. End-to-end sensor-to-ADC characterization in Fig. 14.5.5 shows a resolution of 320 lux and solid linearity with R²0.9994.

From Fig. 14.5.5, net positive RF power is harvested for an RF ambient power >-6 dBm (battery-charging/harvesting sensitivity). The system draws an average 4.2µA current due to always-on battery monitoring, battery charger and battery leakage, while sensing at a 30 acquisition/hour rate (i.e., every 2 minutes). From Fig. 14.5.6, a wireless transmission bit error rate of 10⁻³ is achieved at 3.5m tag-to-router and tag-to-PID distance, assuming a commodity receiver with -86dBm sensitivity. System operation is sustained at 3.5m distance from the PID (27dBm in a 7m x 7m room), even in the pessimistic case with no other RF signal available in the building. The adopted 7.5mAh LiPo battery at 10% discharge depth is restored to full charge in around half an hour. Figure 14.5.7 shows the backscattered spectrum and time-domain OOK waveforms.

Comparison of the proposed system (Fig. 14.5.7) with prior flex light sensors in Fig. 14.5.6 shows the highest level of integration including sensing, harvesting, battery charging, system power management and wireless communications on the same flex integrated circuit (only the RTC off-chip, apart from flex battery and printed antenna). The sticker-like system treated with silicone conformal coating (enabled by thin/flexible nature) is washable as necessary in indoor environments (Fig. 14.5.7). Its 2.02mm² flex IC and 10cm x 10cm overall form factor (with flex battery, antennas) allow unobtrusive and ubiquitous placement, while having sustainable operation with 10-year life at an ambient RF power of -6dBm or higher.

Acknowledgement:

The authors acknowledge the support of the Singapore NRF and DSO National Laboratories under SHINE program, the Singapore MOE (T2EP50223-0040 grant), as well as the Piano di Incentivi per la Ricerca di Ateneo 2024/2026 (Pia.ce.ri) Linea 1 (B3C), University of Catania.

References:

[1] H. Çeliker *et al.*, "Flex6502: A Flexible 8b Microprocessor in 0.8μm Metal-Oxide Thin-Film Transistor Technology Implemented with a Complete Digital Design Flow Running Complex Assembly Code," *ISSCC*, pp. 272-274, Feb. 2022. https://doi.org/10.1109/ISSCC42614.2022.9731790

[2] J. Biggs *et al.*, "A natively flexible 32-bit Arm microprocessor", *Nature*, pp. 532–53, 2021. https://doi.org/10.1038/s41586-021-03625-w

[3] M. Fattori *et al.*, "A Fully-Printed Organic Smart Temperature Sensor for Cold Chain Monitoring Applications," *IEEE CICC*, pp. 1-4, 2020. https://doi.org/10.1109/CICC48029.2020.9075908

[4] H. Ozaki *et al.*, "20-μW operation of an a-IGZO TFT-based RFID chip using purely NMOS "active" load logic gates with ultra-low-consumption power," *IEEE Symp. VLSI Circuits*, pp. 54-55, 2011. https://doi.org/978-1-61284-175-5

[5] L. Huang *et al.*, "A super-regenerative radio on plastic based on thin-film transistors and antennas on large flexible sheets for distributed communication links," *ISSCC*, pp. 458-459, Feb. 2013. https://doi.org/10.1109/ISSCC.2013.6487814

[6] K. Myny, "A flexible ISO14443-A compliant 7.5mW 128b metal-oxide NFC barcode tag with direct clock division circuit from 13.56MHz carrier," *ISSCC*, pp. 258-259, Feb. 2017. https://doi.org/10.1109/ISSCC.2017.7870359

[7] A. Rahaman *et al.*, "A high performance operational amplifier using coplanar dual gate a-IGZO TFTs," *IEEE JEDS*, pp. 655–661, 2019. https://doi.org/10.1109/JEDS.2019.2923208

[8] N. Papadopoulos *et al.*, "In-Panel 31.17dB 140kHz 87μW Unipolar Dual-Gate In-Ga-Zn-O Charge-Sense Amplifier for 500dpi Sensor Array on Flexible Displays," *IEEE ESSCIRC*, pp. 194-197, 2018. https://doi.org/0.1109/ESSCIRC.2018.8494260

[9] F. Meng *et al.*, "A Performance Optimized Operational Amplifier Using Transconductance Enhancement Topology Based on a-IGZO TFTs," *IEEE JEDS*, pp. 159–164, 2024. https://doi.org/10.1109/JEDS.2024.3366554

[10] M. Dandekar *et al.*, "An a-IGZO TFT based Op-Amp with 57 dB DC-Gain, 311 KHz Unity-gain Freq., 75 deg. Phase Margin and 2.43 mW Power on Flexible Substrate," *IEEE ESSCIRC*, pp. 407-410, 2021. https://doi.org/10.1109/ESSCIRC53450.2021.9567794

[11] C. Garripoli *et al.*, "A Fully Integrated 11.2-mm² a-IGZO EMG Front-End Circuit on Flexible Substrate Achieving Up to 41-dB SNR and 29-MΩ Input Impedance," *IEEE SSCL*, pp. 142-145, June 2018. https://doi.org/10.1109/LSSC.2018.2878184

[12] T. Moy *et al.*, "A flexible EEG acquisition and biomarker extraction system based on thin-film electronics," *ISSCC*, pp. 294-295, Feb. 2016. https://doi.org/10.1109/ISSCC.2016.7418023

[13] W. Wu *et al.*, "A 1.25-V Power Source Based on Solid-State Battery and Low Dropout Regulator Integrated With E-Mode a-IGZO and D-Mode ITO Thin-Film Transistors," *IEEE TED*, pp. 7551-7556, 2024. https://doi.org/10.1109/TED.2024.3488686.

[14] M. D. Alea *et al.*, "DERMIS: A Flexible Fully-Integrated 600μm -Resolution Per-Taxel Slip-to-Spikes Tactile Sensor Readout on A-IGZO TFT for Large-Area High-Density Electronic Skins," *IEEE Symp. VLSI Circuits*, pp. 1-3, 2025. https://doi.org/10.23919/VLSITechnologyandCir65189.2025.11074978

[15] N. Papadopoulos *et al.*, "Flexible 16nJ/c.s. 134S/s 6b MIM C-2C ADC using Dual Gate Self-aligned Unipolar Metal-Oxide TFTs," *IEEE CICC*, pp. 1-4, 2019. https://doi.org/10.1109/CICC.2019.8780121

[16] N. Papadopoulos *et al.*, "Toward Temperature Tracking With Unipolar Metal-Oxide Thin-Film SAR C-2C ADC on Plastic," *IEEE JSSC*, pp. 2263-2272, 2018. https://doi.org/10.1109/JSSC.2018.2831211

[17] C. Garripoli *et al.*, "An a-IGZO asynchronous deltasigma modulator on foil achieving up to 43dB SNR and 40dB SNDR in 300Hz bandwidth," *ISSCC*, pp. 260–261, Feb. 2017. https://doi.org/10.1109/ISSCC.2017.7870360

[18] D. Raiteri *et al.*, "An organic VCO-based ADC for quasi-static signals achieving 1LSB INL at 6b resolution," *ISSCC*, pp. 108-109, Feb. 2013. https://doi.org/10.1109/ISSCC.2013.6487658

[19] A. Dey and D. R. Allee, "Amorphous silicon 5 bit flash analog to digital converter," *IEEE CICC*, pp. 1-4, 2012. https://doi.org/10.1109/CICC.2012.6330647

[20] M. Dandekar and K. Myny, "An N-Type-Only a-IGZO Thin-Film-Transistor Based Nyquist-Rate 8-bit CDAC+SAR ADC Consuming 1.7mW at 32ksps and Achieving 44dB SNDR," *IEEE Symp. VLSI Circuits*, pp. 1-3, 2025. https://doi.org/10.23919/VLSITechnologyandCir65189.2025.11075081

[21] Y. Hu *et al.*, "3D multi-gesture sensing system for large areas based on pixel self-capacitance readout using TFT scanning and frequency-conversion circuits," *IEEE CICC*, pp. 1-4, 2014. https://doi.org/10.1109/CICC.2014.6946090

[22] G. Maiellaro *et al.*, "Ambient Light Organic Sensor in a Printed Complementary Organic TFT Technology on Flexible Plastic Foil," *IEEE TCAS-I*, pp. 1036-1043, 2014. https://doi.org/10.1109/TCSI.2013.2286031

[23] Y. Afsar *et al.*, "Large-scale acquisition of large-area sensors using an array of frequency-hopping ZnO thin-film-transistor oscillators", *ISSCC*, pp. 256-257, Feb. 2017. https://doi.org/10.1109/ISSCC.2017.7870358

[24] D. Geng *et al.*, "Touch Sensor Array With Integrated Drivers and Comparator Using a-IGZO TFTs," *IEEE LED*, pp. 391-394, 2017. https://doi.org/10.1109/LED.2017.2661405

[25] S. Elsaegh *et al.*, "A 1.6μW tunable organic transimpedance amplifier for photodetector applications based on gain-boosted common-gate input stage and voltage-controlled resistor with ±0.5% nonlinearity," *IEEE ESSCIRC*, pp. 75-78, 2017. https://doi.org/10.1109/ESSCIRC.2017.8094529

[26] S. Elsaegh *et al.*, "Low-Power Organic Light Sensor Array Based on Active-Matrix Common-Gate Transimpedance Amplifier on Foil for Imaging Applications," *IEEE JSSC*, pp. 2553-2566, 2020. https://doi.org/10.1109/JSSC.2020.2993732

[27] J. Pelgrims *et al.*, "A 24V Thin-Film Ultrasonic Driver for Haptic Feedback in Metal-Oxide Thin-Film Technology using Hybrid DLL Locking Architecture," *IEEE ESSCIRC*, pp. 69-72, 2022. https://doi.org/10.1109/ESSCIRC55480.2022.9911408

[28] K. Kim *et al.*, "a-InGaZnO Thin-Film Transistor-Based Operational Amplifier for an Adaptive DC–DC Converter in Display Driving Systems," *IEEE TED*, pp. 1189-1194, 2015. https://doi.org/10.1109/TED.2015.2402684

[29] H. Marien *et al.*, "A Fully Integrated ΔΣ ADC in Organic Thin-Film Transistor Technology on Flexible Plastic Foil," *IEEE JSSC*, pp. 276-284, 2011. https://doi.org/10.1109/JSSC.2010.2073230

[30] W. -X. Xu *et al.*, "High-Speed Ring Oscillator Using Skewed Delay Scheme Integrated by Metal-Oxide TFTs," *IEEE TED*, pp. 5526-5531, 2020. https://doi.org/10.1109/TED.2020.3029539

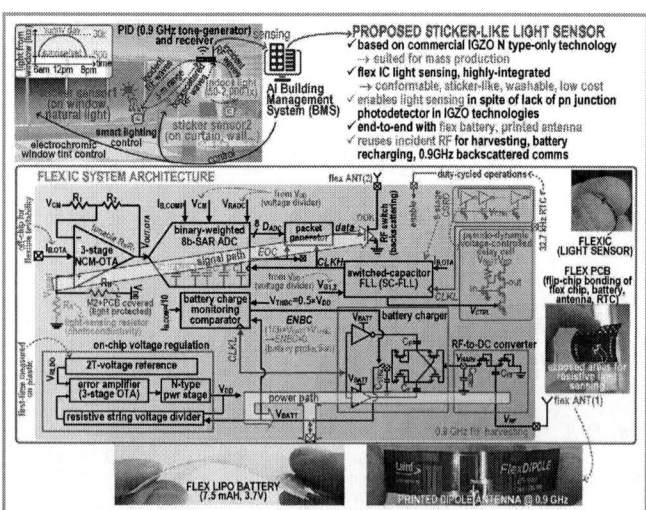

Figure 14.5.1: Sticker-like flexible highly-integrated light sensing for smart lighting system in energy-efficient buildings (top), sensing system architecture in IGZO technology (bottom).

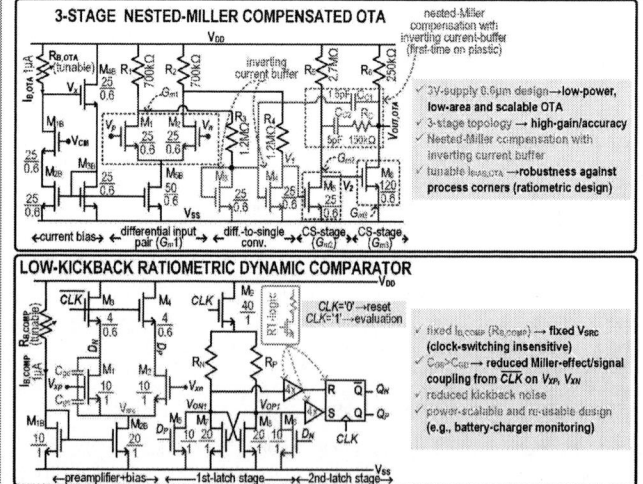

Figure 14.5.2: 3-stage flex OTA with nested-Miller compensation (top), low-kickback noise ratiometric dynamic comparator (gain is ratio of resistors and corner-independent bottom).

ISSCC 2026 / SESSION 14 / UNUSUAL INTERCONNECTS AND OTHER USES FOR LIGHT / 14.5

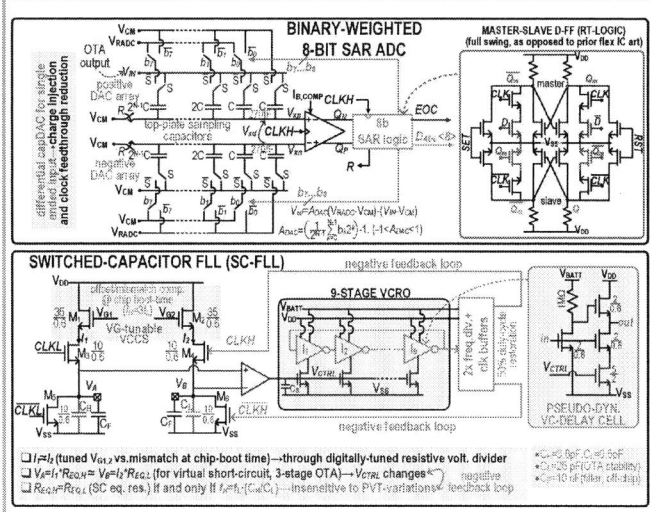

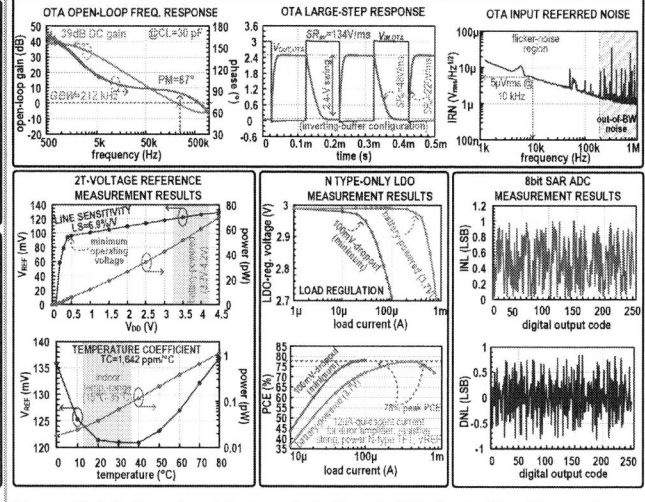

Figure 14.5.3: Single-input differential capDAC SAR ADC (top), switched capacitor FLL (SC-FLL) employing pseudo-dynamic voltage-controlled delay cells (bottom).

Figure 14.5.4: Experimental characterization of OTA (top), 2T-voltage reference (bottom-left), N-type only LDO (bottom-center), SAR ADC INL/DNL (bottom-right).

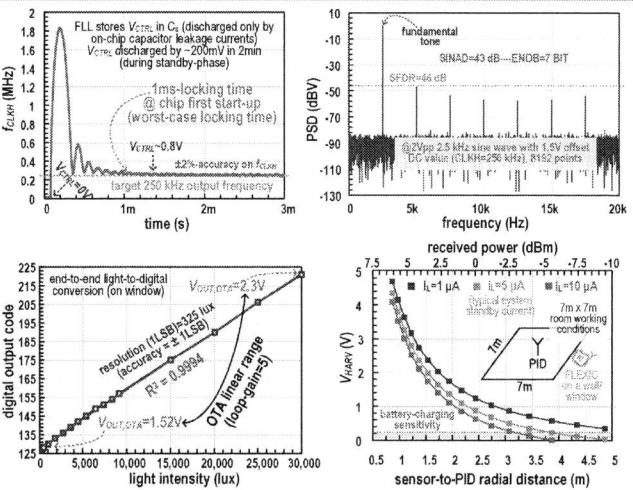

Figure 14.5.5: SC-FLL output frequency (top-left), SAR ADC dynamic performance at 250kHz (top-right), end-to-end light-to-digital conversion for light sensing on window (bottom-left), RF-to-DC vs radial distance/current (bottom-right).

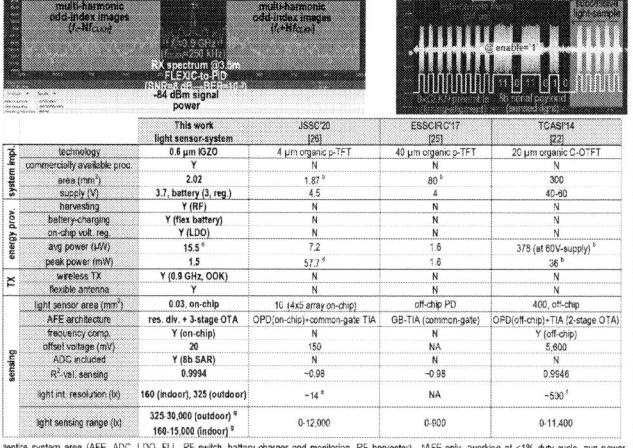

Figure 14.5.6: Backscattered OOK wireless spectrum (top-left), transmitted envelope waveforms (top-right), comparison with prior flex light sensing systems (bottom).

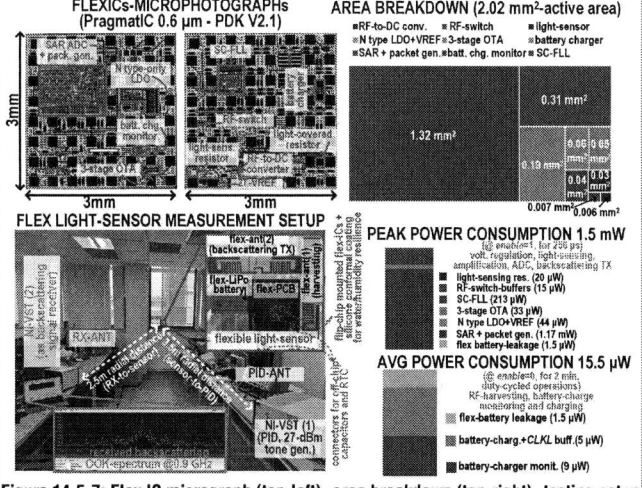

Figure 14.5.7: Flex IC micrograph (top-left), area breakdown (top-right), testing setup (bottom-left, front-end/back-end functions in boards A and B for signals testing convenience), peak and average duty-cycled power breakdown (bottom-right).

• 2026 IEEE International Solid-State Circuits Conference

979-8-3315-8937-0/26 $31.00 © 2026 IEEE

ISSCC 2026 / SESSION 14 / UNUSUAL INTERCONNECTS AND OTHER USES FOR LIGHT / 14.6

14.6 Self-Programmable Twin PUFs via Photovoltaic Energy Harvesting During the Pre-Wafer-Dicing Stage

Eunseok Lee, Jaehong Jung, Maitreyi Ashok, Anantha P. Chandrakasan, Ruonan Han

Massachusetts Institute of Technology, Cambridge, MA

Abstract

We present a CMOS self-programmable twin PUF which can be formed at the pre-wafer-dicing stage via photovoltaic harvesting. Two adjacent PUFs share entropy during oxide breakdown, enabling mutual authentication without pre-stored secrets. The prototype is implemented with twin PUF macros with a 128×8 array and achieves 98.4% pairwise consistency with 0.4% native BER. Using a low-cost LED, we demonstrate photovoltaic twin-PUF generation, confirming wafer-level batch processing feasibility.

Physical Unclonable Functions (PUFs) offer hardware-intrinsic variations to enable secure key storage and authentication without reliance on a dedicated non-volatile memory process [1,2]. After fabrication, the challenge–response pairs (CRPs) of the PUF device are measured and pre-stored for future authentications. Such an enrollment process, however, increases the risk of secret leaks, which is particularly the case for the PUF device since a third-party server is often used for the enrollment (Fig. 14.6.1) [3,4]. The authentication process routing through the server also involves additional communication overhead. Many PUF devices, such as those based on cryptography with internal entropy source [5], support very large number of CRPs by itself, but only a limited portion of those can be enrolled, which reduces the level of security.

Suppose each PUF has one and only one designated counterpart sharing identical CRPs (i.e. twin PUF). While this appears to contradict the unclonability premise, such a deliberately matched twin eliminates the need for verified stored CRPs. This is similar to tearing a paper sheet to two pieces, the shapes of the halves are unique yet matched; if distributed to two parties, the match becomes their "shared unclonable randomness". In critical systems with non-interchangeable device pairs, twin PUFs can provide lightweight mutual authentication without a third-party server. For instance, an ingestible pill and its paired wearable patch can authenticate each other directly, without any server involvement. On the other hand, the randomness-sharing step must be strictly one-time and non-repeatable, and it should be carried out under a controlled environment – preferably at a trusted foundry, as shown in Fig. 14.6.1. Recent studies have explored the matching and distribution of entropy between chips [6–8]. In [6], the concept of duplicating PUF responses is presented using the randomness of RRAM pairs between two chips, but it has not been experimentally prototyped. [7] demonstrates this idea by transferring SRAM entropy to RRAMs using organic transistors. However, these methods [6,7] rely on non-volatile memories and require external power to access and duplicate the entropy, involving temporary electrical connections to the wafer before it is diced. In [8], carbon nanotube (CNT) array-based twin PUFs are presented, utilizing CNT alignment between two adjacent dies on a wafer as the common entropy source during device fabrication. However, these approaches [6–8] remain incompatible with standard CMOS processes, making it difficult to integrate with other standard IP blocks.

In this paper, we present a CMOS-based, self-programmable twin PUF generated via photovoltaic energy harvesting during the pre-wafer-dicing stage. The conceptual twin PUF generation flow at a trusted foundry is illustrated in Fig. 14.6.1. Two PUFs are placed along the edges of two chips and are connected by metal layers within the wafer for sharing entropy. After fabrication in the trusted foundry, the wafer is illuminated with light, and harvested energy activates the twin-PUF generation circuits on each die. Once the wafer is diced, the paired two chips still retain shared unclonable randomness and can perform mutual authentication without any enrollment and communication with a third-party verifier, as shown in Fig. 14.6.1. Among various harvesting methods, photovoltaic harvesting is chosen because broad illumination easily scales to wafer-level batch processing and photodiodes are readily available in CMOS. In contrast, RF powering [9] requires bulky on-chip wave-coupling passives, and mm-wave/sub-THz powering [10–11] only enables small powered area causing long processing time for an entire wafer. Plasma-based harvesting during the fabrication could be explored in the future [12], though practical implementation depends strongly on specific foundry processes.

Figure 14.6.2 shows the twin PUF unit cell and its pairing process. We use an oxide-breakdown PUF [12,13], which is generated after chip fabrication by applying high-voltage stress. Unlike methods such as hot carrier injection that change device characteristics [14] and demand mA-scale current per unit, this approach runs at µA levels per unit [12,13], making it compatible with photovoltaic harvesting. We adopt a 4-transistor (4T) scheme with two transistors per die (Chip1 and Chip2). During twin PUF generation, top/bottom resistors self-protect remaining unbroken transistors by pulling their gate voltages low and raising their source/drain voltages. The detailed mechanism is illustrated in Fig. 14.6.2. When high-voltage stress V_{HV} is applied to M_1-M_4, V_{T1} and V_{T2} initially remain near V_{HV}, while V_{B1} and V_{B2} remain near ground. Each transistor has a different time-to-breakdown (t_{BD}). Assume M_1 breaks first at $t = t_{BD1}$ (i.e., $min(t_{BD1}, t_{BD2}, t_{BD3}, t_{BD4})$), its resistance collapses, lowering V_{T1} and raising V_{B1}. As a result, V_{GS} for M_2 and M_3 falls below V_{HV} (self-protected), while M_4 still sees nearly V_{HV} across its gate–to–source bias and remains unprotected. It therefore breaks next at $t=t_{BD4}$, and generates complementary outputs. Alternatively, a 2-transistor (2T) scheme—used in prior oxide-breakdown PUFs [12,13]—could, in principle, realize a twin PUF by placing one transistor on each die. However, the ~100s-µm inter-die spacing can introduce die-to-die gradients that skew the per-die t_{BD} distributions, with one die's transistors tending to break first under high-voltage stress, thereby biasing the PUF response. In contrast, the 4T scheme mitigates die-to-die t_{BD} gradients because the within-die pairs (M_1/M_2 and M_3/M_4) are co-located and experience similar t_{BD} distributions. Even if one die breaks globally faster (M_1 and M_2 tend to break before M_3 and M_4), the breakdown probability between M_1 and M_2 remains ~50/50. Whichever of M_1/M_2 fails first then determines which of M_3/M_4 receives full stress and breaks next, producing complementary bits, thereby making unbiased PUF responses.

Figure 14.6.3 presents the block diagram of the twin PUF macros with a 128×8 array. Chip1 and Chip2 are symmetric except for the twin PUF generation logic. During twin PUF generation, on-chip photodiodes harvest energy to drive two DC-DC converters and a high-voltage generator that supply V_{DDL} (~1V) and V_{DD_PUF} (~4V), respectively. For the CMOS process used in this work, a gate voltage of ~4V with duration on the order of seconds induces oxide breakdown [15]. Harvesting circuits are placed in the dicing area and eliminated during dicing, thereby avoiding chip area overhead. During the twin PUF generation, upon breakdown, the collapsed resistance induces a µA-level current surge on the V_{DD_PUF}. To support the power-limited energy-harvesting condition, all 1024 PUF cells are generated sequentially, one bit at a time (Fig. 14.6.3). Each breakdown raises the bottom-resistor nodes (V_{B1} and V_{B2}), then triggers the comparator-based breakdown detectors (BD_1, BD_2). Once both signals assert, the counter increments and advances to the next bit. The main-circuit clock is derived from the DC-DC converter. In our tests, the power consumption of the entire circuit is 1.3µW before any breakdown, and rises to 4.0-to-6.1µW when there are breakdowns during the generation mode. The top/bottom resistors limit the maximum current to ensure reliable harvesting operation. To minimize the area penalty of top/bottom resistors (~MΩ), resistors are shared across PUF columns. Circuit schematic of the unit column and the implemented unit cell are shown in Fig. 14.6.3. After dicing, the sense amplifiers (SAs) sense the voltage difference across the bottom resistors to read out the generated twin PUF in each chip.

The chip is fabricated using a 65nm standard CMOS process (Fig. 14.6.7). All empty spaces are filled with photodiodes to increase the available power during PUF generation. The twin-PUF cell concept was first validated and electrically characterized under external power supply and stress voltage (~4V) using test cells. The transient waveform (Fig. 14.6.4) shows that upon power-up, the first breakdown lowers V_{B2} and raises V_{T2}, reducing V_{GS2} and V_{GS3} and thereby protecting M_2 and M_3, which is consistent with the mechanism described in Fig. 14.6.2. Post-breakdown voltage fluctuations are caused by random telegraph noise arising from soft breakdown [16]. Measured normalized t_{BD}, across stress voltages, shows that 0.2V difference in stress voltage leads to >10x in t_{BD}. It should be noted that higher voltage stress shortens t_{BD} and total twin PUF generation time but increases required power, thereby imposing a burden on the harvester or power source. During read mode, the SAs compare the post-breakdown voltages at the V_{BL} and V_{BLB} to produce the output bit. Fig. 14.6.4 shows the normalized V_{diff} distributions for Chip1 and Chip2, defined as V_{diff} ($=\Delta(V_{BL}, V_{BLB})/Avg(V_{BL}, V_{BLB})$). If the twin PUF is formed correctly, the two distributions exhibit opposite polarities. For 256 PUF test unit cells across four samples, twin pairs successfully demonstrate opposite normalized V_{diff} polarity at $V_{DD\_PUF}=2.6V$, validating the twin PUF concept. The post-breakdown absolute normalized V_{diff} shows a mean (µ) of 1.51 and a deviation (σ) of 0.23, indicating a robust differential margin for stable PUF readout.

The twin PUFs with the entire 128×8 arrays are characterized by SAs. Measured example twin PUF data are shown in Fig. 14.6.5. Chip1 and the complement of Chip2 exhibit 98.8% pairwise consistency (opposite breakdown polarity). For 10 measured samples, average consistency of 98.4% is achieved. The residual inconsistency can be addressed using fault-tolerant cryptography [8]. Across 10 samples, the normalized inter-die Hamming distance is µ=50% (σ=4.48%), the normalized Hamming weight is µ=50.9% (σ=4.6%), and the average native bit-error rate is 0.4%. For Hamming distance/weight calculations, each 1024-bit response is split into 128-bit blocks. Next, twin PUF generation with photovoltaic

250 • 2026 IEEE International Solid-State Circuits Conference

979-8-3315-8937-0/26 $31.00 © 2026 IEEE

harvesting is demonstrated using a low-cost LED, as shown in Fig. 14.6.5. When the LED illuminates the chip, V_{DDL} and V_{DD_PUF} reach ~1V and ~4V, respectively. Two breakdown sequences (BD_1 and BD_2) are observed, confirming correct twin-PUF generation using only harvested energy. For a 12-inch wafer, we estimate several hundred watts of LED power may be required for illumination, though the exact amount depends on the number and efficiency of the photodiodes. The 10 measured twin PUFs were evaluated using the NIST SP 800-22 test suite [17], and passed all tests (Fig. 14.6.6), confirming their statistical soundness. A performance summary and comparison with the prior twin PUF prototypes are presented in Fig. 14.6.6.

Acknowledgement:
The authors thank Lockheed Martin, the MIT School of Engineering MathWorks Fellowship, and the Korea Foundation for Advanced Studies Fellowship for their support.

References:
[1] Y. Gao *et al.*, "Physical unclonable functions," *Nature Electron.*, vol. 3, pp. 81–91, 2020. http://doi.org/10.1038/s41928-020-0372-5
[2] C. Herder *et al.*, "Physical Unclonable Functions and Applications: A Tutorial," *Proceedings of the IEEE*, vol. 102, no. 8, pp. 1126-1141, Aug. 2014. http://doi.org/10.1109/JPROC.2014.2320516
[3] V. Suresh *et al.*, "A 0.26% BER, 1028 Challenge-Response Machine-Learning Resistant Strong-PUF in 14nm CMOS Featuring Stability-Aware Adversarial Challenge Selection," *IEEE Symposium on VLSI Circuits*, 2020, pp. 1-2. http://doi.org/10.1109/VLSICircuits18222.2020.9162890
[4] Y. Zheng *et al.*, "PUF-Based Mutual Authentication and Key Exchange Protocol for Peer-to-Peer IoT Applications," *IEEE Transactions on Dependable and Secure Computing*, vol. 20, no. 4, pp. 3299-3316, 1 July-Aug. 2023. http://doi.org/10.1109/TDSC.2022.3193570
[5] K. Liu *et al.*, "A 100-Bit-Output Modeling Attack-Resistant SPN Strong PUF with Uniform and High-Randomness Response," *2023 IEEE Custom Integrated Circuits Conference (CICC)*, 2023, pp. 1-2. https://doi.org/10.1109/CICC57935.2023.10121278
[6] T. Sato *et al.*, "Clonable PUF: on the Design of PUFs That Share Equivalent Responses," *IEEE International Symposium on Circuits and Systems (ISCAS)*, 2021, pp. 1-5. http://doi.org/10.1109/ISCAS51556.2021.9401345
[7] K. Oshima *et al.*, "Design of Aging-Robust Clonable PUF Using an Insulator-Based ReRAM for Organic Circuits," *29th Asia and South Pacific Design Automation Conference (ASP-DAC)*, 2024, pp. 165-170. http://doi.org/10.1109/ASP-DAC58780.2024.10473868
[8] D. Zhong *et al.*, "Twin physically unclonable functions based on aligned carbon nanotube arrays," *Nature Electron.*, vol. 5, no. 7, pp. 424–432, Jul. 2022. https://doi.org/10.1038/s41928-022-00787-x
[9] B. Zhao *et al.*, "A 5.8GHz power-harvesting 116μmx116μm "dielet" near-field radio with on-chip coil antenna," *IEEE International Solid-State Circuits Conference - (ISSCC)*, Feb. 2018, pp. 456-458. http://doi.org/10.1109/ISSCC.2018.8310381
[10] M. Tabesh *et al.*, "A Power-Harvesting Pad-Less mm-Sized 24/60GHz Passive Radio with On-Chip Antennas," *IEEE Symp. VLSI Circuits*, pp. 1-2, 2014. http://doi.org/10.1109/VLSIC.2014.6858380
[11] M. I. W. Khan *et al.*, "A Dual-Antenna, 263-GHz Energy Harvester in CMOS for Ultra-Miniaturized Platforms with 13.6% RF-to-DC Conversion Efficiency at –8 dBm Input Power," *IEEE Radio Frequency Integrated Circuits Symposium (RFIC)*, 2022, pp. 291-294. http://doi.org/10.1109/RFIC54546.2022.9863171

[12] K. Naruse *et al.*, "A Self-Programming PUF Harvesting the High-Energy Plasma During Fabrication," *IEEE International Solid-State Circuits Conference (ISSCC)*, Feb. 2023, pp. 218-220. http://doi.org/10.1109/ISSCC42615.2023.10067576
[13] K. -H. Chuang *et al.*, "A Physically Unclonable Function Using Soft Oxide Breakdown Featuring 0% Native BER and 51.8 fJ/bit in 40-nm CMOS," *IEEE Journal of Solid-State Circuits*, vol. 54, no. 10, pp. 2765-2776, Oct. 2019. http://doi.org/10.1109/JSSC.2019.2920714
[14] K. Liu *et al.*, "A 0.5-V Hybrid SRAM Physically Unclonable Function Using Hot Carrier Injection Burn-In for Stability Reinforcement," *IEEE Journal of Solid-State Circuits*, vol. 56, no. 7, pp. 2193-2204, July 2021. http://doi.org/10.1109/JSSC.2020.3035207
[15] M. Babaie *et al.*, "A study of RF oscillator reliability in nanoscale CMOS," *2013 European Conference on Circuit Theory and Design (ECCTD)*, 2013, pp. 1-4. http://doi.org/10.1109/10.1109/ECCTD.2013.6662205
[16] A. Cester *et al.*, "Soft breakdown current noise in ultra-thin gate oxides," *Solid-State Electron.*, vol. 46, no. 7, pp. 1019–1025, 2002. https://doi.org/10.1016/S0038-1101(02)00036-9
[17] A. Rukhin *et al.*, "A Statistical Test Suite for Random and Pseudorandom Number Generators for Cryptographic Applications," vol. 800. Gaithersburg, MD, USA: National Institute of Standards and Technology, 2010, no. 22-r1a. https://nvlpubs.nist.gov/nistpubs/legacy/sp/nistspecialpublication800-22r1a.pdf

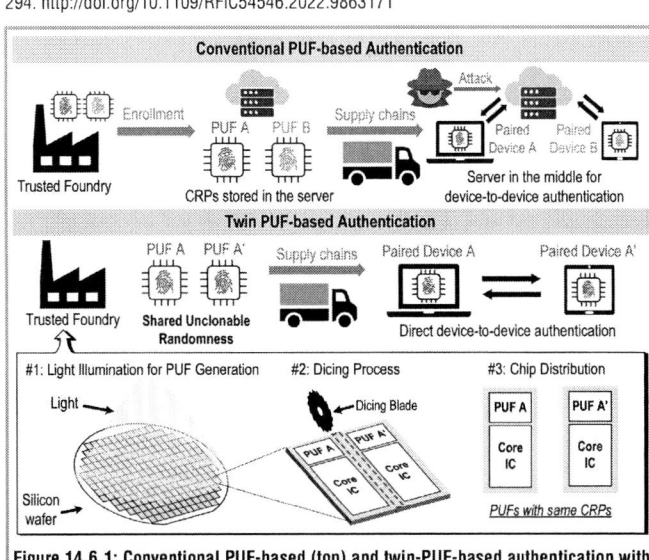

Figure 14.6.1: Conventional PUF-based (top) and twin-PUF-based authentication with conceptual fabrication scenario (bottom).

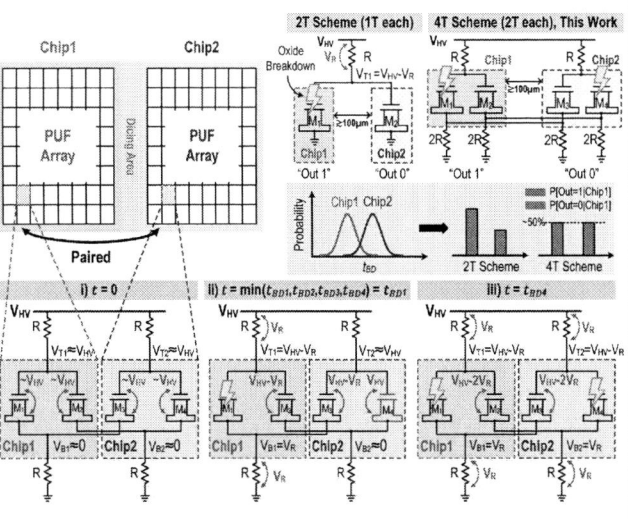

Figure 14.6.2: Concepts of the twin PUF unit cell and arrays with generation principle.

ISSCC 2026 / SESSION 14 / UNUSUAL INTERCONNECTS AND OTHER USES FOR LIGHT / 14.6

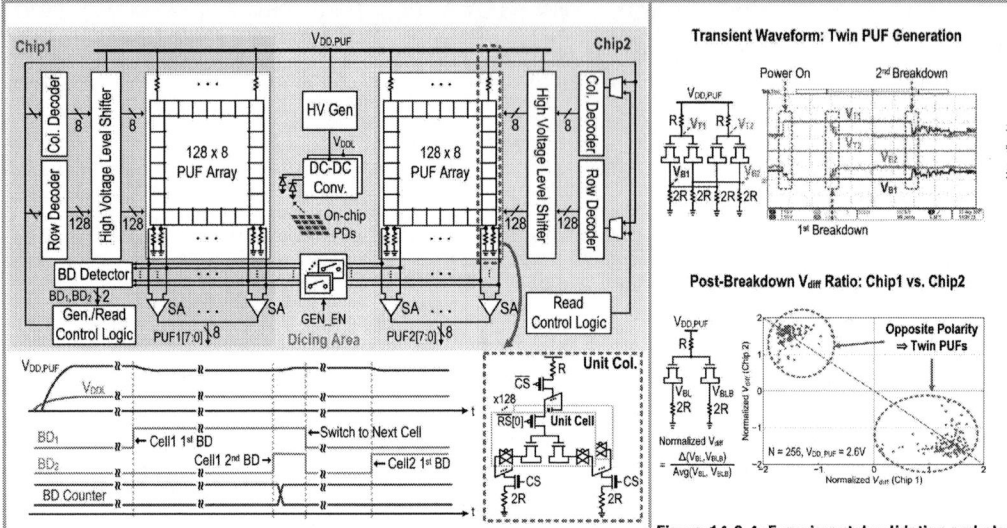

Figure 14.6.3: System diagram and operation of the twin PUF generation circuitry.

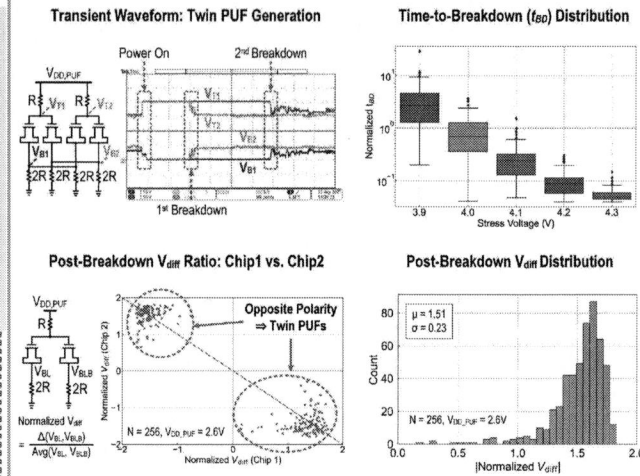

Figure 14.6.4: Experimental validation and electrical characterization of the twin PUF unit cells.

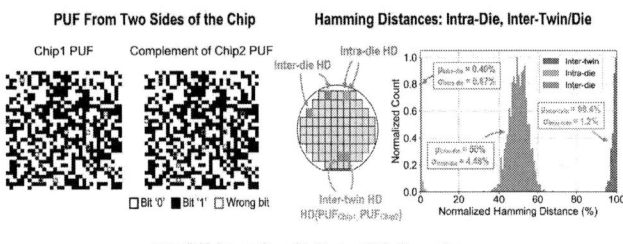

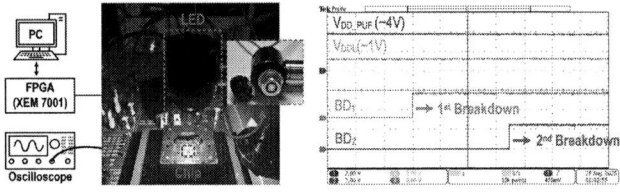

Figure 14.6.5: Performance of twin PUFs, Photovoltaic twin-PUF generation with LED setup and transient breakdown waveform.

NIST 800-22 Statistical Test Results

#	NIST Test	Pass % (10 runs)	Average P-Value	Pass?
1	Frequency	100	0.3166	Yes
2	Block Frequency	100	0.4423	Yes
3	Runs	100	0.5091	Yes
4	Longest runs	100	0.2892	Yes
5	FFT	100	0.4111	Yes
6	Nonoverlapping Template (m=7)	100	0.5236	Yes
7	Serial (m=5)	100	0.4275	Yes
8	Approx. entropy (m=4)	100	0.3598	Yes
9	Cumulative sum	100	0.3317	Yes

Performance Summary and Comparison

	Nature Elec.'22 [7]	This Work
Process/ Technology	Carbon nanotube (CNT)	Standard 65nm CMOS
Entropy Source	Random CNT alignment	Time-dependent gate-oxide breakdown
Twin Pairing Mechanism	Shared nanotube	Shared metal wires, self-protected transistor breakdown
Pairing stage	Frontend fabrication	Pre-dicing-wafer stage
Number of Bits for Each PUF	1120	1024
Bit Error Rate	~0%	0.4% (Native)
Normalized Inter HD	50.1% (binary)	50.0%
Twin-PUF Consistency	95%	98.4%

Figure 14.6.6: NIST SP 800-22 test results with 10 samples, and performance summary.

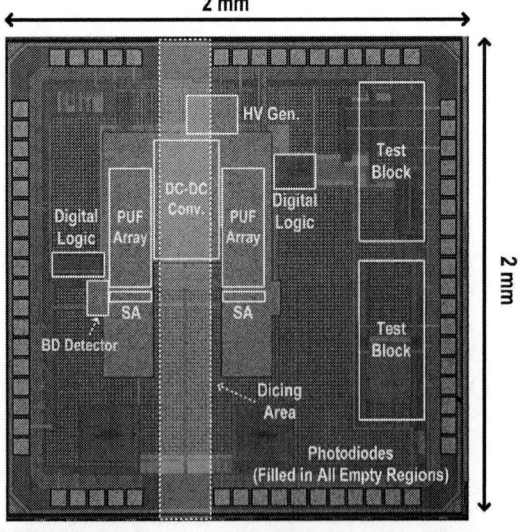

Figure 14.6.7: Die micrograph.

- 2026 IEEE International Solid-State Circuits Conference

979-8-3315-8937-0/26 $31.00 © 2026 IEEE

ISSCC 2026 / SESSION 15 / DRAM, SRAM, AND NON-VOLATILE MEMORIES / OVERVIEW

Session 15 Overview: *DRAM, SRAM, and Non-Volatile Memories*

MEMORY SUBCOMMITTEE

Session Chair: Hidehiro Shiga
KIOXIA
Yokohama, Japan

Session Co-Chair: Gunther Lehmann
Infineon Technologies AG
Neubiberg, Germany

DRAM is increasingly becoming the essential memory for AI and IoT devices, while SRAM continues to be an indispensable memory for high-performance computing due to its overwhelming speed; non-volatile memory solutions supporting these devices are also becoming increasingly important. In this session, the highest density NAND Flash memory and eMRAM are introduced, along with SRAM operating at low-voltage as well as various types of state-of-the-art DRAM solutions.

8:00 AM

15.1 A 2Tb 4b/Cell 6-Plane 3D-Flash Memory with 37.6Gb/mm² Bit Density and >85MB/s Write Throughput
Jayanth M. Thimmaiah, Sandisk, Bengaluru, India
In Paper 15.1, Sandisk/KIOXIA describe a 2Tb 4b/cell NAND flash memory with 37.6Gb/mm² bit density and 85MB/s write throughput.

8:25 AM

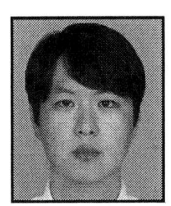

15.2 A 350mV Single-Rail SRAM Using a Custom-Logic-Bitcell in 2nm-CMOS-Nanosheet Technology for Mobile and Edge-AI Applications
Manish Trivedi, MediaTek, Bengaluru, India
In Paper 15.2, MediaTek presents a logic-cell-based 2-port SRAM fabricated in 2nm nanosheet technology.

8:50 AM

15.3 An 8nm eMRAM for Auto-G1 with a 125MHz Read Speed at 0.6V and 19.94Mb/mm² Density
Hyunjin Shin, Samsung Semiconductor, Gyunggido Kiheung, Korea
In Paper 15.3, Samsung shows an eMRAM in 8nm process that reaches the highest-yet density of 19.94Mb/mm².

252 • 2026 IEEE International Solid-State Circuits Conference

979-8-3315-8937-0/26 $31.00 © 2026 IEEE

ISSCC 2026 / February 17, 2026 / 8:00 AM

9:15 AM

15.4 A 16nm 168Mb Embedded STT-MRAM with 0.0249μm² Bit-Cell, Dual-Port Access, and 51.2Gb/s Read Throughput for Automotive and Edge AI Applications

Hon-Jarn Lin, TSMC Design Technology, Hsinchu, Taiwan

In Paper 15.4, TSMC presents a 16nm dual-port eMRAM achieving a very high read throughput of 51.2Gb/s.

9:45 AM

15.5 A 3nm 0.167fJ/b 5.27Mb/mm² Configurable TCAM with Macro-Wise Pipelined Search Methods for Automotive Applications

Yohei Sawada, Renesas Electronics, Tokyo, Japan

In Paper 15.5, Renesas introduces a configurable TCAM that achieves the lowest search energy of 0.167fJ/b and the highest density of 5.27Mb/mm².

10:05 AM

15.6 A 36GB 3.3TB/s HBM4 DRAM with Per-Channel TSV RDQS Auto Calibration and Fully-Programmable MBIST

Sunghwan Joo, Samsung Electronics, Hwaseong, Korea

In Paper 15.6, Samsung describes a 36GB 3.3TB/s 2048 I/Os HBM4 with 12-high 6th-generation 10nm DRAM process core dies and a 4nm Fin-FET logic process base die.

15

10:30 AM

15.7 A 1cnm 14.4Gb/s/pin 16Gb LPDDR6 SDRAM with Efficiency Mode, LDO-Based WCK Tree, Dynamic Write NT-ODT, Fast CS Control and System Meta Mode

Jungtaek You, SK hynix, Icheon, Korea

In Paper 15.7, SK hynix reveals a 14.4Gb/s/pin 16Gb LPDDR6 DRAM fabricated with 1cnm process.

10:55 AM

15.8 A 16Gb 12.8Gb/s LPDDR6 SDRAM with 12-DQ/Sub-Channel Wide NRZ Signaling and Enhanced Reliability by Per-Row Activation Counting and Meta-Data Scheme

Kiheung Kim, Samsung Electronics, Hwaseong, Korea

In Paper 15.8, Samsung also presents an LPDDR6 DRAM that reduces active and standby I/O power.

11:20 AM

15.9 A 48Gb/s 24Gb GDDR7 DRAM for Mid-Range Inference AI with Symmetric 2CH-Mode Operation, Clock-Path Optimization, and RAS Features

Hyunsu Park, SK hynix Semiconductor, Icheon, Korea

In Paper 15.9, SK hynix shows a 24Gb GDDR7 DRAM that achieves 48Gb/s with a supply voltage of 1.2V.

11:35 AM

15.10 A Vertical-Cell-Transistor-Based 4F² DRAM with Cell-on-Peripheral Architecture Using Wafer-to-Wafer Hybrid Copper Bonding

Hyunchul Yoon, Samsung Electronics, Hwaseong, Korea

In Paper 15.10, Samsung presents a 16Gb DRAM with vertical cell transistor based on a 4F² cell and a cell-on-peripheral architecture.

DIGEST OF TECHNICAL PAPERS • 253

979-8-3315-8937-0/26 $31.00 © 2026 IEEE

ISSCC 2026 / SESSION 15 / DRAM, SRAM, AND NON-VOLATILE MEMORIES / 15.1

15.1 A 2Tb 4b/Cell 6-Plane 3D-Flash Memory with 37.6Gb/mm² Bit Density and >85MB/s Write Throughput

Jayanth M. Thimmaiah[1], Ryuji Yamashita[2], In-Soo Yoon[3], Jason Li[3], Cynthia Hsu[3], Takuya Ariki[2], Naoki Ookuma[2], Yosuke Kato[2], Koichiro Hayashi[2], Kazuki Yamauchi[2], Indra K V[1], Masahiro Kano[2], Sirisha Bhamidipati[1], Sneha Bhatia[1], Seema Malhotra[1], Naoki Ojima[2], Ella Wu[3], Zhiyong Yang[3], Frank W. Tsai[3], Mathias Bayle[2], Naoyuki Minami[2], Yasuyuki Fujihara[2], Kei Kitamura[2], Tomofumi Kitani[2], Takuyo Kodama[4], Takaya Handa[4], Naoaki Kanagawa[4], Yuki Ishizaki[4], Susumu Fujimura[4], Yoshinao Suzuki[4], Mario Sako[4], Yumi Higashi[4], Yoshihisa Watanabe[4], Toshiyuki Kouchi[4], Aravinth V[1], Chin-Yi Chen[3], Xiang Yang[3], Guirong Liang[3], Jenny Wang[3]

[1]Sandisk, Bengaluru, India, [2]Sandisk, Yokohama, Japan, [3]Sandisk, Milpitas, CA, [4]KIOXIA, Yokohama, Japan

Abstract

A 332-WL-layer stacked 2Tb 4b/Cell 6-plane 3D Flash memory achieves an areal density of 37.6Gb/mm². An 8kB-physical and 16kB-logical page architecture realizes a <65µs read latency and a >85MB/s program throughput. Write throughput is enhanced by 4.2% through voltage-level sensing based on SEN2–LBUS coupling, and an additional 8% by reduced verify operations using 4 check points. A fast SLC burst programming technique reduces program time by ~11%.

The exponential growth of data and expanding AI applications necessitate high-capacity storage solutions with performance that scales with capacity. 3D Flash memory storing 4b/cell offers high density and a lower bit cost, making it well-suited for such applications.

In this work, a 332-WL layer stacked 2Tb 4b/cell 3D-Flash memory achieves an areal density >37Gb/mm². The design incorporates an 8kB physical and 16kB logical page structure, enabling a <65µs read latency and >85MB/s write throughput. Compared to prior 4b/cell designs [1], this work improves density >30% by using a CMOS directly-bonded-to-array (CBA) structure and a 6-plane configuration. 6-plane configuration enhances performance via increased parallelism: supporting asynchronous read operations across all 6 planes. The CBA structure, illustrated in Fig. 15.1.1(bottom right) which shows a peel-back view, also improves power delivery via additional low-resistance metal layers. A novel voltage-level (VL) sensing scheme based on SEN2-LBUS coupling, boosts write throughput and contributes to overall system performance in high-density high-capacity Flash memory systems. To further enhance write throughput, a reduced verify with 4 check points is implemented. Additionally, the fast self-adjusting read (FSAR) mode, which tracks threshold voltage (V_{th}) valley with quick search capability, improves data retention and V_{th}-valley tracking. FSAR achieves faster read performance (t_{READ}) and reduces read retries compared to conventional methods. A fast SLC burst-programming method is introduced to enhance SLC cache-program performance.

This work enables multi-die stacking to support high-capacity storage and leverages CBA technology to overcome limited power pad availability while enhancing power delivery. A new bus-idle sleep (BIS) mode reduces idle currents to near-standby levels, addressing challenges posed by an increased die count. Separate command-address (SCA) protocol is supported for multi-die stack configuration.

Figure 15.1.1(top left) shows the plane architecture of this work; the chip is comprised of 6 physical planes: the 16kB WL is arranged in two groups of 8kB WL, one on each side of the connection region to the stacked WLs. The planes are arranged in a BL direction in a 1×6 configuration. This arrangement enables a ~2.1% area reduction: compared to the 2×3 configuration [1, 5] shown in Fig. 15.1.1(top right). Area reduction is achieved via a smaller pad region that is placed in chip's bottom side and outside of the memory-cell region. A 1×6 configuration is constrained by the number of power pads and vertical power tracks in the x direction; however, the CBA architecture provides the flexibility to change CMOS wafer process/layers for circuit convenience. Hence, by adding another lower-sheet-resistance metal layer to realize a stronger power mesh to overcome the limitation. In a multi-die scenario, large currents can be injected into package ground, which can result in significant ground bounce that adversely affects the die, performing sense operations. To mitigate ground bounce, we incorporate additional ground pads, as illustrated in Fig. 15.1.1(top left). The main charge-pump circuit, which we have identified as the dominant source of ground bounce, is located on the right side of the chip. We placed additional ground pads on the right side of it to reduce induced ground bounce: even under high-current conditions across the 6-plane configuration. Additionally, a dedicated quiet-ground pad is included to provide an isolated ground connection for circuits that are sensitive to ground noise. These two mitigation strategies enabled the chip to perform asynchronous plane-read operations in a 1×6 configuration.

The need for faster program time (t_{PROG}) is always a benchmark for 3D Flash memory. A lower t_{PROG} is important for 4b/cell flash memories as programming involves verifying 16 states, which can result in a longer t_{PROG}. To reduce SA's x-pitch (to implement CBL), a local clock method was introduced [6]. The sense amplifier is shown in Fig. 15.1.2(top). Utilizing this structure, we propose a new sensing: LBUS coupling sense, which uses SEN2-to-LBUS coupling to directly transfer sense data to a latch without intermediate steps. A typical program verify uses two verify voltage levels per state: a low (V_L) and high (V_H) level [7]. Each verify operation consists of the following four steps: CLK_UP, sensing, CLK_DOWN, and strobe. (1) **CLK_UP** - to obtain a sufficient sense margin, especially for source-biasing sense, we clock up the SEN2 & SEN nodes using LBUS & VLOP; (2) Sensing - next, we sense the data by turning on XXL; (3) **CLK_DOWN** - SEN is clocked down with VLOP; (4) **Strobe** - we then strobe the SEN data onto the DL1 node. Previous overlapping sense schemes on non-clocking sense amplifiers have been discussed [8]; however, for deeper negative sense schemes it was not applicable due to existing CLK up/down pulses between the two senses. The LBUS coupling-sense scheme allows for an overlapping deep negative sense, which provides a 4.2% t_{PROG} improvement.

For a conventional method, as shown in Fig. 15.1.2(bottom left), step(1) to step(4) are done for VL sensing followed by VH sensing. In the proposed LBUS coupling sense, back-to-back V_L & V_H sensing is done, as shown in Fig. 15.1.2(bottom right). The sense amplifier is shown in Fig. 15.1.2(top). Once V_L sense completes, we transfer SEN2 onto DL1 using SEN2-to-LBUS coupling. For highly conductive cells, T_EN turns on and DL1 flips. For low-conductive cells, T_EN stays off and DL1 doesn't flip. For V_H sensing, we adjust the sense timing for good margin between on cell and off cell. We then clock down the SEN node with VLOP and strobe the data onto DL2 node via A_EN. This proposal needs careful tuning of gate voltages (T_EN) to transfer the information from LBUS to DL1 and to differentiate the on and off cells correctly. In a nutshell, LBUS coupling sense consists of parallel operation (overlap V_L STB and V_H sensing), skipping steps: CLKdown ~ CLKup.

The adequately-reduced-verify mechanism is implemented to optimize program operations in 4b/cell Flash memory: the number of program-verify steps are reduced to maintaining accurate cell-state programming, while improving overall performance and power efficiency. Normally, program-verify operations are performed for all the 15 programmable states. Instead of verifying all the program states, only 4 key states, 4 check points (CP) are verified. Figure 15.1.3(top) shows example CP states and the skipped verify states of the program loop, with the bias condition shown in Fig. 15.1.3(bottom). The program loop number and bias strength are dynamically adjusted based on each check point's verify result. Upper states S12-15 (after CP4 pass) are considered for data retention to optimize bias loop. Compared to all CP, we could get 8% performance gain without losing V_{th} margin.

The proposed FSAR mode addresses the limitations of the conventional valley-tracking-read (VTR) mode by significantly reducing the number of read-retry cycles and by improving read latency (t_{READ}); thereby, enhancing data retention (DR) performance. Compared to the baseline normal mode, the conventional VTR incurs ~7× impact to t_{READ}. In FSAR, quick-search and fine-search modes incur the highest latency (+89%), while quick-search-only and fine-search-only modes offer intermediate performance gains of +35% and +54% respectively. The FSAR technique involves two searching method: the quick-search-mode shown in Fig. 15.1.4(top) and fine-search-mode shown in Fig. 15.1.4(bottom). The supported five read modes are (1) quick -search-only, (2) fine-search-only using the quick-search result, (3) fine-search-only using fine-search result, (4) quick-search and fine-search, and (5) normal read using the quick-search result. Quick-search utilizes bit-count (BC) thresholds and read-level-voltage control adjustments via the following 3-step approach, a 2-point detection read with BC followed by a converting BC to shift value, 2-point-detection read-level adjustments are made using plus and minus shifts from default read level and finally performing page by page read with shift value. Fine-search mode uses a 3-point read method with Δ1 and Δ2 comparisons to determine the optimal read levels that involves a 2-step approach. Step 1 proceeds with a 3-point read followed by BC calculation and the storing of the read result in DLs. Step 2 calculates a threshold value and selects the best-read result as per the calculation. Fine-search is to get more accurate read result and the stored fine-search result reflects the next fine-search to save t_{READ} with accurate offset. Quick-search results can be utilized for a fine-search and a normal read without t_{READ} impact. The method supports asynchronous 6-plane read and suspend-read operations for both quick-search and fine-search modes.

To enhance SLC-cache program performance, a fast SLC burst programming method is introduced. The charge pump setup and discharge operations are time consuming for cache-program operations (t_{CPROG}). To reduce charge pump setup and recovery time, a pump set/reset-skip method is proposed. Figure 15.1.5(top) illustrates the pump set/reset-skip scheme. Charge pumps are enabled one time and kept enabled for subsequent program operations as shown in Fig. 15.1.5(bottom). Final discharge is enabled at the end of a burst program sequence. t_{PROG} can save around 11% for each cache program.

As storage density increases, stacking more dies as shown in Fig. 15.1.6(top left) becomes essential. In multi-die architectures, the idle current from unselected dies is approaching active-current levels. To minimize idle current, the data path must be fully shut down to imitate standby mode. An additional feature, BIS mode, is introduced to reduce the idle current from 10's of mA to 100's of µA for each die when enabled. The controller sends a LUNSEL command followed by the address of the target die. All of the stacked die with common CEnx that do not match the address issued will go into BIS mode: where an internal chip-enable signal, CEni, as shown in Fig. 15.1.6(top right), is disabled; thereby, disabling the complete data path and voltage generators, behaving as if the die was in standby mode. To exit BIS mode, the controller only drives CEnx to logic-high, which resets all gating, as shown in Fig. 15.1.6 (bottom).

The die micrograph is shown in Fig. 15.1.7(bottom) and a comparison of this work with prior 4b/cell and 3b/cell 3D Flash memory is given in Fig. 15.1.7(top).

References:
[1] W. Cho et al., "A 321-Layer 2Tb 4b/cell 3D-NAND-Flash Memory with a 75MB/s Program Throughput," *ISSCC*, pp. 512–514, 2025.
 https://doi.org/10.1109/ISSCC49661.2025.10904748
[2] J. Wontaeck et al., "A 280-Layer 1Tb 4b/cell 3D-NAND Flash Memory with a 28.5Gb/mm2 Areal Density and a 3.2GB/s High-Speed IO Rate" *ISSCC*, pp. 236-237, 2024. https://doi.org/10.1109/ISSCC49657.2024.10454343
[3] K. Yanagidaira et al., "A 1Tb 3b/cell 3D-Flash Memory with a 29%-Improved-Energy-Efficiency Read Operation and 4.8Gb/s Power-Isolated Low-Tapped-Termination I/O's" *ISSCC*, pp. 506-507, 2025. https://doi.org/10.1109/ISSCC49661.2025.10904509
[4] S.-S. Park et al., "A 28Gb/mm² 4XX-Layer 1Tb 3b/cell WF-Bonding 3D-NAND Flash with 5.6Gb/s/pin IOs," *ISSCC*, pp. 504-505, 2025.
 https://doi.org/10.1109/ISSCC49661.2025.10904543
[5] K. Kawai et al., "A 1Tb Density 3b/Cell 3D-NAND Flash on a 2YY-Tier Technology with a 300MB/s Write Throughput", *ISSCC*, pp. 244-246, 2024.
https://doi.org/10.1109/ISSCC49657.2024.10454296
[6] H. Maejima et al., "Crossed Bit Line (CBL) Architecture in 3D Flash Memory CMOS Directly Bonded to Array (CBA) Structure," *IEEE Int. Mem. Workshop*, 2025. https://doi.org/10.1109/IMW61990.2025.11026939
[7] B. Kim et al., "A High-Performance 1Tb 3b/Cell 3D-NAND Flash with a 194MB/s Write Throughput on over 300 Layers," *ISSCC*, pp. 402–403, 2023.
https://doi.org/10.1109/ISSCC42615.2023.10067666
[8] W. Cho et al., "A 1-Tb, 4b/Cell, 176-Stacked-WL 3D-NAND Flash Memory with Improved Read Latency and a 14.8Gb/mm2 Density," *ISSCC*, pp. 134-135, 2022. https://doi.org/10.1109/ISSCC42614.2022.9731785

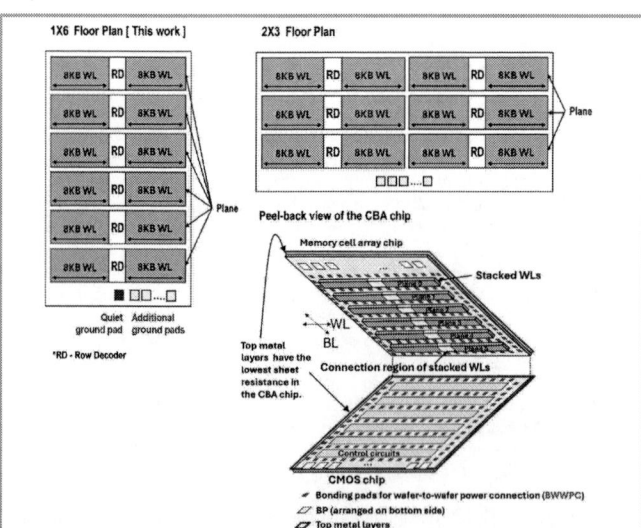

Figure 15.1.1: Floor plan options for 1×6 (top left) and 2×3 (top right) configurations. Peel-back view of the CBA chip (bottom right).

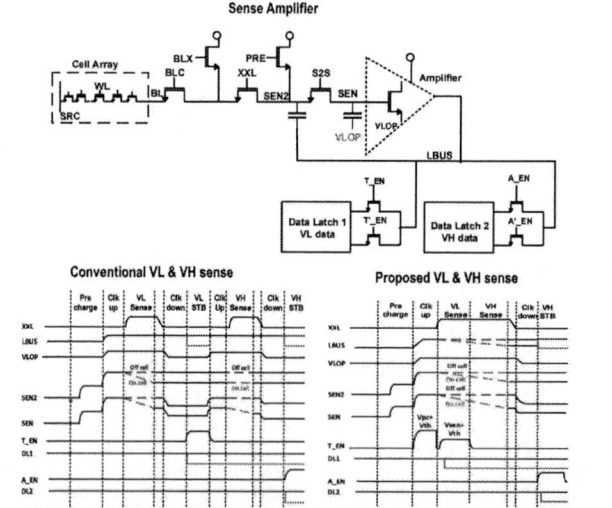

Figure 15.1.2: Sense amplifier schematic (top). Timing diagrams for a conventional VL & VH sense (bottom left) and the proposed (bottom right) VL & VH sense schemes.

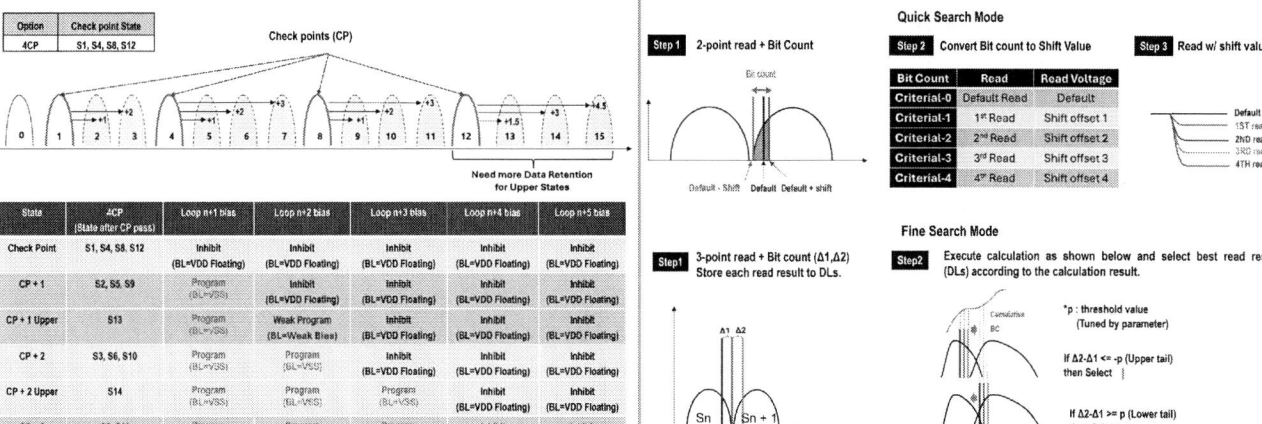

Figure 15.1.3: Adequately-reduced program verify with 4 check points. Example check-points states (top) and program loop with bias condition (bottom).

Figure 15.1.4: Fast self-adjusting read (FSAR) with quick-search mode (top), and fine-search mode (bottom).

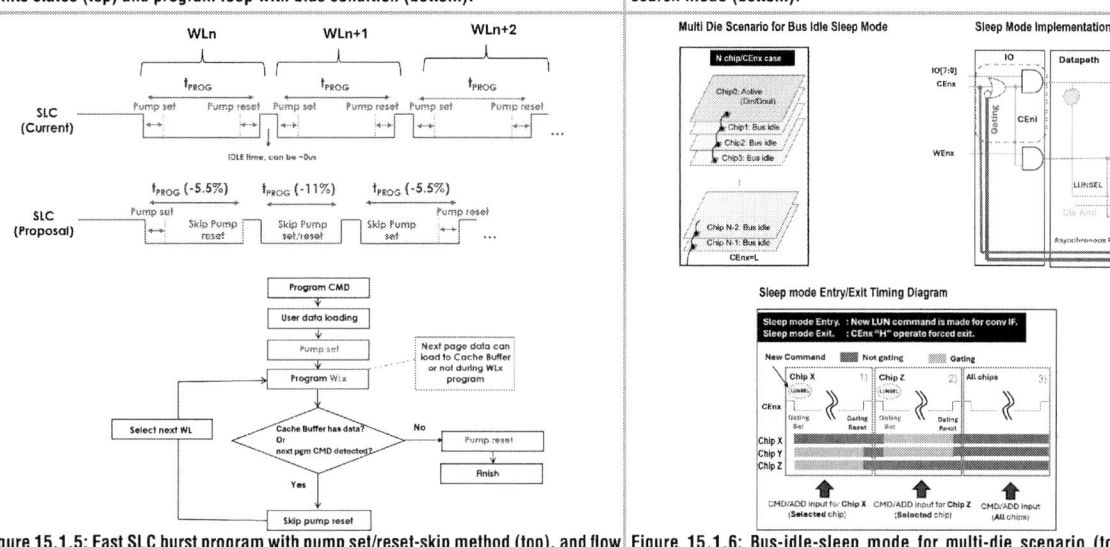

Figure 15.1.5: Fast SLC burst program with pump set/reset-skip method (top), and flow diagram for enabling the charge pump (bottom).

Figure 15.1.6: Bus-idle-sleep mode for multi-die scenario (top-left), sleep-mode implementation (top right), and entry/exit timing diagram (bottom).

Key-feature summary and comparison to prior work.

	This Work	[1] ISSCC'25	[2] ISSCC'24	[3] ISSCC'25	[4] ISSCC'25	[5] ISSCC'24
Technology	CBA, 3D Flash with 332 stacked WL Layers	3D NAND with 321 stacked WL Layers	3D NAND with 280 stacked WL Layers	CBA, 3D Flash with 332 stacked WL Layers	3D NAND with >400 stacked WL Layers	3D NAND with 2YY stacked WL Layers
Bits Per Cell	4	4	4	3	3	3
Capacity	2Tb	2Tb	1Tb	1Tb	1Tb	1Tb
# of Planes	6	6	4	4	4	6
Page Size	16 KB	16 KB	16 KB	16KB	16 KB	16 KB
Bit Density	37.6 Gb/mm²	28.8 Gb/mm²	28.5 Gb/mm²	29 Gb/mm²	>28.2 Gb/mm²	> 26 Gb/mm²
IO Bandwidth	4.8 Gbps	3.2 Gbps	3.2 Gbps	4.8 Gbps	5.6 Gbps	3.6 Gbps
t_{READ}	< 65uS	80uS	85 uS		38 uS	32 uS
Write Throughput	> 85 MB/s	75 MB/s	41 MB/s		231 MB/s	300 MB/s

Figure 15.1.7: Key-feature comparison table and die micrograph.

ISSCC 2026 / SESSION 15 / DRAM, SRAM, AND NON-VOLATILE MEMORIES / 15.2

15.2 A 350mV Single-Rail SRAM Using a Custom-Logic-Bitcell in 2nm-CMOS-Nanosheet Technology for Mobile and Edge-AI Applications

Manish Trivedi[1], Sandipan Sinha[1], Ramesh Halli[1], Girishankar Gurumurthy[1], Jaswinder Singh[1], Chun-Yuan Cheng[2], Linchien Chen[2], Jeff Lin[2], Hugh Mair[3]

[1]MediaTek, Bengaluru, India, [2]MediaTek, Hsinchu, Taiwan, [3]MediaTek, Austin, TX

Abstract

This work presents a single-rail, 21.04Mb/mm² logic-bitcell 2-port SRAM in a 2nm nanosheet technology for CPU, GPU and NPU caches. The implemented xBIT cell in a 2R×1C configuration uses dual BL for balanced NMOS/PMOS devices to form a rectangular array and maximize area density. Optimized for dynamic power, it operates reliably from 0.35 - 1.2V, achieving an industry-leading 102ps read-access time to support 4GHz processor and making it ideal for shallow-depth caches.

SRAM is critical in modern compute systems for optimizing system-level power, performance, and area (PPA). For CPUs, GPUs and NPUs it provides low-latency memory access in the form of caches, register files or scratchpads; thereby, accelerating data access and storing weights to support high-speed processing. In addition, these systems use voltage-frequency scaling to improve efficiency, requiring SRAM to function across a wide voltage range: 0.45 - 1.2V. 6-transistor high-current-bitcell (6T-HC) SRAMs [1-3] commonly use per-column sense amplifiers (SAs) to increase read performance; however, an SA's high dynamic power due to inherent capacitance makes them less suitable at ultra-high speeds (3.5GHz+). Thus, large-signal sensing and segmented BL schemes (e.g., 32 rows per subarray) are preferred, as they reduce dynamic power while improving performance.

To increase the per cycled data throughput, modern compute architectures utilize multi-port SRAMs (e.g. 1R1W and 2R1W) that enable simultaneous read and write operations, while minimizing access conflicts [4,5]. Consequently, SRAM cache designs are increasingly using the 8T bitcell for 1R1W and 2R1W support, along with hierarchical segmented BLs and dual-rail configuration that enables lower peripheral voltage for power savings, while maintaining adequate bitcell voltage [6,7]. However, speed- and power-critical caches are typically moderate in size (16 - 128kb) with shallow word depths (≤512) and high bit widths (128–256). For these cases, a segmented-BL multi-bank architecture using 6T or 8T bitcells incurs a significant area overhead from separator cells and the required logic-to-memory spacing, as shown in Fig. 15.2.1(bottom), increasing the effective implementation area of SRAM by 20 - 30%. Dual-rail architectures also complicate the implementation by requiring additional power connections, hindering interconnect optimization, as well diminishing low-voltage power efficiency by the need of maintaining the bitcells at a higher voltage. Moreover, foundry bitcells are not scaling in newer technology nodes, compared to logic-based storage cells, like multibit flipflops (MBFF), as shown in Fig. 15.2.1(top). To address these issues, single-rail logic-bitcell SRAMs have been implemented [8-10]. While logic-bitcells are larger than foundry bitcells, they allow for the elimination of separator cells and enable direct abutment to logic; thereby, reduces effective implementation area of SRAM [9].

Logic-bitcell SRAMs follow technology scaling, but the traditional 10T logic-bitcell [8] is not area efficient in advanced nodes due to its NMOS-PMOS imbalance in transistor count, leading to wasted area due to the two additional NMOS transistors in the bitcell. The 12T logic-bitcell [9] achieves NMOS-PMOS balance in transistor count, but its higher transistor count increases leakage and area.

This work achieves low dynamic power and minimal area, for SRAM caches ≤128kb, in a 2nm technology by introducing a single-rail logic-bitcell, which enables aggressive voltage scaling and seamless logic integration. To maximize density, we introduce a new bitcell, xBIT, that pairs an NMOS-dominated NBIT (6N/4P) and PMOS-dominated PBIT (4N/6P) forming a rectangular array of balanced NMOS and PMOS devices, see Fig. 15.2.2. Both NBIT and PBIT feature a dedicated read-access transistor stack, for read disturb immunity, and a complementary controlled latch for contention-free writes. The xBIT cell uses two BLs: rbln for NBIT and rblp for PBIT that are pre-charged high and low prior to read, respectively. Read timing diagrams for NBIT and PBIT cells are shown in Fig. 15.2.2. For NBIT, a stored 0 discharges rbln low, while a stored 1 leaves it high. For PBIT, a stored 0 keeps rblp low, while a stored 1 charges it high.

Figure 15.2.3 shows the xBIT SRAM architecture with four bank arrays that are managed by a bank controller (BNKCTRL slice), with split IO arrays at the top and bottom of the BNKCTRL slice to reduce the interconnect delay. Each BNKCTRL has an array of WL drivers (WLDV-ARY) for row selection and local control (LCTRL) to generate bank specific read/write signals. Each IO array slice contains two xBIT subarrays in a butterfly configuration, connected to a local IO (LIO) circuit. The sub-array row count (16 - 64) is determined by speed and power requirements; the bank count is determined by the word size. IO array slices repeat to match bit-width, and global IO (GIO) receives/drives data from/to the SRAM instance.

Figure 15.2.3 also illustrates the write data path from GIO to the bitcell. Input Data Din is latched inside GIO and is sent to the upper and lower banks as global-BL signals: wblbtop and wblbot. For each bank, a local control generates write-enable clocks (went for the upper and wend for the lower subarrays), distributing them to all LIO slices. Each LIO mixes these signals with the global BL using NOR-inverter logic to drive local write BL (wblu for upper and wbld for lower), shared by all NBIT and PBIT cells in the subarray. This local bank clock mixing reduces toggling, optimizing dynamic write power. Write timing diagram is shown in Fig. 15.2.3, write is identical for NBIT and PBIT. Data is written by asserting a single-row WL (wwl0 high and wwlb0 low for NBIT; wwl1 high and wwlb1 low for PBIT) to make the latch transparent, enabling reliable and contention-free writes even at ultra-low voltages. Note that V_{MIN} is no longer limited by write, hence smaller devices can be used in the write path to reduce area and leakage power.

Figure 15.2.4 shows the read path from the xBIT cell to DO, passing through the array, LIO and GIO circuits. Each upper and lower xBIT subarray uses separate BLs for NBITs (rblnu) and PBITs (rblpu), so each BL serves only half the rows. The LIO uses P-type devices to pre-charge NBIT BLs (rblnu/rblnd for upper and lower subarrays) and N-type devices to pre-discharge PBIT BLs (rblpu/rblpd for upper and lower subarrays). It combines rblnu/rblnd with a NAND-inverter for gblnu, and rblpu/rblpd with a NOR-inverter for gblpu, sending these to the GIO. GIO processes global BLs from multiple bank arrays using a read select mux (RSM, as an AOI gate) and a controlled output latch. RSM receives select signals (seln for NBIT, selp for PBIT) from global control (GCTRL) circuit, which determines the cell type based on the input address, NBIT for even addresses and PBIT for odd addresses. With opposite read conventions for NBIT and PBIT, default gblnu state is high (stored 1 leaves it high and a stored 0 discharges it low); gblpu state is low (stored 0 keeps it low and stored 1 charges it high). RSM unifies these outputs by combining signals from the top and bottom banks with a NAND gate, then sends the result to a controlled latch to generate DO.

Figure 15.2.4 (right) shows the read timing. For NBIT reads, pnu disables precharge and read WL (rwln) is asserted high. If storing 0, rblnu discharges, LIO drives gblnu low, and with seln high, RSM keeps gblu high; when the latch is transparent (mx high and mxb low), DO reads 0. For stored 1, rblnu and gblnu stay high, and DO reads 1. PBIT reads utilize the discharge control (dpu) and active low read WL (rwlp), following the polarity as illustrated in Fig. 15.2.4. To prevent glitches, mx/mxb are asserted only after gblu crosses threshold, ensuring correct detection, and avoiding premature latch opening. Signals mx/mxb reset before input changes, closing the latch before transitions to maintain data integrity.

The write mask function blocks write to selected data bits using the BYTE signal, which gates the WL signals (wwl and wwlb), as shown in Fig. 15.2.5. If BYTE is disabled for a given bit, corresponding WLs are not asserted, and no write occurs. In caches, masking is usually done for halfwords (Bits/2), quadwords (Bits/4), or groups of 8, 16, or 32 bits, not for single bits; this allows gating logic overhead to be shared across group as shown in Fig. 15.2.5 (upper-right). Write mask area overhead is 10% for 8-bit groups, 2.5% for 32-bit groups, and under 0.5% for quadwords or halfwords. This group-based method is especially efficient for xCPU caches, offering much lower PPA impact versus methods like read-modify-write [9] and 16T cells [10].

While normal logic performance averages NMOS and PMOS due to its alternating rising/falling nature, the xBIT read performance can be dominated by either NMOS or PMOS read paths, creating a maximum vs. average response to process asymmetry. Figure 15.2.5 shows the 3-σ read delays for each path, indicating that NBIT reads are faster at higher voltages. However, at lower voltages, layout-dependent effects on N-devices slow NBIT down, increasing read delay by 12% at 0.35V. Measured silicon V_{MIN} is set by this worst-case scenario.

The xBIT SRAM is fabricated in 2nm nanosheet technology. Figure 15.2.7 shows the test chip micrograph, implementing a total of 0.5Mb as eight 64kb SRAM blocks in a 256×256 2-Port (1R1W) configuration in standard cell architecture. Each xBIT subarray has 32 rows (16N + 16P). This work uses wide 19n-uLVTLL devices in the read stack to reduce access delay, and narrow 13n-LVT devices in the write latch to minimize leakage. Total 232 test chip samples are tested across a wide voltage and temperature range. The SRAM achieves

4GHz at 0.95V, with 100% yield from 15°C to 65°C, and from 0.35V to 1.2V, as shown in Fig. 15.2.6. It delivers a V_{MIN} of 0.35V and a top density among 2-Port SRAMs of 21.04Mb/mm², see Fig. 15.2.6. It achieves 30% power reduction as compared to conventional 8T SRAM across similar process condition and SRAM size, see Fig. 15.2.6. Compared to conventional 8T SRAM in 2nm, this design delivers higher density across cache-relevant configurations (≤128kb), making it well-suited for CPU, GPU, and NPU caches.

This work introduces a low-power and area-efficient logic-bitcell SRAM architecture that overcomes scaling limitations of conventional foundry SRAM cells. Area efficiency is improved by using dual read polarities to balance the NMOS and PMOS devices across pairs of bitcells. The logic-bitcell enables seamless integration with logic blocks, eliminating the need for separator cells. The single supply-rail operation supports deep voltage scaling, specifically <0.75V where standard SRAMs fail. Connecting each BL to half of the rows per subarray halves BL loading, resulting in faster reads and lower read power. As the number of rows per BL is halved, the fight between the leakage current from unselected bitcells and read current from selected bitcell is reduced, enabling the use of ultra-low-V_t and wider-channel transistor devices in the read transistor-stack path to reduce read-access time. The above traits make logic-bitcell SRAM well-suited for caches with tight read-access time requirements, while offering power and area advantages over foundry-provided 6T/8T bitcell SRAM. By combining cell innovation with circuit efficiency, this architecture delivers a high-density low-power SRAM suitable for next-generation processors.

References:
[1] M. Clinton et al., "A low-power and high-performance 10nm SRAM architecture for mobile applications," *ISSCC*, pp. 210-211, 2017. doi: C10.1109/ISSCC.2017.7870335. https://doi.org/10.1109/ISSCC.2017.7870335
[2] Y. Aoyagi et al., "A 3-nm 27.6-Mbit/mm² Self-timed SRAM Enabling 0.48 - 1.2 V Wide Operating Range with Far-end Pre-charge and Weak-Bit Tracking," *IEEE Symp. on VLSI Tech. & Circuits*, 2023. https://doi.org/10.23919/VLSITechnologyandCir57934.2023.10185429
[3] M. Clinton et al., "A 5GHz 7nm L1 cache memory compiler for high-speed computing and mobile applications," *ISSCC*, pp. 200-201, 2018. doi: 10.1109/ISSCC.2018.8310253. https://doi.org/10.1109/ISSCC.2018.8310253
[4] J. Keane et al., "5.6Mb/mm² 1R1W 8T SRAM arrays operating down to 560mV utilizing small-signal sensing with charge-shared bitline and asymmetric sense amplifier in 14nm FinFET CMOS technology," *ISSCC*, pp. 308-309. 2016 https://doi.org/10.1109/ISSCC.2016.7418030
[5] A. Fritsch et al., "A 6.2 GHz Single Ended Current Sense Amplifier (CSA) Based Compliable 8T SRAM in 7nm FinFET Technology," *ISSCC*, pp. 334-336, 2021. https://doi.org/10.1109/ISSCC42613.2021.9365812
[6] R. Mathur et al., "5GHz SRAM for High-Performance Compute Platform in 5nm CMOS," *IEEE CICC*, 2022. https://doi.org/10.1109/CICC53496.2022.9772840
[7] R. Mathur et al., "A 7GHz High-Bandwidth 1R-1RW SRAM for Arm HPC Processor in 3nm Technology," *IEEE Symp. on VLSI Tech. & Circuits*, 2024. https://doi.org/10.1109/VLSITechnologyandCir46783.2024.10631403

[8] S. Jain et al., "A 280mV-to-1.2V wide-operating-range IA-32 processor in 32nm CMOS," *ISSCC*, pp. 66-68, 2012. https://doi.org/10.1109/ISSCC.2012.6176932
[9] M. E. Sinangil et al., "A 290MV Ultra-Low Voltage One-Port SRAM Compiler Design Using a 12T Write Contention and Read Upset Free Bit-Cell in 7NM FinFET Technology," *IEEE Symp. on VLSI Circuits*, pp. 13-14, 2018. https://doi.org/10.1109/VLSIC.2018.8502419
[10] H. Fujiwara et al., "A 5nm 5.7GHz@1.0V and 1.3GHz@0.5V 4kb Standard-Cell- Based Two-Port Register File with a 16T Bitcell with No Half-Selection Issue," *ISSCC*, pp. 340-341, 2021. https://doi.org/10.1109/ISSCC42613.2021.9366000

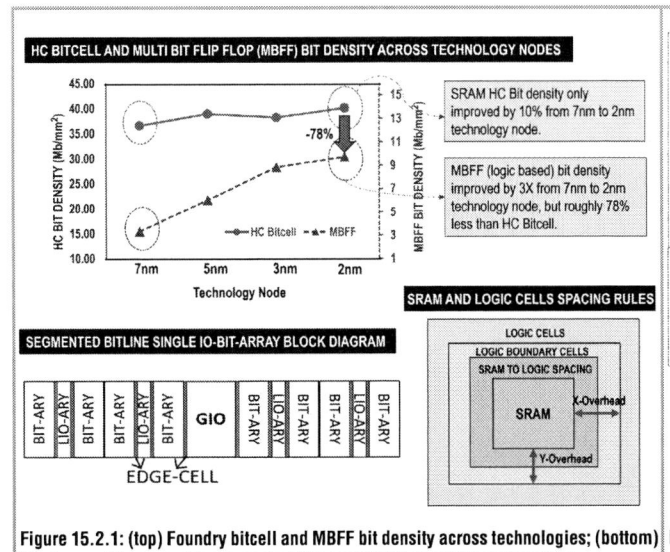

Figure 15.2.1: (top) Foundry bitcell and MBFF bit density across technologies; (bottom) IO slice architecture with segmented BLs and SRAM-to-logic spacing rules.

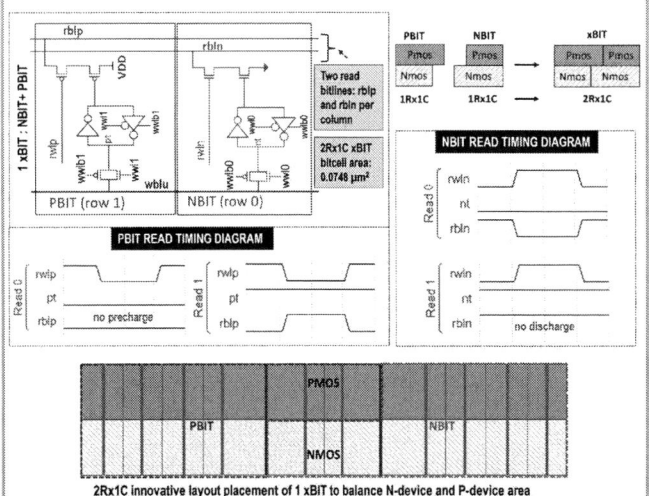

Figure 15.2.2: (top) xBIT schematic arranged in 2 Row×1 Column (2R×1C) scheme, read0/read1 timing waveform; (bottom) xBIT layout.

ISSCC 2026 / SESSION 15 / DRAM, SRAM, AND NON-VOLATILE MEMORIES / 15.2

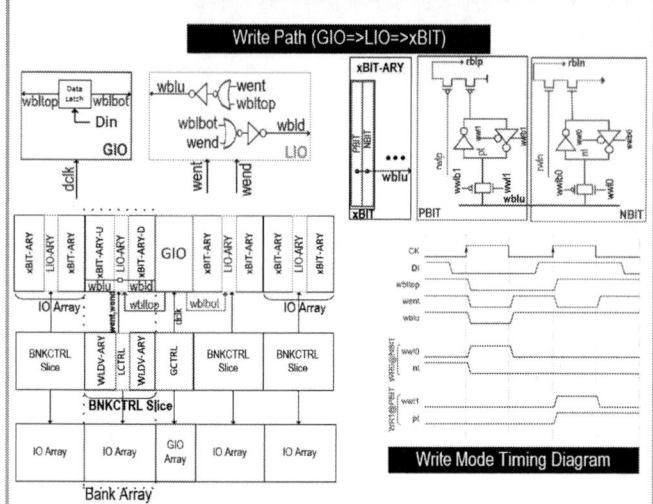

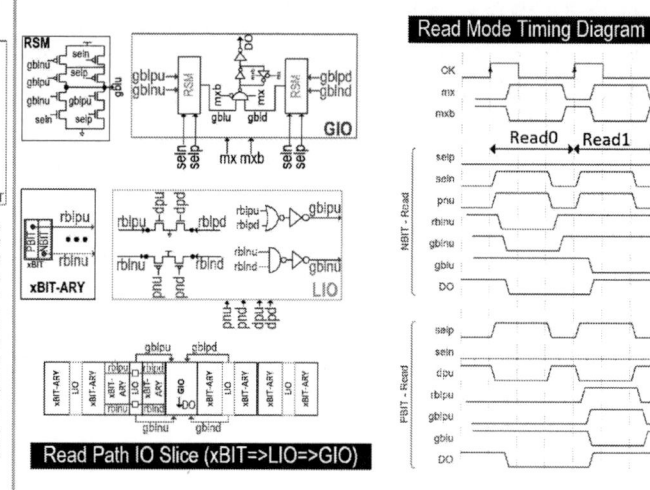

Figure 15.2.3: xBIT SRAM architecture with 4 banks, Data in-to-bitcell write circuit and signal timing diagram shown.

Figure 15.2.4: Read path circuits and signal timing diagram for reading 1 and 0 from NBIT and PBIT.

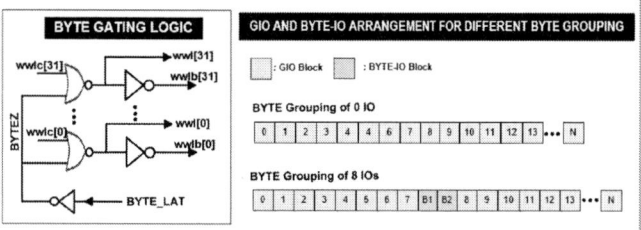

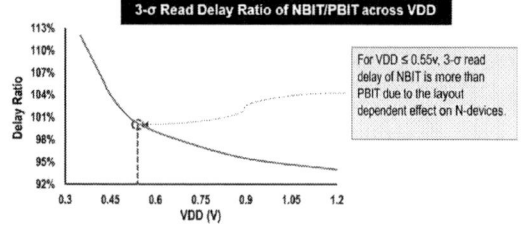

Figure 15.2.5: (top) Byte gating circuit and grouping for different number of IOs shown; (bottom) 3σ read delay ratio of NBIT and PBIT for different V_{DD} voltages.

Density Comparison (Mux:1, Rows/BL:32)
xBIT vs. Foundry 8T Bitcell

Tech: 2nm	2-Port (1R1W)	
WordsxBits	8T (Mb/mm²)	This Work (xBIT) (Mb/mm²)
32x256	9.29	15.15
64x256	13.23	18.51
128x256	14.73	20.07
256x256	17.16	21.04

Energy Comparison (Mux:1, Rows/BL:32)
xBIT vs. Foundry 8T Bitcell

Tech: 2nm	2-Port (1R1W)	
Corner	Instance: 256x256	%(Power Reduction wrt. 8T)
TT_0.5v_25C	Leakage (uW)	-29%
	Dynamic (pJ)	-31%
TT_0.9v_25C	Leakage (uW)	-9%
	Dynamic (pJ)	-36%

Comparison with Prior Works

Reference	ISSCC'16 [4]	ISSCC'21[5]	CICC'22[6]	VLSI'24 [7]	This Work
Technology	14nm	7nm	5nm	3nm	2nm
Size (Kb)	64 Kb	1B Kb	6 Kb	5 kb	64 Kb
Density (Mb/mm²)	5.6	6.1	6.8	11.2	21.04
Rows/Bitline	256	128	64	64	16N+16P
Sensing Scheme	Small Signal	Small Signal	Large Signal	Large/Small	Large Signal
Type	1R1W	1R1W	1R1W	1R1RW	1R1W
F_{MAX}	2.21GHz (1.0V)	6.2GHz (1.0V)	5.22GHz (1.0V)	7.74GHz (0.945V)	4.0GHz (0.95V)
T_{CLK-Q}		161ps (1.0V)	112ps (1.0V)	108ps (0.75V)	102ps (0.95V)
Vmin	0.56V (0.4GHz)	0.5V (1.97GHz)	0.46V (0.57GHz)	0.39V (-)	0.35V (0.1GHz)

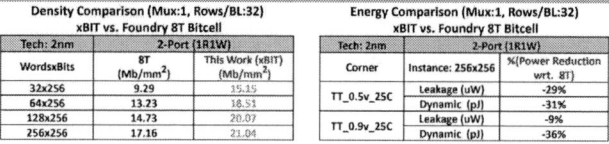

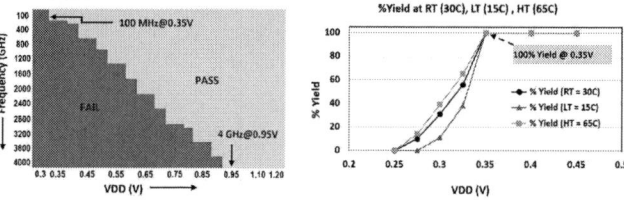

Figure 15.2.6: Density and energy comparison to foundry 8T bitcell, comparison table to prior work, FV Shmoo and yield plots.

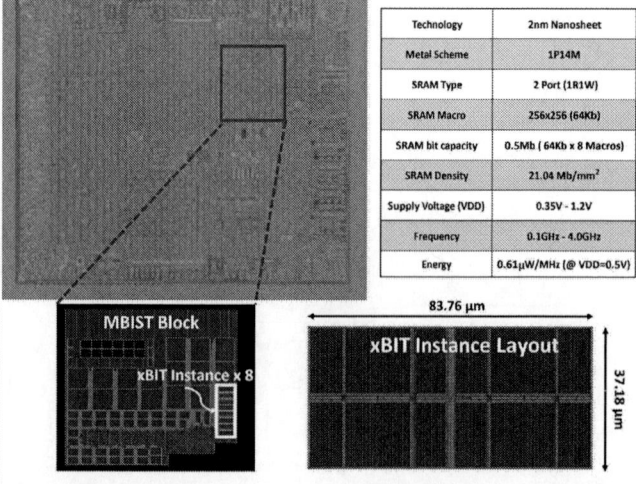

Technology	2nm Nanosheet
Metal Scheme	1P14M
SRAM Type	2 Port (1R1W)
SRAM Macro	256x256 (64Kb)
SRAM bit capacity	0.5Mb (64Kb x 8 Macros)
SRAM Density	21.04 Mb/mm²
Supply Voltage (VDD)	0.35V - 1.2V
Frequency	0.1GHz - 4.0GHz
Energy	0.61µW/MHz (@ VDD=0.5V)

Figure 15.2.7: Chip micrograph, MBIST block, xBIT SRAM instance layout and key technology details.

ISSCC 2026 / SESSION 15 / DRAM, SRAM, AND NON-VOLATILE MEMORIES / 15.3

15.3 An 8nm eMRAM for Auto-G1 with a 125MHz Read Speed at 0.6V and 19.94Mb/mm² Density

Hyunjin Shin[1], Gyuseong Kang[1], Yeseul Kim[1], Dohui Kim[1], Kyuseong Kim[1], Sunkyu Lee[1], Hangil Lee[1], Sanggyeong Won[1], Mijoung Kim[1], Jaeho Lee[1], Seungpil Ko[2], Jaechul Shim[2], Shinhee Han[1], Kiseok Suh[1], Sohee Hwang[1], Hyunchang Lee[1], Jonghoon Jung[1], Sanghoon Baek[1]

[1]Samsung Semiconductor, Gyunggido Kiheung, Korea, [2]Samsung Semiconductor, Gyunggido Hwasung, Korea

Abstract

This paper proposes techniques to improve read margin and write performance for an 8nm eMRAM: temperature compensation, resistance-mismatch improvement and margin-verify-write MVW. These techniques enable 8ns read and 300ns write operations for 283b at a time with a 128Mb density in a temperature range of -40 - 150°C and from a 0.60V supply. A 19.94Mb/mm² memory macro is implemented and silicon validation shows a mass-production yield is achieved.

Embedded non-volatile memories (eNVM) are widely adopted in various applications: such as IoT, AI, and micro controller units (MCU). eNVMs are one of the most essential IPs in system on chip (SoC) implementations. eNVMs act as data loggers and are used for AI-weight storage and to store boot code [1]. Spin-transfer-torque magnetic random-access memory (STT-MRAM) offers several advantages over other next-generation NVMs: high endurance, retention, fast write and read speeds, and low power consumption. Moreover, STT-MRAM cell-area efficiency improves with bit-cell scaling beyond the FinFET process node; thus, making it an efficient on-chip embedded memory solution for AI and high-performance MCU applications [2]. STT-MRAM faces scaling challenges as well, specifically with an increasing back-end-of-line (BEOL) metal resistance that degrades access speed and margins, which currently require precise trimming techniques to ensure reliability. To be cost competitive STT-MRAM needs to have high area efficiency, consume low power, achieve high performance and endurance. This paper introduces techniques to meet automotive grade 1-level reliability requirements, while minimizing impact to area, to achieve fast write and read performance with high scalability and logic process compatibility.

Figure 15.3.1 shows a diagram of the test chip including a built-in self-test (BIST) engine and a controller to evaluate the 128Mb eMRAM compiler. The 128Mb eMRAM macro consists of 64 slices, with a slice storing a minimum of 2Mb. The test chip includes 128Mb together with 64Mb and 16Mb composed of 2Mb slices.

MRAM is a non-volatile memory that utilizes the resistance change of a magnetic-tunnel junction (MTJ) for data storage. The read margin of STT-MRAM is degraded across temperature due to decreasing tunnel magnetoresistance (TMR), unselected-cell leakage, and metal RC mismatch between data and reference cells. The MTJ's P-state resistance decreases due to unselected-cell leakage, while unselected-cell leakage and TMR degradation reduces AP-state resistance. In addition, the temperature-coefficient-of-resistance (TCR) mismatch, due to different metal layers between data and reference bit cells, further reduces read margins. Figure 15.3.2 shows the temperature auto-tracking reference (TATR), which is a temperature compensation circuit designed to address the reduced read margin at high temperature. The proposed TATR combines PTAT CTAT and voltages to generate a temperature-dependent voltage. This enables the reference resistor to be dynamically adjusted based on temperature, thereby aligning the thermal characteristics of the data path and the reference path. The temperature compensation circuit ensures a consistent read margin across operating temperatures by reducing metal-layer mismatches between data and reference paths, as well as compensating for the reduction in TMR of the MTJ at high temperatures. Furthermore, TATR includes a process-trimming circuit to provide robust temperature compensation against process variations. Unselected-cell leakage is also controlled by negative gate-biasing (NGB) cell transistors, thereby reducing leakage.

Figure 15.3.3 shows the proposed address-detected assist circuit (ADAC), designed to reduce address-induced offset. While a conventional local-read assist circuit (LRA) reduces the read path resistance by adding a discharge path for the read current, mismatched metal layers between the data and reference paths cause address-dependent resistance mismatch due to uneven resistance reduction. ADAC uses transistors of sizes according to the row address, it enables the application of an appropriate pull-downs according to the resistance; thereby, ensuring a consistent read margin across all addresses. ADAC achieves a 68% reduction in resistance mismatch compared to the previous.

Due to PVT variations, differences in the read window among IOs during read operations are unavoidable. However, read operations are performed with a global reference value only due to complexity and search-time overhead of the per-IO reference searching logic in the BIST logic. In Fig. 15.3.4, it shows that each IO has its own optimal reference value, resulting in overall read window degradation. Especially as the temperature increases, the overall read window degrades. To enhance the read window, this paper proposes the automatic fine-trim (AFT) scheme for BIST that finds the optimal reference value for each IO. Still, a global reference value can be used for read operations, the fine-trim scheme searches for each IO's optimal reference starting from the global reference value. After fine trim, different reference value offsets are applied to individual IOs, allowing the read windows to be optimally aligned, as shown in Fig, 15.3.4. This resulted in a read margin improvement of up to 33%.

The margin-verify-write (MVW) scheme is proposed to guarantee write characteristics by verifying the read confirmation before and after writing for a one-time-level-write speed guarantee and bit cells with weak characteristics. Figure 15.3.5 shows the MVW function to ensure reliable write operations. This function performs read operations before write operations and skips the write if the stored data matches the intended write data, thereby reducing energy consumption from unnecessary writes as well as minimizing cell stress caused by redundant write operations. However, it regenerates false verification due to issues such as soft read fail and bitcell meta-stability, which can lead to unintentional skipping of necessary writes. To address these errors, this paper proposes adjusting the read reference resistors based on the write data values during compare reads. For instance, when writing 0, the reference resistor is adjusted toward 0, increasing the read margin for a stored 1, and vice versa. These approaches minimize soft errors and ensures reliable write operations.

As shown in Fig. 15.3.6, the 128Mb MRAM achieves a 27% read-speed improvement from a 0.6V supply, compared to prior work on Auto-G1. This is the Shmoo plot for the proposed 128Mb MRAM with a total 283b: 256 data and 27 parity bits. And it was confirmed that write operations can operate at a minimum 1.5V IO voltage at a speed of 300ns. This is a result with write time improved by 62.5% compared to previous work. Although the application of these technologies increased the read power, the write power was able to maintain an equivalent level compared to the previous work.

The MRAM macro designed based on the technology presented in this paper was successfully demonstrated with 8ns read cycle operation from -40 to 150°C from a 0.6V supply; operation up to 150°C was also confirmed. The 128Mb memory macro area is 6.42mm², achieving a 19.94Mb/mm² size efficiency and a high-density implementation along with a robust read margin for automotive grade-1 applications. To our knowledge, this is the first MRAM implemented in an 8nm process. In comparison to prior work, the MRAM macro introduced in this paper has the best competitiveness considering implementation capacity, operating voltage and size, as tabulated in Fig. 15.3.7.

References:
[1] K. Lee et al., "28nm CIS-Compatible Embedded STT-MRAM for Frame Buffer Memory", *IEEE IEDM*, pp. 2.1.1-2.1.4, 2021.
https://ieeexplore.ieee.org/stamp/stamp.jsp?tp=&arnumber=9720537
[2] Y.-J. Mii, "Semiconductor Industry Outlook and New Technology Frontiers," *IEEE IEDM*, 2024.
https://ieeexplore.ieee.org/stamp/stamp.jsp?tp=&arnumber=10873484
[3] G. Kang et al., "A 14nm 128Mb eMRAM Implemented with 17.88 Mb/mm² at 0.60 V for Auto-G1 Applications, *IEEE VLSI Tech. and Circuits*, 2024.
https://ieeexplore.ieee.org/stamp/stamp.jsp?tp=&arnumber=10631510
[4] T. Ogawa et al., "A 22nm 10.8Mb embedded STT-MRAM macro achieving over 200MHz random-read access and a 10.4 Mb/s write throughput with an in-field programmable 0.3 Mb MTJ-OTP for high-end MCUsN," *ISSCC*, pp. 290-292, 2024.
https://ieeexplore.ieee.org/stamp/stamp.jsp?tp=&arnumber=10454409
[5] P.-H. Lee et al., "A 16nm 32Mb embedded STT-MRAM with a 6ns read-access time, a 1M-cycle write endurance, 20-year retention at 150C and MTJ-OTP solutions for magnetic immunity," *ISSCC*, pp. 494-496, 2023.
https://ieeexplore.ieee.org/stamp/stamp.jsp?tp=&arnumber=10067837

ISSCC 2026 / February 17, 2026 / 8:50 AM

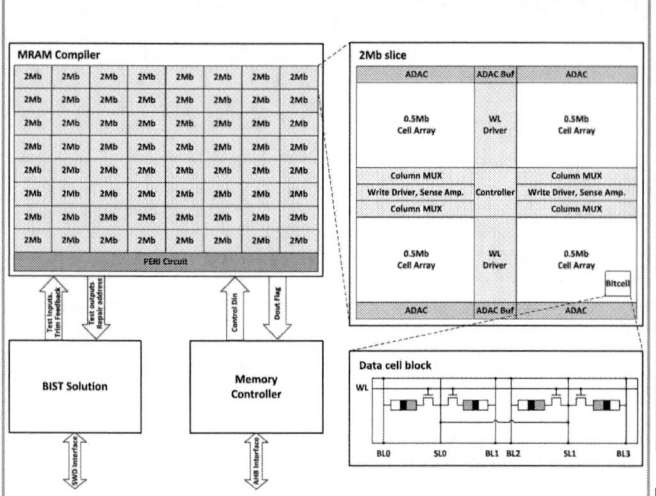

Figure 15.3.1: eMRAM macro test-chip diagram and BIST/Controller solution.

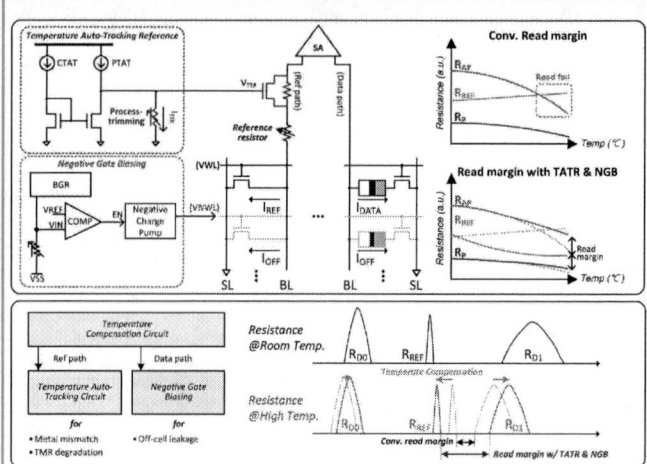

Figure 15.3.2: Challenges to reducing read margin at high temperatures, and the structure and improvements due to the proposed temperature auto-tracking reference (TATR) and negative gate biasing (NGB).

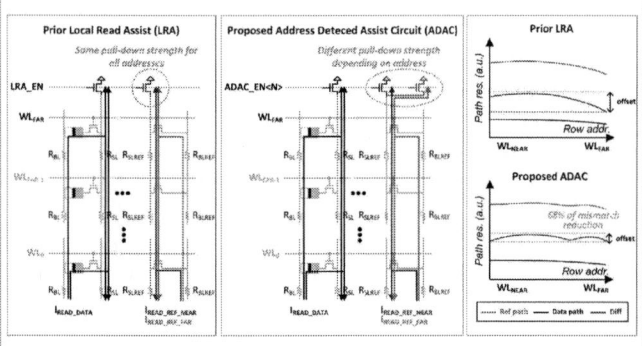

Figure 15.3.3: Prior local-read-assist (LRA) circuit and the proposed address-detected assist circuit (ADAC), a comparison of resistance improvement results.

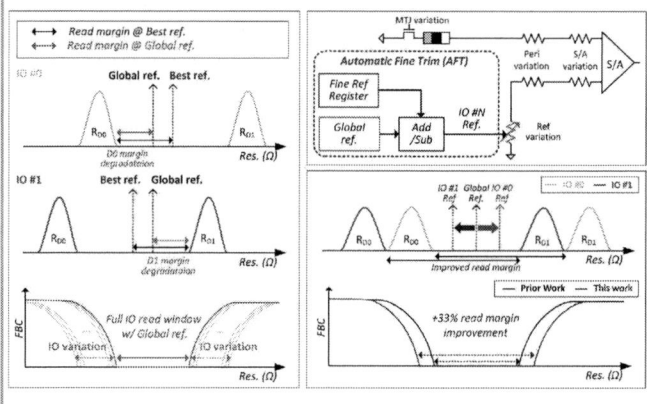

Figure 15.3.4: Read window differences between using a global reference (left) and automatic fine-trim (AFT) scheme (right).

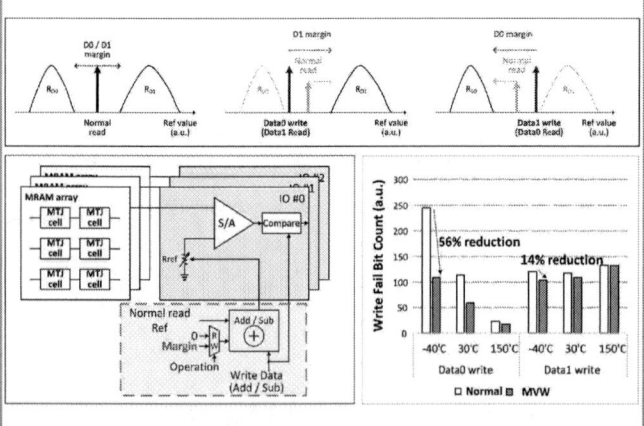

Figure 15.3.5: MVW function (top). Difference between normal read and D0/D1 MVW (left). The proposed MVW function (right) write FBC improvement by MVW function.

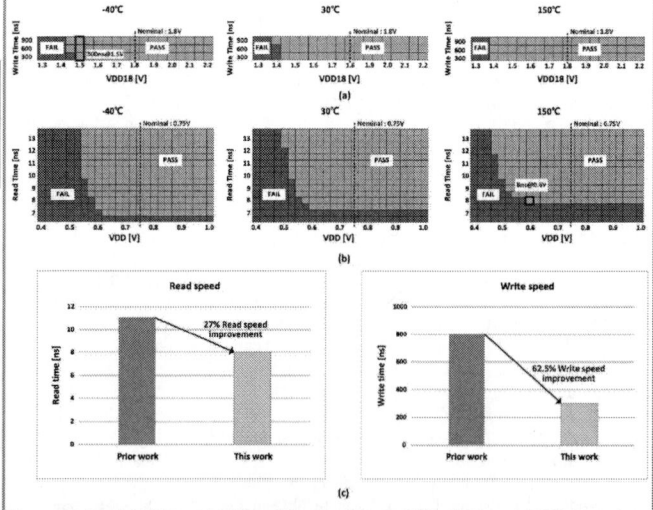

Figure 15.3.6: (a) Write and (b) Read Shmoo plots at -40°C, 30°C, and 150°C for various V_{DD18} and V_{DD} voltages (c) speed improvement level.

DIGEST OF TECHNICAL PAPERS • 259

979-8-3315-8937-0/26 $31.00 © 2026 IEEE

ISSCC 2026 / SESSION 15 / DRAM, SRAM, AND NON-VOLATILE MEMORIES / 15.3

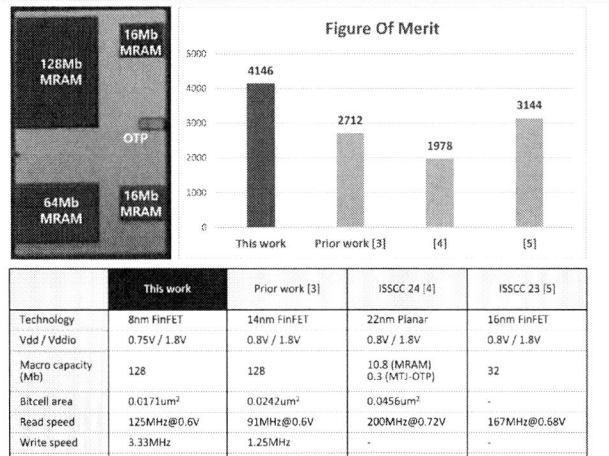

	This work	Prior work [3]	ISSCC 24 [4]	ISSCC 23 [5]
Technology	8nm FinFET	14nm FinFET	22nm Planar	16nm FinFET
Vdd / Vddio	0.75V / 1.8V	0.8V / 1.8V	0.8V / 1.8V	0.8V / 1.8V
Macro capacity (Mb)	128	128	10.8 (MRAM) 0.3 (MTJ-OTP)	32
Bitcell area	0.0171um²	0.0242um²	0.0456um²	-
Read speed	125MHz@0.6V	91MHz@0.6V	200MHz@0.72V	167MHz@0.68V
Write speed	3.33MHz	1.25MHz	-	-
FOM [Freq x Mb /mm2 / Vdd]	4146	2712	1978	3144

Figure 15.3.7: Die micrograph of eMRAM compiler including 128Mb and figure-of-merit comparison table.

ISSCC 2026 / SESSION 15 / DRAM, SRAM, AND NON-VOLATILE MEMORIES / 15.4

15.4 A 16nm 168Mb Embedded STT-MRAM with 0.0249µm² Bit-Cell, Dual-Port Access, and 51.2Gb/s Read Throughput for Automotive and Edge AI Applications

Po-Hao Lee[1], Chia-Fu Lee[1], Hon-Jarn Lin[1], Cheng-Han Lu[1], Yen-An Chang[1], Pranata W. Sanjaya[1], Chia-Jung Tsen[1], Kuan-Chun Chen[1], Ming-Chieh Lin[1], Chao-Jung Hung[1], Tan-Li Chou[1], Chih-Hui Weng[2], Chia-Yu Wang[2], J.J. Wu[2], Harry Chuang[2], Yih Wang[1], Yu-Der Chih[1], Tsung-Yung Jonathan Chang[1]

[1]TSMC Design Technology, Hsinchu, Taiwan, [2]TSMC, Hsinchu, Taiwan

Abstract

A 16nm 168Mb embedded STT-MRAM with a 0.0249µm² bit cell that overcomes density and performance challenges of competing NVM technologies is presented. The design features a modular architecture supporting customizable macros, a dual-port access scheme, and an interleaving read architecture achieving a 51.2Gb/s throughput. Fabricated in a 16nm FinFET process, the chip demonstrates a 20-year data retention at 150°C, 1M-cycle write endurance, and is suitable for automotive and edge AI applications.

Spin-transfer torque magnetoresistive random access memory (STT-MRAM) has emerged as a leading candidate for next-generation memory technologies, offering a compelling combination of non-volatility, excellent high-temperature data retention, high endurance, and fast write speeds [1,2]. Beyond its traditional role as a random-access memory for code and data storage, STT-MRAM can be adapted to function as one-time programmable (OTP) memory, which is irreversible and resistant to magnetic interference, making it particularly suitable for storing sensitive information such as encryption keys and security codes [3]. However, for STT-MRAM to be a universal embedded non-volatile memory replacement, it must achieve bitcell densities that are competitive with other non-volatile memory (NVM) technologies: most notably resistive RAM (ReRAM) [4]. This paper presents a 16nm STT-MRAM technology that reduces the bitcell area from 0.033 to 0.0249µm² through design-technology co-optimization (DTCO) and addresses critical read/write design-margin limitations in scaled MRAM bitcells, while directly overcoming density benchmarks set by competing technologies. Additionally, a modular architecture is developed, offering the flexibility to generate customizable embedded MRAM (eMRAM) macros ranging from 8 - 128Mb. This modular design transforms the core MRAM technology into a scalable platform suitable for diverse application requirements. The design also incorporates a dual-port-access scheme optimized to support over-the-air (OTA) software updates, critical for automotive applications, alongside a high-throughput interleaving-read architecture tailored for data-intensive workloads, such as edge AI applications. A 168Mb STT-MRAM test chip, based on the proposed approaches, is fabricated using a 16nm FinFET CMOS process. Measurements confirm its performance and manufacturability, demonstrating its suitability for enabling high-capacity embedded applications in automotive, industrial, and edge AI applications.

Figure 15.4.1 illustrates the 1T1MTJ bitcell structure, its TEM cross-section, and block diagrams for the two sub-bank architectures. Each 0.0249µm² 1T1MTJ bitcell is 180×144nm² with a metal-3 WL, a metal-6 BL, and a metal-2 common source line (CSL). To mitigate read-window degradation caused by larger on-chip variation in the scaled bitcells. This work utilizes a merged-reference sensing scheme with a local trimmable-reference array architecture [5]. The local-WL location-tracking reference scheme provides natural flexibility to overcome parasitic BL/SL resistance mismatches and enables support for varying BL lengths, which is essential for MRAM compiler design. To enable density scaling, two BL length configurations are implemented, a 612 and a 153kb array, that together enable a flexible MRAM macro architecture. The 612kb array is designed with 512b/BL and 1224b/WL. A large sub-bank configuration, for a 2Mb memory block, consists of four 612kb arrays organized in a butterfly-array structure. Similarly, the smaller sub-bank configuration, for a 0.5Mb memory block, consists of four 153kb arrays, with each array featuring 128b/BL and 1224b/WL, and also arranged in a butterfly-array structure.

Figure 15.4.2 illustrates the organization of the 16, 8 and 2Mb modules designed for MRAM macro integration. The 16Mb module consists of eight 2Mb sub-banks that share local control circuits and local analog circuits. Similarly, the 8 and 2Mb modules consist of four 2 and 0.5Mb sub-banks, which also share local control circuits and local analog circuits. The local analog circuits include write drivers and internal power switches to support dual-port access functionality; thereby, enabling simultaneous read and write operations within a single macro. The scalable MRAM macro employs a modular architecture, which integrates both global analog and far-side analog modules at the boundaries of the design. The analog modules include a bandgap-reference voltage generator, a charge pump, low-dropout regulators, and decoupling capacitors to support a trimmable bias voltage for read and write operations. At the center of the architecture are several scalable MRAM modules, including 16, 8 and 2Mb configurations. The scalable MRAM macro design supports 8 - 128Mb densities within a single macro, which makes it adaptable to different application requirements.

In the automotive industry, software-defined vehicles emphasize the need for frequent and reliable OTA software updates to deploy new features, fix security vulnerabilities, and optimize performance throughout the vehicle's lifecycle. Traditional embedded non-volatile memories, such as eFlash, present limitations for OTA updates due to its slow write process, which involves time-consuming erase-before-write cycles. To overcome these challenges, this work introduces a dual-port-access functionality that enables simultaneous read and write operations to a single MRAM macro. Unlike conventional designs that require two separate macros for OTA, this approach shares analog modules, reducing area usage and improving efficiency. Figure 15.4.3 shows the dual-port-access block diagram: each MRAM module uses an input selection signal to control its internal controller and bias power switches, determining read or write mode. To minimize write soft errors, the design includes a verification step after each write operation to decide if a re-write is required. A separate data output path (DOUT and DOUT_VFY) enables simultaneous read and write-verify operations.

To support simultaneous read and write-verify operations, each MRAM module integrates an individual controller and read timer within the module to generate read-control signals required for both read and write-verify operations. This individual read-control architecture is leveraged to enable interleaved-read functionality, significantly enhancing read throughput. Figure 15.4.4(top) illustrates the block diagram for the interleaved-read interface. Each module operates with its own read clock (CLK) and data output multiplexer (DOUT MUX) control signals. This architecture allows read operation to be initiated on one module while another read operation is concurrently in progress in another module. The DOUT MUX is sequentially controlled to output interleaved read data, functioning like a pipeline and increasing throughput. Figure 15.4.4(bottom) presents the access time Shmoo plot, demonstrating a 7.5ns read access time at 0.8V over a -40 - 150°C temperature range, for a macro density of 84Mb. Interleaved-read operations are performed by toggling CLK[N:0] across multiple MRAM modules in a pipelined sequence boosting the read frequency up to 200MHz and achieving a 51.2Gb/s read throughput.

To enable dual-port access and interleaved-read functionality in wide memory density MRAM macros, reliable analog and mixed-signal circuit designs are essential. During MRAM read and write-verify operations, a global clamping-voltage generator is employed to supply V_{clamp} for the sense amplifier (SA), as shown in Fig. 15.4.5(top): V_{clamp} maintains a stable BL voltage during reads, thereby mitigating read-disturb issues. The SA control waveform is shown in Fig. 15.4.5(bottom). During a read, the SA sensing nodes, Q and QB, are pre-charged to V_{DD} before the sensing phase and are discharged during the sense-evaluation phase. However, the SA's parasitic capacitances couple V_{clamp} to the sensing nodes, when Q/QB are pre-charged or discharged. To counteract voltage coupling a local kick-back capacitor is integrated into the SA design. This capacitor effectively cancels the coupling offset, ensuring a stable V_{clamp} voltage from cycle to cycle. After sense-evaluation phase, Q/QB are pre-charged high and the kick-back capacitor node KICK transitions from high to low, mitigating coupling. This design ensures accurate and reliable operation of the sense amplifier, enabling robust simultaneous dual-port access and interleaved-read functionality in wide-density MRAM macros.

As the MRAM bitcell is scaled from 0.033 to 0.0249µm², DTCO becomes critical to maintaining a sufficient margin between write-induced soft errors and hard errors. This margin is essential for meeting the stringent reliability requirements of automotive applications. In this work, we developed a per-die write-bias-trimming approach with adjustable write bias temperature coefficients to effectively track variations in process, voltage, and temperature. We used built-in self-test capabilities during wafer-sort phase to determine the optimal bias settings. Figure 15.4.6 illustrates the reliability results: the hard-error rate after 1M cycles at -40°C is less than 0.01ppm, and the errors can be corrected by error-correcting-code (ECC). Furthermore, the read-disturb rate at the user read voltage is lower than 10^{-22}ppm at 150°C, demonstrating a sufficient read margin.

Figure 15.4.7 highlights the key metrics of this work, relative to prior work[5], and shows the 168Mb MRAM test chip photo. The test chip consists of two 84Mb macros, fabricated using a 16nm FinFET logic process. Each 84Mb macro contains five 16Mb modules and two 2Mb modules. The 84Mb macro achieves a 7.5ns read-access time at 0.8V and a 200 MHz interleaved-read frequency. The test chip passes a 1M-write-cycle endurance test, reflow testing, and achieves a 20-year data retention at 150°C.

References:
[1] Y.-D. Chih et al., "Design Challenges and Solutions of Emerging Nonvolatile Memory for Embedded Applications," *IEDM*, pp. 2.4.1-2.4.4, 2021.
https://doi.org/10.1109/IEDM19574.2021.9720557
[2] O. Glowinski et al., "MRAM as Embedded Non-Volatile Memory Solution for 22FFL FinFET Technology," *IEDM*, pp. 18.1.1-18.1.4, 2018.
https://doi.org/10.1109/IEDM.2018.8614620
[3] Y.C. Ong et al., "Design-Technology-Reliability Co-Optimization for MRAM-OTP Integration – A Methodological Approach," *IEEE IRPS*, 2025.
https://doi.org/10.1109/IRPS48204.2025.10982744
[4] Y.C. Huang et al., "A 32Mb RRAM in a 12nm FinFet Technology with a 0.0249μm² Bit-Cell, a 3.2GB/S Read Throughput, a 10KCycle Write Endurance and a 10-Year Retention at 105°C," *ISSCC*, pp. 288-290, 2024.
https://doi.org/10.1109/ISSCC49657.2024.10454367
[5] P.H. Lee et al., "A 16nm 32Mb Embedded STT-MRAM with a 6ns Read-Access Time, a 1M-Cycle Write Endurance, 20-Year Retention at150°C and MTJ-OTP Solutions for Magnetic Immunity," *ISSCC*, pp. 494-496, 2023,
https://doi.org/10.1109/ISSCC42615.2023.10067837

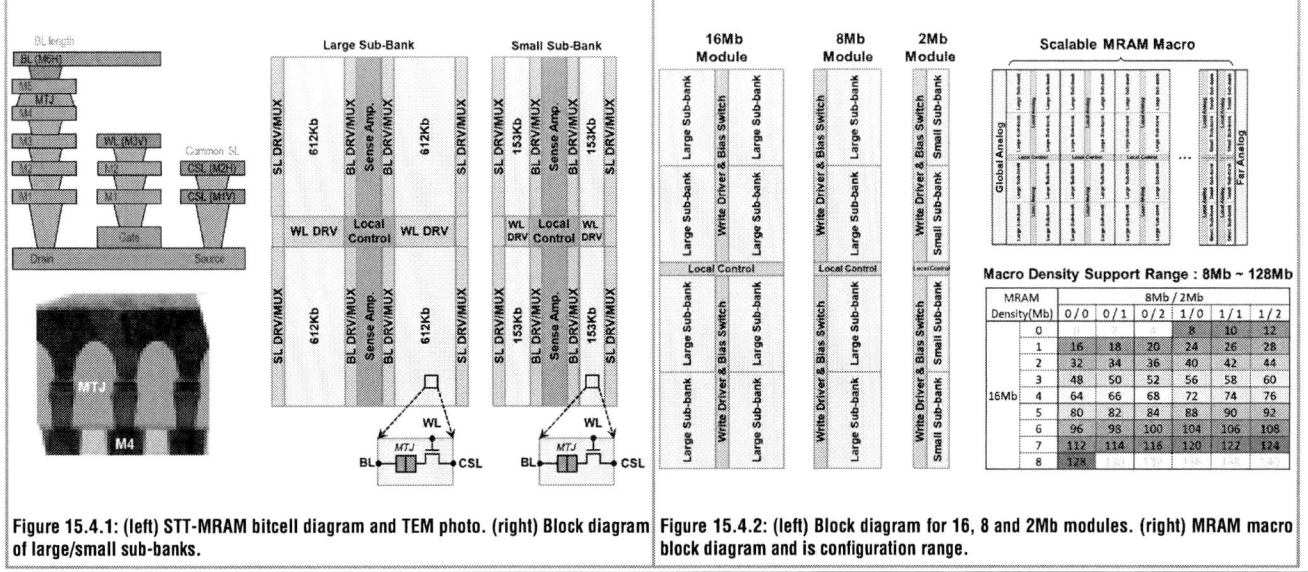

Figure 15.4.1: (left) STT-MRAM bitcell diagram and TEM photo. (right) Block diagram of large/small sub-banks.

Figure 15.4.2: (left) Block diagram for 16, 8 and 2Mb modules. (right) MRAM macro block diagram and is configuration range.

ISSCC 2026 / SESSION 15 / DRAM, SRAM, AND NON-VOLATILE MEMORIES / 15.4

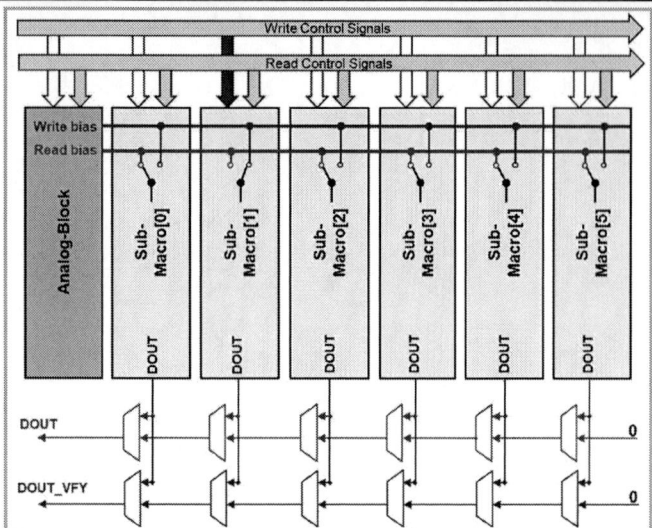

Figure 15.4.3: Dual-port-access interface for simultaneous read and write-verify operations block diagram.

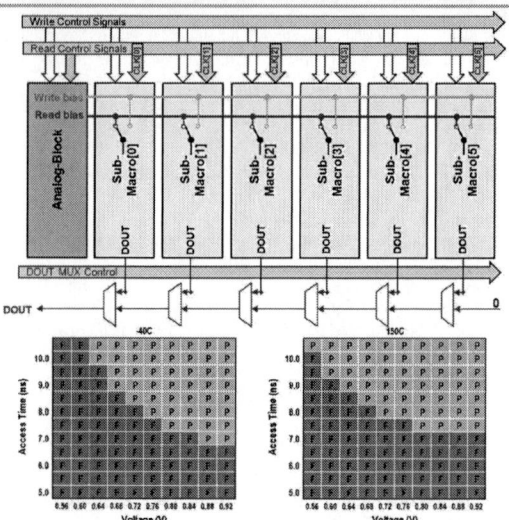

Figure 15.4.4: (top) Interleaved-read interface block diagram. (Bottom) 84Mb read Shmoo at -40 and 150°C.

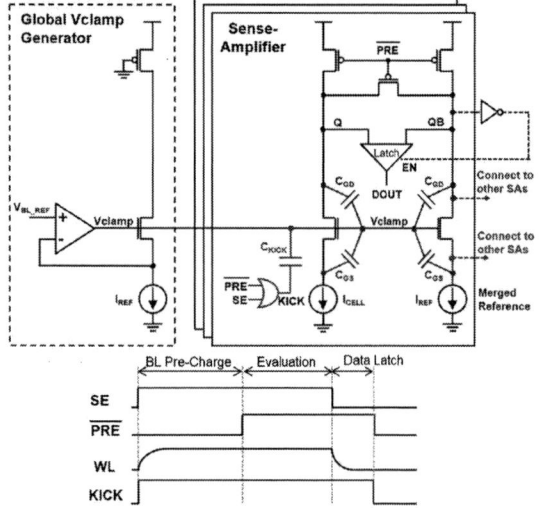

Figure 15.4.5: (top) Read waveform sense amplifier and global read-clamp-bias generator. (bottom) Sense amplifier's control signals.

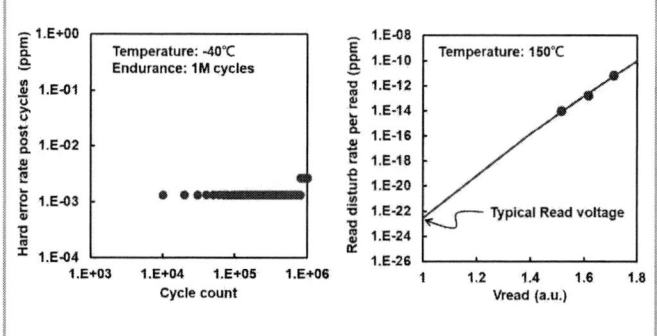

Figure 15.4.6: (left) 1M-cycle MRAM endurance results at -40C, worst-case temperature. (right) Read disturb at a typical read voltage is lower than 10⁻²²ppm at 150°C.

	ISSCC 2023 [4]	This work
Technology	N16 MRAM	N16 MRAM
Physical Cell size (um²)	0.033	0.0249
VDD (V)	0.72~0.88	0.72~0.88
VDIO (V)	1.62~1.98	1.62~1.98
Operation Temperature (Tj, °C)	-40°C ~ 150°C	-40°C ~ 150°C
Macro Density (Mb/mm2)	12.8 (32Mb)	16.0 (32Mb) 14.8 (84Mb) 18.1 (128Mb)
Redundancy	Row-repair Column-repair	Row-repair Column-repair
ECC	DECTED	TECQED
Chip MRAM Density	128K x 275	672K x 284
Macro Read Speed (ns)	6ns (32Mb)	7.5ns (84Mb)
4Mb Bank Read Access Time (ns)	-	3.97ns (4Mb)
Interleaving Read	N/A	200MHz @ 0.72V
Read Throughput (Gb/s)	32	51.2
Dual Port Access Feature	N/A	Yes
Endurance (cycle) (Code/Data)	1M	1M
Retention (yr/°C)	20/150	20/150
Reflow (120s @ 260°C)	Yes	Yes

Figure 15.4.7: (left) Comparison table of key MRAM metrics. (right) 168Mb MRAM test chip die photo.

• 2026 IEEE International Solid-State Circuits Conference

ISSCC 2026 / SESSION 15 / DRAM, SRAM, AND NON-VOLATILE MEMORIES / 15.5

15.5 A 3nm 0.167fJ/b 5.27Mb/mm² Configurable TCAM with Macro-Wise Pipelined Search Methods for Automotive Applications

Shunya Nagata, Yohei Sawada, Kenichiro Takiguchi, Masao Morimoto, Shinji Tanaka, Naoya Fujita, Tomohiro Miura, Daisuke Nakamura, Takashi Ito

Renesas Electronics, Tokyo, Japan

Abstract

This paper presents a TCAM design that includes a bank-less small-grain TCAM-macro compiler, a configurable soft-macro generator, selective macro-wise pipelined search techniques, and a split data-bus architecture for ECC. The proposed TCAM achieves 0.167fJ/b search energy and 5.27Mb/mm² density with a 512-entry and 256b search key configuration in a 3nm bulk Fin-FET process technology for automotive network applications.

TCAM is widely used in various networking devices, and recent technological advances - such as the deployment of 5G networks and the rise of cloud and edge computing - have led to a dramatic increase in network traffic, necessitating larger-capacity TCAMs. TCAMs are required to support search key widths of up to 256b and entry depths of up to 4096 [1], as illustrated in Fig. 15.5.1 (top). Moreover, customers frequently demand multiple entry-depth and search-key-width configurations within a single chip. These trends make it difficult to rely solely on hard-macro TCAMs, as increasing banks and repeaters raises design complexity, complicates timing closure, and expands peripheral area. Moreover, simulating large-scale netlists significantly drives up EDA tool operating costs. Deep entries, such as a 4096-entry, require extensive banks and repeaters, increasing complexity. Deep search keys, such as 256b, also cause huge power consumption because multiple long match lines activate simultaneously. The pipelined search method has been proposed as a counter measure [2], wherein typically half of the search key (e.g., 110b of 220b) is used to search half of the array during the first stage of a search. However, the overall power savings of the TCAM array are limited to about half, even when all comparisons of whole entries in the first stage result in mismatches. While power reduction remains a key concern, another challenge is the growing demand for TCAMs in diverse applications. They are widely used in industrial and consumer domains and are now expanding into automotive communication systems, driven by the increasing volume of high-speed data exchanged between sensors and processors. In automotive networks, safety standards (e.g., ISO26262) make functional safety a primary requirement.

To address these challenges, this work proposes a bank-less small-grain TCAM hard macro generated by a memory compiler, a synthesized soft macro created using an RTL generator, selective macro-wise pipelined search techniques, and a data-bus split architecture for ECC. Using these techniques, the proposed TCAM achieves a 0.167fJ/b search energy and a 5.27Mb/mm² density for a 512-entry TCAM with a 256b search key in a 3nm bulk Fin-FET process technology; thus, achieving a multi-purpose TCAM suitable for automotive networking application. Figure 15.5.1 (bottom) shows an overview of the proposed TCAM. SRAM-based 16T TCAM bitcells support three storage states – 0, 1, and X – depending on the stored data combination in the T- and B-node of each bitcell. The TCAM hard macro supports an entry depth range of 32–128 and a search-key width range of 8–64b, to accommodate various configurations. For entry depths or key widths exceeding the compiler range, the synthesized soft macro is integrated with the hard macro as a single macro. The area overhead of the soft macro is only 1.6% for a 512-entry 256b configuration. This integrated approach as a single macro to realize the 512-entry 256b configuration, reduces the bank-level area overhead seen in conventional designs [2], and improves overall macro density compared with prior work [2-7].

Figure 15.5.2 shows the block diagram of the proposed TCAM hard macro. The write and read blocks are identical to those in conventional designs. The compare block outputs the search results of all entries on the match line outputs (MLO), while the proposed all-mismatch detection circuit determines whether all entries in the hard macro mismatch. AMNO is the output pin that indicates the all-mismatch detection result to the subsequent second search: AMNO is 0 if all entries mismatch. AMNI is the input pin that receives the all-mismatch result from the first search result. If AMNI is 0, then the search operation is disabled: by stopping the search line driver and match amplifier. The proposed all-mismatch detection circuit consists of a NOR-NAND cascade, which reduces fluctuations in the detector's output timing caused by local variations, compared to a wired-OR detector. When a match line (ML) is discharged to V_{SS}, due to a mismatch between the bitcell data and the search key, and then the match amplifier is activated by the match amplifier activation signals (MA=1 and MAB=0), the mismatch result is transferred to the mismatch transfer node (STRN) as 0. If all STRN signals are 0, then the AMNO output becomes 0: indicating that all entries in the TCAM hard macro mismatch. Simulated waveform for AMNO=0 case is shown in the bottom-right corner of Fig. 15.5.2.

This work uses a 2-way pipelined search method, which is illustrated in Fig. 15.5.3. The first partial search determines whether the second search is required. For a no-split search key case (key width ≤ 64b), the macro row-wise pipelined search (MRPS) method is applied. If the first search result is a hit, the priority encoder, placed after the MLOs to send the address of the first matching entry, outputs this address within the first-stage region, even though additional matches may exist in subsequent TCAM regions. Since these additional

matches in later regions can never be the first matching entry, a second search across the remaining TCAMs is unnecessary. In this no split search key case, AMNO=1 from the first search is inverted to 0 by an inverter, and this signal is propagated to the AMNI pins of the remaining TCAMs, disabling their comparison operations. For the search key split case (key width > 64b), the macro column-wise pipelined-search (MCPS) method is applied. If the first search results in an all-mismatch condition, then AMNO outputs 0, which is sent to the AMNI pins of the remaining TCAMs to disable their comparison operations. Figure 15.5.3 (bottom) illustrates examples of outcomes obtained by applying MRPS and MCPS, including the full activation case - where no power reduction occurs - and cases demonstrating power savings achieved through these methods.

By applying these pipelined methods, search power is reduced across the entire range of entry and search-key configurations, which integrate TCAM hard macros with a synthesized soft macro, in scenarios where the second-stage search becomes unnecessary. Figure 15.5.4 presents the search-power reduction achieved compared to full activation without applying these methods, assuming the scenario in which all TCAM hard macros return mismatched results on all match lines. The 512-entry 256b configuration is evaluated on silicon test chip, and the other configurations are calculated based on this result. For the MRPS method, search power is reduced by 65.3% for a 512-entry 64b configuration when the first search results in a hit. As the entry depth increases, the number of disabled TCAM hard macros in the second stage also increases, enabling further power reduction. If frequently searched data is allocated to the TCAM used for the first search, then the hit probability during the first search increases, making the MRPS method more effective as shown in Fig. 15.5.4 (top left). For the MCPS method, power savings increase as the number of TCAM hard macros with all-mismatches for the first search increases. For the case where all macros in the first search result in mismatches, the search power is reduced by 71.1% for a 512-entry 256b configuration. If the data distribution within each TCAM hard macro and each entry used in the first search exhibits high bit-wise similarity – where a large proportion of bits share identical values across entries, forming clusters of nearly identical patterns – and it differs from the search key, then the likelihood of an all-mismatch condition during the first search increases, thereby enhancing the effectiveness of the MCPS method, as shown in Fig. 15.5.4 (top right). Key width dependency of power saving from full activation and search energy is shown in Fig.15.5.4 (bottom). The MRPS method achieves power savings, exceeding 60%, for 48 - 64b keys. The MCPS method achieves power saving ranging from 40.7% for 66b key to 71.1% for 256b key, indicating the greater search key widths result in higher power savings. Regarding pipelined search methods reduce search power across configurations, the lowest value for MRPS is 0.078fJ/b for a 4096-entry 64b configuration, while the lowest value for MCPS is 0.167fJ/b for a 512-entry 256b configuration.

For automotive applications, improving ECC robustness is essential to ensure functional safety. Figure 15.5.5 shows the proposed split data-bus architecture for the ECC module, which is highly effective in preventing various types of failures in memory macros. In a conventional TCAM (Fig. 15.5.5), both user data and ECC parity bits are stored in the same array, similar to SRAM. However, unlike SRAM, TCAM cannot use column multiplexing because all bitcells connected to a match line must be used for comparisons. In SRAM, column multiplexing separates bitcells with the same address, reducing the chance of double-bit errors even though multi-cell upsets occur, which SECDED can correct. In TCAM, bitcells with the same address are adjacent, so if neighboring bitcells experience soft errors, a double-bit error occurs that SECDED cannot fix. To mitigate this, user data and ECC parity bits are split into odd and even groups, each with its own ECC module. Another concern is incorrect word-line selection caused by address faults, such as corruption from a soft error in an address input flip-flop or latch during a write operation. In conventional TCAMs, where user data and ECC parity bits share one array, reading from a wrong address still returns a consistent combination, making such faults undetectable. In the proposed design, ECC parity bits are stored in a dedicated SRAM, separate from the TCAMs. Since TCAMs and SRAM have independent address decoders, the likelihood of both selecting the same wrong address simultaneously is negligible. Therefore, if an address fault occurs in only one hard macro, the combined readout becomes invalid, and the ECC check can detect the error. By applying these techniques, the proposed architecture significantly improves functional safety coverage.

This work demonstrates a 3nm Fin-FET small-grain TCAM hard macro combined with a configurable soft macro, which addresses automotive functional safety requirements. Figure 15.5.6 compares the proposed design against recent TCAMs and includes the measurement Shmoo plot. At 0.75V, TT, 25°C, the proposed TCAM achieves a search energy of 0.167fJ/b and a memory density of 5.27Mb/mm² for a 512-entry 256b search-key configuration: including the soft macro area overhead. The Shmoo plot in Fig. 15.5.6 shows 1.7GHz search operation at 0.75V, TT, 25°C. The memory-density × search-speed ÷ search-energy FoM is 53.8: the highest among prior work [2–7], showing superior overall PPA.

Figure 15.5.7 shows a die photograph of a 128kb TCAM test chip using a 3nm FinFET technology and a 8kb (128-entry 64b) TCAM hard macro layout, whose physical size is 36.72 x 39.728μm².

References:
[1] P. Saggurti, "Understanding the Role and Functionality of TCAMs," *Synopsys*, Apr. 2020. [Online]. Available: https://www.synopsys.com/articles/introduction-to-tcam.html#3. https://www.synopsys.com/articles/introduction-to-tcam.html#3
[2] S. Kumar et al., "A 3nm FinFET 2.2Gsearch/s 0.305fJ/b TCAM with Dynamically Gated Search Lines for Data-Center ASICs," *ISSCC*, pp. 496-498, 2025. https://doi.org/10.1109/ISSCC49661.2025.10904633
[3] Z. Yue et al., "A 0.795fJ/bit Physically-Unclonable Function-Protected TCAM for a Software-Defined Networking Switch," *ISSCC*, pp. 276-278, 2024. https://doi.org/10.1109/ISSCC49657.2024.10454312
[4] I. Arsovski et al., "1.4Gsearch/s 2Mb/mm² TCAM using two-phase-precharge ML sensing and power-grid preconditioning to reduce Ldi/dt power-supply noise by 50%," ISSCC, pp. 212-213, 2017. https://doi.org/10.1109/ISSCC.2017.7870336
[5] M. Yabuuchi et al., "12-NM Fin-FET 3.0G-Search/s 80-Bit × 128-Entry Dual-Port Ternary CAM," *IEEE VLSI*, pp. 19-20, 2018. https://doi.org/10.1109/VLSIC.2018.8502345
[6] M. Yabuuchi et al., "A 7nm Fin-FET 4.04-Mb/mm² TCAM with Improved Electromigration Reliability Using Far-Side Driving Scheme and Self-Adjust Reference Match-Line Amplifier," *IEEE VLSI*, 2020. https://doi.org/10.1109/VLSICircuits18222.2020.9162775
[7] C. Deshpande et al., "A 5nm Fin-FET 2G-search/s 512-entry x 220-bit TCAM with Single Cycle Entry Update Capability for Data Center ASICs," *IEEE VLSI*, 2021. https://doi.org/10.23919/VLSICircuits52068.2021.9492464

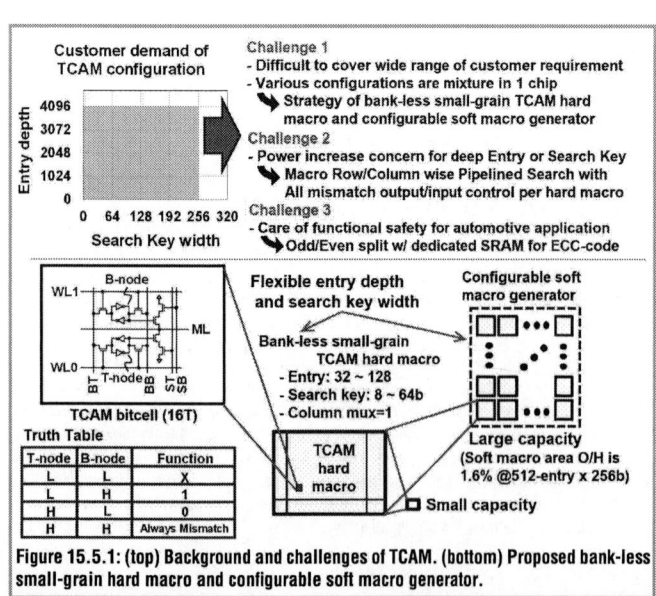

Figure 15.5.1: (top) Background and challenges of TCAM. (bottom) Proposed bank-less small-grain hard macro and configurable soft macro generator.

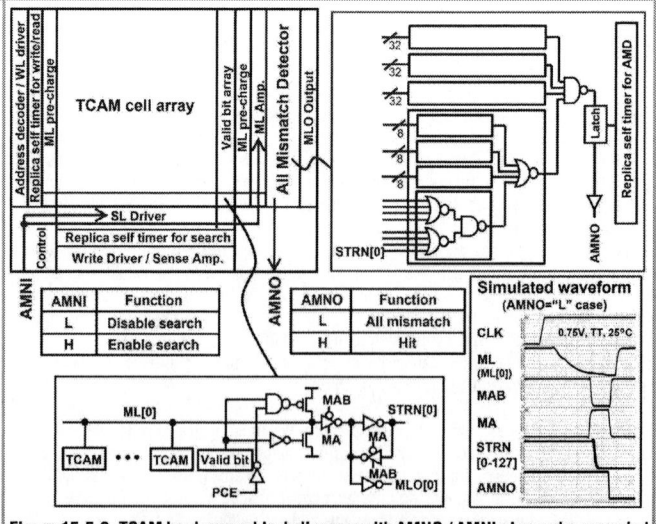

Figure 15.5.2. TCAM hard macro block diagram with AMNO / AMNI pins and a cascaded all-mismatch detector, and the simulated waveform of the detector.

ISSCC 2026 / SESSION 15 / DRAM, SRAM, AND NON-VOLATILE MEMORIES / 15.5

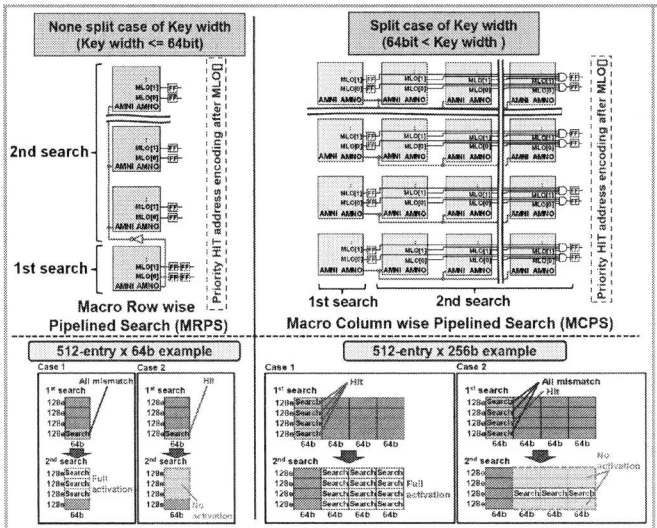

Figure 15.5.3: Macro-wise pipeline-search methods with AMNO / AMNI control, with a search example of each method.

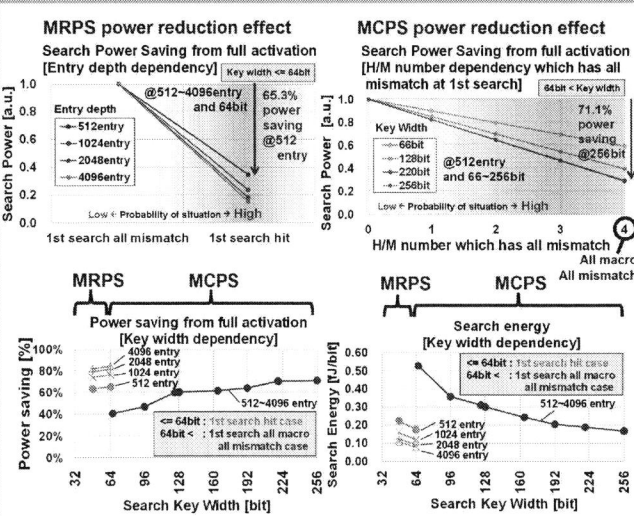

Figure 15.5.4: Search-power reduction with full activation, and search-key width dependency for power saving and search energy.

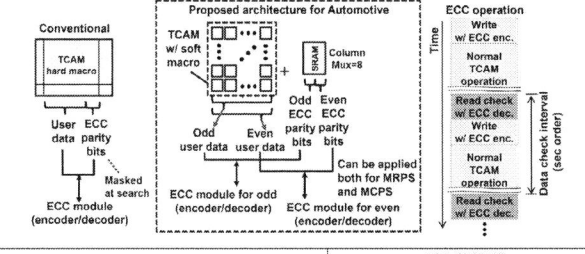

Figure 15.5.5: Proposed split data bus architecture, ECC-check flow, and safety-coverage improvement for each failure mode.

	ISSCC '24 [3]	ISSCC '17 [4]	VLSI '18 [5]	VLSI '20 [6]	VLSI '21 [7]	ISSCC '25 [2]	This work
Process	28nm	14nm	12nm	7nm	5nm	3nm	3nm
Entry Depth	64	256	128	256	512	512	512 *1)
Search Key Width	64	160	80	80	220	220	256 *1)
Search Energy (fJ/bit)	0.795	10.1	2.197	0.586	NA	0.305	0.167 *2)
Memory Density (Mb/mm²)	3.25	2.01	1.08	4.04	3.48	4.97	5.27 *3)
Search Speed (G-Search/s)	0.33	1.4	3.0 *4)	1.6	2.0	2.2	1.7
FoM (Mb/mm²)×(G-Search/s)／(fJ/bit)	1.3	0.3	1.5	11.0	NA	35.8	53.8

*1) Entry depth and search key width are configured by 128-entry and 64b TCAM hard macro with soft macro
*2) Employing MCPS method with all mismatch detection in all macro at 1st search
*3) Soft macro area is included
*4) Dual search rate by Dual port TCAM

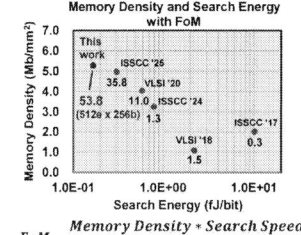

$$FoM = \frac{Memory\ Density * Search\ Speed}{Search\ Energy}$$

Figure 15.5.6: Comparison table to prior work, test chip Shmoo plot, and plot comparing this and prior work with respect to memory density and search energy, with annotated FoM.

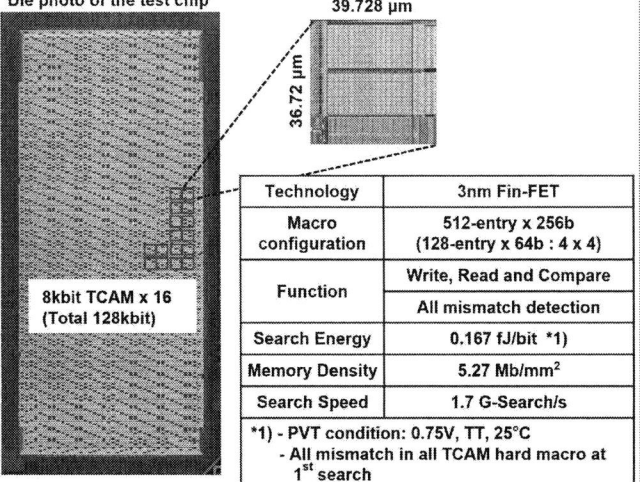

Technology	3nm Fin-FET
Macro configuration	512-entry x 256b (128-entry x 64b : 4 x 4)
Function	Write, Read and Compare
	All mismatch detection
Search Energy	0.167 fJ/bit *1)
Memory Density	5.27 Mb/mm²
Search Speed	1.7 G-Search/s

*1) - PVT condition: 0.75V, TT, 25°C
- All mismatch in all TCAM hard macro at 1st search

Figure 15.5.7: Test chip die photograph showing hard macro layout as insert. Key metric summary table.

• 2026 IEEE International Solid-State Circuits Conference

979-8-3315-8937-0/26 $31.00 © 2026 IEEE

ISSCC 2026 / SESSION 15 / DRAM, SRAM, AND NON-VOLATILE MEMORIES / 15.6

15.6 A 36GB 3.3TB/s HBM4 DRAM with Per-Channel TSV RDQS Auto Calibration and Fully-Programmable MBIST

Sunghwan Joo, Jinyeon Kim, Yongsun Lee, Ji-Young Kim, Youngsik Lee, Yong-Min Kim, ChiSung Oh, Kyu-Ha Shim, Haesuk Lee, Young-Yong Byun, ChangHyun Bae, Joohwan Kim, Je-Min Ryu, Shin-haeng Kang, Jaehoon Lee, Young-Uk Chang, JaeKyung Lee, JongTae Hwang, Daehwan Seo, Ki-Heon Na, Young Guen Song, Daihyun Lim, Kyung-Soo Ha, Young-Soo Sohn, Sang-Joon Hwang

Samsung Electronics, Hwaseong, Korea

Abstract

LLM-based generative AI requires high-capacity and high-bandwidth memory. We propose an HBM4 with 2048 I/Os, 3.3TB/s bandwidth, and 36GB cube density. Core die uses the 6th-gen. DRAM and the base die uses a 4nm Fin-FET process. TSV count and data window are doubled for high-frequency margin. t_{CCDR} is improved via per-channel TSV-RDQS auto calibration, enhancing rank-to-rank access. PMBIST enables at-speed testing in CoW and SiP. The on-die WDQS skew cancellation ensures robust timing, enabling reliable high-speed operation and improved yield.

Large language models (LLM)-based generative AIs are employed across diverse applications: including image and video generation, text composition, and information retrieval. As generative-AI performance continues to improve, high-capacity and high-bandwidth memory becomes essential to handle the increasing model parameters at high speed. To meet this demand, we propose a high-bandwidth memory-4 (HBM4) DRAM. Compared to existing HBM generations, HBM3 [1,2] and HBM3E [3], the proposed HBM4 provides 2048 I/Os and a bandwidth of 3.3TB/s with a cube density of 36GB (using a 12-high stack). HBM4 incorporates several novel schemes and features: the core die uses the 6th-generation 10nm DRAM process, while the base die is implemented using a 4nm Fin-FET logic process. By exploiting the enhanced area efficiency of the new process, the number of through-silicon via (TSVs) DQs is doubled, thereby doubling the TSV data window, ensuring sufficient data margin for high-frequency operation. To enhance overall GPU system performance, a per-channel TSV read-data-strobe (RDQS) timing auto-calibration scheme is adopted to improve the rank-to-rank access delay (t_{CCDR}) between different SIDs. To maximize the programmability of at-speed testing at chip-on-wafer (CoW) and system-in-package (SiP) testing, a programmable-memory built-in self-test (PMBIST) is developed; leveraging the logic processes of the base die. To maintain robust HBM4 timing margins, an on-die at-speed scheme is added in the DWORD write-data strobe (WDQS) distribution, which monitors the 4-phase skew and cancels it using an internal clock. This enables wafer-level outlier screening and allows the per-DWORD trim that is determined at wafer test to be applied in CoW or SiP stages. Consequently, defective dies are identified before costly downstream processing, test time is reduced by the fully automatic on-die calibration, and final-product yield is increased.

In prior HBM generations, both the base die and the core die are fabricated using the DRAM processes. To address AI demands, which requires high bandwidth and fast data rates, we implemented the base die using Samsung Foundry's 4nm Fin-FET logic process for HBM4. Compared to the DRAM processes, the 4nm Fin-FET logic process enables higher operation speed and lower power consumption, while improving area efficiency via smaller transistors and the availability of a larger metal-layer stack. As a result, the HBM4 base die achieves substantial performance improvement while satisfying full JEDEC [4] specifications.

Leveraging the 4nm Fin-FET logic base die and the 6th-generation 10nm DRAM core die improves area efficiency, enables the implementation of double the number of DQ TSVs. Figure 15.6.1 shows the TSV configuration for HBM4: the total TSV count is quadrupled compared to HBM3E, due to 2× the channel count and 2× the DQ TSVs, for enhanced performance. Additionally, adaptive body-bias (ABB) control is used to minimize process-variation effects between core dies. ABB reduces delay variation across core dies, leading to improved t_{CCDR} performance. The use of 2× number of DQ TSVs and ABB control for the core dies enhances the TSV data window and reduces delay variation, thereby improving t_{CCDR} and enabling HBM4 to achieve a maximum operating frequency of up to 13Gb/s.

The minimum required time interval, t_{CCDR}, for consecutive READ commands issued between different stack IDs (SID), is the key timing parameter that enhances HBM performance for high-speed AI workloads. However, for HBMs constructed by stacking multiple core die slices, the system performance is inherently limited by timing variation between stacked core die slices. In addition, as the number of channels increases, the local variation in each core die worsens the per-channel timing variation, further impacting t_{CCDR}. Although the core dies use ABB to mitigate core-die process variation, the remaining delay mismatch and local variations still limit t_{CCDR}.

To address t_{CCDR}, we propose a per-channel TSV RDQS timing-calibration scheme that automatically detects inter-slice delay variation in core dies and compensates for it to increase the maximum data rate while improving t_{CCDR} (Fig. 15.6.2). This calibration scheme operates after power-up: determining the optimal delay code for each delay compensation circuit (DCDL). This scheme considers global variation between core-die slices and local variation between channels.

For each SID channel, a reference signal, derived from each SID's readiness signals, simultaneously propagates through the normal RDQS path, including a replica RDQS tree, to accurately model the timing variations between each channel. These timing differences

are then quantized by a time-to-digital converter (TDC), then fed back into each channel's DCDL. The delay mismatches of each channel are thereby compensated.

The maximum data rate, satisfying $t_{CCDR} = 2n_{CK}$, is further improved from 7.8 to 9.4Gb/s by employing the proposed TSV t_{CCDR} RDQS-timing auto-calibration scheme.

PMBIST is a versatile memory-test pattern generator that embeds programmable logic in an HBM base die; supporting the full JEDEC HBM row/column command set and can issue commands on any clock edge. In traditional DRAM processes, power and area budgets limit the confined MBIST in HBM3E to only a few predefined patterns. The shift to a logic-process in HBM4 makes a fully programmable test methodology feasible, allowing direct programming of complex algorithms. Pattern programming and testing utilize the IEEE 1500 interface, enabling CoW and SiP verification. A dedicated software development kit (SDK) converts high-level test descriptors into binary microcode: speeding up measurement and debugging of critical AC parameters.

Figure 15.6.3 illustrates the PMBIST structure. SDK patterns are loaded into the pattern-storage unit via the IEEE 1500 interface as a set of PMBIST reduced-instruction-set instructions which (i) modify registers, (ii) issue commands/DQ based on register values, and (iii) perform loops. Multiple instructions are fetched and processed in parallel, then queued in a FIFO for sequential, at-speed issuance, enabling any pattern to be generated without impacting speed.

Figure 15.6.4 compares HBM4 PMBIST with HBM3E MBIST in terms of address-looping flexibility. MBIST uses a fixed looping sequence of column, row, bank, pseudo-channel, and SID. On the other hand, PMBIST allows the address-looping order to be freely configured, enabling t_{CCDR} and t_{CCDS} testing to perform SID and PC looping first.

Figure 15.6.5 shows the base-die DWORD WDQS clock-distribution network (CDN) integrated with the on-die scheme, which monitors the 4-phase clock skew and automatically cancels it along the signal path. In HBM4, timing margins at the high-speed physical layer (PHY) data input/output (I/O) become jitter limited as the I/O count and lane density increase; thinner interposer wiring increases loss/coupling and timing variation. This work uses a quarter-rate 4-phase WDQS scheme (I/IB/Q/QB). Therefore, any imbalance among the four phases shortens the one quarter interval and reduces data-interface eye width. Early detection and cancellation of such skew at the silicon bring-up stage is essential.

Consequently, the entire on-chip calibration procedure is performed at the target data rate. An internal clock (a ring oscillator in DWORD or a phase-locked loop in the direct-access region) is self-injected into WDQS_T/C, so no external injection or high-bandwidth probing across thousands of bumps is required. Before quadrature generation, the WDQS_T/C duty cycle is calibrated to 50% to prevent duty-induced quadrature error; the signal is then divided by 2 to produce quadrature phases. The read path on the WDQS CDN is observed, and the network is calibrated to minimize 4-phase skew to a few picoseconds: that is, phase spacing is corrected at the transmitter.

A key benefit is wafer-level screening, since during wafer test, devices that require correction beyond the available trim range (i.e., the trim setting saturates) are screened out early to avoid further test and packaging costs. The samples within the correction range have their per-DWORD trim values logged (IEEE 1500 accessible); these values can then be applied at later manufacturing stages (CoW or SiP), or the on-die trim can be re-run in situ if conditions change. This provides the per-DWORD trims logged at wafer and applied in CoW or SiP stages, avoiding manual per-DWORD tuning.

On the receiver side, write is not actively trimmed in this silicon because its clock routing was already prioritized (short paths, low-resistance metals, minimal vias), yielding intrinsically smaller skew and sufficient margin. The same internal-clocking and access points could be used to trim write if ever required.

Measured CoW proxy-package results at 10Gb/s confirm the effectiveness of this approach. When the wafer-test trims were applied during CoW evaluation, RDQS_T/C deterministic

ISSCC 2026 / February 17, 2026 / 10:05 AM

jitter (DJ) (dual-Dirac) decreased from 10.6 to 4.5ps, with a corresponding improvement in the read-data eye. In summary, adding on-die 4-phase skew monitoring and automatic cancellation to the WDQS CDN enables at-speed self-test at wafer, early outlier screening, and reusable per-DWORD trims that preserve WDQS alignment from wafer to final product.

Figure 15.6.6(top) shows the measured t_{CK} Shmoo plot. The results show that the fabricated HBM4 DRAM stacked in 12-high configuration achieves 13Gb/s/pin, corresponding to 3.3TB/s bandwidth with 32-channel operation. Figure 15.6.6(bottom) compares the key features between HBM3E and HBM4. In summary, the proposed HBM4 DRAM achieves a 260% bandwidth improvement while reducing V_{DDQ} supply voltage by 32%. The chip micrograph of the fabricated core and base dies with TSVs is shown in Fig. 15.6.7.

References:
[1] M.-J. Park *et al.*, "A 192-Gb 12-High 896-GB/s HBM3 DRAM With a TSV Auto-Calibration Scheme and Machine-Learning-Based Layout Optimization," in *IEEE JSSC*, vol. 58, no. 1, pp. 256-269, Jan. 2023, doi: 10.1109/JSSC.2022.3193354.
[2] Y. Ryu *et al.*, "A 16 GB 1024 GB/s HBM3 DRAM With Source-Synchronized Bus Design and On-Die Error Control Scheme for Enhanced RAS Features," in *IEEE JSSC*, vol. 58, no. 4, pp. 1051-1061, April 2023, doi: 10.1109/JSSC.2022.3232096.
[3] J. Lee *et al.*, "13.4 A 48GB 16-High 1280GB/s HBM3E DRAM with All-Around Power TSV and a 6-Phase RDQS Scheme for TSV Area Optimization," *ISSCC*, pp. 238-240, 2024, doi: 10.1109/ISSCC49657.2024.10454440.
[4] JESD270-4: JEDEC Standard High Bandwidth Memory (HBM) DRAM Specification, Apr. 2025. https://www.jedec.org/standards-documents/docs/jesd270-4

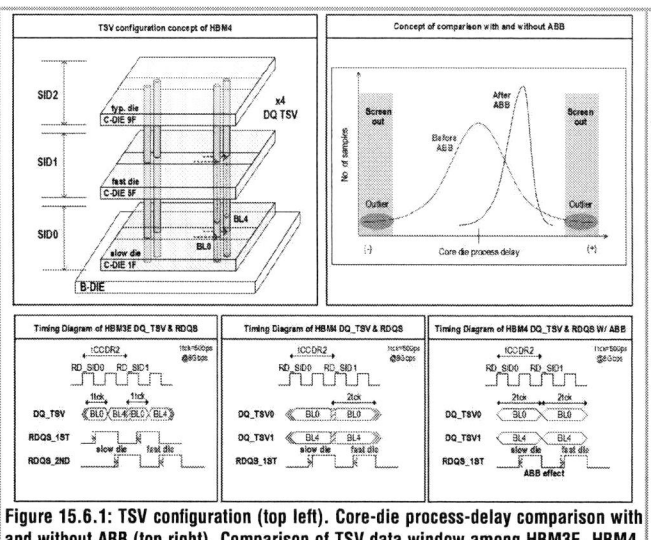

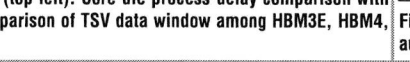

Figure 15.6.1: TSV configuration (top left). Core-die process-delay comparison with and without ABB (top right). Comparison of TSV data window among HBM3E, HBM4, and HBM4 with ABB (bottom).

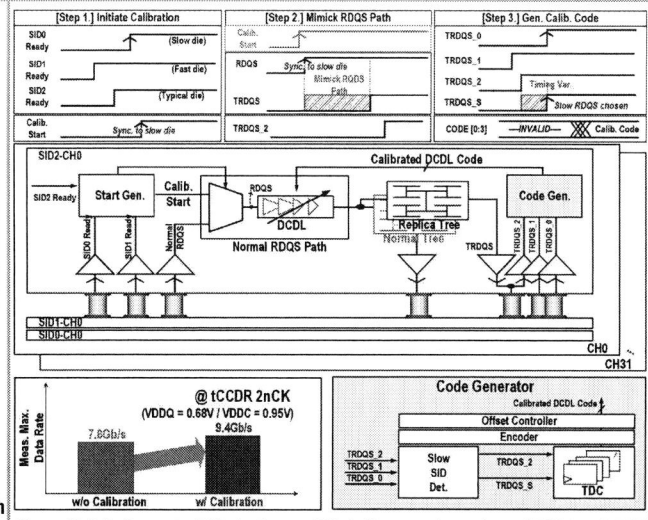

Figure 15.6.2: Conceptual illustration and block diagram of per-channel TSV t_{CCDR} RDQS auto-calibration scheme.

DIGEST OF TECHNICAL PAPERS • 265

979-8-3315-8937-0/26 $31.00 © 2026 IEEE

ISSCC 2026 / SESSION 15 / DRAM, SRAM, AND NON-VOLATILE MEMORIES / 15.6

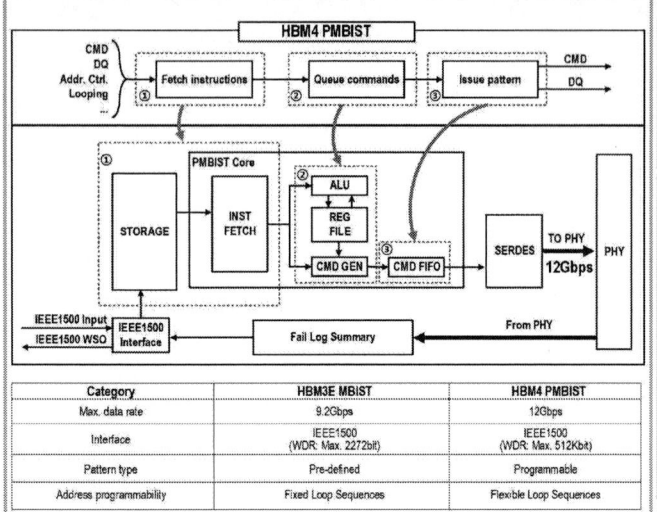

Category	HBM3E MBIST	HBM4 PMBIST
Max. data rate	9.2Gbps	12Gbps
Interface	IEEE1500 (WDR: Max. 2272bit)	IEEE1500 (WDR: Max. 512Kbit)
Pattern type	Pre-defined	Programmable
Address programmability	Fixed Loop Sequences	Flexible Loop Sequences

Figure 15.6.3: HBM4 PMBIST structure, and comparison to HBM3E MBIST.

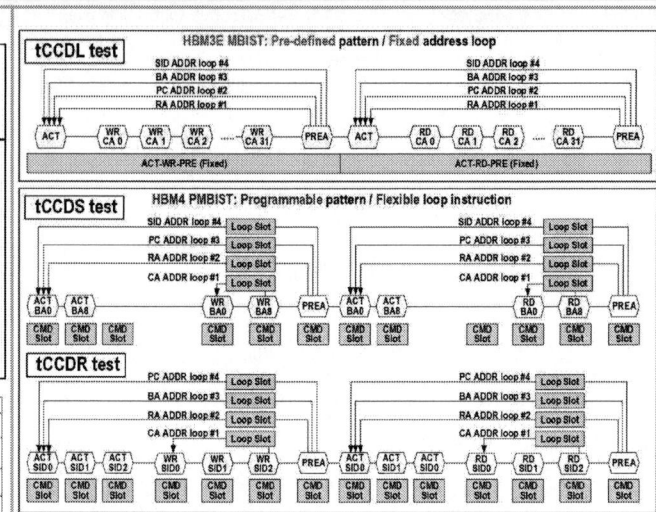

Figure 15.6.4: Address-looping flexibility comparison between HBM4 PMBIST and HBM3E MBIST.

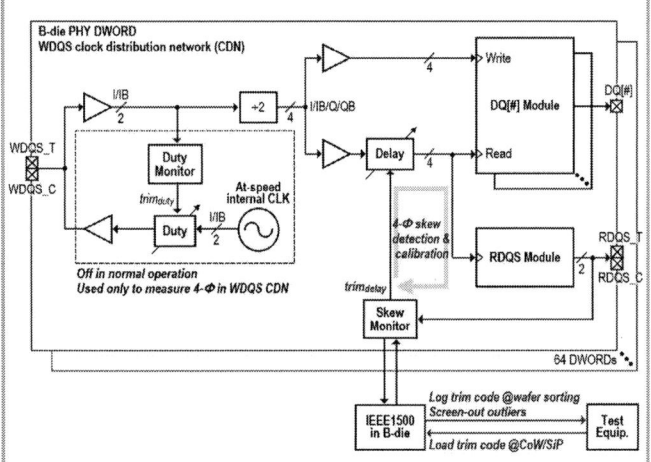

Figure 15.6.5: On-die at-speed 4-phase skew monitoring and cancellation for DWORD WDQS clock distribution, which enables wafer-level outlier screening.

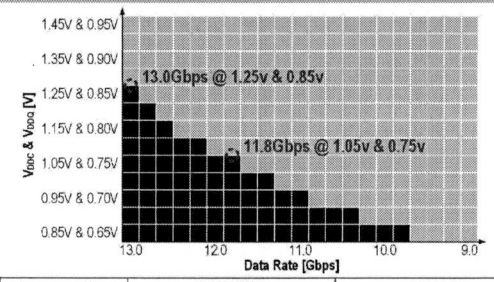

Generation	HBM3E	HBM4
Process	C&B-DIE: 4th Gen 10nm DRAM	C-DIE: 6th Gen 10nm DRAM B-DIE: 4nm Fin-FET
Supply Voltage	VDDC=1.1V, VDDQ=1.1V VDDQL=0.4V, VPPE=1.8V	VDDC=1.05V, VDDQ=0.75V VDDQL=0.4V, VPPE=1.8V
Max Data Rate	10.0Gb/s/pin	13.0Gb/s/pin
Organization	16 channel x 2PCH X 32I/O	32 channel x 2PCH X 32I/O
Bandwidth	1.25 TB/s per Cube	3.25 TB/s per Cube
Max Density	24Gb x 16-High = 48GB	24Gb x 12-High = 36GB
Microbump ballmap	7.08 mm x 8.82 mm	10.77 mm x 8.82 mm
Microbump pitch	96 μm x 110 μm	70 μm x 110μm
Chip Size	11 mm x 11 mm	12.8 mm x 11 mm

Figure 15.6.6: Measured t_{ck} Shmoo plot, and HBM3E and HBM4 key-feature summary table.

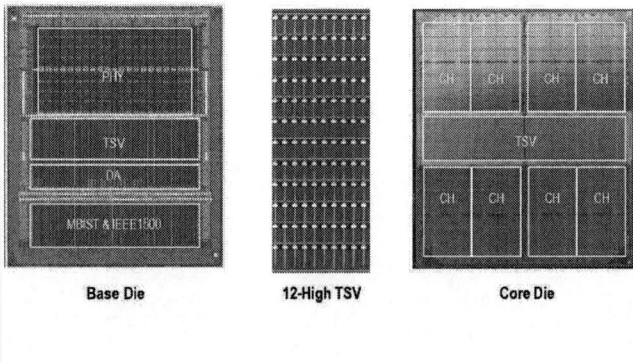

Figure 15.6.7: Chip micrograph of base die (left) and core die (right), with a cross-view of the 12-high TSV (middle).

• 2026 IEEE International Solid-State Circuits Conference

979-8-3315-8937-0/26 $31.00 © 2026 IEEE

ISSCC 2026 / SESSION 15 / DRAM, SRAM, AND NON-VOLATILE MEMORIES / 15.7

15.7 A 1cnm 14.4Gb/s/pin 16Gb LPDDR6 SDRAM with Efficiency Mode, LDO-Based WCK Tree, Dynamic Write NT-ODT, Fast CS Control and System Meta Mode

Jungtaek You*, Mino Kim*, Daehan Kwon, Sangsic Yoon, Joonhong Park, Hyungrok Do, Jeongje Park, Yooseok Jung, Seongseop Lee, Kyuyoung Kim, Hyeongsoo Jeong, Jaeyeol Kang, Hanbyeol Kwon, Minwook Oh, Jieun Kim, Jinseok Oh, Gi-Moon Hong, Dongyoon Ka, Seungho Lee, Heewoong Song, Minchang Kim, Daeho Yun, Sooncheol Kwon, Dongbeom Lee, Taekyun Shin, Jongmyeong Lee, Jungmin Yoon, Kyeongpil Kang, Sanghyuk Heo, Dongkyun Kim, Jieun Jang, Hyungsoo Kim

SK hynix, Icheon, Korea
*Equally Credited Authors (ECAs)

Abstract

This paper presents a 16Gb LPDDR6 DRAM in a 1cnm process, compliant with the JEDEC LPDDR6 specification. In implementing LPDDR6 functions, measures were taken to limit increase in power: including dynamic WR NT-ODT, LDO regulator-based WCK distribution, adding efficiency mode with power gating techniques, using fast CS control and system meta mode. These techniques enhance per-channel bandwidth by more than 50%, achieve 14.4Gb/s at 1.025V and improve power efficiency by over 20% compared to LPDDR5.

LPDDR DRAM is primarily used in power-efficient devices such as smartphones and edge devices; recently, expanding to automotive and AI computing. These applications require higher performance, lower power consumption, and improved reliability, which the next-generation LPDDR6 specification that was finalized this year [1], addresses. This paper introduces a 16Gb LPDDR6 DRAM fabricated in a 1cnm DRAM process, which fully complies with the LPDDR6 JEDEC specification.

In implementing the new LPDDR6 functions, various measures were used to limit increase in power consumption. Specifically, the dynamic WR NT-ODT function that enhances DQ-channel signal integrity (SI), incorporates a Command and address (CA) buffer and dedicated control logic to reduce power consumption. To address WCK jitter, a low drop-out (LDO) regulator-based WCK distribution network is implemented.

In efficiency mode, power gating is used for most circuit blocks, excluding essential circuits in the secondary sub-channel. Additionally, I_{DD2N} consumption is minimized by selectively enabling the CA buffer based on the operating frequency and suppressing CA signal toggles using the Chip Selection (CS) signal. Furthermore, a carving-based system meta mode in LPDDR6 [1] is implemented by placing metadata register adjacent to the DQ Tx/Rx circuitry. This optimization minimizes current consumption during metadata read and write operations.

Through these techniques, the per-channel bandwidth is enhanced by 50% compared to existing LPDDR5 [2], achieving a 14.4Gb/s data-transfer rate at 1.025V (V_{DD2C}), while improving power efficiency by over 20%.

Figure 15.7.1 illustrates that LPDDR6 adopts a dual-sub-channel-per-die configuration, supporting two operational modes: a normal mode, where each sub-channel functions independently, and an efficiency mode, in which both sub-channel banks are managed via the primary sub-channel (SC0). In efficiency mode, the banks of the two sub-channels operate as a pseudo-channel, configured as a unified set of 32 DRAM banks, thereby providing more flexible bank accessibility. To optimize power efficiency in efficiency mode, this work incorporates two key features. (1) to reduce power overhead during interleaved accesses between the two sub-channels, the clock-driven circuits including the command decoder, clock control, and latency control are centralized in the SC0; thereby, eliminating duplicate logic in the secondary sub-channel (SC1). (2) the inactive SC1 control circuits, shown in the gray area in Fig. 15.7.1, are powered off to minimize standby current. As a result, in efficiency mode using a 3.2GHz clock frequency with a 12.8Gb/s data bandwidth achieves standby (I_{DD2N}) and operation (I_{DD4R}) currents that are 87.3% and 81.1% of those in normal mode at 1.6GHz. The proposed architecture provides a power-efficient alternative to frequency scaling in normal mode, enabling the system to halve the data bandwidth between the DRAM and the controller by reducing the I/O count in efficiency mode, rather than by lowering the clock frequency.

Figure 15.7.2 shows LPDDR6's WCK tree architecture and the LDO regulator for the global WCK distribution block. In DRAM operations, the WCK tree delay for read commands is approximately 0.4 - 1ns longer than for write commands; resulting in a significant delay drift at a comparable voltage drop and adversely impacting the read DQ signal's valid window. To mitigate this issue, an LDO is integrated into the global WCK distribution block, targeting the predominant sources of the read WCK tree delay. This LDO has three main features: (1) the response time for DET_OUT is shortened by reducing the amplifier gain and increasing its bandwidth, which facilitates rapid response characteristics during WCK on/off cycles. (2) the implementation of a clocked multi-bit control scheme that is synchronized to the internal oscillator's output ensures the adjustment of PASS TR driver strength. This adjustment is made in accordance with the current load of the WCK tree by updating the register code in sync with the oscillator's output clock. (3) the internal oscillator and shift register in V_{WCK} LDO turn on and off in an event-driven manner. During WCK toggling, the amplifier quickly detects voltage drops in the WCK tree, causing DET_OUT to be driven to a high level, which in turn activates the internal oscillator and shift register.

During WCK non-toggling, all PASS TR control bits are turned off as soon as DET_OUT goes low, preventing overshoot and without adversely impacting the subsequent WCK operation. The proposed LDO reduces WCK jitter by 30%, compared to LPDDR5's WCK-distribution scheme [3].

The maximum operating frequency of the CA bus in LPDDR6 increased approximately threefold, compared to LPDDR5: ranging from 1.6 to 3.6GHz. Using a CA input buffer that operates across a wide frequency range leads to decreased power efficiency at low-frequency operation. The operational frequency range of the CA input buffer is divided into three sets, as shown in Fig. 15.7.3, thereby reducing the operational range. Within each of these operating ranges, the CA input buffer is designed to achieve optimal performance while considering power efficiency. The CA input buffers are selectively activated based on the speed value set in the mode register, which is adjusted according to changes in the CLK frequency.

In a multi-rank structure, CA pins are shared between ranks, resulting in standby current consumption during non-target operations. With increasing demand for high-capacity DRAM, multi-rank configurations are becoming more common, making it increasingly important to reduce CA input standby power consumption. When the memory controller issues a command to the target die, CS and CA are driven as shown in Fig. 15.7.3. The valid window of CS is $0.25 \times t_{CK}$ earlier than that of CA, as CS operates in SDR mode while CA operates in DDR mode. Additionally, within DRAM, there is a delay for adjusting the setup/hold time for each input CA signal. The fast CS control scheme utilizes the earliest-received CS signal to manage whether valid CA inputs pass through the setup/hold delay between the input buffer and the latch. As the $0.25 \times t_{CK}$ timing gap between CS and CA inputs depends on the operating frequency, the control position for valid CA inputs is changed by the mode register setting. At lower operating frequencies, a greater reduction in CA input standby current is achieved by moving the control point closer to the input buffer. Through the implementation of three CA input buffers optimized for narrow frequency ranges and the fast CS control scheme, the I_{DD2N} in normal mode is reduced by 42% at low-frequency and 19% at middle-frequency operation.

In previous LPDDR5 implementations, non-target on die termination (NT-ODT) has effectively mitigated the signal reflection noise caused by the shared DQ structure in a multi-rank configuration [4]. However, as shown in Fig. 15.7.4, for 12.8Gb/s data rates and beyond, the optimal NT-ODT resistance for write and read differs, which limits the use of a conventional fixed-resistance NT-ODT. To address this issue, LPDDR6 supports dynamic write NT-ODT (DWNT-ODT), which is used to control the non-target die and achieve an optimal NT-ODT resistance during write operations. In a multi-rank DRAM package, the CA input pins are shared across all ranks. Therefore, when a write is issued to the target die and successive CS input phases are simultaneously asserted high and low on the non-target die, the non-target die can recognize write command and burst length using the input CA pins, allowing control over the timing and duration of the DWNT-ODT operation. Although the DWNT-ODT feature contributes to enhanced signal integrity (SI) performance, DRAM must maintain DWNT-ODT operation in the standby state and in the power-down state, which inevitably leads to increased power consumption. To reduce the power consumption during the power-down state while DWNT-ODT is enabled, we designed a dedicated ODT control block, which operates using only the CA[0] and CS signals; thereby, eliminating the need for the command decoder, which typically requires all CA pins to distinguish the various commands. Through DWNT-ODT control, we achieved a 21% improvement in the data eye at 12.8Gb/s.

Figure 15.7.5 illustrates the block diagram of the two types of bank architecture and control circuitry within a carving-based system meta mode. Each bank incorporates a 256b metadata register (MDR), from which 16b are selectively accessed during read and write operations. In carving-based system meta mode, meta-read is required to read metadata from the cell array and store it in the metadata register before performing read and write operations. Additionally, upon completion of a write, meta-write is necessary to store

266 • 2026 IEEE International Solid-State Circuits Conference

979-8-3315-8937-0/26 $31.00 © 2026 IEEE

metadata from the metadata register back into the cell array. In Fig. 15.7.5 the left illustration depicts the MDR placed adjacent to each bank; whereas the right illustration shows the MDR positioned closer to the DQ Tx/Rx circuitry. The location of the MDR is crucial in minimizing power consumption during metadata accesses. Placing the MDR adjacent to each bank results in higher power consumption due to the increased data transfer distance between the DQ Tx/Rx circuitry and the MDR, including metadata bits. Consequently, positioning the MDR closer to the DQ Tx/Rx circuitry reduces power consumption during read and write operations. However, the opposite effect occurs during meta-read and meta-write operations. Given that reads and writes occur much more frequently than meta-read and meta-write operations, the MDR's location is determined based on this usage frequency. The chosen location for the MDR may introduce a timing skew between data and metadata accesses; to mitigate this, a dedicated metadata address FIFO is implemented.

Figure 15.7.6 shows the measured t_{WCK} Shmoo plot for various voltages and clock periods and for read and write operations. The implemented design achieves 14.4Gb/s at a 1.025V V_{DD2C} and a 0.875V V_{DD2D}. Figure 15.7.7 shows the chip micrograph of the fabricated DRAM in a 1cnm process.

Acknowledgement:
This paper was result of the research project supported by SK hynix Inc.

References:
[1] LPDDR6 SDRAM Specification (JESD209-6), JEDEC Standard, *JEDEC Solid-State Technology Association*, July 2025.
 https://www.jedec.org/standards-documents/docs/jesd209-6
[2] J.-H. Baek et al., "A 16Gb 12.7Gb/s/pin LPDDR5-Ultra-Pro DRAM with 4-Phase Self-Calibration and AC-Coupled Transceiver Equalization in a 5th-Generation 10nm DRAM Process," *ISSCC*, pp. 510-512, 2025.
https://doi.org/10.1109/ISSCC49661.2025.10904794
[3] Y. Seo et al., "A 1a-nm 1.05V 10.5Gb/s/pin 16Gb LPDDR5 Turbo DRAM with WCK Correction Strategy, a Voltage-Offset-Calibrated Receiver and Parasitic Capacitance Reduction," *ISSCC*, pp. 246-248, 2024.
https://doi.org/10.1109/ISSCC49657.2024.10454381
[4] K.-S. Ha et al., "A 7.5Gb/s/pin LPDDR5 SDRAM with WCK Clocking and Non-Target ODT for High Speed and with DVFS, Internal Data Copy, and Deep-Sleep Mode for Low Power," *ISSCC*, pp. 378-380, 2019. https://doi.org/10.1109/ISSCC.2019.8662509

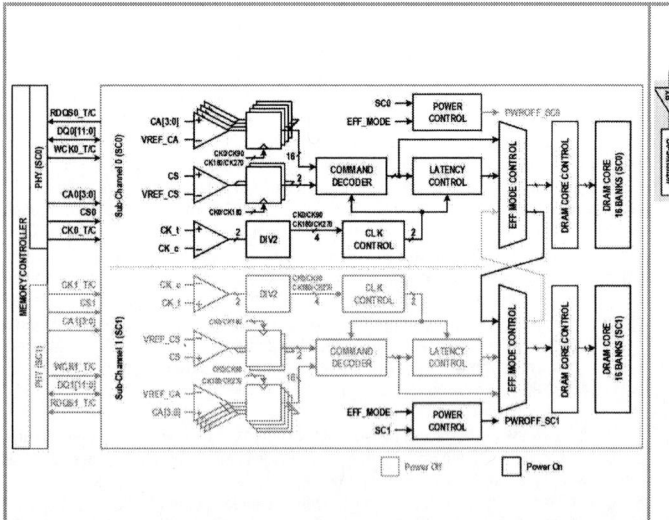

Figure 15.7.1: The proposed LPDDR6 architecture that incorporates an efficiency-mode operation.

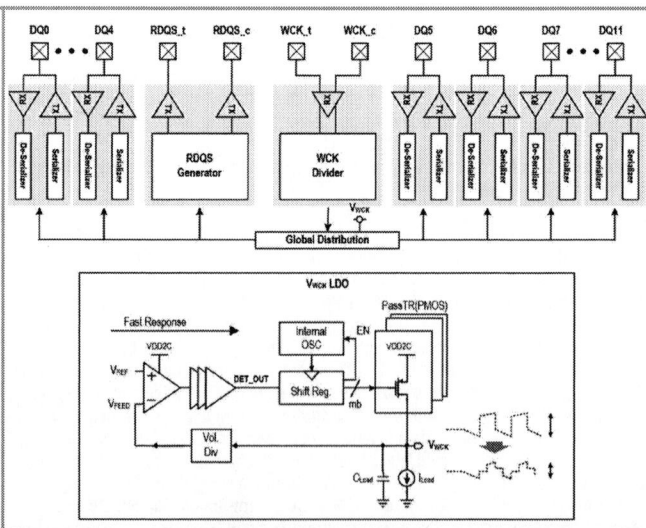

Figure 15.7.2: Proposed WCK distribution network for sub-channels with an LDO-generated V_{WCK}.

ISSCC 2026 / SESSION 15 / DRAM, SRAM, AND NON-VOLATILE MEMORIES / 15.7

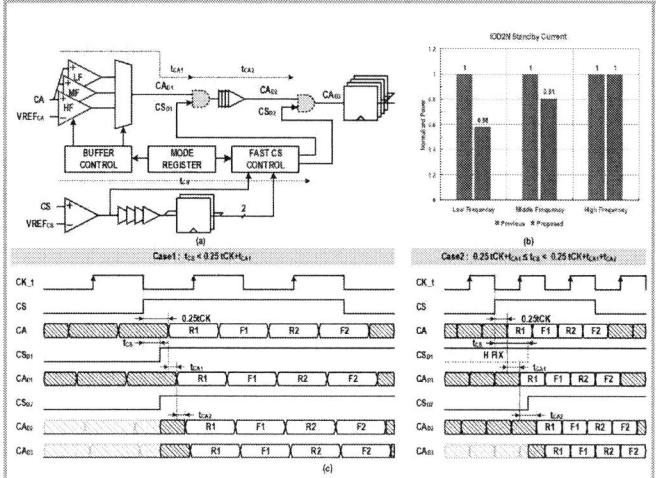

Figure 15.7.3: (a) Fast-CS control scheme block diagram, with frequency-adaptive CA buffers. (b) I_DD2N comparison across three frequency ranges. (c) Timing diagrams for low and middle frequencies.

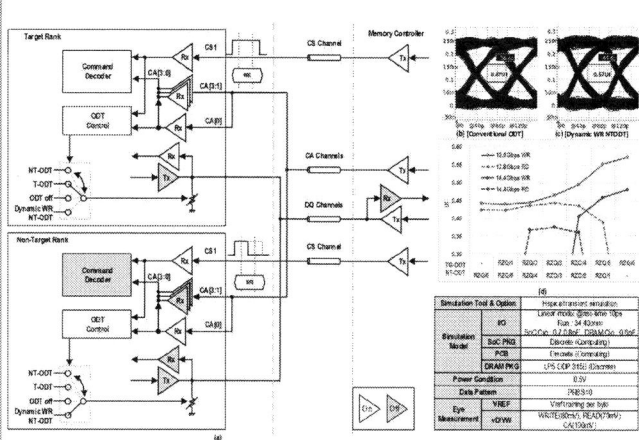

Figure 15.7.4: (a) Block diagram of a 2-rank system with dynamic WR NT-ODT. DQ write eye diagram with (b) conventional ODT and (c) dynamic WR NT-ODT. (d) Simulation results of optimal ODT combination at 12.8 and 14.4Gb/s.

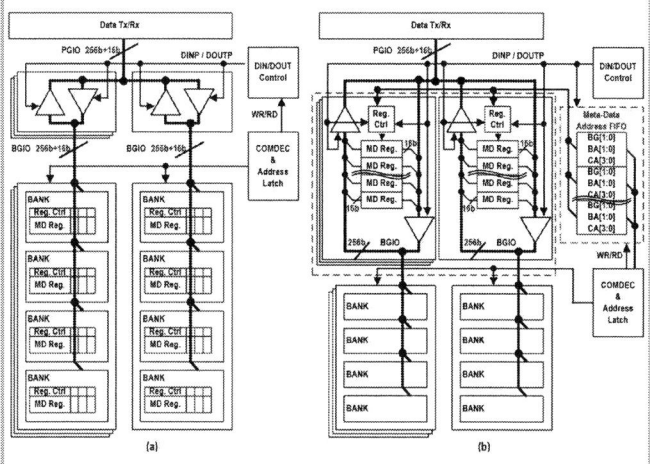

Figure 15.7.5: Bank architecture and control circuitry for carving-based system meta mode. (a) MDR placed adjacent to each bank. (b) Proposed MDR placement: adjacent to DQ transceivers.

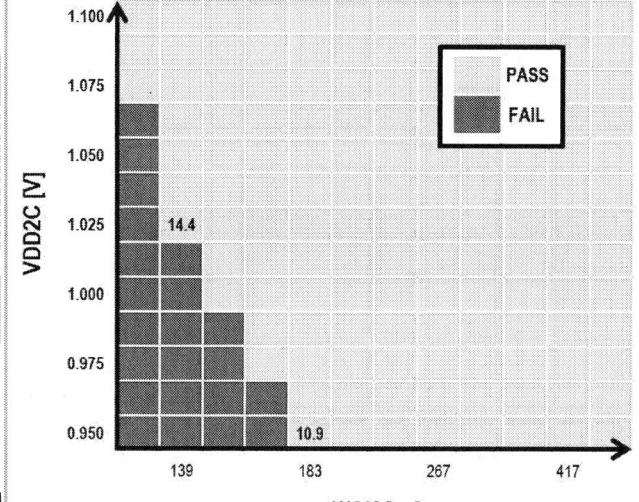

Figure 15.7.6: Measured t_CK Shmoo plot.

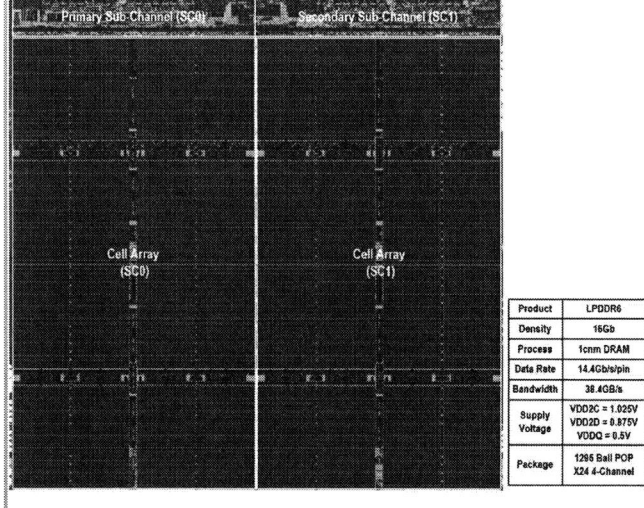

Product	LPDDR6
Density	16Gb
Process	1cnm DRAM
Data Rate	14.4Gb/s/pin
Bandwidth	38.4GB/s
Supply Voltage	VDD2C = 1.025V VDD2D = 0.875V VDDQ = 0.5V
Package	1295 Ball POP X24 4-Channel

Figure 15.7.7: Chip micrograph and summary table of 1cnm 16Gb LPDDR6.

• 2026 IEEE International Solid-State Circuits Conference

979-8-3315-8937-0/26 $31.00 © 2026 IEEE

ISSCC 2026 / SESSION 15 / DRAM, SRAM, AND NON-VOLATILE MEMORIES / 15.8

15.8 A 16Gb 12.8Gb/s LPDDR6 SDRAM with 12-DQ/Sub-Channel Wide NRZ Signaling and Enhanced Reliability by Per-Row Activation Counting and Meta-Data Scheme

Kiheung Kim*, Hong-Joo Song*, Youngdo Um, Jinhun Jang, Seungjun Lee, Hoseok Seol, Hye-Jung Kwon, Ji-Hwan Seol, Hoon Shin, Jaehyung Lee, JongMoon Choi, Sung-Woo Yoon, Sang Yun Kim, Dongkeon Lee, Minsoo Jang, Daehyun Kwon, Sangyong Lee, Jinguk Kim, Jonghyuk Kim, In Jung, Taeha Song, Chang Won Kim, Seung Ho Baek, Jinyong Choi, Young-Hun Seo, Won Ho Choi, Changsik Yoo, SangJoon Hwang

Samsung Electronics, Hwaseong, Korea
*Equally Credited Authors (ECAs)

Abstract

This paper presents a 12.8Gb/s/pin LPDDR6 SDRAM implemented in a 10-nm class process. It shows 33.3% higher bandwidth and 21.0% better energy efficiency than LPDDR5x by using wide NRZ signaling (12 DQ/sub-channel with two sub-channels) and optimum allocation of two supply rails (1.0 and 0.875V) across functional blocks. The dynamic and static efficiency modes provide better energy efficiency and higher capacity, respectively by dynamically or statically turning off the I/O circuitry of a sub-channel.

The demand for high-bandwidth low-power DRAM is accelerating for various applications, such as edge AI, hyperscale data-center platform , and advanced driver assistance system (ADAS), where the workload requires sustained throughput and strict power/thermal limits. This paper presents a 1st-generation 10nm-process LPDDR6 SDRAM that delivers 33.3% higher bandwidth (up to 12.8Gb/s) and 21.0% lower energy per bit than LPDDR5X, and integrates a wide-NRZ scheme with an efficiency mode for flexible power/density scaling. This paper also addresses the increased DQ pin count and the increased CA/CK frequency, by using a compact placement and layout of DQ/CA units, CK duty-cycle correction (DCC), and CA pseudo-DFE scheme. Power consumption is reduced by using V_{DD2D} for most peripheral circuits and via dynamic voltage-frequency scaling (DVFS) for I/O blocks, with a selective V_{DD2C}/V_{DD2D} switching scheme in DQ Tx/Rx, combined with usage of an additional low-power path for DQ/CA signals at low frequencies. Per-row activation counting (PRAC) is implemented with two proposed write-compensation techniques: resilient CSL selection and enhanced ECC. Reliability, availability, and serviceability (RAS) capabilities are improved by metadata support for system-level ECC.

Figure 15.8.1 shows the top-level block diagram of this work. The LPDDR6 SDRAM consists of two independent sub-channels: each with 16 banks. To secure a sufficient data-eye margin for a low-voltage swing-terminated-logic (LVSTL) driver, LPDDR6 adopts a wide-NRZ scheme instead of PAM3 signalling as used in GDDR7. The CA pin count is reduced to four and the DMI pin is removed to compensate for the increased DQ pin count resulting from the wide-NRZ scheme. CK input signals are divided by 2 to mitigate the internal timing margins in the DRAM. Each sub-channel has 12 DQs and a 24-burst-length (BL24) prefetch structure, producing 288b per DRAM access: 256b of normal data, 16b of metadata, 16b of DBI. LPDDR6 supports a dynamic-efficiency mode to reduce standby power and a static-efficiency mode to expand capacity. In efficiency mode, sub-channel 0 becomes the primary channel, controlling sub-channel 1 and enabling data access via sub-channel flag bit in commands. Data from sub-channel 1 passes through a bus repeater before reaching sub-channel 0 DQs, incurring an additional latency.

Achieving both high-frequency operation and low-power consumption is challenging due to their inherent trade-off relationship. LPDDR6 has adopted two constant-power domains, V_{DD2D} (0.875V) and V_{DD2C} (1.0V), while LPDDR5/5X optionally supports two power rails, V_{DD2L} (0.9V) and V_{DD2H} (1.05V) [1], which limits the usage of two power rails. This work leverages V_{DD2D} and V_{DD2C} to overcome the trade-off between speed and power consumption as shown in Fig. 15.8.2. LPDDR6 applies V_{DD2C} to the DQ/WCK, CA/CK I/O, and a portion of the latency logic, and uses V_{DD2D} for all other core and peripheral circuits to minimize power consumption. High-performance transistors help to enable high-frequency operation in the V_{DD2D} domain, while reverse body bias (RBB) and enhanced power gating are applied in the fast corner to suppress leakage. LPDDR6 further extends power gating to the high-frequency DQ region and the data bus; thereby, achieving a 10% reduction in I_{DD2P}. In addition, a lower internal data supply ($V_{INTDATA}$) lower than VDD2D is introduced to minimize toggle- power during read and write operation. LPDDR6 adopts WCK clocking to secure stable high-speed I/O timing and sampling margin, as in LPDDR5X; therefore, WCK-to-CK synchronization is essential [2]. To ensure that voltage variation does not impact synchronization, a delay matched path construction between WCK and CK is required to ensure timing alignment across voltage and temperature variation. WCK is confined to the V_{DD2C} domain, while a portion of the CK latency path operates in the V_{DD2D} domain. To align voltage variation tracking with WCK, the CK path incorporates a V_{DD2D}-to-V_{DD2C} level shifter followed by an additional flip-flop, thereby referencing V_{DD2C} variation while retaining power benefits of V_{DD2D}. Due to these implementations, the read-power and write-power decrease by 27% and 22% relative to LPDDR5X. DQ Tx/Rx utilize internal power switching between V_{DD2C} and V_{DD2D} to guarantee performance at high frequencies, while minimizing power consumption at low frequencies. In power-down mode, they are also used to power gate DQ Tx/Rx. Power switching decreases read (I_{DD4R}) and write power (I_{DD4W}) by 9% and 5% at 3.2Gb/s. To further reduce power consumption at low frequencies, separate low-frequency (LF) paths are employed for DQ Tx and CA Rx. A DQ Tx LF path is implemented after a serializer to prevent increased WCK loading and the path is optimized to balance between jitter injection and power consumption. A CA Rx LF path consists of an additional Rx buffer, which consumes less static current than a normal path and this buffer significantly reduces standby current. When LF paths are enabled, I_{DD4R} and standby current (I_{DD2N}) decrease by 3% and 12% at 3.2Gb/s.

The number of DQ pins has increased to support a wide data-bus NRZ in LPDDR6; thereby, increasing WCK loading and degrading margin of DQ Tx/Rx. As shown in Fig. 15.8.3, a re-distribution layer (RDL) is utilized to place DQ Tx/Rx units as close to WCK as possible, while considering their insertion loss and crosstalk. This placement decreases WCK-to-DQ delay, as well as its voltage and temperature variations; therefore, results in an improvement to the write and read eye margin, while saving operating power of WCK and DQ circuits. The WCK-to-CK ratio is fixed at 2:1 in LPDDR6 to enhance the efficiency of command/address; hence, the frequency of CA has increased more than twice as fast as the frequency of LPDDR5X. This increases challenges the CA system budget in multi-rank configuration because part of the die-to-die CA skew cannot be cancelled by training. The demand to reduce this CA-to-CA skew requires that CA/CK units are placed more closely, via RDL as shown in Fig. 15.8.3, to remove the CK clock-tree that was used to match delays between CK and each CA pin. The placement with RDL minimizes the number of CA/CK stages and achieves reduced CA/CK delay variation due to voltage and temperature variation. In addition, various features to support the increased CA/CK frequencies are employed. Figure 15.8.3 shows the overall configuration of the CK and CA path including a divider, DCC, a skew adjuster for CK and CA signals, and a pseudo DFE. CA-timing margin is improved by minimizing the duty-cycle error: the DCC in the CK buffer corrects the duty-cycle error of the input CK signals [3] and the remaining duty-cycle error, including the internal 4-phase skew, is adjusted by the CK quadrature-skew adjuster. A pseudo 1-tap DFE is adopted for CA to make the CA buffer immune to ISI. The 1-tap DFE is implemented using a current-steering DAC, using cross-coupled transistors. The gates of the cross-coupled transistors are connected to the outputs of the first-stage buffer, enabling timing-mismatch compensation between the main cursor and the feed-back decisions, without the use of a strobe clock.

In DRAM, repeated activations of adjacent rows can disturb charge in neighboring cells and induce data retention failure: this is known as the row-hammer phenomenon; various probabilistic refresh schemes have been proposed to reduce its effects [4]. Since these schemes rely on statistical coverage, new vulnerabilities continue to be exposed as malicious attack scenarios. The proposed PRAC scheme embeds counters on a per-WL basis inside the DRAM array, enabling deterministic tracking of all row activations and thereby improving robustness against adversarial hammering. Implementing PRAC requires a read–modify–write (RMW) operation during each ACT-to-PRE interval, in which the PRAC counter value for the activated row is read, incremented, and written back. Long RMW sequences may degrade system performance; hence, reducing the write time, which dominates the RMW duration, is critical. As shown in Fig. 15.8.4, this paper proposes a resilient WL-selection method that chooses the CSL with the shortest write time, thereby minimizing timing overhead. In addition, errors arising from shortened write cycles are recovered by an ECC capable of correcting small multi-bit errors within the PRAC cell array. As illustrated in Fig. 15.8.4, the PRAC mechanism increments the activation counter (AC) on every row operations. When the counter exceeds a programmable threshold, the current row address is stored in the row-address log register. Since the number of log registers is limited, PRAC replaces the entry whose stored activation count is the minimum among them and writes the new row address into that register. During the mitigation phase, the register retrieves the address with the highest count, resets its counter value, and refreshes its adjacent victim rows according to the configured algorithm. Unlike LPDDR5X, the PRE command in LPDDR6 does not immediately trigger an internal pre-charge but instead begins with an internal read operation, which modifies the READ-to-PRE and WRITE-to-PRE timing relationships. Detailed timing adjustments are described in the following section. Emulation results confirm that PRAC improves row-hammer resilience by approximately 4–5× compared to conventional mitigation schemes.

Figure 15.8.5 illustrates the metadata-scheme block diagram. LPDDR6 supports access to 32bytes of normal data and 2bytes of metadata in a single DRAM read/write. To reduce

DRAM overhead, a column-carved-out scheme is used, where column addresses C[5:0] ≥ 0x3C are reserved for meta data. When meta mode is disabled, these column addresses are used for regular data. Each bank contains a 32bytes meta register positioned between the DINPAR block (bank write-data path) and the DOUTPAR block (bank read-data path). This placement allows metadata to be accessed directly through the carved-out column addresses without activating the main cell array, thereby reducing both access latency and energy for metadata operations. During write operations, the 2bytes metadata is delivered via the metadata bus and stored in the metadata register. After write operations, the 32bytes in the metadata register is written to the 0x3C–0x3F region via a metadata write operation. For read operations, the 32bytes meta data in the 0x3C–0x3F region is first loaded into the metadata register through a metadata read operation, and subsequent read commands output the corresponding 2bytes meta data along with the 32bytes normal data. By providing column-carved-out metadata, LPDDR6 supports system-level RAS feature enhancement.

Figure 15.8.6 shows the measured t_{CK} vs V_{DD2C} Shmoo, and the read/write eye Shmoo 12.8Gb/s operations. The maximum data-rate is 12.8Gb/s at V_{DD2C} = 0.97V: the JEDEC specified minimum voltage. The valid read/write windows are 0.69 and 0.72UI. Figure 15.8.7 shows the fabricated 16Gb DRAM micrograph in a 10nm-class DRAM process: occupying 44.5mm².

References:
[1] D. Kim et al., "A 16Gb 9.5 Gb/S/pin LPDDR5X SDRAM With Low-Power Schemes Exploiting Dynamic Voltage-Frequency Scaling and Offset-Calibrated Readout Sense Amplifiers in a Fourth Generation 10nm DRAM Process," *ISSCC*, pp. 448-450, 2022. https://doi.org/10.1109/ISSCC42614.2022.9731537
[2] K.-S. Ha et al., "A 7.5Gb/s/pin LPDDR5 SDRAM with WCK Clocking and Non-Target ODT for High Speed and with DVFS, Internal Data Copy, and Deep-Sleep Mode for Low Power," *ISSCC*, pp. 378-379, 2019. https://doi.org/10.1109/ISSCC.2019.8662509
[3] H.-J. Chi et al., "An 8.5Gb/s/pin 12Gb-LPDDR5 SDRAM with a Hybrid-Bank Architecture Using Skew-Tolerant, Low-Power and Speed-Boosting Techniques in a 2nd Generation 10nm DRAM Process," *ISSCC*, pp. 382-384, 2020. https://doi.org/10.1109/ISSCC19947.2020.9062914
[4] W. Kim et al., "A 1.1V 16Gb DDR5 DRAM with Probabilistic-Aggressor Tracking, Refresh-Management Functionality, Per-Row Hammer Tracking, a Multi-Step Precharge, and Core-Bias Modulation for Security and Reliability Enhancement," *ISSCC*, 2023. https://doi.org/10.1109/ISSCC42615.2023.10067805

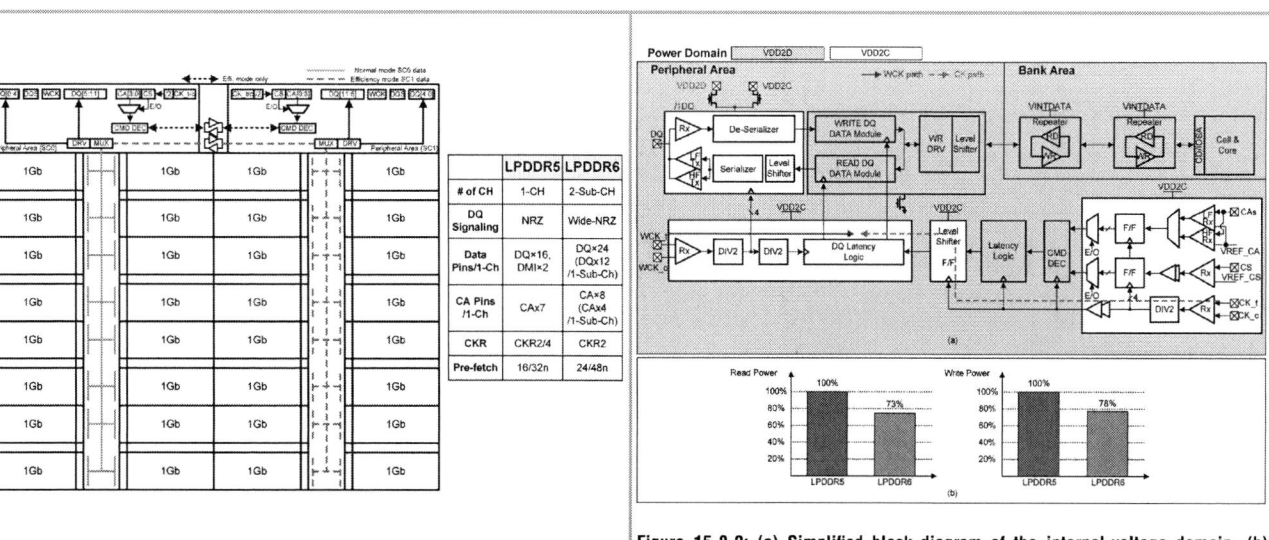

Figure 15.8.1: Proposed LPDDR6 architecture.

Figure 15.8.2: (a) Simplified block diagram of the internal-voltage domain, (b) comparison of read-/write-power.

ISSCC 2026 / SESSION 15 / DRAM, SRAM, AND NON-VOLATILE MEMORIES / 15.8

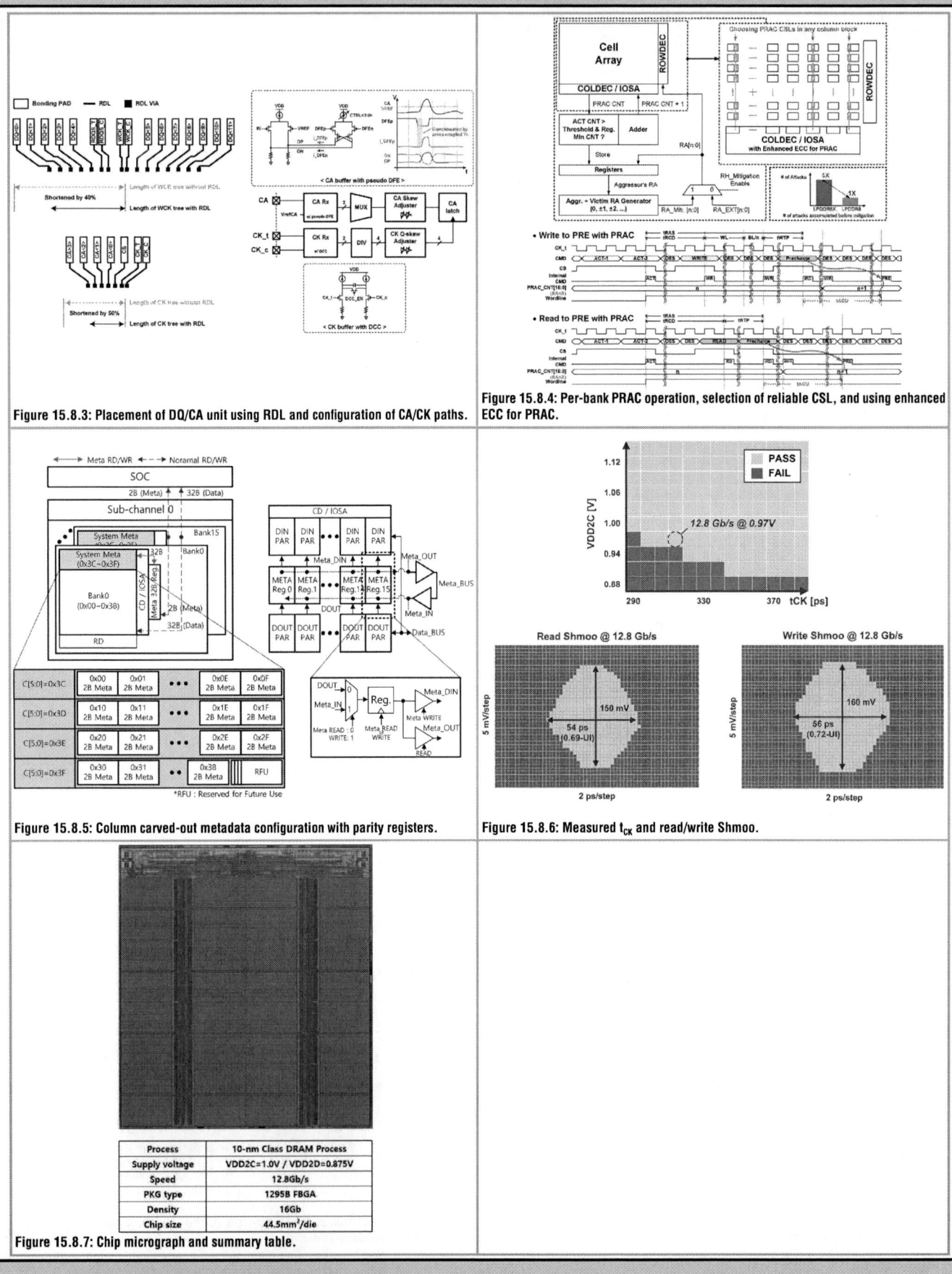

Figure 15.8.3: Placement of DQ/CA unit using RDL and configuration of CA/CK paths.

Figure 15.8.4: Per-bank PRAC operation, selection of reliable CSL, and using enhanced ECC for PRAC.

Figure 15.8.5: Column carved-out metadata configuration with parity registers.

Figure 15.8.6: Measured t_{CK} and read/write Shmoo.

Figure 15.8.7: Chip micrograph and summary table.

Process	10-nm Class DRAM Process
Supply voltage	VDD2C=1.0V / VDD2D=0.875V
Speed	12.8Gb/s
PKG type	1295B FBGA
Density	16Gb
Chip size	44.5mm²/die

• 2026 IEEE International Solid-State Circuits Conference

ISSCC 2026 / SESSION 15 / DRAM, SRAM, AND NON-VOLATILE MEMORIES / 15.9

15.9 A 48Gb/s 24Gb GDDR7 DRAM for Mid-Range Inference AI with Symmetric 2CH-Mode Operation, Clock-Path Optimization, and RAS Features

Hyunsu Park*, Sungkwon Lee*, Kyunghoon Kim, Jinyoup Cha, Hyeongjun Ko, Jaehyeok Yang, Seongjin Kim, Youngtaek Kim, Minsoo Park, Gangsik Lee, Keonho Lee, Sanghoon Lee, Jiseok Lee, Gyunam Jeon, Sera Jeong, Yongsuk Joo, Jaehoon Cha, Seonwoo Hwang, Seulgi Kim, Eunji Song, Junghwan Ji, Byungjun Kang, Boram Kim, Sangyeon Byeon, Hankyu Chi, Hyungsoo Kim, Jonghwan Kim

SK hynix Semiconductor, Icheon, Korea
*Equally Credited Authors (ECAs)

Abstract

This paper introduces a GDDR7 DRAM for mid-range inference AI. A center GIO architecture is used to remove the speed penalty of the 2CH mode, which doubles memory density and reduces the clock path. This work supports ECC, CRC, and DPP to meet the RAS features required in server systems. Additionally, energy efficiency is enhanced for on-device AI applications through a low-V_{DDQ} mode. For nominal conditions, with a 1.2V supply voltage, a 48Gb/s data transmission speed is achieved.

Artificial intelligence (AI) has been rapidly advancing alongside the development of graphic processing units (GPUs) and memory hardware. Currently, high-bandwidth-memory (HBM) dominates the high-end machine-learning-AI applications, due to its superior bandwidth. However, its high cost and limited packaging flexibility impede its application for mid-range and on-device inference AI. There is a growing demand for cost-effective package solutions for AI applications. In this context, GDDR7 emerges as a cost-effective alternative that offers high-speed advantages, making it suitable for future AI applications [1-2]. However, as training-data volume increases, there is an ongoing demand for higher GDDR7 DRAM density. GDDR7 addresses this with its 2-channel (2CH) mode, which doubles memory density [3]. While developing high-capacity DRAM products requires significant development costs and time, the 2CH mode can effectively reduce the costs associated with an increased memory capacity. Nevertheless, addressing the inherent speed degradation in the 2CH mode is essential because the DRAM bandwidth is another key parameter for AI applications. Furthermore, for on-device AI applications, issues related to power consumption and heat dissipation must be addressed. GDDR7 supports a low-V_{DDQ} mode, which reduces system power consumption. Additionally, it is important to implement circuitry supporting low-V_{DDQ} mode while minimizing speed degradation. By overcoming these challenges, GDDR7 can establish itself as a cost-effective and energy-efficient solution for AI.

This paper proposes the center-GIO architecture to improve the speed of the 2CH mode, as shown in Fig. 15.9.1. GDDR7 features independent memory-bus channels and supports the 2CH mode, which connects two adjacent channels to increase the memory capacity per channel. However, in the conventional DRAM architecture, the application of the 2CH mode for increased memory density results in a longer data access time (t_{AA}), than that of the 4CH mode. For the 2CH mode of a conventional DRAM architecture, the global-input/output (GIO) bus, which connects adjacent channels, transmits data across the command and address (CA) area. This additional asynchronous delay causes domain crossing issues in the region where the data from adjacent channels is combined, such as when data is transmitted from Ch. B to Ch. A. In previous GDDR7 devices, this delay is compensated via a replica delay; however, at increased speeds, ensuring reliability during massive data transfers between GIO buses, becomes challenging [1,2]. To mitigate the speed limitations associated with the 2CH mode, the center-GIO architecture is adopted. In this structure, the GIO bus is positioned centrally between two adjacent channels and is controlled through a 4CH:2CH multiplexer (MUX). CA circuits are relocated to the edges of the DRAM to create space for the centralized GIO bus. This symmetric structure equalizes the t_{AA} for both 4CH and 2CH modes; thereby, minimizing the speed degradations between each mode. However, the center-GIO structure causes GIO-bus congestion in the data path, because all the GIO for read and write operations should be connected to the center MUX. Due to the dual-mode support of NRZ and PAM-3 modes, the congestion of the GIO bus is highest near the PAM-3 encoder and decoder blocks. These blocks are located near the GIO MUX, among the two encoders and decoders divided into left and right parts considering the DQ placement [3]. To address this issue, the dual mode MUX is embedded in the PAM-3 encoders and decoders, as shown in Fig. 15.9.1. The embedded MUX reduces the GIO BUS congestion by 17.9%.

To increase memory speed and ensure low-power operation, it is necessary to guarantee the bandwidth of clocking schemes and effectively minimize phase skew. The center-GIO structure reduces the length of the clock-distribution network's path by approximately 20%, as shown in Fig. 15.9.2. If the GIO bus from the banks is relocated to the center, the gap between the global clock circuitry and the local DQ unit blocks can be eliminated because the minimized gap resolves issues with the power-distribution network and coupling induced by the GIO bus. As a result, the bandwidth of the CML-based multi-drop clock driver is improved by 15%, without additional power consumption. To mitigate phase error caused by local variation, internal quadrature-phase clocks are generated with a local clock divider. Despite these efforts, the local clock's phase error remains a critical factor at increased speeds. To address this, a quadrature phase-skew correction (QPSC) technique is adopted, as shown in Fig. 15.9.2. The differential-phase skew between signals such as ICLK (0°) and IBCLK (180°) is corrected using a cross-coupled inverter, while the phase skew between 90° clocks, such as ICLK (0°) and QCLK (90°), is averaged using the feedback network and programmable tri-state buffers with optimized delay for a specific target frequency range and 2b resolution. This ensures that the four-phase write clocks have a very-low-skew error at high speeds. As in the previous clock-distribution network, the divided clock, D_ICK, is used in the local clock generator to synchronize the divided clock phases with the command clock [1]. In this work, D_ICK is converted into a differential clock to enable faster generation of the four-phase clock.

As the speed of DRAM increases, new features have been added to the DRAM transmitter and receiver to ensure signal integrity as shown in Fig. 15.9.3. In the transmitter, a 3-tap feed-forward equalizer (FFE) with pre- and post-cursor cancellation is implemented. The pre-cursor tap, in particular, was added to the DRAM side as relying solely on the continuous-time linear equalizer (CTLE) on the GPU side could amplify high-frequency noise such as crosstalk. In addition, a capacitive pre-emphasis scheme is adopted to compensate for large channel attenuation, which supports the FFE. The input signal of the DRAM receiver is amplified through two independent analog equalizers: CTLEH and CTLEL. This structure that independently equalizes and amplifies the upper and lower eyes is adopted with considerations for testability and power efficiency; however, a common-mode voltage difference occurs between the upper and lower paths (CTLEH and CTLEL) in the single-ended design. This voltage difference causes delays in each amplifier's output signal, which degrades the sampling margin of the upper and lower PAM-3 eyes at the equalizer output stage. In the proposed receiver, the RX sampling clocks for the lower eyes are delayed to mitigate the distorted timing window. The delay offset between the upper and power paths is compensated in the de-serializer stages, which operate on the lower clock speed.

The use of GDDR7 in servers necessitates a variety of reliability, availability, and serviceability (RAS) features to detect or prevent potential errors during operation, as shown in Fig. 15.9.4. It is particularly important to reduce the BER for commands that operate at high speeds, as well as to employ various methods to protect command paths of high-speed input/output data. In GDDR7, the CA signal integrity is improved via a CA DFE, since the CA speed has doubled compared to GDDR6. The proposed GDDR7 includes a 1-tap decision-feedback equalizer (DFE) to ensure the eye-opening of the CA input signal, which is distorted by ISI on the channel. The four mode-register-set (MRS) bits adjust the V_{REFCA} of the CA DFE by shifting it in steps of approximately ±0.5% of V_{DDQ}: ±6mV. Additionally, QPSC is implemented to correct the 4-phase error. Since the CA clock is derived from WCK, without an external reference clock source, a long clock path from WCK to the CA clock not only makes it vulnerable to noise but also causes delay drift due to temperature and voltage variations. The CA clock-path delay is monitored using an internal replicated oscillator for the clock delay drift. To further improve reliability, the DRAM internally computes CA parity. All commands are stored in a pipeline and undergo parity computation for error detection, enabling the DRAM to accurately identify commands containing errors and block subsequent command inputs. If an error is detected, the host is notified, and command execution is halted. The system then re-runs CA-link training. This allows the host to clearly understand which command requires re-execution, eliminating any ambiguous period around the erroneous command. For example, data written after a CA parity error may contain error; therefore, rewrites must be performed. Furthermore, the transmitter of the ERR signal, which outputs CA parity error is located close to the parity computation block and utilizes the same clock as the CA RX to reduce the asynchronous delay of the CA parity error [1]. This configuration enables the ERR signal to be delivered in the shortest possible time without introducing additional latency for domain crossing.

On-die ECC in GDDR7 provides cell reliability that is comparable to server-grade DRAM [4]. GDDR7 is the first graphics DRAM to apply SECDED coding: providing single-bit error-correction and random double-bit error-detection capabilities for the 272b stored data. The designed H-matrix of the on-die ECC also enables the correction of adjacent 2b errors and provides an average 99.3% detection capability for multi-bit errors involving three or more bits.

The data-path protection (DPP) scheme can monitor the parity error in the internal paths that are not covered by CRC or ECC, particularly in blocks such as the PAM-3 decoder and encoder. Basically, DRAMs containing errors within the data path are filtered out through a functional test process. Nevertheless a 1b parity enables error detection for these paths, even in the case of very low-probability events, allowing the controller to appropriately handle the affected data at the right time.

Figure 15.9.5 shows the measurement results across varying speed and supply voltages (V_{DD} and V_{DDQ}). In the normal mode where V_{DD} and V_{DDQ} are at the same level, the maximum data rate at 1.2V exceeds the equipment measurement limit of 48Gb/s. In the V_{DDQ}-tCK Shmoo plot for the low-V_{DDQ} mode, with V_{DD} fixed at 1.05V, the maximum data rate is measured at 30.3Gb/s when V_{DDQ} is 0.9V. A level-shifter-less V_{DD}-to-V_{DDQ} domain-crossing scheme is adopted to mitigate the speed degradation in both normal and low-V_{DDQ} modes. As shown in Fig. 15.9.6, by enabling QPSC at 46Gb/s the average and standard deviation of the 4-phase error is improved by >40%. Power consumption is also improved compared to the previous GDDR7 device [1]: for read and write operations the energy transfer efficiency per bit decreases by 7.4% and 19.3%. The active and idle power consumption, normalized to the operating speed, is decreased by approximately 20% or more. When applying heterogeneous power sources (V_{DD}/V_{DDQ}), the energy efficiency of write and read operations improved by approximately 25.4% and 31%, and the Idle-related current can also be additionally reduced by more than 20%.

Acknowledgement:
This paper is the result of the research supported by SK hynix Inc.

References:
[1] J. Yang et al., "A 35.4-Gb/s/pin 16-Gb GDDR7 with a low-power clocking architecture and PAM3 IO circuitry," *ISSCC*, pp. 232-233, 2023.
https://doi.org/10.1109/ISSCC49657.2024.10454560
[2] S.-H. Kim et al., "A 24Gb 42.5Gb/s GDDR7 DRAM with low-power WCK distribution, R/C optimized dual-emphasis TX, and voltage/time margin enhancement with power reduction," *ISSCC*, pp. 508-509, Feb., 2025.
https://doi.org/10.1109/ISSCC49661.2025.10904689
[3] Graphics Double Data Rate 7 SGRAM Standard (GDDR7), Standard JESD239.01, *JEDEC*, Feb. 2024.
https://www.jedec.org/standards-documents/docs/jesd239c
[4] S. Kaneda and E. Fujiwara, "Single byte error correcting—double byte error detecting codes for memory systems," *IEEE Trans. on Computers*, vol. C-31, no. 7, pp. 596-602, Jul., 1982. https://doi.org/10.1109/TC.1982.1676056

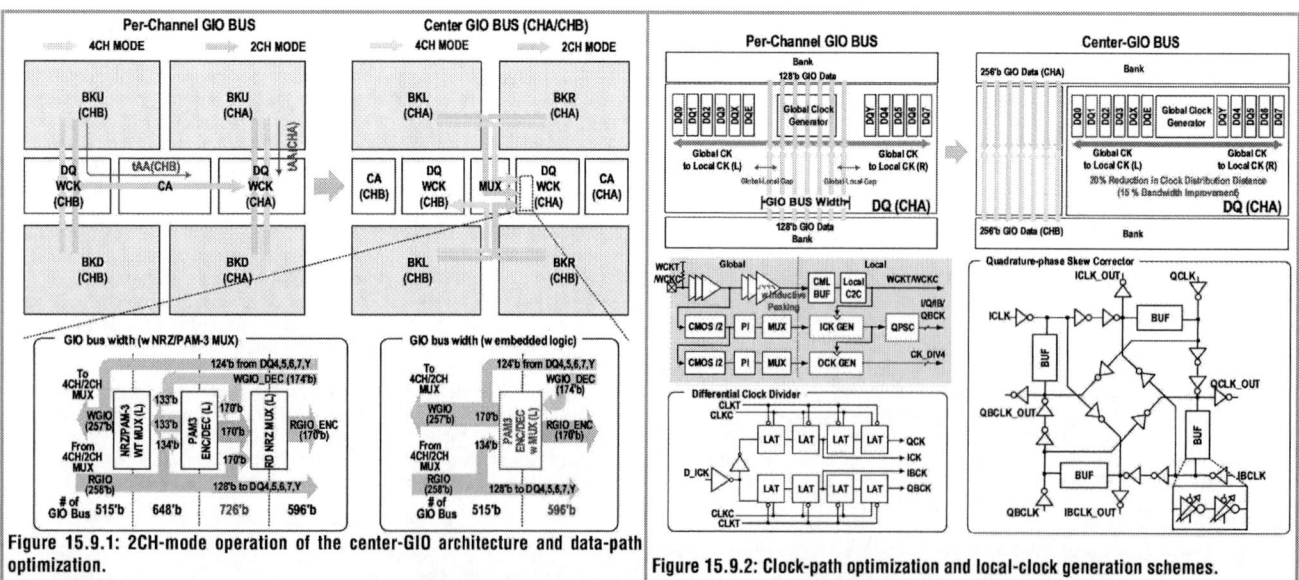

Figure 15.9.1: 2CH-mode operation of the center-GIO architecture and data-path optimization.

Figure 15.9.2: Clock-path optimization and local-clock generation schemes.

ISSCC 2026 / SESSION 15 / DRAM, SRAM, AND NON-VOLATILE MEMORIES / 15.9

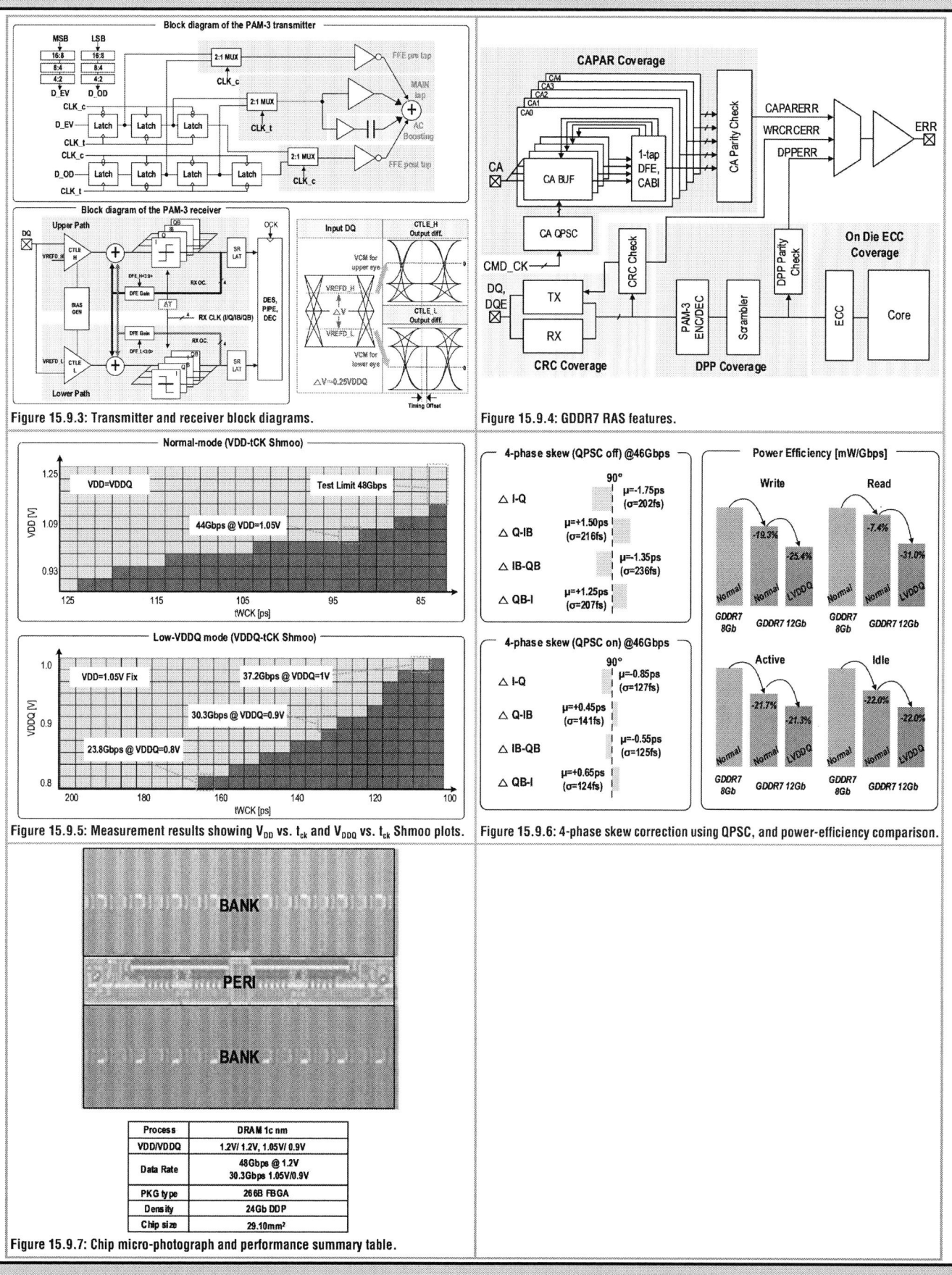

Figure 15.9.3: Transmitter and receiver block diagrams.

Figure 15.9.4: GDDR7 RAS features.

Figure 15.9.5: Measurement results showing V_{DD} vs. t_{ck} and V_{DDQ} vs. t_{ck} Shmoo plots.

Figure 15.9.6: 4-phase skew correction using QPSC, and power-efficiency comparison.

Figure 15.9.7: Chip micro-photograph and performance summary table.

• 2026 IEEE International Solid-State Circuits Conference

979-8-3315-8937-0/26 $31.00 © 2026 IEEE

ISSCC 2026 / SESSION 15 / DRAM, SRAM, AND NON-VOLATILE MEMORIES / 15.10

15.10 A Vertical-Cell-Transistor-Based 4F² DRAM with Cell-on-Peripheral Architecture Using Wafer-to-Wafer Hybrid Copper Bonding

Hyunchul Yoon, Youngseok Park, Tae Jin Park, Suk Lae Kim, Seungjae Jung, Daesun Kim, Kyunghwan Kim, Yongjun Kim, KyuChang Kang, Bokyeon Won, Sang-Hoon Jung, Seunghan Woo, Donggeon Kim, Jonghyuk Kim, In Jung, Junsoo Kim, Jae-Joon Song, InCheol Nam, Young-Hun Seo, Sungsoo Yim, Jemin Park, Changsik Yoo, SangJoon Hwang

Samsung Electronics, Hwaseong, Korea

Abstract

The paper presents a 16Gb SDRAM with a vertical-channel-transistor-based 4F² cell and cell-on-peripheral (COP) architecture. Wafer-to-wafer hybrid copper bonding is applied to COP, which increases the total die count per wafer by more than 20% with the same design rules. Fabricated in a 10nm-class DRAM process, the test chip performs read and write operations successfully, opening the possibility for further DRAM-technology scaling.

The scalability of dynamic random-access memory (DRAM), which integrates capacitors and transistors within a 2-dimensional environment, is facing physical limitations. On the other hand, industry trends consistently demand higher density and bandwidth at a reduced energy-per-bit. To address these challenges, new DRAM architectures are being actively investigated [1,2,3]. In particular, the vertical channel transistor (VCT), in which the source and drain are separated along different planes in a vertical configuration, has been investigated for the DRAM cell [1]. However, the device is fabricated in a monolithic process, hence it is difficult to form well-defined buried bit-line (BL); VCT also suffer from the floating-body effect [2]. In addition, recent research has proposed a 4F² DRAM cell using VCT and wafer-bonding techniques for next-generation DRAM to overcome monolithic-integration limitations [3]. However, the vertical interface to form inter-wafer contacts requires additional area, since each VCT's BL and WL must be connect to the BL sense amplifier (BLSA), which detects the data polarity stored in a DRAM cell, and a sub-WL driver (SWD) to drive the WL, following the wafer-to-wafer bonding process. Meanwhile, as Moore's law slows down the application of heterogeneous 3D integrated circuits is being developed with hybrid copper bonding (HCB). The wafer-to-wafer HCB process has already been commercialized for CMOS image sensor and NAND Flash products [4,5]. HCB is a good alternative for VCT DRAM, which requires wafer bonding techniques.

This paper proposes a new DRAM architecture to replace conventional DRAM for memory-process scaling extension. With sub-10nm VCT, wafer-to-wafer HCB achieves a gross die gain of over 20%, compared to a conventional architecture, by separating the cell and the peripheral circuits (PERI) area, via a cell-over-PERI (COP) architecture. With the adoption of a vertical cell structure, the parasitic capacitance of the BL is reduced. In addition, improved design flexibility for the BLSA and SWD allows for the number of cells on each WL and BL to be freely adjusted, resulting in a 10% and 30% reduction in bank active-precharge current I_{DD0} and burst refresh current I_{DD5}, without compromising additional area.

Device scaling inherently reduces the capacitance of the charge-storage capacitor (C_S), making reliable data retention challenging. Furthermore, as the space between adjacent cells diminishes, the row-hammer (RH) effect, which causes charge loss when adjacent WLs are frequently accessed, becomes more pronounced. Figure 15.10.1 (top-left) shows a VCT cell structure to address challenges faced in DRAM cell shrinkage. The vertical channel mitigates short-channel effects, as the gate length is vertical height of the channel, without compromising area. Moreover, its bulk-less architecture prevents charge loss through the body when adjacent cells are operating, unlike conventional structure that suffer from RH. Figure 15.10.1(bottom-right) shows a top view of the simplified 4 cells. In the VCT architecture, spatial decoupling of the storage node from the BL leads to a significant reduction in BL parasitic capacitance (C_{BL}). Figure 15.10.1(bottom-left) shows an overview of the VCT DRAM architecture, which consists of cell and PERI wafers, indicating the wafer bonding process. Figure 15.10.1 (top-right) illustrates a comparison of the overall vertical structure, highlighting the COP architecture connected via HCB. The HCB process, which fabricates cell and PERI wafers independently and bonds the two wafers via a Cu pad, can connect every BL and WL to the corresponding BLSA and SWD.

The most critical change, due the COP architecture, is the placement of the DRAM cells and PERI on different wafers, followed by the process required to bond two wafers into a single device. The major factor of COP architecture strongly depends on how effectively the DRAM core circuits, represented by BLSA and SWD, are rearranged when adopting the HCB process. Figure 15.10.2 shows a variety of arrangements using the COP architecture. Figure 15.10.2(top-left) shows the conventional architecture where the BLSA and SWD are positioned adjacent to the cell array, aligned respectively with the directions of the BL and WL. However, with a COP architecture, the core circuits can be placed above the cell array since cell and PERI are located on different wafers. Figure 15.10.2(top-right) shows two COP architectures with BLSAs and SWDs respectively overlapped on the cell array. Since the BLSA must detect small signal differences, the strict requirements for sensitivity and reliability pose significant challenges to continued scaling [6,7]. Moreover, typically BLSAs occupy area that is more than double of the SWD, thus it is advantageous to overlay the BLSAs on the cell array. Meanwhile, the SWDs also must fit within a restricted area defined by the fixed cell array length (or width) to directly connect WLs for minimum routing length, which makes it prone to short-channel effects and reliability issues due to reducing the cell

size results in a shorter transistor's length. However, it is difficult to extend SWD's area as it directly affects chip size. Therefore, it is optimal to arrange both the BLSAs and SWD circuits to maximize the utilization of the cell array area without consuming additional space. A structural approach is proposed to address the spatial limitations of the core circuitry. As shown Fig. 15.10.2(bottom-left), the windmill and checkerboard configurations arrange the orthogonal BLSAs and SWDs to avoid overlaps, whereas the zigzag placement of the core circuits results in complex connectivity with PERI. In addition, it is not possible for both the BLSAs and SWDs to overlap at the outermost edge of cell array, hence there is a significant loss in chip area. Alternatively, the sandwich structure, shown in Fig. 15.10.2(bottom-right), independently isolates the BLSAs and SWDs area, resulting in a more straightforward routing flow. Furthermore, SWDs can fully overlay with the cell array even at the edge region, and the BLSAs can overlap by up to half. Therefore, the prototype adopts the sandwich architecture.

Thanks to the COP architecture, the size of the core transistors has a negligible impact on the entire chip size, which allows for high flexibility in applying the number of cells per BL and cells per WL. Traditionally, many cells share one BL, decreasing the required number of BLSAs required; thus, improving cost effectiveness. The greater the number of cells connected to a single BL leads to a degradation of the BLSA's read/write margin, which is a major obstacle to DRAM scaling. Figure 15.10.3(top) shows the trade-off relationship between chip size and BLSA performance for different cells/BL. The conventional approach has a strong negative correlation, while the COP architecture exhibits a relationship that significantly reduced correlation compared to conventional one. Meanwhile, a smaller cell/BL ratio, without compromising additional area for BLSAs, can decrease I_{DD0} and greatly diminishes I_{DD5}, consumed by the DRAM cell's data retention. Furthermore, core circuits represented by BLSA and SWD are designed to occupy an area equivalent to the cell region. As shown in Fig. 15.10.3 (bottom), the approximately 4:1 ratio between cell and core areas enables the expansion of the core transistor's size resulting in 2× improved sensitivity and about 10 times improved on-off current ratio of BLSA's transistors.

Although the COP architecture, using HCB, offers numerous advantages, it inherently possesses certain drawbacks. The interconnection between cells and core circuits requires additional wires resulting in increasing parasitic capacitance and resistance. Figure 15.10.4(left) shows the representative additional wires of the COP architectures. In contrast to the windmill configuration, the sandwich architecture exhibits a longer WL route, while shortening the BL additional wiring to less than half of the windmill's BLSA area. In general, BLSA performance is more critical compared to SWD, thus the sandwich topology is appropriate. In addition, the face-to-face bonding of wafers requires electrical connections passing through every layer of the top and bottom wafers as shown in Fig. 15.10.4(right). It is estimated that amount of the layer consumption accounts for about 20% to 30% of each metal layer on the bottom wafer containing the PERI transistors depending on its design rules. To address this issue, circuit-level efforts are required to reduce the number of interconnects needed to access of cell array. Figure 15.10.5 illustrates circuit techniques used to minimize the number of routes required for cell array access. First, to reduce the input signals of the SWD for WL driving, the number of input signals can be minimized by 75%, as shown in Fig 15.10.5(top). While additional transistors are required for the NOR-type SWD topology, the 4× increase in SWD area within the COP architecture is sufficient to accommodate the placement of additional transistors. Next, as shown Fig. 15.10.5(bottom), the technique achieves a 50% reduction in the number of column address select lines (CSL) by using 2:1-column multiplexers located within the enlarged core area: using the LSB of the column address to choose between two CSLs. However, a series-connected 2:1 multiplexers increase the resistance of the corresponding signal path, leading to degradation of read/write performance. Nonetheless, these technologies are essential for obtaining a sufficient back-end-of-line (BEOL) budget required for HCB.

The prototype DRAM is fabricated in a sub-10nm DRAM process. The array size is 1296cell/BL×1056cell/WL, corresponding to a typical array configuration for commercial DRAM products, see Fig. 15.10.7. A performance variations when data is written to the cell, at different temperatures, are shown in Fig 15.10.6(top). As for conventional DRAM cells, the number of failing bits significantly increases at -25° compared to 95°C, however, this failure level is manageable and repairable. The measured failed bit count per chip versus data retention time is shown Fig. 15.10.6(bottom). In general, 1s tend to be more vulnerable to off-state leakage, compared to 0s in conventional DRAMs; whereas, for VCT 0s are more susceptible due to the floating-body effect. Nevertheless, the overall quantity of failed bits versus retention time shows comparable results to that of a conventional DRAM. These measurement results prove that VCT- based DRAM cells with sub-10nm scaling are competitive in comparison to conventional DRAM; indicating the possibility of replacing conventional DRAM for memory process scaling extension.

References:

[1] K, Kim, "From the future Si technology perspective: Challenges and opportunities," *IEEE IEDM*, pp. 1.1.1-1.1.9, 2010.
https://doi.org/10.1109/IEDM.2010.5703274
[2] H. Chung et al., "Novel 4F2 DRAM cell with Vertical Pillar Transistor (VPT)," *ESSDERC*, pp. 211-214, 2011.
https://doi.org/10.1109/ESSDERC.2011.6044197
[3] J. Park et al., "4F2 DRAM Integration with Vertical Gate (VG) Cell Transistor and Peri-Under-Cell (PUC) Architecture," *IEEE Symp. VLSI Tech. & Circuits*, 2025.
https://doi.org/10.23919/VLSITechnologyandCir65189.2025.11075066
[4] Y. Ouyang et al., "Excellent Reliability of Xtacking™ Bonding Interface," *IEEE IRPS*, 2021. https://doi.org/10.1109/IRPS46558.2021.9405115
[5] Y. Kagawa et al., "Novel stacked CMOS image sensor with advanced Cu2Cu hybrid bonding," *IEEE IEDM*, pp. 8.4.1-8.4.4, 2016. https://doi.org/10.1109/IEDM.2016.7838375
[6] K. Nam.et al., "An Offset Compensated Charge Transfer Pre-Sensing Bitline Sense Amplifier," *IEEE JSSC*, vol. 60, no. 4, pp. 1359-1367, April 2025.
https://doi.org/10.1109/JSSC.2025.3531904
[7] C. Lee et al., "A Single-Ended Offset-Compensating Bit-Line Sense-Amplifier With Ground Precharge and Charge Transfer Pre Sensing for Sub-1V DRAM," *IEEE Solid-State Circuits Letters*, vol. 8, pp. 145-148, 2025. https://doi.org/10.1109/LSSC.2025.3561280

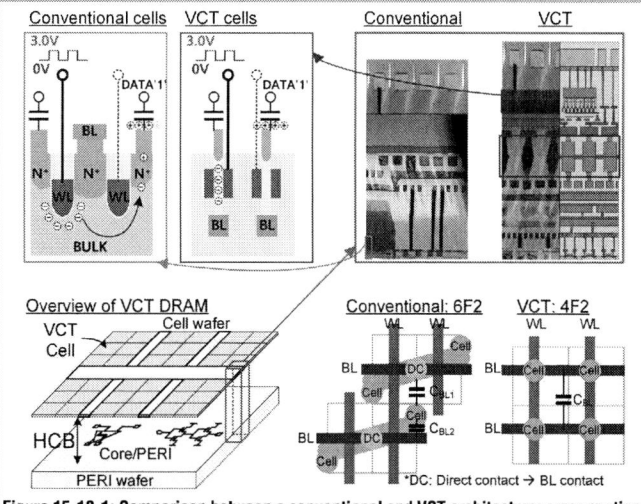

Figure 15.10.1: Comparison between a conventional and VCT architecture: cross section of unit cells (top-left), full integration (top-right), and top view of unit cells (bottom-right).

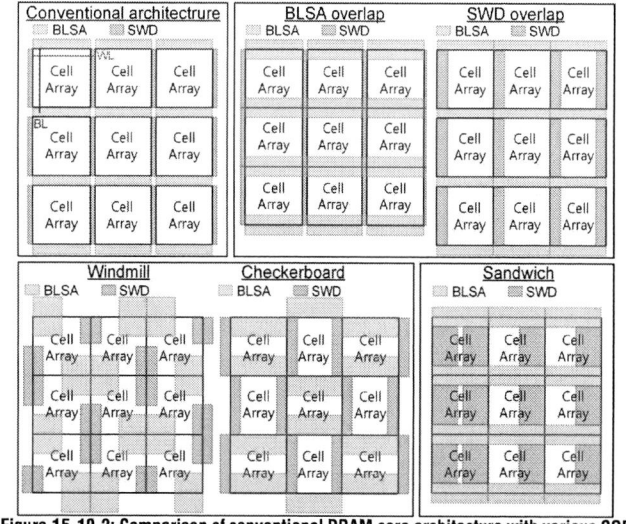

Figure 15.10.2: Comparison of conventional DRAM core architecture with various COP architectures.

ISSCC 2026 / SESSION 15 / DRAM, SRAM, AND NON-VOLATILE MEMORIES / 15.10

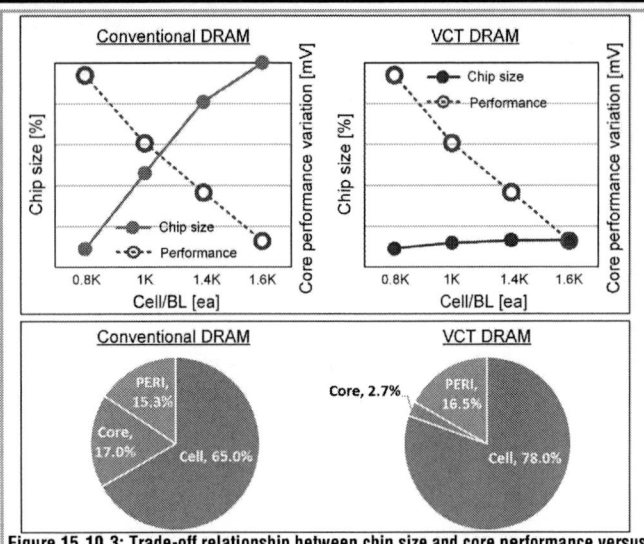

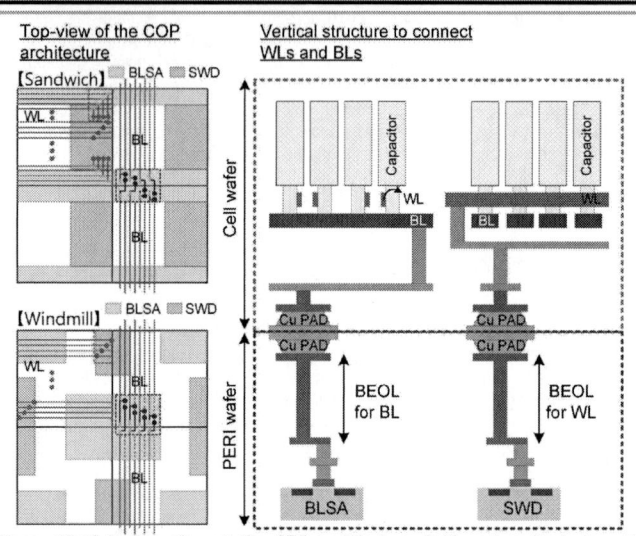

Figure 15.10.3: Trade-off relationship between chip size and core performance versus the different cell/BL (top) and comparison of the chip size portion (bottom).

Figure 15.10.4: Top view of the COP architecture (left) and vertical view of interconnection between top and bottom wafer using HCB (right).

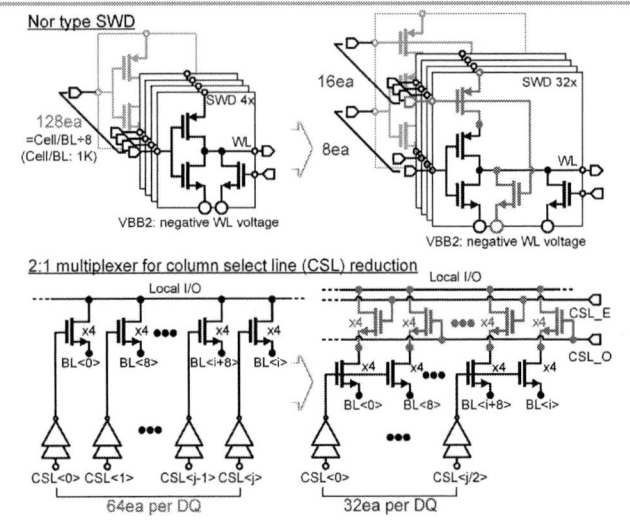

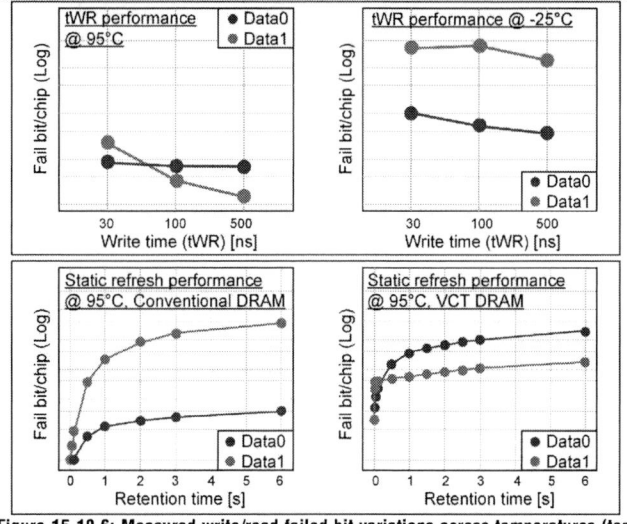

Figure 15.10.5: Circuit technology to reduce interconnection on array: NOR type SWD scheme (top) and 2:1 multiplexer for column-select reduction (bottom).

Figure 15.10.6: Measured write/read failed bit variations across temperatures (top) and static refresh failed bits for conventional and VCT DRAM (bottom).

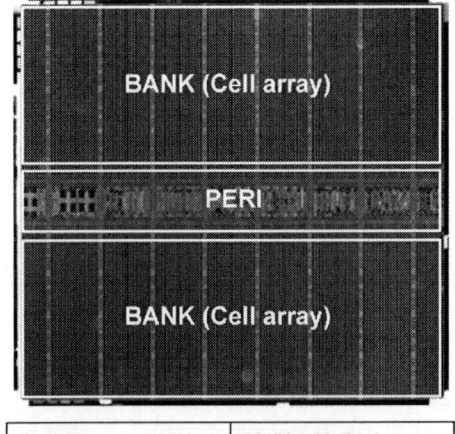

Density	16 Gb, 32 Bank
Cell/BL per unit array	1296 Cell/BL
Cell/WL per unit array	1024 Cell/WL

Figure 15.10.7: Chip micro photograph and array configuration summary.

- 2026 IEEE International Solid-State Circuits Conference

ISSCC 2026 / SESSION 16 / ENERGY HARVESTING, PIEZO AND CHARGERS / OVERVIEW

Session 16 Overview: *Energy Harvesting, Piezo and Chargers*
POWER MANAGEMENT SUBCOMMITTEE

Session Chair: Sung-Wan Hong
Sogang University, Seoul, Korea

Session Co-Chair: Kousuke Miyaji
Shinshu University, Nagano, Japan

Recent innovations in energy-harvesting and charging circuits are driving higher efficiency and autonomy in compact IoT, wearable, and biomedical systems. Advances in photovoltaic, piezoelectric, thermoelectric, and RF harvesters have enabled broader operating ranges and improved robustness through smarter power management. This session highlights ten papers presenting new circuit topologies, MPPT schemes, and energy-recycling techniques that minimize losses and silicon area. Collectively, these works illustrate the progress toward fully self-powered and highly integrated energy solutions.

8:00 AM

16.1 PV Energy-Harvesting Interface Using Reconfigurable Self-Clamp CSCR Converter Achieving 3.83× High-Efficiency VCR Ratio and Open-Voltage-Sense-Free MPPT

Ziyang Zhong, University of Macau, Macau, China

In Paper 16.1, University of Macau presents a self-clamp CSCR switched-capacitor converter for PV energy harvesting, reducing capacitor stress and achieving 17.1mW/mm² power density with 90% peak PCE. A reconfigurable self-clamp extends the efficient VCR ratio to 3.83 using 13 flying capacitors, and improves accuracy of a Sensor-Free OCV MPPT under V_{IN} variations.

8:25 AM

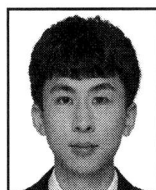

16.2 A Bias-Flip-Based Piezoelectric Energy Harvesting Interface with a Digital Track-and-Lock MPPT Achieving Sampling-Free Operation and 99.8% MPPT Efficiency

Rui Zhang, Sun Yat-Sen University, Guangzhou, China

In Paper 16.2, Sun Yat-Sen University and Chinese University of Hong Kong present a fully digital MPPT for bias-flip PEH that stores and locks MPP parameters to maintain optimal operation without repeated tracking even under varying or missing vibrations. By reusing the ZCD signal without dedicated sampling, it achieves 99.8% peak efficiency with 17nA quiescent current.

8:50 AM

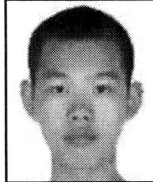

16.3 A Fully Integrated Piezoelectric Energy-Harvesting Interface with Single-Stage Bias-Flip and MPPT Achieving 5.63× Maximum Output Power Improving Rate

Xiaonan Wu, University of Science and Technology of China, Hefei, China

In Paper 16.3, University of Science and Technology of China presents a fully integrated PEH interface that boosts output power with minimal on-chip capacitor area. By reusing the piezoelectric and rectifier capacitors as flying capacitors and employing dense MOS+MIM as DC capacitors, parasitic losses are removed, achieving 5.63× power improvement and 0.09mm²/nF area.

274 • 2026 IEEE International Solid-State Circuits Conference

979-8-3315-8937-0/26 $31.00 © 2026 IEEE

ISSCC 2026 / February 17, 2026 / 8:00 AM

9:15 AM

16.4 A One-Stage Bidirectional Rectifier with Pre-Charge-Based MPPT for Triboelectric Energy Harvesting with 93% MPPT Efficiency and 8.86× Power Enhancement

Wenyu Peng, Delft University of Technology, Delft, The Netherlands

In Paper 16.4, Delft University of Technology presents a one-stage bidirectional rectifier with a V_{PC}-based P&O MPPT for triboelectric energy harvesting. By accounting for circuit non-idealities, the V_{PC}-based MPPT accurately tracks the true MPP with a simpler structure, achieving 93% MPPT efficiency and 8.86× higher harvested energy than a full-bridge rectifier.

9:30 AM

16.5 A Single-Inductor Multi-Channel Thermoelectric Energy Harvesting Interface Realizing Uneven Temperature MPPT with 39.6% Efficiency Enhancement and 62mV Tapped-Inductor-Oscillator-Based Start-Up

Yunzhe Yang, Delft University of Technology, Delft, The Netherlands

In Paper 16.5, Delft University of Technology presents a multi-channel TEG EH system to handle uneven temperatures across multiple TEGs, improving MPPT efficiency compared to conventional series or parallel connections. The system employs a tapped-inductor oscillator for start-up, achieving operation from only 62mV without extra power switches or off-chip components.

10:05 AM

16.6 A Parameter-Free Runtime-Energy-Loss Optimizer Achieving 2.15% Error in Energy-Recycling Duty-Cycled Systems

Jianxin Yang, University of Macau, Macau, China

In Paper 16.6, University of Macau proposes a runtime E_{loss} optimizer that requires no predefined parameters for use in energy-recycling duty-cycled systems. By using a 3-level converter to recycle output-capacitance energy, it achieves <2.15% deviation from optimal E_{loss} with high robustness, 15nA I_Q, and 80% efficiency at 1µA load, reducing energy loss by up to 60%.

16

10:30 AM

16.7 A 90.7%-Efficiency Piezoelectric Resonator-Based Sigma Converter with 6-Phase 2-DoF On-Chip Regulation and Zero-Standby Sigma Mode for Transient and Output Power Enhancement

Shunmin Jiang, Delft University of Technology, Delft, The Netherlands

In Paper 16.7, Delft University of Technology presents an on-chip Sigma regulation system for precise and power-efficient control of piezoelectric-resonator-based DC–DC converters. A zero-standby sigma mode enhances transient response and output power. The PR and PR-sigma converters achieve peak efficiencies of 91.2% and 90.7%, respectively.

10:55 AM

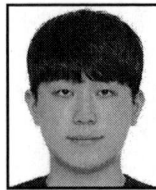

16.8 A Battery Charger Based On Mesh-Connection 2×CF Continuously-Scalable-Conversion-Ratio Converter Achieving 3.2W/mm³ Power Density

Yuanfei Wang, University of Macau, Macau, China

In Paper 16.8, University of Macau proposes a compact high-power-density charger for smart watches using a 2-capacitor CSC with a mesh-connected topology, minimizing routing loss and switch count. A CBS-free bootstrap driver and shared EA enable 92% peak efficiency, smooth CC–CV transition, and 16.8W output, making it highly suitable for low-profile wearable applications.

11:20 AM

16.9 A 96.6% Single-Mode Hybrid Dual-Path Buck-Boost Converter with Conduction Loss Reduction through Conversion-Ratio-Based Adaptive 3-Phase Control

Minsu Kim, Korea University, Seoul, Korea

In Paper 16.9, Korea University and Samsung Electronics present a 3-phase dual-path buck-boost converter to reduce conduction loss through dual-path topology and adaptive 3-phase control. The scheme minimizes inductor current ripple, enabling a low-DCR and small inductor design. Fabricated in 0.13µm, it achieves 96.6% peak efficiency using a 1.92mm³ inductor and 0402-size flying capacitors.

11:35 AM

16.10 Fully Integrated mm-Scale 5G RF MIMO Harvester with -40dBm Sensitivity and Spatial MPPT via Hybrid Transformer-Based Combining/Shifting

Andrea Ballo, University of Catania, Catania, Italy

In Paper 16.10, University of Catania and National University of Singapore propose a 28GHz MIMO RF harvester with spatial MPPT to self-align its high-gain direction to maximum power. Using transformer-based hybrid RF/DC combining and phase shifting without off-chip capacitors, the 22nm chip achieves –40dBm sensitivity and 56.7% PCE at 0dBm, 1.3× better than prior work.

DIGEST OF TECHNICAL PAPERS • 275

979-8-3315-8937-0/26 $31.00 © 2026 IEEE

ISSCC 2026 / SESSION 16 / ENERGY HARVESTING, PIEZO AND CHARGERS / 16.1

16.1 PV Energy-Harvesting Interface Using Reconfigurable Self-Clamp CSCR Converter Achieving 3.83× High-Efficiency VCR Ratio and Open-Voltage-Sense-Free MPPT

Ziyang Zhong, Rui P. Martins, Mo Huang

University of Macau, Macau, China

Abstract

This work presents a PV energy harvesting interface using a self-clamp CSCR SC converter. It reduces V_{CF} stress without additional SC stages, enabling low-voltage C_Fs, 17.1mW/mm² power density and 90% peak PCE. A reconfigurable self-clamp scheme extends the high-

PCE VCR ratio to 3.83 using only 13 C_Fs, outperforming prior works. An open-circuit MPPT scheme reuses the converter as a slope sensor, eliminating the open-voltage sensor, while reconfiguration enhances MPPT accuracy under V_{IN} variations.

Rising demand in smart cities, healthcare, and industrial automation drives the rapid growth of IoT edge sensors [1], featuring smart sensing, embedded processing, and real-time communication (Fig. 16.1.1) [2]. Self-power techniques [3], such as harvesting energy from photovoltaic (PV) panels, are essential for IoT nodes, avoiding frequent battery replacement and maintenance. Tight volume constraints on IoT nodes impose the extensive use of small Li-ion batteries (~2.8V) [4,5], while the energy harvester (EH) must adopt low-profile circuit implementations. To enable maximum power point tracking (MPPT), a DC-DC converter cascaded with the PV panel continuously tunes the input resistance (R_{IN}). With wide variations in PV voltage (V_{IN}), the converter must support a broad voltage conversion ratio (VCR) range while maintaining high power conversion ratio (PCE). Conventional inductive boost converters, limited by bulky inductors [6], are unsuitable for low-profile designs. Switched-capacitor (SC) converters offer higher energy density and the potential for full integration; however, conventional SC topologies suffer severe PCE degradation when operating away from nominal VCRs [7], conflicting with the requirement for continuous R_{IN} tuning.

Continuously Scalable-Conversion Ratio (CSCR) SC Converters [8] mitigate this trade-off, maintaining moderate PCE across a wide VCR range. These topologies employ multiple sub-cells, each comprised of a flying capacitor (C_F) and several switches. The voltage on each C_F (V_{CF}) experiences a step change of ΔV_T at the top plate and ΔV_B at the bottom plate (Fig. 16.1.1), with a full swing amplitude of the CSCR output voltage V_{OUT}. To reduce power loss, both ΔV_T and ΔV_B must be minimized, requiring many switches and C_Fs [9,10], thereby introducing a complexity-loss trade-off. In addition, the V_{OUT} stress on C_F forces the use of high-voltage integrated MOS capacitors, which exhibit much lower density (e.g., 8fF/um² and 3fF/um² for 2V and 5V capacitors, respectively). CSCR can be used as input power sensing in the perturb and observe (P&O) MPPT algorithm [11]. However, accurate sensing requires $\Delta V_T = \Delta V_B$ [4], a condition that holds only at specific VCRs. As the VCR varies, this condition is violated, thus degrading MPPT accuracy (Fig. 16.1.1).

[12] proposed a 4-rail CSCR (Fig. 16.1.1) that clamps the V_{CF} swing between V_{OUT} and ($V_{SC}-V_{IN}$), where V_{SC} is an intermediate DC node generated by a pre-stage SC network. While this approach mitigates the complexity-loss trade-off of conventional CSCRs, it still relies on high-voltage capacitors which swing between $V_{SC}-V_{IN}$ and V_{OUT}. Meanwhile, the SC pre-stage must employ large flying capacitors (C_{SC}) [12], reducing the area available for C_Fs in the CSCR stage, thus limiting output power (P_{OUT}).

This work proposes a self-clamp CSCR (Fig. 16.1.1), where each C_F also serves as a clamping capacitor in specific phases, e.g., C_{F1} and C_{F2} clamp the remaining CSCR sub-cells in phase 1 and phase 2, while subsequently returning to normal operation of CSCR. These actions form a virtual clamping voltage plane V_{clamp} throughout the full CSCR cycles. The proposed architecture preserves the benefits of [12], while reducing the V_{CF} swing between 0-to-V_{clamp} and thus enabling the use of low-voltage, high-density, on-chip MOS capacitors. Reconfigurable clamping supports adaptive adjustment of V_{clamp} to extend the VCR range with high PCE and suppresses $|\Delta V_B-\Delta V_T|$ mismatch against VCR variations to improve MPPT accuracy.

Figure 16.1.2 showcases the working principle of the proposed 13-subcell self-clamp CSCR, featuring two intermediate nodes at the top plate (V_{T1-2}) and three at the bottom plate (V_{B1-3}). Each sub-cell operates in one of four modes: B (top plate fixed, bottom plate connects virtual nodes V_{B1-B3}), T (top plate connects V_{T1-T2}, bottom plate fixed), C (clamp) and D (cornerstone). Each C_F enters a phase with a two-phase delay relative to the previous C_F, achieving out-phasing that splits each virtual node into two voltage levels [8], thereby yielding 26 effective operation phases.

In the ×2T clamp case ($V_{IN} < V_{OUT}/2$), V_{CF1} (in sub-cell 1) rises from phase 1 (B mode) and reaches $V_{OUT}/2$ at phase 13 (D mode). From phases 14 to 15 (clamp mode), C_{F1}'s top plate connects to V_{OUT} while its bottom plate connects to the top of sub-cells 2-13, thereby acting as a clamping capacitor and establishing $V_{clampT} = V_{OUT} - V_{CF1} = V_{OUT}/2$. Once C_{F1} resumes normal CSCR operation, clamping is seamlessly handed over to C_{F2}, and then to subsequent C_Fs, ensuring a continuous plane V_{clampT}.

The proposed topology supports multiple clamp configurations to optimize performance across varying input conditions (Fig. 16.1.2). When $V_{IN} > V_{OUT}/2$, the bottom plates of C_Fs sequentially connect to V_{SS} while their top plates serve as the ground nodes for other sub-cells during clamp mode. We refer to this configuration as ×2B-clamp mode, yielding a clamping plane $V_{clampB} = V_{OUT}/2$. For $V_{IN} < V_{OUT}/3$, ×3T-clamp mode further lowers V_{CF} stress by stacking two C_Fs in series and generating two planes $V_{clampT2} = 2V_{OUT}/3$ and $V_{clampT1} = V_{OUT}/3$. Simulation results (Fig. 16.1.2) show that although each clamp alone achieves high PCE within a limited VCR range, their combination extends the high-PCE range beyond that of conventional CSCR. Furthermore, like stage out-phasing in [13,14], this topology splits the virtual nodes into finer steps through the interaction of the clamping and non-clamping C_Fs (Fig. 16.1.2), referred to as clamp stage out-phasing (CSO). Specifically, the ×2B clamp splits V_T nodes, while the ×2T and ×3T clamps split V_B nodes. Compared to the results without using CSO (Fig. 16.1.2), this out-phasing improves PCE across the entire VCR range.

For MPPT, this work adopts an open-circuit method, offering lower circuit complexity than P&O method. To sample the open-circuit voltage, conventional open-circuit MPPT requires a high-resolution ADC or SC sampler [15,16], which is eliminated here (Fig. 16.1.3). The output power of a PV panel (P_{PV1}) under a given light intensity (LI) follows a high-order polynomial expressed in eq(1) [17], where the coefficients a_1-d_1 correspond to LI_1. At low V_{IN}, eq(1) simplifies to $P_{PV1} \approx d_1 V_{IN}$, with d_1 denoting the asymptotic slope. The MPP voltage for LI_1 (V_{MPP1}) occurs at the intersection of P_{PV1} with a reference line of slope 0.9×d_1, requiring accurate sense of d_1. We reuse the CSCR itself as the sensor. Its input power (P_{IN}) is written in eq(2) [4], where k is a proportional constant, f_{CLK} is the switching frequency, N_{ph} is the number of phases/cycle, and ΔV_{CF} is the voltage swing across C_F. P_{IN} reduces to $kf_{CLK}V_{IN}$ when $\Delta V_T = \Delta V_B$ (ideal conditions). By adjusting f_{CLK} to f_{REF1} such that P_{IN} intersects P_{PV1} at a low voltage V_{REF}, we achieve $d_1 = kf_{REF1}$. Then, the target MPP frequency is set as $f_{MPP1} = 0.9×f_{REF1}$ (Fig. 16.1.3).

In practice, $\Delta V_B-\Delta V_T$ varies with V_{IN} in conventional CSCRs, degrading MPPT accuracy. The proposed reconfigurable self-clamp mitigates the error, expressed as $|\Delta V_B-\Delta V_T|/\Delta V_{CF}$. Adapting the clamping ratio to VCR reduces this error greatly compared to conventional designs (Fig. 16.1.3), maintaining high accuracy across a wide V_{IN} range.

Figure 16.1.3 plots the operating waveforms of MPPT. When LI_1 transitions to LI_2, f_{CLK} initially remains at f_{MPP1}, causing V_{IN} to deviate from the new MPP voltage V_{MPP2}. A new regulation adjusts V_{IN} back to the same V_{REF}, sensing f_{REF2} for LI_2. Subsequently, open-loop operation resumes with f_{CLK} tuned to $f_{MPP2} = 0.9×f_{REF2}$. This regulation is periodically triggered every 2s.

Figure 16.1.3 presents the schematic of the prototype self-clamp CSCR which is comprised of 13 sub-cells. In addition to the switches required for conventional CSCR operation, auxiliary switches (S_{CT1-3} and S_{CB1-3}) are used to implement the three clamping voltage planes. V_{IN} regulation is achieved by integrating the error between V_{IN} and V_{REF} through an error amplifier (EA) and a PI compensator [13]. The output drives a VCO to generate f_{REF}, which is further utilized to derive the target f_{MPP}. A multiplexer selects between f_{MPP} and f_{REF}, producing f_{CLK} which generates control signals for the switches [9]. A clamp-mode selection block is included to determine the appropriate self-clamp ratio based on the VCR.

Figure 16.1.7 shows the prototype chip fabricated in a 180nm CMOS process, where it occupies an active area of 3.52mm². Figure 16.1.4 showcases the measured steady-state waveforms at $V_{OUT} = 2.4V$ and $V_{IN} = 0.8V$, 0.45V and 1.8V. Clamping modes of ×2T, ×3T and ×2B yield V_{CF} swings of 1.22V, 0.81V and 1.21V, respectively, with the corresponding clamping voltage planes observed as analyzed. Additional V_{CF} steps (e.g., in V_{B2}) introduced by clamp stage out-phasing are also evident, consistent with predictions. Figure 16.1.4 also presents the measured PCE versus VCR, demonstrating that the reconfigured self-clamping extends the high-PCE range, maintaining PCE >70% across a VCR span of 1.2-to-4.6, denoted as VCR_{MIN} and VCR_{MAX}.

Figure 16.1.5 displays the measured PCE versus output current (I_{OUT}) under various V_{IN} and V_{OUT} conditions, achieving a peak PCE of 90% at $V_{IN} = 2V$ and $V_{OUT} = 2.4V$. A peak P_{OUT} of

60.2mW is obtained at V_{IN} = 1.8V and V_{OUT} = 3.8V. During the LI_1-to-LI_2 transition, the initial V_{IN} is V_{MPP1} = 940mV, which is then regulated to V_{REF} = 400mV for around 5ms in order to determine f_{REF2}. Subsequently, f_{MPP2} is derived and applied, resulting in V_{MPP2} = 990mV. Figure 16.1.5 compares the measured MPPT accuracy of the fixed ×2T clamp with the proposed reconfigurable clamp scheme. The reconfiguration achieves >96.7% accuracy across 2k-40k lux light intensity, outperforming the fixed-mode clamp due to the reduced $|\Delta V_B - \Delta V_T|$ variations across VCRs.

Figure 16.1.6 benchmarks this work against prior step-up CSCR designs. The proposed design reduces MPPT complexity through an open-circuit method, reusing CSCR as a slope sensor. Reconfigurable clamping reduces $|\Delta V_B - \Delta V_T|$ variation across VCRs, attaining >96.7% MPPT accuracy across wide light intensity range. Moreover, the self-clamp method enables a comparable peak PCE at a higher V_{OUT} of 3.8V with only 13× C_F (N = 13). By lowering V_{CF} stress to $V_{OUT}/2$ or $V_{OUT}/3$, this work allows the use of low-voltage MOS capacitors, achieving the highest power density among compared works. The reconfigurable self-clamp extends the high-PCE (e.g., >70%) VCR range, achieving a VCR ratio (= VCR_{MAX}/VCR_{MIN}) of 3.83. Since this ratio scales with N for ideal CSCR designs (Fig. 16.1.6), a normalized VCR ratio (raw ratio divided by the coefficient in eq(3) [9]) is adopted for fair comparison, where this work advances prior arts.

Acknowledgement:
This work was supported in part by the Science and Technology Development Fund, Macau SAR (0029/2025/AMJ, 0042/2025/RIB1, 004/2023/SKL and 001/2024/COP), in part by the Guangdong-HongKong-Macao Joint Laboratories (2025B1212150003), in part by the Guangdong Basic and Applied Basic Research Foundation (2023B1515130001), in part by the Research Committee of University of Macau (MYRG-GRG2024-00041-IME).

References:
[1] Z. Chen et al., "An Energy Harvesting Interface Based on Reconfigurable Piezoelectric Harvester Array," in IEEE Transactions on Power Electronics, vol. 40, no. 10, pp. 15949-15958, Oct. 2025. https://doi.org/10.1109/TPEL.2025.3577696
[2] Y. -S. Noh, J. -I. Seo, H. -S. Kim and S. -G. Lee, "A Reconfigurable DC-DC Converter for Maximum Thermoelectric Energy Harvesting in a Battery-Powered Duty-Cycling Wireless Sensor Node," in IEEE Journal of Solid-State Circuits, vol. 57, no. 9, pp. 2719-2730, Sept. 2022. https://doi.org/10.1109/JSSC.2022.3152261
[3] M. Pathak, C. Huang and R. Kumar, "Start-up Circuit for Synchronous Switched Energy Extraction from Triboelectric Energy Harvesters," 2023 IEEE 16th Dallas Circuits and Systems Conference (DCAS), Denton, TX, USA, 2023, pp. 1-2. https://doi.org/10.1109/DCAS57389.2023.10130239
[4] M. Kim, M. Jeong, D. Seo, Y. Lee and I. Lee, "A Continuously-Scalable-Conversion-Ratio SC Energy-Harvesting Interface with Exponential DCO for Wide Range Output Power Tracking," 2024 IEEE European Solid-State Electronics Research Conference (ESSERC), Bruges, Belgium, 2024, pp. 548-551. https://doi.org/10.1109/ESSERC62670.2024.10719400
[5] Y. Wang, M. Huang, R. P. Martins and Y. Lu, "A SIDO/DISO VCF-Step-Reconfigurable Continuously Scalable-Conversion-Ratio SC Converter Achieving 91.4%/92.6% Peak Efficiency and Almost-lossless Channel Switching," 2024 IEEE International Solid-State Circuits Conference (ISSCC), San Francisco, CA, USA, 2024, pp. 506-508. https://doi.org/10.1109/ISSCC49657.2024.10454453

[6] S. Bandyopadhyay, P. P. Mercier, A. C. Lysaght, K. M. Stankovic and A. P. Chandrakasan, "A 1.1 nW Energy-Harvesting System with 544 pW Quiescent Power for Next-Generation Implants," in IEEE Journal of Solid-State Circuits, vol. 49, no. 12, pp. 2812-2824, Dec. 2014. https://doi.org/10.1109/JSSC.2014.2350260
[7] H. -P. Le, S. R. Sanders and E. Alon, "Design Techniques for Fully Integrated Switched-Capacitor DC-DC Converters," in IEEE Journal of Solid-State Circuits, vol. 46, no. 9, pp. 2120-2131, Sept. 2011. https://doi.org/10.1109/JSSC.2011.2159054
[8] N. Butzen and M. Steyaert, "Design of Single-Topology Continuously Scalable-Conversion-Ratio Switched- Capacitor DC–DC Converters," in IEEE Journal of Solid-State Circuits, vol. 54, no. 4, pp. 1039-1047, April 2019. https://doi.org/10.1109/JSSC.2018.2884351
[9] Y. Wang, M. Huang, Q. Chen, R. P. Martins and Y. Lu, "A VCF-Step-Reconfigurable Continuously Scalable-Conversion-Ratio Switched- Capacitor Converter," in IEEE Journal of Solid-State Circuits, vol. 60, no. 2, pp. 626-637, Feb. 2025. https://doi.org/10.1109/JSSC.2024.3414448
[10] Y. Yang, et al, "Matryoshka CSCR: A Reconfigurable Matryoshka-Stacked Continuous-Scalable-Conversion-Ratio Switched-Capacitor DC-DC Converter with 0.1-to-1.7V Input," 2025 Symposium on VLSI Technology and Circuits (VLSI Technology and Circuits), Kyoto, Japan, 2025, pp. 1-3. https://doi.org/10.23919/VLSITechnologyandCir65189.2025.11075046
[11] Y. Yoon et al., "A Continuously-Scalable-Conversion-Ratio Step-Up/Down SC Energy-Harvesting Interface With MPPT Enabled by Real-Time Power Monitoring With Frequency-Mapped Capacitor DAC," in IEEE Transactions on Circuits and Systems I: Regular Papers, vol. 69, no. 4, pp. 1820-1831, April 2022. https://doi.org/10.1109/TCSI.2021.3139708
[12] A. Guo, W. Peng, Y. Yang, X. Hu, D. Muratore and S. Du, "A Fully Integrated SC Converter Hybridizing Dickson and Continuously-Scalable-Conversion-Ratio Topologies with a Wide Bipolar VCR Range for Energy Harvesting," 2025 Symposium on VLSI Technology and Circuits (VLSI Technology and Circuits), Kyoto, Japan, 2025, pp. 1-3. https://doi.org/10.23919/VLSITechnologyandCir65189.2025.11074934
[13] Y. Wang, Z. Zhang, Z. Zhong, Y. Zhang, R. P. Martins and M. Huang, "An SC-first Hybrid SCVR with 4xCF Continuously Scalable-Conversion Ratio SC Achieving 92.5% Peak Efficiency," 2025 IEEE Custom Integrated Circuits Conference (CICC), Boston, MA, USA, 2025, pp. 1-3. https://doi.org/10.1109/CICC63670.2025.10983039
[14] N. Butzen et al., "A Monolithic 12.7 W/mm², 92% Peak-Efficiency Switched-Capacitor DC-DC Converter Using CSCR-First Topology," in IEEE Journal of Solid-State Circuits, vol. 59, no. 12, pp. 4114-4123, Dec. 2024. https://doi.org/10.1109/JSSC.2024.3465388
[15] Q. Fang, F. Li, R. P. Martins and M. -K. Law, "A 91.25% Peak Power-Conversion-Efficiency Capacitive Power-Management IC Supporting up to 5.68mJ Burst Energy Delivery Using a Single External Capacitor for mm-Scale IoT Applications," 2025 IEEE International Solid-State Circuits Conference (ISSCC), San Francisco, CA, USA, 2025, pp. 524-526. https://doi.org/10.1109/ISSCC49661.2025.10904683
[16] P. -C. Huang and T. -H. Kuo, "A 100-pA Adaptive-FOCV MPPT Circuit with >99.6% Tracking Efficiency for Indoor Light Energy Harvesting," 2019 IEEE Asian Solid-State Circuits Conference (A-SSCC), Macau, Macao, 2019, pp. 185-188. https://doi.org/10.1109/A-SSCC47793.2019.9056929
[17] T. Nukala and A. K. Panchal, "Maximum Power Point Tracking of PV Module Based on New Explicit I-V Relation," 2017 IEEE 44th Photovoltaic Specialist Conference (PVSC), Washington, DC, USA, 2017, pp. 3061-3066. https://doi.org/10.1109/PVSC.2017.8366697

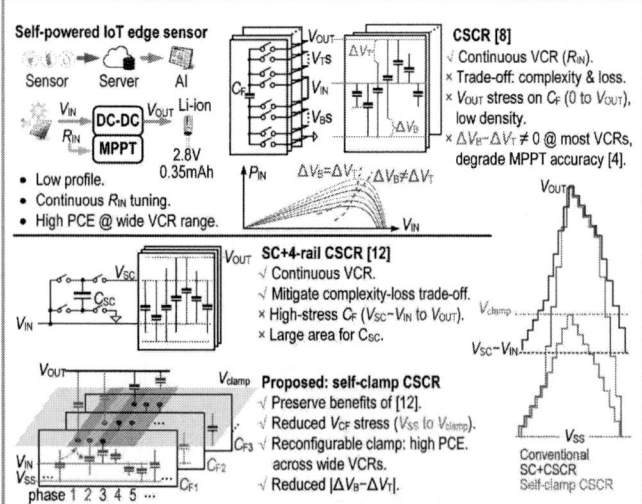

Figure 16.1.1: Motivations, and drawbacks of conventional CSCR in [8] (top), partially addressed by SC+4-rail CSCR in [12] (middle), and proposed scheme (bottom).

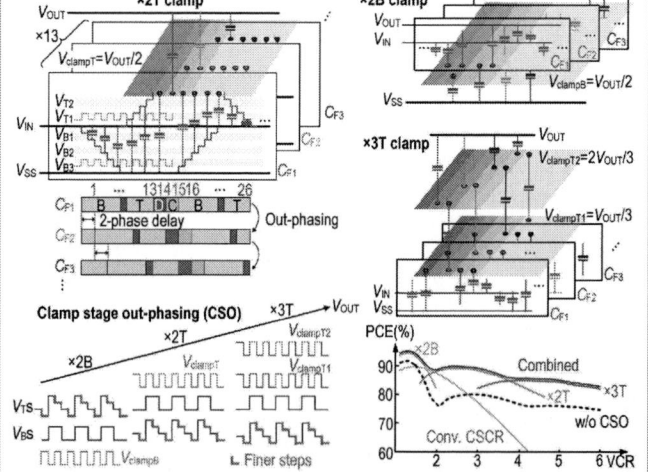

Figure 16.1.2: Operating principles ×2T clamp mode (left top), waveforms of clamp stage out-phasing (left bottom), ×2B clamp and ×3T clamp modes (right top), simulated PCE vs. VCR (right bottom).

ISSCC 2026 / SESSION 16 / ENERGY HARVESTING, PIEZO AND CHARGERS / 16.1

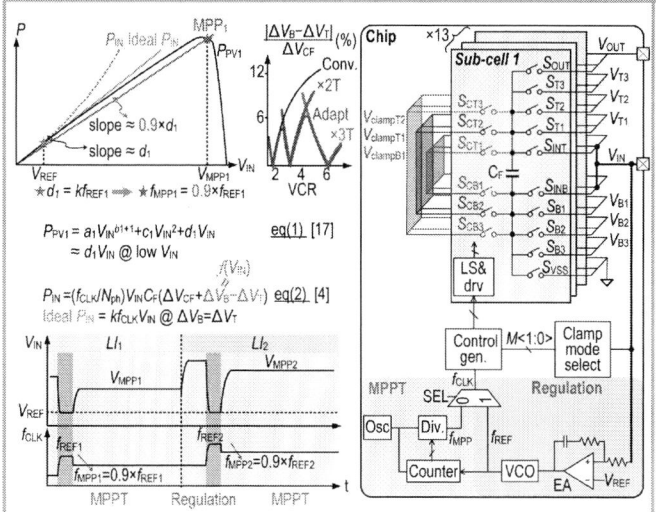

Figure 16.1.3: Proposed MPPT method and calculated $|\Delta V_B - \Delta V_T|$ error vs. VCR (left top), MPPT waveforms in LI transitions (left bottom), schematic (right).

Figure 16.1.4: Measured steady-state waveforms at three clamp modes (top and bottom left), measured PCE versus V_{IN} at V_{OUT}=2.4V (bottom right).

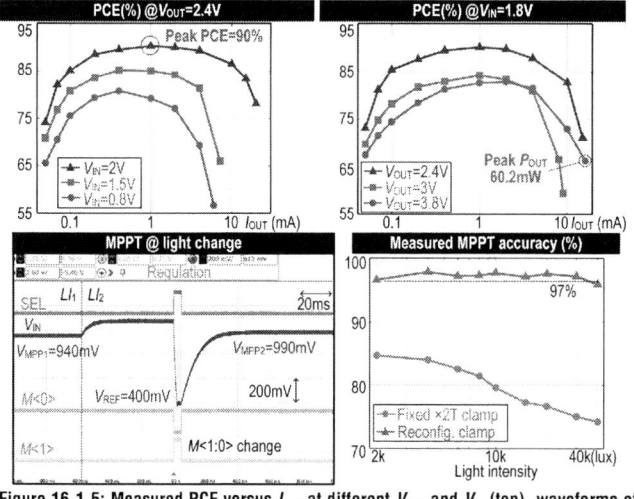

Figure 16.1.5: Measured PCE versus I_{OUT} at different V_{OUT} and V_{IN} (top), waveforms of MPP Tracking (bottom left), MPPT accuracy at fixed and reconfigurable clamp modes (bottom right).

	[11] TCAS-I22	[4] ESSERC24	[9] JSSC25	[12] VLSI25	[15] ISSCC25	This work
Technology	28nm	180nm	65nm	180nm	180nm	180nm
Chip area (mm²)	4.8	7.6	4.76	3.57	4.62	3.52
Topology	CSCR	CSCR	Reconfig. CSCR	Reconfig. SC + CSCR	SC + Reconfig. CSCR	Reconfig. Self-clamp CSCR
Buck or boost	Buck-boost*	Boost	Buck-boost*	Boost	Buck-boost*	Boost
V_{IN} (V)	0.65 - 1.5	0.9 - 1.8	0.4 - 2	0.13 - 0.54	1 - 1.8	0.4 - 2
V_{OUT} (V)	1 - 2	3.6	0.5 - 2	1.2	1.8	2.4 - 3.8
MPPT method	Perturb & Observe	Perturb & Observe	Open-circuit	N.A.	Open-circuit	Open-circuit, reuse CSCR as sensor
MPPT accuracy	> 96.9%	> 96.6%**	N.A.	> 95%**	N.A.	> 96.7%
Total flying cap (nF)	14.88	19.66	19.8	12	13.6	15.6
V_{CF} stress	0 to V_{OUT}	0 to V_{OUT}	0 to V_{OUT}	(V_{SC}-V_{IN}) to V_{OUT}	0 to V_{OUT}	0 to (V_{OUT}/2 or /3)
C_F count (N)	48×	96×	33×	17×	13×	13×
Peak PCE	88.9%	91.5%	90%	84%	90.5%**	90%
Max. P_{OUT} (mW)	2.794	6.056	39.7	10.8	36**	60.2
Power density (mW/mm²)	0.582	0.797	8.34	3.025	7.79	17.1
70%-PCE VCR ratio	3.08	2	2.67**	2.87**	1.8	3.83
Normalized VCR ratio***	0.17	0.061	0.196	0.326	0.237	0.504

* Only boost performance.   ** Estimated from figures.

$$ \text{*** Normalized VCR ratio} = \frac{\text{VCR ratio}}{(N-1)(1-\text{PCE})+4} \quad [9] \quad eq(3) $$

Figure 16.1.6: Performance comparison with state-of-the-art works.

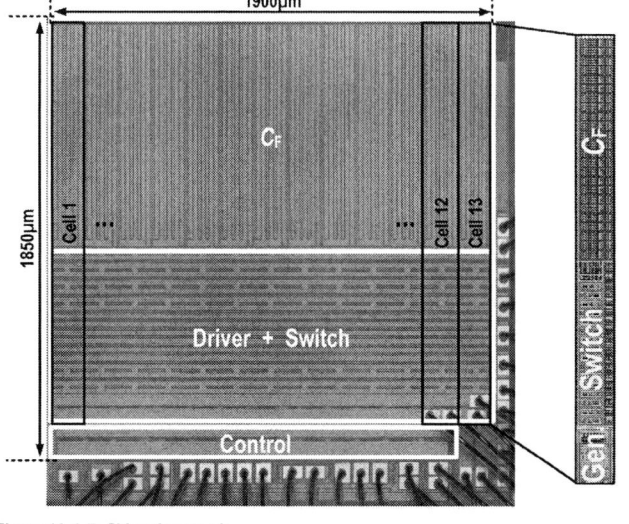

Figure 16.1.7: Chip micrograph.

ISSCC 2026 / SESSION 16 / ENERGY HARVESTING, PIEZO AND CHARGERS / 16.2

16.2 A Bias-Flip-Based Piezoelectric Energy Harvesting Interface with a Digital Track-and-Lock MPPT Achieving Sampling-Free Operation and 99.8% MPPT Efficiency

Rui Zhang[1], Chuhui Wang[2], Zhongxin Ye[1], Ka Nang Leung[2], Jianping Guo[1]

[1]Sun Yat-Sen University, Guangzhou, China, [2]Chinese University of Hong Kong, Hong Kong, China

Abstract

A fully digital MPPT is proposed for a bias-flip piezoelectric energy harvester. Once the MPP is identified, its parameters are stored and locked. Even if the vibration amplitude or frequency changes or disappears entirely, it directly operates at the pre-recorded MPP state. There is no need for repeated tracking, thus ensuring continuous maximum power output.

Piezoelectric energy harvesters (PEHs) are perceived as a vital battery-substituted power-supply solution for self-sustainable wireless sensor nodes (WSNs) [1]. A rectifier is required to extract energy from the vibrating piezoelectric transducer (PT), where the vibration amplitude and frequency are not fixed as the ambient environment changes during operation, as shown in Fig. 16.2.1 (top-left). The output power of a PEH rectifier typically fluctuates under varying environmental and load conditions [2,3], making it challenging to ensure optimal energy extraction. A maximum power point tracking (MPPT) circuit is essential for a rectifier to achieve impedance matching, ensuring it consistently harvests the maximum power from the PT. Fractional open-circuit voltage (FOCV) [4,5], perturb & observe (P&O) [6,7], and duty-cycle-based (DCB) [8,9] are predominant MPPT techniques for PEH rectifiers. The FOCV algorithm samples the open voltage (V_{OC}) of PT, which causes the PT to be disconnected from the rectifier and wastes energy during sampling. P&O and DCB techniques are both continuous stepwise MPPT algorithms. The P&O technique enhances MPPT accuracy through an output power evaluation algorithm, which demands complex sampling and power-hungry circuits. The DCB technique is usually employed for parallel synchronized switch harvesting on inductor (P-SSHI) rectifiers, which is based on the vibration cut-off duty cycle approach. Although independent of V_{OC}, it remains sensitive to frequency variations. Additionally, continuous multiple MPPT operations are required for the P&O and DCB techniques during the vibration process.

In this paper, a P-SSHI rectifier with a digital track and lock (T&L) MPPT algorithm is proposed. The MPP is identified in the start-up process and then locked in the subsequent vibration process. The locked MPP is independent of the vibration frequency (f_{PT}) and V_{OC}, so that no further tracking is necessary after the completion of the initial tracking, even when f_{PT} and V_{OC} change. Additionally, when the vibration shuts down and restarts, proposed rectifier inherently maintains operation at the MPP. Since the digitized MPP information has already been stored through the previous operation, the need for periodic retracking is eliminated. Furthermore, the output power approach of the proposed MPPT is rectifier independent, making it compatible with various rectifiers.

Figure 16.2.1 (middle-left) shows the optimal load (R_{LOAD}) of the rectifiers based on different bias-flip techniques, i.e., P-SSHI, series SSHI (S-SSHI), and synchronized switch harvesting on capacitor (SSHC). The optimal R_{LOAD} is determined by f_{PT}, the piezoelectric capacitance (C_P), and the flipping factor defined for each approach (λ_P, λ_S, λ_C). The buck-boost converter is often connected as a load after the rectifier to achieve impedance matching. For a buck-boost converter working in discontinuous conduction mode (DCM), its input resistance is $2L/(f_{MPPT}T_H^2)$, where f_{MPPT} is the working frequency of the buck-boost converter, and T_H is the inductor magnetizing time. Assuming the input resistance equals the optimal R_{LOAD}, the optimal T_H can thus be derived. Furthermore, if f_{PT}/f_{MPPT} is kept as a constant, k, the optimal T_H becomes a constant, independent of the V_{OC} and f_{PT} input conditions. The locked T_H offers the advantage that the algorithm can adaptively lock the MPP in a single round, regardless of changes in the input conditions. In the proposed PEH system, a battery is used to store energy, whose voltage, V_{BAT}, is assumed to be constant. The output power can then be attained by detecting the output current I_{OUT}. For a buck-boost converter working in DCM, I_{OUT} can be measured by evaluating the inductor demagnetization time (T_L) when V_{BAT} is a constant. Figure 16.2.1 (bottom-right) shows different rectified voltages (V_{REC}) versus different T_Ls, which indicates that only when T_L is at its maximum, the rectifier works at the MPP, i.e., the output power is the highest.

Figure 16.2.2 (top) shows the system architecture of the proposed PEH system, including a P-SSHI rectifier, a buck-boost converter with a controller, and an MPPT controller. The P-SSHI rectifier consists of a negative voltage converter (NVC), an active diode, a bias-flip controller, and an off-chip inductor shared with the buck-boost converter. The buck-boost controller contains a T_H&T_L generator module as well as a control logic module, which provides driving signals (D_H, D_L) for the power stage. A dynamic zero-current detector (ZCD) without an external clock is used for the DC-DC switching control rather than a static comparator to reduce power consumption. Furthermore, the output signal of the ZCD (Q_{CMP}) is also used for T&L MPPT, avoiding the need to sample the output voltage or current and simplifying the MPPT circuit design. The proposed MPPT controller is divided into a track module and a lock module. In the track state, the successive approximation register (SAR) logic module is activated to search for optimal T_L values during different V_{REC} voltages. The

digital code ($Q_{SAR[7:0]}$), that corresponds to the optimal T_L is stored in the registers Q_{NOW} and Q_{LAST}. The digital comparator compares Q_{NOW} with Q_{LAST} to determine whether the MPP has reached. When $Q_{LAST} > Q_{NOW}$, this indicates that the MPP is stored in Q_{LAST}. At this point, the signal EN_L is set high, and the MPP is locked. Q_{LAST} is then transmitted to the integral logic module, and the circuit turns to the lock state, which keeps T_L at its optimum value. Figure 16.2.2 (bottom-left) shows the flowchart of the proposed T&L MPPT algorithm. When the circuit starts tracking, T_H is initialized first. Then the ZCD searches for the optimal T_L during this T_H. The 8-bit SAR logic is used to speed up the search process. After a convergence cycle from $Q_{SAR[7]}$ to $Q_{SAR[0]}$, the optimal T_L is saved to Q_{NOW}. Then, the digital comparator compares Q_{NOW} and Q_{LAST}. If $Q_{NOW} > Q_{LAST}$, this indicates that T_L is increasing and so T_H continues to decrease to track the MPP. If $Q_{NOW} < Q_{LAST}$, it indicates that the previous state corresponds to the MPP. Thus, T_H reverts to its previous state, and the tracking process is completed. According to the previous analysis, the optimal T_H is a constant, implying that the MPP is locked. Figure 16.2.2 (bottom-right) shows the control logic module for buck-boost control. The frequency divider divides the frequency of D_F by k, such that $f/f_{MPPT} = k$, which corresponds to the locking principle shown in Fig. 16.2.1 (middle-left). EN_L tunes the increment or decrement of the 4-bit counter and subsequently controls the reset signal of the DFF. Therefore, D_H is tuned by EN_L. D_L is controlled by the signal $Q_{CAP[7:0]}$, which comes from the T_H&T_L generator module.

Figure 16.2.3 (top) presents the detailed operation principle during the track and lock states. In the track state, the 8-bit SAR logic operates to rapidly converge T_L to its optimal value. In each cycle, each bit of $Q_{SAR[j]}$ is first set high and then updated by the Q_{CMP} result. Once convergence is achieved, $Q_{SAR[7:0]}$ is stored in Q_{NOW}, while the last result is stored in Q_{LAST}. In this circuit, the dynamic ZCD compares the SW node voltage with the ground after M_{N1} turns off, which can lower the requirements for comparator accuracy. If $I_L < 0$ when M_{N1} turns off, V_{SW} will rise above ground. Then Q_{CMP} will output low, causing $Q_{SAR[j]}$ to turn low; otherwise, Q_{CMP} will output high, causing $Q_{SAR[j]}$ to remain high. In the lock state, $Q_{T[7:0]}$ is first set to the final value of Q_{LAST}. Then, $Q_{T[7:0]}$ is incremented or decremented by one, based on the Q_{CMP} output. $Q_{T[7:0]}$ repeats this cycle during the steady lock state. Figure 16.2.3 (bottom-left) shows the schematic of the T_H&T_L generator, which includes a binary-weighted 8-bit capacitor array. The schematic of the dynamic ZCD is given in Fig. 16.2.3 (bottom-right), where no dc bias current is needed, and dynamic power consumption exists only during the transition periods. The ZCD not only controls D_L, but also provides the MPP signal. Consequently, this proposed algorithm does not require a power-hungry sampling circuit.

The proposed interface circuit was fabricated in a 180nm BCD process and occupies an active area of 0.42mm² (Fig. 16.2.7). Figure 16.2.4 (top) shows the proposed MPPT startup waveform. In the track state, V_{REC} increases step by step until it crosses the MPP. A 470μH/1mH off-chip inductor is shared by the P-SSHI rectifier and the buck-boost converter. Figure 16.2.4 (middle) shows the adaptive adjustments of the transient V_{OC} during the locked state, where T_H remains constant amid the transient V_{OC} variations, indicating the MPP is independent of V_{OC}. Figure 16.2.4 (bottom) illustrates how the dynamic ZCD detects I_L. Detection is achieved by reflecting whether I_L crosses zero through V_{SW}. Figure 16.2.5 (top-left) shows that even when the vibration source ceases input, the data of MPP is still stored in the integral logic. When the vibration source restarts, the MPP can be directly locked, and the rectifier directly enters MPP without retracking. Figure 16.2.5 (bottom-left) shows the MPP lock effect during different vibration frequencies, where T_H remains at 9.3μs from 28-to-68Hz, indicating the MPP is independent of f_{PT}. Figure 16.2.5 (top-right) shows the MPPT efficiency versus V_{OC} and f_{PT}, which is obtained by comparing P_{OUT} when MPPT is enabled with manually tuned maximum P_{OUT} with an 18ns T_H step. The MPPT efficiency is higher than 98% across the whole range, and the peak MPPT efficiency is 99.8%. Measurements of the output power from a full-bridge rectifier (FBR) and the proposed P-SSHI rectifier with different inductors at V_{OC} = 1.7V are shown in Fig. 16.2.5 (bottom-right). The proposed rectifier achieves a peak output power of 27.1μW, representing a 4.9× enhancement compared to an FBR (5.52μW).

Figure 16.2.6 compares the proposed T&L MPPT circuit with state-of-the-art works. The proposed rectifier only requires a single round of tracking to lock to the MPP, whereas previous works need endless tracking of the MPP during vibration cycles. By reusing the

By reusing the zero-current detector used for switching control, no dedicated sampling operation is required for MPPT. The measured I_Q of the MPPT circuit is 17nA, and the peak MPPT efficiency is 99.8%.

278 • 2026 IEEE International Solid-State Circuits Conference

979-8-3315-8937-0/26 $31.00 © 2026 IEEE

ZCD signal in the buck-boost converter, this circuit achieves sampling-free, fully digital, V_{OC} and f_{PT} independent MPPT, thereby enhancing robustness and reducing power consumption. It achieves the smallest chip area of 0.42mm² compared with the state-of-the-art, and a current of only 17nA for the MPPT circuit.

Acknowledgement:
This work was partly supported by the National Natural Science Foundation of China under Grant 62274189, and the Industrial Core and Key Technique Research Project of Zhuhai under Grant 2320004002564. Corresponding author: Jianping Guo.

References:
[1] Y. K. Ramadass and A. P. Chandrakasan, "An Efficient Piezoelectric Energy Harvesting Interface Circuit Using a Bias-Flip Rectifier and Shared Inductor," in *IEEE JSSC*, vol. 45, no. 1, pp. 189-204, Jan. 2010. https://doi.org/10.1109/JSSC.2009.2034442.
[2] S. Du and A. A. Seshia, "An Inductorless Bias-Flip Rectifier for Piezoelectric Energy Harvesting," in *IEEE JSSC*, vol. 52, no. 10, pp. 2746-2757, Oct. 2017. https://doi.org/10.1109/JSSC.2017.2725959.
[3] C. Wang et al., "A Bridge-Less Hybrid SSHI Rectifier With High Efficiency Over Wide Load Range for Piezoelectric Energy Harvesting," *IEEE JSSC*, vol. 60, no. 9, pp. 3294-3304, Sept. 2025. https://doi.org/10.1109/JSSC.2024.3520175.
[4] Z. Chen et al., "A Piezoelectric Energy-Harvesting Interface Using Split-Phase Flipping Capacitor Rectifier and Capacitor Reuse Multiple-VCR SC DC-DC Achieving 9.3× Energy-Extraction Improvement," *ISSCC*, pp. 424–425, Feb. 2019. https://doi.org/10.1109/ISSCC.2019.8662323.
[5] S. Li et al., "A 32nA Fully Autonomous Multi-Input Single-Inductor Multi-Output Energy-Harvesting and Power-Management Platform with 1.2×105 Dynamic Range, Integrated MPPT, and Multi-Modal Cold Start-Up," *ISSCC*, pp. 472–473, Feb. 2022. https://doi.org/10.1109/ISSCC42614.2022.9731732.
[6] S. Li et al., "A Piezoelectric Energy-Harvesting System with Parallel-SSHI Rectifier and Integrated MPPT Achieving 417% Energy-Extraction Improvement and 97% Tracking Efficiency," *IEEE Symp. VLSI Circuits*, pp. C324–C325, June 2019. https://doi.org/10.1109/LSSC.2019.2951394.
[7] A. Morel, et al., "Self-Tunable Phase-Shifted SECE Piezoelectric Energy-Harvesting IC with a 30nW MPPT Achieving 446% Energy-Bandwidth Improvement and 94% Efficiency," *ISSCC*, pp. 488–489, Feb. 2020. https://doi.org/10.1109/ISSCC19947.2020.9062972.
[8] X. Yue et al., "A Bias-Flip Rectifier with a Duty-Cycle-Based MPPT Algorithm for Piezoelectric Energy Harvesting with 98% Peak MPPT Efficiency and 738% Energy-Extraction Enhancement," *ISSCC*, pp. 442-443, Feb. 2023. https://doi.org/10.1109/JSSC.2023.3313733.
[9] X. Yue et al., "A Single-Stage Bias-Flip Regulating Rectifier With Fully Digital Duty-Cycle-Based MPPT for Piezoelectric Energy Harvesting," *IEEE JSSC*, pp. 1–11, 2024. https://doi.org/10.1109/JSSC.2024.3495232.
[10] J. Lee et al., "A Biased-SECE Interface for Piezoelectric Energy Harvesting with Geometric-Mean-Computational MPPT Achieving 99.9% MPPT Efficiency, 8.75Cycles/ΔVOC Tracking, and 9.3x Energy Extraction," *ISSCC*, 2025, pp. 1-3. https://doi.org/10.1109/ISSCC49661.2025.10904519.

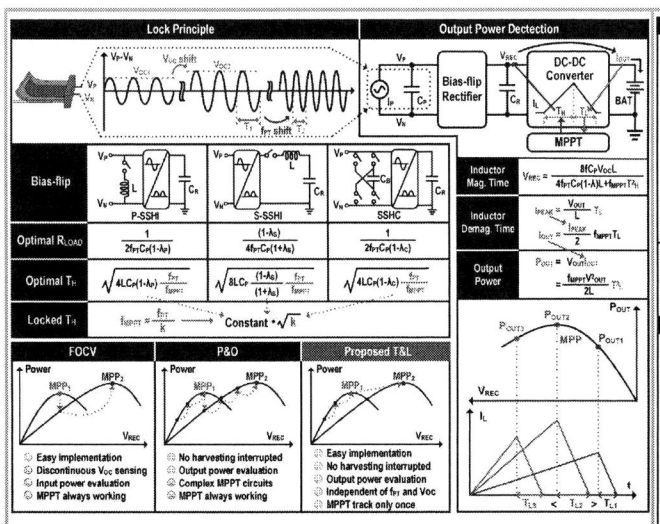

Figure 16.2.1: Conventional MPPT algorithm and the principle of the proposed T&L MPPT algorithm.

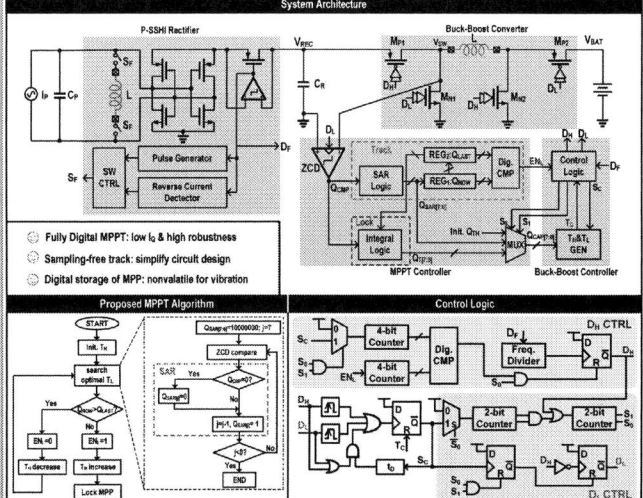

Figure 16.2.2: Block diagram of the proposed PEH system (top); proposed MPPT algorithm (bottom-left); control logic diagram (bottom-right).

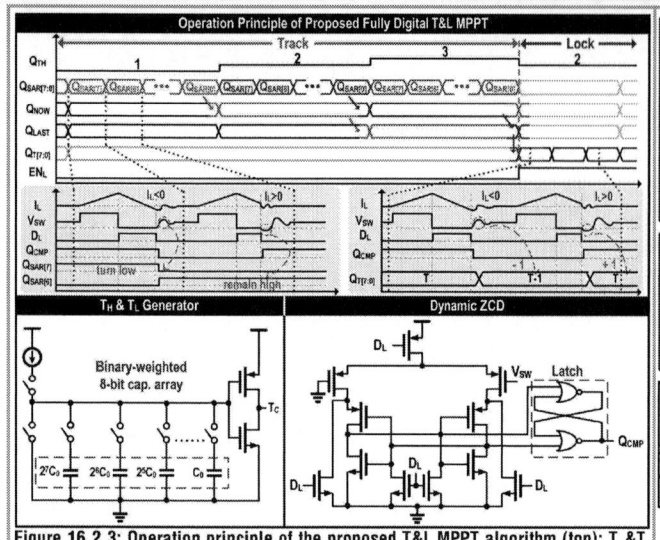

Figure 16.2.3: Operation principle of the proposed T&L MPPT algorithm (top); T_H&T_L generator diagram (bottom-left); dynamic ZCD diagram (bottom-right).

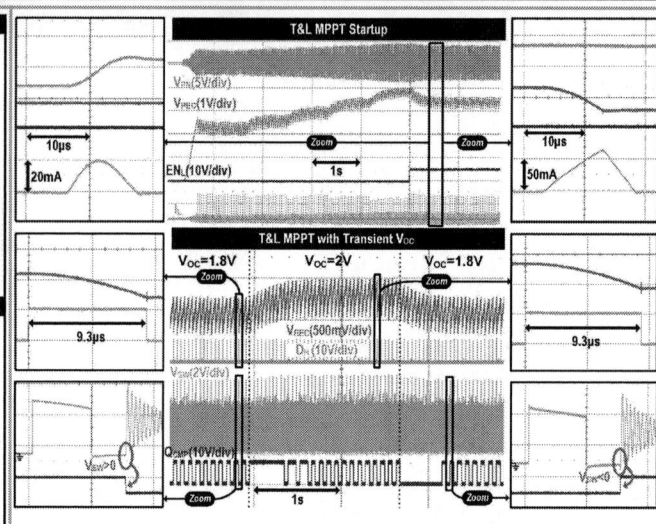

Figure 16.2.4: Measured waveforms of MPPT startup (top) and transient V_{OC} (bottom).

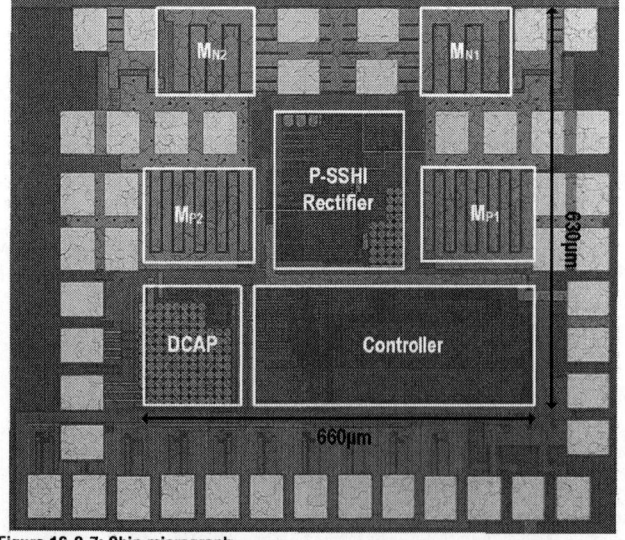

Figure 16.2.5: Measured waveforms of MPP storage (top-left) and under different frequency (bottom-left); measured MPPT efficiency versus V_{OC} and frequency (top-right); measured output power of proposed P-SSHI rectifier (bottom-right).

	This work	[6] VLSI'19	[7] ISSCC'20	[5] ISSCC'22	[8] ISSCC'23	[9] JSSC'24	[10] ISSCC'25
Technology (nm)	180	130	600	65	180	180	180
Active area (mm²)	0.42	1.07	14	3.11	0.46	0.9	4.03
Inductor (µH)	470/1000	3300	220	22	27-120	220	1500
C_P (nF)	32	20&100	24	N/R*	42	23/116	40
Freq (Hz)	28-74	100-180	56	N/R*	230	133/120	260&265
MPPT Algorithm	T&L	P&O	P&O	FOCV	DCB	Stabilized DCB	GMC
Output-based MPPT	Yes	Yes	No	No	No	No	No
MPPT Circuit	Digital	Analog	Analog&Digital	Analog&Digital	Analog	Digital	Analog
Sampling free?	Yes	No/Time	No/Voltage	No/Voltage	Yes	Yes	No/Voltage
V_{OC}/Frequency independent?	Yes	No	No	No	V_{OC}	V_{OC}	No
Once/always MPPT operation?	Once	Always	Always	Always	Always	Always	Always
MPPT efficiency	99.8%	97%	94%	80%	98%	99.9%	99.9%
I_Q of MPPT (nA)	17	N/R*	23**	N/R*	N/R*	8.66	157
FoM	4.9	4.17	3.28	3.2	7.38	9.3	9.29

*N/R: not reported
**Estimated from the paper

Figure 16.2.6: Comparison of the T&L MPPT circuit and state-of-the-art piezoelectric energy harvesting circuits.

Figure 16.2.7: Chip micrograph.

ISSCC 2026 / SESSION 16 / ENERGY HARVESTING, PIEZO AND CHARGERS / 16.3

16.3 A Fully Integrated Piezoelectric Energy-Harvesting Interface with Single-Stage Bias-Flip and MPPT Achieving 5.63× Maximum Output Power Improving Rate

Xiaonan Wu[1], Jinyi Yuan[1], Minghao Shang[1], Baochuang Wang[1], Sijun Du[2], Lin Cheng[1]

[1]University of Science and Technology of China, Hefei, China, [2]TU Delft, Delft, The Netherlands

Abstract

This paper presents a fully integrated PEH interface that enhances output power while minimizing on-chip capacitor area. The proposed design employs a SC structure that reuses both the piezoelectric capacitor and the rectifier capacitor as flying capacitors. Meanwhile, on-chip capacitors serve as DC capacitors, thereby eliminating parasitic-capacitance losses. As a result, the interface achieves a maximum output power improvement of 5.63× and a normalized on-chip capacitor area of 0.09 mm²/nF.

Piezoelectric energy harvesting (PEH) has emerged as a promising solution for scavenging vibration energy to power miniaturized Internet-of-Things (IoT) devices. A fundamental PEH interface consists of two parts (Fig. 16.3.1 top left): a Bias-Flip (BF) AC-DC stage that rapidly reverses the piezoelectric transducer (PT) voltage (V_{PT}) when the PT current (I_P) crosses zero, minimizing energy loss on the piezoelectric capacitor (C_P), and a DC-DC converter that regulates the rectifier capacitor (C_R) voltage (V_R) for maximum-power-point tracking (MPPT). Previous works have achieved high-efficiency BF and MPPT operation using off-chip resonance inductors [1] or capacitor arrays [2]. However, these off-chip components hinder miniaturization. Fully-integrated interface in [3,4] target PTs with C_Ps of a few hundred pF and reply on a fixed flip count, resulting in switched-capacitor values comparable to C_P, which limits their applicability to commercial PTs whose C_P is in the tens of nF. To address this limitation, an arbitrary number of flip counts is implemented in [5], allowing for a much smaller switched capacitor and enabling a fully-integrated interface for commercial PTs. However, this technique relies on an on-chip switched capacitor (SC) (Fig. 16.3.1 top right), whose maximum efficiency is typically constrained by k_{Par}, which is the ratio of the parasitic capacitance ($C_{Par,On\text{-}chip}$) of the flying capacitor (C_{Fly}) to C_{Fly}. For a 2:1 SC, if a metal-insulator-metal (MIM) capacitor with k_{Par} = 1.5% and a MOS capacitor with k_{Par} = 7% are used, their theoretical efficiency is limited to 89% and 79%, respectively [6].

To solve this problem, this paper proposes an Input-Output-Flying (IOF) SC topology, as shown in Fig. 16.3.1 bottom. During BF operation, the off-chip input capacitor C_P and output capacitor C_R, which initially function as fixed grounded DC capacitors for rectification, are reused as flying capacitors between on-chip DC capacitor arrays and the battery. In addition, to regulate V_R and achieve MPPT, the BF operation is designed to operate in two distinct modes: Self-Flip Mode (SM) and Charge-Flip Mode (CM). In SM, energy is exchanged only between C_P and C_R, leaving V_R unchanged. In CM, a charging path to the battery is introduced to decrease V_R after the flip. During non-BF operation, the system works in Rectification Mode (RM), where C_R is charged by the PT through the rectifier, and V_R increases. By alternating among these three modes, V_R is regulated at the maximum power point for MPPT operation while avoiding both the parasitic-capacitor loss of on-chip flying capacitors and the cascade loss of a separated DC-DC stage. Since all on-chip capacitors are DC capacitors that do not incur the parasitic-capacitor loss of flying capacitors, high-density MOS capacitors with significant parasitic capacitance can be utilized. Moreover, the minimal parasitic capacitance of C_P (~400fF) and C_R (~50fF) further ensure high efficiency in the IOF-SC topology.

Figure 16.3.2 (top) illustrates the MPPT operating principle of the proposed IOF-SC-based interface. Prior to each BF operation, the system compares V_R with the maximum power point voltage $V_{R,MPP}$. If V_R is lower than $V_{R,MPP}$, the BF operates in SM, maintaining a constant V_R and continuing to charge C_R in the subsequent RM. After sufficient C_R charging raises V_R above $V_{R,MPP}$, the BF switches to CM, causing V_R to decrease and transferring charge to the battery. By alternating between SM and CM, V_R is effectively regulated at $V_{R,MPP}$. Both SM and CM are divided into nine configurations (cfg_i, i = +4~-4) realized by four on-chip capacitors. Each cfg_i implements a switched-capacitor configuration with a voltage conversion ratio of V_{PT}:V_R = i:5. Among them, cfg_0 represents the direct short-circuit reset of C_P, while the other cfg_is are realized through several sub-phases. The bottom-left of Fig. 16.3.2 shows the typical voltage waveforms and energy transfer in SM. In cfg_{+4}, V_{PT} decreases from V_R to $4/5V_R$, during which the charge stored in C_P is transferred to C_R; in cfg_{-4}, V_{PT} further decreases from $-3/5V_R$ to $-4/5V_R$, with the required charge supplied by C_R. Because C_P experiences the same voltage variation in in cfg_{+4} and cfg_{-4} and the conversion ratios are identical, the charge delivered to C_R in cfg_{+4} cancels the charge released in cfg_{-4}. The same cancellation holds for each (+i, -i) pair, so V_R remains constant over the full BF operation. Over the entire flip, V_{PT} transitions from V_R to $-4/5V_R$, leading to a flip efficiency ($\eta_{Flip} = |V_{PT,after\ Flip} / V_{PT,before\ BF}|$) of 80%. The bottom-right of Fig. 16.3.2 shows the corresponding process in CM. During the cfg_{+4} to cfg_{+1} flipping process, an additional energy transfer path is introduced to deliver the energy of C_P to the battery, BAT. To enable this, certain phases of this SC are reconfigured for the BAT charging process, the details of which are illustrated in Fig. 16.3.3. Compared with SM, the charge transferred to C_R during CM is reduced. Conversely, during the cfg_{-1} to cfg_{-4} transition, the energy extraction from C_R to BAT increases, causing C_R to release more charge than in SM, thereby leading to a net decrease in V_R after the flip. As the vibration increases, the charge harvested by C_R during RM also grows, shifting the BF operation progressively toward CM until SM completely vanishes. Then, the output power is maximized for the current $V_{R,MPP}$. With further vibration growth, V_R will continue to increase; however, because the output charge delivered by C_R also scales with V_R, the voltage does not increase without bound but eventually stabilizes above $V_{R,MPP}$. Subsequently, $V_{R,MPP}$ is updated to a higher value in response to the stronger excitation and V_R settles again around the new $V_{R,MPP}$, maintaining MPPT operation.

Figure 16.3.3 left shows the system architecture, which consists of a high-voltage domain powered by V_{BAT}, including an active-rectifier (AR) and IOF-SC with four 1.2nF on-chip capacitors, and a low-voltage domain supplied by an on-chip LDO, containing the switch controller, config and sub-phase timer, and PWM generator. In the IOF-SC, switches $S_{PT,RT}$ and $S_{PB,RT}$ are implemented by 5V PMOS transistors to handle voltages from V_R to $2V_R$, while other switches are 5V NMOS working from 0 to max(V_R, V_{BAT}). The top-right of Fig. 16.3.3 illustrates the capacitor connection scheme of each sub-phase, $\varphi_{i,j}$, (j = 1 ~ 5) in SM, using cfg_{+3} as an example. C_P continuously discharges to complete the BF operation and transfers 3/5 of its charge to C_R, thereby forming a SC with a voltage conversion ratio of 3:5 from C_P to C_R. During this process, $C_{DC,1-4}$ always maintain their bottom plates connected to ground, thereby eliminating parasitic-capacitor losses. The bottom-right of Fig. 16.3.3 presents the configuration construction algorithm in CM along with an example. When the BF operation is in cfg_i (i > 0), if during a sub-phase $\varphi_{i,j}$, the condition $V_{DC,k} + V_{PT} = V_{BAT}$ is satisfied, the BAT can replace the series connection of $C_{DC,k}$ and C_{PT}, thereby enabling this phase to be used for battery charging. However, if only this single phase is reconfigured, $C_{DC,k}$ would be discharged in the preceding phase $\varphi_{i,j-1}$ without subsequent recharging, which breaks its charge balance and disrupts the normal operation of this SC. To avoid this issue, all phases from $\varphi_{i,1}$ to $\varphi_{i,j}$ can be reconfigured as battery charging phases. In this way, the capacitors in the earlier phases no longer participate in the charge/discharge process, thereby maintaining the steady-state operation of the system. Similarly, when the BF operation is in cfg_i (i < 0), if during a sub-phase $\varphi_{i,j}$,the condition $V_{DC,k} + V_R - |V_{PT}| = V_{BAT}$ is satisfied, then sub-phases $\varphi_{i,j}$ to $\varphi_{i,5}$ are merged into a battery-supplying sub-phase $\varphi_{i,BAT}$, where $C_{DC,k}$, C_P, and C_R are connected in series to supply the battery.

The proposed PEH interface was fabricated in a 0.18μm BCD process and evaluated with a commercial PT (MIDE PPA-1021, C_P = 21nF) excited at f_{PT} = 100 Hz. Figure 16.3.4 left shows the operation waveforms of C_P and C_R at V_{OC} (= $I_P/2\pi f_{PT}C_P$) = 1V, V_R:V_{BAT} = 5:5. This demonstrates that IOF-SC achieves a 79.4% η_{Flip} after the nine-phase operation, with C_P and C_R serving as the flying capacitors. Figure 16.3.4 top right shows the transient waveforms of the system at V_{BAT} = 2.8V and V_R:V_{BAT} = 5:7, undergoing excitation enhancement and the transition of $V_{R,MPP}$ and V_R:V_{BAT}. The dashed vertical line marks the excitation enhancement, where V_{OC} rises from 0.6-to-1V, while V_R remains regulated at $V_{R,MPP}$ = 2V. The solid vertical line marks the moment when the system receives the S_{MPPT} signal, triggering the transition of V_R:V_{BAT} from 5:7 to 5:5. V_R then gradually increases to $V_{R,MPP}$ = 2.8V, completing the MPPT operation. Figure 16.3.4 bottom right shows the details of V_R regulation. When the controller detects that V_R is below V_{BAT}, S_{PWM} is set to low, and so the system operates in SM. Once the controller detects that V_R is above V_{BAT}, S_{PWM} is set to high, and the system operates in CM. By alternating between SM and CM, V_R is ultimately regulated at $V_{R,MPP}$ = 2V.

Figure 16.3.5 top shows the measured output power versus V_{BAT} for different V_{OC} when V_R:V_{BAT} = 5:5. The maximum output power is 16.99μW with a 5.63× maximum output power improving rate (MOPIR) at V_{OC} = 1.2V. Figure 16.3.5 bottom shows that by configuring V_R:V_{BAT}, the system maintains a >4.5× MOPIR across a V_{OC} range of 0.8-to-1.9V. Figure 16.3.6 summarizes and compares the performance of the proposed converter with state-of-the-art. Thanks to the IOF-SC with plate-parasitic loss elimination, the proposed PEH interface achieves the highest η_{Flip} while significantly reducing the normalized capacitor area, which is the ratio of the on-chip capacitor area to the capacitance of C_P. Moreover, the high-efficiency IOF-SC can also ensure that the MOPIR remains superior compared to non-fully integrated designs. A chip micrograph and PCB prototype are shown in Fig. 16.3.7. Since this design does not employ any additional off-chip flying capacitors, the PCB layout is significantly simplified, resulting in a compact footprint of only 10.2mm × 16.5mm. Moreover, the miniaturized PCB is directly soldered onto the PT to reduce the parasitic capacitance caused by wiring, thereby maximizing the reduction of the plate-parasitic

979-8-3315-8937-0/26 $31.00 © 2026 IEEE

capacitance $C_{PAR,P}$ of C_P. Similarly, by mounting C_R directly on top of the chip, the parasitic capacitance introduced by the PCB pads is avoided, thus reducing the plate-parasitic capacitance $C_{PAR,R}$ of C_R.

Acknowledgement:
This work was supported in part by the National Natural Science Foundation of China under Grant No. 62261160647. Corresponding author: Lin Cheng (eecheng@ustc.edu.cn).

References:
[1] S. Li, A. Roy and B. H. Calhoun, "A Piezoelectric Energy-Harvesting System with Parallel-SSHI Rectifier and Integrated MPPT Achieving 417% Energy-Extraction Improvement and 97% Tracking Efficiency," *2019 Symposium on VLSI Circuits*, Kyoto, Japan, 2019, pp. C324-C325. https://doi.org/10.23919/VLSIC.2019.8778144
[2] Z. Chen, Y. Jiang, M. -K. Law, P. -I. Mak, X. Zeng and R. P. Martins, "A Piezoelectric Energy-Harvesting Interface Using Split-Phase Flipping-Capacitor Rectifier and Capacitor Reuse Multiple-VCR SC DC-DC Achieving 9.3× Energy-Extraction Improvement," *2019 IEEE International Solid-State Circuits Conference - (ISSCC)*, San Francisco, CA, USA, 2019, pp. 424-426. https://doi.org/10.1109/ISSCC.2019.8662323
[3] Z. Chen, M. -K. Law, P. -I. Mak, W. -H. Ki and R. P. Martins, "A 1.7mm² inductorless fully integrated flipping-capacitor rectifier (FCR) for piezoelectric energy harvesting with 483% power-extraction enhancement," *2017 IEEE International Solid-State Circuits Conference (ISSCC)*, San Francisco, CA, USA, 2017, pp. 372-373. https://doi.org/10.1109/ISSCC.2017.7870416
[4] S. Du and A. A. Seshia, "A fully integrated split-electrode synchronized-switch-harvesting-on-capacitors (SE-SSHC) rectifier for piezoelectric energy harvesting with between 358% and 821% power-extraction enhancement," *2018 IEEE International Solid-State Circuits Conference - (ISSCC)*, San Francisco, CA, USA, 2018, pp. 152-154. https://doi.org/10.1109/ISSCC.2018.8310229
[5] P. Angelov and M. Nielsen-Lönn, "A Fully Integrated Multilevel Synchronized-Switch-Harvesting-on-Capacitors Interface for Generic PEHs," in *IEEE Journal of Solid-State Circuits*, vol. 55, no. 8, pp. 2118-2128, Aug. 2020. https://doi.org/10.1109/JSSC.2020.2979178
[6] N. Butzen and M. Steyaert, "A 94.6%-efficiency fully integrated switched-capacitor DC-DC converter in baseline 40nm CMOS using scalable parasitic charge redistribution," *2016 IEEE International Solid-State Circuits Conference (ISSCC)*, San Francisco, CA, USA, 2016, pp. 220-221. https://doi.org/10.1109/ISSCC.2016.7417986
[7] Z. Li et al., "Piezoelectric Energy Harvesting Interface Using Self-Bias-Flip Rectifier and Switched-PEH DC–DC for MPPT," in *IEEE Journal of Solid-State Circuits*, vol. 59, no. 7, pp. 2248-2259, July 2024. https://doi.org/10.1109/JSSC.2023.3341865

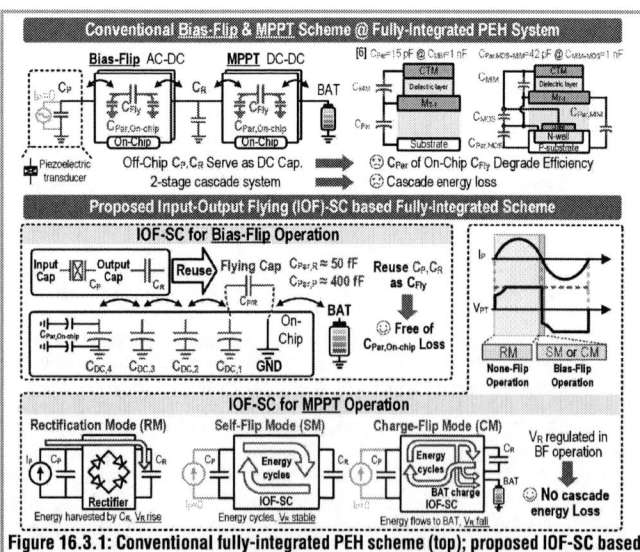

Figure 16.3.1: Conventional fully-integrated PEH scheme (top); proposed IOF-SC based fully-integrated scheme for eliminating plate-parasitic losses and cascade energy loss.

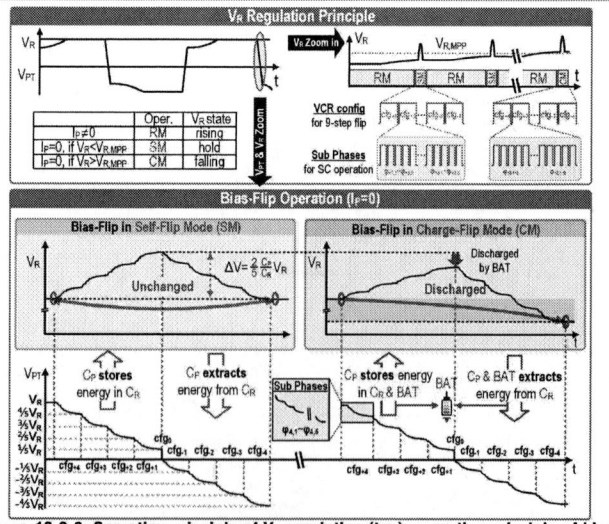

Figure 16.3.2: Operation principle of V_R regulation (top); operation principle of bias-flip operation in self-flip mode and charge-flip mode.

ISSCC 2026 / SESSION 16 / ENERGY HARVESTING, PIEZO AND CHARGERS / 16.3

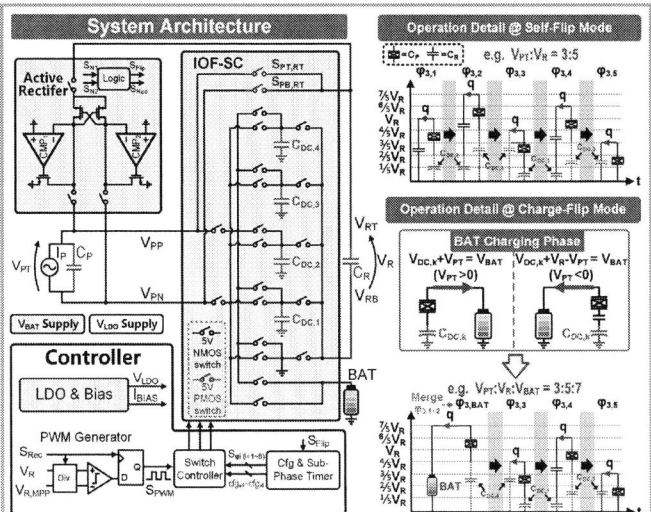

Figure 16.3.3: System architecture of the proposed IOF-SC based PEH interface circuit (left), operation details in self-flip mode and charge-flip mode (right).

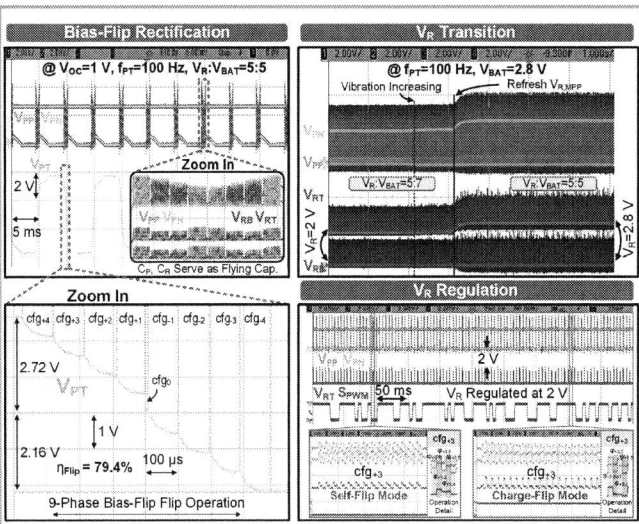

Figure 16.3.4: Measured results of the bias-flip rectification waveforms (left); Transient waveforms of V_R transition and regulation (right).

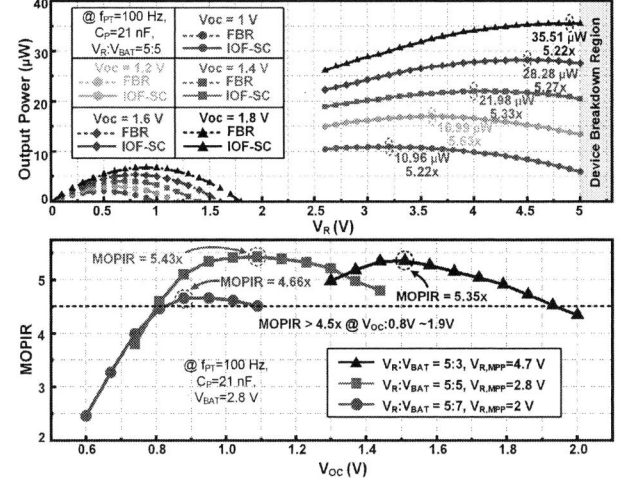

Figure 16.3.5: Output power comparison of the proposed IOF-SC-based interface circuit and ideal full-bridge-rectifier-based interface circuit according to V_{OC} and V_R (top); relationship between MOPIR and V_{OC} under different $V_{R,MPP}$ (bottom).

	ISSCC'17 [3]	ISSCC'18 [4]	VLSI'19 [1]	JSSC'20 [5]	JSSC'24 [7]	This work
Technology	0.18 μm	0.18 μm	0.18 μm	0.18 μm	0.18 μm	**0.18 μm**
Technique	FCR	SE-SSHC	P-SSHI	ML-SSHC	SBFRR	**IOF-SC**
Piezoelectric Transducer	Piezo P5A4E	Custom MEMS	MIDE PPA1021	Vernon S.A. VH1504C-2	MIDE PPA1021	**MIDE PPA1021**
C_P	80 pF	1.94 nF	20 nF	6 nF	4*22 nF	**21 nF**
Excitation Frequency	110 kHz	219 Hz	100-180 Hz	22 Hz	100 Hz	**100 Hz**
Implemented Features	Bias-Flip	Bias-Flip	Bias-Flip +MPPT	Bias-Flip	Bias-Flip +MPPT	**Bias-Flip +MPPT**
Energy Reservoir	On-chip MIM Cap. (1.44 nF)	On-chip MIM Cap. (4 nF)	Off-chip Inductor (3.3 mH)	On-chip MIM Cap. (1.5 nF)	None	**On-chip MIM+MOS Cap. (4*1.2 nF)**
Normalized Cap. Area[a,b]	18.75 mm²/nF	0.67 mm²/nF	-	0.125 mm²/nF	-	**0.09 mm²/nF**
η_{Flip}[c]	70%[a]	71.2%[a]	72%[a]	75%	60%[a]	**79.4%**
MOPIR[d]	4.83x (V_D=0 V)	3.58-8.21x (V_D=0.25 V)	4.17x (V_D=0 V)	7.01x[e] (V_D=0 V)	4.88x (V_D=0 V)	**8.78x (V_D=0.12 V) 5.63x (V_D=0 V)**

(a) Estimated from the paper. (b) Cap. Area/C_P. (c) η_{Flip}=$V_{PT\ after\ flip}$/$V_{PT\ before\ flip}$.
(d) MOPIR=$P_{out,max}$ of proposed interface/$P_{out,max}$ of full bridge rectifier
(e) Exclude an 4.5 mW microcontroller

Figure 16.3.6: Performance summary and comparison with the prior arts.

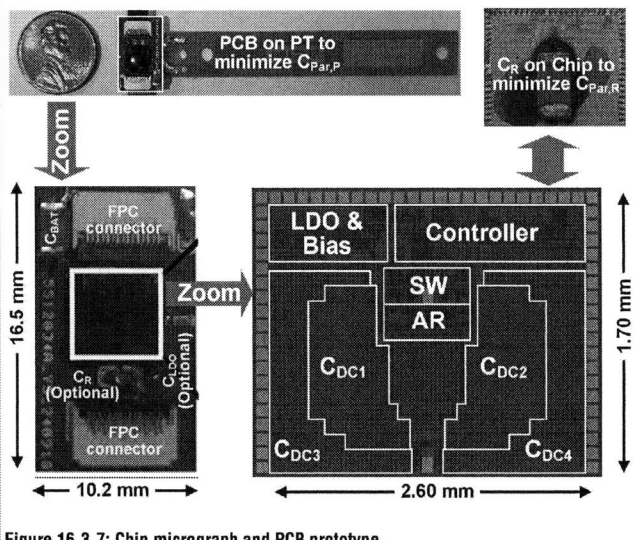

Figure 16.3.7: Chip micrograph and PCB prototype.

- 2026 IEEE International Solid-State Circuits Conference

ISSCC 2026 / SESSION 16 / ENERGY HARVESTING, PIEZO AND CHARGERS / 16.4

16.4 A One-Stage Bidirectional Rectifier with Pre-Charge-Based MPPT for Triboelectric Energy Harvesting with 93% MPPT Efficiency and 8.86× Power Enhancement

Wenyu Peng, Jida Peng, Willem van Driel, Guoqi Zhang, Sijun Du

Delft University of Technology, Delft, The Netherlands

Abstract

This work proposes a one-stage bidirectional rectifier with a pre-charge voltage (V_{PC}) based perturb-and-observe (P&O) maximum power point tracking (MPPT) algorithm for triboelectric energy harvesting. The proposed V_{PC}-based MPPT dynamically locates the true MPP while accounting for circuit non-idealities, using a simplified implementation compared to conventional P&O methods. Consequently, it achieves 93% MPPT efficiency and 8.86× energy enhancement compared to the full-bridge rectifier.

With the growth of Internet-of-Things (IoT) and edge computing, wireless sensor networks (WSNs) are widely deployed at the end where battery life is critical. Energy harvesting (EH) has been a promising solution to eliminate batteries or extend the battery life. Among various EH techniques, triboelectric nanogenerators (TENGs) harvest irregular mechanical energy but produce a high-voltage (HV) AC output (V_T) with a time-varying intrinsic capacitance (C_T) [1,2], as shown in Fig.16.4.1 (top), thus requiring specialized rectifiers to deliver usable sub-5V DC power. Prior interfaces such as bias-flip (BF) [3-7] and electrostatic charge boosting (ECB) [8] rely on an HV storage capacitor (C_{REC}) and a separate DC-DC stage, incurring leakage, slow startup, and cascaded loss.

This work proposes a one-stage bidirectional rectifier with a pre-charge voltage (V_{PC})-based perturb-and-observe (P&O) maximum power point tracking (MPPT) algorithm (Fig. 16.4.1 bottom). The one-stage topology removes the HV-capacitor leakage and directly steps the TENG voltage down to sub-5 V. The proposed V_{PC}-based MPPT dynamically locates the true MPP while accounting for circuit non-idealities, using a simplified implementation compared to conventional P&O methods. Within each kinetic period (T_k), the rectifier alternates between harvest (blue) and invest (red) phases. In the positive half-cycle (Φ_P), phase I starts, where V_T is pre-charged from V_{OUT} to a given pre-charge voltage, $V_{PC,p}$. Phase II maintains V_T near the harvesting voltage (V_{HAR}) (typically the process limit for high power, e.g., 65V) using multi-shot conversions to extract power form the TENG to V_{OUT}. Phase III extracts the remaining energy from C_T to prevent self-discharge. The negative half-cycle (Φ_N) includes phases IV – VI, which mirror phases I – III but operate with reversed polarity and a separate pre-charge voltage $V_{PC,n}$. Among these six phases, $V_{PC,p}$ and $V_{PC,n}$ determine the energy invested from V_{OUT} to the TENG. Since Φ_P typically delivers much higher energy than Φ_N, the MPPT algorithm focuses on tracking the optimal $V_{PC,p}$ (i.e., $V_{PC,MPP}$), while the $V_{PC,n}$ is externally set to reduce system complexity.

The power stage of the proposed rectifier (Fig. 16.4.2) employs seven on-chip 65-V n-type LDMOS transistors (S_{1-7}) to integrate both rectification and bidirectional power conversion. Node V_X, connected to V_{OUT} via an inductor, exhibits a tens-of-pF parasitic capacitance (C_{para}) from the HV switches, inductor, and packaging, which complicates precise MPP tracking. Figure 16.4.2 (bottom) illustrates the waveforms of V_T, V_X, and the inductor current (I_L) across phases I – III. The corresponding operation steps ($OP_\#$) are shown in Fig. 16.4.2 (top-right). In phase I, V_T is pre-charged to $V_{PC,p}$ via two operations: OP_1, which charges the inductor, and OP_{3P}, which transfers energy to the TENG. In phase II, energy extraction begins when V_T reaches 65V. To minimize conduction loss and reduce the quality factor requirements of the inductor, multiple low-current energy conversion shots are used [9]. This allows the use of a compact 470μH inductor. In each shot, C_{para} is first charged through LC resonance by OP_1 and OP_{3P} to suppress the hard-charging loss and V_T ripple caused by C_{para}. The OP_1 duration is adaptively tuned to avoid overcharging. Once V_X peaks during OP_{3P}, OP_{2P} is triggered to let the TENG briefly charge the inductor. A short dead time during OP_0 follows, allowing residual C_{para} energy to enter the inductor. When V_X falls below 0V, OP_1 resumes to transfer I_L efficiently to V_{OUT}. Phase II continues until V_T starts decreasing due to the polarity change of I_T, marking the start of phase III, which uses the same operation sequence as phase II but with different durations. As V_T declines, the duration of OP_{2P} is extended to maintain energy delivery. By the end of phase III, V_T is fully discharged to 0V. Phases IV – VI repeat the same operation pattern during Φ_N with opposite polarities and the replacement of OP_{2P} and OP_{3P} with OP_{2N} and OP_{3N}.

Figure 16.4.3 (left) illustrates the mechanism and implementation of the proposed V_{PC}-based MPPT algorithm. In each cycle, the system must determine an optimal $V_{PC,p}$ that maximizes the net output energy. To do so, it evaluates the energy balance during phase I (investment) and phase II (harvesting) by analyzing each power conversion shot. Each shot is composed of two time periods: the pre-charge period (T_{PC}) when V_{OUT} charges the inductor, and the harvesting period (T_{HAR}) when energy is delivered from the inductor to V_{OUT}. As I_L rises linearly (slope = V_{OUT}/L), the net transferred energy per shot is proportional to $T_{CONV}^2 - T_{PC}^2$, or equivalently, $(T_{CONV} - T_{PC})(T_{CONV} + T_{PC})$. To simplify computation, the term $T_{CONV} - T_{PC}$ is kept constant across all shots, enabling linear rescaling of the energy input and output using only time measurements. Based on this, the system estimates the invested energy $E_{I,k}$ and harvested energy $E_{II,k}$ during the kth T_k. Then, in the $(k+1)^{th}$ cycle, the MPPT controller perturbs $V_{PC,p}$ by injecting two additional pre-charging shots in phase I (i.e., $M_{k+1} = M_k + 2$). The resulting change in harvested energy, $\Delta E = E_{II,k+1} - E_{II,k}$, reflects the system's position relative to the MPP: if ΔE exceeds 150% of the extra injected energy, it implies that $V_{PC,p}$ is below the optimum, and the next perturbation increases to +3 shots; otherwise, if ΔE is smaller, the perturbation is reversed (–3 shots) to reduce $V_{PC,p}$, and in this case, the energy comparison step is skipped to save computation. This pre-charge-based P&O method offers real-time tracking with low hardware complexity, requiring only simple time counters and avoiding multipliers or analog computation.

Figure 16.4.3 (right) shows the architecture of the proposed system. The LDMOS switches in the power stage are driven by level shifters and bootstrap gate drivers. A peak voltage detector (PVD) connected to V_P identifies the transition from positive to negative semi-cycle (Φ_P to Φ_N), thereby synchronizing the control phases. At V_N, a HV sample and hold (S&H) block and a 6-bit successive approximation register (SAR) ADC are used to define $V_{PC,n}$. However, since the inherent capacitance C_T increases during Φ_N and the available harvested energy is very limited, implementing dynamic MPPT for $V_{PC,n}$ is not cost-effective. To reduce system complexity and power overhead, $V_{PC,n}$ is set externally and kept constant. The system also incorporates voltage sensors at both V_P and V_N to monitor V_T. Once V_T reaches 65V, these sensors trigger energy harvesting actions to regulate V_T and protect the circuit. V_X is connected to a second PVD, a zero current detector, and a hard-charging detector. These auxiliary sensors track inductor and capacitor dynamics in real-time and coordinate the switching between internal operations such as OP_1-OP_3.

The proposed chip is fabricated in a 0.18μm BCD process. The TENG used for validation is in-house fabricated, featuring copper-nickel tape electrodes, and triboelectric layers composed of nylon and PTFE films, as shown in Fig. 16.4.4 (top-left). This device achieves an open-circuit voltage exceeding 70V during Φ_N (V_{OC-}), with a C_T swing from approximately 50-to-500pF. Figure 16.4.4 illustrates the measured waveforms when the TENG operates with a 45Hz kinetic excitation. Each operational cycle is divided into six distinct phases (I – VI), each further split into multiple fine-grained energy conversion shots to reduce conduction loss. In phase I, the number of pre-charging shots is dynamically determined by the V_{PC}-based MPPT algorithm. As shown in Fig. 16.4.4 (top-right), $V_{PC,p}$ is perturbed and adjusted across cycles, enabling real-time tracking of net energy gain under different pre-charge voltages. During phase II, once V_P reaches 65V, the system initiates the controlled energy extraction via multi-shot conversions from the TENG to V_{OUT}. The averaged voltage drop during this phase is around 6V, allowing V_T to remain near V_{HAR} for efficient harvesting. At the end of Φ_P, when V_P begins to fall, phase III is initiated to complete charge extraction from C_T, driving V_P toward 0V using a series of small shots. The Φ_N half cycle mirrors this behavior: in phase IV, V_N is pre-charged to ~10V (with $V_{PC,n}$, set manually). In phase V, V_N may reach up to 65V due to the high V_{OC-}, prompting the system to apply several power conversion shots to cap V_N under 65V. In phase VI, the system fully extracts the remaining energy to V_{OUT}. More shots are required here than in phase III, due to the larger C_T during this interval. Finally, Fig. 16.4.5 (top-left) shows the detailed waveforms of V_P, V_N, and V_X during a single energy conversion shot. OP_{3P} charges the parasitic capacitor at V_X up to ~55V via soft charging, effectively reducing the hard-charging loss.

Figure 16.4.5 (top-right) presents the energy-extraction performance of the proposed rectifier. Leveraging the high open-circuit voltage of the in-house TENG, the system achieves maximum harvested energy when V_{HAR} is set to 65V, reaching a peak extracted power of 101μW. This corresponds to an 8.86× improvement over an ideal full-bridge rectifier (FBR), validating the effectiveness of the proposed synchronized harvesting strategy and one-stage architecture. The power conversion efficiency (PCE) is shown in Fig. 16.4.5 (bottom-left), which shows small variations across V_{OUT}, and a peak efficiency of 60.3% at V_{OUT} = 4.7V, demonstrating robust operation under various load conditions. Figure 16.4.5 (bottom-right) shows the V_{PC}-based MPPT performance. The maximum power output is measured when $V_{PC,p}$ is equal to 16V. When the MPPT algorithm is activated, $V_{PC,p}$ is perturbed between 10V and 25V, resulting in a measured extracted power of 93.8μW. Therefore, the MPPT efficiency is calculated as 93%, showing that the proposed V_{PC}-based MPPT algorithm effectively optimizes the energy extraction from the triboelectric energy harvesting system. The comparison table is presented in Fig. 16.4.6. This work exhibits a one-stage bidirectional rectifier for triboelectric energy harvesting. Compared with other works, it requires zero HV off-chip capacitors. Moreover, a novel V_{PC}-based MPPT algorithm is proposed, which

282 • 2026 IEEE International Solid-State Circuits Conference

979-8-3315-8937-0/26 $31.00 © 2026 IEEE

ISSCC 2026 / February 17, 2026 / 9:15 AM

achieves systematic MPPT in triboelectric energy harvesting and a high tracking efficiency of 93%. Lastly, the proposed system improves the power output by 8.86×, indicating a high energy extraction efficiency. Figure 16.4.7 shows the chip micrograph.

References:

[1] Y. Zhou, P. Zhang, J. Li, and X. Mao, "Recent progress of triboelectric nanogenerator-based power management and information processing circuit," *Materials Today Sustainability,* vol. 23, p. 100426, 2023/09/01/ 2023. https://doi.org/10.1016/j.mtsust.2023.100426

[2] W. Peng and S. Du, "The Advances in Conversion Techniques in Triboelectric Energy Harvesting: A Review," in *IEEE Transactions on Circuits and Systems I: Regular Papers,* vol. 70, no. 7, pp. 3049-3062, July 2023. https://doi.org/10.1109/TCSI.2023.3261780

[3] I. Kara, M. Becermis, M. A. -A. Kamar, M. Aktan, H. Dogan and S. Mutlu, "A 70-to-2 V Triboelectric Energy Harvesting System Utilizing Parallel-SSHI Rectifier and DC-DC Converters," in *IEEE Transactions on Circuits and Systems I: Regular Papers,* vol. 68, no. 1, pp. 210-223, Jan. 2021. https://doi.org/10.1109/TCSI.2020.3025468

[4] J. Lee, S. -H. Lee, G. -G. Kang, J. -H. Kim, G. -H. Cho and H. -S. Kim, "A 130V Triboelectric Energy-Harvesting Interface in .18\mu\mathrm{m}$ BCD with Scalable Multi-Chip-Stacked Bias-Flip and Daisy-Chained Synchronous Signaling Technique," *2022 IEEE International Solid-State Circuits Conference (ISSCC),* San Francisco, CA, USA, 2022, pp. 474-476. https://doi.org/10.1109/ISSCC42614.2022.9731605

[5] S. -H. Lee, Y. -W. Jeong, S. -J. Park and S. -U. Shin, "A Rectifier-Reusing Bias-Flip Energy Harvesting Interface Circuit With Adaptively Reconfigurable SC Converter for Wind-Driven Triboelectric Nanogenerator," in *IEEE Transactions on Industrial Electronics,* vol. 70, no. 8, pp. 8022-8031, Aug. 2023. https://doi.org/10.1109/TIE.2022.3220848

[6] W. Peng, X. Yue, W. van Driel, G. Zhang and S. Du, "A 70-V Fully Integrated Dual-SSHC Rectifier for Triboelectric Energy Harvesting with Full-Digital Duty-Cycle-Based MPPT Achieving 598% Power Extraction Enhancement," *2024 IEEE Custom Integrated Circuits Conference (CICC),* Denver, CO, USA, 2024, pp. 1-2. https://doi.org/10.1109/CICC60959.2024.10529024

[7] W. Jung, H. -M. Lee and H. -P. Le, "A Reconfigurable Multi-Level AC-DC/DC-DC Ocean Energy Harvester IC Achieving 77.7% End-to-End Power Efficiency for Triboelectric Nanogenerators," *2025 Symposium on VLSI Technology and Circuits (VLSI Technology and Circuits),* Kyoto, Japan, 2025, pp. 1-3. https://doi.org/10.23919/VLSITechnologyandCir65189.2025.11074895

[8] W. Peng, X. Yue, L. Pakula and S. Du, "A Capacitor-Based Bias-Flip Rectifier with Electrostatic Charge Boosting for Triboelectric Energy Harvesting Achieving Auto-MPPT at Breakdown Voltage and 14× Power Extraction Improvement," *2024 IEEE International Solid-State Circuits Conference (ISSCC),* San Francisco, CA, USA, 2024, pp. 516-518. https://doi.org/10.1109/ISSCC49657.2024.10454538

[9] Y. -W. Jeong, S. -J. Lee, J. -H. Kim, M. -J. Cho, H. -S. Kim and S. -U. Shin, "A Scalable N-Step Equal Split SSHI Piezoelectric Energy Harvesting Circuit Achieving 1170% Power Extraction Improvement and 22nA Quiescent Current with a 1µH–to–10µH Low Q Inductor," *2023 IEEE International Solid-State Circuits Conference (ISSCC),* San Francisco, CA, USA, 2023, pp. 438-440. https://doi.org/10.1109/ISSCC42615.2023.10067389

[10] I. Park, J. Maeng, M. Shim, J. Jeong and C. Kim, "A High-Voltage Dual-Input Buck Converter Achieving 52.9% Maximum End-to-End Efficiency for Triboelectric Energy-Harvesting Applications," in *IEEE Journal of Solid-State Circuits,* vol. 55, no. 5, pp. 1324-1336, May 2020. https://doi.org/10.1109/JSSC.2019.2942370

[11] J. Maeng, I. Park, M. Shim, J. Jeong and C. Kim, "A High-Voltage Dual-Input Buck Converter With Bidirectional Inductor Current for Triboelectric Energy-Harvesting Applications," in *IEEE Journal of Solid-State Circuits,* vol. 56, no. 2, pp. 541-553, Feb. 2021. https://doi.org/10.1109/JSSC.2020.3012991

[12] S. -Y. Moon, A. Shafique, S. C. Chandrarathna and J. -W. Lee, "A High-Voltage PMIC Using an Efficient Perturb and Observe Technique for Energy Harvesting of Triboelectric Nanogenerators," in *IEEE Transactions on Power Electronics,* vol. 40, no. 2, pp. 3225-3239, Feb. 2025. https://doi.org/10.1109/TPEL.2024.3484455

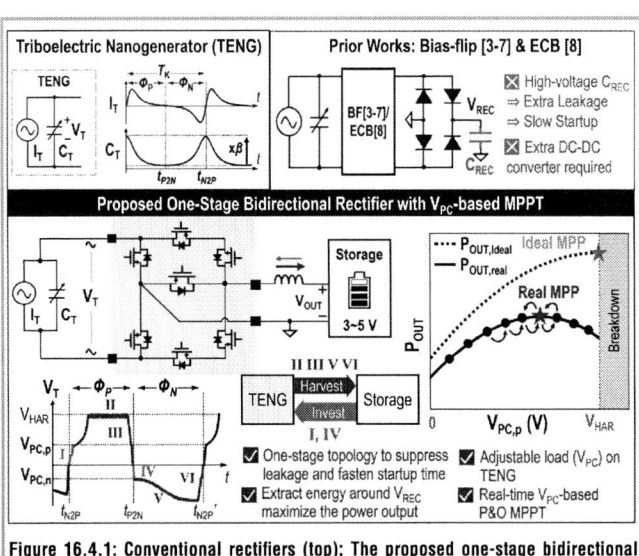

Figure 16.4.1: Conventional rectifiers (top); The proposed one-stage bidirectional rectifier with V_PC-based MPPT (bottom).

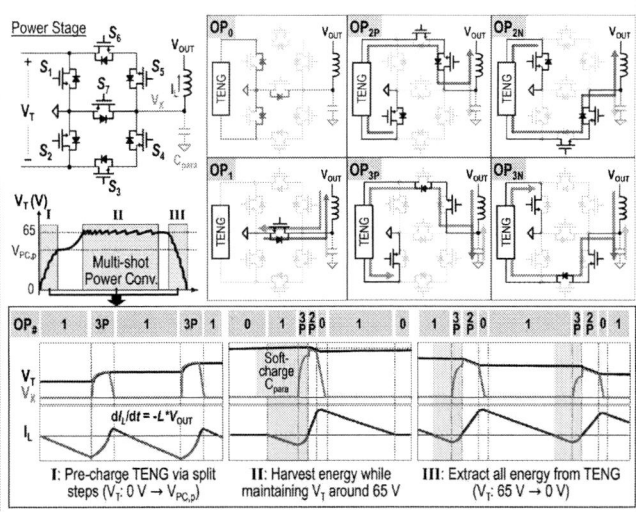

Figure 16.4.2: The operations in each shot in energy harvesting and investing phases.

ISSCC 2026 / SESSION 16 / ENERGY HARVESTING, PIEZO AND CHARGERS / 16.4

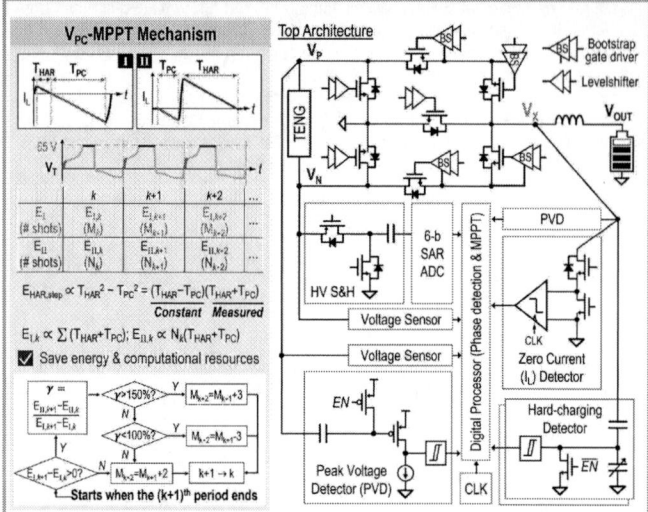

Figure 16.4.3: V_{PC}-MPPT algorithm and logical workflow (left); Top architecture of the proposed energy harvesting system (right).

Figure 16.4.4: The structure of TENG (top-left); The measured transient waveform of the proposed rectifier.

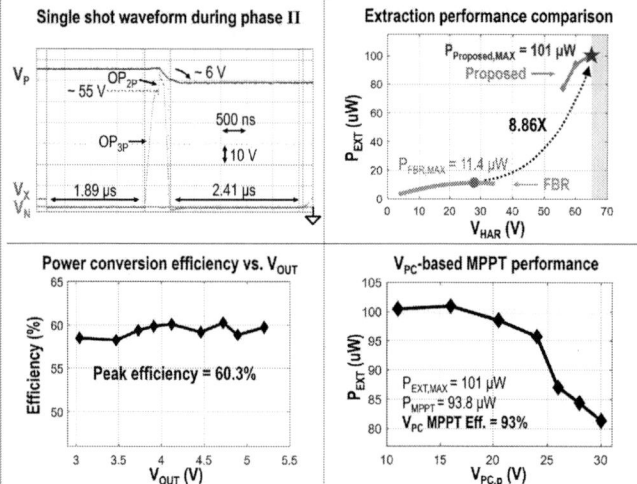

Figure 16.4.5: The measured waveform of a harvesting shot @ phase II (left-top); The measured energy extraction and MPPT performance.

Figure 16.4.6: The comparison table between the proposed work and the state-of-the-art.

	JSSC'20 [10]	JSSC'21 [11]	ISSCC 22 [4]	ISSCC'24 [8]	CICC'24 [6]	TPE'25 [12]	VLSI'25 [7]	This work
Process	0.18-µm BCD	0.18-µm BCD	0.18-µm BCD	0.18-µm BCD	0.18-µm BCD	0.18-µm BCD	0.18-µm BCD	0.18-µm BCD
Transducer	TENG	TENG	Wind-TENG	TENG	TENG	TENG	TENG	TENG
Inherent Capacitance	Varying	Varying	Constant	Varying	Varying	Varying	Constant	Varying
Frequency (Hz)	20-50	40-55	250	150	70	5	2-50	7-80
Topology	Dual-Output FBR + Dual-Input Buck	Dual-Output FBR + Dual-Input Buck	MCS-BF + SC DC-DC Converter	ECB + Buck	D-SSHC + Buck	FBR + Buck	Multi-level BF + Dual-Input Three-Level Buck	One-Stage Bidirectional Rectifier
Max. Input Voltage (V)	70	70	130/195	70	70	70	70	65
HV (>5V) Capacitors	2	2	2/3	1	1	1	3	0
V_{REC}-MPPT	FOCV	FOCV	N/A	Auto-MPPT @Breakdown Voltage	DCB	P&O	N/A	N/A
V_{PC}-MPPT	N/A	N/A	N/A	N/A	N/A	N/A	N/A	P&O
Peak PCE (%)	51.1	84.7	70.7*	N/A	N/A	85.3	83.7	60.3
Chip area (mm²)	2.03	2.34	4.16(main)/ 1.77(follower)	4.34	2.47	2.75	3.5	1
External Inductor	1 mH	10 mH	10 mH	1 µH	1 µH	20 mH	1 mH	470 µH
Peak Output Power (µW)	20.7	7.2	1211	127.6	85.4	153.5	2008	101
V_{PC}-MPPT Eff. (%)	N/A	N/A	N/A	N/A	N/A	N/A	N/A	93
P_{EXT}/P_{FBR}	N/A	N/A	3.14×	14.0×	5.96×	N/A	6.63×	8.86×

* End-to-end efficiency

Figure 16.4.7: Chip micrograph.

ISSCC 2026 / SESSION 16 / ENERGY HARVESTING, PIEZO AND CHARGERS / 16.5

16.5 A Single-Inductor Multi-Channel Thermoelectric Energy Harvesting Interface Realizing Uneven Temperature MPPT with 39.6% Efficiency Enhancement and 62mV Tapped-Inductor-Oscillator-Based Start-Up

Yunzhe Yang, Sijun Du

Delft University of Technology, Delft, The Netherlands

Abstract

This paper presents a thermoelectric energy harvesting (TEH) interface that realizes: 1) a multi-channel TEH for maximum power point tracking (MPPT) under uneven temperature conditions, improving efficiency by up to 39.6%; and 2) a tapped-inductor start-up functionality, which can begin operating at 62mV with no additional inductors or series switches. The overall design achieves a 90.51% peak end-to-end efficiency.

Energy harvesting (EH) from ambient sources has been widely adopted to power wearable and IoT devices as a replacement for batteries [1-5]. In such applications, improving harvesting efficiency and minimizing system size are both critical but challenging. Thermoelectric generators (TEGs) convert temperature differences (ΔT) across their two sides into DC voltage, with larger surface areas providing higher potential energy [6-9]. However, in the real world, the temperature distribution on a large surface is not even. For example, as shown in Fig. 16.5.1 (top-left), even a human hand's skin exhibits a more than 2 °C temperature difference, and more variations can be observed on larger surfaces. When using multiple TEGs, the output voltage (V_{TEG}) from each individual TEG can differ a lot. Conventional multi-TEG EH systems [10-12] simply connect TEGs in series or parallel, treating them as a single source. Such approaches suffer from poor maximum power point tracking (MPPT) efficiency when these V_{TEG}s are different. Besides, the TEG EH interfaces typically require a sub-100mV oscillator (OSC) with enough driving capability for start-up due to the TEGs' low output voltage [13-16]. However, even the start-of-the-art OSCs have shortcomings. As shown in Fig. 16.5.1 (top-right), ring OSCs require no off-chip components but they have a poor drive capability for charging the following stages. Consequently, they either start the system from a relatively high voltage level [17-20] or insert a series switch at the power stage to prevent the weak current from flowing to V_{OUT} too early [21-23]. In contrast, LC OSCs have a stronger drive capability, but they need an extra off-chip inductor [24,25], which occupies more space.

To address the large-area MPPT challenge and reduce the inductor count, this paper proposes a multi-channel TEG EH interface (Fig. 16.5.1 bottom), which enables: 1) multi-channel MPPT that tracks the MPP of each TEG individually, achieving optimized large-area tracking efficiency; and 2) reuse of a single inductor coil for both power conversion and start-up, achieving a low start-up voltage, requiring no additional series power switches, and simultaneously reducing off-chip components. As shown in Fig. 16.5.1 (bottom-left), for a four-TEG system, the proposed 4-channel scheme consistently outperforms conventional series and parallel connections [10-12] under temperature gradients. This advantage becomes even more significant in systems with more TEGs. In Fig. 16.5.1 (bottom-right), tapping the midpoint of the main power inductor transforms it into a transformer, which enables the simple implementation of a transformer-based OSC, without employing extra passive components. Thanks to its good driving capability, no extra series power switch is needed, while achieving ultra-low-voltage start-up.

Figure 16.5.2 (top-left) shows the system architecture of the proposed interface. Four TEGs (TEG_{1-4}) are connected to the chip via a multi-channel input switch matrix, which includes four switches that select which input channel is active for energy extraction. Four small off-chip capacitors (C_{IN1-4}) are connected in parallel with the four TEGs. The main power inductor (L_{BOOST}) functions as a boost inductor during steady-state energy harvesting operations and serves as a transformer during start-up. The MPPT block uses a single-switch fractional open circuit voltage (FOCV) approach where only one open-circuit switch (S_{MPPT}) is inserted to perform MPPT for all of the channels. The MPPT block also generates a clock signal (CLK) with an adaptively tuned period (T_{CLK}), which determines the operating frequency of the boost converter to realize MPPT. The system also integrates zero-current sensing (ZCS), V_{OUT} regulation, and control logic to generate non-overlapping driving signals for M_{L1} and M_{H1}.

Figure 16.5.2 (top-right) shows the inductor current (I_L) and output voltage (V_{OUT}) during power conversion. All input channels are sequentially enabled once for the same duration. The on-time of M_{L1} (t_{ON}) is $4T_{CLK}$. The working principle of the single-switch FOCV multi-channel MPPT is illustrated in Fig. 16.5.2 (middle-right). When employing four identical TEGs, they typically have the same internal resistance (R_{TEG}), which is quite insensitive to temperature. Hence, their maximum power points (MPPs) occur at the same t_{ON} when the total period T is the same across channels via the input matrix. As a result, MPPT can be performed on a single representative channel, minimizing open-circuit loss from S_{MPPT}. Figure 16.5.2 (bottom) shows the topology and operation flow of the MPPT block. When it is enabled by the EN_{MPPT} signal, S_{MPPT} turns off, and a 2:1 voltage divider samples half of the open-circuit voltage, $V_{TEG3}/2$. Then, a comparator determines if V_{IN3} is higher than $V_{TEG3}/2$. If so, T_{CLK} is extended; otherwise, it is shortened to increase V_{IN3}. The V_{TEG} sampling is triggered once every 64 periods to save power consumption from sampling and comparison.

Figure 16.5.3 (top) shows the circuit diagram and working waveform of the tapped-inductor-OSC-based start-up block. It includes three on-chip auxiliary capacitors, C_{1-3}, to help charge C_{OUT}. The reused-coil transformer-based OSC_1 serves as the first stage to initiate operation from the low-voltage input, V_{TEG}. During $0-t_1$, C_1 and C_2 are charged by associate OSCs and charge pumps. Once V_{C2} reaches V_{TH1} at t_1, C_3 starts being charged by C_2, and a negative voltage generator activates, outputting the OFF_{OSC_1} signal to disable the transformer-based OSC_1. After t_2, the coil acts as the inductor L_{BOOST}, and forms a dual-output boost converter, consisting of four switches, M_{L2} and M_{H2-4}. This stage charges C_{OUT} while maintaining adequate voltage on C_2 and C_3 as supplies. After V_{OUT} reaches a preset level, the system is considered started, and the start-up process completes. During normal operation, C_{OUT} is charged by the main boost converter formed by M_{L1} and M_{H1}, with the help of the MPPT. Figure 16.5.3 (bottom) shows the circuit diagram of the tapped-inductor OSC_1 that is used in the proposed start-up block, along with its current flow diagram during and after start-up. The transformer and the native MOSFET transistors, M_{1a} and M_{1b}, form an OSC capable of operating with an input voltage of 62mV. Before and during start-up, the OFF_{OSC_1} signal is held at 0V, enabling depletion-mode switches M_2, M_{3a}, and M_{3b}, which complete the oscillation loop. After start-up, OFF_{OSC_1} is pulled to a negative voltage, turning off the depletion switches to stop the oscillation. Additionally, a tie-up block is activated to pull the gate voltages of M_{1a} and M_{1b} to 0V, preventing current from leaking through these transistors during normal operation.

The proposed interface is fabricated in a 0.18μm CMOS process (Fig. 16.5.7). Figure 16.5.4 (top) shows the start-up waveforms after applying a 62mV V_{TEG}, with a 20Ω R_{TEG}. The auxiliary capacitor C_1 is first charged to 0.4V, followed by proper charging of auxiliary capacitors C_{2-3} as designed. Once C_3 is charged, the disable signal, OFF_{OSC_1}, is pulled below −1V, ensuring the depletion switches in OSC_1 remain off. After start-up, V_{OUT} is regulated to 1.2V by the on-chip regulation block immediately. Figure 16.5.4 (bottom) shows the steady-state waveforms when setting the V_{TEG1-4} to 0mV, 100mV, 200mV, and 300mV, respectively, with R_{TEG} at 20Ω. The output of the input matrix, V_A, standing for the four channels' input voltage, proves that all the channels' voltages are close to their corresponding $V_{TEG}/2$, and the MPPT efficiency is higher than 99.7%. The L_{BOOST} current, I_L, confirms that the power switch M_{H1} is turned off at the right time, validating that the ZCS block operates well for all four channels.

To test efficiency improvements for uneven inputs, the multi-TEG system is measured by setting their input voltages (V_{TEG1-4}) to different deviation voltages (V_D) to an average voltage V_{AVG}, which are: $V_{AVG}-2V_D$, $V_{AVG}-V_D$, $V_{AVG}+V_D$, $V_{AVG}+2V_D$, respectively, across three configurations. In addition to the proposed multi-channel configuration, conventional series and parallel configurations were tested for comparison. As shown in Fig. 16.5.5 (top-left), the chip has a 1-channel mode (S_1 on, S_{2-4} off), which enables measurements with four TEGs in either parallel or series connections using the same interface circuit for fair comparisons. The measured efficiencies under different V_{AVG} values (200mV, 100mV, and 50mV) are shown in the rest of Fig. 16.5.5, with varying V_D. It can be observed that the proposed multi-channel MPPT architecture is able to improve the end-to-end (E2E) efficiency by 28.6%, 35.3%, and 39.6%, respectively, when $V_D = 0.5V_{AVG}$. This demonstrates that more energy can be harvested in application scenarios where the temperature distribution on a surface is uneven, compared to conventional parallel or series configurations. Figure 16.5.6 summarizes the proposed interface's performance and benchmarks it against state-of-the-art designs. Compared with the prior multi-TEG systems [11,12], which lack boost conversion and start-up, this work realizes a complete boost conversion architecture and accurate MPP tracking for uneven temperature. Compared with other systems with single-TEG inputs, this work achieves high E2E efficiency, 62-mV start-up without employing any additional inductors (except the main power inductor) or power switches, enabling a high efficiency with a small form factor.

References:
[1] X. Liu, A. Agrawal, A. Tanaka and B. Calhoun, "A 6nA Fully-Autonomous Triple-Input Hybrid-Inductor-Capacitor Multi-Output Power Management System with Multi-Rail Energy Sharing, All-Rail Cold Startup, and Adaptive Conversion Control for mm-scale Distributed Systems," *2024 IEEE International Solid-State Circuits Conference (ISSCC)*, San Francisco, CA, USA, 2024, pp. 152-154. http://doi.org/10.1109/ISSCC49657.2024.10454435

[2] S. S. Amin and P. P. Mercier, "MISIMO: A multi-input single-inductor multi-output energy harvester employing event-driven MPPT control to achieve 89% peak efficiency and a 60,000x dynamic range in 28nm FDSOI," 2018 IEEE International Solid-State Circuits Conference - (ISSCC), San Francisco, CA, USA, 2018, pp. 144-146. http://doi.org/10.1109/ISSCC.2018.8310225

[3] S. Li, X. Liu and B. H. Calhoun, "A 32nA Fully Autonomous Multi-Input Single-Inductor Multi-Output Energy-Harvesting and Power-Management Platform with 1.2×105 Dynamic Range, Integrated MPPT, and Multi-Modal Cold Start-Up," 2022 IEEE International Solid-State Circuits Conference (ISSCC), San Francisco, CA, USA, 2022, pp. 1-3. http://doi.org/10.1109/ISSCC42614.2022.9731732

[4] Z. -Y. Yang et al., "A 93.2%-Efficiency Multi-Input Bipolar Energy Harvester with 17.9× MPPT Loss Reduction," 2023 IEEE International Solid-State Circuits Conference (ISSCC), San Francisco, CA, USA, 2023. http://doi.org/10.1109/ISSCC42615.2023.10067272

[5] Y. Wang, M. Huang, R. P. Martins and Y. Lu, "A SIDO/DISO VCF-Step-Reconfigurable Continuously Scalable-Conversion-Ratio SC Converter Achieving 91.4%/92.6% Peak Efficiency and Almost-lossless Channel Switching," 2024 IEEE International Solid-State Circuits Conference (ISSCC), San Francisco, CA, USA, 2024, pp. 506-508. http://doi.org/10.1109/ISSCC49657.2024.10454453

[6] Y. -S. Noh et al., "A Reconfigurable DC-DC Converter for Maximum TEG Energy Harvesting in a Battery-Powered Wireless Sensor Node," 2021 IEEE International Solid-State Circuits Conference (ISSCC), San Francisco, CA, USA, 2021, pp. 266-268. http://doi.org/10.1109/ISSCC42613.2021.9365811

[7] E. J. Carlson and J. R. Smith, "A ±0.5-mV-Minimum-Input DC-DC Converter With Stepwise Adiabatic Gate-Drive and Efficient Timing Control for Thermoelectric Energy Harvesting," in IEEE Transactions on Circuits and Systems I: Regular Papers, vol. 70, no. 2, pp. 977-990, Feb. 2023. http://doi.org/10.1109/TCSI.2022.3219402

[8] I. Park, J. Jeon, H. Kim, T. Park, J. Jeong and C. Kim, "A Thermoelectric Energy-Harvesting Interface With Dual-Conversion Reconfigurable DC–DC Converter and Instantaneous Linear Extrapolation MPPT Method," in IEEE Journal of Solid-State Circuits, vol. 58, no. 6, pp. 1706-1718, June 2023. http://doi.org/10.1109/JSSC.2022.3214839

[9] T. Lu et al., "A Thermoelectric Energy Harvesting System Assisted by a Piezoelectric Transducer Achieving 10-mV Cold-Startup and 82.7% Peak Efficiency," in IEEE Transactions on Power Electronics, vol. 39, no. 5, pp. 6352-6363, May 2024. http://doi.org/10.1109/TPEL.2024.3362366

[10] S. Carreon-Bautista, A. Eladawy, A. Nader Mohieldin and E. Sánchez-Sinencio, "Boost Converter With Dynamic Input Impedance Matching for Energy Harvesting With Multi-Array Thermoelectric Generators," in IEEE Transactions on Industrial Electronics, vol. 61, no. 10, pp. 5345-5353, Oct. 2014. http://doi.org/10.1109/TIE.2014.2300035

[11] Q. Wan, Y. -K. Teh, Y. Gao and P. K. T. Mok, "Analysis and Design of a Thermoelectric Energy Harvesting System With Reconfigurable Array of Thermoelectric Generators for IoT Applications," in IEEE Transactions on Circuits and Systems I: Regular Papers, vol. 64, no. 9, pp. 2346-2358, Sept. 2017. http://doi.org/10.1109/TCSI.2017.2708763

[12] Q. Wan and P. K. T. Mok, "A 14-nA, Highly Efficient Triple-Output Thermoelectric Energy Harvesting System Based on a Reconfigurable TEG Array," in IEEE Journal of Solid-State Circuits, vol. 54, no. 6, pp. 1720-1732, June 2019. http://doi.org/10.1109/JSSC.2019.2899973

[13] J. -P. Im, S. -W. Wang, S. -T. Ryu and G. -H. Cho, "A 40 mV Transformer-Reuse Self-Startup Boost Converter With MPPT Control for Thermoelectric Energy Harvesting," in IEEE Journal of Solid-State Circuits, vol. 47, no. 12, pp. 3055-3067, Dec. 2012. http://doi.org/10.1109/JSSC.2012.2225734

[14] Y. -K. Teh and P. K. T. Mok, "Design of Transformer-Based Boost Converter for High Internal Resistance Energy Harvesting Sources With 21 mV Self-Startup Voltage and 74% Power Efficiency," in IEEE Journal of Solid-State Circuits, vol. 49, no. 11, pp. 2694-2704, Nov. 2014. http://doi.org/10.1109/JSSC.2014.2354645

[15] D. Rozgić and D. Marković, "A 0.78mW/cm² autonomous thermoelectric energy-harvester for biomedical sensors," 2015 Symposium on VLSI Circuits (VLSI Circuits), Kyoto, Japan, 2015, pp. C278-C279. http://doi.org/10.1109/VLSIC.2015.7231289

[16] C. Xue et al., "A Compact Dickson Hybrid Boost Converter With 5-mV Input 90.5% Peak Efficiency and On-Chip Cold-Start for Thermoelectric Energy Harvesting," in IEEE Transactions on Circuits and Systems I: Regular Papers, vol. 71, no. 12, pp. 5596-5606, Dec. 2024. http://doi.org/10.1109/TCSI.2024.3383425

[17] P. Cao, Y. Qian, P. Xue, D. Lu, J. He and Z. Hong, "An 84% Peak Efficiency Bipolar-Input Boost/Flyback Hybrid Converter With MPPT and on-Chip Cold Starter for Thermoelectric Energy Harvesting," 2019 IEEE International Solid-State Circuits Conference - (ISSCC), San Francisco, CA, USA, 2019, pp. 420-422. http://doi.org/10.1109/ISSCC.2019.8662368

[18] P. -H. Chen et al., "Startup Techniques for 95 mV Step-Up Converter by Capacitor Pass-On Scheme and V_{TH}-Tuned Oscillator With Fixed Charge Programming," in IEEE Journal of Solid-State Circuits, vol. 47, no. 5, pp. 1252-1260, May 2012. http://doi.org/10.1109/JSSC.2012.2185589

[19] P. -H. Chen et al., "An 80 mV Startup Dual-Mode Boost Converter by Charge-Pumped Pulse Generator and Threshold Voltage Tuned Oscillator With Hot Carrier Injection," in IEEE Journal of Solid-State Circuits, vol. 47, no. 11, pp. 2554-2562, Nov. 2012. http://doi.org/10.1109/JSSC.2012.2210953

[20] A. Shrivastava, N. E. Roberts, O. U. Khan, D. D. Wentzloff and B. H. Calhoun, "A 10 mV-Input Boost Converter With Inductor Peak Current Control and Zero Detection for Thermoelectric and Solar Energy Harvesting With 220 mV Cold-Start and -14.5 dBm, 915 MHz RF Kick-Start," in IEEE Journal of Solid-State Circuits, vol. 50, no. 8, pp. 1820-1832, Aug. 2015. http://doi.org/10.1109/JSSC.2015.2412952

[21] S. Bose, T. Anand and M. L. Johnston, "A 3.5-mV Input Single-Inductor Self-Starting Boost Converter With Loss-Aware MPPT for Efficient Autonomous Body-Heat Energy Harvesting," in IEEE Journal of Solid-State Circuits, vol. 56, no. 6, pp. 1837-1848, June 2021. http://doi.org/10.1109/JSSC.2020.3042962

[22] S. Bose, T. Anand and M. L. Johnston, "Integrated Cold Start of a Boost Converter at 57 mV Using Cross-Coupled Complementary Charge Pumps and Ultra-Low-Voltage Ring Oscillator," in IEEE Journal of Solid-State Circuits, vol. 54, no. 10, pp. 2867-2878, Oct. 2019. http://doi.org/10.1109/JSSC.2019.2930911

[23] J. Goeppert and Y. Manoli, "Fully Integrated Startup at 70 mV of Boost Converters for Thermoelectric Energy Harvesting," in IEEE Journal of Solid-State Circuits, vol. 51, no. 7, pp. 1716-1726, July 2016. http://doi.org/10.1109/JSSC.2016.2563782

[24] B. -M. Lim, J. -I. Seo and S. -G. Lee, "A Colpitts Oscillator-Based Self-Starting Boost Converter for Thermoelectric Energy Harvesting With 40-mV Startup Voltage and 75% Maximum Efficiency," in IEEE Journal of Solid-State Circuits, vol. 53, no. 11, pp. 3293-3302, Nov. 2018. http://doi.org/10.1109/JSSC.2018.2863951

[25] P. -S. Weng, H. -Y. Tang, P. -C. Ku and L. -H. Lu, "50 mV-Input Batteryless Boost Converter for Thermal Energy Harvesting," in IEEE Journal of Solid-State Circuits, vol. 48, no. 4, pp. 1031-1041, April 2013. http://doi.org/10.1109/JSSC.2013.2237998

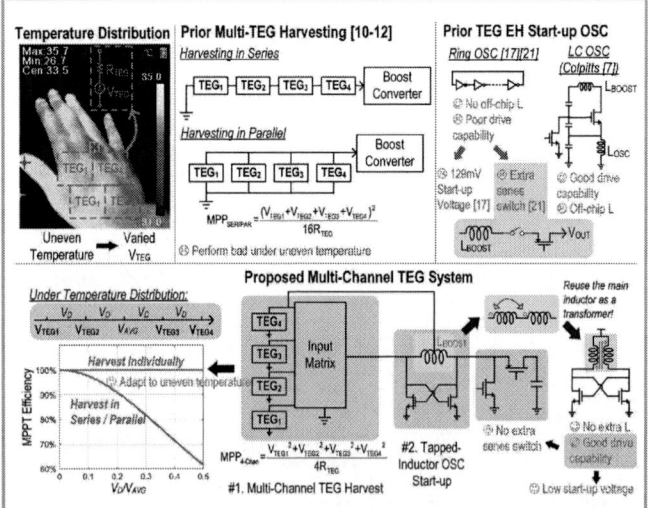

Figure 16.5.1: Uneven distribution of temperature (top left), prior multi-TEG EH (top middle), prior TEG EH start-up oscillators (top right), proposed EH System (bottom).

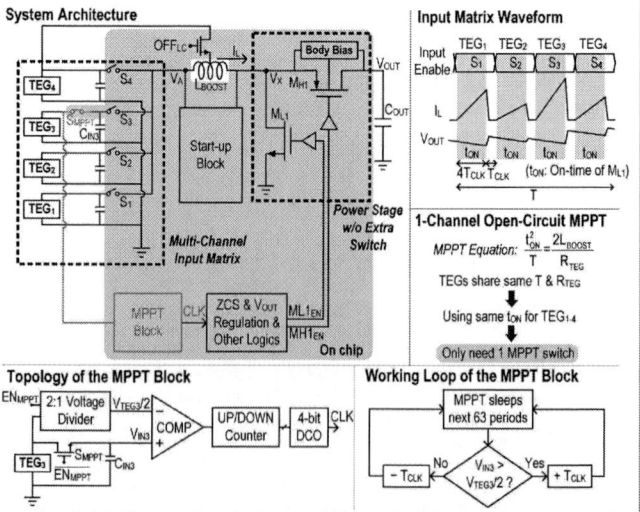

Figure 16.5.2: System architecture (top left), input matrix working sequence (top right), MPPT using single FOCV (middle right), its topology and working loop (bottom).

ISSCC 2026 / SESSION 16 / ENERGY HARVESTING, PIEZO AND CHARGERS / 16.5

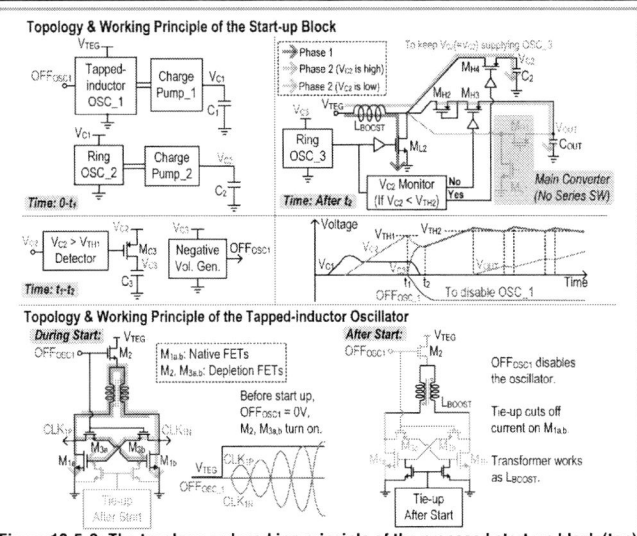

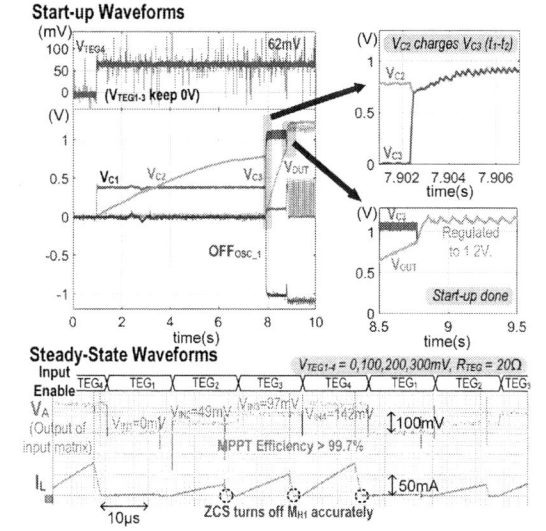

Figure 16.5.3: The topology and working principle of the proposed start-up block (top) and the tapped-inductor oscillator (bottom).

Figure 16.5.4: Measured waveforms in start-up (top) and steady-state (bottom).

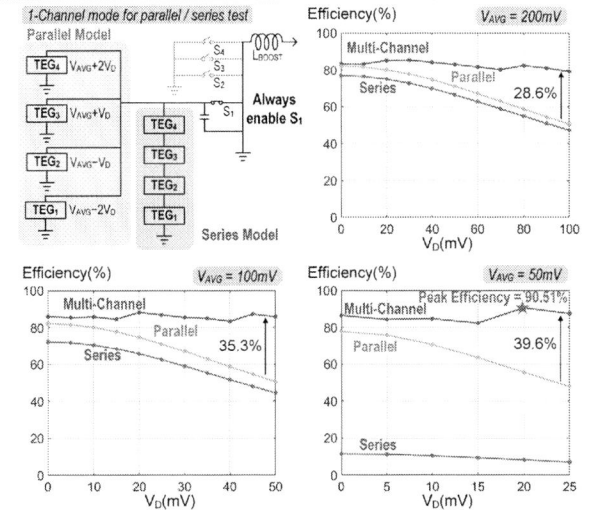

Figure 16.5.5: Measured end-to-end (E2E) efficiency of the multi-channel TEG EH interface versus parallel and series input.

Figure 16.5.6: Performance summary and comparison with the prior state-of-the-art TEG EH interfaces.

	TACSI'17 [11]	JSSC '19 [12]	This Work	ISSCC '19 [17]	JSSC '21 [21]	JSSC '18 [24]	ISSCC '24 [1]
Process	0.35μm	0.35μm	0.18μm	0.18μm	0.18μm	65nm	65nm
Source	TEG	TEG	TEG	TEG	TEG	TEG	TEG,PV
R_{TEG} (Ω)	1500	1500	20	210	5	2.2	N/A
MPPT	Yes	Yes	Yes	Yes	Yes	Yes	Yes
Output Voltage (V)	3-4.2	1.2/2.7/3.6	1.2	0.8	1.2	1.1	0.6/1/1.6
Ouput Regulation	No	Yes	No	Yes	Yes	Yes	Yes
Multiple TEG?	Yes	Yes	Yes	No			
Able to Boost Voltage?	No	No	Yes	Yes	Yes	Yes	Yes
Peak E2E Efficiency	N/A	99%#	90.51%	84%	80%	75%	90.1%
Uneven Temperature MPPT?	No	No	Yes				
Self Start-up	No		Yes	Yes	Yes	Yes	Yes
Magnet Components Number			1	1	1	2	1
Add. Power Switch			0	0	1	0	0
Minimum Start-up Voltage(mV)			62	129	50	40	400*

#: The system does not have boost converter, so only has good efficiency at high (>1V) V_{IN}, which is out of most TEGs working range.

*: Estimated from the paper.

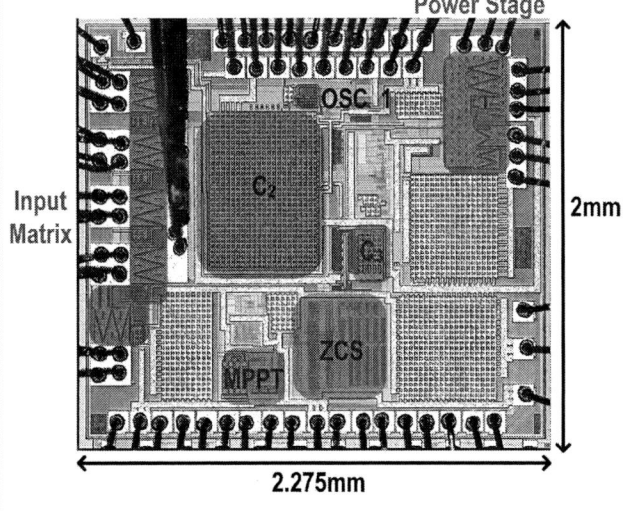

Figure 16.5.7: Chip micrograph.

ISSCC 2026 / SESSION 16 / ENERGY HARVESTING, PIEZO AND CHARGERS / 16.6

16.6 A Parameter-Free Runtime-Energy-Loss Optimizer Achieving 2.15% Error in Energy-Recycling Duty-Cycled Systems

Jianxin Yang, Rui P. Martins, Mo Huang

University of Macau, Macau, China

Abstract

This work proposes a predefined-parameter-free runtime-E_{loss} optimizer for energy-recycling duty-cycled systems. By leveraging a 3-level converter to recycle output-capacitance energy to the flying capacitor, it achieves <2.15% deviation from optimal E_{loss} with increased robustness against battery voltage and leakage current variations. It achieves a quiescent current of 15nA and an 80% efficiency at a 1µA load current, while eliminating sub-harmonic oscillation. It saves energy loss by up to 60% across a wide range of time spent in the sleep state.

Wearable devices impose stringent constraints on volume and cost, driving the use of low-profile batteries and advanced power management techniques to extend battery life. These techniques include low-duty-cycle operation and power gating (PG) in the loading microprocessor (µP). In addition, high power conversion efficiency (PCE) under light-load conditions in the supply buck converter is beneficial. To further reduce the footprint and cost, µPs often integrate the buck converter (Fig. 16.6.1).

Local PG schemes require per-block PG switches (SW_{PG}) that occupy significant silicon area to lower their turn-on resistance [1]. In contrast, global PG schemes [2] eliminate per-block SW_{PG}s by disabling the converter during sleep periods. This causes the output voltage V_0, which is equal to the reference voltage V_{ref} during active periods, to drop from its voltage at the start of the sleep period V_{sleep} to a voltage V_{wake} at the end of the sleep period as the load leakage current I_{lk} dissipates the energy of the output capacitor C_0. Upon wake-up, C_0 recharges (Rchg) to V_{ref}. The total energy loss E_{loss} is evident in the variation of the battery input energy E_{in} during the recharge period. E_{loss} can be expressed as the sum of the energy lost during sleep E_{sleep} and during recharging E_{rchg}.

E_{sleep}, which is proportional to $V_{sleep}^2 - V_{wake}^2$, can dominate the total loss. To reduce it, [3-5] proposed C_0 charge recycling (Recy). The converter is reconfigured as a boost converter, recycling the charge to the battery [4] at the onset of sleep, thereby lowering V_0 to a small V_{sleep} and reducing E_{sleep}. However, E_{loss}, now comprising E_{sleep}, E_{rchg} and energy lost in recycling E_{recy} [eq. (1)], does not monotonically decrease with V_{sleep} (as shown in the energy-V_{sleep} curve of Fig. 16.6.1). A low V_{sleep} increases the total energy processed by the converter, while forcing the converter to operate at high voltage conversion ratios (VCR). These conditions degrade both boost and buck efficiencies η_{boost} and η_{buck}, raising $E_{rchg} + E_{recy}$ and potentially negating the energy savings from recycling. Therefore, it is necessary to select an optimal V_{sleep} to balance E_{sleep} against $E_{recy} + E_{rchg}$. In [4], a runtime E_{loss} optimizer is employed to dynamically determine the optimal V_{sleep} based on eq. (1), where E_{loss} is a function of V_{sleep}, T_{sleep}, I_{lk}, η_{boost} and η_{buck}. Since the latter three parameters are difficult to measure in real time, they require predefinition, which compromises their accuracy under varying operating conditions.

This work proposes a more robust alternative by recycling the charge into a storage capacitor C_{STO}, as shown in Fig. 16.6.1. During the recharge period, C_{STO} recharges the previously recycled energy back to C_0, raising V_0 to an intermediate voltage V_{int} until C_{STO}'s charge balances. Subsequently, V_0 charges from V_{int} to V_{ref} using battery energy. Therefore, E_{loss} is proportional to a simple expression of $V_{ref}^2 - V_{int}^2$ [eq.(2)], which implies that maximizing V_{int} directly minimizes E_{loss}. This approach eliminates the need for the predefined parameters in eq. (1), significantly improving the design simplicity and robustness over [4]. In addition, by designing the C_{STO} voltage V_{STO} to be lower than the battery voltage V_{BAT}, $E_{rchg} + E_{recy}$ is reduced, leading to a lower E_{loss}.

In Fig. 16.6.2 we propose a 3-level converter that recycles C_0 energy to its flying capacitor C_F, which is used as C_{STO}. This topology halves the voltage stress on the four switches (M_1 to M_4) to $V_{BAT}/2$, ensuring they are compatible with standard CMOS transistors rated at 2.5V/3.3V. Compared to a conventional buck converter integrated into the µP, which also requires stacked switches to sustain V_{BAT} (up to 4.2V for Li-ion battery) [6], the total switch count (SW_{CNT}) remains unchanged. Figure 16.6.2 exhibits the converter's working states. In normal mode, performing step down conversion with constant on-time (COT) control, the control sequence is State1, 3, 0, 2, 3, 0, repeated periodically. State1, 2, and 3 follow the operations of a conventional 3-level converter, while State0 deactivates all switches. This sequence guarantees the auto balancing of C_F's voltage, V_{CF}, at $V_{BAT}/2$ [7]. Additionally, the recycle mode employs State4 and 5, while the recharge mode (transferring energy from C_F to C_0) reuses State2 and 3.

Figure 16.6.2 also shows the optimal-E_{loss} search process, starting from a sleep mode interval T_{sleep1}. The search uses a perturbation-observation signal $V_{P\&O}$ as V_{sleep}, producing a corresponding V_{int} after C_F recharges C_0. This V_{int} is then sampled and stored. Subsequently, $\pm1\times V_{step}$ perturbs $V_{P\&O}$, yielding a new V_{int} for comparison with the previously stored value to decide the polarity of the next $V_{P\&O}$. Iterations allow the convergence of V_{sleep} to its optimal value, while causing a steady-state fluctuation between two or three levels.

Depending on the T_{sleep} value, the optimal E_{loss} may occur at the maximum of V_{sleep} (no recycle) or the minimum $V_{sleep} = 0$ (full recycle), leading to a two-level fluctuation. Other cases yield a three-level behavior (Fig. 16.6.2). The simulated maximum search error between the optimal and average E_{loss} values theoretically occurs in two-level scenarios and is 1.67% with $V_{STEP} = 68.75mV$.

To accelerate convergence over a wide T_{sleep} range, we propose a coarse-fine tuning (Fig. 16.6.2). Initially, a lookup table (LUT) stores the simulated optimal V_{sleep} values at discrete T_{sleep} points. When T_{sleep} changes from T_{sleep1} to T_{sleep2}, the LUT provides a coarse estimation of V_{sleep}, followed by a fine-tuning to reduce error.

Figure 16.6.3 presents the schematic of the proposed converter, integrating power switches and control circuits. Switches M_1, M_2 and M_3, M_4 are P-type and N-type 2.5-V devices, controlled by signals S_1-S_4. A switched-capacitor (SC) circuit generates voltages for the gate drivers of M_1 and M_4, as well as the controller supply. An LDO pre-charges C_F during startup.

Conventional COT control uses a comparator (CMP_1) to trigger a short pulse (denoted as ON = '1') when $V_0 < V_{ref}$. This triggers either State1 or 2 (represented by pulses on S_1 or S_2) that energizes the inductor for a fixed duration T_{ON}. Once I_L returns to zero at t_1, which is detected by the zero current detector (ZCD) CMP_3, the converter transitions to State 0. Since CMP_1 remains activated continuously and dominates the quiescent current (I_Q) of low-power converters, it must be highly power-efficient. However, reducing CMP_1's I_Q often induces sub-harmonic oscillation [8]. While the actual crossing of V_0 above V_{ref} occurs at t_2 (Fig. 16.6.3), the delay of CMP_1 extends the ON = '1' pulse to t_3. If $t_3 > t_1$, the controller erroneously forces the inductor to be re-energized, which leads to sub-harmonic oscillations and increased V_0 ripple. The comparator input offset can cause a similar failure. This issue is partially mitigated in [8] by momentarily increasing I_Q during T_{ON}.

To address this issue, a stacked-RAMP scheme is employed (Fig. 16.6.3), where capacitor C_{ramp} charges during T_{ON} and discharges over the remainder of the cycle. The resulting triangular voltage across C_{ramp} is superimposed on V_0, producing the voltage V_{ramp} that is compared with V_{ref} in CMP_1. The large differential input rapidly ends the ON pulse before t_3, eliminating a sub-harmonic oscillation. Additionally, the large differential input greatly reduces CMP_1's I_Q [5], incurring only 8% power loss at load current, $I_{LOAD} = 1µA$ (Fig. 16.6.3). CMP_2 generates an OFF pulse to end the inductor energization, with T_{ON} set by an offset voltage V_{offset}. Both CMP_2 and CMP_3 operate intermittently to save I_Q, while the bias currents of CMP_1 and the C_{ramp} discharging circuit are adaptively scaled with I_{LOAD}. A finite-state machine (FSM) executes the search algorithm for minimal E_{loss}, outputting a code (PO [3:0]) of a digital-to-analog converter (DAC) that produces $V_{P\&O}$. To save I_Q, CMP_1 is reused for $V_{P\&O}$-V_0 comparison.

Fabricated in 65-nm CMOS, the prototype converter occupies an active area of 0.458mm² (Fig. 16.6.7). Figure 16.6.4 shows the measured waveforms for normal, recycle, sleep and recharge modes at $T_{sleep} = 2ms$ and $V_0 = 1.1V$, where V_{int} settles to 0.88V after recharging. V_{CF} is auto balanced at $V_{BAT}/2 = 2.1V$, as predicted. E_{loss} at different V_{sleep} values is manually measured, with a minimum of 1.8µJ identified at $V_{sleep,OPT} = 0.55V$ (Fig. 16.6.4). Due to the P&O behavior in the proposed auto search algorithm, V_{sleep} fluctuates around $V_{sleep,OPT}$, yielding an average E_{loss} of 1.81µJ, slightly above the minimum value. During T_{sleep} transition between 0.5ms and 15ms, coarse-tuning is evident from abrupt $V_{P\&O}$ changes, while fine-tuning achieves convergence within approximately two cycles, in both up and down transitions. In addition, Fig. 16.6.4 compares the measured E_{loss} versus T_{sleep} under different conditions. Across the full T_{sleep} range, the proposed scheme achieves consistently lower E_{loss} than no recycling and fixed-V_{sleep} (e.g., 0.2V and 0.95V) cases. We use the 0.2V V_{sleep} case as the baseline in Fig. 16.6.6. The maximum E_{SAVE} given in [eq.(3)] is around 60%.

Figure 16.6.5 presents the measured E_{loss} error across $V_{BAT} = 2.8V$, 3.6V and 4.2V, and for two load circuits (SoC1 and SoC2, with leakage currents of 4.6mA and 2.2mA at $V_0 = 1.1V$, respectively). The maximum error is 2.15%, validating robustness against variations in the predefined parameters η_{buck}, η_{boost}, and I_{lk} that are required in [4]. Figure 16.6.5 also plots the measured PCE operating at $V_0 = 1.1V$, which peaks at 93.6%. Due to our low-power design (measured 15-nA quiescent current at V_{BAT} = 3.6V), the PCE remains above 80% even with a 1-µA load. Moreover, Fig. 16.6.5 compares the measured waveforms with and without the stacked-RAMP scheme at I_{LOAD} = 150µA. Sub-harmonic oscillations is eliminated when the stacked RAMP is activated, reducing the V_O ripple from 12mV to 6mV.

Figure. 16.6.6 compares the proposed design with state-of-the-art low-power converters. This work achieves a comparable or better peak PCE, while attaining the lowest quiescent current of 15nA. The proposed stacked-RAMP scheme eliminates sub-harmonic oscillations, achieving the lowest V_O ripple. Among designs employing C_O energy recycling, [5] and this work recycle energy to a storage capacitor at $V_{BAT}/2$, reducing energy loss in recycling and recharge modes, yielding comparable E_{save} but at a higher V_{BAT} = 4.2V than in [4] (= 1.8V). In addition, [4] and this work support runtime E_{loss} optimization, significantly reducing error from optimal E_{loss} when compared with the baseline design of leaving V_{sleep} fixed at 0.2V. Unlike [4] that relies on pre-characterization of the design parameters η_{buck}, η_{boost}, and I_{lk}, our approach eliminates these dependencies, reducing the nominal-condition error to 1.91%. Meanwhile, our design maintains strong robustness with a maximum E_{loss} error of 2.15% across V_{BAT} and I_{lk} variations, while [4] is prone to large errors under these variations.

Acknowledgement:
This work was supported in part by the Science and Technology Development Fund, Macau SAR (0029/2025/AMJ, 0042/2025/RIB1 and 004/2023/SKL), in part by the Guangdong-HongKong-Macao Joint Laboratories (2025B1212150003), in part by the Guangdong Basic and Applied Basic Research Foundation (2023B1515130001), in part by the Research Committee of University of Macau (MYRG-GRG2024-00041-IME).

References:
[1] Z. Fan et al., "AIMMI: Audio and Image Multi-Modal Intelligence via a Low-Power SoC With 2-MByte On-Chip MRAM for IoT Devices," in *IEEE Journal of Solid-State Circuits*, vol. 59, no. 10, pp. 3488-3501, Oct. 2024. https://doi.org/10.1109/JSSC.2024.3410306
[2] C. -H. Huang et al., "Regenerative Breaking: Optimal Energy Recycling for Energy Minimization in Duty-Cycled Domains," in *IEEE Journal of Solid-State Circuits*, vol. 58, no. 1, pp. 68-77, Jan. 2023. https://doi.org/10.1109/JSSC.2022.3221143
[3] M. Alioto et al., ""EChO" Reconfigurable Power Management Unit for Energy Reduction in Sleep-Active Transitions," in *IEEE Journal of Solid-State Circuits*, vol. 48, no. 8, pp. 1921-1932, Aug. 2013. https://doi.org/10.1109/JSSC.2013.2258816
[4] C. -H. Huang et al., "Energy Minimization of Duty-Cycled Systems Through Optimal Stored-Energy Recycling from Idle Domains," *2022 IEEE International Solid-State Circuits Conference (ISSCC)*, San Francisco, CA, USA, Feb. 2022. https://doi.org/10.1109/ISSCC42614.2022.9731611
[5] J. Yang et al., "25-nA Modified Hybrid Ladder Converter with Efficient Output-Capacitor Charge Recycling and 90% Battery Lifetime Extension," *2025 IEEE Custom Integrated Circuits Conference (CICC)*, Boston, MA, USA, Apr. 2025. https://doi.org/10.1109/CICC63670.2025.10983162
[6] P. Renz et al., "Switch Stacking in Power Management ICs," in *IEEE Journal of Emerging and Selected Topics in Power Electronics*, vol. 9, no. 3, pp. 3735-3743, Jun. 2021. https://doi.org/10.1109/JESTPE.2020.3012813

[7] J. Xue et al., "A 2 MHz 12–100 V 90% Efficiency Self-Balancing ZVS Reconfigurable Three-Level DC-DC Regulator With Constant-Frequency Adaptive-On-Time V2 Control and Nanosecond-Scale ZVS Turn-On Delay," in *IEEE Journal of Solid-State Circuits*, vol. 51, no. 12, pp. 2854-2866, Dec. 2016. https://doi.org/10.1109/JSSC.2016.2606581
[8] W. Huang et al., "A 240-nA Quiescent Current, 95.8% Efficiency AOT-Controlled Buck Converter With A2-Comparator and Sleep-Time Detector for IoT Application," in *IEEE Transactions on Power Electronics*, vol. 36, no. 11, pp. 12898-12909, Nov. 2021. https://doi.org/10.1109/TPEL.2021.3082896
[9] Y. Lee et al., "A High-Efficiency High-Voltage-Tolerant Buck Converter with Inductor Current Emulator for Battery-Powered IoT Devices," in *IEEE Transactions on Power Electronics*, vol. 38, no. 9, pp. 10917-10932, Sep. 2023. https://doi.org/10.1109/TPEL.2023.3289834

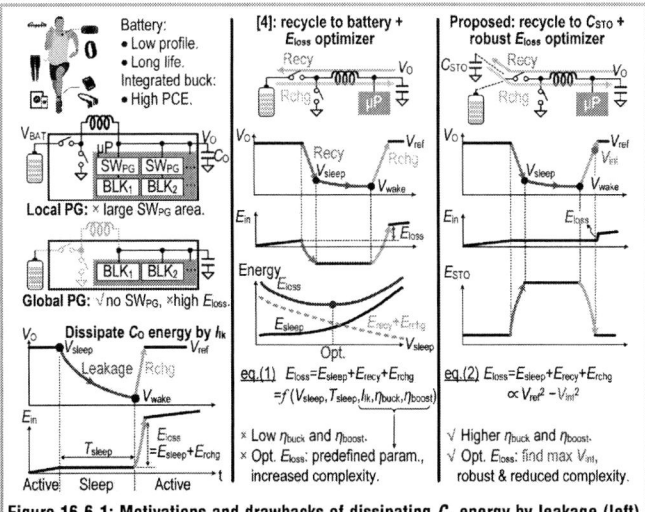

Figure 16.6.1: Motivations and drawbacks of dissipating C_O energy by leakage (left), recycled to battery through a runtime E_{loss} optimizer (middle), this work recycles it to C_{STO} (right).

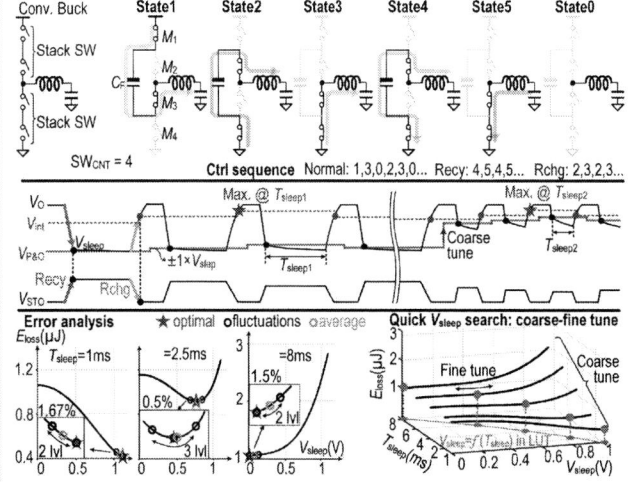

Figure 16.6.2: Operating principles (top), waveforms of auto-search optimal E_{loss} (middle), E_{loss} error analysis (bottom left), coarse-fine tune for quick search (bottom right).

ISSCC 2026 / SESSION 16 / ENERGY HARVESTING, PIEZO AND CHARGERS / 16.6

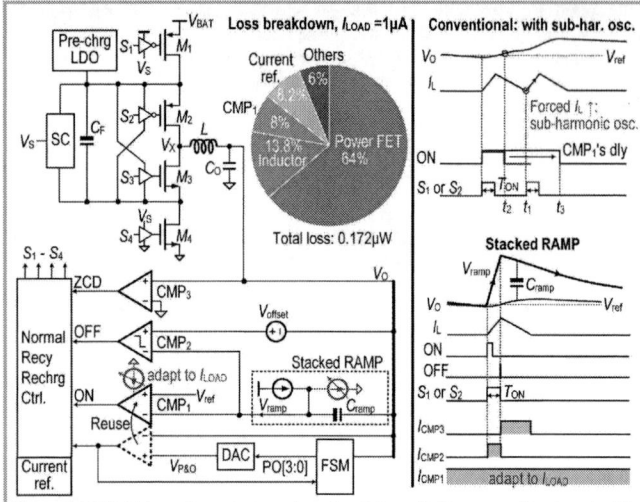

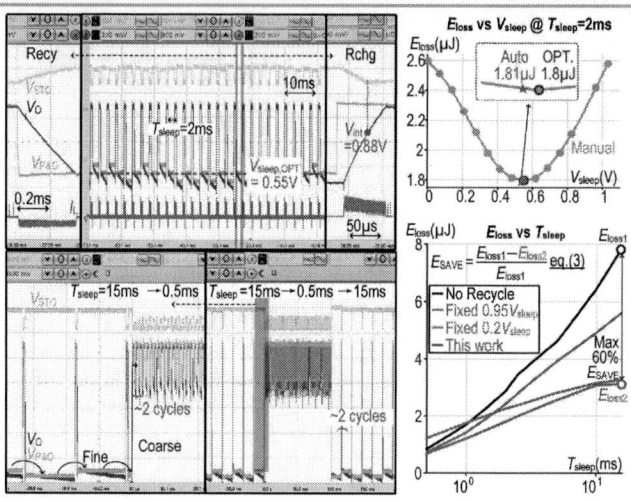

Figure 16.6.3: Schematic and power loss breakdown (left), conventional control has sub-harmonic oscillation (right top), eliminated by the proposed stacked RAMP (right bottom).

Figure 16.6.4: Measured waveforms (top left), E_{sleep} vs. V_{sleep} at T_{sleep} = 2ms (top right), transition between 0.5ms and 15ms T_{sleep} (bottom left), E_{loss} vs. T_{sleep} at different conditions (bottom right).

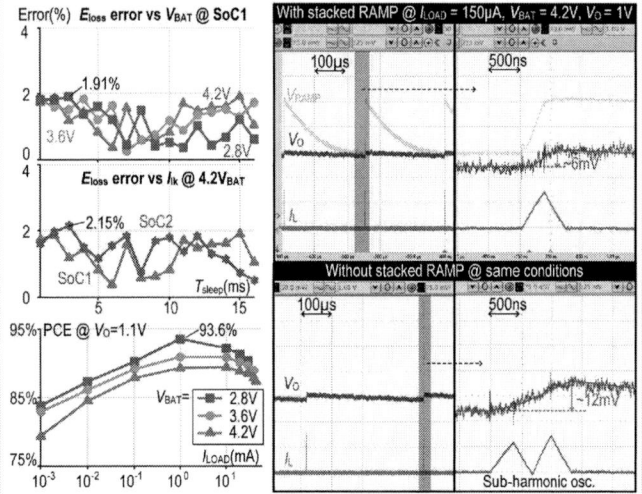

Figure 16.6.5: Measured E_{loss} error vs. V_{BAT} and I_{lk} (left top), PCE vs. I_{LOAD} (left bottom), V_O and I_L waveforms with and without stacked-RAMP scheme (right).

Figure 16.6.6: Performance comparison with state-of-the-art works.

	This work	ISSCC' 22 [4]	JSSC' 13 [3]	CICC' 25 [5]	TPE' 23 [9]
Technology	65nm	65nm	65nm	65nm	28nm
Area	0.458mm²	2.6mm²	0.78mm²	0.47mm²	0.84mm²
Topology	3-level	Buck	SC	Hybrid ladder	Stack Buck
L, C_O	4.7µH, 10µF	10µH, 10µF	N.A., 2nF	4.7µH, 10µF	47µH, 2.2µF
Flying capacitors	4.7µF	0	64×32pF	2×4.7µF	0
V_{IN}	2.8-4.2V	1.8V	1.2V	2.8-5V	2.8-4.2V
V_O	0.5-1.1V	0.6-1V	0.55V	0.5-1V	1V
V_O ripple	6mV	20mV	54mV	10mV	20mV
I_Q	15nA	N.A.	N.A.	25nA	123nA
I_{LOAD} range	1µA-45mA	5mA-50mA	N.A.	1µA-35mA	0.5µA-1mA
Peak PCE	93.6%	91.8%	80%	92.4%	70%
PCE @ light load	83.8%@1µA	79%@5mA	67%@2.5µA	82.9%@1µA	64.8%@1µA
C_O energy recycle to	CAP (V_{BAT}/2)	Battery	Battery	CAP (V_{BAT}/2)	
Max. E_{save} @ V_{BAT}	60%@4.2V	59.4%@1.8V	40%@1.2V	64%@4.2V	
Runtime E_{loss} optimize	No, $0.2V_{sleep}$ (baseline)	Enable	Yes	No	No
Predefined parameters	N.A.	No	I_k, η_{buck}, η_{boost}	N.A.	N.A.
T_{sleep} range	N.A.	0.5ms-15ms	0.2ms-3.2ms	N.A.	N.A.
Max. E_{loss} error @ nominal conditions	80.19%	1.91%	4%	N.A.	
Error under V_{BAT}, I_{lk} variations	N.A.	2.15%	Theoretically large		

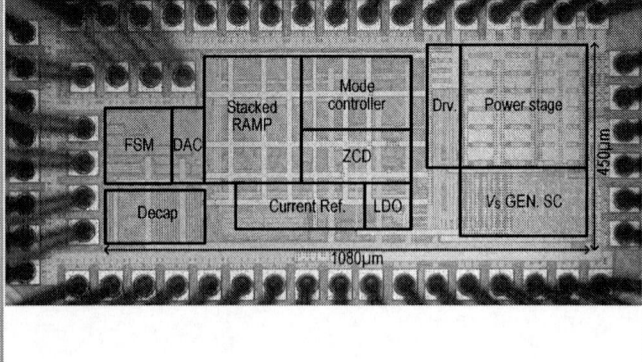

Figure 16.6.7: Chip micrograph.

ISSCC 2026 / SESSION 16 / ENERGY HARVESTING, PIEZO AND CHARGERS / 16.7

16.7 A 90.7%-Efficiency Piezoelectric Resonator-Based Sigma Converter with 6-Phase 2-DoF On-Chip Regulation and Zero-Standby Sigma Mode for Transient and Output Power Enhancement

Shunmin Jiang[1], Tiansu Wang[1], Patrick Mercier[2], Sijun Du[1]

[1]Delft University of Technology, Delft, The Netherlands, [2]University of California, San Diego, CA

Abstract

This paper proposes an on-chip sigma regulation system for the accurate and power-efficient control of piezoelectric resonator (PR)-based DC-DC converters to address the challenges of conventional off-chip controllers. A zero-standby sigma-mode is implemented for transient response/output power improvement. Peak efficiencies of 91.2% and 90.7% are achieved for the PR converter and the PR-sigma converter respectively.

Piezoelectric resonators (PRs) are promising inductor substitutes for compact, magnetically-free DC-DC power conversion. Operating in the inductive band between the resonance (f_r) and anti-resonance (f_{ar}) frequencies, they offer low mechanical impedance and efficient energy transfer (Fig. 16.7.1, top). Various PR-based converter topologies have been developed [1-14]; however, the control mechanisms have been challenging since the undesired vibration harmonics corrupt the voltage/current signals and complicate timing detection, leading to system instability or low performance. As shown in Fig. 16.7.1 (middle-left), open-loop schemes [1-9] struggle with complex switching and amplitude-dependent frequency [15]; while closed-loop designs [10-14] require bulky, power-hungry digital controllers. Moreover, the start-up remains unresolved due to the varying resonance and vibration dynamics. To address these challenges, this work proposes a fully on-chip control system for a PR-based DC-DC converter with a 6-phase 2-degree-of-freedom (2-DoF) control scheme, enabling adaptive frequency tracking, robust output regulation, and static start-up (Fig. 16.7.1 bottom-right). Implemented with four main power switches, S_1-S_4, the converter achieves an adaptive voltage conversion ratio (VCR) up to 0.44, and the scheme is extendable to other PR-based topologies.

While full on-chip control ensures stability and output regulation, solely PR-based converters still suffer from slow transients, since their vibration amplitude must settle before delivering large or small currents, limiting their use with supply-sensitive loads. Pairing an LDO-assist with a DC-DC converter can be an impactful way to improve transient performance [16-18] (Fig. 16.7.1, bottom-left), but this approach has not been used for PR-based converters yet. Cascaded converter-LDO structures ensure fast response, but this incurs cascaded losses and limits output current. In contrast, the sigma converter approach places the LDO in parallel. The sigma converter with a series-input type achieves higher efficiency by splitting V_{IN} into V_{HIGH} and V_{LOW} domains, but its efficiency depends on V_{IN} V_{LOW}>V_{OUT}, limiting the VCR. In addition, the LDO always draws static power, reducing overall efficiency. Other sigma variants suffer from similar drawbacks. To tackle the slow transients of sole-PR DC-DC converters while addressing the challenges in conventional sigma structures, this work proposes a PR-based sigma converter with a zero-standby-current low-side path (Fig. 16.7.1, bottom-right). A 2:1 switched-capacitor (SC) stage is inserted before the LDO. In steady state, the SC output (V_{LOW}) is saturated at V_{IN}/2 by regulating the PR converter's power, forcing the LDO current (I_{LOW}) to near zero. Thus, the overall power conversion efficiency (PCE) equals that of the PR converter and becomes independent of the VCR. During transients, the LDO instantly regulates V_{OUT}, ensuring fast response. At very heavy loads, when PR efficiency drops due to mechanical damping at large vibration amplitudes, the LDO seamlessly supplies the extra current, elegantly extending the maximum output power. The proposed architecture therefore combines a high-efficiency PR path (high side) and an auxiliary zero-standby path (low side), to be detailed in Figs. 16.7.2 and 16.7.3.

Figure 16.7.2 shows the proposed 6-phase 2-DoF PR-based DC-DC converter. The top-left and top-right parts show the control scheme and operating waveforms over two PR periods. During Φ_1, S_1 and S_2 turn on, applying V_{IN}-V_{OUT} across the PR to energize it and charge V_{OUT}. The Φ_1 duration is set by a PWM regulator with a voltage-mode type-III compensator and it determines the output power, which provides the first DoF. In Φ_2, S_1 turns off and the voltage across the PR, V_P, naturally discharges to 0V with the PR open. When V_P reaches 0V, Φ_3 starts, and S_2/S_4 turn on to short the PR to balance the energy, and its end must align with the current zero-crossing (ZC) of the PR current, I_P. A ZC detection (ZCD) block senses this moment using the voltage across S_4. A key challenge for ZCD is that harmonic currents from undesired vibration modes in the MHz range can cause false triggering. Such harmonics often lead to early detection, resulting in PR over-discharging and an incorrect phase-locked loop (PLL) reference in a later phase (Φ_5). To prevent this, an arbiter is introduced to validate the triggering time and filter out spurious short pulses.

After the ZC point, the circuit enters Φ_4, where the PR is open and I_P charges V_P until it reaches V_{OUT}, triggering Φ_5. Phases Φ_5 and Φ_6 must be jointly optimized so that the end of Φ_6 aligns with the ZC of I_P. To prevent hard charging on C_P at the end of Φ_6 (Fig. 16.7.2 top-left), Φ_5 is adjusted such that I_P charges C_P from V_{OUT} to V_{IN}-V_{OUT} precisely within Φ_6. Because I_P is nonlinear, predictive calculation of Φ_5 is impractical. Instead, a digital feedback calibration is adopted. The HALF point (end of Φ_6), indicating the end of a half PR period, is precisely generated by a low-bandwidth charge-pump PLL to suppress harmonic noise.

At this point, the C_P voltage is compared with V_{IN}-V_{OUT}, and a 7-bit delay line tunes the Φ_5 duration cycle-by-cycle, providing the second DoF for charge and energy balance. After Φ_6, the next PR period starts, repeating the six phases.

The bottom-right of Fig. 16.7.2 shows the start-up sequence. From the static state, the PR is excited by a short burst of fixed frequency pulses within its inductive band, building up stable resonance in phases 1 and 2. The circuit then enters phase 3 with the PR shorted for initial lock-in. After a few cycles, the frequency and phase are recorded, and normal six-phase DC-DC operation starts.

Figure 16.7.3 illustrates the architecture and operation of the proposed sigma converter. The top part shows the system diagram: the high-side PR DC-DC converter is combined with a low-side path where the 2:1 SC output voltage (V_{LOW}) serves as the reference for the PR's PWM-based slow-loop regulation. When V_{LOW}<V_{IN}/2, indicating I_{LOW}≠0 and the LDO is supplying current, the PR converter increases its power until V_{LOW}≈V_{IN}/2 and I_{LOW}≈0. Thus, in steady state, the LDO current is nearly zero, and the overall efficiency matches that of a sole-PR converter. Meanwhile, V_{OUT} is also regulated by the LDO fast loop, guaranteeing rapid transient response. The bottom-left part shows the system's transient behavior. For a step-up in load current (I_{LOAD}), the LDO reacts immediately, yielding only a small undershoot at V_{OUT}, while V_{LOW} drops significantly. If I_{LOAD} is lower than the maximum output current capacity of the PR converter ($I_{PR,MAX}$), a slow undershoot regulator gradually restores V_{LOW} to V_{IN}/2, reducing I_{LOW} to zero. If I_{LOAD} exceeds $I_{PR,MAX}$, the LDO automatically provides the surplus current, and V_{LOW} remains below V_{IN}/2, thereby extending the maximum load capacity. During an I_{LOAD} step-down, a slow overshoot regulator feeds V_{OUT} back into the compensator, reducing the current in the PR DC-DC, I_{HIGH}, to suppress overshot. The bottom-right part shows the analog LDO implementation, with a folded-cascode amplifier and a super source-follower stage driving an NMOS pass switch. With its gate driven by V_{IN} and the drain tied to V_{LOW} (≤V_{IN}/2), the NMOS can be fully turned on, delivering sufficient current during step-up transients.

The chip was fabricated in a 180nm BCD process with an active area of 4.2mm². Figure 16.7.4 presents the measured waveforms of the sole-PR converter with the sigma path disabled. The top-left shows start-up and 6-phase operation. From a static state, the PR is excited for 12 cycles near its resonance frequency, then held for phase-locking (bottom-left). Meanwhile, the soft-start reference ramps V_{OUT} from 0-to-1.8V in 3.6ms. Then the converter enters stable 6-phase operation with accurate switching, enabled by the 2-DoF closed-loop control (top-right). For load steps between 10mA and 100mA with 1μs transition times, the sole-PR converter exhibits a 600mV undershoot with 1.2ms settling and a 500mV overshoot with 1.5ms settling.

Fig. 16.7.5 (top) shows the transient performance with the sigma path enabled at V_{IN}=5V and V_{OUT}=1.8V. For 10-to-180mA load steps, assisted by the LDO, an undershoot of only 10mV is achieved with a 30μs settling time, showing orders of magnitude improvement over the sole-PR converter. Since 180mA exceeds the PR's maximum capacity of ≈107mA, the LDO supplies the extra 73mA, preventing V_{LOW} from recovering to 2.5V (V_{IN}/2). During the step-down transition, an overshoot of around 240mV is observed, which is smaller than the overshoot from a lighter down-step shown in Fig. 16.7.4. This is because there is an additional regulation loop associated with the sigma mode (Fig. 16.7.3), and there is around 73mA delivered by the fast-loop LDO. The bottom of Fig. 16.7.5 compares the efficiency of the sole-PR converter and the PR sigma converter. Without the sigma mode, a 91.2% peak efficiency is achieved. With the sigma mode enabled, the peak efficiency slightly drops to 90.7% due to the leakage in I_{LOW} and the maximum I_{LOAD} is doubled or tripled across different VCRs, significantly extending the output power range.

Figure 16.7.6 presents the comparison table. The top part benchmarks this work against prior PR-based converters, highlighting it as the only on-chip closed-loop implementation, which delivers more accurate regulation and much faster transient response. The bottom part compares sigma structures. Despite differences in high-side converters, the proposed topology and regulation method overcome the inherent limitations of prior designs and maintain high efficiency independent of VCR.

Figure 16.7.7 shows the die micrograph and prototype PCB.

288 • 2026 IEEE International Solid-State Circuits Conference

979-8-3315-8937-0/26 $31.00 © 2026 IEEE

References:

[1] W. D. Braun, Z. Tong and J. Rivas-Davila, "Inductorless Soft Switching DC-DC Converter with an Optimized Piezoelectric Resonator," *2020 IEEE Applied Power Electronics Conference and Exposition (APEC)*, New Orleans, LA, USA, 2020, pp. 2272-2278. http://doi.org/10.1109/APEC39645.2020.9124606

[2] J. D. Boles, J. J. Piel and D. J. Perreault, "Enumeration and Analysis of DC-DC Converter Implementations Based on Piezoelectric Resonators," in *IEEE Transactions on Power Electronics*, vol. 36, no. 1, pp. 129-145, Jan. 2021. http://doi.org/10.1109/TPEL.2020.3004147

[3] M. Touhami, G. Despesse and F. Costa, "A New Topology of DC-DC Converter Based on Piezoelectric Resonator," in *IEEE Transactions on Power Electronics*, vol. 37, no. 6, pp. 6986-7000, June 2022. http://doi.org/10.1109/TPEL.2022.3142997

[4] J. D. Boles, J. E. Bonavia, J. H. Lang and D. J. Perreault, "A Piezoelectric-Resonator-Based DC-DC Converter Demonstrating 1 kW/cm Resonator Power Density," in *IEEE Transactions on Power Electronics*, vol. 38, no. 3, pp. 2811-2815, March 2023. http://doi.org/10.1109/TPEL.2022.3217773

[5] W. -C. B. Liu and P. P. Mercier, "A Series/Parallel Magnetic-Less Step-Down Converter based on Piezoelectric Resonators," *2023 IEEE Applied Power Electronics Conference and Exposition (APEC)*, Orlando, FL, USA, 2023, pp. 484-489. http://doi.org/10.1109/APEC43580.2023.10131228

[6] Q. Li, Y. Hou and K. K. Afridi, "Merged Switched-Capacitor Piezoelectric-Resonator Based DC-DC Converter with High Voltage Conversion Ratio," *2023 IEEE 24th Workshop on Control and Modeling for Power Electronics (COMPEL)*, Ann Arbor, MI, USA, 2023, pp. 1-8. http://doi.org/10.1109/COMPEL52896.2023.10221132

[7] W. -C. B. Liu, G. Pillonnet and P. P. Mercier, "An Integrated Dual-side Series/Parallel Piezoelectric Resonator-based 20-to-2.2V DC-DC Converter Achieving a 310% Loss Reduction," *2024 IEEE International Solid-State Circuits Conference (ISSCC)*, San Francisco, CA, USA, 2024, pp. 364-366. http://doi.org/10.1109/ISSCC49657.2024.10454471

[8] B. M. Wanyeki, J. D. Boles, J. H. Lang and D. J. Perreault, "Two-Stage Piezoelectric Resonator / Switched Capacitor DC-DC Converter," *2023 IEEE Energy Conversion Congress and Exposition (ECCE)*, Nashville, TN, USA, 2023, pp. 1297-1304. http://doi.org/10.1109/ECCE53617.2023.10362388

[9] V. Breton, E. Bigot, G. Despesse and F. Costa, "A New Isolated Topology of DC-DC Converter Based on Piezoelectric Resonators," in *IEEE Transactions on Power Electronics*, vol. 38, no. 8, pp. 10012-10025, Aug. 2023. http://doi.org/10.1109/TPEL.2023.3276478

[10] B. Pollet, F. Costa and G. Despesse, "A new inductorless DC-DC piezoelectric flyback converter," *2018 IEEE International Conference on Industrial Technology (ICIT)*, Lyon, France, 2018, pp. 585-590. http://doi.org/10.1109/ICIT.2018.8352243

[11] B. Pollet, G. Despesse and F. Costa, "A New Non-Isolated Low-Power Inductorless Piezoelectric DC-DC Converter," in *IEEE Transactions on Power Electronics*, vol. 34, no. 11, pp. 11002-11013, Nov. 2019. http://doi.org/10.1109/TPEL.2019.2900526

[12] M. Touhami, H. Liew and J. D. Boles, "Phase-Shift Voltage Regulation of DC-DC Converters Based on Piezoelectric Resonators," *2024 IEEE Workshop on Control and Modeling for Power Electronics (COMPEL)*, Lahore, Pakistan, 2024, pp. 1-8. http://doi.org/10.1109/COMPEL57542.2024.10614015

[13] W. -C. B. Liu, G. Pillonnet and P. P. Mercier, "Closed-loop Control of A Dual-side Series/Parallel Piezoelectric-Resonator-based DC-DC Converter," *2025 IEEE Applied Power Electronics Conference and Exposition (APEC)*, Atlanta, GA, USA, 2025, pp. 315-320. http://doi.org/10.1109/APEC48143.2025.10977487

[14] M. Touhami, G. Despesse, F. Costa and B. Pollet, "Implementation of Control Strategy for Step-down DC-DC Converter Based on Piezoelectric Resonator," *2020 22nd European Conference on Power Electronics and Applications (EPE'20 ECCE Europe)*, Lyon, France, 2020, pp. 1-9. http://doi.org/10.23919/EPE20ECCEEurope43536.2020.9215910

[15] L. d. A. Pereira, A. Morel, M. Touhami, T. Lamorelle, G. Despesse and G. Pillonnet, "Operating Frequency Prediction of Piezoelectric DC-DC Converters," in *IEEE Transactions on Power Electronics*, vol. 37, no. 3, pp. 2508-2512, March 2022. http://doi.org/10.1109/TPEL.2021.3115182

[16] X. Yang *et al.*, "A 5V Input 98.4% Peak Efficiency Reconfigurable Capacitive-Sigma Converter With Greater than 90% Peak Efficiency for the Entire 0.4~1.2V Output Range," *2022 IEEE International Solid-State Circuits Conference (ISSCC)*, San Francisco, CA, USA, 2022, pp. 108-110. http://doi.org/10.1109/ISSCC42614.2022.9731550

[17] C. Hu, X. Huang, X. Liu, S. Du, X. Liu and J. Jiang, "A 3.6W 16V-Output 180ns-Response-Time 94%-Efficiency SC Sigma Converter with Output Impedance Compensation and Ripple Mitigation for LiDAR Driver Applications," *2024 IEEE International Solid-State Circuits Conference (ISSCC)*, San Francisco, CA, USA, 2024, pp. 508-510. http://doi.org/10.1109/ISSCC49657.2024.10454359

[18] Y. Li, J. Jin, Y. Guo, J. Zhou and J. Jiang, "A 3-5 V to Sub-1 V DLDO-SC-Sigma Converter With Auxiliary Loop for Efficiency Improvement in High-Density Power Delivery," in *IEEE Journal of Solid-State Circuits*. http://doi.org/10.1109/JSSC.2025.3566089

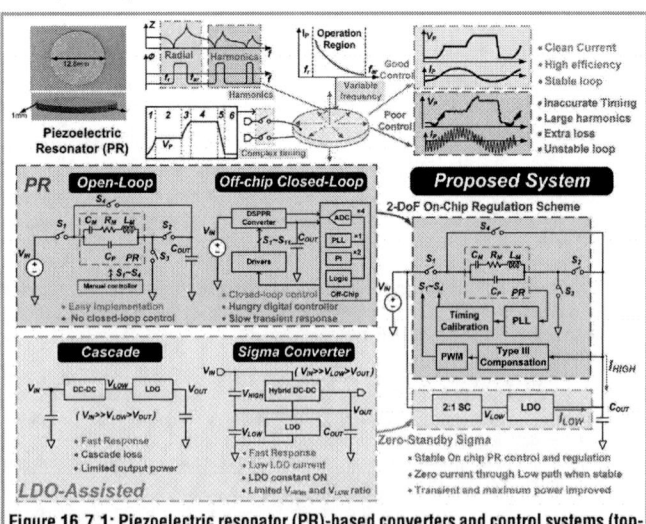

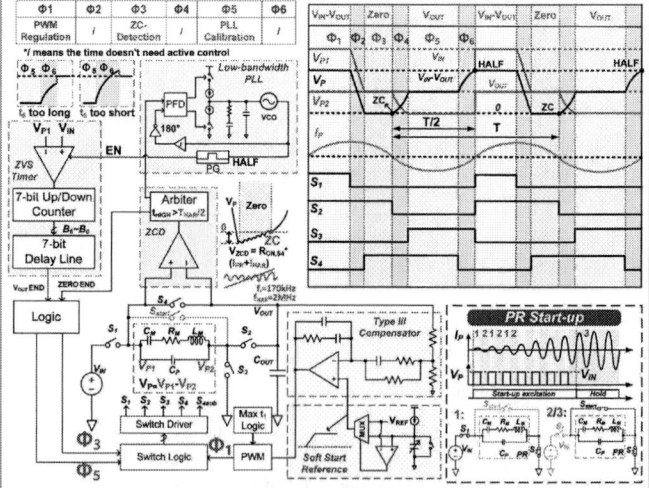

Figure 16.7.1: Piezoelectric resonator (PR)-based converters and control systems (top-left); conventional sigma converters (bottom-left), and the proposed hybrid system (right).

ISSCC 2026 / SESSION 16 / ENERGY HARVESTING, PIEZO AND CHARGERS / 16.7

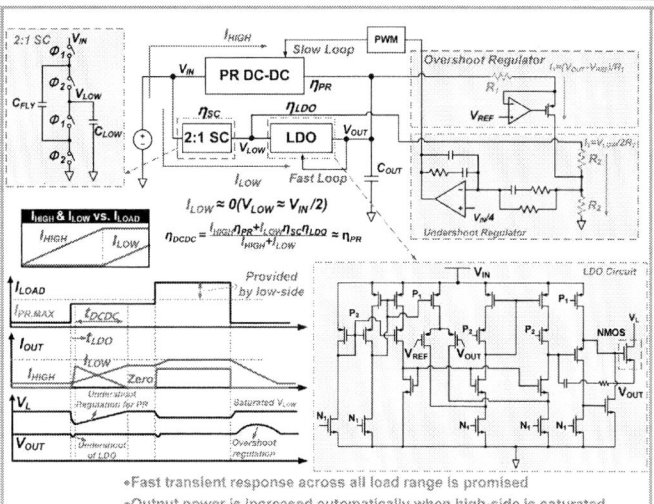

• Fast transient response across all load range is promised
• Output power is increased automatically when high-side is saturated

Figure 16.7.2: Operation principle of the 6-phase PR power converter (top-right), on-chip 2-DoF control systems (left), and static startup operation (bottom-right).

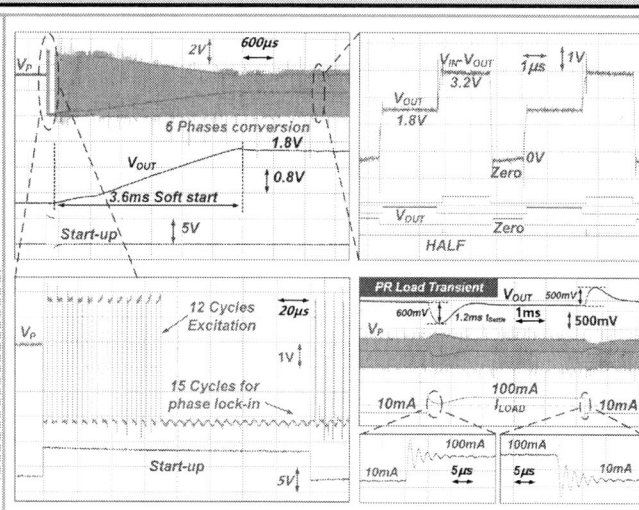

Figure 16.7.3: System architecture of the proposed PR-Sigma power converter (top), with associated waveform under load transients (bottom-left) and LDO circuits (bottom-right).

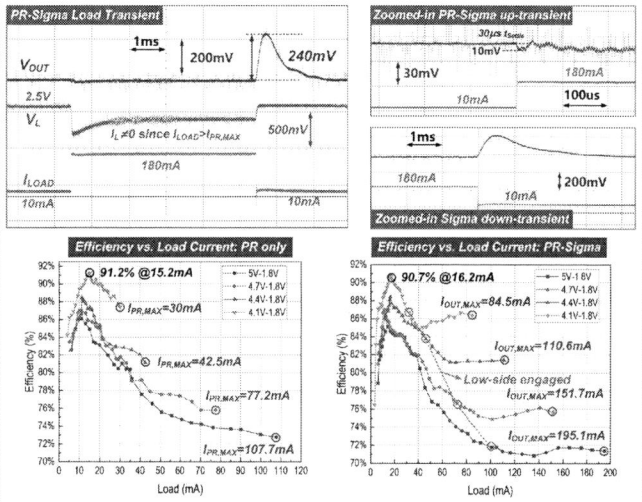

Figure 16.7.4: Measured PR converter's working waveform (top-left); 6-phase working waveform (top-right); start-up figure (bottom-left) and PR load transient (bottom-right).

Parameters	TPEL2023[9]	COMPEL2024[12]	ISSCC 2024[7]	APEC 2025[13]	This work
DC-DC Type	PR	PR	DSPPR	DSPPR	PR-Sigma
PR Material	Fuji Ceramic	APC 1556	PIC181	PIC181	APC1777
Technology	Board	Board	180nm BCD	Board	180nm BCD
Operation frequency	100kHz	90-100kHz	113-129kHz	113-129kHz	170kHz-200kHz
V_{IN}/V_{OUT} range	12-360V/3.3-346V	120V/80-82V	16-20V/1.1-2.2V	36-60V/4-6V	3.8-5/1.8V
Equivalent C_{OUT}	N\R	10uF	N\R	10uF	47uF
$I_{OUT.MAX}$	1.3A	128mA	250mA	500mA	195.1mA
Control methods	Open-Loop	Microcontroller	Digital controller	Digital controller	Integrated
Output regulation	No	Yes	No	Yes	Yes
Peak efficiency	97.2%	N\R	92.9%	94.1	91.2%(PR only)
		Control loss not included (if implemented)			90.7%(PR-Sigma)
Load Up transient Settling time	N\R	26ms@50mA	N\R	40ms@50mA	1.2ms@90mA,PR
					30us@170mA,PR-Sigma
Current density	0.22mA/mm³*	0.33mA/mm³*	0.99mA/mm³*	1.98mA/mm³*	0.376mA/mm³
		Power stage only			(Controller included)

Parameters	ISSCC2022[16]	ISSCC2024[17]	JSSC 2025[18]	This work
Sigma type	Capacitive Sigma	SC-Sigma	Hybrid-Sigma	Zero-Standby PR-Sigma
Sigma requirement	$V_{IN} \gg V_{LOW} > V_{OUT}$	$V_{OUT} > V_{HIGH} \gg V_{LOW}, C_{HIGH} \gg C_{LOW}$	$V_{IN} \gg V_{LOW} > V_{OUT}$	$V_{IN} > V_{LOW} > V_{OUT}$
Theoretical efficiency	$\eta = \frac{V_{HIGH}}{V_{IN}}\eta_{HIGH} + \eta_{LOW}\frac{V_{LOW}}{V_{IN}}$	$\eta = \frac{(V_{HIGH}+V_{LOW})\eta_{LOW}\eta_{HIGH}}{V_{LOW}\eta_{HIGH}+V_{HIGH}\eta_{LOW}}$	$\eta = \frac{V_{HIGH}}{V_{IN}}\eta_{HIGH} + \eta_{LOW}\frac{V_{LOW}}{V_{IN}}$	$\eta \approx \eta_{HIGH}$

N\R : Not Reported
* : Estimated

Figure 16.7.6: Comparison table with other prior works.

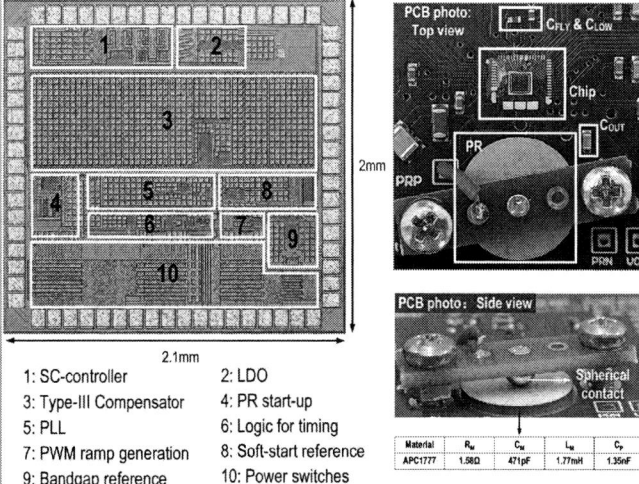

1: SC-controller   2: LDO
3: Type-III Compensator   4: PR start-up
5: PLL   6: Logic for timing
7: PWM ramp generation   8: Soft-start reference
9: Bandgap reference   10: Power switches

Material	R_m	C_m	L_m	C_p
APC1777	1.58Ω	471pF	1.77mH	1.35nF

Figure 16.7.7: Die micrograph and PCB photos.

• 2026 IEEE International Solid-State Circuits Conference

979-8-3315-8937-0/26 $31.00 © 2026 IEEE

ISSCC 2026 / SESSION 16 / ENERGY HARVESTING, PIEZO AND CHARGERS / 16.8

16.8 A Battery Charger Based On Mesh-Connection 2×CF Continuously-Scalable-Conversion-Ratio Converter Achieving 3.2W/mm³ Power Density

Yuanfei Wang[1], Zhiyuan Zhang[1], Yihan Zhang[2], Rui P. Martins[1], Mo Huang[1]

[1]University of Macau, Macau, China, [2]Hong Kong University of Science and Technology, Hong Kong, China

Abstract

This work presents a battery charger for smartwatches based on a mesh-connected continuously scalable-conversion ratio converter that uses only two flying capacitors. The proposed design reduces the number of invalid phases along with the switch and routing resistance, achieving 89.2% efficiency at a 16.8W output and a power density of 3.2 W/mm³.

Smart watches, with stringent low-profile requirements, demand compact chargers while allowing slight compromises in power conversion efficiency (PCE). The chargers typically deliver output power (P_{OUT}) of ~10W from a 12-V input bus to a Li-ion battery whose voltage ranges from 3.5-to-4.2V during charging [1], corresponding to a required voltage conversion ratio (VCR) between 0.3 and 0.35. Conventional inductive chargers achieve high PCE across wide VCRs but are bulky due to high-current inductors. Switched capacitor converters (SC), facilitating flying capacitors (C_F) with >100× density than inductors, have gained increasing attention. However, their PCE peaks only at nominal VCRs and degrades rapidly elsewhere. Multi-VCR SC designs, such as [2] using 4×C_F and 19× switches (Fig. 16.8.1), partially mitigate this issue but suffer heavy-load PCE degradation [3].

Continuously scalable-conversion-ratio converters (CSC) [4], achieving medium PCE over a wide VCR range, represent a promising alternative. Their operation relies on charging and discharging C_F across a large number of phases (N_P is the phase count per full cycle), thereby splitting the high voltage swing (V_{CF}) into fine steps. As a result, conventional CSC designs require many C_Fs and switches, leading to low switch utilization ratio (UR) [5], e.g., 5.7% in a 33×C_F design [6]. Under high-power conditions, their performance is further constrained by the following factors (Fig. 16.8.1). 1) The large C_F count hinders its off-chip implementation, while on-chip C_F density remains limited in most processes. 2) P_{OUT} and power density, as expressed in eq(1) and eq(2), are inversely proportional to N_P, which contradicts the need for high N_P to finely split V_{CF}. 3) The heavy-load PCE is limited by the series resistance (R_S) along C_F charging/discharging power traces [5], consisting of switch resistance (R_{SW}) and routing resistance (R_{route}). For a fixed silicon area for switches, low UR increases R_{SW} per switch, while congested routing raises R_{route}.

Reducing the number of C_Fs in CSC facilitates off-chip implementation and effectively lowers N_P and R_S. For instance, [5] cuts the count to 4× (using off-chip MLCCs), lowering N_P to 32× and switch count to 40× (Fig. 16.8.1). To sustain efficiency, two SC stages clamp V_{CF} to smaller values, while stage out-phasing between SC and CSC [7] further splits V_{CF} steps. However, [5] introduces 7× invalid phases where V_{CF} does not change, reducing P_{OUT}. Moreover, its R_S deserves further improvement.

This work further reduces the CSC C_Fs count to 2×, i.e., C_{F3} and C_{F4}. The CSC employs an intermediate voltage bus V_T connected C_F top plates and a V_B bus at the bottom plate. Preceding stages SC$_1$ and SC$_2$, incorporating C_{F1} and C_{F2} respectively, clamp the CSC stage between the V_H and V_L buses. The conceptual diagram in Fig. 16.8.1 adopts the conventional C_F bus connection, which requires 20× switches in total. However, this connection results in long C_F charging and discharging paths with up to 8× series-connected active switches, substantially increasing R_S. This work proposes a mesh connection (Fig. 16.8.1) that shortens routing and reduces the number of active switches per path to 5×. Note that the mesh connection remains efficient only with low C_F counts. Unlike prior mesh designs [8] with 48×C_F (requiring >1000× switches), the proposed scheme further reduces the switch count to 17×.

Figure 16.8.2 elaborates the generation of invalid phases presented in [5], where SC$_1$ operates at twice the switching frequency of SC$_2$ to enable the stage outphasing among CSC and SCs. For instance, as SC$_1$ transitions between phases from C_{F1} discharging (dchg) to charging (chg) at t_1, a step is generated on V_H. With SC$_2$ connected V_H, this step propagates to V_L and then V_T in CSC, delivering charge to V_0, referred to as a valid phase. However, when SC$_2$ is disconnected from V_H during its own discharging phase, a similar SC$_1$ transition (e.g., at t_2) does not produce any step on V_L or V_T, and thus delivers no charge to V_0 (denoted as an invalid phase). In this work, SC$_1$ and SC$_2$ operate at the same switching frequency, with their transitions from charging to discharging aligning to the phase transitions of C_{F3} and C_{F4} (denoted as T, B and C for the phases with top, bottom and both plates connecting DC nodes). Meanwhile, SC$_1$'s transitions from discharging to charging occur during SC$_2$'s charging phase, thereby eliminating most invalid phases. The only remaining invalid phase occurs when C_{F3} and C_{F4} are in the cornerstone phase (C), where the top plate voltage of C_{F3} or C_{F4} connects to V_H and the SC$_2$ transition does not produce any step on it. This modification reduces the number of SC$_1$ transitions from four to two for generating three V_{CF} steps, as compared to a prior approach illustrated in Fig. 16.8.2, thereby lowering the power loss in SC$_1$.

A shared-error-amplifier scheme ensures a smooth transition between constant current & constant voltage modes. With a peak power conversion efficiency of 92%, this compact and efficient design is well-suited for smartwatch applications.

According to the respective voltage stress, the power stage employs 12-V LDMOS for M_1, 6-V LDMOS for M_{2-5} and 5-V LDMOS for M_{14-17}. To block bidirectional voltages, M_{6-13} use body-switched NMOS devices(Fig. 16.8.2). To minimize R_{SW}, bootstrap (BST) gate driving is necessary. Conventional BST circuits require bootstrap capacitors (C_{BS}) that typically occupy silicon area comparable to the driven power switch. As illustrated in Fig. 16.8.2, this work adopts a C_{BS}-free BST [9], where a Zener diode produces a clamping voltage of ~5.6V (held by a small capacitor of 8pF) for a source follower M_B. The M_B output produces the BST voltage V_{bst}, yielding ~5V boost relative to the source node V_{src} and serving as the high-level V_{gs} for turn-on power switches. The low output impedance of M_B enables quick recharge of V_{bst} during the turn-on intervals. In addition, for complete turn-off of the body-switched NMOS (e.g., M_2 with $V_{src} = V_{T2}$), a lower voltage, such as the bottom plate voltage of C_{F2} (V_{B2}), is employed to generate the low-level V_{gs}. The V_{B2H} node in the driver is bootstrapped from V_{B2}. Moreover, to further save area, switches featuring the same source node (highlighted in Fig. 16.8.2) share one BST circuit. For example, M_2 and M_{6-10}, which have V_{T2} as their common source node, share BST$_1$. This reduces the number of BSTs to seven while driving 17 switches overall.

In battery charging, a smooth transition from constant current (CC) to constant voltage (CV) mode is critical. Both battery current (I_{BAT}) and voltage (V_{BAT}) can be regulated by tuning the VCO frequency (f_{CLK}), with the VCO input (V_{EA}) derived from an error amplifier (EA) with a proportional-integral compensation network (Fig. 16.8.3). For CC regulation, an OTA-shared current sensor mirrors the output currents of M_{14-17} at a ratio of 7000:1. The total sensed current (I_{SEN}) is converted to voltage through a resistor R_{SEN}, then averaged as V_{SEN} by a π network [10]. The EA integrates the error between V_{SEN} and V_{IREF} (representing reference current). For CV regulation, the EA integrates the error between the divided output voltage (V_{DIV}) and the reference voltage (V_{REF}).

Conventional designs use separated EAs for CC and CV loops, producing two outputs voltages ($V_{EA,CV}$ and $V_{EA,CC}$) that may differ in the CC-CV transition. Directly multiplexing them as the VCO input may lead to an abrupt increase in I_{BAT}, leading to V_{BAT} overshoot beyond the cut-off voltage, which may shorten cycle life and raise safety risks. In contrast, this work employs a single shared EA, where V_{SEN} and V_{DIV} are multiplexed as the EA input, ensuring smooth V_{EA} variation during the CC-CV transition. To further expand battery life, we also implement multi-stage constant current charging (MSCC) control between CC and CV modes. A DAC generates a stepwise V_{IREF}, gradually reducing I_{BAT}.

The proposed battery charger was fabricated in a 180-nm BCD process, occupies an effective area of 5.4mm² (Fig. 16.8.7). This design utilizes two 22μF (0603) MLCCs as C_{F1-2} in SC$_1$ and SC$_2$, and two 22μF (0603) as C_{F3-4} in CSC, all mounted on the bottom of the PCB. Figure 16.8.7 also illustrates the layout, where the BST circuits only occupy a small area, leaving most of the area for power switches to reduce R_{SW}. The chip uses a flip-chip package with redistribution layers (RDL) for routing. The power switches are placed in close proximity and connected in a mesh structure, minimizing interconnection length.

Figure 16.8.4 presents measured waveforms under conditions of V_{IN} = 12V and V_0 = 4.2V. The voltage across V_{CF} exhibits seven ramp-up and eight ramp-down steps, with only one invalid phase. From the measured top and bottom plate voltages of C_{F1} and C_{F2}, we observe two transitions in each: transitions (1) and (3) for C_{F1}, and (2) and (3) for C_{F2}. They together form three V_{CF} steps enclosed in a dashed rectangle, as analyzed in Fig. 16.8.2.

Figure 16.8.4 also compares CC to CV transitions with and without the shared EA scheme. We use two series-connected 500F supercapacitors [11] for test. V_{SEN}, reflecting I_{BAT}, remains constant (corresponding to 4A I_{BAT}) while V_0 increases during CC mode. When V_0 reaches the MSCC transition voltage, the controller reduces current from 4A to 0.8A in a stepwise fashion. A smooth CC-CV transition is observed with the shared-EA scheme, avoiding abrupt change in V_{SEN}. In contrast, V_{SEN} spikes during the CC-CV transition with the non-shared-EA scheme, causing V_0 to overshoot the cut-off voltage.

Figure 16.8.5 displays the measured PCE versus I_{BAT} at V_{IN} = 12V and V_0 = 4.2V. A peak PCE of 92% is achieved, with PCE remaining above 90% up to 3.3A. These results are slightly lower than the simulation results. We add a simulated result of a reference design

290 • 2026 IEEE International Solid-State Circuits Conference

979-8-3315-8937-0/26 $31.00 © 2026 IEEE

for comparison, using the same SC-clamp topology but with $4\times C_F$ in the CSC. Although it achieves a marginally higher peak value, its PCE degrades rapidly with I_{BAT}, dropping to 90% at 1.9A, due to the higher R_S. This highlights the advantage of this work. At a lower $V_0 = 3.6V$, PCE decreases by around 5%. From the steady-state waveforms at $I_{BAT} = 4A$ and 1A, f_{CLK} is reduced from 2.6MHz to 630kHz.

Figure 16.8.6 provides a performance summary and comparison with prior works. To our best knowledge, this work achieves the lowest C_F count (=2) in CSC. Compared to prior CSC converters [5,7,16], the reduced C_F count enables the lowest phase counts, which is beneficial for high P_{OUT} capability. In addition, the low C_F count facilitates an efficient mesh connection topology, significantly reducing the switch count and R_S from conventional bus connections, resulting in a highest PCE of 89.2% at a peak P_{OUT} of 16.8W among the state-of-the-art designs in Fig. 16.8.6. When compared to prior inductive [12,13] and hybrid [14,15] chargers, this work demonstrates the highest power density (1.6W/mm² and 3.2W/mm³), due to the inherent high-density in SC converters. Moreover, it maintains comparable PCEs, with a peak PCE of 92% and high PCE at maximum P_{OUT} (Fig. 16.8.5). These features combine to make this design well-suited for smart watch applications.

Acknowledgment:
This work was supported in part by Science and Technology Development Fund, Macau SAR (0029/2025/AMJ, 0042/2025/RIB1 and 004/2023/SKL); in part by Guangdong Basic and Applied Basic Research Foundation (2023B1515130001); in part by Research Committee of University of Macau (MYRG-GRG2024-00041-IME).

References:
[1] S. Schnier et al., "The architecture of a switched-capacitor charger with fast charging and high efficiency," Texas Instruments, 2018. [Online]. Available:. https://www.ti.com/lit/wp/slly002/slly002.pdf
[2] M. Krstic et al., "Analysis and Design of Multiphase, Reconfigurable Switched-Capacitor Converters," *IEEE JESTPE*, vol. 8, no. 4, pp. 4046-4059, Dec. 2020. https://doi.org/10.1109/JESTPE.2019.2960016.
[3] X. Liu et al., "A Highly Efficient Reconfigurable Charge Pump Energy Harvester With Wide Harvesting Range and Two-Dimensional MPPT for Internet of Things," *IEEE JSSC*, vol. 51, no. 5, pp. 1302-1312, May 2016. https://doi.org/10.1109/JSSC.2016.2525822.
[4] N. Butzen et al., "Design of Single-Topology Continuously Scalable-Conversion-Ratio Switched-Capacitor DC-DC Converters," *IEEE JSSC*, vol. 54, no. 4, pp. 1039-1047, Apr. 2019. https://doi.org/10.1109/JSSC.2018.2884351
[5] Y. Wang et al., "An SC-first Hybrid SCVR with 4xCF Continuously Scalable-Conversion Ratio SC Achieving 92.5% Peak Efficiency," *IEEE CICC*, pp. 1-3, Apr. 2025. https://doi.org/10.1109/CICC63670.2025.10983039
[6] Y. Wang et al., "A VCF-Step-Reconfigurable Continuously Scalable-Conversion-Ratio Switched- Capacitor Converter," *IEEE JSSC*, vol. 60, no. 2, pp. 626-637, Feb. 2025,. https://doi.org/10.1109/JSSC.2024.3414448.
[7] N. Butzen et al., "A Monolithic 12.7W/mm² Pmax, 92% Peak-Efficiency CSCR-First Switched-Capacitor DC-DC Converter," *ISSCC*, pp. 462-464, Feb. 2024. https://doi.org/10.1109/ISSCC49657.2024.10454555
[8] Y. Yoon et al., "A Continuously-Scalable-Conversion-Ratio Step-Up/Down SC Energy-Harvesting Interface With MPPT Enabled by Real-Time Power Monitoring With Frequency-Mapped Capacitor DAC," *IEEE TCAS-I*, vol. 69, no. 4, pp. 1820-1831, April 2022. https://doi.org/10.1109/TCSI.2021.3139708.

[9] H. Cao et al., "A 12-Level Series-Capacitor 48-1V DC–DC Converter With On-Chip Switch and GaN Hybrid Power Conversion," *IEEE JSSC*, vol. 56, no. 12, pp. 3628-3638, Dec. 2021. https://doi.org/10.1109/JSSC.2021.3104328.
[10] L. Cheng et al., "A 6.78-MHz Single-Stage Wireless Charger With Constant-Current Constant-Voltage Charging Technique," *IEEE JSSC*, vol. 55, no. 4, pp. 999-1010, April 2020. https://doi.org/10.1109/JSSC.2019.2961852.
[11] CDA Zhifengwei Technology, "CHQ-2R7507R-TW," DigiKey. [Online]. Available:. https://www.digikey.com/en/products/detail/cda-zhifengwei-technology/CHQ-2R7507R-TW/22461627
[12] W. -T. Lin et al., "Dynamic-Charging Current-Scaling Technique with Dual Accurate Current Control and Temperature Loops with Charging-Current Accuracy up to 99.6% for 1.6× Faster Lithium-Ion Battery Charging," *IEEE ISSCC*, pp. 434-436, Feb, 2019. https://doi.org/10.1109/ISSCC.2019.8662335.
[13] Texas Instruments, "BQ25600 I2C controlled 3-A, single-cell, buck battery charger with 40-V overvoltage protection controller and NVDC power path management," *TI.com*. [Online]. Available: https://www.ti.com
[14] C. Hardy et al., "A Flying-Inductor Hybrid DC–DC Converter for 1-Cell and 2-Cell Smart-Cable Battery Chargers," *IEEE JSSC*, vol. 54, no. 12, pp. 3292-3305, Dec. 2019. https://doi.org/10.1109/JSSC.2019.2944837
[15] C. Hardy and H. -P. Le., "A 21W 94.8%-Efficient Reconfigurable Single-Inductor Multi-Stage Hybrid DC-DC Converter," *IEEE ISSCC*, pp. 190-192, Feb, 2023. https://doi.org/10.1109/ISSCC42615.2023.10067316
[16] N. Butzen et al., "A Monolithic 26A/mm² Imax, 88.5% Peak-Efficiency Continuously Scalable Conversion-Ratio Switched-Capacitor DC-DC Converter," *ISSCC*, pp. 232-233, Feb. 2023. https://doi.org/10.1109/ISSCC42615.2023.10067583

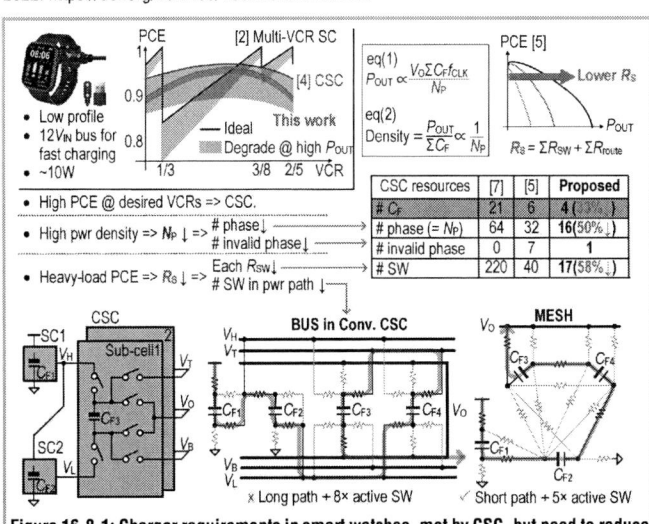

Figure 16.8.1: Charger requirements in smart watches, met by CSC, but need to reduce N_P and R_S (top), achieved by proposed topology (bottom left) and mesh connection (bottom right).

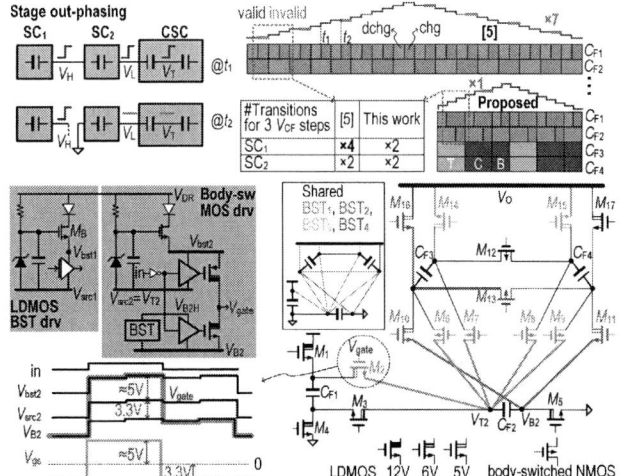

Figure 16.8.2: Invalid phase generation in [5], reduced to one in this work (top), gate drivers for LDMOS and body-switched MOS (bottom left), schematic of power stage and shared bootstraps (bottom right).

ISSCC 2026 / SESSION 16 / ENERGY HARVESTING, PIEZO AND CHARGERS / 16.8

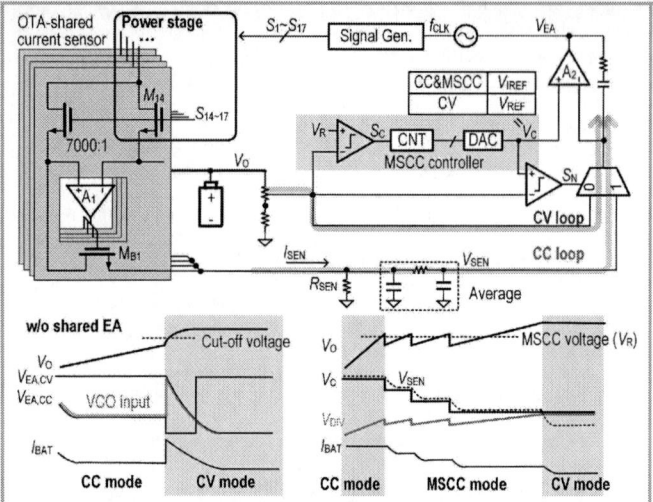

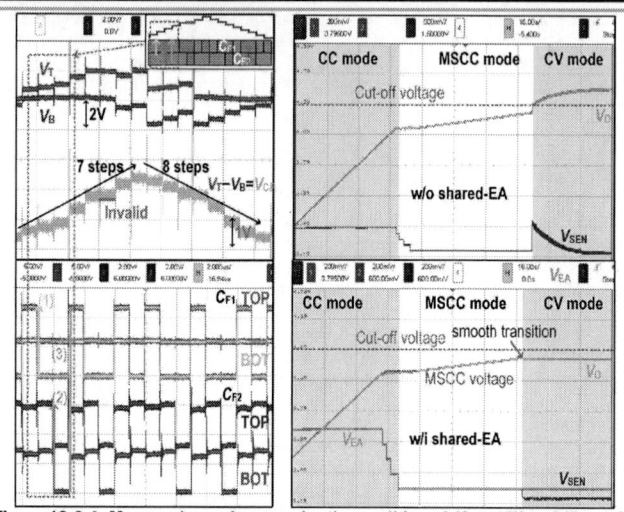

Figure 16.8.3: Control loop with shared EA for CC and CV regulation (top), V_{BAT} overshoot during CC-CV transition without shared EA (bottom left), waveforms in MSCC control (bottom right).

Figure 16.8.4: Measured waveforms under the condition of V_{IN} = 12V and V_O = 4.2V (left). Comparison between CC-to-CV transitions with and without shared EA scheme (right).

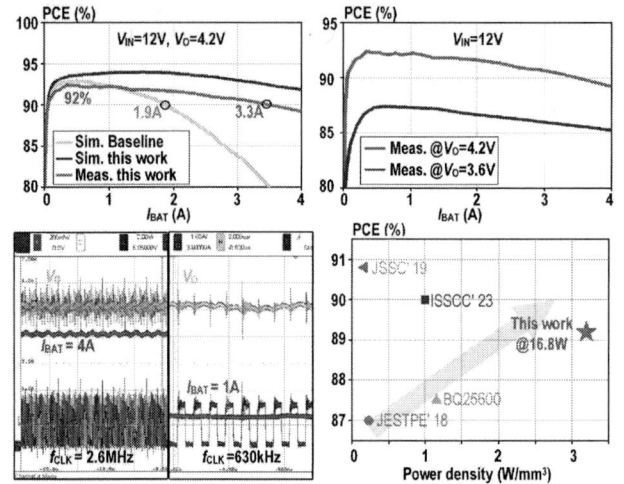

Figure 16.8.5: Measured and simulated PCE versus I_{BAT} at V_{IN} = 12V and V_O = 4.2V (3.6V) of this work and baseline (top). Steady-state waveforms at I_{BAT} = 4A and 1A (bottom left). comparison of PCE versus Power density (bottom right).

Comparison with prior CSC converters.

	ISSCC' 23 [16]	ISSCC' 24 [7]	CICC' 25 [5]	This work
Process	4nm CMOS	16nm CMOS	65nm CMOS	180nm BCD
Connection	Bus	Bus	Bus	Mesh
# C_F	15 in CSC	4 in SC +17 in CSC	2 in SC +4 in CSC	2 in SC +2 in CSC
V_{IN}/V_O	1.3V/0.5 - 0.95V	1.8 - 3V/0.48 - 1.36V	2.5 - 3.3V/0.7 - 1.3V	12V/3.5 - 4.2V
# Phase	30	68	32	16
# SW	165	220	40	17
PCE (%) @ Peak power	60	79	72	89.2

Comparison with prior battery chargers.

		JSSC'19 [14]	ISSCC'19 [12]	ISSCC'23 [15]	BQ25600 [13]	This work
Process		130nm BCD	180nm BCD	180nm BCD	n.a.	180nm BCD
Topology		Hybrid	Inductive	Hybrid	Inductive	CSC
Input voltage (V)		9	12	15	12	12
Output voltage (V)		3 - 4.2	2.5 - 4.2	2.8 - 4.2	2.8 - 4.6	3.5 - 4.2
Max. I_{OUT} (A)		3.4	2.5	5	3	4
Peak PCE (%)		94.3	92	92.7	90	92
Peak P_{OUT} (W)		12.2	10.5	21	11.4	16.8
Total-area (mm²)		23.4*	n.a.	17.4*	9.8*	10.52
Total-volume (mm³)		72.4	n.a.	20.8	9.8	5.26
@ Peak P_{OUT}	PCE (%)	90.8	90.7	90	87.5	89.2
	PD_1(W/mm²)	0.52	n.a.	1.21	1.16	1.6
	PD_2(W/mm³)	0.17	n.a.	1	1.16	3.2

* Active area + inductor/capacitor area.

Figure 16.8.6: Comparison table with prior CSC converters and battery charges.

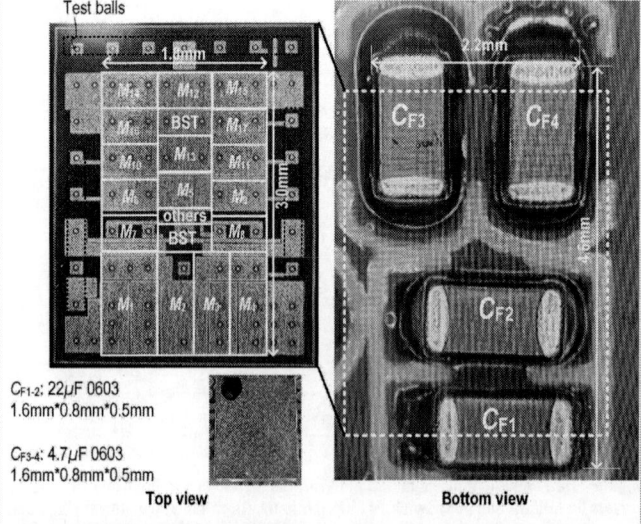

Figure 16.8.7: Chip micrograph and PCB.

ISSCC 2026 / SESSION 16 / ENERGY HARVESTING, PIEZO AND CHARGERS / 16.9

16.9 A 96.6% Single-Mode Hybrid Dual-Path Buck-Boost Converter with Conduction Loss Reduction through Conversion-Ratio-Based Adaptive 3-Phase Control

Minsu Kim[1], Yunho Lee[1], Hyunjun Park[1], Hongseok Kim[1,2], Woojoong Jung[1,2], Minwoo Kim[2], Junhyun Bae[2], Hyoung-Seok Oh[2], Hwayeal Yu[2], Hyung-Min Lee[1]

[1]Korea University, Seoul, Korea, [2]Samsung Electronics, Hwaseong, Korea

Abstract

This paper presents a 3-phase dual-path buck-boost converter that effectively reduces conduction loss by combining a dual-path topology with adaptive 3-phase control. The proposed control technique minimizes inductor current ripple, which allows the use of a low-inductance inductor with low DCR, while reducing the inductor current. Fabricated in a 0.13µm CMOS process, the converter achieves a high peak-efficiency of 96.6% with a small 1.92mm³ inductor and 0402-size flying capacitors.

Buck-boost converters are widely used to provide a stable output voltage (V_O) for powering functional blocks in mobile devices, which are supplied from a Li-ion battery with an input voltage (V_{IN}) range of 2.8- to 4.2V depending on the state of charge, as illustrated in Fig. 16.9.1 (top). Multi-mode buck-boost converters provide optimized performance in each operating mode, but the multi-mode operation causes a significant noise in V_O during mode transitions [1-3]. To fully eliminate mode-transition-related issues, single-mode buck-boost converters have been introduced for noise-sensitive systems [4-9]. In addition, various strategies for reducing substantial conduction loss in the inductor ($P_{CON,L}$) have also been studied to achieve high efficiency in space-constrained devices such as wearables and smartphones, where the parasitic DC resistance (DCR) of the inductor becomes much larger than on-resistance (R_{ON}) of the switches since the use of bulky inductors is not feasible. One of the representative methods to reduce $P_{CON,L}$ is an LC dual-path technique, which lowers the inductor current (I_L) by distributing the input-to-output current through capacitor paths [5-7]. However, although the dual-path converters in [5-7] effectively reduce I_L and mitigate $P_{CON,L}$, the benefit is limited as the large DCR is still a significant source of the loss. Another approach to reduce $P_{CON,L}$ is lowering the voltage swing across the inductor to enable the use of a low-inductance inductor with low DCR while maintaining a similar level of I_L ripple ($I_{L,AC}$) [8,9]. Nevertheless, the converters in [8] and [9] also suffer from a limitation in reducing $P_{CON,L}$, since I_L remains high, being equal to the load current (I_{LOAD}) or even larger in specific operating regions.

To overcome the limitations of the previous dual-path and low-inductance approaches while achieving further reduction in $P_{CON,L}$, this paper proposes a 3-phase dual-path buck-boost (3P-DPBB) converter that effectively combines both a hybrid LC dual-path topology and additional 3-phase control, thereby reducing I_L and enabling to use a low-inductance inductor, as shown in Fig. 16.9.1 (middle). The power stage of the 3P-DPBB converter consists of an inductor (L), two flying capacitors (C_X and C_Y), and six switches (S_1–S_6). In Φ_1, where switches S_1, S_2, and S_6 are turned on, the inductor is charged for a duration of D_1 while delivering the current to the output through the C_X path. In Φ_2, where switches S_3, S_4, and S_5 are turned on, the inductor is discharged for D_2, and the C_Y path additionally supplies the current to the output. In Φ_3, switches S_1, S_2, S_4, and S_5 are turned on, connecting the L, C_X, and C_Y paths to the output for D_3, during which the inductor is either charged or discharged depending on the voltage conversion ratio (M). The switching sequence proceeds in the order of Φ_1, Φ_3 and Φ_2, while Φ_3 acts as a buffer phase that causes no additional switching loss, smooths the slope of I_L to reduce $I_{L,AC}$, and secures sufficient hard-charging time for C_X and C_Y to mitigate capacitive inrush current (I_{CF}). Figure 16.9.1 (bottom) shows the variations in the normalized $I_{L,AC}$, the average I_{CF}s ($I_{CX,AVG}$ and $I_{CY,AVG}$), and the achievable M range as a function of D_3. Assuming the converter requires a 4.7µH inductor when $D_3 = 0$, the same $I_{L,AC}$ is achieved with a 2.2µH inductor when $D_3 = 0.516$, while reducing the average values of I_{CF}s by over 60%. However, as D_3 increases, the minimum duty limitation of either D_1 or D_2 narrows the gap between the upper and lower boundaries of M (M_{UB} and M_{LB}), restricting the operating range of the converter. To address this issue, the 3P-DPBB converter employs an adaptive D_3 control scheme that automatically adjusts D_3 according to the voltage difference between V_{IN} and V_O.

Figure 16.9.2 (left) depicts the adaptive D_3 controller, which consists of op-amp-based circuits that compute the absolute value of the V_{IN}–V_O ($V_{EA,RST}$) and a comparator that generates a timing pulse (V_{RST}) by comparing $V_{EA,RST}$ with a sawtooth signal ($V_{SAW,RST}$). As the voltage difference between V_{IN} and V_O decreases, which means a narrower M range becomes acceptable, $V_{EA,RST}$ decreases, thereby increasing D_3 and leading to a reduction in $I_{L,AC}$. In simulation results, D_3 ranges from 0.424 to 0.688 over the operating range of the converter, exhibiting a similar trend to the calculated results with $K_1 = 1.5$, $K_2 = 4$, $V_{OS1} = V_{OS2} = 0.03V$, and $T_{DEL} = 170$ns. Figure 16.9.2 (right) shows the performance improvement with adaptive D_3 control. When $V_O = 3.3V$, the increase in M_{UB} at low V_{IN} and the decrease in M_{LB} at high V_{IN} allow sufficient M margin across the entire V_{IN} range. In addition, due to the reduction in $I_{CX,AVG}$ and $I_{CY,AVG}$, the sum of conduction loss across all switches ($P_{CON,SW}$) is reduced by 42.9% at $V_{IN} = 2.8V$ and by 27.5% at $V_{IN} = 4.2V$, compared to the case with $D_3 = 0$. Assuming a 4.7µH inductor with 200mΩ DCR for $D_3 = 0$ and a 2.2µH inductor with 110mΩ DCR for the adaptive D_3 case, the simulated $P_{CON,L}$, which is estimated using RMS values of the currents, shows reductions of 28.9% and 18.5% at $V_{IN} = 2.8V$ and 4.2V, respectively.

Figure 16.9.3 compares the 3P-DPBB converter with prior works in terms of the flying capacitor DC bias voltage, $I_{L,AC}$, $P_{CON,L}$, and total conduction loss ($P_{CON,TOTAL}$). The DC bias voltages of flying capacitors C_X and C_Y in the 3P-DPBB converter are $|V_{IN}-V_O|$ and $|V_O-V_{IN}|$, respectively, and remain below 0.9V. The low DC bias voltages mitigate the voltage derating effect of MLCCs and enable the use of capacitors with smaller package sizes. Additionally, despite using a 2.2µH inductor, the 3P-DPBB converter exhibits comparable or even lower $I_{L,AC}$ than prior arts employing a 4.7µH inductor under the same conditions of a 1MHz switching frequency and $V_O = 3.3$ V. Based on this analysis, $P_{CON,L}$ is compared by applying a 200mΩ DCR to the converters using a 4.7µH inductor and a 110mΩ DCR to the converters using a 2.2µH inductor, considering only the DC component of I_L ($I_{L,DC}$). As a result, the 3P-DPBB converter exhibits consistently lower $P_{CON,L}$ due to both lower DCR and reduced $I_{L,DC}$, with reductions of 25.4% and 17.7% at $V_{IN} = 2.8V$ and 4.2V, respectively, compared to the converter in [7]. Finally, the 3P-DPBB converter achieves up to 29.7% reduction in $P_{CON,TOTAL}$, which is the sum of $P_{CON,L}$ and $P_{CON,SW}$, over the prior arts. In this comparison, R_{ON}s of the switches are set using weighted values based on the voltage rating and number of switches as described in [7], and all currents are approximated using average values at $I_{LOAD} = 1A$.

Figure 16.9.4 shows the steady-state operation waveforms measured at $V_O = 3.3V$ and $I_{LOAD} = 1A$ for $V_{IN} = 2.8V$, 3.4V, and 4.2V. V_{X1} and V_{Y1} represent the left and right nodes of the inductor, swinging between V_{IN} and $V_{IN}-V_O$ and between V_O and V_O-V_{IN}, respectively. The 3P-DPBB converter operates in Φ_3 when both V_{X1} and V_{Y1} nodes become high, and D_3 varies from 0.41 to 0.65 as V_{IN} changes, maintaining $I_{L,AC}$ below approximately 440mA. Figure 16.9.5 (left) depicts load transient waveforms measured at each V_{IN}. When I_{LOAD} steps of 1A are applied within 500ns, the undershoot voltages at V_O are measured as 168mV, 172mV, and 143mV at $V_{IN} = 2.8V$, 3.4V, and 4.2V, respectively. Additionally, when V_{IN} is swept from 4.2V to 2.8V to mimic a Li-ion battery discharging as shown in Fig. 16.9.5 (top right), V_O remains well-regulated at 3.3V across the entire V_{IN} range. Figure 16.9.5 (bottom right) depicts the measured power efficiencies with respect to I_{LOAD} at $V_O = 3.3V$ under various V_{IN} conditions. The peak efficiency reaches 96.6% at $V_{IN} = 3.4V$ and $I_{LOAD} = 0.2A$, while a second peak efficiency is 96.2% at $V_{IN} = 3.7V$ and $I_{LOAD} = 0.3A$.

Figure 16.9.6 shows the performance comparison with state-of-the-art converters. With the same inductance, switching frequency, and V_O, the 3P-DPBB converter exhibits the lowest $I_{L,AC}$ averaged across the entire V_{IN} range, which is at least 2.2× lower than that of prior converters using 4.7µH inductors, indicating that I_L ripple performance can be maintained even with a 2.2µH inductor. In addition to its lower DCR, the 3P-DPBB converter can also reduce $I_{L,DC}$ to the levels below I_{LOAD} consistently. Minimizing the average $P_{CON,L}$ to 73.34mW at $I_{LOAD} = 1A$. Consequently, the 3P-DPBB converter achieves a high peak efficiency of 96.6% while using the smallest external components among the compared state-of-the-arts, only with a 1.92mm³-volume inductor and 0402-size capacitors. Figure 16.9.7 presents the die micrograph of the fabricated chip in a 0.13µm process along with the PCB and external components.

Acknowledgement:
This work was supported by Samsung Electronics, Korea.

References:
[1] J. Jin et al., "A 98.6%-Peak-Efficiency 1.47A/mm²-Current-Density Buck-Boost Converter with Always Reduced Conduction Loss," *ISSCC*, pp. 448–450, Feb. 2023. http://doi.org/10.1109/ISSCC42615.2023.10067471
[2] S. Zhao, C. Zhan and Y. Lu, "A Battery-Input Three-Mode Buck–Boost Hybrid DC–DC Converter With 97.6% Peak Efficiency," *IEEE JSSC*, vol. 59, no. 5, pp. 1567–1577, May 2024. http://doi.org/10.1109/JSSC.2023.3320200
[3] M. Kim et al., "A 97% Peak-Efficiency Dual-Channel 3-Fine-Level Buck-Boost Converter with Capacitive Inrush Current Reduction and Seamless Mode Transition Using a Slim 0.65-Thickness Inductor," *IEEE ESSERC*, Sep. 2025. http://doi.org/10.1109/ESSERC66193.2025.11214118
[4] H. Shin et al., "A 96.6%-Efficiency Continuous-Input-Current Hybrid Dual-Path Buck-Boost Converter with Single-Mode Operation and Non-Stopping Output Current Delivery," *IEEE Symp. VLSI Circuits*, pp. 1–2, June 2021. http://doi.org/10.23919/VLSICircuits52068.2021.9492409

292 • 2026 IEEE International Solid-State Circuits Conference

979-8-3315-8937-0/26 $31.00 © 2026 IEEE

[5] D. Cho et al., "A High-Efficiency Single-Mode Dual-Path Buck-Boost Converter With Reduced Inductor Current," *IEEE JSSC*, vol. 58, no. 3, pp. 720–731, Mar. 2023. http://doi.org/10.1109/JSSC.2022.3230424

[6] J. Jin et al., "A 97.3%-Peak-Efficiency Always-Dual-Path Buck-Boost Converter with Single-Mode Operation and Fast Transient Responses," *IEEE CICC*, pp. 1-2, April 2024. http://doi.org/10.1109/CICC60959.2024.10529066

[7] D.-H. Kim and H.-S. Kim, "A Power-Efficient Single-Mode Buck-Boost DC–DC Converter With Bilaterally Symmetrical Hybrid Topology," *IEEE JSSC*, vol. 59, no. 12, pp. 4137-4149, Dec. 2024. http://doi.org/10.1109/JSSC.2024.3452238

[8] H. Park et al., "An Inductor-First Hybrid Buck-Boost Converter Featuring Seamless Single-Mode Operation, 97.2% Peak Efficiency, and 565 mA/mm³ Current Density with Ultra-Compact 1 mm³-Volume Inductor," *IEEE CICC*, pp. 1–3, April 2025. http://doi.org/10.1109/CICC63670.2025.10983352

[9] J. Jin et al., "A 98.3%-Peak-Efficiency Single-Mode Hybrid Buck-Boost Converter with 7mV Maximum Output Ripple for Li-Ion Battery Management," *ISSCC*, pp. 198-200, Feb. 2024. http://doi.org/10.1109/ISSCC49661.2025.10904716

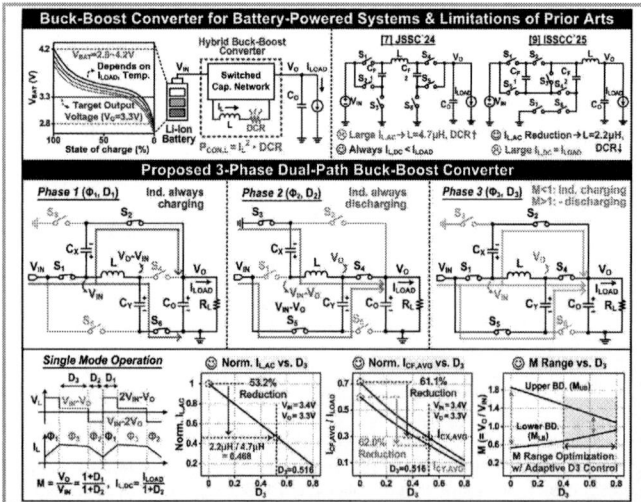

Figure 16.9.1: Buck-boost converter for battery-powered systems and limitations of prior arts (top), the proposed 3-phase dual-path buck-boost converter (middle), and performance change with respect to D_3 (bottom).

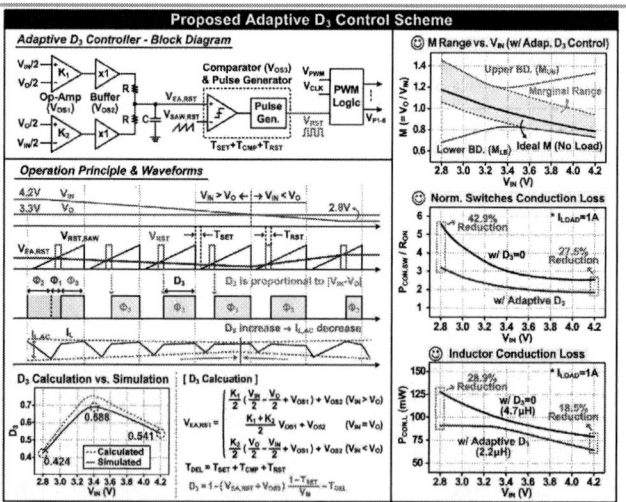

Figure 16.9.2: Proposed adaptive D_3 control scheme: adaptive D_3 controller and its operation principle (left), and simulated performance analysis with D_3 control at V_O = 3.3V and I_{LOAD} = 1A (right).

ISSCC 2026 / SESSION 16 / ENERGY HARVESTING, PIEZO AND CHARGERS / 16.9

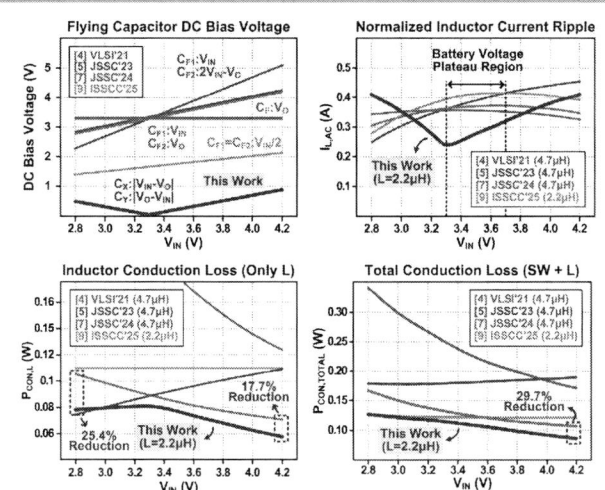

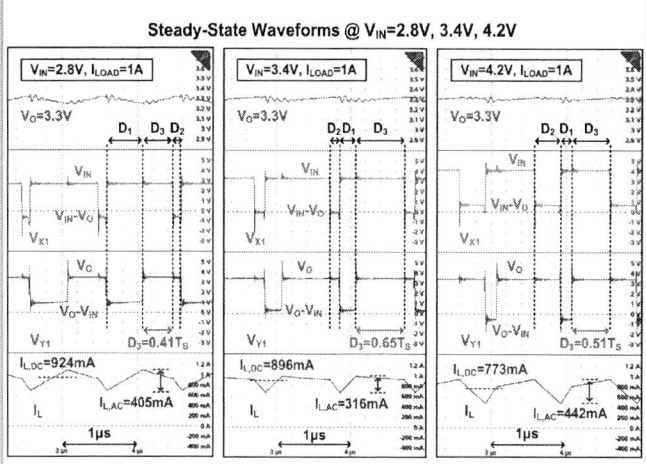

Figure 16.9.3: Comparison of flying capacitor DC bias voltage, inductor current ripple, inductor conduction loss, and total conduction loss of the proposed converter with prior arts.

Figure 16.9.4: Measured steady-state waveforms when V_O = 3.3V, I_{LOAD} = 1A, and V_{IN} = 2.8V, 3.4V, and 4.2V with adaptive D_3 control.

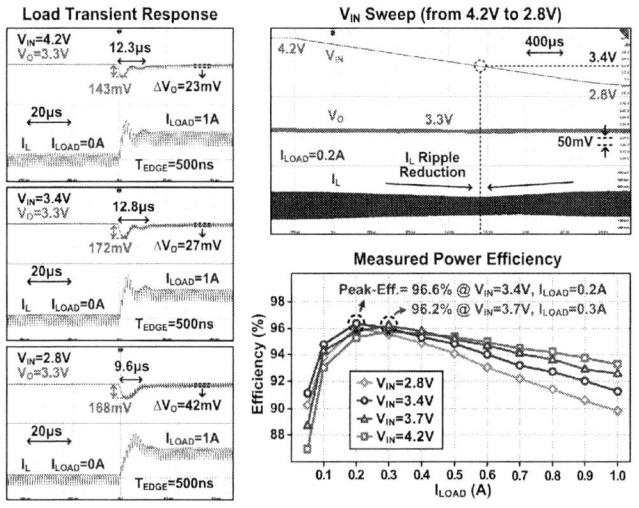

Figure 16.9.5: Measured waveforms of load transient responses (left), I_L ripple variation under V_{IN} sweep (top right), and measured power efficiencies (bottom right).

		[4] VLSI'21	[5] JSSC'23	[7] JSSC'24	[9] ISSCC'25	This Work
Process [μm]		0.18	0.18	0.18	0.18	0.13
Input Voltage [V]		2.8 - 4.2	2.8 - 4.2	2.7 - 4.2	2.8 - 4.2	2.8 - 4.2
Output Voltage [V]		3.3	3.3	3.4	3.3	3.3
Load Current [A]		0.2 - 1	0.2 - 1	0 - 1	0.1 - 1	0.05 - 1
Switching Freq. [MHz]		1	1, 1.5	1	1	1
C_O [μF]		10	4.7	10	10	10
L	Value [μH]	4.7	4.7	4.7	2.2	2.2
	DCR [mΩ]	N.A.	175	250	12	110
	Volume [mm³]	N.A.	2.5	3.84	234.3	1.92
C_F	Value* [μF]	10	10 x 2	10 x 2	4.7 x 2	4.7 x 2
	DC Bias (V_{CF})	V_O	$V_{IN}, 2V_{IN}-V_O$	V_{IN}, V_O	$V_{IN}/2$	$\vert V_{IN}-V_{OUT}\vert$
	Max. V_{CF} [V]	3.3	4.2, 5.1	4.2, 3.4	2.1	0.9
	Package Size (EIA)	N.A.	0805	1210	N.A.	0402
Norm. Value of $I_{L,AC}$**		2.19	2.36	2.21	1.11	1
Always $I_{L,DC} < I_{LOAD}$		No	Yes	Yes	No	Yes
Avg. $P_{CON,L}$**,† [mW]		186.3	93.02	85.48	110	73.34
Peak Efficiency (%)		96.6	93.8	95.5	98.3 (@12mΩ DCR)	96.6

* The value not considering derating due to DC bias voltage ** Averaged value over the entire V_{IN} range
† $P_{CON,L}=I_{L,DC}^2 \cdot DCR$, estimated under the assumption that I_{LOAD}=1A, DCR=200mΩ (4.7μH), 110mΩ (2.2μH)

Figure 16.9.6: Performance summary of the proposed buck-boost converter and comparison with state-of-the-art works.

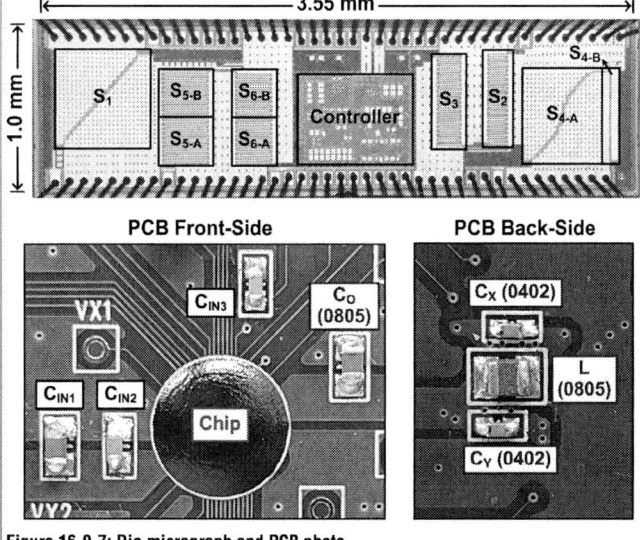

Figure 16.9.7: Die micrograph and PCB photo.

• 2026 IEEE International Solid-State Circuits Conference

ISSCC 2026 / SESSION 16 / ENERGY HARVESTING, PIEZO AND CHARGERS / 16.10

16.10 Fully Integrated mm-Scale 5G RF MIMO Harvester with -40dBm Sensitivity and Spatial MPPT via Hybrid Transformer-Based Combining/Shifting

Andrea Ballo[1,2], Ruiyuan Yang[2], Karim Ali[2], Massimo Bruno Alioto[2]

[1]University of Catania, Catania, Italy, [2]National University of Singapore, Singapore, Singapore

Abstract

A MIMO RF harvester in the 5G 28-GHz band with spatial MPPT is introduced to self-align high-gain harvesting direction with maximum power availability. Spatial scanning is enabled by a hybrid 2-level RF/DC antenna power combining and phase shifting based on transformers, while eliminating conventional off-chip capacitor(s). A 22nm test-chip shows -40dBm harvesting sensitivity, and power conversion efficiency of 56.7% at 0dBm (1.3× better than prior art at 28GHz).

RF harvesting is gaining rapidly growing interest thanks to the ubiquitous availability of RF ambient power as energy source, allowing the elimination of wiring and large batteries, and supporting miniaturization and cost reductions in edge devices [1-23]. The on-going migration to millimeter-wave carrier frequencies in 5G communications (e.g., NR FR2 at 28GHz) represents an opportunity to harvest from additional and densely-available RF sources, while simultaneously reducing system size including the antenna and the energy source [1-20]. Directional RF harvesting and spatial scanning aim to maximize the average harvested power and effectively serve as maximum power point tracking (MPPT) in the spatial domain, in addition to the traditional MPPT at harvesting ports. As key challenges, mm-wave MIMO harvesting architectures require: 1) high power peak conversion efficiency (PCE) and wide input power range with near-peak PCE to maximize the utilization of the wireless environment; 2) a reduction in the power burden of harvesting to extract net positive power even when energy is scarcely available (sensitivity); and 3) an expansion in the angular beam coverage to maximize the harvesting opportunities in any direction.

Recently, a 28-GHz 5G phased-array relay transceiver with vector-summing backscatter and 24-GHz wireless power transfer achieved a PCE of 15% at 0dBm [1], although with limited and sparse angular coverage (-30°, 0°, 30°), a relatively large gain ripple (4dBi) and a limited range of 0.2m. The 24-GHz 4-element harvesting array with a Butler matrix in [2] uses class-F rectifiers for 44% PCE at 4dBm, although only at rigidly fixed directions set at design time. Most of the other RF harvesters are instead based on single antenna (e.g., [4-8]), multi-harvester, or multi-band whose power contributions are combined in DC [9]. Other RF harvesters at 5.8GHz or lower cannot achieve the same level of miniaturization as mm-wave systems due to the antenna and the requirement of off-chip components [4-9].

In this work, a fully-integrated mm-scale 5G RF MIMO harvester is introduced to simultaneously support a competitive 57% peak PCE, a 35% PCE over 48dBm, a state-of-the-art sensitivity of -40dBm, and a broad angular coverage of ±75°. This is achieved by reducing the circuit power burden of harvesting by combining antenna power contributions partly in RF, enabling phase shifting and beamforming capabilities, and partly in DC when further RF power combining would require complex and lossy RF components (e.g., >2-port combiners). The proposed hybrid RF/DC architecture also eliminates the need for any off-chip capacitor.

The proposed harvesting architecture in Fig. 16.10.1 is based on a 3×3 phased antenna array that can be dynamically merged into 3-element sub-arrays (by either row or column) to scan either the azimuth or the elevation plane. Center sub-arrays containing the central element 22 serve as the reference phase for the other elements, hence their power contribution is directly transferred to the rectifier through the 1:1 transformer T2 and the related charge pump for: 1) impedance matching, 2) DC voltage shifting to the battery voltage V_{BAT} via the center tap of the secondary coil, and 3) RF-to-DC conversion. Similarly, the power contribution from the sub-array containing the 11 element (33 element) is connected to transformer T1 (T3), whose output is connected to the RF power combiner with either its original phase or a 180° shift through a double-pole double-throw switch (DPDT in Fig. 16.10.2). The power combiner is comprised of differential quadrature shifters and transformer T4 (Fig. 16.10.1). RF-to-DC conversion of the output power of transformer T4 is carried out by a second charge pump, whose output power is combined in DC with the output of the center sub-array. Mismatch across paths is partially compensated via a differential tuning mechanism that reuses part of the capacitors used for phase shifting. The output of the harvester feeds a conventional battery charger, a power management unit (PMU) to generate auxiliary voltages for the necessary switches and drivers, and provides power to a microcontroller unit (MCU, off-chip) to generate the necessary control signals for the harvester.

The two operating modes are shown in Fig. 16.10.3. In the first one, the system operates with a single antenna (element 22) for higher efficiency at low input power levels, which is typically required during the start-up phase or in the case where the battery is fully charged. In the second mode, MIMO operation is enabled and spatial MPPT occurs through beam steering/scanning in three steps: azimuth/elevation, quadrant, and sector selection. Each step is governed by dedicated sub-blocks managed by the MCU via the <B0...4> digital word that sets the related beam direction scanning the surrounding environment. In detail, initially the word <00000> and <10000> are sequentially generated to respectively combine the antenna elements by row or by column, so that the best plane between azimuth and elevation is identified (i.e., highest harvested voltage as proxy for input power).

Successively, <B1, B2> are set as complements of each other to scan the right quadrant (Fig. 16.10.2). Electrically, the signals at the secondary coil of T1 and T3 are lagged (leaded) by 180° if <B1, B2> is <0,1> (<1,0>), moving the beam to the right (left) side as referred to the azimuth plane. Finally, other combinations of <B3, B4> introduce a further phase shift of M×45°, where M = 0, 1, 2, 3 is the decimal equivalent of <B3, B4>. The proposed spatial MPPT allows a beam resolution of 15°. This is small enough to contain the maximum gain ripple below 1dBi, which is 5× lower than the prior best [1] in a comparable frequency target.

Figure 16.10.3 illustrates the circuit sub-blocks within the top view, comprising the DPDT, the differential tuning, the quadrature shifter and the RF-DC converter. To steer the power flowing through the input port IN1 (IN2) to the output port OUT, transistors M1b and M2a (M1a and M2b) are turned off, whereas M1a and M2b (M1b and M2a) are turned on. The OFF capacitance of transistors M1b and M2a (M1a and M2b) resonates with the primary half-coil inductance of T4 and blocks the path from IN1 port (IN2) to GND port (OUT). The IN1 (IN2) and OUT ports are connected via the ON-resistance of transistor M1a (M2a), whereas the IN2 (IN1) and GND ports are connected by the transistor M2b (M1b) in Fig. 16.10.3. Transistors are sized as a compromise between leakage and isolation. Regarding the differential tuning and the quadrature shifter, the physical design enforces symmetry of the differential path and preserves resonance at the same center frequency for the two coils of T4, regardless of the selected phase shift (beam steering). NMOS-only charge pumps are used for efficient RF-DC conversion. To keep the system efficiency insensitive to the battery state of charge (i.e., its voltage V_{BAT}), transistors within the charge pump are segregated in an isolated well with the n-well ring being connected to V_{BAT}, extending the breakdown voltage of the charge pumps to the breakdown of the parasitic junctions (which is typically larger than 5V).

Figure 16.10.4 shows the measured metrics related to matching and the power conversion efficiency of the test-chip in Fig. 16.10.7, whose active area is 0.62mm[2] and drives an antenna array on a Megtron substrate with a size of 2cm × 2cm. The antenna array S_{11} scattering parameter from measurements under row/column configuration agrees reasonably well with simulations including the effect of the gold stud bumps used for bonding. Their curves are both centered around 28GHz and their magnitude is lower than -15dBm, confirming adequate impedance matching between the test-chip and the antenna array. The plot of PCE versus the input power at max (min) gain direction shows a harvester sensitivity of -40dBm (-30dBm). At higher input power levels, the PCE increases as expected and remains higher than 35% for an input power range of 48dBm. A net output power of 6.47mW is achieved at an input power of 10dBm, which corresponds to a 48% PCE. The nearly-flat behavior of PCE as a function of frequency indicates full coverage of the targeted 5G NR FR2 band (n257) of the local multipoint distribution service (LMDS), allowing unrestricted ability to harvest power from any within-band RF source.

Figure 16.10.5 illustrates the measured output-related parameters across input power and steering angles with distances ranging from 0.65-to-25m. The input-output power characteristics fit the expected line-of-sight behavior with a linear trend. The angular coverage is 83% of the spherical cap, or 75°, allowing the circuit to harvest >440µW at 0dBm input power. In addition, extrapolating the beam pattern from the measured output power versus angle shows a maximum gain of 13.3dBi and a gain ripple of less than 1dBi, which enables the utilization of ambient power in a nearly angle-independent fashion.

The comparison with prior RF harvesters in Fig. 16.10.6 shows that the proposed solution achieves a best-in-class harvesting sensitivity of -40dBm among state-of-the-art harvesters operating from 0.693-to-28GHz. Compared to prior art in a comparable band, this represents a 20-dBm improvement over [2] with fixed beam direction, and a 35-dBm improvement over [1] with some spatial MPPT capability. In addition to the sensitivity improvement, the proposed harvester architecture improves angular coverage to 75° with a 15° step size, which is a 2.5× larger range and a 2× finer step than [1]. The 57% (62.7%) peak PCE at multi-antenna (single-antenna) outperforms prior art at comparable frequencies by 1.3-to-3.8× [1,2]. The sensitivity in single-antenna mode is comparable to prior art at lower frequencies [4,8], and is 11-to-30dBm better than [2,5,6,9]. The input power range at PCE >35% is an improvement of 31dBm over prior art with a comparable frequency [2], and 26dBm over other prior art with lower frequencies [3-9]. The maximum output power of the proposed harvester is 6.47mW, corresponding to an improvement over harvesters at similar frequencies by 2.1-to-5.9× [1,2]. Overall, the proposed harvester architecture with

ISSCC 2026 / February 17, 2026 / 11:35 AM

hybrid RF/DC combining advances the state of the art both at low and high levels of input power, and offers enhanced spatial coverage, thus allowing net-positive power to be harvested over a wider range of scenarios in terms of wireless environment and spatial orientation.

Acknowledgement:
The authors acknowledge the support of the Singapore Ministry of Education (T2EP50125-0009 grant), and GlobalFoundries for chip fabrication.

References:
[1] M. Ide, A. Shirane, K. Yanagisawa, D. You, J. Pang and K. Okada, "A 28-GHz Phased-Array Relay Transceiver for 5G Network Using Vector-Summing Backscatter With 24-GHz Wireless Power and LO Transfer," in *IEEE Journal of Solid-State Circuits*, vol. 57, no. 4, pp. 1211-1223, April 2022, doi: http://doi.org/10.1109/JSSC.2021.3137336.
[2] M. Ghorbanpoor, E. L. Roux, A. M. A. Najafabadi, O. Vorobyov, P. Nussbaum and H. Wang, "A 24-GHz 4-Element Multi-Beam Wireless Energy Harvesting Array with Class-F Rectifiers Achieving 51.5% PCE," *2024 IEEE/MTT-S International Microwave Symposium - IMS 2024*, Washington, DC, USA, 2024, pp. 2-5, doi: http://doi.org/10.1109/IMS40175.2024.10600372.
[3] M. -Y. Huang, T. Chi, F. Wang and H. Wang, "An All-Passive Negative Feedback Network for Broadband and Wide Field-of-View Self-Steering Beam-Forming With Zero DC Power Consumption," in *IEEE Journal of Solid-State Circuits*, vol. 52, no. 5, pp. 1260-1273, May 2017, doi: http://doi.org/10.1109/JSSC.2016.2641947.
[4] K. Ichikawa et al., "A 64.4% Efficiency 5.8GHz RF Wireless Power Transfer Receiver with GaAs E-pHEMT Rectifier and 45.2µs MPPT Time SIDITO Buck-Boost Converter Using VOC Prediction Scheme," *2024 IEEE International Solid-State Circuits Conference (ISSCC)*, San Francisco, CA, USA, 2024, pp. 228-230, doi: http://doi.org/10.1109/ISSCC49657.2024.10454425.
[5] S. Nagaveni, P. Hunasigidad, D. Pathak and A. Dutta, "On-Chip Configurable RF Energy Harvester for Biomedical Implantable Devices," in *IEEE Transactions on Circuits and Systems I: Regular Papers*, vol. 71, no. 11, pp. 5030-5039, Nov. 2024. http://doi.org/10.1109/TCSI.2024.3416252
[6] P. Xu, D. Flandre and D. Bol, "A Self-Gating RF Energy Harvester for Wireless Power Transfer With High-PAPR Incident Waveform," in *IEEE Journal of Solid-State Circuits*, vol. 56, no. 6, pp. 1816-1826, June 2021, doi: http://doi.org/10.1109/JSSC.2021.
[7] K. R. Sadagopan, J. Kang, Y. Ramadass and A. Natarajan, "A 960pW Co-Integrated-Antenna Wireless Energy Harvester for WiFi Backchannel Wireless Powering," 2018 IEEE International Solid-State Circuits Conference - (ISSCC), San Francisco, CA, USA, 2018, pp. 136-138, doi: 10.1109/ISSCC.2018.8310221.
[8] J. Wang, Y. Jiang, J. Dijkhuis, G. Dolmans, H. Gao and P. Baltus, "A 900 MHz RF energy harvesting system in 40 nm CMOS technology with efficiency peaking at 47% and higher than 30% over a 22dB wide input power range," *ESSCIRC 2017 - 43rd IEEE European Solid State Circuits Conference*, Leuven, Belgium, 2017, pp. 299-302, doi: 10.1109/ESSCIRC.2017.8094585
[9] Z. Zhang, C. Zhan, S. Zhao and M. -K. Law, "A High-Efficiency Low-Cost Multi-Antenna Energy Harvesting System With Leakage Suppression," in *IEEE Journal of Solid-State Circuits*, vol. 59, no. 9, pp. 2995-3007, Sept. 2024, doi: 10.1109/JSSC.2024.3387025.
[10] T. Tsutsui, "5G and it's surrounding situations until 2020," *2017 Symposium on VLSI Technology*, Kyoto, Japan, 2017, pp. T2-T6, doi: 10.23919/VLSIT.2017.7998184.

[11] F. Silva, P. Pinho and N. B. Carvalho, "Multibeam Beamforming Technology in Microwave Power Transfer and Harvesting," in *IEEE Journal of Microwaves*, vol. 5, no. 4, pp. 918-938, July 2025, doi: 10.1109/JMW.2025.3575342.
[12] Sheng-Fan Yang, Tzuen-Hsi Huang, Chun-Cheng Chen, Chun-Yi Lu and Pei-Jung Chung, "Beamforming power emitter design with 2×2 antenna array and phase control for microwave/RF-based energy harvesting," *2015 IEEE Wireless Power Transfer Conference (WPTC)*, Boulder, CO, 2015, pp. 1-4, doi: 10.1109/WPT.2015.7140129.
[13] S. -B. Liu, F. -S. Zhang, M. Boyuan, S. -P. Gao and Y. -X. Guo, "Multiband Dual-Polarized Hybrid Antenna With Complementary Beam for Simultaneous RF Energy Harvesting and WPT," in *IEEE Transactions on Antennas and Propagation*, vol. 70, no. 9, pp. 8485-8495, Sept. 2022, doi: 10.1109/TAP.2022.3177484.
[14] S. S. Amin and P. P. Mercier, "MISIMO: A Multi-Input Single-Inductor Multi-Output Energy Harvesting Platform in 28-nm FDSOI for Powering Net-Zero-Energy Systems," in *IEEE Journal of Solid-State Circuits*, vol. 53, no. 12, pp. 3407-3419, Dec. 2018, doi: 10.1109/JSSC.2018.2865467.
[15] N. Shinohara, B. Yang, W. Shao, K. Itoh, N. Sakai and N. Hasegawa, "Novel Energy Harvesting and SWIPT System at 28 GHz with a Simple Phased Array," *2023 IEEE 13th International Conference on RFID Technology and Applications (RFID-TA)*, Aveiro, Portugal, 2023, pp. 209-212, doi: 10.1109/RFID-TA58140.2023.10290371.
[16] O. L. A. López, B. Clerckx and M. Latva-Aho, "Dynamic RF Combining for Multi-Antenna Ambient Energy Harvesting," in *IEEE Wireless Communications Letters*, vol. 11, no. 3, pp. 493-497, March 2022, doi: 10.1109/LWC.2021.3133623.
[17] T. Wu, Y. Zhang and J. Guo, "A Dual-Input Wide-Range RF-DC Rectifier with 72.4% Peak Efficiency for RF Energy Harvesting," *2023 International Conference on Microwave and Millimeter Wave Technology (ICMMT)*, Qingdao, China, 2023, pp. 1-3, doi: 10.1109/ICMMT58241.2023.10277211.
[18] S. Bandyopadhyay, P. P. Mercier, A. C. Lysaght, K. M. Stankovic and A. P. Chandrakasan, "A 1.1 nW Energy-Harvesting System with 544 pW Quiescent Power for Next-Generation Implants," in *IEEE Journal of Solid-State Circuits*, vol. 49, no. 12, pp. 2812-2824, Dec. 2014, doi: 10.1109/JSSC.2014.2350260.
[19] S. S. Amin and P. P. Mercier, "A Fully Integrated Li-Ion-Compatible Hybrid Four-Level DC–DC Converter in 28-nm FDSOI," in *IEEE Journal of Solid-State Circuits*, vol. 54, no. 3, pp. 720-732, March 2019, doi: 10.1109/JSSC.2018.2880183.
[20] X. Li, S. Du, X. Zeng and Z. Chen, "A Globally Optimized 3-D MPPT System for Dual-Band RF Energy Harvesting with Collaborative Source Reconfiguration," *2025 Symposium on VLSI Technology and Circuits (VLSI Technology and Circuits)*, Kyoto, Japan, 2025, pp. 1-3, doi: 10.23919/VLSITechnologyandCir65189.2025.11075023.
[21] H. Yi, W. -H. Yu, P. -I. Mak, J. Yin and R. P. Martins, "A 0.18-V 382-µW Bluetooth Low-Energy Receiver Front-End With 1.33-nW Sleep Power for Energy-Harvesting Applications in 28-nm CMOS," in *IEEE Journal of Solid-State Circuits*, vol. 53, no. 6, pp. 1618-1627, June 2018, doi: 10.1109/JSSC.2018.2815597.
[22] J. Davis, J. Sankman and D. Ma, "An input-powered 1.1-µA Iq 13.56 MHz RF energy harvesting system for biomedical implantable devices," *2015 IEEE 11th International Conference on ASIC (ASICON)*, Chengdu, China, 2015, pp. 1-4, doi: 10.1109/ASICON.2015.7516894.
[23] J. Lee, S. -H. Lee, G. -G. Kang, J. -H. Kim, G. -H. Cho and H. -S. Kim, "A Triboelectric Energy-Harvesting Interface With Scalable Multi-Chip-Stacked Bias-Flip and Daisy-Chained Synchronous Signaling Techniques," in *IEEE Journal of Solid-State Circuits*, vol. 57, no. 12, pp. 3825-3839, Dec. 2022, doi: 10.1109/JSSC.2022.3193738.

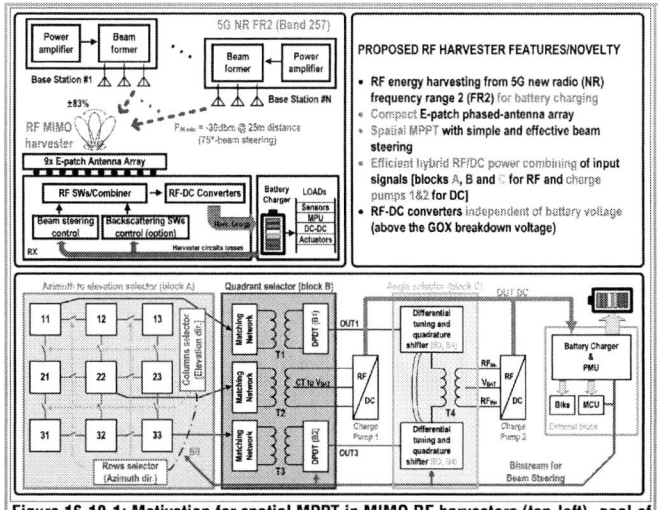

Figure 16.10.1: Motivation for spatial MPPT in MIMO RF harvesters (top-left), goal of this work (top-right) and simplified block diagram of the proposed RF MIMO harvester operating in the 5G NR FR2 band 257 (bottom).

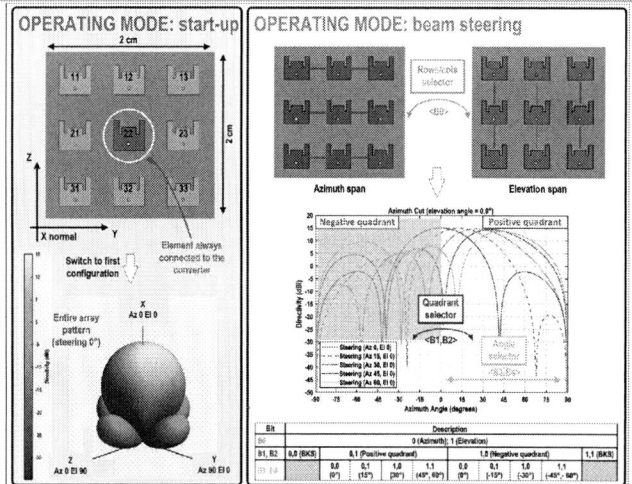

Figure 16.10.2: Adopted antenna array on the Megtron substrate, 3D antenna radiation pattern and start-up mode (left), beam steering strategy and operating modes for focused RF harvesting and spatial MPPT (right).

ISSCC 2026 / SESSION 16 / ENERGY HARVESTING, PIEZO AND CHARGERS / 16.10

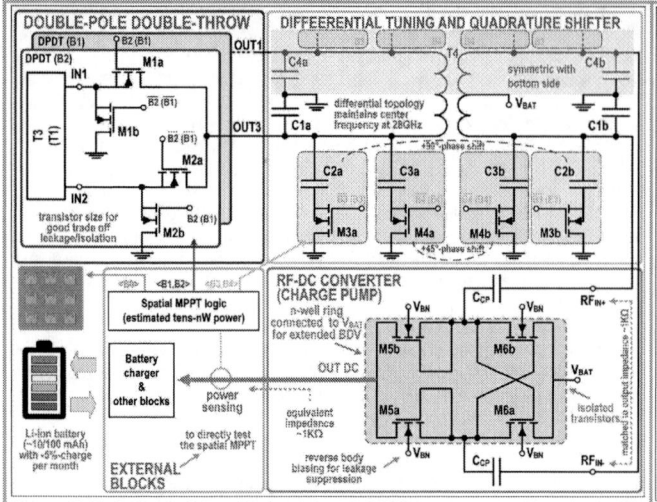

Figure 16.10.3: Key building blocks of the RF MIMO harvester: double-pole double-throw (top-left), differential tuning and quadrature shifter (top-right), RF-DC converter and off-chip blocks for end-to-end demonstration (bottom).

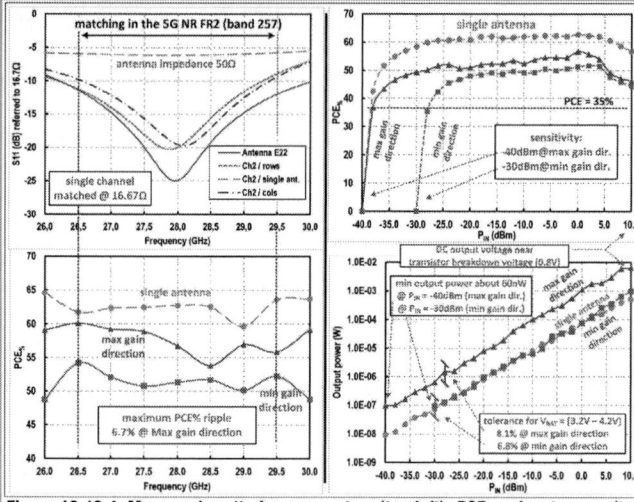

Figure 16.10.4: Measured scattering parameters (top-left), PCE vs. input power (top-right), PCE vs. input signal frequency (bottom-left), output vs. input power (bottom-right).

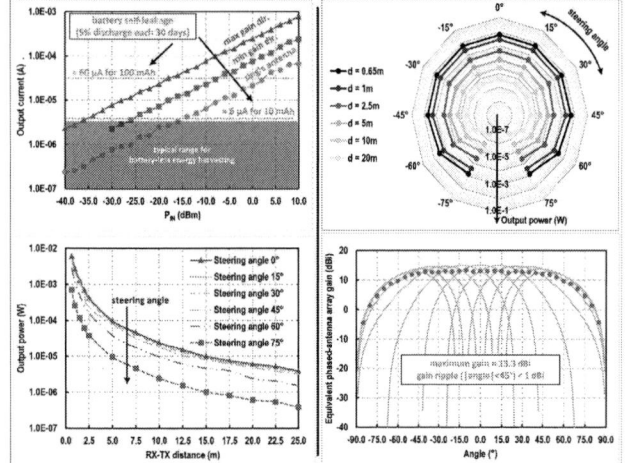

Figure 16.10.5: Output current vs. input power (top-left), output power vs. steering angle for different RX-TX distances (top-right), output power vs. RX-TX distances for different steering angles (bottom-left), phased-antenna array gain vs. angle (bottom-right).

Figure 16.10.6: Comparison with state-of-the-art RF energy harvesters and wireless power transfer systems (best performance or unique feature in bold).

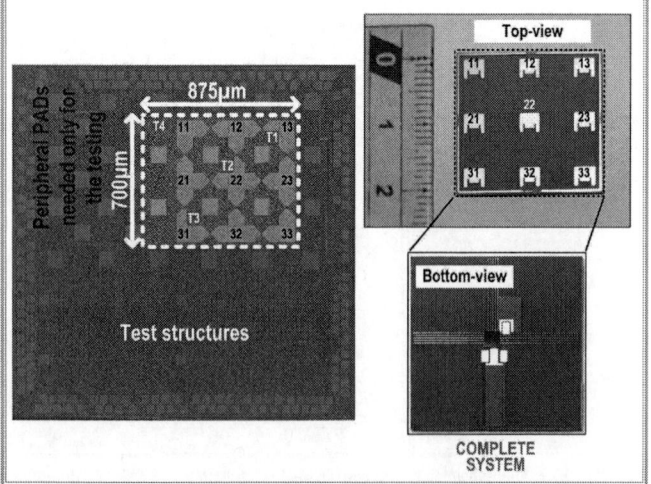

Figure 16.10.7: Chip micrograph (left), complete system top-view (top-right) and bottom-view (bottom-right).

• 2026 IEEE International Solid-State Circuits Conference

Session 17 Overview: *Highlighted Chip Releases for AI*
INVITED INDUSTRY SESSION

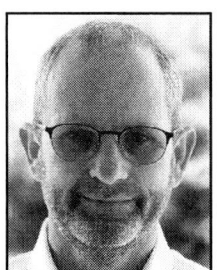

Session Chair: Hugh Mair
MediaTek, Dallas, TX

Session Co-Chair: Sylvain Clerc
CEA LIST, Grenoble, France

Hardware solutions powering the AI revolution are presented in this session, with papers covering a diverse set of use cases. Paper 17.1 targets a new class of desktop AI super-computers, while more traditional datacenter deployment is the target of Paper 17.4. Paper 17.2 targets edge applications while Paper 17.3 demonstrates a scalable approach to power/performance at the edge.

ISSCC 2026 / February 17, 2026 / 1:30 PM

1:30 PM

17.1   NVIDIA GB10: SoC Built for AI Acceleration

A. Skende, Nvidia, Westford, MA

In Paper 17.1, Nvidia details the GB10 SoC that powers the recently launched DGX™ Spark workstation. A dual-die solution, fabricated in TSMC's 3nm process, co-packages a CPU die featuring 20 ARM v9.2 cores alongside a Blackwell iGPU die tuned for desktop performance. The GPU features 5th-generation Tensor Cores, 4th-generation RT cores, and 24MB of L2 cache, achieving 31TFLOPS of FP32 and 1PFLOPS of FP4.

1:55 PM

17.2   The STM32N6 Microcontroller: Enabling Intelligent Edge AI for IoT and Beyond

G. Desoli, STMicroelectronics, Cornaredo, Italy

In Paper 17.2, ST Microelectronics discusses their STM32N6 family of 16nm FinFET microcontrollers powered by an Arm Cortex-M55 CPU and featuring an in-house NPU. Targeting edge applications, the 1GHz Neural-ART NPU delivers 600GOPS and 3TOPS/W. Additional co-processors accelerate graphics/video/imaging and optimize the handling of non-rectangular displays.

17

2:20 PM

17.3   ARIES and REGULUS: A Unified and Scalable Hardware-Software Co-Designed NPU SoC Family for On-Device and On-Premises Multimodal Inference

D. Shin, Mobilint, Seoul, Korea

In Paper 17.3, Mobilint presents their scalable NPU core architecture deployed in the ARIES and REGULUS SoCs. Dual 4-core NPU clusters are implemented in the 181mm^2 Samsung-14nm ARIES, while a single NPU core is deployed on the 50mm^2 TSMC-12nm REGULUS.

2:45 PM

17.4   Maia: A Reticle-Scale AI Accelerator

S. Xu, Microsoft, Mountain View, CA

In Paper 17.4, the architecture and implementation of Microsoft's MAIA AI silicon, a reticle-scale 750W AI SoC, is presented. Innovation across power delivery, thermal management, physical design, validation, and packaging is detailed, resulting in a scalable approach to high-performance AI acceleration.

ISSCC 2026 / SESSION 17 / INVITED INDUSTRY / 17.1

17.1 NVIDIA GB10: SoC Built for AI Acceleration

A. Skende[1], G. Rosseel[2], N. Pinckney[3]

[1]Nvidia, Westford, MA, [2]Nvidia, Santa Clara, CA, [3]Nvidia, Austin, TX

Abstract

This paper details the GB10 SoC that powers the recently launched DGX™ Spark workstation. A dual-die solution, fabricated in TSMC's 3nm process, co-packages a CPU die featuring 20 ARM v9.2 cores alongside a Blackwell iGPU die tuned for desktop performance. The GPU features 5th-generation Tensor Cores, 4th-generation RT cores, and 24MB of L2 cache, achieving 31TFLOPS of FP32 and 1PFLOPS of FP4.

Introduction

NVIDIA recently introduced the DGX Spark workstation (a.k.a. Project DIGITS) powered by the innovative GB10 SoC (Fig. 17.1.1), addressing the growing need for accessible AI development platforms that bridge datacenter capabilities with desktop accessibility. The GB10 SoC enables AI developers, researchers, data scientists, and students to prototype, fine-tune, and run large AI models locally while maintaining seamless workload portability to NVIDIA DGX Cloud or other accelerated cloud infrastructures. This "mini-AI supercomputer" represents a paradigm shift in democratizing access to high-performance AI computing in a small form-factor system based on an innovative multi-dielet SoC architecture enabled through advanced packaging technologies.

The emergence of large language models and sophisticated AI applications has created an unprecedented demand for accessible high-performance computing platforms. Traditional approaches have required researchers and developers to either invest in expensive datacenter-class hardware or rely on cloud-based solutions with associated connectivity and cost constraints. The GB10 SoC addresses this fundamental challenge by delivering the computational capabilities of datacenter AI accelerators in a desktop form factor, powered by a standard wall outlet.

GB10 SoC represents an engineering collaboration between NVIDIA and MediaTek. It combines MediaTek's power efficient implementation of the CPU and memory subsystem with NVIDIA's high performance GPU acceleration.

Multi-Dielet Architecture and System Design

The GB10 introduces a dual-dielet architecture (Fig. 17.1.2), breaking from traditional monolithic chip design. The S-dielet hosts the CPU complex and memory subsystem, incorporating twenty ARM v9.2 cores evenly split across two clusters. Each core is equipped with private L2 cache, while each cluster of cores shares 16MB of L3. A 16MB System Cache, implemented as a memory-side cache, serves as L4 cache for the CPU cores and system cache for other accelerators integrated in the SoC. The G-dielet integrates NVIDIA's Blackwell iGPU, tuned for desktop operation while retaining architectural consistency with the Blackwell datacenter GPU. Its hardware includes 5th-generation Tensor Cores for AI acceleration, 6144 CUDA Cores for graphics processing, 4th-generation RT Cores for ray tracing support, and 24MB of Graphics L2 cache for high-bandwidth local data access.

By integrating heterogeneous computing elements from two industry leaders, the system leverages complementary strengths within a single package. The result is a system that balances efficiency, scalability, and raw performance. Both dielets are fabricated on TSMC's 3nm process. This ensures the density and efficiency required for desktop deployment while delivering compute characteristics approaching workstation and datacenter-class systems. Together, they form the foundation of the GB10's performance-per-watt advantage.

The dual-dielet physical implementation relies on 2.5D packaging technology whose interposer enables die-to-die interconnect density needed for high performance, and at the same time power efficient operation. This approach allows the CPU and GPU dies to communicate at high speed with minimal power overhead, enabling tight integration without the inefficiencies of board-level signaling. At the center of this packaging solution is the NVLINK-C2C interface (Fig. 17.1.3, Fig. 17.1.5), adapted specifically for chip-to-chip requirements and delivering ~600GB/s of aggregate bandwidth.

The design implements hardware-enabled bidirectional coherency between CPU and GPU. Traditionally, heterogeneous systems required explicit cache management operations, introducing overhead. In GB10, coherency is managed at the hardware level, thus improving performance while simplifying the programming model at the driver as well as application level. Hardware-level enabled coherency is made possible through implementation of ATS (Address Translation Services) which enables GPU's MMU to cache translation to system physical address space and GPU's L2 to be physically tagged. This transparency extends to the developer experience. By caching data in system physical address space, the SoC presents a simplified, logically unified memory space. Applications gain the ease of a physical and logical UMA with the efficiency of specialized accelerators.

A 256b LPDDR5x interface provides up to 301GB/s bandwidth at 9400 MTransactions/s, scaling to 128GB of memory capacity. This memory configuration enables inference for models up to 200B parameters and fine-tuning for models up to 70B parameters. By selecting LPDDR5x over HBM, the system balances across bandwidth, capacity and energy efficiency. This decision makes it possible to sustain demanding AI workloads in desktop environments, where power and thermal budgets are far more constrained.

Peripherals

The GB10 SoC integrates broad connectivity options tailored to both workstation and AI workloads. Multiple PCIe interfaces support connectivity to high-capacity SSD storage (up to 4TB) as well as expansion and networking connectivity. Multiple USB3.2 interfaces enable peripheral support. Typical for desktop platform, GB10 enables DGX Spark with integrated Wi-Fi and Bluetooth. The system supports up to three simultaneous displays through two DisplayPort Alt-Mode outputs (4K@120Hz) and one HDMI 2.1a interface (8K@120Hz, SDR/HDR). Together, these capabilities ensure that the GB10 is not only an accelerator but also a complete workstation solution. Developers can store, visualize, and interact with system in a traditional desktop model while performing advanced AI acceleration on a single platform.

Security

The GB10 incorporates enterprise-grade security through a dual-root architecture. A SROOT secure processor enables NVIDIA's hardware-based secure boot and manages cryptographic functions needed for authentication and decryption of system firmware, while an OSROOT processor validates UEFI, OS, and high-level software integrity. This layered approach ensures system trust from the earliest boot stages. SROOT secures proprietary low-level flows and key management, while OSROOT extends verification to higher level drivers and software components. The platform can support both firmware TPM (fTPM) and discrete TPM implementations.

The GB10 GPU extends its utility with robust virtualization support. Through PCIe SR-IOV, it exposes one physical function for bare-metal access and up to 255 virtual functions for fine-grained resource virtualization and partitioning. This makes the system suitable for multi-tenant research environments, enabling multiple concurrent users without performance interference. Hardware-enforced isolation guarantees workload separation and data confidentiality. Support for Address Translation Services (ATS) via the Graphics MMU ensures efficient address translation. Combined with Graphics L2 optimization in system physical address space, virtualized workloads achieve near-native performance. These features transform the GB10 from a personal workstation into a shared compute resource.

GPU Performance and AI Acceleration Capabilities

The Blackwell iGPU in the GB10 delivers 31 TFLOPS of FP32 and 1 PFLOPS of NVFP4 tensor compute. This enables low-latency inference and efficient fine-tuning of mid-scale models, previously feasible only on dedicated datacenter GPUs. Compatibility with GB300 ensures that optimizations and algorithms developed on the GB10 translate seamlessly to larger-scale deployments. This architectural consistency minimizes friction when transitioning workloads between development and production environments.

Key to the GPU's efficiency is the 5th-generation Tensor Cores, specialized for AI matrix operations. These units deliver superior performance per watt compared to general-purpose compute cores, sustaining high throughput without exceeding power budgets.

The inclusion of NVFP4 precision extends this advantage. It maximizes computational density while preserving accuracy, allowing the GB10 to achieve great AI performance within its 140W envelope.

While powerful as a standalone SoC, the GB10 is designed for scale. PCIe Gen5 x8 connectivity enables integration with ConnectX-7 networking (Fig. 17.1.4), allowing multiple DGX Spark systems to be linked into distributed compute clusters. Scaling two systems doubles both compute and memory resources, extending support to models with up to 405B parameters. NVIDIA NCCL enables multi-GPU parallelism, while RDMA and GPU Direct provide low-latency data transfer across the compute nodes. This growth path allows developers to expand capacity incrementally as needs evolve.

Design Implementation

GB10 SoC represents merging of two architecture, MediaTek's low power implementation of CPU and memory subsystem with NVIDIA's high performance GPU design. This was made possible through implementation of industry defined standard interfaces and protocols, as well as a strict adherence to an IP model for NVIDIA accelerators delivered for integration into the MediaTek's die.

A hierarchical verification approach with heavy reliance on BFMs (Bus Functional Models) complemented by end-to-end verification of complex system features was deployed.

Performance modeling also played a critical role, particularly in aligning GPU memory traffic with MediaTek's memory subsystem to achieve the best memory interface efficiency possible. Hardware emulation was fundamental in enabling full-stack verification pre-silicon. This ensured functional issues were identified and addressed before silicon manufacturing. This complex multi prone approach to the design and verification enabled a relatively fast post-silicon validation and allowed moving to mass production using the first revision of the silicon.

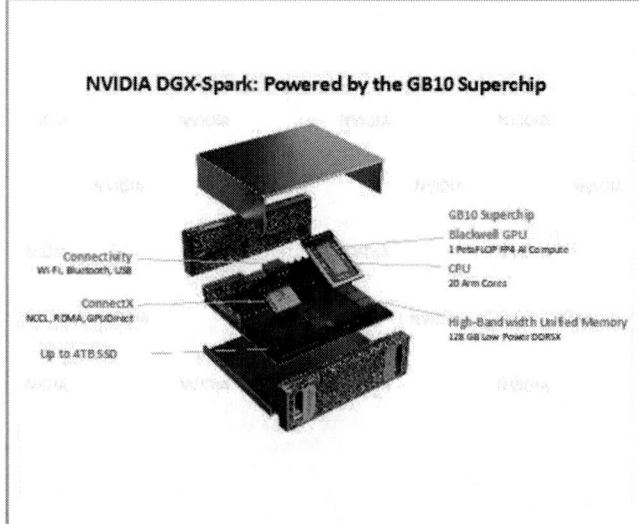

Figure 17.1.1: NVIDIA DGX-Spark Workstation.

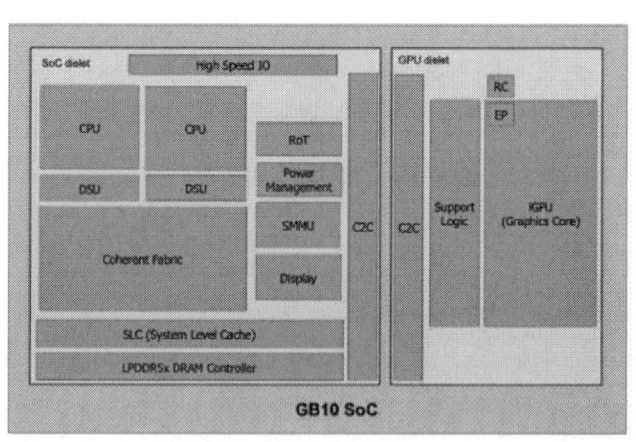

Figure 17.1.2: Dielet-based GB10 SoC.

ISSCC 2026 / SESSION 17 / INVITED INDUSTRY / 17.1

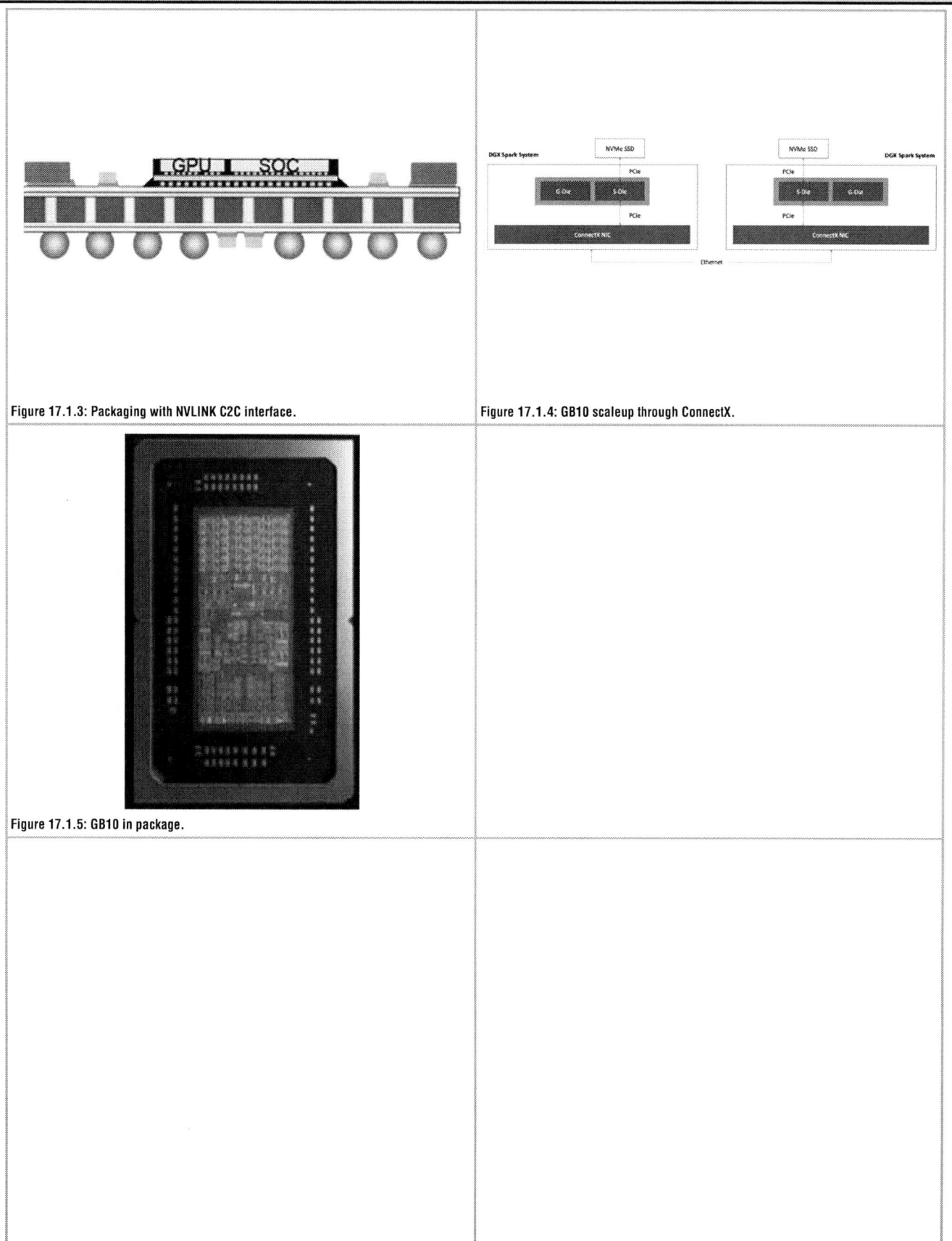

Figure 17.1.3: Packaging with NVLINK C2C interface.

Figure 17.1.4: GB10 scaleup through ConnectX.

Figure 17.1.5: GB10 in package.

ISSCC 2026 / SESSION 17 / INVITED INDUSTRY / 17.2

17.2 The STM32N6 Microcontroller: Enabling Intelligent Edge AI for IoT and Beyond

G. Desoli[1], J-F. Agaësse[2], N. Chawla[3], E. Hilkens[2], T. Boesch[4], V. Taufour[2], M. Ayodhyawasi[3], P. Ravenhill[2], S. Pal-Singh[3], M. Soulie[2]

[1]STMicroelectronics, Cornaredo, Italy, [2]STMicroelectronics, Grenoble, France, [3]STMicroelectronics, Greater Noida, India, [4]STMicroelectronics, Geneva, Switzerland

Abstract

The STM32N6 microcontroller meets the growing need for intelligent edge devices in IoT, wearables, industrial automation, and smart home systems supporting real-time, energy-efficient AI processing at the edge, reducing latency and enhancing data privacy by limiting cloud reliance. At its core, the Neural-ART neural processing unit delivers complex and power efficient AI inference. This paper examines the chip's 16nm FinFET design, performance, computational power and energy efficiency

---

The STM32N6 microcontroller addresses the increasing demand for intelligent edge devices in IoT, wearable technology, industrial automation, and smart-home systems. It enables real-time, energy-efficient AI processing at the edge, reducing latency and enhancing data privacy by minimizing cloud dependency. At its core is the Neural-ART neural processing unit (NPU), a dedicated accelerator that supports complex AI inference with minimal power consumption. This paper explores the chip design, implementation in 16nm FinFET, and performance, highlighting its ability to balance computational power and energy efficiency.

The rapid growth of embedded AI applications in IoT, wearables, and smart systems has increased the demand for energy-efficient, low-latency neural network processors. Traditional microcontrollers often fail to meet the processing requirements of modern algorithms. Consequently, dedicated hardware accelerators have emerged as a viable alternative [1]. Current methods often compromise adaptability for performance by relying on rigid-function components or reducing processing speed to accommodate varied kernel dimensions and layer arrangements.

The STM32N6 chips are based on an Arm® Cortex®-M55 core running at up to 800MHz, while the NPU runs at up to 1GHz, delivering a peak of 600 giga-operations per second (GOPS) and up to three tera-operations per second per watt (TOPS/W). Other coprocessors include a Chrom-ART accelerator for 2D graphics, a Chrom-GRC graphics resource cutter for round and other non-square displays, a 2.5D NeoChrom graphics accelerator, an H.264 video encoder capable of 1080p15 or 720p30, and an image signal processor (ISP) targeting a 5Mpixel camera at 30 frames per second (Fig. 17.2.2).

The chip offers a memory configuration with 4.2MB of contiguous embedded RAM and 64+128KB tightly coupled memories (TCMs), plus external interfaces for multiple memory types, including pseudo-static RAM (PSRAM), synchronous dynamic RAM (SDRAM), and both NOR and NAND flash. It also includes a wide range of external interfaces, such as Gigabit Ethernet and CSI2.0. An Arm® TrustZone® security subsystem is complemented by ST's proprietary SoC security architecture, side-channel-attack-resistant AES acceleration, tenant-aware firewalling, and the goal of achieving SESIP Level 3 and PSA Level 3 security certifications. Designed for edge AI applications, the STM32N6 combines low power consumption, advanced processing capacity, and a feature set tailored to industrial, consumer, and some automotive uses.

The STM32N6 is powered by the Neural-ART NPU, a reconfigurable NN processor designed to address these constraints by way of a streaming data-flow-based NPU that features dynamic kernel configuration, compression, and batch processing, enabling adaptive resource allocation for diverse workloads. A reconfigurable data transfer fabric integrated into the NPU and linked to the on-chip interconnect enhances data reuse and minimizes access to on/off-chip memory, a significant constraint in edge devices [3]. A low-power Arm® Cortex®-M55 microprocessor core with a suite of peripherals and interfaces supports simultaneous processing of data acquisition and transfer, pre- and post-processing, as well as non-hardware-accelerated tasks. The Arm's Helium vector extensions boost machine learning workloads while running on the microcontroller core, reducing idle periods and improving overall system performance.

Recent advances in hardware acceleration for neural networks (NN) focus on optimizing energy efficiency and computational throughput for edge devices. Both research and industrial applications demonstrate that NNs are effective for many sophisticated functionalities; however, their high computational requirements necessitate dedicated hardware. Later designs [1] show that NNs can achieve high accuracy, but their substantial energy demands can make them impractical for embedded applications. In response, several energy-saving accelerators have been proposed. However, these designs often lack flexibility in handling varying kernel sizes and layer configurations. Recent architectures enhance memory efficiency by employing on-chip storage, yet many remain unsuitable for edge devices with limited resources. Conversely, microcontroller-based approaches prioritize energy efficiency but frequently compromise computational capability.

The NPU differentiates itself from previous designs by integrating dynamically configurable acceleration with autonomous layer processing. Unlike rigid ASIC-based architectures [3], this approach dynamically adjusts to the demands of individual layers without increasing power consumption. This adaptability is particularly beneficial for embedded applications, where minimizing latency and maximizing energy efficiency are critical [2]. Consumer and industrial AI algorithms require deeper topologies with many layers, millions of parameters, and varying kernel sizes, resulting in escalating bandwidth, power, and area costs. Meeting these constraints for embedded devices and applications requires efficiency through a hierarchical memory system and effective reuse of local data. In the STM32N6, the NPU relies on unidirectional links transporting data streams via a configurable fully connected switch, including DMAs and accelerators supporting convolutions, pooling, activation, scalar, and other processing kernels. This fabric allows the definition of an arbitrary number of concurrent, virtual processing chains at runtime. A full-featured backpressure mechanism handles data flow control and stream multicasting, enabling the reuse of a data stream at multiple endpoints. Linked lists control the fully autonomous processing of an entire epoch, which can be a fused combination of multiple layers. Multiple accelerators grouped or chained together handle varying feature map sizes and multiple kernels in parallel. The switch topology is defined at compile time and dynamically configured at runtime, with each node in the fabric graph representing a processing element or memory streaming channel.

As illustrated in Fig. 17.2.3 and Fig. 17.2.4, the design supports pipelined operation across multiple tiers while preserving worst-case latency constraints. The configurable accelerator framework (CAF) manages data movement and computation scheduling while automatically adjusting to diverse network layer topologies.

The NPU supports both INT8 and INT16 data formats and dynamically adjusts to variable workloads, reduces memory operations, and minimizes delays, making it particularly suitable for embedded AI implementations. This design-time parametric architecture represents a step toward a paradigm shift from the Von Neumann architecture to a flexible, dedicated dataflow stream processing engine. The instance integrated into the STM32N6 seamlessly connects to the MCU (Microcontroller Unit) backbone via two 64b AXI interfaces and associated DMA streaming channels, while also supporting direct streaming from the imaging sensor pipeline and a NPU data cache for both weights and activations.

A notable feature is the advanced stream ciphering/deciphering capability, which protects the intellectual property of application deployment for both kernel weights and activations without noticeable performance degradation.

The chip is designed in TSMC 16FF (16nm) with a die size of 17.5mm² (Fig. 17.2.7). Fine and coarse clock gating is applied at the logic and processing element level, reducing dynamic inference power per layer and during idle periods. Architectural innovations, such as asynchronous bus nodes, high-end memories, and a network-on-chip backbone, combined with circuit-level optimizations, result in a balanced design suitable for energy-constrained edge deployments.

Figure 17.2.5 illustrates the execution of a YOLOv5 algorithm commonly used in the industry for image object detection and classification. It details performance and power consumption metrics when all layers are mapped onto the NPU. The top charts display the average power consumption of the AI subsystem (NPU and memories) and the execution time for hardware epochs generated by the mapper tool to complete an inference run. The ratio of epoch execution time to power depends on the mapping efficiency of each layer, considering its parameters, hardware capabilities, and memory bandwidth requirements. The network achieves over 100 frames per second at maximum chip performance, enabling reduced frequency and supply voltage in the field to meet a 30 frames per second real-time target. The bottom three charts in Fig. 17.2.5 show the relative power distribution across the AI subsystem for the NPU, bus, and memories, the NPU's internal power breakdown, including weights and activation buffers, and its subunits.

Edge AI with integrated NPUs delivers groundbreaking capabilities to IoT, wearables, healthcare, industrial automation, and smart home systems by providing powerful, low-latency intelligence directly on the device. For the first time, tiny microcontrollers can perform complex computer vision tasks that were once limited to large, costly, and power-hungry microprocessors (MPUs). This breakthrough not only enhances the user's experience of existing products but also enables the development of entirely new solutions that were previously beyond technical and commercial reach. With the help of integrated efficient NPUs, MCUs can now handle diverse vision tasks such as object detection, gesture recognition, and quality inspection, enabling innovative and cost-effective solutions across numerous industries and use cases. In the wearables domain, where a small footprint and low power consumption are crucial, smart glasses can perform real-time analytics on camera images and deliver personalized experiences to users. Smart access control systems, which rely on reliable face recognition algorithms, require high processing power to ensure smooth user experience. Since these systems need to be cost-effective for widespread adoption, MCU-based solutions are preferred. The same applies to automatic license plate recognition systems, which have traditionally been offered only by expensive solutions. STM32 MCUs have been widely adopted by drone manufacturers for flight controllers. With the STM32N6, flight controllers will be equipped with computer vision capabilities to enable features such as follow-me and obstacle avoidance. Figure 17.2.1 shows snapshots of inference running on the device for four different applications, including people detection and pose estimation, hand gesture and pose recognition. Applications like these were previously possible only with much more powerful and costly devices.

In-memory computing (IMC) [6] significantly reduces data transfer with memory, hence power consumption. We are in a mature stage of prototyping this technology for both digital and analog versions, recording up to 8.3× and 16.7× TOPS/W improvements for digital and analog IMC, respectively (Fig. 17.2.6). These technologies support advanced quantization at 1, 2, 4, and 8 bits for further performance improvements and model size reduction while ensuring seamless workflow integration in the continuity of the current NPU.

The STM32N6's adaptability to multiple use cases, varying AI algorithm functionality, and NPU autonomous layer processing capabilities make it particularly suitable for real-time applications in industrial IoT, wearables, and smart agriculture. Its alignment with widely adopted ML frameworks ensures real-world applicability, effectively bridging high-speed acceleration with the limitations of edge devices.

Acknowledgment:
The STM32N6 family is the result of the coordinated effort of many STMicroelectronics engineers and experts, that couldn't be explicitly mentioned but are all deserving of recognition.

References:
[1] C. Silvano et al., "A Survey on Deep Learning Hardware Accelerators for Heterogeneous HPC Platforms," *ACM Comput. Surv.*, vol. 57, no. 11, pp. 1–39, June 2025, doi: 10.1145/3729215.
[2] H. Hua, Y. Li, T. Wang, N. Dong, W. Li, and J. Cao, "Edge Computing with Artificial Intelligence: A Machine Learning Perspective," *ACM Comput. Surv.*, vol. 55, no. 9, pp. 1–35, Jan. 2023, doi: 10.1145/3555802.
[3] S. Alam, C. Yakopcic, Q. Wu, M. Barnell, S. Khan, and T. M. Taha, "Survey of Deep Learning Accelerators for Edge and Emerging Computing," *Electronics*, vol. 13, no. 15, p. 2988, July 2024, doi: 10.3390/electronics13152988.
[4] G. Desoli et al., "A 2.9TOPS/W deep convolutional neural network SoC in FD-SOI 28nm for intelligent embedded systems," *ISSCC*, Feb. 2017.

[5] G. Desoli et al., "A 40-310TOPS/W SRAM-Based All-Digital Up to 4b In-Memory Computing Multi-Tiled NN Accelerator in FD-SOI 18nm for Deep-Learning Edge Applications," *ISSCC*, Feb. 2023.
[6] A. Mehonic et al., "Roadmap to neuromorphic computing with emerging technologies," *APL Materials*, vol. 12, no. 10, Oct. 2024, doi: 10.1063/5.0179424.

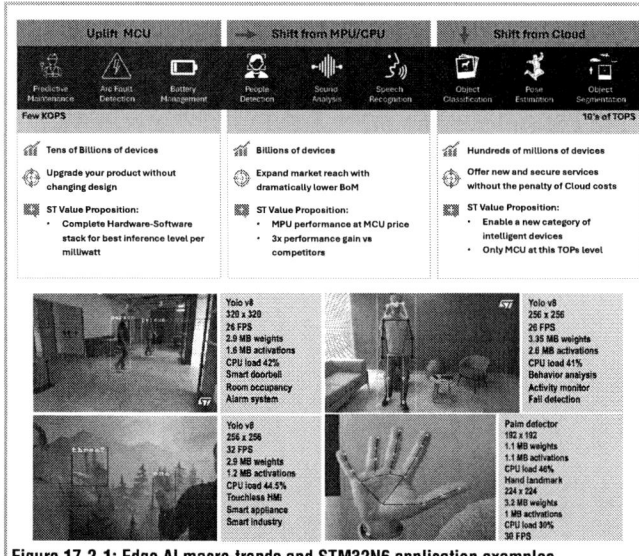

Figure 17.2.1: Edge AI macro-trends and STM32N6 application examples.

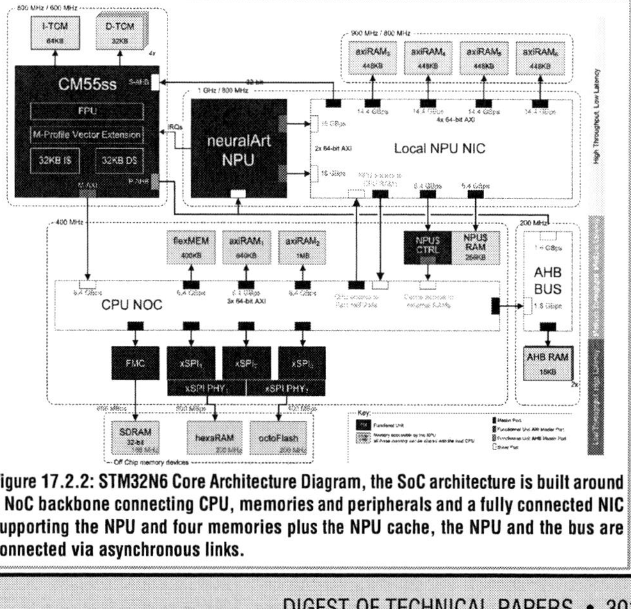

Figure 17.2.2: STM32N6 Core Architecture Diagram, the SoC architecture is built around a NoC backbone connecting CPU, memories and peripherals and a fully connected NIC supporting the NPU and four memories plus the NPU cache, the NPU and the bus are connected via asynchronous links.

ISSCC 2026 / SESSION 17 / INVITED INDUSTRY / 17.2

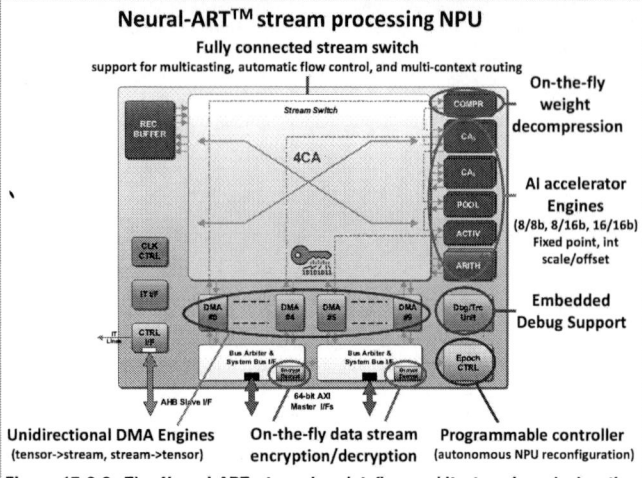

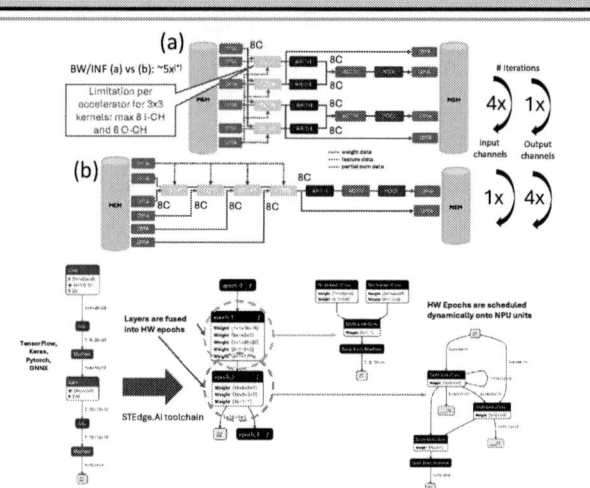

Figure 17.2.3: The Neural-ART streaming dataflow architecture is a design time parametric NPU built around a streaming switch that can be defined with a mix of different accelerators, features and cardinalities to meet specific products requirements.

Figure 17.2.4: The Neural-ART maximizes local data reuse matching bandwidth, buffer space, MAC units, to accelerators that can be rearranged to balance BW and computing for different layer topologies. Iterations on input/output channels can handle arbitrary tensor and kernel dimensions.

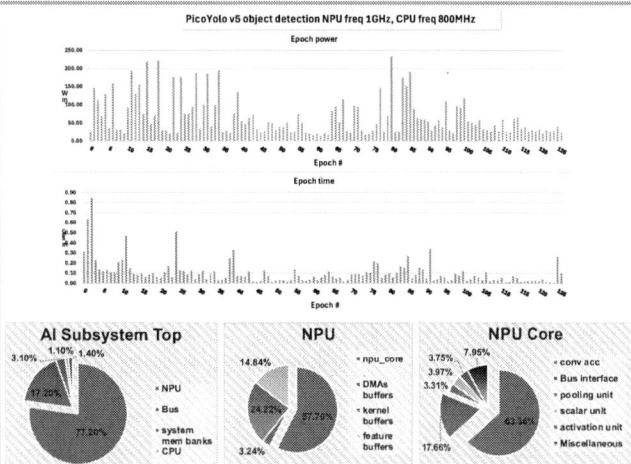

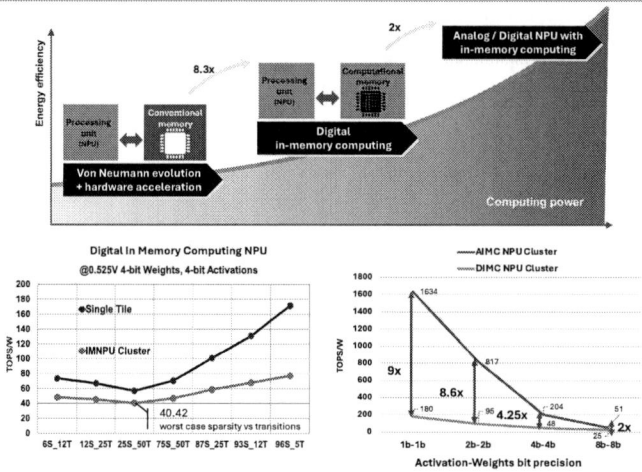

Figure 17.2.5: PicoYolo v5 INT8 inf time 10.59 ms, average AI subsystem power (NPU + memories) 94mW (1.51 TOPS/W), top/middle: average power and compute time for each epoch/ fused layers, bottom left/right: power break down for the whole AI subsystem (NPU,Bus,Mems), NPU+buffers, NPU logic.

Figure 17.2.6: STMicroelectronics roadmap for non-Von Neumann NPUs relies on SRAM-Based 1-2-4-8b In-Memory Computing Multi-Tiled NN Accelerators [2], up to 8.3× and 16.7× for Digital and Analog (approximate) performance increase over STM32N6 peak efficiency measured of 3 TOPS/W in INT8 precision.

Figure 17.2.7: STM32N6 die photo.

• 2026 IEEE International Solid-State Circuits Conference

ISSCC 2026 / SESSION 17 / INVITED INDUSTRY / 17.3

17.3 ARIES and REGULUS: A Unified and Scalable Hardware-Software Co-Designed NPU SoC Family for On-Device and On-Premises Multimodal Inference

D. Shin, H. Yang, S. Jeon, J. Park, S. Han, J. Park, J. Lee, J. Kim, H. Kim, Y. Oh, M. Kim, C. Jung, W. Kim, S. Kim, H. Jeong, G. Kim, K. Lee, G. Song, Y. Min, C. Song, A. Kanybek, Y. Jung, J. Song, S. Cho, H. Na, J. Park, D. Si, B. Lee, B. Park, H. Jeon

Mobilint, Seoul, Korea

Abstract

We present a scalable NPU architecture, proven in two SoCs (ARIES and REGULUS), designed to efficiently execute diverse AI workloads from vision to large language models. The architecture scales seamlessly across cores, clusters, and chips via a unified programming model. Our holistic hardware–software co-design preserves model accuracy through a bound- and distribution-aware mixed-precision quantization scheme, while efficiently handling complex activation functions via a non-uniform piecewise-linear method. Silicon measurements demonstrate near-linear performance scaling and high power and bandwidth efficiency on both vision and language workloads.

The shift of intelligent compute to edge devices requires neural processing units (NPUs) to efficiently execute a wide spectrum of workloads—from compute-bound convolutional neural networks (CNNs) [1] to memory-intensive transformer-based large language models (LLMs) [2,3]—under strict mobile power and bandwidth constraints. This diversity spans models, scales, and serving scenarios, creating architectural pressure to sustain performance across compute- and memory-bound regimes, efficiently realize complex activation functions such as GELU and SiLU, and meet heterogeneous latency–throughput targets. Quantization is essential to mitigate memory bottlenecks and increase effective bandwidth and capacity, but accuracy is often limited by the resolution–range trade-off, particularly in the presence of outliers; in practice, dequantization overheads on hardware datapaths can erode these gains.

To address these challenges, we present a scalable NPU architecture developed via holistic hardware–software co-design and demonstrated in two systems-on-chip (SoCs)—ARIES and REGULUS. Our approach combines: (1) a bound-aware, distribution-aware mixed-precision quantization scheme that preserves accuracy while maximizing memory efficiency; (2) a non-uniform piecewise-linear (PWL) method that computes complex activation functions with high fidelity and low cost; and (3) a unified programming model that orchestrates these techniques on scalable hardware. Together, these elements deliver high performance and energy efficiency across a broad range of models and applications. Figure 17.3.1 summarizes the challenges and the proposed solutions.

Figure 17.3.2 presents block diagrams of the NPU and the SoC variants—ARIES and REGULUS. A single NPU cluster integrates four NPU cores and a shared 16MiB shared scratchpad memory (SPM). ARIES implements two such clusters for a total of eight cores connected by a hybrid topology Network-on-Chip (NoC). REGULUS contains a single core with a 1MiB shared SPM. Both chips share the same core microarchitecture while targeting area and power optimization at different operating points.

Each NPU core consists of a Tensor Compute Engine (TCE), Tensor Manipulation Engine (TME), Vector Compute Engine (VCE), a DMA Engine, a Control Unit (CU), and a local SPM. The TCE performs matrix multiplication (MATMUL) and convolution. It includes a vector feeder (VF), a matrix feeder (MF), a MAC array, an aggregator (AG), and a partial-sum buffer (PB). The MAC array sustains 4K MACs under a weight-stationary dataflow. The VF streams activations, and the MF streams weights and matrix-structured data. The MF incorporates an on-the-fly dequantization pipeline to reduce memory bandwidth. The AG performs INT30 accumulation with fused requantization and activation and writes results directly to the local SPM without extra data movement. The TME handles common shape/layout transforms. It supports fully unaligned data access via flexible strides and offsets, while a dual logical-bank memory organization hides latency to sustain throughput matching the local SPM bandwidth. The VCE executes elementwise and vector operations including depthwise convolution, average/max pooling, nonlinear functions, top-k, and datatype conversion. In transformer workloads, MATMUL map to the TCE, while softmax, normalization, and top-k map to the VCE, forming an inter-engine pipeline. The DMA supports 3D strides, gather/scatter, prefetch, and double buffering between off-chip DRAM and on-chip SRAM. Three independent units for weight-read (WR), activation-read (AR), and activation-write (AW) maximize concurrency with the TCE/TME/VCE. The CU fetches, decodes, and dispatches a topologically ordered command stream. A combination of compile-time events and a runtime scoreboard resolves data and resource dependencies among DMA/TCE/TME/VCE, preserving topological order and memory consistency.

To minimize accuracy loss while achieving high compression, we adopt a hierarchical mixed-precision quantization approach as illustrated in Fig. 17.3.3. Offline analysis first assigns static mixed-precision at the layer level (Level 1) and the group level (Level 2) according to global statistics. For fine-grained control, each group then selects one of two advanced methods based on distribution characteristics. When outliers are prevalent, elementwise mixed-precision (Level 3-1) decomposes values into low-precision inliers and high-precision extremes to preserve accuracy without distorting quantization ranges. When values cluster non-uniformly, a LUT-based non-uniform quantization (Level 3-2) maps low-bit indices to high-precision values, achieving both compression and expressiveness.

Uniform PWL approximation can incur large errors in certain regions. We propose a non-uniform PWL scheme whose segment density adapts to local distribution complexity, implemented with a hierarchical two-level LUT. The L1 stage computes a band index via an affine transform of the input and maps the slope to a band-dependent segment-offset scale. The L2 stage retrieves the PWL coefficients within the selected band. The final output is produced by an affine transform over the segment offset. Concentrating segments in high-curvature regions reduces approximation error at a fixed LUT capacity.

The processor employs a decomposition-aware MATMUL structure to support multiple formats efficiently within a single compute core. A dual scale aligner normalizes partial sums from heterogeneous quantization domains into a common numeric range and leverages a higher-precision previous-scale aligner to minimize normalization loss. Such hardware–software co-design is instrumental in mitigating system-level memory-bandwidth bottlenecks. In particular, quantization substantially reduces the WRDMA time, while the dequantization overhead is overlapped with WRDMA, MAC, and MF.

The NPU core architecture scales uniformly from a single core to multi-core, multi-cluster, and multi-chip systems under a unified programming model. Figure 17.3.4 illustrates the scalability and the performance gains across multi-core, multi-chip systems. At the core level, REGULUS integrates one NPU core. At the cluster level, four local cores and one Global Control Unit (GCU) form a cluster. The GCU provides hardware-level synchronization and performs sharding of weights and activations across tiles. It also orchestrates collective operations such as broadcast, all-gather, and reduce-scatter, supporting parallelization strategies such as tensor, data, and pipeline parallelism. At the multi-cluster level, clusters synchronize via a cluster-signal handshake. At the multi-chip level, diverse parallel execution strategies, as shown in Fig. 17.3.4(a), scale the architecture further.

Under a unified programming model, high utilization is maintained across single-batch/single-model, multi-batch, and multi-model deployments. At the graph level, the compiler partitions workloads to balance compute and memory traffic. Techniques include activation tiling for SPM reuse, output-depth partitioning for fine-grained load balancing, and the selection among tensor, data, and pipeline parallelism per layer. Figure 17.3.4(b, c) illustrates the scalability of ARIES under different execution modes. In throughput-priority (TP) mode, throughput scales almost linearly with the number of cores, while the latency for a single input slightly increases. In contrast, in latency-priority (LP) mode, the throughput scaling is less pronounced, but the latency is effectively reduced as more cores are utilized. This trade-off applies across different batch-size scenarios: LP mode is typically employed in single-batch execution, whereas TP mode is preferable for multi-batch workloads. For multi-model or large models, the compiler applies deterministic partitioning to ensure balanced DMA bandwidth. The same barrier semantics scale from 1 to N cores, clusters, and chips, sustaining near-peak utilization over diverse workloads.

Figure 17.3.5 summarizes the chip-level measurements collected at room temperature under the same software stack. Voltage–frequency shmoo plots are reported for representative vision models, and voltage–performance–efficiency curves show the existence of an operating point with peak energy efficiency. Power breakdown highlights that the NPU is the dominant on-chip consumer during inference, while DRAM, PCIe, and PMIC contribute at the board level.

Figure 17.3.6 compares ARIES to GPUs of similar or higher class across several models. In Fig. 17.3.6(a), GPU baselines are reported in FP16, whereas Fig. 17.3.6(b) uses INT8 baselines. While the GPU can show higher absolute throughput in certain configurations, ARIES demonstrates substantially higher power and bandwidth efficiency. Figure 17.3.6(b) evaluates REGULUS against a representative edge AI platform in a single-batch setting. Despite having lower DRAM bandwidth, REGULUS achieves higher inference speed on key models and better bandwidth efficiency, which is critical in edge and mobile environments.

Tables in Fig. 17.3.6(c) and Fig. 17.3.6(d) indicate that our hardware–software co-designed quantization preserves accuracy close to FP16 on vision tasks and LLMs. For LLMs, we

ISSCC 2026 / February 17, 2026 / 2:20 PM

evaluate our quantization approach on four language models: Llama-3.2-3B/3.1-8B (Instruct) and Qwen-3-4B (Instruct)/3-8B (base; no public Instruct). Following standard practice, we use WikiText2 as our evaluation dataset, with 128 randomly selected sentences of 2048 tokens for calibration and 100 sentences for evaluation. We report perplexity (PPL) as our primary metric to assess model quality after quantization. Our method employs static quantization for activations, with average bit-widths for both activations and weights detailed in Fig. 17.3.6(d). We build upon GPTQ [4] and SmoothQuant (SMQ) [5] as foundational quantization techniques and enhance them by incorporating elementwise mixed-precision (Level 3-1) and LUT-based non-uniform quantization (Level 3-2) to achieve improved performance while maintaining computational efficiency. The experimental results demonstrate that our method achieves superior PPL scores compared to the baseline GPTQ and SMQ methods, validating the effectiveness of our enhanced quantization strategy. Fig. 17.3.7 shows the die photos of ARIES and REGULUS and summarizes key characteristics of each chip.

In summary, this work presents a scalable NPU architecture, proven in two SoCs (ARIES and REGULUS). The core microarchitecture with TCE/TME/VCE, and efficient mixed-precision support accelerates diverse networks effectively. As the number of cores scales from one to eight, system performance scales nearly linearly. Latency or throughput priority mode enables flexible adaptation to a variety of deployment scenarios.

References:

[1] A. Krizhevsky et al., "ImageNet Classification with Deep Convolutional Neural Networks," *Adv. Neural Inf. Process. Syst.* (NeurIPS), vol. 25, pp. 1097-1105, 2012.
[2] A. Vaswani et al., "Attention is All You Need," *Adv. Neural Inf. Process. Syst.* (NeurIPS), vol. 30, 2017.
[3] OpenAI (J. Achiam et al.), "GPT-4 Technical Report," arXiv:2303.08774, 2023.
[4] E. Frantar et al., "GPTQ: Accurate post-training quantization for generative pre-trained transformers," arXiv preprint arXiv:2210.17323, 2022.
[5] G. Xiao et al., "SmoothQuant: Accurate and efficient post-training quantization for large language models," *Int. Conf. on Machine Learning* (ICML), Honolulu, HI, USA, Jul. 23–29, 2023, *Proc. Machine Learning Research*, vol. 202, pp. 38087–38099.
[6] NVIDIA, "GeForce RTX 5080 — Product Page," Available online: https://www.nvidia.com/en-us/geforce/graphics-cards/50-series/rtx-5080/
[7] NVIDIA, "RTX A5000 — Product Page," Available online: https://www.nvidia.com/en-us/products/workstations/rtx-a5000/
[8] NVIDIA, "RTX 2000 Ada Generation — Product Page," Available online: https://www.nvidia.com/en-us/products/workstations/rtx-2000/
[9] NVIDIA, "Jetson Orin Nano Series — Data Sheet (DS-11105-001)," Available online: https://www.mouser.com/pdfDocs/Jetson_Orin_Nano_Series_DS-11105-001_v11.pdf

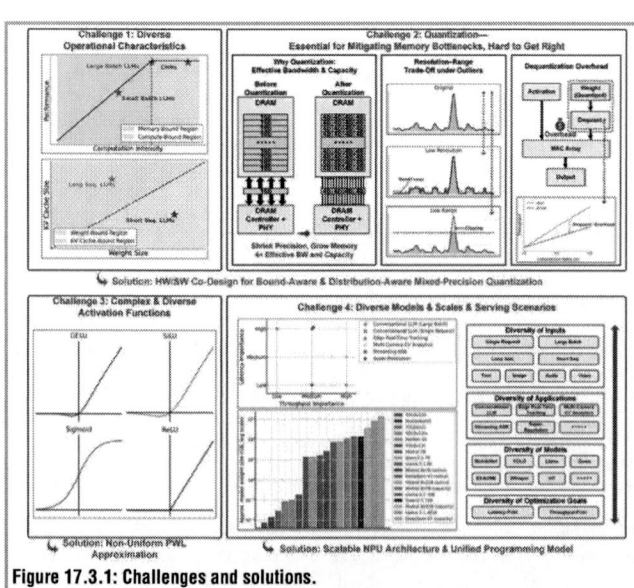

Figure 17.3.1: Challenges and solutions.

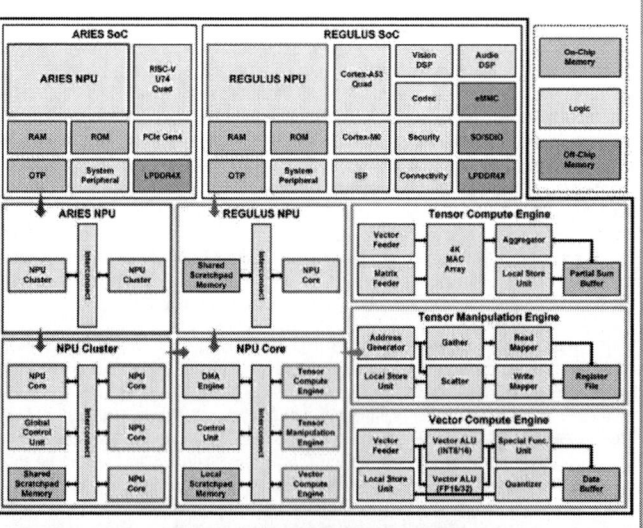

Figure 17.3.2: SoC architecture and NPU microarchitecture diagram.

979-8-3315-8937-0/26 $31.00 © 2026 IEEE

ISSCC 2026 / SESSION 17 / INVITED INDUSTRY / 17.3

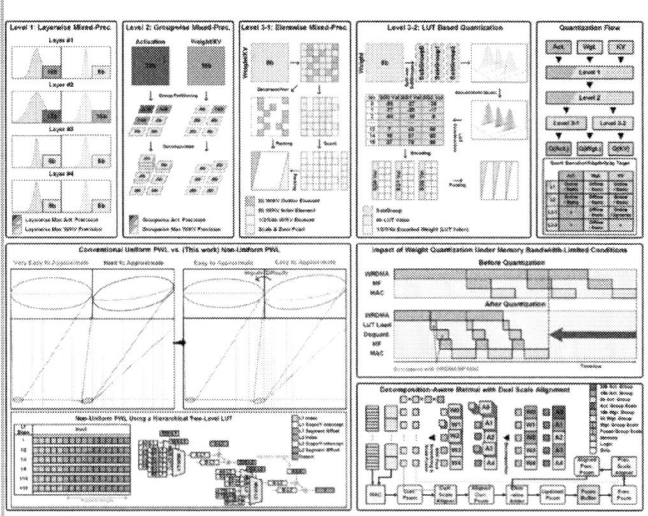

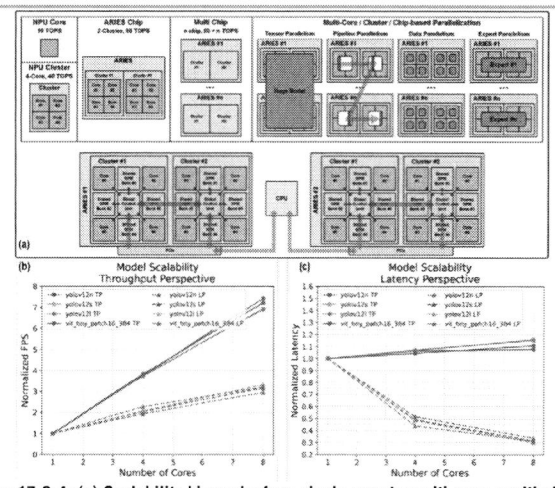

Figure 17.3.3: Mixed-precision execution: INT8 efficiency with FP16 accuracy.

Figure 17.3.4: (a) Scalability hierarchy from single-core to multi-core, multi-cluster, and multi-chip; various parallelism modes on multi-cluster scenario; (b) Normalized FPS scaling in throughput-priority (TP) and latency-priority (LP) modes; (c) normalized scaling in TP and LP modes.

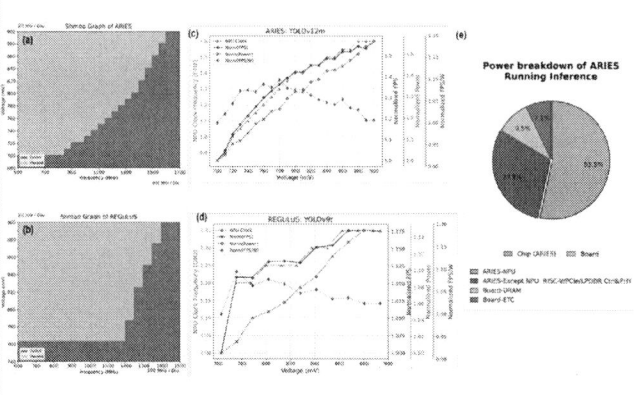

Figure 17.3.5: Room-temperature shmoo plots of (a) ARIES and (b) REGULUS; (c) ARIES running YOLOv12m and (d) REGULUS running YOLOv9t under voltage scaling, showing NPU clock frequency, normalized FPS, normalized power, and normalized efficiency (FPS/W); (e) power breakdown of ARIES during inference.

Figure 17.3.6: (a) Performance and efficiency comparison between ARIES and GPUs; (b) performance and efficiency comparison between REGULUS and Jetson Orin Nano; quantization accuracy comparison of (c) CNN models and (d) language models.

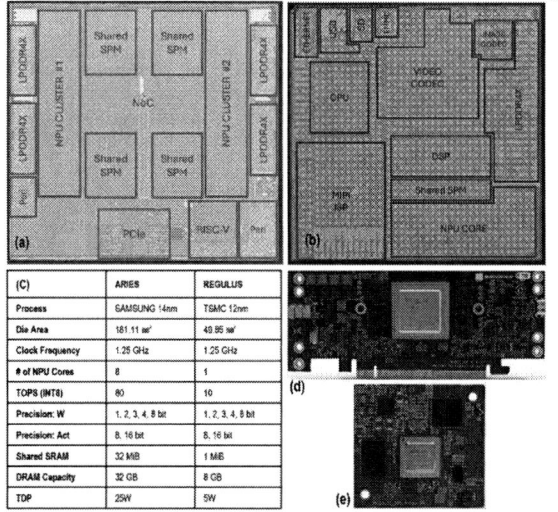

Figure 17.3.7: Die micrograph of (a) ARIES and (b) REGULUS; (c) chip specifications; photographs of the (d) ARIES and (e) REGULUS board and chips.

• 2026 IEEE International Solid-State Circuits Conference

ISSCC 2026 / SESSION 17 / INVITED INDUSTRY / 17.4

17.4 Maia: A Reticle-Scale AI Accelerator

S. Xu[1], G. Mandal[1], S. Jahagirdar[1], S. Tripathy[2], P. Parthasarathy[1], S. Srinivasan[1], A. Levin[1]

[1]Microsoft, Mountain View, CA, [2]Microsoft, Bengaluru, India

Abstract

In Paper 17.4, the architecture and implementation of Microsoft's MAIA AI silicon, a reticle-scale 750W AI SoC, is presented. Innovation across power delivery, thermal management, physical design, validation, and packaging is detailed, resulting in a scalable approach to high-performance AI acceleration.

Introduction: Maia 200 is Microsoft's second-generation custom AI accelerator, engineered for large language model workloads. Compared to Maia 100, Maia 200 achieves over 3× higher HBM bandwidth and compute throughput, while scaling network bandwidth by more than 2× (see Fig. 17.4.1 for detailed specifications). Leveraging a vertically integrated approach across the system, networking, and software, Maia 200 is co-designed to deliver leading performance-per-dollar for inference and synthetic data generation workloads.

Similar to its predecessor, Maia 200 is a reticle-sized die, fabricated on TSMC's 3nm process and integrated using CoWoS-S packaging. With a provisioned 750W SOC TDP power, Maia 200 overcomes significant implementation challenges across power delivery, thermal management, physical design, package integration, and validation—topics that will be addressed in the following sections.

Architecture background: Maia accelerators adopt hierarchical architecture. The basic building block is a "tile." Each tile integrates two primary compute engines—a Tile Tensor Unit (TTU) and a Tile Vector Processor (TVP)—fed by a private, multi-banked Tile SRAM (TSRAM, or L1 SRAM) and a tile DMA subsystem for data movement. The TTU performs high throughput matrix multiply and convolution, which supports a wide range of data types including narrow data types such as FP8/FP6/FP4 that are OCP MX compliant. The TVP is a highly programmable 256-way SIMD unit, optimized for BF16 while supporting many other data formats. Each tile further contains a Tile Control Processor (TCP). The code running on the TCP is the final output of the software stack and delivers work to the tile subsystem. Hardware semaphores are used to synchronize DMA, TVP and TTU operations.

Multiple tiles compose together to form a "cluster." Each cluster contains a large multi-banked Cluster SRAM (CSRAM, or L2 SRAM), a dedicated cluster DMA subsystem for moving data between cluster SRAM and HBM memory, and a dedicated cluster core for control and synchronization. The entire SoC is constructed with multiple instances of clusters. To improve yield, redundancy schemes are implemented for both tiles and SRAM.

Maia accelerators feature a highly optimized data movement infrastructure, centered around its Direct Memory Access (DMA) subsystem and hierarchical Network-on-Chip (NoC), both critical for achieving scalable performance in AI and data-intensive workloads. The DMA engine is architected for multichannel and high-bandwidth data transfers, and it supports 1D, 2D and 3D data strides to optimize ML workloads. It enables asynchronous movement of data between on-chip SRAM, high-bandwidth memory (HBM), and external interfaces, supporting multi-stream operations to maximize memory throughput and minimize compute idle times. The NoC is designed as a hierarchical, scalable interconnect fabric that provides high-bandwidth, low-latency communication across compute clusters and memory subsystems. The NoC supports both unicast and multicast transfers, crucial for distributing large tensor blocks and synchronizing parallel processing elements. To further enhance HBM efficiency, multiple narrow data types are supported as storage formats in both HBM and SRAM. Hardware-based data casting converts these storage types to compute types at line rate, ensuring seamless integration with compute pipelines.

Maia accelerators employ a high-performance, Ethernet-based interconnect to deliver scalable, low-latency communication across multiple accelerator nodes. This interconnect leverages 112G PAM4 SerDes for high-speed data transfer. Unlike many other AI solutions, Maia implements a unified backend network architecture. The Ethernet links are organized into two categories: fixed links, which connect neighboring accelerators directly, and switch links, which interface with an Ethernet switch. This topology minimizes system cost and power consumption while improving link utilization for its target workloads.

Chip Implementation: Maia 200 employs TSMC's 3nm process with a reticle-limited die size (~26×33mm²). The design includes a large number of tiles with high-power density. To address the thermal and power delivery challenges inherent to reticle-scale silicon, the chip resources are partitioned into multiple independent clock and power domains. Dynamic Voltage and Frequency Scaling (DVFS) allows adaptation to workload, minimizing overall power usage. Clock gating is deployed aggressively, both at the register-transfer level for minor blocks and at the macro level, enabling rapid sleep and wake cycles responsive to AI workload demands.

One of the unique features for Maia is the large on-die SRAM. This allows data-intensive tasks to be serviced with minimal off-chip traffic, which helps reduce HBM demand and improve power efficiency. To improve Power-Performance-Area (PPA), customized SRAM macros are used for L1 and L2 SRAM. Given the large size of SRAM, the SRAM supports ECC correction and periodic scrubbing functions.

Maia 200 includes 28 X4 PAM4 channels for interprocessor communication. Each X4 channel is implemented using an X4 SerDes macro. To provide a better channel, each via

and BGA landing were optimized to meet the target impedance as close as possible. The CoWoS-S TSV model was optimized by characterizing the TSVs in a test interposer. Finally, end-to-end SI simulation was done and COM scores were met.

For the host-side interface, Maia 200 uses Gen6 PCIe interface using similar SI practices as described above.

Maia 200 uses six stacks of HBM3e, three on each side of the SoC. Interposer routing was optimized for optimum read- and write-eye with lower power.

Maia 200 is a reticle-size die designed for data center deployments, where high availability is crucial. To achieve this, multiple RAS (Reliability, Availability, and Serviceability) features are integrated: HBM is a major source of potential errors; therefore, advanced on-die ECC and CRC mechanisms are implemented for HBM interfaces. On-demand memory scrubbing is supported for both HBM and on-die SRAM to proactively detect and correct faults. Additionally, the SoC enforces strict error containment policies to prevent error propagation across the system.

Power Design: The optimum power and cooling levels for the SoC are determined by assessing the target workload performance in the Azure data center. The data center control plane configures the power and cooling levels into the SoC at run time. Power consumption is monitored and regulated by the SoC. The compute blocks operate on a dedicated voltage and frequency domain to enable power scaling. This also allows Maia to respond to on-die and system events requiring power consumption reduction. There are on-die thermal and power sensors for runtime overheat detection, power management, as well as telemetry.

Power Delivery: Design a PDN that can supply such a high TDP power was an interesting problem to solve. There were many voltage domains and allocating enough resources to each domain while maintaining the supply noise and ripple specs was challenging to solve. The chip uses 112G Ethernet interface, PCIe6, and HBM3e interfaces. Care was taken to keep sensitive supplies for these interfaces isolated from any other supplies, so that the noise injection from other noisy supplies is low. Also, we made sure there are appropriate amounts of DTCs allocated for these sensitive supplies. And the supply noise was verified in PI analysis.

At the die level, there were several steps taken from the very beginning. For all clock buffers, a concise decision was made to have extra power/ground straps and dedicated decaps. SHDMIM caps were judiciously allocated closer to macros that have higher activity.

Maia 200 uses CoWoS-S packaging technology. For better power delivery Deep Trench Capacitor (DTC) was used for most rails. The availability of DTC helped significantly reduce supply noise.

As the demand current is very high, land side decaps were necessary for power integrity. The decap locations were chosen based on two major factors: first, the BGAs that are removed to make space for decaps, without causing other BGAs to exceed the current limit from reliability perspective; and second, the location of the decaps being close enough to be effective to regions where the change in demand current is high.

The power grid in interposer was done with the priority that each ubump connection to C4 has minimum lateral routing in interposer. To achieve this, we used a custom cell that contains a predefined number of ubumps and is connected to a C4 that is at the center of the ubumps at the other side of interposer. This structure provided a very uniform and well controlled C4-to-ubump resistance.

Finally, at the board level, the PMIC devices were placed close to the SoC package, so that the board level I²R loss could be minimized.

During the design phase, PDN models were generated for SoC, interposer, package, and board. We also included our internally generated PMIC models to simulate the full PDN network to ensure the specification for each rail was met.

Voltage Droop Management: Voltage excursions due to activity changes are a significant consideration due to the implications to power consumption. The SoC's compute can transition between low and high activity levels on short time scales of hundreds of nanoseconds. This can cause significant voltage excursions and power loss. The SoC implements a combination of proactive and reactive excursion management schemes. The compute sub-blocks are turned ON and OFF in a staggered manner to manage the ramp rates. The total amount of compute turning ON and OFF in a given time period is also controlled to limit the ramp rates. The ramp rate controls are tuned to minimize performance impact. Additionally, on-die voltage droop detectors sense and throttle the activity rates if proactive controls are not sufficient to manage the excursions within acceptable limits.

Physical Design: The physical design implementation is a cutting-edge engineering achievement that exemplifies how advanced methodologies can optimize performance, area, power, and manufacturability within the stringent constraints of a reticle-limited scale. As AI accelerators demand unparalleled levels of compute density, memory bandwidth, and energy efficiency, the Maia SoC addresses these challenges with an innovative design approach.

The chip's construction is built on a framework of quadrant-based compute clusters interconnected by a centralized fabric, with memory, PCIe, and scale-out IOs strategically positioned along the periphery along with integration simplification through using abutted design implementation. However, achieving the desired performance metrics posed significant challenges, including handling the multi-hierarchical, multi-initiated module (MIM) designs, routing-intensive and memory-dominated architecture, multi-clock and multi-voltage domain interface crossings, high-speed operation and the integration complexity of diverse IOs such as PCIe, HBM, and scale-out interfaces. Central to the success of the Maia accelerator is its robust Power Distribution Network (PDN), which utilizes 19 metal layers, hierarchical grids, and strategically placed bump pads to ensure reliable current delivery, mitigate IR drop and electromigration, and maintain signal integrity across the die. The chip also employs a custom clocking implementation based on an H-tree topology to minimize clock latency, variation, and skew, enabling high-speed and low-power operation. A customized routing methodology, leveraging all the metal layers to match RC delays across interface bundles and forward clocking implementation by taking advantage of unilateral data transfer, was deployed to optimize both area and latency.

Early-stage thermal simulations played a crucial role in optimizing the placement of high-switching regions, preventing thermal hotspots and improving overall thermal efficiency. Following place-and-route, the SoC undergoes comprehensive signoff flows to validate its functional correctness, performance, and manufacturability. These include multi-mode, multi-corner (MMMC) timing analysis to ensure timing closure across diverse conditions, IR drop and electromigration analysis for power grid robustness, DRC and LVS checks for process compliance and design integrity, and antenna effect mitigation through dummy vias and metal fills to enhance yield and reliability.

These measures ensure that the design is both performant and manufacturable at scale. The integration of these state-of-the-art techniques enables the Maia accelerator to achieve industry-leading performance metrics, including high clock speeds, high compute density, maximum memory fabric bandwidth, and energy efficiency, all while maintaining robust functionality and reliability. As a result, the SoC not only sets a new benchmark in AI accelerator SoC development but also demonstrates how large-scale chip designs can overcome the challenges of reticle-limited architectures, making it a flagship innovation in the field.

Package Technology: Maia 200 package is an enabling technology for scalable, high-performance AI silicon. The package architecture utilizes a silicon interposer, enabling dense interconnect routing and facilitating direct-to-die connections for high-speed, co-packaged memory subsystems.

An advanced silicon interposer forms the electrical and mechanical backbone of the package. The interposer supports ultra-fine pitch micro-bump connections between the main die and adjacent high-bandwidth memory (HBM3E) stacks. The interposer is fabricated with multiple metal layers, each optimized for high-frequency signal integrity and low RC loss, supporting aggregate bandwidths in the multi-terabit-per-second range.

Robust power delivery networks are implemented both within the interposer and the package substrate. Dedicated power and ground planes are assigned to support the high-current demands of the reticle-scale die, minimizing IR drop across the die and HBM. Power/ground bumps are distributed strategically to correspond to areas with the highest current draw, such as tiles and memory controllers. Signal escape routing leverages the interposer's dense wiring capabilities and the package's multi-layer to facilitate reliable, low-crosstalk communication with external system components via high-speed SerDes and PCIe links.

In-package thermal sensors provide real-time temperature monitoring, enabling dynamic control over cooling systems and supporting the chip's adaptive power management features.

Fine-pitch ball grid arrays (BGA) connect the package to the system board, enabling compact footprint and robust mechanical stability. The overall package structure is engineered to manage stress induced by thermal cycling and power transients, with underfill and advanced encapsulants used to protect micro-bump regions and interposer connections. Extensive simulation and reliability testing—including thermal shock, vibration, and electrostatic discharge (ESD)—are conducted to ensure service lifetime in demanding data center operations. The Silicon, Package, Interconnect, board, system mechanics and thermals are all designed and optimized with the view of end-to-end optimization across the stack.

Validation Strategy: Maia 200 employs a staged pre- and post-silicon validation strategy encompassing Field Programmable Gate Arrays (FPGA) prototyping, emulation, wafer-level bring-up, packaged silicon bring-up, silicon validation and qualification, and rack-scale system testing. A standalone FPGA platform enabled early bring-up of SoC firmware and drivers, exercising substantial portions of the chip logic. The FPGA was also linked to silicon bring-up and production boards via adapters to validate PCIe interfaces and FW/SW in a production-like environment—critical for firmware readiness and a seamless transition to silicon.

Emulation complemented these efforts by supporting long-duration functional and performance stress suites that map directly to silicon, covering approximately 95% of validation without relying on the production software stack. Firmware and host drivers were also integrated on emulation, alongside validation of basic kernels and models for both functionality and performance. High-speed I/O initialization sequences were fully verified across simulation, emulation, and FPGA platforms. These measures ensured smooth deployment of the production firmware to stack upon silicon arrival. Stress tests for the production FW/SW stack were likewise exercised on emulation prior to silicon, accelerating readiness.

Conclusion: Maia 200 delivers a reticle-scale AI accelerator optimized for performance, energy efficiency, and manufacturability. Leveraging a vertically integrated approach across system and software design, Maia 200 achieves 3-to-5× performance-per-dollar improvement over Maia 100 on Azure's in-house generative inference workloads. This advancement sets a new benchmark for high-performance AI acceleration in hyperscale data center environments.

	Maia 100	Maia 200
Peak Dense Tensor TOPS	6bit: 3000	FP4: 10145
	9bit: 1500	FP8: 5072
	BF16: 800	BF16: 1268
Peak HBM BW (TB/s)	1.8TB/s	Up to 7TB/s
Num of HBM Stack	4	6
HBM Capacity (GB)	64	216
Host PCIe BW (GB/s)	32	64
PCIe Configuration	Gen4 x16/Gen5 x8	Gen6 x8
Backend Network Bandwidth (GB/s) (bi-directional)	1200	2800
Backend Network Configuration	12x400	28x400

Figure 17.4.1: Maia 100 and Maia 200 performance summary.

ISSCC 2026 / SESSION 18 / TECHNOLOGY AND CIRCUITS FOR DOMAIN-SPECIFIC ACCELERATORS / OVERVIEW

Session 18 Overview: *Technology and Circuits for Domain-Specific Accelerators*

DIGITAL CIRCUITS SUBCOMMITTEE

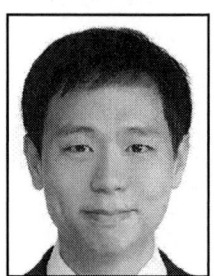

Session Chair: Dongsuk Jeon
Seoul National University
Seoul, Korea

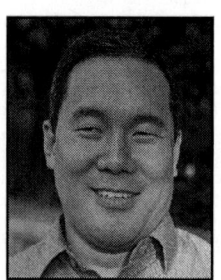

Session Co-Chair: Carlos Tokunaga
Intel
Hillsboro, OR

With machine learning models continuing to evolve both in complexity and applications, there is an increasing need for specialized hardware systems for efficient processing of ML algorithms. This session introduces five papers that use digital circuit innovations to advance the state-of-art in ML hardware acceleration and inter-chip communications. The first paper presents a cutting-edge technology enabling highly efficient chip-to-chip communication critical for multi-die acceleration platforms. The next two papers present well-optimized design examples aimed at multi-speaker automatic speech recognition and LLM processing. A neuromorphic sensing and processing system comes next, with the session concluded by a visual autoregressive accelerator.

306 • 2026 IEEE International Solid-State Circuits Conference

979-8-3315-8937-0/26 $31.00 © 2026 IEEE

ISSCC 2026 / February 17, 2026 / 3:35 PM

3:35 PM

18.1 A 3.19pJ/b Electro-Optical Router with 18ns Setup Frame-Level Routing and 1-to-6 Wavelength-Flexible Link Capacity for Photonic Interposers

Yvain Thonnart, CEA-List, Grenoble, France

In Paper 18.1, CEA-List and CEA-Léti present a dynamically routed 3.19pJ/b electro-optical link for chiplet-interposers in 28nm FD-SOI CMOS, enabling frame-level dynamic routing with an 18ns setup time and 1-to-6 wavelength allocation for photonic interposers. Compact analog drivers and standard-cell-based SerDes and clocking achieve 0.007mm² active area per link.

4:00 PM

18.2 A 22nm 1.87ms/Frame Streaming Multi-Speaker ASR Accelerator Leveraging Contextual-Aware Redundancy Skipping with 2D-Writable Microscaling Compute-in-Memory and Similarity-Aware TCAM Design

Wenjie Ren, Peking University, Beijing, China

In Paper 18.2, Peking University presents a multi-speaker automatic speech recognition (ASR) accelerator in 22nm based on digital compute-in-memory (DCIM) and a similarity-aware TCAM design, achieving 1.87ms/frame and system energy-efficiency of 10.24TFLOPS/W for MXFP8 precision at 582MHz and 1.0V at an area of 5.33mm².

4:25 PM

18.3 SMoLPU: 122.1µJ/Token Sparse MoE-Based Speculative Decoding Language Processing Unit with Adaptive-Offload NPU-CIM Core

Sangwoo Ha, KAIST, Daejeon, Korea

In Paper 18.3, KAIST and MIT introduce a MoE-based speculative decoding LLM processor with an energy-efficient NPU-CIM accelerator for heterogeneous and dynamic workloads, implemented in 28nm FD-SOI technology, achieving 122.1µJ/token and peak energy efficiency of 107.3TOPS/W at 50MHz and 0.7V with an area of 20.25mm².

4:50 PM

18.4 SpikeRAM: A 48.1pW/Synapse/Bit Event-Driven Spiking Compute-Near/In-Memory Processor with Neuromorphic Sensor Enabling Life-Long On-Chip Learning

Haotian Fu, Hong Kong University of Science and Technology, Guangzhou, China

In Paper 18.4, HKUST (Guangzhou), North China Research Institute of Electro-Optics and SynSense describe a 48.1pW/synapse/bit neuromorphic platform integrating a 65nm inference core of 29.9mm² and an 40nm on-chip learning core of 4.3mm², consuming 8.28mW system power operating at 1.1V and 10MHz and achieving 94.3% accuracy for event-based signature verification with 2-shot learning.

5:15 PM

18.5 A 28nm 47.3TFLOPs/W 894mJ/Inference Visual Autoregressive Accelerator with Differential-Amplifier Speculation and Chain-Reaction-Like Parallel Generation

Zhiheng Yue, Tsinghua University, Beijing, China

In Paper 18.5, Tsinghua University presents a 28nm visual accelerator for Visual Autoregressive (VAR) applications that achieves 0.6% Fréchet Inception Distance (FID) loss. The 5.76mm² prototype reaches a peak frequency at 400MHz, 0.9V and accelerates DeiT/ViT VAR at 47.3TFLOPs/W energy efficiency and 2.75TFLOPs/mm² area efficiency.

DIGEST OF TECHNICAL PAPERS • 307

979-8-3315-8937-0/26 $31.00 © 2026 IEEE

ISSCC 2026 / SESSION 18 / TECHNOLOGY AND CIRCUITS FOR DOMAIN-SPECIFIC ACCELERATORS / 18.1

18.1 A 3.19pJ/b Electro-Optical Router with 18ns Setup Frame-Level Routing and 1-to-6 Wavelength-Flexible Link Capacity for Photonic Interposers

Yvain Thonnart[1], Christian Bernard[2], Stéphane Bernabé[2], Myriam Assous[2], Laura Boutafa[2], Benoit Charbonnier[2], Rémi Franiatte[2], César Fuguet[1], Olivier Guille[2], Vincent Josselin[2], Stéphane Malhouitre[2], Laurent Mendizabal[2], André Myko[2], Damien Saint Patrice[2], Rémi Vélard[2], Jean Charbonnier[2]

[1]CEA-List, Grenoble, France, [2]CEA-Léti, Grenoble, France

Abstract

Global interconnect on interposers between chiplets is currently mostly limited to nearest-neighbor communication. We propose an electro-optical router in 28nm on a photonic interposer to handle queuing and routing between dies and optical link drivers capable of 18ns link setup time for frame-level routing, with 1-to-6 lambda flexible wavelength capacity. Compact analog drivers and standard-cell-based SerDes and clocking achieve 0.007mm² active area per link and 3.19pJ/b power.

Chiplet stacking using 3D and 2.5D integration has been a major evolution of large compute chips, using different possible integration schemes on organic, silicon or CMOS interposers. Nevertheless, global interconnects are facing multiple scaling issues with increasing interposer area. Currently, most die-to-die interconnects use short-reach parallel interfaces between neighboring dies, or top and bottom dies, such as UCIe. While these interfaces allow for high aggregated throughput with relatively simple drivers, they are mostly shoreline-limited and are not meant to extend to distant chiplets. Longer distances could be achieved using wireline techniques over longer microstrip or coplanar RF lines working at high datarates with low-latency, but at the cost of expensive SerDes and drivers, with no routing capability and thus limited scaling. Active CMOS interposers [1,2,3] have been proposed to increase flexibility and possibly integrate a complex NoC or switching fabric. However, as regards to communication between distant dies, these solutions still face the issue of global clocking constraints or the need for multiple resynchronization stages, leading to high latency. Photonic interposers have been discussed with architecture proposals for several years [4], but only a very recent realization has been announced [5]. While this solution offers a high datarate and low latency, it comes at the cost of complex drivers, and fixed static routing driven from centralized control using large and costly Mach-Zehnder modulators.

In this paper, we demonstrate a lightweight version of dynamic optical routing on a photonic interposer using an electro-optical chiplet with simple digital control for routing, queuing and arbitration, standard-cell-based SerDes at limited baudrate, and small analog optical drivers controlled by digital routing logic to achieve frame-level dynamic routing from a given transmitter (Tx) in one chiplet to any of seven receivers (Rx) interfacing with distant computing chiplets, such as a CPU core complex or GPU to distant HBM. This routing is achieved using the narrow resonance of microring modulators and filters that also allow for wavelength-division multiplexing. The single-writer (one Tx) multiple-reader (several Rx) link on a given optical waveguide is presented in Fig. 18.1.1. Tx microrings are tuned to the laser wavelength, while Rx microrings are by default tuned slightly off-resonance close to the laser wavelength. When the routing logic in Tx has a frame to route to one specific Rx, it uses low-bandwidth routing signaling metal lines on the interposer to notify the Rx die to dynamically tune its microrings to the laser wavelength. Other off-resonance rings will let light pass through in the waveguide all the way to the appropriate Rx, whose rings will deflect light to photodiodes on the interposer, whose photocurrent is injected in the Rx chain for demodulation.

Figure 18.1.2 presents the detailed architecture of the Rx front-end. Upon receiving a transmit signal from Tx, the p-i-n junction of the microring filter is biased to a configurable resonance shift (using a refractive index change by carrier injection in the ring), so the photodiode receives light from the drop port. The next functional block is a 2-stage push-pull transimpedance amplifier (TIA) made of CMOS inverters with a transimpedance resistance Rtia (1.8kΩ) converting the photocurrent to a small voltage change near the second-stage trip point. In order to tolerate a large range of input optical powers, the DC component of the photocurrent is canceled using a NMOS transistor driven by a RC low-pass filter at the TIA output (Rlpf=1MΩ,Clpf=100fF). After amplification, the TIA output is fed to four interleaved clocked comparators, whose threshold is set dynamically at the beginning of the frame and stored in a memory capacitor Cavg (100fF). Indeed, as we perform dynamic routing, the optical levels are not known beforehand, and can only be determined during the transmission of a frame preamble made of alternating 1 and 0. The output of the comparators is a first 1:4 deserialization, followed by a second 1:4 deserialization using standard-cell latches driven by the appropriate clock divisions and phases generated from the reception of a global optical clock (after TIA amplification) using a cascade of latch rings at each division stage. As a side note, on the Tx side, standard 2-input MUXes are used for a first 4:1 serialization, and a custom 4:1 tristate base MUX standard-cell has been designed for the second 4:1 stage. The bottom part of Fig. 18.1.2 shows the chronograms of electrical and optical signals during frame routing. Electrical signal 'transmit' is received from Tx, and a 'preamble' signal is generated with configurable tp duration equal or similar to the configured preamble length in Tx, so as to sample the average level in the comparators. As the ring is shifted to the laser wavelength λi, optical power on that wavelength Pin(λi) is switched from Pthru(λi) to Pdrop. A specific pattern indicates the end of the preamble and the first bit of data. The end of frame should be

signaled by lowering transmit at least one core clock cycle (1:16) after the last bit of data, to ensure that the whole word is received in the worst case of data/core clock misalignment. Optical power is then rerouted to Pthru(λi), and available for communication with other Rx. During the preamble, the average value of the TIA output is collected on Cavg, using a weak Ron pass-gate, and fixed for the whole frame by opening the pass-gate at the end of the preamble.

Figure 18.1.3 presents frame decoding after deserialization. As our architecture uses an external global optical clock and derives phases and the digital core clock without knowledge of data alignment instead of using clock-data recovery (CDR), the 16b parallel output of each SerDes is often misaligned from the parallel input at Tx. We use two stages of registers to recombine a 32b sub-word in which we identify the end of the preamble. In addition, due to microring resonance variability and thermal difference between Tx and Rx, we do not try to match the ring mapping to laser wavelengths between Tx and Rx, in order to minimize the thermal tuning budget, and tune each active ring to the closest laser wavelength. In order to recombine the wavelength-multiplexed channels, we use a different preamble for the ring carrying the overall word least significant bit than for the other sub-words. End of preamble detection looks for a simultaneous final sequence of 10111000 for this ring and 10000111 for the others. That permits identification of the initial data bit position, the initial ring containing the least significant bit, and the number of active rings. This information is used in the next cycle to rotate data within rings, remap data between rings, and if needed de-sequentialize from 16b, 32b, 48b to 96b (for 1, 2, 3 or 6 active rings). We use the 8 most significant bits of the data word for a checksum using an unfolded CRC with configurable polynomial. Checksum decoding identifys errors in the frame up to very low undetected error rates below 1e-24. The ninth most significant bit in the data word is an end of frame marker to confirm all data has correctly been received. Based on this information, Rx sends back an acknowledgment or error signal to the Tx, for clearing or possible retransmission of the frame queued in a 1024-word memory.

We demonstrate this electro-optical router architecture in a 4mm² chip in STMicroelectronics 28nm FDSOI technology, post-processed for 3D stacking on two variants of photonic interposers, as shown in Fig. 18.1.7. This electro-optical chiplet implements one Tx bank and 7 Rx banks of 6 microring resonators for up to 6 wavelengths. A looped-back version of an interposer using a single electro-optical router to communicate to its own Rx bank allows for standalone tests and intermediate optical measurements. A fully populated 207mm² interposer version instantiates 6 electro-optical routers, with 4 of them connected to computing dies with two 96b parallel short-reach die-to-die interfaces for outbound and inbound traffic, and 2 electro-optical routers interfacing with primary electrical I/O on half-rate 48b interfaces. The two remaining input optical ports of each router are connected to two off-chip optical I/O ports. Four separate power domains are used in the router, with a main digital supply and a quiet Rx supply at 1V, a Tx supply at 2V, and a thermal tuning supply nominally at 1V and up to 1.5V.

We measured the link performance using error counters integrated in each Rx according to checksum validation, and extracted corresponding bit error rates (BER), as presented in Fig. 18.1.4. All tests presented here are performed using wavelength division multiplexing on two wavelengths. The optical power waveforms measured before the Rx bank are collected on a broadband external photodiode summing-up photocurrents on both wavelengths, resulting in a 3-level waveform (0,1 or 2 wavelengths transmitting light). We measured the data BER at various datarates and optical input powers, demonstrating 1e-6 BER up to 4Gbaud. To identify the limiting factor, we also measured clock receiver BER below 1e-6 up to 4.5Gbaud, and back-to-back electrical SerDes with an external LVDS clock up to 7Gbaud. This seems to indicate jitter in the push-pull TIA as the limiting factor. We also analyzed the robustness of our routing mechanism in Fig. 18.1.5, and by tuning the transmit and preamble durations we were able to reduce the preamble length to only 18ns to tune rings and average power levels. We assessed the TIA low-pass filter response by increasing frame length, and measured increasing BER at 1e-4 for 312ns frames, beyond our target 50ns duration. Fig. 18.1.5 also presents the high-speed circuits layout, with a total active area of only 7427μm² for Tx+Rx, of which only 244μm² is for the SerDes, and a power breakdown of 2.09pJ/b for the bare TxRx link to compare to the state of the art, only 0.36pJ/b for the SerDes and clock phase generation, and 0.75pJ/b for all digital logic on

308 • 2026 IEEE International Solid-State Circuits Conference

979-8-3315-8937-0/26 $31.00 © 2026 IEEE

core domain including routing and queuing, for a total of 3.19pJ/b for the overall end-to-end routed transmission.

Compared to recent interposers and optical links, as presented in Fig. 18.1.6, this work combines the flexibility of dynamic routing achieved in active CMOS interposers with improved reach and low latency offered by optical links. It achieves low latency for a reach of around 2.5cm, while being capable of longer reach on an interposer and extension to other dies through optical fiber. Its lightweight digital-intensive implementation makes it small, while preserving power efficiency close to the state of the art.

Acknowledgment:
This research is partly supported by the project NET4EXA, which has received funding from the European High-Performance Computing Joint Undertaking (JU) under grant agreement No 101175702. The JU receives support from the European Union's Horizon Europe research and innovation programme and France (through France 2030), Italy, Greece and Norway.

References:
[1] B. Jiao et al., "SHINSAI: A 586mm² Reusable Active TSV Interposer with Programmable Interconnect Fabric and 512Mb 3D Underdeck Memory," *ISSCC*, pp. 612-613, 2025. https://doi.org/10.1109/ISSCC49661.2025.10904819
[2] A. Smith et al., "AMD InstinctTM MI300 Series Modular Chiplet Package – HPC and AI Accelerator for Exa-Class Systems," *ISSCC*, pp. 490-492, 2024. https://doi.org/10.1109/ISSCC49657.2024.10454441
[3] P. Vivet et al., "IntAct: A 96-Core Processor With Six Chiplets 3D-Stacked on an Active Interposer With Distributed Interconnects and Integrated Power Management," *IEEE JSSC*, vol. 56, no. 1, pp. 79-97, 2021. https://doi.org/10.1109/JSSC.2020.3036341
[4] Y. Thonnart et al., "POPSTAR: a Robust Modular Optical NoC Architecture for Chiplet-based 3D Integrated Systems," *ACM/IEEE DATE*, pp. 1456-1461, 2020. https://doi.org/10.23919/DATE48585.2020.9116214
[5] D. Bunandar, "Passage M1000: 3D photonic interposer for AI," *IEEE Hot Chips Symp.*, 2025. https://doi.org/10.1109/HCS66204.2025.11154385
[6] R. Baghdadi et al., "Monolithically Integrated Microring Transmitter and Receiver for High-Density 3D Co-Packaged Optics," *Optical Fiber Communication Conf.*, 2025. https://doi.org/10.1364/OFC.2025.Tu3J.6
[7] M. Raj et al., "A 0.96pJ/b 7×50Gb/s-per-Fiber WDM Receiver with Stacked 7nm CMOS and 45nm Silicon Photonic Dies," *ISSCC*, pp. 11-13, 2023. https://doi.org/10.1109/ISSCC42615.2023.10067617
[8] P. Bhargava et al., "A 256Gbps Microring-Based WDM Transceiver with Error-Free Wide Temperature Operation for Co-Packaged Optical I/O Chiplets," *IEEE Symp. VLSI Tech.*, 2024. https://doi.org/10.1109/VLSITechnologyandCir46783.2024.10631547
[9] Y. Thonnart et al., "A 10Gb/s Si-Photonic Transceiver with 150µW 120µs-Lock-Time Digitally Supervised Analog Microring Wavelength Stabilization for 1Tb/s/mm² Die-to-Die Optical Networks," *ISSCC*, pp. 350-352, 2018. https://doi.org/10.1109/ISSCC.2018.8310328
[10] D. Saint-Patrice et al., "Process Integration of Photonic Interposer for Chiplet-based 3D Systems," *IEEE Electronic Components and Technology Conf.*, pp. 5-12, 2023. https://doi.org/10.1109/ECTC51909.2023.00009

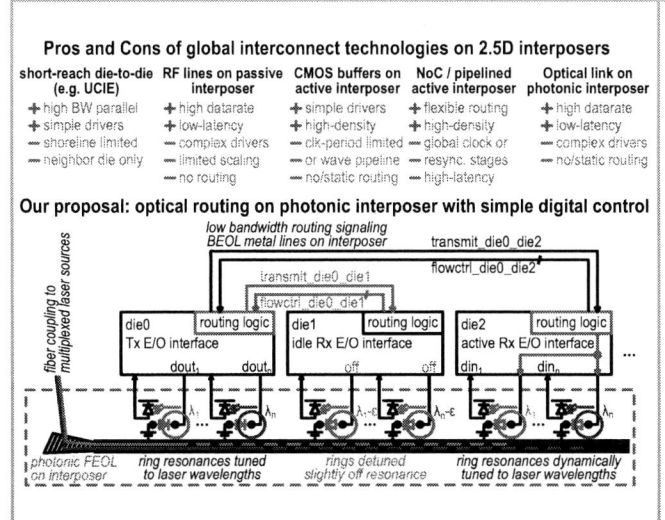

Figure 18.1.1: Scaling perspectives for 3D-stacking technologies and proposed routed optical links on interposer.

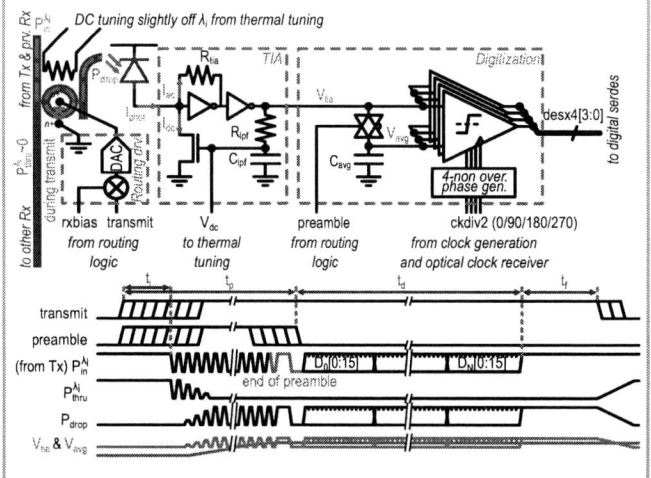

Figure 18.1.2: Rx front-end with routing using resonance shift and dynamic threshold adjustment during preamble.

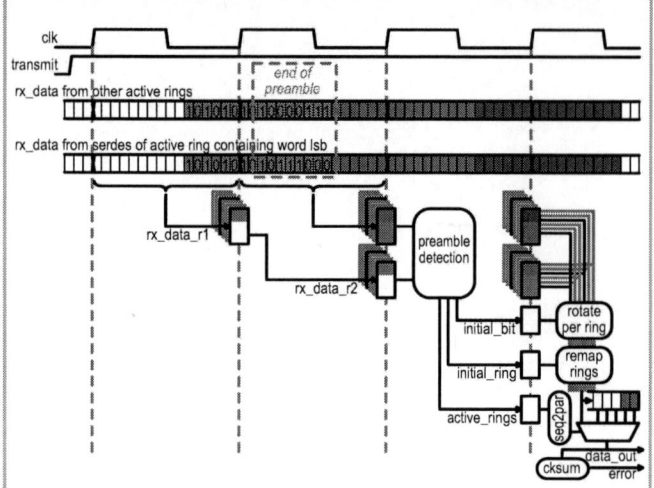

Figure 18.1.3: Frame decoding with data alignment on end of preamble, and dynamic remapping on 1 to 6 wavelengths.

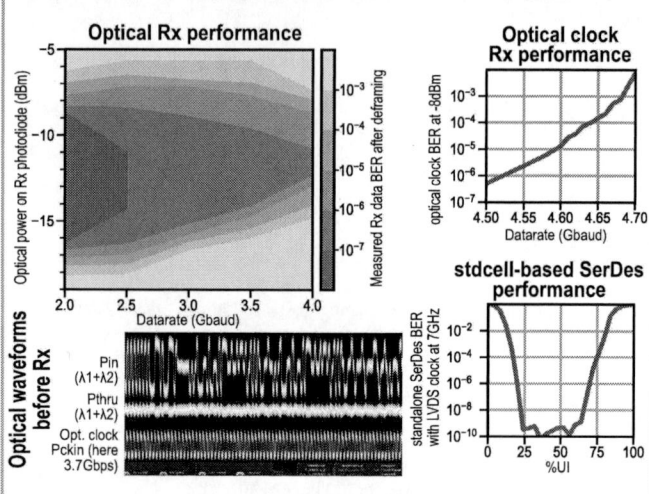

Figure 18.1.4: Performance measurements up to 4Gbaud showing BER for optical data, optical clock and digital SerDes.

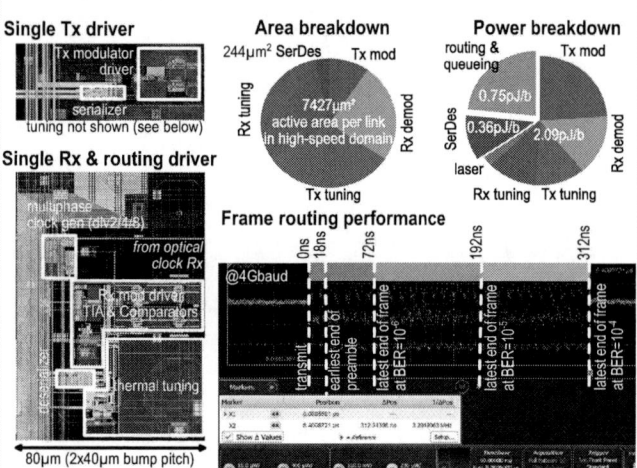

Figure 18.1.5: Tx/Rx cell layout, area and power breakdown, and long-packet routing performance.

Figure 18.1.6: Comparison table.

Metric	this work	ISSCC25 [1]	ISSCC24 [2]	JSSC21 [3]	HotChips25 OFC25 [5,6]	ISSCC23 [7]	VLSI24 [8]	ISSCC18 [9]
Type	Photonic Interposer	Active Interposer	Active Interposer	Active Interposer	Photonic Interposer	Photonic links	Photonic links	Photonic link
Tech node	28nm	28nm	6nm	65nm	45nm	7nm	45nm	65nm
Integration	3D µbump	3D µbump	hybrid bnd.	3D µbump	C4 bumps	Hybrid bnd	3D µbump	3D µbump
3D pitch	40µm	40µm	9µm	20µm	120µm	—	55µm	40µm
Interposer Area	19.2 mm² & 207mm²	586 mm²	4*377mm²	198 mm²	stitched 8*462mm²	—	50 mm²	15mm²
Chiplet Area	3.24mm²	—	—	—	—	~1mm²	—	2mm²
Voltage	1V& 2 V	0.9V	—	0.8-1.3V	1.3V	0.88V (Rx)	1V & 2V	1.2V&2.5V
Reach	28mm & off-chip	24mm	<42mm	25mm	64mm	off-chip	off-chip	5mm & off-chip
Datarate	4Gbaud	500 MHz	2.1GHz	1GHz	56Gbaud	50Gbaud	32Gbaud	10Gbaud
Latency	12-18ns	—	—	15-50ns	—	—	—	—
Routing	dynamic	static+dyn	dynamic	dynamic	static	no	no	no
WDM nb. λ	1-6 dyn.	—	—	—	1-16 static	7	8	4
Link power	2.09pJ/b	—	—	4-13pJ/b	1.51 pJ/b	0.96 pJ/b	3.45 pJ/b	4pJ/b
w. routing	3.19pJ/b	—	—	—	—	—	—	—
Link active area	0.007 mm²	—	—	—	0.006 mm² (no serdes)	0.031 mm²	0.2 mm²	0.019mm²
Link setup time	18ns	—	—	—	—	—	—	—

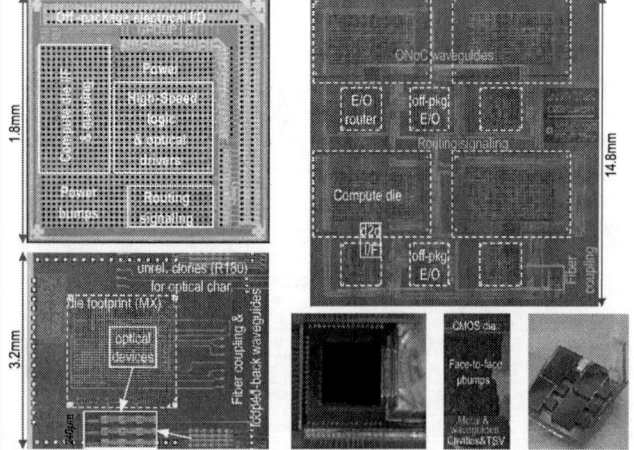

Figure 18.1.7: Die photographs with E/O route, looped-back interposer and 8-port ONoC interposer.

ISSCC 2026 / SESSION 18 / TECHNOLOGY AND CIRCUITS FOR DOMAIN-SPECIFIC ACCELERATORS / 18.2

18.2 A 22nm 1.87ms/Frame Streaming Multi-Speaker ASR Accelerator Leveraging Contextual-Aware Redundancy Skipping with 2D-Writable Microscaling Compute-in-Memory and Similarity-Aware TCAM Design

Wenjie Ren*, Mingxuan Li*, Zhenghao Jin, Yifan Ding, Ruohuang Xu, Xiangjun Ye, Yufei Ma, Tianyu Jia, Le Ye

Peking University, Beijing, China
*Equally Credited Authors (ECAs)

Abstract

This paper presents a DCIM-based accelerator for on-device streaming multi-speaker ASR (MS-ASR), featuring: 1) a context-aware redundancy skipping scheme with online sparse block prediction, 2) a 2D-writable CIM with static pages and dynamic pages, 3) a similarity-aware TCAM for rapid search of similar speakers and N-best semantic results. The test chip achieves a system energy efficiency of 37.50TFLOPS/W in MXFP8 and shows superior performance on MS-ASR tasks with 1.87-7.51ms/frame and 0.158-1.26mJ/frame efficiency.

The rising popularity of AI-assisted devices has sparked heightened interest in real-time automatic speech recognition (ASR). Prior single-pass ASR models, such as connectionist temporal classification (CTC) [1-2], recurrent neural-network transducer (RNN-T) [3-4] and attention-based encoder-decoder (AED) [5-6], exhibit high accuracy in terms of word error rate (WER). However, it is challenging to utilize these models for real-time streaming, multi-speaker (MS) dialogue, where frequent speaker alternation degrades both inference latency and recognition accuracy [7]. Recent two-pass MS-ASR models combine an intrinsic streaming RNN-T and a high accuracy AED model [8-9]. As shown in Fig. 18.2.1, the audio is first encoded by an ASR and a speaker encoder and then attributes to the corresponding speaker by the similarity and clustering module. At the decoding stage, the first RNN-T decoder pass consists of a lightweight joiner and predictor and generates the output sequence simultaneously with the input stream, making it naturally suited for low-latency streaming applications. The second AED pass processes the output list from the RNN-T through an attention mechanism for tokens it generates to achieve higher recognition accuracy.

Previous ASR accelerator chips [10-13] focused on single-speaker ASR – none of them consider real-life MS-ASR. MS-ASR has its unique deployment challenges for efficient real-time streaming. First, the sparse context in real conversation induces block-level computation redundancy, which contains 32-128 tokens in each block. Silent pauses, represented by blank (Φ) tokens, also create computational redundancy that waste significant energy. Moreover, increasing token-level similarity from 45% to 92% is observed as the transformer layer depth increases (i.e., higher block index) due to the continuity characteristic of speech signals. Second, two decoding passes introduces distinct requirements for computation and memory. The streaming RNN-T pass decodes speech embeddings with a light-weight joiner (0.23MB), which has a high invocation frequency consuming 59.2% of memory power [14]. Meanwhile, the attention operator in the transformer-based AED pass requires intense matrix transposes and sparse GEMM, degrading energy efficiency. Third, significant post-processing redundancy arises due to the high overhead of frequent speaker alternation and repetitive evaluations on common words within the N-best rescoring list.

To address the above challenges, this paper presents a digital computing-in-memory (CIM)-based accelerator for on-device streaming MS-ASR, featuring: 1) a context-aware redundancy skipping (CARS) scheme with online sparse block prediction, 2) a 2D-writable CIM with memory-intensive static pages and computation-intensive dynamic pages, 3) a similarity-aware ternary content-addressable memory (TCAM) for rapid search of similar speaker embeddings and N-best semantic results. Combining the above techniques, we prototyped the MS-ASR test chip in 22nm CMOS with a system energy efficiency of 37.50TFLOPS/W in MXFP8 and our chip also shows superior performance on MS-ASR tasks with 1.87-7.51ms/frame and 0.158-1.26mJ/frame compared to prior works [10-11].

Figure 18.2.2 shows the chip microarchitecture. It consists of three CIM-based computing clusters, a similarity-aware TCAM system, a RISC-V CPU, a VPU and peripheral circuits. Each computing cluster contains 5 digital CIM macros and a CARS sparse attention unit. The CARS unit contains a similarity-based skipper, a blank-based skipper to dynamically skip repetitive or non-informative (blank) blocks based on online similarity analysis, while a high accuracy decoding trigger is applied to adaptively schedule a rescoring process based on a confidence score of initial transcription. The debug and function controllers generate control signals for the 5 CIM macros to configure their write direction and computation mode. Each CIM macro contains 17 stationary pages (528×256) to store stationary weights and 2 dynamic pages (64×256) to support 2D update and sparse computation. A similarity-aware TCAM (SA-TCAM) comprising 32 TCAM banks is used to reduce the latency of speaker attribution and the rescoring process. An active speaker finder (ASF) is added into the TCAM unit to support partial activation of match lines (ML) based on search frequencies and the nearest encoder encodes the address of the closest search results. The reuse-aware N-best scheduler reorders computations based on the N-best score in TCAM, which increases the reuse rate of similar words.

Figure 18.2.3 illustrates the workflow of our context-aware redundancy skipping (CARS). A two-level skipping and one trigger mechanism is employed, which includes ① similarity-based token block skipping leveraging sparse attention, ② blank-based token skipping for

blank tokens, and ③ context-aware rescoring trigger for non-confidential RNN-T predictions. The input audio frames are first grouped into blocks, and the in-block similarity $\cos(\theta)$ and compressed attention score (CAS) are evaluated. The in-block similarity is approximated in a co-sim threshold block selector based on inequality A1 (displayed in Fig. 18.2.3 right). In the token compression unit, blocks with high similarity are compressed by summing them across the token length. The resulting compressed q and k vectors are then accumulated across the hidden dimensions to compute the CAS. Redundant blocks can be identified and will be marked with "0" in block mask generator if high similarity and low CAS are observed. According to the block index, the blank-based skipper (②) fetches active blocks and in parallel detects non-blank tokens with a window size of 4. To filter blank frames, a co-sim approximation unit (CAU) is reused to calculate the similarity between the current frame and a reference blank frame, where the left and right part in inequality A2 are generated in adjacent-multiplication and self-multiplication modes of the CAU, respectively. The beam search trigger receives non-blank frames and generates a confidence score for all text candidates. Rescoring is triggered only when there is high ambiguity among the top candidates ($\Delta s < s_{th}$), otherwise the result is accepted directly. Overall, our CARS scheme reduces MACs by 57-62% with negligible impact on WER.

Figure 18.2.4 outlines the design of our 2D-writable MX-DCIM. The CIM macro supports the microscaling (MX) format [15-16] with MX extension modules, which includes an MX block reorder unit, a block max finder for shared scale pre-calculation, and a weight/input aligner for exponent alignment. Since the weights in the joiner module need frequent weight-reload, a 17-static-page (16.5KB) DCIM is adopted to store all weights, reducing memory access by 65% and energy consumption by 32%. Meanwhile, the compute-intensive AED pass contains a transformer model with both matrix transpose (QK^T) and sparse GEMM (PV). Although prior work [17] introduced dual adder trees for transpose MVM, this incurs power and area overhead. Our solution is a 2D-writable MX-CIM with two dynamic pages, providing two key benefits to support (1) native in-memory matrix transpose to avoid costly data reordering of intermediate results (K^T) and (2) column-by-column input configuration to send unique input features to non-zero weights in CIM columns. The two dynamic pages are highly configurable to support four operation modes determined by the SPEN and LW signals. In a standard configuration (SPEN=0, LW=0), it functions as a conventional CIM with row-wise write (R-write) and input broadcast (input-BC). To enable in-memory transpose, the column-wise write (C-write) mode (LW=1) reconfigures the data path, allowing the right DLCs (in Fig. 18.2.4 right) to latch a weight directly from the vertical input feature (IF) ports. Finally, a 'flash input' mode (SPEN=1, LW=1) is designed for sparsity-aware computation, allowing unique input features to be rapidly input into the left DLCs of selected columns. Overall, the 2D-writable MX-DCIM reduces memory access by 72% and saves 43% of energy for the two-pass decoding.

Figure 18.2.5 presents the design of the SA-TCAM. In MS-ASR, speaker embeddings (SimHash compressed) of the same person extracted from different audio clips are often similar but not identical, making the exact matching ineffective. Furthermore, only a few speakers are active in a typical dialogue, making a full search of all stored speaker profiles highly inefficient. The SA-TCAM design supports Hamming distance-based approximate matching and incorporates ASF to dynamically power-gate TCAM entries for inactive speakers. The ASF monitors the search frequency of each speaker by activity down counters, where the entry with minimum distance is selected by the nearest selector (NRS). Once successfully matched, the match signal is asserted by NRS and resets the entry's activity counter to its maximum. Conversely, if the counter expires by reaching zero due to inactivity, the entry's ML is power-gated. This dynamic deactivation of inactive speakers prevents unnecessary searches, yielding a 53% reduction in search power and a 29% reduction in latency. The TCAM is also used to accelerate an N-best rescoring process in full match mode. High-ambiguity candidates from the first pass often exhibit high lexical overlap (i.e., contain many of the same words). Our reuse-aware rescoring (RAR) scheduler leverages this by processing the list from the candidates with highest score to the lowest and storing the decoded key-value pairs of common words in TCAM. When these words reappear (hit) in subsequent candidates, their results are instantly retrieved from the TCAM, bypassing expensive re-computation entirely. The rescoring hit rate is 82% and 71% for single speaker speech and multi-speaker dialogue and this caching mechanism reduces the overall latency by 32%.

310 • 2026 IEEE International Solid-State Circuits Conference

979-8-3315-8937-0/26 $31.00 © 2026 IEEE

Figure 18.2.6 shows the measurement results and comparison table. The chip is fabricated using 22nm and works at 46.4-582MHz with a supply voltage of 0.55-1.0V. The MS-ASR tasks are evaluated using two-pass ASR model trained and quantized in microscaling format, in which the RNN-T pass with a size of 0.234MB is stored in CIM [5, 18]. The CARS scheme, STDY-2D-DCIM and SA-TCAM together obtain 12.9× speedup and 13.6× energy savings for clean, single speaker audio and 12.2× speedup and 18.21× energy savings for 3-speaker mixed audio compared to the baseline, with negligible WER increasement. Our accelerator achieves the system efficiency of 37.50TFLOPS/W in MXFP8 format, which is 1.2× better than prior work [19]. The per-frame latency and energy for MS-ASR are 7.51-1.87ms and 0.158-1.26mJ, respectively, as the SoC voltage scales from 0.55-1.0V, demonstrating 2.39-3.20× speedup and 1.91-3.84× energy reduction compared to a prior ASR accelerator [11]. Fig. 18.2.7 shows the die photo, voltage-frequency scaling curves and specification table.

Acknowledgement:
This work was supported by Grand 2024YFB4505001; NSFC No. U24A20287, 62225401; Grand QYJS-2023-2401-B/QYJS-2023-2402-B. Corresponding authors: Tianyu Jia and Le Ye.

References:
[1] D. Amodei et al., "Deep Speech 2: End-to-End Speech Recognition in English and Mandarin," arXiv:1512.02595, 2015. https://arxiv.org/abs/1512.02595v1
[2] N. Moritz et al., "Streaming End-to-End Speech Recognition with Joint CTC-Attention Based Models," *IEEE ASRU*, pp. 936–943, 2019. http://doi.org/10.1109/ASRU46091.2019.9003920
[3] A. Graves, "Sequence Transduction with Recurrent Neural Networks". https://arxiv.org/abs/1211.3711
[4] X. Wang et al., "Cascade RNN-Transducer: Syllable Based Streaming On-Device Mandarin Speech Recognition with a Syllable-to-Character Converter," arXiv:2011.08469, 2020. https://arxiv.org/abs/2011.08469
[5] W. Chan et al., "Listen, Attend and Spell," arXiv:1508.01211, 2015. https://arxiv.org/abs/1508.01211v2
[6] A. Radford et al., "Robust Speech Recognition via Large-Scale Weak Supervision," arXiv:2212.04356, 2022. https://arxiv.org/abs/2212.04356v1
[7] Y. Lin et al., "Diarization-Aware Multi-Speaker Automatic Speech Recognition via Large Language Models," arXiv:2506.05796, 2025. https://arxiv.org/abs/2506.05796
[8] I. E. Kang et al., "Transformer-based Model for ASR N-Best Rescoring and Rewriting," arXiv:2406.08207, 2024. https://arxiv.org/abs/2406.08207
[9] B. Zhang et al., "WeNet 2.0: More Productive End-to-End Speech Recognition Toolkit," arXiv:2203.15455, 2022. https://arxiv.org/abs/2203.15455
[10] W. Jo et al., "BROCA: A 52.4-to-559.2mW Mobile Social Agent System-on-Chip with Adaptive Bit-Truncate Unit and Acoustic-Cluster Bit Grouping," *ISSCC*, pp. 418–420, 2025. http://doi.org/10.1109/ISSCC49661.2025.10904658
[11] T. Tambe et al., "A 16-nm SoC for Noise-Robust Speech and NLP Edge AI Inference With Bayesian Sound Source Separation and Attention-Based DNNs," *IEEE JSSC*, vol. 58, no. 2, pp. 569–581, 2023. http://doi.org/10.1109/JSSC.2022.3179303
[12] Y.-H. Tsai et al., "A 28-nm 1.3-mW Speech-to-Text Accelerator for Edge AI Devices," *IEEE JSSC*, vol. 59, no. 11, pp. 3816–3826, 2024. http://doi.org/10.1109/JSSC.2024.3389965

[13] S. Zhou et al., "An 8.62µW 75dB-DRSoC End-to-End Spoken-Language-Understanding SoC With Channel-Level AGC and Temporal-Sparsity-Aware Streaming-Mode RNN," *ISSCC*, pp. 238–240, 2025. http://doi.org/10.1109/ISSCC49661.2025.10904788
[14] Y. Li et al., "Breaking Down Power Barriers in On-Device Streaming ASR: Insights and Solutions," arXiv:2402.13076, 2025. https://arxiv.org/abs/2402.13076
[15] B. D. Rouhani et al., "OCP Microscaling Formats (MX) Specification," 2023. https://www.opencompute.org/documents/ocp-microscaling-formats-mx-v1-0-spec-final-pdf
[16] B. D. Rouhani et al., "Microscaling Data Formats for Deep Learning," arXiv:2310.10537, 2023. https://arxiv.org/abs/2310.10537v3
[17] Y. Jing et al., "NeRF-Learner: A 2.79mJ/Frame NeRF-SLAM Processor with Unified Inference/Training Compute-in-Memory for Large-Scale Neural Rendering," *ESSERC*, pp. 145–148, 2024. http://doi.org/10.1109/ESSERC62670.2024.10719471
[18] A. Gulati et al., "Conformer: Convolution-augmented Transformer for Speech Recognition," arXiv:2005.08100, 2020. https://arxiv.org/pdf/2005.08100
[19] Z. Wu et al., "CELLA: A 28nm Compute-Memory Co-Optimized Real-Time Digital CIM-Based Edge LLM Accelerator with 1.78ms-Response in Prefill and 31.32 Token/s in Decoding," *IEEE Symp. VLSI Circuits*, 2025. http://doi.org/10.23919/VLSITechnologyandCir65189.2025.11075101

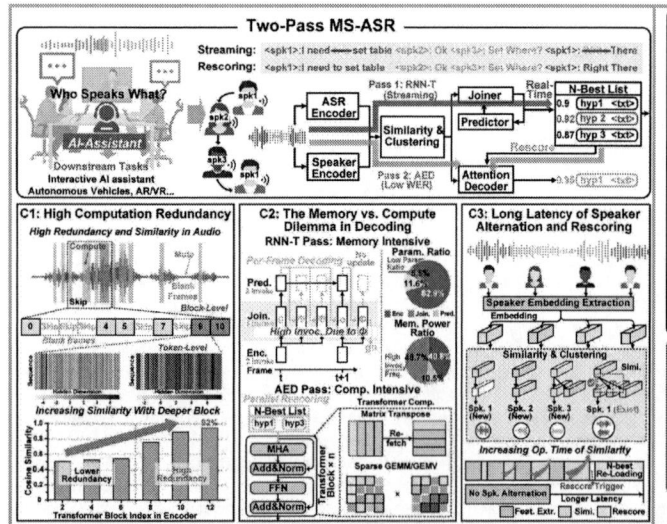

Figure 18.2.1: Overview of two-pass ASR model for multi-speaker and the deployment challenges.

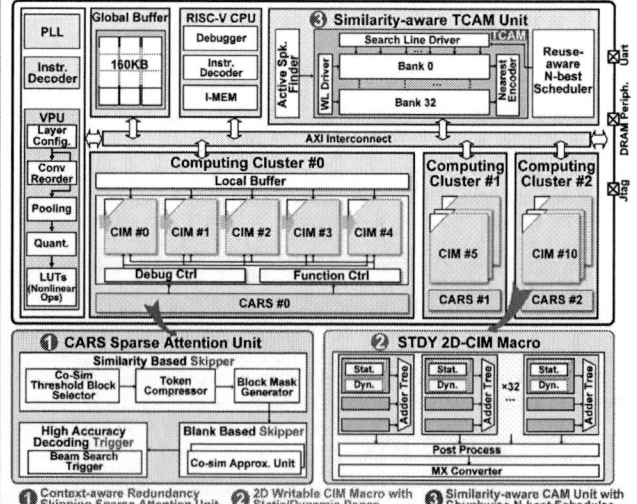

Figure 18.2.2: Overall architecture of the proposed ASR accelerator and three main features.

ISSCC 2026 / SESSION 18 / TECHNOLOGY AND CIRCUITS FOR DOMAIN-SPECIFIC ACCELERATORS / 18.2

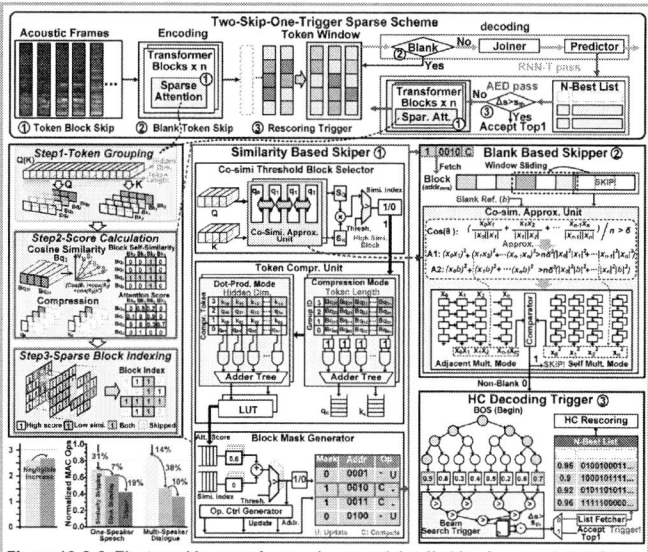

Figure 18.2.3: The two-skip-one-trigger scheme and detailed implementation of CARS.

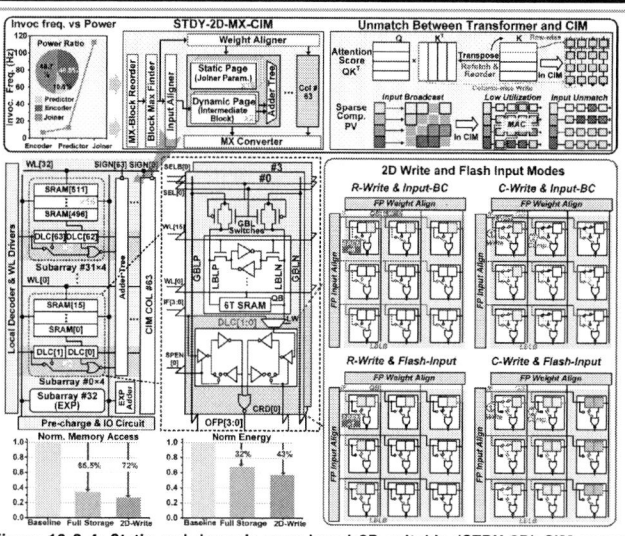

Figure 18.2.4: Static and dynamic page-based 2D writable (STDY-2D) CIM macro supporting microscaling format.

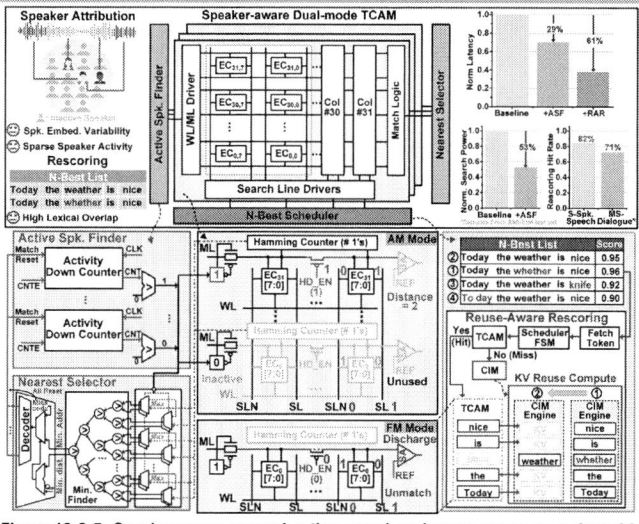

Figure 18.2.5: Speaker-aware approximation search and reuse-aware rescoring with proposed SA-TCAM.

Model	Conformer[10]	Attention based RNN[11]	Multi-Speaker Two-Pass ASR	
Dataset	LibriSpeech	LibriSpeech	LibriSpeech	LibriSpeechMix
Encoder	Conformer	RNN	Transformer	
Decoder	-	RNN	Transducer (0.234 MB) Transformer (3.82 MB)	
Precision	INT8	FP8	MXINT8	
Parameter	10.1 M	3.5 M	17.43 M	
WER[1]	2.8 (+0.1)	10.5	2.98 (+0.08[3])	10.21[3] (+0.2[3])
SA-WER[1]	-	-	-	12.01[3] (+0.24[3])

	ISSCC'25 [10]	JSSC'23 [11]	VLSI'25 [19]	This Work
Tech node (nm)	28	16	28	22
Area (mm²)	20.25	25	12.65	5.33
Architecture	Digital	Digital	Digital CIM+CAM	Digital+DCIM+TCAM
Application	ASR (Agent)	Speech denoising, ASR	LLM (Transformer)	Multi-speaker ASR
Algorithms	Conformer	Attention-based RNN+Bayesian MRF	OPT-1.3B LLaMA2-7B	Hybrid Transformer + Transducer
Parameter (M)	10.1	3.5	-	17.43
Datatype	INT2-8 (A)/INT8 (W)	FP8	FP16-INT8	INT8, FP8, MXINT8, MXFP8
SRAM Cap. (MB)	834KB	9.8 MB	272KB (KV: 256KB)	735.5 KB (CIM: 247.5KB)
Supply Voltage (V)	0.7-1.1	0.55-1.0	0.6-1.0	0.55-1.0
Frequency (MHz)	200	130-775	110-500	46.4-582
Peak Performance[1] (TOPS/TFLOPS)	0.41-1.64	1.17	2.98	2.382[2]
Peak AE[1] (TOPS/ TFLOPS/mm²)	-	FlexASR: 0.13	0.24	0.447[2]
Peak EE[1] (TOPS/TFLOPS/W)	-	FlexASR: 7.8	15.6	10.24-37.50[1] (MXFP8) 16.72-49.17[2] (MXINT8)
Latency per Frame (ms)[5]	32.5-129.9	6[4]-18[4]	-	1.87[2]-7.51[3]
Energy/frame(mJ)[5]	2.5-6.3	0.607[4]-2.414[4]	-	0.158[2]-1.26[3]

Figure 18.2.6: Measurement results and performance comparison table.

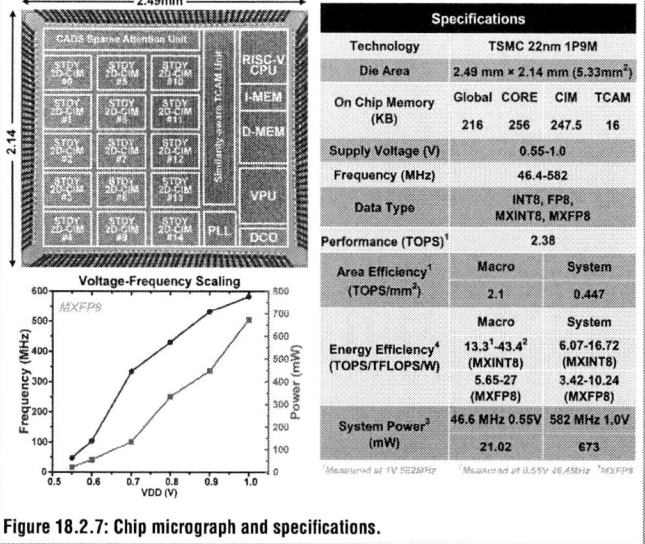

Specifications				
Technology	TSMC 22nm 1P9M			
Die Area	2.49 mm × 2.14 mm (5.33mm²)			
On Chip Memory (KB)	Global	CORE	CIM	TCAM
	216	256	247.5	16
Supply Voltage (V)	0.55-1.0			
Frequency (MHz)	46.4-582			
Data Type	INT8, FP8, MXINT8, MXFP8			
Performance (TOPS)[1]	2.38			
Area Efficiency[1] (TOPS/mm²)	Macro	System		
	2.1	0.447		
Energy Efficiency[4] (TOPS/TFLOPS/W)	Macro	System		
	13.3[1]-43.4[2] (MXINT8)	6.07-16.72 (MXINT8)		
	5.65-27 (MXFP8)	3.42-10.24 (MXFP8)		
System Power[3] (mW)	46.6 MHz 0.55V	582 MHz 1.0V		
	21.02	673		

Figure 18.2.7: Chip micrograph and specifications.

- 2026 IEEE International Solid-State Circuits Conference

ISSCC 2026 / SESSION 18 / TECHNOLOGY AND CIRCUITS FOR DOMAIN-SPECIFIC ACCELERATORS / 18.3

18.3 SMoLPU: 122.1µJ/Token Sparse MoE-Based Speculative Decoding Language Processing Unit with Adaptive-Offload NPU-CIM Core

Sangwoo Ha[1], Jingu Lee[1], Youngjin Moon[1], Sunjoo Whang[1], Wooyoung Jo[1], Gwangtae Park[1], Sangjin Kim[1], Soyeon Um[2], Junha Ryu[1], Yurim Jo[1], Hoi-Jun Yoo[1]

[1]KAIST, Daejeon, Korea, [2]Massachusetts Institute of Technology, Cambridge, MA

Abstract

SMoLPU is an energy-efficient MoE-based speculative decoding LLM processor with an NPU-CIM core. It has 3 features: 1) Token-adaptive expert refinement removes redundant expert activations and schedules expert load order, achieving 2.3x/4.2x energy efficiency improvement in prefill/decode; 2) An adaptive-offload NPU-CIM core and a top scheduling

unit maintain HW utilization under dynamic INT–FP ratios, achieving 3.3x lower latency; 3) A reconfigurable DRAM-based LUT-CIM reduces adder tree power/area, while supporting dynamic input size. The LLM processor achieves 43.5% lower energy per parameter than prior SOTA.

Reasoning strategies such as chain-of-thought (CoT) [1] improve the accuracy of large language models (LLMs) on complex tasks, but lengthen decoding sequences under limited parallelism. The longer sequences repeatedly fetch full model weights from external memory, which inflates external memory access (EMA) and hinders deployment on mobile devices [2-5]. Numerous optimizations have been explored, from conventional methodologies, such as quantization and channel-wise sparsity handling to speculative decoding (SD) [6] and sparse mixture-of-experts (MoE) [7-9] for higher parallelism and reduced EMA. By holistically adopting an MoE-based SD LLM together with the conventional techniques, we can achieve an 89.7% reduction in EMA energy without accuracy loss. However, existing state-of-the-art hardware [2-5] fails to exploit MoE-based SD LLM optimizations. Prior LLM accelerators [2-3] focused on group quantization [10] for low bit-precision and exploiting sparsity for low-power integer (INT) MAC. As group quantization increases the number of floating-point (FP) MACs and exploiting sparsity reduces the number of INT MACs, distinct INT-FP workloads are created. Although NPU with compute-in-memory (NPU-CIM) architectures [11] deliver high energy efficiency on such heterogeneous INT-FP workloads, applying MoE-based SD to NPU-CIM exposes an EMA penalty due to redundancy. As shown in Fig. 18.3.1, the decoding stage suffers from significant weight redundancy, caused by both unnecessary activation of experts for mis-predicted (and thus rejected) tokens and sparsity that arises since the expert outputs are scaled by their routing score (RS). In the prefill stage, sequential loading of 4MB experts requires partial sum (PSUM) caching to aggregate all expert outputs, which enlarges the PSUM footprint and further increases EMA. MoE-based SD LLM also induces channel-wise sparsity that varies over time, creating a dynamically changing ratio of INT and FP MACs across quantization groups. While a prior NPU-CIM architecture supports parallel INT-FP processing with fixed INT-FP ratio, MoE-based SD requires INT MAC results to feed into FP MAC operations such as dequantization, FP scaling, and FP aggregation, with varying INT-FP MAC ratios at runtime. Average hardware utilization drops to 33.7-51.1%, which raises end-to-end latency by 3.3x even on a state-of-the-art NPU-CIM [11]. Condensing sparse workloads into a dense format in CIM can raise INT MAC utilization of CIM, but demands a reconfigurable adder tree (AT) [12, 13] that supports variable input size, increasing power by 3.1x compared with a conventional AT.

To address these issues, we propose SMoLPU, a pipelined integer-floating point NPU-CIM architecture for energy-efficient MoE-based SD LLM inference on mobile devices, with three key features: 1) Token-adaptive Expert Refinement (TaER) with a weight management unit (WMU) and a PSUM management unit (PMU) that eliminates redundant expert fetching and reduces EMA; 2) An adaptive-offload NPU-CIM core (ANC) with a top scheduling unit (TSU) that sustains high utilization and low computing latency under time varying integer to floating-point ratios; 3) A reconfigurable DRAM-based LUT-CIM (RDL-CIM) using bipolar coded DRAM (BCD) that lowers the power overhead of a reconfigurable adder tree. Fig. 18.3.2 shows the overall architecture of SMoLPU, consisting of a memory management unit (MMU), TSU, 4 clusters, and 128KB global memory (GMEM). The MMU comprises WMU for masking redundant experts, and a PMU with 512KB PSUM memory (PSMEM) for eliminating redundant PSUMs. The TSU combines quantization groups into a bundle and forwards indices of bundled groups to the workload allocator of clusters. The cluster consists of 4 ANCs, a workload allocator, a scaling and aggregation unit, and 8KB OAMEM. The ANC consists of an index building unit for the CIM LUT, 4 CIM macros with 4 banks, 4 NPUs with a shift-and-accumulate unit (SAC), a 1.5KB WMEM, and an 8KB scaling factor memory (SFMEM). A CIM bank is composed of 32 local cell arrays (LCAs), each containing a 64×4 6T SRAM cell array with a local decoder, two 10×4 3T1C DRAM LUTs, and read word-line (RWL) switches for refresh.

Figure 18.3.3 presents the proposed TaER algorithm integrated into the MMU architecture to reduce EMA from redundant experts and the PSUM. Since only a fraction of candidate tokens (SD outputs) are accepted and each has inter-token correlation, TaER estimates a per-token acceptance rate (AR) that predicts final acceptance and schedules cache management by exploiting expert-overlapped demands across tokens. The MMU realizes TaER with two-stage dataflow composed of the WMU and the PMU. The WMU masks experts and input channels that contribute little to final accepted tokens, reducing EMA and computation energy. The WMU consists of a redundancy detection unit (RDU) for AR estimation and sorting, and a decision and masking unit (DMU) for mask generation. For

expert masking, the RDU aggregates attention scores per token within each MoE layer to compute AR, and the DMU masks experts with AR below a threshold. Within each expert, the RDU sorts input channels by magnitude scaled with the routing score, and the DMU masks channels below a precomputed top-k threshold. Experts masked by the WMU incur neither EMA nor computation. Masked input channels are skipped and channels masked for all routed tokens are not fetched from external memory, reducing expert EMA and computation energy by 66.3% and 80.0%, respectively. The PMU manages the PSUM cache by evaluating the overlap between the set of currently cached tokens and the set of tokens routed to each expert. It sequences expert load order to fetch the expert that is most commonly required by cached tokens to maximize the PSUM reuse and reduce evictions. The PMU operates in two sequential modes: expert sequencing and PSUM ranking. In expert sequencing, a greedy scheduler computes a reuse score for each expert, defined as the number of cached tokens that route to each expert, and loads the expert with the highest reuse score. In PSUM ranking, the PSMEM capacity miss triggers the eviction of the PSUM with the lowest relevance to the currently loaded expert. This maximizes on-chip PSUM reuse, reducing PSUM EMA by 67.2%. Overall, TaER improves energy efficiency by 2.3x in prefill and 4.2x in decode, while expert-wise pipelined operation keeps latency overhead lower than 3%.

Figure 18.3.4 shows the operation and implementation of the TSU and the ANC, designed to maximize HW utilization under diverse INT-FP ratios. In TaER, fine masking induces sparsity from the RS, which degrades CIM utilization due to finely pruned input channels. Densifying these input channels into bundles improves CIM utilization, but still leads to low NPU utilization when bundles contain fewer quantization groups, leaving FP resources idle during aggregation. We propose an on-the-fly workload allocation scheme that manages CIM-friendly bundles and sparse groups separately, maintaining high utilization on both CIM and NPU. In the TSU, nonzero input channels are fetched and compacted into a group FIFO. A bundling unit then organizes these groups into bundles (up to 4 groups per bundle), which are dispatched to 32 INT MACs and 4 FP MACs, thereby eliminating zero channels. In the ANC, sparse groups are collected into a leftover FIFO and offloaded to the NPU whenever FP resources are available. The hardware autonomously monitors FP MAC availability, enabling offloading without software intervention. For example, when a bundle contains 4 groups, the CIM and NPU operate in a fully pipelined manner. The CIM first performs bit-parallel bit-serial (BPBS) INT MACs, and the outputs are then forwarded to NPU, where the SAC computes a weighted sum before FP MAC. 4 SACs are time-division multiplexed onto a single FP MAC, fully hiding 4-cycle weighted-sum latency. When a bundle has fewer than four groups, leftovers are fed to an underutilized NPU to fill idle FP slots. In this case, the SAC performs INT MACs, and forwards results to FP MAC. This is enabled by SAC reconfigurability, achieved using simple AND gates at its input, which supports INT MAC execution in the NPU with negligible area overhead. Overall, the proposed TSU and ANC enhance HW utilization for INT and FP MAC by 23.1% and 35.2%, respectively, achieving a 3.3x end-to-end execution speedup.

Figure 18.3.5 depicts the RDL-CIM with BCD which reduces the power consumption of the reconfigurable AT. Low-power CIMs [14, 15] adopt a LUT-based AT that replaces multi-level gate switching with a small number of memory read operations. The proposed RDL-CIM utilizes a reconfigurable LUT decoder to support low-power LUT for diverse input sizes. Furthermore, BCD enables the RDL-CIM to overcome the significant area overhead of the LUT. The RDL-CIM reads 4 weights per LCA and supplies them to the local decoder, which activates a single bit-line of the LUT. The entry on the selected bit-line is read out from 3T1C DRAM cells to RWL. While a standard LUT requires a 4b input to generate 16 entries, BCD makes it possible using only 8b entries with bipolar coding. Even though bipolar coding incurs overhead for inverters and MUXes, smaller entries and shared front-end inverters/MUXes across CIM macros make the logic overhead negligible. Physical area is also reduced by using a compact 3T1C DRAM cell for the LUT. Consequently, LUT pitch is reduced by 75.7% and matches precisely the 3.6µm pitch of the SRAM LCA. The local decoder with a router and two 2-to-4 decoders enables reconfigurable LUT-based addition which has 3 different modes: one group, symmetric two group, and asymmetric two group. For example, one group mode decodes W3 and W4 inputs into a single 3-to-8 together with W2 output gating. In the case of asymmetric two group, W3 and W4 inputs decode into 2-to-4 and W1 decodes as 1-to-2. Since any bundle created by the TSU has a maximum of

312 • 2026 IEEE International Solid-State Circuits Conference

979-8-3315-8937-0/26 $31.00 © 2026 IEEE

two groups per 4 weights, all workloads are fully covered by the RDL-CIM. The proposed RDL-CIM reduces power and area by 36.8% and 26.3%, respectively, with only 0.13% refresh power overhead.

Figure 18.3.6 shows measurement results of SMoLPU and a comparison table. SMoLPU is evaluated on MT-bench with GPT-4 as judge (LLM-as-a-judge) [16] using OLMoE [8], DeepSeek-V2 [7], and Qwen3 [9] models. For a sequence of 128 input and 64 output tokens, SMoLPU achieves 34.7-99.3mJ/token with negligible accuracy loss (<0.2), yielding a 42.7~71.1% reduction in total energy consumption. Previous works [2-5] applied various compression techniques to reduce EMA, but were fundamentally limited by degrading model accuracy (3.3 ppl in [3]). In contrast, SMoLPU activates only 1.4B of 15.7B parameters per token, yielding 11.2× EMA reduction, the highest among prior works. SMoLPU also achieves the lowest energy per token per parameter of 7.8μJ/token/parameter, which is 43.5% lower than the state-of-the-art [2] for 1024 input tokens and 1 output token. Its improvement becomes larger with longer output tokens, making the hardware effective for reasoning tasks. SMoLPU is fabricated in 28nm CMOS technology and occupies a 20.25mm² die area, as shown in Fig. 18.3.7. In conclusion, SMoLPU demonstrates a 122.1μJ/token sparse MoE-based speculative decoding LLM inference with a pipelined NPU-CIM architecture for mobile devices.

References:

[1] Wang, Xuezhi et al., "Self-Consistency Improves Chain of Thought Reasoning in Language Models," arXiv preprint arXiv:2203.11171, 2022.
https://doi.org/10.48550/arXiv.2203.11171

[2] S. Kim et al., "Slim-Llama: A 4.69mW Large-Language-Model Processor with Binary/Ternary Weights for Billion-Parameter Llama Model," *ISSCC*, pp. 421-423, 2025.
https://doi.org/10.1109/ISSCC49661.2025.10904761

[3] W. Jo et al., "BROCA: A 52.4-to-559.2mW Mobile Social Agent System-on-Chip with Adaptive Bit-Truncate Unit and Acoustic-Cluster Bit Grouping," *ISSCC*, pp. 418-420, 2025.
https://doi.org/10.1109/ISSCC49661.2025.10904658

[4] Y. Qin et al., "An 88.36TOPS/W Bit-Level-Weight-Compressed Large-Language-Model Accelerator with Cluster-Aligned INT-FP-GEMM and Bi-Dimensional Workflow Reformulation," *ISSCC*, pp. 420-422, 2025.
https://doi.org/10.1109/ISSCC49661.2025.10904774

[5] S. Kim et al., "C-Transformer: A 2.6-18.1μJ/Token Homogeneous DNN-Transformer/Spiking-Transformer Processor with Big-Little Network and Implicit Weight Generation for Large Language Models," *ISSCC*, pp. 368-370, 2024.
https://doi.org/10.1109/ISSCC49657.2024.10454330

[6] Li, Yuhui et al., "EAGLE-3: Scaling Up Inference Acceleration of Large Language Models via Training-Time Test," arXiv preprint arXiv:2503.01840, 2025.
https://doi.org/10.48550/arXiv.2503.01840

[7] Liu, Aixin et al., "DeepSeek-V2: A Strong, Economical, and Efficient Mixture-Of-Experts Language Model," arXiv preprint arXiv:2405.04434, 2024.
https://doi.org/10.48550/arXiv.2405.04434

[8] Muennighoff, Niklas et al., "OLMoE: Open Mixture-Of-Experts Language Models," arXiv preprint arXiv:2409.02060, 2024. https://doi.org/10.48550/arXiv.2409.02060

[9] Yang, An et al., "Qwen3 Technical Report," arXiv preprint arXiv:2505.09388, 2025.
https://doi.org/10.48550/arXiv.2505.09388

[10] Lin, Ji et al., "AWQ: Activation-Aware Weight Quantization for On-Device LLM Compression and Acceleration," *MLSys*, pp. 87-100, 2024.
https://doi.org/10.48550/arXiv.2306.00978

[11] J. Yue et al.,, "A 28nm 16.9-300TOPS/W Computing-in-Memory Processor Supporting Floating-Point NN Inference/Training with Intensive-CIM Sparse-Digital Architecture," *ISSCC*, pp. 252-253, 2023.
https://doi.org/10.1109/ISSCC42615.2023.10067779

[12] Z. Wu et al.,, "CELLA: A 28nm Compute-Memory Co-Optimized Real-Time Digital CIM-Based Edge LLM Accelerator with 1.78ms-Response in Prefill and 31.32 Token/s in Decoding," *IEEE Symp. VLSI Circuits*, 2025.
https://doi.org/10.23919/VLSITechnologyandCir65189.2025.11075101

[13] R. Guo et al.,, "CIMFormer: A Systolic CIM-Array-Based Transformer Accelerator With Token-Pruning-Aware Attention Reformulating and Principal Possibility Gathering," *IEEE JSSC*, vol. 59, no. 10, pp. 3317-3329, Oct. 2024.
https://doi.org/10.1109/JSSC.2024.3402174

[14] Y. Wang et al.,, "LLM-CIM: A 28nm 126.7TOPS/W Input-LUT-Based Digital CIM Macro with Reconfigurable Matrix Multiplication and Nonlinear Operation Modes for LLMs," *IEEE Symp. VLSI Circuits*, 2025.
https://doi.org/10.23919/VLSITechnologyandCir65189.2025.11074939

[15] Y. He et al.,, "A 28nm 2.4Mb/mm2 6.9 - 16.3TOPS/mm2 eDRAM-LUT-Based Digital-Computing-in-Memory Macro with In-Memory Encoding and Refreshing," *ISSCC*, pp. 578-580, 2024. https://doi.org/10.1109/ISSCC49657.2024.10454323

[16] Zheng, Lianmin et al., "Judging LLM-as-a-Judge with MT-bench and Chatbot Arena," *NeurIPS*, pp. 46595-46623, 2023. https://doi.org/10.48550/arXiv.2306.05685

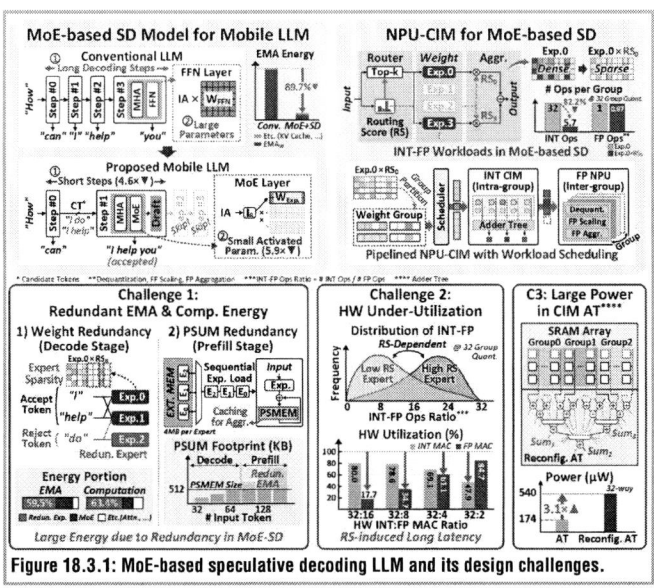

Figure 18.3.1: MoE-based speculative decoding LLM and its design challenges.

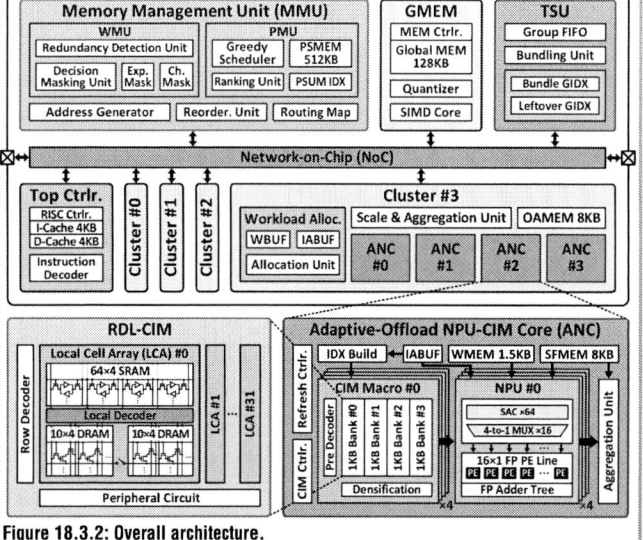

Figure 18.3.2: Overall architecture.

ISSCC 2026 / SESSION 18 / TECHNOLOGY AND CIRCUITS FOR DOMAIN-SPECIFIC ACCELERATORS / 18.3

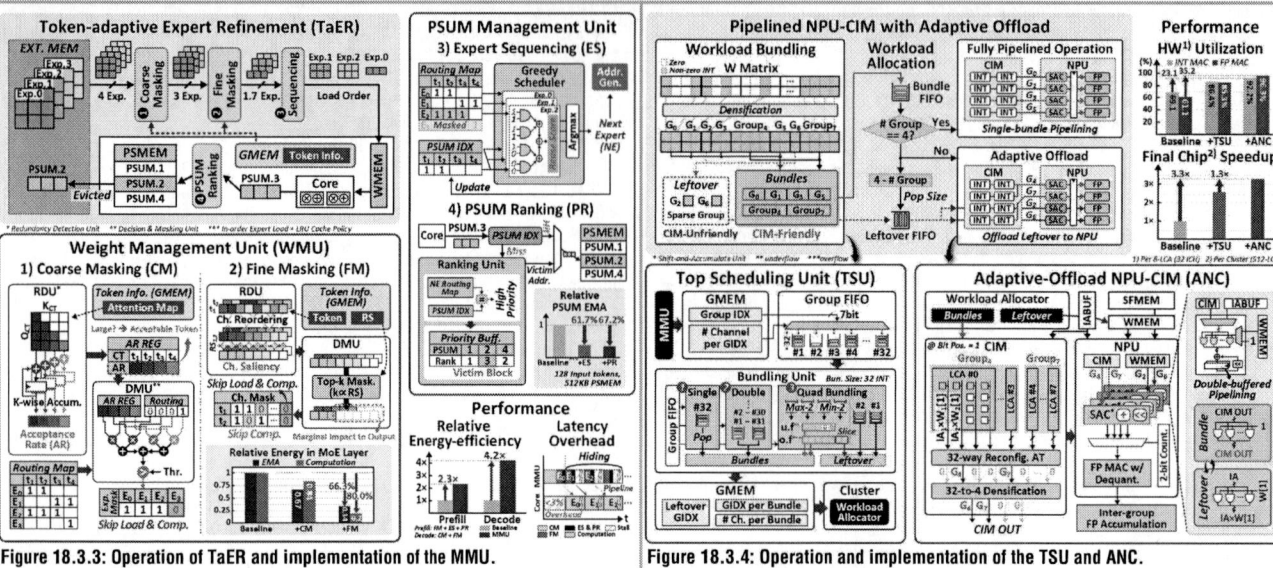

Figure 18.3.3: Operation of TaER and implementation of the MMU.

Figure 18.3.4: Operation and implementation of the TSU and ANC.

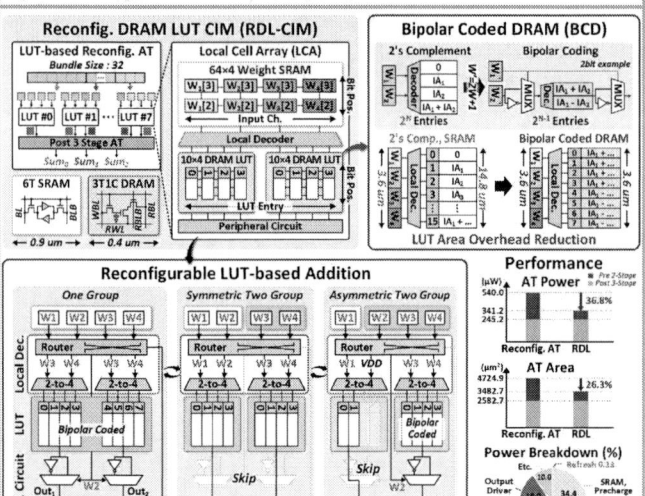

Figure 18.3.5: The RDL-CIM with BCD and operation flow of the reconfigurable LUT.

Figure 18.3.6: Measurement results and performance comparison table.

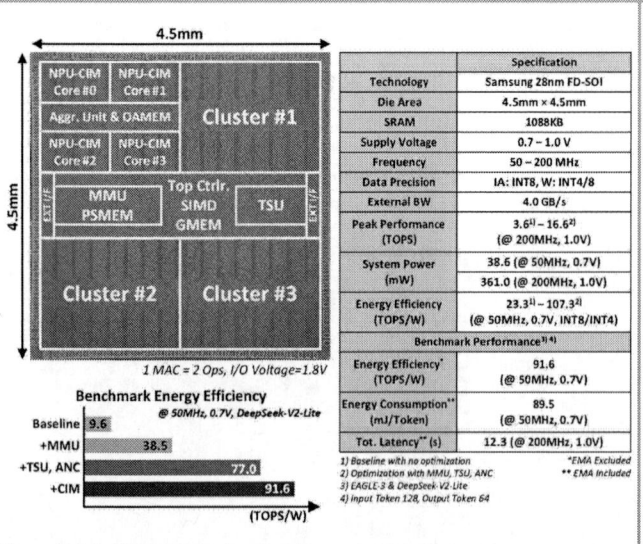

Figure 18.3.7: Chip micrograph and performance summary.

- 2026 IEEE International Solid-State Circuits Conference

979-8-3315-8937-0/26 $31.00 © 2026 IEEE

ISSCC 2026 / SESSION 18 / TECHNOLOGY AND CIRCUITS FOR DOMAIN-SPECIFIC ACCELERATORS / 18.4

18.4 SpikeRAM: A 48.1pW/Synapse/Bit Event-Driven Spiking Compute-Near/In-Memory Processor with Neuromorphic Sensor Enabling Life-Long On-Chip Learning

Haotian Fu[1], Yue Zhou[1], Zhuo Zhang[1], Hongzhao Zheng[1], Renxu Yang[1], Yulong Huang[1], Dezhen Yang[2], Yannan Xing[3], Tugba Demirci[4], Ning Qiao[5], Bojun Cheng[1]

[1]Hong Kong University of Science and Technology, Guangzhou, China, [2]North China Research Institute of Electro-Optics, Beijing, China, [3]SynSense, Chengdu, China, [4]SynSense, Zurich, Switzerland, [5]SynSense, Ningbo, China

Abstract

SpikeRAM is a high efficiency (48.1pW/Synapse/Bit) memory-centric neuromorphic system with perception, computing and on-chip learning, achieving 464M synapses and 8.28mW power in real-time processing, while realizing few-shot learning for event-based signature verification. The EVS-sCNN core enables asynchronous sensing and feature extraction. The sFC-OCL core enables event-driven inference and learning with the e-OTBP algorithm. Gray-code weights and ternary gradients reduce memory write times by over 86%.

Edge perceptual SoCs increasingly demand high energy efficiency, privacy and adaptability for real-world tasks as shown in Fig. 18.4.1 (top) [1–6]. For perception, an event-based vision sensor (EVS) with asynchronous, sparse and edge-only imaging, offers low power and high privacy [7-8]. For computation, the use of a spiking neural network (SNN) further increases energy efficiency through the spike-driven paradigm [9-14]. In addition, on-chip learning (OCL) with emerging non-volatile memories (eNVMs) enables permanent in-situ training without cloud computing, mitigating privacy risks while supporting user-specific and environment adaptation [15-17]. However, to achieve energy efficiency, privacy and adaptability simultaneously, there are three challenges as shown in Fig. 18.4.1 (middle). 1) Inefficient inference arises from data movement due to the separation of sensing, memory and computing, along with frame conversion and redundant multiply-accumulations. 2) High power and memory overhead in spatial-temporal OCL, which scales with the depth of time-window (TW). 3) Excessive memory updates during OCL increase energy and hamper lifetimes for eNVMs with limited writing endurance [18-21].

To tackle these challenges, we introduce the SpikeRAM system comprising an on-chip EVS-integrated spiking convolution neural network (EVS-sCNN) core with near-memory computing (NMC) and a spiking fully connected OCL (sFC-OCL) core featuring compute-in-memory (CIM). Fig. 18.4.1 (bottom) depicts three key features of SpikeRAM: 1) Memory-centric asynchronous near-sensor computing. The SNN core directly processes asynchronous events from on-chip EVS, eliminating frame conversion, reducing data movement and redundant operations. 2) Eligibility one-time backpropagation (e-OTBP) OCL algorithm trains a multi-layer network through a single TW and eligibility trace (ET). It maintains low power, while significantly reducing memory requirements by 73.1% at 32 TWs. 3) Ternary gradients with gray-code weights significantly reduce memory programming times. Combined with the high convergence of e-OTBP, it enhances both durability and energy efficiency. On NMNIST and DVS Gesture, the average programming times are reduced by 89.1% and 88.4%, respectively. To summarize, SpikeRAM is an EVS-based spiking sense-memory-compute system with OCL. It also enables eNVM-based in-situ training and can extend eNVM lifetime by over 10×.

Figure 18.4.2 (top) illustrates the architecture of the SpikeRAM system. In the EVS-sCNN core, raw events from on-chip EVS (3.8MEps and 130.9dB dynamic range) or external sources are denoised and routed to the sCNN processing elements (PEs) for feature extraction through configurable convolution layers, scheduled by a Network-on-Chip (NoC) for diverse tasks. In the sFC-OCL core, extracted feature events are processed through sFC and OCL PEs with configurable filters and neurons for inference and OCL. Fig. 18.4.2 (bottom) shows the workflow of the sCNN PE, which adopts a fully asynchronous design with dual-rail handshake logic to eliminate the clock tree overhead. Each PE performs event-driven convolution, activating kernels only upon event arrival and skipping blank regions to avoid invalid operations, enabling sparsity adaptation. During processing, padding ensures dimensional consistency, and the kernel anchor determines output feature locations. Address sweep generates addresses for Leaky-Integrate-and-Fire (LIF) neurons, where non-zero 8b weights are accumulated into 16b membrane potential (V_{mem}). Once V_{mem} exceeds the threshold, neurons reset and output events. These events are merged via polling, packetized, and routed to the next layer or the sFC-OCL core under NoC control.

Figure 18.4.3 (top) illustrates the system diagram of the CIM-based sFC-OCL core. The first sFC PE contains 8 digital CIM macros, each including an 32Kb RAM, containing 8b weights with sense amplifiers (SAs), gray-code decoders, parallel shift-adders, and 16 configurable neurons. The second sFC PE consists of 10 CIM macros with a total of 10Kb for weights and 10 output neurons. Within each TW, input events are compressed into weighted spikes and routed to a configurable event filter to exploit input sparsity, skipping up to 92.4% of redundant computations in DVS Gesture. Addresses of weighted spikes activate the corresponding word-lines in the first sFC PE and decoded weights are shifted and accumulated into V_{mem} to avoid multiplications. Neurons support LIF, IF and InoF behaviors, allowing dynamic trade-offs between temporal sensitivity and energy efficiency. The generated spikes are routed to the second sFC-PE for subsequent processing. At the end of inference, the OCL PEs based on e-OTBP are activated for backpropagation by reusing the preserved inference states. Fig. 18.4.3 (bottom) shows a handshake-based pipeline that ensures efficient inference and learning in the sFC-OCL core.

Figure 18.4.4 (top) shows the e-OTBP algorithm and ternary gradient in OCL. The e-OTBP performs global learning through a single TW and captures temporal information via both ETs and output V_{mem} across previous TWs, retained in inference and reused in learning. Gradients of layer 2 are computed by errors and corresponding input spikes. Gradients of layer 1 are computed by propagating errors via transposed weights and a surrogate (Surr) function, then combined with ETs. Both gradients are fed into ternary quantization function yielding the ternary value. The computing graph shows that e-OTBP eliminates historical computations compared with conventional backpropagation through time (BPTT) algorithm. For efficient hardware implementation, the Surr function is binarized using two comparators, the ET cell is realized via a cyclic accumulator, and the FC weight transpose is computed via a horizontal adder tree to reuse the weight array. Compared with BPTT, the memory and computation overheads of e-OTBP are 4.3% and 3.1%, respectively.

Figure 18.4.4 (bottom) illustrates the gray-code dataflow. All weights are stored in the array using gray-code encoding. During inference, gray-code decoders implemented by parallel XOR logic are integrated with the SAs to enable binary readout of weights for neuron accumulations. During learning, readout weights are added or subtracted with ternary gradients and updated in gray-code. This combination ensures that each weight is updated by no more than 1b per batch, significantly reducing memory programming times. SpikeRAM can be either standalone or combined with a non-volatile MRAM array to achieve large-scale neural networks and permanent learning. With an external 4Mb MRAM chip, the parameters can be expanded by 5.4× for 101-class N-Caltech101 [23] tasks through time multiplexing.

Figure 18.4.5 (top) shows the event-based signature verification system as a demonstration of real-world private edge learning. Signatures are private data for authentication, leading to cloud training privacy leakage issues. Conventional image-based methods ignore temporal dynamics, reducing accuracy and raising forgery risks. SpikeRAM addresses these by sensing the spatial-temporal features with EVS, followed by SNN-based inference and OCL. Fig. 18.4.5 (top left) shows the system setup. The on-chip EVS with an infrared filter captures the signature signals from an infrared pen at 200Hz within the 850nm band. To suppress interference from sensor and ambient noise, the NeuroPass layer employs a LIF-based sCNN to selectively respond to a specific frequency range, thereby improving the signal-to-noise ratio as shown in Fig. 18.4.5 (top middle). SpikeRAM extracts spatial features while preserving sign speed thanks to the high temporal resolution of EVS as shown in Fig. 18.4.5 (right). Although forgeries appear nearly same in static image, they are easily distinguishable in temporal domain. Based on SpikeRAM, we introduce the EHSV dataset, comprising one class of genuine signatures and four classes of forgeries. With 2-shot OCL from random initialization, SpikeRAM achieves 94.3% accuracy in authenticity verification and classification. Fig. 18.4.5 (middle) shows the measurement results. Static power and dynamic power for real-time processing are 1.26mW and 8.28mW, respectively. On NMNIST, DVS Gesture and DailyDVS, SpikeRAM achieves rapid convergence from random initialization, reaching 96.4%, 92.9% and 90.5% accuracy, respectively, with only 1 OCL epoch.

Figure 18.4.5 (middle right and bottom) shows that integrating ternary gradients, gray-code weights, and the high convergence of e-OTBP significantly reduces MRAM write operations. In DVS Gesture, mean and standard deviation of the update times are reduced by 10.9× and 35.6×, respectively. Across four datasets, the maximum number of memory devices written is reduced by 92.8% and the total write time per sample is reduced by 86.3%, on average. These results indicate that SpikeRAM effectively mitigates MRAM endurance limitations during OCL.

Figure 18.4.6 compares state-of-the-art neuromorphic processors. SpikeRAM improves power density (48.1pW/Synapse/Bit) by 2.7× over [2] through its NMC and CIM hybrid architecture with asynchronous and event-driven sparsity adaptation. SpikeRAM supports 464M synapses and 328K neurons for diverse networks, improving network density (441.9Synapse/Byte) by 24.3× over [10]. Learning efficiency is evaluated by the total energy per learning task. SpikeRAM achieves outstanding learning efficiency and accuracy across multiple event-based tasks, validated with on-chip SRAM and an external 40nm, 4Mb MRAM chip. The MRAM lifetime is extended by 13.9× in OCL, while energy consumption remains comparable. At the system level, SpikeRAM integrates on-chip EVS for real-time sensing, processing and learning, achieving 8.28mW system power in real-world tasks. Fig. 18.4.7 shows the summaries, voltage characteristics, micrographs and test system. SpikeRAM enables both real-time perception and inference, while supporting long-life learning in an eNVM-based system.

Acknowledgement:

This work was supported by 1+1+1: Guangdong S&T program (Project number 2025A0505000036), and National Key R&D Program of China (Project number 2025YFG0100300). The corresponding author is Bojun Cheng (bocheng@hkust-gz.edu.cn).

References:

[1] Y. Yuan et al., "A 28nm 192.3 TFLOPS/W Accurate/Approximate Dual-Mode-Transpose Digital 6T-SRAM CIM Macro for Floating-Point Edge Training and Inference," *ISSCC*, pp. 258-260, Feb. 2025. https://doi.org/10.1109/ISSCC49661.2025.10904659
[2] D. Huo et al., "A 67µW/Channel, 0.13 nW/Synapse/b Nose-on-a-Chip for Noninvasive Diagnosis of Diseases with On-Chip Incremental Learning," *ISSCC*, pp. 350-352, Feb. 2025. https://doi.org/10.1109/ISSCC49661.2025.10904713
[3] J. Liu et al., "A High-Accuracy and Energy-Efficient Zero-Shot-Retraining Seizure-Detection Processor with Hybrid-Feature-Driven Adaptive Processing and Learning-Based Adaptive Channel Selection," *ISSCC*, pp. 542-544, Feb. 2024. https://doi.org/10.1109/ISSCC49657.2024.10454405
[4] J. Zhang et al., "ANP-I: A 28nm 1.5 pJ/SOP Asynchronous Spiking Neural Network Processor Enabling Sub-0.1 µJ/Sample On-Chip Learning for Edge-AI Applications," *ISSCC*, pp. 21-23, Feb. 2023. https://doi.org/10.1109/ISSCC42615.2023.10067650
[5] Y. Zhong et al., "PAICORE: A 1.9-Million-Neuron 5.181-TSOPS/W Digital Neuromorphic Processor With Unified SNN-ANN and On-Chip Learning Paradigm," *IEEE JSSC*, vol. 60, no. 2, pp. 651-671, Feb. 2025. https://doi.org/10.1109/JSSC.2024.3426319
[6] R. Mao et al., "FSNAP: An Ultra-Energy-Efficient Reconfigurable Few-Spikes-Neuron-Based SNN Processor Supporting Unified On-Chip Learning and Adaptive Time-Window Tuning," *IEEE JSSC*, early access, 2025. https://doi.org/10.1109/JSSC.2025.3576485
[7] P. Lichtsteiner et al., "A 128× 128 120 dB 15 µs Latency Asynchronous Temporal Contrast Vision Sensor," *IEEE JSSC*, vol. 43, no. 2, pp. 566-576, Feb. 2008. https://doi.org/10.1109/JSSC.2007.914337
[8] T. Finateu et al., "A 1280×720 Back-Illuminated Stacked Temporal Contrast Event-Based Vision Sensor with 4.86µm Pixels, 1.066GEPS Readout, Programmable Event-Rate Controller and Compressive Data-Formatting Pipeline," *ISSCC*, pp. 112-114, Feb. 2020. https://doi.org/10.1109/ISSCC19947.2020.9063149
[9] C. Frenkel et al., "ReckOn: A 28nm Sub-mm2 Task-Agnostic Spiking Recurrent Neural Network Processor Enabling On-Chip Learning over Second-Long Timescales," *ISSCC*, pp. 1-3, Feb. 2022. https://doi.org/10.1109/ISSCC42614.2022.9731734
[10] Y. Liu et al., "A 22nm 0.26nW/Synapse Spike-Driven Spiking Neural Network Processing Unit Using Time-Step-First Dataflow and Sparsity-Adaptive In-Memory Computing," *ISSCC*, pp. 484-486, Feb. 2024. https://doi.org/10.1109/ISSCC49657.2024.10454472
[11] X. Yang et al., "A 10 000-Inference/s Bio-Inspired Spiking Vision Chip Based on an End-to-End SNN Embedding Image Signal Enhancement," *IEEE JSSC*, early access, 2025. https://doi.org/10.1109/JSSC.2025.3583310

[12] X. Qi et al., "A 0.67-to-5.4 TSOPs/W Spiking Neural Network Accelerator With 128/256 Reconfigurable Neurons and Asynchronous Fully Connected Synapses," *IEEE JSSC*, vol. 59, no. 10, pp. 3366-3377, Oct. 2024. https://doi.org/10.1109/JSSC.2024.3402208
[13] S. Kim et al., "C-Transformer: A 2.6-18.1µJ/Token Homogeneous DNN-Transformer/Spiking-Transformer Processor with Big-Little Network and Implicit Weight Generation for Large Language Models," *ISSCC*, pp. 368-370, Feb. 2024. https://doi.org/10.1109/ISSCC49657.2024.10454330
[14] M. Yao et al., "Spike-Based Dynamic Computing with Asynchronous Sensing-Computing Neuromorphic Chip," *Nature Communications*, vol. 15, article no. 4464, 2024. https://doi.org/10.1038/s41467-024-47811-6
[15] H. -J. Lee et al., "A 13.5µW 35-Keyword End-to-End Keyword Spotting System Featuring Personalized On-Chip Training in 28nm CMOS," *ISSCC*, pp. 620-622, Feb. 2025. https://doi.org/10.1109/ISSCC49661.2025.10904744
[16] X. Li et al., "Federated learning Using a Memristor Compute-in-Memory Chip with In Situ Physical Unclonable Function and True Random Number Generator," *Nature Electronics*, vol. 8, no. 6, pp. 518-528, Jun. 2025. https://doi.org/10.1038/s41928-025-01390-6
[17] W. Zhang et al., "Edge Learning Using a Fully Integrated Neuro-Inspired Memristor Chip," *Science*, vol. 381, no. 6663, pp. 1205-1211, Sep. 2023. https://doi.org/10.1126/science.ade3483
[18] M. Zhao et al., "Characterizing Endurance Degradation of Incremental Switching in Analog RRAM for Neuromorphic Systems," *IEDM*, pp. 20.2.1-20.2.4, Dec. 2018. https://doi.org/10.1109/IEDM.2018.8614664
[19] Y. -D. Chih et al., "Design Challenges and Solutions of Emerging Nonvolatile Memory for Embedded Applications," *IEDM*, pp. 2.4.1-2.4.4, Dec. 2021. https://doi.org/10.1109/IEDM19574.2021.9720557
[20] D. Zeng et al., "Achieving Over 95% Yield of Sub-1 ppm BER with Retention Over 10 years at 125 °C and Endurance of 1×10^{12} Cycles Towards Automotive Non-Volatile RAM Applications," *J. Semicond.*, vol. 46, no. 3, pp. 032301, Mar. 2025. https://doi.org/10.1088/1674-4926/24090037
[21] C. Mu et al., "A 28-nm RRAM/SRAM Collaborative CIM Accelerator Supporting RRAM-Endurance-Latency Awareness for Edge Fine-Tuning," *IEEE JSSC, vol. 60, no. 10, pp. 3626-3638, Oct. 2025.* https://doi.org/10.1109/JSSC.2025.3577335
[22] G. Orchard et al., "Converting Static Image Datasets To Spiking Neuromorphic Datasets Using Saccades," *Frontiers in neuroscience*, vol. 9, no. 437, Nov. 2015. https://doi.org/10.3389/fnins.2015.00437
[23] A. Amir et al., "A Low Power, Fully Event-Based Gesture Recognition System," *IEEE CVPR*, pp. 7388-7397, Jul. 2017. https://doi.org/10.1109/CVPR.2017.781
[24] Q. Liu et al., "Event-based Action Recognition Using Motion Information and Spiking Neural Networks," *IJCAI*, pp. 1743-1749, Aug. 2021. https://doi.org/10.24963/ijcai.2021%2F240

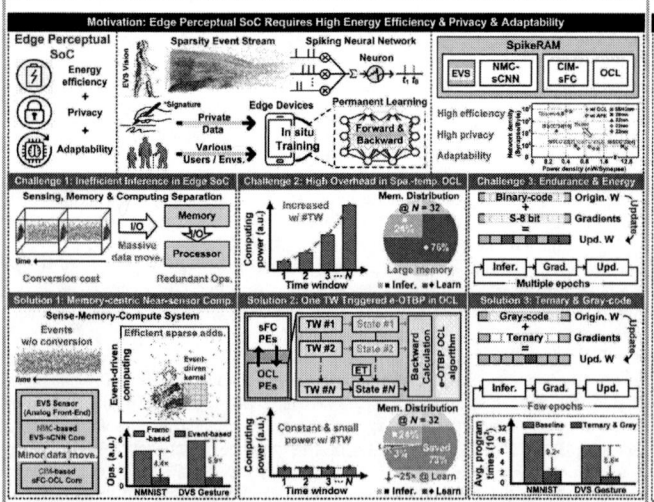

Figure 18.4.1: Motivation of edge perceptual SoC with high energy efficiency, privacy and adaptability (top), challenges (middle), and our solutions in SpikeRAM (bottom).

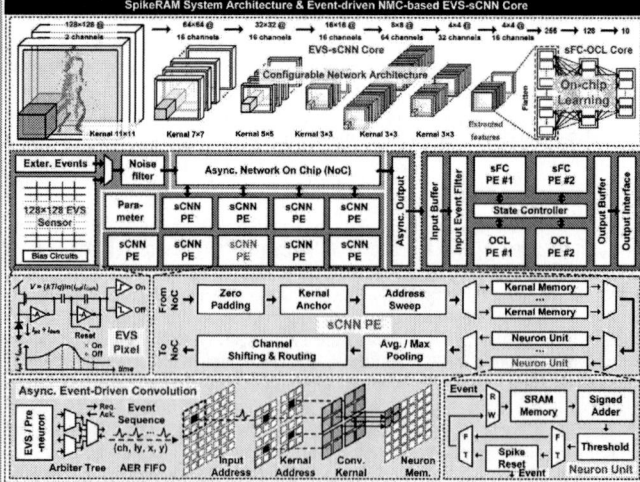

Figure 18.4.2: Architecture of SpikeRAM system (top), and event-driven convolution workflow in EVS-sCNN core (bottom).

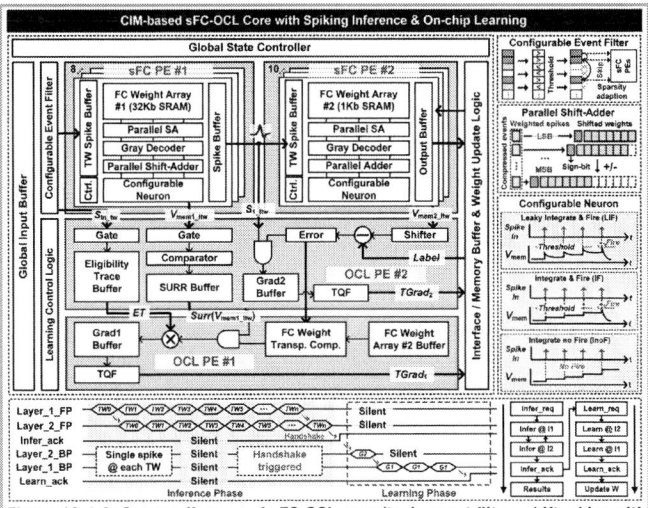

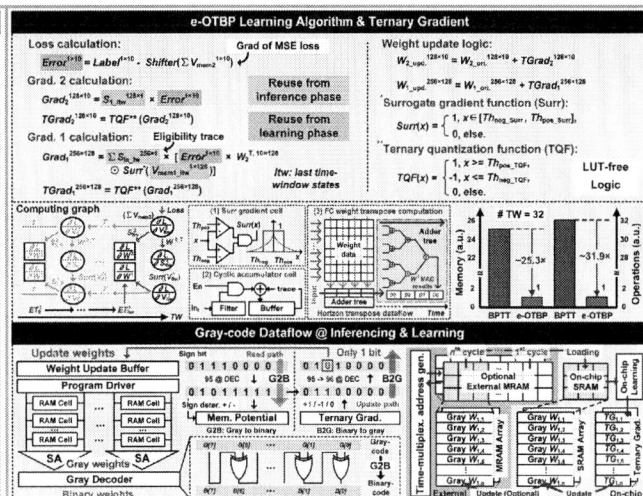

Figure 18.4.3: System diagram of sFC-OCL core (top), event filter, shift-adder with weighed spikes and configurable neurons (top right), and handshake-based pipeline (bottom).

Figure 18.4.4: e-OTBP on-chip learning algorithm and ternary gradient implementations (top), and gray-code dataflow with weight updating and optional learning in MRAM (bottom).

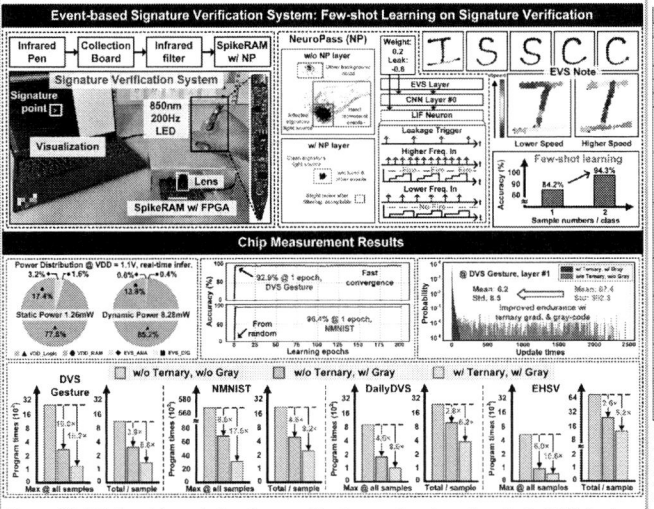

Figure 18.4.5: Event-based signature verification system based on SpikeRAM for few-shot on-chip learning (top), and chip measurement results (bottom).

Figure 18.4.6: Comparison with state-of-the-art neuromorphic processors.

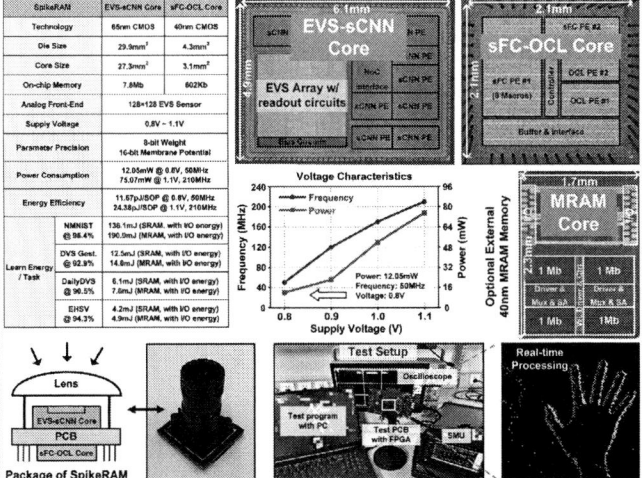

Figure 18.4.7: Summary table (top left), voltage characteristics and chip micrograph (top right), and measurement system (bottom).

ISSCC 2026 / SESSION 18 / TECHNOLOGY AND CIRCUITS FOR DOMAIN-SPECIFIC ACCELERATORS / 18.5

18.5 A 28nm 47.3TFLOPs/W 894mJ/Inference Visual Autoregressive Accelerator with Differential-Amplifier Speculation and Chain-Reaction-Like Parallel Generation

Zhiheng Yue*, Xujiang Xiang*, Jiamu Fu, Shaojun Wei, Yang Wang, Yang Hu, Shouyi Yin

Tsinghua University, Beijing, China
*Equally Credited Authors (ECAs)

Abstract

To accelerate Visual Autoregressive (VAR) applications, this work implements a 28nm VAR accelerator achieving 47.3TFLOPs/W and <0.6% FID loss. A differential visual attention amplifier speculates critical tokens for selective execution; a full-path optimized MXINT PE adapts to biased data distribution; and, a chain reaction-like parallel generation exploits spatial correlation. The 5.76mm^2 chip runs at 400MHz, accelerating DeiT/ViT VAR by 37.6× with 2.75TFLOPs/mm^2 area efficiency.

In the field of generative AI, the Visual Autoregressive (VAR) model exhibits promising intelligence and has surpassed other generative models due to its scalability and versatility [1-4]. The former refers to its general performance scaling across models, and the latter points to the model's adaptability to diverse downstream tasks. These benefits stem from its visual context learning capability, mirroring the large language model (LLM) [5-7]. However, high-dimensional image tasks induce significantly more latency to process (e.g., 332.8s per 512×512 resolution image generation). The latency can be attributed to 3 challenges. At the intra-iteration level: 1) The model evaluates the attention between all tokens, but only the critical token pairs matter. Consequently, much of the attention noise computing is unnecessary. Though, the utility of the attention score is difficult to distinguish what can be ignored. 2) The highly biased data distribution of image context causes redundancy in full-precision and full-width hardware. This is because only a fraction of tokens and a narrowed value range have an impact on the final output. At the inter-iteration level, 3) each pixel is generated by the model iteratively, but the information entropy explored by the model is low because of data correlations. The sequential generation induces numerous iterations, ignoring data redundancy.

To address the model/hardware/data redundancy, we implement a VAR engine with three key features. 1) A differential visual attention amplifier (DVAA) efficiently suppresses attention noise and identifies the significant elements. Then, speculative results are utilized to selectively skip model computing, achieving 1.58× speedup at the system level. 2) A full-path optimized MXINT PE is implemented considering the data distribution of the visual task, including a compression-aware multiplier, a combining-like-term adder tree, and an exponent partitioning hot-cold accumulator, which improves the energy efficiency by 1.70× altogether. 3) A chain-reaction-like parallel generation (CRPG) technique takes advantage of visual spatial correlations and then generates output pixels in parallel, reducing the generation latency by 23.7×.

Figure 18.5.2 shows the overall architecture of the chip. It consists of three key cores and other peripherals. The differential amplifier core (DAC) supports DVAA and guides selective execution in formal generation. The DAC includes 2 Log predictors to evaluate the approximate attention score, then a differentiator outputs the differential attention. Next, a Top-K Sort Core (TSC) produces a Top-K mask based on the differential result. The mask is utilized to accelerate each iteration, i.e., the idea is to only compute the steps that have a meaningful impact on final outputs. The formal acceleration core (FAC) is responsible for guided iteration. It consumes the Top-K mask from the DAC to accelerate execution. The FAC includes 4 PE clusters, each with 8 PE pairs. One pair consists of a left/right PE for 16 parallel multiply-and-accumulation (MAC) operations. The partial results from the PE pair are then accumulated in a shared floating-point (FP) accumulator. Finally, the TSC sorts the outputs of the FAC, yielding Top-K outputs in parallel.

Figure 18.5.3 presents the detailed design of DVAA. To distinguish between essential outputs and noise, the chip integrates a differential amplifier core. By way of analogy to a common-mode amplifier, the core computes the attention and differentiates two halved outputs in the embedding dimension. The attention noise is thus suppressed, making the Top-K elements more obvious. Compared with the non-differential mechanism, the DVAA improves the ratio of $RMS_{Top-64}/RMS_{Non-Top-64}$ by 3.21×. To lower the overhead introduced by DVAA, the Log PE is adopted to approximate the computation of attention scores. As the objective is to identify the Top-K elements rather than to obtain precise values, the Log PE operates solely on the leading-1 of each element. This enables the use of only shift-and-add operations, eliminating multipliers. Moreover, the softmax is substituted with S_2Max, a base-2 implementation that avoids non-linear operations [8]. The approximate scores are differentiated and then sorted by the TSC to generate the Top-K mask.

Guided by the mask, solely the computation associated with the generation of Top-K elements needs to be precisely executed, while the remaining operations can be approximated or bypassed. Specifically, QKV generation can be skipped if the entire row/column of the Top-K mask is 0, as these QKV do not contribute to the final attention score. Next, the mask specifies the valid QKV pairs for matrix multiplication, enabling sparse-aware acceleration. The conventional approach searches for valid data in a row-by-row manner, but incurs redundant memory access. To enhance PE utilization and reduce

memory access, a differential allocator distributes valid data to the PE according to the Top-K mask. First, the leftmost column (K_0) is used as a reference. A differential overlap index (DOI) computation is performed between the reference and other masks. The mask with the highest DOI, exhibiting the greatest difference from the reference, is selected (K_5) and merged to form the new reference (K_{merge}) for subsequent merging. If a conflict arises, the mask table is updated. Two merge iterations are performed, as it leads to >90.1% PE utilization and <49.6% conflict. Finally, in the FFN layer, the number of valid elements in the mask determines the precision of multipliers engaged in the computation to optimize energy efficiency. Overall, the Top-K mask reduces computations by 47.2%, and improves energy efficiency by 1.93×, including the differential speculation.

Figure 18.5.4 illustrates the full-path optimized MXINT PE, which facilitates the MAC during the formal iteration computing. According to the valid mask distribution, a compression-aware Dadda multiplier (CDM) activates different precisions. More valid masks indicate a higher attention score and necessitate more precise computing. Therefore, a triple-mode 4:2 compressor is leveraged to tune its state after comparison with predefined threshold: It dynamically switches among the precise and high/low approximate modes, reducing power consumption by 1.21×. In a conventional Dadda multiplier, after the stages of 4:2 compression, the final stage produces two vectors that must be accumulated by a vector adder and then merged by an adder tree. Since one has a width of 15b and the other 14b, a 15b vector adder is required for accumulation before merging, which results in inefficient bit-width utilization. Therefore, two interleaving combining-like-term adder trees (CATs) separately merge the 14b and 15b vectors from the final stage, improving the area efficiency by 4.7%.

For results generated across different cycles, an FP32 accumulator is required to perform exponent (-126-127) pre-alignment and mantissa shift-accumulation. Empirically, we observed that exponents are concentrated within a limited range ($\mu=0.160$, $\sigma=2.308$), so a wide shifter/adder is seldom activated. To accommodate the distribution, multiple blocks partition the exponent range. During accumulation, the newly generated result is accumulated with the partial sum of the block according to its exponent. Since they are close in value, only a 4b narrow shifter and an integer adder are required. The hot INT adder is activated for each accumulation, while a cold FP adder is activated only in the scenario of overflow or accumulation of different blocks. Instead of including blocks that can handle the full exponent range, a ring buffer dynamically stores 8 blocks, with each covering 4 exponents. When a new maximum exponent (E_{max}) appears, it replaces the block with the minimum exponent, achieving a >32 dynamic exponent interval. As the exponent gap exceeds 32, the resulting error is below 2^{-31}. The partitioning technique lowers the activity of the FP adder by 6.85×, and the power of accumulation by 1.38×. The inactive FP adder can be shared by PE pairs as there are few conflicts, further lowering area overhead.

Figure 18.5.5 details the chain-reaction-like parallel generation. Sequential generation produces one pixel based on all generated pixels. However, image data inherently contains substantial redundancy and correlation. In this work, previously generated pixels serve as guidance to trigger the simultaneous generation of multiple correlated pixels. The key process of parallel generation is selecting the Top-K generated pixels. Compared with full sorting, Top-K sorting only needs to identify elements whose probability ranks within the Top-K set. Accordingly, bit-slice sorting is employed. The highest 2b of Input first compares with the four possible values of a 2b segment, and a counter records the number of matches. Once the number of maximum-value matches exceeds K, the comparison can be terminated early, and a filter selects all qualifying items. The process then iterates to the next lower bit segment until the exact K items are left. This approach exhibits higher efficiency for large K. Moreover, it can be further accelerated by constructing a 'larger first' skewed sequence, so the threshold is reached earlier, triggering early exit.

Owing to spatial correlation, as more pixels are generated within a region, the higher the probability that neighboring pixels belong to the Top-K set. But potential Top-K candidates are widely scattered. Therefore, a skewed parallel generation scheduler is employed to generate a skewed sequence. The scheduler counts the number of generated pixels within each row of divided blocks. The block with the highest confidence is selected first, and the row with the highest confidence is then forwarded to the sorting unit. The block confidence is subsequently updated. With the scheduler's approximate prioritization, a skewed sequence is formed. A Top-K sort unit processes this sequence with parallel comparators and a filter [9]. The latency of Top-K sorting is improved by 1.16× with a skewed scheduler. Owing to the Top-K sorting, the parallel generation accelerates the generation speed by 23.7×.

Figure 18.5.6 shows the measurement results. Owing to the adoption of the differential speculative technique and the parallel generation technique, the chip accelerates the generation by 37.6×, and maintains <0.6% FID loss compared with the baseline. Moreover, the hardware accommodates the data distribution characteristics of the visual tasks, avoiding redundant activities, and achieves a system energy efficiency of 47.3TFLOPs/W and area efficiency of 2.75TFLOPs/mm². The chip is fabricated in 28nm technology and occupies an area of 5.76mm², achieving the peak frequency at 400MHz. The die photo is shown in Fig. 18.5.7.

Acknowledgement:
This work was supported in part by the NSFC under Grant 92464302, Grant 62125403, Grant U24B20164 and Grant 92164301; in part by the National Key Research and Development Program under Grant 2023YFB4403100; in part by Shanghai Municipal Science and Technology Major Project; in part by the Natural Science Foundation of Jiangsu Province Basic Research Program under Grant BK20243042; in part by the Beijing National Research Center for Information Science and Technology; and in part by the Beijing Advanced Innovation Center for Integrated Circuits. Corresponding authors: Shouyi Yin (yinsy@tsinghua.edu.cn) and Yang Wang (wangyang_imec@mail.tsinghua.edu.cn).

References:
[1] S. Kim et al., "EdgeDiff: 418.4mJ/Inference Multi-Modal Few-Step Diffusion Model Accelerator with Mixed-Precision and Reordered Group Quantization," *ISSCC*, pp. 410-412, 2025. https://doi.org/10.1109/ISSCC49661.2025.10904594
[2] S.-W. Hsieh et al., "MAE: A 3nm 0.168mm² 576MAC Mini AutoEncoder with Line-Based Depth-First Scheduling for Generative AI in Vision on Edge Devices," *ISSCC*, pp. 414-416, 2025. https://doi.org/10.1109/ISSCC49661.2025.10904763
[3] D. Han et al., "MEGA.mini: A Universal Generative AI Processor with a New Big/Little Core Architecture for NPU," *ISSCC*, pp. 1-3, 2025. https://doi.org/10.1109/ISSCC49661.2025.10904514
[4] R. Guo et al., "A 28nm 74.34TFLOPS/W BF16 Heterogenous CIM-Based Accelerator Exploiting Denoising-Similarity for Diffusion Models," *ISSCC*, pp. 362-364, 2024. https://doi.org/10.1109/ISSCC49657.2024.10454308
[5] P. Dong et al., "A 28nm 0.22μJ/Token Memory-Compute-Intensity-Aware CNN-Transformer Accelerator with Hybrid-Attention-Based Layer-Fusion and Cascaded Pruning for Semantic-Segmentation," *ISSCC*, pp. 408-410, 2025. https://doi.org/10.1109/ISSCC49661.2025.10904793
[6] S. Kim, et al., "Slim-Llama: A 4.69mW Large-Language-Model Processor with Binary/Ternary Weights for Billion-Parameter Llama Model," *ISSCC*, pp. 421-423, 2025. https://doi.org/10.1109/ISSCC49661.2025.10904761
[7] S. Moon, et al., "T-REX: A 68-to-567μs/Token 0.41-to-3.95μJ/Token Transformer Accelerator with Reduced External Memory Access and Enhanced Hardware Utilization in 16nm FinFET," *ISSCC*, pp. 406-408, 2025. https://doi.org/10.1109/ISSCC49661.2025.10904499

[8] J. R. Stevens, et al., "Softmax: Hardware/Software Co-Design of an Efficient Softmax for Transformers," *IEEE/ACM DAC*, pp. 469-474, 2021. https://doi.org/10.1109/DAC18074.2021.9586134
[9] E. Stehle, et al., "A Memory Bandwidth-Efficient Hybrid Radix Sort On GPUs," *ACM International Conf. on Management of Data*, pp. 417-432, 2017. https://doi.org/10.1145/3035918.3064043

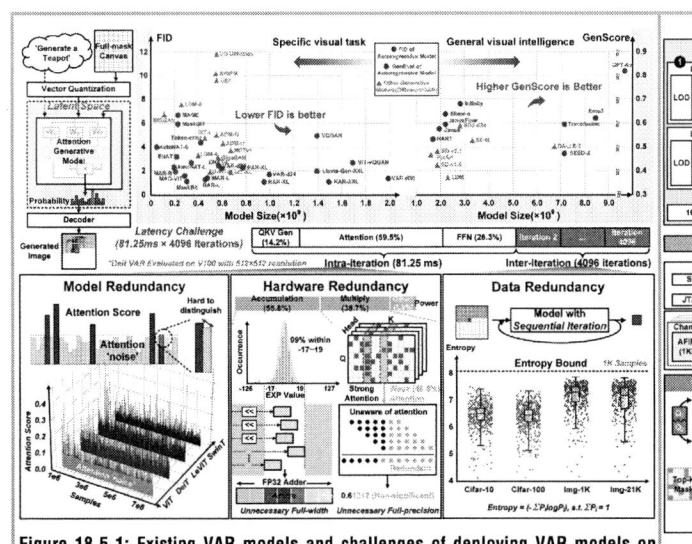

Figure 18.5.1: Existing VAR models and challenges of deploying VAR models on hardware.

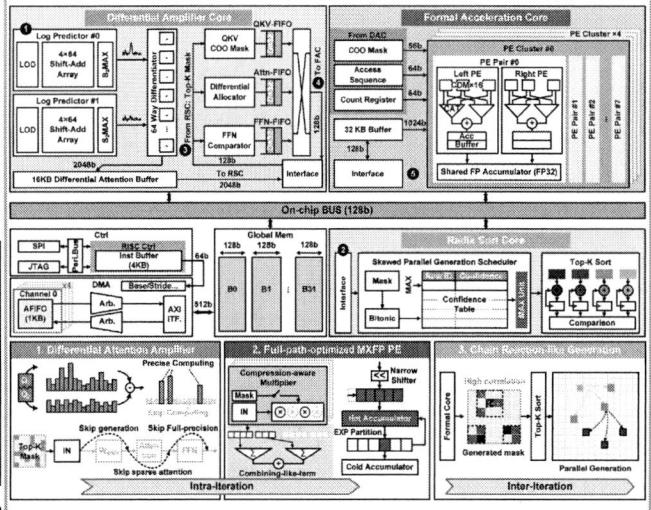

Figure 18.5.2: Overall architecture of the chip and three special features.

ISSCC 2026 / SESSION 18 / TECHNOLOGY AND CIRCUITS FOR DOMAIN-SPECIFIC ACCELERATORS / 18.5

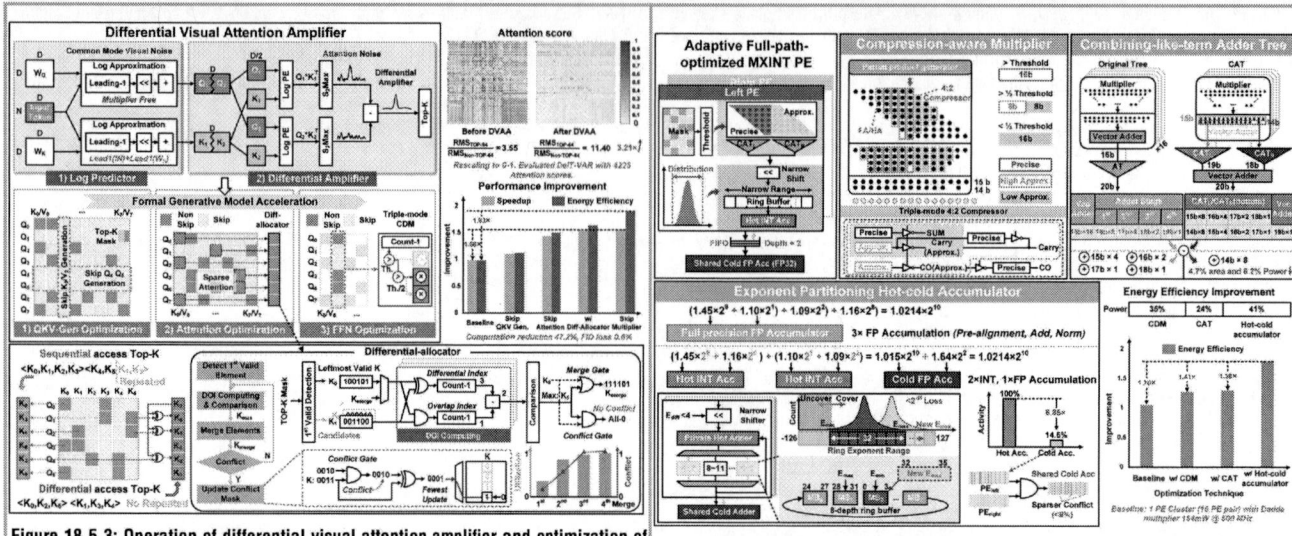

Figure 18.5.3: Operation of differential visual attention amplifier and optimization of model processing.

Figure 18.5.4: Design of data distribution-aware full path optimized MXINT PE.

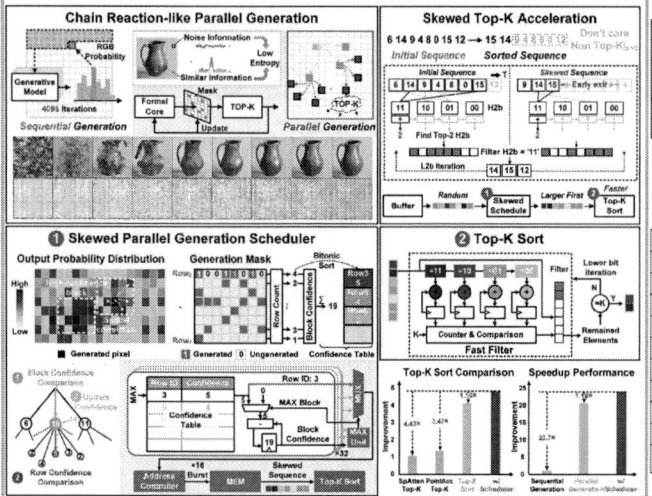

Figure 18.5.5: Implementation of chain-reaction-like parallel generation and skewed sort acceleration.

Figure 18.5.6: Measurement results and comparison table.

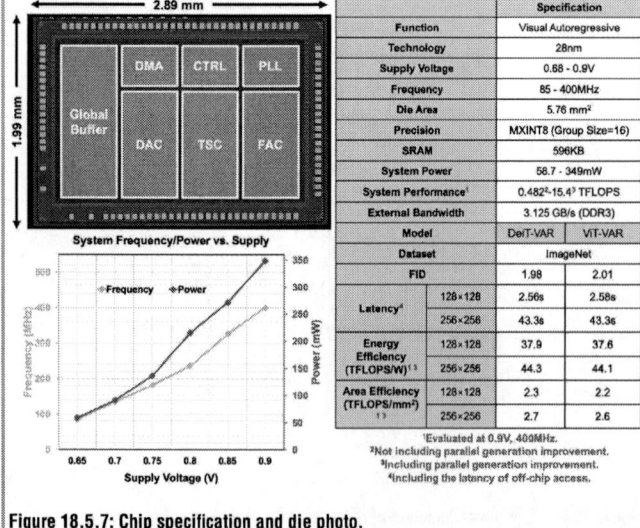

Figure 18.5.7: Chip specification and die photo.

• 2026 IEEE International Solid-State Circuits Conference

ISSCC 2026 / SESSION 19 / HIGH-VOLTAGE, ISOLATED AND DISPLAY POWER / OVERVIEW

Session 19 Overview: *High-Voltage, Isolated and Display Power*

POWER MANAGEMENT SUBCOMMITTEE

Session Chair: Xugang Ke
Zhejiang University
Zhejiang, China

Session Co-Chair: Yuan Gao
Southern University of Science and Technology
Shenzhen, China

High Voltage, Isolated and Display Power are crucial for ensuring high performance, high safety and reliability in harsh environments, such as electric vehicles and industrial automation. This session aims to showcase some of the best work in the high voltage, isolated and display power areas, featuring novel topologies that offer high conversion efficiency, multi-outputs, and advanced power regulation for isolated converters, as well as precise gate control techniques for SiC gate drivers.

1:30 PM

19.1 Piggybacked SC-on-CSCR: A Modular On-Chip Switched-Capacitor Converter for 12-to-60V Input 1.8-to-5V Output Achieving 5.67mW/mm² Power Density and 71.5% Peak Efficiency

Yunzhe Yang, Delft University of Technology, Delft, The Netherlands

In Paper 19.1, Delft University of Technology presents a modular power conversion architecture with SIDO piggybacked SC-on-CSCR module, which eliminates all HV switches and adequately uses the capacitor resources on chip to achieve a large VCR, high efficiency, and a peak power density of 5.67mW/mm² at 12V and 1.77mW/mm² at 48V.

1:55 PM

19.2 A Three-Mode Single-Inductor Four-Quadrant Converter Achieving 94.6% Peak Efficiency with Seamless Zero-Crossing

Peng Cao, Nanjing University, Suzhou, China, Fudan University, Shanghai, China

In Paper 19.2, Nanjing University and Fudan University introduce a three-mode single-inductor four-quadrant converter that enables full four-quadrant operation via buck/boost, inverting buck/boost and auxiliary transition modes. The design achieves 94.6% peak efficiency, with a load current range of −3A to +3A, and seamless zero-crossing and mode switching.

2:20 PM

19.3 A 94.8%-Peak-Efficiency Double Step-Up SIBO Converter Achieving 88% Output Ripple Reduction for AMOLED Display

Hyeon-Ji Choi, Sogang University, Seoul, Korea

In Paper 19.3, Sogang University presents a double step-up single-inductor bipolar-output (DSU-SIBO) converter for AMOLED displays. By employing a 1:2 charge pump, the design reduces both DC and AC components of the inductor current, reducing output ripple by up to 88%, minimizing power loss and achieving a peak efficiency of 94.8%.

318 • 2026 IEEE International Solid-State Circuits Conference

979-8-3315-8937-0/26 $31.00 © 2026 IEEE

2:45 PM

19.4 A Digital-Feedback Active-Gate-Driver IC for 600A 1200V SiC MOSFETs Supporting High- and Low-Side Drive with Simultaneous dV_{ds}/dt Control and V_{ds} Surge Suppression Enabled by Miller Capacitance Calibration

Shusuke Kawai, Toshiba, Kawasaki, Japan

In Paper 19.4, Toshiba presents a digital feedback active gate driver IC applicable to both high- and low-side 600A, 1200V SiC MOSFETs. The design uses a parasitic Miller capacitance calibration circuit and digital feedback for simultaneous turn-on/off, dv/dt control and Vds surge suppression, achieving 58% surge voltage and 66% switching loss reduction.

3:00 PM

19.5 A Binary-Weighted Switched-Capacitor Gate Driver IC for Overcoming Trade-offs Between Driving Loss and Delay Time with Gate-Current Feedback Achieving 85% Driving Loss Reduction

Kohei Horii, Toshiba, Kawasaki, Japan

In Paper 19.5, Toshiba presents a binary-weighted switched-capacitor SiC gate driver IC featuring 9-level gate-voltage segmentation with enhanced charge supply capability and adaptive state-interval control. The design achieves up to an 85% reduction in gate-driving loss.

3:35 PM

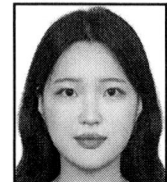

19.6 A 68%-Peak-Efficiency Single-Transformer Multi-Output Isolated DC-DC Converter with a Regulated Negative Rail

Qiao Huang, University of Science and Technology of China, Hefei, China

In Paper 19.6, University of Science and Technology of China, presents a single-transformer multi-output isolated DC-DC converter, demonstrating concurrent generation of +15V, −5V, and +5V outputs using a dual-doubler rectifier with only four power transistors. The converter achieves 68% peak efficiency, and 2.5W maximum output power.

4:00 PM

19.7 A Hybrid Bipolar-Output Isolated Converter with +15V/−5V Outputs for SiC Gate Drivers

Joo-Mi Cho, Sogang University, Seoul, Korea

In Paper 19.7, Sogang University presents a hybrid bipolar-output (+15V/−5V) isolated gate-driver power supply that combines a flyback-based TX with primary-side regulation and a switched-capacitor-based RX. With a compact 1:1 coupled inductor and all-5V-CMOS devices, the design achieves 89.23% peak efficiency at 3W.

4:25 PM

19.8 A Fully Integrated Bidirectional 5-Level Isolated DC-DC Converter with 42.5% Efficiency and 170mW/mm² Transformer Power Density

Kishalay Datta, Dartmouth College, Hanover, NH

In Paper 19.8, Dartmouth College presents a fully-integrated 5kV-isolated DC-DC converter. The design uses a distributed multi-winding transformer to maximize utilization of the standard on-chip metal stackup and a 5-level hybrid full-bridge power stage, achieving a peak efficiency of 42.5% and a system power density of 69mW/mm².

4:50 PM

19.9 A 2.15W 120V/230Vac to 5-to-12Vdc Offline Power Converter with Full-Duty-Cycle Input-Series Dual-Branch Converter Achieving 1088mW/cm³ and 87.2% Peak Efficiency

Gang Liu, Chinese University of Hong Kong, Shenzhen, China, Southern University of Science and Technology, Shenzhen, China

In Paper 19.9, Chinese University of Hong Kong, Southern University of Science and Technology, and Delft University of Technology present a 120V/230Vac to 5-to-12Vdc offline power converter. It employs a full duty cycle input-series dual-branch converter as the second stage, achieving 87.2% peak efficiency, 2.15W output power, and 1088mW/cm³ power density.

ISSCC 2026 / SESSION 19 / HIGH-VOLTAGE, ISOLATED AND DISPLAY POWER / 19.1

19.1 Piggybacked SC-on-CSCR: A Modular On-Chip Switched-Capacitor Converter for 12-to-60V Input 1.8-to-5V Output Achieving 5.67mW/mm² Power Density and 71.5% Peak Efficiency

Yunzhe Yang[1], Xin Zhang[2], Sijun Du[1]

[1]Delft University of Technology, Delft, The Netherlands, [2]IBM T. J. Watson Research Center, Yorktown Heights, NY

Abstract

This paper presents an on-chip switched-capacitor converter, named piggybacked SC-on-CSCR, featuring: 1) a power density of 5.67mW/mm² enabled by eliminating high-voltage switches and jointly utilizing MOM and MOS capacitors; 2) a 12-to-60V input and 1.8-to-5V output achieved through a chip-level modular design; 3) 71.5% peak efficiency enabled by full soft-charging operation; and 4) fast high-voltage startup protection, verified under a 20V/ms power-on slope.

Electric vehicles (EVs) rely heavily on batteries, with typical voltage rails of 12V, 24V, 48V, and 60V [1-3]. Many low-power auxiliary electronics, such as battery and environmental sensors, can be directly powered from these rails without burdening the main converter. Switched-capacitor (SC) DC-DC converters are attractive for this task due to their compactness and full integrability [4-8]. However, conventional SC topologies face challenges at high input voltages, where high-voltage (HV) switches are required, degrading both efficiency and power density [9-11]. A stacked SC converter with single-input-dual-output (SIDO) stages was proposed [12] (Fig. 19.1.1, bottom-left), which reduces per-stage voltage stress and achieves 42V-to-3V conversion. While effective, it still suffers from several drawbacks. At the system level, its input reconfiguration block relies on numerous HV LDMOS switches, reducing the efficiency. Within each SIDO core, switches experience >5V stress and flying capacitors (C_{FLY}) undergo bottom-plate (BP) hard charging, both of which further reduce efficiency. Soft-charging techniques, such as those used in continuously-scalable-conversion-ratio (CSCR) converters [13-18], mitigate BP loss by lowering voltage steps (ΔV) and incrementally shifting the phase. This effectively increases efficiency across a wide voltage conversion ratio (VCR). However, it is challenging for conventional CSCR converters to handle HV inputs, as their switches must withstand the voltage difference between input and output and therefore still require HV devices. Moreover, startup remains critical as a reliable operation must be ensured when the input voltage ramps up with a steep slope at power-on.

In this paper, a modular power conversion architecture based on a SIDO piggybacked SC-on-CSCR module is proposed (Fig. 19.1.2), which eliminates the need for HV switches while efficiently utilizing chip capacitor resources to achieve a large VCR, high efficiency, and high power density. By flexibly configuring chip-level SIDO modules (Fig. 19.1.2, top-left), the system adapts to different input voltages (V_{IN}) without requiring LDMOS devices. To support high V_{IN}, the proposed stacking architecture maintains efficiency even as more stages are added, unlike conventional series cascading, thereby extending the V_{IN} range with a minimal efficiency penalty. As shown in Fig. 19.1.2 (top-middle), the SIDO piggybacked SC-on-CSCR combines HV MOM capacitors and low-voltage (LV) MOS capacitors into two sub-converters: an HV MOM-CSCR stage and a 1:1 LV MOS-SC stage. The MOM-CSCR deploys a dual-output topology, with its higher output V_M serving as the input for the MOS-SC. Compared with using only the MOM-CSCR, adding the MOS-SC stage effectively doubles the voltage step, halves the required number of stacked chips, and improves power density. To realize soft-charging in the MOS-SC without generating additional intermediate rails, the MOS-SC piggybacks on the MOM-CSCR: their top plates are connected, and the MOS-SC follows the same soft-charging steps as the MOM-CSCR, like a child (MOS) riding on a MOM's back (Fig. 19.1.2 top-right). Finally, an HV startup protection scheme is designed to ensure safe operation when the system is connected to a steeply rising high-voltage V_{IN} at power-on, without using any extra hardware.

The diagram and the operating principle of the piggybacked SC-on-CSCR are illustrated in Fig. 19.1.2 (bottom). Each interleaved SC-on-CSCR cell contains both MOM and MOS capacitors. The MOM-CSCR stage (blue) generates intermediate voltage rails (T_{1-6}, B_{1-6}), enabling soft-charging not only for itself but also for the piggybacked MOS-SC stage (red). As a result, all switches operate under 5V stress, eliminating the need for HV devices. Furthermore, leveraging the inherently continuous and wide VCR range of the CSCR topology, the proposed converter supports V_{OUT} from 1.8-to-5V.

Figure 19.1.3 (top-left) shows the architecture of the fully integrated piggybacked SC-on-CSCR chip. It integrates 27 SC-on-CSCR cells, implementing 54-phase interleaving with the outphasing technique [19]. Each cell contains two capacitors (MOM+MOS), 18 switches ($S_{T<0:7>}$, $S_{M<0:1>}$, and $S_{B<0:7>}$), non-overlap (NOP) logic, and level shifters (1.8-to-5V). The chip has two HV domains using V_M and V_{O1} as grounds (referred to as the V_M and V_{O1} domains). The supply generators provide local supplies (V_{DD_H}: V_M+4.5V, V_M+1.6V, V_{O1}+4.5V, and V_{O1}+1.6V), enabling reliable operation of both 5V and 1.8V transistors in each domain. HV level shifters convert CLK and RST signals into V_M and V_{O1} domains, or keep these signals synchronized in the normal LV domain. Their outputs drive finite state machines (FSMs), which generate the phase signals for each SC-on-CSCR cell. Figure 19.1.3 (middle-left) illustrates the HV level shifter [20], consisting of 2 MOM capacitors ($C_{LS\_1,2}$) and 6 inverters. When its input (IN) flips, the currents flowing through C_{LS_1} and C_{LS_2} change the state of the latch in the HV isolation well, thereby flipping the output (OUT). Figure 19.1.3 (bottom-left) shows the implementation of supply generators, which use HV level shifters and NOP generators to output the non-overlapping driving signals for S_{1-4} with two phases (φ_{1-2}). At φ_1, S_1 and S_3 are turned on, allowing V_{SS_H} to charge a capacitor C_{SUP_GEN}. At φ_2, S_2 and S_4 are turned on, and V_{DD_L} together with the stored charge in C_{SUP_GEN} boosts the output node V_{DD_H}.

The system incorporates a dedicated HV startup protection mechanism (Fig. 19.1.3, right) to safeguard transistors during sudden power-on events. The key idea is to create always-on conduction paths from V_I to critical nodes in the V_M and V_{O1} domains whenever overstress risks arise, regardless of the operating phase, acting as dynamic clamps. This chip implements four such fast conduction paths. Path (1): in each SC-on-CSCR cell, when V_I-V_M rises rapidly to about 5.3V, the PMOS in the transmission gate $S_{T<0>}$ is automatically turned on, its pre-stage is supplied by V_M+4.5V. This action charges the capacitor top plates (V_{TOP}). Path (2): at any instant, some cells at phases where $S_{T<7>}$ is on, pulling V_M to V_{TOP}, with $T_{<1:6>}$ similarly pulled. Paths (3) and (4) operate in the same way as paths (1) and (2), ensuring V_{O1} is also pulled up. Together, these conduction paths clamp internal voltages below 5.5V, which is within the safe limit of standard 5V MOSFETs. Functionally, they act like built-in Zener diodes, but without consuming the chip area required by discrete devices.

The proposed chip was fabricated in a 0.18μm BCD process, with 1.48mm² active area (Fig. 19.1.7). Figure 19.1.4 (top-left) shows the measured startup and steady-state waveforms of a single-stage configuration with V_{IN} = 12V and a power-on slope of 20V/ms. In this case, V_M and V_{OUT} settle at 8.8V and 5V, respectively. Figure 19.1.4 (top-right) shows the startup waveforms of a 6-stage configuration with V_{IN} = 48V under the same power-on slope (20V/ms) and a target V_{OUT} of 1.8V. The signals $V_{O1\_1}-V_{O1\_5}$ represent the V_{O1} outputs of chips 1–5 in the stacked system, all of which stabilized at the intended voltage levels without requiring decoupling capacitors. Their zoomed-in waveforms at power-on are shown in Fig. 19.1.4 (bottom-left). Even under the steep V_{IN} ramp, all 5V on-chip devices remain within safe limits, demonstrating the effectiveness of the HV startup protection scheme.

Figure 19.1.4 (bottom-right) shows the efficiency versus V_{OUT} at a 10MHz clock frequency for V_{IN} = 12/24/48/60V with 1-/3-/6-/8-stage SIDO stacking. The system exhibits a relatively flat efficiency curve across the wide input and output ranges, enabling broad applicability. The peak efficiency reaches 71.5% at 12V input (single-chip) and 51.6% at 48V input (six-chip stacking). Figure 19.1.5 (top) further plots efficiency versus V_{IN} (12-to-60V) for 3.3V and 5V outputs under proper stacking configurations at 10MHz. Thanks to the SIDO stacking topology, efficiency degradation remains minimal even with eight stacked chips. Figure 19.1.5 (bottom) shows efficiency versus output current (I_{OUT}) for 12V and 48V inputs with 3.3V and 5V outputs. Without replying on any off-chip components for power conversion, the system achieves a peak power density of 5.67mW/mm² for a single-chip (12V input), and 1.77mW/mm² for a six-chip configuration (48V input).

Figure 19.1.6 summarizes the performance of the proposed SC-on-CSCR converter and compares it with prior SC converters operating in similar voltage ranges [21]. The proposed configurable topology provides wide input and output ranges. In addition, the HV startup protection paths prevent the system from overstressing, which was often overlooked in the prior art. By efficiently leveraging both MOM and MOS capacitors, the chip achieves a high capacitance density of 845.9pF/mm². All switches operate below 5V, eliminating the need for HV devices. This work achieves full soft charging across all capacitors. With these advantages, the proposed converter achieves higher maximum P_{OUT}, peak efficiency, and power density compared with prior related works under comparable operating conditions.

References:
[1] H. -J. Choi et al., "A 92.7% Peak Efficiency 12V-to-60V Input to 1.2V Output Hybrid DC-DC Converter Based on a Series-Parallel-Connected Switched Capacitor," *2024 IEEE International Solid-State Circuits Conference (ISSCC)*, San Francisco, CA, USA, 2024, pp. 156-158. http://doi.org/10.1109/ISSCC49657.2024.10454344
[2] Y. -H. Kao et al., "A 48V-to-5V Buck Converter with Triple EMI Suppression Circuit Meeting CISPR 25 Automotive Standards," *2024 IEEE International Solid-State Circuits Conference (ISSCC)*, San Francisco, CA, USA, 2024, pp. 164-166. http://doi.org/10.1109/ISSCC49657.2024.10454539

[3] M. Ashourloo *et al.*, "A Masterless Fault-Tolerant Hybrid Dickson Converter with 95.3% Peak Efficiency 20V-to-60V Input and 3.3V Output for 48V Multi-Phase Automotive Applications," *2021 IEEE International Solid-State Circuits Conference (ISSCC)*, San Francisco, CA, USA, 2021, pp. 258-260. http://doi.org/10.1109/ISSCC42613.2021.9366016

[4] J. Jiang, X. Liu, W. -H. Ki, P. K. T. Mok and Y. Lu, "Circuit Techniques for High Efficiency Fully-Integrated Switched-Capacitor Converters," in *IEEE Transactions on Circuits and Systems II: Express Briefs*, vol. 68, no. 2, pp. 556-561, Feb. 2021. http://doi.org/10.1109/TCSII.2020.3046514

[5] J. Liu and S. Gregori, "Switched-Capacitor Boost-Buck Ladder Converters With Extended Voltage Range in Standard CMOS," in *IEEE Transactions on Circuits and Systems I: Regular Papers*, vol. 67, no. 12, pp. 4593-4606, Dec. 2020. http://doi.org/10.1109/TCSI.2020.3028268

[6] Y. Lu *et al.*, "A 123-phase DC-DC converter-ring with fast-DVS for microprocessors," *2015 IEEE International Solid-State Circuits Conference - (ISSCC) Digest of Technical Papers*, San Francisco, CA, USA, 2015, pp. 1-3. http://doi.org/10.1109/ISSCC.2015.7063077

[7] T. M. Andersen *et al.*, "A sub-ns response on-chip switched-capacitor DC-DC voltage regulator delivering 3.7W/mm2 at 90% efficiency using deep-trench capacitors in 32nm SOI CMOS," *2014 IEEE International Solid-State Circuits Conference Digest of Technical Papers (ISSCC)*, San Francisco, CA, USA, 2014, pp. 90-91. http://doi.org/10.1109/ISSCC.2014.6757351

[8] H. -P. Le, S. R. Sanders and E. Alon, "Design Techniques for Fully Integrated Switched-Capacitor DC-DC Converters," in *IEEE Journal of Solid-State Circuits*, vol. 46, no. 9, pp. 2120-2131, Sept. 2011. http://doi.org/10.1109/JSSC.2011.2159054

[9] H. Meyvaert, G. Villar Piqué, R. Karadi, H. J. Bergveld and M. S. J. Steyaert, "A Light-Load-Efficient 11/1 Switched-Capacitor DC-DC Converter With 94.7% Efficiency While Delivering 100 mW at 3.3 V," in *IEEE Journal of Solid-State Circuits*, vol. 50, no. 12, pp. 2849-2860, Dec. 2015. http://doi.org/10.1109/JSSC.2015.2461600

[10] R. Rothe *et al.*, "A uW Output Power, >100V, Single-Capacitor Switched DC-DC Up/Down Converter," *2024 IEEE Symposium on VLSI Technology and Circuits (VLSI Technology and Circuits)*, Honolulu, HI, USA, 2024, pp. 1-2. http://doi.org/10.1109/VLSITechnologyandCir46783.2024.10631439

[11] T. Van Daele and F. Tavernier, "Fully Integrating a 400 V-to-12 V DC-DC Converter in High-Voltage CMOS," in *IEEE Journal of Solid-State Circuits*, vol. 58, no. 3, pp. 732-741, March 2023. http://doi.org/10.1109/JSSC.2022.3223900

[12] E. De Pelecijn and M. S. J. Steyaert, "Stacking Isolated SC Cores for High-Voltage Wide Input Range Monolithic DC-DC Conversion," in *IEEE Journal of Solid-State Circuits*, vol. 55, no. 10, pp. 2639-2648, Oct. 2020. http://doi.org/10.1109/JSSC.2020.3005795

[13] N. Butzen and M. Steyaert, "Design of Single-Topology Continuously Scalable-Conversion-Ratio Switched- Capacitor DC-DC Converters," in *IEEE Journal of Solid-State Circuits*, vol. 54, no. 4, pp. 1039-1047, April 2019. http://doi.org/10.1109/JSSC.2018.2884351

[14] N. Butzen *et al.*, "A Monolithic 26A/mm2Imax, 88.5% Peak-Efficiency Continuously Scalable Conversion-Ratio Switched-Capacitor DC-DC Converter," *2023 IEEE International Solid-State Circuits Conference (ISSCC)*, San Francisco, CA, USA, 2023, pp. 232-234. http://doi.org/10.1109/ISSCC42615.2023.10067583

[15] Y. Wang, M. Huang, Q. Chen, R. P. Martins and Y. Lu, "A VCF-Step-Reconfigurable Continuously Scalable-Conversion-Ratio Switched- Capacitor Converter," in *IEEE Journal of Solid-State Circuits*, vol. 60, no. 2, pp. 626-637, Feb. 2025. http://doi.org/10.1109/JSSC.2024.3414448

[16] Y. Wang, M. Huang, R. P. Martins and Y. Lu, "A SIDO/DISO VCF-Step-Reconfigurable Continuously Scalable-Conversion-Ratio SC Converter Achieving 91.4%/92.6% Peak Efficiency and Almost-lossless Channel Switching," *2024 IEEE International Solid-State Circuits Conference (ISSCC)*, San Francisco, CA, USA, 2024, pp. 506-508. http://doi.org/10.1109/ISSCC49657.2024.10454453

[17] A. Guo, W. Peng, Y. Yang, X. Hu, D. Muratore and S. Du, "A Fully Integrated SC Converter Hybridizing Dickson and Continuously-Scalable-Conversion-Ratio Topologies with a Wide Bipolar VCR Range for Energy Harvesting," *2025 Symposium on VLSI Technology and Circuits (VLSI Technology and Circuits)*, Kyoto, Japan, 2025, pp. 1-3. http://doi.org/10.23919/VLSITechnologyandCir65189.2025.11074934

[18] Y. Yang, W. Peng, M. Huang and S. Du, "Matryoshka CSCR: A Reconfigurable Matryoshka-Stacked Continuous-Scalable-Conversion-Ratio Switched-Capacitor DC-DC Converter with 0.1-to-1.7V Input," *2025 Symposium on VLSI Technology and Circuits (VLSI Technology and Circuits)*, Kyoto, Japan, 2025, pp. 1-3. http://doi.org/10.23919/VLSITechnologyandCir65189.2025.11075046

[19] N. Butzen and M. S. J. Steyaert, "Design of Soft-Charging Switched-Capacitor DC–DC Converters Using Stage Outphasing and Multiphase Soft-Charging," in *IEEE Journal of Solid-State Circuits*, vol. 52, no. 12, pp. 3132-3141, Dec. 2017. http://doi.org/10.1109/JSSC.2017.2733539

[20] D. Lutz, A. Seidel and B. Wicht, "A 50V, 1.45ns, 4.1pJ High-Speed Low-Power Level Shifter for High-Voltage DCDC Converters," *ESSDERC 2018 - IEEE 44th European Solid State Circuits Conference (ESSCIRC)*, Dresden, Germany, 2018, pp. 126-129. http://doi.org/10.1109/ESSCIRC.2018.8494292

[21] Y. Ismail, H. Lee, S. Pamarti and C. -K. K. Yang, "A 36-V 49% Efficient Hybrid Charge Pump in Nanometer-Scale Bulk CMOS Technology," in *IEEE Journal of Solid-State Circuits*, vol. 52, no. 3, pp. 781-798, March 2017. http://doi.org/10.1109/JSSC.2016.2636876

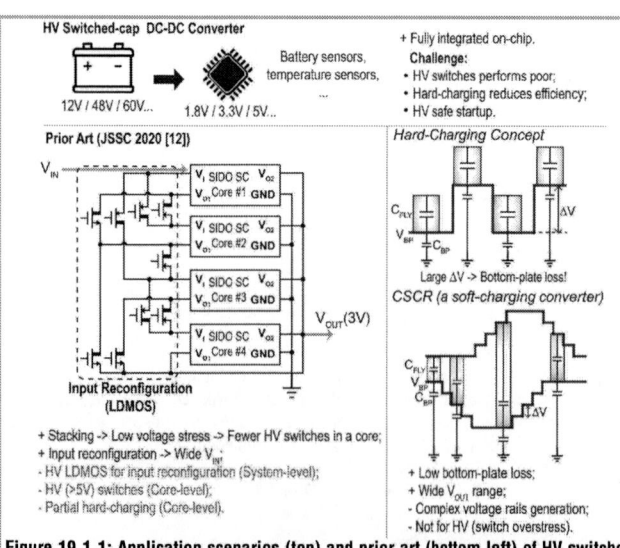

Figure 19.1.1: Application scenarios (top) and prior art (bottom left) of HV switched-capacitor converter, hard-charging and soft-charging concepts (bottom right).

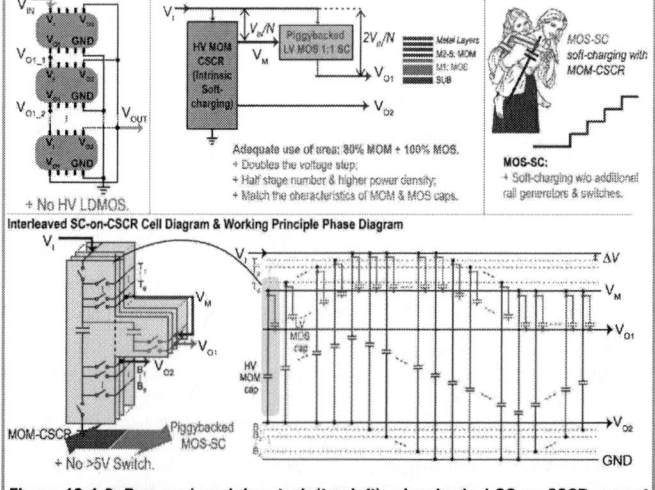

Figure 19.1.2: Proposed modular stack (top left), piggybacked SC-on-CSCR current flowing diagram (top middle) and concept (top right), and working principle (bottom).

ISSCC 2026 / SESSION 19 / HIGH-VOLTAGE, ISOLATED AND DISPLAY POWER / 19.1

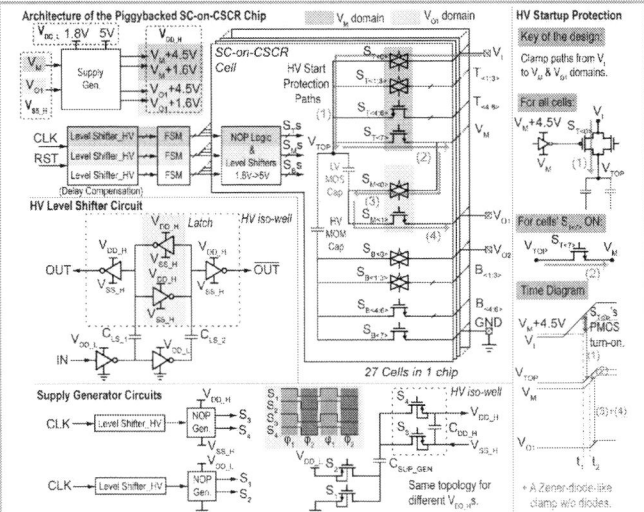

Figure 19.1.3: Architecture of the piggybacked SC-on-CSCR (top left), HV level shifter (middle left), supply generator (bottom left), and HV startup protection (right).

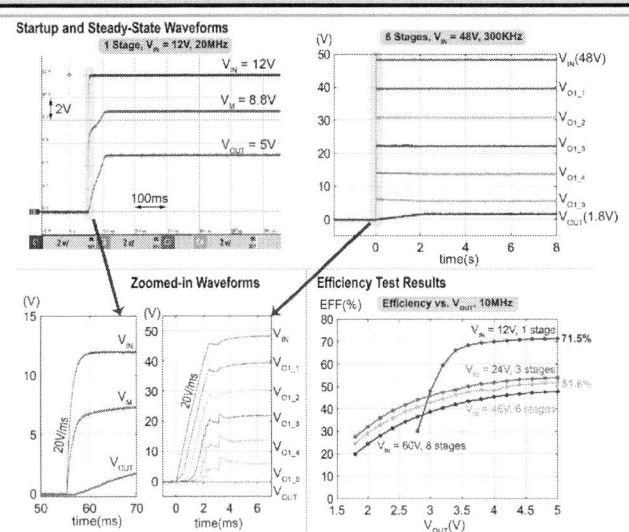

Figure 19.1.4: Measured startup and steady-state waveforms (top), zoomed-in startup waveforms (bottom left), and efficiency versus output voltage (bottom right).

Figure 19.1.5: Measured efficiency versus input voltage (top) and output current (bottom) of the proposed piggybacked SC-on-CSCR converter.

Figure 19.1.6: Performance summary and comparison with prior HV switched-capacitor converters.

	This work	JSSC '20 [12]	JSSC '17 [21]	VLSI '24(i) [10]	VLSI '24(ii) [10]	TCASI '20 [5]
Process	0.18μm	0.35μm	65nm	0.18μm	0.18μm	0.18μm
Area (mm²)	1.48	5.37	2.48*	23.2	23.2	0.31
V_{IN} (V)	12-60	7.5-42	2.25-2.75	30-180	27	1.8
V_{OUT} (V)	1.8-5	3	36	3-23	175	10.8
Has HV Startup Protection?	Yes	No	N/A#	No	No	N/A#
No. of off-chip Caps	0	0	0	1	1	0
On-chip Capacitance (pF)	356 (MOM) + 896 (MOS)	1300	8	260	260	623
Capacitance Density (pF/mm²)	845.9	242.1	3.2	11.2	11.2	2009.6
Using >5V Switches?	No	Yes	No	Yes	Yes	No
Soft-charging	Full	Partial	No	No	No	No
Max. P_{OUT} (mW)	5.67 (12V V_{IN}) / 15.7 (48V V_{IN})	2.15	2.66*	0.00269	0.01625	0.723
Peak Efficiency	71.5% (12V V_{IN}) / 51.6% (48V V_{IN})	68.3% (15V V_{IN}) / 50.3% (42V V_{IN})	49%	42.53%	62.14%	33%
Peak Power Density (mW/mm²)	5.67 (12V V_{IN}) / 1.77 (48V V_{IN})	0.4	1.07*	0.000116	0.0007	2.3

#: Boost converter doesn't need the start protection.
*: Estimated from papers.

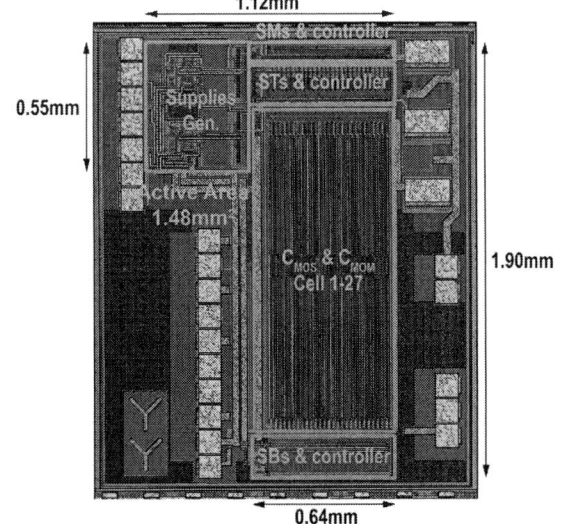

Figure 19.1.7: Chip micrograph.

• 2026 IEEE International Solid-State Circuits Conference

ISSCC 2026 / SESSION 19 / HIGH-VOLTAGE, ISOLATED AND DISPLAY POWER / 19.2

19.2 A Three-Mode Single-Inductor Four-Quadrant Converter Achieving 94.6% Peak Efficiency with Seamless Zero-Crossing

Peng Cao[1,2], Jiawei Xu[2], Zhiliang Hong[2]

[1]Nanjing University, Suzhou, China, [2]Fudan University, Shanghai, China

Abstract

This paper presents a three-mode single-inductor four-quadrant converter with 94.6% peak efficiency for electrochromic smart windows. It supports a 5 to 12V input, a –5 to +5V output and a –3 to 3A load current, achieving full four-quadrant operation by employing buck/boost, inverting buck/boost and auxiliary transition modes. This integrated converter enables seamless zero-crossing and mode switching with a single inductor, thus resulting in a compact, low-cost, and low-complexity system.

A four-quadrant power supply (FQPS) enables bidirectional energy transfer with full control of voltage polarity and current direction, thereby covering all four quadrants of the voltage-current plane (top-left of Fig. 19.2.1). This capability is increasingly indispensable for applications demanding reversible energy flow and dynamic load regulation, including electrochromic (EC) smart windows for automotive and aerospace systems. In EC devices, a dual-polarity voltage is essential to modulate optical transmittance by mediating Li$^+$ migration between the LiCoO$_2$ ion-storage layer and the WO$_3$ electrochromic layer (top-right of Fig. 19.2.1). A positive bias (1–3V) drives Li$^+$ intercalation into WO$_3$, forming Li$_x$WO$_3$ and inducing opacity in the glass. Conversely, reversing the bias facilitates Li$^+$ extraction, reverting the glass to its transparent state. In the absence of an applied voltage, the device's optical transmittance remains unchanged. This reversible electrochemical mechanism highlights the need of compact and efficient FQPS solutions for rapid switching and precise optical modulation in smart transportation systems.

Most conventional DC-DC converters, by contrast, are confined to first-quadrant operation with positive output voltage and current sourcing. While bidirectional topologies have been developed, they only extend operation to Quadrants I and IV, enabling both current sinking and sourcing at positive voltages [1-7]. The four-quadrant controller reported in [8] relies on large positive/negative intermediate voltages, external high-voltage-rated MOSFETs, and bulky coupled inductors. These drawbacks lead to compromised efficiency, significant output voltage ripple, and elevated system cost, which severely limits its suitability for compact EC systems. To date, no existing design achieves practical four-quadrant operation while simultaneously satisfying the stringent requirements on efficiency, integration, and output quality, underscoring the need for a compact and high-performance solution. Conventional bi-directional buck/boost converters inherently operate in Quadrants I/IV, while their inverting counterparts cover Quadrants II/III [9]. Leveraging this complementarity, this work proposes a single-inductor four-quadrant converter (SIFQC) that integrates both topologies and implements three auto-switched operating modes (bottom of Fig. 19.2.1). The proposed SIFQC achieves seamless four-quadrant operation with a reduced component count and compact footprint, providing an efficient and versatile solution for bipolar voltage regulation in EC glass applications.

Figure 19.2.2 illustrates the operating principles of the proposed SIFQC, which adopts three modes: buck/boost, inverting buck/boost, and auxiliary transition, to achieve complete four-quadrant operation. The buck/boost mode governs Quadrants I and IV, where the converter functions as a buck converter in Quadrant I with a conversion ratio (CR) of D, and operates as its current-reversed counterpart in Quadrant IV. Note that in this mode, S$_1$ and S$_3$ remain conducting, ensuring that the flying capacitor (C$_F$) stores the input voltage (V$_{IN}$) and pre-establishes the voltage required for potential mode transitions. The inverting buck/boost mode covers Quadrants II and III. In Quadrant III, the topology operates as an inverting buck converter: during Φ$_1$, switches S$_1$–S$_3$–S$_4$ charge C$_F$ while transferring inductor current to the output, and during Φ$_2$, switches S$_2$–S$_4$ energize the inductor, leading to a CR of D–1 with C$_F$ providing input-voltage inversion. Quadrant II follows the same sequence as Quadrant III but with reversed inductor current.

In both buck/boost and inverting buck/boost modes, the CR is unable to reach zero owing to minimum duty-cycle limitations, potentially causing voltage discontinuities. To address this, an auxiliary transition mode is introduced for operation near zero output voltage (bottom of Fig. 19.2.2). In this mode, the inductor terminal alternates between +V$_{IN}$ and –V$_{IN}$, yielding a CR of 2D–1 and enabling precise polarity control through duty-cycle modulation. Energy transfer is regulated by selectively activating the relevant switches (e.g., S$_1$, S$_3$, S$_5$ in one phase, S$_2$, S$_4$ in the other) to charge the output and flying capacitor while reversing voltage polarity. This mechanism allows continuous operation across all four quadrants, including near-zero output region. Nevertheless, large voltage swings induce significant inductor and output current ripples, alongside energy backflow to the input, which degrade efficiency. Therefore, the auxiliary transition mode is restricted to a narrow output voltage range (–1 to +1V), whereas buck/boost and inverting buck/boost modes are employed for higher voltages to optimize overall performance.

Figure 19.2.3 illustrates the system-level architecture of the proposed SIFQC, along with its timing diagram. The power stage integrates five high-voltage (HV) devices, comprising NMOS transistors M$_1$–M$_4$ and a PMOS transistor M$_5$, together with an external inductor and flying capacitor. The output voltage V$_O$ is regulated through an external control signal V$_{CTRL}$, with V$_O$ = 5(V$_{CTRL}$–1), yielding a tunable V$_O$ range of –5 to +5V for V$_{CTRL}$ ranging from 0 to 2V. In the top right of Fig. 19.2.3, the gate-driver circuitry includes an on-chip LDO that supplies a 5V V$_{DD}$ rail for the controller and the low-side driver of M$_2$, whereas M$_1$ is driven through a conventional bootstrap circuit. For M$_3$, the gate-driving voltage is derived from a Zener-diode-resistor-bias network that generates a reference V$_Z$. This bias drives NMOS M$_{b1}$ to generate V$_{CB2}$ = V$_Z$ – V$_{GS\_Mb1}$, which is shared by M$_3$ and M$_4$ due to their common source connection. The M$_5$ gate driver adopts a similar scheme, employing NMOS M$_{b2}$ to establish V$_{CB3}$ = V$_Z$ – V$_{GS\_Mb2}$.

To optimize the transient response, the proposed SIFQC employs a dual feedback controller with capacitor current sensing (middle of Fig. 19.2.3). The slow loop employs an error amplifier (EA) to regulate the DC output voltage by compensating the difference between feedback and the reference. The fast loop enhances the transient response by incorporating filter-capacitor current feedback. To ensure multi-mode operation, a mode selector compares V$_{CTRL}$ with two reference voltages (0.8V and 1.2V) to generate logic signals M$_P$ and M$_N$ that correspond to three operating modes (Fig. 19.2.3, table). A notable issue in mode transitions with slow output voltage changes is that the converter may experience duty-cycle discontinuities, causing output voltage undershoots or overshoots, or even mode instability. To address these challenges and ensure seamless mode transitions, a mode transition compensator (MTC) is implemented using control-voltage endpoint prediction in the middle-right of Fig. 19.2.3. The MTC employs a state machine that detects mode changes, retrieves preset values from a look-up table (LUT), and drives a DAC to generate a compensation voltage V$_{MC}$ over two switching cycles, thereby pre-setting the inductor current and duty cycle for seamless transitions. For example, as shown in the bottom of Fig. 19.2.3, during the transition from buck/boost mode to the auxiliary transition mode under a slow voltage reduction, the duty cycle approaches 0.5 while the MTC pre-sets the inductor current near its target value, thereby enhancing transient response. The same compensation procedure is applied in the auxiliary transition mode to inverting buck/boost mode transitions. For rapid voltage variations, however, the response remains constrained by the maximum inductor current, which ensures safe operation.

The proposed SIFQC is implemented in a 0.18μm BCD process and supports an input voltage range of 5 to 12V, an output voltage range of –5 to +5V, and a load current range of –3 to +3A, thereby enabling complete four-quadrant operation. Steady-state waveforms under three operating modes are illustrated in Fig. 19.2.4 (top and left). The inductor voltage V$_{LX}$ varies from 0 to +V$_{IN}$ in buck/boost mode, 0 to –V$_{IN}$ in inverting buck/boost mode, and –V$_{IN}$ to +V$_{IN}$ in auxiliary transition mode, aligning with the aforementioned operating principles. Output voltage ripple is measured at 24mV for buck/boost and inverting buck/boost modes and 45mV for the auxiliary transition mode, which is restricted to near-zero output. Fig. 19.2.4 (bottom-right) presents waveforms during output current reversal under V$_{IN}$ = 10V and V$_O$ = +5V. When I$_O$ steps between –1.5A and +1.5A, V$_O$ exhibits an undershoot of 590mV and an overshoot of 620mV, with recovery times of 19μs and 21μs, respectively, demonstrating robust performance during transitions between Quadrants I and IV. Figure 19.2.5 (top-left) shows the output voltage polarity reversal waveforms measured under V$_{IN}$ = 10V and R$_{LOAD}$ = 2.5Ω. As V$_{CTRL}$ varies from 0 to 2V, V$_O$ transitions smoothly from –5 to +5V, demonstrating seamless operation between Quadrants I and III. The auxiliary transition mode ensures uninterrupted continuous zero-crossing. Figure 19.2.5 (top-right) presents the converter's power loss versus load current across all quadrants, while Fig. 19.2.5 (bottom) shows the efficiency curve. Notably, the converter has higher power losses in Quadrants II and III than in I and IV because negative-voltage operation requires inverting the input voltage via C$_F$, adding extra losses to the switched-capacitor circuitry. A peak efficiency of 94.6% is achieved in Quadrant I under V$_{IN}$ = 10V, V$_O$ = +5V, I$_O$ = +900mA.

Compared with prior-art designs (Fig. 19.2.6), the proposed SIFQC achieves four-quadrant operation using a single inductor, significantly reducing system complexity, size, and cost while preserving high efficiency and low output ripple. The fabricated chip occupies an area of 2.17mm × 1.99mm (Fig. 19.2.7), demonstrating the practicality of integrating a high-performance four-quadrant converter in compact form factors. These results highlight the technical novelty of the SIFQC and its potential for broad deployment in applications requiring precise bidirectional power regulation, including EC smart windows and other dynamic load systems.

Acknowledgement:
This work is funded by the National Natural Science Foundation of China under Grant 62404049.

References:
[1] C. Lin et al., "A Wide 0.1-to-10 Conversion-Ratio Symmetric Hybrid Buck-Boost Converter for USB PD Bidirectional Conversion," *2023 IEEE International Solid-State Circuits Conference (ISSCC)*, San Francisco, CA, USA, 2023, pp. 194-196.
http://doi.org/10.1109/ISSCC42615.2023.10067408
[2] T. -W. Wang et al., "Multiple-Phase Accelerated Current Control in Bidirectional Energy Transfer of Automotive High-Voltage and Low-Voltage Batteries," *2023 IEEE International Solid-State Circuits Conference (ISSCC)*, San Francisco, CA, USA, 2023, pp. 308-310.
http://doi.org/10.1109/ISSCC42615.2023.10067291
[3] Y. Lee, H. Park, M. Kim, W. Jung, H. Kim and H. -M. Lee, "A 97.4%-Peak-Efficiency Always-Half-Inductor-Current Hybrid Bidirectional Converter With Adaptive Target Current Tracking for USB-to-2-Cell Bidirectional Power Transfer," *2025 IEEE International Solid-State Circuits Conference (ISSCC)*, San Francisco, CA, USA, 2025, pp. 380-382.
http://doi.org/10.1109/ISSCC49661.2025.10904784
[4] Y. -Y. Lin, Y. -R. Huang, C. -J. Chen, Y. -C. Lin and T. -W. Huang, "A Bidirectional Three-Level Converter Control With Shared Control Circuit and Single-Point Sensing for Flying Capacitor Balance," in *IEEE Transactions on Power Electronics*, vol. 39, no. 1, pp. 1015-1027, Jan. 2024.
http://doi.org/10.1109/TPEL.2023.3321905
[5] W. Hong et al., "A Dual-Input Bidirectional 3-Level Battery Charger with Coarse-Fine VCF Balancing and Wide VCR for Foldable Mobile Applications," *2025 IEEE International Solid-State Circuits Conference (ISSCC)*, San Francisco, CA, USA, 2025, pp. 376-378.
http://doi.org/10.1109/ISSCC49661.2025.10904747
[6] Y. -C. Chiu et al., "Two-Phase Hybrid Buck-Boost Converter With Coupled-Inductors Under ZVS Operation for USB PD Bidirectional Conversion," in *IEEE Transactions on Circuits and Systems I: Regular Papers*, vol. 71, no. 11, pp. 5091-5101, Nov. 2024.
http://doi.org/10.1109/TCSI.2024.3423778
[7] Z. Tong, J. Huang, Y. Lu and R. P. Martins, "A 42W Reconfigurable Bidirectional Power Delivery Voltage-Regulating Cable," *2023 IEEE International Solid-State Circuits Conference (ISSCC)*, San Francisco, CA, USA, 2023, pp. 192-194.
http://doi.org/10.1109/ISSCC42615.2023.10067491
[8] Analog Devices Inc., "LT8714, Bipolar Output Synchronous Controller with Seamless Four Quadrant Operation", Nov. 2015, Available: https://www.analog.com/media/en/technical-documentation/data-sheets/8714f.pdf.
[9] X. Yan et al., "Design and Analysis of a Hybrid Inverting Buck Converter With 5 µs Response Time and 92.9% Efficiency for Micro-LED Displays," in *IEEE Transactions on Power Electronics*, vol. 40, no. 10, pp. 14388-14400, Oct. 2025.
http://doi.org/10.1109/TPEL.2025.3584285

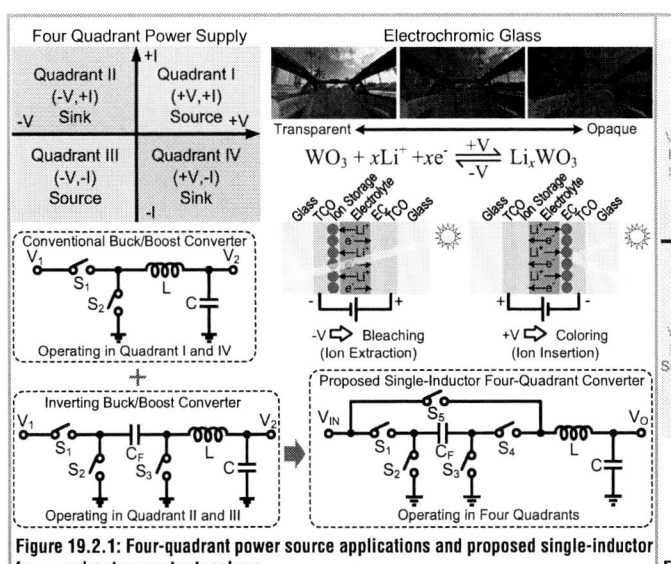

Figure 19.2.1: Four-quadrant power source applications and proposed single-inductor four-quadrant converter topology.

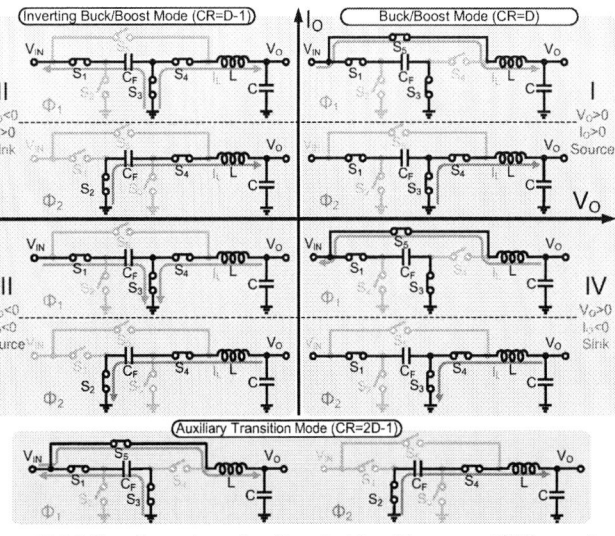

Figure 19.2.2: Operating modes and working principles of the proposed SIFQ converter.

ISSCC 2026 / SESSION 19 / HIGH-VOLTAGE, ISOLATED AND DISPLAY POWER / 19.2

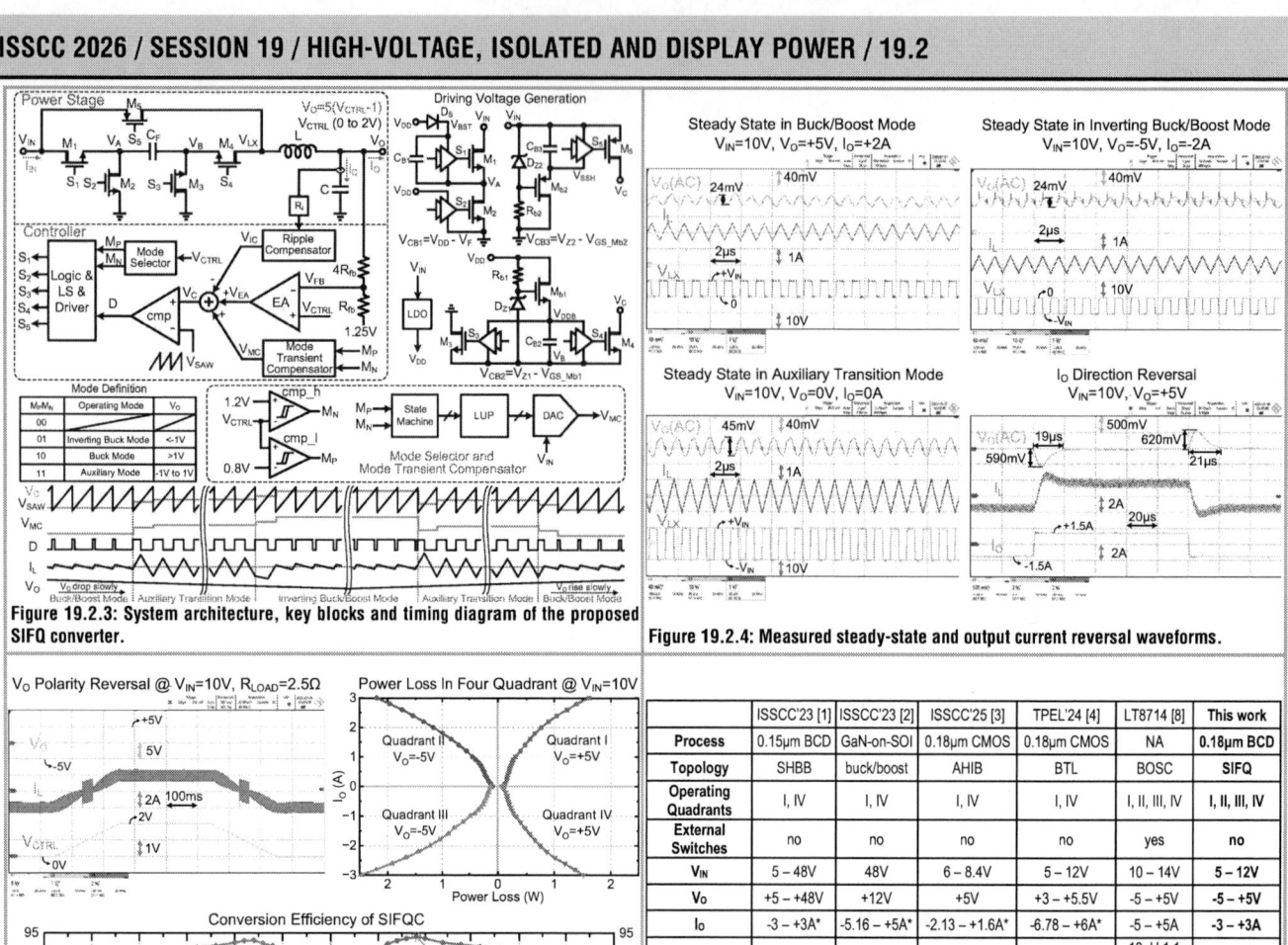

Figure 19.2.3: System architecture, key blocks and timing diagram of the proposed SIFQ converter.

Figure 19.2.4: Measured steady-state and output current reversal waveforms.

Figure 19.2.5: Measured output voltage polarity reversal waveforms, power loss and conversion efficiency.

	ISSCC'23 [1]	ISSCC'23 [2]	ISSCC'25 [3]	TPEL'24 [4]	LT8714 [8]	This work
Process	0.15μm BCD	GaN-on-SOI	0.18μm CMOS	0.18μm CMOS	NA	0.18μm BCD
Topology	SHBB	buck/boost	AHIB	BTL	BOSC	SIFQ
Operating Quadrants	I, IV	I, IV	I, IV	I, IV	I, II, III, IV	I, II, III, IV
External Switches	no	no	no	no	yes	no
V_{IN}	5 – 48V	48V	6 – 8.4V	5 – 12V	10 – 14V	5 – 12V
V_O	+5 – +48V	+12V	+5V	+3 – +5.5V	-5 – +5V	-5 – +5V
I_O	-3 – +3A*	-5.16 – +5A*	-2.13 – +1.6A*	-6.78 – +6A*	-5 – +5A	-3 – +3A
Inductor(L)	0.47μH	2.2μH	4.7μH	470nH	10μH 1:1 Coupled	2.2μH
C_F	8×10μF	no	2×10μF	10μF	2×22μF	1×22μF
C_O	4.7μF	NA	10μF	20μF	4×100μF	10μF
Peak efficiency	95.4%	95.5%	97.4%	96.01%	90.9%**	94.6%
f_s	2MHz	20MHz	1MHz	1.5MHz	0.2MHz	1MHz
Chip area	24.2mm²	3.31mm²	6.48mm²	7.04mm²	NA	4.32mm²

* Estimated value at the point where the output current is uniformly converted to the low-voltage side and normalized.
** Estimated value.

Figure 19.2.6: Performance summary and comparison with the state-of-the-art designs.

Figure 19.2.7: Chip micrograph.

• 2026 IEEE International Solid-State Circuits Conference

ISSCC 2026 / SESSION 19 / HIGH-VOLTAGE, ISOLATED AND DISPLAY POWER / 19.3

19.3 A 94.8%-Peak-Efficiency Double Step-Up SIBO Converter Achieving 88% Output Ripple Reduction for AMOLED Display

Hyeon-Ji Choi, Joo-Mi Cho, Sang-Yun Nam, Sung-Wan Hong

Sogang University, Seoul, Korea

Abstract

This paper presents a double step-up single-inductor bipolar-output (DSU-SIBO) converter for AMOLED displays. By employing a 1:2 charge pump, the proposed design reduces both DC and AC components of the inductor current, thereby minimizing loss and achieving a peak efficiency of 94.8%. With 2-out-of-3-phase energy delivery, each output receives energy during two phases per cycle, reducing output ripple by up to 88%. The prototype is implemented in a 180-nm BCD process.

The growing demand for high-quality visual experiences has necessitated the use of high-performance and large-size AMOLED displays in portable devices, which require both positive and negative supply rails (V_{OP} and V_{ON}). In this application, the power converter generating these supply rails must meet two key requirements: 1) high power efficiency for longer battery time and lower thermal dissipation, and 2) small output voltage ripple to ensure image quality. In modern portable devices, the inductor volume is severely limited, resulting in significant parasitic resistance (R_{PAR}) that includes both DC and AC components. Therefore, it is necessary to minimize the inductor RMS current ($i_{L,RMS}$) for high efficiency. At the same time, voltage ripple ($\Delta V_{OP,ON}$) on the supply rails can cause image fluctuations on the AMOLED display and should be minimized.

Conventional bipolar power delivery typically uses two separate DC-DC converters, such as a boost converter for V_{OP} and an inverting buck-boost converter for V_{ON}, each requiring their own inductor [1], as shown in Fig. 19.3.1 (middle-left). However, the use of two inductors increases both size and cost, making this approach unsuitable for compact and cost-sensitive applications. In contrast, a single-inductor bipolar-output (SIBO) converter can generate both V_{OP} and V_{ON} using only one inductor [2-6], as shown in Fig. 19.3.1 (middle). Nevertheless, in previous SIBO converters, the voltage at the left node of the inductor cannot exceed the input voltage (V_{IN}) during the magnetizing phase, fundamentally limiting the reduction of $i_{L,RMS}$. Therefore, they have limitations in improving the power efficiency.

In addition, SIBO converters suffer from large output voltage ripple, making it difficult to satisfy ripple requirements. Among prior works, [3] and [4] adopt three-phase operation, where the inductor current (i_L) delivers energy (E_{iL}) to the output only during one of the three phases. During the remaining two-thirds of the switching period (T), only the load current (I_O) discharges and charges positive and negative output capacitors (C_{OP} and C_{ON}), respectively. Since the output voltage ripple increases as the E_{iL} delivery duration becomes shorter, [3] and [4] have relatively large $\Delta V_{OP,ON}$. To extend the E_{iL} delivery, [2], [5] and [6] adopt a two-phase operation where E_{iL} is simultaneously delivered to both outputs during one of the two phases. This operation increases the E_{iL} delivery duration compared to [3] and [4], but still has limitations in effectively minimizing $\Delta V_{OP,ON}$.

To overcome these challenges, this paper proposes a double step-up SIBO (DSU-SIBO) converter with 2-out-of-3-phase energy delivery. The proposed DSU-SIBO converter integrates a 1:2 charge pump at the left side of the inductor, boosting the voltage at the left node of the inductor to $2{\times}V_{IN}$ as shown in Fig. 19.3.1 (top-right). With this $2{\times}$ boost operation, the DC component of i_L ($i_{L,DC}$) is reduced for the same output power. In addition, the DSU-SIBO converter reduces the voltage difference across the inductor, resulting in a smaller AC ripple current (Δi_L). Consequently, both the DC and AC components of i_L are reduced, leading to a lower $i_{L,RMS}$ compared to previous designs. As a result, the power loss ($P_{R,L}$) in the R_{PAR} of the inductor is significantly reduced, enabling higher power efficiency than conventional SIBO converters.

At the same time, the DSU-SIBO converter enhances ripple performance. Owing to the effectively doubled V_{IN} by the 1:2 charge pump, the converter modifies its operation sequence to extend the E_{iL} delivery duration. Although the DSU-SIBO converter operates in three phases, each output receives E_{iL} during two of every three phases in a switching cycle. In other words, only one of the three phases lacks E_{iL} delivery to both outputs, which corresponds to the shortest energy-absent duration compared to previous designs. As a result, the DSU-SIBO converter achieves the smallest $\Delta V_{OP,ON}$ among the previous designs, enhancing image quality in AMOLED displays.

Figure 19.3.2 (top-left) shows the DSU-SIBO converter, which consists of an inductor (L), two flying capacitors (C_{F1} and C_{F2}), three PMOS switches (S_1, S_4, and S_8), and five NMOS switches (S_2, S_3, S_5, S_6, and S_7). Since S_1–S_4 and C_{F1}–C_{F2} form the 1:2 charge pump, the voltages across C_{F1} and C_{F2} are maintained at V_{IN} and $2V_{IN}$, respectively. Moreover, all switches are implemented using 5V CMOS because the maximum voltage stress across any switch remains below 5V.

Figure 19.3.2 (right) shows the three operational phases (Φ_1, Φ_2, and Φ_3) of the DSU-SIBO converter, where E_{iL} is delivered to at least one of the two outputs, V_{OP} and V_{ON}, in each phase. In Φ_1, C_{F1} is discharged to simultaneously charge C_{F2} and magnetize L. In this case,

L is magnetized with a slope of $(2V_{IN}\text{-}V_{OP})/L$ and delivers E_{iL} to V_{OP} through i_L. In Φ_2, C_{F1} is charged by V_{IN} while C_{F2} is discharged through i_L. L in this phase is de-magnetized with a slope of $(2V_{IN}+V_{ON})/L$ and delivers E_{iL} to V_{ON}. Lastly, in Φ_3, C_{F1} continues to be charged by V_{IN} while C_{F2} is further discharged through i_L, and L is de-magnetized with a slope of $(2V_{IN}\text{-}V_{OP}+V_{ON})/L$. During this phase, E_{iL} is simultaneously delivered to both V_{OP} and V_{ON}, via i_L. Owing to this operation, the duration during which each output does not receive E_{iL} is shorter than in any previous SIBO converters, as shown Fig. 19.3.2 (right-bottom), resulting in the smallest $\Delta V_{OP,ON}$.

Meanwhile, Fig. 19.3.2 (bottom-left) shows the operational differences between the previous and proposed converters during Φ_1, when L is magnetized. Assuming that all converters operate with the same input power (P_{IN}), it can be expressed as $P_{IN} = V_{IN} \cdot I_{IN}$, where I_{IN} is the current drawn from the input source. In [2] and [3], where only L is connected between V_{IN} and ground, I_{IN} equals i_L. In [4-6], where L and C_F are connected in parallel between V_{IN} and ground, I_{IN} consists of both i_L and the capacitor current (i_C), reducing i_L compared to [2] and [3]. In contrast, the proposed converter effectively boosts V_{IN} to $2V_{IN}$, so the current delivered to L and C_F ($i_L + i_C$) is reduced to $1/2 \cdot I_{IN}$. Consequently, the proposed converter achieves the lowest i_L among the compared designs, resulting in the smallest $P_{R,L}$.

Figure 19.3.3 (top) shows comparative analyses of the power losses caused by the R_{PAR} of L, including prior works and the proposed DSU-SIBO converter. The 1:2 charge pump used in the proposed topology significantly reduces both the DC component ($i_{L,DC}$) and the AC ripple (Δi_L) of i_L. Therefore, compared to the baseline design [2], under the typical load condition of $I_O = 0.2A$, the power loss (P_{DCR}) due to the DC resistance (DCR) is reduced by 85–88%, and the power loss (P_{ACR}) due to AC resistance (ACR) is reduced by 78–93%, assuming ACR is five times larger than DCR. Consequently, the total $P_{R,L}$ is reduced by 84–90%, achieving the lowest value among all compared designs.

In addition, Fig. 19.3.3 (bottom) compares the normalized $\Delta V_{OP,ON}$ values with respect to the baseline design [2] ($\Delta V_{OP,[2]}$ and $\Delta V_{ON,[2]}$), under the typical load condition. The proposed converter achieves a 51–88% reduction in ΔV_{OP} and a 38–79% reduction in ΔV_{ON}, demonstrating the smallest output ripple across the entire V_{IN} range.

Figure 19.3.4 shows the converter waveforms during start-up operation. Since the DSU-SIBO converter uses flying capacitors, these capacitors must be pre-charged to prevent potential damage caused by immediate switching operation after power-up. During power-on, the input voltage ramps up to V_{IN}, gradually charging C_{F1}, C_{F2}, and C_{OP} to V_{IN}. In this mode, C_{F1} is fully charged to its target value of V_{IN}. However, C_{F2}, C_{OP}, and C_{ON} must be further charged to $2V_{IN}$, V_{OP}, and V_{ON}, respectively, which is accomplished during the start-up operation.

The start-up operation is achieved by a simple two-phase open-loop control with a fixed 50% duty cycle. In the first phase (Φ_{S1}), C_{F1} is discharged, while C_{F2} and C_{OP} are charged. Since both C_{F1} and C_{F2} are already pre-charged to V_{IN} during the power-on mode, a significant inrush current would initially flow to charge C_{F2} up to $2V_{IN}$. To mitigate this, the gate voltage of switch S_2 is gradually increased from its threshold level over the entire start-up operation, which smoothly ramps up the charging current into C_{F2}. In the next phase (Φ_{S2}), C_{F1} and C_{OP} are charged, while the C_{F2} and C_{ON} are discharged by i_L.

As these two phases repeat, V_{OP} gradually increases while V_{ON} decreases. Since V_{OP} reaches its target value much faster than V_{ON}, V_{OP} can exceed 5V, which is the maximum allowable voltage of the CMOS devices, before V_{ON} is regulated. Consequently, the drain-to-source voltages of the switch S_6 ($V_{DS,S6}$) may exceed 5V, potentially resulting in device breakdown. To prevent this, additional startup-only switch, which is significantly smaller than the main switches S_1–S_8, is included to discharge excessive charge that causes V_{OP} to exceed its target level. Similarly, additional charge is supplied through this supplementary switch when V_{ON} falls below the target level. As a result, this operation prevents V_{ON} from dropping below the maximum allowable voltage during start-up, thereby avoiding voltage overstress on S_7.

Figure. 19.3.5 shows the measured waveforms under a typical operating condition of V_{IN} = 3.7V, V_{OP} = 4.6V, and V_{ON} = −4.9V. The proposed converter achieves ΔV_{OP} = 8mV and ΔV_{ON} = 10mV at I_O = 0.3A, and ΔV_{OP} = 20mV and ΔV_{ON} = 28mV at I_O = 0.8A. In addition, the load

ISSCC 2026 / February 17, 2026 / 2:20 PM

transient responses when I_O changes between 0.05A and 0.5A within 1µs show undershoots of 64mV and 79mV for V_{OP} and V_{ON}, respectively, and overshoots of 68mV and 62mV.

Figure 19.3.6 (top) shows the measured power efficiency of the DSU-SIBO converter under various I_O conditions, achieving a peak efficiency of 94.8% at I_O = 0.2A.

Figure 19.3.6 (bottom) compares the performance with state-of-the-art designs. The proposed converter achieves the smallest $\Delta V_{OP,ON}$ of the compared designs owing to its extended energy delivery to the outputs. This extended energy delivery also enables the proposed converter to achieve the smallest undershoot and overshoot among the prior works of Fig. 19.3.6, even under wider load transitions. Furthermore, by substantially reducing $P_{R,L}$, the prototype chip achieves a high peak efficiency of 94.8% and maintains efficiency above 86.9% across I_O range from 0.05 to 0.5A, indicating minimal variation in the converter efficiency over a wide I_O range. Benefiting from the proposed topology and its operation, the proposed DSU-SIBO converter attains excellent performance, optimally meeting the requirements of AMOLED display applications.

Figure 19.3.7 shows a die micrograph.

Acknowledgement:
This work was supported by National Research Foundation of Korea (NRF) Granted funded by the Korea government (MSIT) under Grant RS-2023-00207919 and IITP-2025-RS-2023-00260091.

References:
[1] STMicroelectronics, STOD13A, "250 mA dual DC-DC converter for powering AMOLED displays," Dec. 2011, Accessed on Oct. 15, 2024.
https://www.st.com/resource/en/datasheet/stod13a.pdf
[2] Texas Instruments, TPS65136, "Single-Inductor, Multiple-Output (SIMO) Regulator for AMOLED," Apr. 2008, Accessed on Oct. 15, 2024.
https://www.ti.com/lit/ds/symlink/tps65136.pdf
[3] K. -L. Lin et al., "A Single-Inductor Bipolar-Output DC/DC Converter with High Efficiency Over Wide Load Range for Active Matrix OLED," *SID Symp. Dig. Tech. Papers*, pp. 1183-1186, Feb. 2014. https://doi.org/10.1002/j.2168-0159.2014.tb00308.x
[4] S. -W. Wang et al., "High Efficiency Single-Inductor Boost/Buck Inverting Flyback Converter with Hybrid Energy Transfer Media and Multi Level Gate Driving for AM OLED Panel," *IEEE Symp. VLSI Circuits*, pp. 59-60, Jun. 2010.
https://doi.org/10.1109/VLSIC.2010.5560269
[5] S. -W. Hong et al., "A 1.46mm² Simultaneous Energy-Transferring Single-Inductor Bipolar-Output Converter with a Flying Capacitor for Highly Efficient AMOLED Display in 0.5µm CMOS," *ISSCC*, pp. 200-202, Feb. 2020.
https://doi.org/10.1109/ISSCC19947.2020.9063141
[6] J. Jin et al., "A 94.5%-Peak-Efficiency 3.99W/mm²-Power-Density Single-Inductor Bipolar-Output Converter with a Concise PWM Control for AMOLED Displays," *ISSCC*, pp. 144-146, Feb. 2024. https://doi.org/10.1109/ISSCC49657.2024.10454554
[7] B. -C. Kwak et al., "A Highly Power-Efficient Single-Inductor Bipolar-Output DC–DC Converter Using Hysteretic Skipping Control for OLED-on-Silicon Microdisplays," *IEEE TCAS-II*, vol. 65, no. 12, pp. 2017-2021, Dec. 2018.
https://doi.org/10.1109/TCSII.2018.2815994

[8] F. Mao et al., "A Power-Efficient Hybrid Single-Inductor Bipolar-Output DC-DC Converter with Floating Negative Output for AMOLED Displays," *IEEE CICC*, pp. 1-4, Mar. 2020. https://doi.org/10.1109/CICC48029.2020.9075940
[9] F. Mao et al., "A Hybrid Single-Inductor Bipolar-Output DC–DC Converter With Floating Negative Output for AMOLED Displays," *IEEE JSSC*, vol. 56, no. 9, pp. 2760-2769, Sep. 2021. https://doi.org/10.1109/JSSC.2021.3062092
[10] S. -W. Wang et al., "Efficiency enhanced Single-Inductor Boost-Inverting Flyback converter with Dual Hybrid Energy transfer media and a Bifurcation Free Comparator," *IEEE ESSCIRC*, pp. 450-453, Sep. 2010. https://doi.org/10.1109/ESSCIRC.2010.5619740
[11] H. -J. Park et al., "A Simultaneous Energy Transferring SIBO Converter Achieving Low Ripple and High Efficiency for AMOLED Applications," *IEEE JSSC*, vol. 59, no. 5, pp. 1497-1508, May 2024. https://doi.org/10.1109/JSSC.2023.3314834
[12] J. Jin et al., "A Hybrid Single-Inductor Bipolar-Output Converter With a Concise PWM Control for AMOLED Displays," *IEEE JSSC*, vol. 59, no. 12, pp. 4150-4161, Dec. 2024. https://doi.org/10.1109/JSSC.2024.3456190

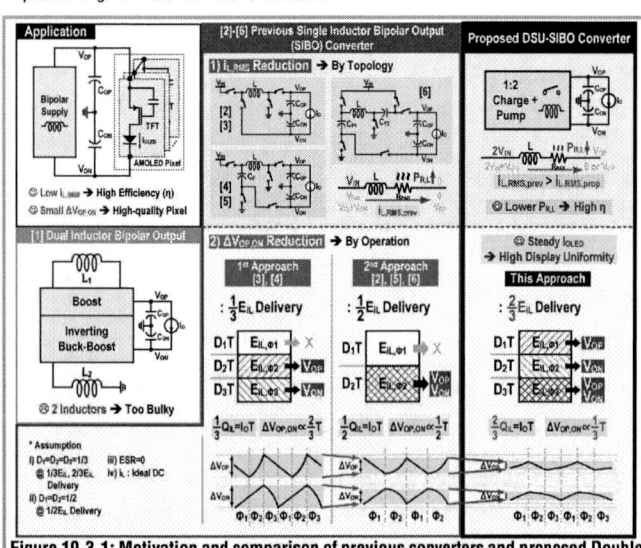

Figure 19.3.1: Motivation and comparison of previous converters and proposed Double Step-Up SIBO (DSU-SIBO) converter with 2-out-of-3-phase energy delivery.

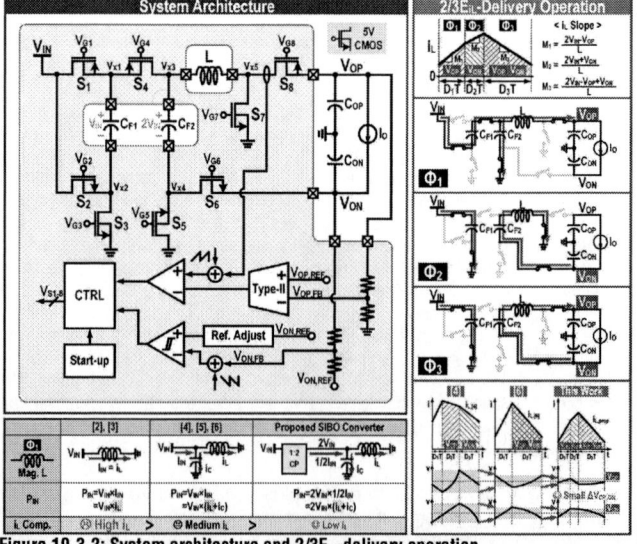

Figure 19.3.2: System architecture and 2/3E_IL-delivery operation.

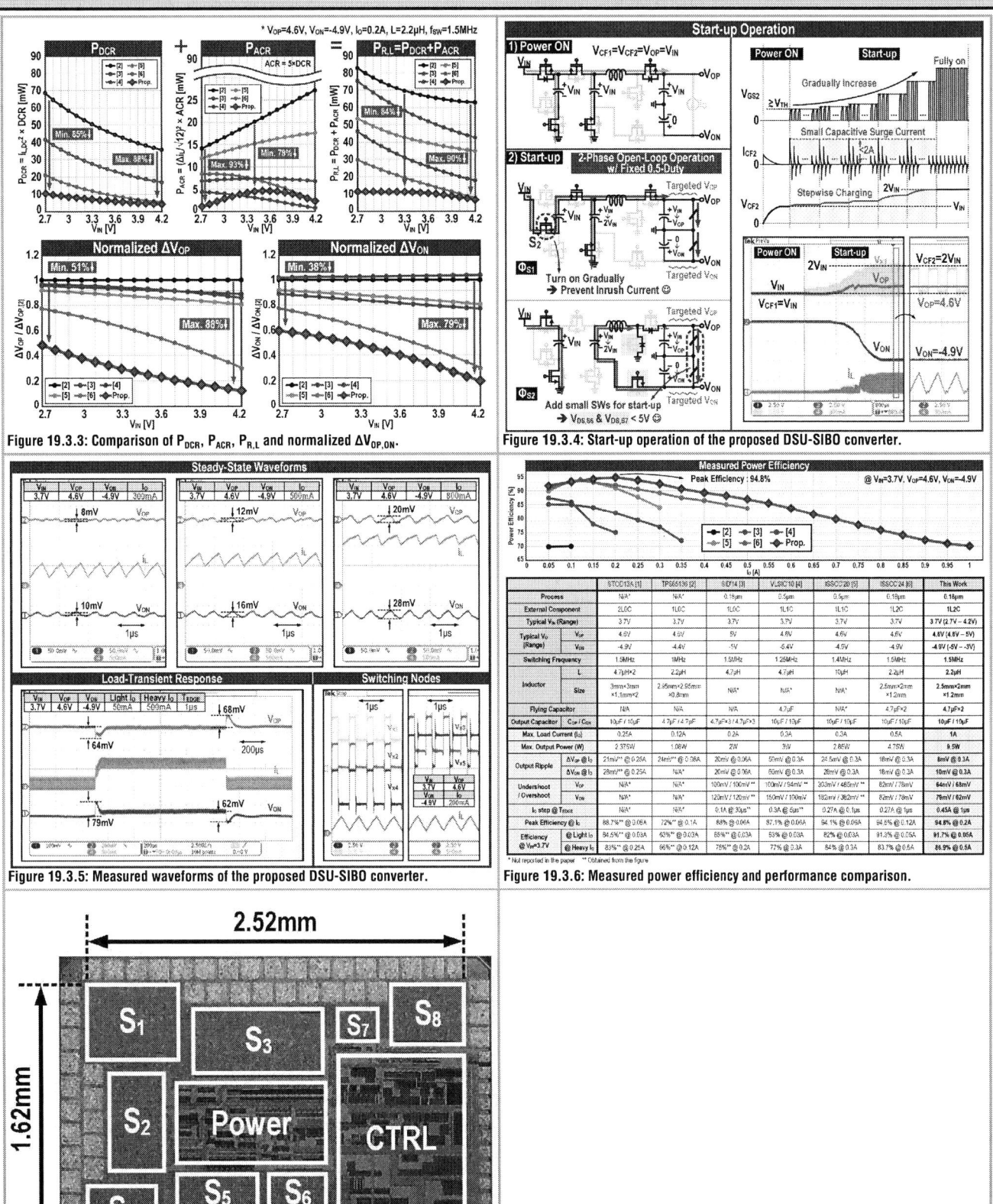

Figure 19.3.3: Comparison of P_{DCR}, P_{ACR}, $P_{R.L}$ and normalized $\Delta V_{OP,ON}$.

Figure 19.3.4: Start-up operation of the proposed DSU-SIBO converter.

Figure 19.3.5: Measured waveforms of the proposed DSU-SIBO converter.

Figure 19.3.6: Measured power efficiency and performance comparison.

Figure 19.3.7: Die micrograph.

19.4 A Digital-Feedback Active-Gate-Driver IC for 600A 1200V SiC MOSFETs Supporting High- and Low-Side Drive with Simultaneous dV_{ds}/dt Control and V_{ds} Surge Suppression Enabled by Miller Capacitance Calibration

Shusuke Kawai, Kohei Horii, Koutaro Miyazaki, Yuto Bushimata, Satoshi Takaya, Hiroaki Ishihara

Toshiba, Kawasaki, Japan

Abstract

A digital feedback active gate driver IC applicable to both high- and low-side 600A, 1200V SiC MOSFETs is proposed and demonstrated on the high-side. High-side voltage sensing uses a parasitic Miller capacitance calibration circuit, while digital feedback enables simultaneous turn-on/off dV_{ds}/dt control and V_{ds} surge suppression. Measurements show 58% surge voltage and 66% switching loss reduction, with dV_{ds}/dt control within 10% error and an additional 29% loss reduction.

Silicon Carbide (SiC) power devices are increasingly being adopted for traction inverters in motor drives of electric vehicles owing to their low-loss and high-speed switching characteristics. However, the drain voltage (V_{ds}) and drain current (I_{ds}) surges generated during high-speed switching act as noise sources, requiring large electromagnetic compatibility (EMC) filters in the inverter. The evolution of SiC power devices targets lower reverse recovery charge (Q_{rr}) and parasitic capacitance [1], suppressing Q_{rr}-induced I_{ds} surges [2], while faster switching enabled by lower capacitance intensifies V_{ds} surges. To suppress V_{ds} surges, the switching speed is reduced, which limits SiC's advantages and results in a trade-off between loss and noise. From an EMC perspective, inverters must comply with the guidelines for the maximum allowable V_{ds} transition rate (dV_{ds}/dt). Maintaining dV_{ds}/dt at its maximum value is desirable for low-loss operation; however, unwanted variations in dV_{ds}/dt due to environmental fluctuations can increase losses.

Active gate driver (AGD) technology effectively mitigates the trade-off between noise and switching loss [3-5]. Analog feedback (FB) techniques enable PVT-robust control but have limited flexibility in gate waveform shaping and cannot control V_{ds} surges and dV_{ds}/dt simultaneously. In [3] and [4], FB is used for V_{ds} surge suppression and dV_{ds}/dt control, but [3] showed loss reduction only at 100V and [4] at 4A, both far below the minimum requirement of 300V/100A for EV applications [5]. The designs in [6,7] focus solely on I_{ds} surge suppression and do not address dV_{ds}/dt or V_{ds} surge control, which are increasingly important with SiC evolution. Digital feedforward (FF) techniques [8] allow flexible waveform design and address various challenges but are sensitive to PVT variations. Since inverters use both low- and high-side devices, AGD must also apply to the high-side. A high-side FF-based AGD [9] ignores load current and PVT variations, and manual tuning for variation compensation increases mass-production cost. While FB techniques are PVT-robust, no prior work implements FB-based dV_{ds}/dt or V_{ds} surge control in high-voltage (>300V) high-side drivers due to difficulty eliminating errors in high-side V_{ds} sensing. A capacitive divider is used for V_{ds} detection since a resistive divider needs a smaller time constant for bandwidth, increasing power loss. In high-side driving as shown in upper part of Fig. 19.4.1, V_{ds} is divided by capacitors C_1 and C_2, and a parasitic capacitance C_p exists between the divided node $V_{ds,div}$ and the power stage ground (GND). Similarly, on the low side, a voltage divider consisting of C_{1L} and C_{2L} and a parasitic capacitance C_{PL} also exist. Since both C_{2L} and C_{PL} share the power stage GND as their reference, the error voltage at $V_{ds,divL}$ caused by C_{PL} is minimal. In contrast, on the high side, C_2 references V_{SW}, which swings by 600V, while C_p references the power stage GND. As a result, C_p behaves like a Miller capacitance. Assuming a typical voltage swing of only a few volts between V_{SW} and $V_{ds,div}$, C_p must handle a charge several hundred times greater than in the low-side case, and as a result, the error at $V_{ds,div}$ becomes non-negligible.

To address these challenges, a digital FB AGD IC that controls both turn-on and turn-off dV_{ds}/dt and V_{ds} surge voltage, with a high-side Miller C_P calibration function, is proposed as shown in lower part of Fig. 19.4.1. The circuit includes a V_{ds} detection circuit with Miller C_p calibration block, a time-to-digital converter (TDC), a 6-bit current output stage, and gate current (I_g) generation logic using FB. The calibration block in the V_{ds} detection circuit compensates for C_p-induced errors, enabling accurate V_{ds} detection. The TDC calculates dV_{ds}/dt from the detected V_{ds} and compares it with the target. Based on the TDC result, the FB I_g generation logic determines the appropriate I_g, achieving both dV_{ds}/dt control and V_{ds} surge suppression. The IC includes a negative-I_{ds} detection circuit to disable AGD feedback under negative-I_{ds} conditions, which is essential for applying AGD to the inverter.

Figure 19.4.2 shows the circuit diagrams of the V_{ds} detection circuit, the TDC, and the negative-I_{ds} detection circuit. Figure 19.4.2 also illustrates FB waveforms enabling V_{ds} surge suppression and dV_{ds}/dt control. The prefix T indicates the timing of I_g changes, DT denotes the digital signals detected by the V_{ds} timing detection circuit corresponding to T, and I_g indicates the gate current amplitude at each timing. V_{ds} surge suppression is achieved by reducing I_g at T_{3Off} when V_{ds} exceeds the steady-state power-stage voltage V_{dc} (600V) and increasing I_g again at T_{4Off} when V_{ds} returns to V_{dc}. T_{3Off} and T_{4Off} are detected by the V_{ds} timing detection comparator, and the detected signals DT_{3Off} and DT_{4Off} are immediately fed back, causing I_g to change within the same switching cycle. For surge control, I_{g1Off} and I_{g3Off} are set to the driver's maximum and minimum current levels. For dV_{ds}/dt control, I_{g2Off} is adjusted based on T_{1Off} and T_{2Off}, corresponding to 10% and 90% of V_{dc}. These timings are also immediately fed back to I_g using the detected signals DT_{1Off} and DT_{2Off}. The TDC converts the interval between DT_{1Off} and DT_{2Off} into an analog voltage $V_{TDCInOff}$. The TDC includes four comparators with reference voltages based on the target dV_{ds}/dt. The comparators compare $V_{TDCInOff}$ with reference values to generate a 4-bit thermometer code, $D_{FB,I2Off}$. The FB I_g data generation logic then uses $D_{FB,I2Off}$ to update the 6-bit data D_{Ig2Off}, which represents the gate current I_{g2Off}. This update occurs after the turn-off has completed and after a short delay, preparing the data for the next switching cycle. The 4-bit output allows for fine adjustments; e.g., $D_{FB,I2Off}$ = 1111 increases I_{g2Off} by a coarse adjustment (4–8 LSBs), while 0111 makes a small adjustment (1–3 LSBs). These coarse and fine adjustment values can be modified. When $D_{FB,I2Off}$ = 0011, I_{g2Off} stays constant, forming a dead zone to suppress dV_{ds}/dt fluctuation. Although the previous example explains turn-off control, dV_{ds}/dt can also be controlled during turn-on. For turn-on, dV_{ds}/dt control uses D_{T1On} and D_{T2On} to adjust I_{g2On}, while I_{g1On} is set to the maximum current the driver can provide.

Since the voltage divided by C_1 and C_2 contains an error caused by C_p, calibration is performed using the circuit shown at the upper left of Fig. 19.4.2. V_{ds} is divided in two steps: first by off-chip capacitors C_1 and C_2, then by on-chip fixed capacitor C_a and variable capacitor C_{att} to generate V_{cmpin}. A voltage V_{Rdiv}, divided by resistors R_1 and R_2, is also provided as a reference for the correct division ratio. R_1 and R_2 do not sense dV_{ds}/dt; instead, the divided voltage is sampled at each 10kHz PWM carrier cycle. This ensures that the correct division ratio is applied to the IC and allows the use of large resistor values to minimize power loss. Comparator CMP_{cal} compares V_{Rdiv} with V_{cmpin} and adjusts the counter until both voltages match. The value of C_{att} is updated according to the counter. Calibration ends when the counter toggles several times. The timing diagram of calibration is shown in the upper-left of Fig. 19.4.3. As mentioned earlier, V_{Rdiv} does not follow the ideal divided V_{ds} (Ideal $V_{ds,div}$) as quickly as V_{cmpin}. After the power device turns off and some time has passed, V_{Rdiv} matches Ideal $V_{ds,div}$, and CMP_{cal} operates at this timing to compare the error. After the device turns on and $V_{ds,div}$ decreases, C_{att} is updated after a short delay. By repeating this process, the error in V_{cmpin} can be corrected.

The upper-right of Fig. 19.4.3 shows the operating waveforms of the negative-I_{ds} detection circuit. In an inverter, negative I_{ds} may flow, and applying AGD can cause malfunction since dV_{ds}/dt cannot be controlled by I_g [9]. When I_{ds} is negative, V_{ds} falls while PWM remains low because the load current I_L freewheels through the diode before the PWM rises. When I_{ds} is positive, V_{ds} falls after the PWM signal goes high. As shown in left-part of Fig. 19.4.2, the negative-I_{ds} detection circuit replaces the sample-and-hold in [10] with a simple comparator and flip-flop design. It samples the PWM signal at the falling edge of V_{ds}, detects the current polarity, and disables FB when negative I_{ds} is detected.

The lower part of Fig. 19.4.3 shows the measured waveform of C_p calibration during high-side operation. CH1, CH2, CH4, and CH8 display V_{gs}, V_{ds}, I_{ds}, and V_{cmpin}. Since both the capacitive divider ratio and oscilloscope scale are 200:1, V_{cmpin} should match V_{ds} if there is no error. In the first switching event, V_{cmpin} and V_{ds} do not match, but the error decreases with repeated switching. The lower-right shows dV_{ds}/dt control for a 3V/ns target: without calibration, the measured dV_{ds}/dt is 2.5V/ns; with calibration, the measured value becomes precisely 3V/ns.

A double-pulse test was performed on a 1200V, 600A 2-in-1 SiC module (BSM600D12P3G001), with FB control applied to the high-side device. The upper-left and upper-right of Fig. 19.4.4 show dV_{ds}/dt control results during turn-on and turn-off, respectively. Without FB, dV_{ds}/dt varies with temperature, I_L, and V_{th} of the power device. When a 3V/ns guideline is specified, operating with dV_{ds}/dt values below 3V/ns increases switching losses for operating conditions with high I_L, at room temperature, and for low-V_{th} devices. Applying FB maintains constant dV_{ds}/dt under these conditions, reducing switching losses. The lower-left of Fig. 19.4.4 shows the rate of loss reduction achieved by FB control: up to 29% during turn-on and 19% during turn-off. The lower-right of Fig. 19.4.4 shows control results for targets of 3, 2, and 1V/ns. With FB control, dV_{ds}/dt variation is suppressed to 7% during turn-on and 10% during turn-off.

The upper-left of Fig. 19.4.5 shows measured V_{ds} surge and switching loss. Regardless of V_{th} or temperature variations of the power device, the trade-off between V_{ds} surge and switching loss improves, achieving up to 66% loss reduction when V_{ds} surge is kept equal. The upper-right of Fig. 19.4.5 shows waveforms with and without V_{ds} suppression, achieving

979-8-3315-8937-0/26 $31.00 © 2026 IEEE

58% surge reduction without extra loss. The lower-left of Fig. 19.4.5 shows a comparison of switching losses with and without FB under an equalized V_{ds} surge of 750V. At $I_L = 140A$, losses drop 49%, and 43% even with a 10% lower V_{dc}. The lower-right of Fig. 19.4.5 shows waveforms where both V_{ds} surge suppression and dV_{ds}/dt FB control during turn-off are applied simultaneously. By individually controlling I_g during the V_{ds} transition and V_{ds} surge occurrence periods, dV_{ds}/dt control at 3V/ns is achieved while keeping V_{ds} surge below 700V.

Figure 19.4.6 compares studies employing AGD on power devices at approximately 300V or higher in EV applications. The proposed IC is measured using a 600A, 1200V SiC-MOSFET under 600V and up to 550A. This work presents the only demonstration of FB control applied on the high-voltage high-side power device to regulate dV_{ds}/dt and suppress V_{ds} surge among the prior arts of Fig. 19.4.6. Figure 19.4.7 shows the chip micrograph and evaluation setup. The chip was fabricated in a 0.6μm 40V HV-CMOS process.

References:
[1] Infineon Technologies AG, "CoolSiC™ 650 V G2 MOSFET Application Note," Application Note, 2023. [Online].
https://www.infineon.com/assets/row/public/documents/24/42/infineon-coolsic-mosfet-650v-g2-applicationnotes-en.pdf
[2] Infineon Technologies AG, "Hard Commutation of Power MOSFET OptiMOS™ FD 200V/250V," Application Note AN 2014-03, Mar. 2014. [Online].
https://www.infineon.com/assets/row/public/documents/24/42/infineon-power-mosfet-optimos-fd-200v-250v-hard-diode-commutation-an-en.pdf
[3] S. Kawai, T. Ueno and K. Onizuka, "A 4.5V/ns Active Slew-Rate-Controlling Gate Driver with Robust Discrete-Time Feedback Technique for 600V Superjunction MOSFETs," *2019 IEEE International Solid-State Circuits Conference - (ISSCC)*, San Francisco, CA, USA, 2019, pp. 252-254. http://doi.org/10.1109/ISSCC.2019.8662534
[4] H. Akiyama, A. Niwa and M. Abe, "Low-Loss Active Gate Driver with Surge Voltage Detection for SiC MOSFET," *PCIM Conference 2025; International Exhibition and Conference for Power Electronics, Intelligent Motion, Renewable Energy and Energy Management*, Nürnberg, Germany, 2025, pp. 482-489.
http://doi.org/10.30420/566541058
[5] C. S. Goli, S. Essakiappan, P. Sahu, M. Manjrekar and N. Shah, "Review of Recent Trends in Design of Traction Inverters for Electric Vehicle Applications," *2021 IEEE 12th International Symposium on Power Electronics for Distributed Generation Systems (PEDG)*, Chicago, IL, USA, 2021, pp. 1-6.
http://doi.org/10.1109/PEDG51384.2021.9494164
[6] D. Zhang, K. Horii, K. Hata and M. Takamiya, "Digital Gate Driver IC with Fully Integrated Automatic Timing Control Function in Stop-and-Go Gate Drive for IGBTs," *2023 IEEE Applied Power Electronics Conference and Exposition (APEC)*, Orlando, FL, USA, 2023, pp. 1225-1231. http://doi.org/10.1109/APEC43580.2023.10131322
[7] Y. Liang, K. Hata and M. Takamiya, "Fully Integrated Closed-Loop Active Gate Driver IC With Real-Time Control of Gate Current Change Timing by Gate Current Sensing," *2025 IEEE Applied Power Electronics Conference and Exposition (APEC)*, Atlanta, GA, USA, 2025, pp. 1084-1089. http://doi.org/10.1109/APEC48143.2025.10977236

[8] S. Kawai *et al.*, "A Load Adaptive Digital Gate Driver IC With Integrated 500 ksps ADC for Drive Pattern Selection and Functional Safety Targeting Dependable SiC Application," in *IEEE Transactions on Power Electronics*, vol. 38, no. 6, pp. 7079-7091, June 2023. http://doi.org/10.1109/TPEL.2023.3244200
[9] Y. Sukita, K. Hata, H. Kondo, K. Watanabe, K. Nagayoshi and M. Takamiya, "Demonstration of Efficiency Increase of 350 V-to-13.3 V Isolated DC-DC Converters for Electric Vehicles by Active Gate Driving," *2025 IEEE Applied Power Electronics Conference and Exposition (APEC)*, Atlanta, GA, USA, 2025, pp. 1102-1107. http://doi.org/10.1109/APEC48143.2025.10977079
[10] H. Akiyama and T. Dewa, "Gate Drive Device," U.S. Patent 11,909,386, Feb. 20, 2024. https://patents.google.com/patent/US11909386B2/en?oq=Patent+11%2c909%2c386
[11] Q. Li, Y. Yang, Y. Wen, G. Zhang and W. Xing, "Active Gate Driver With the Independent Suppression of Overshoot and Oscillation for SiC MOSFET Modules," in *IEEE Transactions on Industrial Electronics*, vol. 72, no. 3, pp. 2325-2335, March 2025. http://doi.org/10.1109/TIE.2024.3433436

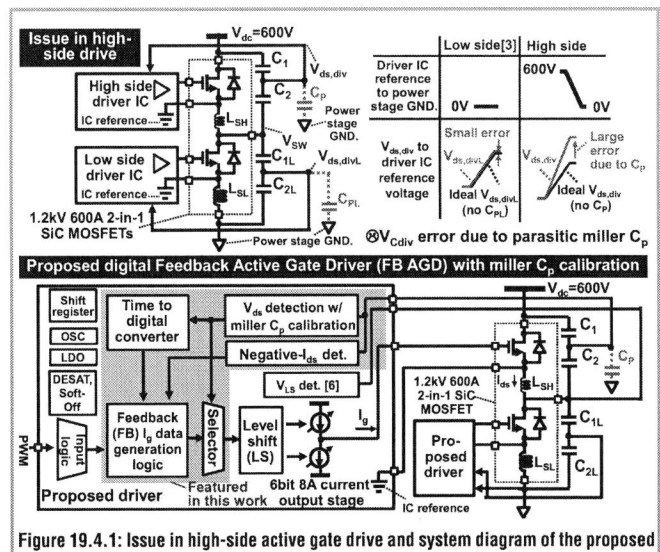

Figure 19.4.1: Issue in high-side active gate drive and system diagram of the proposed digital feedback active gate driver.

Figure 19.4.2: Circuit schematic of the V_{ds} detection circuit, time-to-digital converter, Miller parasitic capacitance (C_p) calibration, and the waveform diagram of proposed IC.

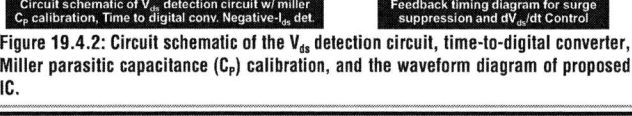

ISSCC 2026 / SESSION 19 / HIGH-VOLTAGE, ISOLATED AND DISPLAY POWER / 19.4

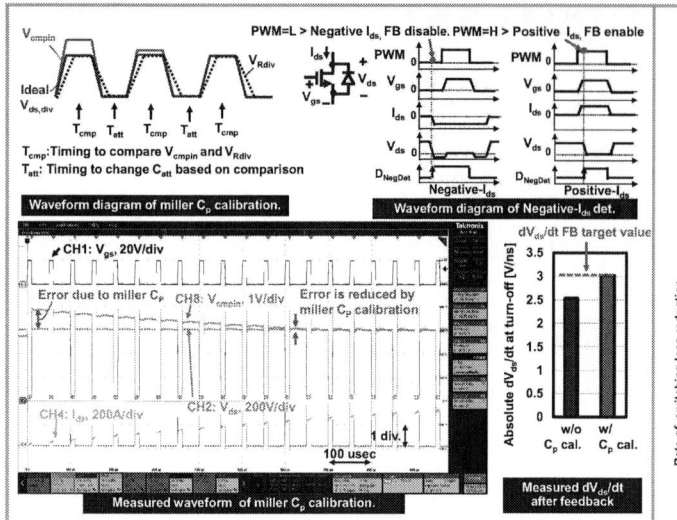

Figure 19.4.3: Timing diagrams for Miller C_p calibration and negative-I_{ds} detection, with measured waveforms of calibration and measured regulated dV_{ds}/dt values.

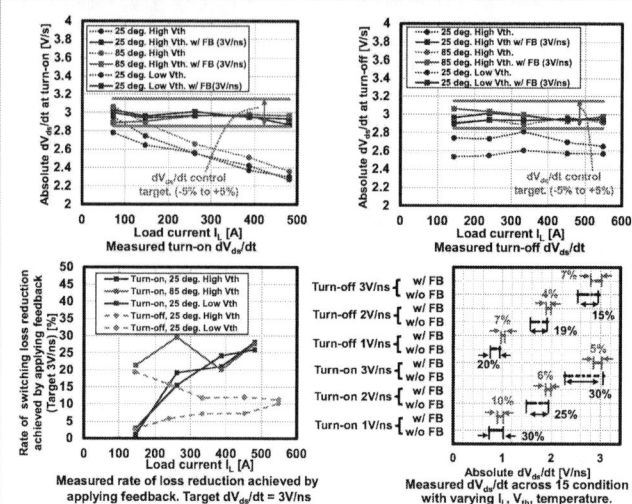

Figure 19.4.4: Measured dV_{ds}/dt at turn-on and turn-off versus load current, with and without feedback, and loss reduction rate achieved by feedback.

Figure 19.4.5: Measured results of switching losses and V_{ds} surges, and measured waveforms of V_{ds} surge suppression and simultaneous V_{ds} suppression with dV_{ds}/dt control.

	ISSCC 2019[3]	APEC 2023[6]	TPE 2023[8]	APEC 2025[9]	TIE 2025[11]	This work
Control method (Feedback/Feedforward)	Feedback	Feedback	Feed-forward (FF)	Feed-forward (FF)	Feedback	Feedback
Power device	SJMOS	IGBT	SiC	SiMOS	SiC	SiC
Measured V_{ds} / I_{ds} of power device	280V / 4A	600V / 80A	400V / 12A	350V / 20A	800V / 120A	600V / 550A
Process of driver IC	0.18um HV-CMOS	0.18um BCD	0.5um HV-CMOS	No chip integration	No chip integration	0.6um HV-CMOS
Gate drive current (I_g)	NA	3A	3.5A	NA	NA	8A
V_{gs} range of driver IC	+18/-0V	+15/-15V	+18/-0V	NA	NA	+18/0V**
Feedback active gate drive with high-side.	No (Low-side)	No (Low-side)	No (Low-side)	No (FF)	No* (Low-side)	Yes
V_{ds} surge suppression	-	33% (I_{ds} surge)	27% (Low side)	27% (I_{ds} surge)	43% (Low side)	58% (High side)
Loss reduction by V_{ds} surge suppression	-	49% (I_{ds} surge)	-	45% (I_{ds} surge)	23% (Low side)	66% (High side)
dV_{ds}/dt control	Yes (Low side)	No	No	No	No	Yes (High side)
Loss reduction by dV_{ds}/dt control	20%(On)	-	-	-	-	29%(On) 19%(Off)
Measurement verification under V_{th}, and temperature variation	V_{th}: Yes Temp.: Yes	V_{th}: No Temp.: Yes	No	No	V_{th}: No Temp.: Yes	V_{th}: Yes Temp.: Yes

*Although circuit diagrams applying active gate drive to high-side devices exist, no measured results are shown.
**The power device in this paper requires a gate voltage of 18V to 0V, whereas the IC supports an output range of 25V to −10V.

Figure 19.4.6: Comparison table.

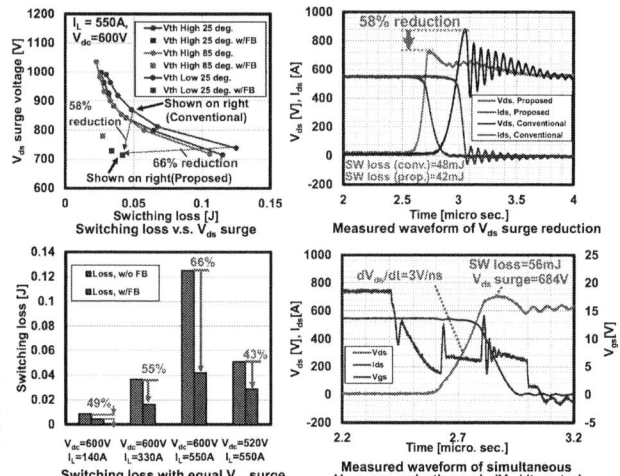

Figure 19.4.7: Chip micrograph and measurement setup.

- 2026 IEEE International Solid-State Circuits Conference

ISSCC 2026 / SESSION 19 / HIGH-VOLTAGE, ISOLATED AND DISPLAY POWER / 19.5

19.5 **A Binary-Weighted Switched-Capacitor Gate Driver IC for Overcoming Trade-offs Between Driving Loss and Delay Time with Gate-Current Feedback Achieving 85% Driving Loss Reduction**

Kohei Horii, Koutaro Miyazaki, Shusuke Kawai, Hiroaki Ishihara

Toshiba, Kawasaki, Japan

Abstract

A binary-weighted switched-capacitor gate driver IC achieving up to 85% reduction in driving loss (E_{drv}) is presented. Four binary-weighted flying capacitors enable nine-level gate-voltage segmentation, enhancing charge supply capability. Adaptive state interval time control based on gate-current feedback mitigates the trade-off between the switching time of the gate-voltage (t_{sw}) and E_{drv}, reducing tsw by 19% at equal E_{drv}, and reducing E_{drv} by 9% at equal t_{sw}.

In recent years, the importance of converters employing high-voltage high-current power devices have grown in response to the rising demand from data center UPSs and onboard chargers for electric vehicles. To maximize converter power density, it is essential to integrate the gate driver IC and minimize the size of the power supply. However, these high-power devices typically exhibit large input capacitance C_{iss} and require a large gate voltage swing (V_g); at the same time, their switching frequency f_{sw} continues to increase to further reduce the size of passive elements. Consequently, the per-cycle gate drive energy ($E_{drv} \approx C_{iss} V_g^2$) and the corresponding power ($P_{drv} = E_{drv} f_{sw}$) both increase significantly. This large power consumption complicates driver IC integration due to thermal constraints, increasing the size of the power supply [1], and further degrading the converters' light load efficiency. Therefore, reducing E_{drv} has become increasingly important.

One method to reduce E_{drv} is by using an LC resonant gate driver [1,2]; however, its inductor is bulky and unsuitable for high power density applications. Another approach is the switched capacitor (SC) gate driver [3], which achieves pseudo-adiabatic charging through gate-voltage segmentation using flying capacitors. In this approach, E_{drv} can be reduced in proportion to the number of gate-voltage segments. Because it uses only capacitors, the SC approach can be far more compact than LC resonant drivers.

However, practical SC implementations require a large total capacitance and a substantial number of flying capacitors. In [3], voltage segmentation is realized by varying the number of series-connected identical flying capacitors with equal applied voltage. This topology is hereafter called a Unary-Weighted Switched-Capacitor (UWSC) gate driver. In a UWSC, the required number of flying capacitors equals the number of voltage segments. Furthermore, the effective capacitance is significantly reduced because all of the flying capacitors are connected in series at the maximum output voltage. Consequently, each capacitor must be much larger than the gate capacitance to ensure sufficient drive strength and charge-supply capability. This requirement makes the UWSC approach impractical for power modules with a very large C_{iss}, which are the focus of this work. Moreover, [3] neglects the strong nonlinearity of C_{iss} during the Miller plateau, which introduces a trade-off between the driving loss and switching time.

To address these challenges, a Binary-Weighted Switched-Capacitor Gate Driver (BWSCGD) is proposed as shown in Fig. 19.5.1. The BWSCGD employs two chips and four flying capacitors (C_{fly1-4}) with binary-weighted capacitances (C, 2C, 4C, 4C). During a pre-drive reset phase, the capacitors are pre-charged to the binary-weighted voltages inversely proportional to their capacitance (10V, 5V, 2.5V, 2.5V), yielding nine gate-voltage levels (0-to-20V in 2.5V step for V_g = 20V). Compared with UWSC, BWSC can increase the equivalent capacitance because fewer flying capacitors are placed in series. The output voltage (V_{out}) level of the BWSCGD is selected using four-bit state signals at the rising edges of the state changing signal (CLK_{state}). The switched-capacitor circuit is composed of C_{fly1-4} and power stages 1 and 2. Switches N1-14 are composed of on-chip bootstrap circuits, level shifters and LDMOS devices to deal with high voltage. In power stage 1, a bidirectional switch is used to prevent unintended turn-on via the LDMOS parasitic diode when the switch is off. The adoption of this switch enables the segmentation of V_g using binary-weighted voltage. As shown in the timing diagram of Fig. 19.5.1, V_{out} is updated on the rising edges of CLK_{state}. The state change timing generator in Controller 2 automatically generates the CLK_{state} based on the edges of the PWM signal and the rising edges of the output of comparators $COMP_{ON}$ and $COMP_{OFF}$. CLK_{state} can also be controlled by external signals experimentally. The gate-current (I_G) feedback control automatically adjusts the state change timing to maintain I_G at the desired level, thereby improving the trade-off between E_{drv} and the switching time of the gate voltage (t_{sw}).

Figure 19.5.2 shows the operation of the BWSCGD. During turn-on, gate driving operations (GDOs) are performed in the order of Φ_0, Φ_1, Φ_2, ..., Φ_9 and V_{out} increases from 0-to-20V at 2.5V steps. During turn-off, the states are traversed in reverse order, Φ_9, Φ_8, Φ_7, ..., Φ_0 and most of the gate charge is recycled from C_{iss} to the flying capacitors. This yields pseudo-adiabatic gate charging in eight steps (nine voltage levels), ideally reducing E_{drv} by 1/8 (= 2.5V/20V). After the turn-on operation is completed, a pull-up state Φ_9 connects the 20V power supply directly to V_{out} to compensate for the residual voltage shortage that the SC network alone cannot supply. The pull-up action increases the amount of charge recycled to the flying capacitor during turn-off, thereby shortening the charge reset operations (CROs) Φ_{R0}-Φ_{R3}.

The CROs precede the GDOs, initializing the flying capacitor charge before the first switching event and replenishing the charge lost during the previous GDOs. During the states Φ_{R0}, Φ_{R1}, and Φ_{R2}, voltage rebalancing is performed to restore the desired binary-weighted voltage ratio across C_{fly1-4}. At Φ_{R3}, a single-phase replenishment from the 20V power supply restores the lost charge to the series-connected C_{fly1-4}. This replenishment is enabled by the prior voltage rebalancing and the binary-weighted capacitance of C_{fly1-4}. After Φ_{R3}, the sequence returns to Φ_0 and GDOs resume.

Figure 19.5.3 shows the trade-off between t_{sw} and E_{drv} during turn-on. In the bottom left of Fig. 19.5.3, reducing t_{sw} requires shorter state intervals. However, if transitions are too fast, the voltage gap between V_{out} and the gate-voltage increases, resulting in higher driving loss and limited E_{drv} reduction. Conversely, as shown in the top-right of Fig. 19.5.3, a longer state interval reduces this voltage difference and E_{drv} but significantly extends t_{sw}. Therefore, the trade-off can be adjusted by tuning the state transition interval. However, the C_{iss} becomes nonlinear in the Miller plateau region, which limits optimization for the trade-off when a fixed interval time is used. Therefore, as shown in the bottom right of Fig. 19.5.3, a state interval time optimization based on I_G feedback control is implemented. During turn-on, a comparator monitors the sensed I_G corresponding voltage V_{Ig} against the threshold current ($I_{G\_ref\_ON}$) corresponding voltage V_{REF_ON} in Fig. 19.5.1. The comparator detects that V_{Ig} has dropped below V_{REF_ON} and advances the state to the next step. With this control method, the system naturally applies longer interval times during the Miller plateau region and shorter interval times elsewhere, approximately maintaining the voltage difference at a near-constant level. As a result, t_{sw} becomes shorter than with fixed-interval slow driving, while E_{drv} becomes lower than with fixed-interval fast driving, thereby mitigating the trade-off. The same feedback strategy is also applied during turn-off.

Figure 19.5.4 shows the measurement results when driving a full SiC module (BSM600D12P3G001) using the BWSCGD. A double-pulse test was performed at 200V and 10A. In the left of Fig. 19.5.4, the E_{drv} of a conventional gate driver (CGD) is 30 µJ, whereas the BWSCGD achieves 4.4 µJ from E_{drv} of the 20V power supply, and 5.5µJ from the 5V power supply including the bootstrap circuit, for a total of 9.9µJ. This corresponds to a 67% reduction compared with a CGD. Considering only the 20V supply, the reduction reaches 85%. The right of Fig. 19.5.4 shows the measurement results of t_{sw} versus E_{drv}, where t_{sw} denotes the sum of the turn-on and turn-off time of V_g. Without I_G feedback control, a trade-off is observed between t_{sw} and E_{drv}. The I_G feedback-based adaptive state interval control reduces t_{sw} by 19% at equal E_{drv} and reduces E_{drv} by 9% at equal t_{sw}. When enabling the feedback control, the 5V power supply incurs a 60mW static loss due to the additional circuitry, including controller 1, 2, the I_G sensor, and the comparators. However, this overhead is negligible compared to the gate-drive power at high switching frequency; for example, E_{drv} = 9.9µJ at f_{sw} = 100kHz corresponds to 990mW.

Figure 19.5.5 shows the switching waveforms corresponding to Fig. 19.5.4. Under fast driving, the timings of the V_{out} increment and decrement are fixed to 160ns by the external CLK_{state}. Under the I_G feedback driving, the I_G sensor triggers state transitions at the V_{Ig} crossover points with V_{REF_ON} and V_{REF_OFF}. In the Miller plateau region, the state interval is extended, thereby suppressing the voltage difference as intended. Conversely, outside this region, transitions occur rapidly, effectively balancing t_{sw} and E_{drv} as shown in Fig. 19.5.3.

Figure 19.5.6 presents a comparison table. The BWSCGD implements a gate driver that applies eight-level segmentation of V_g for driving power modules with extremely large Q_g. The effectiveness of the E_{drv} reduction is experimentally demonstrated. Furthermore, the I_G feedback control resolves the trade-off and eliminates the external CLK source by automating the state transition. Compared with the conventional UWSC, the BWSC architecture offers potential for miniaturization due to its ability to achieve a higher equivalent capacitance with fewer series connections.

Figure 19.5.7 shows a micrograph of the BWSCGD ICs and a board-level photograph of the flying capacitors C_{fly1-4} mounted between chips. The ICs are fabricated using a 0.6µm 40V HV-CMOS process. Each chip is 3.4mm × 6.89mm.

ISSCC 2026 / February 17, 2026 / 3:00 PM

References:

[1] Hao Peng, Han Peng , Ziyue Dang , Yong Kang , Zhiqiang Wang , Maojun He, and Xudan Liu, "A Driving Loss and Speed Co-Optimized Series Resonant Gate Driver with Novel Time Segmented Methodology for High Frequency SiC MOSFETs," in *Proc. 2020 IEEE Applied Power Electronics Conference and Exposition (APEC)*, New Orleans, LA, USA, Jun 2020, pp. 1599-1603. https://doi.org/10.1109/APEC39645.2020.9124508

[2] Wei Liu, Yongzhi Zhu, and Ming Liu, "A High Frequency Active Clamping Source Based SiC Resonant Gate Driving Technology for Multi-kW and Tens of MHz Switched Mode Power Amplifier," *IEEE Transactions on Power Electronics,* vol. 40, Issue. 6, pp. 8380-8394, Feb 2025. https://doi.org/10.1109/TPEL.2025.3540757

[3] Yanqiao Li, Ziyu Xia, Jason T. Stauth, "A Pseudo-Adiabatic Switched-Capacitor Gate Driver for Si and GaN FETs Achieving >5x Power Reduction", in *Proc. 2024 IEEE Custom Integrated Circuits Conference (CICC)*, Denver, CO, USA, May 2024, pp.1-2. https://doi.org/10.1109/CICC60959.2024.10528993

[4] Si-Yi Li, Wei-Chien Hung, Tz-Wun Wang, Ya-Ting Hsu, Ke-Horng Chen, Kuo-Lin Zheng, Ying-Hsi Lin, Shian-Ru Lin, Tsung-Yen Tsai, "High Common-Mode Transient Immunity GaN-on-SOI Gate Driver for High dV/dt SiC Power Switch", in *Proc. 2023 IEEE International Solid-State Circuits Conference (ISSCC)*, San Francisco, CA, USA, Mar 2023, pp. 302-304. https://doi.org/10.1109/ISSCC42615.2023.10067394

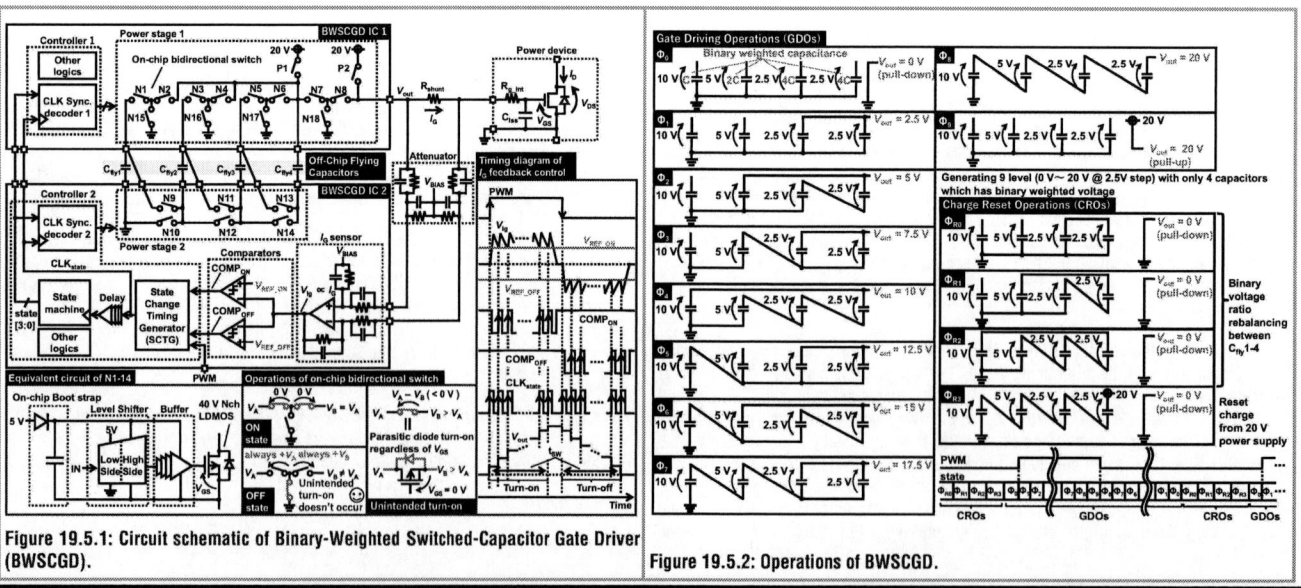

Figure 19.5.1: Circuit schematic of Binary-Weighted Switched-Capacitor Gate Driver (BWSCGD).

Figure 19.5.2: Operations of BWSCGD.

DIGEST OF TECHNICAL PAPERS • 329

979-8-3315-8937-0/26 $31.00 © 2026 IEEE

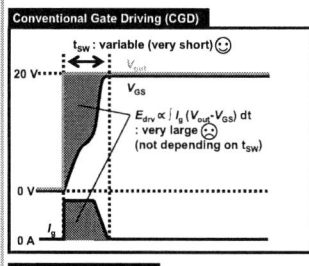

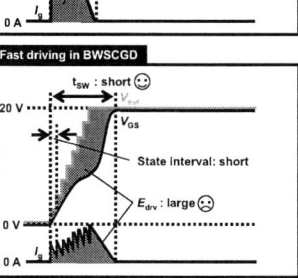

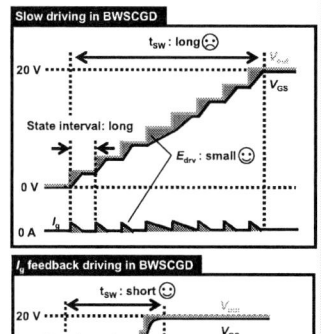

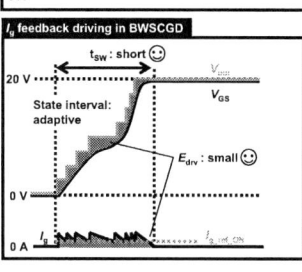

Figure 19.5.3: Switching waveform of conventional gate driving (CGD), fast driving, slow driving and I_G feedback driving in BWSCGD.

Figure 19.5.4: Measured driving loss in each operation mode, improving trade-off between switching time and driving loss.

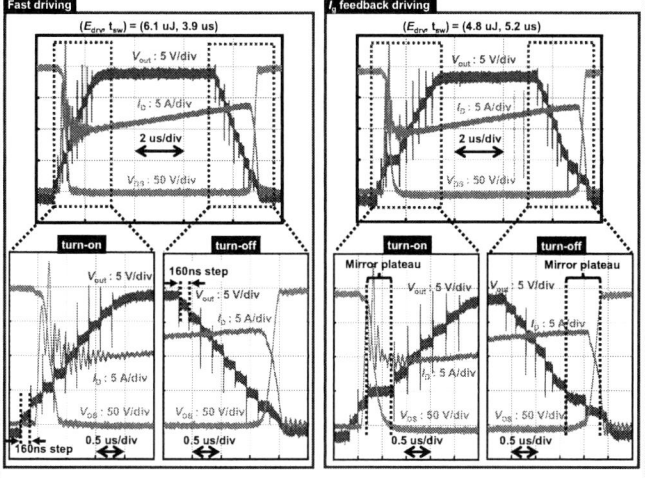

Figure 19.5.5: Measured switching waveform of the fast driving and I_G feedback driving in Fig.19.5.4.

	APEC 2020 [1]	TPEL 2025 [2]	CICC 2024 [3]	ISSCC 2023 [4]	This Work
Process	No chip integration	No chip integration	0.13 um RF-SOI	0.18 um CMOS & GaN	0.6 um 40 V HV-CMOS
Gate driver topology	Resonant-LC	Resonant-LC	UWSC	N/A	BWSC
Target power device	Discreate SiC	Discreate SiC	Discreate GaN & Si	Discreate SiC	Full SiC module
Implementation	PCB	PCB	1 IC	3 IC	2 IC
External passive element	1 inductor (150 nH)	1 capacitor (1.32 nF) 2 inductors (90 nH x 2)	5 capacitors (1 uF x 5)	Not used	4 capacitors (1 uF, 2.2 uF, 4.7 uF x 2)
V_g	25 V	18 V	5 V	24 V	20 V
Number of V_g segment	1	1	5	1	8
$Q_{g, max}$	194 nC	28 nC	49 nC	N/A	1700 nC
Power reduction rating w/o controller	69 %	82.4 %	~80 %	N/A	85 %
Power reduction rating w controller	N/A	N/A	N/A	N/A	67 %

Figure 19.5.6: Comparison table.

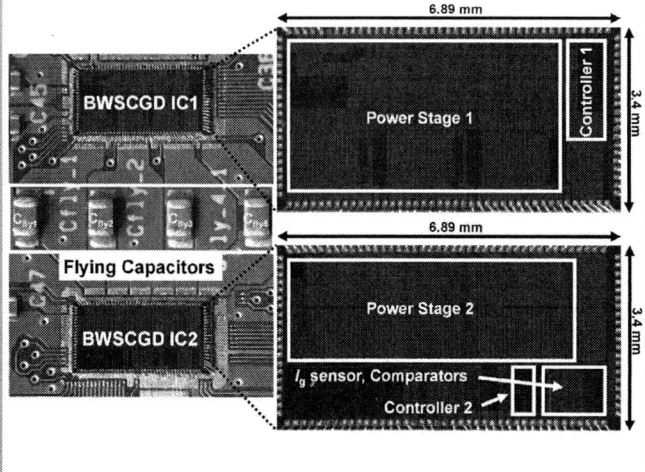

Figure 19.5.7: Micrograph of BWSCGD ICs and flying capacitors.

ISSCC 2026 / SESSION 19 / HIGH-VOLTAGE, ISOLATED AND DISPLAY POWER / 19.6

19.6 A 68%-Peak-Efficiency Single-Transformer Multi-Output Isolated DC-DC Converter with a Regulated Negative Rail

Qiao Huang*[1], Dongfang Pan*[1], Weiwei Xu[2], Yongle Zhang[1], Lin Cheng[1,2]

[1]University of Science and Technology of China, Hefei, China, [2]Hefei CLT Microelectronics, Hefei, China
*Equally Credited Authors (ECAs)

Abstract

This paper presents a single-transformer multi-output isolated DC-DC converter generating three regulated rails (+15V, −5V, and +5V) from wide 12/24V inputs. By sharing a common secondary winding in a dual-doubler rectifier configuration, the design achieves three outputs using only four power transistors. Additionally, a reconfigurable two-mode inverter ensures high efficiency across both input voltage conditions. The converter reaches a peak efficiency of 68% with maximum output power of 2.5W.

SiC and IGBT devices in traction inverters and on-board chargers require isolated positive (V_{DD1}) and negative (V_{EE}) rails (e.g., +15V/−5V) for the isolated gate drivers, with V_{EE} suppressing Miller-induced crosstalk to prevent spurious turn-on. In parallel, a low-voltage (LV) rail V_{DD2} (e.g., 5V) is needed to power monitoring, protection and communication circuits (Fig. 19.6.1). In many commercial products [1,2], V_{DD2} is often derived from V_{DD1} via an additional regulator, which increases cost and reduces efficiency. Therefore, an isolated converter that supplies all rails concurrently is highly desirable. However, in existing multi-output isolated converters, the rectifier in the receiver (RX) provides only two LV rails using six [3] or eight [4] switches and cannot generate the required high-voltage (HV) or negative rails. Replacing the LV switches in [3] and [4] with HV devices to obtain an HV output substantially degrades efficiency. Dual-output topologies and control methods have also been explored in wireless power transfer (WPT) systems [5-7], but they remain restricted to LV outputs and lack a regulated negative rail. Moreover, with a loosely coupled power link and an open-loop transmitter (TX), the design considerations are substantially different.

In this work, a single-transformer multi-output isolated DC-DC converter with a dual-doubler rectifier is proposed, delivering three regulated rails from a 12/24V input voltage (V_{IN}): +15V (V_{DD1}), −5V (V_{EE}), and +5V (V_{DD2}). As shown in Fig. 19.6.1, the RX employs two voltage doublers sharing a common secondary, enabling concurrent generation of HV, LV, and negative domains. A key enabler of this operation is the voltage doubler capacitor C_2, which ensures proper realization of the three outputs. Without C_2, the average RX port voltages at V_{SP} and V_{SN} become equal at steady state, forcing V_{DD2} to follow $V_{DD1}+V_{EE}$. With C_2, the average capacitor voltage equals ($V_{DD1}+V_{EE}−V_{DD2}$)/2, thereby separating the HV and LV domains and decoupling all three rails for independent regulation. In addition, C_2 resonates with the secondary inductance L_2 to suppress circulating current and improve transformer efficiency. Compared with existing designs, this topology offers two key advantages. First, a minimal four-device RX power stage delivers all three outputs without multiple windings or full-bridge rectifiers, thereby reducing conduction loss and gate-drive overhead. The HV branch consists of a 24V Schottky diode (D_1) and a 24V nLDMOS (M_1), while the LV branch employs a 6V nLDMOS (M_2) and a 6V pLDMOS (M_3) forming a complementary pair to drive the LV doubler. Second, in isolated power systems with closed-loop TX control [8-15], maximum link efficiency is achieved when the voltage gain is unity. A 1:1 transformer ratio maximizes the coupling coefficient (k), whereas higher ratios degrade efficiency. To maintain this condition across the wide input range, a reconfigurable half-/full-bridge LLC inverter operating at 16MHz with zero-voltage switching (ZVS) is adopted, with all TX switches implemented using 24V nLDMOS devices. The TX operates in full-bridge+doublers mode at a 12V input voltage and half-bridge+doublers mode at a 24V input voltage, thereby preserving identical voltage gain conditions and sustaining high efficiency.

Figure 19.6.2 illustrates the operating principle of the proposed dual-doubler rectifier. The rectifier operates in four modes depending on the output condition, with Modes 1–3 each containing two charging phases. In Mode 1, the devices conduct alternately: phase Φ_1 charges V_{DD1}, while phase Φ_2 charges V_{DD2} and V_{EE}, supplying both domains within a cycle. In Mode 2, the LV high-side switch M_3 is off while M_2 remains on, so C_{L1} and C_{L2} in the HV domain continue charging while C_{L3} discharges, which corresponds to HV-domain charging. In Mode 3, the HV high-side switch D_1 is off while M_1 remains on, so C_{L3} is charged while C_{L1} discharges, which corresponds to LV-domain charging; here the opposite-phase tank currents cancel at C_{L2}, yielding no net charging, while the load current causes C_{L2} to discharge. In Mode 4, both the TX stage and all RX switches are off, and C_{L1}–C_{L3} supply the loads with stored energy. With the rectifier input equivalently modeled as a fully differential current source [16], the charging behavior of each rail is clearly determined by the phase arrangement. As shown in Fig. 19.6.2, a key feature of the operation is that in Mode 1 both $V_{DD1}−V_{EE}$ and V_{DD2} are charged. Depending on the relative load requirements, the rectifier then smoothly transitions to Mode 2 when I_{L1} exceeds I_{L2} to allocate more power to the HV domain, or to Mode 3 when I_{L1} is smaller to allocate more power to the LV domain. Meanwhile, in Modes 1 and 3, the charging current of V_{DD2} remains identical and independent of the HV rails. As a result, the HV and LV domains are simultaneously regulated, cross-regulation is minimized, and smooth mode transitions are maintained under dynamic loading.

Figure 19.6.3 illustrates the system block diagram of the proposed isolated converter. The dual-doubler rectifier is controlled by two independent type-II error amplifiers (EAs), each comparing its regulated output with a shared RAMP signal to generate PWM_1 and PWM_2. These two signals define the rectifier state, where <PWM_1, PWM_2> = 11, 10, 01, and 00 correspond to Modes 1–4, respectively. For closed-loop coordination, PWM_1 and PWM_2 are combined into PWM_{RX}, encoded as a differential on–off keying (OOK) signal, and transmitted across the isolation barrier to the TX, where it is decoded as PWM_{TX} to regulate V_{DD1}, V_{EE}, and V_{DD2}. V_{EE} serves as the reference ground, while the differential EAs independently sense $V_{DD1}−V_{EE}$ and $V_{DD2}−GND_S$, enabling separate regulation of the HV and LV domains, with a shunt regulator included to maintain balance between V_{DD1} and V_{EE} under extreme conditions. As a result, when the load current I_{L2} of V_{DD2} varies, PWM_2 adjusts its duty cycle and switches the rectifier between Modes 2 and 3 to compensate, while PWM_1 and the $V_{DD1}−V_{EE}$ regulation remain unaffected. Sharing a common RAMP ensures that rectifier transitions follow the orderly sequence (Mode1–2–4 or Mode1–3–4), preventing conflicts between domains. Since the total secondary resonant current is constant across modes, these transitions are inherently smooth, ensuring stable operation under dynamic loading.

The TX and RX were fabricated using a 0.18μm BCD process and assembled with a PCB transformer using 1.5-oz copper and a 230-μm FR4 insulator as a proof of concept, providing at least 5kV isolation. Figure 19.6.4 shows the measured steady-state waveforms of V_{DD1}, V_{DD2}, V_{EE}, and the PWM_{TX} signal at $V_{IN} = 12V$ when both load currents I_{L1} and I_{L2} are 20mA and 80mA. With $V_{IN} = 12V$ and output capacitors $C_{L1} = 3.3\mu F$, $C_{L2} = 10\mu F$, and $C_{L3} = 10\mu F$, the measured ripples of V_{DD1}, V_{DD2}, and V_{EE} are 80mV, 20mV, and 50mV, respectively. Figure 19.6.4 shows the measured load-transient responses: when I_{L1} is fixed at 50mA and I_{L2} steps from 20mA to 80mA, and when I_{L2} is fixed at 50mA and I_{L1} steps from 20mA to 80mA. The corresponding overshoot and undershoot of $V_{DD1}−V_{EE}$ and V_{DD2} are 200mV and 120mV with recovery times of 360μs and 100μs, respectively. These results confirm independent regulation of the HV and LV domains with minimized cross-regulation and stable operation under dynamic loading.

Figure 19.6.5 shows the measured efficiency of the converter under different operating conditions. With V_{DD1}/V_{EE}=+15V/−5V and V_{DD2}=+5V, a peak efficiency of 68% is achieved at V_{IN}=12V/24V when both I_{L1} and I_{L2} are 100mA. Under the 24V V_{IN}, the reconfigurable TX operates in half-bridge mode, which improves the efficiency by 10% compared with full-bridge operation, confirming effective wide-input optimization. Figure 19.6.6 compares the performance of the proposed isolated DC-DC converter with state-of-the-art designs. This work demonstrates concurrent generation of +15V, −5V, and +5V outputs using a dual-doubler rectifier with only four power transistors. Figure 19.6.7 shows the die micrograph of the RX and TX chips along with the parameters of the transformer.

Acknowledgement:
This work was supported in part by the National Natural Science Foundation of China under Grant U23A20353. Corresponding author: Lin Cheng (eecheng@ustc.edu.cn).

References:
[1] Texas Instruments, TIDA-01605, "Automotive Dual-Channel, SiC MOSFET Gate Driver Reference Design With Two-Level Turnoff Protection," Accessed on June. 3, 2020, < https://www.ti.com.cn/lit/ug/tidue55b/tidue55b.pdf>.

[2] MPS, "Designing Multiple Independent Auxiliary Power Supplies with Magentic Isolation," Accessed on August. 8, 2023, <https://media.monolithicpower.com/mps_cms_document/2/0/2023-en-wechat-designing-multiple-independent-auxiliary-power-supplies_r1.0.pdf>.

[3] J. Jiang et al., "A 63% Efficiency 1.29-W Single-Link Multiple-Output (SLiMO) Isolated DC-DC Converter Using FPC Micro-Transformer with Local Voltage and Global Power Regulations," *IEEE JSSC*, vol.59, no.3, pp. 804-816, March. 2024. https://doi.org/10.1109/JSSC.2023.3330173

[4] J. Jiang et al., A Single-Link Multi-Domain-Output (SLiMDO) Isolated DC-DC Converter with Passive Magnetic Flux Sharing for Local Energy Distribution and Rx Behavior Sensing-Based Global Power Modulation," *ISSCC*, pp. 530-531, Feb. 2025. https://doi.org/10.1109/ISSCC49661.2025.10904792

[5] Q. Zhuang et al., "A 6.78MHz Single-Stage Regulating Rectifier with Dual Outputs Simultaneously Charged in a Half Cycle Achieving 92.2% Efficiency and 131mW Output Power," *ISSCC*, pp. 188-120, Feb. 2025. https://doi.org/10.1109/ISSCC49661.2025.10904779

[6] Y. Liu et al., "A 13.56-MHz Single-Input Dual-Output Wireless Power and Data Transfer System for Bio-Implants," *IEEE JSSC*, vol. 59, no. 8, pp. 2557-2567, Aug. 2024. https://doi.org/10.1109/JSSC.2024.3372430

[7] Z. Luo et al., "A 90%-Efficiency 40.68 MHz Single-Stage Dual-Output Regulating Rectifier with ZVS and Synchronous PFM Control for Wireless Powering," *ISSCC*, pp. 454-455, Feb. 2023. https://doi.org/10.1109/ISSCC42615.2023.10067331

[8] W. Qin et al., "An 800mW Fully Integrated Galvanic Isolated Power Transfer System Meeting CISPR 22 Class-B Emission Levels with 6dB Margin," *ISSCC*, pp. 246-247, Feb. 2019. http://doi.org/10.1109/ISSCC.2019.8662349

[9] Y. Zhuo et al., "A 52% Peak-Efficiency >1W Isolated Power Transfer System Using Fully Integrated Magnetic-Core Transformer," *ISSCC*, pp. 244-245, Feb. 2019. http://doi.org/10.1109/ISSCC.2019.8662301

[10] D. Pan et al., "A 1.25W 46.5%-Peak-Efficiency Transformer-in-Package Isolated DCDC Converter Using Glass-Based Fan-Out Wafer-Level Packaging Achieving 50mW/mm2 Power Density," *ISSCC*, pp. 468-469, Feb. 2021. http://doi.org/10.1109/ISSCC42613.2021.9365955

[11] D. Pan et al., "A 1.2W 51%-Peak-Efficiency Isolated DC-DC Converter with a Cross-Coupled Shoot-Through-Free Class-D Oscillator Meeting the CISPR-32 Class-B EMI Standard," *ISSCC*, pp. 240-242, Feb. 2022. http://doi.org/10.1109/ISSCC42614.2022.9731554

[12] D. Pan et al., "An Isolated DC–DC Converter Using a Cross-Coupled Shoot-Through-Free Class-D Oscillator with Low EMI Emissions," *IEEE JSSC*, vol. 59, no. 10, pp. 3457-3467, Oct. 2024. http://doi.org/10.1109/JSSC.2024.3407599

[13] Q. Huang et al., "A Dual-LC-Resonant Isolated DC-DC Converter Achieving 65.4% Peak Efficiency and Inherent Backscattering," *ISSCC*, pp. 534-535, Feb. 2025. http://doi.org/10.1109/ISSCC49661.2025.10904821

[14] D. Pan et al., "A 2W 53.2%-Peak-Efficiency Multi-Core Isolated DC-DC Converter with Embedded Magnetic-Core Transformer Achieving CISPR-32 Class-B EMI Compliance and <5mV Ripple," *ISSCC*, pp. 536-537, Feb. 2025. https://doi.org/10.1109/ISSCC49661.2025.10904708

[15] T. Xia et al., "A 180MHz 45.3%-Peak-Efficiency Isolated Converter Using Q-Downsize Class-D Power Amplifier with Inherent Shoot-Through Current Blocking and High Tolerance for Efficiency Despite Frequency Misalignments," *ISSCC*, pp. 528-529, Feb. 2025. https://doi.org/10.1109/ISSCC49661.2025.10904701

[16] L. Cheng et al., "A 6.78-MHz Single-Stage Wireless Power Receiver Using a 3-Mode Reconfigurable Resonant Regulating Rectifier", *IEEE JSSC*, vol. 52 no. 5, pp. 1412-1423, May. 2017. https://doi.org/10.1109/JSSC.2017.2657603

[17] Texas Instruments, UCC14241-Q1, "Automotive, 2-W, 24-Vin 25-Vout high-density > 5-kVRMS isolated DC/DC module." Apr. 2023, <https://www.ti.com/lit/ds/symlink/ucc14241-q1.pdf>.

[18] D. Pan et al., "A 24–20-V Isolated DC–DC Converter Using a Transformer-Based Supply-Generating Technique", *IEEE JSSC*, Early Access. http://doi.org/10.1109/JSSC.2025.3566939

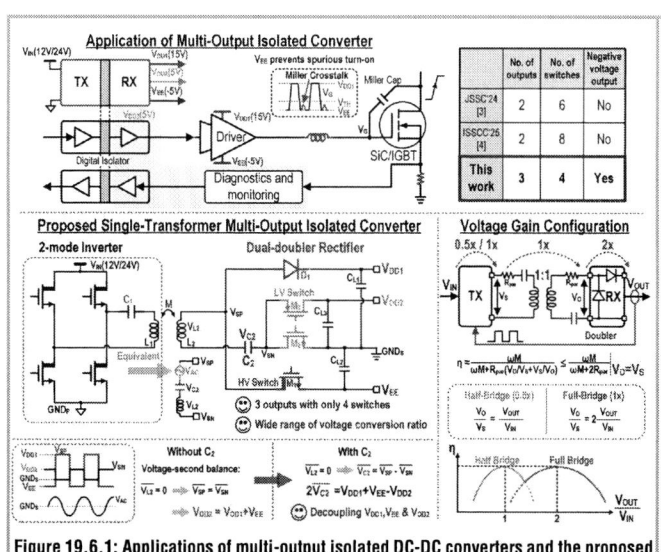

Figure 19.6.1: Applications of multi-output isolated DC-DC converters and the proposed isolated converter with dual-doubler rectifier.

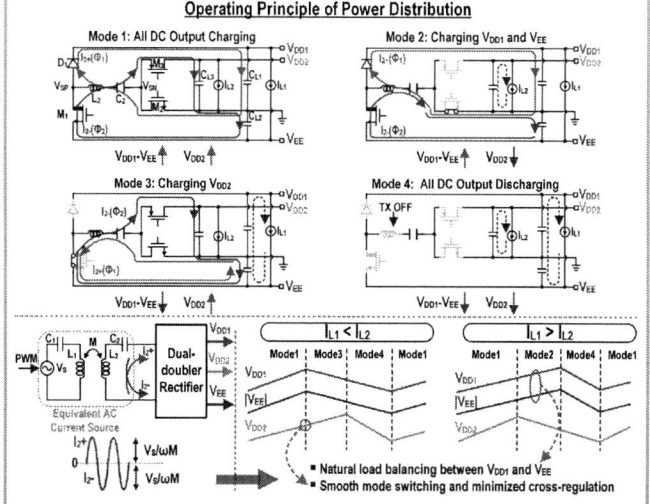

Figure 19.6.2: Operating principle of power distribution in the proposed dual-doubler topology with inherent load-balancing mechanism.

ISSCC 2026 / SESSION 19 / HIGH-VOLTAGE, ISOLATED AND DISPLAY POWER / 19.6

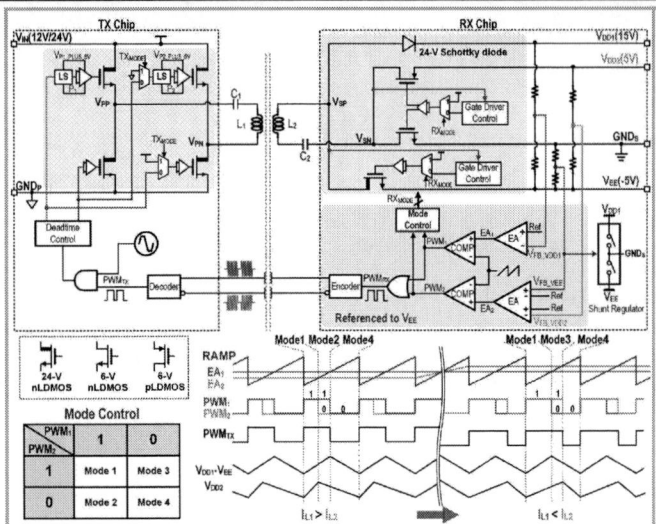

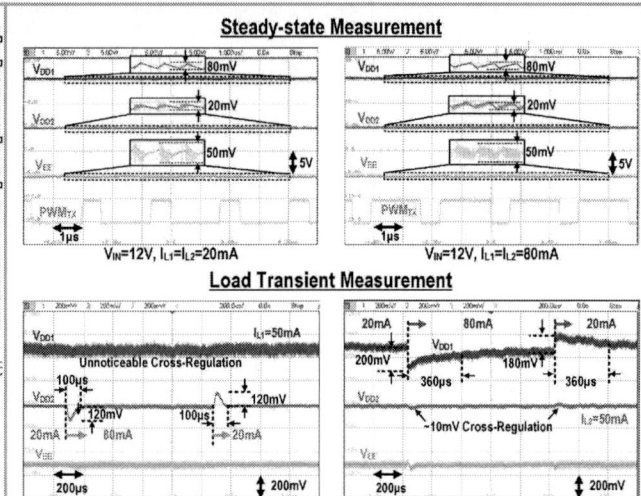

Figure 19.6.3: Block diagram and control scheme of the proposed multi-output isolated DC-DC converter.

Figure 19.6.4: Measured steady-state and load-transient waveforms of the proposed multi-output isolated DC-DC converter.

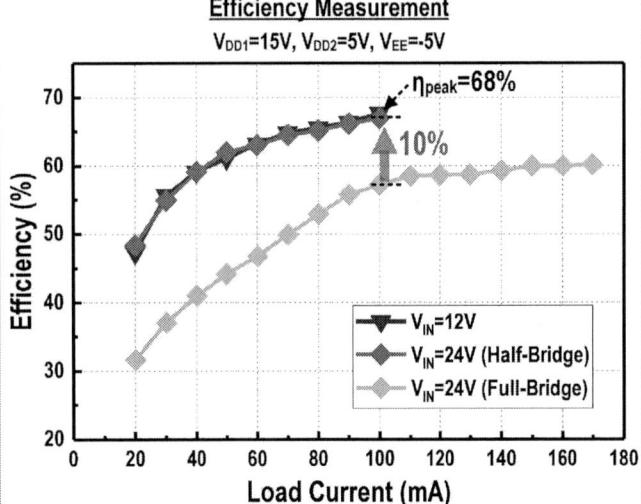

Figure 19.6.5: Measured efficiency of the proposed converter at different conditions.

	ISSCC'22[11]	JSSC'25[18]	JSSC'24[3]	ISSCC'25[4]	TI UCC14241[17]	This work
Technology	0.18μm BCD	0.18μm BCD	0.18μm BCD	0.18μm BCD	N/A	0.18μm BCD
Transformer Type	Coreless (In substrate)	Coreless (In substrate)	Coreless (In substrate)	Coreless (In substrate)	Magnetic-Core (In substrate)	Coreless (In substrate)
Switching Frequency	90MHz	15MHz	6.7-17.2MHz	7.6-23.8MHz	11-15MHz	16MHz
Input Voltage	5V	24V	3.3V	3.3V	24V	12V/24V
Output Voltage	5V	20V	3.3V	3.3V	20V/-5V	15V/5V/-5V
Maximum Load Current	240mA	300mA	460mA	460mA	100mA	200mA
Maximum P_{OUT}	1.2W	6W	1.29W	1.13W	2.5W	2.5W
No. of Outputs	1	1	2	2	2	3
No. of RX Switches	4	4	6	8	4	4
Peak Efficiency	51%	73%	63%	62.6%	60%	68%

Figure 19.6.6: Performance comparison of the proposed isolated DC-DC converter with state-of-the-art designs.

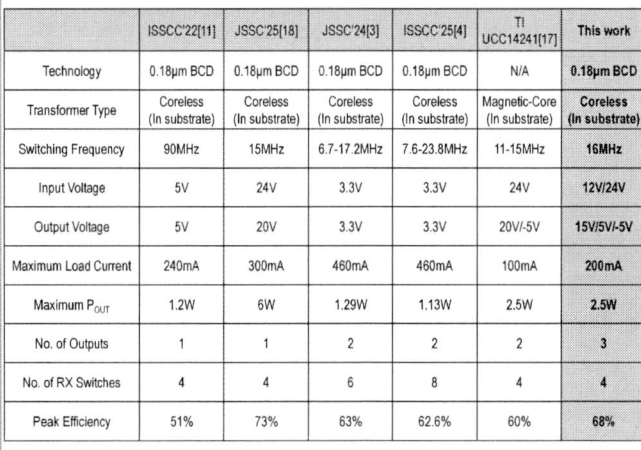

Figure 19.6.7: Die micrograph of the TX and the RX chips and EM simulation results of transformer.

• 2026 IEEE International Solid-State Circuits Conference

979-8-3315-8937-0/26 $31.00 © 2026 IEEE

ISSCC 2026 / SESSION 19 / HIGH-VOLTAGE, ISOLATED AND DISPLAY POWER / 19.7

19.7 A Hybrid Bipolar-Output Isolated Converter with +15V/–5V Outputs for SiC Gate Drivers

Joo-Mi Cho, Jeong-Hun Kim, Yong-Chan Lee, Sung-Wan Hong

Sogang University, Seoul, Korea

Abstract

This paper presents a hybrid bipolar-output isolated (HBO-I) gate-driver power supply (GDPS) that combines a flyback-based TX with primary-side regulation (PSR) and a switched-capacitor-based RX. Even with a compact 1:1 coupled inductor and all 5V CMOS devices, the HBO-I GDPS delivers bipolar outputs of +15V/–5V from a 5V input without additional 5V regulators. Therefore, the HBO-I GDPS attains ≈ 1kV galvanic isolation barrier, robust high CMTI, and 89.23% peak efficiency at 3W.

Silicon Carbide (SiC) power devices, which offer high power delivery capability, have emerged as key candidates to replace conventional silicon-based power devices. To ensure reliable operation of these devices, both a gate driver and the gate driver power supply (GDPS), each requiring a galvanic isolation barrier up to 1kV, are essential, as shown in Fig. 19.7.1 (bottom-left). Typically, a SiC gate driver requires +15V for turn-on. In addition, a negative gate voltage, such as –5V, is required for turn-off to suppress switching off losses caused by parasitic inductance, particularly in high-power applications, as illustrated in Fig. 19.7.1 (top-left) [1]. In this configuration, the source terminal of the SiC device defines the common ground reference (COM), thereby requiring the GDPS to generate bipolar output voltages with respect to COM [2-4].

Figure 19.7.1 (top-right) summarizes previous GDPS designs. Among them, the design in [4] generates bipolar outputs of +15V and –4V from a low input voltage (V_{IN}) of 5V. This approach enables the use of low-voltage (LV) devices on the primary side (TX), thereby reducing power loss. However, it requires a high-turns-ratio transformer (1:3.9) to produce both +15V and –4V, which increases system volume. Moreover, it operates under open-loop control, limiting the accuracy of output voltage (V_{OUT}) regulation. Furthermore, an additional regulator is needed to regulate –4V, and high-voltage (HV) diodes must be used on the secondary side (RX), resulting in additional power loss.

To improve V_{OUT} regulation in a compact size, the design in [5] uses a 1:1 on-board transformer with secondary-side regulation (SSR) feedback. However, due to the 1:1 ratio, this architecture requires a high V_{IN} of 24V to generate a V_{OUT} of 20V. Consequently, an additional auxiliary winding (L_{aux}) must be added on the TX side to supply 5V for TX switches and control circuits, which complicates the overall design. In addition, a separate regulator is also required on the RX side for the 5V SSR feedback circuit, which further reduces efficiency. Moreover, high V_{IN} and V_{OUT} necessitate HV devices on both the TX and RX sides, increasing conduction and switching losses. Since this structure cannot generate a negative voltage, the SiC gate driver devices suffer from large switching off losses. Furthermore, a galvanic isolated capacitor (C_{ISO}) used in the SSR path not only raises process cost but also adds another factor that degrades common-mode transient immunity (CMTI) [6]. As a result, in addition to the inherent CMTI weakness caused by the transformer, the C_{ISO} further weakens the CMTI.

Since SiC gate drivers typically require less than a few watts of power, the flyback converter is suitable for this application. The design in [7] adopts primary-side regulation (PSR) using the reflected V_{OUT}, enabling stable regulation without additional feedback components such as C_{ISO} [5], thereby reducing process cost and improving CMTI robustness. However, with a high V_{IN} of 12V, an additional regulator is still required to generate 5V for the gate driver and controller, which reduces efficiency. In addition, despite using a 1:2 turns-ratio transformer that increases the burden on system volume, this GDPS generates only a single 24V output. Moreover, the use of HV devices on both TX and RX sides further increases power losses.

To overcome these limitations, this work proposes a hybrid bipolar-output isolated (HBO-I) GDPS with a compact 1:1 coupled inductor, as shown in Fig. 19.7.1 (bottom-right). On the flyback-based TX side, the PSR scheme ensures accurate V_{OUT} regulation, and all TX circuits operate with V_{IN} of 5V. On the switched-capacitor (SC)-based RX side, an SC structure generates bipolar outputs of +15V and –5V. Overall, the flyback-based TX and SC-based RX constitute a hybrid flyback-SC topology, in which neither the TX nor RX side requires an additional 5V regulator.

Furthermore, all devices in this structure are rated for 10V. To withstand this voltage rating, all switching devices are implemented in a cascode structure with LV devices, avoiding HV devices that cause large power losses. In addition, the RX side switches operate as active diodes, eliminating the need to transfer a control signal from TX to RX and thereby further simplifying the design [8].

Figure 19.7.2 shows the system architecture of the proposed HBO-I GDPS. The TX and RX are connected through a 1:1 coupled inductor with a coupling coefficient of 0.98.

In the proposed GDPS, V_{OUT} is regulated using peak-current control. On the TX side, the primary-side current (i_P) is sensed from the intermediate-node voltage (V_{CS}) at the main switch (S_M), and V_{OUT} is indirectly sensed through the primary-winding voltage (V_P). The two quantities i_P and V_P are alternately sensed in each operation phase and the TX controller combines them to adjust the on-duty of S_M for V_{OUT} regulation.

On the RX side, the power stage consists of two flying capacitors (C_{F1} and C_{F2}), two NMOS-based active diodes (D_{N1} and D_{N2}), and two PMOS-based active diodes (D_{P1} and D_{P2}). C_{F1} is charged to 5V and C_{F2} to 10V, while a bootstrap circuit provides +10V referenced to COM for driving D_{N2} and D_{P2}. In addition, to prevent malfunctions due to ringing that can occur in discontinuous conduction mode (DCM), a latch-based ringing guard is implemented in the RX controller, as described in detail later.

Figure 19.7.3 illustrates the two-phase operation of the proposed HBO-I GDPS along with the timing diagram under heavy-load conditions in continuous conduction mode (CCM).

In phase Φ_1 (D·T), when the TX switch S_M turns on, i_P splits into the current (i_M) flowing through the magnetizing inductor (L_M) and the current (i_X) flowing through the primary-side inductor (L_P). Due to the 1:1 turns ratio of the coupled inductor, the voltage (V_S) across the secondary-side inductor (L_S) is approximately V_{IN}, and L_S delivers the current (i_S) to the RX side, which is equal to i_X. Therefore, the active diodes of D_{P1} and D_{P2} are turned on. Consequently, i_{CF1} charges C_{F1}, while $i_{OUT,P}$ discharges C_{F2} and charges $C_{OUT,P}$. During this operation, L_M continues to store energy, increasing i_M.

Since C_{F1} is connected in parallel with L_S, it is charged to approximately 5V, which is equal to V_S. In addition, C_{F1} and C_{F2} are connected in series and this series connection is placed in parallel with $C_{OUT,P}$, thereby charging $C_{OUT,P}$ with the sum of V_S and the voltage across C_{F2}.

In phase Φ_2 [(1–D)·T], S_M turns off. Accordingly, L_M releases the energy stored during Φ_1, and i_M flows through L_P in the opposite direction compared to Φ_1. Therefore, L_S draws i_S from the RX side through the active diodes D_{N1} and D_{N2}. In this phase, i_{CF12} charges C_{F2} and discharges C_{F1}, while $i_{OUT,N}$ charges $C_{OUT,N}$. In addition, $C_{OUT,N}$ and C_{F1} are connected in series and this series connection is paralleled with C_{F2}, establishing 10V across C_{F2}. As a result, by repeating these two phases, the voltage across $C_{OUT,P}$ becomes 15V.

Meanwhile, in Φ_2, V_S becomes equal to the voltage across $C_{OUT,N}$. Through the 1:1 coupled inductor, this voltage is reflected to the primary side, so that $V_P = V_{IN} - V_S$, which is approximately 10V. During each switching cycle, the primary side senses this V_P after it stabilizes. This indirect sensing of V_S, which contains V_{OUT} information, enables PSR control.

Figure 19.7.4 (left) shows the operational timing diagram under light-load conditions in DCM, both without and with the DCM ringing guard. Because the active diodes prevent reverse current, LC resonance occurs between L_S and the parasitic capacitance (C_{para}) of the switches under these conditions, causing oscillation of the switching node voltage. As a result, if the HBO-I GDPS operates without the DCM ringing guard, the active diodes repeatedly turn on and off, which increases the amplitude of i_S and eventually destabilizes the system's operation.

In contrast, with the DCM ringing guard, the latch-based structure ensures that active diodes do not erroneously turn on and off, keeping all switching devices in the OFF state after the first trigger. In the next phase, i_S reverses and V_S flips. This transition resets the latch, masking only DCM oscillations and allowing normal detection to resume in the next cycle. Consequently, the LC oscillation is attenuated, and the GDPS can continue operating in the next cycle without any issue.

Figure 19.7.4 (right) shows the start-up operation and waveforms. During start-up, S_M is driven with a limited gate-source voltage to suppress inrush current. When S_M turns on, i_S flows through the body diodes of D_{P1} and D_{P2}, which raises $V_{OUT,P}$. On the other hand, when S_M turns off, i_S flows through the body diodes of D_{N1} and D_{N2}, which lowers $V_{OUT,N}$. By repeating these operations, the voltage across each capacitor gradually increases.

332 • 2026 IEEE International Solid-State Circuits Conference

979-8-3315-8937-0/26 $31.00 © 2026 IEEE

Once the capacitor voltages rise above a predefined threshold, which is sufficient to drive the active diodes, the power-on-reset (POR) circuits enable the driving signals for the active diodes, and the system transitions into its normal operating mode.

Figure 19.7.5 (top) shows transient waveforms which evaluate the CMTI of the HBO-I GDPS when it delivers an output power of 1W. In this test, COM is varied between 0V and 1kV with a slope of approximately 100V/ns. The measured results confirm that both the output voltages and circuit operation remain stable, demonstrating the high CMTI robustness of the proposed GDPS.

Figure 19.7.5 (bottom) shows steady-state waveforms under different load current (I_{LOAD}) conditions. At light I_{LOAD}s of 10mA, 25mA, 50mA, and 75mA, the HBO-I GDPS operates in DCM. At heavy I_{LOAD}s of 100mA, 125mA, and 150mA, the proposed GDPS operates in CCM. These waveforms verify that the HBO-I GDPS operates properly in both DCM and CCM, ensuring stable operation across a wide I_{LOAD} range.

Figure 19.7.6 (top) illustrates the efficiency curve with the peak efficiency reaching 89.23% at the output power of 3W.

Figure 19.7.6 (bottom) compares the performance of the proposed design with previous works. Owing to the hybrid flyback–SC topology, the HBO-I GDPS delivers bipolar outputs of +15V and –5V while operating with a compact 1:1 turns-ratio coupled inductor and a 5V input. In addition, the HBO-I GDPS does not require additional 5V regulators on both the TX and RX sides, because the TX side directly uses the 5V input and the RX side internally generates a 5V rail through its SC structure. Moreover, all switching devices on both the TX and RX sides are implemented using 5V CMOS devices, which reduces both power losses and process cost. Finally, owing to the hybrid flyback–SC topology, the HBO-I GDPS achieves the highest measured efficiency among the compared previous works, with a compact implementation.

Figure 19.7.7 shows the chip micrograph (top) and PCB photograph (bottom).

Acknowledgement:
This work was supported by National Research Foundation of Korea (NRF) Granted funded by the Korea government (MSIT) under Grant RS-2023-00207919 and IITP-2025-RS-2023-00260091

References:
[1] onsemi, AND90103/D, "M1 1200 V SiC MOSFETs & Modules: Characteristics and Driving Recommendations," Rev. 3, June 2022. https://www.onsemi.com/pub/Collateral/AND90103-D.PDF
[2] E. Serban et al., "High-Performance Isolated Gate-Driver Power Supply With Integrated Planar Transformer," *IEEE TPE*, vol. 36, no. 10, pp. 11409-11420, Oct. 2021. https://doi.org/10.1109/TPEL.2021.3070053
[3] Texas Instruments, UCC14240-Q1, "Automotive, 1.5-W, 12-Vin 25-Vout high-density > 5-kVRMS isolated DC/DC module," June. 2023. https://www.ti.com/lit/ds/symlink/ucc14240-q1.pdf
[4] Texas Instruments, TIDA-01605, "Automotive Dual-Channel, SiC MOSFET Gate Driver Reference Design With Two-Level Turnoff Protection," June 2020. https://www.ti.com/lit/ug/tidue55b/tidue55b.pdf

[5] D. Pan et al., "A 24V-to-20V 6W 73.2%-Peak-Efficiency Isolated DC-DC Converter Using a Transformer-Based Supply-Generating Technique," *IEEE CICC*, pp. 1-2, Apr. 2024. https://doi.org/10.1109/CICC60959.2024.10529013
[6] J. Weckbrodt et al., "A Bidirectional Communicating Power Supply Circuit for Smart Gate Driver Boards," *IEEE TPE*, vol. 35, no. 8, pp. 8540-8549, Aug. 2020. https://doi.org/10.1109/TPEL.2019.2960632
[7] L. Zhang et al., "Design Considerations for High-Voltage Insulated Gate Drive Power Supply for 10-kV SiC MOSFET Applied in Medium-Voltage Converter," *IEEE TIE*, vol. 68, no. 7, pp. 5712-5724, Jul. 2021. https://doi.org/10.1109/TIE.2020.3000131
[8] D. A. Philipps et al., "On Dead-Time Optimization and Active Gate Driving in Flyback Converters With Synchronous Rectifiers," *IEEE Access*, vol. 12, pp. 173146-173155, Sep. 2024. https://doi.org/10.1109/ACCESS.2024.3462956
[9] Analog Devices, LT8300, "100V_IN Micropower Isolated Flyback Converter with 150V/260mA Switch," Rev. A, June. 2019. https://www.analog.com/media/en/technical-documentation/data-sheets/lt8300.pdf

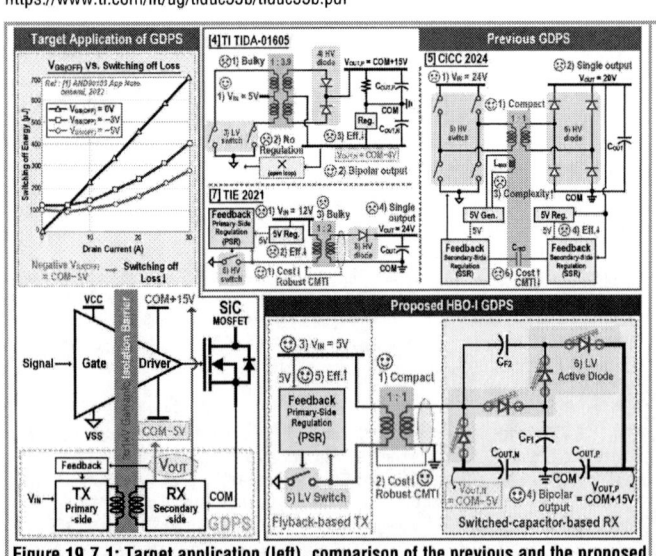

Figure 19.7.1: Target application (left), comparison of the previous and the proposed gate driver power supply (GDPS) (right).

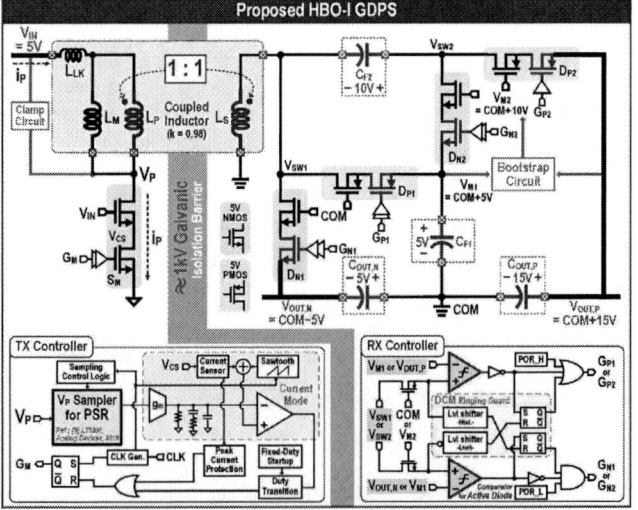

Figure 19.7.2: System architecture of the proposed hybrid bipolar-output isolated (HBO-I) GDPS.

ISSCC 2026 / SESSION 19 / HIGH-VOLTAGE, ISOLATED AND DISPLAY POWER / 19.7

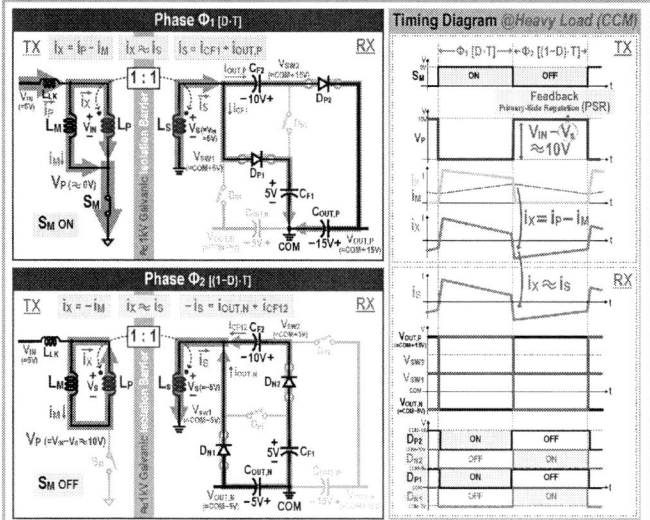

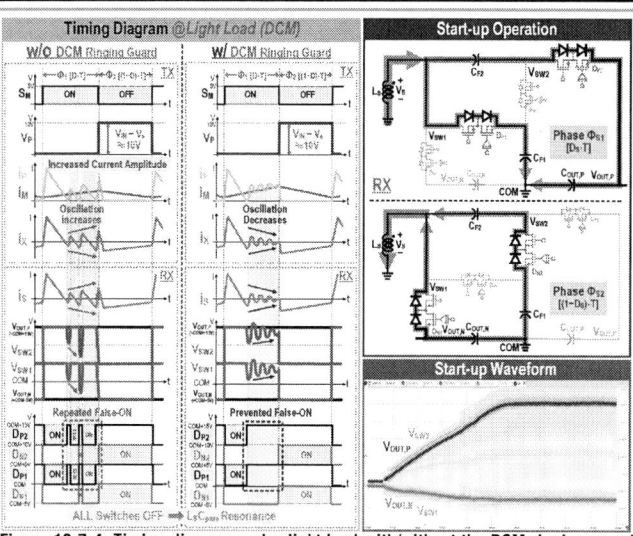

Figure 19.7.3: Operation of the two-phases for the proposed HBO-I GDPS (left) and timing diagram under heavy load (right).

Figure 19.7.4: Timing diagram under light load with/without the DCM ringing guard (left) and start-up operation and waveform (right).

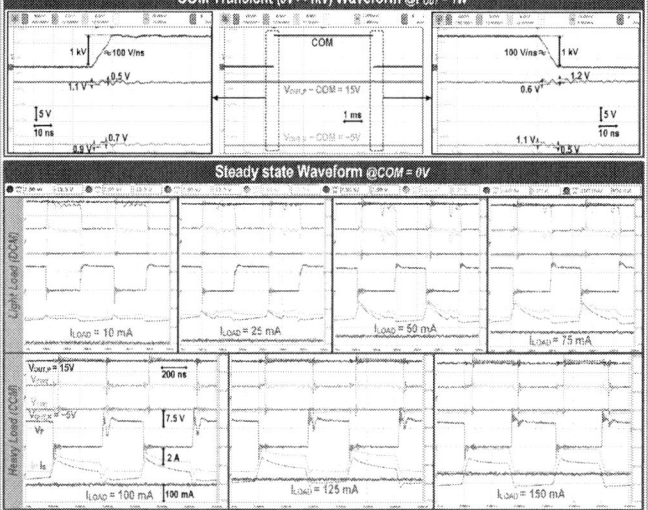

Figure 19.7.5: Measured waveforms of COM transient (top) and steady-state (bottom).

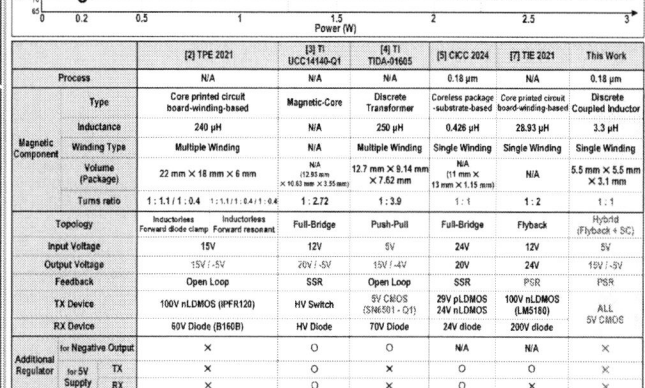

Figure 19.7.6: Efficiency plots (top) and comparison table (bottom).

		[2] TPE 2021	[3] TI UCC14140-Q1	[4] TI TIDA-01605	[5] CICC 2024	[7] TIE 2021	This Work
	Process	N/A	N/A	N/A	0.18 μm	N/A	0.18 μm
	Type	Core printed circuit board-winding-based	Magnetic-Core	Discrete Transformer	Coreless package -substrate-based	Core printed circuit board-winding-based	Discrete Coupled Inductor
Magnetic Component	Inductance	240 μH	N/A	250 μH	0.426 μH	28.93 μH	3.3 μH
	Winding Type	Multiple Winding	N/A	Multiple Winding	Single Winding	Single Winding	Single Winding
	Volume (Package)	22 mm × 18 mm × 6 mm	N/A	12.7 mm × 9.14 mm × 7.62 mm	(11 mm × 13 mm × 1.15 mm)	N/A	5.5 mm × 5.5 mm × 3.1 mm
	Turns ratio	1 : 1.1 / 1 : 0.4	1 : 2.72	1 : 3.9	1 : 1	1 : 2	1 : 1
	Topology	Inductorless Forward diode clamp / Inductorless Forward resonant	Full-Bridge	Push-Pull	Full-Bridge	Flyback	Hybrid (Flyback + SC)
	Input Voltage	15V	12V	5V	24V	12V	5V
	Output Voltage	15V / -5V	20V / -5V	15V / -4V	20V	24V	15V / -5V
	Feedback	Open Loop	SSR	Open Loop	SSR	PSR	PSR
	TX Device	100V nLDMOS (IPFR120)	HV Switch	5V CMOS (SN6501 - Q1)	29V pLDMOS / 24V nLDMOS	100V nLDMOS (LM5180)	ALL 5V CMOS
	RX Device	60V Diode (B160B)	HV Diode	70V Diode	24V diode	200V diode	
Additional Regulator	for Negative Output	×	O	O	N/A	N/A	×
	for 5V Supply TX	×	O	×	O	O	×
	RX	×	O	×	O	×	×
	Max. Output Power	4W	1.5W	1W	6W	2.4W	3W
	Peak Efficiency	85%	87.7%	< 50%	N/A	80%	89.23%

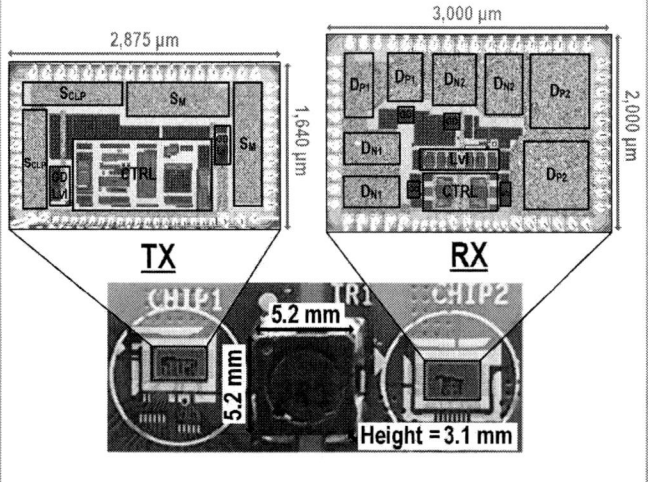

Figure 19.7.7: Chip micrograph (top) and PCB photograph (bottom).

• 2026 IEEE International Solid-State Circuits Conference

ISSCC 2026 / SESSION 19 / HIGH-VOLTAGE, ISOLATED AND DISPLAY POWER / 19.8

19.8 A Fully Integrated Bidirectional 5-Level Isolated DC-DC Converter with 42.5% Efficiency and 170mW/mm² Transformer Power Density

Kishalay Datta, Yue Wu, Charles Sullivan, Jason Stauth

Dartmouth College, Hanover, NH

Abstract

This work presents a bidirectional 5kV-isolated DC-DC converter, fully-integrated in a 180nm SOI CMOS process. The design uses a distributed multi-winding transformer to maximize utilization of the standard on-chip metal stackup. A 5-level hybrid full-bridge power stage allows fundamental and harmonic waveform shaping with current ballasting. The design achieves peak efficiency of 42.5% at over 550mW, for system power density over 69mW/mm² and transformer power density over 170mW/mm².

Power transfer across a galvanic isolation barrier is critical for safety and reliability in applications such as biomedical implants and isolated gate drivers [1,2]. The key challenges in this area are to achieve high power density with galvanic isolation over 5kV while maximizing power-transfer efficiency. Figure 19.8.1 shows past solutions which use separate transmitter (TX) and receiver (RX) chips and typically a transformer (XFMR) implemented on a separate die or interposer, leading to increased system complexity and cost. The XFMR is usually large and realized using advanced process flows [2-12] with thick gold or copper-based metal stacks, which are generally incompatible with regular CMOS technology nodes. While the wide metal traces of interposer-based XFMRs can provide low DC resistance, at high frequencies (10s-100s MHz), skin and proximity effect current crowding can dramatically increase the AC resistance, leading to high loss, low Q, and poor utilization of metal conductance, limiting solution size, frequency, and power-density.

In this work, we demonstrate a 5kV-isolated DC-DC converter that achieves 42.5% efficiency while delivering ~560mW with all circuitry and transformer windings fully integrated in a single die fabricated in a 180nm SOI CMOS process. Unlike previous solutions, this work uses the same integrated circuit for the RX and TX, coupling power through the integrated transformer in a chip-chip stack. This significantly reduces cost and complexity; it also enables the system to operate bidirectionally with the roles of TX and RX swapping as needed. To maximize the utilization of the limited on-chip metal conductance, the design uses multi-winding current ballasting [13,14], where the winding is split into multiple parallel traces, each with their width comparable to an effective skin depth, and driven by a distributed 'unit-cell' converter and resonant ballasting capacitor. Thus, each winding carries an optimal fraction of the total current and is less impacted by high-frequency magnetic-field-induced AC resistance effects, enabling efficient operation at smaller size and higher frequency. The distributed driver is based on a 5-level hybrid full-bridge TX/RX DC-DC converter that uses integrated flying and resonant capacitance to interface with the on-chip transformer windings. Compared to current-mode (class-D/E) drivers [5-9,15,16], voltage mode class-D and LLC approaches [2-4] can have lower peak voltage stress and allow accurate phase and amplitude control, but require active gate driving. Here, 5-level operation helps to improve efficiency and power density by allowing low-voltage devices to interface with a high-voltage supply; it also enables new degrees of waveform shaping to reduce switching harmonics [17], this can reduce power loss and electromagnetic interference.

Shown in Fig. 19.8.2, the 5-level full-bridge (5LFB) is configured in a 'unit-cell' that is distributed around the chip to drive individual windings with each switch rated to half the input voltage $V_{IN}/2$ or a quarter of the peak-peak drive voltage $V_{XPP}/4$ (note that for the symmetric RX, V_{IN} is swapped with V_{OUT}). Each 5LFB unit cell uses two pairs of complementary (merged-interleaved) 3-level buck (3LB) converters; the merged-interleaved structure reduces on-chip bypass capacitance and simplifies layout [18,19]. Each unit cell includes drive circuitry, level shifters, and bootstrap gate driving for the all-NMOS powertrain. On-chip flying and bypass capacitance is a MIM-MOS stack with ~10nF/mm² density ($C_F \approx 64pF$ each, $C_{BP} \approx 100pF$ each). Ballast capacitors C_R use only MIM, have approximately 20pF/unit cell and are tuned separately to resonate with each winding. The converter operates in phases Φ_0 to Φ_7, with $V_X = 0V$ in Φ_0 and Φ_4; $V_X = \pm V_{IN}$ in Φ_2 and Φ_6 respectively, and $V_X = \pm V_{IN}/2$ in $\Phi_{1/3}$ and $\Phi_{5/7}$ respectively. The waveform is synthesized by adjusting the time duration of each of the five levels, allowing voltage regulation by varying the fundamental voltage and current amplitudes and by shaping or eliminating certain harmonics. While even harmonics are generally small for the symmetric waveform, the 3rd, 5th and higher odd harmonics can be significantly reduced, improving efficiency and spectral purity.

Figure 19.8.3 shows the high-level layout and configuration of the unit-cells driving the multi-winding XFMR, which has an outer radius of 0.9mm and is split into eight parallel windings in four groups. Each group includes a thin inner winding (10µm, 12 turns) and a wide outer winding (30µm, 3 turns) which mainly use the top aluminum metal layer (3µm) in a standard 6-metal SOI CMOS stackup. The winding groups are concentric but rotated with quadrilateral symmetry to simplify layout and reduce AC magnetic field effects. Each thin inner winding connects to a single unit cell driver, while the wide outer windings connect to three parallel unit cells. Independent resonant capacitors cancel the reactive component of the winding impedance; the unit cell sizing matches the real impedance of the windings to approximate the optimum current density based on the power transfer versus loss of the individual winding [14]. This is roughly a factor of three for moderate chip-chip spacing

(50-to-100µm) assuming flipped die, surface/sidewall and bondwire resin coating, and 50µm polyester spacer to achieve >5 kV isolation. The unit cells connect to an outer power and control bus which simplifies routing and is designed to match timing delays to within 200ps to synchronize switching.

Figure 19.8.4 shows the implementations for the gate drivers (GD), level shifters (LS) and bootstrap (BS) circuits of each 5LFB leg. The LS uses a pulsed floating amplifier with a 5V cascode to provide symmetric rise and fall propagation delays of ~320ps. Here, a 200ps overlap on-time of M_{L1}-M_{L3} or M_{L2}-M_{L4} provides a high pulse current of ~300µA, which is amplified and rectified in the floating domain; diodes M_{L5}-M_{L6} provide a small leakage current of ~1.3µA for common-mode transient immunity. Due to low latency and tight timing control, complementary powertrain switch pairs S_1/S_3 and S_2/S_4 operate with short deadtime of ~100ps, which trades off overlap loss and diode conduction. A slight 50ps deadtime is also included in M_{PG}/M_{NG} buffer pair to reduce overlap gate drive loss, which can be significant at 75MHz. Gate drivers run off $V_{GD} \approx V_{IN}/2$ to power the NMOS powertrain. The BS is realized using an active nested scheme [20] where C_{B1} is charged from V_{GD}. C_{B1} provides power to C_{B2}, which in turn serves C_{B3}. All BS capacitances are 5.6pF. The active BS switches (M_{B1}-M_{B3}) are themselves driven off of a secondary passive BS network consisting of 5V Schottky diodes, D_{B1}-D_{B3}, and 480fF capacitors, C_{P1}-C_{P3}. For example, when S_2 is off and S_3 is on, C_{B2} is supplying charge to its GD, while C_{P2} is charged by C_{B1}. The diode D_1, connected between V_{GD} and V_{TP}, provides an auxiliary path to ensure C_F is charged to $\sim V_{GD}$ and maintains flying capacitor voltage balance. During startup, flying capacitors pre-charge through D_1 to $V_{GD} \approx V_{IN}/2$ less a diode drop; during normal operation, inherent feedback from the inductor current forces a natural balance of the flying capacitors to a median voltage $V_{IN}/2$, thus there is no current in D_1 in steady state operation.

Also shown in Fig. 19.8.4 are the timing requirements and implementation details. Efficiency is maximized when the currents through the same-side windings are phase aligned to each other, and the TX currents lead the RX currents by 90° [14]. To achieve this, an external reference clock (nominally 150MHz, and which need not be phase aligned or duty cycle accurate) passes through a 32-tap differential delay line (DL) and its last taps Y<31> and YB<31> are passed to a frequency divider of 2× (output nominally 75MHz), such that by tuning the supply of the DL in the RX, the relative phase shift between the GD signals of TX and RX can be adjusted to any desired phase. The divided clock feeds into another copy of the DL composed of 32×200ps unit delay cells. These delayed signals are strategically tapped and passed through three independent 8:1 multiplexers. The multiplexer outputs are combined so that their rising and falling edges can synthesize the GD input signals, which can accomplish any desired 5-level waveform shape independently in both TX and RX. The transmission of external clock signals across the isolation barrier can be accomplished using standard digital isolation architectures and is therefore not included in this work, as the focus is on achieving efficient and compact isolated power transmission.

Figure 19.8.5 shows the measurement results. The peak efficiency is 42.5% at power transfer of ~560mW from TX to RX, operating at 75MHz with V_{IN} = 3.8V and V_{OUT} = 2.4V. This corresponds to power density of 171.6mW/mm² for the on-chip transformer, demonstrating high utilization of the modest on-chip metal, and effective ballasting of the windings. The overall solution power density is 69.6mW/mm² including the total active die stackup and bonding area, shown in Fig. 19.8.7. Efficiency remains high for input voltages down to 1.8V and output voltages down to 0.9V, with load currents from 75mA to 300mA. While the focus of this work is not on active feedback control, the output voltage can be adjusted efficiently by tuning the fundamental of the 5-level waveform on the TX, RX or both. For example, at V_{IN} = 3.6V and with a fixed 9.9Ω resistive load, 5-level tuning can regulate V_{OUT} from 1.75V to 2.25V with efficiencies from 35-to-42%. An important feature of this work is the capability to support bidirectional power flow simply by adjusting the phase shift between the TX and RX. Figure 19.8.5 demonstrates efficient and symmetric bidirectional power transfer. For positive phase shifts, power flows from TX to RX; for negative phase shifts, it flows in the opposite direction with the TX and RX switching roles. It is also confirmed that orthogonal ±90° is optimum for maximum efficiency [14], thus bidirectional control can in principle be initiated by either side of the TX or RX independently.

Figure 19.8.6 shows a comparison table to recent high-performance work. This work introduces the use of a single unique die supporting TX, RX, and transformer functions that meets 5kV AC and DC withstanding. It uses a distributed 5-level full bridge with multi-

334 • 2026 IEEE International Solid-State Circuits Conference

979-8-3315-8937-0/26 $31.00 © 2026 IEEE

winding current ballasting to achieve the highest reported transformer power density and overall system power density of the state-of-the-art works highlighted in Fig. 19.8.6 using a standard power CMOS metal stackup. The design demonstrates efficient bidirectional power transfer and capabilities for waveform and harmonic tuning. Overall, this work highlights a roadmap to achieve higher efficiency and power density with smaller magnetics at higher frequencies using distributed resonant structures, current ballasting, and hybrid multi-level drivers.

Acknowledgement:
This work was supported by the Power Management Integration Center (PMIC), a National Science Foundation (NSF) I/UCRC under IIP 1822140.

References:
[1] B. Chen, "Fully integrated isolated dc-dc converter using micro-transformers," *2008 Twenty-Third Annual IEEE Applied Power Electronics Conference and Exposition*, Austin, TX, USA, 2008, pp. 335-338. https://doi.org/10.1109/APEC.2008.4522743
[2] Y. Zhuo et al., "A 52% Peak Efficiency > 1-W Isolated Power Transfer System Using Fully Integrated Transformer With Magnetic Core," in *IEEE Journal of Solid-State Circuits*, vol. 54, no. 12, pp. 3326-3335, Dec. 2019. https://doi.org/10.1109/JSSC.2019.2940333
[3] J. Jiang, L. Zhao, J. Tang and C. Huang, "A Single-Link Multi-Domain-Output (SLiMDO) Isolated DC-DC Converter with Passive Magnetic Flux Sharing for Local Energy Distribution and Rx Behavior Sensing-Based Global Power Modulation," *2025 IEEE International Solid-State Circuits Conference (ISSCC)*, San Francisco, CA, USA, 2025, pp. 530-532. https://doi.org/10.1109/ISSCC49661.2025.10904792
[4] J. Jiang, J. Tang, L. Zhao, C. Zhan and C. Huang, "A 63% Efficiency 1.29-W Single-Link Multiple-Output (SLiMO) Isolated DC–DC Converter Using FPC Micro-Transformer With Local Voltage and Global Power Regulations," in *IEEE Journal of Solid-State Circuits*, vol. 59, no. 3, pp. 804-816, March 2024. https://doi.org/10.1109/JSSC.2023.3330173
[5] W. Qin et al., "An 800mW Fully Integrated Galvanic Isolated Power Transfer System Meeting CISPR 22 Class-B Emission Levels with 6dB Margin," *2019 IEEE International Solid-State Circuits Conference - (ISSCC)*, San Francisco, CA, USA, 2019. https://doi.org/10.1109/ISSCC.2019.8662349
[6] T. Hu, M. Huang, R. P. Martins and Y. Lu, "A 750mW, 37% Peak Efficiency Isolated DC-DC Converter with 54/18Mb/s Full-Duplex Communication Using a Single Pair of Transformers," *2024 IEEE International Solid-State Circuits Conference (ISSCC)*, San Francisco, CA, USA, 2024, pp. 504-506. https://doi.org/10.1109/ISSCC49657.2024.10454311
[7] Q. Huang, D. Pan, Z. Chen and L. Cheng, "A Dual-LC-Resonant Isolated DC-DC Converter Achieving 65.4% Peak Efficiency and Inherent Backscattering," *2025 IEEE International Solid-State Circuits Conference (ISSCC)*, San Francisco, CA, USA, 2025, pp. 01-03. https://doi.org/10.1109/ISSCC49661.2025.10904821
[8] D. Pan et al., "A 1.25W 46.5%-Peak-Efficiency Transformer-in-Package Isolated DC-DC Converter Using Glass-Based Fan-Out Wafer-Level Packaging Achieving 50mW/mm2 Power Density," *2021 IEEE International Solid-State Circuits Conference (ISSCC)*, San Francisco, CA, USA, 2021, pp. 468-470. https://doi.org/10.1109/ISSCC42613.2021.9365955
[9] D. Pan, W. Xu, L. Zhang, Q. Huang and L. Cheng, "A 2W 53.2%-Peak-Efficiency Multi-Core Isolated DC-DC Converter with Embedded Magnetic-Core Transformer Achieving CISPR-32 Class-B EMI Compliance and <5mV Ripple," *2025 IEEE International Solid-State Circuits Conference (ISSCC)*, San Francisco, CA, USA, 2025, pp. 536-538. https://doi.org/10.1109/ISSCC49661.2025.10904708

[10] L. Li, X. Fang and R. Wu, "An 11MHz Fully Integrated 5kV Isolated DC-DC Converter Without Cross-Isolation-Barrier Feedback," *2020 IEEE International Solid-State Circuits Conference - (ISSCC)*, San Francisco, CA, USA, 2020, pp. 292-294. https://doi.org/10.1109/ISSCC19947.2020.9063050
[11] H. Ishihara and K. Onizuka, "A Fully-Generic-Process Galvanic Isolator for Gate Driver with 123mW 23% Power Transfer and Full-Triplex 21/14/0.5Mb/s Bidirectional Communication Utilizing Reference-Free Dual-Modulation FSK," *2020 IEEE International Solid-State Circuits Conference - (ISSCC)*, San Francisco, CA, USA, 2020, pp. 300-302. https://doi.org/10.1109/ISSCC19947.2020.9063035
[12] T. Xia, Q. Chen, S. Wang, R. P. Martins and M. Huang, "A 180MHz 45.3%-Peak-Efficiency Isolated Converter Using Q-Downsize Class-D Power Amplifier with Inherent Shoot-Through Current Blocking and High Tolerance for Efficiency Despite Frequency Misalignments," *2025 IEEE International Solid-State Circuits Conference (ISSCC)*, San Francisco, CA, USA, 2025, pp. 1-3. https://doi.org/10.1109/ISSCC49661.2025.10904701
[13] K. Datta, P. H. McLaughlin and J. T. Stauth, "A Fully-Integrated Direct-Conversion Resonant Switched Capacitor Converter with Modular Multi-Winding Current Ballasting," *2023 IEEE Custom Integrated Circuits Conference (CICC)*, San Antonio, TX, USA, 2023, pp. 1-2. https://doi.org/10.1109/CICC57935.2023.10121241
[14] K. Datta, Y. Wu, C. R. Sullivan and J. T. Stauth, "Optimal Current Distribution in Multi-Winding Transformers for Isolated and Wireless Power Transfer," *IEEE Open Journal of Power Electronics*, vol. 6, pp. 1296-1309, 2025. https://doi.org/10.1109/OJPEL.2025.3590020
[15] E. Ragonese et al., "A Fully Integrated Galvanically Isolated DC-DC Converter With Data Communication," in *IEEE Transactions on Circuits and Systems I: Regular Papers*, vol. 65, no. 4, pp. 1432-1441, April 2018. https://doi.org/10.1109/TCSI.2017.2742021
[16] P. Lombardo, V. Fiore, E. Ragonese and G. Palmisano, "A fully-integrated half-duplex data/power transfer system with up to 40Mb/s data rate, 23mW output power and on-chip 5kV galvanic isolation," *2016 IEEE International Solid-State Circuits Conference (ISSCC)*, San Francisco, CA, USA, 2016, pp. 300-301. https://doi.org/10.1109/ISSCC.2016.7418026
[17] S. Cochran, C. Zhao and D. Costinett, "Multilevel Switched-Capacitor AC–DC Step-Down Rectifier for Wireless Charging With Reduced Conduction Loss and Harmonic Content," in *IEEE Transactions on Power Electronics*, vol. 37, no. 7, pp. 8669-8681, July 2022. https://doi.org/10.1109/TPEL.2022.3141607
[18] C. Schaef, E. Din and J. T. Stauth, "A digitally controlled 94.8%-peak-efficiency hybrid switched-capacitor converter for bidirectional balancing and impedance-based diagnostics of lithium-ion battery arrays," *2017 IEEE International Solid-State Circuits Conference (ISSCC)*, San Francisco, CA, USA, 2017, pp. 180-181. https://doi.org/10.1109/ISSCC.2017.7870320
[19] Z. Xia and J. Stauth, "A Two-Stage Cascaded Hybrid Switched-Capacitor DC-DC Converter with 96.9% Peak Efficiency Tolerating 0.6V/μs Input Slew Rate During Startup," *2021 IEEE International Solid-State Circuits Conference (ISSCC)*, San Francisco, CA, USA, 2021, pp. 256-258. https://doi.org/10.1109/ISSCC42613.2021.9365763
[20] S. Biswas and D. Reusch, "GaN Based Switched Capacitor Three-Level Buck Converter with Cascaded Synchronous Bootstrap Gate Drive Scheme," *2018 IEEE Energy Conversion Congress and Exposition (ECCE)*, Portland, OR, USA, 2018. https://doi.org/10.1109/ECCE.2018.8557595

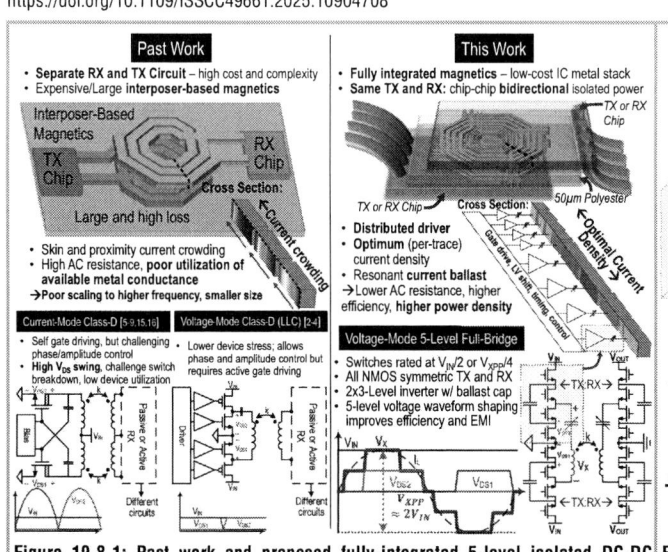

Figure 19.8.1: Past work and proposed fully-integrated 5-level isolated DC-DC converter.

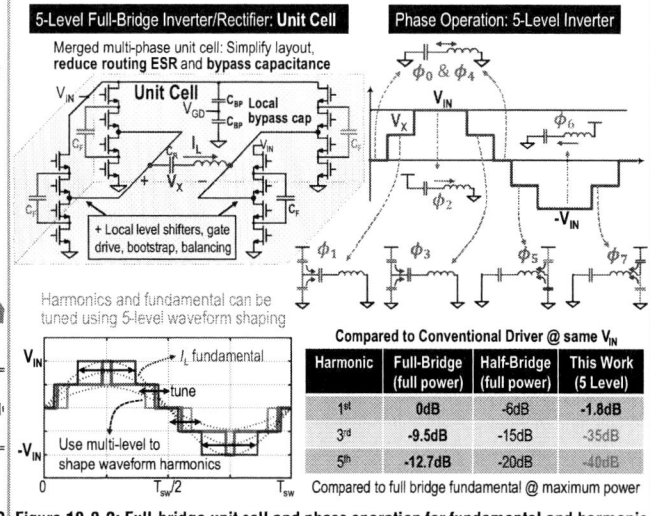

Figure 19.8.2: Full-bridge unit cell and phase operation for fundamental and harmonic control.

ISSCC 2026 / SESSION 19 / HIGH-VOLTAGE, ISOLATED AND DISPLAY POWER / 19.8

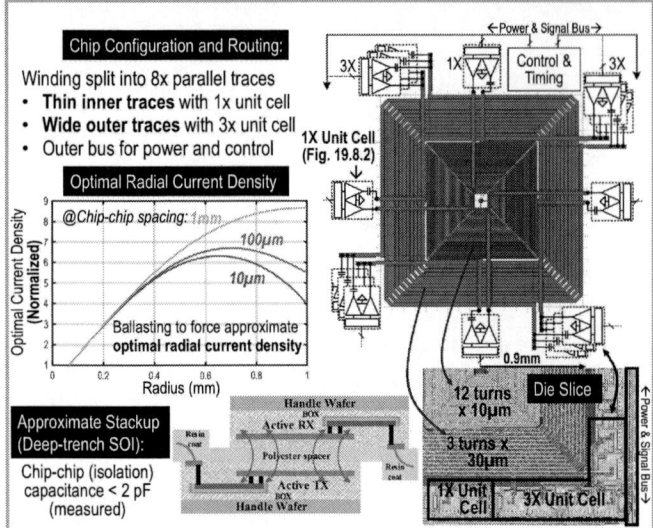

Figure 19.8.3: Chip configuration and routing for optimal radial current density.

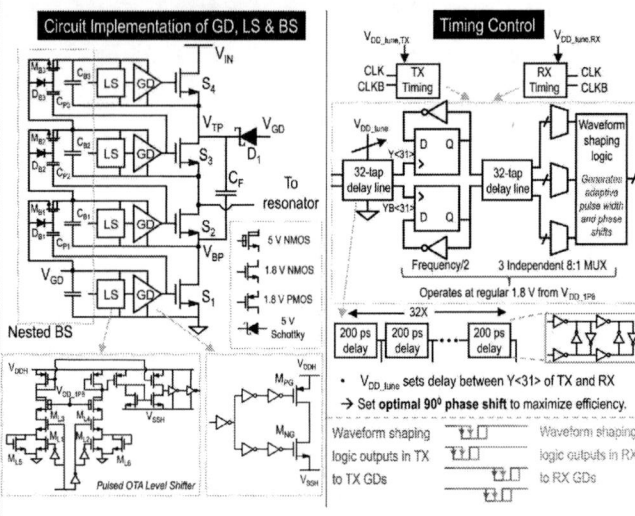

Figure 19.8.4: Circuit implementation of 5-level inverter power train and timing control.

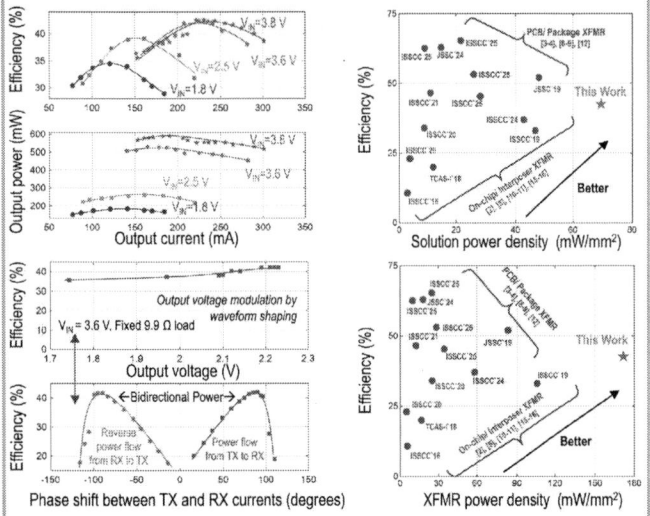

Figure 19.8.5: Measurement results.

	Qin ISSCC 19	Xia ISSCC 25	Pan ISSCC 25	Huang ISSCC 25	Jiang ISSCC 25	This Work
Topology	Class-D oscillator	Q-downsize Class-D PA	Class-D oscillator	Dual-LC	LLC	5-level full bridge
V_{IN} (V)	4.5 – 5.5	4.5 – 5.5	3.3 – 5.5	5	3.3	1 – 3.8
V_{OUT} (V)	3.3 – 5	5	3.3 – 5.5	5	1.8 – 3.3	0.9 – 3.6
Transformer Type	Coreless in Au & polyimide interposer	Coreless in PCB	Embedded Magnetic-Core (in package)	Coreless in PCB substrate	Coreless in FPC substrate	Coreless in standard CMOS metal stack
Technology	350 nm BCD	180 nm BCD	180 nm BCD	180 nm BCD	180 nm BCD	180 nm SOI
Frequency (MHz)	160 - 210	180	80	50	7.6 – 23.8	75
Peak Efficiency @ XP_{den} (mW/mm^2)*, P_{den} (mW/mm^2)**	34% @ 96.2, 42.6	45.3% @ 34.7, 28.2	53.2% @ 28.3, 26	65.4% @ 24.7, 21.7	62.6% @ 9.5, 8.8	42.5% @ 171.6, 69.6
Peak XP_{den} (mW/mm^2)*, Peak P_{den} (mW/mm^2)** @ Efficiency	154, 68.2 @ 33%	69.4, 56.4 @ 42%	35, 32.1 @ 50%	30, 26.3 @ 63%	17.7, 16.4 @ 55.6%	182.2, 73.9 @ 38.5%
Isolation Rating (kV)	5	N.A.	5	N.A.	> 5	> 5
# Unique Dies/ Packages #	3	3	3	2	3	1
Bidirectional Power Flow?	No	No	No	Yes†	No	Yes

* Output power over calculated or estimated XFMR area; **Output power over total solution area: active die(s) plus XFMR and any required off chip passives
# Actual implementation may have one or more of these unique dies. † Capable of bidirectional power flow but not demonstrated, also XFMR is asymmetric.

Figure 19.8.6: Comparison table with prior state-of-the-art works.

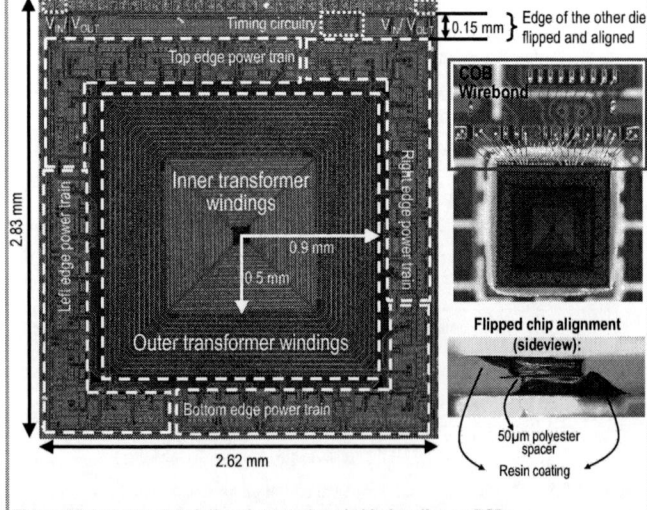

Figure 19.8.7: Annotated die micrograph and chip bonding on PCB.

ISSCC 2026 / SESSION 19 / HIGH-VOLTAGE, ISOLATED AND DISPLAY POWER / 19.9

19.9 A 2.15W 120V/230Vac to 5-to-12Vdc Offline Power Converter with Full-Duty-Cycle Input-Series Dual-Branch Converter Achieving 1088mW/cm³ and 87.2% Peak Efficiency

Gang Liu[1,2], Junmin Jiang[2], Sijun Du[3], Xun Liu[1]

[1]Chinese University of Hong Kong, Shenzhen, China, [2]Southern University of Science and Technology, Shenzhen, China, [3]TU Delft, Delft, The Netherlands

Abstract

This paper presents a 120V/230Vac to 5-to-12Vdc offline power converter consisting of a capacitor-drop AC-DC rectifier and a full-duty-cycle input-series dual-branch DC-DC converter for IoT devices. The proposed converter operates at full-duty-cycle with VCR of D/2 and reduces both the C_{REC} volume and the voltage stress on components, while achieving high efficiency across a broad input and output voltage range. 87.2% peak efficiency, 2.15W output power, and 1088mW/cm³ power density are achieved.

The rapid growth of Internet of Things (IoT) devices and sensors in smart home systems has increased the demand for efficient and compact offline power converters that transfer power from the AC mains to low DC voltages. Traditional offline power converters are typically implemented with isolated flyback [1] or forward [2] topologies, which require bulky transformers to step down the AC voltage (120/230V$_{RMS}$), limiting the power density to typically less than ~500mW/cm³ [1]. In recent years, non-isolated AC-DC converters [3-15] are gaining popularity due to their compact form factor. Direct AC-DC converters [3-5] use high-voltage (HV) low-dropout regulators (LDOs) [3] or DC-DC converters [4,5] to step down the rectified high voltage (>325V) to a low DC voltage of 3.3-to-12V. These extremely high voltage conversion ratios (VCRs) lead to low efficiency and bulky volume. The capacitor-drop AC-DC converter [6-9] uses an X-capacitor as a buffer to withstand most of the high voltage and reduce the rectifiers' input and output voltages. It offers a higher integration level [6], a reduced VCR for the DC-DC stage [7], and higher system efficiency [8,9]; however, it struggles with limited output power, large filtering capacitor (C_{REC}) volume, and low light-load efficiency.

As shown in Fig. 19.9.1, the theoretical output power P_{REC} of a capacitor-drop AC-DC rectifier is $P_{REC} = 4f_{LINE}V_{REC}C_X(\sqrt{2}V_{LINE}-V_{REC})$. The frequency and voltage of the AC mains (f_{LINE} and V_{LINE}) are fixed, and increasing C_X brings more idle power loss. Thus, increasing P_{REC} requires raising the rectifier's output voltage V_{REC}, as demonstrated by prior arts increasing it from 23V [7] and 60V [8] and 70V [9] to achieve up to 1.52W of output power P_{OUT}. However, this necessitates a higher VCR in the subsequent DC-DC stage. In [8] and [9], switched-capacitor (SC) converters with 1/10× and 1/14× VCRs are utilized. The SC converters can only achieve high efficiency within a narrow voltage range near their ideal VCRs. As such, a large-volume filtering capacitor C_{REC} (aluminum electrolytic, 47μF@63V/80V) is required to suppress the ripple voltage of V_{REC}, severely limiting power density. Furthermore, the complex SC structures required for high VCRs result in higher output resistance and switching loss, limiting the maximum output power and light-load efficiency. Consequently, the P_{OUT} in [8] and [9] reached only 33% and 48% of the 3.18W theoretical P_{REC}, respectively. In summary, SC stages with high VCRs have the drawbacks of a narrow voltage range and a large output resistance, which constrains their output power, voltage range, power density, and peak efficiency.

To resolve these issues, we propose an architecture, shown in Fig. 19.9.1, that integrates two key components: 1) a three-mode capacitor-drop AC-DC rectifier [8] to mitigate bridge voltage stress, reduce idle power, and enhance light-load efficiency; and 2) a full-duty-cycle, input-series dual-branch (ISDB) DC-DC converter. The ISDB converter replaces the traditional SC converter and yields a reduced C_{REC} volume, enhanced load capability, and higher efficiency. This architecture leverages the fact that the AC-DC rectifier's output voltage (V_{REC}) and the DC-DC converter's ground are not required to share the same reference. Two series-connected capacitors (C_H and C_L) are placed across V_{REC}, with the DC-DC converter's ground defined at their mid-point. This splits the high V_{REC} voltage domain into two symmetric lower-voltage domains: a positive domain ($V_{REC}/2$) across C_H and a negative domain ($-V_{REC}/2$) across C_L. A buck converter regulates the positive voltage, yielding $V_{OUT} = D \times V_{REC}/2$, while an inverting buck converter regulates the negative voltage with $V_{OUT} = -D \times -V_{REC}/2$. The output currents from both inductors are combined to supply the load. Critically, as each branch now handles only $V_{REC}/2$, the required voltage rating for all components, including filtering capacitors, flying capacitor, power transistors, and isolation rings, is halved. Furthermore, this dual-branch configuration, operating at a full duty cycle with a conversion ratio of D/2, supports a significantly wider input voltage range than a fixed-ratio SC converter. This inherent voltage flexibility relaxes the V_{REC} ripple requirement, allowing the total capacitance (C_H and C_L) to be reduced by approximately 30%. The total inductor plus capacitors' volume in the ISDB converter is nearly identical to that of the prior SC converters. Thereby, the overall converter volume is reduced. Compared to the SC converter, the ISDB converter also sustains much higher efficiency across broad voltage and load ranges, increasing P_{OUT} under the same V_{REC}. To ensure equal voltage allocation between C_H and C_L, a voltage balance scheme is proposed, which guarantees the reliability of the converter and simultaneously balances the inductor currents in the two branches. Finally, a discontinuous conduction mode (DCM) calibration loop is implemented to reduce inductor losses further and improve light-load efficiency.

As shown in Fig. 19.9.2, the ISDB converter can be simplified into two branches. Branch P comprises C_H, S_1, S_2, and L_1, while branch N includes C_L, S_3, S_4, S_5, and L_2. The inputs of the two branches are in series, and the outputs are in parallel. Consequently, both the switches and capacitors withstand only $V_{REC}/2$. Both branches have the same V_{OUT} and can be controlled independently without mutual interference. The ISDB converter extends the duty cycle to 1 while achieving a VCR of D/2. Figure 19.9.2 then shows the operating principle. In the steady state, the voltages across C_H and C_L are typically $V_{REC}/2 = 30V$ and $-V_{REC}/2 = -30V$. In Φ_1, S_1 is turned on and $V_{SW1} = 30V$. S_3 and S_4 are also turned on. The flying capacitor C_F is connected in parallel with C_L, resulting in $V_{CF} = -30V$ and $V_{SW2} = 0$. In Φ_2, S_2 and S_5 are on, $V_{SW1} = 0$ and $V_{SW2} = -C_F = 30V$. In Φ_3, S_2, S_3, and S_4 are on, resulting in $V_{SW1} = V_{SW2} = 0$. In Φ_4, S_1 and S_5 are on, so $V_{SW1} = V_{SW2} = 30V$. To keep two voltages of C_H and C_L equal, a V_{CH} and V_{CL} balance loop is shown in Fig. 19.9.2. The concept is to adjust the duty cycles to control the powers in each branch. When $V_{CH}>V_{CL}$ due to disturbances, the loop makes $D_1>D_2$, and more current is delivered from branch P to the load, reducing V_{CH} and raising V_{CL}, until $V_{CH} = V_{CL}$. Owing to the characteristics of the input-series output-parallel topology [16], the output currents of both branches, which are the inductor currents in the ISDB converter, are proportional to the input voltages, so $I_{L1}/I_{L2} = V_{CH}/V_{CL}$. In this way, the V_{CH} and V_{CL} balance loop also ensures I_{L1} and I_{L2} balance, thus improving efficiency while eliminating complex current-sensing and regulation circuitries.

Figure 19.9.3 shows the system diagram. The power stage of the AC-DC rectifier consists of four 70V-LDMOS (M_{1-4}), two 70V-diodes (D_{1-2}), and five off-chip capacitors C_X, $C_{P1, P2}$, and $C_{H, L}$. C_X is a 310V$_{AC}$ film capacitor and carries most of the AC voltage. C_{P1} and C_{P2} reduce idle power, and 70V low-profile ceramic capacitors are used. Low and high-side transistors in the rectifier bridge are realized with NMOS devises and diodes, respectively. The hysteresis controller generates the gate signals V_{GM1-4} of M_{1-4}. C_H and C_L are implemented using 35V-rated aluminum electrolytic capacitors. For the ISDB DC-DC converter, the switches S_{1-5} are implemented with 40V-LDMOSs. $S_{2,3,4}$ are implemented using NMOSs to reduce on-resistance, and $S_{1,3}$ are PMOS transistors for easy driving. C_F is a 35V-rated ceramic capacitor. A type-III compensation is used for the voltage regulation loop. The V_{CH} and V_{CL} balance loop first senses the mismatch between the voltage of $V_{REC\_P} + V_{REC\_N}$ and GND and then gets the error voltage V_{EA_B} from the error amplifier. V_{EA_B} is used to modify the magnitude of Ramp$_1$ and Ramp$_2$ to modulate the duty cycles of D_1 and D_2 until the balance is achieved. A DCM calibration loop is also designed to enhance the light-load efficiency. Before turning off S_2, the V_{SW1} node and GND voltages are both sampled and then processed by a type-I compensator to generate the error signal V_{EA_DCM}. As illustrated in Fig. 19.9.3, the reverse inductor current reduces V_{EA_DCM}, which advances the turning-off point of S_2, and in this way, negative inductor current is gradually eliminated. Measurement results verify that under light-load DCM conditions with $V_{REC} = 50V$ and $V_{OUT} = 5V$, the efficiency of the ISDB converter is improved by up to 14%. This approach enhances the light-load efficiency of the AC-DC converter from 51% [8] to 78%.

The AC-DC rectifier was fabricated in a 180nm HV SOI process with an area of 1.45mm×0.6mm. To lower the cost, the ISDB converter was implemented in a 180nm BCD process, with an area of 3.03mm×0.55mm. Figure 19.9.4 presents the measured waveforms when $V_{LINE} = 230V@50Hz$ and $V_{OUT} = 5V$. Although V_{REC} varies between ~30V and ~54V, the converter maintains a regulated 5V output, indicating a relaxed V_{REC} ripple requirement. Figure 19.9.4 also shows the measured steady-state waveforms of the ISDB converter under various input and output voltages. At $V_{REC} = 30V$, $V_{OUT} = 5V$, and $I_{OUT} = 0.05A$, the converter operates in DCM. It was observed that the DCM calibration loop effectively eliminates negative currents in I_{L1} and I_{L2}. When the input voltage and output current increase to 60V and 0.4A, the ISDB converter operates in continuous condition mode (CCM). As expected, when duty cycle D = 0.8>0.5, overlapping occurs at $V_{REC} = 30V$ and $V_{OUT} = 12V$, demonstrating support for a wide range of input and output voltages.

Figure 19.9.5 illustrates the measured load transient response waveforms of the ISDB converter for $V_{OUT} = 5V$. When the load current is switched between 50-to-500mA with ~30ns edges, the under- and over-shoots are 400mV and 200mV, respectively. It can be observed that the currents are well equalized by the V_{CH} and V_{CL} balance loop. The reverse inductor currents after transients are eliminated by the DCM calibration loop. Figure 19.9.5 also shows the measured efficiencies of the ISDB converter and the overall AC-DC converter under different AC input and DC output voltages. When V_{LINE} = 230V_{RMS}@50Hz and V_{OUT} = 12V, an 87.2% peak efficiency is achieved. The maximum output power is 2.15W. This work also maintains high efficiency across various output voltages and load currents.

Figure 19.9.6 benchmarks the performance in terms of power density, output power, and efficiency. Figure 19.9.6 also presents the comparison with the state-of-the-art designs. Among the compared designs in Fig. 19.9.6, this work achieves the highest maximum output power of 2.15W, and the highest efficiency of 86.7% at maximum output power. The highest power density of 1088W/cm³ is also achieved. Additionally, the light-load efficiency is improved by at least 18% compared to the prior art in Fig. 19.9.6. Figure 19.9.7 shows the chip micrograph and PCB photos.

Acknowledgment:

This work was supported in part by the National Natural Science Foundation of China (62474153 and 62261160647), by the Guangdong Basic and Applied Basic Research Foundation (2025A1515011313 and 2025A1515010698), by the Shenzhen Science and Technology Program (JCYJ20240813100601003), by the State Key Laboratory of Radio Frequency Heterogeneous Integration (Open Scientific Research Program) (KF2025016), and High-level Special Funds of SUSTech (G03034K007). *Corresponding author: Xun Liu; Junmin Jiang.*

References:

[1] C. Rindfleisch *et al.*, "A Highly-Integrated 20-300V 0.5W Active-Clamp Flyback DCDC Converter with 76.7% Peak Efficiency," *CICC*, pp. 1–2, Apr. 2022. http://doi.org/10.1109/CICC53496.2022.9772834
[2] L. Xu *et al.*, "A Two-Transistor Forward Converter With Variable Turn Ratio Using Dual Optocouplers to Automatically Adjust Crossover Frequency," *IEEE Trans. Circuits Syst. II*, vol. 70, no. 12, pp. 4414–4418, Dec. 2023. http://doi.org/10.1109/TCSII.2023.3288044
[3] M. Pomper *et al.*, "On-Chip Power Supply for 110 V Line Input," *IEEE JSSC*, vol. 13, no. 6, pp. 882–886, Dec. 1978. http://doi.org/10.1109/JSSC.1978.1052063
[4] C. Rindfleisch *et al.*, "A 110/230 V AC and 15–400 V DC 0.3 W Power-Supply IC with Integrated Active Zero-Crossing Buffer," *IEEE JSSC*, vol. 57, no. 12, pp. 3816–3824, Dec. 2022. http://doi.org/10.1109/JSSC.2022.3199653
[5] T. Van Daele *et al.*, "Monolithic 230-V_{RMS}-to-12-V_{DC} AC–DC Converter at 9 mW/mm² Enabled by a 31–325-V_{DC} Input Range Capacitive Multi-Ratio DC–DC Converter," *IEEE JSSC*, vol. 59, no. 4, pp. 1067–1077, Apr. 2024. http://doi.org/10.1109/JSSC.2023.3340976
[6] N. O. Sokal *et al.*, "A Capacitor-Fed, Voltage-Step-Down, Single-Phase, Non-Isolated Rectifier," *APEC*, pp. 208–215, Feb. 1998. http://doi.org/0.1109/APEC.1998.647693
[7] Y. Ramadass *et al.*, "A 120mA Non-Isolated Capacitor-Drop AC/DC Power Supply," *ISSCC*, pp. 290–292, Feb. 2020. http://doi.org/10.1109/ISSCC19947.2020.9062968
[8] G. Liu *et al.*, "An 85-264Vac to 3-4.2Vdc 1.05W Capacitive Power Converter with Idle Power Reduction and 4-Phase 1/10X SC Converter Achieving 5.11mW Quiescent Power and 78.2% Peak Efficiency," *ISSCC*, pp. 512–513, Feb. 2024. http://doi.org/10.1109/ISSCC49657.2024.10454521
[9] F. Song *et al.*, "An 85-to-230V_{AC} to 3.3-to-4.6V_{DC} 1.52W Capacitor-Drop Sigma-Floating-SC AC-DC Converter with 81.3% Peak Efficiency," *ISSCC*, pp. 178–179, Feb. 2025. http://doi.org/10.1109/ISSCC49661.2025.10904504

[10] T. Lu *et al.*, "A 0.49W 120-230V_{RMS} to 8.12V_{DC} Power Converter with Switched-Capacitor Regulation and Rectifier Short Flipping Achieving Maximized Bridge Conduction Time," *CICC*, pp. 1–3, Apr. 2025. http://doi.org/10.1109/CICC63670.2025.10983403
[11] J. Shi *et al.*, "A 9.1mW All-5V-CMOS Series-Capacitor AC-DC Converter with CF Reallocation Operations for 85-230V_{RMS} Mains Achieving 85.6% Efficiency at 858mW/cm³ Density," *VLSI*, pp. 1-3, June 2025. http://doi.org/10.23919/VLSITechnologyandCir65189.2025.11075089
[12] A. A. Tamez *et al.*, "An Integrated 120 Volt AC Mains Voltage Interface in Standard 130 nm CMOS," *ESSCIRC*, pp. 238–241, Sep. 2010. http://doi.org/10.1109/ESSCIRC.2010.5619885
[13] H. Meyvaert *et al.*, "A 265 V_{RMS} Mains Interface Integrated in 0.35 μm CMOS," *IEEE JSSC*, vol. 48, no. 7, pp. 1558–1564, Jul. 2013. http://doi.org/10.1109/JSSC.2013.2253214
[14] D. Lutz *et al.*, "An Integrated 3-mW 120/230-V AC Mains Micropower Supply," *IEEE J. Emerg. Sel. Topics Power Electron.*, vol. 6, no. 2, pp. 581–591, Jun. 2018. http://doi.org/10.1109/JESTPE.2018.2798504
[15] E. De Pelecijn *et al.*, "A Fully Integrated Switched-Capacitor-Based AC–DC Converter for a 120 V_{RMS} Mains Interface," *IEEE JSSC*, vol. 54, no. 7, pp. 2009–2018, Jul. 2019. http://doi.org/10.1109/JSSC.2019.2906750
[16] X. Yang *et al.*, "A 5V Input 98.4% Peak Efficiency Reconfigurable Capacitive-Sigma Converter with Greater than 90% Peak Efficiency for the Entire 0.4~1.2V Output Range," *ISSCC*, pp. 108–110, Feb. 2022. http://doi.org/10.1109/ISSCC42614.2022.9731550.

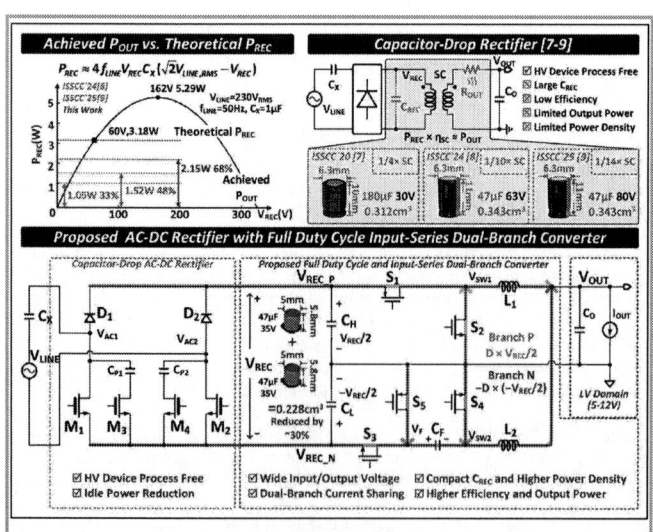

Figure 19.9.1: System diagrams of prior art and the proposed AC-DC converter.

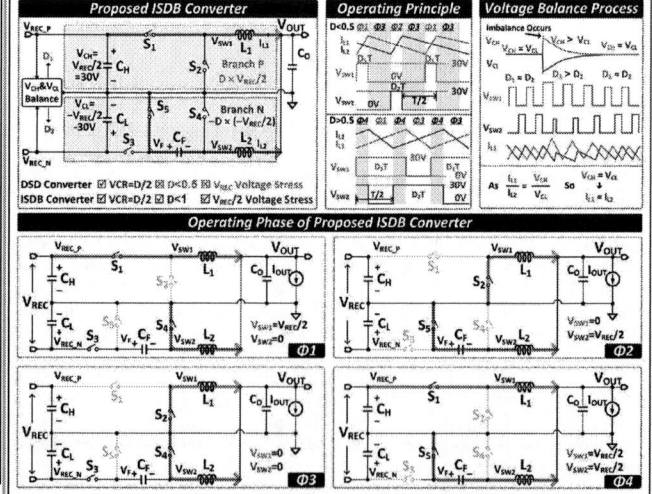

Figure 19.9.2: Comparison of the DSD converter and the proposed ISDB converter, voltage balance process, and the operating principle and phase of the ISDB converter.

ISSCC 2026 / SESSION 19 / HIGH-VOLTAGE, ISOLATED AND DISPLAY POWER / 19.9

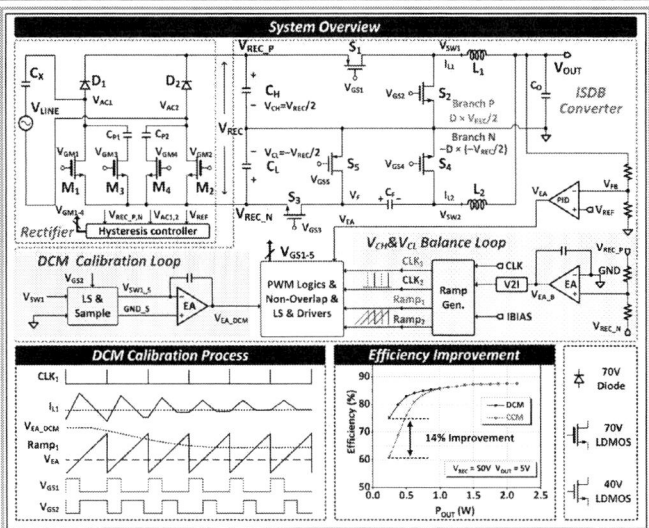

Figure 19.9.3: Schematic and DCM calibration process of the ISDB converter with measurement results of efficiency improvement in DCM.

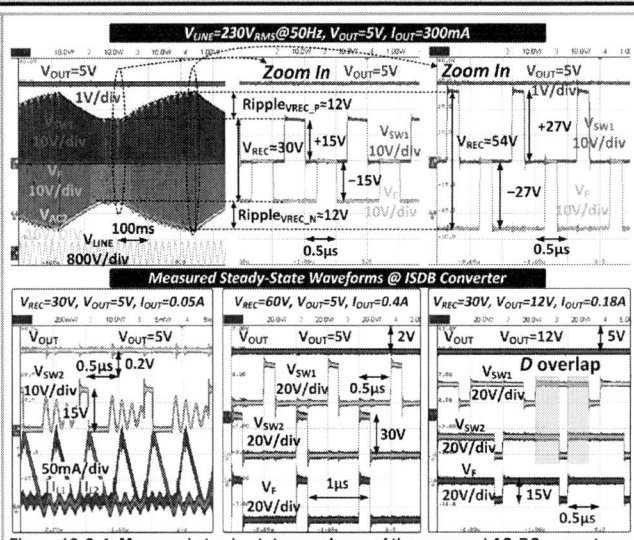

Figure 19.9.4: Measured steady-state waveforms of the proposed AC-DC converter and the ISDB converter.

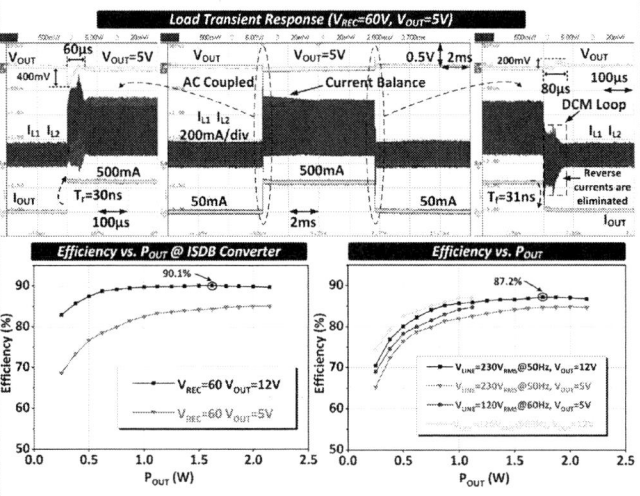

Figure 19.9.5: Measured load transient response, efficiency vs. P_{OUT} under different AC input and DC output voltages.

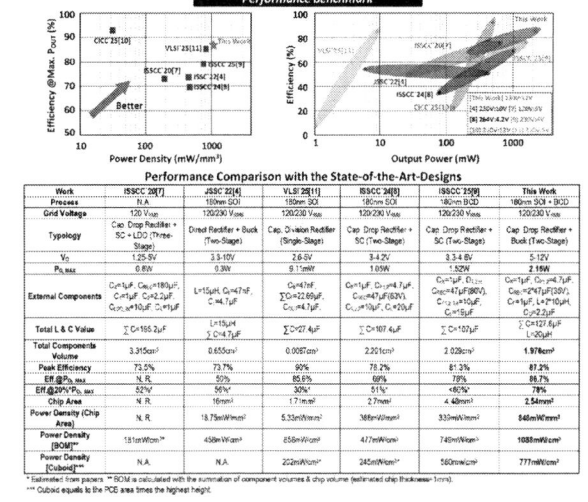

Figure 19.9.6: Performance benchmark and comparison with the state-of-the-art-designs.

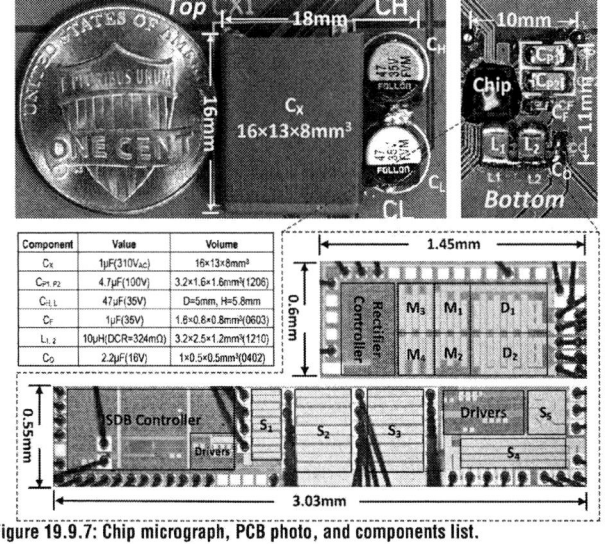

Figure 19.9.7: Chip micrograph, PCB photo, and components list.

- 2026 IEEE International Solid-State Circuits Conference

ISSCC 2026 / SESSION 20 / RF TRANSCEIVER SUBSYSTEMS FROM cm-WAVE TO THz / OVERVIEW

Session 20 Overview: *RF Transceiver Subsystems from cm-Wave to THz*

RF SUBCOMMITTEE

Session Chair: Sudipto Chakraborty
IBM T.J.Watson Research Center
Plano, TX

Session Co-Chair: Mikko Varonen
VTT Technical Research Centre of Finland
Espoo, Finland

This session covers design techniques for high performance RF transceiver subsystems targeting high carrier frequencies, wide bandwidths, and compact and low-cost integrated solutions operating between 10 and 350GHz frequencies and built in Silicon and GaN technologies. The first three papers present innovations in Doherty PA design, including asymmetrical series combination, unequal stacking, and compact footprints for operation in the FR2 and FR3 bands. The fourth paper introduces a VSWR-resilient source-follower PA followed by a self-adaptive load-balance amplifier in GaN. The next two papers describe a high-frequency GaN amplifier operating near 350GHz and a 150GHz bidirectional amplifier, respectively. This is followed by a subharmonic phase-modulating transmitter with a dual-band low-noise amplifier. Finally, the last two papers describe high-efficiency THz radiator arrays enabled by co-design of IC and packaging techniques.

1:30 PM

20.1 An Ultra-Compact Asymmetrically Load-Pulled Series Doherty Power Amplifier in 22nm FDSOI CMOS with 25.3dBm P_{sat} and 29.7% PAE_{6dB} for Ku-Band 6G FR3

Jinglong Xu, ETH Zurich, Zurich, Switzerland

In Paper 20.1, ETH Zurich describes a 6G FR3 ultra-compact asymmetric series Doherty PA developed in 22nm FDSOI CMOS that provides 25.3dBm P_{sat} and 29.7% PAE at 6dB back-off.

1:55 PM

20.2 A High Back-off Efficiency Unequal-Stacked Doherty Power Amplifier Achieving 16.7dBm P_{avg} in a 22nm FDSOI CMOS Technology for 5G FR2 Applications

Seungwon Park, Samsung Electronics, Seoul, Korea

In Paper 20.2, Samsung Electronics describes a 5G FR2 high-back-off-efficiency unequal-stacked Doherty PA developed in 22nm FDSOI CMOS providing 16.7dBm average output power.

2:20 PM

20.3 A mm-Wave Doherty Power Amplifier in a Single-Path Footprint Using Compact Reciprocal Doherty Networks

Lianbo Liu, University of Texas, Austin, TX

In Paper 20.3, UT Austin presents a mm-wave CMOS single-path-footprint Doherty PA that uses compact reciprocal Doherty networks, achieving a core area of only 0.12mm² and supporting 800MHz/400MHz 64-/256-QAM OFDM signals with average efficiencies of 11% and 9%, respectively.

2:45 PM

20.4 An Ultra-Compact, Inherently Resilient FR3 Source-Follower Power Amplifier with 4:1 VSWR Resilience and RIMD Immunity for Large-Scale SATCOM and 6G Phased Arrays

Mohamed Eleraky, ETH Zurich, Zurich, Switzerland

In Paper 20.4, ETH Zurich describes a 4:1 VSWR-resilient source-follower PA operating from 11.5 to 13.8GHz with average power of 10.2dBm and 16.8% average PAE.

ISSCC 2026 / February 17, 2026 / 1:30 PM

3:00 PM

20.5 A 24-to-27.5GHz Self-Adaptive Load-Modulated Balanced Amplifier for Integrated Communication, Sensing, and Power Transfer Scenarios

Luqi Yu, Southeast University, Nanjing, China

In Paper 20.5, Southeast University describes a 24-to-27.5GHz self-adaptive load-modulated balanced amplifier for integrated communication, sensing, and power transfer developed using 150nm GaN to provide 2.4Gbps 64QAM modulation with 26.6dBm P_{avg} and 23.3% average PAE.

3:35 PM

20.6 A 330-to-344GHz GaN Power Amplifier with Maximum-Available-Gain-Boosting Technique and Compact Tandem Coupler Achieving 86mW Output Power at 340GHz

Wenhua Chen, Tsinghua University, Beijing, China

In Paper 20.6, Tianjin University describes a 330-to-344GHz GaN PA based on a 15-stage cascade that demonstrates a peak P_{sat} of 19.2dBm with 0.6% PAE using a maximum-available-gain-boosting technique and a compact tandem coupler.

4:00 PM

20.7 An Ultra-Compact D-Band Transceiver Front-End Based on a Common-Gate Bidirectional Amplifier Achieving 11.2dBm TX P_{sat} and 8.2dB RX Average NF for an Area-Constrained 2D Beamformer AiP

Syed Mohammad Ashab Uddin, Pennsylvania State University, University Park, PA, Texas Instruments, Dallas, TX

In Paper 20.7, Pennsylvania State University describes an ultra-compact D-band (130 to 147GHz) transceiver front-end that uses a common-gate bidirectional amplifier and achieves 11.2dBm P_{sat} and 8.2dB RX NF.

4:25 PM

20.8 A 16-to-256QAM G-Band Subharmonic Phase-Modulating Transmitter for Beyond-5G Communications

Jia Zhou, University of California, Los Angeles, CA

In Paper 20.8, the University of California, Los Angeles presents a G-band (195 to 225GHz) subharmonic phase-modulating transmitter in 28nm CMOS that supports 16-to-256QAM modulation with a peak output power of -2dBm at 210GHz and a total power dissipation of 280 to 350mW.

20

4:50 PM

20.9 A Compact 26/38GHz-Reconfigurable Dual-Band Low-Noise Amplifier Using Transformer-Based Pole-Zero-Inversion Image-Rejection Technique Achieving >39/41dB IRR for 5G Multi-Band Applications

Jiaming Luo, Hangzhou Dianzi University, Hangzhou, China

In Paper 20.9, Hangzhou Dianzi University describes a 65nm-CMOS reconfigurable dual-band LNA that uses transformer-based compact pole-zero inversion to achieve image rejection ratios of 39 and 41dB respectively for 26/38GHz frequency bands with 3.5-to-5.5dB NF to reduce the required LO tuning range.

5:05 PM

20.10 A 214-to-242GHz Miniaturized Co-Packaged PA-Antenna Array with 29dBm Lens-less EIRP in a 0.13µm SiGe Process

Qianqi Meng, Southeast University, Nanjing, China

In Paper 20.10, Southeast University presents a 130nm-SiGe 214-to-242GHz miniaturized co-packaged PA-antenna array to provide 29dBm EIRP and 16.5dBm peak P_{sat} while using folded 10th-order power splitting/combining and staggered-matching method without using any lens.

5:20 PM

20.11 A 215GHz 8×8 Radiator-Oscillator Array with Robust Coupling Achieving 25.5dBm EIRP and 12.9% FTR

Ahmed Elmenshawi, Rensselaer Polytechnic Institute, Troy, NY

In Paper 20.11, Rensselaer Polytechnic Institute describes a 215GHz 8×8 radiator oscillator array with robust coupling that achieves 25.5dBm EIRP and 12.9% frequency tuning range and combines 22nm FDSOI CMOS and horn/glass superstrate.

DIGEST OF TECHNICAL PAPERS • 339

979-8-3315-8937-0/26 $31.00 © 2026 IEEE

ISSCC 2026 / SESSION 20 / RF TRANSCEIVER SUBSYSTEMS FROM CM-WAVE TO THz / 20.1

20.1 An Ultra-Compact Asymmetrically Load-Pulled Series Doherty Power Amplifier in 22nm FDSOI CMOS with 25.3dBm P_{sat} and 29.7% PAE_{6dB} for Ku-Band 6G FR3

Jinglong Xu[1], Tzu-Yuan Huang[1,2], Mohamed Eleraky[1], Islam Tolaib[1], Hua Wang[1]

[1]ETH Zurich, Zurich, Switzerland, [2]ARGUS SPACE AG, Zurich, Switzerland

Abstract

This paper presents an ultra-compact asymmetrically load-pulled series Doherty PA in 22nm CMOS SOI with a single-footprint output combiner. A quantitative analysis of a single transformer as an impedance inverter enables a passive-network transformation to a single-footprint EM structure. In addition, an asymmetric load-pull concept is proposed. The PA achieves 25.3dBm P_{sat}, 1.56W/mm² power density, up to 30% PAE_{6dB}, and wideband modulation for 3GHz QPSK and 2GHz 64QAM.

As 6G continues to redefine the limits of wireless communication, the Frequency Range 3 (FR3) spectrum (7 to 20GHz) is emerging as a leading candidate, with the 10-to-15GHz Ku band offering a more favorable trade-off between coverage and capacity than the existing mm-wave FR2 solutions. While FR2 deployments have struggled due to high installation costs, severe path loss, and limited coverage, the commercial success of Ku-band satellite-communication (SATCOM) constellations like Starlink highlights FR3 practical viability. However, unlike SATCOM systems that can accommodate large antenna arrays with thousands of elements, 6G handsets face stringent constraints in size, cost, and power. This makes power density a critical metric for power-amplifier (PA) design. Moreover, spectrally efficient 6G waveforms impose high peak-to-average power ratios (PAPR), which demand sustained efficiency even at deep power back-off (PBO). These challenges underscore the importance of Doherty architectures [1], particularly compact implementations that enhance back-off efficiency without excessive area overhead. Such designs are, therefore, key enablers for FR3 phased-array front-ends in next-generation mobile devices.

Most existing on-chip differential Doherty PAs, whether parallel or series, typically require at least two output transformers [2-4], doubling the effective channel width in array implementations compared to a single-path Class-AB PA. To improve compactness, recent works have explored alternative layout strategies. Work in [5] proposed a cascaded layout arrangement where the main and auxiliary amplifiers are placed along the signal path, and two λ/4 lines are implemented on either side of the PA before merging into the output transformer. This reduces the combiner size to a quasi-single-transformer footprint. However, this approach introduces feed imbalance between the main and auxiliary paths. Moreover, the lossy λ/4 line remains in the main PA path due to its parallel Doherty architecture, degrading back-off efficiency and ultimately limiting compactness in the dimension perpendicular to the signal path, as shown in [6]. Subsequent works have aimed to further reduce area by adopting a single-driver architecture [7,8]. In these designs, the 90° phase shift is generated after the driver in a compact form, either by inserting a λ/4 line-based phase offset [7] or by employing a stagger-tuned differential quadrature coupler with a shunt L [8]. However, these techniques not only require a large driver but also lack adaptive-bias compatibility in the driving stage, which limits isolation at back-off when the auxiliary path is not fully turned off. In addition, generating the 90° phase shift after the driver typically results in a narrowband circuit.

An alternative approach is reported in [9], where a stacked three-coil output combiner integrates both the impedance-inverting network (IIN) and impedance-scaling network (ISN) functions into a single transformer footprint. This is achieved by eliminating the λ/4-line-based IIN and instead employing a low-k transformer-based IIN. However, the low coupling inherently increases matching loss, and since the parallel architecture places IIN in the main PA path, this loss directly limits efficiency at PBO.

The prior works collectively reveal a common bottleneck: compact Doherty implementations have mainly relied on the parallel Doherty PA architecture, where the lossy IIN remains in the main PA path and limits efficiency at back-off. In contrast, series Doherty PA architectures achieve higher PBO efficiency by shifting the IIN to the auxiliary path. For example, [10] demonstrated a 1.74× enhancement in 6dB power-added efficiency (PAE) over Class-B, but the design occupied 0.37mm² at 28GHz FR2, and the required chip area would be even larger at FR3. Besides the benefits in both high-peak and back-off PAE, the series architectures also present inherently lower load impedance, due to the series power combining, making them more suitable for 6G FR3 that demands higher PA output power than in FR2. In addition, series Doherty PA offers reduced auxiliary loading and a broader load-modulation bandwidth [11]. Building on these advantages, this paper presents an ultra-compact asymmetrically load-pulled series Doherty PA in the Ku band that occupies only 0.217mm², even more compact than the 28GHz series Doherty in [10] despite operating at much lower frequencies, while delivering 25.3dBm output power with the power density of 1.56W/mm² and up to 30% PAE_{6dB}.

A quantitative framework for analyzing a single transformer as an IIN is shown in Fig. 20.1.1. After reflecting C_s in the on-chip transformer model to the primary side, the ideal transformer behaves as an ISN. Overall the IIN behavior is achieved when the passive network highlighted in blue resembles a CLC-π IIN. Solving for C_s shows that, for it to be physically implementable, k<0.707 is required. Unlike [9], which only qualitatively stated that low-k transformers act as IIN and thus represent a limited subset of the generalized condition, our analysis specifies the threshold quantitatively and shows that IIN behavior does not require the extremely low value (e.g., k=0.36) used in [9], where such low coupling inherently degrades efficiency.

This work further proposes an asymmetrically load-pulled Doherty PA in which the main and auxiliary PAs, although equal in size, are load-pulled to different contours. The main PA is optimized for peak PAE, while the auxiliary PA is tuned for output power and linearity. For high-PAPR signals, peak-power events are rare and demand more linearity, whereas the high-probability back-off region benefits from aligning peak PAE with the PDF maximum. This approach achieves both high efficiency in the high-probability back-off region with improved linearity at peak region.

The passive network transformation of the ultra-compact series Doherty combiner is shown in Fig. 20.1.2. Building on the earlier analysis that a single transformer can serve as both ISN and IIN, the λ/4-line-based IIN is replaced with a transformer-based IIN (Steps 1 and 2). Its single-ended equivalent is then derived as in Step 3, and by exploiting series interchangeability (Steps 3 and 4) [12], the Main– and Aux–branches are interchanged while preserving equivalence. This allows the output coil to be folded into a two-turn inductor, with the inner turns coupled to the auxiliary PA as the IIN and the outer turns coupled to the main PA in a distributed manner as the ISN. Notably, the IIN employs k_a=0.7, near the 0.707 upper limit, achieving impedance-inverting behavior with high passive efficiency and showing significant improvement over [9]. At 6dB PBO, the proposed combiner presents ~55Ω to the main PA at 13GHz, and it is modulated to ~35Ω at peak, which is load pulled for the main PA efficiency, confirming the intended Doherty load modulation. The auxiliary PA instead sees ~15Ω, providing the optimal load condition for output power and linearity. This reflects the asymmetric load-pull concept, where a larger impedance inversion ratio is applied to the auxiliary path, while still supporting high passive efficiency, achieving 80.6% at peak and 79.6% at 6dB PBO. The entire Doherty combiner occupies only 0.07mm².

Figure 20.1.3 shows the detailed circuit schematics of the proposed Doherty PA. The PA stage employs a high-voltage extended-drain MOS (EDMOS) device in the GF 22FDX+ process, supporting a 2.4V supply in a two-stack configuration to enable higher output power while preserving reliability and RF performance [13]. Furthermore, common-gate drain-source double neutralization is adopted to improve cascode stability [14]. Adaptive biasing is used to ensure the desired auxiliary path turn-on and support wideband modulation.

Implemented in the GF 22nm CMOS SOI process, the prototype PA achieves an ultra-compact area of just 0.217mm² (Fig. 20.1.7), which is comparable with or even smaller than recently reported 28GHz series Doherty PA designs [10,15], while operating at less than half their frequency.

Figure 20.1.4 presents the measured S-parameters and large-signal performance. The S21 3dB response spans 11.3 to 14.8GHz, corresponding to a 26.8% fractional bandwidth. The input 90° coupler achieves both compactness and wideband operation, maintaining S11 below –10dB across 5.7 to 21.8GHz. Furthermore, the measured k, µ, and µ' stability factors confirm unconditional stability across the entire Ku band. At 12GHz, thanks to the asymmetric load-pull strategy, the PA reaches its maximum PAE of 35.8% at 4.5dB back-off, which is more favorable for high-PAPR signals than peaking at P_{sat}, since the modulated signal rarely operate in the saturation region. Meanwhile, the PA delivers a P_{sat} of 25.3dBm and a close-in OP1dB of 24.5dBm, ensuring strong linearity to handle the infrequent peak-power excursions of high-PAPR signals while still maintaining reasonable PAE. Across the 11-to-15GHz range, the PA achieves a P_{sat} of 24.8 to 25.3dBm and a PAE_{max} of 29.9 to 35.8%. The OP1dB ranges from 23.7 to 25.0dBm, with a corresponding PAE_{1dB} of 26.1 to 27.6%. At 6dB back-off, the PAE remains high at 26.3 to 29.7%, closely approaching the peak values and confirming wideband PBO efficiency enhancement.

ISSCC 2026 / February 17, 2026 / 1:30 PM

In Fig. 20.1.5, the PA is further evaluated under modulated signals using SC QPSK and 64-QAM schemes to demonstrate the benefits of the backed-off peak and its wideband capability. At 12GHz, for QPSK (PAPR = 6.7dB), the PA achieves 20.6dBm P_{avg} with 32.2% PAE_{avg} at a 100MHz bandwidth and $EVM_{rms} < -22dB$. Notably, the modulation bandwidth can be extended up to 2/3GHz while still delivering 22.3/20.4dBm average power with 28.2%/31.4% PAE_{avg}. For 64QAM (PAPR = 8.6dB), the PA achieves 17.1dBm P_{avg} and 22.5% PAE_{avg} at a 100MHz bandwidth with $EVM_{rms} < -25dB$. When the bandwidth is increased to 1/2GHz, it maintains ~16.8/16.9dBm P_{avg} and 22.7%/22.9% PAE_{avg}, demonstrating efficiency enhancement and robust linearity over wide bandwidth.

Figure 20.1.6 summarizes the comparison. Given the limited number of Doherty PAs reported in FR3, the proposed work is also benchmarked with prior-art 28GHz FR2 Doherty PAs [16]. As shown in Fig. 20.1.6, the design achieves one of the highest reported P_{sat} values while occupying a core area comparable to 28GHz PAs, despite operating at less than half their frequency. This leads to a significantly higher power density of 1.56W/mm² than the prior works in Fig. 20.1.6. The PA also maintains competitive efficiency, with particularly strong PAE_{6dB}, and demonstrates wideband modulation with both high P_{avg} and PAE_{avg}. The sustained linearity over wideband modulated signals validates the benefit of a backed-off peak design and highlights the effectiveness of the proposed asymmetric load-pull strategy, making this PA readily integrable into compact, large-scale phased-array systems for 6G FR3.

Acknowledgement:
The authors would like to thank the GlobalFoundries University Program for chip fabrication and the team members of ETH IDEAS group for helpful technical discussions. This work was in part sponsored by the HORIZON-JU-SNS-2023 "6G-REFERENCE" project under Project 101139155, the Swiss State Secretariat for Education, Research, and Innovation (SERI) under the SwissChips initiative, HORIZON-KDT-JU-2023 "SOIL" project with the project number 101139785, and HORIZON-JU-SNS-2023 "X-TREME 6G" project under Project 101192681.

References:
[1] W. H. Doherty, "A New High Efficiency Power Amplifier for Modulated Waves," *Proc. Inst. Radio Eng.*, vol. 24, no. 9, pp. 1163–1182, Sep. 1936. https://doi.org/10.1109/JRPROC.1936.228468
[2] X. Zhang et al., "A 47GHz 4-way Doherty PA with 23.7dBm P1dB and 21.7% / 13.1% PAE at 6 / 12dB Back-off Supporting 2000MHz 5G NR 64-QAM OFDM," *ISSCC*, pp. 520–522, Feb. 2024. https://doi.org/10.1109/ISSCC49657.2024.10454571
[3] X. Zhang et al., "A 24-to-29GHz Compact Transmit/Receive Front-End Module Featuring an Asymmetric Doherty Power Amplifier and 0.22mm² Area," *ISSCC*, pp. 464–465, Feb. 2025. https://doi.org/10.1109/ISSCC49661.2025.10904541
[4] S. Hu et al., "A 28-/37-/39-GHz Linear Doherty Power Amplifier in Silicon for 5G Applications," *IEEE JSSC*, vol. 54, no. 6, pp. 1586–1599, June 2019, https://doi.org/10.1109/JSSC.2019.2902307
[5] H.-C. Park et al., "Single Transformer-Based Compact Doherty Power Amplifiers for 5G RF Phased-Array ICs," *IEEE JSSC*, vol. 57, no. 5, pp. 1267–1279, May 2022. https://doi.org/10.1109/JSSC.2022.3148044

[6] J. Lee et al., "A 22nm FDSOI CMOS-Based Compact 3-Stack Doherty Power Amplifier with a Stacked OPA-Based Bias Scheme Achieving >16.5dBm Pavg for 5G FR2 Applications," *ISSCC*, pp. 96–98, Feb. 2025. https://doi.org/10.1109/ISSCC49661.2025.10904681
[7] H. Oh et al., "A 24.25-to-29.5GHz Extremely Compact Doherty Power Amplifier with Differential-Breaking Phase Offset Achieving 23.7% PAE_{avg} for 5G Base-Station Transceivers," ISSCC, pp. 522–524, Feb. 2024. https://doi.org/10.1109/ISSCC49657.2024.10454406
[8] H. -W. Choi et al., "An Ultra-Compact, >17 dBm P_{OUT}, >30% PAE, Single Transformer-Based Doherty PA in 28-nm CMOS FD-SOI for 5G FR2 UE AiP Products," *IEEE RFIC*, pp. 67-70, June 2025. https://doi.org/10.1109/RFIC61188.2025.11082885
[9] E. Liu and H. Wang, "An Ultra-Compact 28GHz Doherty Power Amplifier with an Asymmetrically-Coupled-Transformer Output Combiner," *ISSCC*, pp. 536–538, Feb. 2024. https://doi.org/10.1109/ISSCC49657.2024.10454274
[10] M. Pashaeifar et al., "A Millimeter-Wave CMOS Series-Doherty Power Amplifier With Post-Silicon Inter-Stage Passive Validation," *IEEE JSSC*, vol. 57, no. 10, pp. 2999–3013, Oct. 2022. https://doi.org/10.1109/JSSC.2022.3175685
[11] M. Pashaeifar et al., "A 24-to-32GHz Series-Doherty PA with Two-Step Impedance Inverting Power Combiner Achieving 20.4dBm P_{sat} and 38%/34% PAE at P_{sat}/6dB PBO for 5G applications," *IEEE ASSCC*, pp. 1–3, Nov. 2021. https://doi.org/10.1109/A-SSCC53895.2021.9634772
[12] H. Xu et al., "A Flip-Chip-Packaged 25.3 dBm Class-D Outphasing Power Amplifier in 32 nm CMOS for WLAN Application," *IEEE JSSC*, vol. 46, no. 7, pp. 1596–1605, July 2011. https://doi.org/10.1109/JSSC.2011.2143930
[13] M. Chang et al., "22FDX EDMOS for 5G mmW Power Amplifier Applications," EuMIC, pp. 301–304, Sep. 2023. https://doi.org/10.23919/EuMIC58042.2023.10288882
[14] J. Xu et al., "A Compact Doubly Neutralized Ku-Band Power Amplifier with 39% Peak PAE and 23-dBm Output Power in 22FDX+ EDMOS for 6G FR3," *IEEE Microw. Wirel. Technol. Lett.*, vol. 35, no. 6, pp. 856–859, June 2025. https://doi.org/10.1109/LMWT.2025.3566279
[15] J. Chen et al., "A 24-to-30 GHz Series-Doherty Power Amplifier with Novel Broadband Combiner Achieving 2.5% Back-off PAE Variation in 65-nm CMOS," ESSCIRC, pp. 453–456, Sep. 2023. https://doi.org/10.1109/ESSCIRC59616.2023.10268766
[16] H. Wang et al., "Power Amplifiers Performance Survey 2000-Present," Accessed: Sept. 3, 2025 [Online]. Available: https://ideas.ethz.ch/Surveys/pa-survey.html

20

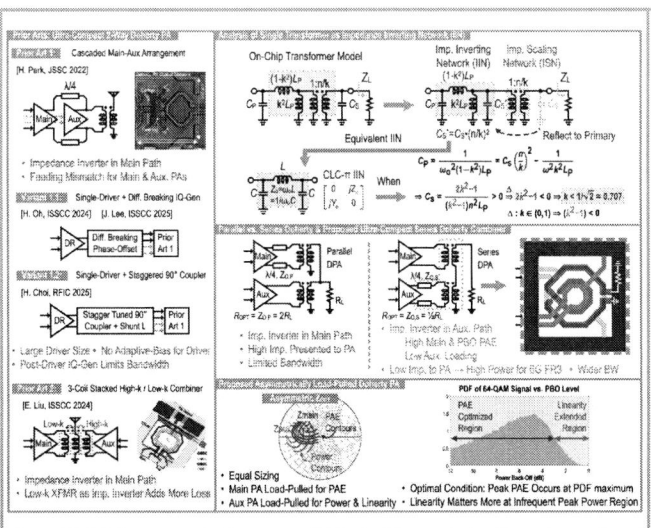

Figure 20.1.1: Summary of prior compact Doherty PAs, analysis of transformer-based IIN, and proposed series Doherty PA with asymmetric load pull.

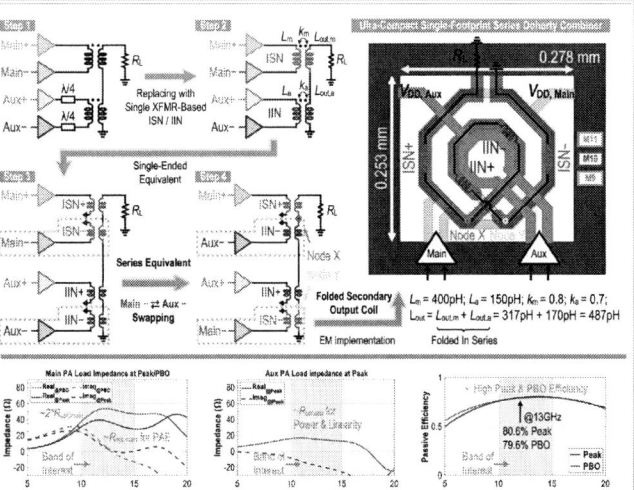

Figure 20.1.2: Passive network transformation of the series Doherty combiner, EM structure, simulated load modulation impedances, and passive efficiency.

DIGEST OF TECHNICAL PAPERS • 341

979-8-3315-8937-0/26 $31.00 © 2026 IEEE

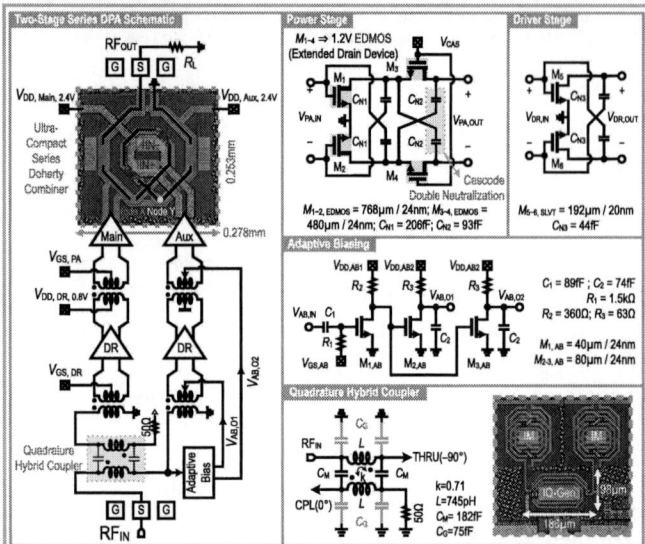

Figure 20.1.3: Detailed implementation and circuit schematics.

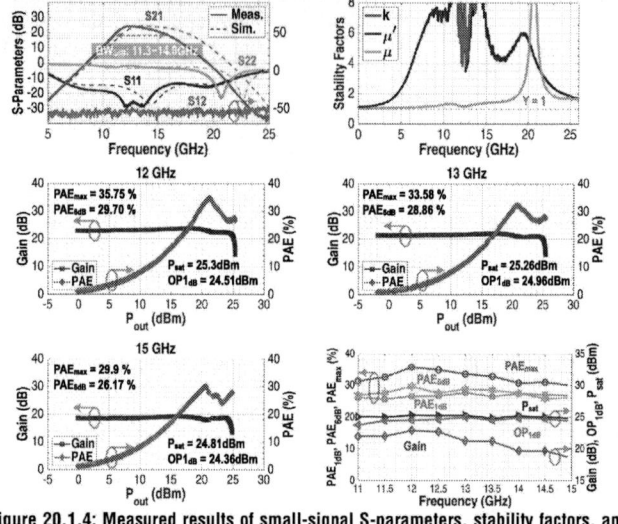

Figure 20.1.4: Measured results of small-signal S-parameters, stability factors, and large-signal CW performance.

Figure 20.1.5: Measured modulation results of 100MHz QPSK/64QAM at 12GHz, and wideband demonstrations up to 3/2GHz for QPSK/64QAM.

Figure 20.1.6: Comparison table of FR3 Doherty and linear PAs, and FR2 series and single-footprint Doherty PAs.

	6G FR3 (10-15GHz) CMOS Doherty / Linear PAs					5G FR2 28GHz CMOS Doherty PAs		
	This Work	N. Rostomyan TMTT 2018	J. Kim RFIC 2024	M. Pashaeifar JSSC 2022	H. Park JSSC 2022	E. Liu ISSCC 2024	H. Oh ISSCC 2024	H. Choi RFIC 2025
Architecture	2-Way, Series Doherty PA	2-Way, Parallel Doherty PA	2-Way, Linear PA	2-Way, Series Doherty PA	2-Way, Parallel Doherty PA	2-Way, Parallel Doherty PA	2-Way, Parallel Doherty PA	2-Way, Parallel Doherty PA
Technology	22nm SOI	45nm SOI	65nm Bulk	40nm Bulk	28nm Bulk	45nm SOI	45nm SOI	28nm SOI
Features	SF + 3-Coil Asymmetric Series Doherty Combiner	Single-Ended 4-Stack	—	Two-Step Imp. Inverter	SF + Cascaded Main & Aux	SF + 3-Coil Stacked	SF + Single Driver + Diff. Breaking	SF + Single Driver + Stagger Tuning
Freq. (GHz)	12	15	13	27	27	28	27	27
3dB BW (GHz) Frac. BW (%)	11.3 to 14.8 (3.5) 26.6%	13.2 to 16.5 (2.6) 19.2%	12.3 to 15.6 (0.3) 23.1%	24 to 32 (8.0) 28.6%	26.3 to 29.5 (4.2) 16.4%	26 to 33 (7.0) 23.7%	24.25 to 29.5 (5.25) 20.3%	26 to 30 (4.0) 14.3%
Supply (V)	0.8 / 2.4	2 / 4.8	1 / 2.2	0.9 / 1.8	0.9 / 1.8	1.0 / 2.0	2.2	1.0 / 2.0
Gain (dB)	22.9	16	24*	17.4	18.5	18.7	22.1	20.2
Psat (dBm)	25.3	25.5	23.4	20.43	18.8	21	22.3	19.8
OP1dB (dBm)	24.51	25.1	22.9	20.2	17.5	20.4	21*	18*
PAEmax (%)	35.8 @ 4.5dB PBO	31.9	46.6	39.1	36.1	38.5*	42.3	31*
PAE 1dB (%)	29.6	28*	45.5	39.1	29*	38	41*	30*
PAE 6dB (%)	29.7	23.5	22.5	34	22.0	29.3	27.8	22*
Core Area (mm²)	0.217	0.44	0.23	0.37	0.161 / 0.21†	0.154	0.14	0.1
Output Doherty Combiner (mm²)	0.070	0.158*	N.A.	0.096*	0.058	0.032	0.073	0.036*
Power Density (W/mm²)	1.56	0.81	0.95	0.30	0.38	0.82	1.21	0.95
Modulation	QPSK / 64QAM	64QAM	64QAM OFDM	64QAM OFDM	64QAM OFDM	64QAM OFDM	256QAM OFDM	QPSK / 64QAM OFDM
PAPR (dB)	6.7 / 8.5	6	11.2	8.5 / 9.7	> 10	> 9	> 10	> 7 / > 10
Bandwidth (MHz)	100 / 3000 / 100 / 2000	200	100	100 / 400	100	100 / 200	100	100 / 100
EVM (dB)	-22.0 / -22.1 / -25.1 / -25.1	-25	-25	-25.1 / -24.5	-25.0	-25.05 / -25.02	< -25	< -22 / < -25
ACPRL/R (dBc)	-19.2 / -33.5 / -34.8 / -25.8 -20.6 / -24.9 / -28.2 / -28.6	-28.8	-30	-36.3 / -28.6 -27.1 / -28.2	-31	-25.9	-29	<-29.5 / <-26.4	<-23.6	<-29.2
Pavg (dBm)	20.56 / 20.4 / 17.06 / 16.92	16.4	17.7	9.8 / 8.8	12.4	11.4	12.05 / 11.06	14.5	13.4	17.0	12.6/13.2^
PAEavg (%)	32.22 / 31.4 / 22.49 / 22.92	15.2	24.1	17.7 / 15	20.2	18.1	21	18.4	25.6	23.8	29.5	20/24^

*: Graphically estimated. SF : Single-Footprint output Doherty combiner. †: Area reported without the input balun. ‡: Estimated area with the input balun. ^: With DPD.

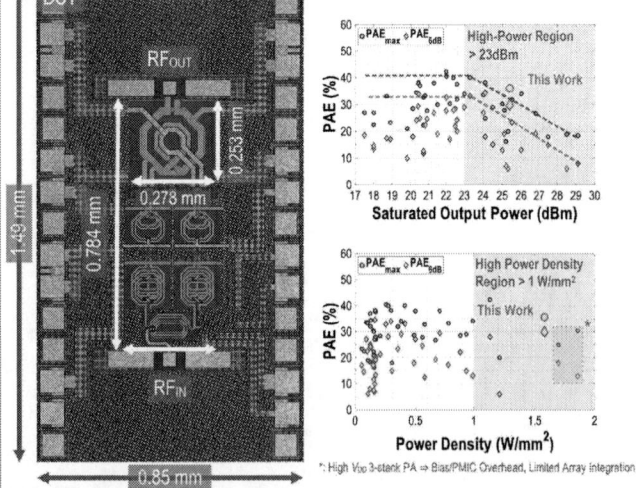

Figure 20.1.7: Die micrograph and comparison scatter plot for recently reported CMOS PAs from 10 to 50GHz.

INDEX TO AUTHORS

A

Abe, Yudai 200
Abolmagd, Hany 588
Adaikkalavan, Ramasamy 42
Adebiyi, Jide Yinka 66
Aflatouni, Firooz 242
Agaësse, J.-F. 300
Agarwal, Akshit 508
Agarwal, Amit 528
Aguirre, Pablo 366
Ahn, Gukchae 44
Akinwande, Deji 66
Akso, Emre 488
Al-Hashimi, Bashir M. 482
Al-Rawhani, Mohammed 112
Albano, Domenico 398
Alcalde, Reinaldo E. 508
Ali, Karim 294, 596
Ali, Tamer 404
Alioto, Massimo 248, 596
Alioto, Massimo Bruno 294
Alizad, Sina Haji 366
Allam, Muhamed Fouad 94
Allen, Mark 490
Alvarez-Fontecilla, Enrique 202
Anand, Tejasvi 630
Anders, Mark A. 528
Antal, Sarthak 428
Aralioglu, Arda 192
Ariki, Takuya 254
Ariyarathna, Viduneth 578
Arthanto, Yashael Faith 44
Artz, Patrick 104
Asci, Cihan 202
Asero, Claudio 398
Aseron, Paolo 424
Ashok, Maitreyi 250
Assmann, Andreas 112
Assous, Myriam 308
Ayesh, Mostafa 592
Ayodhyawasi, M. 300

B

Babaie, Masoud 208, 388, 562
Bae, ChangHyun 264
Bae, Daehee 128
Bae, Jaewan 44
Bae, Jeongyeol 578
Bae, Jooyoung 182
Bae, Junhyun 292
Bae, Junsang 116, 118
Bae, Sung-il 44
Baek, Goeun 480, 578
Baek, Sanghoon 258
Baek, Seung Ho 268
Baek, Seungcheol 44
Baek, Seungjae 342
Bahmani, Faramarz 144
Bai, Fujun 524
Bai, Jyun-Cheng 520, 526
Bajoria, Shagun 192
Baks, Christian W. 386
Ballo, Andrea 248, 294
Banerjee, Utsav 432
Bao, Hongyu Bruce 246
Bardsley, Scott 202
Bartholomew, Kevin 138
Bayle, Mathias 254
Bè, Gabriele 194
Bejarano-Carbo, Andrea 590
Beltran, Francisco Cardenas 354
Benites, Jorge Lagos 198
Bernabé, Stéphane 308
Bernard, Christian 308

Bertulessi, Luca 194
Bhamidipati, Sirisha 254
Bharadia, Dinesh 588
Bhat, Mohmad Aasif 572
Bhat, Savit 600
Bhatia, Sneha 254
Bhattacharya, Tinish 186
Bi, Xiaojun 400
Bijinapally, Kamalakanth 424
Bishop, Robert 202
Blaauw, David 590
Black, Ryan 386
Bo, Xiaochen 180
Boecker, Charles 138
Boesch, T. 300
Bolatkale, Muhammed 192
Bonfanti, Andrea Giovanni 194
Bose, Bella 630
Bose, Soumya 134
Bosi, Alessandro 398
Boutafa, Laura 308
Bowman, Keith 172
Branca, Xavier 112
Breems, Lucien 192
Brian, Michele 120
Brunsilius, Janet 202
Bulzacchelli, John F. 386
Bunsen, Keigo 200
Bushimata, Yuto 326
Byeon, Sangyeon 270
Byun, Sanho 420
Byun, Sung-Jae 128
Byun, Young-Yong 264

C

Cai, Fanxun 582
Cai, Hao 512, 536
Cai, Tianyi 376
Cai, Yuancheng 96
Cammarota, Rosario 424
Cantoni, Adalberto 202
Cao, Chenyao 146
Cao, Hung Van 168
Cao, Jun 174
Cao, Lixuan 566
Cao, Nianzheng 52
Cao, Peng 322
Cao, Qiankai 176
Cao, Yue 524
Cao, Yuluan 434
Casas, Jeremy 424
Catalino, Chris 52
Ceroni, Alessia 194
Cha, Jaehoon 270
Cha, Jinyoup 270
Chae, Heeyoung 44
Chae, Joo-Hyung 634
Chae, Kwanyeob 628
Chakraborty, Sudipto 386
Challagundla, AppaRao 424
Chan, Chi Hou 106
Chan, Chi-Hang 86
Chan, Chun-Kun 56
Chan, Lando 438
Chan, Wei Khuen 594
Chandrakasan, Anantha P. 250
Chang, Cheng-Feng 516
Chang, Chia-Ming 56
Chang, Chin-Yu 490
Chang, Huaichung 170
Chang, Jeongtaek 48, 590
Chang, Ken 398
Chang, Leland 52
Chang, Mau-Chung Frank 354
Chang, Meng-Fan 516, 520, 526

Chang, Min-Hua 64
Chang, Muya 548
Chang, ShenKai 56
Chang, Shih-Chieh 526
Chang, Tsung-Yung Jonathan 260
Chang, Yen-An 260
Chang, Young-Uk 264
Chang, Yung-Chang 56
Chapman, Eric 42
Charbonnier, Benoit 308
Charbonnier, Jean 308
Chatarasi, Prasanth 52
Chaves, Jose M. Rojas 424
Chawla, N. 300
Chen, Allen 144
Chen, Barry 170
Chen, Bu 122
Chen, Candy 42
Chen, Chang-Yuan 516
Chen, Changjin 460
Chen, Chia-Ping 56
Chen, Chin-Yi 254
Chen, Chixiao 240, 632
Chen, Christopher 354
Chen, Dihu 162
Chen, Fuzhan 246
Chen, Gong 576
Chen, Gregory K. 178
Chen, Hao 144
Chen, Hong 606
Chen, Hsinchen 170
Chen, Jhih-Wei 354
Chen, Jiawen 482
Chen, Jinbo 610
Chen, Jinghong 498
Chen, Jixin 348, 358
Chen, Junjie 218
Chen, Ko-Chi 526
Chen, Kuan-Chang 402
Chen, Kuan-Chun 260
Chen, Linchien 256
Chen, Mike Shuo-Wei 592
Chen, Minhan 138
Chen, Peng 348
Chen, Po-Wei 54
Chen, Qin 108
Chen, Tao 170
Chen, Wei-Chih 136
Chen, Weinan 434
Chen, Wenhua 350
Chen, Xi 176, 396, 512
Chen, Xibi 580
Chen, Xin 108
Chen, Xinyang 536
Chen, Xuesong 594
Chen, Yan 350
Chen, Yen-Ming 136
Chen, Yi 542
Chen, Yi-Syuan 56
Chen, Yi-Ting 136
Chen, Yifei 342
Chen, Ying 578
Chen, Yong 442
Chen, Yong-Tai 54
Chen, Yu-Chi 136
Chen, Yutang 162
Chen, Zaize 472
Chen, Zhe 358
Chen, Ziteng 448
Cheng, Bojun 314
Cheng, Chia-Yuan 56
Cheng, Chun-Yuan 256
Cheng, Depeng 108
Cheng, Hsin-Ping 56
Cheng, Jinhui 524

Cheng, Kai 442
Cheng, Kwang-Ting 532
Cheng, Kwang-Ting Tim 610
Cheng, Lin 280, 330, 452, 458, 460
Cheng, Nai-Chen 136
Cheng, Quan 390
Cheng, Sirui 610
Cheng, Ting-Yu 508
Cheng, Xuxu 150, 624
Cheng, Ziyi 606
Chenna, Vinay 234, 236
Chhetri, Ghanshyam 172
Chi, Baoyong 152, 220, 222
Chi, Hankyu 270
Chi, Miock 44
Chi, Taiyun 92, 344
Chiang, Ming-Hsuan 56
Chiang, Po-Han 56
Chien, Shihchieh 578
Chighine, Kevin 500
Chih, Hung-Wei 56
Chih, Yu-Der 260
Chin, Woojin 540
Chiou, Pei-Chen 136
Chiu, Taiwei 228
Cho, Gihwan 128
Cho, Hyun 628
Cho, Hyunjun 430
Cho, Joo-Mi 324, 332
Cho, Nara 44
Cho, S. 302
Cho, Wonhee 424
Choe, Yeounghwan 44
Choi, Dooseok 578
Choi, Haejung 372
Choi, Haidam 618
Choi, Hyeon-Ji 324, 456
Choi, Jaehyuk 116, 118, 124
Choi, Jinho 628
Choi, Jinyong 268
Choi, Jiwon 540
Choi, JongMoon 268
Choi, Juncheol 364
Choi, Junkyeong 44
Choi, Michael 372
Choi, Myunghoon 44
Choi, Sungpill 44
Choi, Won Ho 268
Choi, Won-Jong 364
Choi, Woo-Seok 148
Choi, Yong-Suk 128
Choi, Yoonjae 628
Choi, Youngkil 410
Choi, Yuseon 534
Chon, Hojae 612
Chou, Iris Ying 184
Chou, Kuan-Ting 136
Chou, Lin 64
Chou, Tan-Li 260
Chowdhury, Antroy Roy 404
Chu, Chenyuan 556
Chuang, Harry 260
Chuang, Ming-Han 154
Chun, Jung-Hoon 116, 118, 124
Chung, Jaehyun 578
Clemencon, Vincent 112
Clemons, Jason 548
Clymore, Christopher J. 488
Codega, Nicola 398
Cohen, Matthew 52
Collins, Steven 112
Cooman, Adam 198
Cordoba, Cyril 168
Coudyzer, Gertjan 406
Covington, William 176

• 2026 IEEE International Solid-State Circuits Conference

INDEX TO AUTHORS

Crafton, Brian 520
Craninckx, Jan 198
Criss, Russell 178
Crocherie, Axel 112
Crols, Sander 502
Crumley, Paul 52
Cui, Han 576
Cui, Kai 164
Cui, Yilu 552
Cuskelly, Lachlan 354
Cyrusian, Sasan 398

D

Dai, Jun 378
Dai, Steve 548
Dartizio, Simone Mattia 210, 212
Datsko, Benjamin 232
Datta, Kishalay 334
David, Jean-Baptiste 500
Davidson, Alfred 100
Davies, Andy 386
Davoodi, Mehdi 398
Dawson, Geraldine 494
Dayal, Pranav 578
Dayanik, Batu 144
De, Vivek 436
Dehos, Cédric 500
Demirci, Tugba 314
Demosthenous, Andreas 620
Demsky, Kevin 386
Deng, Heqi 562
Deng, Juncheng 576
Deng, Wei 152, 220, 222
Desai, Shaishav 138
Desoli, G. 300
Devgan, Anirudh 26
Diao, Yumei 204
Dikopoulos, Evangelos 232
Dillon, Chris 202
Ding, Lingke 428
Ding, Shixiang 498
Ding, Yifan 310
Ding, Yifang 96
Ding, Yixiao 490
Ding, Yong 448
Ding, Yuhan 466
Do, Hyungrok 266
Dogiamis, Georgios C. 580
Dong, Jun 512
Dong, Liang 158
Dong, Pingcheng 532
Dou, Chunmeng 514, 522, 524
Driel, Willem van 282
Du, Haoran 536
Du, Jieqiong 354
Du, Li 626
Du, Naike 582
Du, Sijun 68, 72, 160, 280, 282, 284, 288, 320, 336
Du, Xincheng 472
Du, Xingyu 490
Du, Yuan 626
Du, Yucheng 512
Du, Yukan 448
Duan, Zhouchi 146
Duda, Kevin 42
Dunna, Manideep 588
Duong, Cuong Manh 430
Dutta, Barundeb 616
Dutton, Neale A. W. 112

E

Ekman, Jeremy 386
Elbadry, Mohamed Abdelrahman 388

Eleraky, Mohamed 340, 346, 564
Eliezer, Oren 578
Elkholy, Ahmed 174
Elmenshawi, Ahmed 360
ElShater, Ahmed 404
Ema, So 90
Emami, Azita 508
Englund, Dirk 384
Enthoven, Luc 388
Eom, Kyeongho 612, 614
Erickson, Emma 386

F

Fabiano, Ivan 398
Fadila, Ashbir Aviat 476
Fagotti, Damiano 210, 212
Fahimnia, Mehrdad 144
Fakkel, Niels 388, 562
Fan, Qinwen 78, 380
Fang, Eric Jia-Wei 170
Fang, Lele 86
Fang, Wenkai 94
Fang, Xiao 582
Fang, Yidong 344
Fang, Yuan 144
Farhoodfar, Arash 398
Fei, Zheyuan 448
Feng, Lichen 518
Feng, Shuo Sarah 246
Feng, Xiaodi 240
Feng, Xiaoyu 60
Feng, Xinhe 126
Ferrari, Victor 52
Ferrer, Florencia 366
Feygin, Gennady 578
Flynn, Michael P. 232
Fojtik, Matthew 548
Francese, Pier Andrea 386
Franiatte, Rémi 308
Frank, David J. 386
Friedman, Daniel J. 386
Fu, Gaoming 512
Fu, Haotian 314
Fu, Jiamu 316
Fu, Renjie 582
Fu, Yang 398
Fu, Yushen 204
Fu, Zhigang 582
Fuguet, César 308
Fujihara, Yasuyuki 254
Fujimura, Susumu 254
Fujisawa, Yui 506
Fujita, Naoya 262
Fukuoka, Kazuki 168

G

Gaddam, Sudheer 402
Gade, Srinivas Pavan Kumar 42
Gai, Weixin 140
Galbraith, Bob 52
Gallucci, Stefano 210, 212
Gambarelli, Serge 500
Gao, Bin 228
Gao, Hanghang 514
Gao, Hao 358
Gao, Wei 492
Gao, Weichen 60
Garampazzi, Marco 398
Garg, Adesh 174
Garimella, Lakshminarasimha Sastry 100
Gaucher, Brian P. 386
Ge, Yu 458
Gebreyohannes, Fikre 172
Geng, Xiang 374

Geng, Xinli 204
Geng, Xinlin 466
Gerfers, Friedel 104
Geurts, Thijs 616
Ghittori, Nicola 398
Ghorbanpoor, Mohsen 564
Ghosh, Archisman 428
Ghozzy, Sherif 230
Giles, Hope 35
Giorgetti, Daniele 120
Gira, Gabriele 398
Giunco, Fabio 398
Giustolisi, Gianluca 248
Go, Jonghyun 128
Golder, Anupam 424
Gomez, Carrel de 476
Gong, Huijing 424
Gong, Xin-Ce 374
Gonzalez, Chris 52
Gooding, Tom 52
Gourdouparis, Marios 70
Grassi, Alberto 144
Grasso, Alfio Dario 248
Gray, C. Thomas 396
Greensky, James 424
Groen, Eric 138
Gu, Jiangyuan 58, 546
Gu, Jie 176, 602
Gu, Mingyang 196
Gu, Tingyi 244
Gu, Wenxian 84
Gu, Ye 406
Gu, Yucong 582
Guan, Ningzi 498
Gubin, Yaroslav 616
Gui, Xiaoyan 146
Guidry, Matthew 488
Guille, Olivier 308
Guillorn, Michael 52
Guo, An 512
Guo, Hao 92, 106
Guo, Jianping 162, 278
Guo, Luyi 482
Guo, Mingqiang 558
Guo, Ruiqi 58, 546
Guo, Yixin 524
Guo, Zhan 594
Guo, Zijun 106
Gurbaxani, Rishabh 208
Gurumurthy, Girishankar 256
Gutierrez, Christopher N. 424

H

Ha, Gyeongmin 420
Ha, Kyung-Soo 264
Ha, Sangwoo 312
Ha, Sohmyung 618, 620
Hajri, Basma 172
Hall, Drew A. 366
Hall, Duncan 112
Halli, Ramesh 256
Ham, Junhee 44
Hammoud, Ali 232
Han, Chanheum 634
Han, Jeongwon 364
Han, Jinho 402
Han, Jun 426
Han, Ruonan 250, 384, 580
Han, S. 302
Han, Seungjun 364
Han, Shinhee 258
Han, Shousheng 454
Han, Su-Hyun 128
Han, Xiangdong 434
Han, Yang 108

Han, Yiming 66
Han, Yuyang 384
Han, Zhongze 514, 522, 524
Handa, Takaya 254
Hanke, Chris 138
Hao, Zhenqi 228
Hara, Yusaku 168
Hardeman, Gilbert 192
Harris, Isaac B. 384
Hasan, Md Nazmul 588
Hasebe, Kazunori 200
Haselhorst, Tom 386
Hashemi, Hossein 234, 236
Hashimoto, Masanori 390
Hattori, Genma 90
Hayashi, Koichiro 254
He, Long 108
He, Longzhen 114
He, Shitu 152
He, Xiyu 190, 196
He, Yongjun 468
He, Yuchen 632
He, Yukun 146
He, Yuming 70
Hedayati, Hiva 402
Hekmatshoartabari, Bahman 52
Hella, Mona M. 360
Helleputte, Nick Van 502
Hensley, Mike 202
Heo, Hyungseok 44
Heo, Sanghyuk 266
Herbas, Daniel L. 398
Higashi, Yumi 254
Hilkens, E. 300
Hong, Binwen 400
Hong, Dongyeon 440
Hong, Gi-Moon 266
Hong, Kieop 124
Hong, Seongyon 540
Hong, Sung-Wan 324, 332, 456
Hong, Tao 122
Hong, Wei 348, 358
Hong, Zhiliang 76, 322, 368
Honma, Naoki 90
Hoover, Kathy 42
Horii, Kohei 326, 328
Hou, Longxiang 466
Hou, Zhang 434
Hsieh, Chih-Cheng 516, 520, 526
Hsieh, Kenny Cheng-Hsiang 136
Hsieh, Le-Jung 520, 526
Hsieh, Shih-Wei 56
Hsu, Chen-Hsing 64
Hsu, Chen-Kai 202
Hsu, Cynthia 254
Hsu, Hung-Hsi 516, 520, 526
Hsu, Lien-Feng 56
Hsu, Steven K. 528
Hsu, Ting-Hao 520, 526
Hsu, Ying-Tuan 232
Hsueh, Sung S.-Y. 170
Hu, Bo 536
Hu, Deyong 594
Hu, Hao 626
Hu, Hao-Tao 106
Hu, Hongyang 514, 522
Hu, Sanming 96
Hu, Yang 58, 316, 546
Hu, Yaolong 92
Hu, Yizhe 576
Hu, Yong 384
Hu, Yu-Jia 526
Huang, Bo-Jr 170
Huang, Chao-Tsung 54
Huang, Cheng 68, 158, 244

INDEX TO AUTHORS

Huang, Jie-Ren 136
Huang, Junwei 454, 462
Huang, Leilei 498
Huang, Mo 276, 286, 290, 446, 450
Huang, Po-Hao 56
Huang, Qiao 330
Huang, Qijing 548
Huang, Siyu 196
Huang, Stan 56
Huang, Tzu-Yuan 340
Huang, Weiwei 458
Huang, Wen-Hung 136
Huang, Xijie 532
Huang, Yao-Wei 64
Huang, Yen-Che 516
Huang, Yi 184
Huang, Yu-Jie 136
Huang, Yulong 314
Huang, Yunbo 442
Huang, Zebin 606
Huang, Zhangcheng 122
Huang, Zhi 174
Huang, Zhikai 504
Huang, Zhiqiang 470
Huang, Zhiwen 140
Hui, Yiheng 626
Humblet, Alexis 616
Hung, Chao-Jung 260
Huo, Dexuan 606
Hur, Joonhoi 578
Hursey, Josh 52
Hutchinson, George Higgins 186
Hwang, JongTae 264
Hwang, Jung-Hye 124
Hwang, Sang-Joon 264
Hwang, SangJoon 268, 272
Hwang, Seonwoo 270
Hwang, Shih-Arn 170
Hwang, Sohee 258

I

III, Roy H. Olsson 490
III, Thomas H. Greer 396
Iizuka, Tetsuya 468
Ilamurugan, Vinoth 192
Im, Maesoon 612, 614
Inagawa, Takahiro 90
Inoue, Ken 386
Irita, Takahiro 168
Ishibashi, Yushi 506
Ishida, Hideaki 554
Ishihara, Hiroaki 326, 328
Ishizaki, Yuki 254
Ismail, Yousr 174
Ito, Masamichi 128
Ito, Takashi 262
Iyer, Arvindh 144

J

Jacob, Philip 52
Jacquot, Jean-François 500
Jahagirdar, S. 304
Jain, Ajaypat 578
Jain, Radhika 52
Jain, Shubham 52
Jamal, Jismal 484
Jang, Jieun 266
Jang, Jinhun 268
Jang, Minsoo 268
Jang, Yonghwan Harold 578
Jantzi, Stephen 398
Jaussi, James 134
Javadi, Ramin 630
Je, Minkyu 618

Je, Sangeun 44
Jeon, H. 302
Jeon, Jaeho 430
Jeon, Jin-Yong 364
Jeon, S. 302
Jeon, Sehyug 342
Jeon, Taehoon 118
Jeon, Taeyoung 44
Jeon,, Gyunam 270
Jeong, H. 302
Jeong, Hyeongsoo 266
Jeong, Hyun-Woo 456
Jeong, Jaehun 128
Jeong, Ji Yong 120
Jeong, Seokhan 608
Jeong, Sera 270
Jeong, Suheon 542
Ji, Junghwan 270
Ji, Yichao 452, 458, 460
Jia, Haikun 152, 220, 222
Jia, Tianyu 310, 538
Jia, Yaoyao 66
Jiang, Chen 240, 632
Jiang, Dai 620
Jiang, Fuze 504
Jiang, Haijun 524
Jiang, Jianqiang 158, 244
Jiang, Jinghao 524
Jiang, Junmin 336
Jiang, Keyao 538
Jiang, Linrui 66
Jiang, Shunmin 288
Jiang, Wenning 632
Jiang, Xiping 524
Jiang, Xuhao 108
Jiang, Yumeng 606
Jiang, Ziwei 350
Jiao, Junkai 582
Jiao, Tianhui 512
Jie, Lu 190, 196, 378, 552
Jimenez, José Luis Gonzalez 500
Jin, Ji 452
Jin, Jinxuan 400
Jin, Ruibin 398
Jin, Young-Jae 44
Jin, Zhenghao 310
Jing, Zixi 478
Jo, Hyeonsu 578
Jo, Hyunje 44
Jo, Wooyoung 312, 540
Jo, Yun-Rae 420
Jo, Yurim 312
John, Deepesh 42
Joo, Sunghwan 264
Joo, Yongsuk 270
Josselin, Vincent 308
Jou, Yucheun Kevin 56
Ju, Chi-Cheng 56
Juan, Bo-Cheng 64
Junaid, Ammaar 100
Jung, C. 302
Jung, In 268, 272
Jung, Jaehong 250
Jung, Jinwook 52
Jung, Jonghoon 258
Jung, Jueun 46
Jung, Jun Won 144
Jung, Junghoon 128
Jung, Sang-Hoon 272
Jung, Seungjae 272
Jung, Woojoong 292, 372
Jung, Y. 302
Jung, Yooseok 266

K

Ka, Dongyoon 266
Kadaveru, Sreevatsank 588
Kalogerakis, Georgios 396
Kalzhan, Zhamaliddin 540
Kam, Dongyun 544
Kamei, Tatsuya 168
Kanagawa, Naoaki 254
Kaneko, Tohru 554
Kang, Bogyeong 440
Kang, Boyoung 628
Kang, Byungjun 270
Kang, Gyuseong 258
Kang, Hansol 628
Kang, Jaeyeol 266
Kang, Joonghoon 612
Kang, Jubin 124
Kang, Kidong 578
Kang, Kyeongpil 266
Kang, KyuChang 272
Kang, Seonghyeon 578
Kang, Shin-haeng 264
Kang, Sungmoon 44
Kang, Sungsik 628
Kang, Xilong 536
Kankuppe, Anirudh 198
Kano, Masahiro 254
Kanybek, A. 302
Kao, Tony 402
Kao, Yu-Sheng 520, 526
Kapusta, Ron 202
Kar, Monodeep 52
Karahan, Emir Ali 230
Karam, Victor 398
Karmakar, Angshuman 428
Kashirin, Alexander 178
Kashmiri, Mahdi 402
Katayama, Yasushi 200
Kato, Sena 90, 392
Kato, Soichi 200
Kato, Yosuke 254
Kaus, Jonathan 386
Kawai, Shusuke 326, 328
Kazemkhani, Shayan 402
Ke, Han-Tzung 136
Keller, Ben 548
Kelly, Dan 202
Kethareswaran, Lalith 424
Khailany, Brucek 548
Khanna, Devrishi 398
Khellah, Muhammad 176
Khiarak, Mehdi N. 398
Khorami, Ata 588
Khwa, Win-San 516, 520
Ki, Myoungoh 44
Kikkawa, Toshiyuki 200
Kim, Bongjin 182
Kim, Boram 270
Kim, Bumjun 128
Kim, Byeong-Chan 638
Kim, Byeongcheol 534
Kim, Chang Won 268
Kim, Daehoon 44
Kim, Daero 628
Kim, Daesun 272
Kim, Dohui 258
Kim, Donggeon 272
Kim, Donghan 44
Kim, Donghyeon 364
Kim, Donghyuk 116
Kim, Dongkyun 266
Kim, Doyoon 480, 578
Kim, Duhyeong 424
Kim, Eunseo 44
Kim, G. 302

Kim, Geunha 608
Kim, H. 302
Kim, Hangyeol 542
Kim, Hongseok 292
Kim, Hongyun 44
Kim, Hun-Seok 590
Kim, Hyeong-Joon 418
Kim, Hyun-Sik 412, 416, 418
Kim, Hyungsoo 266, 270
Kim, Hyunho 44
Kim, Hyunsung 44
Kim, J. 302
Kim, Jaehee 50
Kim, Jeong-Hun 332
Kim, Ji-Young 264
Kim, Jieun 266
Kim, Jinguk 268
Kim, Jinseok 44
Kim, Jinyeon 264
Kim, Jonghwan 270
Kim, Jonghyuk 268, 272
Kim, Jonghyun 578
Kim, Joo-Young 430, 542
Kim, Joohwan 264
Kim, Joonsuk 578
Kim, Jooseong 372
Kim, JuHwan 44
Kim, Junseong 578
Kim, Junsoo 272
Kim, Kahyun 148
Kim, Kangjoo 410
Kim, Kiheung 268
Kim, Kihyun 412, 418
Kim, Kyu-hyoun 52
Kim, Kyunghoon 270
Kim, Kyunghwan 272, 480, 578
Kim, Kyuseong 258
Kim, Kyuyoung 266
Kim, M. 302
Kim, Mijoung 258
Kim, Minchang 266
Kim, Minju 614
Kim, Minkyung 116
Kim, Mino 266
Kim, Minseo 44
Kim, Minsu 292
Kim, Minwoo 292
Kim, Minwook 636
Kim, Myunggon 628
Kim, S. 302
Kim, Sang Young 402
Kim, Sang Yun 268
Kim, Sangjin 312, 534
Kim, Seong-Jin 116, 118, 124
Kim, Seongjin 270
Kim, Seongjung 578
Kim, Seulgi 270
Kim, Seung-Goo 44
Kim, Seung-Sik 128
Kim, Siwoo 420
Kim, Suk Lae 272
Kim, Suksan 128
Kim, Sungjoo 578
Kim, Sunkwon 410
Kim, Taehyeon 636
Kim, Taewan 342
Kim, Taeyeon 578
Kim, W. 302
Kim, Wan 440
Kim, Wan Jong 578
Kim, Yeseul 258
Kim, Yong-Min 264
Kim, Yongjik 44
Kim, Yongjun 128, 272
Kim, Yoonhyung 628

INDEX TO AUTHORS

Kim, Youngtaek 270
Kimura, Hiroshi 144
Kitamura, Kei 254
Kitani, Tomofumi 254
Knag, Phil C. 178
Ko, Han-Gon 640
Ko, Hyeongjun 270
Ko, Seungpil 258
Ko, Youngwoon 364
Kocaman, Namik 144
Kodama, Takuyo 254
Komma, Demba 590
Kong, Hao 184
Kong, Xiangyu 184
Kong, Zhen 126, 390
Koo, Billy 628
Koo, Si-Gyoung 128
Kornfield, Julia A. 508
Koswatta, Siyu 52
Kota, Kishore 398
Kouchi, Toshiyuki 254
Krishnamurthy, Ram 528
Krishnamurthy, Ram K. 178
Krishnamurthy, Sashank 134
Krishnaswamy, Harish 100
Krithivasan, Sarada 52
Kuang, Honglin 426
Kuang, Jian-Jun 374
Kumar, Limitha 68
Kumar, Neelotpala 66
Kumar, Nitish 64
Kumar, Raghavan 424
Kummari, Shekher 366
Kundu, Suparna 428
Kunihiro, Kazuaki 476
Kuo, Hsin-Hung 136
Kuo, Sheng-Po 56
Kuo, Shih-Kai 588
Kwak, Yong-Sik 440
Kwon, Daehan 266
Kwon, Daehyun 268
Kwon, Dongseok 186
Kwon, Hanbyeol 266
Kwon, Hye-Jung 268
Kwon, Hyukbin 128
Kwon, Kyeongha 364
Kwon, Sooncheol 266
Kwon, Soonhyun 48, 544
Kwon, Taehyun 410
Kwon, Yongil 410

L

L'Bahy, Hassan 202
Lacaita, Andrea Leonardo 194, 210, 212
Lai, Sheng-Tsung 136
Lake, Dan 424
Lalwaney, Poornima 424
Lan, Juntao 220, 222
Lancaster, John David 52
Lang, Tian-Chen 374
Lassalle-Balier, Remy 366
Latham, Alex 366
Lau, Pak Tao Alan 246
Lau, Pak-Kim 578
Law, Duncan 42
Law, Man-Kay 114
Lee, Benjamin G. 396
Lee, Chan-Ho 456
Lee, Chanhee 48
Lee, Chia-Fu 260
Lee, Ching-Yen 414
Lee, Dongbeom 266
Lee, Dongkeon 268
Lee, Dongsoo 544
Lee, Eunseok 250, 384

Lee, Gangsik 270
Lee, Gwangwon 628
Lee, Haesuk 264
Lee, Hangil 258
Lee, Hyun-Su 612, 614
Lee, Hyunchang 258
Lee, Hyung-Min 292, 612, 614
Lee, J. 302
Lee, Jae-kyu 128
Lee, Jae-Yeol 410
Lee, Jae-Youl 420
Lee, Jaebong 44
Lee, Jaeho 258
Lee, Jaehoon 264, 636
Lee, Jaehyung 268
Lee, JaeKyung 264
Lee, Jeonghyeon 640
Lee, Jewon 636
Lee, Jihyun 420
Lee, Jingu 312
Lee, Jinseop 364
Lee, Jiseok 270
Lee, Jongmi 440
Lee, Jongmyeong 266
Lee, Jongwoo 410
Lee, Joonggeun 578
Lee, Joonhee 578
Lee, Jooseok 342
Lee, Joungwoo 44
Lee, Jun-Gi 412
Lee, Junghyup 584, 608
Lee, K. 302
Lee, Keonho 270
Lee, Ki-Soo 634
Lee, Kyongsu 638
Lee, Kyoungtae 584, 608
Lee, Kyoungwon 628
Lee, Kyuho Jason 46
Lee, Minoo 608
Lee, Minseob 578
Lee, Nayeong 540
Lee, Po-Hao 260
Lee, Saekyu 52
Lee, Sang-Gug 364
Lee, Sanggwon 128
Lee, Sangho 46
Lee, Sanghoon 270
Lee, Sangmin 628
Lee, Sangsung 578
Lee, Sangyong 268
Lee, Seon-Kyoo 636, 638
Lee, Seongseop 266
Lee, Seungho 266
Lee, Seungjun 268
Lee, Sunghyuck 372
Lee, Sungjun 578
Lee, Sungkwon 270
Lee, Sunkyu 258
Lee, Woncheol 578
Lee, Woo-Nyoung 420
Lee, Wooram 102, 352
Lee, Yong-Chan 332
Lee, Yongsun 264
Lee, Yoona 148
Lee, Youna 128
Lee, Younggeun 44
Lee, Youngjoo 48, 50, 544
Lee, Youngki 578
Lee, Youngsik 264
Lee, Yunho 292
Lei, Hao 582
Lei, Ka-Meng 80
Lei, Mingqian 204
Lekuch, Scott 386
Lele, Ashwin Sanjay 520

Leon, Ana Sonia 178
Leung, Ka Nang 278
Leung, Michael 398
Levantino, Salvatore 194, 210, 212
Levin, A. 304
Li, B. 302
Li, Bingrui 190, 556
Li, Bo 442
Li, Chang-Yi 136
Li, Chushan 448
Li, Dong 518
Li, Duo 476
Li, Haihua 80
Li, Hanyue 198
Li, Haoran 472
Li, Haoyuan 390
Li, Hongou 538
Li, Humiao 126
Li, Jason 254
Li, Jiamin 126, 390, 604
Li, Jiaming 228
Li, Jiawei 474
Li, Jiaxiang 566
Li, Jiayang 620
Li, Jinben 358
Li, Jinge 472, 474
Li, Liangwei 184
Li, Lin 228
Li, Mao 436
Li, Ming-Che 428
Li, Mingxuan 310
Li, Mingyuan 578
Li, Peng 512
Li, Qiufeng 390
Li, Sensen 344
Li, Shenggao 136
Li, Simon 138
Li, Vincent 172
Li, Weizeng 514, 522
Li, Wuhua 448
Li, Xiao 184
Li, Xiaoyu 576
Li, Xiayang 68
Li, Xueqi 228
Li, Xuyang 498
Li, Yang 214, 224
Li, Yicheng 566
Li, Yida 126
Li, Yiqi 176
Li, Yixi 218
Li, Yuanfang 144
Li, Zhao 202
Li, Zhenghao 150, 624
Li, Zhenhao 158, 244
Li, Zhensheng 558
Li, Zheyi 616
Li, Zhi 514, 522
Li, Zhouzheng 228
Li, Ziyang 582
Li1, Lianming 108
Liang, Dingqi 582
Liang, Guirong 254
Liang, Jin 138
Liang, Luhong 532
Liang, Xiqing 570
Liang, Yitao 180
Liang, Yuan 390
Liao, Yu-Te 64
Liao, Zhipeng 610
Lim, Daihyun 264
Lim, Gyu-Wan 416
Lim, Hyun-Wook 410, 420
Lim, Hyungsun 578
Lim, Kyungtae 128
Lim, Yong 440

Lin, Guan-Yu 64
Lin, Hon-Jarn 260
Lin, I-Ting 600
Lin, Jeff 256
Lin, Jiaming 536
Lin, Lishan 180
Lin, Longyang 126, 390, 604
Lin, Ming-Chieh 260
Lin, Ming-Hung 56
Lin, Minggui 606
Lin, Mu-Shan 136
Lin, Qingxuan 470
Lin, Shu-Ping 64
Lin, Wei-Shuo 136
Lin, Xiaohui 630
Lin, Yalong 400
Lin, Yen-Hua 516
Lin, Yi-Jie 64
Lin, Yifan 162
Lin, Ying-Sheng 414
Lin, Yu-En 526
Lin, Yu-Shiang 178
Lin, Zhen 96
Lin, Zhicheng 106
Lin, Ziyi 152
Linnhoff, Sebastian 104
Lipson, Samuel 42
Liss, Andrew 176
Liu, Bo 512, 536
Liu, Chang 152
Liu, Di 384
Liu, Gang 336
Liu, Hangxing 504
Liu, Jett 56
Liu, Jian 142, 218
Liu, Jiang 70
Liu, Jiayao 368
Liu, Kunyang 438
Liu, Leibo 184, 434
Liu, Lianbo 344
Liu, Liyuan 142, 218
Liu, Mengyao 498
Liu, Ming 122, 240, 514, 522, 524, 632
Liu, Qi 122, 240, 524, 632
Liu, Qian 626
Liu, Qilong 192
Liu, Ren-Shuo 516, 520, 526
Liu, Shen-luan 154
Liu, Shih-Yang 532
Liu, Supeng 594
Liu, Wen 138
Liu, Xiao 126
Liu, Xiaosen 216
Liu, Xing 610
Liu, Xinning 512
Liu, Xuanzhi 524
Liu, Xuejiao 532
Liu, Xun 336
Liu, Yanchao 214, 224
Liu, Yao-Hong 70
Liu, Yen-Jen 64
Liu, Yi 478
Liu, Yichen 216
Liu, Yongpan 60
Liu, Yu 532
Liu, Yunlong 518
Liu, Yuqi 434
Liu, Zeguo 458
Liu, Zezheng 476
Liu, Zhaokai 134
Liu, Zhichao 512
Lo, Chung-Chuan 520, 526
Lo, Wei-Chung 526
Lombard, Christian 500
Lombardo, Domenico Maria 78

INDEX TO AUTHORS

Long, Yucheng 576
Long, Zhijun 204
Lopes, Ward 396
Lopez, Carolina Mora 502, 616
Lou, Liheng 576
Lou, Tsung-Han 520, 526
Lovitt, Travis 398
Lu, Cheng-Han 260
Lu, Cheng-Hsun 50
Lu, Chien-Yu 170
Lu, Hao 66
Lu, Haowei 204
Lu, Ping 138
Lu, Po-Yen 54
Lu, Pong-Fei 52
Lu, Siuchuang Ivan 578
Lu, Tianqi 72, 160
Lu, Xiaoyu 348
Lu, Yan 164, 454, 462
Lu, Yue 498
Lu, Yuri 498
Lu, Zhichao 532
Luo, Chenjie 350
Luo, Hao 204
Luo, Haoyang 180, 190, 556
Luo, Jiaming 356
Luo, Peng 532
Luo, Qing 514, 522
Luo, Xiongshi 150, 624
Luo, Xun 568, 570
Luo, Yunbin 632
Luo, Yuxuan 376
Luo, Zhiren 458
Luong, Howard Cam 478
Lutz, Martin 52
Lv, Hangbing 524
Lyu, Liangjian 84
Lyu, Zhichao 126

M

Ma, He 626
Ma, Heng 380
Ma, Na 606
Ma, Rui 220
Ma, Ruitao Matthew 246
Ma, Songchen 532
Ma, Yuan 100
Ma, Yufei 310
Maangat, Simar 402
Machida, Shiro 168
Magod, Raveesh 176
Mai, Hanning 120
Mair, Hugh 170, 256
Maiyuran, Subramaniam 42
Mak, Pui-In 80, 442, 472, 474
Makinwa, Kofi A. A. 370
Malhotra, Seema 254
Malhouitre, Stéphane 308
Mandal, G. 304
Manetakis, Konstantinos 564
Mannari, Alberto 52
Mao, Junfa 568
Marani, Davide 120
Markuli , Nereo 198
Martens, Ewout 198
Martin, Fabrice 112
Martinelli, Fulvio 398
Martino, Matias Di 494
Martins, Rui 80
Martins, Rui P. 86, 114, 276, 286, 290,
446, 450, 472, 474, 558
Masilamanai, Indu 52
Mathaikutty, Deepak A. 528
Mathew, Sanu 424
Mathew, Sanu K. 436

Matsumoto, Tomohiro 200
Maurel, Vincent 500
Maurice, Lisa 52
Mayeda, Jill 90, 392
Mazzanti, Andrea 484
McGarry, William 602
Mehta, Nandish 396
Mello, Shalini De 548
Mellot, Pascal 112
Mendizabal, Laurent 308
Meng, Qianqi 358
Meng, Xiemei 178
Mercier, Patrick 288, 588
Midoh, Yoshihiro 506
Min, Qingqing 452
Min, Y. 302
Minami, Naoyuki 254
Ming, Xin 374
Miral, Nimesh N. 398
Mishra, Umesh K. 488
Miura, Noriyuki 506
Miura, Tomohiro 262
Miyahara, Masaya 392
Miyazaki, Daisuke 200
Miyazaki, Koutaro 326, 328
Mo, Yaowu 130
Moertl, Daniel 386
Moleri, Riccardo 210, 212
Momtaz, Afshin 144, 174
Monaco, Enrico 398
Mondal, Imon 572
Mondal, Susnata 134
Moon, Byeong-Taek 480, 578
Moon, Ji-Won 636
Moon, Kyoung-Jun 440, 578
Moon, Un-Ku 554
Moon, Youngjin 312
Morimoto, Masao 262
Morioka, Sumio 90
Mourik, Patrick van 192
Mueller, Josef 398
Mukkamala, Raj S. 508
Muller, Rikky 600
Munck, Koen De 616
Muralidharan, Sriram 360
Murata, Kentaro 90
Myko, André 308

N

Na, Daehoon 148
Na, H. 302
Na, Ki-Heon 264
Na, Sewhan 410, 420
Naffziger, Samuel 42
Nagarajan, Amrit 52
Nagata, Makoto 506
Nagata, Shun 200
Nagata, Shunya 262
Nair, Indira 52
Naito, Takahiro 200
Naka, Atsushiro 90
Nakamura, Daisuke 262
Nakano, Hiroyuki 168
Nakhkoob, Behrooz 402
Nallaparaju, Kalyan 138
Nam, InCheol 272
Nam, Sang-Yun 324
Naman, Harshit 428
Namkoong, Jin 402
Nani, Claudio 398
Narasimha, Rahul 590
Narukiyo, Yasuto 392
Naveed, Muhammad Zahid 248
Nazemi, Ali 174
Nedovic, Nikola 396

Nejad, Aboozar Ghorbani 588
Nett, Ryan 52
Neutens, Pieter 616
Newman, Dianne K. 508
Nguyen, Bo 402
Ni, Ronghua 214, 224
Nidhi, Nitin 144
Niitsu, Kiichi 438
Nishi, Yoshi 396
Nisi, Fabrizio De 120
Nitto, Daiki 506
Niu, Gaoqiang 390
Niu, Shengpu 406
Nivarthi, Karthik 424
Nogamida, Takeru 200
Nong, Yuanlin 626
Noori, Ari 386
Nussbaum, Pascal 564

O

O'Callaghan, John 616
Ochiai, Soichi 120
Oh, ChiSung 264
Oh, Hyoung-Seok 292
Oh, Jinseok 266
Oh, Jinwook 44
Oh, Jungju 44
Oh, Jungjun 534
Oh, Kwang-Seok 364
Oh, Minwook 266
Oh, Y. 302
Ojima, Naoki 254
Ok, Jiheon 420
Okada, Kenichi 476
Okamoto, Koichi 120
Okamura, Yusuke 120
Omirzakhov, aisarbek 242
Ommen, Hendrik Benjamin van 388
Oncu, Ahmet 506
Onda, Tomoya 168
Oo, Kaung Myat San 202
Ookuma, Naoki 254
Ou, Tincheng 472
Ouyang, Keqing 204

P

Pabba, Kaustubh 100
Pal-Singh, S. 300
Pan, Dongfang 330
Pan, Quan 150, 624
Pan, Sining 82, 552
Pan, Tao 202
Panahandeh, Mohamadamin 508
Pandey, Aviral 600
Pandey, Vaibhav 402
Pang, Di 532
Parisi, Angelo 198
Park, B. 302
Park, Chan-Hong 480, 578
Park, Chansoo 584
Park, Dongjun 636, 638
Park, Geonho 480, 578
Park, Gunho 544
Park, Gwangtae 312, 534
Park, Ha-Jung 148
Park, Hyun-Chul 480, 578
Park, Hyunjun 292
Park, Hyunsu 270
Park, J. 302
Park, Jemin 272
Park, Jeongbin 636
Park, Jeongje 266
Park, Jihye 372
Park, Joonhong 266

Park, Juho 230
Park, Junchul 420
Park, Jung-June 636, 638
Park, Kwanseo 640
Park, Kyuhyeon 608
Park, Kyungdam 364
Park, Minsoo 270
Park, Minsu 584
Park, Sanggyu 44
Park, Seonghyeok 124
Park, Seonwoo 618
Park, Seungwon 342
Park, Sunghyun 44
Park, Tae Jin 272
Park, Youngseok 272
Park, Yousung 416, 418
Parthasarathy, P. 304
Parthasarathy, Vinay 42
Pasdast, Gerald 134
Pasquo, Alessio Di 398
Pastorelli, Cedric 112
Patil, Nishant 100
Patrice, Damien Saint 308
Peng, Ji 576
Peng, Jida 282
Peng, Wenyu 68, 282
Peng, Yatao 474
Peng, Yimai 172
Perenzoni, Daniele 120
Perenzoni, Matteo 120
Perkins, Gina 66
Piao, Canxing 118
Pinckney, N. 298
Pinckney, Nathaniel 548
Pinkenburg, Jade 600
Piotto, Lorenzo 484
Pirbazari, Mahmoud M. 484
Pirmoradi, Ali 242
Privitera, Marco 248
Proft, Anabel De 616
Pu, Dong 224
Putra, Adiwena 430

Q

Qi, Dekui 130
Qian, He 228
Qian, Liwen 498
Qian, Yun 96
Qiao, Ning 314
Qin, Yubin 58
Qiu, Junyi 134
Qu, Tianxiang 76, 368
Qu, Wanyuan 448
Quarta, Gabriele 120

R

Race, Kylan 424
Rae, Bruce R. 112
Raha, Arnab 528
Rahman, Wahid 404
Ramapragada, Krishna Sai Tarun 432
Ramirez, Daniel 386
Rammohan, Ashwin 600
Ramos, Ernesto Zamora 424
Ran, Liang 126
Randhawa, Kavi 52
Rangarajan, Sundar 42
Ranjan, Ashish 52
Ravenhill, P. 300
Ray, Subhajit 386
Rayudu, Sai Krishna 578
Reddy, Sushmitha 138
Reick, Kevin 52
Rekhi, Angad 396

INDEX TO AUTHORS

Ren, Guanjing 130
Ren, Hongyu 442
Ren, Huanyu 468
Ren, Shengdao 448
Ren, Wenjie 310
Ren, Zhibin 52
Renukaswamy, Pratap 198
Rey-Losada, Daniel 202
Rhew, Ben 372
Rhew, Hyo-Gyuem 628
Ricci, Luca 194
Rider, Scot 52
Riel, Heike 16
Riggelen, Margriet van 388
Rizzini, Daniele Lodi 210, 212
Robinson, Mike 402
Rocca, Francesco Paolo Mattioli Della 120
Rocco, Michele 194
Roewer, Thomas 52
Rogers, Gregory Eric 578
Roh, Hyeonhee 612
Roh, Wonjong 118
Rosno, Pat 386
Rosseel, G. 298
Rossoni, Michele 210, 212, 476
Ruffino, Andrea 386
Rui, Jiaqing 84
Rutten, Robert 192
Ryu, Je-Min 264
Ryu, Junha 312
Ryu, Kiljun 44
Ryu, Seongyoung 410
Ryu, Wonoh 128

S

Sadhu, Bodhisatwa 386
Saeidi, Hooman 94
Saif, Marco 504
Sakai, Hiroyuki 476
Sako, Mario 254
Salik, Marina 138
Salvi, Pietro 210, 212
Samajdar, Ananda 52
Samori, Carlo 194
Sang, Haoyang 114
Sanjaya, Pranata W. 260
Santana, Lucas Moura 198
Sapiro, Guillermo 494
Satterfield, Dave 52
Sawada, Yohei 262
Sawan, Mohamad 610
Sawigun, Chutham 616
Sayeed, Abdullah 428
Schaal, Marcel 52
Schmerbeck, Timothy J. 386
Scholz, Philipp 104
Scouten, Shawn 398
Sebastiano, Fabio 388, 562
Sen, Sanchari 52
Sen, Shreyas 428
Senger, Rob 52
Sengupta, Kaushik 94, 230
Seo, Daehwan 264
Seo, Jin-O 44
Seo, Min-Woong 128
Seo, Young-Hun 268, 272
Seok, Hyun-Gi 578
Seok, Mingoo 436
Seol, Hoseok 268
Seol, Ji-Hwan 268
Seong, Hyeonwoo 636
Serdijn, Wouter 70
Shallal, Aws 138
Shan, Kexin 84
Shan, Weiwei 512

Shan, Xiaoyu 442
Shang, Dechun 350
Shang, Minghao 280, 460
Shao, Yen-Tung 516
Shao, Zijian 94
Sheikh, Kashif M. 202
Shekhar, Sudip 588
Shen, Dawei 190
Shen, Junzhe 514, 522
Shen, Shen 402
Shen, Xiaohan 240
Shen, Xinyu 218
Shen, Yi 590
Shen, Yili 376
Shen, Yizhu 96
Shen, Zhongqiu 76
Sheng, Chaodi 400
Sheng, Maojia 228
Sheng, Yuguo 504
Shi, C.-J. Richard 84
Shi, Chengyao 70
Shi, Chunqi 498
Shi, Dan 80
Shi, Linxu 582
Shih, Ming-En David 56
Shim, Heesung 128
Shim, Jaechul 258
Shim, Kyu-Ha 264
Shim, Yeongseok 128
Shimazaki, Yasuhisa 168
Shin, D. 302
Shin, Deokha 128
Shin, Ho-Joon 640
Shin, Hoon 268
Shin, Hyunjin 258
Shin, Seunghun 182
Shin, Seunghwa 418
Shin, Taekyun 266
Shin, Wongyu 44
Shin, Yeonjae 608
Shinohara, Hirofumi 438
Shiomi, Jun 506
Shirane, Atsushi 90, 392
Shivnraine, Ravi 138
Shokrolahzade, Ehsan 562
Shoobi, Amirreza 242
Shui, Hanyue 434
Shum, Kam Man 106
Shyh-Shyuan, Sheu 526
Si, D. 302
Si, Xin 512, 536
Siby, Emil 192
Sideris, Constantine 506
Silberman, Joel 52
Siligaris, Alexandre 500
Silla, Mark 42
Sim, Jae-Yoon 636, 638
Sin, Sai-Weng 454, 558
Singer, Larry 202
Singh, Abhishek 112
Singh, Jaswinder 256
Singh, Nawab 158
Singh, Teja 42
Singh, Ullas 144
Sinha, Sandipan 256
Siriburanon, Teerachot 482
Skende, A. 298
Smith, Alan 42
Snell, Bryce 386
So, Jeonggyu 540
Sohn, Young-Soo 264
Son, Insang 124
Son, Jihun 584
Son, Juhee 578
Son, Muyoung 542

Son, Sieon 44
Son, Young-Suk 364
Song, C. 302
Song, Eunji 270
Song, G. 302
Song, Heewoong 266
Song, Hong-Joo 268
Song, J. 302
Song, Jae-Joon 272
Song, Jaegeun 628
Song, Jaihyuk 128
Song, Jeongeun 148
Song, Minyoung 584, 608
Song, Sanquan 396
Song, Tae-Gyun 420
Song, Taeha 268
Song, Uijong 410
Song, Young Guen 264
Song, Zhibang 454
Soong, Ruei-Chen 354
Sosio, Marco 398
Soulie, M. 300
Spirito, Marco 562
Srinivasan, S. 304
Srinivasan, Viji 52
Srirambhatla, Vasantha 424
Stanzione, Stefano 70
Staszewski, Robert Bogdan 482
Stauth, Jason 334
Steiner, Michael 424
Stepko, Alexander 94
Still, Greg 52
Stojanovic, Vladimir 404
Stoppa, David 120
Strohman, Ryan 590
Strukov, Dmitri 186
Su, Jian-Wei 526
Su, Yu 146
Su, Yumin 92
Su, Yutong 546
Sugihara, Hiroyuki 128
Suh, Ji-Hoon 618
Suh, Junseuk 578
Suh, Kiseok 258
Sullivan, Charles 334
Sun, Guangyu 512
Sun, Huanfa 146
Sun, Lingling 356
Sun, Mingqian 368
Sun, Nan 196, 378, 552
Sun, Quan 170
Sun, Xuan 158, 244
Sun, Yi 426
Sun, Yiyang 538
Sun, Yuxuan 466
Sundaram, Sriram 42
Surana, Saurabh 144
Suresh, Vikram 424
Suriyasak, Chetphilin 506
Suzuki, Atsuya 200
Suzuki, Yoshinao 254

T

Tagawa, Hironori 438
Takahashi, Yuya 90, 392
Takasaki, Mika 200
Takaya, Satoshi 326
Takehara, Satoshi 554
Takiguchi, Kenichiro 262
Talamala, Bala Prasad 424
Tam, Sai-Wang 354
Tambe, Thierry 548
Taminiau, Tim Hugo 388
Tan, Nick Nianxiong 82
Tan, Yonghao 532

Tan, Zhichao 448
Tanaka, Hirofumi 66
Tanaka, Shinji 262
Taneja, Sachin 424
Tang, Adrian 354
Tang, Dawei 358
Tang, Hongxin 570
Tang, Jianshi 228
Tang, Junyao 158
Tang, Kea-Tiong 516, 520, 526, 606
Tang, Miaoyu 512
Tang, Minzhe 476
Tang, Siyuan 358
Tang, Wei 50, 232
Tang, Xiyuan 180, 190, 556
Tang, Yuxiang 76
Tang, Zhong 82
Tang, Zihao 450
Tangirala, Shankar 138
Tao, Guanren 590
Tao, Weichen 576
Tao, Yunsong 196, 378
Tatani, Keiji 120
Taufour, V. 300
Taupin, Sophie 112
Tavernier, Filip 502
Tell, Stephen G. 396, 548
Tellez, Gustavo 52
Temporiti, Enrico 398
Tessitore, Alex 138
Thasari, Kumar 144
Thimmaiah, Jayanth M 254
Thivin, Mathieu 112
Thonnart, Yvain 308
Thuaire, Herve 112
Tian, Fengshi 610
Tian, Li 204
Tian, Meijun 102
Tien, Chao-Jen 354
Tien, Jen-Chun 516, 520, 526
Tien, Kevin 386
Tilmans, Harrie A. C. 616
Timmerwilke, John 386
Tokunaga, Carlos 178
Tolaib, Islam 340
Tombolan, Giacomo 194
Tomita, Kazutoshi 200
Tong, Lin 512
Tong, Zhiguo 462
Torabi, Mohammadamin 144
Toth, Nandor G. 370
Tran, Trong-Hieu 170
Tripathy, S. 304
Trivedi, Manish 256
Trochut, Severin 112
Trotta, Giovanni Rocco 210
Tsai, C.-J. 170
Tsai, Chien-Chun 136
Tsai, Frank W. 254
Tsai, Ping-Yuan 56
Tsai, Rick 10
Tsai, Tsung-Hsien 136
Tsai, Yao-Hung 154
Tsai, Yueh-Feng 414
Tschanz, James W. 178
Tsen, Chia-Jung 260
Tseng, Mai 520, 526
Tseng, Yu-Cheng 56
Tsui, Chi-Ying 532, 610
Tu, Fengbin 532

U

Uddin, Syed Mohammad Ashab 352
Uggu, Viswanath 56
Um, Soyeon 312

INDEX TO AUTHORS

Um, Youngdo 268
Underwood, Devin 386

V

V, Aravinth 254
V, Indra K 254
Vachon, Robert 172
Vaish, Dhruv 600
Valluri, Sravanth 424
Vamvakos, Socrates 138
Vartak, Adish 424
Vaucher, Cicero S. 208
Vecchi, Federico 484
Vélard, Rémi 308
Vemulapalli, Hanish 42
Venkataramani, Swagath 52
Venkatesan, Rangharajan 548
Veraa, Brian 52
Verbauwhede, Ingrid 428
Verhelst, Marian 502
Vu, Roxanne 138

W

Wadatsumi, Takuya 506
Wakahara, Kohei 168
Wan, Jiapeng 96
Wan, Rentao 436
Wan, Yuting 84
Wang, Adam 504
Wang, Alex 144
Wang, Baochuang 280
Wang, Biao 558
Wang, Bin 184
Wang, Bo 114
Wang, Bohan 514, 522
Wang, Cheng 204, 406
Wang, Chia-Yu 260
Wang, Chih-Ming 56
Wang, Chuhui 278
Wang, Ericbill 170
Wang, Hanning 184, 434
Wang, Hedi 60
Wang, Hua 340, 346, 504, 564
Wang, Huanyu 58, 546
Wang, Hui 606
Wang, Jenny 254
Wang, Jian 578
Wang, Jinchen 384
Wang, Jingpeng 190
Wang, Jingyi 122
Wang, Kaihang 214, 224
Wang, Keping 350
Wang, Lei 626
Wang, Li 442
Wang, Linfang 514, 522
Wang, Nan 368
Wang, Pengcheng 582
Wang, Qian 130
Wang, Quan 626
Wang, Ruohan 498
Wang, Shixuan 582
Wang, Song 426, 524
Wang, Tiansu 288
Wang, Ting-Yu 54
Wang, Wei 140
Wang, Weibo 350
Wang, Weixiao 376
Wang, Wen 424
Wang, Wenqian 476
Wang, Xing 512
Wang, Xu 356
Wang, Yan 216
Wang, Yang 58, 316, 546, 604
Wang, Yih 260

Wang, Yiman 514, 522
Wang, Yiming 76
Wang, Yuan 180, 556
Wang, Yuanfei 290
Wang, Yunfan 590
Wang, Yunzhengmao 632
Wang, Zhao 114
Wang, Zheng 466
Wang, Zhifei 140
Wang, Zhihua 152, 220, 222, 358, 606
Wang, Zijie 582
Wang, Zongnan 556
Watanabe, oshihisa 254
Watanabe, Shunya 90
Wei, Kang 176
Wei, Shangjie 146
Wei, Shaojun 58, 316, 546
Wei, Yijie 602
Wei, Yuhong 498
Wen, Chin-Hua 136
Wen, Jincai 356
Wen, Yixuan 466
Weng, Chih-Hui 260
Whang, Sunjoo 312
Wie, Jeongyoon 608
Wilding, Dominik 104
Wilkerson, Chris 424
Wilkins, Paul 202
Willenborg, Scott M. 386
Winckel, Steven Van 198
Wittenhagen, Enne 104
Won, Bokyeon 272
Won, Sanggyeong 258
Woo, Seunghan 272
Woo, Sunsik 364
Woodward, Sandra 52
Wormald, Luke D. 232
Wu, Chih-Ling 516
Wu, Dabin 228
Wu, Defa 512
Wu, Ella 254
Wu, Hongzhi 150, 624
Wu, Huaqiang 228
Wu, Hui 610
Wu, J.J. 260
Wu, Jing 498
Wu, Lianbo 582
Wu, Lin 518
Wu, Mengze 366
Wu, Nanjian 142, 218
Wu, Peilin 594
Wu, Ping-Sheng 516, 520
Wu, Shuxian 490
Wu, Wanghua 578
Wu, Wei 178
Wu, Weitao 150, 624
Wu, Xiaonan 280
Wu, Xiaoyuan 498
Wu, Xing 84
Wu, Xu 108
Wu, Yihao 182
Wu, Yu 620
Wu, Yue 334, 474
Wu, Yuxia 356
Wu, Zihan 180
Wu, Zuoguo 134

X

Xia, Haiyang 108
Xia, Xiaoyue 106
Xiang, Longxi 368
Xiang, Xujiang 316
Xianyu, Haishu 228
Xiao, Junyue 358
Xiao, Qijing 376

Xie, Feng 566
Xie, Kenan 350
Xie, Qian 466
Xie, Shanshan 178
Xie, Wenao 114
Xing, Chaoyang 552
Xing, Yannan 314
Xiong, Yuang 476
Xu, Aolin 400
Xu, Bocheng 180
Xu, Dacheng 400
Xu, Dingxin 476
Xu, Dongfan 476
Xu, Hongtao 566
Xu, Jianlong 610
Xu, Jiawei 76, 322, 368
Xu, Jinglong 340
Xu, Kai 482
Xu, Li 396
Xu, Ningsheng 240
Xu, Richard 130
Xu, Ruohuang 310
Xu, S. 304
Xu, Shizhe 96
Xu, Shufan 438
Xu, Weiwei 330, 452, 458
Xu, Wende 58
Xu, Yilin 582
Xu, Yiqing 142
Xu, Yiwei 626
Xu, Zhuo 472
Xu, Ziang 400

Y

Yada, Nobuhiro 168
Yada, Satish 178
Yadav, Ajay 402
Yagishita, Yuki 200
Yamada, Shuhei 134
Yamaki, Yuzuru 554
Yamanaka, Sho 168
Yamashita, Ryuji 254
Yamauchi, Kazuki 254
Yan, Xinming 536
Yanaka, Kotaro 392
Yanase, Keishi 506
Yang, Bingzheng 568, 570
Yang, Bohan 434
Yang, Chia-Hsiang 414
Yang, Dezhen 314
Yang, H. 302
Yang, Hanlin 478
Yang, Huazhong 60
Yang, Jaehyeok 270
Yang, Jiachang 556
Yang, Jiacheng 450
Yang, Jianguo 524
Yang, Jianxin 286, 446
Yang, Jiawei 96
Yang, Jiaxin 546
Yang, Jie 218, 610
Yang, Jinjiang 434
Yang, Jun 512, 536
Yang, Jun-Hyeok 372
Yang, Kaiyuan 92
Yang, Renxu 314
Yang, Ruiyuan 294, 596
Yang, Shaoqi 576
Yang, Shiheng 466
Yang, Shu-Chun 136
Yang, Sung-gi 342
Yang, Tianze 566
Yang, Wei 228
Yang, Wen 478
Yang, Xiang 254

Yang, Xiaolin 616
Yang, Xiaolong 58
Yang, Xuecheng 76
Yang, Ya-Tang 520, 526
Yang, Yang 498
Yang, Yi 204, 512
Yang, Ying-Yu 64
Yang, Youming 180
Yang, Yunqi 582
Yang, Yunzhe 284, 320
Yang, Zhen 426
Yang, Zhengke 126, 390
Yang, Zhiyong 254
Yang, Zhizhan 472
Yang, Zunsong 442
Yao, Peng 228
Yao, Shun 490
Yao, Wang 594
Ye, Bingyi 140
Ye, Haifeng 498
Ye, Le 310
Ye, Shen 76
Ye, Tianchen 140
Ye, Xiangjun 310
Ye, Xiuzhu 582
Ye, Zhenkai 400
Ye, Zhongxin 278
Ye, Zonglin 466
Yeck, Mark 386
Yeh, Chao-Yang 170
Yeh, Po-Yu 56
Yeh, Shau-Hua 54
Yeh, Yao-Kai 516, 520, 526
Yennampelli, Nataraj 424
Yeo, Theng Tee 594
Yi, Donghyeon 618
Yi, Shinyoung 628
Yi, Tengyue 632
Yim, Sungsoo 272
Yin, Jun 462, 472, 474, 482
Yin, Shouyi 58, 316, 546
Yin, Shuying 434
Yin, Xin 406
Yin, Yun 566
Yingling, Daniel 172
Yoo, Changsik 268, 272
Yoo, Hoi-Jun 312, 534, 540
Yoo, Jerald 390, 604
Yoo, Junghyun 46
Yoo, Kyungwoo 578
Yoo, Sangmin 440, 578
Yoo, Seungjae 542
Yoo, Sun-Young 128
Yoo, Sungmin 372
Yoo, Sunwoo 544
Yook, Byungho 480, 578
Yoon, Chiweon 636, 638
Yoon, Hyunchul 272
Yoon, In-Soo 254
Yoon, Jae-Sung 44
Yoon, Jong-Hyeok 584, 608
Yoon, Jungmin 266
Yoon, Juyeong 44
Yoon, Sangsic 266
Yoon, Sung-Woo 268
Yoshizawa, Satoshi 200
You, De-Qi 516, 520, 526
You, Jungtaek 266
You, Shujuan 228
You, Xiaohu 108
Youn, Yelim 440
Young, Steve 590
Yu, Chang-Hyo 44
Yu, Chao 348
Yu, Dunshan 140

INDEX TO AUTHORS

Yu, Guodong 350
Yu, Haofeng 126
Yu, Haohong 594
Yu, He 602
Yu, Hwayeal 292
Yu, Luqi 348
Yu, Rui 594
Yu, Taewoo 578
Yu, Xiaohua 214, 224
Yu, Xiaopeng 82
Yu, Xinglong 426
Yu, Yichen 588
Yu, Yucheng 348
Yu, Zhehao 242
Yu, Zhewen 454, 462
Yuan, Chia-Hung 56
Yuan, Fengen 474
Yuan, Jinyi 280, 452
Yuan, Luyao 576
Yuasa, Keito 90
Yue, Chih-Yen 526
Yue, Chik Patrick 246
Yue, Zhiheng 58, 316, 546
Yun, Daeho 266
Yun, Ghangmin 46
Yun, Gichan 618
Yun, Jiwon 388
Yun, Sangbu 48
Yune, Sungwoong 430

Z

Zalani, Vidhi 52
Zanoletti, Gabriele 194
Zeng, Jianping 346
Zeng, Xiaoyang 368
Zeng, Yujie 146
Zha, Wenfeng 514, 522
Zhan, Mingtao 190, 378
Zhan, Xiangxun 472
Zhang, Aodong 378
Zhang, Aoyang 184, 434
Zhang, Bo 374, 516, 520
Zhang, Bodong 478
Zhang, Daxu 476
Zhang, Dong 532
Zhang, Fuyao 446
Zhang, Gaojing 348
Zhang, Guangzhong 582
Zhang, Guohe 146
Zhang, Guoqi 282
Zhang, Hao 378
Zhang, Haoming 468
Zhang, Hongyang 466
Zhang, Hongyong 610
Zhang, Huajun 78, 380
Zhang, Hui 582
Zhang, Jian 216
Zhang, Jilin 606
Zhang, Jing 406
Zhang, Lushuo 606
Zhang, Miao 78
Zhang, Nuo 202
Zhang, Qian 524
Zhang, Qingtian 228
Zhang, Qingyu 620
Zhang, Rui 278
Zhang, Runxi 498
Zhang, Shengxin 130
Zhang, Shiwei 222
Zhang, Shun 108
Zhang, Tantan 594
Zhang, Xiang 348
Zhang, Xin 176, 244, 320
Zhang, Yangyi 150, 624
Zhang, Yibin 350

Zhang, Yihan 290
Zhang, Yongle 330
Zhang, Yuncheng 476
Zhang, Zhao 142, 218, 610
Zhang, Zhaoyu 142
Zhang, Zhengya 48, 50, 232
Zhang, Zhi-Jun 54
Zhang, Zhishuai 196
Zhang, Zhiyuan 290
Zhang, Zhongyuan 96
Zhang, Zhuo 314
Zhang, Ziang 108
Zhao, Bo 160, 376
Zhao, Cankun 434
Zhao, Chen 144
Zhao, Guangshu 114
Zhao, Kangjie 498
Zhao, Lei 158
Zhao, Liang 126, 532
Zhao, Linran 66
Zhao, Menglian 448
Zhao, Weisheng 582
Zhao, Xiaoxi 68
Zhao, Xingpeng 594
Zhao, Yifan 426
Zhao, Yunxia 606
Zhao, Ziyan 498
Zheng, Hongzhao 314
Zhong, Jiajun 204
Zhong, Kangping 246
Zhong, Liwen 102
Zhong, Yi 196, 378, 552
Zhong, Zhiwei 602
Zhong, Ziyang 276
Zhou, Ching 52
Zhou, Feichi 126
Zhou, Jia 354
Zhou, Jiaqi 538
Zhou, Jie 568
Zhou, Jonathan 230
Zhou, Kaiwen 76, 368
Zhou, Peigen 358
Zhou, Qi 80
Zhou, Qiang 92, 344
Zhou, Rui 358
Zhou, Wen 578
Zhou, Yue 314
Zhou, Zeyu 576
Zhou, Zhidao 514, 522
Zhu, Haipeng 204
Zhu, Haiyang 202
Zhu, Jianfeng 184
Zhu, Jinpeng 626
Zhu, Junyu 514, 522
Zhu, Min 96, 434
Zhu, Minji 354
Zhu, Wenping 434
Zhu, Wenshuo 158, 244
Zhu, Wentao 204
Zhu, Xiaoge 204
Zhu, Xiaowei 348
Zhu, Yan 86
Zhu, Yuyang 468
Zhu, Zhangming 82, 518
Zhu, Zhaochen 470
Zhu, Zhongyao 454
Zhuang, Yi-Chang 170
Ziegler, Matthew 52
Zimmer, Brian 396
Zou, Wenjun 610
Zou, Yu 120
Zou, Zihan 536
Zuo, Fengguo 524